CRC Handbook
of
Chemistry and Physics

81st Edition

EDITORIAL ADVISORY BOARD

CRC Handbook
of
Chemistry and Physics

A Ready-Reference Book of Chemical and Physical Data

HANDBOOK OF CHEMISTRY AND PHYSICS

2000-2001

81st

EDITION

CRC PRESS LLC

Editor-in-Chief

David R. Lide, Ph.D.
Former Director, Standard Reference Data
National Institute of Standards and Technology

CRC Press
Boca Raton London New York Washington, D.C.

© 1974, 1975, 1976, 1977, 1978, 1979, 1980, 1981, 1982, 1983, 1984, 1985, 1986, 1987, 1988, 1989, 1990, 1991, 1992, 1993, 1994, 1995, 1996, 1997, 1998, 1999, 2000 by CRC Press LLC

© 1964, 1965, 1966, 1967, 1968, 1969, 1970, 1971, 1972, 1973 by THE CHEMICAL RUBBER CO.
Copyright 1918, 1920 by The Chemical Rubber Company (Copyright renewed 1946, 1948 by The Chemical Rubber Publishing Company)

Copyright 1922 (Copyright renewed 1950), 1925 (Copyright renewed 1953), 1926 (Copyright renewed 1954), 1927 (Copyright renewed 1955), 1929 (Copyright renewed 1957), 1936, 1937 (Copyright renewed 1965 by The Chemical Rubber Co.), 1939, 1940 (Copyright renewed 1968 by the Chemical Rubber Co.), 1941 (Copyright renewed 1969 by The Chemical Rubber Co.), 1942 (Copyright renewed 1970 by The Chemical Rubber Co.), 1943 (Copyright renewed 1971 by The Chemical Rubber Co.), 1944 (Copyright renewed 1972 by The Chemical Rubber Co.), 1945 (Copyright renewed 1973 by The Chemical Rubber Co.), 1947, 1949, 1950, 1951, 1952 (Copyright renewed 1980 by CRC Press, Inc.), 1953 (Copyright renewed 1981a by CRC Press, Inc.), 1954 (Copyright renewed 1982 by CRC Press, Inc.), 1955 (Copyright renewed 1983 by CRC Press, Inc.), 1956 by Chemical Rubber Publishing Company

© 1957, 1958, 1959, 1960, 1962 by Chemical Rubber Publishing Company

© 2000 by CRC Press LLC

No claim to original U.S. Government works
International Standard Book Number 0-8493-0481-4
Library of Congress Card Number 13-11056
Printed in the United States of America 1 2 3 4 5 6 7 8 9 0
Printed on acid-free paper

PREFACE
81st Edition

The 81st Edition of the *CRC Handbook of Chemistry and Physics*, the first to be published in the new millennium, contains several updated and expanded tables, including:

- Fundamental Physical Constants (the new set of CODATA recommended values, replacing the 1986 set)
- The Elements (descriptive texts on the occurrence, properties, history, and uses of all the chemical elements)
- Dissociation Constants of Organic Acids and Bases (expanded by 50%)
- Dipole Moments (revised and expanded)
- Threshold Limits for Airborne Contaminants (including the most recent recommendations)

Among the new topics covered in the 81st Edition are:

- Properties of Liquid Helium
- Electrical Conductivity of Water as a Function of Temperature and Pressure
- Bond Lengths in Organometallic Compounds
- Organic Magnets
- Pressure-Volume-Temperature Relationship for Polymer Melts
- Characteristics of Human Hearing

Finally, in response to requests from a number of users, indexes to melting points and boiling points have been added to the table of Physical Constants of Organic Compounds.

The Editor will appreciate suggestions on new topics for the *Handbook* and notification of any errors. Address all comments to Editor, *Handbook of Chemistry and Physics*, CRC Press LLC, 2000 N. W. Corporate Blvd., Boca Raton, FL 33431. Comments may also be sent by electronic mail to drlide@post.harvard.edu.

The *Handbook of Chemistry and Physics* is dependent on the efforts of many contributors throughout the world. The list of current contributors follows this Preface. I appreciate the valuable suggestions which have come from the Editorial Advisory Board and from many users. I would like to thank Susan Fox and Carole Sweatman at CRC Press for their help, and I am particularly grateful to Karen Feinstein for her cheerful assistance and strong support over the past few years.

David R. Lide
February 15, 2000

This Edition is Dedicated to my Wife, Bettijoyce Breen Lide

Note on the Ordering of Chemical Compounds: The decision on the order in which to list chemical compounds in a table is always difficult. An alphabetical list by name has the disadvantage that several different synonyms are often in common use, with the result that a reader may conclude incorrectly that a compound is not present if he looks it up under the wrong name. Listing by common formula is satisfactory for simple inorganic compounds, but is cumbersome for organics. A listing by molecular formula is attractive because clear rules can be given for locating a compound, but the user may have to go to some effort to determine the molecular formula. In this book the choice is made on pragmatic grounds. The long tables, Physical Constants of Organic Compounds and Physical Constants of Inorganic Compounds, are ordered by systematic name, but indexes to synonyms, formulas, and CAS Registry Numbers are provided. If the table is very short and includes only familiar substances, the listing is usually alphabetical by common formula or name. The remaining tables are ordered by molecular formula using a modification of the Hill convention. In this convention the molecular formula is written with C first, H second, and then all other elements in alphabetical order of their chemical symbols. For tables with organic compounds only, the sequence of entries is determined by the alphabetical order of elements in the molecular formula and the number of atoms of each element, in ascending order, e.g., C_3H_7Cl, C_3H_7N, C_3H_7NO, $C_3H_7NO_2$, etc. (For organic compounds, a quick way to determine the molecular formula is to use the Physical Constants of Organic Compounds table, which starts on Page 3-1, and its synonym index on Page 3-586.) In tables containing non-carbon compounds, those are listed first, followed by a separate listing of compounds that do contain carbon. This is in contrast to the strict Hill convention as followed by Chemical Abstracts Service, where the molecular formulas beginning with A and B precede the formulas for carbon-containing compounds, while those beginning with D... Z follow. For tabular displays, as opposed to an index, it appears more convenient to the user if the non-carbon compounds are listed as a block, rather than being split by the longer list of carbon compounds.

CURRENT CONTRIBUTORS

Douglas Ambrose
9 The Crest
Surbiton
Surrey KT5 8JZ
England

Lev I. Berger
California Institute of Electronics
    and Materials Science
2115 Flame Tree Way
Hemet, California 92545

R.W. Clay
Department of Physics and
    Mathematical Physics
The University of Adelaide
Adelaide, South Australia 5005
Australia

A. K. Covington
Department of Chemistry
University of Newcastle
Newcastle upon Tyne NE1 7RU
England

K.F. Eckerman
Oak Ridge National Laboratory
Oak Ridge, Tennessee 37831

Robert B. Fox
6115 Wiscassett Rd.
Bethesda, Maryland 20816

H. P. R. Frederikse
9625 Dewmar Lane
Kensington, Maryland 20895

J.R. Fuhr
Atomic Physics Division
National Institute of Standards and
    Technology
Gaithersburg, Maryland 20899

Karl A. Gschneidner
Ames Laboratory
Energy and Mineral Resources
    Research Institute
Iowa State University
Ames, Iowa 50011

C. R. Hammond
17 Greystone Rd.
West Hartford, Connecticut 06107

Robert F. Hampson
Chemical Kinetics Division
National Institute of Standards and
    Technology
Gaithersburg, Maryland 20899

Norman E. Holden
National Nuclear Data Center
Brookhaven National Laboratory
Upton, New York 11973

H. Donald Brooke Jenkins
Department of Chemistry
University of Warwick
Coventry CV4 7AL England

Henry V. Kehiaian
ITODYS
1 rue Guy de la Brosse
75005 Paris
France

J. Alistair Kerr
School of Chemistry
University of Birmingham
Birmingham B15 2TT England

Sharon G. Lias
National Institute of Standards and
    Technology
Gaithersburg, Maryland 20899

William C. Martin
Atomic Physics Division
National Institute of Standards and
    Technology
Gaithersburg, Maryland 20899

Joel S. Miller
Department of Chemistry
University of Utah
Salt Lake City, Utah 84112

Thomas M. Miller
Air Force Research Laboratory/VSBP
29 Randolph Rd.
Hanscom AFB, Massachusetts
    01731-3010

Joseph Reader
Atomic Physics Division
National Institute of Standards and
    Technology
Gaithersburg, Maryland 20899

H. K. Roobottom
Department of Chemistry
University of Warwick
Coventry CV4 7AL England

B. L. Sharma
E-1/15 Model Town
Delhi — 110009 India

David W. Stocker
School of Chemistry
University of Leeds
Leeds LS2 9JT England

B. N. Taylor
Physics Laboratory
National Institute of Standards and
    Technology
Gaithersburg, Maryland 20899

Thomas G. Trippe
Particle Data Group
Lawrence Berkeley Laboratory
1 Cyclotron Road
Berkeley, California 94720

Petr Vanýsek
Department of Chemistry
Northern Illinois University
DeKalb, Illinois 60115

Wolfgang L. Wiese
Atomic Physics Division
National Institute of Standards and
    Technology
Gaithersburg, Maryland 20899

Christian Wohlfarth
Institut für Physikalische Chemie
Martin Luther University
D-06217 Merseburg
Germany

TABLE OF CONTENTS

SECTION 1: BASIC CONSTANTS, UNITS, AND CONVERSION FACTORS
Fundamental Physical Constants 1-1
Standard Atomic Weights (1995) 1-12
Atomic Masses and Abundances 1-15
Electron Configuration of Neutral Atoms in the Ground State 1-18
International Temperature Scale of 1990 (ITS-90) 1-20
Conversion of Temperatures from the 1948 and 1968 Scales to ITS-90 1-22
International System of Units (SI) 1-24
Conversion Factors 1-29
Conversion of Temperatures 1-41
Conversion Factors for Energy Units 1-42
Conversion Factors for Pressure Units 1-43
Conversion Factors for Thermal Conductivity Units 1-44
Conversion Factors for Electrical Resistivity Units 1-45
Conversion Factors for Chemical Kinetics 1-46
Conversion Factors for Ionizing Radiation 1-47
Values of the Gas Constant in Different Unit Systems 1-49

SECTION 2: SYMBOLS, TERMINOLOGY, AND NOMENCLATURE
Symbols and Terminology for Physical and Chemical Quantities 2-1
Nomenclature of Chemical Compounds 2-22
Nomenclature for Inorganic Ions and Ligands 2-23
Organic Substituent Groups and Ring Systems 2-25
Scientific Abbreviations and Symbols 2-29
Greek, Russian, and Hebrew Alphabets 2-37
Definitions of Scientific Terms 2-38

SECTION 3: PHYSICAL CONSTANTS OF ORGANIC COMPOUNDS
Physical Constants of Organic Compounds 3-1
Structure Diagrams for Table of Physical Constants 3-331
Synonym Index of Organic Compounds 3-586
Molecular Formula Index of Organic Compounds 3-631
CAS Registry Number Index of Organic Compounds 3-709
Melting Point Index of Organic Compounds 3-737
Boiling Point Index of Organic Compounds 3-751
Diamagnetic Susceptibility of Selected Organic Compounds 3-764

SECTION 4: PROPERTIES OF THE ELEMENTS AND INORGANIC COMPOUNDS
The Elements 4-1
Physical Constants of Inorganic Compounds 4-37
Synonym Index of Inorganic Compounds 4-101
CAS Registry Number Index of Inorganic Compounds 4-107
Physical Properties of the Rare Earth Metals 4-114
Melting, Boiling, and Critical Temperatures of the Elements 4-124
Heat Capacity of the Elements at 25°C 4-125
Vapor Pressure of the Metallic Elements 4-126
Density of Molten Elements and Representative Salts 4-129
Magnetic Susceptibility of the Elements and Inorganic Compounds 4-133
Index of Refraction of Inorganic Liquids 4-139
Physical and Optical Properties of Minerals 4-141
Crystallographic Data on Minerals 4-148

SECTION 5: THERMOCHEMISTRY, ELECTROCHEMISTRY, AND KINETICS

CODATA Key Values for Thermodynamics 5-1
Standard Thermodynamic Properties of Chemical Substances 5-4
Thermodynamic Properties as a Function of Temperature 5-61
Thermodynamic Properties of Aqueous Systems 5-85
Heat of Combustion 5-89
Electrical Conductivity of Water 5-90
Standard KCl Solutions for Calibrating Conductivity Cells 5-91
Molar Conductivity of Aqueous HF, HCl, HBr, and HI 5-92
Equivalent Conductivity of Electrolytes in Aqueous Solution 5-94
Ionic Conductivity and Diffusion at Infinite Dilution 5-95
Activity Coefficients of Acids, Bases, and Salts 5-97
Mean Activity Coefficients of Electrolytes as a Function of Concentration 5-99
Enthalpy of Dilution of Acids 5-104
Enthalpy of Solution of Electrolytes 5-105
Chemical Kinetic Data for Stratospheric Modeling 5-106

SECTION 6: FLUID PROPERTIES

Thermodynamic Properties of Air 6-1
Properties of Water in the Range 0-100°C 6-3
Enthalpy of Vaporization of Water 6-3
Fixed Point Properties of H_2O and D_2O 6-4
Thermal Conductivity of Saturated H_2O and D_2O 6-4
Standard Density of Water 6-5
Properties of Ice and Supercooled Water 6-6
Volumetric Properties of Aqueous Sodium Chloride Solutions 6-8
Density of D_2O 6-9
Vapor Pressure of Ice 6-9
Vapor Pressure of Water from 0 to 370°C 6-10
Boiling Point of Water at Various Pressures 6-12
Melting Point of Ice as a Function of Pressure 6-12
Properties of Water and Steam as a Function of Temperature and Pressure 6-13
Permittivity (Dielectric Constant) of Water at Various Frequencies 6-15
Thermophysical Properties of Fluids 6-16
Virial Coefficients of Selected Gases 6-25
Van der Waals Constants for Gases 6-45
Mean Free Path and Related Properties of Gases 6-46
Influence of Pressure on Freezing Points 6-47
Critical Constants 6-48
Sublimation Pressure of Solids 6-61
Vapor Pressure 6-64
Vapor Pressure of Fluids at Temperatures below 300 K 6-92
Vapor Pressure of Saturated Salt Solutions 6-103
IUPAC Recommended Data for Vapor Pressure Calibration 6-104
Enthalpy of Vaporization 6-105
Enthalpy of Fusion 6-121
Pressure and Temperature Dependence of Liquid Density 6-136
Properties of Cryogenic Fluids 6-138
Properties of Liquid Helium 6-139
Halocarbon Refrigerants 6-140
Density and Specific Volume of Mercury 6-143
Thermal Properties of Mercury 6-144
Surface Tension of Common Liquids 6-145
Permittivity (Dielectric Constant) of Liquids 6-149

Permittivity (Dielectric Constant) of Gases **6**-172
Azeotropic Data for Binary Mixtures **6**-174
Viscosity of Gases **6**-178
Viscosity of Liquids **6**-180
Thermal Conductivity of Gases **6**-185
Thermal Conductivity of Liquids **6**-187
Diffusion in Gases **6**-189
Diffusion of Gases in Water **6**-191
Diffusion Coefficients in Liquids at Infinite Dilution **6**-192

SECTION 7: BIOCHEMISTRY

Properties of Amino Acids **7**-1
Structures of Common Amino Acids **7**-3
Properties of Purine and Pyrimidine Bases **7**-4
The Genetic Code **7**-5
Properties of Fatty Acids **7**-6
Carbohydrate Names and Symbols **7**-7
Biological Buffers **7**-9
Typical pH Values of Biological Materials and Foods **7**-10
Chemical Composition of the Human Body **7**-11

SECTION 8: ANALYTICAL CHEMISTRY

Preparation of Special Analytical Reagents **8**-1
Standard Solutions of Acids, Bases, and Salts **8**-5
Standard Solutions of Oxidation and Reduction Reagents **8**-7
Organic Analytical Reagents for the Determination of Inorganic Substances **8**-8
Acid-Base Indicators **8**-16
Fluorescent Indicators **8**-19
Conversion Formulas for Concentration of Solutions **8**-20
Electrochemical Series **8**-21
Reduction and Oxidation Potentials for Certain Ion Radicals **8**-32
pH Scale for Aqueous Solutions **8**-35
Practical pH Measurements on Natural Waters **8**-41
Buffer Solutions Giving Round Values of pH at 25°C **8**-43
Dissociation Constants of Inorganic Acids and Bases **8**-44
Dissociation Constants of Organic Acids and Bases **8**-46
Concentrative Properties of Aqueous Solutions: Density, Refractive Index, Freezing Point
    Depression, and Viscosity **8**-57
Ion Product of Water Substance **8**-84
Ionization Constant of Normal and Heavy Water **8**-85
Solubility of Selected Gases in Water **8**-86
Solubility of Carbon Dioxide in Water at Various Temperatures and Pressures **8**-90
Aqueous Solubility and Henry's Law Constants of Organic Compounds **8**-91
Aqueous Solubility of Inorganic Compounds at Various Temperatures **8**-102
Solubility Product Constants **8**-111
Solubility of Common Salts at Ambient Temperatures **8**-115
Solubility Chart **8**-116
Reduction of Weighings in Air to Vacuo **8**-119
Volume of One Gram of Water **8**-119
Properties of Carrier Gases for Gas Chromatography **8**-120
Solvents for Ultraviolet Spectrophotometry **8**-121
^{13}C Chemical Shifts of Useful NMR Solvents **8**-122
Mass Spectral Peaks of Common Organic Solvents **8**-123

SECTION 9: MOLECULAR STRUCTURE AND SPECTROSCOPY

Bond Lengths in Crystalline Organic Compounds 9-1
Bond Lengths in Organometallic Compounds 9-15
Bond Lengths and Angles in Gas-Phase Molecules 9-17
Dipole Moments 9-44
Strengths of Chemical Bonds 9-51
Electronegativity 9-75
Force Constants for Bond Stretching 9-76
Fundamental Vibrational Frequencies of Small Molecules 9-77
Spectroscopic Constants of Diatomic Molecules 9-81
Infrared Correlation Charts 9-87
Nuclear Spins, Moments, and Other Data Related to NMR Spectroscopy 9-92
Proton NMR Chemical Shifts for Characteristic Organic Structures 9-95
^{13}C NMR Absorptions of Major Functional Groups 9-96

SECTION 10: ATOMIC, MOLECULAR, AND OPTICAL PHYSICS

Line Spectra of the Elements 10-1
NIST Atomic Transition Probability Tables 10-88
Electron Affinities 10-147
Atomic and Molecular Polarizabilities 10-160
Ionization Potentials of Atoms and Atomic Ions 10-175
Ionization Energies of Gas-Phase Molecules 10-178
X-Ray Atomic Energy Levels 10-196
Electron Binding Energies of the Elements 10-200
Natural Width of X-Ray Lines 10-206
Photon Attenuation Coefficients 10-207
Classification of Electromagnetic Radiation 10-212
Black Body Radiation 10-214
Characteristics of Infrared Detectors 10-216
Refractive Index and Transmittance of Representative Glasses 10-217
Index of Refraction of Water 10-218
Index of Refraction of Liquids for Calibration Purposes 10-219
Index of Refraction of Air 10-220
Characteristics of Laser Sources 10-221
Infrared Laser Frequencies 10-228
Infrared and Far-Infrared Absorption Frequency Standards 10-236

SECTION 11: NUCLEAR AND PARTICLE PHYSICS

Summary Tables of Particle Properties 11-1
Table of the Isotopes 11-50
Neutron Scattering and Absorption Properties 11-159
Cosmic Radiation 11-176

SECTION 12: PROPERTIES OF SOLIDS

Techniques for Materials Characterization 12-1
Symmetry of Crystals 12-7
Ionic Radii in Crystals 12-14
Polarizability of Atoms and Ions in Solids 12-17
Crystal Structures and Lattice Parameters of Allotropes of the Elements 12-19
Lattice Energies 12-22
The Madelung Constant and Crystal Lattice Energy 12-36
Elastic Constants of Single Crystals 12-37
Electrical Resistivity of Pure Metals 12-45
Electrical Resistivity of Selected Alloys 12-48
Permittivity (Dielectric Constant) of Inorganic Solids 12-52

Curie Temperature of Selected Ferroelectric Crystals 12-61
Properties of Antiferroelectric Crystals 12-62
Dielectric Constants of Glasses 12-63
Properties of Superconductors 12-64
High Temperature Superconductors 12-91
Organic Superconductors 12-94
Properties of Semiconductors 12-97
Diffusion Data for Semiconductors 12-108
Properties of Magnetic Materials 12-117
Organic Magnets 12-126
Electron Work Function of the Elements 12-130
Secondary Electron Emission 12-131
Optical Properties of Metals and Semiconductors 12-133
Elasto-optic, Electro-optic, and Magneto-optic Constants 12-157
Nonlinear Optical Constants 12-174
Phase Diagrams 12-178
Heat Capacity of Selected Solids 12-196
Thermal and Physical Properties of Pure Metals 12-197
Thermal Conductivity of Metals and Semiconductors as a Function of Temperature 12-199
Thermal Conductivity of Alloys as a Function of Temperature 12-201
Thermal Conductivity of Crystalline Dielectrics 12-202
Thermal Conductivity of Ceramics and Other Insulating Materials 12-204
Thermal Conductivity of Glasses 12-206
Commercial Metals and Alloys 12-210
Hardness of Minerals and Ceramics 12-211

SECTION 13: POLYMER PROPERTIES

Naming Organic Polymers 13-1
Solvents for Common Polymers 13-3
Glass Transition Temperature for Selected Polymers 13-4
Dielectric Constant of Selected Polymers 13-12
Pressure-Volume-Temperature Relationship for Polymer Melts 13-13

SECTION 14: GEOPHYSICS, ASTRONOMY, AND ACOUSTICS

Astronomical Constants 14-1
Properties of the Solar System 14-2
Satellites of the Planets 14-4
Mass, Dimensions, and other Parameters of the Earth 14-6
Geological Time Scale 14-8
Acceleration Due to Gravity 14-9
Density, Pressure, and Gravity as a Function of Depth within the Earth 14-10
Ocean Pressure as a Function of Depth and Latitude 14-11
Properties of Seawater 14-12
Abundance of Elements in the Earth's Crust and in the Sea 14-14
Solar Spectral Irradiance 14-15
U.S. Standard Atmosphere (1976) 14-16
Geographical and Seasonal Variation in Solar Radiation 14-23
Infrared Absorption by the Earth's Atmosphere 14-24
Atmospheric Concentration of Carbon Dioxide, 1958-1996 14-25
Mean Temperatures in the United States, 1900-1992 14-27
Global Temperature Trend, 1866-1996 14-29
Atmospheric Electricity 14-30
Speed of Sound in Various Media 14-38
Attenuation and Speed of Sound in Air as a Function of Humidity and Frequency 14-40

Speed of Sound in Dry Air 14-41
Musical Scales 14-42
Characteristics of Human Hearing 14-44

SECTION 15: PRACTICAL LABORATORY DATA
Standard ITS-90 Thermocouple Tables 15-1
Properties of Common Laboratory Solvents 15-14
Dependence of Boiling Point on Pressure 15-19
Ebullioscopic Constants for Calculation of Boiling Point Elevation 15-20
Cryogenic Constants for Calculation of Freezing Point Depression 15-21
Freezing Point Lowering by Electrolytes in Aqueous Solution 15-22
Determination of Relative Humidity from Dew Point 15-23
Determination of Relative Humidity from Wet and Dry Bulb Temperatures 15-24
Constant Humidity Solutions 15-25
Standard Salt Solutions for Humidity Calibration 15-26
Low Temperature Baths for Maintaining Constant Temperature 15-26
Wire Tables 15-27
Characteristics of Particles and Particle Dispersoids 15-28
Density of Various Solids 15-29
Dielectric Strength of Insulating Materials 15-30

SECTION 16: HEALTH AND SAFETY INFORMATION
Handling and Disposal of Chemicals in Laboratories 16-1
Flammability of Chemical Substances 16-16
Threshold Limits for Airborne Contaminants 16-32
Octanol-Water Partition Coefficients 16-43
Protection Against Ionizing Radiation 16-48
Annual Limits on Intakes of Radionuclides 16-49
Chemical Carcinogens 16-53

APPENDIX A: MATHEMATICAL TABLES
Miscellaneous Mathematical Constants A-1
Exponential and Hyperbolic Functions and their Common Logarithms A-2
Natural Trigonomic Functions to Four Places A-7
Relation of Angular Functions in Terms of One Another A-11
Derivatives A-12
Integration A-15
Integrals A-20
Differential Equations A-65
Fourier Series A-75
The Fourier Transforms A-80
Series Expansion A-84
Vector Analysis A-87
Moment of Inertial for Various Bodies of Mass A-96
Bessel Functions A-96
The Gamma Function A-99
The Beta Function A-101
The Error Function A-101
Orthogonal Polynomials A-102
Normal Probability Function A-104
Percentage Points, Student's t-Distribution A-105
Percentage Points, Chi-Square Distribution A-105
Percentage Points, F-Distribution A-106

INDEX I-1

Section 1
Basic Constants, Units, and Conversion Factors

Fundamental Physical Constants 1-1
Standard Atomic Weights (1995) 1-12
Atomic Masses and Abundances 1-15
Electron Configuration of Neutral Atoms in the Ground State 1-18
International Temperature Scale of 1990 (ITS-90) 1-20
Conversion of Temperatures from the 1948 and 1968 Scales to ITS-90 1-22
International System of Units (SI) 1-24
Conversion Factors 1-29
Conversion of Temperatures 1-41
Conversion Factors for Energy Units 1-42
Conversion Factors for Pressure Units 1-43
Conversion Factors for Thermal Conductivity Units 1-44
Conversion Factors for Electrical Resistivity Units 1-45
Conversion Factors for Chemical Kinetics 1-46
Conversion Factors for Ionizing Radiation 1-47
Values of the Gas Constant in Different Unit Systems 1-49

FUNDAMENTAL PHYSICAL CONSTANTS

Peter J. Mohr and Barry N. Taylor

These tables give the 1998 self-consistent set of values of the basic constants and conversion factors of physics and chemistry recommended by the Committee on Data for Science and Technology (CODATA) for international use. The 1998 set replaces the previous set of constants recommended by CODATA in 1986; assigned uncertainties have been reduced by a factor of 1/5 to 1/12 (and sometimes even greater) relative to the 1986 uncertainties. The recommended set is based on a least-squares adjustment involving all of the relevant experimental and theoretical data available through December 31, 1998. Full details of the input data and the adjustment procedure are given in Reference 1.

The 1998 adjustment was carried out by P. J. Mohr and B. N. Taylor of the National Institute of Standards and Technology (NIST) under the auspices of the CODATA Task Group on Fundamental Constants. The Task Group was established in 1969 with the aim of periodically providing the scientific and technological communities with a self-consistent set of internationally recommended values of the fundamental physical constants based on all applicable information available at a given point in time. The first set was published in 1973 and was followed by a revised set first published in 1986; the current 1998 set first appeared in 1999. In the future, the CODATA Task Group plans to take advantage of the high level of automation developed for the current set in order to issue a new set of recommended values at least every four years.

At the time of completion of the 1998 adjustment, the membership of the Task Group was as follows:

F. Cabiati, Istituto Elettrotecnico Nazionale "Galileo Ferraris," Italy
E. R. Cohen, Science Center, Rockwell International (retired), United States of America
T. Endo, Electrotechnical Laboratory, Japan
R. Liu, National Institute of Metrology, China (People's Republic of)
B. A. Mamyrin, A. F. Ioffe Physical-Technical Institute, Russian Federation
P. J. Mohr, National Institute of Standards and Technology, United States of America
F. Nez, Laboratoire Kastler-Brossel, France
B. W. Petley, National Physical Laboratory, United Kingdom
T. J. Quinn, Bureau International des Poids et Mesures
B. N. Taylor, National Institute of Standards and Technology, United States of America
V. S. Tuninsky, D. I. Mendeleyev All-Russian Research Institute for Metrology, Russian Federation
W. Wöger, Physikalisch-Technische Bundesanstalt, Germany
B. M. Wood, National Research Council, Canada

REFERENCE

1. Mohr, Peter J., and Taylor, Barry N., *J. Phys Chem. Ref. Data* **28**, No. 6, 1999; *Rev. Mod. Phys.* **72**, No. 2, 2000. The 1998 set of recommended values is also available at the Web site of the Fundamental Constants Data Center of the NIST Physics Laboratory: http://physics.nist.gov/constants.

Fundamental Physical Constants

Quantity	Symbol	Value	Unit	Relative std. uncert. u_r
UNIVERSAL				
speed of light in vacuum	c, c_0	299 792 458	m s^{-1}	(exact)
magnetic constant	μ_0	$4\pi \times 10^{-7}$	N A^{-2}	
		$= 12.566\,370\,614... \times 10^{-7}$	N A^{-2}	(exact)
electric constant $1/\mu_0 c^2$	ε_0	$8.854\,187\,817... \times 10^{-12}$	F m^{-1}	(exact)
characteristic impedance of vacuum $\sqrt{\mu_0/\epsilon_0} = \mu_0 c$	Z_0	$376.730\,313\,461...$	Ω	(exact)
Newtonian constant of gravitation	G	$6.673(10) \times 10^{-11}$	m^3 kg^{-1} s^{-2}	1.5×10^{-3}
	$G/\hbar c$	$6.707(10) \times 10^{-39}$	(GeV/c^2)$^{-2}$	1.5×10^{-3}
Planck constant	h	$6.626\,068\,76(52) \times 10^{-34}$	J s	7.8×10^{-8}
in eV s		$4.135\,667\,27(16) \times 10^{-15}$	eV s	3.9×10^{-8}
$h/2\pi$	$\hbar$	$1.054\,571\,596(82) \times 10^{-34}$	J s	7.8×10^{-8}
in eV s		$6.582\,118\,89(26) \times 10^{-16}$	eV s	3.9×10^{-8}
Planck mass $(\hbar c/G)^{1/2}$	m_P	$2.1767(16) \times 10^{-8}$	kg	7.5×10^{-4}
Planck length $\hbar/m_P c = (\hbar G/c^3)^{1/2}$	l_P	$1.6160(12) \times 10^{-35}$	m	7.5×10^{-4}
Planck time $l_P/c = (\hbar G/c^5)^{1/2}$	t_P	$5.3906(40) \times 10^{-44}$	s	7.5×10^{-4}
ELECTROMAGNETIC				
elementary charge	e	$1.602\,176\,462(63) \times 10^{-19}$	C	3.9×10^{-8}
	e/h	$2.417\,989\,491(95) \times 10^{14}$	A J^{-1}	3.9×10^{-8}
magnetic flux quantum $h/2e$	Φ_0	$2.067\,833\,636(81) \times 10^{-15}$	Wb	3.9×10^{-8}
conductance quantum $2e^2/h$	G_0	$7.748\,091\,696(28) \times 10^{-5}$	S	3.7×10^{-9}
inverse of conductance quantum	G_0^{-1}	$12\,906.403\,786(47)$	Ω	3.7×10^{-9}
Josephson constant[a] $2e/h$	K_J	$483\,597.898(19) \times 10^9$	Hz V^{-1}	3.9×10^{-8}
von Klitzing constant[b] $h/e^2 = \mu_0 c/2\alpha$	R_K	$25\,812.807\,572(95)$	Ω	3.7×10^{-9}
Bohr magneton $e\hbar/2m_e$	μ_B	$927.400\,899(37) \times 10^{-26}$	J T^{-1}	4.0×10^{-8}
in eV T^{-1}		$5.788\,381\,749(43) \times 10^{-5}$	eV T^{-1}	7.3×10^{-9}
	μ_B/h	$13.996\,246\,24(56) \times 10^9$	Hz T^{-1}	4.0×10^{-8}
	μ_B/hc	$46.686\,4521(19)$	m^{-1} T^{-1}	4.0×10^{-8}
	μ_B/k	$0.671\,7131(12)$	K T^{-1}	1.7×10^{-6}
nuclear magneton $e\hbar/2m_p$	μ_N	$5.050\,783\,17(20) \times 10^{-27}$	J T^{-1}	4.0×10^{-8}
in eV T^{-1}		$3.152\,451\,238(24) \times 10^{-8}$	eV T^{-1}	7.6×10^{-9}
	μ_N/h	$7.622\,593\,96(31)$	MHz T^{-1}	4.0×10^{-8}
	μ_N/hc	$2.542\,623\,66(10) \times 10^{-2}$	m^{-1} T^{-1}	4.0×10^{-8}
	μ_N/k	$3.658\,2638(64) \times 10^{-4}$	K T^{-1}	1.7×10^{-6}
ATOMIC AND NUCLEAR				
General				
fine-structure constant $e^2/4\pi\epsilon_0\hbar c$	α	$7.297\,352\,533(27) \times 10^{-3}$		3.7×10^{-9}
inverse fine-structure constant	α^{-1}	$137.035\,999\,76(50)$		3.7×10^{-9}

Fundamental Physical Constants

Quantity	Symbol	Value	Unit	Relative std. uncert. u_r
Rydberg constant $\alpha^2 m_e c/2h$	R_∞	10 973 731.568 549(83)	m^{-1}	7.6×10^{-12}
	$R_\infty c$	$3.289\,841\,960\,368(25) \times 10^{15}$	Hz	7.6×10^{-12}
	$R_\infty hc$	$2.179\,871\,90(17) \times 10^{-18}$	J	7.8×10^{-8}
$R_\infty hc$ in eV		13.605 691 72(53)	eV	3.9×10^{-8}
Bohr radius $\alpha/4\pi R_\infty = 4\pi\epsilon_0\hbar^2/m_e e^2$	a_0	$0.529\,177\,2083(19) \times 10^{-10}$	m	3.7×10^{-9}
Hartree energy $e^2/4\pi\varepsilon_0 a_0 = 2R_\infty hc$				
$= \alpha^2 m_e c^2$	E_h	$4.359\,743\,81(34) \times 10^{-18}$	J	7.8×10^{-8}
in eV		27.211 3834(11)	eV	3.9×10^{-8}
quantum of circulation	$h/2m_e$	$3.636\,947\,516(27) \times 10^{-4}$	$m^2\,s^{-1}$	7.3×10^{-9}
	h/m_e	$7.273\,895\,032(53) \times 10^{-4}$	$m^2\,s^{-1}$	7.3×10^{-9}

Electroweak

Quantity	Symbol	Value	Unit	Relative std. uncert. u_r
Fermi coupling constant[c]	$G_F/(\hbar c)^3$	$1.166\,39(1) \times 10^{-5}$	GeV^{-2}	8.6×10^{-6}
weak mixing angle[d] θ_W (on-shell scheme)				
$\sin^2\theta_W = s_W^2 \equiv 1 - (m_W/m_Z)^2$	$\sin^2\theta_W$	0.2224(19)		8.7×10^{-3}

Electron, e^-

Quantity	Symbol	Value	Unit	Relative std. uncert. u_r
electron mass	m_e	$9.109\,381\,88(72) \times 10^{-31}$	kg	7.9×10^{-8}
in u, $m_e = A_r(e)$ u (electron relative atomic mass times u)		$5.485\,799\,110(12) \times 10^{-4}$	u	2.1×10^{-9}
energy equivalent	$m_e c^2$	$8.187\,104\,14(64) \times 10^{-14}$	J	7.9×10^{-8}
in MeV		0.510 998 902(21)	MeV	4.0×10^{-8}
electron-muon mass ratio	m_e/m_μ	$4.836\,332\,10(15) \times 10^{-3}$		3.0×10^{-8}
electron-tau mass ratio	m_e/m_τ	$2.875\,55(47) \times 10^{-4}$		1.6×10^{-4}
electron-proton mass ratio	m_e/m_p	$5.446\,170\,232(12) \times 10^{-4}$		2.1×10^{-9}
electron-neutron mass ratio	m_e/m_n	$5.438\,673\,462(12) \times 10^{-4}$		2.2×10^{-9}
electron-deuteron mass ratio	m_e/m_d	$2.724\,437\,1170(58) \times 10^{-4}$		2.1×10^{-9}
electron to alpha particle mass ratio	m_e/m_α	$1.370\,933\,5611(29) \times 10^{-4}$		2.1×10^{-9}
electron charge to mass quotient	$-e/m_e$	$-1.758\,820\,174(71) \times 10^{11}$	$C\,kg^{-1}$	4.0×10^{-8}
electron molar mass $N_A m_e$	$M(e), M_e$	$5.485\,799\,110(12) \times 10^{-7}$	$kg\,mol^{-1}$	2.1×10^{-9}
Compton wavelength $h/m_e c$	λ_C	$2.426\,310\,215(18) \times 10^{-12}$	m	7.3×10^{-9}
$\lambda_C/2\pi = \alpha a_0 = \alpha^2/4\pi R_\infty$	$\lambdabar_C$	$386.159\,2642(28) \times 10^{-15}$	m	7.3×10^{-9}
classical electron radius $\alpha^2 a_0$	r_e	$2.817\,940\,285(31) \times 10^{-15}$	m	1.1×10^{-8}
Thomson cross section $(8\pi/3)r_e^2$	σ_e	$0.665\,245\,854(15) \times 10^{-28}$	m^2	2.2×10^{-8}
electron magnetic moment	μ_e	$-928.476\,362(37) \times 10^{-26}$	$J\,T^{-1}$	4.0×10^{-8}
to Bohr magneton ratio	μ_e/μ_B	$-1.001\,159\,652\,1869(41)$		4.1×10^{-12}
to nuclear magneton ratio	μ_e/μ_N	$-1\,838.281\,9660(39)$		2.1×10^{-9}
electron magnetic moment anomaly $|\mu_e|/\mu_B - 1$	a_e	$1.159\,652\,1869(41) \times 10^{-3}$		3.5×10^{-9}
electron g-factor $-2(1 + a_e)$	g_e	$-2.002\,319\,304\,3737(82)$		4.1×10^{-12}
electron-muon magnetic moment ratio	μ_e/μ_μ	206.766 9720(63)		3.0×10^{-8}

Fundamental Physical Constants

Quantity	Symbol	Value	Unit	Relative std. uncert. u_r
electron-proton magnetic moment ratio	μ_e/μ_p	$-658.210\,6875(66)$		1.0×10^{-8}
electron to shielded proton magnetic moment ratio (H$_2$O, sphere, 25 °C)	μ_e/μ_p'	$-658.227\,5954(71)$		1.1×10^{-8}
electron-neutron magnetic moment ratio	μ_e/μ_n	$960.920\,50(23)$		2.4×10^{-7}
electron-deuteron magnetic moment ratio	μ_e/μ_d	$-2\,143.923\,498(23)$		1.1×10^{-8}
electron to shielded helion[e] magnetic moment ratio (gas, sphere, 25 °C)	μ_e/μ_h'	$864.058\,255(10)$		1.2×10^{-8}
electron gyromagnetic ratio $2|\mu_e|/\hbar$	γ_e	$1.760\,859\,794(71) \times 10^{11}$	s^{-1} T^{-1}	4.0×10^{-8}
	$\gamma_e/2\pi$	$28\,024.9540(11)$	MHz T^{-1}	4.0×10^{-8}

Muon, μ^-

Quantity	Symbol	Value	Unit	Relative std. uncert. u_r
muon mass	m_μ	$1.883\,531\,09(16) \times 10^{-28}$	kg	8.4×10^{-8}
in u, $m_\mu = A_r(\mu)$ u (muon relative atomic mass times u)		$0.113\,428\,9168(34)$	u	3.0×10^{-8}
energy equivalent	$m_\mu c^2$	$1.692\,833\,32(14) \times 10^{-11}$	J	8.4×10^{-8}
in MeV		$105.658\,3568(52)$	MeV	4.9×10^{-8}
muon-electron mass ratio	m_μ/m_e	$206.768\,2657(63)$		3.0×10^{-8}
muon-tau mass ratio	m_μ/m_τ	$5.945\,72(97) \times 10^{-2}$		1.6×10^{-4}
muon-proton mass ratio	m_μ/m_p	$0.112\,609\,5173(34)$		3.0×10^{-8}
muon-neutron mass ratio	m_μ/m_n	$0.112\,454\,5079(34)$		3.0×10^{-8}
muon molar mass $N_A m_\mu$	$M(\mu), M_\mu$	$0.113\,428\,9168(34) \times 10^{-3}$	kg mol^{-1}	3.0×10^{-8}
muon Compton wavelength $h/m_\mu c$	$\lambda_{C,\mu}$	$11.734\,441\,97(35) \times 10^{-15}$	m	2.9×10^{-8}
$\lambda_{C,\mu}/2\pi$	$\lambdabar_{C,\mu}$	$1.867\,594\,444(55) \times 10^{-15}$	m	2.9×10^{-8}
muon magnetic moment	μ_μ	$-4.490\,448\,13(22) \times 10^{-26}$	J T^{-1}	4.9×10^{-8}
to Bohr magneton ratio	μ_μ/μ_B	$-4.841\,970\,85(15) \times 10^{-3}$		3.0×10^{-8}
to nuclear magneton ratio	μ_μ/μ_N	$-8.890\,597\,70(27)$		3.0×10^{-8}
muon magnetic moment anomaly $|\mu_\mu|/(e\hbar/2m_\mu) - 1$	a_μ	$1.165\,916\,02(64) \times 10^{-3}$		5.5×10^{-7}
muon g-factor $-2(1 + a_\mu)$	g_μ	$-2.002\,331\,8320(13)$		6.4×10^{-10}
muon-proton magnetic moment ratio	μ_μ/μ_p	$-3.183\,345\,39(10)$		3.2×10^{-8}

Tau, τ^-

Quantity	Symbol	Value	Unit	Relative std. uncert. u_r
tau mass[f]	m_τ	$3.167\,88(52) \times 10^{-27}$	kg	1.6×10^{-4}
in u, $m_\tau = A_r(\tau)$ u (tau relative atomic mass times u)		$1.907\,74(31)$	u	1.6×10^{-4}
energy equivalent	$m_\tau c^2$	$2.847\,15(46) \times 10^{-10}$	J	1.6×10^{-4}
in MeV		$1\,777.05(29)$	MeV	1.6×10^{-4}

Fundamental Physical Constants

Quantity	Symbol	Value	Unit	Relative std. uncert. u_r
tau-electron mass ratio	m_τ/m_e	3477.60(57)		1.6×10^{-4}
tau-muon mass ratio	m_τ/m_μ	16.8188(27)		1.6×10^{-4}
tau-proton mass ratio	m_τ/m_p	1.89396(31)		1.6×10^{-4}
tau-neutron mass ratio	m_τ/m_n	1.89135(31)		1.6×10^{-4}
tau molar mass $N_A m_\tau$	$M(\tau), M_\tau$	$1.90774(31) \times 10^{-3}$	kg mol^{-1}	1.6×10^{-4}
tau Compton wavelength $h/m_\tau c$	$\lambda_{C,\tau}$	$0.69770(11) \times 10^{-15}$	m	1.6×10^{-4}
$\lambda_{C,\tau}/2\pi$	$\lambdabar_{C,\tau}$	$0.111042(18) \times 10^{-15}$	m	1.6×10^{-4}
		Proton, p		
proton mass	m_p	$1.67262158(13) \times 10^{-27}$	kg	7.9×10^{-8}
in u, $m_p = A_r(p)$ u (proton relative atomic mass times u)		1.00727646688(13)	u	1.3×10^{-10}
energy equivalent	$m_p c^2$	$1.50327731(12) \times 10^{-10}$	J	7.9×10^{-8}
in MeV		938.271998(38)	MeV	4.0×10^{-8}
proton-electron mass ratio	m_p/m_e	1836.1526675(39)		2.1×10^{-9}
proton-muon mass ratio	m_p/m_μ	8.88024408(27)		3.0×10^{-8}
proton-tau mass ratio	m_p/m_τ	0.527994(86)		1.6×10^{-4}
proton-neutron mass ratio	m_p/m_n	0.99862347855(58)		5.8×10^{-10}
proton charge to mass quotient	e/m_p	$9.57883408(38) \times 10^7$	C kg^{-1}	4.0×10^{-8}
proton molar mass $N_A m_p$	$M(p), M_p$	$1.00727646688(13) \times 10^{-3}$	kg mol^{-1}	1.3×10^{-10}
proton Compton wavelength $h/m_p c$	$\lambda_{C,p}$	$1.321409847(10) \times 10^{-15}$	m	7.6×10^{-9}
$\lambda_{C,p}/2\pi$	$\lambdabar_{C,p}$	$0.2103089089(16) \times 10^{-15}$	m	7.6×10^{-9}
proton magnetic moment	μ_p	$1.410606633(58) \times 10^{-26}$	J T^{-1}	4.1×10^{-8}
to Bohr magneton ratio	μ_p/μ_B	$1.521032203(15) \times 10^{-3}$		1.0×10^{-8}
to nuclear magneton ratio	μ_p/μ_N	2.792847337(29)		1.0×10^{-8}
proton g-factor $2\mu_p/\mu_N$	g_p	5.585694675(57)		1.0×10^{-8}
proton-neutron magnetic moment ratio	μ_p/μ_n	$-1.45989805(34)$		2.4×10^{-7}
shielded proton magnetic moment (H$_2$O, sphere, 25 °C)	μ_p'	$1.410570399(59) \times 10^{-26}$	J T^{-1}	4.2×10^{-8}
to Bohr magneton ratio	μ_p'/μ_B	$1.520993132(16) \times 10^{-3}$		1.1×10^{-8}
to nuclear magneton ratio	μ_p'/μ_N	2.792775597(31)		1.1×10^{-8}
proton magnetic shielding correction $1 - \mu_p'/\mu_p$ (H$_2$O, sphere, 25 °C)	σ_p'	$25.687(15) \times 10^{-6}$		5.7×10^{-4}
proton gyromagnetic ratio $2\mu_p/\hbar$	γ_p	$2.67522212(11) \times 10^8$	s^{-1} T^{-1}	4.1×10^{-8}
	$\gamma_p/2\pi$	42.5774825(18)	MHz T^{-1}	4.1×10^{-8}
shielded proton gyromagnetic ratio $2\mu_p'/\hbar$ (H$_2$O, sphere, 25 °C)	γ_p'	$2.67515341(11) \times 10^8$	s^{-1} T^{-1}	4.2×10^{-8}
	$\gamma_p'/2\pi$	42.5763888(18)	MHz T^{-1}	4.2×10^{-8}
		Neutron, n		

Fundamental Physical Constants

Quantity	Symbol	Value	Unit	Relative std. uncert. u_r
neutron mass	m_n	$1.674\,927\,16(13) \times 10^{-27}$	kg	7.9×10^{-8}
in u, $m_n = A_r(n)$ u (neutron relative atomic mass times u)		$1.008\,664\,915\,78(55)$	u	5.4×10^{-10}
energy equivalent	$m_n c^2$	$1.505\,349\,46(12) \times 10^{-10}$	J	7.9×10^{-8}
in MeV		$939.565\,330(38)$	MeV	4.0×10^{-8}
neutron-electron mass ratio	m_n/m_e	$1\,838.683\,6550(40)$		2.2×10^{-9}
neutron-muon mass ratio	m_n/m_μ	$8.892\,484\,78(27)$		3.0×10^{-8}
neutron-tau mass ratio	m_n/m_τ	$0.528\,722(86)$		1.6×10^{-4}
neutron-proton mass ratio	m_n/m_p	$1.001\,378\,418\,87(58)$		5.8×10^{-10}
neutron molar mass $N_A m_n$	$M(n), M_n$	$1.008\,664\,915\,78(55) \times 10^{-3}$	kg mol^{-1}	5.4×10^{-10}
neutron Compton wavelength $h/m_n c$	$\lambda_{C,n}$	$1.319\,590\,898(10) \times 10^{-15}$	m	7.6×10^{-9}
$\lambda_{C,n}/2\pi$	$\bar\lambda_{C,n}$	$0.210\,019\,4142(16) \times 10^{-15}$	m	7.6×10^{-9}
neutron magnetic moment	μ_n	$-0.966\,236\,40(23) \times 10^{-26}$	J T^{-1}	2.4×10^{-7}
to Bohr magneton ratio	μ_n/μ_B	$-1.041\,875\,63(25) \times 10^{-3}$		2.4×10^{-7}
to nuclear magneton ratio	μ_n/μ_N	$-1.913\,042\,72(45)$		2.4×10^{-7}
neutron g-factor $2\mu_n/\mu_N$	g_n	$-3.826\,085\,45(90)$		2.4×10^{-7}
neutron-electron magnetic moment ratio	μ_n/μ_e	$1.040\,668\,82(25) \times 10^{-3}$		2.4×10^{-7}
neutron-proton magnetic moment ratio	μ_n/μ_p	$-0.684\,979\,34(16)$		2.4×10^{-7}
neutron to shielded proton magnetic moment ratio (H_2O, sphere, 25 °C)	μ_n/μ_p'	$-0.684\,996\,94(16)$		2.4×10^{-7}
neutron gyromagnetic ratio $2\lvert\mu_n\rvert/\hbar$	γ_n	$1.832\,471\,88(44) \times 10^8$	s^{-1} T^{-1}	2.4×10^{-7}
	$\gamma_n/2\pi$	$29.164\,6958(70)$	MHz T^{-1}	2.4×10^{-7}
Deuteron, d				
deuteron mass	m_d	$3.343\,583\,09(26) \times 10^{-27}$	kg	7.9×10^{-8}
in u, $m_d = A_r(d)$ u (deuteron relative atomic mass times u)		$2.013\,553\,212\,71(35)$	u	1.7×10^{-10}
energy equivalent	$m_d c^2$	$3.005\,062\,62(24) \times 10^{-10}$	J	7.9×10^{-8}
in MeV		$1\,875.612\,762(75)$	MeV	4.0×10^{-8}
deuteron-electron mass ratio	m_d/m_e	$3\,670.482\,9550(78)$		2.1×10^{-9}
deuteron-proton mass ratio	m_d/m_p	$1.999\,007\,500\,83(41)$		2.0×10^{-10}
deuteron molar mass $N_A m_d$	$M(d), M_d$	$2.013\,553\,212\,71(35) \times 10^{-3}$	kg mol^{-1}	1.7×10^{-10}
deuteron magnetic moment	μ_d	$0.433\,073\,457(18) \times 10^{-26}$	J T^{-1}	4.2×10^{-8}
to Bohr magneton ratio	μ_d/μ_B	$0.466\,975\,4556(50) \times 10^{-3}$		1.1×10^{-8}
to nuclear magneton ratio	μ_d/μ_N	$0.857\,438\,2284(94)$		1.1×10^{-8}
deuteron-electron magnetic moment ratio	μ_d/μ_e	$-4.664\,345\,537(50) \times 10^{-4}$		1.1×10^{-8}
deuteron-proton magnetic moment ratio	μ_d/μ_p	$0.307\,012\,2083(45)$		1.5×10^{-8}

Fundamental Physical Constants

Quantity	Symbol	Value	Unit	Relative std. uncert. u_r
deuteron-neutron magnetic moment ratio	μ_d/μ_n	$-0.448\,206\,52(11)$		2.4×10^{-7}

Helion, h

Quantity	Symbol	Value	Unit	Relative std. uncert. u_r
helion mass[e]	m_h	$5.006\,411\,74(39) \times 10^{-27}$	kg	7.9×10^{-8}
in u, $m_h = A_r(h)$ u (helion relative atomic mass times u)		$3.014\,932\,234\,69(86)$	u	2.8×10^{-10}
energy equivalent	$m_h c^2$	$4.499\,538\,48(35) \times 10^{-10}$	J	7.9×10^{-8}
in MeV		$2\,808.391\,32(11)$	MeV	4.0×10^{-8}
helion-electron mass ratio	m_h/m_e	$5495.885\,238(12)$		2.1×10^{-9}
helion-proton mass ratio	m_h/m_p	$2.993\,152\,658\,50(93)$		3.1×10^{-10}
helion molar mass $N_A m_h$	$M(h), M_h$	$3.014\,932\,234\,69(86) \times 10^{-3}$	kg mol^{-1}	2.8×10^{-10}
shielded helion magnetic moment (gas, sphere, 25 °C)	μ'_h	$-1.074\,552\,967(45) \times 10^{-26}$	J T^{-1}	4.2×10^{-8}
to Bohr magneton ratio	μ'_h/μ_B	$-1.158\,671\,474(14) \times 10^{-3}$		1.2×10^{-8}
to nuclear magneton ratio	μ'_h/μ_N	$-2.127\,497\,718(25)$		1.2×10^{-8}
shielded helion to proton magnetic moment ratio (gas, sphere, 25 °C)	μ'_h/μ_p	$-0.761\,766\,563(12)$		1.5×10^{-8}
shielded helion to shielded proton magnetic moment ratio (gas/H$_2$O, spheres, 25 °C)	μ'_h/μ'_p	$-0.761\,786\,1313(33)$		4.3×10^{-9}
shielded helion gyromagnetic ratio $2|\mu'_h|/\hbar$ (gas, sphere, 25 °C)	γ'_h	$2.037\,894\,764(85) \times 10^8$	s^{-1} T^{-1}	4.2×10^{-8}
	$\gamma'_h/2\pi$	$32.434\,1025(14)$	MHz T^{-1}	4.2×10^{-8}

Alpha particle, α

Quantity	Symbol	Value	Unit	Relative std. uncert. u_r
alpha particle mass	m_α	$6.644\,655\,98(52) \times 10^{-27}$	kg	7.9×10^{-8}
in u, $m_\alpha = A_r(\alpha)$ u (alpha particle relative atomic mass times u)		$4.001\,506\,1747(10)$	u	2.5×10^{-10}
energy equivalent	$m_\alpha c^2$	$5.971\,918\,97(47) \times 10^{-10}$	J	7.9×10^{-8}
in MeV		$3\,727.379\,04(15)$	MeV	4.0×10^{-8}
alpha particle to electron mass ratio	m_α/m_e	$7\,294.299\,508(16)$		2.1×10^{-9}
alpha particle to proton mass ratio	m_α/m_p	$3.972\,599\,6846(11)$		2.8×10^{-10}
alpha particle molar mass $N_A m_\alpha$	$M(\alpha), M_\alpha$	$4.001\,506\,1747(10) \times 10^{-3}$	kg mol^{-1}	2.5×10^{-10}

PHYSICO-CHEMICAL

Quantity	Symbol	Value	Unit	Relative std. uncert. u_r
Avogadro constant	N_A, L	$6.022\,141\,99(47) \times 10^{23}$	mol^{-1}	7.9×10^{-8}
atomic mass constant $m_u = \frac{1}{12}m(^{12}C) = 1$ u $= 10^{-3}$ kg mol$^{-1}/N_A$	m_u	$1.660\,538\,73(13) \times 10^{-27}$	kg	7.9×10^{-8}
energy equivalent	$m_u c^2$	$1.492\,417\,78(12) \times 10^{-10}$	J	7.9×10^{-8}
in MeV		$931.494\,013(37)$	MeV	4.0×10^{-8}
Faraday constant[g] $N_A e$	F	$96\,485.3415(39)$	C mol^{-1}	4.0×10^{-8}

Fundamental Physical Constants

Quantity	Symbol	Value	Unit	Relative std. uncert. u_r
molar Planck constant	$N_A h$	$3.990\,312\,689(30) \times 10^{-10}$	J s mol^{-1}	7.6×10^{-9}
	$N_A hc$	$0.119\,626\,564\,92(91)$	J m mol^{-1}	7.6×10^{-9}
molar gas constant	R	$8.314\,472(15)$	J mol^{-1} K^{-1}	1.7×10^{-6}
Boltzmann constant R/N_A	k	$1.380\,6503(24) \times 10^{-23}$	J K^{-1}	1.7×10^{-6}
in eV K^{-1}		$8.617\,342(15) \times 10^{-5}$	eV K^{-1}	1.7×10^{-6}
	k/h	$2.083\,6644(36) \times 10^{10}$	Hz K^{-1}	1.7×10^{-6}
	k/hc	$69.503\,56(12)$	m^{-1} K^{-1}	1.7×10^{-6}
molar volume of ideal gas RT/p				
$T = 273.15$ K, $p = 101.325$ kPa	V_m	$22.413\,996(39) \times 10^{-3}$	m^3 mol^{-1}	1.7×10^{-6}
Loschmidt constant N_A/V_m	n_0	$2.686\,7775(47) \times 10^{25}$	m^{-3}	1.7×10^{-6}
$T = 273.15$ K, $p = 100$ kPa	V_m	$22.710\,981(40) \times 10^{-3}$	m^3 mol^{-1}	1.7×10^{-6}
Sackur-Tetrode constant (absolute entropy constant)^h $\frac{5}{2} + \ln[(2\pi m_u kT_1/h^2)^{3/2}kT_1/p_0]$				
$T_1 = 1$ K, $p_0 = 100$ kPa	S_0/R	$-1.151\,7048(44)$		3.8×10^{-6}
$T_1 = 1$ K, $p_0 = 101.325$ kPa		$-1.164\,8678(44)$		3.7×10^{-6}
Stefan-Boltzmann constant $(\pi^2/60)k^4/\hbar^3 c^2$	σ	$5.670\,400(40) \times 10^{-8}$	W m^{-2} K^{-4}	7.0×10^{-6}
first radiation constant $2\pi hc^2$	c_1	$3.741\,771\,07(29) \times 10^{-16}$	W m^2	7.8×10^{-8}
first radiation constant for spectral radiance $2hc^2$	c_{1L}	$1.191\,042\,722(93) \times 10^{-16}$	W m^2 sr^{-1}	7.8×10^{-8}
second radiation constant hc/k	c_2	$1.438\,7752(25) \times 10^{-2}$	m K	1.7×10^{-6}
Wien displacement law constant $b = \lambda_{max}T = c_2/4.965\,114\,231...$	b	$2.897\,7686(51) \times 10^{-3}$	m K	1.7×10^{-6}

[a] See the "Adopted values" table for the conventional value adopted internationally for realizing representations of the volt using the Josephson effect.

[b] See the "Adopted values" table for the conventional value adopted internationally for realizing representations of the ohm using the quantum Hall effect.

[c] Value recommended by the Particle Data Group, Caso et al., Eur. Phys. J. C 3(1-4), 1-794 (1998).

[d] Based on the ratio of the masses of the W and Z bosons m_W/m_Z recommended by the Particle Data Group (Caso et al., 1998). The value for $\sin^2\theta_W$ they recommend, which is based on a particular variant of the modified minimal subtraction ($\overline{\text{MS}}$) scheme, is $\sin^2\hat{\theta}_W(M_Z) = 0.231\,24(24)$.

[e] The helion, symbol h, is the nucleus of the ^{3}He atom.

[f] This and all other values involving m_τ are based on the value of $m_\tau c^2$ in MeV recommended by the Particle Data Group, Caso et al., Eur. Phys. J. C 3(1-4), 1-794 (1998), but with a standard uncertainty of 0.29 MeV rather than the quoted uncertainty of −0.26 MeV, +0.29 MeV.

[g] The numerical value of F to be used in coulometric chemical measurements is $96\,485.3432(76)$ [7.9×10^{-8}] when the relevant current is measured in terms of representations of the volt and ohm based on the Josephson and quantum Hall effects and the internationally adopted conventional values of the Josephson and von Klitzing constants K_{J-90} and R_{K-90} given in the "Adopted values" table.

[h] The entropy of an ideal monoatomic gas of relative atomic mass A_r is given by $S = S_0 + \frac{3}{2}R \ln A_r - R \ln(p/p_0) + \frac{5}{2}R \ln(T/\text{K})$.

Fundamental Physical Constants — Adopted values

Quantity	Symbol	Value	Unit	Relative std. uncert. u_r
molar mass of ^{12}C	$M(^{12}C)$	12×10^{-3}	kg mol^{-1}	(exact)
molar mass constant[a] $M(^{12}C)/12$	M_u	1×10^{-3}	kg mol^{-1}	(exact)
conventional value of Josephson constant[b]	K_{J-90}	$483\,597.9$	GHz V^{-1}	(exact)
conventional value of von Klitzing constant[c]	R_{K-90}	$25\,812.807$	Ω	(exact)
standard atmosphere		$101\,325$	Pa	(exact)
standard acceleration of gravity	g_n	$9.806\,65$	m s^{-2}	(exact)

[a] The relative atomic mass $A_r(X)$ of particle X with mass $m(X)$ is defined by $A_r(X) = m(X)/m_u$, where $m_u = m(^{12}C)/12 = M_u/N_A = 1$ u is the atomic mass constant, N_A is the Avogadro constant, and u is the atomic mass unit. Thus the mass of particle X in u is $m(X) = A_r(X)$ u and the molar mass of X is $M(X) = A_r(X)M_u$.

[b] This is the value adopted internationally for realizing representations of the volt using the Josephson effect.

[c] This is the value adopted internationally for realizing representations of the ohm using the quantum Hall effect.

Energy Equivalents

	J	kg	m⁻¹	Hz
1 J	$(1\,J) =$ $1\,J$	$(1\,J)/c^2 =$ $1.112\,650\,056 \times 10^{-17}\,kg$	$(1\,J)/hc =$ $5.034\,117\,62(39) \times 10^{24}\,m^{-1}$	$(1\,J)/h =$ $1.509\,190\,50(12) \times 10^{33}\,Hz$
1 kg	$(1\,kg)c^2 =$ $8.987\,551\,787 \times 10^{16}\,J$	$(1\,kg) =$ $1\,kg$	$(1\,kg)c/h =$ $4.524\,439\,29(35) \times 10^{41}\,m^{-1}$	$(1\,kg)c^2/h =$ $1.356\,392\,77(11) \times 10^{50}\,Hz$
1 m⁻¹	$(1\,m^{-1})hc =$ $1.986\,445\,44(16) \times 10^{-25}\,J$	$(1\,m^{-1})h/c =$ $2.210\,218\,63(17) \times 10^{-42}\,kg$	$(1\,m^{-1}) =$ $1\,m^{-1}$	$(1\,m^{-1})c =$ $299\,792\,458\,Hz$
1 Hz	$(1\,Hz)h =$ $6.626\,068\,76(52) \times 10^{-34}\,J$	$(1\,Hz)h/c^2 =$ $7.372\,495\,78(58) \times 10^{-51}\,kg$	$(1\,Hz)/c =$ $3.335\,640\,952 \times 10^{-9}\,m^{-1}$	$(1\,Hz) =$ $1\,Hz$
1 K	$(1\,K)k =$ $1.380\,6503(24) \times 10^{-23}\,J$	$(1\,K)k/c^2 =$ $1.536\,1807(27) \times 10^{-40}\,kg$	$(1\,K)k/hc =$ $69.503\,56(12)\,m^{-1}$	$(1\,K)k/h =$ $2.083\,6644(36) \times 10^{10}\,Hz$
1 eV	$(1\,eV) =$ $1.602\,176\,462(63) \times 10^{-19}\,J$	$(1\,eV)/c^2 =$ $1.782\,661\,731(70) \times 10^{-36}\,kg$	$(1\,eV)/hc =$ $8.065\,544\,77(32) \times 10^5\,m^{-1}$	$(1\,eV)/h =$ $2.417\,989\,491(95) \times 10^{14}\,Hz$
1 u	$(1\,u)c^2 =$ $1.492\,417\,78(12) \times 10^{-10}\,J$	$(1\,u) =$ $1.660\,538\,73(13) \times 10^{-27}\,kg$	$(1\,u)c/h =$ $7.513\,006\,658(57) \times 10^{14}\,m^{-1}$	$(1\,u)c^2/h =$ $2.252\,342\,733(17) \times 10^{23}\,Hz$
1 E_h	$(1\,E_h) =$ $4.359\,743\,81(34) \times 10^{-18}\,J$	$(1\,E_h)/c^2 =$ $4.850\,869\,19(38) \times 10^{-35}\,kg$	$(1\,E_h)/hc =$ $2.194\,746\,313\,710(17) \times 10^7\,m^{-1}$	$(1\,E_h)/h =$ $6.579\,683\,920\,735(50) \times 10^{15}\,Hz$

Derived from the relations $E = mc^2 = hc/\lambda = h\nu = kT$, and based on the 1998 CODATA adjustment of the values of the constants; $1\,eV = (e/C)\,J$, $1\,u = m_u = \frac{1}{12}m(^{12}C) = 10^{-3}\,kg\,mol^{-1}/N_A$, and $E_h = 2R_\infty hc = \alpha^2 m_e c^2$ is the Hartree energy (hartree).

Energy Equivalents

	K	eV	u	E_h
1 J	$(1\,\mathrm{J})/k =$ $7.242\,964(13) \times 10^{22}$ K	$(1\,\mathrm{J}) =$ $6.241\,509\,74(24) \times 10^{18}$ eV	$(1\,\mathrm{J})/c^2 =$ $6.700\,536\,62(53) \times 10^{9}$ u	$(1\,\mathrm{J}) =$ $2.293\,712\,76(18) \times 10^{17}\ E_\mathrm{h}$
1 kg	$(1\,\mathrm{kg})c^2/k =$ $6.509\,651(11) \times 10^{39}$ K	$(1\,\mathrm{kg})c^2 =$ $5.609\,589\,21(22) \times 10^{35}$ eV	$(1\,\mathrm{kg}) =$ $6.022\,141\,99(47) \times 10^{26}$ u	$(1\,\mathrm{kg})c^2 =$ $2.061\,486\,22(16) \times 10^{34}\ E_\mathrm{h}$
1 m⁻¹	$(1\,\mathrm{m}^{-1})hc/k =$ $1.438\,7752(25) \times 10^{-2}$ K	$(1\,\mathrm{m}^{-1})hc =$ $1.239\,841\,857(49) \times 10^{-6}$ eV	$(1\,\mathrm{m}^{-1})h/c =$ $1.331\,025\,042(10) \times 10^{-15}$ u	$(1\,\mathrm{m}^{-1})hc =$ $4.556\,335\,252\,750(35) \times 10^{-8}\ E_\mathrm{h}$
1 Hz	$(1\,\mathrm{Hz})h/k =$ $4.799\,2374(84) \times 10^{-11}$ K	$(1\,\mathrm{Hz})h =$ $4.135\,667\,27(16) \times 10^{-15}$ eV	$(1\,\mathrm{Hz})h/c^2 =$ $4.439\,821\,637(34) \times 10^{-24}$ u	$(1\,\mathrm{Hz})h =$ $1.519\,829\,846\,003(12) \times 10^{-16}\ E_\mathrm{h}$
1 K	$(1\,\mathrm{K}) =$ 1 K	$(1\,\mathrm{K})k =$ $8.617\,342(15) \times 10^{-5}$ eV	$(1\,\mathrm{K})k/c^2 =$ $9.251\,098(16) \times 10^{-14}$ u	$(1\,\mathrm{K})k =$ $3.166\,8153(55) \times 10^{-6}\ E_\mathrm{h}$
1 eV	$(1\,\mathrm{eV})/k =$ $1.160\,4506(20) \times 10^{4}$ K	$(1\,\mathrm{eV}) =$ 1 eV	$(1\,\mathrm{eV})/c^2 =$ $1.073\,544\,206(43) \times 10^{-9}$ u	$(1\,\mathrm{eV}) =$ $3.674\,932\,60(14) \times 10^{-2}\ E_\mathrm{h}$
1 u	$(1\,\mathrm{u})c^2/k =$ $1.080\,9528(19) \times 10^{13}$ K	$(1\,\mathrm{u})c^2 =$ $931.494\,013(37) \times 10^{6}$ eV	$(1\,\mathrm{u}) =$ 1 u	$(1\,\mathrm{u})c^2 =$ $3.423\,177\,709(26) \times 10^{7}\ E_\mathrm{h}$
1 E_h	$(1\,E_\mathrm{h})/k =$ $3.157\,7465(55) \times 10^{5}$ K	$(1\,E_\mathrm{h}) =$ $27.211\,3834(11)$ eV	$(1\,E_\mathrm{h})/c^2 =$ $2.921\,262\,304(22) \times 10^{-8}$ u	$(1\,E_\mathrm{h}) =$ $1\ E_\mathrm{h}$

Derived from the relations $E = mc^2 = hc/\lambda = h\nu = kT$, and based on the 1998 CODATA adjustment of the values of the constants; $1\,\mathrm{eV} = (e/\mathrm{C})$ J, $1\,\mathrm{u} = m_\mathrm{u} = \frac{1}{12}m(^{12}\mathrm{C}) = 10^{-3}$ kg mol$^{-1}/N_\mathrm{A}$, and $E_\mathrm{h} = 2R_\infty hc = \alpha^2 m_e c^2$ is the Hartree energy (hartree).

STANDARD ATOMIC WEIGHTS (1995)

This table of atomic weights is reprinted from the 1995 report of the IUPAC Commission on Atomic Weights and Isotopic Abundances. The Standard Atomic Weights apply to the elements as they exist naturally on Earth, and the uncertainties take into account the isotopic variation found in most laboratory samples. Further comments on the variability are given in the footnotes.

The number in parentheses following the atomic weight value gives the uncertainty in the last digit. An entry in brackets indicates the mass number of the longest-lived isotope of an element that has no stable isotopes and for which a Standard Atomic Weight cannot be defined because of wide variability in isotopic composition (or complete absence) in nature.

REFERENCE

IUPAC Commission on Atomic Weights and Isotopic Abundances, Atomic Weights of the Elements, 1995, *Pure Appl. Chem.*, 68, 2339, 1996.

Name	Symbol	At. no.	Atomic Weight	Footnotes		
Actinium	Ac	89	[227]			
Aluminum	Al	13	26.981538(2)			
Americium	Am	95	[243]			
Antimony	Sb	51	121.760(1)	g		
Argon	Ar	18	39.948(1)	g		r
Arsenic	As	33	74.92160(2)			
Astatine	At	85	[210]			
Barium	Ba	56	137.327(7)			
Berkelium	Bk	97	[247]			
Beryllium	Be	4	9.012182(3)			
Bismuth	Bi	83	208.98038(2)			
Bohrium	Bh	107	[264]			
Boron	B	5	10.811(7)	g	m	r
Bromine	Br	35	79.904(1)			
Cadmium	Cd	48	112.411(8)	g		
Calcium	Ca	20	40.078(4)	g		
Californium	Cf	98	[251]			
Carbon	C	6	12.0107(8)	g		r
Cerium	Ce	58	140.116(1)	g		
Cesium	Cs	55	132.90545(2)			
Chlorine	Cl	17	35.4527(9)		m	
Chromium	Cr	24	51.9961(6)			
Cobalt	Co	27	58.933200(9)			
Copper	Cu	29	63.546(3)			r
Curium	Cm	96	[247]			
Dubnium	Db	105	[262]			
Dysprosium	Dy	66	162.50(3)	g		
Einsteinium	Es	99	[252]			
Erbium	Er	68	167.26(3)	g		
Europium	Eu	63	151.964(1)	g		
Fermium	Fm	100	[257]			
Fluorine	F	9	18.9984032(5)			
Francium	Fr	87	[223]			
Gadolinium	Gd	64	157.25(3)	g		
Gallium	Ga	31	69.723(1)			
Germanium	Ge	32	72.61(2)			
Gold	Au	79	196.96655(2)			
Hafnium	Hf	72	178.49(2)			
Hassium	Hs	108	[269]			
Helium	He	2	4.002602(2)	g		r
Holmium	Ho	67	164.93032(2)			
Hydrogen	H	1	1.00794(7)	g	m	r
Indium	In	49	114.818(3)			
Iodine	I	53	126.90447(3)			

Name	Symbol	At. no.	Atomic Weight	Footnotes		
Iridium	Ir	77	192.217(3)			
Iron	Fe	26	55.845(2)			
Krypton	Kr	36	83.80(1)	g	m	
Lanthanum	La	57	138.9055(2)	g		
Lawrencium	Lr	103	[262]			
Lead	Pb	82	207.2(1)	g		r
Lithium	Li	3	6.941(2)	g	m	r
Lutetium	Lu	71	174.967(1)	g		
Magnesium	Mg	12	24.3050(6)			
Manganese	Mn	25	54.938049(9)			
Meitnerium	Mt	109	[268]			
Mendelevium	Md	101	[258]			
Mercury	Hg	80	200.59(2)			
Molybdenum	Mo	42	95.94(1)	g		
Neodymium	Nd	60	144.24(3)	g		
Neon	Ne	10	20.1797(6)	g	m	
Neptunium	Np	93	[237]			
Nickel	Ni	28	58.6934(2)			
Niobium	Nb	41	92.90638(2)			
Nitrogen	N	7	14.00674(7)	g		r
Nobelium	No	102	[259]			
Osmium	Os	76	190.23(3)	g		
Oxygen	O	8	15.9994(3)	g		r
Palladium	Pd	46	106.42(1)	g		
Phosphorus	P	15	30.973761(2)			
Platinum	Pt	78	195.078(2)			
Plutonium	Pu	94	[244]			
Polonium	Po	84	[209]			
Potassium	K	19	39.0983(1)	g		
Praseodymium	Pr	59	140.90765(2)			
Promethium	Pm	61	[145]			
Protactinium	Pa	91	231.03588(2)			
Radium	Ra	88	[226]			
Radon	Rn	86	[222]			
Rhenium	Re	75	186.207(1)			
Rhodium	Rh	45	102.90550(2)			
Rubidium	Rb	37	85.4678(3)	g		
Ruthenium	Ru	44	101.07(2)	g		
Rutherfordium	Rf	104	[261]			
Samarium	Sm	62	150.36(3)	g		
Scandium	Sc	21	44.955910(8)			
Seaborgium	Sg	106	[266]			
Selenium	Se	34	78.96(3)			
Silicon	Si	14	28.0855(3)			r
Silver	Ag	47	107.8682(2)	g		
Sodium	Na	11	22.989770(2)			
Strontium	Sr	38	87.62(1)	g		r
Sulfur	S	16	32.066(6)	g		r
Tantalum	Ta	73	180.9479(1)			
Technetium	Tc	43	[98]			
Tellurium	Te	52	127.60(3)	g		
Terbium	Tb	65	158.92534(2)			
Thallium	Tl	81	204.3833(2)			
Thorium	Th	90	232.0381(1)	g		
Thulium	Tm	69	168.93421(2)			
Tin	Sn	50	118.710(7)	g		
Titanium	Ti	22	47.867(1)			
Tungsten	W	74	183.84(1)			

STANDARD ATOMIC WEIGHTS (1995) (continued)

Name	Symbol	At. no.	Atomic Weight	Footnotes	
Uranium	U	92	238.0289(1)	g	m
Vanadium	V	23	50.9415(1)		
Xenon	Xe	54	131.29(2)	g	m
Ytterbium	Yb	70	173.04(3)	g	
Yttrium	Y	39	88.90585(2)		
Zinc	Zn	30	65.39(2)		
Zirconium	Zr	40	91.224(2)	g	

g   geological specimens are known in which the element has an isotopic composition outside the limits for normal material. The difference between the atomic weight of the element in such specimens and that given in the table may exceed the stated uncertainty.

m   modified isotopic compositions may be found in commercially available material because it has been subjected to an undisclosed or inadvertent isotopic fractionation. Substantial deviations in atomic weight of the element from that given the table can occur.

r   range in isotopic composition of normal terrestrial material prevents a more precise atomic weight being given; the tabulated atomic weight value should be applicable to any normal material.

ATOMIC MASSES AND ABUNDANCES

This table lists the mass (in atomic mass units, symbol u) and the natural abundance (in percent) of the stable nuclides and a few important radioactive nuclides. A complete table of all nuclides may be found in Section 11 ("Table of the Isotopes").

The atomic masses were taken from the 1995 evaluation of Audi and Wapstra (References 2, 3). Mass values were rounded in accordance with the stated uncertainty. The number in parentheses following the mass value is the uncertainty (specifically, the rounded value of the estimated standard deviation) in the last digit given.

Natural abundance values are also followed by uncertainties in the last digit(s) of the stated values. This uncertainty includes both the estimated measurement uncertainty and the reported range of variation in different terrestrial sources of the element. A * in the Abundance column indicates a radioactive nuclide not present in nature or an element whose isotopic composition varies so widely that a meaningful natural abundance cannot be defined. See Reference 5 for the most recent updates and further details on natural abundance.

REFERENCES

1. Holden, N. E., "Table of the Isotopes", in Lide, D. R., Ed., *CRC Handbook of Chemistry and Physics, 81st Ed.*, CRC Press, Boca Raton, FL, 2000.
2. Audi, G. and Wapstra, A. H., *Nucl. Phys.*, A595, 409, 1995.
3. Audi, G. and Wapstra, A. H., Atomic Mass Data Center, World Wide Web site, <http://csnwww.in2p3.fr>. Note that the print version in Reference 2 gives mass excess in energy units but not the actual atomic mass. The latter is included in the electronic version.
4. IUPAC Commission on Atomic Weights and Isotopic Abundances, *Pure Appl. Chem.*, 63, 991, 1991.
5. Rosman, K.J.R. and Taylor, P.D.P., *J. Phys. Chem. Ref. Data*, 27, 1275 (1998).

Z	Isotope	Mass in u	Abundance in %	Z	Isotope	Mass in u	Abundance in %
1	^{1}H	1.007 825 032(1)	99.985(1)	18	^{38}Ar	37.962 732 2(5)	0.063(1)
1	^{2}H	2.014 101 778(1)	0.015(1)	18	^{39}Ar	38.964 313(5)	*
1	^{3}H	3.016 049 268(1)	*	18	^{40}Ar	39.962 383 123(3)	99.600(3)
2	^{3}He	3.016 029 310(1)	0.000137(3)	19	^{39}K	38.963 706 8(3)	93.2581(44)
2	^{4}He	4.002 603 250(1)	99.999863(3)	19	^{40}K	39.963 998 7(3)	0.0117(1)
3	^{6}Li	6.015 122 3(5)	7.5(2)	19	^{41}K	40.961 826 0(3)	6.7302(44)
3	^{7}Li	7.016 004 0(5)	92.5(2)	20	^{40}Ca	39.962 591 1(3)	96.941(18)
4	^{9}Be	9.012 182 1(4)	100	20	^{42}Ca	41.958 618 3(4)	0.647(9)
5	^{10}B	10.012 937 0(3)	19.9(2)	20	^{43}Ca	42.958 766 8(5)	0.135(6)
5	^{11}B	11.009 305 5(5)	80.1(2)	20	^{44}Ca	43.955 481 1(9)	2.086(12)
6	^{12}C	12 by definition	98.90(3)	20	^{46}Ca	45.953 693(3)	0.004(3)
6	^{13}C	13.003 354 838(1)	1.10(3)	20	^{48}Ca	47.952 534(4)	0.187(4)
6	^{14}C	14.003 241 988(4)	*	21	^{45}Sc	44.955 910(1)	100
7	^{14}N	14.003 074 005(1)	99.634(9)	22	^{46}Ti	45.952 629(1)	8.0(1)
7	^{15}N	15.000 108 898(1)	0.366(9)	22	^{47}Ti	46.951 764(1)	7.3(1)
8	^{16}O	15.994 914 622(2)	99.762(15)	22	^{48}Ti	47.947 947(1)	73.8(1)
8	^{17}O	16.999 131 5(2)	0.038(3)	22	^{49}Ti	48.947 871(1)	5.5(1)
8	^{18}O	17.999 160 4(9)	0.200(12)	22	^{50}Ti	49.944 792(1)	5.4(1)
9	^{19}F	18.998 403 21(8)	100	23	^{50}V	49.947 163(1)	0.250(2)
10	^{20}Ne	19.992 440 176(2)	90.48(3)	23	^{51}V	50.943 964(1)	99.750(2)
10	^{21}Ne	20.993 846 74(4)	0.27(1)	24	^{50}Cr	49.946 050(1)	4.345(13)
10	^{22}Ne	21.991 385 5(2)	9.25(3)	24	^{52}Cr	51.940 512(2)	83.789(18)
11	^{23}Na	22.989 769 7(2)	100	24	^{53}Cr	52.940 654(2)	9.501(17)
12	^{24}Mg	23.985 041 9(2)	78.99(3)	24	^{54}Cr	53.938 885(1)	2.365(7)
12	^{25}Mg	24.985 837 0(2)	10.00(1)	25	^{55}Mn	54.938 050(1)	100
12	^{26}Mg	25.982 593 0(2)	11.01(2)	26	^{54}Fe	53.939 615(1)	5.8(1)
13	^{27}Al	26.981 538 4(1)	100	26	^{56}Fe	55.934 942(1)	91.72(30)
14	^{28}Si	27.976 926 533(2)	92.23(1)	26	^{57}Fe	56.935 399(1)	2.1(1)
14	^{29}Si	28.976 494 72(3)	4.67(1)	26	^{58}Fe	57.933 280(1)	0.28(1)
14	^{30}Si	29.973 770 22(5)	3.10(1)	27	^{59}Co	58.933 200(2)	100
15	^{31}P	30.973 761 5(2)	100	28	^{58}Ni	57.935 348(2)	68.077(9)
16	^{32}S	31.972 070 7(1)	95.02(9)	28	^{60}Ni	59.930 791(2)	26.223(8)
16	^{33}S	32.971 458 5(1)	0.75(4)	28	^{61}Ni	60.931 060(2)	1.140(1)
16	^{34}S	33.967 866 8(1)	4.21(8)	28	^{62}Ni	61.928 349(2)	3.634(2)
16	^{36}S	35.967 080 9(3)	0.02(1)	28	^{64}Ni	63.927 970(2)	0.926(1)
17	^{35}Cl	34.968 852 71(4)	75.77(7)	29	^{63}Cu	62.929 601(2)	69.17(3)
17	^{37}Cl	36.965 902 60(5)	24.23(7)	29	^{65}Cu	64.927 794(2)	30.83(3)
18	^{36}Ar	35.967 546 3(3)	0.337(3)	30	^{64}Zn	63.929 147(2)	48.6(3)
18	^{37}Ar	36.966 775 9(3)	*	30	^{66}Zn	65.926 037(2)	27.9(2)

Z	Isotope	Mass in u	Abundance in %	Z	Isotope	Mass in u	Abundance in %
30	^{67}Zn	66.927 131(2)	4.1(1)	46	^{108}Pd	107.903 894(4)	26.46(9)
30	^{68}Zn	67.924 848(2)	18.8(4)	46	^{110}Pd	109.905 15(1)	11.72(9)
30	^{70}Zn	69.925 325(4)	0.6(1)	47	^{107}Ag	106.905 093(6)	51.839(7)
31	^{69}Ga	68.925 581(3)	60.108(9)	47	^{109}Ag	108.904 756(3)	48.161(7)
31	^{71}Ga	70.924 705(2)	39.892(9)	48	^{106}Cd	105.906 458(6)	1.25(4)
32	^{70}Ge	69.924 250(2)	21.23(4)	48	^{108}Cd	107.904 183(6)	0.89(2)
32	^{72}Ge	71.922 076(2)	27.66(3)	48	^{110}Cd	109.903 006(3)	12.49(12)
32	^{73}Ge	72.923 459(2)	7.73(1)	48	^{111}Cd	110.904 182(3)	12.80(8)
32	^{74}Ge	73.921 178(2)	35.94(2)	48	^{112}Cd	111.902 757(3)	24.13(14)
32	^{76}Ge	75.921 403(2)	7.44(2)	48	^{113}Cd	112.904 401(3)	12.22(8)
33	^{75}As	74.921 596(2)	100	48	^{114}Cd	113.903 358(3)	28.73(28)
34	^{74}Se	73.922 477(2)	0.89(2)	48	^{116}Cd	115.904 755(3)	7.49(12)
34	^{76}Se	75.919 214(2)	9.36(11)	49	^{113}In	112.904 061(4)	4.3(2)
34	^{77}Se	76.919 915(2)	7.63(6)	49	^{115}In	114.903 878(5)	95.7(2)
34	^{78}Se	77.917 310(2)	23.78(9)	50	^{112}Sn	111.904 821(5)	0.97(1)
34	^{80}Se	79.916 522(2)	49.61(10)	50	^{114}Sn	113.902 782(3)	0.65(1)
34	^{82}Se	81.916 700(2)	8.73(6)	50	^{115}Sn	114.903 346(3)	0.34(1)
35	^{79}Br	78.918 338(2)	50.69(7)	50	^{116}Sn	115.901 744(3)	14.53(1)
35	^{81}Br	80.916 291(3)	49.31(7)	50	^{117}Sn	116.902 954(3)	7.68(7)
36	^{78}Kr	77.920 386(2)	0.35(2)	50	^{118}Sn	117.901 606(3)	24.23(11)
36	^{80}Kr	79.916 378(4)	2.25(2)	50	^{119}Sn	118.903 309(3)	8.59(4)
36	^{82}Kr	81.913 485(3)	11.6(1)	50	^{120}Sn	119.902 197(3)	32.59(10)
36	^{83}Kr	82.914 136(3)	11.5(1)	50	^{122}Sn	121.903 440(3)	4.63(3)
36	^{84}Kr	83.911 507(3)	57.0(3)	50	^{124}Sn	123.905 275(1)	5.79(5)
36	^{86}Kr	85.910 610(1)	17.3(2)	51	^{121}Sb	120.903 818(2)	57.36(8)
37	^{85}Rb	84.911 789(3)	72.165(20)	51	^{123}Sb	122.904 216(2)	42.64(8)
37	^{87}Rb	86.909 183(3)	27.835(20)	52	^{120}Te	119.904 02(1)	0.096(2)
38	^{84}Sr	83.913 425(4)	0.56(1)	52	^{122}Te	121.903 047(2)	2.603(4)
38	^{86}Sr	85.909 262(2)	9.86(1)	52	^{123}Te	122.904 273(2)	0.908(2)
38	^{87}Sr	86.908 879(2)	7.00(1)	52	^{124}Te	123.902 819(2)	4.816(6)
38	^{88}Sr	87.905 614(2)	82.58(1)	52	^{125}Te	124.904 425(2)	7.139(6)
39	^{89}Y	88.905 848(3)	100	52	^{126}Te	125.903 306(2)	18.95(1)
40	^{90}Zr	89.904 704(2)	51.45(3)	52	^{128}Te	127.904 461(2)	31.69(1)
40	^{91}Zr	90.905 645(2)	11.22(4)	52	^{130}Te	129.906 223(2)	33.80(1)
40	^{92}Zr	91.905 040(2)	17.15(2)	53	^{127}I	126.904 468(4)	100
40	^{94}Zr	93.906 316(3)	17.38(4)	54	^{124}Xe	123.905 896(2)	0.10(1)
40	^{96}Zr	95.908 276(3)	2.80(2)	54	^{126}Xe	125.904 269(1)	0.09(1)
41	^{93}Nb	92.906 378(2)	100	54	^{128}Xe	127.903 530(2)	1.91(3)
42	^{92}Mo	91.906 810(4)	14.84(4)	54	^{129}Xe	128.904 779 4(9)	26.4(6)
42	^{94}Mo	93.905 088(2)	9.25(3)	54	^{130}Xe	129.903 508(1)	4.1(1)
42	^{95}Mo	94.905 841(2)	15.92(5)	54	^{131}Xe	130.905 082(1)	21.2(4)
42	^{96}Mo	95.904 679(2)	16.68(5)	54	^{132}Xe	131.904 154(1)	26.9(5)
42	^{97}Mo	96.906 021(2)	9.55(3)	54	^{134}Xe	133.905 394 5(9)	10.4(2)
42	^{98}Mo	97.905 408(2)	24.13(7)	54	^{136}Xe	135.907 220(8)	8.9(1)
42	^{100}Mo	99.907 477(6)	9.63(3)	55	^{133}Cs	132.905 447(3)	100
43	^{98}Tc	97.907 216(4)	*	56	^{130}Ba	129.906 310(7)	0.106(2)
43	^{99}Tc	98.906 255(2)	*	56	^{132}Ba	131.905 056(3)	0.101(2)
44	^{96}Ru	95.907 598(8)	5.52(6)	56	^{134}Ba	133.904 503(3)	2.417(27)
44	^{98}Ru	97.905 287(7)	1.88(6)	56	^{135}Ba	134.905 683(3)	6.592(18)
44	^{99}Ru	98.905 939(2)	12.7(1)	56	^{136}Ba	135.904 570(3)	7.854(36)
44	^{100}Ru	99.904 220(2)	12.6(1)	56	^{137}Ba	136.905 821(3)	11.23(4)
44	^{101}Ru	100.905 582(2)	17.0(1)	56	^{138}Ba	137.905 241(3)	71.70(7)
44	^{102}Ru	101.904 350(2)	31.6(2)	57	^{138}La	137.907 107(4)	0.0902(2)
44	^{104}Ru	103.905 430(4)	18.7(2)	57	^{139}La	138.906 348(3)	99.9098(2)
45	^{103}Rh	102.905 504(3)	100	58	^{136}Ce	135.907 14(5)	0.19(1)
46	^{102}Pd	101.905 608(3)	1.02(1)	58	^{137}Ce	136.907 78(5)	*
46	^{104}Pd	103.904 035(5)	11.14(8)	58	^{138}Ce	137.905 99(1)	0.25(1)
46	^{105}Pd	104.905 084(5)	22.33(8)	58	^{139}Ce	138.906 647(8)	*
46	^{106}Pd	105.903 483(5)	27.33(3)	58	^{140}Ce	139.905 434(3)	88.48(10)

Z	Isotope	Mass in u	Abundance in %	Z	Isotope	Mass in u	Abundance in %
58	^{141}Ce	140.908 271(3)	*	72	^{180}Hf	179.946 549(3)	35.100(7)
58	^{142}Ce	141.909 240(4)	11.08(10)	73	^{180}Ta	179.947 466(3)	0.012(2)
59	^{141}Pr	140.907 648(3)	100	73	^{181}Ta	180.947 996(3)	99.988(2)
60	^{142}Nd	141.907 719(3)	27.13(12)	74	^{180}W	179.946 706(5)	0.13(4)
60	^{143}Nd	142.909 810(3)	12.18(6)	74	^{182}W	181.948 206(3)	26.3(2)
60	^{144}Nd	143.910 083(3)	23.80(12)	74	^{183}W	182.950 224(3)	14.3(1)
60	^{145}Nd	144.912 569(3)	8.30(6)	74	^{184}W	183.950 933(3)	30.67(15)
60	^{146}Nd	145.913 112(3)	17.19(9)	74	^{186}W	185.954 362(3)	28.6(2)
60	^{148}Nd	147.916 889(3)	5.76(3)	75	^{185}Re	184.952 956(3)	37.40(2)
60	^{150}Nd	149.920 887(4)	5.64(3)	75	^{187}Re	186.955 751(3)	62.60(2)
61	^{143}Pm	142.910 928(4)	*	76	^{184}Os	183.952 491(3)	0.02(1)
61	^{145}Pm	144.912 744(4)	*	76	^{186}Os	185.953 838(3)	1.58(30)
61	^{147}Pm	146.915 134(3)	*	76	^{187}Os	186.955 748(3)	1.6(3)
62	^{144}Sm	143.911 995(4)	3.1(1)	76	^{188}Os	187.955 836(3)	13.3(7)
62	^{147}Sm	146.914 893(3)	15.0(2)	76	^{189}Os	188.958 145(3)	16.1(8)
62	^{148}Sm	147.914 818(3)	11.3(1)	76	^{190}Os	189.958 445(3)	26.4(12)
62	^{149}Sm	148.917 180(3)	13.8(1)	76	^{192}Os	191.961 479(4)	41.0(8)
62	^{150}Sm	149.917 271(3)	7.4(1)	77	^{191}Ir	190.960 591(3)	37.3(5)
62	^{152}Sm	151.919 728(3)	26.7(2)	77	^{193}Ir	192.962 924(3)	62.7(5)
62	^{154}Sm	153.922 205(3)	22.7(2)	78	^{190}Pt	189.959 930(7)	0.01(1)
63	^{151}Eu	150.919 846(3)	47.8(15)	78	^{192}Pt	191.961 035(4)	0.79(6)
63	^{153}Eu	152.921 226(3)	52.2(15)	78	^{194}Pt	193.962 664(3)	32.9(6)
64	^{152}Gd	151.919 788(3)	0.20(1)	78	^{195}Pt	194.964 774(3)	33.8(6)
64	^{154}Gd	153.920 862(3)	2.18(3)	78	^{196}Pt	195.964 935(3)	25.3(6)
64	^{155}Gd	154.922 619(3)	14.80(5)	78	^{198}Pt	197.967 876(5)	7.2(2)
64	^{156}Gd	155.922 120(3)	20.47(4)	79	^{197}Au	196.966 552(3)	100
64	^{157}Gd	156.923 957(3)	15.65(3)	80	^{196}Hg	195.965 815(4)	0.15(1)
64	^{158}Gd	157.924 101(3)	24.84(12)	80	^{198}Hg	197.966 752(3)	9.97(8)
64	^{160}Gd	159.927 051(3)	21.86(4)	80	^{199}Hg	198.968 262(3)	16.87(10)
65	^{159}Tb	158.925 343(3)	100	80	^{200}Hg	199.968 309(3)	23.10(16)
66	^{156}Dy	155.924 278(7)	0.06(1)	80	^{201}Hg	200.970 285(3)	13.18(8)
66	^{158}Dy	157.924 405(4)	0.10(1)	80	^{202}Hg	201.970 626(3)	29.86(20)
66	^{160}Dy	159.925 194(3)	2.34(6)	80	^{204}Hg	203.973 476(3)	6.87(4)
66	^{161}Dy	160.926 930(3)	18.9(2)	81	^{203}Tl	202.972 329(3)	29.524(14)
66	^{162}Dy	161.926 795(3)	25.5(2)	81	^{205}Tl	204.974 412(3)	70.476(14)
66	^{163}Dy	162.928 728(3)	24.9(2)	82	^{204}Pb	203.973 029(3)	1.4(1)
66	^{164}Dy	163.929 171(3)	28.2(2)	82	^{206}Pb	205.974 449(3)	24.1(1)
67	^{165}Ho	164.930 319(3)	100	82	^{207}Pb	206.975 881(3)	22.1(1)
68	^{162}Er	161.928 775(4)	0.14(1)	82	^{208}Pb	207.976 636(3)	52.4(1)
68	^{164}Er	163.929 197(4)	1.61(2)	83	^{209}Bi	208.980 383(3)	100
68	^{166}Er	165.930 290(3)	33.6(2)	84	^{209}Po	208.982 416(3)	*
68	^{167}Er	166.932 045(3)	22.95(15)	85	^{210}At	209.987 131(9)	*
68	^{168}Er	167.932 368(3)	26.8(2)	86	^{211}Rn	210.990 585(8)	*
68	^{170}Er	169.935 460(3)	14.9(2)	86	^{222}Rn	222.017 570(3)	*
69	^{169}Tm	168.934 211(3)	100	87	^{223}Fr	223.019 731(3)	*
70	^{168}Yb	167.933 894(5)	0.13(1)	88	^{223}Ra	223.018 497(3)	*
70	^{170}Yb	169.934 759(3)	3.05(6)	88	^{225}Ra	225.023 604(3)	*
70	^{171}Yb	170.936 322(3)	14.3(2)	88	^{226}Ra	226.025 403(3)	*
70	^{172}Yb	171.936 378(3)	21.9(3)	89	^{227}Ac	227.027 747(3)	*
70	^{173}Yb	172.938 207(3)	16.12(21)	90	^{229}Th	229.031 755(3)	*
70	^{174}Yb	173.938 858(3)	31.8(4)	90	^{232}Th	232.038 050(2)	100
70	^{176}Yb	175.942 568(3)	12.7(2)	91	^{231}Pa	231.035 879(3)	*
71	^{175}Lu	174.940 768(3)	97.41(2)	92	^{234}U	234.040 946(2)	0.0055(5)
71	^{176}Lu	175.942 682(3)	2.59(2)	92	^{235}U	235.043 923(2)	0.7200(12)
72	^{174}Hf	173.940 040(3)	0.162(3)	92	^{238}U	238.050 783(2)	99.2745(60)
72	^{176}Hf	175.941 402(3)	5.206(5)	93	^{237}Np	237.048 167(2)	*
72	^{177}Hf	176.943 220(3)	18.606(4)	94	^{239}Pu	239.052 157(2)	*
72	^{178}Hf	177.943 698(3)	27.297(4)	94	^{244}Pu	244.064 198(5)	*
72	^{179}Hf	178.945 815(3)	13.629(6)				

ELECTRON CONFIGURATION OF NEUTRAL ATOMS IN THE GROUND STATE

Atomic no.	Element	K 1 s	L 2 s	L 2 p	M 3 s	M 3 p	M 3 d	N 4 s	N 4 p	N 4 d	N 4 f	O 5 s	O 5 p	O 5 d	O 5 f	P 6 s	P 6 p	P 6 d	Q 7 s
1	H	1																	
2	He	2																	
3	Li	2	1																
4	Be	2	2																
5	B	2	2	1															
6	C	2	2	2															
7	N	2	2	3															
8	O	2	2	4															
9	F	2	2	5															
10	Ne	2	2	6															
11	Na	2	2	6	1														
12	Mg	2	2	6	2														
13	Al	2	2	6	2	1													
14	Si	2	2	6	2	2													
15	P	2	2	6	2	3													
16	S	2	2	6	2	4													
17	Cl	2	2	6	2	5													
18	Ar	2	2	6	2	6													
19	K	2	2	6	2	6		1											
20	Ca	2	2	6	2	6		2											
21	Sc	2	2	6	2	6	1	2											
22	Ti	2	2	6	2	6	2	2											
23	V	2	2	6	2	6	3	2											
24	Cr	2	2	6	2	6	5*	1											
25	Mn	2	2	6	2	6	5	2											
26	Fe	2	2	6	2	6	6	2											
27	Co	2	2	6	2	6	7	2											
28	Ni	2	2	6	2	6	8	2											
29	Cu	2	2	6	2	6	10*	1											
30	Zn	2	2	6	2	6	10	2											
31	Ga	2	2	6	2	6	10	2	1										
32	Ge	2	2	6	2	6	10	2	2										
33	As	2	2	6	2	6	10	2	3										
34	Se	2	2	6	2	6	10	2	4										
35	Br	2	2	6	2	6	10	2	5										
36	Kr	2	2	6	2	6	10	2	6										
37	Rb	2	2	6	2	6	10	2	6			1							
38	Sr	2	2	6	2	6	10	2	6			2							
39	Y	2	2	6	2	6	10	2	6	1		2							
40	Zr	2	2	6	2	6	10	2	6	2		2							
41	Nb	2	2	6	2	6	10	2	6	4*		1							
42	Mo	2	2	6	2	6	10	2	6	5		1							
43	Tc	2	2	6	2	6	10	2	6	5		2							
44	Ru	2	2	6	2	6	10	2	6	7		1							
45	Rh	2	2	6	2	6	10	2	6	8		1							
46	Pd	2	2	6	2	6	10	2	6	10*									
47	Ag	2	2	6	2	6	10	2	6	10		1							
48	Cd	2	2	6	2	6	10	2	6	10		2							
49	In	2	2	6	2	6	10	2	6	10		2	1						
50	Sn	2	2	6	2	6	10	2	6	10		2	2						
51	Sb	2	2	6	2	6	10	2	6	10		2	3						
52	Te	2	2	6	2	6	10	2	6	10		2	4						
53	I	2	2	6	2	6	10	2	6	10		2	5						
54	Xe	2	2	6	2	6	10	2	6	10		2	6						
55	Cs	2	2	6	2	6	10	2	6	10		2	6			1			
56	Ba	2	2	6	2	6	10	2	6	10		2	6			2			

Atomic no.	Element	K 1	L 2		M 3			N 4				O 5				P 6			Q 7
$n =$		s	s	p	s	p	d	s	p	d	f	s	p	d	f	s	p	d	s
57	La	2	2	6	2	6	10	2	6	10		2	6	1		2			
58	Ce	2	2	6	2	6	10	2	6	10	1*	2	6	1		2			
59	Pr	2	2	6	2	6	10	2	6	10	3	2	6			2			
60	Nd	2	2	6	2	6	10	2	6	10	4	2	6			2			
61	Pm	2	2	6	2	6	10	2	6	10	5	2	6			2			
62	Sm	2	2	6	2	6	10	2	6	10	6	2	6			2			
63	Eu	2	2	6	2	6	10	2	6	10	7	2	6			2			
64	Gd	2	2	6	2	6	10	2	6	10	7	2	6	1		2			
65	Tb	2	2	6	2	6	10	2	6	10	9*	2	6			2			
66	Dy	2	2	6	2	6	10	2	6	10	10	2	6			2			
67	Ho	2	2	6	2	6	10	2	6	10	11	2	6			2			
68	Er	2	2	6	2	6	10	2	6	10	12	2	6			2			
69	Tm	2	2	6	2	6	10	2	6	10	13	2	6			2			
70	Yb	2	2	6	2	6	10	2	6	10	14	2	6			2			
71	Lu	2	2	6	2	6	10	2	6	10	14	2	6	1		2			
72	Hf	2	2	6	2	6	10	2	6	10	14	2	6	2		2			
73	Ta	2	2	6	2	6	10	2	6	10	14	2	6	3		2			
74	W	2	2	6	2	6	10	2	6	10	14	2	6	4		2			
75	Re	2	2	6	2	6	10	2	6	10	14	2	6	5		2			
76	Os	2	2	6	2	6	10	2	6	10	14	2	6	6		2			
77	Ir	2	2	6	2	6	10	2	6	10	14	2	6	7		2			
78	Pt	2	2	6	2	6	10	2	6	10	14	2	6	9		1			
79	Au	2	2	6	2	6	10	2	6	10	14	2	6	10		1			
80	Hg	2	2	6	2	6	10	2	6	10	14	2	6	10		2			
81	Tl	2	2	6	2	6	10	2	6	10	14	2	6	10		2	1		
82	Pb	2	2	6	2	6	10	2	6	10	14	2	6	10		2	2		
83	Bi	2	2	6	2	6	10	2	6	10	14	2	6	10		2	3		
84	Po	2	2	6	2	6	10	2	6	10	14	2	6	10		2	4		
85	At	2	2	6	2	6	10	2	6	10	14	2	6	10		2	5		
86	Rn	2	2	6	2	6	10	2	6	10	14	2	6	10		2	6		
87	Fr	2	2	6	2	6	10	2	6	10	14	2	6	10		2	6		1
88	Ra	2	2	6	2	6	10	2	6	10	14	2	6	10		2	6		2
89	Ac	2	2	6	2	6	10	2	6	10	14	2	6	10		2	6	1	2
90	Th	2	2	6	2	6	10	2	6	10	14	2	6	10		2	6	2	2
91	Pa	2	2	6	2	6	10	2	6	10	14	2	6	10	2*	2	6	1	2
92	U	2	2	6	2	6	10	2	6	10	14	2	6	10	3	2	6	1	2
93	Np	2	2	6	2	6	10	2	6	10	14	2	6	10	4	2	6	1	2
94	Pu	2	2	6	2	6	10	2	6	10	14	2	6	10	6*	2	6		2
95	Am	2	2	6	2	6	10	2	6	10	14	2	6	10	7	2	6		2
96	Cm	2	2	6	2	6	10	2	6	10	14	2	6	10	7*	2	6	1	2
97	Bk	2	2	6	2	6	10	2	6	10	14	2	6	10	9	2	6		2
98	Cf	2	2	6	2	6	10	2	6	10	14	2	6	10	10	2	6		2
99	Es	2	2	6	2	6	10	2	6	10	14	2	6	10	11	2	6		2
100	Fm	2	2	6	2	6	10	2	6	10	14	2	6	10	12	2	6		2
101	Md	2	2	6	2	6	10	2	6	10	14	2	6	10	13	2	6		2
102	No	2	2	6	2	6	10	2	6	10	14	2	6	10	14	2	6		2
103	Lr	2	2	6	2	6	10	2	6	10	14	2	6	10	14	2	6	1	2
104	Rf	2	2	6	2	6	10	2	6	10	14	2	6	10	14	2	6	2	2

* Note irregularity.

REFERENCE

W. L. Wiese and G. A. Martin, in *A Physicist's Desk Reference*, American Institute of Physics, New York, 1989, 94.

INTERNATIONAL TEMPERATURE SCALE OF 1990 (ITS-90)

B. W. Mangum

A new temperature scale, the International Temperature Scale of 1990 (ITS-90), was officially adopted by the Comité International des Poids et Mesures (CIPM), meeting 26—28 September 1989 at the Bureau International des Poids et Mesures (BIPM). The ITS-90 was recommended to the CIPM for its adoption following the completion of the final details of the new scale by the Comité Consultatif de Thermométrie (CCT), meeting 12—14 September 1989 at the BIPM in its 17th Session. The ITS-90 became the official international temperature scale on 1 January 1990. The ITS-90 supersedes the present scales, the International Practical Temperature Scale of 1968 (IPTS-68) and the 1976 Provisional 0.5 to 30 K Temperature Scale (EPT-76).

The ITS-90 extends upward from 0.65 K, and temperatures on this scale are in much better agreement with thermodynamic values that are those on the IPTS-68 and the EPT-76. The new scale has subranges and alternative definitions in certain ranges that greatly facilitate its use. Furthermore, its continuity, precision, and reproducibility throughout its ranges are much improved over that of the present scales. The replacement of the thermocouple with the platinum resistance thermometer at temperatures below 961.78°C resulted in the biggest improvement in reproducibility.

The ITS-90 is divided into four primary ranges:

1. Between 0.65 and 3.2 K, the ITS-90 is defined by the vapor pressure-temperature relation of ^{3}He, and between 1.25 and 2.1768 K (the λ point) and between 2.1768 and 5.0 K by the vapor pressure-temperature relations of ^{4}He. T_{90} is defined by the vapor pressure equations of the form:

$$T_{90}/K = A_0 + \sum_{i=1}^{9} A_i \left[\left(\ln(p/\mathrm{Pa}) - B \right)/C \right]^i$$

   The values of the coefficients A_i, and of the constants A_o, B, and C of the equations are given below.

2. Between 3.0 and 24.5561 K, the ITS-90 is defined in terms of a ^{3}He or ^{4}He constant volume gas thermometer (CVGT). The thermometer is calibrated at three temperatures — at the triple point of neon (24.5561 K), at the triple point of equilibrium hydrogen (13.8033 K), and at a temperature between 3.0 and 5.0 K, the value of which is determined by using either ^{3}He or ^{4}He vapor pressure thermometry.

3. Between 13.8033 K (–259.3467°C) and 1234.93 K (961.78°C), the ITS-90 is defined in terms of the specified fixed points given below, by resistance ratios of platinum resistance thermometers obtained by calibration at specified sets of the fixed points, and by reference functions and deviation functions of resistance ratios which relate to T_{90} between the fixed points.

4. Above 1234.93 K, the ITS-90 is defined in terms of Planck's radiation law, using the freezing-point temperature of either silver, gold, or copper as the reference temperature.

Full details of the calibration procedures and reference functions for various subranges are given in:

The International Temperature Scale of 1990, *Metrologia*, 27, 3, 1990; errata in *Metrologia*, 27, 107, 1990.

Defining Fixed Points of the ITS-90

Material[a]	Equilibrium state[b]	Temperature	
		T_{90} (K)	t_{90} (°C)
He	VP	3 to 5	–270.15 to –268.15
e-H_2	TP	13.8033	–259.3467
e-H_2 (or He)	VP (or CVGT)	≈17	≈ –256.15
e-H_2 (or He)	VP (or CVGT)	≈20.3	≈ –252.85
Ne[c]	TP	24.5561	–248.5939
O_2	TP	54.3584	–218.7916
Ar	TP	83.8058	–189.3442
Hg[c]	TP	234.3156	–38.8344
H_2O	TP	273.16	0.01
Ga[c]	MP	302.9146	29.7646
In[c]	FP	429.7485	156.5985
Sn	FP	505.078	231.928
Zn	FP	692.677	419.527
Al[c]	FP	933.473	660.323
Ag	FP	1234.93	961.78
Au	FP	1337.33	1064.18
Cu[c]	FP	1357.77	1084.62

Defining Fixed Points of the ITS-90 (continued)

a e-H_2 indicates equilibrium hydrogen, that is, hydrogen with the equilibrium distribution of its ortho and para states. Normal hydrogen at room temperature contains 25% para hydrogen and 75% ortho hydrogen.

b VP indicates vapor pressure point; CVGT indicates constant volume gas thermometer point; TP indicates triple point (equilibrium temperature at which the solid, liquid, and vapor phases coexist); FP indicates freezing point, and MP indicates melting point (the equilibrium temperatures at which the solid and liquid phases coexist under a pressure of 101 325 Pa, one standard atmosphere). The isotopic composition is that naturally occurring.

c Previously, these were secondary fixed points.

Values of Coefficients in the Vapor Pressure Equations for Helium

Coef.or constant	^{3}He 0.65—3.2 K	^{4}He 1.25—2.1768 K	^{4}He 2.1768—5.0 K
A_0	1.053 447	1.392 408	3.146 631
A_1	0.980 106	0.527 153	1.357 655
A_2	0.676 380	0.166 756	0.413 923
A_3	0.372 692	0.050 988	0.091 159
A_4	0.151 656	0.026 514	0.016 349
A_5	–0.002 263	0.001 975	0.001 826
A_6	0.006 596	–0.017 976	–0.004 325
A_7	0.088 966	0.005 409	–0.004 973
A_8	–0.004 770	0.013 259	0
A_9	–0.054 943	0	0
B	7.3	5.6	10.3
C	4.3	2.9	1.9

CONVERSION OF TEMPERATURES FROM THE 1948 AND 1968 SCALES TO ITS-90

This table gives temperature corrections from older scales to the current International Temperature Scale of 1990 (see the preceding table for details on ITS-90). The first part of the table may be used for converting Celsius temperatures in the range -180 to 4000°C from IPTS-68 or IPTS-48 to ITS-90. Within the accuracy of the corrections, the temperature in the first column may be identified with either t_{68}, t_{48}, or t_{90}. The second part of the table is designed for use at lower temperatures to convert values expressed in kelvins from EPT-76 or IPTS-68 to ITS-90.

The references give analytical equations for expressing these relations. Note that Reference 1 supersedes Reference 2 with respect to corrections in the 630 to 1064°C range.

REFERENCES

1. Burns, G. W. et al., in *Temperature: Its Measurement and Control in Science and Industry,* Vol. 6, Schooley, J. F., Ed., American Institute of Physics, New York, 1993.
2. Goldberg, R. N. and Weir, R. D., *Pure and Appl. Chem.*, 1545, 1992.

$t/°C$	$t_{90}-t_{68}$	$t_{90}-t_{48}$	$t/°C$	$t_{90}-t_{68}$	$t_{90}-t_{48}$	$t/°C$	$t_{90}-t_{68}$	$t_{90}-t_{48}$
-180	0.008	0.020	270	-0.039	0.028	720	0.00	0.45
-170	0.010	0.017	280	-0.039	0.030	730	0.02	0.49
-160	0.012	0.007	290	-0.039	0.032	740	0.03	0.53
-150	0.013	0.000	300	-0.039	0.034	750	0.03	0.56
-140	0.014	0.001	310	-0.039	0.035	760	0.04	0.60
-130	0.014	0.008	320	-0.039	0.036	770	0.05	0.63
-120	0.014	0.017	330	-0.040	0.036	780	0.05	0.66
-110	0.013	0.026	340	-0.040	0.037	790	0.05	0.69
-100	0.013	0.035	350	-0.041	0.036	800	0.05	0.72
-90	0.012	0.041	360	-0.042	0.035	810	0.05	0.75
-80	0.012	0.045	370	-0.043	0.034	820	0.04	0.76
-70	0.011	0.045	380	-0.045	0.032	830	0.04	0.79
-60	0.010	0.042	390	-0.046	0.030	840	0.03	0.81
-50	0.009	0.038	400	-0.048	0.028	850	0.02	0.83
-40	0.008	0.032	410	-0.051	0.024	860	0.01	0.85
-30	0.006	0.024	420	-0.053	0.022	870	0.00	0.87
-20	0.004	0.016	430	-0.056	0.019	880	-0.02	0.87
-10	0.002	0.008	440	-0.059	0.015	890	-0.03	0.89
0	0.000	0.000	450	-0.062	0.012	900	-0.05	0.90
10	-0.002	-0.006	460	-0.065	0.009	910	-0.06	0.92
20	-0.005	-0.012	470	-0.068	0.007	920	-0.08	0.93
30	-0.007	-0.016	480	-0.072	0.004	930	-0.10	0.94
40	-0.010	-0.020	490	-0.075	0.002	940	-0.11	0.96
50	-0.013	-0.023	500	-0.079	0.000	950	-0.13	0.97
60	-0.016	-0.026	510	-0.083	-0.001	960	-0.15	0.97
70	-0.018	-0.026	520	-0.087	-0.002	970	-0.16	0.99
80	-0.021	-0.027	530	-0.090	-0.001	980	-0.18	1.00
90	-0.024	-0.027	540	-0.094	0.000	990	-0.19	1.02
100	-0.026	-0.026	550	-0.098	0.002	1000	-0.20	1.04
110	-0.028	-0.024	560	-0.101	0.007	1010	-0.22	1.05
120	-0.030	-0.023	570	-0.105	0.011	1020	-0.23	1.07
130	-0.032	-0.020	580	-0.108	0.018	1030	-0.23	1.10
140	-0.034	-0.018	590	-0.112	0.025	1040	-0.24	1.12
150	-0.036	-0.016	600	-0.115	0.035	1050	-0.25	1.14
160	-0.037	-0.012	610	-0.118	0.047	1060	-0.25	1.17
170	-0.038	-0.009	620	-0.122	0.060	1070	-0.25	1.19
180	-0.039	-0.005	630	-0.125	0.075	1080	-0.26	1.20
190	-0.039	-0.001	640	-0.11	0.12	1090	-0.26	1.20
200	-0.040	0.003	650	-0.10	0.15	1100	-0.26	1.2
210	-0.040	0.007	660	-0.09	0.19	1200	-0.30	1.4
220	-0.040	0.011	670	-0.07	0.24	1300	-0.35	1.5
230	-0.040	0.014	680	-0.05	0.29	1400	-0.39	1.6
240	-0.040	0.018	690	-0.04	0.32	1500	-0.44	1.8
250	-0.040	0.021	700	-0.02	0.37	1600	-0.49	1.9
260	-0.040	0.024	710	-0.01	0.41	1700	-0.54	2.1

CONVERSION OF TEMPERATURES FROM THE 1948 AND 1968 SCALES TO ITS-90 (continued)

$t/°C$	$t_{90}-t_{68}$	$t_{90}-t_{48}$	T/K	$T_{90}-T_{76}$	$T_{90}-T_{68}$	T/K	$T_{90}-T_{76}$	$T_{90}-T_{68}$
1800	-0.60	2.2	28		-0.005	77		0.008
1900	-0.66	2.3	29		-0.006	78		0.008
2000	-0.72	2.5	30		-0.006	79		0.008
2100	-0.79	2.7	31		-0.007	80		0.008
2200	-0.85	2.9	32		-0.008	81		0.008
2300	-0.93	3.1	33		-0.008	82		0.008
2400	-1.00	3.2	34		-0.008	83		0.008
2500	-1.07	3.4	35		-0.007	84		0.008
2600	-1.15	3.7	36		-0.007	85		0.008
2700	-1.24	3.8	37		-0.007	86		0.008
2800	-1.32	4.0	38		-0.006	87		0.008
2900	-1.41	4.2	39		-0.006	88		0.008
3000	-1.50	4.4	40		-0.006	89		0.008
3100	-1.59	4.6	41		-0.006	90		0.008
3200	-1.69	4.8	42		-0.006	91		0.008
3300	-1.78	5.1	43		-0.006	92		0.008
3400	-1.89	5.3	44		-0.006	93		0.008
3500	-1.99	5.5	45		-0.007	94		0.008
3600	-2.10	5.8	46		-0.007	95		0.008
3700	-2.21	6.0	47		-0.007	96		0.008
3800	-2.32	6.3	48		-0.006	97		0.009
3900	-2.43	6.6	49		-0.006	98		0.009
4000	-2.55	6.8	50		-0.006	99		0.009
			51		-0.005	100		0.009
T/K	$T_{90}-T_{76}$	$T_{90}-T_{68}$	52		-0.005	110		0.011
			53		-0.004	120		0.013
5	-0.0001		54		-0.003	130		0.014
6	-0.0002		55		-0.002	140		0.014
7	-0.0003		56		-0.001	150		0.014
8	-0.0004		57		0.000	160		0.014
9	-0.0005		58		0.001	170		0.013
10	-0.0006		59		0.002	180		0.012
11	-0.0007		60		0.003	190		0.012
12	-0.0008		61		0.003	200		0.011
13	-0.0010		62		0.004	210		0.010
14	-0.0011	-0.006	63		0.004	220		0.009
15	-0.0013	-0.003	64		0.005	230		0.008
16	-0.0014	-0.004	65		0.005	240		0.007
17	-0.0016	-0.006	66		0.006	250		0.005
18	-0.0018	-0.008	67		0.006	260		0.003
19	-0.0020	-0.009	68		0.007	270		0.001
20	-0.0022	-0.009	69		0.007	273.16		0.000
21	-0.0025	-0.008	70		0.007	300		-0.006
22	-0.0027	-0.007	71		0.007	400		-0.031
23	-0.0030	-0.007	72		0.007	500		-0.040
24	-0.0032	-0.006	73		0.007	600		-0.040
25	-0.0035	-0.005	74		0.007	700		-0.055
26	-0.0038	-0.004	75		0.008	800		-0.089
27	-0.0041	-0.004	76		0.008	900		-0.124

INTERNATIONAL SYSTEM OF UNITS (SI)

The International System of units (SI) was adopted by the 11th General Conference on Weights and Measures (CGPM) in 1960. It is a coherent system of units built from seven *SI base units*, one for each of the seven dimensionally independent base quantities: they are the meter, kilogram, second, ampere, kelvin, mole, and candela, for the dimensions length, mass, time, electric current, thermodynamic temperature, amount of substance, and luminous intensity, respectively, The definitions of the SI base units are given below. The *SI derived units* are expressed as products of powers of the base units, analogous to the corresponding relations between physical quantities but with numerical factors equal to unity.

In the International System there is only one SI unit for each physical quanity. This is either the appropriate SI base unit itself or the appropriate SI derived unit. However, any of the approved decimal prefixes, called *SI prefixes*, may be used to construct decimal multiples or submultiples of SI units.

It is recommended that only SI units be used in science and technology (with SI prefixes where appropriate). Where there are special reasons for making an exception to this rule, it is recommended always to define the units used in terms of SI units. This section was reprinted with the permission of IUPAC.

Definitions of SI Base Units

Meter — the meter is the length of path travelled by light in vacuum during a time interval of 1/299 792 458 of a second (17th CGPM, 1983).

Kilogram — The kilogram is the unit of mass; it is equal to the mass of the international prototype of the kilogram (3rd CGPM, 1901).

Second — The second is the duration of 9 192 631 770 periods of the radiation corresponding to the transition between the two hyperfine levels of the ground state of the cesium-133 atom (13th CGPM, 1967).

Ampere — The ampere is that constant current which, if maintained in two straight parallel conductors of infinite length, of negligible circular cross-section, and placed 1 meter apart in vacuum, would produce between these conductors a force equal to 2×10^{-7} newton per meter of length (9th CGPM, 1948).

Kelvin — The kelvin, unit of thermodynamic temperature, is the fraction 1/273.16 of the thermodynamic temperature of the triple point of water (13th CGPM, 1967)

Mole — The mole is the amount of substance of a system which contains as many elementary entities as there are atoms in 0.012 kilogram of carbon-12. When the mole is used, the elementary entities must be specified and may be atoms, molecules, ions, electrons, other particles, or specified groups of such particles (14th CGPM, 1971).

Examples of the use of the mole:

1 mol of H_2 contains about 6.022×10^{23} H_2 molecules, or 12.044×10^{23} H atoms

1 mol of $HgCl$ has a mass of 236.04 g

1 mol of Hg_2Cl_2 has a mass of 472.08 g

1 mol of Hg_2^{2+} has a mass of 401.18 g and a charge of 192.97 kC

1 mol of $Fe_{0.91}S$ has a mass of 82.88 g

1 mol of e^- has a mass of 548.60 µg and a charge of –96.49 kC

1 mol of photons whose frequency is 10^{14} Hz has energy of about 39.90 kJ

Candela — The candela is the luminous intensity, in a given direction, of a source that emits monochromatic radiation of frequency 540×10^{12} hertz and that has a radiant intensity in that direction of (1/683) watt per steradian (16th CGPM, 1979).

Names and Symbols for the SI Base Units

Physical quantity	Name of SI unit	Symbol for SI unit
length	meter	m
mass	kilogram	kg
time	second	s
electric current	ampere	A
thermodynamic temperature	kelvin	K
amount of substance	mole	mol
luminous intensity	candela	cd

SI Derived Units with Special Names and Symbols

Physical quantity	Name of SI unit	Symbol for SI unit	Expression in terms of SI base units	
frequency[1]	hertz	Hz	s^{-1}	
force	newton	N	$m\ kg\ s^{-2}$	
pressure, stress	pascal	Pa	$N\ m^{-2}$	$=m^{-1}\ kg\ s^{-2}$
energy, work, heat	joule	J	$N\ m$	$=m^2\ kg\ s^{-2}$
power, radiant flux	watt	W	$J\ s^{-1}$	$=m^2\ kg\ s^{-3}$
electric charge	coulomb	C	$A\ s$	
electric potential, electromotive force	volt	V	$J\ C^{-1}$	$=m^2\ kg\ s^{-3}\ A^{-1}$
electric resistance	ohm	Ω	$V\ A^{-1}$	$=m^2\ kg\ s^{-3}\ A^{-2}$
electric conductance	siemens	S	Ω^{-1}	$=m^{-2}\ kg^{-1}\ s^3\ A^2$
electric capacitance	farad	F	$C\ V^{-1}$	$=m^{-2}\ kg^{-1}\ s^4\ A^2$
magnetic flux density	tesla	T	$V\ s\ m^{-2}$	$=kg\ s^{-2}\ A^{-1}$
magnetic flux	weber	Wb	$V\ s$	$=m^2\ kg\ s^{-2}\ A^{-1}$
inductance	henry	H	$V\ A^{-1}\ s$	$=m^2\ kg\ s^{-2}\ A^{-2}$
Celsius temperature[2]	degree Celsius	°C	K	
luminous flux	lumen	lm	cd sr	
illuminance	lux	lx	$cd\ sr\ m^{-2}$	
activity[3] (radioactive)	becquerel	Bq	s^{-1}	
absorbed dose[3] (of radiation)	gray	Gy	$J\ kg^{-1}$	$=m^2\ s^{-2}$
dose equivalent[3] (dose equivalent index)	sievert	Sv	$J\ kg^{-1}$	$=m^2\ s^{-2}$
plane angle[4]	radian	rad	1	$=m\ m^{-1}$
solid angle[4]	steradian	sr	1	$=m^2\ m^{-2}$

(1) For radial (circular) frequency and for angular velocity the unit rad s^{-1}, or simply s^{-1}, should be used, and this may not be simplified to Hz. The unit Hz should be used only for frequency in the sense of cycles per second.

(2) The Celsius temperature θ is defined by the equation:

$$\theta/°C = T/K - 273.15$$

The SI unit of Celsius temperature interval is the degree Celsius, °C, which is equal to the kelvin, K. °C should be treated as a single symbol, with no space between the ° sign and the letter C. (The symbol °K, and the symbol °, should no longer be used.)

(3) The units gray and sievert are admitted for reasons of safeguarding human health.

(4) The units radian and steradian are described as 'SI supplementary units'. However, in chemistry, as well as in physics, they are usually treated as dimensionless derived units, and this was recognized by CIPM in 1980. Since they are then of dimension 1, this leaves open the possibility of including them or omitting them in expressions of SI derived units. In practice this means that rad and sr may be used when appropriate and may be omitted if clarity is not lost thereby.

SI Prefixes

To signify decimal multiples and submultiples of SI units the following prefixes may be used.

Factor	Prefix	Symbol	Factor	Prefix	Symbol
10^{24}	yotta	Y	10^{-1}	deci	d
10^{21}	zetta	Z	10^{-2}	centi	c
10^{18}	exa	E	10^{-3}	milli	m
10^{15}	peta	P	10^{-6}	micro	μ
10^{12}	tera	T	10^{-9}	nano	n
10^{9}	giga	G	10^{-12}	pico	p
10^{6}	mega	M	10^{-15}	femto	f
10^{3}	kilo	k	10^{-18}	atto	a
10^{2}	hecto	h	10^{-21}	zepto	z
10^{1}	deka	da	10^{-24}	yocto	y

Prefix symbols should be printed in roman (upright) type with no space between the prefix and the unit symbol.

Example kilometer, km

When a prefix is used with a unit symbol, the combination is taken as a new symbol that can be raised to any power without the use of parentheses.

Examples

$$1 \text{ cm}^3 = (0.01 \text{ m})^3 = 10^{-6} \text{ m}^3$$
$$1 \text{ } \mu\text{s}^{-1} = (10^{-6} \text{ s})^{-1} = 10^6 \text{ s}^{-1}$$
$$1 \text{ V/cm} = 100 \text{ V/m}$$
$$1 \text{ mmol/dm}^3 = \text{mol m}^{-3}$$

A prefix should never be used on its own, and prefixes are not to be combined into compound prefixes.

Example pm, not $\mu\mu$m

The names and symbols of decimal multiples and sub-multiples of the SI base unit of mass, the kg, which already contains a prefix, are constructed by adding the appropriate prefix to the word gram and symbol g.

Examples mg, not μkg; Mg, not kkg

The SI prefixes are not to be used with °C.

Units in Use Together with the SI

These units are not part of the SI, but it is recognized that they will continue to be used in appropriate contexts. SI prefixes may be attached to some of these units, such as milliliter, ml; millibar, mbar; megaelectronvolt, MeV; kilotonne, ktonne.

Physical quantity	Name of unit	Symbol for unit	Value in SI units
time	minute	min	60 s
time	hour	h	3600 s
time	day	d	86 400 s
plane angle	degree	°	$(\pi/180)$ rad
plane angle	minute	'	$(\pi/10\,800)$ rad
plane angle	second	"	$(\pi/648\,000)$ rad
length	ångström[1]	Å	10^{-10} m
area	barn	b	10^{-28} m^2
volume	litre	l, L	dm^3 $= 10^{-3}$ m^3
mass	tonne	t	Mg $= 10^3$ kg
pressure	bar[1]	bar	10^5 Pa $= 10^5$ N m^{-2}
energy	electronvolt[2]	eV $(= e \times$ V$)$	$\approx 1.60218 \times 10^{-19}$ J
mass	unified atomic mass unit[2,3]	u $(= m_A(^{12}\text{C})/12)$	$\approx 1.66054 \times 10^{-27}$ kg

(1) The ångström and the bar are approved by CIPM for 'temporary use with SI units', until CIPM makes a further recommendation, However, they should not be introduced where they are not used at present.

(2) The values of these units in terms of the corresponding SI units are not exact, since they depend on the values of the physical constants e (for the electronvolt) and N_A (for the unified atomic mass unit), which are determined by experiment.

(3) The unified atomic mass unit is also sometimes called the dalton, with symbol Da, although the name and symbol have not been approved by CGPM.

Atomic Units

For the purposes of quantum mechanical calculations of electronic wavefunctions, it is convenient to regard certain fundamental constants (and combinations of such constants) as though they were units. They are customarily called *atomic units* (abbreviated: au), and they may be regarded as forming a coherent system of units for the calculation of electronic properties in theoretical chemistry, although there is no authority from CGPM for treating them as units. The first five atomic units in the table below have special names and symbols. Only four of these are independent; all others may be derived by multiplication and division in the usual way, and the table includes a number of examples.

The relation of atomic units to the corresponding SI units involves the values of the fundamental physical constants, and is therefore not exact. The numerical values in the table are based on the 1986 CODATA values of the fundamental constants. The numerical results of calculations in theoretical chemistry are frequently quoted in atomic units, or as numerical values in the form (*physical quantity*)/(*atomic unit*), so that the reader may make the conversion using the current best estimates of the physical constants.

Physical quantity	Name of unit	Symbol for unit	Definition and value of unit in SI
mass	electron rest mass	m_e	$m_e \approx 9.1095 \times 10^{-31}$ kg
charge	elementary charge	e	$e \approx 1.6022 \times 10^{-19}$ C
action	Planck constant/2π	$\hbar$	$\hbar = h/2\pi \approx 1.0546 \times 10^{-34}$ J s
length	bohr	a_0	$4\pi\varepsilon_0\hbar^2/m_e e^2 \approx 5.2918 \times 10^{-11}$ m
energy	hartree	E_h	$\hbar^2/m_e a_0^2 \approx 4.3598 \times 10^{-18}$ J
time	au of time	$\hbar/E_h$	$\approx 2.4189 \times 10^{-17}$ s
velocity[1]	au of velocity	$a_0 E_h/\hbar$	$\approx 2.1877 \times 10^6$ m s^{-1}
force	au of force	E_h/a_0	$\approx 8.2389 \times 10^{-8}$ N
momentum, linear	au of momentum	$\hbar/a_0$	$\approx 1.9929 \times 10^{-24}$ N s
electric current	au of current	$eE_h/\hbar$	$\approx 6.6236 \times 10^{-3}$ A
electric field	au of electric field	E_h/ea_0	$\approx 5.1422 \times 10^{11}$ V m^{-1}
electric dipole moment	au of electric dipole	ea_0	$\approx 8.4784 \times 10^{-30}$ C m
magnetic flux density	au of magnetic flux density	$\hbar/ea_0^2$	$\approx 2.3505 \times 10^5$ T
magnetic dipole moment[2]	au of magnetic dipole	$e\hbar/m_e$	$= 2\mu_B \approx 1.8548 \times 10^{-23}$ J T^{-1}

(1) The numerical value of the speed of light, when expressed in atomic units, is equal to the reciprocal of the fine structure constant α; c/(au of velocity) $= c\hbar/a_0 E_h = \alpha^{-1}$ 137.04.

(2) The atomic unit of magnetic dipole moment is twice the Bohr magneton, μ_B.

REFERENCES

Quantities, Units and Symbols in Physical Chemistry, Ian Mills, Editor, Blackwell Scientific Publications, Oxford, 1987.

Symbols, Units, Nomenclature and Fundamental Constants in Physics, E. R. Cohen and P. Giacomo, Editors, International Union of Pure and Applied Physics Document 25, 1987; also appears in Physica 146A, 1—8, 1987.

ISO Standards Handbook 2, Units of Measurement, International Organization for Standardization, Geneva, 1982.

CONVERSION FACTORS

The following table gives conversion factors from various units of measure to SI units. It is reproduced from NIST Special Publication 811, *Guide for the Use of the International System of Units (SI)*. The table gives the factor by which a quantity expressed in a non-SI unit should be multiplied in order to calculate its value in the SI. The SI values are expressed in terms of the base, supplementary, and derived units of SI in order to provide a coherent presentation of the conversion factors and facilitate computations (see the table "International System of Units" in this Section). If desired, powers of ten can be avoided by using SI Prefixes and shifting the decimal point if necessary.

Conversion from a non-SI unit to a different non-SI unit may be carried out by using this table in two stages, e.g.,

$$1 \text{ cal}_{th} = 4.184 \text{ J}$$

$$1 \text{ Btu}_{IT} = 1.055056 \text{ E}+03 \text{ J}$$

Thus,

$$1 \text{ Btu}_{IT} = (1.055056 \text{ E}+03 \div 4.184) \text{ cal}_{th} = 252.164 \text{ cal}_{th}$$

Conversion factors are presented for ready adaptation to computer readout and electronic data transmission. The factors are written as a number equal to or greater than one and less than ten with six or fewer decimal places. This number is followed by the letter E (for exponent), a plus or a minus sign, and two digits which indicate the power of 10 by which the number must be multiplied to obtain the correct value. For example:

$$3.523\,907 \text{ E}-02 \text{ is } 3.523\,907 \times 10^{-2}$$

or

$$0.035\,239\,07$$

Similarly:

$$3.386\,389 \text{ E}+03 \text{ is } 3.386\,389 \times 10^{3}$$

or

$$3\,386.389$$

A factor in boldface is exact; i.e., all subsequent digits are zero. All other conversion factors have been rounded to the figures given in accordance with accepted practice. Where less than six digits after the decimal point are shown, more precision is not warranted.

It is often desirable to round a number obtained from a conversion of units in order to retain information on the precision of the value. The following rounding rules may be followed:

(1) If the digits to be discarded begin with a digit less than 5, the digit preceding the first discarded digit is not changed.

Example: 6.974 951 5 rounded to 3 digits is 6.97

(2) If the digits to be discarded begin with a digit greater than 5, the digit preceding the first discarded digit is increased by one.

Example: 6.974 951 5 rounded to 4 digits is 6.975

(3) If the digits to be discarded begin with a 5 and at least one of the following digits is greater than 0, the digit preceding the 5 is increased by 1.

Example: 6.974 851 rounded to 5 digits is 6.974 9

(4) If the digits to be discarded begin with a 5 and all of the following digits are 0, the digit preceding the 5 is unchanged if it is even and increased by one if it is odd. (Note that this means that the final digit is always even.)

Examples: 6.974 951 5 rounded to 7 digits is 6.974 952
6.974 950 5 rounded to 7 digits is 6.974 950

REFERENCE

Taylor, B. N., *Guide for the Use of the International System of Units (SI)*, NIST Special Publication 811, 1995 Edition, Superintendent of Documents, U.S. Government Printing Office, Washington, DC 20402, 1995.

To convert from	to		Multiply by
abampere	ampere (A)	1.0	**E+01**
abcoulomb	coulomb (C)	1.0	**E+01**
abfarad	farad (F)	1.0	**E+09**
abhenry	henry (H)	1.0	**E−09**
abmho	siemens (S)	1.0	**E+09**
abohm	ohm (Ω)	1.0	**E−09**
abvolt	volt (V)	1.0	**E−08**
acceleration of free fall, standard (g_n)	meter per second squared (m/s^2)	**9.806 65**	**E+00**
acre (based on U.S. survey foot)[9]	square meter (m^2)	4.046 873	E+03
acre foot (based on U.S. survey foot)[9]	cubic meter (m^3)	1.233 489	E+03
ampere hour (A · h)	coulomb (C)	**3.6**	**E+03**
ångström (Å)	meter (m)	**1.0**	**E−10**
ångström (Å)	nanometer (nm)	**1.0**	**E−01**
are (a)	square meter (m^2)	**1.0**	**E+02**
astronomical unit (AU)	meter (m)	1.495 979	E+11
atmosphere, standard (atm)	pascal (Pa)	**1.013 25**	**E+05**
atmosphere, standard (atm)	kilopascal (kPa)	**1.013 25**	**E+02**
atmosphere, technical (at)[10]	pascal (Pa)	**9.806 65**	**E+04**
atmosphere, technical (at)[10]	kilopascal (kPa)	**9.806 65**	**E+01**
bar (bar)	pascal (Pa)	**1.0**	**E+05**
bar (bar)	kilopascal (kPa)	**1.0**	**E+02**
barn (b)	square meter (m^2)	**1.0**	**E−28**
barrel [for petroleum, 42 gallons (U.S.)](bbl)	cubic meter (m^3)	1.589 873	E−01
barrel [for petroleum, 42 gallons (U.S.)](bbl)	liter (L)	1.589 873	E+02
biot (Bi)	ampere (A)	**1.0**	**E+01**
British thermal unit$_{IT}$ (Btu$_{IT}$)[11]	joule (J)	1.055 056	E+03
British thermal unit$_{th}$ (Btu$_{th}$)[11]	joule (J)	1.054 350	E+03
British thermal unit (mean) (Btu)	joule (J)	1.055 87	E+03
British thermal unit (39 °F) (Btu)	joule (J)	1.059 67	E+03
British thermal unit (59 °F) (Btu)	joule (J)	1.054 80	E+03
British thermal unit (60 °F) (Btu)	joule (J)	1.054 68	E+03
British thermal unit$_{IT}$ foot per hour square foot degree Fahrenheit [Btu$_{IT}$ · ft/(h · ft^2 · °F)]	watt per meter kelvin [W/(m · K)]	1.730 735	E+00
British thermal unit$_{th}$ foot per hour square foot degree Fahrenheit [Btu$_{th}$ · ft/(h · ft^2 · °F)]	watt per meter kelvin [W/(m · K)]	1.729 577	E+00
British thermal unit$_{IT}$ inch per hour square foot degree Fahrenheit [Btu$_{IT}$ · in/(h · ft^2 · °F)]	watt per meter kelvin [W/(m · K)]	1.442 279	E−01
British thermal unit$_{th}$ inch per hour square foot degree Fahrenheit [Btu$_{th}$ · in/(h · ft^2 · °F)]	watt per meter kelvin [W/(m · K)]	1.441 314	E−01
British thermal unit$_{IT}$ inch per second square foot degree Fahrenheit [Btu$_{IT}$ · in/(s · ft^2 · °F)]	watt per meter kelvin [W/(m · K)]	5.192 204	E+02

---

[9] The U.S. survey foot equals (1200/3937) m. 1 international foot = 0.999998 survey foot.

[10] One technical atmosphere equals one kilogram-force per square centimeter (1 at = 1 kgf/cm^2).

[11] The Fifth International Conference on the Properties of Steam (London, July 1956) defined the International Table calorie as 4.1868 J. Therefore the exact conversion factor for the International Table Btu is 1.055 055 852 62 kJ. Note that the notation for International Table used in this listing is subscript "IT". Similarly, the notation for thermochemical is subscript "th." Further, the thermochemical Btu, Btu$_{th}$, is based on the thermochemical calorie, cal$_{th}$, where cal$_{th}$ = 4.184 J exactly.

To convert from	to	Multiply by	
British thermal unit$_{th}$ inch per second square foot degree Fahrenheit [Btu$_{th}$ · in/(s · ft^2 · °F)]	watt per meter kelvin [W/(m · K)]	5.188 732	E+02
British thermal unit$_{IT}$ per cubic foot (Btu$_{IT}$/ft^3)	joule per cubic meter (J/m^3)	3.725 895	E+04
British thermal unit$_{th}$ per cubic foot (Btu$_{th}$/ft^3)	joule per cubic meter (J/m^3)	3.723 403	E+04
British thermal unit$_{IT}$ per degree Fahrenheit (Btu$_{IT}$/°F)	joule per kelvin (J/k)	1.899 101	E+03
British thermal unit$_{th}$ per degree Fahrenheit (Btu$_{th}$/°F)	joule per kelvin (J/k)	1.897 830	E+03
British thermal unit$_{IT}$ per degree Rankine (Btu$_{IT}$/°R)	joule per kelvin (J/k)	1.899 101	E+03
British thermal unit$_{th}$ per degree Rankine (Btu$_{th}$/°R)	joule per kelvin (J/k)	1.897 830	E+03
British thermal unit$_{IT}$ per hour (Btu$_{IT}$/h)	watt (W)	2.930 711	E−01
British thermal unit$_{th}$ per hour (Btu$_{th}$/h)	watt (W)	2.928 751	E−01
British thermal unit$_{IT}$ per hour square foot degree Fahrenheit [Btu$_{IT}$/(h · ft^2 · °F)]	watt per square meter kelvin [W/(m^2 · K)]	5.678 263	E+00
British thermal unit$_{th}$ per hour square foot degree Fahrenheit [Btu$_{th}$/(h · ft^2 · °F)]	watt per square meter kelvin [W/(m^2 · K)]	5.674 466	E+00
British thermal unit$_{th}$ per minute (Btu$_{th}$/min)	watt (W)	1.757 250	E+01
British thermal unit$_{IT}$ per pound (Btu$_{IT}$/lb)	joule per kilogram (J/kg)	**2.326**	**E+03**
British thermal unit$_{th}$ per pound (Btu$_{th}$/lb)	joule per kilogram (J/kg)	2.324 444	E+03
British thermal unit$_{IT}$ per pound degree Fahrenheit [Btu$_{IT}$/(lb · °F)]	joule per kilogram kelvin (J/(kg · K)]	**4.1868**	**E+03**
British thermal unit$_{th}$ per pound degree Fahrenheit [Btu$_{th}$/(lb · °F)]	joule per kilogram kelvin [J/(kg · K)]	**4.184**	**E+03**
British thermal unit$_{IT}$ per pound degree Rankine [Btu$_{IT}$/(lb · °R)]	joule per kilogram kelvin [J/(kg · K)]	**4.1868**	**E+03**
British thermal unit$_{th}$ per pound degree Rankine [Btu$_{th}$/(lb · °R)]	joule per kilogram kelvin [J/(kg · K)]	**4.184**	**E+03**
British thermal unit$_{IT}$ per second (Btu$_{IT}$/s)	watt (W)	1.055 056	E+03
British thermal unit$_{th}$ per second (Btu$_{th}$/s)	watt (W)	1.054 350	E+03
British thermal unit$_{IT}$ per second square foot degree Fahrenheit [Btu$_{IT}$/(s · ft^2 · °F)]	watt per square meter kelvin [W/(m^2 · K)]	2.044 175	E+04
British thermal unit$_{th}$ per second square foot degree Fahrenheit [Btu$_{th}$/(s · ft^2 · °F)]	watt per square meter kelvin [W/(m^2 · K)]	2.042 808	E+04
British thermal unit$_{IT}$ per square foot (Btu$_{IT}$/ft^2)	joule per square meter (J/m^2)	1.135 653	E+04
British thermal unit$_{th}$ per square foot (Btu$_{th}$/ft^2)	joule per square meter (J/m^2)	1.134 893	E+04
British thermal unit$_{IT}$ per square foot hour [(Btu$_{IT}$/(ft^2 · h)]	watt per square meter (W/m^2)	3.154 591	E+00
British thermal unit$_{th}$ per square foot hour [Btu$_{th}$/(ft^2 · h)]	watt per square meter (W/m^2)	3.152 481	E+00
British thermal unit$_{th}$ per square foot minute [Btu$_{th}$/(ft^2 · min)]	watt per square meter (W/m^2)	1.891 489	E+02
British thermal unit$_{IT}$ per square foot second [(Btu$_{IT}$/(ft^2 · s)]	watt per square meter (W/m^2)	1.135 653	E+04
British thermal unit$_{th}$ per square foot second [Btu$_{th}$/(ft^2 · s)]	watt per square meter (W/m^2)	1.134 893	E+04
British thermal unit$_{th}$ per square inch second [Btu$_{th}$/(in^2 · s)]	watt per square meter (W/m^2)	1.634 246	E+06

To convert from	to	Multiply by	
bushel (U.S.) (bu)	cubic meter (m^3)	3.523 907	E−02
bushel (U.S.) (bu)	liter (L)	3.523 907	E+01
calorie$_{IT}$ (cal$_{IT}$)[11]	joule (J)	**4.1868**	**E+00**
calorie$_{th}$ (cal$_{th}$)[11]	joule (J)	**4.184**	**E+00**
calorie (cal) (mean)	joule (J)	4.190 02	E+00
calorie (15 °C) (cal$_{15}$)	joule (J)	4.185 80	E+00
calorie (20 °C) (cal$_{20}$)	joule (J)	4.181 90	E+00
calorie$_{IT}$, kilogram (nutrition)[12]	joule (J)	**4.1868**	**E+03**
calorie$_{th}$, kilogram (nutrition)[12]	joule (J)	**4.184**	**E+03**
calorie (mean), kilogram (nutrition)[12]	joule (J)	4.190 02	E+03
calorie$_{th}$ per centimeter second degree Celsius [cal$_{th}$/(cm · s · °C)]	watt per meter kelvin [W/(m · K)]	**4.184**	**E+02**
calorie$_{IT}$ per gram (cal$_{IT}$/g)	joule per kilogram (J/kg)	**4.1868**	**E+03**
calorie$_{th}$ per gram (cal$_{th}$/g)	joule per kilogram (J/kg)	**4.184**	**E+03**
calorie$_{IT}$ per gram degree Celsius [cal$_{IT}$/(g · °C)]	joule per kilogram kelvin [J/(kg · K)]	**4.1868**	**E+03**
calorie$_{th}$ per gram degree Celsius [cal$_{th}$/(g · °C)]	joule per kilogram kelvin [J/(kg · K)]	**4.184**	**E+03**
calorie$_{IT}$ per gram kelvin [cal$_{IT}$/(g · K)]	joule per kilogram kelvin [J/(kg · K)]	**4.1868**	**E+03**
calorie$_{th}$ per gram kelvin [cal$_{th}$/(g · K)]	joule per kilogram kelvin [J/(kg · K)]	**4.184**	**E+03**
calorie$_{th}$ per minute (cal$_{th}$/min)	watt (W)	6.973 333	E−02
calorie$_{th}$ per second (cal$_{th}$/s)	watt (W)	**4.184**	**E+00**
calorie$_{th}$ per square centimeter (cal$_{th}$/cm^2)	joule per square meter (J/m^2)	**4.184**	**E+04**
calorie$_{th}$ per square centimeter minute [cal$_{th}$/(cm^2 · min)]	watt per square meter (W/m^2)	6.973 333	E+02
calorie$_{th}$ per square centimeter second [cal$_{th}$/(cm^2 · s)]	watt per square meter (W/m^2)	**4.184**	**E+04**
candela per square inch (cd/in^2)	candela per square meter (cd/m^2)	1.550 003	E+03
carat, metric	kilogram (kg)	**2.0**	**E−04**
carat, metric	gram (g)	**2.0**	**E−01**
centimeter of mercury (0 °C)[13]	pascal (Pa)	1.333 22	E+03
centimeter of mercury (0 °C)[13]	kilopascal (kPa)	1.333 22	E+00
centimeter of mercury, conventional (cmHg)[13]	pascal (Pa)	1.333 224	E+03
centimeter of mercury, conventional (cmHg)[13]	kilopascal (kPa)	1.333 224	E+00
centimeter of water (4 °C)[13]	pascal (Pa)	9.806 38	E+01
centimeter of water, conventional (cmH$_2$O)[13]	pascal (Pa)	**9.806 65**	**E+01**
centipoise (cP)	pascal second (Pa · s)	**1.0**	**E−03**
centistokes (cSt)	meter squared per second (m^2/s)	**1.0**	**E−06**
chain (based on U.S. survey foot) (ch)[9]	meter (m)	2.011 684	E+01
circular mil	square meter (m^2)	5.067 075	E−10
circular mil	square millimeter (mm^2)	5.067 075	E−04
clo	square meter kelvin per watt (m^2 · K/W)	1.55	E−01
cord (128 ft^3)	cubic meter (m^3)	3.624 556	E+00
cubic foot (ft^3)	cubic meter (m^3)	2.831 685	E−02
cubic foot per minute (ft^3/min)	cubic meter per second (m^3/s)	4.719 474	E−04
cubic foot per minute (ft^3/min)	liter per second (L/s)	4.719 474	E−01
cubic foot per second (ft^3/s)	cubic meter per second (m^3/s)	2.831 685	E−02

---

[12] The kilogram calorie or "large calorie" is an obsolete term used for the kilocalorie, which is the calorie used to express the energy content of foods. However, in practice, the prefix "kilo" is usually omitted.

[13] Conversion factors for mercury manometer pressure units are calculated using the standard value for the acceleration of gravity and the density of mercury at the stated temperature. Additional digits are not justified because the definitions of the units do not take into account the compressibility of mercury or the change in density caused by the revised practical temperature scale, ITS-90. Similar comments also apply to water manometer pressure units. Conversion factors for conventional mercury and water manometer pressure units are based on ISO 31-3.

To convert from	to	Multiply by	
cubic inch (in^3)[14]	cubic meter (m^3)	1.638 706	E−05
cubic inch per minute (in^3/min)	cubic meter per second (m^3/s)	2.731 177	E−07
cubic mile (mi^3)	cubic meter (m^3)	4.168 182	E+09
cubic yard (yd^3)	cubic meter (m^3)	7.645 549	E−01
cubic yard per minute (yd^3/min)	cubic meter per second (m^3/s)	1.274 258	E−02
cup (U.S.)	cubic meter (m^3)	2.365 882	E−04
cup (U.S.)	liter (L)	2.365 882	E−01
cup (U.S.)	milliliter (mL)	2.365 882	E+02
curie (Ci)	becquerel (Bq)	**3.7**	**E+10**
darcy[15]	meter squared (m^2)	9.869 233	E−13
day (d)	second (s)	**8.64**	**E+04**
day (sidereal)	second (s)	8.616 409	E+04
debye (D)	coulomb meter (C · m)	3.335 641	E−30
degree (angle) (°)	radian (rad)	1.745 329	E−02
degree Celsius (temperature) (°C)	kelvin (K)	$T/\mathrm{K} = t/°\mathrm{C} + 273.15$	
degree Celsius (temperature interval) (°C)	kelvin (K)	**1.0**	**E+00**
degree centigrade (temperature)[16]	degree Celsius (°C)	$t/°\mathrm{C} \approx t/\mathrm{deg. cent.}$	
degree centigrade (temperature interval)[16]	degree Celsius (°C)	1.0	E+00
degree Fahrenheit (temperature) (°F)	degree Celsius (°C)	$t/°\mathrm{C} = (t/°\mathrm{F} - \textbf{32})/\textbf{1.8}$	
degree Fahrenheit (temperature) (°F)	kelvin (K)	$T/\mathrm{K} = (t/°\mathrm{F} + \textbf{459.67})/\textbf{1.8}$	
degree Fahrenheit (temperature interval)(°F)	degree Celsius (°C)	5.555 556	E−01
degree Fahrenheit (temperature interval)(°F)	kelvin (K)	5.555 556	E−01
degree Fahrenheit hour per British thermal unit$_{IT}$ (°F · h/Btu$_{IT}$)	kelvin per watt (K/W)	1.895 634	E+00
degree Fahrenheit hour per British thermal unit$_{th}$ (°F · h/Btu$_{th}$)	kelvin per watt (K/W)	1.896 903	E+00
degree Fahrenheit hour square foot per British thermal unit$_{IT}$ (°F · h · ft^2/Btu$_{IT}$)	square meter kelvin per watt (m^2 · K/W)	1.761 102	E−01
degree Fahrenheit hour square foot per British thermal unit$_{th}$ (°F · h · ft^2/Btu$_{th}$)	square meter kelvin per watt (m^2 · K/W)	1.762 280	E−01
degree Fahrenheit hour square foot per British thermal unit$_{IT}$ inch [°F · h · ft^2/(Btu$_{IT}$ · in)]	meter kelvin per watt (m · K/W)	6.933 472	E+00
degree Fahrenheit hour square foot per British thermal unit$_{th}$ inch [°F · h · ft^2/(Btu$_{th}$ · in)]	meter kelvin per watt (m · K/W)	6.938 112	E+00
degree Fahrenheit second per British thermal unit$_{IT}$ (°F · s/Btu$_{IT}$)	kelvin per watt (K/W)	5.265 651	E−04
degree Fahrenheit second per British thermal unit$_{th}$ (°F · s/Btu$_{th}$)	kelvin per watt (K/W)	5.269 175	E−04
degree Rankine (°R)	kelvin (K)	$T/\mathrm{K} = (T/°\mathrm{R})/\textbf{1.8}$	
degree Rankine (temperature interval) (°R)	kelvin (K)	5.555 556	E−01
denier	kilogram per meter (kg/m)	1.111 111	E−07
denier	gram per meter (g/m)	1.111 111	E−04
dyne (dyn)	newton (N)	**1.0**	**E−05**
dyne centimeter (dyn · cm)	newton meter (N · m)	**1.0**	**E−07**
dyne per square centimeter (dyn/cm^2)	pascal (Pa)	**1.0**	**E−01**
electronvolt (eV)	joule (J)	1.602 177	E−19
EMU of capacitance (abfarad)	farad (F)	**1.0**	**E+09**
EMU of current (abampere)	ampere (A)	**1.0**	**E+01**
EMU of electric potential (abvolt)	volt (V)	**1.0**	**E−08**
EMU of inductance (abhenry)	henry (H)	**1.0**	**E−09**

---

[14] The exact conversion factor is 1.638 706 4 E−05.
[15] The darcy is a unit for expressing the permeability of porous solids, not area.
[16] The centigrade temperature scale is obsolete; the degree centigrade is only approximately equal to the degree Celsius.

To convert from	to		Multiply by
EMU of resistance (abohm)	ohm (Ω)	1.0	**E$-$09**
erg (erg)	joule (J)	1.0	**E$-$07**
erg per second (erg/s)	watt (W)	1.0	**E$-$07**
erg per square centimeter second [[obrkt&1rul/($cm^2 \cdot s$)]]	watt per square meter (W/m^2)	1.0	**E$-$03**
ESU of capacitance (statfarad)	farad (F)	1.112 650	E$-$12
ESU of current (statampere)	ampere (A)	3.335 641	E$-$10
ESU of electric potential (statvolt)	volt (V)	2.997 925	E+02
ESU of inductance (stathenry)	henry (H)	8.987 552	E+11
ESU of resistance (statohm)	ohm (Ω)	8.987 552	E+11
faraday (based on carbon 12)	coulomb (C)	9.648 531	E+04
fathom (based on U.S. survey foot)[9]	meter (m)	1.828 804	E+00
fermi	meter (m)	1.0	**E$-$15**
fermi	femtometer (fm)	1.0	**E+00**
fluid ounce (U.S.) (fl oz)	cubic meter (m^3)	2.957 353	E$-$05
fluid ounce (U.S.) (fl oz)	milliliter (mL)	2.957 353	E+01
foot (ft)	meter (m)	**3.048**	**E$-$01**
foot (U.S. survey) (ft)[9]	meter (m)	3.048 006	E$-$01
footcandle	lux (lx)	1.076 391	E+01
footlambert	candela per square meter (cd/m^2)	3.426 259	E+00
foot of mercury, conventional (ftHg)[13]	pascal (Pa)	4.063 666	E+04
foot of mercury, conventional (ftHg)[13]	kilopascal (kPa)	4.063 666	E+01
foot of water (39.2 °F)[13]	pascal (Pa)	2.988 98	E+03
foot of water (39.2 °F)[13]	kilopascal (kPa)	2.988 98	E+00
foot of water, conventional (ftH_2O)[13]	pascal (Pa)	2.989 067	E+03
foot of water, conventional (ftH_2O)[13]	kilopascal (kPa)	2.989 067	E+00
foot per hour (ft/h)	meter per second (m/s)	8.466 667	E$-$05
foot per minute (ft/min)	meter per second (m/s)	**5.08**	**E$-$03**
foot per second (ft/s)	meter per second (m/s)	**3.048**	**E$-$01**
foot per second squared (ft/s^2)	meter per second squared (m/s^2)	**3.048**	**E$-$01**
foot poundal	joule (J)	4.214 011	E$-$02
foot pound-force (ft · lbf)	joule (J)	1.355 818	E+00
foot pound-force per hour (ft · lbf/h)	watt (W)	3.766 161	E$-$04
foot pound-force per minute (ft · lbf/min)	watt (W)	2.259 697	E$-$02
foot pound-force per second (ft · lbf/s)	watt (W)	1.355 818	E+00
foot to the fourth power (ft^4)[17]	meter to the fourth power (m^4)	8.630 975	E$-$03
franklin (Fr)	coulomb (C)	3.335 641	E$-$10
gal (Gal)	meter per second squared (m/s^2)	1.0	**E$-$02**
gallon [Canadian and U.K. (Imperial)] (gal)	cubic meter (m^3)	**4.546 09**	**E$-$03**
gallon [Canadian and U.K. (Imperial)] (gal)	liter (L)	**4.546 09**	**E+00**
gallon (U.S.) (gal)	cubic meter (m^3)	3.785 412	E$-$03
gallon (U.S.) (gal)	liter (L)	3.785 412	E+00
gallon (U.S.) per day (gal/d)	cubic meter per second (m^3/s)	4.381 264	E$-$08
gallon (U.S.) per day (gal/d)	liter per second (L/s)	4.381 264	E$-$05
gallon (U.S.) per horsepower hour [gal/(hp · h)]	cubic meter per joule (m^3/J)	1.410 089	E$-$09
gallon (U.S.) per horsepower hour [gal/(hp · h)]	liter per joule (L/J)	1.410 089	E$-$06
gallon (U.S.) per minute (gpm)(gal/min)	cubic meter per second (m^3/s)	6.309 020	E$-$05
gallon (U.S.) per minute (gpm)(gal/min)	liter per second (L/s)	6.309 020	E$-$02

---

[17] This is a unit for the quantity second moment of area, which is sometimes called the ''moment of section'' or ''area moment of inertia'' of a plane section about a specified axis.

To convert from	to	Multiply by	
gamma (γ)	tesla (T)	1.0	E−09
gauss (Gs, G)	tesla (T)	1.0	E−04
gilbert (Gi)	ampere (A)	7.957 747	E−01
gill [Canadian and U.K. (Imperial)] (gi)	cubic meter (m³)	1.420 653	E−04
gill [Canadian and U.K. (Imperial)] (gi)	liter (L)	1.420 653	E−01
gill (U.S.) (gi)	cubic meter (m³)	1.182 941	E−04
gill (U.S.) (gi)	liter (L)	1.182 941	E−01
gon (also called grade) (gon)	radian (rad)	1.570 796	E−02
gon (also called grade) (gon)	degree (angle) (°)	9.0	E−01
grain (gr)	kilogram (kg)	6.479 891	E−05
grain (gr)	milligram (mg)	6.479 891	E+01
grain per gallon (U.S.) (gr/gal)	kilogram per cubic meter (kg/m³)	1.711 806	E−02
grain per gallon (U.S.) (gr/gal)	milligram per liter (mg/L)	1.711 806	E+01
gram-force per square centimeter (gf/cm²)	pascal (Pa)	9.806 65	E+01
gram per cubic centimeter (g/cm³)	kilogram per cubic meter (kg/m³)	1.0	E+03
hectare (ha)	square meter (m²)	1.0	E+04
horsepower (550 ft · lbf/s) (hp)	watt (W)	7.456 999	E+02
horsepower (boiler)	watt (W)	9.809 50	E+03
horsepower (electric)	watt (W)	7.46	E+02
horsepower (metric)	watt (W)	7.354 988	E+02
horsepower (U.K.)	watt (W)	7.4570	E+02
horsepower (water)	watt (W)	7.460 43	E+02
hour (h)	second (s)	3.6	E+03
hour (sidereal)	second (s)	3.590 170	E+03
hundredweight (long, 112 lb)	kilogram (kg)	5.080 235	E+01
hundredweight (short, 100 lb)	kilogram (kg)	4.535 924	E+01
inch (in)	meter (m)	2.54	E−02
inch (in)	centimeter (cm)	2.54	E+00
inch of mercury (32 °F)[13]	pascal (Pa)	3.386 38	E+03
inch of mercury (32 °F)[13]	kilopascal (kPa)	3.386 38	E+00
inch of mercury (60 °F)[13]	pascal (Pa)	3.376 85	E+03
inch of mercury (60 °F)[13]	kilopascal (kPa)	3.376 85	E+00
inch of mercury, conventional (inHg)[13]	pascal (Pa)	3.386 389	E+03
inch of mercury, conventional (inHg)[13]	kilopascal (kPa)	3.386 389	E+00
inch of water (39.2 °F)[13]	pascal (Pa)	2.490 82	E+02
inch of water (60 °F)[13]	pascal (Pa)	2.4884	E+02
inch of water, conventional (inH₂O)[13]	pascal (Pa)	2.490 889	E+02
inch per second (in/s)	meter per second (m/s)	2.54	E−02
inch per second squared (in/s²)	meter per second squared (m/s²)	2.54	E−02
inch to the fourth power (in⁴)[17]	meter to the fourth power (m⁴)	4.162 314	E−07
kayser (K)	reciprocal meter (m⁻¹)	1.0	E+02
kelvin (K)	degree Celsius (°C)	$t/°C = T/K - 273.15$	
kilocalorie$_{IT}$ (kcal$_{IT}$)	joule (J)	4.1868	E+03
kilocalorie$_{th}$ (kcal$_{th}$)	joule (J)	4.184	E+03
kilocalorie (mean) (kcal)	joule (J)	4.190 02	E+03
kilocalorie$_{th}$ per minute (kcal$_{th}$/min)	watt (W)	6.973 333	E+01
kilocalorie$_{th}$ per second (kcal$_{th}$/s)	watt (W)	4.184	E+03
kilogram-force (kgf)	newton (N)	9.806 65	E+00
kilogram-force meter (kgf · m)	newton meter (N · m)	9.806 65	E+00

To convert from	to		Multiply by
kilogram-force per square centimeter (kgf/cm²)	pascal (Pa)	**9.806 65**	**E+04**
kilogram-force per square centimeter (kgf/cm²)	kilopascal (kPa)	**9.806 65**	**E+01**
kilogram-force per square meter (kgf/m²)	pascal (Pa)	**9.806 65**	**E+00**
kilogram-force per square millimeter (kgf/mm²)	pascal (Pa)	**9.806 65**	**E+06**
kilogram-force per square millimeter (kgf/mm²)	megapascal (MPa)	**9.806 65**	**E+00**
kilogram-force second squared per meter (kgf · s²/m)	kilogram (kg)	**9.806 65**	**E+00**
kilometer per hour (km/h)	meter per second (m/s)	2.777 778	E−01
kilopond (kilogram-force) (kp)	newton (N)	**9.806 65**	**E+00**
kilowatt hour (kW · h)	joule (J)	**3.6**	**E+06**
kilowatt hour (kW · h)	megajoule (MJ)	**3.6**	**E+00**
kip (1 kip=1000 lbf)	newton (N)	4.448 222	E+03
kip (1 kip=1000 lbf)	kilonewton (kN)	4.448 222	E+00
kip per square inch (ksi) (kip/in²)	pascal (Pa)	6.894 757	E+06
kip per square inch (ksi) (kip/in²)	kilopascal (kPa)	6.894 757	E+03
knot (nautical mile per hour)	meter per second (m/s)	5.144 444	E−01
lambert[18]	candela per square meter (cd/m²)	3.183 099	E+03
langley (cal_th/cm²)	joule per square meter (J/m²)	**4.184**	**E+04**
light year (l.y.)[19]	meter (m)	9.460 73	E+15
liter (L)[20]	cubic meter (m³)	**1.0**	**E−03**
lumen per square foot (lm/ft²)	lux (lx)	1.076 391	E+01
maxwell (Mx)	weber (Wb)	**1.0**	**E−08**
mho	siemens (S)	**1.0**	**E+00**
microinch	meter (m)	**2.54**	**E−08**
microinch	micrometer (μm)	**2.54**	**E−02**
micron (μ)	meter (m)	**1.0**	**E−06**
micron (μ)	micrometer (μm)	**1.0**	**E+00**
mil (0.001 in)	meter (m)	**2.54**	**E−05**
mil (0.001 in)	millimeter (mm)	**2.54**	**E−02**
mil (angle)	radian (rad)	9.817 477	E−04
mil (angle)	degree (°)	**5.625**	**E−02**
mile (mi)	meter (m)	**1.609 344**	**E+03**
mile (mi)	kilometer (km)	**1.609 344**	**E+00**
mile (based on U.S. survey foot) (mi)[9]	meter (m)	1.609 347	E+03
mile (based on U.S. survey foot) (mi)[9]	kilometer (km)	1.609 347	E+00
mile, nautical[21]	meter (m)	**1.852**	**E+03**
mile per gallon (U.S.) (mpg) (mi/gal)	meter per cubic meter (m/m³)	4.251 437	E+05
mile per gallon (U.S.) (mpg) (mi/gal)	kilometer per liter (km/L)	4.251 437	E−01
mile per gallon (U.S.) (mpg) (mi/gal)[22]	liter per 100 kilometer (L/100 km)	divide 235.215 by number of miles per gallon	
mile per hour (mi/h)	meter per second (m/s)	**4.4704**	**E−01**
mile per hour (mi/h)	kilometer per hour (km/h)	**1.609 344**	**E+00**

[18] The exact conversion factor is $10^4/\pi$.

[19] This conversion factor is based on 1 d = 86 400 s; and 1 Julian century = 36 525 d. (See *The Astronomical Almanac for the Year 1995*, page K6, U.S. Government Printing Office, Washington, DC, 1994).

[20] In 1964 the General Conference on Weights and Measures reestablished the name "liter" as a special name for the cubic decimeter. Between 1901 and 1964 the liter was slightly larger (1.000 028 dm³); when one uses high-accuracy volume data of that time, this fact must be kept in mind.

[21] The value of this unit, 1 nautical mile = 1852 m, was adopted by the First International Extraordinary Hydrographic Conference, Monaco, 1929, under the name "International nautical mile."

[22] For converting fuel economy, as used in the U.S., to fuel consumption.

To convert from	to	Multiply by	
mile per minute (mi/min)	meter per second (m/s)	**2.682 24**	**E+01**
mile per second (mi/s)	meter per second (m/s)	**1.609 344**	**E+03**
millibar (mbar)	pascal (Pa)	**1.0**	**E+02**
millibar (mbar)	kilopascal (kPa)	**1.0**	**E−01**
millimeter of mercury, conventional (mmHg)[13]	pascal (Pa)	1.333 224	E+02
millimeter of water, conventional (mmH$_2$O)[13]	pascal (Pa)	**9.806 65**	**E+00**
minute (angle) (')	radian (rad)	2.908 882	E−04
minute (min)	second (s)	**6.0**	**E+01**
minute (sidereal)	second (s)	5.983 617	E+01
oersted (Oe)	ampere per meter (A/m)	7.957 747	E+01
ohm centimeter ($\Omega \cdot$ cm)	ohm meter ($\Omega \cdot$ m)	**1.0**	**E−02**
ohm circular-mil per foot	ohm meter ($\Omega \cdot$ m)	1.662 426	E−09
ohm circular-mil per foot	ohm square millimeter per meter ($\Omega \cdot$ mm^2/m)	1.662 426	E−03
ounce (avoirdupois) (oz)	kilogram (kg)	2.834 952	E−02
ounce (avoirdupois) (oz)	gram (g)	2.834 952	E+01
ounce (troy or apothecary) (oz)	kilogram (kg)	3.110 348	E−02
ounce (troy or apothecary) (oz)	gram (g)	3.110 348	E+01
ounce [Canadian and U.K. fluid (Imperial)] (fl oz)	cubic meter (m^3)	2.841 306	E−05
ounce [Canadian and U.K. fluid (Imperial)] (fl oz)	milliliter (mL)	2.841 306	E+01
ounce (U.S. fluid) (fl oz)	cubic meter (m^3)	2.957 353	E−05
ounce (U.S. fluid) (fl oz)	millimeter (mL)	2.957 353	E+01
ounce (avoirdupois)-force (ozf)	newton (N)	2.780 139	E−01
ounce (avoirdupois)-force inch (ozf $\cdot$ in)	newton meter (N $\cdot$ m)	7.061 552	E−03
ounce (avoirdupois)-force inch (ozf $\cdot$ in)	millinewton meter (mN $\cdot$ m)	7.061 552	E+00
ounce (avoirdupois) per cubic inch (oz/in^3)	kilogram per cubic meter (kg/m^3)	1.729 994	E+03
ounce (avoirdupois) per gallon [Canadian and U.K. (Imperial)] (oz/gal)	kilogram per cubic meter (kg/m^3)	6.236 023	E+00
ounce (avoirdupois) per gallon [Canadian and U.K. (Imperial)] (oz/gal)	gram per liter (g/L)	6.236 023	E+00
ounce (avoirdupois) per gallon(U.S.)(oz/gal)	kilogram per cubic meter (kg/m^3)	7.489 152	E+00
ounce (avoirdupois) per gallon(U.S.)(oz/gal)	gram per liter (g/L)	7.489 152	E+00
ounce (avoirdupois) per square foot (oz/ft^2)	kilogram per square meter (kg/m^2)	3.051 517	E−01
ounce (avoirdupois) per square inch (oz/in^2)	kilogram per square meter (kg/m^2)	4.394 185	E+01
ounce (avoirdupois) per square yard (oz/yd^2)	kilogram per square meter (kg/m^2)	3.390 575	E−02
parsec (pc)	meter (m)	3.085 678	E+16
peck (U.S.) (pk)	cubic meter (m^3)	8.809 768	E−03
peck (U.S.) (pk)	liter (L)	8.809 768	E+00
pennyweight (dwt)	kilogram (kg)	1.555 174	E−03
pennyweight (dwt)	gram (g)	1.555 174	E+00
perm (0 °C)	kilogram per pascal second square meter [kg/(Pa $\cdot$ s $\cdot$ m^2)]	5.721 35	E−11
perm (23 °C)	kilogram per pascal second square meter [kg/(Pa $\cdot$ s $\cdot$ m^2)]	5.745 25	E−11
perm inch (0 °C)	kilogram per pascal second meter [kg/(Pa $\cdot$ s $\cdot$ m)]	1.453 22	E−12
perm inch (23 °C)	kilogram per pascal second meter [kg/(Pa $\cdot$ s $\cdot$ m)]	1.459 29	E−12

To convert from	to	Multiply by	
phot (ph)	lux (lx)	**1.0**	**E+04**
pica (computer) (1/6 in)	meter (m)	4.233 333	E−03
pica (computer) (1/6 in)	millimeter (mm)	4.233 333	E+00
pica (printer's)	meter (m)	4.217 518	E−03
pica (printer's)	millimeter (mm)	4.217 518	E+00
pint (U.S. dry) (dry pt)	cubic meter (m^3)	5.506 105	E−04
pint (U.S. dry) (dry pt)	liter (L)	5.506 105	E−01
pint (U.S. liquid) (liq pt)	cubic meter (m^3)	4.731 765	E−04
pint (U.S. liquid) (liq pt)	liter (L)	4.731 765	E−01
point (computer) (1/72 in)	meter (m)	3.527 778	E−04
point (computer) (1/72 in)	millimeter (mm)	3.527 778	E−01
point (printer's)	meter (m)	3.514 598	E−04
point (printer's)	millimeter (mm)	3.514 598	E−01
poise (P)	pascal second (Pa · s)	**1.0**	**E−01**
pound (avoirdupois) (lb)[23]	kilogram (kg)	4.535 924	E−01
pound (troy or apothecary) (lb)	kilogram (kg)	3.732 417	E−01
poundal	newton (N)	1.382 550	E−01
poundal per square foot	pascal (Pa)	1.488 164	E+00
poundal second per square foot	pascal second (Pa · s)	1.488 164	E+00
pound foot squared (lb · ft^2)	kilogram meter squared (kg · m^2)	4.214 011	E−02
pound-force (lbf)[24]	newton (N)	4.448 222	E+00
pound-force foot (lbf · ft)	newton meter (N · m)	1.355 818	E+00
pound-force foot per inch (lbf · ft/in)	newton meter per meter (N · m/m)	5.337 866	E+01
pound-force inch (lbf · in)	newton meter (N · m)	1.129 848	E−01
pound-force inch per inch (lbf · in/in)	newton meter per meter (N · m/m)	4.448 222	E+00
pound-force per foot (lbf/ft)	newton per meter (N/m)	1.459 390	E+01
pound-force per inch (lbf/in)	newton per meter (N/m)	1.751 268	E+02
pound-force per pound (lbf/lb) (thrust to mass ratio)	newton per kilogram (N/kg)	**9.806 65**	**E+00**
pound-force per square foot (lbf/ft^2)	pascal (Pa)	4.788 026	E+01
pound-force per square inch (psi) (lbf/in^2)	pascal (Pa)	6.894 757	E+03
pound-force per square inch (psi) (lbf/in^2)	kilopascal (kPa)	6.894 757	E+00
pound-force second per square foot (lbf · s/ft^2)	pascal second (Pa · s)	4.788 026	E+01
pound-force second per square inch (lbf · s/in^2)	pascal second (Pa · s)	6.894 757	E+03
pound inch squared (lb · in^2)	kilogram meter squared (kg · m^2)	2.926 397	E−04
pound per cubic foot (lb/ft^3)	kilogram per cubic meter (kg/m^3)	1.601 846	E+01
pound per cubic inch (lb/in^3)	kilogram per cubic meter (kg/m^3)	2.767 990	E+04
pound per cubic yard (lb/yd^3)	kilogram per cubic meter (kg/m^3)	5.932 764	E−01
pound per foot (lb/ft)	kilogram per meter (kg/m)	1.488 164	E+00
pound per foot hour [lb/(ft · h)]	pascal second (Pa · s)	4.133 789	E−04
pound per foot second [lb/(ft · s)]	pascal second (Pa · s)	1.488 164	E+00
pound per gallon [Canadian and U.K. (Imperial)] (lb/gal)	kilogram per cubic meter (kg/m^3)	9.977 637	E+01
pound per gallon [Canadian and U.K. (Imperial)] (lb/gal)	kilogram per liter (kg/L)	9.977 637	E−02
pound per gallon (U.S.) (lb/gal)	kilogram per cubic meter (kg/m^3)	1.198 264	E+02
pound per gallon (U.S.) (lb/gal)	kilogram per liter (kg/L)	1.198 264	E−01
pound per horsepower hour [lb/(hp · h)]	kilogram per joule (kg/J)	1.689 659	E−07
pound per hour (lb/h)	kilogram per second (kg/s)	1.259 979	E−04

---

[23] The exact conversion factor is 4.535 923 7 E−01. All units that contain the pound refer to the avoirdupois pound.

[24] If the local value of the acceleration of free fall is taken as $g_n = 9.806\ 65$ m/s^2 (the standard value), the exact conversion factor is 4.448 221 615 260 5 E+00.

To convert from	to	Multiply by	
pound per inch (lb/in)	kilogram per meter (kg/m)	1.785 797	E+01
pound per minute (lb/min)	kilogram per second (kg/s)	7.559 873	E−03
pound per second (lb/s)	kilogram per second (kg/s)	4.535 924	E−01
pound per square foot (lb/ft^2)	kilogram per square meter (kg/m^2)	4.882 428	E+00
pound per square inch (*not* pound-force) (lb/in^2)	kilogram per square meter (kg/m^2)	7.030 696	E+02
pound per yard (lb/yd)	kilogram per meter (kg/m)	4.960 546	E−01
psi (pound-force per square inch) (lbf/in^2)	pascal (Pa)	6.894 757	E+03
psi (pound-force per square inch) (lbf/in^2)	kilopascal (kPa)	6.894 757	E+00
quad (10^{15} Btu$_{IT}$)[11]	joule (J)	1.055 056	E+18
quart (U.S. dry) (dry qt)	cubic meter (m^3)	1.101 221	E−03
quart (U.S. dry) (dry qt)	liter (L)	1.101 221	E+00
quart (U.S. liquid) (liq qt)	cubic meter (m^3)	9.463 529	E−04
quart (U.S. liquid) (liq qt)	liter (L)	9.463 529	E−01
rad (absorbed dose) (rad)	gray (Gy)	**1.0**	**E−02**
rem (rem)	sievert (Sv)	**1.0**	**E−02**
revolution (r)	radian (rad)	6.283 185	E+00
revolution per minute (rpm) (r/min)	radian per second (rad/s)	1.047 198	E−01
rhe	reciprocal pascal second [(Pa · s)$^{-1}$]	**1.0**	**E+01**
rod (based on U.S. survey foot) (rd)[9]	meter (m)	5.029 210	E+00
roentgen (R)	coulomb per kilogram (C/kg)	**2.58**	**E−04**
rpm (revolution per minute) (r/min)	radian per second (rad/s)	1.047 198	E−01
second (angle) (")	radian (rad)	4.848 137	E−06
second (sidereal)	second (s)	9.972 696	E−01
shake	second (s)	**1.0**	**E−08**
shake	nanosecond (ns)	**1.0**	**E+01**
slug (slug)	kilogram (kg)	1.459 390	E+01
slug per cubic foot (slug/ft^3)	kilogram per cubic meter (kg/m^3)	5.153 788	E+02
slug per foot second [slug/(ft · s)]	pascal second (Pa · s)	4.788 026	E+01
square foot (ft^2)	square meter (m^2)	**9.290 304**	**E−02**
square foot per hour (ft^2/h)	square meter per second (m^2/s)	**2.580 64**	**E−05**
square foot per second (ft^2/s)	square meter per second (m^2/s)	**9.290 304**	**E−02**
square inch (in^2)	square meter (m^2)	**6.4516**	**E−04**
square inch (in^2)	square centimeter (cm^2)	**6.4516**	**E+00**
square mile (mi^2)	square meter (m^2)	2.589 988	E+06
square mile (mi^2)	square kilometer (km^2)	2.589 988	E+00
square mile (based on U.S. survey foot) (mi^2)[9]	square meter (m^2)	2.589 998	E+06
square mile (based on U.S. survey foot) (mi^2)[9]	square kilometer (km^2)	2.589 998	E+00
square yard (yd^2)	square meter (m^2)	8.361 274	E−01
statampere	ampere (A)	3.335 641	E−10
statcoulomb	coulomb (C)	3.335 641	E−10
statfarad	farad (F)	1.112 650	E−12
stathenry	henry (H)	8.987 552	E+11
statmho	siemens (S)	1.112 650	E−12
statohm	ohm (Ω)	8.987 552	E+11
statvolt	volt (V)	2.997 925	E+02
stere (st)	cubic meter (m^3)	**1.0**	**E+00**
stilb (sb)	candela per square meter (cd/m^2)	**1.0**	**E+04**
stokes (St)	meter squared per second (m^2/s)	**1.0**	**E−04**

To convert from	to	Multiply by	
tablespoon	cubic meter (m³)	1.478 676	E−05
tablespoon	milliliter (mL)	1.478 676	E+01
teaspoon	cubic meter (m³)	4.928 922	E−06
teaspoon	milliliter (mL)	4.928 922	E+00
tex	kilogram per meter (kg/m)	1.0	**E−06**
therm (EC)[25]	joule (J)	1.055 06	**E+08**
therm (U.S.)[25]	joule (J)	1.054 804	**E+08**
ton, assay (AT)	kilogram (kg)	2.916 667	E−02
ton, assay (AT)	gram (g)	2.916 667	E+01
ton-force (2000 lbf)	newton (N)	8.896 443	E+03
ton-force (2000 lbf)	kilonewton (kN)	8.896 443	E+00
ton, long (2240 lb)	kilogram (kg)	1.016 047	E+03
ton, long, per cubic yard	kilogram per cubic meter (kg/m³)	1.328 939	E+03
ton, metric (t)	kilogram (kg)	1.0	**E+03**
tonne (called ''metric ton'' in U.S.) (t)	kilogram (kg)	1.0	**E+03**
ton of refrigeration (12 000 Btu_IT/h)	watt (W)	3.516 853	E+03
ton of TNT (energy equivalent)[26]	joule (J)	4.184	**E+09**
ton, register	cubic meter (m³)	2.831 685	E+00
ton, short (2000 lb)	kilogram (kg)	9.071 847	E+02
ton, short, per cubic yard	kilogram per cubic meter (kg/m³)	1.186 553	E+03
ton, short, per hour	kilogram per second (kg/s)	2.519 958	E−01
torr (Torr)	pascal (Pa)	1.333 224	E+02
unit pole	weber (Wb)	1.256 637	E−07
watt hour (W · h)	joule (J)	3.6	**E+03**
watt per square centimeter (W/cm²)	watt per square meter (W/m²)	1.0	**E+04**
watt per square inch (W/in²)	watt per square meter (W/m²)	1.550 003	E+03
watt second (W · s)	joule (J)	1.0	**E+00**
yard (yd)	meter (m)	9.144	**E−01**
year (365 days)	second (s)	3.1536	**E+07**
year (sidereal)	second (s)	3.155 815	E+07
year (tropical)	second (s)	3.155 693	E+07

---

[25] The therm (EC) is legally defined in the Council Directive of 20 December 1979, Council of the European Communities (now the European Union, EU). The therm (U.S.) is legally defined in the Federal Register of July 27, 1968. Although the therm (EC), which is based on the International Table Btu, is frequently used by engineers in the United States, the therm (U.S.) is the legal unit used by the U.S. natural gas industry.

[26] Defined (not measured) value.

CONVERSION OF TEMPERATURES

From	To	
Celsius	Fahrenheit	$t_F/°F = (9/5)\ t/°C + 32$
	Kelvin	$T/K = t/°C + 273.15$
	Rankine	$T/°R = (9/5)\ (t/°C + 273.15)$
Fahrenheit	Celsius	$t/°C = (5/9)\ [(t_F/°F) - 32]$
	Kelvin	$T/K = (5/9)\ [(t_F/°F) - 32] + 273.15$
	Rankine	$T/°R = t_F/°F + 459.67$
Kelvin	Celsius	$t/°C = T/K - 273.15$
	Rankine	$T/°R = (9/5)\ T/K$
Rankine	Fahrenheit	$t_F/°F = T/°R - 459.67$
	Kelvin	$T/K = (5/9)\ T/°R$

Definition of symbols:

T = thermodynamic (absolute) temperature
t = Celsius temperature (the symbol q is also used for Celsius temperature)
t_F = Fahrenheit temperature

DESIGNATION OF LARGE NUMBERS

	U.S.A.	Other Countries
10^6	million	million
10^9	billion	milliard
10^{12}	trillion	billion
10^{15}	quadrillion	billiard
10^{18}	quintillion	trillion

CONVERSION FACTORS FOR ENERGY UNITS

If greater accuracy is required, use the *Energy Equivalents* section of the *Fundamental Physical Constants* table.

	Wavenumber $\bar{\nu}$ cm⁻¹	Frequency ν MHz	Energy E aJ	Energy E eV	Energy E E_h	Molar energy E_m kJ/mol	Molar energy E_m kcal/mol	Temperature T K
$\bar{\nu}$: 1 cm⁻¹	$\doteq$ 1	2.997925×10^4	1.986447×10^{-5}	1.239842×10^{-4}	4.556335×10^{-6}	11.96266×10^{-3}	2.85914×10^{-3}	1.438769
ν: 1 MHz	$\doteq 3.33564 \times 10^{-5}$	1	6.626076×10^{-10}	4.135669×10^{-9}	1.519830×10^{-10}	3.990313×10^{-7}	9.53708×10^{-8}	4.79922×10^{-5}
1 aJ	$\doteq 50341.1$	1.509189×10^9	1	6.241506	0.2293710	602.2137	143.9325	7.24292×10^4
E: 1 eV	$\doteq 8065.54$	2.417988×10^8	0.1602177	1	3.674931×10^{-2}	96.4853	23.0605	1.16045×10^4
E_h	$\doteq 219474.63$	6.579684×10^9	4.359748	27.2114	1	2625.500	627.510	3.15773×10^5
E_m: 1 kJ/mol	$\doteq 83.5935$	2.506069×10^6	1.660540×10^{-3}	1.036427×10^{-2}	3.808798×10^{-4}	1	0.239006	120.272
1 kcal/mol	$\doteq 349.755$	1.048539×10^7	6.947700×10^{-3}	4.336411×10^{-2}	1.593601×10^{-3}	4.184	1	503.217
T: 1 K	$\doteq 0.695039$	2.08367×10^4	1.380658×10^{-5}	8.61738×10^{-5}	3.16683×10^{-6}	8.31451×10^{-3}	1.98722×10^{-3}	1

Examples of the use of this table:
$1 \text{ aJ} \doteq 50341 \text{ cm}^{-1}$
$1 \text{ eV} \doteq 96.4853 \text{ kJ mol}^{-1}$

The symbol $\doteq$ should be read as meaning "corresponds to" or "is equivalent to".
$E = h\nu = hc\bar{\nu} = kT$; $E_m = N_A E$; E_h is the Hartree energy

CONVERSION FACTORS FOR PRESSURE UNITS

	Pa	kPa	MPa	bar	atmos	Torr	μmHg	psi
Pa	1	0.001	0.000001	0.00001	9.8692×10^{-6}	0.0075006	7.5006	0.0001450377
kPa	1000	1	0.001	0.01	0.0098692	7.5006	7500.6	0.1450377
MPa	1000000	1000	1	10	9.8692	7500.6	7500600	145.0377
bar	100000	100	0.1	1	0.98692	750.06	750060	14.50377
atmos	101325	101.325	0.101325	1.01325	1	760	760000	14.69594
Torr	133.322	0.133322	0.000133322	0.00133322	0.00131579	1	1000	0.01933672
μmHg	0.133322	0.000133322	1.33322×10^{-7}	1.33322×10^{-6}	1.31579×10^{-6}	0.001	1	1.933672×10^{-5}
psi	6894.757	6.894757	0.006894757	0.06894757	0.068046	51.7151	51715.1	1

To convert a pressure value from a unit in the left hand column to a new unit, multiply the value by the factor appearing in the column for the new unit. For example:

$1 \text{ kPa} = 9.8692 \times 10^{-3} \text{ atmos}$

$1 \text{ Torr} = 1.33322 \times 10^{-4} \text{ MPa}$

Notes: μmHg is often referred to as "micron"

Torr is essentially identical to mmHg

psi is an abbreviation for the unit pound–force per square inch

psia (as a term for a physical quantity) implies the true (absolute) pressure

psig implies the true pressure minus the local atmospheric pressure

CONVERSION FACTORS FOR THERMAL CONDUCTIVITY UNITS

MULTIPLY
→ by appropriate factor to
OBTAIN→

	Btu_{IT} h^{-1} ft^{-1} $°F^{-1}$	Btu_{IT} in. h^{-1} ft^{-2} $°F^{-1}$	Btu_{th} h^{-1} ft^{-1} $°F^{-1}$	Btu_{th} in. h^{-1} ft^{-2} $°F^{-1}$	cal_{IT} s^{-1} cm^{-1} $°C^{-1}$	cal_{th} s^{-1} cm^{-1} $°C^{-1}$	$kcal_{th}$ h^{-1} m^{-1} $°C^{-1}$	J s^{-1} cm^{-1} K^{-1}	W cm^{-1} K^{-1}	W m^{-1} K^{-1}	mW cm^{-1} K^{-1}
Btu_{IT} h^{-1} ft^{-1} $°F^{-1}$	1	12	1.00067	12.0080	4.13379×10^{-3}	4.13656×10^{-3}	1.48916	1.73073×10^{-2}	1.73073×10^{-2}	1.73073	17.3073
Btu_{IT} in. h^{-1} ft^{-2} $°F^{-1}$	8.33333×10^{-2}	1	8.33891×10^{-2}	1.00067	3.44482×10^{-4}	3.44713×10^{-4}	0.124097	1.44228×10^{-3}	1.44228×10^{-3}	0.144228	1.44228
Btu_{th} h^{-1} ft^{-1} $°F^{-1}$	0.999331	11.9920	1	12	4.13102×10^{-3}	4.13379×10^{-3}	1.48816	1.72958×10^{-2}	1.72958×10^{-2}	1.72958	17.2958
Btu_{th} in. h^{-1} ft^{-2} $°F^{-1}$	8.32776×10^{-2}	0.999331	8.33333×10^{-2}	1	3.44252×10^{-4}	3.44482×10^{-4}	0.124014	1.44131×10^{-3}	1.44131×10^{-3}	0.144131	1.44131
cal_{IT} s^{-1} cm^{-1} $°C^{-1}$	2.41909×10^{2}	2.90291×10^{3}	2.42071×10^{2}	2.90485×10^{3}	1	1.00067	3.60241×10^{2}	4.1868	4.1868	4.1868×10^{2}	4.1868×10^{3}
cal_{th} s^{-1} cm^{-1} $°C^{-1}$	2.41747×10^{2}	2.90096×10^{3}	2.41909×10^{2}	2.90291×10^{3}	0.999331	1	3.6×10^{2}	4.184	4.184	4.184×10^{2}	4.184×10^{3}
$kcal_{th}$ h^{-1} m^{-1} $°C^{-1}$	0.671520	8.05824	0.671969	8.06363	2.77592×10^{-3}	2.77778×10^{-3}	1	1.16222×10^{-2}	1.16222×10^{-2}	1.16222	11.6222
J s^{-1} cm^{-1} K^{-1}	57.7789	6.93347×10^{2}	57.8176	6.93811×10^{2}	0.238846	0.239006	86.0421	1	1	1×10^{2}	1×10^{3}
W cm^{-1} K^{-1}	57.7789	6.93347×10^{2}	57.8176	6.93811×10^{2}	0.238846	0.239006	86.0421	1	1	1×10^{2}	1×10^{3}
W m^{-1} K^{-1}	0.577789	6.93347	0.578176	6.93811	2.38846×10^{-3}	2.39006×10^{-3}	0.860421	1×10^{-2}	1×10^{-2}	1	10
mW cm^{-1} K^{-1}	5.77789×10^{-2}	0.693347	5.78176×10^{-2}	0.693811	2.38846×10^{-4}	2.39006×10^{-4}	8.60421×10^{-2}	1×10^{-3}	1×10^{-3}	0.1	1

CONVERSION FACTORS FOR ELECTRICAL RESISTIVITY UNITS

To convert from
multiply by
appropriate
factor to

Obtain →	abΩ cm	μΩ cm	Ω cm	StatΩ cm	Ω m	Ω cir. mil ft⁻¹	Ω in.	Ω ft
abohm centimeter	1	1×10^{-3}	10^{-9}	1.113×10^{-21}	10^{-11}	6.015×10^{-3}	3.937×10^{-10}	3.281×10^{-11}
microohm centimeter	10^{3}	1	10^{-6}	1.113×10^{-18}	10^{-6}	6.015	3.937×10^{-7}	3.281×10^{-6}
ohm centimeter	10^{8}	10^{6}	1	1.113×10^{-12}	1×10^{-2}	6.015×10^{6}	3.937×10^{-1}	3.281×10^{-2}
statohm centimeter (esu)	8.987×10^{20}	8.987×10^{17}	8.987×10^{11}	1	8.987×10^{9}	5.406×10^{18}	3.538×10^{11}	2.949×10^{10}
ohm meter	10^{11}	10^{8}	10^{2}	1.113×10^{-10}	1	6.015×10^{8}	3.937×10^{1}	3.281
ohm circular mil per foot	1.662×10^{2}	1.662×10^{-1}	1.662×10^{-7}	1.850×10^{-19}	1.662×10^{-9}	1	6.54×10^{-6}	5.45×10^{-9}
ohm inch	2.54×10^{9}	2.54×10^{6}	2.54	2.827×10^{-12}	2.54×10^{-2}	1.528×10^{7}	1	8.3×10^{-2}
ohm foot	3.048×10^{10}	3.048×10^{7}	3.048×10^{-1}	3.3924×10^{-11}	3.048×10^{-1}	1.833×10^{8}	12	1

CONVERSION FACTORS FOR CHEMICAL KINETICS

Equivalent second order rate constants

A \ B	cm³ mol⁻¹ s⁻¹	dm³ mol⁻¹ s⁻¹	m³ mol⁻¹ s⁻¹	cm³ molecule⁻¹ s⁻¹	(mm Hg)⁻¹ s⁻¹	atm⁻¹ s⁻¹	ppm⁻¹ min⁻¹	m² kN⁻¹ s⁻¹
1 cm³ mol⁻¹ s⁻¹ =	1	10^{-3}	10^{-6}	1.66×10^{-24}	$1.604 \times 10^{-5} T^{-1}$	$1.219 \times 10^{-2} T^{-1}$	2.453×10^{-9}	$1.203 \times 10^{-4} T^{-1}$
1 dm³ mol⁻¹ s⁻¹ =	10^{3}	1	10^{-3}	1.66×10^{-21}	$1.604 \times 10^{-2} T^{-1}$	$12.19 T^{-1}$	2.453×10^{-6}	$1.203 \times 10^{-1} T^{-1}$
1 m³ mol⁻¹ s⁻¹ =	10^{6}	10^{3}	1	1.66×10^{-18}	$16.04 T^{-1}$	$1.219 \times 10^{4} T^{-1}$	2.453×10^{-3}	$120.3 T^{-1}$
1 cm³ molecule⁻¹ s⁻¹ =	6.023×10^{23}	6.023×10^{20}	6.023×10^{17}	1	$9.658 \times 10^{18} T^{-1}$	$7.34 \times 10^{21} T^{-1}$	1.478×10^{15}	$7.244 \times 10^{19} T^{-1}$
1 (mm Hg)⁻¹ s⁻¹ =	$6.236 \times 10^{4} T$	$62.36 T$	$6.236 \times 10^{-2} T$	$1.035 \times 10^{-19} T$	1	760	4.56×10^{-2}	7.500
1 atm⁻¹ s⁻¹	$82.06 T$	$8.206 \times 10^{-2} T$	$8.206 \times 10^{-5} T$	$1.362 \times 10^{-22} T$	1.316×10^{-3}	1	6×10^{-5}	9.869×10^{-3}
1 ppm⁻¹ min⁻¹ = at 298 K, 1 atm total pressure	4.077×10^{8}	4.077×10^{5}	407.7	6.76×10^{-16}	21.93	1.667×10^{4}	1	164.5
1 m² kN⁻¹ s⁻¹ =	$8314 T$	$8.314 T$	$8.314 \times 10^{-3} T$	$1.38 \times 10^{-20} T$	0.1333	101.325	6.079×10^{-3}	1

To convert a rate constant from one set of units A to a new set B find the conversion factor for the row A under column B and multiply the old value by it, e.g. to convert cm³ molecule⁻¹ s⁻¹ to m³ mol⁻¹ s⁻¹ multiply by 6.023×10^{17}.

Table adapted from High Temperature Reaction Rate Data No. 5, The University, Leeds (1970).

Equivalent third order rate constants

A \ B	cm⁶ mol⁻² s⁻¹	dm⁶ mol⁻² s⁻¹	m⁶ mol⁻² s⁻¹	cm⁶ molecule⁻² s⁻¹	(mm Hg)⁻² s⁻¹	atm⁻² s⁻¹	ppm⁻² min⁻¹	m⁴ kN⁻² s⁻¹
1 cm⁶ mol⁻² s⁻¹ =	1	10^{-6}	10^{-12}	2.76×10^{-48}	$2.57 \times 10^{-10} T^{-2}$	$1.48 \times 10^{-4} T^{-2}$	1.003×10^{-19}	$1.447 \times 10^{-8} T^{-2}$
1 dm⁶ mol⁻² s⁻¹ =	10^{6}	1	10^{-6}	2.76×10^{-42}	$2.57 \times 10^{-4} T^{-2}$	$148 T^{-2}$	1.003×10^{-13}	$1.447 \times 10^{-2} T^{-2}$
1 m⁶ mol⁻² s⁻¹ =	10^{12}	10^{6}	1	2.76×10^{-36}	$257 T^{-2}$	$1.48 \times 10^{8} T^{-2}$	1.003×10^{-7}	$1.447 \times 10^{4} T^{-2}$
1 cm⁶ molecule⁻² s⁻¹ =	3.628×10^{47}	3.628×10^{41}	3.628×10^{35}	1	$9.328 \times 10^{37} T^{-2}$	$5.388 \times 10^{43} T^{-2}$	3.64×10^{28}	$5.248 \times 10^{39} T^{-2}$
1 (mm Hg)⁻² s⁻¹ =	$3.89 \times 10^{9} T^{2}$	$3.89 \times 10^{3} T^{2}$	$3.89 \times 10^{-3} T^{2}$	$1.07 \times 10^{-38} T^{2}$	1	5.776×10^{5}	3.46×10^{-5}	56.25
1 atm⁻² s⁻¹ =	$6.733 \times 10^{3} T^{2}$	$6.733 \times 10^{-3} T^{2}$	$6.733 \times 10^{-9} T^{2}$	$1.86 \times 10^{-44} T^{2}$	1.73×10^{-6}	1	6×10^{-11}	9.74×10^{-5}
1 ppm⁻² min⁻¹ = at 298 K, 1 atm total pressure	9.97×10^{18}	9.97×10^{12}	9.97×10^{6}	2.75×10^{-29}	2.89×10^{4}	1.667×10^{10}	1	1.623×10^{6}
1 m⁴ kN⁻² s⁻¹ =	$6.91 \times 10^{7} T^{2}$	$6.91 T^{2}$	69.1	$1.904 \times 10^{-40} T^{2}$	0.0178	1.027×10^{4}	6.16×10^{-7}	1

From *J. Phys. Chem. Ref. Data*, 9, 470, 1980, by permission of the authors and the copyright owner, the American Institute of Physics.

CONVERSION FACTORS FOR IONIZING RADIATION

CONVERSION BETWEEN SI AND OTHER UNITS

Quantity	Symbol for quantity	Expression in SI units	Expression in symbols for SI units	Special name for SI units	Symbols using special names	Conventional units	Symbol for conventional unit	Value of conventional unit in SI units
Activity	A	1 per second	s^{-1}	becquerel	Bq	curie	Ci	3.7×10^{10} Bq
Absorbed dose	D	joule per kilogram	$J\ kg^{-1}$	gray	Gy	rad	rad	0.01 Gy
Absorbed dose rate	$\dot{D}$	joule per kilogram second	$J\ kg^{-1}\ s^{-1}$		$Gy\ s^{-1}$	rad	$rad\ s^{-1}$	$0.01\ Gy\ s^{-1}$
Average energy per ion pair	W	joule	J			electronvolt	eV	1.602×10^{-19} J
Dose equivalent	H	joule per kilogram	$J\ kg^{-1}$	sievert	Sv	rem	rem	0.01 Sv
Dose equivalent rate	$\dot{H}$	joule per kilogram second	$J\ kg^{-1}\ s^{-1}$		$Sv\ s^{-1}$	rem per second	$rem\ s^{-1}$	$0.01\ Sv\ s^{-1}$
Electric current	I	ampere	A			ampere	A	1.0 A
Electric potential difference	U, V	watt per ampere	$W\ A^{-1}$	volt	V	volt	V	1.0 A
Exposure	X	coulomb per kilogram	$C\ kg^{-1}$			roentgen	R	$2.58 \times 10^{-4}\ C\ kg^{-1}$
Exposure rate	$\dot{X}$	coulomb per kilogram second	$C\ kg^{-1}\ s^{-1}$			roentgen	$R\ s^{-1}$	$2.58 \times 10^{-4}\ C\ kg^{-1}\ s^{-1}$
Fluence	ϕ	1 per meter squared	m^{-2}			1 per centimeter squared	cm^{-2}	$1.0 \times 10^{4}\ m^{-2}$
Fluence rate	Φ	1 per meter squared second	$m^{-2}\ s^{-1}$			1 per centimeter squared second	$cm^{-2}\ s^{-1}$	$1.0 \times 10^{4}\ m^{-2}\ s^{-1}$
Kerma	K	joule per kilogram	$J\ kg^{-1}$	gray	Gy	rad	rad	0.01 Gy
Kerma rate	$\dot{K}$	joule per kilogram second	$J\ kg^{-1}\ s^{-1}$		$Gy\ s^{-1}$	rad per second	$rad\ s^{-1}$	$0.01\ Gy\ s^{-1}$
Lineal energy	y	joule per meter	$J\ m^{-1}$			kiloelectron volt per micrometer	$keV\ \mu m^{-1}$	$1.602 \times 10^{-10}\ J\ m^{-1}$
Linear energy transfer	L	joule per meter	$J\ m^{-1}$			kiloelectron volt per micrometer	$keV\ \mu m^{-1}$	$1.602 \times 10^{-10}\ J\ m^{-1}$
Mass attenuation coefficient	μ/ρ	meter squared per kilogram	$m^2\ kg^{-1}$			centimeter squared per gram	$cm^2\ g^{-1}$	$0.1\ m^2\ kg^{-1}$
Mass energy transfer coefficient	μ_{tr}/ρ	meter squared per kilogram	$m^2\ kg^{-1}$			centimeter squared per gram	$cm^2\ g^{-1}$	$0.1\ m^2\ kg^{-1}$
Mass energy absorption coefficient	μ_{en}/ρ	meter squared per kilogram	$m^2\ kg^{-1}$			centimeter squared per gram	$cm^2\ g^{-1}$	$0.1\ m^2\ kg^{-1}$
Mass stopping power	S/ρ	joule meter squared per kilogram	$J\ m^2\ kg^{-1}$			MeV centimeter squared per gram	$MeV\ cm^2\ g^{-1}$	$1.602 \times 10^{-14}\ J\ m^2\ kg^{-1}$
Power	P	joule per second	$J\ s^{-1}$	watt	W	watt	W	1.0 W
Pressure	p	newton per meter squared	$N\ m^{-2}$	pascal	Pa	torr	torr	$(101325/760)$Pa
Radiation chemical yield	G	mole per joule	$mol\ J^{-1}$			molecules per 100 electron volts	molecules $(100\ eV)^{-1}$	$1.04 \times 10^{-7}\ mol\ J^{-1}$
Specific energy	z	joule per kilogram	$J\ kg^{-1}$	gray	Gy	rad	rad	0.01 Gy

CONVERSION OF RADIOACTIVITY UNITS FROM MBq TO mCi AND μCi

MBq	mCi	MBq	mCi	MBq	μCi	MBq	μCi
7000	189.	500	13.5	30	810	1	27
6000	162.	400	10.8	20	540	0.9	24
5000	135.	300	8.1	10	270	0.8	21.6
4000	108.	200	5.4	9	240	0.7	18.9
3000	81.	100	2.7	8	220	0.6	16.2
2000	54.	90	2.4	7	189	0.5	13.5
1000	27.	80	2.16	6	162	0.4	10.8
900	24.	70	1.89	5	135	0.3	8.1
800	21.6	60	1.62	4	108	0.2	5.4
700	18.9	50	1.35	3	81	0.1	2.7
600	16.2	40	1.08	2	54		

CONVERSION OF RADIOACTIVITY UNITS FROM mCi AND μCi TO MBq

mCi	MBq	mCi	MBq	μCi	MBq	μCi	MBq
200	7400	10	370	1000	37.0	80	2.96
150	5550	9	333	900	33.3	70	2.59
100	3700	8	296	800	29.6	60	2.22
90	3330	7	259	700	25.9	50	1.85
80	2960	6	222	600	22.2	40	1.48
70	2590	5	185	500	18.5	30	1.11
60	2220	4	148	400	14.8	20	0.74
50	1850	3	111	300	11.1	10	0.37
40	1480	2	74.0	200	7.4	5	0.185
30	1110	1	37.0	100	3.7	2	0.074
20	740			90	3.33	1	0.037

CONVERSION OF RADIOACTIVITY UNITS

100 TBq (10^{14} Bq)	=	2.7 kCi (2.7×10^3 Ci)	100 kBq (10^5 Bq)	=	2.7 μCi (2.7×10^{-6} Ci)
10 TBq (10^{13} Bq)	=	270 Ci (2.7×10^2 Ci)	10 kBq (10^4 Bq)	=	270 nCi (2.7×10^{-7} Ci)
1 TBq (10^{12} Bq)	=	27 Ci (2.7×10^1 Ci)	1 kBq (10^3 Bq)	=	27 nCi (2.7×10^{-8} Ci)
100 GBq (10^{11} Bq)	=	2.7 Ci (2.7×10^0 Ci)	100 Bq (10^2 Bq)	=	2.7 nCi (2.7×10^{-9} Ci)
10 GBq (10^{10} Bq)	=	270 mCi (2.7×10^{-1} Ci)	10 Bq (10^1 Bq)	=	270 pCi (2.7×10^{-10} Ci)
1 GBq (10^9 Bq)	=	27 mCi (2.7×10^{-2} Ci)	1 Bq (10^0 Bq)	=	27 pCi (2.7×10^{-11} Ci)
100 MBq (10^8 Bq)	=	2.7 mCi (2.7×10^{-3} Ci)	100 mBq (10^{-1} Bq)	=	2.7 pCi (2.7×10^{-12} Ci)
10 MBq (10^7 Bq)	=	270 μCi (2.7×10^{-4} Ci)	10 mBq (10^{-2} Bq)	=	270 fCi (2.7×10^{-13} Ci)
1 MBq (10^6 Bq)	=	27 μCi (2.7×10^{-5} Ci)	1 mBq (10^{-3} Bq)	=	27 fCi (2.7×10^{-14} Ci)

CONVERSION OF ABSORBED DOSE UNITS

SI Units		Conventional
100 Gy (10^2 Gy)	=	10,000 rad (10^4 rad)
10 Gy (10^1 Gy)	=	1,000 rad (10^3 rad)
1 Gy (10^0 Gy)	=	100 rad (10^2 rad)
100 mGy (10^{-1} Gy)	=	10 rad (10^1 rad)
10 mGy (10^{-2} Gy)	=	1 rad (10^0 rad)
1 mGy (10^{-3} Gy)	=	100 mrad (10^{-1} rad)
100 μGy (10^{-4} Gy)	=	10 mrad (10^{-2} rad)
10 μGy (10^{-5} Gy)	=	1 mrad (10^{-3} rad)
1 μGy (10^{-6} Gy)	=	100 μrad (10^{-4} rad)
100 nGy (10^{-7} Gy)	=	10 μrad (10^{-5} rad)
10 nGy (10^{-8} Gy)	=	1 μrad (10^{-6} rad)
1 nGy (10^{-9} Gy)	=	100 nrad (10^{-7} rad)

CONVERSION OF DOSE EQUIVALENT UNITS

100 Sv (10^2 Sv)	=	10,000 rem (10^4 rem)
10 Sv (10^1 Sv)	=	1,000 rem (10^3 rem)
1 Sv (10^0 Sv)	=	100 rem (10^2 rem)
100 mSv (10^{-1} Sv)	=	10 rem (10^1 rem)
10 mSv (10^{-2} Sv)	=	1 rem (10^0 rem)
1 mSv (10^{-3} Sv)	=	100 mrem (10^{-1} rem)
100 μSv (10^{-4} Sv)	=	10 mrem (10^{-2} rem)
10 μSv (10^{-5} Sv)	=	1 mrem (10^{-3} rem)
1 μSv (10^{-6} Sv)	=	100 μrem (10^{-4} rem)
100 nSv (10^{-7} Sv)	=	10 μrem (10^{-5} rem)
10 nSv (10^{-8} Sv)	=	1 μrem (10^{-6} rem)
1 nSv (10^{-9} Sv)	=	100 nrem (10^{-7} rem)

VALUES OF THE GAS CONSTANT IN DIFFERENT UNIT SYSTEMS

In SI units the value of the gas constant, R, is:

R = 8.314510 Pa m^3 K^{-1} mol^{-1}

     = 8314.510 Pa L K^{-1} mol^{-1}

     = 0.08314510 bar L K^{-1} mol^{-1}

This table gives the appropriate value of R for use in the ideal gas equation, $PV = nRT$, when the variables are expressed in other units. The following conversion factors for pressure units were used in generating the table:

1 atm = 101325 Pa

1 psi = 6894.757 Pa

1 torr (mmHg) = 133.322 Pa [at 0°C]

1 in Hg = 3386.38 Pa [at 0°C]

1 in H_2O = 249.082 Pa [at 4°C]

1 ft H_2O = 2988.98 Pa [at 4°C]

The advice of Prabir K. Chandra is appreciated.

Units of V, T, n			Units of P						
V	T	n	kPa	atm	psi	mmHg	in Hg	in H_2O	ft H_2O
ft^3	K	mol	0.2936241	0.00289785	0.0425866	2.20237	0.0867074	1.17882	0.0982355
		lb·mol	133.1857	1.31444	19.3169	998.978	39.3298	534.706	44.5589
	°R	mol	0.1631245	0.00160991	0.0236592	1.22354	0.0481708	0.654903	0.0545753
		lb·mol	73.99204	0.730245	10.7316	554.987	21.8499	297.059	24.7549
cm^3	K	mol	8314.510	82.0578	1205.92	62364.1	2455.28	33380.6	2781.72
		lb·mol	3771398	37220.8	546995	28287900	1113700	15141200	1261770
	°R	mol	4619.172	45.5877	669.954	34646.7	1364.04	18544.8	1545.40
		lb·mol	2095221	20678.2	303886	15715500	618720	8411770	700982
L	K	mol	8.314510	0.0820578	1.20592	62.3641	2.45528	33.3806	2.78172
		lb·mol	3771.398	37.2208	546.995	28287.9	1113.70	15141.2	1261.77
	°R	mol	4.619172	0.0455877	0.669954	34.6467	1.36404	18.5448	1.54540
		lb·mol	2095.221	20.6782	303.886	15715.5	618.720	8411.77	700.982
m^3	K	mol	0.008314510	0.0000820578	0.00120592	0.0623641	0.00245528	0.0333806	0.00278172
		lb·mol	3.771398	0.0372208	0.546995	28.2879	1.11370	15.1412	1.26177
	°R	mol	0.004619172	0.0000455877	0.000669954	0.0346467	0.00136404	0.0185448	0.00154540
		lb·mol	2.095221	0.0206782	0.303886	15.7155	0.618720	8.41177	0.700982

Section 2
Symbols, Terminology, and Nomenclature

Symbols and Terminology for Physical and Chemical Quantities 2-1
Nomenclature of Chemical Compounds 2-22
Nomenclature for Inorganic Ions and Ligands 2-23
Organic Substituent Groups and Ring Systems 2-25
Scientific Abbreviations and Symbols 2-29
Greek, Russian, and Hebrew Alphabets 2-37
Definitions of Scientific Terms 2-38

SYMBOLS AND TERMINOLOGY FOR PHYSICAL AND CHEMICAL QUANTITIES

The International Organization for Standardization (ISO), International Union of Pure and Applied Chemistry (IUPAC), and the International Union of Pure and Applied Physics (IUPAP) have jointly developed a set of recommended symbols for physical and chemical quantities. Consistent use of these recommended symbols helps assure unambiguous scientific communication. The list below is reprinted from Reference 1 with permission from IUPAC. Full details may be found in the following references:

1. Ian Mills, Ed., *Quantities, Units, and Symbols in Physical Chemistry*, Blackwell Scientific Publications, Oxford, 1988.
2. E. R. Cohen and P. Giacomo, *Symbols, Units, Nomenclature, and Fundamental Constants in Physics*, Document IUPAP-25, 1987; also published in *Physica*, 146A, 1—68, 1987.
3. *ISO Standards Handbook 2: Units of Measurement*, International Organization of Standardization, Geneva, 1982.

GENERAL RULES

The value of a physical quantity is expressed as the product of a numerical value and a unit, e.g.:

$T = 300$ K
$V = 26.2$ cm^3
$C_p = 45.3$ J mol^{-1} K^{-1}

The symbol for a physical quantity is always given in italic (sloping) type, while symbols for units are given in roman type. Column headings in tables and axis labels on graphs may conveniently be written as the physical quantity symbol divided by the unit symbol, e.g.:

T/K
V/cm^3
C_p/J mol^{-1} K^{-1}

The values in the table or graph axis are then pure numbers.

Subscripts to symbols for physical quantities should be italic if the subscript refers to another physical quantity or to a number, e.g.:

C_p — heat capacity at constant pressure
B_n — nth virial coefficient

Subscripts which have other meanings should be in roman type:

m_p — mass of the proton
E_k — kinetic energy

The following tables give the recommended symbols for the major classes of physical and chemical quantities. The expression in the Definition column is given as an aid in identifying the quantity but is not necessarily the complete or unique definition. The SI Unit gives one (not necessarily unique) expression for the coherent SI unit for the quantity. Other equivalent unit expressions, including those which involve SI prefixes, may be used.

Name	Symbol	Definition	SI unit
SPACE AND TIME			
cartesian space coordinates	x, y, z		m
spherical polar coordinates	r, θ, ϕ		m, 1, 1
generalized coordinate	q, q_i		(varies)
position vector	r	$r = x\boldsymbol{i} + y\boldsymbol{j} + z\boldsymbol{k}$	m
length	l		m
special symbols:			
height	h		
breadth	b		
thickness	d, δ		
distance	d		

Name	Symbol	Definition	SI unit
radius	r		
diameter	d		
path length	s		
length of arc	s		
area	A, A_s, S		m^2
volume	$V, (v)$		m^3
plane angle	$\alpha, \beta, \gamma, \theta, \phi \ldots$	$\alpha = s/r$	rad, 1
solid angle	ω, Ω	$\omega = A/r^2$	sr, 1
time	t		s
period	T	$T = t/N$	s
frequency	v, f	$v = 1/T$	Hz
circular frequency, angular frequency	ω	$\omega = 2\pi v$	rad s^{-1}, s^{-1}
characteristic time interval, relaxation time, time constant	τ, T	$\tau = \lvert dt/d\ln x \rvert$	s
angular velocity	ω	$\omega = d\phi/dt$	rad s^{-1}, s^{-1}
velocity	$v, u, w, c, \dot{r}$	$v = dr/dt$	m s^{-1}
speed	v, u, w, c	$v = \lvert v \rvert$	m s^{-1}
acceleration	$a, (g)$	$a = dv/dt$	m s^{-2}

CLASSICAL MECHANICS

Name	Symbol	Definition	SI unit
mass	m		kg
reduced mass	μ	$\mu = m_1 m_2/(m_1 + m_2)$	kg
density, mass density	ρ	$\rho = m/V$	kg m^{-3}
relative density	d	$d = \rho/\rho^\bullet$	1
surface density	ρ_A, ρ_S	$\rho_A = m/A$	kg m^{-2}
specific volume	v	$v = V/m = 1/\rho$	m^3 kg^{-1}
momentum	p	$p = mv$	kg m s^{-1}
angular momentum, action	L	$L = r \times p$	J s
moment of inertia	I, J	$I = \sum m_i r_i^2$	kg m^2
force	F	$F = dp/dt = ma$	N
torque, moment of a force	$T, (M)$	$T = r \times F$	N m
energy	E		J
potential energy	E_p, V, Φ	$E_p = -\int F \cdot ds$	J
kinetic energy	E_k, T, K	$E_k = \frac{1}{2}mv^2$	J
work	W, w	$W = \int F \cdot ds$	J
Hamilton function	H	$H(q, p) = T(q, p) + V(q)$	J
Lagrange function	L	$L(q, \dot{q}) = T(q, \dot{q}) - V(q)$	J

Name	Symbol	Definition	SI unit
pressure	p, P	$p = F/A$	Pa, N m^{-2}
surface tension	γ, σ	$\gamma = dW/dA$	N m^{-1}, J m^{-2}
weight	$G, (W, P)$	$G = mg$	N
gravitational constant	G	$F = Gm_1 m_2/r^2$	N m^2 kg^{-2}
normal stress	σ	$\sigma = F/A$	Pa
shear stress	τ	$\tau = F/A$	Pa
linear strain, relative elongation	ε, e	$\varepsilon = \Delta l/l$	1
modulus of elasticity, Young's modulus	E	$E = \sigma/\varepsilon$	Pa
shear strain	γ	$\gamma = \Delta x/d$	1
shear modulus	G	$G = \tau/\gamma$	Pa
volume strain, bulk strain	θ	$\theta = \Delta V/V_0$	1
bulk modulus, compression modulus	K	$K = -V_0(dp/dV)$	Pa
viscosity, dynamic viscosity	η, μ	$\tau_{x,z} = \eta(dv_x/dz)$	Pa s
fluidity	ϕ	$\phi = 1/\eta$	m kg^{-1} s
kinematic viscosity	v	$v = \eta/\rho$	m^2 s^{-1}
friction coefficient	$\mu, (f)$	$F_{frict} = \mu F_{norm}$	1
power	P	$P = dW/dt$	W
sound energy flux	P, P_a	$P = dE/dt$	W
acoustic factors,			
reflection factor	ρ	$\rho = P_r/P_0$	1
acoustic absorption factor	$\alpha_a, (\alpha)$	$\alpha_a = 1 - \rho$	1
transmission factor	τ	$\tau = P_{tr}/P_0$	1
dissipation factor	δ	$\delta = \alpha_a - \tau$	1

ELECTRICITY AND MAGNETISM

Name	Symbol	Definition	SI unit
quantity of electricity, electric charge	Q		C
charge density	ρ	$\rho = Q/V$	C m^{-3}
surface charge density	σ	$\sigma = Q/A$	C m^{-2}
electric potential	V, ϕ	$V = dW/dQ$	V, J C^{-1}
electric potential difference	$U, \Delta V, \Delta\phi$	$U = V_2 - V_1$	V
electromotive force	E	$E = \int(F/Q)\cdot ds$.	V
electric field strength	E	$E = F/Q = -\mathrm{grad}\ V$	V m^{-1}
electric flux	Ψ	$\Psi = \int D\cdot dA$	C
electric displacement	D	$D = \varepsilon E$	C m^{-2}
capacitance	C	$C = Q/U$	F, C V^{-1}
permittivity	ε	$D = \varepsilon E$	F m^{-1}

Name	Symbol	Definition	SI unit
permittivity of vacuum	ε_0	$\varepsilon_0 = \mu_0^{-1} c_0^{-2}$	$F\,m^{-1}$
relative permittivity	ε_r	$\varepsilon_r = \varepsilon/\varepsilon_0$	1
dielectric polarization (dipole moment per volume)	P	$P = D - \varepsilon_0 E$	$C\,m^{-2}$
electric susceptibility	χ_e	$\chi_e = \varepsilon_r - 1$	1
electric dipole moment	p, μ	$p = Qr$	$C\,m$
electric current	I	$I = dQ/dt$	A
electric current density	j, J	$I = \int j \cdot dA$	$A\,m^{-2}$
magnetic flux density, magnetic induction	B	$F = Qv \times B$	T
magnetic flux	Φ	$\Phi = \int B \cdot dA$	Wb
magnetic field strength	H	$B = \mu H$	$A\,m^{-1}$
permeability	μ	$B = \mu H$	$N\,A^{-2}, H\,m^{-1}$
permeability of vacuum	μ_0		$H\,m^{-1}$
relative permeability	μ_r	$\mu_r = \mu/\mu_0$	1
magnetization (magnetic dipole moment per volume)	M	$M = B/\mu_0 - H$	$A\,m^{-1}$
magnetic susceptibility	$\chi, \kappa, (\chi_m)$	$\chi = \mu_r - 1$	1
molar magnetic susceptibility	χ_m	$\chi_m = V_m \chi$	$m^3\,mol^{-1}$
magnetic dipole moment	m, μ	$E_p = -m \cdot B$	$A\,m^2, J\,T^{-1}$
electrical resistance	R	$R = U/I$	Ω
conductance	G	$G = 1/R$	S
loss angle	δ	$\delta = (\pi/2) + \phi_I - \phi_U$	1, rad
reactance	X	$X = (U/I)\sin\delta$	Ω
impedance (complex impedance)	Z	$Z = R + iX$	Ω
admittance (complex admittance)	Y	$Y = 1/Z$	S
susceptance	B	$Y = G + iB$	S
resistivity	ρ	$\rho = E/j$	$\Omega\,m$
conductivity	κ, γ, σ	$\kappa = 1/\rho$	$S\,m^{-1}$
self-inductance	L	$E = -L(dI/dt)$	H
mutual inductance	M, L_{12}	$E_1 = L_{12}(dI_2/dt)$	H
magnetic vector potential	A	$B = \nabla \times A$	$Wb\,m^{-1}$
Poynting vector	S	$S = E \times H$	$W\,m^{-2}$

QUANTUM MECHANICS

Name	Symbol	Definition	SI unit
momentum operator	$\hat{p}$	$\hat{p} = -i\hbar\nabla$	$m^{-1}\,J\,s$
kinetic energy operator	$\hat{T}$	$\hat{T} = -(\hbar^2/2m)\nabla^2$	J

Name	Symbol	Definition	SI unit
hamiltonian operator	$\hat{H}$	$\hat{H} = \hat{T} + V$	J
wavefunction, state function	Ψ, ψ, ϕ	$\hat{H}\psi = E\psi$	$(m^{-3/2})$
probability density	P	$P = \psi^*\psi$	(m^{-3})
charge density of electrons	ρ	$\rho = -eP$	$(C\ m^{-3})$
probability current density	S	$S = -i\hbar(\psi^*\nabla\psi - \psi\nabla\psi^*)/2m_e$	$(m^{-2}\ s^{-1})$
electric current density of electrons	j	$j = -eS$	$(A\ m^{-2})$
matrix element of operator $\hat{A}$	$A_{ij}, \langle i|\hat{A}|j\rangle$	$A_{ij} = \int \psi_i^* \hat{A}\psi_j\, d\tau$	(varies)
expectation value of operator $\hat{A}$	$\langle A \rangle, \bar{A}$	$\langle A \rangle = \int \psi^* \hat{A}\psi\, d\tau$	(varies)
hermitian conjugate of $\hat{A}$	$\hat{A}^\dagger$	$(\hat{A}^\dagger)_{ij} = (A_{ji})^*$	(varies)
commutator of $\hat{A}$ and $\hat{B}$	$[\hat{A}, \hat{B}], [\hat{A}, \hat{B}]_-$	$[\hat{A}, \hat{B}] = \hat{A}\hat{B} - \hat{B}\hat{A}$	(varies)
anticommutator	$[\hat{A}, \hat{B}]_+$	$[\hat{A}, \hat{B}]_+ = \hat{A}\hat{B} + \hat{B}\hat{A}$	(varies)
spin wavefunction	$\alpha; \beta$		1
coulomb integral	H_{AA}	$H_{AA} = \int \psi_A^* \hat{H}\psi_A\, d\tau$	J
resonance integral	H_{AB}	$H_{AB} = \int \psi_A^* \hat{H}\psi_B\, d\tau$	J
overlap integral	S_{AB}	$S_{AB} = \int \psi_A^* \psi_B\, d\tau$	1

ATOMS AND MOLECULES

Name	Symbol	Definition	SI unit
nucleon number, mass number	A		1
proton number, atomic number	Z		1
neutron number	N	$N = A - Z$	1
electron rest mass	m_e		kg
mass of atom, atomic mass	m_a, m		kg
atomic mass constant	m_u	$m_u = m_a(^{12}C)/12$	kg
mass excess	Δ	$\Delta = m_a - Am_u$	kg
elementary charge, proton charge	e		C
Planck constant	h		J s
Planck constant/2π	$\hbar$	$\hbar = h/2\pi$	J s
Bohr radius	a_0	$a_0 = 4\pi\varepsilon_0\hbar^2/m_e e^2$	m
Hartree energy	E_h	$E_h = \hbar^2/m_e a_0^2$	J
Rydberg constant	R_∞	$R_\infty = E_h/2hc$	m^{-1}
fine structure constant	α	$\alpha = e^2/4\pi\varepsilon_0\hbar c$	1
ionization energy	E_i		J

Name	Symbol	Definition	SI unit
electron affinity	E_{ea}		J
dissociation energy	E_d, D		J
from the ground state	D_0		J
from the potential minimum	D_e		J
principal quantum number (H atom)	n	$E = -hcR/n^2$	1
angular momentum quantum numbers	see under Spectroscopy		
magnetic dipole moment of a molecule	$\boldsymbol{m}, \boldsymbol{\mu}$	$E_p = -\boldsymbol{m} \cdot \boldsymbol{B}$	$J\,T^{-1}$
magnetizability of a molecule	ξ	$\boldsymbol{m} = \xi \boldsymbol{B}$	$J\,T^{-2}$
Bohr magneton	μ_B	$\mu_B = e\hbar/2m_e$	$J\,T^{-1}$
nuclear magneton	μ_N	$\mu_N = (m_e/m_p)\mu_B$	$J\,T^{-1}$
magnetogyric ratio (gyromagnetic ratio)	γ	$\gamma = \mu/L$	$C\,kg^{-1}$
g factor	g		1
Larmor circular frequency	ω_L	$\omega_L = (e/2m)B$	s^{-1}
Larmor frequency	ν_L	$\nu_L = \omega_L/2\pi$	Hz
longitudinal relaxation time	T_1		s
transverse relaxation time	T_2		s
electric dipole moment of a molecule	$\boldsymbol{p}, \boldsymbol{\mu}$	$E_p = -\boldsymbol{p} \cdot \boldsymbol{E}$	C m
quadrupole moment of a molecule	$Q; \Theta$	$E_p = \frac{1}{2}Q : V'' = \frac{1}{3}\Theta : V''$	C m^2
quadrupole moment of a nucleus	eQ	$eQ = 2\langle \Theta_{zz} \rangle$	C m^2
electric field gradient tensor	q	$q_{\alpha\beta} = -\partial^2 V/\partial\alpha\partial\beta$	V m^{-2}
quadrupole interaction energy tensor	χ	$\chi_{\alpha\beta} = eQq_{\alpha\beta}$	J
electric polarizability of a molecule	α	p (induced) $= \alpha E$	C m^2 V^{-1}
activity (of a radioactive substance)	A	$A = -dN_B/dt$	Bq
decay (rate) constant, disintegration (rate) constant	λ	$A = \lambda N_B$	s^{-1}
half life	$t_{\frac{1}{2}}, T_{\frac{1}{2}}$		s
mean life	τ		s
level width	Γ	$\Gamma = \hbar/\tau$	J

Name	Symbol	Definition	SI unit
disintegration energy	Q		J
cross section (of a nuclear reaction)	σ		m^2

SPECTROSCOPY

Name	Symbol	Definition	SI unit
total term	T	$T = E_{tot}/hc$	m^{-1}
transition wavenumber	$\tilde{\nu}, (\nu)$	$\tilde{\nu} = T' - T''$	m^{-1}
transition frequency	ν	$\nu = (E' - E'')/h$	Hz
electronic term	T_e	$T_e = E_e/hc$	m^{-1}
vibrational term	G	$G = E_{vib}/hc$	m^{-1}
rotational term	F	$F = E_{rot}/hc$	m^{-1}
spin orbit coupling constant	A	$T_{s.o.} = A\langle \hat{L}\cdot\hat{S}\rangle$	m^{-1}
principal moments of inertia	$I_A; I_B; I_C$	$I_A \leqslant I_B \leqslant I_C$	kg m^2
rotational constants,			
in wavenumber	$\tilde{A}; \tilde{B}; \tilde{C}$	$\tilde{A} = h/8\pi^2 c I_A$	m^{-1}
in frequency	$A; B; C$	$A = h/8\pi^2 I_A$	Hz
inertial defect	Δ	$\Delta = I_C - I_A - I_B$	kg m^2
asymmetry parameter	κ	$\kappa = \dfrac{(2B - A - C)}{(A - C)}$	1
centrifugal distortion constants,			
S reduction	$D_J; D_{JK}; D_K; d_1; d_2$		m^{-1}
A reduction	$\Delta_J; \Delta_{JK}; \Delta_K; \delta_J; \delta_K$		m^{-1}
harmonic vibration wavenumber	$\omega_e; \omega_r$		m^{-1}
vibrational anharmonicity constant	$\omega_e x_e; x_{rs}; g_{tt'}$		m^{-1}
vibrational quantum numbers	$v_r; l_t$		1
Coriolis zeta constant	ζ_{rs}^{α}		1
angular momentum quantum numbers	see additional information below		
degeneracy, statistical weight	g, d, β		1
electric dipole moment of a molecule	p, μ	$E_p = -p\cdot E$	C m
transition dipole moment of a molecule	M, R	$M = \int \psi' p \psi'' d\tau$	C m
molecular geometry, interatomic distances,			
equilibrium distance	r_e		m
zero-point average distance	r_z		m
ground state distance	r_0		m

Name	Symbol	Definition	SI unit
substitution structure distance	r_s		m
vibrational coordinates,			
internal coordinates	R_i, r_i, θ_j, etc.		(varies)
symmetry coordinates	S_i		(varies)
normal coordinates			
mass adjusted	Q_r		$kg^{\frac{1}{2}}\,\dot{m}$
dimensionless	q_r		1
vibrational force constants,			
diatomic	$f, (k)$	$f = \partial^2 V/\partial r^2$	$J\,m^{-2}$
polyatomic,			
internal coordinates	f_{ij}	$f_{ij} = \partial^2 V/\partial r_i \partial r_j$	(varies)
symmetry coordinates	F_{ij}	$F_{ij} = \partial^2 V/\partial S_i \partial S_j$	(varies)
dimensionless normal coordinates	$\phi_{rst\ldots}$, $k_{rst\ldots}$		m^{-1}
nuclear magnetic resonance (NMR),			
magnetogyric ratio	γ	$\gamma = \mu/I\hbar$	$C\,kg^{-1}$
shielding constant	σ_A	$B_A = (1 - \sigma_A)B$	1
chemical shift, δ scale	δ	$\delta = 10^6(\nu - \nu_0)/\nu_0$	1
(indirect) spin–spin coupling constant	J_{AB}	$\hat{H}/h = J_{AB}\boldsymbol{I}_A \cdot \boldsymbol{I}_B$	Hz
direct (dipolar) coupling constant	D_{AB}		Hz
longitudinal relaxation time	T_1		s
transverse relaxation time	T_2		s
electron spin resonance, electron paramagnetic resonance (ESR, EPR),			
magnetogyric ratio	γ	$\gamma = \mu/s\hbar$	$C\,kg^{-1}$
g factor	g	$h\nu = g\mu_B B$	1
hyperfine coupling constant,			
in liquids	a, A	$\hat{H}_{hfs}/h = a\boldsymbol{S} \cdot \boldsymbol{I}$	Hz
in solids	T	$\hat{H}_{hfs}/h = \boldsymbol{S} \cdot T \cdot \boldsymbol{I}$	Hz

Angular momentum	Operator symbol	Quantum number symbol Total	Z-axis	z-axis
electron orbital	$\hat{L}$	L	M_L	Λ
one electron only	$\hat{l}$	l	m_l	λ

Angular momentum	Operator symbol	Quantum number symbol		
		Total	Z-axis	z-axis
electron spin	$\hat{S}$	S	M_S	Σ
one electron only	$\hat{s}$	s	m_s	σ
electron orbital+spin	$\hat{L}+\hat{S}$			$\Omega = \Lambda + \Sigma$
nuclear orbital (rotational)	$\hat{R}$	R		K_R, k_R
nuclear spin	$\hat{I}$	I	M_I	
internal vibrational				
spherical top	$\hat{l}$	$l(l\zeta)$		K_l
other	$\hat{j}, \hat{\pi}$			$l(l\zeta)$
sum of $R+L(+j)$	$\hat{N}$	N		K, k
sum of $N+S$	$\hat{J}$	J	M_J	K, k
sum of $J+I$	$\hat{F}$	F	M_F	

ELECTROMAGNETIC RADIATION

Name	Symbol	Definition	SI unit
wavelength	λ		m
speed of light			
in vacuum	c_0		m s^{-1}
in a medium	c	$c = c_0/n$	m s^{-1}
wavenumber in vacuum	$\tilde{\nu}$	$\tilde{\nu} = \nu/c_0 = 1/n\lambda$	m^{-1}
wavenumber (in a medium)	σ	$\sigma = 1/\lambda$	m^{-1}
frequency	ν	$\nu = c/\lambda$	Hz
circular frequency, pulsatance	ω	$\omega = 2\pi\nu$	s^{-1}, rad s^{-1}
refractive index	n	$n = c_0/c$	1
Planck constant	h		J s
Planck constant/2π	$\hbar$	$\hbar = h/2\pi$	J s
radiant energy	Q, W		J
radiant energy density	ρ, w	$\rho = Q/V$	J m^{-3}
spectral radiant energy density			
in terms of frequency	ρ_ν, w_ν	$\rho_\nu = d\rho/d\nu$	J m^{-3} Hz^{-1}
in terms of wavenumber	$\rho_{\tilde{\nu}}, w_{\tilde{\nu}}$	$\rho_{\tilde{\nu}} = d\rho/d\tilde{\nu}$	J m^{-2}
in terms of wavelenglth	ρ_λ, w_λ	$\rho_\lambda = d\rho/d\lambda$	J m^{-4}
Einstein transition probabilities			
spontaneous emission	A_{nm}	$dN_n/dt = -A_{nm}N_n$	s^{-1}
stimulated emission	B_{nm}	$dN_n/dt = -\rho_{\tilde{\nu}}(\tilde{\nu}_{nm}) \times B_{nm}N_n$	s kg^{-1}
stimulated absorption	B_{mn}	$dN_n/dt = \rho_{\tilde{\nu}}(\tilde{\nu}_{nm})B_{mn}N_m$	s kg^{-1}
radiant power, radiant energy per time	Φ, P	$\Phi = dQ/dt$	W

Name	Symbol	Definition	SI unit
radiant intensity	I	$I = \mathrm{d}\Phi/\mathrm{d}\Omega$	$\mathrm{W\ sr^{-1}}$
radiant exitance, (emitted radiant flux)	M	$M = \mathrm{d}\Phi/\mathrm{d}A_{source}$	$\mathrm{W\ m^{-2}}$
irradiance, (radiant flux received)	$E, (I)$	$E = \mathrm{d}\Phi/\mathrm{d}A$	$\mathrm{W\ m^{-2}}$
emittance	ε	$\varepsilon = M/M_{bb}$	1
Stefan–Boltzmann constant	σ	$M_{bb} = \sigma T^4$	$\mathrm{W\ m^{-2}\ K^{-4}}$
first radiation constant	c_1	$c_1 = 2\pi h c_0{}^2$	$\mathrm{W\ m^2}$
second radiation constant	c_2	$c_2 = h c_0/k$	$\mathrm{K\ m}$
transmittance, transmission factor	τ, T	$\tau = \Phi_{tr}/\Phi_0$	1
absorptance, absorption factor	α	$\alpha = \Phi_{abs}/\Phi_0$	1
reflectance, reflection factor	ρ	$\rho = \Phi_{refl}/\Phi_0$	1
(decadic) absorbance	A	$A = -\lg(1-\alpha_i)$	1
napierian absorbance	B	$B = -\ln(1-\alpha_i)$	1
absorption coefficient			
(linear) decadic	a, K	$a = A/l$	$\mathrm{m^{-1}}$
(linear) napierian	α	$\alpha = B/l$	$\mathrm{m^{-1}}$
molar (decadic)	ε	$\varepsilon = a/c = A/cl$	$\mathrm{m^2\ mol^{-1}}$
molar napierian	κ	$\kappa = \alpha/c = B/cl$	$\mathrm{m^2\ mol^{-1}}$
absorption index	k	$k = \alpha/4\pi\tilde{\nu}$	1
complex refractive index	$\hat{n}$	$\hat{n} = n + ik$	1
molar refraction	R, R_m	$R = \dfrac{(n^2-1)}{(n^2+2)} V_m$	$\mathrm{m^3\ mol^{-1}}$
angle of optical rotation	α		1, rad

SOLID STATE

Name	Symbol	Definition	SI unit
lattice vector	R, R_0		m
fundamental translation vectors for the crystal lattice	$a_1; a_2; a_3,$ $a; b; c$	$R = n_1 a_1 + n_2 a_2 + n_3 a_3$	m
(circular) reciprocal lattice vector	G	$G \cdot R = 2\pi m$	$\mathrm{m^{-1}}$
(circular) fundamental translation vectors for the reciprocal lattice	$b_1; b_2; b_3,$ $a^*; b^*; c^*$	$a_i \cdot b_k = 2\pi\delta_{ik}$	$\mathrm{m^{-1}}$
lattice plane spacing	d		m
Bragg angle	θ	$n\lambda = 2d \sin\theta$	1, rad

Name	Symbol	Definition	SI unit
order of reflection	n		1
order parameters			
short range	σ		1
long range	s		1
Burgers vector	b		m
particle position vector	r, R_j		m
equilibrium position vector of an ion	R_0		m
displacement vector of an ion	u	$u = R - R_0$	m
Debye–Waller factor	B, D		1
Debye circular wavenumber	q_D		m^{-1}
Debye circular frequency	ω_D		s^{-1}
Grüneisen parameter	γ, Γ	$\gamma = \alpha V / \kappa C_V$	1
Madelung constant	$\alpha, \mathcal{M}$	$E_{coul} = \dfrac{\alpha N_A z_+ z_- e^2}{4\pi\varepsilon_0 R_0}$	1
density of states	N_E	$N_E = dN(E)/dE$	$J^{-1} m^{-3}$
(spectral) density of vibrational modes	N_ω, g	$N_\omega = dN(\omega)/d\omega$	$s\, m^{-3}$
resistivity tensor	ρ_{ik}	$E = \rho \cdot j$	$\Omega\, m$
conductivity tensor	σ_{ik}	$\sigma = \rho^{-1}$	$S\, m^{-1}$
thermal conductivity tensor	λ_{ik}	$J_q = -\lambda \cdot \mathrm{grad}\, T$	$W\, m^{-1} K^{-1}$
residual resistivity	ρ_R		$\Omega\, m$
relaxation time	τ	$\tau = l/v_F$	s
Lorenz coefficient	L	$L = \lambda/\sigma T$	$V^2\, K^{-2}$
Hall coefficient	A_H, R_H	$E = \rho \cdot j + R_H (B \times j)$	$m^3\, C^{-1}$
thermoelectric force	E		V
Peltier coefficient	Π		V
Thomson coefficient	$\mu, (\tau)$		$V\, K^{-1}$
work function	Φ	$\Phi = E_\infty - E_F$	J
number density, number concentration	$n, (p)$		m^{-3}
gap energy	E_g		J
donor ionization energy	E_d		J
acceptor ionization energy	E_a		J
Fermi energy	E_F, ε_F		J
circular wave vector, propagation vector	k, q	$k = 2\pi/\lambda$	m^{-1}
Bloch function	$u_k(r)$	$\psi(r) = u_k(r)\exp(ik \cdot r)$	$m^{-3/2}$
charge density of electrons	ρ	$\rho(r) = -e\psi^*(r)\psi(r)$	$C\, m^{-3}$

Name	Symbol	Definition	SI unit
effective mass	m^*		kg
mobility	μ	$\mu = v_{\text{drift}}/E$	$m^2\,V^{-1}\,s^{-1}$
mobility ratio	b	$b = \mu_n/\mu_p$	1
diffusion coefficient	D	$dN/dt = -DA(dn/dx)$	$m^2\,s^{-1}$
diffusion length	L	$L = \sqrt{D\tau}$	m
characteristic (Weiss) temperature	θ, θ_W		K
Curie temperature	T_C		K
Néel temperature	T_N		K

STATISTICAL THERMODYNAMICS

Name	Symbol	Definition	SI unit
number of entities	N		1
number density of entities, number concentration	n, C	$n = N/V$	m^{-3}
Avogadro constant	L, N_A		mol^{-1}
Boltzmann constant	k, k_B		$J\,K^{-1}$
gas constant (molar)	R	$R = Lk$	$J\,K^{-1}\,mol^{-1}$
molecular position vector	$r\,(x, y, z)$		m
molecular velocity vector	$c(c_x, c_y, c_z),$ $u(u_x, u_y, u_z)$	$c = dr/dt$	$m\,s^{-1}$
molecular momentum vector	$p(p_x, p_y, p_z)$	$p = mc$	$kg\,m\,s^{-1}$
velocity distribution function (Maxwell)	$f(c_x)$	$f(c_x) = (m/2\pi kT)^{\frac{1}{2}}$ $\times \exp(-mc_x^2/2kT)$	$m^{-1}\,s$
speed distribution function (Maxwell–Boltzmann)	$F(c)$	$F(c) = (m/2\pi kT)^{3/2}$ $\times 4\pi c^2 \exp(-mc^2/2kT)$	$m^{-1}\,s$
average speed	$\bar{c}, \bar{u},$ $\langle c \rangle, \langle u \rangle$	$\bar{c} = \int cF(c)\,dc$	$m\,s^{-1}$
generalized coordinate	q		(m)
generalized momentum	p	$p = \partial L/\partial \dot{q}$	$(kg\,m\,s^{-1})$
volume in phase space	Ω	$\Omega = (1/h)\int p\,dq$	1
probability	P		1
statistical weight, degeneracy	g, d, W, ω, β		1
density of states	$\rho(E)$	$\rho(E) = dN/dE$	J^{-1}
partition function, sum over states, for a single molecule	q, z	$q = \sum_i g_i \exp(-\varepsilon_i/kT)$	1
for a canonical ensemble (system, or assembly)	Q, Z		1

Name	Symbol	Definition	SI unit
microcanonical ensemble	Ω		1
grand (canonical ensemble)	Ξ		1
symmetry number	σ, s		1
reciprocal temperature parameter	β	$\beta = 1/kT$	J^{-1}
characteristic temperature	Θ		K

GENERAL CHEMISTRY

Name	Symbol	Definition	SI unit
number of entities (e.g. molecules, atoms, ions, formula units)	N		1
amount (of substance)	n	$n_B = N_B/L$	mol
Avogadro constant	L, N_A		mol^{-1}
mass of atom, atomic mass	m_a, m		kg
mass of entity (molecule, or formula unit)	m_f, m		kg
atomic mass constant	m_u	$m_u = m_a(^{12}C)/12$	kg
molar mass	M	$M_B = m/n_B$	$kg\,mol^{-1}$
relative molecular mass (relative molar mass, molecular weight)	M_r	$M_{r,B} = m_B/m_u$	1
molar volume	V_m	$V_{m,B} = V/n_B$	$m^3\,mol^{-1}$
mass fraction	w	$w_B = m_B/\Sigma m_i$	1
volume fraction	ϕ	$\phi_B = V_B/\Sigma V_i$	1
mole fraction, amount fraction, number fraction	x, y	$x_B = n_B/\Sigma n_i$	1
(total) pressure	p, P		Pa
partial pressure	p_B	$p_B = y_B p$	Pa
mass concentration (mass density)	γ, ρ	$\gamma_B = m_B/V$	$kg\,m^{-3}$
number concentration, number density of entities	C, n	$C_B = N_B/V$	m^{-3}
amount concentration, concentration	c	$c_B = n_B/V$	$mol\,m^{-3}$
solubility	s	$s_B = c_B$ (saturated solution)	$mol\,m^{-3}$
molality (of a solute)	$m, (b)$	$m_B = n_B/m_A$	$mol\,kg^{-1}$

Name	Symbol	Definition	SI unit
surface concentration	Γ	$\Gamma_B = n_B/A$	$mol\ m^{-2}$
stoichiometric number	v		1
extent of reaction, advancement	ξ	$\Delta\xi = \Delta n_B/v_B$	mol
degree of dissociation	α		1

CHEMICAL THERMODYNAMICS

Name	Symbol	Definition	SI unit
heat	q, Q		J
work	w, W		J
internal energy	U	$\Delta U = q + w$	J
enthalpy	H	$H = U + pV$	J
thermodynamic temperature	T		K
Celsius temperature	θ, t	$\theta/°C = T/K - 273.15$	°C
entropy	S	$dS \geqslant dq/T$	$J\,K^{-1}$
Helmholtz energy, (Helmholtz function)	A	$A = U - TS$	J
Gibbs energy, (Gibbs function)	G	$G = H - TS$	J
Massieu function	J	$J = -A/T$	$J\,K^{-1}$
Planck function	Y	$Y = -G/T$	$J\,K^{-1}$
surface tension	γ, σ	$\gamma = (\partial G/\partial A_s)_{T,p}$	$J\,m^{-2}, N\,m^{-1}$
molar quantity X	X_m	$X_m = X/n$	(varies)
specific quantity X	x	$x = X/m$	(varies)
pressure coefficient	β	$\beta = (\partial p/\partial T)_V$	$Pa\,K^{-1}$
relative pressure coefficient	α_p	$\alpha_p = (1/p)(\partial p/\partial T)_V$	K^{-1}
compressibility,			
isothermal	κ_T	$\kappa_T = -(1/V)(\partial V/\partial p)_T$	Pa^{-1}
isentropic	κ_S	$\kappa_S = -(1/V)(\partial V/\partial p)_S$	Pa^{-1}
linear expansion coefficient	α_l	$\alpha_l = (1/l)(\partial l/\partial T)$	K^{-1}
cubic expansion coefficient	α, α_V, γ	$\alpha = (1/V)(\partial V/\partial T)_p$	K^{-1}
heat capacity,			
at constant pressure	C_p	$C_p = (\partial H/\partial T)_p$	$J\,K^{-1}$
at constant volume	C_V	$C_V = (\partial U/\partial T)_V$	$J\,K^{-1}$
ratio of heat capacities	$\gamma, (\kappa)$	$\gamma = C_p/C_V$	1
Joule–Thomson coefficient	μ, μ_{JT}	$\mu = (\partial T/\partial p)_H$	$K\,Pa^{-1}$
second virial coefficient	B	$pV_m = RT(1 + B/V_m + \cdots)$	$m^3\,mol^{-1}$
compression factor (compressibility factor)	Z	$Z = pV_m/RT$	1

Name	Symbol	Definition	SI unit
partial molar quantity X	$X_B, (X'_B)$	$X_B = (\partial X / \partial n_B)_{T, p, n_{j \neq B}}$	(varies)
chemical potential (partial molar Gibbs energy)	μ	$\mu_B = (\partial G / \partial n_B)_{T, p, n_{j \neq B}}$	$J\,mol^{-1}$
absolute activity	λ	$\lambda_B = \exp(\mu_B / RT)$	1
standard chemical potential	$\mu^\bullet, \mu^\circ$		$J\,mol^{-1}$
standard partial molar enthalpy	$H_B^\bullet$	$H_B^\bullet = \mu_B^\bullet + TS_B^\bullet$	$J\,mol^{-1}$
standard partial molar entropy	$S_B^\bullet$	$S_B^\bullet = -(\partial \mu_B^\bullet / \partial T)_p$	$J\,mol^{-1}\,K^{-1}$
standard reaction Gibbs energy (function)	$\Delta_r G^\bullet$	$\Delta_r G^\bullet = \sum_B \nu_B \mu_B^\bullet$	$J\,mol^{-1}$
affinity of reaction	$A, (\mathscr{A})$	$A = -(\partial G / \partial \xi)_{p, T}$ $= -\sum_B \nu_B \mu_B$	$J\,mol^{-1}$
standard reaction enthalpy	$\Delta_r H^\bullet$	$\Delta_r H^\bullet = \sum_B \nu_B H_B^\bullet$	$J\,mol^{-1}$
standard reaction entropy	$\Delta_r S^\bullet$	$\Delta_r S^\bullet = \sum_B \nu_B S_B^\bullet$	$J\,mol^{-1}\,K^{-1}$
equilibrium constant	$K^\bullet, K$	$K^\bullet = \exp(-\Delta_r G^\bullet / RT)$	1
equilibrium constant, pressure basis	K_p	$K_p = \prod_B p_B^{\nu_B}$	$Pa^{\Sigma \nu}$
concentration basis	K_c	$K_c = \prod_B c_B^{\nu_B}$	$(mol\,m^{-3})^{\Sigma \nu}$
molality basis	K_m	$K_m = \prod_B m_B^{\nu_B}$	$(mol\,kg^{-1})^{\Sigma \nu}$
fugacity	$f, \tilde{p}$	$f_B = \lambda_B \lim\limits_{p \to 0} (p_B / \lambda_B)_T$	Pa
fugacity coefficient	ϕ	$\phi_B = f_B / p_B$	1
activity and activity coefficient referenced to Raoult's law, (relative) activity	a	$a_B = \exp\left[\dfrac{\mu_B - \mu_B^*}{RT}\right]$	1
activity coefficient	f	$f_B = a_B / x_B$	1
activities and activity coefficients referenced to Henry's law, (relative) activity, molality basis	a_m	$a_{m,B} = \exp\left[\dfrac{\mu_B - \mu_B^\bullet}{RT}\right]$	1
concentration basis	a_c	$a_{c,B} = \exp\left[\dfrac{\mu_B - \mu_B^\bullet}{RT}\right]$	1

Name	Symbol	Definition	SI Unit
mole fraction basis	a_x	$a_{x,\,B}=\exp\left[\dfrac{\mu_B-\mu_B^{\bullet}}{RT}\right]$	1
activity coefficient,			
molality basis	γ_m	$a_{m,B}=\gamma_{m,B}\,m_B/m^{\bullet}$	1
concentration basis	γ_c	$a_{c,B}=\gamma_{c,B}\,c_B/c^{\bullet}$	1
mole fraction basis	γ_x	$a_{x,B}=\gamma_{x,B}\,x_B$	1
ionic strength,			
molality basis	$I_m,\,I$	$I_m=\frac{1}{2}\Sigma m_B z_B^2$	$mol\,kg^{-1}$
concentration basis	$I_c,\,I$	$I_c=\frac{1}{2}\Sigma c_B z_B^2$	$mol\,m^{-3}$
osmotic coefficient,			
molality basis	ϕ_m	$\phi_m=(\mu_A^*-\mu_A)/(RTM_A\Sigma m_B)$	1
mole fraction basis	ϕ_x	$\phi_x=(\mu_A-\mu_A^*)/(RT\ln x_A)$	1
osmotic pressure	Π	$\Pi=c_B RT$ (ideal dilute solution)	Pa

(i) *Symbols used as subscripts to denote a chemical process or reaction*
These symbols should be printed in roman (upright) type, without a full stop (period).

vaporization, evaporation (liquid→gas)	vap
sublimation (solid→gas)	sub
melting, fusion (solid→liquid)	fus
transition (between two phases)	trs
mixing of fluids	mix
solution (of solute in solvent)	sol
dilution (of a solution)	dil
adsorption	ads
displacement	dpl
immersion	imm
reaction in general	r
atomization	at
combustion reaction	c
formation reaction	f

(ii) *Recommended superscripts*

standard	$\ominus$, o
pure substance	*
infinite dilution	∞
ideal	id
activated complex, transition state	‡
excess quantity	E

Name	Symbol	Definition	SI unit
CHEMICAL KINETICS			
rate of change of quantity X	$\dot{X}$	$\dot{X} = dX/dt$	(varies)
rate of conversion	$\dot{\xi}$	$\dot{\xi} = d\xi/dt$	$mol\ s^{-1}$
rate of concentration change (due to chemical reaction)	r_B, v_B	$r_B = dc_B/dt$	$mol\ m^{-3}\ s^{-1}$
rate of reaction (based on amount concentration)	v	$v = \dot{\xi}/V$ $= v_B^{-1}dc_B/dt$	$mol\ m^{-3}\ s^{-1}$
partial order of reaction	n_B	$v = k\Pi c_B^{n_B}$	1
overall order of reaction	n	$n = \Sigma n_B$	1
rate constant, rate coefficient	k	$v = k\Pi c_B^{n_B}$	$(mol^{-1}\ m^3)^{n-1}s^{-1}$
Boltzmann constant	k, k_B		$J\ K^{-1}$
half life	$t_{\frac{1}{2}}$	$c(t_{\frac{1}{2}}) = c_0/2$	s
relaxation time	τ	$\tau = 1/(k_1 + k_{-1})$	s
energy of activation, activation energy	E_a, E	$E_a = RT^2\ d\ln k/dT$	$J\ mol^{-1}$
pre-exponential factor	A	$k = A\exp(-E_a/RT)$	$(mol^{-1}\ m^3)^{n-1}s^{-1}$
volume of activation	$\Delta^{\ddagger}V$	$\Delta^{\ddagger}V = -RT\times$ $(\partial\ln k/\partial p)_T$	$m^3\ mol^{-1}$
collision diameter	d	$d_{AB} = r_A + r_B$	m
collision cross-section	σ	$\sigma_{AB} = \pi d_{AB}^2$	m^2
collision frequency	Z_A		s^{-1}
collision number	Z_{AB}, Z_{AA}		$m^{-3}s^{-1}$
collision frequency factor	z_{AB}, z_{AA}	$z_{AB} = Z_{AB}/Lc_Ac_B$	$m^3\ mol^{-1}\ s^{-1}$
standard enthalpy of activation	$\Delta^{\ddagger}H^{\ominus}, \Delta H^{\ddagger}$		$J\ mol^{-1}$
standard entropy of activation	$\Delta^{\ddagger}S^{\ominus}, \Delta S^{\ddagger}$		$J\ mol^{-1}\ K^{-1}$
standard Gibbs energy of activation	$\Delta^{\ddagger}G^{\ominus}, \Delta G^{\ddagger}$		$J\ mol^{-1}$
quantum yield, photochemical yield	ϕ		1
ELECTROCHEMISTRY			
elementary charge (proton charge)	e		C
Faraday constant	F	$F = eL$	$C\ mol^{-1}$

Name	Symbol	Definition	SI unit
charge number of an ion	z	$z_B = Q_B/e$	1
ionic strength	I_c, I	$I_c = \frac{1}{2}\sum c_i z_i^2$	$mol\,m^{-3}$
mean ionic activity	$a_\pm$	$a_\pm = m_\pm \gamma_\pm /m^\bullet$	1
mean ionic molality	$m_\pm$	$m_\pm^{(\nu_+ + \nu_-)} = m_+^{\nu_+} \cdot m_-^{\nu_-}$	$mol\,kg^{-1}$
mean ionic activity coefficient	$\gamma_\pm$	$\gamma_\pm^{(\nu_+ + \nu_-)} = \gamma_+^{\nu_+} \cdot \gamma_-^{\nu_-}$	1
charge number of electrochemical cell reaction	$n, (z)$		1
electric potential difference (of a galvanic cell)	$\Delta V, E, U$	$\Delta V = V_R - V_L$	V
emf, electromotive force	E	$E = \lim_{I \to 0} \Delta V$	V
standard emf, standard potential of the electrochemical cell reaction	$E^\bullet$	$E^\bullet = -\Delta_r G^\bullet /nF$ $= (RT/nF)\ln K^\bullet$	V
standard electrode potential	$E^\bullet$		V
emf of the cell, potential of the electro-chemical cell reaction	E	$E = E^\bullet - (RT/nF)$ $\times \sum \nu_i \ln a_i$	V
pH	pH	$pH \approx -\lg\left[\dfrac{c(H^+)}{mol\,dm^{-3}}\right]$	1
inner electric potential	ϕ	$\nabla \phi = -E$	V
outer electric potential	ψ	$\psi = Q/4\pi\varepsilon_0 r$	V
surface electric potential	χ	$\chi = \phi - \psi$	V
Galvani potential difference	$\Delta\phi$	$\Delta_\alpha^\beta \phi = \phi^\beta - \phi^\alpha$	V
volta potential difference	$\Delta\psi$	$\Delta_\alpha^\beta \psi = \psi^\beta - \psi^\alpha$	V
electrochemical potential	$\tilde{\mu}$	$\tilde{\mu}_B^\alpha = (\partial G/\partial n_B^\alpha)$	$J\,mol^{-1}$
electric current	I	$I = dQ/dt$	A
(electric) current density	j	$j = I/A$	$A\,m^{-2}$
(surface) charge density	σ	$\sigma = Q/A$	$C\,m^{-2}$
electrode reaction rate constant	k	$k_{ox} = I_a/(nFA \prod_i c_i^{n_i})$	(varies)
mass transfer coefficient, diffusion rate constant	k_d	$k_{d,B} = |\nu_B| I_{l,B}/nFcA$	$m\,s^{-1}$
thickness of diffusion layer	δ	$\delta_B = D_B/k_{d,B}$	m

Name	Symbol	Definition	SI unit
transfer coefficient (electrochemical)	α	$\alpha_c = \dfrac{-|\nu|RT}{nF}\dfrac{\partial \ln|I_c|}{\partial E}$	1
overpotential	η	$\eta = E_I - E_{I=0} - IR_u$	V
electrokinetic potential (zeta potential)	ζ		V
conductivity	$\kappa, (\sigma)$	$\kappa = j/E$	$S\,m^{-1}$
conductivity cell constant	K_{cell}	$K_{cell} = \kappa R$	m^{-1}
molar conductivity (of an electrolyte)	Λ	$\Lambda_B = \kappa/c_B$	$S\,m^2\,mol^{-1}$
ionic conductivity, molar conductivity of an ion	λ	$\lambda_B = |z_B|Fu_B$	$S\,m^2\,mol^{-1}$
electric mobility	$u, (\mu)$	$u_B = v_B/E$	$m^2\,V^{-1}\,s^{-1}$
transport number	t	$t_B = j_B/\Sigma j_i$	1
reciprocal radius of ionic atmosphere	κ	$\kappa = (2F^2I/\varepsilon RT)^{\frac{1}{2}}$	m^{-1}

COLLOID AND SURFACE CHEMISTRY

Name	Symbol	Definition	SI unit
specific surface area	a, a_s, s	$a = A/m$	$m^2\,kg^{-1}$
surface amount of B, adsorbed amount of B	n_B^s, n_B^a		mol
surface excess of B	n_B^σ		mol
surface excess concentration of B	$\Gamma_B, (\Gamma_B^\sigma)$	$\Gamma_B = n_B^\sigma/A$	$mol\,m^{-2}$
total surface excess concentration	$\Gamma, (\Gamma^\sigma)$	$\Gamma = \sum_i \Gamma_i$	$mol\,m^{-2}$
area per molecule	a, σ	$a_B = A/N_B^\sigma$	m^2
area per molecule in a filled monolayer	a_m, σ_m	$a_{m,B} = A/N_{m,B}$	m^2
surface coverage	θ	$\theta = N_B^\sigma/N_{m,B}$	1
contact angle	θ		1, rad
film thickness	t, h, δ		m
thickness of (surface or interfacial) layer	τ, δ, t		m
surface tension, interfacial tension	γ, σ	$\gamma = (\partial G/\partial A_s)_{T,p}$	$N\,m^{-1}, J\,m^{-2}$
film tension	Σ_f	$\Sigma_f = 2\gamma_f$	$N\,m^{-1}$
reciprocal thickness of the double layer	κ	$\kappa = [2F^2 I_c/\varepsilon RT]^{\frac{1}{2}}$	m^{-1}
average molar masses			
number-average	M_n	$M_n = \Sigma n_i M_i/\Sigma n_i$	$kg\,mol^{-1}$
mass-average	M_m	$M_m = \Sigma n_i M_i^2/\Sigma n_i M_i$	$kg\,mol^{-1}$

Name	Symbol	Definition	SI unit
Z-average	M_Z	$M_Z = \Sigma n_i M_i^3 / \Sigma n_i M_i^2$	kg mol^{-1}
sedimentation coefficient	s	$s = v/a$	s
van der Waals constant	λ		J
retarded van der Waals constant	β, B		J
van der Waals–Hamaker constant	A_H		J
surface pressure	π^s, π	$\pi^s = \gamma^0 - \gamma$	N m^{-1}

TRANSPORT PROPERTIES

Name	Symbol	Definition	SI unit
flux (of a quantity X)	J_X, J	$J_X = A^{-1} \, dX/dt$	(varies)
volume flow rate	$q_V, \dot{V}$	$q_v = dV/dt$	$\text{m}^3 \, \text{s}^{-1}$
mass flow rate	$q_m, \dot{m}$	$q_m = dm/dt$	kg s^{-1}
mass transfer coefficient	k_d		m s^{-1}
heat flow rate	ϕ	$\phi = dq/dt$	W
heat flux	J_q	$J_q = \phi/A$	W m^{-2}
thermal conductance	G	$G = \phi/\Delta T$	W K^{-1}
thermal resistance	R	$R = 1/G$	K W^{-1}
thermal conductivity	λ, k	$\lambda = J_q/(dT/dl)$	$\text{W m}^{-1} \text{K}^{-1}$
coefficient of heat transfer	$h, (k, K, \alpha)$	$h = J_q/\Delta T$	$\text{W m}^{-2} \text{K}^{-1}$
thermal diffusivity	a	$a = \lambda/\rho c_p$	$\text{m}^2 \, \text{s}^{-1}$
diffusion coefficient	D	$D = J_n/(dc/dl)$	$\text{m}^2 \, \text{s}^{-1}$

The following symbols are used in the definitions of the dimensionless quantities: mass (m), time (t), volume (V), area (A), density (ρ), speed (v), length (l), viscosity (η), pressure (p), acceleration of free fall (g), cubic expansion coefficient (α), temperature (T), surface tension (γ), speed of sound (c), mean free path (λ), frequency (f), thermal diffusivity (a), coefficient of heat transfer (h), thermal conductivity (k), specific heat capacity at constant pressure (c_p), diffusion coefficient (D), mole fraction (x), mass transfer coefficient (k_d), permeability (μ), electric conductivity (κ), and magnetic flux density (B).

Name	Symbol	Definition	SI unit
Reynolds number	Re	$Re = \rho v l/\eta$	1
Euler number	Eu	$Eu = \Delta p/\rho v^2$	1
Froude number	Fr	$Fr = v/(lg)^{\frac{1}{2}}$	1
Grashof number	Gr	$Gr = l^3 g \alpha \Delta T \rho^2/\eta^2$	1
Weber number	We	$We = \rho v^2 l/\gamma$	1
Mach number	Ma	$Ma = v/c$	1
Knudsen number	Kn	$Kn = \lambda/l$	1
Strouhal number	Sr	$Sr = lf/v$	1
Fourier number	Fo	$Fo = at/l^2$	1
Péclet number	Pe	$Pe = vl/a$	1
Rayleigh number	Ra	$Ra = l^3 g \alpha \Delta T \rho/\eta a$	1
Nusselt number	Nu	$Nu = hl/k$	1
Stanton number	St	$St = h/\rho v c_p$	1

Name	Symbol	Definition	SI units
Fourier number for mass transfer	Fo^*	$Fo^* = Dt/l^2$	1
Péclet number for mass transfer	Pe^*	$Pe^* = vl/D$	1
Grashof number for mass transfer	Gr^*	$Gr^* = l^3 g \left(\dfrac{\partial \rho}{\partial x}\right)_{T,p} \left(\dfrac{\Delta x \rho}{\eta}\right)$	1
Nusselt number for mass transfer	Nu^*	$Nu^* = k_d l/D$	1
Stanton number for mass transfer	St^*	$St^* = k_d/v$	1
Prandtl number	Pr	$Pr = \eta/\rho a$	1
Schmidt number	Sc	$Sc = \eta/\rho D$	1
Lewis number	Le	$Le = a/D$	1
magnetic Reynolds number	Rm, Re_m	$Rm = v\mu\kappa l$	1
Alfvén number	Al	$Al = v(\rho\mu)^{\frac{1}{2}}/B$	1
Hartmann number	Ha	$Ha = Bl(\kappa/\eta)^{\frac{1}{2}}$	1
Cowling number	Co	$Co = B^2/\mu\rho v^2$	1

NOMENCLATURE OF CHEMICAL COMPOUNDS

The International Union of Pure and Applied Chemistry (IUPAC) maintains several commissions that deal with the naming of chemical substances. In general, the approach of IUPAC is to present rules for arriving at names in a systematic manner, rather than recommending a unique name for each compound. Thus there are often several alternative "IUPAC names", depending on which nomenclature system is used, each of which may have advantages in specific applications. However, each of these names will be unambiguous.

Organizations such as the Chemical Abstacts Service and the Beilstein Institute that prepare indexes to the chemical literature must adopt a system for selecting unique names in order to avoid excessive cross referencing. Chemical Abstracts Service uses a system which groups together compounds derived from a single parent compound. Thus most index names are inverted (e.g., Benzene, bromo rather than bromobenzene; Acetic acid, sodium salt rather than sodium acetate). In this *Handbook* the CAS Index Names are used only in the table "Physical Constants of Organic Compounds". Other tables use more familiar names which, with a few possible exceptions, conform to one of the IUPAC naming systems.

Recommended names for the most common substituent groups, ligands, ions, and organic rings are given in the two following tables, "Nomenclature for Inorganic Ions and Ligands" and "Organic Substituent Groups and Ring Systems". For the basics of macromolecular nomenclature, see "Naming Organic Polymers" in Section 13.

Some of the most useful recent guides to chemical nomenclature, prepared by IUPAC and other organizations such as the International Union of Biochemistry and Molecular Biology (IUBMB) and the American Chemical Society are listed below . These books contain citations to the more detailed nomenclature documents in each area.

Inorganic Chemistry

International Union of Pure and Applied Chemistry, *Nomenclature of Inorganic Chemistry, Recommendations 1990,* edited by Leigh, G.J., Blackwell Scientific Publications, Oxford, 1990.

Block, B.P., Powell, W.H., and Fernelius, W.C., *Inorganic Chemical Nomenclature, Principles and Practice*, American Chemical Society, Washington, 1990.

Organic Chemistry

International Union of Pure and Applied Chemistry, *A Guide to IUPAC Nomenclature of Organic Compounds, Recommendations 1993,* edited by Panico, R., Powell, W.H., and Richer, J.-C., Blackwell Scientific Publications, Oxford, 1993.

International Union of Pure and Applied Chemistry, *Glossary of Class Names of Organic Compounds and Reactive Intermediates Based on Structure,* edited by Moss, G.P., Smith, P.A.S., and Tavernier, D., *Pure & Appl. Chem*, 67, 1307, 1995.

Rhodes, P.H., *The Organic Chemist's Desk Reference*, Chapman & Hall, London, 1995.

International Union of Pure and Applied Chemistry, *Basic Terminology of Stereochemistry*, edited by Moss, G.P., *Pure & Applied Chemistry*, 68, 2193, 1966.

Macromolecular Chemistry

International Union of Pure and Applied Chemistry, *Compendium of Macromolecular Nomenclature,* edited by, Metanomski, W.V., Blackwell Scientific Publications, Oxford, 1991.

International Union of Pure and Applied Chemistry, *Glossary of Basic Terms in Polymer Science,* edited by Jenkins, A.D., Kratochvil, P., Stepto, R.F.T., and Suter, U.W., *Pure & Appl. Chem*, 68, in press.

Biochemistry

International Union of Biochemistry and Molecular Biology, *Biochemical Nomenclature and Related Documents*, 2nd Edition, 1992, Portland Press, London, 1993; includes recommendations of the IUPAC-IUBMB Joint Commission on Biochemical Nomenclature.

International Union of Biochemistry and Molecular Biology, *Enzyme Nomenclature, 1992*, Academic Press, Orlando, FL, 1992.

IUPAC-IUBMB Joint Commission on Biochemical Nomenclature, *Nomenclature of Carbohydrates, Recommendations 1996*, edited by McNaught, A.D., *Pure & Appl. Chem.*, 68, 1919, 1996.

General

Chemical Abstracts Service, *Naming and Indexing Chemical Substances for Chemical Abstracts, Appendix IV, Chemical Abstracts 1994 Index Guide.*

NOMENCLATURE FOR INORGANIC IONS AND LIGANDS

See the table *Nomenclature of Chemical Compounds* for references. The assistance of Warren H. Powell in preparing this list is gratefully acknowledged.

Group	As cation	As anion	As ligand	As prefix in organic compounds
H	hydrogen	hydride	hydrido	
F	fluorine	fluoride	fluoro	fluoro
Cl	chlorine	chloride	chloro	chloro
Br	bromine	bromide	bromo	bromo
I	iodine	iodide	iodo	iodo
ClO	chlorosyl	hypochlorite	hypochlorito	chlorosyl
ClO_2	chloryl	chlorite	chlorito	chloryl
ClO_3	perchloryl	chlorate	chlorato	perchloryl
ClO_4		perchlorate		
IO	iodosyl	hypoiodite		iodoso
IO_2	iodyl			iodyl; iodoxy
O		oxide	oxo	oxo
O_2		peroxide (O_2^{2-})	peroxo	peroxy
		hyperoxide (O_2^-)		
HO		hydroxide	hydroxo	hydroxy
HO_2		hydrogen peroxide	hydrogen peroxo	hydroperoxy
S		sulfide	thio; sulfido	thio; thioxo
HS		hydrogen sulfide	thiolo	mercapto
S_2		disulfide	disulfido	
SO	sulfinyl; thionyl			sulfinyl
SO_2	sulfonyl; sulfuryl	sulfoxylate		sulfonyl
SO_3		sulfite	sulfito	
HSO_3		hydrogen sulfite	hydrogen sulfito	
S_2O_3		thiosulfate	thiosulfato	
SO_4		sulfate	sulfato	
Se		selenide	seleno	seleno; selenoxo
SeO	seleninyl			seleninyl
SeO_2	selenonyl			selenonyl
SeO_3		selenite	selenito	
SeO_4		selenate	selenato	
Te		telluride	telluro	telluro; telluroxo
CrO_2	chromyl			
UO_2	uranyl			
NpO_2	neptunyl			
PuO_2	plutoryl			
AmO_2	americyl			
N		nitride	nitrido	
N_3		azide	azido	
NH		imide	imido	imino
NH_2		amide	amido	amino
NHOH		hydroxylamide	hydroxylamido	hydroxyamino
N_2H_3		hydrazide	hydrazido	hydrazino; diazanyl
NO	nitrosyl		nitrosyl	nitroso
NO_2	nitryl		nitro	nitro
ONO		nitrite	nitrito	
NS	thionitrosyl			
NO_3		nitrate	nitrato	
N_2O_3		hyponitrite	hyponitrito	
P		phosphide	phosphido	phosphinidyne
PO	phosphoryl			phosphoroso; phosphinylidyne
PO_2	phospho			
PS	thiophosphoryl			phosphinothioylidyne; thiophosphorozo
PH_2O_3		hypophosphite	hypophosphito	
PHO_3		phosphite	phosphito	

NOMENCLATURE FOR INORGANIC IONS AND LIGANDS (continued)

Group	As cation	As anion	As ligand	As prefix in organic compounds
PO_4		phosphate	phosphato	
AsO_4		arsenate	arsenato	
VO	vanadyl			
CO	carbonyl		carbonyl	carbonyl
CS	thiocarbonyl			thiocarbonyl
CH_3O		methanolate	methoxo	methoxy
C_2H_5O		ethanolate	ethoxo	ethoxy
CH_3S		methanethiolate	methanethiolato	methylthio
C_2H_5S		ethanethiolate	ethanethiolato	ethylthio
CN	cyanogen	cyanide	cyano	cyano
OCN		cyanate	cyanato	cyanato
SCN		thiocyanate	thiocyanato	thiocyanato
$SeCN$		selenocyanate	selenocyanato	selenocyanato
$TeCN$		tellurocyanate	tellurocyanato	tellurocyanato
CO_3		carbonate	carbonato	
HCO_3		hydrogen carbonate	hydrogen carbonato	carboxycarbonyl
C_2O_4		oxalate	oxalato	

ORGANIC SUBSTITUENT GROUPS AND RING SYSTEMS

The first part of this table lists substituent groups and their line formulas. A substituent group is defined by IUPAC as a group that replaces one or more hydrogen atoms attached to a parent structure. Such groups are sometimes called radicals, but IUPAC now reserves the term radical for a free molecular species with unpaired electrons. IUPAC does not recommend some of these names, which are marked here with asterisks (e.g., amyl*), but they are included in this list because they are often encountered in the older literature. Substituent group names which are formed by systematic rules (e.g., methyl from methane, ethyl from ethane, etc.) are included here only for the first few members of a homologous series.

In the second part of the table a number of common organic ring compounds are shown, with the conventional numbering of the ring positions indicated.

The help of Warren H. Powell in preparing this table is greatly appreciated. Pertinent references may be found in the table *Nomenclature of Chemical Compounds*.

SUBSTITUENT GROUPS

acetamido (acetylamino)	CH_3CONH-	cyanamido (cyanoamino)	$NCNH-$
acetoacetyl	CH_3COCH_2CO-	cyanato	$NCO-$
acetonyl	CH_3COCH_2-	cyano	$NC-$
acetyl	CH_3CO-	decanedioyl	$-OC(CH_2)_8CO-$
acryloyl* (1-oxo-2-propenyl)	$CH_2=CHCO-$	decanoyl	$CH_3(CH_2)_8CO-$
alanyl (from alanine)	$CH_3CH(NH_2)CO-$	diazo	$N_2=$
β-alanyl	$H_2N(CH_2)_2CO-$	diazoamino	$-NHN=N-$
allyl (2-propenyl)	$CH_2=CHCH_2-$	disilanyl	H_3SiSiH_2-
allylidene (2-propenylidene)	$CH_2=CHCH=$	disiloxanyloxy	$H_3SiOSiH_2O-$
amidino (aminoiminomethyl)	$H_2NC(=NH)-$	disulfinyl	$-S(O)S(O)-$
amino	H_2N-	dithio	$-SS-$
amyl* (pentyl)	$CH_3(CH_2)_4-$	enanthoyl* (heptanoyl)	$CH_3(CH_2)_5CO-$
anilino (phenylamino)	C_6H_5NH-	epoxy	$-O-$
anisidino	$CH_3OC_6H_4NH-$	ethenyl (vinyl)	$CH_2=CH-$
anthranoyl (2-aminobenzoyl)	$2-H_2NC_6H_4CO-$	ethynyl	$HC≡C-$
arsino	AsH_2-	ethoxy	C_2H_5O-
azelaoyl (from azelaic acid)	$-OC(CH_2)_7CO-$	ethyl	CH_3CH_2-
azido	N_3-	ethylene	$-CH_2CH_2-$
azino	$=N-N=$	ethylidene	$CH_3CH=$
azo	$-N=N-$	ethylthio	C_2H_5S-
azoxy	$-N(O)=N-$	formamido (formylamino)	$HCONH-$
benzal* (benzylidene)	$C_6H_5CH=$	formyl	$HCO-$
benzamido (benzoylamino)	C_6H_5CONH-	furmaroyl (from fumaric acid)	$-OCCH=CHCO-$
benzhydryl (diphenylmethyl)	$(C_6H_5)_2CH-$	furfuryl (2-furanylmethyl)	$OC_4H_3CH_2-$
benzoxy* (benzoyloxy)	C_6H_5COO-	furfurylidene (2-furanylmethylene)	$OC_4H_3CH=$
benzoyl	C_6H_5CO-	glutamoyl (from glutamic acid)	$-OC(CH_2)_2CH(NH_2)CO-$
benzyl	$C_6H_5CH_2-$	glutaryl (from glutaric acid)	$-OC(CH_2)_3CO-$
benzylidene	$C_6H_5CH=$	glycylamino	H_2NCH_2CONH-
benzylidyne	$C_6H_5C=$	glycoloyl; glycolyl (hydroxyacetyl)	$HOCH_2CO-$
biphenylyl	$C_6H_5C_6H_5-$	glycyl (aminoacetyl)	H_2NCH_2CO-
biphenylene	$-C_6H_4-C_6H_4-$	glyoxyloyl; glyoxylyl (oxoacetyl)	$HCOCO-$
butoxy	C_4H_9O-	guanidino	$H_2NC(=NH)NH-$
sec-butoxy (1-methylpropoxy)	$C_2H_5CH(CH_3)O-$	guanyl (aminoiminomethyl)	$H_2NC(=NH)-$
tert-butoxy (1,1-dimethylethoxy)	$(CH_3)_3CO-$	heptadecanoyl	$CH_3(CH_2)_{15}CO-$
butyl	$CH_3(CH_2)_3-$	heptanamido	$CH_3(CH_2)_5CONH-$
sec-butyl (1-methylpropyl)	$CH_3CH_2CH(CH_3)-$	heptanedioyl	$-OC(CH_2)_5CO-$
tert-butyl (1,1-dimethylethyl)	$(CH_3)_3C-$	heptanoyl	$CH_3(CH_2)_5CO-$
butyryl (1-oxobutyl)	$CH_3(CH_2)_2CO-$	hexadecanoyl	$CH_3(CH_2)_{14}CO-$
caproyl* (hexanoyl)	$CH_3(CH_2)_4CO-$	hexamethylene (1,6-hexanediyl)	$-(CH_2)_6-$
capryl* (decanoyl)	$CH_3(CH_2)_8CO-$	hexanedioyl	$-OC(CH_2)_4CO-$
capryloyl* (octanoyl)	$CH_3(CH_2)_6CO-$	hippuryl (N-benzoylglycyl)	$C_6H_5CONHCH_2CO-$
carbamido (carbamoylamino)	$H_2NCONH-$	hydrazino	H_2NNH-
carbamoyl (aminocarbonyl)	H_2NCO-	hydrazo	$-HNNH-$
carbamyl (aminocarbonyl)	H_2NCO-	hydrocinnamoyl	$C_6H_5(CH_2)_2CO-$
carbazoyl (hydrazinocarbonyl)	$H_2NNHCO-$	hydroperoxy	$HOO-$
carbethoxy (ethoxycarbonyl)	C_2H_5OCO-	hydroxyamino	$HONH-$
carbonyl	$=C=O$	hydroxy	$HO-$
carboxy	$HOOC-$	imino	$HN=$
cetyl* (hexadecyl)	$CH_3(CH_2)_{15}-$	iodoso* (iodosyl)	$OI-$
chloroformyl (chlorcarbonyl)	$ClCO-$	iodyl	O_2I-
cinnamoyl	$C_6H_5CH=CHCO-$	isoamyl* (isopentyl; 3-methylbutyl)	$(CH_3)_2CH(CH_2)_2-$
cinnamyl (3-phenyl-2-propenyl)	$C_6H_5CH=CHCH_2-$	isobutenyl (2-methyl-1-propenyl)	$(CH_3)_2C=CH-$
cinnamylidene	$C_6H_5CH=CHCH=$	isobutoxy (2-methylpropoxy)	$(CH_3)_2CHCH_2O-$
cresyl* (hydroxymethylphenyl)	$HO(CH_3)C_6H_4-$	isobutyl (2-methylpropyl)	$(CH_3)_2CHCH_2-$
crotonoyl	$CH_3CH=CHCO-$	isobutylidene (3-methylpropylidene)	$(CH_3)_2CHCH=$
crotyl (2-butenyl)	$CH_3CH=CHCH_2-$	isobutyryl (2-methyl-1-oxopropyl)	$(CH_3)_2CHCO-$

isocyanato	OCN-
isocyano	CN-
isohexyl (4-methylpentyl)	$(CH_3)_2CH(CH_2)_3$-
isoleucyl (from isoleucine)	$C_2H_5CH(CH_3)CH(NH_2)CO$-
isonitroso* (hydroxyamino)	HON=
isopentyl (3-methylbutyl)	$(CH_3)_2CH(CH_2)_2$-
isopentylidene (3-methylbutylidene)	$(CH_3)_2CHCH_2CH$=
isopropenyl (1-methylethenyl)	$CH_2=C(CH_3)$-
isopropoxy (1-methylethoxy)	$(CH_3)_2CHO$-
isopropyl (1-methylethyl)	$(CH_3)_2CH$-
isopropylidene (1-methylethylidene)	$(CH_3)_2C$=
isothiocyanato (isothiocyano)	SCN-
isovaleryl* (3-methyl-1-oxobutyl)	$(CH_3)_2CHCH_2CO$-
lactoyl (from lactic acid)	$CH_3CH(OH)CO$-
lauroyl (from lauric acid)	$CH_3(CH_2)_{10}CO$-
lauryl (dodecyl)	$CH_3(CH_2)_{11}$-
leucyl (from leucine)	$(CH_3)_2CHCH_2CH(NH_2)CO$-
levulinoyl (from levulinic acid)	$CH_3CO(CH_2)_2CO$-
malonyl (from malonic acid)	$-OCCH_2CO$-
mandeloyl (from mandelic acid)	$C_6H_5CH(OH)CO$-
mercapto	HS-
mesityl	$2,4,6-(CH_3)_3C_6H_2$-
methacryloyl (from methacrylic acid)	$CH_2=C(CH_3)CO$-
methallyl (2-methyl-2-propenyl)	$CH_2=C(CH_3)CH_2$-
methionyl (from methionine)	$CH_3SCH_2CH_2CH(NH_2)CO$-
methoxy	CH_3O-
methyl	H_3C-
methylene	H_2C=
methylthio	CH_3S-
myristoyl (from myristic acid)	$CH_3(CH_2)_{12}CO$-
myristyl (tetradecyl)	$CH_3(CH_2)_{13}$-
naphthyl	$(C_{10}H_7)$-
naphthylene	$-(C_{10}H_6)$-
neopentyl (2,2-dimethylpropyl)	$(CH_3)_3CCH_2$-
nitramino (nitroamino)	O_2NNH-
nitro	O_2N-
nitrosamino (nitrosoamino)	ONNH-
nitrosimino (nitrosoimino)	ONN=
nitroso	ON-
nonanoyl (from nonanoic acid)	$CH_3(CH_2)_7CO$-
oleoyl (from oleic acid)	$CH_3(CH_2)_7CH=CH(CH_2)_7CO$-
oxalyl (from oxalic acid)	-OCCO-
oxo	O=
palmitoyl (from palmitic acid)	$CH_3(CH_2)_{14}CO$-
pentamethylene (1,5-pentanediyl)	$-(CH_2)_5$-
pentyl	$CH_3(CH_2)_4$-
tert-pentyl	$CH_3CH_2C(CH_3)_2$-
phenacyl	$C_6H_5COCH_2$-
phenacylidene	C_6H_5COCH=
phenethyl (2-phenylethyl)	$C_6H_5CH_2CH_2$-
phenoxy	C_6H_5O-
phenyl	C_6H_5-
phenylene (benzenediyl)	$-C_6H_4$-
phosphino* (phosphanyl)	H_2P-
phosphinyl* (phosphinoyl)	$H_2P(O)$-
phospho	O_2P-
phosphono	$(HO)_2P(O)$-
phthaloyl (from phthalic acid)	$1,2-C_6H_4(CO-)_2$
picryl (2,4,6-trinitrophenyl)	$2,4,6-(NO_2)_3C_6H_2$-
pimeloyl (from pimelic acid)	$-OC(CH_2)_5CO$-
piperidino (1-piperidinyl)	$C_5H_{10}N$-
pivaloyl (from pivalic acid)	$(CH_3)_3CCO$-
prenyl (3-methyl-2-butenyl)	$(CH_3)_2C=CHCH_2$-
propargyl (2-propynyl)	$HC≡CCH_2$-
1-propenyl	$-CH=CHCH_2$
2-propenyl (allyl)	$CH_2=CHCH_2$-
propionyl* (propanyl)	CH_3CH_2CO-
propoxy	$CH_3CH_2CH_2O$-
propyl	$CH_3CH_2CH_2$-
propylidene	CH_3CH_2CH=
pyrryl (pyrrolyl)	C_3H_4N-
salicyloyl (2-hydroxybenzoyl)	$2-HOC_6H_4CO$-
selenyl* (selanyl; hydroseleno)	HSe-
seryl (from serine)	$HOCH_2CH(NH_2)CO$-
siloxy	H_3SiO-
silyl	H_3Si-
silylene	H_2Si=
sorboyl (from sorbic acid)	$CH_3CH=CHCH=CHCO$-
stearoyl (from stearic acid)	$CH_3(CH_2)_{14}CO$-
stearyl (octadecyl)	$CH_3(CH_2)_{17}$-
styryl (2-phenylethenyl)	$C_6H_5CH=CH$-
suberoyl (from suberic acid)	$-OC(CH_2)_6CO$-
succinyl (from succinic acid)	$-OCCH_2CH_2CO$-
sulfamino (sulfoamino)	$HOSO_2NH$-
sulfamoyl (sulfamyl)	H_2NSO_2-
sulfanilyl [(4-aminophenyl)sulfonyl]	$4-H_2NC_6H_4SO_2$-
sulfeno	HOS-
sulfhydryl (mercapto)	HS-
sulfinyl	OS=
sulfo	HO_3S-
sulfonyl (sulfuryl)	$-SO_2$-
terephthaloyl	$1,4-C_6H_4(CO-)_2$
tetramethylene	$-(CH_2)_4$-
thienyl (from thiophene)	(C_4H_3S)-
thiocarbonyl (carbothionyl)	=CS
thiocarboxy	HOSC-
thiocyanato (thiocyano)	NCS-
thionyl* (sulfinyl)	-SO-
threonyl (from threonine)	$CH_3CH(OH)CH(NH_2)CO$-
toluidino [(methylphenyl)amino]	$CH_3C_6H_4NH$-
toluoyl (methylbenzoyl)	$CH_3C_6H_4CO$-
tolyl (methylphenyl)	$CH_3C_6H_4$-
α-tolyl (benzyl)	$C_6H_5CH_2$-
tolylene (methylphenylene)	$-(CH_3C_6H_3)$-
tosyl [(4-methylphenyl) sulfonyl)]	$4-CH_3C_6H_4SO_2$-
triazano	H_2NNHNH-
trimethylene (1,3-propanediyl)	$-(CH_2)_3$-
trityl (triphenylmethyl)	$(C_6H_5)_3C$-
valeryl* (pentanoyl)	$CH_3(CH_2)_3CO$-
valyl (from valine)	$(CH_3)_2CHCH(NH_2)CO$-
vinyl (ethenyl)	$CH_2=CH$-
vinylidene (ethenylidene)	$CH_2=C$=
xylidino [(dimethylphenyl)amino]	$(CH_3)_2C_6H_3NH$-
xylyl (dimethylphenyl)	$(CH_3)_2C_6H_3$-
xylylene [phenelenebis(methylene)]	$-CH_2C_6H_4CH_2$-

ORGANIC RING COMPOUNDS

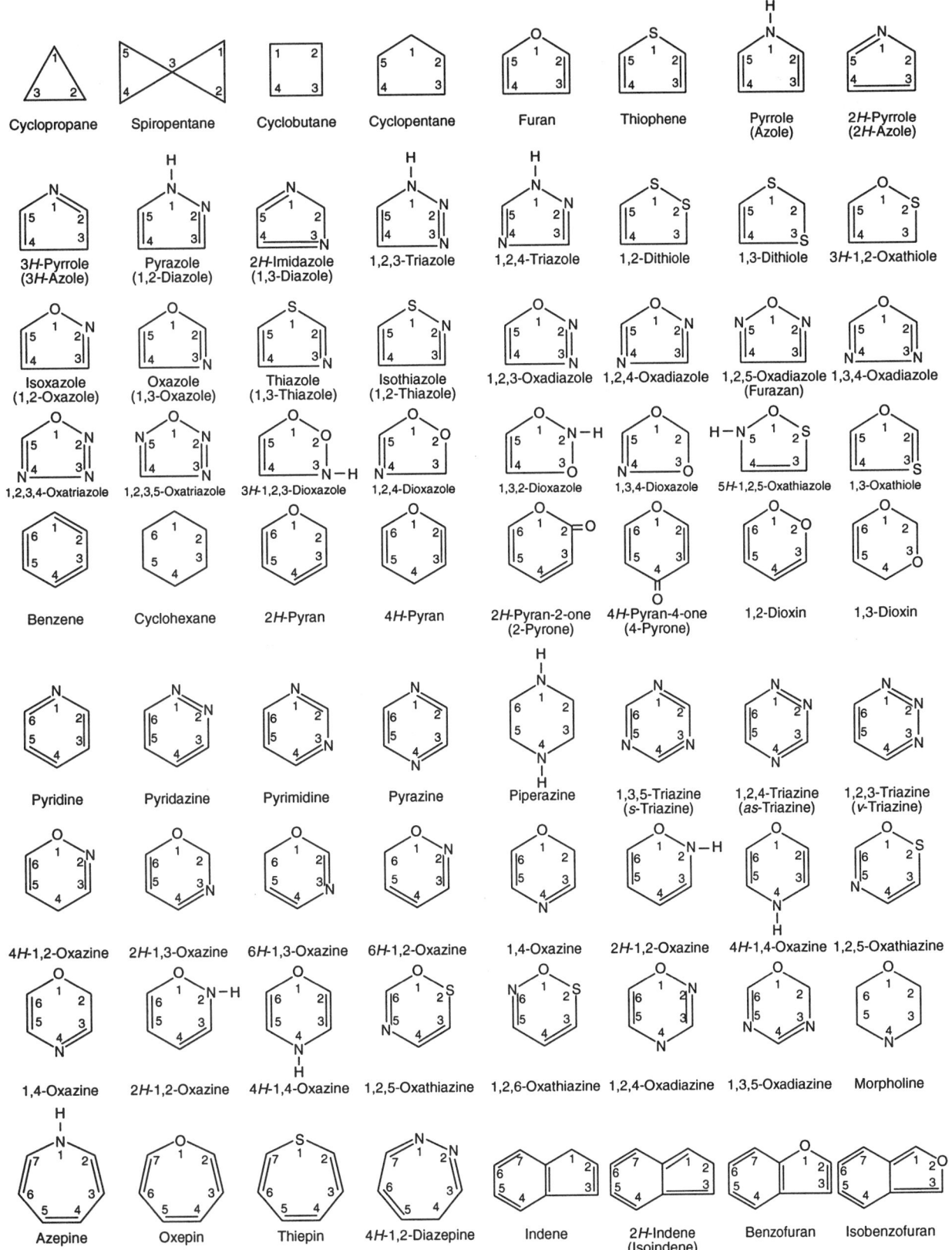

Cyclopropane   Spiropentane   Cyclobutane   Cyclopentane   Furan   Thiophene   Pyrrole (Azole)   2H-Pyrrole (2H-Azole)

3H-Pyrrole (3H-Azole)   Pyrazole (1,2-Diazole)   2H-Imidazole (1,3-Diazole)   1,2,3-Triazole   1,2,4-Triazole   1,2-Dithiole   1,3-Dithiole   3H-1,2-Oxathiole

Isoxazole (1,2-Oxazole)   Oxazole (1,3-Oxazole)   Thiazole (1,3-Thiazole)   Isothiazole (1,2-Thiazole)   1,2,3-Oxadiazole   1,2,4-Oxadiazole   1,2,5-Oxadiazole (Furazan)   1,3,4-Oxadiazole

1,2,3,4-Oxatriazole   1,2,3,5-Oxatriazole   3H-1,2,3-Dioxazole   1,2,4-Dioxazole   1,3,2-Dioxazole   1,3,4-Dioxazole   5H-1,2,5-Oxathiazole   1,3-Oxathiole

Benzene   Cyclohexane   2H-Pyran   4H-Pyran   2H-Pyran-2-one (2-Pyrone)   4H-Pyran-4-one (4-Pyrone)   1,2-Dioxin   1,3-Dioxin

Pyridine   Pyridazine   Pyrimidine   Pyrazine   Piperazine   1,3,5-Triazine (s-Triazine)   1,2,4-Triazine (as-Triazine)   1,2,3-Triazine (v-Triazine)

4H-1,2-Oxazine   2H-1,3-Oxazine   6H-1,3-Oxazine   6H-1,2-Oxazine   1,4-Oxazine   2H-1,2-Oxazine   4H-1,4-Oxazine   1,2,5-Oxathiazine

1,4-Oxazine   2H-1,2-Oxazine   4H-1,4-Oxazine   1,2,5-Oxathiazine   1,2,6-Oxathiazine   1,2,4-Oxadiazine   1,3,5-Oxadiazine   Morpholine

Azepine   Oxepin   Thiepin   4H-1,2-Diazepine   Indene   2H-Indene (Isoindene)   Benzofuran   Isobenzofuran

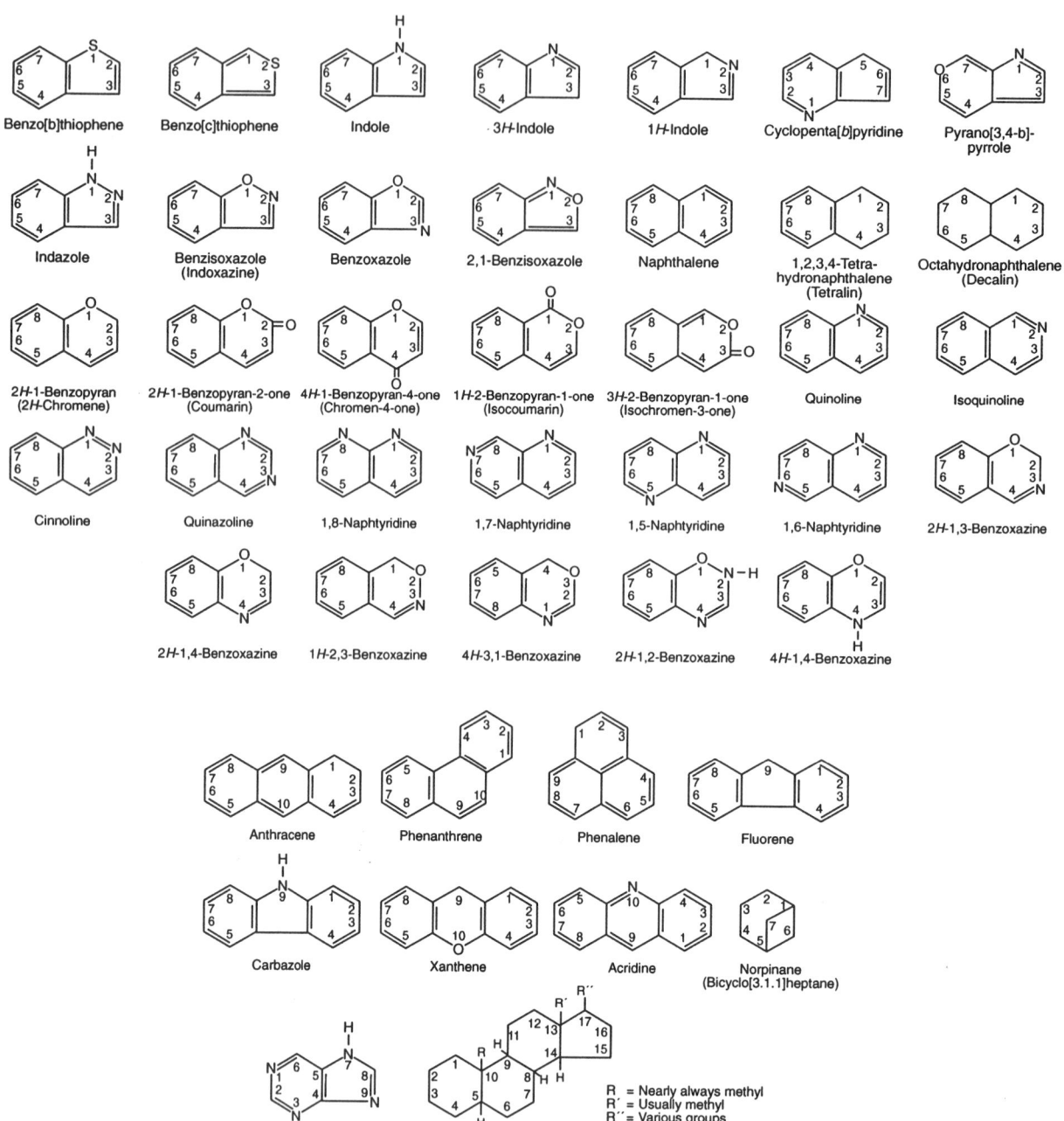

Benzo[b]thiophene

Benzo[c]thiophene

Indole

3H-Indole

1H-Indole

Cyclopenta[b]pyridine

Pyrano[3,4-b]-pyrrole

Indazole

Benzisoxazole (Indoxazine)

Benzoxazole

2,1-Benzisoxazole

Naphthalene

1,2,3,4-Tetra-hydronaphthalene (Tetralin)

Octahydronaphthalene (Decalin)

2H-1-Benzopyran (2H-Chromene)

2H-1-Benzopyran-2-one (Coumarin)

4H-1-Benzopyran-4-one (Chromen-4-one)

1H-2-Benzopyran-1-one (Isocoumarin)

3H-2-Benzopyran-1-one (Isochromen-3-one)

Quinoline

Isoquinoline

Cinnoline

Quinazoline

1,8-Naphtyridine

1,7-Naphtyridine

1,5-Naphtyridine

1,6-Naphtyridine

2H-1,3-Benzoxazine

2H-1,4-Benzoxazine

1H-2,3-Benzoxazine

4H-3,1-Benzoxazine

2H-1,2-Benzoxazine

4H-1,4-Benzoxazine

Anthracene

Phenanthrene

Phenalene

Fluorene

Carbazole

Xanthene

Acridine

Norpinane (Bicyclo[3.1.1]heptane)

7H-Purine

Steroid ring system

R = Nearly always methyl
R′ = Usually methyl
R″ = Various groups

SCIENTIFIC ABBREVIATIONS AND SYMBOLS

This table lists some symbols, abbreviations, and acronyms encountered in the physical sciences. Most entries in italic type are symbols for physical quantities; for more details on these, see the table "Symbols and Terminology for Physical and Chemical Quantities" in this section. Additional information on units may be found in the table "International System of Units" in Section 1. Many of the terms to which these abbreviations refer are included in the tables "Definitions of Scientific Terms" in Section 2 and "Techniques for Materials Characterization" in Section 12.

Publication practices vary with regard to the use of capital or lower case letters for many abbreviations. An effort has been made to follow the most common practices in this table, but much variation is found in the literature. Likewise, policies on the use of periods in an abbreviation vary considerably. Periods are generally omitted in this table unless they are necessary for clarity. Periods should never appear in SI units. The SI prefixes (m, k ,M, etc.) are not listed here, since they should never be used alone, but selected combinations with SI units (e.g., mg, kV, MW) are included.

Abbreviations are listed in alphabetical order without regard to case. Entries beginning with Greek letters fall at the end of the table.

a	absorption coefficient, acceleration, activity		$as, asym$	asymmetrical (as chemical descriptor)
a_0	Bohr radius		ASCII	American National Standard Code for Information Interchange
A	ampere, adenine (in genetic code)		ASE	aromatic stabilization energy
Å	ångstrom		Asn	asparagine
A	absorbance, area, Helmholtz energy, mass number		Asp	aspartic acid
A_H	Hall coefficient		at	atomization
A_r	atomic weight (relative atomic mass)		atm	standard atmosphere
AAS	atomic absorption spectroscopy		ATP	adenosine 5'-triphosphate
Abe	abequose		ATR	attenuated total internal reflection
abs	absolute		at.wt.	atomic weight
ac	alternating current		AU	astronomical unit
Ac	acetyl		av	average
AcOH	acetic acid		avdp	avoirdupois
ACT	activated complex theory		b	barn
ACTH	adrenocorticotropic hormone		B	magnetic flux density, second virial coefficient, susceptance
Ade	adenine		bar	bar (pressure unit)
ADP	adenosine diphosphate		bbl	barrel
ads	adsorption		bcc	body centered cubic
ae	eon (10^9 years)		BCS	Bardeen-Cooper-Schrieffer (theory)
AES	atomic emission spectroscopy, Auger electron spectroscopy		BDE	bond dissociation energy
AF	audio frequency		Bé	Baumé
AFM	atomic force microscopy		BET	Brunauer-Emmett-Teller (method)
AI	artificial intelligence		BeV	billion electronvolt
AIM	atoms in molecules		Bhn	Brinell hardness number
Al	Alfen number		Bi	biot
Ala	alanine		BN	bond number
alc	alcohol		BNS	nuclear backscattering spectroscopy
aliph.	aliphatic		BO	bond order, Born-Oppenheimer (approximation)
alk.	alkaline		BOD	biochemical oxygen demand
All	allose		bp	boiling point
Alt	altrose		bpy	2,2'-bipyridine
am	amorphous solid		Bq	becquerel
Am	amyl		BRE	bond resonance energy
AM	amplitude modulation		BSSE	basis set superposition error
AMP	adenosine 5'-monophosphate		Btu	British thermal unit
amu	atomic mass unit (recommended symbol is u)		bu	bushel
anh, anhyd	anhydrous		Bu	butyl
antilog	antilogarithm		Bz	benzoyl
AO	atomic orbital		Bzl	benzyl
AOM	angular overlap model		c	combustion reaction
Api	apiose		c	amount concentration, specific heat, velocity
APS	appearance potential spectroscopy		c_0	speed of light in vacuum
APW	augmented plane wave		C	coulomb, cytosine (in genetic code)
aq	aqueous		°C	degree Celsius
Ar	aryl		C	capacitance, heat capacity, number concentration
Ara	arabinose		ca.	approximately
Ara-ol	arabinitol		cal	calorie
Arg	arginine			

calc	calculated
CARS	coherent anti-Stokes Raman spectroscopy
CAS RN	Chemical Abstracts Service Registry Number
CAT	clear-air turbulence, computerized axial tomography
CBS	complete basis set
cc	cubic centimeter
CCD	charge-coupled device
cd	candela, condensed phase
c.d.	current density
CD	circular dichroism
CDP	cytidine 5'-diphosphate
CEPA	couplet electron pair approximation
cf.	compare
cfm	cubic feet per minute
cgs	centimeter-gram-second system
CHF	coupled Hartree-Fock (method)
Ci	curie
CI	configuration interaction, chemical ionization
CIDEP	chemically induced dynamic electron polarization
CIDNP	chemically induced dynamic nuclear polarization
cir	circular
CKFF	Cotton-Kraihanzel force field
CL	cathode luminescence
cm	centimeter
c.m.	center of mass
c.m.c.	critical micelle concentration
CMO	canonical molecular orbital
CMP	cytidine 5'-monophosphate
CN	coordination number
CNDO	complete neglect of differential overlap
Co	Cowling number
COD	chemical oxygen demand
conc	concentrated, concentration
const	constant
cos	cosine
cosh	hyperbolic cosine
COSY	correlation spectroscopy (in NMR)
cot	cotangent
coth	hyperbolic cotangent
cp	candle power
cP	centipoise
Cp	cyclopentadienyl
CP	chemically pure
CPA	coherent potential approximation
cpd	contact potential difference
cps	cycles per second
CPT	charge conjugation-space inversion-time reversal (theorem)
CPU	central processing unit
cr, cryst	crystalline (phase)
CRU	constitutional repeating unit
csc	cosecant
ct	carat
CT	charge transfer
CTEM	conventional transmission electron microscopy
CTP	cytidine 5'-triphosphate
CTR	controlled thermonuclear reaction
cu	cubic
CV	cyclic voltammetry
CVD	chemical vapor deposition
cw	continuous wave
cwt	hundredweight (112 pounds)
Cy	cyclohexyl
cyl	cylinder
Cys	cysteine
d	day, deuteron
d	distance, density, dextrorotatory
D	debye unit
D	diffusion coefficient, dissociation energy, electric displacement
Da	dalton
DA	donor-acceptor (complex)
dB	decibel
dc	direct current
DE	delocalization energy
dec	decomposes
deg	degree
den	density
det	determinant
dev	deviation
DFT	density functional theory
diam	diameter
dil	dilute, dilution
DIM	diatomics in molecules
dm	decimeter
dmf, DMF	*N,N*-dimethylformamide
dmso, DMSO	dimethylsulfoxide
DNA	deoxyribonucleic acid
DNase	deoxyribonuclease
DNMR	dynamic nuclear magnetic resonance
DOS	density of states
doz	dozen
d.p.	degree of polymerization
dpl	displacement
dpm	disintegrations per minute
dps	disintegrations per second
dr	dram
dRib	2-deoxyribose
DRIFT	diffuse reflectance infrared Fourier transform
DRS	diffuse reflectance spectroscopy
DSC	differential scanning calorimetry
DTA	differential thermal analysis
dyn	dyne
e	electron, base of natural logarithms
e	elementary charge, linear strain
E	electric field strength, electromotive force, energy, modulus of elasticity, entgegen (*trans* configuration)
E_h	Hartree energy
EA	electron affinity
EAN	effective atomic number
ECP	effective core potential
ECR	electron cyclotron resonance
ED	electron diffraction, effective dose
EDS	energy dispersive X-ray spectroscopy
EDTA	ethylenediaminetetraacetic acid
EELS	electron energy loss spectroscopy
EFFF	energy factored force field
EHMO	extended Hückel molecular orbital
EHT	extended Hückel theory
emf	electromotive force
emu	electromagnetic unit system
en	ethylenediamine

ENDOR	electron-nuclear double resonance		G	gauss, guanine (in genetic code)
EOS	equation of state		G	electrical conductance, Gibbs energy, gravitational constant, sheer modulus
EPMA	electron probe microanalysis		gal	gallon
EPR	electron paramagnetic (spin) resonance		Gal	gal, galileo, galactose
eq, eqn	equation		GalN	galactosamine
eqQ	quadrupole coupling constant		GC	gas chromatography
erf	error function		GC-MS	gas chromatography-mass spectrometry
erg	erg		GDMS	glow discharge mass spectroscopy
ESCA	electron spectroscopy for chemical analysis		gem	geminal (on the same carbon atom)
e.s.d.	estimated standard deviation		GeV	gigaelectronvolt
ESD	electron stimulated desorption		GIAO	gauge invariant atomic orbital
ESR	electron spin resonance		gl	glacial
est	estimate, estimated		GLC	gas-liquid chromatography
esu	electrostatic unit system		Glc	glucose
Et	ethyl		GlcN	glucosamine
ET	electron transfer, ephemeris time		Glc-ol	glucitol
Et_2O	diethyl ether		Gln	glutamine
e.u.	entropy unit		Glu	glutamic acid
Eu	Euler number		Gly	glycine
eV	electronvolt		GMP	guanosine 5'-triphosphate
EWG	electron withdrawing group		GMT	Greenwich mean time
EXAFS	extended x-ray absorption fine structure		gpm	gallons per minute
EXELFS	extended energy loss fine structure		gps	gallons per second
exp	exponential function		gr	grain
expt	experimental		Gr	Grashof number
ext	external		GTO	gaussian type atomic orbital
f	formation reaction		Gua	guanine
f	activity coefficient, aperture ratio, focal length, force constant, frequency, fugacity		Gul	gulose
			GUT	grand unified theory
F	farad		GVB	generalized valence bond
°F	degree Fahrenheit		GWS	Glashow-Weinberg-Salam (theory)
F	Faraday constant, force, angular momentum		Gy	gray, gigayear
FAD	flavin adenine dinucleotide		h	helion, hour
fcc	face centered cubic		h	Planck constant
FEL	free electron laser		H	henry
FEM	field emission microscopy		H	enthalpy, Hamiltonian function, magnetic field
FEMO	free electron molecular orbital		H_0	Hubble constant
FET	field effect transistor		ha	hectare
fid	free induction decay		Ha	Hartmann number
FIM	field ion microscopy		Hacac	acetylacetone
FIR	far infrared		HAM	hydrogenic atoms in molecules
fl	fluid (phase)		hav	haversine
FM	frequency modulation		Hb	hemoglobin
Fo	Fourier number		hcp	hexagonal closed packed
fp	freezing point		Hea	ethanolamine
fpm	feet per minute		HEIS	high energy ion scattering
fps	feet per second, foot-pound-second system		HEP	high energy physics
Fr	franklin		HF	high frequency
Fr	Froude number		hfs	hyperfine structure
Fru	fructose		Him	imidazole
FSGO	floating spherical Gaussian orbital		His	histidine
ft	foot		HMO	Hückel molecular orbital
ft-lb	foot pound		HOMO	highest occupied molecular orbital
FT	Fourier transform		hp	horsepower
FTIR	Fourier transform infrared spectroscopy		HPLC	high-performance liquid chromatography
Fuc	fucose		Hpz	pyrazole
Fuc-ol	fucitol		hr	hour
fus	fusion (melting)		HREELS	high resolution electron energy loss spectroscopy
g	gram, gas		HREM	high resolution electron microscopy
g	acceleration due to gravity, degeneracy, statistical weight, Landé g-factor		HSAB	hard-soft acid-base (theory)

HSE	homodesmotic stabilization energy
Hz	hertz
i	square root of minus one
I	electric current, ionic strength, moment of inertia, nuclear spin angular momentum, radiant intensity
IAT	international atomic time
i-Bu	isobutyl
IC	integrated circuit
ICP	inductively coupled plasma
ICR	ion cyclotron resonance
id	ideal (solution)
ID	inside diameter
Ido	idose
IDP	inosine 5'-diphosphate
IE	ionization energy
i.e.p.	isoelectric point
IEPA	independent electron pair approximation
IF	intermediate frequency
IGLO	individual gauge for localized orbitals
Ile	isoleucine
Im	imaginary part
imm	immersion
IMPATT	impact ionization avalanche transit time
in.	inch
INDO	intermediate neglect of differential overlap
INS	inelastic neutron scattering, ion neutralization spectroscopy
int	internal
I/O	input/output
IP	ionization potential
IPN	interpenetrating polymer network
i-Pr	isopropyl
IPR	isotopic perturbation of resonance
IPTS	International Practical Temperature Scale
IR	infrared
IRAS	reflection-absorption infrared spectroscopy
IRC	intrinsic reaction coordinate
isc	intersystem crossing
ISE	isodesmic stabilization energy
ISS	ion scattering spectroscopy
ITP	inosine 5'-triphosphate
ITS	International Temperature Scale (1990)
IU	international unit
j	angular momentum, electric current density
J	joule
J	angular momentum, electric current density, flux, Massieu function
k	absorption index, Boltzmann constant, rate constant, thermal conductivity, wave vector
K	kelvin
K	absorption coefficient, bulk modulus, equilibrium constant, kinetic energy
kb	kilobar, kilobase (DNA or RNA)
kcal	kilocalorie
KE	kinetic energy
keV	kiloelectronvolt
kg	kilogram
kgf	kilogram force
kJ	kilojoule
km	kilometer
Kn	Knudsen number

kPa	kilopascal
kt	karat
kV	kilovolt
kva	kilovolt ampere
kW	kilowatt
kwh	kilowatt hour
l	liquid, liter
l	angular momentum, length, levorotatory
L	liter, lambert
L	Avogadro constant, inductance, Lagrange function
lat.	latitude
lb	pound
lbf	pound force
lc	liquid crystal
LC	liquid chromatography
LCAO	linear combination of atomic orbitals
LD	lethal dose
Le	Lewis function
LED	light emitting diode
LEED	low-energy electron diffraction
LEIS	low energy ion scattering
Leu	leucine
LFER	linear free energy relationship
lim	limit
LIMS	laser ionization mass spectroscopy, laboratory information management system
liq	liquid
lm	lumen
ln	logarithm (natural)
LNDO	local neglect of differential overlap
log	logarithm (common)
long.	longitude
LST	local sidereal time
LT	local time
LTE	local thermodynamic equilibrium
LUMO	lowest unoccupied molecular orbital
lut	lutidine
lx	lux
ly	langley
l.y.	light year
Lys	lysine
Lyx	lyxose
m	meter, molal (as in 0.1 m solution), metastable (isotope)
m	magnetic dipole moment, mass, molality, angular momentum component, *meta* (as chemical descriptor)
M	molar (as in 0.1 M solution), metal (in chemical formulas)
M	magnetization, molar mass, mutual inductance, torque, angular momentum component
M_r	molecular weight (relative molar mass)
Ma	Mach number
Man	mannose
MASNMR	magic angle spinning nuclear magnetic resonance
max	maximum
MBE	molecular beam epitaxy
MBPT	many body perturbation theory
MC	Monte Carlo (method)
MCD	magnetic circular dichroism
MCPF	modified couple pair functional
MCSCF	multi-configurational self-consistent field

MD	molecular dynamics		NEXAFS	near-edge x-ray absorption fine structure
Me	methyl		ng	nanogram
MEP	molecular electrostatic potential		NIR	near infrared
MERP	minimum energy reaction path		nm	nanometer
Mes	mesityl		NMR	nuclear magnetic resonance
MESFET	metal-semiconductor field-effect transistor		NNDO	neglect of nonbonded differential overlap
Met	methionine		NO	natural orbital
meV	millielectronvolt		NOE	nuclear Overhauser effect
MeV	megaelectronvolt		NPA	natural population analysis
MF	molecular formula		NQR	nuclear quadrupole resonance
mg	milligram		NRA	prompt nuclear reaction analysis
MHD	magnetohydrodynamics		ns	nanosecond
mi	mile		NTP	normal temperature and pressure
MIM	molecules-in-molecules		Nu	nucleophile
min	minimum, minute		Nu	Nusselt number
MINDO	modified intermediate neglect of differential overlap		o	$ortho$ (as chemical descriptor)
MIR	mid-infrared		obs, obsd	observed
misc	miscible		OD	optical density, outside diameter
MKS	meter-kilogram-second system		Oe	oersted
MKSA	meter-kilogram-second-ampere system		ORD	optical rotatory dispersion
mL, ml	milliliter		oz	ounce
mm	millimeter		p	proton
MM	molecular mechanics		p	dielectric polarization, electric dipole moment, momentum, pressure, $para$ (as chemical descriptor)
mmf	magnetomotive force			
mmHg	millimeter of mercury		P	poise
MO	molecular orbital		P	power, pressure, probability, sound energy flux
mol	mole		Pa	pascal
mol.wt.	molecular weight		PA	proton affinity
mon	monomeric form		PAS	photoacoustic spectroscopy
MOS	metal-oxide semiconductor		pc	parsec
MOSFET	metal-oxide semiconductor field-effect transistor		PCR	polymerase chain reaction
mp	melting point		PD	potential difference
MPa	megapascal		pdl	poundal
MPA	Mulliken population analysis		pe	probable error
Mpc	megaparsec		Pe	Péclet number
MRI	magnetic resonance imaging		PES	photoelectron spectroscopy
mRNA	messenger RNA		PET	positron emission tomography
ms	millisecond		peth	petroleum ether
MS	mass spectroscopy		pf	power factor
MSL	mean sea level		pg	picogram
Mur	muramic acid		pH	negative log of hydrogen ion concentration
mV	millivolt		Ph	phenyl
mW	milliwatt		Phe	phenylalanine
MW	megawatt, microwave, molecular weight		pI	isoelectric point
Mx	maxwell		pip	piperidine
n	neutron		pK	negative log of ionization constant
n	amount of substance, number density, principal quantum number, refractive index, normal (in chemical formulas)		pm	picometer
			PMO	perturbational molecular orbital
			PNDO	partial neglect of differential overlap
N	newton		PNRA	prompt nuclear reaction analysis
N	angular momentum, neutron number		pol	polymeric form
N_A	Avogadro constant		ppb	parts per billion
N_E	density of states		ppm	parts per million
NAA	neutron activation analysis		PPP	Pariser-Parr-Pople (method)
NAD	nicotinamide adenine dinucleotide		ppt	parts per thousand, precipitate
NADH	reduced NAD		Pr	propyl
NADP	nicotinamide adenine dinucleotide phosphate		Pr	Prandtl number
NAO	natural atomic orbital		PRDDO	partial retention of diatomic differential overlap
NBO	natural bond order		Pro	proline
nbp	normal boiling point		ps	picosecond
Neu	neuraminic acid		PS	photoelectron spectroscopy

PSD	photon stimulated desorption
Psi	psicose
psi	pounds per square inch
psia	pounds per square inch absolute
psig	pounds per square inch gage
pt	pint
PVT	pressure-volume-temperature
py	pyridine
q	electric field gradient, flow rate, heat, wave vector (phonons)
Q	electric charge, heat, partition function, quadrupole moment, radiant energy, vibrational normal coordinate
QCD	quantum chromodynamics
QED	quantum electrodynamics
Q.E.D.	quod erat demonstrandum (which was to be proved)
QSAR	quantitative structure-activity relationship
QSO	quasi-stellar object (quasar)
qt	quart
quad	quadrillion Btu (= 1.05510^{18} J)
Qui	quinovose
q.v.	quod vide (which you should see)
r	reaction
r	position vector, radius
R	roentgen, alkyl radical (in chemical formulas)
°R	degree Rankine
R	electrical resistance, gas constant, molar refraction, Rydberg constant
RA	right ascension
rad	radian
RAIRS	reflection-absorption infrared spectroscopy
RAM	random access memory
RBS	Rutherford backscattering spectroscopy
RE	resonance energy
Re	real part
RED	radial electron distribution
REM	reflection electron microscopy
rem	roentgen equivalent man
RF	radiofrequency
Rha	rhamnose
RHEED	reflection high-energy electron diffraction
RHF	restricted Hartree-Fock (theory)
RIA	radioimmunoassay
Rib	ribose
Ribulo	ribulose
rms	root mean square
RNA	ribonucleic acid
RNase	ribonuclease
rRNA	ribosomal RNA
ROHF	restricted open shell Hartree-Fock
ROM	read only memory
RPA	random phase approximation
rpm	revolutions per minute
rps	revolutions per second
RRK	Rice-Ramsperger-Kassel (theory)
RRKM	Rice-Ramsperger-Kassel-Marcus (theory)
RRS	resonance Raman spectroscopy
RS	Raman spectroscopy
Ry	rydberg
s	second, solid
s	path length, solubility, spin angular momentum, symmetry number, symmetrical (as stereochemical descriptor)

S	siemens
S	area, entropy, probability current density, Poynting vector, symmetry coordinate, spin angular momentum
SALC	symmetry adapted linear combinations
SALI	surface analysis by laser ionization
SAM	scanning Auger microscopy
SANS	small angle neutron scattering
Sar	sarcosine
sat, satd	saturated
SAXS	small angle x-ray scattering
s-Bu	sec-butyl
Sc	Schmidt number
SCE	saturated calomel electrode
SCF	self-consistent field
SCR	silicon-controlled rectifier
sd	standard deviation
sec	secant, second
sec	secondary (in chemical name)
SEELFS	surface sensitive energy loss fine structure
SEM	scanning electron microscope
sepn	separation
Ser	serine
SERS	surface-enhanced Raman spectroscopy
SET	single electron transfer
SEXAFS	surface extended x-ray absorption fine structure
Sh	Sherwood number
SI	International System of Units
SIMS	secondary ion mass spectroscopy
sin	sine
SINDO	symmetrically orthogonalized INDO method
sinh	hyperbolic sine
SIPN	semi-interpenetrating polymer network
SLAM	scanning laser acoustic microscopy
sln	solution
SMO	semiempirical molecular orbital
SMOW	Standard Mean Ocean Water
SNMS	sputtered neutral mass spectroscopy
SNU	solar neutrino unit
SO	spin orbital
sol	soluble, solution
soln	solution
SOMO	singly occupied molecular orbital
Sor	sorbose
sp gr	specific gravity
SPM	scanned probe microscopy
sq	square
sr	steradian
Sr	Strouhal number
SSMS	spark source mass spectroscopy
St	stoke
St	Stanton number
std, stnd	standard (state)
STEM	scanning transmission electron microscope
STM	scanning tunneling microscopy
STO	Slater type orbital
STP	standard temperature and pressure
sub	sublimation, sublimes
Sv	sievert
t	metric tonne, triton
t	Celsius temperature, thickness, time, transport number
T	tesla

T	kinetic energy, period, term value, temperature (thermodynamic), torque, transmittance
Tag	tagatose
Tal	talose
tan	tangent
tanh	hyperbolic tangent
t-Bu	*tert*-butyl
TCA	trichloroacetic acid
TCE	trichloroethylene
tcne	tetracyanoethylene
TCSCF	two configuration self-consistent field
TE	transverse electric
TED	transmission electron diffraction, transferred electron device
TEM	transmission electron microscopy, transverse electromagnetic
temp	temperature
tert	tertiary (in chemical name)
TFD	Thomas-Fermi-Dirac (method)
TGA	thermo-gravimetric analysis
theor	theoretical
thf, THF	tetrahydrofuran
Thr	threonine
Thy	thymine
TL	thermoluminescence
TLC	thin-layer chromatography
TM	transverse magnetic
Tol	tolyl
Torr	torr
tRNA	transfer RNA
Trp	tryptophan
trs	transition
TS	transition state
tsp	teaspoon
Tyr	tyrosine
u	unified atomic mass unit
u	Bloch function, electric mobility, velocity
U	uracil (in genetic code)
U	electric potential difference, internal energy
UDP	uridine 5'-diphosphate
UHF	ultrahigh frequency, unrestricted Hartree-Fock (theory)
UMP	uridine 5'-monophosphate
uns, unsym	unsymmetrical (as chemical descriptor)
UPES	ultraviolet photoelectron spectroscopy
UPS	ultraviolet photoelectron spectroscopy
ur	urea
Ura	uracil
USP	United States Pharmacopeia
UT	universal time
UTP	uridine 5'-triphosphate
UV	ultraviolet
v	reaction rate, specific volume, velocity, vibrational quantum number, vicinal (as chemical descriptor)
V	volt
V	electric potential, potential energy, volume
Val	valine
vap	vaporization
VB	valence band, valence bond
VCD	vibrational circular dichroism
VHF	very high frequency

vic	vicinal (on adjacent carbon atoms)
VIS	visible region of the spectrum
vit	vitreous
VSEPR	valence shell electron pair repulsion
VSLI	very large scale integrated (circuit)
VUV	vacuum ultraviolet
v/v	volume per volume (volume of solute divided by volume of solution, expressed as percent)
w	energy density, mass fraction, velocity, work
W	watt
W	radiant energy, statistical weight, work
WAXS	wide angle x-ray scattering
Wb	weber
We	Weber number
WKB	Wentzel-Kramers-Brillouin (method)
wt	weight
w/v	weight per volume (mass of solute divided by volume of solution, generally expressed as g/100 mL)
w/w	weight per weight (mass of solute divided by mass of solution, expressed as percent)
x	mole fraction
X	X unit, halogen (in chemical formula)
X	reactance
XAFS	x-ray absorption fine structure
XANES	x-ray absorption near-edge structure
XPES	x-ray photoelectron spectroscopy
XPS	x-ray photoelectron spectroscopy
XRD	x-ray diffraction
XRF	x-ray fluorescence
XRS	x-ray spectroscopy
Xyl	xylose
y, yr	year
Y	admittance, Planck function, Young's modulus
yd	yard
z	charge number (of an ion), collision frequency factor
Z	atomic number, compression factor, collision number, impedance, partition function, zusammen (*cis* configuration)
ZDO	zero differential overlap
ZPE, ZPVE	zero point vibrational energy
ZULU	Greenwich mean time
α	alpha particle
α	absorption coefficient, degree of dissociation, electric polarizability, expansion coefficient, fine structure constant
β	beta particle
γ	photon
γ	activity coefficient, conductivity, magnetogyric ratio, mass concentration, ratio of heat capacities, surface tension
Γ	Gruneisin parameter, level width, surface concentration
δ	chemical shift, Dirac delta function, Kronecker delta, loss angle
Δ	inertia defect, mass defect
ε	emittance, Levi-Civita symbol, linear strain, molar absorption coefficient, permittivity
η	overpotential, viscosity
θ	Bragg angle, temperature, scattering angle, surface coverage
Θ	quadrupole moment

κ	compressibility, conductivity, magnetic susceptibility, molar absorption coefficient, transmission coefficient
λ	absolute activity, radioactive decay constant, thermal conductivity, wavelength
Λ	angular momentum, ionic conductivity
μ	muon
μ	chemical potential, electric dipole moment, electric mobility, friction coefficient, Joule-Thompson coefficient, magnetic dipole moment, mobility, permeability
μF	microfarad
μg	microgram
μm	micrometer
μs	microsecond
ν	frequency, kinematic velocity, stoichiometric number, wavenumber
ν_e	neutrino
π	pion
Π	osmotic pressure, Peltier coefficient
ρ	density, reflectance, resistivity
σ	electrical conductivity, cross section, normal stress, shielding constant (NMR), Stefan-Boltzmann constant, surface tension
τ	transmittance, chemical shift, shear stress, relaxation time
φ	electrical potential, fugacity coefficient, osmotic coefficient, quantum yield, volume fraction, wavefunction
Φ	magnetic flux, potential energy, radiant power, work function
χ	magnetic susceptibility, electronegativity
χ_e	electric susceptibility
ψ	wavefunction
ω	circular frequency, angular velocity, harmonic vibration wavenumber, statistical weight
Ω	ohm
Ω	axial angular momentum, solid angle

GREEK, RUSSIAN, AND HEBREW ALPHABETS

The following table presents the Hebrew, Greek, and Russian alphabets, their letters, the names of the letters, and the English equivalents.

HEBREW[1,3]			GREEK[4]			RUSSIAN	
א	aleph	' [2]	A α	alpha	a	А а	a
ב	beth	b, bh	B β	beta	b	Б б	b
						В в	v
ג	gimel	g, gh	Γ γ	gamma	g, n	Г г	g
ד	daleth	d, dh	Δ δ	delta	d	Д д	d
ה	he	h	E ε	epsilon	e	Е е	e
						Ж ж	zh
ו	waw	w	Z ζ	zeta	z	З з	z
ז	zayin	z	Η η	eta	ē	И и Й й	i, ĭ
ח	heth	ḥ	Θ θ	theta	th	К к	k
ט	teth	ṭ	I ι	iota	i	Л л	l
י	yodh	y	K κ	kappa	k	М м	m
כ ך	kaph	k, kh	Λ λ	lambda	l	Н н	n
ל	lamedh	l	M μ	mu	m	О о	o
מ ם	mem	m	Ξ ξ	xi	x	П п	p
נ ן	nun	n	O o	omicron	o	Р р	r
ס	samekh	s	Π π	pi	p	С с	s
ע	ayin	'	P ρ	rho	r, rh	Т т	t
פ ף	pe	p, ph	Σ σ ς	sigma	s	У у	u
צ ץ	sadhe	ṣ	T τ	tau	t	Ф ф	f
ק	qoph	q	Υ υ	upsilon	y, u	Х х	kh
ר	resh	r	Φ φ	phi	ph	Ц ц	ts
שׂ	sin	ś	X χ	chi	ch	Ч ч	ch
שׁ	shin	sh	Ψ ψ	psi	ps	Ш ш	sh
ת	taw	t, th	Ω ω	omega	ō	Щ щ	shch
						Ъ ъ [5]	''
						Ы ы	y
						Ь ь [6]	'
						Э э	e
						Ю ю	yu
						Я я	ya

[1]   Where two forms of a letter are given, the second one is the form used at the end of a word.

[2]   Not represented in transliteration when initial.

[3]   The Hebrew letters are primarily consonants; a few of them are also used secondarily to represent certain vowels, when provided at all, is by means of a system of dots or strokes adjacent to the consonated characters.

[4]   The letter gamma is transliterated "n" only before velars; the letter upsilon is transliterated "u" only as the final element in diphthongs.

[5]   This sign indicates that the immediately preceding consonant is not palatized even though immediately followed by a palatized vowel.

[6]   This sign indicates that the immediately preceding consonant is palatized even though not immediately followed by a palatized vowel.

DEFINITIONS OF SCIENTIFIC TERMS

Brief definitions of selected terms of importance in chemistry, physics, and related fields of science are given in this section. The selection process emphasizes the following types of terms:

- Physical quantities
- Units of measure
- Classes of chemical compounds and materials
- Important theories, laws, and basic concepts.

Individual chemical compounds are not included.

Definitions have taken wherever possible from the recommendations of international or national bodies, especially the International Union of Pure and Applied Chemistry (IUPAC) and International Organization for Standardization (ISO). For physical quantities and units, the recommended symbol is also given. The source of such definitions is indicated by the reference number in brackets following the definition. In many cases these official definitions have been edited in the interest of stylistic consistency and economy of space. The user is referred to the original source for further details.

* An asterisk following a term indicates that further information can be found by consulting the index of this handbook under the entry for that term.

REFERENCES

1. *ISO Standards Handbook 2, Units of Measurement*, International Organization for Standardization, Geneva, 1992.
2. *Quantities, Units, and Symbols in Physical Chemistry, Second Edition*, International Union of Pure and Applied Chemistry, Blackwell Scientific Publications, Oxford, 1993.
3. *Compendium of Chemical Terminology*, International Union of Pure and Applied Chemistry, Blackwell Scientific Publications, Oxford, 1987.
4. *A Guide to IUPAC Nomenclature of Organic Compounds*, International Union of Pure and Applied Chemistry, Blackwell Scientific Publications, Oxford, 1993.
5. *Glossary of Class Names of Organic Compounds and Reactive Intermediates Based on Structure, Pure and Applied Chemistry*, 67, 1307, 1995.
6. *Compendium of Analytical Nomenclature*, International Union of Pure and Applied Chemistry, Blackwell Scientific Publications, Oxford, 1987.
7. *Nomenclature of Inorganic Chemistry*, International Union of Pure and Applied Chemistry, Blackwell Scientific Publications, Oxford, 1990.
8. *Glossary of Basic Terms in Polymer Science, Pure and Applied Chemistry*, 68, 2287, 1996.
9. *The International Temperature Scale of 1990, Metrologia*, 27, 107, 1990.
10. *Compilation of ASTM Standard Definitions*, American Society of Testing and Materials, Philadelphia, 1990.
11. *ASM Metals Reference Book*, American Society for Metals, Metals Park, OH, 1983.

***Ab initio* method** - An approach to quantum-mechanical calculations on molecules which starts with the Schrödinger equation and carries out a complete integration, without introducing empirical factors derived from experimental measurement.

Absorbance (A) - Defined as $-\log(1-\alpha) = \log(1/\tau)$, where α is the absorptance and τ the transmittance of a medium through which a light beam passes. [2]

Absorbed dose (D) - For any ionizing radiation, the mean energy imparted to an element of irradiated matter divided by the mass of that element. [1]

Absorptance (α) - Ratio of the radiant or luminous flux in a given spectral interval absorbed in a medium to that of the incident radiation. Also called absorption factor. [1]

Absorption coefficient (a) - The relative decrease in the intensity of a collimated beam of electromagnetic radiation, as a result of absorption by a medium, during traversal of an infinitesimal layer of the medium, divided by the length traversed. [1]

Absorption coefficient, molar (ε) - Absorption coefficient divided by amount-of-substance concentration of the absorbing material in the sample solution ($\varepsilon = a/c$). The SI unit is m^2/mol. Also called extinction coefficient, but usually in units of $mol^{-1}dm^3cm^{-1}$. [2]

Acceleration - Rate of change of velocity with respect to time.

Acceleration due to gravity (g)* - The standard value (9.80665 m/s^2) of the acceleration experienced by a body in the earth's gravitational field. [1]

Acenes - Polycyclic aromatic hydrocarbons consisting of fused benzene rings in a rectilinear arrangement. [5]

Acid - Historically, a substance that yields an H^+ ion when it dissociates in solution, resulting in a pH<7. In the Brönsted definition, an acid is a substance that donates a proton in any type of reaction. The most general definition, due to G.N. Lewis, classifies any chemical species capable of accepting an electron pair as an acid.

Acid dissociation constant (K_a)* - The equilibrium constant for the dissociation of an acid HA through the reaction $HA + H_2O \rightleftharpoons A^- + H_3O^+$. The quantity $pK_a = -\log K_a$ is often used to express the acid dissociation constant.

Actinides - The elements of atomic number 89 through 103, e.g., Ac, Th, Pa, U, Np, Pu, Am, Cm, Bk, Cf, Es, Fm, Md, No, Lr. [7]

Activation energy* - In general, the energy that must be added to a system in order for a process to occur, even though the process may already be thermodynamically possible. In chemical kinetics, the activation energy is the height of the potential barrier separating the products and reactants. It determines the temperature dependence of the reaction rate.

Activity - For a mixture of substances, the absolute activity λ of substance B is defined as $\lambda_B = \exp(\mu_B/RT)$, where μ_B is the chemical potential of substance B, R the gas constant, and T the thermodynamic temperature. The relative activity a is defined as $a_B = \exp[(\mu_B - \mu_B^\circ)/RT]$, where μ_B° designates the chemical potential in the standard state. [2]

Activity coefficient (γ)* - Ratio of the activity a_B of component B of a mixture to the concentration of that component. The value of γ depends on the method of stating the composition. For mole fraction x_B, the relation is $a_B = \gamma_B x_B$; for molarity c_B, it is $a_B = \gamma_B c_B/c^\circ$, where c° is the standard state composition (typically chosen as 1 mol/L); for molality

m_B, it is $a_B = \gamma_B m_B/m°$, where $m°$ is the standard state molality (typically 1 mol/kg). [2]

Activity, of radioactive substance (A) - The average number of spontaneous nuclear transitions from a particular energy state occurring in an amount of a radionuclide in a small time interval divided by that interval. [1]

Acyl groups - Groups formed by removing the hydroxy groups from oxoacids that have the general structure RC(=O)(OH) and replacement analogues of such acyl groups. [5]

Adiabatic process - A thermodynamic process in which no heat enters or leaves the system.

Admittance (Y) - Reciprocal of impedance. $Y = G + iB$, where G is conductance and B is susceptance. [1]

Adsorption - A process in which molecules of gas, of dissolved substances in liquids, or of liquids adhere in an extremely thin layer to surfaces of solid bodies with which they are in contact. [10]

Albedo* - The ratio of the light reflected or scattered from a surface to the intensity of incident light. The term is often used in reference to specific types of terrain or to entire planets.

Alcohols - Compounds in which a hydroxy group, -OH, is attached to a saturated carbon atom. [5]

Aldehydes - Compounds RC(=O)H, in which a carbonyl group is bonded to one hydrogen atom and to one R group. [5]

Aldoses - Aldehydic parent sugars (polyhydroxyaldehydes H[CH(OH)]$_n$C(=O)H, $n>1$) and their intramolecular hemiacetals. [5]

Aldoximes - Oximes of aldehydes: RCH=NOH. [5]

Alfvén number (Al) - A dimensionless quantity used in plasma physics, defined by $Al = v(\rho\mu)^{1/2}/B$, where ρ is density, v is velocity, μ is permeability, and B is magnetic flux density. [2]

Alfven waves - Very low frequency waves which can exist in a plasma in the presence of a uniform magnetic field. Also called magnetohydrodynamic waves.

Alicyclic compounds - Aliphatic compounds having a carbocyclic ring structure which may be saturated or unsaturated, but may not be a benzenoid or other aromatic system. [5]

Aliphatic compounds - Acyclic or cyclic, saturated or unsaturated carbon compounds, excluding aromatic compounds. [5]

Alkali metals - The elements lithium, sodium, potassium, rubidium, cesium, and francium.

Alkaline earth metals - The elements calcium, strontium, barium, and radium. [7]

Alkaloids - Basic nitrogen compounds (mostly heterocyclic) occurring mostly in the plant kingdom (but not excluding those of animal origin). Amino acids, peptides, proteins, nucleotides, nucleic acids, and amino sugars are not normally regarded as alkaloids. [5]

Alkanes - Acyclic branched or unbranched hydrocarbons having the general formula C_nH_{2n+2}, and therefore consisting entirely of hydrogen atoms and saturated carbon atoms. [5]

Alkenes - Acyclic branched or unbranched hydrocarbons having one carbon-carbon double bond and the general formula C_nH_{2n}. Acyclic branched or unbranched hydrocarbons having more than one double bond are alkadienes, alkatrienes, etc. [5]

Alkoxides - Compounds, ROM, derivatives of alcohols, ROH, in which R is saturated at the site of its attachment to oxygen and M is a metal or other cationic species. [5]

Alkyl groups - Univalent groups derived from alkanes by removal of a hydrogen atom from any carbon atom: C_nH_{2n+1}-. The groups derived by removal of a hydrogen atom from a terminal carbon atom of unbranched alkanes form a subclass of normal alkyl (n-alkyl) groups. The groups RCH$_2$-, R$_2$CH-, and R$_3$C- (R not equal to H) are primary, secondary, and tertiary alkyl groups, respectively. [5]

Alkynes - Acyclic branched or unbranched hydrocarbons having a carbon-carbon triple bond and the general formula C_nH_{2n-2}, RC≡CR′.

Acyclic branched or unbranched hydrocarbons having more than one triple bond are known as alkadiynes, alkatriynes, etc. [5]

Allotropy - The occurrence of an element in two or more crystalline forms.

Allylic groups - The group CH$_2$=CHCH$_2$- (allyl) and derivatives formed by substitution. The term 'allylic position' or 'allylic site' refers to the saturated carbon atom. A group, such as -OH, attached at an allylic site is sometimes described as "allylic". [5]

Amagat volume unit - A non-SI unit previously used in high pressure science. It is defined as the molar volume of a real gas at one atmosphere pressure and 273.15 K. The approximate value is 22.4 L/mol.

Amides - Derivatives of oxoacids RC(=O)(OH) in which the hydroxy group has been replaced by an amino or substituted amino group. [5]

Amine oxides - Compounds derived from tertiary amines by the attachment of one oxygen atom to the nitrogen atom: R_3N^+-O$^-$. By extension the term includes the analogous derivatives of primary and secondary amines. [5]

Amines - Compounds formally derived from ammonia by replacing one, two, or three hydrogen atoms by hydrocarbyl groups, and having the general structures RNH$_2$ (primary amines), R$_2$NH (secondary amines), R$_3$N (tertiary amines). [5]

Amino acids* - Compounds containing both a carboxylic acid group (-COOH) and an amino group (-NH$_2$). The most important are the α-amino acids, in which the -NH$_2$ group in attached to the C atom adjacent to the -COOH group. In the ß-amino acids, there is an intervening carbon atom. [4]

Ampere (A)* - The SI base unit of electric current. [1]

Ampere's law - The defining equation for the magnetic induction B, viz., $dF = Idl \times B$, where dF is the force produced by a current I flowing in an element of the conductor dl pointing in the direction of the current.

Ångström (Å) - A unit of length used in spectroscopy, crystallography, and molecular structure, equal to 10^{-10} m.

Angular momentum (L) - The angular momentum of a particle about a point is the vector product of the radius vector from this point to the particle and the momentum of the particle; i.e., $L = r \times p$. [1]

Angular velocity (ω) - The angle through which a body rotates per unit time.

Anilides - Compounds derived from oxoacids R(C=O)(OH) by replacing the -OH group by the -NHPh group or derivative formed by ring substitution. Also used for salts formed by replacement of a nitrogen-bound hydrogen of aniline by a metal. [5]

Anion - A negatively charged atomic or molecular particle.

Antiferroelectricity* - An effect analogous to antiferromagnetism in which electric dipoles in a crystal are ordered in two sublattices that are polarized in opposite directions, leading to zero net polarization. The effect vanishes above a critical temperature.

Antiferromagnetism* - A type of magnetism in which the magnetic moments of atoms in a solid are ordered into two antiparallel aligned sublattices. Antiferromagnets are characterized by a zero or small positive magnetic susceptibility. The susceptibility increases with temperature up to a critical value, the Néel temperature, above which the material becomes paramagnetic.

Antiparticle - A particle having the same mass as a given elementary particle and a charge equal in magnitude but opposite in sign.

Appearance potential* - The lowest energy which must be imparted to the parent molecule to cause it to produce a particular specified parent ion. This energy, usually stated in eV, may be imparted by electron impact, photon impact, or in other ways. More properly called appearance energy. [3]

Appearance potential spectroscopy (APS) - See Techniques for Materials Characterization, page **12**-1.

Are (a) - A unit of area equal to 100 m². [1]

Arenes - Monocyclic and polycyclic aromatic hydrocarbons. See aromatic compounds. [5]

Aromatic compounds - Compounds whose structure includes a cyclic delocalized π-electron system. Historical use of the term implies a ring containing only carbon (e.g., benzene, naphthalene), but it is often generalized to include heterocyclic structures such as pyridine and thiophene. [5]

Arrhenius equation - A key equation in chemical kinetics which expresses the rate constant k as $k = A\exp(-E_a/RT)$, where E_a is the activation energy, R the molar gas constant, and T the temperature. A is called the preexponential factor and, for simple gas phase reactions, may be identified with the collision frequency.

Arsines - AsH_3 and compounds derived from it by substituting one, two or three hydrogen atoms by hydrocarbyl groups. $RAsH_2$, R_2AsH, R_3As (R not equal to H) are called primary, secondary and tertiary arsines, respectively. [5]

Aryl groups - Groups derived from arenes by removal of a hydrogen atom from a ring carbon atom. Groups similarly derived from heteroarenes are sometimes subsumed in this definition. [5]

Astronomical unit (AU)* - The mean distance of the earth from the sun, equal to $1.49597870 \times 10^{11}$ m.

Atomic absorption spectroscopy (AAS) - See Techniques for Materials Characterization, page **12-1**.

Atomic emission spectroscopy (AES) - See Techniques for Materials Characterization, page **12-1**.

Atomic force microscopy (AFM) - See Techniques for Materials Characterization, page **12-1**.

Atomic mass* - The mass of a nuclide, normally expressed in unified atomic mass units (u).

Atomic mass unit (u)* - A unit of mass used in atomic, molecular, and nuclear science, defined as the mass of one atom of ^{12}C divided by 12. Its approximate value is 1.66054×10^{-27} kg. Also called the unified atomic mass unit. [1]

Atomic number (Z) - A characteristic property of an element, equal to the number of protons in the nucleus.

Atomic weight (A_r)* - The ratio of the average mass per atom of an element to 1/12 of the mass of nuclide ^{12}C. An atomic weight can be defined for a sample of any given isotopic composition. The standard atomic weight refers to a sample of normal terrestrial isotopic composition. The term relative atomic mass is synonymous with atomic weight. [2]

Attenuated total reflection (ATR) - See Techniques for Materials Characterization, page **12-1**.

Auger effect - An atomic process in which an electron from a higher energy level fills a vacancy in an inner shell, transferring the released energy to another electron which is ejected.

Aurora - An atmospheric phenomenon in which streamers of light are produced when electrons from the sun are guided into the thermosphere by the earth's magnetic field. It occurs in the polar regions at altitudes of 95—300 km.

Avogadro constant (N_A)* - The number of elementary entities in one mole of a substance.

Azeotrope - A liquid mixture in a state where the variation of vapor pressure with composition at constant temperature (or, alternatively, the variation of normal boiling point with composition) shows either a maximum or a minimum. Thus when an azeotrope boils the vapor has the same composition as the liquid.

Azides - Compounds bearing the group $-N_3$, viz. $-N=N^+=N^-$; usually attached to carbon, e.g. PhN_3, phenyl azide or azidobenzene. Also used for salts of hydrazoic acid, HN_3, e.g. NaN_3, sodium azide. [5]

Azines - Condensation products, $R_2C=NN=CR_2$, of two moles of a carbonyl compound with one mole of hydrazine. [5]

Azo compounds - Derivatives of diazene (diimide), $HN=NH$, wherein both hydrogens are substituted by hydrocarbyl groups, e.g., $PhN=NPh$, azobenzene or diphenyldiazene. [5]

Balmer series - The series of lines in the spectrum of the hydrogen atom which corresponds to transitions between the state with principal quantum number $n = 2$ and successive higher states. The wavelengths are given by $1/\lambda = R_H(1/4 - 1/n^2)$, where $n = 3,4,...$ and R_H is the Rydberg constant for hydrogen. The first member of the series ($n = 2 \rightleftharpoons 3$), which is often called the H_α line, falls at a wavelength of 6563 Å.

Bar (bar) - A unit of pressure equal to 10^5 Pa.´

Bardeen-Cooper-Schrieffer (BCS) theory - A theory of superconductivity which is based upon the formation of electron pairs as a result of an electron-lattice interaction. The theory relates the superconducting transition temperature to the density of states and the Debye temperature.

Barn (b) - A unit used for expressing cross sections of nuclear processes, equal to 10^{-28} m^2.

Barrel - A unit of volume equal to 158.9873 L.

Baryon - Any elementary particle built up from three quarks. Examples are the proton, neutron, and various short-lived hyperons. Baryons have odd half-integer spins.

Base - Historically, a substance that yields an OH^- ion when it dissociates in solution, resulting in a pH>7. In the Brönsted definition, a base is a substance capable of accepting a proton in any type of reaction. The more general definition, due to G.N. Lewis, classifies any chemical species capable of donating an electron pair as a base.

Becquerel (Bq)* - The SI unit of radioactivity (disintegrations per unit time), equal to s^{-1}. [1]

Beer's law - An approximate expression for the change in intensity of a light beam that passes through an absorbing medium, viz., $\log(I/I_0) = -\varepsilon cl$, where I_0 is the incident intensity, I is the final intensity, ε is the molar (decadic) absorption coefficient, c is the molar concentration of the absorbing substance, and l is the path length. Also called the Beer-Lambert law

Binding energy* - A generic term for the energy required to decompose a system into two or more of its constituent parts. In nuclear physics, the binding energy is the energy difference between a nucleus and the separated nucleons of which it is composed (the energy equivalent of the mass defect). In atomic physics, it is the energy required to remove an electron from an atom.

Biot (Bi) - A name sometimes used for the unit of current in the emu system.

Birefringence - A property of certain crystals in which two refracted rays result from a single incident light ray. One, the ordinary ray, follows the normal laws of refraction, while the other, the extraordinary ray, exhibits a variable refractive index which depends on the direction in the crystal.

Black body radiation* - The radiation emitted by a perfect black body, i.e., a body which absorbs all radiation incident on it and reflects none. The wavelength dependence of the radiated energy density ρ (energy per unit volume per unit wavelength range) is given by the Planck formula

$$\rho = \frac{8\pi hc}{\lambda^5 \left(e^{hc/\lambda kT} - 1 \right)}$$

where λ is the wavelength, h is Planck's constant, c is the speed of light, k is the Boltzmann constant, and T is the temperature.

Black hole - A very dense object, formed in a supernova explosion, whose gravitational field is so large that no matter or radiation can escape from the object.

Bloch wave function - A solution of the Schrödinger equation for an electron moving in a spatially periodic potential; used in the band theory of solids.

Bohr magneton (μ_B)* - The atomic unit of magnetic moment, defined as

$eh/4\pi m_e$, where h is Planck's constant, m_e the electron mass, and e the elementary charge. It is the moment associated with a single electron spin.

Bohr, bohr radius (a_0)* - The radius of the lowest orbit in the Bohr model of the hydrogen atom, defined as $\varepsilon_0 h^2/\pi m_e e^2$, where ε_0 is the permittivity of a vacuum, h is Planck's constant, m_e the electron mass, and e the elementary charge. It is customarily taken as the unit of length when using atomic units.

Boiling point - The temperature at which the liquid and gas phases of a substance are in equilibrium at a specified pressure. The normal boiling point is the boiling point at normal atmospheric pressure (101.325 kPa).

Boltzmann constant (k)* - The molar gas constant R divided by Avogadro's constant.

Boltzmann distribution - An expression for the equilibrium distribution of molecules as a function of their energy, in which the number of molecules in a state of energy E is proportional to $\exp(-E/kT)$, where k is the Boltzmann constant and T is the temperature.

Bond strength - See Dissociation energy.

Born-Haber cycle* - A thermodynamic cycle in which a crystalline solid is converted to gaseous ions and then reconverted to the solid. The cycle permits calculation of the lattice energy of the crystal.

Bose-Einstein distribution - A modification of the Boltzmann distribution which applies to a system of particles that are bosons. The number of particles of energy E is proportional to $[e^{(E-\mu)/kT}-1]^{-1}$, where μ is a normalization constant, k is the Boltzmann constant, and T is the temperature.

Boson - A particle that obeys Bose-Einstein Statistics; specifically, any particle with spin equal to zero or an integer. This includes the photon, pion, deuteron, and all nuclei of even mass number.

Boyle's law - The empirical law, exact only for an ideal gas, which states that the volume of a gas is inversely proportional to its pressure at constant temperature.

Bragg angle (θ) - Defined by the equation $n\lambda = 2d\sin\theta$, which relates the angle θ between a crystal plane and the diffracted x-ray beam, the wavelength λ of the x-rays, the crystal plane spacing d, and the diffraction order n (any integer).

Bravais lattices* - The 14 distinct crystal lattices that can exist in three dimensions. They include three in the cubic crystal system, two in the tetragonal, four in the orthorhombic, two in the monoclinic, and one each in the triclinic, hexagonal, and trigonal systems.

Breakdown voltage - The potential difference at which an insulating substance undergoes a physical or chemical change that causes it to become a conductor, thus allowing current to flow through the sample.

Bremsstrahlung - Electromagnetic radiation generated when the velocity of a charged particle is reduced (literally, "braking radiation"). An example is the x-ray continuum resulting from collisions of electrons with the target in an x-ray tube.

Brewster angle - The angle of incidence for which the maximum degree of plane polarization occurs when a beam of unpolarized light is incident on the surface of a medium of refractive index n. At this angle, the angle between the reflected and refracted beams is 90°. The value of the Brewster angle is $\tan^{-1}n$.

Brillouin scattering - The scattering of light by acoustic phonons in a solid or liquid.

Brillouin zone - A region of allowed wave vectors and energy levels in a crystalline solid, which plays a part in the propagation of waves through the lattice.

British thermal unit (Btu) - A non-SI unit of energy, equal to approximately 1055 J. Several values of the Btu, defined in slightly different ways, have been used.

Brownian motion - The random movements of small particles suspended in a fluid, which arise from collisions with the fluid molecules.

Brunauer-Emmett-Teller method (BET) - See Techniques for Materials Characterization, page 12-1.

Buffer* - A solution designed to maintain a constant pH when small amounts of a strong acid or base are added. Buffers usually consist of a fairly weak acid and its salt with a strong base. Suitable concentrations are chosen so that the pH of the solution remains close to the pK_a of the weak acid.

Calorie (cal) - A non-SI unit of energy, originally defined as the heat required to raise the temperature of 1 g of water by 1°C. Several calories of slightly different values have been used. The thermochemical calorie is now defined as 4.184 J.

Candela (cd)* - The SI base unit of luminous intensity. [1]

Capacitance (C) - Ratio of the charge acquired by a body to the change in potential. [1]

Carbamates - Salts or esters of carbamic acid, $H_2NC(=O)OH$, or of N-substituted carbamic acids: $R_2NC(=O)OR'$, (R' = hydrocarbyl or a cation). The esters are often called urethanes or urethans, a usage that is strictly correct only for the ethyl esters. [5]

Carbenes - The electrically neutral species H_2C: and its derivatives, in which the carbon is covalently bonded to two univalent groups of any kind or a divalent group and bears two nonbonding electrons, which may be spin-paired (singlet state) or spin-non-paired (triplet state). [5]

Carbinols - An obsolete term for substituted methanols, in which the name carbinol is synonymous with methanol. [5]

Carbohydrates - Originally, compounds such as aldoses and ketoses, having the stoichiometric formula $C_n(H_2O)_n$ (hence "hydrates of carbon"). The generic term carbohydrate now includes mono-, oligo-, and polysaccharides, as well as their reaction products and derivatives. [5]

Carboranes - A contraction of carbaboranes. Compounds in which a boron atom in a polyboron hydride is replaced by a carbon atom with maintenance of the skeletal structure. [5]

Carboxylic acids - Oxoacids having the structure $RC(=O)OH$. The term is used as a suffix in systematic name formation to denote the $-C(=O)OH$ group including its carbon atom. [5]

Carnot cycle - A sequence of reversible changes in a heat engine using a perfect gas as the working substance, which is used to demonstrate that entropy is a state function. The Carnot cycle also provides a means to calculate the efficiency of a heat engine.

Catalyst - A substance that participates in a particular chemical reaction and thereby increases its rate but without a net change in the amount of that substance in the system. [3]

Catenanes, catena compounds - Hydrocarbons having two or more rings connected in the manner of links of a chain, without a covalent bond. More generally, the class catena compounds embraces functional derivatives and hetero analogues. [5]

Cation - A positively charged atomic or molecular particle.

Centipoise (cP) - A common non-SI unit of viscosity, equal to mPa s.

Centrifugal distortion - An effect in molecular spectroscopy in which rotational levels are lowered in energy, relative to the values of a rigid rotor, as the rotational angular momentum increases. The effect may be understood classically as a stretching of the bonds in the molecule as it rotates faster, thus increasing the moment of inertia.

Ceramic - A nonmetallic material of very high melting point.

Cerenkov radiation - Light emitted when a beam of charged particles travels through a medium at a speed greater than the speed of light in the medium. It is typically blue in color.

Cgs system of units - A system of units based upon the centimeter, gram, and second. The cgs system has been supplanted by the International System (SI).

Chalcogens - The Group VIA elements (oxygen, sulfur, selenium, tellurium, and polonium). Compounds of these elements are called chalcogenides. [7]

Chaotic system - A complex system whose behavior is governed by deterministic laws but whose evolution can vary drastically when small changes are made in the initial conditions.

Charge - See Electric charge.

Charles' law - The empirical law, exact only for an ideal gas, which states that the volume of a gas is directly proportional to its temperature at constant pressure.

Charm - A quantum number introduced in particle physics to account for certain properties of elementary particles and their reactions.

Chelate - A compound characterized by the presence of bonds from two or more bonding sites within the same ligand to a central metal atom. [3]

Chemical potential - For a mixture of substances, the chemical potential of constituent B is defined as the partial derivative of the Gibbs energy G with respect to the amount (number of moles) of B, with temperature, pressure, and amounts of all other constituents held constant. Also called partial molar Gibbs energy. [2]

Chemical shift* - A small change in the energy levels (and hence in the spectra associated with these levels) resulting from the effects of chemical binding in a molecule. The term is used in fields such as NMR, Mössbauer, and photoelectron spectroscopy, where the energy levels are determined primarily by nuclear or atomic effects.

Chiral molecule - A molecule which cannot be superimposed on its mirror image. A common example is an organic molecule containing a carbon atom to which four different atoms or groups are attached. Such molecules exhibit optical activity, i.e., they rotate the plane of a polarized light beam.

Chlorocarbons - Compounds consisting solely of chlorine and carbon. [5]

Chromatography* - A method for separation of the components of a sample in which the components are distributed between two phases, one of which is stationary while the other moves. In gas chromatography the gas moves over a liquid or solid stationary phase. In liquid chromatography the liquid mixture moves through another liquid, a solid, or a gel. The mechanism of separation of components may be adsorption, differential solubility, ion-exchange, permeation, or other mechanisms. [6]

Clapeyron equation - A relation between pressure and temperature of two phases of a pure substance that are in equilibrium, viz., $dp/dT = \Delta_{trs}S/\Delta_{trs}V$, where $\Delta_{trs}S$ is the difference in entropy between the phases and $\Delta_{trs}V$ the corresponding difference in volume.

Clathrates - Inclusion compounds in which the guest molecule is in a cage formed by the host molecule or by a lattice of host molecules. [5]

Clausius (Cl) - A non-SI unit of entropy or heat capacity defined as cal/K = 4.184 J/K. [2]

Clausius-Clapeyron equation - An approximation to the Clapeyron equation applicable to liquid-gas and solid-gas equilibrium, in which one assumes an ideal gas with volume much greater than the condensed phase volume. For the liquid-gas case, it takes the form $d(\ln p)/dT = \Delta_{vap}H/RT^2$, where R is the molar gas constant and $\Delta_{vap}H$ is the molar enthalpy of vaporization. For the solid-gas case, $\Delta_{vap}H$ is replaced by the molar enthalpy of sublimation, $\Delta_{sub}H$.

Clausius-Mosotti equation - A relation between the dielectric constant ε_r at optical frequencies and the polarizability α:

$$\frac{\varepsilon_r - 1}{\varepsilon_r + 2} = \frac{\rho N_A \alpha}{3M\varepsilon_0}$$

where ρ is density, N_A is Avogadro's number, M is molar mass, and ε_0 is the permittivity of a vacuum.

Clebsch-Gordon coefficients - A set of coefficients used to describe the vector coupling of angular momenta in atomic and nuclear physics.

Codon - A set of three bases, chosen from the four primary bases found in the DNA molecule (uracil, cytosine, adenine, and guanine), which specifies the production of a particular amino acid or carries some other genetic instruction. For example, the codon UCA specifies the amino acid serine, CAG specifies glutamine, etc. There are a total of 64 codons.

Coercive force - The magnetizing force at which the magnetic flux density is equal to zero. [10]

Coercivity* - The maximum value of coercive force that can be attained when a magnetic material is symmetrically magnetized to saturation induction. [10]

Coherent anti-Stokes Raman spectroscopy (CARS) - See Techniques for Materials Characterization, page **12-1**.

Colloid - Molecules or polymolecular particles dispersed in a medium that have, at least in one direction, a dimension roughly between 1 nm and 1 µm. [3]

Color center - A defect in a crystal that gives rise to optical absorption, thus changing the color of the material. A common type is the F-center, which results when an electron occupies the site of a negative ion.

Compressibility (κ)* - The fractional change of volume as pressure is increased, viz., $\kappa = -(1/V)(dV/dp)$. [1]

Compton wavelength (λ_C)* - In the scattering of electromagnetic radiation by a free particle (e.g., electron, proton), $\lambda_C = h/mc$ is the increase in wavelength, at a 90° scattering angle, corresponding to the transfer of energy from radiation to particle. Here h is Planck's constant, c the speed of light, and m the mass of the particle.

Conductance (G)* - For direct current, the reciprocal of resistance. More generally, the real part of admittance. [1]

Conductivity, electrical (σ)* - The reciprocal of the resistivity. [1]

Conductivity, thermal - See Thermal conductivity.

Congruent transformation - A phase transition (melting, vaporization, etc.) in which the substance preserves its exact chemical composition.

Constitutional repeating unit (CRU) - In polymer science, the smallest constitutional unit, the repetition of which constitutes a regular macromolecule, i.e., a macromolecule with all units connected identically with respect to directional sense. [8]

Copolymer - A polymer derived from more than one species of monomer. [8]

Coriolis effect - The deviation from simple trajectories when a mechanical system is described in a rotating coordinate system. It affects the motion of projectiles on the earth and in molecular spectroscopy leads to an important interaction between the rotational and vibrational motions. The effect may be described by an additional term in the equations of motion, called the Coriolis force.

Cosmic rays* - High energy nuclear particles, electrons, and photons, originating mostly outside the solar system, which continually bombard the earth's atmosphere.

Coulomb (C)* - The SI unit of electric charge, equal to A s. [1]

Coulomb's law - The statement that the force F between two electrical charges q_1 and q_2 separated by a distance r is $F = (4\pi\varepsilon_0)^{-1}q_1q_2/r^2$, where ε_0 is the permittivity of a vacuum.

Covalent bond - A chemical bond between two atoms whose stability results from the sharing of two electrons, one from each atom.

Cowling number (Co) - A dimensionless quantity used in plasma physics, defined by $Co = B^2/\mu\rho v^2$, where ρ is density, v is velocity, μ is permeability, and B is magnetic flux density. [2]

CPT theorem - A theorem in particle physics which states that any local Lagrangian theory that is invariant under proper Lorentz transformations is also invariant under the combined operations of charge conjugation, C, space inversion, P, and time reversal, T, taken in any order.

Critical point* - In general, the point on the phase diagram of a two-phase system at which the two coexisting phases have identical properties and therefore represent a single phase. At the liquid-gas critical point of a pure substance, the distinction between liquid and gas vanishes, and the vapor pressure curve ends. The coordinates of this point are called the critical temperature and critical pressure. Above the critical temperature, it is not possible to liquefy the substance.

Cross section (σ)* - A measure of the probability of collision (or other interaction) between a beam of particles and a target which it encounters. In rough terms it is the effective area the target particles present to the incident ones; however, the precise definition depends on the nature of the interaction. A general definition of σ is the number of encounters per unit time divided by nv, where n is the concentration of incident particles and v their velocity.

Crosslink - In polymer science, a small region in a macromolecule from which at least four chains emanate, and formed by reactions involving sites or groups on existing macromolecules or by interactions between existing macromolecules. [8]

Crown compounds - Macrocyclic polydentate compounds, usually uncharged, in which three or more coordinating ring atoms (usually oxygen or nitrogen) are or may become suitably close for easy formation of chelate complexes with metal ions or other cationic species. [5]

Crust* - The outer layer of the solid earth, above the Mohorovicic discontinuity. Its thickness averages about 35 km on the continents and about 7 km below the ocean floor.

Cryoscopic constant (E_f)* - The constant that expresses the amount by which the freezing point T_f of a solvent is lowered by a non-dissociating solute, through the relation $\Delta T_f = E_f m$, where m is the molality of the solute.

Curie (Ci) - A non-SI unit of radioactivity (disintegrations per unit time), equal to 3.7×10^{10} s^{-1}.

Curie temperature (T_C)* - For a ferromagnetic material, the critical temperature above which the material becomes paramagnetic. Also applied to the temperature at which the spontaneous polarization disappears in a ferroelectric solid. [1]

Cyanohydrins - Alcohols substituted by a cyano group, most commonly, but not limited to, examples having a CN and an OH group attached to the same carbon atom. They are formally derived from aldehydes or ketones by the addition of hydrogen cyanide. [5]

Cycloalkanes - Saturated monocyclic hydrocarbons (with or without side chains). See alicyclic compounds. Unsaturated monocyclic hydrocarbons having one endocyclic double or one triple bond are called cycloalkenes and cycloalkynes, respectively. [5]

Cyclotron resonance - The resonant absorption of energy from a system in which electrons or ions that are orbiting in a uniform magnetic field are subjected to radiofrequency or microwave radiation. The resonance frequency is given by $v = eH/2\pi m^*c$, where e is the elementary charge, H is the magnetic field strength, m^* is the effective mass of the charged particle, and c is the speed of light. The effect occurs in both solids (involving electrons or holes) and in low pressure gasses (involving ions)

Dalton (Da) - A name sometimes used in biochemistry for the unified atomic mass unit (u).

De Broglie wavelength - The wavelength associated with the wave representation of a moving particle, given by h/mv, where h is Planck's constant, m the particle mass, and v the velocity.

De Haas-Van Alphen effect - An effect observed in certain metals and semiconductors at low temperatures and high magnetic fields, characterized by a periodic variation of magnetic susceptibility with field strength.

Debye equation* - The relation between the relative permittivity (dielectric constant) ε_r, polarizability α, and permanent dipole moment μ in a dielectric material whose molecules are free to rotate. It takes the form

$$\frac{\varepsilon_r - 1}{\varepsilon_r + 2} = \frac{\rho N_A}{3M\varepsilon_0}\left(\alpha + \frac{\mu^2}{3kT}\right)$$

where ρ is density, N_A is Avogadro's number, M is molar mass, and ε_0 is the permittivity of a vacuum.

Debye length - In the Debye-Hückel theory of ionic solutions, the effective thickness of the cloud of ions of opposite charge which surrounds each given ion and shields the Coulomb potential produced by that ion.

Debye temperature (θ_D)* - In the Debye model of the heat capacity of a crystalline solid, $\theta_D = hv_D/k$, where h is Planck's constant, k is the Boltzmann constant, and v_D is the maximum vibrational frequency the crystal can support. For $T \ll \theta_D$, the heat capacity is proportional to T^3.

Debye unit (D) - A non-SI unit of electric dipole moment used in molecular physics, equal to 3.335641×10^{-30} C m.

Debye-Waller factor (D) - The factor by which the intensity of a diffraction line is reduced because of lattice vibrations. [1]

Defect - Any departure from the regular structure of a crystal lattice. A Frenkel defect results when an atom or ion moves to an interstitial position and leaves behind a vacancy. A Schottky defect involves either a vacancy where the atom has moved to the surface or a structure where a surface atom has moved to an interstitial position.

Degree of polymerization - The number of monomeric units in a macromolecule or an oligomer molecule. [8]

Dendrite - A tree-like crystalline pattern often observed, for example, in ice crystals and alloys in which the crystal growth branches repeatedly.

Density (ρ)* - In the most common usage, mass density or mass per unit volume. More generally, the amount of some quantity (mass, charge, energy, etc.) divided by a length, area, or volume.

Density of states (N_E, ρ) - The number of one-electron states in an infinitesimal interval of energy, divided by the range of that interval and by volume. [1]

Dew point* - The temperature at which liquid begins to condense as the temperature of a gas mixture is lowered. In meteorology, it is the temperature at which moisture begins to condense on a surface in contact with the air.

Diamagnetism - A type of magnetism characterized by a negative magnetic susceptibility, so that the material, when placed in an external magnetic field, becomes weakly magnetized in the direction opposite to the field. This magnetization is independent of temperature.

Diazo compounds - Compounds having the divalent diazo group, $=N^+=N^-$, attached to a carbon atom, e.g., $CH_2=N_2$ diazomethane. [5]

Dielectric constant (ε)* - Ratio of the electric displacement in a medium to the electric field strength. Also called permittivity. [1]

Dienes - Compounds that contain two fixed double bonds (usually assumed to be between carbon atoms). Dienes in which the two double-bond units are linked by one single bond are termed conjugated. [5]

Differential scanning calorimetry (DSC) - See Techniques for Materials Characterization, page 12-1.

Differential thermal analysis (DTA) - See Techniques for Materials Characterization, page 12-1.

Diffusion* - The migration of atoms, molecules, ions, or other particles as a result of some type of gradient (concentration, temperature, etc.).

Diopter - A unit used in optics, formally equal to m^{-1}. It is used in expressing dioptic power, which is the reciprocal of the focal length of a lens.

Dipole moment, electric (p,μ)* - For a distribution of equal positive and negative charge, the magnitude of the dipole moment vector is the positive charge multiplied by the distance between the centers of positive and negative charge distribution. The direction is given by the line from the center of negative charge to the center of positive charge.

Dipole moment, magnetic (m,μ) - Formally defined in electromagnetic theory as a vector quantity whose vector product with the magnetic flux density equals the torque. The magnetic dipole generated by a current I flowing in a small loop of area A has a magnetic moment of magnitude IA. In atomic and nuclear physics, a magnetic moment is associated with the angular momentum of a particle; e.g., an electron with orbital angular momentum l exhibits a magnetic moment of $-el/2m_e$ where e is the elementary charge and m_e the mass of the electron. [1]

Disaccharides - Compounds in which two monosaccharides are joined by a glycosidic bond. [5]

Dislocation - An extended displacement of a crystal from a regular lattice. An edge dislocation results when one portion of the crystal has partially slipped with respect to the other, resulting in an extra plane of atoms extending through part of the crystal. A screw dislocation transforms successive atomic planes into the surface of a helix.

Dispersion - Splitting of a beam of light (or other electromagnetic radiation) of mixed wavelengths into the constituent wavelengths as a result of the variation of refractive index of the medium with wavelength.

Dissociation constant* - The equilibrium constant for a chemical reaction in which a compound dissociates into its constituent parts.

Dissociation energy (D_e)* - For a diatomic molecule, the difference between the energies of the free atoms at rest and the minimum in the potential energy curve. The term bond dissociation energy (D_0), which can be applied to polyatomic molecules as well, is used for the difference between the energies of the fragments resulting when a bond is broken and the energy of the original molecule in its lowest energy state. The term bond strength implies differences in enthalpy rather than energy.

Domain - A small region of a solid in which the magnetic or electric moments of the individual units (atoms, molecules, or ions) are aligned in the same direction.

Domain wall - The transition region between adjacent ferromagnetic domains, generally a layer with a thickness of a few hundred ångström units. Also called Bloch wall.

Doppler effect - The change in the apparent frequency of a wave (sound, light, or other) when the source of the wave is moving relative to the observer.

Dose equivalent (H) - The product of the absorbed dose of radiation at a point of interest in tissue and various modifying factors which depend on the type of tissue and radiation. [1]

Drift velocity - The velocity of charge carriers (electrons, ions, etc.) moving under the influence of an electric field in a medium which subjects the carriers to some frictional force.

Dyne (dyn) - A non-SI (cgs) unit of force, equal to 10^{-5} N.

Ebullioscopic constant (E_b)* - The constant that expresses the amount by which the boiling point T_b of a solvent is raised by a non-dissociating solute, through the relation $\Delta T_b = E_b\, m$, where m is the molality of the solute.

Eddy currents - Circulating currents set up in conducting bulk materials or sheets by varying magnetic fields.

Effinghausen effect - The appearance of a temperature gradient in a current carrying conductor that is placed in a transverse magnetic field. The direction of the gradient is perpendicular to the current and the field.

Eigenvalue - An allowed value of the constant a in the equation $Au = au$, where A is an operator acting on a function u (which is called an eigenfunction). In quantum mechanics, the outcome of any observation is an eigenvalue of the corresponding operator. Also called characteristic value.

Einstein - A non-SI unit used in photochemistry, equal to one mole of photons.

Einstein temperature (θ_V) - In the Einstein theory of the heat capacity of a crystalline solid, $\theta_V = h\nu/k$, where h is Planck's constant, k is the Boltzmann constant, and ν is the vibrational frequency of the crystal.

Einstein transition probability - A constant in the Einstein relation $A_{ij} + B_{ij}\rho$ for the probability of a transition between two energy levels i and j in a radiation field of energy density ρ. The A_{ij} coefficient describes the probability of spontaneous emission, while B_{ij} and B_{ji} govern the probability of stimulated emission and absorption, respectively ($B_{ij} = B_{ji}$).

Elastic limit - The greatest stress which a material is capable of sustaining without any permanent strain remaining after complete release of the stress. [10]

Elastic modulus - See Young's modulus.

Electric charge (Q) - The quantity of electricity; i.e., the property that controls interactions between bodies through electrical forces.

Electric current (I) - The charge passing through a circuit per unit time. [1]

Electric displacement (D) - A vector quantity whose magnitude equals the electric field strength multiplied by the permittivity of the medium and whose direction is the same as that of the field strength.

Electric field strength (E) - The force exerted by an electric field on a point charge divided by the electric charge. [1]

Electric potential (V) - A scalar quantity whose gradient is equal to the negative of the electric field strength.

Electrical conductance - See Conductance

Electrical resistance - See Resistance

Electrical resistivity - See Resistivity.

Electrochemical series* - An arrangement of reactions which produce or consume electrons in an order based on standard electrode potentials. A common arrangement places metals in decreasing order of their tendency to give up electrons.

Electrode potential* - The electromotive force of a cell in which the electrode on the left is the standard hydrogen electrode and that on the right is the electrode in question. [2]

Electrolysis - The decomposition of a substance as a result of passing an electric current between two electrodes immersed in the sample.

Electromotive force (emf) - The energy supplied by a source divided by the charge transported through the source. [1]

Electron* - An elementary particle in the family of leptons, with negative charge and spin of 1/2.

Electron affinity* - The energy difference between the ground state of a gas-phase atom or molecule and the lowest state of the corresponding negative ion.

Electron cyclotron resonance (ECR) - See Techniques for Materials Characterization, page **12-1**.

Electron energy loss spectroscopy (EELS) - See Techniques for Materials Characterization, page **12-1**.

Electron nuclear double resonance (ENDOR) - See Techniques for Materials Characterization, page **12-1**.

Electron paramagnetic resonance (EPR) - See Techniques for Materials Characterization, page **12-1**.

Electron probe microanalysis (EPMA) - See Techniques for Materials Characterization, page **12-1**.

Electron spectroscopy for chemical analysis (ESCA) - See Techniques for Materials Characterization, page **12-1**.

Electron spin (s) - The quantum number, equal to 1/2, that specifies the intrinsic angular momentum of the electron.

Electron stimulated desorption (ESD) - See Techniques for Materials Characterization, page **12**-1.

Electron volt (eV)* - A non-SI unit of energy used in atomic and nuclear physics, equal to approximately 1.602177×10^{-19} J. The electron volt is defined as the kinetic energy acquired by an electron upon acceleration through a potential difference of 1 V. [1]

Electronegativity* - A parameter originally introduced by Pauling which describes, on a relative basis, the power of an atom or group of atoms to attract electrons from the same molecular entity. [3]

Electrophoresis - The motion of macromolecules or colloidal particles in an electric field. [3]

Emissivity (ε)* - Ratio of the radiant flux emitted per unit area to that of an ideal black body at the same temperature. Also called emittance. [1]

Emu - The electromagnetic system of units, based upon the cm, g, and s plus the emu of current (sometimes called the abampere).

Enantiomers - A chiral molecule and its non-superposable mirror image. The two forms rotate the plane of polarized light by equal amounts in opposite directions. Also called optical isomers.

Energy (E, U)* - The characteristic of a system that enables it to do work.

Energy gap* - In the theory of solids, the region between two energy bands, in which no bound states can occur.

Enols, alkenols - The term refers specifically to vinylic alcohols, which have the structure $HOCR'=CR_2$. Enols are tautomeric with aldehydes ($R' = H$) or ketones (R' not equal to H). [5]

Enthalpy (H)* - A thermodynamic function, especially useful when dealing with constant-pressure processes, defined by $H = E + PV$, where E is energy, P pressure, and V volume. [1]

Enthalpy of combustion* - The enthalpy change in a combustion reaction. Its negative is the heat released in combustion.

Enthalpy of formation, standard* - The enthalpy change for the reaction in which a substance is formed from its constituent elements, each in its standard reference state (normally refers to 1 mol, sometimes to 1 g, of the substance).

Enthalpy of fusion* - The enthalpy change in the transition from solid to liquid state.

Enthalpy of sublimation - The enthalpy change in the transition from solid to gas state.

Enthalpy of vaporization* - The enthalpy change in the transition from liquid to gas state.

Entropy (S)* - A thermodynamic function defined such that when a small quantity of heat dQ is received by a system at temperature T, the entropy of the system is increased by dQ/T, provided that no irreversible change takes place in the system. [1]

Entropy unit (e.u.) - A non-SI unit of entropy, equal to 4.184 J/K mol.

Ephemeris time - Time measured in tropical years from January 1, 1900.

Epoxy compounds - Compounds in which an oxygen atom is directly attached to two adjacent or non-adjacent carbon atoms of a carbon chain or ring system; thus cyclic ethers. [5]

Equation of continuity - Any of a class of equations that express the fact that some quantity (mass, charge, energy, etc.) cannot be created or destroyed. Such equations typically specify that the rate of increase of the quantity in a given region of space equals the net current of the quantity flowing into the region.

Equation of state* - An equation relating the pressure, volume, and temperature of a substance or system.

Equilibrium constant (K)* - For a chemical reaction $aA + bB \rightleftharpoons cC + dD$, the equilibrium constant is defined by:

$$K = \frac{a_C^{\,c} \cdot a_D^{\,d}}{a_A^{\,a} \cdot a_B^{\,b}}$$

where a_i is the activity of component i. To a certain approximation, the activities can be replaced by concentrations. The equilibrium constant is related to $\Delta_r G°$, the standard Gibbs energy change in the reaction, by $RT \ln K = -\Delta_r G°$.

Equivalent conductance - See Conductivity, electrical

Erg (erg) - A non-SI (cgs) unit of energy, equal to 10^{-7} J.

Esters - Compounds formally derived from an oxoacid RC(=O)(OH) and an alcohol, phenol, heteroarenol, or enol by linking, with formal loss of water from an acidic hydroxy group of the former and a hydroxy group of the latter. [5]

Esu - The electrostatic system of units, based upon the cm, g, and s plus the esu of charge (sometimes called the statcoulomb or franklin).

Ethers - Compounds with formula ROR, where R is not equal to H. [5]

Euler number (Eu) - A dimensionless quantity used in fluid mechanics, defined by $Eu = \Delta p / \rho v^2$, where p is pressure, ρ is density, and v is velocity. [2]

Eutectic - The point on a two-component solid-liquid phase diagram which represents the lowest melting point of any possible mixture. A liquid having the eutectic composition will freeze at a single temperature without change of composition.

Excitance (M) - Radiant energy flux leaving an element of a surface divided by the area of that element. [1]

Exciton - A localized excited state consisting of a bound electron-hole pair in a molecular or ionic crystal. The exciton can propagate through the crystal.

Exosphere - The outermost part of the earth's atmosphere, beginning at about 500 to 1000 km above the surface. It is characterized by densities so low that air molecules can escape into outer space.

Expansion coefficient - See thermal expansion coefficient.

Extended electron energy loss fine structure (EXELFS) - See Techniques for Materials Characterization, page **12**-1.

Extended x-ray absorption fine structure (EXAFS) - See Techniques for Materials Characterization, page **12**-1.

Extinction coefficient - See Absorption coefficient, molar

F-Center - See Color center

Fahrenheit temperature ($°F$) - The temperature scale based on the assignment of $32°F = 0°C$ and a temperature interval of $°F = (5/9)°C$; i.e., $t/°F = (9/5)t/°C + 32$.

Farad (F)* - The SI unit of electric capacitance, equal to C/V. [1]

Faraday constant (F)* - The electric charge of 1 mol of singly charged positive ions; i.e., $F = N_A e$, where N_A is Avogadro's constant and e is the elementary charge. [1]

Faraday effect* - The rotation of the plane of plane-polarized light by a medium placed in a magnetic field parallel to the direction of the light beam. The effect can be observed in solids, liquids, and gasses.

Fatty acids - Aliphatic monocarboxylic acids derived from or contained in esterified form in an animal or vegetable fat, oil, or wax. Natural fatty acids commonly have a chain of 4 to 28 carbons (usually unbranched and even-numbered), which may be saturated or unsaturated. By extension, the term is sometimes used to embrace all acyclic aliphatic carboxylic acids. [5]

Fermat's principle - The law that a ray of light traversing one or more media will follow a path which minimizes the time required to pass between two given points.

Fermi (f) - Name sometimes used in nuclear physics for the femtometer.

Fermi level - The highest energy of occupied states in a solid at zero temperature. Sometimes called Fermi energy. The Fermi surface is the surface in momentum space formed by electrons occupying the Fermi level.

Fermi resonance - An effect observed in vibrational spectroscopy when an overtone of one fundamental vibration closely coincides in energy with another fundamental of the same symmetry species. It leads to a splitting of vibrational bands.

Fermi-Dirac distribution - A modification of the Boltzmann distribution which takes into account the Pauli exclusion principle. The number of particles of energy E is proportional to $[e^{(E-\mu)/kT}+1]^{-1}$, where μ is a normalization constant, k the Boltzmann constant, and T the temperature. The distribution is applicable to a system of fermions.

Fermion - A particle that obeys Fermi-Dirac statistics. Specifically, any particle with spin equal to an odd multiple of 1/2. Examples are the electron, proton, neutron, muon, etc.

Ferrimagnetism* - A type of magnetism in which the magnetic moments of atoms in a solid are ordered into two nonequivalent sublattices with unequal magnetic moments, leading to a nonzero magnetic susceptibility.

Ferrite - A ferrimagnetic material of nominal formula MFe_2O_4, where M is a divalent metal; widely used in microwave switches and other solid state devices.

Ferroelectricity* - The retention of electric polarization by certain materials after the external field that produced the polarization has been removed.

Ferromagnetism* - A type of magnetism in which the magnetic moments of atoms in a solid are aligned within domains which can in turn be aligned with each other by a weak magnetic field. Some ferromagnetic materials can retain their magnetization when the external field is removed, as long as the temperature is below a critical value, the Curie temperature. They are characterized by a large positive magnetic susceptibility.

Fick's law - The statement that the flux J of a diffusing substance is proportional to the concentration gradient, i.e., $J = -D(dc/dx)$, where D is called the diffusion coefficient.

Field - A mathematical construct which describes the interaction between particles resulting from gravity, electromagnetism, or other physical phenomena. In classical physics a field is described by equations. Quantum field theory introduces operators to represent the physical observables.

Field emission microscopy (FEM) - See Techniques for Materials Characterization, page **12-1**.

Field ion microscopy (FIM) - See Techniques for Materials Characterization, page **12-1**.

Fine structure - The splitting in spectral lines that results from interactions of the electron spin with the orbital angular momentum.

Fine structure constant (α)* - Defined as $e^2/2hc\varepsilon_0$, where e is the elementary charge, h Planck's constant, c the speed of light, and ε_0 the permittivity of a vacuum. It is a measure of the strength of the electromagnetic interaction between particles.

First radiation constant (c_1)* - Constant ($=2\pi hc^2$) in the equation for the radiant excitance M_λ of a black body:

$$M_\lambda = \frac{c_1\lambda^{-5}\Delta\lambda}{e^{c_2/\lambda T}-1}$$

where λ is the wavelength, T is the temperature, and $c_2 = hc/k$ is the second radiation constant.

Flash point - The lowest temperature at which vapors above a volatile combustible substance will ignite in air when exposed to a flame. [10]

Fluence (F) - Term used in photochemistry to specify the energy per unit area delivered in a given time interval, for example by a laser pulse. [2]

Fluorocarbons - Compounds consisting solely of fluorine and carbon. [5]

Fluxoid - The quantum of magnetic flux in superconductivity theory, equal to $hc/2e$, where h is Planck's constant, c the velocity of light, and e the elementary charge.

Force (F) - The rate of change of momentum with time. [1]

Force constants (f, k)* - In molecular vibrations, the coefficients in the expression of the potential energy in terms of atom displacements from their equilibrium positions. In a diatomic molecule, $f = d^2V/dr^2$, where $V(r)$ is the potential energy and r is the interatomic distance. [2]

Fourier number (Fo) - A dimensionless quantity used in fluid mechanics, defined by $Fo = at/l^2$, where a is thermal diffusivity, t is time, and l is length. [2]

Fourier transform infrared spectroscopy (FTIR) - A technique for obtaining an infrared spectrum by use of an interferometer in which the path length of one of the beams is varied. A Fourier transformation of the resulting interferogram yields the actual spectrum. The technique is also used for NMR and other types of spectroscopy.

Fractals - Geometrical objects that are self-similar under a change of scale; i.e., they appear similar at all levels of magnification. They can be considered to have fractional dimensionality. Examples occur in diverse fields such as geography (rivers and shorelines), biology (trees), and solid state physics (amorphous materials).

Franck-Condon principle - An important principle in molecular spectroscopy which states that the nuclei in a molecule remain essentially stationary while an electronic transition is taking place. The physical interpretation rests on the fact that the electrons move much more rapidly than the nuclei because of their much smaller mass.

Franklin (Fr) - Name sometimes given to the unit of charge in the esu system.

Fraunhofer diffraction - Diffraction of light in situations where the source and observation point are so far removed that the wave surfaces may be considered planar.

Fraunhofer lines - Sharp absorption lines in the spectrum of sunlight, caused by absorption of the solar blackbody radiation by atoms near the sun's surface.

Free radical - See Radicals. The term "free radical" is often used more broadly for molecules that have a paramagnetic ground state (e.g., O_2) and sometimes for any transient or highly reactive molecular species.

Freezing point - See Melting point

Frequency (ν)* - Number of cycles of a periodic phenomenon divided by time. [1]

Fresnel diffraction - Diffraction of light in a situation where the source and observation point are sufficiently close together that the curvature of the wave surfaces must be taken into account.

Froude number (Fr) - A dimensionless quantity used in fluid mechanics, defined by $Fr = v/(lg)^{1/2}$, where v is velocity, l is length, and g is acceleration due to gravity. [2]

Fugacity (f_B) - For a gas mixture, the fugacity of component B is defined as the absolute activity λ_B times the limit, as the pressure p approaches zero at constant temperature, of p_B/λ_B. [2]

Fullerenes - Compounds composed solely of an even number of carbon atoms, which form a cage-like fused-ring polycyclic system with twelve five-membered rings and the rest six-membered rings. The archetypal example is [60]fullerene, where the atoms and bonds delineate a truncated icosahedron. The term has been broadened to include any closed cage structure consisting entirely of three-coordinate carbon atoms. [5]

Fulvalenes - The hydrocarbon fulvalene and its derivatives formed by substitution (and by extension, analogues formed by replacement of one or more carbon atoms of the fulvalene skeleton by a heteroatom). [5]

Fulvenes - The hydrocarbon fulvene and its derivatives formed by substitution (and by extension, analogues formed by replacement of one or more carbon atoms of the fulvene skeleton by a heteroatom). [5]

Fundamental vibrational frequencies* - In molecular spectroscopy, the characteristic vibrational frequencies obtained when the vibrational energy is expressed in normal coordinates. They determine the primary features of the infrared and Raman spectra of the molecule.

γ - Name sometimes used for microgram.

γ-rays* - Electromagnetic radiation (photons) with energy greater than about 0.1 MeV (wavelength less than about 1 pm).

g-Factor of the electron* - The proportionality factor in the equation

DEFINITIONS OF SCIENTIFIC TERMS (continued)

relating the magnetic moment μ of an electron to its total angular momentum quantum number J, i.e., $\mu = -g\mu_B J$, where μ_B is the Bohr magneton. Also called Landé factor.

Gal - A non-SI unit of acceleration, equal to 0.01 m/s. Also called galileo.

Gallon (US) - A unit of volume equal to 3.785412 L.

Gallon (UK, Imperial) - A unit of volume equal to 4.546090 L.

Gauss (G) - A non-SI unit of magnetic flux density (B) equal to 10^{-4} T.

Gaussian system of units - A hybrid system used in electromagnetic theory, which combines features of both the esu and emu systems.

Gel - A colloidal system with a finite, but usually rather small, yield stress (the sheer stress at which yielding starts abruptly). [3]

Genetic code* - The set of relations between each of the 64 codons of DNA and a specific amino acid (or other genetic instruction).

Gibbs energy (G)* - An important function in chemical thermodynamics, defined by $G = H-TS$, where H is the enthalpy, S the entropy, and T the thermodynamic temperature. Sometimes called Gibbs free energy and, in older literature, simply "free energy". [2]

Gibbs phase rule - The relation $F = C - P + 2$, where C is the number of components in a mixture, P is the number of phases, and F is the degrees of freedom, i.e., the number of intensive variables that can be changed independently without affecting the number of phases.

Glass transition temperature* - The temperature at which an amorphous polymer is transformed, in a reversible way, from a viscous or rubbery condition to a hard and relatively brittle one. [10]

Glow discharge mass spectroscopy (GDMS) - See Techniques for Materials Characterization, page 12-1.

Gluon - A hypothetical particle postulated to take part in the binding of quarks, in analogy to the role of the photon in electromagnetic interactions.

Glycerides - Esters of glycerol (propane-1,2,3-triol) with fatty acids, widely distributed in nature. They are by long-established custom subdivided into triglycerides, 1,2- or 1,3-diglycerides, and 1- or 2-monoglycerides, according to the number and positions of acyl groups. [5]

Glycols - Dihydric alcohols in which two hydroxy groups are on different carbon atoms, usually but not necessarily adjacent. Also called diols. [5]

Grain (gr) - A non-SI unit of mass, equal to 64.79891 mg.

Grain boundary - The interface between two regions of different crystal orientation.

Grashof number (Gr) - A dimensionless quantity used in fluid mechanics, defined by $Gr = l^3 g\alpha\Delta T\rho^2/\eta^2$, where T is temperature, ρ is density, l is length, η is viscosity, α is cubic expansion coefficient, and g is acceleration of gravity. [2]

Gravitational constant (G)* - The universal constant in the equation for the gravitational force between two particles, $F = Gm_1m_2/r^2$, where r is the distance between the particles and m_1 and m_2 are their masses. [1]

Gray (Gy)* - The SI unit of absorbed dose of radiation, equal to J/kg. [1]

Gregorian calendar - The modification of the Julian calendar introduced in 1582 by Pope Gregory XII which specified that a year divisible by 100 is a leap year only if divisible by 400.

Grignard reagents - Organomagnesium halides, RMgX, having a carbon-magnesium bond (or their equilibrium mixtures in solution with $R_2Mg + MgX_2$). [5]

Gruneisen parameter (γ) - Defined by $\gamma = \alpha_V/\kappa\ c_V\ \rho$, where α_V is the cubic thermal expansion coefficient, κ is the isothermal compressibility, c_V is the specific heat capacity at constant volume, and ρ is the mass density. γ is independent of temperature for most crystalline solids. [1]

Gyromagnetic ratio (γ) - Ratio of the magnetic moment of a particle to its angular momentum. Also called magnetogyric ratio.

Hadron - Any elementary particle that can take part in the strong interaction. Hadrons are subdivided into baryons, with odd half integer spins, and mesons, which have zero or integral spin.

Hall effect* - The development of a transverse potential difference V in a conducting material when subjected to a magnetic field H perpendicular to the direction of the current. The potential difference is given by $V = R_H BJt$, where B is the magnetic induction, J the current density, t the thickness of the specimen in the direction of the potential difference, and R_H is called the Hall coefficient.

Halocarbon - A compound containing no elements other than carbon, hydrogen, and one or more halogens. In common practice, the term is used mainly for compounds of no more than four or five carbon atoms.

Halogens - The elements F, Cl, Br, I, and At. Compounds of these elements are called halogenides or halides. [7]

Hamiltonian (H) - An expression for the total energy of a mechanical system in terms of the momenta and positions of constituent particles. In quantum mechanics, the Hamiltonian operator appears in the eigenvalue equation $H\psi = E\psi$, where E is an energy eigenvalue and ψ the corresponding eigenfunction.

Hardness* - The resistance of a material to deformation, indentation, or scratching. Hardness is measured on various scales, such as Mohs, Brinell, Knoop, Rockwell, and Vickers. [10]

Hartmann number (Ha) - A dimensionless quantity used in plasma physics, defined by $Ha = Bl(\kappa/\eta)^{1/2}$, where B is magnetic flux density, l is length, κ is electric conductivity, and η is viscosity. [2]

Hartree (E_h)* - An energy unit used in atomic and molecular science, equal to approximately $4.3597482 \times 10^{-18}$ J.

Hartree-Fock method - A iterative procedure for solving the Schrödinger equation for an atom or molecule in which the equation is solved for each electron in an initial assumed potential from all the other electrons. The new potential that results is used to repeat the calculation and the procedure continued until convergence is reached. Also called self-consistent field (SCF) method.

Heat capacity* - Defined in general as dQ/dT, where dQ is the amount of heat that must be added to a system to increase its temperature by a small amount dT. The heat capacity at constant pressure is $C_p = (\partial H/\partial T)_p$; that at constant volume is $C_V = (\partial E/\partial T)_V$, where H is enthalpy, E is internal energy, p is pressure, V is volume, and T is temperature. An upper case C normally indicates the molar heat capacity, while a lower case c is used for the specific (per unit mass) heat capacity. [1]

Heat of formation, vaporization, etc. - See corresponding terms under Enthalpy.

Hectare (ha) - A unit of area equal to 10^4 m^2. [1]

Heisenberg uncertainty principle - The statement that two observable properties of a system that are complementary, in the sense that their quantum-mechanical operators do not commute, cannot be specified simultaneously with absolute precision. An example is the position and momentum of a particle; according to this principle, the uncertainties in position Δq and momentum Δp must satisfy the relation $\Delta p\Delta q \geq h/4\pi$, where h is Planck's constant.

Heitler-London model - An early quantum-mechanical model of the hydrogen atom which introduced the concept of the exchange interaction between electrons as the primary reason for stability of the chemical bond.

Helicon - A low-frequency wave generated when a metal at low temperature is exposed to a uniform magnetic field and a circularly polarized electric field.

Helmholz energy (A) - A thermodynamic function defined by $A = E-TS$, where E is the energy, S the entropy, and T the thermodynamic temperature. [2]

Hemiacetals - Compounds having the general formula $R_2C(OH)OR'$ (R′ not equal to H). [5]

Henry (H)* - The SI unit of inductance, equal to Wb/A. [1]

DEFINITIONS OF SCIENTIFIC TERMS (continued)

Henry's law * - An expression which applies to an ideal dilute solution in which one or more gasses are dissolved, viz., $p_i = H_i x_i$, where p_i is the partial pressure of component i above the solution, x_i is its mole fraction in the solution, and H_i is the Henry's law constant (a characteristic of the given gas and solvent, as well as the temperature).

Hermitian operator - An operator A that satisfies the relation $\int u_m *Au_n dx = (\int u_n *Au_m \, dx)*$, where * indicates the complex conjugate. The eigenvalues of Hermitian operators are real, and eigenfunctions belonging to different eigenvalues are orthogonal.

Hertz (Hz) - The SI unit of frequency, equal to s^{-1}. [1]

Heterocyclic compounds - Cyclic compounds having as ring members atoms of at least two different elements, e.g., quinoline, 1,2-thiazole, bicyclo[3.3.1]tetrasiloxane. [5]

Heusler alloys - Alloys of manganese, copper, aluminum, nickel, and sometimes other metals which find important uses as permanent magnets.

Holography - A technique for creating a three-dimensional image of a object by recording the interference pattern between a light beam diffracted from the object and a reference beam. The image can be reconstructed from this pattern by a suitable optical system.

Homopolymer - A polymer derived from one species of (real, implicit, or hypothetical) monomer. [8]

Hooke's law - The statement that the ratio of stress to strain is a constant in a totally elastic medium.

Horse power - A non-SI unit of energy, equal to approximately 746 W.

Hubble constant - The ratio of the recessional velocity of an extragalactic object to the distance of that object. Its value is about 2×10^{-18} s^{-1}.

Huckel theory - A simple approximation for calculating the energy of conjugated molecules in which only the resonance integrals between neighboring bonds are considered. Also called CNDO method (complete neglect of differential overlap).

Hume-Rothery rules - A set of empirical rules for predicting the occurrence of solid solutions in metallic systems. The rules involve size, crystal structure, and electronegativity.

Hund's rules - A series of rules for predicting the sequence of energy states in atoms and molecules. One of the important results is that when two electrons exist in different orbitals, the state with their spins parallel (triplet state) lies at lower energy than the state with antiparallel spins (singlet).

Hydrazines - Hydrazine (diazane), H_2NNH_2, and its hydrocarbyl derivatives. When one or more substituents are acyl groups, the compound is a hydrazide. [5]

Hydrocarbon - A compound containing only carbon and hydrogen. [5]

Hydrolysis - A reaction occurring in water in which a chemical bond is cleaved and a new bond formed with the oxygen atom of water.

Hyperfine structure - Splitting of energy levels and spectral lines into several closely spaced components as a result of interaction of nuclear spin angular momentum with other angular momenta in the atom or molecule.

Hysteresis* - An irreversible response of a system (parameter A) as a function of an external force (parameter F), usually symmetric with respect to the origin of the A vs. F graph after the initial application of the force. A common example is magnetic induction vs. magnetic field strength in a ferromagnet.

Ideal gas law - The equation of state $pV = RT$, which defines an ideal gas, where p is pressure, V molar volume, T temperature, and R the molar gas constant.

Ideal solution - A solution in which solvent-solvent and solvent-solute interactions are identical, so that properties such as volume and enthalpy are exactly additive. Ideal solutions follow Raoult's law, which states that the vapor pressure p_i of component i is $p_i = x_i p_i*$, where x_i is the mole fraction of component i and p_i* the vapor pressure of the pure substance i.

Ignition temperature* - The lowest temperature at which combustion of a material will occur spontaneously under specified conditions. Sometimes called autoignition temperature, kindling point. [10]

Imides - Diacyl derivatives of ammonia or primary amines, especially those cyclic compounds derived from diacids. Also used for salts having the anion RN_2^-. [5]

Impedence (Z) - The complex representation of potential difference divided by the complex representation of current. In terms of reactance X and resistance R, the impedance is given by $Z = R + iX$. [1]

Index of refraction (n)* - For a non-absorbing medium, the ratio of the velocity of electromagnetic radiation *in vacuo* to the phase velocity of radiation of a specified frequency in the medium. [1]

Inductance - The ratio of the electromagnetic force induced in a coil by a current to the rate of change of the current.

Inductive coupled plasma mass spectroscopy (ICPMS) - See Techniques for Materials Characterization, page 12-1.

Inertial defect - In molecular spectroscopy, the quantity I_c-I_a-I_b for a molecule whose equilibrium configuration is planar, where I_a, I_b, and I_c are the effective principal moments of inertia. The inertial defect for a rigid planar molecule would be zero, but vibration-rotation interactions in a real molecule lead to a positive inertial defect.

Insulator - A material in which the highest occupied energy band (valence band) is completely filled with electrons, while the next higher band (conduction band) is empty. Solids with an energy gap of 5 eV or more are generally considered as insulators at room temperature. Their conductivity is less than 10^{-6} S/m and increases with temperature.

Intercalation compounds - Compounds resulting from reversible inclusion, without covalent bonding, of one kind of molecule in a solid matrix of another compound, which has a laminar structure. The host compound, a solid, may be macromolecular, crystalline, or amorphous. [5]

International System of Units (SI)* - The unit system adopted by the General Conference on Weights and Measures in 1960. It consists of seven base units (meter, kilogram, second, ampere, kelvin, mole, candela), plus derived units and prefixes. [1]

International Temperature Scale (ITS-90)* - The official international temperature scale adopted in 1990. It consists of a set of fixed points and equations which enable the thermodynamic temperature to be determined from operational measurements. [9]

Ion - An atomic or molecular particle having a net electric charge. [3]

Ion exchange - A process involving the adsorption of one or several ionic species accompanied by the simultaneous desorption (displacement) of one or more other ionic species. [3]

Ion neutralization spectroscopy (INS) - See Techniques for Materials Characterization, page 12-1.

Ionic strength (I) - A measure of the total concentration of ions in a solution, defined by $I = 1/2\Sigma_i z_i^2 m_i$, where z_i is the charge of ionic species i and m_i is its molality. For a 1-1 electrolyte at molality m, $I = m$.

Ionization constant* - The equilibrium constant for a reaction in which a substance in solution dissociates into ions.

Ionization potential* - The minimum energy required to remove an electron from an isolated atom or molecule (in its vibrational ground state) in the gaseous phase. More properly called ionization energy. [3]

Irradiance (E) - The radiant energy flux incident on an element of a surface, divided by the area of that element. [1]

Isentropic process - A thermodynamic process in which the entropy of the system does not change.

Ising model - A model describing the coupling between two atoms in a ferromagnetic lattice, in which the interaction energy is proportional to the negative of the product of the spin components along a specified axis.

DEFINITIONS OF SCIENTIFIC TERMS (continued)

Isobar - A line connecting points of equal pressure on a graphical representation of a physical system.

Isochore - A line or surface of constant volume on a graphical representation of a physical system.

Isoelectric point* - The pH of a solution or dispersion at which the net charge on the macromolecules or colloidal particles is zero. In electrophoresis there is no motion of the particles in an electric field at the isoelectric point.

Isomers - In chemistry, compounds that have identical molecular formulas but differ in the nature or sequence of bonding of their atoms or in the arrangement of their atoms in space. In physics, nuclei of the same atomic number Z and mass number A but in different energy states. [3]

Isomorphs - Substances of different chemical nature but having the same crystal structure.

Isotactic macromolecule - A tactic macromolecule, essentially comprising only one species of repeating unit which has chiral or prochiral atoms in the main chain in a unique arrangement with respect to its adjacent constitutional units. [8]

Isotherm - A line connecting points of equal temperature on a graphical representation of a physical system.

Isothermal process - A thermodynamic process in which the temperature of the system does not change.

Isotones - Nuclides having the same neutron number N but different atomic number Z. [3]

Isotopes - Two or more nuclides with the same atomic number Z but different mass number A. The term is sometimes used synonymously with nuclide, but it is preferable to reserve the word nuclide for a species of specific Z and A. [3]

Jahn-Teller effect - An interaction of vibrational and electronic motions in a nonlinear molecule which removes the degeneracy of certain electronic energy levels. It can influence the spectrum, crystal structure, and magnetic properties of the substance.

Johnson noise - Electrical noise generated by random thermal motion of electrons in a conductor or semiconductor. Also called thermal noise.

Josephson effect - The tunneling of electron pairs through a thin insulating layer which separates two superconductors. When a potential difference is applied to the superconductors, an alternating current is generated whose frequency is precisely proportional to the potential difference. This effect has important applications in metrology and determination of fundamental physical constants.

Joule (J)* - The SI unit of energy, equal to N m. [1]

Joule-Thomson coefficient (μ) - A parameter which describes the temperature change when a gas expands adiabatically through a nozzle from a high pressure to a low pressure region. It is defined by $\mu = (\partial T/\partial p)_H$, where H is enthalpy.

Julian calendar - The calendar introduced by Julius Caeser in 46 B.C. which divided the year into 365 days with a leap year of 366 days every fourth year.

Julian date (JD) - The number of days elapsed since noon Greenwich Mean Time on January 1, 4713 B.C. Thus January 1, 2000, 0h (midnight) will be JD 2,451,543.5. This dating system was introduced by Joseph Scaliger in 1582.

Kaon - One of the elementary particles in the family of mesons. Kaons have a spin of zero and may be neutral or charged.

Kelvin (K)* - The SI base unit of thermodynamic temperature. [1]

Kepler's laws - The three laws of planetary motion, which established the elliptical shape of planetary orbits and the relation between orbital dimensions and the period of rotation.

Kerr effect* - An electrooptical effect in which birefringence is induced in a liquid or gas when a strong electric field is applied perpendicular to the direction of an incident light beam. The Kerr constant k is given by $n_1-n_2 = k\lambda E^2$, where λ is the wavelength, E is the electric field strength, and n_1 and n_2 are the indices of refraction of the ordinary and extraordinary rays, respectively.

Ketenes - Compounds in which a carbonyl group is connected by a double bond to an alkylidene group: $R_2C=C=O$. [5]

Ketones - Compounds in which a carbonyl group is bonded to two carbon atoms: $R_1R_2C=O$ (neither R may be H). [5]

Kilogram (kg)* - The SI base unit of mass. [1]

Kinetic energy (E_k, T) - The energy associated with the motion of a system of particles in a specified reference frame. For a single particle of mass m moving at velocity v, $E_k = 1/2mv^2$.

Kirchhoff's laws - Basic rules for electric circuits, which state (a) the algebraic sum of the currents at a network node is zero and (b) the algebraic sum of the voltage drops around a closed path is zero.

Klein-Gordon equation - A relativistic extension of the Schrödinger equation.

Klein-Nishima formula - An expression for the scattering cross section of a photon by an unbound electron, based upon the Dirac electron theory.

Knight shift - The change in magnetic resonance frequency of a nucleus in a metal relative to the same nucleus in a diamagnetic solid. The effect is due to the polarization of the conduction electrons in the metal.

Knudsen number (Kn) - A dimensionless quantity used in fluid mechanics, defined by $Kn = \lambda/l$, where λ is mean free path and l is length. [2]

Kondo effect - A large increase in electrical resistance observed at low temperatures in certain dilute alloys of a magnetic metal in a nonmagnetic material.

Kramers-Kronig relation - A set of equations relating the real and imaginary parts of the index of refraction of a medium

Lactams - Cyclic amides of amino carboxylic acids, having a 1-azacycloalkan-2-one structure, or analogues having unsaturation or heteroatoms replacing one or more carbon atoms of the ring. [5]

Lactones - Cyclic esters of hydroxy carboxylic acids, containing a 1-oxacycloalkan-2-one structure, or analogues having unsaturation or heteroatoms replacing one or more carbon atoms of the ring. [5]

Lagrangian function (L) - A function used in classical mechanics, defined as the kinetic energy minus the potential energy for a system of particles.

Lamb shift - The small energy difference between the $^2S_{1/2}$ and $^2P_{1/2}$ levels in the hydrogen atom, which results from interactions between the electron and the radiation field.

Laminar flow - Smooth, uniform, non-turbulent flow of a gas or liquid in parallel layers, with little mixing between layers. It is characterized by small values of the Reynolds number.

Landé g-factor - See g-Factor of the electron

Langevin function - The mathematical function $L(x) = (e^x+e^{-x})/(e^x-e^{-x})-1/x$, which occurs in the expression for the average dipole moment of a group of rotating polar molecules in an electric field: $\mu_{av} = \mu L(\mu E/kT)$, where μ is the electric dipole moment of a single molecule, E is the electric field strength, k is the Boltzmann constant, and T is the temperature.

Lanthanides - The elements of atomic number 57 through 71, which share common chemical properties: La, Ce, Pr, Nd, Pm, Sm, Eu, Gd, Tb, Dy, Ho, Er, Tm, Yb, Lu. [7]

Larmor frequency (ν_L) - The precession frequency of a magnetic dipole in an applied magnetic field. In particular, a nucleus in a magnetic field of strength B has a Larmor frequency of $\gamma B/2\pi$, where γ is the magnetogyric ratio of the nucleus.

Laser* - A device in which an optical cavity is filled with a medium where a population inversion can be produced by some means. When the resonant frequency of the cavity bears the proper relation to the separation of the inverted energy levels, stimulated emission occurs, producing a highly monochromatic, coherent beam of light.

Laser ionization mass spectroscopy (LIMS) - See Techniques for Materials Characterization, page **12**-1.

Lattice constants* - Parameters specifying the dimensions of a unit cell in a crystal lattice, specifically the lengths of the cell edges and the angles between them.

Lattice energy* - The energy per ion pair required to separate completely the ions in a crystal lattice at a temperature of absolute zero.

Laue diagram - A diffraction pattern produced when an x-ray beam passes through a thin slice of a crystal and impinges on a detector behind the crystal.

Lenz's law - The statement that the current induced in a circuit by a change in magnetic flux is so directed as to oppose the change in flux

Leonard-Jones potential - A simple but useful function for approximating the interaction between two neutral atoms or molecules separated by a distance r by writing the potential energy as $U(r) = 4\varepsilon\{(r_0/r)^{12} - (r_0/r)^6\}$, where ε and r_0 are adjustable parameters. In this form the depth of the potential well is ε and the minimum occurs at $2^{1/6}r_0$. The $(1/r)^{12}$ term is often replaced by other powers of $1/r$.

Lepton - One of the class of elementary particles that do not take part in the strong interaction. Included are the electron, muon, and neutrino. All leptons have a spin of 1/2.

Lewis number (Le) - A dimensionless quantity used in fluid mechanics, defined by $Le = a/D$, where a is thermal diffusivity and D is diffusion coefficient. [2]

Ligand field theory - A description of the structure of crystals containing a transition metal ion surrounded by nonmetallic ions (ligands). It is based on construction of molecular orbitals involving the d-orbitals of the central metal ion and combinations of atomic orbitals of the ligands.

Light year (l.y.) - A unit of distance used in astronomy, defined as the distance light travels in one year in a vacuum. Its approximate value is 9.46073×10^{15} m.

Lignins - Macromolecular constituents of wood related to lignans, composed of phenolic propylbenzene skeletal units, linked at various sites and apparently randomly. [5]

Ligroin - The petroleum fraction consisting mostly of C_7 and C_8 hydrocarbons and boiling in the range 90-140°C; commonly used as a laboratory solvent.

Lipids - A loosely defined term for substances of biological origin that are soluble in nonpolar solvents. They consist of saponifiable lipids, such as glycerides (fats and oils) and phospholipids, as well as nonsaponifiable lipids, principally steroids. [5]

Lipoproteins - Clathrate complexes consisting of a lipid enwrapped in a protein host without covalent binding, in such a way that the complex has a hydrophilic outer surface consisting of all the protein and the polar ends of any phospholipids. [5]

Liter (L)* - A synonym for cubic decimeter. [1]

Lithosphere* - The outer layer of the solid earth, extending from the base of the mantle to the surface of the crust.

Lorentz contraction - The reduction in length of a moving body in the direction of motion, given by the factor $(1-v^2/c^2)^{1/2}$, where v is the velocity of the body and c the velocity of light. Also known as the FitzGerald-Lorentz contraction.

Lorentz force - The force exerted on a point charge Q moving at velocity v in the presence of external fields E and B. It is given (in SI units) by $F = Q(E + v \times B)$.

Loss angle (δ) - For a dielectric material in an alternating electromagnetic field, δ is the phase difference between the current and the potential difference. The function $\tan \delta$ is a measure of the ratio of the power dissipated in the dielectric to the power stored.

Low energy electron diffraction (LEED) - See Techniques for Materials Characterization, page **12**-1.

Lumen (lm)* - The SI unit of luminous flux, equal to cd sr. [1]

Luminous flux (Φ) - The intensity of light from a source multiplied by the solid angle. The SI unit is lumen. [1]

Lux (lx)* - The SI unit of illuminance, equal to cd sr m^{-2}. [1]

Lyddane-Sachs-Teller relation - A relation between the phonon frequencies and dielectric constants of an ionic crystal which states that $(\omega_T/\omega_L)^2 = \varepsilon(\infty)/\varepsilon(0)$, where ω_T is the angular frequency of transverse optical phonons, ω_L that of longitudinal optical phonons, $\varepsilon(0)$ is the static dielectric constant, and $\varepsilon(\infty)$ the dielectric constant at optical frequencies.

Lyman series - The series of lines in the spectrum of the hydrogen atom which corresponds to transitions between the ground state (principal quantum number $n = 1$) and successive excited states. The wavelengths are given by $1/\lambda = R_H(1-1/n^2)$, where $n = 2,3,4,...$ and R_H is the Rydberg constant for hydrogen. The first member of the series ($n = 1 \leftrightarrow 2$), which is often called the Lyman-α line, falls at a wavelength of 1216 Å, and the series converges at 912 Å, the ionization limit of hydrogen.

Mach number (Ma) - A dimensionless quantity used in fluid mechanics, defined by $Ma = v/c$, where v is velocity and c is the speed of sound. [2]

Macromolecule - A molecule of high relative molecular mass (molecular weight), the structure of which essentially comprises the multiple repetition of units derived, actually or conceptually, from molecules of low relative molecular mass [8]

Madelung constant* - A constant characteristic of a particular crystalline material which gives a measure of the electrostatic energy binding the ions in the crystal.

Magnetic field strength (H) - An axial vector quantity, the curl of which is equal to the current density, including the displacement current. [1]

Magnetic induction (B) - An axial vector quantity such that the force exerted on an element of current is equal to the vector product of this element and the magnetic induction. [1]

Magnetic moment - See Dipole moment, magnetic.

Magnetic susceptibility (χ$_m$, κ)* - Defined by $\chi_m = (\mu-\mu_0)/\mu_0$, where μ is the permeability of the medium and μ_0 the permeability of a vacuum. [1]

Magnetization (M) - Defined by $M = (B/\mu_0)-H$, where B is magnetic induction, H magnetic field strength, and μ_0 the permeability of a vacuum. [1]

Magnetogyric ratio (γ) - Ratio of the magnetic moment of a particle to its angular momentum. Also called gyromagnetic ratio.

Magneton - See Bohr magneton, Nuclear magneton.

Magnetostriction* - The change in dimensions of a solid sample when it is placed in a magnetic field.

Magnon - A quantum of magnetic energy associated with a spin wave in a ferromagnetic or antiferromagnetic crystal.

Mantle - The layer of the earth between the crust and the liquid outer core, which begins about 2900 km below the earth's surface.

Maser - A device in which a microwave cavity is filled with a medium where a population inversion can be produced by some means. When the resonant frequency of the cavity bears the proper relation to the separation of the inverted energy levels, the device can serve as an amplifier or oscillator at that frequency.

Mass (m)* - Quantity of matter. Mass can also be defined as "resistance to acceleration".

Mass defect (B) - Defined by $B = Zm(^1H)+Nm_n-m_a$, where Z is the atomic number, $m(^1H)$ is the mass of the hydrogen atom, N is the neutron number, m_n is the rest mass of the neutron, and m_a is the mass of the atom in question. Thus Bc^2 can be equated to the binding energy of the nucleus if the binding energy of atomic electrons is neglected. [1]

Mass excess (Δ) - Defined by $\Delta = m_a - Am_u$, where m_a is the mass of the atom, A the number of nucleons, and m_u the unified atomic mass constant (m_u = 1 u). [1]

Mass fraction (w_B) - The ratio of the mass of substance B to the total mass of a mixture. [1]

Mass number (A) - A characteristic property of a specific isotope of an element, equal to the sum of the number of protons and neutrons in the nucleus.

Mass spectrometry - An analytical technique in which ions are separated according to the mass/charge ratio and detected by a suitable detector. The ions may be produced by electron impact on a gas, a chemical reaction, energetic vaporization of a solid, etc. [6]

Massieu function - A thermodynamic function defined by $J = -A/T$, where A is the Helmholz energy and T the thermodynamic temperature. [2]

Matthiessen's rule - The statement that the electrical resistivity ρ of a metal can be written as $\rho = \rho_L + \rho_i$, where ρ_L is due to scattering of conduction electrons by lattice vibrations and ρ_i to scattering by impurities and imperfections. If the impurity concentration is small, ρ_i is temperature independent.

Maxwell (Mx)* - A non-SI unit of magnetic field strength (H) equal to 10^{-8} Wb. [1]

Maxwell's equations - The fundamental equations of electromagnetism. In a form appropriate to SI units, they are:

curl $H = \partial D/\partial t + j$

div $B = 0$

curl $E = -\partial B/\partial t$

div $D = \rho$

where H is the magnetic field strength, B the magnetic induction, E the electric field strength, D the electric displacement, j the current density, ρ the charge density, and t is time.

Maxwell-Boltzmann distribution - An expression for the fraction of molecules $f(v)$ in a gas that have velocity v within a specified interval. It takes the form

$$f(v) = 4\pi (M/2\pi RT)^{3/2} v^2 e^{-Mv^2/2RT}$$

where M is the molar mass, R the molar gas constant, and T the temperature.

Mean free path* - The average distance a gas molecule travels between collisions.

Meissner effect - The complete exclusion of magnetic induction from the interior of a superconductor.

Melting point* - The temperature at which the solid and liquid phases of a substance are in equilibrium at a specified pressure (normally taken to be atmospheric unless stated otherwise).

Mercaptans - A traditional term abandoned by IUPAC, synonymous with thiols. This term is still widely used. [5]

Meson - Any elementary particle that has zero or integral spin. Mesons are responsible for the forces between protons and neutrons in the nucleus.

Mesosphere - The part of the earth's atmosphere extending from the top of the stratosphere (about 50 km above the surface) to 80-90 km. It is characterized by a decrease in temperature with increasing altitude.

Metal - A material in which the highest occupied energy band (conduction band) is only partially filled with electrons. The electrical conductivity of metals generally decreases with temperature.

Metallocenes - Organometallic coordination compounds in which one atom of a transition metal such as iron, ruthenium or osmium is bonded to and only to the face of two cyclopentadienyl ligands which lie in parallel planes. [5]

Meter (m)* - The SI base unit of length. [1]

Methine group - In organic compounds, the -C= group. [5]

Mho - An archaic name for the SI unit siemens (reciprocal ohm).

Micelle - A particle formed by the aggregation of surfactant molecules (typically, 10 to 100 molecules) in solution. For aqueous solutions, the hydrophilic end of the molecule is on the surface of the micelle, while the hydrophobic end (often a hydrocarbon chain) points toward the center. At the critical micelle concentration (cmc) the previously dissolved molecules aggregate into a micelle.

Micron (μ) - An obsolete name for micrometer.

Mie scattering - The scattering of light by spherical dielectric particles whose diameter is comparable to the wavelength of the light.

Milky way - The band of light in the night sky resulting from the stars in the galactic plane. The term is also used to denote the galaxy in which the sun is located.

Miller indices (hkl) - A set of indices used to label planes in a crystal lattice. [2]

Millimeter of mercury (mmHg) - A non-SI unit of pressure, equal to 133.322 Pa. The name is generally considered interchangeable with torr.

Mobility (μ)* - In solid state physics, the drift velocity of electrons or holes in a solid divided by the applied electric field strength. The term is used in a similar sense in other fields.

Molality (m) - A measure of concentration of a solution in which one states the amount of substance (i.e., number of moles) of solute per kilogram of solvent. Thus a 0.1 molal solution (often written as 0.1 m) has m = 0.1 mol/kg.

Molar mass - The mass of one mole of a substance. It is normally expressed in units of g/mol, in which case its numerical value is identical with the molecular weight (relative molecular mass). [1]

Molar quantity - It is often convenient to express an extensive quantity (e.g., volume, enthalpy, heat capacity, etc.) as the actual value divided by amount of substance (number of moles). The resulting quantity is called molar volume, molar enthalpy, etc

Molar refraction (R) - A property of a dielectric defined by the equation $R = V_{ml}[(n^2-1)/(n^2+2)]$, where n is the index of refraction of the medium (at optical wavelengths) and V_m the molar volume. It is related to the polarizability α of the molecules that make up the medium by the Lorenz-Lorentz equation, $R = N_A\alpha/3\varepsilon_0$, where N_A is Avogadro's constant and ε_0 is the permittivity of a vacuum.

Molarity (c) - A measure of concentration of a solution in which one states the amount of substance (i.e., number of moles) of solute per liter of solution. Thus a 0.1 molar solution (often referred to as 0.1 M) has a concentration c = 0.1 mol/L.

Mole (mol)* - The SI base unit of amount of substance. [1]

Mole fraction (x_B) - The ratio of the amount of substance (number of moles) of substance B to the total amount of substance in a mixture. [1]

Molecular orbital - See Orbital.

Molecular weight (M_r)* - The ratio of the average mass per molecule or specified entity of a substance to 1/12 of the mass of nuclide ^{12}C. Also called relative molar (or molecular) mass. [1]

Moment of inertia (I) - The moment of inertia of a body about an axis is the sum (or integral) of the products of its elements of mass and the squares of their distances from the axis. [1]

Momentum (p) - The product of mass and velocity. [1]

Monomer - A substance consisting of molecules which can undergo polymerization, thereby contributing constitutional units to the essential structure of a macromolecule. [8]

Monosaccharides - A term which includes aldoses, ketoses, and a wide variety of derivatives. [5]

Mössbauer effect - The recoilless emission of γ-rays from nuclei bound in a crystal under conditions where the recoil energy associated with the γ emission is taken up by the crystal as a whole. This results in a very narrow line width, which can be exploited in various types of precise measurements.

DEFINITIONS OF SCIENTIFIC TERMS (continued)

Muon* - An unstable elementary particle of spin 1/2 and mass about 200 times that of the electron.

Naphtha - The petroleum fraction consisting mostly of C_6 to C_8 hydrocarbons and boiling in the range 80-120°C. Solvents derived from this fraction include ligroin and petroleum ether.

Nautical mile - A non-SI unit of length, equal to exactly 1852 m.

Navier-Stokes equations - A set of complex equations for the motion of a viscous fluid subject to external forces.

Néel temperature (T_N)* - The critical temperature above which an antiferromagnetic substance becomes paramagnetic. [1]

Nernst effect - The production of an electric field in a conductor subject to an applied magnetic field and containing a transverse temperature gradient. The electric field is perpendicular to the magnetic field and the temperature gradient.

Network - In polymer science, a highly ramified macromolecule in which essentially each constitutional unit is connected to each other constitutional unit and to the macroscopic phase boundary by many permanent paths through the macromolecule, the number of such paths increasing with the number of intervening bonds. The paths must on the average be coextensive with the macromolecule. [8]

Neutrino - A stable elementary particle in the lepton family. Neutrinos have zero (or at least near-zero) rest mass and spin 1/2.

Neutron* - An elementary particle on spin 1/2 and zero charge. The free neutron has a mean lifetime of 887 seconds. Neutrons and protons, which are collectively called nucleons, are the constituents of the nucleus.

Neutron activation analysis (NAA) - See Techniques for Materials Characterization, page **12**-1.

Neutron number (N) - A characteristic property of a specific isotope of an element, equal to the number of neutrons in the nucleus.

Newton (N)* - The SI unit of force, equal to m kg s^{-2}. [1]

Nitriles - Compounds having the structure RC≡N; thus C-substituted derivatives of hydrocyanic acid, HC≡N. [5]

Nitrosamines - N-Nitroso amines: compounds of the structure R_2NNO. Compounds RNHNO are not ordinarily isolatable, but they, too, are nitrosamines. The name is a contraction of N-nitrosoamine and, as such, does not require the N locant. [5]

Nuclear magnetic resonance (NMR)* - A widely used technique in which the resonant absorption of radiofrequency radiation by magnetic nuclei in a magnetic field is measured. The results give important information on the local environment of each nucleus.

Nuclear magneton (μ_N)* - The unit of nuclear magnetic moment, defined as $eh/4\pi m_p$, where h is Planck's constant, m_p the proton mass, and e the elementary charge.

Nuclear quadrupole resonance (NQR) - See Techniques for Materials Characterization, page **12**-1.

Nuclear reaction analysis (NRA) - See Techniques for Materials Characterization, page **12**-1.

Nuclear spin (I) - The quantum number that specifies the intrinsic angular momentum of a particular nucleus. The magnitude of the angular momentum is given by $[I(I+1)]^{1/2} h/2\pi$, where h is Planck's constant.

Nucleic acids* - Macromolecules, the major organic matter of the nuclei of biological cells, made up of nucleotide units, and hydrolyzable into certain pyrimidine or purine bases (usually adenine, cytosine, guanine, thymine, uracil), D-ribose or 2-deoxy-D-ribose. [5]

Nucleon - A collective term for the proton and neutron.

Nucleosides - Ribosyl or deoxyribosyl derivatives (rarely, other glycosyl derivatives) of certain pyrimidine or purine bases. They are thus glycosylamines or N-glycosides related to nucleotides by the lack of phosphorylation. [5]

Nucleotides - Compounds formally obtained by esterification of the 3′ or 5′ hydroxy group of nucleosides with phosphoric acid. They are the monomers of nucleic acids and are formed from them by hydrolytic cleavage. [5]

Nuclide - A species of atoms in which each atom has identical atomic number Z and identical mass number A. [3]

Nusselt number (Nu) - A dimensionless quantity used in fluid mechanics, defined by $Nu = hl/k$, where h is coefficient of heat transfer, l is length, and k is thermal conductivity. [2]

Nyquist theorem - An expression for the mean square thermal noise voltage across a resistor, given by $4RkT\Delta f$ where R is the resistance, k the Boltzmann constant, T the temperature, and Δf the frequency band within which the voltage is measured.

Octanol-water partition coefficient (P)* - A measure of the way in which a compound will partition itself between the octanol and water phases in the two-phase octanol-water system, and thus an indicator of certain types of biological activity. Specifically, P is the ratio of the concentration (in moles per liter) of the compound in the octanol phase to that in the water phase at infinite dilution. The quantity normally reported is log P.

Oersted (Oe) - A non-SI unit of magnetic field (H), equal to 79.57747 A/m.

Ohm (Ω)* - The SI unit of electric resistance, equal to V/A. [1]

Ohm's law - A relation among electric current I, potential difference V, and resistance R, viz., $I = V/R$. The resistance is constant at constant temperature to high precision for many materials.

Olefins - Acyclic and cyclic hydrocarbons having one or more carbon-carbon double bonds, apart from the formal ones in aromatic compounds. The class olefins subsumes alkenes and cycloalkenes and the corresponding polyenes. [5]

Oligomer - A substance consisting of molecules of intermediate relative molecular mass (molecular weight), the structure of which essentially comprises the multiple repetition of units derived, actually or conceptually, from molecules of low relative molecular mass. In contrast to a polymer, the properties of an oligomer can vary significantly with the removal of one or a few of its units. [8]

Oligopeptides - Peptides containing from three to nine amino groups. [5]

Onsager relations - An important set of equations in the thermodynamics of irreversible processes. They express the symmetry between the transport coefficients describing reciprocal processes in systems with a linear dependence of flux on driving forces.

Optical rotary power - Angle by which the plane of polarization of a light beam is rotated by an optically active medium, divided by path length and by concentration of the active constituent. Depending on whether mass or molar concentration is used, the modifier "specific" or "molar" is attached. [2]

Orbital - A one-electron wavefunction. Atomic orbitals are classified as s-, p-, d,- or f-orbitals according to whether the angular momentum quantum number $l = 0, 1, 2,$ or 3. Molecular orbitals, which are usually constructed as linear combinations of atomic orbitals, describe the distribution of electrons over the entire molecule.

Oscillator strength (f) - A measure of the intensity of a spectroscopic transition, defined by

$$f = \frac{8\pi^2 Me\nu}{3he^2}\left|\mu_{ij}\right|^2$$

where ν is the frequency, μ_{ij} the transition dipole moment, m_e the mass of the electron, e the elementary charge, and h Planck's constant.

Osmosis - The flow of a solvent in a system in which two solutions of different concentration are separated by a semipermeable membrane which cannot pass solute molecules. The solvent will flow from the side of lower concentration to that of higher concentration, thus tending to equalize the concentrations. The pressure that must be applied to the more concentrated side to stop the flow is called the osmotic pressure.

Osmotic coefficient (φ) - Defined by $\phi = \ln a_A/(M_A \Sigma m_B)$, where M_A is the molar mass of substance A (normally the solvent), a_A is its activity, and the m_B are molalities of the solutes. [1]

Osmotic pressure (Π) - The excess pressure necessary to maintain osmotic equilibrium between a solution and the pure solvent separated by a membrane permeable only to the solvent. In an ideal dilute solution $\Pi = c_B RT$, where c_B is the amount-of-substance concentration of the solute, R is the molar gas constant, and T the temperature. [1,2]

Ostwald dilution law - A relation for the concentration dependence of the molar conductivity Λ of an electrolyte solution, viz.,

$$\frac{1}{\Lambda} = \frac{1}{\Lambda^\circ} + \frac{\Lambda c}{K\left(\Lambda^\circ\right)^2}$$

where c is the solute concentration, K is the equilibrium constant for dissociation of the solute, and Λ° is the conductivity at $c\Lambda = 0$.

Ounce (oz) - A non-SI unit of mass. The avoirdupois once equals 28.34952 g, while the troy ounce equals 31.10348 g.

Overpotential (η) - In an electrochemical cell, the difference between the potential of an electrode and its zero-current value.

Oximes - Compounds of structure $R_2C=NOH$ derived from condensation of aldehydes or ketones with hydroxylamine. Oximes from aldehydes may be called aldoximes; those from ketones may be called ketoximes. [5]

Oxo compounds - Compounds containing an oxygen atom, =O, doubly bonded to carbon or another element. The term thus embraces aldehydes, carboxylic acids, ketones, sulfonic acids, amides and esters. [5]

Ozonides - The 1,2,4-trioxolanes formed by the reaction of ozone at a carbon-carbon double bond, or the analogous compounds derived from acetylenic compounds. [5]

Pair production - A process in which a photon is converted into a particle and its antiparticle (e.g., an electron and positron) in the electromagnetic field of a nucleus.

Paraffins - Obsolescent term for saturated hydrocarbons, commonly but not necessarily acyclic. Still widely used in the petrochemical industry, where the term designates acyclic saturated hydrocarbons, and stands in contradistinction to naphthenes. [5]

Paramagnetism* - A type of magnetism characterized by a positive magnetic susceptibility, so that the material becomes weakly magnetized in the direction of an external field. The magnetization disappears when the field in removed. In the simplest approximation (Curie's law) the susceptibility is inversely proportional to temperature.

Parity - The property of a quantum-mechanical wave function that describes its behavior under the symmetry operation of coordinate inversion. A parity of +1 (or even) is assigned if the wave function does not change sign when the signs of all the coordinates are changed; the parity is -1 (or odd) if the wave function changes sign under this operation.

Parsec (pc) - A unit of distance defined as the distance at which 1 astronomical unit (AU) subtends an angle of 1 second of arc. It is equal to 206264.806 AU or 3.085678×10^{16} m.

Particle induced x-ray emission (PIXE) - See Techniques for Materials Characterization, page **12**-1.

Partition function (q,z) - For a single molecule, $q = \Sigma_i g_i \exp(\varepsilon_i/kT)$, where ε_i is an energy level of degeneracy g_i, k the Boltzmann constant, and T the absolute temperature; the summation extends over all energy states. For a system of N non-interacting molecules which are indistinguishable, as in an ideal gas, the canonical partition function $Q = qN/N!$.

Pascal (Pa)* - The SI unit of pressure, equal to N/m^2. [1]

Paschen series - The series of lines in the spectrum of the hydrogen atom which corresponds to transitions between the state with principal quantum number $n = 3$ and successive higher states. The wavelengths are given by $1/\lambda = R_H(1/9 - 1/n^2)$, where $n = 4,5,6,\ldots$ and R_H is the Rydberg constant. The first member of the series ($n = 3 \leftrightarrow 4$), which is often called the P_α line, falls in the infrared at a wavelength of 1.875 µm.

Paschen-Back effect - In atomic spectroscopy, the decoupling of electron spin from orbital angular momentum as the strength of an external magnetic field is increased.

Pauli exclusion principle - The statement that two electrons in an atom cannot have identical quantum numbers; thus if there are two electrons in the same orbital, their spin quantum numbers must be of opposite sign.

Pearson symbol - A code for designating crystallographic information, including the crystal system, the lattice type, and the number of atoms per unit cell.

Péclet number (Pe) - A dimensionless quantity used in fluid mechanics, defined by $Pe = vl/a$, where v is velocity, l is length, and a is thermal diffusivity. [2]

Peltier effect - The absorption or generation of heat (depending on the current direction) which occurs when an electric current is passed through a junction between two materials.

Peptides - Amides derived from two or more amino carboxylic acid molecules (the same or different) by formation of a covalent bond from the carbonyl carbon of one to the nitrogen atom of another with formal loss of water. [5]

Permeability (μ) - Magnetic induction divided by magnetic field strength; i.e. $\mu = B/H$. The relative permeability $\mu_r = \mu/\mu_0$, where μ_0 is the permeability of a vacuum. [1]

Permittivity (ε) - Ratio of the electric displacement in a medium to the electric field strength. Also called dielectric constant. [1]

Peroxides - Compounds of structure ROOR in which R may be any organic group. In inorganic chemistry, salts of the anion O_2^{-2} [5]

Peroxy acids - Acids in which an acidic -OH group has been replaced by an -OOH group; e.g. $CH_3C(=O)OOH$ peroxyacetic acid, $PhS(=O)_2OOH$ benzeneperoxysulfonic acid. [5]

Petroleum ether - The petroleum fraction consisting of C_5 and C_6 hydrocarbons and boiling in the range 35-60°C; commonly used as a laboratory solvent.

pH* - A convenient measure of the acid-base character of a solution, usually defined by pH = -log $[c(H^+)/mol\ L^{-1})]$, where $c(H^+)$ is the concentration of hydrogen ions. The more precise definition is in terms af activity rather than concentration. [2]

Phenols - Compounds having one or more hydroxy groups attached to a benzene or other arene ring. [5]

Phonon - A quantum of energy associated with a vibrational mode of a crystal lattice.

Phosphines - PH_3 and compounds derived from it by substituting one, two or three hydrogen atoms by hydrocarbyl groups. RPH_2, R_2PH and R_3P (R not equal to H) are called primary, secondary and tertiary phosphines, respectively. [5]

Phosphonium compounds - Salts (and hydroxides) $[R_4P]^+X^-$ containing tetracoordinate phosphonium ion and the associated anion. [5]

Phosphonium ylides - Compounds having the structure $R_3P^+ \text{-} C^- R_2 \rightleftharpoons R_3P=CR_2$. Also known as Wittig reagents. [5]

Phosphoresence - The process by which a molecule is excited by light to a higher electronic state and then undergoes a radiationless transition to a state of different multiplicity from which it decays, after some delay, to the ground state. The emitted light is normally of longer wavelength than the exciting light because vibrational energy has been dissipated.

Photoelectric effect - The complete absorption of a photon by a solid with the emission of an electron.

Photon - An elementary particle of zero mass and spin 1/2. The photon is involved in electromagnetic interactions and is the quantum of electromagnetic radiation.

Photon stimulated desorption (PSD) - See Techniques for Materials Characterization, page 12-1.

Pinacols - Tetra(hydrocarbyl)ethane-1,2-diols, $R_2C(OH)C(OH)R_2$, of which the tetramethyl example is the simplest one and is itself commonly known as pinacol. [5]

Pion - An elementary particle in the family of mesons. Pions have zero spin and may be neutral or charged. They participate in the strong interaction which holds the nucleus together.

pK* - The negative logarithm (base 10) of an equilibrium constant K. For pK_a, see Acid dissociation constant.

Planck constant (h)* - The elementary quantum of action, which relates energy to frequency through the equation $E = h\nu$.

Planck distribution - See Black body radiation

Planck function - A thermodynamic function defined by $Y = -G/T$, where G is Gibbs energy and T thermodynamic temperature. [2]

Plasma - A highly ionized gas in which the charge of the electrons is balanced by the charge of the positive ions, so that the system as a whole is electrically neutral.

Plasmon - A quantum associated with a plasma oscillation in the electron gas of a solid.

Point group* - A group of symmetry operations (rotations, reflections, etc.) that leave a molecule invariant. Every molecular conformation can be assigned to a specific point group, which plays a major role in determining the spectrum of the molecule.

Poise (P) - A non-SI unit of viscosity, equal to 0.1 Pa s.

Poiseuille's equation - A formula for the rate of flow of a viscous fluid through a tube:

$$\frac{dV}{dt} = \frac{\left(p_1{}^2 - p_2{}^2\right)\pi r^4}{16l\eta p_0}$$

where V is the volume as measured at pressure p_0; p_1 and p_2 are the pressures at each end of the tube; r is the radius and l the length of the tube; and η is the viscosity.

Poisson ratio (μ) - The absolute value of the ratio of the transverse strain to the corresponding axial strain resulting from uniformly distributed axial stress below the proportional limit (i.e., where Hooke's law is valid). [10]

Polariton - A quantum associated with the coupled modes of photons and optical phonons in an ionic crystal.

Polarizability (α)* - The change in dipole moment of a molecule produced by an external electric field; specifically, $\alpha_{ab} = \partial p_a / \partial E_b$, where p_a is the dipole moment component on the a axis and E_b is the component of the electric field strength along the b axis. [2]

Polymer - A substance composed of molecules of high relative molecular mass (molecular weight), the structure of which essentially comprises the multiple repetition of units derived, actually or conceptually, from molecules of low relative molecular mass. A single molecule of a polymer is called a macromolecule. [8]

Polypeptides - Peptides containing 10 or more amino acid residues. See also Peptides. [5]

Polysaccharides - Compounds consisting of a large number of monosaccharides linked glycosidically. This term is commonly used only for those containing more than ten monosaccharide residues. Also called glycans. [5]

Porphyrins - Natural pigments containing a fundamental skeleton of four pyrrole nuclei united through the α-positions by four methine groups to form a macrocyclic structure (porphyrin is designated porphine in Chemical Abstracts indexes). [5]

Positron - The antiparticle of the electron. It has the same mass and spin as an electron, and an equal but opposite charge.

Positronium - The hydrogen-like "atom" formed from a positron nucleus and an electron. Its lifetime is very short because of annihilation of the positron and electron.

Potential - See Electric potential

Potential energy (E_p, V, U) - The portion of the energy of a system that is associated with its position in a force field.

Pound (lb) - A non-SI unit of mass, equal to 0.4535924 kg.

Power (P) - Rate of energy transfer. For electrical circuits, this is equal to the product of current and potential difference, $P = IV$. [1]

Poynting vector (S) - For electromagnetic radiation, the vector product of the electric field strength and the magnetic field strength. [1]

Prandtl number (Pr) - A dimensionless quantity used in fluid mechanics, defined by $Pr = \eta/\rho a$, where η is viscosity, ρ is density, and a is thermal diffusivity. [2]

Pressure* - Force divided by area. [1]

Proteins - Naturally occurring and synthetic polypeptides having molecular weights greater than about 10,000 (the limit is not precise). See also Peptides. [5]

Proton* - A stable elementary particle of unit positive charge and spin 1/2. Protons and neutrons, which are collectively called nucleons, are the constituents of the nucleus.

Pulsar - A neutron star which rotates rapidly and emits electromagnetic radiation in regular pulses at a frequency related to the rotation period.

Purine bases* - Purine and its substitution derivatives, especially naturally occurring examples. [5]

Pyrimidine bases* - Pyrimidine and its substitution derivatives, especially naturally occurring examples. [5]

Q-switching - A technique for obtaining very high power from a laser by keeping the Q factor of the laser cavity low while the population inversion builds up, then suddenly increasing the Q to initiate the stimulated emission.

Quad - A unit of energy defined as 10^{15} Btu, equal to approximately 1.055056×10^{18} J.

Quadrupole moment - A coefficient of the third term (after monopole and dipole) in the power series expansion of the electric potential of an array of charges. A nucleus of spin greater than 1/2 has a non-vanishing nuclear quadrupole moment which can interact with the electric field gradient of the surrounding electrons. Molecular quadrupole moments have an influence on intermolecular forces.

Quality factor (Q) - The ratio of the absolute value of the reactance of an electrical system to the resistance; thus a measure of the energy stored per cycle relative to the energy dissipated.

Quantum yield - In photochemistry, the number of moles transformed in a specific process, either physically (e.g., by emission of photons) or chemically, per mole of photons absorbed by the system. [3]

Quark - An elementary entity which has not been directly observed but is considered a constituent of protons, neutrons, and other hadrons.

Quasar - An extragalactic object emitting electromagnetic radiation at a very high power level and showing a very large red shift, thus indicating that the object is receding at a speed approaching the speed of light.

Quasicrystal - A solid having conventional crystalline properties but whose lattice does not display translational periodicity.

Quaternary ammonium compounds - Derivatives of ammonium compounds, $NH_4^+ Y^-$, in which all four of the hydrogens bonded to nitrogen have been replaced with hydrocarbyl groups. Compounds having a carbon-nitrogen double bond (i.e. $R_2C=N^+R_2Y^-$) are more accurately called iminium compounds. [5]

Quinones - Compounds having a fully conjugated cyclic dione structure, such as that of benzoquinones, derived from aromatic compounds by conversion of an even number of -CH= groups into -C(=O)- groups with any necessary rearrangement of double bonds. [5]

Racemic mixture - A mixture of equal amounts of a pair of enantiomers (optical isomers); such a mixture is not optically active.

Rad - A non-SI unit of absorbed dose of radiation, equal to 0.01 Gy.

Radiance (L) - The radiant intensity in a given direction from an element of a surface, divided by the area of the orthogonal projection of this element on a plane perpendicular to the given direction. [1]

Radiant intensity (I) - The radiant energy flux leaving an element of a source within an element of solid angle, divided by that element of solid angle. [1]

Radicals - Molecular entities possessing an unpaired electron, such as $\cdot CH_3$, $\cdot SnH_3$, $\cdot Cl$. (In these formulas the dot, symbolizing the unpaired electron, should be placed so as to indicate the atom of highest spin density, if this is possible.) [5]

Raman effect - The inelastic scattering of light by a molecule, in which the incident photon either gives up to, or receives energy from, one of the internal vibrational modes of the molecule. The scattered light thus has either a lower frequency (Stokes radiation) or higher frequency (anti-Stokes radiation) than the incident light. These shifts provide a measure of the normal vibrational frequencies of the molecule.

Rankine cycle - A thermodynamic cycle which can be used to calculate the ideal performance of a heat engine that uses a condensable vapor as the working fluid (e.g., a steam engine or a heat pump).

Rankine temperature - A thermodynamic temperature scale based on a temperature interval °R = (5/9) K ; i.e., $T/°R = (9/5)T/K = t/°F + 459.67$.

Raoult's law - The expression for the vapor pressure p_i of component i in an ideal solution, viz., $p_i = x_i p_{i0}$, where x_i is the mole fraction of component i and p_{i0} the vapor pressure of the pure substance i.

Rare earth elements - The elements Sc, Y, and the lanthanides (La, Ce, Pr, Nd, Pm, Sm, Eu, Gd, Tb, Dy, Ho, Er, Tm, Yb, Lu). [7]

Rayleigh number (Ra) - A dimensionless quantity used in fluid mechanics, defined by $Ra = l^3 g\alpha\Delta T\rho/\eta a$, where l is length, g is acceleration of gravity, α is cubic expansion coefficient, T is temperature, ρ is density, η is viscosity, and a is thermal diffusivity. [2]

Rayleigh scattering - The scattering of light by particles which are much smaller than the wavelength of the light. It is characterized by a scattered intensity which varies as the inverse fourth power of the wavelength.

Rayleigh wave - A guided elastic wave along the surface of a solid; also called surface acoustic wave.

Reactance (X) - The imaginary part of impedance. For an inductive reactance L and a capacitive reactance C in series, the reactance is $X = L\omega - 1/(C\omega)$, where ω is 2π times the frequency of the current. [1]

Red shift - A displacement of a spectral line toward longer wavelengths. This can occur through the Doppler effect (e.g., in the light from receding galaxies) or, in the general theory of relativity, from the effects of a star's gravitational field.

Reflectance (ρ) - Ratio of the radiant or luminous flux at a given wavelength that is reflected to that of the incident radiation. Also called reflection factor. [1]

Reflection high energy electron diffraction (RHEED) - See Techniques for Materials Characterization, page 12-1.

Relative humidity* - The ratio of the partial pressure of water vapor in air to the saturation vapor pressure of water at the same temperature, expressed as a percentage. [10]

Relative molar mass - See Molecular weight.

Rem - A non-SI unit of dose equivalent, equal to 0.01 Sv.

Resistance (R) - Electric potential difference divided by current when there is no electromotive force in the conductor. This definition applies to direct current. More generally, resistance is defined as the real part of impedance. [1]

Resistivity (ρ) - Electric field strength divided by current density when there is no electromotive force in the conductor. Resistivity is an intrinsic property of a material. For a conductor of uniform cross section with area A and length L, and whose resistance is R, the resistivity is given by $\rho = RA/L$. [1]

Reynolds number (Re) - A dimensionless quantity used in fluid mechanics, defined by $Re = \rho v l/\eta$, where ρ is density, v is velocity, l is length, and η is viscosity. [2]

Rheology - The study of the flow of liquids and deformation of solids. Rheology addresses such phenomena as creep, stress relaxation, anelasticity, nonlinear stress deformation, and viscosity.

Ribonucleic acids (RNA) - Naturally occurring polyribonucleotides. See also nucleic acids, nucleosides, nucleotides, ribonucleotides. [5]

Ribonucleotides - Nucleotides in which the glycosyl group is a ribosyl group. See also nucleotides. [5]

Roentgen (R) - A unit used for expressing the charge (positive or negative) liberated by x-ray or γ radiation in air, divided by the mass of air. A roentgen is defined as 2.58×10^{-4} C/kg.

Rotational constants - In molecular spectroscopy, the constants appearing in the expression for the rotational energy levels as a function of the angular momentum quantum numbers. These constants are proportional to the reciprocals of the principal moments of inertia, averaged over the vibrational motion.

Rutherford back scattering (RBS) - See Techniques for Materials Characterization, page 12-1.

Rydberg constant (R_∞)* - The fundamental constant which appears in the equation for the energy levels of hydrogen-like atoms; i.e., $E_n = hcR_\infty Z^2\mu/n^2$, where h is Planck's constant, c the speed of light, Z the atomic number, μ the reduced mass of nucleus and electron, and n the principal quantum number ($n = 1, 2, \ldots$).

Rydberg series - A regular series of lines in the spectrum of an atom or molecule, with the spacing between successive lines becoming smaller as the frequency increases (wavelength decreases). The series eventually converges to a limit which usually corresponds to the complete removal of an electron from the atom or molecule.

Sackur-Tetrode equation* - An equation for the molar entropy S_m of an ideal monatomic gas: $S_m = R\ln(e^{5/2} V/N_A\Lambda^3)$, where R is the molar gas constant, V is the volume, and N_A is Avogadro's number. The constant Λ is given by $\Lambda = h/(2\pi mkT)^{1/2}$, where h is Planck's constant, m the atomic mass, k the Boltzmann constant, and T the temperature.

Salinity (S)* - A parameter used in oceanography to describe the concentration of dissolved salts in seawater. It is defined in terms of electrical conductivity relative to a standard solution of KCl. When expressed in units of parts per thousand, S may be roughly equated to the concentration of dissolved material in grams per kilogram of seawater.

Salt - An ionic compound formed by the reaction of an acid and a base.

Scanned probe microscopy (SPM) - See Techniques for Materials Characterization, page 12-1.

Scanning electron microscopy (SEM) - See Techniques for Materials Characterization, page 12-1.

Scanning laser acoustic microscopy (SLAM) - See Techniques for Materials Characterization, page 12-1.

Scanning transmission electron microscopy (STEM) - See Techniques for Materials Characterization, page 12-1.

Scanning tunneling microscopy (STM) - See Techniques for Materials Characterization, page 12-1.

Schiff bases - Imines bearing a hydrocarbyl group on the nitrogen atom: $R_2C=NR'$ (R' not equal to H). Considered by many to be synonymous with azomethines. [5]

Schmidt number (*Sc*) - A dimensionless quantity used in fluid mechanics, defined by $Sc = \eta/\rho D$, where η is viscosity, ρ is density, and D is diffusion coefficient. [2]

Schottky barrier - A potential barrier associated with a metal-semiconductor contact. It forms the basis for the rectifying device known as the Schottly diode.

Schrödinger equation - The basic equation of wave mechanics which, for systems not dependent on time, takes the form:

$$-(\hbar/2m)\nabla^2\psi + V\psi = E\psi$$

where ψ is the wavefunction, V is the potential energy expressed as a function of the spatial coordinates, E is an energy eigenvalue, ∇^2 is the Laplacian operator, $\hbar$ is Planck's constant divided by 2π, and m is the mass.

Second (s)* - The SI base unit of time. [1]

Second radiation constant (*c*₂)* - See First radiation constant.

Secondary ion mass spectroscopy (SIMS) - See Techniques for Materials Characterization, page 12-1.

Seebeck effect - The development of a potential difference in a circuit where two different metals or semiconductors are joined and their junctions maintained at different temperatures. It is the basis of the thermocouple.

Selenides - Compounds having the structure RSeR (R not equal to H). They are thus selenium analogues of ethers. Also used for metal salts of H_2Se. [5]

Semicarbazones - Compounds having the structure $R_2C{=}NNHC({=}O)NH_2$, formally derived by condensation of aldehydes or ketones with semicarbazide $[NH_2NHC({=}O)NH_2]$. [5]

Semiconductor - A material in which the highest occupied energy band (valence band) is completely filled with electrons at $T = 0$ K, and the energy gap to the next highest band (conduction band) ranges from 0 to 4 or 5 eV. With increasing temperature electrons are excited into the conduction band, leading to an increase in the electrical conductivity.

Semiquinones - Radical anions having the structure -O-Z-O· where Z is an ortho- or para-arylene group or analogous heteroarylene group; they are formally generated by the addition of an electron to a quinone. [5]

SI units* - The International System of Units adopted in 1960 and recommended for use in all scientific and technical fields. [1]

Siemens (S)* - The SI unit of electric conductance, equal to Ω^{-1}. [1]

Sievert (Sv)* - The SI unit of dose equivalent (of radiation), equal to J/kg. [1]

Silanes - Saturated silicon hydrides, analogues of the alkanes; i.e. compounds of the general formula Si_nH_{2n+2}. Silanes may be subdivided into silane, oligosilanes, and polysilanes. Hydrocarbyl derivatives are often referred to loosely as silanes. [5]

Silicones - Polymeric or oligomeric siloxanes, usually considered unbranched, of general formula $[-OSiR_2-]_n$ (R not equal to H). [5]

Siloxanes - Saturated silicon-oxygen hydrides with unbranched or branched chains of alternating silicon and oxygen atoms (each silicon atom is separated from its nearest silicon neighbors by single oxygen atoms). [5]

Skin effect - The concentration of high frequency alternating currents near the surface of a conductor.

Slater orbital - A particular mathematical expression for the radial part of the wave function of a single electron, which is used in quantum-mechanical calculations of the energy and other properties of atoms and molecules.

Small angle neutron scattering (SANS) - See Techniques for Materials Characterization, page 12-1.

Snell's law - The relation between the angle of incidence i and the angle of refraction r of a light beam which passes from a medium of refractive index n_0 to a medium of index n_1, viz., $\sin i/\sin r = n_1/n_0$.

Solar constant* - The mean radiant energy flux from the sun on a unit surface normal to the direction of the rays at the mean distance of the earth from the sun. The value is approximately 1373 W/m^2.

Solar wind - The stream of high velocity hydrogen and helium ions emitted by the sun which flows through the solar system and beyond.

Soliton - A spatially localized wave in a solid or liquid that can interact strongly with other solitons but will afterwards regain its original form.

Solubility* - A quantity expressing the maximum concentration of some material (the solute) that can exist in another liquid or solid material (the solvent) at thermodynamic equilibrium at specified temperature and pressure. Common measures of solubility include the mass of solute per unit mass of solution (mass fraction), mole fraction of solute, molality, molarity, and others.

Solubility product constant (*K*ₛₚ)* - The equilibrium constant for the dissolution of a sparsely soluble salt into its constituent ions.

Space group* - A group of symmetry operations (reflections, rotations, etc.) that leave a crystal invariant. A total of 230 space groups have been identified.

Spark source mass spectroscopy (SSMS) - See Techniques for Materials Characterization, page 12-1.

Specific gravity - Ratio of the mass density of a material to that of water. Since one must specify the temperature of both the sample and the water to have a precisely defined quantity, the use of this term is now discouraged.

Specific heat - Heat capacity divided by mass. See Heat capacity.

Specific quantity - It is often convenient to express an extensive quantity (e.g., volume, enthalpy, heat capacity, etc.) as the actual value divided by mass. The resulting quantity is called specific volume, specific enthalpy, etc.

Specific rotation [α]$^\theta_\lambda$- For an optically active substance, defined by $[\alpha]^\theta_\lambda = \alpha/\gamma l$, where α is the angle through which plane polarized light is rotated by a solution of mass concentration γ and path length l. Here θ is the Celsius temperature and λ the wavelength of the light at which the measurement is carried out. Also called specific optical rotatory power. [2]

Spin (*s*, *I*)* - A measure of the intrinsic angular momentum of a particle, which it possesses independent of its orbital motion. The symbol s is used for the spin quantum number of an electron, while I is generally used for nuclear spin.

Spiro compounds - Compounds having one atom (usually a quaternary carbon) as the only common member of two rings. [5]

Stacking fault - An error in the normal sequence of layer growth in a crystal.

Standard mean ocean water (SMOW) - A standard sample of pure water of accurately known isotopic composition which is maintained by the International Atomic Energy Agency. It is used for precise calibration of density and isotopic composition measurements.

Standard reduction potential (*E*°) - The zero-current potential of a cell in which the specified reduction reaction occurs at the right-hand electrode and the left-hand electrode is the standard hydrogen electrode. Also called Standard electrode potential.

Standard state - A defined state (specified temperature, pressure, concentration, etc.) for tabulating thermodynamic functions and carrying out thermodynamic calculations. The standard state pressure is usually taken as 100,000 Pa (1 bar), but various standard state temperatures are used. [2]

Stanton number (*St*) - A dimensionless quantity used in fluid mechanics, defined by $St = h/\rho v c_p$, where h is coefficient of heat transfer, ρ is density, v is velocity, and c_p is specific heat capacity at constant pressure. [2]

Stark effect - The splitting of an energy level of an atom or molecule, and hence a splitting of spectral lines arising from that level, as a result of the application of an external electric field.

Statistical weight (g) - The number of distinct states corresponding to the same energy level. Also called degeneracy.

Stefan-Boltzmann constant (σ)* - Constant in the equation for the radiant exitance M (radiant energy flux per unit area) from a black body at thermodynamic temperature T, viz. $M = \sigma T^4$. [1]

Stibines - SbH_3 and compounds derived from it by substituting one, two or three hydrogen atoms by hydrocarbyl groups: R_3Sb. $RSbH_2$, R_2SbH, and R_3Sb (R not equal to H) are called primary, secondary and tertiary stibines, respectively. [5]

Stochastic process - A process which involves random variables and whose outcome can thus be described only in terms of probabilities.

Stoichiometric number (ν) - The number appearing before the symbol for each compound in the equation for a chemical reaction. By convention, it is negative for reactants and positive for products. [2]

Stokes (St) - A non-SI unit of kinematic viscosity, equal to 10^{-4} m^2/s.

Stokes' law - The statement, valid under certain conditions, that the viscous force F experienced by a sphere of radius a moving at velocity v in a medium of viscosity η is given by $F = -6\pi\eta av$.

Strain - The deformation of a body that results from an applied stress.

Stratosphere - The part of the earth's atmosphere extending from the top of the troposphere (typically 10 to 15 km above the surface) to about 50 km. It is characterized by an increase in temperature with increasing altitude.

Stress - Force per unit area (pressure) applied to a body. Tensile stress tends to stretch or compress the body in the direction of the applied force. Sheer stress results from a tangential force which tends to twist the body.

Strong interaction - The short range (order of 1 fm) attractive forces between protons, neutrons, and other hadrons which are responsible for the stability of the nucleus.

Strouhal number (Sr) - A dimensionless quantity used in fluid mechanics, defined by $Sr = lf/v$, where l is length, f is frequency, and v is velocity. [2]

Structure factor - In x-ray crystallography, the sum of the scattering factors of all the atoms in a unit cell, weighted by an appropriate phase factor. The intensity of a given reflection is proportional to the square of the structure factor.

Sublimation pressure - The pressure of a gas in equilibrium with a solid at a specified temperature.

Sulfides - Compounds having the structure RSR (R not equal to H). Such compounds were once called thioethers. In an inorganic sense, salts or other derivatives of hydrogen sulfide. [5]

Sulfones - Compounds having the structure, $RS(=O)_2R$ (R not equal to H), e.g. $C_2H_5S(=O)_2CH_3$, ethyl methyl sulfone. [5]

Sulfonic acids - $HS(=O)_2OH$, sulfonic acid, and its S-hydrocarbyl derivatives. [5]

Sulfoxides - Compounds having the structure $R_2S=O$ (R not equal to H), e.g. $Ph_2S=O$, diphenyl sulfoxide. [5]

Superconductor - A material that experiences a nearly total loss of electrical resistivity below a critical temperature T_c. The effect can occur in pure metals, alloys, semiconductors, organic compounds, and certain inorganic solids.

Superfluid - A fluid with near-zero viscosity and extremely high thermal conductivity. Liquid helium exhibits these properties below 2.186 K (the λ point).

Supernova - A star in the process of exploding because of instabilities which follow the exhaustion of its nuclear fuel.

Surface analysis by laser ionization (SALI) - See Techniques for Materials Characterization, page 12-1.

Surface tension (γ,σ)* - The force per unit length in the plane of the interface between a liquid and a gas, which resists an increase in the area of that surface. It can also be equated to the surface Gibbs energy per unit area.

Surfactant - A substance which lowers the surface tension of the medium in which it is dissolved, and/or the interfacial tension with other phases, and accordingly is positively adsorbed at the liquid-vapor or other interfaces. [3]

Susceptance (B) - Imaginary part of admittance. [1]

Svedberg - A non-SI unit of time, used to express sedimentation coefficients, equal to 10^{-13} s.

Syndiotactic macromolecule - A tactic macromolecule, essentially comprising alternating enantiomeric configurational base units which have chiral or prochiral atoms in the main chain in a unique arrangement with respect to their adjacent constitutional units. In this case the repeating unit consists of two configurational base units that are enantiomeric. [8]

Tacticity - The orderliness of the succession of configurational repeating units of a macromolecule or oligomer molecule. In a tactic macromolecule essentially all the configurational repeating units are identical with respect to directional sense. See Configurational repeating unit, Isotactic, Syndiotactic. [8]

Tautomerism - Isomerism of the general form G-X-Y=Z $\rightleftharpoons$ X=Y-Z-G, where the isomers (called tautomers) are readily interconvertible; the atoms connecting the groups X, Y, Z are typically any of C, H, O, or S, and G is a group which becomes an electrofuge (i.e., a group that does not carry away the bonding electron pair when it leaves its position in the molecule) or nucleofuge (a group that does carry away the bonding electrons when leaving) during isomerization. The commonest case, when the electrofuge is H^+, is also known as prototropy. A common example, written so as to illustrate the general pattern given above, is keto-enol tautomerism, such as

$$H\text{-}O\text{-}C(CH_3)=CH\text{-}CO_2Et \text{ (enol)} \rightleftharpoons (CH_3)C(=O)\text{-}CH_2\text{-}CO_2Et \text{ (keto)}$$

In some cases the interconversion rate between tautomers is slow enough to permit isolation of the separate keto and enol forms. [5]

Tensile strength* - In tensile testing, the ratio of maximum load a body can bear before breaking to original cross-sectional area. Also called ultimate strength. [11]

Terpenes - Hydrocarbons of biological origin having carbon skeletons formally derived from isoprene [$CH_2=C(CH_3)CH=CH_2$]. [5]

Terpenoids - Natural products and related compounds formally derived from isoprene units. They contain oxygen in various functional groups. The skeleton of terpenoids may differ from strict additivity of isoprene units by the loss or shift of a methyl (or other) group. [5]

Tesla (T)* - The SI unit of magnetic flux density (B), equal to V s/m^2. [1]

Thermal conductivity* - Rate of heat flow divided by area and by temperature gradient. [1]

Thermal diffusivity - Thermal conductivity divided by density and by specific heat capacity at constant pressure. [1]

Thermal expansion coefficient (α)* - The linear expansion coefficient is defined by $\alpha_l = (1/l)(dl/dT)$; the volume expansion coefficient by $\alpha_V = (1/V)(dV/dT)$. [1]

Thermionic emission - The emission of electrons from a solid as a result of heat. The effect requires a high enough temperature to impart sufficient kinetic energy to the electrons to exceed the work function of the solid.

Thermodynamic laws - The foundation of the science of thermodynamics:

First law: The internal energy of an isolated system is constant; if energy is supplied to the system in the form of heat dq and work dw, then the change in energy $dU = dq + dw$.

Second law: No process is possible in which the only result is the transfer of heat from a reservoir and its complete conversion to work.

Third law: The entropy of a perfect crystal approaches zero as the thermodynamic temperature approaches zero.

Thermoelectric power - For a bar of a pure material whose ends are at different temperatures, the potential difference divided by the difference in temperature of the ends. See also Seeback effect.

Thermogravimetric analysis (TGA) - See Techniques for Materials Characterization, page 12-1.

Thermosphere - The layer of the earth's atmosphere extending from the top of the mesosphere (typically 80-90 km above the surface) to about 500 km. It is characterized by a rapid increase in temperature with increasing altitude up to about 200 km, followed by a leveling off in the 300-500 km region.

Thiols - Compounds having the structure RSH (R not equal to H). Also known by the term mercaptans (abandoned by IUPAC); e.g. CH_3CH_2SH, ethanethiol. [5]

Thomson coefficient (μ, τ) - The heat power developed in the Thomson effect (whereby heat is evolved in a conductor when a current is flowing in the presence of a temperature gradient), divided by the current and the temperature difference. [1]

Tonne (t) - An alternative name for megagram (1000 kg). [1]

Torque (T) - For a force F that produces a torsional motion, $T = r \times F$, where r is a vector from some reference point to the point of application of the force.

Torr - A non-SI unit of pressure, equal to 133.322 Pa. The name is generally considered interchangeable with millimeter of mercury.

Townsend coefficient - In a radiation counter, the number of ionizing collisions by an electron per unit path length in the direction of an applied electric field.

Transducer - Any device that converts a signal from acoustical, optical, or some other form of energy into an electrical signal (or vice versa) while preserving the information content of the original signal.

Transistor - A voltage amplifier using controlled electron currents inside a semiconductor.

Transition metals - Elements characterized by a partially filled d subshell. The First Transition Series comprises Sc, Ti, V, Cr, Mn, Fe, Co, Ni, Cu. The Second and Third Transition Series include the lanthanides and actinides, respectively. [7]

Transition probability* - See Einstein transition probability.

Transmittance (τ) - Ratio of the radiant or luminous flux at a given wavelength that is transmitted to that of the incident radiation. Also called transmission factor. [1]

Tribology - The study of frictional forces between solid surfaces.

Triple point* - The point in p,T space where the solid, liquid, and gas phases of a substance are in thermodynamic equilibrium. The corresponding temperature and pressure are called the triple point temperature and triple point pressure.

Troposphere - The lowest part of the earth's atmosphere, extending to 10-15 km above the surface. It is characterized by a decrease in temperature with increasing altitude. The exact height varies with latitude and season.

Tunnel diode - A device involving a p-n junction in which both sides are so heavily doped that the Fermi level on the p-side lies in the valence band and on the n-side in the conduction band. This leads to a current-voltage curve with a maximum, so that the device exhibits a negative resistance in some regions.

Ultraviolet photoelectron spectroscopy (UPS) - See Techniques for Materials Characterization, page 12-1.

Umklapp process - A process involving the interaction of three or more waves (lattice or electron) in a solid in which the sum of the wave vectors does not equal zero.

Unified atomic mass unit (u)* - A unit of mass used in atomic, molecular, and nuclear science, defined as the mass of one atom of ^{12}C divided by 12. Its approximate value is 1.66054×10^{-27} kg. [1]

Universal time (t_U, UT) - Mean solar time counted from midnight at the Greenwich meridian. Also called Greenwich mean time (GMT). The interval of mean solar time is based on the average, over one year, of the time between successive transits of the sun across the observer's meridian.

Vacancy - A missing atom or ion in a crystal lattice.

Van Allen belts - Two toroidal regions above the earth's atmosphere containing protons and electrons. The outer belt at about 25,000 km above the surface is probably of solar origin. The inner belt at about 3000 km contains more energetic particles from outside the solar system.

Van der Waals' equation* - An equation of state for fluids which takes the form:

$$pV_m = RT\left(\frac{1}{V_m - b} - \frac{a}{V_m^2}\right)$$

where p is pressure, V_m is molar volume, T is temperature, R is the molar gas constant, and a and b are characteristic parameters of the substance which describe the effect of attractive and repulsive intermolecular forces, respectively.

Van der Waals' force - The weak attractive force between two molecules which arises from electric dipole interactions. It can lead to the formation of stable but weakly bound dimer molecules or clusters.

Van't Hoff equation - The equation expressing the temperature dependence of the equilibrium constant K of a chemical reaction:

$$\frac{d\ln k}{dT} = \frac{\Delta_r H°}{RT^2}$$

where $\Delta_r H°$ is the standard enthalpy of reaction, R the molar gas constant, and T the temperature. Also called van't Hoff isochore.

Vapor pressure* - The pressure of a gas in equilibrium with a liquid (or, in some usage, a solid) at a specified temperature.

Varistor - A device that utilizes the properties of certain metal oxides with small amounts of impurities, which show abrupt nonlinearities at specific voltages where the material changes from a semiconductor to an insulator.

Velocity (v) - Rate of change of distance with time.

Verdet constants (V)* - Angle of rotation of a plane polarized light beam passing through a medium in a magnetic field, divided by the field strength and by the path length.

Virial equation of state* - An equation relating the pressure p, molar volume V_m, and temperature T of a real gas in the form of an expansion in powers of the molar volume, viz., $pV_m = RT(1 + BV_m^{-1} + CV_m^{-2} + \ldots)$, where R is the molar gas constant. B is called the second virial coefficient, C the third virial coefficient, etc. The virial coefficients are functions of temperature.

Viscosity (η)* - The proportionality factor between sheer rate and sheer stress, defined through the equation $F = \eta A(dv/dx)$, where F is the tangential force required to move a planar surface of area A at velocity v relative to a parallel surface separated from the first by a distance x. Sometimes called dynamic or absolute viscosity. The term kinematic viscosity (symbol v) is defined as η divided by the mass density.

Volt (V)* - The SI unit of electric potential, equal to W/A. [1]

Volume fraction (ϕ_j) - Defined as $V_j/\Sigma_i V_i$, where V_j is the volume of the specified component and the V_i are the volumes of all the components of a mixture prior to mixing. [2]

DEFINITIONS OF SCIENTIFIC TERMS (continued)

Watt (W)* - The SI unit of power, equal to J/s. [1]

Wave function - A function of the coordinates of all the particles in a quantum mechanical system (and, in general, of time) which fully describes the state of the system. The product of the wave function and its complex conjugate is proportional to the probability of finding a particle at a particular point in space.

Weak interaction - The weak forces (order of 10^{-12} of the strong interaction) between elementary particles which are responsible for beta decay and other nuclear effects.

Weber (Wb)* - The SI unit of magnetic flux, equal to V s. [1]

Weber number (We) - A dimensionless quantity used in fluid mechanics, defined by $We = \rho v^2 l/\gamma$, where ρ is density, v is velocity, l is length, and γ is surface tension. [2]

Weight - That force which, when applied to a body, would give it an acceleration equal to the local acceleration of gravity. [1]

Wiedeman-Franz law - The law stating that the thermal conductivity k and electrical conductivity σ of a pure metal are related by $k = L\sigma T$, where T is the temperature and L (called the Lorenz ratio) has the approximate value 2.45×10^{-8} V^2/K^2.

Wien displacement law - The relation, which can be derived from the Planck formula for black body radiation, that $\lambda_{max}T = 0.0028978$ m K, where λ_{max} is the wavelength of maximum radiance at temperature T.

Wigner-Seitz method - A method of calculating electron energy levels in a solid using a model in which each electron is subject to a spherically symmetric potential.

Wittig reagents - See phosphonium ylides.

Work (W) - Force multiplied by the displacement in the direction of the force. [1]

Work function (Φ)* - The energy difference between an electron at rest at infinity and an electron at the Fermi level in the interior of a substance. It is thus the minimum energy required to remove an electron from the interior of a solid to a point just outside the surface. [1]

X unit (X) - A unit of length used in x-ray crystallography, equal to approximately 1.002×10^{-13} m.

X-ray photoelectron spectroscopy (XPS) - See Techniques for Materials Characterization, page **12-1**.

Yield strength - The stress at which a material exhibits a specified deviation (often chosen as 0.2% for metals) from proportionality of stress and strain. [11]

Young's modulus (E) - In tension or compression of a body below its elastic limit, the ratio of stress to corresponding strain. Since strain is normally expressed on a fractional basis, Young's modulus has dimensions of pressure. Also called elastic modulus. [11]

Zeeman effect - The splitting of an energy level of an atom or molecule, and hence a splitting of spectral lines arising from that level, as a result of the application of an external magnetic field.

Zener diode - A control device utilizing a p-n junction with a well defined reverse-bias avalanche breakdown voltage.

Zeotrope - A liquid mixture that shows no maximum or minimum when vapor pressure is plotted against composition at constant temperature. See Azeotrope.

Zero-point energy - The energy possessed by a quantum mechanical system as a result of the uncertainty principle even when it is in its lowest energy state; e.g., the difference between the lowest energy level of a harmonic oscillator and the minimum in the potential well.

Zeta potential (ζ) - The electric potential at the surface of a colloidal particle relative to the potential in the bulk medium at a long distance. Also called electrokinetic potential.

Zwitterions - Neutral compounds having formal unit electrical charges of opposite sign. Some chemists restrict the term to compounds with the charges on non-adjacent atoms. Sometimes referred to as inner salts, dipolar ions (a misnomer). [5]

Section 3
Physical Constants of Organic Compounds

Physical Constants of Organic Compounds 3-1
Structure Diagrams for Table of Physical Constants 3-331
Synonym Index of Organic Compounds 3-586
Molecular Formula Index of Organic Compounds 3-631
CAS Registry Number Index of Organic Compounds 3-709
Melting Point Index of Organic Compounds 3-737
Boiling Point Index of Organic Compounds 3-751
Diamagnetic Susceptibility of Selected Organic Compounds 3-764

PHYSICAL CONSTANTS OF ORGANIC COMPOUNDS

The basic physical constants for over 12,000 organic compounds are presented in this table, along with structures and references to other sources of information. An effort has been made to include the compounds most frequently encountered in the laboratory, the workplace, and the environment. The selection was based mainly on the appearance of the compounds in various specialized tables in this *Handbook* and in other widely used reference sources, such as the *Merck Index* and the *DIPPR Database of Pure Compound Properties*. The occurrence of a compound on regulatory lists of hazardous chemicals was also taken into consideration, as was the availability of reliable physical constant data. Clearly, criteria of this type are somewhat subjective, and compounds considered important by some users have undoubtedly been omitted. Suggestions for additional compounds or other improvements are welcomed.

The data in the table have been taken from many sources, including both compilations and the primary literature. Where conflicts were found, the value deemed most reliable was chosen. Some of the useful compilations of physical property data are listed at the end of this introduction.

The table is arranged alphabetically by the primary name, which is generally the Index Name from the 8th or 9th Collective Index of Chemical Abstracts Service (CAS). In a few cases, especially pesticides and pharmaceuticals, the common name is used rather than the more complex systematic name. By convention, CAS Index Names are written in inverted order, e.g., chloromethane is listed as methane, chloro and ethyl acetate as acetic acid, ethyl ester. Furthermore, certain important compounds are listed under Index Names which differ from the names by which they are commonly known (e.g. aniline appears as benzenamine and acetone as 2-propanone). In order to facilitate the location of compounds in the table, three indexes are provided:

- **Synonym Index:** Includes common synonyms, but not the primary name by which the table is arranged.
- **Molecular Formula Index:** Lists compounds by molecular formula in the Hill order (see Preface to this *Handbook*).
- **CAS Registry Number Index:** Lists compounds by Chemical Abstracts Service Registry Number.

Two lines of data appear for each compound. The explanation of the data fields follows.

Top Line:
- **No.:** An identification number used in the indexes and to identify the structure diagrams.
- **Name:** Primary name, generally the CAS Index Name.
- **Mol. Form.:** The molecular formula written in the Hill convention.
- **CAS RN:** The Chemical Abstracts Service Registry Number assigned by CAS as a unique identifier for the compound.
- **Merck No:** Monograph Number in *The Merck Index, Eleventh Edition*. It should be noted that this is not a unique identifier for a single compound, since several derivatives or isomers of a compound may be included in the same Monograph.
- **Beil. Ref:** Citation to the *Beilstein Handbook of Organic Chemistry*. An entry of 5-18-11-01234, for example, indicates that the compound may be found in the 5th Series, Volume 18, Subvolume 11, page 1234.
- **Solubility:** Solubility in common solvents on a relative scale: 1 = insoluble; 2 = slightly soluble; 3 = soluble; 4 = very soluble; 5 = miscible; 6 = decomposes. See List of Abbreviations for the solvent abbreviations.

Bottom line:
- **Synonym:** A synonym in common use. When the primary name is non-systematic, the systematic name appears here.
- **Mol. Wt.:** Molecular weight (relative molar mass) as calculated with the 1991 IUPAC Standard Atomic Weights.
- **mp/°C:** Normal melting point in °C. Although some values are quoted to 0.1°C, uncertainties are typically several degrees Celsius. A value is sometimes followed by "dec", indicating decomposition is observed at the stated temperature (so that it is probably not a true melting point). See the List of Abbreviations for other abbreviations.
- **bp/°C:** Boiling point in °C. When available, the normal boiling point is given first, without a superscript. This is the temperature at which the liquid phase is in equilibrium with the vapor at a pressure of 760 mmHg (101.325 kPa). Boiling point values at reduced pressure are also given in many cases; here the superscript indicates the pressure in mmHg. A "dec" or "exp" following the value indicates decomposition or explosion has been observed at the boiling point. A simple entry of "exp" (sometimes followed by a temperature) indicates explosion may occur on heating, even below the boiling point. An entry of "sub" indicates that no boiling point is available, but measurable vapor (sublimation) pressure has been observed upon heating the solid. A temperature may be given, but no precise meaning can be attached because the pressure is not specified.
- **den/g cm^{-3}:** density (mass per unit volume) in g/cm^3. The superscript indicates the temperature in °C. Values are given only for the liquid and solid phases, and all values are true densities, not specific gravities. The number of decimal places gives a rough estimate of the accuracy of the value.
- **n_D:** Refractive index, at the temperature indicated by the superscript. Unless otherwise indicated, all values refer to a wavelength of 589 nm (sodium D line). Values are given only for liquids and solids.

Structures are given, when available, in the section following the main table, using the **No.** in the first column as the linking identifier.

PHYSICAL CONSTANTS OF ORGANIC COMPOUNDS (continued)

LIST OF ABBREVIATIONS

Ac	Acetyl		EtOH	Ethanol
ace	Acetone		exp	Explodes
acid	Acids		gl	Glacial
AcOEt	Ethyl acetate		HOAc	Acetic acid
alk	Alkalis		hp	Heptane
anh	Anhydrous		hx	Hexane
aq	Aqueous		lig	Ligroin
bz	Benzene		MeOH	Methanol
chl	Chloroform		os	Organic solvents
con	Concentrated		peth	Petroleum ether
ctc	Carbon tetrachloride		$PhNO_2$	Nitrobenzene
cyhex	Cyclohexane		PrOH	1-Propanol
dec	Decomposes		py	Pyridine
dil	Dilute		sub	Sublimes
diox	Dioxane		sulf	Sulfuric acid
DMF	Dimethyl formamide		temp.	Temperature
DMSO	Dimethyl sulfoxide		tfa	Trifluoroacetic acid
DMSO-d_6	Deuterated dimethyl sulfoxide		tol	Toluene
eth	Ethyl ether		xyl	Xylene

REFERENCES

1. Lide, D. R., and Milne, G. W. A., Editors, *Handbook of Data on Organic Compounds, Third Edition*, CRC Press, Boca Raton, FL, 1993.
2. Luckenbach, R., Editor, *Beilstein Handbook of Organic Chemistry*, Springer-Verlag, Heidelberg. Data can also be accessed through the on-line Beilstein Database.
3. Budavari, S., Editor, *The Merck Index, Eleventh Edition*, Merck & Co., Rahway, NJ, 1989.
4. Daubert, T. E., Danner, R. P., Sibul, H. M., and Stebbins, C. C., *Physical and Thermodynamic Properties of Pure Compounds: Data Compilation*, extant 1994 (core with 4 supplements), Taylor & Francis, Bristol, PA (also available as database).
5. *TRC Thermodynamic Tables*, Thermodynamic Research Center, Texas A & M University, College Station, TX.
6. Riddick, J. A., Bunger, W. B., and Sakano, T. K., *Organic Solvents, Fourth Edition*, John Wiley & Sons, New York, 1986.
7. *Physical Constants of Hydrocarbon and Non-Hydrocarbon Compounds*, ASTM Data Series DS 4B, ASTM, Philadelphia, 1988.
8. Stevenson, R. M., and Malanowski, S., *Handbook of the Thermodynamics of Organic Compounds*, Elsevier, New York, 1987.
9. Buckingham, J., Editor, *Dictionary of Organic Compounds, Fifth Edition* (and Supplements), Chapman and Hall, New York, 1982-1993.
10. Lide, D. R., and Kehiaian, H. V., *Handbook of Thermophysical and Thermochemical Data*, CRC Press, Boca Raton, FL, 1994.

No.	Name / Synonym	Mol. Form. / Mol. Wt. / mp/°C	CAS RN / bp/°C	Merck No.	Beil. Ref. / den/g cm^{-3}	Solubility / n_D
1	Abate / Phosphorothioic acid, O,O'-(thiodi-4,1-phenylene) O,O,O',O'-tetramethyl ester	$C_{16}H_{20}O_6P_2S_3$ / 466.48 / 30	3383-96-8	9075	1.32	
2	1-Acenaphthylenamine, 1,2-dihydro- / 1-Acenaphthenamine	$C_{12}H_{11}N$ / 169.23 / 135	40745-44-6 / sub		2-12-00-00764	bz 4; EtOH 4
3	Acenaphthylene / Acenaphthalene	$C_{12}H_8$ / 152.20 / 92.5	208-96-8 / 280; 150[28]		4-05-00-02138 / 0.8987[16]	H_2O 1; EtOH 4; eth 4; bz 4
4	Acenaphthylene, 5-bromo-1,2-dihydro- / 5-Bromoacenaphthene	$C_{12}H_9Br$ / 233.11 / 52	2051-98-1 / 335		4-05-00-01839 / 1.4392[52]	H_2O 1; EtOH 2; chl 2 / 1.6565[54]
5	Acenaphthylene, 5-chloro-1,2-dihydro- / 5-Chloroacenaphthene	$C_{12}H_9Cl$ / 188.66 / 70.5	5209-33-6 / 319; 163[13]		4-05-00-01837 / 1.1954[70]	1.6169[100]
6	Acenaphthylene, 1,2-dihydro- / Acenaphthene	$C_{12}H_{10}$ / 154.21 / 93.4	83-32-9 / 279	23	4-05-00-01834 / 1.222[20]	H_2O 1; EtOH 2; bz 4; chl 2 / 1.6048[95]
7	Acenaphthylene, 1,2-dihydro-5-iodo- / 5-Iodoacenaphthene	$C_{12}H_9I$ / 280.11 / 64.3	6861-64-9		2-05-00-00497 / 1.5000[20]	bz 4; EtOH 4 / 1.6909[65]
8	1,2-Acenaphthylenedione	$C_{12}H_6O_2$ / 182.18 / 261	82-86-0 / sub		4-07-00-02498 / 1.4800[20]	H_2O 1; EtOH 2; bz 2; lig 3
9	Acenaphthylene, 1,2,2a,3,4,5-hexahydro-	$C_{12}H_{14}$ / 158.24 / 12.8	480-72-8 / 249		4-05-00-01566 / 1.0290[20]	chl 3
10	Acephate / Phosphoramidothioic acid, acetyl-, O,S-dimethyl ester	$C_4H_{10}NO_3PS$ / 183.17 / 88	30560-19-1		1.35[20]	
11	Acetaldehyde / Ethanal	C_2H_4O / 44.05 / -123	75-07-0 / 20.1	32	4-01-00-03094 / 0.7834[18]	H_2O 5; EtOH 5; eth 5; bz 5 / 1.3316[20]
12	Acetaldehyde, chloro- / Chloroacetaldehyde	C_2H_3ClO / 78.50 / -16.3	107-20-0 / 85.5	2108	4-01-00-03134	eth 3
13	Acetaldehyde, chloro-, trimer / Chloroacetaldehyde, trimer	$C_6H_9Cl_3O_3$ / 235.49 / 87	65438-35-9 / 142[10]		4-19-00-04718	eth 4
14	Acetaldehyde, ethoxy- / α-Ethoxyacetaldehyde	$C_4H_8O_2$ / 88.11	22056-82-2		3-01-00-03181 / 0.942[20]	H_2O 4; ace 4; EtOH 4 / 1.3956[20]
15	Acetaldehyde, hydroxy- / Glycolaldehyde	$C_2H_4O_2$ / 60.05 / 97	141-46-8		4-01-00-03955 / 1.366[100]	chl 3 / 1.4772[19]
16	Acetaldehyde, methoxy- / Methoxyacetaldehyde	$C_3H_6O_2$ / 74.08 / 92	10312-83-1		4-01-00-03955 / 1.005[25]	H_2O 4; ace 4; eth 4; EtOH 4 / 1.3950[20]
17	Acetaldehyde, oxime / Acetaldoxime	C_2H_5NO / 59.07 / 45	107-29-9 / 115	35	4-01-00-03121 / 0.9656[20]	H_2O 3; EtOH 5; eth 5; chl 3 / 1.4264[20]
18	Acetaldehyde, phenylhydrazone	$C_8H_{10}N_2$ / 134.18 / 99.5	935-07-9 / 150[40]; 135[21]		3-15-00-00079	EtOH 4
19	Acetaldehyde, tetramer / Metaldehyde II (tetramer)	$C_4H_{12}O_4$ / 124.14 / 47	66056-06-2 / 110		4-19-00-05643	ace 4; bz 4; eth 4; EtOH 4
20	Acetaldehyde, tribromo- / Bromal	C_2HBr_3O / 280.74	115-17-3 / 174	1372	4-01-00-03155 / 2.6649[25]	ace 4; eth 4; EtOH 4 / 1.5939[20]
21	Acetaldehyde, trichloro- / Chloral	C_2HCl_3O / 147.39 / -57.5	75-87-6 / 97.8	9538	4-01-00-03142 / 1.512[20]	H_2O 4; EtOH 3; eth 3 / 1.4580[20]
22	Acetamide / Ethanamide	C_2H_5NO / 59.07 / 81	60-35-5 / 222.0	36	4-02-00-00399 / 0.9986[85]	H_2O 4; EtOH 4 / 1.4278
23	Acetamide, N-acetyl-	$C_4H_7NO_2$ / 101.11 / 79	625-77-4 / 223.5		4-02-00-00416	H_2O 3; EtOH 3; eth 3; chl 3
24	Acetamide, N-acetyl-N-ethyl-	$C_6H_{11}NO_2$ / 129.16	1563-83-3 / 197		4-04-00-00352 / 1.0092[20]	EtOH 4 / 1.4513[20]
25	Acetamide, N-acetyl-N-methyl-	$C_5H_9NO_2$ / 115.13 / -25	1113-68-4 / 195; 114.5[61]		4-04-00-00183 / 1.0663[25]	H_2O 5; eth 1 / 1.4502[25]
26	Acetamide, N-acetyl-N-phenyl-	$C_{10}H_{11}NO_2$ / 177.20 / 37.5	1563-87-7 / 200[100]; 142[11]		4-12-00-00385	H_2O 2; EtOH 3; bz 3; tol 3
27	Acetamide, N-(aminocarbonyl)-	$C_3H_6N_2O_2$ / 102.09 / 218	591-07-1 / sub 180		3-03-00-00119	H_2O 2; EtOH 3; eth 2
28	Acetamide, N-(2-aminoethyl)-	$C_4H_{10}N_2O$ / 102.14 / 51	1001-53-2 / 136[27]		4-04-00-01193	H_2O 3; EtOH 3; eth 1; bz 3
29	Acetamide, N-(4-aminophenyl)- / p-Aminoacetanilide	$C_8H_{10}N_2O$ / 150.18 / 166.5	122-80-5 / 267	422	4-13-00-00137	H_2O 3; EtOH 4; eth 4
30	Acetamide, N-[(4-aminophenyl)sulfonyl]- / Sulfacetamide	$C_8H_{10}N_2O_3S$ / 214.25 / 183	144-80-9	8870	4-14-00-02662	H_2O 2; EtOH 3; eth 1; ace 4
31	Acetamide, N-[4-(aminosulfonyl)phenyl]- / Acetylsulfanilamide	$C_8H_{10}N_2O_3S$ / 214.25 / 219.5	121-61-9	97	4-14-00-02704	H_2O 3; EtOH 3; ace 3
32	Acetamide, N-[5-(aminosulfonyl)-1,3,4-thiadiazol-2-yl]- / Acetazolamide	$C_4H_6N_4O_3S_2$ / 222.25 / 260.5	59-66-5	45	4-27-00-08219	H_2O 2
33	Acetamide, N-(aminothioxomethyl)- / Acetylthiourea	$C_3H_6N_2OS$ / 118.16 / 165	591-08-2		4-03-00-00354	H_2O 2; EtOH 3; eth 2
34	Acetamide, N-[4-[[(aminothioxomethyl)hydrazono]methyl]phenyl]- / Thiacetazone	$C_{10}H_{12}N_4OS$ / 236.30 / 225 dec	104-06-3	9218	4-14-00-00075	H_2O 1; os 1; CS_2 1
35	Acetamide, N-[1,1'-biphenyl]-2-yl-	$C_{14}H_{13}NO$ / 211.26 / 121	2113-47-5 / 355		4-12-00-03224	H_2O 1; EtOH 4; eth 4
36	Acetamide, N-bromo- / N-Bromoacetamide	C_2H_4BrNO / 137.96 / 103.5	79-15-2	1386	4-02-00-00417	eth 4

No.	Name / Synonym	Mol. Form. / Mol. Wt.	CAS RN / mp/°C	Merck No. / bp/°C	Beil. Ref. / den/g cm⁻³	Solubility / n_D
37	Acetamide, N-(4-bromophenyl)- / p-Bromoacetanilide	C_8H_8BrNO / 214.06	103-88-8 / 168	1387	4-12-00-01504 / 1.717[25]	H_2O 1; EtOH 3; eth 2; bz 2
38	Acetamide, N-butyl-	$C_6H_{13}NO$ / 115.18	1119-49-9 / / 229		4-04-00-00565 / 0.8960[25]	1.4388[25]
39	Acetamide, N-butyl-N-phenyl-	$C_{12}H_{17}NO$ / 191.27	91-49-6 / 24.5 / 281		4-12-00-00379 / 0.9912[20]	chl 2 / 1.5146[20]
40	Acetamide, N,N'-carbonylbis-	$C_5H_8N_2O_3$ / 144.13	638-20-0 / 154.5 / sub		3-03-00-00121	H_2O 2; EtOH 2; eth 3; ace 3
41	Acetamide, 2-chloro- / Chloroacetamide	C_2H_4ClNO / 93.51	79-07-2 / 121 / 225	2109	4-02-00-00490	H_2O 3; EtOH 4; eth 2
42	Acetamide, 2-chloro-N,N-bis(1-methylethyl)-	$C_8H_{16}ClNO$ / 177.67	7403-66-9 / 48.5 / 163[55]; 86[2.7]		4-04-00-00516	1.4619[25]
43	Acetamide, 2-chloro-N-3-chloroallyl- / Chloroacetamide, N-3-chloroallyl	$C_5H_7Cl_2NO$ / 168.02	100130-25-4 / 52.5 / 112[0.5]		4-04-00-01087	peth 4
44	Acetamide, 2-chloro-N,N-di-2-propenyl- / Allidochlor	$C_8H_{12}ClNO$ / 173.64	93-71-0 / 250 / 116[1]; 92[0.7]		4-04-00-01064	H_2O 2; EtOH 3 / 1.4932[25]
45	Acetamide, 2-chloro-N-hexyl- / α-Chloro-N-hexylacetamide	$C_8H_{16}ClNO$ / 177.67	5326-81-8 / 108.5 / 100[0.2]		4-04-00-00713	peth 4
46	Acetamide, 2-chloro-N-(3-methoxypropyl)- / α-Chloro-N-(3-methoxypropyl)acetamide	$C_6H_{12}ClNO_2$ / 165.62	1709-03-1 / 30 / 88[0.5]		4-04-00-01645	1.4712[25]
47	Acetamide, 2-chloro-N-(1-methylpropyl)-	$C_6H_{12}ClNO$ / 149.62	32322-73-9 / 45.5 / 68[0.7]		4-04-00-00621	peth 4
48	Acetamide, N-(4-chlorophenyl)-	C_8H_8ClNO / 169.61	539-03-7 / 179 / 333		4-12-00-01178 / 1.385[22]	H_2O 1; EtOH 3; eth 4; ctc 2
49	Acetamide, 2-cyano- / Cyanoacetamide	$C_3H_4N_2O$ / 84.08	107-91-5 / 121.5	2695	4-02-00-01891	H_2O 4
50	Acetamide, N,N-diacetyl- / Triacetamide	$C_6H_9NO_3$ / 143.14	641-06-5 / 79		4-02-00-00416	eth 4
51	Acetamide, 2,2-dichloro-	$C_2H_3Cl_2NO$ / 127.96	683-72-7 / 99.4 / 234		4-02-00-00505	H_2O 3; EtOH 3; eth 3; ace 2
52	Acetamide, 2,2-dichloro-N-[2-hydroxy-1-(hydroxymethyl)-2-(4-nitrophenyl)ethyl]-, [R-(R*,R*)]- / Chloramphenicol	$C_{11}H_{12}Cl_2N_2O_5$ / 323.13	56-75-7 / 150.5 / sub	2068	3-13-00-02268	ace 4; EtOH 4; chl 4
53	Acetamide, N,N-diethyl-	$C_6H_{13}NO$ / 115.18	685-91-6 / / 185.5		4-04-00-00349 / 0.9130[17]	H_2O 3; EtOH 3; eth 5; ace 5 / 1.4374[17]
54	Acetamide, 2-(diethylamino)-N-(2,6-dimethylphenyl)- / Lidocaine	$C_{14}H_{22}N_2O$ / 234.34	137-58-6 / 68.5 / 180-2[4]	5359	4-12-00-02538	bz 4; eth 4; EtOH 4; chl 4
55	Acetamide, N,N-dimethyl- / N,N-Dimethylethanamide	C_4H_9NO / 87.12	127-19-5 / -18.59 / 165	3216	4-04-00-00180 / 0.9366[25]	H_2O 5; EtOH 5; eth 5; ace 5 / 1.4380[20]
56	Acetamide, N-(2,4-dimethylphenyl)-	$C_{10}H_{13}NO$ / 163.22	2050-43-3 / 129.3 / 170[10]		4-12-00-02548	EtOH 4; chl 4
57	Acetamide, N,N-diphenyl- / Diphenylacetamide	$C_{14}H_{13}NO$ / 211.26	519-87-9 / 103 / sub	3315	4-12-00-00381	H_2O 2; EtOH 3; eth 2; chl 2
58	Acetamide, N,N-dipropyl-	$C_8H_{17}NO$ / 143.23	1116-24-1 / / 209.5		4-04-00-00476 / 0.8992[17]	EtOH 4 / 1.4419[17]
59	Acetamide, N-(2-ethoxyphenyl)-	$C_{10}H_{13}NO_2$ / 179.22	581-08-8 / 79 / >240		3-13-00-00779	H_2O 1; EtOH 3; eth 3; chl 3
60	Acetamide, N-(4-ethoxyphenyl)- / Phenacetin	$C_{10}H_{13}NO_2$ / 179.22	62-44-2 / 137.5	7155	4-13-00-01092	H_2O 2; EtOH 3; eth 2; ace 3 / 1.571
61	Acetamide, N-ethyl- / N-Ethylacetamide	C_4H_9NO / 87.12	625-50-3 / / 205; 104[18]		4-04-00-00347 / 0.942[4]	H_2O 5; EtOH 5; chl 3; HOAc 3 / 1.4338[20]
62	Acetamide, N-ethyl-N-phenyl-	$C_{10}H_{13}NO$ / 163.22	529-65-7 / 55 / 260		4-12-00-00379 / 0.9938[60]	H_2O 3; eth 3; ctc 3
63	Acetamide, 2-fluoro- / Fluoroacetic acid amide	C_2H_4FNO / 77.06	640-19-7 / 108 / sub	4095	4-02-00-00454	H_2O 3; ace 3; chl 2
64	Acetamide, N-9H-fluoren-2-yl- / 2-(Acetylamino)fluorene	$C_{15}H_{13}NO$ / 223.27	53-96-3 / 193	4083	4-12-00-03373	H_2O 1; EtOH 3; eth 3; HOAc 3
65	Acetamide, N-(2-hydroxyethyl)-	$C_4H_9NO_2$ / 103.12	142-26-7 / 63.5 / 166-7[8]		4-04-00-01535 / 1.1079[25]	H_2O 5; ace 3; bz 2; lig 2 / 1.4674[20]
66	Acetamide, N-(2-hydroxy-1-naphthalenyl)-	$C_{12}H_{11}NO_2$ / 201.22	117-93-1 / 235 dec / sub		4-13-00-02095	ace 4; bz 4; eth 4; EtOH 4
67	Acetamide, N-(4-hydroxyphenyl)- / Acetaminophen	$C_8H_9NO_2$ / 151.17	103-90-2 / 170	40	4-13-00-01091 / 1.293[21]	H_2O 1; EtOH 4
68	Acetamide, 2-mercapto-N-2-naphthalenyl- / Thionalide	$C_{12}H_{11}NOS$ / 217.29	93-42-5 / 111.5	9273	3-12-00-03052	H_2O 1; EtOH 4; os 4
69	Acetamide, N-[2-(5-methoxy-1H-indol-3-yl)ethyl]- / Melatonin	$C_{13}H_{16}N_2O_2$ / 232.28	73-31-4 / 117	5695	5-22-12-00042	
70	Acetamide, N-(2-methoxyphenyl)-	$C_9H_{11}NO_2$ / 165.19	93-26-5 / 87.5 / 304		4-13-00-00833	H_2O 4; EtOH 4; eth 3; ace 3
71	Acetamide, N-(4-methoxyphenyl)- / p-Acetanisidine	$C_9H_{11}NO_2$ / 165.19	51-66-1 / 131	43	4-13-00-01092	ace 4; EtOH 4; chl 4
72	Acetamide, N-methyl-	C_3H_7NO / 73.09	79-16-3 / 28 / 205		4-04-00-00176 / 0.9371[25]	ace 4; bz 4; eth 4; EtOH 4 / 1.4301[20]
73	Acetamide, N-[(methylamino)carbonyl]-	$C_4H_8N_2O_2$ / 116.12	623-59-6 / 180.5 / dec		4-04-00-00207	H_2O 3; EtOH 2; eth 2; chl 3

No.	Name / Synonym	Mol. Form. / Mol. Wt.	CAS RN / mp/°C	Merck No. / bp/°C	Beil. Ref. / den/g cm^{-3}	Solubility / n_D
74	Acetamide, N-(1-methylethyl)-N-phenyl- / N-Isopropylacetanilide	$C_{11}H_{15}NO$ / 177.25	5461-51-8 / 40.5	264	3-12-00-00466	lig 4
75	Acetamide, N-methyl-N-(2-methylphenyl)-	$C_{10}H_{13}NO$ / 163.22	573-26-2 / 55.5	260	3-12-00-01854	EtOH 3; chl 3
76	Acetamide, N-(2-methylphenyl)-	$C_9H_{11}NO$ / 149.19	120-66-1 / 110	296	4-12-00-01755 / 1.168^{15}	H_2O 2; EtOH 3; eth 3; ace 3
77	Acetamide, N-(3-methylphenyl)-	$C_9H_{11}NO$ / 149.19	537-92-8 / 65.5	303	4-12-00-01823 / 1.141^{15}	H_2O 2; EtOH 4; eth 4; chl 3
78	Acetamide, N-(4-methylphenyl)-	$C_9H_{11}NO$ / 149.19	103-89-9 / 152	307	4-12-00-01902 / 1.2120^{15}	eth 4; EtOH 4
79	Acetamide, N-methyl-N-phenyl- / N-Methylacetanilide	$C_9H_{11}NO$ / 149.19	579-10-2 / 103	5931 / 256	4-12-00-00378 / 1.0036^{105}	H_2O 3; EtOH 3; eth 3; chl 3 / 1.576
80	Acetamide, 2-nitro-N-phenyl- / α-Nitroacetanilide	$C_8H_8N_2O_3$ / 180.16	10151-95-8 / 94	$100^{0.1}$	1-12-00-00193 / 1.419^{15}	bz 4; EtOH 4; lig 4
81	Acetamide, N-(2-nitrophenyl)-	$C_8H_8N_2O_3$ / 180.16	552-32-9 / 94	$100^{0.1}$	4-12-00-01574 / 1.419^{15}	H_2O 3; EtOH 3; eth 4; bz 3
82	Acetamide, N-(3-nitrophenyl)-	$C_8H_8N_2O_3$ / 180.16	122-28-1 / 155	$100^{0.0074}$	4-12-00-01595	H_2O 3; EtOH 3; eth 1; chl 3
83	Acetamide, N-(4-nitrophenyl)-	$C_8H_8N_2O_3$ / 180.16	104-04-1 / 216	$100^{0.008}$	4-12-00-01632	H_2O 2; EtOH 3; eth 2; chl 2
84	Acetamide, N-(5-nitro-2-propoxyphenyl)- / 5'-Nitro-2'-propoxyacetanilide	$C_{11}H_{14}N_2O_4$ / 238.24	553-20-8 / 102.5	6550	3-13-00-00881	
85	Acetamide, N-(5-nitro-2-thiazolyl)- / Aminitrozole	$C_5H_5N_3O_3S$ / 187.18	140-40-9 / 264.5	421	4-27-00-04676	alk 3
86	Acetamide, N-phenyl- / Acetanilide	C_8H_9NO / 135.17	103-84-4 / 114.3	42 / 304	4-12-00-00373 / 1.2190^{15}	H_2O 2; EtOH 4; eth 3; ace 4
87	Acetamide, N-(phenylmethyl)-	$C_9H_{11}NO$ / 149.19	588-46-5 / 61	157^2	4-12-00-02230	EtOH 4; eth 4
88	Acetamide, N-phenyl-N-propyl- / N-Propylacetanilide	$C_{11}H_{15}NO$ / 177.25	2437-98-1 / 49	268	4-12-00-00379	eth 4; EtOH 4; lig 4
89	Acetamide, N-3-pyridinyl-	$C_7H_8N_2O$ / 136.15	5867-45-8 / 133	326.5	5-22-09-00017	H_2O 3; EtOH 3; eth 2; bz 2
90	Acetamide, N-4-quinolinyl-	$C_{11}H_{10}N_2O$ / 186.21	32433-28-6 / 176	sub	5-22-10-00250	EtOH 4
91	Acetamide, N,N'-(sulfonyldi-4,1-phenylene)bis- / Acedapsone	$C_{16}H_{16}N_2O_4S$ / 332.38	77-46-3 / 290	16	4-13-00-01361	EtOH 4
92	Acetamide, N-(5,6,7,9-tetrahydro-10-hydroxy-1,2,3-trimethoxy-9-oxobenzo[a]heptalen-7-yl)-, (S)- / Colchiceine	$C_{21}H_{23}NO_6$ / 385.42	477-27-0 / 178.5	2469	4-14-00-00943 / 1.24^{25}	H_2O 2; EtOH 4; eth 1; bz 1
93	Acetamide, N-(5,6,7,9-tetrahydro-1,2,3,10-tetramethoxy-9-oxobenzo[a]heptalen-7-yl)-, (S)- / Colchicine	$C_{22}H_{25}NO_6$ / 399.44	64-86-8 / 156	2470	4-14-00-00946	H_2O 4; EtOH 4
94	Acetamide, 2,2,2-tribromo- / Tribromoacetamide	$C_2H_2Br_3NO$ / 295.76	594-47-8 / 121.5	sub	3-02-00-00485	H_2O 4; eth 4; EtOH 4
95	Acetamide, 2,2,2-trichloro-	$C_2H_2Cl_3NO$ / 162.40	594-65-0 / 142	240	4-02-00-00520	H_2O 2; EtOH 4; eth 4
96	Acetamide, 2,2,2-trichloro-N,N-diethyl-	$C_6H_{10}Cl_3NO$ / 218.51	2430-00-4 / 27	112^{12}	4-04-00-00351	1.4900^{24}
97	Acetamide, 2,2,2-trichloro-N,N-dimethyl-	$C_4H_6Cl_3NO$ / 190.46	7291-33-0 / 12	231 dec	4-04-00-00182 / 1.390^{20}	bz 4; chl 4 / 1.5017^{25}
98	Acetamide, 2,2,2-trichloro-N-phenyl-	$C_8H_6Cl_3NO$ / 238.50	2563-97-5 / 96	169^{19}	4-12-00-00377	EtOH 4
99	Acetamide, 2,2,2-trifluoro-N,N-dimethyl-	$C_4H_6F_3NO$ / 141.09	1547-87-1 /	133	4-04-00-00182 / 1.232^{21}	1.3611^{25}
100	Acetic acid / Ethanoic acid	$C_2H_4O_2$ / 60.05	64-19-7 / 16.6	47 / 117.9	4-02-00-00094 / 1.0446^{25}	H_2O 5; EtOH 5; eth 5; ace 5 / 1.3720^{20}
101	Acetic acid, (acetyloxy)-	$C_4H_6O_4$ / 118.09	13831-30-6 / 67.5	145^{12}	4-03-00-00576	H_2O 4; EtOH 4; eth 4; ace 4
102	Acetic acid, (acetyloxy)-, ethyl ester	$C_6H_{10}O_4$ / 146.14	623-86-9 /	179	4-03-00-00584 / 1.0880^{20}	eth 4; EtOH 4; HOAc 4 / 1.4112^{20}
103	Acetic acid, [2-(aminocarbonyl)phenoxy]- / Salicylamide O-acetic acid	$C_9H_9NO_4$ / 195.17	25395-22-6	8298	4-10-00-00184	
104	Acetic acid, aminooxo- / Oxamic acid	$C_2H_3NO_3$ / 89.05	471-47-6 / 210 dec	6870	4-02-00-01857	H_2O 2; EtOH 1; eth 1
105	Acetic acid, aminooxo-, hydrazide / Semioxamazide	$C_2H_5N_3O_2$ / 103.08	515-96-8 / 221 dec	8397	4-02-00-01868	H_2O 2; EtOH 1; eth 1; alk 4
106	Acetic acid, ammonium salt / Ammonium, acetate	$C_2H_7NO_2$ / 77.08	631-61-8 / 114	514 / dec	1.17^{20}	H_2O 4; EtOH 3; ace 2
107	Acetic acid, anhydride / Acetic anhydride	$C_4H_6O_3$ / 102.09	108-24-7 / -73	48 / 139.5	4-02-00-00386 / 1.082^{20}	H_2O 4; EtOH 3; eth 5; bz 3 / 1.3901^{20}
108	Acetic acid, anhydride with nitric acid / Acetyl nitrate	$C_2H_3NO_4$ / 105.05	591-09-3 /	92 / 22^{70}; exp 60	4-02-00-00435 / 1.24^{15}	
109	Acetic acid, azido-	$C_2H_3N_3O_2$ / 101.06	18523-48-3 / 16	116^{12}; 92^3	4-02-00-00541 / 1.354^{33}	
110	Acetic acid, barium salt	$C_4H_6BaO_4$ / 255.42	543-80-6 / 975		4-02-00-00094 / 2.468^{25}	H_2O 3

No.	Name / Synonym	Mol. Form. / Mol. Wt.	CAS RN / mp/°C	Merck No. / bp/°C	Bell. Ref. / den/g cm^{-3}	Solubility / n_D
111	Acetic acid, bromo- / Bromoacetic acid	$C_2H_3BrO_2$ / 138.95	79-08-3 / 50	1388 / 208	4-02-00-00526 / 1.9335[50]	H_2O 5; EtOH 5; eth 5; ace 3 / 1.4804[50]
112	Acetic acid, bromochloro- / Bromochloroacetic acid	$C_2H_2BrClO_2$ / 173.39	5589-96-8 / 31.5	/ 215; 104[11]	4-02-00-00532 / 1.9848[31]	
113	Acetic acid, bromochloro-, ethyl ester / Ethyl bromochloroacetate	$C_4H_6BrClO_2$ / 201.45	22524-32-9 /	/ 174 dec	4-02-00-00532 / 1.5890[22]	eth 4; EtOH 4 / 1.4639[24]
114	Acetic acid, bromo-, 1,1-dimethylethyl ester	$C_6H_{11}BrO_2$ / 195.06	5292-43-3 /	/ 73-4[25]	3-02-00-00482 /	eth 4; EtOH 4 / 1.4430[20]
115	Acetic acid, bromo-, ethyl ester / Ethyl bromacetate	$C_4H_7BrO_2$ / 167.00	105-36-2 /	/ 168.5	4-02-00-00527 / 1.5032[20]	H_2O 1; EtOH 5; eth 5; ace 5 / 1.4489[20]
116	Acetic acid, bromofluoro- / Bromofluoroacetic acid	$C_2H_2BrFO_2$ / 156.94	359-25-1 / 50	/ 183	4-02-00-00531 /	H_2O 4; EtOH 4; chl 4
117	Acetic acid, bromo-, methyl ester / Methyl 2-bromoacetate	$C_3H_5BrO_2$ / 152.98	96-32-2 /	/ 132	4-02-00-00527 / 1.6350[20]	H_2O 1; EtOH 3; eth 3; ace 3 / 1.4520[20]
118	Acetic acid, bromo-, 2-methylpropyl ester	$C_6H_{11}BrO_2$ / 195.06	59956-48-8 /	/ 188; 74.5[10]	1-02-00-00096 / 1.3269[20]	ace 4; eth 4; EtOH 4
119	Acetic acid, bromo-, phenyl ester	$C_8H_7BrO_2$ / 215.05	620-72-4 / 32	/ 140[20]	3-06-00-00599 /	eth 4; EtOH 4
120	Acetic acid, bromo-, propyl ester / Propyl bromoacetate	$C_5H_9BrO_2$ / 181.03	35223-80-4 /	/ 176	4-02-00-00528 / 1.4099[20]	ace 4; eth 4; EtOH 4 / 1.4518[20]
121	Acetic acid, butyl ester / Butyl acetate	$C_6H_{12}O_2$ / 116.16	123-86-4 / -78	1535 / 126.1	4-02-00-00143 / 0.8825[20]	H_2O 2; EtOH 5; eth 5; ace 3 / 1.3941[20]
122	Acetic acid, calcium salt	$C_4H_6CaO_4$ / 158.17	62-54-4 / dec	1643 /	4-02-00-00094 /	H_2O 3; EtOH 2 / 1.55
123	Acetic acid, chloro- / Chloroacetic acid (β)	$C_2H_3ClO_2$ / 94.50	79-11-8 / 63	2111 / 189.3	4-02-00-00474 / 1.4043[40]	H_2O 4; EtOH 3; eth 3; bz 3 / 1.4351[55]
124	Acetic acid, chloro-, anhydride / Chloroacetic anhydride	$C_4H_4Cl_2O_3$ / 170.98	541-88-8 / 46	2112 / 203	4-02-00-00487 / 1.5497[20]	
125	Acetic acid, chloro-, butyl ester	$C_6H_{11}ClO_2$ / 150.60	590-02-3 /	/ 183	4-02-00-00482 / 1.0704[20]	eth 4; EtOH 4 / 1.4297[20]
126	Acetic acid, chloro-, 2-chloroethyl ester	$C_4H_6Cl_2O_2$ / 157.00	3848-12-2 /	/ 203; 102[20]	4-02-00-00481 / 1.3600[25]	eth 4 / 1.4619[25]
127	Acetic acid, chlorodifluoro- / Chlorodifluoroacetic acid	$C_2HClF_2O_2$ / 130.48	76-04-0 / 25	/ 122	4-02-00-00497 /	chl 3 / 1.3559[20]
128	Acetic acid, chloro-, 1,1-dimethylethyl ester / tert-Butyl chloroacetate	$C_8H_{11}ClO_2$ / 150.60	107-59-5 /	1563 / 150; 50[10]	3-02-00-00444 /	H_2O 6 / 1.4260[20]
129	Acetic acid, chloro-, 1,2-ethanediyl ester	$C_6H_8Cl_2O_4$ / 215.03	6941-69-1 / 45.5	/ 157[8]	3-02-00-00448 /	H_2O 1; eth 4
130	Acetic acid, chloro-, ethyl ester / Ethyl chloroacetate	$C_4H_7ClO_2$ / 122.55	105-39-5 / -21	3741 / 144.3	4-02-00-00481 / 1.1585[20]	H_2O 1; EtOH 5; eth 5; ace 5 / 1.4215[20]
131	Acetic acid, chlorofluoro- / Chlorofluoroacetic acid	$C_2H_2ClFO_2$ / 112.49	471-44-3 /	/ 162	4-02-00-00493 / 1.532[25]	/ 1.4085[25]
132	Acetic acid, chlorofluoro-, butyl ester / Chlorofluoroacetic acid, butyl ester	$C_6H_{10}ClFO_2$ / 168.60	368-34-3 /	/ 165.5	3-02-00-00453 / 1.124[25]	/ 1.4067[25]
133	Acetic acid, chlorofluoro-, ethyl ester	$C_4H_6ClFO_2$ / 140.54	401-56-9 /	/ 129	4-02-00-00493 / 1.225[20]	/ 1.3927[20]
134	Acetic acid, chlorofluoro-, methyl ester / Methyl chlorofluoroacetate	$C_3H_4ClFO_2$ / 126.51	433-52-3 /	/ 116	4-02-00-00493 / 1.323[25]	/ 1.3903[25]
135	Acetic acid, chlorofluoro-, propyl ester / Chlorofluoroacetic acid, propyl ester	$C_5H_8ClFO_2$ / 154.57	348-70-9 /	/ 147	3-02-00-00453 / 1.170[25]	/ 1.3994[25]
136	Acetic acid, chloro-, 2-hydroxyethyl ester	$C_4H_7ClO_3$ / 138.55	35280-53-6 /	/ 240 dec; 86[1.5]	3-02-00-00447 / 1.3300[20]	H_2O 4; EtOH 4 / 1.4609[20]
137	Acetic acid, chloro-, 2-methoxyethyl ester	$C_5H_9ClO_3$ / 152.58	13361-36-9 /	/ 86-6[9]	3-02-00-00447 / 1.2015[20]	eth 4 / 1.4382[20]
138	Acetic acid, chloro-, methyl ester / Chloroacetic acid, methyl ester	$C_3H_5ClO_2$ / 108.52	96-34-4 / -32.1	5965 / 129.5	4-02-00-00480 / 1.236[20]	ace 4; bz 4; eth 4; EtOH 4 / 1.4218[20]
139	Acetic acid, chloro-, 1-methylethyl ester / Isopropyl chloroacetate	$C_5H_9ClO_2$ / 136.58	105-48-6 /	/ 150.5	4-02-00-00482 / 1.0888[20]	eth 4 / 1.4382[20]
140	Acetic acid, chloro-, 4-methylphenyl ester	$C_9H_9ClO_2$ / 184.62	15150-39-7 / 32	/ 254; 162[41]	4-06-00-02112 / 1.1840[35]	eth 4; EtOH 4
141	Acetic acid, (4-chloro-2-methylphenoxy)- / MCPA	$C_9H_9ClO_3$ / 200.62	94-74-6 / 120	5645 /	4-06-00-01991 /	H_2O 2; EtOH 4; eth 4; bz 3
142	Acetic acid, chloro-, 1-methylpropyl ester	$C_6H_{11}ClO_2$ / 150.60	17696-64-9 /	/ 163.5	3-02-00-00443 / 1.060[20]	eth 4; EtOH 4 / 1.4251[19]
143	Acetic acid, chloro-, 2-methylpropyl ester	$C_6H_{11}ClO_2$ / 150.60	13361-35-8 /	/ 170	4-02-00-00483 / 1.0612[20]	ace 4; eth 4 / 1.4255[20]
144	Acetic acid, chlorooxo-, ethyl ester	$C_4H_5ClO_3$ / 136.53	4755-77-5 /	/ 137	4-02-00-01853 / 1.2226[20]	bz 4; eth 4
145	Acetic acid, chlorooxo-, methyl ester	$C_3H_3ClO_3$ / 122.51	5781-53-3 /	/ 119	3-02-00-01583 / 1.3316[20]	/ 1.4189[20]
146	Acetic acid, (4-chlorophenoxy)- / (p-Chlorophenoxy)acetic acid	$C_8H_7ClO_3$ / 186.59	122-88-3 / 156.5	/	4-06-00-00845 /	H_2O 4; chl 2
147	Acetic acid, chloro-, phenyl ester / Phenyl chloroacetate	$C_8H_7ClO_2$ / 170.60	620-73-5 / 44.5	/ 232.5	3-06-00-00598 / 1.2202[44]	H_2O 1; EtOH 4; eth 4 / 1.5146[44]
148	Acetic acid, chloro-, phenylmethyl ester	$C_9H_9ClO_2$ / 184.62	140-18-1 /	/ 147.5[9]; 85[0.4]	4-06-00-02264 / 1.1223[4]	eth 4; EtOH 4 / 1.5426[18]
149	Acetic acid, chloro-, propyl ester / Propyl chloroacetate	$C_5H_9ClO_2$ / 136.58	5396-24-7 /	/ 161	4-02-00-00482 / 1.1022[20]	eth 4 / 1.4261[20]

No.	Name / Synonym	Mol. Form. / Mol. Wt.	CAS RN / mp/°C	Merck No. / bp/°C	Beil. Ref. / den/g cm^{-3}	Solubility / n_D
150	Acetic acid, cyano- / Cyanoacetic acid	$C_3H_3NO_2$ / 85.06	372-09-8 / 66	2696 / 108$^{0.15}$; 160 dec	4-02-00-01888	H_2O 3; EtOH 3; eth 3; chl 2
151	Acetic acid, cyano-, butyl ester	$C_7H_{11}NO_2$ / 141.17	5459-58-5	/ 231; 115^{15}	4-02-00-01891 / 1.0010^{20}	1.4200^{20}
152	Acetic acid, cyano-, ethyl ester / Ethyl cyanoacetate ·	$C_5H_7NO_2$ / 113.12	105-56-6 / -22.5	3744 / 205	4-02-00-01889 / 1.0654^{20}	eth 4; EtOH 4 / 1.4175^{20}
153	Acetic acid, cyano-, hydrazide / Cyacetacide	$C_3H_5N_3O$ / 99.09	140-87-4 / 114.5	2688	4-02-00-01895	H_2O 4; EtOH 4
154	Acetic acid, cyano-, methyl ester / Methyl cyanoacetate	$C_4H_5NO_2$ / 99.09	105-34-0 / -22.5	/ 200.5	4-02-00-01889 / 1.1225^{25}	eth 4; EtOH 4 / 1.4176^{20}
155	Acetic acid, cyclohexyl ester / Cyclohexyl acetate	$C_8H_{14}O_2$ / 142.20	622-45-7	/ 173; 96^{75}	4-06-00-00036 / 0.968^{20}	eth 4; EtOH 4 / 1.442^{20}
156	Acetic acid, cyclohexylidene-, methyl ester / Methyl cyclohexylideneacetate	$C_9H_{14}O_2$ / 154.21	40203-74-5	/ 99^{20}	4-09-00-00124 / 1.0021^{19}	1.4838^{19}
157	Acetic acid, decyl ester / Decyl acetate	$C_{12}H_{24}O_2$ / 200.32	112-17-4 / -15	/ 244	4-02-00-00168 / 0.8671^{20}	H_2O 1; EtOH 3; eth 3; bz 3 / 1.4273^{20}
158	Acetic acid, diazo-, ethyl ester / Diazoacetic ester	$C_4H_6N_2O_2$ / 114.10	623-73-4 / -22	2980 / 140 dec	4-03-00-01495 / 1.0852^{18}	H_2O 2; EtOH 5; eth 5; bz 5 / 1.4605^{20}
159	Acetic acid, dibromo- / Dibromoacetic acid	$C_2H_2Br_2O_2$ / 217.84	631-64-1 / 49	/ 195^{250}; 130^{16}	4-02-00-00533	H_2O 4; EtOH 4; eth 3
160	Acetic acid, dibromo-, ethyl ester / Ethyl dibromoacetate	$C_4H_6Br_2O_2$ / 245.90	617-33-4	/ 194	4-02-00-00533 / 1.8991^{20}	H_2O 1; EtOH 5; eth 5 / 1.5017^{13}
161	Acetic acid, dibromofluoro-, ethyl ester / Ethyl dibromofluoroacetate	$C_4H_5Br_2FO_2$ / 263.89	565-53-7	/ 83^{33}	4-02-00-00534 / 1.894^{25}	1.4633^{25}
162	Acetic acid, dichloro- / Dichloroacetic acid	$C_2H_2Cl_2O_2$ / 128.94	79-43-6 / 13.5	3037 / 194; 102^{20}	4-02-00-00498 / 1.5634^{20}	H_2O 5; EtOH 5; eth 5; ace 3 / 1.4658^{20}
163	Acetic acid, dichloro-, butyl ester / Butyl dichloroacetate	$C_6H_{10}Cl_2O_2$ / 185.05	29003-73-4	/ 193.5	4-02-00-00502 / 1.1820^{20}	eth 4; EtOH 4 / 1.4420^{20}
164	Acetic acid, dichloro-, ethyl ester / Ethyl dichloroacetate	$C_4H_6Cl_2O_2$ / 157.00	535-15-9	/ 155; 56^{10}	4-02-00-00501 / 1.2827^{20}	H_2O 2; EtOH 5; eth 5; ace 3 / 1.4386^{20}
165	Acetic acid, dichloro-, 2-hydroxyethyl ester	$C_4H_6Cl_2O_3$ / 173.00	17704-30-2	/ 81-2$^{0.5}$	3-02-00-00460 / 1.438^{20}	EtOH 4 / 1.4735^{20}
166	Acetic acid, dichloro-, methyl ester / Methyl dichloroacetate	$C_3H_4Cl_2O_2$ / 142.97	116-54-1 / -51.9	/ 142.9	4-02-00-00501 / 1.3774^{20}	H_2O 1; EtOH 3; ctc 3 / 1.4429^{20}
167	Acetic acid, dichloro-, 1-methylethyl ester	$C_5H_8Cl_2O_2$ / 171.02	25006-60-4	/ 163.5	4-02-00-00502 / 1.2053^{20}	eth 4; EtOH 4 / 1.4328^{20}
168	Acetic acid, [2,3-dichloro-4-(2-methylene-1-oxobutyl)phenoxy]- / Ethacrynic acid	$C_{13}H_{12}Cl_2O_4$ / 303.14	58-54-8 / 122.5	3669		
169	Acetic acid, (2,4-dichlorophenoxy)- / 2,4-D	$C_8H_6Cl_2O_3$ / 221.04	94-75-7 / 140.5	2802 / 160$^{0.4}$	4-06-00-00908	H_2O 1; EtOH 3; bz 2
170	Acetic acid, (2,4-dichlorophenoxy)-, 2-butoxyethyl ester / 2,4-D 2-Butoxyethyl ester	$C_{14}H_{18}Cl_2O_4$ / 321.20	1929-73-3	/ 156-62^1	4-06-00-00911 / 1.232^{20}	
171	Acetic acid, (2,4-dichlorophenoxy)-, butyl ester / 2,4-D Butyl ester	$C_{12}H_{14}Cl_2O_3$ / 277.15	94-80-4	2802	3-06-00-00706	
172	Acetic acid, (2,4-dichlorophenoxy)-, methyl ester / 2,4-D Methyl ester	$C_9H_8Cl_2O_3$ / 235.07	1928-38-7 / 119	/ 141^{18}	4-06-00-00909	
173	Acetic acid, (2,4-dichlorophenoxy)-, 1-methylethyl ester / 2,4-D Isopropyl ester	$C_{11}H_{12}Cl_2O_3$ / 263.12	94-11-1 / 5	/ 140^1	4-06-00-00910 / 1.26^{25}	1.5209^{25}
174	Acetic acid, diethoxy-, ethyl ester	$C_8H_{16}O_4$ / 176.21	6065-82-3	/ 199	4-03-00-01494 / 0.985^{25}	1.4100^{20}
175	Acetic acid, difluoro- / Difluoroacetic acid	$C_2H_2F_2O_2$ / 96.03	381-73-7 / -1	/ 133	4-02-00-00455 / 1.526^{25}	1.3470^{20}
176	Acetic acid, difluoro-, ethyl ester / Ethyl difluoroacetate	$C_4H_6F_2O_2$ / 124.09	454-31-9	/ 100	4-02-00-00455 / 1.1893^9	H_2O 1
177	Acetic acid, 1,1-dimethylethyl ester / tert-Butyl acetate	$C_6H_{12}O_2$ / 116.16	540-88-5	1537 / 95.1	4-02-00-00151 / 0.8665^{20}	EtOH 3; eth 3; chl 3; HOAc 3 / 1.3855^{20}
178	Acetic acid, dodecyl ester	$C_{14}H_{28}O_2$ / 228.38	112-66-3 / 1.3	/ 265; 180^{40}	4-02-00-00170 / 0.8652^{22}	1.4439^{20}
179	Acetic acid ethenyl ester / Vinyl acetate	$C_4H_6O_2$ / 86.09	108-05-4 / -93.2	9896 / 72.5	4-02-00-00176 / 0.9317^{20}	H_2O 1; EtOH 5; eth 3; ace 3 / 1.3959^{20}
180	Acetic acid, ethoxy- / 2-Ethoxyacetic acid	$C_4H_8O_3$ / 104.11	627-03-2	/ 206.5	4-03-00-00574 / 1.1021^{20}	H_2O 4; EtOH 4; eth 4; chl 3 / 1.4194^{20}
181	Acetic acid, ethoxy-, ethyl ester	$C_6H_{12}O_3$ / 132.16	817-95-8	/ 158	4-03-00-00581 / 0.9702^{20}	EtOH 3; eth 3; ace 3 / 1.4039^{20}
182	Acetic acid, ethoxy-, methyl ester / Methyl ethoxyacetate	$C_5H_{10}O_3$ / 118.13	17640-26-5	/ 147	4-03-00-00578 / 1.0112^{15}	ace 4; eth 4; EtOH 4
183	Acetic acid, ethoxy-, 5-methyl-2-(1-methylethyl)cyclohexyl ester, (1α,2β,5α)-	$C_{14}H_{26}O_3$ / 242.36	579-94-2	/ 155^{20}	2-06-00-00047 / 0.9545^{20}	eth 4; EtOH 4; chl 4
184	Acetic acid, (ethoxysulfonyl)-, ethyl ester / Acetic acid, sulfo, diethyl ester	$C_6H_{12}O_5S$ / 196.22	59376-54-4	/ 116$^{0.5}$	3-04-00-00058 / 1.232^{16}	1.436^{21}
185	Acetic acid, 2-ethylbutyl ester / 2-Ethylbutyl acetate	$C_8H_{16}O_2$ / 144.21	10031-87-5 / <-100	/ 162.5	4-02-00-00161 / 0.8790^{20}	H_2O 1; EtOH 3; eth 3; ctc 3 / 1.4109^{20}
186	Acetic acid ethyl ester / Ethyl acetate	$C_4H_8O_2$ / 88.11	141-78-6 / -83.6	3713 / 77.11	4-02-00-00127 / 0.9003^{20}	H_2O 3; EtOH 5; eth 5; ace 4 / 1.3723^{20}

No.	Name / Synonym	Mol. Form. / Mol. Wt.	CAS RN / mp/°C	Merck No. / bp/°C	Beil. Ref. / den/g cm^{-3}	Solubility / n_D
187	Acetic acid, 2-ethylhexyl ester / 2-Ethylhexyl acetate	$C_{10}H_{20}O_2$ / 172.27	103-09-3 / -80	6683 / 199	4-02-00-00166 / 0.8718[20]	H_2O 1; EtOH 3; eth 3 / 1.4204[20]
188	Acetic acid, (ethylthio)-	$C_4H_8O_2S$ / 120.17	627-04-3 / -8.5	/ 164[83]; 109[5]	4-03-00-00603 / 1.1497[20]	H_2O 4; EtOH 4; eth 4
189	Acetic acid, fluoro- / Fluoroethanoic acid	$C_2H_3FO_2$ / 78.04	144-49-0 / 35.2	4096 / 168	4-02-00-00446 / 1.3693[36]	H_2O 3; EtOH 3
190	Acetic acid, fluoro-, 1,1-dimethylethyl ester	$C_6H_{11}FO_2$ / 134.15	406-74-6 /	/ 132	4-02-00-00450 / 0.9904[20]	1.386[20]
191	Acetic acid, fluoro-, ethyl ester	$C_4H_7FO_2$ / 106.10	459-72-3 /	/ 120	4-02-00-00447 / 1.098[25]	H_2O 4 / 1.3755[20]
192	Acetic acid, fluoro-, methyl ester / Methyl fluoroacetate	$C_3H_5FO_2$ / 92.07	453-18-9 / -35	/ 104.5	4-02-00-00447 / 1.1613[15]	
193	Acetic acid, heptyl ester / Heptyl acetate	$C_9H_{18}O_2$ / 158.24	112-06-1 / -50.2	/ 193	4-02-00-00162 / 0.8750[15]	H_2O 1; EtOH 3; eth 3; ctc 3 / 1.4150[20]
194	Acetic acid, hexyl ester	$C_8H_{16}O_2$ / 144.21	142-92-7 / -80.9	/ 171.5	4-02-00-00159 / 0.8779[15]	eth 4; EtOH 4 / 1.4092[20]
195	Acetic acid, hydrazide	$C_2H_6N_2O$ / 74.08	1068-57-1 / 67	/ 137[25]	4-02-00-00435	H_2O 3; EtOH 3; eth 2
196	Acetic acid, hydroxy- / Glycolic acid	$C_2H_4O_3$ / 76.05	79-14-1 / 79.5	4394 / 100	4-03-00-00571	H_2O 3; EtOH 3; eth 3
197	Acetic acid, hydroxy-, ethyl ester	$C_4H_8O_3$ / 104.11	623-50-7 /	/ 160	4-03-00-00580 / 1.0826[23]	eth 4; EtOH 4 / 1.4180[20]
198	Acetic acid, hydroxy-, methyl ester	$C_3H_6O_3$ / 90.08	96-35-5 /	/ 149; 52[17]	4-03-00-00578 / 1.1677[18]	H_2O 3; EtOH 5; eth 5
199	Acetic acid, hydroxy-, propyl ester / Hydroxyacetic acid, propyl ester	$C_5H_{10}O_3$ / 118.13	90357-58-7 /	/ 170.5	4-03-00-00588 / 1.0631[18]	1.4231[18]
200	Acetic acid, iodo- / Iodoacetic acid	$C_2H_3IO_2$ / 185.95	64-69-7 / 82.5	4918 / dec	4-02-00-00534	H_2O 3; EtOH 3; eth 2; chl 2
201	Acetic acid, iodo-, ethyl ester	$C_4H_7IO_2$ / 214.00	623-48-3 /	/ 179	4-02-00-00535 / 1.8173[13]	EtOH 3; eth 3 / 1.5079[13]
202	Acetic acid, mercapto- / Thioglycolic acid	$C_2H_4O_2S$ / 92.12	68-11-1 / -16.5	9265 / 120[20]	4-03-00-00600 / 1.3253[20]	H_2O 5; EtOH 5; eth 5; chl 2 / 1.5080[20]
203	Acetic acid, mercapto-, ethyl ester	$C_4H_8O_2S$ / 120.17	623-51-8 /	/ 157	4-03-00-00616 / 1.0964[15]	EtOH 3; eth 3; ctc 2 / 1.4582[20]
204	Acetic acid, mercapto-, methyl ester	$C_3H_6O_2S$ / 106.15	2365-48-2 /	/ 42-3[10]	4-03-00-00614	eth 4; EtOH 4 / 1.4657[20]
205	Acetic acid, methoxy-	$C_3H_6O_3$ / 90.08	625-45-6 /	/ 203.5	4-03-00-00574 / 1.1768[20]	H_2O 3; EtOH 3; eth 3 / 1.4168[20]
206	Acetic acid, methoxy-, ethyl ester / Ethyl methoxyacetate	$C_5H_{10}O_3$ / 118.13	3938-96-3 /	/ 142	3-03-00-00381 / 1.0118[15]	H_2O 2; EtOH 4; eth 4; ctc 3 / 1.4050[20]
207	Acetic acid, methoxy-, methyl ester	$C_4H_8O_3$ / 104.11	6290-49-9 /	/ 131	4-03-00-00578 / 1.0511[20]	H_2O 2; EtOH 4; eth 4; ace 3 / 1.3962[20]
208	Acetic acid, methyl ester / Methyl acetate	$C_3H_6O_2$ / 74.08	79-20-9 / -98	5932 / 56.87	4-02-00-00122 / 0.9342[20]	H_2O 4; eth 4; EtOH 4 / 1.3614[20]
209	Acetic acid, 1-methylethyl ester / Isopropyl acetate	$C_5H_{10}O_2$ / 102.13	108-21-4 / -73.4	5093 / 88.6	4-02-00-00141 / 0.8718[20]	H_2O 3; EtOH 3; eth 5; ace 3 / 1.3773[20]
210	Acetic acid, [[5-methyl-2-(1-methylethyl)cyclohexyl]oxy]-, [1R-(1α,2β,5α)]- / l-Methoxyacetic acid	$C_{12}H_{22}O_3$ / 214.30	40248-63-3 / 54	/ 171[11]	4-06-00-00157	H_2O 2; EtOH 4; eth 3; ace 4
211	Acetic acid, 2-methylphenyl ester / o-Cresyl acetate	$C_9H_{10}O_2$ / 150.18	533-18-6 /	2588 / 208	4-06-00-01960 / 1.0533[15]	eth 4; EtOH 4 / 1.5002[20]
212	Acetic acid, 3-methylphenyl ester / m-Cresyl acetate	$C_9H_{10}O_2$ / 150.18	122-46-3 / 12	2587 / 212	4-06-00-02047 / 1.043[20]	bz 4; eth 4; EtOH 4 / 1.4978[20]
213	Acetic acid, 4-methylphenyl ester	$C_9H_{10}O_2$ / 150.18	140-39-6 /	/ 212.5	4-06-00-02112 / 1.0512[17]	H_2O 2; EtOH 3; eth 3; ctc 2 / 1.5163[22]
214	Acetic acid, 1-methylpropyl ester / sec-Butyl acetate	$C_6H_{12}O_2$ / 116.16	105-46-4 / -98.9	1536 / 112	4-02-00-00148 / 0.8748[20]	H_2O 1; EtOH 3; eth 3; ctc 2 / 1.3888[20]
215	Acetic acid, 2-methylpropyl ester / Isobutyl acetate	$C_6H_{12}O_2$ / 116.16	110-19-0 / -98.8	5014 / 116.5	4-02-00-00149 / 0.8712[20]	H_2O 2; EtOH 5; eth 5; ace 3 / 1.3902[20]
216	Acetic acid, (methylthio)-	$C_3H_6O_2S$ / 106.15	2444-37-3 /	/ 130-1[27]	4-03-00-00603 / 1.221[20]	1.495[20]
217	Acetic acid, (2-naphthalenyloxy)- / 2-Naphthoxyacetic acid	$C_{12}H_{10}O_3$ / 202.21	120-23-0 / 156	6317 /	4-06-00-04274	H_2O 3; EtOH 3; eth 3; HOAc 3
218	Acetic acid, nitro-, ethyl ester / Ethyl nitroacetate	$C_4H_7NO_4$ / 133.10	626-35-7 /	/ 106[25]; 83[6]	4-02-00-00537 / 1.1953[20]	H_2O 2; EtOH 5; eth 4; dil alk 3 / 1.4250[20]
219	Acetic acid, 2-nitrophenyl ester	$C_8H_7NO_4$ / 181.15	610-69-5 / 40.5	/ 253 dec; 141[11]	4-06-00-01256	H_2O 3; EtOH 4; eth 4; ace 4
220	Acetic acid, nonyl ester / Nonyl acetate	$C_{11}H_{22}O_2$ / 186.29	143-13-5 / -26	6597 / 210	4-02-00-00167 / 0.8785[15]	1.426[20]
221	Acetic acid, octadecyl ester	$C_{20}H_{40}O_2$ / 312.54	822-23-1 / 34.5	/ 208[9]	4-02-00-00171 / 0.8510[30]	EtOH 4
222	Acetic acid, octyl ester	$C_{10}H_{20}O_2$ / 172.27	112-14-1 / -38.5	/ 210	4-02-00-00165 / 0.8705[20]	H_2O 1; EtOH 3; eth 3; ctc 2 / 1.4150[20]
223	Acetic acid, oxo- / Glyoxylic acid	$C_2H_2O_3$ / 74.04	298-12-4 / 98	4407 /	4-03-00-01489	H_2O 4; EtOH 2; eth 2; bz 2
224	Acetic acid, oxo(phenylamino)-, ethyl ester	$C_{10}H_{11}NO_3$ / 193.20	1457-85-8 / 65.5	/ 280	3-12-00-00550	ace 4; bz 4; eth 4; EtOH 4

No.	Name Synonym	Mol. Form. Mol. Wt.	CAS RN mp/°C	Merck No. bp/°C	Beil. Ref. den/g cm^{-3}	Solubility n_D
225	Acetic acid, [(4-oxo-2-phenyl-4H-1-benzopyran-7-yl)oxy]-, ethyl ester Efloxate	$C_{19}H_{16}O_5$ 324.33	119-41-5 123.7	3490	4-18-00-00695	chl 3
226	Acetic acid, 2,2'-oxybis-	$C_4H_6O_5$ 134.09	110-99-6 148	dec	4-03-00-00577	H_2O 4; eth 4; EtOH 4
227	Acetic acid, pentyl ester Pentyl acetate	$C_7H_{14}O_2$ 130.19	628-63-7 -70.8	149.2	4-02-00-00152 0.8756[20]	H_2O 2; EtOH 5; eth 5; ctc 3 1.4023[20]
228	Acetic acid, phenoxy- Phenoxyacetic acid	$C_8H_8O_3$ 152.15	122-59-8 98.5	7223 285 dec	4-06-00-00634	H_2O 3; EtOH 4; eth 4; bz 4
229	Acetic acid, phenoxy-, ethyl ester Ethyl phenoxyacetate	$C_{10}H_{12}O_3$ 180.20	2555-49-9	109-12[3]	4-06-00-00635 1.1082[20]	1.5080[20]
230	Acetic acid, phenoxy-, methyl ester Methyl phenoxyacetate	$C_9H_{10}O_3$ 166.18	2065-23-8	245	4-06-00-00635 1.1493[20]	eth 4; EtOH 4 1.5155[20]
231	Acetic acid, phenyl ester Phenyl acetate	$C_8H_8O_2$ 136.15	122-79-2	7238 196; 75[8]	4-06-00-00611 1.0780[20]	H_2O 2; EtOH 5; eth 5; ctc 3 1.5035[20]
232	Acetic acid, 2-phenylethyl ester	$C_{10}H_{12}O_2$ 164.20	103-45-7 -31.1	232.6	4-06-00-03073 1.0883[20]	eth 4; EtOH 4 1.5171[20]
233	Acetic acid, phenylmethyl ester Benzyl acetate	$C_9H_{10}O_2$ 150.18	140-11-4 -51.3	1137 213	4-06-00-02262 1.0550[20]	H_2O 2; EtOH 5; eth 3; ace 3 1.5232[20]
234	Acetic acid, potassium salt Potassium acetate	$C_2H_3KO_2$ 98.14	127-08-2 292	7580	4-02-00-00094 1.57[25]	H_2O 4; EtOH 3; eth 1; ace 1
235	Acetic acid, 2-propenyl ester Allyl acetate	$C_5H_8O_2$ 100.12	591-87-7	103.5	4-02-00-00180 0.9275[20]	H_2O 2; EtOH 5; eth 5; ace 3 1.4049[20]
236	Acetic acid, propyl ester Propyl acetate	$C_5H_{10}O_2$ 102.13	109-60-4 -93	7853 101.54	4-02-00-00138 0.8878[20]	H_2O 2; EtOH 5; eth 5; ctc 3 1.3842[20]
237	Acetic acid, sodium salt Sodium acetate	$C_2H_3NaO_2$ 82.03	127-09-3 324	8513	4-02-00-00094 1.528[25]	H_2O 4; EtOH 2 1.464
238	Acetic acid, sulfo- Sulfoacetic acid	$C_2H_4O_5S$ 140.12	123-43-3 85	8932 245 dec	4-04-00-00102	H_2O 4; ace 4; EtOH 4
239	Acetic acid, 2,2'-sulfonylbis- Sulfonyldiacetic acid	$C_4H_6O_6S$ 182.15	123-45-5 187	8940	3-03-00-00424	H_2O 4; EtOH 4; eth 3; sulf 3
240	Acetic acid, 2,2'-thiobis- Thiodiglycolic acid	$C_4H_6O_4S$ 150.16	123-93-3 129	9260	4-03-00-00612	H_2O 2; EtOH 4; bz 3
241	Acetic acid, tribromo- Tribromoacetic acid	$C_2HBr_3O_2$ 296.74	75-96-7 132	9521 245 dec	4-02-00-00534	H_2O 3; EtOH 3; eth 3
242	Acetic acid, tribromo-, ethyl ester Ethyl tribromoacetate	$C_4H_5Br_3O_2$ 324.79	599-99-5	225	3-02-00-00485 2.2260[20]	EtOH 3; eth 3 1.5438[13]
243	Acetic acid, trichloro- Trichloroacetic acid	$C_2HCl_3O_2$ 163.39	76-03-9 57.5	9539 196.5	4-02-00-00508 1.6126[64]	H_2O 4; EtOH 3; eth 3; ctc 2 1.4603[61]
244	Acetic acid, trichloro-, anhydride Trichloroacetic anhydride	$C_4Cl_6O_3$ 308.76	4124-31-6	223 dec	4-02-00-00518 1.6908[20]	eth 4; HOAc 4
245	Acetic acid, trichloro-, butyl ester Butyl trichloroacetate	$C_6H_9Cl_3O_2$ 219.49	3657-07-6	204	4-02-00-00515 1.2778[20]	ctc 3 1.4525[25]
246	Acetic acid, trichloro-, 2-chloroethyl ester	$C_4H_4Cl_4O_2$ 225.89	4974-21-4	217	4-02-00-00515 1.5357[20]	eth 4 1.4813[20]
247	Acetic acid, trichloro-, 1,1-dimethylethyl ester	$C_6H_9Cl_3O_2$ 219.49	1860-21-5 25.5	54-5[7]	4-02-00-00516 1.2363[25]	eth 4; EtOH 4 1.4398[25]
248	Acetic acid, trichloro-, ethyl ester Ethyl trichloroacetate	$C_4H_5Cl_3O_2$ 191.44	515-84-4 167.5	9539	4-02-00-00514 1.3836[20]	H_2O 1; EtOH 3; eth 3; bz 3 1.4505[20]
249	Acetic acid, trichloro-, 2-hydroxyethyl ester	$C_4H_5Cl_3O_3$ 207.44	33560-17-7	130-4[12]	3-02-00-00474 1.532[20]	EtOH 4 1.4775[20]
250	Acetic acid, trichloro-, 2-methoxyethyl ester 2-Methoxyethyl trichloroacetate	$C_5H_7Cl_3O_3$ 221.47	35449-34-4 14.5	98-9[17]	3-02-00-00474 1.3826[20]	bz 4; eth 4; EtOH 4 1.4563[20]
251	Acetic acid, trichloro-, 1-methylpropyl ester	$C_6H_9Cl_3O_2$ 219.49	4484-80-4	93-4[24]	3-02-00-00472 1.2636[20]	bz 4; eth 4; EtOH 4 1.4483[20]
252	Acetic acid, trichloro-, 2-methylpropyl ester Isobutyl trichloroacetate	$C_6H_9Cl_3O_2$ 219.49	33560-15-5	188	4-02-00-00515 1.2636[20]	bz 4; eth 4; EtOH 4 1.4483[20]
253	Acetic acid, trichloro-, methyl ester Methyl trichloroacetate	$C_3H_3Cl_3O_2$ 177.41	598-99-2 -17.5	153.8	4-02-00-00513 1.4874[20]	H_2O 1; EtOH 4; eth 4; ctc 3 1.4572[20]
254	Acetic acid, trichloro-, 1-methylethyl ester Isopropyl trichloroacetate	$C_5H_7Cl_3O_2$ 205.47	3974-99-0	175; 66[15]	4-02-00-00515 1.3034[20]	bz 4; eth 4; EtOH 4 1.4428[20]
255	Acetic acid, trichloro-, 3-methylbutyl ester	$C_7H_{11}Cl_3O_2$ 233.52	57392-55-9	217	4-02-00-00516 1.2314[20]	eth 4; EtOH 4 1.4521[20]
256	Acetic acid, (2,4,5-trichlorophenoxy)- 2,4,5-T	$C_8H_5Cl_3O_3$ 255.48	93-76-5 153	8999	4-06-00-00973	H_2O 1; EtOH 3; bz 4
257	Acetic acid, (2,4,5-trichlorophenoxy)-, 2-butoxyethyl ester 2,4,5-T Butoxyethyl ester	$C_{14}H_{17}Cl_3O_4$ 355.64	2545-59-7	163-6[1]	4-06-00-00976 1.280[20]	ctc 3
258	Acetic acid, (2,4,5-trichlorophenoxy)-, butyl ester 2,4,5-T Butyl ester	$C_{12}H_{13}Cl_3O_3$ 311.59	93-79-8 28.5		4-06-00-00975	
259	Acetic acid, (2,4,5-trichlorophenoxy)-, 1-methylethyl ester 2,4,5-T Isopropyl ester	$C_{11}H_{11}Cl_3O_3$ 297.56	93-78-7 45		4-06-00-00974	chl 4
260	Acetic acid, (2,4,5-trichlorophenoxy)-, 1-methylpropyl ester 2,4,5-T sec-Butyl ester	$C_{12}H_{13}Cl_3O_3$ 311.59	61792-07-2 39		4-06-00-00975	
261	Acetic acid, trichloro-, propyl ester	$C_5H_7Cl_3O_2$ 205.47	13313-91-2	187	4-02-00-00515 1.3221[20]	eth 4; EtOH 4 1.4501[20]
262	Acetic acid, trichloro-, trichloromethyl ester	$C_3Cl_6O_2$ 280.75	6135-29-1 34	191.5	2-03-00-00015 1.6733[35]	bz 4; eth 4; chl 4; lig 4

No.	Name / Synonym	Mol. Form. / Mol. Wt.	CAS RN / mp/°C	Merck No. / bp/°C	Beil. Ref. / den/g cm^{-3}	Solubility / n_D
263	Acetic acid, trifluoro- / Trifluoroacetic acid	$C_2HF_3O_2$ / 114.02	76-05-1 / -15.2	9595 / 73	4-02-00-00458 / 1.5351[25]	H_2O 3; EtOH 3; eth 3; ace 3
264	Acetic acid, trifluoro-, anhydride	$C_4F_6O_3$ / 210.03	407-25-0 / -65	/ 39.5	4-02-00-00469 / 1.490[25]	1.269[25]
265	Acetic acid, trifluoro-, butyl ester / Butyl trifluoroacetate	$C_6H_9F_3O_2$ / 170.13	367-64-6 /	/ 102	4-02-00-00464 / 1.0268[22]	chl 3 / 1.353[22]
266	Acetic acid, trifluoro-, 1,1-dimethylethyl ester	$C_6H_9F_3O_2$ / 170.13	400-52-2 /	/ 83	4-02-00-00465 /	chl 3 / 1.3300[25]
267	Acetic acid, trifluoro-, ethyl ester	$C_4H_5F_3O_2$ / 142.08	383-63-1 /	/ 61	4-02-00-00463 / 1.194[20]	1.308[20]
268	Acetic acid, trifluoro-, 2-methylpentyl ester / 2-Methylpentyl trifluoroacetate	$C_8H_{13}F_3O_2$ / 198.19	10042-29-2 /	/ 140	4-02-00-00466 / 1.0504[20]	chl 2
269	Acetic acid, trifluoro-, propyl ester / Propyl trifluoroacetate	$C_5H_7F_3O_2$ / 156.10	383-66-4 /	/ 82.5	4-02-00-00464 / 1.1285[25]	chl 3 / 1.3233[22]
270	Acetochlor / Acetamide, 2-chloro-N-(ethoxymethyl)-N-(2-ethyl-6-methylphenyl)-	$C_{14}H_{20}ClNO_2$ / 269.77	34256-82-1 /			
271	Acetonitrile / Methyl cyanide	C_2H_3N / 41.05	75-05-8 / -43.8	62 / 81.6	4-02-00-00419 / 0.7857[20]	H_2O 5; EtOH 5; eth 5; ace 5 / 1.3442[30]
272	Acetonitrile, amino- / Aminoacetonitrile	$C_2H_4N_2$ / 56.07	540-61-4 /	423 / 58[15]	4-04-00-02363 /	EtOH 4
273	Acetonitrile, (butylamino)-	$C_6H_{12}N_2$ / 112.17	3010-04-6 /	/ 85[9]	4-04-00-02382 / 0.891[25]	1.4337[20]
274	Acetonitrile, chloro- / Chloromethyl cyanide	C_2H_2ClN / 75.50	107-14-2 /	/ 126.5	4-02-00-00492 / 1.1930[20]	eth 4; EtOH 4 / 1.4202[25]
275	Acetonitrile, (cyclohexylamino)-	$C_8H_{14}N_2$ / 138.21	1074-58-4 / 18	/ 76[1]	4-12-00-00073 / 0.9657[25]	
276	Acetonitrile, cyclohexylidene- / Cyclohexylideneacetonitrile	$C_8H_{11}N$ / 121.18	4435-18-1 /	/ 107-8[22]	4-09-00-00124 / 0.9483[15]	eth 4; EtOH 4 / 1.4382[25]
277	Acetonitrile, dibromo- / Dibromoacetonitrile	C_2HBr_2N / 198.84	3252-43-5 /	/ 67-9[24]	4-02-00-00533 / 2.296[25]	1.5393[20]
278	Acetonitrile, dichloro- / Dichloroacetonitrile	C_2HCl_2N / 109.94	3018-12-0 /	/ 112.5	4-02-00-00506 / 1.369[20]	MeOH 3 / 1.4391[25]
279	Acetonitrile, dichlorofluoro- / Dichlorofluoroacetonitrile	C_2Cl_2FN / 127.93	353-82-2 / -110	/ 33	4-02-00-00508 / 1.3909[25]	1.3682[20]
280	Acetonitrile, (diethylamino)-	$C_6H_{12}N_2$ / 112.17	3010-02-4 /	/ 170	4-04-00-02378 / 0.8660[20]	H_2O 3 / 1.4260[20]
281	Acetonitrile, (dimethylamino)- / Glycinonitrile, N,N-dimethyl-	$C_4H_8N_2$ / 84.12	926-64-7 /	/ 137.5	4-04-00-02368 / 0.8649[20]	H_2O 4; EtOH 4 / 1.4095[20]
282	Acetonitrile, hydroxy- / Glyconitrile	C_2H_3NO / 57.05	107-16-4 / <-72	/ 183 dec; 119[24]	4-03-00-00598 /	H_2O 4; EtOH 4; eth 4; bz 1 / 1.4117[19]
283	Acetonitrile, iodo- / Iodoacetonitrile	C_2H_2IN / 166.95	624-75-9 /	/ 185	2-02-00-00206 / 2.307[25]	1.5744[20]
284	Acetonitrile, methoxy-	C_3H_5NO / 71.08	1738-36-9 /	/ 119	3-03-00-00399 / 0.9492[20]	H_2O 2; EtOH 3; eth 3; ace 3 / 1.3831[20]
285	Acetonitrile, (methyleneamino)- / Methyleneaminoacetonitrile	$C_3H_4N_2$ / 68.08	109-82-0 / 129	5976 /		
286	Acetonitrile, (methylthio)- / Cyanomethyl methyl sulfide	C_3H_5NS / 87.15	35120-10-6 /	/ 61-3[15]	4-03-00-00630 / 1.039[25]	1.4826[20]
287	Acetonitrile, phenoxy- / Phenoxyacetonitrile	C_8H_7NO / 133.15	3598-14-9 /	/ 239.5	4-06-00-00640 / 1.0991[20]	eth 4; EtOH 4 / 1.5246[20]
288	Acetonitrile, trichloro- / Trichloroacetonitrile	C_2Cl_3N / 144.39	545-06-2 / -42	9540 / 85.7	4-02-00-00524 / 1.4403[25]	H_2O 1 / 1.4409[20]
289	Acetonitrile, trifluoro- / Trifluoroacetonitrile	C_2F_3N / 95.02	353-85-5 /	/ -68.8	4-02-00-00472 /	
290	Acetonitrile, trinitro- / Trinitroacetonitrile	$C_2N_4O_6$ / 176.05	630-72-8 / 41.5	/ exp 220	0-02-00-00229 /	eth 4
291	Acetyl bromide / Acetic acid, bromide	C_2H_3BrO / 122.95	506-96-7 / -96	76 / 76	4-02-00-00398 / 1.6625[16]	eth 5; ace 3; bz 5; chl 5 / 1.4486[20]
292	Acetyl bromide, bromo- / Bromoacetyl bromide	$C_2H_2Br_2O$ / 201.85	598-21-0 /	/ 148.5	4-02-00-00530 / 2.312[22]	ace 3; ctc 3 / 1.5449[20]
293	Acetyl bromide, trichloro- / Trichloroacetyl bromide	C_2BrCl_3O / 226.28	34069-94-8 /	/ 143	3-02-00-00476 / 1.898[15]	bz 4; eth 4
294	Acetyl chloride / Ethanoyl chloride	C_2H_3ClO / 78.50	75-36-5 / -112.8	79 / 50.7	4-02-00-00395 / 1.1051[20]	eth 5; ace 5; bz 5; ctc 3 / 1.3886[20]
295	Acetyl chloride, chloro- / Chloroacetyl chloride	$C_2H_2Cl_2O$ / 112.94	79-04-9 / -22	2054 / 106	4-02-00-00488 / 1.4202[20]	eth 5; ace 3; ctc 3 / 1.4530[20]
296	Acetyl chloride, chlorofluoro- / Chlorofluoroacetyl chloride	C_2HCl_2FO / 130.93	359-32-0 /	/ 69.5	3-02-00-00453 / 1.468[25]	1.3992[25]
297	Acetyl chloride, dichloro- / Dichloroacetyl chloride	C_2HCl_3O / 147.39	79-36-7 /	3040 / 108	4-02-00-00504 / 1.5315[16]	H_2O 6; EtOH 6; eth 5 / 1.4591[20]
298	Acetyl chloride, ethoxy-	$C_4H_7ClO_2$ / 122.55	14077-58-8 /	/ 125	3-03-00-00396 / 1.1170[25]	ace 4; eth 4 / 1.4204[20]
299	Acetyl chloride, methoxy-	$C_3H_5ClO_2$ / 108.52	38870-89-2 /	/ 112.5	3-03-00-00396 / 1.1871[20]	eth 3; ace 3; ctc 3; chl 4 / 1.4199[20]
300	Acetyl chloride, trichloro- / Trichloroacetyl chloride	C_2Cl_4O / 181.83	76-02-8 /	/ 117.9	4-02-00-00519 / 1.6202[20]	eth 5 / 1.4695[20]
301	Acetyl fluoride / Acetic acid, fluoride	C_2H_3FO / 62.04	557-99-3 / -84	/ 20.8	4-02-00-00393 / 1.032[25]	EtOH 5; eth 5; bz 3; chl 3

No.	Name / Synonym	Mol. Form. / Mol. Wt.	CAS RN / mp/°C	Merck No. / bp/°C	Beil. Ref. / den/g cm^{-3}	Solubility / n_D
302	Acetyl iodide	C_2H_3IO 169.95	507-02-8 88	108	4-02-00-00399 2.0673[20]	eth 4 1.5491[20]
303	Acetyl isothiocyanate Isothiocyanic acid, acetyl	C_3H_3NOS 101.13	13250-46-9	132.5	3-03-00-00278 1.1523[13]	eth 3; CS_2 3 1.5231[18]
304	Acifluorfen Benzoic acid, 5-[2-chloro-4-(trifluoromethyl)phenoxy]-2-nitro-	$C_{14}H_7ClF_3NO_5$ 361.66	50594-66-6 150	105		
305	Aconitane-3,8,13,14,15-pentol, 20-ethyl-1,6,16-trimethoxy-4-(methoxymethyl)-, Aconine	$C_{25}H_{41}NO_9$ 499.60	509-20-6 132	110	5-21-06-00310	H_2O 3; EtOH 3; eth 2; chl 3
306	Aconitane-3,8,13,14,15-pentol, 20-ethyl-1,6,16-trimethoxy-4-(methoxymethyl)-, 8-acetate 14-benzoate, Aconitine	$C_{34}H_{47}NO_{11}$ 645.75	302-27-2 204	113	5-21-06-00310	bz 4; EtOH 4; chl 4
307	Aconitane-3,8,13,14-tetrol, 20-ethyl-1,6,16-trimethoxy-4-(methoxymethyl)-, 8-acetate 14-benzoate, (1α,3alp Indaconitine	$C_{34}H_{47}NO_{10}$ 629.75	4491-19-4 202-3 dec	4842	5-21-06-00308	eth 4; EtOH 4; chl 4
308	Aconitane-3,8,13,14-tetrol, 20-ethyl-1,6,16-trimethoxy-4-(methoxymethyl)-, 8-acetate 14-(3,4-dimethoxybenzoate),(1α,3α,6α,14α,16β)- Pseudoaconitine	$C_{36}H_{51}NO_{12}$ 689.80	127-29-7 214	7924	5-21-06-00309	eth 4; EtOH 4
309	Aconitane-4,8,9-triol, 20-ethyl-1,14,16-trimethoxy-, 4-[2-(acetylamino)benzoate], (1α,14α,16β)- Lappaconitine	$C_{32}H_{44}N_2O_8$ 584.71	32854-75-4 217.5	5237	5-21-06-00291	H_2O 1; EtOH 2; eth 2; bz 3
310	Aconitane-8,13,14-triol, 20-ethyl-1,6,16-trimethoxy-4-(methoxymethyl)-, 8-acetate 14-(3,4-dimethoxybenzoate), Bikhaconitine	$C_{36}H_{51}NO_{11}$ 673.80	6078-26-8 164	1234	5-21-06-00303	eth 4; EtOH 4; chl 4
311	Aconitane-8,13,14-triol, 1,6,16-trimethoxy-4-(methoxymethyl)-20-methyl-, 8-acetate 14-benzoate, Delphinine	$C_{33}H_{45}NO_9$ 599.72	561-07-9 199	2867	5-21-06-00302	H_2O 1; chl 3; ace 3; eth 3
312	4-Acridinamine	$C_{13}H_{10}N_2$ 194.24	578-07-4 108	183.5	5-22-11-00008	ace 4; bz 4; eth 4; EtOH 4
313	9-Acridinamine Aminacrine	$C_{13}H_{10}N_2$ 194.24	90-45-9 241	418	5-22-11-00008	EtOH 3; ace 3; dil HCl 4
314	Acridine Dibenzo[b,e]pyridine	$C_{13}H_9N$ 179.22	260-94-6 106(form a); 110(form b)	117 344.86	5-20-08-00199 1.005[20]	H_2O 2; EtOH 4; eth 4; bz 4
315	Acridine, 9-chloro- 9-Chloroacridine	$C_{13}H_8ClN$ 213.67	1207-69-8 121	sub	5-20-08-00210	H_2O 4; EtOH 4
316	3,6-Acridinediamine Proflavine	$C_{13}H_{11}N_3$ 209.25	92-62-6 285	7780	5-22-11-00322	H_2O 3; EtOH 4; eth 2; bz 2
317	3,9-Acridinediamine, 7-ethoxy- Ethacridine	$C_{15}H_{15}N_3O$ 253.30	442-16-0 226	3668	5-22-12-00243	
318	Acridine, 9-phenyl- 9-Phenylacridine	$C_{19}H_{13}N$ 255.32	602-56-2 184	404	5-20-08-00564	H_2O 1; EtOH 2; eth 3; bz 4
319	Adenosine β-D-Ribofuranoside, adenine-9	$C_{10}H_{13}N_5O_4$ 267.24	58-61-7 235.5	143	4-26-00-03598	H_2O 2; EtOH 1
320	5'-Adenylic acid Adenosine 5'-monophosphate	$C_{10}H_{14}N_5O_7P$ 347.22	61-19-8 195 dec	148	4-26-00-03615	H_2O 4; EtOH 3
321	Ajmalan-17,21-diol, (17R,21α)- Ajmaline	$C_{20}H_{26}N_2O_2$ 326.44	4360-12-7 206	184	4-23-00-03212	H_2O 1; EtOH 3; eth 2; bz 2
322	Alachlor Acetamide, 2-chloro-N-(2,6-diethylphenyl)-N-(methoxymethyl)-	$C_{14}H_{20}ClNO_2$ 269.77	15972-60-8 40	193 100[0.02]	1.133[25]	
323	β-Alanine 3-Aminopropanoic acid	$C_3H_7NO_2$ 89.09	107-95-9 200 dec	196	4-04-00-02526 1.437[19]	H_2O 3; EtOH 2; eth 1; ace 1
324	DL-Alanine DL-2-Aminopropanoic acid	$C_3H_7NO_2$ 89.09	302-72-7 300 dec	195 sub 250	4-04-00-02481 1.424[25]	H_2O 3; EtOH 4
325	L-Alanine (S)-2-Aminopropanoic acid	$C_3H_7NO_2$ 89.09	56-41-7 297 dec	195 sub 250	4-04-00-02480 1.432[22]	H_2O 3; EtOH 2; eth 1; ace 1
326	Alanine, 3-amino- 2,3-Diaminopropionic acid	$C_3H_8N_2O_2$ 104.11	515-94-6 110	2962	4-04-00-02581	H_2O 4
327	DL-Alanine, N-benzoyl-	$C_{10}H_{11}NO_3$ 193.20	1205-02-3 165.5	dec	4-09-00-00794	H_2O 3; EtOH 3; eth 2; DMSO 2
328	β-Alanine, N,N-diethyl-, ethyl ester Ethyl 3-(diethylamino)propionate	$C_9H_{19}NO_2$ 173.26	5515-83-3	83-4[12]	4-04-00-02538 0.881[25]	1.4253[20]
329	β-Alanine, N-(2,4-dihydroxy-3,3-dimethyl-1-oxobutyl)-, (R)- Pantothenic acid	$C_9H_{17}NO_5$ 219.24	79-83-4	6964	4-04-00-02569	H_2O 4; bz 4; eth 4
330	β-Alanine, N-(3-ethoxy-3-oxopropyl)-N-methyl-, ethyl ester Propionic acid, 3,3'-iminodi-N-methyl, diethyl ester	$C_{11}H_{21}NO_4$ 231.29	6315-60-2	136-8[4]	3-04-00-01278 1.0172[20]	eth 4; EtOH 4 1.4421[20]
331	Alanine, 2-methyl- α-Aminoisobutyric acid	$C_4H_9NO_2$ 103.12	62-57-7 335	457 sub 280	4-04-00-02616	H_2O 4; EtOH 2; eth 1

No.	Name Synonym	Mol. Form. Mol. Wt.	CAS RN mp/°C	Merck No. bp/°C	Beil. Ref. den/g cm^{-3}	Solubility n_D
332	β-Alanine, N-methyl-, ethyl ester	$C_6H_{13}NO_2$ 131.17	2213-08-3 	 80[21]	3-04-00-01264 1.0064[20]	eth 4; EtOH 4 1.4443[20]
333	L-Alanine, 3-(2-propenylsulfinyl)-, (S)- Alliin	$C_6H_{11}NO_3S$ 177.22	556-27-4 165	251	4-04-00-03147	H_2O 4
334	Aldicarb Propanal, 2-methyl-2-(methylthio)-, O-[(methylamino)carbonyl]oxime	$C_7H_{14}N_2O_2S$ 190.27	116-06-3 99	216	 1.195[25]	
335	Aldoxycarb Propanal, 2-methyl-2-(methylsulfonyl)-, O-[(methylamino)carbonyl]oxime	$C_7H_{14}N_2O_4S$ 222.27	1646-88-4			
336	Allethrin 2-Methyl-4-oxo-3-(2-propenyl)-2-cyclopenten-1-yl 2,2-dimethyl-3-(2-methyl-1-propenyl) cyclopropane carboxylate	$C_{19}H_{26}O_3$ 302.41	584-79-2		 1.010[20]	
337	D-Allose	$C_6H_{12}O_6$ 180.16	2595-97-3 128	279	4-01-00-04299	H_2O 4
338	D-Altrose	$C_6H_{12}O_6$ 180.16	1990-29-0 103.5	319	4-01-00-04300	H_2O 4
339	Aluminum, trimethyl Trimethyl aluminum	C_3H_9Al 72.09	75-24-1 15.4	 130; 20[8]	0.752[20]	
340	Aluminum, tris(2-hydroxypropanoato-O1,O2)- Aluminum lactate	$C_9H_{15}AlO_9$ 294.19	18917-91-4 	350		H_2O 3
341	Ametryn N-Ethyl-N'-(1-methylethyl)-6-(methylthio)1,3,5-triazine-2,4-diamine	$C_9H_{17}N_5S$ 227.33	834-12-8 88	402		
342	Amitraz N-Methylbis(2,4-xylyliminomethyl)amine	$C_{19}H_{23}N_3$ 293.41	33089-61-1 86	503	 1.128[20]	
343	Androsta-1,4-dien-3-one, 17-hydroxy-, (17β)- Boldenone	$C_{19}H_{26}O_2$ 286.41	846-48-0 165	1327		
344	Androsta-1,4-dien-3-one, 17-hydroxy-17-methyl-, (17β)- Methandrostenolone	$C_{20}H_{28}O_2$ 300.44	72-63-9 166	5862	4-08-00-01119	
345	Androstane	$C_{19}H_{32}$ 260.46	24887-75-0 50	668 60[0.003]	4-05-00-01211	ace 4; eth 4; EtOH 4; peth 4
346	Androstane-17-carboxylic acid, (5β,17β)- Etiocholanic acid	$C_{20}H_{32}O_2$ 304.47	438-08-4 228.5	3824 sub 160	3-09-00-02644	
347	Androstan-17-one, 3-hydroxy-, (3β,5α)- Epiandrosterone	$C_{19}H_{30}O_2$ 290.45	481-29-8 178	3562	4-08-00-00641	
348	Androstan-17-one, 3-hydroxy-, (3α,5α)- Androsterone	$C_{19}H_{30}O_2$ 290.45	53-41-8 185	673	4-08-00-00642	H_2O 2; EtOH 3; eth 3; ace 3
349	Androstan-3-one, 17-hydroxy-, (5α,17β)- Stanolone	$C_{19}H_{30}O_2$ 290.45	521-18-6 181	8753 sub 135	4-08-00-00634	
350	Androstan-3-one, 17-hydroxy-17-methyl-, (5α,17β)- Mestanolone	$C_{20}H_{32}O_2$ 304.47	521-11-9 192.5	5816	4-08-00-00656	AcOEt 2
351	Androstan-3-one, 2-methyl-17-(1-oxopropoxy)-, (2α,5α,17β)- Dromostanolone propionate	$C_{23}H_{36}O_3$ 360.54	521-12-0 128	3436	4-08-00-00654	
352	Androst-4-ene-3,17-dione 4-Androstene-3,17-dione	$C_{19}H_{26}O_2$ 286.41	63-05-8 143(form a); 173(form b)	671	4-07-00-02381	
353	Androst-4-ene-3,11,17-trione Adrenosterone	$C_{19}H_{24}O_3$ 300.40	382-45-6 222	165 sub	4-07-00-02796	H_2O 2; EtOH 3; eth 3; ace 3
354	Androst-4-en-3-one, 4-chloro-17-hydroxy-, (17β)- Clostebol	$C_{19}H_{27}ClO_2$ 322.87	1093-58-9 189	2409	4-08-00-00989	
355	Androst-4-en-3-one, 9-fluoro-11,17-dihydroxy-17-methyl-, (11β,17β)- Fluoxymesterone	$C_{20}H_{29}FO_3$ 336.45	76-43-7 270	4113	4-08-00-02057	
356	Androst-4-en-3-one, 17-hydroxy-, (17β)- Testosterone	$C_{19}H_{28}O_2$ 288.43	58-22-0 155	9109	4-08-00-00974	H_2O 1; EtOH 3; eth 3; ace 3
357	Androst-4-en-3-one, 17-hydroxy-7,17-dimethyl-, (7β,17β)- Calusterone	$C_{28}H_{48}O$ 400.69	17021-26-0 157.5	1730	4-08-00-01029	
358	Androst-4-en-3-one, 17-hydroxy-17-methyl-, (17β)- 17-Methyltestosterone	$C_{20}H_{30}O_2$ 302.46	58-18-4 163.5	6044	4-08-00-01010	eth 4; EtOH 4
359	Androst-4-en-3-one, 17-(1-oxopropoxy)-, (17β)- Testosterone-17-propionate	$C_{22}H_{32}O_3$ 344.49	57-85-2 120	9115	3-08-00-00897	eth 4; py 4; EtOH 4
360	Anilazine 2,4-Dichloro-6-(o-chloroanilino)-s-triazine	$C_9H_5Cl_3N_4$ 275.52	101-05-3 160	685	 1.8[20]	
361	Aniline, 3,3'-azobis[N,N-dimethyl- Azobenzene, 3,3'-bis(dimethylamino)	$C_{16}H_{20}N_4$ 268.36	21232-53-1 118	 sub	0-16-00-00305	bz 4; EtOH 4; chl 4
362	Aniline, p-isopentyl-	$C_{11}H_{17}N$ 163.26	104177-72-2 257		4-12-00-02858 0.928[15]	1.5305[20]
363	Anisole, 3-isopropyl-2-methyl- o-Cresol, 3-isopropyl, methyl ether	$C_{11}H_{16}O$ 164.25	31202-12-7 -0.5	 67[3]	4-06-00-03328 0.9540[25]	1.5148[20]
364	Anisole, 2,3,5-trinitro- 2,3,5-Trinitroanisole	$C_7H_5N_3O_7$ 243.13	7539-25-5 107.4		2-06-00-00253 1.618[15]	ace 4; bz 4; EtOH 4

No.	Name Synonym	Mol. Form. Mol. Wt.	CAS RN mp/°C	Merck No. bp/°C	Beil. Ref. den/g cm^{-3}	Solubility n_D
365	2-Anthracenamine	$C_{14}H_{11}N$ 193.25	613-13-8 238.8	sub	4-12-00-03435	H_2O 1; EtOH 3; os 2; con sulf 1
366	Anthracene	$C_{14}H_{10}$ 178.23	120-12-7 215.0	712 339.9	4-05-00-02281 1.28[25]	H_2O 1; EtOH 2; eth 2; ace 2
367	Anthracene, 9,10-bis(phenylmethyl)- 9,10-Dibenzylanthracene	$C_{28}H_{22}$ 358.48	3613-42-1 246.5		3-05-00-02640 1.1787[16]	H_2O 1; eth 2; bz 2; lig 1
368	1-Anthracenecarboxylic acid 1-Anthroic acid	$C_{15}H_{10}O_2$ 222.24	607-42-1 251.5	sub	4-09-00-02670	H_2O 1; EtOH 3; eth 3; bz 2
369	2-Anthracenecarboxylic acid 2-Anthroic acid	$C_{15}H_{10}O_2$ 222.24	613-08-1 281	sub	4-09-00-02671	HOAc 4
370	9-Anthracenecarboxylic acid 9-Anthroic acid	$C_{15}H_{10}O_2$ 222.24	723-62-6 217 dec	sub	4-09-00-02671	H_2O 1; EtOH 3
371	2-Anthracenecarboxylic acid, 9,10-dihydro-4,5-dihydroxy-9,10-dioxo- Rhein	$C_{15}H_8O_6$ 284.22	478-43-3 321	8175 sub	4-10-00-04088	H_2O 2; EtOH 2; eth 2; ace 2
372	2-Anthracenecarboxylic acid, 9,10-dihydro-9,10-dioxo-	$C_{15}H_8O_4$ 252.23	117-78-2 291	sub	4-10-00-03200	EtOH 2; eth 1; ace 3; bz 1
373	2-Anthracenecarboxylic acid, 7-β-D-glucopyranosyl-9,10-dihydro-3,5,6,8-tetrahydroxy-1-methyl-9,10-dioxo- Carminic acid	$C_{22}H_{20}O_{13}$ 492.39	1260-17-9 136 dec	1850	3-10-00-04874	H_2O 3; EtOH 3; eth 2; bz 1
374	Anthracene, 1-chloro- 1-Chloroanthracene	$C_{14}H_9Cl$ 212.68	4985-70-0 83.5		4-05-00-02292 1.1707[100]	H_2O 1; EtOH 3; eth 3; bz 3 1.6959[100]
375	Anthracene, 9,10-dibromo- 9,10-Dibromoanthracene	$C_{14}H_8Br_2$ 336.03	523-27-3 226	3000 sub	4-05-00-02295	H_2O 1; EtOH 2; eth 2; bz 2
376	Anthracene, 9,10-dihydro- 9,10-Dihydroanthracene	$C_{14}H_{12}$ 180.25	613-31-0 111	305	4-05-00-02182 1.215[20]	H_2O 1; EtOH 3; eth 3; bz 3
377	Anthracene, 1,3-dimethyl- 1,3-Dimethylanthracene	$C_{16}H_{14}$ 206.29	610-46-8 83	143[2]	4-05-00-02327	eth 4; EtOH 4
378	Anthracene, 2,10-dimethyl- Anthracene, 9,3-dimethyl	$C_{16}H_{14}$ 206.29	27532-76-9 84	140-5[2]	4-05-00-02328	H_2O 1; EtOH 4; eth 4
379	9,10-Anthracenedione Anthraquinone	$C_{14}H_8O_2$ 208.22	84-65-1 286	717 377	4-07-00-02556 1.438[20]	H_2O 1; EtOH 2; eth 2; bz 2
380	9,10-Anthracenedione, 1-amino- 1-Aminoanthraquinone	$C_{14}H_9NO_2$ 223.23	82-45-1 253.5	428 sub	4-14-00-00429	ace 4; bz 4; EtOH 4; chl 4
381	9,10-Anthracenedione, 2-amino- 2-Aminoanthraquinone	$C_{14}H_9NO_2$ 223.23	117-79-3 304.5	sub	4-14-00-00447	H_2O 1; EtOH 2; eth 1; ace 3
382	9,10-Anthracenedione, 3-amino-1,2-dihydroxy-	$C_{14}H_9NO_4$ 255.23	3963-78-8 >300	sub	2-14-00-00185	aq NH_3 4
383	9,10-Anthracenedione, 2-amino-1-hydroxy- Anthraquinone, 2-amino-1-hydroxy	$C_{14}H_9NO_3$ 239.23	568-99-0 258	sub	4-14-00-00891	bz 4; eth 4; EtOH 4
384	9,10-Anthracenedione, 1-amino-2-methyl- C.I. Solvent Orange 35	$C_{15}H_{11}NO_2$ 237.26	82-28-0 205.5		4-14-00-00496	H_2O 1; EtOH 3; eth 2; bz 3
385	9,10-Anthracenedione, 1-bromo-	$C_{14}H_7BrO_2$ 287.11	632-83-7 186.5	sub	3-07-00-04076	bz 4; EtOH 4
386	9,10-Anthracenedione, 2-bromo-	$C_{14}H_7BrO_2$ 287.11	572-83-8 205.8	sub	3-07-00-04076	bz 4
387	9,10-Anthracenedione, 1-chloro-	$C_{14}H_7ClO_2$ 242.66	82-44-0 163	sub	4-07-00-02559	H_2O 1; EtOH 2; eth 5; bz 3
388	9,10-Anthracenedione, 2-chloro-	$C_{14}H_7ClO_2$ 242.66	131-09-9 211	sub	4-07-00-02559	H_2O 1; EtOH 2; eth 1; bz 2
389	9,10-Anthracenedione, 1,5-diamino-	$C_{14}H_{10}N_2O_2$ 238.25	129-44-2 319	sub	4-14-00-00479	H_2O 1; EtOH 2; eth 2; ace 2
390	9,10-Anthracenedione, 2,7-dibromo- 2,7-Dibromoanthraquinone	$C_{14}H_8Br_2O_2$ 366.01	605-42-5 249	sub	2-07-00-00718	bz 4; HOAc 4
391	9,10-Anthracenedione, 1,2-dihydroxy- Alizarin	$C_{14}H_8O_4$ 240.22	72-48-0 289.5	237	4-08-00-03256	H_2O 2; EtOH 3; eth 3; ace 3
392	9,10-Anthracenedione, 1,4-dihydroxy- Quinizarin	$C_{14}H_8O_4$ 240.22	81-64-1 200	8094	4-08-00-03260	H_2O 3; EtOH 3; eth 3; bz 3
393	9,10-Anthracenedione, 1,5-dihydroxy- Anthrarufin	$C_{14}H_8O_4$ 240.22	117-12-4 280	719 sub	4-08-00-03268	H_2O 1; EtOH 2; eth 2; ace 2
394	9,10-Anthracenedione, 1,7-dihydroxy- 9,10-Anthraquinone, 1,7-dihydroxy	$C_{14}H_8O_4$ 240.22	569-08-4 293.3	sub	4-08-00-03270	H_2O 1; EtOH 3; eth 3; bz 3
395	9,10-Anthracenedione, 1,8-dihydroxy- Danthron	$C_{14}H_8O_4$ 240.22	117-10-2 193	2813 sub	4-08-00-03271	H_2O 1; EtOH 3; eth 3; ace 3
396	9,10-Anthracenedione, 2,7-dihydroxy-	$C_{14}H_8O_4$ 240.22	572-93-0 353.8	sub	4-08-00-03273	H_2O 1; EtOH 3; eth 2; bz 2
397	9,10-Anthracenedione, 1,8-dihydroxy-3-(hydroxymethyl)- Aloe-emodol	$C_{15}H_{10}O_5$ 270.24	481-72-1 223.5	303 sub	4-08-00-03578	bz 4; eth 4; EtOH 4
398	9,10-Anthracenedione, 1,8-dihydroxy-3-methyl- Chrysophanic acid	$C_{15}H_{10}O_4$ 254.24	481-74-3 196	2263 sub	4-08-00-03277 0.92[25]	bz 4; HOAc 4
399	9,10-Anthracenedione, 1,2-dihydroxy-3-nitro- Alizarine Orange	$C_{14}H_7NO_6$ 285.21	568-93-4 244 dec	240 sub	4-08-00-03259	H_2O 2; EtOH 3; bz 3; chl 3
400	9,10-Anthracenedione, 1,2-dihydroxy-4-nitro-	$C_{14}H_7NO_6$ 285.21	2243-71-2 289 dec	sub	4-08-00-03259	bz 4; EtOH 4; chl 4; HOAc 4

No.	Name / Synonym	Mol. Form. Mol. Wt.	CAS RN mp/°C	Merck No. bp/°C	Beil. Ref. den/g cm^{-3}	Solubility n_D
401	9,10-Anthracenedione, 1,8-dihydroxy-2,4,5,7-tetranitro- Chrysamminic acid	C$_{14}$H$_4$N$_4$O$_{12}$ 420.21	517-92-0 exp	2253 dec	1-08-00-00723	eth 4; EtOH 4
402	9,10-Anthracenedione, 1,4-dimethyl-	C$_{16}$H$_{12}$O$_2$ 236.27	1519-36-4 140.5	sub	4-07-00-02583	H$_2$O 1; EtOH 2; bz 3; xyl 3
403	9,10-Anthracenedione, 2,3-dimethyl-	C$_{16}$H$_{12}$O$_2$ 236.27	6531-35-7 210.8	sub	3-07-00-04131	EtOH 3; bz 3; chl 2; xyl 3
404	9,10-Anthracenedione, 2,6-dimethyl-	C$_{16}$H$_{12}$O$_2$ 236.27	3837-38-5 242	sub	3-07-00-04132	EtOH 4; HOAc 4; tol 4
405	9,10-Anthracenedione, 1,5-dinitro-	C$_{14}$H$_8$N$_2$O$_6$ 298.21	82-35-9 385	sub	4-07-00-02562	H$_2$O 1; EtOH 2; eth 2; bz 2
406	9,10-Anthracenedione, 1,2,3,5,6,7-hexahydroxy- Rufigallol	C$_{14}$H$_8$O$_8$ 304.21	82-12-2	8267 sub	4-08-00-03750	H$_2$O 1; EtOH 2; eth 2; ace 3
407	9,10-Anthracenedione, 1-hydroxy-	C$_{14}$H$_8$O$_3$ 224.22	129-43-1 193.8	sub	4-08-00-02586	H$_2$O 1; EtOH 3; eth 3; bz 3
408	9,10-Anthracenedione, 2-hydroxy-	C$_{14}$H$_8$O$_3$ 224.22	605-32-3 306	sub	4-08-00-02598	H$_2$O 1; EtOH 3; eth 3; KOH 3
409	9,10-Anthracenedione, 2-methyl- 2-Methylanthraquinone	C$_{15}$H$_{10}$O$_2$ 222.24	84-54-8 177	5943 sub	4-07-00-02574	bz 4; EtOH 4; HOAc 4
410	9,10-Anthracenedione, 1-nitro-	C$_{14}$H$_7$NO$_4$ 253.21	82-34-8 231.5	270^7	4-07-00-02561	H$_2$O 1; EtOH 2; eth 2; ace 3
411	9,10-Anthracenedione, 2-nitro-	C$_{14}$H$_7$NO$_4$ 253.21	605-27-6 184.5	sub	4-07-00-02561	H$_2$O 1; EtOH 2; eth 2; ace 3
412	9,10-Anthracenedione, 1,2,5,6-tetrahydroxy-	C$_{14}$H$_8$O$_6$ 272.21	632-77-9 340	sub	3-08-00-04288	py 4; EtOH 4; HOAc 4
413	9,10-Anthracenedione, 1,2,5,8-tetrahydroxy- Quinalizarin	C$_{14}$H$_8$O$_6$ 272.21	81-61-8 >275	8059	4-08-00-03683	H$_2$O 2; EtOH 2; eth 2; ace 2
414	9,10-Anthracenedione, 1,3,5,7-tetrahydroxy-	C$_{14}$H$_8$O$_6$ 272.21	632-82-6 360	sub	4-08-00-03683	ace 4; bz 4; EtOH 4; HOAc 4
415	9,10-Anthracenedione, 1,2,3-trihydroxy- Anthragallol	C$_{14}$H$_8$O$_5$ 256.21	602-64-2 313	713 sub 290	4-08-00-03567	H$_2$O 2; EtOH 3; eth 3; CS$_2$ 3
416	9,10-Anthracenedione, 1,2,4-trihydroxy- Purpurin	C$_{14}$H$_8$O$_5$ 256.21	81-54-9 259	7962 sub	4-08-00-03568	H$_2$O 2; EtOH 4; eth 3; bz 4
417	9,10-Anthracenedione, 1,2,6-trihydroxy-	C$_{14}$H$_8$O$_5$ 256.21	82-29-1 330	459	4-08-00-03568	H$_2$O 3; EtOH 3; eth 2; bz 3
418	9,10-Anthracenedione, 1,2,7-trihydroxy-	C$_{14}$H$_8$O$_5$ 256.21	602-65-3 374	462	4-08-00-03569	bz 4; EtOH 4; HOAc 4
419	9,10-Anthracenedione, 1,3,8-trihydroxy-6-methyl- Emodin	C$_{15}$H$_{10}$O$_5$ 270.24	518-82-1 257	3518 sub	4-08-00-03575	eth 4; EtOH 4
420	Anthracene, 9-ethyl- 9-Ethylanthracene	C$_{16}$H$_{14}$ 206.29	605-83-4 60		4-05-00-02326 1.0413^{99}	H$_2$O 1; EtOH 3; eth 3 1.6767^{99}
421	Anthracene, 9-ethyl-9,10-dihydro- 9-Ethyl-9,10-dihydroanthracene	C$_{16}$H$_{16}$ 208.30	605-82-3	321.5	0-05-00-00649 1.048^{18}	bz 4; eth 4; EtOH 4
422	Anthracene, 1,2,3,4,5,6-hexahydro- 1,2,3,4,5,6-Hexahydroanthracene	C$_{14}$H$_{16}$ 184.28	6109-22-4 68.5	160^{15}	3-05-00-01676	bz 4
423	Anthracene, 1-methyl- 1-Methylanthracene	C$_{15}$H$_{12}$ 192.26	610-48-0 85.5	199.5	4-05-00-02310 1.0471^{99}	H$_2$O 1; EtOH 3; eth 3; bz 3 1.6802^{99}
424	Anthracene, 2-methyl- 2-Methylanthracene	C$_{15}$H$_{12}$ 192.26	613-12-7 209	sub	4-05-00-02311 1.80^0	H$_2$O 1; EtOH 2; eth 2; ace 1
425	Anthracene, 9-methyl- 9-Methylanthracene	C$_{15}$H$_{12}$ 192.26	779-02-2 81.5	196-7^{12}	4-05-00-02312 1.065^{99}	EtOH 3; eth 3; ace 3; bz 3 1.6959^{99}
426	Anthracene, 9-nitro- 9-Nitroanthracene	C$_{14}$H$_9$NO$_2$ 223.23	602-60-8 146	275^{17}	4-05-00-02296	H$_2$O 1; EtOH 2; ace 4; chl 2
427	Anthracene, 1,2,3,4,5,6,7,8-octahydro-	C$_{14}$H$_{18}$ 186.30	1079-71-6 78	294	4-05-00-01584 0.9703^{80}	H$_2$O 1; EtOH 3; bz 4; ctc 2 1.5372^{80}
428	Anthracene, 9-phenyl- 9-Phenylanthracene	C$_{20}$H$_{14}$ 254.33	602-55-1 156	417	4-05-00-02628	H$_2$O 1; EtOH 3; eth 3; bz 3
429	1,2,10-Anthracenetriol Anthrarobin	C$_{14}$H$_{10}$O$_3$ 226.23	577-33-3 208	718	4-06-00-07601	H$_2$O 2; EtOH 4; eth 4; ace 4
430	1,8,9-Anthracenetriol Anthralin	C$_{14}$H$_{10}$O$_3$ 226.23	1143-38-0 179	714	4-06-00-07602	H$_2$O 1; EtOH 3; eth 2; ace 3
431	1-Anthracenol	C$_{14}$H$_{10}$O 194.23	610-50-4 158	234^{13}	4-06-00-04928	H$_2$O 1; EtOH 4; eth 4; NaOH 3
432	9-Anthracenol Anthranol	C$_{14}$H$_{10}$O 194.23	529-86-2 152	716	4-06-00-04930	
433	9(10H)-Anthracenone Anthrone	C$_{14}$H$_{10}$O 194.23	90-44-8 155	721	4-06-00-04930	ace 3; bz 3; con sulf 3; dil alk 3
434	5,9,14,18-Anthrazinetetrone, 6,15-dihydro- Indanthrene	C$_{28}$H$_{14}$N$_2$O$_4$ 442.43	81-77-6 470-500 dec	4846	5-24-09-00504	H$_2$O 1; EtOH 1; eth 1; ace 1
435	8'-Apo-β,ψ-carotenal, 3-hydroxy-, (3R)- β-Citraurin	C$_{30}$H$_{40}$O$_2$ 432.65	650-69-1 147	2326	3-08-00-01599	H$_2$O 1; EtOH 4; eth 4; ace 4
436	β-D-Arabinopyranose Arabinose (D)(β)	C$_5$H$_{10}$O$_5$ 150.13	6748-95-4 155.5		4-01-00-04215 1.625^{25}	H$_2$O 4
437	DL-Arginine	C$_6$H$_{14}$N$_4$O$_2$ 174.20	7200-25-1 217 dec	805	3-04-00-01359	H$_2$O 1; EtOH 1; eth 1; bz 1
438	L-Arginine	C$_6$H$_{14}$N$_4$O$_2$ 174.20	74-79-3 244 dec	805	4-04-00-02648	H$_2$O 3; EtOH 2; eth 1

No.	Name Synonym	Mol. Form. Mol. Wt.	CAS RN mp/°C	Merck No. bp/°C	Beil. Ref. den/g cm⁻³	Solubility n_D
439	Arsenic acid, triethyl ester Ethyl arsenate	$C_6H_{15}AsO_4$ 226.10	15606-95-8	236.5	4-01-00-01359 1.3021[20]	1.4343[20]
440	Arsenous acid, triethyl ester Ethyl arsenite	$C_6H_{15}AsO_3$ 210.10	3141-12-6	165.5	4-01-00-01358 1.2239[20]	1.4369[13]
441	Arsine, diethyl- Diethylarsine	$C_4H_{11}As$ 134.05	692-42-2	105	3-04-00-01797 1.1338[24]	ace 4; bz 4; eth 4; EtOH 4 1.4709
442	Arsine, dimethyl- Dimethylarsine	C_2H_7As 106.00	593-57-7 -136.1	36	4-04-00-03665 1.208[29]	ace 4; bz 4; eth 4; EtOH 4
443	Arsine, diphenyl- Diphenylarsine	$C_{12}H_{11}As$ 230.14	829-83-4	174[25]	4-16-00-01138 1.30[25]	eth 4; EtOH 4
444	Arsine, ethyl- Ethylarsine	C_2H_7As 106.00	593-59-9	36	3-04-00-01797 1.214[22]	eth 4; EtOH 4
445	Arsine, ethyldifluoro-	$C_2H_5AsF_2$ 141.98	430-40-0 -38.7	94.3	3-04-00-01799 1.708[17]	
446	Arsine, methyl- Methylarsine	CH_5As 91.97	593-52-2 -143	2	3-04-00-01795	ace 4; eth 4; EtOH 4
447	Arsine, oxophenyl- Oxophenylarsine	C_6H_5AsO 168.03	637-03-6 145	6903	4-16-00-01160	H_2O 1; EtOH 2; eth 1; bz 4
448	Arsine, triethyl- Triethylarsine	$C_6H_{15}As$ 162.11	617-75-4	138.5	4-04-00-03666 1.150[20]	ace 4; eth 4; EtOH 4 1.467[20]
449	Arsine, trimethyl- Trimethylarsine	C_3H_9As 120.03	593-88-4 -87.3	52	4-04-00-03665 1.144[15]	bz 4; eth 4; EtOH 4
450	Arsine, triphenyl- Triphenylarsine	$C_{18}H_{15}As$ 306.24	603-32-7 61	360	4-16-00-01139 1.2634[18]	H_2O 1; EtOH 2; eth 4; bz 4 1.6888[21]
451	Arsinic acid, dimethyl- Cacodylic acid	$C_2H_7AsO_2$ 138.00	75-60-5 195	1603	4-04-00-03681	H_2O 4; EtOH 3; eth 1
452	Arsinous acid, dimethyl-, anhydride Cacodyl oxide	$C_4H_{12}As_2O$ 225.98	503-80-0 -25	150	3-04-00-01814 1.4816[15]	eth 4; EtOH 4 1.5225[9]
453	Arsinous chloride, dimethyl- Dimethylchloroarsine	C_2H_6AsCl 140.44	557-89-1 <-45	109; 77[37]	4-04-00-03670 1.505[12]	EtOH 4 1.5203[12]
454	Arsinous chloride, diphenyl- Chlorodiphenylarsine	$C_{12}H_{10}AsCl$ 264.59	712-48-1 44	337	3-16-00-00944 1.4820[16]	ace 4; bz 4; eth 4; EtOH 4 1.6332[56]
455	Arsonic acid, [4-[(aminocarbonyl)amino]phenyl]- Carbarsone	$C_7H_9AsN_2O_4$ 260.08	121-59-5 174	1788	3-16-00-01112	H_2O 2; EtOH 2; eth 1; chl 1
456	Arsonic acid, (4-aminophenyl)- Arsanilic acid	$C_6H_8AsNO_3$ 217.06	98-50-0 232	818	4-16-00-01190 1.9571[10]	H_2O 3; EtOH 2; eth 3; ace 1
457	Arsonic acid, ethyl- Ethanearsonic acid	$C_2H_7AsO_3$ 154.00	507-32-4 99.5	3677 209-11[12]	4-04-00-03682	H_2O 4; EtOH 4
458	Arsonic acid, methyl- Methanearsonic acid	CH_5AsO_3 139.97	124-58-3 160.5	5864	4-04-00-03682	H_2O 3; EtOH 3
459	Arsonic acid, (4-nitrophenyl)- Nitarsone	$C_6H_6AsNO_5$ 247.04	98-72-6 >310 dec	6483	4-16-00-01185	H_2O 2; EtOH 2; DMSO-d_6 2
460	Arsonic acid, phenyl- Benzenearsonic acid	$C_6H_7AsO_3$ 202.04	98-05-5 158- dec	1075	4-16-00-01183	H_2O 4; EtOH 4
461	Arsonic acid, propyl- 1-Propanearsonic acid	$C_3H_9AsO_3$ 168.02	107-34-6 134.5	7810	4-04-00-03682	H_2O 4; EtOH 4; eth 1
462	Arsonous dichloride, phenyl- Dichlorophenylarsine	$C_6H_5AsCl_2$ 222.93	696-28-6 -19	255	4-16-00-01161 1.6516[20]	bz 4; eth 4; EtOH 4 1.6386[15]
463	Arsonous difluoride, methyl- Difluoromethylarsine	CH_3AsF_2 127.95	420-24-6 -29.7	76.5	3-04-00-01796 1.924[18]	
464	Arsonous diiodide, phenyl-	$C_6H_5AsI_2$ 405.84	6380-34-3 15	205[14] 185[10]	3-16-00-00958 1.6264[15]	
465	DL-Ascorbic acid	$C_6H_8O_6$ 176.13	62624-30-0 168.5		5-18-05-00026	H_2O 4; EtOH 3; eth 1; bz 1
466	L-Ascorbic acid Vitamin C	$C_6H_8O_6$ 176.13	50-81-7 190-2 dec	855	5-18-05-00026 1.65[25]	H_2O 4; EtOH 3; eth 1; bz 1
467	L-Ascorbic acid, 6-deoxy- 6-Desoxy-L-ascorbic acid	$C_6H_8O_5$ 160.13	528-81-4 168	2880 sub 160	4-18-00-02301	H_2O 4; ace 4; EtOH 4
468	L-Asparagine α-Aminosuccinamic acid	$C_4H_8N_2O_3$ 132.12	70-47-3 235	859	4-04-00-03004 1.543[15]	H_2O 3; EtOH 1; eth 1; MeOH 1
469	DL-Asparagine, monohydrate	$C_4H_{10}N_2O_4$ 150.13	69833-18-7 182.5	859 214 dec	2-04-00-00900 1.4540[15]	H_2O 2; EtOH 1; eth 1; bz 1
470	DL-Aspartic acid	$C_4H_7NO_4$ 133.10	617-45-8 275		2-04-00-00900 1.6622[13]	H_2O 2; EtOH 1; eth 1; bz 1
471	L-Aspartic acid L-Aminosuccinic acid	$C_4H_7NO_4$ 133.10	56-84-8 270	862	4-04-00-02998 1.6603[13]	H_2O 2; EtOH 1; eth 1; bz 1
472	L-Aspartic acid, N-[2-[(2-amino-2-oxoethyl)amino]-2-carboxyethyl]- Lycomarasmine	$C_9H_{15}N_3O_7$ 277.23	7611-43-0 227-9 dec	5491	3-04-00-01521	
473	Aspidospermidine, 1-acetyl-17-methoxy- Aspidospermine	$C_{22}H_{30}N_2O_2$ 354.49	466-49-9 208	872 220[2]	5-23-12-00210	H_2O 2; EtOH 3; eth 2; bz 3
474	Asulam Methyl [(4-aminophenyl)sulfonyl]carbamate	$C_8H_{10}N_2O_4S$ 230.24	3337-71-1 144			
475	Atisine Anthorine	$C_{22}H_{33}NO_2$ 343.51	466-43-3 58.5	881		eth 4; EtOH 4; chl 4
476	Atrazine 6-Chloro-N-ethyl-N '-(1-methylethyl)-1,3,5-triazine-2,4-diamine	$C_8H_{14}ClN_5$ 215.69	1912-24-9 173	886		

No.	Name Synonym	Mol. Form. Mol. Wt.	CAS RN mp/°C	Merck No. bp/°C	Bell. Ref. den/g cm^{-3}	Solubility n_D
477	Avermectin B1a Abamectin	$C_{48}H_{72}O_{14}$ 873.09	71751-41-2 152	1		
478	7-Azabicyclo[4.1.0]heptane 2,3-Tetramethyleneaziridine	$C_6H_{11}N$ 97.16	286-18-0 21	150; 50[22]	5-20-04-00304 0.9480[27]	bz 4; eth 4; EtOH 4
479	9-Azabicyclo[3.3.1]nonan-3-one, 9-methyl- Pseudopelletierine	$C_9H_{15}NO$ 153.22	552-70-5 54	7936 246	5-21-07-00059 1.001[100]	H_2O 4; eth 4; EtOH 4 1.4760[100]
480	1-Azabicyclo[2.2.2]octane Quinuclidine	$C_7H_{13}N$ 111.19	100-76-5 158	8109	5-20-04-00335	H_2O 4; ace 4; eth 4; EtOH 4
481	8-Azabicyclo[3.2.1]octane-2-carboxylic acid, 3-(benzoyloxy)-8-methyl-, [1R-(exo,exo)]- Benzoylecgonine	$C_{16}H_{19}NO_4$ 289.33	519-09-5 195	1125	5-22-05-00053	bz 4; EtOH 4
482	8-Azabicyclo[3.2.1]octane-2-carboxylic acid, 3-(benzoyloxy)-8-methyl-, methyl ester, [1R-(exo,exo)]- Cocaine	$C_{17}H_{21}NO_4$ 303.36	50-36-2 98	2450 187[0.1]	5-22-05-00054	H_2O 2; EtOH 4; eth 4; ace 3 1.5022[98]
483	8-Azabicyclo[3.2.1]octane-2-carboxylic acid, 3-hydroxy-8-methyl-, [1R-(exo,exo)]- Ecgonine	$C_9H_{15}NO_3$ 185.22	481-37-8 205	3467	5-22-05-00053	H_2O 4; EtOH 4
484	8-Azabicyclo[3.2.1]octane, 8-methyl- Tropane	$C_8H_{15}N$ 125.21	529-17-9 166	9689	5-20-04-00331 0.9251[15]	
485	1-Azabicyclo[2.2.2]octan-3-ol 3-Quinuclidinol	$C_7H_{13}NO$ 127.19	1619-34-7 221	8110 sub 120	5-21-01-00271	ace 3
486	8-Azabicyclo[3.2.1]octan-3-ol, 8-methyl-, exo- Pseudotropine	$C_8H_{15}NO$ 141.21	135-97-7 109	7937 241	5-21-01-00219	H_2O 4; EtOH 4; eth 2; bz 3
487	8-Azabicyclo[3.2.1]octan-3-ol, 8-methyl-, benzoate (ester), endo-	$C_{15}H_{19}NO_2$ 245.32	19145-60-9 41.5	 177.5	5-21-01-00226	H_2O 4; eth 4; EtOH 4
488	8-Azabicyclo[3.2.1]octan-3-ol, 8-methyl-, benzoate (ester), exo- Tropacocaine	$C_{15}H_{19}NO_2$ 245.32	537-26-8 49	9686 dec	5-21-01-00226 1.0426[100]	bz 4; eth 4; EtOH 4; peth 4 1.5080[100]
489	8-Azabicyclo[3.2.1]octan-3-ol, 8-methyl-, endo- Tropine	$C_8H_{15}NO$ 141.21	120-29-6 64	9693 233	5-21-01-00219 1.016[100]	H_2O 4; eth 4; EtOH 4 1.4811[100]
490	8-Azabicyclo[3.2.1]octan-3-one, 8-methyl-	$C_8H_{13}NO$ 139.20	532-24-1 43	227; 113[25]	5-21-07-00032 1.9872[100]	EtOH 3; eth 3; ace 3; bz 3 1.4598[100]
491	8-Azabicyclo[3.2.1]oct-2-ene-2-carboxylic acid, 8-methyl-, (1R)- Ecgonidine	$C_9H_{13}NO_2$ 167.21	484-93-5 226-30 dec	3466	5-22-01-00446	H_2O 4
492	1H-Azepine, hexahydro- Hexamethylenimine	$C_6H_{13}N$ 99.18	111-49-9		5-20-04-00003 0.8643[22]	H_2O 3; EtOH 4; eth 4 1.4631[20]
				138		
493	2H-Azepin-2-one, hexahydro- Caprolactam	$C_6H_{11}NO$ 113.16	105-60-2 69.3	1762 270	5-21-06-00444	H_2O 4; bz 4; EtOH 4; chl 4
494	Azetidine	C_3H_7N 57.10	503-29-7 	 63	5-20-01-00136 0.8436[20]	ace 4; bz 4; eth 4; EtOH 4 1.4287[25]
495	2-Azetidinecarboxylic acid	$C_4H_7NO_2$ 101.11	2517-04-6 217 dec	923	5-22-01-00015	
496	2-Azetidinone	C_3H_5NO 71.08	930-21-2 73.5	106[15]	5-21-06-00312	eth 4; EtOH 4; chl 4
497	Azinphos-methyl O,O-Dimethyl S-[(4-oxo-1,2,3-benzotriazin-3(4H)-yl)methyl]phosphorodithioate	$C_{10}H_{12}N_3O_3PS_2$ 317.33	86-50-0 73	926	 1.44[20]	
498	Aziridine Ethyleneimine	C_2H_5N 43.07	151-56-4 -77.9	3760 56	5-20-01-00003 0.832[25]	H_2O 5; EtOH 3; eth 4; chl 2
499	1-Aziridineethanol	C_4H_9NO 87.12	1072-52-2 	 168	5-20-01-00014 1.088[25]	 1.4560[20]
500	Aziridine, 1-methyl-	C_3H_7N 57.10	1072-44-2 	 27.5	5-20-01-00007 0.7572[19]	H_2O 5 1.3885[19]
501	Aziridine, 2-methyl- 1,2-Propylenimine	C_3H_7N 57.10	75-55-8 	 67	5-20-01-00150 0.812[16]	
502	Azocine, octahydro-	$C_7H_{15}N$ 113.20	1121-92-2 29	51-3[15]	5-20-04-00156 0.896[25]	 1.4720[20]
503	1H-Azonine, octahydro-	$C_8H_{17}N$ 127.23	5661-71-2 -19	91[24]	5-20-04-00203 0.83[25]	 1.4760[20]
504	Azulene Bicyclo[5.3.0]decapentaene	$C_{10}H_8$ 128.17	275-51-4 99	939 270 dec; 125[10]	4-05-00-01636	H_2O 1; EtOH 3; eth 3; ace 3
505	Azulene, 1,4-dimethyl-7-(1-methylethyl)- Guaiazulene	$C_{15}H_{18}$ 198.31	489-84-9 31.5	4464 167-8[12]	4-05-00-01751 0.973[20]	EtOH 3; eth 3; AcOEt 3
506	Azulene, 4,8-dimethyl-2-(1-methylethyl)-	$C_{15}H_{18}$ 198.31	529-08-8 32.3	122[1.0]	4-05-00-01754 0.9735[19]	eth 4; EtOH 4
507	5-Azulenemethanol, 1,2,3,4,5,6,7,8- octahydro-α,α,3,8-tetramethyl-, Guaiol	$C_{15}H_{26}O$ 222.37	489-86-1 91	4466 288 dec; 165[17]	3-06-00-00412 0.9074[100]	H_2O 1; EtOH 3; eth 3 1.4716[100]
508	Bacitracin	$C_{66}H_{103}N_{17}O_{16}S$ 1422.71	1405-87-4 223	948		H_2O 4; EtOH 4
509	Balan Benzenamine, N-butyl-N-ethyl-2,6-dinitro-4-(trifluoromethyl)-	$C_{13}H_{16}F_3N_3O_4$ 335.28	1861-40-1 66	1048 121[0.5]; 148[7]		

No.	Name Synonym	Mol. Form. Mol. Wt.	CAS RN mp/°C	Merck No. bp/°C	Beil. Ref. den/g cm^{-3}	Solubility n_D
510	Barban Carbamic acid, (3-chlorophenyl)-, 4-chloro-2-butynyl ester	$C_{11}H_9Cl_2NO_2$ 258.10	101-27-9 75	969		
511	Basic copper carbonate Copper, [μ-[carbonato(2-)-O:O']]dihydroxydi-	$CH_2Cu_2O_5$ 221.12	12069-69-1 200 dec	2634	3.8[20]	
512	Bayleton 2-Butanone, 1-(4-chlorophenoxy)-3,3-dimethyl-1-(1H-1,2,4-triazol-1-yl)-	$C_{14}H_{16}ClN_3O_2$ 293.75	43121-43-3 82	9507	1.22[20]	
513	Benalaxyl Methyl N-(2,6-dimethylphenyl)-N-(phenylacetyl)-DL-alaninate	$C_{20}H_{23}NO_3$ 325.41	71626-11-4 79		1.27[25]	
514	Bendiocarb 1,3-Benzodioxol-4-ol, 2,2-dimethyl-, methylcarbamate	$C_{11}H_{13}NO_4$ 223.23	22781-23-3 130	1044	1.25[20]	
515	Benomyl Carbamic acid, [1-[(butylamino)carbonyl]-1H-benzimidazol-2-yl]-, methyl ester	$C_{14}H_{18}N_4O_3$ 290.32	17804-35-2 dec	1053		
516	Bensulfuron-methyl 2-[[[[[(4,6-Dimethoxypyrimidin-2-yl)amino]carbonyl]amino]sulfonyl]methyl]benzoic acid, methyl ester	$C_{16}H_{18}N_4O_7S$ 410.41	83055-99-6 187			
517	Bensulide Phosphorodithioic acid, O,O-bis(1-methylethyl) S-[2-[(phenylsulfonyl)amino]ethyl] ester	$C_{14}H_{244}NO_4PS_3$ 397.52	741-58-2 34.4		1.224 [20]	
518	Bentazon 1H-2,1,3-Benzothiadiazin-4(3H)-one, 3-(1-methylethyl)-, 2,2-dioxide	$C_{10}H_{12}N_2O_3S$ 240.28	25057-89-0 138	1060		
519	Benz[a]anthracene 1,2-Benzanthracene	$C_{18}H_{12}$ 228.29	56-55-3 84	1069	4-05-00-02549	EtOH 4
520	Benz[a]anthracene, 7,12-dimethyl- 9,10-Dimethyl-1,2-benzanthracene	$C_{20}H_{16}$ 256.35	57-97-6 122.5	3224	4-05-00-02587	ace 4; bz 4
521	Benz[a]anthracene, 11-methyl-	$C_{19}H_{14}$ 242.32	6111-78-0 118	360 dec	4-05-00-02568	H_2O 1; EtOH 3; HOAc 3
522	Benz[a]anthracene, 3-methyl- 3-Methylbenz[a]anthracene	$C_{19}H_{14}$ 242.32	2498-75-1 156.5	160[0.1]	4-05-00-02562	H_2O 1; EtOH 3; ace 3; bz 3
523	Benz[a]anthracene, 8-methyl-	$C_{19}H_{14}$ 242.32	2381-31-9 156.5	272[3]; 160[0.1]	4-05-00-02566 1.2310[0]	H_2O 1; EtOH 3; eth 3; bz 3
524	Benzadox Acetic acid, [(benzoylamino)oxy]-	$C_9H_9NO_4$ 195.17	5251-93-4 140			
525	Benzaldehyde Benzenecarboxaldehyde	C_7H_6O 106.12	100-52-7 -26	1065 179.0	4-07-00-00505 1.0415[10]	H_2O 2; EtOH 5; eth 5; ace 4 1.5463[20]
526	Benzaldehyde, 2-bromo-	C_7H_5BrO 185.02	6630-33-7 21.5	230	4-07-00-00578	H_2O 1; EtOH 4; bz 4; ctc 2 1.5925[20]
527	Benzaldehyde, 3-bromo-	C_7H_5BrO 185.02	3132-99-8 234.5		4-07-00-00579	H_2O 1; EtOH 4; eth 4; ctc 2 1.5935[20]
528	Benzaldehyde, 4-bromo-	C_7H_5BrO 185.02	1122-91-4 67	66-8[2]	4-07-00-00579	H_2O 1; EtOH 4; bz 4; chl 2
529	Benzaldehyde, 2-chloro-	C_7H_5ClO 140.57	89-98-5 12.4	211.9	4-07-00-00561 1.2483[20]	H_2O 2; EtOH 3; eth 3; ace 3 1.5662[20]
530	Benzaldehyde, 3-chloro-	C_7H_5ClO 140.57	587-04-2 17.5	213.5	4-07-00-00566 1.2410[20]	H_2O 2; EtOH 3; eth 3; ace 3 1.5650[20]
531	Benzaldehyde, 4-chloro-	C_7H_5ClO 140.57	104-88-1 47.5	213.5	4-07-00-00568 1.196[61]	H_2O 3; EtOH 4; eth 4; ace 3 1.555[61]
532	Benzaldehyde, 3-chloro-4-hydroxy-	$C_7H_5ClO_2$ 156.57	2420-16-8 139	150[14]	4-08-00-00296	eth 4; EtOH 4
533	Benzaldehyde, 5-chloro-2-hydroxy-	$C_7H_5ClO_2$ 156.57	635-93-8 100.3	105[12]	4-07-00-00224	H_2O 1; EtOH 4; eth 3; alk 3
534	Benzaldehyde, 3,5-dibromo-2-hydroxy- 3,5-Dibromosalicylaldehyde	$C_7H_4Br_2O_2$ 279.92	90-59-5 86	3011 sub	3-08-00-00188	bz 4; eth 4; chl 4
535	Benzaldehyde, 2,5-dichloro- 2,5-Dichlorobenzaldehyde	$C_7H_4Cl_2O$ 175.01	6361-23-5 58	232	4-07-00-00576	bz 4; eth 4; EtOH 4; chl 4
536	Benzaldehyde, 3,4-dichloro- 3,4-Dichlorobenzaldehyde	$C_7H_4Cl_2O$ 175.01	6287-38-3 44	247.5	4-07-00-00576	H_2O 1; EtOH 3; eth 3; ctc 2
537	Benzaldehyde, 3,5-dichloro- 3,5-Dichlorobenzaldehyde	$C_7H_4Cl_2O$ 175.01	10203-08-4 65	240	4-07-00-00577	ace 4; bz 4; eth 4; EtOH 4
538	Benzaldehyde, 3,4-diethoxy- 3,4-Diethoxybenzaldehyde	$C_{11}H_{14}O_3$ 194.23	2029-94-9 22	279; 200[50]	4-08-00-01766 1.0100[22]	EtOH 4
539	Benzaldehyde, 4-(diethylamino)-	$C_{11}H_{15}NO$ 177.25	120-21-8 41	172[10]	4-14-00-00066	H_2O 4; EtOH 3; eth 3; bz 3
540	Benzaldehyde, 2,3-dihydroxy-	$C_7H_6O_3$ 138.12	24677-78-9 108	235; 120[16]	3-08-00-01979	ace 4; EtOH 4; HOAc 4
541	Benzaldehyde, 2,4-dihydroxy- β-Resorcylaldehyde	$C_7H_6O_3$ 138.12	95-01-2 135	8159 226[22]	4-08-00-01753	H_2O 3; EtOH 4; eth 4; bz 3
542	Benzaldehyde, 3,4-dihydroxy- Protocatechualdehyde	$C_7H_6O_3$ 138.12	139-85-5 153 dec	7908	4-08-00-01762	H_2O 3; EtOH 4; eth 4
543	Benzaldehyde, 2,4-dimethoxy- 2,4-Dimethoxybenzaldehyde	$C_9H_{10}O_3$ 166.18	613-45-6 72	290; 165[10]	4-08-00-01754	H_2O 1; EtOH 3; eth 3; bz 3
544	Benzaldehyde, 2,5-dimethoxy- 2,5-Dimethoxybenzaldehyde	$C_9H_{10}O_3$ 166.18	93-02-7 52	270; 146[10]	4-08-00-01759	H_2O 2; EtOH 3; eth 3

No.	Name / Synonym	Mol. Form. / Mol. Wt.	CAS RN / mp/°C	Merck No. / bp/°C	Bell. Ref. / den/g cm^{-3}	Solubility / n_D
545	Benzaldehyde, 3,4-dimethoxy- / Veratraldehyde	$C_9H_{10}O_3$ / 166.18	120-14-9 / 43	9852 / 281; 155[10]	4-08-00-01765	H_2O 2; EtOH 4; eth 4; chl 2
546	Benzaldehyde, 3,5-dimethoxy- / 3,5-Dimethoxybenzaldehyde	$C_9H_{10}O_3$ / 166.18	7311-34-4 / 46.3	151[16]	4-08-00-01786	H_2O 2; EtOH 3; bz 3; peth 2
547	Benzaldehyde, 2,4-dimethyl- / 2,4-Dimethylbenzaldehyde	$C_9H_{10}O$ / 134.18	15764-16-6 / -9	218	4-07-00-00706	EtOH 3; eth 3; ace 3; bz 3
548	Benzaldehyde, 2,5-dimethyl- / Isoxylaldehyde	$C_9H_{10}O$ / 134.18	5779-94-2	220	4-07-00-00707 / 0.9500[20]	EtOH 4; eth 3; ace 3; bz 3
549	Benzaldehyde, 3,5-dimethyl-	$C_9H_{10}O$ / 134.18	5779-95-3 / 9	221	4-07-00-00707	ace 4; bz 4; eth 4; EtOH 4
550	Benzaldehyde, 4-(dimethylamino)- / Ehrlich's reagent	$C_9H_{11}NO$ / 149.19	100-10-7 / 74.5	3219 / 176-7[17]	4-14-00-00051 / 1.0254[100]	H_2O 2; EtOH 3; eth 3; ace 3
551	Benzaldehyde, 2,4-dinitro- / 2,4-Dinitrobenzaldehyde	$C_7H_4N_2O_5$ / 196.12	528-75-6 / 72	3265 / 200[15]	3-07-00-00923	H_2O 2; EtOH 3; eth 3; bz 3
552	Benzaldehyde, 2-ethoxy-	$C_9H_{10}O_2$ / 150.18	613-69-4 / 21	248	4-08-00-00180	EtOH 5; eth 5; chl 2
553	Benzaldehyde, 3-ethoxy-	$C_9H_{10}O_2$ / 150.18	22924-15-8	245.5	0-08-00-00060 / 1.0768[20]	EtOH 3; eth 3; bz 3 / 1.5408[20]
554	Benzaldehyde, 4-ethoxy-	$C_9H_{10}O_2$ / 150.18	10031-82-0 / 13.5	249	4-08-00-00254 / 1.08[21]	EtOH 4; eth 4; bz 4
555	Benzaldehyde, 3-ethoxy-4-hydroxy- / Ethylvanillin	$C_9H_{10}O_3$ / 166.18	121-32-4 / 77.5	3815	4-08-00-01765	H_2O 2; EtOH 3; eth 3; bz 3
556	Benzaldehyde, 5-ethoxy-2-hydroxy- / 5-Ethoxy-2-hydroxybenzaldehyde	$C_9H_{10}O_3$ / 166.18	80832-54-8 / 51.5	230	3-08-00-02004	eth 4; EtOH 4; chl 4
557	Benzaldehyde, 4-ethoxy-3-methoxy-	$C_{10}H_{12}O_3$ / 180.20	120-25-2 / 64.5	168[13]	4-08-00-01766	H_2O 2; EtOH 3; eth 3; bz 3
558	Benzaldehyde, 2-ethyl-	$C_9H_{10}O$ / 134.18	22927-13-5	209	4-07-00-00696 / 1.0216[17]	1.5425[17]
559	Benzaldehyde, 2-fluoro-	C_7H_5FO / 124.11	446-52-6 / -44.5	175	4-07-00-00559 / 1.178[25]	1.5234[20]
560	Benzaldehyde, 3-fluoro-	C_7H_5FO / 124.11	456-48-4	173	4-07-00-00559 / 1.17[25]	1.5206[20]
561	Benzaldehyde, 4-fluoro-	C_7H_5FO / 124.11	459-57-4 / -10	181.5	4-07-00-00559 / 1.1810[19]	
562	Benzaldehyde, 2-(β-D-glucopyranosyloxy)- / Helicin	$C_{13}H_{16}O_7$ / 284.27	618-65-5 / 175	4544	5-17-07-00150	H_2O 4; EtOH 4
563	Benzaldehyde, hydrazone / Benzylidene hydrazine	$C_7H_8N_2$ / 120.15	5281-18-5 / 16	140[14]	4-07-00-00530	EtOH 3
564	Benzaldehyde, 2-hydroxy- / Salicylaldehyde	$C_7H_6O_2$ / 122.12	90-02-8 / -7	8295 / 197	4-08-00-00176 / 1.1674[20]	H_2O 2; EtOH 5; eth 5; ace 4 / 1.5740[20]
565	Benzaldehyde, 3-hydroxy-	$C_7H_6O_2$ / 122.12	100-83-4 / 108	240	4-08-00-00240 / 1.1179[130]	H_2O 2; EtOH 3; eth 3; ace 3
566	Benzaldehyde, 4-hydroxy- / p-Hydroxybenzaldehyde	$C_7H_6O_2$ / 122.12	123-08-0 / 117	4741	4-08-00-00251 / 1.129[130]	H_2O 2; EtOH 4; eth 4; ace 2 / 1.5705[130]
567	Benzaldehyde, 2-hydroxy-4,6-dimethoxy-	$C_9H_{10}O_4$ / 182.18	708-76-9 / 70	193[25]; 165[10]	4-08-00-02717	H_2O 1; EtOH 4; eth 4; bz 4
568	Benzaldehyde, 3-hydroxy-4,5-dimethoxy-	$C_9H_{10}O_4$ / 182.18	29865-90-5 / 62.5	179[12]	4-08-00-02718	bz 4; EtOH 4; lig 4
569	Benzaldehyde, 4-hydroxy-3,5-dimethoxy- / Syringaldehyde	$C_9H_{10}O_4$ / 182.18	134-96-3 / 113	8996 / 192-3[14]	4-08-00-02718	H_2O 2; EtOH 4; eth 4; bz 4
570	Benzaldehyde, 2-hydroxy-3-methoxy-	$C_8H_8O_3$ / 152.15	148-53-8 / 44.5	265.5	4-08-00-01747	H_2O 2; EtOH 4; eth 4; ctc 4
571	Benzaldehyde, 2-hydroxy-5-methoxy-	$C_8H_8O_3$ / 152.15	672-13-9 / 4	247.5	4-08-00-01759	eth 4; EtOH 4
572	Benzaldehyde, 3-hydroxy-4-methoxy-	$C_8H_8O_3$ / 152.15	621-59-0 / 114	179[15]	4-08-00-01764 / 1.196[25]	H_2O 2; EtOH 3; eth 3; bz 3
573	Benzaldehyde, 3-hydroxy-5-methoxy- / 3-Hydroxy-5-methoxybenzaldehyde	$C_8H_8O_3$ / 152.15	57179-35-8 / 130.5	sub 110	3-08-00-02073	bz 4; EtOH 4; HOAc 4
574	Benzaldehyde, 4-hydroxy-3-methoxy- / Vanillin	$C_8H_8O_3$ / 152.15	121-33-5 / 81.5	9839 / 285	4-08-00-01763 / 1.056[25]	H_2O 2; EtOH 4; eth 4; ace 4
575	Benzaldehyde, 2-hydroxy-3-methyl- / m-Tolualdehyde, 2-hydroxy	$C_8H_8O_2$ / 136.15	824-42-0 / 17	208.5	4-08-00-00430	eth 4; EtOH 4; chl 4
576	Benzaldehyde, 2-hydroxy-4-methyl-	$C_8H_8O_2$ / 136.15	698-27-1 / 60.5	223	4-08-00-00434	eth 4; EtOH 4; chl 4
577	Benzaldehyde, 2-hydroxy-5-methyl-	$C_8H_8O_2$ / 136.15	613-84-3 / 56	217.5	4-08-00-00432 / 1.091[59]	eth 4; EtOH 4; chl 4 / 1.547[59]
578	Benzaldehyde, 2-hydroxy-6-methyl-	$C_8H_8O_2$ / 136.15	18362-36-2 / 32.3	230	4-08-00-00430	bz 4
579	Benzaldehyde, 2-hydroxy-, oxime / Salicylaldoxime	$C_7H_7NO_2$ / 137.14	94-67-7 / 57	8296	1-08-00-00520	H_2O 2; EtOH 4; eth 4; bz 4
580	Benzaldehyde, 2-iodo-	C_7H_5IO / 232.02	26260-02-6 / 37	129[14]	4-07-00-00582	H_2O 2; ace 3
581	Benzaldehyde, 3-iodo-	C_7H_5IO / 232.02	696-41-3 / 57	125[13]	4-07-00-00583	EtOH 3; ace 3; bz 3
582	Benzaldehyde, 4-iodo-	C_7H_5IO / 232.02	15164-44-0 / 77.5	265	4-07-00-00584	H_2O 2; EtOH 3; bz 3
583	Benzaldehyde, 2-methoxy-	$C_8H_8O_2$ / 136.15	135-02-4 / 37.5	243.5	4-08-00-00180 / 1.1326[20]	H_2O 1; EtOH 3; eth 4; ace 4 / 1.5600[20]
584	Benzaldehyde, 3-methoxy-	$C_8H_8O_2$ / 136.15	591-31-1	231	4-08-00-00241 / 1.1187[20]	H_2O 1; EtOH 3; eth 4; ace 4 / 1.5530[20]

No.	Name / Synonym	Mol. Form. / Mol. Wt.	CAS RN / mp/°C	Merck No. / bp/°C	Beil. Ref. / den/g cm^{-3}	Solubility / n_D
585	Benzaldehyde, 4-methoxy- p-Anisaldehyde	$C_8H_8O_2$ 136.15	123-11-5 0	693 248; 134[12]	4-08-00-00252 1.119[15]	H_2O 1; EtOH 5; eth 5; ace 4 1.5730[20]
586	Benzaldehyde, 4-methoxy-3-methyl-	$C_9H_{10}O_2$ 150.18	32723-67-4 80-5[1]		4-08-00-00432 1.025[25]	1.5670[20]
587	Benzaldehyde, 2-methyl- o-Tolualdehyde	C_8H_8O 120.15	529-20-4 200; 94[10]	9453	4-07-00-00667 1.0328[20]	H_2O 2; EtOH 3; eth 3; ace 4 1.5462[20]
588	Benzaldehyde, 3-methyl-	C_8H_8O 120.15	620-23-5 199		4-07-00-00669 1.0189[21]	H_2O 2; EtOH 5; eth 5; ace 4 1.5413[21]
589	Benzaldehyde, 4-methyl- p-Tolualdehyde	C_8H_8O 120.15	104-87-0 204-5.106[10]		4-07-00-00672 1.0194[17]	H_2O 2; EtOH 5; eth 5; ace 5 1.5454[20]
590	Benzaldehyde, 4-(1-methylethyl)- Cuminaldehyde	$C_{10}H_{12}O$ 148.20	122-03-2 235.5	2623	4-07-00-00723 0.9755[20]	H_2O I; EtOH 3; eth 3; ctc 2 1.5301[20]
591	Benzaldehyde, 2-nitro-	$C_7H_5NO_3$ 151.12	552-89-6 43.5	153[23]	4-07-00-00584 1.2844[20]	H_2O 2; EtOH 4; eth 4; ace 4
592	Benzaldehyde, 3-nitro-	$C_7H_5NO_3$ 151.12	99-61-6 58.5	164[23]	4-07-00-00591 1.2792[20]	H_2O 2; EtOH 3; eth 3; ace 4
593	Benzaldehyde, 4-nitro-	$C_7H_5NO_3$ 151.12	555-16-8 107	sub	4-07-00-00598 1.496[25]	H_2O 2; EtOH 4; eth 2; bz 3
594	Benzaldehyde, oxime, (E)-	C_7H_7NO 121.14	622-31-1 35	119[10]	4-07-00-00527 1.145[20]	H_2O 3; EtOH 4; eth 4
595	Benzaldehyde, oxime, (Z)-	C_7H_7NO 121.14	622-32-2 36.5	200	4-07-00-00527 1.1111[20]	bz 4; eth 4; EtOH 4 1.5908[20]
596	Benzaldehyde, pentafluoro-	C_7HF_5O 196.08	653-37-2 20	167	4-07-00-00561	1.4506[20]
597	Benzaldehyde, 3-phenoxy-	$C_{13}H_{10}O_2$ 198.22	39515-51-0 140[0.1]		4-08-00-00242 1.147[25]	1.5954[20]
598	Benzaldehyde, 2,3,5,6-tetramethyl- Durene aldehyde	$C_{11}H_{14}O$ 162.23	17432-37-0 20	135[11]	3-07-00-01148	1.5560[30]
599	Benzaldehyde, 2,4,6-trimethyl-	$C_{10}H_{12}O$ 148.20	487-68-3 14	238.5	4-07-00-00730 1.0154[25]	H_2O 1; EtOH 3; eth 3; ace 3
600	Benzamide Benzoic acid amide	C_7H_7NO 121.14	55-21-0 129.1	1067 290	4-09-00-00725 1.0792[130]	H_2O 2; EtOH 4; eth 2; bz 2
601	Benzamide, 2-amino-	$C_7H_8N_2O$ 136.15	88-68-6 110.5 dec	300	4-14-00-01010	H_2O 3; EtOH 3; eth 2; bz 2
602	Benzamide, 4-amino-N-[2-(diethylamino)ethyl]-, monohydrochloride Procainamide hydrochloride	$C_{13}H_{22}ClN_3O$ 271.79	614-39-1 167	7762	4-14-00-01154	H_2O 4; EtOH 3; eth 1; bz 1
603	Benzamide, N-[(4-aminophenyl)sulfonyl]- Sulfabenzamide	$C_{13}H_{12}N_2O_3S$ 276.32	127-71-9 181.5	8868	4-14-00-02664	
604	Benzamide, 2-bromo-	C_7H_6BrNO 200.03	4001-73-4 160.5	sub	4-09-00-01013	H_2O 2; EtOH 3; eth 2; tfa 2
605	Benzamide, 3-bromo-	C_7H_6BrNO 200.03	22726-00-7 155.3	sub	4-09-00-01015	H_2O 2; EtOH 3
606	Benzamide, 5-bromo-N,2-dihydroxy- 5-Bromosalicylhydroxamic acid	$C_7H_6BrNO_3$ 232.03	5798-94-7 232 dec	1424	4-10-00-00221	
607	Benzamide, N-butyl-4-chloro-2-hydroxy- Buclosamide	$C_{11}H_{14}ClNO_2$ 227.69	575-74-6 91.5	1450		
608	Benzamide, 5-chloro-N-(3,4-dichlorophenyl)-2-hydroxy- 3',4',5-Trichlorosalicylanilide	$C_{13}H_8Cl_3NO_2$ 316.57	642-84-2 247	9558	4-12-00-01273	
609	Benzamide, 3,5-dibromo-N-(4-bromophenyl)-2-hydroxy- Tribromsalan	$C_{13}H_8Br_3NO_2$ 449.92	87-10-5 227	9529		
610	Benzamide, 3,5-dichloro-N-(3,4-dichlorophenyl)-2-hydroxy- 3,3',4',5-Tetrachlorosalicylanilide	$C_{13}H_7Cl_4NO_2$ 351.01	1154-59-2 161	9127		
611	Benzamide, N,N-diethyl-4-hydroxy-3-methoxy- Ethamivan	$C_{12}H_{17}NO_3$ 223.27	304-84-7 95	3673	4-10-00-01481	chl 3
612	Benzamide, N,N-diethyl-3-methyl- DEET	$C_{12}H_{17}NO$ 191.27	134-62-3 160[19]; 111[1]	2848	4-09-00-01716 0.996[20]	H_2O 4; bz 4; eth 4; EtOH 4 1.5212[20]
613	Benzamide, N,2-dihydroxy- Salicylhydroxamic acid	$C_7H_7NO_3$ 153.14	89-73-6 168	8300 sub	4-10-00-00192	H_2O 2; EtOH 4; eth 4; HOAc 3
614	Benzamide, 3,5-dinitro- Nitromide	$C_7H_5N_3O_5$ 211.13	121-81-3 184	6533	4-09-00-01351	H_2O 4
615	Benzamide, 2-ethoxy- Ethenzamide	$C_9H_{11}NO_2$ 165.19	938-73-8 133	3685	4-10-00-00175	H_2O 2; EtOH 4; eth 4; chl 2
616	Benzamide, 2-hydroxy- Salicylamide	$C_7H_7NO_2$ 137.14	65-45-2 142	8297 181.5[14]	4-10-00-00169 1.175[140]	H_2O 2; EtOH 3; eth 2
617	Benzamide, 2-hydroxy-N-phenyl- Salicylanilide	$C_{13}H_{11}NO_2$ 213.24	87-17-2 136.5	8299	4-12-00-00906	H_2O 3; EtOH 2; eth 2; bz 2
618	Benzamide, 4-methoxy-	$C_8H_9NO_2$ 151.17	3424-93-9 166.5	295	4-10-00-00433	H_2O 4; EtOH 4
619	Benzamide, 2-methyl- o-Toluamide	C_8H_9NO 135.17	527-85-5 147	9454	4-09-00-01701	H_2O 4; EtOH 4; eth 2; bz 2
620	Benzamide, N-methyl-	C_8H_9NO 135.17	613-93-4 82	291; 167[12]	4-09-00-00727	EtOH 3; ace 3
621	Benzamide, N-(2-methylphenyl)-	$C_{14}H_{13}NO$ 211.26	584-70-3 145.5		4-12-00-01757 1.205[15]	H_2O 2; EtOH 3; eth 3

No.	Name Synonym	Mol. Form. Mol. Wt.	CAS RN mp/°C	Merck No. bp/°C	Beil. Ref. den/g cm^{-3}	Solubility n_D
622	Benzamide, N-(3-methylphenyl)- m-Benzotoluidide	$C_{14}H_{13}NO$ 211.26	582-77-4 125		4-12-00-01825 1.170^{15}	EtOH 4
623	Benzamide, N-(4-methylphenyl)-	$C_{14}H_{13}NO$ 211.26	582-78-5 158		4-12-00-01910 1.202^{15}	eth 4; EtOH 4
624	Benzamide, 2-nitro-	$C_7H_6N_2O_3$ 166.14	610-15-1 176.6	317	4-09-00-01049	H_2O 3; EtOH 3; eth 3
625	Benzamide, 3-nitro-	$C_7H_6N_2O_3$ 166.14	645-09-0 142.7	312.5	4-09-00-01061	H_2O 3; EtOH 3; eth 3
626	Benzamide, N-phenyl- Benzanilide	$C_{13}H_{11}NO$ 197.24	93-98-1 163	1068 sub 117	4-12-00-00417 1.315^{25}	H_2O 1; EtOH 2; eth 2; HOAc 2
627	Benzamide, N,N,4-trimethyl-	$C_{10}H_{13}NO$ 163.22	14062-78-3 41	156^{10}	0-09-00-00486	H_2O 2; EtOH 2
628	Benz[b]indeno[1,2-d]pyran-9(6H)-one, 6a,7-dihydro-3,4,6a,10-tetrahydroxy- Hematein	$C_{16}H_{12}O_6$ 300.27	475-25-2 250 dec	4553	4-18-00-03343	H_2O 1; EtOH 2; eth 1; bz 1
629	Benz[b]indeno[1,2-d]pyran-3,4,6a,9,10(6H)-pentol, 7,11b-dihydro-, cis-(+)- Hematoxylin	$C_{16}H_{14}O_6$ 302.28	517-28-2 140	4556	2-17-00-00273	H_2O 2; EtOH 3; eth 2; alk 3
630	Benz[c]acridine 12-Azabenz[a]anthracene	$C_{17}H_{11}N$ 229.28	225-51-4 132		5-20-08-00519	bz 4; eth 4; EtOH 4
631	Benz[e]acephenanthrylene Benzo[b]fluoranthene	$C_{20}H_{12}$ 252.32	205-99-2 168		4-05-00-02686	H_2O 1; bz 5
632	1H-Benz[e]indene, 2,3-dihydro-	$C_{13}H_{12}$ 168.24	4944-94-9 	294.5	4-05-00-01865 1.066^{20}	H_2O 1; HOAc 3 1.6290^{20}
633	Benzenamine Aniline	C_6H_7N 93.13	62-53-3 -6.0	687 184.1	4-12-00-00223 1.0217^{20}	H_2O 3; EtOH 5; eth 5; ace 5 1.5863^{20}
634	Benzenamine, 4-[(4-aminophenyl)(4-imino-2,5-cyclohexadien-1-ylidene)methyl]-2-methyl-, monohydrochloride Magenta I	$C_{20}H_{20}ClN_3$ 337.85	632-99-5 200 dec	5528		H_2O 2; EtOH 2; eth 1
635	Benzenamine, 4,4'-azobis- p-Diaminoazobenzene	$C_{12}H_{12}N_4$ 212.25	538-41-0 250.5	2954	4-16-00-00515	H_2O 2; EtOH 3; bz 4; chl 4
636	Benzenamine, 2,6-bis(1-methylethyl)-	$C_{12}H_{19}N$ 177.29	24544-04-5 -45	257	3-12-00-02764 0.94^{25}	1.5332^{20}
637	Benzenamine, 3,5-bis(trifluoromethyl)-	$C_8H_5F_6N$ 229.12	328-74-5 	85^{15}	3-12-00-02498 1.487^{25}	1.4335^{20}
638	Benzenamine, 2-bromo- o-Bromoaniline	C_6H_6BrN 172.02	615-36-1 32	229	4-12-00-01487 1.578^{20}	H_2O 1; EtOH 3; eth 3 1.6113^{20}
639	Benzenamine, 3-bromo- m-Bromoaniline	C_6H_6BrN 172.02	591-19-5 18.5	251	4-12-00-01491 1.5793^{20}	H_2O 2; EtOH 3; eth 3 1.6260^{20}
640	Benzenamine, 4-bromo- p-Bromoaniline	C_6H_6BrN 172.02	106-40-1 66.4	1392 dec	4-12-00-01497 1.4970^{100}	H_2O 1; EtOH 3; eth 3; chl 2
641	Benzenamine, 2-bromo-4,6-dichloro-	$C_6H_4BrCl_2N$ 240.91	697-86-9 83.5	273	2-12-00-00355	bz 4; EtOH 4; chl 4
642	Benzenamine, 4-bromo-N,N-diethyl-	$C_{10}H_{14}BrN$ 228.13	2052-06-4 38	270	4-12-00-01500	H_2O 1; EtOH 4; eth 4
643	Benzenamine, 2-bromo-N,N-dimethyl-	$C_8H_{10}BrN$ 200.08	698-00-0 	$107-8^{14}$	4-12-00-01487 1.3839^{25}	EtOH 4 1.5748^{25}
644	Benzenamine, 3-bromo-N,N-dimethyl-	$C_8H_{10}BrN$ 200.08	16518-62-0 11	259	4-12-00-01492 1.3651^{20}	EtOH 4; HOAc 4; con acid 3
645	Benzenamine, 4-bromo-N,N-dimethyl-	$C_8H_{10}BrN$ 200.08	586-77-6 55	264	4-12-00-01499 1.3220^{100}	H_2O 1; EtOH 3; eth 4; os 3
646	Benzenamine, 2-bromo-4,6-dinitro-	$C_6H_4BrN_3O_4$ 262.02	1817-73-8 153.5	sub	4-12-00-01734	EtOH 4; ace 4; HOAc 3
647	Benzenamine, 2-bromo-4-methyl-	C_7H_8BrN 186.05	583-68-6 26	240	4-12-00-01992 1.510^{20}	H_2O 1; EtOH 3; eth 3 1.5999^{20}
648	Benzenamine, 2-bromo-5-methyl- 2-Bromo-5-methylaniline	C_7H_8BrN 186.05	53078-85-6 46	129^{15}	3-12-00-01998 1.470^{25}	eth 4; EtOH 4
649	Benzenamine, 3-bromo-4-methyl-	C_7H_8BrN 186.05	7745-91-7 26	255.5	4-12-00-01991	eth 4
650	Benzenamine, 4-bromo-2-methyl-	C_7H_8BrN 186.05	583-75-5 59.5	240	4-12-00-01804	H_2O 2; EtOH 3; eth 4; chl 2
651	Benzenamine, 4-bromo-3-methyl-	C_7H_8BrN 186.05	6933-10-4 81	240	4-12-00-01856	H_2O 1; EtOH 4; chl 2
652	Benzenamine, 5-bromo-2-methyl-	C_7H_8BrN 186.05	39478-78-9 33	255 dec; 165^{25}	3-12-00-01923	eth 4; EtOH 4
653	Benzenamine, 2-bromo-6-nitro- Aniline, 2-bromo-6-nitro-	$C_6H_5BrN_2O_2$ 217.02	59255-95-7 74.5		2-12-00-00402 1.988^{25}	EtOH 4
654	Benzenamine, 4-bromo-2-nitro-	$C_6H_5BrN_2O_2$ 217.02	875-51-4 111.5	sub	3-12-00-01670	EtOH 4
655	Benzenamine, 4-butyl-	$C_{10}H_{15}N$ 149.24	104-13-2 	261	4-12-00-02807 0.945^{20}	ctc 2
656	Benzenamine, N-butyl- N-Butylaniline	$C_{10}H_{15}N$ 149.24	1126-78-9 -14.4	243.5	4-12-00-00256 0.9323^{20}	eth 4; EtOH 4 1.5341^{20}
657	Benzenamine, 4,4'-carbonimidoylbis[N,N-dimethyl- 4,4'-Dimethylaminobenzophenonimide	$C_{17}H_{21}N_3$ 267.37	492-80-8 136		4-14-00-00256	H_2O 1; EtOH 3; eth 2

No.	Name / Synonym	Mol. Form. / Mol. Wt.	CAS RN / mp/°C	Merck No. / bp/°C	Bell. Ref. / den/g cm⁻³	Solubility / n_D
658	Benzenamine, 2-chloro- / o-Chloroaniline	C_6H_6ClN / 127.57	95-51-2 / -14	2118 / 208.8	4-12-00-01115 /	H_2O 1; EtOH 5; eth 3; ace 3 / 1.5895[20]
659	Benzenamine, 3-chloro- / m-Chloroaniline	C_6H_6ClN / 127.57	108-42-9 / -10.4	2118 / 230.5	4-12-00-01137 / 1.2161[20]	H_2O 1; EtOH 5; eth 5; ace 5 / 1.5941[20]
660	Benzenamine, 4-chloro- / p-Chloroaniline	C_6H_6ClN / 127.57	106-47-8 / 72.5	2118 / 232	4-12-00-01166 / 1.429[19]	H_2O 3; EtOH 3; eth 3; chl 3 / 1.5546[87]
661	Benzenamine, 4-chloro-N,N-diethyl-	$C_{10}H_{14}ClN$ / 183.68	2873-89-4 / 45.5	/ 252; 113[6]	4-12-00-01170 /	EtOH 3
662	Benzenamine, 2-chloro-N,N-dimethyl-	$C_8H_{10}ClN$ / 155.63	698-01-1 /	/ 205	4-12-00-01116 / 1.1067[20]	bz 4; EtOH 4 / 1.5578[20]
663	Benzenamine, 4-chloro-N,N-dimethyl-	$C_8H_{10}ClN$ / 155.63	698-69-1 / 35.5	/ 231	4-12-00-01169 / 1.0480[100]	EtOH 3
664	Benzenamine, 2-chloro-6-fluoro- / Aniline, 2-chloro-6-fluoro	C_6H_5ClFN / 145.56	363-51-9 /	/ 91[30]	4-12-00-01238 / 1.316[23]	/ 1.5511[23]
665	Benzenamine, 2-chloro-, hydrochloride / Aniline, 2-chloro, hydrochloride	$C_6H_7Cl_2N$ / 164.03	137-04-2 / 235		3-12-00-01281 / 1.505[18]	H_2O 4
666	Benzenamine, 2-chloro-5-methoxy-	C_7H_8ClNO / 157.60	2401-24-3 / 27	/ 142[18]	4-13-00-00999 /	/ 1.5848[25]
667	Benzenamine, 4-chloro-2-methoxy- / 4-Chloro-o-anisidine	C_7H_8ClNO / 157.60	93-50-5 / 52	/ 260	4-13-00-00885 /	EtOH 3; eth 3; bz 3; chl 3
668	Benzenamine, 2-chloro-4-methyl- / 2-Chloro-p-toluidine	C_7H_8ClN / 141.60	615-65-6 / 7	/ 220	3-12-00-02152 / 1.151[20]	EtOH 2; bz 2 / 1.5748[22]
669	Benzenamine, 2-chloro-5-methyl- / 6-Chloro-m-toluidine	C_7H_8ClN / 141.60	95-81-8 / 29.5	/ 229	4-12-00-01849 /	EtOH 4
670	Benzenamine, 2-chloro-N-methyl-	C_7H_8ClN / 141.60	932-32-1 /	/ 218	4-12-00-01116 / 1.1735[11]	EtOH 3; ace 3; bz 3 / 1.5780[25]
671	Benzenamine, 3-chloro-2-methyl- / 3-Chloro-o-toluidine	C_7H_8ClN / 141.60	87-60-5 / 1	/ 245	4-12-00-01800 /	H_2O 3; EtOH 3; eth 1; bz 1 / 1.5880[20]
672	Benzenamine, 3-chloro-4-methyl- / 3-Chloro-p-toluidine	C_7H_8ClN / 141.60	95-74-9 / 26	/ 243	4-12-00-01985 /	EtOH 3; ctc 2
673	Benzenamine, 4-chloro-2-methyl-	C_7H_8ClN / 141.60	95-69-2 / 30.3	/ 244	4-12-00-01796 /	EtOH 3; ctc 2
674	Benzenamine, 4-chloro-3-methyl- / 4-Chloro-m-toluidine	C_7H_8ClN / 141.60	7149-75-9 / 83.5	/ 241	4-12-00-01850 /	ace 4; bz 4; EtOH 4
675	Benzenamine, 4-chloro-N-methyl-	C_7H_8ClN / 141.60	932-96-7 /	/ 240	4-12-00-01168 / 1.169[11]	EtOH 3; ace 3; bz 3 / 1.5835[20]
676	Benzenamine, 5-chloro-2-methyl- / 5-Chloro-o-toluidine	C_7H_8ClN / 141.60	95-79-4 / 26	/ 239; 140[38]	4-12-00-01794 /	EtOH 4
677	Benzenamine, 5-chloro-2-nitro-	$C_6H_5ClN_2O_2$ / 172.57	1635-61-6 / 127.8	/ sub	4-12-00-01673 /	eth 4; EtOH 4
678	Benzenamine, 3-chloro-N-phenyl-	$C_{12}H_{10}ClN$ / 203.67	101-17-7 /	/ 338	4-12-00-01141 / 1.2[25]	H_2O 1; EtOH 3; eth 3; bz 3 / 1.6513[20]
679	Benzenamine, 4-chloro-N-phenyl-	$C_{12}H_{10}ClN$ / 203.67	1205-71-6 / 74	/ 335	4-12-00-01170 /	bz 4; eth 4; EtOH 4; lig 4
680	Benzenamine, N-cyclohexyl-	$C_{12}H_{17}N$ / 175.27	1821-36-9 / 16	/ 279; 192[73]	4-12-00-00268 / 1.0155[20]	H_2O 1; EtOH 3; eth 3; bz 3 / 1.5610[20]
681	Benzenamine, 2,4-dibromo-	$C_6H_5Br_2N$ / 250.92	615-57-6 / 79.5	/ 156[74]	4-12-00-01532 / 2.260[20]	EtOH 3; eth 3; chl 3; HOAc 3
682	Benzenamine, 2,6-dibromo-	$C_6H_5Br_2N$ / 250.92	608-30-0 / 87.5	/ 263	4-12-00-01535 /	EtOH 4; eth 4; bz 4; chl 4
683	Benzenamine, 3,4-dibromo- / 3,4-Dibromoaniline	$C_6H_5Br_2N$ / 250.92	615-55-4 / 80.5	/ 100	4-12-00-01535 /	EtOH 4
684	Benzenamine, N,N-dibutyl-	$C_{14}H_{23}N$ / 205.34	613-29-6 / -32.2	/ 274.8	4-12-00-00256 / 0.9037[20]	H_2O 1; EtOH 5; eth 5; ace 4 / 1.5186[20]
685	Benzenamine, 2,3-dichloro- / 2,3-Dichloroaniline	$C_6H_5Cl_2N$ / 162.02	608-27-5 / 24	/ 252	0-12-00-00621 /	EtOH 3; eth 4; ace 3; bz 2
686	Benzenamine, 2,4-dichloro- / 2,4-Dichloroaniline	$C_6H_5Cl_2N$ / 162.02	554-00-7 / 63.5	/ 245	4-12-00-01241 / 1.567[20]	H_2O 2; EtOH 3; eth 3; chl 2
687	Benzenamine, 2,5-dichloro- / 2,5-Dichloroaniline	$C_6H_5Cl_2N$ / 162.02	95-82-9 / 50	/ 251	4-12-00-01250 /	H_2O 2; EtOH 3; eth 3; bz 3
688	Benzenamine, 2,6-dichloro- / 2,6-Dichloroaniline	$C_6H_5Cl_2N$ / 162.02	608-31-1 / 39		2-12-00-00337 /	H_2O 2; EtOH 3; eth 3
689	Benzenamine, 3,4-dichloro- / 3,4-Dichloroaniline	$C_6H_5Cl_2N$ / 162.02	95-76-1 / 72	3041 / 272	4-12-00-01257 /	EtOH 3; eth 3; bz 2; chl 2
690	Benzenamine, 3,5-dichloro- / 3,5-Dichloroaniline	$C_6H_5Cl_2N$ / 162.02	626-43-7 / 52	/ 261	4-12-00-01274 /	H_2O 1; EtOH 3; eth 3; ctc 3
691	Benzenamine, 2,6-dichloro-4-ethoxy- / Aniline, 2,6-dichloro-4-ethoxy	$C_8H_9Cl_2NO$ / 206.07	51225-20-8 / 46	/ 275	2-13-00-00276 /	bz 4; eth 4; EtOH 4
692	Benzenamine, 2,6-dichloro-4-nitro- / 2,6-Dichloro-4-nitroaniline	$C_6H_4Cl_2N_2O_2$ / 207.02	99-30-9 / 191		4-12-00-01681 /	EtOH 3; DMSO-d_6 2; acid 3
693	Benzenamine, 3,4-diethoxy- / 3,4-Diethoxyaniline	$C_{10}H_{15}NO_2$ / 181.23	39052-12-5 / 48	/ 132-5[0.45]	4-13-00-02507 /	bz 4; eth 4; EtOH 4
694	Benzenamine, 2,6-diethyl-	$C_{10}H_{15}N$ / 149.24	579-66-8 / 3	/ 243	4-12-00-02841 / 0.906[25]	/ 1.5452[20]
695	Benzenamine, N,N-diethyl- / N,N-Diethylaniline	$C_{10}H_{15}N$ / 149.24	91-66-7 / -38.8	3102 / 216.3	4-12-00-00252 / 0.9307[20]	H_2O 2; EtOH 3; eth 4; ace 3 / 1.5409[20]
696	Benzenamine, N,N-diethyl-2-methyl-	$C_{11}H_{17}N$ / 163.26	606-46-2 / -60	/ 209	4-12-00-01748 / 0.9286[20]	H_2O 2; EtOH 5; eth 5; ctc 3 / 1.5153[20]
697	Benzenamine, N,N-diethyl-4-methyl-	$C_{11}H_{17}N$ / 163.26	613-48-9 /	/ 229	4-12-00-01875 / 0.9242[16]	H_2O 2; EtOH 5; eth 5

No.	Name Synonym	Mol. Form. Mol. Wt.	CAS RN mp/°C	Merck No. bp/°C	Beil. Ref. den/g cm⁻³	Solubility n_D
698	Benzenamine, N,N-diethyl-4-nitro-	$C_{10}H_{14}N_2O_2$ 194.23	2216-15-1 77.5		4-12-00-01618 1.225[25]	EtOH 3; lig 2
699	Benzenamine, N,N-diethyl-4-nitroso- Aniline, N,N-diethyl-4-nitroso	$C_{10}H_{14}N_2O$ 178.23	120-22-9 87.5		4-12-00-01559 1.24[15]	H_2O 2; EtOH 3; eth 3; ace 3
700	Benzenamine, 2,4-difluoro- 2,4-Difluoroaniline	$C_6H_5F_2N$ 129.11	367-25-9 -7.5	3131 170	4-12-00-01112 1.268[25]	1.5063[20]
701	Benzenamine, 2,4-diiodo- 2,4-Diiodoaniline	$C_6H_5I_2N$ 344.92	533-70-0 95.5	3173	3-12-00-01508 2.748[25]	ace 4; bz 4; eth 4; EtOH 4
702	Benzenamine, 2,5-dimethoxy-	$C_8H_{11}NO_2$ 153.18	102-56-7 82.5	270	4-13-00-02548	H_2O 3; EtOH 3; chl 3; lig 3
703	Benzenamine, 2,6-dimethoxy- 2,6-Dimethoxyaniline	$C_8H_{11}NO_2$ 153.18	2734-70-5 76.5	146[23]	4-13-00-02526	bz 4; eth 4; EtOH 4; lig 4
704	Benzenamine, 2,3-dimethyl- 2,3-Xylidine	$C_8H_{11}N$ 121.18	87-59-2 <-15	221.5	4-12-00-02497 0.9931[20]	H_2O 2; EtOH 4; eth 4; ctc 3 1.5684[20]
705	Benzenamine, 2,4-dimethyl- 2,4-Xylidine	$C_8H_{11}N$ 121.18	95-68-1 -14.3	214	4-12-00-02545 0.9723[20]	H_2O 2; EtOH 3; eth 3; bz 3 1.5569[20]
706	Benzenamine, 2,5-dimethyl- 2,5-Xylidine	$C_8H_{11}N$ 121.18	95-78-3 15.5	214	4-12-00-02567 0.9790[21]	H_2O 2; eth 3; ctc 3 1.5591[21]
707	Benzenamine, 2,6-dimethyl- 2,6-Xylidine	$C_8H_{11}N$ 121.18	87-62-7 11.2	215	4-12-00-02521 0.9842[20]	eth 4; EtOH 4 1.5610[20]
708	Benzenamine, 3,4-dimethyl- 3,4-Xylidine	$C_8H_{11}N$ 121.18	95-64-7 51	228	4-12-00-02502 1.076[18]	H_2O 2; eth 3; chl 2; lig 4
709	Benzenamine, 3,5-dimethyl- 3,5-Xylidine	$C_8H_{11}N$ 121.18	108-69-0 9.8	220.5	4-12-00-02561 0.9706[20]	H_2O 2; eth 3; ctc 3 1.5581[20]
710	Benzenamine, N,2-dimethyl-	$C_8H_{11}N$ 121.18	611-21-2 	207.5	4-12-00-01746 0.9709[20]	H_2O 1; EtOH 5; eth 5; ace 3 1.5649[20]
711	Benzenamine, N,3-dimethyl-	$C_8H_{11}N$ 121.18	696-44-6 	206.5	4-12-00-01815 	H_2O 1; EtOH 5; eth 5; ace 3 1.5557[25]
712	Benzenamine, N,4-dimethyl-	$C_8H_{11}N$ 121.18	623-08-5 	210	4-12-00-01874 0.9348[55]	H_2O 1; EtOH 5; eth 5; ace 3 1.5568[20]
713	Benzenamine, N,N-dimethyl- N,N-Dimethylaniline	$C_8H_{11}N$ 121.18	121-69-7 2.4	3223 194.1	4-12-00-00243 0.9557[20]	H_2O 2; EtOH 3; eth 3; ace 3 1.5582[20]
714	Benzenamine, 2-(1,1-dimethylethyl)-	$C_{10}H_{15}N$ 149.24	6310-21-0 	234	4-12-00-02823 0.977[15]	EtOH 4; eth 4; bz 4; acid 3 1.5453[20]
715	Benzenamine, 4-(1,1-dimethylethyl)-	$C_{10}H_{15}N$ 149.24	769-92-6 17	241	4-12-00-02828 0.9525[15]	H_2O 2; EtOH 5; eth 5; bz 4 1.5380[20]
716	Benzenamine, N-(1,1-dimethylethyl)- N-tert-Butylaniline	$C_{10}H_{15}N$ 149.24	937-33-7 	215; 95[19]	4-12-00-00257 	EtOH 3; ace 4; bz 4; chl 4 1.5270[20]
717	Benzenamine, N,N-dimethyl-, hydrochloride	$C_8H_{12}ClN$ 157.64	5882-44-0 90		4-12-00-00243 1.1156[19]	H_2O 4; EtOH 4; chl 4
718	Benzenamine, N,N-dimethyl-2-nitro-	$C_8H_{10}N_2O_2$ 166.18	610-17-3 -20	146[20]	4-12-00-01564 1.1794[20]	H_2O 3; EtOH 4; eth 3; chl 4 1.6102[20]
719	Benzenamine, N,N-dimethyl-3-nitro-	$C_8H_{10}N_2O_2$ 166.18	619-31-8 60.5	282.5	4-12-00-01591 1.313[17]	H_2O 1; EtOH 3; eth 3
720	Benzenamine, N,N-dimethyl-4-nitroso- p-Nitroso-N,N-dimethylaniline	$C_8H_{10}N_2O$ 150.18	138-89-6 92.5	6559	4-12-00-01558 1.145[20]	H_2O 2; EtOH 3; eth 3; chl 3
721	Benzenamine, N,N-dimethyl-4-(phenylazo)- 4-(Dimethylamino)azobenzene	$C_{14}H_{15}N_3$ 225.29	60-11-7 117 dec	3218	3-16-00-00340 	H_2O 1; EtOH 4; eth 3; chl 2
722	Benzenamine, N,N-dimethyl-4-(2- phenylethenyl)-, (E)-	$C_{16}H_{17}N$ 223.32	838-95-9 150	 145[0.1]	4-12-00-03400 	H_2O 1; EtOH 3; chl 3
723	Benzenamine, N,N-dimethyl-4-[[3- (trifluoromethyl)phenyl]azo]- Azobenzene, 4-dimethylamino-3'- trifluoromethyl	$C_{15}H_{14}F_3N_3$ 293.29	328-96-1 80	 76[8]	4-16-00-00487 	chl 3
724	Benzenamine, 2,3-dinitro- 2,3-Dinitroaniline	$C_6H_5N_3O_4$ 183.12	602-03-9 128		4-12-00-01689 1.646[50]	H_2O 1; EtOH 3; eth 2
725	Benzenamine, 2,4-dinitro- 2,4-Dinitroaniline	$C_6H_5N_3O_4$ 183.12	97-02-9 180	3263	4-12-00-01689 1.615[14]	H_2O 1; EtOH 2; ace 2; HCl 2
726	Benzenamine, 2,6-dinitro- 2,6-Dinitroaniline	$C_6H_5N_3O_4$ 183.12	606-22-4 141.5	3264	4-12-00-01729 	H_2O 1; EtOH 2; eth 3; bz 3
727	Benzenamine, 3,5-dinitro- 3,5-Dinitroaniline	$C_6H_5N_3O_4$ 183.12	618-87-1 163		4-12-00-01729 1.601[50]	H_2O 1; EtOH 3; eth 3; ace 2
728	Benzenamine, N,N-diphenyl- Triphenylamine	$C_{18}H_{15}N$ 245.32	603-34-9 127	365	4-12-00-00276 0.774[0]	H_2O 1; EtOH 2; eth 3; bz 3 1.353[16]
729	Benzenamine, 4-(diphenylmethyl)-	$C_{19}H_{17}N$ 259.35	603-38-3 84.5	248[12]	4-12-00-03470 	bz 4; eth 4; lig 4
730	Benzenamine, N,N-dipropyl-	$C_{12}H_{19}N$ 177.29	2217-07-4 	242	4-12-00-00255 0.9104[20]	H_2O 1; EtOH 3; eth 3; ace 3 1.5271[20]
731	Benzenamine, 4,4'-(1,2-ethanediyl)bis-	$C_{14}H_{16}N_2$ 212.29	621-95-4 137 sub		4-13-00-00401 	H_2O 1; EtOH 4
732	Benzenamine, 4,4'-(1,2-ethenediyl)bis-, (E)-	$C_{14}H_{14}N_2$ 210.28	7314-06-9 231 sub		4-13-00-00453 	H_2O 2; bz 2; MeOH 3; CS_2 2
733	Benzenamine, 2-ethenyl-	C_8H_9N 119.17	3867-18-3 	112[20]; 97[8]	4-12-00-02914 1.0181[20]	ace 3; bz 3 1.6124[20]
734	Benzenamine, 3-ethenyl-	C_8H_9N 119.17	15411-43-5 	112-5[12]	4-12-00-02915 1.0198[20]	ace 3; bz 3 1.6069[26]
735	Benzenamine, 4-ethenyl-	C_8H_9N 119.17	1520-21-4 23.5	116[9]	4-12-00-02915 1.010[20]	ace 3; bz 3 1.6250[22]

No.	Name / Synonym	Mol. Form. / Mol. Wt.	CAS RN / mp/°C	Merck No. / bp/°C	Beil. Ref. / den/g cm⁻³	Solubility / n_D
736	Benzenamine, 2-ethoxy- / o-Phenetidine	$C_8H_{11}NO$ / 137.18	94-70-2 / <-21	7187 / 232.5	4-13-00-00807	H_2O 2; EtOH 3; eth 3; ctc 2 / 1.5560[20]
737	Benzenamine, 4-ethoxy- / p-Phenetidine	$C_8H_{11}NO$ / 137.18	156-43-4 / 2.4	7188 / 254	4-13-00-01017 / 1.0652[16]	H_2O 2; EtOH 3; eth 3; chl 3 / 1.5528[20]
738	Benzenamine, 3-ethoxy-N,N-diethyl-	$C_{12}H_{19}NO$ / 193.29	1864-92-2 /	/ 286; 97[0.6]	4-13-00-00970	EtOH 3; bz 3; HOAc 3 / 1.5325[25]
739	Benzenamine, 2-ethoxy-5-nitro- / o-Phenetidine, 5-nitro	$C_8H_{10}N_2O_3$ / 182.18	136-79-8 / 96.5	6538 / 205-6[14]	4-13-00-00896	eth 4; EtOH 4
740	Benzenamine, 2-ethyl- / 2-Ethylaniline	$C_8H_{11}N$ / 121.18	578-54-1 / -43	/ 209.5	4-12-00-02411 / 0.983[22]	H_2O 2; EtOH 4; eth 4; chl 2 / 1.5584[22]
741	Benzenamine, 3-ethyl-	$C_8H_{11}N$ / 121.18	587-02-0 / -64	/ 214; 94[6]	4-12-00-02417	eth 4; EtOH 4
742	Benzenamine, 4-ethyl- / p-Ethylaniline	$C_8H_{11}N$ / 121.18	589-16-2 / -4.8	/ 217.5	4-12-00-02419 / 0.9679[20]	H_2O 2; EtOH 4; eth 4; ctc 2 / 1.5554[20]
743	Benzenamine, N-ethyl- / N-Ethylaniline	$C_8H_{11}N$ / 121.18	103-69-5 / -63.5	3722 / 203.0	4-12-00-00250 / 0.9625[20]	H_2O 1; EtOH 5; eth 3; ace 4 / 1.5559[20]
744	Benzenamine, 2-ethyl-N-(2-ethylphenyl)-	$C_{16}H_{19}N$ / 225.33	64653-59-4 / 29	/ 336[603]; 173[10]	4-12-00-02413	H_2O 1; EtOH 4; eth 4; chl 2 / 1.5550[25]
745	Benzenamine, 2-ethyl-6-methyl-	$C_9H_{13}N$ / 135.21	24549-06-2 / -33	/ 231	4-12-00-02638 / 0.968[25]	/ 1.5525[20]
746	Benzenamine, N-ethyl-2-methyl-	$C_9H_{13}N$ / 135.21	94-68-8 / <-15	/ 216	4-12-00-01747 / 0.948[25]	EtOH 3; eth 3 / 1.5456[20]
747	Benzenamine, N-ethyl-3-methyl-	$C_9H_{13}N$ / 135.21	102-27-2 /	/ 221	4-12-00-01816	EtOH 3; eth 3 / 1.5451[20]
748	Benzenamine, N-ethyl-4-methyl- / N-Ethyl-4-toluidine	$C_9H_{13}N$ / 135.21	622-57-1 /	/ 217	4-12-00-01875 / 0.9391[16]	EtOH 3; eth 3
749	Benzenamine, N-ethyl-N-methyl-	$C_9H_{13}N$ / 135.21	613-97-8 /	/ 204	4-12-00-00252 / 0.92[55]	H_2O 1; EtOH 5; eth 5; ctc 3
750	Benzenamine, N-ethyl-N-nitroso-	$C_8H_{10}N_2O$ / 150.18	612-64-6 /	/ 119-20[15]	4-16-00-00865 / 1.0874[20]	H_2O 1; HOAc 3
751	Benzenamine, N-ethyl-N-phenyl-	$C_{14}H_{15}N$ / 197.28	606-99-5 /	/ 295.5; 147[10]	4-12-00-00276 / 1.0377[20]	eth 4; EtOH 4 / 1.6095[20]
752	Benzenamine, 2-fluoro- / o-Fluoroaniline	C_6H_6FN / 111.12	348-54-9 / -34.6	/ 175; 55[12]	4-12-00-01099 / 1.1513[21]	H_2O 1; EtOH 3; eth 3; ctc 2 / 1.5421[20]
753	Benzenamine, 3-fluoro- / m-Fluoroaniline	C_6H_6FN / 111.12	372-19-0 /	/ 188	4-12-00-01101 / 1.1561[19]	H_2O 2; EtOH 3; eth 3; chl 2 / 1.5436[20]
754	Benzenamine, 4-fluoro- / p-Fluoroaniline	C_6H_6FN / 111.12	371-40-4 / -0.8	4098 / 182; 85[19]	4-12-00-01104 / 1.1725[20]	H_2O 2; EtOH 3; eth 3; ctc 2 / 1.5195[20]
755	Benzenamine, 4-fluoro-2-methyl-	C_7H_8FN / 125.15	452-71-1 / 14.2	/ 94[16]	4-12-00-01786 / 1.1263[18]	eth 3; ace 3; bz 3; ctc 3 / 1.5363[18]
756	Benzenamine, hydrochloride / Aniline, hydrochloride	C_6H_8ClN / 129.59	142-04-1 / 198	687 /	4-12-00-00223 / 1.2215[4]	H_2O 4; EtOH 4; eth 1; chl 1
757	Benzenamine, N-hydroxy- / Phenylhydroxylamine	C_6H_7NO / 109.13	100-65-2 / 83.5	7266 /	4-15-00-00004	bz 4; eth 4; EtOH 4; chl 4
758	Benzenamine, N-hydroxy-4-methyl-	C_7H_9NO / 123.15	623-10-9 / 96	/ 117 dec	4-15-00-00019	eth 4; EtOH 4; chl 4
759	Benzenamine, N-hydroxy-N-nitroso-, ammonium salt / Cupferron	$C_6H_9N_3O_2$ / 155.16	135-20-6 / 163.5	2625 /	4-16-00-00889	DMSO 2
760	Benzenamine, 3-iodo-	C_6H_6IN / 219.02	626-01-7 / 33	/ 145[15]	4-12-00-01543	H_2O 1; EtOH 3; chl 3 / 1.6811[20]
761	Benzenamine, 4-iodo- / p-Iodoaniline	C_6H_6IN / 219.02	540-37-4 / 67.5	4920 /	4-12-00-01544	H_2O 2; EtOH 3; eth 3; os 3
762	Benzenamine, 2-iodo-4-methyl- / 2-Iodo-4-methylaniline	C_7H_8IN / 233.05	29289-13-2 / 40	/ dec	4-12-00-01996	ace 4; bz 4; eth 4; EtOH 4
763	Benzenamine, 5-iodo-2-methyl- / o-Toluidine, 5-iodo	C_7H_8IN / 233.05	83863-33-6 / 48.5	/ 273 dec	1-12-00-00391	EtOH 4
764	Benzenamine, N,N'-methanetetraylbis-	$C_{13}H_{10}N_2$ / 194.24	622-16-2 / 169	/ 331; 175[20]	4-12-00-00866	H_2O 2; EtOH 2; eth 2; bz 3
765	Benzenamine, N,N'-methanetetraylbis[4-methyl-	$C_{15}H_{14}N_2$ / 222.29	726-42-1 / 58.5	/ 221[20]	4-12-00-01938 / 1.1500[20]	
766	Benzenamine, 2-methoxy- / o-Anisidine	C_7H_9NO / 123.15	90-04-0 / 6.2	/ 224	4-13-00-00806 / 1.0923[20]	H_2O 2; EtOH 3; eth 3; ace 3 / 1.5715[10]
767	Benzenamine, 3-methoxy- / m-Anisidine	C_7H_9NO / 123.15	536-90-3 / -1	/ 251	4-13-00-00953 / 1.096[20]	H_2O 2; EtOH 3; eth 3; ace 3 / 1.5794[20]
768	Benzenamine, 4-methoxy- / p-Anisidine	C_7H_9NO / 123.15	104-94-9 / 57.2	/ 243	4-13-00-01015 / 1.071[57]	H_2O 3; EtOH 4; eth 4; ace 3 / 1.5559[80]
769	Benzenamine, 2-methoxy-N,N-dimethyl-	$C_9H_{13}NO$ / 151.21	700-75-4 /	/ 210	4-13-00-00816 / 1.0160[23]	chl 4
770	Benzenamine, 2-methoxy-5-methyl- / 5-Methyl-o-anisidine	$C_8H_{11}NO$ / 137.18	120-71-8 / 53	/ 235	3-13-00-01577	H_2O 2; EtOH 3; eth 3; bz 3
771	Benzenamine, 3-methoxy-4-methyl-	$C_8H_{11}NO$ / 137.18	16452-01-0 / 56.5	/ 251	4-13-00-01666	bz 4; eth 4; EtOH 4; lig 4
772	Benzenamine, 4-methoxy-2-methyl-	$C_8H_{11}NO$ / 137.18	102-50-1 / 29.5	/ 248.5	4-13-00-01698 / 1.065[25]	EtOH 4 / 1.5647[20]
773	Benzenamine, 5-methoxy-2-methyl-	$C_8H_{11}NO$ / 137.18	50868-72-9 / 47	/ 253; 140[20]	4-13-00-01711	eth 4

No.	Name Synonym	Mol. Form. Mol. Wt.	CAS RN mp/°C	Merck No. bp/°C	Beil. Ref. den/g cm^{-3}	Solubility n_D
774	Benzenamine, 2-methoxy-6-methyl-	$C_8H_{11}NO$ 137.18	50868-73-0 31	119[16]	3-13-00-01549	EtOH 4
775	Benzenamine, 2-methoxy-5-nitro- 5-Nitro-o-anisidine	$C_7H_8N_2O_3$ 168.15	99-59-2 118		4-13-00-00896 1.2068[15]	H_2O 3; EtOH 4; eth 3; ace 4
776	Benzenamine, 3-methoxy-4-nitro- m-Anisidine, 4-nitro-	$C_7H_8N_2O_3$ 168.15	16292-88-9 169	sub	4-13-00-01005	ace 4; EtOH 4
777	Benzenamine, 5-methoxy-2-nitro-	$C_7H_8N_2O_3$ 168.15	16133-49-6 131	sub	4-13-00-01003	EtOH 4
778	Benzenamine, 2-methyl- o-Toluidine	C_7H_9N 107.16	95-53-4 -16.3	9462 200.3	4-12-00-01744 0.9984[20]	H_2O 2; EtOH 5; eth 5; ctc 5 1.5725[20]
779	Benzenamine, 3-methyl- m-Toluidine	C_7H_9N 107.16	108-44-1 -31.2	9462 203.3	4-12-00-01813 0.9889[20]	ace 4; bz 4; eth 4; EtOH 4 1.5681[20]
780	Benzenamine, 4-methyl- p-Toluidine	C_7H_9N 107.16	106-49-0 43.7	9462 200.4	4-12-00-01866 0.9619[20]	H_2O 2; EtOH 4; eth 3; ace 3 1.5534[45]
781	Benzenamine, N-methyl- N-Methylaniline	C_7H_9N 107.16	100-61-8 -57	5941 196.2	4-12-00-00241 0.9891[20]	H_2O 1; EtOH 3; eth 3; ctc 3 1.5684[20]
782	Benzenamine, 4-(6-methyl-2-benzothiazolyl)-	$C_{14}H_{12}N_2S$ 240.33	92-36-4 194.8	434	4-27-00-05052	EtOH 2; eth 2; bz 2; HOAc 2
783	Benzenamine, N-(3-methylbutyl)-	$C_{11}H_{17}N$ 163.26	2051-84-5	254.5	4-12-00-00259 0.8912[55]	eth 4; EtOH 4 1.5305[20]
784	Benzenamine, 4,4'-methylenebis- p,p'-Diaminodiphenylmethane	$C_{13}H_{14}N_2$ 198.27	101-77-9 92.5	2958 398; 257[18]	4-13-00-00390	H_2O 2; EtOH 4; eth 4; bz 4
785	Benzenamine, 4,4'-methylenebis[2-chloro- 4,4'-Methylenebis[2-chloroaniline]	$C_{13}H_{12}Cl_2N_2$ 267.16	101-14-4 	5978	1-13-00-00074	ctc 3
786	Benzenamine, 4,4'-methylenebis[N,N-dimethyl- Michler's Base	$C_{17}H_{22}N_2$ 254.38	101-61-1 91.5	6099 390 dec; 183[3]	4-13-00-00390	H_2O 1; EtOH 2; eth 4; bz 4
787	Benzenamine, 2-(1-methylethenyl)-	$C_9H_{11}N$ 133.19	52562-19-3	95[13]	4-12-00-02925 0.976[25]	1.5722[20]
788	Benzenamine, 2-(1-methylethyl)- 2-Isopropylaniline	$C_9H_{13}N$ 135.21	643-28-7	221; 95[13]	4-12-00-02630 0.9760[12]	H_2O 1; eth 3; bz 3; ctc 3
789	Benzenamine, 4-(1-methylethyl)- Cumidine	$C_9H_{13}N$ 135.21	99-88-7	2622 225	4-12-00-02632 0.953[20]	
790	Benzenamine, N-(1-methylethyl)- N-Isopropylaniline	$C_9H_{13}N$ 135.21	768-52-5	203	4-12-00-00255 0.9526[25]	EtOH 3; eth 3; ace 3; bz 3 1.5380[20]
791	Benzenamine, 2-methyl-, hydrochloride o-Toluidine hydrochloride	$C_7H_{10}ClN$ 143.62	636-21-5 215		4-12-00-01744	H_2O 4; EtOH 4
792	Benzenamine, N-methyl-, hydrochloride	$C_7H_{10}ClN$ 143.62	2739-12-0 122.7		4-12-00-00241 1.0660[131]	H_2O 4; EtOH 3; eth 1; bz 1
793	Benzenamine, 2-methyl-5-(1-methylethyl)-	$C_{10}H_{15}N$ 149.24	2051-53-8 -16	241	3-12-00-02733 0.9942[20]	ctc 3; CS_2 3 1.5387[20]
794	Benzenamine, 4-methyl-N-(1-methylethyl)-	$C_{10}H_{15}N$ 149.24	10436-75-6	220	4-12-00-01876 0.9226[20]	1.5332
795	Benzenamine, 2-methyl-N-(2-methylphenyl)-	$C_{14}H_{15}N$ 197.28	617-00-5 52.5	316; 192[23]	4-12-00-01750	H_2O 2
796	Benzenamine, 3-methyl-N-(3-methylphenyl)-	$C_{14}H_{15}N$ 197.28	626-13-1 -12-	319.5	2-12-00-00467	H_2O 2; EtOH 3; eth 3; peth 3
797	Benzenamine, 4-methyl-N-(4-methylphenyl)-	$C_{14}H_{15}N$ 197.28	620-93-9 79.8	330.5	4-12-00-01879	eth 4; peth 4
798	Benzenamine, 2-methyl-4-[(2-methylphenyl)azo]- 2',3-Dimethyl-4-aminoazobenzene	$C_{14}H_{15}N_3$ 225.29	97-56-3 102		4-16-00-00528	eth 4; EtOH 4
799	Benzenamine, 2-methyl-3-nitro-	$C_7H_8N_2O_2$ 152.15	603-83-8 92	305	4-12-00-01811 1.3780[15]	H_2O 2; EtOH 3; eth 3; bz 3
800	Benzenamine, 2-methyl-4-nitro-	$C_7H_8N_2O_2$ 152.15	99-52-5 133.5		4-12-00-01809 1.1586[140]	H_2O 2; EtOH 3; bz 3; HOAc 3
801	Benzenamine, 2-methyl-5-nitro- 5-Nitro-o-toluidine	$C_7H_8N_2O_2$ 152.15	99-55-8 105.5		4-12-00-01807	H_2O 2; EtOH 3; eth 3; ace 3
802	Benzenamine, 2-methyl-6-nitro-	$C_7H_8N_2O_2$ 152.15	570-24-1 96		4-12-00-01807 1.1900[100]	H_2O 2; EtOH 3; eth 3; bz 3
803	Benzenamine, 4-methyl-2-nitro-	$C_7H_8N_2O_2$ 152.15	89-62-3 116.3		4-12-00-02000 1.16[121]	H_2O 2; EtOH 3; chl 3
804	Benzenamine, N-methyl-2-nitro-	$C_7H_8N_2O_2$ 152.15	612-28-2 38	158[18]	4-12-00-01564	H_2O 2; EtOH 3; eth 3; ace 3
805	Benzenamine, N-methyl-4-nitro-	$C_7H_8N_2O_2$ 152.15	100-15-2 152	dec	4-12-00-01616 1.201[155]	H_2O 1; EtOH 3; eth 2; bz 3
806	Benzenamine, N-methyl-N-nitroso-	$C_7H_8N_2O$ 136.15	614-00-6 14.7	225 dec; 121[13]	4-16-00-00864 1.1240[20]	H_2O 1; EtOH 3; eth 3 1.5769[20]
807	Benzenamine, N-methyl-N-phenyl- Methyldiphenylamine	$C_{13}H_{13}N$ 183.25	552-82-9 -7.5	5973 293.5	4-12-00-00275 1.0476[20]	H_2O 1; EtOH 2; ctc 3; MeOH 2 1.6193[20]
808	Benzenamine, 3-methyl-N-phenyl-	$C_{13}H_{13}N$ 183.25	1205-64-7 30	316; 183[17]	4-12-00-01818 1.6350[20]	bz 4; eth 4; EtOH 4
809	Benzenamine, 2-methyl-N-(phenylmethylene)-	$C_{14}H_{13}N$ 195.26	5877-55-4 	314	4-12-00-01751 1.041[20]	H_2O 1; ace 3 1.6310[25]
810	Benzenamine, 3-methyl-N-(phenylmethylene)-	$C_{14}H_{13}N$ 195.26	5877-58-7 31	315	3-12-00-01958 1.0390[25]	ace 3; os 3 1.6353[25]

No.	Name Synonym	Mol. Form. Mol. Wt.	CAS RN mp/°C	Merck No. bp/°C	Beil. Ref. den/g cm⁻³	Solubility n_D
811	Benzenamine, 4-methyl-N-(phenylmethylene)-	$C_{14}H_{13}N$ 195.26	2272-45-9 35	318; 178[11]	2-12-00-00496	ace 4
812	Benzenamine, 2-(1-methylpropyl)-	$C_{10}H_{15}N$ 149.24	55751-54-7 	120-2[16]	4-12-00-02814 0.9574[20]	EtOH 3; ace 3; bz 3; ctc 2
813	Benzenamine, 4-(1-methylpropyl)-	$C_{10}H_{15}N$ 149.24	30273-11-1 	238; 118[15]	4-12-00-02814 0.949[15]	bz 4; eth 4 1.5360[29]
814	Benzenamine, 4-(2-methylpropyl)-	$C_{10}H_{15}N$ 149.24	30090-17-6 	238; 112[11]	4-12-00-02818 0.949[15]	ace 4; bz 4; EtOH 4
815	Benzenamine, 4-methyl-N-propyl- N-Propyl-p-toluidine	$C_{10}H_{15}N$ 149.24	54837-90-0 	235	4-12-00-01876 0.9243[20]	1.5367
816	Benzenamine, N-(1-methylpropyl)- N-sec-Butylaniline	$C_{10}H_{15}N$ 149.24	6068-69-5 	225	4-12-00-00257	chl 2; os 3 1.5333[20]
817	Benzenamine, N-(2-methylpropyl)-	$C_{10}H_{15}N$ 149.24	588-47-6 	232; 113[11]	4-12-00-00257 0.940[18]	H_2O 1; eth 4; bz 4 1.5328[20]
818	Benzenamine, N-methyl-N,2,4,6-tetranitro- Nitramine	$C_7H_5N_5O_8$ 287.15	479-45-8 131.5	6488 exp 180	4-16-00-00895 1.57[10]	H_2O 1; EtOH 2; eth 2; ace 3
819	Benzenamine, 2-(methylthio)- Aniline, 2-methylthio	C_7H_9NS 139.22	2987-53-3 	234	4-13-00-00910 1.111[25]	1.6239[20]
820	Benzenamine, 4-(methylthio)-	C_7H_9NS 139.22	104-96-1 	272.5	4-13-00-01289 1.1379[20]	EtOH 3; eth 3; ace 3; bz 3 1.6395[20]
821	Benzenamine, nitrate Aniline nitrate	$C_6H_8N_2O_3$ 156.14	542-15-4 190 dec	687	4-12-00-00223 1.356[4]	H_2O 4; eth 4; EtOH 4
822	Benzenamine, 2-nitro- o-Nitroaniline	$C_6H_6N_2O_2$ 138.13	88-74-4 71.2	6504 284	4-12-00-01563 1.442[15]	H_2O 2; EtOH 3; eth 4; ace 4
823	Benzenamine, 3-nitro- m-Nitroaniline	$C_6H_6N_2O_2$ 138.13	99-09-2 114	6503 306 dec	4-12-00-01589 0.991[25]	H_2O 2; EtOH 3; eth 3; ace 3
824	Benzenamine, 4-nitro- p-Nitroaniline	$C_6H_6N_2O_2$ 138.13	100-01-6 147	6505 332	4-12-00-01613 1.424[20]	H_2O 1; EtOH 3; eth 3; ace 3
825	Benzenamine, N-nitro-	$C_6H_6N_2O_2$ 138.13	645-55-6 47.3	exp	4-16-00-00889	H_2O 4; bz 4; eth 4; EtOH 4
826	Benzenamine, 4-nitro-N-phenyl-	$C_{12}H_{10}N_2O_2$ 214.22	836-30-6 135.3	211[30]	4-12-00-01619	H_2O 1; EtOH 4; ace 2; HOAc 4
827	Benzenamine, 5-nitro-2-propoxy- Aniline, 5-nitro-2-propoxy	$C_9H_{12}N_2O_3$ 196.21	553-79-7 49	6551	4-13-00-00897	EtOH 4
828	Benzenamine, 2-nitroso-	$C_6H_6N_2O$ 122.13	659-49-4 173.5	dec	4-12-00-01558	H_2O 3; EtOH 3; bz 2
829	Benzenamine, 4-nitroso-N-phenyl- p-Nitrosodiphenylamine	$C_{12}H_{10}N_2O$ 198.22	156-10-5 143	6560	4-12-00-01560	H_2O 2; EtOH 4; eth 4; bz 4
830	Benzenamine, N-nitroso-N-phenyl- N-Nitroso-N-diphenylamine	$C_{12}H_{10}N_2O$ 198.22	86-30-6 66.5		4-16-00-00865	EtOH 2; bz 3; chl 2
831	Benzenamine, 4-octyl-	$C_{14}H_{23}N$ 205.34	16245-79-7 20	310; 138[5]	4-12-00-02887 0.9128[20]	eth 4
832	Benzenamine, 4,4'-oxybis- Bis(p-aminophenyl) ether	$C_{12}H_{12}N_2O$ 200.24	101-80-4 189 dec		4-13-00-01038	
833	Benzenamine, 2,3,4,5,6-pentamethyl-	$C_{11}H_{17}N$ 163.26	2243-30-3 152.5	277.5	3-12-00-02758	eth 4; EtOH 4
834	Benzenamine, 2-phenoxy-	$C_{12}H_{11}NO$ 185.23	2688-84-8 45.8	308; 172[14]	4-13-00-00809	EtOH 3; eth 3; ace 3; bz 3
835	Benzenamine, 3-phenoxy-	$C_{12}H_{11}NO$ 185.23	3586-12-7 37	315; 180[10]	4-13-00-00955 1.1583[25]	EtOH 3; eth 3; ace 3; bz 3
836	Benzenamine, N-phenyl- Diphenylamine	$C_{12}H_{11}N$ 169.23	122-39-4 52.9	3317 302	4-12-00-00271 1.158[22]	H_2O 1; EtOH 4; eth 3; ace 4
837	Benzenamine, 4-(phenylazo)- p-Aminoazobenzene	$C_{12}H_{11}N_3$ 197.24	60-09-3 127	430 >360	2-16-00-00149	H_2O 2; EtOH 3; eth 3; bz 3
838	Benzenamine, 2-(2-phenylethyl)-	$C_{14}H_{15}N$ 197.28	5697-85-8 33	193[19]	4-12-00-03311 1.0430[20]	
839	Benzenamine, 4-(phenylmethyl)-	$C_{13}H_{13}N$ 183.25	1135-12-2 34.5	300	4-12-00-03281 1.038[25]	eth 4; EtOH 4; lig 4
840	Benzenamine, N-(phenylmethylene)- Benzylideneaniline	$C_{13}H_{11}N$ 181.24	538-51-2 54	1152 310	2-12-00-00113 1.038[55]	H_2O 1; EtOH 3; eth 3; chl 2 1.600[100]
841	Benzenamine, N-2-propenyl-	$C_9H_{11}N$ 133.19	589-09-3 	219; 106[12]	4-12-00-00265 0.982[25]	H_2O 2; EtOH 3; eth 5; ace 3 1.563[20]
842	Benzenamine, 2-propyl-	$C_9H_{13}N$ 135.21	1821-39-2 	226; 116[15]	4-12-00-02580 0.9602[20]	H_2O 1; EtOH 3; eth 3 1.5427[20]
843	Benzenamine, N-propyl-	$C_9H_{13}N$ 135.21	622-80-0 	222; 98[11]	4-12-00-00254 0.9443[20]	eth 4; EtOH 4 1.5428[20]
844	Benzenamine, N-sulfinyl-	C_6H_5NOS 139.18	1122-83-4 	200	4-12-00-01075 1.236[25]	1.6270[20]
845	Benzenamine, 4,4'-sulfinylbis- 4,4'-Sulfinyldianiline	$C_{12}H_{12}N_2OS$ 232.31	119-59-5 175 dec	8927	4-13-00-01306	H_2O 3; EtOH 3
846	Benzenamine, 4,4'-sulfonylbis- Dapsone	$C_{12}H_{12}N_2O_2S$ 248.31	80-08-0 175.5	2820	4-13-00-01306	EtOH 3; DMSO-d_6 2
847	Benzenamine, 2,3,4,5-tetramethyl- Aniline, 2,3,4,5-tetramethyl	$C_{10}H_{15}N$ 149.24	2217-45-0 69	259.5	3-12-00-02743	eth 4; EtOH 4; lig 4
848	Benzenamine, 2,3,4,6-tetramethyl-	$C_{10}H_{15}N$ 149.24	488-71-1 23.5	255	4-12-00-02849 0.978[24]	EtOH 4
849	Benzenamine, N,N,2,6-tetramethyl-	$C_{10}H_{15}N$ 149.24	769-06-2 -36	196; 88[20]	4-12-00-02522 0.9147[20]	

No.	Name Synonym	Mol. Form. Mol. Wt.	CAS RN mp/°C	Merck No. bp/°C	Beil. Ref. den/g cm^{-3}	Solubility n_D
850	Benzenamine, 4,4'-thiobis- Bis(4-aminophenyl) sulfide	$C_{12}H_{12}N_2S$ 216.31	139-65-1 108.5		4-13-00-01306	H_2O 2; EtOH 4; eth 4; bz 4
851	Benzenamine, 2,4,6-tribromo- 2,4,6-Tribromoaniline	$C_6H_4Br_3N$ 329.82	147-82-0 122	9522 300	4-12-00-01538 2.35^{20}	H_2O 1; EtOH 2; eth 3; chl 3
852	Benzenamine, 2,3,4-trichloro- 2,3,4-Trichloroaniline	$C_6H_4Cl_3N$ 196.46	634-67-3 73	292	0-12-00-00626	EtOH 4
853	Benzenamine, 2,4,5-trichloro-	$C_6H_4Cl_3N$ 196.46	636-30-6 96.5	270	4-12-00-01277	EtOH 3; eth 3; CS_2 4; lig 2
854	Benzenamine, 2,4,6-trichloro-	$C_6H_4Cl_3N$ 196.46	634-93-5 78.5	262	4-12-00-01281	H_2O 1; EtOH 3; eth 3; chl 3
855	Benzenamine, 2-(trifluoromethyl)- o-(Trifluoromethyl)aniline	$C_7H_6F_3N$ 161.13	88-17-5 35.5	68^{15}	4-12-00-01786 1.282^{25}	1.4810^{20}
856	Benzenamine, 3-(trifluoromethyl)- m-(Trifluoromethyl)aniline	$C_7H_6F_3N$ 161.13	98-16-8 5.5	187; 74-5^{10}	4-12-00-01843 1.3047^{12}	H_2O 2; EtOH 3; eth 3 1.4787^{20}
857	Benzenamine, 4-(trifluoromethyl)-	$C_7H_6F_3N$ 161.13	455-14-1 38	117.5^{60}	4-12-00-01982 1.283^{27}	1.4815^{25}
858	Benzenamine, 2,4,5-trimethyl-	$C_9H_{13}N$ 135.21	137-17-7 68	234.5	4-12-00-02645 0.957^{25}	EtOH 4
859	Benzenamine, 2,4,6-trimethyl- Mesitylamine	$C_9H_{13}N$ 135.21	88-05-1 -5	232.5	4-12-00-02648 0.9633^{25}	ctc 2 1.5495^{20}
860	Benzenamine, N,N,2-trimethyl- N,N-Dimethyl-o-toluidine	$C_9H_{13}N$ 135.21	609-72-3 -60	194.1	4-12-00-01747 0.9286^{20}	eth 4; EtOH 4 1.5152^{20}
861	Benzenamine, N,N,3-trimethyl-	$C_9H_{13}N$ 135.21	121-72-2	212	4-12-00-01815 0.9410^{20}	EtOH 5; eth 5 1.5492^{20}
862	Benzenamine, N,N,4-trimethyl-	$C_9H_{13}N$ 135.21	99-97-8	211	4-12-00-01874 0.9366^{20}	H_2O 1; EtOH 5; eth 5; ctc 3 1.5366^{20}
863	Benzenamine, 2,4,6-trinitro-	$C_6H_4N_4O_6$ 228.12	489-98-5 193.5	exp	4-12-00-01735 1.762^{10}	H_2O 1; EtOH 2; eth 2; ace 3
864	Benzenamine, 2,4,6-trinitro-N-(2,4,6-trinitrophenyl)- Dipicrylamine	$C_{12}H_5N_7O_{12}$ 439.21	131-73-7 244 dec	3340	4-12-00-01737	H_2O 1; EtOH 1; eth 2; ace 2
865	Benzenaminium, N-ethyl-3-hydroxy-N,N-dimethyl-, chloride Ammonium, dimethyl ethyl 3-hydroxyphenyl,chloride	$C_{10}H_{16}ClNO$ 201.70	116-38-1	3487	4-13-00-00969	
866	Benzenaminium, N,N,N-trimethyl-, iodide Phenyl trimethyl ammonium iodide	$C_9H_{14}IN$ 263.12	98-04-4 224	7289	4-12-00-00249	H_2O 4; EtOH 3; ace 2; chl 1
867	Benzene [6]Annulene	C_6H_6 78.11	71-43-2 5.5	1074 80.0	4-05-00-00583 0.8765^{20}	H_2O 2; EtOH 5; eth 5; ace 5 1.5011^{20}
868	Benzeneacetaldehyde Phenylacetaldehyde	C_8H_8O 120.15	122-78-1 33.5	7236 195	4-07-00-00664 1.0272^{20}	H_2O 2; EtOH 5; eth 5; ace 3 1.5255^{20}
869	Benzeneacetaldehyde, 3,4-dimethoxy-	$C_{10}H_{12}O_3$ 180.20	5703-21-9 125$^{1.5}$		4-08-00-01828 1.148^{20}	1.5426^{20}
870	Benzeneacetaldehyde, 2-methoxy-	$C_9H_{10}O_2$ 150.18	33567-59-8 245; 117^{17}		1-08-00-00543 1.0897^{20}	ace 4; EtOH 4 1.5393^{20}
871	Benzeneacetaldehyde, 4-methoxy-	$C_9H_{10}O_2$ 150.18	5703-26-4 255.5		4-08-00-00426 1.096^{20}	1.5359^{20}
872	Benzeneacetaldehyde, α-methyl-	$C_9H_{10}O$ 134.18	93-53-8 203.5		1.0089^{20}	EtOH 4 1.5176^{20}
873	Benzeneacetaldehyde, 2-methyl-	$C_9H_{10}O$ 134.18	10166-08-2 221; 92^{10}		1-07-00-00163 1.0241^{10}	eth 4; EtOH 4; chl 4
874	Benzeneacetaldehyde, 4-methyl-	$C_9H_{10}O$ 134.18	104-09-6 40	221.5	4-07-00-00705 1.0052^{20}	eth 4; EtOH 4; chl 4 1.5255^{20}
875	Benzeneacetaldehyde, 2-nitro-	$C_8H_7NO_3$ 165.15	1969-73-9 22.5	135^5	2-07-00-00229	bz 4; eth 4; EtOH 4
876	Benzeneacetaldehyde, α-oxo-, aldoxime Isonitrosoacetophenone	$C_8H_7NO_2$ 149.15	532-54-7 129	5077	4-07-00-02132	H_2O 2; chl 3
877	Benzeneacetaldehyde, α-phenyl-	$C_{14}H_{12}O$ 196.25	947-91-1	315 dec; 157^7	4-07-00-01400 1.1061^{21}	H_2O 1; EtOH 4; eth 4; bz 4 1.5920^{21}
878	Benzeneacetamide α-Phenylacetamide	C_8H_9NO 135.17	103-81-1 157	7237	4-09-00-01632	H_2O 2; EtOH 3; eth 2; bz 2
879	Benzeneacetamide, N-(aminocarbonyl)- Phenacemide	$C_9H_{10}N_2O_2$ 178.19	63-98-9 215	7154	4-09-00-01636	bz 4; eth 4; EtOH 4
880	Benzeneacetamide, 4-butoxy-N-hydroxy- Bufexamac	$C_{12}H_{17}NO_3$ 223.27	2438-72-4 154	1462		
881	Benzeneacetamide, α-ethyl- α-Phenylbutyramide	$C_{10}H_{13}NO$ 163.22	90-26-6 86	7249 185^{16}	4-09-00-01818	H_2O 3; ace 2; ctc 3
882	Benzeneacetamide, α-(1-methylethyl)-	$C_{11}H_{15}NO$ 177.25	5470-47-3 111.5	180^{14}	3-09-00-02518	eth 4; EtOH 4
883	Benzeneacetic acid Phenylacetic acid	$C_8H_8O_2$ 136.15	103-82-2 76.7	7239 265.5	4-09-00-01614 1.228^6	H_2O 2; EtOH 4; eth 4; ace 3
884	Benzeneacetic acid, α-acetyl-, ethyl ester	$C_{12}H_{14}O_3$ 206.24	5413-05-8 156^{22}		4-10-00-02776 1.0855^{20}	eth 4; EtOH 4 1.5176^{20}
885	Benzeneacetic acid, α-amino- α-Phenylglycine	$C_8H_9NO_2$ 151.17	69-91-0 292 dec	7263 sub 255	4-14-00-01317	alk 3; os 2
886	Benzeneacetic acid, 4-amino- p-Aminophenylacetic acid	$C_8H_9NO_2$ 151.17	1197-55-3 200 dec	475	3-14-00-01182	H_2O 1; EtOH 2; DMSO-d_6 2
887	Benzeneacetic acid, anhydride	$C_{16}H_{14}O_3$ 254.29	1555-80-2 73.3	195^{12}	4-09-00-01629	eth 4; chl 4

No.	Name / Synonym	Mol. Form. / Mol. Wt.	CAS RN / mp/°C	Merck No. / bp/°C	Beil. Ref. / den/g cm⁻³	Solubility / n_D
888	Benzeneacetic acid, 4-bromo-	$C_8H_7BrO_2$ 215.05	1878-68-8 116	sub	3-09-00-02275	H_2O 2; EtOH 4; eth 4; CS_2 4
889	Benzeneacetic acid, 4-bromo-α-hydroxy-, (±)- Acetic acid, (4-bromophenyl)-hydroxy(DL)	$C_8H_7BrO_3$ 231.05	7021-04-7 119	1412	3-10-00-00480	H_2O 4; EtOH 4; eth 4; bz 4
890	Benzeneacetic acid, α-butyl-	$C_{12}H_{16}O_2$ 192.26	24716-09-4	182^{20}; $118^{0.1}$	4-09-00-01897 1.0225^{19}	1.5071^{19}
891	Benzeneacetic acid, α-chloro-, ethyl ester, (S)- Acetic acid, chlorophenyl-, ethyl ester, L-(+)-	$C_{10}H_{11}ClO_2$ 198.65	10606-73-2	162^{45}; 133^{12}	3-09-00-02266 1.1594^{20}	1.5152^{20}
892	Benzeneacetic acid, α-chloro-α-phenyl-, ethyl ester	$C_{16}H_{15}ClO_2$ 274.75	52460-86-3 43.5	185^{14}	4-09-00-02508	eth 4; EtOH 4
893	Benzeneacetic acid, α-cyano-, ethyl ester	$C_{11}H_{11}NO_2$ 189.21	4553-07-5	275 dec; 165^{20}	3-09-00-04262 1.091^{20}	ace 4; bz 4; eth 4; EtOH 4 1.5012^{25}
894	Benzeneacetic acid, 4-[2-(diethylamino)-2-oxoethoxy]-3-methoxy-, propyl ester Acetic acid, amide,(4(carboxy(propyl ester)methyl)-2-methoxyphenoxy)-N,N-diethyl	$C_{18}H_{27}NO_5$ 337.42	1421-14-3	7813 $210-2^{0.7}$		H_2O 1; EtOH 3; chl 3
895	Benzeneacetic acid, 2,5-dihydroxy- Homogentisic acid	$C_8H_8O_4$ 168.15	451-13-8 153	4658	4-10-00-01506	H_2O 4; EtOH 4; eth 4; bz 1
896	Benzeneacetic acid, α,4-dihydroxy-3-methoxy- Vanilmandelic acid	$C_9H_{10}O_5$ 198.18	55-10-7 131-3 dec	9840	3-10-00-02100	H_2O 4; ace 4; eth 4
897	Benzeneacetic acid, α-ethyl-	$C_{10}H_{12}O_2$ 164.20	90-27-7 47.5	271	2-09-00-00356	eth 3; bz 3; ctc 3
898	Benzeneacetic acid, α-ethyl-, 2-(diethylamino)ethyl ester Butethamate	$C_{16}H_{25}NO_2$ 263.38	14007-64-8	1516 $167-9^{11}$	3-09-00-02463	1.4909^{20}
899	Benzeneacetic acid, ethyl ester Ethyl phenylacetate	$C_{10}H_{12}O_2$ 164.20	101-97-3 -29.4	3794 227	4-09-00-01618 1.033^{20}	eth 4; EtOH 4 1.4980^{20}
900	Benzeneacetic acid, α-ethyl-, methyl ester	$C_{11}H_{14}O_2$ 178.23	2294-71-5 77.5	228	3-09-00-02462	eth 4; EtOH 4
901	Benzeneacetic acid, 4-fluoro- (p-Fluorophenyl)acetic acid	$C_8H_7FO_2$ 154.14	405-50-5 86	4105 164^2	4-09-00-01672	
902	Benzeneacetic acid, α-(hydroxymethyl)- 8-methyl-8-azabicyclo[3.2.1]oct-3-yl ester endo-(±)- Atropine	$C_{17}H_{23}NO_3$ 289.37	51-55-8 118.5	891 sub 95	5-21-01-00235	H_2O 4; EtOH 4; eth 1; chl 2
903	Benzeneacetic acid, α-hydroxy-, (S)-	$C_8H_8O_3$ 152.15	17199-29-0 134		4-10-00-00564 1.341^{25}	H_2O 4; eth 4; EtOH 4; chl 4
904	Benzeneacetic acid, 2-hydroxy-	$C_8H_8O_3$ 152.15	614-75-5 148	240	4-10-00-00536	H_2O 2; eth 3; chl 2
905	Benzeneacetic acid, 3-hydroxy-	$C_8H_8O_3$ 152.15	621-37-4 132	190^{11}	4-10-00-00541	H_2O 4; EtOH 4; eth 4; bz 3
906	Benzeneacetic acid, 4-hydroxy-	$C_8H_8O_3$ 152.15	156-38-7 152	sub	4-10-00-00543	H_2O 2; EtOH 4; eth 4
907	Benzeneacetic acid, α-hydroxy-, ethyl ester, (±)-	$C_{10}H_{12}O_3$ 180.20	4358-88-7 35.5	254	4-10-00-00569 1.1258^{20}	EtOH 3; eth 3; ctc 2; CS_2 3
908	Benzeneacetic acid, α-(hydroxymethyl)-, 9-methyl-3-oxa-9-azatricyclo[3.3.1.02,4]non-7-yl ester, [(7S)-(1α,2β,4β,5α,7β)]- Scopolamine	$C_{17}H_{21}NO_4$ 303.36	51-34-3	8361	2-27-00-00064	
909	Benzeneacetic acid, α-hydroxy-, ethyl ester	$C_{10}H_{12}O_3$ 180.20	774-40-3 35	150^{21}	2-10-00-00121 1.1270^{20}	eth 4; EtOH 4
910	Benzeneacetic acid, α-hydroxy-, methyl ester, (±)-	$C_9H_{10}O_3$ 166.18	4358-87-6 58	250 dec; 144^{20}	2-10-00-00120 1.1756^{20}	EtOH 4; chl 4
911	Benzeneacetic acid, α-hydroxy-, 8-methyl-8-azabicyclo[3.2.1]oct-3-yl ester, endo-(±)- Homatropine	$C_{16}H_{21}NO_3$ 275.35	87-00-3 99.5	4649	5-21-01-00234	H_2O 2; EtOH 3; eth 3; ace 3
912	Benzeneacetic acid, α-hydroxy-, 3-methylbutyl ester Mandelic acid isoamyl ester	$C_{13}H_{18}O_3$ 222.28	5421-04-5	5600 172^{11}	4-10-00-00571	
913	Benzeneacetic acid, 4-hydroxy-3-methoxy- Homovanillic acid	$C_9H_{10}O_4$ 182.18	306-08-1 143.5	4662	4-10-00-01509	
914	Benzeneacetic acid, α-hydroxy-α-methyl- Atrolactic acid	$C_9H_{10}O_3$ 166.18	515-30-0 116.5	889	2-10-00-00157	H_2O 4; ace 4; bz 4; EtOH 4
915	Benzeneacetic acid, α-(hydroxymethyl)- Tropic acid	$C_9H_{10}O_3$ 166.18	529-64-6 118	9691	3-10-00-00564	eth 4; EtOH 4
916	Benzeneacetic acid, α-(hydroxymethyl)-, 8-azabicyclo[3.2.1]oct-3-yl ester, [3(S)-endo]- Norhyoscyamine	$C_{16}H_{21}NO_3$ 275.35	537-29-1 140.5	6623	5-21-01-00219	EtOH 4; chl 4
917	Benzeneacetic acid, α-(hydroxymethyl)-, 8-methyl-8-azabicyclo[3.2.1]oct-3-yl ester, [3(S)-endo]- Hyoscyamine	$C_{17}H_{23}NO_3$ 289.37	101-31-5 108.5	4795	5-21-01-00235	H_2O 2; EtOH 4; eth 2; bz 2

No.	Name Synonym	Mol. Form. Mol. Wt.	CAS RN mp/°C	Merck No. bp/°C	Bell. Ref. den/g cm^{-3}	Solubility n_D
918	Benzeneacetic acid, α-hydroxy-α-phenyl- Benzilic acid	$C_{14}H_{12}O_3$ 228.25	76-93-7 151	1089 180 dec	4-10-00-01256	H_2O 2; EtOH 4; eth 4; ace 2
919	Benzeneacetic acid, α-hydroxy-α-phenyl-, ethyl ester	$C_{16}H_{16}O_3$ 256.30	52182-15-7 34	201[21]	4-10-00-01258 1.5620[20]	eth 4; EtOH 4
920	Benzeneacetic acid, α-hydroxy-α-phenyl-, methyl ester	$C_{15}H_{14}O_3$ 242.27	76-89-1 75.8	187[13]	4-10-00-01258	eth 4; EtOH 4
921	Benzeneacetic acid, 2-methoxy-	$C_9H_{10}O_3$ 166.18	93-25-4 124	100-1[2]	4-10-00-00536	H_2O 3; EtOH 4; eth 4; ace 4
922	Benzeneacetic acid, 4-methoxy-	$C_9H_{10}O_3$ 166.18	104-01-8 87	138[2]	4-10-00-00544	H_2O 1; EtOH 4; eth 3; bz 3
923	Benzeneacetic acid, 4-methoxy-, ethyl ester	$C_{11}H_{14}O_3$ 194.23	14062-18-1	138-40[70]	4-10-00-00549 1.097[25]	1.5075[20]
924	Benzeneacetic acid, α-methyl-, (±)-	$C_9H_{10}O_2$ 150.18	2328-24-7 <-20	263	4-09-00-01779 1.1[0]	1.5237[20]
925	Benzeneacetic acid, 3-methyl-	$C_9H_{10}O_2$ 150.18	621-36-3 62	120-3[26]	4-09-00-01792	H_2O 3; chl 3
926	Benzeneacetic acid, 4-methyl-	$C_9H_{10}O_2$ 150.18	622-47-9 93	265	4-09-00-01795	bz 4; eth 4; EtOH 4
927	Benzeneacetic acid, α-methylene- Atropic acid	$C_9H_8O_2$ 148.16	492-38-6 106.5	890 267 dec	4-09-00-02050	H_2O 2; EtOH 3; eth 3; bz 3
928	Benzeneacetic acid, α-methylene-, 8-methyl-8-azabicyclo[3.2.1]oct-3-yl ester, endo- Apoatropine	$C_{17}H_{21}NO_2$ 271.36	500-55-0 62	770	5-21-01-00228	H_2O 2; EtOH 4; eth 4; ace 4
929	Benzeneacetic acid, methyl ester	$C_9H_{10}O_2$ 150.18	101-41-7	216.5	4-09-00-01617 1.0622[16]	H_2O 1; EtOH 5; eth 5; ace 3 1.5075[20]
930	Benzeneacetic acid, α-(1-methylethyl)-	$C_{11}H_{14}O_2$ 178.23	3508-94-9 63	159[14]	3-09-00-02518	eth 4; EtOH 4
931	Benzeneacetic acid, α-methyl-α-phenyl-	$C_{15}H_{14}O_2$ 226.27	5558-66-7 176	sub	4-09-00-02535	H_2O 2; EtOH 3; eth 4; bz 3
932	Benzeneacetic acid, 2-methylpropyl ester	$C_{12}H_{16}O_2$ 192.26	102-13-6	247	4-09-00-01619 0.999[18]	H_2O 1; EtOH 3; eth 3
933	Benzeneacetic acid, 4-nitro- p-Nitrophenylacetic acid	$C_8H_7NO_4$ 181.15	104-03-0 154	6543	4-09-00-01687	H_2O 2; EtOH 3; eth 3; bz 3
934	Benzeneacetic acid, α-oxo-	$C_8H_6O_3$ 150.13	611-73-4 66	163[15]	4-10-00-02737	H_2O 4; EtOH 3; eth 3; ctc 2
935	Benzeneacetic acid, α-oxo-, ethyl ester	$C_{10}H_{10}O_3$ 178.19	1603-79-8	256.5	4-10-00-02738 1.1222[25]	1.5190[25]
936	Benzeneacetic acid, α-phenyl- Diphenylacetic acid	$C_{14}H_{12}O_2$ 212.25	117-34-0 147	3316 194[25]	4-09-00-02492 1.257[15]	H_2O 2; EtOH 4; eth 3; chl 3
937	Benzeneacetic acid, α-phenyl-, anhydride	$C_{28}H_{22}O_3$ 406.48	1760-46-9 98	222[15]	4-09-00-02497	bz 4; chl 4
938	Benzeneacetic acid, α-phenyl-, 2-(diethylamino)ethyl ester, hydrochloride Adiphenine hydrochloride	$C_{20}H_{26}ClNO_2$ 347.88	50-42-0 113.5	151	3-09-00-03297	H_2O 4; EtOH 2; eth 2
939	Benzeneacetic acid, α-phenyl-, ethyl ester	$C_{16}H_{16}O_2$ 240.30	3468-99-3 60	195[25]	4-09-00-02493 1.1860[20]	H_2O 1; EtOH 4; eth 4; CS_2 4
940	Benzeneacetic acid, 2-phenylethyl ester	$C_{16}H_{16}O_2$ 240.30	102-20-5 26.5	177-8[4.5]	3-09-00-02183 1.077[25]	EtOH 4
941	Benzeneacetic acid, α-(phenylmethylene)- α-Phenylcinnamic acid	$C_{15}H_{12}O_2$ 224.26	3368-16-9 172.5	7253 sub	3-09-00-03438	eth 4; EtOH 4
942	Benzeneacetic acid, α-(phenylmethylene)-, ethyl ester	$C_{17}H_{16}O_2$ 252.31	24446-63-7 33.5	214-5[28]	3-09-00-03416 1.0971[18]	1.5972[18]
943	Benzeneacetic acid, α-propyl-, (±)- Valeric acid, 2-phenyl-, (±)-	$C_{11}H_{14}O_2$ 178.23	7782-21-0 58	280	4-09-00-01868	eth 4; EtOH 4
944	Benzeneacetic acid, 2,3,6-trichloro- Chlorfenac	$C_8H_5Cl_3O_2$ 239.48	85-34-7 161	2085	4-09-00-01681	
945	Benzeneacetonitrile Benzyl cyanide	C_8H_7N 117.15	140-29-4 -23.8	1145 233.5	4-09-00-01663 1.0205[15]	1.5211[25]
946	Benzeneacetonitrile, 4-amino-	$C_8H_8N_2$ 132.17	3544-25-0 46	312	4-14-00-01312	H_2O 2; EtOH 3; CS_2 2; os 3
947	Benzeneacetonitrile, α-bromo- α-Bromobenzyl cyanide	C_8H_6BrN 196.05	5798-79-8 29	1400 242 dec; 133[12]	3-09-00-02278 1.539[29]	H_2O 1; EtOH 4; eth 4; ace 4
948	Benzeneacetonitrile, 2-bromo-	C_8H_6BrN 196.05	19472-74-3 1	140[13]	4-09-00-01682	H_2O 1; EtOH 4
949	Benzeneacetonitrile, 2-chloro-	C_8H_6ClN 151.60	2856-63-5 24	251	4-09-00-01674 1.1737[18]	
950	Benzeneacetonitrile, 3-chloro-	C_8H_6ClN 151.60	1529-41-5	134-6[10]	4-09-00-01675 1.283[25]	1.5437[20]
951	Benzeneacetonitrile, α-ethyl-	$C_{10}H_{11}N$ 145.20	769-68-6	241	4-09-00-01822 0.977[14]	H_2O 1; EtOH 3; eth 3; bz 3
952	Benzeneacetonitrile, 2-fluoro-	C_8H_6FN 135.14	326-62-5	232; 102[10]	4-09-00-01671 1.059[25]	1.5009[20]
953	Benzeneacetonitrile, 4-fluoro-	C_8H_6FN 135.14	459-22-3	119-20[18]	4-09-00-01673 1.126[25]	1.5002[20]

No.	Name Synonym	Mol. Form. Mol. Wt.	CAS RN mp/°C	Merck No. bp/°C	Beil. Ref. den/g cm^{-3}	Solubility n_D
954	Benzeneacetonitrile, α-[(6-O-β-D-glucopyranosyl-β-D-glucopyranosyl)oxy]-, (R)- Amygdalin	$C_{20}H_{27}NO_{11}$ 457.43	29883-15-6 224.5	629	5-17-08-00118	H_2O 4; EtOH 2; eth 1; chl 1
955	Benzeneacetonitrile, α-(β-D-glucopyranosyloxy)- Mandelonitrile glucoside	$C_{14}H_{17}NO_6$ 295.29	138-53-4 122	5602		H_2O 4; EtOH 4
956	Benzeneacetonitrile, α-hydroxy-, (±)-	C_8H_7NO 133.15	613-88-7 22	170	3-10-00-00473 1.1165[20]	eth 4; EtOH 4; chl 4
957	Benzeneacetonitrile, 2-methoxy-	C_9H_9NO 147.18	7035-03-2 69.8	141[15]	4-10-00-00538	bz 4
958	Benzeneacetonitrile, 4-methoxy-	C_9H_9NO 147.18	104-47-2	286.5	4-10-00-00555 1.0845[20]	EtOH 3; eth 3; chl 3 1.5309[20]
959	Benzeneacetonitrile, α-methyl-	C_9H_9N 131.18	1823-91-2	231	4-09-00-01783 0.9854[20]	eth 4; EtOH 4 1.5095[25]
960	Benzeneacetonitrile, 2-methyl-	C_9H_9N 131.18	22364-68-7	244	4-09-00-01791 1.0156[22]	bz 4; eth 4; EtOH 4 1.5252[20]
961	Benzeneacetonitrile, 3-methyl-	C_9H_9N 131.18	2947-60-6	248; 133[15]	4-09-00-01792 1.0022[22]	H_2O 1; EtOH 3; eth 3; bz 3 1.5233[20]
962	Benzeneacetonitrile, 4-methyl-	C_9H_9N 131.18	2947-61-7 18	242.5	4-09-00-01796 0.992[25]	H_2O 1; EtOH 3; eth 3; bz 3 1.5190[20]
963	Benzeneacetonitrile, α-(1-methylethyl)-	$C_{11}H_{13}N$ 159.23	5558-29-2	247	4-09-00-01879 0.967[15]	bz 4; EtOH 4 1.5038[25]
964	Benzeneacetonitrile, α-(2-methylpropyl)- Valeronitrile, 4-methyl-2-phenyl-	$C_{12}H_{15}N$ 173.26	5558-31-6	264; 137[15]	4-09-00-01903 0.942[16]	bz 4; eth 4; EtOH 4
965	Benzeneacetonitrile, 2-nitro-	$C_8H_6N_2O_2$ 162.15	610-66-2 84	178[12]; 138[1]	3-09-00-02283	ace 4; bz 4; eth 4; EtOH 4
966	Benzeneacetonitrile, 3-nitro-	$C_8H_6N_2O_2$ 162.15	621-50-1 63	180[15]	4-09-00-01687	bz 4; eth 4; EtOH 4; chl 4
967	Benzeneacetonitrile, 4-nitro- p-Nitrobenzyl cyanide	$C_8H_6N_2O_2$ 162.15	555-21-5 117	6512 195-7[12]	4-09-00-01689	H_2O 2; EtOH 3; eth 3; bz 3
968	Benzeneacetonitrile, α-oxo-	C_8H_5NO 131.13	613-90-1 32.5	206	4-10-00-02743	H_2O 1; EtOH 4; eth 4; chl 2
969	Benzeneacetonitrile, α-phenyl-	$C_{14}H_{11}N$ 193.25	86-29-3 74.3	184[16]	4-09-00-02505	EtOH 3; eth 4; chl 3; lig 2
970	Benzeneacetonitrile, α-propyl-	$C_{11}H_{13}N$ 159.23	5558-78-1	255; 126[13]	4-09-00-01869 0.9425[20]	bz 4; EtOH 4 1.5000[20]
971	Benzeneacetyl chloride Phenylacetyl chloride	C_8H_7ClO 154.60	103-80-0	170[250]; 105[24]	4-09-00-01631 1.1682[20]	eth 4 1.5325[20]
972	Benzeneacetyl chloride, α-(acetyloxy)-	$C_{10}H_9ClO_3$ 212.63	1638-63-7 36	152[33]; 125[10]	1-10-00-00089	eth 4; bz 4; chl 4; lig 4
973	Benzeneacetyl chloride, α-chloro-	$C_8H_6Cl_2O$ 189.04	2912-62-1	120[23]	2-09-00-00308 1.196[25]	1.5440[20]
974	Benzeneacetyl chloride, α-ethyl-	$C_{10}H_{11}ClO$ 182.65	36854-57-6	110-2[10]	4-09-00-01817	ctc 3 1.5169[20]
975	Benzeneacetyl chloride, α-phenyl-	$C_{14}H_{11}ClO$ 230.69	1871-76-7 56.5	170[16]	3-09-00-03300	lig 3
976	Benzene, azido-	$C_6H_5N_3$ 119.13	622-37-7 -27.5	70[11]	4-05-00-00759 1.0860[20]	H_2O 1; EtOH 2; eth 2 1.5589[25]
977	Benzene, 1-azido-4-bromo-	$C_6H_4BrN_3$ 198.02	2101-88-4 20	105[10]; 69[2]	4-05-00-00761 1.5827[25]	H_2O 1; EtOH 2; eth 4; bz 4
978	Benzene, 1-azido-4-chloro-	$C_6H_4ClN_3$ 153.57	3296-05-7 20	96[20]	4-05-00-00760 1.2634[25]	H_2O 1; eth 3
979	Benzene, 1-azido-2-iodo-	$C_6H_4IN_3$ 245.02	54467-95-7	90-100[0.9]	0-05-00-00278 1.8893[25]	ace 4; EtOH 4 1.6631[25]
980	Benzene, (azidomethyl)-	$C_7H_7N_3$ 133.15	622-79-7	108[23]; 74[11]	4-05-00-00877 1.0655[25]	H_2O 1; EtOH 5; eth 5 1.5341[25]
981	Benzene, 1-azido-2-methyl-	$C_7H_7N_3$ 133.15	31656-92-5 <-10	90.5[30]; 76[12]	4-05-00-00876 1.0648[25]	eth 4
982	Benzene, 1-azido-4-methyl-	$C_7H_7N_3$ 133.15	2101-86-2	180 dec; 80[10]	4-05-00-00876 1.0527[23]	eth 4; EtOH 4
983	Benzene, 1,2-bis(bromomethyl)-	$C_8H_8Br_2$ 263.96	91-13-4 95	128-30[4.5]	4-05-00-00929 1.988[25]	H_2O 1; EtOH 3; eth 3; ctc 3
984	Benzene, 1,3-bis(bromomethyl)-	$C_8H_8Br_2$ 263.96	626-15-3 77	135-40[20]	4-05-00-00946 1.959[25]	H_2O 1; EtOH 3; eth 3; chl 3
985	Benzene, 1,4-bis(bromomethyl)-	$C_8H_8Br_2$ 263.96	623-24-5 144.5	245	4-05-00-00970 2.012[25]	H_2O 1; EtOH 4; eth 2; bz 3
986	Benzene, 1,2-bis(chloromethyl)-	$C_8H_8Cl_2$ 175.06	612-12-4 55	239.5	4-05-00-00927 1.393[25]	H_2O 1; EtOH 4; eth 4; ctc 3
987	Benzene, 1,3-bis(chloromethyl)-	$C_8H_8Cl_2$ 175.06	626-16-4 34.2	251.5	4-05-00-00944 1.302[20]	H_2O 1; EtOH 4; eth 4; chl 2
988	Benzene, 1,4-bis(chloromethyl)-	$C_8H_8Cl_2$ 175.06	623-25-6 100	245 dec; 135[16]	4-05-00-00967 1.417[25]	H_2O 1; EtOH 4; eth 4; ace 4
989	Benzene, 1,4-bis(dichloromethyl)-	$C_8H_6Cl_4$ 243.95	7398-82-5 9.5		3-05-00-00857 1.606[25]	H_2O 1; EtOH 3; eth 3; bz 3
990	Benzene, 1,4-bis(1,1-dimethylethyl)- Benzene, 1,4-di-tert-butyl	$C_{14}H_{22}$ 190.33	1012-72-2 79.5	238; 109[15]	4-05-00-01163 0.9850[20]	H_2O 1; EtOH 3; eth 3

No.	Name Synonym	Mol. Form. Mol. Wt.	CAS RN mp/°C	Merck No. bp/°C	Beil. Ref. den/g cm^{-3}	Solubility n_D
991	Benzene, 1,4-bis(2-hydroxybenzylideneamino)-2-nitro Ethionine	$C_{20}H_{15}N_3O_4$ 361.36	13073-35-3 195	3693		
992	Benzene, 1,2-bis(1-methylethyl)-	$C_{12}H_{18}$ 162.27	577-55-9 -57	204	4-05-00-01125 0.8701[20]	H$_2$O 1; EtOH 5; eth 5; ace 5 1.4960[20]
993	Benzene, 1,3-bis(1-methylethyl)-	$C_{12}H_{18}$ 162.27	99-62-7 -63.1	203.2	4-05-00-01125 0.8559[20]	H$_2$O 1; EtOH 5; eth 5; ace 5 1.4883[20]
994	Benzene, 1,4-bis(1-methylethyl)-	$C_{12}H_{18}$ 162.27	100-18-5 -17	210.3	4-05-00-01126 0.8568[20]	H$_2$O 1; EtOH 5; eth 5; ace 5 1.4898[20]
995	Benzene, 1,4-bis(phenylmethyl)-	$C_{20}H_{18}$ 258.36	793-23-7 87.5	225[18]	4-05-00-02512	EtOH 3; eth 2; bz 3; chl 3
996	Benzene, 1,4-bis(trichloromethyl)-	$C_8H_4Cl_6$ 312.84	68-36-0 109	1313	4-05-00-00968	chl 3
997	Benzene, 1,3-bis(trifluoromethyl)-	$C_8H_4F_6$ 214.11	402-31-3	116	4-05-00-00943 1.3790[25]	1.3916[25]
998	Benzene, bromo- Bromobenzene	C_6H_5Br 157.01	108-86-1 -30.6	1394 156.0	4-05-00-00670 1.4950[20]	H$_2$O 1; EtOH 4; eth 4; bz 4 1.5597[20]
999	Benzene, 1-bromo-2-(bromomethyl)-	$C_7H_6Br_2$ 249.93	3433-80-5 31	129[19]	4-05-00-00836	eth 4; EtOH 4; HOAc 4
1000	Benzene, 1-bromo-4-(bromomethyl)- p-Bromobenzyl bromide	$C_7H_6Br_2$ 249.93	589-15-1 63	1397	4-05-00-00836	H$_2$O 2; EtOH 3; eth 4; bz 3
1001	Benzene, (3-bromo-3-butenyl)- 1-Butene, 2-bromo-4-phenyl	$C_{10}H_{11}Br$ 211.10	62692-40-4	118[21]; 90[5]	3-05-00-01209 1.2907[20]	ace 4 1.5450[20]
1002	Benzene, (4-bromo-3-butenyl)- 2-Butene, 1-bromo-4-phenyl	$C_{10}H_{11}Br$ 211.10	40734-75-6	126-30[10.5]	4-05-00-01377 1.2660[20]	eth 4 1.5678[20]
1003	Benzene, (4-bromobutoxy)-	$C_{10}H_{13}BrO$ 229.12	1200-03-9 41	154[18]	4-06-00-00558	EtOH 2; ctc 2
1004	Benzene, 1-bromo-2-chloro-	C_6H_4BrCl 191.45	694-80-4 -12.3	204	4-05-00-00680 1.6387[25]	H$_2$O 1; bz 4; ctc 2 1.5809[20]
1005	Benzene, 1-bromo-3-chloro-	C_6H_4BrCl 191.45	108-37-2 -21.5	196	4-05-00-00680 1.6302[20]	H$_2$O 1; EtOH 4; eth 4 1.5771[20]
1006	Benzene, 1-bromo-4-chloro-	C_6H_4BrCl 191.45	106-39-8 68	196	4-05-00-00681 1.576[71]	H$_2$O 1; EtOH 2; eth 3; bz 3 1.5531[70]
1007	Benzene, 1-bromo-3-(chloromethyl)-	C_7H_6BrCl 205.48	932-77-4 32.5	119[18]	4-05-00-00832	EtOH 4
1008	Benzene, 1-bromo-4-(chloromethyl)- p-Bromobenzyl chloride	C_7H_6BrCl 205.48	589-17-3 50	1398 236	4-05-00-00832	H$_2$O 1; EtOH 4; eth 4; peth 3
1009	Benzene, 1-bromo-4-cyclohexyl-	$C_{12}H_{15}Br$ 239.16	25109-28-8	160[25]	3-05-00-01258 1.283[25]	bz 4; chl 4 1.5584[20]
1010	Benzene, 1-bromo-2,3-dichloro-	$C_6H_3BrCl_2$ 225.90	56961-77-4 60	245	0-05-00-00209	bz 4; eth 4; chl 4
1011	Benzene, 1-bromo-2,4-dichloro-	$C_6H_3BrCl_2$ 225.90	1193-72-2 26	235	3-05-00-00564	bz 4; eth 4; chl 4
1012	Benzene, 1-bromo-3,5-dichloro-	$C_6H_3BrCl_2$ 225.90	19752-55-7 83	232	2-05-00-00162	H$_2$O 1; EtOH 3; eth 3; bz 4
1013	Benzene, 2-bromo-1,3-dichloro-	$C_6H_3BrCl_2$ 225.90	19393-92-1 65	242	0-05-00-00210	bz 4; eth 4; chl 4
1014	Benzene, 2-bromo-1,4-dichloro-	$C_6H_3BrCl_2$ 225.90	1435-50-3 35	226; 119[20]	3-05-00-00564	bz 4; eth 4; EtOH 4; lig 4
1015	Benzene, 4-bromo-1,2-dichloro-	$C_6H_3BrCl_2$ 225.90	18282-59-2 25	237	3-05-00-00564	H$_2$O 1; EtOH 2; eth 4; bz 4
1016	Benzene, 4-bromo-1,2-dimethoxy-	$C_8H_9BrO_2$ 217.06	2859-78-1	254.5	4-06-00-05621 1.702[25]	1.5743[20]
1017	Benzene, 1-bromo-2,3-dimethyl-	C_8H_9Br 185.06	576-23-8	214	4-05-00-00928 1.365[25]	ace 4; bz 4; eth 4 1.5587[20]
1018	Benzene, 1-bromo-2,4-dimethyl-	C_8H_9Br 185.06	583-70-0 -17	205	4-05-00-00945 1.3419[20]	H$_2$O 1; EtOH 4; eth 4; ace 4 1.5501[20]
1019	Benzene, 1-bromo-3,5-dimethyl-	C_8H_9Br 185.06	556-96-7	204	4-05-00-00945 1.362[20]	eth 4; ace 3; bz 3 1.5462[22]
1020	Benzene, 2-bromo-1,3-dimethyl-	C_8H_9Br 185.06	576-22-7	203.5; 100[20]	4-05-00-00945	eth 4; ace 3; bz 3 1.5552[20]
1021	Benzene, 2-bromo-1,4-dimethyl-	C_8H_9Br 185.06	553-94-6 9	199; 88-9[13]	4-05-00-00969 1.3582[18]	H$_2$O 1; EtOH 4; bz 3 1.5514[18]
1022	Benzene, 4-bromo-1,2-dimethyl-	C_8H_9Br 185.06	583-71-1 -0.2	214.5	4-05-00-00928 1.3708[20]	H$_2$O 1; EtOH 4; eth 4 1.5530[20]
1023	Benzene, 1-bromo-4-(1,1-dimethylethyl)-	$C_{10}H_{13}Br$ 213.12	3972-65-4 19	231.5	4-05-00-01050 1.2286[20]	H$_2$O 1; eth 3; bz 3; chl 3 1.5436[20]
1024	Benzene, 1-bromo-2,3-dinitro- 1-Bromo-2,3-dinitrobenzene	$C_6H_3BrN_2O_4$ 247.00	19613-76-4 101.5	320	4-05-00-00749	EtOH 4
1025	Benzene, (1-bromoethenyl)-	C_8H_7Br 183.05	98-81-7 -44	86[14]; 71[3]	4-05-00-01349 1.4025[23]	1.5881[20]
1026	Benzene, 1-bromo-2-ethenyl-	C_8H_7Br 183.05	2039-88-5 -52.8	209.2; 98[20]	4-05-00-01349 1.4160[20]	1.5927[20]
1027	Benzene, 1-bromo-3-ethenyl-	C_8H_7Br 183.05	2039-86-3	90-4[20]	3-05-00-01176 1.4059[20]	1.5933[20]
1028	Benzene, 1-bromo-4-ethenyl-	C_8H_7Br 183.05	2039-82-9 7.7	212; 103[30]	4-05-00-01349 1.3984[20]	H$_2$O 1; chl 4; HOAc 3 1.5947[20]
1029	Benzene, (2-bromoethenyl)-, (E)-	C_8H_7Br 183.05	588-72-7 7	219 dec; 108[20]	4-05-00-01349 1.4269[16]	H$_2$O 1; EtOH 5; eth 5; chl 3 1.6093[20]

No.	Name Synonym	Mol. Form. Mol. Wt.	CAS RN mp/°C	Merck No. bp/°C	Beil. Ref. den/g cm^{-3}	Solubility n_D
1030	Benzene, 1,1'-(bromoethenylidene)bis-	$C_{14}H_{11}Br$ 259.15	13249-58-6 41.5	$175^{11.5}$	4-05-00-02177	EtOH 2; eth 3; ace 3; CS_2 3
1031	Benzene, (2-bromoethenyl)-, (Z)-	C_8H_7Br 183.05	588-73-8 -7	$55-6^2$	4-05-00-01349 1.4322^{10}	1.5990^{22}
1032	Benzene, 1-bromo-4-ethoxy-	C_8H_9BrO 201.06	588-96-5 4	231	4-06-00-01045 1.4071^{25}	H_2O 1; EtOH 4; eth 4; chl 3 1.5517^{20}
1033	Benzene, (2-bromoethoxy)-	C_8H_9BrO 201.06	589-10-6 39	240 dec; 128^{20}	4-06-00-00556 1.3555^{20}	H_2O 1; EtOH 4; eth 4
1034	Benzene, (1-bromoethyl)-, (±)-	C_8H_9Br 185.06	38661-81-3	202-3. 85^{13}	4-05-00-00906 1.3605^{20}	EtOH 3; eth 3; bz 3; ctc 2 1.5612^{20}
1035	Benzene, 1-bromo-2-ethyl-	C_8H_9Br 185.06	1973-22-4 -67.9	199.3	4-05-00-00905 1.3548^{20}	ace 4; bz 4; eth 4; EtOH 4 1.5472^{20}
1036	Benzene, 1-bromo-3-ethyl-	C_8H_9Br 185.06	2725-82-8	202	4-05-00-00906 1.3493^{20}	1.5465^{20}
1037	Benzene, 1-bromo-4-ethyl-	C_8H_9Br 185.06	1585-07-5 -43.5	204	4-05-00-00906 1.3423^{20}	ace 4; bz 4; eth 4; EtOH 4 1.5445^{20}
1038	Benzene, (2-bromoethyl)-	C_8H_9Br 185.06	103-63-9	219	4-05-00-00907 1.3587^{20}	H_2O 1; eth 3; bz 3; ctc 2 1.5372^{20}
1039	Benzene, (1-bromoethyl)-, (R)-	C_8H_9Br 185.06	1459-14-9	203	4-05-00-00906 1.3108^{23}	H_2O 1; EtOH 3; eth 3 1.5612^{20}
1040	Benzene, 1-bromo-2-ethynyl- o-Bromophenylacetylene	C_8H_5Br 181.03	766-46-1	$92-3^{20}$	3-05-00-01351 1.4434^{25}	1.5962^{25}
1041	Benzene, 1-bromo-3-ethynyl- m-Bromophenylacetylene	C_8H_5Br 181.03	766-81-4	$85-6^{16}$	3-05-00-01351 1.4466^{25}	1.5896^{25}
1042	Benzene, 1-bromo-4-ethynyl-	C_8H_5Br 181.03	766-96-1 64.5	89^{16}	4-05-00-01529	chl 3
1043	Benzene, 1-bromo-3-fluoro-	C_6H_4BrF 175.00	1073-06-9	150	4-05-00-00678 1.7081^{20}	ctc 3 1.5257^{20}
1044	Benzene, 1-bromo-4-fluoro-	C_6H_4BrF 175.00	460-00-4 -17.4	151.5	4-05-00-00678 1.593^{15}	H_2O 1; EtOH 3; eth 3; chl 3 1.5310^{15}
1045	Benzene, 1-bromo-4-(hexyloxy)-	$C_{12}H_{17}BrO$ 257.17	30752-19-3	$156-7^{13}$	4-06-00-01047 1.2306^{20}	1.5262^{20}
1046	Benzene, 1-bromo-2-iodo-	C_6H_4BrI 282.91	583-55-1 9.5	257; 120^{15}	4-05-00-00698 2.2570^{25}	H_2O 1; EtOH 2; ace 3; HOAc 2 1.6618^{25}
1047	Benzene, 1-bromo-3-iodo- m-Bromoiodobenzene	C_6H_4BrI 282.91	591-18-4 -9.3	252; 120^{18}	4-05-00-00698	H_2O 1; EtOH 2; HOAc 2
1048	Benzene, 1-bromo-4-iodo-	C_6H_4BrI 282.91	589-87-7 92	252	4-05-00-00698	H_2O 1; EtOH 2; eth 3; chl 2
1049	Benzene, 1-bromo-4-isocyanato- p-Bromophenyl isocyanate	C_7H_4BrNO 198.02	2493-02-9 1418	226	1-12-00-00321	eth 4
1050	Benzene, 1-bromo-2-methoxy-	C_7H_7BrO 187.04	578-57-4 2.5	216	4-06-00-01037 1.5018^{20}	H_2O 1; EtOH 4; eth 4 1.5727^{20}
1051	Benzene, 1-bromo-3-methoxy-	C_7H_7BrO 187.04	2398-37-0	211; 105^{16}	4-06-00-01043	H_2O 1; EtOH 3; eth 3; bz 3 1.5635^{20}
1052	Benzene, 1-bromo-4-methoxy-	C_7H_7BrO 187.04	104-92-7 13.5	215	4-06-00-01044 1.4564^{20}	H_2O 2; EtOH 4; eth 4; ctc 3 1.5642^{20}
1053	Benzene, (bromomethyl)- Benzyl bromide	C_7H_7Br 171.04	100-39-0 -3	1142 201	4-05-00-00829 1.4380^{25}	H_2O 1; EtOH 5; eth 5; ctc 3 1.5752^{20}
1054	Benzene, 1-bromo-2-methyl- o-Bromotoluene	C_7H_7Br 171.04	95-46-5 -27.8	181.7	4-05-00-00825 1.4232^{20}	H_2O 1; EtOH 4; eth 4; bz 4 1.5565^{20}
1055	Benzene, 1-bromo-3-methyl- m-Bromotoluene	C_7H_7Br 171.04	591-17-3 -39.8	183.7	4-05-00-00827 1.4099^{20}	H_2O 1; EtOH 3; eth 5; ace 3 1.5510^{20}
1056	Benzene, 1-bromo-4-methyl- p-Bromotoluene	C_7H_7Br 171.04	106-38-7 28.5	1429 184.3	4-05-00-00827 1.3959^{35}	H_2O 1; EtOH 3; eth 3; ace 3 1.5477^{20}
1057	Benzene, 1-(bromomethyl)-3,5-dimethyl- Mesitylene, α-bromo-	$C_9H_{11}Br$ 199.09	27129-86-8 40	230; 118^{22}	4-05-00-01027	bz 4; eth 4; EtOH 4; chl 4
1058	Benzene, 1,1'-(bromomethylene)bis- Bromodiphenylmethane	$C_{13}H_{11}Br$ 247.13	776-74-9 45	184^{20}; 152^2	4-05-00-01850	EtOH 3; bz 4; chl 3
1059	Benzene, 1-bromo-4-(1-methylethyl)-	$C_9H_{11}Br$ 199.09	586-61-8 -22.5	218.7	4-05-00-00994 1.3145^{20}	H_2O 1; eth 3; bz 3; ctc 2 1.5569^{20}
1060	Benzene, (2-bromo-1-methylethyl)-	$C_9H_{11}Br$ 199.09	1459-00-3 -59.3	188.5	4-05-00-00995 1.3020^{20}	
1061	Benzene, 1,1',1"-(bromomethylidyne)tris-	$C_{19}H_{15}Br$ 323.23	596-43-0 153	230^{15}	4-05-00-02500 1.5500^{20}	
1062	Benzene, 1-(bromomethyl)-2-methyl-	C_8H_9Br 185.06	89-92-9 21	217; 108^{16}	4-05-00-00928 1.3811^{23}	H_2O 1; EtOH 3; eth 3; ace 3 1.5730^{20}
1063	Benzene, 1-(bromomethyl)-3-methyl-	C_8H_9Br 185.06	620-13-3	212.5	4-05-00-00945 1.3711^{23}	H_2O 1; EtOH 4; eth 4; os 3 1.5660^{20}
1064	Benzene, 1-(bromomethyl)-4-methyl-	C_8H_9Br 185.06	104-81-4 35	220	4-05-00-00969 1.324^{25}	H_2O 1; EtOH 3; eth 4; chl 4
1065	Benzene, 2-bromo-1-methyl-4-(1-methylethyl)-	$C_{10}H_{13}Br$ 213.12	2437-76-5	234.3	4-05-00-01063 1.2689^{18}	eth 4; EtOH 4; chl 4 1.5360^{20}
1066	Benzene, 1-bromo-2-methyl-4-nitro-	$C_7H_6BrNO_2$ 216.03	7149-70-4 78	141^{17}	4-05-00-00860	eth 4
1067	Benzene, 1-(bromomethyl)-3-nitro-	$C_7H_6BrNO_2$ 216.03	3958-57-4 59.3	162^{13}	4-05-00-00860	H_2O 1; EtOH 3

No.	Name / Synonym	Mol. Form. / Mol. Wt.	CAS RN / mp/°C	Merck No. / bp/°C	Beil. Ref. / den/g cm^{-3}	Solubility / n_D
1068	Benzene, 1-bromo-3-methyl-2-nitro- / 3-Bromo-2-nitrotoluene	$C_7H_6BrNO_2$ / 216.03	52414-97-8 / 28.5	/ 153[30]; 129[10]	4-05-00-00860	eth 4; EtOH 4
1069	Benzene, 1-bromo-3-methyl-5-nitro- / Toluene, 3-bromo-5-nitro-	$C_7H_6BrNO_2$ / 216.03	52488-28-5 / 83	/ 269.5	3-05-00-00752	eth 4; EtOH 4
1070	Benzene, 1-bromo-2-methyl-3-nitro- / Toluene, 2-bromo-6-nitro	$C_7H_6BrNO_2$ / 216.03	55289-35-5 / 42	/ 143[22]	3-05-00-00751	EtOH 4
1071	Benzene, 2-bromo-1-methyl-3-nitro-	$C_7H_6BrNO_2$ / 216.03	41085-43-2 / 41.5	/ 161[30]	4-05-00-00860	EtOH 4
1072	Benzene, 2-bromo-1-methyl-4-nitro-	$C_7H_6BrNO_2$ / 216.03	7745-93-9 / 78	/ 175[38]; 150[20]	4-05-00-00861	eth 4
1073	Benzene, 2-bromo-4-methyl-1-nitro- / 2-Bromo-4-methyl-1-nitrobenzene	$C_7H_6BrNO_2$ / 216.03	40385-54-4 / 37.8	/ 269; 154[20]	4-05-00-00861	eth 4; EtOH 4
1074	Benzene, 4-bromo-1-methyl-2-nitro- / 4-Bromo-2-nitrotoluene	$C_7H_6BrNO_2$ / 216.03	60956-26-5 / 47	/ 256.5	4-05-00-00860	eth 4; EtOH 4
1075	Benzene, 1-bromo-2-nitro-	$C_6H_4BrNO_2$ / 202.01	577-19-5 / 43	/ 258	4-05-00-00728 / 1.6245[80]	H_2O 1; EtOH 4; eth 3; ace 3
1076	Benzene, 1-bromo-3-nitro-	$C_6H_4BrNO_2$ / 202.01	585-79-5 / 56	/ 265	4-05-00-00729 / 1.7036[20]	H_2O 2; EtOH 3; eth 3; bz 3 / 1.5979[20]
1077	Benzene, 1-bromo-4-nitro- / p-Nitrobromobenzene	$C_6H_4BrNO_2$ / 202.01	586-78-7 / 127	/ 256	4-05-00-00729 / 1.948[25]	H_2O 1; EtOH 3; eth 3; bz 3
1078	Benzene, bromopentafluoro- / Bromopentafluorobenzene	C_6BrF_5 / 246.96	344-04-7 / -31	/ 137	4-05-00-00680 / 1.981[25]	1.4490[20]
1079	Benzene, 1-bromo-4-phenoxy- / 4-Bromophenyl phenyl ether	$C_{12}H_9BrO$ / 249.11	101-55-3 / 18.72	/ 124-8[3-4]	4-06-00-01047 / 1.6088[20]	H_2O 1; eth 3; ctc 3 / 1.6084[20]
1080	Benzene, 1-bromo-2-(2-phenylethenyl)-, (E)- / trans-2-Bromostilbene	$C_{14}H_{11}Br$ / 259.15	54737-45-0 / 34	/ 143[0.2]	4-05-00-02164	H_2O I; EtOH 3; peth 3 / 1.6822[25]
1081	Benzene, 1-bromo-2-(2-phenylethenyl)-, (Z)- / Stilbene, 2-bromo (cis)	$C_{14}H_{11}Br$ / 259.15	4877-77-4	/ 121[0.5]	4-05-00-02164	H_2O 1; EtOH 3; peth 3 / 1.6404[25]
1082	Benzene, 1-bromo-4-(1-propenyl)-	C_9H_9Br / 197.07	4489-23-0 / 35	/ 238.5	2-05-00-00372 / 1.332[38]	1.590[36]
1083	Benzene, 1-bromo-4-(2-propenyl)-	C_9H_9Br / 197.07	2294-43-1	/ 224; 96[12]	4-05-00-01363 / 1.324[15]	1.559[15]
1084	Benzene, (3-bromo-1-propenyl)-	C_9H_9Br / 197.07	4392-24-9 / 34	/ 130[10]	3-05-00-01188 / 1.3428[30]	EtOH 4 / 1.613[20]
1085	Benzene, (3-bromopropoxy)-	$C_9H_{11}BrO$ / 215.09	588-63-6 / 10.7	/ 127[18]	4-06-00-00557 / 1.364[16]	eth 4
1086	Benzene, (1-bromopropyl)-, (±)- / Benzene, (1-bromopropyl) (DL)	$C_9H_{11}Br$ / 199.09	63790-14-7	/ 105[17]	2-05-00-00305 / 1.3098[19]	1.5517[19]
1087	Benzene, (2-bromopropyl)-	$C_9H_{11}Br$ / 199.09	2114-39-8	/ 107-9[16]	4-05-00-00981 / 1.2908[16]	chl 4 / 1.5450[20]
1088	Benzene, (3-bromopropyl)-	$C_9H_{11}Br$ / 199.09	637-59-2	/ 219.5; 117[25]	4-05-00-00982 / 1.3106[25]	H_2O 1; eth 4 / 1.5440[25]
1089	Benzene, 1-bromo-2-(trifluoromethyl)-	$C_7H_4BrF_3$ / 225.01	392-83-6	/ 167.5	4-05-00-00831 / 1.652[25]	1.4817[20]
1090	Benzene, 1-bromo-3-(trifluoromethyl)-	$C_7H_4BrF_3$ / 225.01	401-78-5 / 1	/ 151.5	4-05-00-00831 / 1.613[25]	1.4716[20]
1091	Benzene, 1-bromo-4-(trifluoromethyl)-	$C_7H_4BrF_3$ / 225.01	402-43-7	/ 160	4-05-00-00831 / 1.607[25]	1.4705[25]
1092	Benzene, 1-bromo-2,3,5-trimethyl- / 1-Bromo-2,3,5-trimethylbenzene	$C_9H_{11}Br$ / 199.09	31053-99-3 / <-15	/ 238	4-05-00-01014	bz 4 / 1.5516[20]
1093	Benzene, 1-bromo-2,4,5-trimethyl-	$C_9H_{11}Br$ / 199.09	5469-19-2 / 73	/ 234	4-05-00-01014	H_2O 1; EtOH 4
1094	Benzene, 2-bromo-1,3,5-trimethyl-	$C_9H_{11}Br$ / 199.09	576-83-0 / -1	/ 225	4-05-00-01027 / 1.3191[10]	H_2O 1; eth 4; bz 3; ctc 2 / 1.5510[20]
1095	Benzene, 1,1'-(1,3-butadiene-1,4-diyl)bis-, (E,E)-	$C_{16}H_{14}$ / 206.29	538-81-8 / 154.3	/ 352	4-05-00-02319	bz 4; eth 4; EtOH 4; peth 4
1096	Benzene, 1,1'-(1,3-butadiene-1,4-diyl)bis-, (Z,Z)-	$C_{16}H_{14}$ / 206.29	5807-76-1 / 70.5		4-05-00-02318 / 0.9697[100]	H_2O 1; EtOH 2; eth 4; bz 4 / 1.6183[100]
1097	Benzene, 1,3-butadienyl-, (E)-	$C_{10}H_{10}$ / 130.19	16939-57-4 / 4.5	/ 76[11]	4-05-00-01536 / 0.9286[20]	H_2O I; EtOH 3; eth 3; ace 3 / 1.6089[25]
1098	Benzene, 1,3-butadienyl-, (Z)-	$C_{10}H_{10}$ / 130.19	31915-94-3 / -57	/ 86[11]	4-05-00-01536 / 0.9334[20]	1.6095[20]
1099	Benzene, 1,1'-(1,4-butanediyl)bis- / 1,4-Diphenylbutane	$C_{16}H_{18}$ / 210.32	1083-56-3 / 52.5	/ 317	4-05-00-01937 / 0.9880[20]	H_2O 1; EtOH 3; eth 3; chl 3
1100	Benzenebutanoic acid	$C_{10}H_{12}O_2$ / 164.20	1821-12-1 / 52	/ 290	4-09-00-01811	H_2O 3; EtOH 3; eth 3
1101	Benzenebutanoic acid, 4-[bis(2-chloroethyl)amino]- / Chlorambucil	$C_{14}H_{19}Cl_2NO_2$ / 304.22	305-03-3 / 65	2064	4-14-00-01715	
1102	Benzenebutanoic acid, α,2-diamino-γ-oxo- / Kynurenine	$C_{10}H_{12}N_2O_3$ / 208.22	343-65-7 / 191 dec	5207	4-14-00-02562	
1103	Benzenebutanoic acid, 4-hydroxy-γ-(4-hydroxyphenyl)-γ-methyl- / Diphenolic acid	$C_{17}H_{18}O_4$ / 286.33	126-00-1 / 171.5	3312	4-10-00-01890	H_2O 4; ace 4; EtOH 4

No.	Name / Synonym	Mol. Form. / Mol. Wt.	CAS RN / mp/°C	Merck No. / bp/°C	Bell. Ref. / den/g cm^{-3}	Solubility / n_D
1104	Benzenebutanoic acid, γ-methyl-	$C_{11}H_{14}O_2$ 178.23	16433-43-5 13	210[85]; 165[12]	4-09-00-01866 1.0554[15]	eth 4; EtOH 4 1.5167[20]
1105	Benzenebutanol, 4-methoxy-	$C_{11}H_{16}O_2$ 180.25	52244-70-9 3.5	160[8]; 125[1.5]	4-06-00-06005	1.5267[20]
1106	Benzenebutanol, α-methyl-	$C_{11}H_{16}O$ 164.25	2344-71-0	134-5[16]	3-06-00-01954 0.9643[25]	1.518[19]
1107	Benzenebutanol, β-methyl-, (±)- 1-Butanol, 2-methyl-4-phenyl (DL)	$C_{11}H_{16}O$ 164.25	116783-11-0	135[11]	4-06-00-03380 0.9719[20]	ace 4; bz 4; eth 4; EtOH 4 1.5173[16]
1108	Benzenebutanol, β-phenyl-, (R)- 1-Butanol, 2,4-diphenyl-, (-)-	$C_{16}H_{18}O$ 226.32	17297-04-0 51.5	178[1]; 145[0.1]		chl 3 1.5686[25]
1109	Benzene, 1-butenyl-, (E)-	$C_{10}H_{12}$ 132.21	1005-64-7 -43.1	198.7	4-05-00-01374 0.9019[20]	H$_2$O 1; EtOH 3; eth 3; bz 3 1.5420[20]
1110	Benzene, 1-butenyl-, (Z)-	$C_{10}H_{12}$ 132.21	1560-09-4	189	3-05-00-01205 0.9065[25]	H$_2$O 1; EtOH 3; eth 3; bz 3 1.5390[20]
1111	Benzene, 2-butenyl-	$C_{10}H_{12}$ 132.21	1560-06-1	176	3-05-00-01207 0.8831[20]	1.5101[20]
1112	Benzene, 3-butenyl-	$C_{10}H_{12}$ 132.21	768-56-9 -70	177	4-05-00-01378 0.8831[20]	H$_2$O 1; eth 3; bz 3 1.5059[20]
1113	Benzene, 1,1'-(1-butenylidene)bis-	$C_{16}H_{16}$ 208.30	1726-14-3 -1.2	296	4-05-00-02209 0.9934[20]	1.5604[20]
1114	Benzene, butoxy-	$C_{10}H_{14}O$ 150.22	1126-79-0 -19.4	210	4-06-00-00558 0.9351[20]	eth 3; ace 3 1.4969[20]
1115	Benzene, (butoxymethyl)-	$C_{11}H_{16}O$ 164.25	588-67-0	223	4-06-00-02231 0.9227[20]	ace 4; eth 4; EtOH 4 1.4833[20]
1116	Benzene, 1-butoxy-3-methyl-	$C_{11}H_{16}O$ 164.25	23079-65-4	229.2	4-06-00-02040 0.9406[70]	eth 4 1.4970[20]
1117	Benzene, 1-butoxy-4-methyl-	$C_{11}H_{16}O$ 164.25	10519-06-9	229.5	3-06-00-01354 0.9205[25]	eth 3 1.4970[20]
1118	Benzene, butyl- Butylbenzene	$C_{10}H_{14}$ 134.22	104-51-8 -87.9	183.3	4-05-00-01033 0.8601[20]	H$_2$O 1; EtOH 5; eth 5; ace 5 1.4898[20]
1119	Benzene, (1-butylhexadecyl)- Eicosane, 5-phenyl-	$C_{26}H_{46}$ 358.65	2400-04-6 30.2	197[1]	4-05-00-01223 0.8549[20]	1.4796[20]
1120	Benzene, 1,1'-butylidenebis- 1,1-Diphenylbutane	$C_{16}H_{18}$ 210.32	719-79-9 27	287	4-05-00-01944 0.9928[20]	H$_2$O 1; EtOH 3; eth 3; bz 3 1.5664[20]
1121	Benzene, 1-butyl-2-methyl-	$C_{11}H_{16}$ 148.25	1595-11-5	208	3-05-00-00999 0.8710[20]	ace 4; bz 4; eth 4 1.4960[20]
1122	Benzene, 1-butyl-3-methyl-	$C_{11}H_{16}$ 148.25	1595-04-6	205	3-05-00-00999 0.8590[20]	ace 4; bz 4; eth 4 1.4910[20]
1123	Benzene, 1-butyl-4-methyl-	$C_{11}H_{16}$ 148.25	1595-05-7 -85	207	4-05-00-01094 0.8586[20]	ace 4; bz 4; eth 4 1.4916[20]
1124	Benzene, 3-butynyl-	$C_{10}H_{10}$ 130.19	16520-62-0	190	4-05-00-01535 0.9258[20]	1.5208[20]
1125	Benzenecarbodithioic acid, 2-hydroxy- Dithiosalicylic acid	$C_7H_6OS_2$ 170.26	527-89-9 49	3381	2-10-00-00078	bz 4; eth 4; EtOH 4
1126	Benzenecarboperoxoic acid Perbenzoic acid	$C_7H_6O_3$ 138.12	93-59-4 42	7109 100[14]	4-09-00-00715	ace 4; bz 4; eth 4; EtOH 4
1127	Benzenecarboperoxoic acid, 3-chloro- 3-Chloroperoxybenzoic acid	$C_7H_5ClO_3$ 172.57	937-14-4 92-4 dec		4-09-00-00972	
1128	Benzenecarboperoxoic acid, 1,1-dimethylethyl ester Benzoyl tert-butyl peroxide	$C_{11}H_{14}O_3$ 194.23	614-45-9	75-6[0.2]	4-09-00-00715 1.021[25]	1.4990[20]
1129	Benzenecarbothioamide, N-phenyl-	$C_{13}H_{11}NS$ 213.30	636-04-4 102	dec	4-12-00-00423	H$_2$O 1; EtOH 4; eth 3; bz 3
1130	Benzenecarbothioic acid	C_7H_6OS 138.19	98-91-9 24	85-7[10]	4-09-00-01364 1.28[20]	ace 4; bz 4; eth 4; EtOH 4 1.6040[20]
1131	Benzene, chloro- Chlorobenzene	C_6H_5Cl 112.56	108-90-7 -45.2	2121 131.7	4-05-00-00640 1.1058[20]	H$_2$O 1; EtOH 5; eth 5; bz 4 1.5241[20]
1132	Benzene, 1-chloro-2-(chloromethyl)-	$C_7H_6Cl_2$ 161.03	611-19-8 -17	217	4-05-00-00816 1.2699[0]	H$_2$O 1; EtOH 2; eth 4; bz 4 1.5530[20]
1133	Benzene, 1-chloro-3-(chloromethyl)-	$C_7H_6Cl_2$ 161.03	620-20-2	216; 110[25]	4-05-00-00816 1.2695[15]	EtOH 4 1.5554[20]
1134	Benzene, 1-chloro-4-(chloromethyl)- 4-Chlorobenzyl chloride	$C_7H_6Cl_2$ 161.03	104-83-6 31	223	4-05-00-00816	ctc 2
1135	Benzene, 1-chloro-4-(chloromethyl)-2-nitro-	$C_7H_5Cl_2NO_2$ 206.03	57403-35-7	122[10]	4-05-00-00858 1.32[25]	1.5855[20]
1136	Benzene, 1-chloro-4-[(chloromethyl)thio]-	$C_7H_6Cl_2S$ 193.10	7205-90-5	128-9[12]	4-06-00-01593 1.346[25]	1.6055[20]
1137	Benzene, 1-chloro-2,3-dimethyl-	C_8H_9Cl 140.61	608-23-1	188	4-05-00-00928 1.053[20]	bz 4; chl 4 1.526[20]
1138	Benzene, 1-chloro-2,4-dimethyl-	C_8H_9Cl 140.61	95-66-9 -32.5	184	4-05-00-00943 1.0579[20]	H$_2$O 1; ace 3; bz 4 1.5230[25]
1139	Benzene, 2-chloro-1,3-dimethyl-	C_8H_9Cl 140.61	6781-98-2 -35	186	4-05-00-00943 1.053[20]	H$_2$O 1; ace 3; bz 4; chl 3 1.526[20]
1140	Benzene, 2-chloro-1,4-dimethyl-	C_8H_9Cl 140.61	95-72-7 1.6	187	4-05-00-00965 1.0589[15]	H$_2$O 1; ace 3; bz 4; ctc 3
1141	Benzene, 4-chloro-1,2-dimethyl-	C_8H_9Cl 140.61	615-60-1 -6	194	3-05-00-00816 1.0682[15]	H$_2$O 1; ace 3; bz 4; ctc 3
1142	Benzene, 1-chloro-4-(1,1-dimethylethyl)-	$C_{10}H_{13}Cl$ 168.67	3972-56-3	213	4-05-00-01048 1.0075[18]	1.5123[20]

No.	Name Synonym	Mol. Form. Mol. Wt.	CAS RN mp/°C	Merck No. bp/°C	Beil. Ref. den/g cm^{-3}	Solubility n_D
1143	Benzene, (2-chloro-1,1-dimethylethyl)- Neophyl chloride	$C_{10}H_{13}Cl$ 168.67	515-40-2	6375 223; 105[18]	4-05-00-01048 1.047[20]	ace 4; bz 4; eth 4; EtOH 4 1.5247[20]
1144	Benzene, 1-chloro-2,4-dinitro- 1-Chloro-2,4-dinitrobenzene	$C_6H_3ClN_2O_4$ 202.55	97-00-7 53	2135 315	4-05-00-00744 1.4982[75]	H_2O 1; EtOH 2; eth 3; bz 3 1.5857[60]
1145	Benzene, 2-chloro-1,3-dinitro- 2-Chloro-1,3-dinitrobenzene	$C_6H_3ClN_2O_4$ 202.55	606-21-3 88	2136 315	4-05-00-00744 1.6867[16]	H_2O 1; EtOH 3; eth 3; chl 2
1146	Benzene, 1,1'-(1-chloro-1,2-ethenediyl)bis-, (E)-	$C_{14}H_{11}Cl$ 214.69	948-98-1 53.5	322; 128[2]	4-05-00-02161	H_2O 1; EtOH 3; eth 3; bz 3
1147	Benzene, 1,1'-(1-chloro-1,2-ethenediyl)bis-, (Z)-	$C_{14}H_{11}Cl$ 214.69	948-99-2	160-2[12]	4-05-00-02162	H_2O 1; EtOH 3; eth 3; bz 3 1.6281[19]
1148	Benzene, (1-chloroethenyl)-	C_8H_7Cl 138.60	618-34-8 -23	199	4-05-00-01345 1.1016[18]	eth 4; EtOH 4 1.5612[20]
1149	Benzene, 1-chloro-2-ethenyl- o-Chlorostyrene	C_8H_7Cl 138.60	2039-87-4 -63.1	188.7	4-05-00-01345 1.1000[20]	EtOH 3; eth 3; ace 3; ctc 3 1.5649[20]
1150	Benzene, 1-chloro-3-ethenyl-	C_8H_7Cl 138.60	2039-85-2	62-3[6]	4-05-00-01345 1.1168[20]	H_2O 1; EtOH 3; eth 3 1.5625[20]
1151	Benzene, 1-chloro-4-ethenyl-	C_8H_7Cl 138.60	1073-67-2 15.9	192	4-05-00-01345 1.0868[20]	H_2O 1; EtOH 3; eth 3; ace 5 1.5660[20]
1152	Benzene, 1-(1-chloroethenyl)-2,4-dimethyl- Styrene, α-chloro-2,4-dimethyl	$C_{10}H_{11}Cl$ 166.65	74346-30-8	104-5[19]	2-05-00-00381 1.044[13]	1.5446[13]
1153	Benzene, (2-chloroethenyl)-, (E)-	C_8H_7Cl 138.60	4110-77-4	199	4-05-00-01346 1.1095[18]	H_2O 1; EtOH 3; eth 3; ace 3 1.5648[20]
1154	Benzene, 1,1',1''-(1-chloro-1-ethenyl-2-ylidene)tris[4-methoxy- Chlorotrianisene	$C_{23}H_{21}ClO_3$ 380.87	569-57-3 115	2173	4-06-00-07650	
1155	Benzene, (2-chloroethenyl)-, (Z)-	C_8H_7Cl 138.60	4604-28-8	59[3.5]	4-05-00-01346 1.1046[25]	1.5762[25]
1156	Benzene, 1-chloro-2-ethoxy-	C_8H_9ClO 156.61	614-72-2	210	4-06-00-00785 1.1288[15]	bz 4; eth 4; EtOH 4 1.5284[25]
1157	Benzene, 1-chloro-3-ethoxy- m-Chlorophenyl ethyl ether	C_8H_9ClO 156.61	2655-83-6	207	4-06-00-00811 1.1712[20]	H_2O 1; EtOH 4; eth 4; bz 4
1158	Benzene, 1-chloro-4-ethoxy-	C_8H_9ClO 156.61	622-61-7 21	213	4-06-00-00823 1.1254[20]	EtOH 3; eth 3; bz 4; ctc 2 1.5252[20]
1159	Benzene, (2-chloroethoxy)-	C_8H_9ClO 156.61	622-86-6 28	218.5	4-06-00-00556	H_2O 1; EtOH 4; eth 4; ace 4
1160	Benzene, (1-chloroethyl)-, (±)-	C_8H_9Cl 140.61	38661-82-4	81[17]	4-05-00-00898 1.0620[20]	EtOH 3; eth 4; bz 3 1.5276[20]
1161	Benzene, 1-chloro-2-ethyl-	C_8H_9Cl 140.61	89-96-3 -82.7	178.4	4-05-00-00897 1.0569[20]	H_2O 1; ace 3; bz 3; ctc 3 1.5218[20]
1162	Benzene, 1-chloro-3-ethyl-	C_8H_9Cl 140.61	620-16-6 -55	183.8	4-05-00-00897 1.0529[20]	ace 4; bz 4; eth 4; EtOH 4 1.5195[20]
1163	Benzene, 1-chloro-4-ethyl-	C_8H_9Cl 140.61	622-98-0 -62.6	184.4	4-05-00-00897 1.0455[20]	H_2O 1; EtOH 5; eth 5; ace 5 1.5175[20]
1164	Benzene, (2-chloroethyl)-	C_8H_9Cl 140.61	622-24-2	197.5; 92[20]	4-05-00-00899 1.069[25]	H_2O 1; EtOH 3; eth 3; ace 3 1.5276[20]
1165	Benzene, [(2-chloroethyl)thio]-	C_8H_9ClS 172.68	5535-49-9	117-8[11]	4-06-00-01469 1.1769[25]	eth 4; chl 4 1.5828[20]
1166	Benzene, 1-chloro-2-fluoro-	C_6H_4ClF 130.55	348-51-6 -43	137.6	4-05-00-00652 1.2233[30]	H_2O 1; ace 3; bz 3 1.4918[30]
1167	Benzene, 1-chloro-3-fluoro-	C_6H_4ClF 130.55	625-98-9	127.6	4-05-00-00652 1.221[25]	1.4911
1168	Benzene, 1-chloro-4-fluoro-	C_6H_4ClF 130.55	352-33-0 -26.8	130	4-05-00-00652 1.4990[15]	H_2O 1; EtOH 3; eth 3; bz 3 1.4990[15]
1169	Benzene, 1-chloro-3-fluoro-2-methyl-	C_7H_6ClF 144.58	443-83-4	154	3-05-00-00691 1.191[25]	1.5026[20]
1170	Benzene, 2-chloro-4-fluoro-1-methyl-	C_7H_6ClF 144.58	452-73-3	152.5	3-05-00-00691 1.1972[20]	1.4985[25]
1171	Benzene, 1-chloro-2-iodo-	C_6H_4ClI 238.45	615-41-8 0.7	234.5	4-05-00-00695 1.9515[25]	H_2O 1; ace 3; ctc 2 1.6331[25]
1172	Benzene, 1-chloro-4-iodo-	C_6H_4ClI 238.45	637-87-6 57	227	4-05-00-00696 1.886[27]	H_2O 1; EtOH 3; chl 2
1173	Benzene, 1-chloro-2-iodo-4-methyl- Toluene, 4-chloro-3-iodo-	C_7H_6ClI 252.48	2401-22-1 -10	249	3-05-00-00726 1.7940[15]	1.614[23]
1174	Benzene, 4-chloro-1-iodo-2-methyl- Toluene, 5-chloro-2-iodo	C_7H_6ClI 252.48	23399-70-4 -21	240	3-05-00-00726 1.702[17]	1.616[23]
1175	Benzene, 4-chloro-2-iodo-1-methyl-	C_7H_6ClI 252.48	33184-48-4 -25	242.5	3-05-00-00726 1.8358[15]	1.620[23]
1176	Benzene, 1-chloro-2-isocyanato-	C_7H_4ClNO 153.57	3320-83-0 30.5	200; 115[43]	4-12-00-01129	ctc 2
1177	Benzene, 1-chloro-4-isothiocyanato-	C_7H_4ClNS 169.63	2131-55-7 46	249.5	4-12-00-01214	H_2O 1; EtOH 3
1178	Benzene, 1-chloro-2-methoxy-	C_7H_7ClO 142.58	766-51-8 -26.8	198.5	4-06-00-00785 1.1911[20]	H_2O 1; EtOH 3; eth 3; chl 2 1.5480[20]
1179	Benzene, 1-chloro-3-methoxy-	C_7H_7ClO 142.58	2845-89-8	193.5	4-06-00-00811 1.1759[12]	H_2O 1; EtOH 3; eth 3 1.5365[20]
1180	Benzene, 1-chloro-4-methoxy-	C_7H_7ClO 142.58	623-12-1 <-18	197.5	4-06-00-00822 1.201[20]	H_2O 1; EtOH 4; eth 4; ctc 3 1.5390[20]
1181	Benzene, [(chloromethoxy)methyl]-	C_8H_9ClO 156.61	3587-60-8	103[13]	4-06-00-02252 1.1350[20]	1.5192[20]

No.	Name Synonym	Mol. Form. Mol. Wt.	CAS RN mp/°C	Merck No. bp/°C	Beil. Ref. den/g cm⁻³	Solubility n_D
1182	Benzene, (chloromethyl)- Benzyl chloride	C_7H_7Cl 126.59	100-44-7 -45	1143 179	4-05-00-00809 1.1004[20]	H_2O 1; EtOH 5; eth 5; ctc 2 1.5391[20]
1183	Benzene, 1-chloro-2-methyl- o-Chlorotoluene	C_7H_7Cl 126.59	95-49-8 -35.6	2172 159.0	4-05-00-00805 1.0825[20]	H_2O 1; EtOH 3; eth 5; ace 5 1.5268[20]
1184	Benzene, 1-chloro-3-methyl- m-Chlorotoluene	C_7H_7Cl 126.59	108-41-8 -47.8	2172 161.8	4-05-00-00806 1.075[20]	H_2O 1; EtOH 3; eth 5; bz 3 1.5214[19]
1185	Benzene, 1-chloro-4-methyl- p-Chlorotoluene	C_7H_7Cl 126.59	106-43-4 7.5	2172 162.4	4-05-00-00806 1.0697[20]	H_2O 1; EtOH 3; eth 5; ctc 3 1.5150[20]
1186	Benzene, 1-(chloromethyl)-2,4-dimethyl-	$C_9H_{11}Cl$ 154.64	824-55-5	215.5; 110[20]	4-05-00-01012 1.058[19]	bz 4; eth 4; EtOH 4
1187	Benzene, 1,1'-(chloromethylene)bis-	$C_{13}H_{11}Cl$ 202.68	90-99-3 16	140[3]	4-05-00-01847 1.140[25]	chl 3 1.5951[20]
1188	Benzene, 1-(2-chloro-1-methylethenyl)-4-methyl- Styrene, β-chloro, α-4-dimethyl	$C_{10}H_{11}Cl$ 166.65	30926-60-4	107-8[10]	2-05-00-00381 1.058[20]	1.5549[23]
1189	Benzene, (1-chloro-1-methylethyl)-	$C_9H_{11}Cl$ 154.64	934-53-2	98[1]	4-05-00-00992 1.192[25]	1.5290[25]
1190	Benzene, 1-chloro-2-(1-methylethyl)-	$C_9H_{11}Cl$ 154.64	2077-13-6 -74.4	191.1	4-05-00-00992 1.0341[20]	ace 4; bz 4; eth 4; EtOH 4 1.5168[20]
1191	Benzene, 1-(chloromethyl)-4-ethyl-	$C_9H_{11}Cl$ 154.64	1467-05-6	95-6[15]	4-05-00-01004	bz 4; EtOH 4; chl 4 1.5290[25]
1192	Benzene, 1-chloro-4-(1-methylethyl)-	$C_9H_{11}Cl$ 154.64	2621-46-7 -12.3	198.3	4-05-00-00992 1.0208[20]	H_2O 1; EtOH 5; eth 5; ace 5 1.5117[20]
1193	Benzene, (2-chloro-1-methylethyl)-	$C_9H_{11}Cl$ 154.64	824-47-5	85[13]	4-05-00-00992 1.047[19]	1.5245[19]
1194	Benzene, 1-(chloromethyl)-2-fluoro-	C_7H_6ClF 144.58	345-35-7	86[40]	4-05-00-00813 1.216[25]	1.5150[20]
1195	Benzene, 1-(chloromethyl)-4-fluoro-	C_7H_6ClF 144.58	352-11-4	82[26]	4-05-00-00813 1.207[25]	1.5130
1196	Benzene, 1,1',1''-(chloromethylidyne)tris-	$C_{19}H_{15}Cl$ 278.78	76-83-5 113.5	310	4-05-00-02497	H_2O 1; EtOH 2; eth 4; ace 3
1197	Benzene, 1-(chloromethyl)-4-methoxy-	C_8H_9ClO 156.61	824-94-2 24.5	262.5	4-06-00-02137 1.261[20]	ace 4; bz 4; eth 4 1.580[20]
1198	Benzene, 1-(chloromethyl)-2-methyl-	C_8H_9Cl 140.61	552-45-4	198; 90[20]	4-05-00-00926 1.063[25]	eth 4; EtOH 4 1.5410[25]
1199	Benzene, 1-(chloromethyl)-3-methyl-	C_8H_9Cl 140.61	620-19-9	195.5	4-05-00-00943 1.064[20]	H_2O 1; EtOH 3; eth 3 1.5345[20]
1200	Benzene, 1-(chloromethyl)-4-methyl-	C_8H_9Cl 140.61	104-82-5	201; 90[20]	4-05-00-00966 1.0512[20]	H_2O 1; EtOH 3; eth 5 1.5380
1201	Benzene, 2-chloro-1-methyl-4-(1-methylethyl)-	$C_{10}H_{13}Cl$ 168.67	4395-79-3	216.5	4-05-00-01062 1.0104[25]	H_2O 1; ace 3; bz 3 1.5078[20]
1202	Benzene, 2-chloro-4-methyl-1-(1-methylethyl)-	$C_{10}H_{13}Cl$ 168.67	4395-80-6	215; 113[35]	3-05-00-00956 1.018[15]	1.5180[20]
1203	Benzene, 1-(chloromethyl)-2-nitro-	$C_7H_6ClNO_2$ 171.58	612-23-7 51		4-05-00-00854	H_2O 1; EtOH 3; eth 3; ace 4 1.5557[62]
1204	Benzene, 1-chloro-2-methyl-3-nitro-	$C_7H_6ClNO_2$ 171.58	83-42-1 37.8	238	4-05-00-00854	H_2O 1; EtOH 3 1.5377[69]
1205	Benzene, 1-chloro-2-methyl-4-nitro-	$C_7H_6ClNO_2$ 171.58	13290-74-9 42.5	249	4-05-00-00855	eth 4
1206	Benzene, 1-(chloromethyl)-3-nitro-	$C_7H_6ClNO_2$ 171.58	619-23-8 46	173[34]	4-05-00-00855	ace 4; bz 4; eth 4; EtOH 4 1.5577[62]
1207	Benzene, 1-chloro-3-methyl-5-nitro- 3-Chloro-5-nitrotoluene	$C_7H_6ClNO_2$ 171.58	16582-38-0 61		4-05-00-00855	EtOH 4 1.5404[69]
1208	Benzene, 1-(chloromethyl)-4-nitro- 4-Nitrobenzyl chloride	$C_7H_6ClNO_2$ 171.58	100-14-1 71		4-05-00-00856	H_2O 1; EtOH 3; eth 3; ace 4 1.5647[62]
1209	Benzene, 1-chloro-4-methyl-2-nitro-	$C_7H_6ClNO_2$ 171.58	89-60-1 7	261; 118[11]	4-05-00-00855	H_2O 1; ctc 3 1.5572[20]
1210	Benzene, 2-chloro-1-methyl-4-nitro-	$C_7H_6ClNO_2$ 171.58	121-86-8 66.5	260	4-05-00-00855	H_2O 2; EtOH 3; eth 3; chl 2 1.5470[69]
1211	Benzene, 2-chloro-4-methyl-1-nitro-	$C_7H_6ClNO_2$ 171.58	38939-88-7 24	146[15]	4-05-00-00856	H_2O 1; ctc 3
1212	Benzene, 4-chloro-1-methyl-2-nitro-	$C_7H_6ClNO_2$ 171.58	89-59-8 38	242; 115.5[11]	4-05-00-00853 1.2559[80]	H_2O 1; EtOH 3; eth 3; chl 2
1213	Benzene, 4-chloro-2-methyl-1-nitro-	$C_7H_6ClNO_2$ 171.58	5367-28-2 25.5		4-05-00-00854	H_2O 1; ctc 3 1.5496[65]
1214	Benzene, 1-chloro-2-nitro- o-Chloronitrobenzene	$C_6H_4ClNO_2$ 157.56	88-73-3 32.5	2151 245.5	4-05-00-00721 1.368[242]	H_2O 1; EtOH 3; eth 3; ace 4
1215	Benzene, 1-chloro-3-nitro- m-Chloronitrobenzene	$C_6H_4ClNO_2$ 157.56	121-73-3 44.4	2151 235.5	4-05-00-00722 1.3435[50]	H_2O 1; EtOH 3; eth 3; bz 3 1.5374[80]
1216	Benzene, 1-chloro-4-nitro- p-Chloronitrobenzene	$C_6H_4ClNO_2$ 157.56	100-00-5 83.5	2151 242	4-05-00-00723 1.2979[90]	H_2O I; EtOH 3; eth 3; chl 3 1.5376[100]
1217	Benzene, 1-chloro-2-nitro-4-(trifluoromethyl)- 2-Nitro-4-(trifluoromethyl)-1-chlorobenzene	$C_7H_3ClF_3NO_2$ 225.55	121-17-5 -2.5	222; 95[10]	4-05-00-00857 1.511[25]	1.4893[20]
1218	Benzene, 1-chloro-4-nitro-2-(trifluoromethyl)- 4-Chloro-3-(trifluoromethyl)nitrobenzene	$C_7H_3ClF_3NO_2$ 225.55	777-37-7 22	232	4-05-00-00857 1.527[25]	1.5083[26]
1219	Benzene, 4-chloro-1-nitro-2-(trifluoromethyl)- 4-Chloro-2-(trifluoromethyl)nitrobenzene	$C_7H_3ClF_3NO_2$ 225.55	118-83-2		3-05-00-00748 1.526[25]	1.4980[20]
1220	Benzene, chloropentafluoro- Chloropentafluorobenzene	C_6ClF_5 202.51	344-07-0	117.9	4-05-00-00654 1.568[25]	1.4256[20]

No.	Name Synonym	Mol. Form. Mol. Wt.	CAS RN mp/°C	Merck No. bp/°C	Beil. Ref. den/g cm^{-3}	Solubility n_D
1221	Benzene, 1-chloro-4-phenoxy- 4-Chlorophenyl phenyl ether	$C_{12}H_9ClO$ 204.66	7005-72-3	284.5	4-06-00-00826 1.2026[15]	1.599
1222	Benzene, 1-chloro-2-(2-phenylethenyl)-, (E)-	$C_{14}H_{11}Cl$ 214.69	1657-52-9 39.5	209[30]; 139[2]	4-05-00-02161	H_2O 1; EtOH 3; chl 3
1223	Benzene, 1-chloro-3-(2-phenylethenyl)-, (E)-	$C_{14}H_{11}Cl$ 214.69	14064-43-8 73.5	177[0.2]	4-05-00-02161	H_2O 1; EtOH 3; chl 3
1224	Benzene, 1-chloro-4-(phenylmethyl)-	$C_{13}H_{11}Cl$ 202.68	831-81-2 7.5	299; 147-8[8]	4-05-00-01847 1.1247[20]	ace 4
1225	Benzene, 1-chloro-4-(phenylsulfonyl)- Sulphenone	$C_{12}H_9ClO_2S$ 252.72	80-00-2 94	8969	4-06-00-01587	H_2O 1; EtOH 2; eth 3; ace 4
1226	Benzene, (1-chloro-1-propenyl)-	C_9H_9Cl 152.62	35673-03-1	90.5[0.9]	2-05-00-00372 1.085[20]	ace 4; bz 4 1.5635[15]
1227	Benzene, (2-chloro-1-propenyl)-	C_9H_9Cl 152.62	13099-50-8	120[20]; 61[1]	3-05-00-01186 1.0738[19]	ace 4; bz 4; chl 4 1.5565[19]
1228	Benzene, (3-chloro-2-propenyl)-	C_9H_9Cl 152.62	6268-37-7	211	4-05-00-01363 1.073[13]	H_2O 1; eth 3; ace 3; bz 3 1.545[14]
1229	Benzene, 1-chloro-3-(1-propenyl)-, (E)-	C_9H_9Cl 152.62	23204-80-0 8.5	106.7[13]	1.0926[20]	H_2O 1; EtOH 5; eth 5; ace 4 1.5851[20]
1230	Benzene, (3-chloro-1-propenyl)-, (E)-	C_9H_9Cl 152.62	21087-29-6 8.5	106-7[13]	4-05-00-01360 1.0926[20]	ace 4; bz 4; eth 4; EtOH 4 1.5851[20]
1231	Benzene, 1-(3-chloro-2-propenyl)-4-methoxy- Benzene, 1(3-chloroallyl)-4-methoxy	$C_{10}H_{11}ClO$ 182.65	54644-23-4	120[11]	3-06-00-02416 1.155[8]	1.553[8]
1232	Benzene, 1-chloro-4-(2-propenyloxy)-	C_9H_9ClO 168.62	13997-70-1	106-7[12]	4-06-00-00825 1.131[15]	bz 4; eth 4; EtOH 4 1.5348[25]
1233	Benzene, (3-chloropropoxy)-	$C_9H_{11}ClO$ 170.64	3384-04-1 12	250	4-06-00-00557 1.1167[20]	ace 4; eth 4 1.5235[25]
1234	Benzene, (2-chloropropyl)-, (R)- R-(-)-2-Chloro-1-phenylpropane	$C_9H_{11}Cl$ 154.64	55449-46-2	94[17]	4-05-00-00980 1.038[19]	H_2O 1; ace 3; bz 3; chl 3 1.5198[22]
1235	Benzene, (3-chloropropyl)-	$C_9H_{11}Cl$ 154.64	104-52-9	219.5	4-05-00-00980 1.056[21]	ctc 2 1.5160[25]
1236	Benzene, [(3-chloropropyl)thio]-	$C_9H_{11}ClS$ 186.71	4911-65-3	116-7[4]	3-06-00-00981 1.1536[20]	ace 4; py 4 1.5752[20]
1237	Benzene, 1-chloro-2-(trichloromethyl)- 1-Chloro-2-(trichloromethyl)benzene	$C_7H_4Cl_4$ 229.92	2136-89-2 29.4	264.3	4-05-00-00823 1.5187[20]	H_2O 1; eth 3; ace 3; ctc 2 1.5836[20]
1238	Benzene, 1-chloro-3-(trichloromethyl)- 1-Chloro-3-(trichloromethyl)benzene	$C_7H_4Cl_4$ 229.92	2136-81-4 0.6	255	4-05-00-00823 1.495[14]	H_2O 1; eth 3; ace 3; ctc 2 1.4461[20]
1239	Benzene, 1-chloro-4-(trichloromethyl)- 1-Chloro-4-(trichloromethyl)benzene	$C_7H_4Cl_4$ 229.92	5216-25-1	245	4-05-00-00823 1.4463[20]	ace 4; eth 4
1240	Benzene, 1-chloro-2-(trifluoromethyl)-	$C_7H_4ClF_3$ 180.56	88-16-4 -6	152.2	4-05-00-00814 1.2540[30]	chl 3 1.4513[25]
1241	Benzene, 1-chloro-3-(trifluoromethyl)- Benzotrifluoride, 3-chloro	$C_7H_4ClF_3$ 180.56	98-15-7 -56	137.5	4-05-00-00814 1.3311[25]	1.4438[25]
1242	Benzene, 1-chloro-4-(trifluoromethyl)-	$C_7H_4ClF_3$ 180.56	98-56-6 -33	138.5	4-05-00-00815 1.3340[25]	1.4431[30]
1243	Benzene, 2-chloro-1,3,5-trimethyl-	$C_9H_{11}Cl$ 154.64	1667-04-5 <-20	205	4-05-00-01026 1.0337[30]	eth 4; EtOH 4 1.5212[30]
1244	Benzene, 2-chloro-1,3,5-trinitro- Picryl chloride	$C_6H_2ClN_3O_6$ 247.55	88-88-0 83	7390	4-05-00-00757 1.797[20]	H_2O 1; EtOH 3; eth 2; ace 4
1245	Benzene, cyclobutyl-	$C_{10}H_{12}$ 132.21	4392-30-7	190[155]; 101[41]	4-05-00-01387 0.9378[20]	1.5277[20]
1246	Benzene, 1-cyclohexen-1-yl-	$C_{12}H_{14}$ 158.24	771-98-2 -11	252	4-05-00-01557 0.9939[20]	MeOH 4 1.5718[20]
1247	Benzene, 2-cyclohexen-1-yl-	$C_{12}H_{14}$ 158.24	15232-96-9	115-7[16]	4-05-00-01558 0.9800[20]	1.5530[20]
1248	Benzene, cyclohexyl- Cyclohexylbenzene	$C_{12}H_{16}$ 160.26	827-52-1 7.3	240.1	4-05-00-01424 0.9427[20]	H_2O 1; EtOH 4; eth 3; ctc 2 1.5329[20]
1249	Benzene, (1-cyclohexylethyl)-	$C_{14}H_{20}$ 188.31	4413-16-5	265	4-05-00-01454 0.9773[17]	1.549[17]
1250	Benzene, 1-cyclohexyl-2-methoxy-	$C_{13}H_{18}O$ 190.29	2206-48-6	268	4-06-00-03899 1.007[18]	1.5365[18]
1251	Benzene, 1-cyclohexyl-2-nitro- Cyclohexane, 2-nitrophenyl	$C_{12}H_{15}NO_2$ 205.26	7137-56-6 45	174[16]; 113[0.5]	3-05-00-01259 1.111[23]	1.5472[25]
1252	Benzene, 1-cyclohexyl-4-nitro-	$C_{12}H_{15}NO_2$ 205.26	5458-48-0 58.5	210[25]	4-05-00-01426	bz 4; eth 4
1253	Benzene, (cyclohexyloxy)-	$C_{12}H_{16}O$ 176.26	2206-38-4	261	4-06-00-00565 1.0077[20]	H_2O 1; ace 3; bz 3 1.520[22]
1254	Benzene, cyclopentyl-	$C_{11}H_{14}$ 146.23	700-88-9	219	4-05-00-01409 0.9462[20]	eth 4 1.5280[20]
1255	Benzene, (3-cyclopentylpropyl)-	$C_{14}H_{20}$ 188.31	2883-12-7	272	4-05-00-01455 0.9233[19]	1.5130[19]
1256	Benzene, cyclopropyl- Cyclopropylbenzene	C_9H_{10} 118.18	873-49-4 -31	173.6; 80[37]	4-05-00-01370 0.9317[20]	H_2O 1; eth 3; ace 3; chl 3 1.5285[20]
1257	Benzenedecanoic acid, ι-octyl-	$C_{24}H_{40}O_2$ 360.58	1938-17-6 40.5	199-205[0.09]	2-09-00-00376 0.9310[25]	1.4894[20]
1258	Benzene, decyl- Decylbenzene	$C_{16}H_{26}$ 218.38	104-72-3 -14.4	298	4-05-00-01193 0.8555[20]	ace 4; bz 4; eth 4; EtOH 4 1.4832[20]
1259	1,4-Benzenediacetic acid, dimethyl ester	$C_{12}H_{14}O_4$ 222.24	36076-25-2 51.5	189[15]	2-09-00-00624	EtOH 3; eth 3

No.	Name Synonym	Mol. Form. Mol. Wt.	CAS RN mp/°C	Merck No. bp/°C	Beil. Ref. den/g cm⁻³	Solubility n_D
1260	1,2-Benzenediamine o-Phenylenediamine	$C_6H_8N_2$ 108.14	95-54-5 102.5	7255 257	4-13-00-00038	H_2O 3; EtOH 4; eth 3; bz 3
1261	1,3-Benzenediamine m-Phenylenediamine	$C_6H_8N_2$ 108.14	108-45-2 63.5	7254 285	4-13-00-00079 1.0096[58]	H_2O 4; EtOH 3; eth 3; bz 3 1.6339[58]
1262	1,4-Benzenediamine p-Phenylenediamine	$C_6H_8N_2$ 108.14	106-50-3 146	7256 267	4-13-00-00104	H_2O 2; EtOH 3; eth 3; bz 3
1263	1,4-Benzenediamine, N-(4-aminophenyl)- 4,4'-Diaminodiphenylamine	$C_{12}H_{13}N_3$ 199.26	537-65-5 158 dec	2957	4-13-00-00203	eth 4; EtOH 4
1264	1,2-Benzenediamine, 4-chloro-	$C_6H_7ClN_2$ 142.59	95-83-0 76		4-13-00-00068	H_2O 2; EtOH 4; eth 4; bz 3
1265	1,3-Benzenediamine, N,N-diethyl-	$C_{10}H_{16}N_2$ 164.25	26513-20-2 152	277	2-13-00-00026	EtOH 4
1266	1,4-Benzenediamine, N,N-diethyl- N,N-Diethyl-p-phenylenediamine	$C_{10}H_{16}N_2$ 164.25	93-05-0	261	4-13-00-00109	bz 4
1267	1,2-Benzenediamine, N,N-dimethyl-	$C_8H_{12}N_2$ 136.20	2836-03-5	218; 90[22]	4-13-00-00042 0.995[22]	H_2O 2; EtOH 4; eth 4; ace 4
1268	1,3-Benzenediamine, N,N-dimethyl-	$C_8H_{12}N_2$ 136.20	2836-04-6 <-20	270; 138[10]	4-13-00-00080 0.995[25]	H_2O 2; EtOH 4; eth 4
1269	1,2-Benzenediamine, N,N'-dimethyl- o-Phenylenediamine, N,N'-dimethyl	$C_8H_{12}N_2$ 136.20	3213-79-4 34.5	247	4-13-00-00042	
1270	1,4-Benzenediamine, N,N'-dimethyl- Dimethyl-p-phenylenediamine	$C_8H_{12}N_2$ 136.20	99-98-9 53	3242 263	4-13-00-00106 1.036[20]	H_2O 3; EtOH 4; eth 4; bz 4
1271	1,4-Benzenediamine, N,N'-diphenyl- N,N'-Diphenyl-p-phenylenediamine	$C_{18}H_{16}N_2$ 260.34	74-31-7 144	3331	4-13-00-00116	EtOH 2; eth 2; bz 2; chl 2
1272	1,2-Benzenediamine, 4-ethoxy-	$C_8H_{12}N_2O$ 152.20	1197-37-1 71.5	295	3-13-00-01362	H_2O 4; EtOH 3; eth 3; chl 3
1273	1,2-Benzenediamine, 4-methoxy-	$C_7H_{10}N_2O$ 138.17	102-51-2 51	200[21]; 168[11]	4-13-00-01448	eth 4
1274	1,3-Benzenediamine, 4-methoxy- 4-Methoxy-m-phenylenediamine	$C_7H_{10}N_2O$ 138.17	615-05-4 67.5	5920	4-13-00-01425	EtOH 3; eth 3; DMSO-d₆ 2
1275	1,2-Benzenediamine, 3-methyl-	$C_7H_{10}N_2$ 122.17	2687-25-4 63.5	255	3-13-00-00277	ace 4; bz 4; EtOH 4
1276	1,2-Benzenediamine, 4-methyl- Toluene-3,4-diamine	$C_7H_{10}N_2$ 122.17	496-72-0 89.5	265	4-13-00-00260	H_2O 4; lig 3
1277	1,3-Benzenediamine, 2-methyl- 2,6-Diaminotoluene	$C_7H_{10}N_2$ 122.17	823-40-5 106		3-13-00-00291	H_2O 3; EtOH 3; bz 3
1278	1,3-Benzenediamine, 4-methyl- Toluene-2,4-diamine	$C_7H_{10}N_2$ 122.17	95-80-7 99	292	4-13-00-00235	H_2O 4; EtOH 4; eth 4; bz 4
1279	1,4-Benzenediamine, 2-methyl-	$C_7H_{10}N_2$ 122.17	95-70-5 64	273.5	4-13-00-00246	H_2O 3; EtOH 3; eth 3; bz 2
1280	1,4-Benzenediamine, N-methyl-	$C_7H_{10}N_2$ 122.17	623-09-6 36	258	4-13-00-00106	H_2O 4; EtOH 4; eth 4; ace 4
1281	1,2-Benzenediamine, 4-nitro- 4-Nitro-o-phenylenediamine	$C_6H_7N_3O_2$ 153.14	99-56-9 199.5	6544	4-13-00-00075	acid 3
1282	1,2-Benzenediamine, N-phenyl-	$C_{12}H_{12}N_2$ 184.24	534-85-0 79.5	313	4-13-00-00043	H_2O 2; ace 3; bz 3; chl 3
1283	1,4-Benzenediamine, N-phenyl- p-Aminodiphenylamine	$C_{12}H_{12}N_2$ 184.24	101-54-2 66	354	4-13-00-00113	H_2O 2; EtOH 4; eth 3; chl 2
1284	1,3-Benzenediamine, 4-(phenylazo)-, monohydrochloride Chrysoidine hydrochloride	$C_{12}H_{13}ClN_4$ 248.71	532-82-1 118.5	2262	2-16-00-00203	ace 4
1285	1,3-Benzenediamine, N,N,N',N'-tetraethyl- N,N,N',N'-Tetraethyl-m-phenylenediamine	$C_{14}H_{24}N_2$ 220.36	64287-26-9	148[9]	2-13-00-00026 0.9522[12]	eth 4; EtOH 4 1.5537[12]
1286	1,2-Benzenediamine, N,N,N',N'-tetramethyl-	$C_{10}H_{16}N_2$ 164.25	704-01-8 8.9	215.5	4-13-00-00042 0.9560[20]	
1287	1,4-Benzenediamine, N,N,N',N'-tetramethyl- Tetramethyl-p-phenylenediamine	$C_{10}H_{16}N_2$ 164.25	100-22-1 51	9159 260	4-13-00-00107	H_2O 2; EtOH 4; eth 4; bz 4
1288	Benzene, 1,2-dibromo- o-Dibromobenzene	$C_6H_4Br_2$ 235.91	583-53-9 7.1	225	4-05-00-00682 1.9843[20]	H_2O 1; EtOH 3; eth 5; ace 5 1.6155[20]
1289	Benzene, 1,3-dibromo- m-Dibromobenzene	$C_6H_4Br_2$ 235.91	108-36-1 -7	218	4-05-00-00682 1.9523[20]	H_2O 1; EtOH 3; eth 5 1.6083[17]
1290	Benzene, 1,4-dibromo- p-Dibromobenzene	$C_6H_4Br_2$ 235.91	106-37-6 87.3	3001 218.5	4-05-00-00683 2.261[17]	H_2O 1; EtOH 3; eth 4; ace 4 1.5742
1291	Benzene, 1,3-dibromo-5-(bromomethyl)- 3,5-Dibromobenzyl bromide	$C_7H_5Br_3$ 328.83	56908-88-4 96	173[19]	3-05-00-00719	EtOH 4
1292	Benzene, 1,2-dibromo-3-chloro- 1,2-Dibromo-3-chlorobenzene	$C_6H_3Br_2Cl$ 270.35	104514-49-0 73.5	264	1-05-00-00117	bz 4; eth 4; chl 4
1293	Benzene, 1,2-dibromo-4-chloro- 1,2-Dibromo-4-chlorobenzene	$C_6H_3Br_2Cl$ 270.35	60956-24-3 35.5	256	0-05-00-00212	bz 4; eth 4; chl 4
1294	Benzene, 1,3-dibromo-2-chloro- 1,3-Dibromo-2-chlorobenzene	$C_6H_3Br_2Cl$ 270.35	19230-27-4 72	265	3-05-00-00569	bz 4; eth 4; chl 4
1295	Benzene, 1,4-dibromo-2-chloro- 1,4-Dibromo-2-chlorobenzene	$C_6H_3Br_2Cl$ 270.35	3460-24-0 40.5	259	0-05-00-00212	bz 4; eth 4; chl 4
1296	Benzene, 2,4-dibromo-1-chloro-	$C_6H_3Br_2Cl$ 270.35	29604-75-9 27	258	0-05-00-00212	bz 4; eth 4; chl 4
1297	Benzene, 1,2-dibromo-3,4-dimethyl- 3,4-Dibromo-o-xylene	$C_8H_8Br_2$ 263.96	24932-49-8 7	277	2-05-00-00285 1.782[15]	EtOH 4
1298	Benzene, 1,5-dibromo-2,4-dimethyl-	$C_8H_8Br_2$ 263.96	615-87-2 70	255.5	3-05-00-00839	EtOH 4

No.	Name Synonym	Mol. Form. Mol. Wt.	CAS RN mp/°C	Merck No. bp/°C	Beil. Ref. den/g cm^{-3}	Solubility n_D
1299	Benzene, (1,2-dibromoethyl)-	$C_8H_8Br_2$ 263.96	93-52-7 75	133[19]	4-05-00-00909	EtOH 3; eth 3; bz 3; chl 3
1300	Benzene, 2,4-dibromo-1-fluoro-	$C_6H_3Br_2F$ 253.90	1435-53-6 69	103-5[32]	3-05-00-00569 2.047[20]	1.5840[20]
1301	Benzene, (dibromomethyl)-	$C_7H_6Br_2$ 249.93	618-31-5	156[23]	4-05-00-00836 1.8365[28]	H_2O 1; EtOH 5; eth 5 1.6147[20]
1302	Benzene, 1,2-dibromo-3-methyl- 2,3-Dibromotoluene	$C_7H_6Br_2$ 249.93	61563-25-5 30.5	126-8[19]	4-05-00-00835 1.8234[15]	1.5984[25]
1303	Benzene, 1,2-dibromo-4-methyl- 3,4-Dibromotoluene	$C_7H_6Br_2$ 249.93	60956-23-2 -10	240; 123[19]	4-05-00-00836 1.8197[25]	1.5979[25]
1304	Benzene, 1,3-dibromo-2-methyl- 2,6-Dibromotoluene	$C_7H_6Br_2$ 249.93	69321-60-4 5.5	246	4-05-00-00836 1.812[22]	
1305	Benzene, 1,4-dibromo-2-methyl- 2,5-Dibromotoluene	$C_7H_6Br_2$ 249.93	615-59-8 5.6	236	4-05-00-00835 1.8127[17]	H_2O 1 1.5982[18]
1306	Benzene, 2,4-dibromo-1-methyl-	$C_7H_6Br_2$ 249.93	31543-75-6 -9.7	103-4[11]	4-05-00-00835 1.8176[25]	1.5964[25]
1307	Benzene, 1,2-dibromo-4-nitro- 3,4-Dibromonitrobenzene	$C_6H_3Br_2NO_2$ 280.90	5411-50-7 58.5	296	4-05-00-00732 2.354[8]	bz 4; EtOH 4; HOAc 4 1.9835[111]
1308	Benzene, 1,3-dibromo-2-nitro- 2,6-Dibromonitrobenzene	$C_6H_3Br_2NO_2$ 280.90	13402-32-9 84	sub	3-05-00-00621 2.211[8]	ace 4; bz 4; EtOH 4
1309	Benzene, 1,3-dibromo-5-nitro-	$C_6H_3Br_2NO_2$ 280.90	6311-60-0 106		3-05-00-00621 2.36[8]	bz 4; eth 4; EtOH 4
1310	Benzene, 1,4-dibromo-2-nitro-	$C_6H_3Br_2NO_2$ 280.90	3460-18-2 85.5		4-05-00-00732 2.388[8]	H_2O 1; ace 3; bz 3; chl 2
1311	Benzene, 2,4-dibromo-1-nitro-	$C_6H_3Br_2NO_2$ 280.90	51686-78-3 62		3-05-00-00620 2.356[8]	ace 4; bz 4
1312	Benzene, 1,2-dibromo-3,4,5,6-tetrafluoro-	$C_6Br_2F_4$ 307.87	827-08-7 12	198	2.260[20]	1.5151[25]
1313	Benzene, 2,4-dibromo-1,3,5-trimethyl- Mesitylene, 2,4-dibromo-	$C_9H_{10}Br_2$ 277.99	6942-99-0 65.5	285	3-05-00-00921	ctc 2
1314	Benzene, 1,4-dibutoxy-	$C_{14}H_{22}O_2$ 222.33	104-36-9 45.5	158[15]	4-06-00-05721	ctc 3
1315	1,3-Benzenedicarbonitrile 1,3-Dicyanobenzene	$C_8H_4N_2$ 128.13	626-17-5 162	sub	4-09-00-03297 0.992[40]	H_2O 2; EtOH 4; eth 3; bz 3
1316	1,4-Benzenedicarbonitrile	$C_8H_4N_2$ 128.13	623-26-7 224	sub	4-09-00-03328	H_2O 1; EtOH 2; eth 2; bz 3
1317	1,3-Benzenedicarbonitrile, 2,4,5,6-tetrachloro- Chlorothalonil	$C_8Cl_4N_2$ 265.91	1897-45-6 250	2167 350	1.7[25]	H_2O 1; ace 2; cyhex 2
1318	1,2-Benzenedicarbonyl dichloride Phthaloyl chloride	$C_8H_4Cl_2O_2$ 203.02	88-95-9 15.5	7349 281.1	4-09-00-03261 1.4089[20]	1.5684[20]
1319	1,3-Benzenedicarbonyl dichloride	$C_8H_4Cl_2O_2$ 203.02	99-63-8 43.5	276	4-09-00-03295 1.3880[17]	H_2O 2; EtOH 2; eth 3 1.570[47]
1320	1,4-Benzenedicarbonyl dichloride	$C_8H_4Cl_2O_2$ 203.02	100-20-9 83.5	258; 125[9]	4-09-00-03318	eth 3
1321	1,2-Benzenedicarbonyl difluoride	$C_8H_4F_2O_2$ 170.12	445-69-2 42.5	227	4-09-00-03260 1.3066[50]	H_2O 1; peth 3
1322	1,3-Benzenedicarboxaldehyde	$C_8H_6O_2$ 134.13	626-19-7 89.5	246; 136[13]	4-07-00-02139	H_2O 2; EtOH 4; eth 2; ace 3
1323	1,4-Benzenedicarboxaldehyde	$C_8H_6O_2$ 134.13	623-27-8 117	246	4-07-00-02140	H_2O 2; EtOH 4; eth 3; chl 3
1324	1,2-Benzenedicarboxamide Phthalamide	$C_8H_8N_2O_2$ 164.16	88-96-0 222	7343 dec	4-09-00-03265	H_2O 2; EtOH 2; eth 1
1325	1,2-Benzenedicarboxamide, N,N,N',N'- tetraethyl- N,N,N',N'-Tetraethylphthalamide	$C_{16}H_{24}N_2O_2$ 276.38	83-81-8 36	9137 204[16]	4-09-00-03266	
1326	1,3-Benzenedicarboxamide, N,N,N',N'- tetraethyl- Isophthalamide, N,N,N',N'-tetraethyl	$C_{16}H_{24}N_2O_2$ 276.38	13698-87-8 86	242[12]	4-09-00-03296	ace 4; bz 4; eth 4; EtOH 4
1327	1,2-Benzenedicarboxylic acid Phthalic acid	$C_8H_6O_4$ 166.13	88-99-3 230 dec	7345 dec	4-09-00-03167 2.18[191]	H_2O 2; EtOH 3; eth 2; chl 1
1328	1,3-Benzenedicarboxylic acid Isophthalic acid	$C_8H_6O_4$ 166.13	121-91-5 347	5083 sub	4-09-00-03292	H_2O 2; EtOH 3; eth 1; bz 1
1329	1,4-Benzenedicarboxylic acid Terephthalic acid	$C_8H_6O_4$ 166.13	100-21-0	9093 sub 300	4-09-00-03301	H_2O 1; EtOH 1; eth 1; ctc 2
1330	1,3-Benzenedicarboxylic acid, 2-amino-	$C_8H_7NO_4$ 181.15	39622-79-2 340	sub 267	4-14-00-01899	eth 4; EtOH 4
1331	1,3-Benzenedicarboxylic acid, 5-amino-	$C_8H_7NO_4$ 181.15	99-31-0 360	sub	1-14-00-00636	H_2O 1; EtOH 2
1332	1,2-Benzenedicarboxylic acid, 3-benzoyl-	$C_{15}H_{10}O_5$ 270.24	602-82-4 140.5	dec	3-10-00-04007	bz 4; EtOH 4
1333	1,2-Benzenedicarboxylic acid, bis(2-ethoxyethyl) ester	$C_{16}H_{22}O_6$ 310.35	605-54-9 34	345	2-09-00-00597 1.1229[21]	
1334	1,2-Benzenedicarboxylic acid, bis(2-ethylhexyl) ester Bis(2-ethylhexyl) phthalate	$C_{24}H_{38}O_4$ 390.56	117-81-7 -55	1262 384	4-09-00-03181 0.981[25]	ctc 2 1.4853[20]
1335	1,2-Benzenedicarboxylic acid, bis(3-methylbutyl) ester Diisoamyl phthalate	$C_{18}H_{26}O_4$ 306.40	605-50-5	5006 334 dec	4-09-00-03179 1.0209[16]	EtOH 4 1.4871[20]

No.	Name Synonym	Mol. Form. Mol. Wt.	CAS RN mp/°C	Merck No. bp/°C	Beil. Ref. den/g cm^{-3}	Solubility n_D
1336	1,2-Benzenedicarboxylic acid, bis(2-methylpropyl) ester	$C_{16}H_{22}O_4$ 278.35	84-69-5	296.5; 159[4]	4-09-00-03177 1.0490[15]	ctc 3
1337	1,2-Benzenedicarboxylic acid, bis(phenylmethyl) ester	$C_{22}H_{18}O_4$ 346.38	523-31-9 44.0	277[15]	4-09-00-03219	H_2O 1; EtOH 4; eth 4; lig 1
1338	1,4-Benzenedicarboxylic acid, bis(2-methylpropyl) ester	$C_{16}H_{22}O_4$ 278.35	18699-48-4 55	180[6]	4-09-00-03305	eth 3
1339	1,2-Benzenedicarboxylic acid, dibutyl ester Dibutyl phthalate	$C_{16}H_{22}O_4$ 278.35	84-74-2 -35	1586 340	4-09-00-03175 1.0465[20]	H_2O 1; EtOH 5; eth 5; bz 5 1.4911[20]
1340	1,2-Benzenedicarboxylic acid, 3,5-dichloro- Phthalic acid, 3,5-dichloro-	$C_8H_4Cl_2O_4$ 235.02	25641-98-9 164 dec	sub	4-09-00-03270	ace 4; eth 4; EtOH 4
1341	1,2-Benzenedicarboxylic acid, dicyclohexyl ester	$C_{20}H_{26}O_4$ 330.42	84-61-7 66		4-09-00-03189 1.383[20]	H_2O 1; EtOH 3; eth 3; chl 2 1.431[20]
1342	1,2-Benzenedicarboxylic acid, diethyl ester Diethyl phthalate	$C_{12}H_{14}O_4$ 222.24	84-66-2 -40.5	7345 295	4-09-00-03172 1.232[14]	H_2O I; EtOH 5; eth 5; ace 3 1.5000[21]
1343	1,2-Benzenedicarboxylic acid, dimethyl ester Dimethyl phthalate	$C_{10}H_{10}O_4$ 194.19	131-11-3 5.5	3243 283.7	4-09-00-03170 1.1905[20]	H_2O 1; EtOH 5; eth 5; bz 3 1.5138[20]
1344	1,2-Benzenedicarboxylic acid, dioctyl ester Dioctyl phthalate	$C_{24}H_{38}O_4$ 390.56	117-84-0 25		4-09-00-03180	
1345	1,2-Benzenedicarboxylic acid, diphenyl ester Phenyl phthalate	$C_{20}H_{14}O_4$ 318.33	84-62-8 73	7278 253[14]	4-09-00-03217	H_2O 1; EtOH 2; eth 2; ctc 2
1346	1,3-Benzenedicarboxylic acid, diethyl ester Diethyl isophthalate	$C_{12}H_{14}O_4$ 222.24	636-53-3 11.5	302	4-09-00-03294 1.1239[17]	H_2O 1 1.508[18]
1347	1,3-Benzenedicarboxylic acid, dimethyl ester Dimethyl isophthalate	$C_{10}H_{10}O_4$ 194.19	1459-93-4 67.5	282	4-09-00-03293 1.194[20]	H_2O 2 1.5168[20]
1348	1,3-Benzenedicarboxylic acid, 4,6-dimethyl-	$C_{10}H_{10}O_4$ 194.19	2790-09-2 266	sub	3-09-00-04298	EtOH 4
1349	1,4-Benzenedicarboxylic acid, 2,5-dichloro- 2,5-Dichloroterephthalic acid	$C_8H_4Cl_2O_4$ 235.02	13799-90-1 306	sub	4-09-00-03335	H_2O 2; EtOH 3; eth 4
1350	1,4-Benzenedicarboxylic acid, diethyl ester Diethyl terephthalate	$C_{12}H_{14}O_4$ 222.24	636-09-9 44	302	4-09-00-03304 1.0989[45]	H_2O 1; EtOH 4; eth 3
1351	1,4-Benzenedicarboxylic acid, dimethyl ester Dimethyl terephthalate	$C_{10}H_{10}O_4$ 194.19	120-61-6 141	288	4-09-00-03303 1.075[141]	H_2O 2; EtOH 2; eth 3; chl 3
1352	1,3-Benzenedicarboxylic acid, 4-hydroxy- 4-Hydroxyisophthalic acid	$C_8H_6O_5$ 182.13	636-46-4 310	4758	4-10-00-02091	H_2O 1; EtOH 4; eth 4; chl 1
1353	1,2-Benzenedicarboxylic acid, 3,4,5,6-tetrachloro-, monoethyl ester Phthalic acid, tetrachloro-, monoethyl ester	$C_{10}H_8Cl_4O_4$ 331.97	602-21-1 94.5	250 dec	3-09-00-04206	eth 4; EtOH 4
1354	Benzene, 1,2-dichloro- o-Dichlorobenzene	$C_8H_4Cl_2$ 147.00	95-50-1 -16.7	3044 180	4-05-00-00654 1.3059[20]	H_2O 1; EtOH 3; eth 3; ace 5 1.5515[20]
1355	Benzene, 1,3-dichloro- m-Dichlorobenzene	$C_8H_4Cl_2$ 147.00	541-73-1 -24.8	3043 173	4-05-00-00657 1.2884[20]	H_2O 1; EtOH 3; eth 3; ace 5 1.5459[20]
1356	Benzene, 1,4-dichloro- p-Dichlorobenzene	$C_8H_4Cl_2$ 147.00	106-46-7 52.7	3045 174	4-05-00-00658 1.2475[55]	H_2O 1; EtOH 5; eth 3; ace 5 1.5285[20]
1357	Benzene, 1,2-dichloro-4-(chloromethyl)-	$C_7H_5Cl_3$ 195.47	102-47-6 37.5	241	4-05-00-00820	H_2O 1; EtOH 3; ctc 3; os 3
1358	Benzene, 1,3-dichloro-5-(chloromethyl)-	$C_7H_5Cl_3$ 195.47	3290-06-0 36	60[0.35]	3-05-00-00699	EtOH 4
1359	Benzene, 1,2-dichloro-4-(dichloromethyl)- Benzylidene chloride, 3,4-dichloro	$C_7H_4Cl_4$ 229.92	56961-84-3	257	3-05-00-00702 1.515[22]	bz 4; eth 4; EtOH 4
1360	Benzene, 1,2-dichloro-3,4-dimethyl- o-Xylene, 3,4-dichloro	$C_8H_8Cl_2$ 175.06	68266-67-1 9	231	4-05-00-00926 1.2240[20]	
1361	Benzene, 1,3-dichloro-2,5-dimethyl-	$C_8H_8Cl_2$ 175.06	38204-89-6 15	222	3-05-00-00855 1.2156[20]	
1362	Benzene, 1,4-dichloro-2,5-dimethyl-	$C_8H_8Cl_2$ 175.06	1124-05-6 71	222	3-05-00-00855	chl 3
1363	Benzene, 1,5-dichloro-2,4-dimethyl-	$C_8H_8Cl_2$ 175.06	2084-45-9 68.5	223	3-05-00-00835	bz 4; eth 4; chl 4
1364	Benzene, 2,3-dichloro-1,4-dimethyl-	$C_8H_8Cl_2$ 175.06	34840-79-4 -2	230	3-05-00-00855 1.2310[20]	
1365	Benzene, 1,1'-(1,2-dichloro-1,2-ethenediyl)bis-, (E)-	$C_{14}H_{10}Cl_2$ 249.14	951-86-0 143.5	183[18]	1-05-00-00304	H_2O 1; EtOH 4; eth 4
1366	Benzene, 1,1'-(1,2-dichloro-1,2-ethenediyl)bis-, (Z)-	$C_{14}H_{10}Cl_2$ 249.14	5216-32-0 67.5	179[18]	1-05-00-00304	eth 4; EtOH 4
1367	Benzene, 1,2-dichloro-3-ethenyl- 2,3-Dichlorostyrene	$C_8H_6Cl_2$ 173.04	2123-28-6	92-4[4.5]	3-05-00-01174 1.2826[20]	1.5848[20]
1368	Benzene, 1,2-dichloro-4-ethenyl-	$C_8H_6Cl_2$ 173.04	2039-83-0	95[5]	4-05-00-01347 1.256[20]	1.5857[20]
1369	Benzene, 1,3-dichloro-2-ethenyl- 2,6-Dichlorostyrene	$C_8H_6Cl_2$ 173.04	28469-92-3 89	64-5[3]	3-05-00-01174 1.2631[20]	1.5752[20]
1370	Benzene, 1,3-dichloro-5-ethenyl- 3,5-Dichlorostyrene	$C_8H_6Cl_2$ 173.04	2155-42-2	59[1]	3-05-00-01175 1.225[25]	1.5745[25]
1371	Benzene, 1,4-dichloro-2-ethenyl- 2,5-Dichlorostyrene	$C_8H_6Cl_2$ 173.04	1123-84-8	72-3[2]	4-05-00-01347 1.4045[20]	1.5798[20]
1372	Benzene, (2,2-dichloroethenyl)-	$C_8H_6Cl_2$ 173.04	698-88-4	225	4-05-00-01348 1.2531[20]	ace 3; chl 3 1.5852[20]

No.	Name Synonym	Mol. Form. Mol. Wt.	CAS RN mp/°C	Merck No. bp/°C	Beil. Ref. den/g cm^{-3}	Solubility n_D
1373	Benzene, 2,4-dichloro-1-ethenyl- 2,4-Dichlorostyrene	$C_8H_6Cl_2$ 173.04	2123-27-5 	81[6]	4-05-00-01347 1.243[25]	1.5828[20]
1374	Benzene, 1,1'-(dichloroethenylidene)bis[4-chloro- 2,2-Dichloro-1,1-bis(4-chlorophenyl)ethylene	$C_{14}H_8Cl_4$ 318.03	72-55-9 89		4-05-00-02177	
1375	Benzene, 2,4-dichloro-1-ethoxy-	$C_8H_8Cl_2O$ 191.06	5392-86-9 32	237	4-06-00-00885	H_2O 1; EtOH 3; eth 3; bz 3
1376	Benzene, (1,2-dichloroethyl)-	$C_8H_8Cl_2$ 175.06	1074-11-9 	233.5	4-05-00-00901 1.240[15]	H_2O 1; ace 3; bz 3 1.5544[15]
1377	Benzene, 1,4-dichloro-2-ethyl- 2,5-Dichloroethylbenzene	$C_8H_8Cl_2$ 175.06	54484-63-8 	213.5	4-05-00-00901 1.239[25]	eth 2; bz 3; chl 2; peth 2
1378	Benzene, 1,1'-(2,2-dichloroethylidene)bis[4-chloro- 1,1-Dichloro-2,2-bis(p-chlorophenyl)ethane	$C_{14}H_{10}Cl_4$ 320.04	72-54-8 109.5	3049	4-05-00-01884	chl 2
1379	Benzene, (dichlorofluoromethyl)-	$C_7H_5Cl_2F$ 179.02	498-67-9 -26.8	179	4-05-00-00818 1.3138[11]	EtOH 4 1.5180[11]
1380	Benzene, 1,4-dichloro-2-iodo-	$C_6H_3Cl_2I$ 272.90	29682-41-5 22	253	3-05-00-00580	bz 4; eth 4; EtOH 4; chl 4
1381	Benzene, 2,4-dichloro-1-methoxy-	$C_7H_6Cl_2O$ 177.03	553-82-2 28.5	232; 125[10]	4-06-00-00885	chl 2
1382	Benzene, (dichloromethyl)- Benzal chloride	$C_7H_6Cl_2$ 161.03	98-87-3 -17	1064 205	4-05-00-00817 1.26[25]	eth 4; EtOH 4 1.5502[20]
1383	Benzene, 1,2-dichloro-3-methyl- 2,3-Dichlorotoluene	$C_7H_6Cl_2$ 161.03	32768-54-0 6	207.5	4-05-00-00815 1.2458[20]	bz 4 1.5511[20]
1384	Benzene, 1,2-dichloro-4-methyl- 3,4-Dichlorotoluene	$C_7H_6Cl_2$ 161.03	95-75-0 -15.2	208.9	4-05-00-00815 1.2564[20]	H_2O 1; EtOH 5; eth 5; ace 5 1.5471[20]
1385	Benzene, 1,3-dichloro-2-methyl- 2,6-Dichlorotoluene	$C_7H_6Cl_2$ 161.03	118-69-4 25.8	198	4-05-00-00815 1.2686[20]	H_2O 1; chl 3 1.5507[20]
1386	Benzene, 1,3-dichloro-5-methyl- 3,5-Dichlorotoluene	$C_7H_6Cl_2$ 161.03	25186-47-4 26	201.5	4-05-00-00816	H_2O 1
1387	Benzene, 1,4-dichloro-2-methyl- 2,5-Dichlorotoluene	$C_7H_6Cl_2$ 161.03	19398-61-9 5	200	3-05-00-00694 1.2535[20]	H_2O 1; bz 3 1.5449[20]
1388	Benzene, 2,4-dichloro-1-methyl- 2,4-Dichlorotoluene	$C_7H_6Cl_2$ 161.03	95-73-8 -13.5	201	4-05-00-00815 1.2476[20]	H_2O 1; ctc 3 1.5511[20]
1389	Benzene, 1,1'-(dichloromethylene)bis-	$C_{13}H_{10}Cl_2$ 237.13	2051-90-3 	305 dec; 190[21]	4-05-00-01848 1.235[18]	eth 3; bz 3; ctc 3
1390	Benzene, 1-(dichloromethyl)-4-methyl-	$C_8H_8Cl_2$ 175.06	23063-36-7 48.5	105[18]	3-05-00-00855	EtOH 2; bz 2; chl 2
1391	Benzene, 1,2-dichloro-3-nitro-	$C_6H_3Cl_2NO_2$ 192.00	3209-22-1 61.5	257.5	4-05-00-00725 1.721[14]	H_2O 1; EtOH 3; eth 3; ace 3
1392	Benzene, 1,2-dichloro-4-nitro-	$C_6H_3Cl_2NO_2$ 192.00	99-54-7 43	255.5	4-05-00-00726 1.4558[75]	H_2O 1; EtOH 3; eth 3; ctc 2
1393	Benzene, 1,3-dichloro-2-nitro- 2,6-Dichloronitrobenzene	$C_6H_3Cl_2NO_2$ 192.00	601-88-7 72.5	130[8]	4-05-00-00726 1.603[17]	eth 4; EtOH 4
1394	Benzene, 1,3-dichloro-5-nitro-	$C_6H_3Cl_2NO_2$ 192.00	618-62-2 65.4		4-05-00-00727 1.4000[100]	H_2O 1; EtOH 3; eth 3 1.4000[100]
1395	Benzene, 1,4-dichloro-2-nitro-	$C_6H_3Cl_2NO_2$ 192.00	89-61-2 56	267	4-05-00-00726 1.439[75]	H_2O 1; EtOH 3; eth 3; bz 3 1.439[75]
1396	Benzene, 2,4-dichloro-1-nitro-	$C_6H_3Cl_2NO_2$ 192.00	611-06-3 34	258.5	4-05-00-00726 1.4790[80]	H_2O 1; EtOH 3; eth 3; chl 2 1.5512[70]
1397	Benzene, (3,3-dichloro-2-propenyl)-	$C_9H_8Cl_2$ 187.07	38862-78-1 59	242; 142[30]	4-05-00-01363 1.2028[20]	eth 3; bz 3; chl 3; peth 3
1398	Benzene, 1,2-dichloro-3,4,5,6-tetrafluoro- 1,2-Dichlorotetrafluorobenzene	$C_6Cl_2F_4$ 218.97	1198-59-0 			
1399	Benzene, 1,3-dichloro-2,4,5,6-tetrafluoro- 1,3-Dichlorotetrafluorobenzene	$C_6Cl_2F_4$ 218.97	1198-61-4 			
1400	Benzene, 1,4-dichloro-2,3,5,6-tetrafluoro- 1,4-Dichlorotetrafluorobenzene	$C_6Cl_2F_4$ 218.97	1198-62-5 			
1401	Benzene, 1,2-dichloro-4-(trichloromethyl)-	$C_7H_3Cl_5$ 264.36	13014-24-9 25.8	283.1	4-05-00-00824 1.5913[20]	1.5886[20]
1402	Benzene, 2,4-dichloro-1,3,5-trimethyl- Mesitylene, 2,4-dichloro-	$C_9H_{10}Cl_2$ 189.08	57386-83-1 59	243.5	3-05-00-00919	EtOH 3; eth 2; bz 2
1403	Benzene, 1,2-diethenyl- o-Divinylbenzene	$C_{10}H_{10}$ 130.19	91-14-5 	82[14]	4-05-00-01540 0.9325[22]	ace 3; bz 3 1.5767[20]
1404	Benzene, 1,3-diethenyl- m-Divinylbenzene	$C_{10}H_{10}$ 130.19	108-57-6 -52.3	121[76]; 52[3]	3-05-00-01367 0.9294[20]	ace 3; bz 3 1.5760[20]
1405	Benzene, 1,4-diethenyl- p-Divinylbenzene	$C_{10}H_{10}$ 130.19	105-06-6 31	95[18]; 34[0.2]	4-05-00-01541 0.9134[0]	ace 3; bz 3 1.5835[25]
1406	Benzene, 1,2-diethoxy-	$C_{10}H_{14}O_2$ 166.22	2050-46-6 44	219	4-06-00-05565 1.0075[20]	EtOH 3; eth 4; ctc 3 1.5083[25]
1407	Benzene, 1,3-diethoxy-	$C_{10}H_{14}O_2$ 166.22	2049-73-2 12.4	235	4-06-00-05664 1.0170[15]	H_2O 1; EtOH 3; eth 3
1408	Benzene, 1,4-diethoxy-	$C_{10}H_{14}O_2$ 166.22	122-95-2 72	246	4-06-00-05720	EtOH 4; eth 3; bz 3; ctc 3
1409	Benzene, (diethoxymethoxy)-	$C_{11}H_{16}O_3$ 196.25	14444-77-0 	103-4[10]	4-06-00-00610 1.014[25]	1.4822[20]
1410	Benzene, 1,4-diethoxy-2-methyl- 2,5-Diethoxytoluene	$C_{11}H_{16}O_2$ 180.25	41901-72-8 24.5	248	3-06-00-04499 1.0134[15]	bz 4; eth 4; EtOH 4

No.	Name / Synonym	Mol. Form. / Mol. Wt.	CAS RN	mp/°C / bp/°C	Merck No.	Beil. Ref. / den/g cm⁻³	Solubility / n_D
1411	Benzene, 1-(diethoxymethyl)-4-(1-methylethyl)-	$C_{14}H_{22}O_2$ 222.33	35364-90-0	265; 142[18.5]		2-07-00-00247 0.944[19]	1.484[19]
1412	Benzene, (3,3-dietoxy-1-propynyl)-	$C_{13}H_{16}O_2$ 204.27	6142-95-6	99-100[2]		4-07-00-01173 0.991[25]	1.5170[20]
1413	Benzene, 1,2-diethyl- o-Diethylbenzene	$C_{10}H_{14}$ 134.22	135-01-3	-31.2 / 184		4-05-00-01065 0.8800[20]	H_2O 1; EtOH 5; eth 5; ace 5 1.5035[20]
1414	Benzene, 1,3-diethyl- m-Diethylbenzene	$C_{10}H_{14}$ 134.22	141-93-5	-83.9 / 181.1		4-05-00-01066 0.8602[20]	H_2O 1; EtOH 5; eth 5; ace 5 1.4955[20]
1415	Benzene, 1,4-diethyl- p-Diethylbenzene	$C_{10}H_{14}$ 134.22	105-05-5	-42.83 / 183.7		4-05-00-01067 0.8620[20]	H_2O 1; EtOH 5; eth 5; ace 5 1.4967[20]
1416	Benzene, 1,2-diethyl-3-methyl-	$C_{11}H_{16}$ 148.25	13632-93-4	/ 206.6		4-05-00-01105 0.8910[20]	ace 4; bz 4; eth 4; EtOH 4 1.5105[20]
1417	Benzene, 1,2-diethyl-4-methyl- Toluene, 3,4-diethyl-	$C_{11}H_{16}$ 148.25	13732-80-4	/ 203.6		4-05-00-01106 0.8762[20]	ace 4; bz 4; eth 4; EtOH 4 1.5039[20]
1418	Benzene, 1,3-diethyl-2-methyl- Toluene, 2,6-diethyl-	$C_{11}H_{16}$ 148.25	13632-95-6	/ 208.8		4-05-00-01105 0.8907[20]	ace 4; bz 4; eth 4; EtOH 4 1.5106[20]
1419	Benzene, 1,3-diethyl-5-methyl-	$C_{11}H_{16}$ 148.25	2050-24-0	-74.1 / 205		4-05-00-01106 0.8748[20]	H_2O 1; EtOH 5; eth 5; ace 5 1.5027[20]
1420	Benzene, 1,4-diethyl-2-methyl-	$C_{11}H_{16}$ 148.25	13632-94-5	/ 207.1		4-05-00-01106 0.8758[20]	ace 4; bz 4; eth 4; EtOH 4 1.5034[20]
1421	Benzene, 2,4-diethyl-1-methyl-	$C_{11}H_{16}$ 148.25	1758-85-6	/ 205		4-05-00-01105 0.8748[20]	ace 4; bz 4; eth 4; EtOH 4 1.5027[20]
1422	Benzene, 1,2-difluoro- o-Difluorobenzene	$C_6H_4F_2$ 114.09	367-11-3	-34 / 94		4-05-00-00637 1.1599[18]	H_2O 1; ace 3; bz 3; chl 3 1.4451[18]
1423	Benzene, 1,3-difluoro- m-Difluorobenzene	$C_6H_4F_2$ 114.09	372-18-9	-69.0 / 82.6		4-05-00-00637 1.1572[20]	H_2O 1; ace 3; bz 3 1.4374[20]
1424	Benzene, 1,4-difluoro- p-Difluorobenzene	$C_6H_4F_2$ 114.09	540-36-3	-13 / 89	3132	4-05-00-00637 1.1701[20]	H_2O 1; ace 3; bz 3; ctc 2 1.4422[20]
1425	Benzene, (difluoromethyl)-	$C_7H_6F_2$ 128.12	455-31-2	/ 140		4-05-00-00801 1.1370[19]	H_2O 1; EtOH 3 1.4577[20]
1426	Benzene, 2,4-difluoro-1-nitro-	$C_6H_3F_2NO_2$ 159.09	446-35-5	9.8 / 207		4-05-00-00720 1.4571[14]	chl 2 1.5149[14]
1427	Benzene, 1,2-diiodo-	$C_6H_4I_2$ 329.91	615-42-9	27 / 287; 100[3]		4-05-00-00700 2.54[20]	H_2O 2; EtOH 2 1.7179[20]
1428	Benzene, 1,3-diiodo-	$C_6H_4I_2$ 329.91	626-00-6	40.4 / 285		4-05-00-00700 2.47[25]	eth 4; EtOH 4; chl 4
1429	Benzene, 1,4-diiodo-	$C_6H_4I_2$ 329.91	624-38-4	131.5 / 285		4-05-00-00700	H_2O 1; EtOH 3; eth 4; chl 2
1430	Benzene, 1,3-diisocyanato-2-methyl- Toluene-2,6-diisocyanate	$C_9H_6N_2O_2$ 174.16	91-08-7	18.3			H_2O 6; ace 3; bz 3
1431	Benzene, 2,4-diisocyanato-1-methyl- Toluene 2,4-diisocyanate	$C_9H_6N_2O_2$ 174.16	584-84-9	20.5 / 251	9456	4-13-00-00243 1.2244[20]	ace 4; bz 4; eth 4
1432	Benzene, 1,4-diisothiocyanato- Bitoscanate	$C_8H_4N_2S_2$ 192.27	4044-65-9	132	1318	4-13-00-00174	
1433	1,3-Benzenedimethanamine m-Xylenediamine	$C_8H_{12}N_2$ 136.20	1477-55-0	/ 247		4-13-00-00293 1.052[20]	H_2O 4; eth 4; EtOH 4
1434	1,3-Benzenedimethanol	$C_8H_{10}O_2$ 138.17	626-18-6	57 / 156[13]		4-06-00-05966 1.1610[18]	H_2O 4; eth 4; EtOH 4
1435	1,4-Benzenedimethanol	$C_8H_{10}O_2$ 138.17	589-29-7	117.5 / 138-43[1]		4-06-00-05971	H_2O 4; ace 4; eth 4; EtOH 4
1436	Benzene, 1,2-dimethoxy- Veratrole	$C_8H_{10}O_2$ 138.17	91-16-7	22.5 / 206	9857	4-06-00-05564 1.0810[25]	H_2O 2; EtOH 3; eth 3; ctc 3 1.5827[21]
1437	Benzene, 1,3-dimethoxy-	$C_8H_{10}O_2$ 138.17	151-10-0	-52 / 217.5		4-06-00-05663 1.0521[25]	H_2O 2; EtOH 3; eth 3; bz 3 1.5231[20]
1438	Benzene, 1,4-dimethoxy-	$C_8H_{10}O_2$ 138.17	150-78-7	59 / 212.6		4-06-00-05718 1.0375[55]	H_2O 2; EtOH 3; eth 4; bz 4
1439	Benzene, 1,2-dimethoxy-3-methyl-	$C_9H_{12}O_2$ 152.19	4463-33-6	/ 204		4-06-00-05860 1.0335[20]	eth 4; EtOH 4 1.5121[25]
1440	Benzene, 1,2-dimethoxy-4-methyl-	$C_9H_{12}O_2$ 152.19	494-99-5	24 / 220		4-06-00-05879 1.0509[25]	H_2O 1; ctc 2; os 4 1.5257[25]
1441	Benzene, 1,3-dimethoxy-5-methyl-	$C_9H_{12}O_2$ 152.19	4179-19-5	/ 244		3-06-00-04533 1.0478[15]	bz 4; eth 4; EtOH 4 1.5234[20]
1442	Benzene, 1,4-dimethoxy-2-(1-methylethyl)-	$C_{11}H_{16}O_2$ 180.25	4132-71-2	/ 114-6[15]		4-06-00-05990 1.0129[17]	bz 4; eth 4 1.5103[17]
1443	Benzene, 1-(dimethoxymethyl)-3-nitro-	$C_9H_{11}NO_4$ 197.19	3395-79-7	/ 162-4[19]		4-07-00-00592 1.209[15]	bz 4
1444	Benzene, 1-(dimethoxymethyl)-4-nitro-	$C_9H_{11}NO_4$ 197.19	881-67-4	24 / 294		4-07-00-00600	bz 4
1445	Benzene, 1,2-dimethoxy-4-nitro-	$C_8H_9NO_4$ 183.16	709-09-1	98 / 230[15]		4-06-00-05627 1.1888[133]	H_2O 1; EtOH 4; eth 4; chl 3
1446	Benzene, 1,3-dimethoxy-5-nitro-	$C_8H_9NO_4$ 183.16	16147-07-2	89		3-06-00-04347 1.1693[133]	bz 4; EtOH 4
1447	Benzene, 1,4-dimethoxy-2-nitro-	$C_8H_9NO_4$ 183.16	89-39-4	72.5		4-06-00-05786 1.1666[132]	H_2O 1; EtOH 3; bz 3; chl 3
1448	Benzene, 2,4-dimethoxy-1-nitro- 1,3-Dimethoxy-4-nitrobenzene	$C_8H_9NO_4$ 183.16	4920-84-7	76.5		4-06-00-05691 1.1876[132]	H_2O 1; EtOH 3; os 3; lig 1
1449	Benzene, 1,2-dimethoxy-4-(1-propenyl)-	$C_{11}H_{14}O_2$ 178.23	93-16-3	18 / 270.5		2-06-00-00918 1.0521[20]	1.5616[20]

No.	Name / Synonym	Mol. Form. / Mol. Wt.	CAS RN / mp/°C	Merck No. / bp/°C	Beil. Ref. / den/g cm^{-3}	Solubility / n_D
1450	Benzene, 1,2-dimethoxy-4-(2-propenyl)-	$C_{11}H_{14}O_2$ / 178.23	93-15-2 / -4	/ 254.7	4-06-00-06337 / 1.0396[20]	H_2O 1; EtOH 3; eth 3 / 1.5340[20]
1451	Benzene, 1,2-dimethoxy-4-(1-propenyl)-, (Z)-	$C_{11}H_{14}O_2$ / 178.23	6380-24-1 / 70	/ 270.5	4-06-00-06325 / 1.0521[20]	H_2O 1; ace 3; bz 3 / 1.5616[20]
1452	Benzene, 1,2-dimethyl- / o-Xylene	C_8H_{10} / 106.17	95-47-6 / -25.2	9988 / 144.5	4-05-00-00917 / 0.8802[10]	H_2O 1; EtOH 5; eth 5; ace 5 / 1.5055[20]
1453	Benzene, 1,3-dimethyl- / m-Xylene	C_8H_{10} / 106.17	108-38-3 / -47.8	9988 / 139.1	4-05-00-00932 / 0.8642[20]	H_2O 1; EtOH 5; eth 5; ace 5 / 1.4972[10]
1454	Benzene, 1,4-dimethyl- / p-Xylene	C_8H_{10} / 106.17	106-42-3 / 13.2	9988 / 138.3	4-05-00-00951 / 0.8611[20]	H_2O 1; EtOH 5; eth 5; ace 5 / 1.4958[20]
1455	Benzene, (1,1-dimethylbutyl)- / Pentane, 2-methyl-2-phenyl	$C_{12}H_{18}$ / 162.27	1985-57-5 /	/ 205.5	4-05-00-01118 / 0.8796[10]	/ 1.4955[16]
1456	Benzene, 1,1'-(1,2-dimethyl-1,2-ethenediyl)bis-, (E)-	$C_{16}H_{16}$ / 208.30	782-06-9 / 107	/ 156[16]	4-05-00-02207 / 0.987[20]	H_2O 1; EtOH 2; eth 3; MeOH 2
1457	Benzene, 1,1'-(1,2-dimethyl-1,2-ethenediyl)bis-, (Z)-	$C_{16}H_{16}$ / 208.30	782-05-8 / 67.5		4-05-00-02207 / 1.004[20]	H_2O 1; EtOH 3; HOAc 3; MeOH 3 / 1.5612[78]
1458	Benzene, (1,1-dimethylethoxy)-	$C_{10}H_{14}O$ / 150.22	6669-13-2 / -24	/ 185.5	4-06-00-00559 / 0.9214[20]	
1459	Benzene, (1,1-dimethylethyl)- / tert-Butylbenzene	$C_{10}H_{14}$ / 134.22	98-06-6 / -57.8	1551 / 169.1	4-05-00-01045 / 0.8665[20]	H_2O 1; EtOH 4; eth 4; ace 5 / 1.4927[20]
1460	Benzene, 1-(1,1-dimethylethyl)-3,5-dimethyl-	$C_{12}H_{18}$ / 162.27	98-19-1 / -18	/ 207	4-05-00-01130 / 0.8668[20]	ctc 3
1461	Benzene, 1-(1,1-dimethylethyl)-3,5-dimethyl-2,4,6-trinitro- / 2,4,6-Trinitro-3,5-dimethyl-tert-butylbenzene	$C_{12}H_{15}N_3O_6$ / 297.27	81-15-2 / 110		4-05-00-01132	H_2O 1; EtOH 2; eth 3; chl 3
1462	Benzene, 1-(1,1-dimethylethyl)-4-methoxy-	$C_{11}H_{16}O$ / 164.25	5396-38-3 /	/ 238	4-06-00-03297 / 0.9436[20]	/ 1.5039[20]
1463	Benzene, 1-(1,1-dimethylethyl)-2-methoxy-4-methyl-3,5-dinitro-	$C_{12}H_{16}N_2O_5$ / 268.27	83-66-9 / 85	/ 185[16]	4-06-00-03402	H_2O 1; EtOH 2; eth 3; chl 3
1464	Benzene, 1-(1,1-dimethylethyl)-2-methyl-	$C_{11}H_{16}$ / 148.25	1074-92-6 / -50.3	/ 200.4	4-05-00-01096 / 0.8897[20]	ace 4; bz 4; eth 4; EtOH 4 / 1.5076[20]
1465	Benzene, 1-(1,1-dimethylethyl)-3-methyl-	$C_{11}H_{16}$ / 148.25	1075-38-3 / -41.4	/ 189.3	4-05-00-01096 / 0.8657[20]	ace 4; bz 4; eth 4; EtOH 4 / 1.4944[20]
1466	Benzene, 1-(1,1-dimethylethyl)-4-methyl- / p-tert-Butyltoluene	$C_{11}H_{16}$ / 148.25	98-51-1 / -52	/ 190	4-05-00-01097 / 0.8612[20]	H_2O 1; EtOH 2; eth 4; ace 3 / 1.4918[20]
1467	Benzene, 1-(1,1-dimethylethyl)-3-nitro-	$C_{10}H_{13}NO_2$ / 179.22	23132-52-7 / 2	/ 254; 136[16]	4-05-00-01052 / 1.0643[25]	
1468	Benzene, 1-(1,5-dimethyl-4-hexenyl)-4-methyl- / ar-Curcumene	$C_{15}H_{22}$ / 202.34	644-30-4 /	/ 140[19]	4-05-00-01465 / 0.8805[20]	H_2O 1; bz 3 / 1.4989[20]
1469	Benzene, 1,2-dimethyl-3-(1-methylethyl)-	$C_{11}H_{16}$ / 148.25	22539-65-7 /	/ 202.6	4-05-00-01103 / 0.888[20]	ace 4; bz 4; eth 4; EtOH 4 / 1.508[20]
1470	Benzene, 1,2-dimethyl-4-(1-methylethyl)-	$C_{11}H_{16}$ / 148.25	4132-77-8 /	/ 200	4-05-00-01103 / 0.8710[20]	/ 1.4951[20]
1471	Benzene, 1,3-dimethyl-2-(1-methylethyl)-	$C_{11}H_{16}$ / 148.25	14411-75-7 /	/ 199	4-05-00-01103 / 0.890[20]	ace 4; bz 4; eth 4; EtOH 4 / 1.509[20]
1472	Benzene, 2,4-dimethyl-1-(1-methylethyl)-	$C_{11}H_{16}$ / 148.25	4706-89-2 / -82	/ 197	4-05-00-01104 / 0.8734[20]	/ 1.4998[25]
1473	Benzene, 2,4-dimethyl-1-(1-methylpropyl)-	$C_{12}H_{18}$ / 162.27	1483-60-9 /	/ 84[8]	4-05-00-01129 / 0.8664[20]	/ 1.4939[25]
1474	Benzene, 1,2-dimethyl-3-nitro-	$C_8H_9NO_2$ / 151.17	83-41-0 / 15	/ 240	4-05-00-00930 / 1.1402[20]	H_2O 1; EtOH 3; ctc 3 / 1.5441[20]
1475	Benzene, 1,2-dimethyl-4-nitro- / 4-Nitro-o-xylene	$C_8H_9NO_2$ / 151.17	99-51-4 / 30.5	/ 251; 143[21]	4-05-00-00930 / 1.112[15]	H_2O 1; EtOH 5 / 1.5202[20]
1476	Benzene, 1,3-dimethyl-2-nitro-	$C_8H_9NO_2$ / 151.17	81-20-9 / 15	/ 226	4-05-00-00948 / 1.112[15]	H_2O 1; EtOH 4; ctc 3 / 1.5202[20]
1477	Benzene, 1,3-dimethyl-5-nitro-	$C_8H_9NO_2$ / 151.17	99-12-7 / 75	/ 274	4-05-00-00948	H_2O 1; EtOH 4; eth 4
1478	Benzene, 1,4-dimethyl-2-nitro-	$C_8H_9NO_2$ / 151.17	89-58-7 / -25	/ 240.5	4-05-00-00971 / 1.132[15]	H_2O 1; EtOH 3 / 1.5413[20]
1479	Benzene, 2,4-dimethyl-1-nitro-	$C_8H_9NO_2$ / 151.17	89-87-2 / 9	/ 247; 122[18]	4-05-00-00948 / 1.135[15]	H_2O 1; eth 3; ace 3; bz 3 / 1.5473[25]
1480	Benzene, (1,1-dimethylpropyl)-	$C_{11}H_{16}$ / 148.25	2049-95-8 /	/ 192.4	4-05-00-01090 / 0.8748[20]	/ 1.4958[20]
1481	Benzene, (1,2-dimethylpropyl)-	$C_{11}H_{16}$ / 148.25	4481-30-5 /	/ 187	4-05-00-01092 / 0.8672[16]	/ 1.4972[16]
1482	Benzene, 1,3-dimethyl-2-propyl-	$C_{11}H_{16}$ / 148.25	17059-45-9 /	/ 207.6	4-05-00-01102 / 0.8856[20]	ace 4; bz 4; eth 4; EtOH 4 / 1.5063[20]
1483	Benzene, 1,3-dimethyl-5-propyl-	$C_{11}H_{16}$ / 148.25	3982-64-7 / -59.1	/ 202.2	4-05-00-01102 / 0.8607[20]	ace 4; bz 4; eth 4; EtOH 4 / 1.4952[20]
1484	Benzene, (2,2-dimethylpropyl)-	$C_{11}H_{16}$ / 148.25	1007-26-7 /	/ 186	4-05-00-01092 / 0.8581[18]	/ 1.4884[18]
1485	Benzene, 1,2-dimethyl-3-propyl- / o-Xylene, 3-propyl	$C_{11}H_{16}$ / 148.25	17059-44-8 /	/ 210.7	4-05-00-01101 / 0.8864[20]	ace 4; bz 4; eth 4; EtOH 4 / 1.5075[20]
1486	Benzene, 1,4-dimethyl-2-propyl-	$C_{11}H_{16}$ / 148.25	3042-50-0 /	/ 204.3	4-05-00-01102 / 0.8717[20]	ace 4; bz 4; eth 4; EtOH 4 / 1.4999[20]
1487	Benzene, 2,4-dimethyl-1-propyl- / m-Xylene, 4-propyl-	$C_{11}H_{16}$ / 148.25	61827-85-8 /	/ 206.6	4-05-00-01102 / 0.8723[20]	ace 4; bz 4; eth 4; EtOH 4 / 1.4998[20]

No.	Name Synonym	Mol. Form. Mol. Wt.	CAS RN mp/°C	Merck No. bp/°C	Beil. Ref. den/g cm^{-3}	Solubility n_D
1488	Benzene, 1-(1,1-dimethylpropyl)-4-methoxy-	$C_{12}H_{18}O$ 178.27	2050-03-5	238	4-06-00-03383 0.9436[20]	1.5039[20]
1489	Benzene, 1,4-dimethyl-2,3,5-trinitro- p-Xylene, 2,3,5-trinitro-	$C_8H_7N_3O_6$ 241.16	602-27-7 139.5	exp 410	3-05-00-00864 1.59[19]	EtOH 4
1490	Benzene, 2,4-dimethyl-1,3,5-trinitro-	$C_8H_7N_3O_6$ 241.16	632-92-8 184		4-05-00-00950 1.604[19]	H_2O 1; EtOH 2; eth 2; ace 3
1491	Benzene, 1,2-dinitro- o-Dinitrobenzene	$C_6H_4N_2O_4$ 168.11	528-29-0 118.5	318; 194[30]	4-05-00-00738 1.3119[120]	H_2O 1; EtOH 3; bz 3; chl 3 1.565[17]
1492	Benzene, 1,3-dinitro- m-Dinitrobenzene	$C_6H_4N_2O_4$ 168.11	99-65-0 90	291; 167[14]	4-05-00-00739 1.575[18]	H_2O 2; EtOH 4; eth 3; ace 4
1493	Benzene, 1,4-dinitro- p-Dinitrobenzene	$C_6H_4N_2O_4$ 168.11	100-25-4 174	297; 183[34]	4-05-00-00741 1.625[18]	H_2O 1; EtOH 2; ace 3; bz 3
1494	Benzene, 1,1'-(1,2-dinitro-1,2-ethenediyl)bis-, (Z)-	$C_{14}H_{10}N_2O_4$ 270.24	1796-05-0 108.5	dec	4-05-00-02171	H_2O 1; EtOH 3; eth 3; ace 4
1495	Benzene, 2,4-dinitro-1-phenoxy-	$C_{12}H_8N_2O_5$ 260.21	2486-07-9 71	230[27]	4-06-00-01375	eth 4; EtOH 4
1496	Benzene, 2,4-dinitro-1-(2-phenylethenyl)-, (E)- Stilbene, 2,4-dinitro (trans)	$C_{14}H_{10}N_2O_4$ 270.24	56456-42-9 144	exp 412	4-05-00-02170	EtOH 1; bz 2; chl 3; CS_2 3
1497	1,2-Benzenediol Pyrocatechol	$C_6H_6O_2$ 110.11	120-80-9 105	8009 245	4-06-00-05557 1.1493[22]	H_2O 4; bz 4; eth 4; EtOH 4 1.604
1498	1,3-Benzenediol Resorcinol	$C_6H_6O_2$ 110.11	108-46-3 111	8158 178[16]	4-06-00-05658 1.2717[25]	H_2O 3; EtOH 3; eth 3; bz 2
1499	1,4-Benzenediol p-Hydroquinone	$C_6H_6O_2$ 110.11	123-31-9 172.3	4738 287	4-06-00-05712 1.328[15]	H_2O 3; EtOH 4; eth 3; ace 4
1500	1,2-Benzenediol, 4-(2-amino-1-hydroxyethyl)-, (R)- Norepinephrine	$C_8H_{11}NO_3$ 169.18	51-41-2 217 dec	6612		H_2O 2; EtOH 2; eth 2; alk 4
1501	1,4-Benzenediol, 2,5-bis(1,1-dimethylpropyl)- 2,5-Di-tert-pentylhydroquinone	$C_{16}H_{26}O_2$ 250.38	79-74-3 180	3300	3-06-00-04748	
1502	1,3-Benzenediol, 4-bromo-	$C_6H_5BrO_2$ 189.01	6626-15-9 103	140[12]	4-06-00-05687	H_2O 4; EtOH 2; eth 4; bz 2
1503	1,4-Benzenediol, 2-bromo-	$C_6H_5BrO_2$ 189.01	583-69-7 111.5	sub	4-06-00-05780	H_2O 4; EtOH 4; eth 4; bz 4
1504	1,2-Benzenediol, 3-chloro-	$C_6H_5ClO_2$ 144.56	4018-65-9 48.5	110[11]	4-06-00-05613	lig 4
1505	1,2-Benzenediol, 4-chloro-	$C_6H_5ClO_2$ 144.56	2138-22-9 90.5	139[10.5]	4-06-00-05614	H_2O 4; ace 4; eth 4; EtOH 4
1506	1,3-Benzenediol, 5-chloro-	$C_6H_5ClO_2$ 144.56	52780-23-1 118	sub	4-06-00-05684	H_2O 3; eth 3; ace 3; os 3
1507	1,4-Benzenediol, 2-chloro-	$C_6H_5ClO_2$ 144.56	615-67-8 108	263	4-06-00-05767	H_2O 4; EtOH 3; eth 3; bz 4
1508	1,2-Benzenediol, diacetate	$C_{10}H_{10}O_4$ 194.19	635-67-6 64.5	142-3[9]	4-06-00-05582	H_2O 1; EtOH 4; eth 4; chl 4
1509	1,4-Benzenediol, diacetate	$C_{10}H_{10}O_4$ 194.19	1205-91-0 123.5		4-06-00-05741 0.8731[25]	H_2O 3; EtOH 4; eth 4; chl 4
1510	1,3-Benzenediol, 4,6-dichloro-	$C_6H_4Cl_2O_2$ 179.00	137-19-9 113	254	4-06-00-05685	H_2O 4; EtOH 4; eth 4; ace 4
1511	1,2-Benzenediol, 4,5-dimethyl-	$C_8H_{10}O_2$ 138.17	2785-74-2 87.5	160[26]	4-06-00-05950	H_2O 4; EtOH 4; eth 4; bz 2
1512	1,3-Benzenediol, 2,5-dimethyl-	$C_8H_{10}O_2$ 138.17	488-87-9 163	278.5	4-06-00-05970	H_2O 3; EtOH 3; eth 3
1513	1,3-Benzenediol, 4,5-dimethyl-	$C_8H_{10}O_2$ 138.17	527-55-9 136	284	4-06-00-05948 1.1994[20]	H_2O 3; EtOH 4; eth 3; bz 2
1514	1,3-Benzenediol, 4,6-dimethyl-	$C_8H_{10}O_2$ 138.17	615-89-4 126.5	276	4-06-00-05963	H_2O 3; EtOH 3; eth 3
1515	1,2-Benzenediol, 4,4'-(2,3-dimethyl-1,4-butanediyl)bis- Nordihydroguaiaretic acid	$C_{18}H_{22}O_4$ 302.37	500-38-9 185.5	6610	4-06-00-07771	H_2O 2; EtOH 3; eth 3; ace 3
1516	1,2-Benzenediol, 4-(1,1-dimethylethyl)-	$C_{10}H_{14}O_2$ 166.22	98-29-3 54.3	285; 160[22]	4-06-00-06014	tfa 3
1517	1,3-Benzenediol, 2,4-dinitro- 2,4-Dinitroresorcinol	$C_6H_4N_2O_6$ 200.11	519-44-8 147.5	3278	4-06-00-05696	H_2O 2; EtOH 2
1518	1,3-Benzenediol, 4-ethyl-	$C_8H_{10}O_2$ 138.17	2896-60-8 98.5	160[24]; 131[15]	4-06-00-05924	H_2O 2; EtOH 2; eth 2
1519	1,3-Benzenediol, 4-hexyl- 4-Hexylresorcinol	$C_{12}H_{18}O_2$ 194.27	136-77-6 68	4633 334	4-06-00-06048	ace 4; eth 4; EtOH 4; chl 4
1520	1,4-Benzenediol, 2-(hydroxymethyl)- Gentisyl alcohol	$C_7H_8O_3$ 140.14	495-08-9 100	4292 sub 75	4-06-00-07380	H_2O 4; EtOH 4; chl 4
1521	1,2-Benzenediol, 4-[1-hydroxy-2-(methylamino)ethyl]-, (R)- Epinephrine	$C_9H_{13}NO_3$ 183.21	51-43-4 211.5	3569	4-13-00-02927	H_2O 2; EtOH 1; HOAc 3; min acid 3
1522	1,2-Benzenediol, 4-[1-hydroxy-2-[(1-methylethyl)amino]ethyl]- Ethanol, 1-(3,4-dihydroxyphenyl)-2(isopropylamino)	$C_{11}H_{17}NO_3$ 211.26	7683-59-2 170.5	5105	3-13-00-02387	
1523	1,2-Benzenediol, 4-iodo- 4-Iodocatechol	$C_6H_5IO_2$ 236.01	76149-14-9 92	sub	4-06-00-05625	ace 4; bz 4; eth 4; EtOH 4
1524	1,3-Benzenediol, 5-iodo- Resorcinol, 5-iodo	$C_6H_5IO_2$ 236.01	64339-43-1 92.5	sub	2-06-00-00821	EtOH 4

No.	Name / Synonym	Mol. Form. / Mol. Wt.	CAS RN / mp/°C	Merck No. / bp/°C	Bell. Ref. / den/g cm^{-3}	Solubility / n_D
1525	1,2-Benzenediol, 3-methoxy-	$C_7H_8O_3$ 140.14	934-00-9 42.8	163[48]; 129[10]	4-06-00-07329	chl 3
1526	1,3-Benzenediol, 5-methoxy-	$C_7H_8O_3$ 140.14	2174-64-3 80.3	213[16]	4-06-00-07362	H_2O 2; EtOH 4; eth 4; bz 2
1527	1,2-Benzenediol, 3-methyl-	$C_7H_8O_2$ 124.14	488-17-5 68	241	4-06-00-05860	H_2O 3; EtOH 3; bz 3; chl 3
1528	1,2-Benzenediol, 4-methyl-	$C_7H_8O_2$ 124.14	452-86-8 65	251	4-06-00-05878 1.1287[74]	H_2O 3; EtOH 3; eth 3; ace 3 1.5425[74]
1529	1,3-Benzenediol, 2-methyl-	$C_7H_8O_2$ 124.14	608-25-3 120	264	4-06-00-05877	H_2O 4; bz 4; eth 4; EtOH 4
1530	1,3-Benzenediol, 4-methyl-	$C_7H_8O_2$ 124.14	496-73-1 106	268.5	4-06-00-05864	H_2O 3; EtOH 3; eth 3; bz 2
1531	1,3-Benzenediol, 5-methyl- Orcinol	$C_7H_8O_2$ 124.14	504-15-4 107.5	6819 289.5	4-06-00-05892 1.290[4]	H_2O 3; EtOH 3; eth 3; bz 3
1532	1,4-Benzenediol, 2-methyl-	$C_7H_8O_2$ 124.14	95-71-6 128	163[11]	4-06-00-05866	H_2O 4; EtOH 4; eth 4; ace 3
1533	1,2-Benzenediol, 4-[2-(methylamino)ethyl]- Deoxyepinephrine	$C_9H_{13}NO_2$ 167.21	501-15-5 188.5	2885	4-13-00-02606	
1534	1,3-Benzenediol, 4-(3-methylbutyl)-	$C_{11}H_{16}O_2$ 180.25	15116-17-3 67	177[7]	3-06-00-04700	bz 4; eth 4; EtOH 4
1535	1,2-Benzenediol, 4-(1-methylethyl)-	$C_9H_{12}O_2$ 152.19	2138-43-4 78	271	3-06-00-04632	peth 3
1536	1,2-Benzenediol, 3-methyl-6-(1-methylethyl)-	$C_{10}H_{14}O_2$ 166.22	490-06-2 48	267	4-06-00-06019	ace 4; eth 4
1537	1,3-Benzenediol, 2-[3-methyl-6-(1-methylethenyl)-2-cyclohexen-1-yl]-5-pentyl-, (1R-trans)- Cannabidiol	$C_{21}H_{30}O_2$ 314.47	13956-29-1 67	1750 187-90[2]	3-06-00-05362 1.040[40]	H_2O 1; EtOH 3; eth 3; bz 3 1.5404[20]
1538	1,4-Benzenediol, 2-methyl-5-(1-methylethyl)-	$C_{10}H_{14}O_2$ 166.22	2217-60-9 148	290	4-06-00-06019	eth 4; EtOH 4
1539	1,3-Benzenediol, 4-(2-methylpropyl)- Resorcinol, 4-isobutyl-	$C_{10}H_{14}O_2$ 166.22	18979-62-9 62.5	166[6]	3-06-00-04667	eth 4; EtOH 4
1540	1,2-Benzenediol, monoacetate	$C_8H_8O_3$ 152.15	2848-25-1 57.5	189[102]; 148[25]	4-06-00-05581	H_2O 4; ace 4; EtOH 4; chl 4
1541	1,3-Benzenediol, 4-nitro-	$C_6H_5NO_4$ 155.11	3163-07-3 122	178[11]	4-06-00-05691	EtOH 3; eth 3; bz 3; ctc 2
1542	1,3-Benzenediol, 4-[(4-nitrophenyl)azo]- Magneson	$C_{12}H_9N_3O_4$ 259.22	74-39-5 200	5578	4-16-00-00266	H_2O 1; EtOH 2; bz 2; HOAc 2
1543	1,2-Benzenediol, 3-pentadecyl- 3-Pentadecylcatechol	$C_{21}H_{36}O_2$ 320.52	492-89-7 59.5	7061	4-06-00-06109	bz 4; eth 4; EtOH 4
1544	1,3-Benzenediol, 4-pentyl-	$C_{11}H_{16}O_2$ 180.25	533-24-4 72.5	168[6]	3-06-00-04693	bz 4; eth 4; EtOH 4
1545	1,3-Benzenediol, 5-(2-phenylethenyl)-, (E)- Pinosylvin	$C_{14}H_{12}O_2$ 212.25	22139-77-1 156	7419	3-06-00-05577	ace 4; bz 4; chl 4; HOAc 4
1546	1,4-Benzenediol, 2-(phenylmethyl)-	$C_{13}H_{12}O_2$ 200.24	1706-73-6 105.8	230[13]	3-06-00-05404	eth 4; EtOH 4
1547	1,2-Benzenediol, 3-(2-propenyl)-	$C_9H_{10}O_2$ 150.18	1125-74-2	141-5[16]	4-06-00-06334 1.1241[20]	1.5656[20]
1548	1,2-Benzenediol, 4-propyl-	$C_9H_{12}O_2$ 152.19	2525-02-2 60	177[30]; 111[0.2]	3-06-00-04613 1.100[18]	1.4440[18]
1549	1,3-Benzenediol, 4-propyl-	$C_9H_{12}O_2$ 152.19	18979-60-7 82.5	172[14]	4-06-00-05975	H_2O 3; EtOH 2; eth 3
1550	1,3-Benzenediol, 5-propyl-	$C_9H_{12}O_2$ 152.19	500-49-2 82.8	169[8]; 148[3]	3-06-00-04622	ace 4; bz 4; eth 4; EtOH 4
1551	1,4-Benzenediol, 2,3,5,6-tetrabromo-	$C_6H_2Br_4O_2$ 425.70	2641-89-6 244		3-06-00-04440 3.023[21]	eth 4; EtOH 4
1552	1,4-Benzenediol, 2,3,5,6-tetramethyl- Durohydroquinone	$C_{10}H_{14}O_2$ 166.22	527-18-4 233	3451	4-06-00-06028	EtOH 3; eth 2
1553	1,3-Benzenediol, 2,4,6-trimethyl-	$C_9H_{12}O_2$ 152.19	608-98-0 150.8	274.5	4-06-00-06000	bz 4; eth 4; EtOH 4
1554	1,3-Benzenediol, 2,4,6-trinitro- Styphnic acid	$C_6H_3N_3O_8$ 245.11	82-71-3 175.5	8828 sub	4-06-00-05699	eth 4; EtOH 4
1555	Benzene, 1,2-dipropoxy-	$C_{12}H_{18}O_2$ 194.27	6280-98-4	235.5	4-06-00-05566 0.9554[33]	chl 2 1.4950[27]
1556	Benzene, 1,3-dipropoxy- Resorcinol dipropyl ether	$C_{12}H_{18}O_2$ 194.27	56106-37-7	253	3-06-00-04308 1.033[20]	1.5138[83]
1557	Benzene, 1,4-dipropyl-	$C_{12}H_{18}$ 162.27	4815-57-0	109[23]	4-05-00-01124 0.8563[19]	1.4917[19]
1558	1,3-Benzenedisulfonamide, 4-amino-6-chloro- Chloraminophenamide	$C_6H_8ClN_3O_4S_2$ 285.73	121-30-2 254.5	2067	4-14-00-02810	
1559	1,3-Benzenedisulfonamide, 4,5-dichloro- 1,3-Benzenedisulfonic acid, diamide,4,5-dichloro	$C_6H_6ClN_2O_4$ 205.58	120-97-8 228.7	3067	4-11-00-00555	
1560	1,3-Benzenedisulfonic acid, 4-hydroxy- Phenoldisulfonic acid	$C_6H_6O_7S_2$ 254.24	96-77-5 >100 dec	7207	2-11-00-00139	H_2O 4; EtOH 4
1561	1,2-Benzenedithiol	$C_6H_6S_2$ 142.25	17534-15-5 28.5	238.5	4-06-00-05651	EtOH 4; eth 4; bz 4; AcOEt 3

No.	Name / Synonym	Mol. Form. / Mol. Wt.	CAS RN / mp/°C	Merck No. / bp/°C	Beil. Ref. / den/g cm⁻³	Solubility / n_D
1562	1,3-Benzenedithiol	$C_6H_6S_2$ 142.25	626-04-0 27	245	4-06-00-05705	bz 4; eth 4; EtOH 4
1563	1,3-Benzenedithiol, 4-chloro-	$C_6H_5ClS_2$ 176.69	58593-78-5	145-6[13]	4-06-00-05708 1.393[25]	1.6704[20]
1564	1,2-Benzenedithiol, 4-methyl- Toluene-3,4-dithiol	$C_7H_8S_2$ 156.27	496-74-2 29	9457	4-06-00-05890	chl 3
1565	Benzene, docosyl- Docosane, 1-phenyl-	$C_{28}H_{50}$ 386.71	5634-22-0 51	448	3-05-00-01142 0.8544[20]	
1566	Benzene, dodecyl- 1-Phenyldodecane	$C_{18}H_{30}$ 246.44	123-01-3 3	328	4-05-00-01200 0.8551[20]	H_2O 1 1.4824[20]
1567	Benzene, 1,1'-dodecylidenebis[4-methyl- Dodecane, 1,1-di-(4-tolyl)	$C_{26}H_{38}$ 350.59	55268-62-7	197[1.0]	3-05-00-01925 0.9117[20]	1.5001[20]
1568	Benzene, eicosyl-	$C_{26}H_{46}$ 358.65	2398-68-7 44	429	4-05-00-01222 0.8545[20]	1.4805[20]
1569	Benzeneethanamine Phenylethylamine	$C_8H_{11}N$ 121.18	64-04-0	7186 197.5	4-12-00-02453 0.9580[24]	H_2O 3; eth 4; eth 4; ctc 3 1.5290[25]
1570	Benzeneethanamine, 4-bromo- Phenethylamine, 4-bromo	$C_8H_{10}BrN$ 200.08	73918-56-6	63-72[0.2]	4-12-00-02492 1.29[25]	1.5750[20]
1571	Benzeneethanamine, 3,4-dimethoxy-	$C_{10}H_{15}NO_2$ 181.23	120-20-7	163-5[14]	4-13-00-02604	ctc 3 1.5464[20]
1572	Benzeneethanamine, α,α-dimethyl- Phentermine	$C_{10}H_{15}N$ 149.24	122-09-8	7232 205; 100[21]	4-12-00-02820	
1573	Benzeneethanamine, N,α-dimethyl-, (S)- Methamphetamine	$C_{10}H_{15}N$ 149.24	537-46-2	5859	4-12-00-02589	
1574	Benzeneethanamine, N,β-dimethyl- Phenylpropylmethylamine	$C_{10}H_{15}N$ 149.24	93-88-9	7280 207.5	3-12-00-02689 0.915[25]	bz 4; eth 4; EtOH 4
1575	Benzeneethanamine, N,α-dimethyl-N-(phenylmethyl)-, (+)- Benzphetamine	$C_{17}H_{21}N$ 239.36	156-08-1	1130 127[0.02]	4-12-00-02594	eth 4; EtOH 4; MeOH 4; chl 4 1.5515[19]
1576	Benzeneethanamine, N-ethyl-α-methyl- N-Ethylamphetamine	$C_{11}H_{17}N$ 163.26	457-87-4	3720 105-6[14]	3-12-00-02668	1.4986[25]
1577	Benzeneethanamine, 2-methoxy- Phenethylamine, 2-methoxy	$C_9H_{13}NO$ 151.21	2045-79-6	236.5	3-13-00-01625 1.0400[25]	1.5422[20]
1578	Benzeneethanamine, α-methyl-, (±)- Amphetamine	$C_9H_{13}N$ 135.21	300-62-9	616 203	4-12-00-02587 0.9306[25]	H_2O 2; EtOH 3; eth 2; chl 3 1.518[26]
1579	Benzeneethanamine, β-methyl-	$C_9H_{13}N$ 135.21	582-22-9	210	4-12-00-02634 0.9433[4]	bz 4; eth 4; EtOH 4 1.5255[20]
1580	Benzeneethanamine, 4-methyl-	$C_9H_{13}N$ 135.21	3261-62-9	214	4-12-00-02641 0.93[25]	1.5257[20]
1581	Benzeneethanamine, N-methyl-	$C_9H_{13}N$ 135.21	589-08-2	206	4-12-00-02453 0.93[25]	1.5162[20]
1582	Benzeneethanamine, α-methyl-, (S)-, sulfate (2:1) Dextroamphetamine sulfate	$C_{18}H_{28}N_2O_4S$ 368.50	51-63-8 >300	2932	3-12-00-02665 1.15[25]	H_2O 4
1583	Benzeneethanamine, α-methyl-N-(2,2,2-trichloroethylidene)- Amphecloral	$C_{11}H_{12}Cl_3N$ 264.58	5581-35-1	613 96.[0.5]		1.530
1584	Benzeneethanamine, α-phenyl-	$C_{14}H_{15}N$ 197.28	25611-78-3	311; 175[15]	4-12-00-03313 1.031[15]	eth 4; EtOH 4
1585	Benzeneethanamine, N-(2-phenylethyl)-	$C_{16}H_{19}N$ 225.33	6308-98-1 29	360; 190[15]	4-12-00-02459	eth 4; EtOH 4 1.5550[25]
1586	Benzeneethanamine, 3,4,5-trimethoxy- Mescaline	$C_{11}H_{17}NO_3$ 211.26	54-04-6 35.5	5808 180[12]	4-13-00-02919	H_2O 3; EtOH 3; eth 1; bz 3
1587	Benzene, 1,1'-(1,2-ethanediyl)bis- Dibenzyl	$C_{14}H_{14}$ 182.27	103-29-7 52.5	1219 284	4-05-00-01868 0.9780[25]	H_2O 1; EtOH 3; eth 3; CS_2 3 1.5476[60]
1588	Benzene, 1,1'-[1,2-ethanediylbis(oxy)]bis-	$C_{14}H_{14}O_2$ 214.26	104-66-5 98	182[12]	4-06-00-00573	H_2O 1; EtOH 2; eth 3; chl 3
1589	Benzene, 1,1'-(1,2-ethanediyl)bis[3-methyl-	$C_{16}H_{18}$ 210.32	4662-96-8	298	4-05-00-01943 0.9703[22]	1.5566[22]
1590	Benzene, 1,1'-(1,2-ethanediyl)bis[4-methyl- 1,2-Di-p-tolylethane	$C_{16}H_{18}$ 210.32	538-39-6 85	3386 178[18]	4-05-00-01943	H_2O 1; EtOH 2; bz 3; peth 3
1591	Benzene, 1,1',1'',1'''-(1,2-ethanediylidene)tetrakis-	$C_{26}H_{22}$ 334.46	632-50-8 214.5	360	4-05-00-02746	EtOH 2; bz 3; HOAc 3
1592	Benzeneethanol Phenethyl alcohol	$C_8H_{10}O$ 122.17	60-12-8 -27	7185 218.2	4-06-00-03067 1.0202[20]	H_2O 2; EtOH 5; eth 5 1.5325[20]
1593	Benzeneethanol, β-amino-	$C_8H_{11}NO$ 137.18	7568-92-5 58.3	261; 160[17]	4-13-00-01835	H_2O 4; EtOH 4
1594	Benzeneethanol, 2-amino-	$C_8H_{11}NO$ 137.18	5339-85-5	147-8[3.5]	3-13-00-01679 1.045[25]	H_2O 4 1.5849[20]
1595	Benzeneethanol, 4-bromo-	C_8H_9BrO 201.06	4654-39-1	138[9]	4-06-00-03082 1.436[25]	1.5735[20]
1596	Benzeneethanol, 3-chloro-	C_8H_9ClO 156.61	5182-44-5	133-7[13]	4-06-00-03079 1.181[25]	1.5491[20]
1597	Benzeneethanol, 4-chloro-	C_8H_9ClO 156.61	1875-88-3	259	4-06-00-03079 1.1804[20]	1.5487[20]
1598	Benzeneethanol, 4-chloro-α,α-dimethyl-	$C_{10}H_{13}ClO$ 184.67	5468-97-3 34	126[12]	4-06-00-03291	1.5310[20]
1599	Benzeneethanol, 2,5-dimethoxy-	$C_{10}H_{14}O_3$ 182.22	7417-19-8	176[18]	4-06-00-07393 1.055[25]	1.5395[20]

No.	Name / Synonym	Mol. Form. / Mol. Wt.	CAS RN / mp/°C	Merck No. / bp/°C	Beil. Ref. / den/g cm⁻³	Solubility / n_D
1600	Benzeneethanol, α,α-dimethyl-	$C_{10}H_{14}O$ / 150.22	100-86-7 / 24	215	4-06-00-03290 / 0.9787[16]	1.5173[16]
1601	Benzeneethanol, α,α-diphenyl-	$C_{20}H_{18}O$ / 274.36	4428-13-1 / 89.5	222[11]	4-06-00-05057	H_2O 1; EtOH 4; eth 2; chl 2
1602	Benzeneethanol, β-ethyl-	$C_{10}H_{14}O$ / 150.22	2035-94-1	125[25]; 81[2]	2-06-00-00488 / 0.989[16]	1.43[16]
1603	Benzeneethanol, 2-ethyl- / Phenethyl alcohol, 2-ethyl	$C_{10}H_{14}O$ / 150.22	22545-12-6	249.7; 127[20]	4-06-00-03351 / 0.9972[20]	1.5304[20]
1604	Benzeneethanol, 4-ethyl- / Phenethyl alcohol, 4-ethyl	$C_{10}H_{14}O$ / 150.22	22545-13-7 / 7.9	250	4-06-00-03355 / 0.9907[20]	1.5229[20]
1605	Benzeneethanol, α-ethyl-α-methyl-, (±)- / 2-Butanol, 3-chloro (erythro, DL)	C_4H_9ClO / 108.57	116783-12-1	235; 56.1[30]	3-06-00-01958 / 1.0610[25]	eth 4; EtOH 4; chl 4 / 1.4397[25]
1606	Benzeneethanol, 3-hydroxy-	$C_8H_{10}O_2$ / 138.17	13398-94-2	168-73[4]	4-06-00-05936 / 1.082[25]	1.5643[20]
1607	Benzeneethanol, α-methyl-	$C_9H_{12}O$ / 136.19	698-87-3	125[25]	4-06-00-03192 / 0.991[20]	1.5190[20]
1608	Benzeneethanol, β-methyl-	$C_9H_{12}O$ / 136.19	1123-85-9	105-6[11]	2-06-00-00477 / 0.975[25]	H_2O 1; EtOH 3 / 1.5582[2]
1609	Benzeneethanol, 2-methyl-	$C_9H_{12}O$ / 136.19	19819-98-8 / 2	243.5	4-06-00-03234 / 1.016[25]	1.5355[20]
1610	Benzeneethanol, 4-methyl-	$C_9H_{12}O$ / 136.19	699-02-5	244.5; 94[6]	4-06-00-03244 / 1.0028[20]	1.5267[20]
1611	Benzeneethanol, β-(1-methylethyl)- / 1-Butanol, 3-methyl-2-phenyl	$C_{11}H_{16}O$ / 164.25	90499-41-5	130[15]	4-06-00-03390 / 0.9694[25]	ace 4; bz 4; eth 4; EtOH 4 / 1.5137[20]
1612	Benzeneethanol, 2-nitro-	$C_8H_9NO_3$ / 167.16	15121-84-3 / 2	267	3-06-00-01714 / 1.19[25]	1.5637[20]
1613	Benzeneethanol, α-phenyl-	$C_{14}H_{14}O$ / 198.26	614-29-9 / 67	177[15]	2-06-00-00637 / 1.0360[70]	
1614	Benzeneethanol, α-phenyl-, (S)-	$C_{14}H_{14}O$ / 198.26	5773-56-8 / 67.5	168[10]	4-06-00-04701 / 1.0358[70]	H_2O 1; EtOH 4; eth 4
1615	Benzeneethanol, β-phenyl- / 2,2-Diphenylethanol	$C_{14}H_{14}O$ / 198.26	1883-32-5 / 64.5	193[20]; 144[1]	4-06-00-04724	EtOH 2; eth 3; ace 3; chl 3
1616	Benzeneethanol, α,α,β-trimethyl- / 2-Butanol, 2-methyl-3-phenyl	$C_{11}H_{16}O$ / 164.25	3280-08-8 / 46	197; 106[12]	4-06-00-03389 / 0.9794[20]	ace 4; bz 4; eth 4; EtOH 4 / 1.5193[20]
1617	Benzeneethanol, α,β,β-trimethyl-	$C_{11}H_{16}O$ / 164.25	2977-31-3	197	4-06-00-03387	ace 4; bz 4; eth 4; EtOH 4 / 1.5161[13]
1618	Benzene, 1,1',1'',1'''-(1-ethanyl-2-ylidyne)tetrakis- / 1,1,1,2-Tetraphenylethane	$C_{26}H_{22}$ / 334.46	2294-94-2 / 144.8	278[21]	4-05-00-02747	bz 4
1619	Benzene, 1,1'-(1,2-ethenediyl)bis-, (E)- / trans-Stilbene	$C_{14}H_{12}$ / 180.25	103-30-0 / 123	8774 / 307; 166[12]	4-05-00-02156 / 0.9707[20]	H_2O 1; EtOH 2; eth 4; bz 4 / 1.6264[17]
1620	Benzene, 1,1'-(1,2-ethenediyl)bis-, (Z)- / cis-Stilbene	$C_{14}H_{12}$ / 180.25	645-49-8 / -5	8774 / 141[12]	4-05-00-02155 / 1.0143[20]	H_2O 1; EtOH 3; eth 3; ace 3 / 1.6130[20]
1621	Benzene, 1,1'-(1,2-ethenediyl)bis[4-methoxy-	$C_{16}H_{18}O_2$ / 240.30	4705-34-4 / 214.5	sub	4-06-00-06823	H_2O 1; ace 3; bz 3; HOAc 3
1622	Benzene, 1,1'-(1,2-ethenediyl)bis[2-methyl- / 2,2'-Dimethylstilbene	$C_{16}H_{16}$ / 208.30	10311-74-7 / 83	176-80[10]	4-05-00-02208	eth 4; EtOH 4; MeOH 4
1623	Benzene, 1,1',1'',1'''-(1,2-ethenediylidene)tetrakis-	$C_{26}H_{20}$ / 332.44	632-51-9 / 225	420	4-05-00-02780 / 1.155[0]	H_2O 1; EtOH 2; eth 2; bz 4
1624	Benzene, ethenyl- / Styrene	C_8H_8 / 104.15	100-42-5 / -31	8830 / 145	4-05-00-01334 / 0.9060[20]	H_2O 1; EtOH 3; eth 3; ace 3 / 1.5468[20]
1625	Benzene, (1-ethenyl-3-butenyl)- / 1,5-Hexadiene, 3-phenyl	$C_{12}H_{14}$ / 158.24	1076-66-0	78[8]	4-05-00-01555 / 0.8911[25]	1.5141[25]
1626	Benzene, 1-ethenyl-2,4-dimethyl-	$C_{10}H_{12}$ / 132.21	2234-20-0	79-80[12]	4-05-00-01386 / 0.905[20]	1.539[20]
1627	Benzene, 1-ethenyl-3,5-dimethyl-	$C_{10}H_{12}$ / 132.21	5379-20-4	57-8[4]	4-05-00-01386 / 0.894[25]	1.5382[20]
1628	Benzene, 2-ethenyl-1,4-dimethyl-	$C_{10}H_{12}$ / 132.21	2039-89-6	69[10]	4-05-00-01385 / 0.9072[17]	1.5236[17]
1629	Benzene, 4-ethenyl-1,2-dimethyl-	$C_{10}H_{12}$ / 132.21	27831-13-6	94-6[20]	4-05-00-01385 / 0.9062[5]	1.5463[20]
1630	Benzene, 1-ethenyl-2-ethyl-	$C_{10}H_{12}$ / 132.21	7564-63-8 / -75.5	187.3; 68[12]	3-05-00-01216 / 0.9017[25]	1.5380[20]
1631	Benzene, 1-ethenyl-3-ethyl-	$C_{10}H_{12}$ / 132.21	7525-62-4 / -101	190.0	3-05-00-01217 / 0.8945[20]	1.5351[20]
1632	Benzene, 1-ethenyl-4-ethyl-	$C_{10}H_{12}$ / 132.21	3454-07-7 / -49.7	192.3; 86[20]	4-05-00-01384 / 0.8884[25]	1.5376[20]
1633	Benzene, 1-ethenyl-2-fluoro-	C_8H_7F / 122.14	394-46-7	46[32]	4-05-00-01342 / 1.0282[20]	bz 4; eth 4; EtOH 4 / 1.5200[20]
1634	Benzene, 1-ethenyl-3-fluoro-	C_8H_7F / 122.14	350-51-6	30-1[4]	4-05-00-01343 / 1.0177[20]	H_2O 1; EtOH 3; eth 3; bz 3 / 1.5170[20]
1635	Benzene, 1-ethenyl-4-fluoro-	C_8H_7F / 122.14	405-99-2 / -34.5	67.4[50]; 30[4]	4-05-00-01343 / 1.0220[20]	H_2O 1; EtOH 3; eth 3; bz 3 / 1.5150[20]
1636	Benzene, 1,1'-ethenylidenebis- / 1,1-Diphenylethene	$C_{14}H_{12}$ / 180.25	530-48-3 / 8.2	3323 / 277	4-05-00-02173 / 1.0232[20]	H_2O 1; eth 3; chl 3 / 1.6085[20]
1637	Benzene, 1-ethenyl-2-methoxy-	$C_9H_{10}O$ / 134.18	612-15-7 / 29	197; 83[12]	4-06-00-03771 / 1.0049[17]	ace 4; bz 4; eth 4; EtOH 4 / 1.5388[20]
1638	Benzene, 1-ethenyl-3-methoxy-	$C_9H_{10}O$ / 134.18	626-20-0	90-3[15]	4-06-00-03773 / 0.999[16]	H_2O 1; EtOH 3; eth 3; bz 3 / 1.5586[23]

No.	Name Synonym	Mol. Form. Mol. Wt.	CAS RN mp/°C	Merck No. bp/°C	Beil. Ref. den/g cm⁻³	Solubility n_D
1639	Benzene, 1-ethenyl-4-methoxy-	$C_9H_{10}O$ 134.18	637-69-4	205; 91[13]	4-06-00-03775 1.0001[13]	H_2O 1; EtOH 3; eth 3; bz 3 1.5642[13]
1640	Benzene, 1-ethenyl-2-methyl- o-Methylstyrene	C_9H_{10} 118.18	611-15-4 -68.5	169.8	4-05-00-01367 0.9077[25]	H_2O 1; bz 3; chl 3 1.5437[20]
1641	Benzene, 1-ethenyl-3-methyl- m-Methylstyrene	C_9H_{10} 118.18	100-80-1 -86.3	164	4-05-00-01367 0.9076[25]	H_2O 1; EtOH 3; eth 3; bz 3 1.5411[20]
1642	Benzene, 1-ethenyl-4-methyl- p-Methylstyrene	C_9H_{10} 118.18	622-97-9 -34.1	172.8	4-05-00-01369 0.9173[25]	H_2O 1; bz 3 1.5420
1643	Benzene, 1-ethenyl-4-(1-methylethyl)-	$C_{11}H_{14}$ 146.23	2055-40-5 -44.7	204.1	4-05-00-01408 0.8850[20]	ace 4; bz 4; eth 4; EtOH 4 1.5289[20]
1644	Benzene, 1-ethenyl-3-nitro-	$C_8H_7NO_2$ 149.15	586-39-0 -10	120-1[11]	4-05-00-01351 1.1552[32]	H_2O 1; EtOH 3; eth 3; bz 3 1.5836[20]
1645	Benzene, 1-ethenyl-4-nitro-	$C_8H_7NO_2$ 149.15	100-13-0 29	dec	4-05-00-01351	EtOH 4; eth 4; chl 3; HOAc 3
1646	Benzene, (ethenyloxy)-	C_8H_8O 120.15	766-94-9	155.5	4-06-00-00561 0.9770[20]	H_2O 1; eth 4 1.5224[20]
1647	Benzene, 2-ethenyl-1,3,5-trimethyl-	$C_{11}H_{14}$ 146.23	769-25-5 -37	209	4-05-00-01408 0.9057[20]	ctc 3 1.5296[20]
1648	Benzene, 1,1',1''-(1-ethenyl-2-ylidene)tris-	$C_{20}H_{16}$ 256.35	58-72-0 72.5	220-1[14]	4-05-00-02575 1.0373[78]	H_2O 1; EtOH 3; eth 4; chl 3 1.6292[78]
1649	Benzene, ethoxy- Phenetole	$C_8H_{10}O$ 122.17	103-73-1 -29.5	7189 169.8	4-06-00-00554 0.9651[20]	H_2O 1; EtOH 3; eth 3; ctc 3 1.5076[20]
1650	Benzene, 2-ethoxy-1,3-dinitro- Phenetole, 2,6-dinitro-	$C_8H_8N_2O_5$ 212.16	13027-43-5 60.5	137[3]	3-06-00-00868	eth 4
1651	Benzene, (2-ethoxyethenyl)-	$C_{10}H_{12}O$ 148.20	17655-74-2 0	224.5	0.979[20]	1.5496[20]
1652	Benzene, (2-ethoxyethoxy)- Ethane, 1-ethoxy-2-phenoxy-	$C_{10}H_{14}O_2$ 166.22	19594-02-6	230	4-06-00-00572 1.018[11]	H_2O 1
1653	Benzene, 1-ethoxy-4-ethyl-	$C_{10}H_{14}O$ 150.22	1585-06-4	210	3-06-00-01665 0.9385[17]	ace 4; bz 4; EtOH 4
1654	Benzene, 1-ethoxy-2-fluoro-	C_8H_9FO 140.16	451-80-9 -16.7	171.4	4-06-00-00771 1.0874[17]	H_2O 1; ace 3; bz 3; ctc 2 1.4932[17]
1655	Benzene, 1-ethoxy-3-fluoro-	C_8H_9FO 140.16	458-03-7 -27.5	172; 65.2[15]	3-06-00-00669 1.0716[16]	H_2O 1; bz 3 1.4847[17]
1656	Benzene, 1-ethoxy-4-fluoro-	C_8H_9FO 140.16	459-26-7 -8.5	173; 54[7]	4-06-00-00774 1.0715[18]	H_2O 1; bz 3; ctc 2; chl 3 1.4826[18]
1657	Benzene, 1-ethoxy-4-iodo-	C_8H_9IO 248.06	699-08-1 29	250	4-06-00-01077	bz 4; eth 4; EtOH 4; chl 4
1658	Benzene, (ethoxymethyl)- Benzyl ethyl ether	$C_9H_{12}O$ 136.19	539-30-0	1147 186	4-06-00-02229 0.9478[20]	H_2O 1; EtOH 5; eth 5 1.4955[20]
1659	Benzene, 1-ethoxy-2-methyl-	$C_9H_{12}O$ 136.19	614-71-1	184	4-06-00-01944 0.9592[13]	eth 4; EtOH 4 1.508[13]
1660	Benzene, 1-ethoxy-3-methyl-	$C_9H_{12}O$ 136.19	621-32-9	192	4-06-00-02039 0.949[20]	H_2O 1; EtOH 3; eth 3 1.513[20]
1661	Benzene, 1-ethoxy-4-methyl-	$C_9H_{12}O$ 136.19	622-60-6	188.5	4-06-00-02099 0.9509[18]	H_2O 1; EtOH 3; eth 3; ctc 2 1.5058[18]
1662	Benzene, 1-ethoxy-2-nitro-	$C_8H_9NO_3$ 167.16	610-67-3 2.1	267	4-06-00-01250 1.1903[15]	eth 4; EtOH 4 1.5425[20]
1663	Benzene, 1-ethoxy-3-nitro-	$C_8H_9NO_3$ 167.16	621-52-3 36	264; 169[70]	4-06-00-01271	H_2O 1; EtOH 3; eth 3
1664	Benzene, 1-ethoxy-4-nitro-	$C_8H_9NO_3$ 167.16	100-29-8 60	283	4-06-00-01283 1.1176[100]	H_2O 2; EtOH 2; eth 4; ace 5
1665	Benzene, 1-ethoxy-2-(1-propenyl)- Phenetole, 2-propenyl	$C_{11}H_{14}O$ 162.23	67191-37-1	231; 120[17]	0-06-00-00565 0.9731[24]	1.544[24]
1666	Benzene, (3-ethoxy-1-propenyl)-	$C_{11}H_{14}O$ 162.23	1476-07-9	127-8[22]	3-06-00-02404 0.970[15]	1.547[15]
1667	Benzene, 1-ethoxy-2-propyl- Phenetole, 2-propyl	$C_{11}H_{16}O$ 164.25	101144-90-5	213; 100[16]	0-06-00-00499 0.9240[20]	1.494[20]
1668	Benzene, ethyl- Ethylbenzene	C_8H_{10} 106.17	100-41-4 -94.9	3723 136.1	4-05-00-00885 0.8670[20]	H_2O 1; EtOH 5; eth 5; chl 2 1.4959[20]
1669	Benzene, (1-ethylbutyl)-	$C_{12}H_{18}$ 162.27	4468-42-2 -55.4	209	4-05-00-01117 0.8239[25]	1.4859[20]
1670	Benzene, 1-ethyl-2,3-dimethyl- 3-Ethyl-o-xylene	$C_{10}H_{14}$ 134.22	933-98-2 -49.5	194	4-05-00-01069 0.8881[25]	1.5117[20]
1671	Benzene, 1-ethyl-2,4-dimethyl-	$C_{10}H_{14}$ 134.22	874-41-9 -62.9	188.4	4-05-00-01070 0.8763[20]	ace 4; bz 4; eth 4; EtOH 4 1.5038[20]
1672	Benzene, 1-ethyl-3,5-dimethyl-	$C_{10}H_{14}$ 134.22	934-74-7 -84.3	183.6	4-05-00-01071 0.8608[25]	H_2O 1; EtOH 5; eth 5; ace 5 1.4981[20]
1673	Benzene, 2-ethyl-1,3-dimethyl-	$C_{10}H_{14}$ 134.22	2870-04-4 -16.2	190	4-05-00-01069 0.8864[25]	1.5107[20]
1674	Benzene, 2-ethyl-1,4-dimethyl-	$C_{10}H_{14}$ 134.22	1758-88-9 -53.7	186.9	4-05-00-01070 0.8732[25]	H_2O 1; EtOH 5; eth 5; ace 5 1.5043[20]
1675	Benzene, 4-ethyl-1,2-dimethyl-	$C_{10}H_{14}$ 134.22	934-80-5 -66.9	189.5	4-05-00-01069 0.8706[25]	H_2O 1; EtOH 5; eth 5; ace 5 1.5031[20]
1676	Benzene, 1,1'-(1-ethyl-1,2-ethanediyl)bis-	$C_{16}H_{18}$ 210.32	5223-59-6	290; 152[11]	4-05-00-01939 0.9777[20]	H_2O 1; EtOH 3; eth 3; bz 3 1.5554[20]
1677	Benzene, 1,1'-ethylidenebis- 1,1-Diphenylethane	$C_{14}H_{14}$ 182.27	612-00-0 -17.9	272.6	4-05-00-01880 0.9997[20]	H_2O 1; EtOH 5; eth 5; bz 3 1.5756[20]
1678	Benzene, 1,1'-ethylidenebis[4-chloro- 1,1-Bis(4-chlorophenyl)ethane	$C_{14}H_{12}Cl_2$ 251.15	3547-04-4 56	320	4-05-00-01881	

No.	Name / Synonym	Mol. Form. / Mol. Wt.	CAS RN / mp/°C	Merck No. / bp/°C	Beil. Ref. / den/g cm^{-3}	Solubility / n_D
1679	Benzene, 1,1',1''-ethylidynetris-	$C_{20}H_{18}$ 258.36	5271-39-6 95	207[18]	4-05-00-02510	H_2O 1; EtOH 2; eth 3; bz 3
1680	Benzene, 1-ethyl-2-iodo- o-Ethyliodobenzene	C_8H_9I 232.06	18282-40-1	226	4-05-00-00910 1.6189[10]	ace 4; bz 4 1.5941[22]
1681	Benzene, 1-ethyl-4-iodo-	C_8H_9I 232.06	25309-64-2 -17	209	3-05-00-00801 1.6095[10]	ace 4; bz 4 1.5909[22]
1682	Benzene, 1-ethyl-2-methoxy-	$C_9H_{12}O$ 136.19	14804-32-1	187; 80[14]	3-06-00-01656 0.9636[19]	bz 4; eth 4 1.5142[20]
1683	Benzene, 1-ethyl-3-methoxy-	$C_9H_{12}O$ 136.19	10568-38-4	197; 74[10]	3-06-00-01662 0.9575[18]	H_2O 1; eth 4; bz 3; ctc 3 1.5102
1684	Benzene, 1-ethyl-4-methoxy-	$C_9H_{12}O$ 136.19	1515-95-3	198	4-06-00-03021 0.9624[15]	bz 4; eth 4 1.5120[20]
1685	Benzene, 1-ethyl-2-methyl- o-Ethyltoluene	C_9H_{12} 120.19	611-14-3 -80.8	165.2	4-05-00-00999 0.8807[20]	H_2O 1; EtOH 5; eth 5; ace 5 1.5046[20]
1686	Benzene, 1-ethyl-3-methyl- m-Ethyltoluene	C_9H_{12} 120.19	620-14-4 -95.5	161.3	4-05-00-01001 0.8645[20]	H_2O 1; EtOH 4; eth 4; ace 5 1.4966[20]
1687	Benzene, 1-ethyl-4-methyl- p-Ethyltoluene	C_9H_{12} 120.19	622-96-8 -62.3	162	4-05-00-01003 0.8614[20]	H_2O 1; EtOH 4; eth 4; ace 5 1.4959[20]
1688	Benzene, 1-ethyl-2-(1-methylethyl)- Benzene, 1-ethyl-2-isopropyl	$C_{11}H_{16}$ 148.25	18970-44-0	193	4-05-00-01100 0.888[20]	ace 4; bz 4; eth 4; EtOH 4 1.508[20]
1689	Benzene, 1-ethyl-3-(1-methylethyl)- Benzene, 1-ethyl-3-isopropyl	$C_{11}H_{16}$ 148.25	4920-99-4 <-20	192	3-05-00-01005 0.859[20]	H_2O 1; EtOH 3; eth 3; bz 4 1.4921[20]
1690	Benzene, 1-ethyl-4-(1-methylethyl)- Benzene, 1-ethyl-4-isopropyl	$C_{11}H_{16}$ 148.25	4218-48-8 <-20	196.6	4-05-00-01100 0.8585[20]	H_2O 1; eth 3; bz 4 1.4923[20]
1691	Benzene, (1-ethyl-1-methylpropyl)- Pentane, 3-methyl-3-phenyl	$C_{12}H_{18}$ 162.27	1985-97-3	205	4-05-00-01120 0.8773[15]	1.4972[16]
1692	Benzene, 1-ethyl-2-nitro-	$C_8H_9NO_2$ 151.17	612-22-6 -12.3	232.5	4-05-00-00911 1.1207[20]	H_2O 1; EtOH 4; eth 4; ace 3 1.5356[20]
1693	Benzene, 1-ethyl-3-nitro-	$C_8H_9NO_2$ 151.17	7369-50-8 -37.9	242.5	4-05-00-00911 1.1345[25]	ace 4; eth 4; EtOH 4
1694	Benzene, 1-ethyl-4-nitro-	$C_8H_9NO_2$ 151.17	100-12-9 -12.3	245.5	4-05-00-00912 1.1192[20]	H_2O 1; EtOH 4; eth 4; ace 3 1.5455[20]
1695	Benzene, (ethyloctadecyl)- 3-Phenyleicosane	$C_{26}H_{46}$ 358.65	2400-02-4 29.3	202[1]	4-05-00-01222 0.8546[20]	1.4796[20]
1696	Benzene, 1-ethyl-4-(phenylmethyl)-	$C_{15}H_{16}$ 196.29	620-85-9 -24	297	4-05-00-01926 0.9777[20]	EtOH 3; eth 3; chl 3 1.5616[20]
1697	Benzene, 1,1'-(1-ethyl-1,3-propanediyl)bis- Pentane, 1,3-diphenyl-	$C_{17}H_{20}$ 224.35	838-45-9	304; 113[0.8]	4-05-00-01963 0.9734[21]	1.553[21]
1698	Benzene, 1,1'-(2-ethyl-1,3-propanediyl)bis- Propane, 1,1-dibenzyl-	$C_{17}H_{20}$ 224.35	1520-45-2	304.5	4-05-00-01964 0.9734[21]	1.553[21]
1699	Benzene, (1-ethyl-1-propenyl)- 3-Phenyl-2-pentene	$C_{11}H_{14}$ 146.23	4701-36-4	198; 92[18]	4-05-00-01404 0.9173[14]	1.5266[15]
1700	Benzene, (1-ethyl-2-propenyl)-	$C_{11}H_{14}$ 146.23	19947-22-9	191.5; 71[12]	3-05-00-01240 0.8458[23]	1.5030[21]
1701	Benzene, (1-ethylpropyl)-	$C_{11}H_{16}$ 148.25	1196-58-3	187.5	4-05-00-01090 0.8649[20]	1.4880[20]
1702	Benzene, 1-ethyl-2-propyl-	$C_{11}H_{16}$ 148.25	16021-20-8	203	4-05-00-01099 0.8744[20]	1.4992[20]
1703	Benzene, 1-ethyl-4-propyl- 1-Ethyl-4-propylbenzene	$C_{11}H_{16}$ 148.25	20024-90-2	205	4-05-00-01100 0.8594[20]	ace 4; bz 4; eth 4; EtOH 4 1.4921[21]
1704	Benzene, (ethylsulfonyl)-	$C_8H_{10}O_2S$ 170.23	599-70-2 42	160[12]	4-06-00-01469 1.1410[20]	bz 4; eth 4; EtOH 4; chl 4
1705	Benzene, (ethylthio)-	$C_8H_{10}S$ 138.23	622-38-8	205	4-06-00-01468 1.0211[20]	EtOH 3 1.5670[20]
1706	Benzene, 1-(ethylthio)-3-methyl- Ethyl m-tolyl sulfide	$C_9H_{12}S$ 152.26	34786-24-8	219	3-06-00-01332 0.9987[20]	1.5590[20]
1707	Benzene, 1-(ethylthio)-4-methyl-	$C_9H_{12}S$ 152.26	622-63-9	220	4-06-00-02156 0.9996[20]	1.555[20]
1708	Benzene, 1-ethyl-2,3,4-trimethyl- Benzene, 4-ethyl-1,2,3-trimethyl	$C_{11}H_{16}$ 148.25	61827-86-9	220.4	4-05-00-01107 0.9019[20]	ace 4; bz 4; eth 4; EtOH 4 1.5180[20]
1709	Benzene, 1-ethyl-2,3,5-trimethyl- 1-Ethyl-2,3,5-trimethylbenzene	$C_{11}H_{16}$ 148.25	18262-85-6	213	4-05-00-01107 0.8897[20]	ace 4; bz 4; eth 4; EtOH 4 1.5118[20]
1710	Benzene, 1-ethyl-2,4,5-trimethyl- 1-Ethyl-2,4,5-trimethylbenzene	$C_{11}H_{16}$ 148.25	17851-27-3 -13.5	213	4-05-00-01108 0.883[20]	ace 4; bz 4; eth 4; EtOH 4 1.5075[20]
1711	Benzene, 2-ethyl-1,3,4-trimethyl- Benzene, 3-ethyl-1,2,4-trimethyl	$C_{11}H_{16}$ 148.25	61827-87-0	216.6	4-05-00-01107 0.895[20]	ace 4; bz 4; eth 4; EtOH 4 1.5133[20]
1712	Benzene, 2-ethyl-1,3,5-trimethyl-	$C_{11}H_{16}$ 148.25	3982-67-0 -15.5	212.4	4-05-00-01107 0.883[20]	ace 4; bz 4; eth 4; EtOH 4 1.5074[20]
1713	Benzene, 5-ethyl-1,2,3-trimethyl- 5-Ethyl-1,2,3-trimethylbenzene	$C_{11}H_{16}$ 148.25	31366-00-4	215.8	4-05-00-01107 0.8863[20]	ace 4; eth 4; EtOH 4 1.5101[20]
1714	Benzene, 1,1'-(1,2-ethynediyl)bis- Diphenylacetylene	$C_{14}H_{10}$ 178.23	501-65-5 62.5	300	4-05-00-02276 0.9657[100]	H_2O 1; EtOH 2; eth 4; ctc 2
1715	Benzene, ethynyl- Ethynylbenzene	C_8H_8 102.14	536-74-3 -44.8	143	4-05-00-01525 0.9300[20]	H_2O 1; EtOH 5; eth 5; ace 3 1.5470[20]
1716	Benzene, (ethynyloxy)-	C_8H_8O 118.14	4279-76-9 -36	61-2[25]	4-06-00-00565 1.0614[20]	eth 4; EtOH 4 1.5125[20]
1717	Benzene, fluoro- Fluorobenzene	C_6H_5F 96.10	462-06-6 -42.2	4099 84.7	4-05-00-00632 1.0225[20]	bz 4; eth 4; EtOH 4; lig 4 1.4684[30]
1718	Benzene, 1-fluoro-2,4-dinitro- 2,4-Dinitrophenyl fluoride	$C_6H_3FN_2O_4$ 186.10	70-34-8 25.8	4101 296	4-05-00-00742 1.4718[54]	EtOH 3; chl 2 1.5690[20]

No.	Name Synonym	Mol. Form. Mol. Wt.	CAS RN mp/°C	Merck No. bp/°C	Beil. Ref. den/g cm⁻³	Solubility n_D
1719	Benzene, 1-fluoro-2-iodo-	C₆H₄FI 222.00	348-52-7 -41.5	188.6	4-05-00-00693	ace 3; bz 3; chl 3 1.5910[20]
1720	Benzene, 1-fluoro-4-iodo-	C₆H₄FI 222.00	352-34-1 -27	183	4-05-00-00694 1.9523[15]	H₂O 1; EtOH 3; eth 3; ace 3 1.5270[22]
1721	Benzene, 1-fluoro-3-isothiocyanato-	C₇H₄FNS 153.18	404-72-8	227	3-12-00-01276 1.27[25]	1.6186[20]
1722	Benzene, 1-fluoro-2-methoxy-	C₇H₇FO 126.13	321-28-8 -39	154.5	4-06-00-00771 1.5489[17]	H₂O 1; eth 3; ctc 3 1.4969[17]
1723	Benzene, 1-fluoro-3-methoxy-	C₇H₇FO 126.13	456-49-5 -35	159; 51[14]	4-06-00-00772 1.104[25]	1.4876[20]
1724	Benzene, 1-fluoro-4-methoxy-	C₇H₇FO 126.13	459-60-9 -45	157	4-06-00-00773 1.1781[18]	eth 3 1.4886[18]
1725	Benzene, (fluoromethyl)-	C₇H₇F 110.13	350-50-5 -35	140; 40[14]	4-05-00-00800 1.0228[25]	ctc 3 1.4892[25]
1726	Benzene, 1-fluoro-2-methyl- o-Fluorotoluene	C₇H₇F 110.13	95-52-3 -62	4108 115	4-05-00-00799 1.0041[13]	H₂O 1; EtOH 4; eth 4 1.4704[20]
1727	Benzene, 1-fluoro-3-methyl- m-Fluorotoluene	C₇H₇F 110.13	352-70-5 -87	4108 115	4-05-00-00799 0.9974[20]	H₂O 1; EtOH 4; eth 4 1.4691[20]
1728	Benzene, 1-fluoro-4-methyl- p-Fluorotoluene	C₇H₇F 110.13	352-32-9 -56	4108 116.6	4-05-00-00800 0.9975[20]	H₂O 1; EtOH 4; eth 4 1.4699[20]
1729	Benzene, 1-fluoro-4-methyl-2-nitro- Toluene, 4-fluoro-3-nitro-	C₇H₆FNO₂ 155.13	446-11-7 26.5	241	3-05-00-00743 1.2619[28]	1.5237[28]
1730	Benzene, 2-fluoro-4-methyl-1-nitro- 3-Fluoro-4-nitrotoluene	C₇H₆FNO₂ 155.13	446-34-4 53.2	97[3]	4-05-00-00852 1.4380[25]	
1731	Benzene, 4-fluoro-1-methyl-2-nitro-	C₇H₆FNO₂ 155.13	446-10-6 27	213; 138[83]	4-05-00-00851 1.2686[20]	1.5218[20]
1732	Benzene, 4-fluoro-2-methyl-1-nitro-	C₇H₆FNO₂ 155.13	446-33-3 27.5	217; 97-8[10]	4-05-00-00851 1.272[28]	1.5271[20]
1733	Benzene, 1-fluoro-2-nitro- o-Fluoronitrobenzene	C₆H₄FNO₂ 141.10	1493-27-2 -6	215 dec	4-05-00-00718 1.3285[18]	eth 4; EtOH 4 1.5489[17]
1734	Benzene, 1-fluoro-3-nitro- m-Fluoronitrobenzene	C₆H₄FNO₂ 141.10	402-67-5 41	199; 86[19]	4-05-00-00719 1.3254[19]	H₂O 1; EtOH 3; eth 3; bz 2 1.5262[15]
1735	Benzene, 1-fluoro-4-nitro- p-Fluoronitrobenzene	C₆H₄FNO₂ 141.10	350-46-9 21	205	4-05-00-00719 1.3300[20]	H₂O 1; EtOH 3; eth 3; ctc 2 1.5316[20]
1736	Benzene, 1-fluoro-2-(trichloromethyl)-	C₇H₄Cl₃F 213.47	488-98-2	75[5]	3-05-00-00701 1.453[25]	1.5432[20]
1737	Benzene, 1-fluoro-2-(trifluoromethyl)-	C₇H₄F₄ 164.10	392-85-8	114.5	4-05-00-00804 1.293[25]	1.4040[25]
1738	Benzene, 1-fluoro-3-(trifluoromethyl)-	C₇H₄F₄ 164.10	401-80-9 -81.5	101.5	4-05-00-00804 1.3021[17]	
1739	Benzene, 1-fluoro-4-(trifluoromethyl)-	C₇H₄F₄ 164.10	402-44-8 -41.7	103.5	4-05-00-00804 1.293[25]	1.4025[20]
1740	Benzene, 2-fluoro-1,3,5-trimethyl-	C₉H₁₁F 138.18	392-69-8 -36.7	171.5	4-05-00-01025 0.9745[25]	1.4809[25]
1741	Benzene, heneicosyl- Heneicosylbenzene	C₂₇H₄₈ 372.68	40775-09-5 48	439	4-05-00-01226 0.8545[20]	1.4804[20]
1742	Benzene, heptacosyl- Heptacosylbenzene	C₃₃H₆₀ 456.84	61828-25-9 64	397[100]; 262[1]	4-05-00-01243 0.8543[20]	1.4797[20]
1743	Benzene, heptadecyl- Heptadecane, 1-phenyl-	C₂₃H₄₀ 316.57	14752-75-1 32	397	3-05-00-01123 0.8546[20]	1.4810[20]
1744	Benzene, 1,1'-(1-heptenylidene)bis- 1,1-Diphenyl-1-heptene	C₁₉H₂₂ 250.38	1530-20-7		4-05-00-02252 0.9673[18]	EtOH 2 1.5648[18]
1745	Benzene, heptyl-	C₁₃H₂₀ 176.30	1078-71-3 -48	240; 109[10]	4-05-00-01143 0.8567[20]	H₂O 1; bz 3; chl 3 1.4865[20]
1746	Benzene, 1,1'-heptylidenebis- 1,1-Diphenylheptane	C₁₉H₂₄ 252.40	1530-05-8 14	333.5; 137[1]	4-05-00-01992 0.9497[20]	
1747	Benzene, (heptyloxy)-	C₁₃H₂₀O 192.30	32395-96-3 -33.5	267	4-06-00-00560 0.9170[15]	ace 4; eth 4; EtOH 4 1.4912[20]
1748	Benzenehexacarboxylic acid Mellitic acid	C₁₂H₆O₁₂ 342.17	517-60-2 286-8 dec	5706	4-09-00-03825	H₂O 4; EtOH 3; sulf 3
1749	Benzene, hexachloro- Hexachlorobenzene	C₆Cl₆ 284.78	118-74-1 231.8	4600 325	4-05-00-00670 2.044[23]	H₂O 1; EtOH 2; eth 3; bz 4 1.5691[23]
1750	Benzene, hexacosyl- Hexacosane, 1-phenyl-	C₃₂H₅₈ 442.81	13024-80-1 62	388[100]; 250[1]	4-05-00-01242 0.8543[20]	1.4798[20]
1751	Benzene, hexadecyl-	C₂₂H₃₈ 302.54	1459-09-2 27	385	4-05-00-01216 0.8547[20]	H₂O 1; EtOH 2; eth 4; bz 4 1.4813[20]
1752	Benzene, (hexadecyloxy)-	C₂₂H₃₈O 318.54	35021-70-6 41.8	200[1]	3-06-00-00555 0.8434[82]	1.4556[82]
1753	Benzene, 1,3-hexadienyl- 1-Phenyl-1,3-hexadiene	C₁₂H₁₄ 158.24	41635-77-2	128[16]	3-05-00-01378 0.9253[12]	1.6025[12]
1754	Benzene, 1,5-hexadienyl- 1,5-Hexadiene, 1-phenyl-	C₁₂H₁₄ 158.24	1009-81-0	102-3[8]	4-05-00-01554 0.9005[25]	1.5421[25]
1755	Benzene, 3,5-hexadienyl- 6-Phenyl-1,3-hexadiene	C₁₂H₁₄ 158.24	39669-95-9	100.5[11]	3-05-00-01378 0.9304[13]	1.5446[13]
1756	Benzene, hexaethyl- Hexaethylbenzene	C₁₈H₃₀ 246.44	604-88-6 129	298	4-05-00-01208 0.8305[130]	H₂O 1; EtOH 3; eth 4; bz 4 1.4736[130]
1757	Benzene, hexafluoro- Perfluorobenzene	C₆F₆ 186.06	392-56-3 5.3	80.2	4-05-00-00640 1.6184[20]	1.3777[20]

No.	Name Synonym	Mol. Form. Mol. Wt.	CAS RN mp/°C	Merck No. bp/°C	Beil. Ref. den/g cm^{-3}	Solubility n_D
1758	Benzene, hexamethyl- Mellitene	$C_{12}H_{18}$ 162.27	87-85-4 166.5	263.4	4-05-00-01137 1.0630[25]	H_2O 1; EtOH 3; eth 3; ace 3
1759	Benzenehexanoic acid	$C_{12}H_{16}O_2$ 192.26	5581-75-9 23	206[30]	4-09-00-01895	1.5164[21]
1760	Benzene, 2-hexenyl- 2-Hexene, 1-phenyl	$C_{12}H_{16}$ 160.26	67590-77-6 	108[16]	0-05-00-00501 0.8898[16]	1.5058[16]
1761	Benzene, 3-hexenyl-	$C_{12}H_{16}$ 160.26	35008-86-7 	222	4-05-00-01418 0.8790[20]	1.5039[20]
1762	Benzene, 5-hexenyl-	$C_{12}H_{16}$ 160.26	1588-44-9 	94-5[10]	3-05-00-01251 0.8839[20]	1.5033[20]
1763	Benzene, hexyl-	$C_{12}H_{18}$ 162.27	1077-16-3 -61	226.1	4-05-00-01115 0.8575[20]	H_2O 1; eth 5; bz 3; peth 3 1.4864[20]
1764	Benzene, (1-hexylheptyl)- 7-Phenyltridecane	$C_{19}H_{32}$ 260.46	2400-01-3 	183-4[20]	4-05-00-01210 0.8723[20]	1.4931[18]
1765	Benzene, (hexyloxy)-	$C_{12}H_{18}O$ 178.27	1132-66-7 -19	240	4-06-00-00560 0.9174[20]	H_2O 1; eth 3 1.4921[20]
1766	Benzene, iodo- Iodobenzene	C_6H_5I 204.01	591-50-4 -31.3	4922 188.4	4-05-00-00688 1.8308[20]	H_2O 1; EtOH 3; eth 5; ace 5 1.6200[20]
1767	Benzene, 1-iodo-2,3-dimethoxy- 3-Iodoveratrole	$C_8H_9IO_2$ 264.06	25245-33-4 45.5	144-5[14]	3-06-00-04262 1.7799[20]	1.6127[20]
1768	Benzene, 4-iodo-1,2-dimethoxy-	$C_8H_9IO_2$ 264.06	5460-32-2 35	174[30]	4-06-00-05625	bz 4; EtOH 4
1769	Benzene, 1-iodo-2,3-dimethyl-	C_8H_9I 232.06	31599-60-7 	229	3-05-00-00820 1.6395[20]	ace 4 1.6074[20]
1770	Benzene, 1-iodo-2,4-dimethyl-	C_8H_9I 232.06	4214-28-2 	231 dec; 111[14]	4-05-00-00947 1.6282[16]	H_2O 1; ace 3; bz 3 1.6008[16]
1771	Benzene, 1-iodo-3,5-dimethyl- m-Xylene, 5-iodo-	C_8H_9I 232.06	22445-41-6 	230.5	3-05-00-00840 1.6085[18]	ace 4 1.5967[18]
1772	Benzene, 2-iodo-1,3-dimethyl-	C_8H_9I 232.06	608-28-6 11.2	229.5	4-05-00-00947 1.6158[20]	H_2O 1; ace 3; bz 3 1.6035[20]
1773	Benzene, 2-iodo-1,4-dimethyl-	C_8H_9I 232.06	1122-42-5 	227 dec	4-05-00-00970 1.6168[17]	H_2O 1; ace 3; bz 3 1.5992[17]
1774	Benzene, 4-iodo-1,2-dimethyl- 4-Iodo-o-xylene	C_8H_9I 232.06	31599-61-8 	231.5	3-05-00-00821 1.6334[18]	ace 4 1.6049[18]
1775	Benzene, 1-iodo-2-methoxy- o-Iodoanisole	C_7H_7IO 234.04	529-28-2 	4921 241; 91[2]	4-06-00-01070 1.8[20]	EtOH 4; eth 4; ace 4; bz 4
1776	Benzene, 1-iodo-4-methoxy-	C_7H_7IO 234.04	696-62-8 53	238; 138[25]	4-06-00-01075	EtOH 3; eth 3; chl 3
1777	Benzene, (iodomethyl)-	C_7H_7I 218.04	620-05-3 24.5	93[10]	4-05-00-00842 1.7335[25]	bz 4; eth 4; EtOH 4 1.6334[25]
1778	Benzene, 1-iodo-2-methyl-	C_7H_7I 218.04	615-37-2 	211.5	4-05-00-00838 1.713[20]	H_2O 1; EtOH 5; eth 5 1.6079[20]
1779	Benzene, 1-iodo-3-methyl-	C_7H_7I 218.04	625-95-6 -27.2	213	4-05-00-00839 1.705[20]	H_2O 1; EtOH 5; eth 5 1.6053[20]
1780	Benzene, 1-iodo-4-methyl-	C_7H_7I 218.04	624-31-7 36.5	211	4-05-00-00840 1.678[20]	H_2O 1; EtOH 3; eth 3; chl 2
1781	Benzene, 2-iodo-1-methyl-4-(1-methylethyl)- 2-Iodo-p-cymene	$C_{10}H_{13}I$ 260.12	56739-95-8 	139[23]	3-05-00-00960 1.4205[17]	1.5800[17]
1782	Benzene, 2-iodo-4-methyl-1-(1-methylethyl)- p-Cymene, 3-iodo-	$C_{10}H_{13}I$ 260.12	4395-81-7 	123[13]; 80[5]	3-05-00-00960 1.4113[17]	1.5690[17]
1783	Benzene, 1-iodo-2-nitro-	$C_6H_4INO_2$ 249.01	609-73-4 54	290; 162[18]	4-05-00-00732 1.9186[75]	H_2O 1; EtOH 3; eth 3
1784	Benzene, 1-iodo-3-nitro-	$C_6H_4INO_2$ 249.01	645-00-1 38.5	280	4-05-00-00733 1.9477[50]	H_2O 1; EtOH 3; eth 3
1785	Benzene, 1-iodo-4-nitro-	$C_6H_4INO_2$ 249.01	636-98-6 174.7	288	4-05-00-00734 1.8090[155]	H_2O 1; EtOH 3; HOAc 3
1786	Benzene, isocyanato- Phenyl isocyanate	C_7H_5NO 119.12	103-71-9 	7267 163; 55[13]	4-12-00-00864 1.0956[20]	eth 4; chl 2 1.5368[20]
1787	Benzene, 1-isocyanato-2-methyl-	C_8H_7NO 133.15	614-68-6 	185	4-12-00-01768	H_2O 1; eth 3 1.5282[20]
1788	Benzene, 1-isocyanato-2-nitro-	$C_7H_4N_2O_3$ 164.12	3320-86-3 41	137[18]	3-12-00-01535	bz 4; eth 4; chl 4
1789	Benzene, 1-isocyanato-3-nitro-	$C_7H_4N_2O_3$ 164.12	3320-87-4 51	130[11]	4-12-00-01604	bz 4; eth 4; chl 4
1790	Benzene, 1-isocyanato-4-nitro-	$C_7H_4N_2O_3$ 164.12	100-28-7 57	162[20]; 137[11]	4-12-00-01653	bz 4; eth 4; chl 4
1791	Benzene, (isocyanomethyl)-	C_8H_7N 117.15	10340-91-7 	199 dec; 93[55]	4-12-00-02219 0.972[15]	1.5193[20]
1792	Benzene, isothiocyanato- Phenyl isothiocyanate	C_7H_5NS 135.19	103-72-0 -21	7268 221	4-12-00-00867 1.1303[20]	H_2O 1; EtOH 3; eth 3; ctc 3 1.6492[23]
1793	Benzene, 1-isothiocyanato-2-methoxy-	C_8H_7NOS 165.22	3288-04-8 	131-2[11]	4-13-00-00855 1.1878[20]	1.6458[20]
1794	Benzene, (isothiocyanatomethyl)-	C_8H_7NS 149.22	622-78-6 	243	4-12-00-02276 1.1246[16]	H_2O 1; EtOH 5; eth 3 1.6049[15]
1795	Benzenemethanamine Benzylamine	C_7H_9N 107.16	100-46-9 	1139 185; 90[12]	4-12-00-02155 0.9813[20]	H_2O 5; EtOH 5; eth 5; ace 4 1.5401[20]
1796	Benzenemethanamine, 2-amino-	$C_7H_{10}N_2$ 122.17	4403-69-4 61	269	4-13-00-00265	EtOH 4

No.	Name Synonym	Mol. Form. Mol. Wt.	CAS RN mp/°C	Merck No. bp/°C	Beil. Ref. den/g cm^{-3}	Solubility n_D
1797	Benzenemethanamine, N-[bis(4-methoxyphenyl)methylene]- Benzylimidobis(p-methoxyphenyl)methane	$C_{22}H_{21}NO_2$ 331.41	524-96-9 90	1153	3-12-00-02254	eth 4; chl 4
1798	Benzenemethanamine, N,N-bis(phenylmethyl)-	$C_{21}H_{21}N$ 287.40	620-40-6 91.5	385	4-12-00-02183 0.9912[95]	H_2O 2; EtOH 2; eth 3; ctc 3
1799	Benzenemethanamine, 4-bromo- Benzyl amine, 4-bromo	C_7H_8BrN 186.05	3959-07-7 20	250	3-12-00-02350	eth 4
1800	Benzenemethanamine, N-3-butynyl-N-methyl-	$C_{12}H_{15}N$ 173.26	15240-91-2 127[16]		4-12-00-02171 0.9372[20]	1.5202[20]
1801	Benzenemethanamine, N-(2-chloroethyl)-N-(phenylmethyl)-, hydrochloride N-(2-Chloroethyl)dibenzylamine hydrochloride	$C_{16}H_{19}Cl_2N$ 296.24	55-43-6	2138	4-12-00-02180	
1802	Benzenemethanamine, 3-chloro-N-methyl-	$C_8H_{10}ClN$ 155.63	39191-07-6 88[4]		3-12-00-02342	chl 3 1.5350[25]
1803	Benzenemethanamine, 2,4-dichloro-	$C_7H_7Cl_2N$ 176.04	95-00-1 124-6[13]		4-12-00-02376	chl 3 1.5762[25]
1804	Benzenemethanamine, N,N-diethyl-α-phenyl- N,N-Diethylbenzhydrylamine	$C_{17}H_{21}N$ 239.36	519-72-2 58.5	3103 170[17]	4-12-00-03283	
1805	Benzenemethanamine, 3,4-dimethoxy-	$C_9H_{13}NO_2$ 167.21	5763-61-1 	156[12]; 120[3]	4-13-00-02582 1.143[25]	chl 3
1806	Benzenemethanamine, α,α-dimethyl-	$C_9H_{13}N$ 135.21	585-32-0 	196.5	4-12-00-02633 0.9423[20]	1.5181[25]
1807	Benzenemethanamine, N,N-dimethyl- Dimethylbenzylamine	$C_9H_{13}N$ 135.21	103-83-3 	181	4-12-00-02161 0.915[0]	H_2O 2; EtOH 5; eth 5 1.5011[20]
1808	Benzenemethanamine, 2-ethoxy- 2-Ethoxybenzylamine	$C_9H_{13}NO$ 151.21	37806-29-4 	69-74[0.2]	4-13-00-01676 1.015[25]	1.5326[20]
1809	Benzenemethanamine, α-ethyl-	$C_9H_{13}N$ 135.21	2941-20-0 	206; 100[16]	4-12-00-02582 0.9346[25]	1.5173[23]
1810	Benzenemethanamine, N-ethyl-	$C_9H_{13}N$ 135.21	14321-27-8 	194	4-12-00-02163 0.9342[17]	H_2O 2; EtOH 3; eth 3; bz 3 1.5117[20]
1811	Benzenemethanamine, N-ethyl-N-phenyl- Ethylbenzylaniline	$C_{15}H_{17}N$ 211.31	92-59-1 35	3728 288; 185[22]	4-12-00-02176 1.001[55]	H_2O 1; EtOH 3; eth 3; chl 3 1.5943[23]
1812	Benzenemethanamine, 2-methoxy-	$C_8H_{11}NO$ 137.18	6850-57-3 	229	4-13-00-01675 1.051[25]	1.5475[20]
1813	Benzenemethanamine, 4-methoxy-	$C_8H_{11}NO$ 137.18	2393-23-9 	236.5	4-13-00-01719 1.050[15]	H_2O 2; EtOH 2; eth 2 1.5462[20]
1814	Benzenemethanamine, α-methyl-, (±)-	$C_8H_{11}N$ 121.18	618-36-0 32	187	4-12-00-02424 0.9395[15]	H_2O 3; EtOH 5; eth 5; chl 3 1.5238[25]
1815	Benzenemethanamine, 2-methyl-	$C_8H_{11}N$ 121.18	89-93-0 -30	206; 81[15]	4-12-00-02518 0.9766[19]	1.5436[19]
1816	Benzenemethanamine, 3-methyl-	$C_8H_{11}N$ 121.18	100-81-2 	203.5	4-12-00-02565 0.966[25]	1.5360[20]
1817	Benzenemethanamine, 4-methyl-	$C_8H_{11}N$ 121.18	104-84-7 12.5	195	4-12-00-02574 0.952[20]	1.5340[20]
1818	Benzenemethanamine, N-methyl-	$C_8H_{11}N$ 121.18	103-67-3 	180.5	4-12-00-02161 0.9442[18]	H_2O 4
1819	Benzenemethanamine, 4-(1-methylethyl)-	$C_{10}H_{15}N$ 149.24	4395-73-7 	227	4-12-00-02836	eth 4; EtOH 4 1.5182[17]
1820	Benzenemethanamine, N-(1-methylethyl)-	$C_{10}H_{15}N$ 149.24	102-97-6 	200; 93[10]	4-12-00-02165 0.892[25]	1.5025[20]
1821	Benzenemethanamine, N-(2-methylphenyl)-	$C_{14}H_{15}N$ 197.28	5405-13-0 60	302.5	4-12-00-02177 1.0142[65]	ace 4; EtOH 4; chl 4 1.5861[65]
1822	Benzenemethanamine, α-methyl-N-(1-phenylethyl)-	$C_{16}H_{19}N$ 225.33	10024-74-5 296.5		4-12-00-02428 1.018[15]	1.573
1823	Benzenemethanamine, N-methyl-N-2-propynyl- Pargyline	$C_{11}H_{13}N$ 159.23	555-57-7 96-7[11]	6988	2-12-00-00548 0.944[25]	1.5213[20]
1824	Benzenemethanamine, N-(3-methylphenyl)- m-Toluidine, N-benzyl	$C_{14}H_{15}N$ 197.28	5405-17-4 	312	4-12-00-02178 1.0083[65]	ace 4; bz 4; eth 4 1.5845[65]
1825	Benzenemethanamine, N-(4-methylphenyl)- p-Toluidine, N-benzyl	$C_{14}H_{15}N$ 197.28	5405-15-2 19.5	319; 181[10]	4-12-00-02178 1.0064[65]	bz 4; eth 4; EtOH 4; chl 4 1.5832[65]
1826	Benzenemethanamine, α-phenyl- Benzhydrylamine	$C_{13}H_{13}N$ 183.25	91-00-9 34	1085 304; 176[23]	4-12-00-03282 1.0633[20]	H_2O 2; ace 3 1.5963
1827	Benzenemethanamine, N-phenyl- Benzylaniline	$C_{13}H_{13}N$ 183.25	103-32-2 37.5	1140 306.5	4-12-00-02172 1.0298[65]	eth 4; EtOH 4 1.6118[25]
1828	Benzenemethanamine, N-(phenylmethyl)- Dibenzylamine	$C_{14}H_{15}N$ 197.28	103-49-1 -26	2993 300 dec; 270[250]	4-12-00-02179 1.0256[22]	H_2O 1; EtOH 4; eth 4; ctc 3 1.5781[20]
1829	Benzenemethanamine, N-phenyl-N-(phenylmethyl)-	$C_{20}H_{19}N$ 273.38	91-73-6 69	226[10]	3-12-00-02225 1.0444[80]	H_2O 1; EtOH 3; eth 3; bz 3 1.6065[80]
1830	Benzenemethanamine, α-propyl-	$C_{10}H_{15}N$ 149.24	2941-19-7 	221; 108[16]	3-12-00-02716 0.9366[20]	EtOH 4
1831	Benzene, 1,1',1'',1'''-methanetetrayltetrakis- Tetraphenylmethane	$C_{25}H_{20}$ 320.43	630-76-2 282	431	4-05-00-02741	H_2O 1; EtOH 1; eth 1; bz 3
1832	Benzenemethanethiol Thiobenzyl alcohol	C_7H_8S 124.21	100-53-8 -30	9249 194.5	4-06-00-02632 1.058[20]	H_2O 1; EtOH 4; eth 4; ctc 2 1.5151[20]

No.	Name / Synonym	Mol. Form. / Mol. Wt.	CAS RN / mp/°C	Merck No. / bp/°C	Beil. Ref. / den/g cm^{-3}	Solubility / n_D
1833	Benzenemethanethiol, 4-chloro-	C_7H_7ClS 158.65	6258-66-8 19.5	113[17]	4-06-00-02770 1.202[25]	1.5893[20]
1834	Benzenemethanethiol, α-methyl-, (S)- Ethanethiol, 1-phenyl (1)	$C_8H_{10}S$ 138.23	33877-11-1	199; 83[10]	4-06-00-03062 1.022[20]	bz 4; eth 4; EtOH 4 1.5593[20]
1835	Benzenemethanimine, α-phenyl-	$C_{13}H_{11}N$ 181.24	1013-88-3	282	4-07-00-01365 1.0847[19]	eth 4 1.6191[19]
1836	Benzenemethanol Benzyl alcohol	C_7H_8O 108.14	100-51-6 -15.2	1138 205.3	4-06-00-02222 1.0419[24]	H_2O 3; EtOH 3; eth 3; ace 3 1.5396[20]
1837	Benzenemethanol, 2-amino-	C_7H_9NO 123.15	5344-90-1 83.5	273	3-13-00-01615	H_2O 3; EtOH 3; eth 3; bz 4
1838	Benzenemethanol, α-(1-aminoethyl)-, hydrochloride, (R*,S*)-(±)- Phenylpropanolamine hydrochloride	$C_9H_{14}ClNO$ 187.67	154-41-6 194	7279	2-13-00-00371	H_2O 4; EtOH 3; eth 1; bz 1
1839	Benzenemethanol, α-(aminomethyl)- Phenylethanolamine	$C_8H_{11}NO$ 137.18	7568-93-6 56.5	7258 160[17]	4-13-00-01801	H_2O 4; EtOH 3
1840	Benzenemethanol, α-(1-aminopropyl)- α-(α-Aminopropyl)benzyl alcohol	$C_{10}H_{15}NO$ 165.24	5897-76-7 79.5	483	3-13-00-01791	
1841	Benzenemethanol, 2-[bis(4-hydroxyphenyl)methyl]- Phenolphthalol	$C_{20}H_{18}O_3$ 306.36	81-92-5 201.5	7211	4-06-00-07623	
1842	Benzenemethanol, α,α-bis(1-methylethyl)-	$C_{13}H_{20}O$ 192.30	4397-05-1	229; 158[60]	4-06-00-03462 0.9755[20]	eth 4 1.5239[20]
1843	Benzenemethanol, 5-bromo-2-hydroxy- Bromosaligenin	$C_7H_7BrO_2$ 203.04	2316-64-5 113	1426	3-06-00-04541	bz 4; eth 4; EtOH 4; chl 4
1844	Benzenemethanol, α-(bromomethyl)-	C_8H_9BrO 201.06	2425-28-7	109-10[2]	4-06-00-03053 1.4994[20]	H_2O 1 1.5800[17]
1845	Benzenemethanol, 2-bromo-α-methyl-	C_8H_9BrO 201.06	5411-56-3	109-10[2]	4-06-00-03053 1.4994[20]	1.5800[17]
1846	Benzenemethanol, α-butyl- 1-Phenyl-1-pentanol	$C_{11}H_{16}O$ 164.25	583-03-9	3921 141[25]; 102[3]	2-06-00-00503 0.9655[20]	ace 4; eth 4; EtOH 4 1.4086[25]
1847	Benzenemethanol, α-butyl-α-methyl- 2-Hexanol, 2-phenyl-	$C_{12}H_{18}O$ 178.27	4396-98-9	3-06-00-01995 129-30[4]	0.9522[20]	1.5091[20]
1848	Benzenemethanol, 2-chloro-	C_7H_7ClO 142.58	17849-38-6 73	230	4-06-00-02589	H_2O 2; EtOH 4; eth 4; lig 4
1849	Benzenemethanol, 4-chloro-	C_7H_7ClO 142.58	873-76-7 75	235	4-06-00-02593	bz 4; eth 4; EtOH 4
1850	Benzenemethanol, 4-chloro-α-(4-chlorophenyl)-α-methyl- 1,1-Bis(4-chlorophenyl)ethanol	$C_{14}H_{12}Cl_2O$ 267.15	80-06-8 70	2086	4-06-00-04718	H_2O 1; EtOH 1; eth 3; bz 3
1851	Benzenemethanol, 4-chloro-α-(4-chlorophenyl)-α-(trichloromethyl)- 1,1-Bis(p-chlorophenyl)-2,2,2-trichloroethanol	$C_{14}H_9Cl_5O$ 370.49	115-32-2 77.5	3075	4-06-00-04722	H_2O 1; os 1
1852	Benzenemethanol, α-(chloromethyl)-	C_8H_9ClO 156.61	1674-30-2	128[17]	4-06-00-03045 1.1926[20]	EtOH 3; eth 4 1.5523[20]
1853	Benzenemethanol, 2,4-dichloro- 2,4-Dichlorobenzyl alcohol	$C_7H_6Cl_2O$ 177.03	1777-82-8 59.5	3048 150[25]	4-06-00-02597	chl 3
1854	Benzenemethanol, 2,4-dichloro-α-methyl-	$C_8H_8Cl_2O$ 191.06	1475-13-4	125-6[7]	4-06-00-03048 1.293[25]	1.5605[20]
1855	Benzenemethanol, α,α-diethyl-	$C_{11}H_{16}O$ 164.25	1565-71-5 <-17	223.5	4-06-00-03381 0.9831[20]	EtOH 4 1.5165[20]
1856	Benzenemethanol, 3,4-dimethoxy-	$C_9H_{12}O_3$ 168.19	93-03-8	298; 172[12]	4-06-00-07381 1.178[17]	H_2O 3; EtOH 3 1.555[17]
1857	Benzenemethanol, α,α-dimethyl- α-Cumyl alcohol	$C_9H_{12}O$ 136.19	617-94-7 36	202	4-06-00-03219 0.9735[20]	H_2O 1; EtOH 3; eth 3; bz 3 1.5325[20]
1858	Benzenemethanol, α,3-dimethyl- 1-(3-Methylphenyl)ethyl alcohol	$C_9H_{12}O$ 136.19	7287-81-2	112[12]	3-06-00-01823 0.9974[15]	eth 4; EtOH 4 1.5240[20]
1859	Benzenemethanol, α,4-dimethyl- 1-(4-Methylphenyl)ethanol	$C_9H_{12}O$ 136.19	536-50-5	3226 219	4-06-00-03242 0.9668[25]	H_2O 1; EtOH 4; eth 4 1.5246[20]
1860	Benzenemethanol, 2,4-dimethyl-	$C_9H_{12}O$ 136.19	16308-92-2 22	232; 150[44]	4-06-00-03250 1.0310[20]	1.5339[20]
1861	Benzenemethanol, 3,5-dimethyl-	$C_9H_{12}O$ 136.19	27129-87-9	219.5	0-06-00-00521 0.927[25]	1.5312[20]
1862	Benzenemethanol, 4-(dimethylamino)-	$C_9H_{13}NO$ 151.21	1703-46-4 69	123[1]	4-13-00-01771 1.059[14]	1.5727[14]
1863	Benzenemethanol, α-[1-(dimethylamino)ethyl]-, [R-(R*,S*)]- (1R,2S)-N-Methylephedrine	$C_{11}H_{17}NO$ 179.26	552-79-4 87.5	5987		H_2O 1; EtOH 3; eth 3; MeOH 3
1864	Benzenemethanol, α-(1,1-dimethylethyl)-	$C_{11}H_{16}O$ 164.25	3835-64-1 45	111[15]	1-06-00-00270	eth 4; EtOH 4
1865	Benzenemethanol, 4-(1,1-dimethylethyl)-	$C_{11}H_{16}O$ 164.25	877-65-6	140[20]	4-06-00-03402 0.928[25]	1.5179[20]
1866	Benzenemethanol, α,α-diphenyl- Triphenylmethanol	$C_{19}H_{16}O$ 260.34	76-84-6 164.2	9653 380	4-06-00-05014 1.199[0]	H_2O 1; EtOH 4; eth 4; ace 3
1867	Benzenemethanol, α,α-dipropyl- 4-Heptanol, 4-phenyl-	$C_{13}H_{20}O$ 192.30	4436-96-8	134[20]	2-06-00-00513 0.9470[15]	1.516[10]
1868	Benzenemethanol, α-ethenyl- 1-Phenylallyl alcohol	$C_9H_{10}O$ 134.18	4393-06-0		2-06-00-00529 1.0249[21]	H_2O 2; EtOH 3; eth 3; bz 3 1.5406[20]
1869	Benzenemethanol, α-ethenyl-α-methyl- 3-Buten-2-ol, 2-phenyl-	$C_{10}H_{12}O$ 148.20	6051-52-1	92-5[7]	4-06-00-03839 0.9965[21]	1.5277[21]

No.	Name / Synonym	Mol. Form. / Mol. Wt.	CAS RN / mp/°C	Merck No. / bp/°C	Beil. Ref. / den/g cm^{-3}	Solubility / n_D
1870	Benzenemethanol, α-ethyl- / α-Ethylbenzyl alcohol	$C_9H_{12}O$ / 136.19	93-54-9 /	3727 / 219	4-06-00-03183 / 0.9915[25]	bz 4; eth 4; EtOH 4; MeOH 4 / 1.5169[23]
1871	Benzenemethanol, α-[1-(ethylamino)ethyl]-, (R*,S*)- / N-Ethylnorephedrine	$C_{11}H_{17}NO$ / 179.26	37025-57-3 / 51.5	143[18]	4-13-00-01887 /	bz 4
1872	Benzenemethanol, α-ethyl-3-hydroxy- / 1-(m-Hydroxyphenyl)-1-propanol	$C_9H_{12}O_2$ / 152.19	55789-02-1 / 107	177[13]	3-06-00-04623 /	eth 4; EtOH 4
1873	Benzenemethanol, α-ethyl-4-methoxy-	$C_{10}H_{14}O_2$ / 166.22	5349-60-0 /	143[20]	3-06-00-04624 /	ctc 3 / 1.5277[20]
1874	Benzenemethanol, α-ethyl-α-methyl-, (R)-	$C_{10}H_{14}O$ / 150.22	1006-06-0 / -13	211; 113[23]	/ 0.984[25]	H_2O 1; EtOH 3; eth 3 / 1.5185[20]
1875	Benzenemethanol, α-ethyl-α-propyl- / 3-Hexanol, 3-phenyl	$C_{12}H_{18}O$ / 178.27	20731-93-5 /	134[27]	3-06-00-02000 / 0.964[20]	/ 1.5100[20]
1876	Benzenemethanol, α-ethynyl- / 1-Phenylpropargyl alcohol	C_9H_8O / 132.16	4187-87-5 / 22	114[12]	4-06-00-04066 / 1.0655[20]	/ 1.5508[20]
1877	Benzenemethanol, α-ethynyl-, carbamate / Carfimate	$C_{10}H_9NO_2$ / 175.19	3567-38-2 / 86.5	1845	4-06-00-04067 /	
1878	Benzenemethanol, α-ethynyl-α-methyl-	$C_{10}H_{10}O$ / 146.19	127-66-2 / 52.3	217.5; 102[12]	4-06-00-04073 / 1.0314[20]	
1879	Benzenemethanol, 3-fluoro-	C_7H_7FO / 126.13	456-47-3 /	201; 104[22]	4-06-00-02589 / 1.164[25]	1.5095[20]
1880	Benzenemethanol, 4-fluoro-	C_7H_7FO / 126.13	459-56-3 / 23	210	4-06-00-02589 /	1.5080[20]
1881	Benzenemethanol, α-hexyl- / 1-Phenyl-1-heptanol	$C_{13}H_{20}O$ / 192.30	614-54-0 /	275	4-06-00-03454 / 0.946[25]	1.5024[20]
1882	Benzenemethanol, 2-hydroxy- / Salicyl alcohol	$C_7H_8O_2$ / 124.14	90-01-7 / 87	8294 / sub	4-06-00-05896 / 1.1613[25]	H_2O 3; EtOH 3; eth 3; bz 3
1883	Benzenemethanol, 3-hydroxy-	$C_7H_8O_2$ / 124.14	620-24-6 / 73	300 dec	4-06-00-05907 / 1.161[20]	H_2O 4; EtOH 4; eth 4; chl 2
1884	Benzenemethanol, 4-hydroxy- / p-Hydroxybenzyl alcohol	$C_7H_8O_2$ / 124.14	623-05-2 / 124.5	252	4-06-00-05909 /	H_2O 4; EtOH 4; eth 3; bz 4
1885	Benzenemethanol, 2-hydroxy-3-methoxy-	$C_8H_{10}O_3$ / 154.17	4383-05-5 / 61.5	162[12]	4-06-00-07378 /	chl 2
1886	Benzenemethanol, 4-hydroxy-3-methoxy-	$C_8H_{10}O_3$ / 154.17	498-00-0 / 115	dec	4-06-00-07381 /	H_2O 3; EtOH 3; eth 3; bz 3
1887	Benzenemethanol, 4-hydroxy-α-[(methylamino)methyl]- / Synephrine	$C_9H_{13}NO_2$ / 167.21	94-07-5 / 184.5	8994	2-13-00-00491 /	
1888	Benzenemethanol, 2-iodo-	C_7H_7IO / 234.04	5159-41-1 / 92	147-9[32]	4-06-00-02605 /	1.6349[20]
1889	Benzenemethanol, 2-methoxy-	$C_8H_{10}O_2$ / 138.17	612-16-8 /	249	4-06-00-05896 / 1.0386[25]	H_2O 1; EtOH 3; eth 5 / 1.5455[20]
1890	Benzenemethanol, 3-methoxy-	$C_8H_{10}O_2$ / 138.17	6971-51-3 /	252	4-06-00-05907 / 1.112[25]	1.5440[20]
1891	Benzenemethanol, 4-methoxy- / Anise alcohol	$C_8H_{10}O_2$ / 138.17	105-13-5 / 25	695 / 259.1	4-06-00-05909 / 1.109[26]	H_2O 4; EtOH 4; eth 4; ctc 3 / 1.5420[25]
1892	Benzenemethanol, 4-methoxy-, acetate	$C_{10}H_{12}O_3$ / 180.20	104-21-2 / 84	270; 150[23]	4-06-00-05914 / 1.105[25]	ctc 3
1893	Benzenemethanol, 2-methoxy-α-methyl-	$C_9H_{12}O_2$ / 152.19	13513-82-1 / -85.1	125	3-06-00-04563 / 0.9647[20]	ace 4; bz 4; eth 4; EtOH 4 / 1.4024[20]
1894	Benzenemethanol, 3-methoxy-α-methyl-	$C_9H_{12}O_2$ / 152.19	23308-82-9 /	133[15]	4-06-00-05929 / 1.0781[19]	eth 4; EtOH 4 / 1.5325[20]
1895	Benzenemethanol, 4-methoxy-α-methyl-	$C_9H_{12}O_2$ / 152.19	3319-15-1 /	310 dec; 140[17]	4-06-00-05930 / 1.0794[20]	ctc 3 / 1.5310[25]
1896	Benzenemethanol, α-methyl- / α-Methylbenzyl alcohol	$C_8H_{10}O$ / 122.17	98-85-1 / 20	205	2-06-00-00444 / 1.013[25]	H_2O 1; EtOH 4; eth 4 / 1.5265[20]
1897	Benzenemethanol, 2-methyl- / o-Methylbenzyl alcohol	$C_8H_{10}O$ / 122.17	89-95-2 / 38	224; 118[20]	4-06-00-03109 / 1.023[40]	eth 4; EtOH 4; chl 4
1898	Benzenemethanol, 3-methyl- / m-Methylbenzyl alcohol	$C_8H_{10}O$ / 122.17	587-03-1 / <-20	215.5	4-06-00-03162 / 0.9157[17]	H_2O 2; EtOH 4; eth 4; chl 3
1899	Benzenemethanol, 4-methyl- / p-Methylbenzyl alcohol	$C_8H_{10}O$ / 122.17	589-18-4 / 61.5	217	4-06-00-03171 / 0.978[22]	eth 4; EtOH 4
1900	Benzenemethanol, α-[1-(methylamino)ethyl]-, (R*,S*)-(±)- / (±)-Ephedrine	$C_{10}H_{15}NO$ / 165.24	90-81-3 / 76.5	3561 / 135[12]	4-13-00-01881 / 1.1220[20]	H_2O 3; EtOH 3; eth 3; bz 3
1901	Benzenemethanol, α-(1-methylethyl)- / 1-Phenyl-2-methylpropyl alcohol	$C_{10}H_{14}O$ / 150.22	611-69-8 /	223	4-06-00-03289 / 0.9869[14]	H_2O 1; EtOH 3; ace 3 / 1.5193[14]
1902	Benzenemethanol, 4-(1-methylethyl)- / Cumic alcohol	$C_{10}H_{14}O$ / 150.22	536-60-7 / 28	2621 / 249	4-06-00-03348 / 0.9818[20]	H_2O 1; EtOH 5; eth 5; bz 4 / 1.5210[20]
1903	Benzenemethanol, α-[1-[methyl(3-phenyl-2-propenyl)amino]ethyl]- / Cinnamedrine	$C_{19}H_{23}NO$ / 281.40	90-86-8 / 75	2299	3-13-00-01734 /	
1904	Benzenemethanol, α-methyl-α-propyl-	$C_{11}H_{16}O$ / 164.25	4383-18-0 /	216	2-06-00-00505 / 0.9723[22]	EtOH 4
1905	Benzenemethanol, α-(2-methylpropyl)-	$C_{11}H_{16}O$ / 164.25	1565-86-2 /	236; 112[9]	4-06-00-03380 / 0.9537[19]	ace 4; bz 4; eth 4; EtOH 4 / 1.5080[18]
1906	Benzenemethanol, 2-nitro-	$C_7H_7NO_3$ / 153.14	612-25-9 / 74	270; 168[20]	4-06-00-02608 /	H_2O 2; EtOH 3; eth 3

No.	Name / Synonym	Mol. Form. / Mol. Wt.	CAS RN / mp/°C	Merck No. / bp/°C	Beil. Ref. / den/g cm^{-3}	Solubility / n_D
1907	Benzenemethanol, 3-nitro-	$C_7H_7NO_3$ 153.14	619-25-0 30.5	175-80[3]	4-06-00-02609 1.296[19]	H_2O 3; EtOH 3; eth 3; chl 2
1908	Benzenemethanol, 4-nitro- p-Nitrobenzyl alcohol	$C_7H_7NO_3$ 153.14	619-73-8 96.5	255 dec; 185[12]	4-06-00-02611	H_2O 2; EtOH 3; eth 3; ace 2
1909	Benzenemethanol, α-pentyl-	$C_{12}H_{18}O$ 178.27	4471-05-0	170[50]; 128[5]	4-06-00-03417 0.9497[25]	EtOH 3; eth 3 1.5105[20]
1910	Benzenemethanol, α-phenyl- Benzohydrol	$C_{13}H_{12}O$ 184.24	91-01-0 69	1100 298; 180[20]	4-06-00-04648	H_2O 2; EtOH 4; eth 4; ctc 4
1911	Benzenemethanol, α-2-propenyl-	$C_{10}H_{12}O$ 148.20	936-58-3	228.5	4-06-00-03838 1.004[18]	1.5289[21]
1912	Benzenemethanol, α-propyl-, (R)-	$C_{10}H_{14}O$ 150.22	22144-60-1 16	232	3-06-00-01845 0.9740[20]	eth 4; EtOH 4 1.5139[20]
1913	Benzenemethanol, α-[(2-pyridinylamino)methyl]- Phenyramidol	$C_{13}H_{14}N_2O$ 214.27	553-69-5 83.5	7292	5-22-08-00301	
1914	Benzenemethanol, 3,4,5-trimethoxy- 3,4,5-Trimethoxybenzyl alcohol	$C_{10}H_{14}O_4$ 198.22	3840-31-1	228[25]	4-06-00-07695 1.232[25]	1.5439[20]
1915	Benzene, methoxy- Anisole	C_7H_8O 108.14	100-66-3 -37.5	699 153.7	4-06-00-00548 0.9940[20]	H_2O 1; EtOH 3; eth 3; ace 4 1.5174[20]
1916	Benzene, 1-methoxy-2,3-dimethyl-	$C_9H_{12}O$ 136.19	2944-49-2 29	199	4-06-00-03097 0.9596[40]	H_2O 1; EtOH 4; eth 4; ace 3 1.5120[40]
1917	Benzene, 1-methoxy-2,4-dimethyl-	$C_9H_{12}O$ 136.19	6738-23-4	192	4-06-00-03127 0.9740[16]	H_2O 1; EtOH 3; eth 3; bz 3 1.5190[16]
1918	Benzene, 1-methoxy-3,5-dimethyl-	$C_9H_{12}O$ 136.19	874-63-5	194; 89[15]	4-06-00-03142 0.9627[15]	H_2O 1; EtOH 3; eth 3; bz 3 1.5110[20]
1919	Benzene, 2-methoxy-1,3-dimethyl-	$C_9H_{12}O$ 136.19	1004-66-6	182.5	4-06-00-03114 0.9619[14]	H_2O 1; EtOH 3; eth 3; bz 3 1.5053[14]
1920	Benzene, 2-methoxy-1,4-dimethyl-	$C_9H_{12}O$ 136.19	1706-11-2	194	4-06-00-03165 0.9693[13]	H_2O 1; EtOH 3; eth 3; bz 3 1.5182[15]
1921	Benzene, 4-methoxy-1,2-dimethyl-	$C_9H_{12}O$ 136.19	4685-47-6	202	4-06-00-03100 0.9744[14]	H_2O 1; EtOH 3; eth 3; bz 3 1.5198[14]
1922	Benzene, 1-methoxy-2,3-dinitro- Anisole, 2,3-dinitro-	$C_7H_8N_2O_5$ 198.14	16315-07-4 119		4-06-00-01369 1.23[137]	EtOH 4; lig 4
1923	Benzene, 1-methoxy-2,4-dinitro-	$C_7H_8N_2O_5$ 198.14	119-27-7 94.5	206[12]	4-06-00-01372 1.3364[131]	H_2O 2; EtOH 3; eth 3; ace 3 1.546[15]
1924	Benzene, 1-methoxy-3,5-dinitro- 3,5-Dinitroanisole	$C_7H_8N_2O_5$ 198.14	5327-44-6 105.3		4-06-00-01385 1.558[12]	ace 4; bz 4; MeOH 4
1925	Benzene, 2-methoxy-1,3-dinitro-	$C_7H_8N_2O_5$ 198.14	3535-67-9 118		3-06-00-00868 1.3000[128]	EtOH 4
1926	Benzene, 2-methoxy-1,4-dinitro- Anisole, 2,5-dinitro-	$C_7H_8N_2O_5$ 198.14	3962-77-4 97	136-8[2]	4-06-00-01383 1.476[18]	H_2O 2; EtOH 3; ace 3; bz 3
1927	Benzene, 4-methoxy-1,2-dinitro-	$C_7H_8N_2O_5$ 198.14	4280-28-8 71	168[3]	4-06-00-01384 1.3332[110]	EtOH 4
1928	Benzene, (2-methoxyethenyl)-	$C_9H_{10}O$ 134.18	4747-15-3	211.5	2-06-00-00521 0.9894[23]	1.5620[24]
1929	Benzene, 1-methoxy-2-(3-methoxyphenoxy)-	$C_{14}H_{14}O_3$ 230.26	1655-71-6 54.8	327.5	4-06-00-05670	bz 4; eth 4; EtOH 4
1930	Benzene, (methoxymethyl)- Benzyl methyl ether	$C_8H_{10}O$ 122.17	538-86-3 -52.6	1154 170	4-06-00-02229 0.9634[20]	H_2O 1; EtOH 4; eth 4; bz 3 1.5008[20]
1931	Benzene, 1-methoxy-2-methyl-	$C_8H_{10}O$ 122.17	578-58-5 -34.1	171	4-06-00-01943 0.985[25]	H_2O 1; EtOH 3; eth 3; ace 3 1.5161[20]
1932	Benzene, 1-methoxy-3-methyl-	$C_8H_{10}O$ 122.17	100-84-5 -47	175.5	4-06-00-02039 0.969[25]	H_2O 1; EtOH 3; eth 3; ace 3 1.5130[20]
1933	Benzene, 1-methoxy-4-methyl-	$C_8H_{10}O$ 122.17	104-93-8 -32	175.5	4-06-00-02098 0.969[25]	H_2O 1; EtOH 3; eth 3; chl 3 1.5112[20]
1934	Benzene, 4-methoxy-1-methyl-2-nitro-	$C_8H_9NO_3$ 167.16	17484-36-5 17	266.5	3-06-00-01384 1.207[25]	1.5525[20]
1935	Benzene, 1-methoxy-2-nitro- 2-Nitroanisole	$C_7H_7NO_3$ 153.14	91-23-6 10.5	277; 144[4]	4-06-00-01249 1.2540[20]	H_2O 1; EtOH 5; eth 5; ctc 3 1.5161[20]
1936	Benzene, 1-methoxy-3-nitro-	$C_7H_7NO_3$ 153.14	555-03-3 38.5	258	4-06-00-01270 1.373[18]	H_2O 1; EtOH 3; eth 4
1937	Benzene, 1-methoxy-4-nitro- Anisole, 4-nitro	$C_7H_7NO_3$ 153.14	100-17-4 54	274	4-06-00-01282 1.2192[60]	H_2O 1; EtOH 4; eth 4; ctc 3 1.5070[60]
1938	Benzene, 1-methoxy-2-phenoxy-	$C_{13}H_{12}O_2$ 200.24	1695-04-1 79	288; 91[7]	4-06-00-05571	bz 4; eth 4; EtOH 4
1939	Benzene, 1-methoxy-4-(2-phenylethenyl)-, (E)-	$C_{15}H_{14}O$ 210.28	1694-19-5 136.5	142.5[15]	4-06-00-04855	H_2O 1; EtOH 4; eth 4; ace 4
1940	Benzene, 1-methoxy-4-(1-phenylethyl)-	$C_{15}H_{18}O$ 212.29	2605-18-7	180-2[19]	4-06-00-04711 1.0473[20]	bz 4; eth 4; HOAc 4 1.5725[20]
1941	Benzene, 1-methoxy-2-(1-propenyl)-	$C_{10}H_{12}O$ 148.20	10577-44-3	224; 104[13]	4-06-00-03793 0.9962[15]	1.560[15]
1942	Benzene, 1-methoxy-4-(1-propenyl)- Anethole	$C_{10}H_{12}O$ 148.20	104-46-1 21.3	675 234; 115[12]	2-06-00-00523 0.9882[20]	ace 4; bz 4; eth 4; EtOH 4 1.5615[20]
1943	Benzene, 1-methoxy-4-(2-propenyl)- Estragole	$C_{10}H_{12}O$ 148.20	140-67-0	3657 215.5	4-06-00-03817 0.965[25]	EtOH 4; chl 4 1.5195[20]
1944	Benzene, 1-methoxy-4-(1-propenyl)-, (E)-	$C_{10}H_{12}O$ 148.20	4180-23-8 21.35	234; 115[12]	4-06-00-03796 0.9882[20]	H_2O 2; EtOH 5; eth 5; ace 3 1.5615[20]
1945	Benzene, 1-methoxy-4-propyl-	$C_{10}H_{14}O$ 150.22	104-45-0	211.5	4-06-00-03181 0.9472[20]	H_2O 2; EtOH 3; eth 4; ace 3 1.5045[20]

No.	Name / Synonym	Mol. Form. / Mol. Wt.	CAS RN / mp/°C	Merck No. / bp/°C	Beil. Ref. / den/g cm^{-3}	Solubility / n_D
1946	Benzene, 2-methoxy-1,3,5-trinitro- / Methyl picrate	$C_7H_5N_3O_7$ / 243.13	606-35-9 / 69		4-06-00-01456 / 1.4947[80]	H_2O 1; EtOH 4; eth 3; bz 4
1947	Benzene, methyl- / Toluene	C_7H_8 / 92.14	108-88-3 / -94.9	9455 / 110.6	4-05-00-00766 / 0.8669[20]	H_2O 1; EtOH 5; eth 5; ace 3 / 1.4961[20]
1948	Benzene, 1-methyl-2,4-bis(1-methylethyl)- / Toluene, 2,4-diisopropyl-	$C_{13}H_{20}$ / 176.30	1460-98-6 /	82[7]	4-05-00-01152 / 0.8636[20]	ctc 3; CS_2 3 / 1.4912[20]
1949	Benzene, 1-methyl-3,5-bis(1-methylethyl)-	$C_{13}H_{20}$ / 176.30	3055-14-9 / -60.8	216.5	4-05-00-01152 / 0.8668[20]	1.4950[20]
1950	Benzene, (1-methyl-1-butenyl)- / 2-Phenyl-2-pentene	$C_{11}H_{14}$ / 146.23	53172-84-2 /	202	4-05-00-01402 / 0.8950[26]	1.5196[26]
1951	Benzene, [(3-methylbutoxy)methyl]-	$C_{12}H_{18}O$ / 178.27	122-73-6 /	237; 118[19]	3-06-00-01457 / 0.9098[20]	eth 4; EtOH 4 / 1.4792[20]
1952	Benzene, (1-methylbutyl)-	$C_{11}H_{16}$ / 148.25	2719-52-0 /	199	4-05-00-01088 / 0.8594[21]	H_2O 1; EtOH 3; eth 3 / 1.4875[25]
1953	Benzene, (2-methylbutyl)-	$C_{11}H_{16}$ / 148.25	3968-85-2 /	195.4	2-05-00-00332 / 0.8584[20]	1.4873[20]
1954	Benzene, (3-methylbutyl)-	$C_{11}H_{16}$ / 148.25	2049-94-7 /	195	4-05-00-01089 / 0.856[20]	H_2O 1; EtOH 3; eth 3; bz 4 / 1.4867[10]
1955	Benzene, (1-methylcyclopropyl)-	$C_{10}H_{12}$ / 132.21	2214-14-4 /	91[50]	4-05-00-01388 /	ace 3; bz 3; chl 3 / 1.5160[20]
1956	Benzene, 1-methyl-2,4-dinitro- / 2,4-Dinitrotoluene	$C_7H_6N_2O_4$ / 182.14	121-14-2 / 71	300 dec	4-05-00-00865 / 1.3208[71]	H_2O 1; EtOH 3; eth 3; ace 4 / 1.442
1957	Benzene, 1-methyl-3,5-dinitro-	$C_7H_6N_2O_4$ / 182.14	618-85-9 / 93	sub	4-05-00-00867 / 1.2772[111]	H_2O 2; EtOH 3; eth 3; bz 3
1958	Benzene, 2-methyl-1,3-dinitro- / 2,6-Dinitrotoluene	$C_7H_6N_2O_4$ / 182.14	606-20-2 / 66		4-05-00-00866 / 1.2833[111]	EtOH 3; chl 3 / 1.479
1959	Benzene, 2-methyl-1,4-dinitro-	$C_7H_6N_2O_4$ / 182.14	619-15-8 / 52.5		4-05-00-00866 / 1.282[111]	EtOH 3; bz 3; CS_2 4
1960	Benzene, 4-methyl-1,2-dinitro- / 3,4-Dinitrotoluene	$C_7H_6N_2O_4$ / 182.14	610-39-9 / 58.3		4-05-00-00866 / 1.2594[111]	H_2O 1; EtOH 3; chl 2; CS_2 3
1961	Benzene, 1,1'-methylenebis- / Diphenylmethane	$C_{13}H_{12}$ / 168.24	101-81-5 / 25.2	3329 / 265.0	4-05-00-01841 / 1.001[26]	H_2O 1; EtOH 3; eth 3; chl 3 / 1.5753[20]
1962	Benzene, 1,1'-methylenebis[2-bromo- / Diphenylmethane, 2,2'-dibromo-	$C_{13}H_{10}Br_2$ / 326.03	61592-89-0 /	234-5[40]	3-05-00-01795 / 1.6197[20]	1.6300[20]
1963	Benzene, 1,1'-methylenebis[4-chloro-	$C_{13}H_{10}Cl_2$ / 237.13	101-76-8 / 55.5	186-90[18]	4-05-00-01848 / 1.365[17]	EtOH 3
1964	Benzene, 1,1'-methylenebis[4-methyl-	$C_{15}H_{16}$ / 196.29	4957-14-6 / 28	286	4-05-00-01929 / 0.9800[20]	eth 4; EtOH 4
1965	Benzene, 1,1'-[methylenebis(oxy)]bis[4-chloro- / Bis(p-chlorophenoxy)methane	$C_{13}H_{10}Cl_2O_2$ / 269.13	555-89-5 / 70.5	1258 / 189-94[6]	4-06-00-00833	ace 4; bz 4
1966	Benzene, 1,1'-(1-methylene-1,2-ethanediyl)bis-	$C_{15}H_{14}$ / 194.28	948-97-0 / 48	289	4-05-00-02191 / 1.1014[20]	1.5903[20]
1967	Benzene, (1-methylene-2-propenyl)-	$C_{10}H_{10}$ / 130.19	2288-18-8 /	60-1[17]	4-05-00-01539 / 0.9266[20]	H_2O 1; eth 3; bz 3; chl 3 / 1.5489[20]
1968	Benzene, (1-methylenepropyl)-	$C_{10}H_{12}$ / 132.21	2039-93-2 /	182	4-05-00-01380 / 0.887[25]	1.5288[20]
1969	Benzene, 1,1'-(1-methyl-1,2-ethanediyl)bis-	$C_{15}H_{16}$ / 196.29	5814-85-7 / 52	280.5	4-05-00-01915 / 0.9807[20]	1.5700[20]
1970	Benzene, 1,1'-[(1-methyl-1,2-ethanediyl)bis(oxy)]bis- / Propane, 1,2-diphenoxy	$C_{15}H_{16}O_2$ / 228.29	69813-63-4 / 32	175-8[12]	2-06-00-00151 / 1.0748[33]	ace 4; bz 4; eth 4; EtOH 4 / 1.5542[33]
1971	Benzene, 1,1'-(1-methyl-1,2-ethenediyl)bis-	$C_{15}H_{14}$ / 194.28	779-51-1 / 82	285.5	4-05-00-02189 / 0.9857[17]	EtOH 3; eth 3; bz 3; lig 2 / 1.5635[17]
1972	Benzene, (1-methylethenyl)- / α-Methylstyrene	C_9H_{10} / 118.18	98-83-9 / -23.2	165.4	4-05-00-01364 / 0.9106[20]	H_2O 1; EtOH 3; eth 3; ace 5 / 1.5386[20]
1973	Benzene, 1-(1-methylethenyl)-4-(1-methylethyl)-	$C_{12}H_{16}$ / 160.26	2388-14-9 / -30.6	220.8	4-05-00-01422 / 0.8936[20]	ace 4; bz 4; eth 4; EtOH 4 / 1.5238[20]
1974	Benzene, (1-methylethoxy)-	$C_9H_{12}O$ / 136.19	2741-16-4 / -33	176.8	4-06-00-00557 / 0.9408[25]	H_2O 3; EtOH 3; ace 3; bz 3 / 1.4975[20]
1975	Benzene, (1-methylethyl)- / Cumene	C_9H_{12} / 120.19	98-82-8 / -96.0	2619 / 152.4	4-05-00-00985 / 0.8618[20]	H_2O 1; EtOH 5; eth 5; ace 5 / 1.4915[20]
1976	Benzene, 1,1'-(1-methylethylidene)bis-	$C_{15}H_{16}$ / 196.29	778-22-3 / 29	282.5	4-05-00-01925 / 0.9980[20]	
1977	Benzene, 1-(1-methylethyl)-2-nitro-	$C_9H_{11}NO_2$ / 165.19	6526-72-3 /	103[9]	4-05-00-00997 / 1.101[12]	H_2O 1; ace 3; bz 3; HOAc 3 / 1.5259[20]
1978	Benzene, 1-(1-methylethyl)-4-nitro-	$C_9H_{11}NO_2$ / 165.19	1817-47-6 /	122[9]	4-05-00-00997 / 1.0830[20]	H_2O 1; ace 3; bz 3; lig 3 / 1.5367[20]
1979	Benzene, 1-(1-methylethyl)-4-(phenylmethyl)-	$C_{16}H_{18}$ / 210.32	886-58-0 / 13.4	310	4-05-00-01951 / 1.007[18]	
1980	Benzene, [(1-methylethyl)thio]-	$C_9H_{12}S$ / 152.26	3019-20-3 /	208	4-06-00-01471 / 0.9852[20]	1.5464[20]
1981	Benzene, (1-methylheptyl)- / 2-Phenyloctane	$C_{14}H_{22}$ / 190.33	777-22-0 /	123-5[20]	4-05-00-01157 / 0.8611[20]	1.4837[20]
1982	Benzene, 1,1',1''-methylidynetris- / Triphenylmethane	$C_{19}H_{16}$ / 244.34	519-73-3 / 94	9655 / 359; 200[10]	4-05-00-02495 / 1.014[99]	H_2O 1; EtOH 2; eth 4; bz 3 / 1.5839[99]
1983	Benzene, 1,1',1''-[methylidynetris(oxy)]tris-	$C_{19}H_{16}O_3$ / 292.33	16737-44-3 / 76.5	269[50] dec	4-06-00-00611	eth 4; EtOH 4
1984	Benzene, 1-methyl-2-(1-methylethenyl)-	$C_{10}H_{12}$ / 132.21	7399-49-7 /	175	3-05-00-01214 / 0.9180[15]	1.5112[30]

No.	Name Synonym	Mol. Form. Mol. Wt.	CAS RN mp/°C	Merck No. bp/°C	Beil. Ref. den/g cm^{-3}	Solubility n_D
1985	Benzene, 1-methyl-3-(1-methylethenyl)-	C$_{10}$H$_{12}$ 132.21	1124-20-5	185	4-05-00-01383 0.9010[25]	1.5335[20]
1986	Benzene, 1-methyl-4-(1-methylethenyl)-	C$_{10}$H$_{12}$ 132.21	1195-32-0 -20	185.3	4-05-00-01383 0.8936[23]	1.5283[23]
1987	Benzene, 1-methyl-3-(1-methylethoxy)-	C$_{10}$H$_{14}$O 150.22	19177-04-9	195	4-06-00-02040 0.931[20]	1.4959[20]
1988	Benzene, 1-methyl-2-(1-methylethyl)- o-Cymene	C$_{10}$H$_{14}$ 134.22	527-84-4 -71.5	2770 178.1	4-05-00-01057 0.8766[20]	H$_2$O 1; EtOH 5; eth 5; ace 5 1.5006[20]
1989	Benzene, 1-methyl-3-(1-methylethyl)- m-Cymene	C$_{10}$H$_{14}$ 134.22	535-77-3 -63.7	2770 175.1	4-05-00-01058 0.8610[20]	H$_2$O 1; EtOH 5; eth 5; ace 5 1.4930[20]
1990	Benzene, 1-methyl-4-(1-methylethyl)- p-Cymene	C$_{10}$H$_{14}$ 134.22	99-87-6 -68.9	2770 177.1	4-05-00-01060 0.8573[20]	H$_2$O 1; EtOH 5; eth 5; ace 5 1.4909[20]
1991	Benzene, 1-methyl-4-(1-methylethyl)-2-nitro-	C$_{10}$H$_{13}$NO$_2$ 179.22	943-15-7	126[10]	4-05-00-01064 1.0744[20]	eth 4; EtOH 4 1.5301[20]
1992	Benzene, [2-methyl-1-(1-methylethyl)propyl]-	C$_{13}$H$_{20}$ 176.30	21777-84-4 -24.5	220.7	4-05-00-01146 0.8821[10]	1.512
1993	Benzene, 1-methyl-2-[(4-methylphenyl)thio]- 2,4'-Ditolyl sulfide	C$_{14}$H$_{14}$S 214.33	4279-70-3	173[11]	4-06-00-02172 1.0774[15]	eth 4; EtOH 4
1994	Benzene, 1-methyl-2-(1-methylpropyl)-	C$_{11}$H$_{16}$ 148.25	1595-06-8	196	4-05-00-01094 0.873[20]	ace 4; bz 4; eth 4; EtOH 4 1.497[20]
1995	Benzene, 1-methyl-2-(2-methylpropyl)-	C$_{11}$H$_{16}$ 148.25	36301-29-8 -73.3	196	4-05-00-01095 0.8649[20]	ace 4; bz 4; eth 4; EtOH 4 1.4935[20]
1996	Benzene, 1-methyl-3-(1-methylpropyl)-	C$_{11}$H$_{16}$ 148.25	1772-10-7	194	4-05-00-01095 0.858[20]	ace 4; bz 4; eth 4; EtOH 4 1.490[20]
1997	Benzene, 1-methyl-3-(2-methylpropyl)-	C$_{11}$H$_{16}$ 148.25	5160-99-6	194	3-05-00-01001 0.8536[20]	ace 4; bz 4; eth 4; EtOH 4 1.4888[20]
1998	Benzene, 1-methyl-4-(1-methylpropyl)-	C$_{11}$H$_{16}$ 148.25	1595-16-0	197	4-05-00-01095 0.8640[19]	bz 4; eth 4; chl 4 1.493[20]
1999	Benzene, 1-methyl-4-(2-methylpropyl)-	C$_{11}$H$_{16}$ 148.25	5161-04-6	196	4-05-00-01096 0.8517[20]	ace 4; bz 4; eth 4; EtOH 4 1.4874[20]
2000	Benzene, 1-methyl-2-nitro- o-Nitrotoluene	C$_7$H$_7$NO$_2$ 137.14	88-72-2 -10	6572 222	4-05-00-00845 1.1611[19]	H$_2$O 1; EtOH 5; eth 5; ctc 3 1.5450[20]
2001	Benzene, 1-methyl-3-nitro- m-Nitrotoluene	C$_7$H$_7$NO$_2$ 137.14	99-08-1 15.5	6572 232	4-05-00-00847 1.1581[20]	H$_2$O 1; EtOH 3; eth 5; bz 3 1.5466[20]
2002	Benzene, 1-methyl-4-nitro- p-Nitrotoluene	C$_7$H$_7$NO$_2$ 137.14	99-99-0 51.6	6572 238.3	4-05-00-00848 1.1038[75]	H$_2$O 1; EtOH 3; eth 4; ace 4
2003	Benzene, (1-methyl-1-nitroethyl)-	C$_9$H$_{11}$NO$_2$ 165.19	3457-58-7	224 dec; 126[15]	4-05-00-00998 1.1025[20]	ace 4; bz 4 1.5209[20]
2004	Benzene, 1-methyl-4-(nitromethyl)-	C$_8$H$_9$NO$_2$ 151.17	29559-27-1 11.5	150-1[35]	4-05-00-00972 1.1234[20]	H$_2$O 1; bz 3 1.5278[20]
2005	Benzene, 1-methyl-2-(2-nitrophenoxy)- 2-Nitrophenyl 2-tolyl ether	C$_{13}$H$_{11}$NO$_3$ 229.24	54106-40-0 39.5	194-6[14]	4-06-00-01946 1.195[20]	
2006	Benzene, 1-methyl-4-(4-nitrophenoxy)-	C$_{13}$H$_{11}$NO$_3$ 229.24	3402-74-2 69	224[25]	4-06-00-02102	bz 4; HOAc 4
2007	Benzene, (1-methylnonadecyl)- Eicosane, 2-phenyl-	C$_{26}$H$_{46}$ 358.65	2398-66-5 29.0	204.5[1]	4-05-00-01222 0.8547[20]	1.4795[20]
2008	Benzene, (1-methyl-1-pentenyl)- 2-Hexene, 2-phenyl-	C$_{12}$H$_{16}$ 160.26	20247-89-6	66-8[1]	4-05-00-01419 0.910[20]	1.5200[20]
2009	Benzene, (1-methylpentyl)-	C$_{12}$H$_{18}$ 162.27	6031-02-3	208	4-05-00-01116 0.869[15]	1.492[15]
2010	Benzene, (3-methylpentyl)- (3-Methylpentyl)benzene	C$_{12}$H$_{18}$ 162.27	54410-69-4	220	0-05-00-00444 0.8644[14]	1.4896
2011	Benzene, (4-methylpentyl)-	C$_{12}$H$_{18}$ 162.27	4215-86-5	215	4-05-00-01116 0.8568[16]	bz 4; eth 4; EtOH 4
2012	Benzene, 1-methyl-2-phenoxy-	C$_{13}$H$_{12}$O 184.24	3991-61-5 21.7	263.5	4-06-00-01946 1.0480[25]	
2013	Benzene, 1-methyl-3-phenoxy-	C$_{13}$H$_{12}$O 184.24	3586-14-9	272	4-06-00-02041 1.051[25]	1.5727[20]
2014	Benzene, 1-methyl-2-(phenylmethyl)-	C$_{14}$H$_{14}$ 182.27	713-36-0 6.6	280.5	4-05-00-01892 1.0020[20]	1.5763[20]
2015	Benzene, 1-methyl-3-(phenylmethyl)-	C$_{14}$H$_{14}$ 182.27	620-47-3 -28	279	4-05-00-01893 0.9913[20]	H$_2$O 1; EtOH 3; eth 3; bz 3 1.5712[20]
2016	Benzene, 1-methyl-4-(phenylmethyl)-	C$_{14}$H$_{14}$ 182.27	620-83-7 -30	286	4-05-00-01894 0.9976[20]	eth 4; bz 4; EtOH 4; chl 4 1.5712[20]
2017	Benzene, 1-methyl-2-(phenylthio)-	C$_{13}$H$_{12}$S 200.30	13963-35-4	308; 164[12]	4-06-00-02018 1.0893[20]	H$_2$O 1; ace 3; bz 3
2018	Benzene, 1-methyl-3-(phenylthio)-	C$_{13}$H$_{12}$S 200.30	13865-48-0 -6.5	309.5	4-06-00-02080 1.0937[15]	H$_2$O 1; ace 3; bz 3
2019	Benzene, 1-methyl-4-(phenylthio)-	C$_{13}$H$_{12}$S 200.30	3699-01-2 15.7	317	4-06-00-02169 1.0986[25]	H$_2$O 1; ace 3; bz 3 1.6225[25]
2020	Benzene, 1,1'-(1-methyl-1,3-propanediyl)bis-, (±)- Butane, 1,3-diphenyl (DL)	C$_{16}$H$_{18}$ 210.32	116783-21-2	293	4-05-00-01938 0.9722[20]	1.5525[20]
2021	Benzene, 1,1'-(2-methyl-1,3-propanediyl)bis-	C$_{16}$H$_{18}$ 210.32	1520-46-3 -33.7	303	4-05-00-01939 0.9669[20]	
2022	Benzene, 1,1'-(3-methyl-1-propene-1,3-diyl)bis- 1,3-Diphenyl-1-butene	C$_{16}$H$_{16}$ 208.30	7614-93-9 47.5	311	4-05-00-02205 0.9996[20]	1.590[15]
2023	Benzene, 1-methyl-2-(2-propenyl)-	C$_{10}$H$_{12}$ 132.21	1587-04-8	182.5	3-05-00-01212 0.9005[20]	1.5187[20]

No.	Name Synonym	Mol. Form. Mol. Wt.	CAS RN mp/°C	Merck No. bp/°C	Beil. Ref. den/g cm^{-3}	Solubility n_D
2024	Benzene, (2-methyl-1-propenyl)-	C$_{10}$H$_{12}$ 132.21	768-49-0 	99[43.5]	4-05-00-01380 0.9029[20]	1.5388[20]
2025	Benzene, (1-methyl-1-propenyl)-, (E)-	C$_{10}$H$_{12}$ 132.21	768-00-3 -23.5	194.7	4-05-00-01378 0.9138[25]	H$_2$O 1; bz 3; chl 3 1.5425[20]
2026	Benzene, 1-methyl-2-(2-propenyloxy)-	C$_{10}$H$_{12}$O 148.20	936-72-1 	206.5	4-06-00-01946 0.9698[15]	H$_2$O 1; ctc 2 1.5188[15]
2027	Benzene, 1-methyl-3-(2-propenyloxy)-	C$_{10}$H$_{12}$O 148.20	1758-10-7 	212.5	4-06-00-02041 0.9564[20]	bz 4 1.5179[20]
2028	Benzene, 1-methyl-4-(2-propenyloxy)-	C$_{10}$H$_{12}$O 148.20	23431-48-3 	214.5	4-06-00-02101 0.9719[15]	H$_2$O 1; ace 3; ctc 2 1.5157[24]
2029	Benzene, (1-methyl-1-propenyl)-, (Z)-	C$_{10}$H$_{12}$ 132.21	767-99-7 	194.7; 177[500]	3-05-00-01209 0.9191[25]	H$_2$O 1; bz 3; chl 3 1.5402[25]
2030	Benzene, (1-methylpropoxy)-	C$_{10}$H$_{14}$O 150.22	10574-17-1 	194.5	4-06-00-00559 0.9415[20]	H$_2$O 1; eth 3; ctc 2 1.4926[25]
2031	Benzene, (2-methylpropoxy)-	C$_{10}$H$_{14}$O 150.22	1126-75-6 	197.5	4-06-00-00559 0.9232[24]	1.4932[14]
2032	Benzene, 1-methyl-4-propoxy- Toluene, 4-propoxy	C$_{10}$H$_{14}$O 150.22	5349-18-8 0	210.4	3-06-00-01354 0.9496[0]	
2033	Benzene, [(2-methylpropoxy)methyl]- Benzyl isobutyl ether	C$_{11}$H$_{16}$O 164.25	940-49-8 	212	3-06-00-01457 0.9233[20]	H$_2$O 1; eth 4; chl 4 1.4826[20]
2034	Benzene, 1-(2-methylpropoxy)-2-nitro- 1-Isobutoxy-2-nitrobenzene	C$_{10}$H$_{13}$NO$_3$ 195.22	56245-02-4 	277.5	0-06-00-00218 1.1361[20]	eth 4
2035	Benzene, 1-methylpropyl sec-Butylbenzene	C$_{10}$H$_{14}$ 134.22	135-98-8 -82.7	1550 173.5	0.8580[25]	H$_2$O 1; EtOH 5; eth 5; ace 5 1.4895[20]
2036	Benzene, 1-methyl-2-propyl-	C$_{10}$H$_{14}$ 134.22	1074-17-5 -60.3	185	4-05-00-01056 0.8697[25]	1.4996[20]
2037	Benzene, 1-methyl-3-propyl-	C$_{10}$H$_{14}$ 134.22	1074-43-7 -82.5	182	4-05-00-01056 0.8569[25]	1.4935[20]
2038	Benzene, 1-methyl-4-propyl-	C$_{10}$H$_{14}$ 134.22	1074-55-1 -63.6	183.4	4-05-00-01057 0.8544[25]	H$_2$O 1; EtOH 3; eth 3 1.4922[20]
2039	Benzene, (2-methylpropyl)- Isobutylbenzene	C$_{10}$H$_{14}$ 134.22	538-93-2 -51.4	5018 172.7	4-05-00-01042 0.8532[20]	H$_2$O 1; EtOH 5; eth 5; ace 5 1.4866[20]
2040	Benzene, (methylsulfinyl)-	C$_7$H$_8$OS 140.21	1193-82-4 32.0	263.5; 140[13]	4-06-00-01467	1.5885[20]
2041	Benzene, (methylthio)-	C$_7$H$_8$S 124.21	100-68-5 	193	4-06-00-01466 1.0579[20]	H$_2$O 1; EtOH 3; ace 4; os 3 1.5868[20]
2042	Benzene, [(methylthio)methyl]-	C$_8$H$_{10}$S 138.23	766-92-7 -30	210; 120[48]	4-06-00-02633 1.0274[20]	1.5620[20]
2043	Benzene, 1-(methylthio)-2-nitro-	C$_7$H$_7$NO$_2$S 169.20	3058-47-7 64.5		4-06-00-01661 1.2628[78]	H$_2$O 1; EtOH 3; bz 3; chl 3 1.6246[78]
2044	Benzene, 1-(methylthio)-4-nitro-	C$_7$H$_7$NO$_2$S 169.20	701-57-5 72	135-40[2]	4-06-00-01687 1.2391[80]	H$_2$O 1; ace 3; bz 3 1.6401[20]
2045	Benzene, 1-methyl-4-(trifluoromethyl)-	C$_8$H$_7$F$_3$ 160.14	6140-17-6 	129	1.4320[26]	eth 3; ace 3; bz 3; chl 3 1.4320[26]
2046	Benzene, 1-methyl-2,3,4-trinitro- 2,3,4-Trinitrotoluene	C$_7$H$_5$N$_3$O$_6$ 227.13	602-29-9 112	301	4-05-00-00872 1.62[25]	H$_2$O 1; EtOH 2; eth 3; ace 3
2047	Benzene, 1-methyl-2,4,5-trinitro- 2,4,5-Trinitrotoluene	C$_7$H$_5$N$_3$O$_6$ 227.13	610-25-3 104	exp 300	4-05-00-00872	H$_2$O 1; EtOH 3; eth 3; ace 3
2048	Benzene, 2-methyl-1,3,5-trinitro- 2,4,6-Trinitrotoluene	C$_7$H$_5$N$_3$O$_6$ 227.13	118-96-7 80.1	9643 exp 240	4-05-00-00873 1.654[25]	H$_2$O 1; EtOH 2; eth 3; ace 4
2049	Benzene, nitro- Nitrobenzene	C$_6$H$_5$NO$_2$ 123.11	98-95-3 5.7	6509 210.8	4-05-00-00708 1.2037[20]	H$_2$O 2; EtOH 4; eth 4; ace 4 1.5562[20]
2050	Benzene, (2-nitroethenyl)-, (E)-	C$_8$H$_7$NO$_2$ 149.15	5153-67-3 60	255	4-05-00-01352	H$_2$O 1; EtOH 3; eth 4; ace 3
2051	Benzene, (2-nitroethyl)-	C$_8$H$_9$NO$_2$ 151.17	6125-24-2 -23	250; 137[16]	4-05-00-00912 1.126[24]	1.5407[19]
2052	Benzene, (nitromethyl)-	C$_7$H$_7$NO$_2$ 137.14	622-42-4 	226; 135[25]	4-05-00-00850 1.1596[20]	ace 4; eth 4 1.5323[20]
2053	Benzene, 1-nitro-2-phenoxy-	C$_{12}$H$_9$NO$_3$ 215.21	2216-12-8 <-20	235[60]; 184[8]	4-06-00-01252 1.2539[22]	bz 4; eth 4; EtOH 4; chl 4 1.575[20]
2054	Benzene, 1-nitro-4-phenoxy-	C$_{12}$H$_9$NO$_3$ 215.21	620-88-2 61	320; 225[30]	4-06-00-01287	H$_2$O 1; EtOH 2; eth 3; bz 3
2055	Benzene, 1-nitro-2-(2-phenylethenyl)-, (E)-	C$_{14}$H$_{11}$NO$_2$ 225.25	4264-29-3 73	209[11]	4-05-00-02166	H$_2$O 1; EtOH 3; eth 4
2056	Benzene, 1-nitro-2-(2-phenylethenyl)-, (Z)-	C$_{14}$H$_{11}$NO$_2$ 225.25	52208-62-5 64.8	187[11]	4-05-00-02166	EtOH 4; eth 4; AcOEt 4; HOAc 4
2057	Benzene, 1-nitro-2-(phenylthio)-	C$_{12}$H$_9$NO$_2$S 231.28	4171-83-9 82	210[15]	4-06-00-01663	eth 4; EtOH 4
2058	Benzene, 1-nitro-4-(phenylthio)-	C$_{12}$H$_9$NO$_2$S 231.28	952-97-6 56	288[100]; 240[25]	4-06-00-01694	eth 4; EtOH 4
2059	Benzene, nitroso- Nitrosobenzene	C$_6$H$_5$NO 107.11	586-96-9 68.5	58[18]	4-05-00-00702	H$_2$O 1; EtOH 3; eth 3; bz 3
2060	Benzene, 1-nitro-2-(trifluoromethyl)-	C$_7$H$_4$F$_3$NO$_2$ 191.11	384-22-5 32.5	217; 105[20]	4-05-00-00852	H$_2$O 1; EtOH 4; bz 4; ctc 2
2061	Benzene, 1-nitro-3-(trifluoromethyl)- 3-(Trifluoromethyl)-1-nitrobenzene	C$_7$H$_4$F$_3$NO$_2$ 191.11	98-46-4 -2.4	202.8; 103[40]	4-05-00-00853 1.4357[15]	H$_2$O 1; EtOH 3; eth 3; ctc 2 1.4719[20]
2062	Benzene, nonadecyl-	C$_{25}$H$_{44}$ 344.62	29136-19-4 40	419	3-05-00-01126 0.8545[20]	1.4807[20]

No.	Name Synonym	Mol. Form. Mol. Wt.	CAS RN mp/°C	Merck No. bp/°C	Bell. Ref. den/g cm^{-3}	Solubility n_D
2063	Benzenenonanoic acid, θ-nonyl-	$C_{24}H_{40}O_2$ 360.58	1938-22-3 37	200-4[0.09]	2-09-00-00376 0.9312[25]	1.4891[20]
2064	Benzene, nonyl-	$C_{15}H_{24}$ 204.36	1081-77-2 -24	280.5	3-05-00-01075 0.8584[20]	1.4816[20]
2065	Benzene, octacosyl- Octacosylbenzene	$C_{34}H_{62}$ 470.87	61828-26-0 66	405[100]; 268[1]	4-05-00-01243 0.8543[20]	1.4796[20]
2066	Benzene, octadecyl- Octadecylbenzene	$C_{24}H_{42}$ 330.60	4445-07-2 29	408	4-05-00-01219 0.8546[20]	1.4809[20]
2067	Benzene, octyl-	$C_{14}H_{22}$ 190.33	2189-60-8 -36	264.5	4-05-00-01157 0.8562[20]	H$_2$O 1; eth 5; bz 5 1.4845[20]
2068	Benzene, (1-octyldodecyl)- Eicosane, 9-phenyl-	$C_{26}H_{46}$ 358.65	2398-65-4 17.9	196[1.0]	4-05-00-01223 0.8534[20]	1.4790[20]
2069	Benzene, (octyloxy)-	$C_{14}H_{22}O$ 206.33	1818-07-1 8	285	4-06-00-00560 0.9131[15]	H$_2$O 1; EtOH 3; eth 3 1.4875[20]
2070	Benzene, (3-octylundecyl)- 9-(2-Phenylethyl)heptadecane	$C_{25}H_{44}$ 344.62	5637-96-7 -26.7	189[1.0]	4-05-00-01221 0.8560[20]	1.4806[20]
2071	Benzene, 1,1'-oxybis- Diphenyl ether	$C_{12}H_{10}O$ 170.21	101-84-8 26.8	7259 258.0	4-06-00-00568 1.0661[30]	H$_2$O 1; EtOH 3; eth 3; bz 3 1.5787[25]
2072	Benzene, 1,1'-oxybis[4-bromo-	$C_{12}H_8Br_2O$ 328.00	2050-47-7 60.5	339	4-06-00-01048 1.8[25]	H$_2$O 1; EtOH 3; eth 4; bz 3
2073	Benzene, 1,1'-oxybis[4-chloro-	$C_{12}H_8Cl_2O$ 239.10	2444-89-5 30	313	4-06-00-00826 1.1231[20]	H$_2$O 1 1.6112[20]
2074	Benzene, 1,1'-oxybis[2-ethyl-	$C_{16}H_{18}O$ 226.32	56911-77-4	318.5	1.0141[18]	H$_2$O 1; eth 3; chl 3 1.5488[18]
2075	Benzene, 1,1'-oxybis[2-methoxy-	$C_{14}H_{14}O_3$ 230.26	1655-70-5 79.5	330.5	3-06-00-04223	H$_2$O 1; EtOH 4; eth 4; lig 1
2076	Benzene, 1,1'-oxybis[2-methyl-	$C_{14}H_{14}O$ 198.26	4731-34-4	271	4-06-00-01947 1.047[24]	bz 4; eth 4; EtOH 4
2077	Benzene, 1,1'-oxybis[3-methyl-	$C_{14}H_{14}O$ 198.26	19814-71-2	284; 136[14]	4-06-00-02043 1.0323[21]	H$_2$O 1; EtOH 3; eth 3; bz 3
2078	Benzene, 1,1'-oxybis[4-methyl-	$C_{14}H_{14}O$ 198.26	1579-40-4 51	285	4-06-00-02103	bz 4; eth 4; EtOH 4
2079	Benzene, 1,1'-[oxybis(methylene)]bis- Benzyl ether	$C_{14}H_{14}O$ 198.26	103-50-4 3.6	1146 298	4-06-00-02240 1.0428[20]	H$_2$O 1; EtOH 5; eth 5; ctc 3 1.5168[20]
2080	Benzene, 1,1'-(oxydi-2,1-ethanediyl)bis-	$C_{16}H_{18}O$ 226.32	2396-53-4	318.5; 86[0.5]	4-06-00-03071 1.0141[18]	eth 4; chl 4 1.5488[18]
2081	Benzene, 1,1'-(oxydiethylidene)bis-, (R*,R*)- (±)- Bis-(1-phenylethyl)ether (DL)	$C_{16}H_{18}O$ 226.32	53776-69-5	280.2	4-06-00-03034 1.0058[15]	H$_2$O 1; eth 3; chl 3 1.5454[21]
2082	Benzene, pentabromo- Pentabromobenzene	C_6HBr_5 472.59	608-90-2 160.5	sub	4-05-00-00687	bz 4; chl 4
2083	Benzene, pentabromomethyl-	$C_7H_3Br_5$ 486.62	87-83-2 288		3-05-00-00720 2.97[17]	H$_2$O 1; EtOH 2; bz 3; HOAc 2
2084	Benzene, pentachloro- Pentachlorobenzene	C_6HCl_5 250.34	608-93-5 86	277	4-05-00-00669 1.8342[16]	H$_2$O 1; EtOH 1; eth 2; bz 2
2085	Benzene, pentachloro(dichloromethyl)-	C_7HCl_7 333.25	2136-95-0 119.5	334	4-05-00-00824	EtOH 2
2086	Benzene, pentachloromethyl-	$C_7H_3Cl_5$ 264.36	877-11-2 224.8	301	4-05-00-00823	EtOH 2; eth 2; bz 3; tol 3
2087	Benzene, pentachloronitro- Quintozene	$C_6Cl_5NO_2$ 295.34	82-68-8 144	8108 328 dec	4-05-00-00728 1.718[25]	H$_2$O 1; EtOH 2; bz 3; chl 3
2088	Benzene, pentacosyl- Pentacosylbenzene	$C_{31}H_{56}$ 428.79	61828-06-6 59	380[100]; 244[1]	4-05-00-01241 0.8544[20]	1.4799[20]
2089	Benzene, pentadecyl-	$C_{21}H_{36}$ 288.52	2131-18-2 22	373	3-05-00-01119 0.8548[20]	1.4815[20]
2090	Benzene, pentaethyl- Pentaethylbenzene	$C_{16}H_{26}$ 218.38	605-01-6 <-20	277	4-05-00-01197 0.8971[19]	1.5127[20]
2091	Benzene, pentafluoro- Pentafluorobenzene	C_6HF_5 168.07	363-72-4 -47.3	85.7	4-05-00-00639 1.514[25]	1.3905[20]
2092	Benzene, pentafluoroiodo-	C_6F_5I 293.96	827-15-6 -29	166	4-05-00-00695 2.212[20]	1.4950[25]
2093	Benzene, pentafluoromethoxy-	$C_7H_3F_5O$ 198.09	389-40-2 -37	138.5	4-06-00-00782 1.493[20]	1.4087[20]
2094	Benzene, pentafluoromethyl- 2,3,4,5,6-Pentafluorotoluene	$C_7H_3F_5$ 182.09	771-56-2 -29.8	117.5	1.440[20]	1.4016[25]
2095	Benzene, pentafluoro(trifluoromethyl)- Perfluorotoluene	C_7F_8 236.06	434-64-0 -65.6	104.5	4-05-00-00804	1.3670[20]
2096	Benzene, pentaiodo- Pentaiodobenzene	C_6HI_5 707.60	608-96-8 172	sub	4-05-00-00702	bz 4; chl 4
2097	Benzene, pentamethyl- Pentamethylbenzene	$C_{11}H_{16}$ 148.25	700-12-9 54.5	232	4-05-00-01109 0.917[20]	H$_2$O 1; EtOH 4; bz 4; chl 3 1.527[20]
2098	Benzene, 1,1'-(1,5-pentanediyl)bis- 1,5-Diphenylpentane	$C_{17}H_{20}$ 224.35	1718-50-9 -8	327	4-05-00-01963 0.9812[19]	H$_2$O 1 1.559[19]
2099	Benzenepentanoic acid 5-Phenylvaleric acid	$C_{11}H_{14}O_2$ 178.23	2270-20-4 57.5	190[30]	4-09-00-01864	H$_2$O 2; EtOH 4; os 3
2100	Benzenepentanoic acid, δ-oxo-, methyl ester	$C_{12}H_{14}O_3$ 206.24	1501-04-8 -2	186[19]; 147[8]	4-10-00-02783	H$_2$O 4
2101	Benzenepentanol	$C_{11}H_{16}O$ 164.25	10521-91-2	155[20]	4-06-00-03373 0.9725[20]	eth 4; EtOH 4 1.5156[20]

No.	Name Synonym	Mol. Form. Mol. Wt.	CAS RN mp/°C	Merck No. bp/°C	Beil. Ref. den/g cm⁻³	Solubility n_D
2102	Benzenepentanol, α-ethyl- 7-Phenylheptan-3-ol	$C_{13}H_{20}O$ 192.30	60012-63-7	277.5; 150[20]	3-06-00-02027 0.968[20]	1.5115[20]
2103	Benzenepentanol, α-methyl- 6-Phenyl-2-hexanol	$C_{12}H_{18}O$ 178.27	38487-94-4	148[18]	3-06-00-01995 0.9567[20]	1.5079[20]
2104	Benzene, 1-pentenyl-	$C_{11}H_{14}$ 146.23	826-18-6	217	4-05-00-01401 0.8782[22]	1.5158[20]
2105	Benzene, 3-pentenyl-	$C_{11}H_{14}$ 146.23	1745-16-0	201	4-05-00-01401 0.8884[16]	1.5089[16]
2106	Benzene, 4-pentenyl- 1-Pentene, 5-phenyl	$C_{11}H_{14}$ 146.23	1075-74-7	201; 77[10]	3-05-00-01238 0.8889[20]	1.5065[20]
2107	Benzene, pentyl- Amylbenzene	$C_{11}H_{16}$ 148.25	538-68-1 -75	635 205.4	4-05-00-01085 0.8585[20]	H₂O 1; EtOH 5; eth 5; ace 5 1.4878[20]
2108	Benzene, 1,1'-pentylidenebis- Pentane, 1,1-diphenyl-	$C_{17}H_{20}$ 224.35	1726-12-1 -12	308	4-05-00-01966 0.9659[20]	H₂O 1 1.5511[20]
2109	Benzene, (pentyloxy)- Pentyl phenyl ether	$C_{11}H_{16}O$ 164.25	2050-04-6	200; 111[17]	3-06-00-00552 0.9270[20]	ace 4; EtOH 4 1.4947[20]
2110	Benzene, 1,1'-(phenylmethylene)bis[4-methyl-	$C_{21}H_{20}$ 272.39	603-39-4 56	218[12]	4-05-00-02520	H₂O 1; EtOH 3; eth 4; bz 4
2111	Benzene, 1-(phenylmethyl)-4-propyl-	$C_{16}H_{18}$ 210.32	62155-41-3	152-5[10]	4-05-00-01950 0.9739[18]	1.3552[20]
2112	Benzene, [(phenylmethyl)sulfonyl]-	$C_{13}H_{12}O_2S$ 232.30	3112-88-7 146		4-06-00-02647 1.1261[153]	H₂O 1; EtOH 2; eth 2; bz 2
2113	Benzene, [(phenylmethyl)thio]-	$C_{13}H_{12}S$ 200.30	831-91-4 43.5	197[27]	4-06-00-02644	H₂O 1; EtOH 3; eth 3; con sulf 3
2114	Benzenepropanal Hydrocinnamic aldehyde	$C_9H_{10}O$ 134.18	104-53-0 47	224; 117[28]	4-07-00-00692 1.0190[20]	H₂O 1; EtOH 4; eth 5
2115	Benzenepropanal, 2-methyl- Hydrocinnamaldehyde, o-methyl-	$C_{10}H_{12}O$ 148.20	19564-40-0	120[13]	2-07-00-00246 0.998[19]	1.522[19]
2116	Benzenepropanal, 4-methyl-	$C_{10}H_{12}O$ 148.20	5406-12-2	223	2-07-00-00247 0.999[14]	1.525[14]
2117	Benzenepropanal, α-methyl-4-(1-methylethyl)- 3-p-Cumenyl-2-methylpropionaldehyde	$C_{13}H_{18}O$ 190.29	103-95-7	135[99]; 115[9]	4-07-00-00788 0.951[15]	bz 4; eth 4; EtOH 4 1.5068[20]
2118	Benzenepropanal, α-methyl-β-oxo-	$C_{10}H_{10}O_2$ 162.19	16837-43-7 119.3	155[125]	3-07-00-03497	bz 4; EtOH 4; MeOH 4
2119	Benzenepropanamine, α-methyl- 1-Methyl-3-phenylpropylamine	$C_{10}H_{15}N$ 149.24	22374-89-6 143	223; 101[14]	4-12-00-02811 0.9289[15]	EtOH 4 1.5152[20]
2120	Benzenepropanaminium, γ-(aminocarbonyl)- N-methyl-N,N-bis(1-methylethyl)-γ-phenyl-, iod Ammonium, (3-carbamoyl-3,3-diphenylpropyl) diisopropyl methyl,iodide	$C_{23}H_{33}IN_2O$ 480.43	71-81-8	5090	4-14-00-01853	
2121	Benzene, 1,1'-(1,3-propanediyl)bis- 1,3-Diphenylpropane	$C_{15}H_{16}$ 196.29	1081-75-0 6	300.3	4-05-00-01913 1.007[20]	1.5760[20]
2122	Benzene, 1,1'-[1,3-propanediylbis(oxy)]bis-	$C_{15}H_{16}O_2$ 228.29	726-44-3 62	339	4-06-00-00585	H₂O 1; EtOH 3; eth 3
2123	Benzenepropanenitrile Hydrocinnamonitrile	C_9H_9N 131.18	645-59-0 -1	261; 141[25]	4-09-00-01764 1.0016[20]	EtOH 3; eth 3; chl 2 1.5266[28]
2124	Benzenepropanenitrile, β-oxo- Benzoylacetonitrile	C_9H_7NO 145.16	614-16-4 80.5	160[10]	4-10-00-02757	H₂O 2; EtOH 3; eth 3; bz 3
2125	Benzenepropanethiol	$C_9H_{12}S$ 152.26	24734-68-7	109[10]	4-06-00-03206 1.01[25]	1.5494[20]
2126	Benzenepropanoic acid Hydrocinnamic acid	$C_9H_{10}O_2$ 150.18	501-52-0 107	4707 279.8	4-09-00-01752 1.0712[49]	H₂O 3; EtOH 3; eth 3; bz 4
2127	Benzenepropanoic acid, α-acetyl-α- (phenylmethyl)-, ethyl ester	$C_{20}H_{22}O_3$ 310.39	42597-26-2 57	233	3-10-00-03350	EtOH 2; eth 2
2128	Benzenepropanoic acid, β-(aminomethyl)- 4-Amino-3-phenylbutyric acid	$C_{10}H_{13}NO_2$ 179.22	1078-21-3 250-5 dec	476	4-14-00-01723	
2129	Benzenepropanoic acid, α,α-dimethyl-, methyl ester Methyl α,α-dimethylhydrocinnamate	$C_{12}H_{16}O_2$ 192.26	14248-22-7	121-2[23]	4-09-00-01879 1.0043[25]	1.4945[25]
2130	Benzenepropanoic acid, β,β-dimethyl-	$C_{11}H_{14}O_2$ 178.23	1010-48-6 60	167[10]	4-09-00-01874	H₂O 3; ace 3; peth 3 1.5182[20]
2131	Benzenepropanoic acid, ethyl ester	$C_{11}H_{14}O_2$ 178.23	2021-28-5	247.2	4-09-00-01753 1.0147[20]	eth 4; EtOH 4 1.4954[20]
2132	Benzenepropanoic acid, α-ethyl-β-oxo-, ethyl ester	$C_{13}H_{16}O_3$ 220.27	24346-56-3	152[7]	3-10-00-03065 1.0706[15]	eth 4 1.509[15]
2133	Benzenepropanoic acid, α-hydroxy-, (±)- (±)-3-Phenyllactic acid	$C_9H_{10}O_3$ 166.18	828-01-3 98	149[15]	2-10-00-00154	H₂O 4; ace 4; eth 4; EtOH 4
2134	Benzenepropanoic acid, 4-hydroxy- p-Hydroxyhydrocinnamic acid	$C_9H_{10}O_3$ 166.18	501-97-3 130.8	208-10[14]	4-10-00-00631	H₂O 3; EtOH 3; eth 3; bz 3
2135	Benzenepropanoic acid, 4-hydroxy-3,5-diiodo- α-phenyl- Iodoalphionic acid	$C_{15}H_{12}I_2O_3$ 494.07	577-91-3 164	4919	4-10-00-01291	H₂O 1; EtOH 3; eth 3; bz 2
2136	Benzenepropanoic acid, 4-(4-hydroxy-3- iodophenoxy)-3,5-diiodo- 3,3',5-Triiodothyropropionic acid	$C_{15}H_{11}I_3O_4$ 635.96	51-26-3	9346	4-10-00-00641	
2137	Benzenepropanoic acid, α-methyl- α-Methyldihydrocinnamic acid	$C_{10}H_{12}O_2$ 164.20	1009-67-2 36.5	272; 152[12]	2-09-00-00357	eth 4; EtOH 4

No.	Name / Synonym	Mol. Form. / Mol. Wt.	CAS RN / mp/°C	Merck No. / bp/°C	Beil. Ref. / den/g cm⁻³	Solubility / n_D
2138	Benzenepropanoic acid, β-methyl-, (±)-	$C_{10}H_{12}O_2$ 164.20	772-17-8 46.5	140-5[3]	4-09-00-01814 1.0701[20]	H_2O 2; peth 3 1.5155[20]
2139	Benzenepropanoic acid, methyl ester Methyl dihydrocinnamate	$C_{10}H_{12}O_2$ 164.20	103-25-3	238.5; 91[4]	4-09-00-01753 1.0455[25]	H_2O 1; EtOH 3; eth 3; bz 3
2140	Benzenepropanoic acid, 1-methylethyl ester	$C_{12}H_{16}O_2$ 192.26	22767-95-9	126[11]	2-09-00-00339 0.9860[25]	eth 4; EtOH 4
2141	Benzenepropanoic acid, β-oxo-, ethyl ester Ethyl benzoylacetate	$C_{11}H_{12}O_3$ 192.21	94-02-0 <0	3726 267 dec; 167[20]	4-10-00-02756 1.1202[15]	H_2O 2; EtOH 3; eth 3 1.5317[15]
2142	Benzenepropanoic acid, β-oxo-, methyl ester	$C_{10}H_{10}O_3$ 178.19	614-27-7	265 dec; 151[12]	4-10-00-02755 1.158[29]	ace 4; eth 4; EtOH 4 1.537[20]
2143	Benzenepropanoic acid, α-phenyl-, (±)- Propionic acid, 2,3-diphenyl (DL)	$C_{15}H_{14}O_2$ 226.27	94942-89-9 88(form a); 95(form b)	337.5	4-09-00-02528 1.1495[96]	bz 4; eth 4; EtOH 4
2144	Benzenepropanoic acid, α-(phenylmethyl)-	$C_{16}H_{16}O_2$ 240.30	618-68-8 90	235[18]	4-09-00-02546	bz 4; eth 4; EtOH 4
2145	Benzenepropanoic acid, phenylmethyl ester	$C_{16}H_{16}O_2$ 240.30	22767-96-0	325; 198[20]	4-09-00-01755 1.090[15]	eth 4
2146	Benzenepropanoic acid, β-propyl-	$C_{12}H_{16}O_2$ 192.26	5703-52-6	152[4]	4-09-00-01900 1.025[25]	1.5078
2147	Benzenepropanoic-β-14C acid, α-bromo- Hydrocinnamic acid, α-bromo-	$C_9H_9BrO_2$ 229.07	16503-53-0 52	150[13]	4-09-00-01771 1.4790[25]	ace 4; bz 4; EtOH 4
2148	Benzenepropanol Hydrocinnamyl alcohol	$C_9H_{12}O$ 136.19	122-97-4 <-18	235	4-06-00-03198 0.995[25]	H_2O 3; EtOH 5; eth 5; ctc 3 1.5357[25]
2149	Benzenepropanol, carbamate Phenprobamate	$C_{10}H_{13}NO_2$ 179.22	673-31-4 102	7229		H_2O 1; EtOH 3; chl 3
2150	Benzenepropanol, α,α-dimethyl- Benzyl-tert-butanol	$C_{11}H_{16}O$ 164.25	103-05-9 24.5	121[13]	4-06-00-03380 0.9626[21]	H_2O 1; EtOH 4; eth 4; ace 4 1.5077[21]
2151	Benzenepropanol, β,β-dimethyl-	$C_{11}H_{16}O$ 164.25	13351-61-6 34.5	125[14.5]	2-06-00-00507	eth 4; EtOH 4
2152	Benzenepropanol, α-ethyl-, (S)- 3-Pentanol, 1-phenyl (d)	$C_{11}H_{16}O$ 164.25	71747-37-0 38	143[19]	3-06-00-01953 0.9687[20]	EtOH 4
2153	Benzenepropanol, 4-hydroxy-3-methoxy-	$C_{10}H_{14}O_3$ 182.22	2305-13-7 65	197[15]	4-06-00-07405	eth 4; EtOH 4 1.5545[25]
2154	Benzenepropanol, 4-methoxy-	$C_{10}H_{14}O_2$ 166.22	5406-18-8 28	166[18]	4-06-00-05982	1.5315[20]
2155	Benzenepropanol, α-methyl-	$C_{10}H_{14}O$ 150.22	2344-70-9	239; 123[15]	4-06-00-03275 0.9899[16]	1.517[16]
2156	Benzenepropanol, γ-methyl-	$C_{10}H_{14}O$ 150.22	2722-36-3	139[33]; 110[6]	4-06-00-03286 0.9832[20]	1.4965[20]
2157	Benzenepropanol, γ-methylene- 3-Buten-1-ol, 3-phenyl-	$C_{10}H_{12}O$ 148.20	3174-83-2	123[10]	4-06-00-03840 1.0272[20]	1.5577[20]
2158	Benzenepropanol, β-phenyl-	$C_{15}H_{16}O$ 212.29	3536-29-6 51	310; 190[15]	4-06-00-04754 1.0585[24]	1.5742[19]
2159	Benzenepropanol, γ-propyl- 1-Hexanol, 3-phenyl	$C_{12}H_{18}O$ 178.27	67700-22-5	127[5]	3-06-00-02001 0.955[25]	1.5101[25]
2160	Benzenepropanoyl chloride	C_9H_9ClO 168.62	645-45-4	225 dec; 105[10]	4-09-00-01762 1.135[21]	eth 3; CS_2 3
2161	Benzene, 1,1'-(1-propene-1,3-diyl)bis-	$C_{15}H_{14}$ 194.28	5209-18-7 16	302; 178[15]	4-05-00-02187 1.0120[20]	
2162	Benzene, 1-propenyl-	C_9H_{10} 118.18	637-50-3 -27.1	175.5	3-05-00-01185 0.9019[25]	ace 4; bz 4; eth 4; EtOH 4 1.5508[20]
2163	Benzene, 1-propenyl-, (E)-	C_9H_{10} 118.18	873-66-5 -29.3	178.3	4-05-00-01359 0.9023[25]	H_2O 1; EtOH 5; eth 5; ace 5 1.5506[20]
2164	Benzene, 1-propenyl-, (Z)-	C_9H_{10} 118.18	766-90-5 -61.6	167.5	4-05-00-01359 0.9088[20]	H_2O 1; EtOH 5; eth 5; ace 5 1.5420[20]
2165	Benzene, 2-propenyl-	C_9H_{10} 118.18	300-57-2 -40	156	4-05-00-01362 0.8920[20]	H_2O 1; EtOH 3; eth 3; bz 3 1.5131[20]
2166	Benzene, 1,1'-(1-propenylidene)bis-	$C_{15}H_{14}$ 194.28	778-66-5 52	280-1 149[11]	4-05-00-02192 1.0250[20]	H_2O 1; EtOH 3; bz 3 1.5880[20]
2167	Benzene, (2-propenyloxy)-	$C_9H_{10}O$ 134.18	1746-13-0	191.7	4-06-00-00562 0.9811[20]	H_2O 1; EtOH 3; eth 5; ctc 2 1.5223[20]
2168	Benzene, propoxy-	$C_9H_{12}O$ 136.19	622-85-5 -27	189.9	4-06-00-00556 0.9474[20]	EtOH 3; eth 3 1.5014[20]
2169	Benzene, propyl- Propylbenzene	C_9H_{12} 120.19	103-65-1 -99.5	7856 159.2	4-05-00-00977 0.8620[20]	H_2O 1; EtOH 5; eth 5; ace 5 1.4920[20]
2170	Benzene, (1-propylheptadecyl)- Eicosane, 4-phenyl-	$C_{26}H_{46}$ 358.65	2400-03-5 31.4	199[1]	4-05-00-01223 0.8546[20]	1.4794[20]
2171	Benzene, 1,1'-propylidenebis- 1,1-Diphenylpropane	$C_{15}H_{16}$ 196.29	1530-03-6 13.3	281.6	4-05-00-01920 0.9951[14]	1.5681[14]
2172	Benzene, (propylthio)-	$C_9H_{12}S$ 152.26	874-79-3 -45	220	4-06-00-01470 0.9995[20]	1.5571[20]
2173	Benzene, 1-propynyl-	C_9H_8 116.16	673-32-5	183	4-05-00-01530 0.942[15]	1.563[15]
2174	Benzeneseleninic acid Phenylseleninic acid	$C_6H_6O_2Se$ 189.07	6996-92-5 124.5		4-11-00-00708 1.93[20]	H_2O 2; bz 1; alk 4
2175	Benzene, 1,1'-selenobis-	$C_{12}H_{10}Se$ 233.17	1132-39-4 2.5	301.5	4-06-00-01779 1.351[20]	H_2O 1; EtOH 5; eth 5; bz 3 1.5500[20]

No.	Name / Synonym	Mol. Form. / Mol. Wt.	CAS RN / mp/°C	Merck No. / bp/°C	Beil. Ref. / den/g cm^{-3}	Solubility / n_D
2176	Benzene, 1,1'-selenobis[4-methyl-	$C_{14}H_{14}Se$ 261.23	22077-55-0 69.5	196[16]	4-06-00-02218	H_2O 1; EtOH 3; eth 2; chl 2
2177	Benzeneselenol	C_6H_6Se 157.07	645-96-5	183.6; 84[25]	4-06-00-01777 1.4865[15]	H_2O 1; EtOH 3; eth 4; ctc 4
2178	Benzenesulfenyl chloride, 2,4-dinitro- 2,4-Dinitrobenzenesulfenyl chloride	$C_6H_3ClN_2O_4S$ 234.62	528-76-7 99	3267	4-06-00-01772	bz 4; chl 4; HOAc 4; peth 2
2179	Benzenesulfenyl chloride, 4-nitro-	$C_6H_4ClNO_2S$ 189.62	937-32-6 52	125[0.1]	4-06-00-01717	bz 4
2180	Benzenesulfinic acid	$C_6H_6O_2S$ 142.18	618-41-7 84	dec	4-11-00-00003	H_2O 2; EtOH 3; eth 3; bz 3
2181	Benzenesulfinic acid, 2-methyl- o-Toluenesulfinic acid	$C_7H_8O_2S$ 156.21	13165-77-0 80	dec	4-11-00-00008	H_2O 4; ace 4; eth 4; EtOH 4
2182	Benzenesulfinic acid, 4-methyl- p-Toluenesulfinic acid	$C_7H_8O_2S$ 156.21	536-57-2 86.5	9458	4-11-00-00009	H_2O 3; EtOH 4; eth 4; bz 2
2183	Benzene, 1,1'-sulfinylbis-	$C_{12}H_{10}OS$ 202.28	945-51-7 71.2	340[16]	4-06-00-01489	EtOH 4; eth 4; bz 4; chl 2
2184	Benzene, 1,1'-[sulfinylbis(methylene)]bis-	$C_{14}H_{14}OS$ 230.33	621-08-9 134	210 dec	4-06-00-02651	H_2O 1; EtOH 4; eth 3
2185	Benzenesulfinyl chloride	C_6H_5ClOS 160.62	4972-29-6 38	71-2[1.5]	4-11-00-00005 1.3469[25]	eth 3; chl 3 1.3470[25]
2186	Benzenesulfinyl chloride, 4-methyl-	C_7H_7ClOS 174.65	10439-23-3 57	113[3.5]	4-11-00-00010	chl 4
2187	Benzenesulfonamide, 4-amino- Sulfanilamide	$C_6H_8N_2O_2S$ 172.21	63-74-1 165.5	8898	4-14-00-02658 1.08[25]	H_2O 3; EtOH 3; eth 3; ace 3
2188	Benzenesulfonamide, 4-amino-N-[4-(aminosulfonyl)phenyl]- N4-Sulfanilylsulfanilamide	$C_{12}H_{13}N_3O_4S_2$ 327.39	547-52-4 137	8905	4-14-00-02760	H_2O 2; EtOH 3; eth 3; ace 3
2189	Benzenesulfonamide, 4-amino-N-(aminothioxomethyl)- Sulfathiourea	$C_7H_9N_3O_2S_2$ 231.30	515-49-1 182	8921	4-14-00-02671	H_2O 1; EtOH 2
2190	Benzenesulfonamide, 4-amino-N-(aminocarbonyl)- Sulfanilylurea	$C_7H_9N_3O_3S$ 215.23	547-44-4 147 dec	8906	4-14-00-02667	
2191	Benzenesulfonamide, 4-amino-N-(aminoiminomethyl)- Sulfaguanidine	$C_7H_{10}N_4O_2S$ 214.25	57-67-0 191.5	8879	4-14-00-02668	
2192	Benzenesulfonamide, 4-amino-N-[(butylamino)carbonyl]- Carbutamide	$C_{11}H_{17}N_3O_3S$ 271.34	339-43-5 144.5	1839	4-14-00-02667	
2193	Benzenesulfonamide, 4-amino-N-(6-chloro-3-pyridazinyl)- Sulfachlorpyridazine	$C_{10}H_9ClN_4O_2S$ 284.73	80-32-0 187	8871	4-25-00-02069	
2194	Benzenesulfonamide, 4-amino-N-(2,6-dimethoxy-4-pyrimidinyl)- Sulfadimethoxine	$C_{12}H_{14}N_4O_4S$ 310.33	122-11-2 203.5	8876	4-25-00-03483	
2195	Benzenesulfonamide, 4-amino-N-(3,4-dimethyl-5-isoxazolyl)- Sulfisoxazole	$C_{11}H_{13}N_3O_3S$ 267.31	127-69-5 191	8930	4-27-00-04747	
2196	Benzenesulfonamide, 4-amino-N-(4,6-dimethyl-2-pyrimidinyl)- Sulfamethazine	$C_{12}H_{14}N_4O_2S$ 278.33	57-68-1 198.5	8886	4-25-00-02215	H_2O 3; acid 3; alk 3
2197	Benzenesulfonamide, 4-amino-N-(6-methoxy-3-pyridazinyl)- Sulfamethoxypyridazine	$C_{11}H_{12}N_4O_3S$ 280.31	80-35-3 182.5	8890	4-25-00-03332	
2198	Benzenesulfonamide, 4-amino-N-(5-methyl-3-isoxazolyl)- Sulfamethoxazole	$C_{10}H_{11}N_3O_3S$ 253.28	723-46-6	8889	4-27-00-04685	
2199	Benzenesulfonamide, 4-amino-N-(4-methyl-2-pyrimidinyl)- Sulfamerazine	$C_{11}H_{12}N_4O_2S$ 264.31	127-79-7 236	8884	4-25-00-02159	H_2O 2; EtOH 2; eth 1; ace 2
2200	Benzenesulfonamide, 4-amino-N-(4-methyl-2-thiazolyl)- Sulfamethylthiazole	$C_{10}H_{11}N_3O_2S_2$ 269.35	515-59-3 237	8891	4-27-00-04715	EtOH 4
2201	Benzenesulfonamide, 4-amino-N-(1-phenyl-1H-pyrazol-5-yl)- Sulfaphenazole	$C_{15}H_{14}N_4O_2S$ 314.37	526-08-9	8910	4-25-00-02029	
2202	Benzenesulfonamide, 4-amino-N-pyrazinyl- Sulfapyrazine	$C_{10}H_{10}N_4O_2S$ 250.28	116-44-9 251	8912	4-25-00-02143	H_2O 1; EtOH 1; eth 1; ace 2
2203	Benzenesulfonamide, 4-amino-N-2-pyridinyl- Sulfapyridine	$C_{11}H_{11}N_3O_2S$ 249.29	144-83-2 192	8913	5-22-08-00424	H_2O 1; EtOH 3; bz 1; ctc 1
2204	Benzenesulfonamide, 4-amino-N-2-pyrimidinyl- Sulfadiazine	$C_{10}H_{10}N_4O_2S$ 250.28	68-35-9 255-6 dec	8874	4-25-00-02097	H_2O 2; EtOH 2; ace 2
2205	Benzenesulfonamide, 4-amino-N-2-quinoxalinyl- Sulfaquinoxaline	$C_{14}H_{12}N_4O_2S$ 300.34	59-40-5 247.5	8914	4-25-00-02606	H_2O 2; EtOH 2; ace 2; aq alk 3
2206	Benzenesulfonamide, 4-amino-N-2-thiazolyl- Sulfathiazole	$C_9H_9N_3O_2S_2$ 255.32	72-14-0 175(form a); 202(form b)	8920	4-27-00-04623	H_2O 2; EtOH 2; DMSO-d_6 2

No.	Name / Synonym	Mol. Form. / Mol. Wt.	CAS RN / mp/°C	Merck No. / bp/°C	Beil. Ref. / den/g cm^{-3}	Solubility / n_D
2207	Benzenesulfonamide, N-[(butylamino)carbonyl]-4-methyl- Tolbutamide	$C_{12}H_{18}N_2O_3S$ 270.35	64-77-7 128.5	9432	4-11-00-00396 1.245[25]	H_2O 2; EtOH 3; eth 3; chl 3
2208	Benzenesulfonamide, 4-chloro-N-[(propylamino)carbonyl]- Chloropropamide	$C_{10}H_{13}ClN_2O_3S$ 276.74	94-20-2	2187	4-11-00-00119	
2209	Benzenesulfonamide, 4-[(2,4-diaminophenyl)azo]- Prontosil	$C_{12}H_{14}ClN_5O_2S$ 327.79	103-12-8 249.5	8895	4-16-00-00563	H_2O 2; EtOH 3; ace 3; oils 3
2210	Benzenesulfonamide, N,N-dichloro-4-methyl- Dichloramine-T	$C_7H_7Cl_2NO_2S$ 240.11	473-34-7 83	3034	3-11-00-00301	H_2O 1; EtOH 3; eth 3; bz 3
2211	Benzenesulfonamide, N,4-dimethyl-	$C_8H_{11}NO_2S$ 185.25	640-61-9 78.5		4-11-00-00377 1.340[25]	eth 4; EtOH 4
2212	Benzenesulfonamide, N,4-dimethyl-N-nitroso- p-Tolylsulfonylmethylnitrosamide	$C_8H_{10}N_2O_3S$ 214.25	80-11-5 60	9469	4-11-00-00478	H_2O 1; EtOH; eth 4
2213	Benzenesulfonamide, 4-[(phenylmethyl)amino]- N4-Benzylsulfanilamide	$C_{13}H_{14}N_2O_2S$ 262.33	104-22-3 171	1161	3-14-00-02026	
2214	Benzenesulfonic acid Besylic acid	$C_6H_6O_3S$ 158.18	98-11-3 65	1078	4-11-00-00027	H_2O 4; EtOH 4; eth 1; bz 2
2215	Benzenesulfonic acid, 2-amino- Orthanilic acid	$C_6H_7NO_3S$ 173.19	88-21-1 >320 dec	6835	4-14-00-02638	H_2O 2; EtOH 1; eth 1
2216	Benzenesulfonic acid, 3-amino- Metanilic acid	$C_6H_7NO_3S$ 173.19	121-47-1 dec	5832	4-14-00-02647	H_2O 2; EtOH 1
2217	Benzenesulfonic acid, 4-amino- Sulfanilic acid	$C_6H_7NO_3S$ 173.19	121-57-3 288	8901	4-14-00-02655 1.485[25]	H_2O 2; EtOH 1; eth 1
2218	Benzenesulfonic acid, 2-amino-5-[(4-amino-3-sulfophenyl)(4-imino-3-sulfo-2,5-cyclohexadien-1-ylidene)methyl]-3 Fuchsin, acid	$C_{20}H_{17}N_3Na_2O_9S_3$ 585.55	3244-88-0	103		H_2O 2; EtOH 2
2219	Benzenesulfonic acid, sec-butyl ester Benzenesulfonic acid, chloride	$C_{10}H_{14}O_3S$ 214.29	103275-78-1 14.5	251 dec; 120[10]	4-11-00-00032 1.3830[15]	H_2O 1; EtOH 4; eth 3; ctc 3
2220	Benzenesulfonic acid, 4-chloro- p-Chlorobenzenesulfonic acid	$C_6H_5ClO_3S$ 192.62	98-66-8 67	2122	4-11-00-00107 147-8[25]	H_2O 3; EtOH 3; eth 1; bz 1
2221	Benzenesulfonic acid, 4-chloro-, 4-chlorophenyl ester Ovex	$C_{12}H_8Cl_2O_3S$ 303.17	80-33-1 86.5	6856	4-11-00-00109	H_2O 1; EtOH 2; ace 3
2222	Benzenesulfonic acid, 4-[[4-(dimethylamino)phenyl]azo]-, sodium salt Methyl orange	$C_{14}H_{14}N_3NaO_3S$ 327.34	547-58-0 dec	6019		H_2O 2; EtOH; eth 1; py 2
2223	Benzenesulfonic acid, ethyl ester Ethyl benzenesulfonate	$C_8H_{10}O_3S$ 186.23	515-46-8	1078 156[15]	4-11-00-00030 1.2167[20]	H_2O 2; EtOH 3; eth 4; chl 4 1.5081[20]
2224	Benzenesulfonic acid, 4-hydrazino- p-Sulfonylphenylhydrazine	$C_6H_8N_2O_3S$ 188.21	98-71-5 286	4695	3-15-00-00865	H_2O 2; EtOH 2
2225	Benzenesulfonic acid, 2-hydroxy- o-Phenolsulfonic acid	$C_6H_6O_4S$ 174.18	609-46-1 145 dec		4-11-00-00574	H_2O 4; EtOH 4
2226	Benzenesulfonic acid, 4-hydroxy- p-Phenolsulfonic acid	$C_6H_6O_4S$ 174.18	98-67-9	7212	4-11-00-00582	H_2O 4; EtOH 4
2227	Benzenesulfonic acid, methyl ester Methyl benzenesulfonate	$C_7H_8O_3S$ 172.20	80-18-2	154[20]	4-11-00-00029 1.2730[17]	H_2O 2; EtOH 4; eth 4; chl 4 1.5151[20]
2228	Benzenesulfonic acid, 2-methyl-	$C_7H_8O_3S$ 172.20	88-20-0 67.5	128.8[25]	4-11-00-00228	H_2O 4; EtOH 3; eth 1
2229	Benzenesulfonic acid, 4-methyl- p-Toluenesulfonic acid	$C_7H_8O_3S$ 172.20	104-15-4 104.5	9459 140[20]	4-11-00-00241	H_2O 4; EtOH 4; eth 3
2230	Benzenesulfonic acid, 4-methyl-, butyl ester	$C_{11}H_{16}O_3S$ 228.31	778-28-9	164-6[6]	4-11-00-00250 1.1319[20]	H_2O 1; eth 3; ctc 2 1.5050[20]
2231	Benzenesulfonic acid, 4-methyl-, ethyl ester Ethyl p-toluenesulfonate	$C_9H_{12}O_3S$ 200.26	80-40-0 34.5	3814 173[15]	4-11-00-00248 1.166[48]	H_2O 1; EtOH 3; eth 3; ctc 2
2232	Benzenesulfonic acid, 4-methyl-, methyl ester	$C_8H_{10}O_3S$ 186.23	80-48-8 28.5	292; 186[22]	4-11-00-00247 1.2087[40]	H_2O 1; EtOH 4; eth 3; bz 4
2233	Benzenesulfonic acid, 4-methyl-, propyl ester	$C_{10}H_{14}O_3S$ 214.29	599-91-7 <-20	189[9]	4-11-00-00249 1.144[20]	1.4998[20]
2234	Benzenesulfonic acid, 4-(phenylamino)- N-Phenylsulfanilic acid	$C_{12}H_{11}NO_3S$ 249.29	101-57-5 206	7284	3-14-00-02025	H_2O 4; EtOH 4
2235	Benzenesulfonic acid, propyl ester Propyl benzenesulfonate	$C_9H_{12}O_3S$ 200.26	80-42-2	162-3[15]	4-11-00-00031 1.1804[17]	H_2O 2; EtOH 3; eth 4; chl 4 1.5035[25]
2236	Benzenesulfonothioic acid, 4-chloro-, S-phenyl ester Benzenesulfonic acid, p-chlorothio-, S-phenyl ester	$C_{12}H_9ClO_2S_2$ 284.79	1142-97-8 -86.8	118.5	4-11-00-00223 0.8341[20]	1.4087[14]
2237	Benzene, 1,1'-sulfonylbis- Diphenyl sulfone	$C_{12}H_{10}O_2S$ 218.28	127-63-9 128.5	3336 379	4-06-00-01490 1.252[20]	H_2O 1; EtOH 3; eth 3; bz 3
2238	Benzene, 1,1'-sulfonylbis[4-bromo-	$C_{12}H_8Br_2O_2S$ 376.07	2050-48-8 172		4-06-00-01652 1.88[25]	H_2O 1; EtOH 2; HOAc 3
2239	Benzene, 1,1'-sulfonylbis[4-chloro-	$C_{12}H_8Cl_2O_2S$ 287.17	80-07-9 147.9	250[10]	4-06-00-01587	H_2O 2; EtOH 3; chl 3
2240	Benzene, 1,1'-sulfonylbis[4-methyl-	$C_{14}H_{14}O_2S$ 246.33	599-66-6 159	406	4-06-00-02174	H_2O 2; EtOH 3; eth 2; bz 3
2241	Benzene, 1,1'-[sulfonylbis(methylene)]bis-	$C_{14}H_{14}O_2S$ 246.33	620-32-6 152	290 dec	4-06-00-02651	H_2O 1; EtOH 2; ace 4; bz 3

No.	Name Synonym	Mol. Form. Mol. Wt.	CAS RN mp/°C	Merck No. bp/°C	Beil. Ref. den/g cm^{-3}	Solubility n_D
2242	Benzenesulfonyl chloride Phenylsulfonyl chloride	$C_6H_5ClO_2S$ 176.62	98-09-9 14.5	1080 251 dec	4-11-00-00049 1.3470[15]	H_2O 1; EtOH 4; eth 3; ctc 3
2243	Benzenesulfonyl chloride, 4-(acetylamino)- Acetylsulfanilyl chloride	$C_8H_8ClNO_3S$ 233.68	121-60-8 149	99	4-14-00-02703	EtOH 4; eth 4; bz 3; chl 3
2244	Benzenesulfonyl chloride, 4-bromo- p-Brosyl chloride	$C_6H_4BrClO_2S$ 255.52	98-58-8 76	1395 153[15]	4-11-00-00162	H_2O 1; eth 4; chl 3
2245	Benzenesulfonyl chloride, 4-chloro-	$C_6H_4Cl_2O_2S$ 211.07	98-60-2 51	141[15]	4-11-00-00114	eth 4; bz 4
2246	Benzenesulfonyl chloride, 4-fluoro-	$C_6H_4ClFO_2S$ 194.61	349-88-2 30	106[9]	4-11-00-00102	bz 4; eth 4; chl 4
2247	Benzenesulfonyl chloride, 4-iodo- Pipsyl chloride	$C_6H_4ClIO_2S$ 302.52	98-61-3 85	7458	4-11-00-00169	
2248	Benzenesulfonyl chloride, 2-methoxy-	$C_7H_7ClO_3S$ 206.65	10130-87-7 56	126[0.3]	4-11-00-00575	eth 3; peth 3
2249	Benzenesulfonyl chloride, 2-methyl- o-Toluenesulfonyl chloride	$C_7H_7ClO_2S$ 190.65	133-59-5 10.2	154[36]	4-11-00-00229 1.3383[20]	H_2O 1; EtOH 3; eth 3; bz 3 1.5565[20]
2250	Benzenesulfonyl chloride, 4-methyl- p-Toluenesulfonyl chloride	$C_7H_7ClO_2S$ 190.65	98-59-9 71	9460 145-6[15]	4-11-00-00375	H_2O 1; EtOH 3; eth 3; bz 4
2251	Benzenesulfonyl fluoride Phenylsulfonyl fluoride	$C_6H_5FO_2S$ 160.17	368-43-4 	203.5	4-11-00-00049 1.3286[20]	EtOH 3; eth 3 1.4932[18]
2252	Benzenesulfonyl fluoride, 4-amino- p-Sulfanilyl fluoride	$C_6H_6FNO_2S$ 175.18	98-62-4 68.5	8904	4-14-00-02658	
2253	Benzene, 1,2,3,5-tetrabromo- 1,2,3,5-Tetrabromobenzene	$C_6H_2Br_4$ 393.70	634-89-9 99.5	329	4-05-00-00686	bz 4; eth 4; EtOH 4
2254	Benzene, 1,2,4,5-tetrabromo- 2,3,5,6-Tetrabromobenzene	$C_6H_2Br_4$ 393.70	636-28-2 182	3.1[20]	4-05-00-00687 3.072[20]	eth 4
2255	Benzene, 1,2,3,4-tetrabromo-5,6-dimethyl-	$C_8H_6Br_4$ 421.75	2810-69-7 262	374.5	3-05-00-00820	bz 4
2256	Benzene, 1,2,3,5-tetrabromo-4-methoxy- 2,3,4,6-Tetrabromoanisole	$C_7H_4Br_4O$ 423.72	95970-07-3 113.5	340	3-06-00-00766	EtOH 4
2257	1,2,4,5-Benzenetetracarboxylic acid Pyromellitic acid	$C_{10}H_6O_8$ 254.15	89-05-4 276	8015	4-09-00-03804	H_2O 2; EtOH 3
2258	1,2,4,5-Benzenetetracarboxylic acid, tetramethyl ester	$C_{14}H_{14}O_8$ 310.26	635-10-9 144	 sub	4-09-00-03804	EtOH 4
2259	Benzene, 1,2,3,4-tetrachloro- 1,2,3,4-Tetrachlorobenzene	$C_6H_2Cl_4$ 215.89	634-66-2 47.5	254	4-05-00-00667	H_2O 1; EtOH 2; eth 4; CS_2 4
2260	Benzene, 1,2,4,5-tetrachloro- s-Tetrachlorobenzene	$C_6H_2Cl_4$ 215.89	95-94-3 139.5	244.5	4-05-00-00668 1.858[22]	H_2O 1; EtOH 2; eth 3; bz 3
2261	Benzene, 1,2,3,5-tetrachloro-4,6-dimethyl-	$C_8H_6Cl_4$ 243.95	877-09-8 223		4-05-00-00944 1.703[25]	H_2O 1; EtOH 1; eth 3; bz 3
2262	Benzene, 1,2,3,5-tetrachloro-4-methyl- 2,3,4,6-Tetrachlorotoluene	$C_7H_4Cl_4$ 229.92	875-40-1 94	276.5	3-05-00-00702	bz 4; EtOH 4
2263	Benzene, 1,2,4,5-tetrachloro-3-methyl- 2,3,5,6-Tetrachlorotoluene	$C_7H_4Cl_4$ 229.92	1006-31-1 93.5	sub	3-05-00-00702	eth 4; EtOH 4
2264	Benzene, 1,2,4,5-tetrachloro-3-nitro-	$C_6HCl_4NO_2$ 260.89	117-18-0 99.5	304	4-05-00-00728 1.744[25]	H_2O 1; EtOH 3; bz 3; chl 3
2265	Benzene, tetracosyl- Tetracosylbenzene	$C_{30}H_{54}$ 414.76	61828-05-5 57	373[100]; 238[1]	4-05-00-01239 0.8544[20]	1.4800[20]
2266	Benzene, tetradecyl-	$C_{20}H_{34}$ 274.49	1459-10-5 16	359	4-05-00-01212 0.8549[20]	1.4818[20]
2267	Benzene, 1,1'-tetradecylidenebis-	$C_{26}H_{38}$ 350.59	55268-63-8 17.9	207[1]	4-05-00-02017 0.9187[20]	1.5202[20]
2268	Benzene, 1,2,3,4-tetraethyl- 1,2,3,4-Tetraethylbenzene	$C_{14}H_{22}$ 190.33	642-32-0 11.8	252; 121.7[14]	4-05-00-01168 0.8875[20]	eth 4; EtOH 4 1.5125[20]
2269	Benzene, 1,2,4,5-tetraethyl- 1,2,4,5-Tetraethylbenzene	$C_{14}H_{22}$ 190.33	635-81-4 10	250	4-05-00-01168 0.8788[20]	eth 4; EtOH 4 1.5054[20]
2270	Benzene, 1,2,3,4-tetrafluoro- 1,2,3,4-Tetrafluorobenzene	$C_6H_2F_4$ 150.08	551-62-2 	94.3	4-05-00-00639	1.4054[20]
2271	Benzene, 1,2,3,5-tetrafluoro- 1,2,3,5-Tetrafluorobenzene	$C_6H_2F_4$ 150.08	2367-82-0 -48	84.4	4-05-00-00639 1.319[25]	1.4035[20]
2272	Benzene, 1,2,4,5-tetrafluoro- 1,2,4,5-Tetrafluorobenzene	$C_6H_2F_4$ 150.08	327-54-8 4.5	90.2	4-05-00-00639 1.4255[20]	1.4075[20]
2273	Benzene, 1,2,3,4-tetraiodo- 1,2,3,4-Tetraiodobenzene	$C_6H_2I_4$ 581.70	634-68-4 136	sub	0-05-00-00229	eth 4; EtOH 4
2274	Benzene, 1,2,3,5-tetraiodo- 1,2,3,5-Tetraiodobenzene	$C_6H_2I_4$ 581.70	634-92-4 148	sub	0-05-00-00229	HOAc 4
2275	Benzene, 1,2,4,5-tetraiodo- 1,2,4,5-Tetraiodobenzene	$C_6H_2I_4$ 581.70	636-31-7 254	sub	0-05-00-00229	HOAc 4
2276	Benzene, 1,2,4,5-tetrakis(1-methylethyl)- 1,2,4,5-Tetraisopropylbenzene	$C_{18}H_{30}$ 246.44	635-11-0 118.4	259; 133[17]	4-05-00-01207 0.758[150]	
2277	Benzene, 1,2,3,4-tetramethoxy-5-(2-propenyl)-	$C_{13}H_{18}O_4$ 238.28	15361-99-6 25	145[12]	3-06-00-06691 1.087[25]	1.5146[25]
2278	Benzene, 1,2,3,4-tetramethyl-	$C_{10}H_{14}$ 134.22	488-23-3 -6.2	205	4-05-00-01072 0.9052[20]	H_2O 1; EtOH 5; eth 5; ace 5 1.5203[20]
2279	Benzene, 1,2,3,5-tetramethyl- Isodurene	$C_{10}H_{14}$ 134.22	527-53-7 -23.7	5050 198	4-05-00-01073 0.8903[20]	H_2O 1; EtOH 5; eth 5; ace 5 1.5130[20]
2280	Benzene, 1,2,4,5-tetramethyl- Durene	$C_{10}H_{14}$ 134.22	95-93-2 79.3	3450 196.8	4-05-00-01076 0.8380[81]	H_2O 1; EtOH 5; eth 5; ace 5 1.4790[81]

No.	Name Synonym	Mol. Form. Mol. Wt.	CAS RN mp/°C	Merck No. bp/°C	Beil. Ref. den/g cm^{-3}	Solubility n_D
2281	Benzene, 1,1'-(1,1,4,4-tetramethyl-1,4-butanediyl)bis- Bineophyl	$C_{20}H_{26}$ 266.43	17648-05-4 61.8	146-50[0.5]	4-05-00-01997	chl 3
2282	Benzene, 1,2,4,5-tetramethyl-3,6-dinitro-	$C_{10}H_{12}N_2O_4$ 224.22	5465-13-4 211.5	sub	4-05-00-01080	H_2O 1; EtOH 2; eth 4; bz 3
2283	Benzene, 1,1'-thiobis- Phenyl sulfide	$C_{12}H_{10}S$ 186.28	139-66-2 -25.9	7285 296	4-06-00-01488 1.1136[20]	H_2O 1; EtOH 3; eth 5; bz 5 1.6334[20]
2284	Benzene, 1,1'-thiobis[4-bromo-	$C_{12}H_8Br_2S$ 344.07	3393-78-0 115	268.5[40]	4-06-00-01651 1.84[25]	H_2O 1; EtOH 3; eth 3; ctc 3
2285	Benzene, 1,1'-thiobis[2-methyl-	$C_{14}H_{14}S$ 214.33	4537-05-7 64	285	4-06-00-02019	eth 4; EtOH 4; chl 4
2286	Benzene, 1,1'-thiobis[4-methyl-	$C_{14}H_{14}S$ 214.33	620-94-0 57.3	>300; 175[16]	4-06-00-02173	H_2O 1; EtOH 3; eth 4; ace 3
2287	Benzene, 1,1'-[thiobis(methylene)]bis- Benzyl sulfide	$C_{14}H_{14}S$ 214.33	538-74-9 49.5	1162 dec	4-06-00-02649 1.0583[50]	H_2O 1; EtOH 3; eth 3; CS_2 3
2288	Benzene, 1,1'-thiobis[2,4,6-trinitro- Picryl sulfide	$C_{12}H_4N_6O_{12}S$ 456.26	2217-06-3		4-06-00-01775	
2289	Benzenethiol Phenyl mercaptan	C_6H_6S 110.18	108-98-5 -14.9	9285 169.1	4-06-00-01463 1.0775[20]	H_2O 1; EtOH 3; eth 3; bz 3 1.5893[20]
2290	Benzenethiol, 2-amino-	C_6H_7NS 125.19	137-07-5 26	234	4-13-00-00910	EtOH 3; eth 3 1.4606[20]
2291	Benzenethiol, 4-amino-	C_6H_7NS 125.19	1193-02-8 46	143[17]	4-13-00-01289	H_2O 3; EtOH 3
2292	Benzenethiol, 4-bromo-	C_6H_5BrS 189.08	106-53-6 73	230.5	4-06-00-01650 1.5260[83]	H_2O 2; EtOH 2; eth 4; ctc 4
2293	Benzenethiol, 2-chloro-	C_6H_5ClS 144.62	6320-03-2	205.5	4-06-00-01570 1.2752[10]	H_2O 2; EtOH 2
2294	Benzenethiol, 3-chloro-	C_6H_5ClS 144.62	2037-31-2	206	4-06-00-01576 1.2637[13]	H_2O 1; EtOH 3; eth 3; chl 3
2295	Benzenethiol, 4-chloro-	C_6H_5ClS 144.62	106-54-7 61	206	4-06-00-01581 1.1911[20]	H_2O 1; EtOH 4; eth 4; bz 4 1.5480[20]
2296	Benzenethiol, 4-ethoxy-	$C_8H_{10}OS$ 154.23	699-09-2 1.6	238	4-06-00-05792 1.1070[17]	ace 4; bz 4; eth 4; EtOH 4
2297	Benzenethiol, 3-ethyl- 3-Ethylthiophenol	$C_8H_{10}S$ 138.23	62154-77-2 	211	3-06-00-01663 1.038[20]	1.572[30]
2298	Benzenethiol, 4-ethyl-	$C_8H_{10}S$ 138.23	4946-13-8 	211	4-06-00-03027 1.038[20]	1.572[20]
2299	Benzenethiol, 3-methoxy-	C_7H_8OS 140.21	15570-12-4 	224.5; 114[20]	4-06-00-05701 	chl 3 1.5874[20]
2300	Benzenethiol, 4-methoxy-	C_7H_8OS 140.21	696-63-9 	228	4-06-00-05790 1.1313[25]	EtOH 3; eth 3; bz 3; chl 2 1.5801[25]
2301	Benzenethiol, 2-methyl-	C_7H_8S 124.21	137-06-4 15	195	4-06-00-02014 1.041[20]	H_2O 1; EtOH 3; eth 4 1.570[20]
2302	Benzenethiol, 3-methyl-	C_7H_8S 124.21	108-40-7 -20	195	4-06-00-02079 1.044[20]	H_2O 1; EtOH 3; eth 5 1.572[20]
2303	Benzenethiol, 4-methyl-	C_7H_8S 124.21	106-45-6 43	195	4-06-00-02153 1.0220[51]	H_2O 1; EtOH 3; eth 4; chl 3
2304	Benzenethiol, pentafluoro-	C_6HF_5S 200.13	771-62-0 -24	143	1.501[25]	1.4645[20]
2305	1,2,3-Benzenetriamine 1,2,3-Triaminobenzene	$C_6H_9N_3$ 123.16	608-32-2 103	336	3-13-00-00551	H_2O 4; eth 4; EtOH 4
2306	1,2,4-Benzenetriamine	$C_6H_9N_3$ 123.16	615-71-4 97.3	340	4-13-00-00506	H_2O 4; EtOH 4; chl 4
2307	Benzene, 1,2,3-tribromo- 1,2,3-Tribromobenzene	$C_6H_3Br_3$ 314.80	608-21-9 87.5		3-05-00-00569 2.658[25]	eth 4
2308	Benzene, 1,2,4-tribromo- 1,2,4-Tribromobenzene	$C_6H_3Br_3$ 314.80	615-54-3 44.5	275	4-05-00-00685	H_2O 1; EtOH 3; eth 4; ace 4
2309	Benzene, 1,3,5-tribromo- 1,3,5-Tribromobenzene	$C_6H_3Br_3$ 314.80	626-39-1 122.8	271	4-05-00-00685	H_2O 1; EtOH 2; eth 3; bz 3
2310	Benzene, 1,2,3-tribromo-5-methoxy- Anisole, 3,4,5-tribromo	$C_7H_5Br_3O$ 344.83	73557-60-5 93.3	303	2-06-00-00195	ace 4; bz 4; EtOH 4
2311	Benzene, 1,2,4-tribromo-5-methoxy- 2,4,5-Tribromoanisole	$C_7H_5Br_3O$ 344.83	95970-10-8 105	307	4-06-00-01067	ace 4; bz 4; EtOH 4
2312	Benzene, 1,2,5-tribromo-3-methoxy- Anisole, 2,3,5-tribromo	$C_7H_5Br_3O$ 344.83	73931-44-9 82	308.5	2-06-00-00192	ace 4; bz 4; EtOH 4
2313	Benzene, 1,3,5-tribromo-2-methoxy-	$C_7H_5Br_3O$ 344.83	607-99-8 88	298	4-06-00-01067 2.491[25]	H_2O 2; EtOH 2; ace 4; bz 4
2314	Benzene, 1,3,5-tribromo-2-methyl- 2,4,6-Tribromotoluene	$C_7H_5Br_3$ 328.83	6320-40-7 70	290	3-05-00-00719 2.479[17]	
2315	Benzene, 1,2,3-tribromo-5-nitro- 3,4,5-Tribromonitrobenzene	$C_6H_2Br_3NO_2$ 359.80	3460-20-6 112	sub	4-05-00-00732 2.645[25]	
2316	Benzene, 1,3,5-tribromo-2-nitro-	$C_6H_2Br_3NO_2$ 359.80	3463-40-9 125	177[11]	4-05-00-00732	eth 4; chl 4
2317	1,3,5-Benzenetricarbonyl trichloride	$C_9H_3Cl_3O_3$ 265.48	4422-95-1 36.3	180[16]	4-09-00-03748	chl 3
2318	1,2,3-Benzenetricarboxylic acid Hemimellitic acid	$C_9H_6O_6$ 210.14	569-51-7 200		4-09-00-03745 1.546[20]	eth 4; EtOH 4
2319	1,2,4-Benzenetricarboxylic acid Trimellitic acid	$C_9H_6O_6$ 210.14	528-44-9 219	9616	4-09-00-03746	H_2O 4; eth 4; EtOH 4
2320	1,3,5-Benzenetricarboxylic acid, 2-chloro- 2-Chloro-1,3,5-benzenetricarboxylic acid	$C_9H_5ClO_6$ 244.59	56961-24-1 285	sub	4-09-00-03749	H_2O 4; eth 4; EtOH 4

No.	Name Synonym	Mol. Form. Mol. Wt.	CAS RN mp/°C	Merck No. bp/°C	Beil. Ref. den/g cm^{-3}	Solubility n_D
2321	Benzene, 1,2,3-trichloro- 1,2,3-Trichlorobenzene	$C_6H_3Cl_3$ 181.45	87-61-6 53.5	9542 218.5	4-05-00-00664 1.4533[25]	H_2O 1; EtOH 2; eth 4; bz 4
2322	Benzene, 1,2,4-trichloro- 1,2,4-Trichlorobenzene	$C_6H_3Cl_3$ 181.45	120-82-1 17	9543 213.5	4-05-00-00664 1.459[25]	H_2O 1; EtOH 2; eth 4; ctc 2 1.5717[20]
2323	Benzene, 1,3,5-trichloro- 1,3,5-Trichlorobenzene	$C_6H_3Cl_3$ 181.45	108-70-3 63.5	9544 208	4-05-00-00666	H_2O 1; EtOH 2; eth 4; bz 4
2324	Benzene, 1,2,4-trichloro-5-(chloromethyl)-	$C_7H_4Cl_4$ 229.92	3955-26-8	273	4-05-00-00822 1.547[20]	ace 4; eth 4; EtOH 4
2325	Benzene, 1,2,4-trichloro-5-[(4-chlorophenyl)sulfonyl]- Tetradifon	$C_{12}H_6Cl_4O_2S$ 356.06	116-29-0 146	9132	4-06-00-01636 1.151[20]	
2326	Benzene, 1,2,3-trichloro-4-(dichloromethyl)- Benzylidene chloride, 2,3,4-trichloro	$C_7H_3Cl_5$ 264.36	56961-82-1 84	278	1-05-00-00153	bz 4
2327	Benzene, 1,2,4-trichloro-5-(dichloromethyl)- Benzylidene chloride, 2,4,5-trichloro	$C_7H_3Cl_5$ 264.36	33429-70-8 280.5		4-05-00-00824 1.5956[20]	bz 4 1.5992[20]
2328	Benzene, 1,1'-(2,2,2-trichloroethylidene)bis[4-chloro- DDT	$C_{14}H_9Cl_5$ 354.49	50-29-3 108.5	2832 260; 186[0.05]	4-05-00-01885	H_2O 1; EtOH 2; eth 4; ace 4
2329	Benzene, 1,1'-(2,2,2-trichloroethylidene)bis[4-methoxy- Methoxychlor	$C_{16}H_{15}Cl_3O_2$ 345.65	72-43-5 87	5913	4-06-00-06691 1.41[25]	H_2O 1; EtOH 3; eth 4; bz 4
2330	Benzene, 1,2,4-trichloro-3-methoxy-	$C_7H_5Cl_3O$ 211.47	50375-10-5 45	227	3-06-00-00717	ace 4; EtOH 4
2331	Benzene, 1,2,4-trichloro-5-methoxy-	$C_7H_5Cl_3O$ 211.47	6130-75-2 77.5	254	4-06-00-00963	EtOH 4; ace 4; os 3
2332	Benzene, 1,3,5-trichloro-2-methoxy- 2,4,6-Trichloroanisole	$C_7H_5Cl_3O$ 211.47	87-40-1 61.5	9541 241	4-06-00-01005 1.640[25]	EtOH 3; ace 4; bz 3; chl 3
2333	Benzene, (trichloromethyl)- (Trichloromethyl)benzene	$C_7H_5Cl_3$ 195.47	98-07-7 -5	1120 221	4-05-00-00820 1.3723[20]	H_2O 1; EtOH 3; eth 3; bz 3 1.5580[20]
2334	Benzene, 1,2,3-trichloro-4-methyl- 2,3,4-Trichlorotoluene	$C_7H_5Cl_3$ 195.47	7359-72-0 43.5	244	4-05-00-00819	H_2O 1; EtOH 3; eth 3; ace 3
2335	Benzene, 1,2,3-trichloro-5-methyl- Toluene, 3,4,5-trichloro-	$C_7H_5Cl_3$ 195.47	21472-86-6 45.5	246	0-05-00-00299	H_2O 1; EtOH 3; ace 3; os 3
2336	Benzene, 1,2,4-trichloro-5-methyl- 2,4,5-Trichlorotoluene	$C_7H_5Cl_3$ 195.47	6639-30-1 82.4	231	4-05-00-00819	H_2O 1; EtOH 3; ace 3; os 3
2337	Benzene, 1,2,5-trichloro-3-methyl- 2,3,5-Trichlorotoluene	$C_7H_5Cl_3$ 195.47	56961-86-5 45.5	230	0-05-00-00299	H_2O 1; EtOH 3; ace 3; os 3
2338	Benzene, 1,2,3-trichloro-5-nitro-	$C_6H_2Cl_3NO_2$ 226.45	20098-48-0 72.5		2-05-00-00187 1.807[25]	EtOH 4
2339	Benzene, 1,2,4-trichloro-5-nitro-	$C_6H_2Cl_3NO_2$ 226.45	89-69-0 57.5	288	4-05-00-00728 1.790[23]	H_2O 1; EtOH 2; eth 3; bz 3
2340	Benzene, 1,3,5-trichloro-2,4,6-trifluoro- 1,3,5-Trichloro-2,4,6-trifluorobenzene	$C_6Cl_3F_3$ 235.42	319-88-0 198.4			
2341	Benzene, tricosyl- Tricosylbenzene	$C_{29}H_{52}$ 400.73	61828-04-4 54	365[100]; 231[1]	4-05-00-01237 0.8544[20]	1.4801[20]
2342	Benzene, tridecyl- 1-Phenyltridecane	$C_{19}H_{32}$ 260.46	123-02-4 10	9575 346	3-05-00-01108 0.8550[20]	1.4821[20]
2343	Benzene, 1,3,5-triethoxy- 1,3,5-Triethoxybenzene	$C_{12}H_{18}O_3$ 210.27	2437-88-9 43.5	170[24]	4-06-00-07362	eth 4; EtOH 4
2344	Benzene, 1,2,4-triethyl- 1,2,4-Triethylbenzene	$C_{12}H_{18}$ 162.27	877-44-1	218; 99[15]	4-05-00-01133 0.8738[20]	H_2O 1; EtOH 3; eth 3 1.5024[20]
2345	Benzene, 1,3,5-triethyl- 1,3,5-Triethylbenzene	$C_{12}H_{18}$ 162.27	102-25-0 -66.5	215.9	4-05-00-01133 0.8631[20]	H_2O 1; EtOH 4; eth 4 1.4969[20]
2346	Benzene, 1,2,4-trifluoro-	$C_6H_3F_3$ 132.09	367-23-7	90	4-05-00-00638 1.264[25]	1.4171[20]
2347	Benzene, 1,3,5-trifluoro- 1,3,5-Trifluorobenzene	$C_6H_3F_3$ 132.09	372-38-3 -5.5	75.5	4-05-00-00638 1.277[25]	1.4140[20]
2348	Benzene, (trifluoromethyl)- (Trifluoromethyl)benzene	$C_7H_5F_3$ 146.11	98-08-8 -29.1	1121 102.1	4-05-00-00802 1.1884[20]	EtOH 5; eth 5; ace 5; bz 5 1.4146[20]
2349	Benzene, 1,2,3-triiodo- 1,2,3-Triiodobenzene	$C_6H_3I_3$ 455.80	608-29-7 116	sub	1-05-00-00122	eth 4; EtOH 4; chl 4
2350	Benzene, 1,2,4-triiodo- 1,2,4-Triiodobenzene	$C_6H_3I_3$ 455.80	615-68-9 91.5	sub	3-05-00-00585	EtOH 4; chl 4
2351	Benzene, 1,3,5-triiodo- 1,3,5-Triiodobenzene	$C_6H_3I_3$ 455.80	626-44-8 184.2	sub	4-05-00-00702	H_2O 1; EtOH 2; eth 2; bz 2
2352	Benzene, 1,3,5-triiodo-2-methyl- 2,4,6-Triiodotoluene	$C_7H_5I_3$ 469.83	36994-79-3 118.5	300 dec	1-05-00-00158	EtOH 4
2353	Benzene, 1,2,3-triiodo-5-nitro-	$C_6H_2I_3NO_2$ 500.80	53663-23-3 167		3-05-00-00626 3.256[25]	bz 4; eth 4; EtOH 4; chl 4
2354	Benzene, 1,2,3-trimethoxy-	$C_9H_{12}O_3$ 168.19	634-36-6 48.5	235	4-06-00-07329 1.1009[45]	H_2O 1; EtOH 3; eth 3; bz 3
2355	Benzene, 1,3,5-trimethoxy-	$C_9H_{12}O_3$ 168.19	621-23-8 54.5	255.5	4-06-00-07362	H_2O 1; EtOH 3; eth 3; bz 3
2356	Benzene, 1,2,4-trimethoxy-5-(1-propenyl)-	$C_{12}H_{16}O_3$ 208.26	494-40-6 67	296	4-06-00-07476 1.165[20]	eth 4; EtOH 4; peth 4 1.5683[20]
2357	Benzene, 1,2,3-trimethyl- 1,2,3-Trimethylbenzene	C_9H_{12} 120.19	526-73-8 -25.4	176.1	4-05-00-01007 0.8944[20]	H_2O 1; EtOH 5; eth 5; ace 5 1.5139[20]
2358	Benzene, 1,2,4-trimethyl- Pseudocumene	C_9H_{12} 120.19	95-63-6 -43.8	7929 169.3	4-05-00-01010 0.8758[20]	H_2O 1; EtOH 5; eth 5; ace 5 1.5048[20]

No.	Name Synonym	Mol. Form. Mol. Wt.	CAS RN mp/°C	Merck No. bp/°C	Beil. Ref. den/g cm⁻³	Solubility n_D
2359	Benzene, 1,3,5-trimethyl- Mesitylene	C_9H_{12} 120.19	108-67-8 -44.7	5810 164.7	4-05-00-01016 0.8652^{20}	H₂O 1; EtOH 5; eth 5; ace 5 1.4994^{20}
2360	Benzene, 1,3,5-trimethyl-2,4-dinitro-	$C_9H_{10}N_2O_4$ 210.19	608-50-4 85.3	 exp 418	4-05-00-01029	H₂O 1; EtOH 3
2361	Benzene, 1,2,4-trimethyl-5-(1-methylethyl)-	$C_{12}H_{18}$ 162.27	10222-95-4 21	 221	3-05-00-01038 0.8795^{21}	 1.5065^{21}
2362	Benzene, 1,2,4-trimethyl-5-nitro-	$C_9H_{11}NO_2$ 165.19	610-91-3 71	 265	4-05-00-01015	EtOH 3; chl 3; peth 3
2363	Benzene, 1,3,5-trimethyl-2-nitro-	$C_9H_{11}NO_2$ 165.19	603-71-4 44	 255	4-05-00-01028 1.51^{25}	EtOH 4
2364	Benzene, 1,3,5-trimethyl-2,4,6-trinitro-	$C_9H_9N_3O_6$ 255.19	602-96-0 238.2	 exp 415	4-05-00-01029	H₂O 1; EtOH 2; eth 2; ace 3
2365	Benzene, 1,3,5-trinitro- sym-Trinitrobenzene	$C_6H_3N_3O_6$ 213.11	99-35-4 121.5	9639 315	4-05-00-00755 1.4775^{152}	H₂O 2; EtOH 2; eth 2; ace 4
2366	1,2,3-Benzenetriol Pyrogallol	$C_6H_6O_3$ 126.11	87-66-1 133	8010 309	4-06-00-07327 1.453^4	H₂O 4; EtOH 4; eth 4; ace 3 1.561^{134}
2367	1,2,4-Benzenetriol Hydroxyhydroquinone	$C_6H_6O_3$ 126.11	533-73-3 140.5	1081	4-06-00-07338	H₂O 4; EtOH 4; eth 4; bz 1
2368	1,3,5-Benzenetriol Phloroglucinol	$C_6H_6O_3$ 126.11	108-73-6 218.5	7301 sub	4-06-00-07361 1.46^{25}	H₂O 2; EtOH 4; eth 4; ace 3
2369	1,2,3-Benzenetriol, 5-methyl-	$C_7H_8O_3$ 140.14	609-25-6 129	 sub	4-06-00-07376	bz 4
2370	1,2,4-Benzenetriol, triacetate	$C_{12}H_{12}O_6$ 252.22	613-03-6 99	 300	4-06-00-07342	EtOH 3; chl 3; MeOH 3
2371	1,3,5-Benzenetriol, 2,4,6-trichloro- Trichloro-1,3,5-trihydroxybenzene	$C_6H_3Cl_3O_3$ 229.45	56961-23-0 136	 sub	0-06-00-01104	EtOH 4
2372	Benzene, 1,2,4-tris(1-methylethyl)- 1,2,4-Triisopropylbenzene	$C_{15}H_{24}$ 204.36	948-32-3 	 244	4-05-00-01178 0.8574^{25}	 1.4896^{25}
2373	Benzene, 1,3,5-tris(1-methylethyl)-	$C_{15}H_{24}$ 204.36	717-74-8 -7.4	 238	4-05-00-01178 0.8545^{20}	ace 3; bz 3; chl 3 1.4882^{20}
2374	Benzene, undecyl-	$C_{17}H_{28}$ 232.41	6742-54-7 -5	 316	3-05-00-01100 0.8553^{20}	 1.4828^{20}
2375	9,10[1',2']-Benzenoanthracene, 9,10-dihydro- Triptycene	$C_{20}H_{14}$ 254.33	477-75-8	9663	4-05-00-02639	
2376	1H-Benzimidazole N,N'-Methenyl-o-phenylenediamine	$C_7H_6N_2$ 118.14	51-17-2 170.5	1091 >360	5-23-06-00196	H₂O 2; EtOH 4; eth 2; bz 1
2377	1H-Benzimidazole, 4,5-dichloro-2-(trifluoromethyl)- Chlorflurazole	$C_8H_3Cl_2F_3N_2$ 255.03	3615-21-2 213.5			
2378	1H-Benzimidazole, 1,5-dimethyl- 1,5-Dimethylbenzimidazole	$C_9H_{10}N_2$ 146.19	10394-35-1 95	 300	5-23-06-00380	bz 4; eth 4; EtOH 4
2379	1H-Benzimidazole, 2,5-dimethyl-	$C_9H_{10}N_2$ 146.19	1792-41-2 203	 350	5-23-06-00448	eth 4; EtOH 4
2380	1H-Benzimidazole, 5,6-dimethyl- Dimedazole	$C_9H_{10}N_2$ 146.19	582-60-5 205.5	3225 sub	5-23-06-00454	H₂O 3; EtOH 3; eth 3; chl 3
2381	1H-Benzimidazole, 1-ethyl-2-methyl- 1-Ethyl-2-methylbenzimidazole	$C_{10}H_{12}N_2$ 160.22	5805-76-5 51	 296	5-23-06-00326 1.073^{25}	
2382	1H-Benzimidazole, 1-methyl-	$C_8H_8N_2$ 132.17	1632-83-3 66	 286	5-23-06-00215 1.1254^{20}	peth 3 1.6013^7
2383	1H-Benzimidazole, 1-phenyl-	$C_{13}H_{10}N_2$ 194.24	2622-60-8 97	 221^{25}	5-23-06-00222	H₂O 2; EtOH 3; MeOH 3
2384	1H-Benzimidazole, 2-phenyl- Phenzidole	$C_{13}H_{10}N_2$ 194.24	716-79-0 293	7245	5-23-08-00437	H₂O 2; EtOH 3; bz 2; chl 2
2385	1H-Benzimidazole, 2-(phenylmethyl)- Bendazol	$C_{14}H_{12}N_2$ 208.26	621-72-7 187	1043	5-23-08-00498	bz 4; EtOH 4; gl HOAc 4
2386	2H-Benzimidazole-2-thione, 1,3-dihydro- 2-Benzimidazolethiol	$C_7H_6N_2S$ 150.20	583-39-1 298	1092	5-24-02-00413	EtOH 4
2387	1,2-Benzisothiazole 1-Thia-2-azaindene	C_7H_5NS 135.19	272-16-2 38	 220	2-27-00-00016	ctc 3
2388	1,2-Benzisothiazol-3(2H)-one, 1,1-dioxide Saccharin	$C_7H_5NO_3S$ 183.19	81-07-2 228 dec	8282 sub	4-27-00-02649 0.828^{25}	H₂O 2; EtOH 3; eth 2; ace 3
2389	1,2-Benzisoxazole Indoxazene	C_7H_5NO 119.12	271-95-4 	 100^{26}	4-27-00-01067 1.1727^{21}	eth 4 1.5570^{20}
2390	2,1-Benzisoxazole Anthranil	C_7H_5NO 119.12	271-58-9 <-18	 215	4-27-00-01068 1.1827^{20}	ace 4; EtOH 4 1.5845^{20}
2391	Benz[j]aceanthrylene, 1,2-dihydro- Cholanthrene	$C_{20}H_{14}$ 254.33	479-23-2 174	2200	4-05-00-02638	H₂O 1; EtOH 3; bz 3; HOAc 3
2392	Benz[j]aceanthrylene, 1,2-dihydro-3-methyl- 3-Methylcholanthrene	$C_{21}H_{16}$ 268.36	56-49-5 180	5967 280^{80}	4-05-00-02648 1.28^{20}	H₂O 1
2393	11H-Benzo[a]fluorene	$C_{17}H_{12}$ 216.28	238-84-6 189.5	 405	4-05-00-02473	EtOH 2; eth 3; bz 3; chl 3
2394	Benzo[a]phenazine 7,12-Diazabenz[a]anthracene	$C_{16}H_{10}N_2$ 230.27	225-61-6 142.5	 >360	5-23-09-00494	H₂O 1; EtOH 3; eth 3; ace 3
2395	Benzo[a]pyrene 2,3-Benzopyrene	$C_{20}H_{12}$ 252.32	50-32-8 176.5	1113	4-05-00-02687	chl 4
2396	2H-Benzo[a]quinolizin-2-one, 1,3,4,6,7,11b-hexahydro-9,10-dimethoxy-3-(2-methylpropyl)- 1,2-Benzoquinolizine-2-one, 9,10-dimethoxy-1,2,3,4,6,7,11b-heptahydro-3-isobutyl	$C_{19}H_{27}NO_3$ 317.43	58-46-8 128	9118	5-21-13-00178	chl 3

No.	Name Synonym	Mol. Form. Mol. Wt.	CAS RN mp/°C	Merck No. bp/°C	Beil. Ref. den/g cm^{-3}	Solubility n_D
2397	2H,8H-Benzo[1,2-b:5,4-b']dipyran-2-one, 8,8-dimethyl- Xanthyletin	$C_{14}H_{12}O_3$ 228.25	553-19-5 131.5	9978 140-5[0.1]	5-19-04-00470	EtOH 3; peth 3
2398	2H,8H-Benzo[1,2-b:5,4-b']dipyran-2-one, 5-methoxy-8,8-dimethyl- Xanthoxyletin	$C_{15}H_{14}O_4$ 258.27	84-99-1 133	9975	5-19-06-00039	H_2O 1; EtOH 3; eth 2; ace 3
2399	Benzo[b]thiophene Thianaphthene	C_8H_6S 134.20	95-15-8 32	9232 221	5-17-02-00006 1.1484[32]	H_2O 1; EtOH 4; eth 3; ace 3 1.6374[37]
2400	Benzo[b]thiophene-2-carboxylic acid Thionaphthene-2-carboxylic acid	$C_9H_6O_2S$ 178.21	6314-28-9 240.5	9274	5-18-06-00427	eth 4
2401	Benzo[b]thiophene-2,3-dione	$C_8H_4O_2S$ 164.18	493-57-2 121	247	5-17-11-00251	H_2O 1; EtOH 3; bz 3; HOAc 3
2402	Benzo[b]thiophene, 5-methyl-	C_9H_8S 148.23	14315-14-1 20.5	111-5[12]	5-17-02-00031 1.111[22]	
2403	Benzo[b]thiophene-4-ol	C_8H_6OS 150.20	3610-02-4 79.8	sub	5-17-04-00199	EtOH 4 1.615[22]
2404	Benzo[b]thiophen-3(2H)-one, 2-(3-oxobenzo[b]thien-2(3H)-ylidene)-	$C_{16}H_8O_2S_2$ 296.37	522-75-8 359	sub	5-19-05-00281	H_2O 1; EtOH 1; bz 3; chl 2
2405	7H-Benzo[c]carbazole 7-Aza-7H-benzo[c]fluorene	$C_{16}H_{11}N$ 217.27	205-25-4 134	448	5-20-08-00455	bz 4; EtOH 4; HOAc 4
2406	Benzo[c]phenanthrene Tetrahelicene	$C_{18}H_{12}$ 228.29	195-19-7 68	1129	4-05-00-02552	H_2O 1; EtOH 2; lig 2
2407	5H-Benzocyclohepten-5-one, 6,7,8,9-tetrahydro-	$C_{11}H_{12}O$ 160.22	826-73-3 124[7]; 108[1]		4-07-00-01029 1.0780[20]	EtOH 3 1.5698[20]
2408	5H-Benzocyclohepten-5-one, 2,3,4,6-tetrahydroxy- Purpurogallin	$C_{11}H_8O_5$ 220.18	569-77-7 274-5 dec	7963	4-08-00-03456	
2409	3H-1,4-Benzodiazepin-2-amine, 7-chloro-N-methyl-5-phenyl-, 4-oxide Chlorodiazepoxide	$C_{16}H_{14}ClN_3O$ 299.76	58-25-3 236.2	2082		
2410	2H-1,4-Benzodiazepin-2-one, 7-chloro-1,3-dihydro-1-methyl-5-phenyl- Diazepam	$C_{16}H_{13}ClN_2O$ 284.74	439-14-5 132	2977	5-24-04-00300	
2411	2H-1,4-Benzodiazepin-2-one, 7-chloro-1,3-dihydro-5-phenyl- Nordazepam	$C_{15}H_{11}ClN_2O$ 270.72	1088-11-5 216.5	6608	5-24-04-00291	
2412	1,3,2-Benzodioxaborole	$C_6H_5BO_2$ 119.92	274-07-7 12	88[156]; 50[50]	1.2700[20]	1.5070[20]
2413	1,3,2-Benzodioxaphosphole, 2-chloro-	$C_6H_4ClO_2P$ 174.52	1641-40-3 30	80[20]	4-06-00-05598 1.4650[20]	1.5712[20]
2414	1,4-Benzodioxin, 2,3-dihydro-	$C_8H_8O_2$ 136.15	493-09-4	103[6]	5-19-01-00439 1.142[25]	1.5485[20]
2415	1,3-Benzodioxole	$C_7H_6O_2$ 122.12	274-09-9	172.5; 77[27]	5-19-01-00434 1.064[25]	1.5398[20]
2416	1,3-Benzodioxole-5-acetic acid, methyl ester Phenylacetic acid, 3,4-methylenedioxy, methyl ester	$C_{10}H_{10}O_4$ 194.19	326-59-0 153-5[10]		2-19-00-00295 1.246[20]	1.534
2417	1,3-Benzodioxole-5-carboxaldehyde Piperonal	$C_8H_6O_3$ 150.13	120-57-0 37	7445 263	5-19-04-00225	H_2O 2; EtOH 4; eth 5; ace 3
2418	1,3-Benzodioxole-5-carboxylic acid Piperonylic acid	$C_8H_6O_4$ 166.13	94-53-1 229	7447	5-19-07-00300	
2419	1,3-Benzodioxole-5-carboxylic acid, ethyl ester	$C_{10}H_{10}O_4$ 194.19	6951-08-2 18.5	285.5; 135[6]	5-19-07-00300	eth 4; EtOH 4; peth 4
2420	1,3-Benzodioxole-5-carboxylic acid, methyl ester	$C_9H_8O_4$ 180.16	326-56-7 53	273 dec	5-19-07-00300	eth 4; EtOH 4
2421	1,3-Benzodioxole, 5-(chloromethyl)-	$C_8H_7ClO_2$ 170.60	20850-43-5 20.5	134-5[14]	5-19-01-00443 1.312[25]	1.5660[20]
2422	1,3-Benzodioxole, 4,7-dimethoxy-5-(1-propenyl)-, (E)-	$C_{12}H_{14}O_4$ 222.24	17672-88-7 56	303.5	4-19-00-01033	ace 4; bz 4; eth 4; EtOH 4
2423	1,3-Benzodioxole, 4,7-dimethoxy-5-(2-propenyl)- Apiole	$C_{12}H_{14}O_4$ 222.24	523-80-8 29.5	767 294. 179[35]	5-19-03-00307 1.015[20]	ace 4; bz 4; EtOH 4; lig 4 1.5360[20]
2424	1,3-Benzodioxole-5-ethanamine	$C_9H_{11}NO_2$ 165.19	1484-85-1 166[20]		5-19-08-00407 1.225[20]	1.5620[20]
2425	1,3-Benzodioxole, 5-ethenyl-	$C_9H_8O_2$ 148.16	7315-32-4 224		5-19-01-00544 1.1488[18]	eth 4; EtOH 4; chl 4 1.5802
2426	1,3-Benzodioxole-5-methanamine	$C_8H_9NO_2$ 151.17	2620-50-0 138-9[13]		5-19-08-00395 1.214[25]	1.5635[20]
2427	1,3-Benzodioxole-5-methanol	$C_8H_8O_3$ 152.15	495-76-1 58	157[16]	5-19-02-00543	H_2O 2; EtOH 3; eth 3; bz 3
2428	1,3-Benzodioxole, 4-methoxy-6-(2-propenyl)- Myristicin	$C_{11}H_{12}O_3$ 192.21	607-91-0 <-20	6247 276.5	5-19-02-00631 1.1416[20]	H_2O 1; EtOH 2; eth 3; bz 3
2429	1,3-Benzodioxole, 5-(2-propenyl)- Safrole	$C_{10}H_{10}O_2$ 162.19	94-59-7 11.2	8287 234.5	5-19-01-00553 1.1000[20]	H_2O 1; EtOH 4; eth 5; chl 5 1.5381[20]
2430	1,3-Benzodioxole, 5-(1-propenyl)-, (E)-	$C_{10}H_{10}O_2$ 162.19	4043-71-4 6.8	253	5-19-01-00552 1.1224[20]	H_2O 1; EtOH 5; eth 5; ace 4 1.5782[20]

No.	Name / Synonym	Mol. Form. / Mol. Wt.	CAS RN / mp/°C	Merck No. / bp/°C	Beil. Ref. / den/g cm^{-3}	Solubility / n_D
2431	1,3-Benzodioxole, 5-propyl- / Dihydrosafrole	$C_{10}H_{12}O_2$ / 164.20	94-58-6		5-19-01-00473	ctc 3
2432	[1,3]Benzodioxolo[5,6-c]-1,3-dioxolo[4,5-i]phenanthridinium, 13-methyl- / Sanguinarine	$C_{20}H_{15}NO_5$ / 349.34	2447-54-3 / 266	8320		ace 4; bz 4; eth 4; EtOH 4
2433	[1,3]Benzodioxolo[5,6-c]phenanthridinium, 1,2-dimethoxy-12-methyl- / Chelerythrine	$C_{21}H_{18}NO_5$ / 365.39	34316-15-9	2041 / 207		chl 4
2434	1,4-Benzodithiin	$C_8H_6S_2$ / 166.27	255-50-5	/ 67-70[0.1]	5-19-01-00536 / 1.2799[20]	1.6754[25]
2435	Benzo[e]-1,3-dioxolo[4,5-l][2]benzazecin-12(5H)-one, 4,6,7,13-tetrahydro-9,10-dimethoxy-5-methyl- / Cryptopine	$C_{21}H_{23}NO_5$ / 369.42	482-74-6 / 223	2611	4-27-00-06652 / 1.315[20]	H_2O 1; EtOH 2; eth 2; bz 2
2436	Benzo[e]pyrene / 1,2-Benzpyrene	$C_{20}H_{12}$ / 252.32	192-97-2 / 177.5	1114 / 311	4-05-00-02689	
2437	Benzo[f]quinoline / β-Naphthoquinoline	$C_{13}H_9N$ / 179.22	85-02-9 / 94	1115 / 352; 202-5[8]	5-20-08-00220	H_2O 2; EtOH 4; eth 4; ace 3
2438	Benzofuran / Coumarone	C_8H_6O / 118.14	271-89-6 / <-18	1098 / 174	5-17-02-00003 / 1.0913[25]	H_2O 1; EtOH 3; eth 3 / 1.5615[17]
2439	2-Benzofurancarboxylic acid / Coumarilic acid	$C_9H_6O_3$ / 162.14	496-41-3 / 192.5	2562 / 312.5	5-18-06-00419	EtOH 4
2440	2-Benzofurancarboxylic acid, ethyl ester	$C_{11}H_{10}O_3$ / 190.20	3199-61-9 / 30.5	/ 276; 161[15]	5-18-06-00420 / 1.1656[28]	1.564[27]
2441	Benzofuran, 2-(chloromethyl)-2,3-dihydro- / 2-(Chloromethyl)-2,3-dihydrobenzofuran	C_9H_9ClO / 168.62	53491-32-0 / 41.5	/ 118-9[11]	4-17-00-00421 / 1.2183[7]	ace 4; eth 4 / 1.5620[7]
2442	Benzofuran, 2,3-dihydro- / Coumaran	C_8H_8O / 120.15	496-16-2 / -21.5	2560 / 188.5	5-17-01-00581 / 1.058[25]	eth 4; EtOH 4; chl 4 / 1.5497[20]
2443	Benzofuran, 2,3-dihydro-2-methyl-	$C_9H_{10}O$ / 134.18	1746-11-8 /	/ 197.5	5-17-01-00597 / 1.061[25]	1.5308
2444	Benzofuran, 2,3-dihydro-5-methyl- / 2,3-Dihydro-5-methylbenzofuran	$C_9H_{10}O$ / 134.18	76429-68-0 /	/ 210.5	4-17-00-00423 / 1.0463[19]	1.54[20]
2445	Benzofuran, 2,5-dimethyl- / 2,5-Dimethylbenzofuran	$C_{10}H_{10}O$ / 146.19	29040-46-8 /	/ 220	5-17-02-00045 / 1.031[20]	eth 4 / 1.5534[20]
2446	Benzofuran, 2,6-dimethyl- / 2,6-Dimethylbenzofuran	$C_{10}H_{10}O$ / 146.19	24410-51-3 /	/ 217.5	5-17-02-00046 / 1.051[12]	eth 4 / 1.554[15]
2447	Benzofuran, 2,7-dimethyl- / 2,7-Dimethylfuran	$C_{10}H_{10}O$ / 146.19	59020-74-5 /	/ 216	4-17-00-00512 / 1.036[20]	eth 4 / 1.5546[20]
2448	Benzofuran, 3,5-dimethyl- / 3,5-Dimethylbenzofuran	$C_{10}H_{10}O$ / 146.19	10410-35-2 /	/ 220.5	5-17-02-00047 / 1.036[20]	eth 4 / 1.550[20]
2449	Benzofuran, 3,6-dimethyl- / 3,6-DimethylfuranBenzofuran	$C_{10}H_{10}O$ / 146.19	24410-50-2 /	/ 222	4-17-00-00512 / 1.0456[20]	eth 4 / 1.5505[20]
2450	Benzofuran, 4,6-dimethyl-	$C_{10}H_{10}O$ / 146.19	116668-34-9 /	/ 219	0-17-00-00063 / 1.037[20]	eth 4 / 1.5485[21]
2451	Benzofuran, 4,7-dimethyl- / 4,7-Dimethylbenzofuran	$C_{10}H_{10}O$ / 146.19	28715-26-6 /	/ 216	0-17-00-00063 / 1.041[17]	eth 4 / 1.549[17]
2452	Benzofuran, 5,6-dimethyl- / 5,6-Dimethylbenzofuran	$C_{10}H_{10}O$ / 146.19	24410-52-4 /	/ 221	5-17-02-00048 / 1.060[15]	eth 4 / 1.5516[15]
2453	Benzofuran, 5,7-dimethyl- / 5,7-Dimethylbenzofuran	$C_{10}H_{10}O$ / 146.19	64965-91-9 /	/ 222	0-17-00-00063 / 1.0262[18]	eth 4 / 1.5358[20]
2454	Benzofuran, 6,7-dimethyl- / 6,7-Dimethylfuran	$C_{10}H_{10}O$ / 146.19	35355-36-3 /	/ 218	0-17-00-00063 / 1.038[20]	eth 4 / 1.5478[20]
2455	2-Benzofuranmethanol / 2-Hydroxymethylbenzofuran	$C_9H_8O_2$ / 148.16	55038-01-2 /	/ 149-50[14]	5-17-04-00224 / 1.177[13]	1.5550[11]
2456	Benzofuran, 2-methyl-	C_9H_8O / 132.16	4265-25-2 /	/ 197.5	5-17-02-00022 / 1.0540[20]	eth 4; EtOH 4 / 1.5495[22]
2457	Benzofuran, 3-methyl- / 3-Methylbenzo[b]furan	C_9H_8O / 132.16	21535-97-7 /	/ 197; 86[20]	5-17-02-00025 / 1.0540[25]	H_2O 1; EtOH 3; eth 3 / 1.5536[16]
2458	Benzofuran, 5-methyl- / 5-Methylbenzo[b]furan	C_9H_8O / 132.16	18441-43-5 /	/ 198	5-17-02-00030 / 1.0603[19]	H_2O 1; EtOH 3; eth 3; sulf 3 / 1.5570[19]
2459	Benzofuran, 7-methyl- / 7-Methylbenzo[b]furan	C_9H_8O / 132.16	17059-52-8 / 109	/ 190.5	5-17-02-00032 / 1.0490[19]	H_2O 1; EtOH 3; eth 3; sulf 3 / 1.5525[19]
2460	Benzofuran, 4-methyl-7-(1-methylethyl)- / Benzofuran, 7-isopropyl-4-methyl	$C_{12}H_{14}O$ / 174.24	95835-77-1 /	/ 241.5	0-17-00-00065 / 1.0145[18]	1.5363[16]
2461	2(3H)-Benzofuranone	$C_8H_6O_2$ / 134.13	553-86-6 / 50	/ 249	5-17-10-00006 / 1.2236[14]	
2462	3(2H)-Benzofuranone	$C_8H_6O_2$ / 134.13	7169-34-8 / 102.5	/ 152[15]	5-17-04-00185	bz 4
2463	2(3H)-Benzofuranone, 3-[2-(diethylamino)ethyl]-3-phenyl- / Amolanone	$C_{20}H_{23}NO_2$ / 309.41	76-65-3 / 43.4	603 / 192-4[2.0]	4-18-00-07995	1.5614[25]
2464	5-Benzofuranpropanol, 2-(1,3-benzodioxol-5-yl)-7-methoxy- / Egonol	$C_{19}H_{18}O_5$ / 326.35	530-22-3 / 118	/ 228-30[15]	5-19-09-00583	chl 4
2465	Benzofuran, 4,5,6,7-tetrahydro-3,6-dimethyl-	$C_{10}H_{14}O$ / 150.22	494-90-6 / 86	/ 80[18]	5-17-01-00518 / 0.972[15]	
2466	6H-Benzofuro[3,2-c][1]benzopyran-6-one, 3,9-dihydroxy- / Coumestrol	$C_{15}H_8O_5$ / 268.23	479-13-0 / 385 dec	2565	5-19-06-00405	H_2O 1; EtOH 2; eth 1; ace 2

No.	Name Synonym	Mol. Form. Mol. Wt.	CAS RN mp/°C	Merck No. bp/°C	Beil. Ref. den/g cm^{-3}	Solubility n_D
2467	Benzo[g]-1,3-benzodioxolo[5,6-a]quinolizinium, 5,6-dihydro-9,10-dimethoxy- Berberine	C$_{20}$H$_{19}$NO$_5$ 353.37	2086-83-1 145	1169		eth 4; EtOH 4
2468	5H-Benzo[g]-1,3-benzodioxolo[6,5,4-de]quinoline, 6,7,7a,8-tetrahydro-10,11-dimethoxy-7-methyl-, (S)- Dicentrine	C$_{20}$H$_{21}$NO$_4$ 339.39	517-66-8	3028	4-27-00-06486	chl 3
2469	5H-Benzo[g]-1,3-benzodioxolo[6,5,4-de]quinoline, 6,7,7a,8-tetrahydro-11-methoxy-7-methyl-, (R)- Laureline	C$_{19}$H$_{19}$NO$_3$ 309.36	81-38-9 114	5251	1-27-00-00461	H$_2$O 1; EtOH 3; eth 3; dil acid 3
2470	5H-Benzo[g]-1,3-benzodioxolo[6,5,4-de]quinolin-12-ol, 6,7,7a,8-tetrahydro-11-methoxy-7-methyl-, (S)- Bulbocapnine	C$_{19}$H$_{19}$NO$_4$ 325.36	298-45-3 199.5	1471	2-27-00-00554	H$_2$O 1; EtOH 3; chl 4; os 3
2471	Benzo[ghi]perylene 1,12-Benzperylene	C$_{22}$H$_{12}$ 276.34	191-24-2		4-05-00-02766	
2472	Benzo[g]pteridine-2,4(1H,3H)-dione, 7,8-dimethyl- Lumichrome	C$_{12}$H$_{10}$N$_4$O$_2$ 242.24	1086-80-2	5468	4-26-00-02538	
2473	Benzo[g]quinoline 6,7-Benzoquinoline	C$_{13}$H$_9$N 179.22	260-36-6 114	202[14]	5-20-08-00195	eth 4; EtOH 4
2474	Benzo[h]quinoline	C$_{13}$H$_9$N 179.22	230-27-3 52	339; 233[47]	5-20-08-00215 1.2340[20]	H$_2$O 2; EtOH 3; eth 3; ace 3
2475	Benzoic acid Benzenecarboxylic acid	C$_7$H$_6$O$_2$ 122.12	65-85-0 122.4	1101 249.2	4-09-00-00273 1.265[15]	H$_2$O 2; EtOH 4; eth 4; ace 3 1.504[132]
2476	Benzoic acid, 2-acetyl-	C$_9$H$_8$O$_3$ 164.16	577-56-0 114.5		110-2[2]	4-10-00-02766	H$_2$O 4; eth 4; EtOH 4
2477	Benzoic acid, 3-acetyl-	C$_9$H$_8$O$_3$ 164.16	586-42-5 172	110-2[2]	4-10-00-02769	H$_2$O 3; EtOH 5
2478	Benzoic acid, 4-acetyl-	C$_9$H$_8$O$_3$ 164.16	586-89-0 208	sub	4-10-00-02769	H$_2$O 4
2479	Benzoic acid, 4-acetyl-, methyl ester	C$_{10}$H$_{10}$O$_3$ 178.19	3609-53-8 95	142[4]	3-10-00-03030	H$_2$O 4
2480	Benzoic acid, 2-(acetyloxy)- Aspirin	C$_9$H$_8$O$_4$ 180.16	50-78-2 135	873	4-10-00-00138	H$_2$O 3; EtOH 4; eth 3; bz 2
2481	Benzoic acid, 2-(acetyloxy)-5-bromo- 5-Bromoacetylsalicylic acid	C$_9$H$_7$BrO$_4$ 259.06	1503-53-3 60	1425	2-10-00-00064	H$_2$O 1; EtOH 4; eth 3
2482	Benzoic acid, 2-(acetyloxy)-, methyl ester Methyl acetylsalicylate	C$_{10}$H$_{10}$O$_4$ 194.19	580-02-9 51.5	5934 134-6[9]	4-10-00-00147	eth 4; EtOH 4; chl 4
2483	Benzoic acid, 2-(acetyloxy)-, phenyl ester Phenyl acetylsalicylate	C$_{15}$H$_{12}$O$_4$ 256.26	134-55-4 96	7241	4-10-00-00156	
2484	Benzoic acid, 2-amino- o-Anthranilic acid	C$_7$H$_7$NO$_2$ 137.14	118-92-3 146.5	433 sub	4-14-00-01004 1.412[20]	H$_2$O 3; EtOH 3; eth 3; bz 2
2485	Benzoic acid, 3-amino- Aniline-3-carboxylic acid	C$_7$H$_7$NO$_2$ 137.14	99-05-8 173	432	4-14-00-01092 1.51[25]	H$_2$O 2; EtOH 3; eth 3; ace 4
2486	Benzoic acid, 4-amino- Aniline-4-carboxylic acid	C$_7$H$_7$NO$_2$ 137.14	150-13-0 188.5	434	4-14-00-01126 1.374[20]	H$_2$O 3; EtOH 3; eth 3; ace 2
2487	Benzoic acid, 2-amino-5-bromo- 5-Bromoanthranilic acid	C$_7$H$_6$BrNO$_2$ 216.03	5794-88-7 219.5	1393	4-14-00-01081	DMSO 3
2488	Benzoic acid, 4-amino-, butyl ester Butamben	C$_{11}$H$_{15}$NO$_2$ 193.25	94-25-7 58	1504 173-4[8]	4-14-00-01130	H$_2$O 1; EtOH 3; eth 3; bz 3
2489	Benzoic acid, 4-(aminocarbonyl)-	C$_8$H$_7$NO$_3$ 165.15	6051-43-0 300	sub	4-09-00-03318	H$_2$O 1; EtOH 1; eth 1; bz 1
2490	Benzoic acid, 5-amino-2-chloro- 5-Amino-2-chlorobenzoic acid	C$_7$H$_6$ClNO$_2$ 171.58	89-54-3 188		4-14-00-01116 1.519[15]	EtOH 4
2491	Benzoic acid, 2-amino-3,6-dichloro-	C$_7$H$_5$Cl$_2$NO$_2$ 206.03	3032-32-4 155	sub	4-14-00-01078	ace 4; bz 4; eth 4; EtOH 4
2492	Benzoic acid, 3-amino-2,5-dichloro- Chloramben	C$_7$H$_5$Cl$_2$NO$_2$ 206.03	133-90-4 200	2063		DMSO 2
2493	Benzoic acid, 4-amino-, 2-(diethylamino)ethyl ester Procaine	C$_{13}$H$_{20}$N$_2$O$_2$ 236.31	59-46-1 61	7763	4-14-00-01138	H$_2$O 2; EtOH 3; eth 3; bz 3
2494	Benzoic acid, 4-[[(2-amino-1,4-dihydro-4-oxo-6-pteridinyl)methyl]formylamino]- Rhizopterin	C$_{15}$H$_{12}$N$_6$O$_4$ 340.30	119-20-0 >300	8181	4-26-00-03949	H$_2$O 1; EtOH 1; eth 1; aq alk 3
2495	Benzoic acid, 2-amino-, ethyl ester	C$_9$H$_{11}$NO$_2$ 165.19	87-25-2 13	268	4-14-00-01008 1.1174[20]	eth 4; EtOH 4 1.5646[20]
2496	Benzoic acid, 3-amino-, ethyl ester	C$_9$H$_{11}$NO$_2$ 165.19	582-33-2	294; 160[5]	4-14-00-01093 1.1248[22]	H$_2$O 2; EtOH 4; eth 4; ctc 3 1.5600[22]
2497	Benzoic acid, 4-amino-, ethyl ester Ethyl aminobenzoate	C$_9$H$_{11}$NO$_2$ 165.19	94-09-7 92	3719 310	4-14-00-01129	H$_2$O 1; EtOH 4; eth 4; chl 3
2498	Benzoic acid, 4-amino-2-hydroxy- p-Aminosalicylic acid	C$_7$H$_7$NO$_3$ 153.14	65-49-6 150-1 dec	491	4-14-00-01967	H$_2$O 3; EtOH 3; eth 3; ace 3
2499	Benzoic acid, 5-amino-2-hydroxy- Mesalamine	C$_7$H$_7$NO$_3$ 153.14	89-57-6 283	5807	4-14-00-02058	H$_2$O 2; EtOH 1
2500	Benzoic acid, 4-amino-2-hydroxy-, hydrazide p-Aminosalicylic acid hydrazide	C$_7$H$_9$N$_3$O$_2$ 167.17	6946-29-8 195	492	4-14-00-01995	EtOH 4
2501	Benzoic acid, 3-amino-4-hydroxy-, methyl ester Orthocaine	C$_8$H$_9$NO$_3$ 167.16	536-25-4 143	6836	3-14-00-01477	H$_2$O 1; EtOH 4; eth 3; bz 2

No.	Name Synonym	Mol. Form. Mol. Wt.	CAS RN mp/°C	Merck No. bp/°C	Beil. Ref. den/g cm^{-3}	Solubility n_D
2502	Benzoic acid, 4-amino-2-hydroxy-, phenyl ester Phenyl p-aminosalicylate	C$_{13}$H$_{11}$NO$_3$ 229.24	133-11-9 153	7243	4-14-00-01979	
2503	Benzoic acid, 2-amino-, methyl ester Methyl anthranilate	C$_8$H$_9$NO$_2$ 151.17	134-20-3 24.5	5942 256	4-14-00-01008 1.1682[10]	H$_2$O 2; EtOH 4; eth 4 1.5810
2504	Benzoic acid, 3-amino-, methyl ester	C$_8$H$_9$NO$_2$ 151.17	4518-10-9 39	152-3[11]	4-14-00-01093 1.232[20]	EtOH 4; eth 4; bz 4; chl 4
2505	Benzoic acid, 4-amino-, 2-methylpropyl ester Isobutyl p-aminobenzoate	C$_{11}$H$_{15}$NO$_2$ 193.25	94-14-4 64.5	5017	4-14-00-01130	
2506	Benzoic acid, 2-amino-3-nitro- 3-Nitroanthranilic acid	C$_7$H$_6$N$_2$O$_4$ 182.14	606-18-8 208.5		4-14-00-01087 1.558[15]	eth 4; EtOH 4
2507	Benzoic acid, 4-amino-, propyl ester Risocaine	C$_{10}$H$_{13}$NO$_2$ 179.22	94-12-2 75	8226	4-14-00-01130	bz 4; eth 4; EtOH 4; chl 4
2508	Benzoic acid, 4-(aminosulfonyl)- Carzenide	C$_7$H$_7$NO$_4$S 201.20	138-41-0 290-2 dec	1885	4-11-00-00690	H$_2$O 1; EtOH 4; eth 2; bz 1
2509	Benzoic acid, ammonium salt	C$_7$H$_9$NO$_2$ 139.15	1863-63-4 194.5	515	4-09-00-00273	H$_2$O 3
2510	Benzoic acid, anhydride Benzoic anhydride	C$_{14}$H$_{10}$O$_3$ 226.23	93-97-0 42.5	1102 360	4-09-00-00550 1.989[15]	H$_2$O 1; EtOH 3; eth 3; chl 2 1.5767[15]
2511	Benzoic acid, 3-benzoyl-	C$_{14}$H$_{10}$O$_3$ 226.23	579-18-0 163.3	sub	4-10-00-02982	eth 4; EtOH 4
2512	Benzoic acid, 4-benzoyl-	C$_{14}$H$_{10}$O$_3$ 226.23	611-95-0 199	sub	3-10-00-03305	H$_2$O 2; EtOH 3; eth 3; bz 2
2513	Benzoic acid, 3-(benzoylamino)-	C$_{14}$H$_{11}$NO$_3$ 241.25	587-54-2 252.5		4-14-00-01105 1.510[4]	EtOH 4
2514	Benzoic acid, 4-(benzoylamino)-2-hydroxy- Benzoyl-PAS	C$_{14}$H$_{11}$NO$_4$ 257.25	13898-58-3 260.5	1127	4-14-00-02029	
2515	Benzoic acid, 2-benzoyl-, ethyl ester	C$_{16}$H$_{14}$O$_3$ 254.29	604-61-5 59.5		3-10-00-03291 1.221[64]	H$_2$O 1; EtOH 4; eth 4; chl 2 1.560[64]
2516	Benzoic acid, 2-benzoyl-, methyl ester	C$_{15}$H$_{12}$O$_3$ 240.26	606-28-0 52	351	4-10-00-02977 1.190[19]	H$_2$O 1; EtOH 4; eth 4; sulf 3 1.591[20]
2517	Benzoic acid, 2-[bis(4-hydroxyphenyl)methyl]- Phenolphthalin	C$_{20}$H$_{16}$O$_4$ 320.34	81-90-3 230.5	7210	4-10-00-01932	eth 4; EtOH 4
2518	Benzoic acid, 2-bromo-	C$_7$H$_5$BrO$_2$ 201.02	88-65-3 150	sub	4-09-00-01011 1.929[25]	H$_2$O 2; EtOH 3; eth 3; ace 3
2519	Benzoic acid, 3-bromo-	C$_7$H$_5$BrO$_2$ 201.02	585-76-2 155	>280	4-09-00-01013 1.845[20]	H$_2$O 1; EtOH 3; eth 3
2520	Benzoic acid, 4-bromo- p-Bromobenzoic acid	C$_7$H$_5$BrO$_2$ 201.02	586-76-5 254.5	1396	4-09-00-01017 1.894[20]	H$_2$O 2; EtOH 3; eth 3
2521	Benzoic acid, 2-bromo-, ethyl ester	C$_9$H$_9$BrO$_2$ 229.07	6091-64-1 254.5		4-09-00-01012 1.4438[15]	ace 4; bz 4; eth 4; EtOH 4 1.5455[15]
2522	Benzoic acid, 3-bromo-, ethyl ester	C$_9$H$_9$BrO$_2$ 229.07	24398-88-7 261		4-09-00-01014 1.4308[19]	ace 4; bz 4; eth 4; EtOH 4 1.5430[19]
2523	Benzoic acid, 4-bromo-, ethyl ester	C$_9$H$_9$BrO$_2$ 229.07	5798-75-4 263; 125[15]		3-09-00-01405 1.4332[17]	H$_2$O 2; EtOH 3; eth 3; ace 3 1.5438[17]
2524	Benzoic acid, 5-bromo-2-hydroxy-	C$_7$H$_5$BrO$_3$ 217.02	89-55-4 169.8	sub 100	4-10-00-00217	H$_2$O 2; EtOH 4; eth 4; ace 2
2525	Benzoic acid, 3-bromo-, methyl ester	C$_8$H$_7$BrO$_2$ 215.05	618-89-3 32	125[15]	4-09-00-01014	H$_2$O 2; EtOH 3; eth 3
2526	Benzoic acid, 4-bromo-, methyl ester	C$_8$H$_7$BrO$_2$ 215.05	619-42-1 81		4-09-00-01017 1.689[25]	EtOH 3; eth 3; ace 3; bz 4
2527	Benzoic acid, 2-bromo-4-nitro- 2-Bromo-4-nitrobenzoic acid	C$_7$H$_4$BrNO$_4$ 246.02	16426-64-5 166.5	sub 150	3-09-00-01771	eth 4; EtOH 4
2528	Benzoic acid, 2-bromo-5-nitro- 2-Bromo-5-nitrobenzoic acid	C$_7$H$_4$BrNO$_4$ 246.02	943-14-6 180.5	sub	3-09-00-01771	eth 4; EtOH 4
2529	Benzoic acid, 4-bromo-3-nitro- 4-Bromo-3-nitrobenzoic acid	C$_7$H$_4$BrNO$_4$ 246.02	6319-40-0 203.5	sub	4-09-00-01236	EtOH 4
2530	Benzoic acid, 5-bromo-2-nitro- 2-Nitro-5-bromobenzoic acid	C$_7$H$_4$BrNO$_4$ 246.02	6950-43-2 140		1-09-00-00165 1.920[18]	bz 4; EtOH 4
2531	Benzoic acid, 4-(butylamino)-, 2-(dimethylamino)ethyl ester, monohydrochloride Tetracaine hydrochloride	C$_{15}$H$_{25}$ClN$_2$O$_2$ 300.83	136-47-0 147	9123	4-14-00-01172	
2532	Benzoic acid, butyl ester Butyl benzoate	C$_{11}$H$_{14}$O$_2$ 178.23	136-60-7 -22.4	1552 250.3	4-09-00-00290 1.000[20]	H$_2$O 1; EtOH 5; eth 5; ace 3 1.4940[25]
2533	Benzoic acid, 4,4'-carbonylbis-	C$_{15}$H$_{10}$O$_5$ 270.24	964-68-1 365	sub	4-10-00-03497	HOAc 4
2534	Benzoic acid, 2-chloro- o-Chlorobenzoic acid	C$_7$H$_5$ClO$_2$ 156.57	118-91-2 140.2	2125 sub	4-09-00-00956 1.544[20]	H$_2$O 3; EtOH 4; eth 4; ace 4
2535	Benzoic acid, 3-chloro- m-Chlorobenzoic acid	C$_7$H$_5$ClO$_2$ 156.57	535-80-8 158	2124 sub	4-09-00-00969 1.496[25]	H$_2$O 4; EtOH 3; eth 4; bz 2
2536	Benzoic acid, 4-chloro- p-Chlorobenzoic acid	C$_7$H$_5$ClO$_2$ 156.57	74-11-3 243	2126	4-09-00-00973	H$_2$O 1; EtOH 4; eth 2; ace 2
2537	Benzoic acid, 2-chloro-, ethyl ester	C$_9$H$_9$ClO$_2$ 184.62	7335-25-3 243		4-09-00-00957 1.1942[15]	eth 4; EtOH 4 1.5242[15]
2538	Benzoic acid, 3-chloro-, ethyl ester	C$_9$H$_9$ClO$_2$ 184.62	1128-76-3 243		4-09-00-00970 1.1859[15]	H$_2$O 1; EtOH 3; eth 3 1.5223[20]
2539	Benzoic acid, 4-chloro-, ethyl ester	C$_9$H$_9$ClO$_2$ 184.62	7335-27-5 237.5		4-09-00-00975 1.1873[14]	EtOH 4
2540	Benzoic acid, 3-chloro-4-hydroxy-	C$_7$H$_5$ClO$_3$ 172.57	3964-58-7 171	sub	4-10-00-00463	H$_2$O 2; EtOH 4; eth 4; ace 4

No.	Name / Synonym	Mol. Form. / Mol. Wt.	CAS RN / mp/°C	Merck No. / bp/°C	Beil. Ref. / den/g cm^{-3}	Solubility / n_D
2541	Benzoic acid, 3-chloro-2-hydroxy-, ethyl ester / Salicylic acid, 3-chloro, ethyl ester	$C_9H_9ClO_3$ / 200.62	56961-32-1 / 21	269.5	0-10-00-00101	EtOH 4
2542	Benzoic acid, 5-chloro-2-hydroxy-, methyl ester	$C_8H_7ClO_3$ / 186.59	4068-78-4 / 50	249 dec; 120^{12}	4-10-00-00209	EtOH 4
2543	Benzoic acid, 4-chloro-, methyl ester	$C_8H_7ClO_2$ / 170.60	1126-46-1 / 43.5		4-09-00-00974 / 1.382^{20}	EtOH 4
2544	Benzoic acid, 2-chloro-3-nitro- / 2-Chloro-3-nitrobenzoic acid	$C_7H_4ClNO_4$ / 201.57	3970-35-2 / 183.5		4-09-00-01225 / 1.662^{18}	H_2O 3; EtOH 3
2545	Benzoic acid, 2-chloro-5-nitro-	$C_7H_4ClNO_4$ / 201.57	2516-96-3 / 166.5		4-09-00-01228 / 1.608^{18}	H_2O 2; EtOH 3; eth 3; ace 2
2546	Benzoic acid, 3-chloro-2-nitro- / 2-Nitro-3-chlorobenzoic acid	$C_7H_4ClNO_4$ / 201.57	4771-47-5 / 238		4-09-00-01225 / 1.566^{18}	eth 4; EtOH 4
2547	Benzoic acid, 4-chloro-3-nitro-	$C_7H_4ClNO_4$ / 201.57	96-99-1 / 182.8		4-09-00-01226 / 1.645^{18}	H_2O 1; EtOH 2; ace 2
2548	Benzoic acid, 4-chloro-3-nitro-, methyl ester / Methyl 4-chloro-3-nitrobenzoate	$C_8H_6ClNO_4$ / 215.59	14719-83-6 / 83		4-09-00-01226 / 1.522^{18}	EtOH 4
2549	Benzoic acid, 5-chloro-2-nitro-, methyl ester / Methyl 5-chloro-2-nitrobenzoate	$C_8H_6ClNO_4$ / 215.59	51282-49-6 / 48.5		0-09-00-00401 / 1.453^{18}	MeOH 4
2550	Benzoic acid, 3-cyano-	$C_8H_5NO_2$ / 147.13	1877-72-1 / 219	sub	4-09-00-03296	H_2O 2; EtOH 3; eth 3
2551	Benzoic acid, cyclohexyl ester	$C_{13}H_{16}O_2$ / 204.27	2412-73-9 / <-10	285	4-09-00-00296 / 1.0429^{20}	H_2O 1; EtOH 3; eth 3 / 1.5200^{20}
2552	Benzoic acid, 2,3-diamino- / 2,3-Diaminobenzoic acid	$C_7H_8N_2O_2$ / 152.15	603-81-6 / 201	dec	3-14-00-01172	EtOH 4; HOAc 4
2553	Benzoic acid, 2,4-diamino- / 4-Aminoanthranilic acid	$C_7H_8N_2O_2$ / 152.15	611-03-0 / 140	200 dec	4-14-00-01299	EtOH 4; HOAc 4
2554	Benzoic acid, 3,5-diamino- / 3,5-Diaminobenzoic acid	$C_7H_8N_2O_2$ / 152.15	535-87-5 / 228	2955	4-14-00-01304	H_2O 2; EtOH 3; eth 4; tfa 2
2555	Benzoic acid, 2,4-dibromo- / 2,4-Dibromobenzoic acid	$C_7H_4Br_2O_2$ / 279.92	611-00-7 / 174	sub	4-09-00-01027	eth 4; EtOH 4
2556	Benzoic acid, 2,5-dibromo- / 2,5-Dibromobenzoic acid	$C_7H_4Br_2O_2$ / 279.92	610-71-9 / 157	sub	4-09-00-01027	eth 4; EtOH 4; chl 4
2557	Benzoic acid, 2,6-dibromo- / 2,6-Dibromobenzoic acid	$C_7H_4Br_2O_2$ / 279.92	601-84-3 / 150.5	209^{16}	4-09-00-01028	H_2O 3; EtOH 4; eth 4; ace 3
2558	Benzoic acid, 3,5-dibromo-2-hydroxy- / 3,5-Dibromosalicylic acid	$C_7H_4Br_2O_3$ / 295.91	3147-55-5 / 228	3012	4-10-00-00222	ace 3
2559	Benzoic acid, 3,5-dibromo-2-hydroxy-, methyl ester / Methyl 3,5-dibromosalicylate	$C_8H_6Br_2O_3$ / 309.94	21702-79-4 / 149	181^{12}	4-10-00-00223	eth 4
2560	Benzoic acid, 2,6-dibromo-3,4,5-trihydroxy- / Dibromogallic acid	$C_7H_4Br_2O_5$ / 327.91	602-92-6 / 150	3006	2-10-00-00347	H_2O 4; eth 4; EtOH 4
2561	Benzoic acid, 2,4-dichloro- / 2,4-Dichlorobenzoic acid	$C_7H_4Cl_2O_2$ / 191.01	50-84-0 / 164.2	sub	4-09-00-00998	H_2O 3; EtOH 3; eth 3; ace 2
2562	Benzoic acid, 2,5-dichloro- / 2,5-Dichlorobenzoic acid	$C_7H_4Cl_2O_2$ / 191.01	50-79-3 / 154.4	301	4-09-00-01005	H_2O 2; EtOH 3; eth 3
2563	Benzoic acid, 2,6-dichloro- / 2,6-Dichlorobenzoic acid	$C_7H_4Cl_2O_2$ / 191.01	50-30-6 / 144	sub	4-09-00-01005	H_2O 3; EtOH 3; eth 3; bz 3
2564	Benzoic acid, 3,5-dichloro- / 3,5-Dichlorobenzoic acid	$C_7H_4Cl_2O_2$ / 191.01	51-36-5 / 188	sub	4-09-00-01008	H_2O 3; EtOH 3; eth 3; lig2
2565	Benzoic acid, 4-[(dichloroamino)sulfonyl]- / Halazone	$C_7H_5Cl_2NO_4S$ / 270.09	80-13-7 / 195 dec	4503	2-11-00-00220	H_2O 2; chl 2; HOAc 4; peth 1
2566	Benzoic acid, 3,5-dichloro-2-hydroxy-	$C_7H_4Cl_2O_3$ / 207.01	320-72-9 / 220.5	sub	4-10-00-00213	H_2O 2; EtOH 4; eth 3
2567	Benzoic acid, 3,6-dichloro-2-methoxy- / Dicamba	$C_8H_6Cl_2O_3$ / 221.04	1918-00-9 / 115	3026	1.57^{25}	
2568	Benzoic acid, 2,3-dihydroxy-	$C_7H_6O_4$ / 154.12	303-38-8 / 204		4-10-00-01414 / 1.542^{20}	H_2O 3; EtOH 3; eth 3; ace 2
2569	Benzoic acid, 2,4-dihydroxy- / β-Resorcylic acid	$C_7H_6O_4$ / 154.12	89-86-1 / 225-7 dec	8160	4-10-00-01420	H_2O 3; EtOH 3; eth 3; bz 3
2570	Benzoic acid, 2,5-dihydroxy- / Gentisic acid	$C_7H_6O_4$ / 154.12	490-79-9 / 199.5	4290	4-10-00-01441	H_2O 4; EtOH 4; eth 4; ace 3
2571	Benzoic acid, 3,4-dihydroxy- / Protocatechuic acid	$C_7H_6O_4$ / 154.12	99-50-3 / 200-2 dec	7909	4-10-00-01459 / 1.524^4	H_2O 2; EtOH 4; eth 3; bz 1
2572	Benzoic acid, 2-(3,6-dihydroxy-9H-xanthen-9-yl)- / Fluorescin	$C_{20}H_{14}O_5$ / 334.33	518-44-5 / 126	4087	2-18-00-00307	H_2O 1; EtOH 3; eth 3; ace 3
2573	Benzoic acid, 2,4-dihydroxy-6-methyl- / o-Orsellinic acid	$C_8H_8O_4$ / 168.15	480-64-8 / 176 dec	6834	4-10-00-01526	EtOH 3; eth 3
2574	Benzoic acid, 2,4-dihydroxy-6-methyl-, ethyl ester	$C_{10}H_{12}O_4$ / 196.20	2524-37-0 / 132	sub	4-10-00-01526	eth 4; EtOH 4
2575	Benzoic acid, 3,4-dihydroxy-5-[(3,4,5-trihydroxybenzoyl)oxy]- / Digallic acid	$C_{14}H_{10}O_9$ / 322.23	536-08-3 / 268-70 dec	3135	3-10-00-02086	ace 4; EtOH 4
2576	Benzoic acid, 3,4-dimethoxy- / Veratric acid	$C_9H_{10}O_4$ / 182.18	93-07-2 / 181	9854 / sub	4-10-00-01460	H_2O 1; EtOH 4; eth 4; chl 2
2577	Benzoic acid, 3,5-dimethoxy- / 3,5-Dimethoxybenzoic acid	$C_9H_{10}O_4$ / 182.18	1132-21-4 / 185.5	sub	4-10-00-01501	eth 4; EtOH 4

No.	Name / Synonym	Mol. Form. / Mol. Wt.	CAS RN / mp/°C	Merck No. / bp/°C	Beil. Ref. / den/g cm⁻³ / n_D	Solubility
2578	Benzoic acid, 3,4-dimethoxy-, ethyl ester	$C_{11}H_{14}O_4$ / 210.23	3943-77-9 / 43.5	/ 295.5	4-10-00-01469	eth 4; EtOH 4
2579	Benzoic acid, 3,4-dimethoxy-, methyl ester	$C_{10}H_{12}O_4$ / 196.20	2150-38-1 / 60.8	/ 283	4-10-00-01467	bz 4; eth 4; EtOH 4
2580	Benzoic acid, 2,4-dimethyl- / 4-Carboxy-1,3-dimethylbenzene	$C_9H_{10}O_2$ / 150.18	611-01-8 / 90	/ 268	4-09-00-01801	H_2O 2; EtOH 3; ace 3; bz 3
2581	Benzoic acid, 2,5-dimethyl-	$C_9H_{10}O_2$ / 150.18	610-72-0 / 132	/ sub	4-09-00-01802 / 1.069[21]	H_2O 1; EtOH 3; eth 3; ace 3
2582	Benzoic acid, 2,6-dimethyl- / 2,6-Dimethylbenzoic acid	$C_9H_{10}O_2$ / 150.18	632-46-2 / 116	/ 274.5; 155[17]	4-09-00-01798	H_2O 2; EtOH 3; eth 3; lig 2
2583	Benzoic acid, 3,5-dimethyl- / Mesitylenic acid	$C_9H_{10}O_2$ / 150.18	499-06-9 / 171.1	/ sub	4-09-00-01806	H_2O 2; EtOH 4; eth 4
2584	Benzoic acid, 2-(dimethylamino)-	$C_9H_{11}NO_2$ / 165.19	610-16-2 / 72	/ sub	4-14-00-01017	H_2O 4; eth 4; EtOH 4
2585	Benzoic acid, 4-(dimethylamino)- / p-(Dimethylamino)benzoic acid	$C_9H_{11}NO_2$ / 165.19	619-84-1 / 242.5	3221	4-14-00-01164	EtOH 3; eth 2
2586	Benzoic acid, 6-[[6-[2-(dimethylamino)ethyl]-4-methoxy-1,3-benzodioxol-5-yl]acetyl]-2,3-dimethoxy- / Narceine	$C_{23}H_{27}NO_8$ / 445.47	131-28-2 / 138	6341	5-19-09-00051	
2587	Benzoic acid, 2-[[4-(dimethylamino)phenyl]azo]- / Methyl red	$C_{15}H_{15}N_3O_2$ / 269.30	493-52-7 / 183	6037	4-16-00-00504	H_2O 2; EtOH 3; ace 4; bz 4
2588	Benzoic acid, 4-(1,1-dimethylethyl)- / p-tert-Butylbenzoic acid	$C_{11}H_{14}O_2$ / 178.23	98-73-7 / 164.5		4-09-00-01884	H_2O 1; EtOH 4; bz 4; chl 3
2589	Benzoic acid, 2,4-dimethyl-, methyl ester / Methyl 2,4-dimethylbenzoate	$C_{10}H_{12}O_2$ / 164.20	23617-71-2 / -2	/ 232.5	2-09-00-00350 / 1.5052[20]	
2590	Benzoic acid, 3,4-dinitro- / 3,4-Dinitrobenzoic acid	$C_7H_4N_2O_6$ / 212.12	528-45-0 / 166	3268	4-09-00-01242	H_2O 2; EtOH 4; eth 4
2591	Benzoic acid, 3,5-dinitro- / 3,5-Dinitrobenzoic acid	$C_7H_4N_2O_6$ / 212.12	99-34-3 / 205	3269	4-09-00-01242	EtOH 4; HOAc 4
2592	Benzoic acid, 3,5-dinitro-, butyl ester / Butyl 3,5-dinitrobenzoate	$C_{11}H_{12}N_2O_6$ / 268.23	10478-02-1 / 62.5		4-09-00-01246 / 1.488	EtOH 3; DMSO-d_6 2
2593	Benzoic acid, 3,5-dinitro-, ethyl ester / Ethyl 3,5-dinitrobenzoate	$C_9H_8N_2O_6$ / 240.17	618-71-3 / 94		4-09-00-01243 / 1.295[111] / 1.560	EtOH 3
2594	Benzoic acid, 3,3'-[(1,6-dioxo-1,6-hexanediyl)diimino]bis[2,4,6-triiodo- / Iodipamide	$C_{20}H_{14}I_6N_2O_6$ / 1139.77	606-17-7 / 307 dec	4916	4-14-00-01122	H_2O 1; EtOH 2; eth 2; ace 2
2595	Benzoic acid, 4-[(dipropylamino)sulfonyl]- / Probenecid	$C_{13}H_{19}NO_4S$ / 285.36	57-66-9 / 195	7760	4-11-00-00691	
2596	Benzoic acid, 2-ethoxy-	$C_9H_{10}O_3$ / 166.18	134-11-2 / 20.7	/ 211-2[39]	3-10-00-00098	H_2O 2; EtOH 3; ctc 2
2597	Benzoic acid, 3-ethoxy-	$C_9H_{10}O_3$ / 166.18	621-51-2 / 137	/ sub	3-10-00-00245	H_2O 3; EtOH 3; eth 3; bz 3
2598	Benzoic acid, 2-ethoxy-, ethyl ester / Ethyl 2-ethoxybenzoate	$C_{11}H_{14}O_3$ / 194.23	6290-24-0 /	/ 251	3-10-00-00117 / 1.1005[20]	eth 4; EtOH 4
2599	Benzoic acid, 4-ethoxy-, ethyl ester / Ethyl 4-ethoxybenzoate	$C_{11}H_{14}O_3$ / 194.23	23676-09-7 /	/ 275	4-10-00-00368 / 1.076[12]	H_2O 1; EtOH 4; eth 4; ctc 2
2600	Benzoic acid, 2-ethyl-	$C_9H_{10}O_2$ / 150.18	612-19-1 / 68	/ 259	4-09-00-01789 / 1.0431[100] / 1.5099[100]	eth 4; EtOH 4
2601	Benzoic acid, 3-ethyl-	$C_9H_{10}O_2$ / 150.18	619-20-5 / 47		4-09-00-01791 / 1.042[100] / 1.5345[100]	eth 4; EtOH 4
2602	Benzoic acid, ethyl ester / Ethyl benzoate	$C_9H_{10}O_2$ / 150.18	93-89-0 / -34	3725 / 212	4-09-00-00285 / 1.0511[15] / 1.5007[20]	H_2O 1; EtOH 3; eth 5; ace 3
2603	Benzoic acid, 2-fluoro-	$C_7H_5FO_2$ / 140.11	445-29-4 / 126.5		4-09-00-00950 / 1.460[25]	H_2O 2; EtOH 4; eth 4; bz 1
2604	Benzoic acid, 3-fluoro-	$C_7H_5FO_2$ / 140.11	455-38-9 / 124		4-09-00-00952 / 1.474[25]	H_2O 2; eth 3
2605	Benzoic acid, 4-fluoro- / p-Fluorobenzoic acid	$C_7H_5FO_2$ / 140.11	456-22-4 / 185	4100	4-09-00-00953 / 1.479[25]	H_2O 2; EtOH 3; eth 3; ace 2
2606	Benzoic acid, 4-fluoro-, ethyl ester	$C_9H_9FO_2$ / 168.17	451-46-7 / 26	/ 210	4-09-00-00953 / 1.146[25] / 1.4864[20]	eth 4; EtOH 4
2607	Benzoic acid, 2-formyl-	$C_8H_6O_3$ / 150.13	119-67-5 / 98		4-10-00-02748 / 1.404[25]	H_2O 3; EtOH 4; eth 4
2608	Benzoic acid, 6-formyl-2,3-dimethoxy- / Opianic acid	$C_{10}H_{10}O_5$ / 210.19	519-05-1 / 150	6806	4-10-00-03863	EtOH 3; eth 3
2609	Benzoic acid, 3-formyl-4-hydroxy-	$C_8H_6O_4$ / 166.13	584-87-2 / 244	/ sub	4-10-00-03622	eth 4; EtOH 4
2610	Benzoic acid, hexyl ester	$C_{13}H_{18}O_2$ / 206.28	6789-88-4 /	/ 272; 139[8]	4-09-00-00293 / 0.9793[20]	H_2O 1; EtOH 3; ace 3
2611	Benzoic acid, hydrazide	$C_7H_8N_2O$ / 136.15	613-94-5 / 115	/ 267 dec	4-09-00-00922	H_2O 3; EtOH 3; eth 2; ace 2
2612	Benzoic acid, 2-hydroxy- / Salicylic acid	$C_7H_6O_3$ / 138.12	69-72-7 / 158	8301 / 211[20]	4-10-00-00125 / 1.443[20] / 1.565	H_2O 2; EtOH 4; eth 4; ace 4
2613	Benzoic acid, 4-hydroxy- / p-Salicylic acid	$C_7H_6O_3$ / 138.12	99-96-7 / 214.5	4742	4-10-00-00345 / 1.46[25]	H_2O 2; EtOH 4; eth 3; ace 3
2614	Benzoic acid, 2-hydroxy-, butyl ester	$C_{11}H_{14}O_3$ / 194.23	2052-14-4 / -5.9	/ 271	4-10-00-00153 / 1.0728[20] / 1.5115[20]	ctc 2
2615	Benzoic acid, 4-hydroxy-, butyl ester / Butylparaben	$C_{11}H_{14}O_3$ / 194.23	94-26-8 / 68.5	1583	4-10-00-00375	H_2O 2; EtOH 3; ctc 2

No.	Name / Synonym	Mol. Form. / Mol. Wt.	CAS RN / mp/°C	Merck No. / bp/°C	Beil. Ref. / den/g cm^{-3}	Solubility / n_D
2616	Benzoic acid, 2-hydroxy-, 2-carboxyphenyl ester / Salsalate	$C_{14}H_{10}O_5$ / 258.23	552-94-3 / 147	8307	4-10-00-00165	ace 2
2617	Benzoic acid, 2-hydroxy-3,5-diiodo- / 3,5-Diiodosalicylic acid	$C_7H_4I_2O_3$ / 389.92	133-91-5 / 235.5	3175	4-10-00-00226	H_2O 2; EtOH 4; eth 4; bz 1
2618	Benzoic acid, 4-hydroxy-3,5-diiodo-	$C_7H_4I_2O_3$ / 389.92	618-76-8 / 237	260 dec	4-10-00-00479	H_2O 1; EtOH 4; eth 4; bz 2
2619	Benzoic acid, 2-hydroxy-, ethyl ester / Ethyl salicylate	$C_9H_{10}O_3$ / 166.18	118-61-6 / 45	3804 / 150-1[10]	4-10-00-00149 / 1.1326[20]	H_2O 1; EtOH 5; eth 4; ctc 3 / 1.5296[20]
2620	Benzoic acid, 3-hydroxy-, ethyl ester	$C_9H_{10}O_3$ / 166.18	7781-98-8 / 74		4-10-00-00321 / 1.0680[131]	H_2O 2; EtOH 3; eth 3; chl 2
2621	Benzoic acid, 4-hydroxy-, ethyl ester / Ethylparaben	$C_9H_{10}O_3$ / 166.18	120-47-8 / 117	3792 / 297.5	4-10-00-00367	H_2O 2; EtOH 4; eth 4; chl 2
2622	Benzoic acid, 2-hydroxy-, 2-hydroxyethyl ester / Glycol salicylate	$C_9H_{10}O_4$ / 182.18	87-28-5 / 37	4395 / 173[15]	3-10-00-00138 / 1.2526[15]	H_2O 2; EtOH 4; eth 4; bz 4
2623	Benzoic acid, 4-hydroxy-3-iodo- / 4-Hydroxybenzoic acid, 3-iodo	$C_7H_5IO_3$ / 264.02	37470-46-5 / 174.3	sub	4-10-00-00478	eth 4; EtOH 4
2624	Benzoic acid, 5-hydroxy-2-iodo- / 3-Hydroxybenzoic acid, 6-iodo	$C_7H_5IO_3$ / 264.02	57772-57-3 / 198	sub 160	3-10-00-00261	eth 4; EtOH 4
2625	Benzoic acid, 4-hydroxy-3-methoxy- / Vanillic acid	$C_8H_8O_4$ / 168.15	121-34-6 / 211.5	9838 / sub	4-10-00-01459	H_2O 2; EtOH 4; eth 3; DMSO 3
2626	Benzoic acid, 4-hydroxy-3-methoxy-, ethyl ester	$C_{10}H_{12}O_4$ / 196.20	617-05-0 / 44	292	3-10-00-01413	H_2O 1; EtOH 4; eth 4; chl 3
2627	Benzoic acid, 4-hydroxy-3-methoxy-, methyl ester	$C_9H_{10}O_4$ / 182.18	3943-74-6 / 64	286	3-10-00-01410	EtOH 3; chl 2; peth 3
2628	Benzoic acid, 4-hydroxy-3-methoxy-, 2-methylpropyl ester	$C_{12}H_{16}O_4$ / 224.26	7152-88-7 / 56.5	125[2]	3-10-00-01415	chl 2
2629	Benzoic acid, 2-hydroxy-3-methyl- / o-Cresotic acid	$C_8H_8O_3$ / 152.15	83-40-9 / 165.5	2585	4-10-00-00601	H_2O 2; EtOH 3; eth 3; bz 3
2630	Benzoic acid, 2-hydroxy-4-methyl- / m-Cresotic acid	$C_8H_8O_3$ / 152.15	50-85-1 / 177	2584	4-10-00-00617	H_2O 2; EtOH 3; eth 4; bz 3
2631	Benzoic acid, 2-hydroxy-5-methyl- / p-Cresotic acid	$C_8H_8O_3$ / 152.15	89-56-5 / 151	2586	4-10-00-00610	H_2O 2; EtOH 3; eth 3; bz 3
2632	Benzoic acid, 3-hydroxy-5-methyl-	$C_8H_8O_3$ / 152.15	585-81-9 / 210	sub	4-10-00-00609	H_2O 3; EtOH 4; eth 4; chl 2
2633	Benzoic acid, 2-hydroxy-, 3-methylbutyl ester / Isopentyl salicylate	$C_{12}H_{16}O_3$ / 208.26	87-20-7 / 278; 151[15]	5007	4-10-00-00153 / 1.0535[20]	H_2O 1; EtOH 4; eth 3; ctc 2 / 1.5080[20]
2634	Benzoic acid, 2-hydroxy-, methyl ester / Methyl salicylate	$C_8H_8O_3$ / 152.15	119-36-8 / -8	6038 / 222.9	4-10-00-00143 / 1.181[25]	eth 4; EtOH 4; chl 4 / 1.535[20]
2635	Benzoic acid, 2-hydroxy-, 1-methylethyl ester / Isopropyl salicylate	$C_{10}H_{12}O_3$ / 180.20	607-85-2	238	4-10-00-00152 / 1.0729[20]	H_2O 1; EtOH 5; eth 5 / 1.5065[20]
2636	Benzoic acid, 3-hydroxy-, methyl ester	$C_8H_8O_3$ / 152.15	19438-10-9 / 73	281; 178[17]	4-10-00-00319 / 1.1528[100]	EtOH 4; bz 3; chl 2; peth 3
2637	Benzoic acid, 4-hydroxy-, methyl ester / Methylparaben	$C_8H_8O_3$ / 152.15	99-76-3 / 131	6021 / 275 dec	4-10-00-00360	H_2O 2; EtOH 4; eth 4; ace 4
2638	Benzoic acid, 2-hydroxy-3-methyl-, methyl ester	$C_9H_{10}O_3$ / 166.18	23287-26-5 / 29	235	4-10-00-00602 / 1.1683[25]	1.5354[16]
2639	Benzoic acid, 2-hydroxy-4-methyl-, methyl ester	$C_9H_{10}O_3$ / 166.18	4670-56-8 / 27.5	243	2-10-00-00139 / 1.1483[15]	1.5378[15]
2640	Benzoic acid, 2-hydroxy-5-methyl-, methyl ester	$C_9H_{10}O_3$ / 166.18	22717-57-3 / -1	244.5	4-10-00-00611 / 1.1673[25]	1.5351[15]
2641	Benzoic acid, 2-hydroxy-3-methyl-6-(1-methylethyl)-	$C_{11}H_{14}O_3$ / 194.23	4389-53-1 / 136	sub	4-10-00-00738	eth 4; EtOH 4
2642	Benzoic acid, 2-hydroxy-6-methyl-3-(1-methylethyl)- / o-Thymotic acid	$C_{11}H_{14}O_3$ / 194.23	548-51-6 / 127	9342 / sub	3-10-00-00629	bz 4; eth 4; EtOH 4
2643	Benzoic acid, 2-hydroxy-, 2-methylpropyl ester / Isobutyl salicylate	$C_{11}H_{14}O_3$ / 194.23	87-19-4 / 5.9	261	3-10-00-00121 / 1.0639[20]	H_2O 1; EtOH 3; eth 3; ctc 3 / 1.5087[20]
2644	Benzoic acid, 2-hydroxy-, 1-naphthalenyl ester / 1-Naphthyl salicylate	$C_{17}H_{12}O_3$ / 264.28	550-97-0 / 83	6334	2-10-00-00052	eth 4
2645	Benzoic acid, 2-hydroxy-, 2-naphthalenyl ester / 2-Naphthyl salicylate	$C_{17}H_{12}O_3$ / 264.28	613-78-5 / 95.5	6335	4-10-00-00158 / 1.11[116]	H_2O 1; EtOH 2; eth 3; bz 3
2646	Benzoic acid, 2-hydroxy-3-nitro- / 3-Nitrosalicylic acid	$C_7H_5NO_5$ / 183.12	85-38-1 / 148	6553	4-10-00-00228	H_2O 2; EtOH 4; eth 4; ace 3
2647	Benzoic acid, 2-hydroxy-5-nitro- / 5-Nitrosalicylic acid	$C_7H_5NO_5$ / 183.12	96-97-9 / 229.5	6554	4-10-00-00255 / 1.650[20]	H_2O 2; EtOH 4; eth 4; ace 4
2648	Benzoic acid, 2-hydroxy-5-[(4-nitrophenyl)azo]- / Alizarine Yellow R	$C_{13}H_9N_3O_5$ / 287.23	2243-76-7 / 253-4 dec	241	4-16-00-00372	H_2O 4; EtOH 4
2649	Benzoic acid, 2-hydroxy-, pentyl ester	$C_{12}H_{16}O_3$ / 208.26	2050-08-0	270	4-10-00-00153 / 1.064[15]	H_2O 2; EtOH 5; eth 5 / 1.506[20]
2650	Benzoic acid, 2-hydroxy-, phenyl ester / Phenyl salicylate	$C_{13}H_{10}O_3$ / 214.22	118-55-8 / 130.5	7282 / 173[12]	4-10-00-00154 / 1.2614[30]	H_2O 1; EtOH 4; eth 3; ace 4

No.	Name Synonym	Mol. Form. Mol. Wt.	CAS RN mp/°C	Merck No. bp/°C	Beil. Ref. den/g cm^{-3}	Solubility n_D
2651	Benzoic acid, 2-hydroxy-, phenylmethyl ester Benzyl salicylate	$C_{14}H_{12}O_3$ 228.25	118-58-1	1160 320	4-10-00-00157 1.1799[20]	H_2O 2; EtOH 3; eth 3; ctc 3 1.5805[20]
2652	Benzoic acid, 2-hydroxy-, propyl ester	$C_{10}H_{12}O_3$ 180.20	607-90-9 97	239	4-10-00-00152 1.0979[20]	ctc 3; CS_2 3 1.5161[20]
2653	Benzoic acid, 4-hydroxy-, propyl ester Propylparaben	$C_{10}H_{12}O_3$ 180.20	94-13-3 97	7879	4-10-00-00374 1.0630[102]	H_2O 1; EtOH 3; eth 3; chl 2 1.5050[102]
2654	Benzoic acid, 2-hydroxy-5-[[4-[(2-pyridinylamino)sulfonyl]phenyl]azo]- Sulfasalazine	$C_{18}H_{14}N_4O_5S$ 398.40	599-79-1 220 dec	8917	5-22-08-00433	
2655	Benzoic acid, 2-hydroxy-5-sulfo- Sulfosalicylic acid	$C_7H_6O_6S$ 218.19	97-05-2 120	8944	4-11-00-00702	H_2O 5; EtOH 5; eth 4
2656	Benzoic acid, 2,2'-iminobis- Diphenylamine-2,2'-dicarboxylic acid	$C_{14}H_{11}NO_4$ 257.25	579-92-0 296-7 dec	3318	4-14-00-01058	
2657	Benzoic acid, 2-iodo- o-Iodobenzoic acid	$C_7H_5IO_2$ 248.02	88-67-5 163	4923 exp 233	4-09-00-01030 2.25[25]	H_2O 2; EtOH 4; eth 4; ace 2
2658	Benzoic acid, 3-iodo-	$C_7H_5IO_2$ 248.02	618-51-9 188.3	sub	4-09-00-01033	H_2O 2; EtOH 4; eth 2
2659	Benzoic acid, 4-iodo-	$C_7H_5IO_2$ 248.02	619-58-9 270	sub	4-09-00-01035 2.184[20]	H_2O 2; EtOH 2; eth 3
2660	Benzoic acid, 2-iodo-, methyl ester	$C_8H_7IO_2$ 262.05	610-97-9	280; 146[16]	4-09-00-01031	EtOH 3 1.6052[20]
2661	Benzoic acid, 3-iodo-, methyl ester	$C_8H_7IO_2$ 262.05	618-91-7 54.5	277; 150[18]	4-09-00-01034	H_2O 1; EtOH 3; eth 4; ace 4
2662	Benzoic acid, 4-iodo-, methyl ester	$C_8H_7IO_2$ 262.05	619-44-3 114.8	sub	4-09-00-01036 2.0200[10]	EtOH 3; eth 3
2663	Benzoic acid, 2-mercapto- o-Thiosalicylic acid	$C_7H_6O_2S$ 154.19	147-93-3 168.5	9291 sub	4-10-00-00272	H_2O 3; EtOH 3; eth 3; lig 2
2664	Benzoic acid, 2-methoxy-	$C_8H_8O_3$ 152.15	579-75-9 101	200	4-10-00-00130	H_2O 2; EtOH 4; eth 3; bz 3
2665	Benzoic acid, 3-methoxy-	$C_8H_8O_3$ 152.15	586-38-9 107	170[10]	4-10-00-00316	H_2O 2; EtOH 3; eth 3; bz 3
2666	Benzoic acid, 4-methoxy- p-Anisic acid	$C_8H_8O_3$ 152.15	100-09-4 185	696 276.5	4-10-00-00346	H_2O 1; EtOH 4; eth 4; chl 3
2667	Benzoic acid, 2-(4-methoxybenzoyl)- o-(p-Anisoyl)benzoic acid	$C_{15}H_{12}O_4$ 256.26	1151-15-1 146	702	3-10-00-04429	eth 4; EtOH 4; tol 4
2668	Benzoic acid, 4-methoxy-, butyl ester	$C_{12}H_{16}O_3$ 208.26	6946-35-6	183[40]	3-10-00-00307 1.054[16]	1.5141[16]
2669	Benzoic acid, 2-methoxy-, ethyl ester	$C_{10}H_{12}O_3$ 180.20	7335-26-4	261	4-10-00-00150 1.1124[20]	eth 4; EtOH 4 1.5224[20]
2670	Benzoic acid, 3-methoxy-, ethyl ester	$C_{10}H_{12}O_3$ 180.20	10259-22-0	260.5	4-10-00-00321 1.0993[20]	eth 4; EtOH 4 1.5161[20]
2671	Benzoic acid, 4-methoxy-, ethyl ester	$C_{10}H_{12}O_3$ 180.20	94-30-4 7.5	269.5	4-10-00-00368 1.1038[20]	H_2O 1; EtOH 3; eth 3 1.5254[20]
2672	Benzoic acid, 3-methoxy-2-(methylamino)-, methyl ester Damascenine	$C_{10}H_{13}NO_3$ 195.22	483-64-7 28	2809 271; 147[10]	4-14-00-02073	bz 4; eth 4; EtOH 4; lig 4
2673	Benzoic acid, 2-methoxy-, methyl ester	$C_9H_{10}O_3$ 166.18	606-45-1	246.5	4-10-00-00144 1.1571[19]	H_2O 1; EtOH 3 1.534[19]
2674	Benzoic acid, 3-methoxy-, methyl ester	$C_9H_{10}O_3$ 166.18	5368-81-0	248	4-10-00-00319 1.1310[20]	H_2O 3; EtOH 3 1.5224[20]
2675	Benzoic acid, 4-methoxy-, methyl ester	$C_9H_{10}O_3$ 166.18	121-98-2 49	244	4-10-00-00360	H_2O 1; EtOH 3; eth 3; chl 3
2676	Benzoic acid, 2-methyl- o-Toluic acid	$C_8H_8O_2$ 136.15	118-90-1 103.7	9461 259	4-09-00-01697 1.062[115]	H_2O 1; EtOH 4; eth 4; chl 3 1.512[115]
2677	Benzoic acid, 3-methyl- m-Toluic acid	$C_8H_8O_2$ 136.15	99-04-7 108.7	9461	4-09-00-01712 1.054[112]	H_2O 2; EtOH 4; eth 4; chl 2 1.509
2678	Benzoic acid, 4-methyl- p-Toluic acid	$C_8H_8O_2$ 136.15	99-94-5 179.6	9461	4-09-00-01724	H_2O 1; EtOH 4; eth 4; tfa 2
2679	Benzoic acid, 2-(methylamino)-	$C_8H_9NO_2$ 151.17	119-68-6 180.5	80[0.01]	4-14-00-01015	H_2O 2; EtOH 4; eth 4; bz 4
2680	Benzoic acid, 2-(methylamino)-, ethyl ester Anthranilic acid, N-methyl-, ethyl ester	$C_{10}H_{13}NO_2$ 179.22	35472-56-1 39	266; 172[45]	2-14-00-00213	eth 4
2681	Benzoic acid, 2-(methylamino)-, methyl ester	$C_9H_{11}NO_2$ 165.19	85-91-6 19	255	4-14-00-01016 1.120[15]	H_2O 1; EtOH 3; eth 3 1.5839[15]
2682	Benzoic acid, 2-methyl-, anhydride	$C_{16}H_{14}O_3$ 254.29	607-86-3 38.5	>325; 220[11]	4-09-00-01700	eth 4
2683	Benzoic acid, 3-methyl-, anhydride	$C_{16}H_{14}O_3$ 254.29	21436-44-2 71	230[17]	4-09-00-01715	ace 4; bz 4; eth 4; EtOH 4
2684	Benzoic acid, 2-(4-methylbenzoyl)- 2-(p-Toluyl)benzoic acid	$C_{15}H_{12}O_3$ 240.26	85-55-2 146	9465	4-10-00-02991	H_2O 2; EtOH 4; eth 4; ace 4
2685	Benzoic acid, 3-methyl-, butyl ester Butyl m-toluate	$C_{12}H_{16}O_2$ 192.26	6640-77-3	262.3	3-09-00-00397 1.0040[20]	H_2O 1; EtOH 3; eth 5 1.4950[20]
2686	Benzoic acid, 3,3'-methylenebis[6-hydroxy- 5,5'-Methylenedisalicylic acid	$C_{15}H_{12}O_6$ 288.25	122-25-8 243.5	5984	4-10-00-02337	ace 4; eth 4; EtOH 4
2687	Benzoic acid, methyl ester Methyl benzoate	$C_8H_8O_2$ 136.15	93-58-3 -15	5947 199	4-09-00-00283 1.0933[15]	H_2O 1; EtOH 3; eth 5; ctc 3 1.5164[20]
2688	Benzoic acid, 2-(1-methylethyl)-	$C_{10}H_{12}O_2$ 164.20	2438-04-2 64	160-1[25]	4-09-00-01839	H_2O 2; EtOH 4; eth 4; bz 3
2689	Benzoic acid, 4-(1-methylethyl)- Cumic acid	$C_{10}H_{12}O_2$ 164.20	536-66-3 117.5	2620 sub	4-09-00-01843 1.162[4]	H_2O 2; EtOH 4; eth 4; peth 3

No.	Name / Synonym	Mol. Form. / Mol. Wt.	CAS RN / mp/°C	Merck No. / bp/°C	Beil. Ref. / den/g cm^{-3}	Solubility / n_D
2690	Benzoic acid, 2-methyl-, ethyl ester	$C_{10}H_{12}O_2$ 164.20	87-24-1 <-10	227; 113[18]	4-09-00-01699 1.0325[21]	H_2O 1; EtOH 5; eth 5 1.507[22]
2691	Benzoic acid, 3-methyl-, ethyl ester	$C_{10}H_{12}O_2$ 164.20	120-33-2	234	4-09-00-01713 1.0265[21]	eth 4; EtOH 4 1.5052[22]
2692	Benzoic acid, 4-methyl-, ethyl ester	$C_{10}H_{12}O_2$ 164.20	94-08-6	232	4-09-00-01726 1.0269[18]	H_2O 1; EtOH 5; eth 5 1.5089[18]
2693	Benzoic acid, 1-methylethyl ester	$C_{10}H_{12}O_2$ 164.20	939-48-0	216	4-09-00-00289 1.0163[15]	H_2O 1; EtOH 3; eth 3; ace 3 1.4890[20]
2694	Benzoic acid, 4-(1-methylethyl)-, methyl ester	$C_{11}H_{14}O_2$ 178.23	20185-55-1 126[14]		4-09-00-01843 1.018[19]	1.515[19]
2695	Benzoic acid, 4-(1-methylethyl)-, 2-methylpropyl ester / Benzoic acid, 4-isopropyl, isobutyl ester	$C_{14}H_{20}O_2$ 220.31	6315-03-3 155[14]		2-09-00-00359 0.966[19]	1.497[19]
2696	Benzoic acid, 2-methyl-, methyl ester	$C_9H_{10}O_2$ 150.18	89-71-4 <-50	215	4-09-00-01699 1.068[20]	H_2O 1; EtOH 5; eth 5
2697	Benzoic acid, 3-methyl-, methyl ester	$C_9H_{10}O_2$ 150.18	99-36-5	221	4-09-00-01713 1.061[20]	H_2O 1; EtOH 3; ctc 2
2698	Benzoic acid, 4-methyl-, methyl ester	$C_9H_{10}O_2$ 150.18	99-75-2 33.2	220	4-09-00-01726	H_2O 1; EtOH 4; eth 4
2699	Benzoic acid, 2-methylphenyl ester	$C_{14}H_{12}O_2$ 212.25	617-02-7	309; 155[8.5]	4-09-00-00305 1.114[19]	eth 4; EtOH 4
2700	Benzoic acid, 4-methylphenyl ester	$C_{14}H_{12}O_2$ 212.25	614-34-6 71.5	316	4-09-00-00306	eth 4; EtOH 4
2701	Benzoic acid, 2-methylpropyl ester	$C_{11}H_{14}O_2$ 178.23	120-50-3	242	4-09-00-00291 0.9990[20]	H_2O 1; EtOH 5; eth 5; ace 3
2702	Benzoic acid, 2-[(1-naphthalenylamino)carbonyl]- / Naptalam	$C_{18}H_{13}NO_3$ 291.31	132-66-1 185	6338	3-12-00-02876 1.4[20]	H_2O 1; EtOH 2; ace 2; bz 2
2703	Benzoic acid, 2-nitro- / o-Nitrobenzoic acid	$C_7H_5NO_4$ 167.12	552-16-9 147.5		4-09-00-01046 1.575[20]	H_2O 3; EtOH 4; eth 3; ace 4
2704	Benzoic acid, 3-nitro- / m-Nitrobenzoic acid	$C_7H_5NO_4$ 167.12	121-92-6 141		4-09-00-01055 1.494[20]	H_2O 2; EtOH 4; eth 4; ace 4
2705	Benzoic acid, 4-nitro- / p-Nitrobenzoic acid	$C_7H_5NO_4$ 167.12	62-23-7 242	sub	4-09-00-01072 1.610[20]	ace 4; eth 4; EtOH 4; chl 4
2706	Benzoic acid, 4-nitro-, butyl ester	$C_{11}H_{13}NO_4$ 223.23	120-48-9 35.3	160[8]	3-09-00-01544	bz 4; eth 4
2707	Benzoic acid, 2-nitro-, ethyl ester	$C_9H_9NO_4$ 195.17	610-34-4 30	275; 178[23]	4-09-00-01047	eth 4; EtOH 4
2708	Benzoic acid, 3-nitro-, ethyl ester	$C_9H_9NO_4$ 195.17	618-98-4 47	297	4-09-00-01056	H_2O 1; EtOH 4; eth 4
2709	Benzoic acid, 4-nitro-, ethyl ester	$C_9H_9NO_4$ 195.17	99-77-4 57	186.3; 153[8]	4-09-00-01074	H_2O 1; EtOH 3; eth 3
2710	Benzoic acid, 2-nitro-, methyl ester	$C_8H_7NO_4$ 181.15	606-27-9 -13	275	4-09-00-01047 1.2855[20]	H_2O 1; EtOH 3; eth 3; bz 3
2711	Benzoic acid, 3-nitro-, methyl ester	$C_8H_7NO_4$ 181.15	618-95-1 78	279[60]	4-09-00-01056	H_2O 1; EtOH 2; eth 2; MeOH 2
2712	Benzoic acid, pentachloro- / Perchlorobenzoic acid	$C_7HCl_5O_2$ 294.35	1012-84-6 209.3	sub	4-09-00-01011	EtOH 4; tol 4; HOAc 4
2713	Benzoic acid, pentamethyl- / Permethylbenzoic acid	$C_{12}H_{16}O_2$ 192.26	2243-32-5 210.5	sub	4-09-00-01920	H_2O 1; EtOH 4
2714	Benzoic acid, 2-phenoxy-	$C_{13}H_{10}O_3$ 214.22	2243-42-7 113	355	4-10-00-00132 1.1553[50]	H_2O 1; EtOH 4; eth 4; chl 3
2715	Benzoic acid, 2-(phenylamino)- / N-Phenylanthranilic acid	$C_{13}H_{11}NO_2$ 213.24	91-40-7 183.5	7244	4-14-00-01019	H_2O 1; EtOH 4; eth 2; bz 2
2716	Benzoic acid, phenyl ester / Phenyl benzoate	$C_{13}H_{10}O_2$ 198.22	93-99-2 71	7246 314	4-09-00-00303 1.235[20]	H_2O 1; EtOH 3; eth 3; chl 3
2717	Benzoic acid, 2-phenylhydrazide	$C_{13}H_{12}N_2O$ 212.25	532-96-7 168	314	4-15-00-00169	H_2O 2; EtOH 3; eth 2; bz 3
2718	Benzoic acid, 2-(phenylmethyl)-	$C_{14}H_{12}O_2$ 212.25	612-35-1 118	sub	4-09-00-02516	H_2O 2; EtOH 3; eth 3; bz 3
2719	Benzoic acid, 4-(phenylmethyl)-	$C_{14}H_{12}O_2$ 212.25	620-86-0 158.8	sub	4-09-00-02518	bz 4; eth 4; EtOH 4; chl 4
2720	Benzoic acid, phenylmethyl ester / Benzyl benzoate	$C_{14}H_{12}O_2$ 212.25	120-51-4 21	1141 323.5	4-09-00-00307 1.1121[25]	H_2O 1; EtOH 3; eth 3; ace 3 1.5680[20]
2721	Benzoic acid, 4-[[(phenylmethyl)sulfonyl]amino]- / Carinamide	$C_{14}H_{13}NO_4S$ 291.33	536-95-8 229.5	1163	4-14-00-01264	EtOH 4
2722	Benzoic acid, 2-propenyl ester	$C_{10}H_{10}O_2$ 162.19	583-04-0		4-09-00-00296 1.0569[15]	H_2O 1; EtOH 3; eth 3; ace 3 1.5178[20]
2723	Benzoic acid, o-propyl- / Benzoic acid, 2-propyl-	$C_{10}H_{12}O_2$ 164.20	2438-03-1 58	273; 164[20]	4-09-00-01836 1.0020[15]	eth 4; EtOH 4
2724	Benzoic acid, propyl ester	$C_{10}H_{12}O_2$ 164.20	2315-68-6 -51.6	211	4-09-00-00288 1.0230[20]	H_2O 1; EtOH 5; eth 5 1.5000[20]
2725	Benzoic acid, sodium salt	$C_7H_5NaO_2$ 144.11	532-32-1 >300	8527	4-09-00-00273	H_2O 3
2726	Benzoic acid, 2-[[[4-[(2-thiazolylamino)sulfonyl]phenyl]amino]carbonyl]- / Phthalylsulphathiazole	$C_{17}H_{13}N_3O_5S_2$ 403.44	85-73-4 273	7351	4-27-00-04639	H_2O 1; EtOH 2; eth 1; chl 1

No.	Name / Synonym	Mol. Form. / Mol. Wt.	CAS RN / mp/°C	Merck No. / bp/°C	Beil. Ref. / den/g cm^{-3}	Solubility / n_D
2727	Benzoic acid, 2,3,6-trichloro- 2,3,6-Trichlorobenzoic acid	$C_7H_3Cl_3O_2$ 225.46	50-31-7 124.5		4-09-00-01009	H_2O 2; eth 3
2728	Benzoic acid, 2,4,5-trichloro- 2,4,5-Trichlorobenzoic acid	$C_7H_3Cl_3O_2$ 225.46	50-82-8 166.5	sub	4-09-00-01010	H_2O 1; EtOH 3; eth 3
2729	Benzoic acid, 2-[[3-(trifluoromethyl)phenyl]amino]- Flufenamic acid	$C_{14}H_{10}F_3NO_2$ 281.23	530-78-9 133.5	4060	3-14-00-00905	DMSO 3
2730	Benzoic acid, 2,3,4-trihydroxy-	$C_7H_6O_5$ 170.12	610-02-6 221	sub	4-10-00-01971	H_2O 2; EtOH 3; eth 3; ace 3
2731	Benzoic acid, 3,4,5-trihydroxy- Gallic acid	$C_7H_6O_5$ 170.12	149-91-7 253 dec	4251	3-10-00-02070 1.694[6]	H_2O 2; EtOH 4; eth 2; ace 3
2732	Benzoic acid, 3,4,5-trihydroxy-, propyl ester Propyl gallate	$C_{10}H_{12}O_5$ 212.20	121-79-9 130	7872	4-10-00-02003	H_2O 2
2733	Benzoic acid, 2,4,5-trimethoxy-	$C_{10}H_{12}O_5$ 212.20	490-64-2 145	300	4-10-00-01980	H_2O 4; bz 4; EtOH 4; peth 4
2734	Benzoic acid, 3,4,5-trimethoxy-	$C_{10}H_{12}O_5$ 212.20	118-41-2 172.3	225-7[10]	4-10-00-01996	H_2O 2; EtOH 4; eth 4; chl 4
2735	Benzoic acid, 2,4,6-trinitro- 2,4,6-Trinitrobenzoic acid	$C_7H_3N_3O_8$ 257.12	129-66-8 228 dec	9640	4-09-00-01362	H_2O 2; EtOH 4; eth 3; ace 3
2736	1H,5H-Benzo[ij]quinolizine-1,6(7H)-dione, 2,3-dihydro- Julolidine 1,6-dioxo	$C_{12}H_{11}NO_2$ 201.22	39052-57-8 145.5	190-210[0.3]	4-21-00-05525	EtOH 4
2737	1H,5H-Benzo[ij]quinolizine, 2,3,6,7-tetrahydro- Julolidine	$C_{12}H_{15}N$ 173.26	479-59-4 40	280 dec; 155[17]	5-20-07-00189 1.003[20]	1.568[25]
2738	Benzo[j]fluoranthene Dibenzo[a,jk]fluorene	$C_{20}H_{12}$ 252.32	205-82-3 166		4-05-00-02687	H_2O 1; EtOH 2; HOAc 2
2739	Benzo[k]fluoranthene 2,3,1',8'-Binaphthylene	$C_{20}H_{12}$ 252.32	207-08-9 217	480	4-05-00-02686	H_2O 1; EtOH 3; bz 3; HOAc 3
2740	Benzonitrile Phenyl cyanide	C_7H_5N 103.12	100-47-0 -12.7	1107 191.1	4-09-00-00892 1.0093[15]	H_2O 2; EtOH 5; eth 5; ace 4 1.5289[20]
2741	Benzonitrile, 2-amino-	$C_7H_6N_2$ 118.14	1885-29-6 51	263	4-14-00-01013	H_2O 2; EtOH 4; eth 4; ace 4
2742	Benzonitrile, 3-amino-	$C_7H_6N_2$ 118.14	2237-30-1 54.3	289	4-14-00-01095	H_2O 2; EtOH 4; eth 4; ace 4
2743	Benzonitrile, 5-amino-2-methyl-	$C_8H_8N_2$ 132.17	50670-64-9 88	100-10[22]	3-14-00-01201	EtOH 4
2744	Benzonitrile, 2-bromo-	C_7H_4BrN 182.02	2042-37-7 55.5	252	4-09-00-01013	H_2O 3; EtOH 4; chl 2
2745	Benzonitrile, 3-bromo-	C_7H_4BrN 182.02	6952-59-6 39.5	225	4-09-00-01016	EtOH 4; eth 4; chl 2
2746	Benzonitrile, 4-bromo-	C_7H_4BrN 182.02	623-00-7 114	236	4-09-00-01025	H_2O 3; EtOH 3; eth 3; chl 3
2747	Benzonitrile, 2-chloro-	C_7H_4ClN 137.57	873-32-5 46.3	232	4-09-00-00965	H_2O 2; EtOH 3; eth 3; chl 3
2748	Benzonitrile, 3-chloro-	C_7H_4ClN 137.57	766-84-7 41	100[15]	4-09-00-00973	H_2O 1; EtOH 3; eth 3
2749	Benzonitrile, 4-chloro-	C_7H_4ClN 137.57	623-03-0 95	223; 95[5]	4-09-00-00992 1.1133[17]	H_2O 2; EtOH 3; eth 3; bz 3
2750	Benzonitrile, 2-(chloromethyl)-	C_8H_6ClN 151.60	612-13-5 61	252	4-09-00-01707	EtOH 4
2751	Benzonitrile, 3,5-dibromo-4-hydroxy- Bromoxynil	$C_7H_3Br_2NO$ 276.91	1689-84-5 190	1431	4-10-00-00475	
2752	Benzonitrile, 2,6-dichloro- Dichlobenil	$C_7H_3Cl_2N$ 172.01	1194-65-6 144.5	3029 270	4-09-00-01006	
2753	Benzonitrile, 2,6-dimethoxy- 2,6-Dimethoxybenzonitrile	$C_9H_9NO_2$ 163.18	16932-49-3 119	310	2-10-00-00260	H_2O 2; EtOH 4; eth 2; ace 3
2754	Benzonitrile, 2-ethoxy- o-Ethoxybenzonitrile	C_9H_9NO 147.18	6609-57-0 5	260.5	0-10-00-00097 1.0650[15]	eth 4; EtOH 4; lig 4
2755	Benzonitrile, 4-ethoxy-	C_9H_9NO 147.18	25117-74-2 61.5	258	3-10-00-00344	eth 4; EtOH 4; lig 4
2756	Benzonitrile, 4-fluoro-	C_7H_4FN 121.11	1194-02-1 34.8	188.8	4-09-00-00954 1.1070[55]	chl 2; peth 3 1.4925[55]
2757	Benzonitrile, 3-formyl-	C_8H_5NO 131.13	24964-64-5 76.5	210	4-10-00-02752	H_2O 4; EtOH 4; eth 4; chl 4
2758	Benzonitrile, 4-formyl-	C_8H_5NO 131.13	105-07-7 100.5	133[12]	4-10-00-02754	H_2O 3; EtOH 4; eth 4; chl 4
2759	Benzonitrile, 2-hydroxy-	C_7H_5NO 119.12	611-20-1 98	149[14]	4-10-00-00191 1.1052[100]	H_2O 2; EtOH 4; eth 4; bz 4 1.5372[100]
2760	Benzonitrile, 2-methoxy-	C_8H_7NO 133.15	6609-56-9 24.5	255.5	4-10-00-00191 1.1063[20]	EtOH 3; eth 4
2761	Benzonitrile, 3-methoxy-	C_8H_7NO 133.15	1527-89-5	111-2[13]	4-10-00-00329 1.089[25]	1.5402[20]
2762	Benzonitrile, 4-methoxy-	C_8H_7NO 133.15	874-90-8 61.5	256.5	4-10-00-00441	H_2O 1; EtOH 4; eth 4; bz 3
2763	Benzonitrile, 2-methyl- o-Tolunitrile	C_8H_7N 117.15	529-19-1 -13.5	9463 205	4-09-00-01703 0.9955[20]	H_2O 1; EtOH 5; eth 5; ctc 2 1.5279[20]
2764	Benzonitrile, 3-methyl- m-Tolunitrile	C_8H_7N 117.15	620-22-4 -23	213	4-09-00-01717 1.0316[20]	H_2O 1; EtOH 5; eth 5; ctc 2 1.5252[20]
2765	Benzonitrile, 4-methyl- p-Tolunitrile	C_8H_7N 117.15	104-85-8 29.5	9464 217.0	4-09-00-01738 0.9762[30]	H_2O 1; EtOH 4; eth 4; ctc 2

No.	Name / Synonym	Mol. Form. / Mol. Wt.	CAS RN / mp/°C	Merck No. / bp/°C	Beil. Ref. / den/g cm^{-3}	Solubility / n_D
2766	Benzonitrile, 2-methyl-5-nitro-	$C_8H_8N_2O_2$ 162.15	939-83-3 106	174[18]	3-09-00-02314	ace 4; bz 4; eth 4; EtOH 4
2767	Benzonitrile, pentafluoro-	C_7F_5N 193.08	773-82-0 2.4	162	4-09-00-00956 1.563[20]	1.4402[25]
2768	Benzonitrile, 4-(2-phenylethenyl)-, (Z)- 4-Stilbenecarbonitrile (cis)	$C_{15}H_{11}N$ 205.26	14064-68-7 44.5	130[0.1]	4-09-00-02615	H_2O 2; EtOH 3; chl 3
2769	Benzonitrile, 2-(trifluoromethyl)-	$C_8H_4F_3N$ 171.12	447-60-9 18	205	4-09-00-01705	ctc 3
2770	Benzonitrile, 3-(trifluoromethyl)-	$C_8H_4F_3N$ 171.12	368-77-4 14.5	189	3-09-00-02327 1.2813[20]	1.4508[20]
2771	Benzophenone, 3-(dimethylamino)- 3-(Dimethylamino)benzophenone	$C_{15}H_{15}NO$ 225.29	31766-07-1 47	216[15]	4-14-00-00247	EtOH 3
2772	2H-1-Benzopyran 1,2-Chromene	C_9H_8O 132.16	254-04-6 1112 132[102.] 91[13]		5-17-02-00019 1.0993[16]	H_2O 1 1.5869[24]
2773	1H-2-Benzopyran-3-carboxylic acid, 6-[2-(4,9-dihydro-8-hydroxy-5,7-dimethoxy-4,9-dioxonaphtho[2,3-b]furan-2-yl Collinomycin	$C_{27}H_{20}O_{12}$ 536.45	27267-69-2 281	2478	5-19-08-00331	ace 4; diox 4; chl 4
2774	3H-2-Benzopyran-7-carboxylic acid, 4,6-dihydro-8-hydroxy-3,4,5-trimethyl-6-oxo-, (3R-trans)- Citrinin	$C_{13}H_{14}O_5$ 250.25	518-75-2 178-9 dec	2329	5-18-09-00061	H_2O 1; EtOH 2; eth 2; ace 3
2775	2H-1-Benzopyran-3-carboxylic acid, 2-oxo- Coumarin-3-carboxylic acid	$C_8H_6O_3$ 162.14	531-81-7 190 dec	2564	5-18-08-00323	EtOH 4
2776	1H-2-Benzopyran, 3,4-dihydro- Isochroman	$C_9H_{10}O$ 134.18	493-05-0	90[12]	5-17-01-00595 1.067[25]	1.5444[20]
2777	2H-1-Benzopyran, 3,4-dihydro-	$C_9H_{10}O$ 134.18	493-08-3	215; 98[18]	5-17-01-00593 1.0610[20]	H_2O 3; os 5 1.5444[20]
2778	2H-1-Benzopyran, 3,4-dihydro-2-methyl- 2-Methylchroman	$C_{10}H_{12}O$ 148.20	13030-26-7	224.5	5-17-01-00609 1.0340[20]	bz 4; eth 4; EtOH 4 1.532[13]
2779	2H-1-Benzopyran, 3,4-dihydro-3-methyl- Chroman, 3-methyl	$C_{10}H_{12}O$ 148.20	70401-56-8	102-4[15]	4-17-00-00430 1.002[22]	1.5335[20]
2780	2H-1-Benzopyran, 3,4-dihydro-6-methyl- Chroman, 6-methyl-	$C_{10}H_{12}O$ 148.20	3722-74-5	234	4-17-00-00431 1.0374[14]	1.5392[25]
2781	2H-1-Benzopyran, 3,4-dihydro-7-methyl- Chroman, 7-methyl-	$C_{10}H_{12}O$ 148.20	3722-73-4	141-3[60]	4-17-00-00431 1.028[26]	1.5380[20]
2782	2H-1-Benzopyran, 3,4-dihydro-2-phenyl- Flavan	$C_{15}H_{14}O$ 210.28	494-12-2 46.3	332; 190[16]	5-17-02-00283	bz 4; eth 4; EtOH 4
2783	[1]Benzopyrano[3,4-b]furo[2,3-h][1]benzopyran-6(6aH)-one, 1,2,12,12a-tetrahydro-8,9-dimethoxy-2-(1-methylethen Rotenone	$C_{23}H_{22}O_6$ 394.42	83-79-4 176	8248 210-20[0.5]	5-19-10-00581	H_2O 1; EtOH 3; eth 2; ace 3
2784	[2]Benzopyrano[6,5,4-def][2]benzopyran-1,3,6,8-tetrone	$C_{14}H_4O_6$ 268.18	81-30-1 450	sub 320	5-19-05-00423	H_2O 1; Na2CO$_3$ 3; HOAc 3
2785	2H-1-Benzopyran-6-ol, 3,4-dihydro-2,8-dimethyl-2-(4,8,12-trimethyltridecyl)-, [2R-[2R*(4R*,8R*)]]- δ-Tocopherol	$C_{27}H_{46}O_2$ 402.66	119-13-1	9419 150[0.001]		H_2O 1; EtOH 4; eth 4; ace 4
2786	2H-1-Benzopyran-7-ol, 3,4-dihydro-3-(4-hydroxyphenyl)-, (S)- Equol	$C_{14}H_{14}O_3$ 230.26	531-95-3 189.5	3583	5-17-05-00413	
2787	2H-1-Benzopyran-6-ol, 3,4-dihydro-2,5,7,8-tetramethyl-2-(4,8,12-trimethyltridecyl)-, [2R-[2R*(4R*,8R*)]]- Vitamin E	$C_{29}H_{50}O_2$ 430.71	59-02-9 3.0	9931 210[0.1]	5-17-04-00168 0.950[25]	H_2O 1; EtOH 3; eth 3; ace 3 1.5045[25]
2788	2H-1-Benzopyran-6-ol, 3,4-dihydro-2,5,7,8-tetramethyl-2-(4,8,12-trimethyltridecyl)-, acetate, [2R-[2R*(4R*,8R* Vitamin E acetate	$C_{31}H_{52}O_3$ 472.75	58-95-7 -27.5	9932 184[0.01]	5-17-04-00168 0.9533[21]	H_2O 1; EtOH 2; eth 3; ace 3 1.497[20]
2789	2H-1-Benzopyran-6-ol, 3,4-dihydro-2,5,8-trimethyl-2-(4,8,12-trimethyltridecyl)- β-Tocopherol	$C_{28}H_{48}O_2$ 416.69	148-03-8	9417 200-10[0.1]		ace 4; eth 4; EtOH 4; chl 4
2790	2H-1-Benzopyran-6-ol, 3,4-dihydro-2,7,8-trimethyl-2-(4,8,12-trimethyltridecyl)- γ-Tocopherol	$C_{28}H_{48}O_2$ 416.69	7616-22-0 -3	9418 200-10[0.1]		H_2O 1; EtOH 5; eth 5; ace 5
2791	1H-2-Benzopyran-1-one Isocoumarin	$C_9H_6O_2$ 146.15	491-31-6 47	286	5-17-10-00149	H_2O 1; EtOH 4; eth 4; bz 4
2792	2H-1-Benzopyran-2-one Coumarin	$C_9H_6O_2$ 146.15	91-64-5 71	2563 301.7	5-17-10-00143 0.935[20]	H_2O 3; EtOH 3; eth 4; chl 4
2793	4H-1-Benzopyran-4-one	$C_9H_6O_2$ 146.15	491-38-3 59	sub	5-17-10-00139 1.290[20]	H_2O 2; EtOH 3; eth 3; bz 3
2794	4H-1-Benzopyran-4-one, 7-[[6-O-(6-deoxy-α-L-mannopyranosyl)-β-D-glucopyranosyl]oxy]-2,3-dih Hesperidin	$C_{28}H_{34}O_{15}$ 610.57	520-26-3 262	4591	5-18-05-00218	py 4; EtOH 4; HOAc 4
2795	4H-1-Benzopyran-4-one, 7-[[2-O-(6-deoxy-α-L-mannopyranosyl)-β-D-glucopyranosyl]oxy]-2,3-dihydro-5-hy Naringin	$C_{27}H_{32}O_{14}$ 580.54	10236-47-2	6345	5-18-04-00528	H_2O 2; EtOH 2; eth 1; bz 1

No.	Name / Synonym	Mol. Form. / Mol. Wt.	CAS RN / mp/°C	Merck No. / bp/°C	Beil. Ref. / den/g cm^{-3}	Solubility / n_D
2796	4H-1-Benzopyran-4-one, 3-[(6-deoxy-α-L-mannopyranosyl)oxy]-2-(3,4-dihydroxyphenyl)-5,7-dihydroxy- Quercitrin	$C_{21}H_{20}O_{11}$ 448.38	522-12-3 170	8047	5-18-05-00514	H_2O 1; EtOH 3; eth 1; HOAc 3
2797	1H-2-Benzopyran-1-one, 3,4-dihydro-	$C_9H_8O_2$ 148.16	4702-34-5	176[20]; 112[0.4]	5-17-10-00018 1.197[25]	1.5629[25]
2798	2H-1-Benzopyran-2-one, 3,4-dihydro-	$C_9H_8O_2$ 148.16	119-84-6 25	272	5-17-10-00013 1.169[18]	H_2O 1; EtOH 2; eth 2; ctc 2 1.5563[20]
2799	4H-1-Benzopyran-4-one, 2,3-dihydro- 4-Chromanone	$C_9H_8O_2$ 148.16	491-37-2 36.5	160[50]; 127[13]	5-17-10-00014 1.1291[100]	EtOH 3; eth 4; ace 4; bz 4 1.5750
2800	4H-1-Benzopyran-4-one, 2,3-dihydro-5,7-dihydroxy-2-(3-hydroxy-4-methoxyphenyl)-, (S)- Hesperetin	$C_{16}H_{14}O_6$ 302.28	520-33-2 227.5	4590 sub 205	4-18-00-03215	eth 4; EtOH 4
2801	4H-1-Benzopyran-4-one, 2,3-dihydro-5,7-dihydroxy-3-(3-hydroxy-4-methoxyphenyl)- Flavanone, 4'-methoxy-3',5,7-trihydroxy-	$C_{16}H_{14}O_6$ 302.28	99365-26-1 227.5	sub 205	2-18-00-00204	H_2O 1; EtOH 4; eth 3; bz 2
2802	4H-1-Benzopyran-4-one, 2,3-dihydro-5,7-dihydroxy-2-(4-hydroxyphenyl)-, (S)- Naringenin	$C_{15}H_{12}O_5$ 272.26	480-41-1 251	6344	5-18-04-00524	bz 4; eth 4; EtOH 4
2803	4H-1-Benzopyran-4-one, 2,3-dihydro-6-methyl-	$C_{10}H_{10}O_2$ 162.19	39513-75-2 35	141-3[13.5]	5-17-10-00036 1.1245[57]	1.555[57]
2804	2H-1-Benzopyran-2-one, 6,7-dihydroxy- Esculetin	$C_9H_6O_4$ 178.14	305-01-1 276	3645 sub	5-18-03-00202	H_2O 2; EtOH 3; eth 2; ace 3
2805	2H-1-Benzopyran-2-one, 7,8-dihydroxy- Daphnetin	$C_9H_6O_4$ 178.14	486-35-1 262	2816 sub	5-18-03-00211	H_2O 3; EtOH 3; eth 2; bz 2
2806	4H-1-Benzopyran-4-one, 5,7-dihydroxy-2-(4-hydroxyphenyl)- Apigenin	$C_{15}H_{10}O_5$ 270.24	520-36-5 347.5	763	5-18-04-00574	H_2O 1; EtOH 3; py 3; dil alk 4
2807	4H-1-Benzopyran-4-one, 5,7-dihydroxy-3-(4-hydroxyphenyl)- Genistein	$C_{15}H_{10}O_5$ 270.24	446-72-0 301-2 dec	4281	5-18-04-00594	
2808	2H-1-Benzopyran-2-one, 7,8-dihydroxy-6-methoxy- Fraxetin	$C_{10}H_8O_5$ 208.17	574-84-5 231	4178	5-18-04-00332	EtOH 4
2809	4H-1-Benzopyran-4-one, 5,7-dihydroxy-2-(4-methoxyphenyl)- Acacetin	$C_{16}H_{12}O_5$ 284.27	480-44-4 263	9	5-18-04-00575	EtOH 4
2810	4H-1-Benzopyran-4-one, 5,7-dihydroxy-6-methoxy-2-phenyl- Oroxylin A	$C_{16}H_{12}O_5$ 284.27	480-11-5 231.5	6830	5-18-04-00569	ace 4; eth 4; EtOH 4
2811	4H-1-Benzopyran-4-one, 5,7-dihydroxy-2-phenyl- Chrysin	$C_{15}H_{10}O_4$ 254.24	480-40-0 285.5	2261	5-18-04-00076	H_2O 1; EtOH 3; eth 2; ace 3
2812	4H-1-Benzopyran-4-one, 2-(3,4-dihydroxyphenyl)-2,3-dihydro-5,7-dihydroxy-, (S)- Eriodictyol	$C_{15}H_{12}O_6$ 288.26	552-58-9 267 dec	3616	5-18-05-00214	EtOH 4; HOAc 4
2813	4H-1-Benzopyran-4-one, 2-(3,4-dihydroxyphenyl)-3,7-dihydroxy- Fisetin	$C_{15}H_{10}O_6$ 286.24	528-48-3 330	4026	5-18-05-00291	H_2O 1; EtOH 3; eth 2; ace 3
2814	4H-1-Benzopyran-4-one, 2-(3,4-dihydroxyphenyl)-5,7-dihydroxy- Luteolin	$C_{15}H_{10}O_6$ 286.24	491-70-3 329-30 dec	5483	5-18-05-00296	H_2O 2; EtOH 3; eth 3; alk 3
2815	4H-1-Benzopyran-4-one, 2-(3,4-dihydroxyphenyl)-3,5-dihydroxy-7-methoxy- Rhamnetin	$C_{16}H_{12}O_7$ 316.27	90-19-7 295	8170	5-18-05-00495	H_2O 2; EtOH 3; ace 3; dil alk 4
2816	4H-1-Benzopyran-4-one, 2-(2,4-dihydroxyphenyl)-3,5,7-trihydroxy- Morin	$C_{15}H_{10}O_7$ 302.24	480-16-0 303.5	6179	5-18-05-00492	H_2O 2; EtOH 4; eth 2; bz 3
2817	4H-1-Benzopyran-4-one, 2-(3,4-dihydroxyphenyl)-3,5,7-trihydroxy- Quercetin	$C_{15}H_{10}O_7$ 302.24	117-39-5 316.5	8044 sub	5-18-05-00494	H_2O 2; EtOH 3; eth 2; ace 3
2818	2H-1-Benzopyran-2-one, 5,7-dimethoxy- Limettin	$C_{11}H_{10}O_4$ 206.20	487-06-9 149	5370 200 dec	5-18-03-00199	H_2O 2; EtOH 4; eth 1; ace 4
2819	4H-1-Benzopyran-4-one, 8-[(dimethylamino)methyl]-7-methoxy-3-methyl-2-phenyl- Dimefline	$C_{20}H_{21}NO_3$ 323.39	1165-48-6 109.5	3190	5-18-12-00083	chl 3
2820	2H-1-Benzopyran-2-one, 3-ethyl- Coumarin, 3-ethyl-	$C_{11}H_{10}O_2$ 174.20	66898-39-3 72.3	299	5-17-10-00185	eth 4; EtOH 4
2821	2H-1-Benzopyran-2-one, 7-(β-D-glucopyranosyloxy)- Skimmin	$C_{15}H_{16}O_8$ 324.29	93-39-0 220	8506	5-18-01-00409	H_2O 3; EtOH 3; eth 1; chl 1
2822	2H-1-Benzopyran-2-one, 6-(β-D-glucopyranosyloxy)-7-hydroxy- Esculin	$C_{15}H_{16}O_9$ 340.29	531-75-9 205 dec	3646 230 dec	5-18-03-00208	H_2O 2; EtOH 2; eth 2; chl 3

No.	Name / Synonym	Mol. Form. / Mol. Wt.	CAS RN / mp/°C	Merck No. / bp/°C	Beil. Ref. / den/g cm^{-3}	Solubility / n_D
2823	2H-1-Benzopyran-2-one, 8-(β-D-glucopyranosyloxy)-7-hydroxy-6-methoxy- Fraxin	$C_{16}H_{18}O_{10}$ 370.31	524-30-1 205	4179	5-18-04-00334	
2824	4H-1-Benzopyran-4-one, 3-[4-(β-D-glucopyranosyloxy)phenyl]-5,7-dihydroxy- Sophoricoside	$C_{21}H_{20}O_{10}$ 432.38	152-95-4 274	8675	5-18-04-00597	
2825	2H-1-Benzopyran-2-one, 6-hydroxy-	$C_9H_8O_3$ 162.14	6093-68-1 250		5-18-01-00385 1.25^{25}	EtOH 4
2826	2H-1-Benzopyran-2-one, 7-hydroxy- Umbelliferone	$C_9H_8O_3$ 162.14	93-35-6 230.5	9758 sub	5-18-01-00386	EtOH 4; HOAc 4; chl 4
2827	4H-1-Benzopyran-4-one, 7-hydroxy-3-(4-hydroxyphenyl)- Daidzein	$C_{15}H_{10}O_4$ 254.24	486-66-8 323 dec	2805 sub	5-18-04-00089	EtOH 3; eth 3
2828	4H-1-Benzopyran-4-one, 5-hydroxy-3-(4-hydroxyphenyl)-7-methoxy- Prunetin	$C_{16}H_{12}O_5$ 284.27	552-59-0 239.5	7923	5-18-04-00595	
2829	2H-1-Benzopyran-2-one, 7-hydroxy-6-methoxy- Scopoletin	$C_{10}H_8O_4$ 192.17	92-61-5 204	8363	5-18-03-00203	H_2O 2; EtOH 2; eth 2; bz 1
2830	4H-1-Benzopyran-4-one, 7-hydroxy-3-(4-methoxyphenyl)- Formononetin	$C_{16}H_{12}O_4$ 268.27	485-72-3 256.5	4157	5-18-04-00089	
2831	2H-1-Benzopyran-2-one, 7-hydroxy-4-methyl- Hymecromone	$C_{10}H_8O_3$ 176.17	90-33-5 194.5	4792	5-18-01-00439	H_2O 2; EtOH 3; eth 2; chl 2
2832	2H-1-Benzopyran-2-one, 4-hydroxy-3-(3-oxo-1-phenylbutyl)- Warfarin	$C_{19}H_{16}O_4$ 308.33	81-81-2 161	9950	5-18-04-00162	H_2O 1; EtOH 3; ace 3; diox 3
2833	2H-1-Benzopyran-2-one, 3-methyl- 3-Methylcoumarin	$C_{10}H_8O_2$ 160.17	2445-82-1 91	292.5	5-17-10-00165	H_2O 2; EtOH 3; chl 3; KOH 1
2834	2H-1-Benzopyran-2-one, 5-methyl- 5-Methylcoumarin	$C_{10}H_8O_2$ 160.17	42286-84-0 65.8	173^{12}	5-17-10-00167	H_2O 2; EtOH 4; eth 4; bz 4
2835	2H-1-Benzopyran-2-one, 6-methyl-	$C_{10}H_8O_2$ 160.17	92-48-8 76.5	304; 174^{14}	5-17-10-00168	EtOH 4; eth 4; bz 4; chl 2
2836	2H-1-Benzopyran-2-one, 7-methyl- 7-Methylcoumarin	$C_{10}H_8O_2$ 160.17	2445-83-2 128	171.5^{11}	5-17-10-00169	H_2O 2; EtOH 4; eth 3; HOAc 4
2837	2H-1-Benzopyran-2-one, 8-methyl- 8-Methylcoumarin	$C_{10}H_8O_2$ 160.17	1807-36-9 110.5	178^{20}	5-17-10-00170	EtOH 4; eth 4; bz 4; ctc 4
2838	4H-1-Benzopyran-4-one, 3-methyl- Tricromyl	$C_{10}H_8O_2$ 160.17	85-90-5 	9574	5-17-10-00164	chl 3
2839	2H-1-Benzopyran-2-one, 3,3'-methylenebis[4-hydroxy- Dicumarol	$C_{19}H_{12}O_6$ 336.30	66-76-2 290	3080	5-19-06-00682	
2840	4H-1-Benzopyran-4-one, 2-phenyl- Flavone	$C_{15}H_{10}O_2$ 222.24	525-82-6 100	4030	5-17-10-00552	H_2O 1; EtOH 3; eth 3; ace 3
2841	4H-1-Benzopyran-4-one, 3-phenyl- Isoflavone	$C_{15}H_{10}O_2$ 222.24	574-12-9 148	5057	5-17-10-00562	
2842	4H-1-Benzopyran-4-one, 3,5,7-trihydroxy-2-(2-hydroxyphenyl)- Datiscetin	$C_{15}H_{10}O_6$ 286.24	480-15-9 277.5	2822	4-18-00-03281	ace 4; eth 4; EtOH 4
2843	4H-1-Benzopyran-4-one, 3,5,7-trihydroxy-2-(4-hydroxyphenyl)- Kaempferol	$C_{15}H_{10}O_6$ 286.24	520-18-3 277	5156	5-18-05-00251	H_2O 2; EtOH 4; eth 4; ace 4
2844	4H-1-Benzopyran-4-one, 5,6,7-trihydroxy-2-phenyl- Baicalein	$C_{15}H_{10}O_5$ 270.24	491-67-8 264-5 dec	954	5-18-04-00569	H_2O 2; EtOH 3; eth 3; ace 3
2845	2H-1-Benzopyran-2-one, 6,7,8-trimethoxy-	$C_{12}H_{12}O_5$ 236.22	6035-49-0 104.2	90-100$^{0.2}$	5-18-04-00333	EtOH 3; eth 3
2846	2H-1-Benzopyran-3,5,7-triol, 2-(3,4-dihydroxyphenyl)-3,4-dihydro-, (2S-cis)-	$C_{15}H_{14}O_6$ 290.27	35323-91-2 242		5-17-08-00447 1.344^4	ace 4; EtOH 4
2847	1-Benzopyrylium, 3,5,7-trihydroxy-2-(4-hydroxy-3,5-dimethoxyphenyl)-, chloride Malvidin chloride	$C_{17}H_{15}ClO_7$ 366.75	643-84-5 >300	5596	5-17-08-00514	H_2O 2; EtOH 3; MeOH 3
2848	1-Benzopyrylium, 3,5,7-trihydroxy-2-(4-hydroxyphenyl)-, chloride Pelargonidin chloride	$C_{15}H_{11}ClO_5$ 306.70	134-04-3 >350	7014	1-18-00-00410	H_2O 3; EtOH 4; chl 2; con sulf 3
2849	1-Benzopyrylium, 3,5,7-trihydroxy-2-(3,4,5-trihydroxyphenyl)-, chloride Delphinidin	$C_{15}H_{11}ClO_7$ 338.70	528-53-0 >350	2866	5-17-08-00514	H_2O 4; EtOH 4; MeOH 4; AcOEt 3
2850	Benzo[rst]pentaphene 1,2:7,8-Dibenzopyrene	$C_{24}H_{14}$ 302.38	189-55-9 281.5		4-05-00-02803	
2851	2H-1,2,4-Benzothiadiazine-7-sulfonamide, 3-bicyclo[2.2.1]hept-5-en-2-yl-6-chloro-3,4-dihydro-, 1,1-dioxide Cyclothiazide	$C_{14}H_{16}ClN_3O_4S_2$ 389.88	2259-96-3 234	2760		
2852	2H-1,2,4-Benzothiadiazine-7-sulfonamide, 6-chloro-, 1,1-dioxide Chlorothiazide	$C_7H_6ClN_3O_4S_2$ 295.73	58-94-6 350 dec	2169	4-27-00-08036	

No.	Name / Synonym	Mol. Form. / Mol. Wt.	CAS RN / mp/°C	Merck No. / bp/°C	Beil. Ref. / den/g cm^{-3}	Solubility / n_D
2853	2H-1,2,4-Benzothiadiazine-7-sulfonamide, 6-chloro-3-(cyclopentylmethyl)-3,4-dihydro-, 1,1-dioxide Cyclopenthiazide	$C_{13}H_{18}ClN_3O_4S_2$ 379.89	742-20-1 238	2750		
2854	2H-1,2,4-Benzothiadiazine-7-sulfonamide, 6-chloro-3-(dichloromethyl)-3,4-dihydro-, 1,1-dioxide Trichloromethiazide	$C_8H_8Cl_3N_3O_4S_2$ 380.66	133-67-5 270 dec	9537		H_2O 2; EtOH 3
2855	2H-1,2,4-Benzothiadiazine-7-sulfonamide, 6-chloro-3,4-dihydro-, 1,1-dioxide Hydrochlorothiazide	$C_7H_8ClN_3O_4S_2$ 297.74	58-93-5 274	4704		
2856	2H-1,2,4-Benzothiadiazine-7-sulfonamide, 6-chloro-3,4-dihydro-3-(2-methylpropyl)-, 1,1-dioxide Buthiazide	$C_{11}H_{16}ClN_3O_4S_2$ 353.85	2043-38-1 221.5	1519		
2857	2H-1,2,4-Benzothiadiazine-7-sulfonamide, 6-chloro-3,4-dihydro-2-methyl-3-[[(2,2,2-trifluoroethyl)thio]methyl]- Polythiazide	$C_{11}H_{13}ClF_3N_3O_4S_3$ 439.89	346-18-9 214	7561		
2858	2H-1,2,4-Benzothiadiazine-7-sulfonamide, 3,4-dihydro-6-(trifluoromethyl)-, 1,1-dioxide Hydroflumethiazide	$C_8H_8F_3N_3O_4S_2$ 331.30	135-09-1 270.5	4716	4-27-00-08035	
2859	2-Benzothiazolamine 2-Aminobenzothiazole	$C_7H_6N_2S$ 150.20	136-95-8 132	435	4-27-00-04824	H_2O 2; EtOH 3; eth 3; chl 3
2860	6-Benzothiazolamine 6-Aminobenzothiazole	$C_7H_6N_2S$ 150.20	533-30-2 87	436	4-27-00-04884	H_2O 1; EtOH 3; eth 1
2861	Benzothiazole Benzosulfonazole	C_7H_5NS 135.19	95-16-9 2	1118 231	4-27-00-01069 1.2460[20]	H_2O 2; EtOH 4; eth 4; ace 3 1.6379[20]
2862	Benzothiazole, 2-chloro- 2-Chlorobenzothiazole	C_7H_4ClNS 169.63	615-20-3 24	248	4-27-00-01072 1.3715[10]	ace 4; eth 4; EtOH 4 1.6338[10]
2863	Benzothiazole, 2-methyl- 2-Methylbenzothiazole	C_8H_7NS 149.22	120-75-2 14	238	4-27-00-01080 1.1763[19]	H_2O 1; EtOH 3; chl 3 1.6092[19]
2864	Benzothiazole, 2-phenyl- 2-Phenylbenzothiazole	$C_{13}H_9NS$ 211.29	883-93-2 115	371	4-27-00-01385	H_2O 1; EtOH 3; eth 3; CS_2 3
2865	2-Benzothiazolesulfonamide, 6-ethoxy- Ethoxzolamide	$C_9H_{10}N_2O_3S_2$ 258.32	452-35-7 189	3711	4-27-00-04404	
2866	2(3H)-Benzothiazolethione 2-Mercaptobenzothiazole	$C_7H_5NS_2$ 167.26	149-30-4 181	5759	4-27-00-02709 1.42[20]	H_2O 1; EtOH 3; eth 2; bz 2
2867	2(3H)-Benzothiazolethione, 3-methyl- 	$C_8H_7NS_2$ 181.28	2254-94-6 90	335	4-27-00-02714	H_2O 1; EtOH 2; eth 2; bz 4
2868	2(3H)-Benzothiazolone 	C_7H_5NOS 151.19	934-34-9 139	360	4-27-00-02693	H_2O 1; EtOH 4; eth 4
2869	4H-1-Benzothiopyran-4-one, 2,3-dihydro- 	C_9H_8OS 164.23	3528-17-4 29	154[12]	5-17-10-00016 1.2487[14]	1.6395[20]
2870	1,2,4-Benzotriazine 1,2,4-Triazanaphthalene	$C_7H_5N_3$ 131.14	254-87-5 75.8	237.5	4-26-00-00166	H_2O 4; eth 4; EtOH 4
2871	1H-Benzotriazole 1,2,3-Triaza-1H-indene	$C_8H_5N_3$ 119.13	95-14-7 100	1119 204[15]	4-26-00-00093	H_2O 2; EtOH 3; bz 3; chl 3
2872	1H-Benzotriazole, 1-methyl- 	$C_7H_7N_3$ 133.15	13351-73-0 64.5	270.5	4-26-00-00095	bz 4; EtOH 4; HOAc 4
2873	1H-Benzotriazole, 5-nitro- 5-Nitrobenzotriazole	$C_6H_4N_4O_2$ 164.12	2338-12-7 217		4-26-00-00129	
2874	1,2,3-Benzoxadiazole, 5,7-dinitro- 2,3-Benzoxadiazole, 5,7-dinitro-	$C_6H_2N_4O_5$ 210.11	87-31-0 158			EtOH 4
2875	3H-2,1-Benzoxathiol-3-one, 1,1-dioxide 	$C_7H_4O_4S$ 184.17	81-08-3 129.5	184[18]	5-19-04-00215	bz 4; chl 4
2876	2-Benzoxazolamine, 5-chloro- Zoxazolamine	$C_7H_5ClN_2O$ 168.58	61-80-3 184.5	10098	4-27-00-04819	EtOH 4
2877	Benzoxazole 1-Oxa-3-azaindene	C_7H_5NO 119.12	273-53-0 31	182.5	4-27-00-01069 1.1754[20]	H_2O 1; EtOH 3; sulf 3 1.5594[20]
2878	Benzoxazole, 2-chloro- 2-Chlorobenzoxazole	C_7H_4ClNO 153.57	615-18-9 7	201.5	2-27-00-00017 1.3453[18]	1.5678[20]
2879	Benzoxazole, 2,5-dimethyl- 2,5-Dimethylbenzoxazole	C_9H_9NO 147.18	5676-58-4 	218.5	4-27-00-01101 1.0880[18]	ctc 3 1.5412[20]
2880	Benzoxazole, 2-methyl- 2-Methylbenzoxazole	C_8H_7NO 133.15	95-21-6 9.5	200.5	4-27-00-01078 1.1211[20]	H_2O 1; EtOH 4; eth 5 1.5497[20]
2881	2-Benzoxazolol 2-Hydroxybenzoxazole	$C_7H_5NO_2$ 135.12	69564-68-7 141	230[30]	2-27-00-00233	eth 4; EtOH 4
2882	2(3H)-Benzoxazolone 	$C_7H_5NO_2$ 135.12	59-49-4 138	335; 230[30]	4-27-00-02677	H_2O 2; EtOH 3; eth 3; tfa 3
2883	2(3H)-Benzoxazolone, 5-chloro- Chlorzoxazone	$C_7H_4ClNO_2$ 169.57	95-25-0 191.5	2197	4-27-00-02683	EtOH 4; MeOH 4
2884	Benzoyl azide Benzazide	$C_7H_5N_3O$ 147.14	582-61-6 32	exp	4-09-00-00945 1.1680[35]	eth 4; EtOH 4
2885	Benzoyl bromide Benzoic acid, bromide	C_7H_5BrO 185.02	618-32-6 -24	218.5	4-09-00-00724 1.570[15]	eth 5 1.5868[25]
2886	Benzoyl chloride Benzenecarbonyl chloride	C_7H_5ClO 140.57	98-88-4 -1.0	1124 197.2; 71[9]	4-09-00-00721 1.2120[20]	eth 5; bz 3; ctc 3; CS_2 3 1.5537[20]
2887	Benzoyl chloride, 2-bromo- 	C_7H_4BrClO 219.46	7154-66-7 11	243	4-09-00-01013	ctc 2 1.5963[20]
2888	Benzoyl chloride, 4-bromo- 	C_7H_4BrClO 219.46	586-75-4 42	246; 181[125]	4-09-00-01023	EtOH 4; eth 4; bz 4; lig 4

No.	Name Synonym	Mol. Form. Mol. Wt.	CAS RN	mp/°C	Merck No. bp/°C	Bell. Ref. den/g cm^{-3}	Solubility n_D
2889	Benzoyl chloride, 4-butyl-	$C_{11}H_{13}ClO$ 196.68	28788-62-7		155-6[22]	3-09-00-02520 1.051[25]	1.5351[20]
2890	Benzoyl chloride, 2-chloro-	$C_7H_4Cl_2O$ 175.01	609-65-4	-4	238	4-09-00-00962	ctc 3 1.5726[16]
2891	Benzoyl chloride, 3-chloro- m-Chlorobenzoyl chloride	$C_7H_4Cl_2O$ 175.01	618-46-2		225	4-09-00-00972	1.5677[20]
2892	Benzoyl chloride, 4-chloro-	$C_7H_4Cl_2O$ 175.01	122-01-0	16	222	4-09-00-00984 1.3770[20]	chl 2 1.5756[20]
2893	Benzoyl chloride, 2,4-dichloro- 2,4-Dichlorobenzoyl chloride	$C_7H_3Cl_3O$ 209.46	89-75-8 16.5		150[34]· 117[7.5]	4-09-00-01002	ctc 3 1.5895[20]
2894	Benzoyl chloride, 3,4-dichloro- 3,4-Dichlorobenzoyl chloride	$C_7H_3Cl_3O$ 209.46	3024-72-4 25		242	3-09-00-01379	ctc 2
2895	Benzoyl chloride, 4-(1,1-dimethylethyl)-	$C_{11}H_{13}ClO$ 196.68	1710-98-1		135[20]	4-09-00-01887 1.007[25]	1.5364[20]
2896	Benzoyl chloride, 3,5-dinitro- 3,5-Dinitrobenzoyl chloride	$C_7H_3ClN_2O_5$ 230.56	99-33-2 74		3270 196[12]	4-09-00-01350	eth 3; chl 3
2897	Benzoyl chloride, 2-fluoro-	C_7H_4ClFO 158.56	393-52-2 4		90-2[15]	4-09-00-00951 1.328[25]	1.5365[20]
2898	Benzoyl chloride, 3-fluoro-	C_7H_4ClFO 158.56	1711-07-5 -30		189	4-09-00-00953 1.304[25]	1.5285[20]
2899	Benzoyl chloride, 4-fluoro-	C_7H_4ClFO 158.56	403-43-0 9		82[20]	4-09-00-00954 1.342[25]	1.5296[20]
2900	Benzoyl chloride, 2-hydroxy-	$C_7H_5ClO_2$ 156.57	1441-87-8 19		92[15]	3-10-00-00150 1.3112[20]	eth 4 1.5812[20]
2901	Benzoyl chloride, 4-methoxy- p-Anisoyl chloride	$C_8H_7ClO_2$ 170.60	100-07-2 24.5		703 262.5	4-10-00-00430 1.261[20]	eth 3; ace 3; bz 4; ctc 2 1.580[20]
2902	Benzoyl chloride, 2-methyl-	C_8H_7ClO 154.60	933-88-0		213.5	3-09-00-02304	eth 4; EtOH 4 1.5549[20]
2903	Benzoyl chloride, 3-methyl-	C_8H_7ClO 154.60	1711-06-4 -23		219.5	4-09-00-01716 1.0265[21]	eth 4; EtOH 4 1.505[22]
2904	Benzoyl chloride, 4-methyl-	C_8H_7ClO 154.60	874-60-2 -3		226	4-09-00-01733 1.1686[20]	ctc 3 1.5547[20]
2905	Benzoyl chloride, 3-nitro-	$C_7H_4ClNO_3$ 185.57	121-90-4 36		276.5	4-09-00-01061	eth 4
2906	Benzoyl chloride, 4-nitro- p-Nitrobenzoyl chloride	$C_7H_4ClNO_3$ 185.57	122-04-3 75		6511 203[105.] 151[15]	4-09-00-01191	eth 3
2907	Benzoyl chloride, 4-pentyl-	$C_{12}H_{15}ClO$ 210.70	49763-65-7		121[8]	4-09-00-01910 1.036[25]	1.5300[20]
2908	Benzoyl chloride, 4-(pentyloxy)-	$C_{12}H_{15}ClO_2$ 226.70	36823-84-4		196-200[30]	4-10-00-00431 1.087[25]	1.5434[20]
2909	Benzoyl fluoride Benzoic acid, fluoride	C_7H_5FO 124.11	455-32-3 -28		154.5	4-09-00-00721 1.1400[20]	EtOH 4; eth 4; ctc 3
2910	Benzoyl iodide	C_7H_5IO 232.02	618-38-2 3		128[20]	4-09-00-00725 1.746[18]	eth 4; EtOH 4
2911	Berbaman-12-ol, 6,6',7-trimethoxy-2,2'-dimethyl- Berbamine	$C_{37}H_{40}N_2O_6$ 608.73	478-61-5 198.5		1168	2-27-00-00891	H_2O 2; EtOH 3; eth 3; chl 3
2912	Beryllium, bis(2,4-pentanedionato-O,O')-, (T-4)- Beryllium acetylacetonate	$C_{10}H_{14}BeO_4$ 207.23	10210-64-7 108		1180 270		
2913	Bicyclo[2.2.1]hepta-2,5-diene	C_7H_8 92.14	121-46-0 -19.1		89.5	4-05-00-00879 0.9064[20]	H_2O 1; EtOH 3; eth 3; ace 3 1.4702[20]
2914	Bicyclo[2.2.1]heptane	C_7H_{12} 96.17	279-23-2 87.5		105.3	4-05-00-00258	ace 4; bz 4; eth 4; EtOH 4
2915	Bicyclo[4.1.0]heptane Norcarane	C_7H_{12} 96.17	286-08-8 116.5		6605	4-05-00-00257 0.853[25]	1.4564[20]
2916	Bicyclo[2.2.1]heptane-2-carboxaldehyde 2-Norbornanecarboxaldehyde	$C_8H_{10}O$ 122.17	19396-83-9		70-2[22]	4-07-00-00157 1.0227[19]	eth 4 1.4760[25]
2917	Bicyclo[2.2.1]heptane-2-carboxylic acid, 4,7,7-trimethyl-3-oxo- d-Camphocarboxylic acid	$C_{11}H_{16}O_3$ 196.25	18530-30-8 127.5		1737	3-10-00-02925	bz 4; eth 4; EtOH 4
2918	Bicyclo[2.2.1]heptane, 2-chloro-	$C_7H_{11}Cl$ 130.62	29342-53-8 -5		162; 52[11]	4-05-00-00259 1.0603[20]	chl 4 1.4849[20]
2919	Bicyclo[2.2.1]heptane, 2-chloro-1,7,7-trimethyl- Bornane, 2-chloro-	$C_{10}H_{17}Cl$ 172.70	6120-13-4 132.8		207	4-05-00-00319	eth 4; EtOH 4
2920	Bicyclo[2.2.1]heptane, 2-chloro-1,7,7-trimethyl-, endo- Bornyl chloride	$C_{10}H_{17}Cl$ 172.70	464-41-5 132		1341 207.5	4-05-00-00319	bz 4; eth 4; EtOH 4; peth 4
2921	Bicyclo[2.2.1]heptane, 2,2-dimethyl-3-methylene-, (±)-	$C_{10}H_{16}$ 136.24	565-00-4 51.5		158.5	4-05-00-00462 0.879[20]	H_2O 1; EtOH 4; eth 4 1.4551[54]
2922	Bicyclo[2.2.1]heptane, 2,2-dimethyl-5-methylene-	$C_{10}H_{16}$ 136.24	497-32-5		152	4-05-00-00465 0.8591[20]	1.4645[25]
2923	Bicyclo[2.2.1]heptane, 7,7-dimethyl-2-methylene-, (±)- α-Fenchene (DL)	$C_{10}H_{16}$ 136.24	2623-54-3		155	3-05-00-00389 0.8660[20]	ace 4; eth 4; EtOH 4 1.4705[20]

No.	Name Synonym	Mol. Form. Mol. Wt.	CAS RN mp/°C	Merck No. bp/°C	Beil. Ref. den/g cm^{-3}	Solubility n_D
2924	Bicyclo[3.1.1]heptane, 6,6-dimethyl-2-methylene-, (1R)-	$C_{10}H_{16}$ 136.24	19902-08-0 -50	7415 164.6	4-05-00-00456 0.8654[20]	bz 4; eth 4; EtOH 4; chl 4 1.4789[20]
2925	Bicyclo[3.1.1]heptane, 6,6-dimethyl-2-methylene- β-Pinene	$C_{10}H_{16}$ 136.24	127-91-3 -61.5	7415 166	0.860[25]	H_2O 1; bz 3; EtOH 3; eth 3 1.4768[25]
2926	Bicyclo[2.2.1]heptane-2,3-dione, 1,7,7-trimethyl-, (1R)-	$C_{10}H_{14}O_2$ 166.22	10334-26-6 199	sub	4-07-00-02039	H_2O 3; EtOH 4; eth 4; bz 3
2927	Bicyclo[2.2.1]heptane-1-methanamine, 7,7-dimethyl- α-Camphylamine	$C_{10}H_{19}N$ 153.27	54131-44-1	195; 95[12]	1-12-00-00129 0.8688[20]	1.4728[18]
2928	Bicyclo[3.1.1]heptane-2-methanamine, 6,6-dimethyl-, [1S-(1α,2β,5α)]-	$C_{10}H_{19}N$ 153.27	73522-42-6	94-9[27]	4-06-00-00384 0.915[25]	1.4877[20]
2929	Bicyclo[2.2.1]heptane-1-methanesulfonic acid, 7,7-dimethyl-2-oxo-, (1S)- D-Camphorsulfonic acid	$C_{10}H_{16}O_4S$ 232.30	3144-16-9 195 dec		4-11-00-00642	H_2O 4; eth 1; HOAc 2
2930	Bicyclo[3.1.1]heptane-2-methanol, 6,6-dimethyl-, [1S-(1α,2β,5α)]-	$C_{10}H_{18}O$ 154.25	51152-12-6	70-2[1]	4-06-00-00278 0.986[25]	1.4890[20]
2931	Bicyclo[2.2.1]heptane, 1,3,3-trimethyl-, (+)- Fenchane(d)	$C_{10}H_{18}$ 138.25	116435-30-4 <-15	151	3-05-00-00256 0.8345[20]	H_2O 1; EtOH 3; eth 3; HOAc 2 1.44714[20]
2932	Bicyclo[2.2.1]heptane, 1,7,7-trimethyl-	$C_{10}H_{18}$ 138.25	464-15-3	161	4-05-00-00319	H_2O 1; EtOH 3; eth 3; AcOEt 3
2933	Bicyclo[2.2.1]heptane, 2,2,3-trimethyl-	$C_{10}H_{18}$ 138.25	473-19-8 62.5	166	3-05-00-00263 0.8276[67]	ace 4; EtOH 4
2934	Bicyclo[3.1.1]heptane, 2,6,6-trimethyl-, (1α,2α,5α)-(±)- Pinane (DL)	$C_{10}H_{18}$ 138.25	112456-46-9	164.5	1-05-00-00048 0.8551[20]	bz 4; eth 4 1.4609[20]
2935	Bicyclo[3.1.1]heptane, 2,6,6-trimethyl-, [1R-(1α,2β,5α)]-	$C_{10}H_{18}$ 138.25	4795-86-2 -53	169	4-05-00-00318 0.8560[20]	bz 4; eth 4 1.4629[20]
2936	Bicyclo[4.1.0]heptane, 3,7,7-trimethyl-, [1S-(1α,3β,6α)]-	$C_{10}H_{18}$ 138.25	2778-68-9	169	4-05-00-00316 0.8410[20]	1.4569
2937	Bicyclo[4.1.0]heptane, 3,7,7-trimethyl-, [1S-(1α,3α,6α)]-	$C_{10}H_{18}$ 138.25	6069-97-2	169	4-05-00-00316 0.841[20]	1.456
2938	Bicyclo[2.2.1]heptan-2-ol, 1,7-dimethyl-, (exo,syn)- β-Santenol	$C_9H_{16}O$ 140.23	509-12-6 101.5	192; 60[15]	4-06-00-00242	eth 4
2939	Bicyclo[2.2.1]heptan-2-ol, 1-phenyl-, acetate	$C_{15}H_{18}O_2$ 230.31	71173-15-4 44	159[13]. 128[1.0]	3-06-00-02755	ctc 2
2940	Bicyclo[2.2.1]heptan-2-ol, 1,3,3-trimethyl-, endo-(±)-	$C_{10}H_{18}O$ 154.25	36386-49-9 39	199.5	3-06-00-00288 0.9420[40]	eth 4; EtOH 4
2941	Bicyclo[2.2.1]heptan-2-ol, 1,3,3-trimethyl-, (1S-endo)- l-α-Fenchyl alcohol	$C_{10}H_{18}O$ 154.25	512-13-0 48	94[20]	4-06-00-00278 0.9034[84]	
2942	Bicyclo[2.2.1]heptan-2-ol, 1,7,7-trimethyl-, endo-(±)-	$C_{10}H_{18}O$ 154.25	6627-72-1 210.5	sub	4-06-00-00281 1.011[20]	H_2O 1; EtOH 4; eth 4; bz 4
2943	Bicyclo[2.2.1]heptan-2-ol, 1,7,7-trimethyl-, exo-(±)-	$C_{10}H_{18}O$ 154.25	24393-70-2 212	sub	4-06-00-00282 1.10[20]	H_2O 1; EtOH 4; eth 4; bz 2
2944	Bicyclo[3.1.1]heptan-2-ol, 4,6,6-trimethyl-, [1R-(1α,2α,4α,5α)]-	$C_{10}H_{18}O$ 154.25	515-88-8 58	218	3-06-00-00284 0.940[20]	1.4702[20]
2945	Bicyclo[2.2.1]heptan-2-ol, 1,7,7-trimethyl-, acetate, endo-(±)-	$C_{12}H_{20}O_2$ 196.29	36386-52-4 <-17	223.5	4-06-00-00283 1.4630[20]	ctc 3 1.4630[20]
2946	Bicyclo[2.2.1]heptan-2-ol, 1,7,7-trimethyl-, acetate, exo-(±)-	$C_{12}H_{20}O_2$ 196.29	17283-45-3	116[21]	4-06-00-00283 0.9841[20]	ace 4; EtOH 4 1.4640[20]
2947	Bicyclo[2.2.1]heptan-2-ol, 4,7,7-trimethyl-, acetate, (1R-endo)- Epiborneol, acetate	$C_{12}H_{20}O_2$ 196.29	22621-74-5 <-15	101[11]	2-06-00-00092 0.9872[14]	1.4651[14]
2948	Bicyclo[2.2.1]heptan-2-ol, 1,7,7-trimethyl-, acetate, endo- Borneol, acetate	$C_{12}H_{22}O_2$ 198.31	76-49-3 29	1339 221	3-06-00-00303	
2949	Bicyclo[2.2.1]heptan-2-ol, 1,7,7-trimethyl-, formate, (1R-endo)- Bornyl formate (d)	$C_{11}H_{18}O_2$ 182.26	74219-20-8	90[10]	3-06-00-00301 1.009[22]	1.4708[15]
2950	Bicyclo[2.2.1]heptan-2-one, 3-(acetyl-17O)-1,7,7-trimethyl- Camphor, 3-acetyl	$C_{12}H_{18}O_2$ 194.27	15068-90-3	127-8[15]	2-07-00-00563 1.0314[19]	1.4949[17]

No.	Name / Synonym	Mol. Form. / Mol. Wt.	CAS RN / mp/°C	Merck No. / bp/°C	Beil. Ref. / den/g cm⁻³	Solubility / n_D
2951	Bicyclo[2.2.1]heptan-2-one, 1-(bromomethyl)-7,7-dimethyl-, (1S)- d-Camphor, 10-bromo-	$C_{10}H_{15}BrO$ 231.13	64161-50-8 78	 265 dec	4-07-00-00218	ace 4; bz 4; eth 4; EtOH 4
2952	Bicyclo[2.2.1]heptan-2-one, 3-bromo-1,7,7-trimethyl-, (1R-endo)-	$C_{10}H_{15}BrO$ 231.13	10293-06-8 78	 274 dec	 1.449[25]	bz 4; eth 4; EtOH 4
2953	Bicyclo[2.2.1]heptan-2-one, 3,3-dibromo-1,7,7-trimethyl-, (1R)- α,α'-Dibromo-d-camphor	$C_{10}H_{14}Br_2O$ 310.03	514-12-5 61	3002	4-07-00-00220 1.854[21]	H_2O 1; EtOH 4; eth 4; bz 4
2954	Bicyclo[2.2.1]heptan-2-one, 3,3-dimethyl-, (1S)-	$C_9H_{14}O$ 138.21	6069-71-2 41.9	 193; 76[12]	4-07-00-00177 0.807[20]	eth 4
2955	Bicyclo[3.1.1]heptan-2-one, 6,6-dimethyl-, (1R)-	$C_9H_{14}O$ 138.21	38651-65-9 -1	 209	4-07-00-00176 0.9807[20]	eth 4; EtOH 4
2956	Bicyclo[3.1.1]heptan-3-one, 6,6-dimethyl-2-methylene-, (1S)-	$C_{10}H_{14}O$ 150.22	19890-00-7 -1.8	 67-9[4]	 0.9875[15]	1.4787[20]
2957	Bicyclo[2.2.1]heptan-2-one, 3-hydroxy-1,7,7-trimethyl- 3-Hydroxycamphor	$C_{10}H_{16}O_2$ 168.24	10373-81-6 205.5	4747	4-08-00-00081	eth 4; EtOH 4; chl 4
2958	Bicyclo[2.2.1]heptan-2-one, 3-methylene- 3-Methylene-2-norbornanone	$C_8H_{10}O$ 122.17	5597-27-3	 69-71[11]	4-07-00-00301 0.997[25]	1.4891[20]
2959	Bicyclo[2.2.1]heptan-2-one, 1,4,7,7-tetramethyl-	$C_{11}H_{18}O$ 166.26	10309-50-9 168	 213	2-07-00-00113	eth 4; EtOH 4
2960	Bicyclo[2.2.1]heptan-2-one, 1,3,3-trimethyl-, (±)-	$C_{10}H_{16}O$ 152.24	18492-37-0 -18	3911 193.5	3-07-00-00393 0.9492[15]	H_2O 1; EtOH 4; eth 3; ace 3 1.4702[20]
2961	Bicyclo[2.2.1]heptan-2-one, 1,7,7-trimethyl-, (1R)- (+)-Camphor	$C_{10}H_{16}O$ 152.24	464-49-3 178.8	1738 207.4	4-07-00-00213 0.990[25]	H_2O 1; EtOH 4; eth 4; ace 3 1.5462
2962	Bicyclo[2.2.1]heptan-2-one, 1,7,7-trimethyl-, oxime, (1S)- (-)-Camphor oxime	$C_{10}H_{17}NO$ 167.25	36065-15-3 118	 252 dec	4-07-00-00215 1.01[116]	eth 4; EtOH 4
2963	Bicyclo[2.2.1]heptan-2-one, 4,7,7-trimethyl-, (1S)-	$C_{10}H_{16}O$ 152.24	10292-98-5 186.5	 213	3-07-00-00421	H_2O 2; EtOH 4; eth 4; peth 4
2964	Bicyclo[3.1.1]heptan-2-one, 4,6,6-trimethyl-, [1R-(1α,4α,5α)]-	$C_{10}H_{16}O$ 152.24	515-90-2	 222	4-07-00-00210 0.961[20]	1.4752[20]
2965	Bicyclo[3.1.1]heptan-3-one, 2,6,6-trimethyl-, [1S-(1α,2β,5α)]-	$C_{10}H_{16}O$ 152.24	14575-93-0	 212	4-07-00-00209 0.964[25]	1.4735[25]
2966	Bicyclo[2.2.1]hept-5-ene-2-carbonitrile	C_8H_9N 119.17	95-11-4 13	 82-6[10]	 0.999[25]	1.4885[20]
2967	Bicyclo[2.2.1]hept-5-ene-2-carboxaldehyde	$C_8H_{10}O$ 122.17	5453-80-5	 67-70[12]	2-07-00-00125 1.018[25]	1.4893[20]
2968	Bicyclo[2.2.1]hept-2-ene, 2,3-dimethyl- 2-Norbornene, 2,3-dimethyl-	C_9H_{14} 122.21	529-16-8	 140.5	4-05-00-00427 0.8698[17]	eth 3; ace 3; bz 3 1.4688[17]
2969	Bicyclo[3.1.1]hept-2-ene-2-ethanol, 6,6-dimethyl-	$C_{11}H_{18}O$ 166.26	128-50-7	 235; 110[10]	4-06-00-00398 0.973[25]	chl 3 1.4930[20]
2970	Bicyclo[2.2.1]hept-2-ene, 1,7,7-trimethyl-	$C_{10}H_{16}$ 136.24	464-17-5 113	 146	2-05-00-00105	bz 4; eth 4; EtOH 4
2971	Bicyclo[3.1.1]hept-2-ene, 2,6,6-trimethyl-, (1S)- (-)-α-Pinene	$C_{10}H_{16}$ 136.24	7785-26-4 -64	7414 155	4-05-00-00452 0.859[20]	1.4660[20]
2972	Bicyclo[3.1.1]hept-2-ene, 2,6,6-trimethyl-, (±)- (±)-α-Pinene	$C_{10}H_{16}$ 136.24	2437-95-8 -55	7414 156.2	2-05-00-00097 0.8582[20]	H_2O 1; EtOH 5; eth 5; chl 5 1.4658[20]
2973	Bicyclo[3.1.1]hept-2-ene, 2,6,6-trimethyl 2-Pinene	$C_{10}H_{16}$ 136.24	80-56-8 -64	 155.9	 0.8539[25]	1.4632[25]
2974	Bicyclo[4.1.0]hept-3-ene, 3,7,7-trimethyl-, (1S)-	$C_{10}H_{16}$ 136.24	498-15-7	 171; 123[200]	4-05-00-00450 0.8549[30]	ace 4; bz 4; eth 4 1.469[3]
2975	Bicyclo[3.1.1]hept-3-en-2-ol, 4,6,6-trimethyl-, (1α,2β,5α)-	$C_{10}H_{16}O$ 152.24	1845-30-3 15.5	 90[10]	 0.9684[25]	1.4912[25]
2976	Bicyclo[3.1.1]hept-3-en-2-ol, 4,6,6-trimethyl-, (1α,2α,5α)-	$C_{10}H_{16}O$ 152.24	1820-09-3 24	 92[10]	4-06-00-00383 0.9657[25]	1.4908[25]
2977	Bicyclo[3.1.1]hept-2-en-6-one, 2,7,7-trimethyl- Chrysanthenone	$C_{10}H_{14}O$ 150.22	473-06-3	2256 88-9[12]	4-07-00-00327	EtOH 4 1.4720[22]
2978	Bicyclo[3.1.1]hept-3-en-2-one, 4,6,6-trimethyl-, (1R)- d-Verbenone	$C_{10}H_{14}O$ 150.22	18309-32-5 9.8	9862 227.5	4-07-00-00327 0.9978[20]	H_2O 3; EtOH 3; ace 3; bz 3 1.4993[18]
2979	[Bi-1,4-cyclohexadien-1-yl]-3,3',6,6'-tetrone, 2,2'-dihydroxy-4,4'-dimethyl- Phenicin	$C_{14}H_{10}O_6$ 274.23	128-68-7 230.5	7194	4-08-00-03660	H_2O 2; EtOH 4; chl 4; HOAc 4
2980	Bicyclo[3,1,0]hexane, 1-methyl- 1-Methylbicyclo[3,1,0]hexane	C_7H_{12} 96.17	4625-24-5	 93.1		

No.	Name Synonym	Mol. Form. Mol. Wt.	CAS RN mp/°C	Merck No. bp/°C	Beil. Ref. den/g cm^{-3}	Solubility n_D
2981	Bicyclo[3.1.0]hexane, 4-methylene-1-(1-methylethyl)-, (±)- 4(10)-Thujene, (±)-	C$_{10}$H$_{16}$ 136.24	15826-80-9 	 164; 69^{30}	3-05-00-00365 0.8437^{20}	H$_2$O 1; EtOH 4; eth 3; ace 3 1.4676^{20}
2982	Bicyclo[3.1.0]hexane, 4-methyl-1-(1-methylethyl)-	C$_{10}$H$_{18}$ 138.25	471-12-5 	 157	4-05-00-00317 0.8139^{20}	 1.4376^{20}
2983	Bicyclo[3.1.0]hexan-3-ol, 4-methylene-1-(1-methylethyl)- 4(10)-Thujene-3-ol	C$_{10}$H$_{16}$O 152.24	59905-55-4 	 208	4-06-00-00382 0.9461^{25}	 1.4871^{25}
2984	Bicyclo[3.1.0]hexan-3-ol, 4-methylene-1-(1-methylethyl)-, [1S-(1α,3β,5α)]-	C$_{10}$H$_{16}$O 152.24	471-16-9 	 208	4-06-00-00382 0.9488^{19}	eth 3 1.4871^{25}
2985	Bicyclo[3.1.0]hexan-3-ol, 4-methyl-1-(1-methylethyl)-	C$_{10}$H$_{18}$O 154.25	513-23-5 	 209	3-06-00-00278 0.9210^{20}	 1.4621^{25}
2986	Bicyclo[3.1.0]hex-2-ene, 2-methyl-5-(1-methylethyl)-	C$_{10}$H$_{16}$ 136.24	2867-05-2 	 151	4-05-00-00451 0.8301^{20}	 1.4515^{20}
2987	Bicyclo[3.1.0]hex-3-en-2-one, 4-methyl-1-(1-methylethyl)-	C$_{10}$H$_{14}$O 150.22	24545-81-1 	 220	0-07-00-00159 0.9572^{15}	 1.48325
2988	1,1'-Bicyclohexyl Cyclohexylcyclohexane	C$_{12}$H$_{22}$ 166.31	92-51-3 4	 238		H$_2$O 2; EtOH 3; eth 3
2989	[1,1'-Bicyclohexyl]-2-one	C$_{12}$H$_{20}$O 180.29	90-42-6 -32	 264	4-07-00-00248 0.9696^{25}	 1.4877^{25}
2990	Bicyclo[3.3.1]nonane	C$_9$H$_{16}$ 124.23	280-65-9 145.5	 169.5	4-05-00-00293 	EtOH 4
2991	Bicyclo[4.2.0]octane	C$_8$H$_{14}$ 110.20	278-30-8 	 136	4-05-00-00276 0.8573^{20}	 1.4613^{20}
2992	Bicyclo[2.2.2]octane, 2-methyl- 2-Methylbicyclo[2.2.2]octane	C$_9$H$_{16}$ 124.23	766-53-0 33.5	 159	4-05-00-00295 0.8664^{40}	 1.4608^{40}
2993	Bicyclo[4.2.0]oct-2-ene	C$_8$H$_{12}$ 108.18	13367-29-8 	 138	4-05-00-00413 0.8948^{20}	 1.4810^{20}
2994	[1,1'-Bicyclopentyl]-2-ol Bicyclopentyl, 2-hydroxy	C$_{10}$H$_{18}$O 154.25	4884-25-7 20	 235.5	3-06-00-00261 0.9785^{15}	 1.4884^{17}
2995	[1,1'-Bicyclopentyl]-2-one	C$_{10}$H$_{16}$O 152.24	4884-24-6 -13	 232.5	4-07-00-00198 0.9745^{21}	 1.4763
2996	Bicyclo[7.2.0]undec-4-ene, 4,11,11-trimethyl-8-methylene-, [1R-(1R*,4E,9S*)]- Caryophyllene	C$_{15}$H$_{24}$ 204.36	87-44-5 	1884 122$^{13.5}$	4-05-00-01182 0.9075^{20}	bz 4 1.4986^{20}
2997	Bifenox Benzoic acid, 5-(2,4-dichlorophenoxy)-2-nitro-, methyl ester	C$_{14}$H$_9$Cl$_2$NO$_5$ 342.13	42576-02-3 85	1228		
2998	Bifenthrin 2-Methylbiphenyl-3-ylmethyl (Z)-(1RS,3RS)-3-(2-chloro-3,3,3-trifluoroprop-1-enyl)- 2,2-dimethylcyclopropanecarboxylate	C$_{23}$H$_{22}$ClF$_3$O$_2$ 422.87	82657-04-3 69	1229	 1.2^{125}	
2999	[2,2'-Bi-1H-indene]-1,1',3,3'(2H,2'H)-tetrone, 2,2'-dihydroxy- Hydrindantin	C$_{18}$H$_{10}$O$_6$ 322.27	5103-42-4 250 dec	4698	4-08-00-03701	
3000	21H-Biline-8,12-dipropanoic acid, 2,17-diethenyl-1,10,19,22,23,24-hexahydro-3,7,13,18-tetramethyl-1,19-dioxo- Bilirubin	C$_{33}$H$_{36}$N$_4$O$_6$ 584.67	635-65-4 	1235	4-26-00-03268	H$_2$O 1; EtOH 2; eth 2; bz 3
3001	21H-Biline-8,12-dipropanoic acid, 3,18-diethenyl-1,19,22,24-tetrahydro-2,7,13,17-tetramethyl-1,19-dioxo- Biliverdine	C$_{33}$H$_{34}$N$_4$O$_6$ 582.66	114-25-0 >300	1236	4-26-00-03272	H$_2$O 1; EtOH 3; eth 2; bz 3
3002	[2,2'-Bimorphinan]-3,3',6,6'-tetrol, 7,7',8,8'-tetradehydro-4,5:4',5'-diepoxy-17,17'-dimethyl- Pseudomorphine	C$_{34}$H$_{36}$N$_2$O$_6$ 568.67	125-24-6 282.5	7934	4-27-00-08724	H$_2$O 1; EtOH 1; eth 1; chl 1
3003	Binapacryl 2-Butenoic acid, 3-methyl-, 2-(1-methylpropyl)-4,6-dinitrophenyl ester	C$_{15}$H$_{18}$N$_2$O$_6$ 322.32	485-31-4 70	1237	 1.27^{20}	
3004	1,1'-Binaphthalene 1,1'-Binaphthyl	C$_{20}$H$_{14}$ 254.33	604-53-5 160	 >360; 240^{12}	4-05-00-02634 1.3000^{20}	H$_2$O 1; EtOH 2; eth 3; ace 3
3005	2,2'-Binaphthalene	C$_{20}$H$_{14}$ 254.33	612-78-2 188.3	 452	4-05-00-02636	H$_2$O 1; EtOH 2; eth 3; bz 3
3006	[1,1'-Binaphthalene]-4,4'-diol	C$_{20}$H$_{14}$O$_2$ 286.33	1446-34-0 300	 sub	4-06-00-07021	eth 4; EtOH 4
3007	2,2'-Bioxirane, (R*,S*)- Erythritol anhydride	C$_4$H$_8$O$_2$ 86.09	564-00-1 -19	3621 138	5-19-01-00184 1.113^{18}	H$_2$O 5; EtOH 4 1.4330^{20}
3008	1,1'-Biphenyl Diphenyl	C$_{12}$H$_{10}$ 154.21	92-52-4 69	3314 256.1	4-05-00-01807 1.04^{20}	H$_2$O 1; EtOH 3; eth 3; bz 4 1.588^{77}
3009	[1,1'-Biphenyl]-4-acetic acid Felbinac	C$_{14}$H$_{12}$O$_2$ 212.25	5728-52-9 160.5	3893	4-09-00-02520	
3010	[1,1'-Biphenyl]-2-amine	C$_{12}$H$_{11}$N 169.23	90-41-5 51	 299	4-12-00-03223	H$_2$O 1; EtOH 3; eth 3; bz 3
3011	[1,1'-Biphenyl]-4-amine p-Biphenylamine	C$_{12}$H$_{11}$N 169.23	92-67-1 53.5	1248 302	4-12-00-03241	H$_2$O 2; EtOH 3; eth 3; ace 3
3012	[1,1'-Biphenyl]-4-amine, 4'-chloro- 4-Amino-4'-chlorodiphenyl	C$_{12}$H$_{10}$ClN 203.67	135-68-2 134	445	4-12-00-03269	ace 4; bz 4; eth 4

No.	Name Synonym	Mol. Form. Mol. Wt.	CAS RN mp/°C	Merck No. bp/°C	Beil. Ref. den/g cm^{-3}	Solubility n_D
3013	[1,1'-Biphenyl]-2-amine, 6-methyl- Biphenyl, 2-amino-6-methyl	$C_{13}H_{13}N$ 183.25	76472-83-8 43.5	355; 144[2]	3-12-00-03234 1.9060[0]	bz 4; eth 4; EtOH 4
3014	[1,1'-Biphenyl]-4-amine, 3-methyl-	$C_{13}H_{13}N$ 183.25	63019-98-7 43	190[15]	3-12-00-03238	bz 4; eth 4; EtOH 4
3015	[1,1'-Biphenyl]-4-amine, 4'-methyl- Biphenyl, 4-amino-4'-methyl	$C_{13}H_{13}N$ 183.25	1204-78-0 99	190[18]	2-12-00-00771	ace 4; bz 4; eth 4; EtOH 4
3016	1,1'-Biphenyl, 2-bromo-	$C_{12}H_9Br$ 233.11	2052-07-5 1.5	297	4-05-00-01818 1.2175[26]	eth 4; EtOH 4 1.6248[25]
3017	1,1'-Biphenyl, 3-bromo-	$C_{12}H_9Br$ 233.11	2113-57-7	300; 171[17]	4-05-00-01818	H_2O 1 1.6411[20]
3018	1,1'-Biphenyl, 4-bromo-	$C_{12}H_9Br$ 233.11	92-66-0 91.5	310	4-05-00-01819 0.9327[25]	H_2O 1; EtOH 3; eth 3; bz 3
3019	[1,1'-Biphenyl]-4-butanoic acid	$C_{16}H_{16}O_2$ 240.30	6057-60-9 119.5	190[0.3]	4-09-00-02558	eth 4; EtOH 4
3020	[1,1'-Biphenyl]-2-carbonitrile	$C_{13}H_9N$ 179.22	24973-49-7 41	175[13]	3-09-00-03269	eth 4; EtOH 4
3021	[1,1'-Biphenyl]-2-carboxylic acid	$C_{13}H_{10}O_2$ 198.22	947-84-2 114.3	343.5	4-09-00-02472	H_2O 1; EtOH 4; bz 4; HOAc 4
3022	[1,1'-Biphenyl]-4-carboxylic acid	$C_{13}H_{10}O_2$ 198.22	92-92-2 228	sub	4-09-00-02479	H_2O 1; EtOH 3; eth 3; bz 3
3023	1,1'-Biphenyl, 2-chloro-	$C_{12}H_9Cl$ 188.66	2051-60-7 34	274	4-05-00-01816 1.1499[32]	eth 4; EtOH 4; lig 4
3024	1,1'-Biphenyl, 3-chloro-	$C_{12}H_9Cl$ 188.66	2051-61-8 16	284.5	4-05-00-01816 1.1579[25]	ace 4; eth 4; EtOH 4 1.6181[25]
3025	1,1'-Biphenyl, 4-chloro-	$C_{12}H_9Cl$ 188.66	2051-62-9 78.8	292.9; 146[10]	4-05-00-01816	H_2O 1; EtOH 3; eth 3; lig 3
3026	Biphenyl, 2-chloro-2'-methyl- 2-Chloro-2'-methylbiphenyl	$C_{13}H_{11}Cl$ 202.68	19493-31-3 17	276; 174[50]	3-05-00-01799	1.588[25]
3027	1,1'-Biphenyl, 2,2',3,3',4,4',5,5',6,6'-decafluoro-	$C_{12}F_{10}$ 334.12	434-90-2 67.5	206	4-05-00-01816 1.785[20]	
3028	[1,1'-Biphenyl]-2,2'-diamine	$C_{12}H_{12}N_2$ 184.24	1454-80-4 81	162[4]	4-13-00-00356 1.3090[20]	H_2O 3; ace 3; bz 3
3029	[1,1'-Biphenyl]-2,4'-diamine 2,4'-Biphenyldiamine	$C_{12}H_{12}N_2$ 184.24	492-17-1 54.5	1249 363	4-13-00-00360	H_2O 1; EtOH 3; eth 3
3030	[1,1'-Biphenyl]-3,3'-diamine	$C_{12}H_{12}N_2$ 184.24	2050-89-7 93.5	205[0.001]	4-13-00-00363	H_2O 2; eth 4; bz 3
3031	[1,1'-Biphenyl]-4,4'-diamine Benzidine	$C_{12}H_{12}N_2$ 184.24	92-87-5 120	1086 401	4-13-00-00364	H_2O 2; EtOH 3; eth 2; DMSO 2
3032	[1,1'-Biphenyl]-4,4'-diamine, 2,2'-dichloro- o-Dichlorobenzidine	$C_{12}H_{10}Cl_2N_2$ 253.13	84-68-4 165	3046	4-13-00-00384	eth 4; EtOH 4
3033	[1,1'-Biphenyl]-4,4'-diamine, 3,3'-dimethoxy- o-Dianisidine	$C_{14}H_{16}N_2O_2$ 244.29	119-90-4 137	2970	4-13-00-02834	H_2O 1; EtOH 3; eth 3; ace 3
3034	[1,1'-Biphenyl]-4,4'-diamine, 3,3'-dimethyl- o-Tolidine	$C_{14}H_{16}N_2$ 212.29	119-93-7 131.5	9437	4-13-00-00419	H_2O 2; EtOH 4; eth 4; chl 2
3035	[1,1'-Biphenyl]-4,4'-diamine, N,N'-diphenyl- N,N'-Diphenylbenzidine	$C_{24}H_{20}N_2$ 336.44	531-91-9 247	3319	4-13-00-00368	H_2O 1; EtOH 2; eth 2; bz 2
3036	[1,1'-Biphenyl]-4,4'-diamine, tetramethyl-	$C_{16}H_{20}N_2$ 240.35	34314-06-2 198	>360	3-13-00-00429	bz 4; chl 4
3037	1,1'-Biphenyl, 4,4'-dibromo-	$C_{12}H_8Br_2$ 312.00	92-86-4 164	357.5	4-05-00-01820	H_2O 1; EtOH 2; bz 3
3038	[1,1'-Biphenyl]-2,2'-dicarboxylic acid o,o'-Diphenic acid	$C_{14}H_{10}O_4$ 242.23	482-05-3 233.5	3309 sub	4-09-00-03552	H_2O 1; EtOH 3; eth 3; os 3
3039	[1,1'-Biphenyl]-2,2'-dicarboxylic acid, dimethyl ester	$C_{16}H_{14}O_4$ 270.28	5807-64-7 74	204-6[14]	4-09-00-03553	H_2O 1; EtOH 4; eth 3; bz 3
3040	1,1'-Biphenyl, 3,3'-dichloro-	$C_{12}H_8Cl_2$ 223.10	2050-67-1 29	320	3-05-00-01739	bz 4; eth 4; EtOH 4
3041	1,1'-Biphenyl, 4,4'-dichloro-	$C_{12}H_8Cl_2$ 223.10	2050-68-2 149.3	317	4-05-00-01817 1.4420[0]	H_2O 1; EtOH 2; bz 3; chl 2
3042	1,1'-Biphenyl, 2,2'-difluoro-	$C_{12}H_8F_2$ 190.19	388-82-9 118.5		4-05-00-01815 1.393[20]	H_2O 1; EtOH 3; xyl 3
3043	1,1'-Biphenyl, 3,3'-difluoro-	$C_{12}H_8F_2$ 190.19	396-64-5 8	130[14]	4-05-00-01815 1.192[25]	ctc 2 1.5678[20]
3044	1,1'-Biphenyl, 4,4'-difluoro- 4,4'-Difluorodiphenyl	$C_{12}H_8F_2$ 190.19	398-23-2 94.5	3133 254.5	4-05-00-01816	H_2O 1; EtOH 4; eth 3; ace 3
3045	1,1'-Biphenyl, 2,2'-dimethoxy- 2,2'-Dimethoxybiphenyl	$C_{14}H_{14}O_2$ 214.26	4877-93-4 155	307.5	4-06-00-06645 1.268[25]	H_2O 1; EtOH 4; eth 2; bz 4
3046	1,1'-Biphenyl, 3,3'-dimethoxy- 3,3'-Dimethoxybiphenyl	$C_{14}H_{14}O_2$ 214.26	6161-50-8 36	328	4-06-00-06650	H_2O 1; EtOH 4; eth 4; ace 3
3047	1,1'-Biphenyl, 4,4'-dimethoxy-	$C_{14}H_{14}O_2$ 214.26	2132-80-1 175	sub	4-06-00-06651	H_2O 1; EtOH 2; eth 2; bz 4
3048	1,1'-Biphenyl, 2,2'-dimethoxy-5,5'-dimethyl- Biphenyl, 2,2'-dimethoxy-5,5'-dimethyl	$C_{16}H_{18}O_2$ 242.32	7168-55-0 71	188[12]	3-06-00-05447	ace 4; bz 4; eth 4; EtOH 4
3049	1,1'-Biphenyl, 2,5-dimethoxy-2',5-dimethyl- Biphenyl, 2,5-dimethoxy-2',5-dimethyl	$C_{16}H_{18}O_2$ 242.32	72935-12-7 66	168[4]	4-06-00-06704	bz 4; peth 4
3050	1,1'-Biphenyl, 2,2'-dimethyl-	$C_{14}H_{14}$ 182.27	605-39-0 19.5	256	4-05-00-01897 0.9906[20]	H_2O 1; EtOH 4; eth 4; ace 3 1.5752[20]
3051	1,1'-Biphenyl, 2,3'-dimethyl-	$C_{14}H_{14}$ 182.27	611-43-8	270	4-05-00-01902 0.9924[20]	H_2O 1; EtOH 4; eth 4; ace 3 1.5810[20]

No.	Name Synonym	Mol. Form. Mol. Wt.	CAS RN mp/°C	Merck No. bp/°C	Beil. Ref. den/g cm^{-3}	Solubility n_D
3052	1,1'-Biphenyl, 2,3-dimethyl- 2,3-Dimethylbiphenyl	C$_{14}$H$_{14}$ 182.27	3864-18-4 42	141[14]; 98[2]	4-05-00-01897 	H$_2$O 1; eth 3; peth 3 1.5845[23]
3053	1,1'-Biphenyl, 2,4'-dimethyl-	C$_{14}$H$_{14}$ 182.27	611-61-0 	277.2	4-05-00-01903 0.9924[20]	H$_2$O 1; EtOH 4; eth 4; ace 3 1.5826[20]
3054	1,1'-Biphenyl, 2,4-dimethyl- 2,4-Dimethylbiphenyl	C$_{14}$H$_{14}$ 182.27	4433-10-7 	273; 144[20]	4-05-00-01897 0.9947[20]	1.5844[20]
3055	1,1'-Biphenyl, 2,5-dimethyl-	C$_{14}$H$_{14}$ 182.27	7372-85-2 	140[14.5]	4-05-00-01897 0.9931[20]	1.5819[20]
3056	1,1'-Biphenyl, 2,6-dimethyl-	C$_{14}$H$_{14}$ 182.27	3976-34-9 -5	262.5	4-05-00-01897 0.9907[20]	1.5745[20]
3057	1,1'-Biphenyl, 3,3'-dimethyl-	C$_{14}$H$_{14}$ 182.27	612-75-9 9	280	4-05-00-01903 0.9995[20]	H$_2$O 1; EtOH 4; eth 4; ace 3 1.5946[20]
3058	1,1'-Biphenyl, 3,4'-dimethyl-	C$_{14}$H$_{14}$ 182.27	7383-90-6 14.5	289; 153[15]	4-05-00-01905 0.9978[20]	H$_2$O 1; bz 3 1.5968[20]
3059	1,1'-Biphenyl, 3,4-dimethyl- 3,4-Dimethylbiphenyl	C$_{14}$H$_{14}$ 182.27	4433-11-8 29.5	282	4-05-00-01903 1.0087[20]	H$_2$O 1; bz 3; chl 3 1.6036[20]
3060	1,1'-Biphenyl, 3,5-dimethyl- 3,5-Dimethylbiphenyl	C$_{14}$H$_{14}$ 182.27	17057-88-4 22.5	274.5	4-05-00-01903 0.9990[20]	1.5952[20]
3061	1,1'-Biphenyl, 4,4'-dimethyl-	C$_{14}$H$_{14}$ 182.27	613-33-2 125	295	4-05-00-01906 0.917[121]	H$_2$O 1; EtOH 2; eth 3; ace 3
3062	1,1'-Biphenyl, 2,2'-dinitro-	C$_{12}$H$_8$N$_2$O$_4$ 244.21	2436-96-6 126	305	4-05-00-01826 1.45[25]	H$_2$O 1; EtOH 4; eth 3; ace 2
3063	1,1'-Biphenyl, 2,4-dinitro-	C$_{12}$H$_8$N$_2$O$_4$ 244.21	606-81-5 93.5		4-05-00-01827 1.474[25]	H$_2$O 1; EtOH 4; eth 3; bz 3
3064	[1,1'-Biphenyl]-2,2'-diol	C$_{12}$H$_{10}$O$_2$ 186.21	1806-29-7 109	320	4-06-00-06645 1.3420[20]	H$_2$O 3; EtOH 3; eth 3; ace 3
3065	[1,1'-Biphenyl]-2,4'-diol	C$_{12}$H$_{10}$O$_2$ 186.21	611-62-1 162.5	342	3-06-00-05387	eth 4
3066	[1,1'-Biphenyl]-3,3'-diol	C$_{12}$H$_{10}$O$_2$ 186.21	612-76-0 124.8	247[18]	3-06-00-05388	H$_2$O 2; EtOH 3; eth 3; bz 3
3067	[1,1'-Biphenyl]-2,2'-diol, 3,3'-dimethyl-	C$_{14}$H$_{14}$O$_2$ 214.26	32750-14-4 113	sub	3-06-00-05445	bz 4; eth 4; EtOH 4
3068	[1,1'-Biphenyl]-2,2'-diol, 5,5'-dimethyl-	C$_{14}$H$_{14}$O$_2$ 214.26	15519-73-0 153.5	sub	4-06-00-06705	H$_2$O 2; EtOH 3; eth 4; ace 3
3069	[1,1'-Biphenyl]-4,4'-diol, 3,3',5,5'-tetramethyl-	C$_{16}$H$_{18}$O$_2$ 242.32	2417-04-1 221.8	sub	4-06-00-06744	EtOH 2; bz 2; gl HOAc 2; tol 2
3070	[1,1'-Biphenyl]-4,4'-disulfonic acid 4,4'-Biphenyldisulfonic acid	C$_{12}$H$_{10}$O$_6$S$_2$ 314.34	5314-37-4 72.5	>200	3-11-00-00472	H$_2$O 4
3071	1,1'-Biphenyl, 3-ethoxy- 3-Ethoxybiphenyl	C$_{14}$H$_{14}$O 198.26	54852-73-2 35	305	3-06-00-03313	EtOH 3; eth 3; ace 3; bz 3
3072	1,1'-Biphenyl, 2-ethyl-	C$_{14}$H$_{14}$ 182.27	1812-51-7 -6	266; 128[11]	4-05-00-01896 0.9966[20]	1.5758[25]
3073	1,1'-Biphenyl, 3-ethyl-	C$_{14}$H$_{14}$ 182.27	5668-93-9 -27.5	283.5	4-05-00-01896 1.048[25]	1.5859[25]
3074	1,1'-Biphenyl, 2-fluoro-	C$_{12}$H$_9$F 172.20	321-60-8 73.5	248	4-05-00-01815 1.2452[25]	EtOH 3; eth 3; chl 3; lig 2
3075	1,1'-Biphenyl, 4-fluoro-	C$_{12}$H$_9$F 172.20	324-74-3 74.2	253	4-05-00-01815 1.247[25]	EtOH 2; eth 3; gl HOAc 3
3076	1,1'-Biphenyl, 2,2',4,4',6,6'-hexamethyl-	C$_{18}$H$_{22}$ 238.37	4482-03-5 103.5	297	4-05-00-01989 1.023[50]	H$_2$O 1; EtOH 2; eth 3; bz 3
3077	1,1'-Biphenyl, 2-iodo-	C$_{12}$H$_9$I 280.11	2113-51-1 	190[36]; 158[6]	4-05-00-01820 1.5991[25]	H$_2$O 1; EtOH 3; eth 3; bz 3 1.6620[20]
3078	1,1'-Biphenyl, 3-iodo-	C$_{12}$H$_9$I 280.11	20442-79-9 26.5	188[16]	4-05-00-01821 1.5967[25]	
3079	1,1'-Biphenyl, 4-iodo-	C$_{12}$H$_9$I 280.11	1591-31-7 113.5	320; 183[11]	4-05-00-01821	H$_2$O 1; EtOH 3; eth 3; bz 3
3080	1,1'-Biphenyl, 2-methoxy-	C$_{13}$H$_{12}$O 184.24	86-26-0 29	274	4-06-00-04580 1.0233[99]	H$_2$O 1; EtOH 3; ctc 2; peth 3 1.5641[99]
3081	1,1'-Biphenyl, 4-methoxy-	C$_{13}$H$_{12}$O 184.24	613-37-6 90	157[10]	4-06-00-04600 1.0278[100]	H$_2$O 1; EtOH 3; eth 3 1.5744[100]
3082	1,1'-Biphenyl, 2-methyl-	C$_{13}$H$_{12}$ 168.24	643-58-3 -0.2	255.5	4-05-00-01855 1.010[22]	H$_2$O 1; EtOH 3; eth 3 1.5914[20]
3083	1,1'-Biphenyl, 3-methyl-	C$_{13}$H$_{12}$ 168.24	643-93-6 4.5	272.7	4-05-00-01858 1.0182[17]	H$_2$O 1; EtOH 3; eth 3; ctc 3 1.5972[20]
3084	1,1'-Biphenyl, 4-methyl-	C$_{13}$H$_{12}$ 168.24	644-08-6 49.5	267.5	4-05-00-01860 1.015[27]	H$_2$O 1; EtOH 3; eth 3; ctc 2
3085	1,1'-Biphenyl, 2-nitro- o-Nitrobiphenyl	C$_{12}$H$_9$NO$_2$ 199.21	86-00-0 37.2	6513 320	4-05-00-01823 1.44[25]	H$_2$O 1; EtOH 3; eth 3; chl 3
3086	1,1'-Biphenyl, 3-nitro-	C$_{12}$H$_9$NO$_2$ 199.21	2113-58-8 62	227[35]; 143[9]	4-05-00-01823	H$_2$O 1; EtOH 3; eth 3; HOAc 3
3087	1,1'-Biphenyl, 4-nitro- p-Nitrobiphenyl	C$_{12}$H$_9$NO$_2$ 199.21	92-93-3 114	6514 340	4-05-00-01823	H$_2$O 1; EtOH 2; eth 3; bz 3
3088	[1,1'-Biphenyl]-2-ol o-Phenylphenol	C$_{12}$H$_{10}$O 170.21	90-43-7 59	7276 286	4-06-00-04579 1.213[25]	H$_2$O 1; EtOH 3; eth 4; ace 3
3089	[1,1'-Biphenyl]-3-ol m-Phenylphenol	C$_{12}$H$_{10}$O 170.21	580-51-8 78	>300	4-06-00-04597	H$_2$O 2; EtOH 4; eth 4; bz 4
3090	[1,1'-Biphenyl]-4-ol p-Phenylphenol	C$_{12}$H$_{10}$O 170.21	92-69-3 166	7277 305	4-06-00-04600	H$_2$O 2; EtOH 4; eth 4; chl 4

No.	Name / Synonym	Mol. Form. / Mol. Wt.	CAS RN / mp/°C	Merck No. / bp/°C	Bell. Ref. / den/g cm⁻³	Solubility / n_D
3091	[1,1'-Biphenyl]-4-ol, acetate	$C_{14}H_{12}O_2$ 212.25	148-86-7 89	196[13]	3-06-00-03326	H₂O 2; EtOH 2; eth 3; bz 3
3092	[1,1'-Biphenyl]-2-ol, 3-chloro- / 2-Phenyl-6-chlorophenol	$C_{12}H_9ClO$ 204.66	85-97-2 6	7251 317 dec	4-06-00-04586 1.24[25]	H₂O 1; EtOH 3; eth 3; ace 3 1.6237[30]
3093	1,1'-Biphenyl, 2-(phenylmethyl)-	$C_{19}H_{16}$ 244.34	606-97-3 55	283[110]	3-05-00-02323	bz 4; eth 4; EtOH 4
3094	1,1'-Biphenyl, 4-(phenylmethyl)-	$C_{19}H_{16}$ 244.34	613-42-3 85	285-6[110]	4-05-00-02502 1.171[0]	H₂O 1; EtOH 3; eth 4; bz 4
3095	1,1'-Biphenyl, 2,2',4,4'-tetramethyl-	$C_{16}H_{18}$ 210.32	3976-36-1 41	287	3-05-00-01891	EtOH 3; eth 2; bz 2; chl 3
3096	1,1'-Biphenyl, 2,2',5,5'-tetramethyl-	$C_{18}H_{18}$ 210.32	3075-84-1 51.5	285	3-05-00-01892	eth 3; bz 3
3097	[1,1'-Biphenyl]-3,3',5,5'-tetrol / Diresorcinol	$C_{12}H_{10}O_4$ 218.21	531-02-2 310	3361	2-06-00-01129	H₂O 4; eth 4; EtOH 4
3098	2,3'-Bipiperidine / 2,3'-Bipiperidyl	$C_{10}H_{20}N_2$ 168.28	2467-09-6 68.5	269.5	4-23-00-00512	H₂O 4; EtOH 4
3099	2,2'-Bipyridine / α,α'-Dipyridyl	$C_{10}H_8N_2$ 156.19	366-18-7 72	3355 273.5	5-23-08-00016	H₂O 2; EtOH 4; eth 4; bz 4
3100	2,3'-Bipyridine / 2,3'-Bipyridyl	$C_{10}H_8N_2$ 156.19	581-50-0	295.5	5-23-08-00026 1.140[20]	H₂O 1; EtOH 4; eth 4; bz 4 1.6223[20]
3101	2,4'-Bipyridine	$C_{10}H_8N_2$ 156.19	581-47-5 61.5	281	5-23-08-00027	H₂O 2; EtOH 4; eth 4; chl 4
3102	3,3'-Bipyridine / 3,3'-Dipyridyl	$C_{10}H_8N_2$ 156.19	581-46-4 68	291.5	5-23-08-00027 1.1614[20]	H₂O 4; EtOH 4; eth 2
3103	3,4'-Bipyridine / 3,4'-Bipyridyl	$C_{10}H_8N_2$ 156.19	4394-11-0 62	297; 144[15]	5-23-08-00028	EtOH 4; peth 4
3104	4,4'-Bipyridine / γ,γ'-Dipyridyl	$C_{10}H_8N_2$ 156.19	553-26-4 111	3356 305	5-23-08-00028	H₂O 2; EtOH 4; eth 3; bz 4
3105	2,3'-Bipyridine, 1,2,3,6-tetrahydro-, (S)- / Anatabine	$C_{10}H_{14}N_2$ 160.22	581-49-7	661 145-6[10]	5-23-06-00507 1.091[19]	H₂O 5; EtOH 3; eth 3; bz 3 1.5676[20]
3106	[5,5'-Bipyrimidine]-2,2',4,4',6,6'(1H,1'H,3H,3'H,5H,5'H)-hexone-5,5'-dihydroxy Alloxantin	$C_8H_6N_4O_8$ 286.16	76-24-4 253-5 dec	282	4-26-00-02782	H₂O 2; EtOH 2; eth 2
3107	2,3'-Biquinoline / 2,3'-Biquinolyl	$C_{18}H_{12}N_2$ 256.31	612-81-7 175.8	>400	5-23-10-00009	bz 4; eth 4; EtOH 4; chl 4
3108	Bis[1,3]benzodioxolo[4,5-c:5',6'-g]azecin-13(5H)-one, 4,6,7,14-tetrahydro-5,14-dimethyl-, (R)- / Corycavamine	$C_{21}H_{21}NO_5$ 367.40	521-85-7 149	2540	2-27-00-00621	EtOH 4; chl 4
3109	Bis[1,3]benzodioxolo[4,5-c:5',6'-g]azecin-13(5H)-one, 4,6,7,14-tetrahydro-5-methyl- / Protopine	$C_{20}H_{19}NO_5$ 353.37	130-86-9 208	7912	4-27-00-06881	H₂O 1; EtOH 2; eth 2; bz 2
3110	3H-Bis[1]benzopyrano[3,4-b:6',5'-e]pyran-7(7aH)-one, 13,13a-dihydro-7a-hydroxy-9,10-dimethoxy-3,3-dimethyl / Tephrosin	$C_{23}H_{22}O_7$ 410.42	76-80-2 198	9082	4-19-00-05271	ace 4; eth 4; chl 4
3111	Bismuthine, triphenyl- / Triphenylbismuth	$C_{18}H_{15}Bi$ 440.30	603-33-8 77.6	242[14]	4-16-00-01226 1.715[75]	EtOH 2; eth 3; ace 3; bz 3 1.7040[75]
3112	2,2'-Bithiophene	$C_8H_6S_2$ 166.27	492-97-7 33	260	5-19-01-00537	H₂O 1; EtOH 4; eth 3; ctc 3
3113	Boranamine, N,N,1,1-tetramethyl-	$C_4H_{12}BN$ 84.96	1113-30-0 -92	65	4-04-00-04367	eth 4; ace 4; os 5
3114	Boranediamine, 1-fluoro-N,N,N',N'-tetramethyl-	$C_4H_{12}BFN_2$ 117.96	383-90-4 -44.3	106	4-04-00-00303	ace 4; eth 4
3115	Borane, difluoro(4-methylphenyl)-	$C_7H_7BF_2$ 139.94	768-39-8	128	4-16-00-01674 1.055[25]	bz 4; eth 4 1.4535[25]
3116	Borane, difluorophenyl-	$C_6H_5BF_2$ 125.91	368-98-9 -36.2	98	4-16-00-01661 1.087[25]	bz 4; eth 4 1.4441[25]
3117	Borane, dimethoxy- / Dimethoxyborane	$C_2H_7BO_2$ 73.89	4542-61-4 -130.6	25.9	4-01-00-01269	H₂O 6
3118	Borane, triethyl / Triethylborane	$C_6H_{15}B$ 98.00	97-94-9 -93	95	0.70[23]	EtOH 3; eth 3 1.3971
3119	Borane, trimethyl- / Trimethylborane	C_3H_9B 55.92	593-90-8 -161.5	-20.2		
3120	Borane, trimethyl-, monoammoniate / Aminetrimethylboron	$C_3H_{12}BN$ 72.95	1830-95-1 73.5			
3121	Borane, tripropyl-	$C_9H_{21}B$ 140.08	1116-61-6 -56	159	4-04-00-04360 0.7204[25]	1.4135[22]
3122	Borane, tris(3-methylbutyl)- / Borane, triisopentyl-	$C_{15}H_{33}B$ 224.24	3062-81-5	119[14]	2-04-00-01023 0.7600[25]	ace 4; eth 4; EtOH 4 1.4321
3123	Borane, tris(2-methylpropyl)-	$C_{12}H_{27}B$ 182.16	1116-39-8	188; 86[20]	4-04-00-04361 0.7380[25]	bz 4; eth 4; EtOH 4 1.4188[23]
3124	Boric acid, tributyl ester	$C_{12}H_{27}BO_3$ 230.16	688-74-4	234	4-01-00-01544 0.8567[20]	EtOH 3; eth 4; bz 3; MeOH 4 1.4106[18]
3125	Boric acid, triethyl ester	$C_6H_{15}BO_3$ 145.99	150-46-9 -84.8	120	4-01-00-01365 0.8546[20]	EtOH 5; eth 5 1.3749[20]
3126	Boric acid, trimethyl ester / Trimethyl borate	$C_3H_9BO_3$ 103.91	121-43-7 -29.3	9626 67.5	4-01-00-01269 0.915[25]	eth 4; MeOH 4 1.3568[20]

No.	Name Synonym	Mol. Form. Mol. Wt.	CAS RN mp/°C	Merck No. bp/°C	Beil. Ref. den/g cm^{-3}	Solubility n_D
3127	Boric acid, tripropyl ester	$C_9H_{21}BO_3$ 188.07	688-71-1 	 179.5	4-01-00-01436 0.8576[20]	EtOH 4; eth 5; PrOH 3 1.3948[20]
3128	Boric acid, tris(1-methylethyl) ester Isopropyl borate	$C_9H_{21}BO_3$ 188.07	5419-55-6 	 140; 75[76]	4-01-00-01488 0.8251[20]	EtOH 4; eth 4; bz 4; PrOH 4 1.3777[20]
3129	Borinic acid, dimethylthio-, methyl ester	C_3H_9BS 87.98	19163-05-4 -84	 71	4-04-00-04368 	ace 4; eth 4
3130	Borinic acid, diphenyl-	$C_{12}H_{11}BO$ 182.03	2622-89-1 57.5	 150-5[20]	4-16-00-01639 1.0740[20]	EtOH 3; bz 3; peth 3
3131	3-Bornanecarboxylic acid Camphane-3-carboxylic acid	$C_{11}H_{18}O_2$ 182.26	91965-23-0 90.5	153[13] 	3-09-00-00244 	EtOH 4; HOAc 4
3132	Boron, (N,N-dimethylmethanamine)trihydro-, (T-4)- Trimethylamineborane	$C_3H_{12}BN$ 72.95	75-22-9 94	 172	4-04-00-00134 0.792[25]	eth 4; EtOH 4
3133	Boronic acid, ethyl-	$C_2H_7BO_2$ 73.89	4433-63-0 40	 dec	4-04-00-04379 	H_2O 4; eth 4; EtOH 4
3134	Boronic acid, (2-methylphenyl)-	$C_7H_9BO_2$ 135.96	16419-60-6 166.5	 dec	3-16-00-01277 	bz 4; eth 4; EtOH 4
3135	Boronic acid, phenyl- Benzeneboronic acid	$C_6H_7BO_2$ 121.93	98-80-6 219	1076 	4-16-00-01654 	H_2O 2; EtOH 3; eth 3; bz 3
3136	Boronic acid, propyl-	$C_3H_9BO_2$ 87.91	17745-45-8 107	 dec	3-04-00-01964 	H_2O 4; eth 4; EtOH 4
3137	Boron, trifluoro[1,1'-oxybis[ethane]]-, (T-4)- Boron trifluoride etherate	$C_4H_{10}BF_3O$ 141.93	109-63-7 -60.4	1352 125.5	4-01-00-01314 1.3572[20]	eth 4; EtOH 4 1.4447[20]
3138	Boron, trifluoro[oxybis[methane]]-, (T-4)- Boron trifluoride compd. with methyl ether	$C_2H_6BF_3O$ 113.88	353-42-4 -14	 127 dec	4-01-00-01245 1.2410[20]	 1.302[20]
3139	Bromacil 2,4(1H,3H)-Pyrimidinedione, 5-bromo-6-methyl-3-(1-methylpropyl)-	$C_9H_{13}BrN_2O_2$ 261.12	314-40-9 158	1370 	 1.55[25]	
3140	Bromopropylate Benzeneacetic acid, 4-bromo-α-(4-bromophenyl)-α-hydroxy-, 1-methylethyl ester	$C_{17}H_{16}Br_2O_3$ 428.12	18181-80-1 77	1422 	 1.59[20]	
3141	Bufa-20,22-dienolide, 16-(acetyloxy)-3,14-dihydroxy-, (3β,5β,16β)- Bufotalin	$C_{26}H_{36}O_6$ 444.57	471-95-4 223 dec	1466 	5-18-04-00443 	H_2O 1; EtOH 3; chl 3
3142	Bufa-4,20,22-trienolide, 6-(acetyloxy)-3-(β-D-glucopyranosyloxy)-8,14-dihydroxy-, (3β,6β)- Scilliroside	$C_{32}H_{44}O_{12}$ 620.69	507-60-8 169	8356 dec	4-18-00-03178 	H_2O 2; EtOH 4; eth 1; ace 2
3143	sec-Bumeton 1,3,5-Triazine-2,4-diamine, N-ethyl-6-methoxy-N'-(1-methylpropyl)-	$C_{10}H_{19}N_5O$ 225.29	26259-45-0 87			
3144	Butachlor Acetamide, N-(butoxymethyl)-2-chloro-N-(2,6-diethylphenyl)-	$C_{17}H_{26}ClNO_2$ 311.85	23184-66-9 <-5	1498 156-5	 1.070[25]	
3145	1,2-Butadiene Methylallene	C_4H_6 54.09	590-19-2 -136.2	 10.9	4-01-00-00975 0.676[0]	H_2O 1; EtOH 5; eth 5; bz 4 1.4205[1]
3146	1,3-Butadiene Divinyl	C_4H_6 54.09	106-99-0 -108.9	1500 -4.4	4-01-00-00976 *0.6149[25]	H_2O 1; EtOH 3; eth 3; ace 4 1.4292[-25]
3147	1,2-Butadiene, 4-bromo- 4-Bromo-1,2-butadiene	C_4H_5Br 132.99	18668-68-3 	 110	3-01-00-00929 1.4255[20]	ace 4 1.5248[20]
3148	1,3-Butadiene, 1-bromo- 1-Bromo-1,3-butadiene	C_4H_5Br 132.99	21890-35-7 	 93	4-01-00-00989 1.416[0]	eth 4; EtOH 4; peth 4
3149	1,3-Butadiene, 2-bromo-	C_4H_5Br 132.99	1822-86-2 	 42-3[165]	4-01-00-00989 1.397[20]	eth 4; EtOH 4 1.4988[20]
3150	1,2-Butadiene, 4-chloro-	C_4H_5Cl 88.54	25790-55-0 	 88	4-01-00-00975 0.9891[20]	ace 4; bz 4; eth 4 1.4775[20]
3151	1,3-Butadiene, 1-chloro-	C_4H_5Cl 88.54	627-22-5 	 68	3-01-00-00949 0.9606[20]	eth 4; EtOH 4; chl 4 1.4712[20]
3152	1,3-Butadiene, 2-chloro- 2-Chloro-1,3-butadiene	C_4H_5Cl 88.54	126-99-8 -130	 59.4	4-01-00-00984 0.956[20]	H_2O 2; eth 5; ace 5; bz 5 1.4583[20]
3153	1,3-Butadiene, 1-chloro-2-methyl- Chloromethylbutadiene	C_5H_7Cl 102.56	35383-51-8 	 107	3-01-00-00974 0.9710[20]	ace 4; EtOH 4 1.4702[20]
3154	1,3-Butadiene, 1-chloro-3-methyl- 4-Chloroisoprene	C_5H_7Cl 102.56	51034-46-9 	 100.7	4-01-00-01005 0.9543[20]	ace 4; eth 4; EtOH 4; chl 4 1.4719[20]
3155	1,3-Butadiene, 2-chloro-3-methyl-	C_5H_7Cl 102.56	1809-02-5 	 93	4-01-00-01004 0.9593[20]	H_2O 1; EtOH 3; eth 4; ace 3 1.4686[20]
3156	1,3-Butadiene, 1,1-dichloro- 1,1-Dichloro-1,3-butadiene	$C_4H_4Cl_2$ 122.98	6061-06-9 	 42-3[90]	4-01-00-00985 1.1831[20]	H_2O 1; eth 4; bz 4; chl 4 1.5022[20]
3157	1,3-Butadiene, 1,2-dichloro- 1,2-Dichloro-1,3-butadiene	$C_4H_4Cl_2$ 122.98	3574-40-1 	 62[105]; 35[40]	4-01-00-00985 1.1991[20]	ctc 4 1.4960[20]
3158	1,3-Butadiene, 2,3-dichloro-	$C_4H_4Cl_2$ 122.98	1653-19-6 	 98	4-01-00-00986 1.1829[20]	chl 4 1.4890[20]
3159	1,3-Butadiene, 2,3-dimethyl- Diisopropenyl	C_6H_{10} 82.15	513-81-5 -76	3228 68.8	4-01-00-01023 0.7222[25]	ctc 3 1.4394[20]
3160	1,3-Butadiene, 1-ethoxy-	$C_6H_{10}O$ 98.14	5614-32-4 	 110.5	4-01-00-02222 0.8154[20]	ace 4; bz 4; eth 4; EtOH 4 1.4529[20]
3161	1,3-Butadiene, 2-ethoxy-	$C_6H_{10}O$ 98.14	4747-05-1 	 94.5	4-01-00-02223 0.8177[20]	ace 4; bz 4; eth 4; EtOH 4 1.4400[20]
3162	1,3-Butadiene, 2-fluoro- Fluoroprene	C_4H_5F 72.08	381-61-3 	 12	4-01-00-00982 0.843[4]	 1.400[4]
3163	1,3-Butadiene, 1,1,2,3,4,4-hexachloro- Hexachloro-1,3-butadiene	C_4Cl_6 260.76	87-68-3 -21	 215	4-01-00-00988 1.556[25]	H_2O 1; EtOH 3; eth 3 1.5542[20]

No.	Name / Synonym	Mol. Form. / Mol. Wt.	CAS RN / mp/°C	Merck No. / bp/°C	Beil. Ref. / den/g cm^{-3}	Solubility / n_D
3164	1,3-Butadiene, 1,1,2,3,4,4-hexafluoro- 1,3-Butadiene, hexafluoro	C$_4$F$_6$ 162.03	685-63-2 -132	6	4-01-00-00983 1.553^{-20}	1.378^{-20}
3165	1,3-Butadiene, 2-iodo-	C$_4$H$_5$I 179.99	19221-28-4	112	3-01-00-00956 1.7278^{20}	1.5616
3166	1,2-Butadiene, 4-iodo- 4-Iodo-1,2-butadiene	C$_4$H$_5$I 179.99	67885-08-9	130	3-01-00-00929 1.7129^{20}	1.5709^{20}
3167	1,2-Butadiene, 4-methoxy-	C$_5$H$_8$O 84.12	36678-06-5	88	4-01-00-02221 0.8286^{20}	EtOH 4 1.435^{12}
3168	1,3-Butadiene, 1-methoxy-	C$_5$H$_8$O 84.12	3036-66-6	91.5	3-01-00-01975 0.8296^{20}	H$_2$O 3; EtOH 3 1.4594^{20}
3169	1,3-Butadiene, 2-methoxy-	C$_5$H$_8$O 84.12	3588-30-5	75	4-01-00-02223 0.8272^{20}	ace 4; bz 4; eth 4; EtOH 4 1.4442^{20}
3170	1,2-Butadiene, 3-methyl-	C$_5$H$_8$ 68.12	598-25-4 -113.6	40.83	4-01-00-01006 0.6806^{25}	ace 4; bz 4; eth 4; EtOH 4 1.4203^{20}
3171	1,3-Butadiene, 2-methyl- Isoprene	C$_5$H$_8$ 68.12	78-79-5 -145.9	5087 34.0	4-01-00-01001 0.679^{20}	H$_2$O 1; EtOH 5; eth 5; ace 5 1.4219^{20}
3172	1,3-Butadiene, 1,1,2,4,4-pentafluoro-3-(trifluoromethyl)- 1,3-Butadiene, pentafluoro-2-(trifluoromethyl)-	C$_5$F$_8$ 212.04	384-04-3	39	1.527^0	ace 4; bz 4; eth 4 1.3000^0
3173	1,3-Butadiene, 1,2,3,4-tetrachloro-	C$_4$H$_2$Cl$_4$ 191.87	1637-31-6 -4	188	4-01-00-00987 1.515^{15}	H$_2$O 1; EtOH 3; eth 3; ace 3 1.5455^{20}
3174	1,3-Butadiene, 1,2,3-trichloro- 1,2,3-Trichloro-1,3-butadiene	C$_4$H$_3$Cl$_3$ 157.43	1573-58-6	33-4^7	3-01-00-00954 1.4060^{20}	eth 4; chl 4 1.5262^{20}
3175	2,3-Butadien-1-ol	C$_4$H$_6$O 70.09	18913-31-0	128.5	4-01-00-02221 0.9164^{20}	ace 4; bz 4; eth 4; EtOH 4 1.4759^{20}
3176	1,3-Butadien-1-ol, acetate	C$_6$H$_8$O$_2$ 112.13	1515-76-0	60-1^{40}	3-02-00-00295 0.945^{25}	1.4690^{20}
3177	1,3-Butadiyne	C$_4$H$_2$ 50.06	460-12-8 -36.4	10.3	4-01-00-01116 0.7364^0	H$_2$O 4; EtOH 3; eth 4; ace 4 1.4189^5
3178	Butanal Butyraldehyde	C$_4$H$_8$O 72.11	123-72-8 -99	1591 74.8	4-01-00-03229 0.8016^{20}	H$_2$O 3; EtOH 5; eth 5; ace 4 1.3843^{20}
3179	Butanal, 2-bromo-	C$_4$H$_7$BrO 151.00	24764-97-4	33^{17}	4-01-00-03241 1.469^{20}	eth 3; ace 3; bz 3; chl 3 1.4683^{20}
3180	Butanal, 2-chloro-	C$_4$H$_7$ClO 106.55	28832-55-5	107	4-01-00-03238 1.1072^{17}	1.4262^{17}
3181	Butanal, 4-chloro-	C$_4$H$_7$ClO 106.55	6139-84-0	50-1^{13}	4-01-00-03240 1.106^8	ace 4; eth 4; EtOH 4 1.4466^8
3182	Butanal, 2,3-dichloro-	C$_4$H$_6$Cl$_2$O 141.00	55775-41-2	58-60^{20}	4-01-00-03240 1.2666^{21}	ace 4; eth 4; EtOH 4; chl 4 1.4618^{21}
3183	Butanal, 2-ethyl- Diethylacetaldehyde	C$_6$H$_{12}$O 100.16	97-96-1	117-9^{160}	4-01-00-03310 0.8110^{20}	H$_2$O 2; EtOH 5; eth 5; ctc 2 1.4025^{20}
3184	Butanal, 3-hydroxy- Aldol	C$_4$H$_8$O$_2$ 88.11	107-89-1 217	83^{20}	4-01-00-03984 1.103^{20}	H$_2$O 5; EtOH 5; eth 3; ace 4 1.4238^{20}
3185	Butanal, 2-methyl-, (±)-	C$_5$H$_{10}$O 86.13	57456-98-1	92.5	4-01-00-03285 0.8029^{20}	ace 4; eth 4; EtOH 4 1.3869^{20}
3186	Butanal, 3-methyl- Isovaleraldehyde	C$_5$H$_{10}$O 86.13	590-86-3 -51	5118 92.5	4-01-00-03291 0.7977^{20}	H$_2$O 2; EtOH 3; eth 3 1.3902^{20}
3187	Butanal, 3-methyl-, oxime Butyraldehyde, 3-methyl, oxime	C$_5$H$_{11}$NO 101.15	626-90-4 48.5	161.3	4-01-00-03293 0.8934^{20}	ace 4; eth 4; EtOH 4 1.4367^{20}
3188	Butanal, oxime Butyraldehyde, oxime	C$_4$H$_9$NO 87.12	110-69-0 -29.5	154	4-01-00-03234 0.923^{20}	H$_2$O 4; EtOH 5; eth 5; ace 4
3189	Butanal, phenylhydrazone Butyraldehyde, phenylhydrazone	C$_{10}$H$_{14}$N$_2$ 162.23	940-54-5 94	192^{80}; 152^{14}	3-15-00-00080	ctc 2
3190	Butanal, 2-(phenylmethylene)-	C$_{11}$H$_{12}$O 160.22	28467-92-7	157-8^5	4-07-00-01024 1.0201^{22}	1.578^{20}
3191	Butanal, 2,2,3-trichloro- 2,2,3-Trichlorobutyraldehyde	C$_4$H$_5$Cl$_3$O 175.44	76-36-8 164	9545	4-01-00-03241 1.3956^{20}	H$_2$O 4; eth 4; EtOH 4 1.4755^{20}
3192	Butanal, 2,3,4-trihydroxy-, [R-(R*,R*)]- D-Erythrose	C$_4$H$_8$O$_4$ 120.11	583-50-6	3637	4-01-00-04172	H$_2$O 3; EtOH 4
3193	Butanal, 2,3,4-trihydroxy-, [S-(R*,S*)]- Threose (D-)	C$_4$H$_8$O$_4$ 120.11	95-43-2 129	9317	4-01-00-04173	
3194	Butanamide Butyramide	C$_4$H$_9$NO 87.12	541-35-5 114.8	1592 216	4-02-00-00804 0.8850^{120}	H$_2$O 2; EtOH 3; eth 2; bz 1 1.4087^{130}
3195	Butanamide, N-(aminocarbonyl)-2-bromo-2-ethyl- Carbromal	C$_7$H$_{13}$BrN$_2$O$_2$ 237.10 118	77-65-6	1837	4-03-00-00117 1.544^{25}	H$_2$O 2; ace 3; bz 3; chl 2
3196	Butanamide, N-(aminocarbonyl)-2-bromo-3-methyl- Bromisovalum	C$_8$H$_{11}$BrN$_2$O$_2$ 223.07 154	496-67-3	1385 sub	3-03-00-00123 1.56^{15}	ace 4; bz 4; eth 4; EtOH 4
3197	Butanamide, N-[5-(aminosulfonyl)-1,3,4-thiadiazol-2-yl]- Butazolamide	C$_6$H$_{10}$N$_4$O$_3$S$_2$ 250.30	16790-49-1	1511	4-27-00-08223	
3198	Butanamide, 2-bromo-2-ethyl- Diethylbromoacetamide	C$_6$H$_{12}$BrNO 194.07	511-70-6 67	3105	3-02-00-00755	H$_2$O 2; EtOH 4; eth 4; bz 4
3199	Butanamide, 2-bromo-2-ethyl-3-methyl- Ibrotamide	C$_7$H$_{14}$BrNO 208.10	466-14-8 50.5	4809 sub	4-02-00-00979	ace 4; bz 4; eth 4; EtOH 4
3200	Butanamide, N,N-diethyl-	C$_8$H$_{17}$NO 143.23	1114-76-7	206	4-04-00-00354	H$_2$O 4; EtOH 4 1.4403^{25}
3201	Butanamide, N,N-diethyl-3-methyl- Isovaleryl diethylamide	C$_9$H$_{19}$NO 157.26	533-32-4	5122 211	4-04-00-00356 0.8764^{20}	eth 4; EtOH 4 1.4422^{20}

No.	Name Synonym	Mol. Form. Mol. Wt.	CAS RN mp/°C	Merck No. bp/°C	Beil. Ref. den/g cm^{-3}	Solubility n_D
3202	Butanamide, 2,4-dihydroxy-N-(3-hydroxypropyl)-3,3-dimethyl-, (R)- Dexpanthenol	$C_9H_{19}NO_4$ 205.25	81-13-0	2924 dec	4-04-00-01652 1.20^{20}	H_2O 4; EtOH 4; eth 2; MeOH 4 1.497^{20}
3203	Butanamide, N,N-dimethyl-	$C_6H_{13}NO$ 115.18	760-79-2 -40	186; 125^{100}	4-04-00-00185 0.9064^{25}	ace 4; bz 4; eth 4; EtOH 4 1.4391^{25}
3204	Butanamide, N,N'-(3,3'-dimethyl[1,1'-biphenyl]-4,4'-diyl)bis[3-oxo- N,N'-Bis(acetoacetyl)-3,3'-dimethylbenzidine	$C_{22}H_{24}N_2O_4$ 380.44	91-96-3 212		3-13-00-00490	DMSO 2
3205	Butanamide, N-(4-hydroxyphenyl)- 4'-Hydroxybutyranilide	$C_{10}H_{13}NO_2$ 179.22	101-91-7 139.5	4745	4-13-00-01109	H_2O 4; EtOH 4
3206	Butanamide, 3-methyl- Isovaleramide	$C_5H_{11}NO$ 101.15	541-46-8 137	5119 226	4-02-00-00902	H_2O 3; EtOH 3; eth 3; peth 4
3207	Butanamide, 3-oxo-N-phenyl- Acetoacetanilide	$C_{10}H_{11}NO_2$ 177.20	102-01-2 86	50	4-12-00-00955	H_2O 2; EtOH 3; eth 3; bz 3
3208	Butanamide, N-phenyl-	$C_{10}H_{13}NO$ 163.22	1129-50-6 97	189^{15}	4-12-00-00387 1.134^{25}	H_2O 1; EtOH 4; eth 4; chl 2
3209	1-Butanamine Butylamine	$C_4H_{11}N$ 73.14	109-73-9 -49.1	1543 77.0	4-04-00-00540 0.7414^{20}	H_2O 5; EtOH 3; eth 3 1.4031^{20}
3210	2-Butanamine, (±)- sec-Butylamine, (±)-	$C_4H_{11}N$ 73.14	33966-50-6 <-72	1544 63.5	4-04-00-00618 0.7246^{20}	H_2O 3; EtOH 5; eth 5; ace 4 1.3932^{20}
3211	1-Butanamine, N-butyl- Dibutylamine	$C_8H_{19}N$ 129.25	111-92-2 -62	3019 159.6	4-04-00-00550 0.7670^{20}	H_2O 3; EtOH 5; eth 4; ace 3 1.4177^{20}
3212	1-Butanamine, N-butyl-N-nitroso- Dibutylnitrosamine	$C_8H_{18}N_2O$ 158.24	924-16-3	105^8	4-04-00-03389	
3213	1-Butanamine, N,N-dibutyl- Tributylamine	$C_{12}H_{27}N$ 185.35	102-82-9 -70	9530 216.5	4-04-00-00554 0.7770^{20}	H_2O 2; EtOH 4; eth 4; ace 3 1.4299^{20}
3214	1-Butanamine, 4,4-diethoxy-	$C_8H_{19}NO_2$ 161.24	6346-09-4	196	4-04-00-01928 0.933^{25}	1.4275^{20}
3215	1-Butanamine, N,N-dimethyl-	$C_8H_{15}N$ 101.19	927-62-8	95	4-04-00-00546 0.7206^{20}	H_2O 5; EtOH 5; eth 5; ace 5 1.3970^{20}
3216	2-Butanamine, 2,3-dimethyl-	$C_8H_{15}N$ 101.19	4358-75-2 -70.5	104.5	4-04-00-00733 0.7683^0	1.4096^{17}
3217	2-Butanamine, 3,3-dimethyl-	$C_8H_{15}N$ 101.19	3850-30-4 -20	102	4-04-00-00730 0.7668^{20}	H_2O 4 1.4105^{25}
3218	1-Butanamine, N-ethyl- Butylethylamine	$C_8H_{15}N$ 101.19	13360-63-9	107.5	4-04-00-00547 0.7398^{20}	EtOH 5; eth 5; ace 5; bz 5 1.4040^{20}
3219	2-Butanamine, N-ethyl-, (±)- sec-Butyl ethyl amine (DL)	$C_8H_{15}N$ 101.19	116724-10-8 -104.3	98	2-04-00-00636 0.7358^{20}	ace 4; bz 4; eth 4; EtOH 4
3220	1-Butanamine, 3-methyl- Isoamylamine	$C_5H_{13}N$ 87.16	107-85-7	4994 96	4-04-00-00696 0.7505^{20}	H_2O 5; EtOH 5; eth 5; ace 3 1.4083^{20}
3221	1-Butanamine, N-methyl- Butylmethylamine	$C_5H_{13}N$ 87.16	110-68-9	91	4-04-00-00546 0.7637^{15}	
3222	2-Butanamine, 2-methyl-	$C_5H_{13}N$ 87.16	594-39-8 -105	77	4-04-00-00694 0.731^{25}	H_2O 4; ace 4; eth 4; EtOH 4 1.3954^{25}
3223	2-Butanamine, 3-methyl-	$C_5H_{13}N$ 87.16	598-74-3 -50	85.5	2-04-00-00644 0.7574^{19}	H_2O 4; EtOH 3 1.4096^{18}
3224	1-Butanamine, 2-methyl-N,N-bis(2-methylbutyl)-	$C_{15}H_{33}N$ 227.43	620-43-9	232	0-04-00-00179 0.91^{13}	ace 4; bz 4; eth 4; EtOH 4 1.4330^{20}
3225	1-Butanamine, 3-methyl-N,N-bis(3-methylbutyl)-	$C_{15}H_{33}N$ 227.43	645-41-0	235	4-04-00-00700 0.7848^{20}	H_2O 1; EtOH 4; eth 5; bz 5 1.4331^{20}
3226	1-Butanamine, 3-methyl-N-(3-methylbutyl)- Diisoamylamine	$C_{10}H_{23}N$ 157.30	544-00-3 -44	3178 188	4-04-00-00699 0.7672^{21}	H_2O 1; EtOH 3; eth 5 1.4235^{20}
3227	2-Butanamine, N-(1-methylpropyl)-	$C_8H_{19}N$ 129.25	626-23-3	135	4-04-00-00620 0.7534^{20}	H_2O 4; EtOH 3; os 3 1.4162^{20}
3228	1-Butanamine, 1,1,2,2,3,3,4,4,4-nonafluoro-N,N-bis(nonafluorobutyl)-	$C_{12}F_{27}N$ 671.10	311-89-7	178	4-02-00-00819 1.884^{25}	ace 3 1.291^{25}
3229	Butane	C_4H_{10} 58.12	106-97-8 -138.2	1507 -0.5	4-01-00-00236 *0.573^{25}	H_2O 3; EtOH 4; eth; chl 4 1.3326^{20}
3230	Butane, 2,2-bis(ethylsulfonyl)- Sulfonethylmethane	$C_8H_{18}O_4S_2$ 242.36	76-20-0 76	8937 dec	3-01-00-02790 1.199^{85}	chl 3
3231	Butane, 1-bromo- Butyl bromide	C_4H_9Br 137.02	109-65-9 -112.4	1553 101.6	4-01-00-00258 1.2758^{20}	H_2O 1; EtOH 5; eth 5; ace 5 1.4401^{20}
3232	Butane, 2-bromo-, (±)- (±)-sec-Butyl bromide	C_4H_9Br 137.02	5787-31-5 -111.9	1554 91.2	4-01-00-00261 1.2585^{20}	ace 4; eth 4; chl 4 1.4366^{20}
3233	Butane, 1-bromo-4-chloro-	C_4H_8BrCl 171.46	6940-78-9	175; 63^{10}	4-01-00-00264 1.488^{20}	H_2O 1; EtOH 3; eth 3; ctc 2 1.4885^{20}
3234	Butane, 2-bromo-1-chloro- 2-Bromo-1-chlorobutane	C_4H_8BrCl 171.46	79504-01-1	146.5	4-01-00-00264 1.468^{20}	bz 4; eth 4; EtOH 4; chl 4 1.4880^{20}
3235	Butane, 1-bromo-3,3-dimethyl-	$C_6H_{13}Br$ 165.07	1647-23-0	138	3-01-00-00409 1.1556^{20}	eth 4; EtOH 4; chl 4 1.4440^{20}
3236	Butane, 2-bromo-2,3-dimethyl- 2-Bromo-2,3-dimethylbutane	$C_6H_{13}Br$ 165.07	594-52-5 24.5	133; 87^{180}	4-01-00-00374 1.1772^{10}	eth 4; chl 4 1.4517
3237	Butane, 1-bromo-4-fluoro- 1-Bromo-4-fluorobutane	C_4H_8BrF 155.01	462-72-6	135	4-01-00-00263	eth 4; EtOH 4 1.4370^{25}
3238	Butane, 1-bromo-2-methyl-, (S)- d-Amyl bromide	$C_5H_{11}Br$ 151.05	534-00-9 121.6	636	4-01-00-00327 1.2234^{20}	H_2O 1; EtOH 3; eth 3; chl 4 1.4451^{20}
3239	Butane, 1-bromo-3-methyl- Isoamyl bromide	$C_5H_{11}Br$ 151.05	107-82-4 -112	4996 120.4	4-01-00-00328 1.2071^{20}	H_2O 1; EtOH 3; eth 3; ctc 2 1.4420^{20}

No.	Name Synonym	Mol. Form. Mol. Wt.	CAS RN mp/°C	Merck No. bp/°C	Beil. Ref. den/g cm^{-3}	Solubility n_D
3240	Butane, 2-bromo-2-methyl- tert-Amyl bromide	$C_5H_{11}Br$ 151.05	507-36-8	638 108	4-01-00-00327 1.197[18]	⁻1.4421
3241	Butane, 1-chloro- Butyl chloride	C_4H_9Cl 92.57	109-69-3 -123.1	1560 78.6	4-01-00-00246 0.8862[20]	H_2O 1; EtOH 5; eth 5; ctc 2 1.4021[20]
3242	Butane, 2-chloro-, (±)- sec-Butyl chloride (DL)	C_4H_9Cl 92.57	53178-20-4 -131.3	1561 68.2	4-01-00-00248 0.8732[20]	bz 4; eth 4; EtOH 4; chl 4 1.3971[20]
3243	Butane, 3-chloro-1,1-diethoxy- Butyraldehyde, 3-chloro-, diethyl acetal	$C_8H_{17}ClO_2$ 180.67	51786-70-0	70-1[12]	4-01-00-03239 0.9709[20]	bz 4; eth 4; EtOH 4 1.4195[25]
3244	Butane, 3-chloro-1,1-dimethoxy- Butyraldehyde, 3-chloro-, dimethyl acetal	$C_6H_{13}ClO_2$ 152.62	50710-40-2	40-4[7]	4-01-00-03239 1.0024[20]	1.4160[25]
3245	Butane, 1-chloro-2,3-dimethyl- 1-Chloro-2,3-dimethylbutane	$C_6H_{13}Cl$ 120.62	600-06-6	124	4-01-00-00373 0.887[22]	eth 4; EtOH 4; chl 4
3246	Butane, 1-chloro-3,3-dimethyl-	$C_6H_{13}Cl$ 120.62	2855-08-5	115	4-01-00-00369 0.8670[20]	eth 4; EtOH 4; chl 4 1.4161[20]
3247	Butane, 2-chloro-2,3-dimethyl-	$C_6H_{13}Cl$ 120.62	594-57-0 -10.4	112	4-01-00-00373 0.8780[20]	ace 4; EtOH 4 1.4191[20]
3248	Butane, 3-chloro-2,2-dimethyl-	$C_6H_{13}Cl$ 120.62	5750-00-5 0.9	111	4-01-00-00369 0.8767[20]	eth 4 1.4182[20]
3249	Butane, 1-(2-chloroethoxy)-	$C_6H_{13}ClO$ 136.62	10503-96-5	154 dec; 50[11]	4-01-00-01519 0.9335[20]	eth 4 1.4155[20]
3250	Butane, 1-chloro-4-fluoro- 1-Chloro-4-fluorobutane	C_4H_8ClF 110.56	462-73-7	114.7	4-01-00-00249 1.0627[25]	eth 4; EtOH 4 1.4020[25]
3251	Butane, 1-chloro-2-methyl-, (±)- Butane, 1-chloro-2-methyl (DL)	$C_5H_{11}Cl$ 106.60	114180-21-1	99.9	4-01-00-00324 0.8810[15]	EtOH 3; eth 3 1.4102[25]
3252	Butane, 1-chloro-3-methyl- Isoamyl chloride	$C_5H_{11}Cl$ 106.60	107-84-6 -104.4	4998 98.9	4-01-00-00325 0.8750[20]	H_2O 2; EtOH 5; eth 5; chl 4 1.4084[20]
3253	Butane, 2-chloro-2-methyl-	$C_5H_{11}Cl$ 106.60	594-36-5 -73.5	85.6	4-01-00-00324 0.8653[20]	H_2O 2; EtOH 3; eth 3; ctc 3 1.4055[20]
3254	Butane, 2-chloro-2,3,3-trimethyl-	$C_7H_{15}Cl$ 134.65	918-07-0 136	sub	4-01-00-00411	eth 4
3255	Butane, decafluoro- Perfluorobutane	C_4F_{10} 238.03	355-25-9 -128.2	-1.9	4-01-00-00245 1.6484[25]	bz 3; chl 3
3256	Butanedial	$C_4H_6O_2$ 86.09	638-37-9	170 dec; 58[9]	4-01-00-03642 1.064[20]	H_2O 4; ace 4; eth 4; EtOH 4 1.4262[18]
3257	Butanediamide Succinamide	$C_4H_8N_2O_2$ 116.12	110-14-5 268 dec	8837 sub 125	4-02-00-01922	
3258	1,4-Butanediamine Putrescine	$C_4H_{12}N_2$ 88.15	110-60-1 27.5	7964 158.5	4-04-00-01283 0.877[25]	H_2O 3 1.4969[20]
3259	1,4-Butanediamine, N,N'-bis(3-aminopropyl)- Spermine	$C_{10}H_{26}N_4$ 202.34	71-44-3 29	8699 150[5]	4-04-00-01301	
3260	1,4-Butanediamine, dihydrochloride Butane, 1,4-diamino, dihydrochloride	$C_4H_{14}Cl_2N_2$ 161.07	333-93-7 280 dec	sub	4-04-00-01283	H_2O 4; EtOH 4; eth 1; bz 1
3261	2,3-Butanediamine, (R*,R*)-(±)-	$C_4H_{12}N_2$ 88.15	20699-48-3 -22	57-8[60]	4-04-00-01302 0.8499[25]	1.4408[25]
3262	1,4-Butanediamine, N,N,N',N'-tetraethyl- N,N,N',N'-Tetraethyl-1,4-butanediamine	$C_{12}H_{28}N_2$ 200.37	69704-44-5	85-6[5]	4-04-00-01286	H_2O 4; eth 4; EtOH 4 1.4383[25]
3263	1,4-Butanediamine, N,N,N',N'-tetramethyl- N,N,N',N'-Tetramethyl-1,4-diaminobutane	$C_8H_{20}N_2$ 144.26	111-51-3 168	9157	4-04-00-01284 0.7942[15]	H_2O 5; EtOH 3; eth 3 1.4621[25]
3264	Butane, 1,1-dibromo- 1,1-Dibromobutane	$C_4H_8Br_2$ 215.92	62168-25-6	91[101]; 72[43]	4-01-00-00266 1.80[25]	1.4980[25]
3265	Butane, 1,2-dibromo- α-Butylene dibromide	$C_4H_8Br_2$ 215.92	533-98-2 -65.4	1565 166.3	4-01-00-00266 1.7915[20]	H_2O 1; eth 3; chl 3 1.4025[20]
3266	Butane, 1,3-dibromo- 1,3-Dibromobutane	$C_4H_8Br_2$ 215.92	107-80-2	174	4-01-00-00266 1.800[20]	H_2O 1; eth 3; ctc 2; chl 3 1.507[20]
3267	Butane, 1,4-dibromo- 1,4-Dibromobutane	$C_4H_8Br_2$ 215.92	110-52-1 -16.5	197	4-01-00-00267 1.8270[20]	H_2O 1; ctc 2; chl 3 1.5190[20]
3268	Butane, 2,3-dibromo- 2,3-Dibromobutane	$C_4H_8Br_2$ 215.92	5408-86-6 -24	161	3-01-00-00296 1.7893[22]	H_2O 1; eth 3 1.5133[22]
3269	Butane, 1,2-dibromo-2,3-dimethyl- 1,2-Dibromo-2,3-dimethylbutane	$C_6H_{12}Br_2$ 243.97	29916-45-8	80[17]	3-01-00-00414 1.6033[20]	1.5105[20]
3270	Butane, 1,4-dibromo-2,3-dimethyl- 1,4-Dibromo-2,3-dimethylbutane	$C_6H_{12}Br_2$ 243.97	54462-70-3	106-8[22]	4-01-00-00375 1.620[20]	1.5128[20]
3271	Butane, 2,3-dibromo-2,3-dimethyl- 1,2-Dibromo-1,1,2,2-tetramethylethane	$C_6H_{12}Br_2$ 243.97	594-81-0 171 dec	78[10]	4-01-00-00375 1.8110[21]	
3272	Butane, 1,2-dibromo-2-methyl- 1,2-Dibromo-2-methylbutane	$C_5H_{10}Br_2$ 229.94	10428-64-5	47-8[9]	4-01-00-00330 1.6711[20]	1.5088[20]
3273	Butane, 1,4-dibromo-2-methyl-, (R)- Butane, 1,4-dibromo-2-methyl (d)	$C_5H_{10}Br_2$ 229.94	69498-28-8	78-9[12]	4-01-00-00330 1.695[17]	1.5128[20]
3274	Butane, 2,3-dibromo-2-methyl- 1,2-Dibromo-1,1,2-trimethylethane	$C_5H_{10}Br_2$ 229.94	594-51-4 7	61-4[17]	4-01-00-00330 1.6711[20]	1.5729[25]
3275	Butane, 1,4-dibromo-1,1,2,2,3,3,4,4-octafluoro- 1,4-Dibromooctafluorobutane	$C_4Br_2F_8$ 359.84	335-48-8	97		
3276	Butane, 1,1-dichloro- Butylidene chloride	$C_4H_8Cl_2$ 127.01	541-33-3	1571 113.8	4-01-00-00250 1.0863[20]	H_2O 1; chl 3 1.4355[20]
3277	Butane, 1,2-dichloro- 1,2-Dichlorobutane	$C_4H_8Cl_2$ 127.01	616-21-7	124.1	4-01-00-00250 1.1116[25]	H_2O 1; eth 3; ctc 2; chl 3 1.4450[20]
3278	Butane, 1,3-dichloro- 1,3-Dichlorobutane	$C_4H_8Cl_2$ 127.01	1190-22-3	134	4-01-00-00250 1.1158[20]	H_2O 1; eth 3; ctc 2; chl 3 1.4445[20]

No.	Name Synonym	Mol. Form. Mol. Wt.	CAS RN mp/°C	Merck No. bp/°C	Beil. Ref. den/g cm^{-3}	Solubility n_D
3279	Butane, 1,4-dichloro- 1,4-Dichlorobutane	$C_4H_8Cl_2$ 127.01	110-56-5 -37.3	161	4-01-00-00250 1.1408[20]	chl 4 1.4542[20]
3280	Butane, 2,2-dichloro- 2,2-Dichlorobutane	$C_4H_8Cl_2$ 127.01	4279-22-5 -74	104	4-01-00-00251 1.1048[25]	H_2O 1; chl 3 1.4295
3281	Butane, 2,3-dichloro-, (R*,R*)-(±)-	$C_4H_8Cl_2$ 127.01	2211-67-8 -80	118; 53.2[80]	4-01-00-00251 1.063[25]	1.4409[25]
3282	Butane, 1,1-dichloro-3-methyl- 1,1-Dichloro-3-methylbutane	$C_5H_{10}Cl_2$ 141.04	625-66-1	130	4-01-00-00326 1.0473[20]	eth 4; EtOH 4 1.4344[20]
3283	Butane, 1,2-dichloro-2-methyl- 1,2-Dichloro-2-methylbutane	$C_5H_{10}Cl_2$ 141.04	23010-04-0	134	3-01-00-00360 1.0785[20]	eth 4; EtOH 4; chl 4 1.4432[21]
3284	Butane, 1,3-dichloro-3-methyl-	$C_5H_{10}Cl_2$ 141.04	624-96-4	146.7	4-01-00-00326 1.0654[20]	eth 4; chl 4 1.4455[20]
3285	Butane, 1,4-dichloro-2-methyl- 1,4-Dichloro-2-methylbutane	$C_5H_{10}Cl_2$ 141.04	623-34-7	168.5	3-01-00-00360 1.1003[25]	chl 4 1.4562[21]
3286	Butane, 2,3-dichloro-2-methyl- Amylene dichloride	$C_5H_{10}Cl_2$ 141.04	507-45-9 645	129	4-01-00-00325 1.0696[15]	eth 4; EtOH 4 1.4450[18]
3287	Butane, 2,3-dichloro-1,1,1,2,3,4,4,4-octafluoro-	$C_4Cl_2F_8$ 270.94	355-20-4 -68	63	4-01-00-00252 1.6801[20]	1.3100[20]
3288	Butane, 1,4-diiodo-	$C_4H_8I_2$ 309.92	628-21-7 5.8	125[15] dec	4-01-00-00276 2.3659[15]	H_2O 1; ctc 2; os 3 1.6239[15]
3289	Butane, 2,2-dimethyl- Neohexane	C_6H_{14} 86.18	75-83-2 -99	49.7	4-01-00-00367 0.6444[25]	H_2O 1; EtOH 3; eth 3; ace 4 1.3688[20]
3290	Butane, 2,3-dimethyl- 2,3-Dimethylbutane	C_6H_{14} 86.18	79-29-8 -128.8	57.9	4-01-00-00371 0.6616[20]	H_2O 1; EtOH 3; eth 3; ace 4 1.3750[20]
3291	Butanedinitrile Succinonitrile	$C_4H_4N_2$ 80.09	110-61-2 54.5	8843 266	4-02-00-01923 0.9867[60]	H_2O 4; EtOH 3; eth 2; ace 3 1.4173[60]
3292	Butanedinitrile, tetramethyl- Tetramethylsuccinonitrile	$C_8H_{12}N_2$ 136.20	3333-52-6 170.5		4-02-00-02054 1.070[25]	EtOH 3
3293	Butane, 1,4-dinitro- 1,4-Dinitrobutane	$C_4H_8N_2O_4$ 148.12	4286-49-1 33.5	176[13]	4-01-00-00280	H_2O 1; EtOH 2; eth 3; bz 3
3294	Butanedioic acid Succinic acid	$C_4H_6O_4$ 118.09	110-15-6 188	8840 235 dec	4-02-00-01908 1.572[25]	H_2O 3; EtOH 3; eth 3; ace 3 1.450
3295	Butanedioic acid, acetyl-, diethyl ester	$C_{10}H_{16}O_5$ 216.23	1115-30-6 	255; 133[17]	4-03-00-01826 1.081[20]	H_2O 1; EtOH 3; eth 3; bz 3 1.4346[20]
3296	Butanedioic acid, bis(1,1-dimethylethyl) ester Di-tert-butyl succinate	$C_{12}H_{22}O_4$ 230.30	926-26-1 36.5	3024 109-10[9]	4-02-00-01916	
3297	Butanedioic acid, bis(1-methylpropyl) ester Di-sec-butyl succinate	$C_{12}H_{22}O_4$ 230.30	626-31-3	256	3-02-00-01665 0.9735[20]	bz 4; eth 4; EtOH 4 1.4238[25]
3298	Butanedioic acid, bis(phenylmethyl) ester	$C_{18}H_{18}O_4$ 298.34	103-43-5 49.5	245[15]	4-06-00-02271 1.256[25]	H_2O 1; EtOH 3; eth 3; bz 3 1.596
3299	Butanedioic acid, bromo-, (±)-	$C_4H_5BrO_4$ 196.99	584-98-5 161		4-02-00-01929 2.073[25]	H_2O 3; EtOH 3; HOAc 2
3300	Butanedioic acid, bromo- Bromosuccinic acid	$C_4H_5BrO_4$ 196.99	923-06-8 161	1427	1-02-00-00268 2.073[25]	EtOH 4
3301	Butanedioic acid, dibutyl ester Dibutyl succinate	$C_{12}H_{22}O_4$ 230.30	141-03-7 -29.2	3023 274.5	4-02-00-01916 0.9752[20]	H_2O 1; EtOH 3; eth 3; bz 3 1.4299[20]
3302	Butanedioic acid, 2,3-dichloro-, (±)- Succinic acid, 2,3-dichloro-, (±)-	$C_4H_4Cl_2O_4$ 186.98	1114-09-6 175 dec		4-02-00-01928 1.844[15]	ace 4; eth 4; chl 4
3303	Butanedioic acid, diethyl ester Diethyl succinate	$C_8H_{14}O_4$ 174.20	123-25-1 -21	8840 217.7	4-02-00-01914 1.0402[20]	H_2O 1; EtOH 5; eth 5; ace 3 1.4201[20]
3304	Butanedioic acid, 2,3-dihydroxy-, (R*,S*)- meso-Tartaric acid	$C_4H_6O_6$ 150.09	147-73-9 147	9040	4-03-00-01218 1.666[20]	H_2O 4; EtOH 4
3305	Butanedioic acid, 2,3-dihydroxy-, bis(1-methylethyl) ester, (R*,R*)-(±)- Tartaric acid, diisopropyl ester (DL)	$C_{10}H_{18}O_6$ 234.25	58167-01-4 34	275; 154[12]	2-03-00-00337 1.1166[20]	ace 4; eth 4; EtOH 4
3306	Butanedioic acid, 2,3-dihydroxy- [R-(R*,R*)]-, bis(phenylmethyl) ester	$C_{18}H_{18}O_6$ 330.34	622-00-4 50	250-70[4]	3-06-00-01537 1.2036[72]	py 4; EtOH 4
3307	Butanedioic acid, 2,3-dihydroxy-, diethyl ester, (R*,R*)-(±)- Diethyl DL-tartrate	$C_8H_{14}O_6$ 206.20	57968-71-5 18.7	281; 158[14]	4-03-00-01232 1.2046[20]	H_2O 2; EtOH 5; eth 5; ace 3 1.4438[20]
3308	Butanedioic acid, 2,3-dihydroxy-, dimethyl ester, (R*,R*)-(±)-	$C_6H_{10}O_6$ 178.14	608-69-5 90	280	4-03-00-01232 1.2604[90]	H_2O 4; ace 4; eth 4; EtOH 4
3309	Butanedioic acid, 2,3-dihydroxy- [R-(R*,R*)]- L-Tartaric acid	$C_4H_6O_6$ 150.09	87-69-4 169	9039	4-03-00-01219	
3310	Butanedioic acid, 2,3-dihydroxy- [R-(R*,R*)]-, dipropyl ester Tartaric acid, dipropyl ester, (+)-	$C_{10}H_{18}O_6$ 234.25	2217-14-3	303	2-03-00-00337 1.1390[20]	H_2O 4; ace 4; eth 4; EtOH 4
3311	Butanedioic acid, 2,3-dihydroxy-, (R*,R*)-(±)- DL-Tartaric acid	$C_4H_6O_6$ 150.09	133-37-9 206	9038	4-03-00-01229 1.788[25]	H_2O 3; EtOH 3; eth 2; bz 1
3312	Butanedioic acid, 2,3-dihydroxy-, [S-(R*,R*)]- D-Tartaric acid	$C_4H_6O_6$ 150.09	147-71-7 172.5	9037	4-03-00-01229 1.7598[20]	DMSO 2 1.4955[20]
3313	Butanedioic acid, 2,3-dihydroxy- [R-(R*,R*)]-, dibutyl ester	$C_{12}H_{22}O_6$ 262.30	87-92-3 22	320	4-03-00-01232 1.0909[20]	H_2O 4; ace 4; EtOH 4 1.4451[20]
3314	Butanedioic acid, 2,3-dihydroxy- [R-(R*,R*)]-, disodium salt Tartaric acid, disodium salt	$C_4H_4Na_2O_6$ 194.05	868-18-8	8640		H_2O 3

No.	Name Synonym	Mol. Form. Mol. Wt.	CAS RN mp/°C	Merck No. bp/°C	Beil. Ref. den/g cm^{-3}	Solubility n_D
3315	Butanedioic acid, 2,3-dihydroxy- [R-(R*,R*)]-, monoethyl ester Tartaric acid, ethyl ester	$C_8H_{10}O_6$ 178.14	608-89-9 90	3812	3-03-00-01028	H_2O 4; EtOH 4
3316	Butanedioic acid, [(dimethoxyphosphinothioyl)thio]-, diethyl ester Malathion	$C_{10}H_{19}O_6PS_2$ 330.36	121-75-5 2.8	5582 156[0.7] dec	4-03-00-01136 1.2076[20]	H_2O 2; EtOH 3; eth 3; bz 3 1.4960[20]
3317	Butanedioic acid, 2,3-dimethyl-	$C_6H_{10}O_4$ 146.14	13545-04-5 132 dec	87-90[2]	4-02-00-01998	ace 4; eth 4
3318	Butanedioic acid, dimethyl ester	$C_6H_{10}O_4$ 146.14	106-65-0 19	196.4	4-02-00-01913 1.1198[20]	H_2O 2; EtOH 3; eth 4; ace 3 1.4197[20]
3319	Butanedioic acid, dipentyl ester	$C_{14}H_{26}O_4$ 258.36	645-69-2 -10.8	171.5[16]	3-02-00-01665 0.9616[20]	
3320	Butanedioic acid, diphenyl ester Diphenyl succinate	$C_{16}H_{14}O_4$ 270.28	621-14-7 121	330; 222.5[15]	3-06-00-00605	H_2O 1; EtOH 3; eth 3; ace 3
3321	Butanedioic acid, dipropyl ester	$C_{10}H_{18}O_4$ 202.25	925-15-5 -5.9	250.8	4-02-00-01916 1.0020[20]	ace 4; bz 4; eth 4 1.4250[20]
3322	Butanedioic acid, ethyl-, (R)-	$C_6H_{10}O_4$ 146.14	4074-24-2 96	181.5	4-02-00-01995 1.0017[20]	H_2O 4; eth 4; EtOH 4
3323	Butanedioic acid, ethyl methyl ester	$C_7H_{12}O_4$ 160.17	627-73-6 <-20	208.2	0-02-00-00609 1.076[20]	eth 4; EtOH 4
3324	Butanedioic acid, formyl-, diethyl ester	$C_9H_{14}O_5$ 202.21	5472-38-8	130-4[15]	4-03-00-01819	eth 4; EtOH 4 1.4486[25]
3325	Butanedioic acid, methyl-, (±)-	$C_5H_8O_4$ 132.12	636-60-2 115	dec	4-02-00-01948 1.4200[0]	H_2O 4; EtOH 4; eth 3; chl 2 1.4303
3326	Butanedioic acid, methyl-, diethyl ester	$C_9H_{16}O_4$ 188.22	4676-51-1	217.5	3-02-00-01696 1.012[25]	1.4199[20]
3327	Butanedioic acid, methyl-, dimethyl ester	$C_7H_{12}O_4$ 160.17	1604-11-1	196	3-02-00-01696 1.076[25]	1.4200[20]
3328	Butanedioic acid, methylene- Itaconic acid	$C_5H_6O_4$ 130.10	97-65-4 175	5130 dec	4-02-00-02228 1.632[25]	H_2O 3; EtOH 3; eth 2; ace 3
3329	Butanedioic acid, methylene-, diethyl ester	$C_9H_{14}O_4$ 186.21	2409-52-1 58.5	228	4-02-00-02230 1.0467[20]	EtOH 5; eth 3; ace 4; bz 3 1.4377[20]
3330	Butanedioic acid, methylene-, dimethyl ester	$C_7H_{10}O_4$ 158.15	617-52-7 38	208	4-02-00-02229 1.1241[18]	EtOH 3; eth 3; ace 4; MeOH 3 1.4457[20]
3331	Butanedioic acid, (1-methylethylidene)-	$C_7H_{10}O_4$ 158.15	584-27-0 160.5	174 dec	4-02-00-02251	eth 4; EtOH 4
3332	Butanedioic acid, (1-methylpropyl)-, diethyl ester Succinic acid, sec-butyl, diethyl ester	$C_{12}H_{22}O_4$ 230.30	69248-35-7	75-6[0.2]	3-02-00-01779 0.9745[25]	1.4293[25]
3333	Butanedioic acid, monobutyl ester	$C_8H_{14}O_4$ 174.20	5150-93-6 8.6	136.5[3]	3-02-00-01665 1.0732[20]	1.4360[20]
3334	Butanedioic acid, mono(2,2-dimethylhydrazide) Daminozide	$C_6H_{12}N_2O_3$ 160.17	1596-84-5 154.5	2810		
3335	Butanedioic acid, monoethyl ester Succinic acid, ethyl ester	$C_6H_{10}O_4$ 146.14	1070-34-4 8	172[42]; 119[3]	4-02-00-01914 1.1466[20]	H_2O 4; eth 4; EtOH 4 1.4327[20]
3336	Butanedioic acid, monomethyl ester	$C_5H_8O_4$ 132.12	3878-55-5 58	151[20]; 122[4]	4-02-00-01913	H_2O 3
3337	Butanedioic acid, monopentyl ester Succinic acid, pentyl ester	$C_9H_{16}O_4$ 188.22	97479-78-2 17.2	147[3]	2-02-00-00551 1.0460[20]	1.4378[20]
3338	Butanedioic acid, monopropyl ester Succinic acid, monopropyl ester	$C_7H_{12}O_4$ 160.17	6946-88-9 15	126[3]	3-02-00-01664 1.1071[20]	1.4343[20]
3339	Butanedioic acid, oxo- Oxalacetic acid	$C_4H_4O_5$ 132.07	328-42-7 161 dec	6863	4-03-00-01808	
3340	Butanedioic acid, oxo-, diethyl ester Diethyl oxalacetate	$C_8H_{12}O_5$ 188.18	108-56-5 131-2[24]	3791	4-03-00-01809 1.131[20]	H_2O 1; EtOH 5; eth 5; ace 4 1.4561[17]
3341	Butanedioic acid, phenyl-, (±)-	$C_{10}H_{10}O_4$ 194.19	10424-29-0 168	dec	4-09-00-03352	H_2O 2; EtOH 4; eth 4; ace 4
3342	Butanedioic acid, tetrahydroxy- Dihydroxytartaric acid	$C_4H_6O_8$ 182.09	76-30-2 114.5	3172	3-03-00-01419	
3343	Butanedioic acid, tetramethyl-	$C_8H_{14}O_4$ 174.20	630-51-3 200	sub	4-02-00-02054 1.30[25]	bz 4; EtOH 4; chl 4
3344	1,2-Butanediol, (±)-	$C_4H_{10}O_2$ 90.12	26171-83-5	190.5	4-01-00-02507 1.0024[20]	H_2O 3; EtOH 3; ace 3 1.4378[20]
3345	1,3-Butanediol 1,3-Butylene glycol	$C_4H_{10}O_2$ 90.12	107-88-0 <-50	1566 207.5	0-01-00-00477 1.0053[20]	1.4401[20]
3346	1,4-Butanediol Tetramethylene glycol	$C_4H_{10}O_2$ 90.12	110-63-4 20.1	235	4-01-00-02515 1.0171[20]	H_2O 5; EtOH 3; eth 2; DMSO 2 1.4460[20]
3347	2,3-Butanediol, (R*,R*)-(±)-	$C_4H_{10}O_2$ 90.12	6982-25-8 7.6	182.5	4-01-00-02525 1.0033[20]	H_2O 5; EtOH 5; eth 3; ace 3 1.4310[25]
3348	1,4-Butanediol, diacetate	$C_8H_{14}O_4$ 174.20	628-67-1 12	229	4-02-00-00224 1.0479[15]	1.4251[15]
3349	2,3-Butanediol, 1,4-dichloro- 1,4-Dichloro-2,3-butanediol	$C_4H_8Cl_2O_2$ 159.01	2419-73-0 126.5	150[30]	4-01-00-02531	EtOH 4
3350	2,3-Butanediol, 2,3-dimethyl- Pinacol	$C_6H_{14}O_2$ 118.18	76-09-5 43	7408 174.4	4-01-00-02575	H_2O 2; EtOH 4; eth 4; CS$_2$ 2
3351	2,3-Butanediol, 2,3-dimethyl-, hexahydrate	$C_6H_{26}O_8$ 226.27	6091-58-3 45.4	172	2-01-00-00553 0.967[15]	eth 4; EtOH 4

No.	Name / Synonym	Mol. Form. / Mol. Wt.	CAS RN / mp/°C	Merck No. / bp/°C	Beil. Ref. / den/g cm⁻³ / n_D	Solubility
3352	1,2-Butanediol, 3-methyl- / 3-Methyl-1,2-butanediol	$C_5H_{12}O_2$ / 104.15	50468-22-9 /	206	4-01-00-02549 / 0.9987[0]	eth 4; EtOH 4
3353	1,3-Butanediol, 2-methyl- / 2-Methyl-1,3-butanediol	$C_5H_{12}O_2$ / 104.15	684-84-4 /	200; 115[20]	4-01-00-02546 / 0.9912[20]	H_2O 3; EtOH 3; eth 3
3354	1,3-Butanediol, 3-methyl-	$C_5H_{12}O_2$ / 104.15	2568-33-4 /	202.5	4-01-00-02549 / 0.9448[20] / 1.4452[20]	H_2O 3; EtOH 3
3355	2,3-Butanediol, 2-methyl- / 2,3-Dihydroxy-2-methylbutane	$C_5H_{12}O_2$ / 104.15	5396-58-7 /	175	4-01-00-02547 / 0.9920[25] / 1.4375[20]	H_2O 4; eth 4; EtOH 4
3356	1,2-Butanediol, 2-phenyl- / 2-Phenyl-1,2-butanediol	$C_{10}H_{14}O_2$ / 166.22	90925-48-7 / 56	165[23]	3-06-00-04667	H_2O 4; eth 4; EtOH 4
3357	1,3-Butanediol, 1-phenyl- / 1-Phenylbutane-1,3-diol	$C_{10}H_{14}O_2$ / 166.22	65469-88-7 / 73.5	175[21]; 130[2]	4-06-00-06006 / 1.0720[25]	bz 4; EtOH 4; lig 4
3358	1,3-Butanediol, 3-phenyl-	$C_{10}H_{14}O_2$ / 166.22	7133-68-8 /	150[2]; 109[0.5]	4-06-00-06009 / 1.0865[20] / 1.5341[20]	
3359	2,3-Butanedione / Diacetyl	$C_4H_8O_2$ / 86.09	431-03-8 / -2.4	2946 / 88	4-01-00-03644 / 0.9808[18] / 1.3951[20]	H_2O 4; EtOH 5; eth 5; ace 4
3360	2,3-Butanedione, dioxime / Dimethylglyoxime	$C_4H_8N_2O_2$ / 116.12	95-45-4 / 245.5	3235 / sub 234	3-01-00-03105	H_2O 1; EtOH 4; eth 4; bz 2
3361	2,3-Butanedione, monooxime	$C_4H_7NO_2$ / 101.11	57-71-6 / 76.8	185.5	4-01-00-03646	H_2O 2; EtOH 4; eth 4; chl 4
3362	1,3-Butanedione, 1-phenyl-	$C_{10}H_{10}O_2$ / 162.19	93-91-4 / 56	261.5	4-07-00-02151 / 1.059[74] / 1.567[78]	H_2O 1; eth 3; chl 2
3363	1,3-Butanedione, 1-(2-pyridinyl)-	$C_9H_9NO_2$ / 163.18	40614-52-6 / 50	140[15]	4-21-00-04750	bz 4; eth 4; EtOH 4; chl 4
3364	1,3-Butanedione, 1-(3-pyridinyl)-	$C_9H_9NO_2$ / 163.18	3594-37-4 / 85	171[15]	5-21-10-00180	H_2O 2; EtOH 3; eth 3; ace 3
3365	1,3-Butanedione, 1-(4-pyridinyl)-	$C_9H_9NO_2$ / 163.18	75055-73-1 / 62	145[18]	5-21-10-00180	bz 4; eth 4; EtOH 4
3366	Butanedioyl dichloride / Succinyl chloride	$C_4H_4Cl_2O_2$ / 154.98	543-20-4 / 20	8844 / 193.3	4-02-00-01921 / 1.3748[20] / 1.4683[20]	eth 3; ace 3; bz 3
3367	Butanedioyl dichloride, methylene-	$C_5H_4Cl_2O_2$ / 166.99	1931-60-8 /	89[17]	3-02-00-01934 / 1.4919[20]	ace 3
3368	1,4-Butanedithiol / Tetramethylenedithiol	$C_4H_{10}S_2$ / 122.26	1191-08-8 / -53.9	195.5	4-01-00-02523 / 1.0021[0] / 1.5290[20]	H_2O 1; EtOH 4; ctc 2
3369	Butane, 1-(ethenyloxy)- / Butyl vinyl ether	$C_6H_{12}O$ / 100.16	111-34-2 / -92	94	4-01-00-02052 / 0.7888[20] / 1.4026[20]	H_2O 1; EtOH 4; eth 5; ace 4
3370	Butane, 2-(ethenyloxy)-	$C_6H_{12}O$ / 100.16	1888-85-3 /	81	4-01-00-02054 / 1.3970[20]	
3371	Butane, 1-(ethenyloxy)-3-methyl- / Isopentyl vinyl ether	$C_7H_{14}O$ / 114.19	39782-38-2 /	112.5	4-01-00-02055 / 0.7826[20] / 1.4072[20]	eth 4; EtOH 4
3372	Butane, 1-ethoxy- / Butyl ethyl ether	$C_6H_{14}O$ / 102.18	628-81-9 / -124	92.3	4-01-00-01518 / 0.7495[20] / 1.3818[20]	H_2O 1; EtOH 5; eth 5; ace 4
3373	Butane, 2-ethoxy- / sec-Butyl ethyl ether	$C_6H_{14}O$ / 102.18	2679-87-0 /	81	4-01-00-01572 / 0.7503[20] / 1.3802[20]	H_2O 1; EtOH 4; eth 4
3374	Butane, 1-ethoxy-3-methyl- / Ethyl isopentyl ether	$C_7H_{16}O$ / 116.20	628-04-6 /	112.5	4-01-00-01681 / 0.7688[21]	eth 4; EtOH 4
3375	Butane, 2-ethoxy-2-methyl- / Ethyl tert-amyl ether	$C_7H_{16}O$ / 116.20	919-94-8 /	102	3-01-00-01626 / 0.751[18]	eth 4; EtOH 4
3376	Butane, 1-(ethylthio)- / Butyl ethyl sulfide	$C_6H_{14}S$ / 118.24	638-46-0 / -95.1	144.3	4-01-00-01558 / 0.8376[20] / 1.4492[10]	EtOH 4; chl 3
3377	Butane, 1-(ethynyloxy)-	$C_6H_{10}O$ / 98.14	3329-56-4 /	104	4-01-00-02213 / 0.8200[20] / 1.4067	eth 4; EtOH 4
3378	Butane, 1-fluoro- / Butyl fluoride	C_4H_9F / 76.11	2366-52-1 / -134	32.5	4-01-00-00244 / 0.7789[20] / 1.3396[20]	EtOH 4
3379	Butane, 2-fluoro- / sec-Butyl fluoride	C_4H_9F / 76.11	359-01-3 / -121.4	25.1	4-01-00-00244 / 0.7559[25]	
3380	Butane, 1,1,2,2,3,4,4-heptachloro-	$C_4H_3Cl_7$ / 299.24	34973-41-6 /	137.5[13.5]	4-01-00-00257 / 1.739[20] / 1.5407[20]	ace 4; ctc 4
3381	Butane, 1-iodo- / Butyl iodide	C_4H_9I / 184.02	542-69-8 / -103	1572 / 130.6	4-01-00-00271 / 1.6154[20] / 1.5001[20]	H_2O 1; EtOH 5; eth 5; chl 4
3382	Butane, 2-iodo-, (±)- / Butane, 2-iodo (DL)	C_4H_9I / 184.02	52152-71-3 / -104.2	1573 / 120	4-01-00-00272 / 1.5920[20] / 1.4991[20]	H_2O 1; EtOH 5; eth 5; chl 4
3383	Butane, 1-iodo-2-methyl-, (S)- / Butane, 1-iodo-2-methyl (d)	$C_5H_{11}I$ / 198.05	29394-58-9 /	148; 132.6[500]	3-01-00-00366 / 1.5253[20] / 1.4977[20]	eth 4; EtOH 4
3384	Butane, 1-iodo-3-methyl- / Isoamyl iodide	$C_5H_{11}I$ / 198.05	541-28-6 /	5002 / 147	4-01-00-00331 / 1.5118[20] / 1.4939[20]	H_2O 2; EtOH 5; eth 5; ctc 2
3385	Butane, 2-iodo-2-methyl-	$C_5H_{11}I$ / 198.05	594-38-7 /	124.5	4-01-00-00331 / 1.4937[20] / 1.4981[20]	H_2O 1; EtOH 5; eth 5
3386	Butane, 1-isocyanato-3-methyl-	$C_6H_{11}NO$ / 113.16	1611-65-0 /	137	0-04-00-00186 / 0.806[20] / 1.406[20]	H_2O 1; EtOH 5; eth 5
3387	Butane, 1-isocyano- / Butyl isocyanide	C_5H_9N / 83.13	2769-64-4 /	120	4-04-00-00562 / 0.78[20]	eth 4; EtOH 4
3388	Butane, 1-isothiocyanato-	C_5H_9NS / 115.20	592-82-5 /	168	4-04-00-00598 / 0.9546[20] / 1.501[20]	eth 4; EtOH 4
3389	Butane, 2-isothiocyanato-, (±)- / sec-Butyl isothiocyanate (DL)	C_5H_9NS / 115.20	116724-11-9 /	159.5	4-04-00-00624 / 0.944[12]	eth 4; EtOH 4
3390	Butane, 1-isothiocyanato-3-methyl-	$C_6H_{11}NS$ / 129.23	628-03-5 /	183	4-04-00-00707 / 0.9419[17]	eth 4; EtOH 4
3391	Butane, 1-methoxy- / Butyl methyl ether	$C_5H_{12}O$ / 88.15	628-28-4 / -115.5	70.1	4-01-00-01518 / 0.7443[20] / 1.3736[20]	H_2O 1; EtOH 5; eth 5; ace 3

No.	Name Synonym	Mol. Form. Mol. Wt.	CAS RN	mp/°C	Merck No. bp/°C	Beil. Ref. den/g cm⁻³	Solubility n_D
3392	Butane, 2-methoxy-, (S)- sec-Butyl methyl ether (d)	$C_5H_{12}O$ 88.15	66610-39-7		60.7	4-01-00-01571 0.7400[25]	
3393	Butane, 2-methoxy-, (±)- sec-Butyl methyl ether (DL)	$C_5H_{12}O$ 88.15	116783-23-4		60	4-01-00-01571 0.7415[20]	ace 4; eth 4; EtOH 4 1.3680[25]
3394	Butane, 1-methoxy-2-methyl- Methyl (2-methylbutyl) ether	$C_6H_{14}O$ 102.18	62016-48-2		90	4-01-00-01667 0.754[18]	1.3849[20]
3395	Butane, 1-methoxy-3-methyl-	$C_6H_{14}O$ 102.18	626-91-5		91	4-01-00-01681 0.7517[20]	eth 4; EtOH 4 1.3830[20]
3396	Butane, 2-methoxy-2-methyl-	$C_6H_{14}O$ 102.18	994-05-8		86.3	4-01-00-01671 0.7703[20]	eth 4; EtOH 4 1.3855[20]
3397	Butane, 2-methyl- Isopentane	C_5H_{12} 72.15	78-78-4	-159.9	27.8	4-01-00-00320 0.6201[20]	H₂O 1; EtOH 5; eth 5 1.3537[20]
3398	Butane, 1,1'-[methylenebis(oxy)]bis-	$C_9H_{20}O_2$ 160.26	2568-90-3	-58.1	179.2	4-01-00-03029 0.8339[20]	1.4072[17]
3399	Butane, 1-(1-methylethoxy)- Butyl isopropyl ether	$C_7H_{16}O$ 116.20	1860-27-1		109	4-01-00-01519 0.7594[15]	H₂O 1; EtOH 3; eth 3; ace 3 1.3870[15]
3400	Butane, 3-methyl-1-nitro- 3-Methyl-1-nitrobutane	$C_5H_{11}NO_2$ 117.15	627-67-8	97	163.5; 72[27]	4-01-00-00332 0.9458[25]	
3401	Butane, 1-(2-methylpropoxy)- Butyl isobutyl ether	$C_8H_{18}O$ 130.23	17071-47-5		151	4-01-00-01594 0.7980[22]	ace 4; eth 4; EtOH 4 1.4077[21]
3402	Butane, 1-(methylthio)- Butyl methyl sulfide	$C_5H_{12}S$ 104.22	628-29-5	-97.8	123.5	4-01-00-01557 0.8426[20]	EtOH 4; MeOH 4 1.4477[20]
3403	Butanenitrile Propyl cyanide	C_4H_7N 69.11	109-74-0	-111.9	1597 117.6	4-02-00-00806 0.7936[20]	H₂O 2; EtOH 5; eth 5; bz 3 1.3842[20]
3404	Butanenitrile, 4-bromo-	C_4H_6BrN 148.00	5332-06-9		206	4-02-00-00836 1.4967[20]	EtOH 3; eth 3; chl 3 1.4818[20]
3405	Butanenitrile, 4-chloro-	C_4H_6ClN 103.55	628-20-6		192	4-02-00-00827 1.0934[15]	H₂O 1; EtOH 3; eth 3; ctc 2 1.4413[20]
3406	Butanenitrile, 4,4-diethoxy-	$C_8H_{15}NO_2$ 157.21	18381-45-8		104-6[10]	3-03-00-01210 0.937[25]	H₂O 2; EtOH 5; eth 5 1.4186[20]
3407	Butanenitrile, 2-ethyl-	$C_6H_{11}N$ 97.16	617-80-1		145.5	4-02-00-00953 	eth 4; EtOH 4 1.3891[24]
3408	Butanenitrile, 2-hydroxy-3-methyl-2-(1-methylethyl)- Diisopropyl ketone, cyanohydrin	$C_8H_{15}NO$ 141.21	4390-75-4	59	111[18]	4-03-00-00884 	ace 4; bz 4; eth 4; EtOH 4
3409	Butanenitrile, 2-methyl- 2-Methylbutanenitrile	C_5H_9N 83.13	18936-17-9		125	4-02-00-00892 0.7913[15]	eth 4; EtOH 4 1.3933[20]
3410	Butanenitrile, 3-methyl- Isobutyl cyanide	C_5H_9N 83.13	625-28-5	-101	127.5	4-02-00-00902 0.7914[20]	H₂O 2; EtOH 5; eth 5; ace 4 1.3927[20]
3411	Butanenitrile, 4-phenoxy-	$C_{10}H_{11}NO$ 161.20	2243-43-8	45.5	288	4-06-00-00646 	eth 3
3412	Butane, 1-nitro- 1-Nitrobutane	$C_4H_9NO_2$ 103.12	627-05-4		153	4-01-00-00277 0.970[25]	H₂O 2; EtOH 5; eth 5; alk 3 1.4303[20]
3413	Butane, 2-nitro-, (±)-	$C_4H_9NO_2$ 103.12	116781-85-2	-132	140	4-01-00-00278 0.9854[17]	chl 3 1.4044[20]
3414	Butane, 1,1,1,2,3,4,4,4-octafluoro-2,3-bis(trifluoromethyl)- Perfluoro-2,3-dimethylbutane	C_6F_{14} 338.04	354-96-1	-15	59.8		
3415	Butane, 1,1'-oxybis- Dibutyl ether	$C_8H_{18}O$ 130.23	142-96-1	-95.2	1568 140.2	4-01-00-01520 0.7684[20]	H₂O 1; EtOH 5; eth 5; ace 4 1.3992[20]
3416	Butane, 2,2'-oxybis-, (±)- sec-Butyl ether, (±)-	$C_8H_{18}O$ 130.23	17226-28-7		120.5	3-01-00-01533 0.756[25]	ace 4; eth 4; EtOH 4 1.393[25]
3417	Butane, 1,1'-[oxybis(2,1-ethanediyloxy)]bis-	$C_{12}H_{26}O_3$ 218.34	112-73-2	-60	256	4-01-00-02395 0.885[20]	1.4235[20]
3418	Butane, 1,1'-oxybis[3-methyl- Diisoamyl ether	$C_{10}H_{22}O$ 158.28	544-01-4		5000 172.5	4-01-00-01682 0.7777[20]	ace 4; EtOH 4; chl 4 1.4120[20]
3419	Butane, 1,1,2,3,4-pentachloro- Butane, 1,1,2,3,4-pentachloro (liquid)	$C_4H_5Cl_5$ 230.35	77753-24-3		95.5[11]	4-01-00-00256 1.561[18]	EtOH 4; ctc 4 1.5141[18]
3420	Butane, 1,2,2,3,4-pentachloro- 1,2,2,3,4-Pentachlorobutane	$C_4H_5Cl_5$ 230.35	2431-52-9		85[10]	4-01-00-00256 1.5543[20]	ace 4; chl 4 1.5157[20]
3421	Butane, 1-propoxy- Butyl propyl ether	$C_7H_{16}O$ 116.20	3073-92-5		118.1	4-01-00-01519 0.7772[0]	H₂O 1; EtOH 4; eth 4
3422	Butane, 1,1'-sulfinylbis-	$C_8H_{18}OS$ 162.30	2168-93-6	32.6	dec	4-01-00-01561 0.8317[23]	H₂O 1; EtOH 3; eth 3 1.4669[20]
3423	Butane, 1,1'-sulfonylbis-	$C_8H_{18}O_2S$ 178.30	598-04-9	45	291	4-01-00-01561 0.9885[47]	H₂O 1; EtOH 3; eth 3
3424	Butane, 1,1,4,4-tetrabromo-	$C_4H_6Br_4$ 373.71	116779-77-2		138-45[10]	3-01-00-00298 2.529[20]	bz 4; eth 4; chl 4 1.6077[20]
3425	Butane, 1,2,2,3-tetrabromo-	$C_4H_6Br_4$ 373.71	116779-78-3	-2	97.5[7]	3-01-00-00298 2.510[20]	
3426	Butane, 1,2,3,4-tetrabromo-, DL- Butane, 1,2,3,4-tetrabromo (DL)	$C_4H_6Br_4$ 373.71	2657-65-0	118	260; 180[60]	4-01-00-00271 	ace 4; eth 4; EtOH 4; lig 4
3427	Butane, 1,1,1,2-tetrachloro- 1,1,1,2-Tetrachlorobutane	$C_4H_6Cl_4$ 195.90	39966-95-5		135	4-01-00-00253 1.3907[20]	1.4920[25]
3428	Butane, 1,2,2,3-tetrachloro- 1,2,2,3-Tetrachlorobutane	$C_4H_6Cl_4$ 195.90	79630-70-9	-48	182	3-01-00-00286 1.4276[18]	ace 4; chl 4 1.491[20]
3429	Butane, 1,2,3,3-tetrachloro- 1,2,3,3-Tetrachlorobutane	$C_4H_6Cl_4$ 195.90	13138-51-7		90[32]; 56[10]	4-01-00-00254 1.4204[20]	ace 4; eth 4; EtOH 4 1.4958[20]
3430	Butane, 2,2,3,3-tetramethyl- 2,2,3,3-Tetramethylbutane	C_8H_{18} 114.23	594-82-1	100.7	106.4	4-01-00-00447 0.8242[20]	H₂O 1; eth 3; chl 3 1.4695[20]

No.	Name Synonym	Mol. Form. Mol. Wt.	CAS RN mp/°C	Merck No. bp/°C	Beil. Ref. den/g cm^{-3}	Solubility n_D
3431	1,2,3,4-Butanetetrol, (R*,S*)- Erythritol	$C_4H_{10}O_4$ 122.12	149-32-6 121.5	3620 330.5	4-01-00-02807 1.451^{20}	H_2O 3; eth 1; bz 1
3432	1,2,3,4-Butanetetrol, tetranitrate, (R*,S*)- Erythrityl tetranitrate	$C_4H_6N_4O_{12}$ 302.11	7297-25-8 61	3622	4-01-00-02809	EtOH 4
3433	Butane, 1,1'-thiobis- Dibutyl sulfide	$C_8H_{18}S$ 146.30	544-40-1 -79.7	1590 185	4-01-00-01559 0.8386^{20}	eth 4; EtOH 4; chl 4 1.4530^{20}
3434	Butane, 2,2'-thiobis-	$C_8H_{18}S$ 146.30	626-26-6	165	4-01-00-01586 0.8348^{20}	H_2O 1; EtOH 4; eth 4 1.4506^{20}
3435	Butane, 1,1'-thiobis[2-methyl- Bis(2-methylbutyl) sulfide	$C_{10}H_{22}S$ 174.35	96034-00-3	165; 95^{13}	0-01-00-00387 0.8348^{20}	eth 4; EtOH 4 1.4506^{20}
3436	Butane, 1,1'-thiobis[3-methyl- Isoamyl sulfide	$C_{10}H_{22}S$ 174.35	544-02-5 -74.6	211	4-01-00-01689 0.8323^{20}	H_2O 1; EtOH 5; eth 4 1.4520^{20}
3437	1-Butanethiol Butyl mercaptan	$C_4H_{10}S$ 90.19	109-79-5 -115.7	1575 98.5	4-01-00-01555 0.8416^{20}	H_2O 2; EtOH 4; eth 4; chl 2 1.4440^{20}
3438	2-Butanethiol, (±)- 2-Butanethiol (DL)	$C_4H_{10}S$ 90.19	91840-99-2 -165	1576 85	4-01-00-01584 0.8295^{20}	EtOH 3; eth 3; bz 3; ctc 2 1.4366^{20}
3439	1-Butanethiol, 2-methyl-, (S)- (+)-2-Methyl-1-butanethiol	$C_5H_{12}S$ 104.22	20089-07-0	118.2	4-01-00-01668 0.8420^{20}	1.4440^{20}
3440	1-Butanethiol, 3-methyl- Isopentyl mercaptan	$C_5H_{12}S$ 104.22	541-31-1	120	4-01-00-01688 0.8350^{20}	H_2O 1; EtOH 5; eth 5; ctc 3 1.4418^{20}
3441	2-Butanethiol, 2-methyl- 2-Methyl-2-butanethiol	$C_5H_{12}S$ 104.22	1679-09-0	99.1	4-01-00-01674 0.8120^{20}	1.4385^{20}
3442	Butane, 1,1,2-tribromo- 1,1,2-Tribromocyclobutane	$C_4H_7Br_3$ 294.81	3675-68-1	216.2	2-01-00-00084 2.1835^{20}	eth 4; EtOH 4; chl 4 1.5626^{17}
3443	Butane, 1,2,2-tribromo- 1,2,2-Tribromobutane	$C_4H_7Br_3$ 294.81	3675-69-2	213.8	2-01-00-00085 2.1691^{20}	eth 4; EtOH 4; chl 4 1.5624^{10}
3444	Butane, 1,2,3-tribromo- 1,2,3-Tribromobutane	$C_4H_7Br_3$ 294.81	632-05-3 -19	220	4-01-00-00271 2.1907^{20}	eth 4; EtOH 4; chl 4 1.5680^{20}
3445	Butane, 1,2,4-tribromo- 1,2,4-Tribromobutane	$C_4H_7Br_3$ 294.81	38300-67-3 -18	215	4-01-00-00271 2.170^{20}	eth 4; EtOH 4; chl 4 1.5608^{20}
3446	Butane, 1,3,3-tribromo- 1,3,3-Tribromobutane	$C_4H_7Br_3$ 294.81	62127-46-2	202.5	3-01-00-00298 2.1445^{20}	eth 4; EtOH 4; chl 4 1.5564^{20}
3447	Butane, 2,2,3-tribromo- 2,2,3-Tribromobutane	$C_4H_7Br_3$ 294.81	62127-47-3 1.8	206	3-01-00-00298 2.1723^{20}	H_2O 1; EtOH 3; eth 3; ctc 2 1.5602^{20}
3448	Butane, 1,1,3-trichloro- 1,1,3-Trichlorobutane	$C_4H_7Cl_3$ 161.46	13279-87-3	152	4-01-00-00252 1.317^{15}	H_2O 1; EtOH 3; eth 3; chl 4 1.4600^{15}
3449	Butane, 1,2,3-trichloro- 1,2,3-Trichlorobutane	$C_4H_7Cl_3$ 161.46	18338-40-4	168; 63^{20}	3-01-00-00285 1.3164^{20}	H_2O 1; EtOH 3; eth 3; chl 4 1.4790^{20}
3450	Butane, 1,2,4-trichloro- 1,2,4-Trichlorobutane	$C_4H_7Cl_3$ 161.46	1790-22-3	61-3^{10}	4-01-00-00253 1.3175^{20}	H_2O 1; bz 3; ctc 3; chl 4 1.4820^{20}
3451	Butane, 2,2,3-trichloro- 2,2,3-Trichlorobutane	$C_4H_7Cl_3$ 161.46	10403-60-8	144	3-01-00-00285 1.2699^{20}	H_2O 1; chl 4 1.4645^{20}
3452	Butane, 2,2,3-trichloro-1,1,1,3,4,4,4-heptafluoro-	$C_4Cl_3F_7$ 287.39	335-44-4 4	98	4-01-00-00253 1.7484^{20}	1.3530^{20}
3453	Butane, 1,2,3-trichloro-2-methyl- Butane, 2-methyl-1,2,3-trichloro	$C_5H_9Cl_3$ 175.48	62521-69-1	184	3-01-00-00361 1.2527^{20}	chl 4
3454	Butane, 2,2,3-trichloro-3-methyl- Butane, 2-methyl-2,3,3-trichloro	$C_5H_9Cl_3$ 175.48	98070-91-8	182.5	3-01-00-00361 1.215^{15}	chl 4; HOAc 4 1.473^{21}
3455	Butane, 1,1,1-trimethoxy-	$C_7H_{16}O_3$ 148.20	10138-89-3	157	4-01-00-03985 0.935^{25}	1.4032^{20}
3456	Butane, 1,3,3-trimethoxy-	$C_7H_{16}O_3$ 148.20	6607-66-5	61-3^{20}	4-01-00-03995 0.94^{25}	1.4096^{20}
3457	Butane, 2,2,3-trimethyl- 2,2,3-Trimethylbutane	C_7H_{16} 100.20	464-06-2 -25	80.8	4-01-00-00410 0.6901^{20}	H_2O 1; EtOH 3; eth 3; ace 4 1.3864^{20}
3458	1,2,3-Butanetriol 1-Methylglycerol	$C_4H_{10}O_3$ 106.12	4435-50-1	170^{20}	4-01-00-02774	H_2O 4; EtOH 4 1.4462^{20}
3459	1,2,4-Butanetriol	$C_4H_{10}O_3$ 106.12	3068-00-6	172-4^{12}	2-01-00-00596 1.018^{20}	H_2O 4; EtOH 4 1.4688^{20}
3460	Butanoic acid Butyric acid	$C_4H_8O_2$ 88.11	107-92-6 -5.7	1593 163.7	4-02-00-00779 0.9577^{20}	H_2O 5; EtOH 5; eth 5; ctc 2 1.3980^{20}
3461	Butanoic acid, 2-acetyl-3-methyl-, ethyl ester	$C_9H_{16}O_3$ 172.22	1522-46-9	201; 97-8^{20}	4-03-00-01608 0.9648^{18}	H_2O 1; EtOH 5; eth 5 1.4256^{18}
3462	Butanoic acid, 2-acetyl-3-oxo-, ethyl ester	$C_8H_{12}O_4$ 172.18	603-69-0	210	4-03-00-01781 1.1045^{20}	bz 4; eth 4; EtOH 4 1.4690^{20}
3463	Butanoic acid, 2-amino-, (±)-	$C_4H_9NO_2$ 103.12	2835-81-6 305 dec	sub	3-04-00-01296 1.2300^{20}	H_2O 4; EtOH 2; eth 1; bz 1
3464	Butanoic acid, 4-amino- γ-Aminobutyric acid	$C_4H_9NO_2$ 103.12	56-12-2 203 dec	441	4-04-00-02600	H_2O 4; EtOH 2; eth 1; ace 2
3465	Butanoic acid, 3-amino-4-hydroxy- γ-Hydroxy-β-aminobutyric acid	$C_4H_9NO_3$ 119.12	589-44-6 216	455	4-04-00-03216	
3466	Butanoic acid, 4-amino-2-hydroxy-4-oxo-	$C_4H_7NO_4$ 133.10	66398-52-5 147.5		3-03-00-00918 1.577^{18}	H_2O 3; EtOH 2; eth 1; MeOH 2
3467	Butanoic acid, anhydride Butyric anhydride	$C_8H_{14}O_3$ 158.20	106-31-0 -75	1594 200	4-02-00-00802 0.9668^{20}	eth 3; ctc 2 1.4070^{20}
3468	Butanoic acid, 2-bromo-, (±)- DL-α-Bromobutyric acid	$C_4H_7BrO_2$ 167.00	2385-70-8 -4	1401 217 dec; 181^{250}	4-02-00-00833 1.5641^{20}	H_2O 3; EtOH 3; eth 3
3469	Butanoic acid, 2-bromo-, ethyl ester	$C_8H_{11}BrO_2$ 195.06	533-68-6	177; 43^5	4-02-00-00834 1.3273^{20}	H_2O 1; EtOH 5; eth 5; chl 3 1.4475^{20}

No.	Name / Synonym	Mol. Form. / Mol. Wt.	CAS RN mp/°C	Merck No. bp/°C	Beil. Ref. den/g cm⁻³	Solubility n_D
3470	Butanoic acid, 4-bromo-, ethyl ester	$C_6H_{11}BrO_2$ 195.06	2969-81-5	82^{10}	4-02-00-00836 1.363^{25}	1.4559^{20}
3471	Butanoic acid, 2-bromo-3-methyl-, (±)-	$C_5H_9BrO_2$ 181.03	10323-40-7 44	230 dec; 138^{25}	3-02-00-00706 1.459^{20}	H_2O 2; EtOH 4; eth 3; ace 3
3472	Butanoic acid, 3-bromo-3-methyl- β-Bromoisovaleric acid	$C_5H_9BrO_2$ 181.03	5798-88-9 74	1410	4-02-00-00905	bz 4; eth 4; EtOH 4
3473	Butanoic acid, 2-bromo-, methyl ester	$C_5H_9BrO_2$ 181.03	3196-15-4	168	4-02-00-00834 1.4528^{20}	EtOH 4 1.4029^{25}
3474	Butanoic acid, 4-bromo-, methyl ester	$C_5H_9BrO_2$ 181.03	4897-84-1	186.5	4-02-00-00835 1.4^{25}	EtOH 4 1.4567^{25}
3475	Butanoic acid, 2-bromo-3-methyl-, ethyl ester	$C_7H_{13}BrO_2$ 209.08	609-12-1	186	4-02-00-00906 1.2760^{20}	eth 4; EtOH 4 1.4496^{20}
3476	Butanoic acid, 2-bromo-3-methyl-, methyl ester	$C_6H_{11}BrO_2$ 195.06	26330-51-8	177	3-02-00-00706 1.352^{13}	eth 4; EtOH 4 1.4530^{20}
3477	Butanoic acid, 2-bromo-3-oxo-, ethyl ester	$C_6H_9BrO_3$ 209.04	609-13-2	210 dec; 107^{15}	4-03-00-01551 1.4294^{16}	eth 4; EtOH 4 1.463^{14}
3478	Butanoic acid, 4-bromo-3-oxo-, ethyl ester	$C_6H_9BrO_3$ 209.04	13176-46-0	115^{14}	4-03-00-01551 1.4840^{20}	eth 4; EtOH 4 1.5281^{20}
3479	Butanoic acid, butyl ester Butyl butanoate	$C_8H_{16}O_2$ 144.21	109-21-7 -91.5	1556 166	4-02-00-00789 0.8700^{20}	H_2O 1; EtOH 5; eth 5; ctc 3 1.4075^{20}
3480	Butanoic acid, 2-chloro-	$C_4H_7ClO_2$ 122.55	4170-24-5	189^{627}; 101^{15}	4-02-00-00821 1.1796^{20}	H_2O 2; EtOH 4; eth 4 1.441^{20}
3481	Butanoic acid, 3-chloro-, (±)-	$C_4H_7ClO_2$ 122.55	625-68-3 16	116^{22}	4-02-00-00823 1.1898^{20}	EtOH 3; eth 4; ctc 2 1.4221^{20}
3482	Butanoic acid, 4-chloro-	$C_4H_7ClO_2$ 122.55	627-00-9 16	196^{22}; $68^{0.2}$	4-02-00-00825 1.2236^{20}	EtOH 4 1.4642^{20}
3483	Butanoic acid, 2-chloro-, ethyl ester	$C_6H_{11}ClO_2$ 150.60	7425-45-8	163.5	4-02-00-00822 1.056^{18}	eth 4; EtOH 4 1.4180^{25}
3484	Butanoic acid, 3-chloro-, ethyl ester	$C_6H_{11}ClO_2$ 150.60	7425-48-1		4-02-00-00824 1.0517^{20}	EtOH 4 1.4246^{20}
3485	Butanoic acid, 4-chloro-, ethyl ester	$C_6H_{11}ClO_2$ 150.60	3153-36-4	184	4-02-00-00825 1.0756^{20}	ace 4; eth 4; EtOH 4 1.4311^{20}
3486	Butanoic acid, 2-chloro-2-methyl- Butyric acid, 2-chloro-2-methyl-	$C_5H_9ClO_2$ 136.58	73758-54-0	200 dec	3-02-00-00688 1.1204^{20}	eth 4; EtOH 4 1.4445^{20}
3487	Butanoic acid, 2-chloro-3-methyl-	$C_5H_9ClO_2$ 136.58	921-08-4 21	210; 126^{32}	4-02-00-00904 1.135^{13}	H_2O 1; EtOH 3; eth 3 1.4450^{11}
3488	Butanoic acid, 2-chloro-, methyl ester	$C_5H_9ClO_2$ 136.58	26464-32-4	146	3-02-00-00622 1.0979^{14}	eth 4; EtOH 4 1.4247^{20}
3489	Butanoic acid, 4-chloro-, methyl ester	$C_5H_9ClO_2$ 136.58	3153-37-5	175; 55^4	4-02-00-00825 1.1201^{20}	H_2O 1; EtOH 4; eth 4; ace 3 1.4321^{20}
3490	Butanoic acid, 2-chloro-3-methyl-, ethyl ester Ethyl 2-chloro-3-methylbutanoate	$C_7H_{13}ClO_2$ 164.63	91913-99-4	179	0-02-00-00316 1.021^{13}	eth 4; EtOH 4
3491	Butanoic acid, 2-chloro-2-methyl-, ethyl ester Butyric acid, 2-chloro-2-methyl, ethyl ester	$C_7H_{13}ClO_2$ 164.63	58190-94-6	176	3-02-00-00688 1.069^{14}	eth 4; EtOH 4 1.4388^{11}
3492	Butanoic acid, 2-chloro-2-methyl-3-oxo-, ethyl ester Butyric acid, 2-chloro-2-methyl-3-oxo, ethyl ester	$C_7H_{11}ClO_3$ 178.62	37935-39-0	$116-7^{75}$	2-03-00-00433 1.157^{18}	EtOH 4 1.4382^{18}
3493	Butanoic acid, 4-(4-chloro-2-methylphenoxy)- 4-(4-Chloro-2-methylphenoxy)butyric acid	$C_{11}H_{13}ClO_3$ 228.68	94-81-5 100		4-06-00-01996	
3494	Butanoic acid, 4-chloro-3-oxo-, ethyl ester	$C_6H_9ClO_3$ 164.59	638-07-3 -8	220 dec; 115^{14}	4-03-00-01550 1.218^{25}	1.4520^{20}
3495	Butanoic acid, 2-cyano-2-ethyl-, ethyl ester Cyanoacetic acid, diethyl, ethyl ester	$C_9H_{15}NO_2$ 169.22	1619-56-3	214.5	3-02-00-01761	eth 4; EtOH 4 1.4200^{27}
3496	Butanoic acid, cyclohexyl ester	$C_{10}H_{18}O_2$ 170.25	1551-44-6	213	4-06-00-00037 0.9572^0	H_2O 1; EtOH 3; ctc 2; os 3
3497	Butanoic acid, 2,3-dibromo-	$C_4H_6Br_2O_2$ 245.90	600-30-6 87	$100-10^{20}$	4-02-00-00837	H_2O 2; EtOH 4; eth 4; bz 4
3498	Butanoic acid, 2,3-dibromo-, ethyl ester	$C_6H_{10}Br_2O_2$ 273.95	609-11-0 58.5	113^{30}	3-02-00-00634 1.6800^{20}	H_2O 2; EtOH 3; eth 3; ctc 2
3499	Butanoic acid, 2,4-dibromo-, ethyl ester	$C_6H_{10}Br_2O_2$ 273.95	36847-51-5	$149-50^{52}$	4-02-00-00838 1.6987^{20}	H_2O 1; EtOH 3; eth 3 1.4960^{20}
3500	Butanoic acid, 2,2-dichloro- Butyric acid, 2,2-dichloro-	$C_4H_6Cl_2O_2$ 157.00	13023-00-2	$107-10^{14}$	2-02-00-00254 1.389^{20}	H_2O 2; EtOH 4; eth 4
3501	Butanoic acid, 3,4-dichloro-, butyl ester Butyric acid, 3,4-dichloro, butyl ester	$C_8H_{14}Cl_2O_2$ 213.10	116723-94-5	$127-8^{15}$	4-02-00-00829 1.143^{25}	1.454^{24}
3502	Butanoic acid, 3,4-dichloro-, methyl ester	$C_5H_8Cl_2O_2$ 171.02	819-93-2	90^{15}	4-02-00-00829 1.278^{23}	1.459^{23}
3503	Butanoic acid, 2,2-dichloro-3-oxo-, ethyl ester	$C_6H_8Cl_2O_3$ 199.03	6134-66-3	206; 91^{12}	2-03-00-00427 1.291^{16}	EtOH 4 1.4492^{17}
3504	Butanoic acid, 2,2-diethyl-3-oxo-, ethyl ester	$C_{10}H_{18}O_3$ 186.25	1619-57-4	216; 64^3	4-03-00-01625 0.9717^{18}	H_2O 1; EtOH 5; eth 5; ctc 3 1.4326^{17}
3505	Butanoic acid, 2,3-dihydroxypropyl ester	$C_7H_{14}O_4$ 162.19	557-25-5	280; 117^{10}	4-02-00-00798 1.129^{18}	H_2O 4; EtOH 4 1.4531^{20}
3506	Butanoic acid, 2,2-dimethyl-	$C_6H_{12}O_2$ 116.16	595-37-9 -14	186	4-02-00-00954 0.9276^{20}	H_2O 2; EtOH 3; eth 3 1.4145^{20}

No.	Name / Synonym	Mol. Form. / Mol. Wt.	CAS RN / mp/°C	Merck No. / bp/°C	Bell. Ref. / den/g cm^{-3}	Solubility / n_D
3507	Butanoic acid, 2,3-dimethyl- / Butyric acid, 2,3-dimethyl	$C_8H_{12}O_2$ / 116.16	14287-61-7 / -1.5	/ 191.7	4-02-00-00958 / 0.9275[20]	eth 4; EtOH 4 / 1.4146[20]
3508	Butanoic acid, 3,3-dimethyl- / tert-Butylacetic acid	$C_8H_{12}O_2$ / 116.16	1070-83-3 / 6.5	1538 / 190	4-02-00-00955 / 0.9124[20]	EtOH 3; eth 3 / 1.4096[20]
3509	Butanoic acid, 1,1-dimethylethyl ester	$C_8H_{16}O_2$ / 144.21	2308-38-5 /	/ 146	4-02-00-00790 /	ace 4; eth 4; EtOH 4 / 1.4007[17]
3510	Butanoic acid, 3,3-dimethyl-2-oxo-	$C_6H_{10}O_3$ / 130.14	815-17-8 / 90.5	/ 189; 80[15]	4-03-00-01595 /	H_2O 2; eth 3; bz 3; chl 3
3511	Butanoic acid, 2,2-dimethyl-3-oxo-, ethyl ester	$C_8H_{14}O_3$ / 158.20	597-04-6 /	/ 182	4-03-00-01594 / 0.9759[20]	eth 4; EtOH 4 / 1.4180[20]
3512	Butanoic acid, 1,2-ethanediyl ester	$C_{10}H_{18}O_4$ / 202.25	105-72-6 / -80	/ 240	4-02-00-00796 / 1.0005[20]	eth 4; EtOH 4 / 1.4262[20]
3513	Butanoic acid, ethenyl ester / Vinyl butyrate	$C_6H_{10}O_2$ / 114.14	123-20-6 /	/ 116.7; 64[130]	4-02-00-00792 / 0.9006[20]	
3514	Butanoic acid, 2-ethyl- / Diethylacetic acid	$C_6H_{12}O_2$ / 116.16	88-09-5 / -31.8	3099 / 194	4-02-00-00950 / 0.9239[20]	H_2O 2; EtOH 5; eth 5; ctc 2 / 1.4132[20]
3515	Butanoic acid, ethyl ester / Ethyl butanoate	$C_6H_{12}O_2$ / 116.16	105-54-4 / -98	3733 / 121.5	4-02-00-00787 / 0.8844[15]	H_2O 2; EtOH 3; eth 3; ctc 2 / 1.4000[20]
3516	Butanoic acid, 2-ethyl-2-methyl-	$C_7H_{14}O_2$ / 130.19	19889-37-3 / <-20	/ 207	4-02-00-00981 /	EtOH 4 / 1.4250[20]
3517	Butanoic acid, 2-ethyl-, methyl ester	$C_7H_{14}O_2$ / 130.19	816-11-5 /	/ 137	4-02-00-00950 / 0.8886[12]	1.4067[12]
3518	Butanoic acid, 2-ethyl-3-oxo-, ethyl ester	$C_8H_{14}O_3$ / 158.20	607-97-6 /	/ 198.0; 80[10]	4-03-00-01592 / 0.9847[16]	EtOH 5; eth 5 / 1.4214[25]
3519	Butanoic acid, 2-ethyl-3-oxo-, methyl ester	$C_7H_{12}O_3$ / 144.17	51756-08-2 /	/ 182	4-03-00-01592 / 0.995[14]	ace 4; eth 4; EtOH 4
3520	Butanoic acid, 2-furanylmethyl ester	$C_9H_{12}O_3$ / 168.19	623-21-2 /	/ 212.5	4-17-00-01247 / 1.0530[20]	H_2O 2; EtOH 3; eth 5
3521	Butanoic acid, heptafluoro-	$C_4HF_7O_2$ / 214.04	375-22-4 / -17.5	/ 121	4-02-00-00810 / 1.651[20]	H_2O 3; eth 3; tol 3; peth 1 / 1.295[25]
3522	Butanoic acid, heptafluoro-, anhydride	$C_8F_{14}O_3$ / 410.06	336-59-4 / -43	/ 106.5	4-02-00-00817 / 1.665[20]	1.285[20]
3523	Butanoic acid, heptafluoro-, ethenyl ester	$C_6H_3F_7O_2$ / 240.08	356-28-5 /	/ 79	4-02-00-00815 / 1.418[20]	1.309
3524	Butanoic acid, heptafluoro-, ethyl ester	$C_6H_5F_7O_2$ / 242.09	356-27-4 /	/ 95	4-02-00-00813 / 1.394[20]	H_2O 2; eth 3; ace 3 / 1.3011[20]
3525	Butanoic acid, heptafluoro-, methyl ester	$C_5H_3F_7O_2$ / 228.07	356-24-1 / -86	/ 80	4-02-00-00812 / 1.483[20]	H_2O 2; eth 3; ace 3 / 1.295[20]
3526	Butanoic acid, heptafluoro-, 1-methylethyl ester / Isopropyl heptafluorobutyrate	$C_7H_7F_7O_2$ / 256.12	425-23-0 /	/ 106	4-02-00-00813 / 1.324[20]	1.310[20]
3527	Butanoic acid, heptyl ester	$C_{11}H_{22}O_2$ / 186.29	5870-93-9 / -57.5	/ 225.8	4-02-00-00791 / 0.8637[20]	EtOH 4 / 1.4231[20]
3528	Butanoic acid, hexyl ester	$C_{10}H_{20}O_2$ / 172.27	2639-63-6 / -78	/ 208	4-02-00-00791 / 0.8652[20]	H_2O 1; EtOH 3; chl 2; os 3 / 1.4160[15]
3529	Butanoic acid, hydrazide / Butyrhydrazide	$C_4H_{10}N_2O$ / 102.14	3538-65-6 / 45.5	/ 138[20]; 120[10]	4-02-00-00807 /	H_2O 4; eth 4; EtOH 4
3530	Butanoic acid, 2-hydroxy-, (±)-	$C_4H_8O_3$ / 104.11	600-15-7 / 44.2	/ 260 dec; 140[14]	3-03-00-00561 / 1.125[20]	H_2O 3; EtOH 3; eth 3
3531	Butanoic acid, 3-hydroxy-, (±)-	$C_4H_8O_3$ / 104.11	625-71-8 / 49	/ 130[12]; 94[0.1]	4-03-00-00761 /	H_2O 4; EtOH 4; eth 4; bz 1 / 1.4424[20]
3532	Butanoic acid, 2-hydroxy-, ethyl ester, (±)- / Butyric acid, 2-hydroxy, ethyl ester (DL)	$C_6H_{12}O_3$ / 132.16	68057-83-0 /	/ 167	4-03-00-00756 / 1.0069[20]	EtOH 4 / 1.4179[20]
3533	Butanoic acid, 3-hydroxy-, ethyl ester, (±)-	$C_6H_{12}O_3$ / 132.16	35608-64-1 /	/ 185; 76[15]	4-03-00-00762 / 1.017[20]	H_2O 3; EtOH 3; ctc 2 / 1.4182[20]
3534	Butanoic acid, 3-hydroxy-, ethyl ester	$C_6H_{12}O_3$ / 132.16	5405-41-4 /	/ 185; 76[15]	4-03-00-00762 / 1.017[20]	H_2O 4; EtOH 4 / 1.4182[20]
3535	Butanoic acid, 3-hydroxy-3-methyl-	$C_5H_{10}O_3$ / 118.13	625-08-1 / <-32	/ 162[12]	4-03-00-00827 / 0.9384[20]	H_2O 4; eth 4; EtOH 4 / 1.5081[20]
3536	Butanoic acid, 2-hydroxy-, 2-methylpropyl ester / Butyric acid, 2-hydroxy, isobutyl ester	$C_8H_{16}O_3$ / 160.21	116723-95-6 /	/ 196	0-03-00-00302 / 0.944[15]	1.4182
3537	Butanoic acid, 3-mercapto-	$C_4H_8O_2S$ / 120.17	26473-49-4 /	/ 110[10]; 88[2.5]	4-03-00-00769 / 1.1371[20]	1.4782[20]
3538	Butanoic acid, 2-methyl-, (±)- / (±)-2-Methylbutyric acid	$C_5H_{10}O_2$ / 102.13	600-07-7 / <-80	/ 177	4-02-00-00889 / 0.934[20]	H_2O 2; EtOH 5; eth 5; chl 3 / 1.4051[20]
3539	Butanoic acid, 3-methyl- / Isovaleric acid	$C_5H_{10}O_2$ / 102.13	503-74-2 / -29.3	5120 / 176.5	4-02-00-00895 / 0.931[20]	H_2O 3; EtOH 5; eth 5; chl 5 / 1.4033[20]
3540	Butanoic acid, 2-methyl-, 10-(acetyloxy)-9,10-dihydro-8,8-dimethyl-2-oxo-2H,8H-benzo[1,2-b:3,4-b']dipyran-9-yl / Visnadine	$C_{21}H_{24}O_7$ / 388.42	477-32-7 / 85.5	9915 /	5-19-06-00272 /	H_2O 1; EtOH 3; eth 3
3541	Butanoic acid, 3-(methylamino)-, ethyl ester / Butyric acid, 3-methylamino-, ethyl ester	$C_7H_{15}NO_2$ / 145.20	68384-70-3 /	/ 72[12.5]	3-04-00-01313 / 0.9282[20]	1.4250[20]
3542	Butanoic acid, 3-methyl-, anhydride	$C_{10}H_{18}O_3$ / 186.25	1468-39-9 /	/ 215	4-02-00-00901 / 0.9327[20]	eth 4 / 1.4043[20]
3543	Butanoic acid, 3-methyl-, butyl ester / Butyl isovalerate	$C_9H_{18}O_2$ / 158.24	109-19-3 /	/	3-02-00-00698 /	1.4058[25]

No.	Name Synonym	Mol. Form. Mol. Wt.	CAS RN mp/°C	Merck No. bp/°C	Beil. Ref. den/g cm^{-3}	Solubility n_D
3544	Butanoic acid, 3-methylbutyl ester Isopentyl butanoate	$C_9H_{18}O_2$ 158.24	106-27-4	4997 179	4-02-00-00791 0.865[19]	H_2O 1; EtOH 4; eth 4 1.4110[20]
3545	Butanoic acid, 2-methyl-, butyl ester Butyric acid, (2-methylbutyl) ester	$C_9H_{18}O_2$ 158.24	15706-73-7	179	0-02-00-00304 0.8620[20]	1.4135[20]
3546	Butanoic acid, 3-methyl-, 1,1-dimethylpropyl ester tert-Amyl isovalerate	$C_{10}H_{20}O_2$ 172.27	542-37-0	647 173.5	0-02-00-00312 0.8729[0]	EtOH 4
3547	Butanoic acid, methyl ester Methyl butanoate	$C_5H_{10}O_2$ 102.13	623-42-7 -85.8	5957 102.8	4-02-00-00786 0.8984[20]	H_2O 2; EtOH 5; eth 5; ctc 2 1.3878[20]
3548	Butanoic acid, 3-methyl-, ethyl ester Ethyl isovalerate	$C_7H_{14}O_2$ 130.19	108-64-5 -99.3	3772 135.0	4-02-00-00898 0.8656[20]	EtOH 4; eth 4 1.3962[20]
3549	Butanoic acid, 1-methylethyl ester Isopropyl butyrate	$C_7H_{14}O_2$ 130.19	638-11-9	130.5	4-02-00-00789 0.8588[20]	H_2O 1; EtOH 3 1.3936[20]
3550	Butanoic acid, 2-methyl-, ethyl ester, (S)-	$C_7H_{14}O_2$ 130.19	10307-61-6	133; 35[16]	4-02-00-00890 0.8689[25]	H_2O 1; EtOH 3; bz 3 1.3964[20]
3551	Butanoic acid, 3-methyl-, 3-methylbutyl ester Isoamyl isovalerate	$C_{10}H_{20}O_2$ 172.27	659-70-1	5003 190.4	4-02-00-00899 0.8583[19]	1.4130[19]
3552	Butanoic acid, 3-methyl-, methyl ester Methyl isovalerate	$C_6H_{12}O_2$ 116.16	556-24-1	6006 116.5	4-02-00-00897 0.8808[20]	H_2O 1; EtOH 4; eth 4; ace 4 1.3927[20]
3553	Butanoic acid, 3-methyl-, 1-methylethyl ester Butyric acid, 3-methyl, isopropyl ester	$C_8H_{16}O_2$ 144.21	32665-23-9	142; 70[55]	3-02-00-00698 0.8538[17]	ace 4; eth 4; EtOH 4 1.3960[20]
3554	Butanoic acid, 3-methyl-, 5-methyl-2-(1-methylethyl)cyclohexyl ester	$C_{15}H_{28}O_2$ 240.39	16409-46-4	129[9]	2-06-00-00043 0.9089[15]	H_2O 1; EtOH 3; ace 3 1.4486[20]
3555	Butanoic acid, 3-methyl-, 2-methylpropyl ester Isobutyl isovalerate	$C_9H_{18}O_2$ 158.24	589-59-3	5029 168.5	4-02-00-00898 0.853[20]	H_2O 1; EtOH 5; eth 5; ace 4 1.4057[20]
3556	Butanoic acid, 3-methyl-2-oxo-	$C_5H_8O_3$ 116.12	759-05-7 31.5	170.5	4-03-00-01577 0.9968[20]	H_2O 3; EtOH 3; eth 3 1.3850[16]
3557	Butanoic acid, 2-methyl-3-oxo-, ethyl ester	$C_7H_{12}O_3$ 144.17	609-14-3	187	4-03-00-01573 0.9941[20]	H_2O 2; EtOH 3; eth 3; ace 4 1.4185[20]
3558	Butanoic acid, 2-methyl-3-oxo-, methyl ester	$C_6H_{10}O_3$ 130.14	17094-21-2	177.4	3-03-00-01225 1.0217[25]	eth 4; EtOH 4 1.416[24]
3559	Butanoic acid, 1-methylpentyl ester, (S)- Butyric acid (1-methylpentyl) ester (d)	$C_{10}H_{20}O_2$ 172.27	116723-96-7	85[20]	1-02-00-00120 0.8744[21]	eth 4; EtOH 4
3560	Butanoic acid, 3-methyl-, phenylmethyl ester	$C_{12}H_{16}O_2$ 192.26	103-38-8	245; 136[25]	4-06-00-02266 0.9983[15]	1.4884[20]
3561	Butanoic acid, 3-methyl-, 1,2,3-propanetriyl ester	$C_{18}H_{32}O_6$ 344.45	620-63-3	332.5	4-02-00-00900 0.9984[20]	eth 4; EtOH 4 1.4354[20]
3562	Butanoic acid, 3-methyl-, propyl ester Propyl isovalerate	$C_8H_{16}O_2$ 144.21	557-00-6	155.9	4-02-00-00898 0.8617[20]	eth 4; EtOH 4 1.4031[20]
3563	Butanoic acid, 2-methylpropyl ester Isobutyl butanoate	$C_8H_{16}O_2$ 144.21	539-90-2	5020 156.9	4-02-00-00790 0.8364[18]	H_2O 2; EtOH 5; eth 5 1.4032[20]
3564	Butanoic acid, 1-methylpropyl ester, (S)- Butyric acid, sec-butyl ester (d)	$C_8H_{16}O_2$ 144.21	116836-55-6 -91.5	152; 54[18]	3-02-00-00600 0.8737[13]	EtOH 4; bz 3; CS_2 3; py 3 1.4011[20]
3565	Butanoic acid, 3-methyl-, 1,7,7-trimethylbicyclo[2.2.1]hept-2-yl ester, (1R-endo)- d-Bornyl isovalerate	$C_{15}H_{26}O_2$ 238.37	53022-14-3	1342 257.5	2-06-00-08488 0.955[25]	eth 4; EtOH 4
3566	Butanoic acid, octyl ester	$C_{12}H_{24}O_2$ 200.32	110-39-4 -55.6	244.1	4-02-00-00791 0.8629[20]	EtOH 4 1.4267[15]
3567	Butanoic acid, 2-oxo-	$C_4H_6O_3$ 102.09	600-18-0 33	80-2[16]	4-03-00-01524 1.200[17]	H_2O 4; EtOH 4; eth 2 1.3972[20]
3568	Butanoic acid, 3-oxo- Acetoacetic acid	$C_4H_6O_3$ 102.09	541-50-4 36.5	51 100 dec	4-03-00-01527	H_2O 4; eth 4; EtOH 4
3569	Butanoic acid, 3-oxo-, butyl ester	$C_8H_{14}O_3$ 158.20	591-60-6 -35.6	127[50]; 85[8]	4-03-00-01536 0.9671[25]	H_2O 2; EtOH 5; bz 5; lig 5 1.4137[20]
3570	Butanoic acid, 3-oxo-, 2-chloroethyl ester	$C_6H_9ClO_3$ 164.59	54527-68-3	198; 107[14]	4-03-00-01534 1.190[20]	bz 4; eth 4; EtOH 4 1.4430[20]
3571	Butanoic acid, 3-oxo-, ethyl ester Ethyl acetoacetate	$C_6H_{10}O_3$ 130.14	141-97-9 -45	3714 180.8	4-03-00-01528 1.0368[10]	H_2O 4; EtOH 5; eth 5; bz 3 1.4171[20]
3572	Butanoic acid, 4-oxo-, ethyl ester	$C_6H_{10}O_3$ 130.14	10138-10-0	162; 85[12]	4-03-00-01555 1.0490[22]	chl 2 1.4287[20]
3573	Butanoic acid, 3-oxo-, methyl ester Methyl acetoacetate	$C_5H_8O_3$ 116.12	105-45-3 27.5	5933 171.7	4-03-00-01527 1.0762[20]	H_2O 4; EtOH 5; eth 5; ctc 3 1.4184[20]
3574	Butanoic acid, 3-oxo-, 1-methylethyl ester Isopropyl acetoacetate	$C_7H_{12}O_3$ 144.17	542-08-5 -27.3	5094 186	4-03-00-01535 0.9835[20]	eth 4; EtOH 4; lig 4 1.4173[20]
3575	Butanoic acid, 4-oxo-4-(phenylamino)- Succinanilic acid	$C_{10}H_{11}NO_3$ 193.20	102-14-7 148.5	8839	4-12-00-00468	H_2O 2; EtOH 3; eth 4
3576	Butanoic acid, 3-oxo-2-(phenylmethylene)-, ethyl ester	$C_{13}H_{14}O_3$ 218.25	620-80-4 60.5	296; 180[17]	3-10-00-03158	H_2O 1; EtOH 2; eth 2; bz 2
3577	Butanoic acid, 3-oxo-, 2-propenyl ester	$C_7H_{10}O_3$ 142.15	1118-84-9 -85	196; 66.5[14]	4-03-00-01538 1.0366[20]	H_2O 3; EtOH 5; bz 5; lig 3 1.4398[20]
3578	Butanoic acid, 4-oxo-4-[[4-[(2-thiazolylamino)sulfonyl]phenyl]amino]- Succinylsulphathiazole	$C_{13}H_{13}N_3O_5S_2$ 355.40	116-43-8 193.5	8850	4-27-00-04637	H_2O 1; EtOH 2; eth 1; ace 2
3579	Butanoic acid, pentyl ester Amyl butyrate	$C_9H_{18}O_2$ 158.24	540-18-1 -73.2	639 186.4	4-02-00-00790 0.8713[15]	H_2O 1; EtOH 4; eth 4 1.4123[20]
3580	Butanoic acid, 2-phenoxy-	$C_{10}H_{12}O_3$ 180.20	13794-14-4 98	258	3-06-00-00616	H_2O 2; EtOH 3; eth 3; ace 3

No.	Name / Synonym	Mol. Form. / Mol. Wt.	CAS RN / mp/°C	Merck No. / bp/°C	Bell. Ref. / den/g cm^{-3}	Solubility / n_D
3581	Butanoic acid, 4-phenoxy-, ethyl ester / Ethyl 4-phenoxybutyrate	$C_{12}H_{16}O_3$ / 208.26	2364-59-2 /	/ 170-3^{25}	2-06-00-00159 / 1.042^{35}	eth 4; EtOH 4 / 1.491^{35}
3582	Butanoic acid, 2-phenoxyethyl ester	$C_{12}H_{16}O_3$ / 208.26	23511-70-8 /	/ 251; 88^{2}	3-06-00-00571 / 1.0388^{21}	ace 4; eth 4; EtOH 4
3583	Butanoic acid, 2-phenoxy-, phenyl ester / 2-Phenoxybutyric acid, phenyl ester	$C_{16}H_{16}O_3$ / 256.30	116836-19-2 / 48.5	/ 202-3^{25}	4-06-00-00645 / 1.135^{15}	
3584	Butanoic acid, phenyl ester / Phenyl butyrate	$C_{10}H_{12}O_2$ / 164.20	4346-18-3 /	/ 225	4-06-00-00615 / 1.0382^{15}	H$_2$O 1; EtOH 3; eth 3
3585	Butanoic acid, phenylmethyl ester	$C_{11}H_{14}O_2$ / 178.23	103-37-7 /	/ 239	4-06-00-02266 / 1.0111^{20}	H$_2$O 1; EtOH 4; eth 4; ctc 3 / 1.4920^{20}
3586	Butanoic acid, 1,2,3-propanetriyl ester / Tributyrin	$C_{15}H_{26}O_6$ / 302.37	60-01-5 / -75	9532 / 307.5	4-02-00-00799 / 1.0350^{20}	H$_2$O 1; EtOH 3; eth 4; ace 3 / 1.4359^{20}
3587	Butanoic acid, 2-propenyl ester	$C_7H_{12}O_2$ / 128.17	2051-78-7 /	/ 142; 44.5^{15}	4-02-00-00793 / 0.9017^{20}	H$_2$O 1; EtOH 5; eth 5; ctc 2 / 1.4158^{20}
3588	Butanoic acid, propyl ester / Propyl butanoate	$C_7H_{14}O_2$ / 130.19	105-66-8 / -95.2	7858 / 143.0	4-02-00-00788 / 0.8730^{20}	H$_2$O 2; EtOH 5; eth 5 / 1.4001^{20}
3589	Butanoic acid, 3-thioxo-, ethyl ester / Butyric acid, 3-thioxo, ethyl ester	$C_6H_{10}O_2S$ / 146.21	7740-33-2 /	/ 75^{15}	4-03-00-01553 / 1.0554^{31}	eth 4; EtOH 4 / 1.4712^{26}
3590	Butanoic acid, 2,2,3-trichloro- / 2,2,3-Trichlorobutyric acid	$C_4H_5Cl_3O_2$ / 191.44	5344-55-8 / 60	/ 237	3-02-00-00629 /	eth 4
3591	Butanoic acid, 2,2,3-trichloro-, ethyl ester / Butyric acid, 2,2,3-trichloro, ethyl ester	$C_6H_9Cl_3O_2$ / 219.49	116723-97-8 /	/ 212	0-02-00-00281 / 1.3114^{20}	eth 4; EtOH 4
3592	Butanoic acid, 4,4,4-trifluoro-3-oxo-, ethyl ester	$C_6H_7F_3O_3$ / 184.11	372-31-6 /	/ 133	4-03-00-01548 / 1.2586^{15}	EtOH 3; eth 3 / 1.3783^{15}
3593	1-Butanol / Butyl alcohol	$C_4H_{10}O$ / 74.12	71-36-3 / -89.8	1540 / 117.7	4-01-00-01506 / 0.8098^{20}	H$_2$O 3; EtOH 5; eth 5; ace 4 / 1.3993^{20}
3594	2-Butanol / sec-Butyl alcohol	$C_4H_{10}O$ / 74.12	78-92-2 / -114.7	1541 / 99.5	2-01-00-00400 / 0.8063^{20}	H$_2$O 4; EtOH 5; eth 5; ace 4 / 1.3978^{20}
3595	1-Butanol, 2-amino-, (±)-	$C_4H_{11}NO$ / 89.14	13054-87-0 / -2	/ 178	4-04-00-01705 / 0.9162^{20}	H$_2$O 5; EtOH 5; eth 5; chl 2 / 1.4489^{25}
3596	1-Butanol, 3-amino-	$C_4H_{11}NO$ / 89.14	2867-59-6 /	/ 82-5^{19}	4-04-00-01710 /	H$_2$O 4; EtOH 4 / 1.4534^{25}
3597	1-Butanol, 4-amino-	$C_4H_{11}NO$ / 89.14	13325-10-5 /	/ 205; 125^{34}	4-04-00-01711 / 0.967^{12}	H$_2$O 3; EtOH 3; eth 1 / 1.4625^{20}
3598	2-Butanol, 3-amino-	$C_4H_{11}NO$ / 89.14	42551-55-3 / 19	/ 160; 79^{20}	3-04-00-00779 / 0.9399^{20}	H$_2$O 4; eth 4; EtOH 4 / 1.4445^{25}
3599	1-Butanol, 2-amino-3-methyl-, (±)-	$C_5H_{13}NO$ / 103.18	16369-05-4 /	/ 75-7^{8}	4-04-00-01777 / 0.93^{25}	1.4543^{20}
3600	1-Butanol, 3-bromo-	C_4H_9BrO / 153.02	6089-12-9 /	/ 69-71^{20}	4-01-00-01553 / 1.7107^{18}	1.5010^{21}
3601	2-Butanol, 3-bromo-, (R*,S*)-(±)- / 2-Butanol, 3-bromo-, erythro-(±)-	C_4H_9BrO / 153.02	19246 39-0 /	/ 154; 53^{13}	4-01-00-01580 / 1.4550^{20}	eth 4; EtOH 4 / 1.4780^{20}
3602	1-Butanol, 2-chloro-	C_4H_9ClO / 108.57	26106-95-6 /	/ 74-6^{25}	4-01-00-01549 / 1.062^{25}	eth 4; EtOH 4 / 1.4438^{20}
3603	1-Butanol, 3-chloro-	C_6H_9ClO / 120.58	2203-35-2 /	/ 175; 66^{15}	3-01-00-01517 / 1.0883^{20}	EtOH 3; eth 3; ace 3; ctc 2 / 1.4518^{20}
3604	1-Butanol, 4-chloro-	C_4H_9ClO / 108.57	928-51-8 /	/ 84-5^{16}	4-01-00-01550 / 1.0883^{20}	eth 4; EtOH 4 / 1.4518^{20}
3605	2-Butanol, 1-chloro- / α-Butylene chlorohydrin	C_4H_9ClO / 108.57	1873-25-2 /	/ 141	4-01-00-01578 / 1.068^{25}	EtOH 3; eth 3 / 1.4400^{20}
3606	2-Butanol, 3-chloro- / β-Butylene chlorohydrin	C_4H_9ClO / 108.57	563-84-8 /	/ 139	3-01-00-01538 / 1.0669^{20}	EtOH 4; eth 4 / 1.4432^{20}
3607	2-Butanol, 4-chloro-	C_4H_9ClO / 108.57	2203-34-1 /	/ 67^{20}	4-01-00-01578 /	eth 4; EtOH 4 / 1.4408^{20}
3608	2-Butanol, 1-chloro-2,3-dimethyl- / 1-Chloro-2,3-dimethyl-2-butanol	$C_6H_{13}ClO$ / 136.62	66235-62-9 /	/ 163	4-01-00-01730 / 1.049^{20}	1.4459^{20}
3609	2-Butanol, 1-chloro-3,3-dimethyl- / 1-Chloro-3,3-dimethyl-2-butanol	$C_6H_{13}ClO$ / 136.62	36402-31-0 / 13	/ 157.5	4-01-00-01728 / 1.0063^{25}	1.4432^{25}
3610	2-Butanol, 1-chloro-2-methyl-	$C_5H_{11}ClO$ / 122.59	74283-48-0 /	/ 151	4-01-00-01673 / 1.0161^{20}	EtOH 4 / 1.4469^{20}
3611	2-Butanol, 3-chloro-2-methyl-	$C_5H_{11}ClO$ / 122.59	21326-62-5 /	/ 141.5	3-01-00-01627 / 1.0295^{20}	eth 4; EtOH 4 / 1.4436^{20}
3612	1-Butanol, 2,3-dibromo- / 2,3-Dibromo-1-butanol	$C_4H_8Br_2O$ / 231.91	4021-75-4 / 32	/ 99.5^{10}	4-01-00-01554 / 1.9475^{20}	1.5442^{20}
3613	1-Butanol, 3,4-dibromo- / 3,4-Dibromobutanol	$C_4H_8Br_2O$ / 231.91	87018-30-2 /	/ 114^{11}	3-01-00-01518 / 1.98^{15}	1.548^{15}
3614	2-Butanol, 1,4-dibromo- / 2-Hydroxy-1,4-dibromobutane	$C_4H_8Br_2O$ / 231.91	19398-47-1 /	/ 114-5^{15}	4-01-00-01581 / 1.953^{20}	1.544^{20}
3615	1-Butanol, 2,2-dimethyl- / 2,2-Dimethyl-1-butanol	$C_6H_{14}O$ / 102.18	1185-33-7 / <-15	/ 136.5	4-01-00-01726 / 0.8283^{20}	H$_2$O 2; EtOH 3; eth 3 / 1.4208^{20}
3616	1-Butanol, 2,3-dimethyl-, (±)-	$C_6H_{14}O$ / 102.18	20281-85-0 /	/ 144.5	4-01-00-01729 / 0.8297^{20}	EtOH 3; eth 3; ace 3 / 1.4195^{20}
3617	1-Butanol, 3,3-dimethyl-	$C_6H_{14}O$ / 102.18	624-95-3 / -60	/ 143	4-01-00-01729 / 0.844^{15}	H$_2$O 2; EtOH 3; eth 3; ace 3 / 1.4323^{15}
3618	2-Butanol, 2,3-dimethyl-	$C_6H_{14}O$ / 102.18	594-60-5 / -14	/ 118.4	4-01-00-01729 / 0.8236^{20}	H$_2$O 3; EtOH 5; eth 5 / 1.4176^{20}
3619	2-Butanol, 3,3-dimethyl-, (±)- / 3,3-Dimethylbutan-2-ol (DL-)	$C_6H_{14}O$ / 102.18	20281-91-8 / 5.6	/ 120.4	4-01-00-01727 / 0.8122^{25}	H$_2$O 2; EtOH 3; eth 5 / 1.4148^{20}
3620	1-Butanol, 2-ethyl- / 2-Ethyl-1-butanol	$C_6H_{14}O$ / 102.18	97-95-0 / <-15	/ 147	4-01-00-01725 / 0.8326^{20}	H$_2$O 2; EtOH 3; eth 3; chl 3 / 1.4220^{20}

No.	Name Synonym	Mol. Form. Mol. Wt.	CAS RN mp/°C	Merck No. bp/°C	Beil. Ref. den/g cm^{-3}	Solubility n_D
3621	1-Butanol, 4-fluoro- 4-Fluoro-1-butanol	C$_4$H$_9$FO 92.11	372-93-0	58[15]	4-01-00-01546	ace 4; eth 4; EtOH 4 1.3942[15]
3622	1-Butanol, 2,2,3,3,4,4,4-heptafluoro-	C$_4$H$_3$F$_7$O 200.06	375-01-9	95	4-01-00-01547 1.600[20]	EtOH 3; ace 3 1.294[20]
3623	2-Butanol, 1-[(2-hydroxyethyl)amino]- 2-Hydroxyethyl 2-hydroxybutyl amine	C$_6$H$_{15}$NO$_2$ 133.19	6967-43-7	137[9]	4-04-00-01725 1.0310[20]	H$_2$O 4; ace 4; EtOH 4 1.4690[30]
3624	2-Butanol, 3,3'-iminobis- Bis-(3-hydroxy-2-butyl)amine	C$_8$H$_{19}$NO$_2$ 161.24	6959-06-4	112-5[3]	3-04-00-00781 0.9775[20]	H$_2$O 4; ace 4; EtOH 4 1.4162[20]
3625	1-Butanol, 3-methoxy- 3-Methoxy-1-butanol	C$_5$H$_{12}$O$_2$ 104.15	2517-43-3	157	4-01-00-02509 0.923[23]	EtOH 4; eth 3; ace 4; chl 2 1.4148[25]
3626	1-Butanol, 2-methyl- 2-Methyl-1-butanol	C$_5$H$_{12}$O 88.15	137-32-6 128	5952	4-01-00-01666 0.8150[25]	H$_2$O 2; EtOH 5; eth 5; ace 4 1.4092[20]
3627	1-Butanol, 3-methyl- Isopentyl alcohol	C$_5$H$_{12}$O 88.15	123-51-3 -117.2	5081 131.1	4-01-00-01677 0.8104[20]	ace 4; eth 4; EtOH 4 1.4053[20]
3628	2-Butanol, 2-methyl- tert-Amyl alcohol	C$_5$H$_{12}$O 88.15	75-85-4 -8.8	7096 102.4	4-01-00-01668 0.8096[20]	H$_2$O 3; EtOH 5; eth 5; ace 4 1.4052[20]
3629	2-Butanol, 3-methyl-, (±)- (±)-3-Methyl-2-butanol	C$_5$H$_{12}$O 88.15	70116-68-6 112.9	5953	4-01-00-01675 0.8180[20]	H$_2$O 2; EtOH 5; eth 5; ace 4 1.4089[20]
3630	1-Butanol, 2-methyl-, acetate	C$_7$H$_{14}$O$_2$ 130.19	624-41-9 140		4-02-00-00156 0.8740[20]	ace 4; eth 4; EtOH 4 1.4040[20]
3631	1-Butanol, 3-methyl-, acetate Isopentyl acetate	C$_7$H$_{14}$O$_2$ 130.19	123-92-2 -78.5	4993 142.5	4-02-00-00157 0.876[15]	H$_2$O 2; EtOH 5; eth 5; ace 3 1.4000[20]
3632	1-Butanol, 3-methyl-, benzoate Isoamyl benzoate	C$_{12}$H$_{16}$O$_2$ 192.26	94-46-2 261	4995	4-09-00-00292 0.993[15]	EtOH 4
3633	1-Butanol, 3-methyl-, carbamate	C$_6$H$_{13}$NO$_2$ 131.17	543-86-2 59	220	4-03-00-00058 0.9438[71]	eth 4; EtOH 4 1.4175[20]
3634	2-Butanol, 2-methyl-, carbamate tert-Amyl carbamate	C$_6$H$_{13}$NO$_2$ 131.17	590-60-3 86	641	1-03-00-00014	ace 4; bz 4
3635	1-Butanol, 3-methyl-, carbonate (2:1)	C$_{11}$H$_{22}$O$_3$ 202.29	2050-95-5	234; 122[16]	3-03-00-00011 0.9067[20]	H$_2$O 1 1.4174[20]
3636	1-Butanol, 3-methyl-, formate Isopentyl formate	C$_6$H$_{12}$O$_2$ 116.16	110-45-2 -93.5	5001 123.5	4-02-00-00030 0.877[20]	H$_2$O 2; EtOH 3; eth 5; ctc 2 1.3967[20]
3637	1-Butanol, 3-methyl-, nitrate Isoamyl nitrate	C$_5$H$_{11}$NO$_3$ 133.15	543-87-3 148	5004	4-01-00-01683 0.996[22]	1.4122[21]
3638	1-Butanol, 3-methyl-2-nitro-	C$_5$H$_{11}$NO$_3$ 133.15	77392-54-2	111[10]	3-01-00-01645 1.0889[25]	1.4430[20]
3639	1-Butanol, 3-methyl-, propanoate Isopentyl propanoate	C$_8$H$_{16}$O$_2$ 144.21	105-68-0 160.2		4-02-00-00709 0.8697[20]	eth 4; EtOH 4 1.4069[20]
3640	1-Butanol, 2-nitro- 2-Nitro-1-butanol	C$_4$H$_9$NO$_3$ 119.12	609-31-4 -47	105[10]	4-01-00-01555 1.1332[25]	H$_2$O 3; EtOH 5; eth 5; ace 3 1.4390[20]
3641	2-Butanol, 1-nitro- 2-Hydroxy-1-nitrobutane	C$_4$H$_9$NO$_3$ 119.12	3156-74-9	204; 75[2]	4-01-00-01582 1.1353[20]	H$_2$O 3; EtOH 4; eth 4; ace 3 1.4435[20]
3642	2-Butanol, 3-nitro- 3-Nitro-2-butanol	C$_4$H$_9$NO$_3$ 119.12	6270-16-2	112[38]; 55[0.5]	4-01-00-01583 1.1166[7]	1.4414[20]
3643	2-Butanol, 1,1,1-trichloro- Ethyl(trichloromethyl)carbinol	C$_4$H$_7$Cl$_3$O 177.46	6111-61-1	171; 83[22]	4-01-00-01579 1.3630[25]	ace 4; bz 4; eth 4; EtOH 4 1.4800[20]
3644	2-Butanol, 2,3,3-trimethyl-	C$_7$H$_{16}$O 116.20	594-83-2 17	131	4-01-00-01755 0.8380[25]	ace 4; eth 4; EtOH 4 1.4233[22]
3645	2-Butanone Methyl ethyl ketone	C$_4$H$_8$O 72.11	78-93-3 -86.65	5991 79.59	4-01-00-03243 0.7999[25]	H$_2$O 4; EtOH 5; eth 5; ace 5 1.3788[20]
3646	2-Butanone, 1-chloro- Chloromethyl ethyl ketone	C$_4$H$_7$ClO 106.55	616-27-3 137.5		4-01-00-03255 1.0850[20]	MeOH 4 1.4372[20]
3647	2-Butanone, 3-chloro-	C$_4$H$_7$ClO 106.55	4091-39-8 115		4-01-00-03256 1.0554[25]	1.4219[20]
3648	2-Butanone, 4-chloro- 1-Chloro-3-butanone	C$_4$H$_7$ClO 106.55	6322-49-2 120 dec		4-01-00-03256 1.0680[23]	eth 4; EtOH 4 1.4284[23]
3649	2-Butanone, 3-chloro-3-methyl- 3-Chloro-3-methyl-2-butanone	C$_5$H$_9$ClO 120.58	5950-19-6 117.2		3-01-00-02818 1.0083[20]	eth 4; EtOH 4 1.4204[20]
3650	1-Butanone, 4-chloro-1-phenyl-	C$_{10}$H$_{11}$ClO 182.65	939-52-6 19.5	130-3[4]	4-07-00-00711 1.137[25]	1.5459[20]
3651	1-Butanone, 4-[4-(4-chlorophenyl)-4-hydroxy- 1-piperidinyl]-1-(4-fluorophenyl)- Haloperidol	C$_{21}$H$_{23}$ClFNO$_2$ 375.87	52-86-8 151.5	4511	5-21-02-00377	
3652	1-Butanone, 1-cyclohexyl-	C$_{10}$H$_{18}$O 154.25	1462-27-7	94[13]	4-07-00-00080 0.903[20]	1.4537[20]
3653	2-Butanone, 3,4-dibromo- 1,2-Dibromoethyl methyl ketone	C$_4$H$_6$Br$_2$O 229.90	25109-57-3	80-1[10]	3-01-00-02788 1.9662[15]	1.5314[15]
3654	2-Butanone, 1,3-dichloro- 1,3-Dichloro-2-butanone	C$_4$H$_6$Cl$_2$O 141.00	16714-77-5 166.5		3-01-00-02786 1.3116[20]	ace 4; bz 4; eth 4; EtOH 4 1.4686[20]
3655	2-Butanone, 3,4-dichloro-	C$_4$H$_6$Cl$_2$O 141.00	58625-77-7	65-70[16]	3-01-00-02787 1.293[18]	1.4628[18]
3656	2-Butanone, 4-(diethylamino)-	C$_8$H$_{17}$NO 143.23	3299-38-5	84[30]	4-04-00-01930 0.8630[20]	ace 4; bz 4; eth 4; EtOH 4 1.4333[24]
3657	2-Butanone, 3,3-dimethyl- Pinacolone	C$_6$H$_{12}$O 100.16	75-97-8 -52.5	7409 106.1	4-01-00-03310 0.7229[25]	H$_2$O 3; EtOH 3; eth 3; ace 3 1.3952[20]
3658	2-Butanone, 3,3-diphenyl- 2,2-Diphenyl-3-butanone	C$_{16}$H$_{16}$O 224.30	2575-20-4 41	310.5	4-07-00-01451 1.069[20]	H$_2$O 1; EtOH 3; eth 4; chl 4 1.5748[20]
3659	2-Butanone, 4-(2-furanyl)-	C$_8$H$_{10}$O$_2$ 138.17	699-17-2	203	4-17-00-04559 1.0361[19]	1.4696[17]
3660	2-Butanone, 1-hydroxy-	C$_4$H$_8$O$_2$ 88.11	5077-67-8	160; 78[60]	4-01-00-03989 1.0272[20]	H$_2$O 4; EtOH 4; eth 4 1.4189[20]

No.	Name / Synonym	Mol. Form. / Mol. Wt.	CAS RN / mp/°C	Merck No. / bp/°C	Beil. Ref. / den/g cm^{-3}	Solubility / n_D
3661	2-Butanone, 3-hydroxy- / Acetoin	$C_4H_8O_2$ / 88.11	513-86-0 / 15	55 / 143	2-01-00-00870 / 1.0044[20]	H_2O 4; ace 4; EtOH 4 / 1.4171[20]
3662	2-Butanone, 4-hydroxy-	$C_4H_8O_2$ / 88.11	590-90-9 /	/ 109[30]; 90[11]	4-01-00-03994 / 1.1089[24]	H_2O 5; EtOH 5; eth 5; ace 4 / 1.4585[14]
3663	2-Butanone, 4-(4-hydroxy-3-methoxyphenyl)- / Zingerone	$C_{11}H_{14}O_3$ / 194.23	122-48-5 / 40.5	10072 / 187-8[14]	4-08-00-01866 /	eth 4
3664	2-Butanone, 4-hydroxy-3-methyl-	$C_5H_{10}O_2$ / 102.13	3393-64-4 /	/ 90-5[15]	4-01-00-04009 / 0.993[25]	1.4340[20]
3665	1-Butanone, 1-(2-hydroxyphenyl)-	$C_{10}H_{12}O_2$ / 164.20	2887-61-8 / 10	/ 124-6[14]	4-08-00-00497 / 1.0683[24]	1.5379[25]
3666	1-Butanone, 2-hydroxy-1-phenyl- / Butyrophenone, 2-hydroxy-	$C_{10}H_{12}O_2$ / 164.20	16183-46-3 /	/ 132-3[12]	4-08-00-00502 / 1.077[19]	1.529[19]
3667	1-Butanone, 4-hydroxy-1-phenyl-	$C_{10}H_{12}O_2$ / 164.20	39755-03-8 / 30	/ 187-8[9]	3-08-00-00449 /	EtOH 4
3668	2-Butanone, 3-methyl- / Methyl isopropyl ketone	$C_5H_{10}O$ / 86.13	563-80-4 / -92	/ 94.3	4-01-00-03287 / 0.8051[20]	H_2O 2; EtOH 5; eth 5; ace 4 / 1.3880[20]
3669	1-Butanone, 3-methyl-1-(2-methylphenyl)- / Isobutyl 2-tolyl ketone	$C_{12}H_{16}O$ / 176.26	58138-81-1 /	/ 248; 85-6[3]	4-07-00-00765 / 0.9578[20]	1.5104[20]
3670	1-Butanone, 3-methyl-1-(4-methylphenyl)- / Isobutyl 4-tolyl ketone	$C_{12}H_{16}O$ / 176.26	61971-91-3 /	/ 259; 137[18]	4-07-00-00765 / 0.9557[20]	1.5081[20]
3671	1-Butanone, 3-methyl-1-phenyl-	$C_{11}H_{14}O$ / 162.23	582-62-7 /	/ 236.5	4-07-00-00739 / 0.9701[16]	H_2O 1; EtOH 5; eth 5; ace 4 / 1.5139[15]
3672	1-Butanone, 1-(4-methylphenyl)- / Butyrophenone, 4-methyl-	$C_{11}H_{14}O$ / 162.23	4160-52-5 / 12	/ 251.5	4-07-00-00743 / 0.9745[20]	eth 4; EtOH 4 / 1.5232[20]
3673	2-Butanone, (1-methylpropylidene)hydrazone	$C_8H_{18}N_2$ / 140.23	5921-54-0 /	/ 171.5	4-01-00-03252 / 0.8404[20]	1.4511[20]
3674	1-Butanone, 1-(1-naphthalenyl)- / 1-Butyronaphthone	$C_{14}H_{14}O$ / 198.26	2876-62-2 /	/ 317	4-07-00-01305 / 1.0861[25]	bz 4; eth 4; EtOH 4 / 1.596[27]
3675	2-Butanone, oxime	C_4H_9NO / 87.12	96-29-7 / -29.5	/ 152.5	4-01-00-03250 / 0.9232[20]	H_2O 3; EtOH 5; eth 5; chl 3 / 1.4410[20]
3676	1-Butanone, 1-phenyl-	$C_{10}H_{12}O$ / 148.20	495-40-9 / 12	/ 228.5	4-07-00-00709 / 0.988[20]	H_2O 1; EtOH 5; eth 5; ace 4 / 1.5203[20]
3677	2-Butanone, 1-phenyl-	$C_{10}H_{12}O$ / 148.20	1007-32-5 /	/ 230; 111[16]	4-07-00-00712 / 1.002[20]	H_2O 1; EtOH 3; eth 5; ace 4
3678	2-Butanone, 4-phenyl-	$C_{10}H_{12}O$ / 148.20	2550-26-7 / -13	/ 233.5	4-07-00-00713 / 0.9849[22]	H_2O 1; EtOH 3; eth 3; ace 4 / 1.511[22]
3679	2-Butanone, 1,3,4-trihydroxy-, (S)- / L-Erythrulose	$C_4H_8O_4$ / 120.11	533-50-6 / dec	3641 / dec	4-01-00-04176 /	H_2O 4; EtOH 4
3680	Butanoyl bromide / Butyryl bromide	C_4H_7BrO / 151.00	5856-82-6 /	/ 129	4-02-00-00804 / 1.4162[17]	1.596[17]
3681	Butanoyl chloride / Butyryl chloride	C_4H_7ClO / 106.55	141-75-3 / -89	1598 / 102	4-02-00-00803 / 1.0277[20]	eth 5 / 1.4121[20]
3682	Butanoyl chloride, 2-bromo-	C_4H_6BrClO / 185.45	22118-12-3 /	/ 151	4-02-00-00834 / 1.5320[20]	eth 4 / 1.5320[20]
3683	Butanoyl chloride, 2-chloro-	$C_4H_6Cl_2O$ / 141.00	7623-11-2 /	/ 130.5	3-02-00-00623 / 1.2360[17]	eth 4 / 1.4475[20]
3684	Butanoyl chloride, 3-chloro- / 3-Chlorobutyryl chloride	$C_4H_6Cl_2O$ / 141.00	1951-11-7 /	/ 40-1[12]	4-02-00-00824 / 1.2163[20]	1.4509[20]
3685	Butanoyl chloride, 4-chloro-	$C_4H_6Cl_2O$ / 141.00	4635-59-0 /	/ 173.5	4-02-00-00826 / 1.2581[20]	eth 3 / 1.4616[20]
3686	Butanoyl chloride, 2-chloro-3-methyl- / Butyryl chloride, 2-chloro-3-methyl	$C_5H_8Cl_2O$ / 155.02	35383-59-6 /	/ 148.5; 56[20]	4-02-00-00904 / 1.135[13]	eth 4
3687	Butanoyl chloride, 4-chloro-3-oxo- / Butyryl chloride, 4-chloro-3-oxo	$C_4H_4Cl_2O_2$ / 154.98	41295-64-1 /	/ 117-9[17]	3-03-00-01207 / 1.4397[20]	bz 4 / 1.4860[20]
3688	Butanoyl chloride, 2,2-dimethyl-	$C_6H_{11}ClO$ / 134.61	5856-77-9 /	/ 132	4-02-00-00955 / 0.9810[20]	eth 4 / 1.4245[20]
3689	Butanoyl chloride, 2,3-dimethyl- / Butyryl chloride, 2,3-dimethyl	$C_6H_{11}ClO$ / 134.61	51760-90-8 /	/ 136; 38[18]	3-02-00-00761 / 0.9795[20]	eth 4
3690	Butanoyl chloride, 3,3-dimethyl-	$C_6H_{11}ClO$ / 134.61	7065-46-5 /	/ 130; 68[100]	4-02-00-00956 / 0.9696[20]	eth 4 / 1.4210[20]
3691	Butanoyl chloride, 2-ethyl-	$C_6H_{11}ClO$ / 134.61	2736-40-5 /	/ 140	4-02-00-00952 / 0.9825[20]	eth 4 / 1.4234[20]
3692	Butanoyl chloride, heptafluoro-	C_4ClF_7O / 232.48	375-16-6 /	/ 38.5	4-02-00-00818 / 1.55[20]	1.288[20]
3693	Butanoyl chloride, 3-methyl- / Isovaleryl chloride	C_5H_9ClO / 120.58	108-12-3 /	5121 / 114	4-02-00-00901 / 0.9844[20]	eth 3 / 1.4149[20]
3694	Butanoyl chloride, 2-methyl-, (±)- / Butyryl chloride, 2-methyl (DL)	C_5H_9ClO / 120.58	57526-28-0 /	/ 116	4-02-00-00891 / 0.9917[20]	1.4170[20]
3695	2-Butenal, (E)- / trans-Crotonaldehyde	C_4H_6O / 70.09	123-73-9 / -76	2599 / 102.2	4-01-00-03447 / 0.8516[20]	H_2O 3; EtOH 4; eth 4; ace 4 / 1.4366[20]
3696	2-Butenal, 2-bromo-	C_4H_5BrO / 148.99	24247-53-8 /	/ 63[15]	3-01-00-02981 / 1.5797[13]	1.5184[20]
3697	2-Butenal, 2-chloro- / α-Chlorocrotonaldehyde	C_4H_5ClO / 104.54	53175-28-3 /	/ 147.5	3-01-00-02981 / 1.1404[23]	H_2O 2; EtOH 3; eth 3; ctc 3 / 1.4780[25]
3698	2-Butenal, 2-methyl-	C_5H_8O / 84.12	1115-11-3 /	/ 117; 64[110]	3-01-00-02990 / 0.8710[20]	H_2O 3; EtOH 5; eth 5 / 1.4475[20]
3699	2-Butenal, 2-methyl-, (E)-	C_5H_8O / 84.12	497-03-0 /	/ 117; 64[119]	4-01-00-03464 / 0.8710[20]	H_2O 4; EtOH 4 / 1.4475[20]
3700	2-Butenal, 3-methyl- / Senecialdehyde	C_5H_8O / 84.12	107-86-8 /	8399 / 134	4-01-00-03464 / 0.8722[20]	H_2O 3; EtOH 3; eth 3 / 1.4528[20]

No.	Name Synonym	Mol. Form. Mol. Wt.	CAS RN mp/°C	Merck No. bp/°C	Beil. Ref. den/g cm^{-3}	Solubility n_D
3701	2-Butenal, 3-phenyl-	$C_{10}H_{10}O$ 146.19	1196-67-4 95	124[12]; 109[4]	4-07-00-01011	1.5876[23]
3702	2-Butenamide, (E)-	C_4H_7NO 85.11	625-37-6 161.5	sub 140	2-02-00-00392 0.9461[120]	H_2O 2; EtOH 3; eth 2; bz 3 1.4420[165]
3703	3-Butenamide, 3-methyl-4-phenyl- β-Benzalbutyramide	$C_{11}H_{13}NO$ 175.23	7236-47-7 133	1063		
3704	1-Butene 1-Butylene	C_4H_8 56.11	106-98-9 -185.3	1513 -6.2	4-01-00-00765 *0.588[25]	H_2O 1; EtOH 4; eth 4; bz 3 1.3962[20]
3705	2-Butene, (E)- trans-2-Butene	C_4H_8 56.11	624-64-6 -105.5	1514 0.8	4-01-00-00781 *0.599[25]	bz 3 1.3848[-25]
3706	2-Butene, (Z)- cis-2-Butene	C_4H_8 56.11	590-18-1 -138.9	1514 3.7	4-01-00-00778 *0.616[25]	H_2O 1; EtOH 4; eth 4; bz 3 1.3931[-25]
3707	1-Butene, 1-bromo-, (E)- trans-1-Butenyl bromide	C_4H_7Br 135.00	32620-08-9 -100.3	94.7	1.3209[15]	H_2O 1; eth 3; ace 3; bz 3 1.4527[20]
3708	1-Butene, 1-bromo-, (Z)-	C_4H_7Br 135.00	31849-78-2	86.1	1.3265[15]	H_2O 1; eth 3; ace 3; bz 3 1.4536[20]
3709	1-Butene, 2-bromo- 2-Bromo-1-butene	C_4H_7Br 135.00	23074-36-4 -133.4	88	4-01-00-00775 1.3209[15]	H_2O 1; eth 3; ace 3; bz 3 1.4527[20]
3710	1-Butene, 3-bromo- α-Methallyl bromide	C_4H_7Br 135.00	22037-73-6	7[10]	4-01-00-00775 1.2998[25]	1.4618[20]
3711	1-Butene, 4-bromo-	C_4H_7Br 135.00	5162-44-7	98.5	4-01-00-00775 1.3230[20]	bz 4; eth 4; EtOH 4 1.4622[20]
3712	2-Butene, 1-bromo-	C_4H_7Br 135.00	4784-77-4	104.5	4-01-00-00789 1.3371[25]	H_2O 1; EtOH 3; eth 3; bz 4 1.4822[20]
3713	2-Butene, 2-bromo-, (E)-	C_4H_7Br 135.00	3017-71-8 -114.6	85.6	4-01-00-00790 1.3323[15]	H_2O 1; EtOH 3; eth 3; bz 4 1.4602[16]
3714	2-Butene, 2-bromo-, (Z)-	C_4H_7Br 135.00	3017-68-3 -111.5	93.9	4-01-00-00790 1.3416[15]	H_2O 1; EtOH 3; eth 3; bz 4 1.4631[19]
3715	1-Butene, 4-bromo-3-chloro-3,4,4-trifluoro- 4-Bromo-3-chloro-3,4,4-trifluoro-1-butene	$C_4H_3BrClF_3$ 223.42	374-25-4	99.5	4-01-00-00776 1.678[25]	1.4092[25]
3716	1-Butene, 2-bromo-3-methyl- 2-Bromo-3-methyl-1-butene	C_5H_9Br 149.03	31844-96-9	105	3-01-00-00800 1.2328[20]	bz 4; eth 4; chl 4 1.4504[20]
3717	2-Butene, 1-bromo-3-methyl-	C_5H_9Br 149.03	870-63-3	131 dec; 50[40]	4-01-00-00824 1.2817[20]	ace 4; bz 4; eth 4; EtOH 4 1.4930[15]
3718	2-Butene, 2-bromo-3-methyl-	C_5H_9Br 149.03	3017-70-7	119.5	4-01-00-00824 1.2773[20]	ace 4; eth 4; chl 4 1.4738[20]
3719	2-Butene, 2-bromo-3-phenyl- 2-Bromo-3-phenyl-2-butene	$C_{10}H_{11}Br$ 211.10	90841-14-8	120-30[11]	4-05-00-01379 1.3348[20]	bz 4; eth 4; chl 4 1.5811[20]
3720	1-Butene, 1-chloro-, (E)- trans-1-Chloro-1-butene	C_4H_7Cl 90.55	7611-87-2	68	3-01-00-00723 0.9205[15]	H_2O 1; EtOH 5; eth 4; ace 3 1.4223[15]
3721	1-Butene, 1-chloro-, (Z)- cis-1-Chloro-1-butene	C_4H_7Cl 90.55	7611-86-1	63.5	3-01-00-00723 0.9153[15]	ace 4; eth 4; EtOH 4; chl 4 1.4194[15]
3722	1-Butene, 2-chloro- 2-Chloro-1-butene	C_4H_7Cl 90.55	2211-70-3	58.5	4-01-00-00769 0.9107[15]	ace 4; bz 4; eth 4; EtOH 4 1.4165[21]
3723	1-Butene, 3-chloro- 3-Chloro-1-butene	C_4H_7Cl 90.55	563-52-0	2131 64.5	4-01-00-00769 0.8978[20]	eth 4; ace 4; chl 3 1.4149[20]
3724	1-Butene, 4-chloro-	C_4H_7Cl 90.55	927-73-1	75	4-01-00-00771 0.9211[20]	ace 4; eth 4; chl 4 1.4233[20]
3725	2-Butene, 1-chloro-, (E)-	C_4H_7Cl 90.55	4894-61-5	85	4-01-00-00784 0.9295[20]	H_2O 1; ace 3; chl 3 1.4350[20]
3726	2-Butene, 1-chloro-, (Z)-	C_4H_7Cl 90.55	4628-21-1	84.1	4-01-00-00783 0.9426[20]	H_2O 1; EtOH 3; ace 3; chl 3 1.4390[20]
3727	2-Butene, 2-chloro-, (E)-	C_4H_7Cl 90.55	2211-68-9 -105.8	62.8	4-01-00-00785 0.9138[20]	H_2O 1; EtOH 5; ace 3; chl 3 1.4190[20]
3728	2-Butene, 2-chloro-, (Z)-	C_4H_7Cl 90.55	2211-69-0 -117.3	70.6	4-01-00-00785 0.9239[20]	H_2O 1; EtOH 5; ace 3; chl 3 1.4240[20]
3729	1-Butene, 3-chloro-2-(chloromethyl)- 3-Chloro-2-(chloromethyl)-1-butene	$C_5H_8Cl_2$ 139.02	69295-21-2	155	3-01-00-00787 1.1233[20]	ace 4; chl 4 1.4724[20]
3730	2-Butene, 2-chloro-1,1-dimethoxy- Crotonaldehyde-2-chloro, dimethylacetal	$C_6H_{11}ClO_2$ 150.60	108365-83-9	58[13]	3-01-00-02981 1.074[18]	1.4466[18]
3731	2-Butene, 1-chloro-2,3-dimethyl- 1-Chloro-2,3-dimethyl-2-butene	$C_6H_{11}Cl$ 118.61	37866-06-1	112	4-01-00-00856 0.9355[20]	ace 4; eth 4; chl 4 1.4605[20]
3732	1-Butene, 1-chloro-2-methyl- 1-Chloro-2-methyl-1-butene	C_5H_9Cl 104.58	23378-11-2	96.5	3-01-00-00787 0.9170[20]	ace 4; eth 4 1.4141[20]
3733	1-Butene, 1-chloro-3-methyl- 1-Chloro-3-methyl-1-butene	C_5H_9Cl 104.58	23010-00-6	87	3-01-00-00799 	ace 4; eth 4; chl 4 1.4229[20]
3734	1-Butene, 3-chloro-2-methyl- 3-Chloro-2-methyl-1-butene	C_5H_9Cl 104.58	5166-35-8	94	4-01-00-00819 0.9088[20]	ace 4; eth 4; chl 4 1.4304[20]
3735	2-Butene, 1-chloro-2-methyl- 1-Chloro-2-methyl-2-butene	C_5H_9Cl 104.58	13417-43-1	110	4-01-00-00822 0.9327[20]	ace 4; eth 4; EtOH 4 1.4481[20]
3736	2-Butene, 1-chloro-3-methyl-	C_5H_9Cl 104.58	503-60-6	109	4-01-00-00823 0.9273[20]	ace 4; eth 4; EtOH 4; chl 4 1.4485[20]
3737	2-Butene, 2-chloro-3-methyl- 2-Chloro-3-methyl-2-butene	C_5H_9Cl 104.58	17773-65-8	96	3-01-00-00794 0.9324[20]	ace 4; eth 4; EtOH 4; chl 4 1.4320[20]
3738	1-Butene, 1,1-dibromo- 1,1-Dibromo-1-butene	$C_4H_6Br_2$ 213.90	73383-24-1	53-5[22]	3-01-00-00727 1.8315[20]	1.5168[20]
3739	1-Butene, 1,2-dibromo- 1,2-dibromo-1-butene	$C_4H_6Br_2$ 213.90	55030-56-3 -49.5	150	0-01-00-00204 1.887[25]	eth 4; EtOH 4
3740	1-Butene, 3,4-dibromo- 1,2-Dibromo-3-butene	$C_4H_6Br_2$ 213.90	10463-48-6	52[10]	4-01-00-00776 1.865[24]	1.541[21]

No.	Name Synonym	Mol. Form. Mol. Wt.	CAS RN mp/°C	Merck No. bp/°C	Beil. Ref. den/g cm^{-3}	Solubility n_D
3741	2-Butene, 1,3-dibromo- 2,4-Dibromo-2-butene	$C_4H_6Br_2$ 213.90	64930-16-1	168 dec	3-01-00-00746 1.8768[20]	1.5485[20]
3742	2-Butene, 1,4-dibromo-, (E)-	$C_4H_6Br_2$ 213.90	821-06-7 53.4	203; 74[14]	4-01-00-00791	H_2O 2; EtOH 4; ace 3; chl 2
3743	1-Butene, 1,3-dichloro- 1,3-Dichloro-1-butene	$C_4H_6Cl_2$ 125.00	52497-07-1	125	4-01-00-00772 1.1341[24]	ace 4; eth 4; EtOH 4; chl 4 1.4647[20]
3744	1-Butene, 2,3-dichloro- 2,3-Dichloro-1-butene	$C_4H_6Cl_2$ 125.00	7013-11-8	112	4-01-00-00772 1.1340[20]	ace 4; eth 4; chl 4 1.4580[20]
3745	1-Butene, 3,4-dichloro- 1,2-Dichloro-3-butene	$C_4H_6Cl_2$ 125.00	760-23-6 -61	116	4-01-00-00772 1.1170[20]	H_2O 1; EtOH 3; eth 3; bz 4 1.4641[20]
3746	2-Butene, 1,1-dichloro- 1,1-Dichloro-2-butene	$C_4H_6Cl_2$ 125.00	56800-09-0	126	3-01-00-00741 1.1310[20]	1.466[18]
3747	2-Butene, 1,2-dichloro- 1,2-Dichloro-2-butene	$C_4H_6Cl_2$ 125.00	13602-13-6	130.5	3-01-00-00741 1.1601[20]	ace 4; eth 4; chl 4 1.4734[20]
3748	2-Butene, 1,3-dichloro-, (E)-	$C_4H_6Cl_2$ 125.00	7415-31-8	131; 53[50]	4-01-00-00786 1.1585[20]	ace 4; bz 4; eth 4; EtOH 4 1.4719[20]
3749	2-Butene, 1,3-dichloro-, (Z)-	$C_4H_6Cl_2$ 125.00	10075-38-4	131; 34[20]	4-01-00-00786 1.1605[20]	ace 4; bz 4; eth 4; EtOH 4 1.4735[20]
3750	2-Butene, 1,4-dichloro-, (E)- 1,4-Dichloro-trans-2-butene	$C_4H_6Cl_2$ 125.00	110-57-6 2.0	155.4	4-01-00-00787 1.183[25]	ace 4; bz 4; eth 4; EtOH 4 1.4871[25]
3751	2-Butene, 1,4-dichloro-, (Z)-	$C_4H_6Cl_2$ 125.00	1476-11-5 -48	152.5	4-01-00-00787 1.188[25]	ace 4; bz 4; eth 4; EtOH 4 1.4887[25]
3752	2-Butene, 2,3-dichloro-, (E)-	$C_4H_6Cl_2$ 125.00	1587-29-7	102	4-01-00-00787 1.1416[20]	H_2O 1; EtOH 3; eth 3; ace 3 1.4582[20]
3753	2-Butene, 2,3-dichloro-, (Z)-	$C_4H_6Cl_2$ 125.00	1587-28-4	125.5	4-01-00-00787 1.1618[20]	H_2O 1; EtOH 3; eth 3; ace 3 1.4590[20]
3754	2-Butene, 2,3-dichloro-1,1,1,4,4,4-hexafluoro- 2,3-Dichlorohexafluoro-2-butene	$C_4Cl_2F_6$ 232.94	303-04-8 -67.3	68.5	4-01-00-00787 1.6233[20]	1.3459[20]
3755	1-Butene, 3,3-dichloro-2-methyl- 3,3-Dichloro-2-methyl-1-butene	$C_5H_8Cl_2$ 139.02	42101-38-2		3-01-00-00787 1.1276[18]	ace 4; eth 4; chl 4 1.4737[18]
3756	2-Butene, 1,3-dichloro-2-methyl- 1,3-Dichloro-2-methyl-2-butene	$C_5H_8Cl_2$ 139.02	25148-87-2	150	3-01-00-00795 1.1293[20]	ace 4; chl 4
3757	2-Butene, 1,4-dichloro-2-methyl- 1,4-Dichloro-2-methyl-2-butene	$C_5H_8Cl_2$ 139.02	29843-58-1	93[50]; 56[10]	3-01-00-00795 1.1526[20]	ace 4; chl 4 1.4932[20]
3758	2-Butene, 1,1-diethoxy- Crotonaldehyde, diethyl acetal	$C_8H_{16}O_2$ 144.21	10602-34-3	147.5	3-01-00-02980 0.8473[18]	H_2O 1; EtOH 5; eth 5; ace 5 1.4097[20]
3759	1-Butene, 2,3-dimethyl- 2,3-Dimethyl-1-butene	C_6H_{12} 84.16	563-78-0 -157.3	55.6	4-01-00-00852 0.6803[20]	H_2O 1; EtOH 3; eth 3; ace 3 1.3995[20]
3760	1-Butene, 3,3-dimethyl-	C_6H_{12} 84.16	558-37-2 -115.2	41.2	4-01-00-00850 0.6529[20]	H_2O 1; EtOH 3; eth 3; ctc 3 1.3763[20]
3761	2-Butene, 2,3-dimethyl- 2,3-Dimethyl-2-butene	C_6H_{12} 84.16	563-79-1 -74.6	73.3	4-01-00-00853 0.7080[20]	H_2O 1; EtOH 3; eth 3; ace 3 1.4122[20]
3762	2-Butenedinitrile, (E)-	$C_4H_2N_2$ 78.07	764-42-1 96.8	186	4-02-00-02219 0.9416[111]	H_2O 3; EtOH 3; eth 3; ace 3 1.4349[111]
3763	2-Butenedinitrile, 2-chloro-, (Z)- Fumaronitrile, chloro-	C_4HClN_2 112.52	71200-79-8 60.5	172; 68[25]	3-02-00-01909 1.2499[20]	1.4957[20]
3764	2-Butenedioic acid (E)- Fumaric acid	$C_4H_4O_4$ 116.07	110-17-8 287 dec	4200 sub 165	3-02-00-01890 1.635[20]	H_2O 2; EtOH 3; eth 2; ace 2
3765	2-Butenedioic acid (Z)- Maleic acid	$C_4H_4O_4$ 116.07	110-16-7 130.5	5585	4-02-00-02199 1.590[20]	H_2O 4; EtOH 4; eth 3; ace 4
3766	2-Butenedioic acid (E)-, bis(1-methylethyl) ester Diisopropyl fumarate	$C_{10}H_{16}O_4$ 200.23 2.1	7283-70-7	280	4-02-00-02209	ace 4; eth 4; EtOH 4
3767	2-Butenedioic acid (E)-, bis(2-methylpropyl) ester Diisobutyl fumarate	$C_{12}H_{20}O_4$ 228.29	7283-69-4	170[160]; 122[5]	4-02-00-02211 0.9760[20]	H_2O 1; EtOH 3; eth 3; ace 3 1.4432[20]
3768	2-Butenedioic acid (E)-, bis(phenylmethyl) ester Benzyl fumarate	$C_{18}H_{16}O_4$ 296.32	538-64-7 59	1149 210-11[5]	4-06-00-02275	eth 4; EtOH 4; chl 4
3769	2-Butenedioic acid, 2-bromo-, (E)-	$C_4H_3BrO_4$ 194.97	584-99-6 141	dec	4-02-00-02224	H_2O 4; EtOH 4; eth 4
3770	2-Butenedioic acid, 2-bromo-, (Z)- 2-Bromofumaric acid	$C_4H_3BrO_4$ 194.97	644-80-4 184.3	200 dec	4-02-00-02224	H_2O 4; EtOH 4
3771	2-Butenedioic acid, 2-chloro-, diethyl ester, (E)- Diethyl chloromaleate	$C_8H_{11}ClO_4$ 206.63	626-10-8	235 dec; 125[19]	3-02-00-01928 1.1741[20]	ace 4; eth 4; EtOH 4
3772	2-Butenedioic acid, 2-chloro-, diethyl ester, (Z)-	$C_8H_{11}ClO_4$ 206.63	10302-94-0	250 dec; 127[10]	3-02-00-01909 1.1880[20]	eth 4; EtOH 4 1.4571[20]
3773	2-Butenedioic acid, 2-chloro-, dimethyl ester, (E)- Dimethyl chloromaleate	$C_6H_7ClO_4$ 178.57	19393-45-4	224; 100[17]	3-02-00-01928 1.2899[25]	eth 4; EtOH 4 1.4720[18]
3774	2-Butenedioic acid (E)-, dibutyl ester	$C_{12}H_{20}O_4$ 228.29	105-75-9	150[4]	4-02-00-02210 0.9869[20]	H_2O 1; ace 3; chl 3 1.4469[20]
3775	2-Butenedioic acid (E)-, diethyl ester	$C_8H_{12}O_4$ 172.18 1.5	623-91-6	214	4-02-00-02207 1.0452[20]	H_2O 1; ace 3; chl 3 1.4412[20]
3776	2-Butenedioic acid (Z)-, diethyl ester Diethyl maleate	$C_8H_{12}O_4$ 172.18 -8.8	141-05-9	3113 223	4-02-00-02207 1.0662[20]	H_2O 1; EtOH 3; eth 3; chl 2 1.4416[20]

No.	Name / Synonym	Mol. Form. / Mol. Wt.	CAS RN / mp/°C	Merck No. / bp/°C	Beil. Ref. / den/g cm^{-3}	Solubility / n_D
3777	2-Butenedioic acid, 2,3-dihydroxy-, (Z)- Dihydroxymaleic acid	$C_4H_4O_6$ 148.07	526-84-1 155 dec	3170	4-03-00-01975	H_2O 2; EtOH 3; eth 2; MeOH 2
3778	2-Butenedioic acid (E)-, dimethyl ester	$C_6H_8O_4$ 144.13	624-49-7 103.5	193	4-02-00-02205 1.37[20]	H_2O 1; ace 3; chl 3 1.4062[111]
3779	2-Butenedioic acid (Z)-, dimethyl ester	$C_6H_8O_4$ 144.13	624-48-6 -19	202	4-02-00-02204 1.1606[20]	H_2O 1; eth 3; ctc 3; lig 1 1.4416[20]
3780	2-Butenedioic acid (Z)-, diphenyl ester	$C_{16}H_{12}O_4$ 268.27	7242-17-3 73	226[15]	4-06-00-00628	ace 4; bz 4; eth 4; EtOH 4
3781	2-Butenedioic acid (E)-, di-2-propenyl ester Diallyl fumarate	$C_{10}H_{12}O_4$ 196.20	2807-54-7 	140[3]	3-02-00-01906 1.060[25]	ace 4; bz 4; eth 4; EtOH 4 1.4670[25]
3782	2-Butenedioic acid (Z)-, di-2-propenyl ester Diallyl maleate	$C_{10}H_{12}O_4$ 196.20	999-21-3 	108-10[3]	4-02-00-02214 1.0773[20]	chl 3 1.4699[20]
3783	2-Butenedioic acid (E)-, dipropyl ester Dipropyl fumarate	$C_{10}H_{16}O_4$ 200.23	14595-35-8 	110[5]	4-02-00-02209 1.0129[20]	EtOH 3; eth 3 1.4435[20]
3784	2-Butenedioic acid (Z)-, dipropyl ester	$C_{10}H_{16}O_4$ 200.23	2432-63-5 	126[12]	4-02-00-02209 1.0245[20]	H_2O 1; EtOH 3; eth 3; ace 3 1.4434[20]
3785	2-Butenedioic acid, 2-methyl-, (E)- Mesaconic acid	$C_5H_6O_4$ 130.10	498-24-8 204.5	5806 sub	4-02-00-02231 1.466[20]	H_2O 2; EtOH 4; eth 3; bz 2
3786	2-Butenedioic acid, 2-methyl-, (Z)- Citraconic acid	$C_5H_6O_4$ 130.10	498-23-7 93.5	2323	4-02-00-02230 1.617[25]	H_2O 4; eth 2; bz 1; chl 2
3787	2-Butenedioic acid, 2-methyl-, diethyl ester, (E)- Diethyl mesaconate	$C_9H_{14}O_4$ 186.21	2418-31-7 	229	4-02-00-02232 1.0434[20]	ace 4; bz 4; eth 4; EtOH 4 1.4488[20]
3788	2-Butenedioic acid, 2-methyl-, diethyl ester, (Z)-	$C_9H_{14}O_4$ 186.21	691-83-8 	228; 120[20]	4-02-00-02232 1.0491[20]	eth 4; EtOH 4; HOAc 4 1.4467[20]
3789	2-Butenedioic acid, 2-methyl-, dimethyl ester, (E)-	$C_7H_{10}O_4$ 158.15	617-53-8 	203.5	4-02-00-02232 1.0914[20]	ace 4; eth 4; EtOH 4 1.4512[20]
3790	2-Butenedioic acid, 2-methyl-, dimethyl ester, (Z)-	$C_7H_{10}O_4$ 158.15	617-54-9 	210.5	4-02-00-02232 1.1153[20]	ace 4; eth 4; EtOH 4 1.4473[20]
3791	2-Butenedioic acid (E)-, monoethyl ester	$C_6H_8O_4$ 144.13	2459-05-4 70	147[16]	4-02-00-02206 1.1109[87]	EtOH 3; ace 3; chl 2
3792	2-Butene-1,4-diol, (E)- trans-2-Butene-1,4-diol	$C_4H_8O_2$ 88.11	821-11-4 25	131[13]	4-01-00-02660 1.0700[20]	H_2O 4; EtOH 4 1.4755[20]
3793	2-Butene-1,4-diol, (Z)- cis-2-Butene-1,4-diol	$C_4H_8O_2$ 88.11	6117-80-2 4	235	4-01-00-02660 1.0698[20]	H_2O 3; EtOH 4 1.4782[20]
3794	3-Butene-1,2-diol	$C_4H_8O_2$ 88.11	497-06-3 	196.5	4-01-00-02658 1.0470[20]	H_2O 3; EtOH 3 1.4628[21]
3795	2-Butenedioyl dichloride, (E)- Fumaric acid dichloride	$C_4H_2Cl_2O_2$ 152.96	627-63-4 	159	4-02-00-02217 1.408[20]	1.5004[18]
3796	2-Butenedioyl dichloride, 2-chloro-, (Z)-	$C_4HCl_3O_2$ 187.41	17096-37-6 	185 dec	3-02-00-01909 1.564[20]	eth 4; HOAc 4 1.5206[20]
3797	1-Butene, 2-ethyl-3-methyl- 2-Ethyl-3-methylbut-1-ene	C_7H_{14} 98.19	7357-93-9 	89	4-01-00-00871 0.7150[20]	H_2O 1; eth 3; ace 3; bz 3 1.410[20]
3798	2-Butene, 1,1,2,3,4,4-hexachloro- 2-Butene, 1,1,2,3,4,4-hexachloro (liquid)	$C_4H_2Cl_6$ 262.78	13045-99-3 -19	97-8[10]	4-01-00-00789 1.650[15]	H_2O 1; chl 3 1.5331
3799	1-Butene, 2-methyl- 2-Methyl-1-butene	C_5H_{10} 70.13	563-46-2 -137.5	31.2	4-01-00-00818 0.6504[20]	H_2O 1; EtOH 3; eth 3; bz 3 1.3778[20]
3800	1-Butene, 3-methyl- 3-Methyl-1-butene	C_5H_{10} 70.13	563-45-1 -168.5	20.1	4-01-00-00825 0.6213[25]	H_2O 1; EtOH 5; eth 5; bz 3 1.3643[20]
3801	2-Butene, 2-methyl- 2-Methyl-2-butene	C_5H_{10} 70.13	513-35-9 -133.7	644 38.5	4-01-00-00820 0.6623[20]	H_2O 1; EtOH 3; eth 3; bz 3 1.3874[20]
3802	2-Butenenitrile, (E)-	C_4H_5N 67.09	627-26-9 -51.5	120	4-02-00-01507 0.8239[20]	eth 3; ace 3 1.4225[20]
3803	3-Butenenitrile Allyl cyanide	C_4H_5N 67.09	109-75-1 -87	288 119	4-02-00-01491 0.8341[20]	H_2O 2; EtOH 5; eth 5 1.4060[20]
3804	2-Butene, 1,1,1,2,3,4,4,4-octafluoro- Perfluoro-2-butene	C_4F_8 200.03	360-89-4 -129	1.5	4-01-00-00783 1.5297[25]	
3805	2-Butene, 1,1,1,4,4-pentachloro- 1,1,1,4,4-Pentachloro-2-butene	$C_4H_3Cl_5$ 228.33	77753-21-0 	78-80[11]	3-01-00-00744 1.609[21]	EtOH 4; chl 4 1.5538[21]
3806	1-Butene, 1,3,4,4-tetrachloro- 1,3,4,4-Tetrachloro-1-butene	$C_4H_4Cl_4$ 193.89	2984-40-9 	88[20]	4-01-00-00774 1.476[20]	chl 4 1.4773[20]
3807	1-Butene, 2,3,3,4-tetrachloro- 2,3,3,4-Tetrachloro-1-butene	$C_4H_4Cl_4$ 193.89	64346-45-8 	41-2[7]	3-01-00-00726 1.4602[20]	ace 4; chl 4 1.5135[20]
3808	1-Butene, 1,3,4,4-tetrachloro-1,2,3,4-tetrafluoro- 1-Butene, 1,3,4,4-tetrachloro-tetrafluoro	$C_4Cl_4F_4$ 265.85	357-20-0 	140.8	4-01-00-00774 1.6902[25]	1.429[20]
3809	2-Butene, 2,3,4-trichloro- 2,3,4-Trichloro-1-butene	$C_4H_5Cl_3$ 159.44	2431-50-7 	60[20]	3-01-00-00725 1.3430[20]	ace 4; chl 4 1.4944[20]
3810	2-Butene, 1,2,4-trichloro- 1,2,4-Trichloro-2-butene	$C_4H_5Cl_3$ 159.44	2431-54-1 	67-9[10]	3-01-00-00744 1.3843[20]	bz 4; eth 4; EtOH 4; chl 4 1.5175[20]
3811	2-Butene, 1,1,3-triethoxy- 2-Butenal, 3-ethoxy, diethyl acetal	$C_{10}H_{20}O_3$ 188.27	69190-65-4 	192.5	4-01-00-04082 0.908[21]	EtOH 4 1.430[21]
3812	1-Butene, 2,3,3-trimethyl-	C_7H_{14} 98.19	594-56-9 -109.9	77.9	4-01-00-00873 0.7050[20]	H_2O 1; eth 3; bz 3; chl 3 1.4025[20]
3813	3-Butenoic acid	$C_4H_6O_2$ 86.09	625-38-7 -35	169	4-02-00-01491 1.0091[20]	H_2O 3; EtOH 5; eth 5 1.4239[20]

No.	Name / Synonym	Mol. Form. / Mol. Wt.	CAS RN mp/°C	Merck No. bp/°C	Beil. Ref. den/g cm^{-3}	Solubility n_D
3814	2-Butenoic acid, (E)- / trans-Crotonic acid	$C_4H_6O_2$ / 86.09	107-93-7 / 72	2600 / 184.7	4-02-00-01498 / 0.9604[77]	H$_2$O 4; EtOH 4; eth 3; ace 3 / 1.4249[77]
3815	2-Butenoic acid, (Z)- / cis-Crotonic acid	$C_4H_6O_2$ / 86.09	503-64-0 / 15	5048 / 169	4-02-00-01497 / 1.0267[20]	H$_2$O 4; EtOH 3 / 1.4483[14]
3816	2-Butenoic acid, 3-amino-, ethyl ester, (E)-	$C_6H_{11}NO_2$ / 129.16	41867-20-3 / 34	/ 212 dec	3-03-00-01199 / 1.0219[19]	H$_2$O 1; EtOH 3; eth 3; bz 3 / 1.4988[22]
3817	2-Butenoic acid, 4-amino-4-oxo-, (Z)- / Maleamic acid	$C_4H_5NO_3$ / 115.09	557-24-4 / 172.5	5583	4-02-00-02218	H$_2$O 4; EtOH 4
3818	2-Butenoic acid, anhydride / Crotonic acid anhydride	$C_8H_{10}O_3$ / 154.17	623-68-7	/ 247; 129[19]	4-02-00-01505 / 1.0397[20]	eth 4 / 1.4745[20]
3819	2-Butenoic acid, 4-bromo-, ethyl ester, (E)-	$C_6H_9BrO_2$ / 193.04	37746-78-4	/ 97-8[15]	4-02-00-01517 / 1.402[16]	EtOH 4 / 1.4925[20]
3820	2-Butenoic acid, 3-chloro-, (E)-	$C_4H_5ClO_2$ / 120.54	6214-28-4 / 61	/ 195	4-02-00-01510 / 1.1969[69]	H$_2$O 2; EtOH 4; chl 3; CS$_2$ 4
3821	2-Butenoic acid, 3-chloro-, (Z)-	$C_4H_5ClO_2$ / 120.54	6213-90-7 / 61		4-02-00-01510 / 1.1905[66]	H$_2$O 2; EtOH 4; peth 4 / 1.4704[66]
3822	2-Butenoic acid, 4-chloro- / γ-Chlorocrotonic acid	$C_4H_5ClO_2$ / 120.54	16197-90-3 / 83	/ 117[13]	4-02-00-01511	eth 4; HOAc 4
3823	2-Butenoic acid, 2-chloro-, ethyl ester / Ethyl β-chlorocrotonate	$C_6H_9ClO_2$ / 148.59	56216-12-7	/ 176	4-02-00-01510 / 1.1133[20]	1.453[20]
3824	2-Butenoic acid, 2-chloro-, ethyl ester, (E)- / 2-Butenoic acid, 2-chloro, ethyl ester (trans)	$C_6H_9ClO_2$ / 148.59	77825-53-7	/ 177; 75[30]	4-02-00-01510 / 1.1135[20]	eth 4; EtOH 4 / 1.4538[20]
3825	2-Butenoic acid, 2-chloro-, ethyl ester, (Z)- / 2-Butenoic acid, 2-chloro, ethyl ester (cis)	$C_6H_9ClO_2$ / 148.59	77825-54-8	/ 75[30]	4-02-00-01510 / 1.1021[18]	eth 4; EtOH 4
3826	2-Butenoic acid, 3-chloro-, ethyl ester / 3-Chlorocrotonic acid ethyl ester	$C_6H_9ClO_2$ / 148.59	38624-62-3	/ 180; 75[14]	4-02-00-01511 / 1.1062[20]	1.459[20]
3827	2-Butenoic acid, 3-chloro-, ethyl ester, (E)-	$C_6H_9ClO_2$ / 148.59	6127-92-0	/ 184	4-02-00-01511 / 1.1062[20]	eth 4; EtOH 4 / 1.4592[20]
3828	2-Butenoic acid, 3-chloro-, ethyl ester, (Z)-	$C_6H_9ClO_2$ / 148.59	6127-93-1	/ 161.4	4-02-00-01511 / 1.0860[20]	eth 4; EtOH 4 / 1.4542[19]
3829	2-Butenoic acid, 4-chloro-, ethyl ester / Crotonic acid, 4-chloro-, ethyl ester	$C_6H_9ClO_2$ / 148.59	15333-22-9	/ 93[10]	4-02-00-01511 / 1.090[18]	1.462[18]
3830	2-Butenoic acid, 2-chloro-, methyl ester, (E)-	$C_5H_7ClO_2$ / 134.56	22038-57-9	/ 161.5; 63[17]	4-02-00-01510 / 1.160[20]	eth 4 / 1.4569[23]
3831	2-Butenoic acid, 3-chloro-, methyl ester, (E)-	$C_5H_7ClO_2$ / 134.56	6372-01-6	/ 64-7[14]	4-02-00-01510 / 1.157[20]	H$_2$O 2; EtOH 4; eth 4 / 1.4630[20]
3832	2-Butenoic acid, 3-chloro-, methyl ester, (Z)-	$C_5H_7ClO_2$ / 134.56	6214-25-1	/ 142.4; 70[24]	4-02-00-01500 / 1.138[20]	H$_2$O 2; eth 4; MeOH 4 / 1.4573[19]
3833	2-Butenoic acid, 4-chloro-, methyl ester / Methyl γ-chlorocrotonate	$C_5H_7ClO_2$ / 134.56	15320-72-6	/ 68[11]	4-02-00-01511 / 1.169[18]	1.467[18]
3834	2-Butenoic acid, 3-cyclohexyl- / Cicrotoic acid	$C_{10}H_{16}O_2$ / 168.24	25229-42-9 / 85.5	2272		
3835	3-Butenoic acid, 2,2-dimethyl- / 2,2-Dimethyl-3-butenoic acid	$C_6H_{10}O_2$ / 114.14	10276-09-2 / -6	/ 185	4-02-00-01574 / 0.9567[15]	1.4305[15]
3836	2-Butenoic acid, 2,3-dimethyl-, ethyl ester / Ethyl 2,3-dimethyl-2-butenoate	$C_8H_{14}O_2$ / 142.20	13979-28-7	/ 155	4-02-00-01576 / 0.9072[19]	1.430[19]
3837	2-Butenoic acid, 2-ethyl-, (E)- / 2-Ethylcrotonic acid	$C_6H_{10}O_2$ / 114.14	1187-13-9 / 45.5	/ 209	3-02-00-01329 / 0.9578[50]	eth 4; EtOH 4 / 1.4475[50]
3838	2-Butenoic acid, ethyl ester	$C_6H_{10}O_2$ / 114.14	10544-63-5	/ 136.5	2-02-00-00394 / 0.9175[20]	eth 4; EtOH 4 / 1.4243[20]
3839	2-Butenoic acid, ethyl ester, (E)- / Ethyl trans-crotonate	$C_6H_{10}O_2$ / 114.14	623-70-1 / 2600	/ 138	4-02-00-01500 / 0.9175[20]	H$_2$O 1; EtOH 3; eth 3 / 1.4243[20]
3840	2-Butenoic acid, ethyl ester, (Z)- / cis-Ethyl crotonate	$C_6H_{10}O_2$ / 114.14	6776-19-8 / 2600	/ 136	4-02-00-01497 / 0.9182[20]	ace 4; eth 4; EtOH 4 / 1.4242[20]
3841	3-Butenoic acid, ethyl ester / Ethyl 3-butenoate	$C_6H_{10}O_2$ / 114.14	1617-18-1	/ 119	4-02-00-01491 / 0.9122[20]	EtOH 3 / 1.4105[20]
3842	3-Butenoic acid, 2-hydroxy- / Vinylglycolic acid	$C_4H_6O_3$ / 102.09	600-17-9 / 43	/ 124[9]	3-03-00-00685 / 1.2020[21]	H$_2$O 4; eth 4; EtOH 4; chl 4
3843	3-Butenoic acid, 2-hydroxy-, ethyl ester / Ethyl 2-hydroxy-3-butenoate	$C_6H_{10}O_3$ / 130.14	91890-87-8	/ 173 dec; 68[15]	3-03-00-00685 / 1.0470[15]	H$_2$O 4; eth 4; EtOH 4 / 1.436[13]
3844	2-Butenoic acid, 2-methyl-, (E)- / Tiglic acid	$C_5H_8O_2$ / 100.12	80-59-1 / 64.5	9365 / 198.5	4-02-00-01552 / 0.9641[78]	eth 4; EtOH 4 / 1.4330[76]
3845	2-Butenoic acid, 2-methyl-, (Z)- / Angelic acid	$C_5H_8O_2$ / 100.12	565-63-9 / 45.5	678 / 185	4-02-00-01551 / 0.9834[49]	H$_2$O 2; EtOH 3; eth 4 / 1.4434[47]
3846	2-Butenoic acid, 3-methyl-	$C_5H_8O_2$ / 100.12	541-47-9 / 69.5	/ 197	4-02-00-01555 / 1.0062[24]	
3847	2-Butenoic acid, methyl ester, (E)- / Methyl trans-crotonate	$C_5H_8O_2$ / 100.12	623-43-8 / -42	2600 / 121	4-02-00-01500 / 0.9444[20]	H$_2$O 1; EtOH 4; eth 4 / 1.4242[20]
3848	2-Butenoic acid, methyl ester, (Z)- / Methyl cis-crotonate	$C_5H_8O_2$ / 100.12	4358-59-2	5048 / 118	4-02-00-01500	1.4175[20]
3849	2-Butenoic acid, 2-methyl-, ethyl ester, (E)-	$C_7H_{12}O_2$ / 128.17	5837-78-5	/ 156	4-02-00-01553 / 0.9200[20]	1.4340[20]
3850	2-Butenoic acid, 3-methyl-, ethyl ester	$C_7H_{12}O_2$ / 128.17	638-10-8	/ 153.5	4-02-00-01556 / 0.9199[21]	1.4345[20]
3851	2-Butenoic acid, 2-methyl-, methyl ester, (E)-	$C_6H_{10}O_2$ / 114.14	6622-76-0	/ 139	4-02-00-01553 / 0.9498[20]	1.4370[20]
3852	2-Butenoic acid, 3-methyl-, methyl ester	$C_6H_{10}O_2$ / 114.14	924-50-5 / 114	/ 138.5	3-02-00-01313 / 0.9337[20]	1.432[20]

No.	Name / Synonym	Mol. Form. / Mol. Wt.	CAS RN / mp/°C	Merck No. / bp/°C	Beil. Ref. / den/g cm^{-3}	Solubility / n_D
3853	2-Butenoic acid, 4-oxo-4-(phenylamino)-, (Z)- / Maleanilic acid	$C_{10}H_9NO_3$ / 191.19	555-59-9 / 192 dec	5584	3-12-00-00571 / 1.418[30]	
3854	2-Butenoic acid, 3-phenyl-, (Z)-	$C_{10}H_{10}O_2$ / 162.19	704-79-0 / 131.5	170[14]	4-09-00-02057	bz 3; CS_2 3; peth 2
3855	3-Butenoic acid, 4-phenyl- / Styrylacetic acid	$C_{10}H_{10}O_2$ / 162.19	2243-53-0 / 87	302	4-09-00-02054	H_2O 1; EtOH 4; eth 4; CS_2 4
3856	2-Butenoic acid, 2-propenyl ester / Allyl crotonate	$C_7H_{10}O_2$ / 126.16	20474-93-5	88-9[70]	4-02-00-01503 / 0.9440[20]	1.4465[20]
3857	2-Butenoic acid, 2-propenyl ester, (E)-	$C_7H_{10}O_2$ / 126.16	5453-44-1	88-9[70]	4-02-00-01503 / 0.9440[20]	1.4465[20]
3858	3-Buten-2-ol, (±)-	C_4H_8O / 72.11	6118-14-5 / <-100	97.3	4-01-00-02102 / 0.8318[20]	1.4137[20]
3859	2-Buten-1-ol, (E)- / trans-Crotyl alcohol	C_4H_8O / 72.11	504-61-0 / <-30	121.2	4-01-00-02107 / 0.8521[20]	H_2O 4; EtOH 5; eth 5; chl 3 / 1.4288[20]
3860	2-Buten-1-ol, (Z)- / cis-Crotyl alcohol	C_4H_8O / 72.11	4088-60-2	123	4-01-00-02107 / 0.8662[20]	1.4342[25]
3861	3-Buten-1-ol	C_4H_8O / 72.11	627-27-0	113.5	4-01-00-02105 / 0.8424[20]	H_2O 3; EtOH 5; eth 5; ace 3 / 1.4224[20]
3862	2-Buten-1-ol, acetate	$C_6H_{10}O_2$ / 114.14	628-08-0	132	3-02-00-00283 / 0.9192[20]	bz 4; eth 4; EtOH 4 / 1.4181[20]
3863	2-Buten-1-ol, 2-chloro-	C_4H_7ClO / 106.55	116723-93-4	159	3-01-00-01900 / 1.1180[20]	H_2O 4; EtOH 4 / 1.4682[20]
3864	2-Buten-1-ol, 3-chloro- / γ-Chlorocrotyl alcohol	C_4H_7ClO / 106.55	40605-42-3 / -42.9	161.5	4-01-00-02108 / 1.095[23]	1.4652[20]
3865	2-Buten-1-ol, 4-chloro- / 4-Chloro-2-buten-1-ol	C_4H_7ClO / 106.55	7523-44-6	64-5[2]	4-01-00-02110	eth 4; EtOH 4 / 1.4845[20]
3866	3-Buten-2-ol, 1-chloro- / 1-Chloro-3-buten-2-ol	C_4H_7ClO / 106.55	671-56-7	145.5	3-01-00-01893 / 1.111[20]	chl 4 / 1.4643[20]
3867	3-Buten-1-ol, 2-chloro-	C_4H_7ClO / 106.55	75455-41-3	66-7[30]	4-01-00-02106 / 1.1044[20]	eth 4; EtOH 4 / 1.4665[20]
3868	3-Buten-2-ol, 3-chloro- / 3-Chloro-3-buten-2-ol	C_4H_7ClO / 106.55	6498-47-1	53-7[19]	3-01-00-01893 / 1.1138[23]	eth 4; EtOH 4
3869	2-Buten-1-ol, 3-methyl-	$C_5H_{10}O$ / 86.13	556-82-1	140	4-01-00-02129 / 0.848[25]	1.4412[20]
3870	3-Buten-2-ol, 2-methyl-	$C_5H_{10}O$ / 86.13	115-18-4 / -28	97	4-01-00-02132 / 0.82[20]	
3871	3-Buten-2-ol, 3-methyl- / Methyl isopropenyl carbinol	$C_5H_{10}O$ / 86.13	10473-14-0	114	4-01-00-02127 / 0.8531[17]	1.4288[17]
3872	2-Buten-1-ol, 3-phenyl- / 2-Butene-1-ol, 3-phenyl	$C_{10}H_{12}O$ / 148.20	1504-54-7	145[15]	4-06-00-03839 / 1.035[20]	ctc 3 / 1.5678[20]
3873	3-Buten-2-ol, 4-(2,6,6-trimethyl-1-cyclohexen-1-yl)- / β-Ionol	$C_{13}H_{22}O$ / 194.32	22029-76-1	131[15]; 80[0.7]	3-06-00-00401 / 0.9243[20]	EtOH 3; eth 3; ace 3 / 1.4969[20]
3874	3-Buten-2-ol, 4-(2,6,6-trimethyl-2-cyclohexen-1-yl)- / α-Ionol	$C_{13}H_{22}O$ / 194.32	25312-34-9	127[15]	3-06-00-00402 / 0.9474[20]	1.4735[20]
3875	3-Buten-2-one / Methyl vinyl ketone	C_4H_6O / 70.09	78-94-4 / 81.4	6052	4-01-00-03444 / 0.864[20]	H_2O 3; EtOH 3; eth 4; ace 4 / 1.4081[20]
3876	2-Buten-1-one, 1,3-diphenyl- / Dypnone	$C_{16}H_{14}O$ / 222.29	495-45-4 / 342.5	3460	4-07-00-01682 / 1.1080[15]	eth 4; EtOH 4 / 1.6343[20]
3877	1-Buten-1-one, 2-ethyl- / Diethylketene	$C_6H_{10}O$ / 98.14	24264-08-2 / 88.5		4-01-00-03477 / 0.831[20]	1.4135[11]
3878	3-Buten-2-one, 4-(2-furanyl)-	$C_8H_8O_2$ / 136.15	623-15-4 / 39.5	229 dec; 113[10]	5-17-09-00578 / 1.0496[57]	H_2O 1; EtOH 4; eth 4; chl 4 / 1.5788[45]
3879	3-Buten-2-one, 3-methyl- / Isopropenyl methyl ketone	C_5H_8O / 84.12	814-78-8 / -54	98	4-01-00-03462 / 0.8527[20]	EtOH 4 / 1.4220[20]
3880	2-Buten-1-one, 1-phenyl-	$C_{10}H_{10}O$ / 146.19	495-41-0 / 20.5	111-2[9]	3-07-00-01410 / 1.025[15]	1.5626[18]
3881	3-Buten-2-one, 4-phenyl- / Benzylideneacetone	$C_{10}H_{10}O$ / 146.19	122-57-6 / 41.5	1151 / 261	2-07-00-00287 / 1.0368[15]	bz 4; eth 4; EtOH 4; chl 4
3882	3-Buten-2-one, 4-phenyl-, (E)-	$C_{10}H_{10}O$ / 146.19	1896-62-4 / 42	262	4-07-00-01003 / 1.0097[45]	H_2O 1; EtOH 4; eth 3; ace 3 / 1.5836[45]
3883	3-Buten-2-one, 4-(2,5,6,6-tetramethyl-1-cyclohexen-1-yl)- / β-Irone	$C_{14}H_{22}O$ / 206.33	79-70-9 / 85-90[0.1]	4978	4-07-00-00377 / 0.9434[21]	H_2O 2; EtOH 4; eth 4; bz 4 / 1.5017[20]
3884	3-Buten-2-one, 4-(2,5,6,6-tetramethyl-2-cyclohexen-1-yl)- / α-Irone	$C_{14}H_{22}O$ / 206.33	79-69-6 / 87-92[0.4]	4977	/ 0.9362[20]	1.5002[20]
3885	3-Buten-2-one, 4-(2,6,6-trimethyl-1-cyclohexen-1-yl)-, (E)-	$C_{13}H_{20}O$ / 192.30	79-77-6 / 140[18]; 73[0.1]		4-07-00-00361 / 0.9462[20]	H_2O 2; EtOH 5; eth 5; chl 3 / 1.5198[20]
3886	3-Buten-2-one, 4-(2,6,6-trimethyl-2-cyclohexen-1-yl)-, (E)-(±)-	$C_{13}H_{20}O$ / 192.30	30685-95-1 / 146[28]; 75[0.15]		4-07-00-00363 / 0.9298[21]	ace 4; eth 4; EtOH 4 / 1.5041[20]
3887	2-Butenoyl chloride	C_4H_5ClO / 104.54	10487-71-5 / 124.5		2-02-00-00394 / 1.0905[20]	ace 4 / 1.460[18]
3888	2-Butenoyl chloride, 3-methyl-	C_5H_7ClO / 118.56	3350-78-5 / 146		4-02-00-01558 / 1.065[25]	1.4770[20]

No.	Name / Synonym	Mol. Form. / Mol. Wt.	CAS RN / mp/°C	Merck No. / bp/°C	Bell. Ref. / den/g cm⁻³	Solubility / n_D
3889	1-Buten-3-yne / Vinylacetylene	C_4H_4 / 52.08	689-97-4 /	/ 5.1	4-01-00-01083 / 0.7094[0]	H_2O 1; bz 3 / 1.4161[1]
3890	1-Buten-3-yne, 4-chloro-	C_4H_3Cl / 86.52	40589-38-6 /	/ 56	4-01-00-01085 / 1.0022[20]	chl 4 / 1.4656[20]
3891	1-Buten-3-yne, 1-methoxy-	C_5H_6O / 82.10	2798-73-4 /	/ 123 dec	4-01-00-02301 / 0.906[20]	H_2O 1; chl 3; os 4 / 1.4818[20]
3892	1-Buten-3-yne, 2-methyl- / Isopropenylacetylene	C_5H_6 / 66.10	78-80-8 / -113	/ 32	4-01-00-01089 / 0.6801[11]	chl 3 / 1.4140[20]
3893	Butralin / Benzenamine, 4-(1,1-dimethylethyl)-N-(1-methylpropyl)-2,6-dinitro-	$C_{14}H_{21}N_3O_4$ / 295.34	33629-47-9 / 60	1532 / 135[0.5]		
3894	N-sec-Butyl-2,6-dinitrobenzene / Benzeneamine, N-(1-methylpropyl)-2,6-dinitro-	$C_{10}H_{13}N_3O_4$ / 239.23	55702-49-3 /			
3895	2-Butynal / Tetrolaldehyde	C_4H_4O / 68.08	1119-19-3 / -26	/ 106.5	4-01-00-03540 / 0.9264[17]	eth 3; ace 3; chl 3 / 1.446[19]
3896	3-Butyn-2-amine, 2-methyl-	C_5H_9N / 83.13	2978-58-7 / 18	/ 79.5	4-04-00-01145 / 0.79[25]	/ 1.4235[20]
3897	1-Butyne / Ethylacetylene	C_4H_6 / 54.09	107-00-6 / -125.7	/ 8.0	4-01-00-00969 / 0.6783[0]	H_2O 1; EtOH 3; eth 3 / 1.3962[20]
3898	2-Butyne / Dimethylacetylene	C_4H_6 / 54.09	503-17-3 / -32.3	/ 26.9	4-01-00-00971 / 0.6910[20]	H_2O 1; EtOH 3; eth 3; ctc 3 / 1.3921[20]
3899	1-Butyne, 3-chloro-	C_4H_5Cl / 88.54	21020-24-6 /	/ 68.5	4-01-00-00970 / 1.4218[25]	/ 1.4218[25]
3900	2-Butyne, 1-chloro-	C_4H_5Cl / 88.54	3355-17-7 /	/ 103	4-01-00-00973 / 1.0152[20]	ace 4; eth 4; EtOH 4 / 1.4581[20]
3901	1-Butyne, 3-chloro-3-methyl-	C_5H_7Cl / 102.56	1111-97-3 / -61	/ 76	4-01-00-01000 / 0.9061[20]	
3902	2-Butynediamide / Cellocidin	$C_4H_4N_2O_2$ / 112.09	543-21-5 / 217 dec	1958 /	4-02-00-02295 /	H_2O 2; EtOH 2; eth 2; chl 2
3903	2-Butyne-1,4-diamine	$C_4H_8N_2$ / 84.12	53878-96-9 / 44	/ 96[16]	4-04-00-01392 /	H_2O 4; EtOH 4; MeOH 4
3904	2-Butyne, 1,4-dibromo- / 1,4-Dibromo-2-butyne	$C_4H_4Br_2$ / 211.88	2219-66-1 /	/ 92[15]	4-01-00-00974 / 2.014[18]	eth 3; ace 3; chl 4 / 1.588[18]
3905	2-Butyne, 1,4-dichloro-	$C_4H_4Cl_2$ / 122.98	821-10-3 /	/ 165.5	4-01-00-00973 / 1.258[20]	eth 3; ace 3; ctc 2; chl 4 / 1.5058[20]
3906	2-Butyne, 1,4-diiodo-	$C_4H_4I_2$ / 305.88	116529-73-8 / 54.5	/ 70[0.1]	4-01-00-00975 /	H_2O 2; EtOH 3; eth 3; ace 3
3907	1-Butyne, 3,3-dimethyl- / tert-Butylacetylene	C_6H_{10} / 82.15	917-92-0 / -78.2	/ 37.7	4-01-00-01022 / 0.6623[25]	/ 1.3736[20]
3908	2-Butynedinitrile	C_4N_2 / 76.06	1071-98-3 / 20.5	/ 76.5	4-02-00-02295 / 0.9708[25]	/ 1.4647[25]
3909	2-Butynedioic acid, diethyl ester	$C_8H_{10}O_4$ / 170.17	762-21-0 / 1.5	/ 184[200]	4-02-00-02294 / 1.0075[20]	EtOH 3; eth 3; ctc 3 / 1.4425[20]
3910	2-Butynedioic acid, dimethyl ester	$C_6H_6O_4$ / 142.11	762-42-5 /	/ 197 dec; 98[20]	4-02-00-02291 / 1.1564[20]	EtOH 3; eth 3; ctc 3 / 1.4434[20]
3911	2-Butyne-1,4-diol / Bis(hydroxymethyl)acetylene	$C_4H_6O_2$ / 86.09	110-65-6 / 50	/ 238	4-01-00-02687 /	H_2O 4; EtOH 4; eth 2; ace 4 / 1.4804[20]
3912	3-Butyne-1,2-diol / 1-Butyne-3,4-diol	$C_4H_6O_2$ / 86.09	616-28-4 / 40	/ 64[0.2]	4-01-00-02689 /	H_2O 4; EtOH 4
3913	2-Butyne-1,4-diol, diacetate / 1,4-Diacetoxy-2-butyne	$C_8H_{10}O_4$ / 170.17	1573-17-7 /	/ 122-3[10]	4-02-00-00244 /	ctc 3 / 1.4611[20]
3914	2-Butyne, 1,1,1,4,4,4-hexafluoro-	C_4F_6 / 162.03	692-50-2 / -117.4	/ -24.6	4-01-00-00972 /	EtOH 3; eth 3; ace 3; ctc 3
3915	2-Butyne, 1-methoxy-	C_5H_8O / 84.12	2768-41-4 /	/ 99.5	3-01-00-01973 / 0.8496[20]	bz 4; eth 4; EtOH 4 / 1.4262[20]
3916	1-Butyne, 3-methyl-	C_5H_8 / 68.12	598-23-2 / -89.7	/ 26.3	4-01-00-00999 / 0.6660[20]	H_2O 1; EtOH 5; eth 5 / 1.3723[20]
3917	2-Butynoic acid	$C_4H_4O_2$ / 84.07	590-93-2 / 78	/ 203	4-02-00-01690 / 0.9641[20]	H_2O 4; eth 4; EtOH 4; chl 4
3918	2-Butynoic acid, ethyl ester	$C_6H_8O_2$ / 112.13	4341-76-8 /	/ 163	4-02-00-01691 / 0.9641[20]	/ 1.4372[20]
3919	2-Butyn-1-ol	C_4H_6O / 70.09	764-01-2 / -2.2	/ 148	4-01-00-02220 / 0.9370[20]	eth 4; EtOH 4 / 1.4530[20]
3920	3-Butyn-1-ol	C_4H_6O / 70.09	927-74-2 / -63.6	/ 129	4-01-00-02219 / 0.9257[20]	H_2O 4; EtOH 4 / 1.4409[20]
3921	3-Butyn-2-ol	C_4H_6O / 70.09	2028-63-9 / -3	/ 106.5	4-01-00-02218 / 0.8618[20]	H_2O 4; eth 4; EtOH 4 / 1.4207[20]
3922	3-Butyn-2-ol, 2-methyl- / 1,1-Dimethylpropargyl alcohol	C_5H_8O / 84.12	115-19-5 / 3	5956 / 104	4-01-00-02229 / 0.8618[20]	/ 1.4207[20]
3923	3-Butyn-2-one / Ethynyl methyl ketone	C_4H_4O / 68.08	1423-60-5 /	/ 84	4-01-00-03539 / 0.8860[15]	/ 1.4070[20]
3924	3-Butyn-2-one, 4-phenyl-	$C_{10}H_8O$ / 144.17	1817-57-8 /	/ 75-6[0.8]	4-07-00-01175 / 0.99[25]	/ 1.5762[20]
3925	Butyrac 118 / Butanoic acid, 4-(2,4-dichlorophenoxy)-	$C_{10}H_{10}Cl_2O_3$ / 249.09	94-82-6 / 118	2828 /		
3926	Butyric acid, 3,4-dichloro- / 3,4-Dichlorobutyric acid	$C_4H_6Cl_2O_2$ / 157.00	29653-38-1 / 49	/ 123[8]	4-02-00-00828 /	eth 4; chl 4; HOAc 4
3927	Butyric acid, 4-fluoro- / 4-Fluorobutyric acid	$C_4H_7FO_2$ / 106.10	462-23-7 /	/ 76-8[5]	4-02-00-00809 /	eth 4; EtOH 4 / 1.3993[25]

No.	Name Synonym	Mol. Form. Mol. Wt.	CAS RN mp/°C	Merck No. bp/°C	Beil. Ref. den/g cm^{-3}	Solubility n_D
3928	Butyric acid, 2-iodo-, ethyl ester	$C_8H_{11}IO_2$ 242.06	7425-47-0	100[21]; 46[1]	4-02-00-00839 1.570[17]	1.4923[20]
3929	Cadmium, dimethyl- Dimethyl cadmium	C_2H_6Cd 142.48	506-82-1 -4.5	3229 105.5; exp 150	1.9846[18]	peth 3 1.5488
3930	Calcium cyanide Calcyan	C_2CaN_2 92.11	592-01-8	1664		
3931	Camphoric acid, trans-(±)- Isocamphoric acid, (DL)	$C_{10}H_{16}O_4$ 200.23	560-08-7 195		3-09-00-03879 1.249[25]	eth 4
3932	Carbamic acid, (aminocarbonyl)-, ethyl ester	$C_4H_8N_2O_3$ 132.12	626-36-8 196.5	dec	4-03-00-00127	H_2O 1; EtOH 2; eth 1; bz 2
3933	Carbamic acid, butyl-, 2-[[(aminocarbonyl)oxy]methyl]-2-methylpentyl ester 2-Methyl-2-propyltrimethylene butylcarbamate carbamate	$C_{13}H_{26}N_2O_4$ 274.36	4268-36-4	9737		
3934	Carbamic acid, butyl-, butyl ester	$C_9H_{19}NO_2$ 173.26	13105-52-7	88[3]	4-04-00-00577 0.9221[20]	eth 4; EtOH 4 1.4359[20]
3935	Carbamic acid, butyl ester Butyl carbamate	$C_5H_{11}NO_2$ 117.15	592-35-8 53	204 dec; 108[14]	4-03-00-00054	EtOH 4; chl 2
3936	Carbamic acid, butyl-, ethyl ester	$C_7H_{15}NO_2$ 145.20	591-62-8 -22	202; 100[15]	4-04-00-00577 0.9434[26]	1.4278[26]
3937	Carbamic acid, butyl-, methyl ester	$C_6H_{13}NO_2$ 131.17	2594-21-0 -18	92[15]	0-04-00-00158 0.9689[23]	1.4289[23]
3938	Carbamic acid, N-butyl-N-nitro,butyl ester	$C_9H_{18}N_2O_4$ 218.25	110795-27-2	98[3]	1.046[20]	EtOH 5; eth 5 1.448[20]
3939	Carbamic acid, (3-chlorophenyl)-, 1-methylethyl ester Chloropropham	$C_{10}H_{12}ClNO_2$ 213.66	101-21-3 41	2188 149[2]	4-12-00-01149 1.18[30]	1.5388[20]
3940	Carbamic acid, dichloro-, ethyl ester	$C_3H_5Cl_2NO_2$ 157.98	13698-16-3	55-6[15]	3-03-00-00050 1.349[25]	1.4595[20]
3941	Carbamic acid, diethyl- Carbamic acid, N,N-diethyl	$C_5H_{11}NO_2$ 117.15	24579-70-2 -15	171 dec	3-04-00-00222 0.9276[20]	H_2O 4; eth 4; EtOH 4 1.4206[20]
3942	Carbamic acid, (1,1-dimethylethyl)-, ethyl ester	$C_7H_{15}NO_2$ 145.20	1611-50-3 21.5	72[16]	4-04-00-00665 0.943[15]	
3943	Carbamic acid, (1,1-dimethylethyl)-, methyl ester	$C_6H_{13}NO_2$ 131.17	27701-01-5 27	63.3[17]	3-04-00-00325 0.966[15]	
3944	Carbamic acid, (1,1-dimethylethyl)nitro-, ethyl ester	$C_7H_{14}N_2O_4$ 190.20	55696-02-1	56[2] dec	4-04-00-03409 1.049[20]	eth 4; EtOH 4 1.4331[20]
3945	Carbamic acid, dimethyl-, 3-methyl-1-phenyl-1H-pyrazol-5-yl ester Pyrolan	$C_{13}H_{15}N_3O_2$ 245.28	87-47-8 50	8013 160-2[0.2]	5-23-10-00530	ctc 3; CS_2 3
3946	Carbamic acid, diphenyl-, ethyl ester	$C_{15}H_{15}NO_2$ 241.29	603-52-1 72	360	4-12-00-00850	H_2O 4; eth 4; bz 4
3947	Carbamic acid, ethyl-, butyl ester Carbamic acid, N-ethyl, butyl ester	$C_7H_{15}NO_2$ 145.20	16246-07-4	66[3]	0.9396[20]	eth 4; EtOH 4 1.4301[20]
3948	Carbamic acid, ethyl ester Urethan	$C_3H_7NO_2$ 89.09	51-79-6 49	9789 185	4-03-00-00040 0.9862[21]	H_2O 4; EtOH 4; eth 4; bz 4 1.4144[51]
3949	Carbamic acid, ethyl-, ethyl ester	$C_5H_{11}NO_2$ 117.15	623-78-9	176	4-04-00-00365 0.9813[20]	H_2O 4; eth 4; EtOH 4 1.4215[20]
3950	Carbamic acid, ethylnitro-, ethyl ester Carbamic acid, N-ethyl-N-nitro, ethyl ester	$C_5H_{10}N_2O_4$ 162.15	6274-16-4	107[31]	4-04-00-03404 1.161[20]	eth 4; EtOH 4 1.4432[20]
3951	Carbamic acid, ethylnitro-, methyl ester Carbamic acid, N-ethyl-N-nitro, ethyl ester	$C_7H_{14}N_2O_4$ 190.20	6162-79-4	79[3]	4-04-00-03403 1.089[20]	H_2O 4; eth 4; EtOH 4 1.4455[20]
3952	Carbamic acid, (3-methylbutyl)-, ethyl ester Urethane, isopentyl	$C_8H_{17}NO_2$ 159.23	1611-52-5	218	2-04-00-00648 0.9322[20]	1.4326[20]
3953	Carbamic acid, methyl ester Methyl carbamate	$C_2H_5NO_2$ 75.07	598-55-0 54	5958 177	4-03-00-00037 1.1361[56]	H_2O 4; EtOH 4; eth 3 1.4125[56]
3954	Carbamic acid, methyl-, ethyl ester	$C_4H_9NO_2$ 103.12	105-40-8	170	4-04-00-00200 1.0115[20]	H_2O 4; EtOH 4 1.4183[20]
3955	Carbamic acid, 1-methylethyl ester Isopropyl carbamate	$C_4H_9NO_2$ 103.12	1746-77-6 93	183	4-03-00-00053 0.9951[66]	
3956	Carbamic acid, (1-methylethyl)-, ethyl ester	$C_6H_{13}NO_2$ 131.17	2594-20-9	79[15]	4-04-00-00520 0.9531[20]	eth 4; EtOH 4 1.4229[20]
3957	Carbamic acid, (1-methylethyl)nitro-, ethyl ester Carbamic acid, N-isopropyl-N-nitro, ethyl ester	$C_6H_{12}N_2O_4$ 176.17	62261-05-6	72[7]	4-04-00-03407 1.110[20]	eth 4; EtOH 4 1.4381[20]
3958	Carbamic acid, methylphenyl-, methyl ester	$C_9H_{11}NO_2$ 165.19	28685-60-1 44	235	2-12-00-00235 1.2960[19]	eth 4; lig 4
3959	Carbamic acid, 2-methylpropyl ester Isobutyl carbamate	$C_5H_{11}NO_2$ 117.15	543-28-2 67	5021 207	4-03-00-00056	eth 4; EtOH 4 1.4098[76]
3960	Carbamic acid, (1-methylpropyl)-, ethyl ester Urethane, sec-butyl	$C_7H_{15}NO_2$ 145.20	10212-74-5 -14	194	4-04-00-00622 0.9404[26]	1.4267[26]
3961	Carbamic acid, (2-methylpropyl)-, ethyl ester Isobutyl urethane	$C_7H_{15}NO_2$ 145.20	539-89-9 <-65	5037 110[30]	4-04-00-00647 0.9432[20]	eth 4; EtOH 4 1.4288[20]

No.	Name Synonym	Mol. Form. Mol. Wt.	CAS RN mp/°C	Merck No. bp/°C	Beil. Ref. den/g cm^{-3}	Solubility n_D
3962	Carbamic acid, (1-methylpropyl)-, methyl ester Carbamic acid, N-sec-butyl, methyl ester	$C_6H_{13}NO_2$ 131.17	39076-02-3 	83[16]	0-04-00-00162 0.9651[25]	1.4263[25]
3963	Carbamic acid, nitro-, ethyl ester	$C_3H_6N_2O_4$ 134.09	626-37-9 64	140 dec	4-03-00-00247 1.0074[20]	H_2O 4; EtOH 4; eth 4; lig 2
3964	Carbamic acid, nitropropyl-, ethyl ester Carbamic acid, N-propyl-N-nitro, ethyl ester	$C_6H_{12}N_2O_4$ 176.17	13855-77-1 	66[3]	4-04-00-03406 1.121[20]	eth 4; EtOH 4 1.4431[20]
3965	Carbamic acid, phenyl-, ethyl ester Phenylurethane	$C_9H_{11}NO_2$ 165.19	101-99-5 53	7291 237 dec	4-12-00-00619 1.1064[30]	H_2O 1; EtOH 4; eth 4; bz 3 1.5376[30]
3966	Carbamic acid, phenyl-, 1-methylethyl ester Propham	$C_{10}H_{13}NO_2$ 179.22	122-42-9 90	7828	4-12-00-00620 1.09[20]	bz 4; EtOH 4 1.4989[91]
3967	Carbamic acid, (phenylmethyl)-, ethyl ester	$C_{10}H_{13}NO_2$ 179.22	2621-78-5 49	230	4-12-00-02250	bz 4; eth 4; EtOH 4; chl 4
3968	Carbamic acid, phenyl-, 2-methylpropyl ester	$C_{11}H_{15}NO_2$ 193.25	2291-80-7 86	216	3-12-00-00614	bz 4; eth 4; EtOH 4
3969	Carbamic acid, propyl ester	$C_4H_9NO_2$ 103.12	627-12-3 60	196	4-03-00-00052	ace 4; eth 4; EtOH 4
3970	Carbamic acid, propyl-, ethyl ester	$C_6H_{13}NO_2$ 131.17	623-85-8 	192	4-04-00-00480 0.9921[15]	EtOH 4
3971	Carbamic chloride Carbamyl chloride	CH_2ClNO 79.49	463-72-9 	1785 62 dec	3-03-00-00065	
3972	Carbamic chloride, diethyl- Carbamoyl chloride, N,N-diethyl	$C_5H_{10}ClNO$ 135.59	88-10-8 	186	4-04-00-00379	
3973	Carbamic chloride, dimethyl- Dimethylcarbamoyl chloride	C_3H_6ClNO 107.54	79-44-7 -33	167	4-04-00-00224 1.168[25]	1.4540[20]
3974	Carbamic chloride, methylphenyl-	C_8H_8ClNO 169.61	4285-42-1 88.5	280	4-12-00-00839	eth 4; EtOH 4
3975	Carbamodithioic acid, dimethyl-, 2,4-dinitrophenyl ester	$C_9H_9N_3O_4S_2$ 287.32 152.5	89-37-2 		4-06-00-01760 1.54[20]	H_2O 1; EtOH 3; ace 3; bz 3
3976	Carbamothioic acid, butylethyl-, S-propyl ester Pebulate	$C_{10}H_{21}NOS$ 203.35	1114-71-2 	7007 142[20]	0.9458[20]	ace 4; bz 4; MeOH 4 1.4752[20]
3977	Carbamothioic acid, diphenyl-, S-[2-(diethylamino)ethyl] ester Phencarbamide	$C_{19}H_{24}N_2OS$ 328.48 48.5	3735-90-8 	7178 120-6[0.01]		eth 4; chl 4; MeOH 4; peth 4
3978	Carbamothioic acid, O-ethyl ester	C_3H_7NOS 105.16	625-57-0 41	dec	4-03-00-00294 1.069[20]	eth 4; EtOH 4; chl 4 1.520[20]
3979	Carbamothioic acid, S-ethyl ester	C_3H_7NOS 105.16	637-98-9 109	dec	4-03-00-00294	eth 4; EtOH 4
3980	Carbamothioic chloride, dimethyl-	C_3H_6ClNS 123.61	16420-13-6 42.5	98[10]	3-04-00-00147	eth 4; chl 3; peth 3
3981	9H-Carbazole Dibenzo[b,d]pyrrole	$C_{12}H_9N$ 167.21	86-74-8 246.2	1792 354.69	5-20-08-00009	H_2O 1; EtOH 2; eth 2; ace 3
3982	9H-Carbazole-9-acetic acid (N-Carbazolyl)acetic acid	$C_{14}H_{11}NO_2$ 225.25	524-80-1 215	1793	5-20-08-00033	eth 4; EtOH 4; chl 4; HOAc 4
3983	9H-Carbazole, 9-acetyl-	$C_{14}H_{11}NO$ 209.25	574-39-0 69	190[6]	5-20-08-00030 1.158[100]	1.640[100]
3984	9H-Carbazole, 9-butyl-	$C_{16}H_{17}N$ 223.32	1484-08-8 58	218-9[19]	5-20-08-00018	eth 4
3985	9H-Carbazole, 9-ethyl-	$C_{14}H_{13}N$ 195.26	86-28-2 68	190[10]	5-20-08-00016 1.059[80]	H_2O 1; EtOH 4; eth 4 1.6394[80]
3986	9H-Carbazole, 3-methyl-	$C_{13}H_{11}N$ 181.24	4630-20-0 208.5	365	5-20-08-00076	bz 4; eth 4
3987	9H-Carbazole, 9-methyl-	$C_{13}H_{11}N$ 181.24	1484-12-4 89.34	343.64; 195[12]	5-20-08-00015	eth 4
3988	9H-Carbazole, 9-(phenylmethyl)-	$C_{19}H_{15}N$ 257.33	19402-87-0 119	267[24]	5-20-08-00023	bz 4
3989	1H-Carbazole, 2,3,4,9-tetrahydro-	$C_{12}H_{13}N$ 171.24	942-01-8 120	327.5	5-20-07-00468	H_2O 1; EtOH 3; eth 4; bz 4
3990	Carbendazim Carbamic acid, 1H-benzimidazol-2-yl-, methyl ester	$C_9H_9N_3O_2$ 191.19	10605-21-7 300 dec	1794	1.45	
3991	Carbofuran 7-Benzofuranol, 2,3-dihydro-2,2-dimethyl-, methylcarbamate	$C_{12}H_{15}NO_3$ 221.26	1563-66-2 151	1810	1.18	
3992	Carbon dioxide Carbonic anhydride	CO_2 44.01	124-38-9 -56.57 (5.11 atm)	1816 sub -78.5 (1 atm)	0-03-00-00004 *0.720[25]	
3993	Carbon disulfide Carbon bisulfide	CS_2 76.14	75-15-0 -111.5	1818 46	4-03-00-00395 1.2632[20]	H_2O 3; EtOH 5; eth 5; chl 3 1.6319[20]
3994	Carbonic acid, bis(1,1-dimethylethyl) ester	$C_9H_{18}O_3$ 174.24	34619-03-9 40	158	4-03-00-00009	EtOH 4
3995	Carbonic acid, bis(1-methylethyl) ester	$C_7H_{14}O_3$ 146.19	6482-34-4 	147	3-03-00-00009 0.9162[20]	EtOH 4 1.3932[20]
3996	Carbonic acid, bis(2-methylphenyl) ester	$C_{15}H_{14}O_3$ 242.27	617-09-4 60	144[0.5]	3-06-00-01256	HOAc 4
3997	Carbonic acid, bis(2-methylpropyl) ester	$C_9H_{18}O_3$ 174.24	539-92-4 	190	4-03-00-00009 0.9138[20]	H_2O 1; EtOH 5; eth 5 1.4072[20]
3998	Carbonic acid, dibutyl ester Dibutyl carbonate	$C_9H_{18}O_3$ 174.24	542-52-9 	1558 207	4-03-00-00008 0.9251[20]	H_2O 1; EtOH 3; eth 3 1.4117[20]

No.	Name Synonym	Mol. Form. Mol. Wt.	CAS RN mp/°C	Merck No. bp/°C	Beil. Ref. den/g cm^{-3}	Solubility n_D
3999	Carbonic acid, diethyl ester Diethyl carbonate	C$_5$H$_{10}$O$_3$ 118.13	105-58-8 -43	3738 126	4-03-00-00005 0.9692[25]	H$_2$O 1; EtOH 3; eth 3; chl 3 1.3845[20]
4000	Carbonic acid, dimethyl ester Methyl carbonate	C$_3$H$_6$O$_3$ 90.08	616-38-6 0.5	5960 90.5	4-03-00-00003 1.0636[25]	H$_2$O 1; EtOH 3; eth 3; ctc 2 1.3687[20]
4001	Carbonic acid, diphenyl ester Phenyl carbonate	C$_{13}$H$_{10}$O$_3$ 214.22	102-09-0 83	7250 306	4-06-00-00629 1.1215[87]	H$_2$O 1; EtOH 3; eth 3; ctc 3
4002	Carbonic acid, dipropyl ester Dipropyl carbonate	C$_7$H$_{14}$O$_3$ 146.19	623-96-1 	 168	4-03-00-00006 0.9435[20]	H$_2$O 2; EtOH 5; eth 5 1.4008[20]
4003	Carbonic acid, ethyl methyl ester Ethyl methyl carbonate	C$_4$H$_8$O$_3$ 104.11	623-53-0 -14	 107.5	4-03-00-00004 1.012[20]	eth 4; EtOH 4 1.3778[20]
4004	Carbonic acid monosodium salt	CHNaO$_3$ 84.01	144-55-8 	8528 		H$_2$O 3; EtOH 1
4005	Carbonic chloride fluoride Carbonyl chloride fluoride	CClFO 82.46	353-49-1 	 	4-03-00-00031	
4006	Carbonic dibromide Carbonyl bromide	CBr$_2$O 187.82	593-95-3 	 64.5	2.52[15]	
4007	Carbonic dichloride Phosgene	CCl$_2$O 98.92	75-44-5 -127.9	7310 8	4-03-00-00031 *1.3719[25]	bz 3; ctc 3; chl 3; tol 3
4008	Carbonic difluoride Carbonyl fluoride	CF$_2$O 66.01	353-50-4 -111.2	1826 -84.5	4-03-00-00021 1.139[25]	
4009	Carbonic dihydrazide Carbohydrazide	CH$_6$N$_4$O 90.09	497-18-7 154	1811 	4-03-00-00240 1.616[20]	H$_2$O 4; EtOH 4
4010	Carbonic dihydrazide, 2,2'-diphenyl- sym-Diphenylcarbazide	C$_{13}$H$_{14}$N$_4$O 242.28	140-22-7 170 dec	3321 	4-15-00-00182	H$_2$O 2; EtOH 3; eth 2; ace 3
4011	Carbon monoxide Carbon oxide	CO 28.01	630-08-0 -205	1820 -191.5	0-01-00-00720 0.7909[-19]	H$_2$O 2; bz 3; HOAc 3
4012	Carbonochloridic acid, butyl ester Butyl chloroformate	C$_5$H$_9$ClO$_2$ 136.58	592-34-7 	 142	4-03-00-00025 1.074[25]	eth 5; ace 3; ctc 2 1.4114[20]
4013	Carbonochloridic acid, 2-chloroethyl ester 2-Chloroethyl chloroformate	C$_3$H$_4$Cl$_2$O$_2$ 142.97	627-11-2 	 155	4-03-00-00024 1.3847[20]	H$_2$O 1; EtOH 3; eth 3; ace 3 1.4483[20]
4014	Carbonochloridic acid, chloromethyl ester Chloromethyl chloroformate	C$_2$H$_2$Cl$_2$O$_2$ 128.94	22128-62-7 	 107	4-03-00-00030 1.465[15]	ace 4; eth 4 1.4286[22]
4015	Carbonochloridic acid, 3-chloropropyl ester	C$_4$H$_6$Cl$_2$O$_2$ 157.00	628-11-5 	 177	3-03-00-00025 1.2926[25]	H$_2$O 1 1.4456[20]
4016	Carbonochloridic acid, dichloromethyl ester	C$_2$HCl$_3$O$_2$ 163.39	22128-63-8 	 110.5	4-03-00-00031 1.5600[15]	ace 4
4017	Carbonochloridic acid, 2-ethoxyethyl ester	C$_5$H$_9$ClO$_3$ 152.58	628-64-8 	 67.2[14]	3-03-00-00029 1.1341[25]	EtOH 4 1.4169[25]
4018	Carbonochloridic acid, ethyl ester Ethyl chloroformate	C$_3$H$_5$ClO$_2$ 108.52	541-41-3 -80.6	3742 95	4-03-00-00023 1.1352[20]	bz 4; eth 4; chl 4 1.3974[20]
4019	Carbonochloridic acid, 2-methoxyethyl ester	C$_4$H$_7$ClO$_3$ 138.55	628-12-6 	 58.7[13]	3-03-00-00029 1.1905[25]	eth 4 1.4163[20]
4020	Carbonochloridic acid, methyl ester Methyl chlorocarbonate	C$_2$H$_3$ClO$_2$ 94.50	79-22-1 	5966 70.5	4-03-00-00023 1.2231[20]	EtOH 5; eth 5; bz 3; ctc 3 1.3868[20]
4021	Carbonochloridic acid, 3-methylbutyl ester	C$_6$H$_{11}$ClO$_2$ 150.60	628-50-2 	 154.3	4-03-00-00026 1.0288[17]	eth 4; EtOH 4 1.4176[20]
4022	Carbonochloridic acid, 1-methylethenyl ester Isopropenyl chloroformate	C$_4$H$_5$ClO$_2$ 120.54	57933-83-2 	 102	3-03-00-00028 1.101[20]	eth 4
4023	Carbonochloridic acid, 1-methylethyl ester Isopropyl chloroformate	C$_4$H$_7$ClO$_2$ 122.55	108-23-6 	 105	4-03-00-00024 	eth 4 1.4013[20]
4024	Carbonochloridic acid, 2-methylpropyl ester Isobutyl chlorocarbonate	C$_5$H$_9$ClO$_2$ 136.58	543-27-1 	5023 128.8	4-03-00-00026 1.0426[18]	EtOH 3; eth 5; bz 3; chl 3 1.4071[18]
4025	Carbonochloridic acid, pentyl ester	C$_6$H$_{11}$ClO$_2$ 150.60	638-41-5 	 60-2[15]	4-03-00-00026 	eth 3 1.4181[18]
4026	Carbonochloridic acid, phenyl ester Phenyl chloroformate	C$_7$H$_5$ClO$_2$ 156.57	1885-14-9 	 71[9]	4-06-00-00629	
4027	Carbonochloridic acid, phenylmethyl ester Carbobenzoxy chloride	C$_8$H$_7$ClO$_2$ 170.60	501-53-1 	1807 103[20]	4-06-00-02278 1.195[25]	eth 3; ace 3; bz 3 1.5190[20]
4028	Carbonochloridic acid, propyl ester Propyl chlorocarbonate	C$_4$H$_7$ClO$_2$ 122.55	109-61-5 	7860 115.2	4-03-00-00024 1.0901[20]	EtOH 5; eth 5 1.4035[20]
4029	Carbonochloridic acid, trichloromethyl ester Diphosgene	C$_2$Cl$_4$O$_2$ 197.83	503-38-8 -57	3339 128	4-03-00-00033 1.6525[14]	eth 4; EtOH 4 1.4566[22]
4030	Carbonocyanidic acid, ethyl ester	C$_4$H$_5$NO$_2$ 99.09	623-49-4 	 115.5	4-02-00-01862 1.003[25]	H$_2$O 1; EtOH 3; eth 3; ctc 3 1.3820[20]
4031	Carbonodithioic acid, S,S-diethyl ester	C$_5$H$_{10}$OS$_2$ 150.27	623-80-3 	 197	3-03-00-00339 1.085[20]	eth 4; EtOH 4 1.5237[18]
4032	Carbonothioic dichloride Thiophosgene	CCl$_2$S 114.98	463-71-8 	 73	4-03-00-00281 1.508[15]	H$_2$O 1; EtOH 3; eth 3 1.5442[20]
4033	Carbonothioic dihydrazide 1,3-Diamino-2-thiourea	CH$_6$N$_4$S 106.15	2231-57-4 170 dec	 	4-03-00-00388	H$_2$O 4
4034	Carbonotrithioic acid Trithiocarbonic acid	CH$_2$S$_3$ 110.22	594-08-1 -30	9671 57 dec	4-03-00-00428 1.47[17]	EtOH 4; tol 4; chl 4
4035	Carbon oxide sulfide Carbonyl sulfide	COS 60.08	463-58-1 -138.8	 -50	4-03-00-00271 1.028[17]	H$_2$O 2; EtOH 3; KOH 4 1.24[-87]
4036	Carbon selenide Carbon diselenide	CSe$_2$ 169.93	506-80-9 -45.5	1817 125.5	4-03-00-00436 2.6823[20]	 1.8454[20]
4037	Carbophenothion Phosphorodithioic acid, S-[[(4- chlorophenyl)thio]methyl] O,O-diethyl ester	C$_{11}$H$_{16}$ClO$_2$PS$_3$ 342.87	786-19-6 	1827 82[0.01]	 1.271[20]	

No.	Name / Synonym	Mol. Form. / Mol. Wt.	CAS RN / mp/°C	Merck No. / bp/°C	Bell. Ref. / den/g cm^{-3}	Solubility / n_D
4038	Carbosulfan / 2,3-Dihydro-2,2-dimethyl-7-benzofuranyl (dibutylamino)thio methylcarbamate	C$_{20}$H$_{32}$N$_2$O$_3$S / 380.55	55285-14-8	126	1.056^{20}	
4039	Carboxin / 1,4-OxathiiN-3-carboxamide, 5,6-dihydro-2-methyl-N-phenyl-	C$_{12}$H$_{13}$NO$_2$S / 235.31	5234-68-4 / 94	1832		
4040	Card-20(22)-enolide, 3-[(6-deoxy-α-L-mannopyranosyl)oxy]-1,5,11,14,19-pentahydroxy-, (1β,3β / Ouabain	C$_{29}$H$_{44}$O$_{12}$ / 584.66	630-60-4 / 200	6854	5-18-05-00625	H$_2$O 2; EtOH 4
4041	Card-20(22)-enolide, 3-[(O-2,6-dideoxy-β-D-ribo-hexopyranosyl-(1-4)-O-2,6-dideoxy-β-D-ribo-hexopyranosyl-(1-4)-2,6-dideoxy-β-D, 12,14-dihydroxy / Digoxin	C$_{41}$H$_{64}$O$_{14}$ / 780.95	20830-75-5 / 248-50 dec	3150	5-18-04-00381	EtOH 4
4042	Card-20(22)-enolide, 3-[(O-2,6-dideoxy-β-D-ribo-hexopyranosyl-(1-4)-O-2,6-dideoxy-β-D-ribo-hexopyranosyl-(1-4)-2,6-dideoxy-β-D, 14,16-dihydroxy / Gitoxin	C$_{41}$H$_{64}$O$_{14}$ / 780.95	4562-36-1 / 285 dec	4328	5-18-04-00388	
4043	Card-20(22)-enolide, 3-[(O-2,6-dideoxy-β-D-ribo-hexopyranosyl-(1-4)-O-2,6-dideoxy-β-D-ribo-hexopyranosyl-(1-4)-2,6-dideoxy-β-D, 14-dihydroxy / Digitoxin	C$_{41}$H$_{64}$O$_{13}$ / 764.95	71-63-6 / 255.5	3146	5-18-03-00354	H$_2$O 2; EtOH 4; eth 3; chl 3
4044	Card-20(22)-enolide, 3,14-dihydroxy-, (3β,5β) / Digitoxigenin	C$_{23}$H$_{34}$O$_4$ / 374.52	143-62-4 / 253	3145	5-18-03-00348	EtOH 3; MeOH 4
4045	Card-20(22)-enolide, 3-[(6-O-β-D-glucopyranosyl-β-D-glucopyranosyl)oxy]-14-hydroxy-, / Uzarin	C$_{35}$H$_{54}$O$_{14}$ / 698.81	20231-81-6 / 269	9809	4-18-00-01497	
4046	Card-20(22)-enolide, 3,11,14-trihydroxy-, (3β,5β,11α)- / Sarmentogenin	C$_{23}$H$_{34}$O$_5$ / 390.52	76-28-8 / 280	8334	5-18-04-00378	H$_2$O 1; EtOH 3; eth 1; ace 2
4047	Card-20(22)-enolide, 3,12,14-trihydroxy-, (3β,5β,12β)- / Digoxigenin	C$_{23}$H$_{34}$O$_5$ / 390.52	1672-46-4 / 222	3149	5-18-04-00380	EtOH 4; chl 2; MeOH 4
4048	Card-20(22)-enolide, 3,14,16-trihydroxy-, (3β,5β,16β)- / Gitoxigenin	C$_{23}$H$_{34}$O$_5$ / 390.52	545-26-6 / 234	4327	5-18-04-00385	H$_2$O 1; eth 2; chl 3
4049	Card-20(22)-enolide, 3,5,14-trihydroxy-19-oxo-, (3β,5β)- / Strophanthidin	C$_{23}$H$_{32}$O$_6$ / 404.50	66-28-4 / 173 dec	8818	5-18-05-00126	H$_2$O 1; EtOH 3; eth 1; ace 3
4050	4-Carene, (1S,3R,6R)-(-)- / Δ4-Carene (l)	C$_{10}$H$_{16}$ / 136.24	5208-49-1	169; 64^{20}	4-05-00-00451 / 0.8441^{30}	ace 4; bz 4; eth 4 / 1.4740^{30}
4051	β,β-Carotene / β-Carotene, all-trans-	C$_{40}$H$_{56}$ / 536.88	7235-40-7 / 183	1860	4-05-00-02617 / 1.00^{20}	H$_2$O 1; EtOH 2; eth 3; ace 3
4052	β,ψ-Carotene / γ-Carotene	C$_{40}$H$_{56}$ / 536.88	472-93-5 / 178	1861	4-05-00-02616	H$_2$O 1; EtOH 1; eth 2; bz 3
4053	ψ,ψ-Carotene / trans-Lycopene	C$_{40}$H$_{56}$ / 536.88	502-65-8 / 175	5492	4-01-00-01165	EtOH 2; eth 3; bz 4; chl 4
4054	β,ε-Carotene, (6'R)- / α-Carotene	C$_{40}$H$_{56}$ / 536.88	7488-99-5 / 187.5	1859	4-05-00-02620 / 1.00^{20}	bz 4; eth 4; chl 4
4055	β,β-Carotene, 3'-(acetyloxy)-6',7'-didehydro-5,6-epoxy-5,5',6,6',7,8-hexahydro-3,5'-dihydroxy-8-oxo-, / Fucoxanthin	C$_{40}$H$_{60}$O$_6$ / 636.91	3351-86-8 / 168	4195	5-18-04-00673	eth 4; EtOH 4
4056	β,β-Carotene-3,3'-diol, 5,6:5',6'-diepoxy-5,5',6,6'-tetrahydro- / Violaxanthin	C$_{40}$H$_{56}$O$_4$ / 600.88	126-29-4 / 208	9902	5-19-03-00464	EtOH 3; eth 3; CS$_2$ 3; peth 1
4057	β,β-Carotene-3,3'-diol, (3R,3'R)- / Zeaxanthin	C$_{40}$H$_{56}$O$_2$ / 568.88	144-68-3 / 215.5	10019 / 226-9$^{0.06}$	4-06-00-07017	H$_2$O 1; EtOH 2; eth 3; ace 3
4058	β,ε-Carotene-3,3'-diol, (3R,3'R,6'R)- / Xanthophyll	C$_{40}$H$_{56}$O$_2$ / 568.88	127-40-2 / 196	9972	3-06-00-05871	bz 4; eth 4; EtOH 4; peth 4
4059	β,ψ-Caroten-16'-oic acid, 3',4'-didehydro- / Torularhodin	C$_{40}$H$_{52}$O$_2$ / 564.85	514-92-1 / 211	9476	4-09-00-02767	py 4; chl 4; CS$_2$ 4
4060	ψ,ψ-Caroten-16-ol / Lycoxanthin	C$_{40}$H$_{56}$O / 552.88	19891-74-8 / 168	5499	4-01-00-02368	H$_2$O 1; EtOH 2; bz 3; CS$_2$ 3
4061	β,β-Caroten-3-ol, (3R)- / Cryptoxanthin	C$_{40}$H$_{56}$O / 552.88	472-70-8 / 160	2612	4-06-00-05111	bz 4; chl 4
4062	β,ψ-Caroten-3-ol, (3R)- / Rubixanthin	C$_{40}$H$_{56}$O / 552.88	3763-55-1 / 160	8265	4-06-00-05110	EtOH 2; bz 3; chl 3; peth 2
4063	β,κ-Caroten-6'-one, 3,3'-dihydroxy-, (3R,3'S,5'R)- / Capsanthin	C$_{40}$H$_{56}$O$_3$ / 584.88	465-42-9 / 176	1768	4-08-00-02657	
4064	Carpaine / 2,15-Dioxa-12,25-diazatricyclo[22.2.2.211,14]triacontane, carpaine deriv.	C$_{28}$H$_{50}$N$_2$O$_4$ / 478.72	3463-92-1 / 121	1866	2-27-00-00209	ace 4; bz 4; eth 4; EtOH 4

No.	Name / Synonym	Mol. Form. / Mol. Wt.	CAS RN / mp/°C	Merck No. / bp/°C	Beil. Ref. / den/g cm^{-3}	Solubility / n_D
4065	CDEC / Carbamodithioic acid, diethyl-, 2-chloro-2-propenyl ester	$C_8H_{14}ClNS_2$ / 223.79	95-06-7	8882 / 129[1]	1.088	
4066	Cedrene / 1H-3a,7-Methanoazulene, cedrene deriv.	$C_{15}H_{24}$ / 204.36	11028-42-5	/ 262.5	3-05-00-01095 / 0.9342[20]	H_2O 1; bz 3; lig 3 / 1.5034[20]
4067	Cevane-3,4,12,14,16,17,20-heptol, 3-acetate, (3β,4α,16β)- / Sabadine	$C_{29}H_{47}NO_8$ / 537.69	124-80-1 / 258	8281	4-21-00-02883	ace 4; EtOH 4
4068	Cevane-3,4,7,14,15,16,20-heptol, 4,9-epoxy-, (3β,4α,7α,15α,16β)- / Germine	$C_{27}H_{43}NO_8$ / 509.64	508-65-6 / 220	4307	5-21-13-00706	bz 3; MeOH 3; alk 3; acid 3
4069	Cevane-3,4,12,14,16,17,20-heptol, 4,9-epoxy-, 3-(2-methyl-2-butenoate), [3β(Z),4α,16β]- / Cevadine	$C_{32}H_{49}NO_9$ / 591.74	62-59-9 / 213-4 dec	2025	4-21-00-06820	
4070	Cevane-3,4,6,7,14,15,16,20-octol, 4,9-epoxy-, (3β,4α,6α,7α,15α,16β) / Protoverine	$C_{27}H_{43}NO_9$ / 525.64	76-45-9 / 221	7916	4-21-00-06841	H_2O 1; EtOH 3; bz 3; aq acid 3
4071	Chelidonine / Stylophorine	$C_{20}H_{19}NO_5$ / 353.37	476-32-4 / 135.5	2043 / 220[0.002]	2-27-00-00615	H_2O 1; EtOH 3; eth 3; chl 3
4072	Chinomethionat / Carbonic acid, dithio-, cyclic S,S-(6-methyl-2,3-quinoxalinediyl) ester	$C_{10}H_6N_2OS_2$ / 234.30	2439-01-2 / 170	6933		
4073	DL-chiro-Inositol	$C_6H_{12}O_6$ / 180.16	18685-70-6 / 253		4-06-00-07920 / 1.752[15]	H_2O 4; EtOH 2; eth 1; bz 1
4074	D-chiro-Inositol, 2-deoxy- / D-Quercitol	$C_6H_{12}O_5$ / 164.16	488-73-3 / 236	8046	4-06-00-07883 / 1.5845[13]	H_2O 4
4075	Chlorbenside / Benzene, 1-chloro-4-[[(4-chlorophenyl)methyl]thio]-	$C_{13}H_{10}Cl_2S$ / 269.19	103-17-3 / 75	2074	1.4210[20]	
4076	Chlorbromuron / Urea, N'-(4-bromo-3-chlorophenyl)-N-methoxy-N-methyl-	$C_9H_{10}BrClN_2O_2$ / 293.55	13360-45-7 / 96		1.69[20]	
4077	Chlordane / 4,7-Methano-1H-indene, 1,2,4,5,6,7,8,8-octachloro-2,3,3a,4,7,7a-hexahydro-	$C_{10}H_6Cl_8$ / 409.78	57-74-9 / 106	2079 / 175[1]	1.60[25]	
4078	Chlorfenvinphos / Phosphoric acid, 2-chloro-1-(2,4-dichlorophenyl)ethenyl diethyl ester	$C_{12}H_{14}Cl_3O_4P$ / 359.57	470-90-6 /	2087 / 170[0.05]		
4079	Chlorflurecol / 9H-Fluorene-9-carboxylic acid, 2-chloro-9-hydroxy-	$C_{14}H_9ClO_3$ / 260.68	2464-37-1		1.496[20]	
4080	Chloridazon / 3(2H)-Pyridazinone, 5-amino-4-chloro-2-phenyl-	$C_{10}H_8ClN_3O$ / 221.65	1698-60-8 / 205			
4081	Chlorimuron / 2-(4-Chloro-6-methoxypyrimidin-2-ylcarbamoylsulfamoyl) benzoic acid, ethyl ester	$C_{15}H_{15}ClN_4O_6S$ / 414.83	90982-32-4 / 186	2092		
4082	Chlorobenzilate / Benzeneacetic acid, 4-chloro-α-(4-chlorophenyl)-α-hydroxy-, ethyl ester	$C_{16}H_{14}Cl_2O_3$ / 325.19	510-15-6 / 37	/ 157[0.07]	1.2816[20]	
4083	5-Chloro-3-methyl-4-nitro-1H-pyrazole / 1H-Pyrazole, 3-chloro-5-methyl-4-nitro-	$C_4H_4ClN_3O_2$ / 161.55	6814-58-0 / 111			
4084	Chloroneb / Benzene, 1,4-dichloro-2,5-dimethoxy-	$C_8H_8Cl_2O_2$ / 207.06	2675-77-6 / 134	/ 268		
4085	Chlorosulfuric acid, ethyl ester	$C_2H_5ClO_3S$ / 144.58	625-01-4 /	/ 152.5; 93[100]	4-01-00-01326 / 1.3502[25]	eth 4; chl 4; lig 4 / 1.416[20]
4086	Chlorosulfuric acid, methyl ester	CH_3ClO_3S / 130.55	812-01-1 / -70	/ 134	4-01-00-01252 / 1.4805[25]	ace 4; bz 4; eth 4 / 1.4138[18]
4087	Chlorosulfurous acid, ethyl ester	$C_2H_5ClO_2S$ / 128.58	6378-11-6 /	/ 52.5[44]; 32[16]	3-01-00-01316 / 1.2766[25]	eth 4 / 1.4550[25]
4088	Chloroxuron / Urea, N'-[4-(4-chlorophenoxy)phenyl]-N,N-dimethyl-	$C_{15}H_{15}ClN_2O_2$ / 290.75	1982-47-4 / 151			
4089	Chlorpyrifos / Phosphorothioic acid, O,O-diethyl O-(3,5,6-trichloro-2-pyridinyl) ester	$C_9H_{11}Cl_3NO_3PS$ / 350.59	2921-88-2 / 42	2190		
4090	Chlorpyrifos-methyl / Phosphorothioic acid, O,O-dimethyl O-(3,5,6-trichloro-2-pyridinyl) ester	$C_7H_7Cl_3NO_3PS$ / 322.54	5598-13-0 / 43			
4091	Chlorsulfuron / 2-Chloro-N-[[(4-methoxy-6-methyl-1,3,5-triaziN-2-yl)amino]carbonyl]benzenesulfonamide	$C_{12}H_{12}ClN_5O_4S$ / 357.78	64902-72-3 / 176	2192		
4092	Chlorthiophos / Phosphorothioic acid, O-[2,5-dichloro-4-(methylthio)phenyl] O,O-diethyl ester	$C_{11}H_{15}Cl_2O_3PS_2$ / 361.25	21923-23-9 /	/ 150[0.001]		

No.	Name Synonym	Mol. Form. Mol. Wt.	CAS RN mp/°C	Merck No. bp/°C	Beil. Ref. den/g cm^{-3}	Solubility n_D
4093	Cholane 1H-Cyclopenta[a]phenanthrene, hexadecahydro-10,13-dimethyl-17-(1-methylbutyl)-, [8R-[8α,9β,10α,1	C$_{24}$H$_{42}$ 330.60	548-98-1 90	2198 190$^{0.001}$	4-05-00-01220	
4094	Cholan-24-oic acid Cholanic acid	C$_{24}$H$_{40}$O$_2$ 360.58	25312-65-6 163.5	2199	4-09-00-01992	EtOH 3; chl 3; HOAc 3
4095	Cholan-24-oic acid, 3,12-dihydroxy-, (3α,5β,12α)- Deoxycholic acid	C$_{24}$H$_{40}$O$_4$ 392.58	83-44-3 177	2881	4-10-00-01608	
4096	Cholan-24-oic acid, 3,6-dihydroxy-, (3α,5β,6α)- Hyodeoxycholic acid	C$_{24}$H$_{40}$O$_4$ 392.58	83-49-8 198.5	4794	4-10-00-01600	H$_2$O 2; EtOH 3; eth 2; ace 2
4097	Cholan-24-oic acid, 3,7-dihydroxy-, (3α,5β,7α)- Chenodiol	C$_{24}$H$_{40}$O$_4$ 392.58	474-25-9 119	2044	4-10-00-01604	H$_2$O 1; EtOH 4; eth 3; ace 4
4098	Cholan-24-oic acid, 3,7-dihydroxy-, (3α,5β,7β)- Ursodiol	C$_{24}$H$_{40}$O$_4$ 392.58	128-13-2 203	9801	4-10-00-01604	EtOH 4; eth 2
4099	Cholan-24-oic acid, 3-hydroxy-, (3α,5β)- Lithocholic acid	C$_{24}$H$_{40}$O$_3$ 376.58	434-13-9 186	5423	4-10-00-00785	H$_2$O 1; EtOH 3; eth 2; chl 3
4100	Cholan-24-oic acid, 3,7,12-trihydroxy-, (3α,5β,7α,12α)- Cholic acid	C$_{24}$H$_{40}$O$_5$ 408.58	81-25-4 198	2206	4-10-00-02072	H$_2$O 2; EtOH 3; eth 4; ace 3
4101	Cholan-24-oic acid, 3,7,12-trioxo-, (5β)- Dehydrocholic acid	C$_{24}$H$_{34}$O$_5$ 402.53	81-23-2 237	2858	4-10-00-03478	H$_2$O 1; EtOH 2; eth 1; ace 3
4102	Cholesta-3,5-diene	C$_{27}$H$_{44}$ 368.65	747-90-0 80	260^{13}	4-05-00-01620 0.925^{100}	H$_2$O 1; EtOH 3; eth 5; bz 5
4103	Cholesta-5,7-dien-3-ol, (3β)- 7-Dehydrocholesterol	C$_{27}$H$_{44}$O 384.65	434-16-2 150.5	2857	4-06-00-04153	H$_2$O 1; EtOH 2; eth 3; ace 3
4104	Cholesta-8,24-dien-3-ol, (3β,5α)-	C$_{27}$H$_{44}$O 384.65	128-33-6 110	160$^{0.001}$	3-06-00-02828	ace 3; chl 3; MeOH 3
4105	Cholestane, (5α)- 28,29,30-Trinorlanostane	C$_{27}$H$_{48}$ 372.68	481-21-0 80	2202 250^1	4-05-00-01227 0.9090^{88}	H$_2$O 1; EtOH 2; eth 4; bz 4 1.4887^{88}
4106	Cholestane, (5β)- Coprostane	C$_{27}$H$_{48}$ 372.68	481-20-9 72	2518	4-05-00-01226 0.9119^{87}	eth 4; chl 4 1.4884^{88}
4107	Cholestanol Dihydrocholesterol	C$_{27}$H$_{48}$O 388.68	80-97-7 141.5	2203	4-06-00-03577	eth 4; chl 4
4108	Cholestan-3-ol, (3α,5α)- Epicholestanol	C$_{27}$H$_{48}$O 388.68	516-95-0 185.5	3564	4-06-00-03577	chl 3
4109	Cholestan-3-ol, (3β,5β)- Coprosterol	C$_{27}$H$_{48}$O 388.68	360-68-9 102	2519	4-06-00-03576	H$_2$O 1; EtOH 4; eth 4; bz 4
4110	Cholest-5-en-3-ol (3β)- Cholesterol	C$_{27}$H$_{46}$O 386.66	57-88-5 148.5	2204 360 dec; 233$^{0.5}$	4-06-00-04000 1.067^{20}	H$_2$O 1; EtOH 2; eth 4; ace 2
4111	Cholest-4-en-3-ol, (3β)- Allocholesterol	C$_{27}$H$_{46}$O 386.66	517-10-2 132	253	3-06-00-02604	H$_2$O 1; EtOH 3; eth 4; ace 4
4112	Cholest-5-en-3-ol, (3α)- Epicholesterol	C$_{27}$H$_{46}$O 386.66	474-77-1 141.5	3565	4-06-00-04001	EtOH 2
4113	6-Chrysenamine 6-Aminochrysene	C$_{18}$H$_{13}$N 243.31	2642-98-0 210.5	2258	4-12-00-03492	
4114	Chrysene 1,2-Benzophenanthrene	C$_{18}$H$_{12}$ 228.29	218-01-9 258.2	2259 448	4-05-00-02554 1.274^{20}	H$_2$O 1; EtOH 2; eth 2; ace 2
4115	Chrysene, 5,6-dimethyl- 5,6-Dimethylchrysene	C$_{20}$H$_{16}$ 256.35	3697-27-6 129.3	200$^{0.5}$	4-05-00-02590	H$_2$O 1; EtOH 3; CS$_2$ 3; HOAc 3
4116	5,6-Chrysenedione	C$_{18}$H$_{10}$O$_2$ 258.28	2051-10-7 239.5	sub	4-07-00-02646	H$_2$O 1; EtOH 3; eth 2; bz 3
4117	Chrysene, 1-methyl- 1-Methylchrysene	C$_{19}$H$_{14}$ 242.32	3351-28-8 256.5	sub 140	4-05-00-02572	H$_2$O 1; EtOH 3
4118	Chrysene, octadecahydro-	C$_{18}$H$_{30}$ 246.44	2090-14-4 115	353	3-05-00-01107	EtOH 4
4119	Cinchonan-6',9-diol, (8α,9R)- Cupreine	C$_{19}$H$_{22}$N$_2$O$_2$ 310.40	524-63-0 202	2626	4-23-00-03260	EtOH 4
4120	Cinchonan-9-ol, (9S)- Cinchonine	C$_{19}$H$_{22}$N$_2$O 294.40	118-10-5 265	2289	5-23-12-00406	
4121	Cinchonan-9-ol, 10,11-dihydro-, (8α,9R)- Hydrocinchonidine	C$_{19}$H$_{24}$N$_2$O 296.41	485-64-3 229	4705		EtOH 4
4122	Cinchonan-9-ol, 10,11-dihydro-, (9S)- Hydrocinchonine	C$_{19}$H$_{24}$N$_2$O 296.41	485-65-4 268.5	4706		H$_2$O 3; EtOH 2; eth 1
4123	Cinchonan-9-ol, 10,11-dihydro-6'-methoxy-, (8α,9R)- Hydroquinine	C$_{20}$H$_{26}$N$_2$O$_2$ 326.44	522-66-7 172.5	4737	5-23-13-00340	ace 4; eth 4; EtOH 4; chl 4
4124	Cinchonan-9-ol, 10,11-dihydro-6'-methoxy-, (9S)- Hydroquinidine	C$_{20}$H$_{26}$N$_2$O$_2$ 326.44	1435-55-8 168.5	4736	5-23-13-00340	EtOH 3; eth 3; ace 3; chl 3
4125	Cinchonan-9-ol, 6'-methoxy-, (9R)- Epiquinidine	C$_{20}$H$_{24}$N$_2$O$_2$ 324.42	572-59-8 113	3570	0-23-00-00505	EtOH 4; eth 3
4126	Cinchonan-9-ol, 6'-methoxy-, (9S)- Quinidine	C$_{20}$H$_{24}$N$_2$O$_2$ 324.42	56-54-2 174	8072	5-23-13-00395	H$_2$O 2; EtOH 3; eth 2; bz 3

No.	Name / Synonym	Mol. Form. / Mol. Wt.	CAS RN / mp/°C	Merck No. / bp/°C	Beil. Ref. / den/g cm⁻³	Solubility / n_D
4127	Cinchonan-9-ol, 6'-methoxy-, (8α,9R)- Quinine	$C_{20}H_{24}N_2O_2$ 324.42	130-95-0 57	8075	5-23-13-00395	H_2O 2; EtOH 4; eth 3; ace 2 1.625[15]
4128	Cinchonan-9-ol, 6'-methoxy-, carbonate (2:1) (esteR), (8α,9R)- Quinine carbonate	$C_{41}H_{48}N_4O_5$ 674.84	146-06-5 189	8077	4-23-00-03282	
4129	Cinchonan-9-ol, 6'-methoxy-, (8α,9R)-, monoformate (salt) Quinine, formate	$C_{21}H_{26}N_2O_4$ 370.45	130-90-5 149.5	8081	4-23-00-03274	H_2O 4; EtOH 4; chl 4
4130	Cinchonan-9-ol, 6'-methoxy-, monohydrochloride, (8α,9R)- Quinine hydrochloride	$C_{20}H_{25}ClN_2O_2$ 360.88	130-89-2 159	8085		H_2O 4; EtOH 4; chl 4
4131	Cinchonan-9-ol, 6'-methoxy-, (8α,9R)-, mono(2-hydroxybenzoate) (salt) Quinine salicylate	$C_{27}H_{32}N_2O_6$ 480.56	750-90-3 195	8088	4-23-00-03277	EtOH 4; chl 4
4132	Cinchonan-9-ol, 6'-methoxy-, (8α,9R)-, sulfate (2:1) (salt) Quinine sulfate	$C_{40}H_{50}N_4O_8S$ 748.93	804-63-7 235.2	8089		EtOH 4
4133	Cinchonan-9-ol, (8α,9R)- Cinchonidine	$C_{19}H_{22}N_2O$ 294.40	485-71-2 210.5	2288 sub	5-23-12-00406	H_2O 1; EtOH 3; eth 2; bz 1
4134	Cinchonan-9-one, 6'-methoxy-, (8α)- Quininone	$C_{20}H_{22}N_2O_2$ 322.41	84-31-1 108	8093	4-25-00-00204	bz 4; eth 4; EtOH 4
4135	Cinnamic acid, β-propyl- 2-Hexenoic acid, 3-phenyl-	$C_{12}H_{14}O_2$ 190.24	4362-01-0 94	183-4[14]	3-09-00-02811	bz 4
4136	Cinnoline 1,2-Benzodiazine	$C_8H_6N_2$ 130.15	253-66-7 38	2309 114[0.3]	5-23-07-00124	eth 4; EtOH 4
4137	Ciodrin 2-Butenoic acid, 3-[(dimethoxyphosphinyl)oxy]-, 1-phenylethyl ester, (E)-	$C_{14}H_{19}O_6P$ 314.28	7700-17-6 	2603 135[0.03]	1.19[25]	
4138	Clethodim	$C_{17}H_{26}ClNO_3S$ 359.92	99129-21-2		1.14[20]	
4139	Clofentezine 3,6-Bis(2-chlorophenyl)-1,2,4,5-tetrazine	$C_{14}H_8Cl_2N_4$ 303.15	74115-24-5 182	2373		
4140	Clomazone 2-(2-Chlorobenzyl)-4,4-dimethyl-1,2-oxazolidin-3-one	$C_{12}H_{14}ClNO_2$ 239.70	81777-89-1	3203	1.192[20]	
4141	Cloprop Propanoic acid, 2-(3-chlorophenoxy)-	$C_9H_9ClO_3$ 200.62	101-10-0			
4142	Clopyralid 3,6-Dichloro-2-pyridinecarboxylic acid	$C_6H_3Cl_2NO_2$ 192.00	1702-17-6 151	2398		
4143	Con-5-enin-3-amine, N,N-dimethyl-, (3β)- Conessine	$C_{24}H_{40}N_2$ 356.60	546-06-5 125.5	2492 165-7[0.1]		H_2O 2; chl 3; HOAc 3
4144	Conhydrine, N-methyl- N-Methylconhydrine	$C_9H_{19}NO$ 157.26	94826-08-1 	94-5[11]	2-21-00-00011 0.9400[20]	1.4708[19]
4145	Coronene	$C_{24}H_{12}$ 300.36	191-07-1 437.3	525	4-05-00-02830 1.371[25]	H_2O 1; bz 2; con sulf 1
4146	Corynan-16-carboxylic acid, 18,17,18,19-tetradehydro-17-methoxy-, methyl ester, (16E)- Corynantheine	$C_{22}H_{26}N_2O_3$ 366.46	18904-54-6 165.5	2546	4-25-00-01255	EtOH 4
4147	Coumaphos Coumarin, 3-chloro-7-hydroxy-4-methyl, O-ester with O,O-diethylphosphorothioate	$C_{14}H_{16}ClO_5PS$ 362.77	56-72-4 93	2559	1.474	
4148	o-Cresol, 4-(3-methylpentyl)- Phenol, 2-methyl-4-(3-methyl pentyl)	$C_{13}H_{20}O$ 192.30	882-25-7 35	145-6[11]		ace 4 1.5200[25]
4149	Crotonaldehyde, 2-bromo-, diethyl acetal 2-Butenal, 2-bromo, diethyl acetal	$C_8H_{15}BrO_2$ 223.11	98551-14-5 	86[15]	1-01-00-00380 1.2255[21]	1.4565[21]
4150	Crotonic acid, 2-chloro- 2-Chlorocrotonic acid	$C_4H_5ClO_2$ 120.54	600-13-5 99.8	212; 111[14]	4-02-00-01509	eth 4; EtOH 4
4151	Crotonic acid, 4-chloro-, (E)- 2-Butenoic acid, 4-chloro (trans)	$C_4H_5ClO_2$ 120.54	26340-58-9 83	117-8[13]	4-02-00-01511	eth 4
4152	Crotonic acid, 4,4-dichloro- 4,4-Dichlorocrotonic acid	$C_4H_4Cl_2O_2$ 154.98	99980-00-4 100.5	130[18]	4-02-00-01512 1.3331[99]	eth 4; EtOH 4; chl 4 1.4597[99]
4153	Crufomate Phosphoramidic acid, methyl-, 2-chloro-4-(1,1-dimethylethyl)phenyl methyl ester	$C_{12}H_{19}ClNO_3P$ 291.71	299-86-5 60	2607 118[0.01]		
4154	Curan-17-ol, (16α)- Geissoschizoline	$C_{19}H_{26}N_2O$ 298.43	18397-07-4 135 dec	4274	5-23-12-00211	H_2O 1; EtOH 4; eth 4; chl 4
4155	Cyanamide Cyanogenamide	CH_2N_2 42.04	420-04-2 44	2691 140[19]	4-03-00-00145 1.282[20]	H_2O 4; EtOH 4; eth 3; ace 3 1.4418[48]
4156	Cyanamide, diethyl-	$C_5H_{10}N_2$ 98.15	617-83-4 -80.6	188	4-04-00-00381 0.854[20]	H_2O 1; EtOH 3; eth 3 1.4126[25]
4157	Cyanamide, dimethyl-	$C_3H_6N_2$ 70.09	1467-79-4 	163.5	4-04-00-00226	ace 4; eth 4; EtOH 4 1.4089[19]
4158	Cyanamide, diphenyl- Diphenylcyanamide	$C_{13}H_{10}N_2$ 194.24	27779-01-7 73.5	237[60]	4-12-00-00857	EtOH 4; lig 4
4159	Cyanamide, di-2-propenyl- Diallylcyanamide	$C_7H_{10}N_2$ 122.17	538-08-9 	2952 142[90]; 95[9]	4-04-00-01078	EtOH 3; eth 2; ctc 2
4160	Cyanamide, methylphenyl-	$C_8H_8N_2$ 132.17	18773-77-8 32	137[12]	4-12-00-00840 1.0040[55]	

No.	Name / Synonym	Mol. Form. / Mol. Wt.	CAS RN / mp/°C	Merck No. / bp/°C	Beil. Ref. / den/g cm^{-3}	Solubility / n_D
4161	Cyanazine / Propanenitrile, 2-[[4-chloro-6-(ethylamino)-1,3,5-triazin-2-yl]amino]-2-methyl-	$C_9H_{13}ClN_6$ 240.70	21725-46-2 168	2692		
4162	Cyanic acid / Hydrogen cyanate (HOCN)	CHNO 43.03	420-05-3 -86	2693 23	4-03-00-00080 1.140[20]	H_2O 4; bz 4; eth 4; chl 4
4163	Cyanic acid, ethyl ester / Ethyl cyanate	C_3H_5NO 71.08	627-48-5	162 dec; 30[12]	0.89[20]	EtOH 4 1.3788[25]
4164	Cyanic acid, sodium salt / Sodium cyanate	CNNaO 65.01	917-61-3 550	8552	1.893[20]	H_2O 3; EtOH 2; eth 1
4165	Cyanogen bromide / Bromine cyanide	CBrN 105.92	506-68-3 52	2700 61.5	4-03-00-00092 2.015[20]	H_2O 3; EtOH 3; eth 3
4166	Cyanogen chloride / Chlorine cyanide	CClN 61.47	506-77-4 -6.5	2701 13	4-03-00-00090 1.186[20]	H_2O 3; EtOH 3; eth 4
4167	Cyanogen fluoride / Fluorine cyanide	CFN 45.02	1495-50-7 -82		-46	
4168	Cycloate / Carbamothioic acid, cyclohexylethyl-, S-ethyl ester	$C_{11}H_{21}NOS$ 215.36	1134-23-2 11.5	145[10]	1.0156[30]	
4169	Cyclobutanamine / Aminocyclobutane	C_4H_9N 71.12	2516-34-9	82	4-12-00-00003 0.8328[20]	1.4363[19]
4170	Cyclobutane / Tetramethylene	C_4H_8 56.11	287-23-0 -90.6	2720 12.6	4-05-00-00006 0.6890[25]	H_2O 1; EtOH 4; eth 5; ace 4 1.365[290]
4171	Cyclobutaneacetic acid, 3-(ethoxycarbonyl)-2,2-dimethyl-, ethyl ester / Pinic acid diethyl ester	$C_{13}H_{22}O_4$ 242.32	28664-03-1	142-6[10]	2-09-00-00529 1.0104[20]	1.4496[70]
4172	Cyclobutane, 1,2-bis(methylene)-	C_6H_8 80.13	14296-80-1	64	4-05-00-00389 0.7698[20]	1.4232[20]
4173	Cyclobutanecarbonitrile / Cyanocyclobutane	C_5H_7N 81.12	4426-11-3	149.6		
4174	Cyclobutanecarboxylic acid / Cyclobutylcarboxylic acid	$C_5H_8O_2$ 100.12	3721-95-7 -2	190; 74-5[2]	4-09-00-00006 1.0599[20]	H_2O 2; EtOH 5; eth 5 1.4400[20]
4175	Cyclobutanecarboxylic acid, ethyl ester / Ethyl cyclobutanecarboxylate	$C_7H_{12}O_2$ 128.17	14924-53-9	159	3-09-00-00007 0.928[25]	ctc 2 1.4261[20]
4176	Cyclobutanecarboxylic acid, methyl ester / Methyl cyclobutanecarboxylate	$C_6H_{10}O_2$ 114.14	765-85-5	135.5		
4177	Cyclobutane, 1-chloro-1,2,2,3,3,4,4-heptafluoro- / Chloroheptafluorocyclobutane	C_4ClF_7 216.49	377-41-3			
4178	Cyclobutane, 1,1-dibromo- / 1,1-Dibromocyclobutane	$C_4H_6Br_2$ 213.90	33742-81-3	160	0-05-00-00017 1.930[20]	bz 4; EtOH 4; lig 4 1.5362[20]
4179	Cyclobutane, 1,2-dibromo- / 1,2-Dibromocyclobutane	$C_4H_6Br_2$ 213.90	89033-70-5 2.5	172.5	0-05-00-00017 1.972[0]	
4180	1,1-Cyclobutanedicarboxylic acid, diethyl ester / Diethyl 1,1-cyclobutanedicarboxylate	$C_{10}H_{16}O_4$ 200.23	3779-29-1	224	4-09-00-02788 1.0456[20]	EtOH 4; ctc 2 1.4330[26]
4181	1,2-Cyclobutanedicarboxylic acid, dimethyl ester, trans- / Dimethyl trans-1,2-cyclobutanedicarboxylate	$C_8H_{12}O_4$ 172.18	7371-67-7	225; 118[24]	3-09-00-03799 1.1276[20]	1.4450[20]
4182	1,3-Cyclobutanedicarboxylic acid, cis-	$C_6H_8O_4$ 144.13	2398-16-5 143.5	252	3-09-00-03801	H_2O 4; EtOH 4
4183	1,3-Cyclobutanedicarboxylic acid, 2,2-dimethyl-, cis- / cis-Norpinic acid	$C_8H_{12}O_4$ 172.18	3211-48-1 175	sub 100	4-09-00-02813	eth 4; chl 4
4184	1,3-Cyclobutanedicarboxylic acid, 2,2-dimethyl-, dimethyl ester / Norpinic acid, dimethyl ester	$C_{10}H_{16}O_4$ 200.23	91057-79-3	229; 113[11]	3-09-00-03829 1.070[14]	1.4459[17]
4185	1,3-Cyclobutanedicarboxylic acid, trans-	$C_6H_8O_4$ 144.13	7439-33-0 171	sub	3-09-00-03802	H_2O 4; EtOH 4
4186	Cyclobutane, 1,2-diethenyl-, cis-	C_8H_{12} 108.18	16177-46-1	38[38]	4-05-00-00412 0.8010[20]	1.4563[20]
4187	Cyclobutane, 1,2-diethenyl-, trans- / trans-1,2-Divinylcyclobutane	C_8H_{12} 108.18	6553-48-6	112.5	4-05-00-00412 0.7817[20]	1.4451[20]
4188	Cyclobutane, 1,2-dimethyl-, trans- / trans-1,2-Dimethylcyclobutane	C_6H_{12} 84.16	15679-02-4 -122.5	58	4-05-00-00088 0.7032[20]	1.395[20]
4189	Cyclobutane, 1,3-dimethyl-, cis- / cis-1,3-Dimethylcyclobutane	C_6H_{12} 84.16	2398-09-6	60.5	4-05-00-00089 0.7060[25]	1.3933[20]
4190	Cyclobutane, 1,3-dimethyl-, trans- / trans-1,3-Dimethylcyclobutane	C_6H_{12} 84.16	2398-10-9	57.5	4-05-00-00089 0.6970[25]	1.3896[20]
4191	Cyclobutane, ethyl- / Ethylcyclobutane	C_6H_{12} 84.16	4806-61-5 -142.9	70.8	4-05-00-00087 0.7284[20]	H_2O 1; EtOH 5; eth 5; ace 3 1.4020[20]
4192	Cyclobutane, methyl- / Methylcyclobutane	C_5H_{10} 70.13	598-61-8 -161.5	36.3	4-05-00-00021 0.6884[20]	H_2O 1; EtOH 5; eth 5; ace 3 1.3866[20]
4193	Cyclobutane, methylene- / exo-Methylenecyclobutane	C_5H_8 68.12	1120-56-5 -134.7	42.2	4-05-00-00216 0.7401[20]	1.4210[20]
4194	Cyclobutane, octafluoro- / Perfluorocyclobutane	C_4F_8 200.03	115-25-3 -40.1	6668 -5.9	4-05-00-00008 *1.500[25]	eth 3
4195	Cyclobutane, 1,2,3,4-tetrabromo- / 1,2,3,4-Tetrabromocyclobutane	$C_4H_4Br_4$ 371.69	101257-79-8	110[1.5]	4-05-00-00012 2.5672[20]	1.6303[20]
4196	Cyclobutanol / Hydroxycyclobutane	C_4H_8O 72.11	2919-23-5	124	4-06-00-00003 0.9218[15]	1.4371[20]

No.	Name Synonym	Mol. Form. Mol. Wt.	CAS RN mp/°C	Merck No. bp/°C	Beil. Ref. den/g cm^{-3}	Solubility n_D
4197	Cyclobutanone	C_4H_6O 70.09	1191-95-3 -50.9	99	4-07-00-00003 0.9547[0]	H_2O 3; EtOH 4; eth 3; bz 3 1.4215[20]
4198	Cyclobutene	C_4H_6 54.09	822-35-5	2	4-05-00-00207 0.733[0]	ace 4; bz 3; peth 3
4199	Cyclobutene, 1,2-dichloro-3,3,4,4-tetrafluoro-	$C_4Cl_2F_4$ 194.94	377-93-5 -43.4	67	4-05-00-00209 1.534[25]	1.3699[25]
4200	Cyclobutene, hexafluoro-	C_4F_6 162.03	697-11-0 -60	5.5	4-05-00-00208 1.602[-20]	1.298[-20]
4201	Cyclobutene, 1-methyl- 1-Methylcyclobutene	C_5H_8 68.12	1489-60-7	37	4-05-00-00216 0.7244[20]	1.4088[18]
4202	Cyclodecane	$C_{10}H_{20}$ 140.27	293-96-9 10	202	4-05-00-00144 0.8538[25]	1.4716[20]
4203	1,2-Cyclodecanedione Sebacil	$C_{10}H_{16}O_2$ 168.24	96-01-5 40.5	8370 104-5[10]	4-07-02009	
4204	Cyclodecanol	$C_{10}H_{20}O$ 156.27	1502-05-2 40.5	125[12]	4-06-00-00138 0.9606[20]	EtOH 3 1.4926[20]
4205	Cyclodecanone	$C_{10}H_{18}O$ 154.25	1502-06-3 28	106-7[13]	4-07-00-00076 0.9654[20]	bz 4; eth 4; chl 4 1.4806[20]
4206	Cyclodecanone, 2-hydroxy- Sebacoin	$C_{10}H_{18}O_2$ 170.25	96-00-4 38.5	8371 134-8[14]	4-08-00-00037	
4207	Cyclodecene, (E)-	$C_{10}H_{18}$ 138.25	2198-20-1	68-70[10]	4-05-00-00296 0.8672[20]	H_2O 1; bz 3; chl 3 1.4822[20]
4208	Cyclodecene, (Z)-	$C_{10}H_{18}$ 138.25	935-31-9	195	4-05-00-00295 0.8770[20]	1.4854[20]
4209	Cyclodecyne	$C_{10}H_{16}$ 136.24	3022-41-1	204	4-05-00-00430 0.8975[20]	1.4903[20]
4210	Cyclododecane	$C_{12}H_{24}$ 168.32	294-62-2 60.4	247	4-05-00-00169 0.82[80]	
4211	Cyclododecanone Cyclodoecanone	$C_{12}H_{22}O$ 182.31	830-13-7 59	126-8[12]	4-07-00-00100 0.9059[66]	1.4571[60]
4212	1,5,9-Cyclododecatriene CDT	$C_{12}H_{18}$ 162.27	4904-61-4 -17	240	4-05-00-01114 0.84[100]	
4213	1,5,9-Cyclododecatriene, (Z,E,E)- 1,5,9-Cyclododecatriene, (cis, trans, trans)	$C_{12}H_{18}$ 162.27	53859-78-2 -18	100-1[11]	4-05-00-01115 0.8910[20]	bz 4; chl 4 1.5058[20]
4214	1,5,9-Cyclododecatriene, (Z,Z,Z)-	$C_{12}H_{18}$ 162.27	4736-48-5 -1.5	108[17]		bz 4; chl 4 1.5100[25]
4215	Cyclododecene, (E)-	$C_{12}H_{22}$ 166.31	1486-75-5	112-5[17]	4-05-00-00332 0.8[20]	bz 4; chl 4 1.4850[20]
4216	Cyclododecene, (Z)-	$C_{12}H_{22}$ 166.31	1129-89-1	133[35]; 71[2]	4-05-00-00332	bz 4; chl 4 1.4840[20]
4217	9-Cycloheptadecen-1-one 9-Cycloheptadecene-1-one	$C_{17}H_{30}O$ 250.42	74244-64-7 32.5	343; 159[2]	3-07-00-00524 0.9170[33]	H_2O 2; EtOH 3; bz 3 1.4830[33]
4218	9-Cycloheptadecen-1-one, (Z)- Civetone	$C_{17}H_{30}O$ 250.42	542-46-1 32.5	2337 343; 159[2]	3-07-00-00524	
4219	1,3-Cycloheptadiene	C_7H_{10} 94.16	4054-38-0 -110.4	120.5	4-05-00-00390 0.868[25]	1.4978[20]
4220	2,4-Cycloheptadien-1-one, 2,6,6-trimethyl- Eucarvone	$C_{10}H_{14}O$ 150.22	503-93-5	105[22]; 88[16]	4-07-00-00311 0.9490[20]	eth 3; ace 3 1.5087[20]
4221	Cycloheptane	C_7H_{14} 98.19	291-64-5 -8.0	118.4	4-05-00-00092 0.8098[20]	H_2O 1; EtOH 4; eth 4; bz 3 1.4436[20]
4222	Cycloheptane, bromo- Cycloheptyl bromide	$C_7H_{13}Br$ 177.08	2404-35-5	101.5[40]; 75[12]	4-05-00-00093 1.2887[22]	H_2O 1; eth 4; chl 4 1.4996[20]
4223	Cycloheptanecarboxylic acid Cycloheptylcarboxylic acid	$C_8H_{14}O_2$ 142.20	1460-16-8	259; 130[8]	4-09-00-00038 1.0423[20]	ace 3 1.4753[20]
4224	Cycloheptanecarboxylic acid, ethyl ester Ethyl cycloheptanecarboxylate	$C_{10}H_{18}O_2$ 170.25	32777-26-7	74[3]	4-09-00-00038 0.9515[20]	1.4482[20]
4225	1,2-Cycloheptanedione	$C_7H_{10}O_2$ 126.16	3008-39-7 -40	107-9[17]	4-07-00-01989 1.0583[22]	EtOH 3 1.4689[22]
4226	Cycloheptane, methyl- Methylcycloheptane	C_8H_{16} 112.22	4126-78-7	134	4-05-00-00114 0.800[20]	bz 4; eth 4; EtOH 4; peth 4 1.439[18]
4227	Cycloheptane, 1,1,2-trimethyl- 1,1,2-Trimethylcycloheptane	$C_{10}H_{20}$ 140.27	35099-89-9	104-5[100]	3-05-00-00127 0.8243[20]	1.4527[20]
4228	Cycloheptane, 1,1,4-trimethyl- 1,1,4-Trimethylcycloheptane	$C_{10}H_{20}$ 140.27	2158-55-6	164	4-05-00-00146 0.8011[20]	1.4420[20]
4229	Cycloheptanol	$C_7H_{14}O$ 114.19	502-41-0 2	185	4-06-00-00094 0.9554[20]	H_2O 2; EtOH 4; eth 4 1.40705[20]
4230	Cycloheptanol, 1-methyl- 1-Methyl-1-cycloheptanol	$C_8H_{16}O$ 128.21	3761-94-2	183.5	4-06-00-00114 0.9392[21]	1.4677[22]
4231	Cycloheptanol, 2-methyl- 2-Methylcycloheptanol	$C_8H_{16}O$ 128.21	59777-92-3	191	3-06-00-00083 0.9492[15]	1.4762[15]
4232	Cycloheptanol, 4-methyl-	$C_8H_{16}O$ 128.21	90200-61-6	105-6[39]	3-06-00-00084 0.9134[31]	1.4574
4233	Cycloheptanol, 2,6,6-trimethyl- 3,3,7-Trimethylcycloheptanol	$C_{10}H_{20}O$ 156.27	33515-82-1	216	1-06-00-00018 0.9096[24]	1.4369
4234	Cycloheptanone Suberone	$C_7H_{12}O$ 112.17	502-42-1	2728 178.5	4-07-00-00039 0.9508[20]	H_2O 1; EtOH 4; eth 3 1.4608[20]
4235	Cycloheptanone, 2-methyl- 2-Methylcycloheptanone	$C_8H_{14}O$ 126.20	932-56-9	185	4-07-00-00051 0.9395[18]	1.461[18]
4236	Cycloheptanone, 2-phenyl- 2-Phenylcycloheptanone	$C_{13}H_{16}O$ 188.27	14996-78-2 22.2	141[5]; 125[2]	4-07-00-01061	1.5395[20]

No.	Name / Synonym	Mol. Form. / Mol. Wt.	CAS RN / mp/°C	Merck No. / bp/°C	Beil. Ref. / den/g cm^{-3}	Solubility / n_D
4237	Cycloheptanone, 2,2,6-trimethyl-	$C_{10}H_{18}O$ 154.25	1686-41-5	191	3-07-00-00138 0.878[14]	1.4421[16]
4238	Cycloheptanone, 2,6,6-trimethyl- Tetrahydroeucarvone	$C_{10}H_{18}O$ 154.25	4436-59-3	208.5	4-07-00-00079 0.9095[18]	1.4568[18]
4239	Cycloheptasiloxane, tetradecamethyl- Tetradecamethylcycloheptasiloxane	$C_{14}H_{42}O_7Si_7$ 519.08	107-50-6 -26	154[20]	4-04-00-04130 0.9703[20]	1.4040[20]
4240	1,3,5-Cycloheptatriene Tropilidene	C_7H_8 92.14	544-25-2 -79.5	117; 60.5[122]	4-05-00-00765 0.8875[19]	H_2O 1; EtOH 3; eth 3; bz 4 1.5343[20]
4241	1,3,6-Cycloheptatriene-1-carboxylic acid, 5,5-dimethyl- Thujic acid	$C_{10}H_{12}O_2$ 164.20	499-89-8	9325	4-09-00-01810	
4242	2,4,6-Cycloheptatrien-1-one	C_7H_6O 106.12	539-80-0 -7	113[15]; 84[6]	4-07-00-00501 1.095[22]	bz 4; chl 4 1.6172[22]
4243	2,4,6-Cycloheptatrien-1-one, 2-hydroxy-	$C_7H_6O_2$ 122.12	533-75-5 50.8	sub 40	4-08-00-00159	H_2O 3; eth 3; ace 3; os 3
4244	Cycloheptene	C_7H_{12} 96.17	628-92-2 -56	115	4-05-00-00244 0.8228[20]	H_2O 1; EtOH 3; eth 3; bz 3 1.4552[20]
4245	Cycloheptene, 1-methyl- 1-Methylcycloheptene	C_8H_{14} 110.20	1453-25-4	136	4-05-00-00265 0.8243[22]	1.4581[20]
4246	Cycloheptene, 5-methyl- 5-Methylcycloheptene	C_8H_{14} 110.20	2505-06-8	69-70[38]	3-05-00-00211 0.7606[31]	1.4202[31]
4247	Cycloheptene, 1,4,4-trimethyl- 1,4,4-Trimethyl-1-cycloheptene	$C_{10}H_{18}$ 138.25	4755-36-6	165	3-05-00-00232 0.8185[20]	1.4561[20]
4248	1,3-Cyclohexadiene	C_6H_8 80.13	592-57-4 -89	80.5	4-05-00-00382 0.8405[20]	H_2O 1; EtOH 3; eth 4; bz 3 1.4755[20]
4249	1,4-Cyclohexadiene 1,4-Dihydrobenzene	C_6H_8 80.13	628-41-1 -49.2	85.5	4-05-00-00385 0.8471[20]	H_2O 1; EtOH 5; eth 5; bz 3 1.4725[20]
4250	1,3-Cyclohexadiene-1-carboxaldehyde, 2,6,6-trimethyl- Safranal	$C_{10}H_{14}O$ 150.22	116-26-7	8286 70[1]	4-07-00-00318 0.9734[19]	EtOH 4; peth 4 1.5281[19]
4251	1,4-Cyclohexadiene-1,2-dicarbonitrile, 4,5-dichloro-3,6-dioxo- 2,3-Dichloro-5,6-dicyanobenzoquinone	$C_8Cl_2N_2O_2$ 227.01	84-58-2 214.5	3052	0-10-00-00902	bz 4; HOAc 4; diox 4
4252	1,3-Cyclohexadiene, 1,3-dimethyl- 1,3-Dimethyl-1,3-cyclohexadiene	C_8H_{12} 108.18	4573-05-1	135.5	3-05-00-00325 0.8373[20]	1.4776[20]
4253	1,3-Cyclohexadiene, 1,4-dimethyl- 1,4-dimethyl-1,3-cyclohexadiene	C_8H_{12} 108.18	26120-52-5	136.5	4-05-00-00409 0.830[20]	1.4792[19]
4254	1,3-Cyclohexadiene, 1,5-dimethyl- 1,5-Dimethyl-1,3-cyclohexadiene	C_8H_{12} 108.18	1453-17-4	130	1-05-00-00063 0.821[20]	1.471[20]
4255	1,3-Cyclohexadiene, 2,5-dimethyl- 2,5-Dimethyl-1,3-cyclohexadiene	C_8H_{12} 108.18	2050-33-1	127	3-05-00-00325 0.8245[18]	1.4631[18]
4256	1,3-Cyclohexadiene, 2,6-dimethyl- 1,3-Dimethyl-3,5-cyclohexadiene	C_8H_{12} 108.18	2050-32-0	131	0-05-00-00119 0.8225[20]	1.4675[20]
4257	2,5-Cyclohexadiene-1,4-dione p-Benzoquinone	$C_6H_4O_2$ 108.10	106-51-4 115.7	8103 sub	4-07-00-02065 1.318[20]	H_2O 2; EtOH 3; eth 3; chl 3
4258	2,5-Cyclohexadiene-1,4-dione, 2,5-dichloro-3,6-dihydroxy- Chloranilic acid	$C_6H_2Cl_2O_4$ 208.98	87-88-7 283.5	2072	4-08-00-02707	H_2O 3
4259	2,5-Cyclohexadiene-1,4-dione, 2,5-dihydroxy-3,6-dinitro- Nitranilic acid	$C_6H_2N_2O_8$ 230.09	479-22-1 100 dec	6489 exp 170	3-08-00-03351	H_2O 4; EtOH 4; eth 1
4260	2,5-Cyclohexadiene-1,4-dione, 2,5-dihydroxy-3-methoxy-6-methyl- Spinulosin	$C_8H_8O_5$ 184.15	85-23-4 202.5	8706 sub 120	3-08-00-03983	H_2O 2; alk 3
4261	2,5-Cyclohexadiene-1,4-dione, 2,5-dihydroxy-3-undecyl- Embelin	$C_{17}H_{26}O_4$ 294.39	550-24-3 142.5	3513	4-08-00-02769	bz 4; eth 4; EtOH 4
4262	2,5-Cyclohexadiene-1,4-dione, 2,6-dimethoxy- 2,6-Dimethoxy-p-quinone	$C_8H_8O_4$ 168.15	530-55-2 256	3214 sub	4-08-00-02710	H_2O 2; EtOH 2; eth 2; tfa 3
4263	2,5-Cyclohexadiene-1,4-dione, 2,3-dimethyl-	$C_8H_8O_2$ 136.15	526-86-3 55	sub	4-07-00-02093	H_2O 2; EtOH 3; eth 3; chl 3
4264	2,5-Cyclohexadiene-1,4-dione, 2,6-dimethyl-	$C_8H_8O_2$ 136.15	527-61-7 72.5	sub	4-07-00-02095 1.0479[28]	chl 3
4265	2,5-Cyclohexadiene-1,4-dione, 3-hydroxy-2-methoxy-5-methyl- Fumigatin	$C_8H_8O_4$ 168.15	484-89-9 116	4201	3-08-00-03374	ace 4; bz 4; eth 4; EtOH 4
4266	2,5-Cyclohexadiene-1,4-dione, 2-methyl-	$C_7H_6O_2$ 122.12	553-97-9 69	sub	4-07-00-02088 1.08[75]	H_2O 2; EtOH 3; eth 3
4267	2,5-Cyclohexadiene-1,4-dione, 2-methyl-5-(1-methylethyl)-	$C_{10}H_{12}O_2$ 164.20	490-91-5 45.5	232	4-07-00-02100	chl 3
4268	2,5-Cyclohexadiene-1,4-dione, 2,3,5,6-tetrachloro- Chloranil	$C_6Cl_4O_2$ 245.88	118-75-2 290	2071 sub	4-07-00-02083	H_2O 1; EtOH 2; eth 3; chl 2
4269	2,5-Cyclohexadiene-1,4-dione, 2,3,5,6-tetrahydroxy- Tetroquinone	$C_6H_4O_6$ 172.09	319-89-1	9177	4-08-00-03604	H_2O 2; EtOH 4; eth 2; ctc 2
4270	2,5-Cyclohexadiene-1,4-dione, 2,3,5,6-tetramethyl- Duroquinone	$C_{10}H_{12}O_2$ 164.20	527-17-3 111.5	3452	4-07-00-02102	H_2O 1; EtOH 3; eth 3; ace 3
4271	1,3-Cyclohexadiene, 2-methyl- 4,5-Dihydrotoluene	C_7H_{10} 94.16	1489-57-2	107.5	3-06-00-00211 0.8260[18]	1.4662[18]

No.	Name / Synonym	Mol. Form. / Mol. Wt.	CAS RN / mp/°C	Merck No. / bp/°C	Beil. Ref. / den/g cm⁻³	Solubility / n_D
4272	1,3-Cyclohexadiene, 5-methyl-, (±)- 1,3-Cyclohexadiene, 5-methyl (DL)	C_7H_{10} 94.16	116781-86-3 	 101.5	3-05-00-00318 0.8354[20]	bz 4; eth 4; EtOH 4; lig 4 1.4763[20]
4273	1,3-Cyclohexadiene, 1-methyl-4-(1-methylethyl)-	$C_{10}H_{16}$ 136.24	99-86-5 	 175	4-05-00-00435 0.8375[19]	H_2O 1; EtOH 5; eth 5 1.477[19]
4274	1,3-Cyclohexadiene, 2-methyl-5-(1-methylethyl)-, (±)-	$C_{10}H_{16}$ 136.24	13811-01-3 	 175.5	3-05-00-00342 0.8410[22]	H_2O 1; EtOH 1; eth 3 1.4772[19]
4275	1,4-Cyclohexadiene, 1-methyl-4-(1-methylethyl)-	$C_{10}H_{16}$ 136.24	99-85-4 	 183	4-05-00-00436 0.849[20]	 1.4765[14]
4276	1,3-Cyclohexadiene, 1,2,3,4,5,5,6,6-octafluoro-	C_6F_8 224.05	377-70-8 	 63.5	4-05-00-00384 1.601[20]	 1.3149[20]
4277	1,4-Cyclohexadiene, 1,2,3,3,4,5,6,6-octafluoro-	C_6F_8 224.05	775-51-9 	 57.5	4-05-00-00386 1.6010[25]	 1.318[18]
4278	2,5-Cyclohexadien-1-one, 4-[bis(4-hydroxyphenyl)methylene]- Aurin	$C_{19}H_{14}O_3$ 290.32	603-45-2 308-10 dec	899 	4-08-00-02646 	H_2O 1; EtOH 3; eth 2; bz 1
4279	2,5-Cyclohexadien-1-one, 2,6-dibromo-4-(chloroimino)- 2,6-Dibromoquinone-4-chlorimide	$C_6H_2Br_2ClNO$ 299.35	537-45-1 83	3010 	4-07-00-02086 	EtOH 4
4280	2,5-Cyclohexadien-1-one, 2,6-dichloro-4-(chloroimino)- Gibbs' reagent	$C_6H_2Cl_3NO$ 210.45	101-38-2 66	4315 	3-07-00-03377 	
4281	2,4-Cyclohexadien-1-one, 3,5-dihydroxy-2,6,6-tris(3-methyl-2-butenyl)-4-(3-methyl-1-oxobutyl)- Lupulon	$C_{26}H_{38}O_4$ 414.59	468-28-0 93	5482 	3-07-00-04753 	H_2O 1; EtOH 3; peth 3; hx 3
4282	2,4-Cyclohexadien-1-one, 3,5,6-trihydroxy-4,6-bis(3-methyl-2-butenyl)-2-(3-methyl-1-oxobutyl)-, (R)- Humulon	$C_{21}H_{30}O_5$ 362.47	26472-41-3 66.5	4673 	4-08-00-03410 	H_2O 2; EtOH 3; eth 3; ace 3
4283	Cyclohexanamine Cyclohexylamine	$C_6H_{13}N$ 99.18	108-91-8 -17.7	2735 134	4-12-00-00008 0.819[20]	H_2O 3; EtOH 4; eth 5; ace 5 1.4372[20]
4284	Cyclohexanamine, N-cyclohexyl- Dicyclohexylamine	$C_{12}H_{23}N$ 181.32	101-83-7 -0.1	3085 256 dec; 114[9]	4-12-00-00022 0.9123[20]	H_2O 2; EtOH 3; eth 3; bz 3 1.4842[20]
4285	Cyclohexanamine, N,N-diethyl-	$C_{10}H_{21}N$ 155.28	91-65-6 	 193; 85[20]	4-12-00-00019 0.872[0]	EtOH 3; ctc 2
4286	Cyclohexanamine, N,N-dimethyl- Cyclohexyldimethylamine	$C_8H_{17}N$ 127.23	98-94-2 	 162	4-12-00-00018 	
4287	Cyclohexanamine, 2-ethyl- 2-Ethylcyclohexylamine	$C_8H_{17}N$ 127.23	6850-36-8 	 170.5	3-12-00-00093 0.8744[20]	 1.4682[20]
4288	Cyclohexanamine, N-ethyl-	$C_8H_{17}N$ 127.23	5459-93-8 	 164	4-12-00-00018 0.868[0]	H_2O 2; EtOH 5; eth 5; ctc 2
4289	Cyclohexanamine, N-(2-ethylhexyl)-	$C_{14}H_{29}N$ 211.39	5432-61-1 	 270; 140[14]	3-12-00-00018 0.8473[20]	ctc 2
4290	Cyclohexanamine, 1-ethynyl-	$C_8H_{13}N$ 123.20	30389-18-5 	 65-6[20]	4-12-00-00214 0.913[25]	 1.4817[20]
4291	Cyclohexanamine, N,N'-methanetetraylbis- Dicyclohexylcarbodiimide	$C_{13}H_{22}N_2$ 206.33	538-75-0 34.5	3086 123[6]; 99[0.5]	4-12-00-00072 	
4292	Cyclohexanamine, 2-methyl-, cis-	$C_7H_{15}N$ 113.20	2164-19-4 	 153.5	4-12-00-00118 0.8778[20]	 1.4688[20]
4293	Cyclohexanamine, 2-methyl-, trans- trans-2-Methylcyclohexylamine	$C_7H_{15}N$ 113.20	931-10-2 	 150	4-12-00-00118 0.8685[20]	 1.4650[20]
4294	Cyclohexanamine, 3-methyl-, cis-	$C_7H_{15}N$ 113.20	1193-16-4 	 153	4-12-00-00121 0.8552[20]	 1.4538[20]
4295	Cyclohexanamine, 3-methyl-, trans-	$C_7H_{15}N$ 113.20	1193-17-5 	 151.5	4-12-00-00121 0.8572[20]	 1.4547[20]
4296	Cyclohexanamine, 4-methyl-, cis-	$C_7H_{15}N$ 113.20	2523-56-0 	 153.5	4-12-00-00122 0.8587[20]	 1.4559[20]
4297	Cyclohexanamine, 4-methyl-, trans-	$C_7H_{15}N$ 113.20	2523-55-9 	 152	4-12-00-00122 0.8543[20]	 1.4550[20]
4298	Cyclohexanamine, N-methyl-	$C_7H_{15}N$ 113.20	100-60-7 	 147	4-12-00-00017 0.8660[23]	H_2O 2; EtOH 4; eth 5; chl 3 1.4560[20]
4299	Cyclohexanamine, 4,4'-methylenebis-	$C_{13}H_{26}N_2$ 210.36	1761-71-3 15	 320	 0.92[75]	
4300	Cyclohexanamine, 4,4'-methylenebis-, [cis(cis)]-	$C_{13}H_{26}N_2$ 210.36	6693-31-8 61	 141[2]	4-13-00-00031 	H_2O 1 1.5014[27]
4301	Cyclohexanamine, 4,4'-methylenebis-, [trans(cis)]-	$C_{13}H_{26}N_2$ 210.36	6693-30-7 36.5	 127-8[1.2]	4-13-00-00031 0.9608[25]	H_2O 1 1.5046[25]
4302	Cyclohexanamine, 4,4'-methylenebis-, [trans(trans)]-	$C_{13}H_{26}N_2$ 210.36	6693-29-4 64.7	 131[0.8]	4-13-00-00031 	H_2O 1 1.5032[25]
4303	Cyclohexanamine, N-(1-methylethyl)-	$C_9H_{19}N$ 141.26	1195-42-2 	 60-5[12]	4-12-00-00019 0.859[25]	 1.4480[20]

No.	Name / Synonym	Mol. Form. / Mol. Wt.	CAS RN / mp/°C	Merck No. / bp/°C	Beil. Ref. / den/g cm⁻³	Solubility / n_D
4304	Cyclohexanamine, N-2-propenyl-	$C_9H_{17}N$ 139.24	6628-00-8	65-6[12]	4-12-00-00021 0.962[25]	1.4664[20]
4305	Cyclohexane Hexahydrobenzene	C_6H_{12} 84.16	110-82-7 6.59	2729 80.73	4-05-00-00027 0.7739[25]	H_2O 1; EtOH 5; eth 5; ace 5 1.4266[20]
4306	Cyclohexaneacetic acid Cyclohexylacetic acid	$C_8H_{14}O_2$ 142.20	5292-21-7 33	245	4-09-00-00038 1.0423[18]	H_2O 2; eth 3; ace 3; os 3 1.4775[20]
4307	Δ1,α-Cyclohexaneacetic acid, 3,3-dimethyl- Allocyclogeranic acid	$C_{10}H_{16}O_2$ 168.24	6790-47-2	145[14]	4-09-00-00161 0.9921[17]	1.4765[17]
4308	Cyclohexaneacetic acid, ethyl ester	$C_{10}H_{18}O_2$ 170.25	5452-75-5	211	4-09-00-00038 0.9537[14]	1.451[14]
4309	Cyclohexaneacetic acid, α-ethyl-1-hydroxy- Cyclobutyrol	$C_{10}H_{18}O_3$ 186.25	512-16-3 81.5	2721 164[24]	4-10-00-00049 1.0010[18]	ace 4; eth 4; EtOH 4; chl 4 1.4680[18]
4310	Cyclohexaneacetic acid, methyl ester Methyl cyclohexaneacetate	$C_9H_{16}O_2$ 156.22	14352-61-5	201	4-09-00-00038 0.9896[24]	1.459[14]
4311	Cyclohexaneacetic acid, 2-methyl- Cyclohexanacetic acid, 2-methyl	$C_9H_{16}O_2$ 156.22	6617-04-5	145-7[13]	2-09-00-00013 1.012[25]	1.4656[25]
4312	Cyclohexaneacetic acid, 3-methyl- Cyclohexaneacetic acid, 3-methyl	$C_9H_{16}O_2$ 156.22	67451-76-7	148[18]	2-09-00-00013 0.9847[20]	1.495[20]
4313	Cyclohexaneacetonitrile	$C_8H_{13}N$ 123.20	4435-14-7	105[22]	4-09-00-00040 0.9473[21]	eth 4; EtOH 4 1.4787[21]
4314	Cyclohexane, 1,2-bis(methylene)-	C_8H_{12} 108.18	2819-48-9	125; 60-1[90]	4-05-00-00409 0.8229[25]	H_2O 3; EtOH 3; eth 3; ace 4 1.4718[25]
4315	Cyclohexane, bromo- Cyclohexyl bromide	$C_6H_{11}Br$ 163.06	108-85-0 -56.5	2736 166.2	4-05-00-00067 1.3359[20]	H_2O 1; EtOH 5; eth 5; ace 5 1.4957[20]
4316	Cyclohexane, (2-bromoethyl)-	$C_8H_{15}Br$ 191.11	1647-26-3 -57	212	4-05-00-00116 1.2357[20]	1.4899[20]
4317	Cyclohexane, 1-bromo-2-methoxy-	$C_7H_{13}BrO$ 193.08	24618-31-3	72-4[8]	3-06-00-00043 1.3257[20]	1.4871[20]
4318	Cyclohexane, (bromomethyl)-	$C_7H_{13}Br$ 177.08	2550-36-9	76-7[26]	4-05-00-00100 1.2763[25]	bz 4; eth 4; chl 4 1.4907[30]
4319	Cyclohexane, 1-bromo-1-methyl-	$C_7H_{13}Br$ 177.08	931-77-1	158	4-05-00-00100 1.2510[20]	H_2O 1; EtOH 3; chl 3 1.4866[20]
4320	Cyclohexane, 1-bromo-3-methyl- 3-Methylcyclohexyl bromide	$C_7H_{13}Br$ 177.08	13905-48-1	181; 60[11]	3-05-00-00076 1.275[25]	H_2O 1; eth 4; bz 3 1.4979[20]
4321	Cyclohexane, 1-bromo-4-methyl-1-(1-methylethyl)- p-Menthane, 4-bromo	$C_{10}H_{19}Br$ 219.16	116836-10-3 98-9[11]		0-05-00-00051 1.165[20]	1.4872[20]
4322	Cyclohexane, (2-bromo-2-propenyl)- Propylene, 2-bromo-3-cyclohexyl	$C_9H_{15}Br$ 203.12	53608-85-8	90[15]	3-05-00-00224 1.215[17]	eth 4 1.495[17]
4323	Cyclohexane, 1,1'-(1,4-butanediyl)bis- 1,4-Dicyclohexylbutane	$C_{16}H_{30}$ 222.41	6165-44-2 12	305	4-05-00-00357 0.8771[21]	EtOH 2; eth 2; bz 2; chl 2 1.475[21]
4324	Cyclohexanebutanol	$C_{10}H_{20}O$ 156.27	4441-57-0	103-4[4]	4-06-00-00139 0.902[25]	1.4660[20]
4325	Cyclohexane, butyl- Butylcyclohexane	$C_{10}H_{20}$ 140.27	1678-93-9 -74.7	180.9	4-05-00-00146 0.7902[20]	H_2O 1 1.4408[20]
4326	Cyclohexane, 1,1'-butylidenebis- Butane, 1,1-dicyclohexyl-	$C_{16}H_{30}$ 222.41	54890-00-5 -10.5	293	4-05-00-00359 0.8841[16]	1.485[16]
4327	Cyclohexane, 2-butyl-1,1,3-trimethyl- 2-Butyl-1,1,3-trimethylcyclohexane	$C_{13}H_{26}$ 182.35	54676-39-0	94-5[10]	4-05-00-00176 0.8292[19]	1.4563[19]
4328	Cyclohexanecarbonitrile	$C_7H_{11}N$ 109.17	766-05-2			
4329	Cyclohexanecarbonitrile, 1-hydroxy-	$C_7H_{11}NO$ 125.17	931-97-5 35	132[20]	4-10-00-00021 1.0172[20]	H_2O 4; eth 4 1.4693[20]
4330	Cyclohexanecarbonyl chloride	$C_7H_{11}ClO$ 146.62	2719-27-9	180	4-09-00-00026 1.0962[15]	1.4711[29]
4331	Cyclohexanecarboxaldehyde	$C_7H_{12}O$ 112.17	2043-61-0	159.3	4-07-00-00045 0.9035[20]	H_2O 3; eth 3 1.4496[20]
4332	Cyclohexanecarboxylic acid Hexahydrobenzoic acid	$C_7H_{12}O_2$ 128.17	98-89-5 31.5	2730 232.5	4-09-00-00016 1.0334[22]	H_2O 2; EtOH 4; bz 4; ctc 2 1.4530[20]
4333	Cyclohexanecarboxylic acid, 4-(aminomethyl)-, trans- Tranexamic acid	$C_8H_{15}NO_2$ 157.21	1197-18-8 >300	9487	3-14-00-00868	H_2O 4
4334	Cyclohexanecarboxylic acid, ethyl ester	$C_9H_{16}O_2$ 156.22	3289-28-9	196	4-09-00-00018 0.9362[20]	ace 4; eth 4; EtOH 4; chl 4 1.4501[15]
4335	Cyclohexanecarboxylic acid, methyl ester	$C_8H_{14}O_2$ 142.20	4630-82-4	183	4-09-00-00017 0.9954[15]	H_2O 1; EtOH 3; eth 3; ace 3 1.4433[20]
4336	Cyclohexanecarboxylic acid, 3-methyl-4-oxo- Cyclohexanone-4-carboxylic acid, 2-methyl	$C_8H_{12}O_3$ 156.18	101567-36-6 93.5	195[20]	4-10-00-02617	bz 4; EtOH 4; chl 4
4337	Cyclohexanecarboxylic acid, propyl ester Propyl cyclohexanecarboxylate	$C_{10}H_{18}O_2$ 170.25	6739-34-0	215.5	4-09-00-00018 0.9530[15]	ace 4; eth 4; EtOH 4; chl 4 1.4486[15]
4338	Cyclohexanecarboxylic acid, 1,3,4,5-tetrahydroxy-, [1R-(1α,3α,4α,5β)]- Quinic acid	$C_7H_{12}O_6$ 192.17	77-95-2 162.5	8071	4-10-00-02257 1.64[25]	H_2O 4; EtOH 4; HOAc 4
4339	Cyclohexane, chloro- Cyclohexyl chloride	$C_6H_{11}Cl$ 118.61	542-18-7 -44	2738 142	4-05-00-00048 1.000[20]	H_2O 1; EtOH 5; eth 5; ace 5 1.4626[20]
4340	Cyclohexane, cyclopentyl-	$C_{11}H_{20}$ 152.28	1606-08-2	215.1	4-05-00-00328 0.8758[20]	1.4725[20]
4341	Cyclohexane, (3-cyclopentylpropyl)- Propane, 1-cyclohexyl-3-cyclopentyl-	$C_{14}H_{26}$ 194.36	2883-07-0	270	4-05-00-00347 0.8751[20]	1.4765[20]

No.	Name / Synonym	Mol. Form. / Mol. Wt.	CAS RN / mp/°C	Merck No. / bp/°C	Beil. Ref. / den/g cm^{-3}	Solubility / n_D
4342	Cyclohexane, 1,1,2,2,3,3,4,4,5,6-decafluoro-5,6-bis(trifluoromethyl)- Cyclohexane, 1,2-dimethyl, perfluoro	C_8F_{16} 400.06	306-98-9 -56	101.5	3-05-00-00098 1.829[25]	1.283[25]
4343	Cyclohexane, 1,1,2,2,3,4,4,5,5,6-decafluoro-3,6-bis(trifluoromethyl)-	C_8F_{16} 400.06	374-77-6	100.5	4-05-00-00124 1.8503[20]	1.2897[20]
4344	Cyclohexane, decyl-	$C_{16}H_{32}$ 224.43	1795-16-0 -1.7	299	4-05-00-00187 0.8186[20]	1.4534[20]
4345	Cyclohexane, 1,1-dibromo- 1,1-Dibromocyclohexane	$C_6H_{10}Br_2$ 241.95	10489-97-1	87[14]	4-05-00-00070 1.7349[20]	1.5392[25]
4346	Cyclohexane, 1,2-dibromo-, cis- cis-1,2-Dibromocyclohexane	$C_6H_{10}Br_2$ 241.95	19246-38-9 9.7	115[14]	4-05-00-00070 1.798[25]	ace 4; bz 4; eth 4; lig 4 1.5514[25]
4347	Cyclohexane, 1,2-dibromo-, trans-(±)-	$C_6H_{10}Br_2$ 241.95	5183-77-7 -4	145[100.] 105[20]	4-05-00-00071 1.7759[20]	ace 4; bz 4; eth 4; EtOH 4 1.5445[19]
4348	Cyclohexane, 1,3-dibromo-, trans- trans-1,3-Dibromocyclohexane	$C_6H_{10}Br_2$ 241.95	29624-17-7 1	116[16]	4-05-00-00072	bz 4; EtOH 4 1.5480[20]
4349	Cyclohexane, 1,4-dibromo-, cis- cis-1,4-Dibromocyclohexane	$C_6H_{10}Br_2$ 241.95	16661-99-7	137-8[25]	4-05-00-00072 1.7834[20]	eth 4 1.5531[20]
4350	1,2-Cyclohexanedicarboxylic acid, diethyl ester, trans-(±)- Cyclohexane-1,2-dicarboxylic acid, diethyl ester (trans, DL)	$C_{12}H_{20}O_4$ 228.29	96836-97-4	133[11]	4-09-00-02803 1.040[20]	eth 4 1.4522[13]
4351	1,3-Cyclohexanedicarboxylic acid, diethyl ester, cis- Cyclohexane-1,3-dicarboxylic acid, diethyl ester (cis)	$C_{12}H_{20}O_4$ 228.29	62059-56-7	288; 151[15]	3-09-00-03817 1.045[20]	1.452[20]
4352	1,3-Cyclohexanedicarboxylic acid, dimethyl ester, cis- Dimethyl cis-1,3-cyclohexanedicarboxylate	$C_{10}H_{16}O_4$ 200.23	6998-82-9	263	4-09-00-02806 1.0997[20]	1.4568[20]
4353	1,3-Cyclohexanedicarboxylic acid, dimethyl ester, trans- Cyclohexane-1,3-dicarboxylic acid, dimethyl ester (trans)	$C_{10}H_{16}O_4$ 200.23	10021-92-8	139[20]	4-09-00-02806 1.1095[20]	1.4577[25]
4354	1,4-Cyclohexanedicarboxylic acid, trans-	$C_8H_{12}O_4$ 172.18	619-82-9 312.5	sub 300	4-09-00-02807	H_2O 2; EtOH 4; eth 2; ace 3
4355	1,4-Cyclohexanedicarboxylic acid, diethyl ester, cis- Cyclohexane-1,4-dicarboxylic acid, diethyl ester (cis)	$C_{12}H_{20}O_4$ 228.29	116724-15-3	151[13]	2-09-00-00524 1.015[20]	eth 4
4356	1,4-Cyclohexanedicarboxylic acid, diethyl ester, trans- Cyclohexane-1,4-dicarboxylic acid, diethyl ester (trans)	$C_{12}H_{20}O_4$ 228.29	19145-96-1 43.5		2-09-00-00524 1.0110[20]	eth 4 1.4337[64]
4357	1,4-Cyclohexanedicarboxylic acid, dimethyl ester, cis- cis-Dimethyl 1,4-cyclohexanedicarboxylate	$C_{10}H_{16}O_4$ 200.23	3399-21-1 14	132.5[10]	4-09-00-02808 1.1112[20]	
4358	Cyclohexane, 1,1-dichloro- 1,1-Dichlorocyclohexane	$C_6H_{10}Cl_2$ 153.05	2108-92-1 -47	171	4-05-00-00050 1.1559[20]	1.4803[20]
4359	Cyclohexane, 1,2-dichloro-, cis- cis-1,2-Dichlorocyclohexane	$C_6H_{10}Cl_2$ 153.05	10498-35-8 -1.5	206.9	4-05-00-00050 1.2021[20]	bz 4 1.4967[20]
4360	Cyclohexane, 1,2-dichloro-, trans-(±)-	$C_6H_{10}Cl_2$ 153.05	5183-79-9 -6.3	189	4-05-00-00051 1.1839[20]	1.4902[20]
4361	Cyclohexane, 1,4-dichloro-, cis- cis-1,4-Dichlorocyclohexane	$C_6H_{10}Cl_2$ 153.05	16749-11-4 18	80.3[25]	4-05-00-00051 1.1900[20]	1.4942[20]
4362	1,3-Cyclohexanedimethanamine Cyclohexane, 1,3-bis(aminomethyl)	$C_8H_{18}N_2$ 142.24	2579-20-6 <-70	220	0.945[20]	H_2O 4; eth 4; EtOH 4
4363	1,3-Cyclohexanedimethanol Cyclohexane, 1,3-bis(hydroxymethyl)	$C_8H_{16}O_2$ 144.21	3971-28-6	105.5	4-06-00-05236 1.036[25]	1.4912[20]
4364	Cyclohexane, 1,1-dimethyl- 1,1-Dimethylcyclohexane	C_8H_{16} 112.22	590-66-9 -33.3	119.6	4-05-00-00117 0.7809[20]	H_2O 1; EtOH 3; eth 3; ace 3 1.4290[20]
4365	Cyclohexane, 1,2-dimethyl-, cis- cis-1,2-Dimethylcyclohexane	C_8H_{16} 112.22	2207-01-4 -49.9	129.8	4-05-00-00118 0.7963[20]	H_2O 1; EtOH 3; eth 5; ace 5 1.4360[20]
4366	Cyclohexane, 1,2-dimethyl, trans- trans-1,2-Dimethylcyclohexane	C_8H_{16} 112.22	6876-23-9 -90	123.5	4-05-00-00118 0.7760[20]	H_2O 1; EtOH 3; eth 3; ace 5 1.4270[20]
4367	Cyclohexane, 1,3-dimethyl-, cis- cis-1,3-Dimethylcyclohexane	C_8H_{16} 112.22	638-04-0 -75.6	120.1	4-05-00-00121 0.7660[20]	H_2O 1; EtOH 5; eth 5; ace 5 1.4229[20]
4368	Cyclohexane, 1,4-dimethyl-, cis- cis-1,4-Dimethylcyclohexane	C_8H_{16} 112.22	624-29-3 -87.4	124.4	4-05-00-00122 0.7829[20]	H_2O 1; EtOH 5; eth 5; ace 5 1.4230[20]
4369	Cyclohexane, (1,1-dimethylethyl)-	$C_{10}H_{20}$ 140.27	3178-22-1 -41.2	171.5	4-05-00-00147 0.8127[20]	H_2O 1 1.4469[20]
4370	Cyclohexane, 1,3-dimethyl-, trans- trans-1,3-Dimethylcyclohexane	C_8H_{16} 112.22	2207-03-6 -90.1	124.5	0.79[15]	1.4284[25]
4371	Cyclohexane, 1,4-dimethyl-, trans- trans-1,4-Dimethylcyclohexane	C_8H_{16} 112.22	2207-04-7 -36.9	119.4	0.77[15]	1.4185[25]
4372	1,2-Cyclohexanediol, cis-	$C_6H_{12}O_2$ 116.16	1792-81-0 100	120[15]	4-06-00-05193 1.0297[101]	EtOH 3; ace 3; bz 3; chl 2
4373	1,2-Cyclohexanediol, trans-(±)-	$C_6H_{12}O_2$ 116.16	54383-22-1 105	117[13]	4-06-00-05194 1.147[24]	H_2O 4; bz 4; eth 4; EtOH 4
4374	1,4-Cyclohexanediol, trans- trans-1,4-Dihydroxycyclohexane	$C_6H_{12}O_2$ 116.16	6995-79-5 143		4-06-00-05209 1.18[20]	H_2O 3; EtOH 3; eth 1; ace 2

No.	Name / Synonym	Mol. Form. / Mol. Wt.	CAS RN / mp/°C	Merck No. / bp/°C	Beil. Ref. / den/g cm^{-3}	Solubility / n_D
4375	1,2-Cyclohexanedione / 1,2-Dioxocyclohexane	$C_8H_8O_2$ / 112.13	765-87-7 / 40	/ 194	4-07-00-01982 / 1.1187[21]	H_2O 3; EtOH 3; eth 3; bz 3 / 1.4995[20]
4376	1,3-Cyclohexanedione	$C_8H_8O_2$ / 112.13	504-02-9 / 105.5	3161	4-07-00-01985 / 1.0861[91]	H_2O 3; EtOH 3; eth 2; ace 3 / 1.4576[102]
4377	1,4-Cyclohexanedione / Tetrahydroquinone	$C_6H_8O_2$ / 112.13	637-88-7 / 78	/ 132[20]	4-07-00-01986 / 1.0861[91]	H_2O 3; EtOH 3; eth 3; ace 3
4378	1,3-Cyclohexanedione, 5,5-dimethyl- / 5,5-Dimethyldihydroresorcinol	$C_8H_{12}O_2$ / 140.18	126-81-8 / 149-51 dec	3231	4-07-00-01999	H_2O 2; eth 2; ace 3; ctc 3
4379	1,2-Cyclohexanedione, dioxime / Nioxime	$C_6H_{10}N_2O_2$ / 142.16	492-99-9 / 192	6477	4-07-00-01982	H_2O 3; ace 3; chl 3; tfa 2
4380	Cyclohexane, docosyl- / Docosane, 1-cyclohexyl-	$C_{28}H_{56}$ / 392.75	61828-07-7 / 54.4	/ 449	3-05-00-00167 / 0.8334[20]	1.4632[20]
4381	Cyclohexane, dodecafluoro- / Perfluorocyclohexane	C_6F_{12} / 300.05	355-68-0 / 62.5 (trip. pt.)	/ 52.8 (subl. pt.)	4-05-00-00048	
4382	Cyclohexane, dodecyl-	$C_{18}H_{36}$ / 252.48	1795-17-1 / 12.5	/ 331	4-05-00-00191 / 0.8223[20]	1.4559[20]
4383	Cyclohexane, dotriacontyl- / Dotriacontylcyclohexane	$C_{38}H_{76}$ / 533.02	61828-17-9 / 74.8	/ 527	4-05-00-00204 / 0.8388[20]	1.4668[20]
4384	Cyclohexane, eicosyl- / 1-Cyclohexyleicosane	$C_{26}H_{52}$ / 364.70	4443-55-4 / 48.5	/ 430	4-05-00-00198 / 0.8318[20]	1.4622[20]
4385	Cyclohexaneethanamine, N,α-dimethyl- / Propylhexedrine	$C_{10}H_{21}N$ / 155.28	101-40-6 /	7873 / 205; 82[10]	3-12-00-00108 / 0.8501[20]	EtOH 4 / 1.4600[20]
4386	Cyclohexane, 1,1'-(1,2-ethanediyl)bis- / 1,2-Dicyclohexylethane	$C_{14}H_{26}$ / 194.36	3321-50-4 / 11.5	/ 272.5	4-05-00-00344 / 0.8728[24]	1.4745[24]
4387	Cyclohexaneethanol	$C_8H_{16}O$ / 128.21	4442-79-9 /	/ 208	4-06-00-00119 / 0.9229[20]	EtOH 3; eth 3; bz 3 / 1.4641[20]
4388	Cyclohexane, ethenyl-	C_8H_{14} / 110.20	695-12-5 /	/ 128	4-05-00-00268 / 0.8166[19]	1.455[19]
4389	Cyclohexane, 1-ethenyl-1-methyl-2,4-bis(1-methylethenyl)-, (1α,2β,4β)- / β-Elemene	$C_{15}H_{24}$ / 204.36	33880-83-0 /	/ 117-24[15.5]	4-05-00-01180 / 0.8749[20]	1.4935[20]
4390	Cyclohexane, ethyl- / Ethylcyclohexane	C_8H_{16} / 112.22	1678-91-7 / -111.3	/ 131.9	4-05-00-00115 / 0.7880[20]	H_2O 1; EtOH 3; eth 3; ace 3 / 1.4330[20]
4391	Cyclohexane, ethylidene- / Ethylidenecyclohexane	C_8H_{14} / 110.20	1003-64-1 /	/ 136	4-05-00-00267 / 0.822[25]	1.4618[20]
4392	Cyclohexane, 1,1'-ethylidenebis- / 1,1-Dicyclohexylethane	$C_{14}H_{26}$ / 194.36	2319-61-1 / -20.9	/ 256; 112[7]	4-05-00-00345 / 0.9070[25]	1.4887[25]
4393	Cyclohexane, 1-ethylidene-2-methyl- / Cyclohexane, 2-methyl-1-ethylidene	C_9H_{16} / 124.23	40514-70-3 /	/ 158	0-05-00-00078 / 0.823[25]	1.47
4394	Cyclohexane, 1-ethyl-1-methyl-	C_9H_{18} / 126.24	4926-90-3 /	/ 152.2	4-05-00-00135 / 0.8025[25]	1.4419[20]
4395	Cyclohexane, 1-ethyl-2-methyl-, cis-	C_9H_{18} / 126.24	4923-77-7 /	/ 156	4-05-00-00136 / 0.8059[25]	1.4456[20]
4396	Cyclohexane, 1-ethyl-2-methyl-, trans- / Cyclohexane, 1-methyl-2-ethyl (trans)	C_9H_{18} / 126.24	4923-78-8 /	/ 151.7	4-05-00-00136 / 0.790[25]	1.4381[20]
4397	Cyclohexane, 1-ethyl-3-methyl-, cis- / cis-1-Ethyl-3-methylcyclohexane	C_9H_{18} / 126.24	19489-10-2 /	/ 156	4-05-00-00136 / 0.8094[20]	1.4432[20]
4398	Cyclohexane, 1-ethyl-3-methyl-, trans-	C_9H_{18} / 126.24	4926-76-5 /	/ 150	4-05-00-00136 / 0.7972[20]	1.4382[20]
4399	Cyclohexane, 1-ethyl-4-methyl-, cis- / cis-1-Ethyl-4-methylcyclohexane	C_9H_{18} / 126.24	4926-78-7 /	/ 152.6	4-05-00-00136 / 0.7969[20]	1.4374[20]
4400	Cyclohexane, 1-ethyl-4-methyl-, trans- / trans-1-Methyl-4-ethylcyclohexane	C_9H_{18} / 126.24	6236-88-0 / -80.8	/ 149	4-05-00-00136 / 0.7798[20]	1.4304[20]
4401	Cyclohexane, 1-ethyl-1-methyl-2,4-bis(1-methylethyl)-, [1R-(1α,2β,4β)]-	$C_{15}H_{30}$ / 210.40	515-12-8 /	/ 115-9[10]	4-05-00-00185 / 0.8509[20]	bz 4; eth 4; peth 4 / 1.4640[20]
4402	Cyclohexane, 1,1'-(2-ethyl-1,3-propanediyl)bis- / Propane, 1,3-dicyclohexyl-2-ethyl	$C_{17}H_{32}$ / 236.44	54833-34-0 /	/ 306	4-05-00-00363 / 0.8845[20]	1.483[21]
4403	Cyclohexane, fluoro- / Cyclohexyl fluoride	$C_6H_{11}F$ / 102.15	372-46-3 / 13	/ 101	4-05-00-00044 / 0.9279[20]	H_2O 1; py 3 / 1.4146[20]
4404	Cyclohexane, heptadecyl- / Heptadecane, 1-cyclohexyl-	$C_{23}H_{46}$ / 322.62	19781-73-8 / 38	/ 397	3-05-00-00161 / 0.8290[20]	1.4603[20]
4405	Cyclohexane, heptyl-	$C_{13}H_{26}$ / 182.35	5617-41-4 / -30	/ 244	4-05-00-00174 / 0.8109[20]	1.4484[20]
4406	Cyclohexane, 1,2,3,4,5,6-hexachloro-, (1α,2β,3α,4β,5α,6β)- / β-Lindane	$C_6H_6Cl_6$ / 290.83	319-85-7 /	/ 60[0.50]	4-05-00-00061 / 1.89[19]	H_2O 1; EtOH 2; bz 2; chl 2
4407	Cyclohexane, 1,2,3,4,5,6-hexachloro-, (1α,2α,3α,4β,5α,6β)- / δ-Lindane	$C_6H_6Cl_6$ / 290.83	319-86-8 / 141.5	/ 60[0.36]	4-05-00-00056	
4408	Cyclohexane, 1,2,3,4,5,6-hexachloro-, (1α,2α,3β,4α,5β,6β)-(±)- / Cyclohexane, 1,2,3,4,5,6-hexachloro, (α, DL)	$C_6H_6Cl_6$ / 290.83	60291-32-9 / 159.5	/ 288	4-05-00-00060 / 1.87[20]	H_2O 1; EtOH 4; bz 3; chl 3
4409	Cyclohexane, 1,2,3,4,5,6-hexachloro-, (1α,2α,3β,4α,5α,6β)- / Lindane	$C_6H_6Cl_6$ / 290.83	58-89-9 / 112.5	5379 / 323.4	4-05-00-00058	ace 4; bz 4
4410	Cyclohexane, hexacosyl- / Hexacosylcyclohexane	$C_{32}H_{64}$ / 448.86	61828-11-3 / 64	/ 484	4-05-00-00203 / 0.8359[20]	1.4649[20]

No.	Name / Synonym	Mol. Form. / Mol. Wt.	CAS RN / mp/°C	Merck No. / bp/°C	Beil. Ref. / den/g cm⁻³	Solubility / n_D
4411	Cyclohexane, hexadecyl- / Hexadecane, 1-cyclohexyl-	$C_{22}H_{44}$ / 308.59	6812-38-0 / 33.6	385	4-05-00-00196 / 0.8279[20]	1.4596[20]
4412	Cyclohexanehexanoic acid / Caproic acid, 6-cyclohexyl	$C_{12}H_{22}O_2$ / 198.31	4354-56-7 / 34	180[11]	4-09-00-00095 / 0.9626[20]	eth 3; chl 2 / 1.4750[20]
4413	Cyclohexane, hexatriacontyl- / Hexatriacontylcyclohexane	$C_{42}H_{84}$ / 589.13	61828-21-5 / 80	551	4-05-00-00205 / 0.8402[20]	1.4678[20]
4414	Cyclohexane, hexyl-	$C_{12}H_{24}$ / 168.32	4292-75-5 / -43	224	4-05-00-00170 / 0.8076[20]	1.4462[20]
4415	Cyclohexane, iodo- / Cyclohexyl iodide	$C_6H_{11}I$ / 210.06	626-62-0	180 dec; 81[20]	4-05-00-00078 / 1.6244[20]	H_2O 1; EtOH 3; eth 3; ace 3 / 1.5477[20]
4416	Cyclohexane, isocyanato- / Cyclohexyl isocyanate	$C_7H_{11}NO$ / 125.17	3173-53-3	172	4-12-00-00071 / 0.98[25]	1.4551[20]
4417	Cyclohexane, isothiocyanato-	$C_7H_{11}NS$ / 141.24	1122-82-3	220	4-12-00-00073 / 0.996[25]	H_2O 1; EtOH 3; eth 3; ctc 2 / 1.5375[20]
4418	Cyclohexanemethanamine	$C_7H_{15}N$ / 113.20	3218-02-8	160	4-12-00-00123 / 0.87[25]	1.4630[20]
4419	Cyclohexanemethanol / Cyclohexylcarbinol	$C_7H_{14}O$ / 114.19	100-49-2 / -43	2737 / 183	4-06-00-00106 / 0.9297[20]	eth 4; EtOH 4 / 1.4644[20]
4420	Cyclohexanemethanol, α-methyl-	$C_8H_{16}O$ / 128.21	1193-81-3	189	1-06-00-00012 / 0.928[25]	EtOH 4; eth 4; ctc 2 / 1.4656[20]
4421	Cyclohexanemethanol, α-methyl-, (S)-	$C_8H_{16}O$ / 128.21	3113-98-2	105[35]; 82[12]	4-06-00-00117 / 0.9254[20]	1.4635[25]
4422	Cyclohexanemethanol, 2-methyl-, cis- / Cyclohexane, 1-hydroxymethyl-2-methyl (cis)	$C_8H_{16}O$ / 128.21	3937-45-9	188.5	4-06-00-00122 / 0.9342[20]	1.4689[20]
4423	Cyclohexanemethanol, 2-methyl-, trans- / Cyclohexane, 1-hydroxymethyl-2-methyl (trans)	$C_8H_{16}O$ / 128.21	3937-46-0	192	4-06-00-00122 / 0.9224[20]	1.4665[20]
4424	Cyclohexanemethanol, 4-methyl-	$C_8H_{16}O$ / 128.21	34885-03-5	75[2.5]	2-06-00-00030 / 0.9074[20]	1.4617[20]
4425	Cyclohexane, methoxy-	$C_7H_{14}O$ / 114.19	931-56-6 / -74.4	133	4-06-00-00026 / 0.8756[20]	eth 4; EtOH 4 / 1.4355[20]
4426	Cyclohexane, methyl- / Methylcyclohexane	C_7H_{14} / 98.19	108-87-2 / -126.6	100.9	4-05-00-00094 / 0.7694[20]	H_2O 1; EtOH 3; eth 3; ace 5 / 1.4231[20]
4427	Cyclohexane, (3-methylbutyl)-	$C_{11}H_{22}$ / 154.30	54105-76-9	196.5	3-05-00-00143 / 0.8023[20]	bz 4; lig 4 / 1.4420[20]
4428	Cyclohexane, methylene- / Methylenecyclohexane	C_7H_{12} / 96.17	1192-37-6 / -106.7	102.5	4-05-00-00250 / 0.8074[20]	H_2O 1; eth 3; bz 3; chl 3 / 1.4523[20]
4429	Cyclohexane, 1,1'-methylenebis- / Dicyclohexylmethane	$C_{13}H_{24}$ / 180.33	3178-23-2	110-5[18]	4-05-00-00340 / 0.8750[20]	1.4752[20]
4430	Cyclohexane, 1,1'-(1-methyl-1,2-ethanediyl)bis- / 1,2-Dicyclohexylpropane	$C_{15}H_{28}$ / 208.39	41851-34-7 / -22	284.5	4-05-00-00350 / 0.8724[21]	1.479[21]
4431	Cyclohexane, (1-methylethyl)- / Isopropylcyclohexane	C_9H_{18} / 126.24	696-29-7 / -89.4	154.8	4-05-00-00134 / 0.8023[20]	H_2O 1; EtOH 4; eth 4; ace 5 / 1.4410[20]
4432	Cyclohexane, 1,1'-(1-methylethylidene)bis- / Propane, 2,2-dicyclohexyl	$C_{15}H_{28}$ / 208.39	54934-90-6 / 15.6	286	4-05-00-00350 / 0.9001[23]	1.490[23]
4433	Cyclohexane, 1,1',1''-methylidynetris- / Methane, tricyclohexyl-	$C_{19}H_{34}$ / 262.48	1610-24-8 / 48	325.5	4-05-00-00504 / 0.9273[50]	H_2O 1; EtOH 2; eth 3; bz 3 / 1.4986[40]
4434	Cyclohexane, 1-methyl-4-methylene-	C_8H_{14} / 110.20	2808-80-2	122	4-05-00-00271 / 0.7910[19]	1.4465[18]
4435	Cyclohexane, 1-methyl-3-(1-methylethenyl)- / m-Menth-8-ene	$C_{10}H_{18}$ / 138.25	16605-36-0	170	3-05-00-00235 / 0.8178[20]	1.4546
4436	Cyclohexane, 1-methyl-2-(1-methylethyl)-	$C_{10}H_{20}$ / 140.27	16580-23-7	171	4-05-00-00150 / 0.8134[22]	1.447[21]
4437	Cyclohexane, 1-methyl-3-(1-methylethyl)-	$C_{10}H_{20}$ / 140.27	16580-24-8	166.5	3-05-00-00132 / 0.7963[20]	1.44[24]
4438	Cyclohexane, 1-methyl-4-(1-methylethyl)-	$C_{10}H_{20}$ / 140.27	99-82-1 / -87.6	170.7	4-05-00-00151 / 0.797[25]	1.4373[20]
4439	Cyclohexane, 1-methyl-4-(1-methylethyl)-, cis-	$C_{10}H_{20}$ / 140.27	6069-98-3 / -89.9	172	4-05-00-00151 / 0.8039[20]	H_2O 1; EtOH 4; eth 4; bz 3 / 1.4431[20]
4440	Cyclohexane, 1-methyl-4-(1-methylethyl)-, trans-	$C_{10}H_{20}$ / 140.27	1678-82-6 / -86.3	170.6	4-05-00-00151 / 0.7928[20]	bz 4; eth 4; EtOH 4; lig 4 / 1.4366[20]
4441	Cyclohexane, 1-methyl-3-(1-methylethylidene)-	$C_{10}H_{18}$ / 138.25	13828-34-7	174	1-05-00-00043 / 0.8214[20]	1.4670[20]
4442	Cyclohexane, 1-methyl-4-(1-methylethylidene)-	$C_{10}H_{18}$ / 138.25	1124-27-2	173	4-05-00-00302 / 0.819[21]	1.4568[21]
4443	Cyclohexane, 1-methyl-1-nitro- / 1-Methyl-1-nitrocyclohexane	$C_7H_{13}NO_2$ / 143.19	59368-15-9 / -71	118-20[50]	4-05-00-00103 / 1.0384[20]	1.4598[20]
4444	Cyclohexane, 1-methyl-2-nitro- / 2-Methylnitrocyclohexane	$C_7H_{13}NO_2$ / 143.19	74221-86-6	100[20]	4-05-00-00103 / 1.046[25]	1.4608[25]
4445	Cyclohexane, 1-methyl-4-nitro- / 1-Methyl-4-nitrocyclohexane	$C_7H_{13}NO_2$ / 143.19	89895-45-4	102[20]	4-05-00-00103 / 1.037[25]	1.4567[23]
4446	Cyclohexane, 1-methyl-2-pentyl- / 1-Methyl-2-pentylcyclohexane	$C_{12}H_{24}$ / 168.32	54411-01-7	217.5	4-05-00-00171 / 0.815[20]	1.4487[20]
4447	Cyclohexane, 1,1'-(2-methyl-1,3-propanediyl)bis- / 1,3-Dicyclohexyl-2-methylpropane	$C_{16}H_{30}$ / 222.41	2883-08-1 / 0.6	295.2	4-05-00-00358 / 0.8715[20]	1.4756[20]

No.	Name / Synonym	Mol. Form. / Mol. Wt.	CAS RN / mp/°C	Merck No. / bp/°C	Beil. Ref. / den/g cm^{-3}	Solubility / n_D
4448	Cyclohexane, (1-methylpropyl)- sec-Butylcyclohexane	$C_{10}H_{20}$ 140.27	7058-01-7	179.3	4-05-00-00146 0.8131[20]	H_2O 1; ace 3 1.4467[20]
4449	Cyclohexane, 1-methyl-1-propyl- 1-Methyl-1-propylcyclohexane	$C_{10}H_{20}$ 140.27	4258-93-9	172	4-05-00-00148 0.8101[20]	1.4440[20]
4450	Cyclohexane, 1-methyl-2-propyl- 1-Methyl-2-propylcyclohexane	$C_{10}H_{20}$ 140.27	4291-79-6	176; 56[13]	4-05-00-00148 0.8130[19]	1.4468[19]
4451	Cyclohexane, 1-methyl-3-propyl- 1-Methyl-3-propylcyclohexane	$C_{10}H_{20}$ 140.27	4291-80-9	164.5	4-05-00-00149 0.7895[21]	1.4377[20]
4452	Cyclohexane, 1-methyl-4-propyl- 1-Methyl-4-propylcyclohexane	$C_{10}H_{20}$ 140.27	4291-81-0	175.5	4-05-00-00149 0.797[20]	1.4393[20]
4453	Cyclohexane, (2-methylpropyl)-	$C_{10}H_{20}$ 140.27	1678-98-4 -95	171.3	4-05-00-00147 0.7952[20]	H_2O 1; EtOH 3; eth 4; ace 3 1.4386[20]
4454	Cyclohexane, nitro- Nitrocyclohexane	$C_6H_{11}NO_2$ 129.16	1122-60-7 -34	205; 95[22]	4-05-00-00081 1.0610[20]	H_2O 1; EtOH 3; lig 3 1.4612[19]
4455	Cyclohexane, (nitromethyl)- (Nitromethyl)cyclohexane	$C_7H_{13}NO_2$ 143.19	2625-30-1	123[40]; 98[10]	4-05-00-00103 1.0482[20]	1.4705[20]
4456	Cyclohexane, nonadecyl-	$C_{25}H_{50}$ 350.67	22349-03-7 45	420	4-05-00-00197 0.8310[20]	1.4616[20]
4457	Cyclohexane, nonyl-	$C_{15}H_{30}$ 210.40	2883-02-5 -10	282	4-05-00-00183 0.8163[20]	1.4519[20]
4458	Cyclohexane, octacosyl- Octacosylcyclohexane	$C_{34}H_{68}$ 476.91	61828-13-5 68	499	4-05-00-00203 0.8370[20]	1.4656[20]
4459	Cyclohexane, octadecyl- Octadecane, 1-cyclohexyl-	$C_{24}H_{48}$ 336.65	4445-06-1 41.6	409; 175[1]	4-05-00-00197 0.8300[20]	1.4610[20]
4460	Cyclohexane, octyl-	$C_{14}H_{28}$ 196.38	1795-15-9 -20	264	4-05-00-00178 0.8138[20]	1.4503[20]
4461	Cyclohexane, 1,1'-oxybis-	$C_{12}H_{22}O$ 182.31	4645-15-2 -36	242.5	3-06-00-00019 0.9227[20]	1.4741[20]
4462	Cyclohexane, pentacosyl- Pentacosylcyclohexane	$C_{31}H_{62}$ 434.83	61828-10-2 62	476	4-05-00-00202 0.8353[20]	1.4645[20]
4463	Cyclohexane, pentadecyl-	$C_{21}H_{42}$ 294.56	6006-95-7 29	373	4-05-00-00195 0.8267[20]	1.4588[20]
4464	Cyclohexane, 1,1'-(1,5-pentanediyl)bis- Pentane, 1,5-dicyclohexyl-	$C_{17}H_{32}$ 236.44	54833-31-7 -13.6	325	4-05-00-00362 0.8718[21]	1.479[21]
4465	Cyclohexane, pentyl- Pentylcyclohexane	$C_{11}H_{22}$ 154.30	4292-92-6 -57.5	203.7	4-05-00-00164 0.8037[20]	1.4437[20]
4466	Cyclohexane, 1,1'-(1,3-propanediyl)bis- 1,3-Dicyclohexylpropane	$C_{15}H_{28}$ 208.39	3178-24-3 -17	291.5	4-05-00-00349 0.8728[24]	1.4736[24]
4467	Cyclohexanepropanenitrile, 2-oxo-	$C_9H_{13}NO$ 151.21	4594-78-9	138-42[10]	4-10-00-02623 1.0181[20]	ctc 2 1.4755[20]
4468	Cyclohexanepropanoic acid	$C_9H_{16}O_2$ 156.22	701-97-3 16	276.5	4-09-00-00056 0.912[25]	H_2O 3; eth 3; ctc 2 1.4638[20]
4469	Cyclohexanepropanol, α-methyl- 3-Cyclohexyl-1-methylpropanol	$C_{10}H_{20}O$ 156.27	10528-67-3	112[16]	3-06-00-00123 0.903[21]	1.464[21]
4470	Cyclohexane, 2-propenyl-	C_9H_{16} 124.23	2114-42-3	132	4-05-00-00283 0.8135[20]	EtOH 4; eth 3; ace 3; bz 3 1.4500[20]
4471	Cyclohexane, propyl- Propylcyclohexane	C_9H_{18} 126.24	1678-92-8 -94.9	156.7	4-05-00-00134 0.7936[20]	H_2O 1; EtOH 5; eth 3; ace 5 1.4370[20]
4472	Cyclohexane, 1,1'-propylidenebis- Propane, 1,1-dicyclohexyl-	$C_{15}H_{28}$ 208.39	54934-91-7 -23.5	283.6	4-05-00-00350 0.8886[23]	1.485[23]
4473	Cyclohexane, 2-propynyl-	C_9H_{14} 122.21	17715-00-3	157.5	4-05-00-00419 0.8449[20]	eth 3 1.4605[20]
4474	Cyclohexane, tetracosyl- Tetracosylcyclohexane	$C_{30}H_{60}$ 420.81	61828-09-9 59.5	467	4-05-00-00202 0.8347[20]	1.4641[20]
4475	Cyclohexane, tetradecyl-	$C_{20}H_{40}$ 280.54	1795-18-2 24	360	4-05-00-00194 0.8254[20]	1.4579[20]
4476	Cyclohexane, 1,1'-tetradecylidenebis- 1,1-Dicyclohexyltetradecane	$C_{26}H_{50}$ 362.68	55334-08-2 37.6	406	4-05-00-00374 0.8735[20]	1.4799[20]
4477	Cyclohexane, 1,1,3,4-tetramethyl- 1,1,3,4-Tetramethylcyclohexane	$C_{10}H_{20}$ 140.27	24612-75-7	160	4-05-00-00159 0.7976[20]	1.4380[20]
4478	Cyclohexane, 1,1,3,5-tetramethyl-, cis- cis-1,1,3,5-Tetramethylcyclohexane	$C_{10}H_{20}$ 140.27	50876-32-9	152.5	4-05-00-00159 0.7813[20]	1.4319[20]
4479	Cyclohexane, tetratriacontyl- Tetratriacontylcyclohexane	$C_{40}H_{80}$ 561.08	61828-19-1 78	540	4-05-00-00205 0.8395[20]	1.4673[20]
4480	Cyclohexanethiol Cyclohexyl mercaptan	$C_6H_{12}S$ 116.23	1569-69-3	158.9	4-06-00-00072 0.9782[20]	ace 4; bz 4; eth 4; EtOH 4 1.4921[20]
4481	Cyclohexanethione Thiocyclohexanone	$C_6H_{10}S$ 114.21	2720-41-4	82[17]; 74[11]	3-07-00-00039	EtOH 3; eth 3; ace 3 1.5375[20]
4482	Cyclohexane, triacontyl- Triacontylcyclohexane	$C_{36}H_{72}$ 504.97	61828-15-7 71.6	514	4-05-00-00204 0.8379[20]	1.4662[20]
4483	Cyclohexane, tricosyl- Tricosylcyclohexane	$C_{29}H_{58}$ 406.78	61828-08-8 57	459	4-05-00-00202 0.8341[20]	1.4637[20]
4484	Cyclohexane, tridecyl- Tridecane, 1-cyclohexyl-	$C_{19}H_{38}$ 266.51	6006-33-3 18.5	346	4-05-00-00192 0.8239[20]	1.4570[20]
4485	Cyclohexane, 1,1,2-trimethyl- 1,1,2-Trimethylcyclohexane	C_9H_{18} 126.24	7094-26-0 -29	145.2	4-05-00-00137 0.7963[25]	1.4382[20]
4486	Cyclohexane, 1,1,3-trimethyl- Cyclogeraniolane	C_9H_{18} 126.24	3073-66-3 -65.7	136.6	4-05-00-00137 0.7749[25]	1.4295[20]
4487	Cyclohexane, 1,1,4-trimethyl- 1,1,4-Trimethylcyclohexane	C_9H_{18} 126.24	7094-27-1	135	4-05-00-00137 0.7685[25]	1.4251[20]

No.	Name Synonym	Mol. Form. Mol. Wt.	CAS RN mp/°C	Merck No. bp/°C	Beil. Ref. den/g cm^{-3}	Solubility n_D
4488	Cyclohexane, 1,2,3-trimethyl- 1,2,3-Trimethylcyclohexane	C_9H_{18} 126.24	1678-97-3	142	4-05-00-00137 0.7898[20]	1.4346[20]
4489	Cyclohexane, 1,2,3-trimethyl-, (1α,2β,3α)-	C_9H_{18} 126.24	1678-81-5 -66.9	144	4-05-00-00137 0.777[25]	1.430[20]
4490	Cyclohexane, 1,2,3-trimethyl-, (1α,2α,3α)-	C_9H_{18} 126.24	1839-88-9 -85	151.7	4-05-00-00137 0.7989[25]	1.4405[20]
4491	Cyclohexane, 1,2,3-trimethyl-, (1α,2α,3β)-	C_9H_{18} 126.24	7667-55-2 -85.7	151.2	4-05-00-00137 0.7995[25]	1.4399[20]
4492	Cyclohexane, 1,2,4-trimethyl-, (1α,2α,4α)-	C_9H_{18} 126.24	1678-80-4 -77.5	146.6	4-05-00-00138 0.783[25]	1.4340[20]
4493	Cyclohexane, 1,2,4-trimethyl-, (1α,2β,4α)-	C_9H_{18} 126.24	7667-59-6 -86	142.9	4-05-00-00138 0.7683[25]	1.4266[20]
4494	Cyclohexane, 1,2,4-trimethyl-, (1α,2α,4β)-	C_9H_{18} 126.24	7667-58-5 -91.8	146.7	4-05-00-00138 0.7870[25]	1.4345[20]
4495	Cyclohexane, 1,2,4-trimethyl-, (1α,2β,4β)-	C_9H_{18} 126.24	7667-60-9 -83.5	142.9	4-05-00-00138 0.7870[25]	1.4341[20]
4496	Cyclohexane, 1,3,5-trimethyl-, (1α,3α,5α)-	C_9H_{18} 126.24	1795-27-3 -49.7	138.5	4-05-00-00138 0.7708[20]	H_2O 1; eth 3; bz 3; ctc 3 1.4269[20]
4497	Cyclohexane, 1,3,5-trimethyl-, (1α,3α,5β)- trans-1,3,5-Trimethylcyclohexane	C_9H_{18} 126.24	1795-26-2 -107.4	140.5	4-05-00-00138 0.7794[20]	bz 4; eth 4; lig 4 1.4307[20]
4498	Cyclohexane, undecafluoro- Undecafluorocyclohexane	C_6HF_{11} 282.06	308-24-7	62.0		
4499	Cyclohexane, undecafluoro(trifluoromethyl)- Perfluoro(methylcyclohexane)	C_7F_{14} 350.05	355-02-2 -44.7	76.3	4-05-00-00097 1.7878[25]	ace 3; bz 3; ctc 3; tol 3 1.285[17]
4500	Cyclohexanol Cyclohexyl alcohol	$C_6H_{12}O$ 100.16	108-93-0 25.4	2731 160.8	4-06-00-00020 0.9624[20]	H_2O 3; EtOH 3; eth 3; ace 3 1.4641[20]
4501	Cyclohexanol, 2-amino-, trans-(±)- Cyclohexanol, 2-amino (trans, DL)	$C_6H_{13}NO$ 115.18	33092-82-9 68	219; 105[10]	4-13-00-00540	bz 4; chl 4; HOAc 4
4502	Cyclohexanol, 3-(aminomethyl)-3,5,5-trimethyl- 1-Hydroxy-3-aminomethyl-3,5,5-trimethylcyclohexane	171.28	$C_{10}H_{21}NO$ 15647-11-7 45.5	265	0.969[25]	1.4904[20]
4503	Cyclohexanol, 2-bromo- 2-Bromo-1-cyclohexanol	$C_6H_{11}BrO$ 179.06	24796-87-0	85[10]; 60[1]	4-06-00-00063 1.4604[20]	1.5169[20]
4504	Cyclohexanol, 2-butyl-, trans- trans-2-Butylcyclohexanol	$C_{10}H_{20}O$ 156.27	35242-05-8	111-2[16]	3-06-00-00122 0.9020[20]	H_2O 1; eth 3; ace 3; bz 3 1.4641[20]
4505	Cyclohexanol, 2-chloro-, trans- trans-2-Chlorocyclohexanol	$C_6H_{11}ClO$ 134.61	6628-80-4 29	93[20]	4-06-00-00064 1.146[16]	bz 4; eth 4; EtOH 4; chl 4 1.4899[20]
4506	Cyclohexanol, 2-chloro-, cis-(±)-	$C_6H_{11}ClO$ 134.61	116783-28-9 36.5	93-4[26]	4-06-00-00064 1.1261[25]	bz 4; EtOH 4; chl 4 1.4894[25]
4507	Cyclohexanol, 4-chloro-, trans- trans-4-Chlorocyclohexanol	$C_6H_{11}ClO$ 134.61	29538-77-0 82.5	106[14]	4-06-00-00068 1.1435[13]	EtOH 3; eth 3; bz 3; chl 3 1.4930[17]
4508	Cyclohexanol, 1,2-dimethyl-, cis- Cyclohexanol, 1,2-dimethyl-, stereoisomer	$C_8H_{16}O$ 128.21	19879-11-9 23.2	82.8[25]	4-06-00-00121 0.9250[20]	1.4625[20]
4509	Cyclohexanol, 1,2-dimethyl-, trans- Cyclohexanol, 1,2-dimethyl-, stereoisomer	$C_8H_{16}O$ 128.21	19879-12-0 13.2	74[25]	4-06-00-00121 0.9187[20]	1.4590[20]
4510	Cyclohexanol, 1,3-dimethyl-, cis- cis-1,3-Dimethylcyclohexanol	$C_8H_{16}O$ 128.21	15466-94-1 27.5	84[25]	3-06-00-00092 0.9022[20]	1.4575[20]
4511	Cyclohexanol, 1,3-dimethyl-, trans- trans-1,3-Dimethylcyclohexanol	$C_8H_{16}O$ 128.21	15466-93-0 14.5	77.6[25]	3-06-00-00093 0.8894[30]	1.4507[20]
4512	Cyclohexanol, 1,4-dimethyl-, cis-	$C_8H_{16}O$ 128.21	16980-61-3 24	83.5[25]	3-06-00-00098 0.9011[30]	1.4564[20]
4513	Cyclohexanol, 2,2-dimethyl- 2,2-Dimethylcyclohexanol	$C_8H_{16}O$ 128.21	1193-46-0 8	177	0.9225[20]	1.4648[20]
4514	Cyclohexanol, 2,4-dimethyl- 2,4-Dimethylcyclohexanol	$C_8H_{16}O$ 128.21	69542-91-2	177.5	4-06-00-00124 0.900[20]	1.4500[20]
4515	Cyclohexanol, 2,6-dimethyl- 2,6-Dimethyl-1-cyclohexanol	$C_8H_{16}O$ 128.21	5337-72-4 32.5	176	2-06-00-00028 0.944[25]	1.4600[20]
4516	Cyclohexanol, 3,3-dimethyl- 3,3-Dimethylcyclohexanol	$C_8H_{16}O$ 128.21	767-12-4 11.5	185; 99.5[35]	3-06-00-00089 0.9128[14]	1.4606[15]
4517	Cyclohexanol, 3,4-dimethyl- 3,4-Dimethylcyclohexyl alcohol	$C_8H_{16}O$ 128.21	5715-23-1	189	3-06-00-00091 0.9073[16]	1.458[16]
4518	Cyclohexanol, 3,5-dimethyl- 3,5-Dimethylcyclohexanol	$C_8H_{16}O$ 128.21	5441-52-1 11.6	187	2-06-00-00029 0.898[20]	1.4550[20]
4519	Cyclohexanol, 4,4-dimethyl- 4,4-Dimethylcyclohexanol	$C_8H_{16}O$ 128.21	932-01-4 16	186	4-06-00-00120 0.9250[20]	1.4613[20]
4520	Cyclohexanol, 3-(dimethylamino)-	$C_8H_{17}NO$ 143.23	6890-03-5 73	231	3-13-00-00719 0.9737[25]	EtOH 4 1.4852[20]
4521	Cyclohexanol, 2-(1,1-dimethylethyl)-	$C_{10}H_{20}O$ 156.27	13491-79-7 45	139[95]	3-06-00-00125 0.902[25]	
4522	Cyclohexanol, 1-ethyl- 1-Ethylcyclohexanol	$C_8H_{16}O$ 128.21	1940-18-7 34.5	166	4-06-00-00115 0.9227[25]	H_2O 2; bz 3; ctc 2; peth 3 1.4633[20]
4523	Cyclohexanol, 2-ethyl-, trans-(±)- Cyclohexanol, 2-ethyl (trans, DL)	$C_8H_{16}O$ 128.21	89886-25-9	79[12]	4-06-00-00117 0.9193[21]	ace 4; bz 4; eth 4; peth 4 1.4640[21]
4524	Cyclohexanol, 2-ethyl-, cis-(±)-	$C_8H_{16}O$ 128.21	116697-35-9	181	3-06-00-00085 0.9274[21]	ace 4; bz 4; eth 4; peth 4 1.4655[21]
4525	Cyclohexanol, 4-ethyl- 4-Ethylcyclohexanol	$C_8H_{16}O$ 128.21	4534-74-1	84[10]	2-06-00-00026 0.889[25]	1.4625[20]
4526	Cyclohexanol, 1-ethyl-2-methyl- 1-Ethyl-2-methylcyclohexanol	$C_9H_{18}O$ 142.24	102370-18-3	182; 82[73]	4-06-00-00133 0.9235[20]	1.458[20]

No.	Name / Synonym	Mol. Form. / Mol. Wt.	CAS RN / mp/°C	Merck No. / bp/°C	Beil. Ref. / den/g cm⁻³	Solubility / n_D
4527	Cyclohexanol, 1-ethyl-3-methyl- / 1-Ethyl-3-methylcyclohexanol	$C_9H_{18}O$ / 142.24	62067-44-1	88[20]	2-06-00-00035 / 0.9013[20]	1.459
4528	Cyclohexanol, 1,1'-(1,2-ethynediyl)bis-	$C_{14}H_{22}O_2$ / 222.33	78-54-6 / 112.5	182[13]	4-06-00-06076	H_2O 1; EtOH 4; eth 4; ace 4
4529	Cyclohexanol, 1-ethynyl-	$C_8H_{12}O$ / 124.18	78-27-3 / 31.5	174	4-06-00-00348 / 0.9873[20]	H_2O 1; EtOH 3; bz 3; chl 2 / 1.4822[20]
4530	Cyclohexanol, 1-methyl- / 1-Methylcyclohexanol	$C_7H_{14}O$ / 114.19	590-67-0 / 25	155; 70[25]	4-06-00-00095 / 0.9194[20]	H_2O 1; EtOH 3; bz 3; chl 3 / 1.4595[20]
4531	Cyclohexanol, 2-methyl-, cis-(±)- / (±)-cis-2-Methylcyclohexanol	$C_7H_{14}O$ / 114.19	615-38-3 / 7	165	4-06-00-00100 / 0.9360[20]	EtOH 4 / 1.4640[20]
4532	Cyclohexanol, 2-methyl-, trans-(±)- / (±)-trans-2-Methylcyclohexanol	$C_7H_{14}O$ / 114.19	615-39-4 / -4	167.5	4-06-00-00100 / 0.9247[20]	eth 4; EtOH 4 / 1.4616[20]
4533	Cyclohexanol, 3-methyl-, cis-(±)- / cis-3-Methylcyclohexanol	$C_7H_{14}O$ / 114.19	5454-79-5 / -5.5	168; 94[12]	0.9155[20]	eth 4; EtOH 4 / 1.4752[20]
4534	Cyclohexanol, 3-methyl-, trans-(±)- / trans-3-Methylcyclohexanol	$C_7H_{14}O$ / 114.19	7443-55-2 / -0.5	167; 84[13]	0.9214[30]	eth 4; EtOH 4 / 1.4580[20]
4535	Cyclohexanol, 4-methyl-, cis- / cis-4-Methylcyclohexanol	$C_7H_{14}O$ / 114.19	7731-28-4 / -9.2	173	4-06-00-00105 / 0.9170[20]	eth 4; EtOH 4 / 1.4614[20]
4536	Cyclohexanol, 4-methyl-, trans- / trans-4-Methylcyclohexanol	$C_7H_{14}O$ / 114.19	7731-29-5	174	4-06-00-00105 / 0.9118[21]	H_2O 2; EtOH 5; eth 3 / 1.4561[20]
4537	Cyclohexanol, 2-methylene-	$C_7H_{12}O$ / 112.17	4065-80-9	82-4[13]	4-06-00-00208 / 0.955[20]	1.4843[20]
4538	Cyclohexanol, 1-(1-methylethyl)-	$C_9H_{18}O$ / 142.24	3552-01-0	76-9[10]	4-06-00-00131 / 0.928[19]	1.4683[19]
4539	Cyclohexanol, 2-(1-methylethyl)-, cis- / cis-2-Isopropylcyclohexanol	$C_9H_{18}O$ / 142.24	10488-25-2 / 52.5	77[13]	4-06-00-00131 / 0.9223[25]	1.4665[25]
4540	Cyclohexanol, 4-(1-methylethyl)-, cis- / cis-4-Isopropylcyclohexanol	$C_9H_{18}O$ / 142.24	22900-08-9 / 72[1.9]		3-06-00-00111 / 0.9195[20]	1.4671[20]
4541	Cyclohexanol, 4-(1-methylethyl)-, trans- / trans-4-Isopropylcyclohexanol	$C_9H_{18}O$ / 142.24	15890-36-5 / 79.5[1.8]		3-06-00-00112 / 0.9140[20]	1.4658[20]
4542	Cyclohexanol, 1-methyl-4-(1-methylethenyl)- / β-Terpineol	$C_{10}H_{18}O$ / 154.25	138-87-4 / 32.5	210; 90[10]	4-06-00-00254 / 0.917[20]	1.4747[20]
4543	Cyclohexanol, 2-methyl-5-(1-methylethenyl)-, [1S-(1α,2β,5α)]-	$C_{10}H_{18}O$ / 154.25	22567-21-1	225; 107[14]	3-06-00-00256 / 0.9274[20]	1.4780[20]
4544	Cyclohexanol, 5-methyl-2-(1-methylethenyl)-, [1R-(1α,2β,5α)]-	$C_{10}H_{18}O$ / 154.25	89-79-2 / 78	93[14]	3-06-00-00257 / 0.911[20]	H_2O 2; EtOH 3; eth 3 / 1.4723[20]
4545	Cyclohexanol, 5-methyl-2-(1-methylethenyl)-, acetate, [1R-(1α,2β,5α)]-	$C_{12}H_{20}O_2$ / 196.29	57576-09-7 / 85	112-4[8]	4-06-00-00255 / 0.925[25]	1.4566[20]
4546	Cyclohexanol, 1-methyl-4-(1-methylethyl)-	$C_{10}H_{20}O$ / 158.27	21129-27-1	208.5	4-06-00-00147 / 0.90[20]	1.4619[20]
4547	Cyclohexanol, 2-methyl-5-(1-methylethyl)-, [1R-(1α,2α,5β)]-	$C_{10}H_{20}O$ / 156.27	5563-78-0	217.5; 102[18]	4-06-00-00148 / 0.9012[20]	eth 4; EtOH 4 / 1.4632[20]
4548	Cyclohexanol, 5-methyl-2-(1-methylethyl)-, (1α,2β,5α)-(±)- / (±)-Menthol	$C_{10}H_{20}O$ / 156.27	15356-70-4 / 38	5723 / 216	4-06-00-00151 / 0.903[15]	H_2O 1; EtOH 4; eth 4; ace 4 / 1.4615[20]
4549	Cyclohexanol, 5-methyl-2-(1-methylethyl)-, acetate, [1R-(1α,2β,5α)]-	$C_{12}H_{22}O_2$ / 198.31	2623-23-6	109[10]	4-06-00-00153 / 0.9185[20]	1.4469[20]
4550	Cyclohexanol, 2-methyl-1-propyl- / 2-Methyl-1-propylcyclohexanol	$C_{10}H_{20}O$ / 156.27	24580-48-1	97-8[34]	3-06-00-00127 / 0.919[20]	1.48[20]
4551	Cyclohexanol, 3-methyl-1-propyl- / 3-Methyl-1-propylcyclohexanol	$C_{10}H_{20}O$ / 156.27	24580-52-7	199; 96-8[20]	1-06-00-00018 / 0.8903[21]	1.4566[24]
4552	Cyclohexanol, 1-phenyl- / 1-Phenylcyclohexan-1-ol	$C_{12}H_{16}O$ / 176.26	1589-60-2 / 63.5	157[28]; 112[5]	4-06-00-03908 / 1.035[16]	bz 4; eth 4; EtOH 4; lig 4 / 1.5415[16]
4553	Cyclohexanol, 2-phenyl-, cis-(±)- / Cyclohexanol, 2-phenyl (cis, DL)	$C_{12}H_{16}O$ / 176.26	40960-73-4 / 41.5	140-1[16]	4-06-00-03909 / 1.035[16]	1.5415[16]
4554	Cyclohexanol, 2-phenyl-, trans-(±)- / Cyclohexanol, 2-phenyl (trans, DL)	$C_{12}H_{16}O$ / 176.26	40960-69-8 / 56.5	153[16]	4-06-00-03909	EtOH 4; chl 4
4555	Cyclohexanol, 1-(2-propenyl)-	$C_9H_{16}O$ / 140.23	1123-34-8	190	4-06-00-00235 / 0.9341[22]	1.4756[22]
4556	Cyclohexanol, 2-(2-propenyl)-, trans- / Cyclohexanol, 2-allyl-, trans-	$C_9H_{16}O$ / 140.23	24844-28-8	94-6[15]	4-06-00-00235 / 0.947[20]	HOAc 4 / 1.4778[20]
4557	Cyclohexanol, 1-propyl- / 1-Hydroxy-1-propylcyclohexane	$C_9H_{18}O$ / 142.24	5445-24-9	180 dec; 85[20]	4-06-00-00129 / 0.934[12]	1.468[12]
4558	Cyclohexanol, 2-propyl-, cis- / cis-2-Propylcyclohexanol	$C_9H_{18}O$ / 142.24	5857-86-3	84[10]	4-06-00-00130 / 0.9247[11]	1.4688[11]
4559	Cyclohexanol, 2-propyl-, trans- / trans-2-Propylcyclohexanol	$C_9H_{18}O$ / 142.24	5846-43-5	90[14]	4-06-00-00130 / 0.9162[11]	1.4668[11]
4560	Cyclohexanol, 4-propyl-, trans- / trans-4-Propylcyclohexanol	$C_9H_{18}O$ / 142.24	77866-58-1	208; 103[14]	3-06-00-00106 / 0.9072[17]	1.4613[17]
4561	Cyclohexanol, 4-propyl-, cis-	$C_9H_{18}O$ / 142.24	98790-22-8	99[17]	3-06-00-00105 / 0.9102[16]	1.4652[16]
4562	Cyclohexanol, 1,2,2-trimethyl- / 1,2,2-Trimethylcyclohexanol	$C_9H_{18}O$ / 142.24	31720-69-1	81[20]; 76[16]	3-06-00-00114 / 0.9230[20]	bz 4; eth 4; EtOH 4 / 1.4680[20]

No.	Name / Synonym	Mol. Form. / Mol. Wt.	CAS RN / mp/°C	Merck No. / bp/°C	Beil. Ref. / den/g cm^{-3}	Solubility / n_D
4563	Cyclohexanol, 1,2,6-trimethyl- / 1,2,6-Trimethylcyclohexanol	$C_9H_{18}O$ / 142.24	96244-13-2 /	/ 78[22]	1-06-00-00017 / 0.9126[15]	ace 4; bz 4; eth 4; EtOH 4 / 1.4598[15]
4564	Cyclohexanol, 1,3,5-trimethyl- / 1,3,5-Trimethylcyclohexanol	$C_9H_{18}O$ / 142.24	90760-75-1 /	/ 181	3-06-00-00117 / 0.8876[17]	eth 4; EtOH 4; chl 4 / 1.454[16]
4565	Cyclohexanol, 2,2,5-trimethyl- / 2,2,5-Trimethylcyclohexanol	$C_9H_{18}O$ / 142.24	73210-25-0 /	/ 188	0-06-00-00022 / 0.8955[22]	EtOH 4 / 1.4569[20]
4566	Cyclohexanol, 2,2,6-trimethyl- / 1,1,3-Trimethyl-2-cyclohexanol	$C_9H_{18}O$ / 142.24	10130-91-3 / 51	187; 87[28]	4-06-00-00135 / 0.9128[20]	eth 4; EtOH 4; chl 4 / 1.4600[20]
4567	Cyclohexanol, 2,3,3-trimethyl-	$C_9H_{18}O$ / 142.24	116724-17-5 / 28	197	1-06-00-00016 / 0.9002[25]	ace 4; EtOH 4 / 1.4572[25]
4568	Cyclohexanol, 2,3,6-trimethyl- / 2,3,6-Trimethylcyclohexanol	$C_9H_{18}O$ / 142.24	58210-03-0 /	/ 195	0-06-00-00022 / 0.9117[17]	EtOH 4; chl 4
4569	Cyclohexanol, 3,3,5-trimethyl-, cis- / cis-3,3,5-Trimethylcyclohexanol	$C_9H_{18}O$ / 142.24	933-48-2 / 37.3	202; 92[12]	4-06-00-00135 / 0.9006[16]	H_2O 1; EtOH 3; eth 3; chl 3 / 1.4550[16]
4570	Cyclohexanol, 3,3,5-trimethyl-, trans- / trans-3,3,5-Trimethylcyclohexanol	$C_9H_{18}O$ / 142.24	767-54-4 / 55.8	189.2	4-06-00-00135 / 0.8631[60]	H_2O 1; EtOH 3; eth 3; chl 3
4571	Cyclohexanone / Pimelic ketone	$C_6H_{10}O$ / 98.14	108-94-1 / -31	2732 / 155.4	4-07-00-00015 / 0.9478[20]	H_2O 3; EtOH 3; eth 3; ace 3 / 1.4507[20]
4572	Cyclohexanone, 2-acetyl-	$C_8H_{12}O_2$ / 140.18	874-23-7 /	/ 111-2[18]	4-07-00-01997 / 1.0782[25]	ctc 3 / 1.5138[20]
4573	Cyclohexanone, 2,6-bis(phenylmethylene)-	$C_{20}H_{18}O$ / 274.36	897-78-9 / 117.5	185-95[20]	3-07-00-02661	EtOH 2; bz 3; HOAc 3
4574	Cyclohexanone, 2-bromo-	C_6H_9BrO / 177.04	822-85-5 /	/ 89-90[14]	4-07-00-00034 / 1.340[25]	/ 1.5085[25]
4575	Cyclohexanone, 2-butyl- / 2-Butylcyclohexanone	$C_{10}H_{18}O$ / 154.25	1126-18-7 /	/ 70[2]	4-07-00-00079 / 0.905[20]	H_2O 1; ctc 2; os 4 / 1.4545[20]
4576	Cyclohexanone, 2-butylidene- / 2-Butylidenecyclohexanone	$C_{10}H_{16}O$ / 152.24	7153-14-2 /	/ 98-100[10]	4-07-00-00181 / 0.935[20]	ace 4; bz 4; eth 4; EtOH 4 / 1.4800[20]
4577	Cyclohexanone, 2-chloro- / 2-Chlorocyclohexanone	C_6H_9ClO / 132.59	822-87-7 / 23	82[15]	4-07-00-00032 / 1.160[20]	eth 3; bz 3; ctc 2; diox 3 / 1.4825[20]
4578	Cyclohexanone, 4-chloro- / 4-Chlorocyclohexanone	C_6H_9ClO / 132.59	21299-26-3 /	/ 95[17]	4-07-00-00032	eth 4 / 1.4867[20]
4579	Cyclohexanone, 2-(2-chlorophenyl)-2-(methylamino)-, (±)- / Ketamine	$C_{13}H_{16}ClNO$ / 237.73	6740-88-1 / 92.5	5174		
4580	Cyclohexanone, 2,2-dimethyl- / 2,2-Dimethyl-1-cyclohexanone	$C_8H_{14}O$ / 126.20	1193-47-1 / -20.5	172	4-07-00-00055 / 0.9145[20]	1.4486[20]
4581	Cyclohexanone, 2,3-dimethyl-, cis- / cis-2,3-Dimethylcyclohexanone	$C_8H_{14}O$ / 126.20	1551-88-8 /	178.5; 99[57]	3-07-00-00088 / 0.9159[20]	1.4505[20]
4582	Cyclohexanone, 2,3-dimethyl-, trans- / trans-2,3-Dimethylcyclohexanone	$C_8H_{14}O$ / 126.20	1551-89-9 /	96[57]	4-07-00-00057 / 0.908[22]	1.4511[25]
4583	Cyclohexanone, 2,4-dimethyl-, (2S-trans)- / Cyclohexanone, 2,4-dimethyl (trans, d)	$C_8H_{14}O$ / 126.20	93921-42-7 /	179; 69[17]	3-07-00-00093 / 0.9004[16]	ace 4; bz 4; eth 4 / 1.4488[22]
4584	Cyclohexanone, 2,4-dimethyl-, cis-	$C_8H_{14}O$ / 126.20	116783-29-0 /	177	3-07-00-00093 / 0.910[15]	1.4493[15]
4585	Cyclohexanone, 2,5-dimethyl-, trans-(±)- / trans-(±)-2,5-Dimethylcyclohexanone	$C_8H_{14}O$ / 126.20	66395-23-1 /	172	4-07-00-00059 / 0.9025[20]	eth 4; EtOH 4 / 1.4446[20]
4586	Cyclohexanone, 2,6-dimethyl- / 2,6-Dimethylcyclohexanone	$C_8H_{14}O$ / 126.20	2816-57-1 /	175	2-07-00-00025 / 0.925[25]	1.4460[20]
4587	Cyclohexanone, 3,3-dimethyl- / 3,3-Dimethylcyclohexanone	$C_8H_{14}O$ / 126.20	2979-19-3 /	180; 72-3[25]	4-07-00-00056 / 0.909[15]	1.4482[17]
4588	Cyclohexanone, 3,4-dimethyl- / 3,4-Dimethylcyclohexanone	$C_8H_{14}O$ / 126.20	5465-09-8 /	187	3-07-00-00089 / 0.906[20]	1.4507[20]
4589	Cyclohexanone, 3,5-dimethyl- / 3,5-Dimethylcyclohexanone	$C_8H_{14}O$ / 126.20	2320-30-1 /	182.5	4-07-00-00058 / 0.896[20]	1.4407[20]
4590	Cyclohexanone, 4,4-dimethyl- / 4,4-Dimethylcyclohexanone	$C_8H_{14}O$ / 126.20	4255-62-3 / 39	73[14]	4-07-00-00056 / 0.932[20]	1.4537[24]
4591	Cyclohexanone, 2-[(dimethylamino)methyl]- / 2-[(Dimethylamino)methyl]cyclohexanone	$C_9H_{17}NO$ / 155.24	15409-60-6 /	92[10.5]	4-14-00-00008 / 0.9504[20]	H_2O 2; EtOH 3; eth 3 / 1.4672[20]
4592	Cyclohexanone, 4-(1,1-dimethylpropyl)-	$C_{11}H_{20}O$ / 168.28	16587-71-6 /	124-5[16]	4-07-00-00096 / 0.920[25]	1.4677[20]
4593	Cyclohexanone, 4-ethyl- / 4-Ethylcyclohexanone	$C_8H_{14}O$ / 126.20	5441-51-0 /	193	3-07-00-00083 / 0.895[25]	1.4515[20]
4594	Cyclohexanone, 2-ethylidene- / 2-Ethylidenecyclohexanone	$C_8H_{12}O$ / 124.18	1122-25-4 /	92[20]	4-07-00-00146 / 0.962[20]	H_2O 1; os 4 / 1.4882[20]
4595	Cyclohexanone, 2-ethyl-2-methyl- / 2-Ethyl-2-methylcyclohexanone	$C_9H_{16}O$ / 140.23	17206-52-9 /	195	2-07-00-00031 / 0.9037[17]	1.4515[17]
4596	Cyclohexanone, 2-methyl-, (±)- / (±)-2-Methylcyclohexanone	$C_7H_{12}O$ / 112.17	24965-84-2 / -13.9	165	4-07-00-00041 / 0.9250[20]	H_2O 1; EtOH 3; eth 3 / 1.4483[25]
4597	Cyclohexanone, 3-methyl-, (±)-	$C_7H_{12}O$ / 112.17	625-96-7 / -73.5	169; 65[15]	4-07-00-00043 / 0.9136[20]	H_2O 1; EtOH 3; eth 3 / 1.4456[20]
4598	Cyclohexanone, 4-methyl- / 4-Methylcyclohexanone	$C_7H_{12}O$ / 112.17	589-92-4 / -40.6	170	4-07-00-00044 / 0.9138[20]	H_2O 1; EtOH 3; eth 3; ctc 2 / 1.4451[20]
4599	Cyclohexanone, 2-(1-methylethyl)- / 2-Isopropylcyclohexanone	$C_9H_{16}O$ / 140.23	1004-77-9 / 72.5	83[17]	4-07-00-00066 / 0.922[16]	H_2O 1; EtOH 3; eth 3; ace 3 / 1.4564[15]
4600	Cyclohexanone, 4-(1-methylethyl)- / 4-Isopropylcyclohexanone	$C_9H_{16}O$ / 140.23	5432-85-9 /	139-40[100]	4-07-00-00066 / 0.9158[25]	1.4552[25]
4601	Cyclohexanone, 2-methyl-5-(1-methylethenyl)-, (2S-trans)-	$C_{10}H_{16}O$ / 152.24	619-02-3 / -11	221.5	0.9253[20]	1.4717[20]

No.	Name Synonym	Mol. Form. Mol. Wt.	CAS RN mp/°C	Merck No. bp/°C	Beil. Ref. den/g cm^{-3}	Solubility n_D
4602	Cyclohexanone, 5-methyl-2-(1-methylethenyl)-	C$_{10}$H$_{16}$O 152.24	529-00-0	100[18]	4-07-00-00190 0.9198[20]	1.4675[20]
4603	Cyclohexanone, 5-methyl-2-(1-methylethenyl)-, trans-	C$_{10}$H$_{16}$O 152.24	29606-79-9	100[18]	4-07-00-00190 0.9198[20]	1.4675[20]
4604	Cyclohexanone, 2-methyl-5-(1-methylethyl)-, (2S-trans)-	C$_{10}$H$_{18}$O 154.25	13163-73-0	220; 96[15]	4-07-00-00085 0.904[20]	ace 4; EtOH 4; chl 4 1.4553[20]
4605	Cyclohexanone, 5-methyl-2-(1-methylethyl)-, (2R-cis)- Menthone(d)	C$_{10}$H$_{18}$O 154.25	1196-31-2	204; 85[14]	4-07-00-00087 0.8947[20]	H$_2$O 2; EtOH 3; eth 3; ace 3 1.4503[20]
4606	Cyclohexanone, 5-methyl-2-(1-methylethyl)-, (2S-trans)- (-)-Menthone	C$_{10}$H$_{18}$O 154.25	14073-97-3 -6	5724 207	4-07-00-00188 0.8954[20]	H$_2$O 2; EtOH 5; eth 5; ace 3 1.4505[20]
4607	Cyclohexanone, 5-methyl-2-(1-methylethylidene)-	C$_{10}$H$_{16}$O 152.24	15932-80-6	95-6[11]	4-07-00-00188 0.9367[20]	1.4869[20]
4608	Cyclohexanone, 5-methyl-2-(1-methylethylidene)-, (R)- Pulegone	C$_{10}$H$_{16}$O 152.24	89-82-7	7955 224	4-07-00-00188 0.9346[45]	H$_2$O 1; EtOH 5; eth 5; ctc 3 1.4894[20]
4609	Cyclohexanone, 4-(2-methylpropyl)-	C$_{10}$H$_{18}$O 154.25	42061-72-3	218; 98[12]	4-07-00-00082 0.8914[24]	1.4502[24]
4610	Cyclohexanone, oxime (Hydroxyimino)cyclohexane	C$_6$H$_{11}$NO 113.16	100-64-1 90	206	4-07-00-00021	H$_2$O 3; EtOH 3; eth 3; chl 2
4611	Cyclohexanone, 2-(1-piperidinylmethyl)- Pimeclone	C$_{12}$H$_{21}$NO 195.30	534-84-9	7399 118-20[14]	5-20-02-00330	
4612	Cyclohexanone, 2-propyl- 2-Propylcyclohexanone	C$_9$H$_{16}$O 140.23	94-65-5	197	4-07-00-00064 0.927[20]	H$_2$O 1; EtOH 3; eth 4; ace 3 1.4538[20]
4613	Cyclohexanone, 4-propyl-	C$_9$H$_{16}$O 140.23	40649-36-3	213; 97.5[15]	3-07-00-00116 0.9072[19]	1.4530[19]
4614	Cyclohexanone, 2,2,3-trimethyl- 2,2,3-Trimethylcyclohexanone	C$_9$H$_{16}$O 140.23	39257-08-4	66-7[10]	4-07-00-00069 0.9234[15]	1.4569[15]
4615	Cyclohexanone, 2,2,4-trimethyl- 2,2,4-Trimethylcyclohexanone	C$_9$H$_{16}$O 140.23	35413-38-8	86-7[12]	3-07-00-00125 0.8914[20]	1.4520[20]
4616	Cyclohexanone, 2,2,5-trimethyl- 2,2,5-Trimethylcyclohexanone	C$_9$H$_{16}$O 140.23	933-36-8	183	3-07-00-00126 0.8871[24]	1.4432[24]
4617	Cyclohexanone, 2,2,6-trimethyl-	C$_9$H$_{16}$O 140.23	2408-37-9 -31.8	178.5	4-07-00-00069 0.9043[18]	1.4470[20]
4618	Cyclohexanone, 2,3,6-trimethyl- 2,3,6-Trimethylcyclohexanone	C$_9$H$_{16}$O 140.23	42185-47-7	190	3-07-00-00126 0.9129[18]	1.4464[21]
4619	Cyclohexanone, 2,4,4-trimethyl- 2,4,4-Trimethylcyclohexanone	C$_9$H$_{16}$O 140.23	2230-70-8	191	4-07-00-00070 0.902[20]	1.4493[20]
4620	Cyclohexanone, 2,4,5-trimethyl- 2,4,5-Trimethylcyclohexanone	C$_9$H$_{16}$O 140.23	51299-55-9	193	4-07-00-00072 0.897[20]	1.4479[20]
4621	Cyclohexanone, 2,4,6-trimethyl- 2,4,6-Trimethylcyclohexanone	C$_9$H$_{16}$O 140.23	90645-54-8	185	3-07-00-00127 0.8992[20]	1.4458[20]
4622	Cyclohexanone, 3,3,5-trimethyl- Dihydroisophorone	C$_9$H$_{16}$O 140.23	873-94-9	189	4-07-00-00070 0.8919[19]	1.4454[15]
4623	Cyclohexanone, 3,4,4-trimethyl- 3,4,4-Trimethylcyclohexanone	C$_9$H$_{16}$O 140.23	40441-35-8	80-1[13]	3-07-00-00122 0.911[25]	1.4552[20]
4624	Cyclohexasiloxane, dodecamethyl- Dodecamethylcyclohexasiloxane	C$_{12}$H$_{36}$O$_6$Si$_6$ 444.93	540-97-6 -3	3400 245	4-04-00-04129 0.9672[25]	H$_2$O 1 1.4015[20]
4625	Cyclohexasiloxane, 2,4,6,8,10,12-hexamethyl- Hexamethylcyclohexasiloxane	C$_6$H$_{24}$O$_6$Si$_6$ 360.77	6166-87-6 -79		4-04-00-04100 1.006[20]	H$_2$O 1 1.3944[20]
4626	Cyclohexene Tetrahydrobenzene	C$_6$H$_{10}$ 82.15	110-83-8 -103.5	2733 82.9	4-05-00-00218 0.8110[20]	H$_2$O 1; EtOH 5; eth 5; ace 5 1.4465[20]
4627	Cyclohexene, 1-bromo-	C$_6$H$_9$Br 161.04	2044-08-8	165	4-05-00-00236 1.3901[20]	eth 3; ace 4; bz 3; chl 4 1.5134[20]
4628	Cyclohexene, 3-bromo-	C$_6$H$_9$Br 161.04	1521-51-3	80-2[40]	4-05-00-00237 1.3890[20]	H$_2$O 1; eth 3; bz 3; chl 3 1.5320[20]
4629	3-Cyclohexene-1-carbonitrile	C$_7$H$_9$N 107.16	100-45-8	39[1.5]	4-09-00-00116	ctc 3 1.4716[20]
4630	1-Cyclohexenecarbonitrile 1-Cyanocyclohexene	C$_7$H$_9$N 107.16	1855-63-6	81[12]		
4631	1-Cyclohexene-1-carboxaldehyde	C$_7$H$_{10}$O 110.16	1192-88-7	72[15]	4-07-00-00132 0.9694[20]	EtOH 3; eth 3 1.5005[20]
4632	3-Cyclohexene-1-carboxaldehyde	C$_7$H$_{10}$O 110.16	100-50-5 2	105	4-07-00-00134 0.9692[20]	ace 3; ctc 2; MeOH 3
4633	1-Cyclohexene-1-carboxaldehyde, 4-(1-methylethenyl)- Perillaldehyde	C$_{10}$H$_{14}$O 150.22	2111-75-3	7119	2-07-00-00130	
4634	1-Cyclohexene-1-carboxaldehyde, 4-(1-methylethenyl)-, (R)-	C$_{10}$H$_{14}$O 150.22	5503-12-8	238; 99[7]	4-07-00-00316 0.953[20]	1.5058[20]
4635	1-Cyclohexene-1-carboxaldehyde, 4-(1-methylethyl)- Phellandral	C$_{10}$H$_{16}$O 152.24	21391-98-0	223	4-07-00-00187 0.93[20]	1.4911[20]
4636	1-Cyclohexene-1-carboxaldehyde, 2,6,6-trimethyl- β-Cyclocitral	C$_{10}$H$_{16}$O 152.24	432-25-7	95-100[15]	4-07-00-00192 0.959[15]	1.4971[15]

No.	Name Synonym	Mol. Form. Mol. Wt.	CAS RN mp/°C	Merck No. bp/°C	Beil. Ref. den/g cm^{-3}	Solubility n_D
4637	1-Cyclohexene-1-carboxylic acid Cyclohexene-1-carboxylic acid	$C_7H_{10}O_2$ 126.16	636-82-8 38	241	4-09-00-00111 1.109[20]	H_2O 2; EtOH 3; ace 3 1.4902[20]
4638	3-Cyclohexene-1-carboxylic acid	$C_7H_{10}O_2$ 126.16	4771-80-6 17	234.5	4-09-00-00114 1.0820[20]	H_2O 4; EtOH 3; ace 3 1.4814[20]
4639	3-Cyclohexene-1-carboxylic acid, ethyl ester	$C_9H_{14}O_2$ 154.21	15111-56-5	194.5	4-09-00-00114 0.9688[20]	1.4578[20]
4640	3-Cyclohexene-1-carboxylic acid, methyl ester	$C_8H_{12}O_2$ 140.18	6493-77-2	80[20]	4-09-00-00114 0.995[25]	1.4610[20]
4641	2-Cyclohexene-1-carboxylic acid, 2,6-dimethyl-4-oxo-, ethyl ester	$C_{11}H_{18}O_3$ 196.25	6102-15-4	157-8[18]	4-10-00-02674 1.0493[20]	ace 3 1.4773[20]
4642	Cyclohexene, 1-chloro- 1-Chlorocyclohexene	C_6H_9Cl 116.59	930-66-5	142.5	4-05-00-00230 1.0361[19]	eth 3; ace 3; ctc 3; chl 3 1.4797[20]
4643	Cyclohexene, decafluoro- Perfluorocyclohexene	C_6F_{10} 262.05	355-75-9	52.0	4-05-00-00229 1.6650[25]	1.293[20]
4644	1-Cyclohexene-1,2-dicarboxylic acid, diethyl ester Cyclohexene-1,2-dicarboxylic acid, diethyl ester	$C_{12}H_{18}O_4$ 226.27	92687-41-7	160[14]	3-09-00-03940 1.0803[19]	1.4747[19]
4645	1-Cyclohexene-1,3-dicarboxylic acid, diethyl ester Diethyl 1-cyclohexene-1,3-dicarboxylate	$C_{12}H_{18}O_4$ 226.27	38511-07-8	150[12]	3-09-00-03950 1.0772[20]	1.4722[20]
4646	1-Cyclohexene-1,4-dicarboxylic acid, dimethyl ester Cyclohexene-1,4-dicarboxylic acid, dimethyl ester	$C_{10}H_{14}O_4$ 198.22	22646-79-3 39	153[20]	4-09-00-02897	ace 4; eth 4; EtOH 4
4647	Cyclohexene, 1,2-dichloro-3,3,4,4,5,5,6,6-octafluoro-	$C_6Cl_2F_8$ 294.96	336-19-6 -70	113	4-05-00-00231 1.719[20]	1.375[20]
4648	Cyclohexene, 1,2-dimethyl-	C_8H_{14} 110.20	1674-10-8 -84.1	138	4-05-00-00268 0.8220[25]	1.4620[20]
4649	Cyclohexene, 1,3-dimethyl- 1,3-Dimethyl-1-cyclohexene	C_8H_{14} 110.20	2808-76-6	127	4-05-00-00270 0.799[25]	1.449[20]
4650	Cyclohexene, 1,4-dimethyl- 1,4-Dimethyl-1-cyclohexene	C_8H_{14} 110.20	2808-79-9 -59.4	128.5	4-05-00-00270 0.8005[20]	1.4457[20]
4651	Cyclohexene, 1,5-dimethyl- Cyclohexene, 2,4-dimethyl-	C_8H_{14} 110.20	2808-77-7	126	3-05-00-00216 0.8009[25]	1.450[20]
4652	Cyclohexene, 1,6-dimethyl-	C_8H_{14} 110.20	1759-64-4	131	4-05-00-00269 0.815[20]	1.454[20]
4653	Cyclohexene, 3,3-dimethyl- 3,3-Dimethylcyclohexene	C_8H_{14} 110.20	695-28-3	119	4-05-00-00268 0.804[20]	1.445[20]
4654	Cyclohexene, 3,5-dimethyl- 3,5-Dimethylcyclohexene	C_8H_{14} 110.20	823-17-6	124.5	3-05-00-00216 0.8005[18]	1.443[18]
4655	Cyclohexene, 3,6-dimethyl- 3,6-Dimethylcyclohexene	C_8H_{14} 110.20	19550-40-4	125	4-05-00-00271 0.802[25]	1.4433[25]
4656	Cyclohexene, 4,4-dimethyl-	C_8H_{14} 110.20	14072-86-7 -74.4	117	4-05-00-00268 0.7996[20]	1.4420[20]
4657	Cyclohexene, 1-ethenyl-	C_8H_{12} 108.18	2622-21-1	145	4-05-00-00405 0.8623[15]	H_2O 1; eth 3; bz 3; MeOH 4 1.4915[20]
4658	Cyclohexene, 4-ethenyl-	C_8H_{12} 108.18	100-40-3 -108.9	128	4-05-00-00406 0.8299[20]	H_2O 1; eth 3; bz 3; peth 3 1.4639[20]
4659	Cyclohexene, 6-ethenyl-6-methyl-1-(1-methylethyl)-3-(1-methylethylidene)-, (S)-	$C_{15}H_{24}$ 204.36	5951-67-7	120-30[8]	3-05-00-01083 0.8782[20]	ace 4; bz 4 1.5130[26]
4660	Cyclohexene, 1-ethyl-	C_8H_{14} 110.20	1453-24-3 -109.9	137	4-05-00-00266 0.8176[25]	1.4567[20]
4661	Cyclohexene, 3-ethyl- 3-Ethyl-1-cyclohexene	C_8H_{14} 110.20	2808-71-1	134	4-05-00-00266 0.814[20]	1.451[20]
4662	Cyclohexene, 4-ethyl- 4-Ethyl-1-cyclohexene	C_8H_{14} 110.20	3742-42-5	133	4-05-00-00267 0.810[20]	1.449[20]
4663	Cyclohexene, 1-ethyl-4-methyl- 1-Ethyl-4-methylcyclohexene	C_9H_{16} 124.23	62088-36-2	149	4-05-00-00286 0.8169[16]	1.453[16]
4664	1-Cyclohexene-1-methanol, 4-(1-methylethenyl)-	$C_{10}H_{16}O$ 152.24	536-59-4	244; 12.5[12]	2-06-00-00102 0.9690[20]	1.5005[20]
4665	1-Cyclohexene-1-methanol, 2,4,4-trimethyl- β-Cyclolavandulol	$C_{10}H_{18}O$ 154.25	103985-40-6	105-6[15]	4-06-00-00259 0.9153[20]	1.4814[20]
4666	1-Cyclohexene-1-methanol, 2,6,6-trimethyl- Cyclogeranyl alcohol	$C_{10}H_{18}O$ 154.25	472-20-8 43.5		3-06-00-00259 0.945[19]	1.487[19]
4667	2-Cyclohexene-1-methanol, 2,6,6-trimethyl- α-Cyclogeraniol	$C_{10}H_{18}O$ 154.25	6627-74-3	215	4-06-00-00257 0.9382[20]	1.4843[20]
4668	3-Cyclohexene-1-methanol, α,α,4-trimethyl-, (±)-	$C_{10}H_{18}O$ 154.25	2438-12-2 40.5	220	4-06-00-00252 0.9337[20]	ace 4; bz 4; eth 4; EtOH 4 1.4831[20]
4669	3-Cyclohexene-1-methanol, α,α,4-trimethyl-, acetate	$C_{12}H_{20}O_2$ 196.29	80-26-2	140[40]; 105[11]	4-06-00-00253 0.9659[21]	H_2O 1; EtOH 3; eth 3; bz 3 1.4689[21]
4670	Cyclohexene, 1-methyl-	C_7H_{12} 96.17	591-49-1 -120.4	110.3	4-05-00-00245 0.8102[20]	H_2O 1; eth 3; bz 3; ctc 3 1.4503[20]

No.	Name Synonym	Mol. Form. Mol. Wt.	CAS RN mp/°C	Merck No. bp/°C	Beil. Ref. den/g cm^{-3}	Solubility n_D
4671	Cyclohexene, 3-methyl-, (±)- (±)-3-Methylcyclohexene	C_7H_{12} 96.17	56688-75-6 -115.5	104	4-05-00-00247 0.7990[20]	bz 4; eth 4; chl 4; peth 4 1.4414[20]
4672	Cyclohexene, 4-methyl-	C_7H_{12} 96.17	591-47-9 -115.5	102.7	4-05-00-00248 0.7991[20]	H_2O 1; EtOH 3; eth 3 1.4414[20]
4673	Cyclohexene, 3-methylene-6-(1-methylethyl)- β-Phellandrene	$C_{10}H_{18}$ 136.24	555-10-2	7152 171.5	2-05-00-00087 0.8520[20]	H_2O 1; EtOH 1; eth 3 1.4788[20]
4674	Cyclohexene, 3-methylene-6-(1-methylethyl)-, (+)-	$C_{10}H_{18}$ 136.24	6153-16-8	171.5	4-05-00-00436 0.8520[20]	eth 4 1.4788[20]
4675	Cyclohexene, 4-methylene-1-(1-methylethyl)-	$C_{10}H_{18}$ 136.24	99-84-3	173.5	4-05-00-00437 0.838[22]	1.4754[22]
4676	Cyclohexene, 1-methyl-3-(1-methylethenyl)-	$C_{10}H_{18}$ 136.24	38738-60-2	175	4-05-00-00434 0.8479[18]	1.4760[18]
4677	Cyclohexene, 1-methyl-4-(1-methylethenyl)- Limonene	$C_{10}H_{18}$ 136.24	138-86-3 -95	5371 178	4-05-00-00440 0.8402[21]	H_2O 1; EtOH 5; eth 5; ctc 3 1.4727[19]
4678	Cyclohexene, 1-methyl-5-(1-methylethenyl)-	$C_{10}H_{18}$ 136.24	13898-73-2	176; 95[20]	4-05-00-00435 0.8481[20]	eth 4; EtOH 4 1.4804[20]
4679	Cyclohexene, 5-methyl-1-(1-methylethenyl)-, (±)- Cyclohexene, 1-isopropenyl-5-methyl (DL)	$C_{10}H_{16}$ 136.24	105065-24-5	184	3-05-00-00336 0.8594[20]	1.4975
4680	Cyclohexene, 1-methyl-4-(1-methylethyl)-	$C_{10}H_{18}$ 138.25	5502-88-5	174.5	4-05-00-00300 0.8457[15]	1.4735[20]
4681	Cyclohexene, 1-methyl-4-(1-methylethyl)-, (R)-	$C_{10}H_{18}$ 138.25	1195-31-9	173	4-05-00-00299 0.8246[18]	bz 4; EtOH 4; peth 4 1.4563[18]
4682	Cyclohexene, 3-methyl-6-(1-methylethyl)-, (3R-trans)-	$C_{10}H_{18}$ 138.25	5113-93-9	167; 55-6[12]	4-05-00-00300 0.824[20]	1.461[20]
4683	Cyclohexene, 4-methyl-1-(1-methylethyl)-, (R)-	$C_{10}H_{18}$ 138.25	619-52-3	167.5	4-05-00-00301 0.8078[20]	H_2O 1; EtOH 3; eth 3; bz 3 1.4503[15]
4684	Cyclohexene, 1-methyl-4-(1-methylethylidene)- Terpinolene	$C_{10}H_{16}$ 136.24	586-62-9	186	4-05-00-00437 0.8632[15]	H_2O 1; EtOH 5; eth 5; bz 3 1.4883[20]
4685	Cyclohexene, 1-methyl-4-(5-methyl-1-methylene-4-hexenyl)-, (S)-	$C_{15}H_{24}$ 204.36	495-61-4	129-30[10.5]	4-05-00-01174 0.8673[20]	1.4880[20]
4686	5-Cyclohexene-1,2,3,4-tetrone, 5,6-dihydroxy- Rhodizonic acid	$C_6H_2O_6$ 170.08	118-76-3 248 dec	sub	4-08-00-03609	EtOH 3
4687	Cyclohexene, 1,2,3-trimethyl- 1,2,3-Trimethylcyclohexene	C_9H_{16} 124.23	72312-48-2	150	1-05-00-00040 0.8347[11]	1.463[11]
4688	Cyclohexene, 1,3,5-trimethyl- 1,3,5-Trimethylcyclohexene	C_9H_{16} 124.23	3643-64-9	141	3-05-00-00226 0.8025[11]	1.449[13]
4689	Cyclohexene, 1,4,4-trimethyl- 1,4,4-Trimethylcyclohexene	C_9H_{16} 124.23	3419-71-4	140	1-05-00-00039 0.8032[19]	1.444[23]
4690	Cyclohexene, 1,4,5-trimethyl- 1,4,5-Trimethylcyclohexene	C_9H_{16} 124.23	20030-32-4	145	2-05-00-00050 0.805[20]	1.4482[20]
4691	Cyclohexene, 1,5,5-trimethyl- α-Cyclogeraniolene	C_9H_{16} 124.23	503-46-8	140	4-05-00-00286 0.7981[23]	1.4461[21]
4692	Cyclohexene, 1,6,6-trimethyl- 1,6,6-Trimethylcyclohexene	C_9H_{16} 124.23	69745-49-9	147	3-05-00-00225 0.8217[20]	1.456[20]
4693	Cyclohexene, 1,5,6-trimethyl-	C_9H_{16} 124.23	116724-18-6	140	3-05-00-00225 0.829[25]	1.4572[25]
4694	2-Cyclohexen-1-ol	$C_6H_{10}O$ 98.14	822-67-3	164	4-06-00-00196 0.9923[15]	EtOH 3; ace 3 1.4790[25]
4695	3-Cyclohexen-1-ol	$C_6H_{10}O$ 98.14	822-66-2	164; 68[16]	4-06-00-00200 0.9845[23]	H_2O 2; eth 3; ace 3 1.4851[22]
4696	2-Cyclohexen-1-ol, 2-methyl-5-(1-methylethenyl)-	$C_{10}H_{16}O$ 152.24	99-48-9	228	2-06-00-00102 0.9484[25]	1.4942[25]
4697	2-Cyclohexen-1-ol, 3-methyl-6-(1-methylethyl)-	$C_{10}H_{18}O$ 154.25	491-04-3	97-8[15.5]	2-06-00-00065 0.9119[25]	1.4729[25]
4698	2-Cyclohexen-1-ol, 6-methyl-3-(1-methylethyl)- Carvenol	$C_{10}H_{18}O$ 154.25	586-27-6	220	4-06-00-00253 0.925[22]	1.479
4699	3-Cyclohexen-1-ol, 4-methyl-1-(1-methylethyl)-	$C_{10}H_{18}O$ 154.25	562-74-3	209	4-06-00-00250 0.926[20]	1.4785[19]
4700	2-Cyclohexen-1-ol, 1,3,5-trimethyl- Cyclohexen-3-ol, 1,3,5-trimethyl	$C_9H_{16}O$ 140.23	33843-55-9 46	87-90[17]	1-06-00-00036 0.9132[20]	1.4735[19]
4701	2-Cyclohexen-1-ol, 2,4,4-trimethyl- 2,4,4-Trimethyl-2-cyclohexen-1-ol	$C_9H_{16}O$ 140.23	73741-61-4	193	1-06-00-00036 0.9310[25]	bz 4; EtOH 4
4702	2-Cyclohexen-1-ol, 3,5,5-trimethyl- Isophorol	$C_9H_{16}O$ 140.23	470-99-5	69[5]	3-06-00-00230 0.9128[20]	1.4717[20]
4703	2-Cyclohexen-1-one	C_6H_8O 96.13	930-68-7 -53	170	4-07-00-00124 0.9620[20]	EtOH 4; ace 3 1.4883[20]
4704	3-Cyclohexen-1-one	C_6H_8O 96.13	4096-34-8	170	4-07-00-00126 0.9620[25]	ace 4; EtOH 4 1.4842[18]
4705	2-Cyclohexen-1-one, 2,3-dimethyl-	$C_8H_{12}O$ 124.18	1122-20-9	93-6[20]	4-07-00-00148 0.9695[20]	EtOH 3; eth 3 1.4995[20]
4706	2-Cyclohexen-1-one, 2,5-dimethyl- 2,5-Dimethyl-2-cyclohexenone	$C_8H_{12}O$ 124.18	14845-35-3	189.5	1-07-00-00051 0.938[22]	eth 4; EtOH 4 1.4753[22]
4707	2-Cyclohexen-1-one, 3,5-dimethyl-	$C_8H_{12}O$ 124.18	1123-09-7	208.5	4-07-00-00149 0.9400[20]	EtOH 3; eth 3 1.4812[20]

3-130

No.	Name Synonym	Mol. Form. Mol. Wt.	CAS RN mp/°C	Merck No. bp/°C	Beil. Ref. den/g cm^{-3}	Solubility n_D
4708	2-Cyclohexen-1-one, 3,6-dimethyl-	$C_8H_{12}O$ 124.18	15329-10-9	75[19]	4-07-00-00150 1.007[18]	H_2O 1; EtOH 3; eth 3 1.4805[18]
4709	2-Cyclohexen-1-one, 2-hydroxy-3-methyl-6-(1-methylethyl)- Diosphenol	$C_{10}H_{16}O_2$ 168.24	490-03-9 83	3292 109[10]		
4710	2-Cyclohexen-1-one, 2-methyl-	$C_7H_{10}O$ 110.16	1121-18-2	178.5	4-07-00-00131 0.966[20]	bz 3 1.4833[20]
4711	2-Cyclohexen-1-one, 3-methyl-	$C_7H_{10}O$ 110.16	1193-18-6 -21	201	4-07-00-00130 0.9693[20]	H_2O 5; bz 3 1.49475[20]
4712	2-Cyclohexen-1-one, 5-methyl- 5-Methyl-2-cyclohexenone	$C_7H_{10}O$ 110.16	7214-50-8	185	4-07-00-00133 0.947[25]	ace 4; bz 4; EtOH 4 1.4739[25]
4713	3-Cyclohexen-1-one, 3-methyl-	$C_7H_{10}O$ 110.16	31883-98-4 -21	201	4-07-00-00131 0.9693[20]	bz 4 1.4947[20]
4714	3-Cyclohexen-1-one, 4-methyl-	$C_7H_{10}O$ 110.16	5259-65-4	171; 74[17]	4-07-00-00131 0.9551[20]	EtOH 3; ace 3; bz 3 1.4652[20]
4715	2-Cyclohexen-1-one, 2-methyl-5-(1-methylethenyl)-, (S)- (+)-Carvone	$C_{10}H_{14}O$ 150.22	2244-16-8 <15	1883 231	4-07-00-00315 0.965[20]	H_2O 2; EtOH 4; eth 3; chl 3 1.4989[20]
4716	2-Cyclohexen-1-one, 3-methyl-6-(1-methylethyl)-, (±)- (±)-Piperitone	$C_{10}H_{16}O$ 152.24	6091-52-7 -19	7443 232.5	4-07-00-00186 0.9331[20]	ace 4; EtOH 4 1.4845[20]
4717	2-Cyclohexen-1-one, 6-methyl-3-(1-methylethyl)-, (±)- Carvenone (DL)	$C_{10}H_{16}O$ 152.24	23733-68-8	235.5	4-07-00-00187 0.9263[20]	ace 4 1.4826[20]
4718	2-Cyclohexen-1-one, 3-methyl-6-(1-methylethylidene)- Piperitenone	$C_{10}H_{14}O$ 150.22	491-09-8	120-2[14]	4-07-00-00314 0.9774[20]	EtOH 4; eth 4 1.5294[20]
4719	2-Cyclohexen-1-one, 3,4,4-trimethyl- Cyclohexen-3-one, 1,6,6-trimethyl	$C_9H_{14}O$ 138.21	17299-41-1	217; 98[13]	4-07-00-00164 0.944[25]	1.4889[25]
4720	2-Cyclohexen-1-one, 3,4,6-trimethyl- Cyclohexen-3-one, 1,4,6-trimethyl	$C_9H_{14}O$ 138.21	7474-10-4	213	4-07-00-00170 0.943[17]	1.484[17]
4721	2-Cyclohexen-1-one, 3,5,5-trimethyl- Isophorone	$C_9H_{14}O$ 138.21	78-59-1 -8.1	215.2	4-07-00-00165 0.9255[20]	1.4766[18]
4722	2-Cyclohexen-1-one, 3,6,6-trimethyl- 3,6,6-Trimethyl-2-cyclohexenone	$C_9H_{14}O$ 138.21	23438-77-9	208; 86-8[15]	3-07-00-00286 0.927[24]	1.4798[20]
4723	Cyclohexylamine, N,4-dimethyl- Cyclohexane, 1-methylamino-4-methyl	$C_8H_{17}N$ 127.23	90226-23-6 -15	196	0.8188[22]	1.4175[22]
4724	Cyclononane	C_9H_{18} 126.24	293-55-0 11	178.4	4-05-00-00133 0.8463[25]	1.4666[20]
4725	Cyclononanone	$C_9H_{16}O$ 140.23	3350-30-9 34	148[24]; 94[12]	4-07-00-00062 0.9560[20]	EtOH 3 1.4729[20]
4726	Cyclononasiloxane, octadecamethyl- Octadecamethylcyclononasiloxane	$C_{18}H_{54}O_9Si_9$ 667.39	556-71-8	188[20]	3-04-00-01886 	bz 4; lig 4 1.4070[20]
4727	Cyclononene, (E)-	C_9H_{16} 124.23	3958-38-1	94-6[30]	4-05-00-00280 0.8615[20]	1.4799[20]
4728	Cyclononene, (Z)-	C_9H_{16} 124.23	933-21-1	168	4-05-00-00280 0.8671[20]	bz 4 1.4805[20]
4729	Cyclononyne	C_9H_{14} 122.21	6573-52-0	178	4-05-00-00419 0.8972[20]	1.4890[20]
4730	2,4(1H)-Cyclo-3,4-secoakuammilanium, 3,17-dihydroxy-16-(methoxycarbonyl)-4-methyl-, (3β,16R)- Echitamine	$C_{22}H_{30}N_2O_5$ 402.49	6871-44-9 206	3474		H_2O 3; EtOH 3; eth 3; chl 3
4731	1,3-Cyclooctadecadiene	$C_{18}H_{32}$ 248.45	6568-58-7	115[3]	4-05-00-00503 0.8814[25]	1.4899[20]
4732	1,4-Cyclooctadiene	C_8H_{12} 108.18	1073-07-0 -53	145	0.8754[20]	
4733	1,5-Cyclooctadiene	C_8H_{12} 108.18	111-78-4 -56.4	150.8	3-05-00-00321 0.8818[25]	bz 4 1.4905[25]
4734	1,5-Cyclooctadiene, (Z,Z)-	C_8H_{12} 108.18	1552-12-1 -70	151; 51-2[25]	4-05-00-00403 0.8818[25]	bz 4 1.4905[25]
4735	Cyclooctanamine Aminocyclooctane	$C_8H_{17}N$ 127.23	5452-37-9 -48	190	4-12-00-00128 0.928[25]	1.4804[20]
4736	Cyclooctane	C_8H_{16} 112.22	292-64-8 14.8	149	4-05-00-00111 0.8349[20]	H_2O 1; bz 3; lig 3 1.4586[20]
4737	Cyclooctane, butyl- Butylcyclooctane	$C_{12}H_{24}$ 168.32	16538-93-5	90-1[10]	4-05-00-00170 0.8260[25]	1.4585[25]
4738	Cyclooctanecarboxaldehyde Formylcyclooctane	$C_9H_{16}O$ 140.23	6688-11-5	96[15]	3-07-00-00112 0.94[25]	1.4748[20]
4739	Cyclooctanol	$C_8H_{16}O$ 128.21	696-71-9 25.1	99[16]	4-06-00-00113 0.974[20]	EtOH 3 1.4871[20]
4740	Cyclooctanone	$C_8H_{14}O$ 126.20	502-49-8 29	196	4-07-00-00049 0.9581[20]	H_2O 3; EtOH 3; ace 3; bz 3 1.4694[20]
4741	Cyclooctasiloxane, hexadecamethyl- Hexadecamethylcyclooctasiloxane	$C_{16}H_{48}O_8Si_8$ 593.24	556-68-3 31.5	290	3-04-00-01886 1.177[25]	bz 4; lig 4 1.4060[20]
4742	1,3,5,7-Cyclooctatetraene [8]Annulene	C_8H_8 104.15	629-20-9 -4.7	140.5	4-05-00-01331 0.9206[20]	EtOH 3; eth 3; ace 3; bz 3 1.5381[20]
4743	1,3,5,7-Cyclooctatetraene, 1-bromo-	C_8H_7Br 183.05	7567-22-8	52-3[18]	4-05-00-01334 1.4206[25]	ace 3 1.5870[25]
4744	1,3,5,7-Cyclooctatetraene, 1-butyl-	$C_{12}H_{16}$ 160.26	13402-37-4	98[20]	4-05-00-01417 0.8876[25]	1.5083[25]

No.	Name / Synonym	Mol. Form. / Mol. Wt.	CAS RN mp/°C	Merck No. bp/°C	Beil. Ref. den/g cm⁻³	Solubility n_D
4745	1,3,5,7-Cyclooctatetraene, 1-chloro- / Chlorocyclooctatetraene	C_8H_7Cl 138.60	29554-49-2 50-1$^{5.5}$		4-05-00-01334 1.1199^{25}	H_2O 1; ace 3; bz 3; chl 3 1.5542^{25}
4746	1,3,5,7-Cyclooctatetraene, 1-ethyl-	$C_{10}H_{12}$ 132.21	13402-35-2 81^{37}		4-05-00-01373 0.8996^{25}	1.5187^{25}
4747	1,3,5,7-Cyclooctatetraene, 1-methyl-	C_9H_{10} 118.18	2570-12-9 84.5^{67}		4-05-00-01358 0.8978^{25}	H_2O 1; EtOH 3; eth 3; bz 3 1.5249^{25}
4748	1,3,5,7-Cyclooctatetraene, propyl- / Cyclooctatetraene, propyl	$C_{11}H_{14}$ 146.23	13402-36-3 73^9		4-05-00-01400 0.8870^{25}	1.5131^{25}
4749	1,3,5-Cyclooctatriene	C_8H_{10} 106.17	1871-52-9 -83	145.5	4-05-00-00882 0.8971^{25}	1.5035^{25}
4750	1,3,6-Cyclooctatriene	C_8H_{10} 106.17	3725-30-2 -62	68^{60}	4-05-00-00883 0.8940^{25}	
4751	Cyclooctene, (E)-	C_8H_{14} 110.20	931-89-5 -59	143	4-05-00-00263 0.8483^{20}	EtOH 3; ctc 2; chl 3 1.4741^{25}
4752	Cyclooctene, (Z)-	C_8H_{14} 110.20	931-87-3 -12	138	3-05-00-00209 0.8472^{20}	EtOH 3; eth 3; ctc 3 1.4698^{20}
4753	2-Cycloocten-1-ol	$C_8H_{14}O$ 126.20	3212-75-7 74^2		4-06-00-00217 0.9756^{25}	1.4959^{25}
4754	3-Cycloocten-1-one	$C_8H_{12}O$ 124.18	4734-90-1 200.5		4-07-00-00142 0.990^{20}	ace 4; eth 4; chl 4 1.4953^{25}
4755	Cyclooctyne	C_8H_{12} 108.18	1781-78-8 158		4-05-00-00401 0.868^{20}	1.4850^{20}
4756	15H-Cyclopenta[a]phenanthrene, 16,17-dihydro- / 1,2-Cyclopentenophenanthrene	$C_{17}H_{14}$ 218.30	482-66-6 135.5	2749	4-05-00-02423	H_2O 1; EtOH 3; peth 3
4757	Cyclopenta[c]furo[3',2':4,5]furo[2,3-h][1]benzopyran-1,11-dione, 2,3,6a,8,9,9a-hexahydro-4-methoxy-, (6aR- / Aflatoxin B2	$C_{17}H_{14}O_6$ 314.29	7220-81-7 287.5	168	5-19-10-00552	
4758	Cyclopenta[c]pyran-1(4aH)-one, 5,6,7,7a-tetrahydro-4,7-dimethyl- / Nepetalactone	$C_{10}H_{14}O_2$ 166.22	490-10-8 71-2$^{0.05}$	6383	4-17-00-04622 1.0663^{25}	1.4859^{25}
4759	Cyclopentadecane	$C_{15}H_{30}$ 210.40	295-48-7 61.3		4-05-00-00181 0.8364^{61}	1.4592^{61}
4760	Cyclopentadecanol / Exaltol	$C_{15}H_{30}O$ 226.40	4727-17-7 80.5	177^{11}; 145$^{0.3}$	4-06-00-00185 0.930^{20}	1.4555^{98}
4761	Cyclopentadecanone	$C_{15}H_{28}O$ 224.39	502-72-7 63	120$^{0.3}$	4-07-00-00113 0.8895^{25}	H_2O 2; EtOH 3; ace 3 1.4637^{60}
4762	Cyclopentadecanone, 2-methyl- / 2-Methylcyclopentadecanone	$C_{16}H_{30}O$ 238.41	52914-66-6 171-3^{12}		3-07-00-00208 0.9213^{16}	1.4812^{16}
4763	Cyclopentadecanone, 3-methyl- / Muscone	$C_{16}H_{30}O$ 238.41	541-91-3 329; 130$^{0.5}$	6222	4-07-00-00118 0.9221^{17}	ace 4; eth 4; EtOH 4 1.4802^{17}
4764	1,3-Cyclopentadiene / Pyropentylene	C_5H_6 66.10	542-92-7 -85	2744 41	4-05-00-00377 0.8021^{20}	H_2O 1; EtOH 5; eth 5; ace 3 1.4440^{20}
4765	1,3-Cyclopentadiene, 5-ethylidene-	C_7H_8 92.14	3839-50-7 32^{17}		4-05-00-00878 0.865^{18}	1.526^{20}
4766	1,3-Cyclopentadiene, 1,2,3,4,5,5-hexachloro- / Perchlorocyclopentadiene	C_5Cl_6 272.77	77-47-4 -9	239; 48$^{0.3}$	4-05-00-00381 1.7019^{25}	1.5658^{20}
4767	1,3-Cyclopentadiene, 5-methylene- / Fulvene	C_6H_6 78.11	497-20-1 7-8^{56}		4-05-00-00764 0.8241^{20}	H_2O 1; bz 3; chl 3 1.4920^{20}
4768	1,3-Cyclopentadiene, 5-(1-methylethylidene)-	C_8H_{10} 106.17	2175-91-9 49-50^{11}		4-05-00-00974 0.881^{20}	1.5474^{20}
4769	1,3-Cyclopentadiene, 5-(2-propenylidene)-, (E)- / Fulvene, 6-vinyl(trans)	C_8H_8 104.15	116862-65-8 -35	45^{12} 20$^{0.15}$	0.898^{20}	H_2O 1; bz 3; chl 3
4770	1,3-Cyclopentadiene, 1,2,3,4-tetrachloro-5,5-dimethoxy-	$C_7H_6Cl_4O_2$ 263.93	2207-27-4 109		4-07-00-00286 1.501^{25}	1.5282^{20}
4771	Cyclopentanamine / Cyclopentylamine	$C_5H_{11}N$ 85.15	1003-03-8 -82.7	108	4-12-00-00004 0.8689^{20}	ace 3; bz 3; chl 3 1.4728^{25}
4772	Cyclopentanamine, 1-methyl-	$C_6H_{13}N$ 99.18	40571-45-7 121		4-12-00-00113 0.8025^{25}	H_2O 4 1.4408^{25}
4773	Cyclopentanamine, 2-methyl-	$C_6H_{13}N$ 99.18	41223-14-7 123		0-12-00-00007 0.801^{20}	H_2O 4
4774	Cyclopentane / Pentamethylene	C_5H_{10} 70.13	287-92-3 -93.8	2746 49.3	4-05-00-00014 0.7457^{20}	H_2O I; EtOH 5; eth 5; ace 5 1.4065^{20}
4775	Cyclopentaneacetic acid / 2-Cyclopentylacetic acid	$C_7H_{12}O_2$ 128.17	1123-00-8 13.5	228	4-09-00-00033 1.0216^{18}	1.4523^{18}
4776	Cyclopentaneacetic acid, 3-methyl- / 3-Methylcyclopentaneacetic acid	$C_8H_{14}O_2$ 142.20	63370-69-4 237		3-09-00-00060 0.9818^{20}	1.4504^{20}
4777	Cyclopentane, bromo- / Cyclopentyl bromide	C_5H_9Br 149.03	137-43-9 137.5		4-05-00-00019 1.3873^{20}	ctc 2 1.4886^{20}
4778	Cyclopentane, butyl- / Butylcyclopentane	C_9H_{18} 126.24	2040-95-1 -108	156.6	4-05-00-00139 0.7846^{20}	ace 4; bz 4; eth 4; EtOH 4 1.4316^{20}
4779	Cyclopentanecarbonitrile / Cyanocyclopentane	C_6H_9N 95.14	4254-02-8			
4780	Cyclopentanecarboxaldehyde	$C_6H_{10}O$ 98.14	872-53-7 133.5		4-07-00-00037 0.9371^{20}	H_2O 4; eth 4; EtOH 4 1.4432^{20}
4781	Cyclopentanecarboxylic acid / Cyclopentanoic acid	$C_6H_{10}O_2$ 114.14	3400-45-1 -7	212; 104^{11}	4-09-00-00011 1.0527^{20}	H_2O 2; ctc 2; MeOH 3 1.4532^{20}

No.	Name / Synonym	Mol. Form. / Mol. Wt.	CAS RN / mp/°C	Merck No. / bp/°C	Beil. Ref. / den/g cm^{-3}	Solubility / n_D
4782	Cyclopentanecarboxylic acid, 1-amino- / Cycloleucine	$C_6H_{11}NO_2$ / 129.18	52-52-8 / 330 dec	2740	4-14-00-00974	
4783	Cyclopentanecarboxylic acid, 1-methyl- / 1-Methylcyclopentanecarboxylic acid	$C_7H_{12}O_2$ / 128.17	5217-05-0 /	219	4-09-00-00035 / 1.0218[20]	eth 4; EtOH 4 / 1.4529[20]
4784	Cyclopentanecarboxylic acid, 2-methyl-5-(1-methylethyl)- / Puleganic acid	$C_{10}H_{18}O_2$ / 170.25	528-23-4 / -18	152[25]; 139[12]	4-09-00-00080 / 0.9642[20]	1.4524[24]
4785	Cyclopentanecarboxylic acid, 2-oxo-, ethyl ester	$C_8H_{12}O_3$ / 156.18	611-10-9 /	221; 110[16]	4-10-00-02602 / 1.0781[21]	eth 3; bz 3 / 1.4519[20]
4786	Cyclopentanecarboxylic acid, 1,2,2,3-tetramethyl- / Campholic acid	$C_{10}H_{18}O_2$ / 170.25	464-88-0 / 106	254; 146[12]	2-09-00-00019	H_2O 2; EtOH 3; eth 3
4787	Cyclopentanecarboxylic acid, 3-isopropyl-1-methyl-	$C_{12}H_{22}O_2$ / 198.31	104068-42-0 / 18.5	257	4-09-00-00079 / 0.9698[19]	1.4563[20]
4788	Cyclopentanecarboxylic acid, 2-methyl-5-(1-methylethyl)-, ethyl ester / Cyclopentanecarboxylic acid, 2-isopropyl-5-methyl, ethyl ester	$C_{12}H_{22}O_2$ / 198.31	116530-99-5 /	145[4]	1-09-00-00017 / 0.9178[12]	1.4405[12]
4789	Cyclopentane, chloro- / Cyclopentyl chloride	C_5H_9Cl / 104.58	930-28-9 /	114	4-05-00-00018 / 1.0051[20]	H_2O 1; eth 3; ace 3; bz 3 / 1.4510[20]
4790	Cyclopentane, decyl- / Decane, 1-cyclopentyl-	$C_{15}H_{30}$ / 210.40	1795-21-7 / -22	279	4-05-00-00186 / 0.8110[20]	ace 4; bz 4; eth 4; EtOH 4 / 1.4486[20]
4791	1,1-Cyclopentanedicarboxylic acid, diethyl ester	$C_{11}H_{18}O_4$ / 214.26	4167-77-5 /	210	4-09-00-02793 / 1.0186[20]	1.4370[25]
4792	1,2-Cyclopentanedicarboxylic acid, dimethyl ester, trans-(±)- / Cyclopentane-1,2-dicarboxylic acid, dimethyl ester (DL, trans)	$C_9H_{14}O_4$ / 186.21	80656-12-8 /	118-9[17]	4-09-00-02794 / 1.1130[19]	1.4498[17]
4793	1,3-Cyclopentanedicarboxylic acid, cis-	$C_7H_{10}O_4$ / 158.15	876-05-1 / 121	300 dec	4-09-00-02795	H_2O 3; EtOH 3; eth 3; ace 3
4794	1,3-Cyclopentanedicarboxylic acid, 1,2,2-trimethyl-, (1R-cis)-	$C_{10}H_{18}O_4$ / 200.23	124-83-4 / 189.1		4-09-00-02851 / 1.186[20]	H_2O 2; EtOH 4; eth 4; ace 3
4795	1,3-Cyclopentanedicarboxylic acid, 1,2,2-trimethyl-, dimethyl ester, (1R-cis)- / Camphoric acid, dimethyl ester, (+)-	$C_{12}H_{20}O_4$ / 228.29	15797-21-4 / <-16	264	4-09-00-02851 / 1.015[20]	eth 4; EtOH 4 / 1.4627[19]
4796	Cyclopentane, 1,1-diethyl- / 1,1-Diethylcyclopentane	C_9H_{18} / 126.24	2721-38-2 / -95.6	150.5	4-05-00-00141 / 0.7988[25]	1.4388[20]
4797	Cyclopentane, 1,2-diethyl-, cis- / cis-1,2-Diethylcyclopentane	C_9H_{18} / 126.24	932-39-8 / -118	153.6	4-05-00-00141 / 0.7920[25]	1.4355[20]
4798	Cyclopentane, 1,2-diethyl-, trans- / trans-1,2-Diethylcyclopentane	C_9H_{18} / 126.24	932-40-1 / -95	147.5	4-05-00-00141 / 0.7792[25]	bz 4; eth 4; peth 4 / 1.4295[20]
4799	Cyclopentane, 1,3-diethyl- / 1,3-Diethylcyclopentane	C_9H_{18} / 126.24	19398-75-5 /	150	0-05-00-00045 / 0.783[25]	1.430[20]
4800	Cyclopentane, 1,1-dimethyl- / 1,1-Dimethylcyclopentane	C_7H_{14} / 98.19	1638-26-2 / -69.8	87.5	4-05-00-00105 / 0.7499[25]	1.4136[20]
4801	Cyclopentane, 1,2-dimethyl-, cis- / cis-1,2-Dimethylcyclopentane	C_7H_{14} / 98.19	1192-18-3 / -54	99.5	4-05-00-00106 / 0.7680[25]	1.4222[20]
4802	Cyclopentane, 1,2-dimethyl-, trans- / trans-1,2-Dimethylcyclopentane	C_7H_{14} / 98.19	822-50-4 / -117.6	91.9	4-05-00-00107 / 0.7468[25]	1.4120[20]
4803	Cyclopentane, 1,3-dimethyl-, cis- / cis-1,3-Dimethylcyclopentane	C_7H_{14} / 98.19	2532-58-3 / -133.7	90.8	4-05-00-00107 / 0.7402[25]	1.4089[20]
4804	Cyclopentane, 1,3-dimethyl-, trans- / trans-1,3-Dimethylcyclopentane	C_7H_{14} / 98.19	1759-58-6 / -134	91.7	4-05-00-00108 / 0.7443[25]	1.4107[20]
4805	Cyclopentane, (1,1-dimethylethyl)-	C_9H_{18} / 126.24	3875-52-3 / -95.8	144.9	3-05-00-00124 / 0.7870[25]	1.4338[20]
4806	1,3-Cyclopentanediol, cis-	$C_5H_{10}O_2$ / 102.13	16326-97-9 / 31	105[5]; 90[1]	3-06-00-04057 / 1.099[15]	1.4792[20]
4807	1,3-Cyclopentanediol, trans- / Cyclopentane-1,3-diol (trans)	$C_5H_{10}O_2$ / 102.13	16326-98-0 / 40	80-5[0.1]	3-06-00-04057 / 1.094[25]	1.4830[20]
4808	Cyclopentane, docosyl- / Docosylcyclopentane	$C_{27}H_{54}$ / 378.73	62016-55-1 / 45	433	4-05-00-00201 / 0.8295[20]	bz 4; eth 4 / 1.4608[20]
4809	Cyclopentane, dodecyl-	$C_{17}H_{34}$ / 238.46	5634-30-0 / -5	312	3-05-00-00156 / 0.8158[20]	ace 4; bz 4; eth 4; EtOH 4 / 1.4518[20]
4810	Cyclopentaneethanamine, N,α-dimethyl- / Cyclopentamine	$C_9H_{19}N$ / 141.26	102-45-4 /	2745 / 171	4-12-00-00137	1.4500[20]
4811	Cyclopentaneethanol	$C_7H_{14}O$ / 114.19	766-00-7 /	183; 96[24]	4-06-00-00110 / 0.9180[20]	eth 4 / 1.4577[20]
4812	Cyclopentane, ethyl- / Ethylcyclopentane	C_7H_{14} / 98.19	1640-89-7 / -138.4	103.5	4-05-00-00104 / 0.7665[20]	H_2O 1; EtOH 5; eth 5; ace 5 / 1.4198[20]
4813	Cyclopentane, ethylidene- / Ethylidenecyclopentane	C_7H_{12} / 96.17	2146-37-4 / -129.5	115	4-05-00-00253 / 0.8020[20]	1.4481[20]
4814	Cyclopentane, 1-ethyl-1-methyl- / 1-Ethyl-1-methylcyclopentane	C_8H_{16} / 112.22	16747-50-5 / -143.8	121.6	4-05-00-00126 / 0.7767[25]	ace 4; bz 4; eth 4; EtOH 4 / 1.4272[20]
4815	Cyclopentane, 1-ethyl-2-methyl-, cis- / cis-1-Methyl-2-ethylcyclopentane	C_8H_{16} / 112.22	930-89-2 / -106	128	4-05-00-00126 / 0.7852[20]	1.4293[20]

No.	Name Synonym	Mol. Form. Mol. Wt.	CAS RN mp/°C	Merck No. bp/°C	Beil. Ref. den/g cm^{-3}	Solubility n_D
4816	Cyclopentane, 1-ethyl-2-methyl-, *trans*-	C_8H_{16} 112.22	930-90-5 -105.9	121.2	4-05-00-00126 0.7649[25]	1.4219[20]
4817	Cyclopentane, 1-ethyl-3-methyl-, *cis*- 1-Methyl-*cis*-3-ethylcyclopentane	C_8H_{16} 112.22	2613-66-3 	121	4-05-00-00127 0.7724[20]	ace 4; bz 4; eth 4; EtOH 4 1.4203[20]
4818	Cyclopentane, 1-ethyl-3-methyl-, *trans*- 1-Methyl-*trans*-3-ethylcyclopentane	C_8H_{16} 112.22	2613-65-2 -108	121	4-05-00-00127 0.7619[20]	1.4186[20]
4819	Cyclopentane, heneicosyl-	$C_{26}H_{52}$ 364.70	6703-82-8 42	420	4-05-00-00200 0.8286[20]	1.4602[20]
4820	Cyclopentane, heptyl- Heptane, 1-cyclopentyl-	$C_{12}H_{24}$ 168.32	5617-42-5 -53	224	4-05-00-00173 0.8010[20]	ace 4; bz 4; eth 4; EtOH 4 1.4421[20]
4821	Cyclopentane, hexacosyl- Hexacosylcyclopentane	$C_{31}H_{82}$ 434.83	62016-58-4 56	468	4-05-00-00203 0.8326[20]	1.4628[20]
4822	Cyclopentane, hexadecyl- Hexadecane, 1-cyclopentyl-	$C_{21}H_{42}$ 294.56	6812-39-1 21	368	4-05-00-00195 0.8228[20]	ace 4; bz 4; eth 4; EtOH 4 1.4564[20]
4823	Cyclopentane, hexyl-	$C_{11}H_{22}$ 154.30	4457-00-5 -73	203	4-05-00-00168 0.7965[20]	ace 4; bz 4; eth 4; EtOH 4 1.4392[20]
4824	Cyclopentane, iodo- Cyclopentyl iodide	C_5H_9I 196.03	1556-18-9 	166.5	4-05-00-00020 1.7096[20]	H_2O 1; eth 3; bz 3; ctc 2 1.5447[20]
4825	Cyclopentanemethanol	$C_6H_{12}O$ 100.16	3637-61-4 	163	4-06-00-00092 0.9332[20]	1.4579[20]
4826	Cyclopentane, methyl- Methylcyclopentane	C_6H_{12} 84.16	96-37-7 -142.5	71.8	4-05-00-00084 0.7486[20]	H_2O 1; EtOH 5; eth 5; ace 5 1.4097[20]
4827	Cyclopentane, methylene- Methylenecyclopentane	C_6H_{10} 82.15	1528-30-9 	75.5	4-05-00-00241 0.7787[20]	bz 3; chl 3 1.4355[20]
4828	Cyclopentane, (1-methylethyl)- Isopropylcyclopentane	C_8H_{16} 112.22	3875-51-2 -111.4	126.5	4-05-00-00125 0.7765[20]	H_2O 1; EtOH 5; eth 3; ace 5 1.4258[20]
4829	Cyclopentane, 1-methyl-2-(1-methylethyl)- Cyclopentane, 1-isopropyl-2-methyl-	C_9H_{18} 126.24	89223-57-4 	143	4-05-00-00141 0.7832[15]	1.4279[20]
4830	Cyclopentane, 1-methyl-1-nitro- 1-Methyl-1-nitrocyclopentane	$C_6H_{11}NO_2$ 129.16	30168-50-4 181; 92[40]		4-05-00-00087 1.0395[20]	1.4504[20]
4831	Cyclopentane, 1-methyl-2-nitro- 1-Methyl-2-nitrocyclopentane	$C_6H_{11}NO_2$ 129.16	102153-88-8 185 dec; 98[40]		1-05-00-00011 1.0381[22]	1.4488[22]
4832	Cyclopentane, (1-methylpropyl)-	C_9H_{18} 126.24	4850-32-2 	154.4	3-05-00-00124 0.7905[25]	1.4357[20]
4833	Cyclopentane, 1-methyl-1-propyl- 1-Methyl-1-propylcyclopentane	C_9H_{18} 126.24	16631-63-3 	146	4-05-00-00140 0.795[25]	1.437[20]
4834	Cyclopentane, 1-methyl-2-propyl-, *cis*- *cis*-1-Methyl-2-propylcyclopentane	C_9H_{18} 126.24	932-43-4 -104	152.6	4-05-00-00140 0.7881[25]	1.4343[20]
4835	Cyclopentane, 1-methyl-2-propyl-, *trans*- *trans*-1-Methyl-2-propylcyclopentane	C_9H_{18} 126.24	932-44-5 -123	146.4	4-05-00-00140 0.7735[25]	1.4274[20]
4836	Cyclopentane, 1-methyl-3-propyl-, *cis*- *cis*-1-Methyl-3-propylcyclopentane	C_9H_{18} 126.24	2443-04-1 	148	3-05-00-00125 0.776[25]	1.426[20]
4837	Cyclopentane, 1-methyl-3-propyl-, *trans*- *trans*-1-Methyl-3-propylcyclopentane	C_9H_{18} 126.24	2443-03-0 	148	3-05-00-00125 0.776[25]	1.426[20]
4838	Cyclopentane, (2-methylpropyl)- Isobutylcyclopentane	C_9H_{18} 126.24	3788-32-7 -115.2	148	4-05-00-00140 0.7769[25]	1.4298[20]
4839	Cyclopentane, nitro- Nitrocyclopentane	$C_5H_9NO_2$ 115.13	2562-38-1 90-1[40]		4-05-00-00020 1.0776[23]	H_2O l; bz 3; chl 3 1.4538[20]
4840	Cyclopentane, (nitromethyl)- (Nitromethyl)cyclopentane	$C_6H_{11}NO_2$ 129.16	2625-31-2 110[35]		1-05-00-00011 1.0713[20]	1.4587[20]
4841	Cyclopentane, nonyl-	$C_{14}H_{28}$ 196.38	2882-98-6 -29	262	4-05-00-00181 0.8081[20]	ace 4; bz 4; eth 4; EtOH 4 1.4467[20]
4842	Cyclopentane, octadecyl- Octadecane, 1-cyclopentyl-	$C_{23}H_{46}$ 322.62	62016-53-9 30	391	4-05-00-00196 0.8254[20]	ace 4; bz 4; eth 4; EtOH 4 1.4581[20]
4843	Cyclopentane, octyl- Octane, 1-cyclopentyl-	$C_{13}H_{26}$ 182.35	1795-20-6 -44	243	4-05-00-00176 0.8048[20]	1.4446[20]
4844	Cyclopentane, pentadecyl- Pentadecane, 1-cyclopentyl-	$C_{20}H_{40}$ 280.54	4669-01-6 17	355	4-05-00-00195 0.8213[20]	ace 4; bz 4; eth 4; EtOH 4 1.4554[20]
4845	Cyclopentane, pentatriacontyl- Pentatriacontylcyclopentane	$C_{40}H_{80}$ 561.08	61827-98-3 74	531	4-05-00-00205 0.8374[20]	1.4660[20]
4846	Cyclopentane, pentyl-	$C_{10}H_{20}$ 140.27	3741-00-2 -83	180	4-05-00-00160 0.7912[20]	ace 4; bz 4; eth 4; EtOH 4 1.4356[20]
4847	Cyclopentanepropanoic acid	$C_8H_{14}O_2$ 142.20	140-77-2 130-2[12]		4-09-00-00046 0.996[25]	1.4570[20]
4848	Cyclopentane, 2-propenyl-	C_8H_{14} 110.20	3524-75-2 -110.7	125	4-05-00-00272 0.793[25]	chl 3 1.4412[20]
4849	Cyclopentane, propyl- Propylcyclopentane	C_8H_{16} 112.22	2040-96-2 -117.3	131	4-05-00-00125 0.7763[20]	H_2O 1; EtOH 5; eth 5; ace 5 1.4266[20]
4850	Cyclopentane, tetradecyl- Tetradecane, 1-cyclopentyl-	$C_{19}H_{38}$ 266.51	1795-22-8 9	341	3-05-00-00158 0.8196[20]	1.4543[20]
4851	Cyclopentane, 1,1,3,3-tetramethyl- 1,1,3,3-Tetramethylcyclopentane	C_9H_{18} 126.24	50876-33-0 -88.4	118	4-05-00-00143 0.7469[25]	1.4125[20]
4852	Cyclopentanethiol Cyclopentyl mercaptan	$C_5H_{10}S$ 102.20	1679-07-8 	132.1	4-06-00-00015 0.9550[20]	
4853	Cyclopentane, triacontyl- Triacontylcyclopentane	$C_{35}H_{70}$ 490.94	61827-93-8 65	498	4-05-00-00204 0.8350[20]	1.4644[20]
4854	Cyclopentane, tridecyl- Tridecane, 1-cyclopentyl-	$C_{18}H_{36}$ 252.48	6006-34-4 5	327	4-05-00-00192 0.8178[20]	ace 4; bz 4; eth 4; EtOH 4 1.4531[20]
4855	Cyclopentane, 1,1,2-trimethyl- Camphocean	C_8H_{16} 112.22	4259-00-1 -21.6	114; 53[100]	4-05-00-00127 0.7660[20]	1.4199[20]

No.	Name Synonym	Mol. Form. Mol. Wt.	CAS RN mp/°C	Merck No. bp/°C	Beil. Ref. den/g cm^{-3}	Solubility n_D
4856	Cyclopentane, 1,1,3-trimethyl- 1,1,3-Trimethylcyclopentane	C_8H_{16} 112.22	4516-69-2 -142.4	104.9	3-05-00-00111 0.7439[25]	1.4112[20]
4857	Cyclopentane, 1,2,3-trimethyl-, (1α,2β,3α)- Cyclopentane, 1,2,3-trimethyl-, trans-1,2,cis-1,3-	C_8H_{16} 112.22	19374-46-0 -112.7	110.4	4-05-00-00128 0.7492[25]	1.4138[20]
4858	Cyclopentane, 1,2,3-trimethyl-, (1α,2α,3α)- Cyclopentane, 1,2,3-trimethyl-, cis-1,2,cis-1,3-	C_8H_{16} 112.22	2613-69-6 -116.4	123	4-05-00-00128 0.7751[25]	1.4262[20]
4859	Cyclopentane, 1,2,3-trimethyl-, (1α,2α,3β)-	C_8H_{16} 112.22	15890-40-1 -112	117.5	4-05-00-00128 0.7661[25]	1.4218[20]
4860	Cyclopentane, 1,2,4-trimethyl-, (1α,2β,4α)-	C_8H_{16} 112.22	16883-48-0 -130.8	109.3	4-05-00-00129 0.7430[25]	1.4106[20]
4861	Cyclopentane, 1,2,4-trimethyl-, (1α,2α,4α)- Cyclopentane, 1,2,4-trimethyl-, cis-1,2,cis-1,4-	C_8H_{16} 112.22	2613-72-1 -132.3	116.8	4-05-00-00129 0.758[25]	1.4186[20]
4862	Cyclopentane, 1,2,4-trimethyl-, (1α,2α,4β)-	C_8H_{16} 112.22	4850-28-6 -132.6	116.7	4-05-00-00129 0.7592[25]	1.4186[20]
4863	Cyclopentane, undecyl-	$C_{16}H_{32}$ 224.43	6785-23-5 -10	296	4-05-00-00189 0.8135[20]	ace 4; bz 4; eth 4; EtOH 4 1.4503[20]
4864	Cyclopentanol Cyclopentyl alcohol	$C_5H_{10}O$ 86.13	96-41-3 -17.5	2747 140.4	4-06-00-00005 0.9488[20]	H_2O 2; EtOH 3; eth 3; ace 3 1.4530[20]
4865	Cyclopentanol, acetate Cyclopentyl acetate	$C_7H_{12}O_2$ 128.17	933-05-1	51-2[12]	4-06-00-00007 0.9522[16]	
4866	Cyclopentanol, 1-ethynyl- 1-Ethynyl-1-cyclopentanol	$C_7H_{10}O$ 110.16	17356-19-3 27	157.5	4-06-00-00339 0.962[25]	1.4751[20]
4867	Cyclopentanol, 1-methyl-	$C_6H_{12}O$ 100.16	1462-03-9 36	136; 53[30]	4-06-00-00086 0.9044[23]	1.4429[23]
4868	Cyclopentanol, 2-methyl-, cis- cis-2-Methylcyclopentanol	$C_6H_{12}O$ 100.16	25144-05-2	148.5	3-06-00-00055 0.9379[16]	1.4504[16]
4869	Cyclopentanol, 2-methyl-, trans- trans-2-Methylcyclopentanol	$C_6H_{12}O$ 100.16	25144-04-1	150.5	4-06-00-00091 0.9248[16]	1.4499[16]
4870	Cyclopentanol, 3-methyl- 3-Methylcyclopentanol	$C_6H_{12}O$ 100.16	18729-48-1	148.5	2-06-00-00015 0.9158[16]	eth 4; EtOH 4 1.4487[16]
4871	Cyclopentanol, 1-phenyl-	$C_{11}H_{14}O$ 162.23	10487-96-4	132-3[18]	4-06-00-03884 1.0530[20]	1.5479[20]
4872	Cyclopentanol, 1-propyl-	$C_8H_{16}O$ 128.21	1604-02-0 -37.5	173.5	3-06-00-00101 0.9040[25]	1.4502[25]
4873	Cyclopentanol, 2-propyl-, cis- cis-2-Propylcyclopentanol	$C_8H_{16}O$ 128.21	25172-42-3	79-80[12]	3-06-00-00101 0.9163[9]	1.4600[9]
4874	Cyclopentanol, 2-propyl-, trans- trans-2-Propylcyclopentanol	$C_8H_{16}O$ 128.21	25172-43-4	78-9[10]	3-06-00-00101 0.9016[9]	1.4565[9]
4875	Cyclopentanol, 1,2,2-trimethyl- 1,2,2-Trimethylcyclopentanol	$C_8H_{16}O$ 128.21	1121-95-5	156; 60[15]	1-06-00-00014 0.9101[20]	1.4513[20]
4876	Cyclopentanol, 1,2,4-trimethyl- 1,2,4-Trimethylcyclopentanol	$C_8H_{16}O$ 128.21	34103-97-4	158	1-06-00-00014 0.8850[21]	1.4424[21]
4877	Cyclopentanol, 1,2,5-trimethyl- 1,2,5-Trimethylcyclopentanol	$C_8H_{16}O$ 128.21	33840-38-9	56-60[8]	1-06-00-00014 0.9121[15]	1.4554[16]
4878	Cyclopentanone Adipic ketone	C_5H_8O 84.12	120-92-3 -51.3	2748 130.5	4-07-00-00005 0.9487[20]	H_2O 1; EtOH 3; eth 5; ace 3 1.4366[20]
4879	Cyclopentanone, 2-acetyl- 2-Acetylcyclopentanone	$C_7H_{10}O_2$ 126.16	1670-46-8	72-5[20]	4-07-00-01993 1.0431[25]	1.4906[20]
4880	Cyclopentanone, 2-chloro-	C_5H_7ClO 118.56	694-28-0	72-4[12]	4-07-00-00012 1.185[25]	1.4750[20]
4881	Cyclopentanone, 2-cyclopentylidene-	$C_{10}H_{14}O$ 150.22	825-25-2	135[25]	4-07-00-00320 1.0179[18]	1.5215[18]
4882	Cyclopentanone, 2,5-dimethyl- 2,5-Dimethylcyclopentanone	$C_7H_{12}O$ 112.17	4041-09-2	146.5	4-07-00-00048 0.882[25]	1.4310[20]
4883	Cyclopentanone, 2-ethyl-5-methyl- 2-Ethyl-5-methylcyclopentanone	$C_8H_{14}O$ 126.20	51686-60-3	164.5	3-06-00-02552 0.900[12]	1.4360[22]
4884	Cyclopentanone, 4-isopropyl-2-methyl- Cyclopentanone, 3-isopropyl-5-methyl	$C_9H_{16}O$ 140.23	90645-61-7	191.5	2-07-00-00034 0.8862[20]	1.4413[19]
4885	Cyclopentanone, 2-methyl-	$C_6H_{10}O$ 98.14	1120-72-5 -75	139.5	4-07-00-00036 0.9139[20]	H_2O 3; EtOH 4; eth 4; ace 4 1.4364[20]
4886	Cyclopentanone, 3-methyl-, (±)- (±)-3-Methylcyclopentanone	$C_6H_{10}O$ 98.14	6195-92-2 -58.4	144	4-07-00-00037 0.913[22]	H_2O 3; EtOH 4; eth 4; ace 4 1.4329[20]
4887	Cyclopentanone, 2-(1-methylethyl)- 2-Isopropylcyclopentanone	$C_8H_{14}O$ 126.20	14845-55-7	175	4-07-00-00060 0.9105[20]	1.4425[15]
4888	Cyclopentanone, 3-methyl-2-(3-methylbutyl)- Cyclopentanone, 2-isopentyl-3-methyl	$C_{11}H_{20}O$ 168.28	52033-97-3	98-9[8]	3-07-00-00181 0.8938[20]	1.4537[20]
4889	Cyclopentanone, 2-methyl-2-(1-methylethyl)- Cyclopentanone, 2-isopropyl-2-methyl-	$C_9H_{16}O$ 140.23	32116-65-7	97.5[45]	3-07-00-00130 0.9067[16]	1.4495[16]
4890	Cyclopentanone, 2-methyl-5-(1-methylethyl)-	$C_9H_{16}O$ 140.23	6784-18-5	182	3-07-00-00130 0.889[20]	1.4402[20]
4891	Cyclopentanone, 4-methyl-2-(1-methylethyl)- Cyclopentanone, 2-isopropyl-4-methyl	$C_9H_{16}O$ 140.23	69770-98-5	183	1-07-00-00026 0.8850[20]	1.4392[20]
4892	Cyclopentanone, oxime	C_5H_9NO 99.13	1192-28-5 57.8	196	4-07-00-00008	H_2O 4; bz 4
4893	Cyclopentanone, 2-(phenylmethyl)- 2-Benzylcyclopentanone	$C_{12}H_{14}O$ 174.24	2867-63-2	152[18]; 84[0.3]	4-07-00-01043 1.038[20]	1.534[13]
4894	Cyclopentanone, 2-propyl- 2-Propylcyclopentanone	$C_8H_{14}O$ 126.20	1193-70-0 -68.3	183	4-07-00-00060 0.9017[20]	1.4429[20]

No.	Name Synonym	Mol. Form. Mol. Wt.	CAS RN mp/°C	Merck No. bp/°C	Beil. Ref. den/g cm^{-3}	Solubility n_D
4895	Cyclopentanone, 3-propyl- 3-Propylcyclopentanone	$C_8H_{14}O$ 126.20	82322-93-8	190.5	3-07-00-00099 0.9041[20]	1.4456[12]
4896	Cyclopentanone, 2,2,4-trimethyl- 2,2,4-Trimethylcyclopentanone	$C_8H_{14}O$ 126.20	28056-54-4 -40.6	158	4-07-00-00061 0.8775[25]	1.4300[20]
4897	Cyclopentanone, 2,2,5-trimethyl- 2,2,5-Trimethylcyclopentanone	$C_8H_{14}O$ 126.20	4573-09-5	152	3-07-00-00107 0.8781[20]	1.4306[20]
4898	Cyclopentanone, 2,3,5-trimethyl- 2,3,5-Trimethylcyclopentanol	$C_8H_{14}O$ 126.20	90112-26-8	158	1-07-00-00021 0.8778[19]	1.4316[19]
4899	Cyclopentanone, 2,4,4-trimethyl- 2,4,4-Trimethylcyclopentanone	$C_8H_{14}O$ 126.20	4694-12-6 -25.6	160.5	4-07-00-00061 0.8785[18]	1.433[18]
4900	1H-Cyclopentapyrimidine-2,4(3H,5H)-dione, 3-cyclohexyl-6,7-dihydro- Lenacil	$C_{13}H_{18}N_2O_2$ 234.30	2164-08-1 290	5318	5-24-07-00375 1.32[25]	py 4
4901	Cyclopentasiloxane, decamethyl- Decamethylcyclopentasiloxane	$C_{10}H_{30}O_5Si_5$ 370.77	541-02-6 -38	2841 210	4-04-00-04128 0.9593[20]	H_2O 1 1.3982[20]
4902	Cyclopentasiloxane, 2,4,6,8,10-pentamethyl-	$C_5H_{20}O_5Si_5$ 300.64	6166-86-5 -108	169	4-04-00-04099 0.9985[20]	1.3912[20]
4903	Cyclopentene	C_5H_8 68.12	142-29-0 -135.1	44.2	4-05-00-00209 0.7720[20]	H_2O 1; EtOH 3; eth 3; bz 3 1.4225[20]
4904	3-Cyclopentene-1-acetic acid	$C_7H_{10}O_2$ 126.16	767-03-3 19	93-4[2.5]	1.047[25]	1.4676[20]
4905	1-Cyclopentene-1-acetonitrile	C_7H_9N 107.16	22734-04-9	124[100]	3-09-00-00151 0.951[25]	1.4670[20]
4906	Cyclopentene, 1-butyl- 1-Butyl-1-cyclopentene	C_9H_{16} 124.23	2423-01-0 -88.4	157	4-05-00-00289 0.8035[25]	1.4486[20]
4907	1-Cyclopentenecarbonitrile 1-Cyanocyclopentene	C_6H_7N 93.13	3047-38-9			
4908	1-Cyclopentene-1-carboxaldehyde	C_6H_8O 96.13	6140-65-4 -32	146	4-07-00-00127 0.970[21]	1.4872[17]
4909	1-Cyclopentene-1-carboxylic acid	$C_6H_8O_2$ 112.13	1560-11-8 123.5	210	4-09-00-00109	H_2O 3; EtOH 3; eth 3; peth 3
4910	1-Cyclopentene-1-carboxylic acid, 2,3,3-trimethyl-	$C_9H_{14}O_2$ 154.21	5587-63-3 135	255.5	4-09-00-00150	eth 4; EtOH 4
4911	1-Cyclopentene-1-carboxylic acid, 2,3,3-trimethyl-, methyl ester	$C_{10}H_{16}O_2$ 168.24	1460-88-4 203.5		0-09-00-00058 0.9656[20]	1.4571[20]
4912	2-Cyclopentene-1-carboxylic acid, 1,2,3-trimethyl- Laurolenic acid	$C_9H_{14}O_2$ 154.21	6894-69-5 6.5	192[100]	2-09-00-00040 1.0318[25]	1.4766
4913	3-Cyclopentene-1-carboxylic acid, 2,2,3-trimethyl- α-Campholytic acid	$C_9H_{14}O_2$ 154.21	6709-22-4 40.5	241.5; 162[45]	4-09-00-00150 1.0145[18]	1.4712[17]
4914	Cyclopentene, 3-chloro-	C_5H_7Cl 102.56	96-40-2 25-31[30]		4-05-00-00212 1.0577[15]	eth 4; EtOH 4; chl 4 1.4708[26]
4915	1-Cyclopentene-1,2-dicarboxylic acid, diethyl ester Diethyl 1-cyclopentene-1,2-dicarboxylate	$C_{11}H_{16}O_4$ 212.25	70202-92-5 140[16]		3-09-00-03938 1.0805[20]	1.4652[20]
4916	1-Cyclopentene-1,3-dicarboxylic acid, diethyl ester Cyclopentene-1,3-dicarboxylic acid, diethyl ester	$C_{11}H_{16}O_4$ 212.25	30689-41-9 168[21]		3-09-00-03939 1.1121[24]	1.4564
4917	Cyclopentene, 1,2-dichloro-3,3,4,4,5,5-hexafluoro-	$C_5Cl_2F_6$ 244.95	706-79-6 -105.8	90.7	4-05-00-00213 1.6546[20]	1.3676[20]
4918	Cyclopentene, 1,2-dimethyl-	C_7H_{12} 96.17	765-47-9 -90.4	105.8	4-05-00-00254 0.7928[25]	1.4448[20]
4919	Cyclopentene, 1,3-dimethyl- 1,3-Dimethylcyclopentene	C_7H_{12} 96.17	62184-82-1	92	4-05-00-00255 0.766[20]	1.428[20]
4920	Cyclopentene, 1,4-dimethyl- 1,4-Dimethylcyclopentene	C_7H_{12} 96.17	19550-48-2	93.2	4-05-00-00255 0.779[20]	1.4283[20]
4921	Cyclopentene, 1,5-dimethyl-	C_7H_{12} 96.17	16491-15-9 -118	99	4-05-00-00254 0.780[20]	1.4331[20]
4922	Cyclopentene, 4,4-dimethyl-	C_7H_{12} 96.17	19037-72-0	88	4-05-00-00254 0.766[25]	1.423[20]
4923	Cyclopentene, 1,5-dimethyl-4-(1-methylethyl)- Cyclopentene, 2,3-dimethyl-4-isopropyl	$C_{10}H_{18}$ 138.25	6912-05-6	163	1-05-00-00046 0.8085[22]	1.4503[22]
4924	Cyclopentene, 1-ethyl-	C_7H_{12} 96.17	2146-38-5 -118.5	106.3	4-05-00-00252 0.7936[25]	1.4412[20]
4925	Cyclopentene, 3-ethyl-	C_7H_{12} 96.17	694-35-9	97.8	4-05-00-00252 0.7784[25]	1.4315[20]
4926	Cyclopentene, 4-ethyl- 4-Ethylcyclopentene	C_7H_{12} 96.17	3742-38-9	98.3	4-05-00-00253 0.778[25]	1.431[20]
4927	Cyclopentene, 4-isopropyl-1-methyl- 4-Isopropyl-1-methylcyclopentene	C_9H_{16} 124.23	90769-70-3	142.5	1-05-00-00041 0.7945[21]	1.4403[21]
4928	Cyclopentene, 1-methyl-	C_6H_{10} 82.15	693-89-0 -126.5	75.5	4-05-00-00239 0.7748[25]	1.4322[20]
4929	Cyclopentene, 3-methyl- 3-Methylcyclopentene	C_6H_{10} 82.15	1120-62-3 64.9		4-05-00-00240 0.7572[25]	1.4216[20]
4930	Cyclopentene, 4-methyl-	C_6H_{10} 82.15	1759-81-5 -160.8	65.7	4-05-00-00240 0.7634[25]	1.4209[20]

No.	Name Synonym	Mol. Form. Mol. Wt.	CAS RN	mp/°C	Merck No. bp/°C	Beil. Ref. den/g cm^{-3}	Solubility n_D
4931	Cyclopentene, 3-methyl-1-(1-methylethyl)- Pulegene	C_9H_{16} 124.23	51115-02-7		138.5	1-05-00-00040 0.791[22]	1.4380[23]
4932	Cyclopentene, octachloro- Perchlorocyclopentene	C_5Cl_8 343.68	706-78-5	40	283	4-05-00-00214 1.8200[50]	H_2O 1; EtOH 4 1.5660[50]
4933	1-Cyclopentene-1-pentanoic acid, δ-hydroxy-3,5-bis(1-methylpropyl)-β-oxo-	$C_{18}H_{30}O_4$ 310.43	53109-18-5	183		3-10-00-04194 1.269[20]	eth 4; EtOH 4
4934	Cyclopentene, 1-pentyl-	$C_{10}H_{18}$ 138.25	4291-98-9	-83	179	4-05-00-00306 0.8085[25]	1.4516[20]
4935	Cyclopentene, 1-propyl- 1-Propyl-1-cyclopentene	C_8H_{14} 110.20	3074-61-1	-100.3	131.2	4-05-00-00271 0.7978[25]	1.4452[20]
4936	2-Cyclopentene-1-tridecanoic acid, (S)- Chaulmoogric acid	$C_{18}H_{32}O_2$ 280.45	29106-32-9	2037 68.5	247-8[20]	4-09-00-00216	eth 4; chl 4
4937	Cyclopentene, 1,2,3-trimethyl-	C_8H_{14} 110.20	473-91-6		121.6	3-05-00-00219 0.8039[15]	1.4464[16]
4938	Cyclopentene, 1,5,5-trimethyl- 1,5,5-Trimethylcyclopentene	C_8H_{14} 110.20	62184-83-2		109	3-05-00-00219 0.7824[20]	1.4324[20]
4939	4-Cyclopentene-1,2,3-trione, 4,5-dihydroxy- Croconic acid	$C_5H_2O_5$ 142.07	488-86-8	150	dec	4-08-00-03342	H_2O 4; EtOH 3
4940	2-Cyclopentene-1-undecanoic acid, (R)- Hydnocarpic acid	$C_{16}H_{28}O_2$ 252.40	459-67-6	60.5	4679	4-09-00-00215	EtOH 4; chl 4; peth 4
4941	2-Cyclopenten-1-one Cyclopenten-3-one	C_5H_6O 82.10	930-30-3		136; 40[12]	4-07-00-00122 0.989[15]	eth 4; EtOH 4 1.4629[15]
4942	2-Cyclopenten-1-one, 2,3-dihydroxy- Reductic acid	$C_5H_8O_3$ 114.10	80-72-8	212	8134	4-08-00-01714	H_2O 3; EtOH 3; eth 2; ace 2
4943	2-Cyclopenten-1-one, 2,4-dimethyl-	$C_7H_{10}O$ 110.16	23048-13-7		165	4-07-00-00138 0.9423[20]	1.4655[20]
4944	2-Cyclopenten-1-one, 2-methyl-	C_6H_8O 96.13	1120-73-6		157	4-07-00-00127 0.9808[16]	1.4762[15]
4945	2-Cyclopenten-1-one, 3-methyl- 3-Methyl-2-cyclopentenone	C_6H_8O 96.13	2758-18-1		157.5	4-07-00-00126 0.9712[20]	1.4714[20]
4946	2-Cyclopenten-1-one, 5-methyl- Cyclopenten-3-one, 4-methyl	C_6H_8O 96.13	14963-40-7		140	3-07-00-00229 0.942[18]	1.4460[18]
4947	2-Cyclopenten-1-one, 3-(1-methylethyl)-	$C_8H_{12}O$ 124.18	1619-28-9		214	2-07-00-00061 0.9378[20]	1.4788[20]
4948	2-Cyclopenten-1-one, 5-methyl-2-(1-methylethyl)-	$C_9H_{14}O$ 138.21	5587-79-1		189	3-07-00-00291 0.9143[20]	1.4660[20]
4949	2-Cyclopenten-1-one, 3-methyl-2-(2-pentenyl)-, (Z)- Jasmone	$C_{11}H_{16}O$ 164.25	488-10-8		5141 258; 134[12]	4-07-00-00337 0.9437[22]	H_2O 2; EtOH 3; eth 3; ctc 3 1.4979[22]
4950	2-Cyclopenten-1-one, 3-methyl-2-pentyl-	$C_{11}H_{18}O$ 166.26	1128-08-1		143[22]; 116[12]	4-07-00-00230 0.9165[18]	1.4767[20]
4951	2-Cyclopenten-1-one, 3-phenyl-	$C_{11}H_{10}O$ 158.20	3810-26-2	-23	234.2	4-07-00-01180 0.9711[20]	EtOH 3; eth 2; ace 3; chl 3 1.5440[20]
4952	Cycloprate Cyclopropanecarboxylic acid, hexadecyl ester	$C_{20}H_{38}O_2$ 310.52	54460-46-7		154[0.05]		
4953	Cyclopropanamine Cyclopropylamine	C_3H_7N 57.10	765-30-0	-35.4	50.5	4-12-00-00003 0.8240[20]	H_2O 5; EtOH 3; eth 3; chl 3 1.4210[20]
4954	Cyclopropa[d]naphthalene, 1,1a,4,4a,5,6,7,8-octahydro-2,4a,8,8-tetramethyl-, [1aS-(1aα,4aβ,8aR*)]- Thujopsene	$C_{15}H_{24}$ 204.36	470-40-6		9327	4-05-00-01189	
4955	Cyclopropane Trimethylene	C_3H_6 42.08	75-19-4	-127.4	2755 -32.8	4-05-00-00003 *0.617[25]	H_2O 3; EtOH 4; eth 4; bz 3 1.3799[-42]
4956	Cyclopropanecarbonitrile Cyclopropyl cyanide	C_4H_5N 67.09	5500-21-0		135.1	4-09-00-00005 0.8946[20]	eth 3; ctc 2; hx 3 1.4229[20]
4957	Cyclopropanecarbonyl chloride, 2-phenyl-, trans- trans-2-Phenylcyclopropanecarbonyl chloride	$C_{10}H_9ClO$ 180.63	939-87-7		108-10[2]	3-09-00-02774 1.163[25]	1.5560[20]
4958	Cyclopropanecarboxylic acid	$C_4H_6O_2$ 86.09	1759-53-1	18.5	183	4-09-00-00003 1.0885[20]	H_2O 3; EtOH 3; eth 3; ctc 2 1.4390[20]
4959	Cyclopropanecarboxylic acid, 2,2-dimethyl-3-(2-methyl-1-propenyl)-, (1R-cis)-	$C_{10}H_{16}O_2$ 168.24	26771-11-9	41	135[12]; 95[0.1]	4-09-00-00168	eth 4; EtOH 4; chl 4
4960	Cyclopropanecarboxylic acid, 2,2-dimethyl-3-(2-methyl-1-propenyl)-, 2-methyl-4-oxo-3-(2,4-pentadienyl)-2-cyclopenten-1-yl ester, [1R-[1α[S*(Z)],3β]]- Pyrethrin 1	$C_{21}H_{28}O_3$ 328.45	121-21-1.		7978 170[0.1] dec	3-09-00-00215 1.5192[18]	H_2O 1; EtOH 3; eth 3; ctc 3 1.5192[18]
4961	Cyclopropanecarboxylic acid, ethyl ester Ethyl cyclopropylcarboxylate	$C_6H_{10}O_2$ 114.14	4606-07-9		134	4-09-00-00003 0.9608[15]	1.4190[20]
4962	Cyclopropanecarboxylic acid, 3-(3-methoxy-2-methyl-3-oxo-1-propenyl)-2,2-dimethyl-, 2-methyl-4-oxo-3-(2,4-pentadienyl)-2-cyclopenten-1-yl ester [1R-[1α[S*(Z)],3β(E)]]- Pyrethrin 2	$C_{22}H_{28}O_5$ 372.46	121-29-9		7978 200[0.1] dec	3-09-00-03988	H_2O 1; EtOH 3; eth 3; ctc 3 1.5258[20]
4963	Cyclopropanecarboxylic acid, methyl ester Methyl cyclopropanecarboxylate	$C_5H_8O_2$ 100.12	2868-37-3		114.9	4-09-00-00003 0.9848[20]	ace 3; chl 3 1.4144[19]

No.	Name / Synonym	Mol. Form. / Mol. Wt.	CAS RN / mp/°C	Merck No. / bp/°C	Beil. Ref. / den/g cm⁻³	Solubility / n_D
4964	Cyclopropanecarboxylic acid, 2-methyl- / 2-Methylcyclopropanecarboxylic acid	$C_5H_8O_2$ / 100.12	29555-02-0	/ 191; 96.5[14]	2-09-00-00006 / 1.0267[20]	/ 1.4441[20]
4965	Cyclopropanecarboxylic acid, 2,2-dimethyl-3-(2-methyl-1-propenyl)-, methyl ester, trans-(±)- / Chrysanthemum monocarboxylic acid, methyl ester (DL, trans)	$C_{11}H_{18}O_2$ / 182.26	15543-70-1	/ 108[20]; 95[13]	4-09-00-00169 / 0.9274[25]	/ 1.4614[25]
4966	Cyclopropane, (chloromethyl)-	C_4H_7Cl / 90.55	5911-08-0 / -90.9	88	4-05-00-00013 / 0.98[25]	/ 1.4350[20]
4967	1,1-Cyclopropanedicarboxylic acid, diethyl ester	$C_9H_{14}O_4$ / 186.21	1559-02-0	/ 216; 100[12]	4-09-00-02786 / 1.0535[25]	EtOH 4; eth 4 / 1.4345[18]
4968	1,2-Cyclopropanedicarboxylic acid, diethyl ester, cis- / Cyclopropane-1,2-dicarboxylic acid, diethyl ester (cis)	$C_9H_{14}O_4$ / 186.21	710-43-0		2-09-00-00513 / 1.062[12]	eth 4; EtOH 4 / 1.4450[20]
4969	1,2-Cyclopropanedicarboxylic acid, dimethyl ester / Cyclopropane-1,2-dicarboxylic acid, dimethyl ester	$C_7H_{10}O_4$ / 158.15	702-28-3	/ 219; 110[3]	/ 1.1584[16]	eth 4; EtOH 4 / 1.4472[14]
4970	1,2-Cyclopropanedicarboxylic acid, trans-(±)- / Cyclopropane-1,2-dicarboxylic acid (trans, DL)	$C_5H_6O_4$ / 130.10	58616-95-8 / 175	210[30]	4-09-00-02786 / 1.4600[20]	H_2O 4; eth 4; EtOH 4
4971	Cyclopropane, 1,1-dimethyl- / gem-Dimethylcyclopropane	C_5H_{10} / 70.13	1630-94-0 / -109	20.6	4-05-00-00024 / 0.6604[20]	H_2O 1; EtOH 3; eth 4; sulf 4 / 1.3668[20]
4972	Cyclopropane, 1,2-dimethyl-, cis- / cis-1,2-Dimethylcyclopropane	C_5H_{10} / 70.13	930-18-7 / -140.9	37.0	4-05-00-00025 / 0.6889[20]	H_2O 1; EtOH 3; eth 4; ctc 2 / 1.3829[20]
4973	Cyclopropane, 1,2-dimethyl-, trans- / trans-1,2-Dimethylcyclopropane	C_5H_{10} / 70.13	2402-06-4 / -149.6	28.2	4-05-00-00026 / 0.6648[25]	eth 4; EtOH 4 / 1.3713[20]
4974	Cyclopropane, ethyl- / Ethylcyclopropane	C_5H_{10} / 70.13	1191-96-4 / -149.2	35.9	4-05-00-00023 / 0.6790[25]	/ 1.3786[20]
4975	Cyclopropane, 1-ethyl-1-methyl- / Cyclopropane, 1-methyl-1-ethyl	C_6H_{12} / 84.16	53778-43-1 / -130.2	56.8	4-05-00-00091 / 0.6968[25]	/ 1.3887[20]
4976	Cyclopropane, 1-ethyl-2-methyl-, cis- / cis-1-Methyl-2-ethylcyclopropane	C_6H_{12} / 84.16	19781-68-1	58.7	3-05-00-00063 / 0.6935[20]	/ 1.3846[20]
4977	Cyclopropane, 1-ethyl-2-methyl-, trans- / trans-1-Methyl-2-ethylcyclopropane	C_6H_{12} / 84.16	19781-69-2	59	3-05-00-00063 / 0.7146[20]	ctc 3 / 1.3953[20]
4978	Cyclopropanemethanol, 2,2-dimethyl-3-(2-methyl-1-propenyl)-	$C_{10}H_{18}O$ / 154.25	5617-92-5	66-9[0.07]	4-06-00-00264 / 0.888[25]	/ 1.4757[20]
4979	Cyclopropanemethanol, α-methyl-	$C_5H_{10}O$ / 86.13	765-42-4 / -32.1	123.5	4-06-00-00020 / 0.8805[20]	/ 1.4316[20]
4980	Cyclopropane, methoxy- / Cyclopropyl methyl ether	C_4H_8O / 72.11	540-47-6 / -119	2756 / 44.7	3-06-00-00003 / 0.8100[20]	H_2O 4; bz 4; eth 4; EtOH 4 / 1.3802[20]
4981	Cyclopropane, methyl- / Methylcyclopropane	C_4H_8 / 56.11	594-11-6 / -177.3	0.7	4-05-00-00013 / 0.6912[-20]	eth 4; EtOH 4
4982	Cyclopropane, (1-methylethenyl)-	C_6H_{10} / 82.15	4663-22-3	70	4-05-00-00243 / 0.7500[20]	/ 1.4252[20]
4983	Cyclopropane, (1-methylethyl)-	C_6H_{12} / 84.16	3638-35-5 / -112.9	58.3	4-05-00-00090 / 0.6936[25]	/ 1.3865[20]
4984	Cyclopropane, propyl- / Propane, 1-cyclopropyl-	C_6H_{12} / 84.16	2415-72-7	69.1	4-05-00-00090 / 0.7062[25]	/ 1.3930[20]
4985	Cyclopropane, 1,1,2-trimethyl- / 1,1,2-Trimethylcyclopropane	C_6H_{12} / 84.16	4127-45-1 / -138.2	54	4-05-00-00091 / 0.6897[25]	/ 1.3864[20]
4986	Cyclopropane, 1,2,3-trimethyl-, (1α,2α,3α)-	C_6H_{12} / 84.16	4806-58-0	66	3-05-00-00063 / 0.7130[25]	/ 1.3970[20]
4987	Cyclopropane, 1,2,3-trimethyl-, (1α,2α,3β)-	C_6H_{12} / 84.16	4806-59-1	59.7	3-05-00-00063 / 0.6929[25]	/ 1.3873[20]
4988	Cyclopropanone	C_3H_4O / 56.06	5009-27-8 / stable only at low temp.			
4989	1H-Cycloprop[e]azulen-4-ol, decahydro-1,1,4,7-tetramethyl-, (1aR,4R,4aS,7R,7aS,7bS)- / Ledol	$C_{15}H_{26}O$ / 222.37	577-27-5 / 105	5313 / 292	/ 0.9078[100]	ace 4; eth 4; EtOH 4 / 1.4667[110]
4990	Cyclopropene	C_3H_4 / 40.06	2781-85-3	-36 dec		
4991	Cyclotetrasiloxane, ethenylheptamethyl-	$C_9H_{24}O_4Si_4$ / 308.63	3763-39-1	84[20]	/ 0.9505[20]	ctc 3; CS_2 3 / 1.4034[20]
4992	Cyclotetrasiloxane, heptamethyl- / Heptamethylcyclotetrasiloxane	$C_7H_{22}O_4Si_4$ / 282.59	15721-05-8 / -27	165	4-04-00-04125 / 0.9583[20]	H_2O I / 1.3965[20]
4993	Cyclotetrasiloxane, octamethyl- / Octamethylcyclotetrasiloxane	$C_8H_{24}O_4Si_4$ / 296.62	556-67-2 / 17.5	6668 / 175.8	4-04-00-04125 / 0.9561[20]	H_2O I; ctc 3 / 1.3968[20]
4994	Cyclotetrasiloxane, octaphenyl-	$C_{48}H_{40}O_4Si_4$ / 793.19	546-56-5 / 200.5	/ 330[1]	4-16-00-01530	H_2O I; EtOH 2; bz 3; chl 3
4995	Cyclotetrasiloxane, 2,4,6,8-tetrabutyl-2,4,6,8-tetramethyl- / Siloxane, butylmethyl(cyclic tetramer)	$C_{20}H_{48}O_4Si_4$ / 464.94	14685-29-1	291	4-04-00-04173 / 0.9230[20]	/ 1.4300[20]
4996	Cyclotetrasiloxane, 2,4,6,8-tetraethenyl-2,4,6,8-tetramethyl-	$C_{12}H_{24}O_4Si_4$ / 344.66	2554-06-5 / -43.5	224; 111[12]	4-04-00-04184 / 0.9875[20]	ctc 3; CS_2 3

No.	Name / Synonym	Mol. Form. / Mol. Wt.	CAS RN / mp/°C	Merck No. / bp/°C	Beil. Ref. / den/g cm^{-3}	Solubility / n_D
4997	Cyclotetrasiloxane, 2,4,6,8-tetraethyl-2,4,6,8-tetramethyl- Cyclotetrasiloxane, 2,4,6,8-tetramethyl-2,4,6,8-tetraethyl	$C_{12}H_{32}O_4Si_4$ 352.73	7623-01-0 -43.5	245; 111[12]	4-04-00-04154 0.9600[20]	ctc 3; CS_2 3
4998	Cyclotetrasiloxane, 2,4,6,8-tetraethyl-2,4,6,8-tetraphenyl- Cyclotetrasiloxane, 1,3,5,7-tetraethyl-1,3,5,7-tetraphenyl	$C_{32}H_{40}O_4Si_4$ 601.01	18758-34-4 106	212[0.1]	3-16-00-01211 1.1000[20]	1.5430[25]
4999	Cyclotetrasiloxane, 2,4,6,8-tetramethyl-	$C_4H_{16}O_4Si_4$ 240.51	2370-88-9 -65	134.5	4-04-00-04099 0.9912[20]	H_2O I 1.3870[20]
5000	Cyclotetrasiloxane, 2,4,6,8-tetramethyl-2,4,6,8-tetraphenyl-	$C_{28}H_{32}O_4Si_4$ 544.90	77-63-4 99	237[115]	3-16-00-01211 1.1183[20]	H_2O 1; ace 5; hp 5 1.5461[20]
5001	Cyclotrisilazane, 2,2,4,4,6,6-hexamethyl- 1,1,3,3,5,5-Hexamethylcyclotrisilazane	$C_6H_{21}N_3Si_3$ 219.51	1009-93-4 -10	188	4-04-00-04114 0.9196[20]	1.448[20]
5002	Cyclotrisiloxane, hexamethyl- Dimethylsiloxane cyclic trimer	$C_6H_{18}O_3Si_3$ 222.46	541-05-9 64.5	134	4-04-00-04123 1.1200[20]	H_2O 1
5003	Cyclotrisiloxane, 2,4,6-triethyl-2,4,6-trimethyl- 2,4,6-Triethyl-2,4,6-trimethylcyclotrisiloxane	$C_9H_{24}O_3Si_3$ 264.54	15901-49-2 -3	199	4-04-00-04154	ctc 3; CS_2 3
5004	Cyclotrisiloxane, 2,4,6-triethyl-2,4,6-triphenyl- 2,4,6-Triethyl-2,4,6-triphenylcyclotrisiloxane	$C_{24}H_{30}O_3Si_3$ 450.76	546-33-8 177.5	166[0.025]	3-16-00-01211 1.0952[25]	H_2O 1 1.5402[25]
5005	Cyclotrisiloxane, 2,4,6-trimethyl-2,4,6-triphenyl-	$C_{21}H_{24}O_3Si_3$ 408.68	546-45-2 100	190[1.5]	3-16-00-01211 1.1062[20]	1.5397[20]
5006	1,4,8-Cycloundecatriene, 2,6,6,9-tetramethyl-, (E,E,E)- Humulene	$C_{15}H_{24}$ 204.36	6753-98-6	4672 123[10]	4-05-00-01171 0.8905[20]	1.5038[20]
5007	7,20-Cycloveatchane-1,12,15-triol, 21-ethyl-4-methyl-16-methylene-, (1α,12α,15β)- Luciculine	$C_{22}H_{35}NO_3$ 361.52	5008-52-6 149	6286 165[0.02]	4-21-00-02584	EtOH 4
5008	Cyfluthrin Cyano(4-fluoro-3-phenoxyphenyl)methyl 3-(2,2-dichloroethenyl)-2,2-dimethylcyclopropanecarboxylate	$C_{22}H_{18}Cl_2FNO_3$ 434.29	68359-37-5 60	2764		
5009	Cygon Phosphorodithioic acid, O,O-dimethyl S-[2-(methylamino)-2-oxoethyl] ester	$C_5H_{12}NO_3PS_2$ 229.26	60-51-5 52	3209 117[0.1]	1.277[65]	
5010	Cyhalothrin	$C_{23}H_{19}ClF_3NO_3$ 449.86	91465-08-6 49.2			
5011	Cyhexatin Stannane, tricyclohexylhydroxy-	$C_{18}H_{34}OSn$ 385.18	13121-70-5 196	2767		
5012	Cypermethrin Cyclopropanecarboxylic acid, 3-(2,2-dichloroethenyl)-2,2-dimethyl-, cyano(3-phenoxy-phenyl)methyl ester	$C_{22}H_{19}Cl_2NO_3$ 416.30	52315-07-8 70	2775	1.25[20]	
5013	Cyprazine 1,3,5-Triazine-2,4-diamine, 6-chloro-N-cyclopropyl-N'-(1-methylethyl)-	$C_9H_{14}ClN_5$ 227.70	22936-86-3 167			
5014	L-Cysteine Propanoic acid, 2-amino-3-mercapto-, (R)-	$C_3H_7NO_2S$ 121.16	52-90-4 240 dec	2787	4-04-00-03144	H_2O 4; ace 4; EtOH 4
5015	L-Cysteine, N-acetyl- Acetylcysteine	$C_5H_9NO_3S$ 163.20	616-91-1 109.5	82	4-04-00-03160	
5016	L-Cysteine, S-(2-amino-2-carboxyethyl)-, (R)- L-Lanthionine	$C_6H_{12}N_2O_4S$ 208.24	922-55-4 293-5 dec	5234	3-04-00-01593	
5017	L-Cysteine, S-(carboxymethyl)- Carbocysteine	$C_5H_9NO_4S$ 179.20	638-23-3 206	1809	4-04-00-03150	
5018	L-Cysteine, S,S'-methylenebis- Djenkolic acid	$C_7H_{14}N_2O_4S_2$ 254.33	498-59-9 300-50 dec	3393	3-04-00-01591	
5019	L-Cysteine, methyl ester, hydrochloride Methylcysteine hydrochloride	$C_4H_{10}ClNO_2S$ 171.65	18598-63-5 140.5	5668	3-04-00-01601	
5020	L-Cystine 3,3'-Dithiobis(2-aminopropanoic acid)	$C_6H_{12}N_2O_4S_2$ 240.30	56-89-3 260-1 dec	2788	4-04-00-03155 1.677[25]	H_2O 2; EtOH 1; eth 1; bz 1
5021	Cytidine 4-Amino-1-β-D-ribofuranosyl-2(1H)-pyrimidinone	$C_9H_{13}N_3O_5$ 243.22	65-46-3 230-1 dec	2792	4-25-00-03667	H_2O 4; EtOH 2
5022	3'-Cytidylic acid Cytidine 3'-monophosphate	$C_9H_{14}N_3O_8P$ 323.20	84-52-6 233 dec	2794	4-25-00-03669	H_2O 3; EtOH 3
5023	1,3-Decadiene 1-Hexyl-1,3-butadiene	$C_{10}H_{18}$ 138.25	2051-25-4 169		4-01-00-01056 0.752[30]	bz 4
5024	1,9-Decadiene α,ω-Decadiene	$C_{10}H_{18}$ 138.25	1647-16-1 167		4-01-00-01056 0.75[25]	1.4325[20]
5025	2,4-Decadienoic acid, (E,E)- 2,4-Decadienoic acid (trans-2, trans-4)	$C_{10}H_{16}O_2$ 168.24	30361-33-2 49.5	121[0.3]	4-02-00-01730	1.5058[31]
5026	2,4-Decadienoic acid, methyl ester, (E,E)-	$C_{11}H_{18}O_2$ 182.26	7328-33-8 70[0.2]		4-02-00-01731 0.982[22]	1.4918[22]
5027	2,4-Decadienoic acid, methyl ester, (E,Z)-	$C_{11}H_{18}O_2$ 182.26	4493-42-9 71[0.15]		4-02-00-01731 0.9128[22]	1.4874[22]

No.	Name / Synonym	Mol. Form. / Mol. Wt.	CAS RN / mp/°C	Merck No. / bp/°C	Beil. Ref. / den/g cm^{-3}	Solubility / n_D
5028	2,4-Decadienoic acid, methyl ester, (Z,E)- / 2,4-Decadienoic acid, methyl ester (cis-2, trans-4)	$C_{11}H_{18}O_2$ / 182.26	108965-84-0 /	/ 76-80[0.5]	4-02-00-01731 / 0.9131[23]	/ 1.4876[23]
5029	2,4-Decadienoic acid, methyl ester, (Z,Z)- / 2,4-Decadienoic acid, methyl ester (cis-2, cis-4)	$C_{11}H_{18}O_2$ / 182.26	108965-86-2 /	/ 78[0.6]	4-02-00-01731 / 0.9095[23]	/ 1.4830[23]
5030	Decanal / Capraldehyde	$C_{10}H_{20}O$ / 156.27	112-31-2 / -5	/ 208.5	4-01-00-03366 / 0.830[15]	H_2O 1; EtOH 3; eth 3; ace 3 / 1.4287[20]
5031	Decanamide / Capramide	$C_{10}H_{21}NO$ / 171.28	2319-29-1 / 108		4-02-00-01050 / 0.822[130]	ace 4; eth 4; EtOH 4 / 1.4621[110]
5032	1-Decanamine	$C_{10}H_{23}N$ / 157.30	2016-57-1 / 17	/ 220.5	4-04-00-00783 / 0.7936[20]	H_2O 2; EtOH 5; eth 5; ace 5 / 1.4369[20]
5033	Decane / n-Decane	$C_{10}H_{22}$ / 142.28	124-18-5 / -29.7	/ 174.1	4-01-00-00464 / 0.7300[20]	H_2O 1; EtOH 5; eth 3; ctc 2 / 1.4102[20]
5034	Decane, 1-bromo-	$C_{10}H_{21}Br$ / 221.18	112-29-8 / -29.2	/ 240.6	4-01-00-00470 / 1.0702[20]	H_2O 1; eth 4; ctc 3; chl 4 / 1.4557[20]
5035	Decane, 1-bromo-10-fluoro- / 1-Bromo-10-fluorodecane	$C_{10}H_{20}BrF$ / 239.17	334-61-2 /	/ 131-2[11]	4-01-00-00471 / 1.152[20]	eth 4; EtOH 4 / 1.4512[25]
5036	Decane, 1-chloro-	$C_{10}H_{21}Cl$ / 176.73	1002-69-3 / -31.3	/ 223.4	4-01-00-00469 / 0.8705[20]	H_2O 1; eth 4; ctc 3; chl 4 / 1.4379[20]
5037	Decane, 1-chloro-10-fluoro- / 1-Chloro-10-fluorodecane	$C_{10}H_{20}ClF$ / 194.72	334-62-3 /	/ 115[9]	4-01-00-00469 / 0.957[20]	eth 4; EtOH 4 / 1.4333[25]
5038	Decane, 1,10-dibromo- / Decamethylene dibromide	$C_{10}H_{20}Br_2$ / 300.08	4101-68-2 / 28	/ 161[9]; 128[4]	4-01-00-00471 / 1.335[30]	H_2O 1; EtOH 2; eth 3 / 1.4927[25]
5039	Decane, 5,6-dibromo- / 5,6-Dibromodecane	$C_{10}H_{20}Br_2$ / 300.08	77928-86-0 /	/ 119[9]	4-01-00-00472 / 1.3484[20]	/ 1.4912[20]
5040	Decane, 1,10-dichloro- / Decamethylene dichloride	$C_{10}H_{20}Cl_2$ / 211.17	2162-98-3 / 15.6	/ 167-8[28]	4-01-00-00469 / 0.9945[25]	/ 1.4586[25]
5041	Decanedinitrile	$C_{10}H_{16}N_2$ / 164.25	1871-96-1 /	/ 204[16]	4-02-00-02089 / 0.9313[20]	H_2O 1; chl 3 / 1.4474[20]
5042	Decanedioic acid / Sebacic acid	$C_{10}H_{18}O_4$ / 202.25	111-20-6 / 130.8	8369 / 295[100.] 232[10]	4-02-00-02078 / 1.2705[20]	H_2O 2; EtOH 3; eth 3; bz 1 / 1.422[133]
5043	Decanedioic acid, bis(2-ethylhexyl) ester / Bis(2-ethylhexyl) sebacate	$C_{26}H_{50}O_4$ / 426.68	122-62-3 / -48	1263 / 256[5]	4-02-00-02083 / 0.912[25]	ace 4; bz 4; EtOH 4 / 1.451[25]
5044	Decanedioic acid, dibutyl ester / Dibutyl sebacate	$C_{18}H_{34}O_4$ / 314.47	109-43-3 / -10	/ 344.5	4-02-00-02081 / 0.9405[15]	H_2O 1; eth 3; ctc 3 / 1.4433[15]
5045	Decanedioic acid, diethyl ester	$C_{14}H_{26}O_4$ / 258.36	110-40-7 / 5	/ 305; 188[19]	4-02-00-02080 / 0.9646[20]	H_2O 2; EtOH 3; ace 3; bz 1 / 1.4306[20]
5046	Decanedioic acid, dimethyl ester	$C_{12}H_{22}O_4$ / 230.30	106-79-6 / 38	/ 175[20]; 144[5]	4-02-00-02080 / 0.9882[28]	H_2O 1; EtOH 3; eth 3; ace 3 / 1.4355[28]
5047	1,10-Decanediol / Decamethylene glycol	$C_{10}H_{22}O_2$ / 174.28	112-47-0 / 74	2842 / 192[20]	4-01-00-02613 /	H_2O 2; EtOH 4; eth 2; DMSO 3
5048	2,9-Decanediol	$C_{10}H_{22}O_2$ / 174.28	14021-92-2 / 33	/ 114[5]	3-01-00-02232 /	/ 1.4505[21]
5049	Decanedioyl dichloride	$C_{10}H_{16}Cl_2O_2$ / 239.14	111-19-3 / -2.5	/ 220[75]; 165[11]	4-02-00-02088 / 1.1212[20]	/ 1.4684[18]
5050	Decane, docosafluoro- / Perfluorodecane	$C_{10}F_{22}$ / 538.07	307-45-9 /	/ 144.2		
5051	Decane, 1-(ethenyloxy)- / Decyl vinyl ether	$C_{12}H_{24}O$ / 184.32	765-05-9 / -41	/ 101[10]	4-01-00-02057 / 0.812[20]	/ 1.4346[20]
5052	Decane, 1-fluoro- / Decyl fluoride	$C_{10}H_{21}F$ / 160.28	334-56-5 / -35	/ 186.2	4-01-00-00468 / 0.8194[20]	eth 4 / 1.4085
5053	Decane, 1-iodo-	$C_{10}H_{21}I$ / 268.18	2050-77-3 / -16.3	/ 132[15]	4-01-00-00472 / 1.2546[20]	H_2O 1; EtOH 3; eth 3; ctc 3 / 1.4858[20]
5054	Decane, 2-methyl- / 2-Methyldecane	$C_{11}H_{24}$ / 156.31	6975-98-0 / -48.9	/ 189.3	4-01-00-00491 / 0.7368[20]	/ 1.4154[20]
5055	Decane, 3-methyl- / 3-Methyldecane	$C_{11}H_{24}$ / 156.31	13151-34-3 / -92.9	/ 188.1	4-01-00-00491 / 0.7422[20]	/ 1.4177[20]
5056	Decane, 4-methyl- / 4-Methyldecane	$C_{11}H_{24}$ / 156.31	2847-72-5 / -77.5	/ 187	/	/ 1.4352[20]
5057	Decanenitrile / Caprinitrile	$C_{10}H_{19}N$ / 153.27	1975-78-6 / -17.9	/ 243. 106[10]	4-02-00-01051 / 0.8199[20]	ace 4; eth 4; EtOH 4; chl 4 / 1.4296[20]
5058	1-Decanethiol / Decyl mercaptan	$C_{10}H_{22}S$ / 174.35	143-10-2 / -26	/ 240.6	4-01-00-01821 / 0.8443[20]	H_2O 1; EtOH 3; eth 3 / 1.4509[20]
5059	Decanoic acid / Capric acid	$C_{10}H_{20}O_2$ / 172.27	334-48-5 / 31.9	1759 / 268.7	4-02-00-01041 / 0.8858[40]	ace 4; bz 4; eth 4; EtOH 4 / 1.4288[40]
5060	Decanoic acid, anhydride	$C_{20}H_{38}O_3$ / 326.52	2082-76-0 / 24.7		4-02-00-01049 / 0.8865[25]	eth 4; EtOH 4 / 1.400[25]
5061	Decanoic acid, 2-bromo-	$C_{10}H_{19}BrO_2$ / 251.16	2623-95-2 / 4	/ 140-1[2]	4-02-00-01054 / 1.1912[24]	eth 4 / 1.4595[24]
5062	Decanoic acid, decyl ester	$C_{20}H_{40}O_2$ / 312.54	1654-86-0 / 9.7	/ 219[15]	4-02-00-01045 / 0.8586[20]	eth 4 / 1.4423[20]
5063	Decanoic acid, ethyl ester / Ethyl caprate	$C_{12}H_{24}O_2$ / 200.32	110-38-3 / -20	3734 / 241.5	4-02-00-01044 / 0.8650[20]	eth 4; EtOH 4; chl 4 / 1.4256[20]
5064	Decanoic acid, 10-fluoro- / Capric acid, 10-fluoro	$C_{10}H_{19}FO_2$ / 190.26	334-59-8 / 49	/ 135-8[10]	4-02-00-01051 /	eth 4; EtOH 4; lig 4
5065	Decanoic acid, 2-hexyl- / 2-Hexyldecanoic acid	$C_{16}H_{32}O_2$ / 256.43	25354-97-6 / 4630	/ 140-50[0.02]	4-02-00-01189 /	/ 1.4432[24]

No.	Name Synonym	Mol. Form. Mol. Wt.	CAS RN mp/°C	Merck No. bp/°C	Beil. Ref. den/g cm^{-3}	Solubility n_D
5066	Decanoic acid, methyl ester	$C_{11}H_{22}O_2$ 186.29	110-42-9 -18	224	4-02-00-01044 0.8730[20]	H_2O 1; EtOH 4; eth 4; ctc 2 1.4259[20]
5067	Decanoic acid, 1-methylethyl ester	$C_{13}H_{26}O_2$ 214.35	2311-59-3 	121[10]	4-02-00-01045 0.8543[20]	1.4221[25]
5068	Decanoic acid, 2-octyl-	$C_{18}H_{36}O_2$ 284.48	619-39-6 38.5	215[13]	4-02-00-01254 0.844[70]	eth 4; EtOH 4
5069	Decanoic acid, propyl ester	$C_{13}H_{26}O_2$ 214.35	30673-60-0 	128.5[10]	4-02-00-01045 0.8623[20]	1.4280[20]
5070	1-Decanol Capric alcohol	$C_{10}H_{22}O$ 158.28	112-30-1 6.9	2847 231.1	4-01-00-01815 0.8297[20]	H_2O 1; EtOH 5; eth 5; ace 5 1.4372[20]
5071	2-Decanol, (±)- 2-Decanol (DL)	$C_{10}H_{22}O$ 158.28	74742-10-2 -2.4	211	4-01-00-01823 0.8250[20]	EtOH 3; eth 5; ace 5; bz 3 1.4326[25]
5072	4-Decanol 1-Propylheptyl alcohol	$C_{10}H_{22}O$ 158.28	2051-31-2 -11	210.5	4-01-00-01824 0.8261[20]	H_2O 1; EtOH 3; ctc 3 1.4320[20]
5073	1-Decanol, 10-chloro-	$C_{10}H_{21}ClO$ 192.73	51309-10-5 12.5	185-9[15]	4-01-00-01821 0.9630[25]	eth 4; EtOH 4 1.4578[20]
5074	1-Decanol, 10-fluoro- 10-Fluoro-1-decanol	$C_{10}H_{21}FO$ 176.27	334-64-5 22	136-7[15]	4-01-00-01821 0.919[20]	eth 4; EtOH 4 1.4322[25]
5075	2-Decanone Methyl octyl ketone	$C_{10}H_{20}O$ 156.27	693-54-9 14	210; 96[12]	4-01-00-03367 0.8248[20]	H_2O 1; EtOH 3; eth 3; ctc 2 1.4255[20]
5076	3-Decanone Ethyl heptyl ketone	$C_{10}H_{20}O$ 156.27	928-80-3 2.5	203	4-01-00-03368 0.8251[20]	EtOH 3; eth 3; ctc 3 1.4252[20]
5077	4-Decanone Hexyl propyl ketone	$C_{10}H_{20}O$ 156.27	624-16-8 -9	206.5	4-01-00-03368 0.824[20]	H_2O 1; EtOH 5; eth 5 1.4240[21]
5078	Decanoyl chloride Caprinoyl chloride	$C_{10}H_{19}ClO$ 190.71	112-13-0 -34.5	95	4-02-00-01050 0.919[25]	eth 3; ctc 3 1.4410[20]
5079	Decasiloxane, docosamethyl- Decasiloxane, dicosamethyl	$C_{22}H_{66}O_9Si_{10}$ 755.62	556-70-7 	183.[4]	3-04-00-01881 0.925[20]	bz 4; lig 4 1.3988[20]
5080	2-Decenal	$C_{10}H_{18}O$ 154.25	3913-71-1 	230	4-01-00-03511 0.845[17]	1.4533[17]
5081	3-Decenal	$C_{10}H_{18}O$ 154.25	58474-80-9 	93-4[14]	4-01-00-03512 0.850[15]	1.4462[15]
5082	1-Decene	$C_{10}H_{20}$ 140.27	872-05-9 -66.3	170.5	3-01-00-00858 0.7408[20]	H_2O 1; EtOH 5; eth 5 1.4215[20]
5083	4-Decene	$C_{10}H_{20}$ 140.27	19689-18-0 	170.6	4-01-00-00902 0.7404[20]	1.4243[20]
5084	5-Decene, (E)-	$C_{10}H_{20}$ 140.27	7433-56-9 -73	171	4-01-00-00902 0.7401[20]	H_2O 1; EtOH 5; eth 5; ctc 2 1.4243[20]
5085	5-Decene, (Z)-	$C_{10}H_{20}$ 140.27	7433-78-5 -112	171; 73[20]	3-01-00-00859 0.7445[20]	H_2O 1; EtOH 5; eth 5; ctc 2 1.4258[20]
5086	1-Decene, 2-bromo- 2-Bromo-1-decene	$C_{10}H_{19}Br$ 219.16	3017-67-2 	115-6[22]	3-01-00-00859 1.0844[20]	1.4629[20]
5087	2-Decene, 1-bromo- 1-Bromo-2-decene	$C_{10}H_{19}Br$ 219.16	14304-30-4 	121[17]	4-01-00-00902 1.074[18]	lig 4 1.4716[18]
5088	2-Decenoic acid Δ 2-Decenoic acid	$C_{10}H_{18}O_2$ 170.25	3913-85-7 12	165[15]	4-02-00-01606 0.9280[18]	1.4616[20]
5089	3-Decenoic acid	$C_{10}H_{18}O_2$ 170.25	15469-77-9 18	154-63[11]	4-02-00-01606 0.914[15]	1.4510[18]
5090	4-Decenoic acid Deconic acid telomer	$C_{10}H_{18}O_2$ 170.25	26303-90-2 	149[13]	4-02-00-01607 0.9197[20]	bz 4; eth 4 1.4497[20]
5091	9-Decenoic acid Caproleic acid	$C_{10}H_{18}O_2$ 170.25	14436-32-9 	158[21]; 142[4]	4-02-00-01605 0.9238[15]	eth 4; EtOH 4 1.4507[15]
5092	9-Decen-1-ol Decylenic alcohol	$C_{10}H_{20}O$ 156.27	13019-22-2 	236	4-01-00-02184 0.876[25]	1.4480[20]
5093	3-Decen-2-one Heptylidene acetone	$C_{10}H_{18}O$ 154.25	10519-33-2 	102-3[15.3]	4-01-00-03512 0.8473[20]	1.4480[20]
5094	1-Decen-3-yne	$C_{10}H_{16}$ 136.24	33622-26-3 	76[20]	4-01-00-01105 0.7873[20]	1.4620[20]
5095	1-Decen-4-yne	$C_{10}H_{16}$ 136.24	24948-66-1 	73-4[22]	3-01-00-01049 0.7880[20]	1.445[20]
5096	2-Decen-4-yne	$C_{10}H_{16}$ 136.24	116668-40-7 	55[5]	3-01-00-01049 0.7850[25]	1.4609[25]
5097	1-Decyne Octylacetylene	$C_{10}H_{18}$ 138.25	764-93-2 -44	174	4-01-00-01054 0.7655[20]	H_2O 1; EtOH 3; eth 3; os 3 1.4265[20]
5098	3-Decyne	$C_{10}H_{18}$ 138.25	2384-85-2 	177	4-01-00-01055 0.7619[25]	1.4315[20]
5099	4-Decyne	$C_{10}H_{18}$ 138.25	2384-86-3 	74.5[19]	3-01-00-01017 0.772[17]	1.436[17]
5100	5-Decyne Dibutylacetylene	$C_{10}H_{18}$ 138.25	1942-46-7 -73	177; 78.8[25]	4-01-00-01055 0.7690[20]	H_2O 1; EtOH 3; eth 3 1.4331[20]
5101	4-Decyne, 3,3-dimethyl- 3,3-Dimethyl-4-decyne	$C_{12}H_{22}$ 166.31	70732-45-5 	86[20]	3-01-00-01026 0.7731[20]	1.4399[20]
5102	Deltamethrin Cyano(3-phenoxyphenyl)methyl-3-(2,2-dibromoethenyl)-2,2-dimethylcyclopropanecarboxylate	$C_{22}H_{19}Br_2NO_3$ 505.21	52918-63-5 99	2869		
5103	Demeton S methyl Phosphorothioic acid, S-[2-(ethylthio)ethyl] O,O-dimethyl ester	$C_6H_{15}O_3PS_2$ 230.29	919-86-8 	89[0.15]; 118[1]	1.207[20]	

No.	Name / Synonym	Mol. Form. / Mol. Wt.	CAS RN / mp/°C	Merck No. / bp/°C	Beil. Ref. / den/g cm^{-3}	Solubility / n_D
5104	Desmedipham Carbamic acid, [3-[[(phenylamino)carbonyl]oxy]phenyl]-, ethyl ester	$C_{16}H_{16}N_2O_4$ 300.31	13684-56-5 120			
5105	Dialifor Phosphorodithioic acid, S-[2-chloro-1-(1,3-dihydro-1,3-dioxo-2H-isoindol-2-yl)ethyl]	$C_{14}H_{17}ClNO_4PS_2$ 393.85	10311-84-9 68	2949		
5106	Diallate Carbamothioic acid, bis(1-methylethyl)-, S-(2,3-dichloro-2-propenyl) ester	$C_{10}H_{17}Cl_2NOS$ 270.22	2303-16-4 150[9]	2950		
5107	8,8'-Diapo-ψ,ψ-carotenedioic acid Crocetin (trans)	$C_{20}H_{24}O_4$ 328.41	27876-94-4 286	2592	3-02-00-02018	H_2O 2; EtOH 2; eth 1; bz 1
5108	6,6'-Diapo-ψ,ψ-carotenedioic acid, monomethyl ester, 9-cis- Bixin	$C_{25}H_{30}O_4$ 394.51	6983-79-5 198	1320	3-02-00-02021	H_2O 1; EtOH 3; eth 2; ace 3
5109	Diarsine, tetraethyl- Tetraethyldiarsenic	$C_8H_{20}As_2$ 266.09	612-08-8 186		3-04-00-01832 1.1388[24]	eth 4; EtOH 4 1.4709
5110	Diarsine, tetramethyl- Cacodyl	$C_4H_{12}As_2$ 209.98	471-35-2 -6	1602 165	4-04-00-03683 1.447[15]	eth 4; EtOH 4
5111	1,4-Diazabicyclo[2.2.2]octane Triethylenediamine	$C_6H_{12}N_2$ 112.17	280-57-9 159	9584	5-23-03-00487	chl 3
5112	Diazene, bis(2-ethoxyphenyl)- Azobenzene, 2,2'-diethoxy-	$C_{16}H_{18}N_2O_2$ 270.33	613-43-4 131	240 dec	4-16-00-00138	H_2O 1; EtOH 3; HCl 3
5113	Diazene, bis(4-ethoxyphenyl)-	$C_{16}H_{18}N_2O_2$ 270.33	588-52-3 162	dec	4-16-00-00174	H_2O 1; EtOH 2; eth 3; bz 3
5114	Diazene, bis(2-methylphenyl)-	$C_{14}H_{14}N_2$ 210.28	584-90-7 55.5		2-16-00-00019 1.0215[65]	H_2O 1; EtOH 3; eth 4; bz 4 1.6180[65]
5115	Diazene, bis(3-methylphenyl)-, (E)-	$C_{14}H_{14}N_2$ 210.28	51437-67-3 54		4-16-00-00070 1.0123[66]	EtOH 4; eth 4; bz 4 1.6152[66]
5116	Diazene, bis(2-methylphenyl)-, 1-oxide, (Z)- Azoxybenzene, 2,2'-dimethyl (trans)	$C_{14}H_{14}N_2O$ 226.28	51284-68-5 60		4-16-00-00064 1.0215[65]	bz 4; eth 4; EtOH 4 1.6180[65]
5117	Diazene, bis(2-methylphenyl)-, 1-oxide, (E)- Azoxybenzene, 2,2'-dimethyl(trans)	$C_{14}H_{14}N_2O$ 226.28	116723-89-8		2-16-00-00318 1.0215[65]	H_2O 1; EtOH 3; eth 3; bz 3 1.6180[65]
5118	Diazene, bis(3-methylphenyl)-, 1-oxide, (Z)- Azoxybenzene, 3,3'-dimethyl (trans)	$C_{14}H_{14}N_2O$ 226.28	71297-97-7 39		4-16-00-00071 1.0123[66]	bz 4; eth 4; EtOH 4; lig 4 1.6152[66]
5119	Diazene, bis(3-methylphenyl)-, 1-oxide, (E)- Azoxybenzene, 3,3'-dimethyl(trans)	$C_{14}H_{14}N_2O$ 226.28	116723-90-1 39		2-16-00-00319 1.0123[66]	H_2O 1; EtOH 3; eth 3; bz 3 1.6152[66]
5120	Diazene, (4-bromophenyl)phenyl-, 1-oxide Azoxybenzene, 4-bromo-	$C_{12}H_9BrN_2O$ 277.12	16109-68-5 93.5		3-16-00-00584 1.4138[100]	H_2O 1; EtOH 3; bz 3; MeOH 3
5121	Diazenecarbothioic acid, phenyl-, 2-phenylhydrazide Dithizone	$C_{13}H_{12}N_4S$ 256.33	60-10-6 165-9 dec	3383	4-16-00-00018	H_2O 1; EtOH 2; eth 2; chl 3
5122	Diazenecarboxylic acid, phenyl-, 2-phenylhydrazide Diphenylcarbazone	$C_{13}H_{12}N_4O$ 240.26	538-62-5 157 dec	3322	4-16-00-00017	H_2O 1; EtOH 4; bz 4; chl 4
5123	Diazene, dibutyl Azobutane	$C_8H_{18}N_2$ 142.24	2159-75-3 60[18]			
5124	Diazenedicarboxamide Azodicarbonamide	$C_2H_4N_4O_2$ 116.08	123-77-3 212 dec	932	3-03-00-00234	
5125	Diazene, dimethyl- Azomethane	$C_2H_6N_2$ 58.08	503-28-6 -78	1.5	3-04-00-01747 0.743[0]	ace 4; eth 4; EtOH 4 1.4199[19]
5126	Diazene, dimethyl-, (E)-	$C_2H_6N_2$ 58.08	4143-41-3 -78	1.5	4-04-00-03366 0.743[0]	ctc 3; hp 3
5127	Diazene, di-1-naphthalenyl- 1,1'-Azobisnaphthalene	$C_{20}H_{14}N_2$ 282.34	487-10-5 190	sub	2-16-00-00026	H_2O 1; EtOH 2; ace 3; bz 4
5128	Diazene, di-2-naphthalenyl- 2,2'-Azonaphthalene	$C_{20}H_{14}N_2$ 282.34	582-08-1 208	sub	2-16-00-00026	H_2O 1; EtOH 2; eth 2; bz 3
5129	Diazene, diphenyl-, (E)- trans-Azobenzene	$C_{12}H_{10}N_2$ 182.22	17082-12-1 68.5	930 293	4-16-00-00007 1.203[20]	H_2O 2; EtOH 3; eth 3; bz 3 1.6266[78]
5130	Diazene, diphenyl-, (Z)- cis-Azobenzene	$C_{12}H_{10}N_2$ 182.22	1080-16-6 71	930	4-16-00-00007	H_2O 2; EtOH 3; eth 3; bz 3
5131	Diazene, diphenyl-, 1-oxide, (Z)- cis-Azoxybenzene	$C_{12}H_{10}N_2O$ 198.22	20972-43-4 87	937	4-16-00-00010 1.166[20]	1.633[20]
5132	Diazene, dipropyl Azopropane	$C_6H_{14}N_2$ 114.19	821-67-0 114			
5133	Diazene, (4-ethoxyphenyl)phenyl-	$C_{14}H_{14}N_2O$ 226.28	7466-38-8 85	339.5	2-16-00-00040 1.0400[100]	EtOH 4; eth 4; ace 3; bz 4 1.6419[100]
5134	Diazene, ethylphenyl-	$C_8H_{10}N_2$ 134.18	935-08-0 177		4-16-00-00007 0.9628[22]	bz 4; eth 4; EtOH 4 1.5579
5135	Diazene, (2-methoxyphenyl)phenyl-	$C_{13}H_{12}N_2O$ 212.25	6319-21-7 41	195-7[14]	3-16-00-00081 1.0728[100]	H_2O 1; EtOH 3; ace 3; os 4
5136	Diazene, (3-methoxyphenyl)phenyl-	$C_{13}H_{12}N_2O$ 212.25	34238-81-8 33.3	193[15]	4-16-00-00155 1.1023[53]	H_2O 1; EtOH 3; ace 3; os 4
5137	Diazene, (4-methoxyphenyl)phenyl-	$C_{13}H_{12}N_2O$ 212.25	2396-60-3 56	340	2-16-00-00040 1.12[75]	H_2O 1; EtOH 3; eth 3; ace 3
5138	Diazene, (3-methylphenyl)phenyl- Azobenzene, 3-methyl	$C_{13}H_{12}N_2$ 196.25	17478-66-9 18.5	175[19]	3-16-00-00047 1.065[20]	
5139	Diazene, (4-methylphenyl)phenyl- p-Methylazobenzene	$C_{13}H_{12}N_2$ 196.25	949-87-1 71.5	312	3-16-00-00049	bz 4; eth 4; chl 4

No.	Name Synonym	Mol. Form. Mol. Wt.	CAS RN mp/°C	Merck No. bp/°C	Beil. Ref. den/g cm^{-3}	Solubility n_D
5140	1H-1,4-Diazepine, 1-[(4-chlorophenyl)phenylmethyl]hexahydro-4-methyl- Homochlorocyclizine	C$_{19}$H$_{23}$ClN$_2$ 314.86	848-53-3	4653	5-23-03-00248	
5141	1H-1,4-Diazepine, hexahydro-1-methyl-	C$_6$H$_{14}$N$_2$ 114.19	4318-37-0	154	5-23-03-00241 0.9111[20]	1.4769[20]
5142	Dibenz[a,h]anthracene 1,2:5,6-Dibenzanthracene	C$_{22}$H$_{14}$ 278.35	53-70-3 269.5	2989	4-05-00-02772	H$_2$O 1; EtOH 2; ace 3; bz 3
5143	Dibenz[a,j]acridine 7-Azadibenz[a,j]anthracene	C$_{21}$H$_{13}$N 279.34	224-42-0 216		5-20-08-00656	
5144	5H-Dibenz[b,f]azepine-5-carboxamide Carbamazepine	C$_{15}$H$_{12}$N$_2$O 236.27	298-46-4 190.2	1783	5-20-08-00247	
5145	Dibenz[c,e]oxepin-5,7-dione	C$_{14}$H$_8$O$_3$ 224.22	6050-13-1 217	sub	5-17-11-00525	H$_2$O 1; eth 2
5146	10H-Dibenzo[a,d]cyclohepten-10-one, 5,11-dihydro-	C$_{15}$H$_{12}$O 208.26	6374-70-5 30	203-4[7]	4-07-00-01675 1.1635[20]	1.6324[20]
5147	5H-Dibenzo[a,d]cyclohepten-5-one, 10,11-dihydro-	C$_{15}$H$_{12}$O 208.26	1210-35-1 30	203-4[7]	4-07-00-01674 1.1635[20]	1.6324[20]
5148	13H-Dibenzo[a,g]fluorene	C$_{21}$H$_{14}$ 266.34	207-83-0 174.5	195-200[0.5]	4-05-00-02693	H$_2$O 1; EtOH 2; eth 2; bz 3
5149	6H-Dibenzo[a,g]quinolizine, 5,8,13,13a-tetrahydro-2,3,9,10-tetramethoxy-13-methyl-, (13S-trans)- Corydaline (d)	C$_{22}$H$_{27}$NO$_4$ 369.46	518-69-4 136	2543	5-21-06-00173	bz 4; eth 4; EtOH 4; chl 4
5150	6H-Dibenzo[a,g]quinolizin-2-ol, 5,8,13,13a-tetrahydro-3,9,10-trimethoxy-13-methyl-, (13S-trans)- Isocorybulbine	C$_{21}$H$_{25}$NO$_4$ 355.43	22672-74-8 187.5	5045	5-21-06-00173 1.045[20]	H$_2$O 1; EtOH 3; chl 3; acid 3
5151	6H-Dibenzo[a,g]quinolizin-3-ol, 5,8,13,13a-tetrahydro-2,9,10-trimethoxy-13-methyl-, (13S-trans)- Corybulbine	C$_{21}$H$_{25}$NO$_4$ 355.43	518-77-4 237.5	2539	5-21-06-00173	H$_2$O 1; EtOH 2; eth 2; ace 3
5152	13H-Dibenzo[a,h]fluoren-13-one 1,2:6,7-Dibenzo-9-fluorenone	C$_{21}$H$_{12}$O 280.33	4599-94-4 214	sub 200	3-07-00-02897	
5153	11H-Dibenzo[b,e][1,4]dioxepin-11-one, 2,4,7,9-tetrachloro-3-hydroxy-8-methoxy-1,6-dimethyl- Diploicin	C$_{16}$H$_{10}$Cl$_4$O$_5$ 424.06	527-93-5 232	3345	5-19-06-00364	
5154	6H-Dibenzo[b,d]pyran-1-ol, 6,6,9-trimethyl-3-pentyl- Cannabinol	C$_{21}$H$_{26}$O$_2$ 310.44	521-35-7 77	1751 185[0.05]	5-17-04-00567	H$_2$O 1; EtOH 3; eth 3; ace 3
5155	4H-Dibenzo[de,g]quinoline-10,11-diol, 5,6,6a,7-tetrahydro-6-methyl-, (R)- Apomorphine	C$_{17}$H$_{17}$NO$_2$ 267.33	58-00-4 195 dec	776	5-21-05-00333	H$_2$O 2; EtOH 3; ace 3
5156	4H-Dibenzo[de,g]quinoline-2,9-diol, 5,6,6a,7-tetrahydro-1,10-dimethoxy-6-methyl-, (S)- Boldine	C$_{19}$H$_{21}$NO$_4$ 327.38	476-70-0 163	1328	5-21-06-00118	EtOH 4; chl 4
5157	4H-Dibenzo[de,g]quinoline, 5,6,6a,7-tetrahydro-1,2,9,10-tetramethoxy-6-methyl-, (S)- d-Glaucine	C$_{21}$H$_{25}$NO$_4$ 355.43	475-81-0 120	4332	5-21-06-00122	ace 4; EtOH 4; chl 4
5158	4H-Dibenzo[de,g]quinolin-1-ol, 5,6,6a,7-tetrahydro-2,11-dimethoxy-6-methyl-, (S)- Isothebaine	C$_{19}$H$_{21}$NO$_3$ 311.38	568-21-8 203.5	5116	5-21-05-00658	H$_2$O 1; EtOH 5; eth 2; chl 5
5159	4H-Dibenzo[de,g]quinolin-10-ol, 5,6,6a,7-tetrahydro-1,2-dimethoxy-, (R)- Tsuduranine	C$_{18}$H$_{19}$NO$_3$ 297.35	517-97-5 204	9710	5-21-05-00655	ace 4; eth 4; EtOH 4
5160	4H-Dibenzo[de,g]quinolin-11-ol, 5,6,6a,7-tetrahydro-10-methoxy-6-methyl-, (R)- Apocodeine	C$_{18}$H$_{19}$NO$_2$ 281.35	641-36-1 123.5	771	5-21-05-00334	EtOH 2; eth 3; ace 3; bz 3
5161	4H-Dibenzo[de,g]quinolin-11-ol, 5,6,6a,7-tetrahydro-1,2,10-trimethoxy-6-methyl-, (S)- Isocorydine	C$_{20}$H$_{23}$NO$_4$ 341.41	475-67-2 185	5046	5-21-06-00132	chl 4
5162	4H-Dibenzo[de,g]quinolin-1-ol, 5,6,6a,7-tetrahydro-2,10,11-trimethoxy-6-methyl-, (S)- Corydine	C$_{20}$H$_{23}$NO$_4$ 341.41	476-69-7 149	2545	5-21-06-00134	eth 4; EtOH 4; chl 4
5163	Dibenzofuran 2,2'-Biphenylene oxide	C$_{12}$H$_8$O 168.19	132-64-9 86.5	287	5-17-02-00234 1.0886[99]	H$_2$O 2; EtOH 3; eth 4; ace 3 1.6079[99]
5164	Dibenzofuran, 2-bromo- 2-Bromodibenzofuran	C$_{12}$H$_7$BrO 247.09	86-76-0 110	220[40]	5-17-02-00238	EtOH 3; HOAc 2
5165	Dibenzofuran, 3-bromo- 3-Bromodibenzofuran	C$_{12}$H$_7$BrO 247.09	26608-06-0 121	220[40]	4-17-00-00588	H$_2$O 1; EtOH 3; eth 3
5166	Dibenzofuran, 3-nitro- 3-Nitrodibenzofuran	C$_{12}$H$_7$NO$_3$ 213.19	5410-97-9 181.5	180-5[3]	5-17-02-00239	H$_2$O 1; EtOH 2; eth 2; HOAc 3
5167	Dibenzothiophene	C$_{12}$H$_8$S 184.26	132-65-0 99.5	332.5	5-17-02-00239	H$_2$O 3; EtOH 4; bz 4; chl 3
5168	Diborane(6), mu-(methylthio)- Diborane, methylthio	CH$_8$B$_2$S 73.76	91572-15-5 -101.5	53	4-01-00-01287	ace 4; bz 4

No.	Name Synonym	Mol. Form. Mol. Wt.	CAS RN mp/°C	Merck No. bp/°C	Beil. Ref. den/g cm^{-3}	Solubility n_D
5169	Dicarbonic acid, diethyl ester Pyrocarbonic acid diethyl ester	$C_6H_{10}O_5$ 162.14	1609-47-8	8008 93-4[18]	4-03-00-00018 1.120[20]	ace 4; EtOH 4; lig 4 1.3960[20]
5170	Dichlorvos Phosphoric acid, 2,2-dichloroethenyl dimethyl ester	$C_4H_7Cl_2O_4P$ 220.98	62-73-7	3069 140[20]; 84[1]	1.415[25]	
5171	Diclofop-methyl Propanoic acid, 2-[4-(2,4-dichlorophenoxy)phenoxy]-, methyl ester	$C_{16}H_{14}Cl_2O_4$ 341.19	51338-27-3 40	3072 176[0.1]		
5172	Dicrotophos Phosphoric acid, 3-(dimethylamino)-1-methyl-3-oxo-1-propenyl dimethyl ester, (E)-	$C_8H_{16}NO_5P$ 237.19	141-66-2	3077 400; 130[0.1]	1.216[15]	
5173	Diethatyl-ethyl Glycine, N-(chloroacetyl)-N-(2,6-diethylphenyl)-, ethyl ester	$C_{16}H_{22}ClNO_3$ 311.81	38727-55-8			
5174	Difenoconazole cis,trans-3-Chloro-4-[4-methyl-2-(1H-1,2,4-triazol-1-ylmethyl)-1,3-dioxolan-2-yl]phenyl 4-chlorophenyl ether	$C_{19}H_{17}Cl_2N_3O_3$ 406.27	119446-68-3 76	220[0.03]		
5175	Difenzoquat methyl sulfate 1H-Pyrazolium, 1,2-dimethyl-3,5-diphenyl-, methyl sulfate	$C_{17}H_{17}N_2 \cdot CH_3O_4S$ 360.43	43222-48-6 157			
5176	Diflubenzuron Benzamide, N-[[(4-chlorophenyl)amino]carbonyl]-2,6-difluoro-	$C_{14}H_9ClF_2N_2O_2$ 310.69	35367-38-5 239	3128		
5177	1,4:5,8-Dimethanonaphthalene, 1,2,3,4,10,10-hexachloro-1,4,4a,5,8,8a-hexahydro-, Aldrin	$C_{12}H_8Cl_6$ 364.91	309-00-2 104	219	3-05-00-01385	H_2O 1; EtOH 3; eth 3; ace 3
5178	2,7:3,6-Dimethanonaphth[2,3-b]oxirene, 3,4,5,6,9,9-hexachloro-1a,2,2a,3,6,6a,7,7a-octahydro-,(1aα,2β,2aα,3β,6β,6aα,7β,7aα)- Dieldrin	$C_{12}H_8Cl_6O$ 380.91	60-57-1 175.5	3093	5-17-02-00087 1.75[25]	H_2O 1; EtOH 2; ace 3; bz 3
5179	Dimethipin 2,3-Dihydro-5,6-dimethyl-1,4-dithiin, 1,1,4,4-tetraoxide	$C_6H_{10}O_4S_2$ 210.28	55290-64-7 165			
5180	Dimethyl tetrachloroterephthalate 1,4-Benzenedicarboxylic acid, 2,3,5,6-tetrachloro-, dimethyl ester	$C_{10}H_6Cl_4O_4$ 331.97	1861-32-1 155	2830		
5181	Dinitramine 1,3-Benzenediamine, N3,N3-diethyl-2,4-dinitro-6-(trifluoromethyl)-	$C_{11}H_{13}F_3N_4O_4$ 322.24	29091-05-2 98			
5182	Dinocap 2-Butenoic acid, 2-(1-methylheptyl)-4,6-dinitrophenyl ester	$C_{18}H_{24}N_2O_6$ 364.40	6119-92-2	136[0.01]		
5183	Dinoseb Phenol, 2-(1-methylpropyl)-4,6-dinitro-	$C_{10}H_{12}N_2O_5$ 240.22	88-85-7 40	3282	1.265[45]	
5184	2,3-Dioxabicyclo[2.2.2]oct-5-ene, 1-methyl-4-(1-methylethyl)- Ascaridole	$C_{10}H_{16}O_2$ 168.24	512-85-6 3.3	852 115[15]; 39[0.2]	5-19-01-00319 1.0103[20]	H_2O 1; EtOH 3; ace 3; bz 3 1.4769[20]
5185	1,7-Dioxadispiro[4.0.4.2]dodeca-3,9-diene-2,8-dione, trans- Anemonin	$C_{10}H_8O_4$ 192.17	508-44-1 158	674	5-19-05-00101	chl 4
5186	1,3-Dioxane 1,3-Dioxacyclohexane	$C_4H_8O_2$ 88.11	505-22-6 -45	106.1	5-19-01-00011 1.0342[20]	H_2O 5; EtOH 5; eth 5; ace 5 1.4165[20]
5187	1,4-Dioxane 1,4-Dioxacyclohexane	$C_4H_8O_2$ 88.11	123-91-1 11.8	3294 101.5	5-19-01-00016 1.0337[20]	H_2O 5; EtOH 5; eth 5; ace 5 1.4224[20]
5188	1,4-Dioxane, 2,3-dichloro-	$C_4H_6Cl_2O_2$ 157.00	95-59-0 30	80-2[10]	4-19-00-00030 1.468[20]	H_2O 1; eth 4; ace 4; bz 4 1.4928[20]
5189	1,3-Dioxane, 2,4-dimethyl-	$C_6H_{12}O_2$ 116.16	766-20-1	116.5	5-19-01-00072 0.9392[20]	H_2O 2; acid 4 1.4136[20]
5190	1,3-Dioxane-4,6-dione, 2,2-dimethyl- Meldrum's acid	$C_6H_8O_4$ 144.13	2033-24-1 94	5696	5-19-05-00008	
5191	1,3-Dioxane-2-ethanol, 4-hydroxy-α,6-dimethyl- Paraldol	$C_8H_{11}O_4$ 171.17	19404-07-0 90	 90[15]	4-01-00-03987 1.116[20]	H_2O 4; eth 4; EtOH 4 1.4610[20]
5192	1,4-Dioxane, 2,2,3,3,5,5,6-heptachloro- 1,4-Dioxane, heptachloro-	$C_4HCl_7O_2$ 329.22	6629-96-5 55	123-8[8]	4-19-00-00032	H_2O 1; eth 4; ace 4; bz 4
5193	1,3-Dioxane, 4-methyl-	$C_5H_{10}O_2$ 102.13	1120-97-4 -44.5	114	5-19-01-00054 0.9758[20]	H_2O 2; os 4 1.4159[20]
5194	1,3-Dioxane, 4-methyl-4-phenyl- m-Dioxane, 4-methyl-4-phenyl-	$C_{11}H_{14}O_2$ 178.23	1200-73-3 37.5	256	5-19-01-00481 1.0864[20]	H_2O 1; chl 2; os 3 1.5240[20]
5195	1,3-Dioxane, 2-phenyl-	$C_{10}H_{12}O_2$ 164.20	772-01-0 41	253	5-19-01-00459 1.6053[60]	EtOH 4; eth 4
5196	1,3-Dioxane, 4-phenyl-	$C_{10}H_{12}O_2$ 164.20	772-00-9	247	5-19-01-00464 1.1038[20]	H_2O 1; os 3 1.5306[18]
5197	1,3-Dioxan-5-ol, 2-methyl-	$C_5H_{10}O_3$ 118.13	3774-03-6	176; 79[18]	4-19-00-00624 1.0705[17]	H_2O 4 1.4375[17]
5198	1,3,2-Dioxathiane, 4-methyl-, 2-oxide 1,3-Butanediol, sulfite	$C_4H_8O_3S$ 136.17	4426-51-1 5	185	4-01-00-02510 1.2352[20]	EtOH 4; eth 4; ace 4; bz 4 1.4661[20]

No.	Name Synonym	Mol. Form. Mol. Wt.	CAS RN mp/°C	Merck No. bp/°C	Beil. Ref. den/g cm⁻³	Solubility n_D
5199	1,3,2-Dioxathiolane, 2,2-dioxide	$C_2H_4O_4S$ 124.12	1072-53-3 99	sub	3-01-00-02110	ace 4; bz 4; eth 4; EtOH 4
5200	1,3,2-Dioxathiolane, 2-oxide	$C_2H_4O_3S$ 108.12	3741-38-6 -11	173	4-01-00-02409 1.4402[20]	H_2O 4; EtOH 4; eth 4; ace 4 1.4463[20]
5201	Dioxathion Phosphorodithioic acid, S,S'-1,4-dioxane-2,3-diyl O,O,O',O'-tetraethyl ester	$C_{12}H_{26}O_6P_2S_4$ 456.55	78-34-2 -20	3296	1.257[26]	
5202	1,4-Dioxin	$C_4H_4O_2$ 84.07	290-67-5	76	5-19-01-00275 1.115[20]	eth 3; ace 3; bz 3 1.4350[20]
5203	1,4-Dioxin, 2,3-dihydro-	$C_4H_6O_2$ 86.09	543-75-9	94.1	5-19-01-00181 1.0836[20]	ctc 3 1.4372[20]
5204	1,3-Dioxolane 1,3-Dioxacyclopentane	$C_3H_6O_2$ 74.08	646-06-0 -95	78	5-19-01-00006 1.060[20]	H_2O 5; EtOH 3; eth 3; ace 3 1.3974[20]
5205	1,3-Dioxolane, 2-(3-bromophenyl)-	$C_9H_9BrO_2$ 229.07	17789-14-9	132.5	5-19-01-00448 1.514[25]	1.5627[20]
5206	1,3-Dioxolane-4-carboxaldehyde, 2,2-dimethyl-	$C_6H_{10}O_3$ 130.14	5736-03-8	74[50]	5-19-04-00041	H_2O 4 1.4189[25]
5207	1,3-Dioxolane-4-methanol, 2,2-dimethyl- Isopropylidene glycerol	$C_6H_{12}O_3$ 132.16	100-79-8	5101 82[10]	5-19-02-00362 1.064[20]	1.4383[20]
5208	1,3-Dioxolane-4-methanol, 2-methyl-	$C_5H_{10}O_3$ 118.13	3773-93-1	187	5-19-02-00350 1.1243[17]	EtOH 4 1.4413[17]
5209	1,3-Dioxolane-4-methanol, 2-phenyl-	$C_{10}H_{12}O_3$ 180.20	1708-39-0	143-4[2]	5-19-02-00562 1.1916[17]	1.5389[17]
5210	1,3-Dioxolane, 2-methyl-	$C_4H_8O_2$ 88.11	497-26-7	81.5	5-19-01-00042 0.9811[20]	H_2O 4; EtOH 5; eth 5 1.4035[17]
5211	1,3-Dioxolan-2-one Ethylene glycol carbonate	$C_3H_4O_3$ 88.06	96-49-1 36.4	248	5-19-04-00006 1.3214[39]	H_2O 5; EtOH 5; eth 5; bz 5 1.4148[50]
5212	1,3-Dioxolan-4-one, 2,5-bis(trichloromethyl)- Chloralide	$C_5H_2Cl_6O_3$ 322.79	554-21-2 116	272.5	4-19-00-01571	eth 4; HOAc 4
5213	1,3-Dioxolan-2-one, 4-methyl-	$C_4H_6O_3$ 102.09	108-32-7 -48.8	242	5-19-04-00021 1.2047[20]	H_2O 4; EtOH 4; eth 4; ace 4 1.4189[20]
5214	1,3-Dioxolo[4,5-g]isoquinoline, 5,6,7,8-tetrahydro-4-methoxy-6-methyl- Hydrocotarnine	$C_{12}H_{15}NO_3$ 221.26	550-10-7 56	4715	4-27-00-06395	H_2O 1; EtOH 3; eth 3; ace 3
5215	1,3-Dioxolo[4,5-g]isoquinoline, 5,6,7,8-tetrahydro-6-methyl- Hydrohydrastinine	$C_{11}H_{13}NO_2$ 191.23	494-55-3 66	4732 303	4-27-00-06306	ace 4; bz 4; eth 4; EtOH 4
5216	1,3-Dioxolo[4,5-g]isoquinolin-5-ol, 5,6,7,8-tetrahydro-4-methoxy-6-methyl- Cotarnine	$C_{12}H_{15}NO_4$ 237.26	82-54-2 132-3 dec	2551	4-27-00-06448	H_2O 2; EtOH 3; eth 3; bz 3
5217	1,3-Dioxolo[4,5-g]isoquinolin-5-ol, 5,6,7,8-tetrahydro-6-methyl- Hydrastinine	$C_{11}H_{13}NO_3$ 207.23	6592-85-4 116.5	4689	4-27-00-06396	H_2O 3; EtOH 4; eth 4; chl 4
5218	1,3-Dioxolo[4,5-h]quinolin-8(9H)-one, 6-methoxy-9-methyl- Casimiroin	$C_{12}H_{11}NO_4$ 233.22	477-89-4	1894	4-27-00-06618	chl 2
5219	Diphenamid Benzeneacetamide, N,N-dimethyl-α-phenyl-	$C_{16}H_{17}NO$ 239.32	957-51-7 135	3305	1.17[23.3]	
5220	Diphosphoramide, octamethyl- Schradan	$C_8H_{24}N_4O_3P_2$ 286.25	152-16-9 17	8351 188-22[0.3]	4-04-00-00288 1.1343[25]	H_2O 4; EtOH 4; chl 4 1.462[25]
5221	Diphosphoric acid, tetraethyl ester Tetraethyl pyrophosphate	$C_8H_{20}O_7P_2$ 290.19	107-49-3 170 dec	9138 155[3]	4-01-00-01342 1.1847[20]	H_2O 5; EtOH 5; eth 5; ace 5 1.4180[20]
5222	Dipropetryn 1,3,5-Triazine-2,4-diamine, 6-(ethylthio)-N,N'-bis(1-methylethyl)-	$C_{11}H_{21}N_5S$ 255.39	4147-51-7 105	3349		
5223	5H,10H-Dipyrrolo[1,2-a:1',2'-d]pyrazine-5,10-dione Pyrocoll	$C_{10}H_6N_2O_2$ 186.17	484-73-1 268	sub	5-24-08-00263	eth 4; EtOH 4; chl 4
5224	Diquat dibromide Dipyrido[1,2-a:2',1'-c]pyrazinediium, 6,7-dihydro-, dibromide	$C_{12}H_{12}N_2 2Br$ 344.05	85-00-7 337	3359	1.24[20]	
5225	Diselenide, diphenyl Diphenyl diselenide	$C_{12}H_{10}Se_2$ 312.13	1666-13-3 63.5	202[11]	4-06-00-01781 1.557[80]	EtOH 3; eth 3; xyl 3; MeOH 3 1.743[20]
5226	Disilane, 1,2-dichloro-1,1,2,2-tetramethyl-	$C_4H_{12}Cl_2Si_2$ 187.22	4342-61-4	49-50[18]	4-04-00-04280 1.010[20]	1.4548[20]
5227	Disilane, 1,2-difluoro-1,1,2,2-tetramethyl- 1,2-Difluorotetramethyldisilane	$C_4H_{12}F_2Si_2$ 154.31	661-68-7	94	4-04-00-04279 0.9120[20]	1.3837[20]
5228	Disilane, hexamethyl- Permethyldisilane	$C_6H_{18}Si_2$ 146.38	1450-14-2 13.5	113.5	4-04-00-04277 0.7247[22]	H_2O 1; eth 3; ace 3; bz 3 1.4229[20]
5229	Disiloxane, 1,3-bis(bromomethyl)-1,1,3,3-tetramethyl- 1,3-Bis(bromomethyl)tetramethyldisiloxane	$C_6H_{16}Br_2OSi_2$ 320.17	2351-13-5	102-4[15]	4-04-00-04027 1.3918[25]	1.4719[25]
5230	Disiloxane, 1,3-bis(chloromethyl)-1,1,3,3-tetramethyl-	$C_6H_{16}Cl_2OSi_2$ 231.27	2362-10-9 -90	204; 92[21]	4-04-00-04026 1.0381[20]	1.4398[20]
5231	Disiloxane, 1,3-bis(dichloromethyl)-1,1,3,3-tetramethyl- 1,3-Bis-(dichloromethyl)tetramethyl disiloxane	$C_6H_{14}Cl_4OSi_2$ 300.16	2943-70-6	149-50[50]	4-04-00-04031 1.2213[20]	1.4660[20]
5232	Disiloxane, 1,3-dichloro-1,1,3,3-tetramethyl-	$C_4H_{12}Cl_2OSi_2$ 203.22	2401-73-2 -37.5	138	4-04-00-04119 1.038[20]	

No.	Name / Synonym	Mol. Form. / Mol. Wt.	CAS RN / mp/°C	Merck No. / bp/°C	Bell. Ref. / den/g cm^{-3}	Solubility / n_D
5233	Disiloxane, 1,3-diethenyl-1,1,3,3-tetramethyl-	$C_8H_{18}OSi_2$ / 186.40	2627-95-4 / -99.7	39	4-04-00-04080 / 0.811[20]	1.4123[20]
5234	Disiloxane, 1,3-dimethoxy-1,1,3,3-tetramethyl- / Bis(methoxydimethylsilyl) oxide	$C_8H_{18}O_3Si_2$ / 194.38	18187-24-1	139	4-04-00-04117 / 0.9048[20]	1.3835[20]
5235	Disiloxane, hexaethyl- / 1,1,1,3,3,3-Hexaethyldisiloxane	$C_{12}H_{30}OSi_2$ / 246.54	994-49-0	233; 129[30]	4-04-00-04055 / 0.8589[0]	1.4340[20]
5236	Disiloxane, hexamethyl- / Hexamethyldisiloxane	$C_6H_{18}OSi_2$ / 162.38	107-46-0 / -66	99	4-04-00-04018 / 0.7638[20]	H_2O 1 / 1.3774[20]
5237	Disiloxane, 1,1,3,3-tetramethyl-	$C_4H_{14}OSi_2$ / 134.33	3277-26-7	72	4-04-00-03991 / 0.7572[20]	1.3700[20]
5238	Disiloxane, 1,1,3,3-tetramethyl-1,3-diphenyl-	$C_{16}H_{22}OSi_2$ / 286.52	56-33-7	155-8[13]	4-16-00-01475 / 0.9763[20]	ctc 3 / 1.5176[20]
5239	Disiloxane, 1,1,1-trimethyl-3,3,3-triphenyl- / 1,1,1-Trimethyl-3,3,3-triphenyldisiloxane	$C_{21}H_{24}OSi_2$ / 348.59	799-53-1 / 51	349	4-16-00-01486 / 1.0320[25]	ctc 3; CS_2 3
5240	Dispiro[5.1.5.1]tetradecane / Dispiro[5,1,5,1]tetradecane	$C_{14}H_{24}$ / 192.34	184-97-4 / 10	131[10]	4-05-00-00491 / 0.872[26]	1.4735[27]
5241	Distannoxane, hexabutyl- / Bis(tributyltin) oxide	$C_{24}H_{54}OSn_2$ / 596.11	56-35-9 / 45	180[2]		
5242	Disulfide, bis(1,1-dimethylethyl)	$C_8H_{18}S_2$ / 178.36	110-06-5 / -5	88[21]	4-01-00-01638 / 0.9226[20]	1.4899[20]
5243	Disulfide, bis(3-methylbutyl)	$C_{10}H_{22}S_2$ / 206.42	2051-04-9	247	4-01-00-01689 / 0.9192[20]	1.4864[20]
5244	Disulfide, bis(1-methylethyl)	$C_6H_{14}S_2$ / 150.31	4253-89-8 / -69	177	4-01-00-01503 / 0.9435[20]	1.4916[20]
5245	Disulfide, bis(3-methylphenyl)	$C_{14}H_{14}S_2$ / 246.40	20333-41-9 / -21	177[3]	4-06-00-02085	EtOH 3; eth 3; ace 3
5246	Disulfide, bis(4-methylphenyl)	$C_{14}H_{14}S_2$ / 246.40	103-19-5 / 47.5	210-15[20]	4-06-00-02206 / 1.114[51]	H_2O 1; EtOH 3; eth 4; ace 3
5247	Disulfide, bis(3-nitrophenyl) / Nitrophenide	$C_{12}H_8N_2O_4S_2$ / 308.34	537-91-7 / 84	6539	4-06-00-01686	EtOH 2; eth 3; chl 2
5248	Disulfide, bis(phenylmethyl) / Dibenzyl disulfide	$C_{14}H_{14}S_2$ / 246.40	150-60-7 / 71.5	2995	4-06-00-02760	H_2O 2; EtOH 3; eth 3; bz 3
5249	Disulfide, bis(trifluoromethyl)	$C_2F_6S_2$ / 202.14	372-64-5	34.6	4-03-00-00278	EtOH 4; peth 4
5250	Disulfide, diacetyl / Acetyl disulfide	$C_4H_6O_2S_2$ / 150.22	592-22-3 / 20	105[18]	4-02-00-00564	eth 4; EtOH 4
5251	Disulfide, dibenzoyl	$C_{14}H_{10}O_2S_2$ / 274.36	644-32-6 / 134.5	dec	3-09-00-01977	H_2O 1; EtOH 2; eth 2; CS_2 3
5252	Disulfide, dibutyl	$C_8H_{18}S_2$ / 178.36	629-45-8	117[20]; 85[3]	4-01-00-01562 / 0.9371[20]	H_2O 1; EtOH 5; eth 5 / 1.4923[20]
5253	Disulfide, diethyl / Diethyl disulfide	$C_4H_{10}S_2$ / 122.26	110-81-6 / -101.5	154.1	4-01-00-01397 / 0.9931[20]	H_2O 2; EtOH 5; eth 5 / 1.5073[20]
5254	Disulfide, dimethyl / Dimethyl disulfide	$C_2H_6S_2$ / 94.20	624-92-0 / -85	109.8	4-01-00-01281 / 1.0625[20]	H_2O 1; EtOH 5; eth 5 / 1.5289[20]
5255	Disulfide, di-1-naphthalenyl	$C_{20}H_{14}S_2$ / 318.46	39178-11-5 / 91		4-06-00-04245 / 1.144[20]	H_2O 1; EtOH 2; eth 4; chl 2
5256	Disulfide, di-2-naphthalenyl	$C_{20}H_{14}S_2$ / 318.46	5586-15-2 / 139.5		4-06-00-04320 / 1.144[145]	H_2O 1; EtOH 4; eth 4; lig 1 / 1.4555[20]
5257	Disulfide, dipentyl	$C_{10}H_{22}S_2$ / 206.42	112-51-6	119[7]	4-01-00-01654 / 0.9221[20]	H_2O 1; eth 3 / 1.4889[20]
5258	Disulfide, diphenyl	$C_{12}H_{10}S_2$ / 218.34	882-33-7 / 62	310	4-06-00-01560 / 1.353[20]	H_2O 1; EtOH 3; eth 3; bz 3
5259	Disulfide, dipropyl	$C_6H_{14}S_2$ / 150.31	629-19-6 / -85.6	193.5	4-01-00-01454 / 0.9599[20]	1.4981[20]
5260	Disulfoton / Phosphorodithioic acid, O,O-diethyl S-[2-(ethylthio)ethyl] ester	$C_8H_{19}O_2PS_3$ / 274.41	298-04-4 / -25	3371 / 108[0.01]; 128[1]	1.144[20]	
5261	1,2-Dithiane	$C_4H_8S_2$ / 120.24	505-20-4 / 32.5	80[14]; 60[5]	5-19-01-00010	eth 3; bz 3; chl 3 / 1.5981[25]
5262	1,3-Dithiane	$C_4H_8S_2$ / 120.24	505-23-7 / 54	89[14]	5-19-01-00013	bz 4; eth 4; chl 4 / 1.5981[25]
5263	1,4-Dithiane	$C_4H_8S_2$ / 120.24	505-29-3 / 112.3	199.5	5-19-01-00036	H_2O 2; EtOH 3; eth 3; ctc 3
5264	4H-1,3,5-Dithiazine, dihydro-5-methyl- / 1,3,5-Dithiazine, 4,5-dihydro-5-methyl	$C_4H_9NS_2$ / 135.25	6302-94-9 / 65	185	4-27-00-06285	eth 4; EtOH 4; HOAc 4
5265	4H-1,3,5-Dithiazine, dihydro-2,4,6-trimethyl-, (2α,4α,6α)-	$C_6H_{13}NS_2$ / 163.31	638-17-5 / 46	dec	2-27-00-00525 / 1.0613[50]	eth 4; EtOH 4
5266	1,3-Dithiolane / 1,3-Dithiacyclopentane	$C_3H_6S_2$ / 106.21	4829-04-3 / -50	175	5-19-01-00008 / 1.259[17]	EtOH 3; eth 3; xyl 3 / 1.5975[15]
5267	Dithiopyr / S,S'-Dimethyl-2-difluoromethyl-4-isobutyl-6-trifluoromethylpyridine-3,5-dicarbothioate	$C_{15}H_{16}F_5NO_2S_2$ / 401.42	97886-45-8 / 65			
5268	Docosane / n-Docosane	$C_{22}H_{46}$ / 310.61	629-97-0 / 44.4	368.6	4-01-00-00572 / 0.7944[20]	H_2O 1; EtOH 3; eth 4; chl 3 / 1.4455[20]
5269	Docosane, 5-butyl- / 5-n-Butyldocosane	$C_{26}H_{54}$ / 366.71	55282-16-1 / 208	244[10]	4-01-00-00584 / 0.8058[20]	1.4503[20]
5270	Docosane, 7-butyl- / 7-Butyldocosane	$C_{26}H_{54}$ / 366.71	55282-15-0 / 3.2	241[10]	4-01-00-00584 / 0.8046[20]	1.4499[20]

No.	Name Synonym	Mol. Form. Mol. Wt.	CAS RN mp/°C	Merck No. bp/°C	Beil. Ref. den/g cm^{-3}	Solubility n_D
5271	Docosane, 9-butyl- 9-Butyldocosane	$C_{26}H_{54}$ 366.71	55282-14-9 1.3	242.5[10]	4-01-00-00584 0.8044[20]	1.4498[20]
5272	Docosane, 11-butyl- 11-Butyldocosane	$C_{26}H_{54}$ 366.71	13475-76-8	242.5[10]	4-01-00-00584 0.8041[20]	1.4499[20]
5273	Docosane, decyl- Docosane, 11-decyl	$C_{32}H_{66}$ 450.88	28261-97-4 1.0	286[10]	4-01-00-00597 0.8131[20]	1.4546[20]
5274	Docosane, 7-hexyl- 7-Hexyldocosane	$C_{28}H_{58}$ 394.77	55373-86-9 19.3	258[10]	4-01-00-00590 0.8080[20]	1.4517[20]
5275	Docosane, 9-octyl- 9-Octyldocosane	$C_{30}H_{62}$ 422.82	55319-83-0 8.6	272.5[10]	4-01-00-00594 0.8114[20]	1.4537[20]
5276	Docosanoic acid Behenic acid	$C_{22}H_{44}O_2$ 340.59	112-85-6 81	1033 306[60]	4-02-00-01290 0.8223[90]	H_2O 2; EtOH 2; eth 2 1.4270[100]
5277	Docosanoic acid, ethyl ester	$C_{24}H_{48}O_2$ 368.64	5908-87-2 50	240[10]	4-02-00-01292	eth 4; EtOH 4
5278	Docosanoic acid, methyl ester	$C_{23}H_{46}O_2$ 354.62	929-77-1 54		4-02-00-01291	eth 4; EtOH 4 1.4339[60]
5279	1-Docosanol	$C_{22}H_{46}O$ 326.61	661-19-8 72.5	180[0.22]	4-01-00-01906	H_2O 2; EtOH 4; eth 2; chl 3
5280	1-Docosene	$C_{22}H_{44}$ 308.59	1599-67-3 38	367	3-01-00-00883 0.794[25]	
5281	13-Docosenoic acid	$C_{22}H_{42}O_2$ 338.57	1072-39-5 34.7	242[5]	4-02-00-01676 0.8532[20]	1.4444[70]
5282	13-Docosenoic acid, (E)- Brassidic acid	$C_{22}H_{42}O_2$ 338.57	506-33-2 61.9	1359 282[30]; 256[10]	4-02-00-01677 0.8585[57]	1.4347[100]
5283	13-Docosenoic acid, (Z)- Erucic acid	$C_{22}H_{42}O_2$ 338.57	112-86-7 33.5	3619 265[15]	4-02-00-01676 0.860[55]	H_2O 1; EtOH 3; eth 4; ctc 3 1.4758[20]
5284	13-Docosen-1-ol, (Z)- Erucyl alcohol	$C_{22}H_{44}O$ 324.59	629-98-1 34.7	240[10]	4-01-00-02209 0.8416[33]	bz 4; EtOH 4; peth 4
5285	1,11-Dodecadiene	$C_{12}H_{22}$ 166.31	5876-87-9	208.5	4-01-00-01067 0.7702[20]	1.4400[20]
5286	3,9-Dodecadiyne	$C_{12}H_{18}$ 162.27	61827-89-2			
5287	5,7-Dodecadiyne Dibutylbutadiyne	$C_{12}H_{18}$ 162.27	1120-29-2	103[8]	4-01-00-01130	
5288	Dodecanal Lauraldehyde	$C_{12}H_{24}O$ 184.32	112-54-9 44.5	185[100]; 100[2.5]	4-01-00-03380 0.8352[15]	H_2O 1; EtOH 2; eth 3 1.435[22]
5289	Dodecanamide	$C_{12}H_{25}NO$ 199.34	1120-16-7 110	199[12]	4-02-00-01103	H_2O 1; EtOH 3; eth 2; ace 3 1.4287[110]
5290	Dodecanamide, N,N-diethyl-	$C_{16}H_{33}NO$ 255.44	3352-87-2	166-7[2]	4-04-00-00357 0.847[25]	chl 3 1.4545[20]
5291	1-Dodecanamine	$C_{12}H_{27}N$ 185.35	124-22-1 28.3	259	4-04-00-00794 0.8015[20]	H_2O 2; EtOH 5; eth 5; bz 5 1.4421[20]
5292	Dodecane	$C_{12}H_{26}$ 170.34	112-40-3 -9.6	216.3	4-01-00-00498 0.7487[20]	H_2O 1; EtOH 4; eth 4; ace 4 1.4216[20]
5293	Dodecane, 1-bromo- Lauryl bromide	$C_{12}H_{25}Br$ 249.23	143-15-7 -9.5	5260 276	4-01-00-00502 1.0399[20]	H_2O 1; EtOH 3; eth 3; ace 5 1.4583[20]
5294	Dodecane, 1-bromo-12-fluoro- 1-Bromo-12-fluorododecane	$C_{12}H_{24}BrF$ 267.22	353-29-7	85.8[0.15]	4-01-00-00502	ace 4; eth EtOH 4 1.4524[25]
5295	Dodecane, 1-chloro-	$C_{12}H_{25}Cl$ 204.78	112-52-7 -9.3	260	4-01-00-00501 0.8687[20]	H_2O 1; EtOH 4; ace 5; bz 3 1.4433[20]
5296	Dodecane, 1,12-dibromo-	$C_{12}H_{24}Br_2$ 328.13	3344-70-5 41	215[15]	4-01-00-00503	H_2O 1; EtOH 4; eth 3; chl 4
5297	Dodecane, 1,1-dimethoxy- Lauraldehyde, dimethyl acetal	$C_{14}H_{30}O_2$ 230.39	14620-52-1	132-4[5]	4-01-00-03381	eth 4; EtOH 4 1.4310[25]
5298	Dodecane, 1,12-dimethoxy- 1,12-Dodecanediol, dimethyl ether	$C_{14}H_{30}O_2$ 230.39	73120-52-2 11.5	155-6[15]	4-01-00-02627 0.8563[22]	1.436[22]
5299	Dodecanedioic acid	$C_{12}H_{22}O_4$ 230.30	693-23-2 128	222[25]	4-02-00-02126 1.15[25]	tfa 3
5300	Dodecanedioic acid, dimethyl ester	$C_{14}H_{26}O_4$ 258.36	1731-79-9 31.9	170[10]	4-02-00-02126	H_2O 1; AcOEt 3
5301	1,12-Dodecanediol 1,12-Dihydroxydodecane	$C_{12}H_{26}O_2$ 202.34	5675-51-4 81.3	189[12]	4-01-00-02627	tfa 3
5302	Dodecane, 1-iodo-	$C_{12}H_{25}I$ 296.23	4292-19-7 0.3	298.2	4-01-00-00503 1.1999[20]	H_2O 1; EtOH 3; eth 5; ace 5 1.4840[20]
5303	Dodecanenitrile Lauronitrile	$C_{12}H_{23}N$ 181.32	2437-25-4 4	277; 198[100]	4-02-00-01104 0.8240[20]	H_2O 1; EtOH 5; eth 5; ace 5 1.4361[20]
5304	1-Dodecanethiol	$C_{12}H_{26}S$ 202.40	112-55-0	142-5[15]	4-01-00-01851 0.8435[20]	H_2O 1; EtOH 3; eth 3; chl 3 1.4589[20]
5305	Dodecanoic acid Lauric acid	$C_{12}H_{24}O_2$ 200.32	143-07-7 43.2	5254 91.4	4-02-00-01082 0.8679[50]	H_2O 1; EtOH 4; eth 4; ace 3 1.4183[82]
5306	Dodecanoic acid, anhydride	$C_{24}H_{46}O_3$ 382.63	645-66-9 41.8		4-02-00-01101 0.8533[70]	EtOH 4 1.4292[70]
5307	Dodecanoic acid, 2-bromo-	$C_{12}H_{23}BrO_2$ 279.22	111-56-8 32	157-9[2]	4-02-00-01106 1.1474[74]	bz 4; eth 4; EtOH 4; lig 4 1.4585[24]
5308	Dodecanoic acid, 2-bromoethyl ester	$C_{14}H_{27}BrO_2$ 307.27	6309-50-8	176[12]; 145[3]	4-02-00-01092 1.088[25]	1.4547[25]
5309	Dodecanoic acid, 2-bromo-, methyl ester	$C_{13}H_{25}BrO_2$ 293.24	617-60-7	169-71[12]	4-02-00-01107 1.113[15]	1.4572[25]

No.	Name / Synonym	Mol. Form. / Mol. Wt.	CAS RN / mp/°C	Merck No. / bp/°C	Beil. Ref. / den/g cm⁻³	Solubility / n_D
5310	Dodecanoic acid, 2,3-dihydroxypropyl ester, (±)-	$C_{15}H_{30}O_4$	40738-26-9		4-02-00-01096	EtOH 2; eth 4; ace 4; bz 3
	Glycerol, 1-laurate (DL)	274.40	63	186[2]	0.9248[97]	1.4350[86]
5311	Dodecanoic acid, 1,2-ethanediyl ester	$C_{26}H_{50}O_4$	624-04-4	4393	4-02-00-01094	eth 4; EtOH 4
	Ethylene glycol dilaurate	426.68	56.6	188[20]		
5312	Dodecanoic acid, ethyl ester	$C_{14}H_{28}O_2$	106-33-2	3774	4-02-00-01092	H_2O 1; EtOH 4; eth 5; ctc 2
	Ethyl laurate	228.38	-10	271; 154[15]	0.8618[20]	1.4311[20]
5313	Dodecanoic acid, 2-(2-hydroxyethoxy)ethyl ester	$C_{16}H_{32}O_4$	141-20-8	3110		EtOH 5; eth 5; ace 5; bz 3
	Diethylene glycol monolaurate	288.43	17.5	>270	0.96[25]	
5314	Dodecanoic acid, 2-methyl-	$C_{13}H_{26}O_2$	2874-74-0		4-02-00-01121	
		214.35	22	153[1]	0.890[18]	
5315	Dodecanoic acid, methyl ester	$C_{13}H_{26}O_2$	111-82-0		4-02-00-01090	H_2O 1; EtOH 5; eth 5; ace 5
	Methyl laurate	214.35	5.2	267	0.8702[20]	1.4319[20]
5316	Dodecanoic acid, 1-methylethyl ester	$C_{15}H_{30}O_2$	10233-13-3		4-02-00-01092	eth 4; EtOH 4
		242.40		196[60]; 117[2]	0.8536[20]	1.4280[25]
5317	Dodecanoic acid, phenyl ester	$C_{18}H_{28}O_2$	4228-00-6		4-06-00-00618	ace 4; eth 4; EtOH 4
	Phenyl laurate	276.42	24.5	210[15]	0.9354[30]	
5318	Dodecanoic acid, phenylmethyl ester	$C_{19}H_{30}O_2$	140-25-0		4-06-00-02267	bz 4; eth 4; EtOH 4; peth 4
		290.45	8.5	209-11[12]	0.9429[25]	1.4812[24]
5319	Dodecanoic acid, 1,2,3-propanetriyl ester	$C_{39}H_{74}O_6$	538-24-9		4-02-00-01098	H_2O 1; EtOH 3; eth 3; ace 4
		639.01			0.8986[55]	1.4404[60]
5320	Dodecanoic acid, propyl ester	$C_{15}H_{30}O_2$	3681-78-5		4-02-00-01092	
	Propyl laurate	242.40		205[60]; 124[2]	0.8600[20]	1.4335[20]
5321	1-Dodecanol	$C_{12}H_{26}O$	112-53-8	3402	4-01-00-01844	H_2O 1; EtOH 3; eth 3; bz 2
	Lauryl alcohol	186.34	24	259	0.8309[24]	
5322	2-Dodecanol	$C_{12}H_{26}O$	10203-28-8		3-01-00-01793	
		186.34	19	252	0.8286[20]	1.4400[20]
5323	3-Dodecanol	$C_{12}H_{26}O$	10203-30-2		4-01-00-01854	
		186.34	25	130[15]	0.8223[32]	
5324	6-Dodecanol	$C_{12}H_{26}O$	6836-38-0		3-01-00-01794	eth 4; EtOH 4
		186.34	30	225; 119[9]	0.8201[40]	
5325	2-Dodecanone	$C_{12}H_{24}O$	6175-49-1		4-01-00-03382	H_2O 1; EtOH 3; eth 3; ace 3
	Decyl methyl ketone	184.32	21	246.5	0.8198[20]	1.4330[20]
5326	6-Dodecanone	$C_{12}H_{24}O$	6064-27-3		4-01-00-03383	
	Amyl hexyl ketone	184.32	10	112[9]		1.4302[20]
5327	1-Dodecanone, 1-phenyl-	$C_{18}H_{28}O$	1674-38-0		4-07-00-00847	H_2O 1; ace 3; ctc 2
		260.42	47	201[9]; 181[5]	0.8794[18]	1.4700[18]
5328	Dodecanoyl chloride	$C_{12}H_{23}ClO$	112-16-3		4-02-00-01103	eth 4
		218.77	-17	145[18]	0.9169[25]	1.4458[20]
5329	1,3,6,10-Dodecatetraene, 3,7,11-trimethyl-, (E,E)-	$C_{15}H_{24}$	502-61-4	3883	3-01-00-01067	H_2O 1; eth 3; ace 3; peth 5
	α-Farnesene	204.36		129-32[12]	0.8410[20]	1.4836[20]
5330	2,6,10-Dodecatrienal, 3,7,11-trimethyl-	$C_{15}H_{24}O$	19317-11-4		4-01-00-03603	
		220.35		172-4[14]	0.893[18]	1.4995
5331	1,6,10-Dodecatriene, 7,11-dimethyl-3-methylene-, (E)-	$C_{15}H_{24}$	18794-84-8	3884	4-01-00-01133	ace 4; eth 4; chl 4
	β-Farnesene	204.36		121-2[9]	0.8363[20]	1.4899[20]
5332	1,6,10-Dodecatrien-3-ol, 3,7,11-trimethyl-, [S-(Z)]-	$C_{15}H_{26}O$	142-50-7	6388	4-01-00-02336	EtOH 4; eth 3; ace 3; os 3
		222.37		276; 70[0.1]	0.8778[20]	1.4898[20]
5333	2,6,10-Dodecatrien-1-ol, 3,7,11-trimethyl-, (E,E)-	$C_{15}H_{26}O$	106-28-5		4-01-00-02335	H_2O 1; EtOH 4; eth 3; ace 3
		222.37		160[10]	0.8846[20]	1.4877[20]
5334	2,6,10-Dodecatrien-1-ol, 3,7,11-trimethyl-, (Z,E)-	$C_{15}H_{26}O$	3790-71-4		4-01-00-02335	ace 4; eth 4; EtOH 4
		222.37		156[12]; 120[0.3]	0.8908[20]	1.4877[20]
5335	1-Dodecene	$C_{12}H_{24}$	112-41-4		4-01-00-00914	H_2O 1; EtOH 3; eth 3; ace 3
		168.32	-35.2	213.8	0.7584[20]	1.4300[20]
5336	2-Dodecenedioic acid, (E)-	$C_{12}H_{20}O_4$	6402-36-4	9493	4-02-00-02279	eth 4; EtOH 4; chl 4
	Traumatic acid	228.29	165.5			
5337	2-Dodecenoic acid	$C_{12}H_{22}O_2$	4412-16-2		4-02-00-01619	
		198.31	17.1	155[3]; 127[0.15]	0.9265[20]	1.4629[25]
5338	4-Dodecenoic acid	$C_{12}H_{22}O_2$	505-92-0		4-02-00-01619	bz 4; eth 4; chl 4
	Linderic acid	198.31	1.3	171[13]	0.9081[15]	1.4529[20]
5339	5-Dodecenoic acid	$C_{12}H_{22}O_2$	2761-84-4		4-02-00-01619	
		198.31	1.3	170-2[13]	0.9081[20]	
5340	11-Dodecenoic acid	$C_{12}H_{22}O_2$	65423-25-8		4-02-00-01618	
		198.31	20	171[13]; 144[3]	0.9014[20]	1.4510[20]
5341	11-Dodecenoic acid, methyl ester	$C_{13}H_{24}O_2$	29972-79-0		4-02-00-01618	
	Methyl 11-dodecenoate	212.33		138-9[13]	0.8789[22]	1.4414[20]
5342	1-Dodecen-3-yne	$C_{12}H_{20}$	74744-36-8		4-01-00-01112	
		164.29		78[4]	0.7858[25]	1.4510[25]
5343	1-Dodecyne	$C_{12}H_{22}$	765-03-7		4-01-00-01066	
	Decylacetylene	166.31	-19	215	0.7788[20]	1.4340[20]
5344	2-Dodecyne	$C_{12}H_{22}$	629-49-2		0-01-00-00261	
		166.31	-9	105[15]	0.7917[15]	1.4828[20]
5345	3-Dodecyne	$C_{12}H_{22}$	6790-27-8		3-01-00-01025	ace 4; eth 4
		166.31		95[12]	0.7871[20]	1.4442[20]

No.	Name Synonym	Mol. Form. Mol. Wt.	CAS RN mp/°C	Merck No. bp/°C	Beil. Ref. den/g cm^{-3}	Solubility n_D
5346	6-Dodecyne	$C_{12}H_{22}$ 166.31	6975-99-1	210; 100[14]	4-01-00-01067 0.7871[20]	ace 4; eth 4; EtOH 4 1.4442[20]
5347	Dodine Guanidine, dodecyl-, monoacetate	$C_{13}H_{29}N_3 \cdot C_2H_4O_2$ 287.45	2439-10-3 136	3406		
5348	Dotriacontane Bicetyl	$C_{32}H_{66}$ 450.88	544-85-4 69.7	467	4-01-00-00595 0.8124[20]	H_2O 1; EtOH 2; eth 3; bz 4 1.4550[20]
5349	Eburnamenine-14-carboxylic acid, 14,15-dihydro-14-hydroxy-, methyl ester, (3α,14β,16α)- Vincamine	$C_{21}H_{26}N_2O_3$ 354.45	1617-90-9 231.5	9888	4-25-00-01252	
5350	Eicosane	$C_{20}H_{42}$ 282.55	112-95-8 36.8	343	4-01-00-00563 0.7886[20]	H_2O 1; eth 3; ace 4; bz 3 1.4425[20]
5351	Eicosane, 1-cyclopentyl- Cyclopentane, eicosyl	$C_{25}H_{50}$ 350.67	22331-38-0 38	413	4-05-00-00198 0.8276[20]	ace 4; bz 4; eth 4; EtOH 4 1.4595[20]
5352	Eicosanedioic acid 1,18-Octadecanedicarboxylic acid	$C_{20}H_{38}O_4$ 342.52	2424-92-2 125.5	233-4[2]	4-02-00-02185	eth 3
5353	Eicosanedioic acid, diethyl ester	$C_{24}H_{46}O_4$ 398.63	42235-39-2 54.5	240[12]; 230[2]	3-02-00-01881	eth 4; EtOH 4
5354	Eicosane, 9-octyl- Eicosane, 9-ocytl	$C_{28}H_{58}$ 394.77	13475-77-9 0.5	257[10]; 199[0.5]	4-01-00-00590 0.8075[20]	 1.4515[20]
5355	Eicosanoic acid Arachidic acid	$C_{20}H_{40}O_2$ 312.54	506-30-9 76.5	791 328 dec; 204[1]	4-02-00-01275 0.8240[100]	H_2O 1; EtOH 2; eth 4; bz 3 1.425[100]
5356	Eicosanoic acid, ethyl ester	$C_{22}H_{44}O_2$ 340.59	18281-05-5 50	295[100]; 186[2]	3-02-00-01067	H_2O 1; EtOH 3; eth 3; bz 3
5357	Eicosanoic acid, methyl ester	$C_{21}H_{41}O_2$ 325.56	1120-28-1 54.5	215[10]	4-02-00-01276	bz 4; eth 4; EtOH 4; chl 4 1.4317[60]
5358	1-Eicosanol Arachic alcohol	$C_{20}H_{42}O$ 298.55	629-96-9 66.1	309; 222[3]	4-01-00-01900 0.8405[20]	H_2O 1; EtOH 2; ace 4; bz 3 1.4350[20]
5359	2-Eicosanol 1-Methyl-1-nonadecanol	$C_{20}H_{42}O$ 298.55	4340-76-5 63.5	357	4-01-00-01902 0.8378[20]	ace 4; bz 4 1.4912[80]
5360	5,8,11,14-Eicosatetraenoic acid 5,8,11,14-Eicosotetraenoic acid	$C_{20}H_{32}O_2$ 304.47	7771-44-0 -49.5	169[0.15]	4-02-00-01802 0.9219[20]	H_2O 1; EtOH 3; eth 3; ace 3 1.4824[20]
5361	5,8,11,14-Eicosatetraenoic acid, (all-Z)- Arachidonic acid	$C_{20}H_{32}O_2$ 304.47	506-32-1 -49.5	792 163[1]	4-02-00-01802 0.9082[20]	ace 4; eth 4; EtOH 4; peth 4 1.4824[20]
5362	1-Eicosene	$C_{20}H_{40}$ 280.54	3452-07-1 28.5	341; 151[2]	4-01-00-00934 0.7882[30]	H_2O 1; bz 3; peth 3 1.4440[30]
5363	11-Eicosenoic acid Δ 11-Eicosenoic acid	$C_{20}H_{38}O_2$ 310.52	2462-94-4 24	267[15]	4-02-00-01673 0.8826[25]	EtOH 4; MeOH 4
5364	9-Eicosenoic acid	$C_{20}H_{38}O_2$ 310.52	506-31-0 23	220[6]; 170[0.1]	4-02-00-01672 0.8882[25]	 1.4597[25]
5365	9-Eicosenoic acid, (Z)-	$C_{20}H_{38}O_2$ 310.52	29204-02-2 24.5	220[6]	4-02-00-01672 0.8882[25]	
5366	1-Eicosyne	$C_{20}H_{38}$ 278.52	765-27-5 36	340	4-01-00-01077 0.8073[20]	bz 4; peth 4 1.4501[20]
5367	Emetan-6'-ol, 7',10,11-trimethoxy- Cephaeline	$C_{28}H_{38}N_2O_4$ 466.62	483-17-0 115.5	1970	5-23-13-00611	ace 4; EtOH 4; MeOH 4; chl 4
5368	Emetan, 6',7',10,11-tetramethoxy- Emetine	$C_{29}H_{40}N_2O_4$ 480.65	483-18-1 74	3517	4-23-00-03422	H_2O 1; EtOH 3; eth 3; ace 3
5369	Endosulfan 5,9-Methano-2,4,3-benzodioxathiepin, 6,7,8,9,10,10-hexachloro-1,5,5a,6,9,9a-hexahydro-, 3-oxide	$C_9H_6Cl_6O_3S$ 406.93	115-29-7 106	3529 106[0.7]	 1.745[20]	
5370	Endothall, disodium salt 7-Oxabicyclo[2.2.1]heptane-2,3-dicarboxylic acid, disodium salt	$C_8H_{10}Na_2O_5$ 232.14	145-73-3 144	3530	 1.431[20]	
5371	4,7-Epoxyisobenzofuran-1,3-dione, hexahydro-3a,7a-dimethyl-, (3aα,4β,7β,7aα)- Cantharidin	$C_{10}H_{12}O_4$ 196.20	56-25-7 218	1755 sub 84	5-19-05-00051	H_2O 1; EtOH 2; eth 2; ace 2
5372	EPTC Carbamothioic acid, dipropyl-, S-ethyl ester	$C_9H_{19}NOS$ 189.32	759-94-4 127[20]	3580	 0.9546[30]	
5373	Ergoline-8-carboxamide, 9,10-didehydro-N-(2-hydroxy-1-methylethyl)-6-methyl-, [8α(S)]- Ergometrinine	$C_{19}H_{23}N_3O_2$ 325.41	479-00-5 195-7 dec	3599	4-25-00-00950	chl 4
5374	Ergoline-8-carboxamide, 9,10-didehydro-6-methyl-, (8β)- Lysergamide	$C_{16}H_{17}N_3O$ 267.33	478-94-4 137.5	5505	4-25-00-00937	EtOH 2; ace 2; os 2
5375	Ergoline-8-carboxylic acid, 9,10-didehydro-6-methyl-, (8α)- Isolysergic acid	$C_{16}H_{16}N_2O_2$ 268.32	478-95-5 218 dec	5065	4-25-00-00935	H_2O 2; EtOH 2; py 3
5376	Ergoline-8-carboxylic acid, 9,10-didehydro-6-methyl-, (8β)- Lysergic acid	$C_{16}H_{16}N_2O_2$ 268.32	82-58-6 240 dec	5506	4-25-00-00934	H_2O 2; EtOH 3; eth 2; bz 2
5377	Ergostane, (5α)-	$C_{28}H_{50}$ 386.71	511-20-6 85	3602	4-05-00-01234	ace 4; eth 4; chl 4
5378	Ergostane, (5β)- Coproergostane	$C_{28}H_{50}$ 386.71	511-21-7 64	2516	3-05-00-01143	eth 4; chl 4

No.	Name Synonym	Mol. Form. Mol. Wt.	CAS RN mp/°C	Merck No. bp/°C	Beil. Ref. den/g cm⁻³	Solubility n_D
5379	Ergostan-3-ol, (3β,5α)- Ergostanol	$C_{28}H_{50}O$ 402.70	6538-02-9 144.5	3603	4-06-00-03602	H_2O 1; eth 3; chl 3
5380	Ergosta-5,7,9(11),22-tetraen-3-ol, (3β,22E)- Dehydroergosterol	$C_{28}H_{42}O$ 394.64	516-85-8 146	2861 $230^{0.5}$	4-06-00-04838	ace 4; bz 4; eth 4; EtOH 4
5381	Ergosta-5,7,22-trien-3-ol, (3β,10α,22E)- Pyrocalciferol	$C_{28}H_{44}O$ 396.66	128-27-8 94	8007	3-06-00-03098	H_2O 1; EtOH 3; chl 3; MeOH 3
5382	Ergosta-5,7,22-trien-3-ol, (3β,9β,10α,22E)- Lumisterol	$C_{28}H_{44}O$ 396.66	474-69-1 118	5471	4-06-00-04408	H_2O 1; EtOH 3; eth 4; ace 4
5383	Ergosta-5,7,22-trien-3-ol, (3β,22E)- Ergosterol	$C_{28}H_{44}O$ 396.66	57-87-4 170	3607 $250^{0.01}$	4-06-00-04407	H_2O 1; EtOH 2; eth 2; bz 3
5384	Ergost-7-en-3-ol, (3β,5α)- γ-Ergostenol	$C_{28}H_{48}O$ 400.69	516-78-9 146	3606	4-06-00-04029	eth 3
5385	Ergost-8(14)-en-3-ol, (3β,5α)- α-Ergostenol	$C_{28}H_{48}O$ 400.69	632-32-6 131	3604	4-06-00-04031	EtOH 2; eth 3; bz 3; chl 3
5386	Ergost-5-en-3-ol, (3β,24R)- Campesterol	$C_{28}H_{48}O$ 400.69	474-62-4 157.5	1735	3-06-00-02680	
5387	Ergotaman-3',6',18-trione, 12'-hydroxy-2',5'-bis(1-methylethyl)-, Ergocornine	$C_{31}H_{39}N_5O_5$ 561.68	564-36-3 182-4 dec	3591	4-25-00-00963	H_2O 1; EtOH 3; ace 3; bz 3
5388	Ergotaman-3',6',18-trione, 12'-hydroxy-2',5'-bis(1-methylethyl)-, Ergocorninine	$C_{31}H_{39}N_5O_5$ 561.68	564-37-4 228 dec	3592	4-25-00-00963	ace 4; bz 4; EtOH 4; chl 4
5389	Ergotaman-3',6',18-trione, 12'-hydroxy-2'-(1-methylethyl)-5'-(2-methylpropyl)- Ergocryptine	$C_{32}H_{41}N_5O_5$ 575.71	511-09-1 212-4 dec	3595	4-25-00-00964	H_2O 1; EtOH 3; chl 3
5390	Ergotaman-3',6',18-trione, 12'-hydroxy-2'-(1-methylethyl)-5'-(2-methylpropyl)-, (5'α,8α)- Ergocryptinine	$C_{33}H_{41}N_5O_5$ 587.72	511-10-4 245 dec	3596	4-25-00-00964	ace 4; chl 4
5391	Ergotaman-3',6',18-trione, 12'-hydroxy-2'-(1-methylethyl)-5'-(phenylmethyl)- Ergocristine	$C_{35}H_{39}N_5O_5$ 609.73	511-08-0 175 dec	3593	4-25-00-00966	H_2O 1; EtOH 3; ace 3; chl 3
5392	Ergotaman-3',6',18-trione, 12'-hydroxy-2'-(1-methylethyl)-5'-(phenylmethyl)- Ergocristinine	$C_{35}H_{39}N_5O_5$ 609.73	511-07-9 237-8 dec	3594	4-25-00-00967	H_2O 1; EtOH 2; ace 2; chl 2
5393	Ergotaman-3',6',18-trione, 12'-hydroxy-2'-methyl-5'-(2-methylpropyl)-, Ergosine	$C_{30}H_{37}N_5O_5$ 547.65	561-94-4 228 dec	3601	4-25-00-00961	ace 3; chl 3; MeOH 2
5394	Ergotaman-3',6',18-trione, 12'-hydroxy-2'-methyl-5'-(phenylmethyl)-, (5'α)- Ergotamine	$C_{33}H_{55}N_5O_5$ 601.83	113-15-5 213-4 dec	3609	4-25-00-00964	bz 4; eth 4; chl 4
5395	Ergotaman-3',6',18-trione, 12'-hydroxy-2'-methyl-5'-(phenylmethyl)-, Ergotaminine	$C_{33}H_{35}N_5O_5$ 581.67	639-81-6 252 dec	3610	4-25-00-00966	H_2O 1; EtOH 2; ace 2; bz 2
5396	Erythromycin Propiocine	$C_{37}H_{67}NO_{13}$ 733.94	114-07-8 191	3626	5-18-10-00398	ace 4; eth 4; EtOH 4; chl 4
5397	Estra-1,3,5,7,9-pentaen-17-one, 3-hydroxy- Equilenin	$C_{18}H_{18}O_2$ 266.34	517-09-9 258.5	3581 sub 170	4-08-00-01420	EtOH 2; ace 2; chl 2
5398	Estra-1,3,5(10),7-tetraen-17-one, 3-hydroxy- Equilin	$C_{18}H_{20}O_2$ 268.36	474-86-2 239	3582 sub 170	4-08-00-01366	H_2O 2; EtOH 3; ace 3; diox 3
5399	Estra-1,3,5(10)-triene-3,17-diol (17β)- Estradiol	$C_{18}H_{24}O_2$ 272.39	50-28-2 178.5	3653	4-06-00-06611	ace 4; EtOH 4; Diox 4
5400	Estra-1,3,5(10)-triene-3,17-diol, (17α)- α-Estradiol	$C_{18}H_{24}O_2$ 272.39	57-91-0 221.5	3654	4-06-00-06611	H_2O 1; EtOH 3; eth 2; ace 3
5401	Estra-1,3,5(10)-triene-3,17-diol, (8α,17β)- Isoestradiol	$C_{18}H_{24}O_2$ 272.39	517-04-4 181	5051	4-06-00-06611	EtOH 3; diox 3
5402	Estra-5,7,9-triene-3,17-diol, (3β,17β)- Hexahydroequilenin	$C_{18}H_{24}O_2$ 272.39	517-07-7 168.5	4607 sub 150	4-06-00-06618	
5403	Estra-1,3,5(10)-triene-3,17-diol (17β)-, 3-benzoate Estradiol benzoate	$C_{25}H_{28}O_3$ 376.50	50-50-0 196	3655	4-09-00-00406	
5404	Estra-1,3,5(10)-triene-3,16,17-triol, (16α,17β)- Estriol	$C_{18}H_{24}O_3$ 288.39	50-27-1 288 dec	3659	4-06-00-07550 1.27²⁵	EtOH 3; eth 2; bz 2; tfa 2
5405	Estra-1,3,5(10)-triene-3,16,17-triol, (16β,17β)- 16-Epiestriol	$C_{18}H_{24}O_3$ 288.39	547-81-9 290	3587	4-06-00-07549	
5406	Estra-1,3,5(10)-trien-17-one, 3-hydroxy- Estrone	$C_{18}H_{22}O_2$ 270.37	53-16-7 260.2	3660	3-08-00-01171 1.236²⁵	H_2O 1; EtOH 2; eth 2; ace 3
5407	Estra-1,3,5(10)-trien-17-one, 3-hydroxy-, (8α)- 8-Isoestrone	$C_{18}H_{22}O_2$ 270.37	517-06-6 247	5052	3-08-00-01170	eth 4; Diox 4
5408	Estr-4-en-3-one, 17-(1-oxo-3-phenylpropoxy)-, (17β)- Nandrolone phenpropionate	$C_{27}H_{34}O_3$ 406.57	62-90-8	6283	4-09-00-01758	
5409	Ethalfluralin Benzenamine, N-ethyl-N-(2-methyl-2-propenyl)-2,6-dinitro-4-(trifluoromethyl)-	$C_{13}H_{14}F_3N_3O_4$ 333.27	55283-68-6 57	3671 256 dec		
5410	Ethanamine Ethylamine	C_2H_7N 45.08	75-04-7 -80.5	3718 16.5	4-04-00-00307 *0.677²⁵	H_2O 5; EtOH 5; eth 5 1.3663²⁰
5411	Ethanamine, 2-[(3-butyl-1-isoquinolinyl)oxy]-N,N-dimethyl- Dimethisoquin	$C_{17}H_{24}N_2O$ 272.39	86-80-6 146	3207 155-7³	5-21-03-00414	H_2O 3; EtOH 3 1.5486²⁰

No.	Name / Synonym	Mol. Form. / Mol. Wt.	CAS RN / mp/°C	Merck No. / bp/°C	Beil. Ref. / den/g cm^{-3}	Solubility / n_D
5412	Ethanamine, 2-chloro-N,N-bis(2-chloroethyl)- 2,2',2"-Trichlorotriethylamine	$C_6H_{12}Cl_3N$ 204.53	555-77-1 -4	9560 143-4[15]	4-04-00-00447	bz 4; eth 4; EtOH 4
5413	Ethanamine, 2-chloro-N-(2-chloroethyl)-N-methyl- Mechlorethamine	$C_5H_{11}Cl_2N$ 156.05	51-75-2 -60	5655 87[18]; 64[5]		H_2O 2; ctc 5; DMF 5
5414	Ethanamine, 2,2-diethoxy-	$C_6H_{15}NO_2$ 133.19	645-36-3 -78	163	4-04-00-01918 0.9159[25]	H_2O 4; eth 4; EtOH 4; chl 4 1.4123[25]
5415	Ethanamine, 2,2-diethoxy-N,N-diethyl-	$C_{10}H_{23}NO_2$ 189.30	3616-57-7	194.5	4-04-00-01919 0.863[25]	H_2O 3; EtOH 3; eth 3; ctc 2 1.4189[20]
5416	Ethanamine, 2,2-diethoxy-N,N-dimethyl-	$C_8H_{19}NO_2$ 161.24	3616-56-6	170.5	3-04-00-00870 0.885[7]	H_2O 4; EtOH 3; eth 3; ace 3 1.4129[20]
5417	Ethanamine, N,N-diethyl- Triethylamine	$C_6H_{15}N$ 101.19	121-44-8 -114.7	9582 89	4-04-00-00322 0.7275[20]	H_2O 3; EtOH 3; eth 3; ace 4 1.4010[20]
5418	Ethanamine, N,N-diethyl-, hydrochloride	$C_6H_{16}ClN$ 137.65	554-68-7 260 dec	sub 245	4-04-00-00322 1.0689[21]	H_2O 4; EtOH 4; chl 4
5419	Ethanamine, 2,2-dimethoxy-	$C_4H_{11}NO_2$ 105.14	22483-09-6	135-9[95]	4-04-00-01918 0.965[25]	1.4170[20]
5420	Ethanamine, 2,2-dimethoxy-N-methyl-	$C_5H_{13}NO_2$ 119.16	122-07-6	140	2-04-00-00759 0.928[25]	1.4115[20]
5421	Ethanamine, N,N-dimethyl-	$C_4H_{11}N$ 73.14	598-56-1 -140	36.5	4-04-00-00312 0.675[20]	1.3705[25]
5422	Ethanamine, 2-(diphenylmethoxy)-N,N-dimethyl- Diphenhydramine	$C_{17}H_{21}NO$ 255.36	58-73-1	3308	4-06-00-04658	
5423	Ethanamine, 2-ethoxy-	$C_4H_{11}NO$ 89.14	110-76-9	108	4-04-00-01411 0.8512[20]	H_2O 5; EtOH 5; eth 5; ace 3 1.4101[20]
5424	Ethanamine, 2-ethoxy-N,N-dimethyl-	$C_6H_{15}NO$ 117.19	26311-17-1	121	4-04-00-01425 0.806[20]	1.406[20]
5425	Ethanamine, 2-ethoxy-N-(2-ethoxyethyl)- Bis-(2-ethoxyethyl) amine	$C_8H_{19}NO_2$ 161.24	124-21-0 -50	194	4-04-00-01516 0.883[25]	1.4213[20]
5426	Ethanamine, 2-ethoxy-N-methyl- Methyl(2-ethoxyethyl)amine	$C_5H_{13}NO$ 103.16	38256-94-9	115	3-04-00-00647 0.8363[20]	ace 4; bz 4; eth 4; EtOH 4 1.4147[20]
5427	Ethanamine, N-ethyl- Diethylamine	$C_4H_{11}N$ 73.14	109-89-7 -49.8	3100 55.5	4-04-00-00313 0.7056[20]	H_2O 4; EtOH 5; eth 3; ctc 3 1.3864[20]
5428	Ethanamine, N-ethyl-, hydrochloride	$C_4H_{12}ClN$ 109.60	660-68-4 228.5		4-04-00-00313 1.0477[22]	H_2O 4; EtOH 4
5429	Ethanamine, N-ethyl-N-hydroxy-	$C_4H_{11}NO$ 89.14	3710-84-7 10	133	4-04-00-03304 0.8669[20]	1.4195[20]
5430	Ethanamine, N-ethyl-N-methyl-	$C_5H_{13}N$ 87.16	616-39-7 -196	66	4-04-00-00321 0.703[25]	H_2O 4; EtOH 4; eth 4 1.3879[25]
5431	Ethanamine, N-ethyl-N-nitro-	$C_4H_{10}N_2O_2$ 118.14	7119-92-8	206.5	4-04-00-03403 1.057[15]	eth 4; EtOH 4
5432	Ethanamine, N-ethyl-N-nitroso- N-Nitrosodiethylamine	$C_4H_{10}N_2O$ 102.14	55-18-5	6557 176.9	4-04-00-03386 0.9422[20]	H_2O 3; EtOH 3; eth 3; chl 2 1.4386[20]
5433	Ethanamine, hydrochloride	C_2H_8ClN 81.54	557-66-4 109.5	3718	4-04-00-00307 1.2160[20]	H_2O 4; EtOH 4
5434	Ethanamine, N-hydroxy- N-Ethylhydroxylamine	C_2H_7NO 61.08	624-81-7 59-60 dec		4-04-00-03304 0.9079[20]	H_2O 4; EtOH 4 1.4152[66]
5435	Ethanamine, N-methyl-, hydrochloride Ethyl methyl amine, hydrochloride	$C_3H_{10}ClN$ 95.57	624-60-2 128		2-04-00-00589 1.0874[20]	H_2O 4; EtOH 4; eth 1; chl 3
5436	Ethanamine, 1,1,2,2,2-pentafluoro-N,N-bis(pentafluoroethyl)-	$C_6F_{15}N$ 371.05	359-70-6	70.3	4-02-00-00471 1.736[20]	1.262[25]
5437	Ethanamine, N-(phenylmethylene)-	$C_9H_{11}N$ 133.19	6852-54-6	196	4-07-00-00518 0.937[20]	H_2O 1; EtOH 3; eth 3 1.5378[15]
5438	Ethanaminium, 2-(acetyloxy)-N,N,N-trimethyl-, bromide Acetyl choline bromide	$C_7H_{16}BrNO_2$ 226.11	66-23-9 146	80	4-04-00-01446	H_2O 4
5439	Ethanaminium, 2-(acetyloxy)-N,N,N-trimethyl-, chloride Acetyl choline chloride	$C_7H_{16}ClNO_2$ 181.66	60-31-1 150	81	4-04-00-01446	H_2O 3; EtOH 3; eth 1
5440	Ethanaminium, 2-[(aminocarbonyl)oxy]-N,N,N-trimethyl-, chloride Carbamoylcholine chloride	$C_6H_{15}ClN_2O_2$ 182.65	51-83-2 210 dec	1780	4-04-00-01455	H_2O 4; EtOH 2; eth 1; chl 1
5441	Ethanaminium, 2,2',2"-[1,2,3-benzenetriyltris(oxy)]tris[N,N,N-triethyl-, tr Gallamine triethiodide	$C_{30}H_{60}I_3N_3O_3$ 891.54	65-29-2 147.5	4249	4-06-00-07333	H_2O 4; EtOH 4; eth 2; ace 2
5442	Ethanaminium, 2-chloro-N,N,N-trimethyl-, chloride	$C_5H_{13}Cl_2N$ 158.07	999-81-5 239 dec	2103	4-04-00-00445	
5443	Ethanaminium, 2-[(cyclohexylhydroxyphenylacetyl)oxy]-N,N-diethyl-N-methyl-, bromide Oxyphenonium bromide	$C_{21}H_{34}BrNO_3$ 428.41	50-10-2 191.5	6928	3-10-00-00911	H_2O 4; EtOH 2
5444	Ethanaminium, N-[4-[[4-(diethylamino)phenyl]phenylmethylene]-2,5-cyclohexadien-1-ylidene]-N-ethyl-, sulfate (1 Brilliant green	$C_{27}H_{34}N_2O_4S$ 482.64	633-03-4	1367	4-13-00-02281	H_2O 4; EtOH 4
5445	Ethanaminium, N,N-diethyl-N-methyl-2-[(9H-xanthen-9-ylcarbonyl)oxy]-, bromide Methantheline bromide	$C_{21}H_{26}BrNO_3$ 420.35	53-46-3 174.5	5869	5-18-06-00590	H_2O 3; EtOH 3; eth 1; chl 3

No.	Name / Synonym	Mol. Form. / Mol. Wt.	CAS RN / mp/°C	Merck No. / bp/°C	Beil. Ref. / den/g cm⁻³	Solubility / n_D
5446	Ethanaminium, 2-hydroxy-N,N,N-trimethyl-, chloride Choline chloride	$C_5H_{14}ClNO$ 139.63	67-48-1 305-5 dec	2208	4-04-00-01443	H_2O 4; EtOH 4
5447	Ethanaminium, N,N,N-triethyl-, bromide Tetraethylammonium bromide	$C_8H_{20}BrN$ 210.16	71-91-0 285-7 dec	9133	4-04-00-00331 1.3970^{20}	H_2O 4; EtOH 4; chl 4; MeOH 4
5448	Ethane	C_2H_6 30.07	74-84-0 -182.8	3676 -88.6	4-01-00-00108 0.5446^{-89}	bz 4
5449	Ethane, 1,2-bis(2-chloroethoxy)-	$C_6H_{12}Cl_2O_2$ 187.07	112-26-5	232	4-01-00-02379 1.195^{20}	ctc 3 1.4592^{25}
5450	Ethane, 1,1-bis(ethylthio)-	$C_6H_{14}S_2$ 150.31	14252-42-7	184	4-01-00-03162 0.9706^{20}	eth 3; ace 3; chl 3 1.5025^{20}
5451	Ethane, 1,2-bis(ethylthio)-	$C_6H_{14}S_2$ 150.31	5395-75-5 180	217; 108^{27}	4-01-00-02452 0.9815^{20}	EtOH 3; eth 3 1.5118^{20}
5452	Ethane, 1,2-bis(methylthio)-	$C_4H_{10}S_2$ 122.26	6628-18-8	183; 79^{11}	4-01-00-02451 1.0371^{20}	H_2O 4; ace 4; eth 4; EtOH 4 1.5292^{20}
5453	Ethanebis(thioic) acid, S,S-diethyl ester	$C_6H_{10}O_2S_2$ 178.28	615-85-0 27	235	1-02-00-00244 1.0565^{21}	eth 4
5454	Ethane, bromo- Ethyl bromide	C_2H_5Br 108.97	74-96-4 -118.6	3730 38.5	4-01-00-00150 1.4604^{20}	H_2O 2; EtOH 5; eth 5; chl 5 1.4239^{20}
5455	Ethane, 1-bromo-1-chloro- 1-Bromo-1-chloroethane	C_2H_4BrCl 143.41	593-96-4	83	4-01-00-00155 1.667^{10}	1.4660^{2-}
5456	Ethane, 1-bromo-2-chloro-	C_2H_4BrCl 143.41	107-04-0 -16.7	107	4-01-00-00155 1.7392^{20}	H_2O 3; EtOH 3; eth 3; chl 3 1.4908^{20}
5457	Ethane, 1-bromo-2-chloro-1,1,2-trifluoro- 1-Bromo-2-chloro-1,1,2-trifluoroethane	$C_2HBrClF_3$ 197.38	354-06-3 52.5		4-01-00-00156 1.8636^{25}	1.3738^{20}
5458	Ethane, 2-bromo-2-chloro-1,1,1-trifluoro- Halothane	$C_2HBrClF_3$ 197.38	151-67-7 50.2; 20^{243}	4517	1.871^{20}	H_2O 2; peth 3 1.3697^0
5459	Ethane, 2-bromo-1,1-diethoxy-	$C_6H_{13}BrO_2$ 197.07	2032-35-1 $66-7^{18}$		4-01-00-03151 1.310^{25}	EtOH 3; eth 3 1.4387^{20}
5460	Ethane, 2-bromo-1,1-dimethoxy-	$C_4H_9BrO_2$ 169.02	7252-83-7 149		3-01-00-02672 1.430^{20}	eth 3; ace 3; chl 3 1.4450^{20}
5461	Ethane, 1-bromo-2-ethoxy- 2-Bromoethyl ethyl ether	C_4H_9BrO 153.02	592-55-2 127.5		4-01-00-01386 1.3852^0	H_2O 2; EtOH 5; eth 5 1.4447^{20}
5462	Ethane, 1-bromo-2-fluoro-	C_2H_4BrF 126.96	762-49-2 71.5		4-01-00-00154 1.7044^{25}	eth 4; EtOH 4 1.4236^{20}
5463	Ethane, 1-bromo-2-iodo- 1-Bromo-2-iodoethane	C_2H_4BrI 234.86	590-16-9 28	163	4-01-00-00168 2.516^{29}	
5464	Ethane, (bromomethoxy)-	C_3H_7BrO 138.99	53588-92-4	110	3-01-00-02594 1.4402^{20}	eth 4 1.4515^{20}
5465	Ethane, 1-bromo-2-methoxy-	C_3H_7BrO 138.99	6482-24-2	110 dec	4-01-00-01386 1.4623^{20}	1.44753^{20}
5466	Ethane, 2-bromo-1,1,1-trifluoro-	$C_2H_2BrF_3$ 162.94	421-06-7 -93.9	26	4-01-00-00154 1.7881^{20}	1.3331^{20}
5467	Ethane, chloro- Ethyl chloride	C_2H_5Cl 64.51	75-00-3 -138.7	3740 12.3	4-01-00-00124 $*0.8902^{25}$	H_2O 2; EtOH 4; eth 5; chl 2 1.3676^{20}
5468	Ethane, 1-chloro-1-(2-chloroethoxy)- 1-Chloroethyl 2-chloroethyl ether	$C_4H_8Cl_2O$ 143.01	1462-34-6 56^{17}		3-01-00-02655 1.1867^{20}	eth 4; EtOH 4; chl 4 1.4473^{20}
5469	Ethane, 2-chloro-1,1-diethoxy-	$C_6H_{13}ClO_2$ 152.62	621-62-5 157.4		4-01-00-03134 1.0180^{20}	H_2O 2; EtOH 5; eth 5; ctc 2 1.4170^{20}
5470	Ethane, 1-chloro-1,1-difluoro- 1-Chloro-1,1-difluoroethane	$C_2H_3ClF_2$ 100.50	75-68-3 -130.8	-9.7	1.107^{25}	H_2O 1; bz 3
5471	Ethane, 2-chloro-1,1-difluoro- 1-Chloro-2,2-difluoroethane	$C_2H_3ClF_2$ 100.50	338-65-8 35.1			
5472	Ethane, 2-chloro-1,1-dimethoxy-	$C_4H_9ClO_2$ 124.57	97-97-2 127.5		4-01-00-03134 1.068^{20}	EtOH 2; eth 2; bz 2; ctc 2 1.4150^{20}
5473	Ethane, 1-chloro-1-ethoxy-	C_4H_9ClO 108.57	7081-78-9 93.5		4-01-00-03119 0.9655^{20}	1.4053^{20}
5474	Ethane, 1-chloro-2-ethoxy-	C_4H_9ClO 108.57	628-34-2 107.5		4-01-00-01375 0.9895^{20}	H_2O 2; eth 5; chl 3 1.4113^{20}
5475	Ethane, 1-chloro-1-fluoro- 1-Chloro-1-fluoroethane	C_2H_4ClF 82.50	1615-75-4 16.2		4-01-00-00127	
5476	Ethane, 1-chloro-2-fluoro-	C_2H_4ClF 82.50	762-50-5 59		4-01-00-00127 1.1747^{20}	eth 4; EtOH 4 1.3775^{20}
5477	Ethane, 1-chloro-2-iodo-	C_2H_4ClI 190.41	624-70-4 -15.6	140	4-01-00-00167 2.1644^{25}	
5478	Ethane, (chloromethoxy)- Chloromethyl ethyl ether	C_3H_7ClO 94.54	3188-13-4 83		4-01-00-03047 1.0188^{15}	1.4040^{20}
5479	Ethane, 1-chloro-1-methoxy-	C_3H_7ClO 94.54	1538-87-0 73		4-01-00-03119 0.9902^{20}	eth 4 1.4004^{20}
5480	Ethane, 1-chloro-2-methoxy-	C_3H_7ClO 94.54	627-42-9 92.5		4-01-00-01375 1.0345^{20}	H_2O 4; eth 4 1.4111^{20}
5481	Ethane, 1-chloro-2-(methylthio)-	C_3H_7ClS 110.61	542-81-4 60^{30}; 44^{20}		4-01-00-01406 1.1225^{20}	EtOH 3; eth 3; ace 3 1.4902^{20}
5482	Ethane, 1-chloro-1-nitro- 1-Chloro-1-nitroethane	$C_2H_4ClNO_2$ 109.51	598-92-5 124.5		4-01-00-00172 1.2837^{20}	H_2O 1; EtOH 3; ctc 3; alk 3 1.4224^{20}
5483	Ethane, 1-chloro-2-nitro- 2-Chloro-1-nitroethane	$C_2H_4ClNO_2$ 109.51	625-47-8 173.5		4-01-00-00173 1.3550^{20}	1.407^7
5484	Ethane, chloropentafluoro- Chloropentafluoroethane	C_2ClF_5 154.47	76-15-3 -99.4	-37.9	4-01-00-00129 1.5678^{-42}	H_2O 1; EtOH 3; eth 3 1.2678^{-42}

No.	Name / Synonym	Mol. Form. / Mol. Wt.	CAS RN / mp/°C	Merck No. / bp/°C	Bell. Ref. / den/g cm^{-3}	Solubility / n_D
5485	Ethane, 1-chloro-1,1,2,2-tetrafluoro- / 1-Chloro-1,1,2,2-tetrafluoroethane	C_2HClF_4 / 136.48	354-25-6 / -117	-10	4-01-00-00129	
5486	Ethane, 1-chloro-1,2,2,2-tetrafluoro- / 1-Chloro-1,2,2,2-tetrafluoroethane	C_2HClF_4 / 136.48	2837-89-0	-12		
5487	Ethane, 1-chloro-1,1,2-trifluoro / 1-Chloro-1,1,2-trifluoroethane	$C_2H_2ClF_3$ / 118.49	421-04-5	12		
5488	Ethane, 1-chloro-1,2,2-trifluoro / 1-Chloro-1,2,2-trifluoroethane	$C_2H_2ClF_3$ / 118.49	431-07-2	17		
5489	Ethane, 2-chloro-1,1,1-trifluoro- / 2-Chloro-1,1,1-trifluoroethane	$C_2H_2ClF_3$ / 118.49	75-88-7 / -105.5	6.1	4-01-00-00128 / 1.389[0]	1.3090[0]
5490	Ethane, 1-chloro-1,1,2-trifluoro-2-iodo- / Ethane, 1-chloro-2-iodo-1,1,2-trifluoro	C_2HClF_3I / 244.38	354-26-7	77[630]	4-01-00-00167 / 2.181[20]	1.4320[20]
5491	Ethanedial / Glyoxal	$C_2H_2O_2$ / 58.04	107-22-2 / 15	4405 / 50.4	4-01-00-03625 / 1.14[20]	H_2O 4; EtOH 3; eth 3 1.3826[20]
5492	Ethanedial, dioxime	$C_2H_4N_2O_2$ / 88.07	557-30-2 / 178 dec	sub	3-01-00-03084	H_2O 4; EtOH 4; eth 4
5493	Ethanediamide / Oxamide	$C_2H_4N_2O_2$ / 88.07	471-46-5 / 350 dec	6871	4-02-00-01860 / 1.667[20]	H_2O 2; EtOH 2; eth 1
5494	Ethanediamide, N,N'-diethyl-	$C_6H_{12}N_2O_2$ / 144.17	615-84-9 / 175		4-04-00-00359 / 1.169[4]	EtOH 4
5495	Ethanediamide, N,N'-diphenyl-	$C_{14}H_{12}N_2O_2$ / 240.26	620-81-5 / 254	>360	4-12-00-00463	bz 4
5496	1,2-Ethanediamine / Ethylenediamine	$C_2H_8N_2$ / 60.10	107-15-3 / 11.1	3752 / 117	4-04-00-01166 / 0.8979[20]	H_2O 4; EtOH 5; eth 1; bz 1 1.4565[20]
5497	1,2-Ethanediamine, N-(2-aminoethyl)- / Diethylenetriamine	$C_4H_{13}N_3$ / 103.17	111-40-0 / -39	207	4-04-00-01238 / 0.9569[20]	H_2O 5; EtOH 5; eth 1; lig 3 1.4810[25]
5498	1,2-Ethanediamine, N-(2-aminoethyl)-N'-[2-[(2-aminoethyl)amino]ethyl]- / Tetraethylenepentamine	$C_8H_{23}N_5$ / 189.30	112-57-2 / 341.5		4-04-00-01244 / 1.5042[20]	H_2O 3
5499	1,2-Ethanediamine, N,N'-bis(2-aminoethyl)- / Trientine	$C_6H_{18}N_4$ / 146.24	112-24-3 / 12	9579 / 266.5	4-04-00-01242 / 1.4971[20]	H_2O 3; EtOH 3; acid 3
5500	1,2-Ethanediamine, N,N'-bis(phenylmethyl)- / Benzathine	$C_{16}H_{20}N_2$ / 240.35	140-28-3 / 26	1072 / 195[4]	4-12-00-02321 / 1.024[20]	bz 4; eth 4; EtOH 4 1.5635[20]
5501	1,2-Ethanediamine, N,N'-diethyl-	$C_6H_{16}N_2$ / 116.21	111-74-0	146	4-04-00-01174 / 0.8280[20]	H_2O 4; eth 4; EtOH 4; tol 4 1.4340[20]
5502	1,2-Ethanediamine, N,N-diethyl- / N,N-Diethylethylenediamine	$C_6H_{16}N_2$ / 116.21	100-36-7	144	4-04-00-01175 / 0.8280[20]	H_2O 5; EtOH 3; eth 3; ctc 3 1.4340[20]
5503	1,2-Ethanediamine, N,N'-dimethyl-	$C_4H_{12}N_2$ / 88.15	110-70-3	120	4-04-00-01171 / 0.828[15]	EtOH 3; eth 3; dil HCl 3
5504	1,2-Ethanediamine, N,N-dimethyl-N'-(phenylmethyl)-	$C_{11}H_{18}N_2$ / 178.28	103-55-9	122-4[11]	4-12-00-02320 / 0.922[25]	1.5089[20]
5505	1,2-Ethanediamine, N,N-dimethyl-N'-(phenylmethyl)-N'-2-pyridinyl- / Tripelennamine	$C_{16}H_{21}N_3$ / 255.36	91-81-6	9651 / 167-72[0.1]	5-22-08-00378	
5506	1,2-Ethanediamine, N,N-dimethyl-N'-phenyl-N'-(phenylmethyl)- / Phenbenzamine	$C_{17}H_{22}N_2$ / 254.38	961-71-7	7176 / 179-80[7]	4-12-00-02324 / 1.016[25]	1.5794[25]
5507	1,2-Ethanediamine, N,N-dimethyl-N'-2-pyridinyl-N'-(2-thienylmethyl)- / Methapyrilene	$C_{14}H_{19}N_3S$ / 261.39	91-80-5	5871 / 173-5[3]	5-22-08-00401	1.5915[20]
5508	1,2-Ethanediamine, N,N'-diphenyl- / 1,2-Dianilinoethane	$C_{14}H_{16}N_2$ / 212.29	150-61-8 / 74	2969 / 229[12]; 178[2]	4-12-00-00986	H_2O 1; EtOH 3; eth 3; tfa 2
5509	1,2-Ethanediamine, N-ethyl-	$C_4H_{12}N_2$ / 88.15	110-72-5	129	4-04-00-01174 / 0.837[25]	1.4385[20]
5510	1,2-Ethanediamine, N'-ethyl-N,N-dimethyl- / Ethylenediamine, N,N-dimethyl-N'-ethyl	$C_6H_{15}N_2$ / 115.20	123-83-1	134.5	4-04-00-01174 / 0.738[25]	1.4222[20]
5511	1,2-Ethanediamine, N-[(4-methoxyphenyl)methyl]-N',N'-dimethyl-N-2-pyridinyl- / Pyrilamine	$C_{17}H_{23}N_3O$ / 285.39	91-84-9	7996 / 201[5]	5-22-08-00381	
5512	1,2-Ethanediamine, N-methyl-	$C_3H_{10}N_2$ / 74.13	109-81-9	115	4-04-00-01170 / 0.841[25]	1.4395[20]
5513	1,2-Ethanediamine, monohydrate	$C_2H_{10}N_2O$ / 78.11	6780-13-8 / 10	118	4-04-00-01166 / 0.964[20]	H_2O 4 1.4500[20]
5514	1,2-Ethanediamine, N-1-naphthalenyl- / N-(1-Naphthyl)ethylenediamine	$C_{12}H_{14}N_2$ / 186.26	551-09-7	6329 / 204[9]	4-12-00-03101 / 1.114[25]	1.6648[25]
5515	1,2-Ethanediamine, N,N,N',N'-tetraethyl-	$C_{10}H_{24}N_2$ / 172.31	150-77-6	192	4-04-00-01177 / 0.808[25]	1.4343[20]
5516	1,2-Ethanediamine, N,N,N',N'-tetramethyl- / N,N,N',N'-Tetramethyldiaminoethane	$C_6H_{16}N_2$ / 116.21	110-18-9 / -55	121	4-04-00-01172 / 0.77[25]	1.4179[20]
5517	Ethane, 1,1-dibromo- / Ethylidene dibromide	$C_2H_4Br_2$ / 187.86	557-91-5 / -63	109	4-01-00-00157 / 2.0554[20]	H_2O 1; EtOH 3; eth 4; ace 3 1.5128[20]
5518	Ethane, 1,2-dibromo- / Ethylene dibromide	$C_2H_4Br_2$ / 187.86	106-93-4 / 9.9	3753 / 131.6	4-01-00-00158 / 2.1791[20]	ace 4; bz 4; eth 4; EtOH 4 1.5387[20]
5519	Ethane, 1,2-dibromo-1-chloro-1,2,2-trifluoro- / 1,2-Dibromo-1-chloro-1,2,2-trifluoroethane	$C_2Br_2ClF_3$ / 276.28	354-51-8 / 50	93		
5520	Ethane, 1,2-dibromo-1,1-dichloro-	$C_2H_2Br_2Cl_2$ / 256.75	75-81-0 / -26	195	4-01-00-00161 / 2.135[20]	ace 4; bz 4; eth 4; EtOH 4 1.5662[20]
5521	Ethane, 1,2-dibromo-1,2-dichloro-	$C_2H_2Br_2Cl_2$ / 256.75	683-68-1 / -26	195	4-01-00-00161 / 2.135[20]	H_2O 1; EtOH 3; eth 3; ace 3 1.5662[20]

No.	Name / Synonym	Mol. Form. / Mol. Wt.	CAS RN / mp/°C	Merck No. / bp/°C	Beil. Ref. / den/g cm⁻³	Solubility / n_D
5522	Ethane, 1,2-dibromo-1,1-difluoro- / Genetron 132b-B2	$C_2H_2Br_2F_2$ / 223.84	75-82-1 / -61.3	92.5	4-01-00-00160 / 2.2238[20]	1.4456[20]
5523	Ethane, 1,2-dibromo-1-ethoxy-	$C_4H_8Br_2O$ / 231.91	2983-26-8 /	80[20]	4-01-00-03152 / 1.7320[20]	EtOH 4; chl 4 / 1.5044[20]
5524	Ethane, 1,2-dibromo-1,1,2,2-tetrachloro-	$C_2Br_2Cl_4$ / 325.64	630-25-1 / 202.5		3-01-00-00190 / 2.713[25]	
5525	Ethane, 1,2-dibromo-1,1,2,2-tetrafluoro- / 1,2-Dibromotetrafluoroethane	$C_2Br_2F_4$ / 259.82	124-73-2 / -110.4	47.3	4-01-00-00160 / 2.149[25]	1.361[25]
5526	Ethane, 1,2-dibromo-1,1,2-trifluoro- / Halon 2302	$C_2HBr_2F_3$ / 241.83	354-04-1 /	76	4-01-00-00160 / 2.274[27]	1.4191[24]
5527	Ethane, 1,2-dibutoxy- / Ethylene glycol dibutyl ether	$C_{10}H_{22}O_2$ / 174.28	112-48-1 / -69.1	203.3	0.8319[25]	
5528	Ethane, 1,1-dichloro- / Ethylidene dichloride	$C_2H_4Cl_2$ / 98.96	75-34-3 / -96.9	3766 / 57.4	4-01-00-00130 / 1.1757[20]	H_2O 2; EtOH 4; eth 4; ace 3 / 1.4164[20]
5529	Ethane, 1,2-dichloro- / Ethylene dichloride	$C_2H_4Cl_2$ / 98.96	107-06-2 / -35.5	3754 / 83.5	4-01-00-00131 / 1.2351[20]	H_2O 2; EtOH 4; eth 5; ace 3 / 1.4448[20]
5530	Ethane, 1,1-dichloro-2,2-diethoxy-	$C_6H_{12}Cl_2O_2$ / 187.07	619-33-0 /	183.5	4-01-00-03140 / 1.1383[14]	EtOH 3; eth 3; ctc 2
5531	Ethane, 1,2-dichloro-1,1-difluoro-	$C_2H_2Cl_2F_2$ / 134.94	1649-08-7 / -101.2	46.8	4-01-00-00135 / 1.4163[20]	1.36193[20]
5532	Ethane, 1,2-dichloro-1,2-difluoro-	$C_2H_2Cl_2F_2$ / 134.94	431-06-1 / -101.2	46.8	4-01-00-00135 / 1.4163[20]	1.3619[20]
5533	Ethane, 1,2-dichloro-1-ethoxy-	$C_4H_8Cl_2O$ / 143.01	623-46-1 /	145	4-01-00-03136 / 1.1370[20]	chl 2 / 1.4435[20]
5534	Ethane, 1,1-dichloro-1-fluoro- / 1,1-Dichloro-1-fluoroethane	$C_2H_3Cl_2F$ / 116.95	1717-00-6 / -103.5	32	4-01-00-00134 / 1.250[10]	1.3600[10]
5535	Ethane, 1,2-dichloro-1-fluoro-	$C_2H_3Cl_2F$ / 116.95	430-57-9 / -60	73.7	4-01-00-00134 / 1.3814[20]	1.4132[20]
5536	Ethane, 1,1-dichloro-1-nitro- / Ethide	$C_2H_3Cl_2NO_2$ / 143.96	594-72-9 /	123.5	4-01-00-00173 /	ctc 3
5537	Ethane, 1,1-dichloro-1,2,2,2-tetrafluoro- / 1,1-Dichlorotetrafluoroethane	$C_2Cl_2F_4$ / 170.92	374-07-2 / -56.6	4	4-01-00-00136 / *1.455[25]	bz 4; eth 4; EtOH 4 / 1.3092[0]
5538	Ethane, 1,2-dichloro-1,1,2,2-tetrafluoro- / Refrigerant 114	$C_2Cl_2F_4$ / 170.92	76-14-2 / -94	2608 / 3.8	4-01-00-00137 / *1.455[25]	eth 4; EtOH 4 / 1.3092[0]
5539	Ethane, 1,2-dichloro-1,1,2-trifluoro-	$C_2HCl_2F_3$ / 152.93	354-23-4 / -78	28	4-01-00-00136 / 1.50[25]	
5540	Ethane, 2,2-dichloro-1,1,1-trifluoro- / 2,2-Dichloro-1,1,1-trifluoroethane	$C_2HCl_2F_3$ / 152.93	306-83-2 / -107	27.82	4-01-00-00135 / 1.4638[25]	
5541	Ethane, 1,1-diethoxy- / Acetal	$C_6H_{14}O_2$ / 118.18	105-57-7 / -100	31 / 102.2	4-01-00-03103 / 0.8254[20]	H_2O 3; EtOH 5; eth 5; ace 4 / 1.3834[20]
5542	Ethane, 1,2-diethoxy- / Ethylene glycol diethyl ether	$C_6H_{14}O_2$ / 118.18	629-14-1 /	119.4	4-01-00-02379 / 0.8484[20]	ace 4; bz 4; eth 4; EtOH 4 / 1.3860[20]
5543	Ethane, 1,1-diethoxy-2-iodo-	$C_6H_{13}IO_2$ / 244.07	51806-20-3 /	132[90;] 100[10]	4-01-00-03156 / 1.489[22]	1.4734[22]
5544	Ethane, 1,1-difluoro-	$C_2H_4F_2$ / 66.05	75-37-6 / -117	-24.9	4-01-00-00120 / *0.896[20]	1.3011[-72]
5545	Ethane, 1,2-difluoro- / Freon 152	$C_2H_4F_2$ / 66.05	624-72-6 /	30.7	4-01-00-00121 /	bz 4; eth 4; chl 4
5546	Ethane, 1,1-diiodo- / 1,1-Diiodoethane	$C_2H_4I_2$ / 281.86	594-02-5 /	179.5	4-01-00-00169 / 2.84[25]	ace 4; eth 4; EtOH 4; chl 4 / 1.673[20]
5547	Ethane, 1,2-diiodo-	$C_2H_4I_2$ / 281.86	624-73-7 / 83	200	4-01-00-00169 / 3.325[20]	H_2O 2; EtOH 3; eth 3; ace 3 / 1.871[20]
5548	Ethane, 1,1-dimethoxy- / Dimethyl acetal	$C_4H_{10}O_2$ / 90.12	534-15-6 / -113.2	3215 / 64.5	4-01-00-03103 / 0.8501[20]	H_2O 3; EtOH 3; eth 3; ace 4 / 1.3668[20]
5549	Ethane, 1,2-dimethoxy- / Ethylene glycol dimethyl ether	$C_4H_{10}O_2$ / 90.12	110-71-4 / -69.20	3213 / 85	4-01-00-02376 / 0.8691[20]	H_2O 3; EtOH 3; eth 3; ace 3 / 1.3796[20]
5550	Ethanedinitrile / Cyanogen	C_2N_2 / 52.04	460-19-5 / -27.9	2698 / -21.1	4-02-00-01863 / 0.9537[-21]	H_2O 3; EtOH 3; eth 3
5551	Ethane, 1,1-dinitro- / 1,1-Dinitroethane	$C_2H_4N_2O_4$ / 120.06	600-40-8 /	185.5	4-01-00-00174 / 1.349[24]	H_2O 2; EtOH 3; eth 3
5552	Ethane, 1,2-dinitro- / 1,2-Dinitroethane	$C_2H_4N_2O_4$ / 120.06	7570-26-5 / 39.5	94-6[5]	4-01-00-00175 / 1.4597[20]	eth 4; EtOH 4 / 1.4468[20]
5553	Ethanedioic acid / Oxalic acid	$C_2H_2O_4$ / 90.04	144-62-7 / 189.5 dec	6865 / sub 157	4-02-00-01819 / 1.900[17]	H_2O 3; EtOH 4; eth 2; bz 1
5554	Ethanedioic acid, bis(2-chloroethyl) ester	$C_6H_8Cl_2O_4$ / 215.03	7208-92-6 / 45	150[4]	4-02-00-01849 /	bz 4; EtOH 4
5555	Ethanedioic acid, bis(3-methylbutyl) ester	$C_{12}H_{22}O_4$ / 230.30	2051-00-5 / -9	267.5	4-02-00-01851 / 0.968[11]	eth 4; EtOH 4
5556	Ethanedioic acid, bis(1-methylethyl) ester	$C_8H_{14}O_4$ / 174.20	615-81-6 /	191	4-02-00-01850 / 1.001[20]	eth 4; EtOH 4 / 1.4100[20]
5557	Ethanedioic acid, bis(2-methylpropyl) ester / Diisobutyl oxalate	$C_{10}H_{18}O_4$ / 202.25	2050-61-5 /	229	4-02-00-01850 / 0.9737[20]	ace 4; eth 4; EtOH 4 / 1.4180[20]
5558	Ethanedioic acid, dibutyl ester	$C_{10}H_{18}O_4$ / 202.25	2050-60-4 / -30.5	241; 96[2]	4-02-00-01850 / 0.9873[20]	H_2O 1; EtOH 3; eth 3 / 1.4234[20]
5559	Ethanedioic acid, dicyclohexyl ester / Dicyclohexyl oxalate	$C_{14}H_{22}O_4$ / 254.33	620-82-6 / 43.5	190[73]	3-06-00-00026 /	H_2O 1; EtOH 4; eth 4; MeOH 3
5560	Ethanedioic acid, diethyl ester / Diethyl oxalate	$C_6H_{10}O_4$ / 146.14	95-92-1 / -40.6	3115 / 185.7	4-02-00-01848 / 1.0785[20]	H_2O 2; EtOH 5; eth 5; ace 5 / 1.4101[20]

No.	Name Synonym	Mol. Form. Mol. Wt.	CAS RN mp/°C	Merck No. bp/°C	Bell. Ref. den/g cm^{-3}	Solubility n_D
5561	Ethanedioic acid, dihydrate Oxalic acid dihydrate	$C_2H_6O_6$ 126.07	6153-56-6 101.5		4-02-00-01819 1.653[18]	H_2O 3; EtOH 3; eth 2
5562	Ethanedioic acid, dihydrazide	$C_2H_6N_4O_2$ 118.10	996-98-5 243 dec		4-02-00-01868 1.458[22]	H_2O 3; EtOH 2; eth 2; bz 2
5563	Ethanedioic acid, dimethyl ester Dimethyl oxalate	$C_4H_6O_4$ 118.09	553-90-2 54.3	6020 163.5	4-02-00-01847 1.1716[60]	H_2O 2; EtOH 3; eth 3; ace 3 1.379[82]
5564	Ethanedioic acid, di-2-propenyl ester Diallyl oxalate	$C_8H_{10}O_4$ 170.17	615-99-6 217		4-02-00-01851 1.1582[20]	H_2O 1; EtOH 3; ace 3; bz 3 1.4481[20]
5565	Ethanedioic acid, dipropyl ester	$C_8H_{14}O_4$ 174.20	615-98-5 -44.3	211	4-02-00-01849 1.0188[20]	H_2O 2; EtOH 5; eth 3 1.4158[20]
5566	Ethanedioic acid, disodium salt Oxalic acid, disodium salt	$C_2Na_2O_4$ 134.00	62-76-0 260 dec	8601	4-02-00-01819 2.34[25]	H_2O 2; EtOH 1; eth 1
5567	Ethanedioic acid, ethyl methyl ester	$C_5H_8O_4$ 132.12	615-52-1 173.7		0-02-00-00535 1.1555[0]	eth 4; EtOH 4
5568	Ethanedioic acid, lead (II) salt Lead oxalate	C_2O_4Pb 295.22	814-93-7 300 dec	5294	4-02-00-01819 5.28	H_2O I; EtOH 1; HNO_3 3
5569	1,2-Ethanediol Ethylene glycol	$C_2H_6O_2$ 62.07	107-21-1 -13	3755 197.3	4-01-00-02369 1.1135[20]	H_2O 5; EtOH 5; eth 3; ace 5 1.4318[20]
5570	1,1-Ethanediol, diacetate Ethylidene diacetate	$C_6H_{10}O_4$ 146.14	542-10-9 18.9	3767 169	4-02-00-00282 1.070[25]	eth 4; EtOH 4 1.3985[25]
5571	1,2-Ethanediol, diacetate Ethylene glycol diacetate	$C_6H_{10}O_4$ 146.14	111-55-7 -31	3756 190	4-02-00-00217 1.1043[20]	H_2O 4; EtOH 5; eth 5; ace 5 1.4159[20]
5572	1,2-Ethanediol, dibenzoate	$C_{16}H_{14}O_4$ 270.28	94-49-5 73.5	360 dec	4-09-00-00356	H_2O 1; eth 3; chl 3
5573	1,2-Ethanediol, diformate	$C_4H_6O_4$ 118.09	629-15-2 174		4-02-00-00037 1.193[0]	H_2O 2; EtOH 3; eth 3 1.3580
5574	1,2-Ethanediol, dinitrate Ethylene glycol dinitrate	$C_2H_4N_2O_6$ 152.06	628-96-6 -22.3	198.5	4-01-00-02413 1.4918[20]	eth 4; EtOH 4
5575	1,2-Ethanediol, 1,2-diphenyl-, (R^*,R^*)-(±)-	$C_{14}H_{14}O_2$ 214.26	655-48-1 122.5	>300	4-06-00-06682	H_2O 1; EtOH 4; eth 4; ace 3
5576	1,2-Ethanediol, dipropanoate	$C_8H_{14}O_4$ 174.20	123-80-8 211		4-02-00-00715 1.020[15]	H_2O 2; EtOH 5; eth 5; ace 4
5577	1,2-Ethanediol, monoacetate Ethylene glycol, monoacetate	$C_4H_8O_3$ 104.11	542-59-6 188	3757	4-02-00-00214 1.108[15]	H_2O 5; EtOH 5; eth 5
5578	1,2-Ethanediol, monobenzoate	$C_9H_{10}O_3$ 166.18	94-33-7 45	150[10]	4-09-00-00354 1.1101[30]	EtOH 4
5579	1,2-Ethanediol, 1-phenyl- Styrene glycol	$C_8H_{10}O_2$ 138.17	93-56-1 67.5	8831 273	4-06-00-05939	H_2O 4; eth 4; bz 4; EtOH 4
5580	1,2-Ethanediol, 1,1,2,2-tetraphenyl- Benzopinacol	$C_{26}H_{22}O_2$ 366.46	464-72-2 182	1110	4-06-00-07053	H_2O 1; EtOH 2; eth 3; ace 3
5581	1,1-Ethanediol, 2,2,2-tribromo- Acetaldehyde, tribromo, hydrate	$C_2H_3Br_3O_2$ 298.76	507-42-6 53.5 dec	1373	4-01-00-03155 2.5661[40]	eth 4; EtOH 4
5582	1,1-Ethanediol, 2,2,2-trichloro- Chloral hydrate	$C_2H_3Cl_3O_2$ 165.40	302-17-0 57	2061 96 dec	4-01-00-03143 1.9081[20]	H_2O 4; bz 4; eth 4; EtOH 4
5583	Ethanedione, di-2-furanyl-, dioxime α-Furildioxime	$C_{10}H_8N_2O_4$ 220.18	522-27-0 167	4218	4-19-00-02008	EtOH 2; eth 2; bz 2; lig 2
5584	Ethanedione, diphenyl- Benzil	$C_{14}H_{10}O_2$ 210.23	134-81-6 94.8	1087 347	4-07-00-02502 1.084[102]	H_2O 1; EtOH 4; eth 4; ace 3
5585	Ethanedioyl dichloride Oxalyl chloride	$C_2Cl_2O_2$ 126.93	79-37-8 -16	6867 63.5	4-02-00-01853 1.4785[20]	eth 3 1.4316[20]
5586	Ethane, 1,1-diproxy- 1,2-Dipropoxyethane	$C_8H_{18}O_2$ 146.23	18854-56-3			
5587	1,2-Ethanedisulfonic acid Ethylenedisulfonic acid	$C_2H_6O_6S_2$ 190.20	110-04-3 173	3678	4-04-00-00078	Diox 4
5588	Ethanedithioamide Rubeanic acid	$C_2H_4N_2S_2$ 120.20	79-40-3 170 dec	8256	4-02-00-01871	H_2O 2; EtOH 2; con sulf 3
5589	Ethane(dithioic) acid	$C_2H_4S_2$ 92.19	594-03-6 66[85]; 37[15]		4-02-00-00572 1.24[20]	H_2O 3; EtOH 4; eth 4; ace 4
5590	1,2-Ethanedithiol Ethylenedimercaptan	$C_2H_6S_2$ 94.20	540-63-6 -41.2	3679 146	4-01-00-02450 1.234[20]	H_2O 1; EtOH 3; eth 3; ace 3 1.5590[20]
5591	Ethane, 1-ethoxy-2-methoxy- 1-Ethoxy-2-methoxyethane	$C_5H_{12}O_2$ 104.15	5137-45-1 102.1		3-01-00-02078 0.8529[20]	1.3868[20]
5592	Ethane, 1,1'-[ethylidenebis(oxy)]bis[2-chloro-	$C_6H_{12}Cl_2O_2$ 187.07	14689-97-5 195		4-01-00-03104 1.1737[20]	eth 4; EtOH 4 1.4527[20]
5593	Ethane, fluoro- Ethyl fluoride	C_2H_5F 48.06	353-36-6 -143.2	-37.6	4-01-00-00120 0.7182[20]	H_2O 3; EtOH 4; eth 4 1.2656[20]
5594	Ethane, hexabromo- Hexabromoethane	C_2Br_6 503.45	594-73-0 200 dec		3-01-00-00193 3.823[20]	EtOH 2; eth 2; CS_2 2 1.863
5595	Ethane, hexachloro- Hexachloroethane	C_2Cl_6 236.74	67-72-1 187 (triple point)	4601	4-01-00-00148 2.091[20]	H_2O 1; EtOH 4; eth 4; bz 3
5596	Ethane, hexafluoro- Perfluoroethane	C_2F_6 138.01	76-16-4 -100.7	-78.1	4-01-00-00123 1.590[-78]	H_2O 1; EtOH 2; eth 2
5597	Ethane, iodo- Ethyl iodide	C_2H_5I 155.97	75-03-6 -111.1	3769 72.5	4-01-00-00163 1.9358[20]	H_2O 2; EtOH 5; eth 3; chl 3 1.5133[20]
5598	Ethane, isocyanato- Ethyl isocyanate	C_3H_5NO 71.08	109-90-0 60		4-04-00-00402 0.9031[20]	H_2O 1; EtOH 5; eth 5 1.3808[20]
5599	Ethane, isocyano-	C_3H_5N 55.08	624-79-3 <-66 79		4-04-00-00342 0.7402[20]	H_2O 4; EtOH 5; eth 5; ace 3 1.3622[20]
5600	Ethane, isothiocyanato- Ethyl isothiocyanate	C_3H_5NS 87.15	542-85-8 -5.9	3771 131.5	4-04-00-00403 0.9990[20]	H_2O 1; EtOH 5; eth 5 1.5130[20]

No.	Name / Synonym	Mol. Form. / Mol. Wt.	CAS RN / mp/°C	Merck No. / bp/°C	Beil. Ref. / den/g cm^{-3}	Solubility / n_D
5601	Ethane, 1,1',1'',1'''-[methanetetrayltetrakis(oxy)]tetrakis-	$C_9H_{20}O_4$ 192.26	78-09-1	159.5	4-03-00-00006 0.9186[20]	EtOH 5; eth 5; ctc 3 1.3905[25]
5602	Ethane, methoxy- Ethyl methyl ether	C_3H_8O 60.10	540-67-0 -113	3783 7.4	4-01-00-01314 0.7251[0]	H_2O 3; EtOH 5; eth 5; ace 3 1.3420[4]
5603	Ethane, 1,1'-[methylenebis(oxy)]bis- Diethoxymethane	$C_5H_{12}O_2$ 104.15	462-95-3 -66.5	88	4-01-00-03027 0.8319[20]	H_2O 3; EtOH 5; eth 5; ace 4 1.3748[18]
5604	Ethane, 1,1',1''-[methylidynetris(oxy)]tris- Ethyl orthoformate	$C_7H_{16}O_3$ 148.20	122-51-0	143; 60[20]	4-02-00-00025 0.8909[20]	EtOH 3; eth 3 1.3922[20]
5605	Ethane, 1,1',1''-[methylidynetris(thio)]tris-	$C_7H_{16}S_3$ 196.40	6267-24-9	235 dec; 127[12]	4-02-00-00093 1.053[20]	eth 4; EtOH 4 1.5410[15]
5606	Ethane, (methylthio)- Ethyl methyl sulfide	C_3H_8S 76.16	624-89-5 -105.9	66.7	4-01-00-01392 0.8422[20]	H_2O 1; EtOH 5; eth 3; chl 3 1.4404[20]
5607	Ethane, nitro- Nitroethane	$C_2H_5NO_2$ 75.07	79-24-3 -89.5	6518 114.0	4-01-00-00170 1.0448[25]	H_2O 2; EtOH 5; eth 5; ace 3 1.3917[20]
5608	Ethane, 1,1'-oxybis- Diethyl ether	$C_4H_{10}O$ 74.12	60-29-7 -116.3	3762 34.5	4-01-00-01314 0.7138[20]	H_2O 2; EtOH 5; eth 5; ace 4 1.3526[20]
5609	Ethane, 1,1'-oxybis[2-bromo-	$C_4H_8Br_2O$ 231.91	5414-19-7	115[32]	4-01-00-01386 1.8222[27]	1.5131[27]
5610	Ethane, 1,1'-oxybis[1-chloro-	$C_4H_8Cl_2O$ 143.01	6986-48-7	116.5	4-01-00-03120 1.1060[25]	eth 4; EtOH 4; chl 4 1.4186[25]
5611	Ethane, 1,1'-oxybis[2-chloro- Bis(2-chloroethyl) ether	$C_4H_8Cl_2O$ 143.01	111-44-4 -51.9	3055 178.5	4-01-00-01375 1.22[20]	H_2O 1; EtOH 3; eth 3; ace 3 1.451[20]
5612	Ethane, 1,1'-oxybis[2-ethoxy- Diethylene glycol diethyl ether	$C_8H_{18}O_3$ 162.23	112-36-7 -45	3108 188	4-01-00-02394 0.9063[20]	H_2O 4; EtOH 4; eth 3; os 4 1.4115[20]
5613	Ethane, 1,1'-oxybis[2-methoxy- Diethylene glycol dimethyl ether	$C_6H_{14}O_3$ 134.18	111-96-6 -68	3148 162	4-01-00-02393 0.9434[20]	H_2O 5; EtOH 5; eth 5 1.4097[20]
5614	Ethane, 1,1'-[oxybis(methyleneoxy)]bis- Bis(ethoxymethyl)ether	$C_6H_{14}O_3$ 134.18	5648-29-3	140.6		
5615	Ethane, 1,1'-oxybis[1,2,2,2-tetrachloro- Bis(1,2,2,2-tetrachloroethyl) ether	$C_4H_2Cl_8O$ 349.68	41284-12-2 40.2	188; 130[11]	2-01-00-00681	bz 4; peth 4
5616	Ethane, pentabromo- 1,1,1,2,2-Pentabromoethane	C_2HBr_5 424.55	75-95-6 56.5	210[300]	3-01-00-00193 3.312[20]	H_2O 1; EtOH 3; eth 4; chl 3
5617	Ethane, pentachloro- Pentachloroethane	C_2HCl_5 202.29	76-01-7 -29	7058 159.8	4-01-00-00147 1.6796[20]	H_2O 1; EtOH 5; eth 5 1.5025[20]
5618	Ethane, pentachlorofluoro-	C_2Cl_5F 220.28	354-56-3 101.3	135	4-01-00-00148 1.74[25]	H_2O 1; EtOH 3; eth 3
5619	Ethaneperoxoic acid Peroxyacetic acid	$C_2H_4O_3$ 76.05	79-21-0 -0.2	7107 110	4-02-00-00390 1.226[15]	H_2O 4; EtOH 4; eth 4; sulf 4 1.3974[20]
5620	Ethane, 1,1'-selenobis-	$C_4H_{10}Se$ 137.08	627-53-2 55	108	4-01-00-01411 1.2300[20]	1.4768[20]
5621	Ethane, 1,1'-sulfinylbis-	$C_4H_{10}OS$ 106.19	70-29-1 14	104[25]; 90[15]	4-01-00-01395 1.0092[22]	H_2O 4; eth 4; EtOH 4
5622	Ethanesulfonic acid Ethylsulfonic acid	$C_2H_6O_3S$ 110.13	594-45-6 -17	123[1]	4-04-00-00033 1.3341[25]	H_2O 4; EtOH 4 1.4335[20]
5623	Ethanesulfonic acid, 2-amino- Taurine	$C_2H_7NO_3S$ 125.15	107-35-7 328	9043	4-04-00-03289	H_2O 4
5624	Ethanesulfonic acid, 2-(methylamino)- N-Methyltaurine	$C_3H_9NO_3S$ 139.18	107-68-6 241.5	6043	3-04-00-01699	H_2O 4; EtOH 1; eth 1
5625	Ethanesulfonic acid, 2-[[(3α,5β,7α,12α)-3,7,12-trihydroxy-24-oxocholan-24-yl]amino]- Taurocholic acid	$C_{28}H_{45}NO_7S$ 515.71	81-24-3 125 dec	9044	4-10-00-02078	H_2O 4; EtOH 4; eth 2; AcOEt 2
5626	Ethane, 1,1'-sulfonylbis-	$C_4H_{10}O_2S$ 122.19	597-35-3 73.5	248	4-01-00-01396 1.357[20]	H_2O 3; eth 3; bz 4; peth 1
5627	Ethanesulfonyl chloride	$C_2H_5ClO_2S$ 128.58	594-44-5	174	4-04-00-00034 1.357[22]	eth 4; CS_2 3 1.4531[20]
5628	Ethanesulfonyl chloride, 2-bromo- 2-Bromoethanesulfonyl chloride	$C_2H_4BrClO_2S$ 207.48	54429-56-0	102[13]	4-04-00-00037 1.921[20]	1.5242[20]
5629	Ethanesulfonyl chloride, 2-chloro-	$C_2H_4Cl_2O_2S$ 163.02	1622-32-8	201.5	4-04-00-00036 1.555[20]	1.4920[20]
5630	Ethane, 1,1'-tellurobis-	$C_4H_{10}Te$ 185.72	627-54-3	137.5	4-01-00-01412 1.599[15]	EtOH 4 1.5182[15]
5631	Ethane, 1,1,1,2-tetrabromo- 1,2,2,2-Tetrabromoethane	$C_2H_2Br_4$ 345.65	630-16-0 0	112[18] dec	4-01-00-00162 2.8747[20]	ace 4; bz 4; eth 4; EtOH 4 1.6277[20]
5632	Ethane, 1,1,2,2-tetrabromo- Acetylene tetrabromide	$C_2H_2Br_4$ 345.65	79-27-6 0	9121 243.5; 151[54]	4-01-00-00162 2.9655[20]	H_2O 1; EtOH 5; eth 5; ace 3 1.6353[20]
5633	1,1,2,2-Ethanetetracarboxylic acid, tetraethyl ester	$C_{14}H_{22}O_8$ 318.32	632-56-4 77	305 dec	4-02-00-02415 1.064[80]	EtOH 3 1.4105[80]
5634	Ethane, 1,1,1,2-tetrachloro- 1,1,1,2-Tetrachloroethane	$C_2H_2Cl_4$ 167.85	630-20-6 -70.2	130.5	4-01-00-00143 1.5406[20]	H_2O 2; EtOH 5; eth 5; ace 3 1.4821[20]
5635	Ethane, 1,1,2,2-tetrachloro- Acetylene tetrachloride	$C_2H_2Cl_4$ 167.85	79-34-5 -43.8	9125 146.5	4-01-00-00144 1.5953[20]	H_2O 2; EtOH 5; eth 5; ace 3 1.4940[20]
5636	Ethane, 1,1,1,2-tetrachloro-2,2-difluoro- 1,1-Difluoroperchloroethane	$C_2Cl_4F_2$ 203.83	76-11-9 40.6	91.5	4-01-00-00146 1.649[25]	H_2O 1; EtOH 3; eth 3; chl 3
5637	Ethane, 1,1,2,2-tetrachloro-1,2-difluoro- Tetrachloro-1,2-difluoroethane	$C_2Cl_4F_2$ 203.83	76-12-0 26	93	4-01-00-00146 1.6447[25]	H_2O 1; EtOH 3; eth 3; chl 3 1.4130[25]
5638	Ethane, 1,1,2,2-tetrachloro-1-fluoro-	C_2HCl_4F 185.84	354-14-3 -82.6	116	4-01-00-00145 1.5497[17]	1.4390[20]

No.	Name / Synonym	Mol. Form. / Mol. Wt.	CAS RN / mp/°C	Merck No. / bp/°C	Beil. Ref. / den/g cm^{-3}	Solubility / n_D
5639	Ethane, 1,1,1,2-tetrachloro-2-fluoro- / 1,1,1,2-Tetrachloro-2-fluoroethane	C$_2$HCl$_4$F / 185.84	354-11-0 / -95.3	116.5		
5640	Ethane, 1,1,1,2-tetrafluoro- / 1,1,1,2-Tetrafluoroethane	C$_2$H$_2$F$_4$ / 102.03	811-97-2 / -103.3	-26.08	4-01-00-00123 / 1.2072[25]	H$_2$O 1; eth 3
5641	Ethane, 1,1,2,2-tetrafluoro- / 1,1,2,2-Tetrafluoroethane	C$_2$H$_2$F$_4$ / 102.03	359-35-3 / -89	-19.9		
5642	Ethane, 1,1,2,2-tetrafluoro-1,2-dinitro- / Ethane, 1,2-dinitro-1,1,2,2-tetrafluoro	C$_2$F$_4$N$_2$O$_4$ / 192.03	356-16-1 / -41.5	58.5	3-01-00-00203 / 1.6024[25]	H$_2$O 1; ace 3 / 1.3265[25]
5643	Ethanethioamide / Thioacetamide	C$_2$H$_5$NS / 75.13	62-55-5 / 115.5	9246	4-02-00-00565	H$_2$O 4; EtOH 4; eth 2; bz 2
5644	Ethane, 1,1'-thiobis- / Diethyl sulfide	C$_4$H$_{10}$S / 90.19	352-93-2 / -103.9	3809 / 92.1	4-01-00-01394 / 0.8362[20]	H$_2$O 2; EtOH 3; eth 3; ctc 2 / 1.4430[20]
5645	Ethane, 1,1'-thiobis[2-chloro- / Mustard gas	C$_4$H$_8$Cl$_2$S / 159.08	505-60-2 / 13.5	6225 / 216	4-01-00-01407 / 1.2741[20]	1.5313[20]
5646	Ethanethioic acid / Thioacetic acid	C$_2$H$_4$OS / 76.12	507-09-5 / <-17	9247 / 93; 26[35]	4-02-00-00542 / 1.064[20]	H$_2$O 3; EtOH 4; eth 5; ace 4 / 1.4648[20]
5647	Ethanethioic acid, S-ethyl ester	C$_4$H$_8$OS / 104.17	625-60-5	116.4	4-02-00-00543 / 0.9792[20]	H$_2$O 1; EtOH 4; eth 4 / 1.4583[21]
5648	Ethanethiol / Ethyl mercaptan	C$_2$H$_6$S / 62.14	75-08-1 / -147.8	3680 / 35.1	4-01-00-01390 / 0.8315[25]	H$_2$O 2; EtOH 3; eth 3; ace 3 / 1.4310[20]
5649	Ethanethiol, 2-amino- / Cysteamine	C$_2$H$_7$NS / 77.15	60-23-1 / 99.5	2785 / dec	4-04-00-01570	H$_2$O 4; EtOH 4
5650	Ethanethiol, 2-(butylamino)- / 2-(Butylamino)ethanethiol	C$_6$H$_{15}$NS / 133.26	5842-00-2	112-5[10]	4-04-00-01602 / 0.901[25]	1.4711[20]
5651	Ethanethiol, 2-chloro- / 2-Chloroethanethiol	C$_2$H$_5$ClS / 96.58	4325-97-7	113	4-01-00-01406 / 1.1826[20]	eth 4; EtOH 4; Diox 4 / 1.4929[20]
5652	Ethanethiol, 2,2'-thiobis-	C$_4$H$_{10}$S$_3$ / 154.32	3570-55-6	135-6[10]	4-01-00-02455 / 1.183[25]	1.5982[20]
5653	Ethane, 1,1,2-tribromo- / 1,1,2-Tribromoethane	C$_2$H$_3$Br$_3$ / 266.76	78-74-0 / -29.3	188.93	4-01-00-00161 / 2.6210[20]	H$_2$O 1; EtOH 3; eth 3; bz 3 / 1.5933[20]
5654	1,1,2-Ethanetricarboxylic acid, triethyl ester	C$_{11}$H$_{18}$O$_6$ / 246.26	7459-46-3	99[05]	4-02-00-02364 / 1.074[25]	1.4290[20]
5655	Ethane, 1,1,1-trichloro- / Methylchloroform	C$_2$H$_3$Cl$_3$ / 133.40	71-55-6 / -30.4	9549 / 74.0	4-01-00-00138 / 1.3390[20]	H$_2$O 2; EtOH 3; eth 5; chl 3 / 1.4379[20]
5656	Ethane, 1,1,2-trichloro- / 1,1,2-Trichloroethane	C$_2$H$_3$Cl$_3$ / 133.40	79-00-5 / -36.6	9550 / 113.8	4-01-00-00139 / 1.4397[20]	H$_2$O 1; EtOH 3; eth 3; chl 3 / 1.4714[20]
5657	Ethane, 1,1,1-trichloro-2,2-diethoxy- / Chloral, diethyl acetal	C$_6$H$_{11}$Cl$_3$O$_2$ / 221.51	599-97-3 / 197		4-01-00-03144 / 1.266[20]	H$_2$O 2; EtOH 5; eth 5; glycerol 5 / 1.4586[25]
5658	Ethane, 1,1,1-trichloro-2,2-difluoro / 1,1,1-Trichloro-2,2-difluoroethane	C$_2$HCl$_3$F$_2$ / 169.38	354-12-1	73		
5659	Ethane, 1,2,2-trichloro-1,1-difluoro-	C$_2$HCl$_3$F$_2$ / 169.38	354-21-2	71.9	4-01-00-00141 / 1.5447[20]	1.3889[20]
5660	Ethane, 1,2,2-trichloro-1,2-difluoro- / 1,2,2-Trichloro-1,2-difluoroethane	C$_2$HCl$_3$F$_2$ / 169.38	354-15-4	72.5		
5661	Ethane, 1,1,1-trichloro-2-fluoro-	C$_2$H$_2$Cl$_3$F / 151.39	359-28-4	102	4-01-00-00141 / 1.5393[20]	H$_2$O 1 / 1.4390[20]
5662	Ethane, 1,1,1-trichloro-2,2,2-trifluoro- / 1,1,1-Trichlorotrifluoroethane	C$_2$Cl$_3$F$_3$ / 187.38	354-58-5 / 14.2	46.1	4-01-00-00142 / 1.5790[20]	H$_2$O 1; EtOH 3; eth 3; chl 3 / 1.3610[35]
5663	Ethane, 1,1,2-trichloro-1,2,2-trifluoro- / 1,1,2-Trichlorotrifluoroethane	C$_2$Cl$_3$F$_3$ / 187.38	76-13-1 / -35	47.7	4-01-00-00142 / 1.5635[25]	H$_2$O 1; EtOH 3; eth 5; bz 5 / 1.3557[25]
5664	Ethane, 1,1,1-triethoxy-	C$_8$H$_{18}$O$_3$ / 162.23	78-39-7	145	4-02-00-00137 / 0.8847[25]	H$_2$O 1; EtOH 5; eth 5; ctc 5 / 1.3980[20]
5665	Ethane, 1,1,1-trifluoro- / Methylfluoroform	C$_2$H$_3$F$_3$ / 84.04	420-46-2 / -111.3	-47.5	4-01-00-00122	eth 3; chl 3
5666	Ethane, 1,1,2-trifluoro- / 1,1,2-Trifluoroethane	C$_2$H$_3$F$_3$ / 84.04	430-66-0 / -84	5	4-01-00-00123	
5667	Ethane, 1,1,1-trifluoro-2-iodo-	C$_2$H$_2$F$_3$I / 209.94	353-83-3	54.5	4-01-00-00166 / 2.13[25]	1.4009[20]
5668	Ethane, 1,1,1-trifluoro-2-nitro-	C$_2$H$_2$F$_3$NO$_2$ / 129.04	819-07-8	96	4-01-00-00172 / 1.3914[20]	eth 4 / 1.3394[20]
5669	Ethane, 1,1,1-trimethoxy-	C$_5$H$_{12}$O$_3$ / 120.15	1445-45-0	108	4-02-00-00127 / 0.9438[25]	eth 4; EtOH 4 / 1.3859[25]
5670	Ethanimidamide, N,N'-bis(4-ethoxyphenyl)-, monohydrochloride / Holocaine, hydrochloride	C$_{18}$H$_{23}$ClN$_2$O$_2$ / 334.85	620-99-5 / 191	7153	3-13-00-01069	H$_2$O 4; EtOH 4; chl 4
5671	Ethanimidamide, monohydrochloride / Acetamidine, hydrochloride	C$_2$H$_7$ClN$_2$ / 94.54	124-42-5 / 177.5	37	4-02-00-00428	H$_2$O 4; EtOH 4
5672	Ethanol / Ethyl alcohol	C$_2$H$_6$O / 46.07	64-17-5 / -114.1	3716 / 78.2	4-01-00-01289 / 0.7893[20]	H$_2$O 5; EtOH 5; eth 5; ace 5 / 1.3611[20]
5673	Ethanol, 2-[2-(acetyloxy)ethoxy]-	C$_6$H$_{12}$O$_4$ / 148.16	2093-20-1	114-5[7]	2-02-00-00155 / 1.1208[20]	1.4320[20]
5674	Ethanol, 1-amino- / Acetaldehyde ammonia	C$_2$H$_7$NO / 61.08	75-39-8 / 97	33 / 110 dec		H$_2$O 3; eth 2
5675	Ethanol, 2-amino- / Ethanolamine	C$_2$H$_7$NO / 61.08	141-43-5 / 10.5	3681 / 171	4-04-00-01406 / 1.0180[20]	H$_2$O 5; EtOH 5; eth 2; bz 2 / 1.4541[20]
5676	Ethanol, 2-[(2-aminoethyl)amino]- / N-(Aminoethyl)ethanolamine	C$_4$H$_{12}$N$_2$O / 104.15	111-41-1	239; 105[10]	4-04-00-01558 / 1.0286[20]	H$_2$O 5; EtOH 5; ace 3; bz 2 / 1.4863[20]
5677	Ethanol, 2-[(2-amino-2-methylpropyl)amino]-	C$_6$H$_{16}$N$_2$O / 132.21	68750-16-3 / 5	236		chl 3 / 1.4698[25]

No.	Name / Synonym	Mol. Form. / Mol. Wt.	CAS RN / mp/°C	Merck No. / bp/°C	Beil. Ref. / den/g cm⁻³	Solubility / n_D
5678	Ethanol, 2-[2-[2-(3-aminopropoxy)ethoxy]ethoxy]-	$C_9H_{21}NO_4$ 207.27	49542-66-7 -50	184[10]	1.0663[20]	H₂O 4; EtOH 4 1.4668[20]
5679	Ethanol, 2,2'-[(3-aminopropyl)imino]bis- N-(3-Aminopropyl)diethanolamine	$C_7H_{18}N_2O_2$ 162.23	4985-85-7	160[1]	3-04-00-00720	
5680	Ethanol, 1-amino-2,2,2-trichloro- Chloral ammonia	$C_2H_4Cl_3NO$ 164.42	507-47-1 73	2057 100 dec	4-01-00-03147	bz 4; eth 4; EtOH 4
5681	Ethanol, 2-azido-	$C_2H_5N_3O$ 87.08	1517-05-1	75[40]	4-01-00-01389 1.146[24]	H₂O 4
5682	Ethanol, 2-[bis(3-methylbutyl)amino]- Diisopentyl 2-hydroxyethylamine	$C_{12}H_{27}NO$ 201.35	3574-43-4	248	3-04-00-00685 0.8492[20]	eth 4; EtOH 4; chl 4 1.4435[20]
5683	Ethanol, 2-[bis(1-methylethyl)amino]- N,N-Diisopropyl-2-aminoethanol	$C_8H_{19}NO$ 145.25	96-80-0	190	4-04-00-01504 0.826[25]	1.4417[20]
5684	Ethanol, 2-[bis(2-methylpropyl)amino]- Ethanolamine, N,N-diisobutyl	$C_{10}H_{23}NO$ 173.30	4535-66-4	214	4-04-00-01509 0.8407[20]	eth 4; EtOH 4; lig 4; chl 4 1.4355[20]
5685	Ethanol, 2-bromo- Ethylene bromohydrin	C_2H_5BrO 124.97	540-51-2	3749 150; 51[4]	4-01-00-01385 1.7629[20]	H₂O 5; EtOH 5; eth 5; os 4 1.4915[20]
5686	Ethanol, 2-bromo-, acetate	$C_4H_7BrO_2$ 167.00	927-68-4 -13.8	162.5	4-02-00-00136 1.514[20]	H₂O 4; EtOH 5; eth 5; chl 4 1.457[23]
5687	Ethanol, 2-butoxy- Ethylene glycol monobutyl ether	$C_6H_{14}O_2$ 118.18	111-76-2 -74.8	1559 168.4	4-01-00-02380 0.9015[20]	H₂O 5; EtOH 5; eth 5; ctc 2 1.4198[20]
5688	Ethanol, 2-(2-butoxyethoxy)- Diethylene glycol monobutyl ether	$C_8H_{18}O_3$ 162.23	112-34-5 -68	1557 231	4-01-00-02394 0.9553[20]	H₂O 5; EtOH 4; eth 4; ace 4 1.4306[20]
5689	Ethanol, 2-(2-butoxyethoxy)-, acetate	$C_{10}H_{20}O_4$ 204.27	124-17-4 -32	245	3-02-00-00308 0.985[20]	ace 4; eth 4; EtOH 4 1.4262[20]
5690	Ethanol, 2-[2-(2-butoxyethoxy)ethoxy]-	$C_{10}H_{22}O_4$ 206.28	143-22-6	278	4-01-00-02402 0.9890[20]	EtOH 4; MeOH 4 1.4389[20]
5691	Ethanol, 2-(butylamino)-	$C_6H_{15}NO$ 117.19	111-75-1	199; 91-2[11]	3-04-00-00682 0.8907[20]	H₂O 4; EtOH 4; eth 4 1.4437[20]
5692	Ethanol, 2,2'-(butylimino)bis-	$C_8H_{19}NO_2$ 161.24	102-79-4	275; 80[35]	4-04-00-01520 0.9592[20]	chl 3 1.4625[20]
5693	Ethanol, 2-chloro- Ethylene chlorohydrin	C_2H_5ClO 80.51	107-07-3 -67.5	3750 128.6	4-01-00-01372 1.2019[20]	H₂O 5; EtOH 5; eth 2; chl 3 1.4419[20]
5694	Ethanol, 1-chloro-, acetate	$C_4H_7ClO_2$ 122.55	5912-58-3	123	4-02-00-00282 1.110[20]	eth 4 1.409[20]
5695	Ethanol, 2-chloro-, acetate β-Chloroethyl acetate	$C_4H_7ClO_2$ 122.55	542-58-5	2137 145	4-02-00-00136 1.178[20]	H₂O 1; EtOH 5; eth 5; ctc 3 1.4234[20]
5696	Ethanol, 2-chloro-, benzoate	$C_9H_9ClO_2$ 184.62	939-55-9 118	255	4-09-00-00288 1.1789[20]	eth 4; EtOH 4
5697	Ethanol, 2-chloro-, carbonate (2:1) Bis(2-chloroethyl) carbonate	$C_5H_8Cl_2O_3$ 187.02	623-97-2 8	241	4-03-00-00006 1.3506[20]	H₂O 1 1.461[20]
5698	Ethanol, 2-(2-chloroethoxy)-	$C_4H_9ClO_2$ 124.57	628-89-7	79-81[5]	3-01-00-02078 1.18[25]	H₂O 4; EtOH 5; eth 5 1.4529[20]
5699	Ethanol, 2-chloro-, methanesulfonate	$C_3H_7ClO_3S$ 158.61	3570-58-9	125-6[9]	4-04-00-00012 1.39[25]	chl 3 1.4562[20]
5700	Ethanol, 2-[2-[4-[(4-chlorophenyl)phenylmethyl]-1-piperazinyl]ethoxy]- Hydroxyzine	$C_{21}H_{27}ClN_2O_2$ 374.91	68-88-2	4786	5-23-01-00462	
5701	Ethanol, 2-chloro-, phosphate (3:1)	$C_6H_{12}Cl_3O_4P$ 285.49	115-96-8	330; 194[10]	4-01-00-01379 1.39[25]	ctc 3 1.4721[20]
5702	Ethanol, 2-chloro-, phosphite (3:1)	$C_6H_{12}Cl_3O_3P$ 269.49	140-08-9	112-5[2]	4-01-00-01378 1.328[25]	1.4868[20]
5703	Ethanol, 2-(dibutylamino)-	$C_{10}H_{23}NO$ 173.30	102-81-8	114[18]	4-04-00-01506	
5704	Ethanol, 2,2-dichloro- 2,2-Dichloroethanol	$C_2H_4Cl_2O$ 114.96	598-38-9	146	4-01-00-01383 1.4040[25]	H₂O 2; EtOH 3; eth 3; ctc 2 1.4626[25]
5705	Ethanol, 2,2-diethoxy-	$C_6H_{14}O_3$ 134.18	621-63-6	167	4-01-00-03958 0.9[25]	eth 4; EtOH 4 1.4073[20]
5706	Ethanol, 2-(diethylamino)- 2-(Diethylamino)ethanol	$C_6H_{15}NO$ 117.19	100-37-8	3101 163	4-04-00-01471 0.8921[20]	H₂O 5; EtOH 3; eth 3; ace 3 1.4412[20]
5707	Ethanol, 2-[2-(diethylamino)ethoxy]-	$C_8H_{19}NO_2$ 161.24	140-82-9	221.5; 92[7]	4-04-00-01474 0.9421[25]	1.4480[20]
5708	Ethanol, 2,2-difluoro-	$C_2H_4F_2O$ 82.05	359-13-7 -28.2	95.5	4-01-00-01369 1.3084[17]	ace 4; eth 4; EtOH 4 1.3345[11]
5709	Ethanol, 2-(dimethylamino)- Deanol	$C_4H_{11}NO$ 89.14	108-01-0 -59	2834 134	4-04-00-01424 0.8866[20]	H₂O 5; EtOH 5; eth 5; chl 3 1.4300[20]
5710	Ethanol, 2-[(1,1-dimethylethyl)amino]-	$C_6H_{15}NO$ 117.19	4620-70-6 44	176.5; 72[14]	4-04-00-01509 0.8818[20]	
5711	Ethanol, 2,2',2'',2'''-[(4,8-di-1-piperidinylpyrimido[5,4-d]pyrimidine- Dipyridamole	$C_{24}H_{40}N_8O_4$ 504.63	58-32-2 163	3354	4-26-00-03840	
5712	Ethanol, 2-(dipropylamino)-	$C_8H_{19}NO$ 145.25	3238-75-3	196	4-04-00-01502 0.8576[20]	eth 4; EtOH 4 1.4402[20]
5713	Ethanol, 2,2'-[1,2-ethanediylbis(oxy)]bis- Triethylene glycol	$C_6H_{14}O_4$ 150.17	112-27-6 -7	9585 285	4-01-00-02400 1.1274[15]	H₂O 5; EtOH 5; eth 2; bz 5 1.4531[20]
5714	Ethanol, 2,2'-[1,2-ethanediylbis(oxy)]bis-, diacetate Triethylene glycol, diacetate	$C_{10}H_{18}O_6$ 234.25	111-21-7 -50	286	4-02-00-00215 1.1153[20]	H₂O 4; eth 4; EtOH 4
5715	Ethanol, 2,2'-(1,2-ethanediyldiimino)bis-	$C_6H_{16}N_2O_2$ 148.21	4439-20-7 97.5	135-7[1]	4-04-00-01559	H₂O 3

No.	Name / Synonym	Mol. Form. / Mol. Wt.	CAS RN / mp/°C	Merck No. / bp/°C	Beil. Ref. / den/g cm^{-3}	Solubility / n_D
5716	Ethanol, 2-(ethenyloxy)-	$C_4H_8O_2$ / 88.11	764-48-7 /	/ 141.6	4-01-00-02387 / 0.9821[20]	H_2O 3; EtOH 3; eth 3; bz 3 / 1.4564[17]
5717	Ethanol, 2-ethoxy- / Ethylene glycol monoethyl ether	$C_4H_{10}O_2$ / 90.12	110-80-5 / -70	3707 / 135	4-01-00-02377 / 0.9297[20]	H_2O 4; ace 4; eth 4; EtOH 4 / 1.4080[20]
5718	Ethanol, 2-ethoxy-, acetate / Ethylene glycol monoethyl ether acetate	$C_6H_{12}O_3$ / 132.16	111-15-9 / -61.7	3708 / 156.4	4-02-00-00214 / 0.9740[20]	H_2O 4; ace 4; eth 4; EtOH 4 / 1.4054[20]
5719	Ethanol, 2-ethoxy-, carbonate (2:1)	$C_9H_{18}O_5$ / 206.24	2049-74-3 /	/ 245.5	3-03-00-00017 / 1.0439[20]	ace 4; eth 4; EtOH 4 / 1.4227[20]
5720	Ethanol, 2-(2-ethoxyethoxy)- / Diethylene glycol monoethyl ether	$C_6H_{14}O_3$ / 134.18	111-90-0 /	1806 / 196	4-01-00-02393 / 0.9885[20]	H_2O 5; EtOH 5; eth 4; ace 5 / 1.4300[20]
5721	Ethanol, 2-(2-ethoxyethoxy)-, acetate / Diethylene glycol monoethyl ether acetate	$C_8H_{16}O_4$ / 176.21	112-15-2 / -25	/ 218.5	3-02-00-00308 / 1.0096[20]	H_2O 4; ace 4; eth 4; EtOH 4 / 1.4213[20]
5722	Ethanol, 2-(ethylamino)-	$C_4H_{11}NO$ / 89.14	110-73-6 /	/ 169.5	4-04-00-01465 / 0.914[20]	H_2O 4; EtOH 4; eth 4; chl 3 / 1.444[20]
5723	Ethanol, 2,2'-(ethylimino)bis-	$C_6H_{15}NO_2$ / 133.19	139-87-7 / -50	/ 247	3-04-00-00693 / 1.0135[20]	H_2O 4; EtOH 4; eth 2 / 1.4663[20]
5724	Ethanol, 2-[ethyl(3-methylphenyl)amino]-	$C_{11}H_{17}NO$ / 179.26	91-88-3 /	/ 118[1.5]	4-12-00-01819 /	ctc 3 / 1.5540[20]
5725	Ethanol, 2-(ethylsulfonyl)- / Ethylsulfonylethyl alcohol	$C_4H_{10}O_3S$ / 138.19	513-12-2 /	3810 /	4-01-00-02430 /	chl 2
5726	Ethanol, 2-(ethylthio)- / 2-(Ethylthio)ethanol	$C_4H_{10}OS$ / 106.19	110-77-0 / -100	3813 / 184	4-01-00-02430 / 1.0166[20]	H_2O 2; EtOH 3; ace 4 / 1.4867[20]
5727	Ethanol, 2-fluoro- / Ethylene fluorohydrin	C_2H_5FO / 64.06	371-62-0 / -26.4	/ 103.5	4-01-00-01366 / 1.1040[20]	H_2O 5; EtOH 5; eth 5; ace 4 / 1.3647[18]
5728	Ethanol, 2-(hexyloxy)-	$C_8H_{18}O_2$ / 146.23	112-25-4 / -45.1	/ 208	4-01-00-02383 / 0.8878[20]	H_2O 2; EtOH 4; eth 4 / 1.4291[20]
5729	Ethanol, 2-hydrazino- / 2-Hydrazinoethanol	$C_2H_8N_2O$ / 76.10	109-84-2 / -70	4696 / 219; 120[17.5]	4-04-00-03348 / 1.119[25]	H_2O 4; EtOH 4; MeOH 4
5730	Ethanol, 2,2'-iminobis- / Diethanolamine	$C_4H_{11}NO_2$ / 105.14	111-42-2 / 28	3097 / 268.8	4-04-00-01514 / 1.0966[20]	H_2O 4; EtOH 4; eth 2; bz 2 / 1.4776[20]
5731	Ethanol, 2-iodo-	C_2H_5IO / 171.97	624-76-0 /	/ 176 dec	4-01-00-01387 / 2.1967[20]	H_2O 4; eth 4; EtOH 4 / 1.5713[20]
5732	Ethanol, 2-mercapto- / 2-Mercaptoethanol	C_2H_6OS / 78.14	60-24-2 /	5760 / 158; 55[13]	4-01-00-02428 / 1.1143[20]	H_2O 3; EtOH 3; eth 3; bz 3 / 1.4996[20]
5733	Ethanol, 2-methoxy- / Ethylene glycol monomethyl ether	$C_3H_8O_2$ / 76.10	109-86-4 / -85.1	5961 / 124.1	4-01-00-02375 / 0.9647[20]	H_2O 5; EtOH 4; eth 5; ace 3 / 1.4024[20]
5734	Ethanol, methoxy-, acetate / Acetic acid, 2-methoxyethyl ester	$C_5H_{10}O_3$ / 118.13	32718-56-2 /	/ 145; 40[12]	2-02-00-00154 / 1.0074[19]	H_2O 3; EtOH 3; eth 3; ctc 2 / 1.4002[20]
5735	Ethanol, 2-methoxy-, acetate / 2-Methoxyethyl acetate	$C_5H_{10}O_3$ / 118.13	110-49-6 / -70	5962 / 143	4-02-00-00214 / 1.0074[19]	H_2O 4; EtOH 4 / 1.4002[20]
5736	Ethanol, 2-methoxy-, carbonate (2:1) / Bis(2-methoxyethyl) carbonate	$C_7H_{14}O_5$ / 178.19	626-84-6 /	/ 231	3-03-00-00017 / 1.0988[20]	EtOH 3; eth 3; ace 3 / 1.4204[20]
5737	Ethanol, 2-(2-methoxyethoxy)- / 2-(2-Methoxyethoxy)ethanol	$C_5H_{12}O_3$ / 120.15	111-77-3 /	5959 / 193	4-01-00-02392 / 1.035[20]	H_2O 5; EtOH 4; eth 4; ace 5 / 1.4264[20]
5738	Ethanol, 2-(methylamino)- / N-Methyl-2-ethanolamine	C_3H_9NO / 75.11	109-83-1 /	5939 / 158	4-04-00-01422 / 0.937[20]	H_2O 5; EtOH 5; eth 5 / 1.4385[20]
5739	Ethanol, 2-[(3-methylbutyl)amino]- / N-Isopentylethanolamine	$C_7H_{17}NO$ / 131.22	34240-76-1 /	/ 210; 82[5.5]	2-04-00-00728 / 0.8822[20]	H_2O 4; eth 4; EtOH 4 / 1.4447[20]
5740	Ethanol, 2-(1-methylethoxy)- / 2-Isopropoxyethanol	$C_5H_{12}O_2$ / 104.15	109-59-1 /	/ 145	4-01-00-02380 / 0.9030[20]	H_2O 5; EtOH 5; eth 5; ace 3 / 1.4095[20]
5741	Ethanol, 2-[(1-methylethyl)amino]-	$C_5H_{13}NO$ / 103.16	109-56-8 /	/ 173	3-04-00-00681 / 0.8970[20]	H_2O 5; EtOH 5; eth 5 / 1.4395[20]
5742	Ethanol, 2,2'-(methylimino)bis-	$C_5H_{13}NO_2$ / 119.16	105-59-9 / -21	/ 247	4-04-00-01517 / 1.038[25]	H_2O 4 / 1.4685[20]
5743	Ethanol, 2-(methylphenylamino)-	$C_9H_{13}NO$ / 151.21	93-90-3 /	/ 150[14]	4-12-00-00280 / 0.9993[15]	H_2O 3; EtOH 4; eth 4; ace 4
5744	Ethanol, 2-[(2-methylphenyl)amino]-	$C_9H_{13}NO$ / 151.21	136-80-1 /	/ 285.5	4-12-00-01750 / 1.0794[20]	eth 4; EtOH 4 / 1.5675[20]
5745	Ethanol, 2-[(4-methylphenyl)amino]-	$C_9H_{13}NO$ / 151.21	2933-74-6 / 42.5	/ 287	4-12-00-01880 /	bz 4; eth 4; EtOH 4; chl 4
5746	Ethanol, 2-(2-methylpropoxy)-	$C_6H_{14}O_2$ / 118.18	4439-24-1 /	/ 160	4-01-00-02382 / 0.8900[20]	/ 1.4143[20]
5747	Ethanol, 2-[(2-methylpropyl)amino]- / Ethanolamine, N-isobutyl	$C_6H_{15}NO$ / 117.19	17091-40-6 /	/ 199; 90[16]	3-04-00-00683 / 0.8818[20]	H_2O 4; EtOH 4; eth 4 / 1.4402[20]
5748	Ethanol, 2-[(2-methylpropyl)amino]-, 4-aminobenzoate (ester) / Butethamine	$C_{13}H_{20}N_2O_2$ / 236.31	2090-89-3 /	1517 /		
5749	Ethanol, 2-(methylthio)-	C_3H_8OS / 92.16	5271-38-5 /	/ 68-70[20]	4-01-00-02429 / 1.063[20]	H_2O 4; eth 4; EtOH 4 / 1.4861[30]
5750	Ethanol, 2-(2-naphthalenylamino)- / Ethanolamine, N-(β-naphthyl)	$C_{12}H_{13}NO$ / 187.24	36190-77-9 / 52	/ 197-8[3]	4-12-00-03131 /	eth 4
5751	Ethanol, 2,2',2''-nitrilotris- / Triethanolamine	$C_6H_{15}NO_3$ / 149.19	102-71-6 / 20.5	9581 / 335.4	4-04-00-01524 / 1.1242[20]	H_2O 5; EtOH 5; eth 2; bz 2 / 1.4852[20]
5752	Ethanol, 2-nitro- / 2-Nitroethanol	$C_2H_5NO_3$ / 91.07	625-48-9 / -80	/ 194; 102[10]	4-01-00-01388 / 1.270[15]	H_2O 5; EtOH 5; eth 5; bz 1 / 1.4438[19]
5753	Ethanol, 2,2'-oxybis- / Diethylene glycol	$C_4H_{10}O_3$ / 106.12	111-46-6 / -10.4	3109 / 245.8	4-01-00-02390 / 1.1197[15]	H_2O 3; EtOH 3; eth 3; chl 3 / 1.4472[20]
5754	Ethanol, 2,2'-oxybis-, diacetate / Diethylene glycol, diacetate	$C_8H_{14}O_5$ / 190.20	628-68-2 / 18	/ 200	4-02-00-00216 / 1.1068[15]	EtOH 4 / 1.4348[20]
5755	Ethanol, 2,2'-oxybis-, dibenzoate	$C_{18}H_{18}O_5$ / 314.34	120-55-8 / 33.5	/ 280[24]; 250[1]	4-09-00-00356 / 1.1690[15]	H_2O 4; EtOH 4

No.	Name / Synonym	Mol. Form. / Mol. Wt.	CAS RN / mp/°C	Merck No. / bp/°C	Beil. Ref. / den/g cm^{-3}	Solubility / n_D
5756	Ethanol, 2,2'-[oxybis(2,1-ethanediyloxy)]bis- Tetraethylene glycol	$C_8H_{18}O_5$ 194.23	112-60-7 -6.2	328	4-01-00-02403 1.1285[15]	H_2O 4; EtOH 3; eth 3; ctc 3 1.4577[20]
5757	Ethanol, 2-(pentyloxy)-	$C_7H_{16}O_2$ 132.20	6196-58-3	182	4-01-00-02383 0.8918[15]	1.4213[25]
5758	Ethanol, 2-phenoxy- 2-Phenoxyethanol	$C_8H_{10}O_2$ 138.17	122-99-6 14	7226 245	4-06-00-00571 1.102[22]	H_2O 1; EtOH 3; eth 3; chl 3 1.534[20]
5759	Ethanol, 2-(phenylamino)-	$C_8H_{11}NO$ 137.18	122-98-5	279.5; 150[10]	4-12-00-00277 1.0945[20]	H_2O 4; EtOH 4; eth 4; chl 4 1.5760[20]
5760	Ethanol, 2,2'-(phenylimino)bis-	$C_{10}H_{15}NO_2$ 181.23	120-07-0 57	200[10]	4-12-00-00285 1.201[60]	ace 4; bz 4; eth 4; EtOH 4
5761	Ethanol, 2-(phenylmethoxy)-	$C_9H_{12}O_2$ 152.19	622-08-2 <-75	256	4-06-00-02241 1.0640[20]	H_2O 4; eth 4; EtOH 4 1.5233[20]
5762	Ethanol, 2-[(phenylmethyl)amino]-	$C_9H_{13}NO$ 151.21	104-63-2	153-6[12]	4-12-00-02184 1.065[25]	1.5430[20]
5763	Ethanol, 2-(2-propenyloxy)-	$C_5H_{10}O_2$ 102.13	111-45-5	158.5	4-01-00-02388 0.9580[20]	H_2O 5; EtOH 4; bz 3; ctc 3 1.4358[20]
5764	Ethanol, 2-propoxy- 2-Propoxyethanol	$C_5H_{12}O_2$ 104.15	2807-30-9	149.8	4-01-00-02379 0.9112[20]	H_2O 3; EtOH 4; eth 4 1.4133[20]
5765	Ethanol, 2-(propylamino)-	$C_5H_{13}NO$ 103.16	16369-21-4	183	3-04-00-00681 0.9005[20]	1.4428[20]
5766	Ethanol, 2,2'-thiobis- 2,2'-Thiodiethanol	$C_4H_{10}O_2S$ 122.19	111-48-8 -10.2	9259 282	4-01-00-02437 1.1793[25]	H_2O 5; EtOH 5; eth 3; bz 2 1.5211[20]
5767	Ethanol, 2,2,2-tribromo- Tribromoethanol	$C_2H_3Br_3O$ 282.76	75-80-9 81	9525 92-3[10]	3-01-00-01362	bz 4; eth 4; EtOH 4
5768	Ethanol, 2,2,2-trichloro- 2,2,2-Trichloroethanol	$C_2H_3Cl_3O$ 149.40	115-20-8 19	9551 152; 52[11]	4-01-00-01383	H_2O 2; EtOH 5; eth 5; ctc 2 1.4861[20]
5769	Ethanol, 2,2,2-trichloro-1-ethoxy- Chloral alcoholate	$C_4H_7Cl_3O_2$ 193.46	515-83-3 56.5	2056 115.5	4-01-00-03144 1.143[40]	H_2O 3; EtOH 3; eth 3
5770	Ethanol, 2,2,2-trifluoro-	$C_2H_3F_3O$ 100.04	75-89-8 -43.5	74	4-01-00-01370 1.3842[20]	EtOH 4; eth 3; ace 3; bz 3 1.2907[22]
5771	Ethanone, 2-(acetyloxy)-1-phenyl-	$C_{10}H_{10}O_3$ 178.19	2243-35-8 49	270	4-08-00-00368 1.1169[65]	H_2O 1; EtOH 4; eth 4; bz 2 1.5036[65]
5772	Ethanone, 1-(3-aminophenyl)-	C_8H_9NO 135.17	99-03-6 98.5	289.5	4-14-00-00096	H_2O 2; EtOH 3
5773	Ethanone, 1-(4-aminophenyl)-	C_8H_9NO 135.17	99-92-3 106	294; 195[15]	4-14-00-00100	eth 4; EtOH 4
5774	Ethanone, 2-amino-1-phenyl- Phenacylamine	C_8H_9NO 135.17	613-89-8 20	7159 251	4-14-00-00114	H_2O 1; eth 3; ctc 2 1.6160[20]
5775	Ethanone, 1-(2-benzofuranyl)-	$C_{10}H_8O_2$ 160.17	1646-26-0 76	126[11]	5-17-10-00172	H_2O 3
5776	Ethanone, 1-[1,1'-biphenyl]-4-yl-	$C_{14}H_{12}O$ 196.25	92-91-1 121	326	4-07-00-01407 1.2510[0]	H_2O 1; EtOH 4; ace 4; chl 2
5777	Ethanone, 2-bromo-1-(4-bromophenyl)- p-Bromophenacyl bromide	$C_8H_6Br_2O$ 277.94	99-73-0 111	1415	4-07-00-00652	H_2O 1; EtOH 3; eth 3; chl 3
5778	Ethanone, 2-bromo-1-(4-chlorophenyl)- p-Chlorophenacyl bromide	C_8H_6BrClO 233.49	536-38-9 96.5	2153	4-07-00-00651	
5779	Ethanone, 2-bromo-1-(4-methylphenyl)-	C_9H_9BrO 213.07	619-41-0 51	157[14]	4-07-00-00705	eth 4; EtOH 4
5780	Ethanone, 1-(3-bromophenyl)-	C_8H_7BrO 199.05	2142-63-4 7.5	133[19]	4-07-00-00646	H_2O 1; ace 3; bz 3 1.5755[20]
5781	Ethanone, 1-(4-bromophenyl)- p-Bromoacetophenone	C_8H_7BrO 199.05	99-90-1 50.5	1390 257; 130[11]	4-07-00-00647 1.647[25]	H_2O 1; EtOH 3; eth 3; bz 3 1.647
5782	Ethanone, 2-bromo-1-phenyl- ω-Bromacetophenone	C_8H_7BrO 199.05	70-11-1 50.5	1391 135[18]	4-07-00-00649 1.647[20]	H_2O 1; EtOH 3; eth 4; bz 4
5783	Ethanone, 1-(5-bromo-2-thienyl)-	C_6H_5BrOS 205.08	5370-25-2 94.5	103[4]	5-17-09-00392	EtOH 2; ctc 3
5784	Ethanone, 2-chloro-1-(4-chlorophenyl)-	$C_8H_6Cl_2O$ 189.04	937-20-2 101.5	270	4-07-00-00644	EtOH 3; bz 3; MeOH 3
5785	Ethanone, 2-chloro-1,2-diphenyl-	$C_{14}H_{11}ClO$ 230.69	447-31-4 68.5	dec	3-07-00-02106	EtOH 3; chl 2; alk 1
5786	Ethanone, 2-chloro-1-(4-hydroxy-3-methoxyphenyl)-	$C_9H_9ClO_3$ 200.62	6344-28-1 102	208[22]	3-08-00-02114	H_2O 2; EtOH 3
5787	Ethanone, 2-chloro-1-(4-methylphenyl)-	C_9H_9ClO 168.62	4209-24-9 57.5	261.5	4-07-00-00704	eth 4; EtOH 4
5788	Ethanone, 1-(3-chlorophenyl)-	C_8H_7ClO 154.60	99-02-5	244; 129[30]	4-07-00-00638 1.2130[40]	EtOH 3; eth 3; ace 3 1.5494[20]
5789	Ethanone, 1-(4-chlorophenyl)- p-Chloroacetophenone	C_8H_7ClO 154.60	99-91-2 20	2114 232	4-07-00-00639 1.1922[20]	H_2O 1; EtOH 5; eth 5; chl 3 1.5550[20]
5790	Ethanone, 2-chloro-1-phenyl- ω-Chloroacetophenone	C_8H_7ClO 154.60	532-27-4 56.5	2115 247	4-07-00-00641 1.32[15]	H_2O 1; EtOH 4; eth 4; ace 3
5791	Ethanone, 1-(5-chloro-2-thienyl)-	C_6H_5ClOS 160.62	6310-09-4 52	117[17]; 88[4.5]	5-17-09-00391	H_2O 4; eth 4; EtOH 4
5792	Ethanone, 1-cyclobutyl-	$C_6H_{10}O$ 98.14	3019-25-8	138	4-07-00-00037 0.9020[20]	H_2O 2 1.4322[19]
5793	Ethanone, 1-(1-cyclohexen-1-yl)-	$C_8H_{12}O$ 124.18	932-66-1 73	201.5	4-07-00-00143 0.9655[20]	EtOH 3; eth 3 1.4881[20]
5794	Ethanone, 1-cyclohexyl-	$C_8H_{14}O$ 126.20	823-76-7	180.5	4-07-00-00053 0.9176[20]	H_2O 1; eth 3 1.4565[16]
5795	Ethanone, 1-cyclopentyl-	$C_7H_{12}O$ 112.17	6004-60-0	158.5	4-07-00-00047 0.916[20]	eth 3 1.4409[20]

No.	Name / Synonym	Mol. Form. / Mol. Wt.	CAS RN / mp/°C	Merck No. / bp/°C	Beil. Ref. / den/g cm^{-3}	Solubility / n_D
5796	Ethanone, 1-cyclopropyl- / Cyclopropyl methyl ketone	C_5H_8O / 84.12	765-43-5 / -68.3	111.3	4-07-00-00014 / 0.8984[20]	H_2O 4; eth 4; EtOH 4 / 1.4251[20]
5797	Ethanone, 1-(3,5-dibromophenyl)-	$C_8H_6Br_2O$ / 277.94	14401-73-1 / 68	198[15]	3-07-00-00984	bz 4; eth 4; EtOH 4
5798	Ethanone, 2,2-dibromo-1-phenyl-	$C_8H_6Br_2O$ / 277.94	13665-04-8 / 36.5	159[13]	4-07-00-00653	EtOH 3; eth 3; chl 3
5799	Ethanone, 1-(2,4-dichlorophenyl)-	$C_8H_6Cl_2O$ / 189.04	2234-16-4 / 33.5		4-07-00-00643	H_2O 1 / 1.5640[20]
5800	Ethanone, 1-(2,5-dichlorophenyl)-	$C_8H_6Cl_2O$ / 189.04	2476-37-1 / 12	118[12]	4-07-00-00643 / 1.321[30]	1.5595[30]
5801	Ethanone, 1-(3,4-dichlorophenyl)-	$C_8H_6Cl_2O$ / 189.04	2642-63-9 / 76	135[12]	4-07-00-00643	H_2O 1; ctc 3; lig 3
5802	Ethanone, 2,2-dichloro-1-phenyl-	$C_8H_6Cl_2O$ / 189.04	2648-61-5 / 20.5	249	4-07-00-00644 / 1.340[18]	EtOH 3; bz 3; ctc 3 / 1.5686[20]
5803	Ethanone, 1-(2,4-dihydroxyphenyl)- / Resacetophenone	$C_8H_8O_3$ / 152.15	89-84-9 / 146	8143	4-08-00-01792 / 1.18[141]	H_2O 1; EtOH 3; eth 2; bz 2
5804	Ethanone, 1-(3,4-dihydroxyphenyl)-2-(methylamino)- / Adrenalone	$C_9H_{11}NO_3$ / 181.19	99-45-6 / 235-6 dec	161	4-14-00-00832	H_2O 2; EtOH 2; eth 2
5805	Ethanone, 1-(3,4-dimethoxyphenyl)-	$C_{10}H_{12}O_3$ / 180.20	1131-62-0 / 51	287	4-08-00-01815	H_2O 4; bz 4; EtOH 4; chl 4
5806	Ethanone, 1-[3-(dimethylamino)phenyl]-	$C_{10}H_{13}NO$ / 163.22	18992-80-8 / 43	148[13]	4-14-00-00096	ace 4; eth 4
5807	Ethanone, 1-[10-[3-(dimethylamino)propyl]-10H-phenothiazin-2-yl]- / Acepromazine	$C_{19}H_{22}N_2OS$ / 326.46	61-00-7 / 27	230[0.5]	4-27-00-02911	
5808	Ethanone, 1-[4-(1,1-dimethylethyl)phenyl]-	$C_{12}H_{16}O$ / 176.26	943-27-1 /	136-8[20]	4-07-00-00769 / 0.9705[25]	1.518[15]
5809	Ethanone, 1-(2,4-dimethylphenyl)-	$C_{10}H_{12}O$ / 148.20	89-74-7 /	228	4-07-00-00729 / 1.0121[15]	eth 4; EtOH 4 / 1.5340[20]
5810	Ethanone, 1-(2,5-dimethylphenyl)-	$C_{10}H_{12}O$ / 148.20	2142-73-6 / -18.1	232.5	4-07-00-00728 / 0.9963[19]	H_2O 1; EtOH 4; eth 4; bz 4 / 1.5291[20]
5811	Ethanone, 1-(3,4-dimethylphenyl)-	$C_{10}H_{12}O$ / 148.20	3637-01-2 / -3	246.5	4-07-00-00728 / 1.0090[14]	H_2O 1; EtOH 4; eth 4; bz 4 / 1.5413[15]
5812	Ethanone, 1,2-diphenyl-	$C_{14}H_{12}O$ / 196.25	451-40-1 / 60	320	4-07-00-01393 / 1.201[0]	H_2O 2; EtOH 3; eth 3; ctc 3
5813	Ethanone, 2-ethoxy-1,2-diphenyl-	$C_{16}H_{16}O_2$ / 240.30	574-09-4 / 62	194-5[20]	3-08-00-01279 / 1.1016[17]	bz 4; eth 4; EtOH 4; lig 4 / 1.5727[17]
5814	Ethanone, 2-ethoxy-1-phenyl- / 2-Ethoxyacetophenone	$C_{10}H_{12}O_2$ / 164.20	14869-39-7 / 43	243.5	4-08-00-00365 / 1.0036[78]	eth 4; EtOH 4; lig 4
5815	Ethanone, 1-(4-fluorophenyl)-	C_8H_7FO / 138.14	403-42-9 / -45	196	4-07-00-00635 / 1.1382[25]	H_2O 1; bz 3; chl 3 / 1.5081[25]
5816	Ethanone, 2-fluoro-1-phenyl-	C_8H_7FO / 138.14	450-95-3 / 29	90-1[12]	4-07-00-00636 / 1.152[20]	1.5200[20]
5817	Ethanone, 1-(2-furanyl)-	$C_6H_6O_2$ / 110.11	1192-62-7 / 33	175	5-17-09-00381 / 1.098[20]	H_2O 1; EtOH 3; eth 3 / 1.5017[20]
5818	Ethanone, 1-[4-(β-D-glucopyranosyloxy)phenyl]- / Picein	$C_{14}H_{18}O_7$ / 298.29	530-14-3 / 195.5	7367	5-17-07-00151	H_2O 2; EtOH 3; eth 3; chl 1
5819	Ethanone, 1-(1H-indol-3-yl)-	$C_{10}H_9NO$ / 159.19	703-80-0 / 192.3	144[10]	5-21-08-00297	EtOH 4
5820	Ethanone, 1-(1H-pyrrol-2-yl)-	C_6H_7NO / 109.13	1072-83-9 / 90	220	5-21-07-00204	H_2O 3; EtOH 3; eth 3
5821	Ethanone, 1-(1-hydroxycyclohexyl)-	$C_8H_{14}O_2$ / 142.20	1123-27-9 /	125.5; 91[11]	4-08-00-00023 / 1.0248[25]	eth 4; EtOH 4 / 1.4670[25]
5822	Ethanone, 2-hydroxy-1,2-diphenyl-, (±)- / (±)-Benzoin	$C_{14}H_{12}O_2$ / 212.25	579-44-2 / 137	1103 / 344; 194[12]	4-08-00-01279 / 1.310[20]	EtOH 4; chl 4
5823	Ethanone, 1-(2-hydroxy-4-methoxyphenyl)-	$C_9H_{10}O_3$ / 166.18	552-41-0 / 52.5	158[20]	4-08-00-01793 / 1.3102[81]	bz 4; eth 4; EtOH 4; chl 4 / 1.5452[81]
5824	Ethanone, 1-(4-hydroxy-3-methoxyphenyl)- / Apocynin	$C_9H_{10}O_3$ / 166.18	498-02-2 / 115	772 / 297; 234[15]	4-08-00-01814	H_2O 2; EtOH 3; eth 4; ace 3
5825	Ethanone, 1-[6-hydroxy-2-(1-methylethenyl)-5-benzofuranyl]- / Euparin	$C_{13}H_{12}O_3$ / 216.24	532-48-9 / 121.5	3857	5-18-02-00029	eth 3; bz 3; chl 3; NaOH 2
5826	Ethanone, 1-(2-hydroxy-4-methylphenyl)-	$C_9H_{10}O_2$ / 150.18	6921-64-8 / 21	245	4-08-00-00480 / 1.1012[10]	1.5527[13]
5827	Ethanone, 1-(2-hydroxy-5-methylphenyl)-	$C_9H_{10}O_2$ / 150.18	1450-72-2 / 50	210; 120[20]	4-08-00-00472 / 1.0797[53]	bz 4; eth 4; EtOH 4; chl 4
5828	Ethanone, 1-(4-hydroxy-2-methylphenyl)-	$C_9H_{10}O_2$ / 150.18	875-59-2 / 128	313	4-08-00-00469 / 1.0592[135]	eth 4; EtOH 4 / 1.5369[135]
5829	Ethanone, 1-(1-hydroxy-2-naphthalenyl)-	$C_{12}H_{10}O_2$ / 186.21	711-79-5 / 101	325 dec	4-08-00-01175	bz 4; HOAc 4
5830	Ethanone, 1-(2-hydroxyphenyl)-	$C_8H_8O_2$ / 136.15	118-93-4 / 5	218	4-08-00-00320 / 1.1307[20]	eth 4; EtOH 4; HOAc 4 / 1.5584[20]
5831	Ethanone, 1-(3-hydroxyphenyl)-	$C_8H_8O_2$ / 136.15	121-71-1 / 96	296; 153[5]	4-08-00-00334 / 1.0992[109]	H_2O 2; EtOH 4; eth 4; bz 4 / 1.5348[109]
5832	Ethanone, 1-(4-hydroxyphenyl)-	$C_8H_8O_2$ / 136.15	99-93-4 / 109.5	147-8[3]	4-08-00-00339 / 1.1090[109]	H_2O 2; EtOH 4; eth 4 / 1.5577[109]
5833	Ethanone, 2-hydroxy-1-phenyl-	$C_8H_8O_2$ / 136.15	582-24-1 / 90	125[12]; 56[1]	4-08-00-00365 / 1.0963[99]	H_2O 3; EtOH 3; eth 3; chl 3

No.	Name / Synonym	Mol. Form. / Mol. Wt.	CAS RN / mp/°C	Merck No. / bp/°C	Beil. Ref. / den/g cm^{-3}	Solubility / n_D
5834	Ethanone, 2-[6-(2-hydroxy-2-phenylethyl)-1-methyl-2-piperidinyl]-1-phenyl-, [2R-[2α,6α(S*)]]- Lobeline	C$_{22}$H$_{27}$NO$_2$ 337.46	90-69-7 130.5	5432	5-21-12-00627	H$_2$O 2; EtOH 3; eth 3; ace 4
5835	Ethanone, 1-(3-hydroxy-2-thienyl)-	C$_6$H$_6$O$_2$S 142.18	5556-07-0 51.5	47[0.2]	5-18-01-00121 1.5000[20]	EtOH 3 1.5795[20]
5836	Ethanone, 1-(3-iodophenyl)-	C$_8$H$_7$IO 246.05	14452-30-3	129[8]; 117[4]	4-07-00-00654	bz 3 1.622[20]
5837	Ethanone, 1-(4-iodophenyl)-	C$_8$H$_7$IO 246.05	13329-40-3 86	153[18]	4-07-00-00654	EtOH 3; eth 2; bz 3; HOAc 3
5838	Ethanone, 2-iodo-1-phenyl-	C$_8$H$_7$IO 246.05	4636-16-2 34.4	154[15]	4-07-00-00654	ace 3
5839	Ethanone, 2-methoxy-1,2-diphenyl-	C$_{15}$H$_{14}$O$_2$ 226.27	3524-62-7 49.5	188-9[15]	4-08-00-01280 1.1278[14]	bz 4; eth 4; EtOH 4
5840	Ethanone, 1-(3-methoxyphenyl)-	C$_9$H$_{10}$O$_2$ 150.18	586-37-8 95.5	240	4-08-00-00334 1.0343[19]	H$_2$O 3; EtOH 3; ace 3; ctc 3 1.5410[20]
5841	Ethanone, 1-(4-methoxyphenyl)-	C$_9$H$_{10}$O$_2$ 150.18	100-06-1 38.5	258	4-08-00-00340 1.0818[41]	H$_2$O 2; EtOH 3; eth 3; ace 3 1.547[41]
5842	Ethanone, 2-methoxy-1-phenyl-	C$_9$H$_{10}$O$_2$ 150.18	4079-52-1 8	245; 125[19]	4-08-00-00365 1.0897[20]	H$_2$O 2; EtOH 3; ace 3 1.5393[20]
5843	Ethanone, 1-(1-methylcyclohexyl)-	C$_9$H$_{16}$O 140.23	2890-62-2	186.2	4-07-00-00067 0.9504[20]	1.4543[20]
5844	Ethanone, 1-(3-methylcyclohexyl)-	C$_9$H$_{16}$O 140.23	7193-78-4	197	3-07-00-00120 0.912[19]	1.4517[19]
5845	Ethanone, 1-(4-methylcyclohexyl)-	C$_9$H$_{16}$O 140.23	1879-06-7	196	3-07-00-00121 0.9055[18]	1.4509[18]
5846	Ethanone, 1-(1-methylcyclopentyl)-	C$_8$H$_{14}$O 126.20	13388-93-7	48.4[10]	3-07-00-00102 0.9104[20]	1.4430[20]
5847	Ethanone, 1-(2-methylcyclopentyl)-	C$_8$H$_{14}$O 126.20	1601-00-9	170.5; 59.6[15]	3-07-00-00103 0.9222[4]	1.4434[16]
5848	Ethanone, 1-[4-(1-methylethyl)phenyl]-	C$_{11}$H$_{14}$O 162.23	645-13-6	254	4-07-00-00747 0.9753[15]	1.5235[20]
5849	Ethanone, 1-(1-methyl-1H-pyrrol-2-yl)-	C$_7$H$_9$NO 123.15	932-16-1	201[252]; 93[22]	5-21-07-00205 1.0445[15]	EtOH 3; bz 3; chl 3 1.5403[15]
5850	Ethanone, 1-[2-methyl-5-(1-methylethyl)phenyl]-	C$_{12}$H$_{16}$O 176.26	1202-08-0 <-20	249.5	4-07-00-00772 0.956[20]	1.5181[20]
5851	Ethanone, 1-(2-methylphenyl)-	C$_9$H$_{10}$O 134.18	577-16-2	214	4-07-00-00697 1.026[20]	1.5276[20]
5852	Ethanone, 1-(3-methylphenyl)-	C$_9$H$_{10}$O 134.18	585-74-0	220	4-07-00-00699 1.0165[0]	EtOH 3; eth 3; ace 3; ctc 2 1.533[15]
5853	Ethanone, 1-(4-methylphenyl)-	C$_9$H$_{10}$O 134.18	122-00-9 28	226; 93.5[7]	4-07-00-00701 1.0051[20]	bz 4; eth 4; EtOH 4; chl 4 1.5335[20]
5854	Ethanone, 2,2'-(1-methyl-2,6-piperidinediyl)bis[1-phenyl-, cis- Lobelanine	C$_{22}$H$_{25}$NO$_2$ 335.45	579-21-5 99	5430	5-21-11-00432	ace 4; bz 4; EtOH 4; chl 4
5855	Ethanone, 1-(6-methyl-3-pyridinyl)-	C$_8$H$_9$NO 135.17	36357-38-7 17.6	144[50]	5-21-07-00425 1.0168[25]	H$_2$O 4 1.5302[25]
5856	Ethanone, 1-(5-methyl-2-thienyl)-	C$_7$H$_8$OS 140.21	13679-74-8 27.5	232.5	5-17-09-00425 1.1185[25]	eth 3; ace 3; bz 3; chl 2 1.5604
5857	Ethanone, 1-(1-naphthalenyl)-	C$_{12}$H$_{10}$O 170.21	941-98-0 34	297	4-07-00-01292 1.1171[21]	H$_2$O 1; EtOH 3; eth 3; ace 3 1.6280[22]
5858	Ethanone, 1-(2-naphthalenyl)-	C$_{12}$H$_{10}$O 170.21	93-08-3 56	302	4-07-00-01294	EtOH 2; ctc 2
5859	Ethanone, 1-(1-naphthalenyl)-2-phenyl- Benzyl 1-naphthyl ketone	C$_{18}$H$_{14}$O 246.31	605-85-6 66.5	194-6[0.05]	4-07-00-01790	eth 4
5860	Ethanone, 1-(3-nitrophenyl)-	C$_8$H$_7$NO$_3$ 165.15	121-89-1 81	202; 167[18]	4-07-00-00656	H$_2$O 4; EtOH 2; eth 4; chl 2
5861	Ethanone, 1-(4-nitrophenyl)-	C$_8$H$_7$NO$_3$ 165.15	100-19-6 81.8	165[5]	4-07-00-00657	eth 4; EtOH 4
5862	Ethanone, 2-nitro-1-phenyl-	C$_8$H$_7$NO$_3$ 165.15	614-21-1 106	158[16]; 142[10]	4-07-00-00659	eth 4; EtOH 4 1.5468[30]
5863	Ethanone, 1-(2-nitrophenyl)- Acetophene,2-nitro	C$_8$H$_7$N$_3$O 161.16	577-59-3 28.5	178[32]; 158[16]	4-07-00-00655 1.2370[25]	H$_2$O 1; EtOH 4; eth 4; chl 4 1.5468[20]
5864	Ethanone, 1-phenyl- Acetophenone	C$_8$H$_8$O 120.15	98-86-2 20	65 202	4-07-00-00619 1.0281[20]	H$_2$O 1; EtOH 3; eth 3; ace 3 1.5372[20]
5865	Ethanone, 1,1'-(1,3-phenylene)bis-	C$_{10}$H$_{10}$O$_2$ 162.19	6781-42-6 32	152[15]	4-07-00-02156	H$_2$O 2; EtOH 3; bz 3; chl 3
5866	Ethanone, 1-phenyl-, oxime	C$_8$H$_9$NO 135.17	613-91-2 60	245	3-07-00-00954 1.0515[78]	EtOH 3; eth 3; ctc 2; HOAc 3
5867	Ethanone, 1-(2-pyridinyl)-	C$_7$H$_7$NO 121.14	1122-62-9	192	5-21-07-00385 1.077[25]	EtOH 3; eth 3; ctc 2; HOAc 3 1.5203[20]
5868	Ethanone, 1-(3-pyridinyl)- Methyl 3-pyridyl ketone	C$_7$H$_7$NO 121.14	350-03-8 13.5	6034 220	5-21-07-00394 1.5341[20]	H$_2$O 3; EtOH 3; eth 3; acid 3
5869	Ethanone, 1-(4-pyridinyl)-	C$_7$H$_7$NO 121.14	1122-54-9 16	212	5-21-07-00400 1.097[25]	EtOH 2; eth 2; acid 2 1.5282[25]

No.	Name / Synonym	Mol. Form. / Mol. Wt.	CAS RN / mp/°C	Merck No. / bp/°C	Beil. Ref. / den/g cm⁻³	Solubility / n_D
5870	Ethanone, 1-(2-thienyl)-	C_8H_6OS 126.18	88-15-3 10.5	213.5	5-17-09-00387 1.1679^{20}	H₂O 2; EtOH 5; eth 5; ctc 3 1.5667^{20}
5871	Ethanone, 2,2,2-trichloro-1-phenyl-	$C_8H_5Cl_3O$ 223.49	2902-69-4	256.5	4-07-00-00645 1.425^{16}	eth 4; EtOH 4
5872	Ethanone, 2,2,2-trifluoro-1-phenyl-	$C_8H_5F_3O$ 174.12	434-45-7 -40	153	4-07-00-00637 1.279^{20}	1.4583^{20}
5873	Ethanone, 1-(2,3,4-trihydroxyphenyl)- Gallacetophenone	$C_8H_8O_4$ 168.15	528-21-2 173	4248	4-08-00-02721	H₂O 3; EtOH 4; eth 3; ace 4
5874	Ethanone, 1-(2,3,4-trimethoxyphenyl)-	$C_{11}H_{14}O_4$ 210.23	13909-73-4 15.8	296	4-08-00-02721	1.5384^{20}
5875	Ethanone, 1-(2,4,5-trimethylphenyl)-	$C_{11}H_{14}O$ 162.23	2040-07-5 10.5	246.5	4-07-00-00751 1.0039^{15}	H₂O 2; EtOH 4; eth 4; bz 4 1.541^{15}
5876	Ethanone, 1-(2,4,6-trimethylphenyl)-	$C_{11}H_{14}O$ 162.23	1667-01-2	241; 120^{12}	4-07-00-00749 0.9754^{20}	H₂O 1; EtOH 3; eth 3; ace 3 1.5175^{20}
5877	Ethanone, 1-(3,4,5-trimethylphenyl)-	$C_{11}H_{14}O$ 162.23	2047-21-4 4.7	101.5^3	4-07-00-00749 1.0037^{25}	1.5420^{25}
5878	Ethenaminium, N,N,N-trimethyl-, hydroxide Neurine	$C_5H_{13}NO$ 103.16	463-88-7	6393	3-04-00-00442	H₂O 4; eth 4; EtOH 4
5879	Ethene Ethylene	C_2H_4 28.05	74-85-1 -169	3748 -103.7	4-01-00-00677 0.5678^{-104}	H₂O 1; EtOH 2; eth 3; ace 2 1.363^{100}
5880	Ethene, bromo- Vinyl bromide	C_2H_3Br 106.95	593-60-2 -137.8	15.8	4-01-00-00718 1.4933^{20}	H₂O 1; EtOH 3; eth 3; ace 3 1.4380^{20}
5881	Ethene, 1-bromo-2-chloro- Ethylene, 1-bromo-2-chloro-	C_2H_2BrCl 141.39	3018-09-5 -86.7	84.6	3-01-00-00671 1.797^{15}	
5882	Ethene, 2-bromo-1,1-dichloro-	C_2HBrCl_2 175.84	5870-61-1 -88.5	107.5	4-01-00-00720 1.9053^{15}	1.4982
5883	Ethene, 1-bromo-1,2-difluoro- 1-Bromo-1,2-difluoroethylene	C_2HBrF_2 142.93	358-99-6	19	0-01-00-00189 1.8434^{25}	1.3846^0
5884	Ethene, chloro- Vinyl chloride	C_2H_3Cl 62.50	75-01-4 -153.7	9898 -13.3	4-01-00-00700 0.9106^{20}	H₂O 2; EtOH 3; eth 4 1.3700^{20}
5885	Ethene, 2-chloro-1,1-difluoro- 1,1-Difluoro-2-chloroethylene	C_2HClF_2 98.48	359-10-4 -138.5	-18.5	4-01-00-00703	
5886	Ethene, 1-chloro-2-ethoxy- 2-Chlorovinyl ethyl ether	C_4H_7ClO 106.55	928-56-3	120	4-01-00-02061 1.0386^{20}	1.4385^{20}
5887	Ethene, (2-chloroethoxy)- 2-Chloroethyl vinyl ether	C_4H_7ClO 106.55	110-75-8 -70	2139 108	4-01-00-02051 1.0495^{20}	EtOH 4; eth 4; chl 2 1.4378^{20}
5888	Ethene, 1-chloro-2-iodo-	C_2H_2ClI 188.40	20244-71-7 -38.2	119	3-01-00-00674 2.2298^{25}	
5889	Ethene, chlorotrifluoro- Chlorotrifluoroethylene	C_2ClF_3 116.47	79-38-9 -158	-27.8	4-01-00-00704 1.54^{-60}	bz 3; chl 3 1.38^0
5890	Ethene, 1,1-dibromo-	$C_2H_2Br_2$ 185.85	593-92-0	92	4-01-00-00720 2.1779^{21}	EtOH 3; eth 3; ace 3; bz 3
5891	Ethene, 1,2-dibromo-, (E)- trans-1,2-Dibromoethylene	$C_2H_2Br_2$ 185.85	590-12-5 -6.5	108	4-01-00-00721 2.2308^{20}	H₂O 1; EtOH 4; eth 4; ace 3 1.5505^{18}
5892	Ethene, 1,2-dibromo-, (Z)- cis-1,2-Dibromoethylene	$C_2H_2Br_2$ 185.85	590-11-4 -53	112.5	4-01-00-00720 2.2464^{20}	H₂O 1; EtOH 4; eth 4; ace 3 1.5428^{20}
5893	Ethene, 1,1-dibromo-2-ethoxy- Ethylene, 1,1-dibromo-2-ethoxy	$C_4H_6Br_2O$ 229.90	77295-79-5	172; 74^{15}	2-01-00-00473 1.7697^{18}	eth 4
5894	Ethene, 1,1-dichloro- Vinylidene chloride	$C_2H_2Cl_2$ 96.94	75-35-4 -122.5	9900 31.6	4-01-00-00706 1.213^{20}	H₂O 1; EtOH 3; eth 4; ace 3 1.4249^{20}
5895	Ethene, 1,2-dichloro-, (E)- trans-1,2-Dichloroethylene	$C_2H_2Cl_2$ 96.94	156-60-5 -49.8	86 48.7	4-01-00-00709 1.2565^{20}	H₂O 2; EtOH 5; eth 5; ace 5 1.4454^{20}
5896	Ethene, 1,2-dichloro-, (Z)- cis-1,2-Dichloroethylene	$C_2H_2Cl_2$ 96.94	156-59-2 -80	86 60.1	4-01-00-00707 1.2837^{20}	H₂O 2; EtOH 5; eth 5; ace 5 1.4490^{20}
5897	Ethene, 1,1-dichloro-2,2-difluoro-	$C_2Cl_2F_2$ 132.92	79-35-6 -116	19	4-01-00-00711 1.555^{-20}	1.383^{-20}
5898	Ethene, 1,2-dichloro-1,2-difluoro-	$C_2Cl_2F_2$ 132.92	598-88-9 -130.5	21.1	1.4950^0	1.3777^0
5899	Ethene, 1,2-dichloro-1-ethoxy-	$C_4H_6Cl_2O$ 141.00	42345-82-4	128.2	3-01-00-02950 1.1972^{25}	1.4558^{17}
5900	Ethene, 1,1-dichloro-2-fluoro- 1,1-Dichloro-2-fluoroethylene	C_2HCl_2F 114.93	359-02-4 -108.8	37.5	4-01-00-00711 1.3732^{16}	1.4031^{16}
5901	Ethene, 1,1-diethoxy-	$C_6H_{12}O_2$ 116.16	2678-54-8	68^{100}	4-01-00-03420 0.7932^{20}	1.3643^{21}
5902	Ethene, 1,1-difluoro- Vinylidene fluoride	$C_2H_2F_2$ 64.03	75-38-7 -144	-85.7	4-01-00-00696	eth 4; EtOH 4
5903	Ethene, 1,2-difluoro-, (Z)- cis-1,2-Difluoroethylene	$C_2H_2F_2$ 64.03	1630-77-9			
5904	Ethene, 1,2-diiodo-, (Z)-	$C_2H_2I_2$ 279.85	590-26-1 -14	72.5^{16}	4-01-00-00724 3.0625^{20}	eth 3; chl 3
5905	Ethene, ethoxy- Ethyl vinyl ether	C_4H_8O 72.11	109-92-2 -115.8	35.5	4-01-00-02049 0.7589^{20}	H₂O 2; EtOH 3; eth 5; ctc 2 1.3767^{20}
5906	Ethene, fluoro- Vinyl fluoride	C_2H_3F 46.04	75-02-5 -160.5	-72	4-01-00-00694	H₂O 1; EtOH 3; ace 3
5907	Ethene, iodo- Iodoethylene	C_2H_3I 153.95	593-66-8	56	4-01-00-00722 2.037^{20}	eth 4; EtOH 4 1.5385^{20}
5908	Ethene, methoxy- Methyl vinyl ether	C_3H_6O 58.08	107-25-5 -122	5.5	4-01-00-02049 0.7725^0	H₂O 2; EtOH 4; eth 4; ace 4 1.3730^0
5909	Ethene, (methylsulfonyl)-	$C_3H_6O_2S$ 106.15	3680-02-2	$122-4^{24}$	4-01-00-02065 1.2117^{20}	eth 3; ace 3 1.4636^{20}

No.	Name Synonym	Mol. Form. Mol. Wt.	CAS RN mp/°C	Merck No. bp/°C	Beil. Ref. den/g cm^{-3}	Solubility n_D
5910	Ethene, (methylthio)-	C_3H_6S 74.15	1822-74-8	69.5	4-01-00-02065 0.9026[20]	eth 3; ace 3; chl 3 1.4837[20]
5911	Ethene, nitro-	$C_2H_3NO_2$ 73.05	3638-64-0 -55.5	98.5	4-01-00-00725 1.2212[14]	EtOH 4; eth 4; ace 4; bz 4 1.4282[20]
5912	Ethene, 1,1'-oxybis- Divinyl ether	C_4H_6O 70.09	109-93-3 -101	9899 28.3	4-01-00-02058 0.773[20]	H_2O 1; EtOH 5; eth 5; ace 5 1.3989[20]
5913	Ethene, 1,1'-sulfonylbis-	$C_4H_6O_2S$ 118.16	77-77-0 -26	234.5	4-01-00-02068 1.177[25]	1.4765[20]
5914	Ethene, tetrabromo- Tetrabromoethylene	C_2Br_4 343.64	79-28-7 56.5	226	4-01-00-00722	H_2O 1; EtOH 3; eth 3; ace 3
5915	Ethenetetracarbonitrile Tetracyanoethylene	C_6N_4 128.09	670-54-2 199	9129 223	4-02-00-02450 1.348[25]	eth 2; ace 3; bz 2; ctc 2 1.560[25]
5916	Ethenetetracarboxylic acid, tetraethyl ester	$C_{14}H_{20}O_8$ 316.31	6174-95-4 58	328 dec; 203[13]	4-02-00-02450	H_2O 1; EtOH 4; eth 4
5917	Ethene, tetrachloro- Tetrachloroethylene	C_2Cl_4 165.83	127-18-4 -22.3	9126 121.3	4-01-00-00715 1.6227[20]	H_2O 1; EtOH 5; eth 5; bz 5 1.5053[20]
5918	Ethene, tetrafluoro- Tetrafluoroethylene	C_2F_4 100.02	116-14-3 -142.5	-75.9	4-01-00-00698 1.519[-76]	H_2O 1
5919	Ethene, tetraiodo- Tetraiodoethylene	C_2I_4 531.64	513-92-8 187	9151 sub	4-01-00-00724 2.983[20]	bz 4; chl 4
5920	Ethene, 1,1'-thiobis-	C_4H_6S 86.16	627-51-0 20	84	4-01-00-02068 0.9174[15]	H_2O 2; EtOH 5; eth 5; ace 3
5921	Ethene, tribromo-	C_2HBr_3 264.74	598-16-3	164	4-01-00-00722 2.708[20]	H_2O 2; EtOH 4; eth 3; ace 3 1.6045[16]
5922	Ethene, trichloro- Trichloroethylene	C_2HCl_3 131.39	79-01-6 -84.7	9552 87.2	4-01-00-00712 1.4642[20]	H_2O 2; EtOH 5; eth 5; ace 3 1.4773[20]
5923	Ethene, trichlorofluoro-	C_2Cl_3F 149.38	359-29-5 -108.9	71	4-01-00-00715 1.5460[20]	chl 3 1.4379[20]
5924	Ethene, trifluoro- Trifluoroethylene	C_2HF_3 82.03	359-11-5	-51	4-01-00-00697 1.26[-70]	H_2O 1; EtOH 2; eth 3
5925	Ethene, trifluoroiodo-	C_2F_3I 207.92	359-37-5	30	4-01-00-00723 2.284[25]	1.4143[0]
5926	1,4-Etheno-3H,7H-benzo[1,2-c:3,4-c']dipyran-3,7-dione, 9-(3-furanyl)-1,4,4a,5,6,6a,9,10,10a,10b-decahydro-4-hydroxy-4a,10a-dimethyl-, [1R-(1α,4β,4aα,6aβ,9β,10aα,10b)]- Columbin	$C_{20}H_{22}O_6$ 358.39	546-97-4	2489	5-19-10-00546	
5927	Ethenone Ketene	C_2H_2O 42.04	463-51-4 -151	5177 -49.8	4-01-00-03418	eth 2; ace 2
5928	Ethenone, diphenyl- Diphenylketene	$C_{14}H_{10}O$ 194.23	525-06-4	3327 267.5	4-07-00-01653 1.1107[13]	1.615[14]
5929	Ethephon Phosphonic acid, (2-chloroethyl)-	$C_2H_6ClO_3P$ 144.49	16672-87-0 74	3686	1.2	
5930	Ethiolate Carbamothioic acid, diethyl-, S-ethyl ester	$C_7H_{15}NOS$ 161.27	2941-55-1	52[0.5]; 142[87]	0.9791[30]	
5931	Ethion Phosphorodithioic acid, S,S'-methylene O,O,O',O'-tetraethyl ester	$C_9H_{22}O_4P_2S_4$ 384.48	563-12-2 -13	3691 165[0.3]	1.22[20]	
5932	Ethirimol 4(1H)-Pyrimidinone, 5-butyl-2-(ethylamino)-6-methyl-	$C_{11}H_{19}N_3O$ 209.29	23947-60-6 160	3695	1.21[25]	
5933	Ethofumesate 2-Ethoxy-2,3-dihydro-3,3-dimethyl-5-benzofuranyl methanesulfonate	$C_{13}H_{18}O_5S$ 286.35	26225-79-6 71	3697	1.14	
5934	Ethoprop Phosphorodithioic acid, O-ethyl S,S-dipropyl ester	$C_8H_{19}O_2PS_2$ 242.34	13194-48-4	3702 88[0.2]	1.094[20]	
5935	Ethylene, 1-bromo-1,2-dichloro-, trans- Ethylene, 1-bromo-1,2-dichloro (dis)	C_2HBrCl_2 175.84	6795-75-1 -83.5	113.8	3-01-00-00671 1.9133[15]	1.5218[15]
5936	Ethyne Acetylene	C_2H_2 26.04	74-86-2 -80.7 (trip. pt.)	84 -84.7 (subl. point)	4-01-00-00939 *0.377[25]	H_2O 2; EtOH 2; ace 3; bz 3
5937	Ethyne, bromo- Bromoacetylene	C_2HBr 104.93	593-61-3	4.7	3-01-00-00919	eth 4
5938	Ethyne, chloro- Chloroacetylene	C_2HCl 60.48	593-63-5 -126	-30	4-01-00-00957	EtOH 2
5939	Ethyne, dibromo-	C_2Br_2 183.83	624-61-3 -24	exp 76	3-01-00-00919	ace 4; bz 4; eth 4; EtOH 4
5940	Ethyne, dichloro- Dichloroacetylene	C_2Cl_2 94.93	7572-29-4 -66	33	4-01-00-00957 1.261[20]	EtOH 3; eth 3; ace 3 1.42790[20]
5941	Ethyne, diiodo- Diiodoacetylene	C_2I_2 277.83	624-74-8 81.5	exp	4-01-00-00958	ace 4; bz 4; eth 4; EtOH 4
5942	Ethyne, ethoxy-	C_4H_6O 70.09	927-80-0	50	4-01-00-02211 0.800[20]	1.3796[20]
5943	Ethyne, fluoro- Fluoroacetylene	C_2HF 44.03	2713-09-9		4-01-00-00957	
5944	Ethyne, methoxy-	C_3H_4O 56.06	6443-91-0		4-01-00-02211 0.8001[20]	eth 4; EtOH 4 1.3812[20]

No.	Name / Synonym	Mol. Form. / Mol. Wt.	CAS RN / mp/°C	Merck No. / bp/°C	Beil. Ref. / den/g cm^{-3}	Solubility / n_D
5945	Etrimfos / Phosphorothioic acid, O-(6-ethoxy-2-ethyl-4-pyrimidinyl) O,O-dimethyl ester	$C_{10}H_{17}N_2O_4PS$ / 292.30	38260-54-7 / -3.35	3847	1.195[20]	
5946	Famphur / Phosphorothioic acid, O-[4-[(dimethylamino)sulfonyl]phenyl] O,O-dimethyl ester	$C_{10}H_{16}NO_5PS_2$ / 325.35	52-85-7 / 53	3882		
5947	Fenamiphos / Phosphoramidic acid, (1-methylethyl)-, ethyl 3-methyl-4-(methylthio)phenyl ester	$C_{13}H_{22}NO_3PS$ / 303.36	22224-92-6 / 49	3901	1.15[20]	
5948	Fenarimol / 5-Pyrimidinemethanol, α-(2-chlorophenyl)-α-(4-chlorophenyl)-	$C_{17}H_{12}Cl_2N_2O$ / 331.20	60168-88-9 / 118	3903		
5949	Fenbuconazole / 4-(4-Chlorophenyl)-2-phenyl-2-(1H-1,2,4-triazol-1-ylmethyl)butyronitrile	$C_{20}H_{19}ClN_4$ / 350.85	114369-43-6 / 125			
5950	Fenbutatin oxide / Distannoxane, hexakis(2-methyl-2-phenylpropyl)-	$C_{60}H_{78}OSn_2$ / 1052.70	13356-08-6 / 138	3907		
5951	Fenitrothion / Phosphorothioic acid, O,O-dimethyl O-(3-methyl-4-nitrophenyl) ester	$C_9H_{12}NO_5PS$ / 277.24	122-14-5	3922 / 118[0.05]; 164[1]	1.3227[25]	
5952	Fenoxycarb / Ethyl 2-(4-phenoxyphenoxy)ethylcarbamate	$C_{17}H_{19}NO_4$ / 301.34	79127-80-3 / 53			
5953	Fenpropathrin / Cyano(3-phenoxyphenyl)methyl 2,2,3,3-tetramethylcyclopropanecarboxylate	$C_{22}H_{23}NO_3$ / 349.43	64257-84-7 / 47		1.15[25]	
5954	Fensulfothion / Phosphorothioic acid, O,O-diethyl O-[4-(methylsulfinyl)phenyl] ester	$C_{11}H_{17}O_4PS_2$ / 308.36	115-90-2	3943 / 140[0.01]	1.202[20]	
5955	Fenthion / Phosphorothioic acid, O,O-dimethyl O-[3-methyl-4-(methylthio)phenyl] ester	$C_{10}H_{15}O_3PS_2$ / 278.33	55-38-9 / 7.5	3945 / 87[0.01]	1.246[20]	
5956	Fenvalerate / Benzeneacetic acid, 4-chloro-α-(1-methylethyl)-, cyano(3-phenoxyphenyl)methyl ester	$C_{25}H_{22}ClNO_3$ / 419.91	51630-58-1 / dec	3952	1.15[25]	
5957	Ferbam / Iron, tris(dimethylcarbamodithioato-S,S')-, (OC-6-11)-	$C_9H_{18}FeN_3S_6$ / 416.51	14484-64-1 / 180 dec	3954		
5958	Ferrate(2-), chloro[7,12-diethenyl-3,8,13,17-tetramethyl-21H,23H-porphine-2,18-dipropanoato(4-)-N21,N22,N23, / Hemin	$C_{34}H_{32}ClFeN_4O_4$ / 651.95	16009-13-5 / >300	4563		
5959	Ferrate(2-), [7,12-diethenyl-3,8,13,17-tetramethyl-21H,23H-porphine-2,18-dipropanoato(4-)-N21,N22,N23,N24 / Hematin	$C_{34}H_{33}FeN_4O_5$ / 633.51	15489-90-4 / >200	4554	4-26-00-03047	H_2O 1; EtOH 3; eth 1; alk 3
5960	Ferrocene / Bis(cyclopentadienyl)iron	$C_{10}H_{10}Fe$ / 186.04	102-54-5 / 172.5	3985 / 249		H_2O 1
5961	Ferrocene, 1,1'-bis(1,1,3,3-tetramethyl-3-phenyldisilyloxanyl)- / Disiloxane, 1-(cyclopentadienyl)-1,1,3,3-tetramethyl-3-phenyl-, Fe deriv.	$C_{30}H_{42}FeO_2Si_4$ / 602.85	12321-04-9 / 18.7	200-5[0.03]	1.1063[25]	1.5473[25]
5962	Fluazipop-butyl / (±)-2-[4-[[5-(Trifluoromethyl)-2-pyridinyl]oxy]phenoxy]propanoic acid, butyl ester	$C_{19}H_{20}F_3NO_4$ / 383.37	69806-50-4 / 13	4049 / 170[0.5]	1.21[20]	
5963	Fluazipop-P-butyl / (R)-2-[4-(5-Trifluoromethyl-2-pyridyloxy)phenoxy]propionic acid, butyl ester	$C_{19}H_{20}F_3NO_4$ / 383.37	79241-46-6 / 5			
5964	Fluchloralin / Benzenamine, N-(2-chloroethyl)-2,6-dinitro-N-propyl-4-(trifluoromethyl)-	$C_{12}H_{13}ClF_3N_3O_4$ / 355.70	33245-39-5 / 42	4052		
5965	Flucythrinate / Cythrin	$C_{26}H_{23}F_2NO_4$ / 451.47	70124-77-5 / 108[0.35]	4055	1.189[22]	
5966	Fluoranthene / 1,2-(1,8-Naphthylene)benzene	$C_{16}H_{10}$ / 202.26	206-44-0 / 107.8	384	4-05-00-02463 / 1.252[0]	H_2O 1; EtOH 3; eth 3; bz 3
5967	9H-Fluorene / 2,2'-Methylenebiphenyl	$C_{13}H_{10}$ / 166.22	86-73-7 / 114.8	4081 / 295	4-05-00-02142 / 1.203[0]	H_2O 1; EtOH 2; eth 3; ace 3
5968	9H-Fluorene, 2-bromo- / 2-Bromofluorene	$C_{13}H_9Br$ / 245.12	1133-80-8 / 113.5	185[135]	4-05-00-02147	H_2O 1; EtOH 3; chl 4; HOAc 3
5969	9H-Fluorene-2-carboxylic acid, 9-oxo-	$C_{14}H_8O_3$ / 224.22	784-50-9 / 341	sub 340	4-10-00-03040	H_2O 1; EtOH 3; HOAc 3
5970	9H-Fluorene, 9-[(2-chlorophenyl)methylene]- / Fluorene, 9-(2-chlorobenzylidene)	$C_{20}H_{13}Cl$ / 288.78	1643-49-8 / 69.5	180[0.7]	4-05-00-02631	EtOH 4; HOAc 4
5971	9H-Fluorene-2,7-diamine / 2,7-Diaminofluorene	$C_{13}H_{12}N_2$ / 196.25	525-64-4 / 166	4082	4-13-00-00449	H_2O 1; EtOH 3; chl 3
5972	9H-Fluorene, 2,7-dichloro- / 2,7-Dichlorofluorene	$C_{13}H_8Cl_2$ / 235.11	7012-16-0 / 128	sub	4-05-00-02147	bz 4; chl 4

No.	Name Synonym	Mol. Form. Mol. Wt.	CAS RN mp/°C	Merck No. bp/°C	Beil. Ref. den/g cm^{-3}	Solubility n_D
5973	1H-Fluorene, dodecahydro-	$C_{13}H_{22}$ 178.32	5744-03-6	260	4-05-00-00489 0.92[25]	1.5012[20]
5974	9H-Fluorene, 9-methyl- 9-Methylfluorene	$C_{14}H_{12}$ 180.25	2523-37-7 46.5	154-6[15]	4-05-00-02185 1.0263[66]	H_2O 1; EtOH 3; eth 3; ace 3 1.610[66]
5975	9H-Fluoren-9-one	$C_{13}H_8O$ 180.21	486-25-9 84	341.5	4-07-00-01629 1.1300[99]	H_2O 1; EtOH 3; eth 4; ace 3 1.6309[99]
5976	9H-Fluoren-9-one, 2-chloro-	$C_{13}H_7ClO$ 214.65	3096-47-7 125.5	sub	4-07-00-01634	EtOH 4
5977	9H-Fluoren-9-one, 2-nitro-	$C_{13}H_7NO_3$ 225.20	3096-52-4 224.3	sub	4-07-00-01636	EtOH 2; ace 3; sulf 3; HOAc 3
5978	9H-Fluoren-9-one, 1,2,3,4-tetrahydro- Phentydrone	$C_{13}H_{12}O$ 184.24	634-19-5 81.5	7235 139-40[0.05]	4-07-00-01302	
5979	9H-Fluoren-9-one, 2,4,7-trinitro- 2,4,7-Trinitrofluorenone	$C_{13}H_5N_3O_7$ 315.20	129-79-3 176	9641	4-07-00-01638	H_2O 2; ace 4; bz 4; chl 4
5980	Fluorodifen Benzene, 2-nitro-1-(4-nitrophenoxy)-4-(trifluoromethyl)-	$C_{13}H_7F_3N_2O_5$ 328.20	15457-05-3 94			
5981	Fluridone 4(1H)-Pyridinone, 1-methyl-3-phenyl-5-[3-(trifluoromethyl)phenyl]-	$C_{19}H_{14}F_3NO$ 329.32	59756-60-4 155			
5982	Fluroxypyr [(4-Amino-3,5-dichloro-6-fluoro-2-pyridyl)oxy]acetic acid	$C_7H_5Cl_2FN_2O_3$ 255.03	69377-81-7 232	4128		
5983	Fluvalinate 2-[[2-Chloro-4-(trifluoromethyl)phenyl]amino]-3-methylbutanoic acid, (±)-cyano(3-phenoxyphenyl)methyl ester	$C_{26}H_{22}ClF_3N_2O_3$ 502.92	102851-06-9 	>450	1.29[25]	
5984	Folpet 1H-Isoindole-1,3(2H)-dione, 2-[(trichloromethyl)thio]-	$C_9H_4Cl_3NO_2S$ 296.56	133-07-3 177	4142		
5985	Fomesafen 5-[2-Chloro-4-(trifluoromethyl)phenoxy]-N-(methylsulfonyl)-2-nitrobenzamide	$C_{15}H_{10}ClF_3N_2O_6S$ 438.77	72178-02-0 220		1.28[20]	
5986	Fonofos Phosphonodithioic acid, ethyl-, O-ethyl S-phenyl ester	$C_{10}H_{15}OPS_2$ 246.33	944-22-9 	4147 130[0.1]	1.16[25]	
5987	Formaldehyde Methanal	CH_2O 30.03	50-00-0 -92	4148 -19.1	4-01-00-03017 0.815[-20]	H_2O 3; EtOH 3; eth 5; ace 5
5988	Formaldehyde, oxime	CH_3NO 45.04	75-17-2 2.5	109[15]	4-01-00-03055 1.133[25]	H_2O 3; EtOH 4; eth 4
5989	Formamide Methanamide	CH_3NO 45.04	75-12-7 2.55	4151 220	4-02-00-00045 1.1334[20]	H_2O 5; EtOH 5; eth 2; ace 3 1.4472[20]
5990	Formamide, N-cyclohexyl-	$C_7H_{13}NO$ 127.19	766-93-8 39.5	260; 113[0.7]	4-12-00-00038 1.0123[17]	
5991	Formamide, N,N-diethyl-	$C_5H_{11}NO$ 101.15	617-84-5 	177.5	4-04-00-00346 0.9080[19]	H_2O 5; EtOH 4; eth 4; ace 5 1.4321[25]
5992	Formamide, N,N-dimethyl- N,N-Dimethylformamide	C_3H_7NO 73.09	68-12-2 -60.4	3232 153	4-04-00-00171 0.944[25]	H_2O 5; EtOH 5; eth 5; ace 5 1.4305[20]
5993	Formamide, N,N-diphenyl-	$C_{13}H_{11}NO$ 197.24	607-00-1 73.5	337.5; 189[13]	4-12-00-00370	H_2O 1; EtOH 3; eth 3; bz 3
5994	Formamide, N-ethyl- N-Ethylformamide	C_3H_7NO 73.09	627-45-2 	198	4-04-00-00346 0.9552[20]	H_2O 5; EtOH 5; eth 5 1.4320[20]
5995	Formamide, N-methyl- N-Methylformamide	C_2H_5NO 59.07	123-39-7 -3.8	199.5	4-04-00-00170 1.011[19]	H_2O 4; ace 4; EtOH 4 1.4319[20]
5996	Formamide, N-(2-methylphenyl)-	C_8H_9NO 135.17	94-69-9 62	288	3-12-00-01852 1.086[55]	H_2O 3; EtOH 4
5997	Formamide, N-(3-methylphenyl)-	C_8H_9NO 135.17	3085-53-8 <-18	290 dec; 176[17]	3-12-00-01962	H_2O 3; NaOH 3
5998	Formamide, N-methyl-N-phenyl-	C_8H_9NO 135.17	93-61-8 14.5	243	4-12-00-00370 1.0948[20]	H_2O 2; EtOH 3; ace 3; ctc 2 1.5589[20]
5999	Formamide, N-phenyl- Formanilide	C_7H_7NO 121.14	103-70-8 46	4152 271	4-12-00-00388 1.1186[50]	H_2O 3; EtOH 4; eth 3; bz 3
6000	Formamide, N-(2,2,2-trichloro-1-hydroxyethyl)- Chloral formamide	$C_3H_4Cl_3NO_2$ 192.43	515-82-2 120	2060	3-02-00-00059	ace 4; eth 4; EtOH 4
6001	Formetanate hydrochloride Methanimidamide, N,N-dimethyl-N'-[3-[[(methylamino)carbonyl]oxy]phenyl]-, monohydrochloride	$C_{11}H_{16}ClN_3O_2$ 257.72	23422-53-9			
6002	Formic acid Methanoic acid	CH_2O_2 46.03	64-18-6 8.3	4153 101	4-02-00-00003 1.220[20]	H_2O 5; EtOH 5; eth 5; ace 4 1.3714[20]
6003	Formic acid, ammonium salt Ammonium formate	CH_5NO_2 63.06	540-69-2	548	4-02-00-00003	
6004	Formic acid, butyl ester Butyl formate	$C_5H_{10}O_2$ 102.13	592-84-7 -91.5	106.1	4-02-00-00028 0.8885[20]	H_2O 2; EtOH 5; eth 5; ace 3 1.3912[20]
6005	Formic acid, cyclohexyl ester	$C_7H_{12}O_2$ 128.17	4351-54-6 	163	4-06-00-00035 1.0057[0]	H_2O 1; EtOH 3; eth 4; HCOOH 3 1.4430[20]

No.	Name Synonym	Mol. Form. Mol. Wt.	CAS RN mp/°C	Merck No. bp/°C	Beil. Ref. den/g cm^{-3}	Solubility n_D
6006	Formic acid, ethenyl ester Vinyl formate	$C_3H_4O_2$ 72.06	692-45-5		4-02-00-00033	
6007	Formic acid, ethyl ester Ethyl formate	$C_3H_6O_2$ 74.08	109-94-4 -79.6	3763 54.4	4-02-00-00023 0.9168[20]	H_2O 3; EtOH 5; eth 5; ace 4 1.3598[10]
6008	Formic acid, heptyl ester Heptyl formate	$C_8H_{16}O_2$ 144.21	112-23-2	178.1	4-02-00-00031 0.8784[20]	H_2O 1; EtOH 5; eth 5 1.4140[20]
6009	Formic acid, hexyl ester Hexyl formate	$C_7H_{14}O_2$ 130.19	629-33-4 -62.6	155.5	4-02-00-00031 0.8813[20]	H_2O 1; EtOH 5; eth 5 1.4071[20]
6010	Formic acid, methyl ester Methyl formate	$C_2H_4O_2$ 60.05	107-31-3 -99	5994 31.7	4-02-00-00020 0.9742[20]	H_2O 4; EtOH 5; eth 3; chl 3 1.3433[20]
6011	Formic acid, 1-methylethyl ester Isopropyl formate	$C_4H_8O_2$ 88.11	625-55-8 68.2		4-02-00-00027 0.8728[20]	H_2O 2; EtOH 5; eth 5; ace 4 1.3678[20]
6012	Formic acid, 1-methylpropyl ester	$C_5H_{10}O_2$ 102.13	589-40-2 97		4-02-00-00029 0.8846[20]	H_2O 2; EtOH 5; eth 5; ace 3 1.3865[20]
6013	Formic acid, 2-methylpropyl ester Isobutyl formate	$C_5H_{10}O_2$ 102.13	542-55-2 -95.8	5026 98.2	4-02-00-00029 0.8776[20]	H_2O 2; EtOH 5; eth 5; ace 4 1.3857[20]
6014	Formic acid, octyl ester	$C_9H_{18}O_2$ 158.24	112-32-3 -39.1	198.8	4-02-00-00031 0.8744[20]	H_2O 1; EtOH 3; eth 5; ctc 2 1.4208[15]
6015	Formic acid, pentyl ester Pentyl formate	$C_6H_{12}O_2$ 116.16	638-49-3 -73.5	130.4	4-02-00-00030 0.8853[20]	H_2O 2; EtOH 5; eth 5 1.3992[20]
6016	Formic acid, phenylmethyl ester Benzyl formate	$C_8H_8O_2$ 136.15	104-57-4 203; 84[10]	1148	4-06-00-02262 1.081[20]	H_2O 1; EtOH 3; eth 5; ace 3 1.5154[20]
6017	Formic acid, 2-propenyl ester Allyl formate	$C_4H_6O_2$ 86.09	1838-59-1 83.6		3-02-00-00046 0.9460[20]	H_2O 2; EtOH 3; eth 5
6018	Formic acid, propyl ester Propyl formate	$C_4H_8O_2$ 88.11	110-74-7 -92.9	7871 80.9	4-02-00-00026 0.9058[20]	H_2O 2; EtOH 5; eth 5; ctc 2 1.3779[20]
6019	Formyl fluoride Fluoroformaldehyde	CHFO 48.02	1493-02-3 -142.2	4162 -26.5	4-02-00-00042 1.1950[-30]	
6020	Fosetyl-Al Ethyl hydrogen phosphonate, aluminum salt	$C_6H_{18}AlO_9P_3$ 354.11	39148-24-8 >300	4167		
6021	Fosthiazate (RS)-S-sec-Butyl O-ethyl 2-oxo-1,3-thiazolidin-3-ylphosphonothioate	$C_9H_{18}NO_3PS_2$ 283.35	98886-44-3			
6022	D-Fructopyranose, 3-O-α-D-glucopyranosyl- D-Turanose	$C_{12}H_{22}O_{11}$ 342.30	547-25-1 168	9728	5-17-07-00213	H_2O 4; EtOH 3; MeOH 3
6023	β-D-Fructose β-Levulose	$C_6H_{12}O_6$ 180.16	53188-23-1 103 dec	4185	4-01-00-04401 1.60[20]	H_2O 4; EtOH 3; ace 4; MeOH 3
6024	Fulminic acid Carbyloxime	CHNO 43.03	506-85-4 unstable in pure form			eth 3
6025	Furan Oxacyclopentadiene	C_4H_4O 68.08	110-00-9 -85.6	4206 31.5	5-17-01-00291 0.9514[20]	H_2O 1; EtOH 4; eth 4; ace 3 1.4214[20]
6026	2-Furanacetic acid	$C_6H_6O_3$ 126.11	2745-26-8 68.5	102[0.4]	5-18-06-00206	H_2O 3; bz 3; MeOH 3; peth 3
6027	2-Furanacetic acid, butyl ester 2-Furylacetic acid, butyl ester	$C_{10}H_{14}O_3$ 182.22	4915-23-5 110-11[13]		4-18-00-04062 1.0232[25]	1.4558[25]
6028	2-Furanacetic acid, ethyl ester	$C_8H_{10}O_3$ 154.17	4915-21-3 88[15]		5-18-06-00206 1.0763[25]	
6029	2-Furanacetic acid, methyl ester	$C_7H_8O_3$ 140.14	4915-22-4 87-8[21]		5-18-06-00206 1.1250[20]	1.4571[25] 1.4638[25]
6030	3-Furanacetic acid, tetrahydro-2,2-dimethyl-5-oxo- Terpenylic acid	$C_8H_{12}O_4$ 172.18	26754-48-3 90	9100	5-18-08-00032	H_2O 4
6031	2-Furanacetonitrile 2-Furylacetonitrile	C_6H_5NO 107.11	2745-25-7 75-80[20]		5-18-06-00207 1.0854[25]	eth 4; EtOH 4 1.4693[20]
6032	Furan, 2,5-bis(1,1-dimethylethyl)-	$C_{12}H_{20}O$ 180.29	4789-40-6 210		5-17-01-00439 0.837[20]	H_2O 1; EtOH 3; eth 3 1.4369[20]
6033	Furan, 2-bromo-	C_4H_3BrO 146.97	584-12-3 103		5-17-01-00295 1.6500[20]	H_2O 2; EtOH 3; eth 3; ace 3 1.4980[20]
6034	Furan, 3-bromo- 3-Bromofuran	C_4H_3BrO 146.97	22037-28-1 103		5-17-01-00295 1.6606[20]	ace 4; bz 4; eth 4; EtOH 4 1.4958[20]
6035	Furan, 2-(bromomethyl)-	C_5H_5BrO 161.00	4437-18-7 32-4[2]		4-17-00-00268 1.557[20]	1.5380[20]
6036	Furan, 2-(bromomethyl)tetrahydro-	C_5H_9BrO 165.03	1192-30-9 170; 70[22]		5-17-01-00080 1.3653[20]	EtOH 3; eth 3 1.4850[20]
6037	Furan, 2-(butoxymethyl)-	$C_9H_{14}O_2$ 154.21	56920-82-2 190		2-17-00-00114 0.9516[20]	eth 4; EtOH 4 1.4522[20]
6038	Furan, 2,2'-sec-butylidenedi- Butane, 2,2-di(2-furyl)	$C_{12}H_{14}O_2$ 190.24	100121-73-1 64-6[1]		4-19-00-00287 1.0330[20]	eth 4; EtOH 4 1.4970[20]
6039	2-Furancarbonitrile	C_5H_3NO 93.08	617-90-3 147		5-18-06-00122 1.0822[20]	EtOH 3; eth 3 1.4798[20]
6040	2-Furancarbonyl chloride	$C_5H_3ClO_2$ 130.53	527-69-5 -2 173		5-18-06-00110 1.324[25]	H_2O 1; eth 3; ctc 2; chl 3 1.5310[20]
6041	2-Furancarboxaldehyde Furfural	$C_5H_4O_2$ 96.09	98-01-1 -36.5	4214 161.7	5-17-09-00292 1.1594[20]	H_2O 3; EtOH 4; eth 5; ace 4 1.5261[20]
6042	3-Furancarboxaldehyde	$C_5H_4O_2$ 96.09	498-60-2 145; 70-2[43]		5-17-09-00372 1.109[20]	1.4945[20]
6043	2-Furancarboxaldehyde, 5-bromo-	$C_5H_3BrO_2$ 174.98	1899-24-7 83.5 201; 112[16]		5-17-09-00320	eth 4; EtOH 4

No.	Name Synonym	Mol. Form. Mol. Wt.	CAS RN mp/°C	Merck No. bp/°C	Beil. Ref. den/g cm^{-3}	Solubility n_D
6044	2-Furancarboxaldehyde, 5-(hydroxymethyl)- 5-(Hydroxymethyl)-2-furaldehyde	$C_6H_8O_3$ 126.11	67-47-0 31.5	4764 114-6[1]	5-18-01-00130 1.2062[25]	H_2O 3; EtOH 3; eth 2; bz 3 1.5627[18]
6045	2-Furancarboxaldehyde, 5-methyl-	$C_6H_6O_2$ 110.11	620-02-0	187; 89[26]	5-17-09-00404 1.1072[18]	H_2O 3; EtOH 4; eth 5; ctc 2 1.5264[20]
6046	2-Furancarboxaldehyde, 5-nitro-	$C_5H_3NO_4$ 141.08	698-63-5 35.5	128-32[10]	5-17-09-00323	H_2O 2; peth 3
6047	2-Furancarboxaldehyde, oxime, (E)-	$C_5H_5NO_2$ 111.10	620-03-1	205 dec; 98[9]	5-17-09-00306 1.1550[80]	H_2O 3; EtOH 4; eth 4; bz 4
6048	2-Furancarboxaldehyde, oxime, (Z)-	$C_5H_5NO_2$ 111.10	1450-58-4 91	100[12]	5-17-09-00306 1.3800[20]	H_2O 2; EtOH 4; eth 4; bz 4
6049	2-Furancarboxaldehyde, tetrahydro-	$C_5H_8O_2$ 100.12	7681-84-7	142; 46[29]	5-17-09-00027 1.0727[20]	H_2O 4; eth 4 1.4366[20]
6050	2-Furancarboxylic acid 2-Furoic acid	$C_5H_4O_3$ 112.08	88-14-2 133.5	4219 231	5-18-06-00102	H_2O 3; EtOH 3; eth 4; ace 2
6051	3-Furancarboxylic acid	$C_5H_4O_3$ 112.08	488-93-7 122.5	sub 105	5-18-06-00196	H_2O 2; EtOH 3; eth 4; AcOEt 3
6052	2-Furancarboxylic acid, butyl ester	$C_9H_{12}O_3$ 168.19	583-33-5	233	4-18-00-03918 1.0555[20]	H_2O 1; EtOH 3; eth 3; bz 3 1.4740
6053	2-Furancarboxylic acid, 5-chloro-, ethyl ester Ethyl 5-chloro-2-furancarboxylate	$C_7H_7ClO_3$ 174.58	4301-39-7 2	217; 108[23]	5-18-06-00131 1.2418[25]	
6054	3-Furancarboxylic acid, 4,5-dihydro-5-oxo-	$C_5H_4O_4$ 128.08	585-68-2 164	dec	4-18-00-05333	H_2O 4; EtOH 3; MeOH 3
6055	3-Furancarboxylic acid, 2,5-dimethyl-, ethyl ester	$C_9H_{12}O_3$ 168.19	29113-63-1 208.5; 96[19]		5-18-06-00233 1.0478[23]	1.4686[20]
6056	2-Furancarboxylic acid, ethyl ester	$C_7H_8O_3$ 140.14	614-99-3 34.5	196.8	5-18-06-00104 1.1174[21]	H_2O 1; EtOH 5; eth 5; ace 5 1.4797[21]
6057	2-Furancarboxylic acid, 2-furanylmethyl ester	$C_{10}H_8O_4$ 192.17	615-11-2 27.5	122[2]	4-18-00-03936 1.2347[25]	ace 4; bz 4; eth 4; EtOH 4 1.5280[20]
6058	2-Furancarboxylic acid, hexyl ester α-Furoic acid, hexyl ester	$C_{11}H_{16}O_3$ 196.25	39251-86-0	105-7[1]	2-18-00-00267 1.0170[20]	H_2O 1; EtOH 3; eth 3; bz 5
6059	2-Furancarboxylic acid, methyl ester	$C_6H_6O_3$ 126.11	611-13-2	181.3	5-18-06-00103 1.1786[21]	H_2O 1; EtOH 3; eth 3; bz 3 1.4860[20]
6060	2-Furancarboxylic acid, 5-methyl-	$C_6H_6O_3$ 126.11	1917-15-3 109.5	105[1]	5-18-06-00217	H_2O 2; EtOH 3; eth 3; bz 2
6061	2-Furancarboxylic acid, 1-methylethyl ester	$C_8H_{10}O_3$ 154.17	6270-34-4	198.5	0-18-00-00275 1.0655[24]	H_2O 1; EtOH 3; eth 3; ace 3 1.4682[24]
6062	2-Furancarboxylic acid, 2-methylpropyl ester	$C_9H_{12}O_3$ 168.19	20279-53-2	222	2-18-00-00266 1.0388[20]	ace 4; bz 4; eth 4; EtOH 4 1.4676[20]
6063	3-Furancarboxylic acid, methyl ester	$C_6H_6O_3$ 126.11	13129-23-2	160	5-18-06-00196 1.1733[15]	H_2O 1; EtOH 3; ace 3 1.4676[20]
6064	3-Furancarboxylic acid, 2-methyl-, ethyl ester	$C_8H_{10}O_3$ 154.17	28921-35-9	85-7[20]	5-18-06-00214 1.0102[25]	eth 4 1.4620[25]
6065	3-Furancarboxylic acid, 5-methyl- 3-Furoic acid, 5-methyl-	$C_6H_6O_3$ 126.11	21984-93-0 119	sub	5-18-06-00216	H_2O 4; eth 4; EtOH 4
6066	2-Furancarboxylic acid, 5-nitro-	$C_5H_3NO_5$ 157.08	645-12-5 186	sub	5-18-06-00139	H_2O 3; EtOH 3; eth 3; ace 2
6067	2-Furancarboxylic acid, octyl ester α-Furoic acid, octyl ester	$C_{13}H_{20}O_3$ 224.30	39251-88-2	126-7[1]	2-18-00-00267 0.9885[20]	H_2O 1; EtOH 3
6068	2-Furancarboxylic acid, pentyl ester 2-Furoic acid, pentyl ester	$C_{10}H_{14}O_3$ 182.22	4996-48-9	95-7[1]	2-18-00-00266 1.0335[20]	H_2O 1; EtOH 5
6069	2-Furancarboxylic acid, phenylmethyl ester	$C_{12}H_{10}O_3$ 202.21	5380-40-5	179-81[18]	4-18-00-03921 1.1623[22]	ace 4; eth 4 1.5550[20]
6070	2-Furancarboxylic acid, 2-propenyl ester	$C_8H_8O_3$ 152.15	4208-49-5	207.5	5-18-06-00105 1.115[25]	eth 3; ace 3; ctc 2 1.4945[20]
6071	2-Furancarboxylic acid, propyl ester	$C_8H_{10}O_3$ 154.17	615-10-1	210.9; 112[20]	5-18-06-00104 1.0745[20]	H_2O 1; EtOH 3; eth 3; ace 3 1.4787[20]
6072	2-Furancarboxylic acid, tetrahydro-	$C_5H_8O_3$ 116.12	16874-33-2 21	145[25]	5-18-06-00010 1.1912[20]	H_2O 4 1.4612[20]
6073	Furan, 2-chloro- 2-Chlorofuran	C_4H_3ClO 102.52	3187-94-8	78	5-17-01-00294 1.1923[20]	ace 4; eth 4; EtOH 4 1.4569[20]
6074	Furan, 3-chloro- 3-Chlorofuran	C_4H_3ClO 102.52	50689-17-3	80	5-17-01-00295 1.2094[20]	ace 4; eth 4 1.4601[20]
6075	Furan, 2-(chloromethyl)-	C_5H_5ClO 116.55	617-88-9	49[26]	5-17-01-00323 1.1783[20]	bz 4; eth 4; EtOH 4 1.4941[20]
6076	Furan, 2-(cyclohexyloxy)- Cyclohexyl 2-furyl ether	$C_{10}H_{14}O_2$ 166.22	99172-57-3	118-9[28]	4-17-00-01219 1.0200[28]	ace 4; eth 4; EtOH 4 1.4861[28]
6077	Furan, 2,3-dibromo- 2,3-Dibromofuran	$C_4H_2Br_2O$ 225.87	30544-34-4	167	5-17-01-00295 1.9938[25]	1.5458[25]
6078	Furan, 2,5-dibromo- 2,5-Dibromofuran	$C_4H_2Br_2O$ 225.87	32460-00-7 9.5	164.5	5-17-01-00295 2.27[20]	1.5455[20]
6079	2,3-Furandicarboxylic acid Isocitrate lactone	$C_6H_4O_5$ 156.09	4282-24-0 226	sub	5-18-07-00043	H_2O 4; EtOH 4; eth 2; AcOEt 4
6080	2,4-Furandicarboxylic acid	$C_6H_4O_5$ 156.09	4282-28-4 266	sub	5-18-07-00045	H_2O 2; EtOH 4; ace 4; chl 2
6081	2,5-Furandicarboxylic acid Dehydromucic acid	$C_6H_4O_5$ 156.09	3238-40-2 342	sub	5-18-07-00047 1.7400[20]	H_2O 2; EtOH 2

No.	Name / Synonym	Mol. Form. / Mol. Wt.	CAS RN / mp/°C	Merck No. / bp/°C	Beil. Ref. / den/g cm^{-3}	Solubility / n_D
6082	3,4-Furandicarboxylic acid, diethyl ester / 2,3-Bis(ethoxycarbonyl)furan	$C_{10}H_{12}O_5$ / 212.20	30814-77-8 /	/ 155[13]	5-18-07-00053 / 1.14[25]	/ 1.4717[20]
6083	2,5-Furandicarboxylic acid, dimethyl ester / Dimethyl 2,5-furandicarboxylate	$C_8H_8O_5$ / 184.15	4282-32-0 / 112	/ 154[15]	5-18-07-00048 /	eth 4; EtOH 4; chl 4
6084	Furan, 2-(diethoxymethyl)-	$C_9H_{14}O_3$ / 170.21	13529-27-6 /	/ 191.5	5-17-09-00299 / 0.9976[20]	EtOH 4 / 1.4451[20]
6085	Furan, 2,2-diethyltetrahydro- / 2,2-Diethyltetrahydrofuran	$C_8H_{16}O$ / 128.21	1193-35-7 /	/ 146	4-17-00-00107 / 0.8703[20]	ace 4; bz 4; eth 4; EtOH 4 / 1.4317[20]
6086	Furan, 2,3-dihydro- / 2,3-Dihydrofuran	C_4H_6O / 70.09	1191-99-7 /	/ 54.5	5-17-01-00172 / 0.927[25]	/ 1.4239[20]
6087	Furan, 2,5-dihydro-2,5-dimethoxy- / 2,5-Dimethoxy-2,5-dihydrofuran	$C_6H_{10}O_3$ / 130.14	332-77-4 /	/ 161	5-17-05-00073 / 1.073[25]	/ 1.4339[20]
6088	Furan, 2,4-dimethyl- / 3,5-Dimethylfuran	C_6H_8O / 96.13	3710-43-8 /	/ 94	5-17-01-00343 / 0.8993[20]	/ 1.4371[20]
6089	Furan, 2,5-dimethyl- / 2,5-Dimethylfuran	C_6H_8O / 96.13	625-86-5 / -62.8	/ 93.5	5-17-01-00344 / 0.8883[20]	H_2O 1; EtOH 3; eth 3; ace 3 / 1.4363[20]
6090	Furan, 2-(1,1-dimethylethyl)- / 2-tert-Butylfuran	$C_8H_{12}O$ / 124.18	7040-43-9 /	/ 119.5	5-17-01-00371 / 0.869[20]	ace 4; eth 4; EtOH 4 / 1.4373[20]
6091	2,5-Furandione / Maleic anhydride	$C_4H_2O_3$ / 98.06	108-31-6 / 52.8	5586 / 202	5-17-11-00055 / 1.314[60]	H_2O 3; eth 3; ace 3; chl 3
6092	2,5-Furandione, 3-chloro-	C_4HClO_3 / 132.50	96-02-6 / 33	/ 196	5-17-11-00063 / 1.5375[25]	/ 1.4980[20]
6093	2,5-Furandione, dihydro- / Succinic anhydride	$C_4H_4O_3$ / 100.07	108-30-5 / 119	8841 / 261	5-17-11-00006 / 1.2[20]	H_2O 1; EtOH 3; eth 2; chl 3
6094	2,5-Furandione, dihydro-3-methyl-	$C_5H_6O_3$ / 114.10	4100-80-5 / 34	/ 239	5-17-11-00011 / 1.22[25]	
6095	2,5-Furandione, dihydro-3-methylene-	$C_5H_4O_3$ / 112.08	2170-03-8 / 69	/ 139[30.] 114[18.]	5-17-11-00066 /	eth 2; chl 4
6096	2,5-Furandione, dihydro-3-phenyl-, (±)- / Succinic anhydride, phenyl (DL)	$C_{10}H_8O_3$ / 176.17	112489-85-7 / 54	/ 204[22]	5-17-11-00275 /	ace 4; bz 4; eth 4; EtOH 4
6097	2,5-Furandione, 3,4-dimethyl-	$C_6H_6O_3$ / 126.11	766-39-2 / 96	/ 223	5-17-11-00069 / 1.107[100]	H_2O 2; EtOH 4; eth 4; bz 4
6098	2,5-Furandione, 3-methyl-	$C_5H_4O_3$ / 112.08	616-02-4 / 7.5	/ 213.5	5-17-11-00065 / 1.2469[16]	ace 4; eth 4; EtOH 4 / 1.4710[21]
6099	2,5-Furandione, 3,3,4,4-tetrafluorodihydro-	$C_4F_4O_3$ / 172.04	699-30-9 /	/ 54.5	5-17-11-00007 / 1.6209[20]	/ 1.3240[20]
6100	Furan, 2,5-diphenyl-	$C_{16}H_{12}O$ / 220.27	955-83-9 / 91	/ 344	5-17-02-00398 /	H_2O 1; EtOH 4; eth 4; ace 3
6101	Furan, 2,2'-[dithiobis(methylene)]bis-	$C_{10}H_{10}O_2S_2$ / 226.32	4437-20-1 / 10	/ 167[13.] 112[0.5]	4-17-00-01258 /	EtOH 4
6102	Furan, 2-ethenyl-	C_6H_6O / 94.11	1487-18-9 / -94	/ 99.5	5-17-01-00477 / 0.9445[19]	/ 1.4992[19]
6103	Furan, 2-ethoxy- / Ethyl 2-furyl ether	$C_6H_8O_2$ / 112.13	5809-07-4 /	/ 125.5	4-17-00-01219 / 0.9849[23]	/ 1.4500[23]
6104	Furan, 2-(ethoxymethyl)-	$C_7H_{10}O_2$ / 126.16	6270-56-0 /	/ 149	5-17-03-00340 / 0.9844[20]	H_2O 1; EtOH 3; eth 3 / 1.4523[20]
6105	Furan, 2-ethyl- / 2-Ethylfuran	C_6H_8O / 96.13	3208-16-0 /	/ 92	5-17-01-00338 / 0.911[15]	EtOH 3; eth 3; ace 3; bz 3 / 1.4466[23]
6106	Furan, 2-ethyl-5-methyl- / 2-Methyl-5-ethylfuran	$C_7H_{10}O$ / 110.16	1703-52-2 /	/ 118.5	5-17-01-00357 / 0.8883[20]	bz 4; eth 4; EtOH 4 / 1.4473[20]
6107	Furan, 2-ethyltetrahydro- / 2-Ethyltetrahydrofuran	$C_6H_{12}O$ / 100.16	1003-30-1 /	/ 109	5-17-01-00095 / 0.8570[19]	ace 4; bz 4; eth 4; EtOH 4 / 1.4147[19]
6108	Furan, 2-ethynyl-	C_6H_4O / 92.10	18649-64-4 /	/ 105.5	4-17-00-00396 / 0.9919[20]	/ 1.5055[20]
6109	Furan, 2,2,3,3,4,4,5-heptafluorotetrahydro-5-(nonafluorobutyl)- / Perfluoro-2-butyltetrahydrofuran	$C_8F_{16}O$ / 416.06	335-36-4 /	/ 102.6	/	
6110	Furan, 2-iodo- / 2-Furyl iodide	C_4H_3IO / 193.97	54829-48-0 /	/ 43-5[15]	5-17-01-00295 / 2.024[20]	eth 4 / 1.5661[20]
6111	Furan, 3-iodo- / 3-Iodofuran	C_4H_3IO / 193.97	29172-20-1 /	/ 133; 37-8[22]	5-17-01-00296 / 2.045[20]	H_2O 1; eth 3 / 1.5610[20]
6112	Furan, 2-isopropoxy- / 2-Furyl isopropyl ether	$C_7H_{10}O_2$ / 126.16	98272-34-5 /	/ 135.5	4-17-00-01219 / 0.9689[20]	/ 1.4419[20]
6113	2-Furanmethanamine / Furfurylamine	C_5H_7NO / 97.12	617-89-0 /	/ 145.5	5-18-09-00541 / 1.0995[20]	H_2O 5; EtOH 5; eth 3; chl 3 / 1.4908[20]
6114	2-Furanmethanamine, N-(2-furanylmethyl)- / Difurfurylamine	$C_{10}H_{11}NO_2$ / 177.20	18240-50-1 /	/ 135-42[15]	4-18-00-07092 / 1.1045[20]	eth 4 / 1.5168[20]
6115	2-Furanmethanamine, N-methyl-	C_6H_9NO / 111.14	4753-75-7 /	/ 149	5-18-09-00541 / 0.989[25]	/ 1.4729[20]
6116	2-Furanmethanamine, tetrahydro- / Tetrahydrofurfurylamine	$C_5H_{11}NO$ / 101.15	4795-29-3 /	/ 153	5-18-09-00488 / 0.9752[20]	H_2O 4; eth 4; EtOH 4 / 1.4551[20]
6117	2-Furanmethanethiol	C_5H_6OS / 114.17	98-02-2 /	/ 157	5-17-03-00351 / 1.1319[20]	H_2O 1; chl 2 / 1.5329[20]
6118	2-Furanmethanol / Furfuryl alcohol	$C_5H_6O_2$ / 98.10	98-00-0 / -31	4215 / 171	5-17-03-00338 / 1.1296[20]	H_2O 5; EtOH 4; eth 4; chl 3 / 1.4869[20]
6119	2-Furanmethanol, acetate	$C_7H_8O_3$ / 140.14	623-17-6 /	/ 179	5-17-03-00344 / 1.1175[20]	H_2O 1; EtOH 3; eth 3 / 1.4327[20]
6120	2-Furanmethanol, 5-(aminomethyl)tetrahydro-	$C_6H_{13}NO_2$ / 131.17	589-14-0 / <-70	/ 103[0.3]	1.1021[25]	/ 1.4870[25]

No.	Name / Synonym	Mol. Form. / Mol. Wt.	CAS RN / mp/°C	Merck No. / bp/°C	Beil. Ref. / den/g cm^{-3}	Solubility / n_D
6121	2-Furanmethanol, α-methyl-	$C_6H_8O_2$ / 112.13	4208-64-4	162.5	5-17-03-00374 / 1.0739[25]	1.4827[15]
6122	2-Furanmethanol, 5-methyl-	$C_6H_8O_2$ / 112.13	3857-25-8	195 dec; 81[23]	5-17-03-00387 / 1.0769[20]	eth 4; EtOH 4 / 1.4853[20]
6123	2-Furanmethanol, propanoate	$C_8H_{10}O_3$ / 154.17	623-19-8	195	2-17-00-00115 / 1.1085[20]	H_2O 2; EtOH 3; eth 5; ace 3
6124	2-Furanmethanol, tetrahydro- / Tetrahydrofurfuryl alcohol	$C_5H_{10}O_2$ / 102.13	97-99-4 / <-80	9146 / 178	5-17-03-00115 / 1.0524[20]	ace 4; eth 4 / 1.4520[20]
6125	2-Furanmethanol, tetrahydro-, acetate	$C_7H_{12}O_3$ / 144.17	637-64-9	193; 89[18]	5-17-03-00119 / 1.0624[20]	H_2O 4; eth 4; EtOH 4; chl 4 / 1.4350[25]
6126	2-Furanmethanol, tetrahydro-, propanoate	$C_8H_{14}O_3$ / 158.20	637-65-0	205.5	5-17-03-00119 / 1.044[20]	eth 4; EtOH 4; chl 4
6127	Furan, 2-methoxy- / 2-Methoxyfuran	$C_5H_6O_2$ / 98.10	25414-22-6	110.5	5-17-03-00283 / 1.0646[25]	1.4468[25]
6128	Furan, 2-(methoxymethyl)-	$C_6H_8O_2$ / 112.13	13679-46-4	132	5-17-03-00340 / 1.0163[20]	H_2O 1; EtOH 3; eth 4 / 1.4570[20]
6129	Furan, 2-(methoxymethyl)-5-nitro- / 2-(Methoxymethyl)-5-nitrofuran	$C_6H_7NO_4$ / 157.13	586-84-5	5917 / 104-5[3]	4-17-00-01254 / 1.281[20]	EtOH 4 / 1.5325[20]
6130	Furan, 2-methyl- / 2-Methylfuran	C_5H_6O / 82.10	534-22-5 / -87.5	65	5-17-01-00322 / 0.9132[20]	H_2O 2; EtOH 3; eth 3; ctc 2 / 1.4342[20]
6131	Furan, 3-methyl- / 3-Methylfuran	C_5H_6O / 82.10	930-27-8	66	5-17-01-00330 / 0.923[18]	H_2O 1; EtOH 3; eth 3 / 1.4330[19]
6132	Furan, 2,2'-methylenebis-	$C_9H_8O_2$ / 148.16	1197-40-6	94[22]	5-19-01-00545 / 1.102[20]	H_2O 1; EtOH 4; eth 4; ace 4 / 1.5049[20]
6133	Furan, 3-(4-methyl-3-pentenyl)-	$C_{10}H_{14}O$ / 150.22	539-52-6	185.5	5-17-01-00515 / 0.9017[20]	1.4705[21]
6134	Furan, 2-nitro-	$C_4H_3NO_3$ / 113.07	609-39-2 / 30	134[123]; 84[13]	5-17-01-00296	H_2O 3; EtOH 3; eth 3
6135	Furan, 2-(octyloxy)- / 2-Furyl octyl ether	$C_{12}H_{20}O_2$ / 196.29	100314-90-7	129-30[18]	4-17-00-01219 / 0.9214[28]	eth 4 / 1.4520[20]
6136	2-Furanol, 3,4-bis(1,3-benzodioxol-5-ylmethyl)tetrahydro- / Cubebin	$C_{20}H_{20}O_6$ / 356.38	18423-69-3 / 131.5	2616	5-19-12-00048	eth 4; EtOH 4; chl 4
6137	3-Furanol, tetrahydro-	$C_4H_8O_2$ / 88.11	453-20-3	181	5-17-03-00064 / 1.09[25]	1.4500[20]
6138	2-Furanol, tetrahydro-, benzoate	$C_{11}H_{12}O_3$ / 192.21	3333-44-6	302; 139[2]	5-17-03-00060 / 1.137[20]	eth 4; EtOH 4
6139	2(3H)-Furanone, 3-acetyldihydro- / α-Acetylbutyrolactone	$C_6H_8O_3$ / 128.13	517-23-7 / 77	107-8[5]	5-17-11-00016 / 1.1846[20]	H_2O 4 / 1.4585[20]
6140	2(3H)-Furanone, 5-butyldihydro-	$C_8H_{14}O_2$ / 142.20	104-50-7	132-3[20]	5-17-09-00073 / 0.9796[19]	EtOH 3; ctc 2 / 1.4451[19]
6141	2(3H)-Furanone, 5-(chloromethyl)dihydro- / Valeric acid, 5-chloro-4-hydroxy-, γ-lactone	$C_5H_7ClO_2$ / 134.56	39928-72-8	132-5[12]	5-17-09-00025 / 1.625[18]	EtOH 2; eth 2; bz 2
6142	2(3H)-Furanone, dihydro- / γ-Butyrolactone	$C_4H_8O_2$ / 86.09	96-48-0 / -43.3	1596 / 204	5-17-09-00007 / 1.1284[16]	ace 4; bz 4; eth 4; EtOH 4 / 1.4341[20]
6143	2(3H)-Furanone, dihydro-3-hydroxy-4,4-dimethyl-, (R)- / Pantolactone	$C_6H_{10}O_3$ / 130.14	599-04-2 / 92	6963	5-18-01-00022	
6144	2(3H)-Furanone, dihydro-3-(hydroxyphenylmethyl)-4-[(1-methyl-1H-imidazol-5-yl)methyl]-, / Isopilosine	$C_{16}H_{18}N_2O_3$ / 286.33	491-88-3 / 187	5085	1-27-00-00612	EtOH 4
6145	2(3H)-Furanone, dihydro-3-methyl-	$C_5H_8O_2$ / 100.12	1679-47-6	78-81[10]	5-17-09-00028 / 1.063[25]	1.4325[20]
6146	2(3H)-Furanone, dihydro-5-methyl-, (±)-	$C_5H_8O_2$ / 100.12	57129-69-8 / -31	206	5-17-09-00024 / 1.0465[25]	H_2O 5; EtOH 3; ace 3; ctc 2 / 1.4328[20]
6147	2(3H)-Furanone, dihydro-3-methylene- / α-Methylene butyrolactone	$C_5H_6O_2$ / 98.10	547-65-9	5981 / 85-6[10]	5-17-09-00127 / 1.1206[20]	H_2O 3; EtOH 4; eth 3; ace 3 / 1.4650[20]
6148	2(3H)-Furanone, dihydro-5-methylene-	$C_5H_6O_2$ / 98.10	10008-73-8	85-6[10]	5-17-09-00124 / 1.1206[20]	ace 4; bz 4; eth 4; EtOH 4 / 1.4650[20]
6149	2(3H)-Furanone, dihydro-5-pentyl- / 4-Hydroxynonanoic acid lactone	$C_9H_{16}O_2$ / 156.22	104-61-0		5-17-09-00084	
6150	2(3H)-Furanone, 5-ethyldihydro-	$C_6H_{10}O_2$ / 114.14	695-06-7 / -18	215.5	5-17-09-00040 / 1.0261[20]	H_2O 4; EtOH 4 / 1.4495[20]
6151	2(3H)-Furanone, 3-ethyldihydro-4-[(1-methyl-1H-imidazol-5-yl)methyl]-, (3S-cis)- / Pilocarpine	$C_{11}H_{16}N_2O_2$ / 208.26	92-13-7 / 34	7395 / 260[5]	2-27-00-00694	H_2O 3; EtOH 3; eth 2; bz 2
6152	2(3H)-Furanone, 5-heptyldihydro- / 4-Hydroxyundecanoic acid lactone	$C_{11}H_{20}O_2$ / 184.28	104-67-6	286	5-17-09-00098 / 0.9494[20]	EtOH 4 / 1.4512[20]
6153	2(3H)-Furanone, 5-methyl-	$C_5H_6O_2$ / 98.10	591-12-8 / 18	56[12]	5-17-09-00120 / 1.084[20]	H_2O 3; EtOH 3; eth 3; ctc 2 / 1.4476[20]
6154	2(5H)-Furanone, 5-methyl-	$C_5H_6O_2$ / 98.10	591-11-7 / <-17	209; 98[15]	5-17-09-00121 / 1.0810[20]	H_2O 5; EtOH 3; eth 3 / 1.4454[20]
6155	2(5H)-Furanone, 5-methylene- / Protoanemonin	$C_5H_4O_2$ / 96.09	108-28-1	7907 / 73[11]	5-17-09-00371	H_2O 2; chl 3
6156	Furan, 2,2'-[oxybis(methylene)]bis-	$C_{10}H_{10}O_3$ / 178.19	4437-22-3	101[2]	5-17-03-00347 / 1.1405[20]	H_2O 1 / 1.5088[20]
6157	Furan, 2-phenoxy- / 2-Furyl phenyl ether	$C_{10}H_8O_2$ / 160.17	60698-31-9	105-6[18]	4-17-00-01219 / 1.1010[23]	eth 3 / 1.5418[23]

No.	Name Synonym	Mol. Form. Mol. Wt.	CAS RN mp/°C	Merck No. bp/°C	Bell. Ref. den/g cm^{-3}	Solubility n_D
6158	Furan, 2-(phenoxymethyl)- Furfuryl phenyl ether	$C_{11}H_{10}O_2$ 174.20	4437-23-4	139-40[20]	5-17-03-00340 1.123[17]	1.5535[17]
6159	Furan, 2-phenyl- 2-Phenylfuran	$C_{10}H_8O$ 144.17	17113-33-6	107-8[18]	5-17-02-00139 1.083[20]	ace 4; bz 4 1.5920[20]
6160	Furan, 2-(phenylthio)- 2-Furyl phenyl sulfide	$C_{10}H_8OS$ 176.24	16003-14-8	119-20[8]	5-17-03-00285 1.134[26]	eth 4; EtOH 4 1.5976[20]
6161	2-Furanpropanoic acid	$C_7H_8O_3$ 140.14	935-13-7 58	229	5-18-06-00224	H_2O 4; eth 4; chl 4
6162	2-Furanpropanoic acid, methyl ester	$C_8H_{10}O_3$ 154.17	37493-31-5	89[15]	5-18-06-00224 1.088[20]	1.4662[20]
6163	2-Furanpropanoic acid, β-oxo-, ethyl ester	$C_9H_{10}O_4$ 182.18	615-09-8	142-3[10]	5-18-08-00143 1.165[17]	H_2O 1; EtOH 3; eth 3
6164	2-Furanpropanoic acid, tetrahydro- Propionic acid, 3-(2-tetrahydrofuryl)	$C_7H_{12}O_3$ 144.17	935-12-6 263		5-18-06-00033 1.1135[20]	ace 4; EtOH 4 1.4578[25]
6165	2-Furanpropanoic acid, tetrahydro-, ethyl ester	$C_9H_{16}O_3$ 172.22	4525-36-4	222; 73[2]	5-18-06-00034 1.023[7]	ace 4; EtOH 4 1.440[20]
6166	2-Furanpropanol, tetrahydro-α-[2-(tetrahydro-2-furanyl)ethyl]- 1,5-Bis(2-tetrahydrofuryl)-3-pentanol	$C_{13}H_{24}O_3$ 228.33	6265-26-5	116[0.05]	4-19-00-00704	chl 2 1.4541[25]
6167	Furan, 2-propyl- 2-Propylfuran	$C_7H_{10}O$ 110.16	4229-91-8	115	5-17-01-00354 0.8876[20]	EtOH 3; eth 3; ace 3 1.4549[20]
6168	Furan, tetrahydro- Tetrahydrofuran	C_4H_8O 72.11	109-99-9 -108.3	9144 65	5-17-01-00027 0.8892[20]	H_2O 3; EtOH 4; eth 4; ace 4 1.4050[20]
6169	Furan, tetrahydro-2,5-dimethoxy-	$C_6H_{12}O_3$ 132.16	696-59-3	145.7	5-17-05-00004 1.02[25]	1.4180[20]
6170	Furan, tetrahydro-2-(methoxymethyl)-	$C_6H_{12}O_2$ 116.16	19354-27-9	140	5-17-03-00116 0.952[25]	1.4270[20]
6171	Furan, tetrahydro-2-methyl- 2-Methyltetrahydrofuran	$C_5H_{10}O$ 86.13	96-47-9	78	5-17-01-00078 0.8552[20]	H_2O 3; EtOH 4; eth 4; ace 4 1.4059[21]
6172	Furan, tetrahydro-3-methyl- 3-Methyltetrahydrofuran	$C_5H_{10}O$ 86.13	13423-15-9	86.5	5-17-01-00081 0.8642[20]	ace 4; bz 4; eth 4; EtOH 4 1.4122[20]
6173	Furan, tetrahydro-2-propyl-	$C_7H_{14}O$ 114.19	3208-22-8	136	5-17-01-00116 0.8547[20]	1.4242[20]
6174	Furan, tetrahydro-2,2,4,4-tetramethyl- 2,2,4,4-Tetramethyltetrahydrofuran	$C_8H_{16}O$ 128.21	3358-28-9	121	4-17-00-00109	H_2O 1; EtOH 5; eth 5 1.4074[20]
6175	Furan, tetraphenyl- 2,3,4,5-Tetraphenylfuran	$C_{28}H_{20}O$ 372.47	1056-77-5 175	220	4-17-00-00810	H_2O 1; EtOH 3; eth 3; ace 3
6176	Furazan, dimethyl-	$C_4H_6N_2O$ 98.10	4975-21-7 -7	156	4-27-00-07096 1.0528[14]	eth 4; EtOH 4 1.4327[20]
6177	3aH-Furo[2,3-b]indol-3a-ol, 8a-(5-ethenyl-1-azabicyclo[2.2.2]oct-2-yl)-2,3,8,8a-tetrahydro- Conquinamine	$C_{19}H_{24}N_2O_2$ 312.41	464-86-8 123	2505	2-27-00-00667	H_2O 2; EtOH 3; eth 3; chl 3
6178	3aH-Furo[2,3-b]indol-3a-ol, 8a-(5-ethenyl-1-azabicyclo[2.2.2]oct-2-yl)-2,3,8,8a-tetrahydro- Quinamine	$C_{19}H_{24}N_2O_2$ 312.41	464-85-7 185.5	8060	2-27-00-00667	H_2O 1; EtOH 4; eth 3; ace 3
6179	Furo[2,3-b]quinoline, 4,8-dimethoxy- Fagarine	$C_{13}H_{11}NO_3$ 229.24	524-15-2 142	3879	4-27-00-02211	H_2O 2; EtOH 3; eth 3; bz 3
6180	Furo[2,3-b]quinoline, 4-methoxy- Dictamnine	$C_{12}H_9NO_2$ 199.21	484-29-7 133.5	3079	4-27-00-02030	H_2O 2; EtOH 4; eth 3; chl 3
6181	Furo[2,3-b]quinoline, 4,7,8-trimethoxy- Skimmianine	$C_{14}H_{13}NO_4$ 259.26	83-95-4 177	8505	4-27-00-02296	H_2O 1; EtOH 3; eth 2; chl 3
6182	Furo[2,3-b]quinolin-4(2H)-one, 3,9-dihydro-8-methoxy-9-methyl-2-(1-methylethyl)-, (R)- Lunacrine	$C_{16}H_{19}NO_3$ 273.33	82-40-6	5473	4-27-00-03566	chl 3
6183	4H-Furo[3,2-c]pyran-2(6H)-one, 4-hydroxy- Patulin	$C_7H_8O_4$ 154.12	149-29-1 111	7002	5-18-03-00005	H_2O 3; EtOH 3; eth 3; ace 3
6184	1H,12H-Furo[3',2':4,5]furo[2,3-h]pyrano[3,4-c][1]benzopyran-1,12-dione, 3,4,7a,10a-tetrahydro-5-meth Aflatoxin G1	$C_{17}H_{12}O_7$ 328.28	1165-39-5 245		5-19-12-00081	
6185	5H-Furo[3,2-g][1]benzopyran-5-one, 4,9-dimethoxy-7-methyl- Khellin	$C_{14}H_{12}O_5$ 260.25	82-02-0 154-5 dec	5189 180-200[0.05]	5-19-06-00320	H_2O 1; EtOH 3; eth 2; ace 3
6186	5H-Furo[3,2-g][1]benzopyran-5-one, 4-methoxy-7-methyl- Visnagin	$C_{13}H_{10}O_4$ 230.22	82-57-5 144.5	9916	5-19-06-00030	H_2O 2; EtOH 2; chl 4
6187	7H-Furo[3,2-g][1]benzopyran-7-one, 9-methoxy- Methoxsalen	$C_{12}H_8O_4$ 216.19	298-81-7 148	5911	5-19-06-00015	H_2O 2; EtOH 4; eth 2; ace 2
6188	7H-Furo[3,2-g][1]benzopyran-7-one, 3-methoxy-2-(1-methylethyl)- Peucedanin	$C_{15}H_{14}O_4$ 258.27	133-26-6 85	7144 276-81[17]	5-19-06-00042	H_2O 2; EtOH 3; eth 3; bz 2
6189	7H-Furo[3,2-g][1]benzopyran-7-one, 9-[(3-methyl-2-butenyl)oxy]- Imperatorin	$C_{16}H_{14}O_4$ 270.28	482-44-0 102	4839	5-19-06-00016	H_2O 2; EtOH 3; eth 3; bz 3
6190	3-Furoic acid, 5-bromo-, ethyl ester β-Furoic acid, 5-bromo, ethyl ester	$C_7H_7BrO_3$ 219.03	32460-20-1 17	235; 135[34]	5-18-06-00198 1.528[20]	eth 4; EtOH 4
6191	Furo[3',4':6,7]naphtho[2,3-d]-1,3-dioxol-6(5aH)-one, 5,8,8a,9-tetrahydro-9-hydroxy-5-(3,4,5-trimethoxyphenyl)- Picropodophyllin	$C_{22}H_{22}O_8$ 414.41	477-47-4 228	7385	5-19-10-00665	ace 4; bz 4; eth 4; EtOH 4

No.	Name / Synonym	Mol. Form. / Mol. Wt.	CAS RN / mp/°C	Merck No. / bp/°C	Beil. Ref. / den/g cm⁻³	Solubility / n_D
6192	Furo[3',4':6,7]naphtho[2,3-d]-1,3-dioxol-6(5aH)-one, 5,8,8a,9-tetrahydro-9-hydroxy-5-(3,4,5-trimetho Podophyllotoxin	$C_{22}H_{22}O_8$ 414.41	518-28-5 183	7520	5-19-10-00666	H_2O 2; EtOH 4; eth 1; ace 3
6193	Galactaric acid Mucic acid	$C_6H_{10}O_8$ 210.14	526-99-8 255 dec	4237	4-03-00-01292	
6194	Galactitol Dulcose	$C_6H_{14}O_6$ 182.17	608-66-2 189.5	4238 275-80[1]	4-01-00-02844 1.47[20]	H_2O 3; EtOH 2; eth 1; bz 1
6195	β-D-Galactopyranoside, (2α,3β,5α,15β,25R)-2,15-dihydroxyspirostan-3-yl Digitonin	$C_{56}H_{92}O_{29}$ 1229.33	11024-24-1 237.5	3144	5-19-03-00603	
6196	β-D-Galactopyranoside, (3β,5α,22β,25S)-spirosolan-3-yl O-β- Tomatine	$C_{50}H_{83}NO_{21}$ 1034.20	17406-45-0 270	9471	4-27-00-01954	EtOH 4; diox 4
6197	D-Galactose	$C_6H_{12}O_6$ 180.16	59-23-4 170	4241	4-01-00-04336	H_2O 4; EtOH 2; eth 1; bz 1
6198	Galactose, 6-deoxy-3-O-methyl- Digitalose	$C_7H_{14}O_5$ 178.19	4481-08-7 119	3142	4-01-00-04270	H_2O 4
6199	D-Galacturonic acid	$C_6H_{10}O_7$ 194.14	685-73-4 166	4242	1-03-00-00306	H_2O 3; EtOH 3; eth 1
6200	Galanthamine Lycoremine	$C_{17}H_{21}NO_3$ 287.36	357-70-0 126.5	4245	4-27-00-02184	ace 4; EtOH 4; chl 4
6201	Galanthan-1,2-diol, 3,12-didehydro-9,10-[methylenebis(oxy)]-, (1α,2β)- Lycorine	$C_{16}H_{17}NO_4$ 287.32	476-28-8 280	5498 sub	2-27-00-00547	H_2O 1; EtOH 2; eth 2; chl 2
6202	Gardona Phosphoric acid, 2-chloro-1-(2,4,5-trichlorophenyl)ethenyl dimethyl ester	$C_{10}H_9Cl_4O_4P$ 365.96	961-11-5			
6203	Gelsemine Spiro[3,5,8-ethanylylidene-1H-pyrano[3,4-c]pyridine-10,3'-[3H]indol]-2'(1'H)-one, 5-ethenyl	$C_{20}H_{22}N_2O_2$ 322.41	509-15-9 178	4277	4-27-00-07526	ace 4; bz 4; eth 4; EtOH 4
6204	Gibb-3-ene-1,10-dicarboxylic acid, 2,4a,7-trihydroxy-1-methyl-8-methylene-, 1,4a-lactone, Gibberellic acid	$C_{19}H_{22}O_6$ 346.38	77-06-5 234	4313	5-18-09-00269	ace 4; EtOH 4; MeOH 4
6205	D-Glucaric acid D-Tetrahydroxyadipic acid	$C_6H_{10}O_8$ 210.14	87-73-0 125.5	4344	4-03-00-01291	H_2O 4; EtOH 4; eth 2; chl 2
6206	D-Glucitol Sorbitol	$C_6H_{14}O_6$ 182.17	50-70-4 111	8680 295[3.5]	4-01-00-02839 1.489[20]	H_2O 4; ace 4 1.3330[20]
6207	D-Glucitol, 1-amino-1-deoxy- Glucamine	$C_6H_{15}NO_5$ 181.19	488-43-7 127	4343	4-04-00-01913	H_2O 4; EtOH 4
6208	D-Glucitol, 1-deoxy-1-(methylamino)- N-Methylglucamine	$C_7H_{17}NO_5$ 195.22	6284-40-8 128.5	5995	4-04-00-01914	H_2O 3
6209	D-Gluconic acid Gluconic acid	$C_6H_{12}O_7$ 196.16	526-95-4 131	4350	4-03-00-01255	H_2O 3; EtOH 2; eth 1; bz 1
6210	D-Gluconic acid, 4-O-β-D-galactopyranosyl- Lactobionic acid	$C_{12}H_{22}O_{12}$ 358.30	96-82-2	5219	5-17-07-00436	H_2O 4; EtOH 2; eth 1; HOAc 2
6211	Gluconic acid, β-lactone Oxetane, gluconic acid deriv.	$C_6H_{10}O_6$ 178.14	3087-62-5 153		2-18-00-00190 1.610[-5]	H_2O 2; EtOH 2; eth 1; ace 1
6212	D-Gluconic acid, δ-lactone δ-D-Gluconolactone	$C_6H_{10}O_6$ 178.14	90-80-2	4351	5-18-05-00011	
6213	D-Gluconic acid, monosodium salt Gluconic acid, sodium salt	$C_6H_{11}NaO_7$ 218.14	527-07-1	8569	4-03-00-01255	H_2O 3
6214	β-D-Glucopyranose, 4-O-β-D-galactopyranosyl-	$C_{12}H_{22}O_{11}$ 342.30	5965-66-2 254		5-17-07-00196 1.59[20]	H_2O 4; EtOH 2; eth 1; chl 1
6215	α-L-Glucopyranose, 2-(methylamino)-2-deoxy- N-Methyl-α-L-glucosamine	$C_7H_{15}NO_5$ 193.20	42852-95-9	5996		
6216	α-D-Glucopyranose, pentaacetate	$C_{16}H_{22}O_{11}$ 390.34	604-68-2 113.3	sub	5-17-07-00318	H_2O 2; EtOH 2; eth 3; chl 3
6217	β-D-Glucopyranose, pentaacetate	$C_{16}H_{22}O_{11}$ 390.34	604-69-3 134	sub	5-17-07-00319 1.274[20]	H_2O 1; EtOH 2; eth 2; bz 3
6218	α-D-Glucopyranoside, β-D-fructofuranosyl Sucrose	$C_{12}H_{22}O_{11}$ 342.30	57-50-1 185.5	8855	5-17-08-00399 1.5805[17]	H_2O 3; EtOH 2; eth 1; py 3 1.5376
6219	α-D-Glucopyranoside, β-D-fructofuranosyl O-α-D-galactopyranosyl-(1 Raffinose	$C_{18}H_{32}O_{16}$ 504.44	512-69-6 80	8120	4-17-00-03801 1.465[25]	H_2O 4; EtOH 2; eth 1; MeOH 4
6220	α-D-Glucopyranoside, α-D-glucopyranosyl Trehalose	$C_{12}H_{22}O_{11}$ 342.30	99-20-7 203	9496	1.58[24]	H_2O 4; EtOH 3; eth 1; bz 1
6221	α-D-Glucopyranoside, O-α-D-glucopyranosyl-(13)-β-D-fructofuranosyl Melezitose	$C_{18}H_{32}O_{16}$ 504.44	597-12-6 153	5698	5-17-08-00414 1.5565[25]	H_2O 4
6222	β-D-Glucopyranoside, 2-(hydroxymethyl)phenyl Salicin	$C_{13}H_{18}O_7$ 286.28	138-52-3 207	8293 240 dec	5-17-07-00113 1.434[20]	H_2O 4; EtOH 4; HOAc 4
6223	β-D-Glucopyranoside, 2-(hydroxymethyl)phenyl, 6-benzoate Populin	$C_{20}H_{22}O_8$ 390.39	99-17-2 180	7571	5-17-07-00343	

No.	Name Synonym	Mol. Form. Mol. Wt.	CAS RN mp/°C	Merck No. bp/°C	Beil. Ref. den/g cm⁻³	Solubility n_D
6224	β-D-Glucopyranoside, 4-hydroxyphenyl Arbutin	$C_{12}H_{16}O_7$ 272.25	497-76-7 199.5	799	5-17-07-00110	H_2O 4; EtOH 3; eth 2; bz 1
6225	β-D-Glucopyranoside, 4-(3-hydroxy-1-propenyl)-2,6-dimethoxyphenyl Syringin	$C_{17}H_{24}O_9$ 372.37	118-34-3 192	8997	4-17-00-03002	EtOH 4
6226	β-D-Glucopyranoside, 4-(3-hydroxy-1-propenyl)-2-methoxyphenyl Coniferin	$C_{16}H_{22}O_8$ 342.35	531-29-3 186	2498	4-17-00-02999	H_2O 3; EtOH 2; eth 1; py 3
6227	α-D-Glucopyranoside, methyl α-Methylglucoside	$C_7H_{14}O_6$ 194.18	97-30-3 168	5997 200^0.2	5-17-07-00010 1.46^30	H_2O 4
6228	α-D-Glucopyranoside, methyl 2,3,4,6-tetra-O-methyl- α-D-Glucose, pentamethyl ether	$C_{11}H_{22}O_6$ 250.29	605-81-2 180^0.4		5-17-07-00029 1.0944^20	H_2O 4; ace 4; eth 4; EtOH 4 1.4466^20
6229	α-D-Glucopyranoside, 1,3,4,6-tetra-O-acetyl-β-D-fructofuranosyl, tetraacetate Octaacetyl sucrose	$C_{28}H_{38}O_{19}$ 678.60	126-14-7 86.5	8856 250^1	5-17-08-00410 1.27^16	H_2O 2; EtOH 3; eth 3; ace 3 1.4660
6230	α-D-Glucopyranosiduronic acid, (3β,20β)-20-carboxy-11-oxo-30-norolean-12-en-3-yl 2-O- Glycyrrhizic acid	$C_{42}H_{62}O_{16}$ 822.94	1405-86-3 220 dec	4401	4-18-00-05156	H_2O 4
6231	β-D-Glucopyranosiduronic acid, 2,2,2-trichloroethyl Urochloralic acid	$C_8H_{11}Cl_3O_7$ 325.53	97-25-6 142	9798	4-18-00-05113	H_2O 4; EtOH 4
6232	β-D-Glucose	$C_6H_{12}O_6$ 180.16	28905-12-6 150		4-01-00-04306 1.5620^18	H_2O 4; EtOH 2; eth 1; py 3
6233	D-Glucose, 2-amino-2-deoxy- D-Glucosamine	$C_6H_{13}NO_5$ 179.17	3416-24-8	4352	4-04-00-02017	H_2O 4
6234	D-Glucose, 6-O-α-L-arabinopyranosyl- Vicianose	$C_{11}H_{20}O_{10}$ 312.27	14116-69-9 210 dec	9879	5-17-06-00112	H_2O 4
6235	D-Glucose, 6-deoxy- Quinovose	$C_6H_{12}O_5$ 164.16	7658-08-4 139.5	8106	4-01-00-04260	H_2O 4; EtOH 4
6236	D-Glucose, 6-O-(6-deoxy-α-L-mannopyranosyl)- Rutinose	$C_{12}H_{22}O_{10}$ 326.30	90-74-4 189-92 dec	8277	5-17-06-00269	H_2O 4; EtOH 4
6237	D-Glucose, 6-O-β-D-xylopyranosyl- Primeverose	$C_{11}H_{20}O_{10}$ 312.27	26531-85-1 210	7752	4-17-00-02447	H_2O 4; MeOH 4
6238	D-Glucuronic acid	$C_6H_{10}O_7$ 194.14	6556-12-3 165	4360	4-03-00-01996	H_2O 4; EtOH 4
6239	D-Glucuronic acid, γ-lactone D-Glucuronolactone	$C_6H_8O_6$ 176.13	32449-92-6 177.5	4362	5-18-05-00033 1.76^20	H_2O 3; EtOH 2; bz 1; MeOH 2
6240	DL-Glutamic acid	$C_5H_9NO_4$ 147.13	617-65-2 199 dec		4-04-00-03028 1.4601^20	H_2O 2; EtOH 1; eth 2; CS_2 1
6241	L-Glutamic acid (S)-2-Aminopentanedioic acid	$C_5H_9NO_4$ 147.13	56-86-0 160 dec	4363 sub 175	4-04-00-03028 1.538^20	
6242	L-Glutamic acid, N-(4-aminobenzoyl)- N-(p-Aminobenzoyl)-L(+)-glutamic acid	$C_{12}H_{14}N_2O_5$ 266.25	4271-30-1 173	437	4-14-00-01153	
6243	L-Glutamic acid, N-[4-[[(2-amino-1,4-dihydro-4-oxo-6-pteridinyl)methyl]amino]benzoyl]- Folic acid	$C_{19}H_{19}N_7O_6$ 441.40	59-30-3 250 dec	4140	4-26-00-03944	py 4; EtOH 4; HOAc 4
6244	L-Glutamic acid, N-[4-[[(2-amino-5-formyl-1,4,5,6,7,8-hexahydro-4-oxo-6-pteridinyl)methyl]amino]benzoyl]- Folinic acid	$C_{20}H_{23}N_7O_7$ 473.45	58-05-9 245 dec	4141	4-26-00-03881	H_2O 2
6245	L-Glutamic acid, 5-ethyl ester γ-Ethyl L-glutamate	$C_7H_{13}NO_4$ 175.18	1119-33-1 191	4364	4-04-00-03033	H_2O 2
6246	DL-Glutamic acid, 3-hydroxy-	$C_5H_9NO_5$ 163.13	5985-23-9 209 dec		4-04-00-03253	H_2O 4; EtOH 1; eth 1; bz 1
6247	L-Glutamic acid, 5-[2-[4-(hydroxymethyl)phenyl]hydrazide] Agaritine	$C_{12}H_{17}N_3O_4$ 267.28	2757-90-6 205-9 dec	175		H_2O 4
6248	L-Glutamic acid, monosodium salt Glutamic acid, monosodium salt	$C_5H_8NNaO_4$ 169.11	142-47-2	6165		H_2O 3
6249	L-Glutamine 2-Aminoglutaramic acid	$C_5H_{10}N_2O_3$ 146.15	56-85-9 185-6 dec	4365	4-04-00-03038	H_2O 3; EtOH 1; eth 1; bz 1
6250	Glycine Aminoacetic acid	$C_2H_5NO_2$ 75.07	56-40-6 290 dec	4386	4-04-00-02349 1.161^20	H_2O 4; EtOH 1; eth 1; ace 2
6251	Glycine, N-acetyl- Aceturic acid	$C_4H_7NO_3$ 117.10	543-24-8 206	74	4-04-00-02399	H_2O 4; ace 4; EtOH 4
6252	Glycine, N-(4-aminobenzoyl)- p-Aminohippuric acid	$C_9H_{10}N_2O_3$ 194.19	61-78-9 198.5	454	4-14-00-01152	ace 4; bz 4; EtOH 4
6253	Glycine, N-(aminoiminomethyl)- Glycocyamine	$C_3H_7N_3O_2$ 117.11	352-97-6 282	4391	4-04-00-02414	H_2O 2; EtOH 2; eth 2
6254	Glycine, N-(aminoiminomethyl)-N-methyl- Creatine	$C_4H_9N_3O_2$ 131.13	57-00-1 303 dec	2570	4-04-00-02426 1.33^25	H_2O 3; EtOH 2; eth 1
6255	Glycine, N-benzoyl- Hippuric acid	$C_9H_9NO_3$ 179.18	495-69-2 191.5	4636	4-09-00-00778 1.371^20	H_2O 3; EtOH 3; eth 2; bz 2
6256	Glycine, N,N-bis(carboxymethyl)- Nitrilotriacetic acid	$C_6H_9NO_6$ 191.14	139-13-9 242 dec	6499	4-04-00-02441	H_2O 2; EtOH 3; DMSO-d_6 2

No.	Name / Synonym	Mol. Form. / Mol. Wt.	CAS RN / mp/°C	Merck No. / bp/°C	Bell. Ref. / den/g cm^{-3}	Solubility / n_D
6257	Glycine, N,N-bis[2-[bis(carboxymethyl)amino]ethyl]- Pentetic acid	$C_{14}H_{23}N_3O_{10}$ 393.35	67-43-6	7083	4-04-00-02454	
6258	Glycine, N,N-bis(ethoxycarbonyl)-, ethyl ester Aminoacetic acid, N,N-(dicarboethoxy), ethyl ester	$C_{10}H_{17}NO_6$ 247.25	127665-32-1 36.5	152^{10}	0-04-00-00365	bz 4; eth 4; EtOH 4
6259	Glycine, N,N-bis(2-hydroxyethyl)- Bicine	$C_6H_{13}NO_4$ 163.17	150-25-4 193-5 dec	1221	4-04-00-02390	H_2O 4; EtOH 1
6260	Glycine, N-(carboxymethyl)- Iminodiacetic acid	$C_4H_7NO_4$ 133.10	142-73-4 247.5	4832	4-04-00-02428	H_2O 2; EtOH 1; eth 1
6261	Glycine, N,N-dimethyl- N,N-Dimethylglycine	$C_4H_9NO_2$ 103.12	1118-68-9 185.5	3233	4-04-00-02365	H_2O 4; EtOH 3; eth 3; ace 3
6262	Glycine, N,N'-1,2-ethanediylbis[N-(carboxymethyl)- Edetic acid	$C_{10}H_{16}N_2O_8$ 292.25	60-00-4 245 dec	3484	4-04-00-02449	
6263	Glycine, ethyl ester	$C_4H_9NO_2$ 103.12	459-73-4	149; 58^{18}	4-04-00-02355 1.0275^{10}	H_2O 5; EtOH 5; eth 5; ace 5 1.4242^{10}
6264	Glycine, N-(N-L-γ-glutamyl-L-cysteinyl)- Glutatione	$C_{10}H_{17}N_3O_6S$ 307.33	70-18-8 195	4369	4-04-00-03165	H_2O 4; EtOH 1; eth 1; DMF 3
6265	Glycine, N-glycyl- 2-(Aminoacetamido)acetic acid	$C_4H_8N_2O_3$ 132.12	556-50-3 215 dec	4399	4-04-00-02459	H_2O 3
6266	Glycine, hydrazide Aminoacetic acid hydrazide	$C_2H_7N_3O$ 89.10	14379-80-7 80.5	150 dec	3-04-00-01121	EtOH 2; eth 1; chl 3
6267	Glycine, N-(4-hydroxyphenyl)- N-(4-Hydroxyphenyl)glycine	$C_8H_9NO_3$ 167.16	122-87-2 245-7 dec	4771	4-13-00-01210	H_2O 2; EtOH 1; eth 1; ace 2
6268	Glycine, N-methyl- Sarcosine	$C_3H_7NO_2$ 89.09	107-97-1 212-3 dec	8331	4-04-00-02363	
6269	Glycine, N-methyl-N-(1-oxododecyl)-, sodium salt Gardol	$C_{15}H_{28}NNaO_3$ 293.38	137-16-6	4267		H_2O 2
6270	Glycine, N-phenyl- Phenylaminoacetic acid	$C_8H_9NO_2$ 151.17	103-01-5 127.5	7262	4-12-00-00872	H_2O 4; EtOH 4
6271	Glycine, N-phenyl-, ethyl ester	$C_{10}H_{13}NO_2$ 179.22	2216-92-4 58	273.5	4-12-00-00873	eth 4; EtOH 4
6272	Glycine, N-(phenylmethyl)-, ethyl ester	$C_{11}H_{15}NO_2$ 193.25	6436-90-4	175-9^{50}	4-12-00-02277	EtOH 4; eth 4; bz 4 1.5041^{20}
6273	Glycine, N-[(3α,5β,7α,12α)-3,7,12-trihydroxy-24-oxocholan-24-yl]- Glycocholic acid	$C_{26}H_{43}NO_6$ 465.63	475-31-0 166.5	4390	4-10-00-02077	H_2O 2; EtOH 4; eth 2
6274	Glycogen	$(C_6H_{10}O_5)_x$ 162.14	9005-79-2	4392		H_2O 4; EtOH 1; eth 1
6275	Glyodin 1H-Imidazole, 2-heptadecyl-4,5-dihydro-, monoacetate	$C_{20}H_{40}N_2 \cdot C_2H_4O_2$ 368.60	556-22-9	4404	1.035^{20}	
6276	Glyphosate Glycine, N-(phosphonomethyl)-	$C_3H_8NO_5P$ 169.07	1071-83-6 230 dec	4408		
6277	Glyphosine Glycine, N,N-bis(phosphonomethyl)-	$C_4H_{11}NO_8P_2$ 263.08	2439-99-8	4409		
6278	Guanidine Aminomethanamidine	CH_5N_3 59.07	113-00-8 50	4475	4-03-00-00148	H_2O 4; EtOH 4
6279	Guanidine, N,N'-bis(2-methylphenyl)-	$C_{15}H_{17}N_3$ 239.32	97-39-2 179		4-12-00-01764 1.10^{20}	H_2O 2; EtOH 2; eth 4; chl 3
6280	Guanidine, cyano- Dicyanodiamide	$C_2H_4N_4$ 84.08	461-58-5 211	3082	4-03-00-00160 1.404^{14}	H_2O 3; EtOH 3; eth 1; ace 3
6281	Guanidine, N,N'-dimethyl-N''-(phenylmethyl)- Bethanidine	$C_{10}H_{15}N_3$ 177.25	55-73-2 196	1208		
6282	Guanidine, N,N'-diphenyl- 1,3-Diphenylguanidine	$C_{13}H_{13}N_3$ 211.27	102-06-7 150	3325 170 dec	4-12-00-00769 1.13^{20}	H_2O 2; EtOH 3; eth 4; ctc 3
6283	Guanidine, (3-methyl-2-butenyl)- Galegine	$C_6H_{13}N_3$ 127.19	543-83-9 62.5 dec	4246	3-04-00-00465	H_2O 4; EtOH 4
6284	Guanidine, N-methyl-N'-nitro-N-nitroso- N-Methyl-N'-nitro-N-nitrosoguanidine	$C_2H_5N_5O_3$ 147.09	70-25-7	6017	4-04-00-03386	DMSO 3
6285	Guanidine, monohydrochloride	CH_6ClN_3 95.53	50-01-1 182.3		4-03-00-00148 1.354^{20}	H_2O 4; EtOH 4
6286	Guanidine, mononitrate	$CH_6N_4O_3$ 122.08	506-93-4 217 dec		4-03-00-00148	H_2O 4; EtOH 4
6287	Guanidine, nitro- Nitroguanidine	$CH_4N_4O_2$ 104.07	556-88-7 239 dec	6529	4-03-00-00249	H_2O 2; EtOH 2; eth 1; alk 4
6288	Guanidine, N,N',N''-triphenyl-	$C_{19}H_{17}N_3$ 287.36	101-01-9 146.5 dec		4-12-00-00866 1.163^{20}	H_2O 2; EtOH 3
6289	Guanosine 2-Amino-1,9-dihydro-9-β-D-ribofuranosyl-6H-purin-6-one	$C_{10}H_{13}N_5O_5$ 283.24	118-00-3 239 dec	4480	4-26-00-03901	H_2O 2; EtOH 1; eth 1; HOAc 4
6290	D-Gulose Gulose, D-	$C_6H_{12}O_6$ 180.16	4205-23-6 dec	4490	4-01-00-04333	H_2O 4
6291	Haloxyfop-(2-ethoxyethyl) (RS)-2-[4-(3-Chloro-5-trifluoromethyl-2-pyridyloxy)phenoxy]propionic acid, ethyl ester	$C_{17}H_{15}ClF_3NO_4$ 389.76	87237-48-7 60		1.34	

No.	Name Synonym	Mol. Form. Mol. Wt.	CAS RN mp/°C	Merck No. bp/°C	Beil. Ref. den/g cm^{-3}	Solubility n_D
6292	7H-Benz[de]anthracen-7-one Benzanthrone	$C_{17}H_{10}O$ 230.27	82-05-3 170	1070		bz 2
6293	Helenine	$C_{15}H_{20}O_2$ 232.32	1407-14-3 78	275; 192[10]	4-17-00-05030	bz 4; eth 4; EtOH 4
6294	Heneicosane	$C_{21}H_{44}$ 296.58	629-94-7 40.5	356.5	4-01-00-00569 0.7919[20]	H_2O 1; EtOH 2; peth 3 1.4441[20]
6295	Heneicosane, 11-decyl-	$C_{31}H_{64}$ 436.85	55320-06-4 10.0	282.0[10]	4-01-00-00595 0.8116[20]	1.4540[20]
6296	Heneicosane, 11-(2,2-dimethylpropyl)-	$C_{26}H_{54}$ 366.71	55282-10-5 -21	238.0[10]	4-01-00-00585 0.8031[20]	1.4491[20]
6297	Heneicosane, 11-phenyl- Benzene, (1-decylundecyl)-	$C_{27}H_{48}$ 372.68	6703-80-6 20.8	205[1.0]	4-05-00-01226 0.8531[20]	1.4788[20]
6298	10-Heneicosene, 11-phenyl-	$C_{27}H_{46}$ 370.66	6703-78-2 48.2	203[1.0] 190[0.5]	4-05-00-01506 0.8638[20]	1.4922[20]
6299	Hentriacontane Untriacontane	$C_{31}H_{64}$ 436.85	630-04-6 67.9	458	4-01-00-00594 0.781[68]	EtOH 2; eth 2; bz 2; chl 2 1.4278[90]
6300	16-Hentriacontanone	$C_{31}H_{62}O$ 450.83	502-73-8 83.8	250[0.1]	4-01-00-03413 0.7947[91]	H_2O 1; EtOH 2; eth 3; ace 2 1.4297[94]
6301	Heptachlor epoxide 2,5-Methano-2H-indeno[1,2-b]oxirene, 2,3,4,5,6,7,7-heptachloro-1a,1b,5,5a,6,6a-hexahydro-, (1aα,1bβ,2α,5α,5aβ,6β,6aα)-	$C_{10}H_5Cl_7O$ 389.32	1024-57-3 160			
6302	Heptacosane	$C_{27}H_{56}$ 380.74	593-49-7 59.5	442	4-01-00-00586 0.7796[60]	H_2O 1; EtOH 1; eth 2 1.4345[65]
6303	14-Heptacosanone	$C_{27}H_{54}O$ 394.73	542-50-7 77.5	238[0.1]	4-01-00-03411 0.7986[81]	
6304	Heptadecanal Margaric aldehyde	$C_{17}H_{34}O$ 254.46	629-90-3 36	204[26]	4-01-00-03395	bz 4; eth 4
6305	1-Heptadecanamine	$C_{17}H_{37}N$ 255.49	4200-95-7 49	336	4-04-00-00824 0.8510[20]	H_2O 1; EtOH 3; eth 3 1.4510[20]
6306	Heptadecane	$C_{17}H_{36}$ 240.47	629-78-7 22	302.0	4-01-00-00548 0.7780[20]	H_2O 1; EtOH 2; eth 3; ctc 2 1.4369[20]
6307	Heptadecane, 1-bromo- 1-Bromoheptadecane	$C_{17}H_{35}Br$ 319.37	3508-00-7 32	349	4-01-00-00549 0.9916[20]	H_2O 1; chl 4 1.4625[20]
6308	Heptadecane, 1,17-dibromo-	$C_{17}H_{34}Br_2$ 398.26	81726-82-1 38	208[3]	3-01-00-00564	chl 4
6309	Heptadecane, 9-hexyl- 9-Hexylheptadecane	$C_{23}H_{48}$ 324.63	55124-79-3 -19.4	213[10]; 157[0.5]	4-01-00-00578 0.7976[20]	1.4465[20]
6310	Heptadecane, 2-methyl- 2-Methylheptadecane	$C_{18}H_{38}$ 254.50	1560-89-0 5.7	311	4-01-00-00556 0.7838[15]	1.4394[15]
6311	Heptadecane, 9-methyl- 9-Methylheptadecane	$C_{18}H_{38}$ 254.50	18869-72-2	172-4[10]	3-01-00-00568 0.7810[20]	1.4388[20]
6312	Heptadecanenitrile	$C_{17}H_{33}N$ 251.46	5399-02-0 34	349	4-02-00-01195 0.8315[20]	H_2O 1; EtOH 2; eth 4; chl 2 1.4467[20]
6313	Heptadecane, 9-octyl-	$C_{25}H_{52}$ 352.69	7225-64-1 -13.8	231.5[10]	4-01-00-00583 0.8020[20]	1.4487[20]
6314	Heptadecanoic acid Margaric acid	$C_{17}H_{34}O_2$ 270.46	506-12-7 61.3	5634 227[100]	4-02-00-01193 0.8532[60]	EtOH 2; eth 3; ace 3; bz 3 1.4342[60]
6315	Heptadecanoic acid, ethyl ester	$C_{19}H_{38}O_2$ 298.51	14010-23-2	185[5]	4-02-00-01194 0.8517[30]	ace 4; bz 4; eth 4; EtOH 4
6316	Heptadecanoic acid, methyl ester	$C_{18}H_{36}O_2$ 284.48	1731-92-6 30	185[9]; 152[0.05]	4-02-00-01194	H_2O 1; EtOH 3; eth 4; ace 3
6317	1-Heptadecanol Heptadecyl alcohol	$C_{17}H_{36}O$ 256.47	1454-85-9 53.8	333	4-01-00-01884 0.8475[20]	H_2O 1; EtOH 3; eth 3
6318	2-Heptadecanol	$C_{17}H_{36}O$ 256.47	16813-18-6 54	155[1]	4-01-00-01885	eth 4; EtOH 4 1.4407[37]
6319	9-Heptadecanol	$C_{17}H_{36}O$ 256.47	624-08-8 61	174[9]	4-01-00-01886	H_2O 1; EtOH 3; eth 3; ace 3 1.4262[80]
6320	2-Heptadecanone Pentadecyl methyl ketone	$C_{17}H_{34}O$ 254.46	2922-51-2 48	320	4-01-00-03395 0.8049[48]	H_2O 1; EtOH 2; eth 4; ace 3
6321	9-Heptadecanone	$C_{17}H_{34}O$ 254.46	540-08-9 53	251.5; 142[1.5]	4-01-00-03396 0.8140[48]	EtOH 2; MeOH 3
6322	1-Heptadecene Hexahydroaplotaxene	$C_{17}H_{34}$ 238.46	6765-39-5 11.5	300	4-01-00-00929 0.7852[20]	H_2O 1; eth 4; bz 3; lig 5 1.4432[20]
6323	8-Heptadecene, 9-octyl-	$C_{25}H_{50}$ 350.67	24306-18-1	227[10]	4-01-00-00936 0.8086[20]	1.4554[20]
6324	9-Heptadecenoic acid	$C_{17}H_{32}O_2$ 268.44	10136-52-4 12.5	175[0.5]	4-02-00-01632 0.8942[20]	1.4598[20]
6325	2,4-Heptadienal, (E,E)-	$C_7H_{10}O$ 110.16	4313-03-5	84.5	4-01-00-03549 0.881[25]	1.5315[20]
6326	1,2-Heptadiene Butylallene	C_7H_{12} 96.17	2384-90-9	105.5	4-01-00-01027 0.7306[18]	1.432[18]
6327	1,4-Heptadiene	C_7H_{12} 96.17	5675-22-9	93	3-01-00-00999 0.7270[20]	bz 4; eth 4; peth 4 1.4370[20]
6328	1,5-Heptadiene 1-Methyldiallyl	C_7H_{12} 96.17	1541-23-7	94	3-01-00-00999 0.7186[20]	ace 4; bz 4; eth 4; EtOH 4 1.4200[20]

No.	Name Synonym	Mol. Form. Mol. Wt.	CAS RN mp/°C	Merck No. bp/°C	Beil. Ref. den/g cm^{-3}	Solubility n_D
6329	2,4-Heptadiene	C_7H_{12} 96.17	628-72-8	108	4-01-00-01029 0.7384[20]	H_2O 1; EtOH 3; eth 3; ace 3 1.4578[20]
6330	1,3-Heptadiene, 2,6-dimethyl-	C_9H_{16} 124.23	2436-84-2	141	4-01-00-01051 0.7648[10]	1.4606[22]
6331	1,5-Heptadiene, 2,6-dimethyl-	C_9H_{16} 124.23	6709-39-3 -70	143	4-01-00-01051 0.7648[25]	
6332	2,4-Heptadiene, 2,4-dimethyl-	C_9H_{16} 124.23	74421-05-9	138	4-01-00-01051 0.7750[4]	1.4587[4]
6333	2,4-Heptadiene, 3,5-dimethyl-	C_9H_{16} 124.23	101935-28-8	142	4-01-00-01052 0.7728[20]	1.4487[20]
6334	1,6-Heptadien-3,5-dione, 1,7-bis(4-hydroxy-3-methoxyphenyl)-, (E,E)- Curcumin	$C_{21}H_{20}O_6$ 368.39	458-37-7 183	2681	4-08-00-03697	EtOH 4; HOAc 4
6335	1,5-Heptadiene, 3-methyl- 3-Methyl-1,5-heptadiene	C_8H_{14} 110.20	4894-62-6 -57	111.	4-01-00-01040 0.7250[25]	
6336	1,5-Heptadien-4-ol	$C_7H_{12}O$ 112.17	5638-26-6	156; 68[24]	4-01-00-02248 0.8598[20]	ace 4 1.4510[25]
6337	1,6-Heptadien-4-ol 4-Hydroxy-1,6-heptadiene	$C_7H_{12}O$ 112.17	2883-45-6	151	4-01-00-02249 0.864[25]	1.4505[20]
6338	1,6-Heptadien-4-ol, 4-methyl- 1,6-Heptadiene-4-ol, 4-methyl	$C_8H_{14}O$ 126.20	25201-40-5	158.4	4-01-00-02260 0.8611[20]	ace 4; EtOH 4 1.4500[23]
6339	3,5-Heptadien-2-one	$C_7H_{10}O$ 110.16	3916-64-1	88[28]	4-01-00-03549 0.8946[19]	eth 4 1.5177[19]
6340	2,5-Heptadien-4-one, 2,6-dimethyl- Phorone	$C_9H_{14}O$ 138.21	504-20-1 28	7307 197.5	4-01-00-03564 0.8850[20]	H_2O 2; EtOH 3; eth 3; ace 3 1.4998[20]
6341	1,6-Heptadien-3-yne 1,6-Heptadiene-3-yne	C_7H_8 92.14	5150-80-1	110	3-01-00-01061 0.787[25]	bz 3; peth 3 1.4694[25]
6342	1,5-Heptadiyne	C_7H_8 92.14	764-56-7	26[30]	2-01-00-00247 0.8100[21]	bz 4; peth 4 1.4521[21]
6343	1,6-Heptadiyne	C_7H_8 92.14	2396-63-6 -85	112	4-01-00-01121 0.8164[17]	H_2O 1; bz 3; HOAc 3 1.451[17]
6344	Heptanal Heptaldehyde	$C_7H_{14}O$ 114.19	111-71-7 -43.3	4578 152.8	4-01-00-03314 0.8132[25]	H_2O 2; EtOH 5; eth 5; ctc 2 1.4113[20]
6345	Heptanal, 5-methyl-	$C_8H_{16}O$ 128.21	75579-88-3	72[25]	4-01-00-03345 0.8164[25]	1.4144[20]
6346	Heptanal, oxime Enanthaldoxime	$C_7H_{15}NO$ 129.20	629-31-2 57.5	195	4-01-00-03316 0.8583[55]	H_2O 2; EtOH 3; eth 3 1.4210[20]
6347	Heptanal, 2-(phenylmethylene)-	$C_{14}H_{18}O$ 202.30	122-40-7 80	174-5[20]	2-07-00-00310 0.9711[20]	H_2O 1; ace 3; ctc 3 1.5381[20]
6348	Heptanamide	$C_7H_{15}NO$ 129.20	628-62-6 96	254	4-02-00-00963 0.8521[110]	H_2O 3; EtOH 3; eth 3; chl 2 1.4217[110]
6349	1-Heptanamine Heptylamine	$C_7H_{17}N$ 115.22	111-68-2 -18	156	4-04-00-00734 0.7754[20]	H_2O 2; EtOH 5; eth 5; chl 2 1.4251[20]
6350	2-Heptanamine Tuaminoheptane	$C_7H_{17}N$ 115.22	123-82-0	9712 142	3-04-00-00374 0.7665[19]	H_2O 2; EtOH 3; eth 3; chl 2 1.4199[19]
6351	4-Heptanamine	$C_7H_{17}N$ 115.22	16751-59-0	139.5	4-04-00-00745 0.767[20]	1.4172[20]
6352	1-Heptanamine, N-heptyl- Diheptylamine	$C_{14}H_{31}N$ 213.41	2470-68-0 31.5	271; 135[9]	4-04-00-00736 0.7956[21]	H_2O 2; EtOH 3; eth 4
6353	2-Heptanamine, 6-methyl-, (±)-	$C_8H_{19}N$ 129.25	5984-58-7	6678 155	4-04-00-00764 0.767[25]	1.4209[20]
6354	2-Heptanamine, 6-methyl-N-(3-methylbutyl)- Octamylamine	$C_{13}H_{29}N$ 199.38	502-59-0	6671 100-1[7]	3-04-00-00387	
6355	Heptane	C_7H_{16} 100.20	142-82-5 -90.6	4580 98.5	4-01-00-00376 0.6837[20]	H_2O 1; EtOH 4; eth 5; ace 5 1.3878[20]
6356	Heptane, 1-bromo- Heptyl bromide	$C_7H_{15}Br$ 179.10	629-04-9 -58	179	4-01-00-00391 1.1400[20]	H_2O 1; EtOH 4; eth 4; ctc 2 1.4502[20]
6357	Heptane, 2-bromo- 2-Heptyl bromide	$C_7H_{15}Br$ 179.10	1974-04-5 47	166	3-01-00-00431 1.1277[20]	H_2O 1; bz 4; ctc 3; chl 3 1.4503[20]
6358	Heptane, 4-bromo- 4-Heptyl bromide	$C_7H_{15}Br$ 179.10	998-93-6	84.6[72]; 60[18]	4-01-00-00392 1.1351[20]	H_2O 1; bz 3; ctc 3; chl 3 1.4495[20]
6359	Heptane, 1-bromo-7-fluoro-	$C_7H_{14}BrF$ 197.09	334-42-9	85[11]	4-01-00-00392	eth 4; EtOH 4 1.4463[20]
6360	Heptane, 1-chloro- Heptyl chloride	$C_7H_{15}Cl$ 134.65	629-06-1 -69.5	159	4-01-00-00389 0.8758[20]	H_2O 1; EtOH 5; eth 5; ctc 2 1.4256[20]
6361	Heptane, 2-chloro- 2-Chloroheptane	$C_7H_{15}Cl$ 134.65	1001-89-4	46[19]	4-01-00-00390 0.8672[20]	H_2O 1; eth 4; bz 3; chl 3 1.4221[20]
6362	Heptane, 3-chloro- 3-Chloroheptane	$C_7H_{15}Cl$ 134.65	999-52-0	144; 48.3[20]	4-01-00-00390 0.8960[20]	bz 4; eth 4 1.4228[20]
6363	Heptane, 4-chloro- 4-Chloroheptane	$C_7H_{15}Cl$ 134.65	998-95-8	144	4-01-00-00390 0.8710[20]	bz 4; eth 4 1.4237[20]
6364	Heptane, 3-chloro-3-ethyl-	$C_9H_{19}Cl$ 162.70	28320-89-0	46[3]	3-01-00-00510 0.8856[20]	EtOH 4; chl 4 1.4400[20]
6365	Heptane, 1-chloro-7-fluoro-	$C_7H_{14}ClF$ 152.64	334-43-0	70[10]	4-01-00-00390 0.993[20]	bz 4; eth 4; EtOH 4; chl 4 1.422[25]
6366	Heptane, 2-chloro-2-methyl- 2-Chloro-2-methylheptane	$C_8H_{17}Cl$ 148.68	4325-49-9	50[15]	4-01-00-00428 0.8568[25]	bz 4; eth 4; EtOH 4; chl 4 1.4240[25]
6367	Heptane, 2-chloro-6-methyl- 2-Chloro-6-methylheptane	$C_8H_{17}Cl$ 148.68	2350-19-8	74[35]	3-01-00-00472	bz 4; eth 4; EtOH 4; chl 4 1.4260[15]
6368	Heptane, 3-(chloromethyl)-	$C_8H_{17}Cl$ 148.68	123-04-6	172	4-01-00-00430 0.8769[20]	H_2O 1; EtOH 3; eth 3; ace 3 1.4319[20]

No.	Name / Synonym	Mol. Form. / Mol. Wt.	CAS RN	mp/°C	Merck No. bp/°C	Beil. Ref. den/g cm^{-3}	Solubility n_D
6369	Heptane, 3-chloro-3-methyl- / 3-Methyl-3-chloroheptane	$C_8H_{17}Cl$ / 148.68	5272-02-6		64[27]	4-01-00-00430 / 0.8764[20]	bz 4; eth 4; EtOH 4; chl 4 / 1.4317[20]
6370	Heptane, 4-chloro-4-methyl- / 2-Chloro-2-propylpentane	$C_8H_{17}Cl$ / 148.68	61764-94-1		50[12]	3-01-00-00477 / 0.8690[20]	bz 4; eth 4; EtOH 4; chl 4 / 1.4310[15]
6371	1,7-Heptanediamine	$C_7H_{18}N_2$ / 130.23	646-19-5	28.9	224	4-04-00-01354	EtOH 3; eth 3; ace 3
6372	Heptane, 1,2-dibromo- / 1,2-Dibromoheptane	$C_7H_{14}Br_2$ / 258.00	42474-21-5		228	3-01-00-00432 / 1.5086[20]	1.4986[20]
6373	Heptane, 1,5-dibromo- / 1,5-Dibromoheptane	$C_7H_{14}Br_2$ / 258.00	1622-10-2		113-5[11]	4-01-00-00393 / 1.536[15]	1.5041[15]
6374	Heptane, 1,7-dibromo- / Heptamethylene dibromide	$C_7H_{14}Br_2$ / 258.00	4549-31-9	41.7	263	4-01-00-00393 / 1.5306[20]	H_2O 1; eth 3; ace 3; bz 3 / 1.5034[20]
6375	Heptane, 2,3-dibromo- / 2,3-Dibromoheptane	$C_7H_{14}Br_2$ / 258.00	21266-88-6		101[17]	3-01-00-00432 / 1.5139[20]	1.4992[20]
6376	Heptane, 3,4-dibromo- / 3,4-Dibromoheptane	$C_7H_{14}Br_2$ / 258.00	21266-90-0		107[24]	3-01-00-00433 / 1.5182[20]	1.5010[20]
6377	Heptane, 1,1-dichloro- / 1,1-Dichloroheptane	$C_7H_{14}Cl_2$ / 169.09	821-25-0		191	4-01-00-00390 / 1.011[20]	H_2O 1; eth 3; bz 3; ctc 3 / 1.4440[20]
6378	Heptane, 1,2-dichloro- / 1,2-Dichloroheptane	$C_7H_{14}Cl_2$ / 169.09	10575-87-8		68-72[7]	3-01-00-00430 / 1.064[20]	bz 4; eth 4; chl 4 / 1.4490[20]
6379	Heptane, 1,7-dichloro- / 1,7-Dichloroheptane	$C_7H_{14}Cl_2$ / 169.09	821-76-1		124-5[35]	4-01-00-00390 / 1.0408[25]	1.4565[25]
6380	Heptane, 2,2-dichloro- / 2,2-Dichloroheptane	$C_7H_{14}Cl_2$ / 169.09	65786-09-6		77[25]	3-01-00-00430 / 1.012[20]	bz 4; eth 4; chl 4 / 1.4440[20]
6381	Heptane, 4,4-dichloro-	$C_7H_{14}Cl_2$ / 169.09	89796-76-9		86[27]	2-01-00-00117 / 1.008[17]	eth 4; chl 4 / 1.448[17]
6382	Heptane, 2,6-dichloro-2,6-dimethyl-	$C_9H_{18}Cl_2$ / 197.15	35951-36-1	43	93[16]	3-01-00-00513	bz 4; chl 4
6383	Heptane, 2,2-dimethyl- / 2,2-Dimethylheptane	C_9H_{20} / 128.26	1071-26-7	-113	132.7	4-01-00-00457 / 0.7105[20]	H_2O 1; eth 3; ace 4; bz 5 / 1.4016[20]
6384	Heptane, 2,3-dimethyl- / 2,3-Dimethylheptane	C_9H_{20} / 128.26	3074-71-3	-116	140.5	4-01-00-00457 / 0.7260[20]	H_2O 1; EtOH 5; eth 5; ace 5 / 1.4088[20]
6385	Heptane, 2,4-dimethyl- / 2,4-Dimethylheptane	C_9H_{20} / 128.26	2213-23-2		132.9	4-01-00-00457 / 0.7115[25]	H_2O 1; EtOH 5; eth 5; ace 5 / 1.4034[20]
6386	Heptane, 2,5-dimethyl- / 3,6-Dimethylheptane	C_9H_{20} / 128.26	2216-30-0		136	4-01-00-00457 / 0.7198[20]	ace 4; bz 4; eth 4; EtOH 4 / 1.4033[20]
6387	Heptane, 2,6-dimethyl- / 2,6-Dimethylheptane	C_9H_{20} / 128.26	1072-05-5	-102.9	135.2	4-01-00-00458 / 0.7089[20]	chl 2 / 1.4011[20]
6388	Heptane, 3,3-dimethyl- / 3,3-Dimethylheptane	C_9H_{20} / 128.26	4032-86-4		137.3	4-01-00-00458 / 0.7254[20]	H_2O 1; EtOH 5; eth 3; ace 4 / 1.4087[20]
6389	Heptane, 3,4-dimethyl- / 3,4-Dimethylheptane	C_9H_{20} / 128.26	922-28-1		140.6	3-01-00-00514 / 0.7314[20]	H_2O 1; eth 3; ace 4; bz 5 / 1.4108[20]
6390	Heptane, 3,5-dimethyl- / 3,5-Dimethylheptane	C_9H_{20} / 128.26	926-82-9		136	4-01-00-00458 / 0.7225[20]	H_2O 1; eth 3; ace 4; bz 5 / 1.4083[20]
6391	Heptane, 4,4-dimethyl- / 4,4-Dimethylheptane	C_9H_{20} / 128.26	1068-19-5		135.2	4-01-00-00458 / 0.7221[20]	H_2O 1; eth 3; ace 4; bz 5 / 1.4076[20]
6392	Heptanedinitrile	$C_7H_{10}N_2$ / 122.17	646-20-8	-31.4	155[14]	4-02-00-02006 / 0.949[18]	H_2O 1; EtOH 5; eth 5; chl 5 / 1.4472[20]
6393	Heptanedioic acid / Pimelic acid	$C_7H_{12}O_4$ / 160.17	111-16-0	106	7401 / 272[100]; 212[10]	4-02-00-02003 / 1.329[15]	H_2O 3; EtOH 3; eth 3; bz 1
6394	Heptanedioic acid, diethyl ester / Diethyl pimelate	$C_{11}H_{20}O_4$ / 216.28	2050-20-6	-24	254; 140[15]	4-02-00-02004 / 0.9945[20]	eth 4; EtOH 4 / 1.4305[20]
6395	Heptanedioic acid, dimethyl ester	$C_9H_{16}O_4$ / 188.22	1732-08-7	-21	120[10]; 80[1]	4-02-00-02004 / 1.0625[20]	H_2O 2; EtOH 3; eth 3; bz 3 / 1.4309[20]
6396	Heptanedioic acid, monoethyl ester	$C_9H_{16}O_4$ / 188.22	33018-91-6	10	182[18]; 160[4]	3-02-00-01742	1.4415[20]
6397	1,4-Heptanediol	$C_7H_{16}O_2$ / 132.20	40646-07-9		243; 127[4]	4-01-00-02579 / 0.9542[20]	1.4510[25]
6398	1,5-Heptanediol	$C_7H_{16}O_2$ / 132.20	60096-09-5		135-6[11]	3-01-00-02213 / 0.9705[20]	eth 4; EtOH 4; chl 4 / 1.4571[22]
6399	1,7-Heptanediol	$C_7H_{16}O_2$ / 132.20	629-30-1	22.5	262	4-01-00-02580 / 0.9569[25]	eth 4; EtOH 4 / 1.4520[25]
6400	2,4-Heptanediol	$C_7H_{16}O_2$ / 132.20	20748-86-1		107-8[8]	4-01-00-02580 / 0.926[20]	1.4386[25]
6401	2,4-Heptanediol, 3-methyl- / 3-Methylheptan-2,4-diol	$C_8H_{18}O_2$ / 146.23	6964-04-1		115[3]	0-01-00-00491 / 0.928[20]	H_2O 1; EtOH 3; ctc 2 / 1.4459[20]
6402	2,3-Heptanedione / Acetyl valeryl	$C_7H_{12}O_2$ / 128.17	96-04-8		46[13]	4-01-00-03697 / 0.919[18]	1.4150[18]
6403	2,6-Heptanedione	$C_7H_{12}O_2$ / 128.17	13505-34-5	33.5	222; 97[11]	4-01-00-03699 / 0.9399[37]	bz 4; eth 4; EtOH 4 / 1.4277[37]
6404	3,5-Heptanedione, 2,2,6,6-tetramethyl- / Dipivaloylmethane	$C_{11}H_{20}O_2$ / 184.28	1118-71-4		72-3[6]	4-01-00-03729 / 0.883[25]	ctc 2 / 1.4589[20]
6405	Heptane, 3-[(ethenyloxy)methyl]-	$C_{10}H_{20}O$ / 156.27	103-44-6	-100	76[32]	4-01-00-02056 / 0.8108[20]	1.4247[25]
6406	Heptane, 1-ethoxy-	$C_9H_{20}O$ / 144.26	1969-43-3	-68.3	166	4-01-00-01733 / 0.790[16]	
6407	Heptane, 3-ethyl- / 3-Ethylheptane	C_9H_{20} / 128.26	15869-80-4	-114.9	143.0	4-01-00-00456 / 0.7225[25]	1.4093[20]
6408	Heptane, 4-ethyl- / 4-Ethylheptane	C_9H_{20} / 128.26	2216-32-2		141.2	4-01-00-00457 / 0.7241[25]	H_2O 1; EtOH 5; eth 3; ace 5 / 1.4096[20]

No.	Name Synonym	Mol. Form. Mol. Wt.	CAS RN mp/°C	Merck No. bp/°C	Beil. Ref. den/g cm^{-3}	Solubility n_D
6409	Heptane, 3-ethyl-2-methyl- Heptane, 2-methyl-3-ethyl	$C_{10}H_{22}$ 142.28	14676-29-0	163	4-01-00-00480 0.7398[25]	1.4174[20]
6410	Heptane, 3-ethyl-3-methyl-	$C_{10}H_{22}$ 142.28	17302-01-1	163.8	4-01-00-00481 0.7463[25]	1.4208[20]
6411	Heptane, 3-ethyl-4-methyl- 3-Ethyl-4-methylheptane	$C_{10}H_{22}$ 142.28	52896-91-0	165	4-01-00-00481 0.7468[25]	1.4207[20]
6412	Heptane, 3-ethyl-5-methyl- 3-Ethyl-5-methylheptane	$C_{10}H_{22}$ 142.28	52896-90-9	158.2	4-01-00-00481 0.7368[25]	1.4164[20]
6413	Heptane, 4-ethyl-2-methyl- 4-Ethyl-2-methylheptane	$C_{10}H_{22}$ 142.28	52896-88-5	158	4-01-00-00481 0.7322[25]	1.4137[20]
6414	Heptane, 4-ethyl-3-methyl- 4-Ethyl-3-methylheptane	$C_{10}H_{22}$ 142.28	52896-89-6	165	4-01-00-00481 0.7466[25]	1.4206[20]
6415	Heptane, 4-ethyl-4-methyl- 4-Ethyl-4-methylheptane	$C_{10}H_{22}$ 142.28	17302-04-4	160.8	4-01-00-00481 0.7472[25]	1.4210[25]
6416	Heptane, 5-ethyl-2-methyl-	$C_{10}H_{22}$ 142.28	13475-78-0	159.7	4-01-00-00481 0.7318[25]	1.4134[20]
6417	Heptane, 1-fluoro-	$C_7H_{15}F$ 118.19	661-11-0 -73	117.9	4-01-00-00387 0.8062[20]	H$_2$O 1; eth 3; ace 3; bz 3 1.3854[20]
6418	Heptane, hexadecafluoro- Perfluoroheptane	C_7F_{16} 388.05	335-57-9 -78	82.5	4-01-00-00388 1.7333[20]	ace 4; eth 4; EtOH 4; chl 4 1.2618[20]
6419	Heptane, 1-iodo-	$C_7H_{15}I$ 226.10	4282-40-0 -48.2	204	4-01-00-00393 1.3791[20]	H$_2$O 1; EtOH 3; eth 3; ace 3 1.4904[20]
6420	Heptane, 2-iodo- 2-Iodoheptane	$C_7H_{15}I$ 226.10	18589-29-2	98[50]	4-01-00-00393 1.304[20]	H$_2$O 1; ace 3; bz 3; ctc 3 1.4826
6421	Heptane, 1-methoxy-	$C_8H_{18}O$ 130.23	629-32-3	151	3-01-00-01682 0.7862[15]	ace 4; eth 4; EtOH 4 1.4073[20]
6422	Heptane, 2-methyl- 2-Methylheptane	C_8H_{18} 114.23	592-27-8 -108.9	117.6	4-01-00-00428 0.6980[20]	H$_2$O 1; EtOH 5; eth 3; ace 5 1.3949[20]
6423	Heptane, 3-methyl-, (S)-	C_8H_{18} 114.23	6131-25-5 -120	118.9	4-01-00-00429 0.7075[16]	H$_2$O 1; EtOH 3; eth 3; ace 5 1.4002[18]
6424	Heptane, 4-methyl- 4-Methylheptane	C_8H_{18} 114.23	589-53-7 -121	117.7	4-01-00-00431 0.7046[20]	H$_2$O 1; EtOH 5; eth 3; ace 5 1.3979[20]
6425	Heptane, 3-methylene-	C_8H_{16} 112.22	1632-16-2	120	4-01-00-00884 0.7270[20]	H$_2$O 1; eth 4; bz 4; peth 4 1.4157[20]
6426	Heptane, 4-methylene- 2-Propyl-1-pentene	C_8H_{16} 112.22	15918-08-8	117.7	4-01-00-00885 0.7198[25]	1.4136[20]
6427	Heptane, 4-(1-methylethyl)- 4-Isopropylheptane	$C_{10}H_{22}$ 142.28	52896-87-4	158.9	4-01-00-00481 0.7354[25]	1.4153[20]
6428	Heptanenitrile	$C_7H_{13}N$ 111.19	629-08-3	183; 70-2[10]	4-02-00-00963 0.8106[20]	H$_2$O 1; eth 3; ace 3; bz 3 1.4104[30]
6429	Heptane, 1-nitro- 1-Nitroheptane	$C_7H_{15}NO_2$ 145.20	693-39-0	194	4-01-00-00395 0.9476[17]	eth 4; EtOH 4
6430	Heptane, 1,1'-oxybis-	$C_{14}H_{30}O$ 214.39	629-64-1	258.5	4-01-00-01733 0.8008[20]	eth 4; EtOH 4 1.4275[20]
6431	Heptane, 3,3'-[oxybis(methylene)]bis- Ether, dihexyl,2,2'-diethyl	$C_{18}H_{34}O$ 242.45	10143-60-9	269.4	4-01-00-01785	ctc 2 1.4325[20]
6432	Heptane, 1,1,1,2,2,3,3,4,4,5,5,6,6,7,7-pentadecafluoro- 1H-Pentadecafluoroheptane	C_7HF_{15} 370.06	375-83-7	96.0	4-01-00-00388 1.725[25]	1.2690[25]
6433	Heptane, 2,2,4,6,6-pentamethyl- 2,2,4,6,6-Pentamethylheptane	$C_{12}H_{26}$ 170.34	13475-82-6 -67	177.8	4-01-00-00510 0.7463[20]	1.4440[20]
6434	Heptane, 4-propyl- 4-Propylheptane	$C_{10}H_{22}$ 142.28	3178-29-8	157.5	4-01-00-00480 0.7321[25]	1.4135[20]
6435	Heptane, 1,1'-thiobis-	$C_{14}H_{30}S$ 230.46	629-65-2 70	298	4-01-00-01739 0.8416[20]	H$_2$O 1; eth 3 1.4606[20]
6436	1-Heptanethiol Heptyl mercaptan	$C_7H_{16}S$ 132.27	1639-09-4 -43	177	4-01-00-01738 0.8427[20]	H$_2$O 1; EtOH 5; eth 5; chl 3 1.4521[20]
6437	Heptane, 2,2,3-trimethyl- 2,2,3-Trimethylheptane	$C_{10}H_{22}$ 142.28	52896-92-1	157.8	4-01-00-00481 0.7385[25]	1.4168[20]
6438	Heptane, 2,2,4-trimethyl- 2,2,4-Trimethylheptane	$C_{10}H_{22}$ 142.28	14720-74-2	148.3	4-01-00-00481 0.7237[25]	H$_2$O 1; bz 3; ctc 3; chl 3 1.4092[20]
6439	Heptane, 2,2,5-trimethyl- 2,2,5-Trimethylheptane	$C_{10}H_{22}$ 142.28	20291-95-6	150.8	4-01-00-00482 0.7243[25]	1.4101[20]
6440	Heptane, 2,2,6-trimethyl- 2,2,6-Trimethylheptane	$C_{10}H_{22}$ 142.28	1190-83-6 -105	148.9	4-01-00-00482 0.7200[25]	1.4078[20]
6441	Heptane, 2,3,3-trimethyl- 2,3,3-Trimethylheptane	$C_{10}H_{22}$ 142.28	52896-93-2	160.2	4-01-00-00482 0.7450[25]	1.4202[20]
6442	Heptane, 2,3,4-trimethyl- 2,3,4-Trimethylheptane	$C_{10}H_{22}$ 142.28	52896-95-4	161	4-01-00-00482 0.7447[25]	1.4195[20]
6443	Heptane, 2,3,5-trimethyl- 2,3,5-Trimethylheptane	$C_{10}H_{22}$ 142.28	20278-85-7	160.7	4-01-00-00482 0.7413[25]	1.4169[20]
6444	Heptane, 2,3,6-trimethyl- 2,3,6-Trimethylheptane	$C_{10}H_{22}$ 142.28	4032-93-3	156.0	4-01-00-00482 0.7305[25]	1.4131[20]
6445	Heptane, 2,4,4-trimethyl- 2,4,4-Trimethylheptane	$C_{10}H_{22}$ 142.28	4032-92-2	151.0	4-01-00-00482 0.7308[25]	1.4142[20]
6446	Heptane, 2,4,5-trimethyl- 2,4,5-Trimethylheptane	$C_{10}H_{22}$ 142.28	20278-84-6	156.5	4-01-00-00483 0.7373[25]	1.4160[20]
6447	Heptane, 2,4,6-trimethyl- 2,4,6-Trimethylheptane	$C_{10}H_{22}$ 142.28	2613-61-8	147.6	4-01-00-00483 0.7190[25]	1.4071[20]
6448	Heptane, 2,5,5-trimethyl- 2,5,5-Trimethylheptane	$C_{10}H_{22}$ 142.28	1189-99-7	152.8	4-01-00-00483 0.7362[25]	1.4149[20]

No.	Name Synonym	Mol. Form. Mol. Wt.	CAS RN mp/°C	Merck No. bp/°C	Beil. Ref. den/g cm^{-3}	Solubility n_D
6449	Heptane, 3,3,4-trimethyl- 3,3,4-Trimethylheptane	C$_{10}$H$_{22}$ 142.28	20278-87-9 	 161.9	4-01-00-00483 0.7527[25]	 1.4236[20]
6450	Heptane, 3,3,5-trimethyl- 3,3,5-Trimethylheptane	C$_{10}$H$_{22}$ 142.28	7154-80-5 	 155.7	4-01-00-00483 0.7248[20]	H$_2$O 1; bz 3; ctc 3; chl 3 1.4170[20]
6451	Heptane, 3,4,4-trimethyl- 3,4,4-Trimethylheptane	C$_{10}$H$_{22}$ 142.28	20278-88-0 	 161.1	4-01-00-00484 0.7535[25]	 1.4235[20]
6452	Heptane, 3,4,5-trimethyl- 3,4,5-Trimethylheptane	C$_{10}$H$_{22}$ 142.28	20278-89-1 	 162.5	4-01-00-00484 0.7519[25]	 1.4229[20]
6453	1,4,7-Heptanetriol	C$_7$H$_{16}$O$_3$ 148.20	3920-53-4 -35	231[25]; 146[1]	4-01-00-02787 1.075[18]	H$_2$O 3; EtOH 3; ace 3 1.4725[20]
6454	2,4,6-Heptanetrione	C$_7$H$_{10}$O$_3$ 142.15	626-53-9 49	121[10]	4-01-00-03783 1.0599[40]	H$_2$O 4; eth 4; EtOH 4 1.4930[20]
6455	Heptanoic acid Enanthylic acid	C$_7$H$_{14}$O$_2$ 130.19	111-14-8 -7.5	4581 222.2	4-02-00-00958 0.9181[20]	H$_2$O 2; EtOH 3; eth 3; ace 3 1.4170[20]
6456	Heptanoic acid, anhydride	C$_{14}$H$_{26}$O$_3$ 242.36	626-27-7 -12.4	 269.5	4-02-00-00962 0.9321[20]	H$_2$O 1; EtOH 3; eth 3 1.4335[15]
6457	Heptanoic acid, 2-bromo- 2-Bromoheptanoic acid	C$_7$H$_{13}$BrO$_2$ 209.08	2624-01-3 	250 dec; 147[12]	4-02-00-00967 1.319[15]	ace 4; eth 4 1.471[18]
6458	Heptanoic acid, 7-bromo-	C$_7$H$_{13}$BrO$_2$ 209.08	30515-28-7 31	280; 165[12]	4-02-00-00968 	ace 4; bz 4; eth 4; EtOH 4
6459	Heptanoic acid, butyl ester Butyl enanthate	C$_{11}$H$_{22}$O$_2$ 186.29	5454-28-4 -67.5	 226.2	3-02-00-00768 0.8638[20]	ace 4; bz 4; eth 4; EtOH 4 1.4204[20]
6460	Heptanoic acid, 2,2-dimethyl-6-oxo- Geronic acid	C$_9$H$_{16}$O$_3$ 172.22	461-11-0 	278; 169[12]	3-03-00-01264 1.0211[20]	eth 4; EtOH 4 1.4488[20]
6461	Heptanoic acid, 4,4-dimethyl-6-oxo- Isogeronic acid	C$_9$H$_{16}$O$_3$ 172.22	471-04-5 50	178[20]; 162[10]	3-03-00-01266 	H$_2$O 4; bz 4; eth 4; EtOH 4
6462	Heptanoic acid, ethyl ester Ethyl oenanthate	C$_9$H$_{18}$O$_2$ 158.24	106-30-9 -66.1	3790 187	4-02-00-00960 0.8817[20]	H$_2$O 2; EtOH 3; eth 3; ctc 2 1.4100[20]
6463	Heptanoic acid, 7-fluoro-	C$_7$H$_{13}$FO$_2$ 148.18	334-28-1 	133[10]	4-02-00-00964 1.039[20]	 1.4207[25]
6464	Heptanoic acid, heptyl ester Heptyl heptanoate	C$_{14}$H$_{28}$O$_2$ 228.38	624-09-9 -33	 277	4-02-00-00961 0.8649[20]	H$_2$O 1; EtOH 3; eth 3 1.4320[20]
6465	Heptanoic acid, hexyl ester Hexyl heptanoate	C$_{13}$H$_{26}$O$_2$ 214.35	1119-06-8 -48	 261	3-02-00-00768 0.8611[20]	ace 4; bz 4; eth 4; EtOH 4 1.429[15]
6466	Heptanoic acid, methyl ester Methyl heptanoate	C$_8$H$_{16}$O$_2$ 144.21	106-73-0 -56	4581 174	4-02-00-00960 0.8815[20]	H$_2$O 2; EtOH 3; eth 3; ace 2 1.4152[20]
6467	Heptanoic acid, 2-methylpropyl ester	C$_{11}$H$_{22}$O$_2$ 186.29	7779-80-8 	 208	1-02-00-00145 0.8593[20]	ace 4; bz 4; eth 4; EtOH 4
6468	Heptanoic acid, octyl ester Octyl enanthate	C$_{15}$H$_{30}$O$_2$ 242.40	5132-75-2 -22.5	 290	3-02-00-00768 0.8596[20]	ace 4; bz 4; eth 4; EtOH 4 1.4349[15]
6469	Heptanoic acid, 6-oxo-	C$_7$H$_{12}$O$_3$ 144.17	3128-07-2 40.2	251[280]; 135[1]	4-03-00-01598 	H$_2$O 4; ace 4; eth 4; EtOH 4 1.4306[25]
6470	Heptanoic acid, pentyl ester Amyl heptanoate	C$_{12}$H$_{24}$O$_2$ 200.32	7493-82-5 -50	 245.4	4-02-00-00960 0.8623[20]	ace 4; bz 4; eth 4; EtOH 4 1.4263[15]
6471	Heptanoic acid, propyl ester Propyl heptanoate	C$_{10}$H$_{20}$O$_2$ 172.27	7778-87-2 -63.5	 208	3-02-00-00767 0.8641[15]	ace 4; bz 4; eth 4; EtOH 4 1.4183[15]
6472	1-Heptanol Heptyl alcohol	C$_7$H$_{16}$O 116.20	111-70-6 -34	4582 176.4	4-01-00-01731 0.8219[20]	H$_2$O 2; EtOH 5; eth 5; ctc 2 1.4249[20]
6473	2-Heptanol, (±)-	C$_7$H$_{16}$O 116.20	52390-72-4 	4583 159	4-01-00-01740 0.8167[20]	H$_2$O 2; EtOH 3; eth 3; ctc 2 1.4210[20]
6474	3-Heptanol, (S)-	C$_7$H$_{16}$O 116.20	26549-25-7 -70	157; 66[18]	2-01-00-00444 0.8227[20]	H$_2$O 2; EtOH 3; eth 3; ctc 2 1.4201[20]
6475	4-Heptanol Dipropylcarbinol	C$_7$H$_{16}$O 116.20	589-55-9 -41.2	 156	4-01-00-01743 0.8183[20]	H$_2$O 1; EtOH 3; eth 3 1.4205[20]
6476	4-Heptanol, acetate	C$_9$H$_{18}$O$_2$ 158.24	5921-84-6 	 171	4-02-00-00163 0.8741[0]	eth 4 1.4105[19]
6477	1-Heptanol, 7-chloro- Heptamethylene chlorohydrin	C$_7$H$_{15}$ClO 150.65	55944-70-2 11	150[20]	4-01-00-01738 0.9998[15]	EtOH 4; peth 4 1.4537[25]
6478	2-Heptanol, 1-chloro- 1-Chloro-2-heptanol	C$_7$H$_{15}$ClO 150.65	53660-21-2 	93[13]	4-01-00-01741 0.9885[20]	EtOH 3; eth 3; ace 3 1.4499[20]
6479	2-Heptanol, 2,6-dimethyl- 2,6-Dimethyl-2-heptanol	C$_9$H$_{20}$O 144.26	13254-34-7 	 173	2-01-00-00457 0.8186[20]	 1.4242[20]
6480	3-Heptanol, 2,3-dimethyl- Butylisopropylmethylcarbinol	C$_9$H$_{20}$O 144.26	19549-71-4 	173; 75[18]	3-01-00-01752 0.8395[20]	 1.4355[20]
6481	3-Heptanol, 2,6-dimethyl- 2,6-Dimethyl-3-heptanol	C$_9$H$_{20}$O 144.26	19549-73-6 	 175	3-01-00-01753 0.8212[20]	 1.4246[20]
6482	3-Heptanol, 3,5-dimethyl-	C$_9$H$_{20}$O 144.26	19549-74-7 	95-7[50]	4-01-00-01811 0.8177[25]	 1.4251[20]
6483	3-Heptanol, 3,6-dimethyl- 3,6-Dimethylheptan-3-ol	C$_9$H$_{20}$O 144.26	1573-28-0 	 173	4-01-00-01810 0.8285[16]	eth 4; EtOH 4; lig 4 1.4326[16]
6484	4-Heptanol, 2,4-dimethyl- 2,4-Dimethyl-4-heptanol	C$_9$H$_{20}$O 144.26	19549-77-0 	171; 70[26]	4-01-00-01810 0.8215[20]	 1.4292[20]
6485	4-Heptanol, 2,6-dimethyl- Diisobutylcarbinol	C$_9$H$_{20}$O 144.26	108-82-7 	 174.5	4-01-00-01810 0.8114[20]	H$_2$O 1; EtOH 3; eth 3; ctc 2 1.4242[20]
6486	4-Heptanol, 3,5-dimethyl- 3,5-Dimethyl-4-heptanol	C$_9$H$_{20}$O 144.26	19549-79-2 	 186	4-01-00-01811 0.836[18]	 1.4283[20]
6487	4-Heptanol, 4-ethyl-	C$_9$H$_{20}$O 144.26	597-90-0 	 182	4-01-00-01809 0.8350[20]	eth 4; EtOH 4 1.4332[20]

No.	Name / Synonym	Mol. Form. / Mol. Wt.	CAS RN / mp/°C	Merck No. / bp/°C	Beil. Ref. / den/g cm^{-3}	Solubility / n_D
6488	3-Heptanol, 3-ethyl-2-methyl-	$C_{10}H_{22}O$ 158.28	66719-37-7	193	1-01-00-00214 0.8455[20]	1.4378
6489	3-Heptanol, 3-ethyl-6-methyl-	$C_{10}H_{22}O$ 158.28	66719-40-2	83-6[15]	0-01-00-00427 0.852[25]	1.4409
6490	3-Heptanol, 5-ethyl-4-methyl-	$C_{10}H_{22}O$ 158.28	66731-94-0	86-8[14]	3-01-00-01770 0.865[17]	1.4565[17]
6491	1-Heptanol, 7-fluoro-	$C_7H_{15}FO$ 134.19	408-16-2	98-9[12]	4-01-00-01736 0.956[20]	eth 4; EtOH 4 1.4197[25]
6492	1-Heptanol, 3-methyl- 3-Methyl-1-heptanol	$C_8H_{18}O$ 130.23	1070-32-2	101[20]	4-01-00-01782 0.824[24]	1.4295[25]
6493	1-Heptanol, 4-methyl- 4-Methyl-1-heptanol	$C_8H_{18}O$ 130.23	817-91-4	188	4-01-00-01789 0.8065[25]	EtOH 4 1.4253[25]
6494	1-Heptanol, 6-methyl- 6-Methyl-1-heptanol	$C_8H_{18}O$ 130.23	1653-40-3 -106	188; 95.8[20]	4-01-00-01782 0.8176[25]	H_2O 1; EtOH 3; eth 3 1.4251[25]
6495	2-Heptanol, 2-methyl- 2-Methyl-2-heptanol	$C_8H_{18}O$ 130.23	625-25-2 -50.4	156	4-01-00-01780 0.8142[20]	H_2O 1; EtOH 3; eth 3 1.4250[20]
6496	2-Heptanol, 3-methyl- 3-Methyl-2-heptanol	$C_8H_{18}O$ 130.23	31367-46-1	166.1	0.8177[25]	H_2O 1; EtOH 3; eth 3; ctc 3 1.4199[25]
6497	2-Heptanol, 5-methyl- 5-Methyl-2-heptanol	$C_8H_{18}O$ 130.23	54630-50-1 -61	170	4-01-00-01783 0.8174[21]	
6498	2-Heptanol, 6-methyl-	$C_8H_{18}O$ 130.23	4730-22-7 -105	174	3-01-00-01728 0.8218[20]	1.4238[10]
6499	3-Heptanol, 2-methyl-, (±)-	$C_8H_{18}O$ 130.23	100296-26-2 -85	167.5	4-01-00-01781 0.8235[20]	H_2O 2; EtOH 3; eth 3; ctc 3 1.4265[20]
6500	3-Heptanol, 3-methyl- 2-Ethyl-2-hexanol	$C_8H_{18}O$ 130.23	5582-82-1 -83	163	4-01-00-01783 0.8282[20]	H_2O 1; EtOH 3; eth 3; ctc 3 1.4279[20]
6501	3-Heptanol, 4-methyl- 4-Methyl-3-heptanol	$C_8H_{18}O$ 130.23	14979-39-6 -123	170	4-01-00-01789 0.827[25]	1.4300[20]
6502	3-Heptanol, 5-methyl-	$C_8H_{18}O$ 130.23	18720-65-5 -91.2	172	4-01-00-01783 0.8425[25]	1.433[24]
6503	4-Heptanol, 2-methyl- 6-Methyl-4-heptanol	$C_8H_{18}O$ 130.23	21570-35-4 -81	164	4-01-00-01781 0.8207[20]	eth 4; EtOH 4 1.4203
6504	4-Heptanol, 3-methyl-, (R*,S*)-(±)- 4-Heptanol, 3-methyl(DL)	$C_8H_{18}O$ 130.23	92737-91-2	162	3-01-00-01731 0.8335[25]	H_2O 2; EtOH 3; eth 3; ctc 3 1.4211[25]
6505	4-Heptanol, 4-methyl- 4-Hydroxy-4-methylheptane	$C_8H_{18}O$ 130.23	598-01-6 -82	161	4-01-00-01789 0.8248[20]	H_2O 1; EtOH 3; eth 3; ctc 3 1.4258[20]
6506	2-Heptanol, 6-methyl-, acetate Acetic acid, (6-methyl-2-heptyl) ester	$C_{10}H_{20}O_2$ 172.27	67952-57-2	187	0-02-00-00135 0.8474[20]	EtOH 4 1.413[20]
6507	3-Heptanol, 6-methyl-, acetate Acetic acid, (6-methyl-3-heptyl) ester	$C_{10}H_{20}O_2$ 172.27	32764-34-4	184.5	0-02-00-00135 0.8554[20]	EtOH 4 1.4160[20]
6508	4-Heptanol, 4-propyl-	$C_{10}H_{22}O$ 158.28	2198-72-3	191	4-01-00-01831 0.8337[21]	bz 4; eth 4; EtOH 4 1.4355[21]
6509	4-Heptanol, 2,4,6-trimethyl-	$C_{10}H_{22}O$ 158.28	60836-07-9	181; 79[12]	4-01-00-01833 0.823[21]	1.4334[18]
6510	2-Heptanone Methyl pentyl ketone	$C_7H_{14}O$ 114.19	110-43-0 -35	4584 151.0	4-01-00-03318 0.8111[20]	H_2O 4; EtOH 3; eth 3 1.4088[20]
6511	3-Heptanone Ethyl butyl ketone	$C_7H_{14}O$ 114.19	106-35-4 -39	147	4-01-00-03321 0.8183[20]	H_2O 1; EtOH 5; eth 5; ctc 2 1.4057[20]
6512	4-Heptanone Dipropyl ketone	$C_7H_{14}O$ 114.19	123-19-3 -33	3351 144	4-01-00-03323 0.8174[20]	H_2O 1; EtOH 5; eth 5; ctc 3 1.4069[20]
6513	2-Heptanone, 1-chloro- 1-Chloro-2-heptanone	$C_7H_{13}ClO$ 148.63	41055-92-9	83[16]	4-01-00-03320 0.802[20]	eth 4; EtOH 4 1.4371[20]
6514	4-Heptanone, 2,6-dimethyl- Diisobutyl ketone	$C_9H_{18}O$ 142.24	108-83-8 -41.5	169.4	4-01-00-03360 0.8062[20]	H_2O 1; EtOH 5; eth 5; ctc 3 1.412[21]
6515	4-Heptanone, 3,5-dimethyl- Di-sec-butyl ketone	$C_9H_{18}O$ 142.24	19549-84-9	162	3-01-00-02893 0.826[14]	1.4193[14]
6516	3-Heptanone, 6-(dimethylamino)-4,4-diphenyl-, hydrochloride Methadone hydrochloride	$C_{21}H_{28}ClNO$ 345.91	1095-90-5 235	5852	3-14-00-00279	H_2O 4; EtOH 4
6517	4-Heptanone, 2-hydroxy-	$C_7H_{14}O_2$ 130.19	54862-92-9	95[12]	4-01-00-04034 0.9296[19]	1.4357[19]
6518	2-Heptanone, 4-hydroxy-3-methyl- 4-Heptanol-2-one, 3-methyl	$C_8H_{18}O_2$ 144.21	56072-27-6	110[16]	3-01-00-03249 0.9238[20]	1.442[20]
6519	3-Heptanone, 5-hydroxy-5-methyl- 3-Heptanol-5-one, 3-methyl	$C_8H_{18}O_2$ 144.21	39121-37-4	85[15]	4-01-00-04045 0.9315[15]	1.4367[15]
6520	2-Heptanone, 3-methyl- 3-Methyl-2-heptanone	$C_8H_{16}O$ 128.21	2371-19-9	164	3-01-00-02878 0.8218[20]	H_2O 2; EtOH 3; eth 3; ace 3 1.4172[20]
6521	2-Heptanone, 6-methyl-	$C_8H_{16}O$ 128.21	928-68-7	167	4-01-00-03344 0.8151[20]	H_2O 2; EtOH 4; eth 4; ace 5 1.4162[20]
6522	3-Heptanone, 2-methyl- Butyl isopropyl ketone	$C_8H_{16}O$ 128.21	13019-20-0	158	4-01-00-03343 0.8163[20]	H_2O 1; EtOH 3; eth 3; ace 4 1.4115[20]
6523	3-Heptanone, 6-methyl-	$C_8H_{16}O$ 128.21	624-42-0	164	4-01-00-03344 0.8304[20]	H_2O 1; EtOH 3; eth 3; bz 3 1.4209[20]
6524	4-Heptanone, 2-methyl- Isobutyl propyl ketone	$C_8H_{16}O$ 128.21	626-33-5	155	4-01-00-03343 0.813[22]	H_2O 1; EtOH 3; eth 3
6525	4-Heptanone, 3-methyl- 3-Methyl-4-heptanone	$C_8H_{16}O$ 128.21	15726-15-5	154	4-01-00-03344 0.817[20]	1.4103[25]
6526	1-Heptanone, 1-phenyl-	$C_{13}H_{18}O$ 190.29	1671-75-6 16.4	283.3	4-07-00-00780 0.9516[20]	ace 4; eth 4; EtOH 4 1.5060[20]
6527	Heptanoyl chloride	$C_7H_{13}ClO$ 148.63	2528-61-2 -83.8	125.2	4-02-00-00963 0.9590[20]	eth 3; ctc 2; lig 4 1.4345[18]

No.	Name Synonym	Mol. Form. Mol. Wt.	CAS RN mp/°C	Merck No. bp/°C	Beil. Ref. den/g cm^{-3}	Solubility n_D
6528	Heptasiloxane, hexadecamethyl- Hexadecamethylheptasiloxane	$C_{16}H_{48}O_6Si_7$ 533.15	541-01-5 -78	270	4-04-00-04129 0.9012[20]	bz 4; lig 4 1.3965[20]
6529	1,3,5-Heptatriene	C_7H_{10} 94.16	2196-23-8 23	113.5; 31[25]	4-01-00-01097 0.764[20]	1.5079
6530	2-Heptenal Butylacrolein	$C_7H_{12}O$ 112.17	2463-63-0	166	4-01-00-03478 0.864[17]	1.4468[17]
6531	3-Heptenal	$C_7H_{12}O$ 112.17	89896-73-1	151	4-01-00-03480 0.851[15]	eth 3; ace 3; bz 3 1.4348[15]
6532	5-Hepten-2-amine, N,6-dimethyl- Isometheptene	$C_8H_{19}N$ 141.26	503-01-5	5070 177	3-04-00-00467	eth 4; EtOH 4
6533	1-Heptene	C_7H_{14} 98.19	592-76-7 -119.7	93.6	4-01-00-00857 0.6970[20]	H$_2$O 1; EtOH 3; eth 3; ctc 2 1.3998[20]
6534	2-Heptene, (E)-	C_7H_{14} 98.19	14686-13-6 -109.5	98	4-01-00-00860 0.7012[20]	H$_2$O 1; EtOH 3; eth 3; ace 3 1.4045[20]
6535	2-Heptene, (Z)- cis-2-Heptene	C_7H_{14} 98.19	6443-92-1	98.4	3-01-00-00824 0.708[20]	H$_2$O 1; EtOH 3; eth 3; ace 3 1.406[20]
6536	3-Heptene, (E)- trans-3-Heptene	C_7H_{14} 98.19	14686-14-7 -136.6	95.7	4-01-00-00861 0.6981[20]	H$_2$O 1; EtOH 3; eth 3; ace 3 1.4043[20]
6537	3-Heptene, (Z)- cis-3-Heptene	C_7H_{14} 98.19	7642-10-6 -136.6	95.8	4-01-00-00861 0.7030[20]	H$_2$O 1; EtOH 3; eth 3; ace 3 1.4059[20]
6538	1-Heptene, 1-chloro- 1-Chloro-1-heptene	$C_7H_{13}Cl$ 132.63	2384-75-0	155	3-01-00-00823 0.8948[20]	ace 4; bz 4; eth 4; chl 4 1.4380[20]
6539	1-Heptene, 2-chloro-	$C_7H_{13}Cl$ 132.63	65786-11-0	139	3-01-00-00823 0.8895[20]	eth 4; EtOH 4; chl 4 1.4349[20]
6540	3-Heptene, 4-chloro-	$C_7H_{13}Cl$ 132.63	2431-24-5	139	3-01-00-00827 0.883[14]	eth 4; EtOH 4; chl 4 1.437[14]
6541	2-Heptene, 6-chloro-2-methyl-	$C_8H_{15}Cl$ 146.66	80325-37-7	60-1[15]	2-01-00-00200 0.8931[18]	ace 4; bz 4; eth 4; EtOH 4 1.4458[18]
6542	2-Heptene, 2,3-dimethyl- 2,3-Dimethyl-2-heptene	C_9H_{18} 126.24	3074-64-4 -108.5	145.2	4-01-00-00897 0.731[25]	1.4519[20]
6543	1-Heptene, 2-methyl- 2-Methyl-1-heptene	C_8H_{16} 112.22	15870-10-7 -90	119.3	4-01-00-00881 0.7104[25]	1.4123[20]
6544	1-Heptene, 3-methyl- 3-Methyl-1-heptene	C_8H_{16} 112.22	4810-09-7	111	4-01-00-00883 0.707[25]	1.406[20]
6545	1-Heptene, 4-methyl- 4-Methyl-1-heptene	C_8H_{16} 112.22	13151-05-8	112.8	4-01-00-00884 0.713[25]	1.410[20]
6546	1-Heptene, 5-methyl- 5-Methyl-1-heptene	C_8H_{16} 112.22	13151-04-7	113.3	4-01-00-00884 0.7125[25]	1.4094[20]
6547	1-Heptene, 6-methyl- 6-Methyl-1-heptene	C_8H_{16} 112.22	5026-76-6	113.2	4-01-00-00883 0.7079[25]	1.4070[20]
6548	2-Heptene, 2-methyl- 2-Methyl-2-heptene	C_8H_{16} 112.22	627-97-4	122.6	4-01-00-00882 0.7200[25]	H$_2$O 1; eth 3; bz 3; ctc 3 1.4170[20]
6549	2-Heptene, 3-methyl-, (E)- 3-Methyl-trans-2-heptene	C_8H_{16} 112.22	22768-20-3	122	4-01-00-00883 1.419[20]	
6550	2-Heptene, 3-methyl-, (Z)- 2-Heptene, 3-methyl (cis)	C_8H_{16} 112.22	22768-19-0	122	3-01-00-00843 0.725[25]	1.419[20]
6551	2-Heptene, 4-methyl-, (E)- 4-Methyl-trans-2-heptene	C_8H_{16} 112.22	66225-17-0	114	4-01-00-00885 0.712[25]	1.410[20]
6552	2-Heptene, 4-methyl-, (Z)- 2-Heptene, 4-methyl (cis)	C_8H_{16} 112.22	66225-16-9	114	4-01-00-00885 0.712[25]	1.408[20]
6553	2-Heptene, 6-methyl- 2-Methyl-5-heptene	C_8H_{16} 112.22	73548-72-8	117	3-01-00-00843 0.714[25]	1.410[20]
6554	3-Heptene, 2-methyl- 2-Methyl-3-heptene	C_8H_{16} 112.22	17618-76-7 -107.5	112	4-01-00-00882 0.702[25]	1.407[20]
6555	3-Heptene, 3-methyl- 2-Ethyl-2-hexene	C_8H_{16} 112.22	7300-03-0	121	3-01-00-00843 0.724[25]	1.418[20]
6556	3-Heptene, 4-methyl- 4-Methyl-3-heptene	C_8H_{16} 112.22	4485-16-9	121	1-01-00-00094 0.721[25]	1.417[20]
6557	3-Heptene, 5-methyl-	C_8H_{16} 112.22	13172-91-3	112	4-01-00-00884 0.709[25]	1.410[20]
6558	3-Heptene, 6-methyl- 2-Methyl-4-heptene	C_8H_{16} 112.22	3404-57-7	115	4-01-00-00882 0.709[25]	1.410[20]
6559	3-Heptene, 4-propyl- 4-Propyl-3-heptene	$C_{10}H_{20}$ 140.27	4485-13-6	160.5	4-01-00-00906 0.7518[17]	1.4302[18]
6560	1-Heptene, 1,1,2,3,3,4,4,5,5,6,6,7,7,7-tetradecafluoro- Perfluoro-1-heptene	C_7F_{14} 350.05	355-63-5	81.0	4-01-00-00859	
6561	2-Heptenoic acid	$C_7H_{12}O_2$ 128.17	18999-28-5 -19	226.5	4-02-00-01577 0.9575[20]	1.4488[20]
6562	4-Heptenoic acid	$C_7H_{12}O_2$ 128.17	35194-37-7	98-100[5]	4-02-00-01578 0.949[15]	1.4418[15]
6563	5-Heptenoic acid	$C_7H_{12}O_2$ 128.17	3593-00-8	223	4-02-00-01578 0.9496[20]	1.4444[20]
6564	6-Heptenoic acid	$C_7H_{12}O_2$ 128.17	1119-60-4 -6.5	226	4-02-00-01576 0.9515[14]	1.4404[14]
6565	2-Heptenoic acid, 6-methyl-	$C_8H_{14}O_2$ 142.20	90112-75-7 3	239.5	1-02-00-00194 0.938[20]	1.4511[10]
6566	1-Hepten-4-ol	$C_7H_{14}O$ 114.19	3521-91-3	152.1	3-01-00-01936 0.8384[22]	1.4337[20]
6567	2-Hepten-1-ol, (E)-	$C_7H_{14}O$ 114.19	33467-76-4	178; 75[10]	3-01-00-01936 0.8516[20]	EtOH 3; ace 3 1.4460[20]

No.	Name / Synonym	Mol. Form. / Mol. Wt.	CAS RN / mp/°C	Merck No. / bp/°C	Beil. Ref. / den/g cm^{-3}	Solubility / n_D
6568	2-Hepten-1-ol, (Z)- / cis-2-Hepten-1-ol	$C_7H_{14}O$ / 114.19	55454-22-3 /	178; 82[20]	4-01-00-02155 / 0.8421[20]	1.4410[20]
6569	2-Hepten-4-ol, (±)-	$C_7H_{14}O$ / 114.19	115113-98-9 /	153	4-01-00-02155 / 0.8445[20]	EtOH 3; eth 3 / 1.4373[20]
6570	3-Hepten-2-ol	$C_7H_{14}O$ / 114.19	98991-54-9 /	66-7[16]	4-01-00-02156 / 0.834[17]	1.4391[18]
6571	6-Hepten-2-ol	$C_7H_{14}O$ / 114.19	24395-10-6 /	64-5[13]	2-01-00-00488 / 0.8484[18]	1.4387[20]
6572	6-Hepten-3-ol / 1-Heptene-4-ol	$C_7H_{14}O$ / 114.19	19781-77-2 /	60-2[11]	4-01-00-02154 / 0.8447[18]	1.4369[18]
6573	1-Hepten-4-ol, 4-methyl- / 1-Heptene-4-ol, 4-methyl	$C_8H_{16}O$ / 128.21	1186-31-8 /	159.5	3-01-00-01945 / 0.8345[20]	eth 4; EtOH 4 / 1.4479[18]
6574	2-Hepten-4-ol, 5-methyl-	$C_8H_{16}O$ / 128.21	4048-31-1 /	69-70[18]	2-01-00-00490 / 0.8473[20]	1.4411[20]
6575	2-Hepten-4-ol, 6-methyl- / 6-Methyl-2-hepten-4-ol	$C_8H_{16}O$ / 128.21	4798-62-3 /	75[21]; 67[11]	2-01-00-00490 / 0.835[15]	1.4378[20]
6576	3-Hepten-2-ol, 2-methyl-	$C_8H_{16}O$ / 128.21	116668-45-2 /	62-3[14]	2-01-00-00490 / 0.8398[10]	1.4416[10]
6577	4-Hepten-2-ol, 2-methyl-	$C_8H_{16}O$ / 128.21	116668-44-1 / -28	63-4[20]	3-01-00-01944 / 0.8424[14]	1.4407[14]
6578	4-Hepten-3-ol, 3-methyl-	$C_8H_{16}O$ / 128.21	4048-30-0 /	103[75]; 62[15]	2-01-00-00490 / 0.8477[17]	1.4465[20]
6579	4-Hepten-3-ol, 4-methyl-	$C_8H_{16}O$ / 128.21	81280-12-8 /	66[17]	4-01-00-02170 / 0.8525[18]	1.4479[18]
6580	5-Hepten-2-ol, 6-methyl-	$C_8H_{16}O$ / 128.21	1569-60-4 /	175	1-01-00-00230 / 0.8545[20]	1.4505[20]
6581	6-Hepten-1-ol, 3-methyl-	$C_8H_{16}O$ / 128.21	4048-32-2 /	97-9[22]	2-01-00-00491 / 0.8562[19]	1.4470[14]
6582	6-Hepten-2-ol, 2-methyl- / 1,1-Dimethyl-5-hexen-1-ol	$C_8H_{16}O$ / 128.21	77437-98-0 /	65-6[16]	3-01-00-01945 / 0.8393[16]	1.4393[16]
6583	1-Hepten-3-one / Butyl vinyl ketone	$C_7H_{12}O$ / 112.17	2918-13-0 /	42-3[15]	4-01-00-03477 / 0.8434[20]	1.4305[19]
6584	4-Hepten-2-one	$C_7H_{12}O$ / 112.17	24332-22-7 /	61-2[20]	3-01-00-03004 / 0.8618[21]	1.4290[21]
6585	6-Hepten-2-one / Methyl 4-pentenyl ketone	$C_7H_{12}O$ / 112.17	21889-88-3 /	41-3[10]	4-01-00-03478 / 0.8673[18]	1.4350[18]
6586	6-Hepten-3-one / 1-Hepten-5-one	$C_7H_{12}O$ / 112.17	2565-39-1 /	142	4-01-00-03478 / 0.8487[18]	1.4254[18]
6587	3-Hepten-2-one, (E)-	$C_7H_{12}O$ / 112.17	5609-09-6 /	62[15]	4-01-00-03481 / 0.8496[20]	H_2O 1; EtOH 4; eth 4 / 1.4436[20]
6588	3-Hepten-2-one, (Z)- / 3-Hepten-2-one (cis)	$C_7H_{12}O$ / 112.17	69668-88-8 /	34-9[11]	4-01-00-03481 / 0.8440[20]	1.4325[20]
6589	5-Hepten-2-one, (E)-	$C_7H_{12}O$ / 112.17	1071-94-9 /	153	4-01-00-03480 / 0.8445[20]	1.4309[20]
6590	3-Hepten-2-one, 3-ethyl-4-methyl-	$C_{10}H_{18}O$ / 154.25	54244-90-5 /	190; 83[14]	3-01-00-03028 / 0.8559[20]	1.4535[20]
6591	4-Hepten-2-one, 3-ethyl-4-methyl-	$C_{10}H_{18}O$ / 154.25	53252-08-7 /	69[11]	3-01-00-03028 / 0.8450[20]	1.4405[20]
6592	4-Hepten-3-one, 5-ethyl-4-methyl-	$C_{10}H_{18}O$ / 154.25	22319-28-4 /	80[10]	4-01-00-03517 / 0.8622[10]	1.4545[19]
6593	5-Hepten-2-one, 4-ethyl-6-methyl- / 2-Hepten-6-one, 4-ethyl-2-methyl	$C_{10}H_{18}O$ / 154.25	104035-66-7 /	74-5[13]	4-01-00-03517 / 0.8430[20]	1.4546[20]
6594	5-Hepten-3-one, 5-ethyl-4-methyl-	$C_{10}H_{18}O$ / 154.25	74764-56-0 / 74.5	135-9[3]	4-01-00-03517 / 0.8564[21]	1.4452[21]
6595	1-Hepten-3-one, 5-methyl-	$C_8H_{14}O$ / 126.20	44829-76-7 /	40-1[7]	3-01-00-03014 / 0.8479[19]	1.4360[19]
6596	2-Hepten-4-one, 5-methyl- / 5-Methyl-2-hepten-4-one	$C_8H_{14}O$ / 126.20	81925-81-7 /	171	4-01-00-03496 / 0.848[25]	1.4400[25]
6597	2-Hepten-4-one, 6-methyl-	$C_8H_{14}O$ / 126.20	49852-35-9 /	170	4-01-00-03494 / 0.841[25]	1.4388[25]
6598	3-Hepten-2-one, 6-methyl-	$C_8H_{14}O$ / 126.20	2009-74-7 /	179	3-01-00-03011 / 0.8443[17]	1.4427[10]
6599	4-Hepten-2-one, 6-methyl- / 2-Methyl-3-hepten-6-one	$C_8H_{14}O$ / 126.20	39273-81-9 /	163	1-01-00-00385 / 0.8345[20]	1.4315
6600	4-Hepten-3-one, 5-methyl- / Homomesityl oxide	$C_8H_{14}O$ / 126.20	1447-26-3 /	166.5	3-01-00-03012 / 0.8591[20]	1.4476[20]
6601	5-Hepten-2-one, 3-methyl- / 3-Methyl-5-hepten-2-one	$C_8H_{14}O$ / 126.20	38552-72-6 /	62-4[20]	4-01-00-03496 / 0.8463[18]	1.4345[18]
6602	5-Hepten-2-one, 5-methyl-	$C_8H_{14}O$ / 126.20	10339-67-0 /	84[45]	4-01-00-03494 / 0.8647[16]	1.4434[16]
6603	5-Hepten-2-one, 6-methyl-	$C_8H_{14}O$ / 126.20	110-93-0 /	173.5	4-01-00-03493 / 0.8546[16]	eth 4; EtOH 4 / 1.4445[20]
6604	5-Hepten-3-one, 2-methyl-	$C_8H_{14}O$ / 126.20	77958-21-5 /	161.5	1-01-00-00385 / 0.842[20]	1.4310[20]
6605	5-Hepten-3-one, 5-methyl- / 5-Methylhept-5-en-3-one	$C_8H_{14}O$ / 126.20	1190-34-7 /	63[19]	3-01-00-03011 / 0.8524[21]	1.4367[21]
6606	1-Hepten-3-yne	C_7H_{10} / 94.16	2384-73-8 /	110	4-01-00-01097 / 0.7603[20]	ace 4; bz 4; EtOH 4; peth 4 / 1.4520[25]
6607	1-Hepten-4-yne, 6,6-dimethyl-	C_9H_{14} / 122.21	31508-08-4 /	125	/ 0.758[20]	ace 4; bz 4; EtOH 4; peth 4 / 1.4312[20]

No.	Name Synonym	Mol. Form. Mol. Wt.	CAS RN mp/°C	Merck No. bp/°C	Beil. Ref. den/g cm^{-3}	Solubility n_D
6608	6-Hepten-4-yn-3-ol, 3-ethyl- / Diethylvinylethynylcarbinol	C$_9$H$_{14}$O / 138.21	3142-84-5	62[4]	3-01-00-02034 / 0.8875[20]	eth 4; EtOH 4 / 1.4800[20]
6609	1-Heptyne	C$_7$H$_{12}$ / 96.17	628-71-7 / -81	99.7	4-01-00-01025 / 0.7328[20]	H$_2$O 2; EtOH 5; eth 5; bz 3 / 1.4087[20]
6610	2-Heptyne / 1-Methyl-2-butylacetylene	C$_7$H$_{12}$ / 96.17	1119-65-9	112	4-01-00-01026 / 0.744[25]	H$_2$O 1; EtOH 5; eth 5; bz 3 / 1.4230[20]
6611	3-Heptyne / 1-Ethyl-2-propylacetylene	C$_7$H$_{12}$ / 96.17	2586-89-2 / -130.5	107.2	4-01-00-01027 / 0.7336[25]	H$_2$O 1; EtOH 5; eth 5; bz 3 / 1.4189[20]
6612	1-Heptyne, 1-bromo- / 1-Bromo-1-heptyne	C$_7$H$_{11}$Br / 175.07	19821-84-2	164; 69[25]	3-01-00-00998 / 1.2120[22]	ace 4; eth 4; EtOH 4; chl 4 / 1.4678[22]
6613	2-Heptyne, 1-bromo- / 2-Heptynyl bromide	C$_7$H$_{11}$Br / 175.07	18495-26-6	104[55]; 84[20]	4-01-00-01026 /	ace 4; eth 4; EtOH 4 / 1.4878[25]
6614	1-Heptyne, 1-chloro- / 1-Chloro-1-heptyne	C$_7$H$_{11}$Cl / 130.62	51556-10-6	141	3-01-00-00997 / 0.9250[24]	eth 4; EtOH 4 / 1.4411[24]
6615	2-Heptyne, 7-chloro- / 1-Chloro-5-heptyne	C$_7$H$_{11}$Cl / 130.62	70396-13-3	166	3-01-00-00998	eth 4; EtOH 4 / 1.4507[25]
6616	3-Heptyne, 7-chloro- / 4-Heptynyl chloride	C$_7$H$_{11}$Cl / 130.62	51575-85-0	164	3-01-00-00999	eth 4; EtOH 4 / 1.4517[20]
6617	3-Heptyne, 2,6-dimethyl-	C$_9$H$_{16}$ / 124.23	19549-97-4	130.6	3-01-00-01015 / 0.785[20]	ace 4; eth 4
6618	3-Heptyne, 5,5-dimethyl- / 5,5-Dimethyl-3-heptyne	C$_9$H$_{16}$ / 124.23	23097-98-5	69[100]	3-01-00-01015 / 0.7610[20]	ace 4; eth 4 / 1.4360[20]
6619	3-Heptyne, 5-ethyl-5-methyl- / 5-Ethyl-5-methyl-3-heptyne	C$_{10}$H$_{18}$ / 138.25	61228-10-2	88[100]	3-01-00-01021 / 0.7714[20]	ace 4; eth 4 / 1.4386[20]
6620	1-Heptyne, 1-iodo- / 1-Iodo-1-heptyne	C$_7$H$_{11}$I / 222.07	54573-13-6	90-2[17]	3-01-00-00998 / 1.470[19]	1.5123[19]
6621	Hexacene	C$_{26}$H$_{16}$ / 328.41	258-31-1 / 380	sub	4-05-00-02834	H$_2$O 1; EtOH 1
6622	Hexacosane	C$_{26}$H$_{54}$ / 366.71	630-01-3 / 56.4	412.2	4-01-00-00583 / 0.7783[60]	bz 4; lig 4; chl 4 / 1.4357[60]
6623	Hexacosane, 13-dodecyl-	C$_{38}$H$_{78}$ / 535.04	55517-73-2 / 13.7	272[1]	4-01-00-00600 / 0.8188[20]	1.4577[20]
6624	Hexacosanoic acid / Cerotic acid	C$_{26}$H$_{52}$O$_2$ / 396.70	506-46-7 / 88.5		4-02-00-01310 / 0.8198[100]	H$_2$O 1; EtOH 4; eth 4 / 1.4301[100]
6625	Hexacosanoic acid, methyl ester / Methyl hexacosanoate	C$_{27}$H$_{54}$O$_2$ / 410.72	5802-82-4 / 63.8	286[15]; 237[2]	4-02-00-01311 /	chl 3 / 1.4301[80]
6626	1-Hexacosanol	C$_{26}$H$_{54}$O / 382.71	506-52-5 / 80	305 [20] dec	4-01-00-01912	H$_2$O 1; EtOH 3; eth 3
6627	6,10-Hexadecadiyne	C$_{16}$H$_{26}$ / 218.38	10160-99-3	157[10]	3-01-00-01067 / 0.7907[20]	1.4523[20]
6628	Hexadecanal	C$_{16}$H$_{32}$O / 240.43	629-80-1 / 35	200[29]	4-01-00-03393	H$_2$O 1; EtOH 3; eth 3; ace 3
6629	Hexadecanamide	C$_{16}$H$_{33}$NO / 255.44	629-54-9 / 107	236[12]	4-02-00-01182 / 1.0000[20]	H$_2$O 1; EtOH 2; eth 2; ace 2
6630	Hexadecanamide, N-phenyl-	C$_{22}$H$_{37}$NO / 331.54	6832-98-0 / 91.8	283[17]; 132[10]	4-12-00-00400	ace 4; bz 4; EtOH 4; chl 4
6631	1-Hexadecanamine	C$_{16}$H$_{35}$N / 241.46	143-27-1 / 46.8	322.5	4-04-00-00818 / 0.8129[20]	H$_2$O 1; EtOH 4; eth 4; ace 3 / 1.4496[20]
6632	Hexadecane / Cetane	C$_{16}$H$_{34}$ / 226.45	544-76-3 / 18.1	286.8	4-01-00-00537 / 0.7733[20]	H$_2$O 1; EtOH 2; eth 5; ctc 3 / 1.4345
6633	Hexadecane, 1-bromo-	C$_{16}$H$_{33}$Br / 305.34	112-82-3 / 18	336	4-01-00-00542 / 0.9991[20]	H$_2$O 1; eth 3 / 1.4618[25]
6634	Hexadecane, 1-chloro-	C$_{16}$H$_{33}$Cl / 260.89	4860-03-1 / 17.9	322	4-01-00-00542 / 0.8652[20]	H$_2$O 1 / 1.4505[20]
6635	Hexadecane, 1,16-dibromo- / Hexadecamethylene dibromide	C$_{16}$H$_{32}$Br$_2$ / 384.24	45223-18-5 / 56	204[4]	2-01-00-00138	chl 4
6636	Hexadecane, 1,1-dimethoxy- / Palmitaldehyde, dimethyl acetal	C$_{18}$H$_{38}$O$_2$ / 286.50	2791-29-9 / 10	144[2]	4-01-00-03393 / 0.8542[20]	ace 4; eth 4; EtOH 4 / 1.4382[25]
6637	Hexadecane, 6,11-dipentyl-	C$_{26}$H$_{54}$ / 366.71	15874-03-0 / -16.2	231[10]	4-01-00-00586 / 0.8072[20]	1.4502[20]
6638	Hexadecane, 1-(ethenyloxy)-	C$_{18}$H$_{36}$O / 268.48	822-28-6 / 16	160[2]	4-01-00-02057 / 0.821[27]	1.4444[25]
6639	Hexadecane, 1-fluoro- / 1-Fluorohexadecane	C$_{16}$H$_{33}$F / 244.44	408-38-8 / 18	289	4-01-00-00542 / 0.8321[20]	eth 4; lig 4 / 1.4317[20]
6640	Hexadecane, 1-iodo-	C$_{16}$H$_{33}$I / 352.34	544-77-4 / 22	212[15]	4-01-00-00543 / 1.123[25]	H$_2$O 1; EtOH 2; eth 3; ace 3 / 1.4806[20]
6641	Hexadecanenitrile	C$_{16}$H$_{31}$N / 237.43	629-79-8 / 31	333	4-02-00-01183 / 0.8303[20]	H$_2$O 1; EtOH 4; eth 4; ace 4 / 1.4450[20]
6642	Hexadecane, 1,1'-oxybis-	C$_{32}$H$_{66}$O / 466.88	4113-12-6 / 55	270 dec	4-01-00-01878 / 0.978[19]	H$_2$O 1; EtOH 3; eth 3; chl 3
6643	1-Hexadecanethiol / Cetyl mercaptan	C$_{16}$H$_{34}$S / 258.51	2917-26-2 / 19	123-8[0.5]	4-01-00-01881	H$_2$O 1; EtOH 2; eth 3; ctc 2
6644	Hexadecanoic acid / Palmitic acid	C$_{16}$H$_{32}$O$_2$ / 256.43	57-10-3 / 63.1	6947 / 351.5	4-02-00-01157 / 0.8527[62]	H$_2$O 1; EtOH 5; eth 5; ace 3 / 1.43345[60]
6645	Hexadecanoic acid, aluminum salt	C$_{48}$H$_{93}$AlO$_6$ / 793.24	555-35-1	361	4-02-00-01157	H$_2$O 1; EtOH 3; peth 3; bz 3
6646	Hexadecanoic acid, anhydride	C$_{32}$H$_{62}$O$_3$ / 494.84	623-65-4 / 64		4-02-00-01181 / 0.8388[83]	eth 4 / 1.4364[68]
6647	Hexadecanoic acid, butyl ester	C$_{20}$H$_{40}$O$_2$ / 312.54	111-06-8 / 16.9		4-02-00-01167	H$_2$O 1; EtOH 3; eth 3 / 1.4312[50]

No.	Name / Synonym	Mol. Form. / Mol. Wt.	CAS RN / mp/°C	Merck No. / bp/°C	Beil. Ref. / den/g cm⁻³	Solubility / n_D
6648	Hexadecanoic acid, 2-[(dichloroacetyl)amino]-3-hydroxy-3-(4-nitrophenyl)propyl ester, [R-(R*,R*)]- Chloramphenicol palmitate	$C_{27}H_{42}Cl_2N_2O_6$ 561.55	530-43-8 90	2069	4-13-00-02753	bz 4; eth 4; EtOH 4
6649	Hexadecanoic acid, 1,2-ethanediyl ester	$C_{34}H_{66}O_4$ 538.90	624-03-3 72		4-02-00-01169 0.8594[78]	H_2O 1; EtOH 1; eth 3; ace 4
6650	Hexadecanoic acid, ethyl ester	$C_{18}H_{36}O_2$ 284.48	628-97-7 24	191[10]	4-02-00-01165 0.8577[25]	H_2O 1; EtOH 3; eth 3; ace 3 1.4347[34]
6651	Hexadecanoic acid, hexadecyl ester Cetyl palmitate	$C_{32}H_{64}O_2$ 480.86	540-10-3 54	2023	4-02-00-01168 0.989[20]	eth 4; EtOH 4 1.4398[70]
6652	Hexadecanoic acid, 2-hydroxyethyl ester	$C_{18}H_{36}O_3$ 300.48	4219-49-2 51	173-4[3]	4-02-00-01168 0.8768[60]	EtOH 4
6653	Hexadecanoic acid, methyl ester	$C_{17}H_{34}O_2$ 270.46	112-39-0 30	417; 148[2]	4-02-00-01165 0.8247[75]	H_2O 1; EtOH 4; eth 3; ace 4
6654	Hexadecanoic acid, 1-methylethyl ester	$C_{19}H_{38}O_2$ 298.51	142-91-6 13.5	160[2]	4-02-00-01167 0.8404[38]	ace 4; bz 4; eth 4; EtOH 4 1.4364[25]
6655	Hexadecanoic acid, phenylmethyl ester Benzyl palmitate	$C_{23}H_{38}O_2$ 346.55	41755-60-6 36		3-06-00-01481 0.9109[35]	bz 4; eth 4; EtOH 4; chl 4 1.4689[50]
6656	Hexadecanoic acid, 1,2,3-propanetriyl ester Tripalmitin	$C_{51}H_{98}O_6$ 807.34	555-44-2 66.5	9648 315	4-02-00-01176 0.8752[70]	H_2O 1; EtOH 2; eth 4; bz 3 1.4381[80]
6657	Hexadecanoic acid, propyl ester	$C_{19}H_{38}O_2$ 298.51	2239-78-3 20.4	190[12]	4-02-00-01167 0.8455[33]	1.4392[25]
6658	Hexadecanoic acid, 3,7,11,15-tetramethyl- Phytanic acid	$C_{20}H_{40}O_2$ 312.54	14721-66-5 -65			
6659	1-Hexadecanol Cetyl alcohol	$C_{16}H_{34}O$ 242.45	36653-82-4 49.3	2020 334	4-01-00-01876 0.8187[50]	H_2O 1; EtOH 2; eth 4; ace 3 1.4283[79]
6660	2-Hexadecanol	$C_{16}H_{34}O$ 242.45	14852-31-4 44	314	4-01-00-01882 0.8338[20]	1.4479[20]
6661	3-Hexadecanol	$C_{16}H_{34}O$ 242.45	593-03-3 50	152[4]	4-01-00-01882 0.8000[67]	
6662	1-Hexadecanol, acetate	$C_{18}H_{36}O_2$ 284.48	629-70-9 -18.5	220-5[205]	4-02-00-00171 0.8574[25]	H_2O 1; EtOH 2; ctc 3 1.4438[20]
6663	3-Hexadecanone	$C_{16}H_{32}O$ 240.43	18787-64-9 43	184[17]; 140[2]	4-01-00-03394	chl 3
6664	1-Hexadecanone, 1-phenyl-	$C_{22}H_{36}O$ 316.53	6697-12-7 59	251[15]	4-07-00-00860 0.8692[76]	1.4675[76]
6665	Hexadecanoyl chloride	$C_{16}H_{31}ClO$ 274.87	112-67-4 12	199[20]	4-02-00-01182 0.9016[25]	eth 4 1.4514[20]
6666	6,10,14-Hexadecatrienoic acid Hiragonic acid	$C_{16}H_{26}O_2$ 250.38	4444-12-6	180-90[15]	3-02-00-01505 0.9296[20]	eth 4; EtOH 4 1.4850[50]
6667	1-Hexadecene 1-Cetene	$C_{16}H_{32}$ 224.43	629-73-2 4.1	284.9	4-01-00-00927 0.7811[20]	H_2O 1; EtOH 3; eth 3; ctc 3 1.4412[20]
6668	7-Hexadecenoic acid Dinoroleic acid	$C_{16}H_{30}O_2$ 254.41	1191-75-9 33	230[10]	0-02-00-00460	EtOH 4; eth 4
6669	9-Hexadecenoic acid	$C_{16}H_{30}O_2$ 254.41	2091-29-4 0.5	181[1]	4-02-00-01629	bz 4; eth 4; peth 4
6670	9-Hexadecenoic acid, (Z)- Palmitoleic acid	$C_{16}H_{30}O_2$ 254.41	373-49-9 -0.1	131[0.06]	4-02-00-01629	
6671	1-Hexadecen-3-ol, 3,7,11,15-tetramethyl- Isophytol	$C_{20}H_{40}O$ 296.54	505-32-8	5084 107-10[0.01]	2-01-00-00504 0.8519[20]	bz 4; eth 4; EtOH 4 1.4571[20]
6672	2-Hexadecen-1-ol, 3,7,11,15-tetramethyl-, [R-[R*,R*-(E)]]- Phytol	$C_{20}H_{40}O$ 296.54	150-86-7	7362 203[10]	4-01-00-02208 0.8497[25]	1.4595[25]
6673	1-Hexadecyne	$C_{16}H_{30}$ 222.41	629-74-3 15	284	4-01-00-01073 0.7965[20]	bz 4 1.4440[20]
6674	2-Hexadecyne	$C_{16}H_{30}$ 222.41	629-75-4 20	160[15]	3-01-00-01028 0.8039[20]	
6675	2,4-Hexadienal, (E,E)-	C_6H_8O 96.13	142-83-6	174; 76[30]	4-01-00-03545 0.898[32]	1.5384[20]
6676	1,2-Hexadiene Propylallene	C_6H_{10} 82.15	592-44-9	76	4-01-00-01011 0.7149[20]	eth 4; chl 4 1.4282[20]
6677	1,4-Hexadiene	C_6H_{10} 82.15	592-45-0	65	4-01-00-01013 0.7000[20]	H_2O 1; eth 4 1.4150[20]
6678	1,5-Hexadiene Biallyl	C_6H_{10} 82.15	592-42-7 -140.7	59.4	4-01-00-01013 0.6878[25]	H_2O 1; EtOH 3; eth 3; bz 3 1.4042[20]
6679	2,3-Hexadiene Methylethylallene	C_6H_{10} 82.15	592-49-4	78	4-01-00-01016 0.684[15]	1.395[20]
6680	2,4-Hexadiene	C_6H_{10} 82.15	592-46-1 -79	80	3-01-00-00985 0.7196[20]	H_2O 1; EtOH 3; eth 3; chl 3 1.4500[20]
6681	1,3-Hexadiene, (E)-	C_6H_{10} 82.15	20237-34-7 -102.4	73.2	4-01-00-01011 0.6995[25]	1.4406[20]
6682	1,3-Hexadiene, (Z)-	C_6H_{10} 82.15	14596-92-0	73.1	4-01-00-01011 0.7033[25]	1.4379[20]
6683	1,4-Hexadiene, (E)-	C_6H_{10} 82.15	7319-00-8 -138.7	65.0	0.695[25]	1.4104[20]
6684	1,4-Hexadiene, (Z)-	C_6H_{10} 82.15	7318-67-4	66.3	0.695[25]	eth 4 1.4049[20]
6685	2,4-Hexadiene, (E,E)-	C_6H_{10} 82.15	5194-51-4 -44.9	82.2	4-01-00-01016 0.7101[25]	1.4510[20]
6686	2,4-Hexadiene, (E,Z)-	C_6H_{10} 82.15	5194-50-3 -96.1	83.5	4-01-00-01016 0.7185[25]	1.4560[20]

No.	Name Synonym	Mol. Form. Mol. Wt.	CAS RN mp/°C	Merck No. bp/°C	Beil. Ref. den/g cm^{-3}	Solubility n_D
6687	2,4-Hexadiene, (Z,Z)-	C_6H_{10} 82.15	6108-61-8	85	0.7298[25]	1.4606[20]
6688	1,3-Hexadiene, 3-chloro-	C_6H_9Cl 116.59	101870-06-8	68[117]	4-01-00-01012 0.9390[20]	eth 4; chl 4 1.4770[20]
6689	1,5-Hexadiene, 1,1,2,3,3,4,4,5,6,6-decachloro-	C_6Cl_{10} 426.59	29030-84-0 49	121[0.03]	4-01-00-01016 1.905[52]	ace 4; eth 4; chl 4 1.6012[51]
6690	2,4-Hexadiene, 1,3-dichloro-	$C_6H_8Cl_2$ 151.03	73454-83-8	80-2[17]	4-01-00-01017 1.1456[20]	bz 4; chl 4 1.5271[20]
6691	1,5-Hexadiene, 2,5-dimethyl-	C_8H_{14} 110.20	627-58-7 -75.6	114.3	4-01-00-01042 0.743[20]	H_2O 1; ace 3; chl 3 1.43995[21]
6692	2,4-Hexadiene, 2,5-dimethyl-	C_8H_{14} 110.20	764-13-6 14	134.5	4-01-00-01043 0.7577[25]	H_2O 1; EtOH 3; eth 3; bz 3 1.4785[20]
6693	1,5-Hexadiene-3,4-diol, (R*,R*)-(±)-	$C_6H_{10}O_2$ 114.14	19700-97-1 21.7	90-1[8]	1.017[19]	H_2O 3; EtOH 3; eth 3; chl 3 1.4790[19]
6694	1,5-Hexadiene, 3-ethyl-2-methyl- 1,5-Hexadiene, 2-methyl-3-ethyl	C_9H_{16} 124.23	73398-15-9 -70	145	0-01-00-00260 0.7594[25]	
6695	1,5-Hexadiene, 2-methyl-	C_7H_{12} 96.17	4049-81-4 -128.8	88.1	4-01-00-01030 0.7153[25]	1.4183[20]
6696	2,4-Hexadiene, 2-methyl-	C_7H_{12} 96.17	28823-41-8 -74.2	111.5	4-01-00-01030 0.7411[25]	
6697	2,4-Hexadiene, 1,3,4,6-tetrachloro-	$C_6H_6Cl_4$ 219.92	100367-45-1	84-9[2]	3-01-00-00987 1.4013[20]	MeOH 4; chl 4 1.5465[20]
6698	2,4-Hexadienoic acid Sorbic acid	$C_6H_8O_2$ 112.13	110-44-1 134.5	8677 228 dec; 153[50]	2-02-00-00452 1.204[19]	H_2O 3; EtOH 3; eth 4; chl 3
6699	2,4-Hexadienoic acid, ethyl ester	$C_8H_{12}O_2$ 140.18	110318-09-7	195.5 85[20]	2-02-00-00452 0.9506[20]	H_2O 1; EtOH 3; eth 3; chl 3 1.4951[20]
6700	2,4-Hexadienoic acid, ethyl ester, (E,E)-	$C_8H_{12}O_2$ 140.18	2396-84-1	195.5	4-02-00-01703 0.9506[20]	eth 4; EtOH 4; chl 4 1.4951[20]
6701	2,5-Hexadienoic acid, 3-methoxy-5-methyl-4-oxo- Penicillic acid	$C_8H_{10}O_4$ 170.17	90-65-3 83	7033	3-03-00-01467	H_2O 3; EtOH 4; eth 4; ace 3
6702	2,4-Hexadienoic acid, methyl ester Methyl 2,4-hexadienoate	$C_7H_{10}O_2$ 126.16	1515-80-6 15	180; 70[20]	2-02-00-00452 0.9777[20]	H_2O 1; EtOH 3; eth 3 1.5025[22]
6703	2,4-Hexadienoic acid, methyl ester, (E,E)-	$C_7H_{10}O_2$ 126.16	689-89-4 15	180	4-02-00-01703 0.9777[20]	eth 4; EtOH 4 1.5025[22]
6704	2,4-Hexadienoic acid, potassium salt, (E,E)-	$C_6H_7KO_2$ 150.22	24634-61-5 >270 dec	7661	1.361[25]	H_2O 4; EtOH 3
6705	1,4-Hexadien-3-ol	$C_6H_{10}O$ 98.14	1070-14-0	50-2[15]	4-01-00-02238 0.859[23]	1.4502[23]
6706	1,5-Hexadien-3-ol 3-Hydroxy-1,5-hexadiene	$C_6H_{10}O$ 98.14	924-41-4	133.5	4-01-00-02239 0.8596[25]	1.4464[25]
6707	2,4-Hexadien-1-ol 1-Hydroxy-2,4-hexadiene	$C_6H_{10}O$ 98.14	111-28-4 30.5	8678 76[12]	4-01-00-02239 0.8967[23]	H_2O 1; EtOH 3; eth 3 1.4981[20]
6708	3,5-Hexadien-2-ol 1,3-Hexadien-5-ol	$C_6H_{10}O$ 98.14	3280-51-1	77-8[26]	4-01-00-02237 0.8678[20]	EtOH 4 1.4896[20]
6709	2,4-Hexadien-1-ol, (E,E)-	$C_6H_{10}O$ 98.14	17102-64-6	76[12]	4-01-00-02239 0.8967[20]	eth 4; EtOH 4 1.4981[20]
6710	3,5-Hexadien-2-one, 6-phenyl-	$C_{12}H_{12}O$ 172.23	4173-44-8 68	171[15.] 155[12]	4-07-00-01183	bz 4; eth 4; EtOH 4; chl 4
6711	2,4-Hexadienoyl chloride, (E,E)-	C_6H_7ClO 130.57	2614-88-2	78[15]	3-02-00-01460 1.0666[19]	ace 4 1.5545[20]
6712	1,3-Hexadien-5-yne	C_6H_8 78.11	10420-90-3 -81	83.5	4-01-00-01120 0.7806[20]	bz 4 1.5095[20]
6713	1,5-Hexadien-3-yne Divinylacetylene	C_6H_8 78.11	821-08-9 -88	85	4-01-00-01120 0.7851[20]	H_2O 1; bz 3 1.5035[20]
6714	1,5-Hexadien-3-yne, 2,5-dimethyl- Diisopropenylacetylene	C_8H_{10} 106.17	3725-05-1	123	4-01-00-01124 0.7863[25]	bz 4; chl 4 1.4845[20]
6715	1,4-Hexadiyne	C_6H_8 78.11	10420-91-4 <-80	81.7	3-01-00-01057 0.825[0]	bz 4; chl 4
6716	1,5-Hexadiyne Bipropargyl	C_6H_8 78.11	628-16-0 -6	86	4-01-00-01118 0.8049[20]	H_2O 1; EtOH 3; eth 3; ace 3 1.4380[23]
6717	2,4-Hexadiyne Dimethyldiacetylene	C_6H_8 78.11	2809-69-0 67.8	129.5	4-01-00-01119	EtOH 4; eth 4
6718	Hexanal Caproaldehyde	$C_6H_{12}O$ 100.16	66-25-1 -56	1761 131	4-01-00-03296 0.8335[20]	H_2O 2; EtOH 4; eth 4; ace 3 1.4039[20]
6719	Hexanal, 2-ethyl- 2-Ethylhexanal	$C_8H_{16}O$ 128.21	123-05-7 <-100	163	4-01-00-03345 0.8540[20]	H_2O 3; EtOH 3; eth 3; ctc 2 1.4142[20]
6720	Hexanal, 3-methyl- Caproaldehyde, 3-methyl	$C_7H_{14}O$ 114.19	19269-28-4	143	2-01-00-00756 0.8203[20]	H_2O 1; EtOH 3; eth 3 1.4122[20]
6721	Hexanamide	$C_6H_{13}NO$ 115.18	628-02-4 101	255	4-02-00-00929 0.999[20]	bz 4; eth 4; EtOH 4; chl 4 1.4200[110]
6722	Hexanamide, N-phenyl-	$C_{12}H_{17}NO$ 191.27	621-15-8 95		4-12-00-00392 1.112[25]	eth 4; EtOH 4
6723	1-Hexanamine Hexylamine	$C_6H_{15}N$ 101.19	111-26-2 -22.9	132.8	4-04-00-00709 0.7660[20]	H_2O 2; EtOH 5; eth 5; chl 3 1.4180[20]
6724	2-Hexanamine, (±)-	$C_6H_{15}N$ 101.19	68107-05-1 -19	117.5	4-04-00-00721 0.7533[20]	EtOH 3; eth 3 1.4080[25]

No.	Name Synonym	Mol. Form. Mol. Wt.	CAS RN mp/°C	Merck No. bp/°C	Bell. Ref. den/g cm^{-3}	Solubility n_D
6725	1-Hexanamine, N,N-dihexyl-	C$_{18}$H$_{39}$N 269.51	102-86-3	261.7	4-04-00-00711 0.7976[21]	H$_2$O 1; EtOH 4; eth 4; acid 3
6726	1-Hexanamine, 2-ethyl- 2-Ethyl-1-hexylamine	C$_8$H$_{19}$N 129.25	104-75-6		4-04-00-00766	
6727	1-Hexanamine, N-hexyl-	C$_{12}$H$_{27}$N 185.35	143-16-8 -13.1	236; 75[1]	4-04-00-00711 0.7889[20]	EtOH 3; eth 3 1.4339[20]
6728	1-Hexanamine, 3-methyl- Hexylamine, 3-methyl-	C$_7$H$_{17}$N 115.22	65530-93-0	149; 67[45]	2-04-00-00653 0.772[26]	1.4249[25]
6729	1-Hexanamine, 4-methyl- Hexylamine, 4-methyl-	C$_7$H$_{17}$N 115.22	34263-68-8	153	2-04-00-00654 0.7802[20]	1.4238[20]
6730	2-Hexanamine, 4-methyl- Methylhexaneamine	C$_7$H$_{17}$N 115.22	105-41-9	6000 132.5	4-04-00-00747 0.7655[20]	H$_2$O 2; EtOH 4; eth 4; chl 4 1.4150[25]
6731	Hexane	C$_6$H$_{14}$ 86.18	110-54-3 -95.3	4613 68.7	4-01-00-00338 0.6548[25]	H$_2$O 1; EtOH 4; eth 3; chl 3 1.3749[20]
6732	Hexane, 1-bromo- Hexyl bromide	C$_6$H$_{13}$Br 165.07	111-25-1 -84.7	155.3	4-01-00-00352 1.1744[20]	H$_2$O 1; EtOH 5; eth 5; ace 3 1.4478[20]
6733	Hexane, 2-bromo- 2-Bromohexane	C$_6$H$_{13}$Br 165.07	3377-86-4	144; 78[90]	4-01-00-00353 1.1658[20]	H$_2$O 1; EtOH 3; eth 3; ace 3 1.4832[25]
6734	Hexane, 3-bromo- 3-Bromohexane	C$_6$H$_{13}$Br 165.07	3377-87-5	142	4-01-00-00353 1.1799[20]	ace 4; eth 4; EtOH 4; chl 4 1.4472[20]
6735	Hexane, 1-bromo-6-fluoro- 1-Bromo-6-fluorohexane	C$_6$H$_{12}$BrF 183.06	373-28-4	67-8[11]	4-01-00-00353 1.293[20]	ace 4; eth 4; EtOH 4; chl 4 1.4436[25]
6736	Hexane, 1-bromo-6-methoxy-	C$_7$H$_{15}$BrO 195.10	50592-87-5	112[35]; 98[19]	4-01-00-01704 1.1887[25]	1.4469[25]
6737	Hexane, 1-chloro- Hexyl chloride	C$_6$H$_{13}$Cl 120.62	544-10-5 -94	2144 135	4-01-00-00349 0.8785[20]	H$_2$O 1; EtOH 3; eth 3; ace 3 1.4199[20]
6738	Hexane, 2-chloro- 2-Hexyl chloride	C$_6$H$_{13}$Cl 120.62	638-28-8	122.5	4-01-00-00349 0.8694[21]	ace 4; bz 4; eth 4; EtOH 4 1.4142[22]
6739	Hexane, 3-chloro- 3-Chlorohexane	C$_6$H$_{13}$Cl 120.62	2346-81-8	123	4-01-00-00349 0.8684[20]	ace 4; bz 4; eth 4; EtOH 4 1.4163[20]
6740	Hexane, 2-chloro-2,5-dimethyl- 2-Chloro-2,5-dimethylhexane	C$_8$H$_{17}$Cl 148.68	29342-44-7	86[100]	4-01-00-00434 0.8476[18]	ace 4; bz 4; eth 4; EtOH 4 1.4232[20]
6741	Hexane, 3-chloro-2,3-dimethyl-	C$_8$H$_{17}$Cl 148.68	101654-30-2	41-3[12]	3-01-00-00481 0.8869[20]	EtOH 4 1.4333[25]
6742	Hexane, 1-chloro-6-fluoro- 1-Chloro-6-fluorohexane	C$_6$H$_{12}$ClF 138.61	1550-09-0	168; 62[15]	4-01-00-00350 1.015[20]	eth 4; EtOH 4; chl 4 1.4168[25]
6743	Hexane, 1-chloro-3-methyl-	C$_7$H$_{15}$Cl 134.65	101257-63-0	151	2-01-00-00119 0.8766[20]	eth 4; EtOH 4; chl 4 1.4274[207]
6744	Hexane, 2-chloro-2-methyl-	C$_7$H$_{15}$Cl 134.65	4398-65-6	135 dec; 60[52]	4-01-00-00398 0.8635[20]	eth 4; EtOH 4; chl 4 1.4200[20]
6745	Hexane, 2-chloro-5-methyl- 2-Chloro-5-methylhexane	C$_7$H$_{15}$Cl 134.65	58766-17-9	138 dec	3-01-00-00436 0.863[20]	eth 4; EtOH 4; chl 4
6746	Hexane, 3-chloro-3-methyl-	C$_7$H$_{15}$Cl 134.65	43197-78-0	135	3-01-00-00438 0.8787[20]	eth 4; EtOH 4; chl 4 1.4250[20]
6747	Hexane, 3-chloro-2,2,3-trimethyl-	C$_9$H$_{19}$Cl 162.70	102449-95-6	64-5[13]	3-01-00-00515 0.9010[20]	eth 4; EtOH 4; chl 4 1.4465[20]
6748	Hexanedial	C$_6$H$_{10}$O$_2$ 114.14	1072-21-5 -8	92-4[9]	4-01-00-03686 1.003[19]	bz 4; eth 4; EtOH 4 1.4350[20]
6749	1,6-Hexanediamine Hexamethylenediamine	C$_6$H$_{16}$N$_2$ 116.21	124-09-4 41.5	4614 205	4-04-00-01320	H$_2$O 4; EtOH 3; bz 3
6750	1,6-Hexanediamine, N,N'-dibutyl- N,N'-Dibutylhexamethylenediamine	C$_{14}$H$_{32}$N$_2$ 228.42	4835-11-4	137-40[3-4]	4-04-00-01327	1.4470[25]
6751	2,5-Hexanediamine, 2,5-dimethyl-	C$_8$H$_{20}$N$_2$ 144.26	23578-35-0	63-4[8]	4-04-00-01363 0.832[25]	1.4459[20]
6752	1,6-Hexanediamine, N,N,N',N'-tetramethyl-	C$_{10}$H$_{24}$N$_2$ 172.31	111-18-2	209.5	4-04-00-01322 0.806[25]	1.4359[20]
6753	1,6-Hexanediaminium, N,N'-di-9H-fluoren-9-yl-N,N,N',N'-tetramethyl-dibromide Hexafluorenium bromide	C$_{36}$H$_{42}$Br$_2$N$_2$ 662.55	317-52-2 153.5	4606	4-12-00-03397	
6754	Hexane, 1,2-dibromo- 1,2-Dibromohexane	C$_6$H$_{12}$Br$_2$ 243.97	624-20-4	104[36]; 82[12]	3-01-00-00392 1.5774[20]	bz 4; eth 4; chl 4 1.5024[20]
6755	Hexane, 1,4-dibromo- 1,4-Dibromohexane	C$_6$H$_{12}$Br$_2$ 243.97	25118-28-9	106-8[15]	3-01-00-00392 1.602[15]	1.5084[15]
6756	Hexane, 1,5-dibromo- 1,5-Dibromohexane	C$_6$H$_{12}$Br$_2$ 243.97	627-96-3	153[100]; 103[9]	4-01-00-00354 1.5989[20]	1.5072[15]
6757	Hexane, 1,6-dibromo-	C$_6$H$_{12}$Br$_2$ 243.97	629-03-8 -2.3	245.5	4-01-00-00354 1.5948[15]	H$_2$O 1; eth 3; ace 3; ctc 2 1.5037[20]
6758	Hexane, 2,3-dibromo- 2,3-Dibromohexane	C$_6$H$_{12}$Br$_2$ 243.97	6423-02-5	90[16]	3-01-00-00393 1.5812[20]	1.5025[20]
6759	Hexane, 2,5-dibromo-, (R*,R*)-(±)- Hexane, 2,5-dibromo(DL)	C$_6$H$_{12}$Br$_2$ 243.97	54462-68-9 -44.64	108[30] 94[13]	3-01-00-00393 1.5788[20]	H$_2$O 1; eth 3; ace 3; ctc 3 1.5007[20]
6760	Hexane, 3,4-dibromo-	C$_6$H$_{12}$Br$_2$ 243.97	89583-12-0	80-1[13]	3-01-00-00394 1.6027[20]	1.5043[20]
6761	Hexane, 1,2-dichloro- 1,2-Dichlorohexane	C$_6$H$_{12}$Cl$_2$ 155.07	2162-92-7	173	4-01-00-00350 1.085[15]	eth 4; chl 4
6762	Hexane, 1,6-dichloro- 1,6-Dichlorohexane	C$_6$H$_{12}$Cl$_2$ 155.07	2163-00-0	204	4-01-00-00350 1.0677[20]	H$_2$O 1; eth 3; ctc 3; chl 3 1.4572[20]
6763	Hexane, 2,2-dichloro- 2,2-Dichlorohexane	C$_6$H$_{12}$Cl$_2$ 155.07	42131-89-5	68[49]	3-01-00-00390 1.0150[25]	eth 4; chl 4 1.4353[25]

No.	Name Synonym	Mol. Form. Mol. Wt.	CAS RN mp/°C	Merck No. bp/°C	Bell. Ref. den/g cm^{-3}	Solubility n_D
6764	Hexane, 2,3-dichloro- 2,3-Dichlorohexane	C$_6$H$_{12}$Cl$_2$ 155.07	54305-87-2	163.5	4-01-00-00350 1.0527[11]	eth 4; chl 4
6765	Hexane, 2,5-dichloro-2,5-dimethyl- 2,5-Dimethyl-2,5-dichlorohexane	C$_8$H$_{16}$Cl$_2$ 183.12	6223-78-5 67.5		4-01-00-00435 0.9543[20]	bz 4; eth 4; EtOH 4; chl 4
6766	Hexane, 2,5-dichloro-, (R^*,R^*)-(±)- (±)-2,5-Dichlorohexane	C$_6$H$_{12}$Cl$_2$ 155.07	41761-12-0 -38	177; 106[91]	4-01-00-00350 1.0474[20]	chl 4 1.4491[20]
6767	Hexane, 3,3-diethyl-	C$_{10}$H$_{22}$ 142.28	17302-02-2	166.3	4-01-00-00484 0.7575[25]	1.4258[20]
6768	Hexane, 3,4-diethyl- 3,4-Diethylhexane	C$_{10}$H$_{22}$ 142.28	19398-77-7	163.9	4-01-00-00484 0.7472[25]	1.4190[20]
6769	Hexane, 1,6-diiodo- Hexamethylene diiodide	C$_6$H$_{12}$I$_2$ 337.97	629-09-4 9.5	163[17]; 141[10]	4-01-00-00356 2.0416[15]	H$_2$O 1; EtOH 4; eth 4 1.5899[15]
6770	Hexane, 1,6-diisocyanato- Hexamethylene diisocyanate	C$_8$H$_{12}$N$_2$O$_2$ 168.20	822-06-0	122[10]; 94[1]	4-04-00-01349 1.0528[20]	1.4585[20]
6771	Hexane, 2,2-dimethyl- 2,2-Dimethylhexane	C$_8$H$_{18}$ 114.23	590-73-8 -121.1	106.8	4-01-00-00432 0.6953[20]	ace 4; bz 4; eth 4; EtOH 4 1.3935[20]
6772	Hexane, 2,3-dimethyl- 2,3-Dimethylhexane	C$_8$H$_{18}$ 114.23	584-94-1	115.6	4-01-00-00432 0.6912[25]	ace 4; bz 4; EtOH 4; lig 4 1.4011[20]
6773	Hexane, 2,4-dimethyl- 2,4-Dimethylhexane	C$_8$H$_{18}$ 114.23	589-43-5	109.5	0.6962[25]	1.3929[25]
6774	Hexane, 2,5-dimethyl- 2,5-Dimethylhexane	C$_8$H$_{18}$ 114.23	592-13-2 -91	109.1	4-01-00-00434 0.6901[25]	H$_2$O 1; EtOH 5; eth 3; ace 5 1.3925[20]
6775	Hexane, 3,3-dimethyl- 3,3-Dimethylhexane	C$_8$H$_{18}$ 114.23	563-16-6 -126.1	111.9	4-01-00-00435 0.7100[20]	H$_2$O 1; EtOH 5; eth 4; ace 4 1.4001[20]
6776	Hexane, 3,4-dimethyl- 3,4-Dimethylhexane	C$_8$H$_{18}$ 114.23	583-48-2	117.7	4-01-00-00436 0.7151[25]	H$_2$O 1; EtOH 5; eth 3; ace 5 1.4041[20]
6777	Hexanedinitrile Adiponitrile	C$_6$H$_8$N$_2$ 108.14	111-69-3 1	295	4-02-00-01975 0.9676[20]	H$_2$O 2; EtOH 3; eth 2; chl 3 1.4380[20]
6778	Hexane, 1,6-dinitro- 1,6-Dinitrohexane	C$_6$H$_{12}$N$_2$O$_4$ 176.17	4286-47-9 37.5	100-3[0.3]	4-01-00-00357	H$_2$O 1; HOAc 3
6779	Hexanedioic acid Adipic acid	C$_6$H$_{10}$O$_4$ 146.14	124-04-9 152 153.2	 337.5	4-02-00-01956 1.360[25]	H$_2$O 2; EtOH 4; eth 3; HOAc 1
6780	Hexanedioic acid, bis(2-ethylbutyl) ester Adipic acid, di-2-ethylbutyl ester	C$_{18}$H$_{34}$O$_4$ 314.47	10022-60-3 -15	200[19]	4-02-00-01964 0.934[25]	ace 4; EtOH 4 1.4434[20]
6781	Hexanedioic acid, bis(2-ethylhexyl) ester Bis(2-ethylhexyl) adipate	C$_{22}$H$_{42}$O$_4$ 370.57	103-23-1 -67.8	214[5]	4-02-00-01964 0.922[25]	ace 4; eth 4; EtOH 4 1.4474[20]
6782	Hexanedioic acid, bis(1-methylethyl) ester	C$_{12}$H$_{22}$O$_4$ 230.30	6938-94-9 -1.1	120[6.5]	4-02-00-01961 0.9569[20]	ace 4; eth 4; EtOH 4 1.4247[20]
6783	Hexanedioic acid, bis(2-methylpropyl) ester	C$_{14}$H$_{26}$O$_4$ 258.36	141-04-8	186-8[15]	4-02-00-01962 0.9543[19]	1.4301[20]
6784	Hexanedioic acid, dibutyl ester	C$_{14}$H$_{26}$O$_4$ 258.36	105-99-7 -32.4	165[10]	4-02-00-01961 0.9613[20]	H$_2$O 1; EtOH 5; eth 5 1.4369[20]
6785	Hexanedioic acid, diethyl ester	C$_{10}$H$_{18}$O$_4$ 202.25	141-28-6 -19.8	245	4-02-00-01960 1.0076[20]	H$_2$O 1; EtOH 3; eth 3 1.4272[20]
6786	Hexanedioic acid, dimethyl ester	C$_8$H$_{14}$O$_4$ 174.20	627-93-0 10.3	115[13]	4-02-00-01959 1.0600[20]	H$_2$O 1; EtOH 3; eth 3; ctc 3 1.4283[20]
6787	Hexanedioic acid, dipropyl ester	C$_{12}$H$_{22}$O$_4$ 230.30	106-19-4 -15.7	151[11]	4-02-00-01961 0.9790[20]	eth 4; EtOH 4; chl 4 1.4314[20]
6788	Hexanedioic acid, 2-methyl-	C$_7$H$_{12}$O$_4$ 160.17	626-70-0 64	209[13]	4-02-00-02010	H$_2$O 4; eth 4; EtOH 4; chl 4
6789	Hexanedioic acid, 3-methyl-, (±)- Adipic acid, 3-methyl (DL)	C$_7$H$_{12}$O$_4$ 160.17	81177-02-8 97	230[30]	4-02-00-02014	H$_2$O 4; EtOH 4; eth 4; ace 4
6790	Hexanedioic acid, 3-methyl-, diethyl ester Diethyl 3-methyladipate	C$_{11}$H$_{20}$O$_4$ 216.28	55877-01-5	258; 126.5[10]	4-02-00-02015 0.9948[20]	1.4335
6791	Hexanedioic acid, monoethyl ester	C$_8$H$_{14}$O$_4$ 174.20	626-86-8 29	285	4-02-00-01960 0.9796[20]	EtOH 3; eth 3; peth 3 1.4311[20]
6792	Hexanedioic acid, monomethyl ester	C$_7$H$_{12}$O$_4$ 160.17	627-91-8 9	158[10]	3-02-00-01711 1.0623[20]	EtOH 3 1.4283[20]
6793	1,3-Hexanediol	C$_6$H$_{14}$O$_2$ 118.18	21531-91-9	123[13]	4-01-00-02555 0.958[22]	1.4461[22]
6794	1,4-Hexanediol	C$_6$H$_{14}$O$_2$ 118.18	16432-53-4	134-5[18.5]	4-01-00-02556 0.982[16]	eth 4; EtOH 4 1.4530[16]
6795	1,5-Hexanediol	C$_6$H$_{14}$O$_2$ 118.18	928-40-5	237; 90[0.5]	4-01-00-02556 0.981[25]	1.4511[20]
6796	1,6-Hexanediol Hexamethylene glycol	C$_6$H$_{14}$O$_2$ 118.18	629-11-8 42.8	4610 208	4-01-00-02556	H$_2$O 3; EtOH 3; eth 2; ace 3 1.4579[25]
6797	2,3-Hexanediol	C$_6$H$_{14}$O$_2$ 118.18	617-30-1 60	205	4-01-00-02561 0.9900[15]	H$_2$O 4; eth 4; EtOH 4 1.4510[15]
6798	2,4-Hexanediol	C$_6$H$_{14}$O$_2$ 118.18	19780-90-6	211; 105[9]	4-01-00-02562 0.9516[21]	1.4418[21]
6799	2,5-Hexanediol Diisopropanol	C$_6$H$_{14}$O$_2$ 118.18	2935-44-6 43	218; 86[1]	4-01-00-02562 0.9610[20]	H$_2$O 3; EtOH 3; eth 3; ctc 2 1.4475[20]
6800	3,4-Hexanediol, 3,4-diethyl- 3,4-Diethyl-3,4-hexanediol	C$_{10}$H$_{22}$O$_2$ 174.28	6931-71-1 28	230	4-01-00-02621 0.9602[13]	eth 4; EtOH 4 1.467[13]
6801	2,5-Hexanediol, 2,5-dimethyl- 1,1,4,4-Tetramethyl-1,4-butanediol	C$_8$H$_{18}$O$_2$ 146.23	110-03-2 92	214	4-01-00-02600. 0.898[20]	H$_2$O 3; EtOH 4; bz 4; chl 4
6802	1,3-Hexanediol, 2-ethyl- Ethohexadiol	C$_8$H$_{18}$O$_2$ 146.23	94-96-2 -40	3699 244	4-01-00-02597 0.9325[22]	H$_2$O 2; EtOH 3; eth 3 1.4497[20]

3-187

No.	Name / Synonym	Mol. Form. / Mol. Wt.	CAS RN / mp/°C	Merck No. / bp/°C	Beil. Ref. / den/g cm^{-3}	Solubility / n_D
6803	2,4-Hexanediol, 2-methyl-	$C_7H_{16}O_2$ / 132.20	66225-35-2	121[30]	2-01-00-00554 / 0.9321[18]	1.4407[18]
6804	2,4-Hexanedione / Propionylacetone	$C_8H_{10}O_2$ / 114.14	3002-24-2	160	4-01-00-03687 / 0.959[20]	1.4516[20]
6805	2,5-Hexanedione / Acetonylacetone	$C_6H_{10}O_2$ / 114.14	110-13-4 / -5.5	63 / 194	4-01-00-03688 / 0.7370[20]	H_2O 4; bz 4; eth 4; EtOH 4 / 1.4232[20]
6806	3,4-Hexanedione / Bipropionyl	$C_6H_{10}O_2$ / 114.14	4437-51-8 / -10	130	4-01-00-03690 / 0.941[21]	1.4130[21]
6807	3,4-Hexanedione, 2,5-dimethyl- / Biisobutyryl	$C_8H_{14}O_2$ / 142.20	4388-87-8	144.5	4-01-00-03713 / 0.9232[20]	1.4206[20]
6808	2,5-Hexanedione, 3-hydroxy- / 3-Hydroxy-2,5-hexanedione	$C_6H_{10}O_3$ / 130.14	61892-85-1	62-7[0.5]	3-01-00-03317	bz 4; eth 4 / 1.4497[25]
6809	2,3-Hexanedione, 5-methyl- / 2-Methylhexa-4,5-dione	$C_7H_{12}O_2$ / 128.17	13706-86-0	138	4-01-00-03701 / 0.908[22]	1.4119[20]
6810	2,5-Hexanedione, 3-methyl- / 3-Methyl-2,5-hexanedione	$C_7H_{12}O_2$ / 128.17	4437-50-7	196; 71[10]	3-01-00-03137 / 0.9527[20]	eth 4; EtOH 4 / 1.4260[20]
6811	3,4-Hexanedione, 2,2,5,5-tetramethyl- / Pivalil	$C_{10}H_{18}O_2$ / 170.25	4388-88-9 / -2	169	4-01-00-03726 / 0.8776[20]	H_2O 1; eth 5 / 1.4157[20]
6812	1,6-Hexanedithiol	$C_6H_{14}S_2$ / 150.31	1191-43-1	118-9[15]	4-01-00-02559 / 0.983[25]	1.5110[20]
6813	Hexane, 1-(ethenyloxy)- / Hexyl vinyl ether	$C_8H_{16}O$ / 128.21	5363-64-4	143	4-01-00-02055 / 0.7966[20]	1.4171[20]
6814	Hexane, 1-ethoxy-	$C_8H_{18}O$ / 130.23	5756-43-4	143	4-01-00-01697 / 0.7722[20]	eth 4; EtOH 4 / 1.4008[20]
6815	Hexane, 3-ethyl- / 3-Ethylhexane	C_8H_{18} / 114.23	619-99-8	118.6	4-01-00-00431 / 0.7136[20]	H_2O 1; EtOH 5; eth 5; ace 5 / 1.4018[20]
6816	Hexane, 3-ethyl-2,2-dimethyl- / 3-Ethyl-2,2-dimethylhexane	$C_{10}H_{22}$ / 142.28	20291-91-2	156.1	4-01-00-00484 / 0.7447[25]	1.4197[20]
6817	Hexane, 3-ethyl-2,3-dimethyl- / 3-Ethyl-2,3-dimethylhexane	$C_{10}H_{22}$ / 142.28	52897-00-4	166	4-01-00-00484 / 0.7599[25]	1.4270[207]
6818	Hexane, 3-ethyl-2,4-dimethyl- / 3-Ethyl-2,4-dimethylhexane	$C_{10}H_{22}$ / 142.28	7220-26-0	162	4-01-00-00484 / 0.7514[25]	1.4225[20]
6819	Hexane, 3-ethyl-2,5-dimethyl- / 3-Ethyl-2,5-dimethylhexane	$C_{10}H_{22}$ / 142.28	52897-04-8	154.1	4-01-00-00484 / 0.7368[25]	1.4157[20]
6820	Hexane, 3-ethyl-3,4-dimethyl- / 3-Ethyl-3,4-dimethylhexane	$C_{10}H_{22}$ / 142.28	52897-06-0	162.1	4-01-00-00485 / 0.7596[25]	1.4267[207]
6821	Hexane, 4-ethyl-2,2-dimethyl- / Hexane, 2,2-dimethyl-4-ethyl	$C_{10}H_{22}$ / 142.28	52896-99-8	147	4-01-00-00485 / 0.7330[20]	1.4131[20]
6822	Hexane, 4-ethyl-2,3-dimethyl- / 4-Ethyl-2,3-dimethylhexane	$C_{10}H_{22}$ / 142.28	52897-01-5	162	4-01-00-00485 / 0.7516[25]	1.4226[20]
6823	Hexane, 4-ethyl-2,4-dimethyl- / 4-Ethyl-2,4-dimethylhexane	$C_{10}H_{22}$ / 142.28	52897-03-7	161.1	4-01-00-00485 / 0.7525[25]	1.4235[20]
6824	Hexane, 4-ethyl-3,3-dimethyl- / 4-Ethyl-3,3-dimethylhexane	$C_{10}H_{22}$ / 142.28	52897-05-9	162.9	4-01-00-00485 / 0.7598[25]	1.4269[20]
6825	Hexane, 3-ethyl-2-methyl- / 3-Ethyl-2-methylhexane	C_9H_{20} / 128.26	16789-46-1	138	4-01-00-00459 / 0.7310[20]	1.4106[20]
6826	Hexane, 3-ethyl-3-methyl- / 3-Ethyl-3-methylhexane	C_9H_{20} / 128.26	3074-76-8	140.6	4-01-00-00459 / 0.7371[25]	1.4140[20]
6827	Hexane, 3-ethyl-4-methyl- / 4-Ethyl-3-methylhexane	C_9H_{20} / 128.26	3074-77-9	140	3-01-00-00514 / 0.7420[20]	1.4134[20]
6828	Hexane, 4-ethyl-2-methyl- / 4-Ethyl-2-methylhexane	C_9H_{20} / 128.26	3074-75-7	133.8	4-01-00-00459 / 0.7195[25]	1.4063[20]
6829	Hexane, 1-fluoro- / Hexyl fluoride	$C_6H_{13}F$ / 104.17	373-14-8 / -103	91.5	4-01-00-00348 / 0.7995[20]	eth 3; bz 3 / 1.3738[20]
6830	Hexane, 2-fluoro- / 2-Fluorohexane	$C_6H_{13}F$ / 104.17	372-54-3	86	4-01-00-00348 / 0.7916[20]	1.3693[20]
6831	Hexane, 1-iodo- / Hexyl iodide	$C_6H_{13}I$ / 212.07	638-45-9 / -74.2	181	4-01-00-00355 / 1.4397[20]	H_2O 1 / 1.4929[20]
6832	Hexane, 2-iodo-, (R)- / Hexane, 2-iodo (l)	$C_6H_{13}I$ / 212.07	59654-13-6	90[70]; 45[9]	4-01-00-00355 / 1.4354[17]	H_2O 1; ace 3; chl 3 / 1.4878[25]
6833	Hexane, 1-methoxy- / Hexyl methyl ether	$C_7H_{16}O$ / 116.20	4747-07-3	126.1		
6834	Hexane, 2-methyl- / 2-Methylhexane	C_7H_{16} / 100.20	591-76-4 / -118.2	90.0	4-01-00-00397 / 0.6787[20]	H_2O 1; EtOH 3; eth 5; ace 5 / 1.3848[20]
6835	Hexane, 3-methyl-, (S)- / (+)-3-Methylhexane	C_7H_{16} / 100.20	6131-24-4 / -119	92	4-01-00-00399 / 0.6880[20]	H_2O 1; EtOH 3; eth 5; ace 5 / 1.3887[20]
6836	Hexane, 2-methyl-3-methylene- / 2-Isopropyl-1-pentene	C_8H_{16} / 112.22	16746-02-4	113	4-01-00-00887 / 0.721[25]	1.414[20]
6837	Hexane, 2-methyl-4-methylene- / 1-Pentene, 2-ethyl-4-methyl-	C_8H_{16} / 112.22	3404-80-6	110.3	4-01-00-00888 / 0.7152[25]	1.4105[20]
6838	Hexane, 3-methyl-4-methylene-	C_8H_{16} / 112.22	3404-67-9	112.5	4-01-00-00890 / 0.725[25]	1.4142[20]
6839	Hexane, 2-methyl-3-(1-methylethyl)- / Hexane, 3-isopropyl-2-methyl-	$C_{10}H_{22}$ / 142.28	62016-13-1	165	4-01-00-00484 / 0.7436[25]	1.4195[20]
6840	Hexanenitrile / Capronitrile	$C_6H_{11}N$ / 97.16	628-73-9 / -80.3	163.6	4-02-00-00930 / 0.8051[20]	H_2O 1; EtOH 3; eth 3; chl 2 / 1.4068[20]
6841	Hexane, 1-nitro- / 1-Nitrohexane	$C_6H_{13}NO_2$ / 131.17	646-14-0	193; 84[21]	4-01-00-00356 / 0.9396[20]	H_2O 1; EtOH 3; eth 3; ace 3 / 1.4270[20]
6842	Hexane, 1,1'-oxybis-	$C_{12}H_{26}O$ / 186.34	112-58-3	223	4-01-00-01698 / 0.7936[20]	H_2O 1; eth 3; ctc 2 / 1.4204[20]

No.	Name Synonym	Mol. Form. Mol. Wt.	CAS RN mp/°C	Merck No. bp/°C	Beil. Ref. den/g cm⁻³	Solubility n_D
6843	Hexane, tetradecafluoro- Perfluorohexane	C_6F_{14} 338.04	355-42-0 -87.1	56.6	4-01-00-00348 1.6995[20]	H_2O 1; eth 3; bz 3; chl 3 1.2515[20]
6844	Hexane, 2,2,3,3-tetramethyl- 2,2,3,3-Tetramethylhexane	$C_{10}H_{22}$ 142.28	13475-81-5 -54	160.3	4-01-00-00485 0.7609[25]	1.4282[20]
6845	Hexane, 2,2,3,4-tetramethyl- 2,2,3,4-Tetramethylhexane	$C_{10}H_{22}$ 142.28	52897-08-2	157	4-01-00-00485 0.7513[25]	1.4216[20]
6846	Hexane, 2,2,3,5-tetramethyl- 2,2,3,5-Tetramethylhexane	$C_{10}H_{22}$ 142.28	52897-09-3	148.4	4-01-00-00485 0.7336[25]	1.4142[20]
6847	Hexane, 2,2,4,4-tetramethyl- 2,2,4,4-Tetramethylhexane	$C_{10}H_{22}$ 142.28	51750-65-3	153.8	4-01-00-00486 0.7424[25]	1.4208[20]
6848	Hexane, 2,2,4,5-tetramethyl- 2,2,4,5-Tetramethylhexane	$C_{10}H_{22}$ 142.28	16747-42-5	147.9	4-01-00-00486 0.7316[25]	1.4132[20]
6849	Hexane, 2,2,5,5-tetramethyl- 2,2,5,5-Tetramethylhexane	$C_{10}H_{22}$ 142.28	1071-81-4 -12.6	137.4	4-01-00-00486 0.7148[25]	1.4055[20]
6850	Hexane, 2,3,3,4-tetramethyl- 2,3,3,4-Tetramethylhexane	$C_{10}H_{22}$ 142.28	52897-10-6	164.6	4-01-00-00486 0.7656[25]	1.4298[20]
6851	Hexane, 2,3,3,5-tetramethyl- 2,3,3,5-Tetramethylhexane	$C_{10}H_{22}$ 142.28	52897-11-7	153.1	4-01-00-00486 0.7449[25]	1.4196[20]
6852	Hexane, 2,3,4,4-tetramethyl- 2,3,4,4-Tetramethylhexane	$C_{10}H_{22}$ 142.28	52897-12-8	161.6	4-01-00-00487 0.7586[25]	1.4267[20]
6853	Hexane, 2,3,4,5-tetramethyl- 2,3,4,5-Tetramethylhexane	$C_{10}H_{22}$ 142.28	52897-15-1	156.2	4-01-00-00487 0.7456[25]	1.4204[20]
6854	Hexane, 3,3,4,4-tetramethyl- 3,3,4,4-Tetramethylhexane	$C_{10}H_{22}$ 142.28	5171-84-6	170.0	3-01-00-00534 0.7789[25]	1.4368[20]
6855	Hexane, 1,1'-thiobis-	$C_{12}H_{26}S$ 202.40	6294-31-1	230; 136[20]	4-01-00-01706 0.8411[20]	1.4586[20]
6856	1-Hexanethiol Hexyl mercaptan	$C_6H_{14}S$ 118.24	111-31-9 -81	151	4-01-00-01705 0.8424[20]	H_2O 1; EtOH 4; eth 4 1.4496[20]
6857	2-Hexanethiol	$C_6H_{14}S$ 118.24	1679-06-7 -147	142	4-01-00-01711 0.8345[20]	H_2O 1; EtOH 3; eth 3; bz 3 1.4451[20]
6858	3-Hexanethiol 3-Mercaptohexane	$C_6H_{14}S$ 118.24	1633-90-5	57[25]	4-01-00-01713 0.9206[20]	EtOH 3 1.4496[20]
6859	Hexane, 2,2,3-trimethyl- 2,2,3-Trimethylhexane	C_9H_{20} 128.26	16747-25-4	133.6	4-01-00-00459 0.7257[25]	1.4106[20]
6860	Hexane, 2,2,4-trimethyl- 2,2,4-Trimethylhexane	C_9H_{20} 128.26	16747-26-5 -120	126.5	4-01-00-00459 0.7110[20]	1.4033[20]
6861	Hexane, 2,2,5-trimethyl- 2,2,5-Trimethylhexane	C_9H_{20} 128.26	3522-94-9 -105.7	124.0	4-01-00-00460 0.7072[20]	H_2O 1; EtOH 4; eth 4; ace 4 1.3997[20]
6862	Hexane, 2,3,3-trimethyl- 2,3,3-Trimethylhexane	C_9H_{20} 128.26	16747-28-7 -116.8	137.7	4-01-00-00461 0.7345[25]	1.4141[20]
6863	Hexane, 2,3,4-trimethyl- 2,3,4-Trimethylhexane	C_9H_{20} 128.26	921-47-1	139.1	4-01-00-00461 0.7354[25]	1.4144[20]
6864	Hexane, 2,3,5-trimethyl- 2,3,5-Trimethylhexane	C_9H_{20} 128.26	1069-53-0 -127.9	131.4	4-01-00-00461 0.7218[20]	1.4051[20]
6865	Hexane, 2,4,4-trimethyl- 2,4,4-Trimethylhexane	C_9H_{20} 128.26	16747-30-1 -113.4	130.7	4-01-00-00461 0.7201[20]	1.4074[20]
6866	Hexane, 3,3,4-trimethyl- 3,3,4-Trimethylhexane	C_9H_{20} 128.26	16747-31-2 -101.2	140.5	4-01-00-00462 0.7414[25]	1.4178[20]
6867	1,2,5-Hexanetriol	$C_6H_{14}O_3$ 134.18	10299-30-6	181[10]; 149[3]	4-01-00-02784 1.1012[20]	H_2O 4; EtOH 4
6868	1,2,6-Hexanetriol 1,2,6-Trihydroxyhexane	$C_6H_{14}O_3$ 134.18	106-69-4	170[3]	4-01-00-02784 1.03[20]	1.58[20]
6869	2,3,4-Hexanetriol	$C_6H_{14}O_3$ 134.18	93972-93-1 47	256.5	3-01-00-02349	H_2O 4; EtOH 4
6870	Hexanoic acid Caproic acid	$C_6H_{12}O_2$ 116.16	142-62-1 1760	205.2	4-02-00-00917 0.9274[20]	H_2O 1; EtOH 3; eth 3; chl 3 1.4163[20]
6871	Hexanoic acid, 6-(acetylamino)- ε-Acetamidocaproic acid	$C_8H_{15}NO_3$ 173.21	57-08-9 104.5	38	4-04-00-02700	
6872	Hexanoic acid, 2-acetyl-, ethyl ester	$C_{10}H_{18}O_3$ 186.25	1540-29-0	221.5	4-03-00-01616 0.9523[20]	ace 4; eth 4 1.4301[20]
6873	Hexanoic acid, 6-amino- ε-Aminocaproic acid	$C_6H_{13}NO_2$ 131.17	60-32-2 205	442	4-04-00-02695	H_2O 4; EtOH 1; MeOH 2
6874	Hexanoic acid, anhydride	$C_{12}H_{22}O_3$ 214.30	2051-49-2 -41	255 dec	4-02-00-00928 0.9240[15]	eth 4; EtOH 4 1.4297[20]
6875	Hexanoic acid, 2-bromo-, (±)-	$C_6H_{11}BrO_2$ 195.06	2681-83-6 4	240; 140[23]	3-02-00-00736 1.2810[33]	EtOH 3; eth 3
6876	Hexanoic acid, 6-bromo-	$C_6H_{11}BrO_2$ 195.06	4224-70-8 35	165-70[20]	4-02-00-00940	peth 4
6877	Hexanoic acid, 2-bromo-, ethyl ester, (±)-	$C_8H_{15}BrO_2$ 223.11	63927-44-6	208.7	4-02-00-00938 1.2210[25]	EtOH 3
6878	Hexanoic acid, 5-bromo-, ethyl ester Caproic acid, 5-bromo,ethyl ester	$C_8H_{15}BrO_2$ 223.11	90202-08-7	121[21]; 70[3]	3-02-00-00737 1.1908[25]	1.4525[25]
6879	Hexanoic acid, 6-bromo-, ethyl ester	$C_8H_{15}BrO_2$ 223.11	25542-62-5 33	126-7[21]	4-02-00-00940 1.238[23]	1.4566[21]
6880	Hexanoic acid, butyl ester	$C_{10}H_{20}O_2$ 172.27	626-82-4 -64.3	208	4-02-00-00922 0.8653[20]	H_2O 1; EtOH 3; eth 5 1.4152[20]
6881	Hexanoic acid, 2-cyano-, ethyl ester 2-Cyanocaproic acid, ethyl ester	$C_9H_{15}NO_2$ 169.22	7391-39-1	105[9]	3-02-00-01746 0.988[15]	1.4248[20]
6882	Hexanoic acid, 2-ethyl- 2-Ethylhexanoic acid	$C_8H_{16}O_2$ 144.21	149-57-5	228; 120[13]	3-02-00-00804 0.9031[25]	H_2O 3; EtOH 2; eth 3; ctc 3 1.4241[20]

No.	Name Synonym	Mol. Form. Mol. Wt.	CAS RN mp/°C	Merck No. bp/°C	Beil. Ref. den/g cm^{-3}	Solubility n_D
6883	Hexanoic acid, 3-ethyl- 3-Ethylhexanoic acid	C$_8$H$_{16}$O$_2$ 144.21	41065-91-2	106[5]	4-02-00-01008 0.911[30]	1.4287[25]
6884	Hexanoic acid, ethyl ester Ethyl hexanoate	C$_8$H$_{16}$O$_2$ 144.21	123-66-0 -67	3735 167	4-02-00-00921 0.873[20]	eth 4; EtOH 4 1.4073[20]
6885	Hexanoic acid, 2-ethyl-, ethyl ester	C$_{10}$H$_{20}$O$_2$ 172.27	2983-37-1	90[28]	4-02-00-01004 0.8586[25]	1.4123[25]
6886	Hexanoic acid, 2-ethyl-3-methyl- 2-Ethyl-3-methylhexanoic acid	C$_9$H$_{18}$O$_2$ 158.24	74581-94-5	231	3-02-00-00832 0.9060[25]	1.4302[25]
6887	Hexanoic acid, 3-ethyl-3-methyl- 3-Ethyl-3-methylhexanoic acid	C$_9$H$_{18}$O$_2$ 158.24	50902-82-4	139-40[22]	4-02-00-01036 0.9205[25]	1.4375[25]
6888	Hexanoic acid, 6-fluoro- Caproic acid, 6-fluoro	C$_6$H$_{11}$FO$_2$ 134.15	373-05-7	138[28]; 67[0.6]	4-02-00-00932	eth 4; EtOH 4 1.4166[25]
6889	Hexanoic acid, heptyl ester	C$_{13}$H$_{26}$O$_2$ 214.35	6976-72-3 -34.4	261	4-02-00-00923 0.8611[20]	ace 4; bz 4; eth 4; EtOH 4 1.4293[15]
6890	Hexanoic acid, hexyl ester	C$_{12}$H$_{24}$O$_2$ 200.32	6378-65-0 -55	246	4-02-00-00922 0.865[18]	ace 4; bz 4; eth 4; EtOH 4 1.4264[15]
6891	Hexanoic acid, 2-hydroxy-, (±)-	C$_6$H$_{12}$O$_3$ 132.16	636-36-2 61.3	270	4-03-00-00838	H$_2$O 4; EtOH 4; eth 3; chl 4
6892	Hexanoic acid, 3-hydroxy-2,2-dimethyl-, β-lactone Caproic acid, 2,2-dimethyl-3-hydroxy, β-lactone	C$_8$H$_{14}$O$_2$ 142.20	90112-82-6 143.5	93-100[14]		DMSO 2 1.4260[20]
6893	Hexanoic acid, 2-methyl-, (±)- Caproic acid, 2-methyl (DL)	C$_7$H$_{14}$O$_2$ 130.19	22160-12-9	215.5	4-02-00-00969 0.9612[20]	H$_2$O 5; EtOH 5; eth 5; ace 5 1.4195[20]
6894	Hexanoic acid, 3-methyl-	C$_7$H$_{14}$O$_2$ 130.19	3780-58-3	213; 116[15]	2-02-00-00298 0.9187[20]	1.4222[20]
6895	Hexanoic acid, 4-methyl-, (±)-	C$_7$H$_{14}$O$_2$ 130.19	22160-41-4 -80	217.5	4-02-00-00974 0.9215[20]	H$_2$O 2; EtOH 3; eth 3; ace 3 1.4211[20]
6896	Hexanoic acid, 5-methyl-	C$_7$H$_{14}$O$_2$ 130.19	628-46-6 <-25	216	4-02-00-00970 0.9138[21]	ace 4; bz 4; eth 4; EtOH 4 1.4220[20]
6897	Hexanoic acid, 3-methylbutyl ester	C$_{11}$H$_{22}$O$_2$ 186.29	2198-61-0	225.5	3-02-00-00727 0.861[20]	H$_2$O 1; EtOH 3; eth 3
6898	Hexanoic acid, methyl ester Methyl caproate	C$_7$H$_{14}$O$_2$ 130.19	106-70-7 -71	149.5	4-02-00-00921 0.8846[20]	H$_2$O 1; EtOH 4; eth 4; ace 3 1.4049[20]
6899	Hexanoic acid, 3-methyl-, ethyl ester Ethyl 3-methylhexanoate	C$_9$H$_{18}$O$_2$ 158.24	41692-47-1	177	2-02-00-00298 0.8679[20]	1.4119[20]
6900	Hexanoic acid, 4-methyl-, ethyl ester Ethyl 4-methylhexanoate	C$_9$H$_{18}$O$_2$ 158.24	1561-10-0	180	4-02-00-00974 0.8708[20]	1.4051[20]
6901	Hexanoic acid, 4-methyl-5-oxo-	C$_7$H$_{12}$O$_3$ 144.17	6818-07-1	160[15]; 140[5]	4-03-00-01601 0.9803[20]	1.4320[20]
6902	Hexanoic acid, 4-methyl-5-oxo-, ethyl ester Valeric acid, 4-acetyl, ethyl ester	C$_9$H$_{16}$O$_3$ 172.22	53068-88-5	114[15]	4-03-00-01601 0.9803[20]	1.4282[25]
6903	Hexanoic acid, 4-methyl-5-oxo-, methyl ester Valeric acid, 4-acetyl, methyl ester	C$_8$H$_{14}$O$_3$ 158.20	36045-56-4	113-7[16]	4-03-00-01601 1.0988[20]	1.4288[20]
6904	Hexanoic acid, octyl ester	C$_{14}$H$_{28}$O$_2$ 228.38	4887-30-3 -28	275	3-02-00-00727 0.8603[20]	ace 4; bz 4; eth 4; EtOH 4 1.4326[15]
6905	Hexanoic acid, 5-oxo-	C$_6$H$_{10}$O$_3$ 130.14	3128-06-1 13.5	274.5	4-03-00-01583 1.09[25]	H$_2$O 3; EtOH 3; eth 3; ctc 2 1.4451[20]
6906	Hexanoic acid, 5-oxo-, ethyl ester Butyric acid, 4-acetyl, ethyl ester	C$_8$H$_{14}$O$_3$ 158.20	13984-57-1	221.5	4-03-00-01583 0.989[25]	1.4277[20]
6907	Hexanoic acid, pentyl ester Amyl caproate	C$_{11}$H$_{22}$O$_2$ 186.29	540-07-8 -47	640 226	4-02-00-00922 0.8612[25]	EtOH 3; eth 3; ace 3; ctc 2 1.4202[25]
6908	Hexanoic acid, 1,2,3-propanetriyl ester	C$_{21}$H$_{38}$O$_6$ 386.53	621-70-5 -60	>200	4-02-00-00926 0.9867[20]	H$_2$O 1; EtOH 5; eth 5; ace 4 1.4427[20]
6909	Hexanoic acid, propyl ester	C$_9$H$_{18}$O$_2$ 158.24	626-77-7 -68.7	187	4-02-00-00922 0.8672[20]	eth 4; EtOH 4 1.4170[20]
6910	Hexanoic acid, 3,5,6-trichloro-2,2,3,4,4,5,6,6-octafluoro- Hexanoic acid, 2,2,3,4,4,5,6,6-octafluoro-3,5,6-trichloro	C$_6$HCl$_3$F$_8$O$_2$ 363.42	2106-54-9	105-6[5]	4-02-00-00937 1.860[20]	1.3903[20]
6911	1-Hexanol Caproyl alcohol	C$_6$H$_{14}$O 102.18	111-27-3 -44.6	4615 157.6	4-01-00-01694 0.8136[20]	H$_2$O 2; EtOH 3; eth 5; ace 3 1.4178[20]
6912	2-Hexanol, (±)-	C$_6$H$_{14}$O 102.18	20281-86-1	140	4-01-00-01708 0.8159[20]	H$_2$O 2; EtOH 3; eth 3; ctc 2 1.4144[20]
6913	2-Hexanol, (R)-	C$_6$H$_{14}$O 102.18	26549-24-6	137	3-01-00-01663 0.8178[18]	H$_2$O 2; EtOH 3; eth 3
6914	3-Hexanol, (±)-	C$_6$H$_{14}$O 102.18	17015-11-1	135	4-01-00-01711 0.8182[20]	H$_2$O 2; EtOH 3; eth 5; ace 3 1.4167[20]
6915	3-Hexanol, (R)- 3-Hexanol (l)	C$_6$H$_{14}$O 102.18	13471-42-6	135	0.8213[20]	H$_2$O 2; EtOH 3; eth 5; ace 3 1.4140[20]
6916	1-Hexanol, 6-chloro-	C$_6$H$_{13}$ClO 136.62	2009-83-8	107[12]	4-01-00-01704 1.0241[20]	H$_2$O 2; EtOH 4; eth 4 1.4550[20]
6917	2-Hexanol, 1-chloro- 1-Chloro-2-hexanol	C$_6$H$_{13}$ClO 136.62	52802-07-0	73-5[12]	4-01-00-01710 1.0139[20]	ace 4; eth 4; EtOH 4 1.4478[20]
6918	3-Hexanol, 1-chloro- 1-Chloro-3-hexanol	C$_6$H$_{13}$ClO 136.62	52418-81-2	120[35]; 78[6]	4-01-00-01712 1.003[25]	ace 4; bz 4; eth 4; EtOH 4 1.446[25]
6919	3-Hexanol, 5-chloro-	C$_6$H$_{13}$ClO 136.62	58588-28-6	78-9[13]	4-01-00-01712 1.0012[15]	ace 4; bz 4; eth 4; EtOH 4 1.4433[19]
6920	1-Hexanol, 3,5-dimethoxy-	C$_8$H$_{18}$O$_3$ 162.23	90952-10-6	114[13]	4-01-00-02785 0.9631[25]	1.4329[25]

No.	Name / Synonym	Mol. Form. / Mol. Wt.	CAS RN / mp/°C	Merck No. / bp/°C	Beil. Ref. / den/g cm⁻³	Solubility / n_D
6921	1-Hexanol, 2,5-dimethyl-	$C_8H_{18}O$ 130.23	6886-16-4	179.5	0-01-00-00422 0.8280[20]	1.5095[17]
6922	2-Hexanol, 2,3-dimethyl- 2,3-Dimethyl-2-hexanol	$C_8H_{18}O$ 130.23	19550-03-9	160; 62[10]	3-01-00-01737 0.8365[20]	1.4335[20]
6923	2-Hexanol, 2,5-dimethyl- 2,5-Dimethyl-2-hexanol	$C_8H_{18}O$ 130.23	3730-60-7	154.5	4-01-00-01792 0.8227[20]	1.4208[20]
6924	2-Hexanol, 3,4-dimethyl-	$C_8H_{18}O$ 130.23	19550-05-1	165.5	3-01-00-01740 0.840[15]	1.4325[15]
6925	3-Hexanol, 2,2-dimethyl- Propyl-tert-butylcarbinol	$C_8H_{18}O$ 130.23	4209-90-9	156	3-01-00-01737 0.8342[20]	1.4275[20]
6926	3-Hexanol, 2,3-dimethyl-	$C_8H_{18}O$ 130.23	4166-46-5	61-2[18]	4-01-00-01791 0.8371[20]	1.4309[20]
6927	3-Hexanol, 2,5-dimethyl- 2,5-Dimethyl-3-hexanol	$C_8H_{18}O$ 130.23	19550-07-3	157.5	4-01-00-01793 0.8212[20]	1.4246[20]
6928	3-Hexanol, 3,5-dimethyl-, (±)- 3-Hexanol, 3,5-dimethyl (DL)	$C_8H_{18}O$ 130.23	19113-78-1	152	4-01-00-01792 0.8373[20]	bz 4; eth 4; EtOH 4 1.4300[20]
6929	3-Hexanol, 4,4-dimethyl-	$C_8H_{18}O$ 130.23	19550-09-5	160.4	4-01-00-01793 0.8341[20]	1.4345[20]
6930	1-Hexanol, 2-ethyl- 2-Ethyl-1-hexanol	$C_8H_{18}O$ 130.23	104-76-7 -70	3764 184.6	4-01-00-01783 0.8319[25]	H_2O 1; EtOH 3; eth 3; ace 3 1.4300[20]
6931	3-Hexanol, 3-ethyl-	$C_8H_{18}O$ 130.23	597-76-2	160	4-01-00-01790 0.8373[20]	H_2O 1; EtOH 3; eth 3; ace 3 1.4300[20]
6932	1-Hexanol, 2-ethyl-3-methyl-	$C_9H_{20}O$ 144.26	66794-04-5	193; 83[10]	4-02-00-02107 0.8486[20]	1.4455[20]
6933	2-Hexanol, 3-ethyl-2-methyl-	$C_9H_{20}O$ 144.26	66794-02-3	71.5[14]	4-01-00-01812 0.8334[25]	1.4331[25]
6934	3-Hexanol, 3-ethyl-2-methyl-	$C_9H_{20}O$ 144.26	66794-03-4	177.5	4-01-00-01811 0.8445[25]	1.4369[25]
6935	3-Hexanol, 3-ethyl-5-methyl-	$C_9H_{20}O$ 144.26	597-77-3	172; 68[18]	3-01-00-01755 0.8396[22]	eth 4; EtOH 4 1.4346[13]
6936	1-Hexanol, 6-fluoro-	$C_8H_{13}FO$ 120.17	373-32-0	85-6[14]	4-01-00-01703 0.975[20]	eth 4; EtOH 4 1.4141[25]
6937	1-Hexanol, 2-methyl-, (±)- 1-Hexanol, 2-methyl (DL)	$C_7H_{16}O$ 116.20	111768-04-8	164; 71[15]	4-01-00-01745 0.8270[20]	eth 4; EtOH 4 1.4226[20]
6938	1-Hexanol, 2-methyl-, (R)- (R)-2-Methyl-1-hexanol	$C_7H_{16}O$ 116.20	66050-98-4	164; 71[15]	0.8313[13]	eth 4; EtOH 4 1.4245[17]
6939	1-Hexanol, 3-methyl-, (±)-	$C_7H_{16}O$ 116.20	111768-08-2	169; 91[13]	4-01-00-01748 0.8258[20]	H_2O 1; EtOH 4; eth 4; ace 3 1.4245[20]
6940	1-Hexanol, 4-methyl-, (±)- 1-Hexanol, 4-methyl (DL)	$C_7H_{16}O$ 116.20	111768-05-9	173	4-01-00-01749 0.8239[20]	ace 4; bz 4; eth 4; EtOH 4 1.4219[20]
6941	1-Hexanol, 5-methyl- 5-Methyl-1-hexanol	C_7H_{16} 100.20	627-98-5	170; 53-5[15]	4-01-00-01748 0.8119[20]	eth 4; EtOH 4 1.4175[20]
6942	2-Hexanol, 2-methyl-	$C_7H_{16}O$ 116.20	625-23-0	143	4-01-00-01745 0.8119[20]	H_2O 2; EtOH 5; eth 5 1.4175[20]
6943	2-Hexanol, 3-methyl- 3-Methyl-2-hexanol	$C_7H_{16}O$ 116.20	2313-65-7	79-81[52]	0.8820[25]	H_2O 1; EtOH 4; eth 4; ace 3 1.4198[18]
6944	2-Hexanol, 5-methyl- 5-Methyl-2-hexanol	$C_7H_{16}O$ 116.20	627-59-8	151; 78[28]	4-01-00-01747 0.814[20]	H_2O 1; EtOH 3; eth 3 1.4180[20]
6945	3-Hexanol, 2-methyl-, (±)- (±)-2-Methyl-3-hexanol	$C_7H_{16}O$ 116.20	100295-82-7	143	4-01-00-01746 0.8407[20]	H_2O 1; EtOH 3; eth 3; ctc 3 1.4149[20]
6946	3-Hexanol, 3-methyl- 2-Ethyl-2-pentanol	$C_7H_{16}O$ 116.20	597-96-6	143	4-01-00-01749 0.8233[20]	H_2O 2; EtOH 3; eth 3; ctc 3 1.4231[20]
6947	3-Hexanol, 5-methyl-, (±)- (±)-5-Methyl-3-hexanol	$C_7H_{16}O$ 116.20	100295-83-8	148	4-01-00-01747 0.827[25]	H_2O 1; EtOH 3; eth 3; ctc 2 1.4128[20]
6948	1-Hexanol, 5-methyl-2-(1-methylethyl)- 2-Isopropyl-5-methyl-1-hexanol	$C_{10}H_{22}O$ 158.28	2051-33-4	211	4-01-00-01833 0.8345[20]	H_2O 1; EtOH 3; eth 4 1.4369[20]
6949	3-Hexanol, 1-phenyl-	$C_{12}H_{18}O$ 178.27	2180-43-0 34	170[50.] 146[16]	4-06-00-03418 0.9525[20]	
6950	3-Hexanol, 2,2,5,5-tetramethyl-	$C_{10}H_{22}O$ 158.28	55073-86-4 52.5	170	3-01-00-01772	ace 4; eth 4; EtOH 4; peth 4
6951	1-Hexanol, 3,5,5-trimethyl-	$C_9H_{20}O$ 144.26	3452-97-9	194	4-01-00-01813 0.8236[25]	1.4300[25]
6952	2-Hexanol, 2,3,4-trimethyl-	$C_9H_{20}O$ 144.26	21102-13-6	57[5]	3-01-00-01756 0.853[15]	ace 4; eth 4; EtOH 4; peth 4 1.4415[15]
6953	3-Hexanol, 2,2,3-trimethyl-	$C_9H_{20}O$ 144.26	5340-41-0	170	3-01-00-01755 0.8474[20]	EtOH 3; eth 3; ace 3 1.4402[20]
6954	3-Hexanol, 2,3,5-trimethyl-	$C_9H_{20}O$ 144.26	65927-60-8	72[21]	4-01-00-01813 0.8256[20]	ace 4; eth 4; EtOH 4 1.4321[20]
6955	3-Hexanol, 2,4,4-trimethyl-	$C_9H_{20}O$ 144.26	66793-92-8	170	3-01-00-01756 0.8489[20]	EtOH 3; eth 3; ace 3 1.4395[20]
6956	3-Hexanol, 2,5,5-trimethyl-	$C_9H_{20}O$ 144.26	66793-72-4	77[32]	3-01-00-01756 0.8250[20]	ace 4; eth 4; EtOH 4 1.4286[20]
6957	3-Hexanol, 3,4,4-trimethyl-	$C_9H_{20}O$ 144.26	66793-74-6	165.5	0-01-00-00425 0.8322[21]	EtOH 3; eth 3; ace 3 1.4341[21]
6958	3-Hexanol, 3,5,5-trimethyl- 3,5,5-Trimethylhexan-3-ol	$C_9H_{20}O$ 144.26	66810-87-5	62[14]	3-01-00-01755 0.8350[20]	ace 4; eth 4; EtOH 4 1.4352[20]
6959	2-Hexanone Butyl methyl ketone	$C_8H_{12}O$ 100.16	591-78-6 -55.5	5955 127.6	4-01-00-03298 0.8113[20]	H_2O 2; EtOH 5; eth 5; ace 3 1.4007[20]
6960	3-Hexanone Ethyl propyl ketone	$C_8H_{12}O$ 100.16	589-38-8 -55.5	123.5	4-01-00-03301 0.8118[20]	H_2O 2; EtOH 5; eth 5; ace 3 1.4004[20]

No.	Name / Synonym	Mol. Form. / Mol. Wt.	CAS RN / mp/°C	Merck No. / bp/°C	Beil. Ref. / den/g cm^{-3}	Solubility / n_D
6961	2-Hexanone, 6-bromo-	$C_6H_{11}BrO$ 179.06	10226-29-6	216; 136[90]	3-01-00-02829 1.3494[0]	eth 4; EtOH 4
6962	1-Hexanone, 1-(2,4-dihydroxyphenyl)-	$C_{12}H_{16}O_3$ 208.26	3144-54-5 57	343	4-08-00-01888	H_2O 1; EtOH 3; eth 3; ace 3
6963	2-Hexanone, 3,3-dimethyl- 3,3-Dimethyl-2-hexanone	$C_8H_{16}O$ 128.21	26118-38-7	149	4-01-00-03350 0.838[0]	ace 4; eth 4; EtOH 4 1.4098[20]
6964	2-Hexanone, 3,4-dimethyl- 3,4-Dimethyl-2-hexanone	$C_8H_{16}O$ 128.21	19550-10-8	158	3-01-00-02882 0.8295[22]	EtOH 3; eth 3; ace 3 1.4193[20]
6965	3-Hexanone, 2,2-dimethyl- tert-Butyl propyl ketone	$C_8H_{16}O$ 128.21	5405-79-8	147	4-01-00-03347 0.8105[25]	EtOH 3; eth 3; ace 3 1.4119[20]
6966	3-Hexanone, 2,5-dimethyl-	$C_8H_{16}O$ 128.21	1888-57-9	147.5	4-01-00-03349 0.8269[0]	ace 4; eth 4; EtOH 4 1.4049[20]
6967	3-Hexanone, 4,4-dimethyl-	$C_8H_{16}O$ 128.21	19550-14-2	148	4-01-00-03350 0.8285[20]	bz 4; EtOH 4; chl 4 1.4208[25]
6968	3-Hexanone, 6-(dimethylamino)-4,4-diphenyl- Normethadone	$C_{20}H_{25}NO$ 295.42	467-85-6	6628 164-7[3]	4-14-00-00297	
6969	2-Hexanone, 5-hydroxy- 5-Hydroxy-2-hexanone	$C_6H_{12}O_2$ 116.16	56745-61-0	203[270]; 80[10]	4-01-00-04019 0.9626[25]	H_2O 4; eth 4; EtOH 4 1.4312[25]
6970	2-Hexanone, 6-hydroxy-	$C_6H_{12}O_2$ 116.16	21856-89-3	227	4-01-00-04020 0.9886[15]	eth 4; EtOH 4 1.4494
6971	3-Hexanone, 4-hydroxy-	$C_6H_{12}O_2$ 116.16	4984-85-4	133[227]; 73[20]	4-01-00-04021 0.956[21]	ace 4; EtOH 4 1.4340[21]
6972	3-Hexanone, 5-hydroxy- 5-Hydroxy-3-hexanone	$C_6H_{12}O_2$ 116.16	33683-44-2	75-8[12]	4-01-00-04021 0.950[20]	1.4280[25]
6973	3-Hexanone, 2-hydroxy-2-methyl- 2-Hexanol-3-one, 2-methyl-	$C_7H_{14}O_2$ 130.19	18905-91-4	60-2[15]	4-01-00-04035 0.899[18]	1.419[18]
6974	3-Hexanone, 5-hydroxy-5-methyl- 2-Hexanol-4-one, 5-methyl-	$C_7H_{14}O_2$ 130.19	59357-07-2	72-3[9]	4-01-00-04035 0.929[20]	1.4278[20]
6975	3-Hexanone, 4-hydroxy-2,2,5,5-tetramethyl- Pivaloin	$C_{10}H_{20}O_2$ 172.27	815-66-7 81	80[10]	4-01-00-04060	eth 4
6976	2-Hexanone, 3-methyl- 3-Methyl-2-hexanone	$C_7H_{14}O$ 114.19	2550-21-2	143.5	3-01-00-02863 0.828[25]	ace 4; bz 4; eth 4; EtOH 4 1.4035[20]
6977	2-Hexanone, 4-methyl- Methyl 2-methylbutyl ketone	$C_7H_{14}O$ 114.19	105-42-0	144.5	4-01-00-03329 0.8130[20]	H_2O 2; EtOH 4; eth 4; ace 4 1.4081[24]
6978	2-Hexanone, 5-methyl- Methyl isopentyl ketone	$C_7H_{14}O$ 114.19	110-12-3	144	4-01-00-03329 0.888[20]	H_2O 2; EtOH 5; eth 5; ace 4 1.4062[20]
6979	3-Hexanone, 2-methyl-	$C_7H_{14}O$ 114.19	7379-12-6	135	4-01-00-03328 0.8091[20]	EtOH 3; eth 3; ace 4; chl 3 1.4042[20]
6980	3-Hexanone, 4-methyl-	$C_7H_{14}O$ 114.19	17042-16-9	134.5	4-01-00-03329 0.8162[20]	ace 4; bz 4; eth 4; EtOH 4 1.4069[20]
6981	3-Hexanone, 5-methyl-	$C_7H_{14}O$ 114.19	623-56-3	135	4-01-00-03328 0.8090[20]	H_2O 1; EtOH 5; eth 5; ctc 3 1.4047[20]
6982	2-Hexanone, 5-methyl-, oxime	$C_7H_{15}NO$ 129.20	624-44-2	195.5	0-01-00-00701 0.8881[20]	chl 2 1.4448[20]
6983	1-Hexanone, 1-phenyl-	$C_{12}H_{16}O$ 176.26	942-92-7 27	265	4-07-00-00754 0.9576[20]	H_2O 2; EtOH 3; eth 3; ace 3 1.5027[25]
6984	2-Hexanone, 3-phenyl-	$C_{12}H_{16}O$ 176.26	6306-30-5	235.5; 114[13]	3-07-00-01160 0.970[0]	1.5020[20]
6985	3-Hexanone, 2-phenyl- 1-Phenylethyl propyl ketone	$C_{12}H_{16}O$ 176.26	65248-43-3	241	2-07-00-00258 0.941[25]	1.4988[25]
6986	Hexanoyl chloride Caproyl chloride	$C_6H_{11}ClO$ 134.61	142-61-0 -87	1763 153	4-02-00-00928 0.9784[20]	eth 3; ace 3 1.4264[20]
6987	Hexanoyl chloride, 2-ethyl-	$C_8H_{15}ClO$ 162.66	760-67-8	67-8[11]	3-02-00-00805 0.939[25]	1.4335[20]
6988	Hexanoyl chloride, 3-methyl- Caproyl chloride, 3-methyl	$C_7H_{13}ClO$ 148.63	57323-93-0	163; 82[50]	3-02-00-00777 0.967[20]	bz 4 1.4293[25]
6989	Hexanoyl chloride, 4-methyl- Caproyl chloride, 4-methyl	$C_7H_{13}ClO$ 148.63	50599-73-0	167	2-02-00-00299 0.9677[20]	eth 4
6990	Hexasiloxane, tetradecamethyl- Tetradecamethylhexasiloxane	$C_{14}H_{42}O_5Si_6$ 459.00	107-52-8 -59	9131 245.5	4-04-00-04127 0.8910[20]	bz 4 1.3948[20]
6991	Hexatriacontane	$C_{36}H_{74}$ 506.98	630-06-8 76.5	298.4[3]	4-01-00-00599 0.7803[80]	1.4397[80]
6992	1,3,5-Hexatriene, (E)-	C_6H_8 80.13	821-07-8 -12	78.5	4-01-00-01093 0.7369[15]	H_2O 1; EtOH 3; ace 3; chl 3 1.5135[20]
6993	1,3,5-Hexatriene, (Z)-	C_6H_8 80.13	2612-46-6 -12	78	0.7175[20]	H_2O 1; EtOH 3; ace 3; chl 3 1.4577[20]
6994	1,3,5-Hexatriene, 2,5-dimethyl-	C_8H_{12} 108.18	4916-63-6 -9	146	4-01-00-01102 0.7822[20]	H_2O 1; ace 3; MeOH 3; lig 3 1.5122[20]
6995	Hexazinone 1,3,5-Triazine-2,4(1H,3H)-dione, 3-cyclohexyl-6-(dimethylamino)-1-methyl-	$C_{12}H_{20}N_4O_2$ 252.32	51235-04-2 99	4617 dec	1.25	
6996	2-Hexenal, (E)-	$C_6H_{10}O$ 98.14	6728-26-3	146.5; 50[20]	4-01-00-03468 0.8491[20]	1.4480[20]
6997	3-Hexenal, (E)-	$C_6H_{10}O$ 98.14	69112-21-6	42-3[28]	4-01-00-03469 0.8455[22]	H_2O 1; eth 3; ace 3 1.4275[21]
6998	3-Hexenal, (Z)-	$C_6H_{10}O$ 98.14	6789-80-6	121	4-01-00-03469 0.8533[22]	1.4300[21]
6999	1-Hexene	C_6H_{12} 84.16	592-41-6 -139.7	63.4	4-01-00-00828 0.6731[20]	bz 4; eth 4; EtOH 4; peth 4 1.3837[20]

No.	Name Synonym	Mol. Form. Mol. Wt.	CAS RN mp/°C	Merck No. bp/°C	Bell. Ref. den/g cm^{-3}	Solubility n_D
7000	2-Hexene, (E)- trans-2-Hexene	C$_6$H$_{12}$ 84.16	4050-45-7 -133	67.9	4-01-00-00834 0.6732[25]	H$_2$O 1; EtOH 3; eth 3; bz 3 1.3936[20]
7001	2-Hexene, (Z)- cis-2-Hexene	C$_6$H$_{12}$ 84.16	7688-21-3 -141.1	68.8	4-01-00-00833 0.6869[20]	H$_2$O 1; EtOH 3; eth 3; bz 3 1.3979[20]
7002	3-Hexene, (E)-	C$_6$H$_{12}$ 84.16	13269-52-8 -115.4	67.1	4-01-00-00837 0.6772[20]	H$_2$O 1; EtOH 3; eth 3; bz 3 1.3943[20]
7003	3-Hexene, (Z)-	C$_6$H$_{12}$ 84.16	7642-09-3 -137.8	66.4	4-01-00-00837 0.6796[20]	H$_2$O 1; EtOH 3; eth 3; bz 3 1.3947[20]
7004	1-Hexene, 1-chloro- 1-Chloro-1-hexene	C$_6$H$_{11}$Cl 118.61	22922-67-4	121	4-01-00-00831 0.8872[22]	ace 4; bz 4; eth 4; chl 4 1.4300[22]
7005	1-Hexene, 2-chloro- 2-Chloro-1-hexene	C$_6$H$_{11}$Cl 118.61	10124-73-9	113	3-01-00-00803 0.8886[25]	ace 4; bz 4; chl 4 1.4278[25]
7006	1-Hexene, 5-chloro- 5-Chloro-1-hexene	C$_6$H$_{11}$Cl 118.61	927-54-8	120.7	4-01-00-00832 0.8891[25]	ace 4; bz 4; eth 4; chl 4 1.4305[20]
7007	2-Hexene, 4-chloro- 4-Chloro-2-hexene	C$_6$H$_{11}$Cl 118.61	6734-98-1	123	4-01-00-00835 0.8934[20]	ace 4; bz 4; eth 4; chl 4 1.4400[20]
7008	3-Hexene, 1-chloro- 3-Hexen-1-yl chloride	C$_6$H$_{11}$Cl 118.61	62706-16-5	61[60]	4-01-00-00838 0.900[24]	ace 4; bz 4; eth 4; chl 4 1.435[24]
7009	3-Hexene, 3-chloro-, (Z)- 3-Hexene, 3-chloro (cis)	C$_6$H$_{11}$Cl 118.61	17226-34-5	117	4-01-00-00838 0.9009[20]	ace 4; bz 4; eth 4; chl 4 1.4360[20]
7010	3-Hexene, 1-chloro-4-ethyl-	C$_8$H$_{15}$Cl 146.66	82507-04-8	173	2-01-00-00201 0.9102[20]	bz 4; chl 4 1.4524[20]
7011	1-Hexene, 1,2-dichloro-, (E)-	C$_6$H$_{10}$Cl$_2$ 153.05	59697-51-7	63-5[22]	3-01-00-00803 1.1167[25]	bz 4; chl 4 1.4576[25]
7012	1-Hexene, 1,2-dichloro-, (Z)-	C$_6$H$_{10}$Cl$_2$ 153.05	59697-55-1	88[30]	3-01-00-00803 1.0812[25]	bz 4; chl 4 1.4631[25]
7013	1-Hexene, 2,3-dimethyl- 2,3-Dimethyl-1-hexene	C$_8$H$_{16}$ 112.22	16746-86-4	110.5	3-01-00-00845 0.7172[25]	1.4113[20]
7014	1-Hexene, 2,4-dimethyl- 2,4-Dimethyl-1-hexene	C$_8$H$_{16}$ 112.22	16746-87-5	111.2	4-01-00-00887 0.716[25]	1.409[20]
7015	1-Hexene, 2,5-dimethyl- 2,5-Dimethyl-1-hexene	C$_8$H$_{16}$ 112.22	6975-92-4	111.6	3-01-00-00846 0.7129[25]	1.4105[20]
7016	1-Hexene, 3,3-dimethyl- 3,3-Dimethyl-1-hexene	C$_8$H$_{16}$ 112.22	3404-77-1	104	4-01-00-00889 0.7099[25]	1.4070[20]
7017	1-Hexene, 3,4-dimethyl- 3,4-Dimethyl-1-hexene	C$_8$H$_{16}$ 112.22	16745-94-1	112	4-01-00-00889 0.720[25]	1.413[20]
7018	1-Hexene, 3,5-dimethyl-	C$_8$H$_{16}$ 112.22	7423-69-0	104	3-01-00-00846 0.712[25]	1.404[20]
7019	1-Hexene, 4,4-dimethyl- 4,4-Dimethyl-1-hexene	C$_8$H$_{16}$ 112.22	1647-08-1	107.2	3-01-00-00847 0.7157[25]	1.4102[20]
7020	1-Hexene, 4,5-dimethyl- 4,5-Dimethyl-1-hexene	C$_8$H$_{16}$ 112.22	16106-59-5	109	4-01-00-00887 0.724[25]	1.414[20]
7021	1-Hexene, 5,5-dimethyl- 5,5-Dimethyl-1-hexene	C$_8$H$_{16}$ 112.22	7116-86-1	104	4-01-00-00886 0.705[25]	1.4049[20]
7022	2-Hexene, 2,3-dimethyl- 2,3-Dimethyl-2-hexene	C$_8$H$_{16}$ 112.22	7145-20-2 -115.1	121.8	4-01-00-00887 0.7366[25]	1.4268[20]
7023	2-Hexene, 2,4-dimethyl- 2,4-Dimethyl-2-hexene	C$_8$H$_{16}$ 112.22	14255-23-3	110.6	4-01-00-00887 0.7171[25]	1.4118[20]
7024	2-Hexene, 2,5-dimethyl- 2,5-Dimethyl-2-hexene	C$_8$H$_{16}$ 112.22	3404-78-2	112.2	4-01-00-00888 0.7182[20]	1.4140[20]
7025	2-Hexene, 4,5-dimethyl- 2,3-Dimethyl-4-hexene	C$_8$H$_{16}$ 112.22	73548-71-7	110	3-01-00-00846 0.721[25]	1.413[20]
7026	2-Hexene, 5,5-dimethyl-, (E)-	C$_8$H$_{16}$ 112.22	39782-43-9	104.1	4-01-00-00886 0.7023[25]	1.4055[20]
7027	2-Hexene, 5,5-dimethyl-, (Z)- cis-5,5-Dimethyl-2-hexene	C$_8$H$_{16}$ 112.22	39761-61-0	106.9	4-01-00-00886 0.7125[25]	1.4113[20]
7028	3-Hexene, 2,2-dimethyl-, (E)-	C$_8$H$_{16}$ 112.22	690-93-7	100.8	4-01-00-00885 0.6995[25]	1.4063[20]
7029	3-Hexene, 2,2-dimethyl-, (Z)-	C$_8$H$_{16}$ 112.22	690-92-6 -137.4	105.5	4-01-00-00885 0.7086[25]	1.4099[20]
7030	3-Hexene, 2,4-dimethyl-, (E)- 2,4-Dimethyl-trans-3-hexene	C$_8$H$_{16}$ 112.22	61847-78-7	107.6	4-01-00-00887 0.7101[25]	1.4126[20]
7031	3-Hexene, 2,4-dimethyl-, (Z)- 3-Hexene, 2,4-dimethyl (cis)	C$_8$H$_{16}$ 112.22	37549-89-6	109	4-01-00-00887 0.7135[25]	1.4140[20]
7032	1-Hexene, 1,1,2,3,3,4,4,5,5,6,6,6-dodecafluoro- Perfluoro-1-hexene	C$_6$F$_{12}$ 300.05	755-25-9	57.0	4-01-00-00831	chl 4
7033	1-Hexene, 3-ethyl- 3-Propyl-1-pentene	C$_8$H$_{16}$ 112.22	3404-58-8	110.3	4-01-00-00885 0.711[25]	1.407[20]
7034	3-Hexene, 3-ethyl-	C$_8$H$_{16}$ 112.22	16789-51-8	116	3-01-00-00845 0.725[25]	1.418[20]
7035	3-Hexene, 1,2,3,4,5,6-hexachloro- 1,2,3,4,5,6-Hexachloro-3-hexene	C$_6$H$_6$Cl$_6$ 290.83	1725-74-2 58.5	112[2]	3-01-00-00808	MeOH 4; chl 4
7036	1-Hexene, 2-methyl- 2-Methyl-1-hexene	C$_7$H$_{14}$ 98.19	6094-02-6 -102.8	92	4-01-00-00863 0.7000[20]	1.4035[20]
7037	1-Hexene, 3-methyl- 3-Methyl-1-hexene	C$_7$H$_{14}$ 98.19	3404-61-3	83.9	4-01-00-00865 0.6871[25]	1.3965[20]
7038	1-Hexene, 4-methyl- 4-Methyl-1-hexene	C$_7$H$_{14}$ 98.19	3769-23-1 -141.5	86.7	3-01-00-00830 0.6942[25]	1.4000[20]
7039	1-Hexene, 5-methyl- 5-Methyl-1-hexene	C$_7$H$_{14}$ 98.19	3524-73-0	85.3	4-01-00-00865 0.6877[25]	1.3967[20]

No.	Name / Synonym	Mol. Form. / Mol. Wt.	CAS RN / mp/°C	Merck No. / bp/°C	Beil. Ref. / den/g cm^{-3}	Solubility / n_D
7040	2-Hexene, 2-methyl- / 2-Methyl-2-hexene	C_7H_{14} / 98.19	2738-19-4 / -130.4	95.4	4-01-00-00863 / 0.7038[25]	1.4106[20]
7041	2-Hexene, 3-methyl-, (Z)-	C_7H_{14} / 98.19	10574-36-4 / -118.5	95.6	4-01-00-00866 / 0.712[20]	1.4126[20]
7042	2-Hexene, 4-methyl-, (E)-	C_7H_{14} / 98.19	3683-22-5 / -125.7	87.6	4-01-00-00867 / 0.6925[25]	1.4025[20]
7043	2-Hexene, 4-methyl-, (Z)-	C_7H_{14} / 98.19	3683-19-0	86.3	4-01-00-00867 / 0.6952[25]	1.4026[20]
7044	2-Hexene, 5-methyl-, (E)-	C_7H_{14} / 98.19	7385-82-2 / -124.3	88.1	4-01-00-00865 / 0.6883[25]	1.4006[20]
7045	2-Hexene, 5-methyl-, (Z)-	C_7H_{14} / 98.19	13151-17-2	89.5	4-01-00-00865 / 0.697[25]	1.404[20]
7046	3-Hexene, 2-methyl-, (E)-	C_7H_{14} / 98.19	692-24-0 / -141.6	85.9	4-01-00-00864 / 0.6853[25]	1.4001[20]
7047	3-Hexene, 2-methyl-, (Z)-	C_7H_{14} / 98.19	15840-60-5	86	4-01-00-00864 / 0.690[25]	1.401[20]
7048	1-Hexene, 1,1,2-trichloro-	$C_8H_9Cl_3$ / 187.50	53977-99-4	90-3[10]	3-01-00-00803 / 1.125[25]	eth 4 / 1.4760[25]
7049	2-Hexenoic acid	$C_8H_{10}O_2$ / 114.14	1191-04-4 / 36.5	216.5	4-02-00-01563 / 0.965[20]	eth 4 / 1.4460[40]
7050	3-Hexenoic acid / Hydrosorbic acid	$C_8H_{10}O_2$ / 114.14	4219-24-3 / 12	208	4-02-00-01566 / 0.9640[23]	1.4935[20]
7051	5-Hexenoic acid / 5-Hexanoic acid	$C_8H_{10}O_2$ / 114.14	1577-22-6 / -37	203	4-02-00-01562 / 0.9610[20]	eth 3; EtOH 4 / 1.4343[20]
7052	4-Hexenoic acid, 6-(1,3-dihydro-4-hydroxy-6-methoxy-7-methyl-3-oxo-5-isobenzofuranyl)-4-methyl-, (E)- / Mycophenolic acid	$C_{17}H_{20}O_6$ / 320.34	24280-93-1 / 141	6238	4-18-00-06513	H_2O 1; EtOH 4; eth 4; bz 2
7053	2-Hexenoic acid, ethyl ester / Ethyl 2-hexenoate	$C_8H_{14}O_2$ / 142.20	1552-67-6	174.5; 80[22]	3-02-00-01318 / 0.8986[20]	1.4348[20]
7054	3-Hexenoic acid, ethyl ester / Hydrosorbic acid, ethyl ester	$C_8H_{14}O_2$ / 142.20	2396-83-0	166.5	4-02-00-01567 / 0.8957[20]	1.4255[20]
7055	2-Hexenoic acid, 2-methyl-, (E)-	$C_7H_{12}O_2$ / 128.17	97961-66-5	205; 118[11]	4-02-00-01581 / 0.9627[20]	eth 3; ace 3 / 1.4601[20]
7056	2-Hexenoic acid, 4-methyl- / 4-Methyl-2-hexenoic acid	$C_7H_{12}O_2$ / 128.17	37549-83-0	125[13]	3-02-00-01336 / 0.9441[20]	1.4526[20]
7057	2-Hexenoic acid, 5-methyl- / 5-Methyl-2-hexenoic acid	$C_7H_{12}O_2$ / 128.17	41653-96-7 / 16.5	226.5	2-02-00-00411 / 0.942[20]	1.4425[17]
7058	3-Hexenoic acid, 2-methyl- / Hydrosorbic acid, 2-methyl-	$C_7H_{12}O_2$ / 128.17	73513-50-5	122[24]	4-02-00-01582 / 0.9353[20]	1.4379[20]
7059	3-Hexenoic acid, 3-methyl-	$C_7H_{12}O_2$ / 128.17	35205-71-1	113[10]	3-02-00-01335 / 0.9549[20]	1.4469[20]
7060	3-Hexenoic acid, 4-methyl- / 4-Ethyl-3-pentenoic acid	$C_7H_{12}O_2$ / 128.17	55665-79-7	118[12]	3-02-00-01336 / 0.9644[16]	1.4512[16]
7061	4-Hexenoic acid, 5-methyl-	$C_7H_{12}O_2$ / 128.17	5636-65-7 / -28	217	4-02-00-01581 / 0.9862[0]	1.4504
7062	3-Hexenoic acid, methyl ester / Methyl 3-hexenoate	$C_7H_{12}O_2$ / 128.17	2396-78-3	67-8[34]	4-02-00-01566 / 0.9132[25]	1.4240[23]
7063	2-Hexenoic acid, 2-methyl-, ethyl ester	$C_9H_{16}O_2$ / 156.22	26311-33-1	72[10]	3-02-00-01333 / 0.9031[20]	1.4407[20]
7064	3-Hexenoic acid, 2-methyl-, ethyl ester	$C_9H_{16}O_2$ / 156.22	21994-78-5	78[25]	3-02-00-01334 / 0.8778[20]	1.4237[20]
7065	3-Hexenoic acid, 3-methyl-, ethyl ester	$C_9H_{16}O_2$ / 156.22	21994-77-4	84-5[26]	3-02-00-01335 / 0.8961[20]	1.4309[20]
7066	1-Hexen-3-ol	$C_6H_{12}O$ / 100.16	4798-44-1	134	4-01-00-02136 / 0.834[22]	ace 4; eth 4; EtOH 4 / 1.4297[18]
7067	4-Hexen-2-ol	$C_6H_{12}O$ / 100.16	52387-50-5	137.5	4-01-00-02139 / 0.8405[18]	1.4392[20]
7068	5-Hexen-2-ol	$C_6H_{12}O$ / 100.16	626-94-8	139	3-01-00-01926 / 0.842[16]	H_2O 2
7069	2-Hexen-1-ol, (E)-	$C_6H_{12}O$ / 100.16	928-95-0	157	4-01-00-02138 / 0.8490[16]	1.4340[20]
7070	2-Hexen-1-ol, (Z)-	$C_6H_{12}O$ / 100.16	928-94-9	157	4-01-00-02138 / 0.8472[20]	H_2O 3; EtOH 4; eth 3; ace 3 / 1.4397[20]
7071	3-Hexen-1-ol, (Z)-	$C_6H_{12}O$ / 100.16	928-96-1	156.5	4-01-00-02141 / 0.8478[22]	H_2O 3; EtOH 4; eth 3 / 1.4380[20]
7072	4-Hexen-1-ol, (E)-	$C_6H_{12}O$ / 100.16	928-92-7	159	4-01-00-02140 / 0.8513[20]	1.4402[20]
7073	1-Hexen-3-ol, 3,5-dimethyl-	$C_8H_{16}O$ / 128.21	3329-48-4	146.5	4-01-00-02173 / 0.8382[20]	1.4342[20]
7074	4-Hexen-3-ol, 2,2-dimethyl- / 2-Hexen-4-ol, 5,5-dimethyl	$C_8H_{16}O$ / 128.21	37674-67-2	75-6[50]	4-01-00-02171 / 0.8308[25]	1.4369[25]
7075	4-Hexen-3-ol, 2,5-dimethyl- / 2,5-Dimethyl-4-hexen-3-ol	$C_8H_{16}O$ / 128.21	60703-31-3	162	2-01-00-00491 / 0.8444[20]	1.4449[20]
7076	4-Hexen-3-ol, 3,5-dimethyl- / 2-Hexen-4-ol, 2,4-dimethyl	$C_8H_{16}O$ / 128.21	1569-43-3	48-50[3-6]	4-01-00-02172 / 0.8600[17]	1.4460[17]
7077	1-Hexen-3-ol, 5-methyl-	$C_7H_{14}O$ / 114.19	4798-46-3	125; 58[15]	3-01-00-01939 / 0.8306[15]	1.4263[23]
7078	2-Hexen-1-ol, 5-methyl- / 5-Methyl-2-hexenol	$C_7H_{14}O$ / 114.19	77053-92-0	169	4-01-00-02158 / 0.8355[20]	1.4390[20]

No.	Name / Synonym	Mol. Form. / Mol. Wt.	CAS RN / mp/°C	Merck No. / bp/°C	Beil. Ref. / den/g cm^{-3}	Solubility / n_D
7079	3-Hexen-2-ol, 2-methyl-	$C_7H_{14}O$ 114.19	18812-62-9	49[11]	4-01-00-02157 0.8536[18]	1.443[18]
7080	3-Hexen-2-ol, 3-methyl-	$C_7H_{14}O$ 114.19	110383-31-8	89[55]	0-01-00-00447 0.8678[9]	1.4487[9]
7081	4-Hexen-3-ol, 2-methyl-	$C_7H_{14}O$ 114.19	4798-60-1	139.5	3-01-00-01939 0.8426[20]	1.438[20]
7082	4-Hexen-3-ol, 3-methyl-	$C_7H_{14}O$ 114.19	60111-14-0	72-3[60]	4-01-00-02160 0.8471[25]	1.4268[16]
7083	5-Hexen-2-ol, 2-methyl- Dimethyl(vinylethyl)carbinol	$C_7H_{14}O$ 114.19	16744-89-1 -22	142.5	4-01-00-02158 0.8397[15]	1.4349[17]
7084	5-Hexen-3-ol, 3-methyl- Methylethylallylcarbinol	$C_7H_{14}O$ 114.19	1569-44-4	140; 61[35]	4-01-00-02160 0.8432[20]	1.4370[20]
7085	5-Hexen-3-ol, 4-methyl- 4-Methyl-5-hexen-3-ol	$C_7H_{14}O$ 114.19	1838-77-3	140.5	3-01-00-01940 0.8452[22]	1.4365[22]
7086	3-Hexen-2-one 1-Butenyl methyl ketone	$C_6H_{10}O$ 98.14	763-93-9	140	4-01-00-03468 0.8655[20]	ace 4; eth 4; EtOH 4 1.4418[20]
7087	4-Hexen-3-one 2-Hexen-5-one	$C_6H_{10}O$ 98.14	25659-22-7	127	3-01-00-02994 0.8520[16]	1.4300[16]
7088	4-Hexen-3-one	$C_6H_{10}O$ 98.14	2497-21-4	138.5	4-01-00-03468 0.8559[20]	EtOH 3; eth 3; ace 4 1.4388[20]
7089	5-Hexen-2-one	$C_6H_{10}O$ 98.14	109-49-9	129.5	4-01-00-03467 0.833[27]	1.4178[27]
7090	3-Hexen-2-one, (E)-	$C_6H_{10}O$ 98.14	4376-23-2	140	4-01-00-03469 0.8665[20]	EtOH 3; eth 3; ace 4 1.4418[20]
7091	3-Hexen-2-one, 3,4-dimethyl-	$C_8H_{14}O$ 126.20	1635-02-5	158; 65[20]	4-01-00-03500 0.8585[20]	1.4418[20]
7092	4-Hexen-2-one, 3,4-dimethyl- 2-Hexen-5-one, 3,4-dimethyl	$C_8H_{14}O$ 126.20	53252-21-4	154	4-01-00-03500 0.8538[19]	1.4377[19]
7093	1-Hexen-3-one, 5-methyl- Isobutyl vinyl ketone	$C_7H_{12}O$ 112.17	2177-32-4	41[22]; 32[10]	4-01-00-03483 0.8400[15]	1.4293[15]
7094	3-Hexen-2-one, 5-methyl- 2-Oxo-5-methylhex-3-ene	$C_7H_{12}O$ 112.17	5166-53-0	77.5[50]	3-01-00-03005 0.8407[22]	1.4395[22]
7095	4-Hexen-2-one, 5-methyl-	$C_7H_{12}O$ 112.17	28332-44-7	76[50]	4-01-00-03482 0.9012[21]	1.4385[20]
7096	4-Hexen-3-one, 2-methyl-	$C_7H_{12}O$ 112.17	53252-19-0	148	4-01-00-03483 0.843[20]	EtOH 3; eth 3; ace 4 1.4345[20]
7097	5-Hexen-2-one, 3-methyl- 1-Methyl-1-allyl acetone	$C_7H_{12}O$ 112.17	2550-22-3	137	4-01-00-03484 0.845[15]	1.4215[25]
7098	5-Hexen-2-one, 4-methyl- 4-Methyl-5-hexen-2-one	$C_7H_{12}O$ 112.17	61675-14-7	137.5	3-01-00-03006 0.8273[25]	1.4193[22]
7099	5-Hexen-2-one, 5-methyl- 5-Hexene-2-one,5-methyl	$C_7H_{12}O$ 112.17	3240-09-3	150	4-01-00-03482 0.8460[20]	ace 4; eth 4; EtOH 4 1.4348[20]
7100	1-Hexen-3-one, 5-methyl-1-phenyl-	$C_{13}H_{16}O$ 188.27	2892-18-4 43	154[25]	4-07-00-01058 0.9509[46]	H_2O 2; EtOH 3; bz 3; chl 3 1.5523[25]
7101	5-Hexen-2-one, 4-phenyl-	$C_{12}H_{14}O$ 174.24	50552-30-2	85-6[1]	3-07-00-01460 0.9848[25]	1.5193[25]
7102	5-Hexen-2-one, 6-phenyl- 6-Phenyl-5-hexen-2-one	$C_{12}H_{14}O$ 174.24	69371-59-1	154[10]; 98[0.3]	3-07-00-01455 0.9967[25]	1.5458[25]
7103	1-Hexen-3-yne Ethylvinylacetylene	C_6H_8 80.13	13721-54-5	85	4-01-00-01091 0.7492[20]	bz 4; eth 4; peth 4; chl 4 1.4522[20]
7104	1-Hexen-4-yne	C_6H_8 80.13	5009-11-0	87	4-01-00-01092 0.767[14]	1.446[14]
7105	1-Hexen-5-yne Hex-5-en-1-yne	C_6H_8 80.13	14548-31-3	70	4-01-00-01092 0.7650[20]	bz 4; eth 4; peth 4; chl 4 1.4318[20]
7106	2-Hexen-4-yne 2-Hexene-4-yne	C_6H_8 80.13	14092-20-7	88.5	4-01-00-01092 0.7710[20]	1.4918[20]
7107	1-Hexen-3-yne, 5-chloro-5-methyl- Dimethylvinylethynylchloromethane	C_7H_9Cl 128.60	819-44-3	48[28]	4-01-00-01098 0.9375[15]	ace 4; bz 4; eth 4; EtOH 4 1.4778[20]
7108	3-Hexen-1-yne, 3-propyl-	C_9H_{14} 122.21	688-52-8	136	3-01-00-01049 0.7799[25]	bz 4; peth 4; chl 4 1.4432[25]
7109	Hexitol, 2,5-anhydro-3,4-dideoxy- Tetrahydro-2,5-furandimethanol	$C_6H_{12}O_3$ 132.16	104-80-3 <-50	9145 265	4-17-00-02052 1.154[20]	H_2O 4; ace 4; bz 4; EtOH 4
7110	D-arabino-Hexose, 2-deoxy- 2-Deoxy-D-glucose	$C_6H_{12}O_5$ 164.16	154-17-6 146.5	2886	4-01-00-04282	
7111	D-ribo-Hexose, 2,6-dideoxy- Digitoxose	$C_6H_{12}O_4$ 148.16	527-52-6 112	3147	4-01-00-04191	H_2O 4; ace 4; py 3; AcOEt 3
7112	ribo-Hexose, 2,6-dideoxy-3-O-methyl- Cymarose	$C_7H_{14}O_4$ 162.19	579-04-4 101	2769	4-01-00-04193	H_2O 4; ace 4; EtOH 4
7113	3-Hexynal	C_6H_8O 96.13	89533-67-5	42-3[17]	4-01-00-03544 0.9036[23]	1.4498[22]
7114	1-Hexyne Butylacetylene	C_6H_{10} 82.15	693-02-7 -131.9	71.3	4-01-00-01006 0.7155[20]	H_2O 1; EtOH 3; eth 3; bz 3 1.3989[20]
7115	2-Hexyne 1-Methyl-2-propylacetylene	C_6H_{10} 82.15	764-35-2 -89.6	84.5	4-01-00-01009 0.7315[20]	H_2O 1; EtOH 5; eth 5; bz 3 1.4138[20]
7116	3-Hexyne Diethylacetylene	C_6H_{10} 82.15	928-49-4 -103	81	4-01-00-01009 0.7231[20]	H_2O 1; EtOH 3; eth 3; bz 3 1.4115[20]
7117	3-Hexyne, 2,5-dichloro-2,5-dimethyl- 3-Hexyne, 2,5-dimethyl-2,5-dichloro	$C_8H_{12}Cl_2$ 179.09	2431-30-3 29	178; 62[14]	4-01-00-01042 1.0118[20]	chl 4
7118	3-Hexyne-2,5-diol	$C_6H_{10}O_2$ 114.14	3031-66-1	120[11]	3-01-00-02271 1.0180[20]	1.4691[20]

No.	Name / Synonym	Mol. Form. / Mol. Wt.	CAS RN / mp/°C	Merck No. / bp/°C	Beil. Ref. / den/g cm^{-3}	Solubility / n_D
7119	3-Hexyne-2,5-diol, 2,5-dimethyl-	$C_8H_{14}O_2$ / 142.20	142-30-3 / 95	205	4-01-00-02699 / 0.947[20]	H_2O 3; EtOH 4; eth 4; ace 4
7120	1-Hexyne, 5-methyl- / 5-Methyl-1-hexyne	C_7H_{12} / 96.17	2203-80-7 / -125	92	3-01-00-01000 / 0.7274[20]	H_2O 1; EtOH 3; eth 3; bz 3 / 1.4059[-20]
7121	2-Hexyne, 5-methyl-	C_7H_{12} / 96.17	53566-37-3 / -92.9	102.5	4-01-00-01030 / 0.7378[20]	H_2O 1; eth 3; ace 3; bz 3 / 1.4176[20]
7122	3-Hexyne, 2-methyl- / 1-Ethyl-2-isopropylacetylene	C_7H_{12} / 96.17	36566-80-0 / -116.7	95.2	4-01-00-01029 / 0.7263[20]	bz 4; eth 4; chl 4; peth 4 / 1.4120[20]
7123	2-Hexynoic acid / Propiolic acid, propyl-	$C_6H_8O_2$ / 112.13	764-33-0 / 27	126[24]	4-02-00-01699 / 0.9820[20]	eth 4; EtOH 4; lig 4
7124	1-Hexyn-3-ol	$C_6H_{10}O$ / 98.14	105-31-7 / -80	142	4-01-00-02234 / 0.8704[20]	ctc 3 / 1.4340[25]
7125	2-Hexyn-1-ol	$C_6H_{10}O$ / 98.14	764-60-3	100[53]; 80[21]	4-01-00-02235 / 0.8472[20]	1.4341[20]
7126	3-Hexyn-1-ol / 3-Hexynol	$C_6H_{10}O$ / 98.14	1002-28-4	163; 65[12]	4-01-00-02236 / 0.8982[20]	1.4530[20]
7127	5-Hexyn-2-ol	$C_6H_{10}O$ / 98.14	23470-12-4	152	4-01-00-02235 / 0.899[20]	1.4481[20]
7128	1-Hexyn-3-ol, 3-methyl- / Ethynylmethylpropylcarbinol	$C_7H_{12}O$ / 112.17	4339-05-3	137	4-01-00-02252 / 0.8620[20]	H_2O 4; eth 4; EtOH 4 / 1.4338[20]
7129	3-Hexyn-2-ol, 2-methyl- / 2-Methyl-3-hexyn-2-ol	$C_7H_{12}O$ / 112.17	5075-33-2 / -6	146	4-01-00-02250 / 0.962[25]	eth 4; EtOH 4 / 1.4392[25]
7130	3-Hexyn-2-one / 3-Hexyne-2-one	C_6H_8O / 96.13	1679-36-3	77-8[75]	4-01-00-03544 / 0.8804[20]	1.4400[20]
7131	5-Hexyn-2-one	C_6H_8O / 96.13	2550-28-9	149	4-01-00-03543 / 0.9065[20]	1.4366[20]
7132	L-Histidine / Glyoxaline-5-alanine	$C_6H_9N_3O_2$ / 155.16	71-00-1 / 287 dec	4642	4-25-00-04343	H_2O 3; EtOH 2; eth 1; ace 1
7133	L-Histidine, N-β-alanyl- / Carnosine	$C_9H_{14}N_4O_3$ / 226.24	305-84-0 / 260	1857	4-25-00-04381	H_2O 4
7134	Homoserine / (S)-2-Amino-4-hydroxybutanoic acid	$C_4H_9NO_3$ / 119.12	672-15-1 / 203 dec	4661	4-04-00-03187	H_2O 4; EtOH 2; eth 1; bz 1
7135	L-Homoserine, O-[(aminoiminomethyl)amino]- / Canavanine	$C_5H_{12}N_4O_3$ / 176.18	543-38-4	1745		H_2O 4
7136	Hydramethylnon / Tetrahydro-5,5-dimethyl-2-(1H)-pyrimidinone[3[4-(trifluoromethyl-styryl)-cinnamylidene- hydrazone	$C_{25}H_{24}F_6N_4$ / 494.48	67485-29-4 / 190	4684		
7137	Hydrazine, 1,2-bis(1-methylethyl)-	$C_6H_{16}N_2$ / 116.21	3711-34-0	125	4-04-00-03339 / 0.7894[20]	ace 4; bz 4; eth 4; EtOH 4 / 1.4173[20]
7138	Hydrazine, 1,2-bis(3-methylphenyl)-	$C_{14}H_{16}N_2$ / 212.29	621-26-1 / 38	224	4-15-00-01220	H_2O 1; EtOH 3; eth 3; bz 3
7139	Hydrazine, 1,2-bis(4-methylphenyl)-	$C_{14}H_{16}N_2$ / 212.29	637-47-8 / 135		4-15-00-01230 / 0.957[20]	bz 4; eth 4; EtOH 4
7140	Hydrazine, 1,2-bis(2-methylpropyl)-	$C_8H_{20}N_2$ / 144.26	3711-37-3	171; 63[18]	2-04-00-00962 / 0.8002[20]	ace 4; bz 4; eth 4; EtOH 4 / 1.4276
7141	Hydrazine, (4-bromophenyl)- / (p-Bromophenyl)hydrazine	$C_6H_7BrN_2$ / 187.04	589-21-9 / 108	1417	4-15-00-00282	eth 4; EtOH 4; lig 4
7142	Hydrazinecarbothioamide / Thiosemicarbazide	CH_5N_3S / 91.14	79-19-6 / 183	9292	4-03-00-00374	H_2O 4; EtOH 4
7143	Hydrazinecarboxamide, monohydrochloride / Semicarbazide hydrochloride	CH_6ClN_3O / 111.53	563-41-7 / 175-7 dec	8396	4-03-00-00177	H_2O 4
7144	Hydrazinecarboxamide, 2-[(5-nitro-2-furanyl)methylene]- / Nitrofurazone	$C_6H_6N_4O_4$ / 198.14	59-87-0 / 236-40 dec	6521	5-17-09-00335	H_2O 1; EtOH 2; eth 1; DMSO 2
7145	Hydrazinecarboxamide, 2-phenyl- / Phenicarbazide	$C_7H_9N_3O$ / 151.17	103-03-7 / 172	7193	4-15-00-00180	H_2O 2; EtOH 3; eth 2; ace 3
7146	Hydrazinecarboxamide, N-phenyl- / 4-Phenylsemicarbazide	$C_7H_9N_3O$ / 151.17	537-47-3 / 128	7283	3-12-00-00822	H_2O 2; EtOH 4; eth 1; chl 4
7147	Hydrazinecarboximidamide / Aminoguanidine	CH_6N_4 / 74.09	79-17-4 / dec	453	4-03-00-00236	H_2O 4; EtOH 4
7148	Hydrazinecarboxylic acid, ethyl ester	$C_3H_8N_2O_2$ / 104.11	4114-31-2 / 46		198 dec; 93[9]	4-03-00-00174	EtOH 3; eth 3; chl 2
7149	Hydrazinecarboxylic acid, methyl ester	$C_2H_6N_2O_2$ / 90.08	6294-89-9 / 73	108[12]	2-03-00-00078	H_2O 3; EtOH 3; bz 2; peth 1
7150	1,1-Hydrazinedicarboxylic acid, diethyl ester	$C_6H_{12}N_2O_4$ / 176.17	5311-96-6 / 30.5	138[12]	2-03-00-00079	EtOH 4
7151	1,2-Hydrazinedicarboxylic acid, diethyl ester	$C_6H_{12}N_2O_4$ / 176.17	4114-28-7 / 135	250 dec	4-03-00-00175 / 1.324[8]	eth 4; EtOH 4
7152	Hydrazine, 1,1-diethyl- / 1,1-Diethylhydrazine	$C_4H_{12}N_2$ / 88.15	616-40-0	99	4-04-00-03335 / 0.8804[20]	bz 4; eth 4; EtOH 4; chl 4 / 1.4214[20]
7153	Hydrazine, 1,2-diethyl- / 1,2-Diethylhydrazine	$C_4H_{12}N_2$ / 88.15	1615-80-1	85.5	4-04-00-03336 / 0.797[26]	bz 4; eth 4; EtOH 4 / 1.4204[20]
7154	Hydrazine, 1,1-dimethyl- / 1,1-Dimethylhydrazine	$C_2H_8N_2$ / 60.10	57-14-7 / -58	3236 / 63.9	4-04-00-03322 / 0.791[22]	H_2O 4; EtOH 4; eth 4; MeOH 4 / 1.4075[22]
7155	Hydrazine, 1,2-dimethyl- / 1,2-Dimethylhydrazine	$C_2H_8N_2$ / 60.10	540-73-8 / 81	3237	4-04-00-03323 / 0.8274[20]	H_2O 5; EtOH 5; eth 5 / 1.4209[20]
7156	Hydrazine, (2,4-dinitrophenyl)- / (2,4-Dinitrophenyl)hydrazine	$C_6H_6N_4O_4$ / 198.14	119-26-6 / 194	3277	4-15-00-00380	H_2O 1; EtOH 3; eth 2; bz 2

No.	Name Synonym	Mol. Form. Mol. Wt.	CAS RN mp/°C	Merck No. bp/°C	Beil. Ref. den/g cm^{-3}	Solubility n_D
7157	Hydrazine, 1,1-diphenyl- 1,1-Diphenylhydrazine	$C_{12}H_{12}N_2$ 184.24	530-50-7 50.5	3326 220[40]	4-15-00-00055 1.190[16]	bz 4; eth 4; EtOH 4; chl 4
7158	Hydrazine, 1,2-diphenyl- *N,N*'-Diphenylhydrazine	$C_{12}H_{12}N_2$ 184.24	122-66-7 131		4-15-00-00056 1.158[16]	EtOH 4; bz 2; DMSO-d$_6$ 2; HOAc 1
7159	Hydrazine, 1-ethyl-1-phenyl- 1-Phenyl-1-ethylhydrazine	$C_8H_{12}N_2$ 136.20	644-21-3 	 237	3-15-00-00074 1.018[21]	ace 4; bz 4; eth 4; EtOH 4 1.5711[21]
7160	Hydrazine, 1-ethyl-2-phenyl- 1-Ethyl-2-phenylhydrazine	$C_8H_{12}N_2$ 136.20	622-82-2 	241; 110[14]	4-15-00-00054 1.0150[20]	bz 4; eth 4; EtOH 4; chl 4 1.5676[20]
7161	Hydrazine, methyl- Methylhydrazine	CH_6N_2 46.07	60-34-4 -52.4	6001 87.5	4-04-00-03322 	H_2O 3; EtOH 5; eth 3; ctc 3 1.4325[20]
7162	Hydrazine, 1-(3-methylbutyl)-1-phenyl- Hydrazine, 1-isopentyl-1-phenyl	$C_{11}H_{18}N_2$ 178.28	636-10-2 	 260	0-15-00-00121 0.9588[15]	
7163	Hydrazine, 1-methyl-2-(3-methylphenyl)- Hydrazine, 1-methyl-2-(3-tolyl)	$C_8H_{12}N_2$ 136.20	116836-06-7 60		2-15-00-00229 1.0265[100]	bz 4; EtOH 4
7164	Hydrazine, (3-methylphenyl)-	$C_7H_{10}N_2$ 122.17	536-89-0 	 244 dec	4-15-00-01220 1.057[20]	H_2O 2; EtOH 4; eth 4; bz 4
7165	Hydrazine, 1-methyl-1-phenyl-	$C_7H_{10}N_2$ 122.17	618-40-6 	228; 131[35]	4-15-00-00053 1.0404[20]	H_2O 2; EtOH 5; eth 5; bz 5 1.5691[20]
7166	Hydrazine, 1-methyl-2-phenyl- 1-Methyl-2-phenylhydrazine	$C_7H_{10}N_2$ 122.17	622-36-6 	231; 112[14]	4-15-00-00054 1.0320[20]	bz 4; eth 4; EtOH 4; chl 4 1.5733[20]
7167	Hydrazine, 1-naphthalenyl-	$C_{10}H_{10}N_2$ 158.20	2243-55-2 117	203[20]	3-15-00-00728 	bz 4; eth 4; EtOH 4; chl 4
7168	Hydrazine, (4-nitrophenyl)- (4-Nitrophenyl)hydrazine	$C_6H_7N_3O_2$ 153.14	100-16-3 158 dec	6545 	4-15-00-00317 	H_2O 2; EtOH 3; eth 3; bz 3
7169	Hydrazine, phenyl- Phenylhydrazine	$C_6H_8N_2$ 108.14	100-63-0 19.6	7264 243.5	4-15-00-00050 1.0986[20]	H_2O 3; EtOH 5; eth 5; ace 4 1.6084[10]
7170	Hydrazine, (1-phenylethyl)-	$C_8H_{12}N_2$ 136.20	65-64-5 	69[0.65]	4-15-00-01267 0.9672[25]	 1.5436[25]
7171	Hydrazine, phenyl-, hemihydrate	$C_{12}H_{18}N_4O$ 234.30	6152-31-4 24	120[12]	3-15-00-00067 1.0938[25]	 1.6081[20]
7172	Hydrazine, phenyl-, monohydrochloride	$C_6H_9ClN_2$ 144.60	59-88-1 243-6 dec	7265 sub		H_2O 4; EtOH 4
7173	Hydrazine, triphenyl- Triphenylhydrazine	$C_{18}H_{16}N_2$ 260.34	606-88-2 142 dec		4-15-00-00058 0.869[70]	EtOH 2; bz 2
7174	Hydrazobenzene, 3-methyl- Hydrazine, 1-phenyl-2-(3-tolyl)	$C_{13}H_{14}N_2$ 198.27	621-25-0 61		2-15-00-00229 1.0265[100]	bz 4; EtOH 4; lig 4
7175	Hydrocinnamic acid, β-ethyl-β-methyl- Valeric acid, 3-methyl-3-phenyl	$C_{12}H_{16}O_2$ 192.26	105401-59-0 284	174[14]	4-09-00-01905 1.050[25]	H_2O 3 1.5197[25]
7176	Hydrocyanic acid Hydrogen cyanide	CHN 27.03	74-90-8 -13.4	4722 26	4-02-00-00050 0.6876[20]	H_2O 5; EtOH 5; eth 5 1.2614[20]
7177	Hydroperoxide, cyclohexyl Cyclohexyl hydroperoxide	$C_6H_{12}O_2$ 116.16	766-07-4 -20	 42[0.1]	4-06-00-00053 1.019[20]	eth 4; EtOH 4; HOAc 4 1.4645[25]
7178	Hydroperoxide, 1,1-dimethylethyl *tert*-Butyl hydroperoxide	$C_4H_{10}O_2$ 90.12	75-91-2 6	1569 89 dec; 36[17]	4-01-00-01616 0.8960[20]	H_2O 3; EtOH 3; eth 3; ctc 3 1.4015[20]
7179	Hydroperoxide, ethyl Ethyl hydrogen peroxide	$C_2H_6O_2$ 62.07	3031-74-1 -100	 95	4-01-00-01323 0.9332[20]	H_2O 4; bz 4; eth 4; EtOH 4 1.3800[20]
7180	Hydroperoxide, methyl	CH_4O_2 48.04	3031-73-0 	38-40[65]	4-01-00-01249 1.9967[15]	H_2O 4; bz 4; eth 4; EtOH 4 1.3641[15]
7181	Hydroperoxide, 1-methyl-1-phenylethyl Cumene hydroperoxide	$C_9H_{12}O_2$ 152.19	80-15-9 		4-06-00-03221 	
7182	Hydroprene 2,4-Dodecadienoic acid, 3,7,11-trimethyl-, ethyl ester, (*E,E*)-	$C_{17}H_{30}O_2$ 266.42	41096-46-2 	174[19]	 0.8955[20]	
7183	Hypochlorous acid, 1,1-dimethylethyl ester *tert*-Butyl hypochlorite	C_4H_9ClO 108.57	507-40-4 	1570 77.5	4-01-00-01621 0.9583[18]	H_2O 1; eth 4; ace 3; bz 4 1.403[20]
7184	Hypochlorous acid, ethyl ester Ethyl hypochlorite	C_2H_5ClO 80.51	624-85-1 	 37	4-01-00-01324 1.013[-6]	H_2O 1; EtOH 3; eth 5; bz 5
7185	Hypophosphoric acid, tetraethyl ester	$C_8H_{20}O_6P_2$ 274.19	679-37-8 	116-7[2]	4-01-00-01358 1.1283[18]	bz 3; ctc 3 1.4284[20]
7186	Ibogamine, 12-methoxy- Ibogaine	$C_{20}H_{26}N_2O$ 310.44	83-74-9 148	4806 	5-23-12-00284 	chl 3
7187	Imazalil 1H-Imidazole, 1-[2-(2,4-dichlorophenyl)-2-(2- propenyloxy)ethyl]-	$C_{14}H_{14}Cl_2N_2O$ 297.18	35554-44-0 50	3537 dec	 1.243[23]	
7188	Imazapyr 2-(4-Isopropyl-4-methyl-5-oxo-2-imidazolin-2- yl)nicotinic acid	$C_{13}H_{15}N_3O_3$ 261.28	81334-34-1 171			
7189	Imazaquin (*RS*)-2-(4-Isopropyl-4-methyl-5-oxo-2- imidazolin-2-yl)quinoline-3-carboxylic acid	$C_{17}H_{17}N_3O_3$ 311.34	81335-37-7 221	4826 		
7190	Imazethapyr (*RS*)-5-Ethyl-2-(4-isopropyl-4-methyl-5-oxo-2- imidazolin-2-yl)nicotinic acid	$C_{15}H_{19}N_3O_3$ 289.33	81335-77-5 173			
7191	Imidazo[4,5-d]imidazole-2,5(1H,3H)-dione, tetrahydro- Acetyleneurea	$C_4H_6N_4O_2$ 142.12	496-46-8 300 dec	87 	2-26-00-00260 	H_2O 2; EtOH 1; eth 3; HCl 3
7192	1H-Imidazole 1,3-Diazole	$C_3H_4N_2$ 68.08	288-32-4 90.5	4828 257	5-23-04-00191 1.0303[101]	H_2O 4; EtOH 4; eth 3; ace 3 1.4801[101]
7193	1H-Imidazole, 1,1'-carbonylbis- *N,N*'-Carbonyldiimidazole	$C_7H_6N_4O$ 162.15	530-62-1 119	1825 	5-23-04-00245 	

No.	Name Synonym	Mol. Form. Mol. Wt.	CAS RN mp/°C	Merck No. bp/°C	Beil. Ref. den/g cm^{-3}	Solubility n_D
7194	1H-Imidazole-1-carboxylic acid, 2,3-dihydro-3-methyl-2-thioxo-, ethyl ester Carbimazole	$C_7H_{10}N_2O_2S$ 186.23	22232-54-8 123.5	1803	4-24-00-00064	ace 4; chl 4
7195	1H-Imidazole-4,5-dicarboxylic acid	$C_5H_4N_2O_4$ 156.10	570-22-9 290 dec		4-25-00-01051 1.749[25]	H_2O 2; EtOH 1; eth 1; bz 1
7196	1H-Imidazole, 1,2-diethyl- 1,2-Diethylimidazole	$C_7H_{12}N_2$ 124.19	51807-53-5 219		5-23-05-00165 0.9813[25]	H_2O 4; eth 4; EtOH 4
7197	1H-Imidazole, 4,5-dihydro-2-methyl- Lysidine	$C_4H_8N_2$ 84.12	534-26-9 107	5508 196.5	5-23-03-00385	H_2O 4; EtOH 4; eth 1; chl 3
7198	1H-Imidazole, 4,5-dihydro-2-(phenylmethyl)- 2-Imidazolidine, 2-benzyl	$C_{10}H_{12}N_2$ 160.22	59-98-3	9430	5-23-06-00488	
7199	1H-Imidazole, 4,5-dihydro-2,4,5-triphenyl-, cis-￼	$C_{21}H_{18}N_2$ 298.39	573-33-1 134.5	198 dec	5-23-10-00053	bz 4; eth 4; EtOH 4
7200	1H-Imidazole, 1,2-dimethyl-	$C_5H_8N_2$ 96.13	1739-84-0 206		5-23-05-00039 1.0051[11]	H_2O 4; eth 4; EtOH 4
7201	1H-Imidazole, 1,4-dimethyl- 1,4-Dimethylimidazole	$C_5H_8N_2$ 96.13	6338-45-0 25.8	199	5-23-05-00094 0.9960[16]	
7202	1H-Imidazole, 1,2-dimethyl-5-nitro- Dimetridazole	$C_5H_7N_3O_2$ 141.13	551-92-8 138.5	3255	5-23-05-00058	eth 4; EtOH 4
7203	1H-Imidazole-4-ethanamine Histamine	$C_5H_9N_3$ 111.15	51-45-6 83	4640 209[18]	4-25-00-02049	H_2O 3; EtOH 3; eth 2; chl 3
7204	1H-Imidazole-4-ethanaminium, α-carboxy-2,3-dihydro-N,N,N-trimethyl-2-thioxo-, hydroxide, in Ergothioneine	$C_9H_{15}N_3O_2S$ 229.30	497-30-3 290 dec	3611	4-25-00-04487	H_2O 4; EtOH 2; eth 1; ace 2
7205	1H-Imidazole-1-ethanol	$C_5H_8N_2O$ 112.13	1615-14-1 39	203[20]	5-23-04-00309	H_2O 4; EtOH 4
7206	1H-Imidazole-1-ethanol, 2-methyl-5-nitro- Metronidazole	$C_6H_9N_3O_3$ 171.16	443-48-1 160.5	6079	5-23-05-00063	
7207	1H-Imidazole, 1-ethyl-	$C_5H_8N_2$ 96.13	7098-07-9 208		5-23-04-00261 0.999[25]	H_2O 5
7208	1H-Imidazole, 1-methyl-	$C_4H_6N_2$ 82.11	616-47-7 -6	195.5	5-23-04-00256 1.0325[20]	H_2O 4; ace 4; eth 4; EtOH 4 1.4970[20]
7209	1H-Imidazole, 2-methyl-	$C_4H_6N_2$ 82.11	693-98-1 144	267	5-23-05-00035	H_2O 4; EtOH 4
7210	1H-Imidazole, 4-methyl-	$C_4H_6N_2$ 82.11	822-36-6 56	263	5-23-05-00089 1.0416[14]	H_2O 4; EtOH 4 1.5037[14]
7211	1H-Imidazole, 2-nitro- Azomycin	$C_3H_3N_3O_2$ 113.08	527-73-1 287 dec	934	5-23-04-00471	
7212	1H-Imidazole, 1-phenyl-	$C_9H_8N_2$ 144.18	7164-98-9 13	276	5-23-04-00267 1.1397[15]	H_2O 1; eth 4; ace 4; chl 4 1.6025[25]
7213	1H-Imidazole, 2-phenyl-	$C_9H_8N_2$ 144.18	670-96-2 149.3	340	5-23-07-00194	EtOH 4
7214	2H-Imidazole-2-thione, 1,3-dihydro-1-methyl- Methimazole	$C_4H_6N_2S$ 114.17	60-56-0 146	5892 280 dec	5-24-01-00305	H_2O 4; EtOH 3; eth 2; bz 2
7215	1H-Imidazole, 2,4,5-triphenyl-	$C_{21}H_{16}N_2$ 296.37	484-47-9 275	sub	5-23-10-00107	H_2O 1; EtOH 3; eth 3
7216	2,4-Imidazolidinedione Hydantoin	$C_3H_4N_2O_2$ 100.08	461-72-3 220	4678	5-24-05-00188	H_2O 4; EtOH 3; eth 2; alk 3
7217	2,4-Imidazolidinedione, 1,3-dibromo-5,5-dimethyl- Dibromantine	$C_5H_6Br_2N_2O_2$ 285.92	77-48-5 197-9 dec		5-24-05-00374	
7218	2,4-Imidazolidinedione, 5,5-dimethyl-	$C_5H_8N_2O_2$ 128.13	77-71-4 178	sub	5-24-05-00348	H_2O 4; EtOH 4; eth 4; ace 4
7219	2,4-Imidazolidinedione, 5,5-diphenyl- Phenytoin	$C_{15}H_{12}N_2O_2$ 252.27	57-41-0 286	7293	5-24-08-00376	H_2O 1; EtOH 3; eth 2; ace 3
7220	2,4-Imidazolidinedione, 5-(1-ethylpentyl)-3-[(trichloromethyl)thio]- Chlordantoin	$C_{11}H_{17}Cl_3N_2O_2S$ 347.69	5588-20-5	2080		CS_2 3
7221	2,4-Imidazolidinedione, 3-methyl-5-phenyl- 3-Methyl-5-phenylhydantoin	$C_{10}H_{10}N_2O_2$ 190.20	6846-11-3 164.5	6026	5-24-08-00005	chl 3
7222	2,4-Imidazolidinedione, 1-[[(5-nitro-2-furanyl)methylene]amino]- Nitrofurantoin	$C_8H_6N_4O_5$ 238.16	67-20-9 263	6520	5-24-05-00224	
7223	4-Imidazolidinehexanoic acid, 5-methyl-2-oxo-, (4R-cis)- Desthiobiotin	$C_{10}H_{18}N_2O_3$ 214.26	533-48-2 157	2914	4-25-00-01543	H_2O 3
7224	4-Imidazolidinehexanoic acid, 5-methyl-2-oxo-, methyl ester, (4R-cis)- Desthiobiotin, methyl ester	$C_{11}H_{20}N_2O_3$ 228.29	6020-51-5 69.5	194-7[0.03]	4-25-00-01544	EtOH 4
7225	2-Imidazolidinethione Ethylenethiourea	$C_3H_6N_2S$ 102.16	96-45-7 203	3759	5-24-01-00165	H_2O 4; EtOH 3; eth 1; bz 1
7226	Imidazolidinetrione Parabanic acid	$C_3H_2N_2O_3$ 114.06	120-89-8 244 dec	6970 sub 100	5-24-09-00009	H_2O 3; EtOH 4
7227	2-Imidazolidinone Ethyleneurea	$C_3H_6N_2O$ 86.09	120-93-4 131	4830	5-24-01-00022	H_2O 4; EtOH 4; eth 2; chl 2
7228	4-Imidazolidinone, 5,5-diphenyl- Doxenitoin	$C_{15}H_{14}N_2O$ 238.29	3254-93-1 183	3424	5-24-04-00145	
7229	4-Imidazolidinone, 5-(2-methylpropyl)-3-(2-propenyl)-2-thioxo- Albutoin	$C_{10}H_{16}N_2OS$ 212.32	830-89-7 210.5	210	4-24-00-01139	

No.	Name Synonym	Mol. Form. Mol. Wt.	CAS RN mp/°C	Merck No. bp/°C	Beil. Ref. den/g cm^{-3}	Solubility n_D
7230	4H-Imidazol-4-one, 2-amino-1,5-dihydro-1-methyl- Creatinine	$C_4H_7N_3O$ 113.12	60-27-5 300 dec	2571	4-25-00-03543	H_2O 3; EtOH 2; eth 1; ace 1
7231	Imidodicarbonic acid, diethyl ester	$C_6H_{11}NO_4$ 161.16	19617-44-8 50	226	3-03-00-00050	H_2O 4
7232	Imidodicarbonic diamide Biuret	$C_2H_5N_3O_2$ 103.08	108-19-0 190 dec	1319	4-03-00-00141	H_2O 2; EtOH 4; eth 1
7233	Imidodicarbonimidic diamide Biguanide	$C_2H_7N_5$ 101.11	56-03-1 136	1233 142 dec	4-03-00-00162	H_2O 4; EtOH 3; bz 1; chl 1
7234	Imidodicarbonimidic diamide, N-phenyl- Phenyl biguanide	$C_8H_{11}N_5$ 177.21	102-02-3 143	7247	4-12-00-00771	
7235	1H-Indazole 1H-Benzopyrazole	$C_7H_6N_2$ 118.14	271-44-3 148	4848 269	5-23-06-00156	H_2O 3; EtOH 3; eth 3
7236	1H-Indazole, 3-chloro- 3-Chloroindazole	$C_7H_5ClN_2$ 152.58	29110-74-5 149	sub	5-23-06-00173	H_2O 3; EtOH 4; eth 4; bz 4
7237	1H-Indazole, 1,5-dimethyl- 1,5-Dimethylindazole	$C_9H_{10}N_2$ 146.19	70127-93-4 62.5	260	2-23-00-00157	ace 4; eth 4; EtOH 4; MeOH 4
7238	2H-Indazole, 2,3-dimethyl- 2,3-Dimethylindazole	$C_9H_{10}N_2$ 146.19	50407-18-6 79.5	286	5-23-06-00317	eth 4; EtOH 4
7239	1H-Indazole, 1-methyl- 1-Methylindazole	$C_8H_8N_2$ 132.17	13436-48-1 60.5	231	5-23-06-00159 1.0315[99]	H_2O 1; EtOH 3; ace 3; chl 3
7240	1H-Indazole, 3-methyl- 3-Methylindazole	$C_8H_8N_2$ 132.17	3176-62-3 113	280.5	5-23-06-00312	H_2O 3; EtOH 3; eth 3; chl 3
7241	1H-Indazole, 5-methyl- 5-Methylindazole	$C_8H_8N_2$ 132.17	1776-37-0 117	294	5-23-06-00319	H_2O 3; EtOH 3; eth 3; chl 3
7242	2H-Indazole, 2-methyl- 2-Methyl-2H-indazole	$C_8H_8N_2$ 132.17	4838-00-0 56	261	5-23-06-00165 1.0450[99]	H_2O 1; EtOH 3; eth 3; ace 3
7243	1H-Inden-1-amine, 2,3-dihydro-, (±)- (RS)-1-Aminoindan	$C_9H_{11}N$ 133.19	61949-83-5 221; 96[8]		3-12-00-02798 1.038[15]	H_2O 2; eth 3; ace 3; bz 3 1.5613[20]
7244	1H-Inden-5-amine, 2,3-dihydro-	$C_9H_{11}N$ 133.19	24425-40-9 37.5	248; 131[15]	4-12-00-02929	H_2O 2; eth 3; ace 3; bz 3
7245	1H-Indene Indonaphthene	C_9H_8 116.16	95-13-6 -1.8	4851 182	4-05-00-01532 0.9960[25]	H_2O 1; EtOH 5; eth 5; ace 3 1.5768[20]
7246	1H-Indene-1-carboxylic acid 1-Indenecarboxylic acid	$C_{10}H_8O_2$ 160.17	5020-21-3 74	193-5[12]	3-09-00-03068	H_2O 2; EtOH 4; eth 2; bz 2
7247	1H-Indene, 1,2-dibromo-2,3-dihydro- 1,2-Dibromoindan	$C_9H_8Br_2$ 275.97	20357-79-3 32	140[11]	2-05-00-00377 1.7470[25]	
7248	2H-Indene-2,2-dicarboxylic acid, 1,3-dihydro-1,3-dioxo-, diethyl ester Phthaloylmalonic acid, diethyl ester	$C_{15}H_{14}O_6$ 290.27	116836-20-5		1-10-00-00441 1.1896[83]	eth 4; EtOH 4 1.541[84]
7249	1H-Indene, 1,2-dichloro-2,3-dihydro- 1,2-Dichloroindan	$C_9H_8Cl_2$ 187.07	74925-48-7 87-90[2]		3-05-00-01202 1.254[25]	1.5715[23]
7250	1H-Indene, 2,3-dihydro- Indan	C_9H_{10} 118.18	496-11-7 -51.4	4844 177.9	4-05-00-01371 0.9639[20]	H_2O 1; EtOH 5; eth 5; chl 2 1.5378[20]
7251	1H-Indene, 2,3-dihydro-1,1-dimethyl- 1,1-Dimethylindan	$C_{11}H_{14}$ 146.23	4912-92-9	191	4-05-00-01415 0.919[20]	1.5135[25]
7252	1H-Indene, 2,3-dihydro-1,2-dimethyl- 1,2-Dimethylindan	$C_{11}H_{14}$ 146.23	17057-82-8 79-80[10]		4-05-00-01415 0.927[20]	1.5186[20]
7253	1H-Indene, 2,3-dihydro-4,7-dimethyl- 4,7-Dimethylindan	$C_{11}H_{14}$ 146.23	6682-71-9 94-7[10]		4-05-00-01416 0.949[20]	1.5342[20]
7254	1H-Indene, 2,3-dihydro-5,6-dimethyl- 5,6-Dimethylindan	$C_{11}H_{14}$ 146.23	1075-22-5 94[10]		4-05-00-01416 0.9449[20]	1.5360[20]
7255	1H-Indene, 2,3-dihydro-1-methyl- 1-Methylindan	$C_{10}H_{12}$ 132.21	767-58-8 190.6		4-05-00-01397 0.938[25]	H_2O 1 1.5266[20]
7256	1H-Indene, 2,3-dihydro-2-methyl- 2-Methylindan	$C_{10}H_{12}$ 132.21	824-63-5 195		4-05-00-01397 0.940[25]	H_2O 1 1.5220[20]
7257	1H-Indene, 2,3-dihydro-4-methyl- 4-Methylindan	$C_{10}H_{12}$ 132.21	824-22-6 205.3		4-05-00-01397 0.9577[20]	1.5356[20]
7258	1H-Indene, 2,3-dihydro-5-methyl- 5-Methylindan	$C_{10}H_{12}$ 132.21	874-35-1 202		4-05-00-01398 0.944[25]	1.5336[20]
7259	1H-Indene, 2,3-dihydro-2-phenyl- 2-Phenylindan	$C_{15}H_{14}$ 194.28	22253-11-8 162-3[10]		4-05-00-02198 1.0821[16]	1.5955[15]
7260	1H-Indene, 2,3-dihydro-1,1,4,7-tetramethyl- 1,1,4,7-Tetramethylindan	$C_{13}H_{18}$ 174.29	1078-04-2 114.2[15]		3-05-00-01282 0.9343[25]	1.5216[25]
7261	1H-Indene, 2,3-dihydro-1,1,5-trimethyl- 1,1,5-Trimethylindan	$C_{12}H_{16}$ 160.26	40650-41-7 86[10]		4-05-00-01433 0.9119[20]	1.5126[20]
7262	1H-Indene, 2,3-dihydro-1,4,7-trimethyl- 1,4,7-Trimethylindan	$C_{12}H_{16}$ 160.26	54340-87-3 95[10]		4-05-00-01434 0.938[20]	1.5252[20]
7263	1H-Indene, 2,3-dihydro-1,1,3-trimethyl-3-phenyl- Indan, 1-phenyl-1,3,3-trimethyl	$C_{18}H_{20}$ 236.36	3910-35-8 52.5	308.5	4-05-00-02246 1.0009[20]	H_2O 1; EtOH 3; bz 3; MeOH 3 1.5681[20]
7264	1H-Indene-1,3(2H)-dione	$C_9H_6O_2$ 146.15	606-23-5 131-2 dec		4-07-00-02344 1.37[21]	H_2O 2; EtOH 4; eth 3; bz 3
7265	1H-Indene-1,3(2H)-dione, 2-(4-bromophenyl)- Bromindione	$C_{15}H_9BrO_2$ 301.14	1146-98-1 138	1381	4-07-00-02572	
7266	1H-Indene-1,3(2H)-dione, 2-(4-chlorophenyl)- Clorindione	$C_{15}H_9ClO_2$ 256.69	1146-99-2 145.5	2402	4-07-00-02571	bz 4; eth 4; EtOH 4
7267	1H-Indene-1,3(2H)-dione, 2,2-dihydroxy- Ninhydrin	$C_9H_6O_4$ 178.14	485-47-2 241-3 dec	6470	4-07-00-02786	H_2O 4; EtOH 3; eth 2; alk 3

No.	Name / Synonym	Mol. Form. / Mol. Wt.	CAS RN / mp/°C	Merck No. / bp/°C	Beil. Ref. / den/g cm^{-3}	Solubility / n_D
7268	1H-Indene-1,3(2H)-dione, 2-(diphenylacetyl)- / Diphenadione	$C_{23}H_{16}O_3$ 340.38	82-66-6 146.5	3304	4-07-00-02838	ace 4; HOAc 4 1.670
7269	1H-Indene-1,3(2H)-dione, 2-(4-methoxyphenyl)- / Anisindione	$C_{16}H_{12}O_3$ 252.27	117-37-3 156.5	698	3-08-00-02931	
7270	1H-Indene-1,3(2H)-dione, 2-phenyl- / Phenindione	$C_{15}H_{10}O_2$ 222.24	83-12-5 150	7196	4-07-00-02570	H_2O 1; EtOH 3; eth 3; ace 3
7271	1H-Indene, 1,3-diphenyl- / 1,3-Diphenylindene	$C_{21}H_{16}$ 268.36	4467-88-3 85	230[15]	3-05-00-02474	ace 4; eth 4
7272	1H-Indene, 2,3-diphenyl- / 2,3-Diphenylindene	$C_{21}H_{16}$ 268.36	5324-00-5 109.3	237[12]	4-05-00-02641	H_2O 1; EtOH 3; eth 3; ace 3
7273	1H-Indene, 1-ethyl-2,3-dihydro- / 1-Ethylindan	$C_{11}H_{14}$ 146.23	4830-99-3	212	4-05-00-01414 0.9348[25]	1.5121[25]
7274	1H-Indene, 5-hexyl-2,3-dihydro- / 5-Hexylindan	$C_{15}H_{22}$ 202.34	54889-55-3 -34	292.1	4-05-00-01474 0.9114[20]	1.5122[20]
7275	1H-Indene, 1-methyl- / 1-Methylindene	$C_{10}H_{10}$ 130.19	767-59-9	199; 82[15]	3-05-00-01371 0.970[25]	1.5616[20]
7276	1H-Indene, 2-methyl- / 2-Methylindene	$C_{10}H_{10}$ 130.19	2177-47-1 80	208	4-05-00-01545 0.974[25]	H_2O 1; eth 3; ace 3; bz 3 1.5652[20]
7277	1H-Indene, 3-methyl- / 3-Methylindene	$C_{10}H_{10}$ 130.19	767-60-2	198	4-05-00-01545 0.972[25]	H_2O 1; eth 3; ace 3; bz 3 1.5621[20]
7278	1H-Indene, 4-methyl- / 4-Methylindene	$C_{10}H_{10}$ 130.19	7344-34-5	209; 88[13]	2-05-00-00418 0.989[25]	1.568[20]
7279	1H-Indene, 6-methyl- / 6-Methylindene	$C_{10}H_{10}$ 130.19	20232-11-5	207; 90[17]	4-05-00-01545 0.977[25]	1.566[20]
7280	1H-Indene, 7-methyl- / 7-Methylindene	$C_{10}H_{10}$ 130.19	7372-92-1	209; 90[15]	2-05-00-00418 0.989[25]	1.568[20]
7281	1H-Indene, octahydro- / Hexahydroindan	C_9H_{16} 124.23	496-10-6 -53	167	4-05-00-00292 0.876[25]	1.4702[20]
7282	6H-Indeno[1,2-b]oxirene, 1a,6a-dihydro-	C_9H_8O 132.16	768-22-9 24.5	113[20]; 98[6]	5-17-02-00033 1.1255[24]	chl 3
7283	1H-Inden-4-ol, 7-chloro-2,3-dihydro- / Chlorindanol	C_9H_9ClO 168.62	145-94-8 92	2094	4-06-00-03828	
7284	1H-Inden-1-ol, 2,3-dihydro-	$C_9H_{10}O$ 134.18	6351-10-6 54.8	220; 128[12]	4-06-00-03824	bz 4; EtOH 4; chl 4
7285	1H-Inden-5-ol, 2,3-dihydro-	$C_9H_{10}O$ 134.18	1470-94-6 58	253	4-06-00-03829	H_2O 2; EtOH 4; eth 4; sulf 3
7286	1H-Inden-1-one, 2,3-dihydro-	C_9H_8O 132.16	83-33-0 42	243; 129[12]	4-07-00-00999 1.0943[40]	H_2O 2; EtOH 4; eth 4; ace 4 1.561[25]
7287	2H-Inden-2-one, 1,3-dihydro- / 2-Indanone	C_9H_8O 132.16	615-13-4 59	218 dec	4-07-00-01002 1.0712[69]	H_2O 1; EtOH 4; eth 4; ace 4 1.538[67]
7288	1H-Inden-1-one, 2,3-dihydro-3,3-dimethyl-	$C_{11}H_{12}O$ 160.22	26465-81-6	130-1[18]	4-07-00-01035 1.0320[14]	1.5453[14]
7289	1H-Inden-1-one, 2,3-dihydro-2-methyl-	$C_{10}H_{10}O$ 146.19	17496-14-9	250; 125[18]	4-07-00-01019 1.0651[21]	1.553[22]
7290	1H-Inden-1-one, 2,3-dihydro-4-methyl- / 4-Methyl-1-indanone	$C_{10}H_{10}O$ 146.19	24644-78-8 95	144[13]; 120[5]	4-07-00-01020	bz 4; eth 4; EtOH 4
7291	1H-Inden-1-one, 2,3-dihydro-2-phenyl- / 2-Phenyl-1-indanone	$C_{15}H_{12}O$ 208.26	16619-12-8 77.5	344	4-07-00-01673	ace 4; bz 4; EtOH 4
7292	2H-Inden-2-one, octahydro-, cis-	$C_9H_{14}O$ 138.21	5689-04-3 10	225; 113[25]	4-07-00-00174 0.9970[20]	EtOH 3; bz 3; lig 3 1.4830[20]
7293	2H-Inden-2-one, octahydro-, trans-	$C_9H_{14}O$ 138.21	16484-17-6 -12	218	4-07-00-00174 0.9807[17]	EtOH 3; bz 3; lig 3 1.4769[17]
7294	1H-Indole / 2,3-Benzopyrrole	C_8H_7N 117.15	120-72-9 52.5	4869 253.6	5-20-07-00005 1.22[25]	H_2O 3; EtOH 4; eth 4; bz 3
7295	1H-Indole-3-acetic acid / Indoleacetic acid	$C_{10}H_9NO_2$ 175.19	87-51-4 168.5	4870	5-22-03-00065	H_2O 1; EtOH 4; eth 3; ace 3
7296	1H-Indole-3-butanoic acid / Indolebutyric acid	$C_{12}H_{13}NO_2$ 203.24	133-32-4 124.5	4871	5-22-03-00140	bz 4; DMSO 3; peth 1
7297	1H-Indole, 2,3-dihydro-	C_8H_9N 119.17	496-15-1	229	5-20-06-00238 1.069[20]	H_2O 2; eth 3; ace 3; bz 3 1.5923[20]
7298	1H-Indole, 1,3-dimethyl-	$C_{10}H_{11}N$ 145.20	875-30-9 142	258.5	5-20-07-00071	eth 3
7299	1H-Indole, 2,5-dimethyl- / 2,5-Dimethylindole	$C_{10}H_{11}N$ 145.20	1196-79-8 115.3	188[40]	5-20-07-00111	H_2O 1; EtOH 2; eth 3; bz 3
7300	1H-Indole-2,3-dione / Isatin	$C_8H_5NO_2$ 147.13	91-56-5 203 dec	4985	5-21-10-00221	H_2O 3; EtOH 4; eth 2; ace 3
7301	1H-Indole-5,6-dione, 2,3-dihydro-3-hydroxy-1-methyl- / Adrenochrome	$C_9H_9NO_3$ 179.18	54-06-8 125 dec	162	5-21-13-00079	H_2O 4; EtOH 4; eth 1; bz 1
7302	1H-Indole-3-ethanamine / Tryptamine	$C_{10}H_{12}N_2$ 160.22	61-54-1 118	9705 137[0.15]	5-22-10-00045	H_2O 1; EtOH 3; eth 1; ace 3
7303	1H-Indole-3-ethanamine, N,N-dimethyl- / N,N-Dimethyltryptamine	$C_{12}H_{16}N_2$ 188.27	61-50-7 46	3251	5-22-10-00048	
7304	1H-Indole-3-ethanamine, 5-methoxy- / 5-Methoxytryptamine	$C_{11}H_{14}N_2O$ 190.25	608-07-1 121.5	5926	5-22-12-00018	
7305	1H-Indole-3-ethanol / Tryptophol	$C_{10}H_{11}NO$ 161.20	526-55-6 59	9708 174[2]	5-21-03-00061	ace 4; eth 4; EtOH 4; chl 4
7306	1H-Indole-3-ethanol, 2-(5-ethenyl-1-azabicyclo[2.2.2]oct-2-yl)- / Cinchonamine	$C_{19}H_{24}N_2O$ 296.41	482-28-0 186	2287	4-23-00-02735	H_2O 1; EtOH 4; eth 4; bz 3

No.	Name / Synonym	Mol. Form. / Mol. Wt.	CAS RN / mp/°C	Merck No. / bp/°C	Beil. Ref. / den/g cm^{-3}	Solubility / n_D
7307	1H-Indole, 1-ethyl-	$C_{10}H_{11}N$ / 145.20	10604-59-8 / 105	/ 252.5; 122[14]	5-20-07-00014 / 1.2563[15]	
7308	1H-Indole, 3-ethyl-	$C_{10}H_{11}N$ / 145.20	1484-19-1 / 37	/ 279; 138[6]	5-20-07-00099	bz 4; eth 4; EtOH 4
7309	1H-Indole, 3-ethyl-2-methyl-	$C_{11}H_{13}N$ / 159.23	35246-18-5 / 44.5	/ 292; 156[12]	5-20-07-00136 / 1.0414[0]	eth 4; EtOH 4
7310	1H-Indole-3-methanamine, N,N-dimethyl- / Gramine	$C_{11}H_{14}N_2$ / 174.25	87-52-5 / 138.5	4440	5-22-10-00025	H_2O 1; EtOH 3; eth 3; chl 3
7311	1H-Indole, 7-methoxy-	C_9H_9NO / 147.18	3189-22-8	/ 108-10[0.3]	5-21-03-00024 / 1.126[25]	1.6120[20]
7312	1H-Indole, 1-methyl-	C_9H_9N / 131.18	603-76-9	/ 237	5-20-07-00012 / 1.070[25]	H_2O 1; EtOH 3; eth 3; bz 3
7313	1H-Indole, 2-methyl-	C_9H_9N / 131.18	95-20-5 / 61	/ 272	5-20-07-00059 / 1.07[20]	H_2O 2; EtOH 4; eth 4; ace 3
7314	1H-Indole, 3-methyl- / Skatole	C_9H_9N / 131.18	83-34-1 / 97.5	8503 / 266	5-20-07-00069	H_2O 3; EtOH 3; eth 3; ace 3
7315	1H-Indole, 4-methyl-	C_9H_9N / 131.18	16096-32-5 / 5	/ 267	5-20-07-00079 / 1.062[20]	
7316	1H-Indole, 5-methyl-	C_9H_9N / 131.18	614-96-0 / 60	/ 267	5-20-07-00080 / 1.0202[78]	H_2O 3; EtOH 3; eth 3; bz 3
7317	1H-Indole, 7-methyl- / 7-Methylindole	C_9H_9N / 131.18	933-67-5 / 85	/ 266	5-20-07-00081 / 1.0202[100]	
7318	1H-Indole, 3-methyl-2-phenyl-	$C_{15}H_{13}N$ / 207.27	10257-92-8 / 91.5	/ 280-90[120]	5-20-08-00274	H_2O 1; EtOH 3; bz 3
7319	1H-Indole, octahydro-	$C_8H_{15}N$ / 125.21	4375-14-8 / 143	/ 185.5	5-20-04-00358 / 0.9472[20]	bz 4; eth 4; EtOH 4 / 1.4892[20]
7320	1H-Indole, 2-phenyl-	$C_{14}H_{11}N$ / 193.25	948-65-2 / 190.5	/ 250[10]	5-20-08-00232	H_2O 2; eth 3; bz 3; chl 3
7321	1H-Indole, 1-(phenylmethyl)-	$C_{15}H_{13}N$ / 207.27	3377-71-7 / 45.8	/ 172[2]	5-20-07-00016	bz 4; eth 4; EtOH 4; lig 4
7322	1H-Indole, 1,2,3-trimethyl- / 1,2,3-Trimethylindole	$C_{11}H_{13}N$ / 159.23	1971-46-6 / 18	/ 283	5-20-07-00102	H_2O 2; EtOH 2; eth 2; bz 2
7323	1H-Indole, 2,3,5-trimethyl- / 2,3,5-Trimethylindole	$C_{11}H_{13}N$ / 159.23	21296-92-4 / 121.5	/ 297	5-20-07-00143	EtOH 4; chl 4; lig 4
7324	Indolizine	C_8H_7N / 117.15	274-40-8 / 75	/ 205	5-20-07-00047	H_2O 1; EtOH 3
7325	Indolizine, octahydro-	$C_8H_{15}N$ / 125.21	13618-93-4	/ 75[43]	5-20-04-00360 / 0.9074[10]	eth 4; EtOH 4 / 1.4748
7326	1H-Indol-5-ol, 3-[2-(dimethylamino)ethyl]- / Bufotenine	$C_{12}H_{16}N_2O$ / 204.27	487-93-4 / 146.5	1467 / 320[0.1]	5-22-12-00026	eth 4; EtOH 4
7327	2H-Indol-2-one, 1,3-dihydro-	C_8H_7NO / 133.15	59-48-3 / 128	/ 227[23]; 195[17]	5-21-08-00007	H_2O 3; EtOH 3; eth 3
7328	2H-Indol-2-one, 1,3-dihydro-1,3-dimethyl-	$C_{10}H_{11}NO$ / 161.20	24438-17-3 / 55	/ 273; 137[11]	5-21-08-00077	eth 4; EtOH 4
7329	2H-Indol-2-one, 1,3-dihydro-3,3-dimethyl-	$C_{10}H_{11}NO$ / 161.20	19155-24-9 / 154.3	/ 300	5-21-08-00113	bz 4; eth 4; EtOH 4
7330	2H-Indol-2-one, 1,3-dihydro-4,7-dimethyl- / 4,7-Dimethyl-2-indolinone	$C_{10}H_{11}NO$ / 161.20	59022-71-8 / 159	/ 205-10[15]	4-21-00-03676	ace 4; EtOH 4; chl 4
7331	3H-Indol-3-one, 2-(1,3-dihydro-3-oxo-2H-indol-2-ylidene)-1,2-dihydro- / Indigo	$C_{16}H_{10}N_2O_2$ / 262.27	482-89-3 / 390 dec	4855 / sub 300	5-24-08-00503	
7332	2H-Indol-2-one, 1-ethyl-1,3-dihydro-3-methyl- / 2-Indolinone, 1-ethyl-3-methyl-	$C_{11}H_{13}NO$ / 175.23	84258-49-1	/ 283; 105[0.5]	2-21-00-00258	eth 4; EtOH 4 / 1.557[25]
7333	Indolo[2',3':3,4]pyrido[2,1-b]quinazolin-5(7H)-one, 8,13,13b,14-tetrahydro-14-methyl-, (S)- / Evodiamine	$C_{19}H_{17}N_3O$ / 303.36	518-17-2 / 28	3865	4-26-00-00525	
7334	Indolo[2',3':3,4]pyrido[2,1-b]quinazolin-5(7H)-one, 8,13-dihydro- / Rutecarpine	$C_{18}H_{13}N_3O$ / 287.32	84-26-4 / 259.5	8271	4-26-00-00528	EtOH 2; ace 2; bz 2
7335	Inosine / Hypoxanthine riboside	$C_{10}H_{12}N_4O_5$ / 268.23	58-63-9 / 218 dec	4880	4-26-00-02086	EtOH 4
7336	scyllo-Inositol	$C_6H_{12}O_6$ / 180.16	488-59-5 / 353 dec		4-06-00-07920 / 1.659[19]	H_2O 2; EtOH 1; eth 1; bz 1
7337	Iodine cyanide / Cyanogen iodide	CIN / 152.92	506-78-5 / 146.7	2702 / sub 45	4-03-00-00093 / 2.84[18]	eth 4; EtOH 4
7338	Iprodione / 1-Imidazolidinecarboxamide, 3-(3,5-dichlorophenyl)-N-(1-methylethyl)-2,4-dioxo-	$C_{13}H_{13}Cl_2N_3O_3$ / 330.17	36734-19-7 / 136	4964		
7339	Iron carbonyl / Iron pentacarbonyl	C_5FeO_5 / 195.90	13463-40-6 / -20	4980 / 103	/ 1.5[20]	H_2O 1; EtOH 2; bz 3; ace 3 / 1.453[22]
7340	Isazophos / Phosphorothioic acid, O-[2-chloro-I-(I-methylethyl)-IH-imidazol-4-yl] O,O-diethyl ester	$C_9H_{17}ClN_3O_3PS$ / 313.74	67329-04-8	4988 / 170; 100[0.001]	/ 1.22[20]	
7341	5-Isobenzofurancarboxylic acid, 1,3-dihydro-1,3-dioxo- / Trimellitic anhydride	$C_9H_4O_5$ / 192.13	552-30-7 / 162	9617 / 240-3[14]	5-18-08-00562	
7342	1,3-Isobenzofurandione / Phthalic anhydride	$C_8H_4O_3$ / 148.12	85-44-9 / 130.8	7346 / 295	5-17-11-00253 / 1.527[4]	H_2O 2; EtOH 3; eth 2; ace 3

No.	Name / Synonym	Mol. Form. / Mol. Wt.	CAS RN / mp/°C	Merck No. / bp/°C	Beil. Ref. / den/g cm^{-3}	Solubility / n_D
7343	1,3-Isobenzofurandione, 4,5-dichloro- / Phthalic anhydride, 3,4-dichloro	$C_8H_2Cl_2O_3$ / 217.01	56962-07-3 / 121	/ 329	4-17-00-06142	EtOH 4; chl 4
7344	1,3-Isobenzofurandione, 5,6-dichloro- / 4,5-Dichlorophthalic anhydride	$C_8H_2Cl_2O_3$ / 217.01	942-06-3 / 188	/ 313	5-17-11-00260	eth 4; EtOH 4; tol 4
7345	1,3-Isobenzofurandione, hexahydro- / Hexahydrophthalic anhydride	$C_8H_{10}O_3$ / 154.17	85-42-7 / 32	/ 145[18]	5-17-11-00076	
7346	1,3-Isobenzofurandione, 4,5,6,7-tetrachloro- / Tetrachlorophthalic anhydride	$C_8Cl_4O_3$ / 285.90	117-08-8 / 254.5	/ sub	5-17-11-00260 / 1.49[275]	eth 2
7347	1,3-Isobenzofurandione, 4,5,6,7-tetrahydro- / 3,4,5,6-Tetrahydrophthalic anhydride	$C_8H_8O_3$ / 152.15	2426-02-0 / 74		5-17-11-00133 / 1.2[105]	EtOH 3; eth 4; ace 3; chl 3
7348	1,3-Isobenzofurandione, 4,5,6,7-tetraiodo-	$C_8I_4O_3$ / 651.70	632-80-4 / 327.5	/ sub	5-17-11-00266	H_2O 1; EtOH 1; bz 1; HOAc 2
7349	1(3H)-Isobenzofuranone	$C_8H_6O_2$ / 134.13	87-41-2 / 75	/ 290	5-17-10-00007 / 1.1636[99]	H_2O 3; EtOH 4; eth 4; chl 2 / 1.536[99]
7350	1(3H)-Isobenzofuranone, 3,3-bis(3,5-dibromo-4-hydroxyphenyl)- / 3',3",5',5"-Tetrabromophenolphthalein	$C_{20}H_{10}Br_4O_4$ / 633.91	76-62-0 / 296	9122	5-18-04-00190	H_2O 1; EtOH 2; eth 4; alk 3
7351	1(3H)-Isobenzofuranone, 3,3-bis(4-hydroxy-3,5-diiodophenyl)-	$C_{20}H_{10}I_4O_4$ / 821.91	386-17-4 / 308		4-18-00-01949 / 2.0201[22]	H_2O 1; EtOH 1; eth 1; chl 2
7352	1(3H)-Isobenzofuranone, 3,3-bis[4-hydroxy-2-methyl-5-(1-methylethyl)phenyl]- / Thymolphthalein	$C_{28}H_{30}O_4$ / 430.54	125-20-2 / 253	9336	5-18-04-00194	H_2O 1; EtOH 3; eth 3; ace 3
7353	1(3H)-Isobenzofuranone, 3,3-bis(4-hydroxy-3-methylphenyl)- / o-Cresolphthalein	$C_{22}H_{18}O_4$ / 346.38	596-27-0 / 223	2582	5-18-04-00193	EtOH 4
7354	1(3H)-Isobenzofuranone, 3,3-bis(4-hydroxyphenyl)- / Phenolphthalein	$C_{20}H_{14}O_4$ / 318.33	77-09-8 / 262.5	7208	5-18-04-00188 / 1.277[32]	H_2O 1; EtOH 4; eth 3; ace 4
7355	1(3H)-Isobenzofuranone, 6,7-dimethoxy- / Meconin	$C_{10}H_{10}O_4$ / 194.19	569-31-3 / 102.5	5665	5-18-03-00068	H_2O 2; EtOH 3; eth 3; ace 3
7356	1(3H)-Isobenzofuranone, 6,7-dimethoxy-3-(5,6,7,8-tetrahydro-4-methoxy-6-methyl-1,3-dioxolo[4,5-g]isoquinolin-5 / Noscapine	$C_{22}H_{23}NO_7$ / 413.43	128-62-1 / 176	6638	4-27-00-06838	H_2O 1; EtOH 3; eth 2; ace 4
7357	1(3H)-Isobenzofuranone, 6,7-dimethoxy-3-(5,6,7,8-tetrahydro-6-methyl-1,3-dioxolo[4,5-g]isoquinolin-5-yl)-, [S- / Hydrastine	$C_{21}H_{21}NO_6$ / 383.40	118-08-1 / 132	4688	4-27-00-06833	H_2O 1; ace 3; bz 3
7358	1(3H)-Isobenzofuranone, 3,3-diphenyl-	$C_{20}H_{14}O_2$ / 286.33	596-29-2 / 120	235[15]	5-17-10-00735	chl 2
7359	Isofenphos / Benzoic acid, 2-[[ethoxy[(1-methylethyl)amino]phosphinothioyl]oxy]-, 1-methylethyl ester	$C_{15}H_{24}NO_4PS$ / 345.40	25311-71-1 / <-12	5055 / 120[0.01]	1.134[20]	
7360	Isofulminic acid	CHNO / 43.03	51060-05-0 / stable only at low temp.			
7361	2H-Isoindole-2-acetic acid, 1,3-dihydro-1,3-dioxo-, ethyl ester	$C_{12}H_{11}NO_4$ / 233.22	6974-10-3 / 112.5	/ 300	5-21-10-00431	bz 4; eth 4; EtOH 4; chl 4
7362	1H-Isoindole-1,3(2H)-dione / Phthalimide	$C_8H_5NO_2$ / 147.13	85-41-6 / 238	7347	5-21-10-00270	bz 4
7363	1H-Isoindole-1,3(2H)-dione, 2-(2,6-dioxo-3-piperidinyl)- / Thalidomide	$C_{13}H_{10}N_2O_4$ / 258.23	50-35-1 / 270	9182	5-22-13-00224	py 4; diox 4
7364	1H-Isoindole-1,3(2H)-dione, N-ethyl-	$C_{10}H_9NO_2$ / 175.19	5022-29-7 / 79	/ 285.5	5-21-10-00274	EtOH 3; eth 3
7365	1H-Isoindole-1,3(2H)-dione, 2-methyl-	$C_9H_7NO_2$ / 161.16	550-44-7 / 134	/ 286	5-21-10-00273	H_2O 1; EtOH 2
7366	1H-Isoindole-1,3(2H)-dione, 2-phenyl-	$C_{14}H_9NO_2$ / 223.23	520-03-6 / 210	/ sub	5-21-10-00292	H_2O 1; EtOH 2; chl 5
7367	1H-Isoindole-1,3(2H)-dione, 2-(phenylmethyl)-	$C_{15}H_{11}NO_2$ / 237.26	2142-01-0 / 116		5-21-10-00299 / 1.343[18]	EtOH 3; DMSO-d_6 2; HOAc 3
7368	1H-Isoindole-1,3(2H)-dione, 3a,4,7,7a-tetrahydro-2-[(trichloromethyl)thio]- / Captan	$C_9H_8Cl_3NO_2S$ / 300.59	133-06-2 / 172.5	1771	5-21-10-00136 / 1.74[25]	chl 4
7369	1H-Isoindol-1-one, 2,3-dihydro-	C_8H_7NO / 133.15	480-91-1 / 151	338; 103[18]	5-21-08-00018	eth 4; EtOH 4; chl 4
7370	L-Isoleucine / 2-Amino-3-methylpentanoic acid	$C_6H_{13}NO_2$ / 131.17	73-32-5 / 284 dec	5064	4-04-00-02775	H_2O 3; EtOH 1
7371	Isopropalin / Benzenamine, 4-(1-methylethyl)-2,6-dinitro-N,N-dipropyl-	$C_{15}H_{23}N_3O_4$ / 309.37	33820-53-0	5089		
7372	5-Isoquinolinamine / 5-Aminoisoquinoline	$C_9H_8N_2$ / 144.18	1125-60-6 / 128	196[18]	5-22-10-00342	chl 2; lig 3
7373	Isoquinoline / Benzopyridine	C_9H_7N / 129.16	119-65-3 / 26.47	5110 / 243.2	5-20-07-00333 / 1.0910[30]	H_2O 1; EtOH 4; eth 5; ace 4 / 1.6148[20]
7374	Isoquinoline, 4-bromo- / 4-Bromoisoquinoline	C_9H_6BrN / 208.06	1532-97-4 / 41.5	/ 282.5	5-20-07-00363	eth 4
7375	5-Isoquinolinecarbonitrile	$C_{10}H_6N_2$ / 154.17	27655-41-0 / 137	/ sub 100	4-22-00-01208	eth 4; EtOH 4

No.	Name Synonym	Mol. Form. Mol. Wt.	CAS RN mp/°C	Merck No. bp/°C	Beil. Ref. den/g cm⁻³	Solubility n_D
7376	Isoquinoline, 1-chloro- 1-Chloroisoquinoline	C_9H_8ClN 163.61	19493-44-8 37.5	274.5	5-20-07-00361	H_2O 1; bz 3
7377	Isoquinoline, 1-[(3,4-diethoxyphenyl)methyl]-6,7-diethoxy- Isoquinoline, 6,7-diethoxy-1(3,4-diethoxybenzyl)	$C_{24}H_{29}NO_4$ 395.50	486-47-5 100	3682	5-21-06-00183	H_2O 1; EtOH 3; eth 2; chl 2
7378	Isoquinoline, 1-[(3,4-dimethoxyphenyl)methyl]-6,7-dimethoxy- Papaverine	$C_{20}H_{21}NO_4$ 339.39	58-74-2 147.5	6968 sub 135	5-21-06-00182 1.337[20]	H_2O 2; EtOH 4; eth 2; ace 3 1.625
7379	Isoquinoline, 1-[(3,4-dimethoxyphenyl)methyl]-1,2,3,4-tetrahydro-6,7-dimethoxy-2-methyl-, (S)- Laudanosine	$C_{21}H_{27}NO_4$ 357.45	2688-77-9 89	5249	5-21-06-00049	ace 4; eth 4; EtOH 4; chl 4
7380	Isoquinoline, 7-methoxy- 7-Methoxyisoquinoline	$C_{10}H_9NO$ 159.19	39989-39-4 49	194[50.] 184[34]	5-21-03-00326	H_2O 1; EtOH 3; lig 3
7381	Isoquinoline, 1-methyl- Isoquinaldine	$C_{10}H_9N$ 143.19	1721-93-3 10	248	5-20-07-00407 1.077[20]	H_2O 2; eth 3; ace 3; bz 3 1.6095[20]
7382	Isoquinoline, 3-methyl- 3-Methylisoquinoline	$C_{10}H_9N$ 143.19	1125-80-0 68	249	5-20-07-00410	H_2O 2; eth 3; ace 3; chl 2
7383	Isoquinoline, 4-methyl- 4-Methylisoquinoline	$C_{10}H_9N$ 143.19	1196-39-0 <-75	256	5-20-07-00413	H_2O 2; eth 3; ace 3; bz 3
7384	Isoquinoline, 6-methyl- 6-Methylisoquinoline	$C_{10}H_9N$ 143.19	42398-73-2 85.5	265.5	5-20-07-00414	ace 4; bz 4; eth 4; EtOH 4
7385	Isoquinoline, 8-methyl- 8-Methylisoquinoline	$C_{10}H_9N$ 143.19	62882-00-2 205	258	5-20-07-00414	ace 4; bz 4; eth 4
7386	Isoquinoline, 5-nitro- 5-Nitroisoquinoline	$C_9H_6N_2O_2$ 174.16	607-32-9 110	sub	5-20-07-00365	H_2O 2; EtOH 4; eth 4; chl 4
7387	Isoquinoline, 3-(phenylmethyl)- Isoquinoline, 3-benzyl	$C_{16}H_{13}N$ 219.29	90210-56-3 104	311[23]	5-20-08-00383	bz 4; chl 4
7388	Isoquinoline, 1,2,3,4-tetrahydro-	$C_9H_{11}N$ 133.19	91-21-4 <-15	232.5	5-20-06-00320 1.0642[24]	H_2O 1; EtOH 3; bz 3; acid 3 1.5668[20]
7389	Isoquinoline, 5,6,7,8-tetrahydro- 5,6,7,8-Tetrahydroisoquinoline	$C_9H_{11}N$ 133.19	36556-06-6 146.5	218; 106[13]	5-20-06-00319 1.0504[10]	H_2O 1; ace 3; bz 3; chl 3 1.5276[10]
7390	Isoquinoline, 1,2,3,4-tetrahydro-6,7-dimethoxy-1,2-dimethyl-, (±)- Carnegine	$C_{13}H_{19}NO_2$ 221.30	490-53-9 170[1]	1854	5-21-04-00501	H_2O 4; eth 4; EtOH 4
7391	Isoquinoline, 1,2,3,4-tetrahydro-6,7-dimethoxy- Isoquinoline, 6,7-dimethoxy-1,2-dimethyl-1,2,3,4-tetrahydro	$C_{13}H_{19}NO_2$ 221.30	71783-56-7 262.5	170[1]; 100[0.01]	5-21-04-00501	H_2O 3; EtOH 3; eth 3; chl 3
7392	7-Isoquinolinol, 1-[(3,4-dimethoxyphenyl)methyl]-1,2,3,4-tetrahydro-6-methoxy-2-methyl-, (S)- Codamine	$C_{20}H_{25}NO_4$ 343.42	21040-59-5 127	2458	4-21-00-02704	eth 4; EtOH 4; chl 4
7393	8-Isoquinolinol, 1,2,3,4-tetrahydro-6,7-dimethoxy- Anhalamine	$C_{11}H_{15}NO_3$ 209.25	643-60-7 187.5	682	5-21-05-00513	eth 4; EtOH 4
7394	8-Isoquinolinol, 1,2,3,4-tetrahydro-6,7-dimethoxy-1,2-dimethyl-, (S)- Pellotine	$C_{13}H_{19}NO_3$ 237.30	83-14-7 111.5	7018	4-21-00-02525	ace 4; eth 4; EtOH 4; peth 4
7395	8-Isoquinolinol, 1,2,3,4-tetrahydro-6,7-dimethoxy-1-methyl-, (S)- Anhalonidine	$C_{12}H_{17}NO_3$ 223.27	17627-77-9 160.5	683	4-21-00-02524	H_2O 4; EtOH 4
7396	7-Isoquinolinol, 1,2,3,4-tetrahydro-1-[(4-hydroxyphenyl)methyl]-6-methoxy-, (S)- Coclaurine	$C_{17}H_{19}NO_3$ 285.34	486-39-5 220.5	2455	4-21-00-02605	
7397	6-Isoquinolinol, 1,2,3,4-tetrahydro-7-methoxy-1-methyl-, (S)- Salsoline	$C_{11}H_{15}NO_2$ 193.25	89-31-6 221.5	8308	5-21-04-00499	H_2O 2; EtOH 2; eth 1; bz 2
7398	1(2H)-Isoquinolinone, 3,4-dihydro-6,7-dimethoxy- Corydaldine	$C_{11}H_{13}NO_3$ 207.23	493-49-2 175	2542	5-21-13-00086	H_2O 4; bz 4; eth 4; EtOH 4
7399	Isothiocyanic acid	CHNS 59.09	3129-90-6			
7400	DL-Isovaline	$C_5H_{11}NO_2$ 117.15	595-39-1 316.5	sub 300	4-04-00-02655	H_2O 3; EtOH 3; eth 2; bz 1
7401	L-Isovaline 2-Amino-2-methylbutyric acid	$C_5H_{11}NO_2$ 117.15	595-40-4	5124	4-04-00-02655	
7402	Isoxazole 1-Oxa-2-azacyclopentadiene	C_3H_3NO 69.06	288-14-2 95		4-27-00-00959 1.078[20]	H_2O 3 1.4298[17]
7403	Isoxazole, 3,5-dimethyl- 3,5-Dimethylisoxazole	C_5H_7NO 97.12	300-87-8 143		4-27-00-00978 0.99[25]	1.4421[20]
7404	5-Isoxazolemethanol, 3-methyl-	$C_5H_7NO_2$ 113.12	14716-89-3 130[20]		4-27-01-01745 1.132[25]	1.4790[20]
7405	Isoxazole, 5-methyl- 5-Methylisoxazole	C_4H_5NO 83.09	5765-44-6 122		4-27-00-00966 1.023[20]	DMSO 3 1.4386[20]
7406	3-Isoxazolidinone, 4-amino-, (R)- Cycloserine	$C_3H_6N_2O_2$ 102.09	68-41-7 155-6 dec	2758	4-27-00-05549	H_2O 3; MeOH 2
7407	Isoxazolium, 2-ethyl-5-(3-sulfophenyl)-, hydroxide, inner salt Woodward's Reagent K	$C_{11}H_{11}NO_4S$ 253.28	4156-16-5 dec 207	9962		

No.	Name / Synonym	Mol. Form. / Mol. Wt.	CAS RN / mp/°C	Merck No. / bp/°C	Beil. Ref. / den/g cm^{-3}	Solubility / n_D
7408	Kepone 1,3,4-Metheno-2H-cyclobuta[cd]pentalen-2-one, 1,1a,3,3a,4,5,5,5a,5b,6-decachloro-octahydro-	$C_{10}Cl_{10}O$ 490.64	143-50-0 350 dec	2081	1.61[25]	
7409	Ketone, 1-methyl-4-piperidyl phenyl Piperidine, 4-benzoyl-N-methyl	$C_{13}H_{17}NO$ 203.28	92040-00-1 36	190[21]; 133[2]	4-21-00-03689	H_2O 3; EtOH 3; eth 4; ace 3 1.5430[23]
7410	Lactofen (±)-2-Ethoxy-1-ethyl-2-oxoethyl 5-[2-chloro-4-(trifluoromethyl)phenoxy]-2-nitrobenzoate	$C_{19}H_{15}ClF_3NO_7$ 461.78	77501-63-4			
7411	Lanosta-8,24-dien-3-ol, (3β)- Lanosterol	$C_{30}H_{50}O$ 426.73	79-63-0 140.5	5232	4-06-00-04188	eth 4; EtOH 4; chl 4
7412	DL-Leucine	$C_6H_{13}NO_2$ 131.17	328-39-2 293	sub	2-04-00-00870 1.293[18]	H_2O 3; EtOH 2; eth 1
7413	L-Leucine 2-Amino-4-methylpentanoic acid	$C_6H_{13}NO_2$ 131.17	61-90-5 293	5331 sub	4-04-00-02738 1.293[18]	H_2O 2; EtOH 1; eth 1
7414	DL-Leucine, N-glycyl-	$C_8H_{16}N_2O_3$ 188.23	688-14-2 242 dec		4-04-00-02766 1.181[25]	H_2O 4; EtOH 1
7415	Linuron Urea, N'-(3,4-dichlorophenyl)-N-methoxy-N-methyl-	$C_9H_{10}Cl_2N_2O_2$ 249.10	330-55-2 93	5387		
7416	Lup-20(29)-ene-3,28-diol, (3β)- Betulin	$C_{30}H_{50}O_2$ 442.73	473-98-3 250	1212 sub 240	4-06-00-06534	H_2O 1; EtOH 3; eth 3; bz 2
7417	Lup-20(29)-en-3-ol, (3β)- Lupeol	$C_{30}H_{50}O$ 426.73	545-47-1 216	5478	4-06-00-04200 0.9457[218]	H_2O 1; EtOH 4; eth 4; ace 4 1.4910[218]
7418	DL-Lysine 2,6-Diaminohexanoic acid	$C_6H_{14}N_2O_2$ 146.19	70-54-2 224		2-04-00-00858	H_2O 2
7419	L-Lysine 2,6-Diaminohexanoic acid	$C_6H_{14}N_2O_2$ 146.19	56-87-1 224-5 dec	5509	4-04-00-02717	H_2O 4; EtOH 1; eth 1; ace 1
7420	Maneb Manganese, [[1,2-ethanediylbis[carbamodithioato]](2-)]-	$C_4H_6MnN_2S_4$ 265.31	12427-38-2 dec 200	5603		
7421	Manganous dimethyldithiocarbamate Carbamic acid, dimethyldithio-, manganese(2+) salt	$C_6H_{12}MnN_2S_4$ 295.38	15339-36-3			
7422	D-Mannitol Cordycepic acid	$C_6H_{14}O_6$ 182.17	69-65-8 168	5629 295[3.5]	4-01-00-02841 1.489[20]	H_2O 4; EtOH 2; eth 1; py 2 1.3330
7423	D-Mannitol, 1,4:3,6-dianhydro-	$C_6H_{10}O_4$ 146.14	641-74-7 88	274 dec	5-19-03-00201	H_2O 4
7424	D-Mannitol, hexanitrate Mannitol, hexanitrate	$C_6H_8N_6O_{18}$ 452.16	15825-70-4 107	5630 exp	4-01-00-02849 1.8[20]	bz 4; eth 4; EtOH 4
7425	D-Mannose Seminose	$C_6H_{12}O_6$ 180.16	3458-28-4 132 dec	5632	4-01-00-04328 1.539[20]	H_2O 4; EtOH 2; eth 1; bz 1
7426	Matridin-15-one Matrine	$C_{15}H_{24}N_2O$ 248.37	519-02-8 223[6]	5641	5-24-02-00301	H_2O 3; EtOH 4; eth 3; ace 3 1.5286[25]
7427	Mefluidide Acetamide, N-[2,4-dimethyl-5-[[(trifluoromethyl)sulfonyl]amino]phenyl]-	$C_{11}H_{13}F_3N_2O_3S$ 310.30	53780-34-0 184	5684		
7428	m-Menthan-6-ol Cyclohexanol-4-isopropyl-2-methyl	$C_{10}H_{20}O$ 156.27	1490-05-7	119.21[28]	2-06-00-00038 0.9140[20]	eth 4; EtOH 4 1.4666[20]
7429	p-Menth-3-ene, (S)-(-)- Cyclohexene, 1-isopropyl-4-methyl (l)	$C_{10}H_{18}$ 138.25	22564-83-6	167	4-05-00-00301 0.8099[19]	H_2O 1; EtOH 3; eth 3; bz 3 1.4511[20]
7430	Mepiquat chloride Piperidinium, 1,1-dimethyl-, chloride	$C_7H_{16}NCl$ 149.66	24307-26-4 223	5746		
7431	Mercury, (acetato-O)phenyl- Phenylmercuric acetate	$C_8H_8HgO_2$ 336.74	62-38-4 153	7271	4-16-00-01720	chl 3
7432	Mercury, bis(4-methylphenyl)- Bis(p-tolyl)mercury	$C_{14}H_{14}Hg$ 382.86	537-64-4 245.7	3387	4-16-00-01705	
7433	Mercury, dibutyl- Dibutylmercury	$C_8H_{18}Hg$ 314.82	629-35-6	105[10]	4-04-00-04429 1.7779[20]	1.5057[20]
7434	Mercury, diethyl- Diethyl mercury	$C_4H_{10}Hg$ 258.71	627-44-1	159; 57[16]	2.43[20]	eth 3; EtOH 2
7435	Mercury, dimethyl- Dimethyl mercury	C_2H_6Hg 230.66	593-74-8 93	3238	3.17[25]	H_2O 1; EtOH 4; eth 4 1.5452[20]
7436	Mercury, diphenyl- Benzene, mercuriodi-	$C_{12}H_{10}Hg$ 354.80	587-85-9	204[10]	4-16-00-01702 2.318[25]	H_2O 1; EtOH 2; eth 2; bz 3
7437	Mercury, [2-methyl-5-nitrophenolato(2-)-C6,O1]- Nitromersol	$C_7H_5HgNO_3$ 351.71	133-58-4	6531		H_2O 1; ace 2; EtOH 3; alk 3
7438	Merphos Phosphorotrithious acid, S,S,S-tributyl ester	$C_{12}H_{27}PS_3$ 298.52	150-50-5 100	137[0.7]; 176[15]	1.02[20]	
7439	Metalaxyl N-(2,6-Dimethylphenyl)-N-(methoxyacetyl)alanine, methyl ester	$C_{15}H_{21}NO_4$ 279.34	57837-19-1 71	5826		
7440	Metaldehyde Acetaldehyde polymer	$(C_2H_4O)_x$	37273-91-9 246	5827 sub 115	4-19-00-05643	H_2O 1; EtOH 2; eth 2; ace 1
7441	Methamidophos Phosphoramidothioic acid, O,S-dimethyl ester	$C_2H_8NO_2PS$ 141.13	10265-92-6 46	5858	1.31[20]	

PHYSICAL CONSTANTS OF ORGANIC COMPOUNDS (continued)

No.	Name / Synonym	Mol. Form. / Mol. Wt.	CAS RN / mp/°C	Merck No. / bp/°C	Beil. Ref. / den/g cm⁻³	Solubility / n_D
7442	Methanamine / Methylamine	CH₅N / 31.06	74-89-5 / -93.4	5938 / -6.3	4-04-00-00118 / *0.656²⁵	H₂O 4; EtOH 3; eth 5; ace 3
7443	Methanamine, 1,1-diethoxy-N,N-dimethyl-	C₇H₁₇NO₂ / 147.22	1188-33-6 /	/ 129	4-04-00-00174 / 0.859²⁵	1.4007²⁰
7444	Methanamine, N,N-dimethyl- / Trimethylamine	C₃H₉N / 59.11	75-50-3 / -117.1	9625 / 2.8	4-04-00-00134 / *0.627²⁵	H₂O 4; EtOH 3; eth 3; bz 3 / 1.3631⁰
7445	Methanamine, N,N-dimethyl-, hydrochloride	C₃H₁₀ClN / 95.57	593-81-7 / 277.5	/ sub 200	4-04-00-00134	H₂O 4; EtOH 4; chl 4
7446	Methanamine, hydrochloride	CH₆ClN / 67.52	593-51-1 / 227.5	5938 / 227¹⁵	4-04-00-00118	H₂O 4; EtOH 4
7447	Methanamine, N-hydroxy- / Hydroxylamine, N-methyl	CH₅NO / 47.06	593-77-1 / 87.5	/ 62.5¹⁵	4-04-00-03299 / 1.0003²⁰	H₂O 4; EtOH 4 / 1.4164²⁰
7448	Methanamine, N-methyl- / Dimethylamine	C₂H₇N / 45.08	124-40-3 / -92.2	3217 / 6.8	4-04-00-00128 / 0.6804⁰	H₂O 4; EtOH 3; eth 3 / 1.350¹⁷
7449	Methanamine, N-methyl-N-nitro-	C₂H₆N₂O₂ / 90.08	4164-28-7 / 58	/ 187	4-04-00-03401 / 1.1090⁷²	H₂O 4; ace 4; eth 4; EtOH 4 / 1.4462⁷²
7450	Methanamine, N-methyl-N-nitroso- / N-Nitrosodimethylamine	C₂H₆N₂O / 74.08	62-75-9 /	6558 / 154	4-04-00-03384 / 1.0059²⁰	H₂O 3; EtOH 3; eth 3; chl 3 / 1.4358²⁰
7451	Methanamine, N-nitro-	CH₄N₂O₂ / 76.06	598-57-2 / 38	/ 80-5¹⁰	4-04-00-03400 / 1.2433⁴⁹	H₂O 4; EtOH 4; eth 3; bz 4 / 1.4616⁴⁹
7452	Methanamine, N-(phenylmethylene)-	C₈H₉N / 119.17	622-29-7 /	/ 185; 92³⁴	4-07-00-00518 / 0.9671¹⁴	EtOH 3; eth 3; ace 3; chl 3 / 1.5526²⁰
7453	Methanaminium, N-[4-[bis[4-(dimethylamino)phenyl]methylene]-2,5-cyclohexadien-1-ylidene]-N-methyl-, chloride / Crystal violet	C₂₅H₃₀ClN₃ / 407.99	548-62-9 / 215 dec	4287		H₂O 4; chl 4
7454	Methanaminium, 1-carboxy-N,N,N-trimethyl-, hydroxide, inner salt / Betaine	C₅H₁₁NO₂ / 117.15	107-43-7 / 293 dec	1201	4-04-00-02369	H₂O 4; EtOH 4; eth 2; chl 2
7455	Methanaminium, N-[4-[[4-(dimethylamino)phenyl]phenylmethylene]-2,5-cyclohexadien-1-ylidene]-N-methyl-, chlorid / Malachite green	C₂₃H₂₅ClN₂ / 364.92	569-64-2 /	5581		H₂O 4; EtOH 4; MeOH 4
7456	Methanaminium, N,N,N-trimethyl-, bromide / Tetramethylammonium bromide	C₄H₁₂BrN / 154.05	64-20-0 / 230 dec		4-04-00-00145 / 1.56²⁵	H₂O 4; EtOH 2; eth 1; bz 1
7457	Methanaminium, N,N,N-trimethyl-, chloride / Tetramethylammonium chloride	C₄H₁₂ClN / 109.60	75-57-0 / 420 dec		4-04-00-00145 / 1.169²⁰	H₂O 3; EtOH 2; eth 1; bz 1
7458	Methanaminium, N,N,N-trimethyl-, iodide / Tetramethylammonium iodide	C₄H₁₂IN / 201.05	75-58-1 / >230 dec	9156	4-04-00-00145 / 1.829²⁵	H₂O 2; EtOH 2; eth 1; ace 2
7459	Methane	CH₄ / 16.04	74-82-8 / -182.4	5863 / -161.5	4-01-00-00003 / 0.4228⁻¹⁶²	H₂O 3; EtOH 3; eth 3; ace 2
7460	Methane, azido / Methyl azide	CH₃N₃ / 57.06	624-90-8 /	/ 20.5; exp 500	0.869¹⁵	
7461	Methane, bis(methylthio)- / Bis[methylmercapto]methane	C₃H₈S₂ / 108.23	1618-26-4 /	1267 / 148	4-01-00-03088	
7462	Methane, bromo- / Methyl bromide	CH₃Br / 94.94	74-83-9 / -93.7	5951 / 3.5	4-01-00-00068 / 1.6755²⁰	H₂O 2; EtOH 5; eth 5; chl 5 / 1.4218²⁰
7463	Methane, bromochloro- / Bromochloromethane	CH₂BrCl / 129.38	74-97-5 / -87.9	/ 68.0	4-01-00-00074 / 1.9344²⁰	H₂O 1; EtOH 3; eth 3; ace 3 / 1.4838²⁰
7464	Methane, bromochlorodifluoro- / Bromochlorodifluoromethane	CBrClF₂ / 165.36	353-59-3 / -159.5	/ -3.7	4-01-00-00075	
7465	Methane, bromochlorodinitro- / Bromochlorodinitromethane	CBrClN₂O₄ / 219.38	33829-48-0 / 9.3	/ 75-6¹⁵	2-01-00-00044 / 2.0393²⁰	EtOH 4 / 1.4793
7466	Methane, bromochlorofluoro- / Bromochlorofluoromethane	CHBrClF / 147.37	593-98-6 / -115	/ 36	4-01-00-00075 / 1.9771⁰	H₂O 1; eth 3; ace 3; chl 3 / 1.4144²⁵
7467	Methane, bromodichloro- / Bromodichloromethane	CHBrCl₂ / 163.83	75-27-4 / -57	/ 90	4-01-00-00076 / 1.980²⁰	H₂O 1; EtOH 4; eth 4; ace 4 / 1.4964²⁰
7468	Methane, bromodichlorofluoro- / Bromodichlorofluoromethane	CBrCl₂F / 181.82	353-58-2 /		4-01-00-00076	
7469	Methane, bromodifluoro- / Bromodifluoromethane	CHBrF₂ / 130.92	1511-62-2 / -14.5		4-01-00-00072 / 1.55¹⁶	H₂O 3; EtOH 4
7470	Methane, bromofluoro- / Bromofluoromethane	CH₂BrF / 112.93	373-52-4 /	/ 19	4-01-00-00072	EtOH 3; chl 4
7471	Methane, bromoiodo- / Bromoiodomethane	CH₂BrI / 220.84	557-68-6 /	/ 139.5	4-01-00-00095 / 2.926¹⁷	chl 4 / 1.6410²⁰
7472	Methane, bromomethoxy-	C₂H₅BrO / 124.97	13057-17-5 /	/ 87	4-01-00-03051 / 1.5976²⁰	1.4562²⁰
7473	Methane, bromonitro- / Bromonitromethane	CH₂BrNO₂ / 139.94	563-70-2 /	/ 149; 71⁴⁰	4-01-00-00106	EtOH 4 / 1.4880²⁰
7474	Methane, bromotrichloro- / Bromotrichloromethane	CBrCl₃ / 198.27	75-62-7 / -5.7	/ 105	4-01-00-00077 / 2.012²⁵	eth 4; EtOH 4 / 1.5065²⁰
7475	Methane, bromotrifluoro- / Bromotrifluoromethane	CBrF₃ / 148.91	75-63-8 / -172	/ -57.8	4-01-00-00073 / 1.5800²⁰	chl 4
7476	Methane, bromotrinitro- / Bromotrinitromethane	CBrN₃O₆ / 229.93	560-95-2 / 17.5	/ 56¹⁰	3-01-00-00116 / 2.0312²⁰	EtOH 4; chl 4 / 1.4808²⁰
7477	Methane, chloro- / Methyl chloride	CH₃Cl / 50.49	74-87-3 / -97.7	5964 / -24.0	4-01-00-00028 / *0.911²⁵	H₂O 3; EtOH 3; eth 5; ace 5 / 1.3389²⁰
7478	Methane, chlorodifluoro- / Chlorodifluoromethane	CHClF₂ / 86.47	75-45-6 / -157.4	/ -40.7	4-01-00-00032 / 1.4909⁻⁶⁹	H₂O 4; eth 3; ace 3; chl 3

No.	Name Synonym	Mol. Form. Mol. Wt.	CAS RN mp/°C	Merck No. bp/°C	Beil. Ref. den/g cm^{-3}	Solubility n_D
7479	Methane, chlorodiiodo-	CHClI$_2$ 302.28	638-73-3 -4	200 dec; 101^{33}	4-01-00-00097 3.1700^0	ace 4; eth 4; chl 4
7480	Methane, chlorodinitro- Chlorodinitromethane	CHClN$_2$O$_4$ 140.48	921-13-1 34-6^{13}		3-01-00-00115 1.6125^{20}	1.4575^{20}
7481	Methane, chlorofluoro- Chlorofluoromethane	CH$_2$ClF 68.48	593-70-4 -133	-9.1	4-01-00-00032	chl 4
7482	Methane, chloroiodo- Chloroiodomethane	CH$_2$ClI 176.38	593-71-5 109		4-01-00-00094 2.422^{20}	ace 4; bz 4; eth 4; EtOH 4 1.5822^{20}
7483	Methane, chloromethoxy- Chloromethyl methyl ether	C$_2$H$_5$ClO 80.51	107-30-2 -103.5	2146 59.5	4-01-00-03046 1.063^{10}	EtOH 3; eth 3; ace 3; chl 3 1.397^{20}
7484	Methane, chloro(methylthio)-	C$_2$H$_5$ClS 96.58	2373-51-5 105		4-01-00-03079 1.153^{25}	' 1.4963^{20}
7485	Methane, chloronitro- Chloronitromethane	CH$_2$ClNO$_2$ 95.49	1794-84-9 122.5		4-01-00-00106 1.466^{15}	H$_2$O 4
7486	Methane, chlorotrifluoro- Chlorotrifluoromethane	CClF$_3$ 104.46	75-72-9 -181	-81.4	4-01-00-00034	
7487	Methane, chlorotrinitro-	CClN$_3$O$_6$ 185.48	1943-16-4 4.5	134 dec; 56^{40}	3-01-00-00116 1.6769^{20}	eth 4; EtOH 4; chl 4 1.4500^{20}
7488	Methanediamine, 1-(2-furanyl)-N,N'-bis(2-furanylmethylene)- Hydrofuramide	C$_{15}$H$_{12}$N$_2$O$_3$ 268.27	494-47-3 117	4718	5-17-09-00306	H$_2$O 1; EtOH 4; eth 4
7489	Methanediamine, N,N,N',N'-tetraethyl-	C$_9$H$_{22}$N$_2$ 158.29	102-53-4 165.8		4-04-00-00338 0.8000^{20}	1.4420^{25}
7490	Methanediamine, N,N,N',N'-tetramethyl- Tetramethylmethanediamine	C$_5$H$_{14}$N$_2$ 102.18	51-80-9 83		4-04-00-00153 0.7491^{18}	H$_2$O 3
7491	Methane, diazo- Diazomethane	CH$_2$N$_2$ 42.04	334-88-3 -145	2983 -23	4-01-00-03056	eth 4; diox 4
7492	Methane, dibromo- Methylene bromide	CH$_2$Br$_2$ 173.83	74-95-3 -52.5	5980 97	4-01-00-00078 2.4969^{20}	H$_2$O 2; EtOH 5; eth 5; ace 5 1.5420^{20}
7493	Methane, dibromochloro- Chlorodibromomethane	CHBr$_2$Cl 208.28	124-48-1 -20	120	4-01-00-00081 2.451^{20}	H$_2$O 1; EtOH 3; eth 3; ace 3 1.5482^{20}
7494	Methane, dibromochlorofluoro- Dibromochlorofluoromethane	CBr$_2$ClF 226.27	353-55-9 80.3		4-01-00-00082 2.3173^{22}	1.4570^{20}
7495	Methane, dibromodichloro- Dibromodichloromethane	CBr$_2$Cl$_2$ 242.72	594-18-3 38	150.2	4-01-00-00082 2.42^{25}	H$_2$O 1; EtOH 3; eth 3; ace 3
7496	Methane, dibromodifluoro- Dibromodifluoromethane	CBr$_2$F$_2$ 209.82	75-61-6 -110.1	25	4-01-00-00080	H$_2$O 3; eth 3; ace 3; bz 3
7497	Methane, dibromodinitro- Dibromodinitromethane	CBr$_2$N$_2$O$_4$ 263.83	2973-00-4 5.5	158 dec; 77^{21}	3-01-00-00115 2.4439^{20}	EtOH 4 1.5280^{25}
7498	Methane, dibromofluoro- Dibromofluoromethane	CHBr$_2$F 191.83	1868-53-7 -78	64.9	4-01-00-00080 2.421^{20}	H$_2$O 1; EtOH 3; eth 3; ace 3 1.4685^{20}
7499	Methane, dichloro- Methylene chloride	CH$_2$Cl$_2$ 84.93	75-09-2 -95.1	5982 40	4-01-00-00035 1.3266^{20}	H$_2$O 2; EtOH 5; eth 5; ctc 3 1.4242^{20}
7500	Methane, dichlorodifluoro- Dichlorodifluoromethane	CCl$_2$F$_2$ 120.91	75-71-8 -158	3053 -29.8	4-01-00-00040	H$_2$O 3; EtOH 3; eth 3; HOAc 3
7501	Methane, dichlorodinitro- Dichlorodinitromethane	CCl$_2$N$_2$O$_4$ 174.93	1587-41-3 121.5		3-01-00-00115 1.6124^{20}	bz 4; eth 4; EtOH 4; chl 4 1.4575^{20}
7502	Methane, dichlorofluoro- Dichlorofluoromethane	CHCl$_2$F 102.92	75-43-4 -135	8.9	4-01-00-00039 1.405^9	H$_2$O 1; EtOH 3; eth 3; ctc 3 1.3724^9
7503	Methane, dichloroiodo- Dichloroiodomethane	CHCl$_2$I 210.83	594-04-7 132		4-01-00-00095 2.392^{20}	ace 4; bz 4; eth 4; EtOH 4 1.5840^{20}
7504	Methane, difluoro- Methylene difluoride	CH$_2$F$_2$ 52.02	75-10-5 -136	-51.6	4-01-00-00024 *0.960^{25}	H$_2$O 1; EtOH 3
7505	Methane, difluoroiodo- Difluoroiodomethane	CHF$_2$I 177.92	1493-03-4 -122	21.6	4-01-00-00092 3.238^{-19}	
7506	Methane, diiodo- Methylene iodide	CH$_2$I$_2$ 267.84	75-11-6 6.1	5985 182	4-01-00-00096 3.3212^{20}	H$_2$O 2; EtOH 3; eth 3; bz 3 1.7425^{20}
7507	Methane, dimethoxy- Methylal	C$_3$H$_8$O$_2$ 76.10	109-87-5 -104.8	5936 42	4-01-00-03026 0.8593^{20}	ace 4; bz 4; eth 4; EtOH 4 1.3513^{20}
7508	Methane, dinitro- Dinitromethane	CH$_2$N$_2$O$_4$ 106.04	625-76-3 <-15	exp 100	4-01-00-00107	H$_2$O 1; EtOH 3; eth 3
7509	Methanediol, dibenzoate	C$_{15}$H$_{12}$O$_4$ 256.26	5342-31-4 99	225 dec	4-09-00-00448 1.275^{22}	ace 4; bz 4; eth 4
7510	Methanediol, 2-furanyl-, diacetate	C$_9$H$_{10}$O$_5$ 198.18	613-75-2 53.3	220	5-17-09-00300	bz 4; eth 4; EtOH 4
7511	Methanediol, (3-nitrophenyl)-, diacetate (ester) Benzaldehyde, 3-nitro, diacetate	C$_{11}$H$_{11}$NO$_6$ 253.21	29949-19-7 72		3-07-00-00900 1.393^{25}	ace 4; bz 4; eth 4; EtOH 4
7512	Methanediol, phenyl-, diacetate	C$_{11}$H$_{12}$O$_4$ 208.21	581-55-5 46	220	4-07-00-00516 1.11^{20}	bz 4; eth 4; EtOH 4
7513	Methane, diphenyl-m-tolyl- Triphenylmethane, 3-methyl	C$_{20}$H$_{18}$ 258.36	603-26-9 62	357	3-05-00-02332 1.07^{16}	bz 4; eth 4; chl 4; lig 4
7514	Methanedisulfonic acid Methionic acid	CH$_4$O$_6$S$_2$ 176.17	503-40-2 98	5895	4-01-00-03054	H$_2$O 1; HNO$_3$ 3
7515	Methane, fluoro- Methyl fluoride	CH$_3$F 34.03	593-53-3 -141.8	4103 -78.4	4-01-00-00022 *0.566^{25}	H$_2$O 4; EtOH 4; eth 4; bz 3
7516	Methane, fluorodiiodo-	CHFI$_2$ 285.83	1493-01-2 -34.5	100	4-01-00-00097 3.1969^{22}	eth 4; EtOH 4

No.	Name Synonym	Mol. Form. Mol. Wt.	CAS RN mp/°C	Merck No. bp/°C	Beil. Ref. den/g cm^{-3}	Solubility n_D
7517	Methane, fluoroiodo- Fluoroiodomethane	CH$_2$FI 159.93	373-53-5	53.4	4-01-00-00092 2.366[20]	ace 4; bz 4; eth 4; chl 4 1.5256[20]
7518	Methane, iodo- Methyl iodide	CH$_3$I 141.94	74-88-4 -66.4	6002 42.5	4-01-00-00087 2.279[20]	H$_2$O 2; EtOH 5; eth 5; ace 3 1.5380[20]
7519	Methane, iodomethoxy- (Iodomethyl) methyl ether	C$_2$H$_5$IO 171.97	13057-19-7	122	4-01-00-03052 2.030[20]	1.5472[20]
7520	Methane, isocyanato- Methyl isocyanate	C$_2$H$_3$NO 57.05	624-83-9 -45	6004 39.5	4-04-00-00247 0.9230[27]	H$_2$O 4 1.3419[18]
7521	Methane, isothiocyanato- Methyl isothiocyanate	C$_2$H$_3$NS 73.12	556-61-6 36	6005 119	4-04-00-00248 1.0691[37]	H$_2$O 2; EtOH 5; eth 4 1.5258
7522	Methane, nitro- Nitromethane	CH$_3$NO$_2$ 61.04	75-52-5 -28.5	6532 101.1	4-01-00-00100 1.1371[20]	H$_2$O 3; EtOH 3; eth 3; ace 3 1.3817[20]
7523	Methane, oxybis- Dimethyl ether	C$_2$H$_6$O 46.07	115-10-6 -141.5	5990 -24.8	4-01-00-01245	H$_2$O 3; EtOH 3; eth 3; ace 3
7524	Methane, oxybis[bromo-	C$_2$H$_4$Br$_2$O 203.86	4497-29-4 -34	154.5	4-01-00-03052 2.2013[20]	
7525	Methane, oxybis[chloro- Bis(chloromethyl) ether	C$_2$H$_4$Cl$_2$O 114.96	542-88-1 -41.5	3058 106	4-01-00-03051 1.323[15]	EtOH 5; eth 5 1.435[21]
7526	Methane, selenobis-	C$_2$H$_6$Se 109.03	593-79-3	55	4-01-00-01288 1.4077[15]	eth 4; EtOH 4; chl 4
7527	Methanesulfenyl chloride, trichloro- Perchloromethyl mercaptan	CCl$_4$S 185.89	594-42-3	147.5	4-03-00-00290 1.6947[20]	eth 3 1.5484[20]
7528	Methanesulfenyl fluoride, trifluoro- Methyl mercaptan, perfluoro	CF$_4$S 120.07	17742-04-0	147.5		eth 4 1.6947[20] 1.5484[20]
7529	Methane, sulfinylbis- Dimethyl sulfoxide	C$_2$H$_6$OS 78.14	67-68-5 18.5	3247 189	4-01-00-01277 1.1014[20]	H$_2$O 3; EtOH 3; eth 3; ace 3 1.4170[20]
7530	Methanesulfonic acid Methylsulfonic acid	CH$_4$O$_3$S 96.11	75-75-2 20	5865 167[10]	4-04-00-00010 1.4812[18]	H$_2$O 3 1.4317[18]
7531	Methanesulfonic acid, ethyl ester Ethyl methanesulfonate	C$_3$H$_8$O$_3$S 124.16	62-50-0	3782 86[10]	4-04-00-00012	
7532	Methanesulfonic acid, methyl ester Methyl methanesulfonate	C$_2$H$_6$O$_3$S 110.13	66-27-3 20	6010 202.5	4-04-00-00011 1.2943[20]	1.4138[20]
7533	Methanesulfonic acid, trifluoro-, ethyl ester	C$_3$H$_5$F$_3$O$_3$S 178.13	425-75-2	115	4-03-00-00034 1.3740[0]	eth 3
7534	Methane, sulfonylbis- Dimethyl sulfone	C$_2$H$_6$O$_2$S 94.13	67-71-0 109	3246 238	4-01-00-01279 1.1700[110]	H$_2$O 3; EtOH 3; bz 3 1.4226
7535	Methanesulfonyl chloride Methanesulfonic acid chloride	CH$_3$ClO$_2$S 114.55	124-63-0	5866 162; 55[11]	4-04-00-00027 1.4805[18]	H$_2$O 1; EtOH 3; eth 3 1.4573[20]
7536	Methanesulfonyl chloride, trichloro-	CCl$_4$O$_2$S 217.89	2547-61-7 140.5	170	4-03-00-00036	H$_2$O 1; EtOH 3; eth 3; CS$_2$ 3
7537	Methanesulfonyl chloride, trifluoro-	CClF$_3$O$_2$S 168.52	421-83-0	162; 62[18]	4-03-00-00035	H$_2$O 1 1.3344[20]
7538	Methanesulfonyl fluoride Methanesulfonic acid fluoride	CH$_3$FO$_2$S 98.10	558-25-8	123.5	4-04-00-00027	
7539	Methane, tetrabromo- Carbon tetrabromide	CBr$_4$ 331.63	558-13-4 90.1	189.5	4-01-00-00085 2.9608[100]	H$_2$O 1; EtOH 3; eth 3; chl 3 1.5942[100]
7540	Methane, tetrachloro- Carbon tetrachloride	CCl$_4$ 153.82	56-23-5 -23	1822 76.8	4-01-00-00056 1.5940[20]	H$_2$O 1; EtOH 3; eth 5; ace 3 1.4601[20]
7541	Methane, tetrafluoro- Carbon tetrafluoride	CF$_4$ 88.00	75-73-0 -183.5	1823 -128.0	4-01-00-00026 3.034[25]	H$_2$O 2; bz 3; chl 3
7542	Methane, tetraiodo- Carbon tetraiodide	CI$_4$ 519.63	507-25-5 171	1824 130-40[1-2]	4-01-00-00098 4.23[20]	py 4; chl 4
7543	Methane, tetramethoxy-	C$_5$H$_{12}$O$_4$ 136.15	1850-14-2 -5	114	4-03-00-00004 1.023[25]	1.3845[20]
7544	Methane, tetranitro- Tetranitromethane	CN$_4$O$_8$ 196.03	509-14-8 13.8	9164 126.1	4-01-00-00107 1.6380[20]	H$_2$O 1; EtOH 3; eth 3 1.4384[20]
7545	Methane, thiobis- Dimethyl sulfide	C$_2$H$_6$S 62.14	75-18-3 -98.3	6042 37.3	4-01-00-01275 0.8483[20]	H$_2$O 1; EtOH 3; eth 3 1.4438[20]
7546	Methanethiol Methyl mercaptan	CH$_4$S 48.11	74-93-1 -123	5867 5.9	4-01-00-01273 0.8665[20]	H$_2$O 2; EtOH 4; eth 4; chl 2
7547	Methanethione, diphenyl-	C$_{13}$H$_{10}$S 198.29	1450-31-3 53.5	174[14]	4-07-00-01385	EtOH 2; eth 2; bz 4; chl 4
7548	Methane, tribromo- Bromoform	CHBr$_3$ 252.73	75-25-2 8.0	1407 149.1	4-01-00-00082 2.899[15]	H$_2$O 2; EtOH 5; eth 5; bz 3 1.6005[15]
7549	Methane, tribromochloro- Tribromochloromethane	CBr$_3$Cl 287.18	594-15-0 55	158.5	4-01-00-00085 2.71[15]	eth 4
7550	Methane, tribromofluoro- Tribromofluoromethane	CBr$_3$F 270.72	353-54-8 -73.6	108		EtOH 3
7551	Methane, tribromonitro-	CBr$_3$NO$_2$ 297.73	464-10-8 10	127[18]	4-01-00-00106 2.811[12]	H$_2$O 1; EtOH 3; eth 3; ace 4 1.5790[20]
7552	Methanetricarboxylic acid, triethyl ester Triethyl methanetricarboxylate	C$_{10}$H$_{16}$O$_6$ 232.23	6279-86-3 29	253	4-02-00-02360 1.1084[16]	1.4243[20]
7553	Methanetricarboxylic acid, trimethyl ester Trimethyl methanetricarboxylate	C$_7$H$_{10}$O$_6$ 190.15	1186-73-8 46.5	242.7; 128[15]	3-02-00-02023	bz 4; eth 4; EtOH 4; chl 4
7554	Methane, trichloro- Chloroform	CHCl$_3$ 119.38	67-66-3 -63.6	2141 61.1	4-01-00-00042 1.4832[20]	H$_2$O 2; EtOH 5; eth 5; ace 3 1.4459[20]
7555	Methane, trichlorofluoro- Trichlorofluoromethane	CCl$_3$F 137.37	75-69-4 -111.1	9553 23.7	4-01-00-00054	
7556	Methane, trichloroiodo- Trichloroiodomethane	CCl$_3$I 245.27	594-22-9 -7.8	142	4-01-00-00095 2.355[20]	ace 4; bz 4; eth 4; chl 4 1.5854[20]

No.	Name Synonym	Mol. Form. Mol. Wt.	CAS RN mp/°C	Merck No. bp/°C	Beil. Ref. den/g cm^{-3}	Solubility n_D
7557	Methane, trichloronitro- Chloropicrin	CCl_3NO_2 164.37	76-06-2 -64	2156 112	4-01-00-00106 1.6558[20]	H_2O 3; EtOH 5; ace 5; bz 5 1.4611[20]
7558	Methane, trifluoro- Fluoroform	CHF_3 70.01	75-46-7 -155.1	4102 -82.1	4-01-00-00024 *0.673[25]	H_2O 3; EtOH 4; ace 3; bz 3
7559	Methane, trifluoroiodo- Trifluoroiodomethane	CF_3I 195.91	2314-97-8 	 -22.5	4-01-00-00092 2.3607[-32]	 1.3790[-32]
7560	Methane, triiodo- Iodoform	CHI_3 393.73	75-47-8 119	4926 218	4-01-00-00097 4.008[25]	H_2O 1; EtOH 3; eth 3; ace 3
7561	Methane, trimethoxy-	$C_4H_{10}O_3$ 106.12	149-73-5 15	 104	4-02-00-00022 0.9676[20]	EtOH 3; eth 3 1.3793[20]
7562	Methane, trinitro- Trinitromethane	CHN_3O_6 151.04	517-25-9 15	9642 exp	4-01-00-00107 1.479[20]	ace 4; EtOH 4 1.4451[24]
7563	Methanimidamide Formamidine	CH_4N_2 44.06	463-52-5 81	 dec	4-02-00-00082	H_2O 4; EtOH 4
7564	Methanimidamide, N'-(4-chloro-2-methylphenyl)-N,N-dimethyl- Chlordimeform	$C_{10}H_{13}ClN_2$ 196.68	6164-98-3 35	2083 156-7[0.4]	 1.105[25]	bz 4; eth 4; EtOH 4 1.5885[25]
7565	Methanimidamide, N,N'-diphenyl-	$C_{13}H_{12}N_2$ 196.25	622-15-1 142	 >250	4-12-00-00372	H_2O 2; EtOH 3; eth 4; ace 3
7566	Methanimidamide, N-hydroxy-	CH_4N_2O 60.06	624-82-8 114.5	 dec	4-02-00-00084	EtOH 4
7567	Methanimidic acid, N-phenyl, ethyl ester	$C_9H_{11}NO$ 149.19	6780-49-0 	 214; 87[10]	4-12-00-00371 1.0051[20]	eth 3; bz 3 1.5279[20]
7568	1,4-Methanoazulene, decahydro-4,8,8-trimethyl-9-methylene-, [1S-(1α,3aβ,4α,8aβ)]- Longifolene	$C_{15}H_{24}$ 204.36	475-20-7 	5446 258; 126[15]	4-05-00-01192 0.9319[18]	H_2O 1; bz 3 1.5040[20]
7569	1,4-Methanoazulen-9-ol, decahydro-1,5,5,8a-tetramethyl-, [1R-(1α,3aβ,4α,8aβ,9S*)]-	$C_{15}H_{26}O$ 222.37	465-24-7 112	 287 dec	3-06-00-00426 1.0441[20]	 1.519
7570	1H-3a,7-Methanoazulen-6-ol, octahydro-3,6,8,8-tetramethyl-, Cedrol	$C_{15}H_{26}O$ 222.37	77-53-2 86	1919	3-06-00-00424 0.9479[90]	eth 4; EtOH 4; chl 4 1.4824[90]
7571	7,14-Methano-2H,6H-dipyrido[1,2-a:1',2'-e][1,5]diazocine, dodecahydro-, Sparteine	$C_{15}H_{26}N_2$ 234.38	90-39-1 30.5	8692 325; 173[8]	5-23-05-00497 1.0196[20]	eth 4; EtOH 4; chl 4 1.5312[20]
7572	7,14-Methano-2H,6H-dipyrido[1,2-a:1',2'-e][1,5]diazocin-6-one, dodecahydro-, [7R-(7α,7aβ,14α,14aalp Aphylline	$C_{15}H_{24}N_2O$ 248.37	577-37-7 54.5	761 200 dec	5-24-02-00298	ace 4; bz 4; eth 4; EtOH 4
7573	7,14-Methano-4H,6H-dipyrido[1,2-a:1',2'-e][1,5]diazocin-4-one, 7,7a,8,9,10,11,13,14-octahydro-, Anagyrine	$C_{15}H_{20}N_2O$ 244.34	486-89-5 	660 265[12]; 212[4]	5-24-03-00410	H_2O 3; EtOH 4; eth 3; bz 3
7574	7,14-Methano-4H,6H-dipyrido[1,2-a:1',2'-e][1,5]diazocin-4-one, dodecahydro-9-hydroxy-, Hydroxylupanine	$C_{15}H_{24}N_2O_2$ 264.37	15358-48-2 169.5	4760	4-25-00-00082	H_2O 4; EtOH 4; chl 4
7575	2,5-Methano-2H-furo[3,2-b]pyrrol-6-ol, hexahydro-4-methyl-, (2α,3aβ,5α,6β,6aβ Scopoline	$C_8H_{13}NO_2$ 155.20	487-27-4 108.5	8365 248	2-27-00-00061 1.0891[134]	H_2O 3
7576	4,7-Methano-1H-indene, 1,4,5,6,7,8,8-heptachloro-3a,4,7,7a-tetrahydro- Heptachlor	$C_{10}H_5Cl_7$ 373.32	76-44-8 95.5	4576	 1.57[9]	bz 4; eth 4; EtOH 4; lig 4
7577	4,7-Methano-1H-indene, 3a,4,7,7a-tetrahydro-, (3aα,4α,7α,7aα)-	$C_{10}H_{12}$ 132.21	1755-01-7 32	 170 dec; 65[14]	4-05-00-01399 0.9302[35]	eth 4; EtOH 4 1.5050[35]
7578	1,5-Methano-8H-pyrido[1,2-a][1,5]diazocin-8-one, 1,2,3,4,5,6-hexahydro-3-methyl-, (1R)- Caulophylline	$C_{12}H_{18}N_2O$ 204.27	486-86-2 137	1914	5-24-02-00538	H_2O 4; ace 4; bz 4; EtOH 4
7579	3,6-Methano-8H-1,5,7-trioxacyclopenta[ij]cycloprop[a]azulene-4,8(3H)-dione, hexahydro-2a-hydroxy Picrotoxin	$C_{30}H_{34}O_{13}$ 602.59	124-87-8 203.5	7388	4-19-00-05245	py 4; EtOH 4
7580	4,7-Methanoisobenzofuran-1,3-dione, 3a,4,7,7a-tetrahydro-, (3aα,4α,7α,7aα)- Carbic anhydride	$C_9H_8O_3$ 164.16	129-64-6 164.5	1801	5-17-11-00192 1.417[25]	ace 4; bz 4; EtOH 4; chl 4
7581	Methanol Methyl alcohol	CH_4O 32.04	67-56-1 -97.6	5868 64.6	4-01-00-01227 0.7914[20]	H_2O 5; EtOH 5; eth 5; ace 5 1.3288[20]
7582	Methanol, chloro-, acetate	$C_3H_5ClO_2$ 108.52	625-56-9 	 116	4-02-00-00280 1.194[20]	eth 4; EtOH 4 1.409[20]
7583	Methanol, (1,3,5-triazine-2,4,6-triyltriimino)tris- Trimethylolmelamine	$C_6H_{12}N_6O_3$ 216.20	1017-56-7	9631	4-26-00-01273	
7584	Methanol, (1,3,5-triazine-2,4,6-triyltrinitrilo)hexakis- Hexamethylolmelamine	$C_9H_{18}N_6O_6$ 306.28	531-18-0 137	4611	4-26-00-01274	H_2O 4
7585	Methanol, trichloro-, carbonate (2:1)	$C_3Cl_6O_3$ 296.75	32315-10-9 79	 203	4-03-00-00033 1.6290[80]	
7586	1,6-Methanonaphthalen-1(2H)-ol, octahydro-4,8a,9,9-tetramethyl-, Patchouli alcohol	$C_{15}H_{26}O$ 222.37	5986-55-0 56	7001	3-06-00-00426 0.9906[65]	H_2O 1; EtOH 3; eth 3 1.5029[65]

No.	Name / Synonym	Mol. Form. / Mol. Wt.	CAS RN / mp/°C	Merck No. / bp/°C	Beil. Ref. / den/g cm^{-3}	Solubility / n_D
7587	Methanone, 1,3,5-benzenetriyltris[phenyl- / 1,3,5-Tribenzoylbenzene	$C_{27}H_{18}O_3$ / 390.44	25871-69-6 / 120	260[1]	4-07-00-02847	chl 3
7588	Methanone, 2-benzofuranylphenyl-	$C_{15}H_{10}O_2$ / 222.24	6272-40-8 / 91	360	5-17-10-00568	H_2O 1
7589	Methanone, bis(3-aminophenyl)-	$C_{13}H_{12}N_2O$ / 212.25	611-79-0 / 173.5	285[11]	4-14-00-00255	eth 4; EtOH 4
7590	Methanone, bis(4-bromophenyl)-	$C_{13}H_8Br_2O$ / 340.01	3988-03-2 / 177	395	4-07-00-01380	bz 4; HOAc 4; chl 4
7591	Methanone, bis(3-chlorophenyl)-	$C_{13}H_8Cl_2O$ / 251.11	7094-34-0 / 124	163[2]	4-07-00-01376	H_2O 1; EtOH 3; eth 3
7592	Methanone, bis(4-chlorophenyl)-	$C_{13}H_8Cl_2O$ / 251.11	90-98-2 / 147.5	353	4-07-00-01376 / 1.4500[20]	H_2O 1; EtOH 3; eth 4; ace 2
7593	Methanone, bis[4-(dimethylamino)phenyl- / Michler's ketone	$C_{17}H_{20}N_2O$ / 268.36	90-94-8 / 179	6100 / 360 dec	4-14-00-00255	H_2O 1; EtOH 2; eth 1; bz 4
7594	Methanone, bis(2,4-dimethylphenyl)- / 2,2',4,4'-Tetramethylbenzophenone	$C_{17}H_{18}O$ / 238.33	3478-88-4	190[10]	4-07-00-01470 / 1.043[15]	1.5790[25]
7595	Methanone, bis(2-hydroxy-4-methoxyphenyl)- / Benzophenone 6	$C_{15}H_{14}O_5$ / 274.27	131-54-4 / 139.5	1109	4-08-00-03505	
7596	Methanone, bis(2-hydroxyphenyl)-	$C_{13}H_{10}O_3$ / 214.22	835-11-0 / 59.5	333	4-08-00-02446	H_2O 1; EtOH 3; eth 3; chl 3
7597	Methanone, bis(4-hydroxyphenyl)-	$C_{13}H_{10}O_3$ / 214.22	611-99-4 / 210		4-08-00-02452 / 1.133[131]	H_2O 2; EtOH 3; eth 3; ace 3
7598	Methanone, bis(4-iodophenyl)- / 4,4'-Diiodobenzophenone	$C_{13}H_8I_2O$ / 434.01	5630-56-8 / 238.5	281[12]	4-07-00-01381 / 2.2600[20]	
7599	Methanone, bis(4-methylphenyl)-	$C_{15}H_{14}O$ / 210.28	611-97-2 / 96.5	334	4-07-00-01434	ace 4; bz 4; eth 4; EtOH 4
7600	Methanone, (2-bromophenyl)phenyl- / 2-Bromobenzophenone	$C_{13}H_9BrO$ / 261.12	13047-06-8 / 42	345	3-07-00-02079	ace 4; EtOH 4; lig 4
7601	Methanone, (3-bromophenyl)phenyl-	$C_{13}H_9BrO$ / 261.12	1016-77-9 / 81	186[6]	4-07-00-01378	EtOH 4
7602	Methanone, (4-bromophenyl)phenyl-	$C_{13}H_9BrO$ / 261.12	90-90-4 / 82.5	350	4-07-00-01378	H_2O 1; EtOH 2; eth 2; bz 2
7603	Methanone, (2-chlorophenyl)(4-chlorophenyl)-	$C_{13}H_8Cl_2O$ / 251.11	85-29-0 / 67	214-5[22]	4-07-00-01376 / 1.393[14]	EtOH 3; chl 2
7604	Methanone, (4-chlorophenyl)phenyl-	$C_{13}H_9ClO$ / 216.67	134-85-0 / 77.5	332	4-07-00-01375	EtOH 3; eth 3; ace 3; ctc 2
7605	Methanone, cyclobutylphenyl-	$C_{11}H_{12}O$ / 160.22	5407-98-7	260	4-07-00-01027 / 1.0426[25]	1.5472[20]
7606	Methanone, dicyclohexyl-	$C_{13}H_{22}O$ / 194.32	119-60-8	159[20]	4-07-00-00262 / 0.986[0]	eth 3; ace 3; ctc 3 / 1.4860[20]
7607	Methanone, dicyclopropyl- / Dicyclopropyl ketone	$C_7H_{10}O$ / 110.16	1121-37-5	161	4-07-00-00139 / 0.977[25]	1.4670[20]
7608	Methanone, (2,6-dihydroxy-4-methoxyphenyl)phenyl- / Cotoin	$C_{14}H_{12}O_4$ / 244.25	479-21-0 / 130.5	2553	4-08-00-03163	ace 4; bz 4; eth 4; EtOH 4
7609	Methanone, (2,4-dihydroxyphenyl)phenyl- / Benzoresorcinol	$C_{13}H_{10}O_3$ / 214.22	131-56-6 / 144	1117	4-08-00-02442	H_2O 1; EtOH 3; eth 4; bz 2
7610	Methanone, (3,4-dihydroxyphenyl)(2,4,6-trihydroxyphenyl)- / Maclurin	$C_{13}H_{10}O_6$ / 262.22	519-34-6 / 222.5	5519	3-08-00-04226	eth 4; EtOH 4
7611	Methanone, (6,7-dimethoxy-1-isoquinolinyl)(3,4-dimethoxyphenyl)- / Papaveraldine	$C_{20}H_{19}NO_5$ / 353.37	522-57-6 / 210.5	6966	5-21-13-00658	H_2O 1; EtOH 2; eth 2; bz 3
7612	Methanone, (2,4-dimethoxyphenyl)phenyl- / 2,4-Dimethoxybenzophenone	$C_{15}H_{14}O_3$ / 242.27	3555-84-8 / 87.5	218[10]	4-08-00-02442	H_2O 1; EtOH 4; eth 2; chl 4
7613	Methanone, [4-(dimethylamino)phenyl]phenyl- / 4-(Dimethylamino)benzophenone	$C_{15}H_{15}NO$ / 225.29	530-44-9 / 92.5	3222	4-14-00-00248	H_2O 1; EtOH 2; eth 4; chl 3
7614	Methanone, (2,4-dimethylphenyl)phenyl-	$C_{15}H_{14}O$ / 210.28	1140-14-3	321.5	4-07-00-01429 / 1.071[20]	bz 4; eth 4; EtOH 4; chl 4
7615	Methanone, (3,4-dimethylphenyl)phenyl- / 3,4-Dimethylbenzophenone	$C_{15}H_{14}O$ / 210.28	2571-39-3 / 47.5	341	4-07-00-01433	bz 4; eth 4; EtOH 4; chl 4
7616	Methanone, diphenyl- / Benzophenone	$C_{13}H_{10}O$ / 182.22	119-61-9 / 47.8	1108 / 305.4	4-07-00-01357 / 1.6077[19]	H_2O 1; EtOH 4; eth 4; ace 4
7617	Methanone, di-2-thienyl- / Di-(2-thienyl) ketone	$C_9H_6OS_2$ / 194.28	704-38-1 / 90	326	5-19-04-00318	ace 4; eth 4
7618	Methanone, 2-furanylphenyl-	$C_{11}H_8O_2$ / 172.18	2689-59-0 / <-15	285	5-17-10-00289 / 1.1732[20]	eth 4; EtOH 4 / 1.6055[20]
7619	Methanone, (2-hydroxy-4-methoxyphenyl)(2-hydroxyphenyl)- / Dioxybenzone	$C_{14}H_{12}O_4$ / 244.25	131-53-3	3298 / 170-5[1]	4-08-00-03163	
7620	Methanone, (2-hydroxy-4-methoxyphenyl)phenyl- / Oxybenzone	$C_{14}H_{12}O_3$ / 228.25	131-57-7 / 65.5	6907	4-08-00-02442	ctc 3
7621	Methanone, [2-hydroxy-4-(octyloxy)phenyl]phenyl- / Octabenzone	$C_{21}H_{26}O_3$ / 326.44	1843-05-6 / 48.5	6662		
7622	Methanone, (2-hydroxyphenyl)phenyl-	$C_{13}H_{10}O_2$ / 198.22	117-99-7 / 40	250[560]	4-08-00-01246	H_2O 1; EtOH 4; eth 4; bz 4
7623	Methanone, (2-methoxyphenyl)phenyl-	$C_{14}H_{12}O_2$ / 212.25	2553-04-0 / 41	210[27]; 195[18]	4-08-00-01246	H_2O 1; EtOH 3; bz 3; HOAc 3

No.	Name / Synonym	Mol. Form. / Mol. Wt.	CAS RN / mp/°C	Merck No. / bp/°C	Beil. Ref. / den/g cm^{-3}	Solubility / n_D
7624	Methanone, (3-methoxyphenyl)phenyl-	$C_{14}H_{12}O_2$ / 212.25	6136-67-0 / 44	343; 201[17]	4-08-00-01262	H_2O 1; EtOH 4; bz 4; HOAc 4
7625	Methanone, (4-methoxyphenyl)phenyl-	$C_{14}H_{12}O_2$ / 212.25	611-94-9 / 61.5	355; 168[12]	4-08-00-01263	H_2O 1; EtOH 4; eth 4; ace 3
7626	Methanone, [2-(methylamino)phenyl]phenyl-	$C_{14}H_{13}NO$ / 211.26	1859-76-3 / 69	282	4-14-00-00244	bz 4; EtOH 4; HOAc 4
7627	Methanone, (4-methyl-2-nitrophenyl)phenyl- / 4-Methyl-2-nitrobenzophenone	$C_{14}H_{11}NO_3$ / 241.25	100224-75-7 / 126.5	sub	0-07-00-00442	bz 4; EtOH 4; chl 4
7628	Methanone, (2-methylphenyl)(4-methylphenyl)- / 2,4'-Dimethylbenzophenone	$C_{15}H_{14}O$ / 210.28	1140-16-5	317	4-07-00-01432 / 1.074[19]	eth 4; EtOH 4; HOAc 4
7629	Methanone, (2-methylphenyl)-1-naphthalenyl-	$C_{18}H_{14}O$ / 246.31	68723-25-1 / 64	365	3-07-00-02639	eth 4; EtOH 4
7630	Methanone, (4-methylphenyl)(4-nitrophenyl)-	$C_{14}H_{11}NO_3$ / 241.25	5350-47-0 / 124	sub	4-07-00-01406	bz 4; eth 4; EtOH 4; chl 4
7631	Methanone, (2-methylphenyl)phenyl-	$C_{14}H_{12}O$ / 196.25	131-58-8 / <-18	308; 128[12]	4-07-00-01401 / 1.1098[20]	H_2O 1; EtOH 4; os 3
7632	Methanone, (3-methylphenyl)phenyl-	$C_{14}H_{12}O$ / 196.25	643-65-2	317; 170[9]	4-07-00-01403 / 1.088[17]	H_2O 1; EtOH 3; eth 3; bz 3
7633	Methanone, (4-methylphenyl)phenyl-	$C_{14}H_{12}O$ / 196.25	134-84-9 / 59.5	228[70]	4-07-00-01403 / 0.9926[0]	H_2O 1; EtOH 2; eth 3; bz 3
7634	Methanone, 1-naphthalenyl-2-naphthalenyl- / 1,2'-Dinaphthyl ketone	$C_{21}H_{14}O$ / 282.34	605-79-8 / 136.5	235[0.6]	4-07-00-01886	chl 4
7635	Methanone, (3-nitrophenyl)phenyl-	$C_{13}H_9NO_3$ / 227.22	2243-80-3 / 95	234[18]	4-07-00-01382	H_2O 1; EtOH 3
7636	Methanone, (4-nitrophenyl)phenyl-	$C_{13}H_9NO_3$ / 227.22	1144-74-7 / 138		4-07-00-01382 / 1.406[9]	bz 4
7637	Methanone, phenyl-1H-pyrrol-2-yl- / Pyrrole, 2-benzoyl	$C_{11}H_9NO$ / 171.20	7697-46-3 / 79	305	5-21-08-00461	bz 4; EtOH 4; peth 4; HOAc 4
7638	Methanone, phenyl-2-pyridinyl-	$C_{12}H_9NO$ / 183.21	91-02-1 / 42	317	5-21-08-00566 / 1.1556[20]	chl 3
7639	Methanone, phenyl-4-pyridinyl-	$C_{12}H_9NO$ / 183.21	14548-46-0 / 72	315; 170[10]	5-21-08-00576	H_2O 2; EtOH 3; eth 3; bz 3
7640	Methanone, phenyl-4-quinolinyl-	$C_{16}H_{11}NO$ / 233.27	54885-00-6 / 60	222[15]	1-21-00-00320	bz 4; eth 4; EtOH 4
7641	Methanone, phenyl-2-thienyl-	$C_{11}H_8OS$ / 188.25	135-00-2 / 56.5	300	5-17-10-00291 / 1.1890[54]	H_2O 1; EtOH 3; eth 3 / 1.6181[54]
7642	Methanone, phenyl-3-thienyl-	$C_{11}H_8OS$ / 188.25	6453-99-2 / 63.5	129[3]	5-17-10-00295	H_2O 1; EtOH 3; eth 3
7643	Methazole / 1,2,4-Oxadiazolidine-3,5-dione, 2-(3,4-dichlorophenyl)-4-methyl-	$C_9H_6Cl_2N_2O_3$ / 261.06	20354-26-1 / 123	5876	1.24[25]	
7644	1,3,4-Metheno-1H-cyclobuta[cd]pentalene, 1,1a,2,2,3,3a,4,5,5,5a,5b,6-dodecachlorooctahydro- / Mirex	$C_{10}Cl_{12}$ / 545.54	2385-85-5 / 485 dec	6126	3-05-00-01237	bz 4; diox 4
7645	Methidathion / Phosphorodithioic acid, S-[(5-methoxy-2-oxo-1,3,4-thiadiazol-3(2H)-yl)methyl], O,O-dimethyl ester	$C_6H_{11}N_2O_4PS_3$ / 302.34	950-37-8 / 39	5891		
7646	Methiocarb / Phenol, 3,5-dimethyl-4-(methylthio)-, methylcarbamate	$C_{11}H_{15}NO_2S$ / 225.31	2032-65-7 / 120	5893		
7647	DL-Methionine	$C_5H_{11}NO_2S$ / 149.21	59-51-8 / 281 dec		4-04-00-03190 / 1.340[25]	H_2O 4; EtOH 2; eth 1; bz 1
7648	L-Methionine / 2-Amino-4-methylthiobutanoic acid	$C_5H_{11}NO_2S$ / 149.21	63-68-3 / 281 dec	5896	4-04-00-03189	H_2O 3; EtOH 1; eth 1; ace 1
7649	L-Methionine, N-acetyl- / Methionamine	$C_7H_{13}NO_3S$ / 191.25	65-82-7 / 105.5	90	4-04-00-03205	
7650	Methomyl / Acetimidic acid, N-[(methylcarbamoyl)oxy]thio-, methyl ester	$C_5H_{10}N_2O_2S$ / 162.21	16752-77-5 / 78	5905	1.2946[24]	
7651	Methoprene / 2,4-Dodecadienoic acid, 11-methoxy-3,7,11-trimethyl-, 1-methylethyl ester, (E,E)-	$C_{19}H_{34}O_3$ / 310.48	40596-69-8 / 100[0.05]	5906	0.926[20]	
7652	Methylamine, N-(o-chlorobenzylidene)- / Toluene, 2-chloro-α-methylimino	C_8H_8ClN / 153.61	17972-08-6 / 7	78-80[2]	3-07-00-00866	1.5660[20]
7653	Methyl 3,6-dichloro-o-anisate / Benzoic acid, 3,6-dichloro-2-methoxy-, methyl ester	$C_9H_8Cl_2O_3$ / 235.07	6597-78-0	115[2.4]		
7654	Methylene bispropionate / Propanoic acid, methylene ester	$C_7H_{12}O_4$ / 160.17	7044-96-4	98[1]; 191[745]	1.053[20]	
7655	Methylisonitrile / Methyl isocyanate	C_2H_3N / 41.05	593-75-9 / -45	59.6 exp	0.756[4]	
7656	Metobromuron / Urea, 3-(p-bromophenyl)-1-methoxy-1-methyl-	$C_9H_{11}BrN_2O_2$ / 259.10	3060-89-7 / 95	6061	1.60[20]	
7657	Metolachlor / Acetamide, 2-chloro-N-(2-ethyl-6-methylphenyl)-N-(2-methoxy-1-methylethyl)-	$C_{15}H_{22}ClNO_2$ / 283.80	51218-45-2 / 100[0.001]	6067	1.12[20]	

No.	Name / Synonym	Mol. Form. / Mol. Wt.	CAS RN / mp/°C	Merck No. / bp/°C	Beil. Ref. / den/g cm^{-3}	Solubility / n_D
7658	Metribuzin 1,2,4-Triazin-5(4H)-one, 4-amino-6-(1,1-dimethylethyl)-3-(methylthio)-	$C_8H_{14}N_4OS$ 214.29	21087-64-9 126	6076	1.31[20]	
7659	Mevinphos 2-Butenoic acid, 3-[(dimethoxyphosphinyl)oxy]-, methyl ester,(E)-	$C_7H_{13}O_6P$ 224.15	7786-34-7 21 (E), 6.9 (Z)	6089 101[0.3]		
7660	MGK 264 4,7-Methano-1H-isoindole-1,3(2H)-dione, 2-(2-ethylhexyl)-3a,4,7,7a-tetrahydro-	$C_{17}H_{25}NO_2$ 275.39	113-48-4 <-20	157	1.04	
7661	Molinate 1H-Azepine-1-carbothioic acid, hexahydro-, S-ethyl ester	$C_9H_{17}NOS$ 187.31	2212-67-1	202[10]	1.063[20]	
7662	Molybdenum carbonyl Molybdenum hexacarbonyl	C_6MoO_6 264.00	13939-06-5 dec 150	701		os 3
7663	Monocrotophos (E)-Dimethyl 1-methyl-3-(methylamino)-3-oxo-1-propenyl phosphate	$C_7H_{14}NO_5P$ 223.17	6923-22-4 55	6161 125[0.0005]	1.33[20]	
7664	Morphinan, 6,7-didehydro-4,5-epoxy-3,6-dimethoxy-17-methyl-, (5α)- Dihydrothebaine	$C_{19}H_{23}NO_3$ 313.40	561-25-1 162.5	3164	2-27-00-00110	H_2O 1; EtOH 3; bz 3; AcOEt 3
7665	Morphinan-3,6-diol, 7,8-didehydro-4,5-epoxy-, (5α,6α)- Normorphine	$C_{16}H_{17}NO_3$ 271.32	466-97-7 273	6630		
7666	Morphinan-3,6-diol, 7,8-didehydro-4,5-epoxy-17-methyl- (5α,6α)- Morphine	$C_{17}H_{19}NO_3$ 285.34	57-27-2 255	6186 sub 190	4-27-00-02223	H_2O 1; EtOH 2; eth 1; ace 1
7667	Morphinan-3,6-diol, 7,8-didehydro-4,5-epoxy-17-methyl- (5α,6α)-, diacetate (ester) Diacetylmorphine	$C_{21}H_{23}NO_5$ 369.42	561-27-3 173	2948 272-4[12]	4-27-00-02236 1.56[25]	bz 4; chl 4
7668	Morphinan-3,6-diol, 7,8-didehydro-4,5-epoxy-17-methyl- (5α,6α)-, 17-oxide Morphine, N-oxide	$C_{17}H_{19}NO_4$ 301.34	639-46-3 274.5	6192	4-27-00-02249	H_2O 2; EtOH 2; ace 1; bz 1
7669	Morphinan-6-ol, 7,8-didehydro-4,5-epoxy-3-methoxy-17-methyl-, (5α,6α)- Codeine	$C_{18}H_{21}NO_3$ 299.37	76-57-3 157.5	2459 250[22,] 140[1.5]	2-27-00-00136 1.32[25]	H_2O 3; EtOH 4; eth 3; bz 3
7670	Morphinan-6-ol, 7,8-didehydro-4,5-epoxy-3-methoxy-17-methyl-, (5α,6β)-	$C_{18}H_{21}NO_3$ 299.37	509-64-8 171.5	dec	2-27-00-00175 1.87[4]	1.675
7671	Morphinan-6-ol, 8,14-didehydro-4,5-epoxy-3-methoxy-17-methyl-, (5α,6α)- Neopine	$C_{18}H_{21}NO_3$ 299.37	467-14-1 127.5	6376		H_2O 3; EtOH 3; eth 3; bz 3
7672	Morphinan-8-ol, 6,7-didehydro-4,5-epoxy-3-methoxy-17-methyl-, (5α,8β)- Pseudocodeine	$C_{18}H_{21}NO_3$ 299.37	466-96-6 181.5	7927	2-27-00-00112 1.290[80]	1.574
7673	Morphinan-6-ol, 7,8-didehydro-4,5-epoxy-3-methoxy-17-methyl-, (5α,6α)-, phosphate (1:1) (salt) Codeine, phosphate	$C_{18}H_{24}NO_7P$ 397.36	52-28-8 220-35 dec	2462	4-27-00-02228	EtOH 4; chl 4
7674	Morphinan-6-ol, 4,5-epoxy-3-methoxy-17-methyl-, (5α,6α)- Dihydrocodeine	$C_{18}H_{23}NO_3$ 301.39	125-28-0 112.5	3154 248[15]	2-27-00-00103	
7675	Morphinan-6-one, 7,8-didehydro-4,5-epoxy-14-hydroxy-3-methoxy-17-methyl-, (5α)- Hydroxycodeinone	$C_{18}H_{19}NO_4$ 313.35	508-54-3 275 dec	4752		
7676	Morphinan-6-one, 7,8-didehydro-4-hydroxy-3,7-dimethoxy-17-methyl-, (9α,13α,14α)- Sinomenine	$C_{19}H_{23}NO_4$ 329.40	115-53-7 162	8496	5-21-13-00516	H_2O 2; EtOH 3; eth 2; ace 3
7677	Morphinan-6-one, 7,8-didehydro-4-hydroxy-3-methoxy-17-methyl- Thebainone	$C_{18}H_{21}NO_3$ 299.37	467-98-1 151.5	9203	5-21-13-00268	H_2O 2; EtOH 2; eth 2; ace 3
7678	Morphinan-6-one, 4,5-epoxy-3-methoxy-17-methyl-, (5α)- Hydrocodone	$C_{18}H_{21}NO_3$ 299.37	125-29-1 198	4708	4-27-00-03580	H_2O 1; EtOH 3
7679	Morphinan, 6,7,8,14-tetradehydro-4,5-epoxy-3,6-dimethoxy-17-methyl-, (5α)- Thebaine	$C_{19}H_{21}NO_3$ 311.38	115-37-7 193	9202 sub 91	4-27-00-02271 1.305[20]	H_2O 1; EtOH 4; eth 2; bz 3
7680	4-Morpholinamine	$C_4H_{10}N_2O$ 102.14	4319-49-7	166	4-27-00-00623 1.059[25]	1.4772[20]
7681	Morpholine Tetrahydro-1,4-oxazine	C_4H_9NO 87.12	110-91-8 -4.9	6194 128	4-27-00-00015 1.0005[20]	H_2O 5; EtOH 3; eth 3; ace 3 1.4548[20]
7682	Morpholine, 4-acetyl-	$C_6H_{11}NO_2$ 129.16	1696-20-4 14.5	152[50,] 118[12]	4-27-00-00274 1.1145[20]	H_2O 5; EtOH 3; ace 3; ctc 3 1.4827[20]
7683	Morpholine, 4-[3-(4-butoxyphenoxy)propyl]- Pramoxine	$C_{17}H_{27}NO_3$ 293.41	140-65-8 196[6]	7707	4-27-00-00127	
7684	Morpholine, 4-butyl-	$C_8H_{17}NO$ 143.23	1005-67-0 -57.1	213.5	4-27-00-00026 0.9068[20]	H_2O 4; ace 4; bz 4; EtOH 4 1.4451[20]
7685	4-Morpholinecarboxaldehyde	$C_5H_9NO_2$ 115.13	4394-85-8 21	239	4-27-00-00274 1.1520[20]	1.4845[20]

No.	Name / Synonym	Mol. Form. / Mol. Wt.	CAS RN / mp/°C	Merck No. / bp/°C	Beil. Ref. / den/g cm^{-3}	Solubility / n_D
7686	4-Morpholinecarboxamide, N,N'-1,2-ethanediylbis[N-butyl- / Dimorpholamine	C$_{20}$H$_{38}$N$_4$O$_4$ / 398.55	119-48-2 / 41.5	3257 / 229$^{0.4}$	4-27-00-00298	H$_2$O 4
7687	Morpholine, 2,6-dimethyl- / 2,6-Dimethylmorpholine	C$_6$H$_{13}$NO / 115.18	141-91-3 / -88	/ 146.6	4-27-00-00674 / 0.9329^{20}	H$_2$O 5; EtOH 3; ace 3; bz 5 / 1.4460^{20}
7688	Morpholine, 4,4'-dithiobis- / 4,4'-Dithiodimorpholine	C$_8$H$_{16}$N$_2$O$_2$S$_2$ / 236.36	103-34-4 / 124.5	3379	4-27-00-00613	chl 3
7689	4-Morpholineethanamine	C$_6$H$_{14}$N$_2$O / 130.19	2038-03-1 / 25.6	/ 205	4-27-00-00370 / 0.9897^{20}	H$_2$O 5; EtOH 5; ace 3; bz 5 / 1.4715^{20}
7690	Morpholine, 4,4'-(1,2-ethanediyl)bis-	C$_{10}$H$_{20}$N$_2$O$_2$ / 200.28	1723-94-0 / 75	/ 285; 160^{25}	4-27-00-00375	H$_2$O 4; ace 4; bz 4; EtOH 4
7691	4-Morpholineethanol	C$_6$H$_{13}$NO$_2$ / 131.17	622-40-2 / -1.6	/ 227	4-27-00-00056 / 1.0710^{20}	H$_2$O 3; EtOH 3; ctc 2 / 1.4763^{20}
7692	4-Morpholineethanol, α-methyl-	C$_7$H$_{15}$NO$_2$ / 145.20	2109-66-2	/ 92-4^{13}	4-27-00-00119 / 1.0174^{20}	H$_2$O 4; ace 4; bz 4; EtOH 4 / 1.4638^{20}
7693	Morpholine, 4-(2-ethoxyethyl)- / Morpholine, N-(2-ethoxyethyl)	C$_8$H$_{17}$NO$_2$ / 159.23	622-09-3 / -100	/ 206	4-27-00-00057 / 0.963^{20}	H$_2$O 4; ace 4; bz 4; eth 4
7694	Morpholine, 4-ethyl- / N-Ethylmorpholine	C$_6$H$_{13}$NO / 115.18	100-74-3 /	/ 138.5	4-27-00-00023 / 0.8996^{20}	H$_2$O 5; EtOH 5; eth 5; ace 3 / 1.4400^{20}
7695	Morpholine, 4-(2-hydroxybenzoyl)-	C$_{11}$H$_{10}$NO$_3$ / 204.21	3202-84-4	8302	4-27-00-00346	DMSO 3
7696	Morpholine, 4-methyl- / N-Methylmorpholine	C$_5$H$_{11}$NO / 101.15	109-02-4 /	/ 116	4-27-00-00022 / 0.9051^{20}	H$_2$O 3; EtOH 3; eth 3 / 1.4332^{20}
7697	Morpholine, 3-methyl-2-phenyl- / Phenmetrazine	C$_{11}$H$_{15}$NO / 177.25	134-49-6 /	7200 / 139^{12}; 104^1	4-27-00-01050	
7698	Morpholine, 4-(4-methylphenyl)-	C$_{11}$H$_{15}$NO / 177.25	3077-16-5 / 51	/ 167^{30}	2-27-00-00004	eth 4; EtOH 4
7699	Morpholine, 4-nitroso- / N-Nitrosomorpholine	C$_4$H$_8$N$_2$O$_2$ / 116.12	59-89-2 / 29	6561 / 225; 140^{25}		H$_2$O 3
7700	Morpholine, 4-(phenylmethyl)-	C$_{11}$H$_{15}$NO / 177.25	10316-00-4 / 194	/ 260.5	4-27-00-00039 / 1.0387^{20}	ace 4; bz 4 / 1.5302^{20}
7701	4-Morpholinepropanamine / 4-(3-Aminopropyl)morpholine	C$_7$H$_{16}$N$_2$O / 144.22	123-00-2 / -15	/ 220; 134^{50}	4-27-00-00411 / 0.9854^{20}	H$_2$O 5; EtOH 5; ace 3; bz 5 / 1.4762^{20}
7702	Naled / Phosphoric acid, 1,2-dibromo-2,2-dichloroethyl dimethyl ester	C$_4$H$_7$Br$_2$Cl$_2$O$_4$P / 380.78	300-76-5 / 27	6272 / 110$^{0.5}$	1.96^{20}	
7703	Naphthacene / 2,3-Benzanthracene	C$_{18}$H$_{12}$ / 228.29	92-24-0 / 357	6288 / sub	4-05-00-02545	H$_2$O 2; bz 2; con sulf 3; os 2
7704	2-Naphthacenecarboxamide, 7-chloro-4-(dimethylamino)-1,4,4a,5,5a,6,11,12a-octahydro-3,6,10,12,12a-pentahydroxy-6-methyl-1,11-dioxo, [4S-(4α,4aα,5aα,6β,12aα)]- / Chlortetracycline	C$_{22}$H$_{23}$ClN$_2$O$_8$ / 478.89	57-62-5 / 168.5	2193	4-14-00-02629	H$_2$O 1; EtOH 2; eth 1; ace 2
7705	2-Naphthacenecarboxamide, 4-(dimethylamino)-1,4,4a,5,5a,6,11,12a-octahydro-3,5,6,10,12,12a-hexahydroxy-6-methyl-1,11-dioxo-,[4S-(4α,4aα,5aα,6β,12aα)]- / Oxytetracycline	C$_{22}$H$_{24}$N$_2$O$_9$ / 460.44	79-57-2 / 184.5	6931	4-14-00-02633 / 1.634^{20}	
7706	2-Naphthacenecarboxamide, 4-(dimethylamino)-1,4,4a,5,5a,6,11,12a-octahydro-3,6,10,12,12a-pentahydroxy-6-methyl-1,11-dioxo-,[4S-(4α,4aα,5aα,6β,12aα)]- / Tetracycline	C$_{22}$H$_{24}$N$_2$O$_8$ / 444.44	60-54-8 / 170-5 dec	9130	4-14-00-02625	
7707	1,2-Naphthacenedione / 9,10-Naphthacenequinone	C$_{18}$H$_{10}$O$_2$ / 258.28	29276-40-2 / 294	/ sub	3-07-00-04273	EtOH 4
7708	1-Naphthalenamine / 1-Naphthylamine	C$_{10}$H$_9$N / 143.19	134-32-7 / 49.2	6318 / 300.7	4-12-00-03009 / 1.0228^{20}	chl 3 / 1.6140^{20}
7709	2-Naphthalenamine / 2-Naphthylamine	C$_{10}$H$_9$N / 143.19	91-59-8 / 113	6319 / 306.2	4-12-00-03122 / 1.6414^{98}	H$_2$O 3; EtOH 3; eth 3 / 1.6493^{98}
7710	2-Naphthalenamine, N,N-bis(2-chloroethyl)- / Chlornaphazine	C$_{14}$H$_{15}$Cl$_2$N / 268.19	494-03-1 / 55	2107 / 210^5	4-12-00-03126	ace 4; bz 4; eth 4; EtOH 4
7711	1-Naphthalenamine, 5-bromo- / 1-Amino-5-bromonaphthalene	C$_{10}$H$_8$BrN / 222.08	4766-33-0 / 70	/ sub	4-12-00-03111	bz 4; eth 4; EtOH 4; lig 4
7712	1-Naphthalenamine, decahydro-	C$_{10}$H$_{19}$N / 153.27	7250-95-5 / 13.5	/ 227.6	3-12-00-00183 / 0.9508^{21}	
7713	1-Naphthalenamine, N,N-diethyl-	C$_{14}$H$_{17}$N / 199.30	84-95-7 /	/ 285	3-12-00-02855 / 1.013^{20}	EtOH 3; eth 3; bz 3; ctc 2 / 1.5961^{20}
7714	1-Naphthalenamine, 5,8-dihydro- / 1-Naphthylamine, 5,8-dihydro-	C$_{10}$H$_{11}$N / 145.20	32666-56-1 / 37.5	/ 247^{408}	4-12-00-02995	EtOH 4; chl 4
7715	1-Naphthalenamine, N,N-dimethyl- / N,N-Dimethyl-1-naphthylamine	C$_{12}$H$_{13}$N / 171.24	86-56-6 /	3240 / 250; 140^{13}	4-12-00-03013 / 1.0423^{20}	H$_2$O 1; EtOH 3; eth 3; ctc 3 / 1.624^{15}
7716	2-Naphthalenamine, N,N-dimethyl-	C$_{12}$H$_{13}$N / 171.24	2436-85-3 / 52.5	/ 305	4-12-00-03125 / 1.0279^{60}	H$_2$O 1; EtOH 3; eth 3 / 1.6443^{53}
7717	1-Naphthalenamine, N,N-diphenyl- / 1-Naphthylamine, N,N-diphenyl-	C$_{22}$H$_{17}$N / 295.38	61231-45-6 / 142	/ 335-40^{80}	0-12-00-01225	ace 4; bz 4; eth 4; EtOH 4
7718	1-Naphthalenamine, N-ethyl-	C$_{12}$H$_{13}$N / 171.24	118-44-5 /	/ 305; 191^{16}	4-12-00-03013 / 1.060^{20}	eth 4; EtOH 4 / 1.6477^{15}

No.	Name / Synonym	Mol. Form. / Mol. Wt.	CAS RN / mp/°C	Merck No. / bp/°C	Beil. Ref. / den/g cm^{-3}	Solubility / n_D
7719	2-Naphthalenamine, N-ethyl-	C$_{12}$H$_{13}$N / 171.24	2437-03-8 / 236	/ 316.5	4-12-00-03126 / 1.0545[21]	/ 1.6544[21]
7720	1-Naphthalenamine, 4-methyl-	C$_{11}$H$_{11}$N / 157.22	4523-45-9 / 51.5	/ 176[12]	4-12-00-03190	eth 4
7721	1-Naphthalenamine, N-methyl-	C$_{11}$H$_{11}$N / 157.22	2216-68-4 / 174	/ 294.5	4-12-00-03013	eth 4; EtOH 4 / 1.6722[20]
7722	2-Naphthalenamine, N-methyl-	C$_{11}$H$_{11}$N / 157.22	2216-67-3 /	/ 317	4-12-00-03125	eth 4; EtOH 4 / 1.6722[20]
7723	1-Naphthalenamine, N-(2-methylphenyl)- / 1-Naphthyl(2-tolyl)amine	C$_{17}$H$_{15}$N / 233.31	634-41-3 / 94.5	/ 200[9]	3-12-00-02857	bz 4; eth 4; EtOH 4
7724	1-Naphthalenamine, N-(4-methylphenyl)-	C$_{17}$H$_{15}$N / 233.31	634-43-5 / 79	/ 360[528], 230[10]	4-12-00-03016	bz 4; eth 4
7725	2-Naphthalenamine, N-(2-methylphenyl)- / 2-Naphthyl(2-tolyl)amine	C$_{17}$H$_{15}$N / 233.31	644-15-5 / 95.5	/ 402.5	1-12-00-00535	ace 4; bz 4; eth 4; EtOH 4
7726	1-Naphthalenamine, N-1-naphthalenyl-	C$_{20}$H$_{15}$N / 269.35	737-89-3 / 115	/ 311[15]	4-12-00-03018	H$_2$O 1; EtOH 3; eth 3; ace 3
7727	2-Naphthalenamine, N-2-naphthalenyl- / β,β'-Dinaphthylamine	C$_{20}$H$_{15}$N / 269.35	532-18-3 / 172.2	3260 / 471	4-12-00-03131	H$_2$O 1; EtOH 2; eth 3; bz 2
7728	2-Naphthalenamine, N-phenyl- / N-Phenyl-β-naphthylamine	C$_{16}$H$_{13}$N / 219.29	135-88-6 / 108	/ 395.5	4-12-00-03128	H$_2$O 1; EtOH 3; eth 3; bz 3
7729	2-Naphthalenamine, 1-(phenylazo)- / Yellow AB	C$_{16}$H$_{13}$N$_3$ / 247.30	85-84-7 / 103	10003		EtOH 4; HOAc 4
7730	2-Naphthalenamine, N-(phenylmethyl)-	C$_{17}$H$_{15}$N / 233.31	13672-18-9 / 68	/ 405	4-12-00-03130	bz 4; lig 4
7731	1-Naphthalenamine, 5,6,7,8-tetrahydro-	C$_{10}$H$_{13}$N / 147.22	2217-41-6 / 38	/ 279	4-12-00-02934 / 1.0625[16]	H$_2$O 2; EtOH 3; eth 3; acid 3 / 1.5900[20]
7732	Naphthalene	C$_{10}$H$_8$ / 128.17	91-20-3 / 80.2	6289 / 217.9	4-05-00-01640 / 1.0253[20]	H$_2$O 1; EtOH 3; eth 4; ace 4 / 1.5898[25]
7733	1-Naphthaleneacetamide / Acetic acid, amide,1-naphthyl	C$_{12}$H$_{11}$NO / 185.23	86-86-2 /	/ sub 180	4-09-00-02426	H$_2$O 1; eth 3; bz 3; CS$_2$ 3
7734	1-Naphthaleneacetic acid / 1-Naphthylacetic acid	C$_{12}$H$_{10}$O$_2$ / 186.21	86-87-3 / 135	6290 / dec	4-09-00-02424	H$_2$O 2; EtOH 2; eth 4; ace 4
7735	1-Naphthaleneacetic acid, ethyl ester / Ethyl 1-naphthylacetate	C$_{14}$H$_{14}$O$_2$ / 214.26	2122-70-5 / 88.5	/ 222[20], 118[13]	4-09-00-02425	EtOH 3; eth 3
7736	1-Naphthaleneacetonitrile	C$_{12}$H$_9$N / 167.21	132-75-2 / 32.5	/ 192[18], 163[12]	4-09-00-02428	EtOH 3 / 1.6192[20]
7737	2-Naphthaleneacetonitrile	C$_{12}$H$_9$N / 167.21	7498-57-9 / 83	/ 303	3-09-00-03212 / 1.092[25]	
7738	Naphthalene, 1-azido-	C$_{10}$H$_7$N$_3$ / 169.19	6921-40-0 / 12	/ 75[0.001]	/ 1.1713[25]	EtOH 4; eth 5; ace 5 / 1.6550[25]
7739	Naphthalene, 1-bromo- / 1-Naphthyl bromide	C$_{10}$H$_7$Br / 207.07	90-11-9 / -1.8	1413 / 281	4-05-00-01665 / 1.4785[20]	H$_2$O 3; EtOH 5; eth 5; ace 3 / 1.658[20]
7740	Naphthalene, 2-bromo- / 2-Bromonaphthalene	C$_{10}$H$_7$Br / 207.07	580-13-2 / 59	1414 / 281.5	4-05-00-01667 / 1.605[25]	H$_2$O 1; EtOH 3; eth 3; bz 3 / 1.6382[60]
7741	Naphthalene, 1-bromo-4-chloro- / 1-Bromo-4-chloronaphthalene	C$_{10}$H$_6$BrCl / 241.51	53220-82-9 / 66.5	/ 303	3-05-00-01585	eth 4; HOAc 4
7742	Naphthalene, 1-(bromomethyl)-	C$_{11}$H$_9$Br / 221.10	3163-27-7 / 56	/ 183[18], 167[10]	4-05-00-01693	ace 4; bz 4; eth 4; EtOH 4
7743	Naphthalene, 2-(bromomethyl)-	C$_{11}$H$_9$Br / 221.10	939-26-4 / 56	/ 213[100], 167[14]	4-05-00-01698	EtOH 3; eth 3; chl 3; HOAc 3
7744	Naphthalene, 1-bromo-3-nitro-	C$_{10}$H$_6$BrNO$_2$ / 252.07	7499-65-2 / 132.3	/ sub	4-05-00-01678	eth 4; EtOH 4
7745	Naphthalene, 2-bromo-1-nitro- / 2-Bromo-1-nitronaphthalene	C$_{10}$H$_6$BrNO$_2$ / 252.07	4185-62-0 / 104.3	/ >360	4-05-00-01677	bz 4; eth 4; EtOH 4
7746	Naphthalene, 1-butyl-	C$_{14}$H$_{16}$ / 184.28	1634-09-9 / -19.8	/ 289.3	4-05-00-01737 / 0.9738[20]	H$_2$O 1; EtOH 3; eth 3; ace 3 / 1.5819[20]
7747	Naphthalene, 2-butyl- / 2-Butylnaphthalene	C$_{14}$H$_{18}$ / 186.30	1134-62-9 / -5	/ 292	4-05-00-01738 / 0.9673[20]	ace 4; bz 4; EtOH 4 / 1.5777[20]
7748	1-Naphthalenecarbonitrile	C$_{11}$H$_7$N / 153.18	86-53-3 / 37.5	/ 299	4-09-00-02407 / 1.1080[25]	H$_2$O 1; EtOH 4; eth 4; lig 3 / 1.6298[18]
7749	2-Naphthalenecarbonitrile	C$_{11}$H$_7$N / 153.18	613-46-7 / 66	/ 306.5	4-09-00-02419 / 1.0755[60]	H$_2$O 2; EtOH 3; eth 3; chl 2
7750	2-Naphthalenecarbonyl chloride	C$_{11}$H$_7$ClO / 190.63	2243-83-6 / 51	/ 305	4-09-00-02417	bz 4; eth 4; chl 4
7751	1-Naphthalenecarboxaldehyde	C$_{11}$H$_8$O / 156.18	66-77-3 / 33.5	/ 292	4-07-00-01286 / 1.1503[20]	H$_2$O 1; EtOH 3; eth 3; ace 3 / 1.6507[20]
7752	2-Naphthalenecarboxaldehyde	C$_{11}$H$_8$O / 156.18	66-99-9 / 62	/ 160[19]	4-07-00-01288 / 1.0775[99]	H$_2$O 2; EtOH 4; eth 4; ace 3 / 1.6211[99]
7753	1-Naphthalenecarboxaldehyde, 2-ethoxy-	C$_{13}$H$_{12}$O$_2$ / 200.24	19523-57-0 / 115	/ 185-7[25]	3-08-00-01110	EtOH 3; HOAc 3
7754	1-Naphthalenecarboxaldehyde, 2-hydroxy-	C$_{11}$H$_8$O$_2$ / 172.18	708-06-5 / 83	/ 192[27]	4-08-00-01160	H$_2$O 1; EtOH 3; eth 3; aq alk 3
7755	1-Naphthalenecarboxaldehyde, 4-methoxy-	C$_{12}$H$_{10}$O$_2$ / 186.21	15971-29-6 / 34.8	/ 212[40]	4-08-00-01170 / 1.1879[39]	

No.	Name Synonym	Mol. Form. Mol. Wt.	CAS RN mp/°C	Merck No. bp/°C	Beil. Ref. den/g cm^{-3}	Solubility n_D
7756	1-Naphthalenecarboxaldehyde, 4-methyl- 1-Naphthaldehyde, 4-methyl-	$C_{12}H_{10}O$ 170.21	33738-48-6 33.5	175[13]	3-07-00-01972 1.1252[38]	
7757	1-Naphthalenecarboxamide	$C_{11}H_9NO$ 171.20	2243-81-4 205.8	sub	4-09-00-02405	H_2O 2; EtOH 2; con HCl 3; HOAc 3
7758	1-Naphthalenecarboxylic acid 1-Naphthoic acid	$C_{11}H_8O_2$ 172.18	86-55-5 161	6301 >300	4-09-00-02402 1.398[25]	eth 4; EtOH 4; chl 4 1.46
7759	2-Naphthalenecarboxylic acid 2-Naphthoic acid	$C_{11}H_8O_2$ 172.18	93-09-4 185.5	6302 >300	4-09-00-02414 1.077[100]	H_2O 2; EtOH 3; eth 3; chl 3
7760	1-Naphthalenecarboxylic acid, 5-amino-	$C_{11}H_9NO_2$ 187.20	32018-88-5 212.3	sub	4-14-00-01812	EtOH 4; HOAc 4
7761	1-Naphthalenecarboxylic acid, 5-bromo-	$C_{11}H_7BrO_2$ 251.08	16726-67-3 261	sub	3-09-00-03153	H_2O 2; EtOH 2; eth 2; bz 4
7762	1-Naphthalenecarboxylic acid, 5-chloro- 1-Naphthoic acid, 5-chloro-	$C_{11}H_7ClO_2$ 206.63	16650-52-5 245	sub	3-09-00-03149	H_2O 4; EtOH 4; bz 2; HOAc 2
7763	1-Naphthalenecarboxylic acid, 8-chloro-	$C_{11}H_7ClO_2$ 206.63	4537-00-2 171.5	sub	3-09-00-03151	H_2O 3; EtOH 3; ace 4; bz 3
7764	1-Naphthalenecarboxylic acid, 3,4-dihydro- 3,4-Dihydro-1-naphthoic acid	$C_{11}H_{10}O_2$ 174.20	3333-23-1 125	306	4-09-00-02338	H_2O 2; EtOH 4; chl 3; MeOH 3
7765	1-Naphthalenecarboxylic acid, ethyl ester	$C_{13}H_{12}O_2$ 200.24	3007-97-4	310	4-09-00-02403 1.1264[15]	H_2O 1; EtOH 3 1.5966[15]
7766	1-Naphthalenecarboxylic acid, methyl ester	$C_{12}H_{10}O_2$ 186.21	2459-24-7 59.5	168[20]; 101[0.04]	4-09-00-02402 1.1290[20]	bz 4; EtOH 4 1.6086[20]
7767	1-Naphthalenecarboxylic acid, 5-nitro-	$C_{11}H_7NO_4$ 217.18	1975-44-6 241.5	sub	4-09-00-02411	H_2O 1; EtOH 3; eth 2; bz 2
7768	2-Naphthalenecarboxylic acid, 3-amino- 3-Amino-2-naphthoic acid	$C_{11}H_9NO_2$ 187.20	5959-52-4 216.5	464	3-14-00-01341	EtOH 3; eth 3
7769	2-Naphthalenecarboxylic acid, 5-bromo- 5-Bromo-2-naphthoic acid	$C_{11}H_7BrO_2$ 251.08	1013-83-8 270	sub	3-09-00-03196	bz 4; eth 4; EtOH 4; HOAc 4
7770	2-Naphthalenecarboxylic acid, ethyl ester	$C_{13}H_{12}O_2$ 200.24	3007-91-8 32	308.5	4-09-00-02415 1.1143[23]	eth 4; EtOH 4; chl 4; HOAc 4 1.5951[23]
7771	2-Naphthalenecarboxylic acid, 1-hydroxy- 1-Hydroxy-2-naphthoic acid	$C_{11}H_8O_3$ 188.18	86-48-6 195	4766	4-10-00-01181	H_2O 2; EtOH 4; eth 4; bz 3
7772	2-Naphthalenecarboxylic acid, 3-hydroxy- 3-Hydroxy-2-naphthoic acid	$C_{11}H_8O_3$ 188.18	92-70-6 222.5	4767	4-10-00-01184	H_2O 2; EtOH 4; eth 4; bz 3
7773	2-Naphthalenecarboxylic acid, 3-hydroxy-, ethyl ester	$C_{13}H_{12}O_3$ 216.24	7163-25-9 85	291	4-10-00-01186	ace 4; chl 4
7774	2-Naphthalenecarboxylic acid, 3-hydroxy-, hexadecyl ester Hexadecyl 3-hydroxy-2-naphthoate	$C_{27}H_{40}O_3$ 412.61	531-84-0 72.5	4604	3-10-00-01087	bz 4; HOAc 4
7775	2-Naphthalenecarboxylic acid, 3-hydroxy-, methyl ester	$C_{12}H_{10}O_3$ 202.21	883-99-8 75.5	206	4-10-00-01186	H_2O 1; EtOH 3; os 3
7776	2-Naphthalenecarboxylic acid, methyl ester	$C_{12}H_{10}O_2$ 186.21	2459-25-8 77	290	4-09-00-02414	bz 4; eth 4; EtOH 4; chl 4
7777	2-Naphthalenecarboxylic acid, 4,4'- methylenebis[3-hydroxy- Pamoic acid	$C_{23}H_{16}O_6$ 388.38	130-85-8 315	6955	4-10-00-02362	
7778	2-Naphthalenecarboxylic acid, 1,2,3,4- tetrahydro- 1,2,3,4-Tetrahydro-2-naphthoic acid	$C_{11}H_{12}O_2$ 176.22	53440-12-3 97.7	168[15]	3-09-00-02805	EtOH 4; bz 4; EtOH 4; chl 4
7779	2-Naphthalenecarboxylic acid, 5,6,7,8- tetrahydro-	$C_{11}H_{12}O_2$ 176.22	1131-63-1 155.5	216[14]	4-09-00-02096	EtOH 4; bz 3; HOAc 3; peth 1
7780	Naphthalene, 1-chloro- 1-Naphthyl chloride	$C_{10}H_7Cl$ 162.62	90-13-1 -2.5	2149 259; 106.5[5]	4-05-00-01658 1.1938[20]	H_2O 1; EtOH 3; eth 3; bz 3 1.6326[20]
7781	Naphthalene, 2-chloro- 2-Chloronaphthalene	$C_{10}H_7Cl$ 162.62	91-58-7 61	2150 256	4-05-00-01660 1.1377[71]	H_2O 1; EtOH 3; eth 3; bz 3 1.6079[13]
7782	Naphthalene, 2-chlorodecahydro- Decalin, 2-chloro	$C_{10}H_{17}Cl$ 172.70	5597-81-9	112[15]	4-05-00-00313 1.0421[20]	1.5020[20]
7783	Naphthalene, 1-(chloromethyl)-	$C_{11}H_9Cl$ 176.65	86-52-2 32	291.5	4-05-00-01692 1.1813[20]	H_2O 1; EtOH 3; ctc 3; peth 3 1.6380[20]
7784	Naphthalene, 1-chloro-2-methyl- 1-Chloro-2-methylnaphthalene	$C_{11}H_9Cl$ 176.65	5859-45-0	150-2[10]	3-05-00-01632	ctc 3 1.6265[20]
7785	Naphthalene, 2-(chloromethyl)-	$C_{11}H_9Cl$ 176.65	2506-41-4 48.5	169[20]	4-05-00-01697	H_2O 1; EtOH 3; peth 3
7786	Naphthalene, 1-chloro-5-nitro- 1-Chloro-5-nitronaphthalene	$C_{10}H_8ClNO_2$ 207.62	605-63-0 111	>360	3-05-00-01597	EtOH 4
7787	Naphthalene, 1-chloro-8-nitro- 1-Chloro-8-nitronaphthalene	$C_{10}H_8ClNO_2$ 207.62	602-37-9 95.8	175[2]	4-05-00-01676	EtOH 3; bz 3; chl 2; HOAc 3
7788	Naphthalene, 2-chloro-1-nitro- 2-Chloro-1-nitronaphthalene	$C_{10}H_6ClNO_2$ 207.62	4185-63-1 99.5	>360	4-05-00-01676 1.4854[25]	EtOH 4; eth 4; ace 4; bz 4
7789	Naphthalene, decahydro- Decalin	$C_{10}H_{18}$ 138.25	91-17-8 -43	2839 155.5	3-05-00-00245 0.8965[22]	ace 4; bz 4; eth 4; EtOH 4 1.4810[20]
7790	Naphthalene, decahydro-, cis- cis-Decalin	$C_{10}H_{18}$ 138.25	493-01-6 -42.9	2839 195.8	4-05-00-00310 0.8965[20]	H_2O 1; EtOH 5; eth 4; ace 4 1.4810[20]
7791	Naphthalene, decahydro-, trans- trans-Decalin	$C_{10}H_{18}$ 138.25	493-02-7 -30.3	2839 187.3	4-05-00-00311 0.8699[20]	H_2O 1; EtOH 4; eth 4; ace 4 1.4695[20]

No.	Name Synonym	Mol. Form. Mol. Wt.	CAS RN mp/°C	Merck No. bp/°C	Beil. Ref. den/g cm^{-3}	Solubility n_D
7792	Naphthalene, decahydro-2-methylene-, *trans*- Decalin, 2-methylene (*trans*)	C$_{11}$H$_{18}$ 150.26	7787-72-6	201; 82[10]	3-05-00-00397 0.8897[20]	1.4841[22]
7793	Naphthalene, 1-decyl- Decane, 1-(1-naphthyl)-	C$_{20}$H$_{28}$ 268.44	26438-27-7 15	379	4-05-00-01786 0.9322[20]	1.5435[20]
7794	Naphthalene, 2-decyl-	C$_{20}$H$_{28}$ 268.44	14188-79-5 20	387	4-05-00-01786 0.9253[20]	1.5413[20]
7795	1,2-Naphthalenediamine 1,2-Diaminonaphthalene	C$_{10}$H$_{10}$N$_2$ 158.20	938-25-0 98.5	214[13]; 150[0.5]	4-13-00-00337	H$_2$O 2; EtOH 4; eth 4; chl 4
7796	1,4-Naphthalenediamine 1,4-Diaminonaphthalene	C$_{10}$H$_{10}$N$_2$ 158.20	2243-61-0 120		4-13-00-00338	ace 4; bz 4; EtOH 4; chl 4 1.6441[18]
7797	1,5-Naphthalenediamine	C$_{10}$H$_{10}$N$_2$ 158.20	2243-62-1 190	sub	4-13-00-00340 1.4[25]	H$_2$O 3; EtOH 3; eth 3; chl 4
7798	1,8-Naphthalenediamine Naphthalene, 1,8-diamino	C$_{10}$H$_{10}$N$_2$ 158.20	479-27-6 66.5	6291 205[12]	4-13-00-00344 1.1265[90]	eth 4; EtOH 4 1.6828[99]
7799	2,3-Naphthalenediamine 2,3-Diaminonaphthalene	C$_{10}$H$_{10}$N$_2$ 158.20	771-97-1 199		4-13-00-00346 1.0968[26]	H$_2$O 2; EtOH 4; eth 3 1.6392[26]
7800	1,4-Naphthalenediamine, 2-methyl- Vitamin K6	C$_{11}$H$_{12}$N$_2$ 172.23	83-68-1 113.5	9937	3-13-00-00406	
7801	Naphthalene, 1,3-dibromo- 1,3-Dibromonaphthalene	C$_{10}$H$_6$Br$_2$ 285.97	52358-73-3 62.8	290	3-05-00-01586	EtOH 4
7802	Naphthalene, 1,4-dibromo- 1,4-Dibromonaphthalene	C$_{10}$H$_6$Br$_2$ 285.97	83-53-4 83	310	4-05-00-01669	H$_2$O 1; EtOH 3; eth 3; HOAc 2
7803	Naphthalene, 1,5-dibromo- 1,5-Dibromonaphthalene	C$_{10}$H$_6$Br$_2$ 285.97	7351-74-8 131.3	326	4-05-00-01669	eth 4; EtOH 4
7804	1,8-Naphthalenedicarbonyl dichloride Naphthaloyl chloride	C$_{12}$H$_6$Cl$_2$O$_2$ 253.08	6423-29-6 85	197[0.2]	3-09-00-04468	bz 4; chl 4
7805	1,8-Naphthalenedicarboxylic acid Naphthalic acid	C$_{12}$H$_8$O$_4$ 216.19	518-05-8 260	6299	4-09-00-03534	H$_2$O 1; EtOH 2; eth 2
7806	1,8-Naphthalenedicarboxylic acid, diethyl ester Naphthalic acid, diethyl ester	C$_{16}$H$_{16}$O$_4$ 272.30	58618-39-6 59.5	238[19]; 202[11]	4-09-00-03534 1.1399[70]	
7807	Naphthalene, 1,2-dichloro- 1,2-Dichloronaphthalene	C$_{10}$H$_6$Cl$_2$ 197.06	2050-69-3 36	296.5	4-05-00-01661 1.3147[49]	EtOH 3; eth 3 1.5338[49]
7808	Naphthalene, 1,3-dichloro- 1,3-Dichloronaphthalene	C$_{10}$H$_6$Cl$_2$ 197.06	2198-75-6 62.3	291	4-05-00-01661	EtOH 3
7809	Naphthalene, 1,4-dichloro- 1,4-Dichloronaphthalene	C$_{10}$H$_6$Cl$_2$ 197.06	1825-31-6 67.5	288; 147[12]	4-05-00-01661 1.2997[76]	H$_2$O 1; EtOH 2; eth 3; ace 4 1.6228[76]
7810	Naphthalene, 1,5-dichloro- 1,5-Dichloronaphthalene	C$_{10}$H$_6$Cl$_2$ 197.06	1825-30-5 107	sub	4-05-00-01662 1.4900[20]	H$_2$O 1; EtOH 2; eth 3
7811	Naphthalene, 1,7-dichloro- 1,7-Dichloronaphthalene	C$_{10}$H$_6$Cl$_2$ 197.06	2050-73-9 63.5	285.5	4-05-00-01662 1.2611[100]	EtOH 3; eth 3; bz 3; HOAc 3 1.6092[100]
7812	Naphthalene, 1,8-dichloro- 1,8-Dichloronaphthalene	C$_{10}$H$_6$Cl$_2$ 197.06	2050-74-0 89	sub	4-05-00-01662 1.2924[100]	EtOH 3; peth 3 1.6236[100]
7813	Naphthalene, 2,6-dichloro- 2,6-Dichloronaphthalene	C$_{10}$H$_6$Cl$_2$ 197.06	2065-70-5 140.5	285	4-05-00-01662	EtOH 2; eth 3; bz 3; chl 3
7814	Naphthalene, 1,2-dihydro-	C$_{10}$H$_{10}$ 130.19	447-53-0 -8	206.5	4-05-00-01543 0.9974[20]	1.5814[20]
7815	Naphthalene, 1,4-dihydro- Δ 2-Dialin	C$_{10}$H$_{10}$ 130.19	612-17-9 25	211.5	4-05-00-01544 0.9928[33]	1.5577[20]
7816	Naphthalene, 1,2-dihydro-3-methyl-	C$_{11}$H$_{12}$ 144.22	2717-44-4	105[13]; 64[0.6]	4-05-00-01552 0.9819[20]	H$_2$O 1; eth 3; bz 3 1.5751[20]
7817	Naphthalene, 1,2-dihydro-4-methyl-	C$_{11}$H$_{12}$ 144.22	4373-13-1	112[18]; 84[5]	4-05-00-01551 0.9895[20]	H$_2$O 1; eth 3; bz 3 1.5758[20]
7818	Naphthalene, 1,2-dimethyl- 1,2-Dimethylnaphthalene	C$_{12}$H$_{12}$ 156.23	573-98-8 1.6	266.5	4-05-00-01708 1.0179[20]	H$_2$O 1; eth 3; bz 3 1.6166[20]
7819	Naphthalene, 1,3-dimethyl- 1,3-Dimethylnaphthalene	C$_{12}$H$_{12}$ 156.23	575-41-7 -6	263	4-05-00-01708 1.0144[20]	H$_2$O 1; eth 3; bz 3 1.6140[20]
7820	Naphthalene, 1,4-dimethyl- 1,4-Dimethylnaphthalene	C$_{12}$H$_{12}$ 156.23	571-58-4 7.6	268	4-05-00-01709 1.0166[20]	H$_2$O 1; EtOH 4; eth 5; ace 5 1.6127[20]
7821	Naphthalene, 1,5-dimethyl- 1,5-Dimethylnaphthalene	C$_{12}$H$_{12}$ 156.23	571-61-9 82	265	4-05-00-01710	bz 4; eth 4
7822	Naphthalene, 1,6-dimethyl- 1,6-Dimethylnaphthalene	C$_{12}$H$_{12}$ 156.23	575-43-9 -16.9	264	4-05-00-01711 1.0021[20]	H$_2$O 1; eth 3; bz 3 1.6166[20]
7823	Naphthalene, 1,7-dimethyl- 1,7-Dimethylnaphthalene	C$_{12}$H$_{12}$ 156.23	575-37-1 -13.9	263	4-05-00-01711 1.0115[20]	H$_2$O 1; eth 3; bz 3 1.6083[20]
7824	Naphthalene, 1,8-dimethyl- 1,8-Dimethylnaphthalene	C$_{12}$H$_{12}$ 156.23	569-41-5 65	270	4-05-00-01712 1.003[20]	H$_2$O 1; eth 3; bz 3
7825	Naphthalene, 2,3-dimethyl- Guajen	C$_{12}$H$_{12}$ 156.23	581-40-8 105	268	4-05-00-01713 1.003[20]	bz 4; eth 4 1.5060[20]
7826	Naphthalene, 2,6-dimethyl- 2,6-Dimethylnaphthalene	C$_{12}$H$_{12}$ 156.23	581-42-0 112	262	4-05-00-01714 1.003[20]	
7827	Naphthalene, 2,7-dimethyl- 2,7-Dimethylnaphthalene	C$_{12}$H$_{12}$ 156.23	582-16-1 97	265	4-05-00-01715 1.003[20]	
7828	Naphthalene, 1-(1,1-dimethylethyl)-	C$_{14}$H$_{16}$ 184.28	17085-91-5 -11.3	279	4-05-00-01739 0.9629[20]	1.5726[20]
7829	Naphthalene, 2-(1,1-dimethylethyl)-	C$_{14}$H$_{16}$ 184.28	2876-35-9 -4	276; 125[4]	4-05-00-01740 0.9674[20]	ace 4; bz 4; EtOH 4 1.5685[20]

No.	Name Synonym	Mol. Form. Mol. Wt.	CAS RN mp/°C	Merck No. bp/°C	Beil. Ref. den/g cm^{-3}	Solubility n_D
7830	Naphthalene, 1,6-dimethyl-4-(1-methylethyl)- Cadalene	$C_{15}H_{18}$ 198.31	483-78-3	1607 294; 149[10]	4-05-00-01758 0.9667[25]	oils 4 1.5785[25]
7831	Naphthalene, 1,3-dinitro- 1,3-Dinitronaphthalene	$C_{10}H_6N_2O_4$ 218.17	606-37-1 148	sub	4-05-00-01680	H_2O 1; EtOH 3; ace 3
7832	Naphthalene, 1,5-dinitro- 1,5-Dinitronaphthalene	$C_{10}H_6N_2O_4$ 218.17	605-71-0 219	sub	4-05-00-01680 1.5860[20]	H_2O 1; EtOH 2; eth 4; ace 2
7833	Naphthalene, 1,8-dinitro- 1,8-Dinitronaphthalene	$C_{10}H_6N_2O_4$ 218.17	602-38-0 173	445 dec	4-05-00-01681	H_2O 1; EtOH 2; ace 3; bz 2
7834	1,3-Naphthalenediol Naphthoresorcinol	$C_{10}H_8O_2$ 160.17	132-86-5 123.5	6316	4-06-00-06543	H_2O 3; EtOH 3; eth 3; ace 2
7835	1,5-Naphthalenediol	$C_{10}H_8O_2$ 160.17	83-56-7 262 dec	sub	4-06-00-06554	H_2O 2; EtOH 2; eth 4; ace 4
7836	1,6-Naphthalenediol	$C_{10}H_8O_2$ 160.17	575-44-0 138	sub	4-06-00-06557	H_2O 2; EtOH 2; eth 3; ace 3
7837	1,7-Naphthalenediol	$C_{10}H_8O_2$ 160.17	575-38-2 180.5	sub	4-06-00-06559	H_2O 2; EtOH 4; eth 4; bz 3
7838	2,6-Naphthalenediol	$C_{10}H_8O_2$ 160.17	581-43-1 220	sub	4-06-00-06566	H_2O 2; EtOH 3; eth 3; ace 3
7839	2,7-Naphthalenediol	$C_{10}H_8O_2$ 160.17	582-17-2 193	sub	4-06-00-06570	H_2O 3; EtOH 3; eth 3; ace 2
7840	1,8-Naphthalenediol, cyclic sulfite 1-Naphthalenesulfonic acid, 8-hydroxy, sulfone	$C_{10}H_6O_3S$ 206.22	21849-97-8 97	>360	2-06-00-00954	bz 4; chl 4
7841	1,4-Naphthalenediol, 2-methyl-, diacetate Menadiol diacetate	$C_{15}H_{14}O_4$ 258.27	573-20-6 113	5710	4-06-00-06581	EtOH 4
7842	1,2-Naphthalenedione 1,2-Naphthoquinone	$C_{10}H_6O_2$ 158.16	524-42-5 146	6314	4-07-00-02417 1.450[25]	H_2O 3; EtOH 3; eth 3; sulf 3
7843	1,4-Naphthalenedione 1,4-Naphthoquinone	$C_{10}H_6O_2$ 158.16	130-15-4 128.5	6315 sub	4-07-00-02422	H_2O 2; EtOH 4; eth 3; bz 3
7844	1,2-Naphthalenedione, 3-bromo- 3-Bromo-1,2-naphthoquinone	$C_{10}H_5BrO_2$ 237.05	7474-83-1 178	sub	4-07-00-02422	bz 4; EtOH 4
7845	1,4-Naphthalenedione, 2-bromo-3-hydroxy-	$C_{10}H_5BrO_3$ 253.05	1203-39-0 202	sub	4-08-00-02364	ace 4
7846	1,4-Naphthalenedione, 2-bromo-3-methyl-	$C_{11}H_7BrO_2$ 251.08	3129-39-3 152.8	sub 100	4-07-00-02431	ace 4; bz 4; eth 4; EtOH 4
7847	1,4-Naphthalenedione, 5-chloro- 5-Chloro-1,4-naphthoquinone	$C_{10}H_5ClO_2$ 192.60	40242-15-7 163	sub	4-07-00-02425	EtOH 4; lig 4
7848	1,4-Naphthalenedione, 2-chloro-3-hydroxy-	$C_{10}H_5ClO_3$ 208.60	1526-73-4 216.5	sub	4-08-00-02363	bz 4; eth 4; EtOH 4
7849	1,2-Naphthalenedione, 3,4-dichloro-	$C_{10}H_4Cl_2O_2$ 227.05	18398-36-2 184	sub	4-07-00-02422	bz 4; chl 4
7850	1,4-Naphthalenedione, 2,3-dichloro- Dichlone	$C_{10}H_4Cl_2O_2$ 227.05	117-80-6 195	3032	4-07-00-02426	H_2O 1; EtOH 2; eth 2; bz 2
7851	1,4-Naphthalenedione, 5,6-dichloro- 1,4-Naphthoquinone, 5,6-dichloro	$C_{10}H_4Cl_2O_2$ 227.05	56961-93-4 181	sub	0-07-00-00730	eth 4
7852	1,4-Naphthalenedione, 2,3-dihydroxy-	$C_{10}H_6O_4$ 190.16	605-37-8 282	sub	4-08-00-02953	H_2O 2; EtOH 2; eth 2; ace 3
7853	1,4-Naphthalenedione, 5,8-dihydroxy-	$C_{10}H_6O_4$ 190.16	475-38-7 232	sub	4-08-00-02946	H_2O 2; EtOH 2; eth 2; HOAc 3
7854	1,4-Naphthalenedione, 5,8-dihydroxy-6-(1-hydroxy-4-methyl-3-pentenyl)-, (S)- Alkannin	$C_{16}H_{16}O_5$ 288.30	23444-65-7 149	243 sub 140	3-08-00-04089	EtOH 4
7855	1,4-Naphthalenedione, 5,8-dihydroxy-6-methoxy-2-methyl-3-(2-oxopropyl)- Javanicin	$C_{15}H_{14}O_6$ 290.27	476-45-9 208 dec	5143	4-08-00-03646	alk 3
7856	1,4-Naphthalenedione, 3,5-dihydroxy-2-methyl-	$C_{11}H_8O_4$ 204.18	478-40-0 181	sub 100	4-08-00-02958	H_2O 3; EtOH 3; eth 3; peth 3
7857	1,4-Naphthalenedione, 2-ethyl-3,5,6,7,8-pentahydroxy- Echinochrome A	$C_{12}H_{10}O_7$ 266.21	517-82-8 220 dec	3470 sub 120	4-08-00-03714	H_2O 2; EtOH 4; ace 3
7858	1,4-Naphthalenedione, 6-ethyl-2,3,5,7,8-pentahydroxy-	$C_{12}H_{10}O_7$ 266.21	1471-96-1 220 dec	120[0.0004]	4-08-00-03714	ace 4; eth 4; EtOH 4
7859	1,4-Naphthalenedione, 2-hydroxy- Lawsone	$C_{10}H_6O_3$ 174.16	83-72-7 195-6 dec	5263	4-08-00-02360	EtOH 4; eth 1; bz 1; chl 1
7860	1,4-Naphthalenedione, 5-hydroxy- Juglone	$C_{10}H_6O_3$ 174.16	481-39-0 155	5150 sub	4-08-00-02368	H_2O 1; EtOH 3; eth 3; bz 3
7861	1,4-Naphthalenedione, 2-hydroxy-3-methyl- Phthiocol	$C_{11}H_8O_3$ 188.18	483-55-6 173.5	7352 sub	4-08-00-02375	ace 4; eth 4
7862	1,4-Naphthalenedione, 2-hydroxy-3-(3-methyl-2-butenyl)- Lapachol	$C_{15}H_{14}O_3$ 242.27	84-79-7 139.5	5235	4-08-00-02487	H_2O 1; EtOH 3; eth 3; bz 3
7863	1,4-Naphthalenedione, 5-hydroxy-2-methyl- 1,4-Naphthoquinone, 5-hydroxy-2-methyl	$C_{11}H_8O_3$ 188.18	481-42-5 78.5	7511 sub	4-08-00-02376	ace 4; bz 4; eth 4; EtOH 4
7864	1,4-Naphthalenedione, 2-methyl- Menadione	$C_{11}H_8O_2$ 172.18	58-27-5 107	5714	4-07-00-02430	H_2O 1; EtOH 2; eth 3; bz 3
7865	1,4-Naphthalenedione, 2-methyl-3-(3,7,11,15-tetramethyl-2-hexadecenyl)-, [R-[R*,R*-(E)]]- Vitamin K1	$C_{31}H_{46}O_2$ 450.71	84-80-0 -20	9933 140-5[0.001]	4-07-00-02496 0.964[25]	ace 4; bz 4; eth 4; EtOH 4 1.5250[25]

No.	Name Synonym	Mol. Form. Mol. Wt.	CAS RN mp/°C	Merck No. bp/°C	Beil. Ref. den/g cm^{-3}	Solubility n_D
7866	1,4-Naphthalenedione, 2-(phenylamino)-	$C_{16}H_{11}NO_2$ 249.27	6628-97-3 193	sub	4-14-00-00399	bz 4; eth 4
7867	1,5-Naphthalenedisulfonic acid Armstrong's acid	$C_{10}H_8O_6S_2$ 288.30	81-04-9 240-5 dec	812	4-11-00-00561 1.493[25]	H_2O 4; EtOH 3; eth 1
7868	1,6-Naphthalenedisulfonic acid Eiver-pick acid	$C_{10}H_8O_6S_2$ 288.30	525-37-1 125 dec	6292	4-11-00-00562	H_2O 4; EtOH 3; eth 1
7869	2,7-Naphthalenedisulfonic acid Naphthalene-2,7-disulfonic acid	$C_{10}H_8O_6S_2$ 288.30	92-41-1 199	6294	4-11-00-00563	H_2O 3; con HCl 2
7870	1,3-Naphthalenedisulfonic acid, 7-amino- Amido-G-Acid	$C_{10}H_9NO_6S_2$ 303.32	86-65-7 274	412	4-14-00-02811	H_2O 4; EtOH 4
7871	1,3-Naphthalenedisulfonic acid, 7-hydroxy- 2-Naphthol-6,8-disulfonic acid	$C_{10}H_8O_7S_2$ 304.30	118-32-1	6308	4-11-00-00625	H_2O 3
7872	1,6-Naphthalenedisulfonic acid, 4-amino- 1-Naphthylamine-4,7-disulfonic acid	$C_{10}H_9NO_6S_2$ 303.32	85-75-6	6322	2-14-00-00476	H_2O 4
7873	1,7-Naphthalenedisulfonic acid, 4-amino- 1-Naphthylamine-4,6-disulfonic acid	$C_{10}H_9NO_6S_2$ 303.32	85-74-5	6321	2-14-00-00477	H_2O 4; EtOH 4
7874	2,7-Naphthalenedisulfonic acid, 4-amino-5-hydroxy- 1-Amino-8-naphthol-3,6-disulfonic acid	$C_{10}H_9NO_7S_2$ 319.32	90-20-0	6305	4-14-00-02824	H_2O 2; EtOH 2; eth 2
7875	2,7-Naphthalenedisulfonic acid, 4,5-dihydroxy- Chromotropic acid	$C_{10}H_8O_8S_2$ 320.30	148-25-4	2243	4-11-00-00636	H_2O 3; EtOH 1; eth 1; alk 3
7876	2,7-Naphthalenedisulfonic acid, 3-hydroxy- 2-Naphthol-3,6-disulfonic acid	$C_{10}H_8O_7S_2$ 304.30	148-75-4 dec	6307	4-11-00-00624	H_2O 4; EtOH 4
7877	1,3-Naphthalenedisulfonic acid, 6,6'-[(3,3'-dimethyl[1,1'-biphenyl]-4,4'-diyl)bis(azo)]bis[4-amino-5-hydroxy-, C.I. Direct blue 53	$C_{34}H_{24}N_6Na_4O_{14}S_4$ 960.82	314-13-6	3863		H_2O 3; EtOH 3; acid 3
7878	2,7-Naphthalenedisulfonic acid, 4,5-dihydroxy-3-[(4-nitrophenyl)azo]-, disodium salt	$C_{16}H_9N_3Na_2O_{10}S_2$ 513.37	548-80-1	2242		
7879	Naphthalene, 1-dodecyl- Dodecane, 1-(1-naphthyl)-	$C_{22}H_{32}$ 296.50	26438-28-8 27	415	4-05-00-01790 0.9240[20]	1.5364[20]
7880	Naphthalene, 2-dodecyl- Dodecane, 1-(2-naphthyl)-	$C_{22}H_{32}$ 296.50	60899-39-0 26	414	4-05-00-01791 0.9177[20]	1.5343[20]
7881	Naphthalene, 1-ethenyl-	$C_{12}H_{10}$ 154.21	826-74-4	124-5[15]	4-05-00-01833 1.036[18]	1.644[20]
7882	Naphthalene, 2-ethenyl- 2-Vinylnaphthalene	$C_{12}H_{10}$ 154.21	827-54-3 66	135[18]; 95[2]	4-05-00-01833	H_2O 1; EtOH 3; ace 3; bz 3
7883	Naphthalene, 1-ethoxy-	$C_{12}H_{12}O$ 172.23	5328-01-8 5.5	280.5	4-06-00-04212 1.060[20]	H_2O 1; EtOH 4; eth 4 1.5953[25]
7884	Naphthalene, 2-ethoxy- 2-Ethoxynaphthalene	$C_{12}H_{12}O$ 172.23	93-18-5 37.5	282	4-06-00-04257 1.0640[20]	H_2O 1; EtOH 3; eth 3; tol 3 1.5975[36]
7885	Naphthalene, 1-ethyl- 1-Ethylnaphthalene	$C_{12}H_{12}$ 156.23	1127-76-0 -13.9	258.6	4-05-00-01705 1.0082[20]	H_2O 1; EtOH 5; eth 5 1.6062[20]
7886	Naphthalene, 2-ethyl- 2-Ethylnaphthalene	$C_{12}H_{12}$ 156.23	939-27-5 -7.4	258	4-05-00-01707 0.9922[20]	H_2O 1; EtOH 5; eth 5; chl 2 1.5999[20]
7887	Naphthalene, 1-ethyl-5-methyl- 1-Methyl-5-ethylnaphthalene	$C_{13}H_{14}$ 170.25	17057-92-0 40	133[10]	3-05-00-01659	1.600[30]
7888	Naphthalene, 1-ethyl-1,2,3,4-tetrahydro- 1-Ethyltetralin	$C_{12}H_{16}$ 160.26	13556-58-6	241	4-05-00-01429 0.9528[20]	1.5318[20]
7889	Naphthalene, 2-ethyl-1,2,3,4-tetrahydro- 2-Ethyltetralin	$C_{12}H_{16}$ 160.26	32367-54-7 -42.8	235	4-05-00-01429 0.9401[15]	1.5250[15]
7890	Naphthalene, 5-ethyl-1,2,3,4-tetrahydro-	$C_{12}H_{16}$ 160.26	42775-75-7 -44.5	245	4-05-00-01428 0.973[20]	1.540[20]
7891	Naphthalene, 6-ethyl-1,2,3,4-tetrahydro- 6-Ethyltetralin	$C_{12}H_{16}$ 160.26	22531-20-0 -70	244	4-05-00-01429 0.9632[17]	1.5414[16]
7892	Naphthalene, 1-ethynyl-	$C_{12}H_8$ 152.20	15727-65-8 2	143-4[25]	3-05-00-01935 1.0513[20]	1.6360[20]
7893	Naphthalene, 1-fluoro- 1-Fluoronaphthalene	$C_{10}H_7F$ 146.16	321-38-0 -9	215; 80[11]	4-05-00-01657 1.1322[20]	H_2O 1; EtOH 3; eth 3; bz 3 1.5939[20]
7894	Naphthalene, 2-fluoro- 2-Fluoronaphthalene	$C_{10}H_7F$ 146.16	323-09-1 61	212; 90[16]	4-05-00-01658	H_2O 1; EtOH 3; eth 3; bz 3
7895	Naphthalene, 1-heptyl- 1-Heptylnaphthalene	$C_{17}H_{22}$ 226.36	2876-52-0 -8	340	4-05-00-01775 0.9491[20]	1.5582[20]
7896	Naphthalene, 2-heptyl- 2-Heptylnaphthalene	$C_{17}H_{22}$ 226.36	2876-45-1 1	341	4-05-00-01775 0.9410[20]	1.5556[20]
7897	Naphthalene, 1,2,3,4,4a,8a-hexahydro- Naphthalene, 1,2,3,4,9,10-hexahydro	$C_{10}H_{14}$ 134.22	62690-62-4	197	4-05-00-01082 0.934[23]	bz 4; eth 4 1.5260[16]
7898	Naphthalene, 1,2,4a,5,8,8a-hexahydro-4,7-dimethyl-1-(1-methylethyl)-, [1S-(1α,4aβ,8aα)]-	$C_{15}H_{24}$ 204.36	523-47-7	274; 136[11]	3-05-00-01086 0.9230[20]	eth 4; lig 4 1.5059[20]
7899	Naphthalene, 1,2,3,5,6,8a-hexahydro-4,7-dimethyl-1-(1-methylethyl)-, (1S-cis)-	$C_{15}H_{24}$ 204.36	483-76-1	125-6[12]	2-05-00-00348 0.9160[15]	1.5089[15]
7900	Naphthalene, 1-hexyl- 1-Hexylnaphthalene	$C_{16}H_{20}$ 212.33	2876-53-1 -18	322	4-05-00-01767 0.9566[20]	1.5647[20]
7901	Naphthalene, 2-hexyl-	$C_{16}H_{20}$ 212.33	2876-46-2 -3	324	4-05-00-01767 0.9479[20]	1.5620[20]
7902	Naphthalene, 1-iodo-	$C_{10}H_7I$ 254.07	90-14-2 4.2	302	4-05-00-01670 1.7399[20]	H_2O 1; EtOH 5; eth 5; bz 5 1.7026[20]

No.	Name / Synonym	Mol. Form. / Mol. Wt.	CAS RN / mp/°C	Merck No. / bp/°C	Beil. Ref. / den/g cm⁻³	Solubility / n_D
7903	Naphthalene, 2-iodo-	$C_{10}H_7I$ / 254.07	612-55-5 / 54.5	308	4-05-00-01671 / 1.6319^{99}	H_2O 1; EtOH 4; eth 4; HOAc 4 / 1.6662^{99}
7904	Naphthalene, 1-isocyanato- / 1-Naphthyl isocyanate	$C_{11}H_7NO$ / 169.18	86-84-0	6330 / 269	4-12-00-03094 / 1.1774^{20}	eth 3; bz 3
7905	Naphthalene, 1-isopropyl-2,4,7-trimethyl- / Cadalene, 3-methyl	$C_{16}H_{20}$ / 212.33	6995-30-8	$129\text{-}31^{1.5}$	3-05-00-01689 / 0.9721^{30}	1.5840^{30}
7906	Naphthalene, 1-isothiocyanato- / 1-Naphthyl isothiocyanate	$C_{11}H_7NS$ / 185.25	551-06-4 / 58	6331	4-12-00-03094	bz 4; eth 4; EtOH 4; chl 4
7907	2-Naphthalenemethanamine	$C_{11}H_{11}N$ / 157.22	2018-90-8 / 59.5	180^{24}; 148^{12}	4-12-00-03207	eth 4; EtOH 4
7908	1-Naphthalenemethanol	$C_{11}H_{10}O$ / 158.20	4780-79-4 / 64	304; 163^{12}	4-06-00-04332 / 1.1039^{80}	H_2O 2; EtOH 4; eth 4
7909	2-Naphthalenemethanol	$C_{11}H_{10}O$ / 158.20	1592-38-7 / 81.3	178^{12}	4-06-00-04340	H_2O 2; EtOH 3; eth 3
7910	1-Naphthalenemethanol, α,α-diphenyl- / Benzhydrol, α-(1-naphthyl)	$C_{23}H_{18}O$ / 310.40	630-95-5 / 136.5	dec	4-06-00-05134	bz 4; eth 4; EtOH 4
7911	1-Naphthalenemethanol, α-methyl-, (±)- / (±)-1-(α-Naphthyl)ethanol	$C_{12}H_{12}O$ / 172.23	57605-95-5 / 66	178^{15}	4-06-00-04346 / 1.1190^{14}	ace 4; bz 4; EtOH 4; chl 4 / 1.6188^{25}
7912	1-Naphthalenemethanol, α-phenyl-	$C_{17}H_{14}O$ / 234.30	642-28-4 / 87.8	360	4-06-00-04989	H_2O 1; EtOH 3; eth 3; bz 3
7913	Naphthalene, 1-methoxy-	$C_{11}H_{10}O$ / 158.20	2216-69-5 / <-10	269	4-06-00-04211 / 1.0963^{14}	H_2O 1; EtOH 3; eth 3; bz 3 / 1.6940^{25}
7914	Naphthalene, 2-methoxy- / 2-Methoxynaphthalene	$C_{11}H_{10}O$ / 158.20	93-04-9 / 73.5	5918 / 274	4-06-00-04257	bz 4; eth 4; chl 4
7915	Naphthalene, 1-methyl- / 1-Methylnaphthalene	$C_{11}H_{10}$ / 142.20	90-12-0 / -30.4	244.7	4-05-00-01687 / 1.0202^{20}	H_2O 1; EtOH 4; eth 4; bz 3 / 1.6170^{20}
7916	Naphthalene, 2-methyl- / 2-Methylnaphthalene	$C_{11}H_{10}$ / 142.20	91-57-6 / 34.4	241.1	4-05-00-01693 / 1.0058^{20}	H_2O 1; EtOH 4; eth 4; bz 3 / 1.6015^{40}
7917	Naphthalene, 1-(3-methylbutoxy)-	$C_{15}H_{18}O$ / 214.31	20213-30-3	319; 150^3	3-06-00-02925 / 1.0069^{14}	1.5705^{16}
7918	Naphthalene, 2-(3-methylbutoxy)- / Isopentyl β-Naphthyl ether	$C_{15}H_{18}O$ / 214.31	635-88-1 / 26.5	325 dec	4-06-00-04259 / 1.0155^{12}	eth 4; EtOH 4 / 1.5768^{12}
7919	Naphthalene, 1,1'-methylenebis-	$C_{21}H_{16}$ / 268.36	607-50-1 / 109	>360; 270^{14}	4-05-00-02646	H_2O 2; EtOH 2; eth 3; bz 3
7920	Naphthalene, 1-(1-methylethyl)-	$C_{13}H_{14}$ / 170.25	6158-45-8 / -16	268	4-05-00-01722 / 0.9956^{20}	1.5952^{20}
7921	Naphthalene, 2-(1-methylethyl)-	$C_{13}H_{14}$ / 170.25	2027-17-0 / 14.5	268.2	4-05-00-01723 / 0.9753^{20}	H_2O 1; EtOH 4; eth 4; bz 3 / 1.5848^{20}
7922	Naphthalene, 1-methyl-4-(1-methylethyl)-	$C_{14}H_{16}$ / 184.28	1680-53-1	$145\text{-}8^{12}$	3-05-00-01671 / 0.9934^{14}	1.5907^{14}
7923	Naphthalene, 1-methyl-7-(1-methylethyl)-	$C_{14}H_{16}$ / 184.28	490-65-3	140^{11}	4-05-00-01741 / 0.9740^{20}	H_2O 1; eth 3; ace 3; bz 3 / 1.5833^{20}
7924	Naphthalene, 7-methyl-1-(1-methylethyl)- / Naphthalene, 1-isopropyl-7-methyl	$C_{14}H_{16}$ / 184.28	66577-17-1	282	3-05-00-01672 / 0.9833^{20}	1.5884^{20}
7925	Naphthalene, 1-methyl-4-nitro- / 4-Methyl-1-nitronaphthalene	$C_{11}H_9NO_2$ / 187.20	880-93-3 / 71.5	182^{18}	3-05-00-01624	ace 4; eth 4; EtOH 4
7926	Naphthalene, 2-methyl-1-nitro- / 2-Methyl-1-nitronaphthalene	$C_{11}H_9NO_2$ / 187.20	881-03-8 / 81.5	188^{20}	4-05-00-01698	H_2O 1; EtOH 3; ace 4; os 4
7927	Naphthalene, 1-(1-methylpropyl)-	$C_{14}H_{16}$ / 184.28	1680-58-6	273	4-05-00-01738 / 0.9742^{20}	1.5701^{20}
7928	Naphthalene, 1-(2-naphthalenyloxy)- / 1,2'-Bis-naphthyl ether	$C_{20}H_{14}O$ / 270.33	611-49-4 / 81	264^{15}	2-06-00-00600	bz 4; eth 4
7929	Naphthalene, 1-nitro- / 1-Nitronaphthalene	$C_{10}H_7NO_2$ / 173.17	86-57-7 / 61	6535 / 180^{14}	4-05-00-01673 / 1.332^{20}	H_2O 1; EtOH 4; eth 4; bz 4
7930	Naphthalene, 2-nitro- / 2-Nitronaphthalene	$C_{10}H_7NO_2$ / 173.17	581-89-5 / 79	314; 165^{15}	4-05-00-01675	H_2O 1; EtOH 4; eth 4
7931	Naphthalene, 1-nonyl-	$C_{19}H_{26}$ / 254.42	26438-26-6 / 8	366	4-05-00-01784 / 0.9371^{20}	1.5477^{20}
7932	Naphthalene, 2-nonyl-	$C_{19}H_{26}$ / 254.42	61886-67-7 / 12	372	4-05-00-01784 / 0.9298^{20}	1.5454^{20}
7933	Naphthalene, octachloro- / Perchloronaphthalene	$C_{10}Cl_8$ / 403.73	2234-13-1 / 197.5	441^7; $248^{0.5}$	4-05-00-01665	EtOH 2; bz 4; chl 4; lig 4
7934	Naphthalene, octadecafluorodecahydro- / Perfluorodecalin	$C_{10}F_{18}$ / 462.08	306-94-5 / -10	142	3-05-00-00249	
7935	Naphthalene, octafluoro- / Perfluoronaphthalene	$C_{10}F_8$ / 272.10	313-72-4 / 87.5	209	4-05-00-01658	
7936	Naphthalene, 1,2,3,4,4a,5,6,8a-octahydro-7-methyl-4-methylene-1-(1-methylethyl)-, (1α,4aβ,8aα)- / γ-Cadinene	$C_{15}H_{24}$ / 204.36	39029-41-9	$124\text{-}8^{12}$	3-05-00-01088 / 0.9182^{15}	1.3166^{20}
7937	Naphthalene, 1-octyl- / 1-Octylnaphthalene	$C_{18}H_{24}$ / 240.39	2876-51-9 / -2	356	4-05-00-01779 / 0.9427^{20}	1.5526^{20}
7938	Naphthalene, 2-octyl-	$C_{18}H_{24}$ / 240.39	2876-44-0 / 12	357	4-05-00-01779 / 0.9350^{20}	1.5501^{20}
7939	Naphthalene, 1,1'-oxybis- / 1-Naphthyl ether	$C_{20}H_{14}O$ / 270.33	607-52-3 / 110	$280\text{-}5^{22}$	4-06-00-04215	bz 4; eth 4
7940	Naphthalene, 2,2'-oxybis-	$C_{20}H_{14}O$ / 270.33	613-80-9 / 105	250^{19} dec	4-06-00-04260	H_2O 1; EtOH 2; eth 4; bz 4

No.	Name / Synonym	Mol. Form. / Mol. Wt.	CAS RN / mp/°C	Merck No. / bp/°C	Beil. Ref. / den/g cm^{-3}	Solubility / n_D
7941	Naphthalene, 1-pentyl-	$C_{15}H_{18}$ 198.31	86-89-5 -22	307	4-05-00-01755 0.9656[20]	1.5725[20]
7942	Naphthalene, 2-pentyl- 2-Pentylnaphthalene	$C_{15}H_{18}$ 198.31	93-22-1 -4	310	4-05-00-01755 0.9561[20]	1.5694[20]
7943	Naphthalene, 1-(pentyloxy)- 1-Naphthyl pentyl ether	$C_{15}H_{18}O$ 214.31	108438-40-0 30	322	4-06-00-04213	bz 4; eth 4; EtOH 4
7944	Naphthalene, 2-(pentyloxy)- 2-Naphthyl pentyl ether	$C_{15}H_{18}O$ 214.31	31059-19-5 24.5	335	3-06-00-02973	bz 4; eth 4; EtOH 4; chl 4 1.5587[30]
7945	Naphthalene, 1-phenoxy-	$C_{16}H_{12}O$ 220.27	3402-76-4 55.5	349.5	4-06-00-04214	H_2O 1; EtOH 2; eth 3; ace 3
7946	Naphthalene, 2-phenoxy-	$C_{16}H_{12}O$ 220.27	19420-29-2 46	335	3-06-00-02975	H_2O 1; eth 3; chl 3; HOAc 3
7947	Naphthalene, 1-phenyl- 1-Phenylnaphthalene	$C_{16}H_{12}$ 204.27	605-02-7 45	334	4-05-00-02411 1.096[20]	H_2O 1; EtOH 4; eth 4; bz 4 1.6664[20]
7948	Naphthalene, 2-phenyl-	$C_{16}H_{12}$ 204.27	612-94-2 103.5	345.5	4-05-00-02412 1.2180[20]	EtOH 3; eth 4; bz 3; chl 3
7949	Naphthalene, 1-(phenylmethyl)-	$C_{17}H_{14}$ 218.30	611-45-0 59.5	350	4-05-00-02420 1.166[17]	H_2O 1; EtOH 2; eth 4; bz 3
7950	Naphthalene, 2-(phenylmethyl)- 2-Benzylnaphthalene	$C_{17}H_{14}$ 218.30	613-59-2 58	350	4-05-00-02420 1.176[12]	bz 4; eth 4; chl 4
7951	Naphthalene, 1-(phenylthio)-	$C_{16}H_{12}S$ 236.34	7570-98-1 41.8	255[43]; 222[11]	4-06-00-04242 1.167[15]	eth 4; EtOH 4
7952	1-Naphthalenepropanoic acid	$C_{13}H_{12}O_2$ 200.24	3243-42-3 156.5	179[11]	4-09-00-02435	H_2O 3; EtOH 3; bz 3
7953	2-Naphthalenepropanoic acid, 6-hydroxy- Allenolic acid	$C_{13}H_{12}O_3$ 216.24	553-39-9 180.5	247	3-10-00-01113	py 4; EtOH 4; MeOH 4
7954	Naphthalene, 1-(2-propenyl)-	$C_{13}H_{12}$ 168.24	2489-86-3	266	3-05-00-01805 1.0228[20]	EtOH 3; bz 4; chl 3 1.6140[20]
7955	Naphthalene, 1-propoxy-	$C_{13}H_{14}O$ 186.25	20009-26-1	293.5	4-06-00-04213 1.0447[18]	1.5928[18]
7956	Naphthalene, 2-propoxy-	$C_{13}H_{14}O$ 186.25	19718-45-7 41	305	3-06-00-02973	EtOH 4
7957	Naphthalene, 1-propyl-	$C_{13}H_{14}$ 170.25	2765-18-6 -8.6	274.5	4-05-00-01721 0.9897[20]	1.5923[20]
7958	Naphthalene, 2-propyl-	$C_{13}H_{14}$ 170.25	2027-19-2 -3	273.5	4-05-00-01722 0.9770[20]	1.5872[20]
7959	1-Naphthalenesulfonic acid	$C_{10}H_8O_3S$ 208.24	85-47-2 140	6295	4-11-00-00521	H_2O 3; EtOH 3; eth 2
7960	2-Naphthalenesulfonic acid β-Naphthylsulfonic acid	$C_{10}H_8O_3S$ 208.24	120-18-3 91	6296 dec	4-11-00-00527 1.441[25]	H_2O 4; EtOH 4; eth 3; bz 2
7961	1-Naphthalenesulfonic acid, 2-amino- 2-Naphthylamine-1-sulfonic acid	$C_{10}H_9NO_3S$ 223.25	81-16-3	6326	4-14-00-02792	DMSO 3
7962	1-Naphthalenesulfonic acid, 4-amino- 1-Naphthylamine-4-sulfonic acid	$C_{10}H_9NO_3S$ 223.25	84-86-6 dec	6323	4-14-00-02793 1.6703[25]	H_2O 1; EtOH 2; MeOH 3; py 3
7963	1-Naphthalenesulfonic acid, 5-amino- 1-Naphthylamine-5-sulfonic acid	$C_{10}H_9NO_3S$ 223.25	84-89-9	6324	4-14-00-02800	H_2O 3; eth 1
7964	1-Naphthalenesulfonic acid, 7-amino- Badische acid	$C_{10}H_9NO_3S$ 223.25	86-60-2	952	4-14-00-02802	HOAc 4
7965	1-Naphthalenesulfonic acid, 8-amino- 1-Naphthylamine-8-sulfonic acid	$C_{10}H_9NO_3S$ 223.25	82-75-7	6325	4-14-00-02802	gl HOAc 4
7966	1-Naphthalenesulfonic acid, 4-amino-3-hydroxy- 1-Amino-2-naphthol-4-sulfonic acid	$C_{10}H_9NO_4S$ 239.25	116-63-2	466	4-14-00-02825	
7967	2-Naphthalenesulfonic acid, 8-amino- 1,7-Cleve's acid	$C_{10}H_9NO_3S$ 223.25	119-28-8	2349	4-14-00-02805	
7968	1-Naphthalenesulfonic acid, 3,6-bis(1,1-dimethylethyl)-, ethyl ester Ethyl dibunate	$C_{20}H_{28}O_3S$ 348.51	5560-69-0	3745		chl 3
7969	1-Naphthalenesulfonic acid, 3,3'-[[1,1'-biphenyl]-4,4'-diylbis(azo)]bis[4-amino-, disodium salt Congo red	$C_{32}H_{22}N_6Na_2O_6S_2$ 696.68	573-58-0 >360	2493	1-16-00-00342	H_2O 3; EtOH 2; eth 1
7970	1-Naphthalenesulfonic acid, 3,4-dihydro-3,4-dioxo-, sodium salt 1,2-Naphthoquinone-4-sulfonic acid,sodium salt	$C_{10}H_5NaO_5S$ 260.20	521-24-4 287 dec	8597		
7971	1-Naphthalenesulfonic acid, 4-hydroxy- 1-Naphthol-4-sulfonic acid	$C_{10}H_8O_4S$ 224.24	84-87-7 170 dec	6311	4-11-00-00614	H_2O 4; eth 1
7972	1-Naphthalenesulfonic acid, 7-hydroxy- Croceic acid	$C_{10}H_8O_4S$ 224.24	132-57-0	2591	4-11-00-00623	H_2O 3
7973	2-Naphthalenesulfonic acid, 1-hydroxy- 1-Naphthol-2-sulfonic acid	$C_{10}H_8O_4S$ 224.24	567-18-0 >250	6310	4-11-00-00613	H_2O 2; EtOH 3; eth 1; dil HCl 2
7974	2-Naphthalenesulfonic acid, 6-hydroxy- 2-Naphthol-6-sulfonic acid	$C_{10}H_8O_4S$ 224.24	93-01-6 125	6312	4-11-00-00622	H_2O 4; EtOH 4; eth 1; HOAc 3
7975	1-Naphthalenesulfonyl chloride	$C_{10}H_7ClO_2S$ 226.68	85-46-1 68	209[20]; 147[0.9]	3-11-00-00383	bz 4; eth 4; EtOH 4

No.	Name / Synonym	Mol. Form. / Mol. Wt.	CAS RN / mp/°C	Merck No. / bp/°C	Beil. Ref. / den/g cm^{-3}	Solubility / n_D
7976	2-Naphthalenesulfonyl chloride	C$_{10}$H$_7$ClO$_2$S 226.68	93-11-8 81	201^{13}; 148$^{0.5}$	4-11-00-00529	H$_2$O 1; EtOH 3; eth 4; bz 3
7977	1-Naphthalenesulfonyl chloride, 5-(dimethylamino)- Dansyl chloride	C$_{12}$H$_{12}$ClNO$_2$S 269.75	605-65-2 70	2812	4-14-00-02801	
7978	2-Naphthalenesulfonyl chloride, 5,6,7,8-tetrahydro- 6-Tetralinsulfonyl chloride	C$_{10}$H$_{11}$ClO$_2$S 230.71	61551-49-3 58	197^{18}	3-11-00-00375	eth 4
7979	Naphthalene, 1,2,3,4-tetrahydro- Tetralin	C$_{10}$H$_{12}$ 132.21	119-64-2 -35.7	9152 207.6	4-05-00-01388 0.9660^{25}	H$_2$O 1; EtOH 4; eth 4; chl 3 1.5413^{20}
7980	Naphthalene, 1,2,3,4-tetrahydro-1,1-dimethyl- 1,1-Dimethyltetralin	C$_{12}$H$_{16}$ 160.26	1985-59-7	221	4-05-00-01430 0.950^{20}	1.5292^{20}
7981	Naphthalene, 1,2,3,4-tetrahydro-1,2-dimethyl- 1,2,3,4-Tetrahydro-1,2-dimethylnaphthalene	C$_{12}$H$_{16}$ 160.26	5195-40-4	235	3-05-00-01262 0.9470^{20}	1.5286^{20}
7982	Naphthalene, 1,2,3,4-tetrahydro-1,3-dimethyl- Tetralin, 1,3-dimethyl	C$_{12}$H$_{16}$ 160.26	5195-37-9	234; 78^1	3-05-00-01263 0.940^{20}	1.525^{20}
7983	Naphthalene, 1,2,3,4-tetrahydro-1,4-dimethyl- Tetralin, 1,4-dimethyl	C$_{12}$H$_{16}$ 160.26	4175-54-6	226	4-05-00-01430 0.940^{20}	1.525^{20}
7984	Naphthalene, 1,2,3,4-tetrahydro-1,5-dimethyl- Tetralin, 1,5-dimethyl	C$_{12}$H$_{16}$ 160.26	21564-91-0	239	3-05-00-01263 0.941^{20}	1.526^{20}
7985	Naphthalene, 1,2,3,4-tetrahydro-2,2-dimethyl- Tetralin, 2,2-dimethyl	C$_{12}$H$_{16}$ 160.26	13556-55-3	230	4-05-00-01431 0.935^{20}	1.5200^{20}
7986	Naphthalene, 1,2,3,4-tetrahydro-2,3-dimethyl- Tetralin, 2,3-dimethyl	C$_{12}$H$_{16}$ 160.26	21564-92-1 -5		3-05-00-01265 0.940^{20}	1.523^{20}
7987	Naphthalene, 1,2,3,4-tetrahydro-2,5-dimethyl- Tetralin, 2,5-dimethyl	C$_{12}$H$_{16}$ 160.26	25419-37-8	236	3-05-00-01263 0.946^{20}	1.526^{20}
7988	Naphthalene, 1,2,3,4-tetrahydro-2,6-dimethyl- Tetralin, 2,6-dimethyl	C$_{12}$H$_{16}$ 160.26	7524-63-2 20	240; 117^{15}	4-05-00-01432 0.941^{20}	1.526^{20}
7989	Naphthalene, 1,2,3,4-tetrahydro-2,7-dimethyl- Tetralin, 2,7-dimethyl	C$_{12}$H$_{16}$ 160.26	13065-07-1	237	3-05-00-01265 0.941^{20}	1.526^{20}
7990	Naphthalene, 1,2,3,4-tetrahydro-2,8-dimethyl- Tetralin, 2,8-dimethyl	C$_{12}$H$_{16}$ 160.26	25419-36-7	236	3-05-00-01264 0.941^{20}	1.526^{20}
7991	Naphthalene, 1,2,3,4-tetrahydro-5,6-dimethyl- Tetralin, 5,6-dimethyl	C$_{12}$H$_{16}$ 160.26	20027-77-4	252	4-05-00-01430 0.975^{20}	1.552^{20}
7992	Naphthalene, 1,2,3,4-tetrahydro-5,7-dimethyl- Tetralin, 5,7-dimethyl	C$_{12}$H$_{16}$ 160.26	21693-54-9 -6	253	4-05-00-01430 0.9583^{20}	1.5405^{20}
7993	Naphthalene, 1,2,3,4-tetrahydro-5,8-dimethyl- Tetralin, 5,8-dimethyl	C$_{12}$H$_{16}$ 160.26	14108-88-4 -4.1	254	4-05-00-01430 0.967^{20}	1.547^{20}
7994	Naphthalene, 1,2,3,4-tetrahydro-6,7-dimethyl-	C$_{12}$H$_{16}$ 160.26	1076-61-5 10	252	4-05-00-01431 0.954^{20}	1.538^{20}
7995	Naphthalene, 1,2,3,4-tetrahydro-1-methyl-	C$_{11}$H$_{14}$ 146.23	1559-81-5	220.6	4-05-00-01412 0.9583^{20}	1.5353^{20}
7996	Naphthalene, 1,2,3,4-tetrahydro-2-methyl-	C$_{11}$H$_{14}$ 146.23	3877-19-8 -43.1	221	4-05-00-01414 0.952^{20}	1.531^{20}
7997	Naphthalene, 1,2,3,4-tetrahydro-5-methyl-	C$_{11}$H$_{14}$ 146.23	2809-64-5 -23	234	4-05-00-01412 0.9720^{20}	1.5439^{20}
7998	Naphthalene, 1,2,3,4-tetrahydro-6-methyl-	C$_{11}$H$_{14}$ 146.23	1680-51-9 -40	229	4-05-00-01413 0.9537^{20}	1.5357^{20}
7999	Naphthalene, 1,2,3,4-tetrahydro-1,1,2,6-tetramethyl- Irene	C$_{14}$H$_{20}$ 188.31	1681-22-7	120-5^{10}	4-05-00-01460 0.9332^{20}	H$_2$O 1; ace 3; bz 3 1.5217^{20}
8000	Naphthalene, 1,2,3,4-tetrahydro-1,1,6-trimethyl-	C$_{13}$H$_{18}$ 174.29	475-03-6	240; 90^4	4-05-00-01445 0.9356^{20}	EtOH 3; eth 3; bz 3; chl 3 1.5257^{20}
8001	Naphthalene, 1,2,4,7-tetramethyl-	C$_{14}$H$_{18}$ 184.28	16020-17-0 -3	174-5^{12}	4-05-00-01746 1.011^{20}	1.6032^{20}
8002	Naphthalene, 1,2,5,6-tetramethyl-	C$_{14}$H$_{16}$ 184.28	2131-43-3 118	152^{12}	4-05-00-01746	H$_2$O 1; EtOH 3; bz 3; peth 3
8003	Naphthalene, 1,3,6,8-tetramethyl- 1,3,6,8-Tetramethylnaphthalene	C$_{14}$H$_{16}$ 184.28	14558-14-6 84.5	115-6^2	4-05-00-01746	H$_2$O 1; EtOH 3
8004	Naphthalene, 1,3,6,8-tetranitro-	C$_{10}$H$_4$N$_4$O$_8$ 308.16	28995-89-3 207	exp	4-05-00-01683 1.6400^0	bz 4; HOAc 4
8005	Naphthalene, 1,1'-thiobis-	C$_{20}$H$_{14}$S 286.40	607-53-4 110	289^{15}	4-06-00-04243	bz 4; HOAc 4
8006	Naphthalene, 2,2'-thiobis-	C$_{20}$H$_{14}$S 286.40	613-81-0 151	296^{15}	4-06-00-04314	EtOH 1; bz 4; CS$_2$ 4; HOAc 4
8007	1-Naphthalenethiol 1-Naphthyl mercaptan	C$_{10}$H$_8$S 160.24	529-36-2	6297 285 dec; 161^{20}	4-06-00-04241 1.1607^{20}	H$_2$O 2; EtOH 4; eth 4; dil alk 2 1.6802^{20}
8008	2-Naphthalenethiol 2-Naphthyl mercaptan	C$_{10}$H$_8$S 160.24	91-60-1 81	6298 288	4-06-00-04312 1.550^{25}	H$_2$O 2; EtOH 4; eth 4; lig 4
8009	Naphthalene, 1,2,4-trimethyl- 1,2,4-Trimethylnaphthalene	C$_{13}$H$_{14}$ 170.25	2717-42-2 55.5	146^{12}	4-05-00-01725	EtOH 3; eth 3; bz 3; chl 3
8010	Naphthalene, 1,2,5-trimethyl- 1,2,5-Trimethylnaphthalene	C$_{13}$H$_{14}$ 170.25	641-91-8 33.5	140^{12}	4-05-00-01726 1.0103^{22}	H$_2$O 1; EtOH 2; eth 4; bz 4 1.6093^{22}
8011	Naphthalene, 1,2,6-trimethyl- 1,2,6-Trimethylnaphthalene	C$_{13}$H$_{14}$ 170.25	3031-05-8 14	148^{10}	4-05-00-01726	H$_2$O 1; eth 3; bz 3 1.6010^{20}
8012	Naphthalene, 1,2,7-trimethyl-	C$_{13}$H$_{14}$ 170.25	486-34-0	147-8^{16}	4-05-00-01727 1.0087^{20}	H$_2$O 1; eth 3; bz 3 1.6097^{20}

No.	Name / Synonym	Mol. Form. / Mol. Wt.	CAS RN / mp/°C	Merck No. / bp/°C	Beil. Ref. / den/g cm^{-3}	Solubility / n_D
8013	Naphthalene, 1,3,7-trimethyl- 2,4,6-Trimethylnaphthalene	$C_{13}H_{14}$ 170.25	2131-38-6 12	131-3[9]	4-05-00-01728 0.9801[21]	1.5972[15]
8014	Naphthalene, 1,2,5-trimethyl-8-(1-methylethyl)- Cadalene, 5-methyl	$C_{16}H_{20}$ 212.33	6897-76-3	144-5[1.5]	3-05-00-01689 0.9847[30]	1.5926[30]
8015	Naphthalene, 1,3,8-trimethyl-5-(1-methylethyl)- Cadalene, 8-methyl	$C_{16}H_{20}$ 212.33	6897-88-7	135-6[1.5]	4-05-00-01770 0.9800[30]	1.5876[30]
8016	Naphthalene, 1,3,5-trinitro-	$C_{10}H_5N_3O_6$ 263.17	2243-94-9 124	exp 364	4-05-00-01682	ace 4; EtOH 4; chl 4
8017	1-Naphthalenol 1-Naphthol	$C_{10}H_8O$ 144.17	90-15-3 95	6303 288; 184[40]	4-06-00-04208 1.0989[99]	H_2O 1; EtOH 4; eth 4; ace 3 1.6224[99]
8018	2-Naphthalenol 2-Naphthol	$C_{10}H_8O$ 144.17	135-19-3 123	6304 285	4-06-00-04253 1.28[20]	H_2O 1; EtOH 4; eth 4; bz 3
8019	2-Naphthalenol, 6-amino-	$C_{10}H_9NO$ 159.19	4363-04-6 193	212 dec	3-13-00-01902	H_2O 4; EtOH 4
8020	2-Naphthalenol, 8-amino-	$C_{10}H_9NO$ 159.19	118-46-7 206	sub	3-13-00-01907	H_2O 3; EtOH 4; eth 3; bz 2
8021	2-Naphthalenol, benzoate 2-Naphthyl benzoate	$C_{17}H_{12}O_2$ 248.28	93-44-7 107	6328	4-09-00-00339	H_2O 1; EtOH 3; eth 2; HOAc 2
8022	2-Naphthalenol, 1-bromo-	$C_{10}H_7BrO$ 223.07	573-97-7 84	130	4-06-00-04301	H_2O 1; EtOH 3; eth 3; bz 3
8023	2-Naphthalenol, 6-chloro-	$C_{10}H_7ClO$ 178.62	40604-49-7 116.5	sub	4-06-00-04294	bz 4; eth 4; EtOH 4; chl 4
8024	2-Naphthalenol, decahydro-	$C_{10}H_{18}O$ 154.25	825-51-4	109[14]	1-06-00-00044 0.996[25]	1.4992[20]
8025	1-Naphthalenol, 2,3-dichloro- 2,3-Dichloro-1-naphthol	$C_{10}H_6Cl_2O$ 213.06	71284-96-3 101	sub	0-06-00-00612	bz 4; eth 4; EtOH 4; chl 4
8026	1-Naphthalenol, 2,4-dichloro-	$C_{10}H_6Cl_2O$ 213.06	2050-76-2 107.5	180	4-06-00-04233	bz 4; eth 4; EtOH 4
8027	2-Naphthalenol, 1,4-dimethyl- 2-Naphthol, 1,4-dimethyl-	$C_{12}H_{12}O$ 172.23	4705-94-6 135.5	315.5	4-06-00-04351	eth 4; EtOH 4
8028	2-Naphthalenol, 1-[(2,4-dimethylphenyl)azo]- C.I. Solvent Orange 7	$C_{18}H_{16}N_2O$ 276.34	3118-97-6 166		4-16-00-00234	eth 4; EtOH 4
8029	2-Naphthalenol, 1-[(2,5-dimethylphenyl)azo]- 1-(2,5-Xylylazo)-2-naphthol	$C_{18}H_{16}N_2O$ 276.34	85-82-5 153	9997	4-16-00-00234	
8030	1-Naphthalenol, 4-methyl-	$C_{11}H_{10}O$ 158.20	10240-08-1 85.8	166[13]	4-06-00-04330	H_2O 3; EtOH 3; eth 3; ace 3
8031	2-Naphthalenol, 1-methyl-	$C_{11}H_{10}O$ 158.20	1076-26-2 112	160[10]	4-06-00-04329	H_2O 2; EtOH 4; eth 4; ace 4
8032	1-Naphthalenol, methylcarbamate Carbaryl	$C_{12}H_{11}NO_2$ 201.22	63-25-2 145	1789	4-06-00-04219 1.228[25]	ace 4; DMF 4
8033	2-Naphthalenol, 1-[[2-methyl-4-[(2-methylphenyl)azo]phenyl]azo]- Scarlet red	$C_{24}H_{20}N_4O$ 380.45	85-83-6 186 dec	8349 260 dec	4-16-00-00249	chl 4; peth 4
8034	2-Naphthalenol, 1-nitro- 1-Nitro-2-naphthol	$C_{10}H_7NO_3$ 189.17	550-60-7 104	6536 115[0.05]	4-06-00-04307	H_2O 3; EtOH 3; eth 4; chl 2
8035	2-Naphthalenol, 1-nitroso- 1-Nitroso-2-naphthol	$C_{10}H_7NO_2$ 173.17	131-91-9 109.5	6562	4-07-00-02419	bz 4; eth 4
8036	1-Naphthalenol, 2-(phenylazo)-	$C_{16}H_{12}N_2O$ 248.28	3375-23-3 138	sub	4-16-00-00219	EtOH 4; HOAc 4
8037	2-Naphthalenol, 1-(phenylazo)- Azo, benzene 1'-naphthalene,2'-hydroxy	$C_{16}H_{12}N_2O$ 248.28	842-07-9		4-16-00-00228	
8038	2-Naphthalenol, 1-[[4-(phenylazo)phenyl]azo]- Sudan III	$C_{22}H_{16}N_4O$ 352.40	85-86-9 195	8858	4-16-00-00248	H_2O 1; EtOH 3; eth 3; ace 3
8039	1-Naphthalenol, 2-(phenylmethyl)-	$C_{17}H_{14}O$ 234.30	36441-32-4 73.5	237[12]	4-06-00-04990	H_2O 1; bz 3; lig 2
8040	2-Naphthalenol, 1-(phenylmethyl)- 2-Naphthol, 1-benzyl	$C_{17}H_{14}O$ 234.30	36441-31-3 114	247[14]	4-06-00-04988	ace 4; bz 4; eth 4; EtOH 4
8041	1-Naphthalenol, 1,2,3,4-tetrahydro-	$C_{10}H_{12}O$ 148.20	529-33-9	102-4[2]	3-06-00-02457 1.09[25]	1.5638[20]
8042	1-Naphthalenol, 5,6,7,8-tetrahydro-	$C_{10}H_{12}O$ 148.20	529-35-1 70	266; 143[11]	4-06-00-03851 1.0556[75]	
8043	2-Naphthalenol, 1,2,3,4-tetrahydro- 2-Tetralol	$C_{10}H_{12}O$ 148.20	530-91-6 15.5	9153 140[12]	0-06-00-00580	
8044	2-Naphthalenol, 5,6,7,8-tetrahydro-	$C_{10}H_{12}O$ 148.20	1125-78-6 57	275.5	4-06-00-03854 1.0552[65]	
8045	1(2H)-Naphthalenone, 2-acetyl-3,4-dihydro-	$C_{12}H_{12}O_2$ 188.23	17216-08-9 56	135-42[1]	4-07-00-02358	chl 3
8046	1(2H)-Naphthalenone, 3,4-dihydro-	$C_{10}H_{10}O$ 146.19	529-34-0	115[6]	4-07-00-01015 1.0988[16]	1.5672[20]
8047	2(1H)-Naphthalenone, 3,4-dihydro-	$C_{10}H_{10}O$ 146.19	530-93-8 18	237	4-07-00-01018 1.1055[27]	H_2O 1; eth 3; bz 3 1.5598[20]
8048	1(2H)-Naphthalenone, 3,4-dihydro-6-methoxy- 6-Methoxy-α-tetralone	$C_{11}H_{12}O_2$ 176.22	1078-19-9 78	5924 171[11]	4-08-00-00904	
8049	1(2H)-Naphthalenone, 3,4-dihydro-2-methyl-	$C_{11}H_{12}O$ 160.22	1590-08-5	127-31[12]	4-07-00-01033 1.057[25]	1.5535[20]
8050	1(2H)-Naphthalenone, 3,4-dihydro-4-methyl-	$C_{11}H_{12}O$ 160.22	19832-98-5	133-4[12]	4-07-00-01031 1.0779[19]	1.5620[19]

No.	Name / Synonym	Mol. Form. / Mol. Wt.	CAS RN / mp/°C	Merck No. / bp/°C	Beil. Ref. / den/g cm^{-3}	Solubility / n_D
8051	1(2H)-Naphthalenone, 7-ethyl-3,4-dihydro- / 1-Tetralone, 7-ethyl	$C_{12}H_{14}O$ / 174.24	22531-06-2	152-3[12]	4-07-00-01049 / 1.0556[17]	bz 4; eth 4; EtOH 4 / 1.5599[17]
8052	2(3H)-Naphthalenone, 4,4a,5,6,7,8-hexahydro-4,4a-dimethyl-6-(1-methylethylidene)-, (4R-cis)-	$C_{15}H_{22}O$ / 218.34	15764-04-2 / 51.5	144[2]	1.0035[20]	ace 4 / 1.5370[20]
8053	1(2H)-Naphthalenone, octahydro-, cis- / 1-Decalone (cis)	$C_{10}H_{16}O$ / 152.24	32166-40-8 / 2	126[20]	4-07-00-00201 / 1.008[20]	1.4936[20]
8054	1(2H)-Naphthalenone, octahydro-, trans-	$C_{10}H_{16}O$ / 152.24	21370-71-8 / 33	122[20]	4-07-00-00201 / 0.986[20]	1.4849[21]
8055	2(1H)-Naphthalenone, octahydro-	$C_{10}H_{16}O$ / 152.24	4832-17-1	96[2.5]	4-07-00-00203 / 0.979[25]	1.4900[20]
8056	Naphtho[2,3-b]-1,4-dithiin-2,3-dicarbonitrile, 5,10-dihydro-5,10-dioxo- / Dithianone	$C_{14}H_4N_2O_2S_2$ / 296.33	3347-22-6 / 220	3375	5-19-08-00189	
8057	Naphtho[1,2-b]furan-2,8(3H,4H)-dione, 3a,5,5a,9b-tetrahydro-3,5a,9-trimethyl-, / α-Santonin	$C_{15}H_{18}O_3$ / 246.31	481-06-1 / 175	8325	5-17-11-00314 / 1.590[25]	H_2O 2; EtOH 2; eth 2; bz 3
8058	Naphtho[1,2-b]furan-2,8(3H,4H)-dione, 3a,5,5a,9b-tetrahydro-4,5-dihydroxy-9-methoxy-3,5a-dimethyl- / Cedrin	$C_{15}H_{18}O_6$ / 294.30	6040-62-6 / 266	1918		EtOH 4; MeOH 4; chl 4
8059	Naphtho[2,3-b]furan-2(3H)-one, 3a,5,6,7,8,8a,9,9a-octahydro-5,8a-dimethyl-3-methylene-, / Alantolactone	$C_{15}H_{20}O_2$ / 232.32	546-43-0 / 76	198 / 275	5-17-10-00097	bz 4; eth 4; EtOH 4; chl 4
8060	Naphtho[1,2-c]furan-1,3-dione, 4,5-dihydro-	$C_{12}H_8O_3$ / 200.19	37845-14-0 / 126.5	227-30[23]	5-17-11-00444	bz 4; MeOH 4
8061	Naphtho[1,2,3,4-def]chrysene / Dibenzo[a,e]pyrene	$C_{24}H_{14}$ / 302.38	192-65-4 / 233.5		4-05-00-02803	EtOH 2; ace 2; bz 2; tol 3
8062	Naphtho[2,3-f]quinoline-7,12-dione, 5,6-dihydroxy- / Alizarine Blue	$C_{17}H_9NO_4$ / 291.26	568-02-5 / 269	239	5-21-13-00590	bz 4; gl HOAc 4
8063	2-Naphthol, 8-chloro-	$C_{10}H_7ClO$ / 178.62	29921-50-4 / 101	307.5	3-06-00-02992	bz 4; eth 4; EtOH 4; chl 4
8064	2-Naphthylamine, 1,4-dimethyl- / Naphthalene, 2-amino-1,4-dimethyl	$C_{12}H_{13}N$ / 171.24	878-93-3 / 75	333	0-12-00-01317	eth 4; EtOH 4
8065	1,5-Naphthyridine / 1,5-Diazanaphthalene	$C_8H_6N_2$ / 130.15	254-79-5 / 75	112[12]	5-23-07-00156 / 1.2100[20]	
8066	1,8-Naphthyridine-3-carboxylic acid, 1-ethyl-1,4-dihydro-7-methyl-4-oxo-	$C_{12}H_{12}N_2O_3$ / 232.24	389-08-2 / 229.5	6273		EtOH 2; eth 2; chl 3
8067	Napropamide / Propanamide, N,N-diethyl-2-(1-naphthalenyloxy)-	$C_{17}H_{21}NO_2$ / 271.36	15299-99-7 / 75	6336		
8068	Nickel carbonyl / Nickel tetracarbonyl	C_4NiO_4 / 170.74	13463-39-3 / -19.3	6411 / 43	1.31[25]	
8069	Nicosulfuron / 2-(4,6-DimethoxypyrimidiN-2-ylcarbamoylsulfamoyl)-N,N-dimethylnicotinamide	$C_{15}H_{18}N_6O_6S$ / 410.41	111991-09-4 / 172			
8070	Nitralin / Benzenamine, 4-(methylsulfonyl)-2,6-dinitro-N,N-dipropyl-	$C_{13}H_{19}N_3O_6S$ / 345.38	4726-14-1 / 150	6486		
8071	Nitrapyrin / Pyridine, 2-chloro-6-(trichloromethyl)-	$C_8H_3Cl_4N$ / 230.91	1929-82-4 / 63	6490 / 136[11]		
8072	Nitric acid, butyl ester / Butyl nitrate	$C_4H_9NO_3$ / 119.12	928-45-0	133	4-01-00-01524 / 1.0228[30]	H_2O 1; EtOH 3; eth 3; ctc 2 / 1.4013[23]
8073	Nitric acid, decyl ester / Decyl nitrate	$C_{10}H_{21}NO_3$ / 203.28	2050-78-4	128[11]; 88[1]	4-01-00-01819 / 0.951[0]	EtOH 3; eth 3
8074	Nitric acid, ethyl ester / Ethyl nitrate	$C_2H_5NO_3$ / 91.07	625-58-1 / -94.6	87.2	4-01-00-01327 / 1.1084[20]	H_2O 3; EtOH 5; eth 5 / 1.3852[20]
8075	Nitric acid, methyl ester / Methyl nitrate	CH_3NO_3 / 77.04	598-58-3 / -82.5	6015 / 64.6 exp	4-01-00-01254 / 1.2075[20]	H_2O 2; EtOH 3; eth 3 / 1.3748[20]
8076	Nitric acid, 1-methylethyl ester / Isopropyl nitrate	$C_3H_7NO_3$ / 105.09	1712-64-7	100	4-01-00-01475 / 1.034[19]	EtOH 3; eth 3 / 1.3912[16]
8077	Nitric acid, 1-methylpropyl ester	$C_4H_9NO_3$ / 119.12	924-52-7	124	4-01-00-01573 / 1.0264[20]	eth 4; EtOH 4 / 1.4015[20]
8078	Nitric acid, 2-methylpropyl ester / Isobutyl nitrate	$C_4H_9NO_3$ / 119.12	543-29-3 / 123.4	5031	4-01-00-01595 / 1.0152[20]	1.4028[20]
8079	Nitric acid, octyl ester	$C_8H_{17}NO_3$ / 175.23	629-39-0	110-2[20]	4-01-00-01762 / 0.8409[17]	H_2O 2; EtOH 3; eth 3
8080	Nitric acid, propyl ester / Propyl nitrate	$C_3H_7NO_3$ / 105.09	627-13-4	7877 / 110	4-01-00-01424 / 1.0538[20]	H_2O 2; EtOH 3; eth 3; ctc 3 / 1.3973[20]
8081	Nitrofen / Benzene, 2,4-dichloro-1-(4-nitrophenoxy)-	$C_{12}H_7Cl_2NO_3$ / 284.10	1836-75-5 / 70	6519		
8082	Nitrofluorfen / Benzene, 2-chloro-1-(4-nitrophenoxy)-4-(trifluoromethyl)-	$C_{13}H_7ClF_3NO_3$ / 317.65	42874-01-1 / 68			
8083	Nitrous acid, butyl ester / Butyl nitrite	$C_4H_9NO_2$ / 103.12	544-16-1 / 78	1581	4-01-00-01523 / 0.9114[25]	EtOH 5; eth 5 / 1.3762[20]

No.	Name Synonym	Mol. Form. Mol. Wt.	CAS RN mp/°C	Merck No. bp/°C	Beil. Ref. den/g cm⁻³	Solubility n_D
8084	Nitrous acid, decyl ester Decyl nitrite	$C_{10}H_{21}NO_2$ 187.28	1653-57-2	105-8[12]	4-01-00-01819	EtOH 3; eth 3 1.4247^{20}
8085	Nitrous acid, 1,1-dimethylethyl ester tert-Butyl nitrite	$C_4H_9NO_2$ 103.12	540-80-7	1582 63	4-01-00-01622 0.8670^{20}	H_2O 2; EtOH 3; eth 3; chl 3 1.368^{20}
8086	Nitrous acid, 1,2-ethanediyl ester Ethylene glycol, dinitrite	$C_2H_4N_2O_4$ 120.06	629-16-3 <-15	98	4-01-00-02411 1.2156^0	eth 4; EtOH 4
8087	Nitrous acid, ethyl ester Ethyl nitrite	$C_2H_5NO_2$ 75.07	109-95-5	3786 18	4-01-00-01327 0.899^{15}	EtOH 5; eth 5 1.3418^{10}
8088	Nitrous acid, heptyl ester Heptyl nitrite	$C_7H_{15}NO_2$ 145.20	629-43-6	155.8	4-01-00-01735 0.8939^0	eth 4 1.4032^{20}
8089	Nitrous acid, hexyl ester Hexyl nitrite	$C_6H_{13}NO_2$ 131.17	638-51-7	129; 52[44]	4-01-00-01699 0.8778^{20}	H_2O 1; EtOH 3; eth 3 1.3987^{20}
8090	Nitrous acid, 3-methylbutyl ester Isoamyl nitrite	$C_5H_{11}NO_2$ 117.15	110-46-3	5005 99.2	4-01-00-01683 0.8828^{20}	H_2O 2; EtOH 5; eth 5 1.3918^{20}
8091	Nitrous acid, methyl ester Methyl nitrite	CH_3NO_2 61.04	624-91-9 -16	-12	4-01-00-01253 0.991^{15}	EtOH 3; eth 3
8092	Nitrous acid, 1-methylethyl ester Isopropyl nitrite	$C_3H_7NO_2$ 89.09	541-42-4	5104 40	4-01-00-01474 0.8684^{15}	H_2O 1; EtOH 3; eth 3
8093	Nitrous acid, 1-methylpropyl ester	$C_4H_9NO_2$ 103.12	924-43-6	68.5	4-01-00-01573 0.8726^{20}	eth 4; EtOH 4; chl 4 1.3710^{20}
8094	Nitrous acid, 2-methylpropyl ester Isobutyl nitrite	$C_4H_9NO_2$ 103.12	542-56-3	5032 67	4-01-00-01595 0.8699^{22}	H_2O 2; EtOH 3; eth 3 1.3715^{22}
8095	Nitrous acid, octyl ester	$C_8H_{17}NO_2$ 159.23	629-46-9	174.5	4-01-00-01762 0.862^{17}	H_2O 2; EtOH 4; eth 4 1.4127^{20}
8096	Nitrous acid, pentyl ester	$C_5H_{11}NO_2$ 117.15	463-04-7	104.5	4-01-00-01644 0.8817^{20}	H_2O 2; EtOH 5; eth 5 1.3851^{20}
8097	Nitrous acid, phenylmethyl ester Benzyl nitrite	$C_7H_7NO_2$ 137.14	935-05-7	80-3[35]	4-06-00-02568 1.075^{25}	1.4989^{25}
8098	Nitrous acid, propyl ester Propyl nitrite	$C_3H_7NO_2$ 89.09	543-67-9	7878 48	4-01-00-01424 0.886^{20}	H_2O 2; EtOH 3; eth 3 1.3604^{20}
8099	Nonacosane Celidoniol, deoxy-	$C_{29}H_{60}$ 408.80	630-03-5 63.7	440.8	4-01-00-00591 0.8083^{20}	H_2O 1; EtOH 4; eth 4; ace 4 1.4529^{20}
8100	1-Nonacosanol Nonacosanyl alcohol	$C_{29}H_{60}O$ 424.79	6624-76-6 84.5	200[1.0]	4-01-00-01916	EtOH 3
8101	Nonadecane	$C_{19}H_{40}$ 268.53	629-92-5 32.1	329.9	4-01-00-00560 0.7855^{20}	H_2O 1; EtOH 2; eth 3; ace 3 1.4409^{20}
8102	1,2,3-Nonadecanetricarboxylic acid, 2-hydroxy- Agaricic acid	$C_{22}H_{40}O_7$ 416.56	666-99-9 142 dec	174	4-03-00-01284	H_2O 3; EtOH 2; eth 2; bz 1
8103	Nonadecanoic acid	$C_{19}H_{38}O_2$ 298.51	646-30-0 69.4	297[100]; 228[10]	4-02-00-01256 0.8468^{70}	H_2O 1; EtOH 4; eth 4; bz 4
8104	1-Nonadecanol	$C_{19}H_{40}O$ 284.53	1454-84-8 63.3	166[0.3]	4-01-00-01898	eth 3; ace 3 1.4328^{75}
8105	10-Nonadecanone	$C_{19}H_{38}O$ 282.51	504-57-4 65.5	>350; 156[1.1]	4-01-00-03402	H_2O 1; EtOH 2; eth 3; ace 3
8106	2-Nonadecanone	$C_{19}H_{38}O$ 282.51	629-66-3 57	266[110]; 165[2]	4-01-00-03400 0.8108^{56}	H_2O 1; EtOH 2; eth 4; ace 3
8107	1-Nonadecyne	$C_{19}H_{36}$ 264.49	26186-01-6 37.5	327	3-01-00-01030 0.8054^{20}	ace 4; bz 4; eth 4 1.4488^{20}
8108	2,6-Nonadienal Leaf aldehyde violet	$C_9H_{14}O$ 138.21	26370-28-5	85-7[11]	4-01-00-03560 0.8678^{20}	1.4460^{20}
8109	2,4-Nonadienal, (E,E)-	$C_9H_{14}O$ 138.21	5910-87-2	97-8[10]	4-01-00-03560 0.862^{25}	1.5207^{20}
8110	1,8-Nonadiene	C_9H_{16} 124.23	4900-30-5	142.5	4-01-00-01048 0.7511^{20}	1.4302^{20}
8111	2,7-Nonadiene	C_9H_{16} 124.23	51333-70-1 -72.5	151.5	3-01-00-01013 0.7499^{20}	1.4358^{20}
8112	2,6-Nonadien-1-ol	$C_9H_{16}O$ 140.23	7786-44-9	96-100[11]	4-01-00-02266 0.8604^{25}	1.4598^{25}
8113	6,8-Nonadien-2-one, 8-methyl-5-(1-methylethyl)-, [S-(E)]- Solanone	$C_{13}H_{22}O$ 194.32	1937-54-8	8663 60[1]	0.870^{20}	1.4755^{20}
8114	1,4-Nonadiyne	C_9H_{12} 120.19	6088-94-4	83-4[41]	4-01-00-01125 0.8112^{25}	1.4518^{25}
8115	1,8-Nonadiyne	C_9H_{12} 120.19	2396-65-8 -27.3	162	4-01-00-01125 0.8158^{20}	eth 3; ace 3 1.4490^{20}
8116	2,7-Nonadiyne	C_9H_{12} 120.19	31699-35-1 4.3	180	3-01-00-01063 0.8332^{20}	1.4674^{20}
8117	Nonanal Nonaldehyde	$C_9H_{18}O$ 142.24	124-19-6	191	4-01-00-03352 0.8264^{22}	eth 3; chl 3 1.4273^{20}
8118	Nonanamide	$C_9H_{19}NO$ 157.26	1120-07-6 99.5	sub	4-02-00-01023 0.8394^{110}	H_2O 1; EtOH 2; eth 2 1.4248^{110}
8119	1-Nonanamine	$C_9H_{21}N$ 143.27	112-20-9 -1	202.2	4-04-00-00777 0.7886^{20}	H_2O 2; EtOH 3; eth 3; chl 2 1.4336^{20}
8120	Nonane	C_9H_{20} 128.26	111-84-2 -53.5	150.8	4-01-00-00447 0.7176^{20}	H_2O 1; EtOH 4; eth 4; ace 5 1.4054^{20}
8121	Nonane, 1-bromo-	$C_9H_{19}Br$ 207.15	693-58-3	88[4]	4-01-00-00451 1.0165^{20}	1.4533^{20}

No.	Name Synonym	Mol. Form. Mol. Wt.	CAS RN mp/°C	Merck No. bp/°C	Beil. Ref. den/g cm^{-3}	Solubility n_D
8122	Nonane, 5-butyl-	$C_{13}H_{28}$ 184.37	17312-63-9	217.5	4-01-00-00517 0.7635[18]	1.4273[18]
8123	Nonane, 1-chloro-	$C_9H_{19}Cl$ 162.70	2473-01-0 -39.4	203.5	4-01-00-00450 0.8720[20]	H_2O 1; eth 3; chl 3 1.4345[20]
8124	Nonane, 2-chloro- 2-Chlorononane	$C_9H_{19}Cl$ 162.70	2216-36-6	190	0-01-00-00166 0.8790[20]	chl 4 1.4420[20]
8125	Nonane, 5-chloro- 5-Chlorononane	$C_9H_{19}Cl$ 162.70	28123-70-8	85-7[14]	2-01-00-00128 0.8639[15]	eth 4 1.4314[15]
8126	Nonane, 1-chloro-9-fluoro-	$C_9H_{18}ClF$ 180.69	463-23-0	102[11]	4-01-00-00450 0.966[20]	eth 4; EtOH 4 1.4301[25]
8127	Nonane, 1,2-dibromo- 1,2-Dibromononane	$C_9H_{18}Br_2$ 286.05	73642-91-8	141.5[20]	3-01-00-00506 1.3980[20]	1.4942[20]
8128	Nonane, 1,9-dichloro- 1,9-Dichlorononane	$C_9H_{18}Cl_2$ 197.15	821-99-8	260; 138[17]	4-01-00-00451 1.0173[25]	1.4586[25]
8129	Nonanedinitrile	$C_9H_{14}N_2$ 150.22	1675-69-0	198-9[25]	4-02-00-02059 0.9200[19]	eth 4; EtOH 4; bz 4 1.4518[19]
8130	Nonanedioic acid Azelaic acid	$C_9H_{16}O_4$ 188.22	123-99-9 106.5	921 287[100]; 225[10]	4-02-00-02055 1.225[25]	H_2O 2; EtOH 3; eth 2; bz 2 1.4303[111]
8131	Nonanedioic acid, bis(2-ethylbutyl) ester	$C_{21}H_{40}O_4$ 356.55	105-03-3 -45	230[5]	4-01-00-02057 0.928[25]	ace 4; bz 4; EtOH 4 1.443[25]
8132	Nonanedioic acid, bis(2-ethylhexyl) ester	$C_{25}H_{48}O_4$ 412.65	103-24-2 -78	237[5]	3-02-00-01787 0.915[25]	H_2O 1; EtOH 3; ace 3; bz 3 1.446[25]
8133	Nonanedioic acid, diethyl ester	$C_{13}H_{24}O_4$ 244.33	624-17-9 -18.5	291.5	4-02-00-02056 0.9729[20]	H_2O 1; EtOH 3; eth 3 1.4351[20]
8134	Nonanedioic acid, dimethyl ester	$C_{11}H_{20}O_4$ 216.28	1732-10-1	156[20]	4-02-00-02056 1.0082[20]	H_2O 1; EtOH 3; ace 3; bz 3 1.4367[20]
8135	Nonanedioic acid, 3-methyl- Azelaic acid, 3-methyl	$C_{10}H_{18}O_4$ 202.25	76078-85-8 44	168[0.2]	3-02-00-01819	1.4670[19]
8136	1,9-Nonanediol 1,9-Dihydroxynonane	$C_9H_{20}O_2$ 160.26	3937-56-2 45.8	173[20]; 150[3]	4-01-00-02607	H_2O 2; EtOH 4; eth 4; bz 3
8137	2,4-Nonanedione Methyl β-oxoheptyl ketone	$C_9H_{16}O_2$ 156.22	6175-23-1 -18	105-7[22]	4-01-00-03715 0.938[25]	
8138	Nonanedioyl dichloride	$C_9H_{14}Cl_2O_2$ 225.11	123-98-8	166[18]	3-02-00-01789 1.4680[20]	eth 3; bz 4 1.4680[20]
8139	Nonane, icosafluoro Perfluorononane	C_9F_{20} 488.07	375-96-2	125.3	1.800[25]	
8140	Nonane, 2-methyl- 2-Methylnonane	$C_{10}H_{22}$ 142.28	871-83-0 -74.6	167.1	4-01-00-00473 0.7281[20]	H_2O 1; eth 3; bz 3; chl 3 1.4099[20]
8141	Nonane, 3-methyl- 3-Methylnonane	$C_{10}H_{22}$ 142.28	5911-04-6 -84.8	167.9	2-01-00-00130 0.7354[20]	bz 4; eth 4; chl 4 1.4125[20]
8142	Nonane, 4-methyl- 4-Methylnonane	$C_{10}H_{22}$ 142.28	17301-94-9 -99	165.7	4-01-00-00475 0.7323[20]	bz 4; eth 4; chl 4 1.4123[20]
8143	Nonane, 5-methyl- 5-Methylnonane	$C_{10}H_{22}$ 142.28	15869-85-9 -87.7	165.1	4-01-00-00475 0.7326[20]	H_2O 1; eth 3; bz 3; chl 3 1.4116[20]
8144	Nonanenitrile	$C_9H_{17}N$ 139.24	2243-27-8 -34.2	224.4	4-02-00-01024 0.8178[20]	H_2O 1; EtOH 3; eth 3; ctc 2 1.4255[20]
8145	1-Nonanethiol Nonyl mercaptan	$C_9H_{20}S$ 160.32	1455-21-6 -20.1	220	4-01-00-01802 0.842[25]	1.4548[20]
8146	Nonanoic acid Pelargonic acid	$C_9H_{18}O_2$ 158.24	112-05-0 12.3	7013 254.5	4-02-00-01018 0.9052[20]	H_2O 1; EtOH 3; eth 3; chl 3 1.4343[19]
8147	Nonanoic acid, butyl ester Pelargonic acid, butyl ester	$C_{13}H_{26}O_2$ 214.35	50623-57-9 -38	122-4[20]	3-02-00-00820 0.8520[25]	1.4262[25]
8148	Nonanoic acid, 9-cyano-, methyl ester Sebacic acid, monomitrile, methyl ester	$C_{11}H_{19}NO_2$ 197.28	53663-26-6 3.5	178[16]	4-02-00-02089 0.934[20]	eth 4 1.4398[25]
8149	Nonanoic acid, ethenyl ester	$C_{11}H_{20}O_2$ 184.28	6280-03-1	133.5[50]	4-02-00-01020 0.8689[30]	1.4447[30]
8150	Nonanoic acid, ethyl ester Ethyl nonanoate	$C_{11}H_{22}O_2$ 186.29	123-29-5 -36.7	3793 227.0	4-02-00-01019 0.8657[20]	H_2O 1; EtOH 3; eth 3; ace 3 1.4220[20]
8151	Nonanoic acid, 9-fluoro- Pelargonic acid, 9-fluoro	$C_9H_{17}FO_2$ 176.23	463-16-1 18	89[0.2]	4-02-00-01024	eth 4; EtOH 4 1.4289[25]
8152	Nonanoic acid, heptyl ester Pelargonic acid, heptyl ester	$C_{16}H_{32}O_2$ 256.43	71605-85-1 -15.5	210[75]	3-02-00-00821 0.8553[25]	1.4350[20]
8153	Nonanoic acid, 3-methyl- Pelargonic acid, 3-methyl	$C_{10}H_{20}O_2$ 172.27	35205-79-9	147-8[12]	4-02-00-01057 0.9012[20]	1.4342[20]
8154	Nonanoic acid, methyl ester	$C_{10}H_{20}O_2$ 172.27	1731-84-6	213.5	4-02-00-01019 0.8799[15]	H_2O 1; EtOH 3; eth 3; ctc 2 1.4214[20]
8155	Nonanoic acid, 3-methyl-, ethyl ester Pelargonic acid 3-methyl, ethyl ester	$C_{12}H_{24}O_2$ 200.32	86051-37-8	115[13]	3-02-00-00849 0.8653[20]	1.4240[20]
8156	Nonanoic acid, pentyl ester Pelargonic acid, pentyl ester	$C_{14}H_{28}O_2$ 228.38	61531-45-1 -27	130-2[20]	3-02-00-00820 0.8506[25]	1.4318[20]
8157	1-Nonanol Nonyl alcohol	$C_9H_{20}O$ 144.26	143-08-8 -5	6598 213.3	4-01-00-01798 0.8273[20]	H_2O 1; EtOH 3; eth 3; ctc 2 1.4333[20]
8158	2-Nonanol, (±)-	$C_9H_{20}O$ 144.26	74683-66-2 -35	193.5	4-01-00-01803 0.8471[20]	eth 4; EtOH 4 1.4353[20]
8159	3-Nonanol, (±)-	$C_9H_{20}O$ 144.26	74742-08-8 22	195; 93[18]	4-01-00-01803 0.8250[20]	H_2O 1; EtOH 3; eth 3 1.4289[20]
8160	4-Nonanol, (S)-	$C_9H_{20}O$ 144.26	52708-03-9	192.5; 94[18]	3-01-00-01747 0.8282[20]	H_2O 1; EtOH 3; eth 3 1.4197[20]
8161	5-Nonanol Dibutylcarbinol	$C_9H_{20}O$ 144.26	623-93-8	193; 97[20]	4-01-00-01804 0.8356[20]	H_2O 1; EtOH 3 1.4289[20]

No.	Name / Synonym	Mol. Form. / Mol. Wt.	CAS RN / mp/°C	bp/°C	Beil. Ref. / den/g cm⁻³	Solubility / n_D
8162	5-Nonanol, 5-butyl-	$C_{13}H_{28}O$ / 200.38	597-93-3 / 20	232 dec	4-01-00-01863 / 0.8408^{20}	EtOH 4 / 1.4445^{20}
8163	1-Nonanol, 9-chloro- / 9-Chlorononan-1-ol	$C_9H_{19}ClO$ / 178.70	51308-99-7 / 28	147^{14}	4-01-00-01802 /	eth 4; EtOH 4 / 1.4575^{20}
8164	5-Nonanol, 2,8-dimethyl- / 2,8-Dimethylnonan-5-ol	$C_{11}H_{24}O$ / 172.31	19780-96-2 /	105^9	0-01-00-00428 / 0.8305^{12}	1.4380^{12}
8165	1-Nonanol, 9-fluoro-	$C_9H_{19}FO$ / 162.25	463-24-1 /	125-6^{15}	4-01-00-01801 / 0.928^{20}	eth 4; EtOH 4 / 1.4279^{25}
8166	1-Nonanol, 3-methyl- / 3-Hexylbutanol-1	$C_{10}H_{22}O$ / 158.28	22663-64-5 /	114^{14}; 103^9	4-01-00-01825 / 0.8342^{20}	1.4361^{20}
8167	1-Nonanol, 4-methyl-	$C_{10}H_{22}O$ / 158.28	1489-47-0 /	120^{17}	4-01-00-01826 / 0.826^{27}	1.4364^{20}
8168	1-Nonanol, 8-methyl / 8-Methyl-1-nonanol	$C_{10}H_{22}O$ / 158.28	55505-26-5 /	108^{10}		
8169	3-Nonanol, 2-methyl-	$C_{10}H_{22}O$ / 158.28	26533-33-5 /	115^{28}	4-01-00-01825 / 0.8290^{20}	1.4346^{20}
8170	4-Nonanol, 4-methyl- / 1-Methyl-1-propylhexanol	$C_{10}H_{22}O$ / 158.28	23418-38-4 /	92-3^{15}	4-01-00-01826 / 0.8245^{25}	1.4338^{20}
8171	5-Nonanol, 5-methyl- / 2-Butyl-2-hexanol	$C_{10}H_{22}O$ / 158.28	33933-78-7 /	84-5^{10}	4-01-00-01827 / 0.8290^{20}	1.4341^{20}
8172	2-Nonanone / Heptyl methyl ketone	$C_9H_{18}O$ / 142.24	821-55-6 / -7.5	195.3	4-01-00-03353 / 0.8208^{20}	H_2O 1; EtOH 3; eth 3; ace 4 / 1.4210^{20}
8173	3-Nonanone / Ethyl hexyl ketone	$C_9H_{18}O$ / 142.24	925-78-0 / -8	190; 86^{20}	4-01-00-03354 / 0.8241^{20}	H_2O 1; EtOH 3; eth 3; ace 4 / 1.4208^{20}
8174	4-Nonanone / Propyl amyl ketone	$C_9H_{18}O$ / 142.24	4485-09-0 /	187.5	4-01-00-03354 / 0.8190^{25}	H_2O 1; EtOH 3; eth 3; ace 4 / 1.4189^{20}
8175	5-Nonanone / Dibutyl ketone	$C_9H_{18}O$ / 142.24	502-56-7 / -5.9	188.4	4-01-00-03355 / 0.8217^{20}	H_2O 1; EtOH 3; eth 4; chl 4 / 1.4195^{20}
8176	5-Nonanone, 2-methyl- / 2-Methyl-5-nonanone	$C_{10}H_{20}O$ / 156.27	22287-02-1 /	203.5	3-01-00-02900 / 0.8213^{20}	1.4239^{20}
8177	Nonanoyl chloride	$C_9H_{17}ClO$ / 176.69	764-85-2 / -60.5	215.3	4-02-00-01023 / 0.9463^{15}	eth 3; ace 3
8178	Nonasiloxane, eicosamethyl- / Eicosamethylnonasiloxane	$C_{20}H_{60}O_8Si_9$ / 681.46	2652-13-3 / 3494	307.5; 198^{16}	4-04-00-04130 / 0.9173^{20}	bz 4 / 1.3980^{20}
8179	2-Nonenal / 2-Nonen-1-al	$C_9H_{16}O$ / 140.23	2463-53-8 /	56-$8^{0.1}$	4-01-00-03501 / 0.8418^{25}	1.4502^{25}
8180	2-Nonenal, (E)-	$C_9H_{16}O$ / 140.23	18829-56-6 /	6595	4-01-00-03502 /	
8181	6-Nonenamide, N-[(4-hydroxy-3-methoxyphenyl)methyl]-8-methyl-, (E)- / Capsaicin	$C_{18}H_{27}NO_3$ / 305.42	404-86-4 / 65	1767 210-$20^{0.01}$	4-13-00-02588 /	H_2O 1; EtOH 4; eth 3; bz 3
8182	1-Nonene	C_9H_{18} / 126.24	124-11-8 / -81.3	146.9	4-01-00-00894 / 0.7253^{25}	1.4257^{20}
8183	3-Nonene, (E)-	C_9H_{18} / 126.24	20063-92-7 /	147.5	3-01-00-00853 / 0.732^{21}	H_2O 1; eth 3; bz 3; ctc 3 / 1.4181^{21}
8184	4-Nonene, 5-butyl- / 5-Butyl-4-nonene	$C_{13}H_{26}$ / 182.35	7367-38-6 /	215.5	4-01-00-00922 / 0.7745^{20}	1.4375^{20}
8185	1-Nonene, 2-methyl- / 2-Methylnon-1-ene	$C_{10}H_{20}$ / 140.27	2980-71-4 / -64.2	168.4	4-01-00-00903 / 0.7412^{25}	1.4241^{20}
8186	3-Nonene, 2-methyl-	$C_{10}H_{20}$ / 140.27	53966-53-3 /	161.0	4-01-00-00903 / 0.7340^{20}	1.4202^{20}
8187	3-Nonenoic acid	$C_9H_{16}O_2$ / 156.22	4124-88-3 /	106^1	4-02-00-01598 / 0.9254^{20}	1.4454^{25}
8188	8-Nonenoic acid	$C_9H_{16}O_2$ / 156.22	31642-67-8 / 5	116-8^1	4-02-00-01597 / 0.9146^{16}	1.4492^{15}
8189	1-Nonen-3-ol / 1-Vinylheptanol	$C_9H_{18}O$ / 142.24	21964-44-3 /	193.5	3-01-00-01947 / 0.824^{21}	1.4382^{15}
8190	8-Nonen-1-ol	$C_9H_{18}O$ / 142.24	13038-21-6 /	135^{20}	3-01-00-01947 / 0.8394^{25}	1.4450^{23}
8191	5-Nonen-4-one, 6-methyl-	$C_{10}H_{18}O$ / 154.25	7036-98-8 /	90-2^{16}	3-01-00-03022 / 0.8608^{20}	1.4518^{20}
8192	6-Nonen-4-one, 6-methyl-	$C_{10}H_{18}O$ / 154.25	53252-20-3 /	94^{18}	3-01-00-03022 / 0.8413^{21}	1.4429^{21}
8193	1-Nonen-3-yne	C_9H_{14} / 122.21	57223-18-4 /	27-8^4	4-01-00-01103 / 0.7602^{25}	ace 4; eth 4 / 1.4487^{25}
8194	1-Nonen-4-yne	C_9H_{14} / 122.21	31508-12-0 /	58^{22}	4-01-00-01103 / 0.777^{25}	ace 4; eth 4 / 1.4418^{25}
8195	2-Nonen-4-yne	C_9H_{14} / 122.21	90644-60-3 /	70^{20}	3-01-00-01048 / 0.7832^{25}	ace 4; eth 4 / 1.4590^{25}
8196	1-Nonyne / Heptylacetylene	C_9H_{16} / 124.23	3452-09-3 / -50	150.8	4-01-00-01047 / 0.7658^{20}	H_2O 1; eth 3; bz 3; ctc 3 / 1.4217^{20}
8197	2-Nonyne	C_9H_{16} / 124.23	19447-29-1 /	161.9	3-01-00-01013 / 0.7690^{20}	H_2O 1; eth 3; lig 3 / 1.4337^{20}
8198	3-Nonyne	C_9H_{16} / 124.23	20184-89-8 /	157.1	3-01-00-01013 / 0.7616^{20}	H_2O 1; eth 3; lig 3 / 1.4299^{20}
8199	4-Nonyne	C_9H_{16} / 124.23	20184-91-2 /	152	4-01-00-01047 / 0.757^{25}	H_2O 1; eth 3; ace 3 / 1.4296^{25}
8200	1-Nonyne, 1-chloro-	$C_9H_{15}Cl$ / 158.67	90722-14-8 /	75-7^{15}	3-01-00-01012 / 0.9062^{20}	eth 4 / 1.450^{20}
8201	4-Nonyne, 3,3-dimethyl- / 3,3-Dimethyl-4-nonyne	$C_{11}H_{20}$ / 152.28	29022-31-9 /	82^{40}	3-01-00-01024 / 0.7667^{20}	bz 4; eth 4 / 1.4317^{20}

No.	Name Synonym	Mol. Form. Mol. Wt.	CAS RN mp/°C	Merck No. bp/°C	Beil. Ref. den/g cm^{-3}	Solubility n_D
8202	4-Nonyne, 8-methyl- 8-Methyl-4-nonyne	$C_{10}H_{18}$ 138.25	70732-43-3	104.5[97]	3-01-00-01018 0.7681[20]	bz 4; eth 4 1.4311[20]
8203	2-Nonynoic acid	$C_9H_{14}O_2$ 154.21	1846-70-4 -8	154-6[18]	4-02-00-01721 0.9525[12]	1.4605[25]
8204	3-Nonynoic acid	$C_9H_{14}O_2$ 154.21	56630-33-2 17	118[2]	4-02-00-01721	1.4603[25]
8205	5-Nonynoic acid	$C_9H_{14}O_2$ 154.21	56630-34-3 -3	115[1]	4-02-00-01723	1.4558[25]
8206	6-Nonynoic acid	$C_9H_{14}O_2$ 154.21	56630-31-0 14	130[3]	4-02-00-01723	1.4578[25]
8207	8-Nonynoic acid	$C_9H_{14}O_2$ 154.21	30964-01-3 19	100[6]	4-02-00-01721	1.4524[25]
8208	1-Nonyn-3-ol	$C_9H_{16}O$ 140.23	7383-20-2	85[10]; 63[2]	4-01-00-02265 0.8627[20]	1.4440[20]
8209	2-Nonyn-1-ol	$C_9H_{16}O$ 140.23	5921-73-3	109[11]	4-01-00-02265 0.878[16]	1.4576[16]
8210	24-Norcholan-23-oic acid, (5β)- Norcholanic acid	$C_{23}H_{38}O_2$ 346.55	511-18-2 177	6606	3-09-00-02652	
8211	29-Nordammara-1,17(20),24-trien-21-oic acid, 6,16-bis(acetyloxy)-3,7-dioxo-, Helvolic acid	$C_{33}H_{44}O_8$ 568.71	29400-42-8 212 dec	4552	4-10-00-04083	H_2O 2; EtOH 2; eth 3; ace 3
8212	Norea Urea, N,N-dimethyl-N'-(octahydro-4,7-methano-1H-indeN-5-yl)-, (3aα,4α,5α,7α,7aα)-	$C_{13}H_{22}N_2O$ 222.33	18530-56-8 177	6611		
8213	Norflurazon 3(2H)-Pyridazinone, 4-chloro-5-(methylamino)-2-[3-(trifluoromethyl)phenyl]-	$C_{12}H_9ClF_3N_3O$ 303.67	27314-13-2 184	6618		
8214	Norleucine 2-Aminohexanoic acid	$C_6H_{13}NO_2$ 131.17	327-57-1 301 dec	6624	4-04-00-02686	H_2O 4
8215	DL-Norleucine Hexanoic acid, 2-amino(DL)	$C_6H_{13}NO_2$ 131.17	616-06-8 327 dec	6624	3-04-00-01386 1.172[25]	H_2O 3; EtOH 2; eth 1
8216	19-Norpregna-1,3,5(10)-trien-20-yne-3,17-diol, (17α)- 17α-Ethinyl-17β-estradiol	$C_{20}H_{24}O_2$ 296.41	57-63-6	3689	4-06-00-06877	chl 2
8217	DL-Norvaline	$C_5H_{11}NO_2$ 117.15	760-78-1 303 sub	6636	4-04-00-02629	H_2O 3; EtOH 1; eth 1; chl 1
8218	L-Norvaline (S)-2-Aminopentanoic acid	$C_5H_{11}NO_2$ 117.15	6600-40-4 307	6636	4-04-00-02629	H_2O 4
8219	Nuarimol 5-Pyrimidinemethanol, α-(2-chlorophenyl)-α-(4-fluorophenyl)-	$C_{17}H_{12}ClFN_2O$ 314.75	63284-71-9 126			
8220	Octacosane	$C_{28}H_{58}$ 394.77	630-02-4 64.5	431.6	4-01-00-00588 0.8067[20]	H_2O 1; ace 5; bz 3; chl 3 1.4330[70]
8221	Octacosanoic acid Montanic acid	$C_{28}H_{56}O_2$ 424.75	506-48-9 90.9		4-02-00-01318 0.8191[100]	bz 4; chl 4 1.4313[100]
8222	1-Octacosanol Montanyl alcohol	$C_{28}H_{58}O$ 410.77	557-61-9 83.3	6664 200[1]	4-01-00-01915	
8223	9,12-Octadecadienoic acid (Z,Z)- Linoleic acid	$C_{18}H_{32}O_2$ 280.45	60-33-3 -12	5382 229-30[16]	4-02-00-01754 0.9022[20]	ace 4; bz 4; eth 4; EtOH 4 1.4699[20]
8224	10,12-Octadecadienoic acid, (E,E)-	$C_{18}H_{32}O_2$ 280.45	1072-36-2 56.5		4-02-00-01752 0.8686[70]	1.4689[40]
8225	9,12-Octadecadienoic acid	$C_{18}H_{32}I_2$ 502.26	2197-37-7 -5	229-30[16]	4-02-00-01754 0.9022[20]	ace 4; bz 4; eth 4; EtOH 4 1.4699[20]
8226	9,12-Octadecadienoic acid (Z,Z)-, ethyl ester Ethyl linoleate	$C_{20}H_{36}O_2$ 308.50	544-35-4	3776 272[180]; 212[12]	4-02-00-01757 0.8865[20]	eth 4; EtOH 4
8227	9,12-Octadecadienoic acid, ethyl ester	$C_{20}H_{36}O_2$ 308.50	7619-08-1	272[180] 212[12]	2-02-00-00461 0.8865[20]	H_2O 1; EtOH 3; eth 3
8228	9,12-Octadecadienoic acid (Z,Z)-, methyl ester Methyl linoleate	$C_{19}H_{34}O_2$ 294.48	112-63-0 -35	6008 215[20]	4-02-00-01756 0.8886[10]	eth 4; EtOH 4 1.4638[20]
8229	9,12-Octadecadienoic acid, methyl ester Methyl 9,12-octadecadienoate	$C_{19}H_{34}O_2$ 294.48	2462-85-3 -35	215[20] 169[1]	2-02-00-00461 0.8886[18]	H_2O 1; EtOH 3; eth 3; ctc 2 1.4638[20]
8230	7,11-Octadecadiyne	$C_{18}H_{30}$ 246.44	103697-14-9	167-8[7]	3-01-00-01068 0.841[19]	1.4698[19]
8231	Octadecanamide	$C_{18}H_{37}NO$ 283.50	124-26-5 109	250[12]	4-02-00-01240	eth 4; chl 4
8232	Octadecanamide, N-(4-hydroxyphenyl)-	$C_{24}H_{41}NO_2$ 375.60	103-99-1 133.8	239.5[10]	4-13-00-01113	H_2O 1; eth 2; ace 3; bz 2
8233	1-Octadecanamine	$C_{18}H_{39}N$ 269.51	124-30-1 52.9	346.8	4-04-00-00825 0.8618[20]	H_2O 1; EtOH 3; eth 3; ace 2 1.4522[20]
8234	Octadecane	$C_{18}H_{38}$ 254.50	593-45-3 28.2	316.3	4-01-00-00553 0.7768[28]	H_2O 1; EtOH 2; eth 3; ace 3 1.4390[20]
8235	Octadecane, 1-bromo-	$C_{18}H_{37}Br$ 333.40	112-89-0 28.2	210[10]	4-01-00-00555 0.9848[20]	H_2O 1; EtOH 3; eth 3; ctc 2 1.4631[20]
8236	Octadecane, 1-chloro-	$C_{18}H_{37}Cl$ 288.94	3386-33-2 28.6	348	4-01-00-00554 0.8641[20]	H_2O 1; ctc 2 1.4531[20]
8237	Octadecane, 1,1-dimethoxy- Octadecanal, dimethyl acetal	$C_{20}H_{42}O_2$ 314.55	14620-55-4	168-70[3]	4-01-00-03397	eth 4; EtOH 4 1.4410[25]

No.	Name / Synonym	Mol. Form. / Mol. Wt.	CAS RN / mp/°C	Merck No. / bp/°C	Beil. Ref. / den/g cm^{-3}	Solubility / n_D
8238	Octadecanedioic acid, diethyl ester	$C_{22}H_{42}O_4$ 370.57	1472-90-8 54.5	240[12]	4-02-00-02176	eth 4; EtOH 4
8239	Octadecanedioic acid, 9,10-dihydroxy-, (R*,R*)-(±)- Phloionic acid	$C_{18}H_{34}O_6$ 346.46	23843-52-9 126	7297	4-03-00-01250	
8240	Octadecane, 1-(ethenyloxy)-	$C_{20}H_{40}O$ 296.54	930-02-9 30	182[3]	4-01-00-02057 0.8138[40]	chl 2
8241	Octadecane, 3-ethyl-5-(2-ethylbutyl)-	$C_{26}H_{54}$ 366.71	55282-12-7	229.5[10]	4-01-00-00586 0.8115[20]	1.45239[20]
8242	Octadecane, 1-iodo-	$C_{18}H_{37}I$ 380.40	629-93-6 34	383	4-01-00-00556 1.0994[20]	H_2O 1; EtOH 2; eth 2 1.4810[20]
8243	Octadecanenitrile	$C_{18}H_{35}N$ 265.48	638-65-3 41	362	4-02-00-01242 0.8325[20]	H_2O 1; EtOH 3; eth 4; ace 4 1.4389[45]
8244	1-Octadecanethiol Stearyl mercaptan	$C_{18}H_{38}S$ 286.57	2885-00-9 30	204-10[11]	4-01-00-01894 0.8475[20]	eth 4 1.4645[20]
8245	Octadecane, 9-p-tolyl- Toluene, p-(1-octyldecyl)-	$C_{25}H_{44}$ 344.62	4445-08-3	185[10]	4-05-00-01221 0.8549[20]	1.4811[20]
8246	9,11,13-Octadecanetrienoic acid (Z,Z,Z) Eleostearic acid	$C_{18}H_{30}O_2$ 278.44	3884-88-6 48.5			
8247	Octadecanoic acid Stearic acid	$C_{18}H_{36}O_2$ 284.48	57-11-4 68.8	8761 350 dec; 232[15]	4-02-00-01206 0.9408[20]	H_2O 1; EtOH 2; eth 4; ace 3 1.4299[80]
8248	Octadecanoic acid, aluminum salt	$C_{54}H_{105}AlO_6$ 877.41	637-12-7 118	370	4-02-00-01206	H_2O 1; EtOH 3; peth 3
8249	Octadecanoic acid, anhydride	$C_{36}H_{70}O_3$ 550.95	638-08-4 72		4-02-00-01239 0.8365[82]	H_2O 1; EtOH 1; eth 2; bz 2 1.4362[80]
8250	Octadecanoic acid, 18-bromo- Stearic acid, 18-bromo	$C_{18}H_{35}BrO_2$ 363.38	2536-38-1 75.5	240[4]	2-02-00-00361	bz 4; eth 4; EtOH 4
8251	Octadecanoic acid, butyl ester Butyl stearate	$C_{22}H_{44}O_2$ 340.59	123-95-5 27	1589 343	4-02-00-01219 0.854[25]	H_2O 1; EtOH 3; ace 4 1.4328[50]
8252	Octadecanoic acid, calcium salt	$C_{36}H_{70}CaO_4$ 607.03	1592-23-0 179.5	1710	4-02-00-01206	H_2O 1; EtOH 1; eth 1
8253	Octadecanoic acid, cyclohexyl ester Stearic acid, cyclohexyl ester	$C_{24}H_{46}O_2$ 366.63	104-07-4 44		4-06-00-00038 0.889[15]	eth 4
8254	Octadecanoic acid, 9,10-dihydroxy- 9,10-Dihydroxystearic acid	$C_{18}H_{36}O_4$ 316.48	120-87-6 90	3171	4-03-00-01092	H_2O 1; EtOH 2; eth 2
8255	Octadecanoic acid, 2,3-dihydroxypropyl ester, (±)-	$C_{21}H_{42}O_4$ 358.56	22610-63-5 74		4-02-00-01225 0.9841[20]	H_2O 1; EtOH 2; eth 2; lig 3 1.4400[86]
8256	Octadecanoic acid, 1,2-ethanediyl ester	$C_{38}H_{74}O_4$ 595.00	627-83-8 79	241[20]	4-02-00-01223 0.8581[78]	H_2O 1; EtOH 1; eth 4; ace 4
8257	Octadecanoic acid, ethyl ester	$C_{20}H_{40}O_2$ 312.54	111-61-5 33	199[10]	4-02-00-01218 1.057[20]	H_2O 1; EtOH 3; eth 3; ace 4 1.4349[40]
8258	Octadecanoic acid, hexadecyl ester	$C_{34}H_{68}O_2$ 508.91	1190-63-2 57		4-02-00-01220	ace 4; eth 4; chl 4 1.4410[70]
8259	Octadecanoic acid, 2-[2-[2-(2-hydroxyethoxy)ethoxy]ethoxy]ethyl ester	$C_{26}H_{52}O_6$ 460.70	106-07-0 40	328	1.1285[15]	1.4593[20]
8260	Octadecanoic acid, 2-hydroxyethyl ester	$C_{20}H_{40}O_3$ 328.54	111-60-4 60.5	189-91[3]	4-02-00-01222 0.8780[60]	EtOH 2; eth 3 1.4310[60]
8261	Octadecanoic acid, lead (II) salt Lead stearate	$C_{36}H_{70}O_4Pb$ 774.15	7428-48-0 125		4-02-00-01206 1.4	H_2O 1; Hot EtOH 3; eth 1
8262	Octadecanoic acid, 14-methyl- Stearic acid, 14-methyl	$C_{19}H_{38}O_2$ 298.51	94434-64-7 37.5	182[0.4]	4-02-00-01265 0.9400[20]	
8263	Octadecanoic acid, 17-methyl-	$C_{19}H_{38}O_2$ 298.51	2724-59-6 67.5	180[0.3]	4-02-00-01260 0.8420[70]	1.4336[70]
8264	Octadecanoic acid, 9-methyl- Stearic acid, 9-methyl	$C_{19}H_{38}O_2$ 298.51	86073-38-3 40	171[0.1]	4-02-00-01271 0.9980[20]	
8265	Octadecanoic acid, 3-methylbutyl ester Stearic acid, isopentyl ester	$C_{23}H_{46}O_2$ 354.62	627-88-3 25.5	192[2]	2-02-00-00353 0.855[20]	H_2O 1; EtOH 2; eth 3; ace 3 1.433[50]
8266	Octadecanoic acid, methyl ester	$C_{19}H_{38}O_2$ 298.51	112-61-8 39.1	443; 215[15]	4-02-00-01216 0.8498[40]	eth 4; chl 4 1.4367[40]
8267	Octadecanoic acid, 1-methylethyl ester	$C_{21}H_{42}O_2$ 326.56	112-10-7 28	207[6]	4-02-00-01219 0.8403[38]	ace 4; eth 4; EtOH 4; chl 4
8268	Octadecanoic acid, 2-methylpropyl ester Isobutyl stearate	$C_{22}H_{44}O_2$ 340.59	646-13-9 28.9	5034 223[15]	3-02-00-01017 0.8498[20]	eth 4
8269	Octadecanoic acid, 12-oxo-, ethyl ester Stearic acid, 12-oxo, ethyl ester	$C_{20}H_{38}O_3$ 326.52	88472-61-1 38	199[3]	3-03-00-01294	EtOH 4
8270	Octadecanoic acid, pentyl ester Stearic acid, pentyl ester	$C_{23}H_{46}O_2$ 354.62	6382-13-4 30		4-02-00-01220	eth 4; EtOH 4 1.4342[50]
8271	Octadecanoic acid, phenyl ester Stearic acid, phenyl ester	$C_{24}H_{40}O_2$ 360.58	637-55-8 52	267[15]	4-06-00-00618	H_2O 1; EtOH 3; eth 3
8272	Octadecanoic acid, 1,2,3-propanetriyl ester Tristearin	$C_{57}H_{110}O_6$ 891.50	555-43-1	9669	4-02-00-01233 0.8559[90]	H_2O 1; EtOH 1; ace 3; bz 2 1.4395[80]
8273	Octadecanoic acid, propyl ester	$C_{21}H_{42}O_2$ 326.56	3634-92-2 28.9	186.8[2]	4-02-00-01219 0.8452[38]	ace 4; eth 4; EtOH 4 1.4400[30]
8274	Octadecanoic acid, 9,10,12,13-tetrabromo-, methyl ester Stearic acid, 9,10,12,13-tetrabromo, methyl ester	$C_{19}H_{34}Br_4O_2$ 614.09	62080-86-8 63	215[15]	3-02-00-01049	eth 4; EtOH 4; chl 4 1.4346[45]

No.	Name Synonym	Mol. Form. Mol. Wt.	CAS RN mp/°C	Merck No. bp/°C	Beil. Ref. den/g cm^{-3}	Solubility n_D
8275	Octadecanoic acid, (tetrahydro-2-furanyl)methyl ester Stearic acid, tetrahydrofurfuryl ester	$C_{23}H_{44}O_3$ 368.60	6940-09-6 22		4-02-00-01248 0.914[25]	eth 4; EtOH 4
8276	Octadecanoic acid, 9,10,18-trihydroxy-, (R*,R*)- Phloionolic acid	$C_{18}H_{36}O_5$ 332.48	583-86-8 101.5	7298	4-03-00-01119	
8277	Octadecanoic acid, zinc salt	$C_{36}H_{70}O_4Zn$ 632.34	557-05-1 130	10063	4-02-00-01206	H_2O 1; EtOH 1; eth 1
8278	1-Octadecanol Stearyl alcohol	$C_{18}H_{38}O$ 270.50	112-92-5 59.5	210.5[15]	4-01-00-01888 0.8124[59]	H_2O 1; EtOH 3; eth 3; ace 2
8279	3-Octadecanol, (±)- 3-Octadecanol (DL)	$C_{18}H_{38}O$ 270.50	111897-18-8 51	202[13]	4-01-00-01896 0.7858[80]	1.4290[80]
8280	3-Octadecanone	$C_{18}H_{36}O$ 268.48	18261-92-2 51	197[11]	4-01-00-03398	ctc 3
8281	Octadecanoyl chloride	$C_{18}H_{35}ClO$ 302.93	112-76-5 23	215[15]	4-02-00-01240 0.8969[0]	EtOH 2 1.4523[24]
8282	9,11,13-Octadecatrienoic acid, (E,E,E)-	$C_{18}H_{30}O_2$ 278.44	544-73-0 71.5	188[1]	4-02-00-01787 0.8839[80]	EtOH 4 1.5000[80]
8283	9,12,15-Octadecatrienoic acid, ethyl ester, (Z,Z,Z)-	$C_{20}H_{34}O_2$ 306.49	1191-41-9 218[15]		4-02-00-01782 0.8919[20]	eth 4; EtOH 4 1.4694[20]
8284	9,11,13-Octadecatrienoic acid, (E,Z,E)-	$C_{18}H_{30}O_2$ 278.44	506-23-0 49	235[12] dec; 170[1]	4-02-00-01787 0.9028[50]	eth 4; EtOH 4 1.5112[50]
8285	9,12,15-Octadecatrienoic acid, methyl ester, (Z,Z,Z)-	$C_{19}H_{32}O_2$ 292.46	301-00-8 182[3]		4-02-00-01782 0.896[25]	 1.4709[20]
8286	9,11,13-Octadecatrienoic acid, (Z,Z,E)-	$C_{18}H_{30}O_2$ 278.44	544-72-9 62		4-02-00-01787 0.9027[20]	EtOH 4; peth 4 1.5114[50]
8287	9,12,15-Octadecatrienoic acid, (Z,Z,Z)- Linolenic acid	$C_{18}H_{30}O_2$ 278.44	463-40-1 -11	5383 231[17]; 129[0.05]	4-02-00-01781 0.9164[20]	H_2O 1; EtOH 3; eth 3; bz 2 1.4800[20]
8288	9-Octadecenal Octadecenyl aldehyde	$C_{18}H_{34}O$ 266.47	5090-41-5 168-9[3]		4-01-00-03533 0.8509[20]	 1.4558[20]
8289	9-Octadecenal, (Z)-	$C_{18}H_{34}O$ 266.47	2423-10-1 168-9[3]		4-01-00-03533 0.8509[20]	 1.4558[20]
8290	9-Octadecenamide, N-phenyl-, (Z)-	$C_{24}H_{39}NO$ 357.58	5429-85-6 41	143.5[10]	4-12-00-00411	bz 4; eth 4; HOAc 4
8291	1-Octadecene	$C_{18}H_{36}$ 252.48	112-88-9 17.5	179[15]; 145[8]	4-01-00-00930 0.7891[20]	H_2O 1; ace 3; ctc 3 1.4448[20]
8292	9-Octadecene	$C_{18}H_{36}$ 252.48	5557-31-3 -30.5	162[9]	4-01-00-00932 0.7916[20]	1.4470[20]
8293	9-Octadecenenitrile, (Z)-	$C_{18}H_{33}N$ 263.47	112-91-4 -1	332 dec	4-02-00-01668 0.847[17]	EtOH 4 1.4566[20]
8294	9-Octadecenoic acid (Z)- Oleic acid	$C_{18}H_{34}O_2$ 282.47	112-80-1 13.4	6788 286[100]	4-02-00-01641 0.8935[20]	H_2O 1; EtOH 5; eth 5; ace 5 1.4582[20]
8295	11-Octadecenoic acid, (E)- Vaccenic acid	$C_{18}H_{34}O_2$ 282.47	693-72-1 44	9810	4-02-00-01640	ace 3 1.4499[60]
8296	2-Octadecenoic acid, (E)-	$C_{18}H_{34}O_2$ 282.47	2825-79-8 58.5		4-02-00-01632 0.8484[90]	H_2O 1; EtOH 2; eth 3; bz 3
8297	9-Octadecenoic acid, (E)- Elaidic acid	$C_{18}H_{34}O_2$ 282.47	112-79-8 45	3497 288[100]; 234[15]	4-02-00-01647 0.8734[45]	H_2O 1; EtOH 3; eth 3; bz 3 1.4499[45]
8298	9-Octadecenoic acid (Z)-, butyl ester	$C_{22}H_{42}O_2$ 338.57	142-77-8 -26.4	227-8[15]	4-02-00-01653 0.8704[15]	EtOH 4 1.4480[25]
8299	9-Octadecenoic acid (Z)-, 2,3-dihydroxypropyl ester	$C_{21}H_{40}O_4$ 356.55	111-03-5 35	238-40[3]	2-02-00-00439 0.9420[20]	H_2O 1; EtOH 3; eth 3; chl 3 1.4626[20]
8300	6-Octadecenoic acid, 6,7-diiodo- Diiodotariric acid	$C_{18}H_{32}I_2O_2$ 534.26	533-86-8 50.3	dec	4-02-00-01638	bz 4; eth 4; chl 4
8301	9-Octadecenoic acid (Z)-, ethyl ester	$C_{20}H_{38}O_2$ 310.52	111-62-6 207[13]		4-02-00-01651 0.8720[20]	eth 4; EtOH 4 1.4515[20]
8302	9-Octadecenoic acid, ethyl ester, (E)-	$C_{20}H_{38}O_2$ 310.52	6114-18-7 5.8	217-9[15]	4-02-00-01652 0.8664[25]	eth 4; EtOH 4 1.4480[25]
8303	9-Octadecenoic acid, 12-hydroxy-, [R-(Z)]- Ricinoleic acid	$C_{18}H_{34}O_3$ 298.47	141-22-0 227[10]	8213	4-03-00-01026 0.9450[21]	eth 4; EtOH 4 1.4716[21]
8304	9-Octadecenoic acid, 12-hydroxy-, butyl ester, [R-(Z)]-	$C_{22}H_{42}O_3$ 354.57	151-13-3 278[12]		0-03-00-00388 0.9058[22]	eth 4 1.4566[22]
8305	9-Octadecenoic acid, 12-hydroxy-, ethyl ester, [R-(Z)]-	$C_{20}H_{38}O_3$ 326.52	55066-53-0 258[13]		4-03-00-01029 0.9045[22]	 1.4618[22]
8306	9-Octadecenoic acid (Z)-, 3-methylbutyl ester Oleic acid, isopentyl ester	$C_{23}H_{44}O_2$ 352.60	627-89-4 223-4[10]		3-02-00-01414 0.897[15]	eth 4; EtOH 4
8307	9-Octadecenoic acid (Z)-, methyl ester Methyl oleate	$C_{19}H_{36}O_2$ 296.49	112-62-9 -19.9	218.5[20]	4-02-00-01649 0.8739[20]	H_2O 1; EtOH 5; eth 5; chl 3 1.4522[20]
8308	9-Octadecenoic acid, methyl ester, (E)-	$C_{19}H_{36}O_2$ 296.49	1937-62-8 213-5[15]		4-02-00-01651 0.8730[20]	eth 4; EtOH 4 1.4513[20]
8309	9-Octadecenoic acid (Z)-, phenylmethyl ester Oleic acid, benzyl ester	$C_{25}H_{40}O_2$ 372.59	55130-16-0 237[1]		4-06-00-02269 0.9302[25]	eth 4; EtOH 4 1.4875[25]

No.	Name / Synonym	Mol. Form. / Mol. Wt.	CAS RN / mp/°C	Merck No. / bp/°C	Beil. Ref. / den/g cm^{-3}	Solubility / n_D
8310	9-Octadecenoic acid (Z)-, 1,2,3-propanetriyl ester Triolein	$C_{57}H_{104}O_6$ 885.45	122-32-7 -32	9644 235-40[18]	4-02-00-01664 0.8988[40]	H_2O 1; EtOH 2; eth 4; chl 3 1.4621[40]
8311	9-Octadecen-1-ol, (E)-	$C_{18}H_{36}O$ 268.48	506-42-3 36.5	 333	4-01-00-02204 0.8338[40]	ace 4; eth 4; EtOH 4 1.4552[40]
8312	9-Octadecen-1-ol, (Z)- Oleyl alcohol	$C_{18}H_{36}O$ 268.48	143-28-2 6.5	6791 205-10[15]	4-01-00-02204 0.8489[20]	H_2O 1; EtOH 3; eth 3; ctc 2 1.4606[20]
8313	1-Octadecyne	$C_{18}H_{34}$ 250.47	629-89-0 22.5	 313	4-01-00-01075 0.8025[20]	1.4774[20]
8314	2-Octadecyne	$C_{18}H_{34}$ 250.47	61847-97-0 30	 184[15]	4-01-00-01075 0.8016[30]	
8315	9-Octadecyne	$C_{18}H_{34}$ 250.47	35365-59-4 3	 163-4[7]	4-01-00-01076 0.8012[20]	1.4488[25]
8316	9-Octadecynoic acid 9-Stearolic acid	$C_{18}H_{32}O_2$ 280.45	506-24-1 48	 189[1.8]	4-02-00-01751	H_2O 1; EtOH 3; eth 4 1.4510[54]
8317	2,6-Octadienal, 3,7-dimethyl-, (E)-	$C_{10}H_{16}O$ 152.24	141-27-5	 229	4-01-00-03569 0.8888[20]	H_2O 1; EtOH 5; eth 5 1.4898[20]
8318	2,6-Octadienal, 3,7-dimethyl-, (Z)-	$C_{10}H_{16}O$ 152.24	106-26-3	 120[20]	4-01-00-03569 0.8869[20]	H_2O 1; EtOH 5; eth 5 1.4869[20]
8319	1,7-Octadiene α,ω-Octadiene	C_8H_{14} 110.20	3710-30-3	 115.5	4-01-00-01038 0.734[20]	1.4245[20]
8320	2,6-Octadiene 1,6-Dimethyldiallyl	C_8H_{14} 110.20	4974-27-0 -76	 124.5	4-01-00-01039 0.7406[25]	1.4292[17]
8321	2,6-Octadiene, 1-bromo-3,7-dimethyl-, (E)- trans-Geranyl bromide	$C_{10}H_{17}Br$ 217.15	6138-90-5	 101[12]; 47[0.005]	4-01-00-01059 1.0940[22]	1.5027[20]
8322	2,4-Octadiene, 3,7-dimethyl- 3,7-Dimethylocta-2,4-diene	$C_{10}H_{18}$ 138.25	56523-26-3	 165.5	3-01-00-01020 0.7933[20]	1.456[20]
8323	2,5-Octadiene, 2,6-dimethyl-	$C_{10}H_{18}$ 138.25	116668-48-5	 59.5[12]	3-01-00-01018 0.773[20]	1.4500[20]
8324	2,6-Octadiene, 2,6-dimethyl-, (Z)- cis-2,6-Dimethyl-2,6-octadiene	$C_{10}H_{18}$ 138.25	2492-22-0	 168	4-01-00-01058 0.775[21]	ace 4; eth 4; EtOH 4 1.4498[20]
8325	2,6-Octadiene, 2,7-dimethyl- 2,7-Dimethyl-2,6-octadiene	$C_{10}H_{18}$ 138.25	16736-42-8 -74.4	 166	4-01-00-01061 0.7849[20]	1.4481[20]
8326	2,6-Octadiene, 3,6-dimethyl-	$C_{10}H_{18}$ 138.25	116668-49-6	 154	4-01-00-01061 0.7767[20]	1.4445[20]
8327	2,6-Octadiene, 4,5-dimethyl- 4,5-Dimethyl-2,6-octadiene	$C_{10}H_{18}$ 138.25	18476-57-8 -64.8	 153.5	3-01-00-01021 0.7611[25]	1.4375[25]
8328	2,4-Octadiene, 7-methyl-	C_9H_{16} 124.23	2216-70-8	 149	2-01-00-00239 0.7521[18]	1.4543[18]
8329	3,5-Octadiene, 4-methyl- 4-Methyl-3,5-octadiene	C_9H_{16} 124.23	36903-95-4	 149.5	2-01-00-00239 0.7640[25]	1.4628[25]
8330	1,6-Octadiene, 7-methyl-3-methylene- β-Myrcene	$C_{10}H_{16}$ 136.24	123-35-3	6243 167	4-01-00-01108 0.8013[15]	H_2O 1; EtOH 3; eth 3; bz 3 1.4722[20]
8331	2,4-Octadienoic acid, 3,7-dimethyl-	$C_{10}H_{16}O_2$ 168.24	101715-60-0	 151-4[14]	4-02-00-01736 0.959[17]	1.4919
8332	1,6-Octadien-3-ol, 3,7-dimethyl-, (±)- Linalol	$C_{10}H_{18}O$ 154.25	22564-99-4	 198; 86[13]	4-01-00-02278 0.870[15]	1.4627
8333	2,6-Octadien-1-ol, 3,7-dimethyl-, (E)- trans-Geraniol	$C_{10}H_{18}O$ 154.25	106-24-1 <-15	4298 230	4-01-00-02277 0.8894[20]	H_2O 1; EtOH 3; eth 3; ace 3 1.4766[20]
8334	2,6-Octadien-1-ol, 3,7-dimethyl-, (Z)- Nerol	$C_{10}H_{18}O$ 154.25	106-25-2 <-15	6387 225; 125[25]	4-01-00-02276 0.8756[20]	EtOH 4 1.4746[20]
8335	1,6-Octadien-3-ol, 3,7-dimethyl-, acetate, (R)-	$C_{12}H_{20}O_2$ 196.29	16509-46-9	 220; 115[25]	4-02-00-00204 0.8951[20]	eth 4; EtOH 4 1.4544[21]
8336	2,6-Octadien-1-ol, 3,7-dimethyl-, acetate, (Z)-	$C_{12}H_{20}O_2$ 196.29	141-12-8	 134[25]; 93[3]	2-02-00-00153 0.905[15]	1.452[20]
8337	1,6-Octadien-3-ol, 3,7-dimethyl-, formate Linalyl formate	$C_{11}H_{18}O_2$ 182.26	115-99-1	 100-3[10]	4-02-00-00035 0.915[25]	EtOH 4 1.456[20]
8338	2,6-Octadien-1-ol, 3,7-dimethyl-, formate, (E)-	$C_{11}H_{18}O_2$ 182.26	105-86-2	 229 dec; 113[25]	4-02-00-00035 0.9086[25]	H_2O 1; EtOH 4; eth 3; ace 3 1.4659[20]
8339	2,6-Octadien-4-yne, 3,6-diethyl-	$C_{12}H_{18}$ 162.27	100319-48-0	 170	3-01-00-01066 0.8196[20]	ace 4; eth 4 1.4965[20]
8340	2,6-Octadien-4-yne, 3,6-dimethyl- 3,6-Dimethyl-2,6-octadiene-4-yne	$C_{10}H_{14}$ 134.22	3725-07-3 -45	 170	4-01-00-01129 0.8071[22]	ace 4; eth 4 1.4998[20]
8341	1,5-Octadien-3-yne, 5-propyl- 5-Propyl-1,5-octadien-3-yne	$C_{11}H_{16}$ 148.25	70732-44-4	 57-8[6]	4-01-00-01129 0.8047[20]	ace 4; eth 4 1.4949[20]
8342	1,7-Octadiyne	C_8H_{10} 106.17	871-84-1	 135.5; 59[35]	4-01-00-01122 0.8169[21]	eth 3 1.4521[18]
8343	2,6-Octadiyne	C_8H_{10} 106.17	764-73-8 27	 62[19]	4-01-00-01123 0.828[80]	eth 4 1.4658[30]
8344	3,5-Octadiyne	C_8H_{10} 106.17	16387-70-5	 163.5	4-01-00-01123 0.826[0]	eth 3 1.4968[0]
8345	3,5-Octadiyne, 2,7-dimethyl- Diisopropylbutadiyne	$C_{10}H_{14}$ 134.22	14813-68-4	 74[12]	1-01-00-00128 0.8090[20]	eth 4
8346	Octanal Caprylic aldehyde	$C_8H_{16}O$ 128.21	124-13-0	1766 171	4-01-00-03337 0.8211[20]	ace 4; bz 4; eth 4; EtOH 4 1.4217[20]
8347	Octanal, 7-hydroxy-3,7-dimethyl-	$C_{10}H_{20}O_2$ 172.27	107-75-5	 103[3]	4-01-00-04058 0.9220[20]	H_2O 2; EtOH 3; ace 3 1.4494[20]
8348	Octanal, oxime	$C_8H_{17}NO$ 143.23	929-55-5 60	 112[9]	2-01-00-00758	ace 4; EtOH 4

No.	Name Synonym	Mol. Form. Mol. Wt.	CAS RN mp/°C	Merck No. bp/°C	Beil. Ref. den/g cm^{-3}	Solubility n_D
8349	Octanamide	$C_8H_{17}NO$ 143.23	629-01-6 108	239	4-02-00-00992 0.8450^{110}	H_2O 2; EtOH 4; eth 3; ace 3
8350	1-Octanamine	$C_8H_{19}N$ 129.25	111-86-4 0	179.6	4-04-00-00751 0.7826^{20}	H_2O 4; EtOH 4; eth 4; ctc 3 1.4292^{20}
8351	2-Octanamine, (±)-	$C_8H_{19}N$ 129.25	44855-57-4 97	164	4-04-00-00762 0.7744^{20}	eth 4; EtOH 4 1.4232^{25}
8352	2-Octanamine, N-(1-methylheptyl)-	$C_{16}H_{35}N$ 241.46	5412-92-0	283	4-04-00-00764 0.7948^{20}	H_2O 1; EtOH 4; eth 4
8353	1-Octanamine, N-methyl-N-octyl-	$C_{17}H_{37}N$ 255.49	4455-26-9 -30.1	158^{10}	4-04-00-00753	1.4424^{20}
8354	1-Octanamine, N-octyl-	$C_{16}H_{35}N$ 241.46	1120-48-5 35.5	297.5	4-04-00-00753 0.7963^{26}	eth 4; EtOH 4 1.4415^{26}
8355	Octane	C_8H_{18} 114.23	111-65-9 -56.8	6672 125.6	4-01-00-00412 0.6986^{25}	H_2O 1; EtOH 5; eth 3; ace 5 1.3974^{20}
8356	Octane, 1-bromo- Octyl bromide	$C_8H_{17}Br$ 193.13	111-83-1 -55	6684 200	4-01-00-00422 1.108^{25}	H_2O 1; EtOH 5; eth 5; ctc 2 1.4524^{20}
8357	Octane, 2-bromo-, (±)- Octane, 2-bromo (DL)	$C_8H_{17}Br$ 193.13	60251-57-2	188.5	4-01-00-00423 1.0878^{25}	H_2O 1; EtOH 5; eth 5 1.4442^{25}
8358	Octane, 1-bromo-8-fluoro-	$C_8H_{16}BrF$ 211.12	593-12-4	118-20$^{22.5}$	4-01-00-00424	eth 4; EtOH 4 1.4500^{20}
8359	Octane, 1-chloro- Octyl chloride	$C_8H_{17}Cl$ 148.68	111-85-3 -57.8	181.5	4-01-00-00419 0.8738^{20}	H_2O 1; EtOH 4; eth 4; ctc 2 1.4305^{20}
8360	Octane, 2-chloro-	$C_8H_{17}Cl$ 148.68	628-61-5	172; 75^{28}	1-01-00-00060 0.8658^{17}	H_2O 1; EtOH 4; eth 3 1.4273^{21}
8361	Octane, 1-chloro-8-fluoro-	$C_8H_{16}ClF$ 166.67	593-14-6	87^{10}	4-01-00-00420 0.978^{20}	ace 4; eth 4; EtOH 4 1.4266^{25}
8362	Octane, 3-chloro-3-methyl-	$C_9H_{19}Cl$ 162.70	28320-88-9	73-4^{15}	3-01-00-00508 0.8680^{25}	ace 4; bz 4; eth 4; chl 4 1.4351^{20}
8363	Octane, 4-chloro-4-methyl- 4-Chloro-4-methyloctane	$C_9H_{19}Cl$ 162.70	36903-89-6	71$^{14.5}$	3-01-00-00509 0.8723^{20}	eth 4; chl 4 1.4360^{20}
8364	Octanedial	$C_8H_{14}O_2$ 142.20	638-54-0	235 dec; 97^3	4-01-00-03706	H_2O 4; EtOH 4 1.4439^{20}
8365	1,8-Octanediamine	$C_8H_{20}N_2$ 144.26	373-44-4 52	225	3-04-00-00612	H_2O 4; eth 4; EtOH 4
8366	Octane, 1,2-dibromo- 1,2-Dibromooctane	$C_8H_{16}Br_2$ 272.02	6269-92-7	241	3-01-00-00468 1.4580^{20}	1.4970^{20}
8367	Octane, 1,4-dibromo-	$C_8H_{16}Br_2$ 272.02	70690-24-3	125-6^{11}	4-01-00-00424 1.466^{11}	1.5003^{11}
8368	Octane, 1,5-dibromo- 1,5-Dibromooctane	$C_8H_{16}Br_2$ 272.02	17912-17-3	127-8^{11}	4-01-00-00424 1.447^{16}	1.4968^{16}
8369	Octane, 1,8-dibromo- Octamethylene dibromide	$C_8H_{16}Br_2$ 272.02	4549-32-0 15.5	271	4-01-00-00424 1.4594^{25}	H_2O 1; eth 3; ctc 3; chl 3 1.4971^{25}
8370	Octane, 4,5-dibromo-	$C_8H_{16}Br_2$ 272.02	61539-75-1	84$^{4.3}$	3-01-00-00469 1.4569^{20}	1.4981^{20}
8371	Octane, 1,8-dichloro- 1,8-Dichlorooctane	$C_8H_{16}Cl_2$ 183.12	2162-99-4	241	4-01-00-00420 1.0248^{25}	1.4572^{25}
8372	Octane, 2,3-dichloro-	$C_8H_{16}Cl_2$ 183.12	21948-47-0 -55.1	206	4-01-00-00425 0.9712^{20}	1.4523^{20}
8373	Octane, 2,2-dimethyl- 2,2-Dimethyloctane	$C_{10}H_{22}$ 142.28	15869-87-1	155	4-01-00-00476 0.7208^{25}	1.4082^{20}
8374	Octane, 2,3-dimethyl- 2,3-Dimethyloctane	$C_{10}H_{22}$ 142.28	7146-60-3	164.3	4-01-00-00476 0.7377^{20}	1.4146^{20}
8375	Octane, 2,4-dimethyl- 2,4-Dimethyloctane	$C_{10}H_{22}$ 142.28	4032-94-4	156	4-01-00-00477 0.7226^{25}	1.4091^{20}
8376	Octane, 2,5-dimethyl- 2,5-Dimethyloctane	$C_{10}H_{22}$ 142.28	15869-89-3	158.5	4-01-00-00477 0.7264^{25}	1.4112^{20}
8377	Octane, 2,6-dimethyl- 2,6-Dimethyloctane	$C_{10}H_{22}$ 142.28	2051-30-1	160.4	2-01-00-00131 0.7313^{20}	1.4097^{20}
8378	Octane, 2,7-dimethyl-	$C_{10}H_{22}$ 142.28	1072-16-8 -54.9	159.9	4-01-00-00478 0.7202^{25}	eth 3; HOAc 3 1.4086^{20}
8379	Octane, 3,3-dimethyl- 3,3-Dimethyloctane	$C_{10}H_{22}$ 142.28	4110-44-5	161.2	4-01-00-00479 0.7351^{25}	1.4165^{20}
8380	Octane, 3,4-dimethyl- 3,4-Dimethyloctane	$C_{10}H_{22}$ 142.28	15869-92-8	163.4	4-01-00-00479 0.7410^{25}	1.4182^{20}
8381	Octane, 3,5-dimethyl- 3,5-Dimethyloctane	$C_{10}H_{22}$ 142.28	15869-93-9	159.4	4-01-00-00479 0.7329^{25}	1.4139^{20}
8382	Octane, 3,6-dimethyl- 3,6-Dimethyloctane	$C_{10}H_{22}$ 142.28	15869-94-0	160.8	4-01-00-00479 0.7324^{25}	1.4139^{20}
8383	Octane, 4,4-dimethyl- 4,4-Dimethyloctane	$C_{10}H_{22}$ 142.28	15869-95-1	157.5	4-01-00-00480 0.7312^{25}	1.4144^{20}
8384	Octane, 4,5-dimethyl- 4,5-Dimethyloctane	$C_{10}H_{22}$ 142.28	15869-96-2	162.15	4-01-00-00480 0.7432^{25}	1.4190^{20}
8385	Octanedinitrile Suberonitrile	$C_8H_{12}N_2$ 136.20	629-40-3 -3.5	185^{15}	3-02-00-01768 0.954^{25}	1.4436^{20}
8386	Octanedioic acid Suberic acid	$C_8H_{14}O_4$ 174.20	505-48-6 144	8833 219^{20}	4-02-00-02028	H_2O 1; eth 5; bz 5
8387	Octanedioic acid, diethyl ester	$C_{12}H_{22}O_4$ 230.30	2050-23-9 5.9	282.6	4-02-00-02029 0.9811^{20}	H_2O 1; EtOH 3; eth 3; ctc 2 1.4328^{20}
8388	Octanedioic acid, dimethyl ester	$C_{10}H_{18}O_4$ 202.25	1732-09-8 -3.1	268	4-02-00-02029 1.0217^{20}	H_2O 1; EtOH 3; eth 3; ace 3 1.4341^{20}

No.	Name Synonym	Mol. Form. Mol. Wt.	CAS RN mp/°C	Merck No. bp/°C	Bell. Ref. den/g cm⁻³	Solubility n_D
8389	1,8-Octanediol	$C_8H_{18}O_2$ 146.23	629-41-4 63	172[20]	4-01-00-02592	H_2O 2; EtOH 4; eth 2; bz 3
8390	2,4-Octanediol	$C_8H_{18}O_2$ 146.23	90162-24-6	117-8[8]	3-01-00-02219 0.918[25]	1.4422[25]
8391	4,5-Octanediol, (±)-	$C_8H_{18}O_2$ 146.23	22520-40-7 30	110[8]	4-01-00-02593	ctc 2 1.4419[25]
8392	1,7-Octanediol, 3,7-dimethyl-	$C_{10}H_{22}O_2$ 174.28	107-74-4	265	3-01-00-02233 0.937[20]	bz 2; tol 2 1.4599[20]
8393	2,4-Octanedione Valerylacetone	$C_8H_{14}O_2$ 142.20	14090-87-0	79-83[20]	4-01-00-03706 0.9233[20]	1.4559[20]
8394	2,7-Octanedione	$C_8H_{14}O_2$ 142.20	1626-09-1 44	110[13]	4-01-00-03707	EtOH 4
8395	3,6-Octanedione	$C_8H_{14}O_2$ 142.20	2955-65-9 35.5	98[14]	4-01-00-03708	EtOH 4
8396	4,5-Octanedione	$C_8H_{14}O_2$ 142.20	5455-24-3	168	4-01-00-03708 0.934[0]	EtOH 4; eth 4; ace 4
8397	4,5-Octanedione, dioxime Dipropylglyoxime	$C_8H_{16}N_2O_2$ 172.23	61050-68-8 186.5	sub	2-01-00-00846	eth 4; EtOH 4
8398	3,5-Octanedione, 6,6,7,7,8,8,8-heptafluoro-2,2-dimethyl-	$C_{10}H_{11}F_7O_2$ 296.18	17587-22-3 38	46-7[5]	1.273[25]	1.3766[20]
8399	2,3-Octanedione, 3-oxime 3-Oximino-2-octanone	$C_8H_{15}NO_2$ 157.21	584-92-9 59	133[13]	0-01-00-00795	eth 4
8400	3,6-Octanedione, 2,2,7,7-tetramethyl-2,2,7,7-Tetramethyl-3,6-octanedione	$C_{12}H_{22}O_2$ 198.31	27610-88-4 2.5	116[17]; 57[0.5]	4-01-00-03734 0.900[27]	EtOH 4 1.4400[20]
8401	Octane, 1-(ethenyloxy)-	$C_{10}H_{20}O$ 156.27	929-62-4	58[4]	4-01-00-02056 0.8020[20]	1.4268[20]
8402	Octane, 1-ethoxy-	$C_{10}H_{22}O$ 158.28	929-61-3 12.5	186.3	4-01-00-01759 0.7847[20]	EtOH 4 1.4127[20]
8403	Octane, 3-ethyl- 3-Ethyloctane	$C_{10}H_{22}$ 142.28	5881-17-4	166.5	4-01-00-00476 0.7359[25]	1.4156[20]
8404	Octane, 4-ethyl- 4-Ethyloctane	$C_{10}H_{22}$ 142.28	15869-86-0	163.7	4-01-00-00476 0.7343[25]	1.4151[20]
8405	Octane, 1-fluoro- Octyl fluoride	$C_8H_{17}F$ 132.22	463-11-6	142.4	4-01-00-00418 0.8108[20]	1.3935[20]
8406	Octane, 1-iodo-	$C_8H_{17}I$ 240.13	629-27-6 -45.7	225.5	4-01-00-00425 1.3297[20]	EtOH 3; eth 3 1.4889[20]
8407	Octane, 2-iodo-, (±)- (±)-2-Octyl iodide	$C_8H_{17}I$ 240.13	36049-78-2	210; 95[16]	4-01-00-00425 1.3251[20]	H_2O 1; EtOH 3; eth 3; lig 3 1.4896[20]
8408	Octane, 2-methyl- 2-Methyloctane	C_9H_{20} 128.26	3221-61-2 -80.3	143.2	4-01-00-00454 0.7095[25]	H_2O 1; EtOH 3; eth 3; ctc 2 1.4031[20]
8409	Octane, 3-methyl- 3-Methyloctane	C_9H_{20} 128.26	2216-33-3 -107.6	144.2	0.717[25]	1.4040[25]
8410	Octane, 4-methyl- 4-Methyloctane	C_9H_{20} 128.26	2216-34-4 -113.3	142.4	0.716[25]	1.4039[25]
8411	Octanenitrile Caprylnitrile	$C_8H_{15}N$ 125.21	124-12-9 -45.6	205.2	4-02-00-00993 0.8136[20]	eth 4 1.4203[20]
8412	Octane, 1-nitro- 1-Nitrooctane	$C_8H_{17}NO_2$ 159.23	629-37-8 15	208.5	4-01-00-00426 0.9346[20]	1.4322[20]
8413	Octane, 2-nitro-, (+)-	$C_8H_{17}NO_2$ 159.23	116836-12-5	102-5[23]	4-01-00-00426 0.9166[20]	1.4280[20]
8414	Octane, 3-nitro-	$C_8H_7NO_2$ 149.15	4609-92-1	91[25]	4-01-00-00427 0.917[25]	1.4210[25]
8415	Octane, octadecafluoro- Perfluorooctane	C_8F_{18} 438.06	307-34-6	105.9	4-01-00-00418 1.73[20]	1.282[20]
8416	Octane, 1,1'-oxybis-	$C_{16}H_{34}O$ 242.45	629-82-3 -7.6	283	4-01-00-01760 0.8063[20]	H_2O 2; EtOH 3; eth 3; ctc 3 1.4327[20]
8417	Octane, 1,1'-thiobis-	$C_{16}H_{34}S$ 258.51	2690-08-6	180[10]	4-01-00-01768 0.842[25]	1.4610[20]
8418	1-Octanethiol Octyl mercaptan	$C_8H_{18}S$ 146.30	111-88-6 -49.2	199.1	4-01-00-01767 0.8433[20]	EtOH 3; ctc 2 1.4540[20]
8419	2-Octanethiol, (±)- 2-Octanethiol (DL)	$C_8H_{18}S$ 146.30	10435-81-1 -79	186.4	4-01-00-01777 0.8366[20]	bz 4; eth 4; EtOH 4 1.4504[20]
8420	Octane, 2,2,4-trimethyl- Octane, 2,2,4-trimethyl	$C_{11}H_{24}$ 156.31	18932-14-4	168.5	4-01-00-00494 0.7381[20]	1.4155[20]
8421	Octane, 2,4,7-trimethyl 2,4,7-Trimethyloctane	$C_{11}H_{24}$ 156.31	62016-38-0	168.1		
8422	Octanoic acid Caprylic acid	$C_8H_{16}O_2$ 144.21	124-07-2 16.3	1765 239	4-02-00-00982 0.9106[20]	H_2O 2; EtOH 5; chl 5; CH_3CN 5 1.4285[20]
8423	Octanoic acid, 2-amino-, (±)-	$C_8H_{17}NO_2$ 159.23	644-90-6 270	sub	4-04-00-02799	H_2O 2; EtOH 2; eth 2; bz 2
8424	Octanoic acid, anhydride	$C_{16}H_{30}O_3$ 270.41	623-66-5 -1	282.5	3-02-00-00796 0.9065[18]	ace 4; eth 4; EtOH 4 1.4358[18]
8425	Octanoic acid, 2-bromo-	$C_8H_{15}BrO_2$ 223.11	2623-82-7	140[5]; 118[0.5]	4-02-00-01000 1.2785[24]	1.4613[24]
8426	Octanoic acid, 8-bromo-	$C_8H_{15}BrO_2$ 223.11	17696-11-6 38.5	147[2]	4-02-00-01000	bz 4; eth 4; EtOH 4
8427	Octanoic acid, butyl ester	$C_{12}H_{24}O_2$ 200.32	589-75-3 -42.9	240.5	4-02-00-00987 0.8628[20]	ace 4; eth 4; EtOH 4 1.4232[25]

No.	Name / Synonym	Mol. Form. / Mol. Wt.	CAS RN / mp/°C	Merck No. / bp/°C	Beil. Ref. / den/g cm^{-3}	Solubility / n_D
8428	Octanoic acid, ethyl ester / Ethyl octanoate	$C_{10}H_{20}O_2$ / 172.27	106-32-1 / -43.1	3736 / 208.5	4-02-00-00987 / 0.866[18]	H_2O 1; EtOH 4; eth 4; ctc 2 / 1.4178[20]
8429	Octanoic acid, 8-fluoro- / Caprylic acid, 8-fluoro	$C_8H_{15}FO_2$ / 162.20	353-25-3 / 35	/ 132[4]	4-02-00-00994 /	eth 4; EtOH 4
8430	Octanoic acid, heptyl ester	$C_{15}H_{30}O_2$ / 242.40	4265-97-8 / -10.6	/ 290.5	3-02-00-00794 / 0.8596[20]	ace 4; eth 4; EtOH 4 / 1.4340[20]
8431	Octanoic acid, hexyl ester	$C_{14}H_{28}O_2$ / 228.38	1117-55-1 / -30.6	/ 277.4	3-02-00-00794 / 0.8603[20]	H_2O 1; EtOH 3; eth 3; ace 3 / 1.4323[25]
8432	Octanoic acid, 2-hydroxy-	$C_8H_{16}O_3$ / 160.21	617-73-2 / 70	/ 162[10]	3-03-00-00642 /	H_2O 2; EtOH 4; eth 4; chl 2
8433	Octanoic acid, 3-methyl- / Caprylic acid, 3-methyl	$C_9H_{18}O_2$ / 158.24	6061-10-5 /	/ 135[16]	3-02-00-00826 / 0.899[23]	/ 1.4298[25]
8434	Octanoic acid, methyl ester / Methyl caprylate	$C_9H_{18}O_2$ / 158.24	111-11-5 / -40	/ 192.9	4-02-00-00986 / 0.8775[20]	H_2O 1; EtOH 4; eth 4; ctc 2 / 1.4170[20]
8435	Octanoic acid, 1-methylethyl ester	$C_{11}H_{22}O_2$ / 186.29	5458-59-3 /	/ 93.8[10]	4-02-00-00987 / 0.8555[20]	/ 1.4147[25]
8436	Octanoic acid, 2-methyl-3-oxo-, ethyl ester / Caprylic acid, 2-methyl-3-oxo, ethyl ester	$C_{11}H_{20}O_3$ / 200.28	10488-94-5 /	/ 128-9[12]	4-03-00-01629 / 0.963[0]	EtOH 4
8437	Octanoic acid, octyl ester	$C_{16}H_{32}O_2$ / 256.43	2306-88-9 / -18.1	/ 306.8	4-02-00-00988 / 0.8554[20]	ace 4; eth 4; EtOH 4 / 1.4352[20]
8438	Octanoic acid, pentadecafluoro-, methyl ester	$C_9H_3F_{15}O_2$ / 428.10	376-27-2 /	/ 158	4-02-00-00995 / 1.684[20]	/ 1.304[27]
8439	Octanoic acid, pentyl ester	$C_{13}H_{26}O_2$ / 214.35	638-25-5 / -34.8	/ 260.2	3-02-00-00794 / 0.8613[20]	H_2O 1; EtOH 3; eth 3; ace 3 / 1.4262[25]
8440	Octanoic acid, 1,2,3-propanetriyl ester	$C_{27}H_{50}O_6$ / 470.69	538-23-8 / 10	/ 233	4-02-00-00991 / 0.9540[20]	H_2O 1; EtOH 5; eth 4; bz 4 / 1.4482[20]
8441	Octanoic acid, propyl ester	$C_{11}H_{22}O_2$ / 186.29	624-13-5 / -46.2	/ 226.4	4-02-00-00987 / 0.8659[20]	ace 4; eth 4; EtOH 4 / 1.4191[25]
8442	Octanoic acid, 3,5,7,8-tetrachloro-2,2,3,4,4,5,6,6,7,8,8-undecafluoro- / Octanoic acid, 3,5,7,8-tetrachloro-2,2,3,4,4,5,6,6,7,8,8-undecafluoro	$C_8HCl_4F_{11}O_2$ / 479.89	2923-68-4 /	/ 154.0-6.5[10]	4-02-00-00998 / 1.899[20]	/ 1.3980[20]
8443	1-Octanol / Caprylic alcohol	$C_8H_{18}O$ / 130.23	111-87-5 / -15.5	6674 / 195.1	4-01-00-01756 / 0.8262[25]	H_2O 1; EtOH 5; eth 5; ctc 3 / 1.4295[20]
8444	2-Octanol, (±)-	$C_8H_{18}O$ / 130.23	4128-31-8 / -31.6	6675 / 180	4-01-00-01770 / 0.8193[20]	H_2O 2; EtOH 3; eth 3; ace 3 / 1.4203[20]
8445	3-Octanol / Ethylamylcarbinol	$C_8H_{18}O$ / 130.23	589-98-0 / -45	/ 171	3-01-00-01723 / 0.8258[20]	
8446	4-Octanol, (±)-	$C_8H_{18}O$ / 130.23	74778-22-6 / -40.7	/ 176.3	4-01-00-01779 / 0.8186[20]	H_2O 2; EtOH 3; ctc 2 / 1.4248[20]
8447	2-Octanol, acetate, (±)- / Acetic acid, 2-octyl ester(DL)	$C_{10}H_{20}O_2$ / 172.27	74112-36-0 /	/ 195	4-02-00-00165 / 0.8626[14]	eth 4; EtOH 4 / 1.4146[20]
8448	1-Octanol, 8-chloro-	$C_8H_{17}ClO$ / 164.68	23144-52-7 /	/ 139[19]	4-01-00-01766 /	eth 4; EtOH 4 / 1.4563[25]
8449	1-Octanol, 3,7-dimethyl-, (±)- / Geraniol, tetrahydro (DL)	$C_{10}H_{22}O$ / 158.28	59204-02-3 /	/ 212.5; 106[13]	4-01-00-01830 / 0.8308[10]	H_2O 1; EtOH 3; eth 3; ace 3 / 1.4367[20]
8450	2-Octanol, 2,6-dimethyl- / Tetrahydromyrcenol	$C_{10}H_{22}O$ / 158.28	18479-57-7 /	/ 80.5[10]	4-01-00-01829 / 0.8023[25]	/ 1.4220[25]
8451	3-Octanol, 2,3-dimethyl- / Amylisopropylmethylcarbinol	$C_{10}H_{22}O$ / 158.28	19781-10-3 /	/ 69-70[5]	4-01-00-01828 / 0.8401[20]	/ 1.4380[20]
8452	3-Octanol, 2,7-dimethyl-	$C_{10}H_{22}O$ / 158.28	66719-55-9 /	/ 193.5	1-01-00-00214 / 0.8152[20]	/ 1.4302[20]
8453	3-Octanol, 3,6-dimethyl-	$C_{10}H_{22}O$ / 158.28	151-19-9 / -67.5	/ 202.2	1-01-00-00214 / 0.8347[22]	/ 1.4370[20]
8454	3-Octanol, 3,6-dimethyl-(+-)- / Linalool, tetrahydro (DL)	$C_{10}H_{22}O$ / 158.28	57706-88-4 / 31.5	/ 196.5	3-01-00-01767 / 0.8280[20]	EtOH 4 / 1.4335[20]
8455	4-Octanol, 2,4-dimethyl-	$C_{10}H_{22}O$ / 158.28	33933-79-8 /	/ 97[27]	4-01-00-01828 / 0.8238[20]	/ 1.4338[20]
8456	4-Octanol, 2,5-dimethyl-	$C_{10}H_{22}O$ / 158.28	66719-53-7 /	/ 102-4[34]	1-01-00-00213 / 0.8125[25]	/ 1.4260[25]
8457	4-Octanol, 2,6-dimethyl-	$C_{10}H_{22}O$ / 158.28	66719-54-8 /	/ 195	2-01-00-00460 / 0.8114[20]	/ 1.4270[20]
8458	4-Octanol, 2,7-dimethyl- / 2,7-Dimethyloctan-4-ol	$C_{10}H_{22}O$ / 158.28	19781-11-4 /	/ 96[18]	3-01-00-01769 / 0.8139[20]	/ 1.4280[20]
8459	3-Octanol, 3-ethyl-	$C_{10}H_{22}O$ / 158.28	2051-32-3 /	/ 199	3-01-00-01766 / 0.8361[25]	EtOH 4 / 1.4390[20]
8460	1-Octanol, 8-fluoro-	$C_8H_{17}FO$ / 148.22	408-27-5 /	/ 106-7[10]	4-01-00-01766 / 0.945[20]	eth 4; EtOH 4 / 1.4248[25]
8461	1-Octanol, 3-methyl- / 3-Methyl-1-octanol	$C_9H_{20}O$ / 144.26	38514-02-2 /	/ 110[25]	4-01-00-01806 / 0.827[4]	/ 1.4328[25]
8462	1-Octanol, 4-methyl- / 4-Methyl-1-octanol	$C_9H_{20}O$ / 144.26	38514-03-3 /	/ 106[17]	4-01-00-01807 / 0.820[27]	/ 1.4335[25]
8463	1-Octanol, 7-methyl-	$C_9H_{20}O$ / 144.26	2430-22-0 / 64.5	/ 206	4-01-00-01806 / 0.8260[25]	/ 1.4316[25]
8464	2-Octanol, 2-methyl- / 2-Methyl-2-octanol	$C_9H_{20}O$ / 144.26	628-44-4 /	/ 178	4-01-00-01805 / 0.8210[20]	H_2O 1; EtOH 3; eth 3 / 1.4280[20]
8465	2-Octanol, 3-methyl-	$C_9H_{20}O$ / 144.26	27644-49-1 /	/ 75[15]	3-01-00-01748 / 0.831[27]	/ 1.437[27]
8466	3-Octanol, 3-methyl- / Amylethylmethylcarbinol	$C_9H_{20}O$ / 144.26	5340-36-3 /	/ 97.5[50]; 36[3]	4-01-00-01807 / 0.8258[25]	/ 1.4257[25]

No.	Name / Synonym	Mol. Form. / Mol. Wt.	CAS RN / mp/°C	Merck No. / bp/°C	Beil. Ref. / den/g cm^{-3}	Solubility / n_D
8467	3-Octanol, 4-methyl-	$C_9H_{20}O$ 144.26	66793-80-4	132-3[20]	3-01-00-01750 0.8437[25]	1.4372[20]
8468	3-Octanol, 6-methyl- 6-Methyl-3-octanol	$C_9H_{20}O$ 144.26	40225-75-0	81-3[15]	3-01-00-01749 0.8320[28]	1.4372[20]
8469	4-Octanol, 4-methyl- 4-Methyl-4-octanol	$C_9H_{20}O$ 144.26	23418-37-3	180; 78[15]	4-01-00-01807 0.8267[20]	1.4327[20]
8470	4-Octanol, 5-methyl-	$C_9H_{20}O$ 144.26	59734-23-5	74-6[9]	1-01-00-00211 0.8156[25]	1.4262[25]
8471	2-Octanone Hexyl methyl ketone	$C_8H_{16}O$ 128.21	111-13-7 -16	4632 172.5	4-01-00-03339 0.820[20]	H_2O 2; EtOH 5; eth 5 1.4151[20]
8472	3-Octanone Ethyl amyl ketone	$C_8H_{16}O$ 128.21	106-68-3	3721 167.5	4-01-00-03341 0.822[25]	H_2O 1; EtOH 5; eth 5 1.4150[20]
8473	4-Octanone Butyl propyl ketone	$C_8H_{16}O$ 128.21	589-63-9	163	4-01-00-03342 0.8146[25]	H_2O 1; EtOH 5; eth 5; ctc 3 1.4173[14]
8474	4-Octanone, 5-hydroxy- Butyroin	$C_8H_{16}O_2$ 144.21	496-77-5 -10	1595 185	4-01-00-04042 0.9107[16]	1.4345[16]
8475	2-Octanone, 3-methyl- 3-Methyl-2-octanone	$C_9H_{18}O$ 142.24	6137-08-2	64-5[18]	4-01-00-03356 0.832[27]	1.424[27]
8476	3-Octanone, 4-methyl- 4-Methyl-3-octanone	$C_9H_{18}O$ 142.24	6137-15-1	62-4[10]	4-01-00-03357 0.820[25]	1.4186[25]
8477	3-Octanone, 7-methyl- 7-Methyl-3-octanone	$C_9H_{18}O$ 142.24	5408-57-1	182.5	3-01-00-02889 0.8353[25]	1.4748
8478	4-Octanone, 3-methyl-	$C_9H_{18}O$ 142.24	20754-04-5	174	4-01-00-03357 0.829[14]	1.4200[14]
8479	4-Octanone, 7-methyl- 7-Methyl-4-octanone	$C_9H_{18}O$ 142.24	20809-46-5	178	4-01-00-03356 0.8239[20]	H_2O 1; EtOH 3; eth 3 1.4210[20]
8480	1-Octanone, 1-phenyl-	$C_{14}H_{20}O$ 204.31	1674-37-9 22.8	285; 164[15]	4-07-00-00798 0.9360[30]	EtOH 3; eth 3
8481	Octanoyl chloride	$C_8H_{15}ClO$ 162.66	111-64-8 -63	195.6	4-02-00-00992 0.9535[15]	eth 3 1.4335[20]
8482	Octasiloxane, octadecamethyl- Octadecamethyloctasiloxane	$C_{18}H_{54}O_7Si_8$ 607.31	556-69-4	153[5.1]	4-04-00-04130 0.913[25]	bz 4; peth 4; lig 4 1.3970[20]
8483	1,3,5,7-Octatetraene Octatetraene	C_8H_{10} 106.17	1482-91-3 50	sub	4-01-00-01124	peth 3; HOAc 3
8484	2,4,6-Octatrienal Octatrienal	$C_8H_{10}O$ 122.17	17609-31-3 56.8	75[3]	4-01-00-03591 0.8891[60]	eth 4; EtOH 4
8485	1,3,6-Octatriene, 3,7-dimethyl- Ocimene	$C_{10}H_{16}$ 136.24	13877-91-3	6661 177 dec; 73[21]	3-01-00-01052 0.800[20]	bz 4; eth 4; EtOH 4; chl 4 1.4862[20]
8486	1,3,7-Octatriene, 3,7-dimethyl-	$C_{10}H_{16}$ 136.24	502-99-8	177 dec	3-01-00-01050 0.8000[20]	H_2O 1; EtOH 3; eth 3; chl 3 1.4862[20]
8487	2,4,6-Octatriene, 2,6-dimethyl-, (E,E)-	$C_{10}H_{16}$ 136.24	3016-19-1 -35.4	188; 91[20]	4-01-00-01106 0.8118[20]	1.5446[20]
8488	2,4,6-Octatriene, 2,6-dimethyl-, (E,Z)-	$C_{10}H_{16}$ 136.24	7216-56-0 -20.6	89[20]	4-01-00-01106 0.8060[20]	1.5446[20]
8489	2,4,6-Octatriene, (E,E,E)-	C_8H_{12} 108.18	15192-80-0 52	147.5	4-01-00-01101 0.7961[23]	EtOH 3; chl 3; lig 3 1.5131[27]
8490	6-Octenal, 3,7-dimethyl- Resorcinol, 4-acetyl	$C_{10}H_{18}O$ 154.25	106-23-0	2331 47[1]	1-01-00-00387 0.85[20]	H_2O 2; EtOH 3
8491	1-Octene Caprylene	C_8H_{16} 112.22	111-66-0 -101.7	1764 121.2	4-01-00-00874 0.7149[20]	H_2O 1; EtOH 5; eth 3; ace 3 1.4087[20]
8492	2-Octene, (E)-	C_8H_{16} 112.22	13389-42-9 -87.7	125	4-01-00-00879 0.7199[20]	H_2O 1; EtOH 3; eth 3; ace 3 1.4132[20]
8493	2-Octene, (Z)-	C_8H_{16} 112.22	7642-04-8 -100.2	125.6	4-01-00-00878 0.7243[20]	H_2O 1; EtOH 3; eth 3; ace 3 1.4150[20]
8494	3-Octene, (E)-	C_8H_{16} 112.22	14919-01-8 -110	123.3	4-01-00-00880 0.7152[20]	H_2O 1; EtOH 3; eth 3; ace 3 1.4126[20]
8495	3-Octene, (Z)-	C_8H_{16} 112.22	14850-22-7 -126	122.9	4-01-00-00880 0.7159[20]	ace 4; bz 4; eth 4; EtOH 4 1.4135[20]
8496	4-Octene, (E)-	C_8H_{16} 112.22	14850-23-8 -93.8	122.3	4-01-00-00880 0.7141[20]	H_2O 1; EtOH 3; eth 3; ace 3 1.4114[20]
8497	4-Octene, (Z)-	C_8H_{16} 112.22	7642-15-1 -118.7	122.5	4-01-00-00880 0.7212[20]	ace 4; bz 4; eth 4; EtOH 4 1.4148[20]
8498	1-Octene, 2-chloro-	$C_8H_{15}Cl$ 146.66	31283-43-9	169	3-01-00-00840 0.9273[0]	ace 4; bz 4; eth 4
8499	2-Octene, 2-chloro-	$C_8H_{15}Cl$ 146.66	90202-21-4	167.5; 82[50]	4-01-00-00879 0.8914[16]	ace 4; bz 4; eth 4; EtOH 4 1.4424[16]
8500	4-Octene, 4-chloro-, (Z)- 4-Octene, 4-chloro (cis)	$C_8H_{15}Cl$ 146.66	7321-48-4	165.3; 75[50]	4-01-00-00881 0.8912[20]	eth 4 1.4447[20]
8501	2-Octene, 4-chloro-	$C_8H_{15}Cl$ 146.66	116668-50-9	153	3-01-00-00840 0.8924[20]	ace 4; bz 4; eth 4; chl 4 1.4952[20]
8502	1-Octene, 3,7-dimethyl- 2,6-Dimethyl-7-octene	$C_{10}H_{20}$ 140.27	4984-01-4	154	3-01-00-00862 0.7396[20]	1.4212[20]
8503	2-Octene, 2,6-dimethyl- 2,6-Dimethyl-2-octene	$C_{10}H_{20}$ 140.27	4057-42-5	163	3-01-00-00861 0.746[22]	1.425[22]
8504	4-Octene-2,7-dione	$C_8H_{12}O_2$ 140.18	2130-23-6 30.5	72[0.03]	4-01-00-03757	eth 4; EtOH 4
8505	1-Octene, 2-methyl- 2-Methyl-1-octene	C_9H_{18} 126.24	4588-18-5 -77.8	144.8	4-01-00-00896 0.7343[20]	1.4184[20]
8506	2-Octene, 3-methyl-	C_9H_{18} 126.24	113426-22-5	145	3-01-00-00854 0.7409[25]	1.4247[20]

No.	Name / Synonym	Mol. Form. / Mol. Wt.	CAS RN / mp/°C	Merck No. / bp/°C	Beil. Ref. / den/g cm^{-3}	Solubility / n_D
8507	3-Octene, 7-methyl-	C_9H_{18} 126.24	86668-33-9	142	3-01-00-00854 0.7278[20]	1.4168[20]
8508	4-Octene, 2-methyl-	C_9H_{18} 126.24	64501-77-5	139	3-01-00-00854 0.7379[20]	1.4181[20]
8509	1-Octen-4-ol	$C_8H_{12}O$ 124.18	40575-42-6	172; 68[10]	3-01-00-01942 0.8373[20]	1.4383[20]
8510	2-Octen-1-ol	$C_8H_{16}O$ 128.21	22104-78-5	87-9[11]	4-01-00-02164 0.850[20]	1.4470[20]
8511	2-Octen-4-ol	$C_8H_{16}O$ 128.21	4798-61-2	175	3-01-00-01943 0.8393[20]	1.4395
8512	4-Octen-1-ol	$C_8H_{16}O$ 128.21	67700-26-9	95[17]	4-01-00-02166 0.842[20]	1.4462[20]
8513	5-Octen-1-ol	$C_8H_{16}O$ 128.21	90200-83-2	91.5[14]	4-01-00-02166 0.8492[21]	1.4478[21]
8514	6-Octen-1-ol, 3,7-dimethyl-, (±)- (±)-Citronellol	$C_{10}H_{20}O$ 156.27	26489-01-0 2332	224; 99[10]	4-01-00-02188 0.8560[20]	H_2O 2; EtOH 5; eth 5 1.4543[20]
8515	6-Octen-3-ol, 3,7-dimethyl-	$C_{10}H_{20}O$ 156.27	18479-51-1	94[14]	2-01-00-00494 0.8695[15]	1.4569[15]
8516	7-Octen-1-ol, 3,7-dimethyl-, (S)- (-)-α-Citronellol	$C_{10}H_{20}O$ 156.27	6812-78-8 8186	114-5[12]	2-01-00-00497 0.8549[20]	eth 4; EtOH 4 1.4556[20]
8517	1-Octen-4-ol, 4-methyl-	$C_9H_{18}O$ 142.24	62108-06-9	178	2-01-00-00492 0.8440[20]	1.4399
8518	2-Octen-4-ol, 7-methyl-	$C_9H_{18}O$ 142.24	4798-64-5	85[12]	2-01-00-00492 0.8402[14]	1.4449[14]
8519	3-Octen-2-ol, 2-methyl-, (Z)- 3-Octen-2-ol, 2-methyl (cis)	$C_9H_{18}O$ 142.24	18521-07-8	177	4-01-00-02177 0.8378[20]	1.4426[20]
8520	4-Octen-3-ol, 3-methyl-	$C_9H_{18}O$ 142.24	90676-55-4	69-70[10]	4-01-00-02178 0.812[14]	1.4414[14]
8521	5-Octen-4-ol, 5-methyl- 3-Octen-5-ol, 4-methyl	$C_9H_{18}O$ 142.24	36903-92-1	89[16]	2-01-00-00492 0.8468[25]	1.4446[25]
8522	1-Octen-3-one, 5-methyl-	$C_9H_{16}O$ 140.23	116836-13-6	72-3[16]	3-01-00-03018 0.8450[19]	1.4380[19]
8523	5-Octen-4-one, 2-methyl- 2-Methyl-5-octen-4-one	$C_9H_{16}O$ 140.23	17577-93-4	183	4-01-00-03504 0.839[25]	1.4387[25]
8524	5-Octen-4-one, 7-methyl-	$C_9H_{16}O$ 140.23	32064-78-1	68-78[24]	3-01-00-03018 0.901[12]	1.4748
8525	7-Octen-4-one, 2-methyl- 1-Octen-5-one, 7-methyl	$C_9H_{16}O$ 140.23	54298-97-4	62-3[14]	2-01-00-00801 0.8361[12]	1.4288[12]
8526	1-Octen-3-yne	C_8H_{12} 108.18	17679-92-4	62[60]	4-01-00-01100 0.7830[20]	eth 4 1.4592[20]
8527	7-Octen-5-yn-4-ol, 4-methyl- 1-Octen-3-yne-5-ol, 5-methyl	$C_9H_{14}O$ 138.21	39118-35-9	80[13]	3-01-00-02034 0.8851[15]	EtOH 4 1.4785[20]
8528	Octhilinone 3(2H)-Isothiazolone, 2-octyl-	$C_{11}H_{19}NOS$ 213.34	26530-20-1 6677	120[0.01]		
8529	1-Octyne Hexylacetylene	C_8H_{14} 110.20	629-05-0 -79.3	126.3	4-01-00-01034 0.7461[20]	H_2O 1; EtOH 3; eth 3 1.4159[20]
8530	2-Octyne Methylpentylacetylene	C_8H_{14} 110.20	2809-67-8 -61.6	137.6	4-01-00-01035 0.7596[20]	H_2O 1; EtOH 3; eth 3 1.4278[20]
8531	3-Octyne	C_8H_{14} 110.20	15232-76-5 -103.9	133.1	4-01-00-01036 0.7529[20]	H_2O 1; EtOH 3; eth 3 1.4250[20]
8532	4-Octyne Dipropylacetylene	C_8H_{14} 110.20	1942-45-6 -101	131.6	4-01-00-01037 0.7509[20]	H_2O 1; EtOH 3; eth 3 1.4248[20]
8533	1-Octyne, 1-chloro- 1-Chloro-1-octyne	$C_8H_{13}Cl$ 144.64	64531-26-6	61-2[17]	3-01-00-01005 0.912[20]	eth 4; EtOH 4 1.445[20]
8534	3-Octyne, 2-chloro-2-methyl- 2-Chloro-2-methyloct-3-yne	$C_9H_{15}Cl$ 158.67	20599-21-7	68[15]	4-01-00-01048 0.8929[20]	H_2O 1; eth 4 1.4480[20]
8535	3-Octyne, 2,2-dimethyl- 2,2-Dimethyl-3-octyne	$C_{10}H_{18}$ 138.25	19482-57-6	79[60]	3-01-00-01018 0.7491[20]	eth 4 1.4270[20]
8536	2-Octyne, 1-methoxy- Methyl (2-octyn-1-yl) ether	$C_9H_{16}O$ 140.23	18495-23-3	77[19]	3-01-00-01996 0.8370[25]	eth 4 1.4380[20]
8537	3-Octyne, 7-methyl- 7-Methyl-3-octyne	C_9H_{18} 124.23	37050-06-9	87[99]	3-01-00-01014 0.7599[20]	eth 4 1.4280[20]
8538	2-Octynoic acid Pentylpropiolic acid	$C_8H_{12}O_2$ 140.18	5663-96-7 4	148[19]; 114[2]	4-02-00-01714 0.9623[13]	1.4595[20]
8539	3-Octynoic acid	$C_8H_{12}O_2$ 140.18	57074-96-1 18	110[2]	4-02-00-01715	1.4577[25]
8540	5-Octynoic acid	$C_8H_{12}O_2$ 140.18	76469-08-4 8	111[2]	4-02-00-01715 0.9787[27]	1.4540[25]
8541	7-Octynoic acid	$C_8H_{12}O_2$ 140.18	10297-09-3 19	123[2]	4-02-00-01714	1.4502[25]
8542	2-Octynoic acid, methyl ester	$C_9H_{14}O_2$ 154.21	111-12-6	107[20]; 94[10]	4-02-00-01715 0.926[20]	1.4464[20]
8543	2-Octyn-1-ol 2-Octynol	$C_8H_{14}O$ 126.20	20739-58-6 -18	98-9[15]	4-01-00-02256 0.8805[20]	eth 4 1.4556[20]
8544	1-Octyn-3-ol, 3-methyl- 1-Octyne-3-ol, 3-methyl	$C_9H_{16}O$ 140.23	23580-51-0	75[10]	4-01-00-02268 0.863[10]	1.443[10]
8545	Olean-12-en-28-oic acid, 3,23-dihydroxy-, (3β,4α)- Hederagenin	$C_{30}H_{48}O_4$ 472.71	465-99-6 333	4540		

PHYSICAL CONSTANTS OF ORGANIC COMPOUNDS (continued)

No.	Name / Synonym	Mol. Form. / Mol. Wt.	CAS RN / mp/°C	Merck No. / bp/°C	Beil. Ref. / den/g cm⁻³	Solubility / n_D
8546	Olean-12-en-28-oic acid, 3,16-dihydroxy-23-oxo-, (3β,4α,16α)- / Quillaic acid	$C_{30}H_{46}O_5$ 486.69	631-01-6 294	8049	3-10-00-04652	ace 4; eth 4; py 4; EtOH 4
8547	Olean-12-en-28-oic acid, 3-hydroxy-, (3β)- / Oleanolic acid	$C_{30}H_{48}O_3$ 456.71	508-02-1 310 dec	6787 sub 280	4-10-00-01164	H_2O 1; EtOH 2; eth 2; ace 2
8548	Olean-12-en-3-ol, (3β)- / β-Amyrin	$C_{30}H_{50}O$ 426.73	559-70-6 197	654 260⁰⁵	4-06-00-04195	H_2O 1; EtOH 2; eth 3; bz 3
8549	Orcein	$C_{28}H_{24}N_2O_7$ 500.51	1400-62-0	6818	2-06-00-00876	
8550	L-Ornithine / (S)-2,5-Diaminopentanoic acid	$C_5H_{12}N_2O_2$ 132.16	70-26-8 140	6826	4-04-00-02644	H_2O 4; EtOH 4
8551	L-Ornithine, N5-(aminocarbonyl)- / Citrulline	$C_6H_{13}N_3O_3$ 175.19	372-75-8 235.5	2333	4-04-00-02647	H_2O 3; EtOH 1; MeOH 1
8552	Orthoformic acid, triisoamyl ester / Isopentyl alcohol, orthoformate (3:1)	$C_{16}H_{34}O_3$ 274.44	5337-70-2 268 dec		4-02-00-00030 0.8628²⁰	eth 4; EtOH 4 1.4238²⁰
8553	Oryzalin / Benzenesulfonamide, 4-(dipropylamino)-3,5-dinitro-	$C_{12}H_{18}N_4O_6S$ 346.36	19044-88-3 141	6840		
8554	7-Oxabicyclo[4.1.0]heptane	$C_6H_{10}O$ 98.14	286-20-4 <-10		5-17-01-00203 0.9663²⁰	H_2O 1; EtOH 4; eth 4; ace 4 1.4519²⁰
8555	7-Oxabicyclo[4.1.0]heptane, 3-ethenyl- / Cyclohexane, 1,2-epoxy-4-vinyl	$C_8H_{12}O$ 124.18	106-86-5 <-100	131.5 169; 70²⁰	4-17-00-00314 0.9581²⁰	1.4700²⁰
8556	7-Oxabicyclo[2.2.1]heptane, 1-methyl-4-(1-methylethyl)-	$C_{10}H_{18}O$ 154.25	470-67-7 1	173.5	5-17-01-00273 0.8997²⁰	H_2O 2; EtOH 5; eth 5; bz 3 1.4562²⁰
8557	7-Oxabicyclo[4.1.0]heptane, 3-oxiranyl- / 4-Vinyl-1-cyclohexene dioxide	$C_8H_{12}O_2$ 140.18	106-87-6 <-55	227	5-19-01-00295 1.0966²⁰	H_2O 4 1.4787²⁰
8558	6-Oxabicyclo[3.1.0]hexane	C_5H_8O 84.12	285-67-6	102	5-17-01-00190 0.964²⁵	1.4336²⁰
8559	3-Oxabicyclo[3.2.1]octane-2,4-dione, 1,8,8-trimethyl-, (±)-	$C_{10}H_{14}O_3$ 182.22	595-30-2 221	270	5-17-11-00089 1.194²⁰	bz 4
8560	2-Oxabicyclo[2.2.2]octane, 1,3,3-trimethyl- / Eucalyptol	$C_{10}H_{18}O$ 154.25	470-82-6 1.5	3851 176.4	5-17-01-00273 0.9267²⁰	H_2O 1; EtOH 3; eth 3; ctc 2 1.4586²⁰
8561	6-Oxabicyclo[3.2.1]oct-3-ene, 4,7,7-trimethyl-, (±)- / (±)-Pinol	$C_{10}H_{16}O$ 152.24	60761-04-8 183.5		4-17-00-00327 0.9515²⁰	eth 4; EtOH 4 1.4695²⁰
8562	Oxacyclohexadecan-2-one / Exaltolide	$C_{15}H_{28}O_2$ 240.39	106-02-5	3867	5-17-09-00106	
8563	1,3,4-Oxadiazole / 1-Oxa-3,4-diazacyclopentadiene	$C_2H_2N_2O$ 70.05	288-99-3 150	6859		1.4300²⁵
8564	1,2,4-Oxadiazole, 3,5-diphenyl- / 3,5-Diphenyl-1,2,4-oxadiazole	$C_{14}H_{10}N_2O$ 222.25	888-71-1 109	290; 210¹⁷	4-27-00-07210	bz 4; eth 4; EtOH 4
8565	Oxadiazon / 1,2,4-Oxadiazolidin-3,5-dione, 2-tert-butyl-4-(2,4-dichloro-5-isopropoxyphenyl)-	$C_{15}H_{18}Cl_2N_2O_3$ 345.23	19666-30-9 90	6860		
8566	Oxadixyl / N-(2,6-Dimethylphenyl)-2-methoxy-N-(2-oxo-3-oxazolidinyl)acetamide	$C_{14}H_{18}N_2O_4$ 278.31	77732-09-3 104			
8567	16(15H)-Oxaerythrinan-15-one, 1,2,6,7-tetradehydro-14,17-dihydro-3-methoxy-, (3β)- / β-Erythroidine	$C_{16}H_{19}NO_3$ 273.33	466-81-9 99.5	3625	4-27-00-03568	H_2O 3; EtOH 4; eth 3; bz 4
8568	Oxamyl / Ethanimidothioic acid, 2-(dimethylamino)-N-[[(methylamino)carbonyl]oxy]-2-oxo-, methyl ester	$C_7H_{13}N_3O_3S$ 219.26	23135-22-0 109 dec	6873	0.97²⁵	
8569	1,4-Oxathiane	C_4H_8OS 104.17	15980-15-1 147		5-19-01-00033 1.1174²⁰	H_2O 2
8570	1,2-Oxathiolane, 2,2-dioxide / Propane sultone	$C_3H_6O_3S$ 122.14	1120-71-4		5-19-01-00005	chl 3
8571	Oxayohimban-16-carboxylic acid, 16,17-didehydro-19-methyl-, methyl ester, (19α)- / Raubasine	$C_{21}H_{24}N_2O_3$ 352.43	483-04-5 258 dec	8129	4-27-00-07927	H_2O 1; MeOH 3
8572	Oxayohimbanium, 3,4,5,6,16,17-hexadehydro-16-(methoxycarbonyl)-19-methyl-, (19α)- / Serpentine	$C_{21}H_{20}N_2O_3$ 348.40	18786-24-8 175	8415		H_2O 1; EtOH 3; eth 3; ace 3
8573	2H-1,3-Oxazine-2,4(3H)-dione, 5,5-diethyldihydro- / Diethadione	$C_8H_{13}NO_3$ 171.20	702-54-5 97.5	3096	4-27-00-03263	
8574	2-Oxazolamine, 4,5-dihydro-5-phenyl- / Aminorex	$C_9H_{10}N_2O$ 162.19	2207-50-3 137	490		
8575	Oxazole / 1,3-Oxazole	C_3H_3NO 69.06	288-42-6 69.5		4-27-00-00960	1.4285¹⁷
8576	Oxazole, 4,5-dihydro-2-methyl-	C_4H_7NO 85.11	1120-64-5 111		4-27-00-00921 1.005²⁵	1.4340²⁰
8577	Oxazole, 2,4-dimethyl- / 2,4-Dimethyloxazole	C_5H_7NO 97.12	7208-05-1 108		4-27-00-00980 0.9352¹⁵	H_2O 4; eth 4; EtOH 4 1.4166¹⁵
8578	Oxazole, 2,5-dimethyl- / 2,5-Dimethyloxazole	C_5H_7NO 97.12	23012-11-5 117.5		4-27-00-00985 0.9958²¹	H_2O 4 1.4385²¹

3-235

No.	Name / Synonym	Mol. Form. / Mol. Wt.	CAS RN / mp/°C	Merck No. / bp/°C	Beil. Ref. / den/g cm⁻³	Solubility / n_D
8579	Oxazole, 2,4-diphenyl- / 2,4-Diphenyloxazole	$C_{15}H_{11}NO$ / 221.26	838-41-5 / 103	/ 339	4-27-00-01435	EtOH 4; eth 4; bz 4
8580	Oxazole, 2,5-diphenyl-	$C_{15}H_{11}NO$ / 221.26	92-71-7 / 74	/ 360	4-27-00-01435 / 1.0940[100]	H_2O 1; EtOH 4; eth 4; chl 2 / 1.6231[100]
8581	Oxazole, 4,5-diphenyl- / 4,5-Diphenyloxazole	$C_{15}H_{11}NO$ / 221.26	4675-18-7 / 44	/ 193[15]; 132[0.05]	4-27-00-01438	H_2O 2; EtOH 4; eth 4; con acid 3 / 1.6283[100]
8582	Oxazole, 5-methoxy-2-phenyl- / 5-Methoxy-2-phenyloxazole	$C_{10}H_9NO_2$ / 175.19	40527-16-0	/ 140[10]	4-27-00-01958	ctc 3 / 1.584[23]
8583	Oxazole, 4-methyl- / 4-Methyloxazole	C_4H_5NO / 83.09	693-93-6	/ 88	4-27-00-00968 / 1.015[25]	1.4317[20]
8584	2,4-Oxazolidinedione, 5,5-dimethyl- / Dimethadione	$C_5H_7NO_3$ / 129.12	695-53-4 / 76.5	3201	4-27-00-03237	
8585	2,4-Oxazolidinedione, 5,5-dipropyl- / 5,5-Dipropyl-2,4-oxazolidinedione	$C_9H_{15}NO_3$ / 185.22	512-12-9 / 42.5	3352 / 148-50[3]	4-27-00-03269	
8586	2,4-Oxazolidinedione, 5-methyl-3-(2-propenyl)- / Aloxidone	$C_7H_9NO_3$ / 155.15	526-35-2	305 / 138[35]; 86[0.5]	4-27-00-03214	1.4688[25]
8587	2,4-Oxazolidinedione, 3,5,5-trimethyl- / Trimethadione	$C_6H_9NO_3$ / 143.14	127-48-0 / 46	9620 / 78-80[5]	4-27-00-03237	H_2O 3; EtOH 4; eth 4; ace 4
8588	2-Oxazolidinone, 5-[(2-methoxyphenoxy)methyl]- / Mephenoxalone	$C_{11}H_{13}NO_4$ / 223.23	70-07-5 / 144	5739		
8589	2-Oxazolidinone, 3-[[(5-nitro-2-furanyl)methylene]amino]- / Furazolidone	$C_8H_7N_3O_5$ / 225.16	67-45-8 / 255	4210	4-27-00-02530	
8590	Oxepane	$C_6H_{12}O$ / 100.16	592-90-5	/ 119	5-17-01-00090 / 0.89[25]	1.4400[20]
8591	2-Oxepanone / Caprolactone	$C_6H_{10}O_2$ / 114.14	502-44-3 / -18	/ 215	5-17-09-00034 / 1.0693[20]	EtOH 3; eth 3; ace 3 / 1.4611[20]
8592	Oxetane / Trimethylene oxide	C_3H_6O / 58.08	503-30-0 / -97	9630 / 47.6	5-17-01-00011 / 0.8930[25]	H_2O 5; EtOH 5; eth 3; ace 4 / 1.3961[20]
8593	Oxetane, 3,3-dimethyl- / 3,3-Dimethyloxetane	$C_5H_{10}O$ / 86.13	6921-35-3	/ 80.6	5-17-01-00085 / 0.834[25]	1.3965[20]
8594	Oxetane, 2-methyl-	C_4H_8O / 72.11	2167-39-7	/ 59	5-17-01-00054 / 0.841[25]	1.3885[20]
8595	2-Oxetanone / β-Propiolactone	$C_3H_4O_2$ / 72.06	57-57-8 / -33.4	7832 / 162	5-17-09-00003 / 1.1460[20]	eth 5; chl 3 / 1.4105[20]
8596	3-Oxetanone	$C_3H_4O_2$ / 72.06	6704-31-0			
8597	2-Oxetanone, 3,3-dimethyl-4-(1-methylethylidene)-	$C_8H_{12}O_2$ / 140.18	3173-79-3 / -18	/ 170	5-17-09-00181 / 0.947[25]	1.4380[20]
8598	2-Oxetanone, 4-methylene- / Diketene	$C_4H_4O_2$ / 84.07	674-82-8 / -6.5	/ 126.1	5-17-09-00115 / 1.0877[20]	1.4379[20]
8599	Oxirane / Ethylene oxide	C_2H_4O / 44.05	75-21-8 / -111.7	3758 / 10.6	5-17-01-00003 / 0.882[10]	H_2O 3; EtOH 3; eth 3; ace 3 / 1.3579[7]
8600	Oxirane, (bromomethyl)-, (±)- / Propane, 3-bromo-1,2-epoxy (DL)	C_3H_5BrO / 136.98	82584-73-4 / -40	/ 137	5-17-01-00023 / 1.615[14]	H_2O 1; EtOH 3; eth 3; bz 3 / 1.4841[20]
8601	Oxirane, 2,2'-[1,4-butanediylbis(oxymethylene)]bis-	$C_{10}H_{18}O_4$ / 202.25	2425-79-8	/ 266; 155[11]	5-17-03-00023 / 1.1[25]	1.4611[20]
8602	Oxirane, (butoxymethyl)- / Butyl glycidyl ether	$C_7H_{14}O_2$ / 130.19	2426-08-6	/ 169; 75[26]	5-17-03-00011 / 0.918[20]	
8603	Oxiranecarboxaldehyde / Glycidaldehyde	$C_3H_4O_2$ / 72.06	765-34-4 / -62	/ 112.5	5-17-09-00006	1.4265[20]
8604	Oxiranecarboxylic acid, 3-methyl-3-phenyl-, ethyl ester	$C_{12}H_{14}O_3$ / 206.24	77-83-8	/ 273.5	5-18-06-00385 / 1.044[20]	1.5182[20]
8605	Oxirane, (chloromethyl)-, (±)- / DL-α-Epichlorohydrin	C_3H_5ClO / 92.52	13403-37-7 / -26	3563 / 117; 62[100]	5-17-01-00020 / 1.1812[20]	H_2O 5; EtOH 5; eth 5; bz 3 / 1.4361[20]
8606	Oxirane, 2-(chloromethyl)-2-methyl-	C_4H_7ClO / 106.55	598-09-4	/ 122	5-17-01-00059 / 1.1011[20]	H_2O 4; eth 4 / 1.4310[20]
8607	Oxirane, 2,2-dimethyl-	C_4H_8O / 72.11	558-30-5 / 52		5-17-01-00058 / 0.8112[20]	EtOH 3; eth 3 / 1.3712[22]
8608	Oxirane, 2,3-dimethyl-, cis-	C_4H_8O / 72.11	1758-33-4 / -80	/ 60	5-17-01-00061 / 0.8226[25]	eth 4; ace 4; bz 4; os 4 / 1.3802[20]
8609	Oxirane, 2,3-dimethyl-, trans-	C_4H_8O / 72.11	21490-63-1 / -85	/ 56.5	5-17-01-00061 / 0.8010[25]	eth 4; ace 4; bz 4; os 4 / 1.3736[20]
8610	Oxirane, [(1,1-dimethylethoxy)methyl]-	$C_7H_{14}O_2$ / 130.19	7665-72-7 / -70	/ 152	5-17-03-00011 / 0.898[20]	
8611	Oxirane, [[4-(1,1-dimethylethyl)phenoxy]methyl]-	$C_{13}H_{18}O_2$ / 206.28	3101-60-8	/ 165-70[14]	5-17-03-00017 / 1.038[25]	1.5145[20]
8612	Oxirane, 2-(2,2-dimethylpropyl)-2-methyl-	$C_8H_{16}O$ / 128.21	107-48-2 / -64	/ 140.9	4-17-00-00112 / 0.8272[20]	bz 4; eth 4 / 1.4097[20]
8613	Oxirane, ethenyl-	C_4H_6O / 70.09	930-22-3	/ 68	5-17-01-00180 / 0.9006[25]	EtOH 3; eth 3; bz 3; os 3 / 1.4168[20]
8614	Oxirane, (ethoxymethyl)- / 2,3-Epoxypropyl ethyl ether	$C_5H_{10}O_2$ / 102.13	4016-11-9	/ 128	5-17-03-00010 / 0.9700[20]	H_2O 3; EtOH 3; eth 3; ctc 2 / 1.4320[20]

No.	Name / Synonym	Mol. Form. / Mol. Wt.	CAS RN / mp/°C	Merck No. / bp/°C	Beil. Ref. / den/g cm^{-3}	Solubility / n_D
8615	Oxirane, ethyl- 1,2-Epoxybutane	$C_4H_8O_2$ 88.11	106-88-7 -150	63.3	5-17-01-00056 0.8297[20]	EtOH 4; eth 5; ace 4; os 4 1.3851[20]
8616	Oxirane, (fluoromethyl)-	C_3H_5FO 76.07	503-09-3	85.5	5-17-01-00019 1.067[25]	1.3715[20]
8617	Oxirane, (iodomethyl)-	C_3H_5IO 183.98	624-57-7	161; 62[24]	5-17-01-00024 1.982[24]	eth 4; EtOH 4
8618	Oxiranemethanol Glycidol	$C_3H_6O_2$ 74.08	556-52-5	4385 167 dec; 66[2.5]	5-17-03-00009 1.1143[25]	H$_2$O 5; EtOH 5; eth 3; ace 3 1.4287[20]
8619	Oxiranemethanol, 3-phenyl-	$C_9H_{10}O_2$ 150.18	21915-53-7 26.5	134[4]	4-17-00-01349 1.1512[21]	1.5427[27]
8620	Oxirane, (methoxymethyl)-	$C_4H_8O_2$ 88.11	930-37-0	113	5-17-03-00010 0.9890[20]	H$_2$O 4; ace 4; eth 4; EtOH 4 1.4320[20]
8621	Oxirane, methyl- Propylene oxide	C_3H_6O 58.08	75-56-9 -111.9	7869 35	5-17-01-00017 0.859[0]	H$_2$O 4; eth 4; EtOH 4 1.3670[20]
8622	Oxirane, 2-methyl-2-phenyl-	$C_9H_{10}O$ 134.18	2085-88-3	83-5[15]	5-17-01-00591 1.0280[20]	1.5232[20]
8623	Oxirane, (phenoxymethyl)- Phenyl glycidyl ether	$C_9H_{10}O_2$ 150.18	122-60-1	243	5-17-03-00013 1.1109[21]	1.5307[21]
8624	Oxirane, phenyl- Phenylethylene oxide	C_8H_8O 120.15	96-09-3 -35.6	194.1	5-17-01-00577 1.0523[16]	H$_2$O 1; EtOH 3; eth 3; chl 3 1.5342[20]
8625	Oxirane, 2,2'-[1,3-phenylenebis(oxymethylene)]bis- Diglycidyl resorcinol ether	$C_{12}H_{14}O_4$ 222.24	101-90-6 42.5	143-51[0.4]	5-17-03-00024 1.2183[30]	1.5408[20]
8626	Oxirane, [(2-propenyloxy)methyl]- Allyl glycidyl ether	$C_6H_{10}O_2$ 114.14	106-92-3	154	5-17-03-00012 0.9698[20]	1.4332[20]
8627	Oxirane, tetramethyl-	$C_6H_{12}O$ 100.16	5076-20-0	90-4[74.5]	5-17-01-00108 0.8156[16]	H$_2$O 3 1.3984[16]
8628	Oxirane, (trichloromethyl)-	$C_3H_3Cl_3O$ 161.41	3083-23-6	149; 44[13]	5-17-01-00023 1.495[20]	eth 4; chl 3 1.4737[25]
8629	Oxirane, trifluoro(trifluoromethyl)- Perfluoropropylene oxide	C_3F_6O 166.02	428-59-1		5-17-01-00020	
8630	Oxyacanthan, 6',12'-dimethoxy-2,2'-dimethyl-6,7-[methylenebis(oxy)]- Cepharanthine	$C_{37}H_{38}N_2O_6$ 606.72	481-49-2 150	1981	4-27-00-09061	
8631	Oxyacanthan-12'-ol, 6,6',7-trimethoxy-2,2'-dimethyl- Oxyacanthine	$C_{37}H_{40}N_2O_6$ 608.73	548-40-3 216.5	6906	2-27-00-00894	H$_2$O 1; EtOH 3; eth 3; bz 3
8632	Oxydemeton-methyl Phosphorothioic acid, S-[2-(ethylsulfinyl)ethyl] O,O-dimethyl ester	$C_6H_{15}O_4PS_2$ 246.29	301-12-2 <-20	106[0.01]	1.289[20]	
8633	Oxyfluorfen Benzene, 2-chloro-1-(3-ethoxy-4-nitrophenoxy)-4-(trifluoromethyl)-	$C_{15}H_{11}ClF_3NO_4$ 361.70	42874-03-3 84	6916 358.2 dec	1.35[73]	
8634	Paraformaldehyde Formaldehyde polymer	$(CH_2O)_x$	30525-89-4 164 dec	6974	5-19-09-00103	H$_2$O 2; EtOH 1; eth 1; alk 3
8635	Pendimethalin Benzenamine, N-(1-ethylpropyl)-3,4-dimethyl-2,6-dinitro-	$C_{13}H_{19}N_3O_4$ 281.31	40487-42-1 56	7026 dec	1.19[25]	
8636	Pentacene Benzo[b]naphthacene	$C_{22}H_{14}$ 278.35	135-48-8 257	7057	4-05-00-02721	H$_2$O 1; bz 2; PhNO$_2$ 3; os 2
8637	Pentacosane	$C_{25}H_{52}$ 352.69	629-99-2 54	401.9; 282[40]	4-01-00-00582 0.8012[20]	bz 3; chl 3 1.4491[20]
8638	Pentacosane, 13-phenyl- Benzene, (1-dodecyltridecyl)-	$C_{31}H_{56}$ 428.79	6006-90-2 31.7	235[10]	4-05-00-01241 0.8537[20]	1.4787[20]
8639	Pentacosane, 13-undecyl-	$C_{36}H_{74}$ 506.98	55517-89-0 9.7	307[10]	4-01-00-00600 0.8168[20]	1.4567[20]
8640	Pentadecanal Pentadecaldehyde	$C_{15}H_{30}O$ 226.40	2765-11-9 24.5	185[25]	4-01-00-03391	ace 4; eth 4; EtOH 4
8641	1-Pentadecanamine	$C_{15}H_{33}N$ 227.43	2570-26-5 37.3	307.6	4-04-00-00817 0.8104[20]	eth 4; EtOH 4 1.4480[20]
8642	Pentadecane	$C_{15}H_{32}$ 212.42	629-62-9 9.9	270.6	4-01-00-00529 0.7685[20]	H$_2$O 1; EtOH 4; eth 4 1.4315[20]
8643	Pentadecane, 1-bromo-	$C_{15}H_{31}Br$ 291.32	629-72-1 19	322	4-01-00-00531 1.0675[20]	H$_2$O 1; ace 3; chl 4 1.4611[20]
8644	Pentadecane, 2,6,10,14-tetramethyl- Pristane	$C_{19}H_{40}$ 268.53	1921-70-6	7757 296	3-01-00-00570 0.783[20]	bz 4; eth 4; chl 4; peth 4 1.4379[20]
8645	Pentadecanoic acid Pentadecylic acid	$C_{15}H_{30}O_2$ 242.40	1002-84-2 52.3	257[100]; 158[1]	4-02-00-01147 0.8423[80]	H$_2$O 1; EtOH 4; eth 3; ace 4 1.4254[80]
8646	Pentadecanoic acid, methyl ester Methyl pentadecanoate	$C_{16}H_{32}O_2$ 256.43	7132-64-1 18.5	153.5	3-02-00-00936 0.8618[25]	EtOH 3; eth 3; os 3 1.4390[25]
8647	2-Pentadecanol sec-Pentadecyl alcohol	$C_{15}H_{32}O$ 228.42	1653-34-5 35	299	4-01-00-01871 0.8328[20]	1.4463[20]
8648	3-Pentadecanol	$C_{15}H_{32}O$ 228.42	53346-71-7 38.8	388; 163[12]	4-01-00-01871	1.4227[80]
8649	1-Pentadecanol, 2-methyl- 2-Methyl-1-pentadecanol	$C_{16}H_{34}O$ 242.45	25354-98-7 12.5	185-7[15]	3-01-00-01827 0.8320[25]	1.4453[25]
8650	6-Pentadecanol, 6-methyl-	$C_{16}H_{34}O$ 242.45	108836-86-8	199-200[50]	3-01-00-01828 0.8316[25]	1.4446[25]

No.	Name Synonym	Mol. Form. Mol. Wt.	CAS RN mp/°C	Merck No. bp/°C	Beil. Ref. den/g cm^{-3}	Solubility n_D
8651	2-Pentadecanone	$C_{15}H_{30}O$ 226.40	2345-28-0 39.5	294	4-01-00-03391 0.8182[39]	
8652	8-Pentadecanone	$C_{15}H_{30}O$ 226.40	818-23-5 43	291	4-01-00-03392 0.8180[39]	EtOH 3; eth 3; bz 3; ctc 3
8653	1-Pentadecene	$C_{15}H_{30}$ 210.40	13360-61-7 -2.8	268.2	4-01-00-00926 0.7764[20]	H_2O 1; ace 3 1.4389[20]
8654	7-Pentadecene	$C_{15}H_{30}$ 210.40	15430-98-5	114[3.2]	3-01-00-00875 0.7765[20]	1.4420[20]
8655	1-Pentadecen-3-ol	$C_{15}H_{30}O$ 226.40	99814-65-0 27.5	150-2[6]	3-01-00-01961 0.834[32]	1.4481[32]
8656	1-Pentadecyne	$C_{15}H_{28}$ 208.39	765-13-9 10	268	4-01-00-01072 0.7928[20]	ace 4 1.4419[20]
8657	2,4-Pentadienal, 5-phenyl- 5-Phenyl-2,4-pentadienal	$C_{11}H_{10}O$ 158.20	13466-40-5 42.5	160[3]; 133[1.0]	4-07-00-01179	H_2O 1; EtOH 5; eth 4; bz 5
8658	1,2-Pentadiene Ethylallene	C_5H_8 68.12	591-95-7 -137.3	44.9	4-01-00-00993 0.6926[20]	EtOH 5; eth 5; ace 5; bz 5 1.4209[20]
8659	1,4-Pentadiene	C_5H_8 68.12	591-93-5 -148.8	26	4-01-00-00998 0.6608[20]	H_2O 1; EtOH 4; eth 4; ace 4 1.3888[20]
8660	2,3-Pentadiene 1,3-Dimethylallene	C_5H_8 68.12	591-96-8 -125.6	48.2	4-01-00-00999 0.6950[20]	H_2O 1; EtOH 5; eth 5; ace 5 1.4284[20]
8661	1,3-Pentadiene, (E)- trans-Piperylene	C_5H_8 68.12	2004-70-8 -87.4	42	4-01-00-00994 0.6710[25]	1.4301[20]
8662	1,3-Pentadiene, (Z)- cis-Piperylene	C_5H_8 68.12	1574-41-0 -140.8	44.1	4-01-00-00994 0.6910[20]	EtOH 5; eth 5; ace 5; bz 5 1.4363[20]
8663	1,3-Pentadiene, 3-chloro- 3-Chloro-1,3-pentadiene	C_5H_7Cl 102.56	37710-49-9	100	3-01-00-00962 0.9576[20]	ace 4; bz 4; eth 4 1.4785[20]
8664	1,2-Pentadiene, 1-chloro-3-methyl-	C_6H_9Cl 116.59	32337-74-9	68-70[100]	3-01-00-00990 0.9562[20]	eth 4
8665	1,3-Pentadiene, 2,4-dimethyl-	C_7H_{12} 96.17	1000-86-8 -114	93.2	4-01-00-01034 0.7343[23]	1.4390[23]
8666	1,2-Pentadiene, 3-methyl- 1-Ethyl-1-methylallene	C_6H_{10} 82.15	7417-48-3	70	4-01-00-01021 0.710[25]	1.425[20]
8667	1,2-Pentadiene, 4-methyl-	C_6H_{10} 82.15	13643-05-5	70	2-01-00-00232 0.703[25]	1.424[20]
8668	1,3-Pentadiene, 2-methyl-, (Z)- cis-2-Methyl-1,3-pentadiene	C_6H_{10} 82.15	1501-60-6 -117.6	76	4-01-00-01019 0.714[25]	1.446[20]
8669	1,3-Pentadiene, 3-methyl-	C_6H_{10} 82.15	4549-74-0	77	4-01-00-01021 0.730[25]	1.452[20]
8670	1,3-Pentadiene, 4-methyl- 1,1-Dimethylbutadiene	C_6H_{10} 82.15	926-56-7	76.5	4-01-00-01020 0.7181[20]	1.4532[20]
8671	1,4-Pentadiene, 2-methyl- 2-Methyl-1,4-pentadiene	C_6H_{10} 82.15	763-30-4	56	4-01-00-01020 0.689[25]	1.405[20]
8672	2,3-Pentadiene, 2-methyl-	C_6H_{10} 82.15	3043-33-2	72	4-01-00-01020 0.706[25]	1.425[20]
8673	2,4-Pentadienenitrile	C_5H_5N 79.10	1615-70-9 -60	136.5	4-02-00-01692 0.8444[20]	1.4880
8674	2,4-Pentadienenitrile, (E)-	C_5H_5N 79.10	2180-68-9 -43	41[13]	4-02-00-01695 0.8576[20]	eth 3; ace 3 1.4986[20]
8675	2,4-Pentadienenitrile, (Z)-	C_5H_5N 79.10	2180-69-0 -64	49[32]; 32[13]	4-02-00-01695 0.8541[26]	eth 3; ace 3 1.4855[20]
8676	2,4-Pentadienoic acid	$C_5H_6O_2$ 98.10	626-99-3 79	112 dec	3-02-00-01451	H_2O 4; EtOH 4; eth 4; bz 4
8677	2,4-Pentadienoic acid, 5-(1,3-benzodioxol-5-yl)- 2,4-Pentadienoic acid, 5(3,4-metylenedioxyphenyl)	$C_{12}H_{10}O_4$ 218.21	5285-18-7 215	7437	2-19-00-00300	H_2O 1; EtOH 3; eth 2; bz 2
8678	2,4-Pentadienoic acid, 5-(1,3-benzodioxol-5-yl)-, (E,E)-	$C_{12}H_{10}O_4$ 218.21	136-72-1 215.8	sub	5-19-07-00382	EtOH 4
8679	2,4-Pentadienoic acid, 5-(1-hydroxy-2,6,6-trimethyl-4-oxo-2-cyclohexen-1-yl)-3-methyl-, [S-(Z,E)]- Abscisic acid	$C_{15}H_{20}O_4$ 264.32	21293-29-8 160	6 sub 120		ace 4; eth 4; chl 4
8680	2,4-Pentadienoic acid, 5-phenyl-, ethyl ester Ethyl cinnamylideneacetate	$C_{13}H_{14}O_2$ 202.25	1552-95-0 26	149-50[4]	2-09-00-00441 1.0467[20]	eth 4; EtOH 4 1.5768[80]
8681	1,4-Pentadien-3-ol	C_5H_8O 84.12	922-65-6	115.5	4-01-00-02229 0.860[23]	1.4400[17]
8682	1,4-Pentadien-3-one, 1,5-di-2-furanyl-	$C_{13}H_{10}O_3$ 214.22	886-77-1 60.5	181[4]	5-19-04-00459	eth 4; EtOH 4; chl 4
8683	1,4-Pentadien-3-one, 1,5-diphenyl- Dibenzalacetone	$C_{17}H_{14}O$ 234.30	538-58-9 113 dec	2988 dec	4-07-00-01747	H_2O 1; EtOH 2; eth 2; ace 3
8684	1,3-Pentadiyne Methyldiacetylene	C_5H_4 64.09	4911-55-1 -38.5	55	4-01-00-01117 0.7909[20]	H_2O 1; eth 3; bz 3; chl 3 1.4431[21]
8685	1,5-Pentalenedione, hexahydro- Bicyclo[3.3.0]octane-2,6-dione	$C_8H_{10}O_2$ 138.17	62353-69-9 45	86-8[0.2]	3-07-00-03280 1.1290[60]	1.4877[54]
8686	Pentalene, octahydro-, cis-	C_8H_{14} 110.20	1755-05-1 <-80	139	4-05-00-00277 0.8638[25]	H_2O 1; EtOH 3 1.4595[25]
8687	Pentalene, octahydro-, trans-	C_8H_{14} 110.20	5597-89-7 -30	132	4-05-00-00277 0.862[18]	H_2O 1; EtOH 3 1.4625[18]
8688	2(1H)-Pentalenone, hexahydro- Bicyclo[3.3.0]octane-2-one	$C_8H_{12}O$ 124.18	56180-61-1	72[13]	4-07-00-00156 1.0097[20]	ace 4; EtOH 4 1.4790[20]

No.	Name / Synonym	Mol. Form. / Mol. Wt.	CAS RN / mp/°C	Merck No. / bp/°C	Beil. Ref. / den/g cm^{-3}	Solubility / n_D
8689	2(1H)-Pentalenone, hexahydro-, cis- / cis-Bicyclo[3.3.0]octan-3-one	C$_8$H$_{12}$O / 124.18	19915-11-8	50[2.3]	4-07-00-00156 / 1.0660[25]	1.4766[25]
8690	Pentanal / Valeraldehyde	C$_5$H$_{10}$O / 86.13	110-62-3 / -91.5	9813 / 103	4-01-00-03268 / 0.8095[20]	H$_2$O 2; EtOH 3; eth 3 / 1.3944[20]
8691	Pentanal, 3-hydroxy-2-methyl-	C$_6$H$_{12}$O$_2$ / 116.16	615-30-5	94-6[23]	4-01-00-04022 / 0.986[25]	H$_2$O 4; ace 4; eth 4; EtOH 4 / 1.4502[20]
8692	Pentanal, 3-hydroxy-2,2,4-trimethyl-	C$_8$H$_{16}$O$_2$ / 144.21	918-79-6	118-20[14]	4-01-00-04049 / 0.9482[20]	1.4501[20]
8693	Pentanal, 2-methyl- / 2-Methylvaleraldehyde	C$_6$H$_{12}$O / 100.16	123-15-9	117	4-01-00-03304	H$_2$O 3; eth 3; ace 3; ctc 2
8694	Pentanal, 3-methyl-	C$_6$H$_{12}$O / 100.16	15877-57-3	121	3-01-00-02836 / 0.811[25]	1.4026[20]
8695	Pentanal, 2-oxo-	C$_5$H$_8$O$_2$ / 100.12	7332-93-6	112	4-01-00-03658	H$_2$O 4; eth 4; EtOH 4 / 1.4043[25]
8696	Pentanal, 4-oxo-	C$_5$H$_8$O$_2$ / 100.12	626-96-0 / <-21	187 dec	4-01-00-03659 / 1.0134[21]	H$_2$O 4; ace 4; eth 4; EtOH 4 / 1.4257[22]
8697	Pentanal, 2-propyl- / Valeraldehyde, 2-propyl-	C$_8$H$_{16}$O / 128.21	18295-59-5	160	3-01-00-02880 / 0.834[15]	1.4142[15]
8698	Pentanamide	C$_5$H$_{11}$NO / 101.15	626-97-1 / 106	225	4-02-00-00874 / 0.8735[110]	H$_2$O 4; EtOH 4; eth 4; chl 2 / 1.4183[110]
8699	Pentanamide, N,N-dimethyl-	C$_7$H$_{15}$NO / 129.20	6225-06-5 / -51	141[100]	3-04-00-00127 / 0.8962[25]	H$_2$O 4; eth 4; EtOH 4 / 1.4419[25]
8700	1-Pentanamine / Amylamine	C$_5$H$_{13}$N / 87.16	110-58-7 / -55	630 / 104.3	4-04-00-00674 / 0.7544[20]	H$_2$O 5; EtOH 5; eth 5; ace 4 / 1.448[20]
8701	2-Pentanamine	C$_5$H$_{13}$N / 87.16	625-30-9	92	4-04-00-00689 / 0.7384[20]	ace 4; bz 4; eth 4; EtOH 4 / 1.4027[20]
8702	3-Pentanamine	C$_5$H$_{13}$N / 87.16	616-24-0	89	4-04-00-00692 / 0.7487[20]	EtOH 3; chl 2 / 1.4063[20]
8703	1-Pentanamine, N,N-dimethyl-	C$_7$H$_{17}$N / 115.22	26153-88-8	123	4-04-00-00675 / 0.743[20]	eth 4 / 1.4083[20]
8704	2-Pentanamine, 2,4-dimethyl-	C$_7$H$_{17}$N / 115.22	64379-30-2	122	4-04-00-00749 / 0.7719[20]	1.4009[20]
8705	1-Pentanamine, N,N-dipentyl-	C$_{15}$H$_{33}$N / 227.43	621-77-2	242.5	4-04-00-00676 / 0.7907[20]	H$_2$O 1; EtOH 3; eth 3; acid 3 / 1.4366[20]
8706	1-Pentanamine, 3-methyl- / 3-Methylpentylamine	C$_6$H$_{15}$N / 101.19	42245-37-4	122	3-04-00-00366 / 0.767[20]	1.4196[25]
8707	1-Pentanamine, 4-methyl-	C$_6$H$_{15}$N / 101.19	5344-20-7 / -94.4	125	4-04-00-00727 / 0.758[52]	
8708	1-Pentanamine, N-pentyl- / Diamylamine	C$_{10}$H$_{23}$N / 157.30	2050-92-2	202.5	4-04-00-00676 / 0.7771[20]	H$_2$O 2; EtOH 4; eth 5; ace 3 / 1.4272[20]
8709	Pentane	C$_5$H$_{12}$ / 72.15	109-66-0 / -129.7	7072 / 36.0	4-01-00-00303 / 0.6262[20]	H$_2$O 2; EtOH 5; eth 5; ace 5 / 1.3575[20]
8710	Pentane, 1-bromo- / Amyl bromide	C$_5$H$_{11}$Br / 151.05	110-53-2 / -95	637 / 129.8	4-01-00-00312 / 1.2182[20]	H$_2$O 1; EtOH 3; eth 5; bz 3 / 1.4447[20]
8711	Pentane, 2-bromo- / 2-Pentyl bromide	C$_5$H$_{11}$Br / 151.05	107-81-3 / -95.5	117.4	4-01-00-00312 / 1.2075[20]	bz 4; eth 4; EtOH 4; chl 4 / 1.4413[20]
8712	Pentane, 3-bromo- / 3-Pentyl bromide	C$_5$H$_{11}$Br / 151.05	1809-10-5 / -126.2	118.6	4-01-00-00313 / 1.214[20]	H$_2$O 1; EtOH 3; eth 3; bz 3 / 1.4441[20]
8713	Pentane, 1-bromo-5-fluoro- / 1-Bromo-5-fluoropentane	C$_5$H$_{10}$BrF / 169.04	407-97-6	162	4-01-00-00313 / 1.3604[25]	eth 4; EtOH 4 / 1.4406[25]
8714	Pentane, 1-bromo-2-methyl- / 2-Methylpentyl bromide	C$_6$H$_{13}$Br / 165.07	25346-33-2	141	4-01-00-00361 / 1.1624[20]	eth 4; chl 4 / 1.4495[20]
8715	Pentane, 1-bromo-3-methyl- / 3-Methylpentyl bromide	C$_6$H$_{13}$Br / 165.07	51116-73-5	148	4-01-00-00365 / 1.1829[20]	H$_2$O 1; eth 3; chl 4 / 1.4496[20]
8716	Pentane, 1-bromo-4-methyl-	C$_6$H$_{13}$Br / 165.07	626-88-0	145	4-01-00-00361 / 1.1683[20]	eth 4; chl 4 / 1.4490
8717	Pentane, 2-bromo-2-methyl- / 1-Bromo-1,1-dimethylbutane	C$_6$H$_{13}$Br / 165.07	4283-80-1	142.5	4-01-00-00361	eth 4; chl 4 / 1.442[23]
8718	Pentane, 3-bromo-3-methyl- / 3-Bromo-3-methylpentane	C$_6$H$_{13}$Br / 165.07	25346-31-0	130; 76[100]	4-01-00-00365 / 1.1835[20]	eth 4; chl 4 / 1.4525[20]
8719	Pentane, 1-chloro- / Amyl chloride	C$_5$H$_{11}$Cl / 106.60	543-59-9 / -99	642 / 107.8	4-01-00-00309 / 0.8820[20]	H$_2$O 1; EtOH 5; eth 5; bz 3 / 1.4127[20]
8720	Pentane, 2-chloro-, (+)- / Pentane, 2-chloro (d)	C$_5$H$_{11}$Cl / 106.60	29882-57-3 / -137	96.9	4-01-00-00309 / 0.8698[20]	H$_2$O 1; EtOH 3; eth 3; bz 3 / 1.4069[20]
8721	Pentane, 3-chloro- / 1-Ethylpropyl chloride	C$_5$H$_{11}$Cl / 106.60	616-20-6 / -105	97.5	4-01-00-00309 / 0.8731[20]	H$_2$O 1; EtOH 3; eth 3; ace 2 / 1.4082[20]
8722	Pentane, 2-chloro-2,3-dimethyl-	C$_7$H$_{15}$Cl / 134.65	59889-45-1	38-9[20]	4-01-00-00406	bz 4; eth 4 / 1.4264[20]
8723	Pentane, 2-chloro-2,4-dimethyl-	C$_7$H$_{15}$Cl / 134.65	35951-33-8	129; 33[20]	4-01-00-00408 / 0.861[20]	eth 4 / 1.4180[20]
8724	Pentane, 3-chloro-2,3-dimethyl-	C$_7$H$_{15}$Cl / 134.65	595-38-0	136 dec; 41[20]	4-01-00-00406 / 0.882[22]	eth 4; chl 4 / 1.4318[20]
8725	Pentane, 4-chloro-2,2-dimethyl-	C$_7$H$_{15}$Cl / 134.65	33429-72-0	93[250]	4-01-00-00404 / 0.855[20]	eth 4; chl 4 / 1.4180[20]
8726	Pentane, 3-chloro-3-ethyl-	C$_7$H$_{15}$Cl / 134.65	994-25-2	143.5	4-01-00-00403 / 0.8856[20]	eth 4 / 1.4400[20]
8727	Pentane, 3-chloro-3-ethyl-2,2-dimethyl- / Pentane 3-chloro-2,2-dimethyl-3-ethyl	C$_9$H$_{19}$Cl / 162.70	86661-53-2	53[6] dec	4-01-00-00462	eth 4; chl 4 / 1.4528[25]
8728	Pentane, 1-chloro-5-fluoro-	C$_5$H$_{10}$ClF / 124.59	407-98-7	143.2	4-01-00-00309 / 1.0325[25]	eth 4; EtOH 4 / 1.4120[23]

No.	Name Synonym	Mol. Form. Mol. Wt.	CAS RN	mp/°C bp/°C	Merck No.	Beil. Ref. den/g cm^{-3}	Solubility n_D
8729	Pentane, 2-chloro-2-methyl-	C$_6$H$_{13}$Cl 120.62	4325-48-8	110 dec; 36[15]		4-01-00-00360 0.863[20]	eth 4 1.4126[20]
8730	Pentane, 2-chloro-4-methyl- 2-Chloro-4-methylpentane	C$_6$H$_{13}$Cl 120.62	25346-32-1	113		3-01-00-00399 0.8610[20]	eth 4 1.4113[20]
8731	Pentane, 3-(chloromethyl)-	C$_6$H$_{13}$Cl 120.62	4737-41-1	126; 83[202]		3-01-00-00403 0.8914[20]	bz 4; eth 4; chl 4 1.4222[20]
8732	Pentane, 3-chloro-3-methyl-	C$_6$H$_{13}$Cl 120.62	918-84-3	116		4-01-00-00365 0.8900[20]	bz 4; eth 4; chl 4 1.4210[20]
8733	Pentane, 2-chloro-2,4,4-trimethyl-	C$_8$H$_{17}$Cl 148.68	6111-88-2 -26	147 dec; 44[16]		4-01-00-00444 0.8746[20]	EtOH 4 1.4308[20]
8734	Pentanedial Glutaraldehyde	C$_5$H$_8$O$_2$ 100.12	111-30-8	188 dec	4366	4-01-00-03659	H$_2$O 5; EtOH 5; bz 3
8735	1,5-Pentanediamine Cadaverine	C$_5$H$_{14}$N$_2$ 102.18	462-94-2 9	179	1608	4-04-00-01310 0.873[25]	H$_2$O 3; EtOH 3; eth 2 1.463[20]
8736	1,4-Pentanediamine, N1,N1-diethyl- Novoldiamine	C$_9$H$_{22}$N$_2$ 158.29	140-80-7	201	6642	4-04-00-01308 0.814[20]	1.4429[20]
8737	1,5-Pentanediamine, N-(6-methoxy-8- quinolinyl)-N '-(1-methylethyl)-	C$_{18}$H$_{27}$N$_3$O 301.43	86-78-2	165-70[0.02]		5-22-12-00126	dil HCl 4 1.5785[25]
8738	1,5-Pentanediaminium, N,N,N ',N ',N '- hexamethyl-, dibromide Ammonium, pentamethylenebis trimethyl, bromide	C$_{11}$H$_{28}$Br$_2$N$_2$ 348.16	541-20-8 301		7070	4-04-00-01311	H$_2$O 2
8739	Pentane, 1,3-dibromo- 1,3-Dibromopentane	C$_5$H$_{10}$Br$_2$ 229.94	42474-20-4	193; 64[10]		4-01-00-00314 1.6721[20]	1.5048[207]
8740	Pentane, 1,4-dibromo- 1,4-Dibromopentane	C$_5$H$_{10}$Br$_2$ 229.94	626-87-9 -34.4	146[150]; 99[14]		4-01-00-00314 1.6222[20]	1.5086[20]
8741	Pentane, 1,5-dibromo-	C$_5$H$_{10}$Br$_2$ 229.94	111-24-0 -39.5	222.3		4-01-00-00314 1.7018[20]	H$_2$O 1; bz 3; ctc 2; chl 3 1.5126[20]
8742	Pentane, 2,3-dibromo- 2,3-Dibromopentane	C$_5$H$_{10}$Br$_2$ 229.94	5398-25-4 -56	178; 91[50]		3-01-00-00347 1.6865[14]	1.5087[20]
8743	Pentane, 2,4-dibromo- 2,4-Dibromopentane	C$_5$H$_{10}$Br$_2$ 229.94	19398-53-9	60[12]		3-01-00-00348 1.6659[20]	1.4987[20]
8744	Pentane, 1,2-dichloro- 1,2-Dichloropentane	C$_5$H$_{10}$Cl$_2$ 141.04	1674-33-5	148.3		3-01-00-00341 1.0872[20]	H$_2$O 1; EtOH 3; chl 4 1.4485[20]
8745	Pentane, 1,3-dichloro- 1,3-Dichloropentane	C$_5$H$_{10}$Cl$_2$ 141.04	30122-12-4	80.4[60]		3-01-00-00342 1.0834[20]	ace 4; chl 4 1.4485[20]
8746	Pentane, 1,4-dichloro- 1,4-Dichloropentane	C$_5$H$_{10}$Cl$_2$ 141.04	626-92-6	162; 88[60]		4-01-00-00309 1.0840[20]	ace 3; chl 3 1.4503[20]
8747	Pentane, 1,5-dichloro- 1,5-Dichloropentane	C$_5$H$_{10}$Cl$_2$ 141.04	628-76-2 -72.8	179		4-01-00-00310 1.1006[22]	H$_2$O 1; EtOH 3; eth 3; bz 3 1.4564[20]
8748	Pentane, 2,2-dichloro- 2,2-Dichloropentane	C$_5$H$_{10}$Cl$_2$ 141.04	34887-14-4	128.5		2-01-00-00095 1.040[20]	H$_2$O 1; eth 3; bz 3; chl 3 1.434[20]
8749	Pentane, 2,3-dichloro- 2,3-Dichloropentane	C$_5$H$_{10}$Cl$_2$ 141.04	600-11-3 -77.3	139		4-01-00-00310 1.0789[20]	1.4464[20]
8750	Pentane, 2,4-dichloro- 2,4-Dichloropentane	C$_5$H$_{10}$Cl$_2$ 141.04	625-67-2	145		3-01-00-00343 1.0634[15]	H$_2$O 1; eth 4; bz 3; chl 4 1.447[18]
8751	Pentane, 3,3-dichloro- 3,3-Dichloropentane	C$_5$H$_{10}$Cl$_2$ 141.04	21571-91-5	132; 32[14]		3-01-00-00343 1.053[20]	bz 4; chl 4 1.442[20]
8752	Pentane, 1,2-dichloro-4,4-dimethyl- 1,2-Dichloro-4,4-dimethylpentane	C$_7$H$_{14}$Cl$_2$ 169.09	6065-90-3	175; 58[12]		3-01-00-00445 1.0259[20]	bz 4; chl 4 1.4489[20]
8753	Pentane, 1,5-dichloro-3,3-dimethyl- 1,5-Dichloro-3,3-dimethylpentane	C$_7$H$_{14}$Cl$_2$ 169.09	62496-53-1	135[80]; 58[8]		4-01-00-00409 1.0563[20]	chl 4 1.4652[20]
8754	Pentane, 2,4-dichloro-2,4-dimethyl- 2,4-Dichloro-2,4-dimethylpentane	C$_7$H$_{14}$Cl$_2$ 169.09	33553-93-4 23.5	51-7[8]		4-01-00-00408 1.0292[20]	1.4537[20]
8755	Pentane, 1,1-diethoxy-	C$_9$H$_{20}$O$_2$ 160.26	3658-79-5	59[12]		4-01-00-03269 0.829[22]	1.4029[22]
8756	Pentane, 3,3-diethyl- 3,3-Diethylpentane	C$_9$H$_{20}$ 128.26	1067-20-5 -33.1	146.3		4-01-00-00462 0.7536[20]	H$_2$O 1; eth 3; bz 3; os 3 1.4206[20]
8757	Pentane, 3,3-diethyl-2-methyl- 3,3-Diethyl-2-methylpentane	C$_{10}$H$_{22}$ 142.28	52897-16-2	172		4-01-00-00487 0.7755[25]	1.4343[20]
8758	Pentane, 1,5-diiodo- Pentamethylene diiodide	C$_5$H$_{10}$I$_2$ 323.94	628-77-3 9	149[20]; 101[3]		4-01-00-00317 2.1903[15]	H$_2$O 1; eth 3; chl 3 1.6046[15]
8759	Pentane, 2,4-diiodo-	C$_5$H$_{10}$I$_2$ 323.94	66719-29-7	147[34]		0-01-00-00133 2.195[20]	1.600[20]
8760	Pentane, 2,2-dimethyl- 2,2-Dimethylpentane	C$_7$H$_{16}$ 100.20	590-35-2 -123.8	79.2		4-01-00-00403 0.6739[20]	H$_2$O 1; EtOH 3; eth 3; ace 5 1.3822[20]
8761	Pentane, 2,3-dimethyl- 2,3-Dimethylpentane	C$_7$H$_{16}$ 100.20	565-59-3	89.7		3-01-00-00445 0.6951[20]	H$_2$O 1; EtOH 3; eth 3; ace 5 1.3919[20]
8762	Pentane, 2,4-dimethyl- 2,4-Dimethylpentane	C$_7$H$_{16}$ 100.20	108-08-7 -119.9	80.4		4-01-00-00406 0.6727[20]	H$_2$O 1; EtOH 3; eth 3; ace 5 1.3815[20]
8763	Pentane, 3,3-dimethyl- 3,3-Dimethylpentane	C$_7$H$_{16}$ 100.20	562-49-2 -134.9	86.0		4-01-00-00409 0.6936[20]	H$_2$O 1; EtOH 3; eth 3; ace 5 1.3909[20]
8764	Pentane, 2,2-dimethyl-3-methylene-	C$_8$H$_{16}$ 112.22	18231-53-3	110		3-01-00-00848 0.724[25]	1.4159[20]
8765	Pentane, 2,4-dimethyl-3-methylene- 1-Butene, 2-isopropyl-3-methyl-	C$_8$H$_{16}$ 112.22	111823-35-9	104		3-01-00-00851 0.718[25]	1.4086[20]

No.	Name / Synonym	Mol. Form. / Mol. Wt.	CAS RN / mp/°C	Merck No. / bp/°C	Beil. Ref. / den/g cm^{-3}	Solubility / n_D
8766	Pentane, 2,4-dimethyl-3-(1-methylethyl)-	C$_{10}$H$_{22}$ 142.28	13475-79-1 -81.7	157.1	3-01-00-00534 0.7545[25]	1.4246[20]
8767	Pentanedinitrile / Glutaronitrile	C$_5$H$_6$N$_2$ 94.12	544-13-8 -29	4368 286	4-02-00-01941 0.9911[15]	EtOH 4; chl 4 1.4295[20]
8768	Pentanedinitrile, 2-bromo-2-(bromomethyl)- / 1,2-Dibromo-2,4-dicyanobutane	C$_6$H$_6$Br$_2$N$_2$ 265.94	35691-65-7 52	3004		H$_2$O 1; ace 4; bz 4; DMF 4
8769	Pentane, 1,5-dinitro- / 1,5-Dinitropentane	C$_5$H$_{10}$N$_2$O$_4$ 162.15	6848-84-6	134[1.2]	4-01-00-00319	H$_2$O 1; bz 3 1.461[20]
8770	Pentanedioic acid / Glutaric acid	C$_5$H$_8$O$_4$ 132.12	110-94-1 97.8	4367 303 dec	4-02-00-01934 1.429[15]	H$_2$O 4; EtOH 4; eth 4; bz 1 1.4188[106]
8771	Pentanedioic acid, 2-acetyl-, diethyl ester	C$_{11}$H$_{18}$O$_5$ 230.26	1501-06-0	271 dec	4-03-00-01835 1.0712[20]	H$_2$O 2; EtOH 3; eth 3; ctc 2 1.4420[15]
8772	Pentanedioic acid, diethyl ester	C$_9$H$_{16}$O$_4$ 188.22	818-38-2 -24.1	236.5	4-02-00-01937 1.0220[20]	eth 4 1.4241[20]
8773	Pentanedioic acid, 3,3-dimethyl-	C$_7$H$_{12}$O$_4$ 160.17	4839-46-7 103.5	126[415]; 89[2]	4-02-00-02023 1.4278[20]	H$_2$O 4; EtOH 4; eth 4; bz 2
8774	Pentanedioic acid, 3,3-dimethyl-, dimethyl ester	C$_9$H$_{16}$O$_4$ 188.22	19184-67-9	103-4[15]	3-02-00-01758 1.0366[20]	ctc 3
8775	Pentanedioic acid, dimethyl ester	C$_7$H$_{12}$O$_4$ 160.17	1119-40-0 -42.5	214; 109[21]	4-02-00-01937 1.0876[20]	EtOH 4; eth 4; chl 3 1.4242[20]
8776	Pentanedioic acid, diphenyl ester / Diphenyl glutarate	C$_{17}$H$_{16}$O$_4$ 284.31	47172-89-4 54	236.5[15]	4-06-00-00626	ace 4; EtOH 4; lig 4
8777	Pentanedioic acid, 3-ethyl-3-methyl- / 3-Ethyl-3-methylglutaric acid	C$_8$H$_{14}$O$_4$ 174.20	5345-01-7 87	261	3-02-00-01781	H$_2$O 3; EtOH 3; eth 3; bz 3
8778	Pentanedioic acid, 3-hydroxy-, diethyl ester / Diethyl 3-hydroxyglutarate	C$_9$H$_{16}$O$_5$ 204.22	32328-03-3	156-7[23]	4-03-00-01149 1.103[25]	1.4368[20]
8779	Pentanedioic acid, 3-methyl- / Glutaric acid, 3-methyl	C$_6$H$_{10}$O$_4$ 146.14	626-51-7 87	165-7[0.5]	4-02-00-01992	H$_2$O 3; EtOH 3; eth 3; bz 2
8780	Pentanedioic acid, monomethyl ester	C$_6$H$_{10}$O$_4$ 146.14	1501-27-5	150-1[10]	4-02-00-01937 1.169[25]	1.4381[20]
8781	Pentanedioic acid, 2-oxo- / α-Ketoglutaric acid	C$_5$H$_6$O$_5$ 146.10	328-50-7 115.5	5182	4-03-00-01813	H$_2$O 4; EtOH 4; eth 4; ace 3
8782	Pentanedioic acid, 3-oxo- / Acetonedicarboxylic acid	C$_5$H$_6$O$_5$ 146.10	542-05-2 138 dec	60	4-03-00-01816	H$_2$O 3; EtOH 3; eth 2; bz 1
8783	Pentanedioic acid, 3-oxo-, diethyl ester	C$_9$H$_{14}$O$_5$ 202.21	105-50-0	250	4-03-00-01817 1.113[20]	H$_2$O 2; EtOH 5
8784	Pentanedioic acid, 3-oxo-, dimethyl ester	C$_7$H$_{10}$O$_5$ 174.15	1830-54-2	150[25]	4-03-00-01817 1.185[25]	1.4434[20]
8785	1,3-Pentanediol	C$_5$H$_{12}$O$_2$ 104.15	3174-67-2	120-2[16]	4-01-00-02538 0.9863[20]	1.4659[20]
8786	1,4-Pentanediol / 4-Hydroxy-1-pentanol	C$_5$H$_{12}$O$_2$ 104.15	626-95-9	202; 125[10]	3-01-00-02191 0.9883[20]	H$_2$O 4; EtOH 4; chl 4 1.4452[23]
8787	1,5-Pentanediol / Pentamethylene glycol	C$_5$H$_{12}$O$_2$ 104.15	111-29-5 -18	7073 239	4-01-00-02540 0.9914[20]	H$_2$O 3; EtOH 3; eth 2; bz 2 1.4494[20]
8788	2,3-Pentanediol	C$_5$H$_{12}$O$_2$ 104.15	42027-23-6	187.5; 100[17]	4-01-00-02543 0.9798[19]	H$_2$O 3; EtOH 3; eth 2 1.4412[25]
8789	2,4-Pentanediol / 2,4-Amylene glycol	C$_5$H$_{12}$O$_2$ 104.15	625-69-4	199; 97-8[13]	3-01-00-02195 0.9635[20]	H$_2$O 4; EtOH 4 1.4349[20]
8790	1,2-Pentanediol, (±)- / 1,2-Pentanediol (DL)	C$_5$H$_{12}$O$_2$ 104.15	91049-43-3	211	4-01-00-02538 0.9784[20]	1.4412[19]
8791	1,5-Pentanediol, diacetate / Pentamethylene acetate	C$_9$H$_{16}$O$_4$ 188.22	6963-44-6	122-3[3]	4-02-00-00226 1.0296[20]	1.4261[19]
8792	1,3-Pentanediol, 2,2-dimethyl- / 2,2-Dimethyl-1,3-dihydroxypentane	C$_7$H$_{16}$O$_2$ 132.20	2157-31-5 62.3	213	0-01-00-00490	EtOH 4
8793	2,4-Pentanediol, 2-methyl- / Hexylene glycol	C$_6$H$_{14}$O$_2$ 118.18	107-41-5 -50	4631 197.1	4-01-00-02565 0.923[15]	H$_2$O 3; EtOH 3; eth 3; ctc 2 1.4276[20]
8794	2,4-Pentanediol, 3-methyl- / 3-Methyl-2,4-pentanediol	C$_6$H$_{14}$O$_2$ 118.18	5683-44-3	211.5	4-01-00-02572 0.9640[20]	H$_2$O 3; EtOH 3 1.4433[20]
8795	1,3-Pentanediol, 2,2,4-trimethyl- / 2,2,4-Trimethyl-1,3-pentanediol	C$_8$H$_{18}$O$_2$ 146.23	144-19-4 51.5	235; 81-2[1]	4-01-00-02604 0.936[15]	H$_2$O 2; EtOH 4; eth 4; bz 3 1.4513[15]
8796	1,4-Pentanediol, 2,2,4-trimethyl-	C$_8$H$_{18}$O$_2$ 146.23	80864-10-4 86	210	0-01-00-00493	eth 4; EtOH 4
8797	2,3-Pentanedione / Acetylpropionyl	C$_5$H$_8$O$_2$ 100.12	600-14-6	108	4-01-00-03660 0.9565[19]	H$_2$O 3; EtOH 5; eth 5; ace 5 1.4014[19]
8798	2,4-Pentanedione / Acetylacetone	C$_5$H$_8$O$_2$ 100.12	123-54-6 -23	75 138	4-01-00-03662 0.9721[25]	H$_2$O 4; EtOH 5; eth 5; ace 5 1.4494[20]
8799	2,4-Pentanedione, 3,3-dimethyl- / 3,3-Dimethylacetylacetone	C$_7$H$_{12}$O$_2$ 128.17	3142-58-3 19	173	4-01-00-03705 0.9575[20]	eth 4 1.4306[20]
8800	2,3-Pentanedione, dioxime	C$_5$H$_{10}$N$_2$O$_2$ 130.15	4775-86-4 172.5	sub	4-01-00-03661	EtOH 4
8801	2,4-Pentanedione, 3-ethyl-	C$_7$H$_{12}$O$_2$ 128.17	1540-34-7	178.5	4-01-00-03703 0.9531[19]	eth 4; EtOH 4; chl 4 1.4408[19]
8802	2,4-Pentanedione, 1,1,1,5,5,5-hexafluoro- / Acetylacetone, hexafluoro	C$_5$H$_2$F$_6$O$_2$ 208.06	1522-22-1	54.1	4-01-00-03681 1.485[20]	1.3333[20]
8803	2,3-Pentanedione, 3-oxime / 3-Oximino-2-pentanone	C$_5$H$_9$NO$_2$ 115.13	609-29-0 58.5	183	4-01-00-03661	eth 4; EtOH 4; chl 4
8804	1,4-Pentanedione, 1-phenyl-	C$_{11}$H$_{12}$O$_2$ 176.22	583-05-1	162[12]	4-07-00-02160	ace 4 1.5250[30]
8805	2,4-Pentanedione, 3-(2-propenyl)-	C$_8$H$_{12}$O$_2$ 140.18	3508-78-9	199; 92[16]	4-01-00-03759 0.974[15]	1.4698[14]

No.	Name Synonym	Mol. Form. Mol. Wt.	CAS RN mp/°C	Merck No. bp/°C	Beil. Ref. den/g cm^{-3}	Solubility n_D
8806	Pentanedioyl dichloride	$C_5H_6Cl_2O_2$ 169.01	2873-74-7	217	4-02-00-01939 1.324[20]	eth 3; chl 2 1.4728[20]
8807	Pentane, dodecafluoro- Perfluoropentane	C_5F_{12} 288.04	678-26-2 -10	29.2	4-01-00-00308	
8808	Pentane, 1-ethoxy- Ethyl pentyl ether	$C_7H_{16}O$ 116.20	17952-11-3	117.6	3-01-00-01602 0.7622[20]	eth 4; EtOH 4 1.3927[20]
8809	Pentane, 3-ethyl- 3-Ethylpentane	C_7H_{16} 100.20	617-78-7 -118.6	93.5	4-01-00-00402 0.6982[20]	H_2O 1; EtOH 3; eth 3; ace 5 1.3934[20]
8810	Pentane, 3-ethyl-2,2-dimethyl-	C_9H_{20} 128.26	16747-32-3 -99.3	133.8	4-01-00-00462 0.7438[20]	1.4123[20]
8811	Pentane, 3-ethyl-2,3-dimethyl-	C_9H_{20} 128.26	16747-33-4	144.7	4-01-00-00462 0.7508[25]	1.4221[20]
8812	Pentane, 3-ethyl-2,4-dimethyl-	C_9H_{20} 128.26	1068-87-7 -122.4	136.7	4-01-00-00462 0.7365[20]	1.4131[20]
8813	Pentane, 3-ethyl-2-methyl- 3-Ethyl-2-methylpentane	C_8H_{18} 114.23	609-26-7 -114.9	115.6	4-01-00-00437 0.7193[20]	H_2O 1; EtOH 5; eth 3; ace 5 1.4040[20]
8814	Pentane, 3-ethyl-3-methyl- 3-Ethyl-3-methylpentane	C_8H_{18} 114.23	1067-08-9 -90.9	118.2	4-01-00-00438 0.7274[20]	H_2O 1; EtOH 5; eth 3; ace 5 1.4078[20]
8815	Pentane, 3-ethyl-2,2,3-trimethyl- 3-Ethyl-2,2,3-trimethylpentane	$C_{10}H_{22}$ 142.28	52897-17-3 -42.5	169.5	4-01-00-00487 0.7780[25]	1.4420[20]
8816	Pentane, 3-ethyl-2,3,4-trimethyl- 3-Ethyl-2,3,4-trimethylpentane	$C_{10}H_{22}$ 142.28	52897-19-5	169.47	4-01-00-00487 0.7735[25]	1.4333[20]
8817	Pentane, 1-fluoro- Pentyl fluoride	$C_5H_{11}F$ 90.14	592-50-7 -120	62.8	4-01-00-00308 0.7907[20]	eth 4; EtOH 4 1.3591[2-]
8818	Pentane, 1-iodo- Amyl iodide	$C_5H_{11}I$ 198.05	628-17-1 -85.6	155	4-01-00-00315 1.5161[20]	chl 3 1.4959[20]
8819	Pentane, 2-iodo-, (±)- Pentane, 2-iodo (DL)	$C_5H_{11}I$ 198.05	52152-72-4	141	4-01-00-00316 1.5096[20]	H_2O 1; eth 3; ace 3; bz 3 1.4961[20]
8820	Pentane, 3-iodo- 3-Iodopentane	$C_5H_{11}I$ 198.05	1809-05-8	145.5	3-01-00-00349 1.5176[20]	ace 4; bz 4; eth 4 1.4974[20]
8821	Pentane, 1-isocyano- Pentyl isocyanide	$C_6H_{11}N$ 97.16	18971-59-0 -51.1	155.5	3-04-00-00331 0.806[20]	H_2O 1; EtOH 3
8822	Pentane, 1-methoxy- Methyl pentyl ether	$C_6H_{14}O$ 102.18	628-80-8	99	4-01-00-01643 0.759[22]	ace 4; eth 4; EtOH 4 1.3862[22]
8823	Pentane, 2-methyl- Isohexane	C_6H_{14} 86.18	107-83-5 -153.7	60.2	4-01-00-00358 0.650[25]	H_2O 1; EtOH 3; eth 3; ace 5 1.3715[20]
8824	Pentane, 3-methyl- 3-Methylpentane	C_6H_{14} 86.18	96-14-0 -162.9	63.2	4-01-00-00363 0.6598[25]	H_2O 1; EtOH 3; eth 5; ace 5 1.3765[20]
8825	Pentane, 3-methylene- 2-Ethyl-1-butene	C_6H_{12} 84.16	760-21-4 -131.5	64.7	4-01-00-00850 0.6894[20]	H_2O 1; eth 3; ace 3; bz 3 1.3969[20]
8826	Pentane, 1-(methylthio)- Methyl pentyl sulfide	$C_6H_{14}S$ 118.24	1741-83-9 -94	145.1	3-01-00-01608 0.8431[20]	EtOH 3; eth 3; ace 3; bz 3 1.4506[20]
8827	Pentanenitrile Valeronitrile	C_5H_9N 83.13	110-59-8 -96.2	141.3	4-02-00-00875 0.8008[20]	eth 3; ace 3; bz 3; ctc 2 1.3971[20]
8828	Pentanenitrile, 5-bromo-	C_5H_8BrN 162.03	5414-21-1	111[12]	4-02-00-00884 1.386[25]	1.4780[20]
8829	Pentanenitrile, 2-hydroxy-2-propyl-	$C_8H_{15}NO$ 141.21	5699-74-1	119-20[21]	4-03-00-00880 0.9077[18]	EtOH 4 1.4337[18]
8830	Pentanenitrile, 3-imino-2-methyl- Valeronitrile, 3-imino-2-methyl	$C_6H_{10}N_2$ 110.16	95642-49-2 46.3	257.5	0.9525[19]	eth 4; EtOH 4
8831	Pentanenitrile, 4-methyl- Isoamyl cyanide	$C_6H_{11}N$ 97.16	542-54-1 -51	4999 156.5	4-02-00-00946 0.8030[20]	H_2O 1; EtOH 3; eth 5; ctc 2 1.4059[20]
8832	Pentane, 1-nitro- 1-Nitropentane	$C_5H_{11}NO_2$ 117.15	628-05-7	172.5	4-01-00-00317 0.9525[20]	EtOH 3; eth 3; bz 3 1.4175[20]
8833	Pentane, 2-nitro- 2-Nitropentane	$C_5H_{11}NO_2$ 117.15	4609-89-6	149; 88[100]	4-01-00-00318 0.9382[25]	1.4092[25]
8834	Pentane, 3-nitro- 3-Nitropentane	$C_5H_{11}NO_2$ 117.15	551-88-2	6537 154	4-01-00-00318 0.957[0]	ace 4; eth 4; EtOH 4
8835	Pentane, 1,1'-oxybis- Amyl ether	$C_{10}H_{22}O$ 158.28	693-65-2 -69	646 190	4-01-00-01643 0.7833[20]	H_2O 1; EtOH 5; eth 5; chl 3 1.4119[20]
8836	Pentane, 2,2,3,3,4-pentamethyl- 2,2,3,3,4-Pentamethylpentane	$C_{10}H_{22}$ 142.28	16747-44-7 -36.4	166.1	4-01-00-00487 0.7767[25]	1.4361[20]
8837	Pentane, 2,2,3,4,4-pentamethyl- 2,2,3,4,4-Pentamethylpentane	$C_{10}H_{22}$ 142.28	16747-45-8 -38.7	159.3	4-01-00-00487 0.7636[25]	1.4307[20]
8838	Pentane, 1,1,1,5-tetrachloro- 1,1,1,5-Tetrachloropentane	$C_5H_8Cl_4$ 209.93	2467-10-9	112[24]	4-01-00-00311 1.3416[25]	1.4859[25]
8839	Pentane, 2,2,3,3-tetramethyl- 2,2,3,3-Tetramethylpentane	C_9H_{20} 128.26	7154-79-2 -9.8	140.2	4-01-00-00463 0.7530[25]	1.4236[20]
8840	Pentane, 2,2,3,4-tetramethyl- 2,2,3,4-Tetramethylpentane	C_9H_{20} 128.26	1186-53-4 -121.0	133.0	4-01-00-00463 0.7389[20]	1.4147[20]
8841	Pentane, 2,2,4,4-tetramethyl- Di-tert-butylmethane	C_9H_{20} 128.26	1070-87-7 -66.5	122.2	4-01-00-00464 0.7195[20]	H_2O 1; EtOH 4; bz 4; os 4 1.4069[20]
8842	Pentane, 2,3,3,4-tetramethyl- 2,3,3,4-Tetramethylpentane	C_9H_{20} 128.26	16747-38-9 -102.1	141.5	4-01-00-00464 0.7547[20]	1.4222[20]
8843	Pentane, 1,1'-thiobis-	$C_{10}H_{22}S$ 174.35	872-10-6 -51.3	86[3.7]	4-01-00-01654 0.8407[20]	H_2O 1; eth 3 1.4561[20]
8844	1-Pentanethiol Pentyl mercaptan	$C_5H_{12}S$ 104.22	110-66-7 -75.7	648 126.6	4-01-00-01653 0.850[20]	H_2O 1; EtOH 5; eth 5 1.4469[20]
8845	2-Pentanethiol sec-Amylmercaptan	$C_5H_{12}S$ 104.22	2084-19-7 -169	112.9	4-01-00-01662 0.8327[20]	EtOH 3; lig 3 1.4412[20]

No.	Name Synonym	Mol. Form. Mol. Wt.	CAS RN mp/°C	Merck No. bp/°C	Beil. Ref. den/g cm^{-3}	Solubility n_D
8846	3-Pentanethiol 3-Pentyl mercaptan	$C_5H_{12}S$ 104.22	616-31-9 -110.8	105	4-01-00-01665 0.8410[20]	EtOH 3; DMSO-d_6 2 1.4447[20]
8847	Pentane, 2,2,3-trimethyl- 2-*tert*-Butylbutane	C_8H_{18} 114.23	564-02-3 -112.2	110	4-01-00-00438 0.7161[20]	H$_2$O 1; EtOH 5; eth 5; ace 5 1.4030[20]
8848	Pentane, 2,2,4-trimethyl- Isooctane	C_8H_{18} 114.23	540-84-1 -107.3	5079 99.2	4-01-00-00439 0.6877[25]	H$_2$O 1; EtOH 3; eth 5; ace 5 1.3915[20]
8849	Pentane, 2,3,3-trimethyl- 2,3,3-Trimethylpentane	C_8H_{18} 114.23	560-21-4 -100.9	114.8	4-01-00-00445 0.7262[20]	H$_2$O 1; EtOH 4; eth 5; ace 5 1.4075[20]
8850	Pentane, 2,3,4-trimethyl- 2,3,4-Trimethylpentane	C_8H_{18} 114.23	565-75-3 -109.2	113.5	4-01-00-00446 0.7191[20]	H$_2$O 1; EtOH 4; eth 5; ace 5 1.4042[20]
8851	1,2,5-Pentanetriol 1,4,5-Pentanetriol	$C_5H_{12}O_3$ 120.15	14697-46-2 	190-1[13]	4-01-00-02779 1.136[20]	H$_2$O 4; EtOH 4 1.4730[20]
8852	1,3,5-Pentanetriol	$C_5H_{12}O_3$ 120.15	4328-94-3 	188[11]; 139[0.1]	4-01-00-02779 1.1291[20]	H$_2$O 4; EtOH 4; eth 3; ace 3 1.4785[20]
8853	Pentane, 1,1,1,2,2,3,4,4,5,5,5-undecafluoro-3- (trifluoromethyl) Perfluoro-3-methylpentane	C_6F_{14} 338.04	865-71-4 -115	58.4	 	bz 3
8854	Pentane, 1,1,1,2,2,3,3,4,5,5,5-undecafluoro-4- (trifluoromethyl)- Perfluoro(2-methylpentane)	C_6F_{14} 338.04	355-04-4 	57.6	4-01-00-00360 1.7326[20]	H$_2$O 1; bz 3 1.2564[22]
8855	Pentanoic acid Valeric acid	$C_5H_{10}O_2$ 102.13	109-52-4 -34	9815 186.1	4-02-00-00868 0.9391[20]	H$_2$O 3; EtOH 3; eth 3; ctc 2 1.4085[20]
8856	Pentanoic acid, 2-acetyl-, ethyl ester	$C_9H_{16}O_3$ 172.22	1540-28-9 	224; 90[15]	4-03-00-01602 0.9661[20]	eth 4; EtOH 4 1.4255[20]
8857	Pentanoic acid, 5-amino-	$C_5H_{11}NO_2$ 117.15	660-88-8 157-8 dec	dec	4-04-00-02636 	H$_2$O 3; EtOH 2; eth 1; bz 1
8858	Pentanoic acid, anhydride	$C_{10}H_{18}O_3$ 186.25	2082-59-9 -56.1	227	4-02-00-00874 0.924[20]	eth 4; EtOH 4 1.4171[26]
8859	Pentanoic acid, 2-bromo-, (±)-	$C_5H_9BrO_2$ 181.03	2681-92-7 	118[12]	4-02-00-00883 1.381[20]	eth 4; EtOH 4
8860	Pentanoic acid, 2-bromo-, ethyl ester	$C_7H_{13}BrO_2$ 209.08	615-83-8 	191	3-02-00-00681 1.226[18]	H$_2$O 1; EtOH 3; eth 3 1.4496[20]
8861	Pentanoic acid, 5-bromo-, ethyl ester	$C_7H_{13}BrO_2$ 209.08	14660-52-7 	107-8[20]	4-02-00-00884 1.2804[20]	ctc 2 1.4543[20]
8862	Pentanoic acid, butyl ester	$C_9H_{18}O_2$ 158.24	591-68-4 -92.8	185.8	3-02-00-00671 0.8710[15]	H$_2$O 2; EtOH 3; eth 3 1.4128[20]
8863	Pentanoic acid, 2-chloro-, (±)-	$C_5H_9ClO_2$ 136.58	94347-45-2 -15	222	4-02-00-00878 1.141[13]	eth 4; EtOH 4 1.4481[11]
8864	Pentanoic acid, 3-chloro- Valeric acid, 3-chloro	$C_5H_9ClO_2$ 136.58	51637-47-9 33	112[10]	4-02-00-00879 1.1484[20]	eth 4; EtOH 4 1.4462[20]
8865	Pentanoic acid, 4-chloro- 4-Chlorovaleric acid	$C_5H_9ClO_2$ 136.58	32607-54-8 	116[10]	4-02-00-00879 1.1514[20]	eth 4; EtOH 4 1.4458[20]
8866	Pentanoic acid, 5-chloro-	$C_5H_9ClO_2$ 136.58	1119-46-6 18	230	4-02-00-00880 1.3416[25]	eth 4; EtOH 4 1.4555[20]
8867	Pentanoic acid, 4-chloro-, ethyl ester Ethyl 4-chlorovalerate	$C_7H_{13}ClO_2$ 164.63	70786-82-2 	196	4-02-00-00879 1.0393[20]	eth 4; EtOH 4 1.4310[20]
8868	Pentanoic acid, 5-chloro-, ethyl ester	$C_7H_{13}ClO_2$ 164.63	2323-81-1 	205.5	4-02-00-00880 1.0561[20]	H$_2$O 1; EtOH 3; eth 3; chl 2 1.4355[20]
8869	Pentanoic acid, 2-chloroethyl ester	$C_7H_{13}ClO_2$ 164.63	7735-33-3 	185	3-02-00-00670 1.040[12]	eth 4 1.4307[11]
8870	Pentanoic acid, 2,3-dibromo- Valeric acid, 2,3-dibromo-	$C_5H_8Br_2O_2$ 259.93	79912-57-5 57	104-6[13]	3-02-00-00683 	bz 4; CS$_2$ 4
8871	Pentanoic acid, 2,5-dibromo- Valeric acid, 2,5-dibromo-	$C_5H_8Br_2O_2$ 259.93	1450-81-3 	173[15]; 151[5]	3-02-00-00683 1.8629[25]	1.5347[25]
8872	Pentanoic acid, 2,3-dibromo-, ethyl ester Valeric acid, 2,3-dibromo-, ethyl ester	$C_7H_{12}Br_2O_2$ 287.98	79912-55-3 	117[14]	2-02-00-00269 1.6199[15]	1.4953[15]
8873	Pentanoic acid, 2,5-dibromo-, ethyl ester Valeric acid, 2,5-dibromo-, ethyl ester	$C_7H_{12}Br_2O_2$ 287.98	29823-16-3 	133-5[14]	3-02-00-00683 1.6289[25]	1.4947[25]
8874	Pentanoic acid, 2,2-dimethyl-4-oxo-	$C_7H_{12}O_3$ 144.17	470-49-5 76.8	138[15]	3-03-00-01247 	H$_2$O 4; bz 4; eth 4; EtOH 4
8875	Pentanoic acid, 2,4-dioxo-	$C_5H_8O_4$ 130.10	5699-58-1 99.5	130[37]	3-03-00-01331 	H$_2$O 3; EtOH 3; eth 3; ace 3
8876	Pentanoic acid, 2,4-dioxo-, ethyl ester	$C_7H_{10}O_4$ 158.15	615-79-2 18	214	4-03-00-01777 1.1251[20]	eth 4; EtOH 4 1.4757[17]
8877	Pentanoic acid, 2-ethyl-, (±)- Valeric acid, 2-ethyl (DL)	$C_7H_{14}O_2$ 130.19	116908-83-9 	209.2	4-02-00-00975 0.9311[33]	eth 4; EtOH 4
8878	Pentanoic acid, ethyl ester Ethyl valerate	$C_7H_{14}O_2$ 130.19	539-82-2 -91.2	146.1	4-02-00-00872 0.8770[20]	H$_2$O 1; EtOH 5; eth 5; ctc 2 1.4120[20]
8879	Pentanoic acid, 5-fluoro- Valeric acid, 5-fluoro-	$C_5H_9FO_2$ 120.12	407-75-0 	83[2]	4-02-00-00876 	eth 4; EtOH 4 1.4080[25]
8880	Pentanoic acid, 2-furanylmethyl ester Furfuryl valerate	$C_{10}H_{14}O_3$ 182.22	36701-01-6 	228; 82[1]	2-17-00-00115 1.0284[20]	eth 4; EtOH 4
8881	Pentanoic acid, heptyl ester	$C_{12}H_{24}O_2$ 200.32	5451-80-9 -46.4	245.2	4-02-00-00872 0.8623[20]	ace 4; eth 4; EtOH 4 1.4254[15]
8882	Pentanoic acid, hexyl ester	$C_{11}H_{22}O_2$ 186.29	1117-59-5 -63.1	226.3	3-02-00-00671 0.8635[20]	ace 4; eth 4; EtOH 4 1.4228[15]
8883	Pentanoic acid, 2-hydroxy-	$C_5H_{10}O_3$ 118.13	617-31-2 34	sub	4-03-00-00807 	H$_2$O 3; EtOH 3; eth 3
8884	Pentanoic acid, 4-hydroxy-, ethyl ester Valeric acid, 4-hydroxy-, ethyl ester	$C_7H_{14}O_3$ 146.19	6149-46-8 	85-6[2]	3-03-00-00612 0.9504[25]	eth 4; EtOH 4 1.4265[25]

No.	Name / Synonym	Mol. Form. / Mol. Wt.	CAS RN / mp/°C	Merck No. / bp/°C	Beil. Ref. / den/g cm^{-3}	Solubility / n_D
8885	Pentanoic acid, 2-hydroxyethyl ester / Valeric acid, 2-hydroxyethyl ester	C$_7$H$_{14}$O$_3$ / 146.19	16179-36-5 /	190	3-02-00-00672 / 0.9839[20]	1.4220[20]
8886	Pentanoic acid, 2-methoxyphenyl ester / Guaiacol valerate	C$_{12}$H$_{16}$O$_3$ / 208.26	531-39-5 /	4461 / 265	1.05[25]	bz 4; eth 4; EtOH 4
8887	Pentanoic acid, 2-methyl-, (±)-	C$_6$H$_{12}$O$_2$ / 116.16	22160-39-0 /	195.6	4-02-00-00942 / 0.9230[20]	H$_2$O 3; EtOH 3; eth 3; ctc 2 / 1.413[20]
8888	Pentanoic acid, 3-methyl-, (±)- / Valeric acid, 3-methyl (DL)	C$_6$H$_{12}$O$_2$ / 116.16	22160-40-3 / -41.6	197.5	4-02-00-00948 / 0.9262[20]	eth 4; EtOH 4 / 1.4159[20]
8889	Pentanoic acid, 4-methyl-	C$_6$H$_{12}$O$_2$ / 116.16	646-07-1 / -33	200.5	4-02-00-00944 / 0.9225[20]	H$_2$O 2; EtOH 3; eth 3; chl 3 / 1.4144[20]
8890	Pentanoic acid, 2-methylene- / Valeric acid, 2-methylene	C$_6$H$_{10}$O$_2$ / 114.14	5650-75-9 / 24.4	214	0-02-00-00437 / 0.9783[25]	bz 4; eth 4; chl 4
8891	Pentanoic acid, methyl ester / Methyl valerate	C$_6$H$_{12}$O$_2$ / 116.16	624-24-8 /	127.4	4-02-00-00871 / 0.8947[20]	H$_2$O 2; EtOH 5; eth 5; ace 3 / 1.4003[20]
8892	Pentanoic acid, 4-methyl-, ethyl ester	C$_8$H$_{16}$O$_2$ / 144.21	25415-67-2 /	163; 52[10]	4-02-00-00945 / 0.8705[20]	1.4050[20]
8893	Pentanoic acid, 1-methylethyl ester	C$_8$H$_{16}$O$_2$ / 144.21	18362-97-5 /		3-02-00-00671 / 0.8579[20]	H$_2$O 1; EtOH 3; eth 3; ace 3 / 1.4061[20]
8894	Pentanoic acid, 4-methyl-, 2-methylpropyl ester	C$_{10}$H$_{20}$O$_2$ / 172.27	25415-70-7 /	172; 76[10]	4-02-00-00945 / 0.8536[20]	1.4135[20]
8895	Pentanoic acid, 2-methyl-4-oxo-	C$_6$H$_{10}$O$_3$ / 130.14	6641-83-4 /	165[40]	4-03-00-01587 / 1.11[25]	1.4431[20]
8896	Pentanoic acid, 3-methyl-4-oxo-	C$_6$H$_{10}$O$_3$ / 130.14	6628-79-1 / 31.5	241.5	3-03-00-01238 / 1.0932[17]	H$_2$O 4; EtOH 4 / 1.4443[17]
8897	Pentanoic acid, 4-methyl-3-oxo-, ethyl ester	C$_8$H$_{14}$O$_3$ / 158.20	7152-15-0 / -9	173	4-03-00-01589 / 0.98[25]	1.250[20]
8898	Pentanoic acid, 3-methyl-4-oxo-, methyl ester / Levulinic acid, 3-methyl-, methyl ester	C$_7$H$_{12}$O$_3$ / 144.17	25234-83-7 /	200; 82[10]	1-03-00-00240 / 1.022[20]	1.4052[73]
8899	Pentanoic acid, 1-methylpropyl ester, (+)- / Valeric acid, sec-butyl ester (d)	C$_9$H$_{18}$O$_2$ / 158.24	116836-32-9 /	174.5	3-02-00-00671 / 0.8605[20]	bz 4; eth 4; py 4; EtOH 4 / 1.4070[20]
8900	Pentanoic acid, 2-methylpropyl ester	C$_9$H$_{18}$O$_2$ / 158.24	10588-10-0 /	179	3-02-00-00671 / 0.8625[25]	H$_2$O 1; EtOH 5; eth 3; ace 3 / 1.4046[20]
8901	Pentanoic acid, octyl ester / Valeric acid, octyl ester	C$_{13}$H$_{26}$O$_2$ / 214.35	5451-85-4 / -42.3	261.6	3-02-00-00672 / 0.8615[20]	ace 4; eth 4; EtOH 4 / 1.4273[15]
8902	Pentanoic acid, 2-oxo-	C$_5$H$_8$O$_3$ / 116.12	1821-02-9 / 6.5	179	4-03-00-01558 / 1.0970[14]	H$_2$O 2; eth 3; bz 3; chl 3
8903	Pentanoic acid, 4-oxo- / Levulinic acid	C$_5$H$_8$O$_3$ / 116.12	123-76-2 / 33	5352 / 245 dec	4-03-00-01560 / 1.1335[20]	H$_2$O 4; EtOH 4; eth 4; chl 3 / 1.4396[20]
8904	Pentanoic acid, 4-oxo-, butyl ester	C$_9$H$_{16}$O$_3$ / 172.22	2052-15-5 /	237.5	4-03-00-01563 / 0.9735[20]	chl 2 / 1.4290[20]
8905	Pentanoic acid, 2-oxo-, ethyl ester / Valeric acid, 2-oxo-, ethyl ester	C$_7$H$_{12}$O$_3$ / 144.17	50461-74-0 /	182.5; 70[11]	4-03-00-01558 / 0.9985[18]	1.4170[18]
8906	Pentanoic acid, 3-oxo-, ethyl ester	C$_7$H$_{12}$O$_3$ / 144.17	4949-44-4 /	191	4-03-00-01559 / 1.0120[20]	bz 4; eth 4; EtOH 4 / 1.4230[20]
8907	Pentanoic acid, 4-oxo-, ethyl ester / Ethyl levulinate	C$_7$H$_{12}$O$_3$ / 144.17	539-88-8 /	3775 / 205.8	4-03-00-01562 / 1.0111[20]	H$_2$O 4; EtOH 4 / 1.4229[20]
8908	Pentanoic acid, 4-oxo-, methyl ester	C$_6$H$_{10}$O$_3$ / 130.14	624-45-3 /	196	4-03-00-01562 / 1.0511[20]	H$_2$O 2; EtOH 3; eth 5; ace 3 / 1.4233[20]
8909	Pentanoic acid, 4-oxo-, 1-methylethyl ester / Levulinic acid, isopropyl ester	C$_8$H$_{14}$O$_3$ / 158.20	21884-26-4 /	209.3	4-03-00-01563 / 0.9842[20]	ace 4; bz 4; eth 4; EtOH 4 / 1.4420[20]
8910	Pentanoic acid, 4-oxo-, 2-methylpropyl ester	C$_9$H$_{16}$O$_3$ / 172.22	3757-32-2 /	228	4-03-00-01563 / 0.9669[20]	ace 4; eth 4; EtOH 4 / 1.4249[20]
8911	Pentanoic acid, 4-oxo-, phenylmethyl ester / Levulinic acid, benzyl ester	C$_{12}$H$_{14}$O$_3$ / 206.24	6939-75-9 /	132-4[2]	4-06-00-02481 / 1.0935[20]	tol 4 / 1.5090[20]
8912	Pentanoic acid, 4-oxo-, propyl ester / Levulinic acid, propyl ester	C$_8$H$_{14}$O$_3$ / 158.20	645-67-0 /	218	4-03-00-01563 / 0.9896[20]	H$_2$O 2; EtOH 3; eth 3; ace 3 / 1.4258[20]
8913	Pentanoic acid, pentyl ester	C$_{10}$H$_{20}$O$_2$ / 172.27	2173-56-0 / -78.8	203.7	4-02-00-00872 / 0.8638[20]	H$_2$O 2; EtOH 5; eth 5 / 1.4164[20]
8914	Pentanoic acid, propyl ester	C$_8$H$_{16}$O$_2$ / 144.21	141-06-0 / -70.7	167.5	4-02-00-00872 / 0.8699[20]	H$_2$O 1; EtOH 3; eth 3; chl 3 / 1.4065[20]
8915	1-Pentanol / Amyl alcohol	C$_5$H$_{12}$O / 88.15	71-41-0 / -78.9	7074 / 137.9	4-01-00-01640 / 0.8144[20]	H$_2$O 1; EtOH 5; eth 5; ace 3 / 1.4101[20]
8916	2-Pentanol / sec-Amyl alcohol	C$_5$H$_{12}$O / 88.15	6032-29-7 / -73	7075 / 119.3	3-01-00-01609 / 0.8094[20]	H$_2$O 4; EtOH 3; eth 3; ctc 3 / 1.4053[20]
8917	3-Pentanol / Diethyl carbinol	C$_5$H$_{12}$O / 88.15	584-02-1 / -69	7076 / 116.2	4-01-00-01662 / 0.8203[20]	H$_2$O 2; EtOH 3; eth 3; ace 3 / 1.4104[20]
8918	2-Pentanol, acetate, (R)- / sec-Amyl acetate (R)	C$_7$H$_{14}$O$_2$ / 130.19	54638-10-7 /	142	0-02-00-00132 / 0.8803[18]	eth 4; EtOH 4 / 1.4012[20]
8919	3-Pentanol, acetate	C$_7$H$_{14}$O$_2$ / 130.19	620-11-1 /	133	4-02-00-00155 / 0.8712[20]	eth 4; EtOH 4 / 1.4005[20]
8920	1-Pentanol, 5-amino-	C$_5$H$_{13}$NO / 103.16	2508-29-4 / 38.5	221.5	4-04-00-01750 / 0.9488[17]	H$_2$O 5; EtOH 5; ace 5 / 1.4618[17]
8921	3-Pentanol, 2-amino-	C$_5$H$_{13}$NO / 103.16	116836-16-9 /	171	1-04-00-00442 / 0.9289[23]	1.4458
8922	1-Pentanol, 2-amino-4-methyl-, (±)-	C$_6$H$_{15}$NO / 117.19	16369-17-8 /	199	2-04-00-00748 / 0.9173[13]	H$_2$O 4; EtOH 4
8923	1-Pentanol, 5-chloro-	C$_5$H$_{11}$ClO / 122.59	5259-98-3 /	112[12]	4-01-00-01650 /	eth 4; EtOH 4 / 1.4518[20]
8924	2-Pentanol, 1-chloro-	C$_5$H$_{11}$ClO / 122.59	17658-32-1 /	160; 60[14]	3-01-00-01613 / 1.035[20]	eth 4; EtOH 4 / 1.4404[25]

No.	Name / Synonym	Mol. Form. / Mol. Wt.	CAS RN / mp/°C	Merck No. / bp/°C	Beil. Ref. / den/g cm^{-3}	Solubility / n_D
8925	3-Pentanol, 1-chloro- / 1-Chloro-3-hydroxypentane	$C_5H_{11}ClO$ / 122.59	32541-33-6 /	/ 173	4-01-00-01665 / 1.0327[25]	eth 4; EtOH 4 / 1.448[25]
8926	1-Pentanol, 2-(diethylamino)-4-methyl-, 4-aminobenzoate (ester), monomethanesulfonate (salt) / Panthesin	$C_{18}H_{32}N_2O_5S$ / 388.53	135-44-4 / 158	5332 /	4-14-00-01143 /	H_2O 4; EtOH 4
8927	1-Pentanol, 2,4-dimethyl-, (±)- / 1-Pentanol, 2,4-dimethyl (DL)	$C_7H_{16}O$ / 116.20	111768-02-6 /	/ 159	4-01-00-01753 / 0.793[20]	eth 4; EtOH 4 / 1.427[20]
8928	2-Pentanol, 2,4-dimethyl- / Isobutyldimethylcarbinol	$C_7H_{16}O$ / 116.20	625-06-9 / <-20	/ 133.1	4-01-00-01753 / 0.8103[20]	H_2O 1; EtOH 3; eth 3; ctc 3 / 1.4172[20]
8929	3-Pentanol, 2,2-dimethyl- / 2,2-Dimethyl-3-pentanol	$C_7H_{16}O$ / 116.20	3970-62-5 / -5	/ 135	3-01-00-01697 / 0.8253[20]	H_2O 1; EtOH 3; eth 3 / 1.4223[20]
8930	3-Pentanol, 2,3-dimethyl- / 2,3-Dimethyl-3-pentanol	$C_7H_{16}O$ / 116.20	595-41-5 / <-30	/ 139.7	4-01-00-01752 / 0.833[20]	H_2O 1; EtOH 3; eth 3; bz 2 / 1.4287[20]
8931	3-Pentanol, 2,4-dimethyl-	$C_7H_{16}O$ / 116.20	600-36-2 / <-70	/ 138.7	4-01-00-01754 / 0.8288[20]	H_2O 2; EtOH 3; eth 3 / 1.4250[20]
8932	3-Pentanol, 3-ethyl-	$C_7H_{16}O$ / 116.20	597-49-9 / -12.5	/ 142	4-01-00-01750 / 0.8407[22]	H_2O 1; EtOH 3; eth 3 / 1.4294[20]
8933	2-Pentanol, 3-ethyl-2-methyl-	$C_8H_{18}O$ / 130.23	19780-63-3 /	/ 156	4-01-00-01794 / 0.8382[20]	bz 4; eth 4; EtOH 4; lig 4 / 1.4325[20]
8934	3-Pentanol, 3-ethyl-2-methyl- / Diethylisopropylcarbinol	$C_8H_{18}O$ / 130.23	597-05-7 /	160; 56[48] /	4-01-00-01794 / 0.8280[20]	H_2O 1; EtOH 3; eth 3 / 1.4372[10]
8935	1-Pentanol, 5-fluoro-	$C_5H_{11}FO$ / 106.14	592-80-3 /	70-1[11] /	4-01-00-01647 /	eth 4; EtOH 4 / 1.4057[25]
8936	2-Pentanol, 1,1'-iminobis[2-methyl- / Dipentylamine, 2,2'-dimethyl,2,2'-dihydroxy	$C_{12}H_{27}NO_2$ / 217.35	85733-97-7 /	165-70[15] /	0.9264[20]	ace 4; eth 4; EtOH 4 / 1.4584[20]
8937	1-Pentanol, 2-methyl- / 2-Methyl-1-pentanol	$C_6H_{14}O$ / 102.18	105-30-6 /	/ 149	3-01-00-01665 / 0.8263[20]	EtOH 3; eth 3; ace 3; ctc 3 / 1.4182[20]
8938	1-Pentanol, 3-methyl-, (±)- / 1-Pentanol, 3-methyl (DL)	$C_6H_{14}O$ / 102.18	20281-83-8 /	/ 153	4-01-00-01722 / 0.8242[20]	H_2O 1; EtOH 3; eth 3 / 1.4112[23]
8939	1-Pentanol, 4-methyl- / Isohexyl alcohol	$C_6H_{14}O$ / 102.18	626-89-1 /	/ 151.9	4-01-00-01721 / 0.8131[20]	H_2O 1; EtOH 3; eth 3 / 1.4134[25]
8940	2-Pentanol, 2-methyl- / 2-Methyl-2-pentanol	$C_6H_{14}O$ / 102.18	590-36-3 / -103	/ 121.1	4-01-00-01714 / 0.8350[16]	H_2O 2; EtOH 3; eth 3 / 1.4100[20]
8941	2-Pentanol, 3-methyl- / 3-Methyl-2-pentanol	$C_6H_{14}O$ / 102.18	565-60-6 /	/ 134.3	3-01-00-01672 / 0.8307[20]	H_2O 2; EtOH 3; eth 3 / 1.4182[20]
8942	2-Pentanol, 4-methyl- / 4-Methyl-2-pentanol	$C_6H_{14}O$ / 102.18	108-11-2 / -90	/ 131.6	4-01-00-01717 / 0.8075[20]	H_2O 2; EtOH 3; eth 3; ctc 2 / 1.4100[20]
8943	3-Pentanol, 2-methyl- / 2-Methyl-3-pentanol	$C_6H_{14}O$ / 102.18	565-67-3 /	/ 126.5	4-01-00-01716 / 0.8243[20]	H_2O 2; EtOH 5; eth 5 / 1.4175[20]
8944	3-Pentanol, 3-methyl-	$C_6H_{14}O$ / 102.18	77-74-7 / -23.6	/ 122.4	4-01-00-01723 / 0.8286[20]	H_2O 2; EtOH 5; eth 5; ctc 2 / 1.4186[20]
8945	1-Pentanol, 2-methyl-, acetate / Acetic acid, (2-methylpentyl) ester	$C_8H_{16}O_2$ / 144.21	7789-99-3 /	/ 163	4-02-00-00160 / 0.869[25]	eth 4; EtOH 4
8946	2-Pentanol, 2-methyl-, acetate / 1,1-Dimethylbutyl acetate	$C_8H_{16}O_2$ / 144.21	34859-98-8 /	153; 34[10] /	3-02-00-00257 / 0.8798[18]	eth 4; EtOH 4 / 1.4068[10]
8947	2-Pentanol, 4-methyl-, acetate / 1,3-Dimethylbutyl acetate	$C_8H_{16}O_2$ / 144.21	108-84-9 /	/ 147.5	4-02-00-00161 / 0.8805[25]	eth 4; EtOH 4 / 1.3980[20]
8948	3-Pentanol, 2-methyl-, acetate / Acetic acid, (2-methyl-3-pentyl) ester	$C_8H_{16}O_2$ / 144.21	35897-16-6 /	/ 149	4-02-00-00160 / 0.8688[20]	eth 4; EtOH 4
8949	3-Pentanol, 3-methyl-, acetate / Acetic acid, (3-methyl-3-pentyl) ester	$C_8H_{16}O_2$ / 144.21	10250-47-2 /	/ 148	4-02-00-00161 / 0.8818[20]	eth 4; EtOH 4 / 1.4169[18]
8950	1-Pentanol, 2-nitro- / 2-Nitro-1-pentanol	$C_5H_{11}NO_3$ / 133.15	2899-90-3 /	133[28]; 117[10] /	3-01-00-01606 / 1.0818[25]	/ 1.4405[25]
8951	2-Pentanol, 1-nitro-	$C_5H_{11}NO_3$ / 133.15	2224-37-5 /	87-8[1] /	4-01-00-01661 / 1.0968[20]	/ 1.4493[20]
8952	2-Pentanol, 3-nitro- / 3-Nitro-2-pentanol	$C_5H_{11}NO_3$ / 133.15	5447-99-4 /	60[0.5] /	3-01-00-01614 / 1.075[25]	/ 1.4430[20]
8953	1-Pentanol, 2,2,3,3,4,4,5,5-octafluoro- / 2,2,3,3,4,4,5,5-Octafluoro-1-pentanol	$C_5H_4F_8O$ / 232.07	355-80-6 /	6667 / 140.5	4-01-00-01648 / 1.6647[20]	/ 1.3178[20]
8954	2-Pentanol, 2,4,4-trimethyl- / 2,4,4-Trimethyl-2-pentanol	$C_8H_{18}O$ / 130.23	690-37-9 / -20	/ 147.5	4-01-00-01796 / 0.8225[20]	H_2O 1; EtOH 2; eth 3 / 1.4284[20]
8955	3-Pentanol, 2,2,4-trimethyl- / Isopropyl-tert-butylcarbinol	$C_8H_{18}O$ / 130.23	5162-48-1 / -13	/ 150.5	4-01-00-01795 / 0.8297[20]	/ 1.4288[20]
8956	3-Pentanol, 2,3,4-trimethyl-	$C_8H_{18}O$ / 130.23	3054-92-0 /	/ 156.5	4-01-00-01798 / 0.8492[20]	/ 1.4353[20]
8957	2-Pentanone / Methyl propyl ketone	$C_5H_{10}O$ / 86.13	107-87-9 / -76.9	6032 / 102.2	4-01-00-03271 / 0.809[20]	H_2O 2; EtOH 5; eth 5; ctc 2 / 1.3895[20]
8958	3-Pentanone / Diethyl ketone	$C_5H_{10}O$ / 86.13	96-22-0 / -39	3111 / 101.96	4-01-00-03279 / 0.8098[25]	H_2O 4; EtOH 5; eth 5; ctc 3 / 1.3905[25]
8959	2-Pentanone, 4-amino-4-methyl- / Diacetonamine	$C_6H_{13}NO$ / 115.18	625-04-7 /	2942 / 25[0.14]	3-04-00-00894 /	H_2O 3; EtOH 5; eth 5
8960	1-Pentanone, 1-(4-aminophenyl)-	$C_{11}H_{15}NO$ / 177.25	38237-74-0 / 74.5	/ 160-3[3]	4-14-00-00178 /	H_2O 1; EtOH 3; eth 3
8961	2-Pentanone, 5-chloro-	C_5H_9ClO / 120.58	5891-21-4 /	76[34] /	4-01-00-03277 / 1.0523[20]	eth 3; ace 3; ctc 2 / 1.4375[20]
8962	3-Pentanone, 1-chloro-	C_5H_9ClO / 120.58	32830-97-0 /	68[20] /	4-01-00-03284 /	eth 4; EtOH 4 / 1.4361[20]
8963	2-Pentanone, 3,5-dichloro- / 3,5-Dichloro-2-pentanone	$C_5H_8Cl_2O$ / 155.02	58371-98-5 /	99[15]; 58[3] /	4-01-00-03277 / 1.239[15]	/ 1.4632[20]

No.	Name / Synonym	Mol. Form. / Mol. Wt.	CAS RN / mp/°C	Merck No. / bp/°C	Beil. Ref. / den/g cm^{-3}	Solubility / n_D
8964	2-Pentanone, 5-(diethylamino)-	$C_9H_{19}NO$ 157.26	105-14-6	83-5[15]	4-04-00-01935 0.861[25]	1.4350[20]
8965	2-Pentanone, 4,4-dimethyl-	$C_7H_{14}O$ 114.19	590-50-1 -64	126	4-01-00-03331 0.809[25]	1.4036[20]
8966	3-Pentanone, 2,2-dimethyl- 2,2-Dimethyl-3-pentanone	$C_7H_{14}O$ 114.19	564-04-5 -45	125.6	4-01-00-03331 0.8125[20]	H_2O 2; EtOH 3; eth 3; ace 3 1.4065[20]
8967	3-Pentanone, 2,4-dimethyl- Diisopropyl ketone	$C_7H_{14}O$ 114.19	565-80-0 -69	125.4	4-01-00-03334 0.8108[20]	H_2O 1; EtOH 5; eth 5; bz 3 1.3999[20]
8968	3-Pentanone, 2,4-dimethyl-, oxime Diisopropyl ketone oxime	$C_7H_{15}NO$ 129.20	1113-74-2 34	181; 91.5[21]	4-01-00-03335 0.9022[17]	
8969	3-Pentanone, 1,5-diphenyl- Diphenethyl ketone	$C_{17}H_{18}O$ 238.33	5396-91-8 13.5	352	4-07-00-01462 1.0356[0]	
8970	2-Pentanone, 3-ethyl-4-methyl-	$C_8H_{16}O$ 128.21	71172-57-1	154.5	3-01-00-02883 0.812[20]	H_2O 3; EtOH 5; eth 5; bz 5 1.4105[20]
8971	2-Pentanone, 1-hydroxy- 1-Pentanol-2-one	$C_5H_{10}O_2$ 102.13	64502-89-2	152	4-01-00-04004 0.9860[20]	H_2O 4; eth 4; EtOH 4 1.4234[12]
8972	2-Pentanone, 3-hydroxy-	$C_5H_{10}O_2$ 102.13	3142-66-3	147.5	3-01-00-03220 0.9500[20]	H_2O 4; ace 4; bz 4; eth 4 1.4350[10]
8973	2-Pentanone, 4-hydroxy-	$C_5H_{10}O_2$ 102.13	4161-60-8	177	4-01-00-04005 1.0071[20]	H_2O 3; EtOH 3; eth 3 1.4415[18]
8974	2-Pentanone, 5-hydroxy-	$C_5H_{10}O_2$ 102.13	1071-73-4	209; 117[33]	4-01-00-04006 1.0071[20]	H_2O 5; EtOH 3; eth 3 1.4390[20]
8975	3-Pentanone, 2-hydroxy- 2-Pentanol-3-one	$C_5H_{10}O_2$ 102.13	5704-20-1	152.5	4-01-00-04008 0.9742[20]	eth 4; EtOH 4 1.4128[20]
8976	2-Pentanone, 4-hydroxy-4-methyl- Diacetone alcohol	$C_6H_{12}O_2$ 116.16	123-42-2 -44	2944 167.9	4-01-00-04023 0.9387[20]	H_2O 5; EtOH 5; eth 5; chl 3 1.4213[20]
8977	2-Pentanone, 4-imino-	C_5H_9NO 99.13	870-74-6 43	209	4-01-00-03678 0.9427[0]	H_2O 4; eth 4
8978	2-Pentanone, 4-methoxy-4-methyl- Pentoxone	$C_7H_{14}O_2$ 130.19	107-70-0	160	4-01-00-04024 0.8980[25]	1.418[20]
8979	1-Pentanone, 1-(3-methoxyphenyl)-	$C_{12}H_{16}O_2$ 192.26	20359-55-1	196[30]	4-08-00-00533 1.012[20]	1.5242[31]
8980	2-Pentanone, 3-methyl-, (±)- 2-Pentanone, 3-methyl (DL)	$C_6H_{12}O$ 100.16	55156-16-6	118	4-01-00-03309 0.8130[20]	H_2O 2; EtOH 5; eth 5; chl 3 1.4002[20]
8981	2-Pentanone, 4-methyl- Isobutyl methyl ketone	$C_6H_{12}O$ 100.16	108-10-1 -84	5095 116.5	4-01-00-03305 0.7965[25]	H_2O 2; EtOH 5; eth 5; ace 5 1.3962[20]
8982	3-Pentanone, 2-methyl- Ethyl isopropyl ketone	$C_6H_{12}O$ 100.16	565-69-5	113.5	4-01-00-03304 0.814[18]	H_2O 2; EtOH 4; eth 5; ace 5 1.3975[20]
8983	1-Pentanone, 4-methyl-1-phenyl-	$C_{12}H_{16}O$ 176.26	2050-07-9 -2	255.5	3-07-00-01158 0.9623[15]	ace 4; bz 4; eth 4; EtOH 4 1.533[20]
8984	2-Pentanone, 3-methyl-1-phenyl- Benzyl sec-butyl ketone	$C_{12}H_{16}O$ 176.26	27993-42-6	127-8[17]	4-07-00-00758 0.964[20]	1.5019[20]
8985	2-Pentanone, oxime Methyl propyl ketone oxime	$C_5H_{11}NO$ 101.15	623-40-5	168	4-01-00-03274 0.9095[20]	H_2O 4; eth 4; EtOH 4 1.4450[20]
8986	1-Pentanone, 1-phenyl-	$C_{11}H_{14}O$ 162.23	1009-14-9 -9.4	245	4-07-00-00733 0.986[20]	H_2O 1; EtOH 4; eth 4; ctc 2 1.5158[20]
8987	2-Pentanone, 3-(phenylmethylene)- Methyl α-ethylstyryl ketone	$C_{12}H_{14}O$ 174.24	3437-89-6	136-8[12]	3-07-00-01461 1.0005[22]	1.5650[22]
8988	3-Pentanone, 2,2,4,4-tetramethyl-	$C_9H_{18}O$ 142.24	815-24-7 -25.2	152	4-01-00-03365 0.8240[18]	H_2O 1; EtOH 3; eth 3; ace 3 1.4194[20]
8989	3-Pentanone, 2,2,4-trimethyl- tert-Butyl isopropyl ketone	$C_8H_{16}O$ 128.21	5857-36-3	135.1	4-01-00-03351 0.8065[20]	H_2O 1; eth 3; ace 3 1.4060
8990	Pentanoyl chloride Valeroyl chloride	C_5H_9ClO 120.58	638-29-9 -110	109	4-02-00-00874 1.015[15]	1.4200[20]
8991	Pentanoyl chloride, 2-chloro-	$C_5H_8Cl_2O$ 155.02	61589-68-2	156	4-02-00-00879 1.1765[20]	eth 4 1.4465[20]
8992	Pentanoyl chloride, 5-chloro-	$C_5H_8Cl_2O$ 155.02	1575-61-7	83[12]	4-02-00-00881 1.210[18]	eth 4 1.4639[20]
8993	Pentanoyl chloride, 2-methyl-, (±)- Valeryl chloride, 2-methyl (DL)	$C_6H_{11}ClO$ 134.61	116908-84-0	141	3-02-00-00741 0.9781[20]	eth 4 1.4330[27]
8994	Pentanoyl chloride, 3-methyl-, (±)- Valeryl chloride, 3-methyl (DL)	$C_6H_{11}ClO$ 134.61	116908-85-1	141	4-02-00-00949 0.9781[20]	eth 4
8995	2,5,8,11,14-Pentaoxapentadecane Tetraethylene glycol dimethyl ether	$C_{10}H_{22}O_5$ 222.28	143-24-8	9141 275.3	4-01-00-02404 1.0114[20]	H_2O 5; EtOH 3; eth 3; ctc 3
8996	Pentaphene 2,3:6,7-Dibenzphenanthrene	$C_{22}H_{14}$ 278.35	222-93-5 257	2992	4-05-00-02721	H_2O 1; EtOH 2; eth 2; bz 3
8997	Pentasiloxane, dodecamethyl- Dodecamethylpentasiloxane	$C_{12}H_{36}O_4Si_5$ 384.84	141-63-9 -80	3401 232; 105[12]	4-04-00-04124 0.8755[20]	ctc 3; CS_2 3 1.3925[20]
8998	1,2,3,5,6-Pentathiepane Lenthionine	$C_2H_4S_5$ 188.38	292-46-6 60.5	5321	5-19-12-00251	
8999	Pentatriacontane	$C_{35}H_{72}$ 492.96	630-07-9 75	490	4-01-00-00598 0.8157[20]	H_2O 1; eth 2; ace 3 1.4568[20]
9000	18-Pentatriacontanone	$C_{35}H_{70}O$ 506.94	504-53-0 89.0	270[0.1]	4-01-00-03413 0.793[95]	H_2O 1; EtOH 2; eth 2; ace 2
9001	4-Pentenal Pent-4-en-1-al	C_5H_8O 84.12	2100-17-6	99	4-01-00-03459 0.852[20]	H_2O 1; eth 3; ace 3 1.4191[20]
9002	2-Pentenal, 2-methyl-	$C_6H_{10}O$ 98.14	623-36-9	136.5	4-01-00-03471 0.8581[20]	H_2O 1; EtOH 3; eth 3; bz 3 1.4488[20]
9003	4-Pentenamide, 2,2-diethyl- Novonal	$C_9H_{17}NO$ 155.24	512-48-1 75.5	6643	3-02-00-01351	eth 4; EtOH 4

No.	Name Synonym	Mol. Form. Mol. Wt.	CAS RN mp/°C	Merck No. bp/°C	Beil. Ref. den/g cm^{-3}	Solubility n_D
9004	1-Pentene α-Amylene	C_5H_{10} 70.13	109-67-1 -165.2	7079 29.9	4-01-00-00808 0.6405[20]	H_2O 1; EtOH 5; eth 5; bz 3 1.3715[20]
9005	2-Pentene, (E)- trans-2-Pentene	C_5H_{10} 70.13	646-04-8 -140.2	7080 36.3	4-01-00-00814 0.6431[25]	H_2O 1; EtOH 5; eth 5; bz 3 1.3793[20]
9006	2-Pentene, (Z)- cis-2-Pentene	C_5H_{10} 70.13	627-20-3 -151.4	7080 36.9	4-01-00-00814 0.6556[20]	H_2O 1; EtOH 5; eth 5; bz 3 1.3830[20]
9007	1-Pentene, 1-bromo- 1-Bromo-1-pentene	C_5H_9Br 149.03	60468-23-7 	 119	3-01-00-00775 1.2606[20]	bz 4; eth 4; chl 4 1.4572[20]
9008	1-Pentene, 2-bromo- 2-Bromo-1-pentene	C_5H_9Br 149.03	31844-95-8 	 107.5	3-01-00-00775 1.228[20]	bz 4; eth 4; lig 4 1.4535[20]
9009	1-Pentene, 3-bromo- 1-Ethylallyl bromide	C_5H_9Br 149.03	53045-71-9 	 30.5[30]	2-01-00-00183 1.2417[25]	ace 4; bz 4; chl 4 1.4626[25]
9010	1-Pentene, 5-bromo-	C_5H_9Br 149.03	1119-51-3 	 125.5	3-01-00-00775 1.2581[20]	 1.4640[20]
9011	2-Pentene, 1-bromo-	C_5H_9Br 149.03	20599-27-3 	 123.5	3-01-00-00783 1.2545[20]	ace 4; bz 4; chl 4 1.4731[20]
9012	2-Pentene, 2-bromo- 2-Bromo-2-pentene	C_5H_9Br 149.03	80204-19-9 	 111	3-01-00-00784 1.275[20]	ace 4; bz 4; chl 4 1.4580[20]
9013	2-Pentene, 3-bromo- 3-Bromo-2-pentene	C_5H_9Br 149.03	21964-23-8 	 116	3-01-00-00784 1.271[20]	ace 4; bz 4; chl 4 1.4628[20]
9014	2-Pentene, 4-bromo-	C_5H_9Br 149.03	1809-26-3 	117 dec; 22[9]	4-01-00-00816 1.2312[21]	ace 4; bz 4; chl 4 1.4752[21]
9015	2-Pentene, 5-bromo- 3-Pentenyl bromide	C_5H_9Br 149.03	51952-42-2 	 121.7[621]	4-01-00-00816 1.2715[20]	ace 4; bz 4; chl 4 1.4695[20]
9016	1-Pentene, 2-chloro- 2-Chloro-1-pentene	C_5H_9Cl 104.58	42131-85-1 	 96	3-01-00-00774 0.872[5]	eth 4; EtOH 4
9017	1-Pentene, 3-chloro- 1-Ethylallyl chloride	C_5H_9Cl 104.58	24356-00-1 	 93	3-01-00-00774 0.8978[20]	ace 4; eth 4; EtOH 4 1.4254[20]
9018	1-Pentene, 4-chloro- 4-Chloro-1-pentene	C_5H_9Cl 104.58	10524-08-0 	 97	2-01-00-00183 0.934[15]	H_2O 1; eth 4; chl 4 1.417[15]
9019	1-Pentene, 5-chloro-	C_5H_9Cl 104.58	928-50-7 	 105	4-01-00-00810 0.9125[20]	ace 4; eth 4 1.4297[20]
9020	2-Pentene, 1-chloro- γ-Ethylallyl chloride	C_5H_9Cl 104.58	10071-60-0 	 108	3-01-00-00782 0.908[22]	ace 4; eth 4; EtOH 4; chl 4 1.4352[22]
9021	2-Pentene, 2-chloro- 2-Chloro-2-pentene	C_5H_9Cl 104.58	67747-70-0 	 96	3-01-00-00782 0.9067[20]	ace 4; eth 4 1.4261[20]
9022	2-Pentene, 3-chloro- 3-Chloro-2-pentene	C_5H_9Cl 104.58	34238-52-3 	 91	3-01-00-00782 0.9125[20]	ace 4; eth 4 1.423[24]
9023	2-Pentene, 4-chloro-	C_5H_9Cl 104.58	1458-99-7 	103; 47[25]	4-01-00-00816 0.8988[20]	ace 4; eth 4; chl 4 1.4322[20]
9024	2-Pentene, 5-chloro-	C_5H_9Cl 104.58	16435-50-0 	 108	4-01-00-00816 0.9043[20]	ace 4; bz 4; chl 4 1.4310[20]
9025	1-Pentene, 3-chloro-2-methyl-	$C_6H_{11}Cl$ 118.61	4104-01-2 	 122.5	3-01-00-00809 	bz 4; eth 4; chl 4 1.4422[20]
9026	2-Pentene, 5-chloro-2-methyl- 5-Chloro-2-methyl-2-pentene	$C_6H_{11}Cl$ 118.61	7712-60-9 	 133	4-01-00-00843 0.9135[20]	ace 4; chl 4
9027	1-Pentene, 1,1,2,3,3,4,4,5,5,5-decafluoro-	C_5F_{10} 250.04	376-87-4 	 30	4-01-00-00810 	chl 4 1.2571[25]
9028	2-Pentene, 2,5-dichloro- 1,4-Dichloro-3-pentene	$C_5H_8Cl_2$ 139.02	20177-02-0 	 40-1[8]	4-01-00-00816 1.1182[15]	chl 4
9029	1-Pentene, 2,3-dimethyl- 2,3-Dimethyl-1-pentene	C_7H_{14} 98.19	3404-72-6 -134.3	 84.3	3-01-00-00832 0.7051[20]	H_2O 1; EtOH 5; eth 5; dil sulf 4 1.4033[20]
9030	1-Pentene, 2,4-dimethyl- 2,4-Dimethyl-1-pentene	C_7H_{14} 98.19	2213-32-3 -124.1	 81.6	4-01-00-00871 0.6943[20]	H_2O 1; EtOH 5; eth 5; bz 3 1.3986[20]
9031	1-Pentene, 3,3-dimethyl- 3,3-Dimethyl-1-pentene	C_7H_{14} 98.19	3404-73-7 -134.3	 77.5	4-01-00-00873 0.6974[20]	H_2O 1; EtOH 5; eth 5; bz 3 1.3984[20]
9032	1-Pentene, 3,4-dimethyl- 3,4-Dimethyl-1-pentene	C_7H_{14} 98.19	7385-78-6 	 80.8	4-01-00-00871 0.6934[25]	 1.3992[20]
9033	1-Pentene, 4,4-dimethyl-	C_7H_{14} 98.19	762-62-9 -136.6	 72.5	4-01-00-00869 0.6827[20]	H_2O 1; EtOH 5; eth 5; bz 3 1.3818[20]
9034	2-Pentene, 2,3-dimethyl- Ethyltrimethylethylene	C_7H_{14} 98.19	10574-37-5 -118.3	 97.5	4-01-00-00870 0.7277[20]	H_2O 1; EtOH 3; eth 3; bz 3 1.4208[20]
9035	2-Pentene, 2,4-dimethyl- 2,4-Dimethyl-2-pentene	C_7H_{14} 98.19	625-65-0 -127.7	 83.4	4-01-00-00872 0.6954[20]	H_2O 1; EtOH 3; eth 3; bz 3 1.4040[20]
9036	2-Pentene, 3,4-dimethyl- 3,4-Dimethyl-2-pentene	C_7H_{14} 98.19	24910-63-2 -124.2	 89	3-01-00-00833 0.7126[20]	H_2O 1; EtOH 3; eth 3; bz 3 1.4070[20]
9037	2-Pentene, 3,4-dimethyl-, (E)-	C_7H_{14} 98.19	4914-92-5 -124.2	 91.5	4-01-00-00871 0.7124[25]	 1.4128[20]
9038	2-Pentene, 3,4-dimethyl-, (Z)-	C_7H_{14} 98.19	4914-91-4 -113.4	 89.3	4-01-00-00871 0.7092[25]	 1.4104[20]
9039	2-Pentene, 4,4-dimethyl-, (E)-	C_7H_{14} 98.19	690-08-4 -115.2	 76.7	4-01-00-00868 0.6889[20]	H_2O 1; EtOH 3; eth 3; bz 3 1.3982[20]
9040	2-Pentene, 4,4-dimethyl-, (Z)-	C_7H_{14} 98.19	762-63-0 -135.4	 80.4	4-01-00-00868 0.6951[25]	 1.4026[20]
9041	2-Pentenedioic acid, diethyl ester Diethyl glutaconate	$C_9H_{14}O_4$ 186.21	2049-67-4 	 237	 1.0496[20]	eth 4; EtOH 4 1.4411[20]
9042	2-Pentenedioic acid, diethyl ester, (E)- trans-2-Pentenedioic acid diethyl ester	$C_9H_{14}O_4$ 186.21	73178-43-5 	237; 126[14]	4-02-00-02227 1.0496[20]	eth 4; EtOH 4 1.4411[20]

No.	Name Synonym	Mol. Form. Mol. Wt.	CAS RN mp/°C	Merck No. bp/°C	Beil. Ref. den/g cm^{-3}	Solubility n_D
9043	1-Pentene, 2-ethyl- 2-Ethyl-1-pentene	C$_7$H$_{14}$ 98.19	3404-71-5	94	4-01-00-00867 0.7079[20]	bz 4; eth 4; EtOH 4 1.405[20]
9044	1-Pentene, 3-ethyl- 3-Ethyl-1-pentene	C$_7$H$_{14}$ 98.19	4038-04-4 -127.5	84.1	4-01-00-00867 0.6917[25]	1.3982[20]
9045	2-Pentene, 3-ethyl- 3-Ethyl-2-pentene	C$_7$H$_{14}$ 98.19	816-79-5	96	4-01-00-00868 0.7204[20]	H$_2$O 1; EtOH 3; eth 3; bz 3 1.4148[20]
9046	1-Pentene, 3-ethyl-2-methyl-	C$_8$H$_{16}$ 112.22	19780-66-6 -112.9	109.5	4-01-00-00891 0.7262[20]	1.4140[20]
9047	1-Pentene, 3-ethyl-3-methyl- 3-Ethyl-3-methyl-1-pentene	C$_8$H$_{16}$ 112.22	6196-60-7	112	4-01-00-00891 0.7264[20]	1.418[20]
9048	1-Pentene, 3-ethyl-4-methyl- 1-Pentene, 4-methyl-3-ethyl	C$_8$H$_{16}$ 112.22	61847-80-1	107.5	4-01-00-00890 0.7158[25]	1.4097[20]
9049	2-Pentene, 3-ethyl-2-methyl- 3-Ethyl-2-methyl-2-pentene	C$_8$H$_{16}$ 112.22	19780-67-7	115	4-01-00-00890 0.735[25]	1.4247[20]
9050	2-Pentene, 3-ethyl-4-methyl-, (E)- 2-Pentene, 4-methyl-3-ethyl (trans)	C$_8$H$_{16}$ 112.22	42067-49-2	114.3	4-01-00-00890 0.7308[25]	1.4210[20]
9051	2-Pentene, 3-ethyl-4-methyl-, (Z)- 2-Pentene, 4-methyl-3-ethyl (cis)	C$_8$H$_{16}$ 112.22	42067-48-1	116	4-01-00-00890 0.735[25]	1.424[20]
9052	1-Pentene, 2-methyl- 2-Methyl-1-pentene	C$_6$H$_{12}$ 84.16	763-29-1 -135.7	62.1	4-01-00-00841 0.6799[20]	H$_2$O 1; EtOH 3; bz 3; ctc 2 1.3920[20]
9053	1-Pentene, 3-methyl- sec-Butylethene	C$_6$H$_{12}$ 84.16	760-20-3 -153	54.2	3-01-00-00812 0.6675[20]	H$_2$O 1; EtOH 3; bz 3; chl 3 1.3841[20]
9054	1-Pentene, 4-methyl- Isobutylethene	C$_6$H$_{12}$ 84.16	691-37-2 -153.6	53.9	4-01-00-00846 0.6642[20]	H$_2$O 1; EtOH 3; bz 3; chl 3 1.3828[20]
9055	2-Pentene, 2-methyl- 2-Methyl-2-pentene	C$_6$H$_{12}$ 84.16	625-27-4 -135	67.3	4-01-00-00842 0.6863[20]	H$_2$O 1; EtOH 3; bz 3; ctc 3 1.4004[20]
9056	2-Pentene, 3-methyl-, (E)-	C$_6$H$_{12}$ 84.16	616-12-6 -138.5	70.4	4-01-00-00848 0.6930[25]	H$_2$O 1; EtOH 3; bz 3; ctc 3 1.4045[20]
9057	2-Pentene, 3-methyl-, (Z)- (cis)-3-Methyl-2-pentene	C$_6$H$_{12}$ 84.16	922-62-3 -134.8	67.7	4-01-00-00848 0.6886[25]	H$_2$O 1; EtOH 3; bz 3; chl 3 1.4016[20]
9058	2-Pentene, 4-methyl-, (E)- 4-Methyl-trans-2-pentene	C$_6$H$_{12}$ 84.16	674-76-0 -140.8	58.6	4-01-00-00844 0.6686[20]	H$_2$O 1; EtOH 3; bz 3; ctc 2 1.3889[20]
9059	2-Pentene, 4-methyl-, (Z)- cis-4-Methyl-2-pentene	C$_6$H$_{12}$ 84.16	691-38-3 -134.8	56.3	4-01-00-00844 0.6690[20]	H$_2$O 1; EtOH 3; bz 3; chl 3 1.3800[20]
9060	1-Pentene, 3-methylene-	C$_6$H$_{10}$ 82.15	3404-63-5	75	4-01-00-01022 0.712[25]	1.445[20]
9061	3-Pentenenitrile, (E) trans-3-Pentenenitrile	C$_5$H$_7$N 81.12	16545-78-1	142.6		
9062	4-Pentenenitrile	C$_5$H$_7$N 81.12	592-51-8	140	4-02-00-01543 0.8239[24]	H$_2$O 1; EtOH 5; eth 5 1.4213[14]
9063	3-Pentenenitrile, 2-hydroxy-, (Z)-	C$_5$H$_7$NO 97.12	116908-78-2	139[70]	4-03-00-01001 0.9675[15]	bz 4; eth 4; EtOH 4; chl 4 1.4460[21]
9064	Pentene, 2,4,4-trimethyl- Diisobutylene	C$_8$H$_{16}$ 112.22	25167-70-8	101	4-01-00-00892	
9065	1-Pentene, 2,3,3-trimethyl- 2,3,3-Trimethyl-1-pentene	C$_8$H$_{16}$ 112.22	560-23-6 -69	108.3	4-01-00-00894 0.7308[25]	1.4174[20]
9066	1-Pentene, 2,3,4-trimethyl- 2,3,4-Trimethyl-1-pentene	C$_8$H$_{16}$ 112.22	565-76-4	108	4-01-00-00894 0.725[25]	1.415[20]
9067	1-Pentene, 2,4,4-trimethyl- 2,2,4-Trimethyl-4-pentene	C$_8$H$_{16}$ 112.22	107-39-1 -93.5	101.4	4-01-00-00892 0.7150[20]	H$_2$O 1; eth 3; bz 3; ctc 3 1.4086[20]
9068	1-Pentene, 3,3,4-trimethyl-	C$_8$H$_{16}$ 112.22	560-22-5	105	3-01-00-00851 0.725[25]	1.4144[20]
9069	1-Pentene, 3,4,4-trimethyl- 3,4,4-Trimethyl-1-pentene	C$_8$H$_{16}$ 112.22	564-03-4	104	4-01-00-00891 0.715[25]	1.412[20]
9070	2-Pentene, 2,3,4-trimethyl- 2,3,4-Trimethyl-2-pentene	C$_8$H$_{16}$ 112.22	565-77-5 -113.4	116.5	4-01-00-00894 0.7434[20]	1.4274[20]
9071	2-Pentene, 2,4,4-trimethyl- 2,4,4-Trimethyl-2-pentene	C$_8$H$_{16}$ 112.22	107-40-4 -106.3	104.9	4-01-00-00894 0.7218[20]	H$_2$O 1; eth 3; bz 3; ctc 3 1.4160[20]
9072	2-Pentene, 3,4,4-trimethyl- 3,4,4-Trimethyl-2-pentene	C$_8$H$_{16}$ 112.22	598-96-9	112	4-01-00-00891 0.735[25]	EtOH 4 1.423[20]
9073	4-Pentenoic acid Allylacetic acid	C$_5$H$_8$O$_2$ 100.12	591-80-0 -22.5	188.5	4-02-00-01542 0.9809[20]	H$_2$O 2; EtOH 4; eth 4 1.4281[20]
9074	4-Pentenoic acid, 2-acetyl-, ethyl ester	C$_9$H$_{14}$O$_3$ 170.21	610-89-9	208	3-03-00-01314 0.9898[20]	EtOH 5; eth 5; bz 5 1.4388[18]
9075	4-Pentenoic acid, 2,2-dimethyl- 2,2-Dimethylpent-4-enoic acid	C$_7$H$_{12}$O$_2$ 128.17	16386-93-9	116[30]	4-02-00-01585 0.933[20]	1.4338[20]
9076	2-Pentenoic acid, 2-methyl-, (E)- 2-Pentenoic acid, 2-methyl (trans)	C$_6$H$_{10}$O$_2$ 114.14	16957-70-3 24.4	214; 112[12]	4-02-00-01568 0.9751[20]	H$_2$O 2; eth 3; chl 3; CS$_2$ 3 1.4513[20]
9077	2-Pentenoic acid, 3-methyl-, (Z)- 2-Pentenoic acid, 3-methyl (cis)	C$_6$H$_{10}$O$_2$ 114.14	19866-51-4 12	96[5]	4-02-00-01571 0.9830[20]	1.4650
9078	2-Pentenoic acid, 4-methyl- 4,4-Dimethyl-2-butenoic acid	C$_6$H$_{10}$O$_2$ 114.14	10321-71-8 35	217	3-02-00-01324 0.9529[21]	ace 4; eth 4; EtOH 4 1.4489[21]
9079	3-Pentenoic acid, 2-methyl-	C$_6$H$_{10}$O$_2$ 114.14	37674-63-8	199	3-02-00-01324 0.966[15]	1.4402[25]
9080	2-Pentenoic acid, 4-methyl-, ethyl ester Ethyl 4,4-dimethylcrotonate	C$_8$H$_{14}$O$_2$ 142.20	2351-97-5	174; 76-7[25]	3-02-00-01324 0.9048[21]	1.4328[20]
9081	3-Pentenoic acid, 3-methyl-, ethyl ester Ethyl 3-methyl-3-pentenoate	C$_8$H$_{14}$O$_2$ 142.20	86623-80-5	61[13]	3-02-00-01328 0.9183[20]	1.4298[20]
9082	3-Pentenoic acid, 4-methyl-, ethyl ester Ethyl 4-methyl-3-pentenoate	C$_8$H$_{14}$O$_2$ 142.20	6849-18-9	58[11]	3-02-00-01324 0.9134[17]	1.4329[17]

No.	Name Synonym	Mol. Form. Mol. Wt.	CAS RN mp/°C	Merck No. bp/°C	Beil. Ref. den/g cm^{-3}	Solubility n_D
9083	2-Pentenoic acid, 2-methyl-, methyl ester	$C_7H_{12}O_2$ 128.17	10478-12-3 51[11]		4-02-00-01568 0.920[25]	1.4336
9084	2-Pentenoic acid, 3-methyl-, methyl ester, (E)-	$C_7H_{12}O_2$ 128.17	17447-01-7 79[50]		4-02-00-01571 0.9302[20]	1.4446[20]
9085	2-Pentenoic acid, 3-methyl-, methyl ester, (Z)- cis-3-Methyl-2-pentenoic acid methyl ester	$C_7H_{12}O_2$ 128.17	17447-00-6 74[50]		4-02-00-01571 0.9279[20]	1.4420[20]
9086	3-Pentenoic acid, 3-methyl-, methyl ester Methyl 3-methyl-3-pentenoate	$C_7H_{12}O_2$ 128.17	2258-58-4 74[50]		4-02-00-01573 0.9949[20]	1.4306[20]
9087	1-Penten-3-ol	$C_5H_{10}O$ 86.13	616-25-1 115		1-01-00-00227 0.839[20]	H_2O 2; EtOH 5; eth 5 1.4239[20]
9088	4-Penten-1-ol	$C_5H_{10}O$ 86.13	821-09-0 141		4-01-00-02119 0.8457[20]	H_2O 2; eth 3; ctc 2 1.4309[20]
9089	4-Penten-2-ol	$C_5H_{10}O$ 86.13	625-31-0 116		1-01-00-00227 0.8367[20]	H_2O 4; EtOH 5; eth 5 1.4225[20]
9090	2-Penten-1-ol, (E)-	$C_5H_{10}O$ 86.13	1576-96-1 138		4-01-00-02121 0.8471[20]	EtOH 3; eth 3; ace 3 1.4341[20]
9091	2-Penten-1-ol, (Z)-	$C_5H_{10}O$ 86.13	1576-95-0 138		4-01-00-02121 0.8529[20]	EtOH 3; eth 3; ace 3 1.4354[20]
9092	3-Penten-2-ol, (±)-	$C_5H_{10}O$ 86.13	42569-16-4 121.6; 65[70]		4-01-00-02122 0.8328[25]	ace 4; eth 4; EtOH 4 1.4280[20]
9093	2-Penten-1-ol, 5-(2,3-dimethyltricyclo[2.2.1.02,6]hept-3-yl)-2-methyl-, stereoisomer α-Santalol	$C_{15}H_{24}O$ 220.35	115-71-9 301.5	8321	4-06-00-03521 0.9679[20]	H_2O 1; EtOH 3 1.5023[20]
9094	2-Penten-1-ol, 2-methyl-	$C_6H_{12}O$ 100.16	1610-29-3 167.5		4-01-00-02144 0.8501[24]	1.444[24]
9095	2-Penten-1-ol, 4-methyl-	$C_6H_{12}O$ 100.16	5362-55-0 159		4-01-00-02146 0.8489[16]	1.4403[16]
9096	3-Penten-1-ol, 4-methyl- 4-Methyl-3-pentenol	$C_6H_{12}O$ 100.16	763-89-3 157		4-01-00-02145 0.8600[20]	1.4432[21]
9097	3-Penten-2-ol, 2-methyl- 2-Methyl-3-penten-2-ol	$C_6H_{12}O$ 100.16	63468-05-3 112; 65[72]		4-01-00-02145 0.8347[20]	1.4302[20]
9098	3-Penten-2-ol, 3-methyl- 3-Methyl-3-penten-2-ol	$C_6H_{12}O$ 100.16	2747-53-7 140		4-01-00-02149 0.8792[0]	1.4428[17]
9099	3-Penten-2-ol, 4-methyl-	$C_6H_{12}O$ 100.16	4325-82-0 134		4-01-00-02144 0.840[15]	1.9377[15]
9100	4-Penten-2-ol, 2-methyl-	$C_6H_{12}O$ 100.16	624-97-5 -73 119.5		4-01-00-02146 0.8300[20]	H_2O 2; EtOH 3; eth 3; ctc 2 1.4268[20]
9101	4-Penten-2-ol, 3-methyl-	$C_6H_{12}O$ 100.16	1569-59-1 125.5		3-01-00-01933 0.8429[22]	1.4326[22]
9102	4-Penten-2-ol, 4-methyl-	$C_6H_{12}O$ 100.16	2004-67-3 129		4-01-00-02143 0.8436[20]	1.4297[17]
9103	2-Penten-1-ol, 2-methyl-5-(2-methyl-3-methylenebicyclo[2.2.1]hept-2-yl)-, [1S-[1α,2α(Z),4α]] β-Santalol	$C_{15}H_{24}O$ 220.35	77-42-9 167-8[10]	8322	4-06-00-03519 0.9750[20]	1.5115[20]
9104	1-Penten-3-one Ethyl vinyl ketone	C_5H_8O 84.12	1629-58-9 103; 44[90]		4-01-00-03457 0.8468[20]	H_2O 1; EtOH 3; eth 3; ace 3 1.4195[20]
9105	3-Penten-2-one, (E)-	C_5H_8O 84.12	3102-33-8 122		4-01-00-03460 0.8624[20]	H_2O 3; eth 3; ace 3; ctc 3 1.4350[20]
9106	1-Penten-3-one, 4,4-dimethyl-1-phenyl-	$C_{13}H_{16}O$ 188.27	538-44-3 43	154[25]	3-07-00-01490 0.9508[46]	1.5523[25]
9107	1-Penten-3-one, 2-methyl-	$C_6H_{10}O$ 98.14	25044-01-3 -69.5 118.5		4-01-00-03470 1.8530[20]	ace 4; EtOH 4 1.4289[20]
9108	3-Penten-2-one, 4-methyl- Mesityl oxide	$C_6H_{10}O$ 98.14	141-79-7 -59 130	5811	4-01-00-03471 0.8653[20]	H_2O 3; EtOH 5; eth 5; ace 3 1.4440[20]
9109	4-Penten-2-one, 4-methyl-	$C_6H_{10}O$ 98.14	3744-02-3 -72.6 124.2		4-01-00-03470 0.8411[20]	
9110	1-Penten-3-one, 1-phenyl-	$C_{11}H_{12}O$ 160.22	3152-68-9 38.5	142[12]	2-07-00-00298 0.8697[20]	H_2O 2; EtOH 4; eth 4; bz 4 1.5684[20]
9111	1-Penten-3-yne Methylvinylacetylene	C_5H_6 66.10	646-05-9 59.5		4-01-00-01087 0.7401[20]	bz 4; eth 4 1.4496[20]
9112	1-Penten-4-yne	C_5H_6 66.10	871-28-3 42.5		4-01-00-01088 0.738[16]	H_2O 1; eth 3; bz 3 1.4125[16]
9113	3-Penten-1-yne, (E)- trans-3-Penten-1-yne	C_5H_6 66.10	2004-69-5 52.2		4-01-00-01085	
9114	3-Penten-1-yne, (Z)- cis-3-Penten-1-yne	C_5H_6 66.10	1574-40-9 44.6			
9115	3-Penten-1-yne, 3-ethyl-	C_7H_{10} 94.16	14272-54-9 96.5		4-01-00-01099 0.7886[25]	bz 4; eth 4 1.4338[25]
9116	1-Penten-3-yne, 2-methyl- Methylisopropenylacetylene	C_6H_8 80.13	926-55-6 81-2[100]		4-01-00-01095 0.739[20]	bz 4; eth 4 1.4002[20]
9117	3-Penten-1-yne, 3-methyl- 3-Methyl-3-penten-1-yne	C_6H_8 80.13	1574-33-0 66.5		4-01-00-01096 0.739[20]	eth 3; bz 3 1.4332[20]
9118	4-Penten-2-yn-1-ol 1-Hydroxypent-2-yn-4-ene	C_5H_6O 82.10	2919-05-3 57[10]; 44[2.5]		4-01-00-02303 0.9581[20]	1.4986[20]
9119	Pentitol, 1,2-dideoxy- 1,2,3-Pentanetriol	$C_5H_{12}O_3$ 120.15	5371-48-2 192[63]		4-01-00-02778 1.0849[34]	H_2O 4; EtOH 4; eth 4
9120	Pentonic acid, 5-C-[3,5-bis(1-methylpropyl)-1-cyclopenten-1-yl]-4-deoxy-	$C_{18}H_{32}O_5$ 328.45	491-14-5 196		3-10-00-02045 1.292[14]	EtOH 4

No.	Name Synonym	Mol. Form. Mol. Wt.	CAS RN mp/°C	Merck No. bp/°C	Beil. Ref. den/g cm⁻³	Solubility n_D
9121	1-Pentyn-3-amine, 3-ethyl-	$C_7H_{13}N$ 111.19	3234-64-8	71-2[90]	4-04-00-01154 0.828[25]	1.4409[20]
9122	1-Pentyne Propylacetylene	C_5H_8 68.12	627-19-0 -90	40.1	4-01-00-00990 0.6901[20]	H_2O 1; EtOH 4; eth 5; bz 3 1.3852[20]
9123	2-Pentyne	C_5H_8 68.12	627-21-4 -109.3	56.1	4-01-00-00992 0.7058[25]	H_2O 1; EtOH 4; eth 5; bz 3 1.4039[20]
9124	1-Pentyne, 1-bromo- 1-Pentynyl bromide	C_5H_7Br 147.01	14752-60-4	44-6[52]	3-01-00-00958 1.28[13]	1.4579[13]
9125	1-Pentyne, 3-chloro-3-ethyl- 3-Chloro-3-ethyl-1-pentyne	$C_7H_{11}Cl$ 130.62	6080-79-1	73-6[100]	4-01-00-01032 0.9230[19]	bz 4; eth 4; chl 4 1.4437[19]
9126	1-Pentyne, 3-chloro-3-methyl- 3-Chloro-3-methyl-1-pentyne	C_6H_9Cl 116.59	14179-94-3	102.3	4-01-00-01021 0.9163[20]	H_2O 1; eth 3; bz 3; chl 3 1.4330[20]
9127	2-Pentyne, 4-chloro-4-methyl-	C_6H_9Cl 116.59	999-79-1	55[70]	4-01-00-01018	ace 4; bz 4; chl 4 1.4143[20]
9128	1-Pentyne, 3,3-dimethyl- 3,3-Dimethyl-1-pentyne	C_7H_{12} 96.17	918-82-1	70	4-01-00-01034 0.7032[25]	1.3934[20]
9129	1-Pentyne, 4,4-dimethyl- 4,4-Dimethyl-1-pentyne	C_7H_{12} 96.17	13361-63-2 -75.7	76.1	4-01-00-01033 0.7142[20]	bz 4; eth 4; chl 4 1.3983[20]
9130	2-Pentyne, 4,4-dimethyl-	C_7H_{12} 96.17	999-78-0 -82.4	83	4-01-00-01032 0.7176[20]	H_2O 1; eth 3; bz 3; ctc 2 1.4071[20]
9131	1-Pentyne, 3-ethyl- 3-Ethyl-1-pentyne	C_7H_{12} 96.17	21020-26-8	84	3-01-00-01002 0.7246[25]	bz 4; eth 4; chl 4 1.4043[25]
9132	1-Pentyne, 3-ethyl-3-methyl- 3-Ethyl-3-methyl-1-pentyne	C_8H_{14} 110.20	919-12-0	101.5	4-01-00-01046 0.7422[20]	bz 4; eth 4; chl 4 1.4110[20]
9133	1-Pentyne, 1-iodo- 1-Iodo-1-pentyne	C_5H_7I 194.02	14752-61-5	54[23]	4-01-00-00992 1.6127[19]	1.5148[19]
9134	1-Pentyne, 3-methyl- 3-Methyl-1-pentyne	C_6H_{10} 82.15	922-59-8	57.7	4-01-00-01021 0.6992[25]	1.3916[20]
9135	1-Pentyne, 4-methyl-	C_6H_{10} 82.15	7154-75-8 -104.6	61.2	4-01-00-01019 0.7000[25]	H_2O 1; bz 3; chl 3 1.3936[20]
9136	2-Pentyne, 4-methyl-	C_6H_{10} 82.15	21020-27-9 -110.3	73.1	4-01-00-01018 0.7112[25]	bz 4; chl 4 1.4057[20]
9137	2-Pentynoic acid Ethylpropiolic acid	$C_5H_6O_2$ 98.10	5963-77-9 50	122[10]; 81[3]	4-02-00-01693 0.978[20]	H_2O 4 1.4619[20]
9138	4-Pentynoic acid Propargylacetic acid	$C_5H_6O_2$ 98.10	6089-09-4 57.7	110[30]; 102[17]	4-02-00-01693	eth 4; EtOH 4
9139	2-Pentynoic acid, methyl ester Methyl 2-pentynoate	$C_6H_8O_2$ 112.13	24342-04-9	72[10]	4-02-00-01694 0.937[20]	1.4455[20]
9140	1-Pentyn-3-ol Ethylethynylcarbinol	C_5H_8O 84.12	4187-86-4	125	4-01-00-02224 0.8926[15]	1.4347[15]
9141	2-Pentyn-1-ol	C_5H_8O 84.12	6261-22-9	61-2[15]	4-01-00-02226 0.913[17]	1.4518[17]
9142	3-Pentyn-1-ol	C_5H_8O 84.12	10229-10-4	154	4-01-00-02227 0.9002[20]	1.4454[20]
9143	4-Pentyn-1-ol	C_5H_8O 84.12	5390-04-5	154	4-01-00-02225 0.913[20]	1.4414[20]
9144	1-Pentyn-3-ol, 3,4-dimethyl-	$C_7H_{12}O$ 112.17	1482-15-1	134	4-01-00-02255 0.8691[20]	H_2O 4; eth 4; EtOH 4 1.4372[20]
9145	1-Pentyn-3-ol, 3-ethyl-	$C_7H_{12}O$ 112.17	6285-06-9	138	4-01-00-02254 0.8691[20]	H_2O 2
9146	1-Pentyn-3-ol, 3-methyl- Meparfynol	$C_6H_{10}O$ 98.14	77-75-8 30.5	5731 120.5	2-01-00-00506 0.8888[20]	1.4310[20]
9147	2-Pentynoyl chloride, 4,4-dimethyl- 4,4-Dimethyl-2-pentynoyl chloride	C_7H_9ClO 144.60	52324-03-5	52[20]	2-02-00-00454 0.9743[20]	1.4443[20]
9148	Perchloric acid, trichloromethyl ester Trichloro perchlorate	CCl_4O_4 217.82	67632-66-0 -55	exp 40	3-03-00-00037	eth 4
9149	Perfluidone o-Toluidine, 4-(phenylsulfonyl)-N- [(trifluoromethyl)sulfono]-	$C_{14}H_{12}F_3NO_4S_2$ 379.38	37924-13-3 143	7113		
9150	Permethrin Cyclopropanecarboxylic acid, 3-(2,2- dichloroethenyl)-2,2-dimethyl-, (3- phenoxyphenyl), methyl ester	$C_{21}H_{20}Cl_2O_3$ 391.29	52645-53-1 34	7132 200[0.01]	1.23[20]	
9151	Peroxide, acetyl benzoyl Acetozone	$C_9H_8O_4$ 180.16	644-31-5 37	71 130[19]	4-09-00-00717	eth 4
9152	Peroxide, bis(2,4-dichlorobenzoyl)	$C_{14}H_6Cl_4O_4$ 380.01	133-14-2 106		4-09-00-01001	
9153	Peroxide, bis(1,1-dimethylethyl) DTBP	$C_8H_{18}O_2$ 146.23	110-05-4 -40	3446 111	4-01-00-01619 0.704[20]	H_2O 1; ace 5; ctc 3; os 3 1.3890[20]
9154	Peroxide, bis(1,1-dimethylpropyl)	$C_{10}H_{22}O_2$ 174.28	10508-09-5	58.5[14]; 44[10]	4-01-00-01671 0.821[20]	1.4095[20]
9155	Peroxide, bis(1-oxododecyl) Lauroyl peroxide	$C_{24}H_{46}O_4$ 398.63	105-74-8 49		4-02-00-01102	H_2O 1; chl 3
9156	Peroxide, diacetyl Acetyl peroxide	$C_4H_6O_4$ 118.09	110-22-5 30	63[21]	4-02-00-00392	eth 4; EtOH 4
9157	Peroxide, dibenzoyl Benzoyl peroxide	$C_{14}H_{10}O_4$ 242.23	94-36-0 105	1128 exp	4-09-00-00717	H_2O 2; EtOH 3; eth 3; ace 3 1.543
9158	Peroxide, diethyl	$C_4H_{10}O_2$ 90.12	628-37-5 -70	65	4-01-00-01323 0.8240[19]	H_2O 2; EtOH 5; eth 5 1.3715[17]

No.	Name Synonym	Mol. Form. Mol. Wt.	CAS RN mp/°C	Merck No. bp/°C	Beil. Ref. den/g cm^{-3}	Solubility n_D
9159	Peroxide, dimethyl	$C_2H_6O_2$ 62.07	690-02-8 	 14	4-01-00-01249 0.8677[0]	EtOH 2; eth 2; tol 3; HOAc 3 1.3503[0]
9160	Perthane Ethane, 1,1-dichloro-2,2-bis(p-ethylphenyl)-	$C_{18}H_{20}Cl_2$ 307.26	72-56-0 56	3050		
9161	Perylene Dibenz[de,kl]anthracene	$C_{20}H_{12}$ 252.32	198-55-0 274	7137	4-05-00-02689 1.35[25]	H_2O 1; EtOH 2; eth 2; ace 4
9162	9-Phenanthrenamine	$C_{14}H_{11}N$ 193.25	947-73-9 138.3	 sub	4-12-00-03443	eth 2; bz 2; chl 2
9163	Phenanthrene	$C_{14}H_{10}$ 178.23	85-01-8 99.2	7167 340	4-05-00-02297 0.9800[4]	H_2O 1; EtOH 3; eth 3; ace 3 1.5943
9164	Phenanthrene, 9-bromo- 9-Phenanthryl bromide	$C_{14}H_9Br$ 257.13	573-17-1 64.5	 >360	4-05-00-02303 1.4093[10]	H_2O 1; EtOH 3; eth 3; chl 2
9165	9-Phenanthrenecarboxylic acid 9-Phenanthroic acid	$C_{15}H_{10}O_2$ 222.24	837-45-6 257.3	 sub	4-09-00-02674	H_2O 1; EtOH 3; eth 3; bz 3
9166	1-Phenanthrenecarboxylic acid, 1,2,3,4,4a,5,6,9,10,10a-decahydro-1,4a-dimethyl-7-(1-methylethyl)-, Palustric acid	$C_{20}H_{30}O_2$ 302.46	1945-53-5 164.5	6950	4-09-00-02174	
9167	1-Phenanthrenecarboxylic acid, 1,2,3,4,4a,4b,5,6,10,10a-decahydro-1,4a-dimethyl-7-(1-methylethyl)-, Abietic acid	$C_{20}H_{30}O_2$ 302.46	514-10-3 173.5	2 250[9]	4-09-00-02175	ace 4; bz 4; eth 4; EtOH 4
9168	1-Phenanthrenecarboxylic acid, 1,2,3,4,4a,4b,5,6,10,10a-decahydro-1,4a-dimethyl-7-(1-methylethyl)-, methyl ester, [1R-(1α,4aβ,4bα,10aα)]- Methyl abietate	$C_{21}H_{32}O_2$ 316.48	127-25-3 225-6[16]	5930	4-09-00-02176 1.049[20]	H_2O 1; EtOH 3; HOAc 3 1.5344
9169	1-Phenanthrenecarboxylic acid, 1,2,3,4,4a,4b,5,9,10,10a-decahydro-1,4a-dimethyl-7-(1-methylethyl)-, Levopimaric acid	$C_{20}H_{30}O_2$ 302.46	79-54-9 150	5348	4-09-00-02173	
9170	1-Phenanthrenecarboxylic acid, 7-ethenyl-1,2,3,4,4a,4b,5,6,7,9,10,10a-dodecahydro-1,4a,7-trimethyl-, [1R-(1α,4aβ,4bα,7β,10aα)]- Pimaric acid	$C_{20}H_{30}O_2$ 302.46	127-27-5 218.5	7398 282[18]	4-09-00-02179	eth 4; py 4; EtOH 4
9171	Phenanthrene, 9,10-dihydro- 9,10-Dihydrophenanthrene	$C_{14}H_{12}$ 180.25	776-35-2 34.5	 168-9[15]	4-05-00-02183 1.0757[40]	chl 3 1.6415[20]
9172	Phenanthrene, 9,10-dimethyl- 9,10-Phenanthraquinodimethane	$C_{16}H_{14}$ 206.29	604-83-1 144	 sub	4-05-00-02335	EtOH 2; bz 3; chl 3; HOAc 3
9173	3,4-Phenanthrenediol	$C_{14}H_{10}O_2$ 210.23	478-71-7 143	 sub 130	3-06-00-05689	eth 4; EtOH 4
9174	9,10-Phenanthrenedione Phenanthrenequinone	$C_{14}H_8O_2$ 208.22	84-11-7 209	7168	4-07-00-02565 1.405[22]	H_2O 1; EtOH 2; eth 3; bz 2
9175	9,10-Phenanthrenedione, 2-hydroxy-	$C_{14}H_8O_3$ 224.22	4088-81-7 284	 sub	3-08-00-02929	H_2O 4
9176	9,10-Phenanthrenedione, 3-hydroxy- 9,10-Phenanthraquinone, 3-hydroxy	$C_{14}H_8O_3$ 224.22	57404-54-3 330 dec	 sub	2-08-00-00398	H_2O 4; EtOH 4
9177	9,10-Phenanthrenedione, 1-methyl-7-(1-methylethyl)-	$C_{18}H_{16}O_2$ 264.32	5398-75-4 197.5	 sub	4-07-00-02595	eth 4; EtOH 4
9178	Phenanthrene, dodecahydro- Dodecahydrophenanthrene	$C_{14}H_{22}$ 190.33	1322-67-4 	 81-2[1.5]	3-05-00-01074 0.9674[20]	ace 4; bz 4 1.5102[20]
9179	Phenanthrene, 9-ethyl- 9-Ethylphenanthrene	$C_{16}H_{14}$ 206.29	3674-75-7 62.5	 199-200[1]	4-05-00-02332 1.0603[78]	H_2O 1; EtOH 3; bz 3; peth 2 1.6582[78]
9180	Phenanthrene, 3-methyl- 3-Methylphenanthrene	$C_{15}H_{12}$ 192.26	832-71-3 65	 350; 145[6]	4-05-00-02314	H_2O 1; EtOH 3; ace 3; chl 2
9181	Phenanthrene, 4-methyl- 4-Methylphenanthrene	$C_{15}H_{12}$ 192.26	832-64-4 53.5	 175-80[10]	3-05-00-02154	H_2O 1; EtOH 3; ctc 3
9182	Phenanthrene, 9-(1-methylethenyl)- 9-Isopropenylphenanthrene	$C_{17}H_{14}$ 218.30	58873-44-2 38	 163[20]	3-05-00-02240	 1.6765[22]
9183	Phenanthrene, 1-methyl-7-(1-methylethyl)- Retene	$C_{18}H_{18}$ 234.34	483-65-8 101	8162 390	4-05-00-02368 1.035[25]	H_2O 1; EtOH 3; eth 3; bz 3
9184	Phenanthrene, 1,2,3,4,4a,9,10,10a-octahydro- 1,2,3,4,4a,9,10,10a-Octahydrophenanthrene	$C_{14}H_{18}$ 186.30	16306-39-1 	 159[15]	4-05-00-01585 0.9973[32]	 1.5527[19]
9185	Phenanthrene, 1,2,3,4,5,6,7,8-octahydro- 1,2,3,4,5,6,7,8-Octahydrophenanthrene	$C_{14}H_{18}$ 186.30	5325-97-3 16.7	 295	4-05-00-01585 1.028[20]	H_2O 1; ace 3; bz 3; CS_2 3 1.5569[17]
9186	Phenanthrene, tetradecahydro-	$C_{14}H_{24}$ 192.34	5743-97-5 	 86-9[2]	4-05-00-00492 0.9447[20]	H_2O 1; eth 3; ace 3; bz 3 1.5011[20]
9187	Phenanthrene, 1,2,3,4-tetrahydro- 1,2,3,4-Tetrahydrophenanthrene	$C_{14}H_{14}$ 182.27	1013-08-7 33.5	 173[11]	4-05-00-01909 1.0601[40]	H_2O 1; EtOH 3; eth 3; ace 3
9188	Phenanthridine	$C_{13}H_9N$ 179.22	229-87-8 107.4	 349	5-20-08-00223	H_2O 2; EtOH 4; eth 4; ace 3
9189	1,10-Phenanthroline o-Phenanthroline	$C_{12}H_8N_2$ 180.21	66-71-7 117	7169 >300	5-23-08-00419	H_2O 4; EtOH 3; ace 3; bz 3
9190	1,7-Phenanthroline	$C_{12}H_8N_2$ 180.21	230-46-6 78	 360	5-23-08-00417	H_2O 3; EtOH 4; eth 1; bz 1
9191	4,7-Phenanthroline	$C_{12}H_8N_2$ 180.21	230-07-9 177	 sub 100	5-23-08-00430	H_2O 3; EtOH 4; eth 2; bz 2
9192	1,10-Phenanthroline, 2,9-dimethyl- Neocuproine	$C_{14}H_{12}N_2$ 208.26	484-11-7 159.5	6364	5-23-08-00527	
9193	Phenazine Dibenzopyrazine	$C_{12}H_8N_2$ 180.21	92-82-0 176.5	7173	5-23-08-00389	H_2O 2; EtOH 3; eth 2; bz 3

No.	Name Synonym	Mol. Form. Mol. Wt.	CAS RN mp/°C	Merck No. bp/°C	Beil. Ref. den/g cm^{-3}	Solubility n_D
9194	2,3-Phenazinediamine 2,3-Diaminophenazine	C$_{12}$H$_{10}$N$_4$ 210.24	655-86-7 264	2960 sub	4-25-00-03028	bz 4; EtOH 4
9195	Phenazine, 2-methyl- Tolazine	C$_{13}$H$_{10}$N$_2$ 194.24	1016-94-0 118.5	350	5-23-08-00475	eth 4; EtOH 4; chl 4
9196	Phenazinium, 1-hydroxy-5-methyl-, hydroxide, inner salt Pyocyanine	C$_{13}$H$_{10}$N$_2$O 210.24	85-66-5 133 dec	7965	5-24-04-00100	H$_2$O 2; EtOH 3; eth 1; ace 3
9197	1-Phenazinol Hemipyocyanine	C$_{12}$H$_8$N$_2$O 196.21	528-71-2 158	4564 sub	5-23-12-00297	H$_2$O 2; EtOH 2; bz 3; os 4
9198	Phenmedipham Carbamic acid, (3-methylphenyl)-, 3-[(methoxycarbonyl)amino]phenyl ester	C$_{16}$H$_{16}$N$_2$O$_4$ 300.31	13684-63-4 143	7199		
9199	Phenol Hydroxybenzene	C$_6$H$_6$O 94.11	108-95-2 40.9	7206 181.8	4-06-00-00531 1.0545[45]	H$_2$O 3; EtOH 3; eth 4; ace 5 1.5408[41]
9200	Phenol, amino- Hydroxyphenylamine	C$_6$H$_7$NO 109.13	27598-85-2 83.5		2-15-00-00004	H$_2$O 3; EtOH 4; eth 4; bz 4
9201	Phenol, 2-amino- o-Aminophenol	C$_6$H$_7$NO 109.13	95-55-6 174	473 sub 153	4-13-00-00805 1.328[25]	H$_2$O 3; EtOH 4; eth 3; bz 2
9202	Phenol, 3-amino- m-Aminophenol	C$_6$H$_7$NO 109.13	591-27-5 123	472 164[11]	4-13-00-00952	H$_2$O 3; EtOH 4; eth 4; bz 2
9203	Phenol, 4-amino- p-Aminophenol	C$_6$H$_7$NO 109.13	123-30-8 187.5	474 110[0.3]	4-13-00-01014	H$_2$O 2; EtOH 4; bz 1; chl 1
9204	Phenol, 2-amino-4-chloro- 2-Hydroxy-5-chloroaniline	C$_6$H$_6$ClNO 143.57	95-85-2 140		4-13-00-00878	DMSO 2
9205	Phenol, 2-amino-4,6-dichloro- 2-Amino-4,6-dichlorophenol	C$_6$H$_5$Cl$_2$NO 178.02	527-62-8 95.5	448 sub 70	4-13-00-00889	
9206	Phenol, 4-amino-2,6-dichloro-	C$_6$H$_5$Cl$_2$NO 178.02	5930-28-9 168	sub	2-13-00-00274	H$_2$O 1; EtOH 4; eth 4; ace 3
9207	Phenol, 2-amino-4,6-dinitro- Picramic acid	C$_6$H$_5$N$_3$O$_5$ 199.12	96-91-3 169	7379	4-13-00-00909	bz 4; EtOH 4
9208	Phenol, 4-(2-aminoethyl)- Tyramine	C$_8$H$_{11}$NO 137.18	51-67-2 164.5	9743 205-7[25]	4-13-00-01788	H$_2$O 2; EtOH 3; bz 2; xzl 3
9209	Phenol, 2-amino-3-methyl-	C$_7$H$_9$NO 123.15	2835-97-4 150	sub	2-13-00-00324	eth 4
9210	Phenol, 2-amino-4-methyl-	C$_7$H$_9$NO 123.15	95-84-1 136	sub	4-13-00-01712	H$_2$O 2; EtOH 4; eth 3; bz 2
9211	Phenol, 3-amino-4-methyl-	C$_7$H$_9$NO 123.15	2836-00-2 157.8	sub	4-13-00-01710	eth 4
9212	Phenol, 3-amino-5-methyl- m-Cresol, 5-amino-	C$_7$H$_9$NO 123.15	76619-89-1 139	245	4-13-00-01697	EtOH 4
9213	Phenol, 4-amino-2-methyl-	C$_7$H$_9$NO 123.15	2835-96-3 176.5	sub	4-13-00-01672	H$_2$O 2; EtOH 3; eth 3; bz 2
9214	Phenol, 2-amino-3-nitro-	C$_6$H$_6$N$_2$O$_3$ 154.13	603-85-0 216.5	sub	4-13-00-00895	H$_2$O 4
9215	Phenol, 4-(2-aminopropyl)-, (±)- Hydroxyamphetamine	C$_9$H$_{13}$NO 151.21	1518-86-1 125.5	4740	4-13-00-01871	H$_2$O 3; EtOH 3; bz 3; chl 3
9216	Phenol, 2,2'-azobis-	C$_{12}$H$_{10}$N$_2$O$_2$ 214.22	2050-14-8 173	140[0.001]	2-16-00-00033	H$_2$O 1; EtOH 2; eth 4; bz 2
9217	Phenol, 2-(2-benzothiazolyl)-	C$_{13}$H$_9$NOS 227.29	3411-95-8 131	175-93[3]	4-27-00-02081	EtOH 3
9218	Phenol, 2-(2-benzoxazolyl)-	C$_{13}$H$_9$NO$_2$ 211.22	835-64-3 123.5	338	4-27-00-02080	H$_2$O 2; EtOH 4; eth 3; ace 3
9219	Phenol, 2,4-bis(1,1-dimethylethyl)-	C$_{14}$H$_{22}$O 206.33	96-76-4 56.5	263.5	4-06-00-03493	ctc 2; alk 1 1.5080[20]
9220	Phenol, 2,6-bis(1,1-dimethylethyl)-	C$_{14}$H$_{22}$O 206.33	128-39-2 39	161[50;] 133[20]	4-06-00-03492	EtOH 2; ctc 3; alk 1 1.5001[20]
9221	Phenol, 2,6-bis(1,1-dimethylethyl)-4-ethyl-	C$_{16}$H$_{26}$O 234.38	4130-42-1 44	272	4-06-00-03529	alk 1
9222	Phenol, 2,4-bis(1,1-dimethylethyl)-5-methyl- DBMC	C$_{15}$H$_{24}$O 220.35	497-39-2 62.1	2829 282	4-06-00-03510 0.912[80]	H$_2$O 1; EtOH 3; eth 3; ace 3
9223	Phenol, 2,4-bis(1,1-dimethylethyl)-6-methyl-	C$_{15}$H$_{24}$O 220.35	616-55-7 51	269	4-06-00-03510 0.891[80]	alk 1
9224	Phenol, 2,6-bis(1,1-dimethylethyl)-4-methyl- 2,6-Di-tert-butyl-4-methylphenol	C$_{15}$H$_{24}$O 220.35	128-37-0 71	1548 265	4-06-00-03511 0.8937[75]	H$_2$O 1; EtOH 3; ace 3; bz 3 1.4859[75]
9225	Phenol, 2,6-bis(1,1-dimethylpropyl)-4-methyl- p-Cresol, 2,6-di-tert-pentyl-	C$_{17}$H$_{28}$O 248.41	56103-67-4	283	4-06-00-03533 0.931[25]	1.4950[20]
9226	Phenol, 2,6-bis(1-methylethyl)- Propofol	C$_{12}$H$_{18}$O 178.27	2078-54-8 18	7847 256	4-06-00-03435 1.5140[20]	1.5140[20]
9227	Phenol, 2,6-bis(1-methylpropyl)- 2,6-Di-sec-butylphenol	C$_{14}$H$_{22}$O 206.33	5510-99-6 -42	257.5	4-06-00-03490	1.5080[20]
9228	Phenol, 2-bromo-	C$_6$H$_5$BrO 173.01	95-56-7 5.6	194.5	4-06-00-01037 1.4924[20]	H$_2$O 2; EtOH 3; eth 3; chl 2 1.589[20]
9229	Phenol, 3-bromo-	C$_6$H$_5$BrO 173.01	591-20-8 33	236.5	4-06-00-01042	H$_2$O 2; EtOH 4; eth 4; ctc 2
9230	Phenol, 4-bromo-	C$_6$H$_5$BrO 173.01	106-41-2 66.4	238	4-06-00-01043 1.840[15]	H$_2$O 3; EtOH 4; eth 4; chl 3
9231	Phenol, 3-bromo-5-chloro-	C$_6$H$_4$BrClO 207.45	56962-04-0 70	256	2-06-00-00187	EtOH 4; HOAc 4
9232	Phenol, 4-bromo-2-chloro- 4-Bromo-2-chlorophenol	C$_6$H$_4$BrClO 207.45	3964-56-5 50.5	233.5	3-06-00-00750 1.6170[20]	EtOH 4; eth 4; ace 4; bz 4 1.5859[20]

No.	Name Synonym	Mol. Form. Mol. Wt.	CAS RN mp/°C	Merck No. bp/°C	Beil. Ref. den/g cm⁻³	Solubility n_D
9233	Phenol, 2-bromo-4,6-dichloro-	$C_6H_3BrCl_2O$ 241.90	4524-77-0 68	220[200]	3-06-00-00751	bz 4; eth 4; chl 4
9234	Phenol, 2-bromo-4,6-dinitro-	$C_6H_3BrN_2O_5$ 263.00	2316-50-9 119.3	sub	3-06-00-00871	H_2O 2; EtOH 3; eth 4; bz 4
9235	Phenol, 4-bromo-2,6-dinitro-	$C_6H_3BrN_2O_5$ 263.00	40466-95-3 78	sub	3-06-00-00872	bz 4; eth 4; EtOH 4; chl 4
9236	Phenol, 2-bromo-4-methyl-	C_7H_7BrO 187.04	6627-55-0 56.5	213.5	4-06-00-02143 1.5422[25]	H_2O 2; EtOH 3; bz 3; chl 3 1.5772[20]
9237	Phenol, 2-bromo-5-methyl- 6-Bromo-m-cresol	C_7H_7BrO 187.04	14847-51-9 38.8	209	3-06-00-01320	ace 4; eth 4; EtOH 4
9238	Phenol, 3-bromo-2-methyl-	C_7H_7BrO 187.04	7766-23-6 95	55[4]	0-06-00-00360	ace 4; bz 4; eth 4; EtOH 4
9239	Phenol, 3-bromo-4-methyl-	C_7H_7BrO 187.04	60710-39-6 56	246	4-06-00-02143	ace 4; bz 4; eth 4; EtOH 4
9240	Phenol, 3-bromo-5-methyl-	C_7H_7BrO 187.04	74204-00-5 56.5	161[28]	2-06-00-00357	eth 4; EtOH 4
9241	Phenol, 4-bromo-2-methyl-	C_7H_7BrO 187.04	2362-12-1 64.7	235	4-06-00-02006	ace 4; eth 4; EtOH 4
9242	Phenol, 4-bromo-3-methyl-	C_7H_7BrO 187.04	14472-14-1 63.5	142[23]	4-06-00-02072	eth 4; py 4
9243	Phenol, 2-bromo-3-nitro-	$C_6H_4BrNO_3$ 218.01	101935-40-4 147.5	sub	3-06-00-00844	eth 4; EtOH 4
9244	Phenol, 4-bromo-2-nitro- 4-Bromo-2-nitrophenol	$C_6H_4BrNO_3$ 218.01	7693-52-9 92	sub	4-06-00-01363	bz 4; eth 4; EtOH 4; chl 4
9245	Phenol, 2-butoxy-	$C_{10}H_{14}O_2$ 166.22	39075-90-6 	232.5	4-06-00-05566 1.026[25]	1.5113[25]
9246	Phenol, 2-butyl-	$C_{10}H_{14}O$ 150.22	3180-09-4 -20	235	4-06-00-03267 0.975[20]	H_2O 1; EtOH 3; eth 3; alk 3 1.5180[25]
9247	Phenol, 3-butyl-	$C_{10}H_{14}O$ 150.22	4074-43-5 	248	4-06-00-03269 0.974[20]	eth 4; EtOH 4
9248	Phenol, 4-butyl-	$C_{10}H_{14}O$ 150.22	1638-22-8 22	248	4-06-00-03269 0.976[22]	H_2O 1; EtOH 3; eth 3; ctc 2 1.5165[25]
9249	Phenol, 4,4'-butylidenebis-	$C_{16}H_{18}O_2$ 242.32	4731-84-4 137	270[12]	4-06-00-06736	bz 4; EtOH 4; tol 4
9250	Phenol, 2-butyl-4-methyl-	$C_{11}H_{16}O$ 164.25	6891-45-8 19	228; 124[15]	4-06-00-03392 0.9680[20]	
9251	Phenol, 2-chloro- o-Chlorophenol	C_6H_5ClO 128.56	95-57-8 9.8	2154 174.9	4-06-00-00782 1.2634[20]	H_2O 2; EtOH 3; eth 3; bz 4 1.5524[20]
9252	Phenol, 3-chloro- m-Chlorophenol	C_6H_5ClO 128.56	108-43-0 32.6	2154 214	4-06-00-00810 1.245[45]	H_2O 2; EtOH 3; eth 3; bz 4 1.5565[40]
9253	Phenol, 4-chloro- p-Chlorophenol	C_6H_5ClO 128.56	106-48-9 42.7	2154 220	4-06-00-00820 1.265[40]	H_2O 1; EtOH 4; eth 4; bz 4 1.5579[40]
9254	Phenol, 2-chloro-4-cyclohexyl- 2-Chloro-4-cyclohexylphenol	$C_{12}H_{15}ClO$ 210.70	3964-61-2 40.5	160[12]; 122[3]	4-06-00-02508 1.1600[15]	ctc 3; CS_2 3
9255	Phenol, 4-chloro-2-cyclopentyl- Dowicide 9	$C_{11}H_{13}ClO$ 196.68	13347-42-7 	3420 181-5[18]	3-06-00-02478	
9256	Phenol, 2-chloro-3,4-dimethyl-	C_8H_9ClO 156.61	10283-15-5 27	188	2-06-00-00456 1.5538[20]	peth 4
9257	Phenol, 4-chloro-3,5-dimethyl- Chloroxylenol	C_8H_9ClO 156.61	88-04-0 115	2176 246	4-06-00-03152	H_2O 2; EtOH 3; eth 3; bz 2
9258	Phenol, 4-chloro-2,6-dinitro- 4-Chloro-2,6-dinitrophenol	$C_6H_3ClN_2O_5$ 218.55	88-87-9 81		4-06-00-01386 1.74[22]	eth 4; EtOH 4; chl 4
9259	Phenol, 5-chloro-2,4-dinitro- 3-Chloro-4,6-dinitrophenol	$C_6H_3ClN_2O_5$ 218.55	54715-57-0 92		4-06-00-01385 1.74[22]	eth 4; EtOH 4; chl 4; peth 4
9260	Phenol, 2-chloro-3-methyl-	C_7H_7ClO 142.58	608-26-4 56	230	4-06-00-02062	H_2O 2; HOAc 4; peth 3
9261	Phenol, 2-chloro-4-methyl-	C_7H_7ClO 142.58	6640-27-3 	195.5	4-06-00-02135 1.1785[27]	bz 4; eth 4; EtOH 4 1.5200[27]
9262	Phenol, 2-chloro-5-methyl- 6-Chloro-m-cresol	C_7H_7ClO 142.58	615-74-7 45.5	196	3-06-00-01315 1.21[15]	H_2O 4; EtOH 4
9263	Phenol, 2-chloro-6-methyl-	C_7H_7ClO 142.58	87-64-9 	189; 80[20]	4-06-00-01984	eth 3 1.5449[20]
9264	Phenol, 3-chloro-2-methyl-	C_7H_7ClO 142.58	3260-87-5 86	225	4-06-00-02000	bz 4; eth 4; EtOH 4
9265	Phenol, 3-chloro-4-methyl-	C_7H_7ClO 142.58	615-62-3 55.5	228	3-06-00-01374	bz 4; eth 4; EtOH 4
9266	Phenol, 4-chloro-2-methyl- 4-Chloro-o-cresol	C_7H_7ClO 142.58	1570-64-5 51	223	4-06-00-01987	H_2O 2; peth 3
9267	Phenol, 4-chloro-3-methyl- 4-Chloro-m-cresol	C_7H_7ClO 142.58	59-50-7 67	2133 235	4-06-00-02064	H_2O 2; EtOH 3; eth 3; chl 2
9268	Phenol, 4-chloro-5-methyl-2-(1-methylethyl)- Chlorothymol	$C_{10}H_{13}ClO$ 184.67	89-68-9 63	2171 258.5	4-06-00-03344	H_2O 4; EtOH 3; eth 3; bz 3
9269	Phenol, 5-chloro-2-nitro- 5-Chloro-2-nitrophenol	$C_6H_4ClNO_3$ 173.56	611-07-4 41	sub	3-06-00-00836	H_2O 2; EtOH 3; eth 3; HOAc 3
9270	Phenol, 4-chloro-2-(phenylmethyl)- Clorophene	$C_{13}H_{11}ClO$ 218.68	120-32-1 48.5	2403 160-2[3.5]	4-06-00-04636 1.185[58]	ctc 3; CS_2 3
9271	Phenol, 4-chloro-2-(2-propenyl)-	C_9H_9ClO 168.62	13997-73-4 48	130-2[15]	4-06-00-03813 1.171[15]	bz 4; EtOH 4
9272	Phenol, 4-cyclohexyl-	$C_{12}H_{16}O$ 176.26	1131-60-8 133	294; 133[4]	4-06-00-03904	H_2O 1; EtOH 4; eth 4; bz 3

No.	Name Synonym	Mol. Form. Mol. Wt.	CAS RN mp/°C	Merck No. bp/°C	Bell. Ref. den/g cm^{-3}	Solubility n_D
9273	Phenol, 2-cyclohexyl-4,6-dinitro- 4,6-Dinitro-o-cyclohexylphenol	$C_{12}H_{14}N_2O_5$ 266.25	131-89-5	2739	4-06-00-03903	
9274	Phenol, 2,4-diamino- 2,4-Diaminophenol	$C_8H_8N_2O$ 124.14	95-86-3 78-80 dec	2961	4-13-00-01425	H_2O 4; ace 4; EtOH 4
9275	Phenol, 4,4'-(1,2-diarsenediyl)bis[2-amino-, dihydrochloride Salvarsan	$C_{12}H_{20}As_2Cl_2O_5$ 465.04	139-93-5 185-95 dec	839	2-16-00-00560	H_2O 4
9276	Phenol, 2,4-dibromo- 2,4-Dibromophenol	$C_6H_4Br_2O$ 251.91	615-58-7 38	238.5	4-06-00-01061 2.070[20]	H_2O 2; EtOH 4; eth 4; bz 4
9277	Phenol, 2,6-dibromo- 2,6-Dibromophenol	$C_6H_4Br_2O$ 251.91	608-33-3 56.5	255; 162[21]	4-06-00-01064	H_2O 3; EtOH 4; eth 4
9278	Phenol, 3,5-dibromo- 3,5-Dibromophenol	$C_6H_4Br_2O$ 251.91	626-41-5 81	274; 121[3]	4-06-00-01065	H_2O 2; EtOH 4; eth 3; lig 2
9279	Phenol, 2,4-dibromo-6-methyl-	$C_7H_6Br_2O$ 265.93	609-22-3 58	265 dec; 105[4]	3-06-00-01271	chl 3
9280	Phenol, 3,5-dibromo-2-methyl-	$C_7H_6Br_2O$ 265.93	14122-00-0 100.3	285	2-06-00-00334	peth 4
9281	Phenol, 2,4-dibromo-3-methyl-6-(1- methylethyl)- Thymol, 2,6-dibromo	$C_{10}H_{12}Br_2O$ 308.01	52262-38-1 3.7	160-1[16]	2-06-00-00500 1.6568[17]	
9282	Phenol, 2,3-dichloro- 2,3-Dichlorophenol	$C_6H_4Cl_2O$ 163.00	576-24-9 58		4-06-00-00883	EtOH 3; eth 3; bz 3; lig 3
9283	Phenol, 2,4-dichloro- 2,4-Dichlorophenol	$C_6H_4Cl_2O$ 163.00	120-83-2 45	3061 210	4-06-00-00885	H_2O 2; EtOH 3; eth 3; bz 3
9284	Phenol, 2,5-dichloro- 2,5-Dichlorophenol	$C_6H_4Cl_2O$ 163.00	583-78-8 59	211	4-06-00-00942	H_2O 2; EtOH 4; eth 4; bz 3
9285	Phenol, 2,6-dichloro- 2,6-Dichlorophenol	$C_6H_4Cl_2O$ 163.00	87-65-0 68.5	3062 220; 82[4]	4-06-00-00949 1.653[20]	EtOH 4; eth 4; bz 3; peth 3
9286	Phenol, 3,4-dichloro- 3,4-Dichlorophenol	$C_6H_4Cl_2O$ 163.00	95-77-2 68	253	4-06-00-00952	H_2O 2; EtOH 4; eth 4; bz 3
9287	Phenol, 3,5-dichloro- 3,5-Dichlorophenol	$C_6H_4Cl_2O$ 163.00	591-35-5 68	233	4-06-00-00957	H_2O 2; EtOH 4; eth 4; peth 3
9288	Phenol, 2,4-dichloro-, acetate 2,4-Dichlorophenyl acetate	$C_8H_6Cl_2O_2$ 205.04	6341-97-5	244.5	3-06-00-00704	EtOH 3; eth 3; chl 3
9289	Phenol, 2,4-dichloro-, benzenesulfonate Genite	$C_{12}H_8Cl_2O_3S$ 303.17	97-16-5 45.5	4283	4-11-00-00034	ctc 3; CS_2 3
9290	Phenol, 2,4-dichloro-3,5-dimethyl- Dichloroxylenol	$C_8H_8Cl_2O$ 191.06	133-53-9 83	3066	4-06-00-03156	eth 4
9291	Phenol, 2,6-dichloro-3,5-dimethyl-	$C_8H_8Cl_2O$ 191.06	1943-54-0 87.5	105-10[1]	4-06-00-03157	eth 4; chl 4
9292	Phenol, 2,4-dichloro-3-methyl- m-Cresol, 2,4-dichloro	$C_7H_6Cl_2O$ 177.03	17788-00-0 58	236; 77[4]	3-06-00-01319	eth 4; chl 4
9293	Phenol, 2,4-dichloro-5-methyl-	$C_7H_6Cl_2O$ 177.03	1124-07-8 73	234	4-06-00-02069 1.572[20]	chl 4; peth 3
9294	Phenol, 2,6-dichloro-3-methyl- 2,6-Dichloro-m-cresol	$C_7H_6Cl_2O$ 177.03	13481-70-4 28	237; 77[4]	3-06-00-01319	eth 4; chl 4
9295	Phenol, 2,6-dichloro-4-methyl-	$C_7H_6Cl_2O$ 177.03	2432-12-4 39	231; 138[28]	4-06-00-02141	eth 4; EtOH 4; HOAc 4
9296	Phenol, 2,6-dichloro-4-nitro-	$C_6H_3ClNO_3$ 172.55	618-80-4 127 exp		4-06-00-01361 1.822[25]	eth 4; chl 4
9297	Phenol, 3-(diethylamino)-	$C_{10}H_{15}NO$ 165.24	91-68-9 78	276; 170[15]	4-13-00-00969	H_2O 3; EtOH 3; eth 3; CS_2 3
9298	Phenol, 4,4'-(1,2-diethyl-1,2-ethanediyl)bis-, (R*,S*)- Hexestrol	$C_{18}H_{22}O_2$ 270.37	84-16-2 186.5	4621	4-06-00-06761	ace 4; eth 4; EtOH 4
9299	Phenol, 4,4'-(1,2-diethyl-1,2-ethanediyl)bis[2- methyl- Methestrol	$C_{20}H_{26}O_2$ 298.43	130-73-4 145	5888	4-06-00-06786	
9300	Phenol, 4,4'-(1,2-diethyl-1,2-ethanediyl)bis-, (E)- Diethylstilbestrol	$C_{18}H_{20}O_2$ 268.36	56-53-1 170.5	3118	4-06-00-06856	eth 4; EtOH 4; chl 4
9301	Phenol, 4,4'-(1,2-diethyl-1,2-ethanediyl)bis-, dipropanoate, (E)- Diethylstilbestrol dipropionate	$C_{24}H_{28}O_4$ 380.48	130-80-3 104	3119	4-06-00-06857	bz 4; eth 4; EtOH 4
9302	Phenol, 4,4'-(1,2-diethylidene-1,2- ethanediyl)bis- Dienestrol	$C_{18}H_{18}O_2$ 266.34	84-17-3 227.5	3094 sub 130	4-06-00-06916	ace 4; eth 4; EtOH 4
9303	Phenol, 4,4'-(1,2-diethyl-3-methyl-1,3- propanediyl)bis- Benzestrol	$C_{20}H_{26}O_2$ 298.43	85-95-0 164	1082	3-06-00-05538	ace 4; eth 4; EtOH 4; HOAc 4
9304	Phenol, 2,6-diiodo-4-nitro- Disophenol	$C_6H_3I_2NO_3$ 390.90	305-85-1 157	3365	4-06-00-01368	EtOH 4
9305	Phenol, 2,6-dimethoxy-	$C_8H_{10}O_3$ 154.17	91-10-1 56.5	261	4-06-00-07329	eth 4; EtOH 4
9306	Phenol, 3,5-dimethoxy-	$C_8H_{10}O_3$ 154.17	500-99-2 37	199[35]; 170[10]	4-06-00-07362	eth 3; bz 3; lig 2
9307	Phenol, 2,3-dimethyl- 2,3-Xylenol	$C_8H_{10}O$ 122.17	528-75-0 72.8	9989 216.9	4-06-00-03096	H_2O 2; EtOH 3; eth 3 1.5420[20]

PHYSICAL CONSTANTS OF ORGANIC COMPOUNDS (continued)

No.	Name Synonym	Mol. Form. Mol. Wt.	CAS RN mp/°C	Merck No. bp/°C	Bell. Ref. den/g cm⁻³	Solubility n_D
9308	Phenol, 2,4-dimethyl- 2,4-Xylenol	$C_8H_{10}O$ 122.17	105-67-9 24.5	9989 210.9	4-06-00-03126 0.9650[20]	H_2O 2; EtOH 5; eth 5; ctc 3 1.5420[14]
9309	Phenol, 2,5-dimethyl- 2,5-Xylenol	$C_8H_{10}O$ 122.17	95-87-4 74.8	9989 211.1	4-06-00-03164	H_2O 3; EtOH 3; eth 4; chl 2
9310	Phenol, 2,6-dimethyl- 2,6-Xylenol	$C_8H_{10}O$ 122.17	576-26-1 45.7	9989 201.0	4-06-00-03112	H_2O 3; EtOH 3; eth 3; ctc 3
9311	Phenol, 3,4-dimethyl- 3,4-Xylenol	$C_8H_{10}O$ 122.17	95-65-8 60.8	9989 227	4-06-00-03099 0.9830[20]	H_2O 2; EtOH 3; eth 5; ctc 3
9312	Phenol, 3,5-dimethyl- 3,5-Xylenol	$C_8H_{10}O$ 122.17	108-68-9 63.6	9989 221.7	4-06-00-03141 0.9680[20]	H_2O 3; EtOH 3; ctc 3
9313	Phenol, 2,4-dimethyl-, acetate	$C_{10}H_{12}O_2$ 164.20	877-53-2 226		2-06-00-00459 1.0298[15]	eth 4; EtOH 4 1.4990[15]
9314	Phenol, 2,5-dimethyl-, acetate	$C_{10}H_{12}O_2$ 164.20	877-48-5 <-20	237	2-06-00-00467 1.0624[15]	eth 4; EtOH 4
9315	Phenol, 3,4-dimethyl-, acetate	$C_{10}H_{12}O_2$ 164.20	22618-23-1 22	235	4-06-00-03102	bz 4; eth 4; EtOH 4
9316	Phenol, 3,5-dimethyl-, acetate	$C_{10}H_{12}O_2$ 164.20	877-82-7 <-20	237	3-06-00-01757 1.0624[15]	eth 4; EtOH 4
9317	Phenol, 2-(dimethylamino)-	$C_8H_{11}NO$ 137.18	3743-22-4 45	199.5	4-13-00-00815	eth 4; EtOH 4; HOAc 4
9318	Phenol, 3-(dimethylamino)-	$C_8H_{11}NO$ 137.18	99-07-0 86	266.5	4-13-00-00959	H_2O 1; EtOH 3; eth 3; ace 3 1.5895[26]
9319	Phenol, 4-(dimethylamino)-	$C_8H_{11}NO$ 137.18	619-60-3 77	165[30]	4-13-00-01040	H_2O 2; EtOH 3; eth 3
9320	Phenol, 4-[2-(dimethylamino)ethyl]- Hordenine	$C_{10}H_{15}NO$ 165.24	539-15-1 117.5	4666 173[11]	4-13-00-01790	eth 4; EtOH 4; chl 4
9321	Phenol, 2-(1,1-dimethylethyl)-	$C_{10}H_{14}O$ 150.22	88-18-6 -6.8	223	4-06-00-03292 0.9783[20]	EtOH 3; eth 4; ctc 3; alk 3 1.5160[20]
9322	Phenol, 3-(1,1-dimethylethyl)-	$C_{10}H_{14}O$ 150.22	585-34-2 42.3	240	4-06-00-03294	EtOH 3; eth 4; alk 3
9323	Phenol, 4-(1,1-dimethylethyl)- p-tert-Butylphenol	$C_{10}H_{14}O$ 150.22	98-54-4 98	1584 237	4-06-00-03296 0.908[80]	H_2O 3; EtOH 3; eth 3; chl 3 1.4787[114]
9324	Phenol, 2-(1,1-dimethylethyl)-4,5-dimethyl- 3,4-Xylenol, 6-tert-butyl-	$C_{12}H_{18}O$ 178.27	1445-23-4 46	258.5	3-06-00-02019 0.920[80]	1.5222[20]
9325	Phenol, 2-(1,1-dimethylethyl)-4,6-dimethyl-	$C_{12}H_{18}O$ 178.27	1879-09-0 22.3	249	4-06-00-03442 0.917[80]	alk 1 1.5183[20]
9326	Phenol, 4-(1,1-dimethylethyl)-2,5-dimethyl-	$C_{12}H_{18}O$ 178.27	17696-37-6 71.2	264	3-06-00-02019 0.939[80]	alk 3 1.5311[20]
9327	Phenol, 4-(1,1-dimethylethyl)-2,6-dimethyl-	$C_{12}H_{18}O$ 178.27	879-97-0 82.4	248	3-06-00-02019 0.916[80]	alk 3
9328	Phenol, 2-(1,1-dimethylethyl)-4,6-dinitro- 2-tert-Butyl-4,6-dinitrophenol	$C_{10}H_{12}N_2O_5$ 240.22	1420-07-1		4-06-00-03293	
9329	Phenol, (1,1-dimethylethyl)-4-methoxy- Butylated hydroxyanisole	$C_{11}H_{16}O_2$ 180.25	25013-16-5 51	1547 268		H_2O 1; peth 3; EtOH 3
9330	Phenol, 2-(1,1-dimethylethyl)-4-methyl-	$C_{11}H_{16}O$ 164.25	2409-55-4 51.5	237	4-06-00-03397 0.9247[75]	H_2O 2; ace 3; bz 3; chl 3 1.4969[75]
9331	Phenol, 2-(1,1-dimethylethyl)-5-methyl-	$C_{11}H_{16}O$ 164.25	88-60-8 46.5	127[11]	4-06-00-03400 0.922[80]	H_2O 1; EtOH 3; eth 3; ace 3 1.5250[20]
9332	Phenol, 2-(1,1-dimethylethyl)-6-methyl-	$C_{11}H_{16}O$ 164.25	2219-82-1 31	230	4-06-00-03396 0.9240[80]	1.5195[20]
9333	Phenol, 4-(1,1-dimethylethyl)-2-methyl-	$C_{11}H_{16}O$ 164.25	98-27-1 27.5	237; 132[20]	4-06-00-03398 0.965[20]	H_2O 1; eth 3; ace 3; bz 3 1.5230[20]
9334	Phenol, 3,6-dimethyl-2-nitro-	$C_8H_9NO_3$ 167.16	71608-10-1 34.5	236 dec; 150[15]	4-06-00-03169	EtOH 4; eth 3; ace 3; bz 3
9335	Phenol, 2,4-dimethyl-, phosphate (3:1)	$C_{24}H_{27}O_4P$ 410.45	3862-12-2 233.5		0-06-00-00488 1.142[38]	H_2O 1; bz 3; chl 3; hx 3 1.5550[20]
9336	Phenol, 2,5-dimethyl-, phosphate (3:1)	$C_{24}H_{27}O_4P$ 410.45	19074-59-0 79.8	260-5[8]	3-06-00-01773 1.197[25]	H_2O 1; EtOH 2; eth 3; bz 3
9337	Phenol, 2,6-dimethyl-, phosphate (3:1)	$C_{24}H_{27}O_4P$ 410.45	121-06-2 137.8	262-4[6]	4-06-00-03121	H_2O 1; EtOH 2; bz 3; hx 2
9338	Phenol, 3,4-dimethyl-, phosphate (3:1)	$C_{24}H_{27}O_4P$ 410.45	3862-11-1 72	260-3[7]	0-06-00-00482	H_2O 1; EtOH 2; bz 3; chl 2
9339	Phenol, 4-(1,1-dimethylpropyl)- p-tert-Pentylphenol	$C_{11}H_{16}O$ 164.25	80-46-6 95	7098 262.5	4-06-00-03383	
9340	Phenol, 4-(1,1-dimethylpropyl)-2-methyl-	$C_{12}H_{18}O$ 178.27	71745-63-6 15	260	3-06-00-02010 0.969[20]	chl 2 1.5240[20]
9341	Phenol, 2,3-dinitro- 2,3-Dinitrophenol	$C_6H_4N_2O_5$ 184.11	66-56-8 144.5		4-06-00-01369 1.681[20]	H_2O 2; EtOH 4; eth 4; bz 3
9342	Phenol, 2,4-dinitro- 2,4-Dinitrophenol	$C_6H_4N_2O_5$ 184.11	51-28-5 115.5	3274 sub	4-06-00-01369 1.683[24]	H_2O 2; EtOH 3; eth 3; ace 3
9343	Phenol, 2,5-dinitro- 2,5-Dinitrophenol	$C_6H_4N_2O_5$ 184.11	329-71-5 108	3275	4-06-00-01383	bz 4; eth 4
9344	Phenol, 2,6-dinitro- 2,6-Dinitrophenol	$C_6H_4N_2O_5$ 184.11	573-56-8 63.5	3276	4-06-00-01383	H_2O 1; EtOH 4; eth 4; ace 4
9345	Phenol, 3,4-dinitro-	$C_6H_4N_2O_5$ 184.11	577-71-9 134		4-06-00-01384 1.672[25]	bz 4; eth 4; EtOH 4
9346	Phenol, 3,5-dinitro- 3,5-Dinitrophenol	$C_6H_4N_2O_5$ 184.11	586-11-8 125.1		4-06-00-01385 1.702[25]	bz 4; eth 4; EtOH 4; chl 4
9347	Phenol, 2,6-dipropyl- 2,6-Dipropylphenol	$C_{12}H_{18}O$ 178.27	6626-32-0 27	256	3-06-00-02013 1.6450[0]	eth 4

3-255

No.	Name / Synonym	Mol. Form. / Mol. Wt.	CAS RN / mp/°C	Merck No. / bp/°C	Beil. Ref. / den/g cm^{-3}	Solubility / n_D
9348	Phenol, 2-ethenyl-	C_8H_8O 120.15	695-84-1 29.5	 101[14]	4-06-00-03771 1.0609[18]	eth 4; EtOH 4 1.5851[20]
9349	Phenol, 3-ethenyl-	C_8H_8O 120.15	620-18-8 	 114-6[16]	4-06-00-03773 1.0353[21]	 1.5804[21]
9350	Phenol, 2-ethoxy-	$C_8H_{10}O_2$ 138.17	94-71-3 29	 217	4-06-00-05565 1.0903[25]	H_2O 2; EtOH 5; eth 5; ctc 2
9351	Phenol, 3-ethoxy-	$C_8H_{10}O_2$ 138.17	621-34-1 	 131[10]	4-06-00-05664 1.0705[4]	H_2O 1; EtOH 3; eth 3; bz 3
9352	Phenol, 4-ethoxy-	$C_8H_{10}O_2$ 138.17	622-62-8 66.5	 246.5	4-06-00-05719	H_2O 2; EtOH 4; eth 4; chl 3
9353	Phenol, 2-ethyl- o-Ethylphenol	$C_8H_{10}O$ 122.17	90-00-6 18	7302 204.5	4-06-00-03011 1.0146[25]	ace 4; bz 4; eth 4; EtOH 4 1.5367[20]
9354	Phenol, 3-ethyl- m-Ethylphenol	$C_8H_{10}O$ 122.17	620-17-7 -4	 218.4	4-06-00-03016 1.0283[20]	H_2O 2; EtOH 4; eth 4; chl 2
9355	Phenol, 4-ethyl- p-Ethylphenol	$C_8H_{10}O$ 122.17	123-07-9 45.0	 217.9	4-06-00-03020	H_2O 2; EtOH 4; eth 4; ace 3 1.5239[25]
9356	Phenol, 3-(ethylamino)-	$C_8H_{11}NO$ 137.18	621-31-8 62	 176[12]	0-13-00-00408	bz 4; eth 4; EtOH 4
9357	Phenol, 4-ethyl-2-methoxy-	$C_9H_{12}O_2$ 152.19	2785-89-9 -7	 236.5	4-06-00-05927 1.0931[18]	
9358	Phenol, 4-[1-ethyl-2-(4-methoxyphenyl)-1-butenyl]- Mestilbol	$C_{19}H_{22}O_2$ 282.38	7773-60-6 117.5	5818 185-95[0.3]	4-06-00-06856	ace 4; eth 4; EtOH 4
9359	Phenol, 4-[1-ethyl-2-(4-methoxyphenyl)-1-butenyl]-, (E)-	$C_{19}H_{22}O_2$ 282.38	18839-90-2 117	 185-95[0.3]	4-06-00-06856	EtOH 4
9360	Phenol, 2-fluoro-	C_6H_5FO 112.10	367-12-4 16.1	 151.5	4-06-00-00770 1.120[25]	H_2O 3 1.5144[20]
9361	Phenol, 3-fluoro-	C_6H_5FO 112.10	372-20-3 13.7	 178	4-06-00-00772 1.238[25]	 1.5140[20]
9362	Phenol, 4-fluoro-	C_6H_5FO 112.10	371-41-5 48	 185.5	4-06-00-00773 1.1889[56]	H_2O 2; ace 3; peth 3
9363	Phenol, 4,4'-(3H-2,1-benzoxathiol-3-ylidene)bis-, S,S-dioxide Phenolsulfonphthalein	$C_{19}H_{14}O_5S$ 354.38	143-74-8 >300	7213	5-19-03-00457	H_2O 2; EtOH 2; eth 1; ace 2
9364	Phenol, 4,4'-(3H-2,1-benzoxathiol-3-ylidene)bis[2-bromo-6-methyl-, S,S-dioxide Bromcresol purple	$C_{21}H_{16}Br_2O_5S$ 540.23	115-40-2 241.5	1376	5-19-03-00460	
9365	Phenol, 4,4'-(3H-2,1-benzoxathiol-3-ylidene)bis[2-bromo-3-methyl-6-(1-methylethyl)-, S,S-dioxide Bromthymol blue	$C_{27}H_{28}Br_2O_5S$ 624.39	76-59-5 201	1435	5-19-03-00461	eth 4; EtOH 4
9366	Phenol, 4,4'-(3H-2,1-benzoxathiol-3-ylidene)bis[2,6-dibromo-, S,S-dioxide Bromphenol blue	$C_{19}H_{10}Br_4O_5S$ 669.97	115-39-9 279 dec	1434	5-19-03-00458	H_2O 2; EtOH 3; bz 3; HOAc 3
9367	Phenol, 4,4'-(3H-2,1-benzoxathiol-3-ylidene)bis[2,6-dibromo-3-methyl-, S,S-dioxide Bromcresol green	$C_{21}H_{14}Br_4O_5S$ 698.02	76-60-8 218.5	1375	5-19-03-00460	H_2O 2; EtOH 4; eth 4; bz 3
9368	Phenol, 4,4'-(3H-2,1-benzoxathiol-3-ylidene)bis[2-methyl-, S,S-dioxide Cresol red	$C_{21}H_{18}O_5S$ 382.44	1733-12-6 >300	2583	5-19-03-00460	H_2O 4; EtOH 4
9369	Phenol, 4,4'-(3H-2,1-benzoxathiol-3-ylidene)bis[5-methyl-2-(1-methylethyl)-, S,S-dioxide Thymol blue	$C_{27}H_{30}O_5S$ 466.60	76-61-9 221-4 dec	9334	5-19-03-00462	H_2O 2; EtOH 3; ace 2; bz 2
9370	Phenol, 4-(3-hydroxy-1-propenyl)-2-methoxy- Coniferyl alcohol	$C_{10}H_{12}O_3$ 180.20	458-35-5 74	2499 163-5[3]	2-06-00-01093	H_2O 1; EtOH 3; eth 4; alk 3
9371	Phenol, 2-iodo- o-Iodophenol	C_6H_5IO 220.01	533-58-4 43	4928 186[160]; 91[2]	4-06-00-01070 1.8757[80]	H_2O 3; EtOH 4; eth 4; CS_2 4
9372	Phenol, 3-iodo-	C_6H_5IO 220.01	626-02-8 118	 186[100]	4-06-00-01073	H_2O 2; EtOH 3; eth 3
9373	Phenol, 4-iodo- p-Iodophenol	C_6H_5IO 220.01	540-38-5 93.5	4929 139[5] dec	4-06-00-01074 1.8573[112]	H_2O 2; EtOH 4; eth 4
9374	Phenol, 2-iodo-4-methyl- p-Cresol, 2-iodo-	C_7H_7IO 234.04	16188-57-1 35	 117[12]	4-06-00-02147 1.684[25]	 1.5331[25]
9375	Phenol, 2-iodo-6-methyl-	C_7H_7IO 234.04	24885-45-8 15.5	 70-5[2]	4-06-00-02008 1.661[25]	 1.610[25]
9376	Phenol, 2-mercapto-	C_6H_6OS 126.18	1121-24-0 5.5	 217; 89[8]	4-06-00-05633 1.2371[0]	bz 4; eth 4; EtOH 4
9377	Phenol, 3-mercapto-	C_6H_6OS 126.18	40248-84-8 16.5	 188[35]	2-06-00-00827	bz 4; EtOH 4
9378	Phenol, 4-mercapto-	C_6H_6OS 126.18	637-89-8 29.5	 167[45]; 135[11]	4-06-00-05790 1.1285[25]	H_2O 3; EtOH 3; alk 3; con sulf 3 1.5101[25]
9379	Phenol, 2-methoxy- Guaiacol	$C_7H_8O_2$ 124.14	90-05-1 32	4457 205	4-06-00-05563 1.1287[21]	H_2O 2; EtOH 3; eth 3; ctc 3 1.5429[20]
9380	Phenol, 3-methoxy-	$C_7H_8O_2$ 124.14	150-19-6 <-17	 114[5]	4-06-00-05662 1.131[25]	H_2O 2; EtOH 5; eth 5; chl 2 1.5510[20]

No.	Name / Synonym	Mol. Form. / Mol. Wt.	CAS RN / mp/°C	Merck No. / bp/°C	Beil. Ref. / den/g cm^{-3}	Solubility / n_D
9381	Phenol, 4-methoxy- / 4-Methoxyphenol	$C_7H_8O_2$ / 124.14	150-76-5 / 57	/ 243	4-06-00-05717 /	H_2O 3; EtOH 4; eth 4; bz 3 /
9382	Phenol, 2-methoxy-, acetate	$C_9H_{10}O_3$ / 166.18	613-70-7 / 31.5	/ 123-4[13]	4-06-00-05581 / 1.1285[25]	H_2O 1; EtOH 3; eth 3 / 1.5101[25]
9383	Phenol, 2-methoxy-, benzoate / Guaiacol benzoate	$C_{14}H_{12}O_3$ / 228.25	531-37-3 / 57.5	4458 /	4-09-00-00371 /	eth 4; chl 4 /
9384	Phenol, 2-methoxy-, carbonate (2:1) / Guaiacol carbonate	$C_{15}H_{14}O_5$ / 274.27	553-17-3 / 89	4459 /	3-06-00-04233 /	H_2O 1; EtOH 2; eth 3; chl 4 /
9385	Phenol, 2-methoxy-4-methyl- / Creosol	$C_8H_{10}O_2$ / 138.17	93-51-6 / 5.5	2573 / 221	4-06-00-05878 / 1.098[20]	eth 4; EtOH 4 / 1.5353[25]
9386	Phenol, 2-methoxy-3-nitro-	$C_7H_7NO_4$ / 169.14	20734-71-8 / 70	/ sub	3-06-00-04263 /	H_2O 2; EtOH 4; chl 3; peth 3 /
9387	Phenol, 2-methoxy-, phosphate (3:1) / Guaiacol phosphate	$C_{21}H_{21}O_7P$ / 416.37	563-03-1 / 91	4460 / 275-80[3]	3-06-00-04246 /	ace 4; tol 4; chl 4 /
9388	Phenol, 2-methoxy-4-(1-propenyl)- / Isoeugenol	$C_{10}H_{12}O_2$ / 164.20	97-54-1 /	5054 / 266	3-06-00-04993 / 1.080[25]	eth 4; EtOH 4 / 1.5739[19]
9389	Phenol, 2-methoxy-4-(2-propenyl)- / Eugenol	$C_{10}H_{12}O_2$ / 164.20	97-53-0 / -7.5	3855 / 253.2	4-06-00-06337 / 1.0652[20]	H_2O 1; EtOH 5; eth 5; chl 3 / 1.5405[20]
9390	Phenol, 2-methoxy-5-(2-propenyl)-	$C_{10}H_{12}O_2$ / 164.20	501-19-9 / 8.5	/ 253.5	3-06-00-05024 / 1.0613[25]	eth 4; EtOH 4 / 1.5413[20]
9391	Phenol, 2-methoxy-6-(1-propenyl)-	$C_{10}H_{12}O_2$ / 164.20	1076-55-7 / 81	/ 267.5	3-06-00-04991 /	H_2O 2; EtOH 3; eth 3; ace 3 /
9392	Phenol, 2-methoxy-6-(2-propenyl)-	$C_{10}H_{12}O_2$ / 164.20	579-60-2 /	/ 250.5	4-06-00-06334 / 1.062[20]	ace 3; bz 3 / 1.5545[20]
9393	Phenol, 2-methoxy-4-(1-propenyl)-, acetate, (E)-	$C_{12}H_{14}O_3$ / 206.24	5912-87-8 / 80.5	/ 282.5	4-06-00-06330 / 1.0251[100]	eth 4 / 1.5052[100]
9394	Phenol, 2-methoxy-4-(1-propenyl)-, acetate, (Z)- / Isoeugenol, acetate (cis)	$C_{12}H_{14}O_3$ / 206.24	97412-23-2 /	/ 160-2[13]	3-06-00-05007 / 1.0947[19]	H_2O 1; eth 3 / 1.5418[20]
9395	Phenol, 2-methoxy-4-(2-propenyl)-, acetate	$C_{12}H_{14}O_3$ / 206.24	93-28-7 / 30.5	/ 281; 127-8[6]	3-06-00-05029 / 1.0806[20]	H_2O 1; EtOH 3; ctc 2 / 1.5205[20]
9396	Phenol, 2-methoxy-4-(1-propenyl)-, (E)-	$C_{10}H_{12}O_2$ / 164.20	5932-68-3 / 33.5	/ 141-2[13]	4-06-00-06324 / 1.0852[20]	H_2O 2; EtOH 3; eth 3; chl 3 / 1.5784[20]
9397	Phenol, 2-methoxy-4-(1-propenyl)-, (Z)-	$C_{10}H_{12}O_2$ / 164.20	5912-86-7 /	/ 133[13]; 80[0.5]	4-06-00-06324 / 1.0837[20]	H_2O 2; EtOH 3; eth 3 / 1.5726[20]
9398	Phenol, 2-methoxy-5-[(1,2,3,4-tetrahydro-6,7-dimethoxy-2-methyl-1-isoquinolinyl)methyl]-, (±)- / Laudanine	$C_{20}H_{25}NO_4$ / 343.42	85-64-3 / 167	5248 /	5-21-06-00048 / 1.26[20]	H_2O 2; EtOH 2; eth 2; bz 3 /
9399	Phenol, 2-methoxy-5-[(1,2,3,4-tetrahydro-6,7-dimethoxy-2-methyl-1-isoquinolinyl)methyl]-, (R)- / Laudanidine	$C_{20}H_{25}NO_4$ / 343.42	301-21-3 / 184.5	5247 /	5-21-06-00048 /	H_2O 4; bz 4 /
9400	Phenol, 2-methyl- / o-Cresol	C_7H_8O / 108.14	95-48-7 / 29.8	2580 / 191.0	4-06-00-01940 / 1.135[25]	H_2O 3; EtOH 4; eth 4; ace 5 / 1.5361[20]
9401	Phenol, 3-methyl- / m-Cresol	C_7H_8O / 108.14	108-39-4 / 11.8	2579 / 202.2	4-06-00-02035 / 1.0341[20]	H_2O 2; EtOH 5; eth 5; ace 5 / 1.5438[20]
9402	Phenol, 4-methyl- / p-Cresol	C_7H_8O / 108.14	106-44-5 / 35.5	2581 / 201.9	4-06-00-02093 / 1.0185[40]	H_2O 2; EtOH 5; eth 5; ace 5 / 1.5312[20]
9403	Phenol, 4-[2-(methylamino)propyl]- / Pholedrine	$C_{10}H_{15}NO$ / 165.24	370-14-9 / 161	7304 /	3-13-00-01710 /	eth 4; EtOH 4 /
9404	Phenol, 4-(3-methylbutyl)-	$C_{11}H_{16}O$ / 164.25	1805-61-4 / 29	/ 255; 128[7]	4-06-00-03378 / 0.9562[23]	/ 1.505[27]
9405	Phenol, 2-methyl-4,6-dinitro- / Dinitrocresol	$C_7H_6N_2O_5$ / 198.14	534-52-1 / 86.5	3272 /	4-06-00-02014 /	H_2O 2; EtOH 3; eth 3; ace 3 /
9406	Phenol, 3,3'-methylenebis- / Diphenylmethane, 3,3'-dihydroxy	$C_{13}H_{12}O_2$ / 200.24	10193-50-7 / 102.5	/ 235[3]	3-06-00-05411 /	eth 4; EtOH 4; HOAc 4 /
9407	Phenol, 4,4'-methylenebis-	$C_{13}H_{12}O_2$ / 200.24	620-92-8 / 162.5	/ sub	4-06-00-06664 /	EtOH 3; eth 3; chl 3; DMSO-d_6 2 /
9408	Phenol, 2,2'-methylenebis[4-chloro- / Dichlorophene	$C_{13}H_{10}Cl_2O_2$ / 269.13	97-23-4 / 177.5	3059 /	4-06-00-06658 /	H_2O 1; EtOH 3; ace 3 /
9409	Phenol, 2,2'-methylenebis[3,4,6-trichloro- / Hexachlorophene	$C_{13}H_6Cl_6O_2$ / 406.91	70-30-4 / 166.5	4602 /	4-06-00-06659 /	H_2O 1; EtOH 3; eth 3; ace 3 /
9410	Phenol, 2-(1-methylethyl)-	$C_9H_{12}O$ / 136.19	88-69-7 / 15.5	/ 213.5	4-06-00-03210 / 1.012[20]	H_2O 2; EtOH 3; eth 3; bz 3 / 1.5315[20]
9411	Phenol, 3-(1-methylethyl)-	$C_9H_{12}O$ / 136.19	618-45-1 / 26	/ 228	4-06-00-03214 /	eth 4 / 1.5261[20]
9412	Phenol, 4-(1-methylethyl)-	$C_9H_{12}O$ / 136.19	99-89-8 / 62.3	/ 230; 110[10]	4-06-00-03215 / 0.990[20]	H_2O 2; EtOH 3; chl 3 / 1.5228[20]
9413	Phenol, 4,4'-(1-methylethylidene)bis- / Bisphenol A	$C_{15}H_{16}O_2$ / 228.29	80-05-7 / 153	1311 / 220[4]; 222[3]	4-06-00-06717 /	H_2O 1; EtOH 4; eth 4; bz 4 /
9414	Phenol, 2-methyl-4-(1-methylethyl)-	$C_{10}H_{14}O$ / 150.22	1740-97-2 / 8.6	/ 230; 83[3]	4-06-00-03330 / 0.9793[25]	H_2O 3; EtOH 3; bz 3; chl 3 / 1.5253[20]
9415	Phenol, 2-methyl-5-(1-methylethyl)- / Carvacrol	$C_{10}H_{14}O$ / 150.22	499-75-2 / 1	1881 / 237.7	4-06-00-03331 / 0.9772[20]	H_2O 2; EtOH 3; eth 3; ace 4 / 1.5230[20]
9416	Phenol, 2-methyl-6-(1-methylethyl)-	$C_{10}H_{14}O$ / 150.22	3228-04-4 / -14.5	/ 225.5	3-06-00-01882 / 0.9782[25]	EtOH 3; bz 3; chl 3; os 3 / 1.5239[15]
9417	Phenol, 4-methyl-2-(1-methylethyl)-	$C_{10}H_{14}O$ / 150.22	4427-56-9 / 36.5	/ 228.5	4-06-00-03329 / 0.9910[20]	H_2O 2; EtOH 3; bz 3; chl 3 / 1.5275[20]

No.	Name Synonym	Mol. Form. Mol. Wt.	CAS RN mp/°C	Merck No. bp/°C	Beil. Ref. den/g cm^{-3}	Solubility n_D
9418	Phenol, 5-methyl-2-(1-methylethyl)- Thymol	C$_{10}$H$_{14}$O 150.22	89-83-8 51.5	9333 232.5	4-06-00-03334 0.970[25]	H$_2$O 1; EtOH 4; eth 4; chl 4 1.5227[20]
9419	Phenol, 2-methyl-5-(1-methylethyl)-, acetate Carvacrol, acetate	C$_{12}$H$_{16}$O$_2$ 192.26	6380-28-5 	2-06-00-00494 246.5	0.9896[25]	eth 4; EtOH 4 1.4913[28]
9420	Phenol, 5-methyl-2-(1-methylethyl)-, acetate Thymol, acetate	C$_{12}$H$_{16}$O$_2$ 192.26	528-79-0 	4-06-00-03337 245	1.009[9]	bz 4; eth 4; EtOH 4; chl 4
9421	Phenol, 5-methyl-2-(1-methylpropyl)- 6-sec-Butyl-m-cresol	C$_{11}$H$_{16}$O 164.25	29472-95-5 	4-06-00-03394 89-90[2]	0.9612[25]	1.5188[20]
9422	Phenol, 2-methyl-4-nitro-	C$_7$H$_7$NO$_3$ 153.14	99-53-6 96	4-06-00-02011 188[9]; 132[0.2]		H$_2$O 2; EtOH 4; eth 4; bz 4
9423	Phenol, 2-methyl-6-nitro-	C$_7$H$_7$NO$_3$ 153.14	13073-29-5 70	4-06-00-02009 255 dec; 102[9]		H$_2$O 1; EtOH 4; eth 4; peth 3
9424	Phenol, 4-methyl-2-nitro-	C$_7$H$_7$NO$_3$ 153.14	119-33-5 36.5	4-06-00-02149 125[22]	1.2399[20]	ace 4; bz 4; eth 4; EtOH 4 1.5744[40]
9425	Phenol, 5-methyl-2-pentyl- 6-n-Amyl-m-cresol	C$_{12}$H$_{18}$O 178.27	1300-94-3 24	643 137-9[15]	3-06-00-02005	ace 4; eth 4; EtOH 4
9426	Phenol, 4-methyl-2-(phenylazo)-	C$_{13}$H$_{12}$N$_2$O 212.25	952-47-6 108.5	4-16-00-00197 sub		EtOH 3; eth 3; bz 3; chl 3
9427	Phenol, 2-methyl-6-(2-propenyl)-	C$_{10}$H$_{12}$O 148.20	3354-58-3 	4-06-00-03844 232	0.992[25]	1.5381[20]
9428	Phenol, 4-methyl-2-(1-propenyl)-	C$_{10}$H$_{12}$O 148.20	53889-94-4 	4-06-00-03844 126-8[14]	1.019[20]	1.5727[20]
9429	Phenol, 4-methyl-2-(2-propenyl)-	C$_{10}$H$_{12}$O 148.20	6628-06-4 	4-06-00-03845 237	0.96[25]	1.5385[20]
9430	Phenol, 2-(1-methylpropyl)- o-sec-Butylphenol	C$_{10}$H$_{14}$O 150.22	89-72-5 16	3-06-00-01852 228; 116[21]	0.9804[25]	1.5200[25]
9431	Phenol, 4-(1-methylpropyl)-	C$_{10}$H$_{14}$O 150.22	99-71-8 61.5	4-06-00-03279 241	0.986[20]	H$_2$O 1; EtOH 3; eth 4; alk 3 1.5182[21]
9432	Phenol, 4-(2-methylpropyl)-	C$_{10}$H$_{14}$O 150.22	4167-74-2 51.5	4-06-00-03288 237	0.9778[20]	ace 4; eth 4; EtOH 4 1.5319[25]
9433	Phenol, 4,4'-(1-methylpropylidene)bis- Bisphenol B	C$_{16}$H$_{18}$O$_2$ 242.32	77-40-7 120.5	1312	4-06-00-06738	ace 4; MeOH 4
9434	Phenol, 3-methyl-2,4,6-trinitro- 3-Methylpicric acid	C$_7$H$_5$N$_3$O$_7$ 243.13	602-99-3 109.5	4-06-00-02079 exp 150		H$_2$O 2; EtOH 3; eth 3; ace 3
9435	Phenol, 2-nitro- o-Nitrophenol	C$_6$H$_5$NO$_3$ 139.11	88-75-5 44.8	6541 216	4-06-00-01246 1.2942[40]	H$_2$O 2; EtOH 4; eth 4; ace 4 1.5723[50]
9436	Phenol, 3-nitro- m-Nitrophenol	C$_6$H$_5$NO$_3$ 139.11	554-84-7 96.8	6540 194[70]	4-06-00-01269 1.2797[100]	H$_2$O 2; EtOH 4; eth 4; ace 4
9437	Phenol, 4-nitro- p-Nitrophenol	C$_6$H$_5$NO$_3$ 139.11	100-02-7 113.8	6542	4-06-00-01279 1.479[20]	H$_2$O 2; EtOH 4; eth 4; ace 4
9438	Phenol, 4-nitroso- 4-Nitrosophenol	C$_6$H$_5$NO$_2$ 123.11	104-91-6 144 dec	6563	4-07-00-02073	H$_2$O 2; EtOH 3; eth 3; ace 3
9439	Phenol, pentabromo- Pentabromophenol	C$_6$HBr$_5$O 488.59	608-71-9 229.5	sub	3-06-00-00766	H$_2$O 1; EtOH 3; eth 2; bz 3
9440	Phenol, pentachloro- Pentachlorophenol	C$_6$HCl$_5$O 266.34	87-86-5 174	7059 310 dec	4-06-00-01025 1.978[22]	H$_2$O 2; EtOH 4; eth 4; bz 3
9441	Phenol, 3-pentadecyl-	C$_{21}$H$_{36}$O 304.52	501-24-6 53.5	230[8]; 197[1.5]	4-06-00-03556	ace 4; bz 4; EtOH 4
9442	Phenol, pentafluoro- Pentafluorophenol	C$_6$HF$_5$O 184.07	771-61-9 32.8	145.6	4-06-00-00782	1.4263[20]
9443	Phenol, pentamethyl- Pentamethylphenol	C$_{11}$H$_{16}$O 164.25	2819-86-5 128	267	4-06-00-03412	H$_2$O 1; EtOH 3
9444	Phenol, 4-pentyl-	C$_{11}$H$_{16}$O 164.25	14938-35-3 23	250.5	4-06-00-03370 0.960[20]	eth 4; EtOH 4 1.5272[25]
9445	Phenol, 2-(phenylamino)-	C$_{12}$H$_{11}$NO 185.23	644-71-3 69.5	185[20]	4-13-00-00822	H$_2$O 2; EtOH 3; eth 3; bz 2
9446	Phenol, 3-(phenylamino)-	C$_{12}$H$_{11}$NO 185.23	101-18-8 81.5	340	3-13-00-00944	H$_2$O 2; EtOH 4; eth 4; ace 4
9447	Phenol, 4-(phenylamino)-	C$_{12}$H$_{11}$NO 185.23	122-37-2 73	330	4-13-00-01052	H$_2$O 2; EtOH 4; eth 4; bz 4
9448	Phenol, 4-(phenylazo)-	C$_{12}$H$_{10}$N$_2$O 198.22	1689-82-3 155	225[20] dec	2-16-00-00038	H$_2$O 1; EtOH 4; eth 4; bz 3
9449	Phenol, 4-(phenylazo)-, acetate (ester)	C$_{14}$H$_{12}$N$_2$O$_2$ 240.26	13102-31-3 89	360 dec	1-16-00-00236	EtOH 4; lig 4
9450	Phenol, 2-[(phenylimino)methyl]-	C$_{13}$H$_{11}$NO 197.24	779-84-0 49.5	4-12-00-00339 1.087[25]		H$_2$O 1; EtOH 3
9451	Phenol, 2-(phenylmethoxy)-	C$_{13}$H$_{12}$O$_2$ 200.24	6272-38-4 	4-06-00-05572 173-4[13]	1.154[22]	eth 4; EtOH 4 1.5906[18]
9452	Phenol, 3-(phenylmethoxy)-	C$_{13}$H$_{12}$O$_2$ 200.24	3769-41-3 69.2	4-06-00-05669 200[5]		H$_2$O 4
9453	Phenol, 4-(phenylmethoxy)- Monobenzone	C$_{13}$H$_{12}$O$_2$ 200.24	103-16-2 122	6159	4-06-00-05728	H$_2$O 2; EtOH 4; eth 4; ace 3
9454	Phenol, 2-(phenylmethyl)- o-Benzylphenol	C$_{13}$H$_{12}$O 184.24	28994-41-4 21	1158 312	4-06-00-04628	ace 4; bz 4; EtOH 4 1.5994[20]
9455	Phenol, 4-(phenylmethyl)- p-Benzylphenol	C$_{13}$H$_{12}$O 184.24	101-53-1 84	1159 322	4-06-00-04640	H$_2$O 3; EtOH 3; eth 3; bz 3
9456	Phenol, 2-(1-propenyl)-	C$_9$H$_{10}$O 134.18	6380-21-8 -6	3-06-00-02393 220	1.0246[15]	eth 4 1.5181[20]
9457	Phenol, 2-(2-propenyl)-	C$_9$H$_{10}$O 134.18	1745-81-9 -6	4-06-00-03807 220	1.0246[15]	eth 4 1.5181[20]

No.	Name Synonym	Mol. Form. Mol. Wt.	CAS RN mp/°C	Merck No. bp/°C	Beil. Ref. den/g cm^{-3}	Solubility n_D
9458	Phenol, 4-(2-propenyl)- Chavicol	$C_9H_{10}O$ 134.18	501-92-8 15.8	2039 238	4-06-00-03817 1.0203[15]	eth 4; EtOH 4; chl 4 1.5441[18]
9459	Phenol, 2-(1-propenyl)-, (E)-	$C_9H_{10}O$ 134.18	23619-59-2 37.5	230.5	4-06-00-03793 1.0441[16]	eth 4; lig 4 1.5823[21]
9460	Phenol, 2-propoxy-	$C_9H_{12}O_2$ 152.19	6280-96-2	227	4-06-00-05566 1.0523[25]	EtOH 4 1.5176[25]
9461	Phenol, 2-propyl-	$C_9H_{12}O$ 136.19	644-35-9 7	220	4-06-00-03176 1.015[20]	eth 4; EtOH 4
9462	Phenol, 3-propyl-	$C_9H_{12}O$ 136.19	621-27-2 26	228	4-06-00-03180 0.987[20]	H_2O 2; EtOH 3; eth 3 1.5223[20]
9463	Phenol, 4-propyl-	$C_9H_{12}O$ 136.19	645-56-7 22	232.6	4-06-00-03181 1.009[20]	H_2O 2; EtOH 3; ctc 2 1.5379[25]
9464	Phenol, 4,4'-propylidenebis-	$C_{15}H_{16}O_2$ 228.29	1576-13-2 132	275[20]	3-06-00-05457	eth 4; EtOH 4
9465	Phenol, 4,4'-(2-pyridinylmethylene)bis-, diacetate (ester) Bisacodyl	$C_{22}H_{19}NO_4$ 361.40	603-50-9 133.5	1253	5-21-05-00415	
9466	Phenol, 4,4'-sulfonylbis-	$C_{12}H_{10}O_4S$ 250.28	80-09-1 240.5		4-06-00-05809 1.3663[15]	H_2O 1; EtOH 3; eth 3; bz 2
9467	Phenol, 2,3,4,6-tetrabromo- 2,3,4,6-Tetrabromophenol	$C_6H_2Br_4O$ 409.70	14400-94-3 113.5	sub	3-06-00-00766	bz 4; EtOH 4
9468	Phenol, 2,3,4,5-tetrabromo-6-methyl- 3,4,5,6-Tetrabromo-o-cresol	$C_7H_4Br_4O$ 423.72	576-55-6 208 dec	9120	3-06-00-01272	H_2O 1; EtOH 3; eth 3; bz 3
9469	Phenol, 2,3,4,5-tetrachloro- 2,3,4,5-Tetrachlorophenol	$C_6H_2Cl_4O$ 231.89	4901-51-3 116.5	sub	4-06-00-01020	EtOH 4
9470	Phenol, 2,3,4,6-tetrachloro- 2,3,4,6-Tetrachlorophenate	$C_6H_2Cl_4O$ 231.89	58-90-2 70	150[15]	4-06-00-01021	H_2O 1; EtOH 3; bz 3; chl 3
9471	Phenol, 2,3,5,6-tetrachloro- 2,3,5,6-Tetrachlorophenol	$C_6H_2Cl_4O$ 231.89	935-95-5 115		4-06-00-01025	H_2O 2; bz 4; lig 3
9472	Phenol, 2,3,5,6-tetrachloro-4-methoxy- Drosophilin A	$C_7H_4Cl_4O_2$ 261.92	484-67-3 116	3441	3-06-00-04436	
9473	Phenol, 2,3,4,5-tetramethyl- Prehnitenol	$C_{10}H_{14}O$ 150.22	488-70-0 85.3	266	4-06-00-03359	H_2O 2; EtOH 4; eth 4; lig 2
9474	Phenol, 2,3,4,6-tetramethyl-	$C_{10}H_{14}O$ 150.22	3238-38-8 80.5	240	4-06-00-03360	EtOH 3
9475	Phenol, 2,3,5,6-tetramethyl-	$C_{10}H_{14}O$ 150.22	527-35-5 118.5	247	4-06-00-03363	chl 3; peth 3; HOAc 3
9476	Phenol, 2,3,4,6-tetranitro- 2,3,4,6-Tetranitrophenol	$C_6H_2N_4O_9$ 274.10	641-16-7 140 dec	exp	2-06-00-00284	H_2O 4; chl 3
9477	Phenol, 2,2'-thiobis[4-chloro- Fenticlor	$C_{12}H_8Cl_2O_2S$ 287.17	97-24-5 174	3947	4-06-00-05645	H_2O 1; EtOH 3; eth 3; gl HOAc 3
9478	Phenol, 2,2'-thiobis[4,6-dichloro- Bithionol	$C_{12}H_6Cl_4O_2S$ 356.06	97-18-7 188	1316	4-06-00-05646 1.73[25]	ace 4
9479	Phenol, 2,4,6-tribromo- 2,4,6-Tribromophenol	$C_6H_3Br_3O$ 330.80	118-79-6 95.5	9526 286	4-06-00-01067 2.55[20]	H_2O 2; EtOH 4; eth 3; bz 3
9480	Phenol, 2,4,6-tribromo-3-methyl- 2,4,6-Tribromo-m-cresol	$C_7H_5Br_3O$ 344.83	4619-74-3 84	9524	3-06-00-01324	EtOH 3; eth 3; bz 3; chl 2
9481	Phenol, 2,3,4-trichloro- 2,3,4-Trichlorophenol	$C_6H_3Cl_3O$ 197.45	15950-66-0 83.5	sub	3-06-00-00716	EtOH 3; eth 3; bz 3; alk 3
9482	Phenol, 2,3,5-trichloro- 2,3,5-Trichlorophenol	$C_6H_3Cl_3O$ 197.45	933-78-8 62	248-9[250]	3-06-00-00716	eth 4; EtOH 4
9483	Phenol, 2,3,6-trichloro- 2,3,6-Trichlorophenol	$C_6H_3Cl_3O$ 197.45	933-75-5 58		4-06-00-00962	H_2O 2; EtOH 4; eth 4; bz 4
9484	Phenol, 2,4,5-trichloro- 2,4,5-Trichlorophenol	$C_6H_3Cl_3O$ 197.45	95-95-4 69	9555 247	4-06-00-00962	H_2O 2; EtOH 4; eth 4; bz 4
9485	Phenol, 2,4,6-trichloro- 2,4,6-Trichlorophenol	$C_6H_3Cl_3O$ 197.45	88-06-2 69	9556 246	4-06-00-01005 1.4901[75]	H_2O 2; EtOH 3; eth 3; HOAc 3
9486	Phenol, 3,4,5-trichloro- 3,4,5-Trichlorophenol	$C_6H_3Cl_3O$ 197.45	609-19-8 101	275	3-06-00-00729	H_2O 2; eth 3; lig 2
9487	Phenol, 2,3,4-trichloro-6-methyl- 4,5,6-Trichloro-o-cresol	$C_7H_5Cl_3O$ 211.47	551-78-0 77	9548	3-06-00-01268	
9488	Phenol, 2,3,6-trichloro-4-methyl- 2,3,6-Trichloro-p-cresol	$C_7H_5Cl_3O$ 211.47	551-77-9 66.5	9546	0-06-00-00404	EtOH 4
9489	Phenol, 2,4,6-trichloro-3-methyl- 2,4,6-Trichloro-m-cresol	$C_7H_5Cl_3O$ 211.47	551-76-8 46	9547 265	4-06-00-02070	EtOH 4; MeOH 4; chl 4
9490	Phenol, 3,4,6-trichloro-2-nitro- 3,4,6-Trichloro-2-nitrophenol	$C_6H_2Cl_3NO_3$ 242.44	82-62-2 92.5	9554	3-06-00-00842	
9491	Phenol, 3-(trifluoromethyl)-	$C_7H_5F_3O$ 162.11	98-17-9 -1.8	178	4-06-00-02060 1.3418[25]	
9492	Phenol, 2,4,6-triiodo- 2,4,6-Triiodophenol	$C_6H_3I_3O$ 471.80	609-23-4 159.8	sub	4-06-00-01085	H_2O 1; EtOH 2; eth 3; ace 3
9493	Phenol, 2,3,4-trimethyl-	$C_9H_{12}O$ 136.19	526-85-2 81	236	4-06-00-03245	bz 4; eth 4; EtOH 4
9494	Phenol, 2,4,5-trimethyl-	$C_9H_{12}O$ 136.19	496-78-6 72	232	4-06-00-03247	H_2O 1; EtOH 4; eth 4
9495	Phenol, 2,4,6-trimethyl-	$C_9H_{12}O$ 136.19	527-60-6 73	220	4-06-00-03253	eth 4; EtOH 4
9496	Phenol, 2,4,5-trimethyl-, acetate 2,4,5-Trimethylphenyl acetate	$C_{11}H_{14}O_2$ 178.23	69305-42-6 34	245.5	2-06-00-00482	eth 4; EtOH 4

No.	Name Synonym	Mol. Form. Mol. Wt.	CAS RN mp/°C	Merck No. bp/°C	Beil. Ref. den/g cm^{-3}	Solubility n_D
9497	Phenol, 2,4,6-trinitro- Picric acid	$C_6H_3N_3O_7$ 229.11	88-89-1 122.5	7380 exp 300	4-06-00-01388	H_2O 2; EtOH 3; eth 3; ace 4 1.763
9498	Phenol, 2,4,6-tris(1,1-dimethylethyl)-	$C_{18}H_{30}O$ 262.44	732-26-3 131	278	4-06-00-03539 0.864[27]	H_2O 1; EtOH 3; ace 3; ctc 3
9499	10H-Phenothiazine Thiodiphenylamine	$C_{12}H_9NS$ 199.28	92-84-2 187.5	7220 371	4-27-00-01214	ace 4; bz 4; eth 4; EtOH 4
9500	10H-Phenothiazine-2-acetic acid, 10-methyl- Metiazinic acid	$C_{15}H_{13}NO_2S$ 271.34	13993-65-2 144	6057		chl 3
9501	10H-Phenothiazine, 2-chloro-10-[3-(4-methyl-1-piperazinyl)propyl]- Phenothiazine, 2-chloro-10(3(4-methyl-1-piperazinyl)propyl)	$C_{20}H_{24}ClN_3S$ 373.95	58-38-8 228	7768	4-27-00-01307	
9502	10H-Phenothiazine, 10-[(dimethylamino)acetyl]- Ahistan	$C_{16}H_{16}N_2OS$ 284.38	518-61-6 144.5	179	4-27-00-01272	
9503	10H-Phenothiazine-10-ethanamine, N,N-diethyl- Diethazine	$C_{18}H_{22}N_2S$ 298.45	60-91-3	3098 167[0.5]	4-27-00-01240	H_2O 1; dil HCl 3
9504	10H-Phenothiazine-10-ethanamine, N,N,α-trimethyl- Promethazine	$C_{17}H_{20}N_2S$ 284.43	60-87-7 60	7797 190-3[0.5]	4-27-00-01253	H_2O 1; dil HCl 4
9505	10H-Phenothiazine-10-propanamine, 2-chloro-N,N-dimethyl- Chlorpromazine	$C_{17}H_{19}ClN_2S$ 318.87	50-53-3 200-5[0.8]	2186	4-27-00-01300	H_2O 1; EtOH 4; eth 4; bz 4
9506	Phenothiazin-5-ium, 3,7-bis(dimethylamino)-, chloride Methylene blue	$C_{18}H_{18}ClN_3S$ 319.86	61-73-4	5979	4-27-00-05152	H_2O 3; EtOH 3; eth 1; chl 3
9507	10H-Phenoxazine Phenoxazine	$C_{12}H_9NO$ 183.21	135-67-1 156 dec	7222	4-27-00-01209	bz 4; eth 4; EtOH 4
9508	3H-Phenoxazin-3-one, 7-hydroxy-, 10-oxide Resazurin	$C_{12}H_7NO_4$ 229.19	550-82-3 sub	8144	4-27-00-03594	H_2O 1; EtOH 2; eth 1; alk 3
9509	DL-Phenylalanine	$C_9H_{11}NO_2$ 165.19	150-30-1 284-8 dec	7242 sub	4-14-00-01553	H_2O 4; EtOH 1; eth 1; bz 1
9510	L-Phenylalanine (S)-α-Aminobenzenepropanoic acid	$C_9H_{11}NO_2$ 165.19	63-91-2 283 dec	7242	4-14-00-01552	H_2O 4; EtOH 1; eth 1; bz 1
9511	L-Phenylalanine, N-L-α-aspartyl-, 1-methyl ester Aspartame	$C_{14}H_{18}N_2O_5$ 294.31	22839-47-0 246.5	861		
9512	L-Phenylalanine, ethyl ester	$C_{11}H_{15}NO_2$ 193.25	3081-24-1 148[13]		4-14-00-01556 1.065[15]	H_2O 2
9513	Phorate Phosphorodithioic acid, O,O-diethyl S-[(ethylthio)methyl] ester	$C_7H_{17}O_2PS_3$ 260.38	298-02-2 <-15 119[0.8]	7305	1.16[25]	
9514	Phosalone Phosphorodithioic acid, S-[(6-chloro-2-oxo-3(2H)-benzoxazolyl)methyl]O,O-diethyl ester	$C_{12}H_{15}ClNO_4PS_2$ 367.81	2310-17-0 46	7308		
9515	Phosmet Phosphorodithioic acid, S-[(1,3-dihydro-1,3-dioxo-2H-isoindol-2-yl)methyl] O,O-dimethyl ester	$C_{11}H_{12}NO_4PS_2$ 317.33	732-11-6 72 dec	7311		
9516	Phosphine, dicyclohexyl- Dicyclohexylphosphine	$C_{12}H_{23}P$ 198.29	829-84-5 281; 129[8]		4-16-00-00947 0.904[25]	1.5163[20]
9517	Phosphine oxide, triethyl-	$C_6H_{15}OP$ 134.16	597-50-2 48 243		4-04-00-03463	H_2O 4; eth 4; EtOH 4
9518	Phosphine oxide, triphenyl-	$C_{18}H_{15}OP$ 278.29	791-28-6 156.5 >360		4-16-00-01011 1.2124[23]	H_2O 2; EtOH 4; eth 2; bz 4
9519	Phosphine, phenyl- Monophenylphosphine	C_6H_7P 110.10	638-21-1 160.5		4-16-00-00948 1.001[15]	1.5796[20]
9520	Phosphine, tributyl- Tributylphosphine	$C_{12}H_{27}P$ 202.32	998-40-3 150[50]		4-04-00-03436 0.812[25]	1.4619[20]
9521	Phosphine, triethyl- Triethyl phosphine	$C_6H_{15}P$ 118.16	554-70-1 -88 129	9590	4-04-00-03431 0.8006[19]	H_2O 1; EtOH 5; eth 5 1.458[15]
9522	Phosphine, trimethyl- Trimethylphosphine	C_3H_9P 76.08	594-09-2 -85 37.5		4-04-00-03429	H_2O 1; eth 3
9523	Phosphine, triphenyl- Triphenylphosphine	$C_{18}H_{15}P$ 262.29	603-35-0 80 188[1]	9657	4-16-00-00951 1.0749[80]	H_2O 1; EtOH 3; eth 4; bz 3 1.6358[80]
9524	Phosphine, tris(chloromethyl)- Tris(chloromethyl)phosphine	$C_3H_6Cl_3P$ 179.41	13482-62-7 100[7]		3-01-00-02608 1.414[20]	ace 4; bz 4; eth 4; EtOH 4
9525	Phosphinic acid, diethyl- Diethylphosphinic acid	$C_4H_{11}O_2P$ 122.10	813-76-3 18.5 320		4-04-00-03474	H_2O 4; EtOH 4; eth 4
9526	Phosphinic acid, diethyl-, anhydride	$C_8H_{20}O_3P_2$ 226.19	7495-97-8 188[14]		4-04-00-03477 1.1053[20]	1.4647[20]
9527	Phosphinic acid, diethyl-, ethyl ester Ethyl diethylphosphinate	$C_6H_{15}O_2P$ 150.16	4775-09-1 95[14]		4-04-00-03475 0.9908[20]	eth 4; EtOH 4 1.4337[20]
9528	Phosphinic acid, dimethyl- Dimethylphosphinic acid	$C_2H_7O_2P$ 94.05	3283-12-3 92 377		4-04-00-03470	H_2O 4; EtOH 4; eth 4; bz 3
9529	Phosphinic chloride, diethyl-	$C_4H_{10}ClOP$ 140.55	1112-37-4 104[15]		4-04-00-03476 1.1394[20]	1.4647[20]
9530	Phosphonic acid, acetyl-, diethyl ester O,O-Diethyl acetylphosphonate	$C_6H_{13}O_4P$ 180.14	919-19-7 114-5[20]		4-02-00-00440 1.1005[20]	1.4200[26]

No.	Name / Synonym	Mol. Form. / Mol. Wt.	CAS RN / mp/°C	Merck No. / bp/°C	Beil. Ref. / den/g cm^{-3}	Solubility / n_D
9531	Phosphonic acid, bis(1,1-dimethylethyl) ester	$C_8H_{19}O_3P$ / 194.21	13086-84-5	70-2[10]	4-01-00-01623 / 0.975[25]	1.4168[25]
9532	Phosphonic acid, dibutyl ester	$C_8H_{19}O_3P$ / 194.21	1809-19-4	131[19]; 77[1.5]	4-01-00-01525 / 0.9485[20]	1.4220[20]
9533	Phosphonic acid, (1,1-dimethylethyl)-	$C_4H_{11}O_3P$ / 138.10	4923-84-6 / 192	dec	4-04-00-03557	H_2O 4; EtOH 4
9534	Phosphonic acid, diphenyl ester	$C_{12}H_{11}O_3P$ / 234.19	4712-55-4 / 12	218-9[26]	4-06-00-00703 / 1.223[25]	1.5575[20]
9535	Phosphonic acid, ethenyl-, diethyl ester	$C_6H_{13}O_3P$ / 164.14	682-30-4	110[2]	4-04-00-03569 / 1.068[25]	1.4290[20]
9536	Phosphonic acid, ethyl-, diethyl ester / Diethyl ethylphosphonate	$C_6H_{15}O_3P$ / 166.16	78-38-6	198; 90[16]	4-04-00-03524 / 1.0259[20]	H_2O 2; EtOH 3; eth 3 / 1.4163[20]
9537	Phosphonic acid, ethyl-, dimethyl ester	$C_4H_{11}O_3P$ / 138.10	6163-75-3	82[18]	4-04-00-03523 / 1.1029[30]	H_2O 4; bz 4; EtOH 4 / 1.4128[30]
9538	Phosphonic acid, methyl-	CH_5O_3P / 96.02	993-13-5 / 108.5	dec	4-04-00-03498	H_2O 4; EtOH 4; eth 4; bz 1
9539	Phosphonic acid, methyl-, diethyl ester	$C_5H_{13}O_3P$ / 152.13	683-08-9	194	4-04-00-03500 / 1.0406[30]	H_2O 3; EtOH 3; eth 3; bz 1 / 1.4101[30]
9540	Phosphonic acid, methyl-, dimethyl ester	$C_3H_9O_3P$ / 124.08	756-79-6	181; 79.5[20]	4-04-00-03499 / 1.1507[20]	H_2O 3; EtOH 3; eth 3 / 1.4099[30]
9541	Phosphonic acid, methyl-, diphenyl ester / Diphenyl methylphosphonate	$C_{13}H_{13}O_3P$ / 248.22	7526-26-3 / 35	205[13]	4-06-00-00703 / 1.2051[20]	H_2O 1
9542	Phosphonic acid, (1-methylethyl)- / 2-Propanephosphonic acid	$C_3H_9O_3P$ / 124.08	4721-37-3 / 74.5	dec	4-04-00-03548	H_2O 4; eth 4; EtOH 4
9543	Phosphonic acid, methyl-, monoethyl ester	$C_3H_9O_3P$ / 124.08	1832-53-7	106-7[0.1]	4-04-00-03499 / 1.1800[20]	1.4258[20]
9544	Phosphonic acid, (2-methylpropyl)-	$C_4H_{11}O_3P$ / 138.10	4721-34-0 / 119	dec	4-04-00-03556	H_2O 4; eth 4; EtOH 4; xyl 4
9545	Phosphonic acid, (phenylmethyl)-, diethyl ester	$C_{11}H_{17}O_3P$ / 228.23	1080-32-6	110[2]	4-16-00-01093	ctc 3 / 1.4930[20]
9546	Phosphonic acid, 1,3-propanediylbis-, tetraethyl ester	$C_{11}H_{26}O_6P_2$ / 316.27	22401-25-8	185-6[3]	4-04-00-03582 / 1.1227[20]	1.4500[20]
9547	Phosphonic acid, propyl-	$C_3H_9O_3P$ / 124.08	4672-38-2 / 73.8	dec	4-04-00-03542	H_2O 4; eth 4; EtOH 4
9548	Phosphonic dichloride, phenyl-	$C_6H_5Cl_2OP$ / 194.98	824-72-6 / 1	258	4-16-00-01074 / 1.197[25]	DMSO-d_6 2 / 1.5581[25]
9549	Phosphonothioic acid, methyl-, O,O-bis(1-methylethyl) ester / Phosphonic acid, thiono, methyl, diisopropyl ester	$C_7H_{17}O_2PS$ / 196.25	66295-45-2	42[1]	4-04-00-03516 / 0.9885[25]	1.4512[25]
9550	Phosphonothioic acid, methyl-, O,O-dimethyl ester	$C_3H_9O_2PS$ / 140.14	681-06-1	43[4]	4-04-00-03514 / 1.1375[25]	1.4738[25]
9551	Phosphonothioic acid, phenyl-, O-ethyl O-(4-nitrophenyl) ester / O-Ethyl O-p-nitrophenyl benzenethiophosphonate	$C_{14}H_{14}NO_4PS$ / 323.31	2104-64-5 / 36	3576	4-16-00-01089 / 1.27[25]	bz 4; eth 4; EtOH 4 / 1.5978[30]
9552	Phosphonothioic dichloride, phenyl- / Dichlorophenylphosphine sulfide	$C_6H_5Cl_2PS$ / 211.05	3497-00-5	205[130]	4-16-00-01090 / 1.376[13]	
9553	Phosphonous dichloride, (2,5-dimethylphenyl)- / Phosphonous dichloride, 2,5-xylyl-	$C_8H_9Cl_2P$ / 207.04	57150-66-0 / -30	253.5	3-16-00-00848 / 1.25[18]	
9554	Phosphonous dichloride, (4-ethylphenyl)-	$C_8H_9Cl_2P$ / 207.04	5274-50-0	251	4-16-00-00974 / 1.237[20]	1.584[20]
9555	Phosphonous dichloride, [4-(1-methylethyl)phenyl]- / Benzene, 1-chlorophosphino-4-isopropyl	$C_9H_{11}Cl_2P$ / 221.07	82906-78-3	269	3-16-00-00848 / 1.190[12]	ace 4 / 1.5677[25]
9556	Phosphonous dichloride, phenyl- / Dichlorophenylphosphine	$C_6H_5Cl_2P$ / 178.98	644-97-3 / -51	225; 142[57]	4-16-00-00972 / 1.356[20]	bz 4 / 1.6030[20]
9557	Phosphoramidothioic dichloride, dimethyl- / Phosphoric acid, amide,thiono,dichloride,N,N-dimethyl	$C_2H_6Cl_2NPS$ / 178.02	1498-65-3	77.5[10]	4-04-00-00293 / 1.3731[20]	1.5390[20]
9558	Phosphoric acid, diethyl ester	$C_4H_{11}O_4P$ / 154.10	598-02-7	203 dec	4-01-00-01339 / 1.186[25]	eth 4 / 1.4170[20]
9559	Phosphoric acid, dimethyl ester	$C_2H_7O_4P$ / 126.05	813-78-5	174 dec	4-01-00-01259 / 1.335[25]	H_2O 4; ace 4; EtOH 4 / 1.408[25]
9560	Phosphoric acid, 2-ethylhexyl diphenyl ester	$C_{20}H_{27}O_4P$ / 362.41	1241-94-7	232[5]	4-06-00-00718 / 1.090[25]	1.510[25]
9561	Phosphoric acid, monoethyl ester	$C_2H_7O_4P$ / 126.05	1623-14-9	dec	4-01-00-01338 / 1.430[25]	H_2O 4; ace 4; eth 4; EtOH 4 / 1.427
9562	Phosphoric acid tributyl ester / Tributyl phosphate	$C_{12}H_{27}O_4P$ / 266.32	126-73-8	9531 / 289	4-01-00-01531 / 0.9727[25]	H_2O 3; EtOH 5; eth 3; bz 3 / 1.4224[25]
9563	Phosphoric acid, triethyl ester / Ethyl phosphate	$C_6H_{15}O_4P$ / 182.16	78-40-0 / -56.4	9589 / 215.5	4-01-00-01339 / 1.0695[20]	H_2O 3; EtOH 4; eth 3; bz 3 / 1.4053[20]
9564	Phosphoric acid, trimethyl ester	$C_3H_9O_4P$ / 140.08	512-56-1 / -46	197.2	4-01-00-01259 / 1.2144[20]	H_2O 4; EtOH 2; eth 3 / 1.3967[20]
9565	Phosphoric acid, tripentyl ester	$C_{15}H_{33}O_4P$ / 308.40	2528-38-3	225[50.] 167[5]	4-01-00-01645 / 0.9608[20]	H_2O 1; EtOH 3; eth 3; tol 3 / 1.4319[20]
9566	Phosphoric acid, triphenyl ester / Triphenyl phosphate	$C_{18}H_{15}O_4P$ / 326.29	115-86-6 / 50.5	9656 / 245[11]	4-06-00-00720 / 1.2055[50]	H_2O 1; EtOH 3; eth 4; bz 4

No.	Name / Synonym	Mol. Form. / Mol. Wt.	CAS RN / mp/°C	Merck No. / bp/°C	Beil. Ref. / den/g cm^{-3}	Solubility / n_D
9567	Phosphoric acid, tripropyl ester / Tripropyl phosphate	$C_9H_{21}O_4P$ / 224.24	513-08-6	/ 252	4-01-00-01428 / 1.0121[20]	H_2O 2; EtOH 3; eth 3; chl 2 / 1.4165[20]
9568	Phosphoric acid, tris(1-methylethyl) ester	$C_9H_{21}O_4P$ / 224.24	513-02-0	/ 219	4-01-00-01478 / 0.9867[20]	EtOH 4 / 1.4057[20]
9569	Phosphoric acid, tris(2-methylphenyl) ester / Tri-o-tolyl phosphate	$C_{21}H_{21}O_4P$ / 368.37	78-30-8 / 11	9676 / 410	4-06-00-01979 / 1.1955[20]	H_2O 1; EtOH 4; eth 4; ctc 4 / 1.5575[20]
9570	Phosphoric acid, tris(3-methylphenyl) ester	$C_{21}H_{21}O_4P$ / 368.37	563-04-2 / 25.5	/ 260[15]	4-06-00-02057 / 1.150[25]	H_2O 1; EtOH 2; eth 3; ctc 4 / 1.5575[20]
9571	Phosphoric acid, tris(4-methylphenyl) ester	$C_{21}H_{21}O_4P$ / 368.37	78-32-0 / 77.5	/ 224[35]	4-06-00-02130 / 1.247[25]	EtOH 3; eth 3; bz 3; chl 3
9572	Phosphoric acid, tris(2-methylpropyl) ester / Triisobutyl phosphate	$C_{12}H_{27}O_4P$ / 266.32	126-71-6	/ 264	4-01-00-01598 / 0.9681[20]	H_2O 4; bz 4; eth 4; EtOH 4 / 1.4193[20]
9573	Phosphoric triamide, hexamethyl- / Tris(dimethylamino)phosphine oxide	$C_6H_{18}N_3OP$ / 179.20	680-31-9 / 7.0	4568 / 232.5	4-04-00-00284 / 1.03[20]	EtOH 3; eth 3 / 1.4579[20]
9574	Phosphorochloridic acid, diethyl ester / Diethyl chlorophosphate	$C_4H_{10}ClO_3P$ / 172.55	814-49-3 /	/ 93.5	4-01-00-01345 / 1.205[19]	/ 1.4170[20]
9575	Phosphorochloridic acid, diphenyl ester	$C_{12}H_{10}ClO_3P$ / 268.64	2524-64-3	/ 314-6[272]	4-06-00-00737 / 1.296[25]	tfa 3 / 1.5500[20]
9576	Phosphorochloridothioic acid, O,O-diethyl ester / Diethyl chlorothiophosphate	$C_4H_{10}ClO_2PS$ / 188.61	2524-04-1	/ 45[3]	4-01-00-01352	ctc 3
9577	Phosphorodichloridic acid, ethyl ester / Ethyl dichlorophosphate	$C_2H_5Cl_2O_2P$ / 162.94	1498-51-7	/ 60-5[10]	4-01-00-01345	/ 1.4338[20]
9578	Phosphorodichloridothioic acid, O-(2,4,5-trichlorophenyl) ester / Phenol, 2,4,5-trichloro-, O-phosphorodichloridothioate	$C_6H_2Cl_5OPS$ / 330.38	16805-78-0	/ 110[1]	4-06-00-00996 / 1.6653[20]	/ 1.6084[20]
9579	Phosphorodithioic acid, O,O,S-triethyl ester	$C_6H_{15}O_2PS_2$ / 214.29	2524-09-6	/ 128[20]	4-01-00-01402 / 1.1336[0]	/ 1.5043
9580	Phosphorofluoridic acid, bis(1-methylethyl) ester / Isoflurophate	$C_6H_{14}FO_3P$ / 184.15	55-91-4	5060 / 62[9]	4-01-00-01480 / 1.055[25]	H_2O 2; eth 3; oils 4; lig 2 / 1.3830[25]
9581	Phosphorothioic acid, O-(2,4-dichlorophenyl) O,O-diethyl ester / Dichlofenthion	$C_{10}H_{13}Cl_2O_3PS$ / 315.16	97-17-6	3030	4-06-00-00939	ctc 3; CS_2 3
9582	Phosphorothioic acid, O,O-diethyl O-[6-methyl-2-(1-methylethyl)-4-pyrimidinyl] ester / Diazinon	$C_{12}H_{21}N_2O_3PS$ / 304.35	333-41-5	2978 / 85-90[0.05]	5-23-11-00187 / 1.1088[20]	/ 1.4922[20]
9583	Phosphorothioic acid, O,O-diethyl O-(4-nitrophenyl) ester / Parathion	$C_{10}H_{14}NO_5PS$ / 291.26	56-38-2 / 6.1	6983 / 375	4-06-00-01337 / 1.2681[20]	H_2O 1; EtOH 5; eth 3; ace 3 / 1.5370[25]
9584	Phosphorothioic acid, O,O-dimethyl O-(2,4,5-trichlorophenyl) ester / Ronnel	$C_8H_8Cl_3O_3PS$ / 321.55	299-84-3 / 41	8239 / 151-4[0.4]	4-06-00-00994 / 1.44[32]	/ 1.5335[35]
9585	Phosphorothioic acid, O,O,O-triethyl ester	$C_6H_{15}O_3PS$ / 198.22	126-68-1	/ 100[16]	4-01-00-01348 / 1.082[25]	/ 1.4480[20]
9586	Phosphorous acid, bis(phenylmethyl) ester / Dibenzyl phosphite	$C_{14}H_{15}O_3P$ / 262.25	17176-77-1 / -5	2996 / 162[0.1]	4-06-00-02571	/ 1.5521[18]
9587	Phosphorous acid, diethyl ester	$C_4H_{11}O_3P$ / 138.10	123-22-8	/ 87[20]		EtOH 3; eth 3 / 1.4101[20]
9588	Phosphorous acid, dimethyl ester	$C_2H_7O_3P$ / 110.05	868-85-9	/ 170.5	/ 1.2002[20]	EtOH 3; ctc 2; py 3 / 1.4036[20]
9589	Phosphorous acid, 2-ethylhexyl diphenyl ester	$C_{20}H_{27}O_3P$ / 346.41	15647-08-2	/ 148-56[0.15]	4-06-00-00695 / 1.054[20]	/ 1.5207[27]
9590	Phosphorous acid, tributyl ester	$C_{12}H_{27}O_3P$ / 250.32	102-85-2	/ 122[12]	4-01-00-01527 / 0.9259[20]	EtOH 3; eth 4; ctc 2 / 1.4321[19]
9591	Phosphorous acid, triethyl ester / Triethyl phosphite	$C_6H_{15}O_3P$ / 166.16	122-52-1	/ 157.9	4-01-00-01333 / 0.9629[20]	H_2O 1; EtOH 4; eth 4 / 1.4127[20]
9592	Phosphorous acid, trimethyl ester / Trimethyl phosphite	$C_3H_9O_3P$ / 124.08	121-45-9	/ 111.5	4-01-00-01256 / 1.0518[20]	EtOH 4; eth 4; ctc 2 / 1.4095[20]
9593	Phosphorous acid, triphenyl ester	$C_{18}H_{15}O_3P$ / 310.29	101-02-0 / 25	/ 360	4-06-00-00695 / 1.1842[20]	H_2O 1; EtOH 4; os 4 / 1.5900[20]
9594	Phosphorous acid, tripropyl ester / Propyl phosphite	$C_9H_{21}O_3P$ / 208.24	923-99-9	/ 206.5	4-01-00-01426 / 0.9417[20]	eth 4; EtOH 4 / 1.4282[20]
9595	Phosphorous acid, tris(1-methylethyl) ester	$C_9H_{21}O_3P$ / 208.24	116-17-6	/ 60-1[10]	4-01-00-01476 / 0.9685[18]	EtOH 3; eth 3; chl 3 / 1.4085[25]
9596	Phosphorous acid, tris(2-methylphenyl) ester	$C_{21}H_{21}O_3P$ / 352.37	2622-08-4	/ 238[11] 198[2]	4-06-00-01977 / 1.1423[20]	eth 3; chl 2 / 1.5740[28]
9597	Phosphorous acid, tris(4-methylphenyl) ester	$C_{21}H_{21}O_3P$ / 352.37	620-42-8	/ 250-5[10]	4-06-00-02128 / 1.1280[25]	eth 4 / 1.5703[28]
9598	p-Phthalaldehydic acid, 3-hydroxy- / 3-Hydroxybenzoic acid, 4-formyl	$C_8H_6O_4$ / 166.13	619-12-5 / 234	/ sub	4-10-00-03623	eth 4; EtOH 4
9599	Phthalazine / 2,3-Benzodiazine	$C_8H_8N_2$ / 130.15	253-52-1 / 90.5	7344 / 316	5-23-07-00149	H_2O 3; EtOH 3; eth 2; bz 3
9600	1,4-Phthalazinedione, 5-amino-2,3-dihydro- / Luminol	$C_8H_7N_3O_2$ / 177.16	521-31-3 / 330.5	5470	4-25-00-04192	H_2O 1; EtOH 2; eth 2; alk 4
9601	Picene / Benzo[a]chrysene	$C_{22}H_{14}$ / 278.35	213-46-7 / 368	7368 / 519	4-05-00-02724	H_2O 1; EtOH 2; bz 2; chl 2
9602	Piperazine / Diethylenediamine	$C_4H_{10}N_2$ / 86.14	110-85-0 / 106	7431 / 146	5-23-01-00030	H_2O 3; EtOH 3; eth 1; chl 3 / 1.446[113]

No.	Name / Synonym	Mol. Form. / Mol. Wt.	CAS RN / mp/°C	Merck No. / bp/°C	Beil. Ref. / den/g cm⁻³	Solubility / n_D
9603	1-Piperazinecarboxamide, N,N-diethyl-4-methyl- / Diethylcarbamazine	$C_{10}H_{21}N_3O$ / 199.30	90-89-1 / 48	3106 / 109-11[3]	4-23-00-00225	
9604	1-Piperazinecarboxylic acid, ethyl ester	$C_7H_{14}N_2O_2$ / 158.20	120-43-4 /	/ 237	5-23-01-00055	H_2O 4; eth 4; EtOH 4 / 1.4760[25]
9605	Piperazine, 1-[(4-chlorophenyl)phenylmethyl]-4-[(3-methylphenyl)methyl]- / Meclizine	$C_{25}H_{27}ClN_2$ / 390.96	569-65-3 /	5657 / 230	5-23-01-00235	CS_2 3
9606	1,4-Piperazinedicarboxylic acid, diethyl ester / N,N'-Piperazinedicarboxylic acid, diethyl ester	$C_{10}H_{18}N_2O_4$ / 230.26	5470-28-0 / 47	/ 315	4-23-00-00254	eth 4; EtOH 4
9607	Piperazine, dihydrochloride / Diethylenediamine dihydrochloride	$C_4H_{12}Cl_2N_2$ / 159.06	142-64-3 /		5-23-01-00030	H_2O 2; EtOH 1
9608	Piperazine, 1,4-dimethyl-	$C_6H_{14}N_2$ / 114.19	106-58-1 /	/ 131	5-23-01-00128 / 0.8600[20]	H_2O 4; EtOH 4; eth 4 / 1.4474[20]
9609	Piperazine, 2,5-dimethyl-, cis- / cis-2,5-Dimethylpiperazine	$C_6H_{14}N_2$ / 114.19	6284-84-0 / 114	/ 162	5-23-03-00304	H_2O 4; EtOH 4; eth 2; bz 2 / 1.4720[20]
9610	Piperazine, 2,5-dimethyl-, trans- / trans-2,5-Dimethylpiperazine	$C_6H_{14}N_2$ / 114.19	2815-34-1 / 118.5	/ 162	5-23-03-00304	H_2O 4; EtOH 4; eth 2; bz 2
9611	Piperazine, 2,6-dimethyl-, cis- / cis-2,6-Dimethylpiperazine	$C_6H_{14}N_2$ / 114.19	21655-48-1 / 114	/ 160	5-23-03-00326	H_2O 3; EtOH 3; eth 1; bz 2
9612	2,5-Piperazinedione / Glycine anhydride	$C_4H_6N_2O_2$ / 114.10	106-57-0 / 312 dec	7434 / sub 260	5-24-05-00294	H_2O 2; EtOH 2; HCl 3
9613	Piperazine, 1,4-diphenyl-	$C_{16}H_{18}N_2$ / 238.33	613-39-8 / 165	/ 300 dec; 232[12]	5-23-01-00173	H_2O 1; EtOH 2
9614	1,4-Piperazinedipropanamine / 1,4-Bis(3-aminopropyl)piperazine	$C_{10}H_{24}N_4$ / 200.33	7209-38-3 / 15	/ 150-2[2]	5-23-01-00356 / 0.973[25]	1.5015[20]
9615	1-Piperazineethanamine / 1-(2-Aminoethyl)piperazine	$C_6H_{15}N_3$ / 129.21	140-31-8 /	/ 220	5-23-01-00257 / 0.985[25]	1.4983[20]
9616	1-Piperazineethanol	$C_6H_{14}N_2O$ / 130.19	103-76-4 /	/ 246	5-23-01-00406 / 1.061[25]	1.5065[20]
9617	1-Piperazineethanol, 4-[3-(2-chloro-10H-phenothiazin-10-yl)propyl]- / Perphenazine	$C_{21}H_{26}ClN_3OS$ / 403.98	58-39-9 / 97	7135 /	4-27-00-01307	
9618	Piperazine, 1-methyl-	$C_5H_{12}N_2$ / 100.16	109-01-3 /	/ 138	5-23-01-00082	H_2O 4; eth 4; EtOH 4 / 1.4378[20]
9619	Piperazine, 2-methyl- / 2-Methylpiperazine	$C_5H_{12}N_2$ / 100.16	109-07-9 / 62	/ 153	5-23-03-00267	H_2O 4; EtOH 3; eth 3; bz 3
9620	Piperazine, 1-phenyl-	$C_{10}H_{14}N_2$ / 162.23	92-54-6 /	/ 286.5; 161[15]	5-23-01-00160 / 1.0621[20]	H_2O 1; EtOH 5; eth 5; chl 3 / 1.5875[20]
9621	Piperazine, 1-(phenylmethyl)-	$C_{11}H_{16}N_2$ / 176.26	2759-28-6 /	/ 145-7[12]	5-23-01-00193	H_2O 3; EtOH 3; eth 3; chl 2 / 1.5430[28]
9622	Piperazine, 1,2,4-trimethyl- / 1,2,4-Trimethylpiperazine	$C_7H_{16}N_2$ / 128.22	120-85-4 /	/ 149.5	5-23-03-00271	ctc 3 / 1.4433[20]
9623	Piperazinium, 4-(2-cyclohexyl-2-hydroxy-2-phenylethyl)-1,1-dimethyl-, methyl sulfate (salt) / Hexocyclium methyl sulfate	$C_{21}H_{36}N_2O_5S$ / 428.59	115-63-9 /	4627 /		
9624	1-Piperidinamine	$C_5H_{12}N_2$ / 100.16	2213-43-6 /	/ 147	5-20-03-00516 / 0.928[25]	1.4750[20]
9625	1-Piperidinamine, 2,6-dimethyl-	$C_7H_{16}N_2$ / 128.22	39135-39-2 /	/ 65-80[30]	5-20-04-00187 / 0.865[25]	1.4650[20]
9626	3-Piperidinamine, 1-ethyl-	$C_7H_{16}N_2$ / 128.22	6789-94-2 /	/ 155; 82[40]	5-22-08-00026 / 0.923[25]	1.4715[20]
9627	4-Piperidinamine, 1-methyl-N-phenyl-N-(phenylmethyl)- / Bamipine	$C_{19}H_{24}N_2$ / 280.41	4945-47-5 / 115	966 /	5-22-08-00036	
9628	4-Piperidinamine, 2,2,6,6-tetramethyl-	$C_9H_{20}N_2$ / 156.27	36768-62-4 / 17	/ 188.5	5-22-08-00162 / 0.912[25]	1.4706[20]
9629	Piperidine / Azacyclohexane	$C_5H_{11}N$ / 85.15	110-89-4 / -11.03	7438 / 106.2	5-20-02-00003 / 0.8606[20]	H_2O 5; EtOH 5; eth 3; ace 3 / 1.4530[20]
9630	Piperidine, 1-acetyl-	$C_7H_{13}NO$ / 127.19	618-42-8 / -13.4	/ 226.5	5-20-02-00440 / 1.011[9]	H_2O 4; EtOH 4 / 1.4790[25]
9631	Piperidine, 1-acetyl-3-methyl-	$C_8H_{15}NO$ / 141.21	4593-16-2 / -13.6	/ 239	5-20-04-00107 / 0.9684[25]	H_2O 4 / 1.4731[25]
9632	Piperidine, 1-[5-(1,3-benzodioxol-5-yl)-1-oxo-2,4-pentadienyl]-, (E,E)- / Piperine	$C_{17}H_{19}NO_3$ / 285.34	94-62-2 / 131.5	7442 /	5-20-03-00469	H_2O 1; EtOH 3; eth 2; bz 3
9633	Piperidine, 1-[5-(1,3-benzodioxol-5-yl)-1-oxo-2,4-pentadienyl]-, (Z,Z)- / Chavicine	$C_{17}H_{19}NO_3$ / 285.34	495-91-0 /	2038 /	5-20-03-00469	eth 4; EtOH 4; peth 4
9634	Piperidine, 1-benzoyl-	$C_{12}H_{15}NO$ / 189.26	776-75-0 / 49	/ 320.5	5-20-02-00450	H_2O 1; EtOH 3; eth 3; ctc 2
9635	Piperidine, 1-butyl-	$C_9H_{19}N$ / 141.26	4945-48-6 /	/ 176	5-20-02-00029 / 0.8245[20]	1.4467[20]
9636	Piperidine, 3-butyl- / 3-Butylpiperidine	$C_9H_{19}N$ / 141.26	13603-21-9 /	/ 196.5	5-20-04-00230 / 0.8620[18]	eth 4; EtOH 4; chl 4
9637	Piperidine, 4-butyl- / 4-n-Butylpiperidine	$C_9H_{19}N$ / 141.26	24152-39-4 /	/ 194	4-20-00-01635 / 0.879[23]	1.4472[23]
9638	1-Piperidinecarboxaldehyde	$C_6H_{11}NO$ / 113.16	2591-86-8 / -30.8	/ 222.5	5-20-02-00432 / 1.0158[25]	H_2O 5; EtOH 5; eth 5; bz 5 / 1.4805[25]

No.	Name / Synonym	Mol. Form. / Mol. Wt.	CAS RN / mp/°C	Merck No. / bp/°C	Beil. Ref. / den/g cm^{-3}	Solubility / n_D
9639	2-Piperidinecarboxamide, N-(2,6-dimethylphenyl)-1-methyl- Mepivacaine	$C_{15}H_{22}N_2O$ 246.35	96-88-8 	5748 	5-22-01-00223	CS$_2$ 3
9640	2-Piperidinecarboxylic acid Pipecolic acid	$C_6H_{11}NO_2$ 129.16	535-75-1 264	7425 	5-22-01-00220	H$_2$O 3; EtOH 2
9641	3-Piperidinecarboxylic acid Nipecotic acid	$C_6H_{11}NO_2$ 129.16	498-95-3 261 dec	6478 	5-22-01-00235	H$_2$O 4
9642	4-Piperidinecarboxylic acid Isonipecotic acid	$C_6H_{11}NO_2$ 129.16	498-94-2 336	5075 	5-22-01-00244	
9643	4-Piperidinecarboxylic acid, ethyl ester	$C_8H_{15}NO_2$ 157.21	1126-09-6 	 100-1[10]	4-22-00-00128	H$_2$O 4; bz 4; eth 4; EtOH 4 1.4591[20]
9644	4-Piperidinecarboxylic acid, methyl ester Methyl isonipecotate	$C_7H_{13}NO_2$ 143.19	2971-79-1 	 107-10[22]	4-22-00-00128	H$_2$O 4; bz 4; eth 4; EtOH 4 1.4635[25]
9645	4-Piperidinecarboxylic acid, 1-methyl-, methyl ester Methyl N-methylisonipecotinate	$C_8H_{15}NO_2$ 157.21	1690-75-1 	 98[20]; 84[10]	5-22-01-00245	H$_2$O 3; EtOH 3; eth 3; bz 3 1.4539[24]
9646	2,6-Piperidinediethanol, 1-methyl-α,α'-diphenyl-, [2α(R*),6α(S*)]- Lobelanidine	$C_{22}H_{29}NO_2$ 339.48	552-72-7 150	5429 	4-21-00-02386	H$_2$O 1; EtOH 3; eth 2; ace 4
9647	Piperidine, 2,5-diethyl-	$C_9H_{19}N$ 141.26	116836-21-6 	 190	0-20-00-00128 0.8722[25]	eth 4; chl 4
9648	Piperidine, 1,2-dimethyl-, (±)- Piperidine, N,2-dimethyl (DL)	$C_7H_{15}N$ 113.20	2512-81-4 	 127.5	5-20-04-00072 0.824[15]	H$_2$O 4; eth 4; EtOH 4 1.4395[20]
9649	Piperidine, 2,5-dimethyl- 2,5-Dimethylpiperidine	$C_7H_{15}N$ 113.20	34893-50-0 	 137	5-20-04-00179 0.8317[20]	EtOH 4 1.4452[25]
9650	Piperidine, 2,6-dimethyl-	$C_7H_{15}N$ 113.20	504-03-0 	 127	5-20-04-00180 0.8158[25]	H$_2$O 5; EtOH 5; eth 5; ctc 2 1.4377[20]
9651	Piperidine, 3,3-dimethyl- 3,3-Dimethylpiperidine	$C_7H_{15}N$ 113.20	1193-12-0 	 137	5-20-04-00189	EtOH 4 1.4452[25]
9652	Piperidine, 3,5-dimethyl- 3,5-Lupetidine	$C_7H_{15}N$ 113.20	35794-11-7 	 144	5-20-04-00191 0.853[25]	1.4454[20]
9653	Piperidine, 1-(1,1-dimethylethyl)- Piperidine, N-tert-butyl	$C_9H_{19}N$ 141.26	14446-69-6 	 166	5-20-02-00031 0.8465[20]	eth 4; EtOH 4 1.4532[20]
9654	2,4-Piperidinedione, 3,3-diethyl- Piperidione	$C_9H_{15}NO_2$ 169.22	77-03-2 104	7439 	5-21-09-00609	H$_2$O 4; EtOH 4; chl 4; MeOH 4
9655	2,6-Piperidinedione, 4-[2-(3,5-dimethyl-2-oxocyclohexyl)-2-hydroxyethyl]-, Cycloheximide	$C_{15}H_{23}NO_4$ 281.35	66-81-9 119	2734 	5-21-13-00434	EtOH 4
9656	2,6-Piperidinedione, 4-ethyl-4-methyl- Bemegride	$C_8H_{13}NO_2$ 155.20	64-65-3 126.5	1036 sub 100	5-21-09-00601	chl 3
9657	Piperidine, 1-[2-[3-(2,2-diphenylethyl)-1,2,4-oxadiazol-5-yl]ethyl]-, monohydrochloride Prenoxdiazine hydrochloride	$C_{23}H_{28}ClN_3O$ 397.95	982-43-4 186.5	7743 		
9658	Piperidine, 1-(3,3-diphenylpropyl)- Fenpiprane	$C_{20}H_{25}N$ 279.43	3540-95-2 41.5	3934 210-208[8]	5-20-02-00091	
9659	Piperidine, 1,1'-dithiobis-	$C_{10}H_{20}N_2S_2$ 232.41	10220-20-9 64.2	 160.1[11]	5-20-03-00487	acid 1
9660	Piperidine, 1-dodecyl-	$C_{17}H_{35}N$ 253.47	5917-47-5 	 161[5]; 115[0.6]	4-20-00-00321 0.8378[20]	1.4588[20]
9661	Piperidine, 1,1'-(1,2-ethanediyl)bis-	$C_{12}H_{24}N_2$ 196.34	1932-04-3 -0.5	 265	5-20-03-00148 0.9160[25]	1.4853[25]
9662	1-Piperidineethanol	$C_7H_{15}NO$ 129.20	3040-44-6 17.9	 202; 90[12]	5-20-02-00101 0.9703[25]	H$_2$O 5; EtOH 4 1.4749[20]
9663	2-Piperidineethanol 2-(2-Hydroxyethyl)piperidine	$C_7H_{15}NO$ 129.20	1484-84-0 69	 202; 145[36]	5-21-01-00121 1.01[27]	H$_2$O 4
9664	4-Piperidineethanol 4-(2-Hydroxyethyl)piperidine	$C_7H_{15}NO$ 129.20	622-26-4 132.5	 227.5	4-21-00-00093 1.0059[15]	H$_2$O 4; eth 4; EtOH 4 1.4907[20]
9665	1-Piperidineethanol, 2-(diphenylmethyl)- Diphemethoxidine	$C_{20}H_{25}NO$ 295.42	13862-07-2 106.5	3303 180-1[0.1]	5-20-08-00173	
9666	Piperidine, 3-ethenyl- Piperidine, 3-vinyl-	$C_7H_{13}N$ 111.19	57502-49-5 	 153.5	4-20-00-01946 0.9274[25]	1.4731[25]
9667	Piperidine, 1-ethyl- N-Ethylpiperidine	$C_7H_{15}N$ 113.20	766-09-6 	 130.8	5-20-02-00025 0.8237[20]	1.4480[20]
9668	Piperidine, 2-ethyl-, (±)- Piperidine, 2-ethyl (DL)	$C_7H_{15}N$ 113.20	78738-37-1 	 142.5	5-20-04-00169 0.8649[0]	1.4494[21]
9669	Piperidine, 3-ethyl-, (±)- Piperidine, 3-ethyl (DL)	$C_7H_{15}N$ 113.20	59433-08-8 	 152.6	5-20-04-00173 0.8585[23]	ace 4 1.4531[20]
9670	Piperidine, 4-ethyl- 4-Ethylpiperidine	$C_7H_{15}N$ 113.20	3230-23-7 	 154	5-20-04-00174 0.8759[25]	1.4503[25]
9671	Piperidine, 1-(1-ethyl-1-methylpropyl)- Piperidine, N-(3-methyl-3-pentyl)	$C_{11}H_{23}N$ 169.31	14045-26-2 	 214	5-20-02-00033 0.8614[20]	1.4637[20]
9672	Piperidine, 1-hexyl- 1-Hexylpiperidine	$C_{11}H_{23}N$ 169.31	7335-01-5 	 219.2	5-20-02-00032 0.8292[20]	1.4522[20]
9673	4-Piperidinemethanamine 4-(Aminomethyl)piperidine	$C_6H_{14}N_2$ 114.19	7144-05-0 25	 200; 31[10]	5-22-08-00128	1.4900[20]
9674	2-Piperidinemethanol	$C_6H_{13}NO$ 115.18	3433-37-2 69	 104[10]; 80[1]	5-21-01-00098	chl 2
9675	3-Piperidinemethanol	$C_6H_{13}NO$ 115.18	4606-65-9 61	 106-7[3.5]	5-21-01-00107 1.0263[20]	chl 2 1.4964[20]

No.	Name Synonym	Mol. Form. Mol. Wt.	CAS RN mp/°C	Merck No. bp/°C	Beil. Ref. den/g cm⁻³	Solubility n_D
9676	2-Piperidinemethanol, α,α-diphenyl- Pipradrol	$C_{18}H_{21}NO$ 267.37	467-60-7	7455	5-21-04-00070	
9677	2-Piperidinemethanol, α-ethyl-, (R*,S*)-(±)- Conhydrine (DL) (lower m.p.)	$C_8H_{17}NO$ 143.23	3238-62-8 121	2495 226	5-21-01-00138	H_2O 2; bz 4; eth 4; EtOH 4
9678	3-Piperidinemethanol, 1-methyl-	$C_7H_{15}NO$ 129.20	7583-53-1	142.5; 114[20]	5-21-01-00108 1.013[25]	1.4772[20]
9679	Piperidine, 1-methyl- N-Methylpiperidine	$C_6H_{13}N$ 99.18	626-67-5	107	5-20-02-00021 0.8159[20]	H_2O 4; EtOH 5; eth 5; ctc 3 1.4355[20]
9680	Piperidine, 2-methyl-, (±)-	$C_6H_{13}N$ 99.18	3000-79-1 -4.9	118	5-20-04-00071 0.8436[24]	H_2O 4; EtOH 3; eth 3; chl 2 1.4459[20]
9681	Piperidine, 3-methyl-, (±)- Piperidine, 3-methyl (DL)	$C_6H_{13}N$ 99.18	53152-98-0 -24	125.5	5-20-04-00099 0.8446[26]	H_2O 4; chl 2 1.4470[20]
9682	Piperidine, 4-methyl-	$C_6H_{13}N$ 99.18	626-58-4	130	5-20-04-00116 0.8874[25]	H_2O 4; chl 2 1.4458[20]
9683	Piperidine, 1,1'-methylenebis-	$C_{11}H_{22}N_2$ 182.31	880-09-1	122-3[15]	5-20-02-00293 0.915[25]	1.4820[20]
9684	Piperidine, 1-(1-methylethyl)-	$C_8H_{17}N$ 127.23	766-79-0	149.5	5-20-02-00028 0.8389[20]	H_2O 4 1.4491[20]
9685	Piperidine, 1-(2-methylpropyl)-	$C_9H_{19}N$ 141.26	10315-89-6	160.5	5-20-02-00030 0.8161[25]	H_2O 4 1.4382[25]
9686	Piperidine, 1-methyl-2-propyl-, (S)- Methylconiine	$C_9H_{19}N$ 141.26	35305-13-6	5968 174	5-20-04-00206 0.8326[22]	ace 4; EtOH 4 1.4538[12]
9687	Piperidine, 1-nitro-	$C_5H_{10}N_2O_2$ 130.15	7119-94-0 -5.5	245; 121[20]	5-20-03-00531 1.1519[26]	1.4954[26]
9688	Piperidine, 1-nitroso- N-Nitrosopiperidine	$C_5H_{10}N_2O$ 114.15	100-75-4	219; 109[20]	5-20-03-00530 1.063[18]	H_2O 3; HCl 3 1.4933[18]
9689	Piperidine, 1-nonyl-	$C_{14}H_{29}N$ 211.39	30538-80-8	135-7[11]	4-20-00-00319 0.8313[25]	1.4538[25]
9690	Piperidine, 1-octyl-	$C_{13}H_{27}N$ 197.36	7335-02-6	137[13]; 89[1]	5-20-02-00034 0.8324[20]	1.4544[20]
9691	Piperidine, 1,2,2,6,6-pentamethyl- Pempidine	$C_{10}H_{21}N$ 155.28	79-55-0	7022 147	5-20-04-00238 0.8580[0]	1.4550[21]
9692	Piperidine, 1-pentyl-	$C_{10}H_{21}N$ 155.28	10324-58-0	198.2	5-20-02-00031 0.8282[20]	1.4498[20]
9693	Piperidine, 1-phenyl-	$C_{11}H_{15}N$ 161.25	4096-20-2 4.7	258	5-20-02-00057 0.9944[25]	EtOH 4; eth 4; bz 4; chl 4 1.5598[25]
9694	Piperidine, 3-phenyl-	$C_{11}H_{15}N$ 161.25	3973-62-4 14.5	257; 140[19]	5-20-06-00405 1.0010[25]	1.5473[25]
9695	Piperidine, 4-phenyl- 4-Phenylpiperidine	$C_{11}H_{15}N$ 161.25	771-99-3 60.5	257	5-20-06-00408 0.9996[16]	chl 3
9696	Piperidine, 1-(1-phenylcyclohexyl)- Phencyclidine	$C_{17}H_{25}N$ 243.39	77-10-1 46.5	7179 135-7[1.0]	5-20-02-00078	
9697	Piperidine, 1-(1-phenylethyl)-	$C_{13}H_{19}N$ 189.30	7529-63-7 -13	272	5-20-02-00064 0.945[25]	1.5218[25]
9698	Piperidine, 1-(phenylmethyl)-	$C_{12}H_{17}N$ 175.27	2905-56-8	245	5-20-02-00062 0.9625[16]	1.5227[20]
9699	Piperidine, 2-(phenylmethyl)- 2-Benzylpiperidine	$C_{12}H_{17}N$ 175.27	32838-55-4 32	267.5	5-20-06-00438 0.9660[25]	eth 4; EtOH 4
9700	Piperidine, 4-(phenylmethyl)- 4-Benzylpiperidine	$C_{12}H_{17}N$ 175.27	31252-42-3 16.8	270; 150[17]	5-20-06-00439 0.9970[20]	H_2O 1; EtOH 3; eth 3 1.5337[25]
9701	Piperidine, 4,4'-(1,3-propanediyl)bis-	$C_{13}H_{26}N_2$ 210.36	16898-52-5 67.1	329	5-23-04-00101	H_2O 4
9702	Piperidine, 4,4'-(1,3-propanediyl)bis[1-methyl- Propane, 1,3-bis(1-methyl-4-piperidyl)	$C_{15}H_{30}N_2$ 238.42	64168-11-2 13.7	215[50]	4-23-00-00536 0.8962[25]	1.4804[25]
9703	1-Piperidinepropanenitrile	$C_8H_{14}N_2$ 138.21	3088-41-3 -6.8	145[50]	5-20-03-00047 0.9403[25]	1.4676[25]
9704	1-Piperidinepropanoic acid, ethyl ester Piperidine, N-carboxyethyl, ethyl ester	$C_{10}H_{19}NO_2$ 185.27	19653-33-9	139[50]	5-20-03-00042 0.9627[25]	H_2O 4 1.4525[25]
9705	1-Piperidinepropanol, α-cyclohexyl-α-phenyl-, hydrochloride Trihexyphenidyl hydrochloride	$C_{20}H_{32}ClNO$ 337.93	52-49-3 258.5	9607		
9706	Piperidine, 2-(1-propenyl)-, (±)-	$C_8H_{15}N$ 125.21	5913-85-9 6	168	5-20-04-00346 0.8716[15]	eth 4; EtOH 4
9707	Piperidine, 1-propyl-	$C_8H_{17}N$ 127.23	5470-02-0	151.5	5-20-02-00027 0.8231[20]	H_2O 4; eth 4; EtOH 4 1.4446[20]
9708	Piperidine, 2-propyl-, (S)- Coniine	$C_8H_{17}N$ 127.23	458-88-8 -2	2500 166.5	5-20-04-00205 0.8440[20]	H_2O 2; EtOH 5; eth 4; bz 3 1.4512[22]
9709	Piperidine, 4-propyl- 4-Propylpiperidine	$C_8H_{17}N$ 127.23	22398-09-0	173	5-20-04-00208 0.864[22]	EtOH 3; chl 2 1.4465[23]
9710	Piperidine, 2,2,4,6-tetramethyl- 2,2,4,6-Tetramethylpiperidine	$C_9H_{19}N$ 141.26	6292-82-6	156	4-20-00-01638 0.816[25]	1.4374[25]
9711	Piperidine, 2,2,6,6-tetramethyl- Norpempidine	$C_9H_{19}N$ 141.26	768-66-1 28	156	5-20-04-00237 0.8367[16]	eth 4 1.4455[20]
9712	Piperidine, 2,2,4-trimethyl-	$C_8H_{17}N$ 127.23	101257-71-0	148	4-20-00-01624 0.832[15]	eth 4; EtOH 4 1.4458[20]
9713	Piperidine, 2,3,6-trimethyl- 2,3,6-Trimethylpiperidine	$C_8H_{17}N$ 127.23	50402-72-7	36[5]	4-20-00-01625 0.8302[20]	1.4434[20]
9714	Piperidine, 2,4,6-trimethyl- 2,4,6-Trimethylpiperidine	$C_8H_{17}N$ 127.23	21974-48-1 <20	145	5-20-04-00220 0.8315[19]	H_2O 2 1.4412[20]
9715	3-Piperidinol, 1-ethyl- 1-Ethyl-3-piperidinol	$C_7H_{15}NO$ 129.20	13444-24-1 93-5[15]	3799	5-21-01-00039	1.4777[14]

No.	Name Synonym	Mol. Form. Mol. Wt.	CAS RN mp/°C	Merck No. bp/°C	Beil. Ref. den/g cm^{-3}	Solubility n_D
9716	3-Piperidinol, 1-methyl-	$C_6H_{13}NO$ 115.18	3554-74-3	76-8[11]	5-21-01-00029 0.999[25]	1.4735[20]
9717	4-Piperidinol, 1-methyl-	$C_6H_{13}NO$ 115.18	106-52-5 29	200	5-21-01-00052	1.4775[20]
9718	3-Piperidinol, 6-propyl-, (3S-trans)- Pseudoconhydrine	$C_8H_{17}NO$ 143.23	140-55-6 106	7928 236	5-21-01-00138	H$_2$O 4; eth 4; EtOH 4
9719	2-Piperidinone	C_5H_9NO 99.13	675-20-7 39.5	256	5-21-06-00396	H$_2$O 4; EtOH 4; eth 4; dil acid 3
9720	4-Piperidinone, 1-acetyl-	$C_7H_{11}NO_2$ 141.17	32161-06-1	218; 124[0.2]	5-21-06-00426 1.146[25]	1.5026[20]
9721	2-Piperidinone, 1-methyl-	$C_6H_{11}NO$ 113.16	931-20-4	105[12]; 94[9]	5-21-06-00398 1.0283[25]	1.4820[20]
9722	2-Piperidinone, 3-methyl-	$C_6H_{11}NO$ 113.16	3768-43-2 55.3	249.5	5-21-06-00479	bz 4; eth 4; EtOH 4
9723	3-Piperidinone, 1-methyl-	$C_6H_{11}NO$ 113.16	5519-50-6	63-4[13]	5-21-06-00417 0.9655[25]	1.4559[25]
9724	4-Piperidinone, 1-methyl-	$C_6H_{11}NO$ 113.16	1445-73-4	56-8[11]	5-21-06-00419 0.9696[25]	1.4580[25]
9725	4-Piperidinone, 1,2,3,6-tetramethyl- 1,2,5,6-Tetramethyl-4-piperidinone	$C_9H_{17}NO$ 155.24	53381-90-1	86-8[7.5]	4-21-00-03272 0.9499[20]	1.4680[20]
9726	4-Piperidinone, 2,2,6,6-tetramethyl-	$C_9H_{17}NO$ 155.24	826-36-8 36	205	5-21-06-00538	H$_2$O 3; EtOH 3; eth 3; chl 2
9727	Piperonyl butoxide 1,3-Benzodioxole, 5-[[2-(2-butoxyethoxy)ethoxy]methyl]-6-propyl-	$C_{19}H_{30}O_5$ 338.44	51-03-6	7446 180[1]	1.05[25]	
9728	Pirimicarb Carbamic acid, dimethyl-, 2-(dimethylamino)-5,6-dimethyl-4-pyrimidinyl ester	$C_{11}H_{18}N_4O_2$ 238.29	23103-98-2 90.5	7468		
9729	Pirimiphos-ethyl Phosphorothioic acid, O-[2-(diethylamino)-6-methyl-4-pyrimidinyl] O,O-diethyl ester	$C_{13}H_{24}N_3O_3PS$ 333.39	23505-41-1 dec >130	7469	1.14[20]	
9730	Pirimiphos-methyl Phosphorothioic acid, O-[2-(diethylamino)-6-methyl-4-pyrimidinyl] O,O-dimethyl ester	$C_{11}H_{20}N_3O_3PS$ 305.34	29232-93-7 15 dec		1.17[20]	
9731	Plumbane, tetraethyl Tetraethyl lead	$C_8H_{20}Pb$ 323.45	78-00-2 200 dec	9136	1.653[20]	H$_2$O 1; bz 3; EtOH 2 1.5198[20]
9732	Plumbane, tetramethyl Tetramethyl lead	$C_4H_{12}Pb$ 267.34	75-74-1 -27.5 110		1.995[20]	
9733	Plumbane, tetraphenyl-	$C_{24}H_{20}Pb$ 515.62	595-89-1 228.3	126[13]	4-16-00-01615 1.5298[20]	chl 3
9734	21H,23H-Porphine	$C_{20}H_{14}N_4$ 310.36	101-60-0 360	7574 sub 300	4-26-00-01900 1.336[25]	H$_2$O 1; EtOH 2; eth 1; ace 1
9735	21H,23H-Porphine-2,18-dipropanoic acid, 7,12-bis(1-hydroxyethyl)-3,8,13,17-tetramethyl- Hematoporphyrin	$C_{34}H_{38}N_4O_6$ 598.70	14459-29-1 172.5	4555	4-26-00-03157	H$_2$O 1; EtOH 3; eth 2; chl 2
9736	Pregna-4,6-diene-21-carboxylic acid, 17-hydroxy-3-oxo-, γ-lactone, (17α)- Canrenone	$C_{22}H_{28}O_3$ 340.46	976-71-6 150	1753	5-17-11-00476	
9737	Pregna-1,4-diene-3,20-dione, 9-fluoro-11,17,21-trihydroxy-16-methyl-, (11β,16α)- Dexamethasone	$C_{22}H_{29}FO_5$ 392.47	50-02-2 262	2922		
9738	Pregna-1,4-diene-3,20-dione, 11,17,21-trihydroxy-, (11β)- Prednisolone	$C_{21}H_{28}O_5$ 360.45	50-24-8 235	7719	4-08-00-03467	
9739	Pregna-1,4-diene-3,11,20-trione, 17,21-dihydroxy- Prednisone	$C_{21}H_{26}O_5$ 358.43	53-03-2 233-5 dec	7727	4-08-00-03531	
9740	Pregnane, (5α)- Allopregnane	$C_{21}H_{36}$ 288.52	641-85-0 84.5	257	4-05-00-01215	
9741	Pregnane, (5β)- 17β-Ethyletiocholane	$C_{21}H_{36}$ 288.52	481-26-5 83.5	7731	4-05-00-01215 1.032[15]	H$_2$O 1; chl 3; MeOH 3
9742	Pregnane-3,20-diol, (3α,5β,20S)- Pregnanediol	$C_{21}H_{36}O_2$ 320.52	80-92-2 243.5	7732	4-06-00-06111 1.15[25]	EtOH 2; eth 2; ace 3; os 2
9743	Pregnane-3,20-dione, (5β)- 5β-Pregnane-3,20-dione	$C_{21}H_{32}O_2$ 316.48	128-23-4 123	7733	4-07-00-02198	H$_2$O 1; EtOH 4; eth 3; ace 3
9744	Pregnane-11,20-dione, 3,17,21-trihydroxy-, (3α,5β)- Tetrahydrocortisone	$C_{21}H_{32}O_5$ 364.48	53-05-4 190	9143	4-08-00-03395	
9745	Pregnan-20-one, 3-hydroxy-, (3α,5β)- Pregnan-3α-ol-20-one	$C_{21}H_{34}O_2$ 318.50	128-20-1 149.5	7734		EtOH 4
9746	Pregn-4-en-18-al, 11,21-dihydroxy-3,20-dioxo-, (11β)- Aldosterone	$C_{21}H_{28}O_5$ 360.45	52-39-1 166.5	218	4-08-00-03491	
9747	Pregn-4-ene-3,20-dione, 21-(acetyloxy)-11,17-dihydroxy-, (11β)- Hydrocortisone acetate	$C_{23}H_{23}O_6$ 395.43	50-03-3 223 dec	4711	4-08-00-03424 1.289[20]	
9748	Pregn-4-ene-3,20-dione, 11,21-dihydroxy-, (11β)- Corticosterone	$C_{21}H_{30}O_4$ 346.47	50-22-6 181	2532 190[0.01]	4-08-00-02907	H$_2$O 1; EtOH 3; eth 3; ace 3

No.	Name / Synonym	Mol. Form. / Mol. Wt.	CAS RN / mp/°C	Merck No. / bp/°C	Beil. Ref. / den/g cm^{-3}	Solubility / n_D
9749	Pregn-4-ene-3,20-dione, 17,21-dihydroxy-	$C_{21}H_{30}O_4$	152-58-9	2912	4-08-00-02913	ace 4; EtOH 4; chl 4
	11-Deoxy-17-hydroxycorticosterone	346.47	215			
9750	Pregn-4-ene-3,20-dione, 17-hydroxy-	$C_{21}H_{30}O_3$	68-96-2	4773	4-08-00-02189	chl 2
	17α-Hydroxyprogesterone	330.47				
9751	Pregn-4-ene-3,20-dione, 21-hydroxy-	$C_{21}H_{30}O_3$	64-85-7	2882	4-08-00-02195	H_2O 2; EtOH 4; eth 2; ace 4
	Deoxycorticosterone	330.47	141.5			
9752	Pregn-4-ene-3,20-dione, 17-hydroxy-6-methyl-, (6α)-	$C_{22}H_{32}O_3$	520-85-4	5677	4-08-00-02211	chl 4
	Medroxyprogesterone	344.49	214.5			
9753	Pregn-4-ene-3,20-dione, 11,17,21-trihydroxy-, (11β)-	$C_{21}H_{30}O_5$	50-23-7	4710	4-08-00-03422	H_2O 2; EtOH 3; diox 3; HOAc 3
	Hydrocortisone	362.47	220			
9754	Pregn-4-ene-3,11,20-trione, 17,21-dihydroxy-	$C_{21}H_{28}O_5$	53-06-5	2533	4-08-00-03480	H_2O 2; EtOH 3; eth 2; ace 3
	Cortisone	360.45	222			
9755	Pregn-4-ene-3,11,20-trione, 21-hydroxy-	$C_{21}H_{28}O_4$	72-23-1	2859	4-08-00-02998	H_2O 1; EtOH 3; ace 3; bz 3
	11-Dehydrocorticosterone	344.45	183.5			
9756	Pregn-5-en-20-one, 3-hydroxy-, (3β)-	$C_{21}H_{32}O_2$	145-13-1	7739	4-08-00-01019	
	Pregnenolone	316.48	192			
9757	Pregn-4-en-3-one, 11,17,20,21-tetrahydroxy-, (11β,20R)-	$C_{21}H_{32}O_5$	116-58-5	7736	4-08-00-03384	ace 4; EtOH 4
	4-Pregnene-11β,17α,20β,21-tetrol-3-one	364.48	125 dec			
9758	Pregn-4-en-20-yn-3-one, 17-hydroxy-, (17α)-	$C_{21}H_{28}O_2$	434-03-7	3696	4-08-00-01225	
	Ethisterone	312.45	272			
9759	Procyazine	$C_{10}H_{13}ClN_6$	32889-48-8			
	Propanenitrile, 2-[[4-chloro-6-(cyclopropylamino)-1,3,5-triazin-2-yl]amino]-2-methyl-	252.71	170			
9760	Procymidone	$C_{13}H_{11}Cl_2NO_2$	32809-16-8			
	N-(3,5-Dichlorophenyl)-1,2-dimethylcyclopropane-1,2-dicarboximide	284.14	166		1.452^{25}	
9761	Prodiamine	$C_{13}H_{17}F_3N_4O_4$	29091-21-2	7773		
	1,3-Benzenediamine, 2,4-dinitro-N3,N3-dipropyl-6-(trifluoromethyl)-	350.30	124		1.47^{25}	
9762	Profenofos	$C_{11}H_{15}BrClO_3PS$	41198-08-7			
	Phosphorothioic acid, O-(4-bromo-2-chlorophenyl) O-ethyl S-propyl ester	373.63		$110^{0.001}$	1.455^{20}	
9763	Profluralin	$C_{14}H_{16}F_3N_3O_4$	26399-36-0	7781		
	Benzenamine, N-(cyclopropylmethyl)-2,6-dinitro-N-propyl-4-(trifluoromethyl)-	347.29	34			
9764	L-Proline	$C_5H_9NO_2$	147-85-3	7790	5-22-01-00031	H_2O 4; EtOH 2; eth 1; ace 2
	2-Pyrrolidinecarboxylic acid	115.13	221 dec			
9765	DL-Proline	$C_5H_9NO_2$	609-36-9		5-22-01-00033	H_2O 4; EtOH 4
		115.13	205 dec			
9766	L-Proline, 1-acetyl-4-hydroxy-, trans-	$C_7H_{11}NO_4$	33996-33-7	6857	5-22-05-00013	H_2O 4; MeOH 4
	Oxaceprol	173.17	132			
9767	L-Proline, 4-hydroxy-, trans-	$C_5H_9NO_3$	51-35-4	4775	5-22-05-00007	H_2O 4; EtOH 2
	4-Hydroxy-L-proline	131.13	274			
9768	L-Proline, 5-oxo-	$C_5H_7NO_3$	98-79-3	8012	5-22-06-00007	DMSO 3
	L-Pyroglutamic acid	129.12	162			
9769	Promecarb	$C_{12}H_{17}NO_2$	2631-37-0	7794		
	Phenol, 3-methyl-5-(1-methylethyl)-, methylcarbamate	207.27	87	$117^{0.01}$		
9770	Prometryn	$C_{10}H_{19}N_5S$	7287-19-6	7800		
	1,3,5-Triazine-2,4-diamine, N,N'-bis(1-methylethyl)-6-(methylthio)-	241.36	119		1.157^{20}	
9771	Propachlor	$C_{11}H_{14}ClNO$	1918-16-7	7805		
	Acetamide, 2-chloro-N-(1-methylethyl)-N-phenyl-	211.69	77	$110^{0.03}$	1.242^{25}	
9772	1,2-Propadiene	C_3H_4	463-49-0		4-01-00-00966	bz 4; peth 4
	Allene	40.06	-136.2	-34.4	$*0.584^{25}$	1.4168
9773	1,2-Propadiene-1,3-dione	C_3O_2	504-64-3	1821	4-01-00-03764	eth 3; bz 3; CS_2 3
	Carbon suboxide	68.03	-107	6.8	1.114^0	1.4538^0
9774	Propanal	C_3H_6O	123-38-6	7835	4-01-00-03165	H_2O 3; EtOH 5; eth 5
	Propionaldehyde	58.08	-80	48	0.8657^{25}	1.3636^{20}
9775	Propanal, 2-bromo-	C_3H_5BrO	19967-57-8		4-01-00-03177	eth 4
		136.98		109.5	1.592^{20}	1.4813^{20}
9776	Propanal, 2-bromo-2-methyl-	C_4H_7BrO	13206-46-7		4-01-00-03267	
		151.00		113	1.383^{25}	1.4518^{25}
9777	Propanal, 2-chloro-	C_3H_5ClO	683-50-1		3-01-00-02691	bz 4; eth 4
		92.52		86	1.182^{15}	1.431^{17}
9778	Propanal, 3-chloro-	C_3H_5ClO	19434-65-2		4-01-00-03174	eth 4; EtOH 4
		92.52	<20	130; 45^{10}	1.268^{15}	1.475^{25}
9779	Propanal, 2-chloro-2-methyl-	C_4H_7ClO	917-93-1		4-01-00-03267	eth 4; EtOH 4
		106.55		90	1.053^{15}	1.4160^{16}
9780	Propanal, 2,3-dibromo-	$C_3H_4Br_2O$	5221-17-0		4-01-00-03178	eth 4
		215.87		$73-5^{10}$	2.198^{15}	1.5082^{20}
9781	Propanal, 2,3-dichloro-	$C_3H_4Cl_2O$	10140-89-3		4-01-00-03174	
		126.97		73^{50}; 48^{14}	1.400^{20}	1.4762^{20}
9782	Propanal, 2,3-dihydroxy-, (±)-	$C_3H_6O_3$	56-82-6		3-01-00-03282	H_2O 3; EtOH 2; eth 2; bz 1
		90.08	145	$140-50^{0.8}$	1.453^{18}	

No.	Name Synonym	Mol. Form. Mol. Wt.	CAS RN mp/°C	Merck No. bp/°C	Beil. Ref. den/g cm^{-3}	Solubility n_D
9783	Propanal, 2,2-dimethyl- Pivaldehyde	C$_5$H$_{10}$O 86.13	630-19-3 6	77.5	4-01-00-03295 0.7923[17]	EtOH 3; eth 3 1.3791[20]
9784	Propanal, 2-(hydroxyimino)-, oxime	C$_3$H$_6$N$_2$O$_2$ 102.09	1804-15-5 157	sub	4-01-00-03633	H$_2$O 2; EtOH 3; eth 3; bz 3
9785	Propanal, 2-methyl- Isobutanal	C$_4$H$_8$O 72.11	78-84-2 -65.9	5038 64.5	4-01-00-03262 0.7891[20]	H$_2$O 3; eth 3; ace 3; ctc 2 1.3730[20]
9786	Propanal, 3-(methylthio)- 3-(Methylthio)propionaldehyde	C$_4$H$_8$OS 104.17	3268-49-3	62[11]	4-01-00-03974	
9787	Propanal, oxime	C$_3$H$_7$NO 73.09	627-39-4 40	131.5	4-01-00-03170 0.9258[20]	1.4287[20]
9788	Propanal, 2-oxo- Pyruvaldehyde	C$_3$H$_4$O$_2$ 72.06	78-98-8	8030 72	4-01-00-03631 1.0455[20]	EtOH 3; eth 3; bz 3 1.4002[18]
9789	Propanal, 2-oxo-, 1-oxime Isonitrosoacetone	C$_3$H$_5$NO$_2$ 87.08	306-44-5 69	5076 sub	3-01-00-03092 1.0744[67]	H$_2$O 3; eth 3; bz 2; ctc 2
9790	Propanal, 2,2,3-trichloro-	C$_3$H$_3$Cl$_3$O 161.41	7789-90-4	63-5[45]	3-01-00-02693 1.470[25]	eth 4 1.473[25]
9791	Propanamide Propionamide	C$_3$H$_7$NO 73.09	79-05-0 81.3	7836 213	4-02-00-00725 0.9262[110]	H$_2$O 4; EtOH 4; eth 4; chl 4 1.4180[110]
9792	Propanamide, 2-bromo-2-methyl- Isobutyramide, 2-bromo	C$_4$H$_8$BrNO 166.02	7462-74-0 148	145[17]	4-02-00-00863	EtOH 4; chl 4
9793	Propanamide, N,N-diethyl-	C$_7$H$_{15}$NO 129.20	1114-51-8	191	4-04-00-00353 0.8972[20]	EtOH 4 1.4425[20]
9794	Propanamide, N,N-diethyl-2,2-dimethyl- Pivalamide, N,N-diethyl	C$_9$H$_{19}$NO 157.26	24331-72-4	206	3-04-00-00211 0.891[15]	eth 4; EtOH 4
9795	Propanamide, N-(2,3-dihydro-1,5-dimethyl-3-oxo-2-phenyl-1H-pyrazol-4-yl)-2-(dimethylamino)- Aminopropylon	C$_{16}$H$_{22}$N$_4$O$_2$ 302.38	3690-04-8 181	484	4-25-00-03595	H$_2$O 4
9796	Propanamide, N-(4-ethoxyphenyl)-2-hydroxy- p-Lactophenetide	C$_{11}$H$_{15}$NO$_3$ 209.25	539-08-2 118	5220	3-13-00-01135	H$_2$O 3; EtOH 4; eth 2; bz 2
9797	Propanamide, 2-hydroxy-, (±)-	C$_3$H$_7$NO$_2$ 89.09	65144-02-7 75.5		4-03-00-00674 1.1381[80]	H$_2$O 4; EtOH 4
9798	Propanamide, 2-methyl-	C$_4$H$_9$NO 87.12	563-83-7 128	218	4-02-00-00852	chl 3
9799	Propanamide, N-methyl-	C$_4$H$_9$NO 87.12	1187-58-2 -30.9	148	4-04-00-00183 0.9305[25]	1.4345[25]
9800	Propanamide, N-phenyl-	C$_9$H$_{11}$NO 149.19	620-71-3 105.5	222.2	4-12-00-00385 1.175[25]	H$_2$O 2; EtOH 4; eth 4
9801	Propanamide, N-phenyl-N-[1-(2-phenylethyl)-4-piperidinyl]- Fentanyl	C$_{22}$H$_{28}$N$_2$O 336.48	437-38-7 87.5	3944	5-22-08-00049	
9802	Propanamide, N-propyl-	C$_6$H$_{13}$NO 115.18	3217-86-5 154	215; 108[9]	4-04-00-00477 0.8995[25]	H$_2$O 2; eth 2
9803	1-Propanamine Propylamine	C$_3$H$_9$N 59.11	107-10-8 -83	7855 47.2	4-04-00-00464 0.7173[20]	H$_2$O 3; EtOH 4; eth 4; ace 4 1.3870[20]
9804	2-Propanamine Isopropylamine	C$_3$H$_9$N 59.11	75-31-0 -95.1	5097 31.7	4-04-00-00504 0.6891[20]	H$_2$O 5; EtOH 5; eth 5; ace 4 1.3742[20]
9805	1-Propanamine, N,N-dichloro- N,N-Dichloropropylamine	C$_3$H$_7$Cl$_2$N 128.00	69947-01-9	117	0-04-00-00145 1.1454[23]	1.4525[23]
9806	1-Propanamine, 2,2-dimethyl-	C$_5$H$_{13}$N 87.16	5813-64-9	82	4-04-00-00707 0.7455[20]	eth 4 1.4023[20]
9807	1-Propanamine, N,N-dimethyl- Dimethylpropylamine	C$_5$H$_{13}$N 87.16	926-63-6	66	4-04-00-00467 0.7152[20]	bz 4; eth 4; EtOH 4 1.3860[20]
9808	1-Propanamine, N,N-dipropyl- Tripropylamine	C$_9$H$_{21}$N 143.27	102-69-2 -93.5	156	4-04-00-00470 0.7558[20]	eth 4; EtOH 4 1.4181[20]
9809	1-Propanamine, N-ethyl-	C$_5$H$_{13}$N 87.16	20193-20-8	81	4-04-00-00468 0.7204[17]	H$_2$O 2; ace 4; EtOH 4 1.3858[25]
9810	2-Propanamine, N-ethyl- Ethylisopropylamine	C$_5$H$_{13}$N 87.16	19961-27-4	69.6	4-04-00-00508	1.3872[25]
9811	2-Propanamine, N-ethyl-N-(1-methylethyl)-	C$_8$H$_{19}$N 129.25	7087-68-5	126.5	4-04-00-00511 0.742[25]	ctc 3 1.4138[20]
9812	1-Propanamine, N-ethyl-N-propyl-	C$_8$H$_{19}$N 129.25	20634-92-8	138	2-04-00-00622 0.807[24]	ace 4; bz 4; eth 4
9813	1-Propanamine, 1,1,2,2,3,3,3-heptafluoro-N,N-bis(heptafluoropropyl)-	C$_9$F$_{21}$N 521.07	338-83-0	130	4-02-00-00742 1.822[4]	1.279[25]
9814	1-Propanamine, N-hydroxy-N-propyl- N,N-Dipropylhydroxylamine	C$_6$H$_{15}$NO 117.19	7446-43-7 29	154.5	4-04-00-03308	bz 4; eth 4; EtOH 4; chl 4
9815	2-Propanamine, N,N-methanetetraylbis-	C$_7$H$_{14}$N$_2$ 126.20	693-13-0	147	4-04-00-00531 0.806[25]	1.4320[20]
9816	1-Propanamine, 3-methoxy-	C$_4$H$_{11}$NO 89.14	5332-73-0	117.5	4-04-00-01623 0.8727[20]	H$_2$O 3; ace 3; bz 3; ctc 3 1.4391[20]
9817	1-Propanamine, 2-methyl- Isobutylamine	C$_4$H$_{11}$N 73.14	78-81-9 -86.7	5016 67.7	4-04-00-00625 0.724[25]	1.3988[19]
9818	2-Propanamine, 2-methyl- tert-Butylamine	C$_4$H$_{11}$N 73.14	75-64-9 -66.9	1545 44.0	4-04-00-00657 0.6958[20]	H$_2$O 5; EtOH 5; eth 5; chl 3 1.3784[20]
9819	2-Propanamine, N-methyl- Isopropylmethylamine	C$_4$H$_{11}$N 73.14	4747-21-1	50.4		
9820	1-Propanamine, 2-methyl-N,N-bis(2-methylpropyl)- Triisobutylamine	C$_{12}$H$_{27}$N 185.35	1116-40-1 -21.8	191.5	4-04-00-00631 0.7684[20]	eth 4; EtOH 4 1.4252[17]

No.	Name / Synonym	Mol. Form. / Mol. Wt.	CAS RN / mp/°C	Merck No. / bp/°C	Beil. Ref. / den/g cm^{-3}	Solubility / n_D
9821	2-Propanamine, N-(1-methylethyl)- / Diisopropylamine	$C_6H_{15}N$ / 101.19	108-18-9 / -61	3181 / 83.9	4-04-00-00510 / 0.7153[20]	ace 4; bz 4; eth 4; EtOH 4 / 1.3924[20]
9822	2-Propanamine, N-(1-methylethyl)-N-nitroso-	$C_6H_{14}N_2O$ / 130.19	601-77-4 / 48	/ 194.5	4-04-00-03388 / 0.9422[20]	H_2O 2; EtOH 3; eth 3; bz 3
9823	1-Propanamine, 2-methyl-N-(2-methylpropyl)- / Diisobutylamine	$C_8H_{19}N$ / 129.25	110-96-3 / -73.5	/ 139.6	4-04-00-00630 /	H_2O 2; EtOH 3; eth 3; ace 3 / 1.4090[20]
9824	1-Propanamine, N-nitro-	$C_3H_8N_2O_2$ / 104.11	627-07-6 / -21	/ 128-9[40]	4-04-00-03405 / 1.1046[15]	eth 4; EtOH 4 / 1.4610[20]
9825	1-Propanamine, N-nitroso-N-propyl- / Dipropylnitrosamine	$C_6H_{14}N_2O$ / 130.19	621-64-7 /	/ 206; 113[40]	4-04-00-03388 / 0.9163[20]	H_2O 2; EtOH 5; eth 5 / 1.4437[20]
9826	1-Propanamine, N-propyl- / Dipropylamine	$C_6H_{15}N$ / 101.19	142-84-7 / -63	3350 / 109.3	4-04-00-00469 / 0.7400[20]	H_2O 3; EtOH 3; eth 5; ace 4 / 1.4050[20]
9827	2-Propanamine, N-propyl / Isopropylpropylamine	$C_6H_{15}N$ / 101.19	21968-17-2 /	/ 96.9		
9828	1-Propanamine, 3-(triethoxysilyl)-	$C_9H_{23}NO_3Si$ / 221.37	919-30-2 /	/ 122-3[31]	4-04-00-04273 / 0.9506[20]	/ 1.4225[20]
9829	1-Propanamine, N,N,2-trimethyl-	$C_6H_{15}N$ / 101.19	7239-24-9 /	/ 80.5	4-04-00-00627 / 0.7097[20]	H_2O 4 / 1.3907[20]
9830	1-Propanaminium, 2-(acetyloxy)-N,N,N-trimethyl-, chloride / Methacholine chloride	$C_8H_{18}ClNO_2$ / 195.69	62-51-1 / 172	5847 /		H_2O 4; EtOH 4; chl 4
9831	1-Propanaminium, 2-hydroxy-N,N,N-trimethyl-, chloride / (2-Hydroxypropyl)trimethyl ammonium chloride	$C_6H_{16}ClNO$ / 153.65	2382-43-6 / 165	/ dec	3-04-00-00754 /	H_2O 4; EtOH 4
9832	2-Propanaminium, N-methyl-N-(1-methylethyl)-N-[2-[(9H-xanthen-9-ylcarbonyl)oxy]ethyl]-, bromide / Propantheline bromide	$C_{23}H_{30}BrNO_3$ / 448.40	50-34-0 / 159.5	7816 /	5-18-06-00590 /	
9833	1-Propanaminium, N,N,N-tripropyl-, iodide / Tetrapropylammonium iodide	$C_{12}H_{28}IN$ / 313.27	631-40-3 / 280 dec	/	4-04-00-00471 / 1.3138[25]	H_2O 4; EtOH 3; eth 2; chl 4
9834	Propane	C_3H_8 / 44.10	74-98-6 / -187.6	7809 / -42.1	4-01-00-00176 / *0.493[25]	H_2O 3; EtOH 3; eth 4; ace 2
9835	Propane, 2,2-bis(ethylsulfonyl)- / Sulfonmethane	$C_7H_{16}O_4S_2$ / 228.33	115-24-2 / 125.8	8939 / 300 dec	3-01-00-02754 /	bz 4; EtOH 4; chl 4
9836	Propane, 1-bromo- / Propyl bromide	C_3H_7Br / 122.99	106-94-5 / -110	7857 / 71.1	4-01-00-00205 / 1.3537[20]	H_2O 2; EtOH 3; eth 3; ace 3 / 1.4343[20]
9837	Propane, 2-bromo- / Isopropyl bromide	C_3H_7Br / 122.99	75-26-3 / -89	5098 / 59.5	4-01-00-00208 / 1.3140[20]	H_2O 2; EtOH 5; eth 5; ace 3 / 1.4251[20]
9838	Propane, 1-bromo-2-chloro- / 1-Bromo-2-chloropropane	C_3H_6BrCl / 157.44	3017-96-7 /	/ 118	4-01-00-00212 / 1.531[20]	ace 4; bz 4; eth 4; EtOH 4 / 1.4745[20]
9839	Propane, 1-bromo-3-chloro- / 1-Bromo-3-chloropropane	C_3H_6BrCl / 157.44	109-70-6 / -58.9	/ 143.3	4-01-00-00212 / 1.5969[20]	H_2O 1; EtOH 4; eth 4; chl 4 / 1.4864[20]
9840	Propane, 2-bromo-1-chloro- / 2-Bromo-1-chloropropane	C_3H_6BrCl / 157.44	3017-95-6 /	/ 118	4-01-00-00212 / 1.537[20]	H_2O 1; EtOH 4; eth 4; ace 3 / 1.4795[20]
9841	Propane, 2-bromo-2-chloro- / 2-Bromo-2-chloropropane	C_3H_6BrCl / 157.44	2310-98-7 /	/ 95	4-01-00-00213 / 1.474[22]	ace 4; bz 4; eth 4; EtOH 4 / 1.4575[20]
9842	Propane, 1-bromo-3-chloro-2-methyl-	C_4H_8BrCl / 171.46	6974-77-2 /	/ 155	4-01-00-00298 / 1.4839[20]	/ 1.4796[25]
9843	Propane, 1-bromo-2,2-dimethyl-	$C_5H_{11}Br$ / 151.05	630-17-1 /	/ 106	4-01-00-00337 / 1.1997[20]	H_2O 1; EtOH 3; eth 3; ace 3 / 1.4370[20]
9844	Propane, 1-bromo-3-fluoro-	C_3H_6BrF / 140.98	352-91-0 /	/ 101.4	4-01-00-00210 / 1.542[25]	bz 4; eth 4; EtOH 4; chl 4 / 1.4290[25]
9845	Propane, 1-bromo-2-methyl- / Isobutyl bromide	C_4H_9Br / 137.02	78-77-3 / -119	5019 / 91.1	4-01-00-00294 / 1.272[15]	EtOH 4; eth 4; ace 4; bz 4 / 1.4348[20]
9846	Propane, 2-bromo-2-methyl- / tert-Butyl bromide	C_4H_9Br / 137.02	507-19-7 / -16.2	1555 / 73.3	4-01-00-00295 / 1.4278[20]	H_2O 1; ctc 2 / 1.4278[20]
9847	Propane, 2-bromo-2-nitro-	$C_3H_6BrNO_2$ / 167.99	5447-97-2 /	/ 153; 73-5[50]	4-01-00-00233 / 1.6582[25]	eth 4; EtOH 4
9848	Propane, 1-chloro- / Propyl chloride	C_3H_7Cl / 78.54	540-54-5 / -122.8	7859 / 46.5	4-01-00-00189 / 0.8899[20]	H_2O 2; EtOH 5; eth 5; bz 3 / 1.3879[20]
9849	Propane, 2-chloro- / Isopropyl chloride	C_3H_7Cl / 78.54	75-29-6 / -117.2	5099 / 35.7	4-01-00-00191 / 0.8617[20]	H_2O 2; EtOH 5; eth 5; bz 3 / 1.3777[20]
9850	Propane, 3-chloro-1,1-diethoxy-	$C_7H_{15}ClO_2$ / 166.65	35573-93-4 /	/ 84[25]	3-01-00-02592 / 0.9951[19]	ace 4; bz 4 / 1.4268[20]
9851	Propane, 1-chloro-2,2-difluoro- / 2,2-Difluoropropyl chloride	$C_3H_5ClF_2$ / 114.52	420-99-5 / -56.2	/ 55	4-01-00-00193 / 1.2001[20]	eth 3; bz 3; chl 3 / 1.3520[20]
9852	Propane, 3-chloro-1,1-dimethoxy-	$C_5H_{11}ClO_2$ / 138.59	35502-06-8 /	/ 86[100]; 51[19]	3-01-00-02692 / 1.064[15]	/ 1.4163[20]
9853	Propane, 1-chloro-2,2-dimethyl-	$C_5H_{11}Cl$ / 106.60	753-89-9 / -20	/ 84.3	4-01-00-00336 / 0.8660[20]	bz 4; eth 4; EtOH 4; chl 4 / 1.4044[20]
9854	Propane, 1-(1-chloroethoxy)- / 1-Chloroethyl propyl ether	$C_5H_{11}ClO$ / 122.59	692-35-3 /	/ 114	3-01-00-02655 / 0.9322[20]	eth 4 / 1.4013[20]
9855	Propane, 2-[(2-chloroethyl)thio]-2-methyl-	$C_6H_{13}ClS$ / 152.69	4303-44-0 / -49	/ 81-2[30]	3-01-00-01592 / 1.0001[25]	
9856	Propane, 1-chloro-3-fluoro- / 1-Chloro-3-fluoropropane	C_3H_6ClF / 96.53	462-38-4 /	/ 80	4-01-00-00193 /	bz 4; eth 4; EtOH 4; chl 4 / 1.3871[25]
9857	Propane, 1-chloro-3-iodo-	C_3H_6ClI / 204.44	6940-76-7 /	/ 171	4-01-00-00226 / 1.904[20]	H_2O 1; eth 3; bz 3; ctc 2 / 1.5472[20]
9858	Propane, 1-(chloromethoxy)-	C_4H_9ClO / 108.57	3587-57-3 /	/ 109	3-01-00-02589 / 0.9884[20]	eth 4; EtOH 4 / 1.4125[20]

No.	Name Synonym	Mol. Form. Mol. Wt.	CAS RN mp/°C	Merck No. bp/°C	Beil. Ref. den/g cm^{-3}	Solubility n_D
9859	Propane, 1-chloro-2-methoxy-	C_4H_9ClO 108.57	5390-72-7	103.5	3-01-00-01470 1.009[20]	1.4137[20]
9860	Propane, 1-chloro-3-methoxy-	C_4H_9ClO 108.57	36215-07-3	111	4-01-00-01441 1.0013[20]	1.4131[20]
9861	Propane, 2-chloro-1-methoxy-	C_4H_9ClO 108.57	5390-71-6	99	3-01-00-01424 0.9946[20]	1.4075[20]
9862	Propane, 1-chloro-2-methyl- Isobutyl chloride	C_4H_9Cl 92.57	513-36-0 -130.3	5022 68.5	4-01-00-00287 0.8773[20]	eth 3; ace 3; ctc 2; chl 3 1.3984[20]
9863	Propane, 2-chloro-2-methyl- tert-Butyl chloride	C_4H_9Cl 92.57	507-20-0 -26	1562 50.9	4-01-00-00288 0.8420[20]	H_2O 2; EtOH 5; eth 5; bz 3 1.3857[20]
9864	Propane, 1-chloro-3-(methylthio)- 3-Chloropropyl methyl sulfide	C_4H_9ClS 124.63	13012-59-4	71-2[29]	4-01-00-01458 1.0844[20]	1.4833[20]
9865	Propane, 2-chloro-1-(methylthio)- 2-Chloropropyl methyl sulfide	C_4H_9ClS 124.63	19987-13-4	67[37]	4-01-00-01457 1.076[20]	1.4905[20]
9866	Propane, 1-chloro-1-nitro- 1-Chloro-1-nitropropane	$C_3H_6ClNO_2$ 123.54	600-25-9	142	4-01-00-00232 1.207[20]	H_2O 2; EtOH 3; eth 3; chl 2 1.4251[20]
9867	Propane, 1-chloro-2-nitro-	$C_3H_6ClNO_2$ 123.54	2425-66-3	172.5	4-01-00-00233 1.245[22]	eth 4; EtOH 4; chl 4 1.4432[25]
9868	Propane, 1-chloro-3-nitro- 1-Chloro-3-nitropropane	$C_3H_6ClNO_2$ 123.54	16694-52-3	197 dec; 115[40]	3-01-00-00259 1.267[20]	eth 4; EtOH 4
9869	Propane, 2-chloro-1-nitro-	$C_3H_6ClNO_2$ 123.54	503-76-4	172	4-01-00-00232 1.2361[15]	eth 4; EtOH 4 1.4447[20]
9870	Propane, 2-chloro-2-nitro- 2-Chloro-2-nitropropane	$C_3H_6ClNO_2$ 123.54	594-71-8	134 dec; 57[50]	4-01-00-00233 1.230[19]	H_2O 2; EtOH 3; eth 3; ctc 3 1.4378[19]
9871	Propane, 3-chloro-1,1,1-trifluoro-	$C_3H_4ClF_3$ 132.51	460-35-5 -106.5	45.1	4-01-00-00193 1.3253[20]	1.3350[20]
9872	1,2-Propanediamine Propylenediamine	$C_3H_{10}N_2$ 74.13	78-90-0 119.5	7865	2-04-00-00698 0.878[15]	H_2O 4; eth 1 1.4460[20]
9873	1,3-Propanediamine 1,3-Diaminopropane	$C_3H_{10}N_2$ 74.13	109-76-2 139.8		4-04-00-01258 0.884[25]	H_2O 3; EtOH 5; eth 5 1.4600[20]
9874	1,3-Propanediamine, N-(3-aminopropyl)- Bis(3-aminopropyl)amine	$C_6H_{17}N_3$ 131.22	56-18-8 -14	151[50]	4-04-00-01278 0.938[25]	chl 3 1.4810[20]
9875	1,3-Propanediamine, N-(3-aminopropyl)-N-methyl-	$C_7H_{19}N_3$ 145.25	105-83-9	232.5; 112[6]	4-04-00-01279 0.9023[20]	1.4705[25]
9876	1,3-Propanediamine, N,N-diethyl- 3-(Diethylamino)propylamine	$C_7H_{18}N_2$ 130.23	104-78-9	168.5	4-04-00-01260 0.822[20]	1.443[20]
9877	1,3-Propanediamine, N,N-diethyl-N'-(6-methoxy-8-quinolinyl)- Plasmocid	$C_{17}H_{25}N_3O$ 287.41	551-01-9	182[1.0]	5-22-12-00126 1.0569[24]	1.5855[24]
9878	1,2-Propanediamine, 2-methyl-	$C_4H_{12}N_2$ 88.15	811-93-8	123	4-04-00-01306 0.841[25]	ctc 3 1.4410[20]
9879	1,3-Propanediamine, N,N,N',N'-tetramethyl-	$C_7H_{18}N_2$ 130.23	110-95-2	144	4-04-00-01259 0.7837[18]	H_2O 5; EtOH 5; eth 5
9880	Propane, 1,1-dibromo- 1,1-Dibromopropane	$C_3H_6Br_2$ 201.89	598-17-4	133.5	4-01-00-00215 1.982[20]	eth 4; EtOH 4; chl 4 1.5100[20]
9881	Propane, 1,2-dibromo- Propylene dibromide	$C_3H_6Br_2$ 201.89	78-75-1 -55.2	7866 141.9	4-01-00-00215 1.9324[20]	EtOH 3; eth 3; ctc 2; chl 3 1.5201[20]
9882	Propane, 1,3-dibromo- 1,3-Dibromopropane	$C_3H_6Br_2$ 201.89	109-64-8 -34.2	9628 167.3	4-01-00-00216 1.9822[20]	EtOH 3; eth 3; ctc 2; chl 3 1.5232[20]
9883	Propane, 2,2-dibromo- 2,2-Dibromopropane	$C_3H_6Br_2$ 201.89	594-16-1	115	4-01-00-00217 1.7825[20]	eth 4; EtOH 4; chl 4
9884	Propane, 1,3-dibromo-2,2-bis(bromomethyl)-	$C_5H_8Br_4$ 387.73	3229-00-3 163	305.5	4-01-00-00337 2.596[15]	EtOH 3; eth 2; bz 3; chl 2
9885	Propane, 1,2-dibromo-3-chloro- Dibromochloropropane	$C_3H_5Br_2Cl$ 236.33	96-12-8 3003	196	3-01-00-00250 2.093[14]	1.553[14]
9886	Propane, 1,1-dibromo-2,2-dimethyl- Neopentane, 1,1-dibromo	$C_5H_{10}Br_2$ 229.94	2443-91-6	64-5[43]	0-01-00-00141 1.7880[0]	1.5085
9887	Propane, 1,3-dibromo-2,2-dimethyl-	$C_5H_{10}Br_2$ 229.94	5434-27-5	187 dec; 80[28]	4-01-00-00337 1.7049[0]	1.5090
9888	Propane, 1,2-dibromo-3-methoxy-	$C_4H_8Br_2O$ 231.91	5836-66-8	185	4-01-00-01447 1.8320[12]	eth 4 1.5123[20]
9889	Propane, 1,2-dibromo-2-methyl-	$C_4H_8Br_2$ 215.92	594-34-3 10.5	150	4-01-00-00298 1.7827[20]	EtOH 3; eth 3; chl 3 1.5119[20]
9890	Propane, 1,3-dibromo-2-methyl-	$C_4H_8Br_2$ 215.92	28148-04-1	177.5; 62[13]	4-01-00-00299 1.8204[20]	1.5068[20]
9891	Propane, 1,1-dichloro- Propylidene chloride	$C_3H_6Cl_2$ 112.99	78-99-9	7874 88.1	4-01-00-00195 1.1321[20]	EtOH 3; eth 3; bz 3; chl 3 1.4289[20]
9892	Propane, 1,2-dichloro-, (±)- Propane, 1,2-dichloro (DL)	$C_3H_6Cl_2$ 112.99	26198-63-0 -100.4	7887 96.4	4-01-00-00195 1.1560[20]	H_2O 2; EtOH 3; eth 3; bz 3 1.4394[20]
9893	Propane, 1,3-dichloro- Trimethylene dichloride	$C_3H_6Cl_2$ 112.99	142-28-9 -99.5	120.9	4-01-00-00196 1.1876[20]	H_2O 2; EtOH 4; eth 4; bz 3 1.4487[20]
9894	Propane, 2,2-dichloro-	$C_3H_6Cl_2$ 112.99	594-20-7 -33.8	69.3	4-01-00-00196 1.1136[20]	H_2O 1; EtOH 3; eth 5; bz 3 1.4148[20]
9895	Propane, 1,3-dichloro-2,2-bis(chloromethyl)-	$C_5H_8Cl_4$ 209.93	3228-99-7 97	110[12]	4-01-00-00336	H_2O 1; eth 3; ctc 3; chl 3

No.	Name Synonym	Mol. Form. Mol. Wt.	CAS RN mp/°C	Merck No. bp/°C	Beil. Ref. den/g cm^{-3}	Solubility n_D
9896	Propane, 1,3-dichloro-2-(ethoxymethoxy)- Formaldehyde, (2,2'-dichloroisopropyl) ethyl acetal	$C_8H_{12}Cl_2O_2$ 187.07	89583-61-9	96-8[16]	3-01-00-02574 1.181[17]	eth 4 1.4491[17]
9897	Propane, 1,2-dichloro-2-fluoro-	$C_3H_5Cl_2F$ 130.98	420-97-3 -91.7	88.6	4-01-00-00197 1.2624[20]	H_2O 2; ace 3; bz 3 1.4099[20]
9898	Propane, 1,2-dichloro-1,1,2,3,3,3-hexafluoro- 1,2-Dichlorohexafluoropropane	$C_3Cl_2F_6$ 220.93	661-97-2	34.1		
9899	Propane, 1,1-dichloro-2-methyl- Isobutane, 1,1-dichloro	$C_4H_8Cl_2$ 127.01	598-76-5	105.5	4-01-00-00292 1.011[12]	bz 4; eth 4; EtOH 4; chl 4 1.4330[25]
9900	Propane, 1,2-dichloro-2-methyl- Isobutane, 1,2-dichloro	$C_4H_8Cl_2$ 127.01	594-37-6	106.5	4-01-00-00292 1.093[20]	H_2O 1; EtOH 5; eth 5; ace 5 1.4370[20]
9901	Propane, 1,3-dichloro-2-methyl-	$C_4H_8Cl_2$ 127.01	616-19-3	136.4	3-01-00-00319 1.131[20]	H_2O 1; EtOH 3; eth 4; bz 4 1.4488[25]
9902	Propane, 1,1-dichloro-1-nitro- 1,1-Dichloro-1-nitropropane	$C_3H_5Cl_2NO_2$ 157.98	595-44-8	145	3-01-00-00259 1.312[20]	ctc 3
9903	Propane, 1,1-diethoxy-	$C_7H_{16}O_2$ 132.20	4744-08-5	123	4-01-00-03168 0.8232[20]	H_2O 3; EtOH 4; eth 4; ace 3 1.3924[19]
9904	Propane, 2,2-diethoxy-	$C_7H_{16}O_2$ 132.20	126-84-1	114	4-01-00-03200 0.8200[21]	EtOH 3; eth 4; ace 3; bz 3 1.3891[20]
9905	Propane, 1,3-difluoro- 1,3-Difluoropropane	$C_3H_6F_2$ 80.08	462-39-5	41.6	3-01-00-00218 1.0057[25]	bz 4 1.3190[26]
9906	Propane, 2,2-difluoro- Dimethyldifluoromethane	$C_3H_6F_2$ 80.08	420-45-1 -104.8	-0.4	4-01-00-00187 0.9205[20]	
9907	Propane, 1,3-diiodo- Trimethylene diiodide	$C_3H_6I_2$ 295.89	627-31-6 -20	227 dec; 110[19]	4-01-00-00228 2.5754[20]	H_2O 1; eth 3; ctc 3; chl 3 1.6423[20]
9908	Propane, 2,2-diiodo- 2,2-Diiodopropane	$C_3H_6I_2$ 295.89	630-13-7	173; 156[500]	3-01-00-00255 2.5754[20]	eth 4; chl 4 1.651[20]
9909	Propane, 2,2-dimethoxy-	$C_5H_{12}O_2$ 104.15	77-76-9 -47	83	4-01-00-03199 0.847[25]	1.3780[20]
9910	Propane, 2,2-dimethyl- Neopentane	C_5H_{12} 72.15	463-82-1 -16.6	6372 9.4	4-01-00-00333 *0.5852[25]	H_2O 1; EtOH 3; eth 3; ctc 3 1.3476[6]
9911	Propanedinitrile Malononitrile	$C_3H_2N_2$ 66.06	109-77-3 32	5592 218.5	4-02-00-01892 1.1910[20]	H_2O 3; EtOH 4; eth 4; ace 3 1.4146[34]
9912	Propanedinitrile, dichloro-	$C_3Cl_2N_2$ 134.95	13063-43-9	97	3-02-00-01638 1.312[25]	1.4312[20]
9913	Propanedinitrile, dimethyl-	$C_5H_6N_2$ 94.12	7321-55-3 31.5	169.5; 62[22]	3-02-00-01704	H_2O 3; EtOH 2
9914	Propanedinitrile, (ethoxymethylene)-	$C_6H_6N_2O$ 122.13	123-06-8 66	160[12]	4-03-00-01194	EtOH 3; eth 3; chl 2
9915	Propanedinitrile, oxo- Carbonyl cyanide	C_3N_2O 80.05	1115-12-4 -36	65.5	4-03-00-01806 1.124[20]	eth 3; ace 3; ctc 3; chl 3 1.3919[20]
9916	Propanedinitrile, phenyl- Phenylmalononitrile	$C_9H_6N_2$ 142.16	3041-40-5 70.5	152[21]	4-09-00-03341	EtOH 4
9917	Propanedinitrile, (phenylmethyl)-	$C_{10}H_8N_2$ 156.19	1867-37-4 91	174[23]	4-09-00-03359	bz 4; eth 4; EtOH 4
9918	Propane, 1,1-dinitro- 1,1-Dinitropropane	$C_3H_6N_2O_4$ 134.09	601-76-3 -42	184	4-01-00-00234 1.2610[25]	alk 3 1.4339[20]
9919	Propane, 1,3-dinitro- 1,3-Dinitropropane	$C_3H_6N_2O_4$ 134.09	6125-21-9 -21.4	103[1]	4-01-00-00234 1.353[26]	H_2O 1; eth 3 1.4654[20]
9920	Propane, 2,2-dinitro- 2,2-Dinitropropane	$C_3H_6N_2O_4$ 134.09	595-49-3 53	185.5	4-01-00-00234 1.30[25]	H_2O 2
9921	Propanedioic acid Malonic acid	$C_3H_4O_4$ 104.06	141-82-2 135 dec	5591 sub	4-02-00-01874 1.619[10]	H_2O 4; EtOH 3; eth 3; bz 1
9922	Propanedioic acid, (acetylamino)-, diethyl ester	$C_9H_{15}NO_5$ 217.22	1068-90-2 96.3	185[20]	4-04-00-02993	H_2O 2; EtOH 3; eth 2; tfa 3
9923	Propanedioic acid, acetyl-, diethyl ester	$C_9H_{14}O_5$ 202.21	570-08-1	232	4-03-00-01819 1.0834[26]	H_2O 1; ace 3 1.4435[25]
9924	Propanedioic acid, amino-, diethyl ester Diethyl aminomalonate	$C_7H_{13}NO_4$ 175.18	6829-40-9	122-3[16]	4-04-00-02991 1.100[16]	H_2O 4; EtOH 4; eth 4; ace 3 1.4353[16]
9925	Propanedioic acid, bis(1,1-dimethylethyl) ester Di-tert-butyl malonate	$C_{11}H_{20}O_4$ 216.28	541-16-2 -6	3021 113[31]; 66[2]	4-02-00-01884 1.4184[20]	ace 3; chl 3 1.4184[29]
9926	Propanedioic acid, bis(phenylmethyl)-, diethyl ester Diethyl dibenzylmalonate	$C_{21}H_{24}O_4$ 340.42	597-55-7 14	234-5[23]	3-09-00-04552 1.093[20]	eth 4; EtOH 4
9927	Propanedioic acid, bis(phenylmethyl) ester	$C_{17}H_{16}O_4$ 284.31	15014-25-2	188[0.2]	4-06-00-02270 1.137[25]	1.5447[20]
9928	Propanedioic acid, bromo-, diethyl ester	$C_7H_{11}BrO_4$ 239.07	685-87-0 -54	254 dec	4-02-00-01904 1.4022[25]	H_2O 1; EtOH 5; eth 5; ace 3 1.4521[20]
9929	Propanedioic acid, butyl- n-Butylmalonic acid	$C_7H_{12}O_4$ 160.17	534-59-8 104.5	1574	4-02-00-02011	H_2O 4; EtOH 3; eth 3
9930	Propanedioic acid, butyl-, diethyl ester	$C_{11}H_{20}O_4$ 216.28	133-08-4	237.5	4-02-00-02011	EtOH 4; eth 4 1.4250[20]
9931	Propanedioic acid, chloro-, diethyl ester	$C_7H_{11}ClO_4$ 194.61	14064-10-9	222	4-02-00-01903 1.2040[20]	H_2O 1; EtOH 5; eth 5; chl 5 1.4327[20]
9932	Propanedioic acid, (3-chloropropyl)-, diethyl ester	$C_{10}H_{17}ClO_4$ 236.70	18719-43-2	147-9[10]	4-02-00-01992	H_2O 1; EtOH 3; eth 3; ctc 2 1.4429[20]

No.	Name / Synonym	Mol. Form. / Mol. Wt.	CAS RN / mp/°C	Merck No. / bp/°C	Beil. Ref. / den/g cm^{-3}	Solubility / n_D
9933	Propanedioic acid, (2-cyanoethyl)-, diethyl ester	$C_{10}H_{15}NO_4$ 213.23	17216-62-5	115-6[2]	4-02-00-02365 1.078[25]	1.4368[20]
9934	Propanedioic acid, cyclohexyl-, diethyl ester	$C_{13}H_{22}O_4$ 242.32	2163-44-2	163-5[20]	3-09-00-03834 1.0281[19]	ace 4; bz 4; eth 4; EtOH 4 1.4478[25]
9935	Propanedioic acid, 2-cyclopenten-1-yl-, diethyl ester / Malonic acid, 2-cyclopentenyl, diethyl ester	$C_{12}H_{18}O_4$ 226.27	53608-93-8	141[10]	3-09-00-03951 1.0507[20]	H$_2$O 1; eth 3; ace 3 1.4536[20]
9936	Propanedioic acid, cyclopentylidene-, diethyl ester / Malonic acid, cyclopentylidene, diethyl ester	$C_{12}H_{18}O_4$ 226.27	41589-42-8	140[10]	3-09-00-03952 1.0616[20]	ace 4; eth 4 1.4724[20]
9937	Propanedioic acid, dibutyl-, diethyl ester / Diethyl dibutylmalonate	$C_{15}H_{28}O_4$ 272.38	596-75-8	153-4[14]	4-02-00-02119 0.9457[20]	H$_2$O 1; EtOH 3; eth 3; ctc 3 1.4341[20]
9938	Propanedioic acid, dibutyl ester	$C_{11}H_{20}O_4$ 216.28	1190-39-2 -83	251.5	4-02-00-01884 0.9824[20]	H$_2$O 1; EtOH 3; eth 3; ace 3 1.4262[20]
9939	Propanedioic acid, diethyl- / Diethylmalonic acid	$C_7H_{12}O_4$ 160.17	510-20-3 127 dec	3114	4-02-00-02026	H$_2$O 4; EtOH 4; eth 4; bz 2
9940	Propanedioic acid, diethyl-, diethyl ester / Diethyl 2,2-diethylmalonate	$C_{11}H_{20}O_4$ 216.28	77-25-8	3746 230	4-02-00-02026 0.9643[30]	H$_2$O 1; EtOH 5; eth 5; ctc 3 1.4240[20]
9941	Propanedioic acid, diethyl ester / Diethyl malonate	$C_7H_{12}O_4$ 160.17	105-53-3 -50	3779 200	4-02-00-01881 1.0551[20]	H$_2$O 2; EtOH 5; eth 5; ace 4 1.4139[20]
9942	Propanedioic acid, dimethyl-, diethyl ester / Diethyl dimethylmalonate	$C_9H_{16}O_4$ 188.22	1619-62-1 -30.4	197	4-02-00-01955 0.9964[20]	H$_2$O 1; EtOH 5; eth 5; ctc 2 1.4129[20]
9943	Propanedioic acid, dimethyl ester / Methyl malonate	$C_5H_8O_4$ 132.12	108-59-8 -61.9	6009 181.4	4-02-00-01880 1.528[20]	H$_2$O 2; EtOH 5; eth 5; ace 4 1.4135[20]
9944	Propanedioic acid, di-2-propenyl-, diethyl ester	$C_{13}H_{20}O_4$ 240.30	3195-24-2	243.5	3-02-00-02003 0.9943[20]	EtOH 3; eth 3; ace 3; ctc 2 1.4445[22]
9945	Propanedioic acid, dipropyl ester / Dipropyl malonate	$C_9H_{16}O_4$ 188.22	1117-19-7 -77.1	229	3-02-00-01619 1.0097[20]	ace 4; bz 4; eth 4; EtOH 4 1.4206[20]
9946	Propanedioic acid, di-2-propynyl-, diethyl ester / Diethyl dipropargylmalonate	$C_{13}H_{16}O_4$ 236.27	2689-88-5 45.5	129[11]	4-02-00-02340	chl 3
9947	Propanedioic acid, (ethoxymethylene)-, diethyl ester	$C_{10}H_{16}O_5$ 216.23	87-13-8	280 dec; 165[19]	4-03-00-01192	H$_2$O 1; EtOH 3; eth 3; chl 2 1.4600[20]
9948	Propanedioic acid, ethyl-	$C_5H_8O_4$ 132.12	601-75-2 114	180[0.05]	4-02-00-01952	H$_2$O 4; EtOH 3; eth 3; ace 1
9949	Propanedioic acid, ethyl-, diethyl ester	$C_9H_{16}O_4$ 188.22	133-13-1	208; 98[12]	4-02-00-01953 1.0047[20]	H$_2$O 2; EtOH 4; eth 4; ace 3 1.4166[20]
9950	Propanedioic acid, ethylidene-, diethyl ester	$C_9H_{14}O_4$ 186.21	1462-12-0	115-8[17]	4-02-00-02234 1.0194[17]	eth 4; EtOH 4 1.4308[17]
9951	Propanedioic acid, ethyl(1-methylethyl)-, diethyl ester / Malonic acid, ethylisopropyl-, diethyl ester	$C_{12}H_{22}O_4$ 230.30	2049-66-3	234; 109[11]	4-02-00-02052	ace 4; eth 4; EtOH 4 1.4280[25]
9952	Propanedioic acid, ethylphenyl-, diethyl ester	$C_{15}H_{20}O_4$ 264.32	76-67-5	170[19]	4-09-00-03385 1.071[20]	H$_2$O 1; EtOH 3; eth 3; chl 2 1.4896[25]
9953	Propanedioic acid, (formylamino)-, diethyl ester	$C_8H_{13}NO_5$ 203.19	6326-44-9 48.5	173[11]	4-04-00-02992	EtOH 4
9954	Propanedioic acid, hexadecyl-, diethyl ester	$C_{23}H_{44}O_4$ 384.60	41433-81-2	238[14]	4-02-00-02183 0.9118[20]	EtOH 4 1.4433[20]
9955	Propanedioic acid, hexyl-, diethyl ester	$C_{13}H_{24}O_4$ 244.33	5398-10-7	269; 143[15]	3-02-00-01791 0.9577[21]	ace 4; bz 4; eth 4; EtOH 4 1.4278[21]
9956	Propanedioic acid, hydroxy- / Tartronic acid	$C_3H_4O_5$ 120.06	80-69-3 157	9042 sub	4-03-00-01120	H$_2$O 3; EtOH 3; eth 2
9957	Propanedioic acid, hydroxy-, diethyl ester	$C_7H_{12}O_5$ 176.17	13937-08-1 -2.5	221	3-03-00-00905 1.1520[15]	ace 4; bz 4; eth 4; EtOH 4
9958	Propanedioic acid, (hydroxyimino)-, diethyl ester	$C_7H_{11}NO_5$ 189.17	6829-41-0	172[12]	4-03-00-01805 1.1821[18]	ace 4; bz 4; eth 4; EtOH 4 1.4544[18]
9959	Propanedioic acid, (hydroxymethyl)- / β-Isomalic acid	$C_4H_6O_5$ 134.09	4360-96-7 142 dec	170 dec	0-03-00-00441	H$_2$O 4; EtOH 4; eth 4
9960	Propanedioic acid, methyl-	$C_4H_6O_4$ 118.09	516-05-2 135 dec		4-02-00-01932 1.455[20]	H$_2$O 4; EtOH 4; eth 4; bz 2
9961	Propanedioic acid, (3-methylbutyl)-, diethyl ester	$C_{12}H_{22}O_4$ 230.30	5398-08-3	241	3-02-00-01777 0.9580[25]	ace 4; eth 4; EtOH 4 1.4255[25]
9962	Propanedioic acid, methyl-, diethyl ester	$C_8H_{14}O_4$ 174.20	609-08-5	201	4-02-00-01932 1.0225[20]	H$_2$O 2; EtOH 4; eth 4; ace 3 1.4126[20]
9963	Propanedioic acid, methyl-, dimethyl ester	$C_6H_{10}O_4$ 146.14	609-02-9	174	3-02-00-01681 1.0977[20]	ace 4; eth 4; EtOH 4; chl 4 1.4128[20]
9964	Propanedioic acid, (1-methylethyl)-, diethyl ester	$C_{10}H_{18}O_4$ 202.25	759-36-4	215	4-02-00-02001 0.9961[20]	H$_2$O 2; EtOH 4; eth 4; ctc 2 1.4188[21]
9965	Propanedioic acid, (1-methylethylidene)-	$C_6H_8O_4$ 144.13	4441-90-1 170.5	dec	3-02-00-01948	EtOH 4
9966	Propanedioic acid, (1-methylethylidene)-, diethyl ester / Diethyl isopropylidenemalonate	$C_{10}H_{16}O_4$ 200.23	6802-75-1	176.5; 116[14]	4-02-00-02244 1.0282[18]	ace 4; EtOH 4 1.4486[17]

No.	Name Synonym	Mol. Form. Mol. Wt.	CAS RN mp/°C	Merck No. bp/°C	Bell. Ref. den/g cm⁻³	Solubility n_D
9967	Propanedioic acid, (1-methylpropyl)-, diethyl ester	$C_{11}H_{20}O_4$	83-27-2		4-02-00-02019	H_2O 1; EtOH 4; eth 4; ctc 2
		216.28			0.988^{15}	1.4248^{20}
9968	Propanedioic acid, (2-methylpropyl)-	$C_7H_{12}O_4$ 160.17	4361-06-2 112 dec	dec	4-02-00-02023	eth 4; EtOH 4
9969	Propanedioic acid, (2-methylpropyl)-, diethyl ester Diethyl isobutylmalonate	$C_{11}H_{20}O_4$ 216.28	10203-58-4		3-02-00-01756 0.9804^{20}	H_2O 1; EtOH 4; eth 4; chl 3 1.4236^{20}
9970	Propanedioic acid, 1-naphthalenyl-, diethyl ester	$C_{17}H_{18}O_4$ 286.33	6341-60-2 61.3	$176^{1.0}$	4-09-00-03536	EtOH 3; bz 3; chl 2
9971	Propanedioic acid, octyl-	$C_{11}H_{20}O_4$ 216.28	760-55-4 116		4-02-00-02114 1.173^{17}	ace 4; EtOH 4
9972	Propanedioic acid, (1-oxobutyl)-, diethyl ester	$C_{11}H_{18}O_5$ 230.26	21633-79-4	249.5	4-03-00-01833 1.056^{20}	1.4451^{20}
9973	Propanedioic acid, oxo-, diethyl ester	$C_7H_{10}O_5$ 174.15	609-09-6 -30	210; 105^{19}	3-03-00-01356 1.1419^{16}	H_2O 4; EtOH 3; eth 3; chl 3 1.4310^{22}
9974	Propanedioic acid, pentyl-, diethyl ester	$C_{12}H_{22}O_4$ 230.30	6065-59-4	$134-6^{14}$	4-02-00-02034 0.9652^{20}	H_2O 1; EtOH 4; eth 4; ctc 2 1.4253^{20}
9975	Propanedioic acid, phenyl-, diethyl ester	$C_{13}H_{16}O_4$ 236.27	83-13-6 16.5	205 dec; 168^{12}	4-09-00-03339 1.0950^{20}	ace 4; EtOH 4 1.4977^{20}
9976	Propanedioic acid, phenyl-, dimethyl ester	$C_{11}H_{12}O_4$ 208.21	37434-59-6 51.6	147^{13}	4-09-00-03339	eth 4; EtOH 4
9977	Propanedioic acid, (phenylmethyl)-, diethyl ester	$C_{14}H_{18}O_4$ 250.29	607-81-8	300	4-09-00-03357 1.076^{15}	H_2O 1; chl 2 1.4872^{20}
9978	Propanedioic acid, (phenylmethylene)-, diethyl ester	$C_{14}H_{16}O_4$ 248.28	5292-53-5 32	216^{30}; 196^{14}	4-09-00-03461 1.1045^{20}	H_2O 1; EtOH 3; eth 3; ace 3 1.5389^{20}
9979	Propanedioic acid, 2-propenyl-	$C_6H_8O_4$ 144.13	2583-25-7 105	180 dec	3-02-00-01945	H_2O 4; eth 4; EtOH 4
9980	Propanedioic acid, 2-propenyl-, diethyl ester	$C_{10}H_{16}O_4$ 200.23	2049-80-1	222.5; 93^6	4-02-00-02240 1.0098^{20}	H_2O 1; EtOH 4; eth 4; ctc 3 1.4305^{20}
9981	Propanedioic acid, propyl-, diethyl ester	$C_{10}H_{18}O_4$ 202.25	2163-48-6	221; 114^{22}	4-02-00-01991 0.9873^{20}	H_2O 2; EtOH 4; eth 4 1.4197^{20}
9982	1,2-Propanediol 1,2-Propylene glycol	$C_3H_8O_2$ 76.10	57-55-6 -60	7868 187.6	3-01-00-02142 1.0361^{20}	H_2O 5; EtOH 5; eth 3; bz 3 1.4324^{20}
9983	1,3-Propanediol Trimethylene glycol	$C_3H_8O_2$ 76.10	504-63-2 -26.7	9629 214.4	4-01-00-02493 1.0538^{20}	H_2O 5; EtOH 5; eth 4; bz 2 1.4398^{20}
9984	1,2-Propanediol, 3-amino-, (±)- 1,2-Propanediol, 3-amino(DL)	$C_3H_9NO_2$ 91.11	13552-31-3	265 dec; 145^9	3-04-00-00839 1.1752^{20}	H_2O 4; EtOH 3; eth 1; bz 1 1.4910^{25}
9985	1,3-Propanediol, 2-amino-2-ethyl- 2-Amino-2-ethyl-1,3-propanediol	$C_5H_{13}NO_2$ 119.16	115-70-8 37.5	451 $152-3^{10}$	3-04-00-00850 1.099^{20}	H_2O 5 1.490^{20}
9986	1,3-Propanediol, 2-amino-2-(hydroxymethyl)- Tromethamine	$C_4H_{11}NO_3$ 121.14	77-86-1 171.5	9684 $219-20^{10}$	4-04-00-01903	H_2O 4; MeOH 3
9987	1,3-Propanediol, 2-amino-2-methyl- 2-Amino-2-methyl-1,3-propanediol	$C_4H_{11}NO_2$ 105.14	115-69-5 110	460 $151-2^{10}$	4-04-00-01881	H_2O 4; EtOH 3
9988	1,3-Propanediol, 2,2-bis[(acetyloxy)methyl]-, diacetate Pentaerythrityl tetraacetate	$C_{13}H_{20}O_8$ 304.30	597-71-7 83.5	7065	4-02-00-00264 1.273^{18}	H_2O 3; EtOH 4; eth 4
9989	1,3-Propanediol, 2,2-bis(hydroxymethyl)- Pentaerythritol	$C_5H_{12}O_4$ 136.15	115-77-5 260	7062 sub	4-01-00-02812	H_2O 3; eth 1; bz 1 1.548
9990	1,3-Propanediol, 2,2-bis[(nitrooxy)methyl]-, dinitrate (ester) Pentaerythrityl tetranitrate	$C_5H_8N_4O_{12}$ 316.14	78-11-5 140.5	7066	4-01-00-02816 1.773^{20}	H_2O 2; EtOH 2; eth 2; ace 4
9991	1,3-Propanediol, 2-bromo-2-nitro- Bronopol	$C_3H_6BrNO_4$ 199.99	52-51-7 131.5	1437	4-01-00-02501	
9992	1,3-Propanediol, 2-butyl-2-ethyl-	$C_9H_{20}O_2$ 160.26	115-84-4 43.8	262	4-01-00-02611 0.927^{50}	H_2O 2; EtOH 3; ace 2 1.4587^{25}
9993	1,3-Propanediol, 2-butyl-2-methyl-	$C_8H_{18}O_2$ 146.23	3121-83-3 43.8	262	4-01-00-02599 0.927^{50}	EtOH 4 1.4587^{25}
9994	1,2-Propanediol, 3-chloro- α-Chlorohydrin	$C_3H_7ClO_2$ 110.54	96-24-2 213 dec; 116^{11}	2145	4-01-00-02484 1.325^{18}	H_2O 3; EtOH 3; eth 3 1.4809^{20}
9995	1,3-Propanediol, 2-chloro- Glycerol β-chlorohydrin	$C_3H_7ClO_2$ 110.54	497-04-1	146^{18}	4-01-00-02499 1.3219^{20}	H_2O 4; ace 4; EtOH 4 1.4831^{20}
9996	1,2-Propanediol, 3-chloro-, diacetate	$C_7H_{11}ClO_4$ 194.61	869-50-1	245	3-02-00-00313 1.199^{25}	eth 4; EtOH 4 1.4407^{20}
9997	1,2-Propanediol, 3-chloro-, dinitrate Clonitrate	$C_3H_5ClN_2O_6$ 200.54	2612-33-1 192.5	2390	4-01-00-02487 1.5112^9	ace 4; EtOH 4; chl 4
9998	1,2-Propanediol, 3-chloro-2-methyl- Isobutylene glycol, 3-chloro	$C_4H_9ClO_2$ 124.57	597-33-1	$114-7^{20}$	4-01-00-02536 1.2362^{20}	H_2O 4; eth 4; EtOH 4 1.4748^{20}
9999	1,2-Propanediol, 3-(2-chlorophenoxy)-	$C_9H_{11}ClO_3$ 202.64	5112-21-0 71.5	250^{19}	4-06-00-00790	H_2O 3; eth 4; bz 3
10000	1,2-Propanediol, 3-(4-chlorophenoxy)- Chlorphenesin	$C_9H_{11}ClO_3$ 202.64	104-29-0 78	2178 $214-5^{19}$	4-06-00-00831	H_2O 1; EtOH 4; eth 4; bz 3

No.	Name Synonym	Mol. Form. Mol. Wt.	CAS RN mp/°C	Merck No. bp/°C	Beil. Ref. den/g cm^{-3}	Solubility n_D
10001	1,2-Propanediol, 3-(4-chlorophenoxy)-, 1-carbamate Chlorphenesin carbamate	$C_{10}H_{12}ClNO_4$ 245.66	886-74-8 90	2179		ace 4; EtOH 3; diox 4
10002	1,2-Propanediol, diacetate	$C_7H_{12}O_4$ 160.17	623-84-7 190.5		4-02-00-00220 1.059[20]	H_2O 4; EtOH 3; eth 3 1.4173[20]
10003	1,3-Propanediol, diacetate	$C_7H_{12}O_4$ 160.17	628-66-0 209.5		4-02-00-00221 1.070[14]	H_2O 4; EtOH 3 1.4192
10004	1,3-Propanediol, 2,2-diethyl-	$C_7H_{16}O_2$ 132.20	115-76-4 61.5	240.5	4-01-00-02589 1.050[20]	H_2O 4; EtOH 4; eth 4; chl 3 1.4574[25]
10005	1,3-Propanediol, 2,2-dimethyl- Neopentyl glycol	$C_5H_{12}O_2$ 104.15	126-30-7 130	6374 208	4-01-00-02551	H_2O 3; EtOH 4; eth 4; bz 3
10006	1,3-Propanediol, 2-ethyl-2-(hydroxymethyl)- Trimethylolpropane	$C_6H_{14}O_3$ 134.18	77-99-6 58	160[5]	4-01-00-02786	H_2O 4; EtOH 4
10007	1,3-Propanediol, 2-ethyl-2-methyl-	$C_6H_{14}O_2$ 118.18	77-84-9 40.3	225.8	4-01-00-02574	chl 3
10008	1,3-Propanediol, 2-ethyl-2-nitro- 2-Nitro-2-ethyl-1,3-propanediol	$C_5H_{11}NO_4$ 149.15	597-09-1 57.5	dec	4-01-00-02550	H_2O 4; eth 4; EtOH 4
10009	1,2-Propanediol, 3-(hexadecyloxy)-	$C_{19}H_{40}O_3$ 316.52	53584-29-5 64	205[1.0]	3-01-00-02322	ace 4; chl 4; peth 4
10010	1,2-Propanediol, 3-(hexadecyloxy)-, (S)- Chimyl alcohol	$C_{19}H_{40}O_3$ 316.52	506-03-6 64	2049 120[0.005]	3-01-00-02322	ace 4; peth 4; chl 4
10011	1,3-Propanediol, 2-(hydroxymethyl)-2-methyl-	$C_5H_{12}O_3$ 120.15	77-85-0 204	135-7[15]	4-01-00-02780	H_2O 5; EtOH 4; eth 1; bz 1
10012	1,3-Propanediol, 2-(hydroxymethyl)-2-nitro- Tris(hydroxymethyl)nitromethane	$C_4H_9NO_5$ 151.12	126-11-4 165	9667 dec	4-01-00-02777	H_2O 4; eth 4; EtOH 4
10013	1,2-Propanediol, 3-mercapto- Thioglycerol	$C_3H_8O_2S$ 108.16	96-27-5 100-1[1]	9264	3-01-00-02339 1.2455[20]	H_2O 2; EtOH 5; eth 2; ace 4 1.5268[20]
10014	1,2-Propanediol, 3-methoxy-	$C_4H_{10}O_3$ 106.12	623-39-2 220		4-01-00-02755 1.114[20]	H_2O 4; EtOH 4; eth 3; ace 4 1.442[25]
10015	1,3-Propanediol, 2-methoxy- Glycerol, 2-methyl ether	$C_4H_{10}O_3$ 106.12	761-06-8 232		3-01-00-02317 1.124[25]	H_2O 4; ace 4; eth 4; EtOH 4 1.4505[12]
10016	1,2-Propanediol, 3-(2-methoxyphenoxy)- Guaifenesin	$C_{10}H_{14}O_4$ 198.22	93-14-1 78.5	4465 215[19.] 127[0.2]	4-06-00-05576	H_2O 3; EtOH 4; bz 3; chl 3
10017	1,2-Propanediol, 2-methyl-	$C_4H_{10}O_2$ 90.12	558-43-0 176		4-01-00-02533 1.0024[20]	H_2O 4; eth 4; EtOH 4 1.4350[20]
10018	1,3-Propanediol, 2-methyl-2-nitro-	$C_4H_9NO_4$ 135.12	77-49-6 150.1	dec	4-01-00-02537	H_2O 4; EtOH 4; DMSO-d$_6$ 2
10019	1,2-Propanediol, 3-(2-methylphenoxy)- Mephenesin	$C_{10}H_{14}O_3$ 182.22	59-47-2 70 dec	5737	4-06-00-01952	H_2O 2; EtOH 3; eth 2
10020	1,3-Propanediol, 2-methyl-2-propyl-	$C_7H_{16}O_2$ 132.20	78-26-2 62.5	234; 121[10]	4-01-00-02585	H_2O 3; chl 2; os 3; hx 3
10021	1,3-Propanediol, 2-methyl-2-propyl-, dicarbamate Meprobamate	$C_9H_{18}N_2O_4$ 218.25	57-53-4 105	5751	4-03-00-00073	bz 4; eth 4; EtOH 4
10022	1,2-Propanediol, 1-nitrate	$C_3H_7NO_4$ 121.09	20266-65-3 148-9[4]		4-01-00-02476 1.2417[20]	1.4368[20]
10023	1,2-Propanediol, 3-(octadecyloxy)- Batyl alcohol	$C_{21}H_{44}O_3$ 344.58	544-62-7 70.5	1021 217[2]	4-01-00-02758	eth 4
10024	Propanediol, oxybis-	$C_6H_{14}O_5$ 166.17	59113-36-9	267[15]; 225[3]	3-01-00-02327 1.2751[20]	H_2O 4; EtOH 4 1.4890[20]
10025	1,2-Propanediol, 3,3'-oxybis-	$C_6H_{14}O_5$ 166.17	627-82-7	267[15]; 225[3]	3-01-00-02327 1.2751[20]	H_2O 4; EtOH 4 1.4890[20]
10026	1,2-Propanediol, 3-phenoxy- Phenylglyceryl ether	$C_9H_{12}O_3$ 168.19	538-43-2 67.5	7261 200[22]	4-06-00-00589 1.225[20]	H_2O 4; bz 4; eth 4; EtOH 4
10027	1,3-Propanedione, 1,3-diphenyl- Dibenzoylmethane	$C_{15}H_{12}O_2$ 224.26	120-46-7 70.5	2991	4-07-00-02512	EtOH 3; eth 3; chl 3; dil NaOH 3
10028	1,2-Propanedione, 1-phenyl-	$C_9H_8O_2$ 148.16	579-07-7 <20	222; 102[12]	4-07-00-02143 1.1006[20]	H_2O 3; EtOH 3; eth 3 1.537[10]
10029	Propanedioyl dichloride	$C_3H_2Cl_2O_2$ 140.95	1663-67-8 58[26]		4-02-00-01887 1.4509[20]	eth 3; AcOEt 3 1.4639[20]
10030	1,2-Propanedithiol	$C_3H_8S_2$ 108.23	814-67-5 152		4-01-00-02492 1.08[20]	chl 3 1.532[20]
10031	1,3-Propanedithiol Trimethylene dimercaptan	$C_3H_8S_2$ 108.23	109-80-8 -79	7811 172.9	4-01-00-02503 1.0772[20]	H_2O 2; EtOH 5; eth 5; bz 5 1.5392[20]
10032	Propane, 1,1'-[1,2-ethanediylbis(oxy)]bis[2-methyl- Ethylene glycol diisobutyl ether	$C_{10}H_{22}O_2$ 174.28	54922-86-0 171.3		4-01-00-02382 0.8200[20]	1.4021[20]
10033	Propane, 1-(ethenyloxy)-	$C_5H_{10}O$ 86.13	764-47-6 65		4-01-00-02052 0.7674[20]	1.3908[20]
10034	Propane, 2-(ethenyloxy)-	$C_5H_{10}O$ 86.13	926-65-8 -140	55.5	4-01-00-02052 0.7534[20]	ace 4; bz 4; eth 4; EtOH 4 1.3840[20]
10035	Propane, 1-(ethenyloxy)-2-methyl- Isobutyl vinyl ether	$C_6H_{12}O$ 100.16	109-53-5 -112	83	4-01-00-02054 0.7645[20]	H_2O 2; EtOH 4; eth 5; ace 4 1.3966[20]
10036	Propane, 2-(ethenyloxy)-2-methyl-	$C_6H_{12}O$ 100.16	926-02-3 -112	75	4-01-00-02055 0.7691[20]	1.3922[20]
10037	Propane, 1-ethoxy- Ethyl propyl ether	$C_5H_{12}O$ 88.15	628-32-0 -127.5	63.2	4-01-00-01421 0.7386[20]	eth 4; EtOH 4; HOAc 4 1.3695[20]
10038	Propane, 2-ethoxy- Ethyl isopropyl ether	$C_5H_{12}O$ 88.15	625-54-7 54.1		4-01-00-01471 0.720[25]	H_2O 3; EtOH 5; eth 5; ace 3 1.3698[25]

No.	Name / Synonym	Mol. Form. / Mol. Wt.	CAS RN / mp/°C	Merck No. / bp/°C	Beil. Ref. / den/g cm^{-3}	Solubility / n_D
10039	Propane, 1-ethoxy-2-methyl-	$C_6H_{14}O$ 102.18	627-02-1	81	4-01-00-01593 0.7512[20]	H$_2$O 1; EtOH 5; eth 5; ace 3 1.3739[25]
10040	Propane, 2-ethoxy-2-methyl- Ethyl tert-butyl ether	$C_6H_{14}O$ 102.18	637-92-3 -94	3732 72.6	4-01-00-01615 0.736[25]	H$_2$O 1; EtOH 4; eth 4 1.3756[20]
10041	Propane, 1,1'-[ethylidenebis(oxy)]bis[2-methyl-	$C_{10}H_{22}O_2$ 174.28	5669-09-0	171.3	4-01-00-03107 0.8200[20]	1.4021[20]
10042	Propane, 1-(ethylthio)- Ethyl propyl sulfide	$C_5H_{12}S$ 104.22	4110-50-3 -117	118.6	4-01-00-01451 0.8370[20]	EtOH 3 1.4462[20]
10043	Propane, 2-(ethylthio)- Ethyl isopropyl sulfide	$C_5H_{12}S$ 104.22	5145-99-3 -122.2	107.5	4-01-00-01501 0.8246[20]	
10044	Propane, 1-(ethylthio)-2-methyl- Ethyl isobutyl sulfide	$C_6H_{14}S$ 118.24	1613-45-2	134.2	4-01-00-01606 0.8306[20]	chl 3 1.4450[20]
10045	Propane, 1-(ethynyloxy)-	C_5H_8O 84.12	42842-08-0	75	4-01-00-02213 0.8080[20]	eth 4; EtOH 4 1.3935[20]
10046	Propane, 1-fluoro- Propyl fluoride	C_3H_7F 62.09	460-13-9 -159	2.5	4-01-00-00187 0.7956[20]	H$_2$O 2; EtOH 4; eth 4 1.3115[20]
10047	Propane, 2-fluoro- Isopropyl fluoride	C_3H_7F 62.09	420-26-8		4-01-00-00187	
10048	Propane, 1,1,1,2,2,3,3-heptachloro- 1,1,1,2,2,3,3-Heptachloropropane	C_3HCl_7 285.21	594-89-8 29.4	247.5	4-01-00-00205 1.8048[34]	ctc 2; chl 3
10049	Propane, 1,1,1,2,3,3,3-heptachloro- 1,1,1,2,3,3,3-Heptachloropropane	C_3HCl_7 285.21	3849-33-0 11	249	4-01-00-00205 1.7921[34]	chl 4 1.5427[21]
10050	Propane, 1,1,1,2,3,3,3-heptafluoro-2-(trifluoromethyl)- Perfluoroisobutane	C_4F_{10} 238.03	354-92-7	0		
10051	Propane, 1,1,1,2,2,3-hexachloro- 1,1,1,2,2,3-Hexachloropropane	$C_3H_2Cl_6$ 250.77	24425-97-6 26	216	4-01-00-00204 1.7187[25]	1.5282[25]
10052	Propane, 1,1,1,3,3,3-hexachloro- 1,1,1,3,3,3-Hexachloropropane	$C_3H_2Cl_6$ 250.77	3607-78-1 -27	206	4-01-00-00205 1.6800[20]	1.5179[20]
10053	Propane, 1,1,2,2,3,3-hexachloro- 1,1,2,2,3,3-Hexachloropropane	$C_3H_2Cl_6$ 250.77	15600-01-8	218.5	4-01-00-00205 1.7137[34]	1.5262[18]
10054	Propane, 1-iodo- Propyl iodide	C_3H_7I 169.99	107-08-4 -101.3	7875 102.6	4-01-00-00222 1.7489[20]	H$_2$O 2; EtOH 5; eth 5; ctc 2 1.5058[20]
10055	Propane, 2-iodo- Isopropyl iodide	C_3H_7I 169.99	75-30-9 -90	5102 89.5	4-01-00-00223 1.7042[20]	H$_2$O 2; EtOH 5; eth 5; bz 5 1.5028[20]
10056	Propane, 1-iodo-2,2-dimethyl-	$C_5H_{11}I$ 198.05	15501-33-4	128 dec	4-01-00-00338 1.4940[20]	H$_2$O 1; EtOH 3; eth 3 1.4890[20]
10057	Propane, 1-iodo-2-methyl- Isobutyl iodide	C_4H_9I 184.02	513-38-2	5027 121.1	4-01-00-00299 1.6035[20]	1.4959[20]
10058	Propane, 2-iodo-2-methyl- tert-Butyl iodide	C_4H_9I 184.02	558-17-8 -38.2	100.1	4-01-00-00300 1.571[25]	EtOH 5; eth 5 1.4918[20]
10059	Propane, 1-isocyanato- Propyl isocyanate	C_4H_7NO 85.11	110-78-1	83.5	4-04-00-00491 0.908[25]	1.3970[20]
10060	Propane, 2-isocyanato- Isopropyl isocyanate	C_4H_7NO 85.11	1795-48-8	74.5	4-04-00-00530 0.866[25]	1.3825[20]
10061	Propane, 1-isocyanato-2-methyl- Isobutyl isocyanate	C_5H_9NO 99.13	1873-29-6	106	4-04-00-00653	
10062	Propane, 2-isocyanato-2-methyl- tert-Butyl isocyanate	C_5H_9NO 99.13	1609-86-5 85.5		4-04-00-00669 0.8670[7]	1.4061[20]
10063	Propane, 2-isocyano- Isopropyl isocyanide	C_4H_7N 69.11	598-45-8	87	0-04-00-00154 0.7596[25]	H$_2$O 1; EtOH 5; eth 5
10064	Propane, 1-isothiocyanato-	C_4H_7NS 101.17	628-30-8	153	4-04-00-00491 0.9781[16]	H$_2$O 2; EtOH 5; eth 5 1.5085[16]
10065	Propane, 1-isothiocyanato-2-methyl-	C_5H_9NS 115.20	591-82-2	160	4-04-00-00653 0.9631[14]	1.5005[14]
10066	Propane, 2-isothiocyanato-2-methyl- tert-Butyl isothiocyanate	C_5H_9NS 115.20	590-42-1 10.5	140	4-04-00-00669 0.9187[10]	
10067	Propane, 1-isothiocyanato-3-(methylsulfonyl)- Cheirolin	$C_5H_9NO_2S_2$ 179.26	505-34-0 47.5	2040 200[3]	4-04-00-01664	EtOH 4; chl 4
10068	Propane, 1,1',1'',1'''-[methanetetrayltetrakis(oxy)]tetrakis-	$C_{13}H_{28}O_4$ 248.36	597-72-8	224.2	4-03-00-00007 0.897[20]	EtOH 4; eth 4 1.4100[20]
10069	Propane, 1-methoxy- Methyl propyl ether	$C_4H_{10}O$ 74.12	557-17-5	6031 39.1	4-01-00-01421 0.7356[13]	H$_2$O 3; EtOH 5; eth 5; ace 3 1.3579[25]
10070	Propane, 2-methoxy- Isopropyl methyl ether	$C_4H_{10}O$ 74.12	598-53-8	30.7	4-01-00-01471 0.7237[15]	H$_2$O 2; EtOH 5; eth 5 1.3576[20]
10071	Propane, 1-methoxy-2-methyl- Isobutyl methyl ether	$C_5H_{12}O$ 88.15	625-44-5	58.6	4-01-00-01593 0.7311[20]	eth 4; EtOH 4
10072	Propane, 2-methoxy-2-methyl- tert-Butyl methyl ether	$C_5H_{12}O$ 88.15	1634-04-4 -108.6	5954 55.2	4-01-00-01615 0.7405[20]	H$_2$O 3; EtOH 4; eth 4 1.3690[20]
10073	Propane, 2-methyl- Isobutane	C_4H_{10} 58.12	75-28-5 -138.3	-11.7	4-01-00-00282 *0.5510[25]	H$_2$O 2; EtOH 3; eth 3; chl 3 1.3518[-25]
10074	Propane, 1,1'-[methylenebis(oxy)]bis-	$C_7H_{16}O_2$ 132.20	505-84-0 -97.3	140.5	4-01-00-03029 0.8345[20]	ace 4; bz 4; eth 4; EtOH 4 1.3939[19]
10075	Propane, 1,1'-[methylenebis(oxy)]bis[2-methyl-	$C_9H_{20}O_2$ 160.26	2568-91-4	165.5	4-01-00-03030 0.8268[20]	1.4108[17]
10076	Propane, 1-(1-methylethoxy)-	$C_6H_{14}O$ 102.18	627-08-7	83	4-01-00-01471 0.7370[20]	H$_2$O 2; EtOH 4; eth 3; ace 3 1.376[21]
10077	Propane, 1-[(1-methylethyl)thio]- Isopropyl propyl sulfide	$C_6H_{14}S$ 118.24	5008-73-1	132.1	4-01-00-01501 0.8269[20]	

No.	Name / Synonym	Mol. Form. / Mol. Wt.	CAS RN	mp/°C / bp/°C	Merck No.	Beil. Ref. / den/g cm^{-3}	Solubility / n_D
10078	Propane, 1,1',1"-[methylidynetris(oxy)]tris-	$C_{10}H_{22}O_3$ 190.28	621-76-1	191; 93[30]		4-02-00-00027 0.8805[20]	eth 4; EtOH 4 1.4072[20]
10079	Propane, 2,2',2"-[methylidynetris(oxy)]tris-	$C_{10}H_{22}O_3$ 190.28	4447-60-3	167		4-02-00-00028 0.8621[20]	eth 4; EtOH 4 1.4000[20]
10080	Propane, 1,1',1"-[methylidynetris(oxy)]tris[2-methyl- Orthoformic acid, triisobutyl ester	$C_{13}H_{28}O_3$ 232.36	16754-49-7	223		4-02-00-00029 0.8582[20]	eth 4; EtOH 4 1.4120[20]
10081	Propane, 2-methyl-2-(1-methylethoxy)-	$C_7H_{16}O$ 116.20	17348-59-3	-88 87.6		4-01-00-01616 0.7365[25]	chl 3
10082	Propane, 2-methyl-1-(methylthio)-	$C_5H_{12}S$ 104.22	5008-69-5	112.5		4-01-00-01606 0.8335[20]	EtOH 3; eth 3; ace 3 1.4433[20]
10083	Propane, 2-methyl-2-(methylthio)- tert-Butyl methyl sulfide	$C_5H_{12}S$ 104.22	6163-64-0	99		4-01-00-01636	
10084	Propane, 2-methyl-1-nitro-	$C_4H_9NO_2$ 103.12	625-74-1	140.5		4-01-00-00301 0.9597[25]	H_2O 2; EtOH 5; eth 5 1.4066[25]
10085	Propane, 2-methyl-2-nitro-	$C_4H_9NO_2$ 103.12	594-70-7	26.23 127.16		4-01-00-00301 0.9501[28]	EtOH 5; eth 5; ace 5; bz 5 1.4015[20]
10086	Propane, 2-methyl-1-propoxy-	$C_7H_{16}O$ 116.20	15268-49-2	107		4-01-00-01594 0.7549[20]	eth 4; EtOH 4 1.3852[25]
10087	Propane, 2-methyl-2-propoxy-	$C_7H_{16}O$ 116.20	29072-93-3	100		3-01-00-01578 0.7472[25]	1.3830[25]
10088	Propane, 1-(methylthio)- Methyl propyl sulfide	$C_4H_{10}S$ 90.19	3877-15-4	-113 95.6		4-01-00-01450 0.8424[20]	H_2O 3; EtOH 3; eth 3; ace 3 1.4442[20]
10089	Propane, 2-(methylthio)- Isopropyl methyl sulfide	$C_4H_{10}S$ 90.19	1551-21-9	-101.5 84.8		4-01-00-01500 0.8291[20]	EtOH 3; eth 3; ace 3 1.4932[20]
10090	Propanenitrile Ethyl cyanide	C_3H_5N 55.08	107-12-0	-92.8 97.1	7839	4-02-00-00728 0.7818[20]	H_2O 4; EtOH 3; eth 3; ace 3 1.3655[20]
10091	Propanenitrile, 2-(acetyloxy)-	$C_5H_7NO_2$ 113.12	15657-96-2	-75 172.5		4-03-00-00676 1.0278[20]	H_2O 4 1.4027[20]
10092	Propanenitrile, 3-amino- 3-Aminopropionitrile	$C_3H_6N_2$ 70.09	151-18-8	185; 88[20]	481	4-04-00-02530 0.9584[20]	1.4396[20]
10093	Propanenitrile, 2,2'-azobis[2-methyl- 2,2'-Azobis(isobutyronitrile)	$C_8H_{12}N_4$ 164.21	78-67-1		931	4-04-00-03377	H_2O 1; EtOH 2; eth 2
10094	Propanenitrile, 2-bromo-	C_3H_4BrN 133.98	19481-82-4	59[24]		4-02-00-00764 1.5505[20]	ace 4; eth 4 1.4585[20]
10095	Propanenitrile, 3-bromo-	C_3H_4BrN 133.98	2417-90-5	92[25]; 69[7]		4-02-00-00766 1.6152[20]	EtOH 4; eth 4; ctc 2 1.4800[20]
10096	Propanenitrile, 2-bromo-2-methyl-	C_4H_6BrN 148.00	41658-69-9	139.5		4-02-00-00863 1.4796[15]	ace 4; eth 4; EtOH 4 1.4379[15]
10097	Propanenitrile, 3-chloro- β-Chloropropionitrile	C_3H_4ClN 89.52	542-76-7	-51 175.5	2160	4-02-00-00751 1.1573[20]	ctc 2 1.4360[20]
10098	Propanenitrile, 3-chloro-2-hydroxy-2-methyl- Lactonitrile, 3-chloro-2-methyl-	C_4H_6ClNO 119.55	33401-05-7	110[27]		3-03-00-00599 1.2027[15]	H_2O 4; EtOH 4; ace 4 1.4356[11]
10099	Propanenitrile, 2,3-dichloro-	$C_3H_3Cl_2N$ 123.97	2601-89-0	58-9[8]		4-02-00-00757 1.3500[20]	1.4640[20]
10100	Propanenitrile, 2,2-dimethyl- tert-Butyl cyanide	C_5H_9N 83.13	630-18-2	15 106.1		4-02-00-00913 0.7586[25]	1.3774[20]
10101	Propanenitrile, 3-(dimethylamino)- 3-(Dimethylamino)propionitrile	$C_5H_{10}N_2$ 98.15	1738-25-6	173		4-04-00-02533 0.8705[20]	
10102	Propanenitrile, 2-ethoxy- 2-Ethoxypropionitrile	C_5H_9NO 99.13	14631-45-9	136		4-03-00-00675 0.8743[20]	eth 4; EtOH 4 1.3890[22]
10103	Propanenitrile, 3-ethoxy-	C_5H_9NO 99.13	2141-62-0	171		4-03-00-00708 0.9285[15]	eth 4; EtOH 4 1.4068[20]
10104	Propanenitrile, 2-(β-D-glucopyranosyloxy)-2-methyl- Linamarin	$C_{10}H_{17}NO_6$ 247.25	554-35-8	145	5375	5-17-07-00397	ace 4
10105	Propanenitrile, 2-hydroxy- Acetaldehyde, cyanohydrin	C_3H_5NO 71.08	78-97-7	-40 183		3-03-00-00451 0.9877[20]	H_2O 5; EtOH 5; eth 3; chl 3 1.4058[18]
10106	Propanenitrile, 3-hydroxy- Hydracrylonitrile	C_3H_5NO 71.08	109-78-4	-46 221	3751	4-03-00-00708 1.0404[25]	H_2O 5; EtOH 5; eth 2; chl 3 1.4248[20]
10107	Propanenitrile, 3-[2-(2-hydroxyethoxy)ethoxy]-	$C_7H_{13}NO_3$ 159.19	10143-54-1	186[9]		4-03-00-00713 1.089[32]	1.4452[22]
10108	Propanenitrile, 2-hydroxy-2-methyl- Acetone cyanohydrin	C_4H_7NO 85.11	75-86-5	-19 82[23]	59	4-03-00-00785 0.932[19]	H_2O 4; EtOH 4; eth 4; ace 3 1.3992[20]
10109	Propanenitrile, 2-methoxy- 2-Methoxypropionitrile	C_4H_7NO 85.11	33695-59-9	119		4-03-00-00675 0.8928[20]	EtOH 4 1.3818[20]
10110	Propanenitrile, 3-methoxy-	C_4H_7NO 85.11	110-67-8	163		4-03-00-00708 0.9379[20]	EtOH 3; eth 3; chl 3 1.4043[20]
10111	Propanenitrile, 2-methyl- Isobutyronitrile	C_4H_7N 69.11	78-82-0	-71.5 103.9		4-02-00-00853 0.7704[20]	H_2O 2; EtOH 4; eth 4; ace 3 1.3720[20]
10112	Propanenitrile, 3-(methylamino)-	$C_4H_8N_2$ 84.12	693-05-0	101-4[49]		3-04-00-01264 0.8992[20]	H_2O 3; ace 3; bz 3; chl 3 1.4320[20]
10113	Propanenitrile, 2-oxo-	C_3H_3NO 69.06	631-57-2	92.3		4-03-00-01515 0.9745[20]	eth 3; ace 3; CH_3CN 3 1.3764[20]
10114	Propanenitrile, 3,3'-oxybis-	$C_6H_8N_2O$ 124.14	1656-48-0	110-12[0.5]		4-03-00-00720 1.043[25]	1.4405[20]
10115	Propanenitrile, 3,3,3-trichloro-2-hydroxy- Chlorocyanohydrin	$C_3H_2Cl_3NO$ 174.41	513-96-2	61 217 dec	2134	4-03-00-00680	H_2O 4; eth 4; EtOH 4
10116	Propane, 1-nitro- 1-Nitropropane	$C_3H_7NO_2$ 89.09	108-03-2	-108 131.1	6548	4-01-00-00229 0.9961[25]	H_2O 2; EtOH 5; eth 5; chl 3 1.4018[20]

No.	Name Synonym	Mol. Form. Mol. Wt.	CAS RN mp/°C	Merck No. bp/°C	Beil. Ref. den/g cm^{-3}	Solubility n_D
10117	Propane, 2-nitro- 2-Nitropropane	C$_3$H$_7$NO$_2$ 89.09	79-46-9 -91.3	6549 120.2	4-01-00-00230 0.9821[25]	H$_2$O 2; chl 3 1.3944[20]
10118	Propane, octafluoro- Perfluoropropane	C$_3$F$_8$ 188.02	76-19-7 -147.6	-36.6	4-01-00-00189	
10119	Propane, 1,1'-oxybis- Dipropyl ether	C$_6$H$_{14}$O 102.18	111-43-3 -126.1	7870 90.0	4-01-00-01422 0.7466[20]	eth 4; EtOH 4 1.3809[20]
10120	Propane, 2,2'-oxybis- Diisopropyl ether	C$_6$H$_{14}$O 102.18	108-20-3 -86.8	5100 68.5	4-01-00-01471 0.7241[20]	H$_2$O 2; EtOH 5; eth 5; ace 3 1.3679[20]
10121	Propane, 1,1'-oxybis[2-chloro-	C$_6$H$_{12}$Cl$_2$O 171.07	54460-96-7 188		2-01-00-00370 1.109[20]	EtOH 3; eth 3 1.4467[20]
10122	Propane, 1,1'-oxybis[3-chloro-	C$_6$H$_{12}$Cl$_2$O 171.07	629-36-7 216; 90.5[11]		4-01-00-01442 1.138[20]	EtOH 3; eth 3 1.4158[20]
10123	Propane, 2,2'-oxybis[1-chloro- 2,2'-Dichlorodiisopropyl ether	C$_6$H$_{12}$Cl$_2$O 171.07	108-60-1 187		3-01-00-01470 1.103[20]	H$_2$O 1; EtOH 5; eth 5; ace 5 1.4505[20]
10124	Propane, 2,2'-oxybis[2-methyl- Di-*tert*-Butyl ether	C$_8$H$_{18}$O 130.23	6163-66-2 107.2	3020	4-01-00-01616 0.7658[20]	1.3949[20]
10125	Propane, 1,1,1,2,3-pentachloro- 1,1,1,2,3-Pentachloropropane	C$_3$H$_3$Cl$_5$ 216.32	21700-31-2 179.5		4-01-00-00203	eth 4; EtOH 4 1.5130[20]
10126	Propane, 1,1,2,3,3-pentachloro- 1,1,2,3,3-Pentachloropropane	C$_3$H$_3$Cl$_5$ 216.32	15104-61-7 199; 99[20]		4-01-00-00204 1.6086[39]	H$_2$O l; EtOH 4; eth 3; chl 3 1.5131[17]
10127	Propane, 1,1,1,2,3-pentachloro-2-methyl- Isobutane, 1,1,1,2,3-pentachloro	C$_4$H$_5$Cl$_5$ 230.35	4749-31-9 73.5	215	4-01-00-00294 1.5686[25]	chl 4 1.5165[20]
10128	Propane, 1,1,1,2,2-pentafluoro- 1,1,1,2,2-Pentafluoropropane	C$_3$H$_3$F$_5$ 134.05	1814-88-6 -17.4			
10129	Propane, 1,1'-sulfinylbis-	C$_6$H$_{14}$OS 134.24	4253-91-2 22.5	80[2]	4-01-00-01453 0.9654[20]	eth 4; EtOH 4 1.4663[20]
10130	2-Propanesulfonic acid Isopropanesulfonic acid	C$_3$H$_8$O$_3$S 124.16	14159-48-9 -37	159[1.4]	4-04-00-00042 1.187[25]	H$_2$O 4 1.4332[20]
10131	Propane, 1,1'-sulfonylbis-	C$_6$H$_{14}$O$_2$S 150.24	598-03-8 29.5		4-01-00-01453 1.0278[50]	H$_2$O 2; EtOH 3; eth 3 1.4456[30]
10132	Propane, 1,1'-sulfonylbis[2-methyl-	C$_8$H$_{18}$O$_2$S 178.30	10495-45-1 17	265	4-01-00-01607 1.0056[18]	
10133	1-Propanesulfonyl chloride	C$_3$H$_7$ClO$_2$S 142.61	10147-36-1	180 dec; 77[12]	4-04-00-00039 1.2826[15]	1.452[20]
10134	1-Propanesulfonyl chloride, 2-methyl-	C$_4$H$_9$ClO$_2$S 156.63	35432-36-1 190		4-04-00-00049 1.2014[20]	1.4520[25]
10135	Propane, 1,1,1,2-tetrabromo-	C$_3$H$_4$Br$_4$ 359.68	62127-49-5 122[15]		3-01-00-00251 2.679[20]	1.6187[20]
10136	Propane, 1,1,2,2-tetrabromo- 1,1,2,2-Tetrabromopropane	C$_3$H$_4$Br$_4$ 359.68	34570-59-7 105-7[9]		4-01-00-00222 2.687[20]	1.6166[20]
10137	Propane, 1,1,3,3-tetrabromo-	C$_3$H$_4$Br$_4$ 359.68	51525-97-4 154-6[19]		2-01-00-00077 2.702[21]	1.6225[21]
10138	Propane, 1,2,2,3-tetrabromo- 1,2,2,3-Tetrabromopropane	C$_3$H$_4$Br$_4$ 359.68	54268-02-9 10.5	169[80]; 115[9]	3-01-00-00251 2.703[20]	1.6200[20]
10139	1,1,2,3-Propanetetracarboxylic acid, tetraethyl ester Tetraethyl 1,1,2,3-propanetetracarboxylate	C$_{15}$H$_{24}$O$_8$ 332.35	635-03-0 203-4[18]		3-02-00-02077 1.1184[20]	EtOH 4 1.4395[20]
10140	1,1,3,3-Propanetetracarboxylic acid, tetraethyl ester 1,1,3,3-Tetracarbethoxypropane	C$_{15}$H$_{24}$O$_8$ 332.35	2121-66-6 -30	305 dec; 195[8]	4-02-00-02417 1.116[20]	EtOH 3 1.4398[20]
10141	Propane, 1,1,1,2-tetrachloro- 1,1,1,2-Tetrachloropropane	C$_3$H$_4$Cl$_4$ 181.88	812-03-3 -64	152.5	4-01-00-00201 1.473[20]	H$_2$O 1; EtOH 4; eth 3; chl 3 1.4887[20]
10142	Propane, 1,1,1,3-tetrachloro- 1,1,1,3-Tetrachloropropane	C$_3$H$_4$Cl$_4$ 181.88	1070-78-6 157		4-01-00-00201 1.4509[20]	H$_2$O 1; EtOH 4; eth 4; bz 4 1.4825[20]
10143	Propane, 1,1,2,2-tetrachloro- 1,1,2,2-Tetrachloropropane	C$_3$H$_4$Cl$_4$ 181.88	13116-60-4 153		4-01-00-00201 1.47[12]	H$_2$O 1; EtOH 5; eth 5; chl 3 1.4850[25]
10144	Propane, 1,1,2,3-tetrachloro- 1,1,2,3-Tetrachloropropane	C$_3$H$_4$Cl$_4$ 181.88	18495-30-2 179.5		4-01-00-00202 1.513[17]	H$_2$O 1; EtOH 3; eth 4; chl 3 1.5037[17]
10145	Propane, 1,2,2,3-tetrachloro- 1,2,2,3-Tetrachloropropane	C$_3$H$_4$Cl$_4$ 181.88	13116-53-5 165		4-01-00-00202 1.500[18]	H$_2$O 1; EtOH 4; eth 4; chl 3 1.4940[18]
10146	Propane, 1,1,2,3-tetrachloro-2-(chloromethyl)- 1,1,2,3-Tetrachloro-2-(chloromethyl)propane	C$_4$H$_5$Cl$_5$ 230.35	66997-45-3 227; 95[9]		4-01-00-00321 1.5686[25]	bz 4; EtOH 4; chl 4 1.5165[25]
10147	Propane, 1,1,1,2-tetrachloro-2-methyl- Isobutane, 1,1,1,2-tetrachloro	C$_4$H$_6$Cl$_4$ 195.90	7086-07-9 178.5	192[117]	4-01-00-00293	eth 4; EtOH 4; chl 4
10148	Propane, 1,1,2,3-tetrachloro-2-methyl-	C$_4$H$_6$Cl$_4$ 195.90	18963-01-4 -46	190.5	4-01-00-00294 1.4393[25]	bz 4; eth 4; chl 4 1.4963[20]
10149	Propane, 1,2,2,3-tetrachloro-1,1,3,3-tetrafluoro- Propane, 1,2,2,3-tetrachloro-tetrafluoro	C$_3$Cl$_4$F$_4$ 253.84	677-68-9 -42.9	112	4-01-00-00203 1.7199[20]	1.3958[20]
10150	Propane, 1,1,3,3-tetramethoxy-	C$_7$H$_{16}$O$_4$ 164.20	102-52-3 183; 66[12]		4-01-00-03635 0.997[25]	1.4081[20]
10151	Propane, 1,1'-thiobis- Dipropyl sulfide	C$_6$H$_{14}$S 118.24	111-47-7 -102.5	7881 142.9	4-01-00-01452 0.814[17]	H$_2$O 1; EtOH 3; eth 3 1.4487[20]
10152	Propane, 2,2'-thiobis- Diisopropyl sulfide	C$_6$H$_{14}$S 118.24	625-80-9 -78.1	120.1	4-01-00-01502 0.8142[20]	H$_2$O 1; EtOH 3; eth 3 1.4438[20]
10153	Propane, 1,1'-thiobis[3-chloro- Bis(3-chloropropyl) sulfide	C$_6$H$_{12}$Cl$_2$S 187.13	55882-21-8 162[43]		3-01-00-01438 1.1774[25]	eth 4; EtOH 4; tol 4 1.5075[20]
10154	Propane, 1,1'-thiobis[2-methyl- Diisobutyl sulfide	C$_8$H$_{18}$S 146.30	592-65-4 -105.5	5035 171	4-01-00-01607 0.8363[10]	

No.	Name Synonym	Mol. Form. Mol. Wt.	CAS RN mp/°C	Merck No. bp/°C	Beil. Ref. den/g cm^{-3}	Solubility n_D
10155	Propane, 2,2'-thiobis[2-methyl- Di-*tert*-butyl sulfide	C$_8$H$_{18}$S 146.30	107-47-1 -9.0	149.1	4-01-00-01637 0.815^{25}	1.4506^{20}
10156	1-Propanethiol Propyl mercaptan	C$_3$H$_8$S 76.16	107-03-9 -113.3	67.8	4-01-00-01449 0.8411^{20}	H$_2$O 2; EtOH 3; eth 3; ace 3 1.4380^{20}
10157	2-Propanethiol Isopropyl mercaptan	C$_3$H$_8$S 76.16	75-33-2 -130.5	52.6	4-01-00-01498 0.8143^{20}	H$_2$O 2; EtOH 5; eth 5; ace 4 1.4255^{20}
10158	1-Propanethiol, 2-methyl- Isobutyl mercaptan	C$_4$H$_{10}$S 90.19	513-44-0 <-70	5030 88.5	4-01-00-01605 0.8357^{20}	H$_2$O 2; EtOH 4; eth 4; ace 4 1.4387^{20}
10159	2-Propanethiol, 2-methyl- *tert*-Butyl mercaptan	C$_4$H$_{10}$S 90.19	75-66-1 -0.5	1577 64.3	4-01-00-01634 0.7943^{25}	H$_2$O 1; ctc 3; hp 3 1.4232^{20}
10160	Propane, 1,1,1-tribromo-	C$_3$H$_5$Br$_3$ 280.78	62127-61-1	77-83^{25}	4-01-00-00221 2.35^{25}	1.5606^{25}
10161	Propane, 1,1,2-tribromo-	C$_3$H$_5$Br$_3$ 280.78	14602-62-1	200.5	3-01-00-00251 2.354^{720}	H$_2$O 1; EtOH 3; eth 4; chl 3 1.5790^{20}
10162	Propane, 1,2,2-tribromo-	C$_3$H$_5$Br$_3$ 280.78	14476-30-3	190.5	2-01-00-00077 2.2984^{20}	eth 4; EtOH 4; chl 4 1.5670^{20}
10163	Propane, 1,2,3-tribromo- 1,2,3-Tribromopropane	C$_3$H$_5$Br$_3$ 280.78	96-11-7 16.9	9527 222.1	4-01-00-00221 2.4208^{20}	H$_2$O 1; EtOH 4; eth 4; ctc 2 1.5862^{20}
10164	Propane, 1,2,3-tribromo-2-methyl-	C$_4$H$_7$Br$_3$ 294.81	631-28-7	178^{235}; 108^{18}	4-01-00-00299 2.2105^{14}	H$_2$O 1; eth 3; chl 3 1.5701^{21}
10165	1,2,3-Propanetricarboxylic acid Tricarballylic acid	C$_6$H$_8$O$_6$ 176.13	99-14-9 166	9534	4-02-00-02366	H$_2$O 4; EtOH 4; eth 2
10166	1,2,3-Propanetricarboxylic acid, 2-(acetyloxy)-, triethyl ester	C$_{14}$H$_{22}$O$_8$ 318.32	77-89-4	214^{40}	3-03-00-01105 1.135^{25}	1.4380
10167	1,2,3-Propanetricarboxylic acid, 2-hydroxy- Citric acid	C$_6$H$_8$O$_7$ 192.13	77-92-9 153	2328 dec	4-03-00-01272 1.665^{20}	H$_2$O 4; EtOH 4; eth 3; bz 1
10168	1,2,3-Propanetricarboxylic acid, 2-hydroxy-, tributyl ester Butyl citrate	C$_{18}$H$_{32}$O$_7$ 360.45	77-94-1 -20	1564 233^{22}	4-03-00-01276 1.043^{20}	1.4460^{20}
10169	1,2,3-Propanetricarboxylic acid, 2-hydroxy-, triethyl ester	C$_{12}$H$_{20}$O$_7$ 276.29	77-93-0	294	4-03-00-01276 1.1369^{20}	H$_2$O 1; EtOH 3; eth 3; ctc 2 1.4455^{20}
10170	1,2,3-Propanetricarboxylic acid, 2-hydroxy-, trimethyl ester Trimethyl citrate	C$_9$H$_{14}$O$_7$ 234.21	1587-20-8 79.3	285; 176^{16}	4-03-00-01276	eth 4; EtOH 4
10171	Propane, 1,1,1-trichloro- 1,1,1-Trichloropropane	C$_3$H$_5$Cl$_3$ 147.43	7789-89-1	108	4-01-00-00198 1.287^{23}	H$_2$O 1; EtOH 3; eth 4; chl 4
10172	Propane, 1,1,2-trichloro- 1,1,2-Trichloropropane	C$_3$H$_5$Cl$_3$ 147.43	598-77-6	132.0; 117^{500}	4-01-00-00199 1.372^{15}	H$_2$O 1; EtOH 3; eth 4; ctc 2
10173	Propane, 1,1,3-trichloro- 1,1,3-Trichloropropane	C$_3$H$_5$Cl$_3$ 147.43	20395-25-9 -59	145.5	4-01-00-00199 1.3557^{20}	eth 4; EtOH 4; chl 4 1.4718^{20}
10174	Propane, 1,2,2-trichloro- 1,2,2-Trichloropropane	C$_3$H$_5$Cl$_3$ 147.43	3175-23-3	124	4-01-00-00199 1.318^{25}	H$_2$O 1; EtOH 3; eth 3; chl 4 1.4609^{20}
10175	Propane, 1,2,3-trichloro- 1,2,3-Trichloropropane	C$_3$H$_5$Cl$_3$ 147.43	96-18-4 -14.7	157	4-01-00-00199 1.3889^{20}	H$_2$O 2; EtOH 3; eth 3; ctc 2 1.4852^{20}
10176	Propane, 1,2,3-trichloro-2-(chloromethyl)-	C$_4$H$_6$Cl$_4$ 195.90	18963-00-3	211; 87^9	3-01-00-00321 1.5036^{25}	chl 4 1.508^{20}
10177	Propane, 1,1,2-trichloro-2-methyl- Isobutane, 1,1,2-trichloro	C$_4$H$_7$Cl$_3$ 161.46	29559-52-2 -6	163	4-01-00-00293 1.2588^{20}	eth 4; chl 4 1.4666^{20}
10178	Propane, 1,2,3-trichloro-2-methyl-	C$_4$H$_7$Cl$_3$ 161.46	1871-58-5	162.5	4-01-00-00293 1.3012^{25}	chl 4 1.4765^{20}
10179	Propane, 1,1,1-trichloro-2,2,3,3,3-pentafluoro-	C$_3$Cl$_3$F$_5$ 237.38	4259-43-2 -80	74	4-01-00-00200 1.643^{25}	
10180	Propane, 1,2,2-trichloro-1,1,3,3,3-pentafluoro-	C$_3$Cl$_3$F$_5$ 237.38	1599-41-3 -4.3	72.0	3-01-00-00233 1.6681^{20}	1.3519^{20}
10181	Propane, 1,2,3-trichloro-1,1,2,3,3-pentafluoro-	C$_3$Cl$_3$F$_5$ 237.38	76-17-5 -72	73.7	4-01-00-00200 1.6631^{20}	1.3512^{20}
10182	Propane, 1,1,1-triethoxy-	C$_9$H$_{20}$O$_3$ 176.26	115-80-0	171	4-02-00-00707	eth 4; EtOH 4 1.4000^{25}
10183	Propane, 1,1,3-triethoxy-	C$_9$H$_{20}$O$_3$ 176.26	7789-92-6	185 dec; 78^{14}	4-01-00-03971 0.898^{19}	H$_2$O 2; EtOH 3; ctc 3 1.4067^{20}
10184	Propane, 1,1,1-trifluoro-3-nitro- Propane, 1-nitro-3,3,3-trifluoro	C$_3$H$_4$F$_3$NO$_2$ 143.07	461-35-8	135	4-01-00-00232 1.4178^{20}	eth 4 1.3549^{20}
10185	Propane, 1,2,3-trimethoxy- Glycerol trimethyl ether	C$_6$H$_{14}$O$_3$ 134.18	20637-49-4	148	4-01-00-02755 0.9460^{15}	H$_2$O 5; eth 3; ace 4; bz 3 1.4055^{15}
10186	1,2,3-Propanetriol Glycerol	C$_3$H$_8$O$_3$ 92.09	56-81-5 18.2	4379 290	4-01-00-02751 1.2613^{20}	H$_2$O 5; EtOH 5; eth 2; bz 1 1.4746^{20}
10187	1,2,3-Propanetriol, 1-acetate, (±)- Glycerol 1-monoacetate(DL)	C$_5$H$_{10}$O$_4$ 134.13	93713-40-7	158^{165} 130^3	4-02-00-00251 1.2060^{20}	H$_2$O 3; EtOH 3; eth 2; bz 1 1.4157^{20}
10188	1,2,3-Propanetriol, 1-(4-aminobenzoate) Glyceryl *p*-aminobenzoate	C$_{10}$H$_{13}$NO$_4$ 211.22	136-44-7	4382		H$_2$O 1; EtOH 3
10189	1,2,3-Propanetriol, diacetate Diacetin	C$_7$H$_{12}$O$_5$ 176.17	25395-31-7	2941 259	4-02-00-00252 1.184^{16}	H$_2$O 4; eth 4; EtOH 4 1.44^{20}
10190	1,2,3-Propanetriol, 1,3-diacetate 1,3-Diacetin	C$_7$H$_{12}$O$_5$ 176.17	105-70-4	2941 173^{40}	4-02-00-00252 1.179^{15}	H$_2$O 4; EtOH 4; eth 2; CS$_2$ 1 1.4395^{20}

No.	Name / Synonym	Mol. Form. / Mol. Wt.	CAS RN / mp/°C	Merck No. / bp/°C	Beil. Ref. / den/g cm^{-3}	Solubility / n_D
10191	1,2,3-Propanetriol, 1,3-dinitrate	$C_3H_6N_2O_7$ 182.09	623-87-0 26	148[15]; 116[0.6]	2-01-00-00591 1.523[20]	H_2O 4; eth 4; EtOH 4 1.4715[20]
10192	1,2,3-Propanetriol, 1-nitrate	$C_3H_7NO_5$ 137.09	624-43-1 61	157.5	4-01-00-02761 1.4164[20]	H_2O 4; EtOH 4; eth 2 1.4698[20]
10193	1,2,3-Propanetriol, 2-nitrate	$C_3H_7NO_5$ 137.09	620-12-2 54	157.5	2-01-00-00591 1.40[22]	H_2O 4; eth 4; EtOH 4
10194	1,2,3-Propanetriol, 1-phenyl- Glycerol, 1-phenyl	$C_9H_{12}O_3$ 168.19	63157-81-3 100.5	185	4-06-00-07407 1.2213[13]	1.5606[15]
10195	1,2,3-Propanetriol, triacetate Triacetin	$C_9H_{14}O_6$ 218.21	102-76-1 -78	9504 259	4-02-00-00253 1.1583[20]	H_2O 2; EtOH 5; eth 5; ace 4 1.4301[20]
10196	1,2,3-Propanetriol, tribenzoate	$C_{24}H_{20}O_6$ 404.42	614-33-5 76		3-09-00-00666 1.228[12]	H_2O 1; EtOH 3; eth 4; ace 4
10197	1,2,3-Propanetriol, trinitrate Nitroglycerin	$C_3H_5N_3O_9$ 227.09	55-63-0 13.5	6528 93[0.31]; exp 218	4-01-00-02762 1.5931[20]	H_2O 2; EtOH 3; eth 5; ace 4 1.4786[12]
10198	1,2,3-Propanetriol, tripropanoate	$C_{12}H_{20}O_6$ 260.29	139-45-7	175-6[20]	4-02-00-00717 1.099[20]	H_2O 1; EtOH 3; eth 4; chl 3 1.4318[19]
10199	Propanetrione, diphenyl-	$C_{15}H_{10}O_3$ 238.24	643-75-4 70	289[175]; 179[4]	4-07-00-02822	eth 4
10200	Propanil Propanamide, N-(3,4-dichlorophenyl)-	$C_9H_9Cl_2NO$ 218.08	709-98-8 92	7814	1.25[25]	
10201	Propanoic acid Propionic acid	$C_3H_6O_2$ 74.08	79-09-4 -20.7	7837 141.1	4-02-00-00695 0.9930[20]	H_2O 5; EtOH 5; eth 3; chl 2 1.3809[20]
10202	Propanoic acid, 2-(acetyloxy)-, (±)-	$C_5H_8O_4$ 132.12	3853-80-3	167-70[78]	4-03-00-00638 1.1758[20]	bz 4; EtOH 4 1.4240[20]
10203	Propanoic acid, anhydride Propionic anhydride	$C_6H_{10}O_3$ 130.14	123-62-6 -45	7838 170; 67.5[18]	4-02-00-00722 1.0110[20]	eth 5; ctc 2 1.4038[20]
10204	Propanoic acid, 2-bromo-, (±)-	$C_3H_5BrO_2$ 152.98	10327-08-9 25.7	203.5	1.7000[20]	H_2O 4; EtOH 4; eth 4; chl 2 1.4753[20]
10205	Propanoic acid, 3-bromo- β-Bromopropionic acid	$C_3H_5BrO_2$ 152.98	590-92-1 62.5	1421 140-2[45]	4-02-00-00764 1.48[25]	H_2O 3; EtOH 3; eth 3; bz 3
10206	Propanoic acid, 3-bromo-, butyl ester	$C_7H_{13}BrO_2$ 209.08	6973-79-1	130[26]	4-02-00-00765 1.4549[20]	ace 4; eth 4; EtOH 4 1.3051[20]
10207	Propanoic acid, 2-bromo-, ethyl ester Ethyl α-bromopropionate	$C_5H_9BrO_2$ 181.03	535-11-5	3731 160 dec; 71[26]	4-02-00-00762 1.4135[20]	H_2O 1; EtOH 5; eth 5; chl 3 1.4490[20]
10208	Propanoic acid, 3-bromo-, ethyl ester	$C_5H_9BrO_2$ 181.03	539-74-2	179; 65[15]	4-02-00-00765 1.4123[18]	EtOH 3; eth 3; ace 3; ctc 2 1.4516[20]
10209	Propanoic acid, 2-bromo-2-methyl- α-Bromoisobutyric acid	$C_4H_7BrO_2$ 167.00	2052-01-9 48.5	1408 199; 115[24]	4-02-00-00862 1.4969[60]	
10210	Propanoic acid, 3-bromo-, 3-methylbutyl ester Isopentyl 3-bromopropionate	$C_8H_{15}BrO_2$ 223.11	100983-12-8	110-1[11]	2-02-00-00231 1.2217[15]	eth 4; EtOH 4 1.4556[9]
10211	Propanoic acid, 2-bromo-, methyl ester, (±)-	$C_4H_7BrO_2$ 167.00	57885-43-5	144	4-02-00-00762 1.4966[25]	eth 4; EtOH 4 1.4451[22]
10212	Propanoic acid, 3-bromo-, methyl ester	$C_4H_7BrO_2$ 167.00	3395-91-3	105.5[60]	4-02-00-00765 1.4897[15]	EtOH 3; eth 3; ace 3; os 4 1.4542[20]
10213	Propanoic acid, 2-bromo-2-methyl-, ethyl ester	$C_6H_{11}BrO_2$ 195.06	600-00-0	163	4-02-00-00862 1.3263[20]	H_2O 1; EtOH 3; eth 5 1.4446[20]
10214	Propanoic acid, butyl ester Butyl propionate	$C_7H_{14}O_2$ 130.19	590-01-2 -89	1587 146.8	4-02-00-00708 0.8754[20]	H_2O 2; EtOH 5; eth 5; ctc 2 1.4014[20]
10215	Propanoic acid, 2-chloro-	$C_3H_5ClO_2$ 108.52	598-78-7	185	4-02-00-00745 1.2585[20]	H_2O 5; EtOH 5; eth 5; ace 3 1.4380[20]
10216	Propanoic acid, 3-chloro- β-Chloropropionic acid	$C_3H_5ClO_2$ 108.52	107-94-8 41	2159 204 dec	4-02-00-00748	H_2O 3; EtOH 3; eth 5; chl 3
10217	Propanoic acid, 2-chloro-, butyl ester	$C_7H_{13}ClO_2$ 164.63	54819-86-2	184	4-02-00-00746 1.0253[20]	eth 4 1.4263[20]
10218	Propanoic acid, 3-chloro-, butyl ester	$C_7H_{13}ClO_2$ 164.63	27387-79-7	104[22]	4-02-00-00749 1.0370[20]	H_2O 4; eth 4 1.4321[20]
10219	Propanoic acid, 3-chloro-2,2-dimethyl-	$C_5H_9ClO_2$ 136.58	13511-38-1 41.5	110[10]	4-02-00-00914	ctc 4
10220	Propanoic acid, 2-chloro-, ethyl ester Ethyl α-chloropropionate	$C_5H_9ClO_2$ 136.58	535-13-7	3743 147	4-02-00-00746 1.0793[20]	H_2O 1; EtOH 5; eth 5; ctc 2 1.4178[20]
10221	Propanoic acid, 3-chloro-, ethyl ester	$C_5H_9ClO_2$ 136.58	623-71-2	162	4-02-00-00749 1.1086[20]	H_2O 2; EtOH 5; eth 5; os 3 1.4254[20]
10222	Propanoic acid, 2-chloro-, methyl ester Methyl 2-chloropropionate	$C_4H_7ClO_2$ 122.55	17639-93-9	132.5	4-02-00-00746 1.0750[25]	
10223	Propanoic acid, 2-chloro-, 1-methylethyl ester Isopropyl 2-chloropropionate	$C_6H_{11}ClO_2$ 150.60	40058-87-5	151.5	4-02-00-00746 1.0315[20]	H_2O 1; EtOH 3; eth 3 1.4149[20]
10224	Propanoic acid, 2-chloro-, 2-methylpropyl ester, (R)- Propionic acid, 2-chloro, isobutyl ester (d)	$C_7H_{13}ClO_2$ 164.63	114489-96-2	176	0-02-00-00248 1.0312[20]	1.4247[20]
10225	Propanoic acid, 3-chloro-, methyl ester	$C_4H_7ClO_2$ 122.55	6001-87-2	156	4-02-00-00748 1.1861[15]	EtOH 3; ctc 3 1.4263[20]
10226	Propanoic acid, 3-chloro-, 3-methylbutyl ester	$C_8H_{15}ClO_2$ 178.66	62108-70-7	208; 87[12]	4-02-00-00749 1.0171[20]	eth 4; EtOH 4 1.4343[20]
10227	Propanoic acid, 3-chloro-, 2-methylpropyl ester	$C_7H_{13}ClO_2$ 164.63	62108-68-3	191.3	4-02-00-00749 1.0323[20]	eth 4; EtOH 4 1.4295[20]

No.	Name / Synonym	Mol. Form. / Mol. Wt.	CAS RN / mp/°C	Merck No. / bp/°C	Beil. Ref. / den/g cm^{-3}	Solubility / n_D
10228	Propanoic acid, 2-chloro-2-methyl-	$C_4H_7ClO_2$ 122.55	594-58-1 31	118[50]; 93[20]	4-02-00-00858 1.1790[20]	EtOH 4 1.450[20]
10229	Propanoic acid, 2-chloro-2-methyl-, methyl ester	$C_5H_9ClO_2$ 136.58	22421-97-2	135	4-02-00-00859 1.0883[15]	eth 4 1.4122[21]
10230	Propanoic acid, 3-chloro-2-methyl-	$C_4H_7ClO_2$ 122.55	16674-04-7	128-33[50]	4-02-00-00860 1.0153[20]	eth 4; EtOH 4 1.4310[20]
10231	Propanoic acid, 2-(4-chlorophenoxy)-2-methyl-, ethyl ester Clofibrate	$C_{12}H_{15}ClO_3$ 242.70	637-07-0	2374 148-50[20]	4-06-00-00851	
10232	Propanoic acid, 3-chloro-, propyl ester	$C_6H_{11}ClO_2$ 150.60	62108-66-1	180	4-02-00-00749 1.0656[20]	eth 4; EtOH 4 1.4290[20]
10233	Propanoic acid, 2-cyano-	$C_4H_5NO_2$ 99.09	632-07-5 35	143[11]	3-02-00-01683 1.1400[20]	
10234	Propanoic acid, 3-cyano-, methyl ester	$C_5H_7NO_2$ 113.12	4107-62-4	211	3-02-00-01672 1.4250[20]	chl 3
10235	Propanoic acid, cyclohexyl ester	$C_9H_{18}O_2$ 156.22	6222-35-1	193; 93[35]	4-06-00-00037 0.9359[20]	H_2O 1; EtOH 3; eth 3; ace 3 1.4403[20]
10236	Propanoic acid, 2-diazo-, ethyl ester Ethyl 2-diazopropionate	$C_5H_8N_2O_2$ 128.13	6111-99-5	65-8[41]	2-03-00-00403 1.086[12]	1.4472[18]
10237	Propanoic acid, 2-diazo-, methyl ester Methyl 2-diazopropionate	$C_4H_6N_2O_2$ 114.10	34757-14-7	43-5[11]	2-03-00-00403 1.1011[13]	1.4487[20]
10238	Propanoic acid, 2,3-dibromo-	$C_3H_4Br_2O_2$ 231.87	600-05-5 66.5	160[20]; 138[12]	4-02-00-00767	bz 4; eth 4; EtOH 4
10239	Propanoic acid, 2,3-dibromo-, butyl ester Butyl 2,3-dibromopropionate	$C_7H_{12}Br_2O_2$ 287.98	21179-48-6	237.5	4-02-00-00768 1.6107[20]	1.4890[20]
10240	Propanoic acid, 2,3-dibromo-, ethyl ester	$C_5H_8Br_2O_2$ 259.93	3674-13-3	214.5	4-02-00-00767 1.7966[20]	EtOH 3; eth 3 1.5007[20]
10241	Propanoic acid, 2,3-dibromo-, methyl ester	$C_4H_6Br_2O_2$ 245.90	1729-67-5	206	4-02-00-00767 1.9333[20]	EtOH 3 1.5127[20]
10242	Propanoic acid, 2,3-dibromo-2-methyl-, methyl ester	$C_5H_8Br_2O_2$ 259.93	3673-79-8	203	4-02-00-00864 1.805[20]	1.505[20]
10243	Propanoic acid, 2,3-dibromo-, propyl ester Propyl 2,3-dibromopropionate	$C_6H_{10}Br_2O_2$ 273.95	79762-76-8	222.5	3-02-00-00572 1.6799[20]	1.4950[20]
10244	Propanoic acid, 2,2-dichloro- Dalapon	$C_3H_4Cl_2O_2$ 142.97	75-99-0	2806 187.5; 92[14]	4-02-00-00753 1.389[12]	H_2O 4; EtOH 4; eth 3; ctc 3
10245	Propanoic acid, 2,3-dichloro-	$C_3H_4Cl_2O_2$ 142.97	565-64-0 50	210; 130[26]	4-02-00-00756 1.4650[20]	H_2O 3; EtOH 3; eth 3; peth 2
10246	Propanoic acid, 2,3-dichloro-, ethyl ester	$C_5H_8Cl_2O_2$ 171.02	6628-21-3	183.5	4-02-00-00756 1.2401[20]	eth 4; EtOH 4 1.4482[20]
10247	Propanoic acid, 3,3-dichloro-2-hydroxy-2-methyl-	$C_4H_6Cl_2O_3$ 173.00	31340-44-0 82.5	dec	2-03-00-00224	H_2O 3; EtOH 3; eth 3
10248	Propanoic acid, 2,3-dichloro-, methyl ester	$C_4H_6Cl_2O_2$ 157.00	3674-09-7	92[50]	4-02-00-00756 1.3282[20]	ace 4; eth 4; EtOH 4
10249	Propanoic acid, 2,3-dichloro-, 1-methylethyl ester	$C_6H_{10}Cl_2O_2$ 185.05	54774-99-1	61-2[5]	4-02-00-00756 1.2010[25]	1.4470
10250	Propanoic acid, 2-(2,4-dichlorophenoxy)- Dichlorprop	$C_9H_8Cl_2O_3$ 235.07	120-36-5 117.5	3068	3-06-00-00707	H_2O 2; EtOH 3; eth 3; lig 2
10251	Propanoic acid, 2,2-dichloro-3,3,3-trifluoro-, methyl ester Propionic acid, 2,2-dichloro-3,3,3-trifluoro-, methyl ester	$C_4H_3Cl_2F_3O_2$ 210.97	378-68-7	116-7[625]	4-02-00-00758 1.5092[20]	1.3806[20]
10252	Propanoic acid, 2,3-dihydroxy- Glyceric acid	$C_3H_6O_4$ 106.08	473-81-4	4378 dec	4-03-00-01050	H_2O 4; ace 4; EtOH 4
10253	Propanoic acid, 2,3-dihydroxy-, ethyl ester, (±)- Glyceric acid, ethyl ester (DL)	$C_5H_{10}O_4$ 134.13	89300-69-6	235; 120[14]	4-03-00-01052 1.1897[15]	H_2O 4; eth 4; EtOH 4
10254	Propanoic acid, 2,3-dihydroxy-, methyl ester, (±)- Methyl DL-glycerate	$C_4H_8O_4$ 120.11	15909-76-9	241.5; 119[14]	4-03-00-01052 1.2802[15]	H_2O 4; EtOH 4 1.4502[20]
10255	Propanoic acid, 2,3-dihydroxy-2-methyl-, ethyl ester Glyceric acid, 2-methyl, ethyl ester	$C_6H_{12}O_4$ 148.16	67535-07-3	95[10]	3-03-00-00860 1.114[25]	1.4370[25]
10256	Propanoic acid, 2,3-dihydroxy-2-methyl-, methyl ester Glyceric acid, 2-methyl-, methyl ester	$C_5H_{10}O_4$ 134.13	19860-56-1	89[10]	4-03-00-01060 1.185[25]	1.4438[25]
10257	Propanoic acid, 2,3-dihydroxy-, propyl ester	$C_6H_{12}O_4$ 148.16	116435-95-1	132-4[3]	3-03-00-00854 1.1537[20]	1.4503[20]
10258	Propanoic acid, 2,2-dimethyl- Pivalic acid	$C_5H_{10}O_2$ 102.13	75-98-9 35	7482 164	4-02-00-00908 0.905[50]	H_2O 2; EtOH 4; eth 4 1.3931[30]
10259	Propanoic acid, 2,2-dimethyl-, ethyl ester Ethyl 2,2-dimethylpropionate	$C_7H_{14}O_2$ 130.19	3938-95-2 -89.5	118.4	4-02-00-00910 0.856[20]	EtOH 3; eth 3 1.3906[20]
10260	Propanoic acid, 2,2-dimethyl-, methyl ester Methyl 2,2-dimethylpropionate	$C_6H_{12}O_2$ 116.16	598-98-1	101.1	4-02-00-00909 0.891[0]	eth 4; EtOH 4 1.3905[20]
10261	Propanoic acid, ethenyl ester Vinyl propionate	$C_5H_8O_2$ 100.12	105-38-4	91.2	4-02-00-00711	

No.	Name / Synonym	Mol. Form. / Mol. Wt.	CAS RN / mp/°C	Merck No. / bp/°C	Beil. Ref. / den/g cm^{-3}	Solubility / n_D
10262	Propanoic acid, 3-ethoxy-, ethyl ester	$C_7H_{14}O_3$ 146.19	763-69-9	48-9[5]	4-03-00-00697 0.92[25]	1.4065[20]
10263	Propanoic acid, ethyl ester Ethyl propionate	$C_5H_{10}O_2$ 102.13	105-37-3 -73.9	3801 99.1	4-02-00-00705 0.8917[20]	H_2O 2; EtOH 5; eth 5; ace 3 1.3839[20]
10264	Propanoic acid, heptyl ester	$C_{10}H_{20}O_2$ 172.27	2216-81-1 -50.9	210	4-02-00-00710 0.8679[20]	H_2O 1; EtOH 3; eth 3; ace 3 1.4201[15]
10265	Propanoic acid, hexyl ester	$C_9H_{18}O_2$ 158.24	2445-76-3 -57.5	190	4-02-00-00709 0.8698[20]	H_2O 1; EtOH 3; eth 3; ace 3 1.4162[15]
10266	Propanoic acid, 2-hydroxy-, (±)- Lactic acid (DL)	$C_3H_6O_3$ 90.08	598-82-3 18	5215 122[15]	2-03-00-00192 1.2060[21]	H_2O 4; eth 2; EtOH 4 1.4392[20]
10267	Propanoic acid, 3-hydroxy- Hydracrylic acid	$C_3H_6O_3$ 90.08	503-66-2	4681 dec	4-03-00-00689	H_2O 4; EtOH 3; eth 5 1.4489[20]
10268	Propanoic acid, 2-hydroxy-, butyl ester, (R)- Lactic acid, butyl ester (d)	$C_7H_{14}O_3$ 146.19	34451-18-8	77[10]	3-03-00-00450 0.9744[27]	eth 4; EtOH 4
10269	Propanoic acid, 2-hydroxy-, ethyl ester, (±)- Ethyl lactate	$C_5H_{10}O_3$ 118.13	2676-33-7 -26	3773 154.5	1.0328[20]	H_2O 4; eth 4; EtOH 4 1.4124[20]
10270	Propanoic acid, 3-hydroxy-, ethyl ester	$C_5H_{10}O_3$ 118.13	623-72-3	187.5	2-03-00-00213 1.059[20]	1.4271[23]
10271	Propanoic acid, 2-hydroxy-, hexadecyl ester Cetyl lactate	$C_{19}H_{38}O_3$ 314.51	35274-05-6 41	2022 219[10]; 170[1]	4-03-00-00660	1.4410[40]
10272	Propanoic acid, 3-hydroxy-2-(hydroxymethyl)-2-methyl- Dimethylolpropionic acid	$C_5H_{10}O_4$ 134.13	4767-03-7 190	3241	4-03-00-01066	
10273	Propanoic acid, 2-hydroxy-, methyl ester, (±)- Lactic acid, methyl ester (DL)	$C_4H_8O_3$ 104.11	2155-30-8	144.8	4-03-00-00640 1.0928[20]	H_2O 4; eth 4; EtOH 4 1.4141[20]
10274	Propanoic acid, 2-hydroxy-, 3-methylbutyl ester	$C_8H_{16}O_3$ 160.21	19329-89-6	202.4	3-03-00-00483 0.9589[25]	eth 4; EtOH 4 1.4240[25]
10275	Propanoic acid, 2-hydroxy-, 1-methylethyl ester Isopropyl lactate	$C_6H_{12}O_3$ 132.16	617-51-6	167	3-03-00-00479 0.9980[20]	H_2O 4; bz 4; eth 4; EtOH 4 1.4082[25]
10276	Propanoic acid, 3-hydroxy-, methyl ester	$C_4H_8O_3$ 104.11	6149-41-3	179	3-03-00-00525 1.105[16]	H_2O 5; EtOH 5; eth 5 1.43[23]
10277	Propanoic acid, 3-hydroxy-, 1-methylethyl ester Hydracrylic acid, isopropyl ester	$C_6H_{12}O_3$ 132.16	84098-45-3	128.5[82]; 95[12]	1-03-00-00113 1.058[25]	1.4303[23]
10278	Propanoic acid, 2-hydroxy-2-methyl-	$C_4H_8O_3$ 104.11	594-61-6 82.5	212	4-03-00-00782	H_2O 4; EtOH 4; eth 4; bz 2
10279	Propanoic acid, 2-hydroxy-2-methyl-, ethyl ester	$C_6H_{12}O_3$ 132.16	80-55-7	150	4-03-00-00783 0.987[20]	H_2O 5; EtOH 5; ctc 3 1.4080[20]
10280	Propanoic acid, 2-hydroxy-2-methyl-, methyl ester	$C_5H_{10}O_3$ 118.13	2110-78-3	137	4-03-00-00783	H_2O 4; EtOH 4 1.4056[20]
10281	Propanoic acid, 2-hydroxy-, 2-propenyl ester	$C_6H_{10}O_3$ 130.14	5349-55-3	56-60[8]	3-03-00-00486 1.0452[20]	py 3 1.4369[20]
10282	Propanoic acid, 2-iodo-, (±)-	$C_3H_5IO_2$ 199.98	18791-49-6 46	93-6[0.2]	3-02-00-00573 2.073[18]	H_2O 2; EtOH 4; eth 4
10283	Propanoic acid, 3-iodo-, ethyl ester	$C_5H_9IO_2$ 228.03	6414-69-3	202	4-02-00-00770 1.6226[30]	1.4961[25]
10284	Propanoic acid, 3-iodo-, methyl ester	$C_4H_7IO_2$ 214.00	5029-66-3	188	3-02-00-00574 1.8408[7]	EtOH 4
10285	Propanoic acid, 2-mercapto- Thiolactic acid	$C_3H_6O_2S$ 106.15	79-42-5 12	9269 106-7[15]	4-03-00-00682 1.1938[20]	H_2O 4; eth 4; EtOH 4 1.4810[20]
10286	Propanoic acid, 3-mercapto-	$C_3H_6O_2S$ 106.15	107-96-0 18	111[15]; 86[3]	4-03-00-00726 1.218[21]	H_2O 3; EtOH 3; eth 3; ctc 3 1.494[20]
10287	Propanoic acid, 3-mercapto-, methyl ester	$C_4H_8O_2S$ 120.17	2935-90-2	54-5[14]	3-03-00-00553 1.085[25]	1.4640[20]
10288	Propanoic acid, 2-mercapto-2-methyl-	$C_4H_8O_2S$ 120.17	4695-31-2 47	101[15]	4-03-00-00797	H_2O 4
10289	Propanoic acid, 3-methoxy-, methyl ester	$C_5H_{10}O_3$ 118.13	3852-09-3	142.8	4-03-00-00693 1.0139[15]	1.4030[20]
10290	Propanoic acid, 2-methyl- Isobutyric acid	$C_4H_8O_2$ 88.11	79-31-2 -46	5039 154.4	4-02-00-00843 0.9681[20]	H_2O 4; EtOH 5; eth 5; ctc 2 1.3930[20]
10291	Propanoic acid, 2-methyl-, anhydride Isobutyric anhydride	$C_8H_{14}O_3$ 158.20	97-72-3 -53.5	183; 89[32]	4-02-00-00851 0.9535[20]	eth 5; chl 3 1.4061[19]
10292	Propanoic acid, 2-methyl-, 2-chloroethyl ester 2-Chloroethyl isobutyrate	$C_6H_{11}ClO_2$ 150.60	33662-96-3	170	4-02-00-00847 1.062[25]	eth 4; EtOH 4 1.4109[16]
10293	Propanoic acid, 2-methyl-, cyclohexyl ester Isobutyric acid, cyclohexyl ester	$C_{10}H_{18}O_2$ 170.25	1129-47-1	204	3-06-00-00024 0.9489[0]	eth 4; EtOH 4
10294	Propanoic acid, 2-methyl-, 1,1-dimethylethyl ester	$C_8H_{16}O_2$ 144.21	16889-72-8	126.7	3-02-00-00648	ace 4; eth 4; EtOH 4 1.3921[20]
10295	Propanoic acid, 2-methyl-, 3,7-dimethyl-2,6-octadienyl ester, (E)-	$C_{14}H_{24}O_2$ 224.34	2345-26-8	135-7[13]	2-02-00-00261 0.8997[15]	1.4576[20]
10296	Propanoic acid, methyl ester Methyl propionate	$C_4H_8O_2$ 88.11	554-12-1 -87.5	6030 79.8	4-02-00-00704 0.9150[20]	H_2O 2; EtOH 5; eth 5; ace 3 1.3775[20]
10297	Propanoic acid, 2-methyl-, ethenyl ester	$C_6H_{10}O_2$ 114.14	2424-98-8	104.5	4-02-00-00848 0.8932[20]	1.4061[20]

No.	Name / Synonym	Mol. Form. / Mol. Wt.	CAS RN / mp/°C	Merck No. / bp/°C	Beil. Ref. / den/g cm^{-3}	Solubility / n_D
10298	Propanoic acid, 2-methyl-, ethyl ester / Ethyl isobutanoate	$C_6H_{12}O_2$ / 116.16	97-62-1 / -88.2	3770 / 110.1	4-02-00-00846 / 0.868[20]	H_2O 2; EtOH 5; eth 5; ace 3 / 1.3869[18]
10299	Propanoic acid, 1-methylethyl ester / Isopropyl propionate	$C_6H_{12}O_2$ / 116.16	637-78-5 /	/ 109.5	4-02-00-00708 / 0.8660[20]	H_2O 2; EtOH 5; eth 5 / 1.3872[20]
10300	Propanoic acid, 2-methyl-, 2-furanylmethyl ester	$C_9H_{12}O_3$ / 168.19	6270-55-9 /	/ 85-6[15]	2-17-00-00115 / 1.0313[20]	eth 4; EtOH 4
10301	Propanoic acid, 2-methyl-, methyl ester / Methyl isobutyrate	$C_5H_{10}O_2$ / 102.13	547-63-7 / -84.7	6003 / 92.5	4-02-00-00846 / 0.8906[20]	H_2O 2; EtOH 5; eth 5; ace 3 / 1.3840[20]
10302	Propanoic acid, 2-methyl-, 3-methylbutyl ester	$C_9H_{18}O_2$ / 158.24	2050-01-3 /	/ 168.5	3-02-00-00649 / 0.8627[20]	H_2O 2; EtOH 3; eth 3; ace 3
10303	Propanoic acid, 2-methyl-, 1-methylethyl ester / Isopropyl isobutyrate	$C_7H_{14}O_2$ / 130.19	617-50-5 /	/ 120.7	4-02-00-00847 / 0.8471[21]	H_2O 1; EtOH 3; eth 3; ace 3
10304	Propanoic acid, 2-methyl-, 2-methylpropyl ester / Isobutyl isobutanoate	$C_8H_{16}O_2$ / 144.21	97-85-8 / -80.7	5028 / 148.6	4-02-00-00847 / 0.8542[20]	H_2O 2; EtOH 3; eth 5; ace 3 / 1.3999[20]
10305	Propanoic acid, 2-methyl-, 2-nitropentyl ester / Isobutyric acid, (2-nitropentyl) ester	$C_9H_{17}NO_4$ / 203.24	116530-63-3 /	/ 249.5; 121[10]	3-02-00-00648 / 1.0310[20]	ace 4; eth 4 / 1.4315[20]
10306	Propanoic acid, 2-methyl-, 2-phenylethyl ester	$C_{12}H_{16}O_2$ / 192.26	103-48-0 /	/ 122-4[15]	4-06-00-03073 / 0.9950[15]	/ 1.4871[20]
10307	Propanoic acid, 2-methyl-, phenylmethyl ester	$C_{11}H_{14}O_2$ / 178.23	103-28-6 /	/ 114-5[20]	4-06-00-02266 / 1.0075[15]	/ 1.4883[20]
10308	Propanoic acid, 2-methyl-, propyl ester / Propyl isobutyrate	$C_7H_{14}O_2$ / 130.19	644-49-5 /	/ 135.4	4-02-00-00847 / 0.8843[0]	H_2O 2; EtOH 3; eth 4; ace 3 / 1.3955[20]
10309	Propanoic acid, 1-methylpropyl ester	$C_7H_{14}O_2$ / 130.19	591-34-4 /	/ 133	4-02-00-00709 / 0.8657[20]	EtOH 3; eth 3 / 1.3952[20]
10310	Propanoic acid, 2-methylpropyl ester / Isobutyl propionate	$C_7H_{14}O_2$ / 130.19	540-42-1 / -71.4	5033 / 137	4-02-00-00709 / 0.888[0]	H_2O 2; EtOH 4; eth 4; ace 3 / 1.3973[20]
10311	Propanoic acid, 3-(methylthio)-, methyl ester	$C_5H_{10}O_2S$ / 134.20	13532-18-8 /	/ 74-5[13]	4-03-00-00737 / 1.077[25]	/ 1.4650[20]
10312	Propanoic acid, 2-nitro-, ethyl ester	$C_5H_9NO_4$ / 147.13	2531-80-8 /	/ 190.5	4-02-00-00771 /	bz 4; eth 4; EtOH 4 / 1.4210[20]
10313	Propanoic acid, octyl ester	$C_{11}H_{22}O_2$ / 186.29	142-60-9 / -42.6	/ 228	4-02-00-00710 / 0.8663[20]	H_2O 1; EtOH 3; eth 3; bz 3 / 1.4221[15]
10314	Propanoic acid, 2-oxo- / Pyruvic acid	$C_3H_4O_3$ / 88.06	127-17-3 / 13.8	8032 / 165 dec; 54[10]	4-03-00-01505 / 1.2272[20]	H_2O 5; EtOH 5; eth 5; ace 3 / 1.4280[20]
10315	Propanoic acid, 2-oxo-, ethyl ester	$C_5H_8O_3$ / 116.12	617-35-6 / -50	/ 155	4-03-00-01513 / 1.0596[15]	H_2O 2; EtOH 5; eth 5; ace 3 / 1.4052[20]
10316	Propanoic acid, 2-oxo-, methyl ester	$C_4H_6O_3$ / 102.09	600-22-6 /	/ 135.5	4-01-00-04146 / 1.154[0]	H_2O 2; EtOH 5; eth 5; ace 3 / 1.4046[25]
10317	Propanoic acid, pentafluoro-, methyl ester / Pentafluoropropionic acid, methyl ester	$C_4H_3F_5O_2$ / 178.06	378-75-6 /	/ 59.5	4-02-00-00740 / 1.390[25]	/ 1.2869[25]
10318	Propanoic acid, pentyl ester	$C_8H_{16}O_2$ / 144.21	624-54-4 / -73.1	/ 168.6	4-02-00-00709 / 0.8761[25]	H_2O 1; EtOH 5; eth 5; bz 3 / 1.4096[15]
10319	Propanoic acid, 2-phenoxy-, (±)-	$C_9H_{10}O_3$ / 166.18	1912-21-6 / 115.5	/ 265.5	4-06-00-00642 / 1.1865[20]	eth 4; EtOH 4 / 1.5184[20]
10320	Propanoic acid, 3-phenoxy-	$C_9H_{10}O_3$ / 166.18	7170-38-9 / 97.5	/ 240; 188[26]	4-06-00-00643 /	H_2O 3; chl 2; lig 3
10321	Propanoic acid, 2-phenoxy-, ethyl ester	$C_{11}H_{14}O_3$ / 194.23	42412-84-0 /	/ 243.5	4-06-00-00643 / 1.360[17]	H_2O 1; EtOH 3; eth 3 / 1.360[17]
10322	Propanoic acid, 3-phenoxy-, ethyl ester	$C_{11}H_{14}O_3$ / 194.23	22409-91-2 / 24	/ 142[11]	4-06-00-00644 / 1.0745[20]	/ 1.5007[18]
10323	Propanoic acid, phenyl ester	$C_9H_{10}O_2$ / 150.18	637-27-4 / 20	/ 211	4-06-00-00615 / 1.0436[25]	H_2O 1; EtOH 4; eth 4; bz 3 / 1.4980[20]
10324	Propanoic acid, 2-phenylethyl ester / 2-Phenethyl propionate	$C_{11}H_{14}O_2$ / 178.23	122-70-3 /	/	3-06-00-01709 /	
10325	Propanoic acid, 2-propenyl ester	$C_6H_{10}O_2$ / 114.14	2408-20-0 /	/ 124	4-02-00-00711 / 0.9140[20]	EtOH 3; eth 3; ace 3 / 1.4105[20]
10326	Propanoic acid, propyl ester / Propyl propionate	$C_6H_{12}O_2$ / 116.16	106-36-5 / -75.9	7880 / 122.5	4-02-00-00707 / 0.8809[20]	H_2O 2; EtOH 5; eth 5; ace 3 / 1.3935[20]
10327	Propanoic acid, sodium salt	$C_3H_5NaO_2$ / 96.06	137-40-6 /	8623 /	4-02-00-00695 /	H_2O 2
10328	Propanoic acid, 2,3,3,3-tetrafluoro-, ethyl ester / Ethyl 2,3,3,3-tetrafluoropropionate	$C_5H_6F_4O_2$ / 174.10	399-92-8 /	/ 108.5	4-02-00-00737 / 1.289[20]	/ 1.3260[20]
10329	Propanoic acid, 3,3'-thiobis- / 3,3'-Thiodipropionic acid	$C_6H_{10}O_4S$ / 178.21	111-17-1 / 129	9261 /	4-03-00-00735 /	H_2O 4; EtOH 4
10330	Propanoic acid, 3,3'-thiobis-, diethyl ester	$C_{10}H_{18}O_4S$ / 234.32	673-79-0 /	/ 121[2]	4-03-00-00741 / 1.095[25]	/ 1.4655[20]
10331	Propanoic acid, 3,3'-thiobis-, dimethyl ester	$C_8H_{14}O_4S$ / 206.26	4131-74-2 /	/ 148[18]	4-03-00-00738 / 1.198[25]	/ 1.4740[20]
10332	Propanoic acid, 2,2,3-trichloro-	$C_3H_3Cl_3O_2$ / 177.41	3278-46-4 / 65.5	/ 140[40]	4-02-00-00759 /	H_2O 3; EtOH 3; bz 3; CS_2 3
10333	Propanoic acid, 3,3,3-trichloro-2-hydroxy- / Lactic acid, 3,3,3-trichloro-	$C_3H_3Cl_3O_3$ / 193.41	599-01-9 / 125	/ 140[45]	3-03-00-00501 /	H_2O 4; eth 4; EtOH 4; chl 4
10334	1-Propanol / Propyl alcohol	C_3H_8O / 60.10	71-23-8 / -126.1	7854 / 97.2	4-01-00-01413 / 0.7997[25]	H_2O 5; EtOH 5; eth 5; ace 3 / 1.3850[20]
10335	2-Propanol / Isopropyl alcohol	C_3H_8O / 60.10	67-63-0 / -89.5	5096 / 82.3	4-01-00-01461 / 0.7809[25]	H_2O 5; EtOH 5; eth 5; ace 3 / 1.3776[20]

No.	Name Synonym	Mol. Form. Mol. Wt.	CAS RN	mp/°C bp/°C	Merck No.	Beil. Ref. den/g cm^{-3}	Solubility n_D
10336	1-Propanol, 2-amino-, (±)-	C_3H_9NO 75.11	6168-72-5	 174.5		3-04-00-00737	H_2O 4; EtOH 4; eth 4; chl 2 1.4502[20]
10337	1-Propanol, 3-amino- Propanolamine	C_3H_9NO 75.11	156-87-6	11 187.5		4-04-00-01623 0.9824[26]	H_2O 3; EtOH 3; eth 3 1.4617[20]
10338	2-Propanol, 1-amino-, (±)- DL-1-Amino-2-propanol	C_3H_9NO 75.11	1674-56-2	1.7 159.4		4-04-00-01665 0.9611[20]	H_2O 5; EtOH 5; eth 5; ace 5 1.4479[20]
10339	2-Propanol, 1-amino-3-(diethylamino)- 3-(Diethylamino)-2-hydroxypropylamine	$C_7H_{18}N_2O$ 146.23	6322-01-6	 223		3-04-00-00767 0.931[20]	EtOH 3; eth 3; chl 2 1.465[20]
10340	2-Propanol, 1-[(2-aminoethyl)amino]- N-(2-Hydroxypropyl)ethylenediamine	$C_5H_{14}N_2O$ 118.18	123-84-2	450 94[3]		4-04-00-01684 0.9837[25]	 1.4738[20]
10341	1-Propanol, 2-amino-2-methyl- 2-Aminoisobutanol	$C_4H_{11}NO$ 89.14	124-68-5	25.5 165.5	461	4-04-00-01740 0.934[20]	H_2O 5; ctc 3 1.449[20]
10342	2-Propanol, 2-[(2-amino-2-methylpropyl)amino]-2-methyl- 1-Propanol, 2-[2-(2-amino-2-methylpropyl)amino]-2-methyl-	$C_8H_{20}N_2O$ 160.26	72622-74-3	46 231		3-04-00-00788 0.9360[20]	chl 2 1.4635[25]
10343	2-Propanol, 1,3-bis(dimethylamino)-	$C_7H_{18}N_2O$ 146.23	5966-51-8	 181.5		4-04-00-01695 0.8788[20]	H_2O 4 1.4418[20]
10344	1-Propanol, 3-bromo-	C_3H_7BrO 138.99	627-18-9	 105[185]; 62[5]		4-01-00-01446 1.5374[20]	H_2O 3; EtOH 5; eth 5 1.4834[25]
10345	2-Propanol, 1-bromo-	C_3H_7BrO 138.99	19686-73-8	 146.5		3-01-00-01474 1.5585[30]	H_2O 3; EtOH 4; eth 4 1.4801[20]
10346	2-Propanol, 1-butoxy-	$C_7H_{16}O_2$ 132.20	5131-66-8	 171.5; 71[20]		4-01-00-02471 0.882[20]	EtOH 3; eth 3; bz 3; ctc 3 1.4168[20]
10347	1-Propanol, 2-chloro- Propylene chlorohydrin	C_3H_7ClO 94.54	78-89-7	7863 133.5		4-01-00-01440 1.103[20]	H_2O 4; eth 4; EtOH 4 1.4390[20]
10348	1-Propanol, 3-chloro- 3-Chloro-1-propanol	C_3H_7ClO 94.54	627-30-5	 165		4-01-00-01441 1.1309[20]	H_2O 4; EtOH 3; eth 3; ctc 2 1.4459[20]
10349	2-Propanol, 1-chloro- sec-Propylene chlorohydrin	C_3H_7ClO 94.54	127-00-4	7864 127		4-01-00-01490 1.113[20]	H_2O 5; EtOH 5; eth 5; ctc 2 1.4392[20]
10350	1-Propanol, 3-chloro-, acetate	$C_5H_9ClO_2$ 136.58	628-09-1	 164; 62[10]		4-02-00-00140 1.250[19]	H_2O 1; EtOH 3; eth 3; ace 3 1.431[20]
10351	2-Propanol, 1-chloro-, acetate	$C_5H_9ClO_2$ 136.58	623-60-9	 149.5		4-02-00-00142 1.0788[20]	eth 4; EtOH 4 1.4223[20]
10352	1-Propanol, 2-chloro-2-methyl- Isobutyl alcohol, 2-chloro	C_4H_9ClO 108.57	558-38-3	 132 dec; 60[50]		4-01-00-01603 1.0472[20]	 1.4388[20]
10353	1-Propanol, 3-chloro-2-methyl- Isobutyl alcohol, 3-chloro	C_4H_9ClO 108.57	10317-10-9	 76-8[11]		3-01-00-01564 1.083[25]	eth 4; EtOH 4 1.4460[25]
10354	2-Propanol, 1-chloro-2-methyl- Isobutylene chlorohydrin	C_4H_9ClO 108.57	558-42-9	-20 128.5		4-01-00-01628 1.0628[20]	H_2O 4; EtOH 4 1.4380[24]
10355	2-Propanol, 1-chloro-3-(1-methylethoxy)- 2-Propanol, 1-chloro-3-isopropoxy-	$C_6H_{13}ClO_2$ 152.62	4288-84-0	 87[20]		4-01-00-02485 1.050[25]	EtOH 3; eth 3 1.4370[25]
10356	2-Propanol, 1-chloro-3-propoxy-	$C_6H_{13}ClO_2$ 152.62	6943-58-4	 92-5[15]		3-01-00-02153 1.0526[25]	ace 4; eth 4 1.4378[25]
10357	2-Propanol, 1,3-dibromo-	$C_3H_6Br_2O$ 217.89	96-21-9	 219 dec; 105[18]		4-01-00-01496 2.1201[25]	ace 4; eth 4; EtOH 4 1.5495[25]
10358	1-Propanol, 2,3-dibromo-, phosphate (3:1) Tris(2,3-dibromopropyl) phosphate	$C_9H_{15}Br_6O_4P$ 697.61	126-72-7	9665			chl 3
10359	1-Propanol, 2,3-dichloro-	$C_3H_6Cl_2O$ 128.99	616-23-9	 184		4-01-00-01442 1.3607[20]	H_2O 2; EtOH 5; eth 5; ace 5 1.4819[20]
10360	1-Propanol, 3,3-dichloro-	$C_3H_6Cl_2O$ 128.99	83682-72-8	 82-3[20]		3-01-00-01427 1.316[25]	 1.467[20]
10361	2-Propanol, 1,1-dichloro-	$C_3H_6Cl_2O$ 128.99	53894-19-2	 147		2-01-00-00383 1.3334[22]	ace 4; eth 4; EtOH 4
10362	2-Propanol, 1,3-dichloro- 1,3-Dichloro-2-propanol	$C_3H_6Cl_2O$ 128.99	96-23-1	3063 176		4-01-00-01491 1.3506[17]	H_2O 4; EtOH 4; eth 5; ace 3 1.4837[20]
10363	2-Propanol, 1,3-dichloro-, acetate	$C_5H_8Cl_2O_2$ 171.02	3674-10-0	 205		3-02-00-00233 1.281[20]	eth 4; EtOH 4; chl 4 1.4542[20]
10364	2-Propanol, 1,1-dichloro-2-methyl- 2,2-Dichloro-1,1-dimethylethanol	$C_4H_8Cl_2O$ 143.01	4773-53-9	8 152.5		4-01-00-01629 1.2363[19]	bz 4; EtOH 4 1.4598[19]
10365	2-Propanol, 1,3-dichloro-2-methyl- β-Methylglycerol α,γ-dichlorohydrin	$C_4H_8Cl_2O$ 143.01	597-32-0	 174.5		4-01-00-01629 1.2745[20]	H_2O 3; EtOH 4; ace 3 1.4744[21]
10366	1-Propanol, 2-(diethylamino)- 2-Diethylamino-1-propanol	$C_7H_{17}NO$ 131.22	611-12-1	 168; 78[12]		4-04-00-01618 0.8649[27]	EtOH 3; eth 3; ace 3; bz 3 1.4332[20]
10367	1-Propanol, 3-(diethylamino)-	$C_7H_{17}NO$ 131.22	622-93-5	 189.5		4-04-00-01633 0.8600[20]	EtOH 3; eth 3; ace 3; bz 3 1.4439[20]
10368	2-Propanol, 1-(diethylamino)-, (±)- 2-Propanol, 1-diethylamino (DL)	$C_7H_{17}NO$ 131.22	78738-36-0	13.5 159; 63[22]		3-04-00-00759 0.8511[20]	H_2O 4; ace 4; EtOH 4 1.4255[20]
10369	2-Propanol, 1,3-difluoro- 1,3-Difluoro-2-propanol	$C_3H_6F_2O$ 96.08	453-13-4	 54-5[34]		4-01-00-01489 1.24[25]	 1.3725[20]
10370	1-Propanol, 2,3-dimercapto- Dimercaprol	$C_3H_8OS_2$ 124.23	59-52-9	3196 82-4[0.8]		4-01-00-02770 1.2463[20]	EtOH 3; eth 3; chl 2; oils 3 1.5749[20]
10371	2-Propanol, 1,3-dimercapto-	$C_3H_8OS_2$ 124.23	584-04-3	 94[12]		4-01-00-02773 1.2386[20]	 1.5700[20]
10372	2-Propanol, 1,3-dimethoxy- Glycerol 1,3-dimethyl ether	$C_5H_{12}O_3$ 120.15	623-69-8	 169		3-01-00-02318 1.0085[20]	H_2O 5; EtOH 4; eth 3 1.4192[20]
10373	1-Propanol, 2,2-dimethyl- Neopentyl alcohol	$C_5H_{12}O$ 88.15	75-84-3	6373 113.5		4-01-00-01690 0.812[20]	H_2O 2; EtOH 4; eth 4; ctc 3

No.	Name Synonym	Mol. Form. Mol. Wt.	CAS RN mp/°C	Merck No. bp/°C	Beil. Ref. den/g cm⁻³	Solubility n_D
10374	1-Propanol, 2-(dimethylamino)- 2-Dimethylamino-1-propanol	$C_5H_{13}NO$ 103.16	15521-18-3	150.3	4-04-00-01617 0.8820^{26}	H_2O 3
10375	1-Propanol, 3-(dimethylamino)-	$C_5H_{13}NO$ 103.16	3179-63-3	163.5	4-04-00-01628 0.872^{25}	ctc 3 1.4360^{20}
10376	2-Propanol, 1-(dimethylamino)-	$C_5H_{13}NO$ 103.16	108-16-7	124.5	3-04-00-00756 0.837^{25}	ctc 3 1.4193^{20}
10377	2-Propanol, 1,3-diphenoxy-	$C_{15}H_{16}O_3$ 244.29	622-04-8 81.5	224^{17}; 175^2	4-06-00-00590 1.179^{24}	bz 4; eth 4; EtOH 4; chl 4
10378	2-Propanol, 1,3-diphenoxy-, acetate Glycerol, 1,3-diphenyl ether-2-acetate	$C_{17}H_{18}O_4$ 286.33	71159-31-4 33	190^{160}	3-06-00-00583	bz 4; eth 4; EtOH 4; chl 4
10379	2-Propanol, 1,1',1'',1'''-(1,2-ethanediyldinitrilo)tetrakis- ENTPROL	$C_{14}H_{32}N_2O_4$ 292.42	102-60-3	3551	4-04-00-01685 1.030^{25}	chl 2 1.478^{25}
10380	1-Propanol, 2-ethoxy- Propylene glycol β-monoethyl ether	$C_5H_{12}O_2$ 104.15	19089-47-5	140.5	3-01-00-02147 0.9044^{20}	H_2O 3; EtOH 3; eth 3 1.4122^{20}
10381	2-Propanol, 1-ethoxy- 1-Ethoxy-2-propanol	$C_5H_{12}O_2$ 104.15	1569-02-4	131	4-01-00-02471 0.9028^{20}	H_2O 3; EtOH 3; eth 3 1.4075^{20}
10382	1-Propanol, 3-fluoro- 3-Hydroxypropyl fluoride	C_3H_7FO 78.09	462-43-1	127.8	4-01-00-01437 1.0390^{25}	H_2O 4; eth 4; EtOH 4 1.3771^{25}
10383	2-Propanol, 1-[(2-hydroxyethyl)amino]-	$C_5H_{13}NO_2$ 119.16	6579-55-1	$120-3^1$	4-04-00-01678 1.0455^{20}	H_2O 3; ace 3 1.4695^{20}
10384	2-Propanol, 1,1'-iminobis- Diisopropanolamine	$C_6H_{15}NO_2$ 133.19	110-97-4 44.5	250; 151^{23}	3-04-00-00761 0.989^{20}	H_2O 3; EtOH 3; eth 2
10385	1-Propanol, 3-iodo-	C_3H_7IO 185.99	627-32-7	115^{38}; 88^4	4-01-00-01448 1.9976^{20}	1.5585^{20}
10386	1-Propanol, 2-methoxy- 2-Methoxy-1-propanol	$C_4H_{10}O_2$ 90.12	1589-47-5	130	4-01-00-02471 0.938^{20}	1.4070^{20}
10387	2-Propanol, 1-methoxy- 1-Methoxy-2-propanol	$C_4H_{10}O_2$ 90.12	107-98-2	119	3-01-00-02146 0.9620^{20}	1.4034^{20}
10388	1-Propanol, 2-methyl- Isobutyl alcohol	$C_4H_{10}O$ 74.12	78-83-1 -108	5015 107.8	4-01-00-01588 0.8018^{20}	H_2O 2; EtOH 3; eth 3; ace 3 1.3955^{20}
10389	2-Propanol, 2-methyl- tert-Butyl alcohol	$C_4H_{10}O$ 74.12	75-65-0 25.4	1542 82.4	4-01-00-01609 0.7887^{20}	H_2O 5; EtOH 5; eth 5; chl 3 1.3878^{20}
10390	2-Propanol, 1-(1-methylethoxy)- 1-Isopropoxy-2-propanol	$C_6H_{14}O_2$ 118.18	3944-36-3	137.5	4-01-00-02471 0.879^{20}	1.4070^{20}
10391	2-Propanol, 2-methyl-2-nitro-	$C_4H_9NO_3$ 119.12	76-39-1 89.5	94^{10}	4-01-00-01604	H_2O 2; EtOH 4; eth 4; chl 3
10392	1-Propanol, 3-[methyl(phenylmethyl)amino]-	$C_{11}H_{17}NO$ 179.26	5814-42-6	$132-5^4$	4-12-00-02208 1.001^{25}	ctc 2; CS_2 3 1.5230^{20}
10393	2-Propanol, 1,1',1''-nitrilotris-	$C_9H_{21}NO_3$ 191.27	122-20-3 45	$170-80^{10}$	4-04-00-01680 1.0^{20}	H_2O 3; EtOH 3; chl 2
10394	1-Propanol, 2-nitro-	$C_3H_7NO_3$ 105.09	2902-96-7	100^{12}	4-01-00-01448 1.1841^{25}	H_2O 3; EtOH 3; eth 3; chl 2 1.4379^{20}
10395	1-Propanol, 3-nitro-	$C_3H_7NO_3$ 105.09	25182-84-7	$138-40^{32}$	3-01-00-01429 1.173^{13}	H_2O 4; eth 4; EtOH 4
10396	2-Propanol, 1-nitro- 1-Nitro-2-propanol	$C_3H_7NO_3$ 105.09	3156-73-8	112^{13}; 68^1	4-01-00-01496 1.1906^{20}	H_2O 3; EtOH 3; eth 3; bz 3 1.4383^{20}
10397	Propanol, oxybis- Dipropylene glycol	$C_6H_{14}O_3$ 134.18	25265-71-8	230.5	4-01-00-02473 1.0206^{20}	H_2O 5; EtOH 3
10398	1-Propanol, 2-phenoxy-	$C_9H_{12}O_2$ 152.19	4169-04-4	244	4-06-00-00582 0.9801^{25}	EtOH 3; eth 3 1.4760^{25}
10399	2-Propanol, 1-phenoxy-	$C_9H_{12}O_2$ 152.19	770-35-4	134.5^{20}	4-06-00-00582 1.0622^{20}	1.5232^{20}
10400	1-Propanol, 3-(phenylmethoxy)-	$C_{10}H_{14}O_2$ 166.22	4799-68-2	172^{43}	4-06-00-02243 1.0474^{20}	EtOH 3; eth 3
10401	2-Propanol, 1-propoxy- 1,2-Propylene glycol 1-propyl ether	$C_6H_{14}O_2$ 118.18	1569-01-3	150	3-01-00-02147 0.8886^{20}	1.4130^{20}
10402	2-Propanol, 1,1,3,3-tetrachloro- 1,1,3,3-Tetrachloro-2-propanol	$C_3H_4Cl_4O$ 197.87	18992-39-7	$80-90^{14}$	3-01-00-01474 1.612^{20}	eth 3 1.5133^{20}
10403	2-Propanol, 1,1,1,3-tetrachloro-	$C_3H_4Cl_4O$ 197.87	88947-16-4	$95-6^{17}$	3-01-00-01474 1.610^{20}	eth 4 1.5145^{20}
10404	1-Propanol, 2,2,3,3-tetrafluoro- 2,2,3,3-Tetrafluoro-1-propanol	$C_3H_4F_4O$ 132.06	76-37-9 -15	9139 109.5	4-01-00-01438 1.4853^{20}	EtOH 3; ace 3; chl 3 1.3197^{20}
10405	2-Propanol, 1,1,1-tribromo-2-methyl- Tribromo-tert-butyl alcohol	$C_4H_7Br_3O$ 310.81	76-08-4 169	9523 sub	3-01-00-01588	H_2O 2; EtOH 3; eth 3; chl 2
10406	2-Propanol, 1,1,1-trichloro- 1,1,1-Trichloro-2-propanol	$C_3H_5Cl_3O$ 163.43	76-00-6 50.5	9557 163; $53-5^{12}$	3-01-00-01474	ace 4; bz 4; eth 4; EtOH 4
10407	2-Propanol, 1,1,1-trichloro-2-methyl- Chlorobutanol	$C_4H_7Cl_3O$ 177.46	57-15-8 97	2129 167	4-01-00-01629	H_2O 1; EtOH 3; eth 3; ace 3
10408	2-Propanol, 1,1,1-trichloro-3-nitro- 2-Propanol, 1-nitro-3,3,3-trichloro	$C_3H_4Cl_3NO_3$ 208.43	760-40-7 45.5	$105^{3.5}$	4-01-00-01497	eth 4; EtOH 4
10409	2-Propanol, 1,1,1-trifluoro-, (±)- 2-Propanol, 1,1,1-trifluoro(DL)	$C_3H_5F_3O$ 114.07	17556-48-8 -52	78	4-01-00-01489 1.263^{25}	EtOH 4; eth 4; ace 3; bz 3 1.3130^{25}
10410	1-Propanol, 3-(trimethylsilyl)-	$C_6H_{16}OSi$ 132.28	2917-47-7	141; 82^{24}	4-04-00-03953 0.822^{25}	1.4298^{20}
10411	2-Propanone Acetone	C_3H_6O 58.08	67-64-1 -94.8	58 56.05	4-01-00-03180 0.7845^{25}	H_2O 5; EtOH 5; eth 5; ace 5 1.3588^{20}
10412	2-Propanone, 1-(acetyloxy)-	$C_5H_8O_3$ 116.12	592-20-1	171; 63^{11}	4-02-00-00297 1.0757^{20}	H_2O 4; eth 4; EtOH 4 1.4141^{20}
10413	1-Propanone, 1-(4-aminophenyl)- p-Aminopropiophenone	$C_9H_{11}NO$ 149.19	70-69-9 140	482	4-14-00-00139	DMSO 3

No.	Name Synonym	Mol. Form. Mol. Wt.	CAS RN mp/°C	Merck No. bp/°C	Beil. Ref. den/g cm^{-3}	Solubility n_D
10414	1-Propanone, 2-amino-1-phenyl-	$C_9H_{11}NO$ 149.19	5265-18-9 46.5	93[0.8]	3-14-00-00147	ace 4; eth 4; EtOH 4
10415	2-Propanone, 1,3-bis(1-methyl-2-pyrrolidinyl)-, (R*,S*)- Cuscohygrine	$C_{13}H_{24}N_2O$ 224.35	454-14-8	2685 169[23]; 122[2]	5-24-01-00458 0.9733[20]	H_2O 4; bz 4; eth 4; EtOH 4 1.4832[20]
10416	2-Propanone, 1-bromo- Bromoacetone	C_3H_5BrO 136.98	598-31-2 -36.5	1389 138; 31.5[8]	4-01-00-03223 1.634[23]	H_2O 2; EtOH 3; eth 3; ace 3 1.4697[15]
10417	2-Propanone, 1-bromo-3-chloro- 1-Bromo-3-chloroacetone	C_3H_4BrClO 171.42	53535-68-5 34.5	179	4-01-00-03224	eth 4; EtOH 4
10418	1-Propanone, 1-(4-bromophenyl)-	C_9H_9BrO 213.07	10342-83-3 48	169[15]; 137[2]	4-07-00-00684	H_2O 1; EtOH 3; eth 3; ace 3
10419	1-Propanone, 2-bromo-1-phenyl-	C_9H_9BrO 213.07	2114-00-3	247.5	4-07-00-00684 1.4298[20]	H_2O 1; EtOH 3; eth 3; ace 3 1.5720[20]
10420	2-Propanone, 1-chloro- Chloroacetone	C_3H_5ClO 92.52	78-95-5 -44.5	2113 119	4-01-00-03215 1.15[20]	H_2O 3; EtOH 3; eth 3; chl 3
10421	2-Propanone, 1-chloro-1,1,3,3,3-pentafluoro- Chloropentafluoroacetone	C_3ClF_5O 182.48	79-53-8 -133	8		
10422	1-Propanone, 1-(4-chlorophenyl)-	C_9H_9ClO 168.62	6285-05-8 37.3	135[31]; 114[2]	4-07-00-00683	H_2O 1; EtOH 3; chl 2; CS_2 3
10423	1-Propanone, 1-cyclohexyl- Cyclohexyl ethyl ketone	$C_9H_{16}O$ 140.23	1123-86-0	196	4-07-00-00064 0.9105[20]	1.4530[20]
10424	2-Propanone, 1,3-dibromo-	$C_3H_4Br_2O$ 215.87	816-39-7 26	97[22]	4-01-00-03225 2.167[18]	eth 4; CS_2 4
10425	2-Propanone, 1,1-dichloro- 1,1-Dichloroacetone	$C_3H_4Cl_2O$ 126.97	513-88-2	3038 120	4-01-00-03218 1.304[18]	H_2O 2; EtOH 3; eth 5
10426	2-Propanone, 1,3-dichloro- 1,3-Dichloroacetone	$C_3H_4Cl_2O$ 126.97	534-07-6 45	3039 173.4	4-01-00-03219 1.3826[46]	H_2O 3; EtOH 3; eth 3 1.4716[40]
10427	2-Propanone, 1-(diethylamino)-	$C_7H_{15}NO$ 129.20	1620-14-0	157 dec	4-04-00-01925 0.8620[20]	H_2O 5; EtOH 5; eth 5; ctc 3 1.4249[20]
10428	1-Propanone, 2-(diethylamino)-1-phenyl- Diethylpropion	$C_{13}H_{19}NO$ 205.30	90-84-6	3116	4-14-00-00144	
10429	2-Propanone, 1,3-dihydroxy- Dihydroxyacetone	$C_3H_6O_3$ 90.08	96-26-4 90	3166	4-01-00-04119	H_2O 3; EtOH 3; eth 3; ace 3
10430	1-Propanone, 1-(2,4-dihydroxyphenyl)-	$C_9H_{10}O_3$ 166.18	5792-36-9 99	176[6]	4-08-00-01837	bz 4; eth 4; EtOH 4
10431	1-Propanone, 2,2-dimethyl-1-phenyl-	$C_{11}H_{14}O$ 162.23	938-16-9	220	4-07-00-00740 0.963[26]	ace 3 1.5086[19]
10432	2-Propanone, 1,1-diphenyl-	$C_{15}H_{14}O$ 210.28	781-35-1 46	307; 174[10]	4-07-00-01426	EtOH 3; eth 3; bz 3; chl 3 1.5361[16]
10433	2-Propanone, 1,3-diphenyl-	$C_{15}H_{14}O$ 210.28	102-04-5 35	331	4-07-00-01420 1.195[0]	H_2O 1; EtOH 3; eth 3; peth 3
10434	1-Propanone, 3-(3-ethenyl-4-piperidinyl)-1-(6-methoxy-4-quinolinyl)-, (3R-cis)- Viquidil	$C_{20}H_{24}N_2O_2$ 324.42	84-55-9 60	9908	4-25-00-00181	eth 4; EtOH 4; chl 4
10435	1-Propanone, 3-(3-ethenyl-4-piperidinyl)-1-(4-quinolinyl)-, (3R-cis)- Cinchotoxine	$C_{19}H_{22}N_2O$ 294.40	69-24-9 59	2291	4-24-00-00651	H_2O 1; EtOH 4; eth 4; ace 4
10436	1-Propanone, 3-ethoxy-1-phenyl-	$C_{11}H_{14}O_2$ 178.23	34008-71-4 11.5	135[18]	4-08-00-00453 1.0386[20]	1.5190[20]
10437	2-Propanone, 1-fluoro-	C_3H_5FO 76.07	430-51-3	77	4-01-00-03213 1.0288[20]	1.3700[20]
10438	1-Propanone, 1-(2-furanyl)-	$C_7H_8O_2$ 124.14	3194-15-8 28	88[14]	5-17-09-00415 1.0626[28]	eth 3; ctc 2 1.4922[25]
10439	2-Propanone, 1-(2-furanyl)-	$C_7H_8O_2$ 124.14	6975-60-6 29	179.5	5-17-09-00418 1.104[20]	1.5035[20]
10440	2-Propanone, 1,1,1,3,3,3-hexachloro- Hexachloroacetone	C_3Cl_6O 264.75	116-16-5 -2	203	4-01-00-03223 1.743[12]	H_2O 2; ace 3 1.5112[20]
10441	2-Propanone, 1,1,1,3,3,3-hexafluoro- Perfluoroacetone	C_3F_6O 166.02	684-16-2 -125	-27.4	4-01-00-03215	
10442	2-Propanone, 1-(1H-indol-3-yl)- 3-Indolylacetone	$C_{11}H_{11}NO$ 173.21	1201-26-9 116	4873	5-21-08-00327	
10443	2-Propanone, 1-hydroxy-	$C_3H_6O_2$ 74.08	116-09-6 -17	145.5	4-01-00-03977 1.0805[20]	H_2O 4; EtOH 4; eth 4 1.4295[20]
10444	1-Propanone, 1-(2-hydroxy-3-methylphenyl)- 2-Methyl-6-propionylphenol	$C_{10}H_{12}O_2$ 164.20	3338-15-6 22.5	124[13]	4-08-00-00517 1.0900[20]	ctc 3
10445	1-Propanone, 1-(2-hydroxy-4-methylphenyl)- 2'-Hydroxy-4'-methylpropiophenone	$C_{10}H_{12}O_2$ 164.20	2886-52-4 44	115[10]	4-08-00-00519	chl 3
10446	1-Propanone, 1-(2-hydroxy-5-methylphenyl)-	$C_{10}H_{12}O_2$ 164.20	938-45-4 2	129-30[16.5]	4-08-00-00517 1.084[14]	chl 3 1.549[13]
10447	1-Propanone, 1-(2-hydroxy-5-methylphenyl)-2-methyl-	$C_{11}H_{14}O_2$ 178.23	64207-03-0	251; 125[11]	4-08-00-00545 1.0460[16]	1.538[16]
10448	1-Propanone, 1-(4-hydroxy-3-methylphenyl)-2-methyl- o-Cresol, 4-isobutyryl	$C_{11}H_{14}O_2$ 178.23	73206-57-2 122	182[12]	2-08-00-00130	eth 4; EtOH 4
10449	1-Propanone, 1-(2-hydroxyphenyl)-	$C_9H_{10}O_2$ 150.18	610-99-1	150[80]; 115[15]	4-08-00-00436	H_2O 2; EtOH 3; eth 3; ctc 3 1.5501[20]
10450	1-Propanone, 1-(4-hydroxyphenyl)- Paroxypropione	$C_9H_{10}O_2$ 150.18	70-70-2 149	6992	4-08-00-00441	H_2O 2; EtOH 3; eth 3; ace 2

No.	Name Synonym	Mol. Form. Mol. Wt.	CAS RN mp/°C	Merck No. bp/°C	Beil. Ref. den/g cm^{-3}	Solubility n_D
10451	1-Propanone, 2-hydroxy-1-phenyl-	C$_9$H$_{10}$O$_2$ 150.18	5650-40-8	251	4-08-00-00450 1.1085[18]	1.536[23]
10452	1-Propanone, 3-hydroxy-1-phenyl-	C$_9$H$_{10}$O$_2$ 150.18	5650-41-9	98[0.2]	4-08-00-00453 1.1170[20]	1.5450[20]
10453	1-Propanone, 3-(4-hydroxyphenyl)-1-(2,4,6-trihydroxyphenyl)- Phloretin	C$_{15}$H$_{14}$O$_5$ 274.27	60-82-2 263-4 dec	7299	4-08-00-03518	H$_2$O 2; EtOH 5; eth 1; ace 3
10454	2-Propanone, 1-iodo-	C$_3$H$_5$IO 183.98	3019-04-3	62[12]	4-01-00-03226 2.17[15]	EtOH 3
10455	2-Propanone, 1-methoxy-	C$_4$H$_8$O$_2$ 88.11	5878-19-3	116	4-01-00-03977 0.957[25]	1.3970[20]
10456	1-Propanone, 1-(3-methoxyphenyl)- 3-Methoxypropiophenone	C$_{10}$H$_{12}$O$_2$ 164.20	37951-49-8	259	4-08-00-00441 1.0812[25]	eth 4; EtOH 4 1.5230[25]
10457	1-Propanone, 2-methoxy-1-phenyl- Propiophenone, 2-methoxy-	C$_{10}$H$_{12}$O$_2$ 164.20	6493-83-0	79-80[1]	4-08-00-00451 1.0704[20]	1.5225[20]
10458	1-Propanone, 3-methoxy-1-phenyl-	C$_{10}$H$_{12}$O$_2$ 164.20	55563-72-9	125-6[16]	4-08-00-00453 1.0602[20]	1.5250[20]
10459	2-Propanone, 1-(methoxyphenyl)- 2-Propanone, (methoxyphenyl)-	C$_{10}$H$_{12}$O$_2$ 164.20	31116-84-4 <-15	267-9 142[14]	2-08-00-00104 1.0670[18]	H$_2$O 2; EtOH 3; eth 3 1.5253[20]
10460	2-Propanone, 1-(3-methoxyphenyl)-	C$_{10}$H$_{12}$O$_2$ 164.20	3027-13-2	259	4-08-00-00459 1.0812[6]	EtOH 3; eth 3 1.5230[25]
10461	2-Propanone, 1-(4-methoxyphenyl)-	C$_{10}$H$_{12}$O$_2$ 164.20	122-84-9 <-15	268	4-08-00-00460 1.0694[17]	eth 4; EtOH 4 1.5253[20]
10462	1-Propanone, 2-methyl-1,2-di-3-pyridinyl- Metyrapone	C$_{14}$H$_{14}$N$_2$O 226.28	54-36-4 50.5	6083	5-24-04-00048	
10463	2-Propanone, (1-methylethylidene)hydrazone	C$_6$H$_{12}$N$_2$ 112.17	627-70-3 -12.5	133	4-01-00-03207 0.8390[20]	H$_2$O 5; EtOH 5; eth 5; ace 3 1.4535[20]
10464	1-Propanone, 2-methyl-1-(1-naphthalenyl)- Isopropyl 1-naphthyl ketone	C$_{14}$H$_{14}$O 198.26	61838-78-6	307	3-07-00-01985 1.0761[25]	bz 4; eth 4; EtOH 4 1.5948[20]
10465	1-Propanone, 2-methyl-1-(2-naphthalenyl)- 2-Isobutyronaphthone	C$_{14}$H$_{14}$O 198.26	107574-57-2	313	3-07-00-01985 1.0617[25]	bz 4; eth 4; EtOH 4
10466	1-Propanone, 1-(2-methylphenyl)-	C$_{10}$H$_{12}$O 148.20	2040-14-4 -27.6	219.5	4-07-00-00720 1.0119[0]	
10467	1-Propanone, 1-(3-methylphenyl)-	C$_{10}$H$_{12}$O 148.20	51772-30-6	235; 132[31]	4-07-00-00721 1.0059[0]	H$_2$O 1; EtOH 3; eth 3; ace 3
10468	1-Propanone, 1-(4-methylphenyl)-	C$_{10}$H$_{12}$O 148.20	5337-93-9 7.2	236	4-07-00-00722 0.9926[20]	H$_2$O 1; EtOH 3; eth 3; ace 3 1.5278[20]
10469	2-Propanone, 2-methyl-1-phenyl-	C$_{10}$H$_{12}$O 148.20	611-70-1 -1.3	220	4-07-00-00716 0.9863[11]	eth 4; EtOH 4 1.5172[20]
10470	2-Propanone, 1-(1-methyl-2-piperidinyl)-, (±)-	C$_9$H$_{17}$NO 155.24	18747-42-7	96-8[13]	5-21-06-00523 0.9478[20]	H$_2$O 4; lig 4 1.4674[20]
10471	2-Propanone, 1-(1-methyl-2-pyrrolidinyl)-, (R)- Hygrine	C$_8$H$_{15}$NO 141.21	496-49-1	4788 76.5[11]	5-21-06-00511	EtOH 4; chl 4 1.4555[20]
10472	1-Propanone, 2-methyl-1-(2-thienyl)- Isopropyl 2-thienyl ketone	C$_8$H$_{10}$OS 154.23	36448-60-9	229	5-17-09-00439 1.0894[20]	1.5405[20]
10473	1-Propanone, 1-(1-naphthalenyl)-	C$_{13}$H$_{12}$O 184.24	2876-63-3	306	4-07-00-01298 1.0971[20]	1.6108[20]
10474	1-Propanone, 1-(2-naphthalenyl)-	C$_{13}$H$_{12}$O 184.24	6315-96-4 60	313	4-07-00-01299	EtOH 4; chl 4
10475	2-Propanone, 1-nitro-	C$_3$H$_5$NO$_3$ 103.08	10230-68-9 50.3	103[24]	4-01-00-03226	bz 4; eth 4; EtOH 4
10476	2-Propanone, oxime Acetoxime	C$_3$H$_7$NO 73.09	127-06-0 61	68 136; 61[20]	4-01-00-03202 0.9113[62]	H$_2$O 3; EtOH 3; eth 3; chl 3 1.4156[20]
10477	2-Propanone, 1,1,1,3,3-pentabromo- Pentabromoacetone	C$_3$HBr$_5$O 452.56	79-49-2 79.5	7056 sub	4-01-00-03226	H$_2$O 1; EtOH 4; eth 4; ace 4
10478	2-Propanone, 1,1,1,3,3-pentachloro-	C$_3$HCl$_5$O 230.30	1768-31-6 2.1	192[253]; 98[40]	4-01-00-03222 1.69[15]	
10479	2-Propanone, 1-phenoxy-	C$_9$H$_{10}$O$_2$ 150.18	621-87-4	229.5	4-06-00-00604 1.0903[20]	eth 3; ace 3 1.5228[20]
10480	1-Propanone, 1-phenyl- Propiophenone	C$_9$H$_{10}$O 134.18	93-55-0 18.6	7842 217.5	4-07-00-00680 1.0096[20]	H$_2$O 1; EtOH 3; eth 3; chl 3 1.5269[20]
10481	2-Propanone, 1-phenyl- Phenylacetone	C$_9$H$_{10}$O 134.18	103-79-7 -15	7240 216.5	4-07-00-00687 1.0157[20]	H$_2$O 1; EtOH 4; eth 4; bz 5 1.5168[20]
10482	2-Propanone, phenylhydrazone Acetone, phenylhydrazone	C$_9$H$_{12}$N$_2$ 148.21	103-02-6 42	163[50]	4-15-00-00060	EtOH 3; eth 3; dil acid 3
10483	1-Propanone, 1-phenyl-, oxime	C$_9$H$_{11}$NO 149.19	2157-50-8 54.8	245; 165[38]	4-07-00-00682 1.0639[18]	H$_2$O 1; EtOH 3; eth 3; chl 2
10484	1-Propanone, 1-(2-piperidinyl)- Piperidine, 2(1-propionyl)	C$_8$H$_{15}$NO 141.21	97073-23-9	94[18]	5-21-06-00522 0.9380[20]	1.4621[20]
10485	2-Propanone, 1-(2-piperidinyl)-, (R)- (-)-Pelletierine	C$_8$H$_{15}$NO 141.21	2858-66-4	7015 195; 125[100]	5-21-06-00522 0.988[20]	eth 4; EtOH 4; chl 4 1.4683[20]
10486	2-Propanone, 1,1,1,3-tetrachloro-	C$_3$H$_2$Cl$_4$O 195.86	16995-35-0 65	183	4-01-00-03222 1.624[25]	H$_2$O 2; eth 4; ace 4 1.497[18]
10487	2-Propanone, 1,1,3,3-tetrachloro-	C$_3$H$_2$Cl$_4$O 195.86	632-21-3	183	4-01-00-03222	H$_2$O 4; EtOH 4; eth 4; ace 4 1.4944[20]
10488	2-Propanone, 1,1,1-trichloro-	C$_3$H$_3$Cl$_3$O 161.41	918-00-3	149; 28[10]	4-01-00-03221 1.435[20]	H$_2$O 4; EtOH 4; eth 4 1.4635[17]
10489	1-Propanone, 1-(2,4,6-trihydroxyphenyl)- Flopropione	C$_9$H$_{10}$O$_4$ 182.18	2295-58-1 175.5	4039	3-08-00-03413	eth 4; EtOH 4

No.	Name Synonym	Mol. Form. Mol. Wt.	CAS RN mp/°C	Merck No. bp/°C	Beil. Ref. den/g cm^{-3}	Solubility n_D
10490	Propanoyl bromide Propionic acid bromide	C_3H_5BrO 136.98	598-22-1 	 102	4-02-00-00724 1.5210[16]	eth 4 1.4578[16]
10491	Propanoyl bromide, 2-bromo-2-methyl-	$C_4H_6Br_2O$ 229.90	20769-85-1 	 163	3-02-00-00661 1.4067[14]	ace 4; CS_2 4
10492	Propanoyl bromide, 2-methyl-	C_4H_7BrO 151.00	2736-37-0 	 117	4-02-00-00852 1.4067[15]	1.4552[25]
10493	Propanoyl chloride Propionyl chloride	C_3H_5ClO 92.52	79-03-8 -94	7840 80	4-02-00-00724 1.0646[20]	eth 3 1.4032[20]
10494	Propanoyl chloride, 2-(acetyloxy)-, (±)- Lactyl chloride, acetate (DL)	$C_5H_7ClO_3$ 150.56	55057-45-9 	150 dec; 56[11]	4-03-00-00674 1.1920[17]	1.4241[17]
10495	Propanoyl chloride, 2-bromo-	C_3H_4BrClO 171.42	7148-74-5 	 132	4-02-00-00764 1.697[11]	eth 3; ctc 2; chl 3 1.4780[20]
10496	Propanoyl chloride, 2-chloro-, (S)- Propanoic acid, chloride, 2-chloro(d)	$C_3H_4Cl_2O$ 126.97	70110-24-6 	 110	4-02-00-00747 1.2394[7]	1.4400[20]
10497	Propanoyl chloride, 3-chloro-	$C_3H_4Cl_2O$ 126.97	625-36-5 	 144	4-02-00-00750 1.3307[13]	H_2O 2; EtOH 4; eth 4; chl 4 1.4549[20]
10498	Propanoyl chloride, 2-chloro-2-methyl-	$C_4H_6Cl_2O$ 141.00	13222-26-9 	 126.5	4-02-00-00859 	eth 4 1.4369[20]
10499	Propanoyl chloride, 3-chloro-2-methyl- Isobutyryl chloride, 3-chloro	$C_4H_6Cl_2O$ 141.00	7623-10-1 -90	 92	4-02-00-00860 1.0174[20]	eth 4 1.4079[20]
10500	Propanoyl chloride, 2,2-dichloro-	$C_3H_3Cl_3O$ 161.41	26073-26-7 	118; 70[89]	4-02-00-00755 1.4062[20]	1.4524[20]
10501	Propanoyl chloride, 2,3-dichloro-	$C_3H_3Cl_3O$ 161.41	7623-13-4 	52-4[16]	4-02-00-00757 1.4757[20]	1.4764[20]
10502	Propanoyl chloride, 3,3-dichloro- 3,3-Dichloropropionyl chloride	$C_3H_3Cl_3O$ 161.41	17880-86-3 	43-4[10]	3-02-00-00562 1.4557[20]	eth 4; Diox 4 1.4738[20]
10503	Propanoyl chloride, 2,2-dimethyl- Pivalic acid chloride	C_5H_9ClO 120.58	3282-30-2 	 107	4-02-00-00912 1.003[20]	eth 4 1.4139[20]
10504	Propanoyl chloride, 2-methyl- Isobutyric acid chloride	C_4H_7ClO 106.55	79-30-1 -90	 92	4-02-00-00852 	eth 3 1.4079[20]
10505	Propanoyl chloride, 2-phenoxy-	$C_9H_9ClO_2$ 184.62	122-35-0 	147; 116[10]	4-06-00-00643 1.1865[20]	eth 3 1.5178[20]
10506	Propanoyl fluoride Propionyl fluoride	C_3H_5FO 76.07	430-71-7 	 44	4-02-00-00724 0.972[15]	1.329[13]
10507	Propargite Sulfurous acid, 2-[4-(1,1-dimethylethyl)phenoxy]cyclohexyl 2-propynyl ester	$C_{19}H_{26}O_4S$ 350.48	2312-35-8 	7818 	 1.10[25]	
10508	Propazine 1,3,5-Triazine-2,4-diamine, 6-chloro-N,N'-bis(1-methylethyl)-	$C_9H_{16}ClN_5$ 229.71	139-40-2 213	7822 	 1.162[20]	
10509	2-Propenal Acrolein	C_3H_4O 56.06	107-02-8 -87.7	122 52.6	4-01-00-03435 0.840[20]	H_2O 4; EtOH 3; eth 3; ace 3 1.4017[20]
10510	2-Propenal, 3-bromo-3-phenyl-	C_9H_7BrO 211.06	14804-59-2 	144-6[12]	3-07-00-01383 1.492[20]	eth 4
10511	2-Propenal, 2-chloro-	C_3H_3ClO 90.51	683-51-2 	40[30]	4-01-00-03440 1.199[20]	eth 4; EtOH 4 1.463[20]
10512	2-Propenal, 3-(2-furanyl)-	$C_7H_6O_2$ 122.12	623-30-3 54	135[14]	5-17-09-00570 	H_2O 1; EtOH 5; eth 3; chl 2
10513	2-Propenal, 3-(2-furanyl)-, (E)-	$C_7H_6O_2$ 122.12	39511-08-5 50	143[37]; 100[12]	5-17-09-00570 	bz 4; eth 4; EtOH 4 1.5286[20]
10514	2-Propenal, 3-(4-hydroxy-3-methoxyphenyl)-	$C_{10}H_{10}O_3$ 178.19	458-36-6 84	 	4-08-00-01979 1.1562[102]	bz 4; eth 4; EtOH 4
10515	2-Propenal, 2-methyl- Methacrolein	C_4H_6O 70.09	78-85-3 	 68.4	4-01-00-03455 0.840[25]	H_2O 5; EtOH 5; eth 5 1.4144[20]
10516	2-Propenal, 2-methyl-3-phenyl-	$C_{10}H_{10}O$ 146.19	101-39-3 	150[100]	2-07-00-00291 1.0407[17]	1.6057[17]
10517	2-Propenal, 3-(4-methylphenyl)-	$C_{10}H_{10}O$ 146.19	1504-75-2 41.5	154[25]	4-07-00-01013 0.9670[23]	EtOH 4
10518	2-Propenal, 3-phenyl-, (E)- trans-Cinnamaldehyde	C_9H_8O 132.16	14371-10-9 -7.5	253 dec; 127[16]	4-07-00-00984 1.0497[20]	H_2O 2; EtOH 3; eth 3; chl 3 1.6195[20]
10519	2-Propenamide Acrylamide	C_3H_5NO 71.08	79-06-1 84.5	123 192.6	4-02-00-01471 	H_2O 4; EtOH 3; eth 3; ace 3
10520	2-Propenamide, N-(3,4-dichlorophenyl)-2-methyl- Dicryl	$C_{10}H_9Cl_2NO$ 230.09	2164-09-2 128	3078 	 	ace 4; EtOH 4
10521	2-Propenamide, N-(1,1-dimethyl-3-oxobutyl)- Diacetone acrylamide	$C_9H_{15}NO_2$ 169.22	2873-97-4 	2943 	 	chl 3
10522	2-Propen-1-amine Allylamine	C_3H_7N 57.10	107-11-9 -88.2	285 53.3	4-04-00-01057 0.758[20]	H_2O 5; EtOH 5; eth 5; chl 3 1.4205[20]
10523	2-Propen-1-amine, N,N-di-2-propenyl- Triallylamine	$C_9H_{15}N$ 137.22	102-70-5 94	 155.5	4-04-00-01061 0.809[20]	EtOH 3; eth 3; ace 3; bz 3 1.4502[20]
10524	2-Propen-1-amine, N-methyl-	C_4H_9N 71.12	627-37-2 	 65	4-04-00-01058 	H_2O 4; ace 4; eth 4; EtOH 4 1.4065[20]
10525	2-Propen-1-amine, N-(1-methylethyl)-	$C_6H_{13}N$ 99.18	35000-22-7 	 96.5	4-04-00-01059 0.7400[30]	1.4140[25]
10526	2-Propen-1-amine, N-2-propenyl- Diallylamine	$C_6H_{11}N$ 97.16	124-02-7 	2951 111	4-04-00-01060 	EtOH 3; eth 3 1.4387[20]

No.	Name / Synonym	Mol. Form. / Mol. Wt.	CAS RN / mp/°C	Merck No. / bp/°C	Beil. Ref. / den/g cm^{-3}	Solubility / n_D
10527	2-Propen-1-amine, N-propyl-	$C_6H_{13}N$ 99.18	5666-21-7	112	0-04-00-00207 0.7708[18]	H_2O 4 1.4140[25]
10528	1-Propene Propylene	C_3H_6 42.08	115-07-1 -185.2	7862 -47.6	4-01-00-00725 *0.5O5[25]	H_2O 4; EtOH 4; HOAc 4 1.3567[-70]
10529	1-Propene, 1-bromo-, (Z)-	C_3H_5Br 120.98	590-13-6 -113	57.8	4-01-00-00754 1.4291[20]	H_2O 1; eth 3; ace 3; chl 3 1.4560[20]
10530	1-Propene, 2-bromo-	C_3H_5Br 120.98	557-93-7 -126	48.4	4-01-00-00754 1.3965[16]	H_2O 1; eth 3; ace 3; chl 3 1.4467[16]
10531	1-Propene, 3-bromo- Allyl bromide	C_3H_5Br 120.98	106-95-6 -119	286 70.1	4-01-00-00754 1.398[20]	H_2O 1; EtOH 5; eth 5; ctc 3 1.4697[20]
10532	1-Propene, 3-bromo-3,3-difluoro- Propylene, 3-bromo-3,3-difluoro	$C_3H_3BrF_2$ 156.96	420-90-6	42	4-01-00-00756 1.543[25]	1.3773[25]
10533	1-Propene, 1-chloro-, (E)- $trans$-1-Chloropropene	C_3H_5Cl 76.53	16136-85-9 -99	37.4	4-01-00-00737 0.9349[20]	H_2O 1; eth 3; ace 3; bz 3 1.4054[20]
10534	1-Propene, 1-chloro-, (Z)- cis-1-Chloropropene	C_3H_5Cl 76.53	16136-84-8 -134.8	32.8	4-01-00-00737 0.9347[20]	H_2O 1; eth 3; ace 3; bz 3 1.4055[20]
10535	1-Propene, 2-chloro- Isopropenyl chloride	C_3H_5Cl 76.53	557-98-2 -137.4	22.6	4-01-00-00737 0.9017[20]	H_2O 1; eth 3; ace 3; bz 3 1.3973[20]
10536	1-Propene, 3-chloro- Allyl chloride	C_3H_5Cl 76.53	107-05-1 -134.5	287 45.1	4-01-00-00738 0.9376[20]	H_2O 1; EtOH 5; eth 5; ace 5 1.4157[20]
10537	1-Propene, 3-chloro-2-(chloromethyl)-	$C_4H_6Cl_2$ 125.00	1871-57-4 -14	138	4-01-00-00805 1.1782[20]	EtOH 4; chl 4 1.4753
10538	1-Propene, 2-chloro-3-isothiocyanato-	C_4H_4ClNS 133.60	14214-31-4	182	0-04-00-00219 1.27[12]	EtOH 2
10539	1-Propene, 1-chloro-2-methyl- 1-Chloro-2-methylpropene	C_4H_7Cl 90.55	513-37-1	2147 68	4-01-00-00803 0.9186[20]	chl 3 1.4221[20]
10540	1-Propene, 3-chloro-2-methyl- 3-Chloro-2-methylpropene	C_4H_7Cl 90.55	563-47-3	2148 71.5	4-01-00-00803 0.9165[20]	EtOH 5; eth 5; ace 3; chl 4 1.4291[20]
10541	1-Propene, 2-chloro-1,1,3,3,3-pentafluoro-	C_3ClF_5 166.48	2804-50-4 -130	6.8	4-01-00-00741 1.5920[20]	
10542	1-Propene, 1,1-dibromo-	$C_3H_4Br_2$ 199.87	13195-80-7	125	4-01-00-00759 1.9767[20]	H_2O 2; bz 3; ctc 3; chl 3 1.5260[20]
10543	1-Propene, 1,3-dibromo-, (E)- Propylene, 1,3-dibromo ($trans$)	$C_3H_4Br_2$ 199.87	32121-07-6	66[25]	4-01-00-00760 1.9791[20]	1.5495[20]
10544	1-Propene, 1,3-dibromo-, (Z)- Propylene, 1,3-dibromo (cis)	$C_3H_4Br_2$ 199.87	32121-06-5	60[25]	4-01-00-00760 2.0599[20]	1.5550[20]
10545	1-Propene, 2,3-dibromo- 2,3-Dibromopropene	$C_3H_4Br_2$ 199.87	513-31-5	3009 141; 37.7[11]	4-01-00-00760 2.0345[25]	H_2O 1; eth 3; ace 3; chl 3 1.5416[20]
10546	1-Propene, 1,3-dibromo-2-methyl- Isobutylene, 1,3-dibromo	$C_4H_6Br_2$ 213.90	35911-17-2	102[100]	3-01-00-00769 1.8939[4]	1.5396[15]
10547	1-Propene, 1,1-dichloro-	$C_3H_4Cl_2$ 110.97	563-58-6	76.5	4-01-00-00742 1.1864[25]	H_2O 1; eth 3; ace 3; chl 3 1.4430[25]
10548	1-Propene, 1,2-dichloro-, (E)-	$C_3H_4Cl_2$ 110.97	7069-38-7	77	4-01-00-00742 1.1818[20]	H_2O 1; EtOH 4; ctc 4; MeOH 4 1.4471[20]
10549	1-Propene, 1,2-dichloro-, (Z)-	$C_3H_4Cl_2$ 110.97	6923-20-2	93	4-01-0U-00742 1.1818[20]	H_2O 1; ace 3; bz 3; chl 3 1.4549[20]
10550	1-Propene, 1,3-dichloro-, (E)- $trans$-1,3-Dichloropropylene	$C_3H_4Cl_2$ 110.97	10061-02-6	3064 112	4-01-00-00744 1.217[20]	H_2O 1; eth 3; bz 3; chl 3 1.4730[20]
10551	1-Propene, 1,3-dichloro-, (Z)- cis-1,3-Dichloropropylene	$C_3H_4Cl_2$ 110.97	10061-01-5	3064 104.3	4-01-00-00743 1.224[20]	H_2O 1; eth 3; bz 3; chl 3 1.4682[20]
10552	1-Propene, 2,3-dichloro- 2,3-Dichloropropene	$C_3H_4Cl_2$ 110.97	78-88-6 10	94	4-01-00-00744 1.211[20]	H_2O 1; EtOH 5; eth 3; bz 3 1.4603[20]
10553	1-Propene, 3,3-dichloro-	$C_3H_4Cl_2$ 110.97	563-57-5	84.4	4-01-00-00745 1.175[20]	bz 4; eth 4; EtOH 4 1.4510[20]
10554	1-Propene, 1,1-dichloro-2-fluoro- Propylene, 1,1-dichloro-2-fluoro	$C_3H_3Cl_2F$ 128.96	430-95-5	78	4-01-00-00745 1.3026[25]	1.4196[25]
10555	1-Propene, 1,1-dichloro-2-methyl-	$C_4H_6Cl_2$ 125.00	6065-93-6 -43	108.5	4-01-00-00804 1.1447[20]	bz 4; eth 4; chl 4 1.4580[20]
10556	1-Propene, 3,3-dichloro-2-methyl-	$C_4H_6Cl_2$ 125.00	22227-75-4	108	4-01-00-00805 1.136[24]	bz 4; eth 4; chl 4 1.4523[24]
10557	1-Propene, 1,2-dichloro-1,3,3,3-tetrafluoro-	$C_3Cl_2F_4$ 182.93	431-53-8 -137	47.3	4-01-00-00747 1.5468[25]	1.3511[20]
10558	Propene, 1,2-dichloro-3,3,3-trifluoro- Propylene, 1,2-dichloro-3,3,3-trifluoro	$C_3HCl_2F_3$ 164.94	431-27-6 -109.2	53.7	4-01-00-00746 1.4653[20]	1.3670[20]
10559	1-Propene, 1,1-diethoxy-	$C_7H_{14}O_2$ 130.19	21504-43-8	133.5	4-01-00-03434 0.8628[25]	1.4083[25]
10560	1-Propene, 3,3-diethoxy- Acrolein, diethyl acetal	$C_7H_{14}O_2$ 130.19	3054-95-3	123.5	4-01-00-03437 0.8543[15]	H_2O 2; EtOH 5; eth 5 1.4000[20]
10561	1-Propene, 1,1-difluoro 1,1-Difluoro-1-propene	$C_3H_4F_2$ 78.06	430-63-7			
10562	1-Propene, 3,3-dimethoxy-	$C_5H_{10}O_2$ 102.13	6044-68-4	88	4-01-00-03437 0.862[25]	1.3954[20]
10563	2-Propene-1,1-diol, diacetate	$C_7H_{10}O_4$ 158.15	869-29-4 -37.6	180	4-02-00-00291 1.0760[20]	ace 4; bz 4; eth 4; EtOH 4 1.4193[20]
10564	2-Propene-1,1-diol, 2-methyl-, diacetate Methacrolein diacetate	$C_8H_{12}O_4$ 172.18	10476-95-6	191	4-02-00-00292	1.4241[20]
10565	1-Propene, 3-(ethenyloxy)-	C_5H_8O 84.12	3917-15-5	66	4-01-00-02085 0.8050[20]	H_2O 1; eth 3; ace 3; chl 3 1.4062[20]

No.	Name Synonym	Mol. Form. Mol. Wt.	CAS RN mp/°C	Merck No. bp/°C	Beil. Ref. den/g cm^{-3}	Solubility n_D
10566	1-Propene, 3-ethoxy- Allyl ethyl ether	$C_5H_{10}O$ 86.13	557-31-3	291 67.6	4-01-00-02083 0.7651[20]	H_2O 1; EtOH 5; eth 5; ace 3 1.3881[20]
10567	1-Propene, 1-fluoro- (Z) cis-1-Fluoropropene	C_3H_5F 60.07	19184-10-2			
10568	1-Propene, 1-fluoro- (E) trans-1-Fluoropropene	C_3H_5F 60.07	20327-65-5			
10569	1-Propene, 2-fluoro- 2-Fluoropropene	C_3H_5F 60.07	1184-60-7	-24	4-01-00-00733	
10570	1-Propene, 3-fluoro- 3-Fluoropropene	C_3H_5F 60.07	818-92-8	-3	4-01-00-00733	H_2O 2; EtOH 4; eth 4; chl 3
10571	1-Propene, 1,1,2,3,3,3-hexachloro- Hexachloropropylene	C_3Cl_6 248.75	1888-71-7 -72.9	209.5	4-01-00-00753 1.7632[20]	H_2O 1; ctc 3; chl 3 1.5091[20]
10572	1-Propene, 1,1,2,3,3,3-hexafluoro- Perfluoropropene	C_3F_6 150.02	116-15-4 -156.5	-29.6	4-01-00-00735	1.583^{-40}
10573	1-Propene, 3-iodo- Allyl iodide	C_3H_5I 167.98	556-56-9 -99.3	292 103	4-01-00-00761 1.84[12]	H_2O 1; EtOH 3; eth 3; chl 3 1.5540[21]
10574	1-Propene, 3-isocyano- Allyl isocyanide	C_4H_5N 67.09	2835-21-4	98	0-04-00-00208 0.794[17]	eth 4; EtOH 4
10575	1-Propene, 3-isothiocyanato- Allyl isothiocyanate	C_4H_5NS 99.16	57-06-7 -80	293 152	4-04-00-01081 1.0126[20]	bz 4; eth 4; EtOH 4 1.5306[20]
10576	1-Propene, 3-methoxy-	C_4H_8O 72.11	627-40-7	44	4-01-00-02083 0.77[11]	H_2O 1; EtOH 5; eth 5; ace 3 1.3778[20]
10577	1-Propene, 3-methoxy-2-methyl-	$C_5H_{10}O$ 86.13	22418-49-1	68	3-01-00-01904 0.7698[20]	1.3964[20]
10578	1-Propene, 2-methyl- Isobutene	C_4H_8 56.11	115-11-7 -140.4	5024 -6.9	4-01-00-00796 *0.589[25]	H_2O 1; EtOH 4; eth 4; bz 3 1.3926[-25]
10579	1-Propene, 3-(1-methylethoxy)-	$C_6H_{12}O$ 100.18	6140-80-3	83.5	4-01-00-02083 0.7764[20]	ace 4; eth 4; EtOH 4 1.3946[20]
10580	1-Propene, 2-methyl-, tetramer Propene, 2-methyl(tetramer)	$C_{18}H_{32}$ 224.43	15220-85-6 -98	244; 109[15]	3-01-00-00763 0.7944[20]	1.4482[20]
10581	1-Propene, 3-(methylthio)-	C_4H_8S 88.17	10152-76-8	92	4-01-00-02095 0.8767[20]	1.4714[20]
10582	1-Propene, 2-methyl-, trimer Isobutylene (trimer)	$C_{12}H_{24}$ 168.32	7756-94-7 -76	180	3-01-00-00763 0.7590[20]	1.4314[20]
10583	2-Propenenitrile Acrylonitrile	C_3H_3N 53.06	107-13-1 -83.5	125 77.3	4-02-00-01473 0.8060[20]	ace 4; bz 4; eth 4; EtOH 4 1.3911[20]
10584	2-Propenenitrile, 2-chloro-	C_3H_2ClN 87.51	920-37-6 -65	88.5	4-02-00-01484 1.096[25]	1.4290[20]
10585	2-Propenenitrile, 3-(2-furanyl)- 2-Furanacrylonitrile	C_7H_5NO 119.12	7187-01-1 38	4208 96	4-18-00-04152	tol 4 1.5824[25]
10586	2-Propenenitrile, 2-methyl- Methylacrylonitrile	C_4H_5N 67.09	126-98-7 -35.8	5850 90.3	4-02-00-01539 0.8001[20]	H_2O 1; EtOH 5; eth 5; ace 5 1.4003[20]
10587	2-Propenenitrile, 3-phenyl-, (E)-	C_9H_7N 129.16	1885-38-7 22	263.8	4-09-00-02027 1.0304[20]	H_2O 1; EtOH 3; ace 3; ctc 3 1.6013[20]
10588	2-Propenenitrile, 3-phenyl-, (Z)-	C_9H_7N 129.16	24840-05-9 -4.4	249; 139[30]	4-09-00-02027 1.0289[20]	H_2O 1; EtOH 3; bz 4 1.5843[20]
10589	1-Propene, 1-nitro-	$C_3H_5NO_2$ 87.08	3156-70-5	60[34]; 37[10]	3-01-00-00715 1.0661[20]	eth 3; ace 3; chl 3 1.4527[20]
10590	1-Propene, 2-nitro-	$C_3H_5NO_2$ 87.08	4749-28-4	57[100]; 32[30]	4-01-00-00764 1.0643[20]	eth 3; ace 3; chl 3 1.4358[20]
10591	1-Propene, 3-nitro-	$C_3H_5NO_2$ 87.08	625-46-7	127.5	4-01-00-00764 1.051[22]	eth 4; EtOH 4
10592	1-Propene, 3,3'-oxybis- Allyl ether	$C_6H_{10}O$ 98.14	557-40-4 -6	290 94	4-01-00-02086 0.8260[20]	H_2O 1; EtOH 5; eth 5; ace 4 1.4163[20]
10593	1-Propene, 1,1,2,3,3-pentachloro-	C_3HCl_5 214.30	1600-37-9	185	4-01-00-00753 1.6317[34]	eth 4 1.5313[20]
10594	1-Propene, 3-propoxy-	$C_6H_{12}O$ 100.16	1471-03-0	91	4-01-00-02083 0.7764[20]	ace 4; eth 4; EtOH 4 1.3919[20]
10595	2-Propene-1-sulfinothioic acid, S-2-propenyl ester Allicin	$C_6H_{10}OS_2$ 162.28	539-86-6	249 dec	4-04-00-00007 1.112[20]	H_2O 4 1.561[20]
10596	1-Propene, 3,3'-sulfinylbis-	$C_6H_{10}OS$ 130.21	14180-63-3 23.5	114[12]; 89[2.6]	4-01-00-02097 1.0261[20]	EtOH 3; eth 3; ace 3 1.5115[20]
10597	1-Propene, 3,3'-sulfonylbis-	$C_6H_{10}O_2S$ 146.21	16841-48-8	109[3]	4-01-00-02098 1.1213[20]	1.4893[20]
10598	1-Propene, 1,2,3,3-tetrachloro-	$C_3H_2Cl_4$ 179.86	20589-85-9	162	3-01-00-00708 1.537[19]	bz 4; chl 4 1.5121[19]
10599	1-Propene, 3,3'-thiobis- Diallyl sulfide	$C_6H_{10}S$ 114.21	592-88-1 -85	295 138.6	4-01-00-02097 0.8877[27]	eth 4; EtOH 4 1.4870[25]
10600	2-Propene-1-thiol	C_3H_6S 74.15	870-23-5	65	4-01-00-02095 0.925[23]	H_2O 1; EtOH 5; eth 5; chl 3 1.4832[20]
10601	1-Propene-1,2,3-tricarboxylic acid, triethyl ester, (E)- Aconitic acid, triethyl ester (trans)	$C_{12}H_{18}O_6$ 258.27	68077-28-1	275 dec; 159[9]	4-02-00-02405 1.1064[20]	eth 4; EtOH 4
10602	1-Propene, 1,1,2-trichloro-	$C_3H_3Cl_3$ 145.41	21400-25-9	118	4-01-00-00747 1.382[20]	H_2O 1; EtOH 3; eth 3; bz 3 1.4827[20]
10603	1-Propene, 1,2,3-trichloro-	$C_3H_3Cl_3$ 145.41	96-19-5	142	4-01-00-00748 1.412[20]	H_2O 1; EtOH 4; eth 4; bz 3 1.5030[20]
10604	1-Propene, 3,3,3-trichloro-	$C_3H_3Cl_3$ 145.41	2233-00-3 -30	114.5	4-01-00-00749 1.367[20]	H_2O 1; EtOH 3; eth 3; bz 3 1.4827[20]

No.	Name Synonym	Mol. Form. Mol. Wt.	CAS RN mp/°C	Merck No. bp/°C	Beil. Ref. den/g cm^{-3}	Solubility n_D
10605	1-Propene, 1,1,3-trichloro-2-methyl-	$C_4H_5Cl_3$ 159.44	31702-33-7	156	4-01-00-00805 1.346[20]	ace 4; bz 4; chl 4 1.4990[20]
10606	1-Propene, 3,3,3-trichloro-2-methyl-	$C_4H_5Cl_3$ 159.44	4749-27-3	133	4-01-00-00805 1.293[20]	H_2O 1; ace 3; bz 3; chl 4 1.4770[20]
10607	1-Propene, 1,1,2-trichloro-3,3,3-trifluoro-	$C_3Cl_3F_3$ 199.39	431-52-7 -114.6	88.1	4-01-00-00751 1.617[20]	
10608	1-Propene, 3,3,3-trifluoro- 3,3,3-Trifluoropropene	$C_3H_3F_3$ 96.05	677-21-4 -17		4-01-00-00734 0	
10609	2-Propenoic acid Acrylic acid	$C_3H_4O_2$ 72.06	79-10-7 12.3	124 141	4-02-00-01455 1.0511[20]	H_2O 5; EtOH 5; eth 5; ace 3 1.4224[20]
10610	2-Propenoic acid, 2-bromo-	$C_3H_3BrO_2$ 150.96	10443-65-9 71.5	dec	4-02-00-01486	H_2O 4; EtOH 4
10611	2-Propenoic acid, 2-bromoethyl ester	$C_5H_7BrO_2$ 179.01	4823-47-6	156.5; 53[5]	4-02-00-01462 1.4581[25]	1.4660[25]
10612	2-Propenoic acid, 3-bromo-2-methyl- Methacrylic acid, 3-bromo	$C_4H_5BrO_2$ 164.99	89123-63-7 64	229	0-02-00-00424	H_2O 4
10613	2-Propenoic acid, 2-bromo-3-phenyl-, ethyl ester	$C_{11}H_{11}BrO_2$ 255.11	57296-00-1	295	4-09-00-02042 1.3885[25]	1.5845[25]
10614	2-Propenoic acid, 3-bromo-3-phenyl-, ethyl ester, (Z)- Cinnamic acid, β-bromo-, ethyl ester	$C_{11}H_{11}BrO_2$ 255.11	1504-70-7	167[12]	3-09-00-02733 1.4044[10]	1.5841[10]
10615	2-Propenoic acid, 2-bromo-3-phenyl-, methyl ester Methyl α-bromocinnamate	$C_{10}H_9BrO_2$ 241.08	24127-62-6 23	120-1[0.6]	4-09-00-02042 1.4975[20]	
10616	2-Propenoic acid, butyl ester Butyl acrylate	$C_7H_{12}O_2$ 128.17	141-32-2 -64.6	1539 145	4-02-00-01463 0.8998[20]	H_2O 1; EtOH 3; eth 3; ace 3 1.4185[20]
10617	2-Propenoic acid, 3-chloro-, (E)-	$C_3H_3ClO_2$ 106.51	2345-61-1 86	94[18]	4-02-00-01481	EtOH 3; eth 3; chl 2
10618	2-Propenoic acid, 3-chloro-, (Z)-	$C_3H_3ClO_2$ 106.51	1609-93-4 63.5	106[17]	4-02-00-01481	EtOH 3; eth 3
10619	2-Propenoic acid, chloro-, ethyl ester Acrylic acid, chloro-, ethyl ester	$C_5H_7ClO_2$ 134.56	30600-19-2	51-3[18]	3-02-00-01245 1.1404[20]	EtOH 4; eth 4 1.4384[20]
10620	2-Propenoic acid, 2-chloroethyl ester	$C_5H_7ClO_2$ 134.56	2206-89-5	51-3[18]	3-02-00-01225 1.1604[20]	eth 4; EtOH 4 1.4384[20]
10621	2-Propenoic acid, 2-chloro-, methyl ester Methyl 2-chloroacrylate	$C_4H_5ClO_2$ 120.54	80-63-7	52[51]	4-02-00-01482 1.189[20]	eth 4 1.4420[20]
10622	2-Propenoic acid, 2-chloro-3-phenyl-, methyl ester, (E)- Cinnamic acid, α-chloro-, methyl ester, (E)-	$C_{10}H_9ClO_2$ 196.63	14737-94-1 33	108-9[0.5]	4-09-00-02035	ace 4
10623	2-Propenoic acid, 3-chloro-3-phenyl-, methyl ester Cinnamic acid, β-chloro-, methyl ester	$C_{10}H_9ClO_2$ 196.63	87541-87-5 29	113-4[0.5]	4-09-00-02035 1.2248[21]	ace 4 1.5781[21]
10624	2-Propenoic acid, 2-cyano-, methyl ester Mecrylate	$C_5H_5NO_2$ 111.10	137-05-3 47-8[2]	5667	4-02-00-02226 1.1012[20]	1.4430
10625	2-Propenoic acid, 2-cyano-3-phenyl-, ethyl ester	$C_{12}H_{11}NO_2$ 201.22	2025-40-3 51	188[15]	2-09-00-00640 1.1076[25]	ace 4; chl 4 1.5033
10626	2-Propenoic acid, cyclohexyl ester	$C_9H_{14}O_2$ 154.21	3066-71-5	183; 88[20]	4-06-00-00038 1.0275[20]	H_2O 1; EtOH 5; eth 5; chl 3 1.4673[20]
10627	2-Propenoic acid, 2,3-dibromo-, (Z)-	$C_3H_2Br_2O_2$ 229.86	24557-10-6 85.5	148[6]	3-02-00-01249	H_2O 2; EtOH 3; eth 3; bz 2
10628	2-Propenoic acid, 3,3-dibromo- Acrylic acid, 3,3-dibromo-	$C_3H_2Br_2O_2$ 229.86	1578-21-8 86.3	245	4-02-00-01487	eth 4; EtOH 4
10629	2-Propenoic acid, 2,3-dibromo-3-phenyl-, (Z)-	$C_9H_8Br_2O_2$ 305.95	708-82-7 100	124[0.5]	4-09-00-02043	H_2O 1; EtOH 3; eth 3; chl 3
10630	2-Propenoic acid, 3,3-dichloro- Acrylic acid, 3,3-dichloro-	$C_3H_2Cl_2O_2$ 140.95	1561-20-2 77.5	sub	4-02-00-01484	eth 4; chl 4
10631	2-Propenoic acid, 2-(diethylamino)ethyl ester	$C_9H_{17}NO_2$ 171.24	2426-54-2 <-60	81[10]	4-04-00-01477 0.937[20]	1.4376[25]
10632	2-Propenoic acid, 3-(3,4-dihydroxyphenyl)- Caffeic acid	$C_9H_8O_4$ 180.16	331-39-5 225 dec	1634	2-10-00-00294	EtOH 4
10633	2-Propenoic acid, 2-(dimethylamino)ethyl ester	$C_7H_{13}NO_2$ 143.19	2439-35-2 <-60	95[50]	4-04-00-01431 0.938[20]	
10634	2-Propenoic acid, 2-ethoxyethyl ester	$C_7H_{12}O_3$ 144.17	106-74-1 -47	174	4-02-00-01469 0.983[20]	1.4274[20]
10635	2-Propenoic acid, ethyl ester Ethyl acrylate	$C_5H_8O_2$ 100.12	140-88-5 -71.2	3715 99.4	4-02-00-01460 0.9234[20]	H_2O 2; EtOH 5; eth 5; chl 3 1.4068[20]
10636	2-Propenoic acid, 2-ethylhexyl ester 2-Ethylhexyl acrylate	$C_{11}H_{20}O_2$ 184.28	103-11-7 -90	123-7[60]	4-02-00-01467 0.880[25]	1.4332[25]
10637	2-Propenoic acid, 2-fluoro-3-phenyl- Cinnamic acid, α-fluoro-	$C_9H_7FO_2$ 166.15	350-90-3 157.6	290	1-09-00-00237	EtOH 3; eth 3
10638	2-Propenoic acid, 3-(2-furanyl)- 2-Furanacrylic acid	$C_7H_6O_3$ 138.12	539-47-9 141	4207 286	5-18-06-00306	eth 4; EtOH 4
10639	2-Propenoic acid, 3-(2-furanyl)-, ethyl ester	$C_9H_{10}O_3$ 166.18	623-20-1 24.5	232	5-18-06-00307 1.0891[25]	H_2O 1; EtOH 5; eth 5; chl 2 1.5286[20]
10640	2-Propenoic acid, 3-(2-furanyl)-, methyl ester	$C_8H_8O_3$ 152.15	623-18-7 36	227; 117[23]	4-18-00-04144	bz 4; eth 4; EtOH 4 1.4447[20]

No.	Name / Synonym	Mol. Form. / Mol. Wt.	CAS RN / mp/°C	Merck No. / bp/°C	Beil. Ref. / den/g cm^{-3}	Solubility / n_D
10641	2-Propenoic acid, 3-(2-furanyl)-, pentyl ester Acrylic acid, 3-(2-furyl), pentyl ester	$C_{12}H_{16}O_3$ 208.26	2438-19-9 	116-8[2]	4-18-00-04146 1.0322[20]	eth 4; EtOH 4 1.5289[14]
10642	2-Propenoic acid, 3-(2-furanyl)-, propyl ester Acrylic acid, 3-(2-furyl), propyl ester	$C_{10}H_{12}O_3$ 180.20	623-22-3 	91-4[3]	4-18-00-04146 1.0744[20]	bz 4; eth 4; EtOH 4 1.5392[24]
10643	2-Propenoic acid, hexyl ester	$C_9H_{16}O_2$ 156.22	2499-95-8 -45	40[1]	4-02-00-01466 0.878[20]	
10644	2-Propenoic acid, 3-(4-hydroxyphenyl)- p-Coumaric acid	$C_9H_8O_3$ 164.16	7400-08-0 211.5	2561	4-10-00-01005	eth 4; EtOH 4
10645	2-Propenoic acid, 2-isocyano-3-phenyl-, ethyl ester Malononmononitrile, benzylidene,monoethyl ester	$C_{12}H_{11}NO_2$ 201.22	52744-85-1 51	188[15]; 118[0.4]	1.0762[25]	H_2O 1; EtOH 2; eth 3; bz 4 1.5033
10646	2-Propenoic acid, isodecyl ester	$C_{13}H_{24}O_2$ 212.33	1330-61-6 -100	158[50]	0.885[20]	1.4416[20]
10647	2-Propenoic acid, 3-(2-methoxyphenyl)-, methyl ester	$C_{11}H_{12}O_3$ 192.21	15854-58-7 	304; 161-3[3]	3-10-00-00835 1.1366[17]	1.5854[16]
10648	2-Propenoic acid, 2-methyl- Methacrylic acid	$C_4H_8O_2$ 86.09	79-41-4 16	5849 162.5	4-02-00-01518 1.0153[20]	H_2O 3; EtOH 5; eth 5; chl 3 1.4314[20]
10649	2-Propenoic acid, 2-methyl-, anhydride Methacrylic acid anhydride	$C_8H_{10}O_3$ 154.17	760-93-0 	89[5]	4-02-00-01537 1.0153[20]	EtOH 5; eth 5 1.4540[20]
10650	2-Propenoic acid, 2-methyl-, butyl ester Butyl methacrylate	$C_8H_{14}O_2$ 142.20	97-88-1 	160	4-02-00-01525 0.8936[20]	eth 4; EtOH 4 1.4240[20]
10651	2-Propenoic acid, methylbutyl ester Propenoic acid, 2-methyl,butyl ester	$C_8H_{14}O_2$ 142.20	90638-48-5 	160; 71[37]	3-02-00-01286 0.8936[20]	H_2O 1; EtOH 5; eth 5; ctc 2 1.4240[20]
10652	2-Propenoic acid, 2-methylbutyl ester Acrylic acid, (2-methylbutyl) ester	$C_8H_{14}O_2$ 142.20	44914-03-6 	160; 45[10]	3-02-00-01228 0.8936[20]	eth 4; EtOH 4 1.4240[20]
10653	2-Propenoic acid, 2-methyl-, cyclohexyl ester	$C_{10}H_{16}O_2$ 168.24	101-43-9 	210	4-06-00-00039 0.9626[20]	1.4578[20]
10654	2-Propenoic acid, 2-methyl-, 2-(dimethylamino)ethyl ester β-Dimethylaminoethyl methacrylate	$C_8H_{15}NO_2$ 157.21	2867-47-2 	62-5[6]	4-04-00-01432	
10655	2-Propenoic acid, methyl ester Methyl acrylate	$C_4H_8O_2$ 86.09	96-33-3 <-75	5935 80.7	4-02-00-01457 0.9535[20]	H_2O 2; EtOH 3; eth 3; ace 3 1.4040[20]
10656	2-Propenoic acid, 2-methyl-, 1,2-ethanediyl ester	$C_{10}H_{14}O_4$ 198.22	97-90-5 -40	260	4-02-00-01532 1.053[20]	bz 4; EtOH 4; lig 4 1.4532[25]
10657	2-Propenoic acid, 2-methyl-, 1,2-ethanediylbis(oxy-2,1-ethanediyl) ester	$C_{14}H_{22}O_6$ 286.33	109-16-0 		4-02-00-01531 1.072[25]	ace 4; eth 4; EtOH 4; peth 4 1.4595[25]
10658	2-Propenoic acid, 2-methyl-, ethyl ester Ethyl methacrylate	$C_8H_{10}O_2$ 114.14	97-63-2 	117	4-02-00-01523 0.9135[20]	H_2O 2; EtOH 5; eth 5; chl 2 1.4147[20]
10659	2-Propenoic acid, 2-methyl-, 2-ethylhexyl ester 2-Ethylhexyl methacrylate	$C_{12}H_{22}O_2$ 198.31	688-84-6 	110[14]	4-02-00-01528 0.880[25]	1.436[25]
10660	2-Propenoic acid, 2-methyl-, hexyl ester	$C_{10}H_{18}O_2$ 170.25	142-09-6 	162; 86[17]	4-02-00-01527 0.880[25]	ace 4; bz 4; eth 4; EtOH 4 1.429[25]
10661	2-Propenoic acid, 2-methyl-, 2-hydroxyethyl ester	$C_8H_{10}O_3$ 130.14	868-77-9 	67[3.5]	4-02-00-01530 1.034[25]	1.4515[20]
10662	2-Propenoic acid, 2-methyl-, 2-hydroxypropyl ester	$C_7H_{12}O_3$ 144.17	923-26-2 	90[9]	4-02-00-01532 1.066[25]	1.4458[20]
10663	2-Propenoic acid, 2-methyl-, methyl ester Methyl methacrylate	$C_5H_8O_2$ 100.12	80-62-6 -48	5849 100.5	4-02-00-01519 0.9440[20]	H_2O 2; EtOH 5; eth 5; ace 5 1.4142[20]
10664	2-Propenoic acid, 2-methyl-, 1-methylethyl ester	$C_7H_{12}O_2$ 128.17	4655-34-9 	125	4-02-00-01525 0.8847[20]	ace 4; bz 4; eth 4; EtOH 4 1.4122[20]
10665	2-Propenoic acid, 2-methyl-, 1-methylpropyl ester	$C_8H_{14}O_2$ 142.20	2998-18-7 	147; 35[4]	4-02-00-01526 0.886[20]	1.4179[20]
10666	2-Propenoic acid, 2-methyl-, 2-methylpropyl ester Isobutyl methacrylate	$C_8H_{14}O_2$ 142.20	97-86-9 	155	4-02-00-01526 0.8858[20]	H_2O 1; EtOH 5; eth 5 1.4199[20]
10667	2-Propenoic acid, 2-methyl-, octadecyl ester	$C_{22}H_{42}O_2$ 338.57	32360-05-7 	195[6]	4-02-00-01529 0.880[25]	1.429[25]
10668	2-Propenoic acid, 2-methyl-, oxiranylmethyl ester	$C_7H_{10}O_3$ 142.15	106-91-2 	75[10]	5-17-03-00035 1.07[25]	bz 4; eth 4; EtOH 4 1.448[25]
10669	2-Propenoic acid, 2-methyl-, oxydi-2,1-ethanediyl ester	$C_{12}H_{18}O_5$ 242.27	2358-84-1 	>200	4-02-00-01531 1.056[20]	1.4571[25]
10670	2-Propenoic acid, 2-methyl-3-phenyl-, (Z)-	$C_{10}H_{10}O_2$ 162.19	15250-29-0 94	288	4-09-00-02063	H_2O 2; EtOH 4; eth 4; bz 4
10671	2-Propenoic acid, 2-methyl-, 2-propenyl ester	$C_7H_{10}O_2$ 126.16	96-05-9 	55[30]; 60[43]	4-02-00-01529 0.927[20]	1.4360[20]
10672	2-Propenoic acid, 2-methyl-, propyl ester Propyl methacrylate	$C_7H_{12}O_2$ 128.17	2210-28-8 	141	4-02-00-01524 0.9022[20]	H_2O 1; EtOH 5; eth 5 1.4190[20]
10673	2-Propenoic acid, 2-methylpropyl ester Isobutyl acrylate	$C_7H_{12}O_2$ 128.17	106-63-8 -61	132	4-02-00-01465 0.8896[20]	H_2O 2; EtOH 3; eth 3; MeOH 3 1.4150[20]

No.	Name / Synonym	Mol. Form. / Mol. Wt.	CAS RN / mp/°C	Merck No. / bp/°C	Beil. Ref. / den/g cm^{-3}	Solubility / n_D
10674	2-Propenoic acid, 2-methyl-, (tetrahydro-2-furanyl)methyl ester	$C_9H_{14}O_3$ 170.21	2455-24-5	265; 81[4]	4-17-00-01105 1.040[25]	1.4554[25]
10675	2-Propenoic acid, 3-(1-naphthalenyl)-, (E)- Acrylic acid, 3-(1-naphtyl) (trans)	$C_{13}H_{10}O_2$ 198.22	2006-14-6 211.5	sub	4-09-00-02486	H_2O 2; EtOH 3; eth 3; chl 3
10676	2-Propenoic acid, 3-(2-nitrophenyl)-, (E)-	$C_9H_7NO_4$ 193.16	1013-96-3 244.3	sub	4-09-00-02043	H_2O 1; EtOH 3; bz 1
10677	2-Propenoic acid, 3-(2-nitrophenyl)-, methyl ester, (E)- Cinnamic acid, 2-nitro, methyl ester (trans)	$C_{10}H_9NO_4$ 207.19	39228-29-0 73	187[15]	4-09-00-02044	EtOH 4
10678	2-Propenoic acid, 3-(3-nitrophenyl)-, methyl ester, (E)- Cinnamic acid, 3-nitro, methyl ester (trans)	$C_{10}H_9NO_4$ 207.19	659-04-1 125.3	dec	3-09-00-02742	bz 4; eth 4; chl 4
10679	2-Propenoic acid, 3-(4-nitrophenyl)-, methyl ester, (E)-	$C_{10}H_9NO_4$ 207.19	637-57-0 162	283.5	4-09-00-02046	EtOH 4
10680	2-Propenoic acid, oxiranylmethyl ester	$C_8H_8O_3$ 128.13	106-90-1	53[10]	4-17-00-01005 1.1[20]	bz 4 1.4490[20]
10681	2-Propenoic acid, 2-phenoxyethyl ester	$C_{11}H_{12}O_3$ 192.21	48145-04-6	110[2]	3-06-00-00572 1.090[25]	ace 4; eth 4; chl 4
10682	2-Propenoic acid, 3-phenyl-, (E)- trans-Cinnamic acid	$C_9H_8O_2$ 148.16	140-10-3 133	2300 300	4-09-00-02002 1.2475[4]	H_2O 1; EtOH 4; eth 3; ace 3
10683	2-Propenoic acid, 3-phenyl-, (Z)-	$C_9H_8O_2$ 148.16	102-94-3 42		4-09-00-02001	EtOH 4; HOAc 4; lig 4
10684	2-Propenoic acid, 3-phenyl-, anhydride Cinnamic anhydride	$C_{18}H_{14}O_3$ 278.31	538-56-7 136	2300		bz 4
10685	2-Propenoic acid, 3-phenyl-, ethyl ester, (E)-	$C_{11}H_{12}O_2$ 176.22	4192-77-2 10	2300 271.5	4-09-00-02006 1.0491[20]	H_2O 1; EtOH 4; eth 4; ace 4 1.5598[20]
10686	2-Propenoic acid, 3-phenyl-, methyl ester, (E)-	$C_{10}H_{10}O_2$ 162.19	1754-62-7 36.5	2300 261.9	4-09-00-02005 1.042[36]	H_2O 1; EtOH 4; eth 4; ace 3 1.5766[22]
10687	2-Propenoic acid, 3-phenyl-, 1-methylethyl ester, (E)- Cinnamic acid, isopropyl ester (trans)	$C_{12}H_{14}O_2$ 190.24	60512-85-8	269	4-09-00-02008 1.0320[20]	H_2O 1; EtOH 3; eth 3; ace 3 1.5455[20]
10688	2-Propenoic acid, phenylmethyl ester	$C_{10}H_{10}O_2$ 162.19	2495-35-4	228	3-06-00-01481 1.0573[20]	H_2O 1; EtOH 3; eth 3; ace 3 1.5143[20]
10689	2-Propenoic acid, 3-phenyl-, phenyl ester, (E)- Cinnamic acid, phenyl ester (trans)	$C_{15}H_{12}O_2$ 224.26	25695-77-6 74.3	205[15]	4-09-00-02011 1.0650[100]	bz 4; eth 4
10690	2-Propenoic acid, 3-phenyl-, phenylmethyl ester, (E)- Cinnamic acid, benzyl ester (trans)	$C_{16}H_{14}O_2$ 238.29	78277-23-3 39	350 dec; 244[5]	4-09-00-02012 1.109[15]	H_2O 1; EtOH 3; eth 3; bz 2
10691	2-Propenoic acid, 3-phenyl-, 3-phenyl-2-propenyl ester Cinnamyl cinnamate	$C_{18}H_{16}O_2$ 264.32	122-69-0 44	2307	4-09-00-02013 1.1565[4]	H_2O 1; EtOH 3; eth 4; chl 3
10692	2-Propenoic acid, 3-phenyl-, 2-propenyl ester Cinnamic acid, allyl ester	$C_{12}H_{12}O_2$ 188.23	1866-31-5	268 dec; 163[17]	4-09-00-02010 1 048[23]	H_2O 1; EtOH 4; eth 5; ctc 2 1.530[20]
10693	2-Propenoic acid, 3-phenyl-, propyl ester, (E)- Cinnamic acid, propyl ester (trans)	$C_{12}H_{14}O_2$ 190.24	74513-58-9	285	4-09-00-02007 1.0433[0]	H_2O 1
10694	2-Propenoic acid, 2-propenyl ester	$C_6H_8O_2$ 112.13	999-55-3	121	4-02-00-01468 0.9441[20]	H_2O 2; EtOH 3; eth 3; acid 3 1.4320[20]
10695	2-Propenoic acid, (tetrahydro-2-furanyl)methyl ester	$C_8H_{12}O_3$ 156.18	2399-48-6 <-60	96[6]	4-17-00-01104 1.061[20]	
10696	2-Propenoic acid, 2,3,3-trichloro-	$C_3HCl_3O_2$ 175.40	2257-35-4 76	222	4-02-00-01486	H_2O 3; EtOH 3; eth 3; ctc 3 1.5271[18]
10697	2-Propen-1-ol Allyl alcohol	C_3H_6O 58.08	107-18-6 -129	284 97.0	4-01-00-02079 0.8540[20]	H_2O 5; EtOH 5; eth 5; chl 3 1.4135[20]
10698	1-Propen-2-ol, acetate Isopropenyl acetate	$C_5H_8O_2$ 100.12	108-22-5 -92.9	5091 94	4-02-00-00179 0.9090[20]	H_2O 2; EtOH 3; eth 4; ace 3 1.4033[20]
10699	2-Propen-1-ol, 2-bromo-	C_3H_5BrO 136.98	598-19-6	154; 62[11]	4-01-00-02094 1.621[18]	eth 4; chl 4 1.500[18]
10700	2-Propen-1-ol, 2-chloro-	C_3H_5ClO 92.52	5976-47-6	135	4-01-00-02091 1.1618[20]	EtOH 3; ctc 2 1.4588[20]
10701	2-Propen-1-ol, 3-chloro- Allyl alcohol, 3-chloro	C_3H_5ClO 92.52	29560-84-7	147	4-01-00-02090 1.1769[20]	1.4738[20]
10702	2-Propen-1-ol, 2-methyl- Methallyl alcohol	C_4H_8O 72.11	513-42-8	114.5	4-01-00-02114 0.8515[20]	H_2O 4; EtOH 5; eth 5 1.4255[20]
10703	2-Propen-1-ol, 2-methyl-, carbonate (2:1) Bis-(2-methallyl)carbonate	$C_9H_{14}O_3$ 170.21	64057-79-0 201.3	66[3]	4-03-00-00011 0.943[25]	1.4371[20]
10704	2-Propen-1-ol, 3-phenyl-, (E)-	$C_9H_{10}O$ 134.18	4407-36-7 34	257.5	4-06-00-03799 1.0440[20]	H_2O 2; EtOH 4; eth 4; chl 2 1.5819[20]
10705	2-Propen-1-ol, 3-phenyl-, (Z)-	$C_9H_{10}O$ 134.18	4510-34-3 34	257.5	3-06-00-02401 1.0440[20]	eth 4; EtOH 4 1.5819[20]
10706	2-Propen-1-ol, 3-phenyl-, acetate, (E)-	$C_{11}H_{12}O_2$ 176.22	21040-45-9	145[15]; 114[1]	4-06-00-03801 1.0567[20]	H_2O 1; EtOH 3; eth 3; ace 3 1.5425[20]
10707	2-Propen-1-ol, 3-phenyl-, formate	$C_{10}H_{10}O_2$ 162.19	104-65-4 0	252	0-06-00-00571 1.086[25]	
10708	2-Propen-1-one, 1,3-diphenyl- Chalcone	$C_{15}H_{12}O$ 208.26	94-41-7 59	2028 346 dec	4-07-00-01658 1.0712[62]	bz 4; EtOH 4; chl 4

No.	Name Synonym	Mol. Form. Mol. Wt.	CAS RN mp/°C	Merck No. bp/°C	Beil. Ref. den/g cm^{-3}	Solubility n_D
10709	2-Propen-1-one, 1,3-diphenyl-, (E)-	C$_{15}$H$_{12}$O 208.26	614-47-1 59	346 dec	4-07-00-01658 1.0712[62]	H$_2$O 1; EtOH 2; eth 3; bz 3
10710	2-Propen-1-one, 3-(2-furanyl)-1-phenyl-	C$_{13}$H$_{10}$O$_2$ 198.22	717-21-5 47	317	5-17-10-00371 1.1140[20]	EtOH 3; eth 3
10711	2-Propen-1-one, 3-hydroxy-1,3-diphenyl-	C$_{15}$H$_{12}$O$_2$ 224.26	1704-15-0 79.8	220[18]; 140[1.0]	4-07-00-02512	eth 4; EtOH 4; chl 4
10712	2-Propen-1-one, 3-(3-methoxyphenyl)-1-phenyl-	C$_{16}$H$_{14}$O$_2$ 238.29	5470-91-7 65	247[12]	1-08-00-00579	ace 4; bz 4; eth 4; EtOH 4
10713	2-Propen-1-one, 3-(4-methoxyphenyl)-1-phenyl-	C$_{16}$H$_{14}$O$_2$ 238.29	959-33-1 79	187[19]	2-08-00-00218	H$_2$O 1; EtOH 4; eth 3; ctc 3
10714	2-Propenoyl chloride Acrylic acid chloride	C$_3$H$_3$ClO 90.51	814-68-6	75.5	4-02-00-01471 1.1136[20]	chl 4 1.4343[20]
10715	2-Propenoyl chloride, 2-methyl- Methacrylic acid chloride	C$_4$H$_5$ClO 104.54	920-46-7 -60	96	4-02-00-01537 1.0871[20]	eth 3; ace 3; chl 3 1.4435[20]
10716	2-Propenoyl chloride, 3-phenyl- Cinnamoyl chloride	C$_9$H$_7$ClO 166.61	102-92-1 36	2303 257	2-09-00-00395 1.1617[45]	lig 4 1.614[42]
10717	2-Propenoyl chloride, 3-phenyl-, (E)-	C$_9$H$_7$ClO 166.61	17082-09-6 37.5	257.5	2-09-00-00390 1.1617[45]	H$_2$O 1; EtOH 3; ctc 3; lig 3 1.614[42]
10718	2-Propenoyl chloride, 2,3,3-trichloro-	C$_3$Cl$_4$O 193.84	815-58-7	158	4-02-00-01486	bz 4 1.5271[18]
10719	Propetamphos 2-Butenoic acid, 3-[[(ethylamino)methoxyphosphinothioyl]oxy]-, 1-methylethyl ester, (E)-	C$_{10}$H$_{20}$NO$_4$PS 281.31	31218-83-4	7827 88[0.005]	1.1294[20]	
10720	Propiconazole 1-[[2-(2,4-Dichlorophenyl)-4-propyl-1,3-dioxolan-2-yl]methyl]-1H-1,2,4-triazole	C$_{15}$H$_{17}$Cl$_2$N$_3$O$_2$ 342.22	60207-90-1	180[0.1]	1.27[20]	
10721	Propionic acid, 2,2-dichloro-3,3,3-trifluoro-, propyl ester Propanoic acid, 2,2-dichloro-3,3,3-trifluoro, propyl ester	C$_6$H$_7$Cl$_2$F$_3$O$_2$ 239.02	357-49-3	144.5[625]	4-02-00-00758 1.3531[20]	1.3888[20]
10722	Propoxur Phenol, 2-(1-methylethoxy)-, methylcarbamate	C$_{11}$H$_{15}$NO$_3$ 209.25	114-26-1 87	7849 dec	1.12[20]	
10723	2-Propynal Propargyl aldehyde	C$_3$H$_2$O 54.05	624-67-9	60	4-01-00-03537 0.9152[20]	H$_2$O 5; EtOH 3; eth 3; ace 3 1.4033[25]
10724	2-Propynal, 3-phenyl-	C$_9$H$_6$O 130.15	2579-22-8	127[28]; 104[11]	4-07-00-01173 1.0680[12]	1.6079[12]
10725	2-Propyn-1-amine	C$_3$H$_5$N 55.08	2450-71-7	83	4-04-00-01135 0.803[25]	1.4480[20]
10726	2-Propyn-1-amine, N-methyl-	C$_4$H$_7$N 69.11	35161-71-8	83	0-04-00-00228 0.819[25]	1.4332[20]
10727	1-Propyne Methylacetylene	C$_3$H$_4$ 40.06	74-99-7 -102.7	-23.2	4-01-00-00958 *0.607[25]	H$_2$O 2; EtOH 4; bz 3; chl 3 1.3863[-40]
10728	1-Propyne, 3-bromo- Propargyl bromide	C$_3$H$_3$Br 118.96	106-96-7	89	4-01-00-00964 1.579[19]	EtOH 3; eth 3; bz 3; ctc 3 1.4922[20]
10729	1-Propyne, 3-chloro- Propargyl chloride	C$_3$H$_3$Cl 74.51	624-65-7	7820 58	4-01-00-00963 1.030[25]	H$_2$O 1; EtOH 5; eth 5; bz 3 1.4349[20]
10730	1-Propyne, 1,3-dibromo- Propyne, 1,3-dibromo-	C$_3$H$_2$Br$_2$ 197.86	627-16-7	73-4[30]	4-01-00-00965 2.1894[20]	eth 4; chl 4 1.5690[20]
10731	1-Propyne, 3,3-diethoxy-	C$_7$H$_{12}$O$_2$ 128.17	10160-87-9	139	4-01-00-03538 0.8942[22]	ace 4; eth 4; EtOH 4; chl 4 1.4140[20]
10732	1-Propyne, 1-ethoxy-	C$_5$H$_8$O 84.12	14273-06-4	84	3-01-00-01969 0.8276[20]	eth 4; EtOH 4 1.4039[20]
10733	1-Propyne, 3-ethoxy-	C$_5$H$_8$O 84.12	628-33-1	82	4-01-00-02215 0.8326[20]	eth 4; EtOH 4 1.4039[20]
10734	1-Propyne, 1-iodo-	C$_3$H$_3$I 165.96	624-66-8	110	4-01-00-00965 2.08[22]	eth 4; EtOH 4
10735	1-Propyne, 3-iodo-	C$_3$H$_3$I 165.96	659-86-9	115	4-01-00-00965 2.0177[25]	eth 3
10736	1-Propyne, 3-methoxy-	C$_4$H$_6$O 70.09	627-41-8	63	4-01-00-02215 0.83[12]	eth 4; EtOH 4 1.5035[20]
10737	2-Propynenitrile Cyanoacetylene	C$_3$HN 51.05	1070-71-9 5	42.5	4-02-00-01689 0.8167[17]	H$_2$O 2; EtOH 3 1.3868[25]
10738	2-Propynoic acid Propiolic acid	C$_3$H$_2$O$_2$ 70.05	471-25-0 9	7833 144 dec; 72[50]	4-02-00-01687 1.1380[20]	H$_2$O 4; eth 4; EtOH 4; chl 4 1.4306[20]
10739	2-Propynoic acid, ethyl ester	C$_5$H$_6$O$_2$ 98.10	623-47-2	120	4-02-00-01688 0.9560[25]	H$_2$O 1; EtOH 4; eth 4; chl 4 1.4105[20]
10740	2-Propynoic acid, 3-(2-nitrophenyl)- o-Nitrophenylpropiolic acid	C$_9$H$_5$NO$_4$ 191.14	530-85-8 157 dec	6546 exp	4-09-00-02330	eth 4; EtOH 4
10741	2-Propynoic acid, 3-phenyl-, ethyl ester	C$_{11}$H$_{10}$O$_2$ 174.20	2216-94-6	265; 128[1.6]	4-09-00-02328 1.055[25]	eth 3 1.5520[20]
10742	2-Propynoic acid, 3-phenyl-, methyl ester	C$_{10}$H$_8$O$_2$ 160.17	4891-38-7 26	158[48]; 132[16]	4-09-00-02328 1.0830[25]	1.5618[25]

No.	Name Synonym	Mol. Form. Mol. Wt.	CAS RN mp/°C	Merck No. bp/°C	Beil. Ref. den/g cm^{-3}	Solubility n_D
10743	2-Propyn-1-ol Propargyl alcohol	C_3H_4O 56.06	107-19-7 -51.8	7819 113.6	4-01-00-02214 0.9478[20]	H_2O 3; EtOH 5; eth 5; chl 3 1.4322[20]
10744	2-Propyn-1-ol, acetate Propargyl acetate	$C_5H_6O_2$ 98.10	627-09-8 	 121.5	4-02-00-00197 0.9982[20]	H_2O 2; EtOH 3; eth 3 1.4187[20]
10745	2-Propyn-1-ol, 3-phenyl-	C_9H_8O 132.16	1504-58-1 137-8[15]	 	4-06-00-04065 1.07[18]	eth 3; ace 3; bz 3 1.5873[28]
10746	Propyzamide Benzamide, 3,5-dichloro-N-(1,1-dimethyl-2-propynyl)-	$C_{12}H_{11}Cl_2NO$ 256.13	23950-58-5 155	7886 	 	
10747	Pteridine Pyrazino[2,3-d]pyrimidine	$C_6H_4N_4$ 132.12	91-18-9 139.5	7947 sub 125	4-26-00-01770 	H_2O 4; EtOH 2; eth 2; bz 2
10748	2,4(1H,3H)-Pteridinedione Lumazine	$C_6H_4N_4O_2$ 164.12	487-21-8 348.5	5467 	4-26-00-02489 	HOAc 4
10749	4,6-Pteridinedione, 2-amino-1,5-dihydro- Xanthopterin	$C_6H_5N_5O_2$ 179.14	119-44-8 >410 dec	9973 98-100[18]	4-26-00-04000 1.559[25]	H_2O 1; EtOH 2; eth 2; acid 4
10750	2,4,7-Pteridinetriamine, 6-phenyl- Triamterene	$C_{12}H_{11}N_7$ 253.27	396-01-0 	9515 	4-26-00-03837 	
10751	1H-Purin-6-amine Adenine	$C_5H_5N_5$ 135.13	73-24-5 360 dec	141 sub 220	4-26-00-03561 	H_2O 3; EtOH 2; eth 1; chl 1
10752	1H-Purine 7H-Imidazo[4,5-d]pyrimidine	$C_5H_4N_4$ 120.11	120-73-0 216.5	7959 	4-26-00-01736 	H_2O 4; EtOH 4; eth 2; ace 3
10753	1H-Purine, 6-chloro- 6-Chloropurine	$C_5H_3ClN_4$ 154.56	87-42-3 175-7 dec	2161 	4-26-00-01742 	
10754	1H-Purine-2,6-diamine 2,6-Diaminopurine	$C_5H_6N_6$ 150.14	1904-98-9 302	2963 	4-26-00-03785 	
10755	1H-Purine-2,6-dione, 3,7-dihydro- Xanthine	$C_5H_4N_4O_2$ 152.11	69-89-6 dec	9968 sub	4-26-00-02327 	
10756	1H-Purine-2,6-dione, 3,7-dihydro-1,3-dimethyl- Theophylline	$C_7H_8N_4O_2$ 180.17	58-55-9 273	9212 	4-26-00-02331 	H_2O 4; EtOH 2; eth 2; chl 2
10757	1H-Purine-2,6-dione, 3,7-dihydro-3,7-dimethyl- Theobromine	$C_7H_8N_4O_2$ 180.17	83-67-0 357	9209 sub 290	4-26-00-02336 	H_2O 2; EtOH 2; eth 1; bz 1
10758	1H-Purine-2,6-dione, 3,7-dihydro-7-(2-hydroxyethyl)-1,3-dimethyl- Etofylline	$C_9H_{12}N_4O_3$ 224.22	519-37-9 158	3835 	4-26-00-02357 	H_2O 4; EtOH 2; eth 2; bz 2
10759	1H-Purine-2,6-dione, 3,7-dihydro-8-methoxy-1,3,7-trimethyl-	$C_9H_{12}N_4O_3$ 224.22	569-34-6 176	 	4-26-00-02727 1.399[25]	H_2O 3; EtOH 4; bz 3
10760	1H-Purine-2,6-dione, 3,7-dihydro-1,3,7-trimethyl- Caffeine	$C_8H_{10}N_4O_2$ 194.19	58-08-2 238	1635 sub 90	4-26-00-02338 1.23[19]	H_2O 2; EtOH 2; eth 1; ace 2
10761	1H-Purine-2,6-dione, 7-(2,3-dihydroxypropyl)-3,7-dihydro-1,3-dimethyl- Dyphylline	$C_{10}H_{14}N_4O_4$ 254.25	479-18-5 161.5	3459 	4-26-00-02370 	
10762	6H-Purine-6-thione, 2-amino-1,7-dihydro- Thioguanine	$C_5H_5N_5S$ 167.19	154-42-7 >360	9266 	4-26-00-03926 	
10763	6H-Purine-6-thione, 1,7-dihydro- 6-Mercaptopurine	$C_5H_4N_4S$ 152.18	50-44-2 313 dec	5762 	 	H_2O 1; alk 3
10764	1H-Purine-2,6,8(3H)-trione, 7,9-dihydro- Uric acid	$C_5H_4N_4O_3$ 168.11	69-93-2 dec	9790 dec	4-26-00-02619 1.89[25]	H_2O 1; EtOH 1; eth 1; alk 4
10765	1H-Purine-2,6,8(3H)-trione, 4,9-dihydro-3-methyl- 3-Methyluric acid	$C_6H_6N_4O_3$ 182.14	605-99-2 >350	 	4-26-00-02621 1.6104[25]	H_2O 2; EtOH 2; alk 3 1.6334[25]
10766	1H-Purine-2,6,8(3H)-trione, 7,9-dihydro-7-methyl- 7-Methyluric acid	$C_6H_6N_4O_3$ 182.14	612-37-3 370-80 dec	 	4-26-00-02622 1.706[25]	H_2O 2; NaOH 3
10767	2H-Purin-2-one, 6-amino-1,3-dihydro- Isoguanine	$C_5H_5N_5O$ 151.13	3373-53-3 >360	 	 	
10768	6H-Purin-6-one, 2-amino-1,7-dihydro- Guanine	$C_5H_5N_5O$ 151.13	73-40-5 360 dec	4477 sub	4-26-00-03890 	H_2O 1; EtOH 2; eth 2; alk 3
10769	6H-Purin-6-one, 2-amino-1,7-dihydro-7-methyl 7-Methylguanine	$C_6H_7N_5O$ 165.15	578-76-7 370	 	 	
10770	6H-Purin-6-one, 1,7-dihydro- Hypoxanthine	$C_5H_4N_4O$ 136.11	68-94-0 150 dec	4805 	4-26-00-02081 	H_2O 2; alk 3; dil acid 3
10771	Pyran	C_5H_6O 82.10	33941-07-0 	 80	0-17-00-00036 	EtOH 3; eth 3; bz 3 1.4559[20]
10772	2H-Pyran-5-carboxylic acid, 3,4-dihydro-3-(hydroxymethyl)-2-methyl-4-[(9-methyl-9H-pyrido[3,4-b]indol-1-yl)methyl]-, methyl ester, [2S-(2α,3β,4β)]- Alstonidine	$C_{22}H_{24}N_2O_4$ 380.44	25394-75-6 189	313 	4-27-00-07962 	ace 4; EtOH 4
10773	2H-Pyran-5-carboxylic acid, 2-oxo- Coumalic acid	$C_6H_4O_4$ 140.10	500-05-0 205-10 dec	2558 218[120]	5-18-08-00120 	H_2O 2; EtOH 3; eth 2; ace 2
10774	2H-Pyran-4-carboxylic acid, tetrahydro- Pyran-4-carboxylic acid, tetrahydro-	$C_6H_{10}O_3$ 130.14	5337-03-1 88	 144[11]	5-18-06-00023 	H_2O 3; peth 2
10775	4H-Pyran-2,6-dicarboxylic acid, 3-hydroxy-4-oxo- Meconic acid	$C_7H_4O_7$ 200.10	497-59-6 120 dec	5664 120 dec	5-18-09-00319 	H_2O 2; EtOH 3; eth 2; ace 2
10776	4H-Pyran-2,6-dicarboxylic acid, 4-oxo- Chelidonic acid	$C_7H_4O_6$ 184.11	99-32-1 262	2042 	5-18-08-00646 	H_2O 3; EtOH 2

No.	Name / Synonym	Mol. Form. / Mol. Wt.	CAS RN / mp/°C	Merck No. / bp/°C	Beil. Ref. / den/g cm^{-3}	Solubility / n_D
10777	2H-Pyran, 3,4-dihydro- / 3,4-Dihydro-2H-pyran	C_5H_8O / 84.12	110-87-2 /	/ 86	5-17-01-00181 / 0.921[19]	H_2O 3; EtOH 3; chl 2 / 1.4402[19]
10778	2H-Pyran, 3,6-dihydro-4-methyl-	$C_6H_{10}O$ / 98.14	16302-35-5 /	/ 117.5	5-17-01-00194 / 0.912[25]	/ 1.4495[20]
10779	2H-Pyran-2,4(3H)-dione, 3-acetyl-6-methyl- / Dehydroacetic acid	$C_8H_8O_4$ / 168.15	520-45-6 / 109	2855 / 270	4-17-00-06699 /	H_2O 4; EtOH 2; eth 4; chl 2
10780	2H-Pyran-2,6(3H)-dione, dihydro-	$C_5H_6O_3$ / 114.10	108-55-4 / 56.3	/ 158[15]	5-17-11-00009 /	/ 1.4110[20]
10781	2H-Pyran-2,6(3H)-dione, dihydro-3-phenyl-	$C_{11}H_{10}O_3$ / 190.20	2959-96-8 / 95	/ 224[13]	5-17-11-00279 /	EtOH 4
10782	2H-Pyran-2,6(3H)-dione, dihydro-4-phenyl-	$C_{11}H_{10}O_3$ / 190.20	4160-80-9 / 105.6	/ 217[15]	5-17-11-00279 /	bz 4; eth 4; chl 4
10783	2H-Pyran, 2-ethoxy-3,4-dihydro-	$C_7H_{12}O_2$ / 128.17	103-75-3 /	/ 42[16]	5-17-03-00196 / 0.957[25]	/ 1.4394[20]
10784	2H-Pyran-2-methanol, tetrahydro-	$C_6H_{12}O_2$ / 116.16	100-72-1 /	/ 185	5-17-03-00136 / 1.027[25]	/ 1.458[20]
10785	4H,5H-Pyrano[4,3-b]pyran-4,5-dione, 2,3-dihydro-3-hydroxy-2-methyl-7-(1-propenyl)-, [2S-[2α,3β,7(E)]]- / Radicinin	$C_{12}H_{12}O_5$ / 236.22	10088-95-6 / 221.5	8117 /	4-19-00-02746 /	chl 2
10786	Pyrano[3,2-c][2]benzopyran-6(2H)-one, 3,4,4a,10b-tetrahydro-3,4,8,10-tetrahydroxy-2-(hydroxymethyl)-9-methoxy-[2R-(2α,3β,4α,4aα,10bβ)]- / Bergenin	$C_{14}H_{16}O_9$ / 328.28	477-90-7 / 238	1174 /	5-19-07-00132 /	H_2O 4; EtOH 4
10787	2H-Pyran-4-ol, tetrahydro-	$C_5H_{10}O_2$ / 102.13	2081-44-9 /	/ 87[15]	5-17-03-00111 / 1.071[25]	/ 1.4600[20]
10788	2H-Pyran-2-one	$C_5H_4O_2$ / 96.09	504-31-4 / 8.5	/ 207.5	5-17-09-00288 / 1.200[20]	H_2O 5; ace 4 / 1.5270[25]
10789	4H-Pyran-4-one	$C_5H_4O_2$ / 96.09	108-97-4 / 32.5	/ 212.5	5-17-09-00290 / 1.190[25]	H_2O 4; EtOH 3; eth 4; bz 3 / 1.5238
10790	2H-Pyran-2-one, 6-[2-(1,3-benzodioxol-5-yl)ethenyl]-5,6-dihydro-4-methoxy-, [R-(E)]- / Methysticin	$C_{15}H_{14}O_5$ / 274.27	495-85-2 / 137	6056 /	5-19-10-00456 /	
10791	2H-Pyran-2-one, 5,6-dihydro-6-methyl-, (S)- / Parasorbic acid	$C_6H_8O_2$ / 112.13	10048-32-5 / 100[15]	6981 /	5-17-09-00130 / 1.079[18]	H_2O 4; eth 4; EtOH 4 / 1.4736[20]
10792	2H-Pyran-2-one, 4,6-dimethyl-	$C_7H_8O_2$ / 124.14	675-09-2 / 51.5	/ 245	5-17-09-00409 /	H_2O 4; eth 4; EtOH 4
10793	4H-Pyran-4-one, 2,6-dimethyl-	$C_7H_8O_2$ / 124.14	1004-36-0 / 132	/ 251; 140[25]	5-17-09-00410 / 0.9953[137]	H_2O 3; EtOH 3; eth 3; ace 3
10794	2H-Pyran-2-one, 3-hydroxy-	$C_5H_4O_3$ / 112.08	496-64-0 / 93.8	/ 112[20]; 102[15]	5-18-01-00095 /	H_2O 3; EtOH 3; eth 3; bz 3
10795	4H-Pyran-4-one, 5-hydroxy-2-(hydroxymethyl)- / Kojic acid	$C_6H_6O_4$ / 142.11	501-30-4 / 153.5	5197 /	5-18-02-00516 /	H_2O 2; EtOH 3; eth 3; ace 3
10796	4H-Pyran-4-one, 3-hydroxy-2-methyl- / Maltol	$C_6H_6O_3$ / 126.11	118-71-8 / 161.5	5594 / sub 93	5-18-01-00114 /	H_2O 2; eth 2; bz 2; chl 4
10797	4H-Pyran-4-one, 2-methoxy-3,5-dimethyl-6-[tetrahydro-4-[2-methyl-3-(4-nitrophenyl)-2-propenylidene]-2-furanyl] / Aureothin	$C_{22}H_{23}NO_6$ / 397.43	2825-00-5 / 158	897 /	5-19-06-00099 /	ace 4; EtOH 4; chl 4
10798	2H-Pyran-2-one, tetrahydro-	$C_5H_8O_2$ / 100.12	542-28-9 / -12.5	/ 219	5-17-09-00017 / 1.0794[20]	H_2O 3; EtOH 5; eth 5; ctc 2 / 1.4503[20]
10799	4H-Pyran-4-one, tetrahydro-	$C_5H_8O_2$ / 100.12	29943-42-8 /	/ 166.5	5-17-09-00021 / 1.084[25]	/ 1.4520[20]
10800	2H-Pyran, tetrahydro- / Oxane	$C_5H_{10}O$ / 86.13	142-68-7 / -45	9149 / 88	5-17-01-00064 / 0.8814[20]	EtOH 3; eth 3; bz 3; ctc 3 / 1.4200[20]
10801	Pyrazine / 1,4-Diazine	$C_4H_4N_2$ / 80.09	290-37-9 / 55	7971 / 115	5-23-05-00351 / 1.0311[61]	H_2O 3; EtOH 3; eth 3; ace 3 / 1.4953[61]
10802	Pyrazinecarboxamide / Pyrazinamide	$C_5H_5N_3O$ / 123.11	98-96-4 / 192	7970 / sub	4-25-00-00772 /	H_2O 3; EtOH 3
10803	Pyrazinecarboxylic acid / Pyrazinoic acid	$C_5H_4N_2O_2$ / 124.10	98-97-5 / 225 dec	7973 / sub	4-25-00-00771 /	
10804	2,3-Pyrazinedicarboxylic acid / 2,3-Dicarboxypyrazine	$C_6H_4N_2O_4$ / 168.11	89-01-0 / 193 dec	7972 /	4-25-00-01064 /	H_2O 4; EtOH 2; eth 2; ace 3
10805	Pyrazine, 2,3-dimethyl- / 2,3-Dimethylpyrazine	$C_6H_8N_2$ / 108.14	5910-89-4 /	/ 156	5-23-05-00402 / 1.0281[0]	H_2O 3; EtOH 3; eth 3
10806	Pyrazine, 2,5-dimethyl- / 2,5-Dimethylpyrazine	$C_6H_8N_2$ / 108.14	123-32-0 / 15	/ 155	5-23-05-00403 / 0.9887[20]	H_2O 5; EtOH 5; eth 5; ace 3 / 1.4980[20]
10807	Pyrazine, 2,6-dimethyl- / 2,6-Dimethylpyrazine	$C_6H_8N_2$ / 108.14	108-50-9 / 47.5	/ 155.6	5-23-05-00406 / 0.9647[50]	H_2O 3; EtOH 3; eth 3; ctc 2
10808	Pyrazineethanol	$C_6H_8N_2O$ / 124.14	6705-31-3 /	/ 129[10]	5-23-11-00173 / 1.163[20]	/ 1.5378[20]
10809	Pyrazine, 3-ethyl-2,5-dimethyl-	$C_8H_{12}N_2$ / 136.20	13360-65-1 /	/ 180.5	5-23-05-00440 / 0.9657[24]	H_2O 2; EtOH 2; eth 2 / 1.5014[24]
10810	Pyrazine, methyl- / 2-Methylpyrazine	$C_5H_8N_2$ / 94.12	109-08-0 / -29	/ 137	5-23-05-00386 / 1.03[20]	H_2O 5; EtOH 5; eth 5; ace 3 / 1.5042[20]
10811	2(1H)-Pyrazinone, 1-hydroxy-6-(1-methylpropyl)-3-(2-methylpropyl)-, (+)- / Aspergillic acid	$C_{12}H_{20}N_2O_2$ / 224.30	490-02-8 / 98	863 /	4-24-00-00235 /	bz 4; eth 4; EtOH 4

No.	Name Synonym	Mol. Form. Mol. Wt.	CAS RN mp/°C	Merck No. bp/°C	Beil. Ref. den/g cm^{-3}	Solubility n_D
10812	1H-Pyrazol-4-amine 4-Aminopyrazole	$C_3H_5N_3$ 83.09	28466-26-4 81	sub	4-25-00-02029	H_2O 4; EtOH 4; chl 4
10813	1H-Pyrazol-4-amine, 3-methyl-1-phenyl- Pyrazole, 4-amino-3-methyl-1-phenyl-	$C_{10}H_{11}N_3$ 173.22	103095-51-8 88.8	314.5	4-25-00-02033	bz 4; eth 4; EtOH 4; chl 4
10814	1H-Pyrazol-5-amine, 3-methyl-1-phenyl-	$C_{10}H_{11}N_3$ 173.22	1131-18-6 116	333	4-25-00-02034	H_2O 3; EtOH 3; bz 2; chl 3
10815	1H-Pyrazole 1,2-Diazole	$C_3H_4N_2$ 68.08	288-13-1 68	7974 187	5-23-04-00122	H_2O 4; EtOH 3; eth 3; bz 3 1.4203
10816	1H-Pyrazole, 4-bromo-1,3-dimethyl- 4-Bromo-1,3-dimethylpyrazole	$C_5H_7BrN_2$ 175.03	5775-82-6	76-7[10]	5-23-05-00022 1.4975[15]	ace 4; eth 4; EtOH 4 1.5214[15]
10817	1H-Pyrazole, 4-bromo-3-methyl- 4-Bromo-3-methylpyrazole	$C_4H_5BrN_2$ 161.00	13808-64-5 76.5		5-23-05-00022 1.5638[100]	ace 4; eth 4; EtOH 4 1.5182[100]
10818	1H-Pyrazole, 3-chloro-1,5-dimethyl- 3-Chloro-1,5-dimethylpyrazole	$C_5H_7ClN_2$ 130.58	51500-32-4 47.5	211	2-23-00-00049 1.0823[100]	ace 4; bz 4; eth 4; EtOH 4 1.4648[100]
10819	1H-Pyrazole, 4-chloro-3,5-dimethyl-	$C_5H_7ClN_2$ 130.58	15953-73-8 117.5	221	5-23-05-00151	ace 4; bz 4; eth 4; EtOH 4
10820	1H-Pyrazole, 5-chloro-1,3-dimethyl- 5-Chloro-1,3-dimethylpyrazole	$C_5H_7ClN_2$ 130.58	54454-10-3	157.5	5-23-05-00019 1.1367[18]	H_2O 4 1.4877[18]
10821	1H-Pyrazole, 3-chloro-5-methyl-	$C_4H_5ClN_2$ 116.55	15953-45-4 118.5	258; 138[15]	5-23-05-00019	H_2O 4
10822	1H-Pyrazole, 4,5-dihydro- 2-Pyrazoline	$C_3H_6N_2$ 70.09	109-98-8 144	7975	5-23-03-00361 1.0200[17]	H_2O 4; eth 4; EtOH 4 1.4796[17]
10823	1H-Pyrazole, 4,5-dihydro-1-phenyl- 1-Phenyl-2-pyrazoline	$C_9H_{10}N_2$ 146.19	936-53-8 52	273; 151[17]	5-23-03-00363 1.0689[58]	H_2O 1; EtOH 3; eth 3; bz 3 1.6015[58]
10824	1H-Pyrazole, 4,5-dihydro-3-phenyl- 2-Pyrazoline, 3-phenyl	$C_9H_{10}N_2$ 146.19	936-48-1 45.8	164[17]	5-23-06-00396 1.0892[54]	
10825	1H-Pyrazole, 1,3-dimethyl-	$C_5H_8N_2$ 96.13	694-48-4	137	5-23-05-00012 0.9561[17]	H_2O 4 1.4734[15]
10826	1H-Pyrazole, 1,5-dimethyl-	$C_5H_8N_2$ 96.13	694-31-5	149	5-23-05-00006 0.9813[17]	H_2O 4; EtOH 4; eth 4 1.4782[16]
10827	1H-Pyrazole, 3,4-dimethyl-	$C_5H_8N_2$ 96.13	2820-37-3 58	110[11]	5-23-05-00108 0.9325[99]	EtOH 4; eth 3; ace 3; bz 3
10828	1H-Pyrazole, 3,5-dimethyl-	$C_5H_8N_2$ 96.13	67-51-6 107.5	218	5-23-05-00110 0.8839[16]	H_2O 3; EtOH 4; eth 4; ace 3
10829	1H-Pyrazole, 3,5-dimethyl-1-phenyl-	$C_{11}H_{12}N_2$ 172.23	1131-16-4	273; 145[12.5]	5-23-05-00127 1.0566[20]	eth 4; EtOH 4; chl 4 1.5738[19]
10830	1H-Pyrazole, 1,3-diphenyl- 1,3-Diphenylpyrazole	$C_{15}H_{12}N_2$ 220.27	4492-01-7 85.8	341[270]	5-23-07-00178 1.0794[101]	
10831	1H-Pyrazole, 1,5-diphenyl- 1,5-Diphenylpyrazole	$C_{15}H_{12}N_2$ 220.27	6831-89-6 55.5	338; 167[12]	5-23-07-00176 1.0696[100]	bz 4; eth 4; EtOH 4
10832	1H-Pyrazole, 1-ethyl-	$C_5H_8N_2$ 96.13	2817-71-2	139	5-23-04-00135 0.9537[20]	ace 4; bz 4; eth 4; EtOH 4 1.4700[20]
10833	1H-Pyrazole, 1-ethyl-3-methyl- 1-Ethyl-3-methylpyrazole	$C_6H_{10}N_2$ 110.16	30433-57-9 115	152	5-23-05-00013 0.936[20]	
10834	1H-Pyrazole, 1-methyl- 1-Methylpyrazole	$C_4H_6N_2$ 82.11	930-36-9	127	5-23-04-00134 0.9929[13]	1.4787[13]
10835	1H-Pyrazole, 3-methyl-	$C_4H_6N_2$ 82.11	1453-58-3 36.5	204; 108[25]	5-23-05-00004 1.0203[16]	H_2O 5; EtOH 5; eth 5 1.4915[20]
10836	1H-Pyrazole, 4-methyl- 4-Methylpyrazole	$C_4H_6N_2$ 82.11	7554-65-6 10146	206; 95[13]	5-23-05-00031 1.015[20]	
10837	1H-Pyrazole, 1-methyl-3-phenyl-	$C_{10}H_{10}N_2$ 158.20	3463-26-1 56	280.5	5-23-07-00178 1.0232[99]	
10838	1H-Pyrazole, 3-methyl-1-phenyl-	$C_{10}H_{10}N_2$ 158.20	1128-54-7 37	255	5-23-05-00013 1.076[20]	
10839	1H-Pyrazole, 1,3,4-trimethyl- 1,3,4-Trimethylpyrazole	$C_6H_{10}N_2$ 110.16	15802-99-0	160	5-23-05-00109 0.9587[17]	1.4866[17]
10840	1H-Pyrazole, 1,3,5-trimethyl- 1,3,5-Trimethylpyrazole	$C_6H_{10}N_2$ 110.16	1072-91-9	170	5-23-05-00125 0.9130[57]	1.4589[57]
10841	1H-Pyrazole, 1,4,5-trimethyl- 1,4,5-Trimethylpyrazole	$C_6H_{10}N_2$ 110.16	15802-97-8	176.5	5-23-05-00108 0.9685[17]	1.4849[17]
10842	1H-Pyrazole, 3,4,5-trimethyl-	$C_6H_{10}N_2$ 110.16	5519-42-6 138.5	233	5-23-05-00198	bz 4; eth 4; EtOH 4
10843	3,5-Pyrazolidinedione, 1,4-diphenyl- Phenopyrazone	$C_{15}H_{12}N_2O_2$ 252.27	3426-01-5 233.5	7217	5-24-08-00003	
10844	3-Pyrazolidinone, 1-phenyl- 1-Phenyl-3-pyrazolidinone	$C_9H_{10}N_2O$ 162.19	92-43-3 126	7281	5-24-01-00011	eth 1; lig 1
10845	3H-Pyrazol-3-one, 4-amino-1,2-dihydro-1,5-dimethyl-2-phenyl- Ampyrone	$C_{11}H_{13}N_3O$ 203.24	83-07-8 109	625	4-25-00-03554	H_2O 3; EtOH 3; eth 2; bz 3
10846	3H-Pyrazol-3-one, 4-bromo-1,2-dihydro-1,5-dimethyl-2-phenyl- 4-Bromoantipyrine	$C_{11}H_{11}BrN_2O$ 267.13	5426-65-3 122	300[9]	5-24-01-00343 1.5900[20]	H_2O 3; EtOH 4; eth 3; chl 4
10847	3H-Pyrazol-3-one, 1,2-dihydro-1,5-dimethyl-2-phenyl- Antipyrine	$C_{11}H_{12}N_2O$ 188.23	60-80-0 114	748 319	5-24-01-00330	H_2O 4; EtOH 4
10848	3H-Pyrazol-3-one, 2,4-dihydro-5-methyl-4-nitro-2-(4-nitrophenyl)- Picrolonic acid	$C_{10}H_8N_4O_5$ 264.20	550-74-3 116	7383 dec		H_2O 2; EtOH 3; eth 3; MeOH 3

No.	Name Synonym	Mol. Form. Mol. Wt.	CAS RN mp/°C	Merck No. bp/°C	Beil. Ref. den/g cm^{-3}	Solubility n_D
10849	3H-Pyrazol-3-one, 1,2-dihydro-5-methyl-2-phenyl-	$C_{10}H_{10}N_2O$	19735-89-8		5-24-01-00317	H$_2$O 3; EtOH 3; bz 2; peth 1
		174.20	128	287[105.] 191[17]	1.2600[20]	1.637
10850	3H-Pyrazol-3-one, 1,2-dihydro-2-(1-methyl-4-piperidinyl)-5-phenyl-4-(phenylmethyl)- Benzpiperylon	$C_{22}H_{25}N_3O$	53-89-4	1131	5-24-04-00398	
		347.46	181-3 dec			
10851	3H-Pyrazol-3-one, 4-(dimethylamino)-1,2-dihydro-1,5-dimethyl-2-phenyl- Aminopyrine	$C_{13}H_{17}N_3O$	58-15-1	488	4-24-00-00096	H$_2$O 4; bz 4; EtOH 4
		231.30	134.5			
10852	4H-Pyrazolo[3,4-d]pyrimidin-4-one, 1,5-dihydro- Allopurinol	$C_5H_4N_4O$	315-30-0	278	4-26-00-02076	
		136.11	350			
10853	Pyrene Benzo[def]phenanthrene	$C_{16}H_{10}$ 202.26	129-00-0 151.2	7977 404	4-05-00-02467 1.271[23]	H$_2$O 1; EtOH 3; eth 3; bz 3
10854	1-Pyrenecarboxylic acid 3-Pyrenecarboxylic acid	$C_{17}H_{10}O_2$ 246.27	19694-02-1 274	sub	3-09-00-03575	eth 4; chl 4
10855	Pyridazine 1,2-Diazabenzene	$C_4H_4N_2$ 80.09	289-80-5 -8	7982 208	5-23-05-00321 1.1035[23]	H$_2$O 5; EtOH 5; eth 4; ace 4 1.5218[20]
10856	Pyridazine, 3-methyl- 3-Methyl-1,2-diazine	$C_5H_6N_2$ 94.12	1632-76-4 184	214	5-23-05-00371 1.0450[26]	1.5145[20]
10857	2-Pyridinamine 2-Aminopyridine	$C_5H_6N_2$ 94.12	504-29-0 57.5	486 105[20]	5-22-08-00280	EtOH 3; eth 3; ace 3; bz 3
10858	3-Pyridinamine β-Aminopyridine	$C_5H_6N_2$ 94.12	462-08-8 64.5	487 252	5-22-09-00003	H$_2$O 3; EtOH 3; eth 3; lig 2
10859	4-Pyridinamine 4-Aminopyridine	$C_5H_6N_2$ 94.12	504-24-5 158.5	273	5-22-09-00106	H$_2$O 3; EtOH 4; eth 3; bz 3
10860	3-Pyridinamine, 6-butoxy-	$C_9H_{14}N_2O$	539-23-1		4-22-00-05574	EtOH 3; eth 3; chl 2; dil acid 3
		166.22		148-50[12]	1.037[25]	1.5373[20]
10861	2-Pyridinamine, 5-chloro-	$C_5H_5ClN_2$ 128.56	1072-98-6 137	127[11]	5-22-08-00451	H$_2$O 3; EtOH; peth 1; lig 1
10862	3-Pyridinamine, 2,6-dichloro- Pyridine, 3-amino-2,6-dichloro	$C_5H_4Cl_2N_2$ 163.01	62476-56-6 119	110[0.3]	4-22-00-04093	eth 4; EtOH 4
10863	2-Pyridinamine, N,N-dimethyl-	$C_7H_{10}N_2$ 122.17	5683-33-0 182	196	5-22-08-00287 1.0149[14]	EtOH 3; eth 3; bz 3 1.5663[20]
10864	4-Pyridinamine, 2,6-dimethyl-	$C_7H_{10}N_2$ 122.17	3512-80-9 192.5	246	5-22-09-00417	H$_2$O 3; EtOH; eth 2; ace 3
10865	3-Pyridinamine, 6-methoxy-	$C_6H_8N_2O$ 124.14	6628-77-9 30	125[10]; 87[1]	5-22-11-00408	1.5745[20]
10866	2-Pyridinamine, 3-methyl-	$C_6H_8N_2$ 108.14	1603-40-3 33.5	222; 95[8]	5-22-09-00272	H$_2$O 4; EtOH 3; eth 3; ace 3
10867	2-Pyridinamine, 4-methyl- 2-Amino-4-picoline	$C_6H_8N_2$ 108.14	695-34-1 100	478 115-7[11]	5-22-09-00325	H$_2$O 4; EtOH 3; eth 3; ace 3
10868	2-Pyridinamine, 6-methyl-	$C_6H_8N_2$ 108.14	1824-81-3 41	208.5	5-22-09-00210	H$_2$O 4; EtOH 3; eth 3; ace 3
10869	2-Pyridinamine, N-methyl-	$C_6H_8N_2$ 108.14	4597-87-9 15	200.5	5-22-08-00286 1.048[29]	H$_2$O 3; EtOH 4; eth 4; bz 3
10870	3-Pyridinamine, 4-methyl-	$C_6H_8N_2$ 108.14	3430-27-1 106	255	5-22-09-00344	H$_2$O 2; EtOH 2; eth 2; ace 2
10871	2-Pyridinamine, N-2-pyridinyl-	$C_{10}H_9N_3$ 171.20	1202-34-2 90.5	307.5	5-22-08-00415	H$_2$O 2; EtOH 4; eth 4; ace 4
10872	Pyridine Azine	C_5H_5N 79.10	110-86-1 -41.6	7983 115.2	5-20-05-00160 0.9819[20]	H$_2$O 5; EtOH 5; eth 5; ace 5 1.5095[20]
10873	1(4H)-Pyridineacetic acid, 3,5-diiodo-4-oxo-, propyl ester Propyliodone	$C_{10}H_{11}I_2NO_3$	587-61-1	7876	5-21-07-00164	
		447.01	186			
10874	Pyridine, 2,6-bis(1,1-dimethylethyl)- 2,6-Di-tert-butylpyridine	$C_{13}H_{21}N$ 191.32	585-48-8	3022 120-1[20]	5-20-06-00181	
10875	Pyridine, 2-bromo-	C_5H_4BrN 158.00	109-04-6 158.00	193; 75[13]	5-20-05-00422 1.657[15]	H$_2$O 2; EtOH 3; eth 3; ctc 3 1.5734[20]
10876	Pyridine, 3-bromo- 3-Bromopyridine	C_5H_4BrN 158.00	626-55-1 -27.3	173; 69[18]	5-20-05-00426 1.645[0]	H$_2$O 3; EtOH 4; eth 4 1.5694[20]
10877	Pyridine, 4-bromo- 4-Bromopyridine	C_5H_4BrN 158.00	1120-87-2 0.5	28-30[0.4]	5-20-05-00431 1.6450[0]	ace 3; bz 3 1.5694[20]
10878	Pyridine, 2-butyl- 2-Butylpyridine	$C_9H_{13}N$ 135.21	5058-19-5 97	192	5-20-06-00118 0.9135[15]	
10879	Pyridine, 4-butyl-	$C_9H_{13}N$ 135.21	5335-75-1	209; 98[20]	5-20-06-00120 0.9157[20]	1.4946[20]
10880	Pyridine, 4-(1-butylpentyl)-	$C_{14}H_{23}N$ 205.34	2961-47-9	181[50]	5-20-06-00189 0.8878[25]	1.4846[25]
10881	2-Pyridinecarbonitrile	$C_6H_4N_2$ 104.11	100-70-9 29	224.5	5-22-02-00019 1.0810[25]	H$_2$O 3; EtOH 4; eth 4; bz 4 1.5242[25]
10882	3-Pyridinecarbonitrile	$C_6H_4N_2$ 104.11	100-54-9 51	206.9; 170[300]	5-22-02-00115 1.1590[25]	H$_2$O 4; EtOH 4; eth 4; bz 4
10883	3-Pyridinecarbonitrile, 1,2-dihydro-4-methoxy-1-methyl-2-oxo- Ricinine	$C_8H_8N_2O_2$	524-40-3	8212	5-22-07-00318	H$_2$O 3; EtOH 2; bz 2; chl 3
		164.16	201.5	sub 170		

No.	Name / Synonym	Mol. Form. / Mol. Wt.	CAS RN / mp/°C	Merck No. / bp/°C	Beil. Ref. / den/g cm^{-3}	Solubility / n_D
10884	3-Pyridinecarbonitrile, 1,2-dihydro-1-methyl-2-oxo-	C$_7$H$_6$N$_2$O 134.14	767-88-4 140	243[28]	5-22-06-00129	EtOH 4
10885	4-Pyridinecarbothioamide, 2-ethyl- Ethionamide	C$_8$H$_{10}$N$_2$S 166.25	536-33-4 163	3692	5-22-02-00360	
10886	4-Pyridinecarbothioamide, 2-propyl- Protionamide	C$_9$H$_{12}$N$_2$S 180.27	14222-60-7 136.7	7905	5-22-02-00376	
10887	2-Pyridinecarboxaldehyde	C$_6$H$_5$NO 107.11	1121-60-4	180; 62[13]	5-21-07-00293 1.1255[18]	H$_2$O 3; EtOH 3; eth 3; ctc 2 1.5389[18]
10888	3-Pyridinecarboxaldehyde	C$_6$H$_5$NO 107.11	500-22-1	89.5[14]	5-21-07-00334 1.1394[25]	H$_2$O 3; EtOH 3; eth 2; ace 3
10889	4-Pyridinecarboxaldehyde	C$_6$H$_5$NO 107.11	872-85-5	77-8[12]	5-21-07-00351	H$_2$O 3; eth 3; ctc 3 1.5423[20]
10890	3-Pyridinecarboxamide Niacinamide	C$_6$H$_6$N$_2$O 122.13	98-92-0 130	6398 157[0.0005]	5-22-02-00080 1.400[25]	H$_2$O 4; EtOH 4; chl 2; glycerol 4 1.466
10891	3-Pyridinecarboxamide, N,N-diethyl- Nikethamide	C$_{10}$H$_{14}$N$_2$O 178.23	59-26-7 25	6459 280 dec; 175[25]	5-22-02-00084 1.060[25]	DMSO-d$_6$ 2 1.525[20]
10892	4-Pyridinecarboxamide, N,N-diethyl- Isonicotinic acid diethylamide	C$_{10}$H$_{14}$N$_2$O 178.23	530-40-5	5074 119-20[1]	5-22-02-00196	H$_2$O 4; ace 4; eth 4; EtOH 4 1.525[20]
10893	2-Pyridinecarboxylic acid Picolinic acid	C$_6$H$_5$NO$_2$ 123.11	98-98-6 136.5	7375 sub	5-22-02-00003	H$_2$O 2; EtOH 3; eth 1; bz 2
10894	3-Pyridinecarboxylic acid Nicotinic acid	C$_6$H$_5$NO$_2$ 123.11	59-67-6 236.6	6435 sub	5-22-02-00057 1.473[25]	H$_2$O 2; EtOH 2; eth 2
10895	4-Pyridinecarboxylic acid Isonicotinic acid	C$_6$H$_5$NO$_2$ 123.11	55-22-1 315	5073 sub 260	5-22-02-00188	H$_2$O 2; EtOH 2; eth 2; bz 2
10896	2-Pyridinecarboxylic acid, 4-amino-3,5,6-trichloro- Picloram	C$_6$H$_3$Cl$_3$N$_2$O$_2$ 241.46	1918-02-1 218.5	7370	5-22-13-00585	
10897	3-Pyridinecarboxylic acid, 6-amino- 6-Aminonicotinic acid	C$_6$H$_6$N$_2$O$_2$ 138.13	3167-49-5 312	468	4-22-00-06726	
10898	2-Pyridinecarboxylic acid, 5-butyl- Fusaric acid	C$_{10}$H$_{13}$NO$_2$ 179.22	536-69-6 97	4227	5-22-02-00384	
10899	3-Pyridinecarboxylic acid, 1,6-dihydro-6-oxo-	C$_6$H$_5$NO$_3$ 139.11	5006-66-6 310 dec	sub	5-22-06-00119	H$_2$O 2; EtOH 1; eth 1; bz 1
10900	4-Pyridinecarboxylic acid, 1,2-dihydro-6-hydroxy-2-oxo- Citrazinic acid	C$_6$H$_5$NO$_4$ 155.11	99-11-6 >300 dec	2327	5-22-07-00024	H$_2$O 3; alk 3; HCl 2
10901	2-Pyridinecarboxylic acid, ethyl ester	C$_8$H$_9$NO$_2$ 151.17	2524-52-9 1	243	5-22-02-00004 1.1194[20]	H$_2$O 4; eth 4; EtOH 4 1.5104[20]
10902	3-Pyridinecarboxylic acid, ethyl ester	C$_8$H$_9$NO$_2$ 151.17	614-18-6 8.5	224	5-22-02-00060 1.1070[20]	H$_2$O 4; EtOH 4; eth 4; bz 4 1.5024[20]
10903	4-Pyridinecarboxylic acid, ethyl ester	C$_8$H$_9$NO$_2$ 151.17	1570-45-2 23	219.5	5-22-02-00190 1.0091[15]	H$_2$O 2; EtOH 3; eth 4; bz 3 1.5017[20]
10904	4-Pyridinecarboxylic acid, [1-(2-furanyl)ethylidene]hydrazide Furonazide	C$_{12}$H$_{11}$N$_3$O$_2$ 229.24	3460-67-1 202.3	4220	5-22-02-00264	
10905	4-Pyridinecarboxylic acid, hydrazide Isoniazid	C$_6$H$_7$N$_3$O 137.14	54-85-3 171.4	5071	5-22-02-00219	H$_2$O 4; EtOH 4
10906	3-Pyridinecarboxylic acid, methyl ester Methyl nicotinate	C$_7$H$_7$NO$_2$ 137.14	93-60-7 42.5	6014 204	5-22-02-00059	H$_2$O 3; EtOH 3; bz 3
10907	3-Pyridinecarboxylic acid, 1-methylethyl ester	C$_9$H$_{11}$NO$_2$ 165.19	553-60-6	126[30]; 92.5[5]	5-22-02-00060 1.0624[20]	1.4926[20]
10908	4-Pyridinecarboxylic acid, methyl ester	C$_7$H$_7$NO$_2$ 137.14	2459-09-8 16.1	208	5-22-02-00189 1.1599[20]	H$_2$O 2; EtOH 3; eth 3; bz 3 1.5135[20]
10909	3-Pyridinecarboxylic acid, 1-oxide Oxiniacic acid	C$_6$H$_5$NO$_3$ 139.11	2398-81-4 254-5 dec	6895	5-22-02-00140	H$_2$O 4; MeOH 4
10910	4-Pyridinecarboxylic acid, 2-[3-oxo-3-[(phenylmethyl)amino]propyl]hydrazide Nialamide	C$_{16}$H$_{18}$N$_4$O$_2$ 298.34	51-12-7 151.6	6399	5-22-02-00251	
10911	3-Pyridinecarboxylic acid, phenylmethyl ester Nicotinic acid benzyl ester	C$_{13}$H$_{11}$NO$_2$ 213.24	94-44-0	6436 170[3]	5-22-02-00062	
10912	3-Pyridinecarboxylic acid, 1,2,5,6-tetrahydro- Guvacine	C$_6$H$_9$NO$_2$ 127.14	498-96-4 295 dec	4495	5-22-01-00322	H$_2$O 4
10913	3-Pyridinecarboxylic acid, 1,2,5,6-tetrahydro-1-methyl- Arecaidine	C$_7$H$_{11}$NO$_2$ 141.17	499-04-7 232 dec	801	5-22-01-00322	H$_2$O 4; EtOH 1; eth 1; bz 1
10914	3-Pyridinecarboxylic acid, 1,2,5,6-tetrahydro-1-methyl-, methyl ester Arecoline	C$_8$H$_{13}$NO$_2$ 155.20	63-75-2	802 209	5-22-01-00322 1.0485[20]	H$_2$O 5; EtOH 5; eth 5; chl 3 1.486[20]
10915	Pyridine, 2-chloro- 2-Chloropyridine	C$_5$H$_4$ClN 113.55	109-09-1	170	5-20-05-00402 1.205[15]	H$_2$O 2; EtOH 3; eth 3 1.5320[20]
10916	Pyridine, 3-chloro- 3-Chloropyridine	C$_5$H$_4$ClN 113.55	626-60-8	148; 86[100]	5-20-05-00406	H$_2$O 2 1.5304[20]
10917	Pyridine, 4-chloro- 4-Chloropyridine	C$_5$H$_4$ClN 113.55	626-61-9 -43.5	147.5	5-20-05-00410 1.200[25]	H$_2$O 3; EtOH 5
10918	Pyridine, 2-chloro-6-methoxy-	C$_6$H$_6$ClNO 143.57	17228-64-7	185.5	5-21-02-00018 1.207[25]	1.5263[20]
10919	Pyridine, 2-chloro-4-methyl-	C$_6$H$_6$ClN 127.57	3678-62-4 115	192	5-20-05-00579 1.1459[20]	1.5293[8]

PHYSICAL CONSTANTS OF ORGANIC COMPOUNDS (continued)

No.	Name / Synonym	Mol. Form. / Mol. Wt.	CAS RN / mp/°C	Merck No. / bp/°C	Beil. Ref. / den/g cm⁻³	Solubility / n_D
10920	Pyridine, 2-chloro-6-methyl-	C_6H_6ClN / 127.57	18368-63-3 /	/ 183.5; 64[10]	5-20-05-00498 / 1.167[25]	/ 1.5270[20]
10921	Pyridine, 2-[(4-chlorophenyl)methyl]-	$C_{12}H_{10}ClN$ / 203.67	4350-41-8 /	/ 181-3[20]	5-20-07-00558 / 1.39[25]	/ 1.5868[20]
10922	Pyridine, 4-(3-cyclohexen-1-yl)- / 4-(Cyclohex-3-enyl)pyridine	$C_{11}H_{13}N$ / 159.23	70644-46-1 / 22.1	226	4-20-00-03239 / 1.0222[25]	/ 1.5466[25]
10923	2,3-Pyridinediamine	$C_5H_7N_3$ / 109.13	452-58-4 / 120.8	148-50[5]	5-22-11-00241 /	H_2O 3; EtOH 3; bz 3
10924	2,5-Pyridinediamine / 2,5-Diaminopyridine	$C_5H_7N_3$ / 109.13	4318-76-7 / 110.3	182[12]	5-22-11-00252 /	H_2O 4; EtOH 4
10925	2,6-Pyridinediamine	$C_5H_7N_3$ / 109.13	141-86-6 / 121.5	285; 148[5]	5-22-11-00255 /	H_2O 2; ace 2
10926	Pyridine, 3,5-dibromo- / 3,5-Dibromopyridine	$C_5H_3Br_2N$ / 236.89	625-92-3 / 112	222	5-20-05-00436 /	H_2O 2; EtOH 3; eth 3
10927	2,3-Pyridinedicarboxylic acid / Quinolinic acid	$C_7H_5NO_4$ / 167.12	89-00-9 / 228.5	8102	5-22-04-00115 /	H_2O 2; EtOH 1; eth 1; bz 1
10928	2,4-Pyridinedicarboxylic acid	$C_7H_5NO_4$ / 167.12	499-80-9 / 249		5-22-04-00119 / 0.942[25]	H_2O 3; EtOH 3; eth 1; bz 1
10929	2,5-Pyridinedicarboxylic acid / Isocinchomeronic acid	$C_7H_5NO_4$ / 167.12	100-26-5 / 254	5043	5-22-04-00124 /	H_2O 3; EtOH 2; eth 1; bz 1
10930	3,4-Pyridinedicarboxylic acid / Cinchomeronic acid	$C_7H_5NO_4$ / 167.12	490-11-9 / 256	2285 sub	5-22-04-00135 /	H_2O 2; EtOH 2; eth 1; bz 2
10931	3,5-Pyridinedicarboxylic acid	$C_7H_5NO_4$ / 167.12	499-81-0 / 324	sub	5-22-04-00139 /	H_2O 1; eth 2; HCl 3; HOAc 2
10932	3,5-Pyridinedicarboxylic acid, 2,6-dimethyl-, diethyl ester	$C_{13}H_{17}NO_4$ / 251.28	1149-24-2 / 71	301; 208[40]	5-22-04-00162 /	H_2O 1; EtOH 3; eth 3; bz 3
10933	Pyridine, 2,4-diethyl- / 2,4-Diethylpyridine	$C_9H_{13}N$ / 135.21	626-21-1 / 112	187.5	4-20-00-02825 / 0.9338[25]	
10934	3,4-Pyridinedimethanol, 5-hydroxy-6-methyl-, hydrochloride / Pyridoxine hydrochloride	$C_8H_{12}ClNO_3$ / 205.64	58-56-0 / 207	7995 sub	5-21-05-00492 /	H_2O 4
10935	Pyridine, 2,6-dimethoxy- / 2,6-Dimethoxypyridine	$C_7H_9NO_2$ / 139.15	6231-18-1 /	179	5-21-04-00416 / 1.053[25]	/ 1.5029[20]
10936	Pyridine, 2,3-dimethyl- / 2,3-Lutidine	C_7H_9N / 107.16	583-61-9 /	161.2	5-20-06-00015 / 0.9319[25]	H_2O 3; EtOH 3; eth 3 / 1.5057[20]
10937	Pyridine, 2,4-dimethyl- / 2,4-Lutidine	C_7H_9N / 107.16	108-47-4 / -64	158.5	5-20-06-00019 / 0.9309[20]	H_2O 4; EtOH 4; eth 4; ace 3 / 1.5010[20]
10938	Pyridine, 2,5-dimethyl- / 2,5-Lutidine	C_7H_9N / 107.16	589-93-5 / -16	157.0	5-20-06-00027 / 0.9297[20]	H_2O 2; EtOH 4; eth 5; ace 3 / 1.5006[20]
10939	Pyridine, 2,6-dimethyl- / 2,6-Lutidine	C_7H_9N / 107.16	108-48-5 / -6.1	5485 144.1	5-20-06-00032 / 0.9226[20]	H_2O 5; EtOH 2; eth 3; ace 3 / 1.4953[20]
10940	Pyridine, 3,4-dimethyl- / 3,4-Lutidine	C_7H_9N / 107.16	583-58-4 / -11	179.1	5-20-06-00054 / 0.9281[20]	H_2O 2; EtOH 3; eth 3; ace 3 / 1.5096[20]
10941	Pyridine, 3,5-dimethyl- / 3,5-Lutidine	C_7H_9N / 107.16	591-22-0 / -6.6	172	5-20-06-00060 / 0.9419[20]	H_2O 3; EtOH 3; eth 3; ace 3 / 1.5061[20]
10942	Pyridine, 2-(1,1-dimethylethyl)- / 2-tert-Butylpyridine	$C_9H_{13}N$ / 135.21	5944-41-2 / -33	170	5-20-06-00121 /	/ 1.4891[20]
10943	Pyridine, 3-(1,1-dimethylethyl)- / 3-tert-Butylpyridine	$C_9H_{13}N$ / 135.21	38031-78-6 / -44	194.8	5-20-06-00122 /	/ 1.4965[20]
10944	Pyridine, 4-(1,1-dimethylethyl)-	$C_9H_{13}N$ / 135.21	3978-81-2 / -41	196.5	5-20-06-00123 / 0.915[25]	ctc 3; CS_2 3 / 1.4958[20]
10945	Pyridine, 2,6-dimethyl-, 1-oxide	C_7H_9NO / 123.15	1073-23-0 /	115-9[18]	5-20-06-00040 / 1.073[25]	/ 1.5706[20]
10946	2,4(1H,3H)-Pyridinedione, 3,3-diethyl- / Pyrithyldione	$C_9H_{13}NO_2$ / 167.21	77-04-3 / 90.7	8005	5-21-10-00086 /	
10947	Pyridine, 2,6-diphenyl- / 2,6-Diphenylpyridine	$C_{17}H_{13}N$ / 231.30	3558-69-8 / 82	397	5-20-08-00459 /	eth 4; EtOH 4
10948	2-Pyridineethanamine	$C_7H_{10}N_2$ / 122.17	2706-56-1 /	131.5[50]	5-22-09-00368 / 1.0220[25]	/ 1.5335[25]
10949	4-Pyridineethanamine	$C_7H_{10}N_2$ / 122.17	13258-63-4 /	121[10]	4-22-00-04201 / 1.0302[25]	H_2O 4 / 1.5381[25]
10950	2-Pyridineethanol	C_7H_9NO / 123.15	103-74-2 / -7.8	190[200]; 170[100]	5-21-02-00206 / 1.091[25]	H_2O 4; EtOH 4; eth 2; chl 4 / 1.5366[20]
10951	2-Pyridineethanol, α-methyl- / 2-(2-Hydroxypropyl)pyridine	$C_8H_{11}NO$ / 137.18	5307-19-7 / 33	124[20]; 110[10]	5-21-02-00234 /	H_2O 4; EtOH 4; chl 4
10952	Pyridine, 2-ethenyl-	C_7H_7N / 105.14	100-69-6 /	159.5	5-20-06-00211 / 0.9983[20]	H_2O 2; EtOH 4; eth 4; ace 4 / 1.5495[20]
10953	Pyridine, 3-ethenyl-	C_7H_7N / 105.14	1121-55-7 /	162	5-20-06-00212 /	H_2O 2; EtOH 3; eth 3 / 1.5530[20]
10954	Pyridine, 4-ethenyl-	C_7H_7N / 105.14	100-43-6 /	121[150]; 65[15]	5-20-06-00213 / 0.9800[20]	H_2O 3; EtOH 3; eth 2; chl 3 / 1.5449[20]
10955	Pyridine, 5-ethenyl-2-methyl- / 5-Vinyl-2-picoline	C_8H_9N / 119.17	140-76-1 /		5-20-06-00218 /	
10956	Pyridine, 2-ethyl- / 2-Ethylpyridine	C_7H_9N / 107.16	100-71-0 / -63.1	148.6	5-20-06-00003 / 0.9502[25]	H_2O 3; EtOH 5; eth 4; ace 4 / 1.4964[20]
10957	Pyridine, 3-ethyl- / 3-Ethylpyridine	C_7H_9N / 107.16	536-78-7 / -76.9	3802 165	5-20-06-00007 / 0.9539[25]	H_2O 3; EtOH 3; eth 3; ace 4 / 1.5021[20]

No.	Name Synonym	Mol. Form. Mol. Wt.	CAS RN mp/°C	Merck No. bp/°C	Beil. Ref. den/g cm^{-3}	Solubility n_D
10958	Pyridine, 4-ethyl- 4-Ethylpyridine	C_7H_9N 107.16	536-75-4 -90.5	3803 168.3	5-20-06-00010 0.9417[20]	H_2O 3; EtOH 3; eth 3; ace 4 1.5009[20]
10959	Pyridine, 2-ethyl-3,5-dimethyl-	$C_9H_{13}N$ 135.21	1123-96-2	198; 86[15]	5-20-06-00129 0.9338[25]	H_2O 3; EtOH 4; eth 4; ace 4
10960	Pyridine, 3-ethyl-2,6-dimethyl-	$C_9H_{13}N$ 135.21	23580-52-1	83-5[23]	5-20-06-00129 0.9120[18]	H_2O 2
10961	Pyridine, 4-ethyl-2,6-dimethyl-	$C_9H_{13}N$ 135.21	36917-36-9 97	187.5	5-20-06-00130 0.9089[25]	1.4964[25]
10962	Pyridine, 2-ethyl-4-methyl-	$C_8H_{11}N$ 121.18	2150-18-7	175	5-20-06-00086 0.9238[20]	ace 4; eth 4; EtOH 4
10963	Pyridine, 2-ethyl-6-methyl-	$C_8H_{11}N$ 121.18	1122-69-6	160.5	5-20-06-00089 0.9207[25]	H_2O 3; EtOH 3; eth 3; ace 4 1.4920[25]
10964	Pyridine, 3-ethyl-4-methyl- 3-Ethyl-4-picoline	$C_8H_{11}N$ 121.18	529-21-5	3796 198	5-20-06-00090 0.9286[17]	H_2O 2; EtOH 3; eth 3; ace 4
10965	Pyridine, 4-ethyl-2-methyl- 4-Ethyl-2-picoline	$C_8H_{11}N$ 121.18	536-88-9	3797 179	5-20-06-00085 0.9130[25]	ace 4; bz 4; eth 4; EtOH 4
10966	Pyridine, 5-ethyl-2-methyl- 5-Ethyl-2-picoline	$C_8H_{11}N$ 121.18	104-90-5	3798 178.3	5-20-06-00086 0.9202[20]	H_2O 2; EtOH 3; eth 3; ace 4 1.4971[20]
10967	Pyridine, 2-(1-ethylpropyl)-	$C_{10}H_{15}N$ 149.24	7399-50-0	195.4	5-20-06-00145 0.8981[20]	
10968	Pyridine, 4-(1-ethylpropyl)- 4-(3-Pentyl)pyridine	$C_{10}H_{15}N$ 149.24	35182-51-5 125.5	217; 80[12]	5-20-06-00145 0.9085[25]	1.40905[25]
10969	Pyridine, hydrochloride Pyridinium chloride	C_5H_6ClN 115.56	628-13-7 146	222	5-20-05-00160	H_2O 4; EtOH 4; chl 4
10970	Pyridine, 2-iodo- 2-Iodopyridine	C_5H_4IN 205.00	5029-67-4	93[13]	5-20-05-00438 1.9735[20]	EtOH 3; eth 3; ace 3; bz 3 1.6366[20]
10971	2-Pyridinemethanamine	$C_6H_8N_2$ 108.14	3731-51-9	203; 91[17]	5-22-09-00232 1.0525[25]	H_2O 4 1.5431[25]
10972	3-Pyridinemethanamine	$C_6H_8N_2$ 108.14	3731-52-0 -21.1	226	5-22-09-00301 1.064[20]	H_2O 4; eth 4; EtOH 4 1.552[20]
10973	4-Pyridinemethanamine	$C_6H_8N_2$ 108.14	3731-53-1 -7.6	230; 103[11]	5-22-09-00345 1.072[20]	H_2O 4 1.5495[25]
10974	2-Pyridinemethanamine, N-(2-pyridinylmethyl)-	$C_{12}H_{13}N_3$ 199.26	1539-42-0	200[10]	5-22-09-00267 1.1074[25]	1.5757[25]
10975	2-Pyridinemethanol	C_6H_7NO 109.13	586-98-1	112[16]; 102.5[8]	5-21-02-00150 1.1317[20]	H_2O 5; EtOH 4; eth 4; ace 4 1.5444[20]
10976	3-Pyridinemethanol Nicotinyl alcohol	C_6H_7NO 109.13	100-55-0 -6.5	6438 266	5-21-02-00172 1.131[20]	H_2O 4; eth 4 1.5455[20]
10977	4-Pyridinemethanol 4-Picolyl alcohol	C_6H_7NO 109.13	586-95-8 53	140-2[12]	5-21-02-00191	chl 3
10978	3-Pyridinemethanol, 4-(aminomethyl)-5-hydroxy-6-methyl-, dihydrochloride Pyridoxamine dihydrochloride	$C_8H_{14}Cl_2N_2O_2$ 241.12	524-36-7 226-7 dec	7993	5-22-12-00324	H_2O 4; EtOH 2
10979	Pyridine, 2-(2-methoxyethyl)- Metyridine	$C_8H_{11}NO$ 137.18	114-91-0	6084 203; 96[17]	5-21-02-00206 0.988[20]	H_2O 4; EtOH 4 1.4975[20]
10980	Pyridine, 2-methyl- 2-Picoline	C_6H_7N 93.13	109-06-8 -66.7	7372 129.3	5-20-05-00464 0.9443[20]	H_2O 4; EtOH 5; eth 5; ace 4 1.4957[20]
10981	Pyridine, 3-methyl- 3-Picoline	C_6H_7N 93.13	108-99-6 -18.1	7373 144.1	5-20-05-00506 0.9566[20]	H_2O 5; EtOH 5; eth 5; ace 4 1.5040[20]
10982	Pyridine, 4-methyl- 4-Picoline	C_6H_7N 93.13	108-89-4 3.66	7374 145.3	5-20-05-00543 0.9548[20]	H_2O 5; EtOH 5; eth 5; ace 3 1.5037[20]
10983	Pyridine, 2-[2-(1-methylethoxy)ethyl]- Pyridine, 2-(2-isopropoxyethyl)	$C_{10}H_{15}NO$ 165.24	70715-19-4	133[50]	0.9502[25]	H_2O 4 1.4820[25]
10984	Pyridine, 2-(1-methylethyl)-	$C_8H_{11}N$ 121.18	644-98-4 -141	159.8	5-20-06-00080 0.9342[25]	H_2O 2; EtOH 5; eth 5; ace 4 1.4915[20]
10985	Pyridine, 3-(1-methylethyl)- 3-Isopropylpyridine	$C_8H_{11}N$ 121.18	6304-18-3 -45.8	177.5	5-20-06-00081 0.9217[16]	
10986	Pyridine, 4-(1-methylethyl)-	$C_8H_{11}N$ 121.18	696-30-0 -54.9	178	5-20-06-00082 0.9382[25]	H_2O 2; EtOH 5; eth 5; ace 4 1.4962[20]
10987	Pyridine, 3-(1-methyl-1H-pyrrol-2-yl)-	$C_{10}H_{10}N_2$ 158.20	487-19-4	281; 150[16]	5-23-07-00227 1.241[20]	EtOH 4 1.6057[20]
10988	Pyridine, 3-methyl-, 1-oxide	C_6H_7NO 109.13	1003-73-2 39	148[15]	5-20-05-00517	chl 3
10989	Pyridine, 2-(4-methylphenyl)-	$C_{12}H_{11}N$ 169.23	4467-06-5	170-80[20]	5-20-07-00569 0.999[25]	1.6125[20]
10990	Pyridine, 2-methyl-5-phenyl-	$C_{12}H_{11}N$ 169.23	3256-88-0	189[50]	5-20-07-00573 1.0590[25]	1.6055[25]
10991	Pyridine, 3-methyl-2-phenyl- 3-Picoline, 2-phenyl-	$C_{12}H_{11}N$ 169.23	10273-90-2	148-9[16]	5-20-07-00571 1.065[25]	1.6026[20]
10992	Pyridine, 3-(1-methyl-2-pyrrolidinyl)-, (S)- Nicotine	$C_{10}H_{14}N_2$ 162.23	54-11-5 -79	6434 247; 125[18]	5-23-06-00064 1.0097[20]	H_2O 5; EtOH 4; eth 4; chl 4 1.5282[20]
10993	Pyridine, 4-nitro-, 1-oxide 4-Nitropyridine N-oxide	$C_5H_4N_2O_3$ 140.10	1124-33-0 160.5		5-20-05-00448	
10994	Pyridine, 1-oxide Pyridine N-oxide	C_5H_5NO 95.10	694-59-7 65.5	7984 146-7[13]	5-20-05-00217	
10995	Pyridine, pentachloro-	C_5Cl_5N 251.33	2176-62-7 125.5	280	5-20-05-00422	bz 4; EtOH 4; lig 4
10996	Pyridine, 2-(3-pentenyl)-	$C_{10}H_{13}N$ 147.22	2057-43-4	216; 93[12]	5-20-06-00355 0.9234[25]	1.5076[25]

No.	Name / Synonym	Mol. Form. / Mol. Wt.	CAS RN / mp/°C	Merck No. / bp/°C	Beil. Ref. / den/g cm^{-3}	Solubility / n_D
10997	Pyridine, 2-phenyl- / 2-Phenylpyridine	$C_{11}H_9N$ / 155.20	1008-89-5	271	5-20-07-00545 / 1.0833[25]	H_2O 2; EtOH 5; eth 5 / 1.6210[20]
10998	Pyridine, 3-phenyl- / 3-Phenylpyridine	$C_{11}H_9N$ / 155.20	1008-88-4 / 164	272	5-20-07-00548	H_2O 2; EtOH 3; eth 3 / 1.6123[25]
10999	Pyridine, 4-phenyl- / 4-Phenylpyridine	$C_{11}H_9N$ / 155.20	939-23-1 / 77.5	281	5-20-07-00549	H_2O 3; EtOH 3; eth 3
11000	Pyridine, 2-(2-phenylethenyl)-, (E)-	$C_{13}H_{11}N$ / 181.24	538-49-8 / 91.5	325; 194[14]	5-20-08-00057 / 1.0147[100]	H_2O 1; EtOH 3; eth 3; bz 3
11001	Pyridine, 4-(2-phenylethenyl)-, (E)-	$C_{13}H_{11}N$ / 181.24	5097-93-8 / 131	208[33]	5-20-08-00063	H_2O 1; EtOH 3; eth 3; chl 3
11002	Pyridine, 2-(2-phenylethenyl)-, (Z)-	$C_{13}H_{11}N$ / 181.24	1519-59-1 / -50	141[10]	4-20-00-03874	EtOH 3; eth 3; bz 3
11003	Pyridine, 2-(2-phenylethyl)-	$C_{13}H_{13}N$ / 183.25	2116-62-3 / -3	289	5-20-07-00588 / 1.0465[0]	
11004	Pyridine, 2-(phenylmethyl)- / 2-Benzylpyridine	$C_{12}H_{11}N$ / 169.23	101-82-6 / 12.5	277; 149[16]	5-20-07-00556 / 1.067[0]	H_2O 1; EtOH 3; eth 3; chl 3 / 1.5785[20]
11005	Pyridine, 3-(phenylmethyl)- / 3-Benzylpyridine	$C_{12}H_{11}N$ / 169.23	620-95-1 / 34	287	5-20-07-00560 / 1.061[25]	eth 4; EtOH 4
11006	Pyridine, 4-(phenylmethyl)-	$C_{12}H_{11}N$ / 169.23	2116-65-6 / 12.4	288; 180[31]	5-20-07-00561 / 1.0612[20]	H_2O 1; EtOH 3; eth 4; ctc 3 / 1.5818[20]
11007	Pyridine, 4-(3-phenylpropyl)-	$C_{14}H_{15}N$ / 197.28	2057-49-0	322; 150[5]	5-20-07-00607 / 1.5016[25]	bz 4; eth 4; py 4; EtOH 4
11008	Pyridine, 3-(2-piperidinyl)-, (S)- / Anabasine	$C_{10}H_{14}N_2$ / 162.23	494-52-0 / 9	655 / 276	5-23-06-00096 / 1.0436[20]	H_2O 5; EtOH 3; eth 3; bz 3 / 1.5430[20]
11009	Pyridine, 4-(4-piperidinyl)-	$C_{10}H_{14}N_2$ / 162.23	581-45-3 / 79	292	5-23-06-00115	H_2O 2; EtOH 4; eth 3; bz 3
11010	2-Pyridinepropanamine, N,N-dimethyl-γ-phenyl- / Pheniramine	$C_{16}H_{20}N_2$ / 240.35	86-21-5	7198 / 181[13]. 135[0.5]	5-22-10-00487 / 1.0081[25]	bz 4; eth 4; EtOH 4; chl 4 / 1.5519[25]
11011	1(4H)-Pyridinepropanoic acid, α-amino-3-hydroxy-4-oxo-, (S)- / Mimosine	$C_8H_{10}N_2O_4$ / 198.18	500-44-7 / 228 dec	6118	5-21-12-00107	H_2O 2; EtOH 1; eth 1; ace 1
11012	2-Pyridinepropanol / 2-(3-Hydroxypropyl)pyridine	$C_8H_{11}NO$ / 137.18	2859-68-9 / 34	260.2; 116[4]	5-21-02-00236 / 1.060[25]	H_2O 4 / 1.5298[20]
11013	3-Pyridinepropanol	$C_8H_{11}NO$ / 137.18	2859-67-8	284; 130[3]	5-21-02-00237 / 1.063[25]	H_2O 4 / 1.5313[20]
11014	Pyridine, 2-(2-propenyl)-	C_8H_9N / 119.17	2835-33-8	190; 63[15]	5-20-06-00216 / 0.9590[20]	eth 4; EtOH 4
11015	Pyridine, 2-propyl-	$C_8H_{11}N$ / 121.18	622-39-9 / 2	167	5-20-06-00076 / 0.9119[20]	H_2O 2; EtOH 5; eth 5; ace 4 / 1.4925[20]
11016	Pyridine, 4-propyl-	$C_8H_{11}N$ / 121.18	1122-81-2	185	5-20-06-00078 / 0.9381[15]	eth 4; EtOH 4 / 1.4966[20]
11017	Pyridine, 3-(2-pyrrolidinyl)-, (S)- / Nornicotine	$C_9H_{12}N_2$ / 148.21	494-97-3	6631 / 270	4-23-00-00998 / 1.0737[19]	H_2O 4; ace 4; eth 4; EtOH 4 / 1.5378[18]
11018	3-Pyridinesulfonic acid / 3-Pyridylsulfonic acid	$C_5H_5NO_3S$ / 159.17	636-73-7 / 357 dec		5-22-07-00552 / 1.713[25]	H_2O 4; EtOH 2; eth 1
11019	Pyridine, 2,3,5,6-tetrachloro- / 2,3,5,6-Tetrachloropyridine	C_5HCl_4N / 216.88	2402-79-1 / 90.5	250.5	5-20-05-00421	eth 4; EtOH 4; peth 4
11020	Pyridine, 1,2,3,6-tetrahydro- / Δ 3-Piperidine	C_5H_9N / 83.13	694-05-3 / -48	108	5-20-04-00282 / 0.911[25]	chl 3 / 1.4800[20]
11021	Pyridine, 2,3,4,5-tetrahydro-6-propyl- / γ-Coniceine	$C_8H_{15}N$ / 125.21	1604-01-9	2497 / 174	5-20-04-00345 / 0.8753[15]	/ 1.4661[16]
11022	Pyridine, 2,3,4,6-tetramethyl- / 2,3,4,6-Tetramethylpyridine	$C_9H_{13}N$ / 135.21	20820-82-0	203	5-20-06-00132 / 0.9322[25]	eth 4; EtOH 4 / 1.5087[25]
11023	Pyridine, 2-(trichloromethyl)- / 2-(Trichloromethyl)pyridine	$C_8H_4Cl_3N$ / 196.46	4377-37-1 / -10	125-6[25]	5-20-05-00499 / 1.4526[25]	/ 1.5596[25]
11024	Pyridine, 2,3,4-trimethyl- / 2,3,4-Collidine	$C_8H_{11}N$ / 121.18	2233-29-6	192.5	5-20-06-00091 / 0.9127[15]	H_2O 3; EtOH 3; eth 3 / 1.5150[20]
11025	Pyridine, 2,3,5-trimethyl- / 2,3,5-Collidine	$C_8H_{11}N$ / 121.18	695-98-7	184	5-20-06-00092 / 0.9352[19]	H_2O 2; EtOH 3; eth 3; ace 3 / 1.5057[25]
11026	Pyridine, 2,3,6-trimethyl- / 2,3,6-Collidine	$C_8H_{11}N$ / 121.18	1462-84-6	171.6	5-20-06-00092 / 0.9220[25]	H_2O 3; EtOH 3; eth 3; ace 3 / 1.5053[20]
11027	Pyridine, 2,4,5-trimethyl- / 2,4,5-Collidine	$C_8H_{11}N$ / 121.18	1122-39-0	188	5-20-06-00093 / 0.9330[25]	ace 4; bz 4; eth 4; EtOH 4 / 1.5054[25]
11028	Pyridine, 2,4,6-trimethyl- / 2,4,6-Collidine	$C_8H_{11}N$ / 121.18	108-75-8 / -46	9632 / 170.6	5-20-06-00093 / 0.9166[22]	H_2O 3; EtOH 3; eth 3; ace 3 / 1.4959[25]
11029	Pyridinium, 3-carboxy-1-methyl-, hydroxide, inner salt / Trigonelline	$C_7H_7NO_2$ / 137.14	535-83-1	9606	5-22-02-00143	H_2O 4
11030	Pyridinium, 1-hexadecyl-, chloride / Hexadecyl pyridinium chloride	$C_{21}H_{38}ClN$ / 339.99	123-03-5 / 80	2024	5-20-05-00233	H_2O 4; chl 4
11031	2-Pyridinol	C_5H_5NO / 95.10	72762-00-6 / 107.8		5-21-07-00106 / 1.3910[20]	H_2O 4; bz 4; EtOH 4
11032	4-Pyridinol	C_5H_5NO / 95.10	626-64-2 / 149.8	>350; 257[10]	5-21-07-00152	H_2O 3; EtOH 3; eth 1; bz 1
11033	2(1H)-Pyridinone	C_5H_5NO / 95.10	142-08-5 / 107.8	280	5-21-07-00106 / 1.3910[20]	H_2O 3; EtOH 3; eth 2; bz 3
11034	2(1H)-Pyridinone, hydrazone	$C_5H_7N_3$ / 109.13	4930-98-7 / 46.6	185[140]; 90[1]	5-22-14-00486	chl 3

No.	Name / Synonym	Mol. Form. / Mol. Wt.	CAS RN / mp/°C	Merck No. / bp/°C	Beil. Ref. / den/g cm^{-3}	Solubility / n_D
11035	2(1H)-Pyridinone, 1-methyl-	C_6H_7NO 109.13	694-85-9 31	250	5-21-07-00109 1.1120[20]	H_2O 5; peth 2; lig 2
11036	2(1H)-Pyridinone, 3-methyl-	C_6H_7NO 109.13	1003-56-1 141.5	289	5-21-07-00199	H_2O 4; EtOH 4; chl 4
11037	2(1H)-Pyridinone, 4-methyl-	C_6H_7NO 109.13	13466-41-6 130	308	5-21-07-00201	H_2O 3; EtOH 3; eth 2; bz 2
11038	3H-Pyrido[3,4-b]indole, 4,9-dihydro-7-methoxy-1-methyl- Harmaline	$C_{13}H_{14}N_2O_{13}$ 406.26	304-21-2 250 dec	4528	5-23-12-00148	H_2O 2; EtOH 2; eth 2; chl 3
11039	9H-Pyrido[3,4-b]indole, 7-methoxy-1-methyl- Harmine	$C_{13}H_{12}N_2O$ 212.25	442-51-3 273 sub	4531	5-23-12-00237	H_2O 2; EtOH 2; eth 2; chl 2
11040	9H-Pyrido[3,4-b]indole, 1-methyl- Harman	$C_{12}H_{10}N_2$ 182.22	486-84-0 236.5	4530	5-23-08-00261	
11041	9H-Pyrido[3,4-b]indole, 1-[(2-methylphenyl)methyl]-	$C_{19}H_{16}N_2$ 272.35	525-15-5 217.8	150[0.01]	5-23-09-00536	EtOH 3; eth 3; bz 3; chl 3
11042	2-Pyrimidinamine	$C_4H_5N_3$ 95.10	109-12-6 127.5	sub	4-25-00-02071	H_2O 3; chl 2
11043	4-Pyrimidinamine, 2-chloro-N,N,6-trimethyl- Crimidine	$C_7H_{10}ClN_3$ 171.63	535-89-7 87	2590 140-7[4]	4-25-00-02184	EtOH 4
11044	2-Pyrimidinamine, 4,5-dimethyl- 2-Amino-4,5-dimethylpyrimidine	$C_6H_9N_3$ 123.16	1193-74-4 215.8	sub	4-25-00-02203	ace 4; bz 4; EtOH 4; chl 4
11045	4-Pyrimidinamine, 2,6-dimethyl- Kyanmethin	$C_6H_9N_3$ 123.16	461-98-3 183	5205 sub	4-25-00-02192	H_2O 2; EtOH 2; bz 2; chl 2
11046	2-Pyrimidinamine, 4-methyl-	$C_5H_7N_3$ 109.13	108-52-1 160.3	sub	4-25-00-02152	H_2O 3; EtOH 3; chl 2
11047	2-Pyrimidinamine, 5-methyl-	$C_5H_7N_3$ 109.13	50840-23-8 193.5	sub	5-25-10-00191	EtOH 4
11048	4-Pyrimidinamine, 6-methyl-	$C_5H_7N_3$ 109.13	3435-28-7 197	sub	4-25-00-02182	H_2O 3; EtOH 3
11049	Pyrimidine 1,3-Diazine	$C_4H_4N_2$ 80.09	289-95-2 22	7998 123.8	5-23-05-00334 1.4998[20]	H_2O 5; EtOH 3
11050	4-Pyrimidinecarboxylic acid, 1,2,3,6-tetrahydro-2,6-dioxo- Orotic acid	$C_5H_4N_2O_4$ 156.10	65-86-1 345.5	6828	4-25-00-01759	H_2O 2; os 1
11051	2,4-Pyrimidinediamine, 5-(4-chlorophenyl)-6-ethyl- Pyrimethamine	$C_{12}H_{13}ClN_4$ 248.71	58-14-0 233.5	7997	4-25-00-03014	
11052	Pyrimidine, 2,4-dichloro-5-methyl- 2,4-Dichloro-5-methylpyrimidine	$C_5H_4Cl_2N_2$ 163.01	1780-31-0 26	235	5-23-05-00385	H_2O 2; EtOH 4; eth 4; bz 4
11053	Pyrimidine, 2,4-dichloro-6-methyl- 2,6-Dichloro-4-methylpyrimidine	$C_5H_4Cl_2N_2$ 163.01	5424-21-5 46.5	219	5-23-05-00381	bz 4; eth 4; EtOH 4; chl 4
11054	Pyrimidine, 2,4-dichloro-5-nitro- 2,4-Dichloro-5-nitropyrimidine	$C_4HCl_2N_3O_2$ 193.98	49845-33-2 29.3	154[58], 135[17]	5-23-05-00350	EtOH 2; eth 2
11055	Pyrimidine, 2,4-dimethyl- 2,4-Dimethylpyrimidine	$C_6H_8N_2$ 108.14	14331-54-5 -4	150.5	5-23-05-00397 1.168[14]	1.4880[25]
11056	Pyrimidine, 2,5-dimethyl- 2,5-Dimethylpyrimidine	$C_6H_8N_2$ 108.14	22868-76-4 19	59[13]	5-23-05-00399	H_2O 4
11057	Pyrimidine, 4,5-dimethyl- 4,5-Dimethylpyrimidine	$C_6H_8N_2$ 108.14	694-81-5 3	177	5-23-05-00399	H_2O 4
11058	Pyrimidine, 4,6-dimethyl- 4,6-Dimethylpyrimidine	$C_6H_8N_2$ 108.14	1558-17-4 25	159	5-23-05-00400	H_2O 4 1.4880[20]
11059	2,4(1H,3H)-Pyrimidinedione Uracil	$C_4H_4N_2O_2$ 112.09	66-22-8 338	9761	5-24-06-00043	H_2O 2; EtOH 4; eth 4
11060	2,4(1H,3H)-Pyrimidinedione, 6-amino-3-ethyl-1-(2-propenyl)- Aminometradine	$C_9H_{13}N_3O_2$ 195.22	642-44-4 143	463	4-25-00-04113	
11061	2,4(1H,3H)-Pyrimidinedione, 5-[bis(2-chloroethyl)amino]- Uracil mustard	$C_8H_{11}Cl_2N_3O_2$ 252.10	66-75-1 206 dec	9762	4-25-00-04127	H_2O 2
11062	2,4(1H,3H)-Pyrimidinedione, 5-bromo- 5-Bromouracil	$C_4H_3BrN_2O_2$ 190.98	51-20-7 310	1430	5-24-06-00322	
11063	4,6(1H,5H)-Pyrimidinedione, dihydro-5-(2-methylpropyl)-5-(2-propenyl)-2-thioxo-, monosodium salt Barbituric acid, 5-allyl-5-isobutyl-2-thio, sodium salt	$C_{11}H_{15}N_2NaO_2S$ 262.31	510-90-7	1518		H_2O 4; EtOH 2; eth 1; bz 1
11064	4,6(1H,5H)-Pyrimidinedione, 5-ethyldihydro-5-(2-methyl-2-propenyl)-2-thioxo- Methallatal	$C_{10}H_{14}N_2O_2S$ 226.30	115-56-0 160.5	5855	5-24-09-00224	
11065	4,6(1H,5H)-Pyrimidinedione, 5-ethyldihydro-5-(1-methylpropyl)-2-thioxo- Barbituric acid, 5-sec-butyl-5-ethyl-2-thio-	$C_{10}H_{16}N_2O_2S$ 228.32	2095-57-0 169	9250	5-24-09-00164	
11066	4,6(1H,5H)-Pyrimidinedione, 5-ethyldihydro-5-phenyl- Primidone	$C_{12}H_{14}N_2O_2$ 218.26	125-33-7 281.5	7753	5-24-08-00102	
11067	2,4(1H,3H)-Pyrimidinedione, 5-fluoro- 5-Fluorouracil	$C_4H_3FN_2O_2$ 130.08	51-21-8 283 sub 190	4109	5-24-06-00271	
11068	2,4(1H,3H)-Pyrimidinedione, 5-methyl- Thymine	$C_5H_6N_2O_2$ 126.11	65-71-4 316	9332	5-24-07-00055	H_2O 2; EtOH 2; eth 2

No.	Name Synonym	Mol. Form. Mol. Wt.	CAS RN mp/°C	Merck No. bp/°C	Beil. Ref. den/g cm⁻³	Solubility n_D
11069	2,4(1H,3H)-Pyrimidinedione, 6-methyl- 6-Methyluracil	$C_5H_8N_2O_2$ 126.11	626-48-2 270-80 dec	6051	5-24-07-00006	H_2O 3; EtOH 3; eth 2; tfa 2
11070	Pyrimidine, 2-methyl- 2-Methyl-1,3-diazine	$C_5H_8N_2$ 94.12	5053-43-0 -4	138	5-23-05-00376	H_2O 5
11071	Pyrimidine, 4-methyl- 4-Methyl-1,3-diazine	$C_5H_8N_2$ 94.12	3438-46-8 32	142	5-23-05-00379 1.030¹⁶	H_2O 5 1.500²⁰
11072	Pyrimidine, 5-methyl- 5-Methyl-1,3-diazine	$C_5H_8N_2$ 94.12	2036-41-1 30.5	153	5-23-05-00383	H_2O 4
11073	2,4,5,6(1H,3H)-Pyrimidinetetrone Alloxan	$C_4H_2N_2O_4$ 142.07	50-71-5 256 dec	281 sub	5-24-09-00382	H_2O 4; EtOH 3; ace 3; bz 3
11074	2,4,5,6(1H,3H)-Pyrimidinetetrone, 5-oxime Violuric acid	$C_4H_3N_3O_4$ 157.09	87-39-8 203-4 dec	9904	5-24-09-00386	H_2O 2; EtOH 3
11075	Pyrimidine, 2,4,6-trichloro- 2,4,6-Trichloropyrimidine	$C_4HCl_3N_2$ 183.42	3764-01-0 22.5	212.5	5-23-05-00347	1.5700²⁰
11076	2,4,6(1H,3H)-Pyrimidinetrione Barbituric acid	$C_4H_4N_2O_3$ 128.09	67-52-7 248	973 260 dec	5-24-09-00078	H_2O 3; EtOH 2; eth 3
11077	2,4,6(1H,3H,5H)-Pyrimidinetrione, 5-amino- Uramil	$C_4H_5N_3O_3$ 143.10	118-78-5 >400	9763	4-25-00-04228	H_2O 3; eth 1; bz 1; chl 3
11078	2,4,6(1H,3H,5H)-Pyrimidinetrione, 5-(2-bromo-2-propenyl)-5-(1-methylethyl)- Propallylonal	$C_{10}H_{13}BrN_2O_3$ 289.13	545-93-7 181	7807	4-24-00-01994	H_2O 2; EtOH 4; eth 2; ace 4
11079	2,4,6(1H,3H,5H)-Pyrimidinetrione, 5-(2-bromo-2-propenyl)-1-methyl-5-(1-methylethyl)- Narcobarbital	$C_{11}H_{15}BrN_2O_3$ 303.16	125-55-3 115	6342	4-24-00-01995	H_2O 2; EtOH 3; py 3
11080	2,4,6(1H,3H,5H)-Pyrimidinetrione, 5-(2-bromo-2-propenyl)-5-(1-methylpropyl)- Butallylonal	$C_{11}H_{15}BrN_2O_3$ 303.16	1142-70-7 131.5	1503	5-24-09-00231	eth 4; EtOH 4
11081	2,4,6(1H,3H,5H)-Pyrimidinetrione, 5-butyl-1-cyclohexyl- Bucolome	$C_{14}H_{22}N_2O_3$ 266.34	841-73-6 84	1452 185-7⁰·⁸	5-24-09-00132	
11082	2,4,6(1H,3H,5H)-Pyrimidinetrione, 5-butyl-5-ethyl- Butethal	$C_{10}H_{16}N_2O_3$ 212.25	77-28-1 128.5	1515	5-24-09-00161	
11083	2,4,6(1H,3H,5H)-Pyrimidinetrione, 5-(1-cyclohexen-1-yl)-1,5-dimethyl- Hexobarbital	$C_{12}H_{16}N_2O_3$ 236.27	56-29-1 146.5	4625	5-24-09-00253	
11084	2,4,6(1H,3H,5H)-Pyrimidinetrione, 5-(1-cyclohexen-1-yl)-5-ethyl- Cyclobarbital	$C_{12}H_{16}N_2O_3$ 236.27	52-31-3 173	2717	5-24-09-00258	H_2O 1; EtOH 4; eth 3; dil alk 3
11085	2,4,6(1H,3H,5H)-Pyrimidinetrione, 5-(2-cyclopenten-1-yl)-5-(2-propenyl)- Cyclopentobarbital	$C_{12}H_{14}N_2O_3$ 234.25	76-68-6 139.5	2751	5-24-09-00269	H_2O 2; EtOH 4
11086	2,4,6(1H,3H,5H)-Pyrimidinetrione, 5,5-dibromo-	$C_4H_2Br_2N_2O_3$ 285.88	511-67-1 235	dec	4-24-00-01881	eth 4; EtOH 4
11087	2,4,6(1H,3H,5H)-Pyrimidinetrione, 5,5-diethyl- Barbital	$C_8H_{12}N_2O_3$ 184.19	57-44-3 190	972	5-24-09-00137 1.220²⁵	H_2O 2; EtOH 3; eth 3; ace 3
11088	2,4,6(1H,3H,5H)-Pyrimidinetrione, 5,5-diethyl-1-methyl- Metharbital	$C_9H_{14}N_2O_3$ 198.22	50-11-3 150.5	5873	5-24-09-00144	H_2O 3; chl 2
11089	2,4,6(1H,3H,5H)-Pyrimidinetrione, 5,5-diethyl-1-phenyl- Phenetharbital	$C_{14}H_{16}N_2O_3$ 260.29	357-67-5 178	7183	4-24-00-01908	EtOH 4
11090	2,4,6(1H,3H,5H)-Pyrimidinetrione, 5-(2,2-dimethylpropyl)-5-(2-propenyl)- Nealbarbital	$C_{12}H_{18}N_2O_3$ 238.29	561-83-1 156	6350	5-24-09-00238	ace 4; eth 4; EtOH 4
11091	2,4,6(1H,3H,5H)-Pyrimidinetrione, 5,5-di-2-propenyl- Allobarbital	$C_{10}H_{12}N_2O_3$ 208.22	52-43-7 172	252	5-24-09-00245	H_2O 2; EtOH 3; eth 3; bz 3
11092	2,4,6(1H,3H,5H)-Pyrimidinetrione, 5-ethyl-5-(3-methylbutyl)- Amobarbital	$C_{11}H_{18}N_2O_3$ 226.28	57-43-2 157	601	5-24-09-00173	bz 4; EtOH 4; chl 4
11093	2,4,6(1H,3H,5H)-Pyrimidinetrione, 5-ethyl-5-(1-methylethyl)- Probarbital	$C_9H_{14}N_2O_3$ 198.22	76-76-6 203	7759	5-24-09-00159	H_2O 2; EtOH 1; eth 3
11094	2,4,6(1H,3H,5H)-Pyrimidinetrione, 5-ethyl-1-methyl-5-phenyl- Mephobarbital	$C_{13}H_{14}N_2O_3$ 246.27	115-38-8 176	5742	5-24-09-00294	H_2O 2; EtOH 4; eth 2; chl 3
11095	2,4,6(1H,3H,5H)-Pyrimidinetrione, 5-ethyl-5-phenyl- Phenobarbital	$C_{12}H_{12}N_2O_3$ 232.24	50-06-6 174	7201	5-24-09-00286	H_2O 1; EtOH 3; eth 3; bz 1
11096	2,4,6(1H,3H,5H)-Pyrimidinetrione, 5-ethyl-5-(1-piperidinyl)- 5-Ethyl-5-(1-piperidyl)barbituric acid	$C_{11}H_{17}N_3O_3$ 239.27	509-87-5 215	3800	4-25-00-04239	ace 4; eth 4; EtOH 4
11097	2,4,6(1H,3H,5H)-Pyrimidinetrione, 5-[(hexahydro-2,4,6-trioxo-5-pyrimidinyl)imino]-, monoammonium salt Murexide	$C_8H_{10}N_8O_7$ 302.20	3051-09-0	6215	4-25-00-04236	H_2O 2; EtOH 1; eth 1; alk 3
11098	2,4,6(1H,3H,5H)-Pyrimidinetrione, 5-(1-methylethyl)-5-(2-propenyl)- Aprobarbital	$C_{10}H_{14}N_2O_3$ 210.23	77-02-1 141	782	5-24-09-00224	ace 4; eth 4; EtOH 4; chl 4

No.	Name Synonym	Mol. Form. Mol. Wt.	CAS RN mp/°C	Merck No. bp/°C	Beil. Ref. den/g cm^{-3}	Solubility n_D
11099	2,4,6(1H,3H,5H)-Pyrimidinetrione, 1-methyl-5-(1-methylethyl)-5-(2-propenyl)- Enallylpropymal	C$_{11}$H$_{16}$N$_2$O$_3$ 224.26	1861-21-8 56.5	3523 176-8[12]	5-24-09-00225	bz 4; eth 4; EtOH 4; chl 4
11100	2,4,6(1H,3H,5H)-Pyrimidinetrione, 5-methyl-5-phenyl- Phenylmethylbarbituric acid	C$_{11}$H$_{10}$N$_2$O$_3$ 218.21	76-94-8 220	7275	5-24-09-00283	H$_2$O 1; EtOH 3; eth 3; alk 3
11101	2,4,6(1H,3H,5H)-Pyrimidinetrione, 5-(2-methylpropyl)-5-(2-propenyl)- Butalbital	C$_{11}$H$_{16}$N$_2$O$_3$ 224.26	77-26-9 138.5	1502	5-24-09-00231	H$_2$O 2; EtOH 3; eth 3; ace 3
11102	2,4,6(1H,3H,5H)-Pyrimidinetrione, 5-nitro- 5-Nitrobarbituric acid	C$_4$H$_3$N$_3$O$_5$ 173.09	480-68-2 180.5	6507	5-24-09-00105	H$_2$O 3; EtOH 3; eth 1
11103	2,4,6(1H,3H,5H)-Pyrimidinetrione, 5-phenyl-5-(2-propenyl)- Phenallymal	C$_{13}$H$_{12}$N$_2$O$_3$ 244.25	115-43-5 156.5	7162	5-24-09-00334	H$_2$O 2; EtOH 4; eth 4; bz 2
11104	2(1H)-Pyrimidinone, 4-amino- Cytosine	C$_4$H$_5$N$_3$O 111.10	71-30-7 320-5 dec	2801	4-25-00-03654	H$_2$O 3; EtOH 2; eth 1; chl 2
11105	2(1H)-Pyrimidinone, 4-amino-5-(hydroxymethyl)- 5-Hydroxymethylcytosine	C$_5$H$_7$N$_3$O$_2$ 141.13	1123-95-1 >300 dec			
11106	2(1H)-Pyrimidinone, 4-amino-5-methyl- 5-Methylcytosine	C$_5$H$_7$N$_3$O 125.13	554-01-8 270 dec	5970	4-25-00-03727	H$_2$O 3; EtOH 2; eth 1; acid 3
11107	4(1H)-Pyrimidinone, 2,6-diamino-5-(β-D-glucopyranosyloxy)- Vicine	C$_{10}$H$_{16}$N$_4$O$_7$ 304.26	152-93-2 239-42 dec	9880	4-25-00-04285	H$_2$O 2; EtOH 2; acid 4; alk 4
11108	4(1H)-Pyrimidinone, 2,3-dihydro-6-methyl-2-thioxo- Methylthiouracil	C$_5$H$_6$N$_2$OS 142.18	56-04-2 330 dec	6048 sub	5-24-07-00049	H$_2$O 1; EtOH 2; eth 2; bz 2
11109	4(1H)-Pyrimidinone, 2,3-dihydro-6-propyl-2-thioxo- Propylthiouracil	C$_7$H$_{10}$N$_2$OS 170.24	51-52-5 219	7882	5-24-07-00257	H$_2$O 2; EtOH 2; eth 1; bz 1
11110	4(1H)-Pyrimidinone, 2,3-dihydro-2-thioxo- 2-Thiouracil	C$_4$H$_4$N$_2$OS 128.15	141-90-2 >340 dec	9298	5-24-06-00361	H$_2$O 2; EtOH 2; anh HF 3
11111	5H-Pyrimido[4,5-d]thiazolo[3,2-a]pyrimidine-8-ethanol, 2,7-dimethyl- Thiochrome	C$_{12}$H$_{14}$N$_4$OS 262.34	92-35-3 228.8	sub	4-27-00-09599	H$_2$O 3; EtOH 2; eth 2; ace 2
11112	1H-Pyrrole Azole	C$_4$H$_5$N 67.09	109-97-7 -23.42	8025 129.79	5-20-05-00003 0.9698[20]	H$_2$O 2; EtOH 3; eth 3; ace 3 1.5085[20]
11113	1H-Pyrrole, 1-butyl-	C$_8$H$_{13}$N 123.20	589-33-3	170.5	5-20-05-00010 0.8747[20]	1.4727[20]
11114	1H-Pyrrole-2-carboxaldehyde	C$_5$H$_5$NO 95.10	1003-29-8 46.5	218	5-21-07-00169	chl 2; os 4; lig 2 1.5939[16]
11115	1H-Pyrrole-3-carboxylic acid, 2,4-dimethyl-, ethyl ester	C$_9$H$_{13}$NO$_2$ 167.21	2199-51-1 78.5	291	5-22-01-00427	eth 4; EtOH 4
11116	1H-Pyrrole-3-carboxylic acid, 2,5-dimethyl-, ethyl ester	C$_9$H$_{13}$NO$_2$ 167.21	2199-52-2 117.5	291; 130[15]	5-22-01-00432	EtOH 4
11117	1H-Pyrrole, 3,4-diethyl-2-methyl- 3,4-Diethyl-2-methylpyrrole	C$_9$H$_{15}$N 137.22	34874-30-1	202.5	4-20-00-02171 0.9100[17]	1.4988[17]
11118	1H-Pyrrole, 2,5-dihydro- 3-Pyrroline	C$_4$H$_7$N 69.11	109-96-6	8028 90.5	5-20-04-00274 0.9097[20]	H$_2$O 4; ace 4; eth 4; EtOH 4 1.4664[20]
11119	1H-Pyrrole, 2,4-dimethyl- 2,4-Dimethylpyrrole	C$_6$H$_9$N 95.14	625-82-1	168	5-20-05-00063 0.9236[20]	H$_2$O 2; EtOH 4; eth 4; bz 4 1.5048[20]
11120	1H-Pyrrole, 2,5-dimethyl-	C$_6$H$_9$N 95.14	625-84-3	171; 51[8]	5-20-05-00064 0.9353[20]	H$_2$O 1; EtOH 4; eth 4 1.5036[20]
11121	1H-Pyrrole-2,5-dione	C$_4$H$_3$NO$_2$ 97.07	541-59-3 94	sub	5-21-10-00003 1.2493[106]	H$_2$O 3; EtOH 3; eth 3
11122	1H-Pyrrole-2,5-dione, 1-ethyl- N-Ethylmaleimide	C$_6$H$_7$NO$_2$ 125.13	128-53-0 45.5	3778	5-21-10-00006	H$_2$O 2; EtOH 4; eth 4; chl 3
11123	1H-Pyrrole-2,5-dione, 1-phenyl- N-Phenylmaleimide	C$_{10}$H$_7$NO$_2$ 173.17	941-69-5 90.5	7270 162[12]	5-21-10-00009	bz 4; eth 4; EtOH 4
11124	1H-Pyrrole, 1-ethyl-	C$_6$H$_9$N 95.14	617-92-5	129.5	5-20-05-00010 0.9009[20]	EtOH 4 1.4841[20]
11125	1H-Pyrrole, 2-ethyl-	C$_6$H$_9$N 95.14	1551-06-0	164	5-20-05-00061 0.9042[20]	H$_2$O 1; EtOH 3; chl 3 1.4942[20]
11126	1H-Pyrrole, 3-ethyl-2,4-dimethyl-	C$_8$H$_{13}$N 123.20	517-22-6 0	199; 96[16]	5-20-05-00099 0.913[20]	H$_2$O 2; EtOH 3; eth 3; bz 3 1.4961[20]
11127	1H-Pyrrole, 4-ethyl-2,3-dimethyl-	C$_8$H$_{13}$N 123.20	491-18-9 16.5	200; 113[16]	4-20-00-02152 0.915[20]	H$_2$O 4
11128	1H-Pyrrole, 3-ethyl-4-methyl-	C$_7$H$_{11}$N 109.17	488-92-6 3	70[11]	5-20-05-00083 0.9059[20]	eth 4; EtOH 4 1.4913[20]
11129	Pyrrole, 2-isopropyl-	C$_7$H$_{11}$N 109.17	7696-51-7	172	5-20-05-00081 0.908[25]	eth 4; EtOH 4 1.491[25]
11130	1H-Pyrrole, 1-methyl-	C$_5$H$_7$N 81.12	96-54-8 -56.32	112.81	5-20-05-00008 0.9145[15]	H$_2$O 1; EtOH 5; eth 5 1.4875[20]
11131	1H-Pyrrole, 2-methyl-	C$_5$H$_7$N 81.12	636-41-9 -35.6	147.6	5-20-05-00053 0.9446[15]	H$_2$O 1; EtOH 5; eth 5 1.5035[16]
11132	1H-Pyrrole, 3-methyl- 3-Methylpyrrole	C$_5$H$_7$N 81.12	616-43-3 -48.4	142.9; 45[11]	5-20-05-00056	EtOH 5; eth 5 1.4970[20]
11133	1H-Pyrrole, 2,2'-methylenebis-	C$_9$H$_{10}$N$_2$ 146.19	21211-65-4 73	164[12]	5-23-06-00468	bz 4; eth 4; EtOH 4

No.	Name / Synonym	Mol. Form. / Mol. Wt.	CAS RN / mp/°C	Merck No. / bp/°C	Beil. Ref. / den/g cm^{-3}	Solubility / n_D
11134	1H-Pyrrole, 1-phenyl-	C$_{10}$H$_9$N / 143.19	635-90-5 / 62	/ 234	5-20-05-00012	H$_2$O 1; EtOH 3; eth 3; ace 3
11135	1H-Pyrrole, 2-phenyl- / 2-Phenylpyrrole	C$_{10}$H$_9$N / 143.19	3042-22-6 / 129	/ 272	5-20-07-00369	H$_2$O 1; EtOH 4; eth 4; bz 4
11136	1H-Pyrrole, 1-(phenylmethyl)-	C$_{11}$H$_{11}$N / 157.22	2051-97-0 / 15	/ 247	5-20-05-00013 / 1.0183[20]	H$_2$O 1; EtOH 4; eth 4 / 1.5655[24]
11137	1H-Pyrrole-1-propanenitrile	C$_7$H$_8$N$_2$ / 120.15	43036-06-2	/ 132-3[10]	5-20-05-00023 / 1.048[25]	
11138	1H-Pyrrole, 1-propyl-	C$_7$H$_{11}$N / 109.17	5145-64-2	/ 145.5	5-20-05-00010 / 0.8833[20]	eth 4; EtOH 4 / 1.5103[20]
11139	1H-Pyrrole, 3-propyl- / 3-Propylpyrrole	C$_7$H$_{11}$N / 109.17	1551-09-3	/ 90[30]; 49[2]	5-20-05-00081	eth 3; ace 3; chl 2 / 1.4900[25]
11140	1H-Pyrrole, 2,3,4,5-tetraiodo- / Iodopyrrole	C$_4$HI$_4$N / 570.68	87-58-1 / 150 dec	4934	5-20-05-00045	ace 4; eth 4; chl 4
11141	1H-Pyrrole, 2,3,4,5-tetramethyl-	C$_8$H$_{13}$N / 123.20	1003-90-3 / 111	130[7]	5-20-05-00099	EtOH 5; eth 5; ace 5; bz 4
11142	1H-Pyrrole, 1,2,5-trimethyl- / 1,2,5-Trimethylpyrrole	C$_7$H$_{11}$N / 109.17	930-87-0	/ 171	5-20-05-00065 / 0.807[25]	/ 1.4969[20]
11143	Pyrrolidine / Azacyclopentane	C$_4$H$_9$N / 71.12	123-75-1 / -57.8	8026 / 86.5	5-20-01-00162 / 0.8586[20]	H$_2$O 5; EtOH 3; eth 3; bz 2 / 1.4431[20]
11144	3-Pyrrolidineacetic acid, 2-carboxy-4-(1-methylethenyl)-, [2S-(2α,3β,4β)]- / Kainic acid	C$_{10}$H$_{15}$NO$_4$ / 213.23	487-79-6	5157	4-22-00-01523	
11145	1-Pyrrolidineacetic acid, 2,5-dioxo-, ethyl ester / Aminoacetic acid, N-succinyl, ethyl ester	C$_8$H$_{11}$NO$_4$ / 185.18	14181-05-6 / 67	/ 198[32]	5-21-09-00474	H$_2$O 4; EtOH 4
11146	Pyrrolidine, 1-butyl-	C$_8$H$_{17}$N / 127.23	767-10-2	/ 154.5	5-20-01-00170 / 0.816[25]	EtOH 4 / 1.4373[25]
11147	Pyrrolidine, 2-butyl- / 2-Butylpyrrolidine	C$_8$H$_{17}$N / 127.23	3446-98-8	/ 174; 67[18]	5-20-04-00222 / 0.8277[20]	H$_2$O 4; EtOH 4 / 1.4490[20]
11148	1,2-Pyrrolidinedicarboxylic acid, 1-(phenylmethyl) ester, (S)-	C$_{13}$H$_{15}$NO$_4$ / 249.27	1148-11-4 / 77		5-22-01-00084	chl 2 / 1.5310[20]
11149	Pyrrolidine, 2,2-dimethyl- / 2,2-Dimethylpyrrolidine	C$_6$H$_{13}$N / 99.18	35018-15-6	/ 106	4-20-00-01529 / 0.8211[25]	/ 1.4304[25]
11150	Pyrrolidine, 2,4-dimethyl-	C$_6$H$_{13}$N / 99.18	13603-04-8	/ 116	4-20-00-01537 / 0.8297[20]	H$_2$O 4; eth 4; EtOH 4 / 1.4325[20]
11151	Pyrrolidine, 2,5-dimethyl-, cis- / cis-2,5-Dimethylpyrrolidine	C$_6$H$_{13}$N / 99.18	39713-71-8	/ 106.5	5-20-04-00146 / 0.8205[20]	H$_2$O 4; eth 4; EtOH 4 / 1.4299[20]
11152	2,5-Pyrrolidinedione / Succinimide	C$_4$H$_5$NO$_2$ / 99.09	123-56-8 / 126.5	8842 / 287 dec	5-21-09-00438 / 1.418[25]	H$_2$O 3; EtOH 2; eth 2; ace 2
11153	2,5-Pyrrolidinedione, 1-bromo- / N-Bromosuccinimide	C$_4$H$_4$BrNO$_2$ / 177.99	128-08-5 / 174	1428	5-21-09-00543 / 2.098[25]	H$_2$O 2; eth 2; ace 4
11154	2,5-Pyrrolidinedione, 1-chloro- / N-Chlorosuccinimide	C$_4$H$_4$ClNO$_2$ / 133.53	128-09-6 / 150	2165	5-21-09-00543 / 1.65[25]	H$_2$O 2; EtOH 2; ace 3; bz 2
11155	2,5-Pyrrolidinedione, 1-(chloromethyl)-	C$_5$H$_6$ClNO$_2$ / 147.56	54553-14-9 / 58	/ 158[12]	5-21-09-00460	ace 4; bz 4
11156	2,5-Pyrrolidinedione, 1,3-dimethyl-3-phenyl- / Methsuximide	C$_{12}$H$_{13}$NO$_2$ / 203.24	77-41-8 / 52.5	5928 / 121-2[0.1]	5-21-11-00209	
11157	2,5-Pyrrolidinedione, 1-ethyl-	C$_6$H$_9$NO$_2$ / 127.14	2314-78-5 / 26	/ 236	5-21-09-00442	eth 4; EtOH 4
11158	2,5-Pyrrolidinedione, 3-ethyl-3-methyl- / Ethosuximide	C$_7$H$_{11}$NO$_2$ / 141.17	77-67-8 / 64.5	3704	5-21-09-00595	H$_2$O 4
11159	2,5-Pyrrolidinedione, 1-iodo- / N-Iodosuccinimide	C$_4$H$_4$INO$_2$ / 224.99	516-12-1 / 200.5	4936	5-21-09-00544 / 2.245[25]	H$_2$O 4; EtOH 3; eth 2; ace 3
11160	2,5-Pyrrolidinedione, 1-methyl-	C$_5$H$_7$NO$_2$ / 113.12	1121-07-9 / 71	/ 234	5-21-09-00441	H$_2$O 3; EtOH 3; eth 4
11161	2,5-Pyrrolidinedione, 1-methyl-3-phenyl- / Phensuximide	C$_{11}$H$_{11}$NO$_2$ / 189.21	86-34-0 / 72	7231	5-21-11-00188	EtOH 4; MeOH 4
11162	2,5-Pyrrolidinedione, 1-phenyl- / Succinanil	C$_{10}$H$_9$NO$_2$ / 175.19	83-25-0 / 156	8838 / 400	5-21-09-00446 / 1.356[25]	H$_2$O 1; EtOH 3; eth 3
11163	2,5-Pyrrolidinedione, 1-(phenylmethyl)- / N-Benzylsuccinimide	C$_{11}$H$_{11}$NO$_2$ / 189.21	2142-06-5 / 103	/ 393	5-21-09-00449	bz 4; eth 4; EtOH 4; chl 4
11164	1-Pyrrolidineethanamine	C$_6$H$_{14}$N$_2$ / 114.19	7154-73-6	/ 66-70[23]	5-20-01-00405 / 0.901[25]	/ 1.4687[20]
11165	1-Pyrrolidineethanol	C$_6$H$_{13}$NO / 115.18	2955-88-6	/ 79-81[13]	4-20-00-00092 / 0.985[25]	/ 1.4713[20]
11166	Pyrrolidine, 1-ethyl-	C$_6$H$_{13}$N / 99.18	7335-06-0	/ 104	5-20-01-00168 / 0.8156[20]	/ 1.4113[15]
11167	2-Pyrrolidinemethanamine, 1-ethyl-	C$_7$H$_{16}$N$_2$ / 128.22	26116-12-1	/ 58-60[16]	5-22-08-00096 / 0.887[25]	/ 1.4665[20]
11168	Pyrrolidine, 1-methyl- / N-Methylpyrrolidine	C$_5$H$_{11}$N / 85.15	120-94-5	/ 81	5-20-01-00166 / 0.8188[20]	H$_2$O 4; eth 4 / 1.4247[20]
11169	Pyrrolidine, 1-[2-(3-methylbutoxy)-2-phenylethyl]- / Amixetrine	C$_{17}$H$_{27}$NO / 261.41	24622-72-8 / 507	/ 121[2]		/ 1.4978[22]
11170	Pyrrolidine, 1-nitroso- / N-Nitrosopyrrolidine	C$_4$H$_8$N$_2$O / 100.12	930-55-2	6564 / 214	5-20-01-00521 / 1.085[25]	/ 1.4880[25]
11171	Pyrrolidine, 1-phenyl-	C$_{10}$H$_{13}$N / 147.22	4096-21-3	/ 119-20[12]	5-20-01-00182 / 1.0260[25]	eth 3 / 1.5813[20]

No.	Name Synonym	Mol. Form. Mol. Wt.	CAS RN mp/°C	Merck No. bp/°C	Beil. Ref. den/g cm^{-3}	Solubility n_D
11172	Pyrrolidine, 1-(phenylmethyl)-	$C_{11}H_{15}N$ 161.25	29897-82-3	45[0.7]	5-20-01-00184 0.965[25]	1.5270[20]
11173	Pyrrolidine, 1-propyl-	$C_7H_{15}N$ 113.20	7335-07-1 105	132.5	5-20-01-00169 0.8171[20]	1.4389[20]
11174	Pyrrolidine, 3-propyl-	$C_7H_{15}N$ 113.20	116632-47-4	159	4-20-00-01598 0.8450[20]	1.4469[20]
11175	Pyrrolidine, 2,2,4-trimethyl- 2,2,4-Trimethylpyrrolidine	$C_7H_{15}N$ 113.20	35018-28-1	119	5-20-04-00198 0.8063[25]	1.4259[25]
11176	Pyrrolidine, 2,2,5-trimethyl-	$C_7H_{15}N$ 113.20	6496-48-6	112	4-20-00-01603 0.7980[25]	1.4223[25]
11177	Pyrrolidinium, 2-carboxy-1,1-dimethyl-, hydroxide, inner salt, (S)- Stachydrine	$C_7H_{13}NO_2$ 143.19	471-87-4 235	8729		H_2O 4; EtOH 4; EtOH 4
11178	Pyrrolidinium, 2-carboxy-4-hydroxy-1,1-dimethyl-, hydroxide, inner salt, (2S-trans)- Betonicine	$C_7H_{13}NO_3$ 159.19	515-25-3 252 dec	1209		EtOH 4
11179	Pyrrolidinium, 3-[(cyclopentylhydroxyphenylacetyl)oxy]-1,1-dimethyl-, bromide Glycopyrrolate	$C_{19}H_{28}BrNO_3$ 398.34	596-51-0 192.5	4397	5-21-01-00015	
11180	2-Pyrrolidinone 2-Pyrrolidone	C_4H_7NO 85.11	616-45-5 25	8027 251; 133[12]	5-21-06-00317 1.120[20]	H_2O 4; EtOH 4; eth 4; bz 4 1.4806[30]
11181	2-Pyrrolidinone, 1,5-dimethyl-	$C_6H_{11}NO$ 113.16	5075-92-3	217; 87.5[10]	5-21-06-00431	H_2O 4; eth 4 1.4650[20]
11182	2-Pyrrolidinone, 3,3-dimethyl- 3,3-Dimethyl-2-pyrrolidone	$C_6H_{11}NO$ 113.16	4831-43-0 66	237	0-21-00-00242	H_2O 4; ace 4; eth 4; EtOH 4
11183	2-Pyrrolidinone, 1-ethenyl-	C_6H_9NO 111.14	88-12-0 13.5	193[400]; 93[11]	5-21-06-00330 1.04[20]	
11184	2-Pyrrolidinone, 1-(2-hydroxyethyl)-	$C_6H_{11}NO_2$ 129.16	3445-11-2 20	295	5-21-06-00341 1.1435[20]	
11185	2-Pyrrolidinone, 1-methyl- N-Methyl-2-pyrrolidone	C_5H_9NO 99.13	872-50-4 -24	202	5-21-06-00321 1.0230[25]	H_2O 4; eth 3; ace 3; chl 3 1.4684[20]
11186	2-Pyrrolidinone, 5-methyl-	C_5H_9NO 99.13	108-27-0 43	248	5-21-06-00431 1.0458[20]	
11187	3-Pyrrolidinone, 1-methyl-	C_5H_9NO 99.13	68165-06-0	46-7[18]	5-21-06-00389 0.9646[25]	1.4431[25]
11188	2-Pyrrolidinone, 1-[4-(1-pyrrolidinyl)-2-butynyl]- Oxotremorine	$C_{12}H_{18}N_2O$ 206.29	70-22-4 124[0.1]	6904	5-21-06-00370 0.991[25]	1.5160[20]
11189	2-Pyrroline, 2-methyl-	C_5H_9N 83.13	100791-95-5	96	0-20-00-00135 0.8995[22]	H_2O 4; eth 4; EtOH 4
11190	Pyrrolo[1,2-a]pyrimidine, 2,3,4,6,7,8-hexahydro-	$C_7H_{12}N_2$ 124.19	3001-72-7	95-8[7]	5-23-05-00239 1.005[25]	1.5190[20]
11191	Pyrrolo[2,3-b]indol-5-ol, 1,2,3,3a,8,8a-hexahydro-1,3a,8-trimethyl-, methylcarbamate (ester), (3aS-cis)- Physostigmine	$C_{15}H_{21}N_3O_2$ 275.35	57-47-6 105.5	7357	5-23-11-00401	H_2O 2; EtOH 3; eth 3; bz 3
11192	Pyrrolo[2,1-b]quinazolin-3-ol, 1,2,3,9-tetrahydro-, (R)- Vasicine	$C_{11}H_{12}N_2O$ 188.23	6159-55-3 211.5	9842	5-23-12-00040	H_2O 2; EtOH 3; eth 2; ace 3
11193	1,1':2',1":2",1'''-Quaterphenyl	$C_{24}H_{18}$ 306.41	641-96-3 118.5	420; 193[1.0]	4-05-00-02729	H_2O 1; EtOH 2; eth 3; ace 3
11194	1,1':4',1":4",1'''-Quaterphenyl	$C_{24}H_{18}$ 306.41	135-70-6 320	428[18]	4-05-00-02730	H_2O 1; EtOH 1; eth 1; bz 3
11195	Quinazoline 1,3-Benzodiazine	$C_8H_6N_2$ 130.15	253-82-7 48	8062 241	5-23-07-00129	H_2O 4; EtOH 3; eth 3; ace 3
11196	Quinazoline, 3,4-dihydro-3-phenyl- Orexine	$C_{14}H_{12}N_2$ 208.26	612-97-5 96.5	dec	4-23-00-01108 1.290[4]	bz 4; eth 4; EtOH 4; chl 4
11197	6-Quinazolinesulfonamide, 7-chloro-2-ethyl-1,2,3,4-tetrahydro-4-oxo- Quinethazone	$C_{10}H_{12}ClN_3O_3S$ 289.74	73-49-4	8067		tfa 3
11198	4(1H)-Quinazolinone, 2,3-dihydro-1-(4-morpholinylacetyl)-3-phenyl- Moquizone	$C_{20}H_{21}N_3O_3$ 351.41	19395-58-5 136	6174		chl 3
11199	4(3H)-Quinazolinone, 2-methyl-3-(2-methylphenyl)- Methaqualone	$C_{16}H_{14}N_2O$ 250.30	72-44-6 120	5872	5-24-03-00132	eth 4; EtOH 4; chl 4
11200	4(1H)-Quinazolinone, 1-methyl-2-(phenylmethyl)- Glycosine	$C_{16}H_{14}N_2O$ 250.30	6873-15-0 161.5	4398	5-24-04-00363	
11201	Quinclorac 3,7-Dichloroquinoline-8-carboxylic acid	$C_{10}H_5Cl_2NO_2$ 242.06	84087-01-4 274		1.75	
11202	2-Quinolinamine 2-Aminoquinoline	$C_9H_8N_2$ 144.18	580-22-3 131.5	sub	5-22-10-00220	H_2O 4; EtOH 3; eth 3; ace 3
11203	4-Quinolinamine	$C_9H_8N_2$ 144.18	578-68-7 154.8	180[12]	5-22-10-00241	H_2O 3; EtOH 4; eth 4; bz 3
11204	5-Quinolinamine	$C_9H_8N_2$ 144.18	611-34-7 110	310; 184[10]	5-22-10-00297	H_2O 2; EtOH 4; eth 4; bz 3
11205	6-Quinolinamine 6-Aminoquinoline	$C_9H_8N_2$ 144.18	580-15-4 114	187[12]	5-22-10-00303	H_2O 2; EtOH 3; eth 2; NH_3 3

No.	Name Synonym	Mol. Form. Mol. Wt.	CAS RN mp/°C	Merck No. bp/°C	Beil. Ref. den/g cm^{-3}	Solubility n_D
11206	8-Quinolinamine	$C_9H_8N_2$ 144.18	578-66-5 70	157[19]	5-22-10-00316	H_2O 4; EtOH 4
11207	4-Quinolinamine, 3-bromo- 4-Amino-3-bromoquinoline	$C_9H_7BrN_2$ 223.07	36825-36-2 203	sub	4-22-00-04662	eth 4; EtOH 4
11208	2-Quinolinamine, 4-methyl- Lepidine, 2-amino-	$C_{10}H_{10}N_2$ 158.20	27063-27-0 133	320	5-22-10-00375	eth 4; EtOH 4; chl 4
11209	3-Quinolinamine, 2-methyl-	$C_{10}H_{10}N_2$ 158.20	21352-22-7 160	278; 198[16]	5-22-10-00347	bz 4; eth 4; EtOH 4
11210	4-Quinolinamine, 2-methyl-	$C_{10}H_{10}N_2$ 158.20	6628-04-2 169	333	5-22-10-00347	H_2O 2; EtOH 4; eth 4; ace 4
11211	7-Quinolinamine, 2-methyl- 7-Amino-2-methylquinoline	$C_{10}H_{10}N_2$ 158.20	64334-96-9 148	304	5-22-10-00364	ace 4; bz 4; eth 4; EtOH 4
11212	8-Quinolinamine, 6-methyl- 8-Amino-6-methylquinoline	$C_{10}H_{10}N_2$ 158.20	68420-93-9 73	sub	4-22-00-04822	ace 4; bz 4; eth 4; EtOH 4
11213	6-Quinolinamine, N,N,2-trimethyl-	$C_{12}H_{14}N_2$ 186.26	92-99-9 101	319	2-22-00-00361	ctc 3; CS_2 3
11214	Quinoline 1-Azanaphthalene	C_9H_7N 129.16	91-22-5 -14.78	8097 237.1	5-20-07-00276 1.0977[15]	H_2O 1; EtOH 5; eth 5; ace 5 1.6268[20]
11215	Quinoline, 6-amino-3-bromo-	$C_9H_7BrN_2$ 223.07	7101-96-4 106	sub	5-22-10-00312	EtOH 4
11216	Quinoline, 2-[2-(1,3-benzodioxol-5-yl)ethyl]-4-methoxy- Cusparine	$C_{19}H_{17}NO_3$ 307.35	529-92-0 92(form a); 116(form b)	2687	2-27-00-00545	ace 4; bz 4; eth 4; EtOH 4
11217	Quinoline, 3-bromo- 3-Bromoquinoline	C_9H_6BrN 208.06	5332-24-1 13.3	275	5-20-07-00318	chl 3; HOAc 4 1.6641[20]
11218	Quinoline, 5-bromo- 5-Bromoquinoline	C_9H_6BrN 208.06	4964-71-0 52	280; 106[12]	5-20-07-00319	acid 3
11219	Quinoline, 6-bromo- 6-Bromoquinoline	C_9H_6BrN 208.06	5332-25-2 24	281	5-20-07-00319	EtOH 3; eth 3; acid 3
11220	Quinoline, 7-bromo- 7-Bromoquinoline	C_9H_6BrN 208.06	4965-36-0 34	290	4-20-00-03390	EtOH 3; eth 3; acid 4
11221	Quinoline, 6-bromo-5-nitro- 6-Bromo-5-nitroquinoline	$C_9H_5BrN_2O_2$ 253.05	98203-04-4 130	sub	0-20-00-00376	eth 4; EtOH 4
11222	Quinoline, 2-butyl- 2-Butylquinoline	$C_{13}H_{15}N$ 185.27	7661-39-4 	153[14]	5-20-07-00482 1.003[20]	1.5699[20]
11223	2-Quinolinecarbonitrile 2-Cyanoquinoline	$C_{10}H_6N_2$ 154.17	1436-43-7 95	160[23]	5-22-03-00190	H_2O 3; EtOH 4; eth 4; bz 4
11224	4-Quinolinecarbonitrile	$C_{10}H_6N_2$ 154.17	2973-27-5 103.5	242.5	5-22-03-00206	ace 4; bz 4; eth 4; EtOH 4
11225	4-Quinolinecarboxaldehyde Cinchoninaldehyde	$C_{10}H_7NO$ 157.17	4363-93-3 51	122[4]	5-21-08-00443	eth 4; tol 4
11226	4-Quinolinecarboxamide, 2-butoxy-N-[2-(diethylamino)ethyl]-, monohydrochloride Dibucaine hydrochloride	$C_{20}H_{30}ClN_3O_2$ 379.93	61-12-1 94 dec	3016	5-22-05-00278	chl 3
11227	2-Quinolinecarboxylic acid Quinaldic acid	$C_{10}H_7NO_2$ 173.17	93-10-7 156	8054	5-22-03-00183	H_2O 3; bz 4
11228	7-Quinolinecarboxylic acid	$C_{10}H_7NO_2$ 173.17	1078-30-4 249.5	sub	5-22-03-00217	EtOH 4
11229	8-Quinolinecarboxylic acid 8-Carboxyquinoline	$C_{10}H_7NO_2$ 173.17	86-59-9 187	8099 sub	5-22-03-00217	EtOH 4
11230	2-Quinolinecarboxylic acid, 4,8-dihydroxy- Xanthurenic acid	$C_{10}H_7NO_4$ 205.17	59-00-7 289	9977	4-22-00-02513	H_2O 1; EtOH 3; eth 2; bz 2
11231	1(2H)-Quinolinecarboxylic acid, 2-ethoxy-, ethyl ester EEDQ	$C_{14}H_{17}NO_3$ 247.29	16357-59-8 56.5	3488 125-8[0.1]	5-21-03-00028	chl 3
11232	2-Quinolinecarboxylic acid, 4-hydroxy- Kynurenic acid	$C_{10}H_7NO_3$ 189.17	492-27-3 282.5	5206	5-22-06-00280	H_2O 2; EtOH 3; eth 1; os 2
11233	4-Quinolinecarboxylic acid, 3-hydroxy-2-phenyl- Oxycinchophen	$C_{16}H_{11}NO_3$ 265.27	485-89-2 206-7 dec	6910	4-22-00-02383	bz 4; EtOH 4; HOAc 4
11234	4-Quinolinecarboxylic acid, 6-methoxy- Quininic acid	$C_{11}H_9NO_3$ 203.20	86-68-0 285 dec	8092 sub	4-22-00-02297	H_2O 2; EtOH 3; eth 2; bz 2
11235	4-Quinolinecarboxylic acid, 2-phenyl- Cinchophen	$C_{16}H_{11}NO_2$ 249.27	132-60-5 214.5	2290	5-22-03-00484	H_2O 1; EtOH 3; eth 3; ace 2
11236	4-Quinolinecarboxylic acid, 2-phenyl-, 2-propenyl ester	$C_{19}H_{15}NO_2$ 289.33	524-34-5 31.5	262[15]	4-22-00-01361	ace 4; eth 4; EtOH 4
11237	Quinoline, 2-chloro- 2-Chloroquinoline	C_9H_6ClN 163.61	612-62-4 38	266; 153[22]	5-20-07-00312 1.2464[25]	H_2O 1; EtOH 4; eth 4; bz 3 1.6342[25]
11238	Quinoline, 4-chloro- 4-Chloroquinoline	C_9H_6ClN 163.61	611-35-8 34.5	262; 130[15]	5-20-07-00314 1.251[25]	H_2O 2; EtOH 4; eth 4; dil HCl 3
11239	Quinoline, 5-chloro- 5-Chloroquinoline	C_9H_6ClN 163.61	635-27-8 45	256	5-20-07-00314	EtOH 3
11240	Quinoline, 6-chloro- 6-Chloroquinoline	C_9H_6ClN 163.61	612-57-7 43.8	263	5-20-07-00315	1.6110[56]
11241	Quinoline, 7-chloro- 7-Chloroquinoline	C_9H_6ClN 163.61	612-61-3 31.5	267.5	5-20-07-00315 1.2158[58]	H_2O 2; EtOH 4; eth 4; ace 4 1.6108[58]
11242	Quinoline, 8-chloro- 8-Chloroquinoline	C_9H_6ClN 163.61	611-33-6 -20	288.5	5-20-07-00315 1.2834[14]	H_2O 3; EtOH 4; eth 4; ace 4 1.6408[14]

No.	Name Synonym	Mol. Form. Mol. Wt.	CAS RN mp/°C	Merck No. bp/°C	Beil. Ref. den/g cm^{-3}	Solubility n_D
11243	Quinoline, 2-chloro-4-methyl-	$C_{10}H_8ClN$ 177.63	634-47-9 59	296	5-20-07-00396	H_2O 1; EtOH 4; eth 4; chl 4
11244	Quinoline, 2-chloro-8-methyl- 2-Chloro-8-methylquinoline	$C_{10}H_8ClN$ 177.63	4225-85-8 61	287	5-20-07-00407	bz 4; eth 4; EtOH 4; chl 4
11245	Quinoline, 4-chloro-2-methyl-	$C_{10}H_8ClN$ 177.63	4295-06-1 42.5	269.5	5-20-07-00383 0.881[25]	H_2O 3; eth 3; bz 3 1.6224[20]
11246	Quinoline, decahydro-, cis- cis-Decahydroquinoline	$C_9H_{17}N$ 139.24	10343-99-4 -40	206; 90[20]	5-20-04-00405 0.9426[20]	H_2O 2; EtOH 3; eth 3 1.4926[20]
11247	Quinoline, decahydro-, trans-(±)- Quinoline, decahydro (trans, DL)	$C_9H_{17}N$ 139.24	105728-23-2 48	204	5-20-04-00405 0.9610[22]	ace 4; bz 4; eth 4; EtOH 4 1.4692[56]
11248	Quinoline, 2,4-dichloro- 2,4-Dichloroquinoline	$C_9H_5Cl_2N$ 198.05	703-61-7 67.5	281	4-20-00-03382	bz 4; eth 4; EtOH 4; chl 4
11249	Quinoline, 2,7-dichloro- 2,7-Dichloroquinoline	$C_9H_5Cl_2N$ 198.05	613-77-4 120	100[2]	4-20-00-03383	eth 4; EtOH 4
11250	Quinoline, 4,5-dichloro-	$C_9H_5Cl_2N$ 198.05	21617-18-5 118	134[5.5]	4-20-00-03383	eth 4; EtOH 4
11251	Quinoline, 4,7-dichloro- 4,7-Dichloroquinoline	$C_9H_5Cl_2N$ 198.05	86-98-6 93	148[10]	5-20-07-00316	chl 2
11252	Quinoline, 5,8-dichloro-	$C_9H_5Cl_2N$ 198.05	703-32-2 97.5	sub	5-20-07-00317	eth 4; EtOH 4
11253	Quinoline, 2-[2-(3,4-dimethoxyphenyl)ethyl]-4-methoxy- Galipine	$C_{20}H_{21}NO_3$ 323.39	525-68-8 115.5	4247	4-21-00-02656	ace 4; bz 4; eth 4; EtOH 4
11254	Quinoline, 2,3-dimethyl- 2,3-Dimethylquinoline	$C_{11}H_{11}N$ 157.22	1721-89-7 68.5	263	5-20-07-00431 1.1013[25]	eth 4; EtOH 4; lig 4
11255	Quinoline, 2,4-dimethyl- 4-Methylquinaldine	$C_{11}H_{11}N$ 157.22	1198-37-4	265	5-20-07-00431 1.0611[15]	H_2O 2; EtOH 4; eth 4; chl 2 1.6075[20]
11256	Quinoline, 2,5-dimethyl-	$C_{11}H_{11}N$ 157.22	26190-82-9 61	264.5	4-20-00-03524	bz 4; eth 4; EtOH 4
11257	Quinoline, 2,6-dimethyl-	$C_{11}H_{11}N$ 157.22	877-43-0 60	266.5	5-20-07-00433	H_2O 2; EtOH 2; eth 2; bz 4
11258	Quinoline, 2,7-dimethyl- m-Toluquinaldine	$C_{11}H_{11}N$ 157.22	93-37-8 61	264.5	5-20-07-00435	H_2O 2; EtOH 3; eth 3; chl 3
11259	Quinoline, 2,8-dimethyl-	$C_{11}H_{11}N$ 157.22	1463-17-8 27	255.3	5-20-07-00436 1.0394[20]	H_2O 2; EtOH 4; eth 4; ctc 3 1.6022[20]
11260	Quinoline, 3,4-dimethyl- 3,4-Dimethylquinoline	$C_{11}H_{11}N$ 157.22	2436-92-2 73.5	291; 114[4]	5-20-07-00437	eth 4; EtOH 4
11261	Quinoline, 3,8-dimethyl- 3,8-Dimethylquinoline	$C_{11}H_{11}N$ 157.22	20668-29-5	270	5-20-07-00437 1.049[20]	1.6063[20]
11262	Quinoline, 4,6-dimethyl- 6-Methyllepidine	$C_{11}H_{11}N$ 157.22	826-77-7 22	273.5; 140[12]	5-20-07-00437 1.0577[20]	eth 4; EtOH 4 1.6100[20]
11263	Quinoline, 5,8-dimethyl- 5,8-Dimethylquinoline	$C_{11}H_{11}N$ 157.22	2623-50-9 4.5	266	5-20-07-00438 1.070[21]	eth 4; EtOH 4
11264	Quinoline, 6,8-dimethyl- 6,8-Dimethylquinoline	$C_{11}H_{11}N$ 157.22	2436-93-3	268.5	5-20-07-00439 1.0665[4]	eth 4; EtOH 4
11265	Quinoline, 6-ethoxy-1,2-dihydro-2,2,4-trimethyl- Ethoxyquin	$C_{14}H_{19}NO$ 217.31	91-53-2	123-5[2]	5-21-03-00095 1.026[25]	1.569[25]
11266	Quinoline, 2-ethyl- 2-Ethylquinoline	$C_{11}H_{11}N$ 157.22	1613-34-9	245.5; 134[16]	5-20-07-00428 1.050[17]	eth 4; EtOH 4; chl 4 1.5979[23]
11267	Quinoline, 3-ethyl- 3-Ethylquinoline	$C_{11}H_{11}N$ 157.22	1873-54-7	266	5-20-07-00428 1.0508[20]	1.6160[20]
11268	Quinoline, 6-ethyl- 6-Ethylquinoline	$C_{11}H_{11}N$ 157.22	19655-60-8	135-6[14]	5-20-07-00429 1.043[25]	1.6009[25]
11269	Quinoline, 2-ethyl-3-methyl- 2-Ethyl-3-methylquinoline	$C_{12}H_{13}N$ 171.24	27356-52-1 57	269; 140[12]	5-20-07-00456	bz 4; eth 4; EtOH 4
11270	Quinoline, 2-ethyl-4-methyl- Lepidine, 2-ethyl-	$C_{12}H_{13}N$ 171.24	33357-44-7	272.5	5-20-07-00456 1.025[20]	1.5941[20]
11271	Quinoline, 3-ethyl-6-methyl-	$C_{12}H_{13}N$ 171.24	105688-74-2	284.5	4-20-00-03555 1.0280[20]	1.5955[20]
11272	Quinoline, 4-iodo- 4-Iodoquinoline	C_9H_6IN 255.06	16560-43-3 98.5	sub	5-20-07-00322	eth 4; EtOH 4
11273	Quinoline, 6-iodo-	C_9H_6IN 255.06	13327-31-6 91	120[1.0]	5-20-07-00322	eth 4; EtOH 4
11274	Quinoline, 4-methoxy- 4-Methoxyquinoline	$C_{10}H_9NO$ 159.19	607-31-8 41	245	5-21-03-00229	H_2O 1; EtOH 3; eth 3; bz 3
11275	Quinoline, 6-methoxy- 6-Methoxyquinoline	$C_{10}H_9NO$ 159.19	5263-87-6 26.5	306; 153[12]	5-21-03-00244 1.152[20]	EtOH 3; eth 3; chl 3; dil HCl 3
11276	Quinoline, 8-methoxy- 8-Methoxyquinoline	$C_{10}H_9NO$ 159.19	938-33-0 49.5	283; 164[14]	5-21-03-00254 1.034[29]	EtOH 3; eth 3; bz 3; peth 3
11277	Quinoline, 2-methyl- Quinaldine	$C_{10}H_9N$ 143.19	91-63-4 -1.5	8055 246.5	5-20-07-00375 1.06[25]	H_2O 2; EtOH 3; eth 3; ace 3 1.6116[20]
11278	Quinoline, 3-methyl- 3-Methylquinoline	$C_{10}H_9N$ 143.19	612-58-8 16.5	259.8	5-20-07-00388 1.0673[20]	ace 4; eth 4; EtOH 4 1.6171[20]
11279	Quinoline, 4-methyl- Lepidine	$C_{10}H_9N$ 143.19	491-35-0 9.5	5324 262	5-20-07-00389 1.083[20]	H_2O 2; EtOH 3; eth 3; ace 3 1.6200[20]
11280	Quinoline, 5-methyl- 5-Methylquinoline	$C_{10}H_9N$ 143.19	7661-55-4 19	262.7	5-20-07-00399 1.0832[20]	H_2O 2; EtOH 5; eth 5; ace 3 1.6219[20]
11281	Quinoline, 6-methyl- 6-Methylquinoline	$C_{10}H_9N$ 143.19	91-62-3 -22	258.6	5-20-07-00400 1.0654[20]	H_2O 2; EtOH 3; eth 3; ace 3 1.6157[20]

No.	Name Synonym	Mol. Form. Mol. Wt.	CAS RN mp/°C	Merck No. bp/°C	Beil. Ref. den/g cm^{-3}	Solubility n_D
11282	Quinoline, 7-methyl- *m*-Toluquinoline	$C_{10}H_9N$ 143.19	612-60-2 39	257.6	5-20-07-00402 1.0609[20]	H_2O 2; EtOH 3; eth 3; ace 3 1.6150[20]
11283	Quinoline, 8-methyl- 8-Methylquinoline	$C_{10}H_9N$ 143.19	611-32-5 -80	247.5	5-20-07-00405 1.0719[20]	H_2O 2; EtOH 5; eth 5; ace 3 1.6164[20]
11284	Quinoline, 5-nitro- 5-Nitroquinoline	$C_9H_6N_2O_2$ 174.16	607-34-1 74	sub	5-20-07-00325	H_2O 2; EtOH 3; bz 3; chl 2
11285	Quinoline, 6-nitro- 6-Nitroquinoline	$C_9H_6N_2O_2$ 174.16	613-50-3 153.5	170[0.2]	5-20-07-00326	H_2O 3; EtOH 3; eth 2; bz 4
11286	Quinoline, 7-nitro- 7-Nitroquinoline	$C_9H_6N_2O_2$ 174.16	613-51-4 132.5	sub	5-20-07-00327	H_2O 3; EtOH 3; eth 3; chl 4
11287	Quinoline, 4-nitro-, 1-oxide 4-Nitroquinoline oxide	$C_9H_6N_2O_3$ 190.16	56-57-5 154		5-20-07-00324	
11288	Quinoline, 2-phenyl- 2-Phenylquinoline	$C_{15}H_{11}N$ 205.26	612-96-4 86	363; 194[6]	5-20-08-00355	H_2O 2; EtOH 4; eth 4; ace 4
11289	Quinoline, 3-phenyl- 3-Phenylquinoline	$C_{15}H_{11}N$ 205.26	1666-96-2 52	205[12]	5-20-08-00359	H_2O 2; EtOH 3; eth 3; ace 4
11290	Quinoline, 4-phenyl- 4-Phenylquinoline	$C_{15}H_{11}N$ 205.26	605-03-8 61.5	260[77;] 156[0.3]	5-20-08-00360	H_2O 1; EtOH 4; eth 4; ace 4
11291	Quinoline, 6-phenyl- 6-Phenylquinoline	$C_{15}H_{11}N$ 205.26	612-95-3 110	260[77]	5-20-08-00361 1.1945[20]	H_2O 2; EtOH 4; eth 2; ace 4
11292	Quinoline, 2-propyl- 2-Propylquinoline	$C_{12}H_{13}N$ 171.24	1613-32-7 	130-1[10]	5-20-07-00453 1.038[17]	1.5886[23]
11293	5-Quinolinesulfonic acid, 8-ethoxy- Actinoquinol	$C_{11}H_{11}NO_4S$ 253.28	15301-40-3 286 dec	134	5-22-07-00581	alk 3
11294	5-Quinolinesulfonic acid, 8-hydroxy- 8-Hydroxy-5-quinolinesulfonic acid	$C_9H_7NO_4S$ 225.22	84-88-8 322.5	4780	5-22-07-00581	H_2O 2
11295	5-Quinolinesulfonic acid, 8-hydroxy-7-iodo- Ferron	$C_9H_6INO_4S$ 351.12	547-91-1 260 dec	4757	5-22-07-00582	H_2O 2; EtOH 2; eth 1; bz 1
11296	Quinoline, 1,2,3,4-tetrahydro-	$C_9H_{11}N$ 133.19	635-46-1 20	251	5-20-06-00293 1.0588[20]	H_2O 3; EtOH 5; eth 5; chl 3 1.6062[19]
11297	Quinoline, 5,6,7,8-tetrahydro- 2,3-Cyclohexenopyridine	$C_9H_{11}N$ 133.19	10500-57-9 	222	5-20-06-00292 1.0304[13]	H_2O 2; EtOH 3; eth 3; ace 3 1.5435[20]
11298	Quinoline, 1,2,3,4-tetrahydro-6-methoxy-	$C_{10}H_{13}NO$ 163.22	120-15-0 42.5	284; 128[1]	5-21-02-00314 	chl 3 1.5718[20]
11299	Quinoline, 1,2,3,4-tetrahydro-1-methyl-	$C_{10}H_{13}N$ 147.22	491-34-9 	246	5-20-06-00294 1.022[20]	EtOH 4 1.5802[23]
11300	Quinoline, 2,3,6-trimethyl-	$C_{12}H_{13}N$ 171.24	2437-73-2 86.5	285	4-20-00-03557	eth 4; EtOH 4
11301	Quinoline, 2,3,8-trimethyl-	$C_{12}H_{13}N$ 171.24	4945-28-2 	283	4-20-00-03557 1.0069[20]	H_2O 1; EtOH 3; eth 3; bz 2
11302	Quinoline, 2,4,6-trimethyl- 4,6-Dimethylquinaldine	$C_{12}H_{13}N$ 171.24	2243-89-2 68	280	5-20-07-00458	ace 4; bz 4; eth 4; EtOH 4
11303	Quinoline, 2,4,7-trimethyl- 2,4,7-Trimethylquinoline	$C_{12}H_{13}N$ 171.24	71633-43-7 63.5	280.5	5-20-07-00458 1.0337[20]	H_2O 2; EtOH 4; eth 4; ace 4 1.5973[24]
11304	Quinoline, 2,5,7-trimethyl-	$C_{12}H_{13}N$ 171.24	102871-67-0 	287; 107[1.2]	4-20-00-03559	H_2O 1; EtOH 4; eth 4 1.5980[20]
11305	5-Quinolinol 5-Hydroxyquinoline	C_9H_7NO 145.16	578-67-6 226 dec	sub	5-21-03-00240	H_2O 3; EtOH 2; eth 2; bz 3
11306	6-Quinolinol	C_9H_7NO 145.16	580-16-5 195	360	5-21-03-00244	H_2O 1; EtOH 2; eth 2; bz 1
11307	7-Quinolinol 7-Hydroxyquinoline	C_9H_7NO 145.16	580-20-1 239	sub	5-21-03-00249	EtOH 4
11308	8-Quinolinol 8-Hydroxyquinoline	C_9H_7NO 145.16	148-24-3 75.5	4778 267	5-21-03-00252 1.034[20]	H_2O 1; EtOH 4; eth 1; ace 3
11309	8-Quinolinol, benzoate (ester) Benzoxiquine	$C_{16}H_{11}NO_2$ 249.27	86-75-9 	1122	5-21-03-00259	chl 2
11310	8-Quinolinol, 5-bromo-	C_9H_6BrNO 224.06	1198-14-7 124	sub	5-21-03-00288	H_2O 4; EtOH 4; bz 4; chl 4
11311	8-Quinolinol, 5-chloro- Cloxyquin	C_9H_6ClNO 179.61	130-16-5 130	2416	5-21-03-00279	
11312	8-Quinolinol, 5-chloro-7-iodo- Iodochlorhydroxyquin	C_9H_5ClINO 305.50	130-26-7 178.5	4924	5-21-03-00294	EtOH 2; HOAc 3
11313	8-Quinolinol, 5,7-dibromo- Broxyquinoline	$C_9H_5Br_2NO$ 302.95	521-74-4 196	1441 sub	5-21-03-00290	H_2O 1; EtOH 3; eth 2; ace 3
11314	8-Quinolinol, 5,7-dichloro- Chloroxine	$C_9H_5Cl_2NO$ 214.05	773-76-2 179.5	2175	5-21-03-00286	EtOH 2; ace 2; bz 3; chl 2
11315	8-Quinolinol, 5,7-diiodo- Iodoquinol	$C_9H_5I_2NO$ 396.95	83-73-8 210	4935	5-21-03-00296	H_2O 2; EtOH 4; eth 2; bz 2
11316	4-Quinolinol, 2,8-dimethyl-	$C_{11}H_{11}NO$ 173.21	15644-80-1 262.3	sub	5-21-08-00321	EtOH 4
11317	6-Quinolinol, 2,4-dimethyl- 2,4-Dimethyl-6-hydroxyquinoline	$C_{11}H_{11}NO$ 173.21	64165-34-0 214	360 dec	2-21-00-00068	ace 4; EtOH 4
11318	8-Quinolinol, 2,4-dimethyl- Quinoline, 2,4-dimethyl-8-hydroxy	$C_{11}H_{11}NO$ 173.21	115310-98-0 65	281	4-21-00-01292	ace 4; bz 4; eth 4; EtOH 4
11319	6-Quinolinol, 2-methyl-	$C_{10}H_9NO$ 159.19	613-21-8 213	304.5	5-21-03-00339 1.1665[25]	eth 4; EtOH 4
11320	7-Quinolinol, 6-methyl- Quinoline, 7-hydroxy-6-methyl	$C_{10}H_9NO$ 159.19	84583-53-9 244	sub	0-21-00-00111	bz 4; eth 4; EtOH 4
11321	8-Quinolinol, 2-methyl-	$C_{10}H_9NO$ 159.19	826-81-3 73.8	267	5-21-03-00341	H_2O 1; EtOH 3; eth 3; bz 3

No.	Name Synonym	Mol. Form. Mol. Wt.	CAS RN mp/°C	Merck No. bp/°C	Beil. Ref. den/g cm^{-3}	Solubility n_D
11322	8-Quinolinol, 6-methyl-	$C_{10}H_9NO$ 159.19	20984-33-2 95.5	sub	4-21-00-01275	EtOH 4
11323	8-Quinolinol, 5-nitro- Nitroxoline	$C_9H_6N_2O_3$ 190.16	4008-48-4 180	6577	5-21-03-00297	
11324	8-Quinolinol, sulfate (2:1) (salt) 8-Hydroxyquinoline sulfate	$C_9H_8NO_3S$ 210.23	134-31-6 177.5	4779	5-21-03-00252	H_2O 4; EtOH 3; eth 1
11325	2(1H)-Quinolinone Carbostyril	C_9H_7NO 145.16	59-31-4 199.5	1831 sub	5-21-08-00217	H_2O 2; EtOH 4; eth 4
11326	2(1H)-Quinolinone, 4-bromo-	C_9H_6BrNO 224.06	938-39-6 266.5	sub	5-21-08-00229	EtOH 4
11327	2(1H)-Quinolinone, 3,4-dihydro- Hydrocarbostyril	C_9H_9NO 147.18	553-03-7 163.5	4702 201[45]	5-21-08-00057	eth 4; EtOH 4
11328	2(1H)-Quinolinone, 1-methyl-	$C_{10}H_9NO$ 159.19	606-43-9 74	325	5-21-08-00218	H_2O 2; EtOH 3; eth 3; ace 3
11329	2(1H)-Quinolinone, 6-methyl-	$C_{10}H_9NO$ 159.19	4053-34-3 237	240[12]	5-21-08-00289	ace 4; bz 4; eth 4; EtOH 4
11330	4(1H)-Quinolinone, 1-methyl- Echinopsine	$C_{10}H_9NO$ 159.19	83-54-5	3472	5-21-08-00210	H_2O 3; EtOH 4; eth 2; bz 4
11331	2H-Quinolizine-1-methanol, octahydro-, (1R-trans)- Lupinine	$C_{10}H_{19}NO$ 169.27	486-70-4 70	5479 270	5-21-01-00338	H_2O 3; EtOH 3; eth 3; bz 3
11332	6-Quinoxalinamine 6-Aminoquinoxaline	$C_8H_7N_3$ 145.16	6298-37-9 159	sub	4-25-00-02611	H_2O 4; EtOH 4; eth 3; bz 2
11333	Quinoxaline 1,4-Benzodiazine	$C_8H_6N_2$ 130.15	91-19-0 28	8107 229.5	5-23-07-00135 1.1334[48]	H_2O 3; EtOH 5; eth 5; ace 5 1.6231[48]
11334	Quinoxaline, 6-chloro- 6-Chloroquinoxaline	$C_8H_5ClN_2$ 164.59	5448-43-1 64	117[10]	5-23-07-00144	H_2O 2
11335	Quinoxaline, 2,6-dimethyl-	$C_{10}H_{10}N_2$ 158.20	60814-29-1 54	268	0-23-00-00192	H_2O 4; ace 4; bz 4; EtOH 4
11336	Quinoxaline, 6-methoxy- 6-Methoxyquinoxaline	$C_9H_8N_2O$ 160.18	6639-82-3 58.9	128[7]	5-23-11-00461	ace 4; eth 4
11337	Quinoxaline, 2-methyl- 2-Methylquinoxaline	$C_9H_8N_2$ 144.18	7251-61-8 180.5	244	5-23-07-00219	H_2O 5; EtOH 4; eth 5; ace 5
11338	Quinoxaline, 6-methyl- 6-Methylquinoxaline	$C_9H_8N_2$ 144.18	6344-72-5 218.5	249; 141.5[29]	5-23-07-00223 1.1164[20]	H_2O 5; EtOH 5; eth 5; ace 5 1.6211[18]
11339	Quinoxaline, 1,2,3,4-tetrahydro- 1,2,3,4-Tetrahydroquinoxaline	$C_8H_{10}N_2$ 134.18	3476-89-9 99	289	5-23-06-00024	H_2O 3; EtOH 4; eth 4; bz 4
11340	(-)-cis-Resmethrin Cyclopropanecarboxylic acid, 2,2-dimethyl-3-(2-methyl-1-propenyl)-, [5-(phenylmethyl)- 3-furanyl] methyl ester	$C_{22}H_{26}O_3$ 338.45	10453-86-8 75			
11341	Retinol Vitamin A	$C_{20}H_{30}O$ 286.46	68-26-8 63.5	9918 137-8[10-6]	4-06-00-04133	H_2O 1; EtOH 3; eth 3; ace 3
11342	Rheadan, 8-methoxy-16-methyl-2,3:10,11-bis[methylenebis(oxy)]-, (8β)- Rheadine	$C_{21}H_{21}NO_6$ 383.40	2718-25-4 257	8174 sub	4-27-00-06897	
11343	Ribitol Adonitol	$C_5H_{12}O_5$ 152.15	488-81-3 104	157	4-01-00-02832	H_2O 3; EtOH 3; eth 1; lig 1
11344	Riboflavin Riboflavine	$C_{17}H_{20}N_4O_6$ 376.37	83-88-5 280 dec	8201	4-26-00-02542	H_2O 1; EtOH 2; eth 1; ace 1
11345	D-Ribose	$C_5H_{10}O_5$ 150.13	50-69-1 88	8205	4-01-00-04211	H_2O 3; EtOH 2
11346	Sarpagan-17-al Vellosimine	$C_{19}H_{20}N_2O$ 292.38	6874-98-2 305.5	9846 sub 180		
11347	Sarpagan-10,17-diol Sarpagine	$C_{19}H_{22}N_2O_2$ 310.40	482-68-8 320	8336	5-23-13-00401	H_2O 1; EtOH 3
11348	9,10-Secocholesta-5,7,10(19)-trien-3-ol, (3β,5Z,7E)- Vitamin D3	$C_{27}H_{44}O$ 384.65	67-97-0 84.5	9929	4-06-00-04149	H_2O 1; os 3
11349	9,10-Secoergosta-5(10),6,8,22-tetraen-3-ol, (3β,6E,22E)- Tachysterol	$C_{28}H_{44}O$ 396.66	115-61-7	9002	4-06-00-04404	H_2O 1; EtOH 3; eth 3; ace 3
11350	9,10-Secoergosta-5,7,10(19),22-tetraen-3-ol, (3β,5Z,7E,22E)- Vitamin D2	$C_{28}H_{44}O$ 396.66	50-14-6 116.5	9928 sub	4-06-00-04404	H_2O 1; EtOH 3; eth 3; ace 3
11351	16,19-Secostrychnidine-10,16-dione, 4-hydroxy-19-methyl- Vomicine	$C_{22}H_{24}N_2O_4$ 380.44	125-15-5 282	9946	4-27-00-07873	EtOH 2; eth 2; ace 2; chl 4
11352	9,10-Secostrychnidin-10-oic acid, 2,3-dihydro-4-nitro-2,3-dioxo- Cacotheline	$C_{21}H_{21}N_3O_7$ 427.41	561-20-6 >300	1604	4-27-00-08014	H_2O 2
11353	Selenanthrene Selanthrene	$C_{12}H_8Se_2$ 310.12	262-30-6 181	223[11]	5-19-02-00057	EtOH 2; eth 2; CS_2 3
11354	Selenourea Carbamimidoselenoic acid	CH_4N_2Se 123.02	630-10-4 200 dec		4-03-00-00435	H_2O 4
11355	D-Serine	$C_3H_7NO_3$ 105.09	312-84-5 229 dec	dec	4-04-00-03118	H_2O 4; EtOH 1; eth 1; bz 1
11356	DL-Serine	$C_3H_7NO_3$ 105.09	302-84-1 246 dec		4-04-00-03119 1.603[22]	H_2O 3; EtOH 1; eth 1; bz 1
11357	L-Serine (S)-2-Amino-3-hydroxypropanoic acid	$C_3H_7NO_3$ 105.09	56-45-1 228 dec	8411 sub 150	4-04-00-03118 1.6[22]	H_2O 4; EtOH 1; eth 1; bz 1

No.	Name / Synonym	Mol. Form. / Mol. Wt.	CAS RN / mp/°C	Merck No. / bp/°C	Beil. Ref. / den/g cm^{-3}	Solubility / n_D
11358	Sesone / Ethanol, 2-(2,4-dichlorophenoxy)-, hydrogen sulfate, sodium salt	$C_8H_8Cl_2NaO_5S$ / 310.11	136-78-7 / 245 dec	3372		
11359	Sethoxydim / (±)-2-[1-(Ethoxyimino)butyl]-5-[2-(ethylthio)propyl]-3-hydroxy-2-cyclohexen-1-one	$C_{17}H_{29}NO_3S$ / 327.49	74051-80-2 / >90[0.00003]	8424	1.043[25]	
11360	Silacyclohexane, 1,1-diphenyl- / 1,1-Diphenylsilacyclohexane	$C_{17}H_{20}Si$ / 252.43	18002-79-4 /	193-8[5]	4-27-00-09833 / 1.0319[25]	1.5779[20]
11361	Silanamine, pentamethyl-	$C_5H_{15}NSi$ / 117.27	2083-91-2 /	86	4-04-00-04010 / 0.7400[20]	1.4379[24]
11362	Silanamine, 1,1,1-trimethyl-*N*-phenyl-*N*-(trimethylsilyl)-	$C_{12}H_{23}NSi_2$ / 237.49	4147-89-1 / 16.5	98-9[11]	0.8951[23]	1.4855[20]
11363	Silanamine, 1,1,1-trimethyl-*N*-(trimethylsilyl)- / Hexamethyldisilazane	$C_6H_{19}NSi_2$ / 161.39	999-97-3 / 125	10158	4-04-00-04014 / 0.7741[25]	1.4090[20]
11364	Silane, [1,1'-biphenyl]-4-yltrichloro-	$C_{12}H_9Cl_3Si$ / 287.65	18030-61-0 /	209-11[30]	4-16-00-01572 / 1.3190[20]	ctc 3; CS_2 3
11365	Silane, (bromomethyl)chlorodimethyl- / (Bromomethyl)dimethylchlorosilane	$C_3H_8BrClSi$ / 187.54	16532-02-8 / 131		4-04-00-04024 / 1.375[25]	1.4630[25]
11366	Silane, (bromomethyl)trimethyl- / Bromomethyltrimethylsilane	$C_4H_{11}BrSi$ / 167.12	18243-41-9 / 116.5		4-04-00-03878 / 1.170[25]	1.4460[20]
11367	Silane, (4-bromophenoxy)trimethyl-	$C_9H_{13}BrOSi$ / 245.19	17878-44-3 / 126[25]		4-06-00-01056 / 1.2619[20]	1.5145[20]
11368	Silane, bromotriethyl-	$C_6H_{15}BrSi$ / 195.17	1112-48-7 / 66-7[24]		4-04-00-04051 / 1.1403[20]	1.4561[20]
11369	Silane, butylchlorodimethyl-	$C_6H_{15}ClSi$ / 150.72	1000-50-6 / 139		4-04-00-04068 / 0.8751[20]	1.5145[20]
11370	Silane, butyltrichloro- / Trichlorobutylsilane	$C_4H_9Cl_3Si$ / 191.56	7521-80-4 / 148.5		4-04-00-04246 / 1.1606[20]	eth 3; bz 3; tol 3; AcOEt 3 / 1.4363[20]
11371	Silane, butyltrimethyl- / Trimethylbutylsilane	$C_7H_{18}Si$ / 130.31	1000-49-3 / 115		4-04-00-03907 / 0.7353[0]	H_2O 1
11372	Silane, chloro(chloromethyl)dimethyl-	$C_3H_8Cl_2Si$ / 143.09	1719-57-9 / 115.5		4-04-00-04024 / 1.0865[20]	1.4360[20]
11373	Silane, chloro(dichloromethyl)dimethyl- / (Dichloromethyl)dimethylchlorosilane	$C_3H_7Cl_3Si$ / 177.53	18171-59-0 / -48	149	4-04-00-04031 / 1.2369[20]	1.461[20]
11374	Silane, chlorodimethylphenyl-	$C_8H_{11}ClSi$ / 170.71	768-33-2 / 80-4[16]		4-16-00-01475 / 1.032[20]	1.5082[20]
11375	Silane, chlorodimethyl-2-propenyl-	$C_5H_{11}ClSi$ / 134.68	4028-23-3 / 111		4-04-00-04080 / 0.8964[20]	1.4195[20]
11376	Silane, chloroethenyldimethyl-	C_4H_9ClSi / 120.65	1719-58-0 / 83.5		4-04-00-04080 / 0.8744[20]	1.4141[20]
11377	Silane, chloroethyldimethyl-	$C_4H_{11}ClSi$ / 122.67	6917-76-6 / 89.5		4-04-00-04035 / 0.8675[20]	1.4105[20]
11378	Silane, chloromethyl / Chloromethylsilane	CH_5ClSi / 80.59	993-00-0 / -135	7; -45[63]		
11379	Silane, (chloromethyl)dimethylphenyl-	$C_9H_{13}ClSi$ / 184.74	1833-51-8 / 225		4-16-00-01376 / 1.0240[25]	ctc 3; CS_2 3
11380	Silane, chloromethyldiphenyl-	$C_{13}H_{13}ClSi$ / 232.78	144-79-6 / 295		4-16-00-01479 / 1.1277[20]	
11381	Silane, chloromethylphenyl-	C_7H_9ClSi / 156.69	1631-82-9 / 113[100]		4-16-00-01473 / 1.0540[20]	1.5171[20]
11382	Silane, (chloromethyl)trimethyl-	$C_4H_{11}ClSi$ / 122.67	2344-80-1 / 98.5		4-04-00-03877 / 0.879[25]	1.4175[20]
11383	Silane, (4-chlorophenoxy)trimethyl-	$C_9H_{13}ClOSi$ / 200.74	17005-59-3 / 214; 101[14]		4-06-00-00878 / 1.0320[20]	1.4930[20]
11384	Silane, chlorophenyl- / Phenylchlorosilane	C_6H_7ClSi / 142.66	4206-75-1 / 162.5		4-16-00-01472 / 1.0683[20]	1.5340[20]
11385	Silane, (3-chlorophenyl)trimethyl-	$C_9H_{13}ClSi$ / 184.74	4405-42-9 / 206.5		4-16-00-01386 / 1.0071[20]	1.5108[20]
11386	Silane, (3-chloropropyl)trimethyl-	$C_6H_{15}ClSi$ / 150.72	2344-83-4 / 151		4-04-00-03905 / 0.8789[20]	1.4319[20]
11387	Silane, chlorotriethoxy-	$C_6H_{15}ClO_3Si$ / 198.72	4667-99-6 / -51	156	4-01-00-01363 / 1.030[20]	EtOH 4 / 1.3999[20]
11388	Silane, chlorotriethyl-	$C_6H_{15}ClSi$ / 150.72	994-30-9 / 144.5		4-04-00-04051 / 0.8967[20]	1.4314[20]
11389	Silane, chlorotrimethyl- / Trimethylchlorosilane	C_3H_9ClSi / 108.64	75-77-4 / -40	60	4-04-00-04007 / 0.856[25]	1.3870[20]
11390	Silane, dichloro(chloromethyl)methyl- / Dichloro(chloromethyl)methylsilane	$C_2H_5Cl_3Si$ / 163.51	1558-33-4 / 121.5		4-04-00-04135 / 1.2858[20]	1.4500[20]
11391	Silane, dichloro(dichloromethyl)methyl- / Dichloromethyl methyl dichlorosilane	$C_2H_4Cl_4Si$ / 197.95	1558-31-2 / 149		4-04-00-04144 / 1.4116[20]	1.4700[20]
11392	Silane, dichlorodiethoxy-	$C_4H_{10}Cl_2O_2Si$ / 189.11	4667-38-3 / -130	135.9	4-01-00-01364 / 1.1290[20]	EtOH 4
11393	Silane, dichlorodiethyl- / Dichlorodiethylsilane	$C_4H_{10}Cl_2Si$ / 157.11	1719-53-5 / -96.5	129 dec	4-04-00-04158 / 1.0504[20]	1.4309[20]
11394	Silane, dichlorodimethyl- / Dichlorodimethylsilane	$C_2H_6Cl_2Si$ / 129.06	75-78-5 / -16	70.3	4-04-00-04111 / 1.064[25]	H_2O 6; EtOH 6 / 1.4038[20]
11395	Silane, dichlorodiphenyl- / Dichlorodiphenylsilane	$C_{12}H_{10}Cl_2Si$ / 253.20	80-10-4 / 305		4-16-00-01526 / 1.204[25]	EtOH 3; eth 3; ace 3; bz 3 / 1.5800[20]

No.	Name Synonym	Mol. Form. Mol. Wt.	CAS RN mp/°C	Merck No. bp/°C	Beil. Ref. den/g cm^{-3}	Solubility n_D
11396	Silane, dichloroethenylmethyl-	$C_3H_6Cl_2Si$ 141.07	124-70-9	92.5	4-04-00-04184 1.0868[20]	H_2O 6 1.4270[20]
11397	Silane, dichloroethylmethyl- Methylethyldichlorosilane	$C_3H_8Cl_2Si$ 143.09	4525-44-4	101	4-04-00-04152 1.0630[20]	1.4197[20]
11398	Silane, dichloromethyl- Dichloromethylsilane	CH_4Cl_2Si 115.03	75-54-7 -93	41	4-04-00-04096 1.105[25]	
11399	Silane, dichloromethyl(1-methylethyl)-	$C_4H_{10}Cl_2Si$ 157.11	18236-89-0	121.5	4-04-00-04171 1.0385[20]	1.4270[20]
11400	Silane, dichloromethyl(4-methylphenyl)-	$C_8H_{10}Cl_2Si$ 205.16	18236-57-2 57		1.1619[20]	1.5170[20]
11401	Silane, dichloromethylphenyl- Dichloromethylphenylsilane	$C_7H_8Cl_2Si$ 191.13	149-74-6	206.5	4-16-00-01517 1.1866[20]	1.5180[20]
11402	Silane, dichloromethyl-2-propenyl-	$C_4H_8Cl_2Si$ 155.10	1873-92-3	119.5	4-04-00-04188 1.0758[20]	1.4419[20]
11403	Silane, dichlorophenyl- Phenyldichlorosilane	$C_6H_6Cl_2Si$ 177.10	1631-84-1	181	4-16-00-01514 1.221[25]	H_2O 6
11404	Silane, diethenyldiphenyl-	$C_{16}H_{16}Si$ 236.39	17937-68-7	130-1[0.05]	4-16-00-01368 1.0092[25]	1.5350[25]
11405	Silane, diethoxydimethyl- Dimethyldiethoxysilane	$C_6H_{16}O_2Si$ 148.28	78-62-6 -87	114	4-04-00-04101 0.865[25]	ctc 3 1.3811[20]
11406	Silane, diethoxydiphenyl- Diphenyldiethoxysilane	$C_{16}H_{20}O_2Si$ 272.42	2553-19-7	167[15]	4-16-00-01524 1.0329[20]	1.5269[20]
11407	Silane, diethoxymethylphenyl-	$C_{11}H_{18}O_2Si$ 210.35	775-56-4	218	4-16-00-01515 0.9627[20]	1.4690[20]
11408	Silane, diethoxymethyl-2-propenyl- Silane, allyldiethoxymethyl-	$C_8H_{18}O_2Si$ 174.32	18388-45-9	155	4-04-00-04187 0.8572[25]	1.4104[20]
11409	Silane, diethyl- 3-Silapentane	$C_4H_{12}Si$ 88.22	542-91-6 -134.3	3117 57	4-04-00-03894 0.6843[20]	H_2O 1 1.3921[20]
11410	Silane, diethyldifluoro- Diethyldifluorosilane	$C_4H_{10}F_2Si$ 124.21	358-06-5 -78.7	61	4-04-00-04157 0.9348[20]	1.3385[20]
11411	Silane, difluorodiphenyl-	$C_{12}H_{10}F_2Si$ 220.29	312-40-3	157[50]	4-16-00-01525 1.145[17]	1.5221[25]
11412	Silane, dimethoxydimethyl-	$C_4H_{12}O_2Si$ 120.22	1112-39-6	82	4-04-00-04100 0.8646[20]	H_2O 6 1.3708[20]
11413	Silane, dimethoxydiphenyl-	$C_{14}H_{16}O_2Si$ 244.37	6843-66-9	161[15]	4-16-00-01524 1.0771[20]	1.5447[20]
11414	Silane, dimethyl- 2-Silapropane	C_2H_8Si 60.17	1111-74-6 -150	-20	4-04-00-03874 0.68[-80]	
11415	Silane, dimethyldiphenoxy- Diphenoxydimethylsilane	$C_{14}H_{16}O_2Si$ 244.37	3440-02-6 -23	130-2[5]	4-06-00-00764 1.0599[25]	1.5330[20]
11416	Silane, dimethyldiphenyl- Diphenyl(dimethyl)silane	$C_{14}H_{16}Si$ 212.37	778-24-5	277; 173[45]	4-16-00-01366 0.9867[20]	1.5644[20]
11417	Silane, dimethyldi-2-propenyl-	$C_8H_{16}Si$ 140.30	1113-12-8	68[50]	4-04-00-03929 0.7679[20]	1.4420[20]
11418	Silane, dimethylphenyl-	$C_8H_{12}Si$ 136.27	766-77-8	156.5	4-16-00-01360 0.8891[20]	H_2O 1 1.4995[20]
11419	Silanediol, dimethyl-, diacetate	$C_6H_{12}O_4Si$ 176.24	2182-66-3 -12.5	165	4-04-00-04108 1.0540[20]	1.4030[20]
11420	Silane, diphenyl- Diphenylsilane	$C_{12}H_{12}Si$ 184.31	775-12-2	95-7[13]	4-16-00-01366 0.9969[20]	ctc 3; CS_2 3 1.5800[20]
11421	Silane, 1,2-ethenediylbis[trimethyl-, (E)-	$C_8H_{20}Si_2$ 172.42	18178-59-1 -18.5	145.5	4-04-00-03948 0.7589[20]	H_2O 1 1.4310[20]
11422	Silane, ethenyldiethoxymethyl-	$C_7H_{16}O_2Si$ 160.29	5507-44-8	133	4-04-00-04183 0.8620[20]	1.4001[20]
11423	Silane, ethenylethoxydimethyl-	$C_6H_{14}OSi$ 130.26	5356-83-2	99	4-04-00-04079 0.790[20]	1.3983[20]
11424	Silane, ethenyltriethoxy-	$C_8H_{18}O_3Si$ 190.31	78-08-0	148[535]; 62[20]	4-04-00-04256 0.9027[20]	chl 3 1.3960[25]
11425	Silane, ethenyltrimethyl-	$C_5H_{12}Si$ 100.24	754-05-2	55	4-04-00-03922 0.6910[20]	H_2O 1 1.3914[20]
11426	Silane, ethenyltris(1-methylethoxy)-	$C_{11}H_{24}O_3Si$ 232.40	18023-33-1	179.5; 77[20]	4-04-00-04257 0.8627[25]	ctc 3 1.3981[20]
11427	Silane, ethenyltris(2-propenyloxy)- Tris(allyloxy)vinylsilane	$C_{11}H_{18}O_3Si$ 226.35	17988-31-7	90.3[7.5]	4-04-00-04257 0.9435[25]	1.4380[20]
11428	Silane, ethoxytriethyl-	$C_8H_{20}OSi$ 160.33	597-67-1	154.5	4-04-00-04042 0.8160[20]	eth 4; EtOH 4 1.4140[20]
11429	Silane, ethoxytrifluoro- Ethoxytrifluorosilicon	$C_2H_5F_3OSi$ 130.14	460-55-9 -122	-7	3-01-00-01336 1.3200[-63]	
11430	Silane, ethoxytrimethyl-	$C_5H_{14}OSi$ 118.25	1825-62-3	76	4-04-00-03994 0.7573[20]	H_2O 1; EtOH 3; eth 3; ace 3 1.3741[20]
11431	Silane, ethoxytriphenyl-	$C_{20}H_{20}OSi$ 304.46	1516-80-9 65	344	4-16-00-01481	chl 3
11432	Silane, ethyltrifluoro- Trifluoroethylsilane	$C_2H_5F_3Si$ 114.14	353-89-9 -105	-4.4	4-04-00-04226 1.227[-76]	
11433	Silane, ethyltrimethoxy- Ethyltrimethoxysilane	$C_5H_{14}O_3Si$ 150.25	5314-55-6	124.3	4-04-00-04223 0.9488[20]	EtOH 4 1.3838[20]
11434	Silane, 1,2-ethynediylbis[trimethyl-	$C_8H_{18}Si_2$ 170.40	14630-40-1 26	134	4-04-00-03950 0.770[20]	1.413[20]
11435	Silane, ethynyl- Ethynylsilane	C_2H_4Si 56.14	1066-27-9	-22.5		

No.	Name / Synonym	Mol. Form. / Mol. Wt.	CAS RN / mp/°C	Merck No. / bp/°C	Bell. Ref. / den/g cm⁻³	Solubility / n_D
11436	Silane, methoxy / Methyl silyl ether	CH_6OSi / 62.14	2171-96-2 /			
11437	Silane, methyl- / Methylsilane	CH_6Si / 46.14	992-94-9 / -156.5	-57.5	4-04-00-03873	
11438	Silane, methyldiphenyl-	$C_{13}H_{14}Si$ / 198.34	776-76-1 /	93.5[1]	4-16-00-01366 / 0.9973[20]	ctc 3 / 1.5694[20]
11439	Silane, methylenebis-	CH_8Si_2 / 76.25	1759-88-2 /	15	4-01-00-03072 / 0.6979[4]	1.4115[4]
11440	Silane, methylenebis[trichloro-	$CH_2Cl_6Si_2$ / 282.91	4142-85-2 /	183; 64[10]	4-01-00-03077 / 1.5567[20]	eth 4; EtOH 4 / 1.4740[20]
11441	Silane, (2-methylphenoxy)triphenyl-	$C_{25}H_{22}OSi$ / 366.53	18858-65-6 /	192	4-16-00-01482 / 0.9287[20]	1.4830[20]
11442	Silane, methylphenyl-	$C_7H_{10}Si$ / 122.24	766-08-5 /	140	4-16-00-01360 / 0.8895[20]	1.5058[20]
11443	Silane, methyltriphenoxy- / Methyltriphenoxysilane	$C_{19}H_{18}O_3Si$ / 322.44	3439-97-2 /	269[100]; 210[12]	4-06-00-00765 / 1.135[20]	1.5599[20]
11444	Silane, methyltriphenyl- / Triphenylmethylsilane	$C_{19}H_{18}Si$ / 274.44	791-29-7 / 69.5	198[9]	4-16-00-01370 / 1.0880[20]	ctc 3; CS$_2$ 3
11445	Silane, methyltri-p-tolyl- / Silane, methyl tri-(4-tolyl)	$C_{22}H_{24}Si$ / 316.52	18752-92-6 / 92	198[4]	4-16-00-01399	chl 2
11446	Silane, phenyl-	C_6H_8Si / 108.22	694-53-1 /	119	4-16-00-01360 / 0.8681[20]	H_2O 1 / 1.5125[20]
11447	Silane, [1,3-phenylenebis(oxy)]bis[trimethyl-	$C_{12}H_{22}O_2Si_2$ / 254.48	4520-29-0 /	240	4-06-00-05683 / 0.950[20]	1.4748[20]
11448	Silane, phenyltripropyl- / Phenyltripropylsilane	$C_{15}H_{26}Si$ / 234.46	78938-11-1 /	146-7[18]	4-16-00-01362 / 0.8799[20]	1.4950[20]
11449	Silane, tetraethenyl- / Tetravinylsilane	$C_8H_{12}Si$ / 136.27	1112-55-6 /	130.2	4-04-00-03925 / 0.7999[20]	1.4625[20]
11450	Silane, tetraethyl- / Tetraethylsilane	$C_8H_{20}Si$ / 144.33	631-36-7 /	154.7	4-04-00-03895 / 0.7658[20]	H_2O 1 / 1.4268[20]
11451	Silane, tetramethyl- / Tetramethylsilane	$C_4H_{12}Si$ / 88.22	75-76-3 / -99.0	26.6	4-04-00-03875 / 0.648[19]	H_2O 1; EtOH 4; eth 4; sulf 1 / 1.3587[20]
11452	Silane, tetraphenyl- / Tetraphenylsilane	$C_{24}H_{20}Si$ / 336.51	1048-08-4 / 236.5	228[3]	4-16-00-01372 / 1.078[20]	ctc 3; CS$_2$ 3
11453	Silane, tribromomethyl-	CH_3Br_3Si / 282.83	4095-09-4 / -28.4	132	4-04-00-04213 / 2.2130[25]	1.5152[25]
11454	Silane, tributyl- / Tributylsilane	$C_{12}H_{28}Si$ / 200.44	998-41-4 /	221	4-04-00-03908 / 0.7794[20]	1.4380[20]
11455	Silane, tributylphenyl- / Silane, phenyl-tributyl	$C_{18}H_{32}Si$ / 276.54	18510-29-7 /	140-2[4]	4-16-00-01362 / 0.8753[20]	1.4915[20]
11456	Silane, trichloro(chloromethyl)- / (Chloromethyl)trichlorosilane	CH_2Cl_4Si / 183.92	1558-25-4 /	118	4-01-00-03074 / 1.4650[20]	1.4555[20]
11457	Silane, trichloro(4-chlorophenyl)-	$C_6H_4Cl_4Si$ / 245.99	825-94-5 /	115-7[20]	4-16-00-01563 / 1.4316[20]	1.5418[20]
11458	Silane, trichloro(3-chloropropyl)-	$C_3H_6Cl_4Si$ / 211.98	2550-06-3 /	181.5	4-04-00-04243 / 1.3590[20]	1.4668[20]
11459	Silane, trichloro(dichloromethyl)- / (Dichloromethyl)trichlorosilane	$CHCl_5Si$ / 218.37	1558-24-3 /	145	4-02-00-00088 / 1.5518[20]	1.4714[20]
11460	Silane, trichlorododecyl- / Dodecyltrichlorosilane	$C_{12}H_{25}Cl_3Si$ / 303.77	4484-72-4 /	155[10]	4-04-00-04255	1.4581[20]
11461	Silane, trichloroethenyl- / Vinyltrichlorosilane	$C_2H_3Cl_3Si$ / 161.49	75-94-5 / -95	91.5	4-04-00-04258 / 1.2426[20]	chl 4 / 1.4295[20]
11462	Silane, trichloroethoxy-	$C_2H_5Cl_3OSi$ / 179.50	1825-82-7 / -135	101.9	4-01-00-01364 / 1.2274[20]	EtOH 4 / 1.4045[20]
11463	Silane, trichloroethyl- / Ethyltrichlorosilane	$C_2H_5Cl_3Si$ / 163.51	115-21-9 / -105.6	100.5	4-04-00-04227 / 1.2373[20]	ctc 3 / 1.4256[20]
11464	Silane, trichlorohexyl- / Hexyltrichlorosilane	$C_6H_{13}Cl_3Si$ / 219.61	928-65-4 /	190	4-04-00-04251 / 1.1100[20]	H_2O 6
11465	Silane, trichloromethyl- / Methyltrichlorosilane	CH_3Cl_3Si / 149.48	75-79-6 / -90	65.6	4-04-00-04212 / 1.273[20]	H_2O 6; EtOH 6 / 1.4106[20]
11466	Silane, trichloro(1-methylethyl)-	$C_3H_7Cl_3Si$ / 177.53	4170-46-1 / -87.7	121	4-04-00-04244 / 1.1934[20]	H_2O 6 / 1.4319[20]
11467	Silane, trichloro(2-methylphenyl)-	$C_7H_7Cl_3Si$ / 225.58	13835-81-9 / -25.5	226.7	4-16-00-01565 / 1.306[25]	1.5336[25]
11468	Silane, trichloro(3-methylphenyl)-	$C_7H_7Cl_3Si$ / 225.58	13688-75-0 / -26	226.5	4-16-00-01565 / 1.306[25]	1.5336[25]
11469	Silane, trichloro(2-methylpropyl)-	$C_4H_9Cl_3Si$ / 191.56	18169-57-8 /	143.3	4-04-00-04248 / 1.154[20]	H_2O 6
11470	Silane, trichlorooctadecyl- / Octadecyltrichlorosilane	$C_{18}H_{37}Cl_3Si$ / 387.94	112-04-9 /	223[10]	4-04-00-04256 / 0.984[25]	1.4602[20]
11471	Silane, trichlorooctyl- / Octyltrichlorosilane	$C_8H_{17}Cl_3Si$ / 247.67	5283-66-9 /	232	4-04-00-04253	H_2O 6; EtOH 6; ctc 3 / 1.4480[20]
11472	Silane, trichloropentyl- / Amyltrichlorosilane	$C_5H_{11}Cl_3Si$ / 205.59	107-72-2 /	172; 60.5[15]	4-04-00-04249 / 1.1330[20]	1.4503[20]
11473	Silane, trichlorophenyl- / Phenyltrichlorosilane	$C_6H_5Cl_3Si$ / 211.55	98-13-5 /	201	4-16-00-01560 / 1.321[20]	ctc 3; chl 3; CS$_2$ 3 / 1.5230[20]
11474	Silane, trichloro(2-phenylethyl)-	$C_8H_9Cl_3Si$ / 239.60	940-41-0 /	242; 98-9[5]	4-16-00-01568 / 1.2397[20]	1.5185[20]
11475	Silane, trichloro-2-propenyl- / Allyltrichlorosilane	$C_3H_5Cl_3Si$ / 175.52	107-37-9 / 35	117.5	4-04-00-04261 / 1.2011[20]	1.4460[20]

No.	Name / Synonym	Mol. Form. / Mol. Wt.	CAS RN / mp/°C	Merck No. / bp/°C	Beil. Ref. / den/g cm^{-3}	Solubility / n_D
11476	Silane, trichloropropyl- / Propyltrichlorosilane	$C_3H_7Cl_3Si$ / 177.53	141-57-1 /	/ 123.5	4-04-00-04239 / 1.195[20]	/ 1.4310[20]
11477	Silane, triethoxy- / Triethoxysilane	$C_6H_{16}O_3Si$ / 164.28	998-30-1 /	/ 133.5	4-01-00-01359 / 0.8745[20]	
11478	Silane, triethoxyethyl-	$C_8H_{20}O_3Si$ / 192.33	78-07-9 /	/ 158.5	4-04-00-04223 / 0.8963[20]	H_2O 1; EtOH 5; eth 5; chl 3 / 1.3955[20]
11479	Silane, triethoxymethyl-	$C_7H_{18}O_3Si$ / 178.30	2031-67-6 /	/ 142	4-04-00-04204 / 0.8948[25]	/ 1.3832[20]
11480	Silane, triethoxypentyl-	$C_{11}H_{26}O_3Si$ / 234.41	2761-24-2 /	/ 95[13]	4-04-00-04248 / 0.8862[20]	/ 1.4059[20]
11481	Silane, triethoxyphenyl-	$C_{12}H_{20}O_3Si$ / 240.37	780-69-8 /	/ 112-3[10]	4-16-00-01556 / 0.996[25]	/ 1.4604[20]
11482	Silane, triethoxy-2-propenyl-	$C_9H_{20}O_3Si$ / 204.34	2550-04-1 /	/ 100[50]	4-04-00-04261 / 0.9030[20]	/ 1.4072[20]
11483	Silane, triethyl- / Triethylsilane	$C_6H_{16}Si$ / 116.28	617-86-7 /	/ 109	4-04-00-03895 / 0.7302[20]	H_2O 1; sulf 1 / 1.447[20]
11484	Silane, triethylfluoro-	$C_6H_{15}FSi$ / 134.27	358-43-0 /	/ 110	4-04-00-04051 / 0.8354[25]	peth 4 / 1.3900[25]
11485	Silane, triethylphenyl-	$C_{12}H_{20}Si$ / 192.38	2987-77-1 /	/ 236	4-16-00-01362 / 0.8915[20]	/ 1.4999[20]
11486	Silane, trifluorophenyl-	$C_6H_5F_3Si$ / 162.19	368-47-8 / -18	/ 101.5	4-16-00-01559 / 1.2169[20]	bz 4; EtOH 4 / 1.4110[20]
11487	Silane, trimethoxymethyl-	$C_4H_{12}O_3Si$ / 136.22	1185-55-3 /	/ 102.5	4-04-00-04203 / 0.9548[20]	chl 3 / 1.3696[20]
11488	Silane, trimethoxyphenyl-	$C_9H_{14}O_3Si$ / 198.29	2996-92-1 /	/ 130[45]	4-16-00-01556 / 1.064[20]	ctc 3; CS_2 3 / 1.4734[20]
11489	Silane, trimethyl- / Trimethylsilane	$C_3H_{10}Si$ / 74.20	993-07-7 / -135.9	/ 6.7		
11490	Silane, trimethyl(4-methylphenyl)-	$C_{10}H_{16}Si$ / 164.32	3728-43-6 /	/ 192; 73[10]	4-16-00-01396 / 0.860[25]	/ 1.4900[20]
11491	Silane, trimethyl(2-methylpropyl)-	$C_7H_{18}Si$ / 130.31	1118-09-8 /	/ 108.5	1-04-00-00580 / 0.7330[0]	
11492	Silane, trimethylphenoxy-	$C_9H_{14}OSi$ / 166.30	1529-17-5 / -55	/ 119	4-06-00-00762 / 0.8681[20]	/ 1.5125[20]
11493	Silane, trimethylphenyl-	$C_9H_{14}Si$ / 150.30	768-32-1 /	/ 169.5	4-16-00-01361 / 0.8722[20]	ctc 3; CS_2 3 / 1.4907[20]
11494	Silane, trimethyl(phenylmethyl)-	$C_{10}H_{16}Si$ / 164.32	770-09-2 /	/ 190.5	4-16-00-01401 / 0.8933[20]	/ 1.4941[20]
11495	Silane, trimethyl-2-propenyl-	$C_6H_{14}Si$ / 114.26	762-72-1 /	/ 86	4-04-00-03927 / 0.7158[25]	H_2O 1 / 1.4074[20]
11496	Silane, trimethylpropyl- / Trimethylpropylsilane	$C_6H_{16}Si$ / 116.28	3510-70-1 /	/ 89	4-04-00-03901 / 0.7196[0]	/ 1.3929[20]
11497	Silane, trimethyl[4-[(trimethylsilyl)oxy]phenyl]- / Silane, (oxy-p-phenylene)bis[trimethyl-	$C_{12}H_{22}OSi_2$ / 238.48	18036-81-2 /	/ 132[25]	4-16-00-01441 / 0.900[25]	/ 1.4794[25]
11498	Silanetriol, ethenyl-, triacetate	$C_8H_{12}O_6Si$ / 232.27	4130-08-9 /	/ 115[10]	4-04-00-04258 / 1.167[20]	/ 1.4226[20]
11499	Silanetriol, methyl-, triacetate	$C_7H_{12}O_6Si$ / 220.25	4253-34-3 / 40.5	/ 110-12[17]	4-04-00-04208 / 1.1750[20]	/ 1.4083[20]
11500	Silane, tripropyl- / Tripropylsilane	$C_9H_{22}Si$ / 158.36	998-29-8 /	/ 172	4-04-00-03902 / 0.7723[0]	H_2O 1 / 1.4280[20]
11501	Silanol, ethyldimethyl- / Silanol, dimethyl ethyl	$C_4H_{12}OSi$ / 104.22	5906-73-0 /	/ 120; 58[50]	4-04-00-04034 / 0.8332[20]	/ 1.4070[20]
11502	Silanol, methyldiphenyl-	$C_{13}H_{14}OSi$ / 214.34	778-25-6 / 167	/ 184[24]; 148[3]	4-16-00-01478 / 1.0840[25]	ctc 3; CS_2 3
11503	Silanol, triethyl-	$C_6H_{16}OSi$ / 132.28	597-52-4 /	/ 154	4-04-00-04040 / 0.8647[20]	H_2O 1; EtOH 5; eth 5 / 1.4329[20]
11504	Silanol, triphenyl-	$C_{18}H_{16}OSi$ / 276.41	791-31-1 / 154.8		4-16-00-01480 / 1.1777[20]	ctc 3; CS_2 3
11505	Silicic acid (H4SiO4), tetrabutyl ester	$C_{16}H_{36}O_4Si$ / 320.54	4766-57-8 /	/ 120[3]	4-01-00-01541 / 0.8990[20]	/ 1.4128[20]
11506	Silicic acid (H4SiO4), tetraethyl ester / Ethyl silicate	$C_8H_{20}O_4Si$ / 208.33	78-10-4 / -82.5	3805 / 168.8	4-01-00-01360 / 0.9320[20]	H_2O 6 / 1.3928[20]
11507	Silicic acid (H4SiO4), tetrakis(2-ethylbutyl) ester	$C_{24}H_{52}O_4Si$ / 432.76	78-13-7 /		4-01-00-01725 / 0.8920[20]	H_2O 1; EtOH 2; eth 3; bz 3 / 1.4307[20]
11508	Silicic acid (H4SiO4), tetramethyl ester / Methyl silicate	$C_4H_{12}O_4Si$ / 152.22	681-84-5 / -2	/ 121	4-01-00-01266 / 1.0232[20]	EtOH 4 / 1.3683[20]
11509	Silicic acid (H4SiO4), tetraphenyl ester	$C_{24}H_{20}O_4Si$ / 400.51	1174-72-7 / 49	/ 417; 236[1]	4-06-00-00766 / 1.1412[60]	
11510	Silicic acid (H4SiO4), tetrapropyl ester	$C_{12}H_{28}O_4Si$ / 264.44	682-01-9 /	/ 226	4-01-00-01435 / 0.9158[20]	ctc 3; CS_2 3 / 1.4012[20]
11511	Silicic acid (H4SiO4), triethyl phenyl ester	$C_{12}H_{20}O_4Si$ / 256.37	18023-36-4 /	/ 120-2[10]	4-06-00-00765 / 1.0283[20]	/ 1.4525[20]
11512	Silvex / Propanoic acid, 2-(2,4,5-trichlorophenoxy)-	$C_9H_7Cl_3O_3$ / 269.51	93-72-1 / 181.6	8483		
11513	Simazine / 1,3,5-Triazine-2,4-diamine, 6-chloro-N,N'-diethyl-	$C_7H_{12}ClN_5$ / 201.66	122-34-9 / 226	8485	1.302[20]	
11514	Sodium dimethyldithiocarbamate / Carbamodithioic acid, dimethyl-, sodium salt	$C_3H_7NNaS_2$ / 144.22	128-04-1 /			

No.	Name Synonym	Mol. Form. Mol. Wt.	CAS RN mp/°C	Merck No. bp/°C	Beil. Ref. den/g cm^{-3}	Solubility n_D
11515	Solanid-5-ene-3,12-diol, (3β,12α)- Rubijervine	$C_{27}H_{43}NO_2$ 413.64	79-58-3 242	8264	5-21-05-00208	bz 4; EtOH 4; chl 4
11516	Solanid-5-ene-3,18-diol, (3β)- Isorubijervine	$C_{27}H_{43}NO_2$ 413.64	468-45-1 242.5	5111	5-21-05-00208	bz 4; chl 4
11517	Sonar 2-Naphthalenol, 1,1'-methylenebis-, monosodium salt	$C_{21}H_{15}NaO_2$ 323.35	38827-66-6 132			
11518	DL-Sorbose	$C_6H_{12}O_6$ 180.16	65732-90-3 162.5	8681	4-01-00-04410 1.638[17]	H_2O 4; MeOH 4
11519	L-Sorbose L-Sorbinose	$C_6H_{12}O_6$ 180.16	87-79-6 165	8681	4-01-00-04412 1.612[17]	H_2O 3; EtOH 2; eth 2; MeOH 2
11520	Spiro[benzofuran-2(3H),1'-[2]cyclohexene]- 3,4'-dione, 7-chloro-2',4,6-trimethoxy- Griseofulvin, (+)-	$C_{17}H_{17}ClO_6$ 352.77	126-07-8 220	4453	5-18-05-00150	H_2O 1; EtOH 2; eth 2; ace 2
11521	Spiro[4.5]dec-6-en-8-one, 6,10-dimethyl-2-(1-methylethylidene)-, (5R-cis)-	$C_{15}H_{22}O$ 218.34	18444-79-6 44.5	141-2[2]	4-07-00-00822 1.0001[20]	ace 3 1.5309[20]
11522	Spiro[isobenzofuran-1(3H),9'-[9H]xanthen]-3-one, 3',6'-dihydroxy- Fluorescein	$C_{20}H_{12}O_5$ 332.31	2321-07-5 315 dec	4085	5-19-06-00456	H_2O 2; EtOH 2; eth 2; ace 4
11523	Spiro[isobenzofuran-1(3H),9'-[9H]xanthen]-3-one, 3',6'-dihydroxy-4',5'-diiodo- Fluorescein, 4',5'-diiodo	$C_{20}H_{10}I_2O_5$ 584.11	38577-97-8	3174	5-19-06-00466	
11524	Spiro[isobenzofuran-1(3H),9'-[9H]xanthen]-3-one, 3',6'-dihydroxy-2',4',5',7'-tetraiodo-, disodium salt Erythrosin	$C_{20}H_8I_4O_5$ 835.90	16423-68-0	3640	4-19-00-02923	eth 4; EtOH 4
11525	Spiro[isobenzofuran-1(3H),9'-[9H]xanthen]-3-one, 2',4',5',7'-tetrabromo-3',6'-dihydroxy-, disodium salt Eosin	$C_{20}H_8Br_4Na_2O_5$ 691.86	17372-87-1 295.5	3555		EtOH 4
11526	Spiro[isobenzofuran-1(3H),9'-[9H]xanthen]-3-one, 3',4',5',6'-tetrahydroxy- Gallein	$C_{20}H_{12}O_5$ 332.31	2103-64-2 >300	4250	5-19-07-00083	ace 4; EtOH 4
11527	Spiropentane Spiro[2.2]pentane	C_5H_8 68.12	157-40-4 -134.6	39	4-05-00-00218 0.7266[20]	1.4120[20]
11528	Spirosolan-3-ol, (3β,5α,22β,25S)- Tomatidine	$C_{27}H_{45}NO_2$ 415.66	77-59-8 210.5	9470		EtOH 3; eth 3
11529	Spirosol-5-en-3-ol, (3β,22α,25R)- Solasodine	$C_{27}H_{43}NO_2$ 413.64	126-17-0 202	8665	4-27-00-02000	EtOH 3; eth 2; ace 3; bz 3
11530	Spirostan-2,3-diol, (2α,3β,5α,25R)- Gitogenin	$C_{27}H_{44}O_4$ 432.64	511-96-6 271.5	4325	5-19-03-00338	H_2O 1; EtOH 3; eth 2; chl 3
11531	Spirostan-3-ol, (3β,5β,25S)- Sarsasapogenin	$C_{27}H_{44}O_3$ 416.64	126-19-2 200.5	8338	5-19-02-00672	EtOH 3; ace 3; bz 3; chl 3
11532	Spirostan-3-ol, (3β,5α,25R)- Tigogenin	$C_{27}H_{44}O_3$ 416.64	77-60-1 205.5	9367		EtOH 3; eth 3; ace 3; ctc 3
11533	Spirostan-3-ol, (3β,5β,25R)- Smilagenin	$C_{27}H_{44}O_3$ 416.64	126-18-1 185	8508	5-19-02-00673	ace 4; bz 4; EtOH 4
11534	Spirostan-12-one, 3-hydroxy-, (3β,5α,25R)- Hecogenin	$C_{27}H_{42}O_4$ 430.63	467-55-0 266.5	4537	5-19-05-00592	ace 4; eth 4; EtOH 4
11535	Spirostan-2,3,15-triol, (2α,3β,5α,15β,25R)- Digitogenin	$C_{27}H_{44}O_5$ 448.64	511-34-2 281.5	3143	5-19-03-00603	chl 4
11536	Spirost-5-en-3-ol, (3β,25R)- Diosgenin	$C_{27}H_{42}O_3$ 414.63	512-04-9 205.5	3289	5-19-03-00030	EtOH 4
11537	Spiro[5.5]undecane Spir[5.]undecane	$C_{11}H_{20}$ 152.28	180-43-8 207		4-05-00-00328 0.8783[20]	1.4731
11538	Stannane, (acetyloxy)tributyl- Acetoxytributylstannane	$C_{14}H_{30}O_2Sn$ 349.10	56-36-0 84.7			
11539	Stannane, (acetyloxy)triphenyl- Triphenyltin acetate	$C_{20}H_{18}O_2Sn$ 409.07	900-95-8 121.5			
11540	Stannane, chlorotrimethyl-	C_3H_9ClSn 199.27	1066-45-1 38.5	148		H_2O 3; chl 3; os 3
11541	Stannane, chlorotriphenyl- Triphenyltin chloride	$C_{18}H_{15}ClSn$ 385.48	639-58-7 103.5			chl 3
11542	Stannane, chlorotripropyl-	$C_9H_{21}ClSn$ 283.43	2279-76-7 -23.5	123[13]	1.2678[28]	ctc 3; os 3 1.49102[28]
11543	Stannane, tetraethenyl-	$C_8H_{12}Sn$ 226.89	1112-56-7 55-7[17]		4-04-00-04316 1.2651[25]	ctc 3 1.4993[25]
11544	Stannane, tetramethyl- Tetramethylstannane	$C_4H_{12}Sn$ 178.85	594-27-4 -54.8	78	4-04-00-04307 1.314[25]	H_2O 1; ctc 3; CS_2 3; os 3 1.4386
11545	Stannane, tetraphenyl-	$C_{24}H_{20}Sn$ 427.13	595-90-4 228	420	4-16-00-01592	chl 2
11546	Stannane, tetrapropyl-	$C_{12}H_{28}Sn$ 291.06	2176-98-9 -109.1	228	4-04-00-04310 1.1065[20]	1.4745[20]
11547	Stibine, dichlorophenyl-	$C_6H_5Cl_2Sb$ 269.77	5035-52-9 69.5	110-5[10]	4-16-00-01205	ace 4; EtOH 4; HOAc 4
11548	Stibine oxide, dihydroxyphenyl- Benzenestibonic acid	$C_6H_7O_3Sb$ 248.88	535-46-6 139	1077	3-16-00-01178	

No.	Name Synonym	Mol. Form. Mol. Wt.	CAS RN mp/°C	Merck No. bp/°C	Beil. Ref. den/g cm^{-3}	Solubility n_D
11549	Stibine, triethyl- Triethylstibine	$C_6H_{15}Sb$ 208.94	617-85-6 -98	161.4	4-04-00-03685 1.3224[15]	H_2O 1; EtOH 3; eth 3
11550	Stibine, trimethyl- Trimethylstibine	C_3H_9Sb 166.86	594-10-5 -62	80.6	4-04-00-03684 1.523[15]	H_2O 1; EtOH 3; eth 3; CS_2 3 1.42[15]
11551	Stibine, triphenyl-	$C_{18}H_{15}Sb$ 353.07	603-36-1 53.5	>360	4-16-00-01198 1.4343[25]	H_2O 1; EtOH 3; eth 4; ace 4 1.6948[42]
11552	Stibine, tris(3-methylphenyl)-	$C_{21}H_{21}Sb$ 395.15	35569-54-1 72		4-16-00-01199 1.3957[16]	bz 4; eth 4; EtOH 4; chl 4
11553	Stibine, tris(4-methylphenyl)-	$C_{21}H_{21}Sb$ 395.15	5395-43-7 127.5	218[3]	4-16-00-01199 1.3595[16]	bz 4; eth 4; chl 4
11554	Stigmasta-5,7-dien-3-ol, (3β)- 7-Dehydrositosterol	$C_{29}H_{48}O$ 412.70	521-04-0 144.5	2863	3-06-00-02855	bz 4; eth 4; EtOH 4
11555	Stigmasta-5,22-dien-3-ol, (3β,22E)- Stigmasterol	$C_{29}H_{48}O$ 412.70	83-48-7 170	8771	4-06-00-04170	bz 4; eth 4; EtOH 4
11556	Stigmasta-7,24(28)-dien-3-ol, 4-methyl-, (3β,4α,5α,24Z)- α1-Sitosterol	$C_{30}H_{50}O$ 426.73	474-40-8 166	8499	3-06-00-02876	EtOH 4; chl 4
11557	Stigmastan-3-ol, (3β,5α)- Stigmastanol	$C_{29}H_{52}O$ 416.73	83-45-4 144	8770	3-06-00-02172	
11558	Stigmast-5-en-3-ol, (3β)- β-Sitosterol	$C_{29}H_{50}O$ 414.72	83-46-5 140	8500	4-06-00-04037	EtOH 3; eth 3; HOAc 3
11559	Stigmast-5-en-3-ol, (3β,24S)- γ-Sitosterol	$C_{29}H_{50}O$ 414.72	83-47-6	8501		
11560	Stilbene, α-ethyl- 1-Butene, 1,2-diphenyl	$C_{16}H_{16}$ 208.30	22692-70-2 57	296.5	4-05-00-02206 1.0124[18]	1.593[18]
11561	D-Streptamine, O-2-deoxy-2-(methylamino)-α- L-glucopyranosyl-(12)-O-5-deoxy-3-C-formyl- α-L-lyxofuranosyl-(1-4)-N,N ' 55- bis(aminoiminomethyl)-, sulfate (2:3)(salt) Streptomycin, sulfate	$C_{42}H_{84}N_{14}O_{36}S_3$ 1457.40	3810-74-0		5-18-11-00082	
11562	Streptomycin N-Methyl-L- glucosamidinostreptosidostreptidine	$C_{21}H_{39}N_7O_{12}$ 581.58	57-92-1	8786		
11563	Strychnidin-10-one Strychnine	$C_{21}H_{22}N_2O_2$ 334.42	57-24-9 287	8822 270[5]	4-27-00-07537 1.36[20]	H_2O 2; EtOH 2; eth 1; ace 2
11564	Strychnidin-10-one, 2,3-dimethoxy- Brucine	$C_{23}H_{26}N_2O_4$ 394.47	357-57-3 178	1443	4-27-00-07875	H_2O 2; EtOH 4; eth 2; bz 2
11565	Strychnidin-10-one, mononitrate Strychnine nitrate	$C_{21}H_{23}N_3O_5$ 397.43	66-32-0 295	8823	2-27-00-00723 1.627[25]	H_2O 4; MeOH 4
11566	Strychnidin-10-one, sulfate (2:1) Strychnine, sulfate	$C_{42}H_{46}N_4O_8S$ 766.92	60-41-3 200 dec	8826		H_2O 3; EtOH 3; eth 1; chl 2
11567	Succinimide, N-(ethoxymethyl)-	$C_7H_{11}NO_3$ 157.17	98431-97-1 31.5	262	2-21-00-00304	H_2O 4; EtOH 4
11568	Sulfamic acid, cyclohexyl- Cyclamic acid	$C_6H_{13}NO_3S$ 179.24	100-88-9 169.5	2707	4-12-00-00102	alk 4
11569	Sulfamide, tetramethyl-	$C_4H_{12}N_2O_2S$ 152.22	3768-63-6 73	225	4-04-00-00270	EtOH 4
11570	Sulfide, m-tolyl p-tolyl 3,4'-Ditolyl sulfide	$C_{14}H_{14}S$ 214.33	107770-92-3 28	179[11]	4-06-00-02173	eth 4; EtOH 4
11571	Sulfoxylic acid, diethyl ester	$C_4H_{10}O_2S$ 122.19	10297-38-8	118	3-01-00-01314 0.9940[20]	H_2O 3; EtOH 4; eth 4; ace 4 1.4234[20]
11572	Sulfur cyanide	C_2N_2S 84.10	627-52-1 63.5	sub 30	4-03-00-00339	eth 4; EtOH 4
11573	Sulfuric acid, diethyl ester Diethyl sulfate	$C_4H_{10}O_4S$ 154.19	64-67-5 -24	3120 208	4-01-00-01326 1.172[25]	H_2O 1; EtOH 5; eth 5 1.3989[20]
11574	Sulfuric acid, dimethyl ester Dimethyl sulfate	$C_2H_6O_4S$ 126.13	77-78-1 -27	3244 188 dec; 76[15]	4-01-00-01251 1.3322[20]	H_2O 3; EtOH 5; eth 3; bz 3 1.3874[20]
11575	Sulfuric acid, dipropyl ester	$C_6H_{14}O_4S$ 182.24	598-05-0 121[30]		4-01-00-01423 1.1064[20]	peth 4 1.4135[20]
11576	Sulfuric acid, monoethyl ester Ethyl sulfate	$C_2H_6O_4S$ 126.13	540-82-9 280 dec	3808	4-01-00-01324 1.3657[20]	H_2O 4 1.4105[20]
11577	Sulfuric acid, monomethyl ester Methyl sulfate	CH_4O_4S 112.11	75-93-4 <-30	6041 135 dec	4-01-00-01250	H_2O 4; eth 4; EtOH 4
11578	Sulfurous acid, bis(2-methylpropyl) ester	$C_8H_{18}O_3S$ 194.30	18748-27-1 210; 92-4[13]		4-01-00-01595 0.9862[20]	1.4268[20]
11579	Sulfurous acid, 2-chloroethyl 2-[4-(1,1- dimethylethyl)phenoxy]-1-methylethyl ester Aramite	$C_{15}H_{23}ClO_4S$ 334.86	140-57-8 -37.3	794 195[2]	4-06-00-03301 1.143[20]	ace 4; bz 4; eth 4; EtOH 4 1.5100[20]
11580	Sulfurous acid, dibutyl ester Butyl sulfite	$C_8H_{18}O_3S$ 194.30	626-85-7 230		4-01-00-01522 0.9957[20]	EtOH 3; eth 3 1.4310[20]
11581	Sulfurous acid, diethyl ester Ethyl sulfite	$C_4H_{10}O_3S$ 138.19	623-81-4 158; 51[13]		4-01-00-01324 1.1[20]	EtOH 3; eth 3 1.4310[20]
11582	Sulfurous acid, dimethyl ester	$C_2H_6O_3S$ 110.13	616-42-2 126		4-01-00-01250 1.2129[20]	H_2O 3; EtOH 3; eth 3 1.4083[20]
11583	Sulfurous diamide, tetramethyl-	$C_4H_{12}N_2OS$ 136.22	3768-60-3 31	209; 74[13]	4-04-00-00269	eth 4
11584	Sulfur, pentafluoro(trifluoromethyl)-, (OC-6- 21)-	CF_8S 196.06	373-80-8 -87	-20	4-03-00-00035	H_2O 1; EtOH 1; CS_2 3

No.	Name / Synonym	Mol. Form. / Mol. Wt.	CAS RN / mp/°C	Merck No. / bp/°C	Beil. Ref. / den/g cm^{-3}	Solubility / n_D
11585	Sulfuryl chloride isocyanate	CClNO$_3$S 141.53	1189-71-5 -44	107	4-03-00-00086 1.626[25]	1.4467[20]
11586	Sutan Carbamothioic acid, bis(2-methylpropyl)-, S-ethyl ester	C$_{11}$H$_{23}$NOS 217.38	2008-41-5	1546 138[21]	0.9402[25]	
11587	D-Tagatose Tagatose, D-	C$_6$H$_{12}$O$_6$ 180.16	87-81-0 134.5	9005	4-01-00-04414	H$_2$O 4
11588	Tartar emetic Antimonate(2-), bis[μ-[2,3-dihydroxybutanedioato(4-)-O1,O2:O3,O4]]di-, dipotassium, trihydrate, stereoisomer	C$_8$H$_{10}$K$_2$O$_{15}$Sb$_2$ 667.87	28300-74-5	732	2.6	
11589	Tebuconazole (RS)-1-p-Chlorophenyl-4,4-dimethyl-3-(1H-1,2,4-triazol-1-ylmethyl)pentan-3-ol	C$_{16}$H$_{23}$ClN$_3$O 308.83	107534-96-3 102.4			
11590	Tebuthiuron Urea, N-[5-(1,1-dimethylethyl)-1,3,4-thiadiazol-2-yl]-N,N-dimethyl-	C$_9$H$_{16}$N$_4$OS 228.32	34014-18-1 163 dec	9053		
11591	Terbacil 2,4(1H,3H)-Pyrimidinedione, 5-chloro-3-(1,1-dimethylethyl)-6-methyl-	C$_9$H$_{13}$ClN$_2$O$_2$ 216.67	5902-51-2 176	9085 sub 175	1.34[25]	
11592	Terbufos Phosphorodithioic acid, S-[[(1,1-dimethylethyl)thio]methyl] O,O-diethyl ester	C$_9$H$_{21}$O$_2$PS$_3$ 288.44	13071-79-9 -29.2	9088 69[0.01]	1.105[24]	
11593	Terbuthylazine 1,3,5-Triazine-2,4-diamine, 6-chloro-N-(1,1-dimethylethyl)-N'-ethyl-	C$_9$H$_{16}$ClN$_5$ 229.71	5915-41-3 178		1.188[20]	
11594	Terbutryn 1,3,5-Triazine-2,4-diamine, N-(1,1-dimethylethyl)-N'-ethyl-6-(methylthio)-	C$_{10}$H$_{19}$N$_5$S 241.36	886-50-0 104	157[0.06]	1.115[20]	
11595	1,1':2',1"-Terphenyl o-Terphenyl	C$_{18}$H$_{14}$ 230.31	84-15-1 56.2	332	4-05-00-02478	H$_2$O 1; ace 3; bz 3; chl 3
11596	1,1':3',1"-Terphenyl m-Terphenyl	C$_{18}$H$_{14}$ 230.31	92-06-8 87	363	4-05-00-02480	H$_2$O 1; EtOH 3; eth 3; bz 3
11597	1,1':4',1"-Terphenyl p-Terphenyl	C$_{18}$H$_{14}$ 230.31	92-94-4 210.1	376	4-05-00-02483	H$_2$O 1; EtOH 2; eth 3; bz 3
11598	1,1':3',1"-Terphenyl, 5'-phenyl-	C$_{24}$H$_{18}$ 306.41	612-71-5 176	462	4-05-00-02732 1.199[30]	H$_2$O 1; EtOH 3; eth 3; bz 4
11599	Terrazole 1,2,4-Thiadiazole, 5-ethoxy-3-(trichloromethyl)-	C$_5$H$_5$Cl$_3$N$_2$OS 247.53	2593-15-9 19.9	95[1]	1.503[25]	
11600	2,2':5',2"-Terthiophene α-Terthienyl	C$_{12}$H$_8$S$_3$ 248.39	1081-34-1	9106	5-19-09-00226	
11601	1,3,5,7-Tetraazatricyclo[3.3.1.13,7]decane Methenamine	C$_6$H$_{12}$N$_4$ 140.19	100-97-0 >250	5879 sub	4-26-00-01680 1.331[-5]	H$_2$O 4; EtOH 3; eth 2; ace 3
11602	2,6,10,14,18,22-Tetracosahexaene, 2,6,10,14,19,23-hexamethyl-, (all-E)- Squalene	C$_{30}$H$_{50}$ 410.73	111-02-4 <-20	8727 280[17]; 241[4]	3-01-00-01071 0.8584[20]	H$_2$O 1; EtOH 2; eth 3; ace 3 1.4990[20]
11603	Tetracosane	C$_{24}$H$_{50}$ 338.66	646-31-1 54	391.3	4-01-00-00578 0.7991[20]	H$_2$O 1; EtOH 2; eth 4 1.4283[70]
11604	Tetracosane, 11-decyl-	C$_{34}$H$_{70}$ 478.93	55429-84-0 10.8	301[10]	4-01-00-00598 0.8161[20]	1.4556[20]
11605	Tetracosane, 3-ethyl-	C$_{26}$H$_{54}$ 366.71	55282-17-2 30.1	255.5[10]	4-01-00-00584 0.7949[40]	1.4436[40]
11606	Tetracosane, 2,6,10,15,19,23-hexamethyl- Squalane	C$_{30}$H$_{62}$ 422.82	111-01-3 -38	8726 350	4-01-00-00593 0.8115[15]	H$_2$O 1; EtOH 2; eth 3; ace 2 1.4530[15]
11607	Tetracosanoic acid Lignoceric acid	C$_{24}$H$_{48}$O$_2$ 368.64	557-59-5 87.5	5364 272[10]	4-02-00-01301 0.8207[100]	bz 4; eth 4 1.4287[100]
11608	15-Tetracosenoic acid	C$_{24}$H$_{46}$O$_2$ 366.63	14490-79-0 61		4-02-00-01681	EtOH 3; chl 3; ace 3
11609	5,9-Tetradecadien-7-yne, 6,9-dimethyl-	C$_{16}$H$_{26}$ 218.38	101427-54-7	95-8[0.3]	4-01-00-01133 0.8241[20]	bz 4; eth 4 1.4866[20]
11610	6,8-Tetradecadiyne	C$_{14}$H$_{22}$ 190.33	16387-72-7 2	118-9[4]	3-01-00-01067 0.8699[16]	eth 4
11611	Tetradecanamide	C$_{14}$H$_{29}$NO 227.39	638-58-4 104	217[12]	4-02-00-01138	EtOH 4
11612	1-Tetradecanamine	C$_{14}$H$_{31}$N 213.41	2016-42-4 83.1	291.2	4-04-00-00812 0.8079[20]	H$_2$O 1; EtOH 4; eth 4; ace 3 1.4463[20]
11613	Tetradecane	C$_{14}$H$_{30}$ 198.39	629-59-4 5.8	253.5	4-01-00-00520 0.7628[20]	H$_2$O 1; EtOH 4; eth 4; ctc 3 1.4290[20]
11614	Tetradecane, 1-bromo-	C$_{14}$H$_{29}$Br 277.29	112-71-0 5.6	307	4-01-00-00523 1.0170[20]	ace 4; bz 4; EtOH 4 1.4603[20]
11615	Tetradecane, 1-chloro-	C$_{14}$H$_{29}$Cl 232.84	2425-54-9 4.9	292	3-01-00-00550 0.8665[20]	H$_2$O 1; EtOH 3; ace 4; bz 4 1.4473[20]
11616	Tetradecane, 1,14-dibromo- Tetradecamethylene dibromide	C$_{14}$H$_{28}$Br$_2$ 356.18	37688-96-3 50.4	190[8]	4-01-00-00523	eth 4; EtOH 4; chl 4
11617	Tetradecane, 1,1-dimethoxy- Myristaldehyde, dimethyl acetal	C$_{16}$H$_{34}$O$_2$ 258.44	14620-53-2	134-6[4]	4-01-00-03389	eth 4; EtOH 4 1.4342[25]
11618	1,14-Tetradecanediol	C$_{14}$H$_{30}$O$_2$ 230.39	19812-64-7 85.8	200[9]	4-01-00-02631	eth 4; EtOH 4

No.	Name Synonym	Mol. Form. Mol. Wt.	CAS RN mp/°C	Merck No. bp/°C	Beil. Ref. den/g cm^{-3}	Solubility n_D
11619	Tetradecanenitrile Myristonitrile	$C_{14}H_{27}N$ 209.38	629-63-0 19	226^{100}; 119^1	4-02-00-01139 0.8281^{19}	H$_2$O 1; EtOH 5; eth 5; ace 5 1.4392^{23}
11620	1-Tetradecanesulfonic acid Tetradecylsulfonic acid	$C_{14}H_{30}O_3S$ 278.46	7314-37-6 65		4-04-00-00066 0.9996^{25}	H$_2$O 4
11621	1-Tetradecanethiol	$C_{14}H_{30}S$ 230.46	2079-95-0	$176-80^{22}$	4-01-00-01867 0.8469^{20}	H$_2$O 1; EtOH 3; eth 3; ctc 3 1.4597^{20}
11622	Tetradecanoic acid Myristic acid	$C_{14}H_{28}O_2$ 228.38	544-63-8 53.9	6246 250^{100}	4-02-00-01126 0.8622^{54}	H$_2$O 1; EtOH 3; eth 2; ace 3 1.4723^{70}
11623	Tetradecanoic acid, anhydride	$C_{28}H_{54}O_3$ 438.73	626-29-9 53.4		4-02-00-01138 0.8502^{70}	eth 4; EtOH 4 1.4335^{70}
11624	Tetradecanoic acid, ethyl ester	$C_{16}H_{32}O_2$ 256.43	124-06-1 12.3	295	4-02-00-01131 0.8573^{25}	H$_2$O 1; EtOH 3; eth 2; ctc 3 1.4362^{20}
11625	Tetradecanoic acid, methyl ester	$C_{15}H_{30}O_2$ 242.40	124-10-7 19	295; 155^7	4-02-00-01131 0.8671^{20}	H$_2$O 1; EtOH 5; eth 5; ace 5 1.425^{45}
11626	Tetradecanoic acid, 1-methylethyl ester Isopropyl myristate	$C_{17}H_{34}O_2$ 270.46	110-27-0 5103	193^{20}; 140^2	4-02-00-01132 0.8532^{20}	H$_2$O 1; EtOH 3; eth 3; ace 4 1.4325^{25}
11627	Tetradecanoic acid, phenylmethyl ester	$C_{21}H_{34}O_2$ 318.50	31161-71-4 20.5	229.3^{11}	2-06-00-00417 0.9293^{25}	bz 4; eth 4; EtOH 4; chl 4
11628	Tetradecanoic acid, 1,2,3-propanetriyl ester Trimyristin	$C_{45}H_{86}O_6$ 723.17	555-45-3 56.5	9638 311	4-02-00-01135 0.8848^{60}	H$_2$O 2; eth 3; ace 3 1.4428^{60}
11629	Tetradecanoic acid, propyl ester	$C_{17}H_{34}O_2$ 270.46	14303-70-9	147^2	4-02-00-01132 0.8592^{20}	ace 4; bz 4; eth 4; EtOH 4 1.4356^{25}
11630	1-Tetradecanol Tetradecyl alcohol	$C_{14}H_{30}O$ 214.39	112-72-1 39.5	6248 289	4-01-00-01864 0.8236^{38}	H$_2$O 1; EtOH 4; eth 4; ace 4
11631	2-Tetradecanol	$C_{14}H_{30}O$ 214.39	4706-81-4 34	284	4-01-00-01867 0.8315^{20}	1.4444^{20}
11632	3-Tetradecanol	$C_{14}H_{30}O$ 214.39	1653-32-3 31.5	173^{25}; 146^{10}	4-01-00-01868 0.8098^{53}	bz 4; eth 4; EtOH 4 1.4340^{45}
11633	2-Tetradecanone Dodecylmethylketone	$C_{14}H_{28}O$ 212.38	2345-27-9 33.5	205^{100}; 134^{13}	4-01-00-03389	H$_2$O 1; EtOH 3; ace 3; os 3
11634	3-Tetradecanone	$C_{14}H_{28}O$ 212.38	629-23-2 34	152^{16}	4-01-00-03389	H$_2$O 1; EtOH 3; ace 3; os 3
11635	Tetradecanoyl chloride Myristoyl chloride	$C_{14}H_{27}ClO$ 246.82	112-64-1 -1	171^{16}	4-02-00-01138 0.9078^{25}	eth 3
11636	1-Tetradecene	$C_{14}H_{28}$ 196.38	1120-36-1 -12	233	4-01-00-00924 0.7745^{25}	H$_2$O 1; EtOH 4; eth 4; bz 3 1.4351^{20}
11637	4-Tetradecenoic acid Tuduic acid	$C_{14}H_{26}O_2$ 226.36	544-65-0 18.5	$185-8^{13}$	4-02-00-01626 0.9024^{20}	bz 4; peth 4 1.4559^{20}
11638	5-Tetradecenoic acid Physoteric acid	$C_{14}H_{26}O_2$ 226.36	544-66-1 20	$190-5^{15}$	3-02-00-01373 0.9046^{20}	1.4552^{20}
11639	9-Tetradecenoic acid	$C_{14}H_{26}O_2$ 226.36	13147-06-3 -4	$144^{0.6}$	4-02-00-01626 0.9018^{20}	1.4519^{20}
11640	2-Tetradecyne	$C_{14}H_{26}$ 194.36	638-60-8 6.5	252.5	0-01-00-00262 0.8000^{20}	eth 4; EtOH 4
11641	7-Tetradecyne	$C_{14}H_{26}$ 194.36	35216-11-6	144^{30}	3-01-00-01027 0.7991^{20}	eth 4; EtOH 4 1.4330^{25}
11642	2,5,8,11-Tetraoxadodecane Triglyme	$C_8H_{18}O_4$ 178.23	112-49-2 -45	9604 216	4-01-00-02401 0.986^{20}	H$_2$O 4; bz 4 1.4224^{20}
11643	2,4,8,10-Tetraoxaspiro[5.5]undecane	$C_7H_{12}O_4$ 160.17	126-54-5 48.3	147^{53}; 68^1	5-19-11-00342	H$_2$O 4; ace 4; eth 4; EtOH 4
11644	Tetraphenylene Tetrabenzocyclooctatetraene	$C_{24}H_{16}$ 304.39	212-74-8 233	 sub 200	4-05-00-02773	EtOH 3; eth 2; AcOEt 3; PhNO$_2$ 3
11645	Tetraphosphoric acid, hexaethyl ester Ethyl tetraphosphate	$C_{12}H_{30}O_{13}P_4$ 506.26	757-58-4 -40	150 dec	1.2917^{27}	ace 4; bz 4; EtOH 4 1.4273^{27}
11646	Tetrasiloxane, decamethyl- Decamethyltetrasiloxane	$C_{10}H_{30}O_3Si_4$ 310.69	141-62-8 -76	2843 194	4-04-00-04119 0.8536^{25}	H$_2$O 1; EtOH 2; bz 3; peth 3 1.3895^{20}
11647	Tetrasiloxane, 1,1,1,3,5,7,7,7-octamethyl- 1,1,1,3,5,7,7,7-Octamethyltetrasiloxane	$C_8H_{26}O_3Si_4$ 282.63	16066-09-4 	 170	4-04-00-04098 0.8559^{20}	1.3854^{20}
11648	Tetrasul p-Chlorophenyl 2,4,5-trichlorophenyl sulfide	$C_{12}H_6Cl_4S$ 324.06	2227-13-6			
11649	Tetrasulfide, bis(1,1-dimethylethyl) Di-tert-Butyl tetrasulfide	$C_8H_{18}S_4$ 242.49	5943-35-1 2.3	$70^{0.2}$	4-01-00-01638 1.0690^{20}	1.5660^{20}
11650	Tetratriacontane	$C_{34}H_{70}$ 478.93	14167-59-0 72.6	285.4^3	4-01-00-00597 0.7728^{90}	1.4296^{90}
11651	1,2,4,5-Tetrazine sym-Tetrazine	$C_2H_2N_4$ 82.06	290-96-0 99	sub	4-26-00-01710	H$_2$O 3; EtOH 3; eth 3; sulf 3
11652	1H-Tetrazole	CH_2N_4 70.05	288-94-8 157	sub	4-26-00-01652 1.4060^{20}	H$_2$O 2
11653	2H-Tetrazolium, 2,3,5-triphenyl-, chloride Triphenyltetrazolium chloride	$C_{19}H_{15}ClN_4$ 334.81	298-96-4 243 dec	9658	4-26-00-01774	H$_2$O 3; EtOH 1; ace 3
11654	5H-Tetrazolo[1,5-a]azepine, 6,7,8,9- tetrahydro- Pentylenetetrazole	$C_6H_{10}N_4$ 138.17	54-95-5 59.5	7097 194^{12}	4-26-00-01712	H$_2$O 4; EtOH 4; eth 3; ace 4
11655	4-Thia-1-azabicyclo[3.2.0]heptane-2- carboxylic acid, 3,3-dimethyl-7-oxo-6- [(phenylacetyl)amino]- Penicillin G, sodium salt	$C_{16}H_{17}N_2NaO_4S$ 356.38	69-57-8	1157	4-27-00-05861	

No.	Name Synonym	Mol. Form. Mol. Wt.	CAS RN mp/°C	Merck No. bp/°C	Beil. Ref. den/g cm^{-3}	Solubility n_D
11656	Thiabendazole 1H-Benzimidazole, 2-(4-thiazolyl)-	$C_{10}H_7N_3S$ 201.25	148-79-8	9217 sub 305		
11657	1,2,4-Thiadiazole 1-Thia-2,4-diazacyclopentadiene	$C_2H_2N_2S$ 86.12	288-92-6 -33	121	4-27-00-07093 1.3298[20]	
11658	1,2,5-Thiadiazole Piazthiole	$C_2H_2N_2S$ 86.12	288-39-1 -50.1	94	1.268[25]	1.5150[25]
11659	1,3,4-Thiadiazole, 2,5-dimethyl- 2,5-Dimethyl-1,3,4-thiadiazole	$C_4H_6N_2S$ 114.17	27464-82-0 65	202.5	4-27-00-07097	H_2O 2; EtOH 2; eth 2
11660	Thianthrene	$C_{12}H_8S_2$ 216.33	92-85-3 159.3	365	5-19-02-00049 1.4420[20]	H_2O 1; EtOH 2; eth 3; bz 3
11661	Thianthrene, 2,7-dimethyl- Mesulphen	$C_{14}H_{12}S_2$ 244.38	135-58-0 123	5821 184[3]	5-19-02-00072	ace 4; eth 4; peth 4; chl 4
11662	2-Thiazolamine 2-Aminothiazole	$C_3H_4N_2S$ 100.14	96-50-4 93	494 140[11]	4-27-00-04574	H_2O 2; EtOH 2; eth 2; chl 2
11663	2-Thiazolamine, 5-[(4-aminophenyl)sulfonyl]- Thiazolsulfone	$C_9H_9N_3O_2S_2$ 255.32	473-30-3 219-21 dec	9237	4-27-00-05401	ace 4; eth 4; EtOH 4; diox 4
11664	2-Thiazolamine, 4,5-dihydro-	$C_3H_6N_2S$ 102.16	1779-81-3 85.3	dec	4-27-00-04555	H_2O 4; EtOH 4; bz 4; chl 4
11665	2-Thiazolamine, 4-methyl- 2-Amino-4-methylthiazole	$C_4H_6N_2S$ 114.17	1603-91-4 45.5	462 125[20]; 70[0.4]	4-27-00-04687	H_2O 4; EtOH 4; eth 4
11666	2-Thiazolamine, 5-nitro- 2-Amino-5-nitrothiazole	$C_3H_3N_3O_2S$ 145.14	121-66-4 202 dec	469	4-27-00-04675	
11667	Thiazole	C_3H_3NS 85.13	288-47-1 118	9235	4-27-00-00960 1.1998[17]	H_2O 2; EtOH 3; eth 3; ace 3 1.5969[20]
11668	Thiazole, 2-bromo- 2-Bromo-1,3-thiazole	C_3H_2BrNS 164.03	3034-53-5 171		4-27-00-00962 1.82[25]	1.5927[20]
11669	5-Thiazolecarboxylic acid, 4-methyl- 4-Methyl-5-thiazolecarboxylic acid	$C_5H_5NO_2S$ 143.17	20485-41-0 280 dec	sub 250	4-27-00-04008	H_2O 2; EtOH 3; eth 3; bz 2
11670	Thiazole, 5-(2-chloroethyl)-4-methyl- Clomethiazole	C_6H_8ClNS 161.65	533-45-9 92[7]	2382	4-27-00-00990 1.233[25]	
11671	2,4-Thiazolediamine, 5-phenyl- Amiphenazole	$C_9H_9N_3S$ 191.26	490-55-1 163-4 dec	498	4-27-00-05139	
11672	Thiazole, 4,5-dihydro-2-methyl-	C_4H_7NS 101.17	2346-00-1 -101	145	4-27-00-00921 1.067[25]	1.5200[20]
11673	Thiazole, 4,5-dihydro-2,4,4-trimethyl-	$C_6H_{11}NS$ 129.23	4145-94-2 147		4-27-00-00931 0.969[25]	1.4825[25]
11674	Thiazole, 2,4-dimethyl- 2,4-Dimethylthiazole	C_5H_7NS 113.18	541-58-2 146; 71[50]	3249	4-27-00-00981 1.0562[15]	H_2O 2; EtOH 3; eth 3; chl 3 1.5091[20]
11675	Thiazole, 4,5-dimethyl- 4,5-Dimethylthiazole	C_5H_7NS 113.18	3581-91-7 83.5	158	4-27-00-00986 1.0699[20]	eth 4; EtOH 4
11676	Thiazole, 2,4-diphenyl- 2,4-Diphenylthiazole	$C_{15}H_{11}NS$ 237.33	1826-14-8 92.5	>360	4-27-00-01435 1.1554[98]	eth 4; EtOH 4
11677	5-Thiazoleethanol, 4-methyl- 4-Methyl-5-thiazoleethanol	C_6H_9NOS 143.21	137-00-8 135[7]	6046	4-27-00-01754 1.196[24]	H_2O 4; EtOH 3; eth 3; bz 3
11678	Thiazole, 2-methyl- 2-Methylthiazole	C_4H_5NS 99.16	3581-87-1 128.5		4-27-00-00967	H_2O 5; EtOH 3; ace 3 1.510
11679	Thiazole, 4-methyl- 4-Methylthiazole	C_4H_5NS 99.16	693-95-8 133.3		4-27-00-00969 1.112[25]	H_2O 3; EtOH 3; eth 3
11680	2(3H)-Thiazolethione, 4-methyl-	$C_4H_5NS_2$ 131.22	5685-06-3 89.3	188[3]	4-27-00-02584	EtOH 4
11681	Thiazolidine	C_3H_7NS 89.16	504-78-9 164.5		4-27-00-00009 1.131[25]	H_2O 5; EtOH 3; eth 4; ace 4 1.551[20]
11682	4-Thiazolidinecarboxylic acid Timonacic	$C_4H_7NO_2S$ 133.17	444-27-9 196.5	9375		H_2O 4
11683	2,4-Thiazolidinedione	$C_3H_3NO_2S$ 117.13	2295-31-0 128	179[19]	4-27-00-03144	eth 4
11684	4-Thiazolidinone, 5-[[4-(dimethylamino)phenyl]methylene]-2-thioxo- p-(Dimethylamino)benzal-5-rhodanine	$C_{12}H_{12}N_2OS_2$ 264.37	536-17-4 270 dec	3220	4-27-00-05697	H_2O 1; EtOH 2; eth 4; ace 3
11685	4-Thiazolidinone, 2-thioxo- Rhodanine	$C_3H_3NOS_2$ 133.19	141-84-4 170	8184	4-27-00-03188 0.868[25]	H_2O 2; EtOH 4; eth 4
11686	Thiazolium, 3-[(4-amino-2-methyl-5-pyrimidinyl)methyl]-5-(2-hydroxyethyl)-4-methyl- chloride, monohydrochlorid Thiamine hydrochloride	$C_{12}H_{18}Cl_2N_4OS$ 337.27	67-03-8 248 dec	9222	4-27-00-01764	H_2O 4; EtOH 2; eth 1; ace 2
11687	Thidiazuron Urea, N-phenyl-N'-1,2,3-thiadiazol-5-yl-	$C_9H_8N_4OS$ 220.25	51707-55-2 211 dec			
11688	Thieno[2,3-b]thiophene	$C_6H_4S_2$ 140.23	250-84-0 6.5	225	5-19-01-00431	eth 4
11689	1H-Thieno[3,4-d]imidazole-4-pentanoic acid, hexahydro-2-oxo-, [3aS-(3aα,4β,6aα)]- Biotin	$C_{10}H_{16}N_2O_3S$ 244.31	58-85-5 232 dec	1244	4-27-00-07979	H_2O 3; EtOH 3; eth 2; chl 2
11690	Thiepane	$C_6H_{12}S$ 116.23	4753-80-4 0.5	173.5	5-17-01-00090 0.9883[20]	H_2O 1; eth 3; ace 3; chl 3 1.5044[18]
11691	Thietane Trimethylene sulfide	C_3H_6S 74.15	287-27-4 -73.2	95	5-17-01-00014 1.0200[20]	H_2O 1; EtOH 4; ace 3; bz 4 1.5102[20]
11692	Thietane, 1,1-dioxide Trimethylene sulfone	$C_3H_6O_2S$ 106.15	5687-92-3 75.5	91.2[14]	5-17-01-00015	H_2O 3; EtOH 3; eth 2; peth 2 1.5156[20]
11693	Thietane, 2-methyl-	C_4H_8S 88.17	17837-41-1 107		5-17-01-00055 0.9571[20]	H_2O 1; EtOH 3; eth 3; ace 3 1.4852[20]

No.	Name / Synonym	Mol. Form. / Mol. Wt.	CAS RN / mp/°C	Merck No. / bp/°C	Beil. Ref. / den/g cm^{-3}	Solubility / n_D
11694	Thietane, 1-oxide 1,3-Propanesulfone	$C_3H_6O_3S$ 122.14	13153-11-2 31.5	180^{30}	5-17-01-00015 1.392^{25}	
11695	Thiirane	C_2H_4S 60.12	420-12-2 	57 dec	5-17-01-00009 1.0368^0	EtOH 2; eth 2; ace 3; chl 3 1.4935^{20}
11696	Thiirane, methyl-	C_3H_6S 74.15	1072-43-1 -91	72.5	5-17-01-00025 0.941^{20}	chl 3 1.472^{20}
11697	Thiobencarb Carbamothioic acid, diethyl-, S-[(4-chlorophenyl)methyl] ester	$C_{12}H_{16}ClNOS$ 257.78	28249-77-6 3.3	127$^{0.008}$	1.16^{20}	
11698	Thioboric acid, trimethyl ester	$C_3H_9BS_3$ 152.11	997-49-9 5	218.2	4-01-00-01287 1.126^{20}	ace 4; eth 4 1.5788^{20}
11699	Thiocyanic acid, butyl ester Butyl thiocyanate	C_5H_9NS 115.20	628-83-1 	186	4-03-00-00329 0.9563^{15}	H_2O 1; EtOH 3; eth 3 1.4360^{20}
11700	Thiocyanic acid, chloromethyl ester	C_2H_2ClNS 107.56	3268-79-9 	185	2-03-00-00124 1.37^{15}	os 5
11701	Thiocyanic acid, 1,1-dimethylethyl ester tert-Butyl thiocyanate	C_5H_9NS 115.20	37985-18-5 10.5	140 dec; 40^{10}	4-03-00-00330 0.9187^{10}	
11702	Thiocyanic acid, 1,2-ethanediyl ester	$C_4H_4N_2S_2$ 144.22	629-17-4 90	dec	4-03-00-00333 1.4200^0	H_2O 2; EtOH 3; eth 3; ace 4
11703	Thiocyanic acid, ethyl ester Ethyl thiocyanate	C_3H_5NS 87.15	542-90-5 -85.5	146	4-03-00-00328 1.007^{23}	H_2O 1; EtOH 5; eth 5; chl 3 1.4684^{15}
11704	Thiocyanic acid, heptyl ester Heptyl thiocyanate	$C_8H_{15}NS$ 157.28	5416-94-4 	235	4-03-00-00331 0.92^{20}	eth 4; EtOH 4
11705	Thiocyanic acid, methyl ester Methyl thiocyanate	C_2H_3NS 73.12	556-64-9 -5	6047 132.9	4-03-00-00327 1.0678^{25}	H_2O 2; EtOH 5; eth 5; ctc 3 1.4669^{25}
11706	Thiocyanic acid, 1-methylethyl ester Isopropyl thiocyanate	C_4H_7NS 101.17	625-59-2 	153	4-03-00-00329 0.9784^{20}	H_2O 1; EtOH 5; eth 5
11707	Thiocyanic acid, 2-methylpropyl ester Isobutyl thiocyanate	C_5H_9NS 115.20	591-84-4 -59	5036 175.4	4-03-00-00330	eth 4; EtOH 4
11708	Thiocyanic acid, octyl ester	$C_9H_{17}NS$ 171.31	19942-78-0 105	141-2^{19}	4-03-00-00331 0.9149^{25}	H_2O 1; EtOH 4; eth 4; bz 3 1.4649^{20}
11709	Thiocyanic acid, phenyl ester Phenyl thiocyanate	C_7H_5NS 135.19	5285-87-0 	232.5	4-06-00-01536 1.153^{18}	H_2O 1; EtOH 3; eth 3
11710	Thiocyanic acid, phenylmethyl ester	C_8H_7NS 149.22	3012-37-1 43	232	4-06-00-02680	H_2O 1; EtOH 3; eth 3; chl 3
11711	Thiocyanic acid, potassium salt KSCN	CKNS 97.18	333-20-0	7687	4-03-00-00299	
11712	Thiocyanic acid, 2-propenyl ester	C_4H_5NS 99.16	764-49-8 	161	4-03-00-00332 1.056^{15}	eth 4; EtOH 4
11713	Thiodicarb Dimethyl N,N'-[thiobis[[(methylimino)carbonyl]oxyl]]bis[ethanimidothioate]	$C_{10}H_{18}N_4O_4S_3$ 354.48	59669-26-0 173	9258	1.4^{20}	
11714	Thiodicarbonic diamide, tetraethyl- Sulfiram	$C_{10}H_{20}N_2S_3$ 264.48	95-05-6 	8928 225-40^3	4-04-00-00397 1.12^{20}	chl 3
11715	Thiodicarbonic diamide, tetramethyl-	$C_6H_{12}N_2S_3$ 208.37	97-74-5 109.5		4-04-00-00238 1.37^{25}	H_2O 1; EtOH 3; eth 2; ace 3
11716	Thiodiphosphoric acid, tetraethyl ester Sulfotep	$C_8H_{20}O_5P_2S_2$ 322.32	3689-24-5 	8945 136-9^2	4-01-00-01351 1.196^{25}	H_2O 1; EtOH 3 1.4753^{25}
11717	Thiofanox 2-Butanone, 3,3-dimethyl-1-(methylthio)-, O-[(methylamino)carbonyl]oxime	$C_9H_{18}N_2O_2S$ 218.32	39196-18-4 57			
11718	Thioimidodicarbonic diamide 2,4-Dithiobiuret	$C_2H_5N_3S_2$ 135.21	541-53-7 181 dec	3378	4-03-00-00356	ace 4
11719	Thiomorpholine Thiamorpholine	C_4H_9NS 103.19	123-90-0 	9229 175; 110^{100}	4-27-00-00636 1.0882^{20}	H_2O 4; ace 4; eth 4; EtOH 4 1.5386^{20}
11720	Thionazin Phosphorothioic acid, O,O-diethyl O-pyrazinyl ester	$C_8H_{13}N_2O_3PS$ 248.24	297-97-2 -1.7	9275 80		
11721	Thioperoxydicarbonic diamide ([(H2N)C(S)]2S2), tetraethyl- Disulfiram	$C_{10}H_{20}N_2S_4$ 296.55	97-77-8 71.5	3370 117^{17}	4-04-00-00398	H_2O 1; EtOH 3; eth 2; chl 4
11722	Thioperoxydicarbonic diamide ([(H2N)C(S)]2S2), tetramethyl- Thiram	$C_6H_{12}N_2S_4$ 240.44	137-26-8 155.6	9304 129^{20}	4-04-00-00242	chl 4
11723	Thiophanate-methyl Carbamic acid, [1,2-phenylenebis(iminocarbonothioyl)]bis-, dimethyl ester	$C_{12}H_{14}N_4O_4S_2$ 342.40	23564-05-8 172 dec			
11724	Thiophene Thiofuran	C_4H_4S 84.14	110-02-1 -39.4	9283 84.0	5-17-01-00297 1.0649^{20}	EtOH 5; eth 5; ace 5; bz 5 1.5289^{20}
11725	2-Thiopheneacetonitrile	C_6H_5NS 123.18	20893-30-5 	115-20^{22}	5-18-06-00208 1.157^{25}	1.5425^{20}
11726	Thiophene, 2-bromo- 2-Thienyl bromide	C_4H_3BrS 163.04	1003-09-4 	150	5-17-01-00305 1.684^{20}	H_2O 1; eth 4; ace 4; ctc 3 1.5868^{20}
11727	Thiophene, 3-bromo-	C_4H_3BrS 163.04	872-31-1 	159.5	5-17-01-00306 1.735^{20}	H_2O 1; ace 3; bz 3; chl 2 1.5919^{20}
11728	Thiophene, 2-bromo-5-chloro- 2-Bromo-5-chlorothiophene	C_4H_2BrClS 197.48	2873-18-9 -20	70^{18}	5-17-01-00307 1.803^{25}	eth 4 1.5925^{25}

No.	Name Synonym	Mol. Form. Mol. Wt.	CAS RN mp/°C	Merck No. bp/°C	Beil. Ref. den/g cm⁻³	Solubility n_D
11729	Thiophene, 2-bromo-3-methyl- 2-Bromo-3-methylthiophene	C_5H_5BrS 177.06	14282-76-9	176; 27[1.8]	5-17-01-00333 1.5844[18]	bz 4; eth 4 1.5714[20]
11730	Thiophene, 2-bromo-5-methyl-	C_5H_5BrS 177.06	765-58-2	178; 29[1.8]	5-17-01-00327 1.5529[20]	eth 3; bz 3 1.5673[20]
11731	Thiophene, 2-butyl- 2-Butylthiophene	$C_8H_{12}S$ 140.25	1455-20-5	181.5	5-17-01-00369 0.9537[20]	1.5090[20]
11732	2-Thiophenecarbonitrile 2-Cyanothiophene .	C_5H_3NS 109.15	1003-31-2	192	5-18-06-00171 1.172[25]	chl 3 1.5629[20]
11733	2-Thiophenecarboxaldehyde	C_5H_4OS 112.15	98-03-3	197; 85[16]	5-17-09-00349 1.2127[21]	H_2O 1; EtOH 4; eth 3; chl 2 1.5920[20]
11734	3-Thiophenecarboxaldehyde 3-Formylthiophene	C_5H_4OS 112.15	498-62-4	86.7[20]	5-17-09-00373 1.5855[20]	H_2O 1; EtOH 4; eth 3
11735	2-Thiophenecarboxaldehyde, 5-chloro-	C_5H_4ClOS 147.60	7283-96-7	77.5[5]	5-17-09-00360 1.6036[25]	chl 2
11736	2-Thiophenecarboxaldehyde, 5-methyl-	C_6H_6OS 126.18	13679-70-4	114[25]	5-17-09-00406 1.5825[20]	chl 3
11737	2-Thiophenecarboxylic acid 2-Carboxythiophene	$C_5H_4O_2S$ 128.15	527-72-0 129.5	9284 260 dec	5-18-06-00158	H_2O 4; EtOH 4; eth 4; chl 3
11738	3-Thiophenecarboxylic acid 3-Thenoic acid	$C_5H_4O_2S$ 128.15	88-13-1 138	9206	5-18-06-00199	H_2O 3
11739	2-Thiophenecarboxylic acid, ethyl ester	$C_7H_8O_2S$ 156.21	2810-04-0	218	5-18-06-00159 1.1623[16]	EtOH 3; ace 3; ctc 2; os 3 1.5248[20]
11740	Thiophene, 2-chloro- 2-Thienyl chloride	C_4H_3ClS 118.59	96-43-5 -71.9	128.3	5-17-01-00303 1.2863[20]	H_2O 1; EtOH 5; eth 5; chl 2 1.5487[20]
11741	Thiophene, 2-chloro-5-methyl- 2-Chloro-5-methylthiophene	C_5H_5ClS 132.61	17249-82-0	155; 55[19]	5-17-01-00326 1.2147[25]	ace 4; bz 4; eth 4; EtOH 4 1.5372[20]
11742	Thiophene, 2,3-dibromo- 2,3-Dibromothiophene	$C_4H_2Br_2S$ 241.93	3140-93-0 -17.5	218.5; 89[13]	5-17-01-00308	1.6304[22]
11743	Thiophene, 2,4-dibromo- 2,4-Dibromothiophene	$C_4H_2Br_2S$ 241.93	3140-92-9 -26.3	210	5-17-01-00308 2.1488[20]	
11744	Thiophene, 2,5-dibromo- 2,5-Dibromothiophene	$C_4H_2Br_2S$ 241.93	3141-27-3 -6	210.3	5-17-01-00308 2.14[23]	H_2O 1; EtOH 4; eth 4; ctc 3 1.6288[20]
11745	2,5-Thiophenedicarboxylic acid 2,5-Dicarboxythiophene	$C_6H_4O_4S$ 172.16	4282-31-9 359	sub 150	5-18-07-00049	H_2O 2; EtOH 3; eth 3
11746	Thiophene, 2,3-dichloro- 2,3-Dichlorothiophene	$C_4H_2Cl_2S$ 153.03	17249-79-5 -26.2	173.5	5-17-01-00304 1.4605[20]	H_2O 1; EtOH 5; eth 3; ace 3 1.5651[20]
11747	Thiophene, 2,4-dichloro- 2,4-Dichlorothiophene	$C_4H_2Cl_2S$ 153.03	17249-75-1 -34	167.6	5-17-01-00304 1.4553[20]	H_2O 1; EtOH 4; eth 3; ace 3 1.5660[20]
11748	Thiophene, 2,5-dichloro- 5-Chloro-2-thienyl chloride	$C_4H_2Cl_2S$ 153.03	3172-52-9 -40.5	162	5-17-01-00304 1.4422[20]	H_2O 1; EtOH 5; eth 5; ctc 3 1.5626[20]
11749	Thiophene, 3,4-dichloro- 3,4-Dichlorothiophene	$C_4H_2Cl_2S$ 153.03	17249-76-2 1	185	5-17-01-00305 1.4867[20]	H_2O 1; EtOH 4; eth 4; ace 3 1.5821[19]
11750	Thiophene, 2,3-dihydro- 2,3-Dihydrothiophene	C_4H_6S 86.16	1120-59-8	112.1		
11751	Thiophene, 2,5-dihydro- 2,5-Dihydrothiophene	C_4H_6S 86.16	1708-32-3	122.4		
11752	Thiophene, 2,5-dihydro-, 1,1-dioxide 3-Sulfolene	$C_4H_6O_2S$ 118.16	77-79-2 64.5	8935	5-17-01-00177	chl 3
11753	Thiophene, 2,5-diiodo- 2,5-Diiodothiophene	$C_4H_2I_2S$ 335.93	625-88-7 41.5	139[15]	5-17-01-00313	H_2O 1; EtOH 4
11754	Thiophene, 2,3-dimethyl- 2,3-Dimethylthiophene	C_6H_8S 112.20	632-16-6 -49	141.6	5-17-01-00342 1.0021[20]	H_2O 1; EtOH 4; eth 4; bz 3 1.5192[20]
11755	Thiophene, 2,4-dimethyl- 2,4-Dimethylthiophene	C_6H_8S 112.20	638-00-6	140.7	5-17-01-00344 0.9938[20]	H_2O 1; EtOH 3; eth 3; bz 3 1.5104[20]
11756	Thiophene, 2,5-dimethyl- 2,5-Dimethylthiophene	C_6H_8S 112.20	638-02-8 -62.8	136.5	5-17-01-00345 0.9850[20]	H_2O 1; EtOH 3; eth 3; bz 3 1.5129[20]
11757	Thiophene, 3,4-dimethyl- 3,4-Dimethylthiophene	C_6H_8S 112.20	632-15-5	145	5-17-01-00347 0.993[25]	H_2O 1; EtOH 3; eth 4 1.5206[20]
11758	Thiophene, 2,5-dinitro- 2,5-Dinitrothiophene	$C_4H_2N_2O_4S$ 174.14	59434-05-8 81	290	5-17-01-00315	eth 4; EtOH 4
11759	Thiophene, 2-ethyl- 2-Ethylthiophene	C_6H_8S 112.20	872-55-9	134	5-17-01-00339 0.9930[20]	H_2O 1; EtOH 4; eth 4 1.5122[20]
11760	Thiophene, 3-ethyl- 3-Ethylthiophene	C_6H_8S 112.20	1795-01-3 -89.1	136	5-17-01-00341 0.9980[20]	H_2O 1; EtOH 3; eth 4 1.5146[20]
11761	Thiophene, 2-ethyl-3-methyl- 2-Ethyl-3-methylthiophene	$C_7H_{10}S$ 126.22	31805-48-8	161	5-17-01-00357 0.9815[20]	1.5105[20]
11762	Thiophene, 2-ethyl-5-methyl- 2-Ethyl-5-methylthiophene	$C_7H_{10}S$ 126.22	40323-88-4 -68.4	160	5-17-01-00357 0.9663[20]	1.5073[20]
11763	Thiophene, 3-ethyl-4-methyl-	$C_7H_{10}S$ 126.22	66577-03-5	165.5	4-17-00-00301 0.9976[20]	1.5186[20]
11764	Thiophene, 4-ethyl-2-methyl- 5-Methyl-3-ethylthiophene	$C_7H_{10}S$ 126.22	13678-54-1 -60	163	4-17-00-00299 0.9742[20]	1.5098[20]
11765	Thiophene, 2-ethyltetrahydro-	$C_6H_{12}S$ 116.23	1551-32-2	157	5-17-01-00096 0.944[20]	ace 4; bz 4; eth 4; EtOH 4 1.490[20]
11766	Thiophene, 3-ethyltetrahydro- 3-Ethyltetrahydrothiophene	$C_6H_{12}S$ 116.23	62184-67-2	165	5-17-01-00096 0.950[20]	1.491[20]
11767	Thiophene, 2-iodo-	C_4H_3IS 210.04	3437-95-4 -40	181	5-17-01-00309 2.0595[25]	EtOH 4; eth 4; chl 2 1.6465[25]
11768	2-Thiophenemethanol	C_5H_6OS 114.17	636-72-6	207; 86[10]	5-17-03-00358 1.2053[16]	EtOH 3; ace 3 1.5280[20]

No.	Name Synonym	Mol. Form. Mol. Wt.	CAS RN mp/°C	Merck No. bp/°C	Beil. Ref. den/g cm^{-3}	Solubility n_D
11769	Thiophene, 2-methyl- 2-Methylthiophene	C_5H_6S 98.17	554-14-3 -63.4	112.6	5-17-01-00324 1.0193[20]	H_2O 1; EtOH 5; eth 5; ace 5 1.5203[20]
11770	Thiophene, 3-methyl- 3-Thiotolene	C_5H_6S 98.17	616-44-4 -69	115.5	5-17-01-00331 1.0218[20]	H_2O 1; EtOH 5; eth 5; ace 5 1.5204[20]
11771	Thiophene, 2-(3-methylbutyl)- 2-Isopentylthiophene	$C_9H_{14}S$ 154.28	26963-33-7	74-5[11]	4-17-00-00316 0.9481[12]	1.5014[12]
11772	Thiophene, 2-(1-methylethyl)- 2-Isopropylthiophene	$C_7H_{10}S$ 126.22	4095-22-1	153	5-17-01-00356 0.9678[20]	1.5038[20]
11773	Thiophene, 3-(1-methylethyl)- 3-Isopropylthiophene	$C_7H_{10}S$ 126.22	29488-27-5	157	5-17-01-00356 0.9733[20]	1.5052[20]
11774	Thiophene, 2-methyl-5-phenyl- 2-Methyl-5-phenylthiophene	$C_{11}H_{10}S$ 174.27	5069-26-1 51	271	5-17-02-00153	EtOH 4; eth 4; lig 4
11775	Thiophene, 2-nitro- 2-Nitrothiophene	$C_4H_3NO_2S$ 129.14	609-40-5 46.5	224.5	5-17-01-00313 1.3644[43]	H_2O 1; EtOH 4; alk 3; peth 2
11776	Thiophene, 3-nitro- 3-Nitrothiophene	$C_4H_3NO_2S$ 129.14	822-84-4 78.5	225	5-17-01-00313	H_2O 1; EtOH 4; ctc 2; peth 3
11777	Thiophene-2-ol	C_4H_4OS 100.14	17236-58-7 9	218	1.255[20]	H_2O 3; ace 3; chl 3 1.5644[20]
11778	Thiophene, 2-propyl- 2-Propylthiophene	$C_7H_{10}S$ 126.22	1551-27-5	158	5-17-01-00355 0.9683[20]	H_2O 1; EtOH 3; eth 3; bz 3 1.5076[20]
11779	Thiophene, 3-propyl- 3-Propylthiophene	$C_7H_{10}S$ 126.22	1518-75-8	163.2	5-17-01-00355 0.9738[20]	H_2O 1; EtOH 3; eth 3; bz 3 1.5057[20]
11780	2-Thiophenesulfonyl chloride	$C_4H_3ClO_2S_2$ 182.65	16629-19-9 28	100[6]	5-18-09-00426	eth 3
11781	Thiophene, tetrabromo- 2,3,4,5-Tetrabromothiophene	C_4Br_4S 399.73	3958-03-0 117.5	326	5-17-01-00309	H_2O 1; EtOH 3; eth 4
11782	Thiophene, tetrachloro-	C_4Cl_4S 221.92	6012-97-1 30.5	233.4	5-17-01-00305 1.7036[30]	H_2O 1; EtOH 4; eth 5 1.5915[30]
11783	Thiophene, tetrahydro- Thiacyclopentane	C_4H_8S 88.17	110-01-0 -96.1	121.0	5-17-01-00036 0.9987[20]	H_2O 1; EtOH 5; eth 5; ace 5 1.4871[18]
11784	Thiophene, tetrahydro-2,4-dimethyl-, 1,1-dioxide 2,4-Dimethylsulfolane	$C_6H_{12}O_2S$ 148.23	1003-78-7 -3	3245 281	5-17-01-00097 1.1362[20]	lig 4 1.4732[20]
11785	Thiophene, tetrahydro-2,5-dimethyl-, *cis*-	$C_6H_{12}S$ 116.23	5161-13-7 -89	142.3	5-17-01-00098 0.9222[20]	ace 4; bz 4; eth 4; EtOH 4 1.4799[20]
11786	Thiophene, tetrahydro-, 1,1-dioxide Sulfolane	$C_4H_8O_2S$ 120.17	126-33-0 27.6	8934 287.3	5-17-01-00039 1.2723[18]	chl 3 1.4833[18]
11787	Thiophene, tetrahydro-2-methyl-	$C_5H_{10}S$ 102.20	1795-09-1 -100.7	134	5-17-01-00080 0.9564[18]	H_2O 1; EtOH 4; eth 4; ace 4 1.4886[15]
11788	Thiophene, tetrahydro-3-methyl-	$C_5H_{10}S$ 102.20	4740-00-5	139	5-17-01-00082 0.9596[18]	H_2O 1; EtOH 4; eth 4; ace 4 1.4886[18]
11789	Thiophene, tetrahydro-3-methyl-, 1,1-dioxide 3-Methylsulfolane	$C_5H_{10}O_2S$ 134.20	872-93-5 1	276	5-17-01-00082 1.188[25]	1.4772[20]
11790	Thiophene, 2,3,5-tribromo- 2,3,5-Tribromothiophene	C_4HBr_3S 320.83	3141-24-0 29	260	5-17-01-00309	chl 3
11791	Thiophene, 2,3,5-trichloro- 2,3,5-Trichlorothiophene	C_4HCl_3S 187.48	17249-77-3 -16.1	198.7	5-17-01-00305 1.5856[20]	H_2O 1; EtOH 5; eth 5 1.5791[20]
11792	Thiophene, 2,3,4-trimethyl- 2,3,4-Trimethylthiophene	$C_7H_{10}S$ 126.22	1795-04-6 -26.2	162; 75[30]	5-17-01-00359 0.995[20]	1.5208[20]
11793	Thiophene, 2,3,5-trimethyl- 2,3,5-Trimethylthiophene	$C_7H_{10}S$ 126.22	1795-05-7	164.5	5-17-01-00359 0.9753[20]	H_2O 1; EtOH 4; eth 4; ace 4 1.5112[20]
11794	2(3H)-Thiophenone, dihydro-	C_4H_6OS 102.16	1003-10-7	39-40[1]	5-17-09-00012 1.18[25]	1.5230[20]
11795	2H-Thiopyran, tetrahydro- Thiacyclohexane	$C_5H_{10}S$ 102.20	1813-51-0 19	141.8	5-17-01-00068 0.9861[20]	H_2O 1; EtOH 3; eth 3; ace 3 1.5067[20]
11796	2H-Thiopyran, tetrahydro-2-methyl-	$C_6H_{12}S$ 116.23	5161-16-0 -58	153	5-17-01-00093 0.9428[20]	1.4905[20]
11797	2H-Thiopyran, tetrahydro-3-methyl-	$C_6H_{12}S$ 116.23	5258-50-4 -60	158	5-17-01-00094 0.9473[20]	1.4922[20]
11798	2H-Thiopyran, tetrahydro-4-methyl-	$C_6H_{12}S$ 116.23	5161-17-1 -28	158.6	5-17-01-00094 0.9471[20]	1.4923[20]
11799	Thiourea Thiocarbamide	CH_4N_2S 76.12	62-56-6 182	9299	4-03-00-00342 1.405[25]	H_2O 3; EtOH 3; eth 1
11800	Thiourea, N,N'-bis[4-(3-methylbutoxy)phenyl]- Tiocarlide	$C_{23}H_{32}N_2O_2S$ 400.59	910-86-1 146	9382	4-13-00-01187	
11801	Thiourea, (2-chlorophenyl)- 1-(2-Chlorophenyl)thiourea	$C_7H_7ClN_2S$ 186.66	5344-82-1 146		3-12-00-01296	bz 4; EtOH 4
11802	Thiourea, N-(3,5-dichlorophenyl)-N'-(4-fluorophenyl)- Loflucarban	$C_{13}H_9Cl_2FN_2S$ 315.20	790-69-2 163.5	5440	4-12-00-01275	
11803	Thiourea, N,N'-diethyl-	$C_5H_{12}N_2S$ 132.23	105-55-5 78	dec	4-04-00-00375	H_2O 3; EtOH 3; eth 4; ctc 2
11804	Thiourea, N,N'-dimethyl- N,N'-Dimethylthiourea	$C_3H_8N_2S$ 104.18	534-13-4 62	3250	4-04-00-00217	H_2O 4; EtOH 4; eth 2; ace 4
11805	Thiourea, N,N'-diphenyl- *sym*-Diphenylthiourea	$C_{13}H_{12}N_2S$ 228.32	102-08-9 154.5	3337	4-12-00-00810 1.32[25]	H_2O 2; EtOH 4; eth 4; chl 4
11806	Thiourea, (2-methylphenyl)- *o*-Tolylthiourea	$C_8H_{10}N_2S$ 166.25	614-78-8 182		4-12-00-01766	H_2O 4; EtOH 4; eth 2

No.	Name Synonym	Mol. Form. Mol. Wt.	CAS RN mp/°C	Merck No. bp/°C	Beil. Ref. den/g cm⁻³	Solubility n_D
11807	Thiourea, 1-naphthalenyl- ANTU	$C_{11}H_{10}N_2S$ 202.28	86-88-4 198	755	4-12-00-03086	H_2O 1; EtOH 2; eth 2; ace 2
11808	Thiourea, phenyl- Phenylthiourea	$C_7H_8N_2S$ 152.22	103-85-5 154	7286	4-12-00-00804	H_2O 2; EtOH 3; NaOH 3
11809	Thiourea, 2-propenyl- Thiosinamine	$C_4H_8N_2S$ 116.19	109-57-9 78	9293	4-04-00-01072 1.217[20]	H_2O 3; EtOH 3; eth 2; bz 1 1.5936[78]
11810	Thiourea, tetramethyl-	$C_5H_{12}N_2S$ 132.23	2782-91-4 79.3	245	4-04-00-00232	H_2O 3; EtOH 3; eth 2; chl 3
11811	9H-Thioxanthene Dibenzothiapyran	$C_{13}H_{10}S$ 198.29	261-31-4 128.5	9300 341	5-17-02-00254	chl 3
11812	9H-Thioxanthen-9-one Thioxanthone	$C_{13}H_8OS$ 212.27	492-22-8 209	9301 373	5-17-10-00437	H_2O 1; EtOH 2; bz 3; chl 3
11813	L-Threonine [R-(R*,S*)]-2-Amino-3-hydroxybutanoic acid	$C_4H_9NO_3$ 119.12	72-19-5 256 dec	9316	4-04-00-03171	H_2O 3; EtOH 1; eth 1; chl 1
11814	Thujane, (1S,4R,5S)-(+)- Sabinane (d)	$C_{10}H_{18}$ 138.25	7712-66-5 157		4-05-00-00318 0.8139[20]	1.4376[20]
11815	4(10)-Thujene, (+)- 4(10)-Thujene (d)	$C_{10}H_{16}$ 136.24	2514-91-2 164		4-05-00-00451 0.842[20]	ace 4; bz 4; eth 4; EtOH 4 1.4678[20]
11816	Thymidine Thymine 2-desoxyriboside	$C_{10}H_{14}N_2O_5$ 242.23	50-89-5 186.5	9331	5-24-07-00109	H_2O 3; EtOH 3; ace 3; chl 2
11817	Toluene, 3-chloro-2-iodo-	C_7H_6ClI 252.48	5100-98-1 27.5	123[14]	3-05-00-00726	1.608[23]
11818	o-Toluenesulfonic acid, thio-, S-phenyl ester Benzenesulfonic acid, thiolo,2-methyl,phenyl ester	$C_{13}H_{12}O_2S_2$ 264.37	96097-69-7 59.5	241.6	4-11-00-00237	H_2O 2; EtOH 4; eth 3; chl 2 1.53412[20]
11819	Triacontane	$C_{30}H_{62}$ 422.82	638-68-6 65.8	449.7	4-01-00-00592 0.8097[20]	H_2O 1; EtOH 2; eth 3; bz 4 1.4352[70]
11820	Triacontanoic acid	$C_{30}H_{60}O_2$ 452.81	506-50-3 93.6		4-02-00-01321	bz 4; CS_2 4; chl 4 1.4323[100]
11821	1-Triacontanol Myricyl alcohol	$C_{30}H_{62}O$ 438.82	593-50-0 88	9506	4-01-00-01918 0.777[95]	bz 4; eth 4; EtOH 4
11822	Triadimenol Mercury, chloro(2-methoxyethyl)-	C_3H_7ClHgO 295.13	123-88-6			
11823	Triallate Carbamothioic acid, bis(1-methylethyl)-, S-(2,3,3-trichloro-2-propenyl) ester	$C_{10}H_{16}Cl_3NOS$ 304.67	2303-17-5 29	9510 117[0.0003]	1.273[25]	
11824	Triasulfuron 2-(2-Chloroethoxy)-N-[[(4-methoxy-6-methyl-1,3,5-triaziN-2-yl)amino]carbonyl]benzene sulfonamide	$C_{14}H_{16}ClN_5O_5S$ 401.83	82097-50-5 186			
11825	3,5,7-Triaza-1-azoniatricyclo[3.3.1.13,7]decane, 1-(2-propenyl)-, iodide Hexamethylenetetramine, allyl iodide	$C_9H_{17}IN_4$ 308.17	36895-62-2	5880		
11826	1-Triazene, 1,3-diphenyl- Diazoaminobenzene	$C_{12}H_{11}N_3$ 197.24	136-35-6 98	2981	4-16-00-00904	H_2O 1; EtOH 4; eth 4; bz 4
11827	1,3,5-Triazine	$C_3H_3N_3$ 81.08	290-87-9 86	114	4-26-00-00063 1.38[25]	EtOH 3; eth 3
11828	1,3,5-Triazine-2,4-diamine, 6-chloro-N,N,N',N'-tetraethyl-	$C_{11}H_{20}ClN_5$ 257.77	580-48-3 27	154-6[9]	4-26-00-01208 1.0956[20]	bz 4; chl 4; EtOH 4; lig 4 1.5320[20]
11829	1,3,5-Triazine-2,4-diamine, 6-phenyl- Benzoguanamine	$C_9H_9N_5$ 187.20	91-76-9 226.5	1099	4-26-00-01244	EtOH 3; eth 3; tfa 2
11830	1,3,5-Triazine-2,4-diamine, N-phenyl- Amanozine	$C_9H_9N_5$ 187.20	537-17-7 235.5	379	4-26-00-01195	
11831	1,2,4-Triazine-3,5(2H,4H)-dione, 6-methyl- 6-Azathymine	$C_4H_5N_3O_2$ 127.10	932-53-6 211	919	4-26-00-00556	H_2O 3; EtOH 3; ace 3
11832	1,2,4-Triazine-3,5(2H,4H)-dione, 2-β-D-ribofuranosyl- 6-Azauridine	$C_8H_{11}N_3O_6$ 245.19	54-25-1 158	920	4-26-00-00554	H_2O 3
11833	1,3,5-Triazine, hexahydro-1,3,5-trinitro- Cyclonite	$C_3H_6N_6O_6$ 222.12	121-82-4 205.5	2742	4-26-00-00022 1.82[20]	H_2O 1; EtOH 1; eth 2; ace 3
11834	1,3,5-Triazine, hexahydro-1,3,5-triphenyl-	$C_{21}H_{21}N_3$ 315.42	91-78-1 144	185; 60[29]	4-26-00-00007	H_2O 1; EtOH 2; eth 3; ace 3
11835	1,3,5-Triazine-2,4,6-triamine Melamine	$C_3H_6N_6$ 126.12	108-78-1 345 dec	5691 sub	4-26-00-01253 1.573[16]	H_2O 2; EtOH 2; eth 1 1.872[20]
11836	1,3,5-Triazine-2,4,6-tricarbonitrile	C_6N_6 156.11	7615-57-8 119	262; 119[1.0]	1-26-00-00091	bz 4
11837	1,3,5-Triazine, 2,4,6-trichloro- Cyanuric acid trichloride	$C_3Cl_3N_3$ 184.41	108-77-0 154	192	4-26-00-00066	EtOH 4
11838	1,3,5-Triazine-2,4,6(1H,3H,5H)-trione, 1,3-dichloro- Dichlorocyanuric acid	$C_3HCl_2N_3O_3$ 197.96	2782-57-2			
11839	1,3,5-Triazine-2,4,6(1H,3H,5H)-trione, dihydrate Cyanuric acid, dihydrate	$C_3H_7N_3O_5$ 165.11	6202-04-6 >360 dec	dec	4-26-00-00632 2.50[20]	H_2O 3; EtOH 2; eth 1; ace 1 2.500[20]
11840	1,3,5-Triazine-2,4,6(1H,3H,5H)-trione, 1,3,5-trichloro- Symclosene	$C_3Cl_3N_3O_3$ 232.41	87-90-1 246.7 dec	8993	4-26-00-00642	

No.	Name / Synonym	Mol. Form. / Mol. Wt.	CAS RN / mp/°C	Merck No. / bp/°C	Beil. Ref. / den/g cm^{-3}	Solubility / n_D
11841	1,3,5-Triazine-2,4,6(1H,3H,5H)-trione, 1,3,5-trimethyl-	$C_6H_9N_3O_3$ 171.16	827-16-7 176.5	274	4-26-00-00635	H_2O 1; EtOH 3
11842	1,3,5-Triazine-2,4,6(1H,3H,5H)-trione, 1,3,5-tri-2-propenyl-	$C_{12}H_{15}N_3O_3$ 249.27	1025-15-6 20.5	149[4]; 105[0.5]	4-26-00-00637 1.1590[20]	
11843	1,3,5-Triazine, 2,4,6-tri-2-pyridinyl- 2,4,6-Tripyridyl-s-triazine	$C_{18}H_{12}N_6$ 312.33	3682-35-7 210	9664	4-26-00-04192	
11844	1,3,5-Triazine, 2,4,6-tris(1-aziridinyl)- Triethylenemelamine	$C_9H_{12}N_6$ 204.23	51-18-3	9586	4-26-00-01268	
11845	1H-1,2,4-Triazol-3-amine Amitrole	$C_2H_4N_4$ 84.08	61-82-5 159	506	4-26-00-01073	H_2O 4; EtOH 4; eth 1; ace 1
11846	1H-1,2,3-Triazole	$C_2H_3N_3$ 69.07	288-36-8 23	204	4-26-00-00029 1.1861[25]	H_2O 3; eth 3; ace 3; os 3 1.4854[25]
11847	1H-1,2,4-Triazole, 1-butyl 1-Butyl-s-triazole	$C_6H_{11}N_3$ 125.17	6086-22-2			
11848	4H-1,2,4-Triazole, 4-cyclohexyl-3-ethyl- Hexazole	$C_{10}H_{17}N_3$ 179.27	4671-03-8 89.5	4618 227[10]	4-26-00-00049	H_2O 4; bz 4; chl 4
11849	1H-1,2,3-Triazole, 1,5-dimethyl- 1,5-Dimethyl-1,2,3-triazole	$C_4H_7N_3$ 97.12	15922-53-9 -4	255	1-26-00-00005	H_2O 4
11850	1,2,4-Triazolidine-3,5-dione Urazole	$C_2H_3N_3O_2$ 101.06	3232-84-6 249-50 dec	9780	4-26-00-00538	
11851	7H-1,2,3-Triazolo[4,5-d]pyrimidin-7-one, 5-amino-1,4-dihydro- 8-Azaguanine	$C_4H_4N_6O$ 152.12	134-58-7 300	912	4-26-00-04171	
11852	S,S,S-Tributyl phosphorotrithioate S,S,S-Tributyl trithiophosphate	$C_{12}H_{27}OPS_3$ 314.52	78-48-8 <-25	150[0.3]	1.057[20]	
11853	Trichlorfon Phosphonic acid, (2,2,2-trichloro-1-hydroxyethyl)-, dimethyl ester	$C_4H_8Cl_3O_4P$ 257.44	52-68-6 77	9536 100[0.1]	1.73[20]	
11854	Triclopyr Acetic acid, [(3,5,6-trichloro-2-pyridinyl)oxy]-	$C_7H_4Cl_3NO_3$ 256.47	55335-06-3 149	9572 dec 290		
11855	Tricosane	$C_{23}H_{48}$ 324.63	638-67-5 47.6	380	4-01-00-00576 0.7785[48]	H_2O 1; EtOH 2; eth 3; ctc 3 1.4468[20]
11856	Tricosane, 2-methyl- 2-Methyltricosane	$C_{24}H_{50}$ 338.66	1928-30-9 37.6	205[3]	3-01-00-00576 0.7539[90]	1.4201[90]
11857	12-Tricosanone Diundecyl ketone	$C_{23}H_{46}O$ 338.62	540-09-0 70.2		4-01-00-03408 0.8086[69]	bz 4; eth 4; chl 4 1.4283[80]
11858	Tricyclazole 1,2,4-Triazolo[3,4-b]benzothiazole, 5-methyl-	$C_9H_7N_3S$ 189.24	41814-78-2 187			
11859	Tricyclo[3.3.1.1³,⁷]decan-1-amine Amantadine	$C_{10}H_{17}N$ 151.25	768-94-5 180	380		H_2O 2
11860	Tricyclo[3.3.1.1³,⁷]decane Adamantane	$C_{10}H_{16}$ 136.24	281-23-2 268	140 sub	4-05-00-00469 1.07[25]	bz 3; ctc 3 1.568
11861	Tricyclo[4.4.0.0²,⁷]dec-3-ene, 1,3-dimethyl-8-(1-methylethyl)-, stereoisomer Copaene	$C_{15}H_{24}$ 204.36	3856-25-5	2510 248.5	4-05-00-01189 0.8996[20]	H_2O 1; eth 3; ace 3; HOAc 3 1.4894[20]
11862	Tricyclo[2.2.1.0²,⁶]heptane-1-carboxylic acid, 7,7-dimethyl- Tricyclenic acid	$C_{10}H_{14}O_2$ 166.22	512-60-7 151.3	263	4-09-00-00246	bz 4; EtOH 4; peth 4
11863	Tricyclo[2.2.1.0²,⁶]heptane-1-carboxylic acid, 7,7-dimethyl-, ethyl ester Tricyclenic acid, ethyl ester	$C_{12}H_{18}O_2$ 194.27	56694-98-5	100-1[10]	2-09-00-00064 1.0143[20]	1.4230[20]
11864	Tricyclo[2.2.1.0²,⁶]heptane-1-carboxylic acid, 7,7-dimethyl-, methyl ester Tricyclenic acid, methyl ester	$C_{11}H_{16}O_2$ 180.25	56694-97-4 45.5	99[14]	2-09-00-00064 1.0255[42]	1.4695[42]
11865	Tricyclo[2.2.1.0²,⁶]heptane, 3,3-dimethyl-	C_9H_{14} 122.21	473-02-9 42.5	138	4-05-00-00430 0.8710[40]	bz 4; eth 4; EtOH 4 1.4514[40]
11866	Tricyclo[2.2.1.0²,⁶]heptane, 1,3,3-trimethyl- Cyclofenchene	$C_{10}H_{16}$ 136.24	488-97-1	145	3-05-00-00392 0.859[20]	1.4503[22]
11867	Tricyclo[2.2.1.0²,⁶]heptane, 1,7,7-trimethyl-	$C_{10}H_{16}$ 136.24	508-32-7 67.5	152.5	4-05-00-00468 0.8668[80]	1.4296[80]
11868	1,12-Tridecadiyne	$C_{13}H_{20}$ 176.30	38628-39-6 -3	115.5[12]	2-01-00-00248 0.8262[21]	1.454[20]
11869	Tridecanal	$C_{13}H_{26}O$ 198.35	10486-19-8 14	156[13]	4-01-00-03386 0.8356[18]	H_2O 1; EtOH 3; os 4 1.4384[18]
11870	1-Tridecanamine Tridecane, 1-amino	$C_{13}H_{29}N$ 199.38	2869-34-3 27.4	275.8	4-04-00-00810 0.8049[20]	H_2O 2; EtOH 3; eth 3 1.4443[20]
11871	Tridecane	$C_{13}H_{28}$ 184.37	629-50-5 -5.3	235.4	4-01-00-00512 0.7564[20]	H_2O 1; EtOH 4; eth 4; ctc 3 1.4256[20]
11872	Tridecane, 1-bromo-	$C_{13}H_{27}Br$ 263.26	765-09-3 6.2	296	4-01-00-00514 1.0177[20]	H_2O 1; chl 4 1.4593[20]
11873	Tridecane, 1-bromo-12-methyl- Tridecane, 2-methyl-13-bromo	$C_{14}H_{29}Br$ 277.29	111772-90-8	120-2[3]	3-01-00-00551 1.0241[20]	1.4598[20]
11874	Tridecane, 1,13-dibromo- Tridecamethylene dibromide	$C_{13}H_{26}Br_2$ 342.16	31772-05-1 9	188-92[12]	3-01-00-00548 1.276[15]	eth 4; chl 4 1.4880[27]
11875	1,12-Tridecanediol	$C_{13}H_{28}O_2$ 216.36	99706-68-0 60.5	188[8]	2-01-00-00563	EtOH 4
11876	1,13-Tridecanediol	$C_{13}H_{28}O_2$ 216.36	13362-52-2 76.5	195[10]	4-01-00-02630	EtOH 4

No.	Name / Synonym	Mol. Form. / Mol. Wt.	CAS RN / mp/°C	Merck No. / bp/°C	Beil. Ref. / den/g cm⁻³	Solubility / n_D
11877	Tridecane, 2,2,4,10,12,12-hexamethyl-7-(3,5,5-trimethylhexyl)-	$C_{28}H_{58}$ 394.77	3035-75-4	$166^{0.5}$	4-01-00-00590 0.8137^{20}	1.4558^{20}
11878	Tridecane, 7-hexyl-	$C_{19}H_{40}$ 268.53	7225-66-3 -28.3	170.5^{10}	4-01-00-00562 0.7877^{20}	1.4409^{20}
11879	Tridecane, 7-methyl- 7-Methyltridecane	$C_{14}H_{30}$ 198.39	26730-14-3 -37.2	115.5^{10}	4-01-00-00525 0.7634^{20}	1.4291^{20}
11880	Tridecanenitrile	$C_{13}H_{25}N$ 195.35	629-60-7 9.7	293	3-02-00-00906 0.8257^{20}	EtOH 4; eth 4 1.4378^{20}
11881	Tridecanoic acid	$C_{13}H_{26}O_2$ 214.35	638-53-9 41.5	236^{100}; 140^1	4-02-00-01117 0.8458^{80}	H_2O 1; EtOH 4; eth 4; ace 3 1.4286^{60}
11882	Tridecanoic acid, methyl ester Methyl tridecanoate	$C_{14}H_{28}O_2$ 228.38	1731-88-0 6.5	92^1	4-02-00-01118	EtOH 5; ctc 3 1.4405^{20}
11883	1-Tridecanol Tridecyl alcohol	$C_{13}H_{28}O$ 200.36	112-70-9 32.5	152^{14}	4-01-00-01860 0.8223^{31}	H_2O 1; EtOH 3; eth 3
11884	2-Tridecanol 2-Hydroxytridecane	$C_{13}H_{28}O$ 200.36	1653-31-2 23	161^{30}; 145^{12}	4-01-00-01861	bz 4; eth 4; EtOH 4 1.4188^{70}
11885	3-Tridecanol	$C_{13}H_{28}O$ 200.36	10289-68-6 32	139^{12}	4-01-00-01861 0.8139^{46}	
11886	1-Tridecanol, 12-methyl- 12-Methyl-1-tridecanol	$C_{14}H_{30}O$ 214.39	21987-21-3 10.5	$145-50^6$	4-01-00-01869 0.8414^{20}	1.4437^{20}
11887	3-Tridecanol, 2-methyl-	$C_{14}H_{30}O$ 214.39	98930-89-3 20	274	1-01-00-00219 0.8390^{20}	1.4460^{20}
11888	4-Tridecanol, 2-methyl Acetic acid, trifluoro-,(1,2-dimethyl-1-ethylbutyl) ester	$C_{14}H_{30}O$ 214.39	36691-77-7	$147-8^{14}$	4-01-00-01869	chl 2 1.4404^{25}
11889	2-Tridecanone Methyl undecyl ketone	$C_{13}H_{26}O$ 198.35	593-08-8 30.5	263	4-01-00-03386 0.8217^{30}	H_2O 1; EtOH 4; eth 4; ace 4 1.4318^{20}
11890	3-Tridecanone	$C_{13}H_{26}O$ 198.35	1534-26-5 31	140^{17}	4-01-00-03387	H_2O 1; ace 3
11891	7-Tridecanone Dihexyl ketone	$C_{13}H_{26}O$ 198.35	462-18-0 33	261	4-01-00-03315 0.825^{30}	EtOH 3; eth 4; chl 3; lig 3
11892	3-Tridecanone, 2-methyl- Decyl isopropyl ketone	$C_{14}H_{28}O$ 212.38	40239-35-8 47	267; 142^{17}	4-01-00-03390 0.8314^{20}	
11893	1-Tridecene	$C_{13}H_{26}$ 182.35	2437-56-1 -13	232.8	4-01-00-00921 0.7658^{20}	H_2O 1; EtOH 4; eth 4; bz 3 1.4340^{20}
11894	2-Tridecenoic acid	$C_{13}H_{24}O_2$ 212.33	6969-16-0	$167-71^2$	4-02-00-01622 0.8995^{30}	1.4612^{20}
11895	12-Tridecenoic acid (10-Undecenyl)acetic acid	$C_{13}H_{24}O_2$ 212.33	6006-06-0 38.5	192^{20}	2-02-00-00422	bz 4; eth 4; EtOH 4; peth 4
11896	12-Tridecenoic acid, methyl ester	$C_{14}H_{26}O_2$ 226.36	29780-00-5	143^8	4-02-00-01622 0.8803^{20}	1.4438^{20}
11897	1-Tridecyne	$C_{13}H_{24}$ 180.33	26186-02-7	94.5^{25}	4-01-00-01069 0.7729^{20}	bz 4; eth 4 1.4309^{20}
11898	Tridemorph Morpholine, 2,6-dimethyl-4-tridecyl-	$C_{19}H_{39}NO$ 297.52	24602-86-6	9576 $141^{1.3}$	0.86	
11899	Tridiphane (±)-2-(3,5-Dichlorophenyl)-2-(2,2,2-trichloroethyl)oxirane	$C_{10}H_7Cl_5O$ 320.43	58138-08-2 42.8			
11900	Triflumizole (E)-1-[1-[[4-Chloro-2-(trifluoromethyl)phenyl]imino]-2-propoxyethyl]-1H-imidazole	$C_{15}H_{15}ClF_3N_3O$ 345.75	68694-11-1 63.5			
11901	Trifluralin Benzenamine, 2,6-dinitro-N,N-dipropyl-4-(trifluoromethyl)-	$C_{13}H_{16}F_3N_3O_4$ 335.28	1582-09-8 49	9598 $140^{4.2}$		
11902	Triforine Formamide, N,N '-[1,4-piperazinediylbis(2,2,2-trichloroethylidene)]bis-	$C_{10}H_{14}Cl_6N_4O_2$ 434.96	26644-46-2 155 dec	9602		
11903	2,5,7-Trioxabicyclo[2.2.1]heptane, 1,3,3,4,6,6-hexamethyl- Bicyclo[2.2.1]heptane, 1,3,3,4,6,6-hexamethyl-2,5,7-trioxa	$C_{10}H_{18}O_3$ 186.25	7045-89-8	166	5-19-09-00139 0.9702^{17}	H_2O 1
11904	2H,4aH-1,4,5-Trioxadicyclopent[a,hi]indene-7-carboxylic acid, 3-ethylidene-3,3a,7a,9b-tetrahydro-2-oxo-, methyl ester, [3aS-(3E,3aα,4aβ,7aβ,9aR*,9bβ)]- Plumericin	$C_{15}H_{14}O_6$ 290.27	77-16-7	7512	5-19-11-00097	chl 3
11905	1,3,5-Trioxane Formaldehyde, trimer	$C_3H_6O_3$ 90.08	110-88-3 60.2	9646 114.5	5-19-09-00103 1.17^{65}	H_2O 4; EtOH 3; eth 3; bz 3
11906	1,3,5-Trioxane, 2,4,6-triethenyl- Metacrolein	$C_9H_{12}O_3$ 168.19	588-01-2 50	170	5-19-09-00165	eth 4; EtOH 4
11907	1,3,5-Trioxane-2,4,6-triimine Cyamelide	$C_3H_3N_3O_3$ 129.08	462-02-2 dec	2689 dec	3-03-00-00030 1.127^{15}	eth 4; EtOH 4
11908	1,3,5-Trioxane, 2,4,6-trimethyl- Paraldehyde	$C_6H_{12}O_3$ 132.16	123-63-7 12.6	6975 124.3	5-19-09-00112 0.9943^{20}	H_2O 2; EtOH 5; eth 5; chl 5 1.4049^{20}

No.	Name / Synonym	Mol. Form. / Mol. Wt.	CAS RN / mp/°C	Merck No. / bp/°C	Beil. Ref. / den/g cm^{-3}	Solubility / n_D
11909	2,6,7-Trioxa-1-phosphabicyclo[2.2.2]octane, 4-ethyl- / Trimethylolpropane phosphite	$C_6H_{11}O_3P$ / 162.13	824-11-3 / 53.7			chl 3
11910	Triphenylene / 9,10-Benzophenanthrene	$C_{18}H_{12}$ / 228.29	217-59-4 / 199	9654 / 425	4-05-00-02556	H_2O 1; EtOH 3; bz 4; chl 4
11911	Triphenyltin hydroxide / Stannane, hydroxytriphenyl-	$C_{18}H_{16}OSn$ / 367.03	76-87-9 / 119	9659	1.54[20]	
11912	Tripropylamine, eicosafluoro-2-(trifluoromethyl)- / Dipropylisobutylamine, perfluoro	$C_{10}F_{23}N$ / 571.08	307-11-9 /	/ 147	4-02-00-00858 / 1.84[25]	eth 4; EtOH 4 / 1.283[25]
11913	Trisiloxane, 1,5-dichloro-1,1,3,3,5,5-hexamethyl-	$C_6H_{18}Cl_2O_2Si_3$ / 277.37	3582-71-6 / -53	/ 184	4-04-00-04123 / 1.018[20]	H_2O 6
11914	Trisiloxane, octamethyl- / Octamethyltrisiloxane	$C_8H_{24}O_2Si_3$ / 236.53	107-51-7 / -80	6669 / 153; 50-2[17]	4-04-00-04115 / 0.8200[20]	EtOH 2; bz 3; peth 3 / 1.3840[20]
11915	Trisiloxane, 1,1,3,5,5-pentamethyl-1,3,5-triphenyl- / 1,1,3,5,5-Pentamethyl-1,3,5-triphenyltrisiloxane	$C_{23}H_{30}O_2Si_3$ / 422.75	80-14-8 /	/ 169[0.7]	3-16-00-01210 / 1.0227[20]	bz 4; lig 4 / 1.5280[20]
11916	Trisulfide, diethyl	$C_4H_{10}S_3$ / 154.32	3600-24-6 /	/ 96-7[26]	4-01-00-01398 / 1.114[20]	1.5689[13]
11917	Trisulfide, di-2-propenyl	$C_6H_{10}S_3$ / 178.34	2050-87-5 /	/ 112-22[16]	0-01-00-00441 / 1.0845[15]	eth 4
11918	Tritetracontane	$C_{43}H_{88}$ / 605.17	7098-21-7 / 85.5	/ 332[3]	3-01-00-00593 / 0.7812[90]	1.4340[90]
11919	1,3,5-Trithiane	$C_3H_6S_3$ / 138.28	291-21-4 / 220	/ sub	5-19-00-00105 / 1.6374[24]	H_2O 2; EtOH 2; eth 2; bz 3
11920	1,3,5-Trithiane, 2,4,6-trimethyl- / Thioacetaldehyde trimer	$C_6H_{12}S_3$ / 180.36	2765-04-0 /	9245	4-19-00-04718	
11921	1,3,5-Trithiane, 2,4,6-trimethyl-, (2α,4α,6β)-	$C_6H_{12}S_3$ / 180.36	23769-39-3 / 101	/ 246.5	2-19-00-00399	H_2O 1; EtOH 3; eth 3; ace 3
11922	L-Tryptophan / l-α-Aminoindole-3-propionic acid	$C_{11}H_{12}N_2O_2$ / 204.23	73-22-3 / 289	9707	5-22-14-00014	H_2O 2; EtOH 3; eth 1; chl 1
11923	L-Tryptophan, N-methyl- / L-Abrine	$C_{12}H_{14}N_2O_2$ / 218.26	526-31-8 / 295 dec	5	5-22-14-00040	H_2O 2; MeOH 2; eth 1; alk 3
11924	Tubocuraran-7',12'-diol, 6,6'-dimethoxy-2,2'-dimethyl-, (1'α)- / Bebeerine	$C_{36}H_{38}N_2O_6$ / 594.71	477-60-1 / 221	1024	2-27-00-00896	EtOH 3; eth 3; ace 4; chl 4
11925	Tubocuraran-7',12'-diol, 6,6'-dimethoxy-2,2'-dimethyl-, (1β)- / Curine	$C_{36}H_{38}N_2O_6$ / 594.71	436-05-5 / 221	2682	4-27-00-08725	ace 4; bz 4; py 4
11926	L-Tyrosine / L-Phenylalanine, 4-hydroxy-	$C_9H_{11}NO_3$ / 181.19	60-18-4 / 343 dec	9747 / sub	4-14-00-02264	H_2O 2; EtOH 1; eth 1; HOAc 2
11927	L-Tyrosine, 3,5-dibromo- / Tyrosine, 3,5-dibromo(L)	$C_9H_9Br_2NO_3$ / 338.98	300-38-9 / 245	3014	4-14-00-02365	H_2O 2; EtOH 2; eth 1; alk 3
11928	Tyrosine, 3,5-diiodo- / 3,5-Diiodotyrosine	$C_9H_9I_2NO_3$ / 432.98	66-02-4 / 213 dec	3177	2-14-00-00384	H_2O 2; EtOH 1; eth 1; bz 1
11929	L-Tyrosine, 3-hydroxy- / Levodopa	$C_9H_{11}NO_4$ / 197.19	59-92-7 / 277 dec	5344	4-14-00-02492	H_2O 3; EtOH 1; eth 1; ace 1
11930	L-Tyrosine, O-(4-hydroxy-3,5-diiodophenyl)-3,5-diiodo- / Thyroxine	$C_{15}H_{11}I_4NO_4$ / 776.87	51-48-9 / 235.5	9348	4-14-00-02373	H_2O 2; EtOH 1; bz 1
11931	L-Tyrosine, 3-hydroxy-α-methyl- / Methyldopa	$C_{10}H_{13}NO_4$ / 211.22	555-30-6 / 300 dec	5974		
11932	L-Tyrosine, 3-iodo- / 3-Iodotyrosine	$C_9H_{10}INO_3$ / 307.09	70-78-0 / 204-6 dec	4937	4-14-00-02369	
11933	L-Tyrosine, N-methyl- / Surinamine	$C_{10}H_{13}NO_3$ / 195.22	537-49-5 / 293	8988	3-14-00-01513	
11934	1,10-Undecadiyne	$C_{11}H_{16}$ / 148.25	4117-15-1 / -17	/ 83[12]	2-01-00-00248 / 0.8182[21]	ace 4; bz 4 / 1.453[21]
11935	Undecanal	$C_{11}H_{22}O$ / 170.30	112-44-7 / -4	/ 117[18]	4-01-00-03374 / 0.8251[23]	H_2O 1; EtOH 3; eth 3 / 1.4520[20]
11936	Undecanal, 2-methyl-	$C_{12}H_{24}O$ / 184.32	110-41-8 /	/ 114[10]	4-01-00-03383 / 0.830[15]	H_2O 2; EtOH 3; eth 3 / 1.4321[20]
11937	1-Undecanamine	$C_{11}H_{25}N$ / 171.33	7307-55-3 / 17	/ 242	4-04-00-00792 / 0.7979[20]	H_2O 3; EtOH 3; eth 1; ctc 2 / 1.4398[20]
11938	Undecane / Hendecane	$C_{11}H_{24}$ / 156.31	1120-21-4 / -25.6	/ 195.9	4-01-00-00487 / 0.7402[20]	H_2O 1; EtOH 5; eth 5 / 1.4398[20]
11939	Undecane, 1-bromo-11-fluoro- / 1-Bromo-11-fluoroundecane	$C_{11}H_{22}BrF$ / 253.20	463-33-2 /	/ 95[0.6]	4-01-00-00490	eth 4; EtOH 4 / 1.4518[25]
11940	Undecane, 1,11-dibromo- / 1,11-Dibromoundecane	$C_{11}H_{22}Br_2$ / 314.10	16696-65-4 / 166	/ 170-5[15]	3-01-00-00536 / 1.332[15]	
11941	Undecanenitrile / Decyl cyanide	$C_{11}H_{21}N$ / 167.29	2244-07-7 /	/ 253	3-02-00-00860 / 0.8254[30]	H_2O 1; EtOH 3; eth 3; ctc 3 / 1.4293[30]
11942	1-Undecanethiol / Undecyl mercaptan	$C_{11}H_{24}S$ / 188.38	5332-52-5 / -3	/ 257.4	4-01-00-01838 / 0.8448[20]	1.4585[20]
11943	Undecanoic acid	$C_{11}H_{22}O_2$ / 186.29	112-37-8 / 28.6	/ 280	4-02-00-01068 / 0.8907[20]	H_2O 1; EtOH 4; eth 3; ace 4 / 1.4294[45]

No.	Name / Synonym	Mol. Form. / Mol. Wt.	CAS RN / mp/°C	Merck No. / bp/°C	Beil. Ref. / den/g cm^{-3}	Solubility / n_D
11944	Undecanoic acid, 11-bromo- / 11-Bromoundecanoic acid	$C_{11}H_{21}BrO_2$ / 265.19	2834-05-1 / 57	188[18]	4-02-00-01073	ace 4; bz 4; eth 4; EtOH 4
11945	Undecanoic acid, 2-bromo- / 2-Bromoundecanoic acid	$C_{11}H_{21}BrO_2$ / 265.19	2623-84-9 / 10	180[14]	4-02-00-01073 / 1.1586[24]	
11946	Undecanoic acid, ethyl ester / Ethyl undecylate	$C_{13}H_{26}O_2$ / 214.35	627-90-7 / -15	131[14]	3-02-00-00858 / 0.8633[20]	H_2O 1; EtOH 3; eth 3; ace 3 / 1.4285[20]
11947	Undecanoic acid, 11-fluoro-	$C_{11}H_{21}FO_2$ / 204.28	463-17-2 / 36	113[0.25]	4-02-00-01071	eth 4; EtOH 4; lig 4
11948	Undecanoic acid, 11-hydroxy- / 11-Hydroxyundecanoic acid	$C_{11}H_{22}O_3$ / 202.29	3669-80-5 / 65.5	148-50[4]	4-03-00-00903	eth 4; EtOH 4
11949	Undecanoic acid, 11-hydroxy-, methyl ester / Methyl 11-hydroxyhendecanoate	$C_{12}H_{24}O_3$ / 216.32	24724-07-0 / 27.5	168[8]	2-03-00-00243 / 0.9542[20]	bz 4; eth 4; EtOH 4
11950	1-Undecanol	$C_{11}H_{24}O$ / 172.31	112-42-5 / 19	243	4-01-00-01835 / 0.8298[20]	H_2O 1; EtOH 3; eth 4 / 1.4392[20]
11951	2-Undecanol, (±)-	$C_{11}H_{24}O$ / 172.31	113666-64-1 / 12	228	4-01-00-01838 / 0.8268[19]	H_2O 1; EtOH 3; eth 3 / 1.4369[20]
11952	3-Undecanol, (R)-	$C_{11}H_{24}O$ / 172.31	107494-37-1 / 17	229	2-01-00-00462 / 0.8295[20]	bz 4; eth 4; EtOH 4 / 1.4367[20]
11953	5-Undecanol / 1-Butylheptyl alcohol	$C_{11}H_{24}O$ / 172.31	37493-70-2 / -3.5	229	4-01-00-01839 / 0.8292[20]	eth 4; EtOH 4 / 1.4354[24]
11954	6-Undecanol	$C_{11}H_{24}O$ / 172.31	23708-56-7 / 25	228	4-01-00-01839 / 0.8334[20]	H_2O 1; EtOH 3; ace 3 / 1.4374[20]
11955	1-Undecanol, 2-methyl- / 2-Methyl-1-undecanol	$C_{12}H_{26}O$ / 186.34	10522-26-6	129-31[12]	4-01-00-01855 / 0.830[15]	eth 4; EtOH 4 / 1.4382[20]
11956	3-Undecanol, 2-methyl- / 3-Hydroxy-2-methylundecane	$C_{12}H_{26}O$ / 186.34	60671-36-5	135[26]	1-01-00-00217 / 0.8327[20]	1.4405[20]
11957	5-Undecanol, 2-methyl-	$C_{12}H_{26}O$ / 186.34	33978-71-1	130-3[20]	4-01-00-01855 / 0.8235[20]	eth 4; EtOH 4 / 1.4346[20]
11958	2-Undecanone / Methyl nonyl ketone	$C_{11}H_{22}O$ / 170.30	112-12-9 / 15	231.5	4-01-00-03374 / 0.8250[20]	H_2O 1; EtOH 3; eth 3; ace 3 / 1.4291[20]
11959	3-Undecanone / Ethyl octyl ketone	$C_{11}H_{22}O$ / 170.30	2216-87-7 / 12	227	4-01-00-03375 / 0.8272[20]	H_2O 1; EtOH 3; eth 3; ctc 2 / 1.4296[20]
11960	4-Undecanone / Heptyl propyl ketone	$C_{11}H_{22}O$ / 170.30	14476-37-0 / 4.5	106[13]	4-01-00-03375 / 0.8274[25]	eth 4; EtOH 4 / 1.4248[24]
11961	5-Undecanone / Butyl hexyl ketone	$C_{11}H_{22}O$ / 170.30	33083-83-9 / 2	227	4-01-00-03375 / 0.8278[19]	H_2O 1; EtOH 3; eth 3; ctc 3 / 1.4275[18]
11962	6-Undecanone / Diamyl ketone	$C_{11}H_{22}O$ / 170.30	927-49-1 / 14.5	228	4-01-00-03376 / 0.8308[20]	H_2O 1; EtOH 4; eth 4 / 1.4270[20]
11963	4-Undecanone, 7-ethyl-2-methyl-	$C_{14}H_{28}O$ / 212.38	6976-00-7	252.5	3-01-00-02920 / 0.8362[20]	H_2O 1; EtOH 4; eth 4; ace 4 / 1.4370[20]
11964	Undecasiloxane, tetracosamethyl- / Tetracosamethylhendecasiloxane	$C_{24}H_{72}O_{10}Si_{11}$ / 829.77	107-53-9	9128 / 322.8; 202[47]	3-04-00-01881 / 0.9247[25]	bz 4 / 1.3994[20]
11965	3,5,9-Undecatrien-2-one, 6,10-dimethyl- / Pseudoionone	$C_{13}H_{20}O$ / 192.30	141-10-6	7931 / 143-5[12]	3-01-00-03067 / 0.8984[20]	EtOH 3; eth 3; chl 3; MeOH 3 / 1.5335[20]
11966	1-Undecene	$C_{11}H_{22}$ / 154.30	821-95-4 / -49.2	192.7	4-01-00-00910 / 0.7503[20]	H_2O 1; eth 3; chl 3; lig 3 / 1.4261[20]
11967	3-Undecene	$C_{11}H_{22}$ / 154.30	60669-40-1 / -62	193.5	0.7516[20]	1.4290[20]
11968	2-Undecene, (E)-	$C_{11}H_{22}$ / 154.30	693-61-8 / -48.3	192.5	4-01-00-00911 / 0.7528[20]	1.4292[20]
11969	2-Undecene, (Z)-	$C_{11}H_{22}$ / 154.30	821-96-5 / -66.5	196.1	0.7576[20]	
11970	4-Undecene, (E)-	$C_{11}H_{22}$ / 154.30	693-62-9 / -63.7	193	0.7508[20]	1.4285[20]
11971	4-Undecene, (Z)-	$C_{11}H_{22}$ / 154.30	821-98-7 / -97	192.6	0.7541[20]	1.4302[20]
11972	5-Undecene, (E)-	$C_{11}H_{22}$ / 154.30	764-97-6 / -61.1	192	4-01-00-00911 / 0.7497[20]	eth 4; chl 4; lig 4 / 1.4285[20]
11973	5-Undecene, (Z)-	$C_{11}H_{22}$ / 154.30	764-96-5 / -106.5	192.3	4-01-00-00911 / 0.7537[20]	1.4302[20]
11974	10-Undecenoic acid / Undecylenic acid	$C_{11}H_{20}O_2$ / 184.28	112-38-9 / 24.5	9760 / 275	4-02-00-01612 / 0.9072[24]	H_2O 1; EtOH 3; eth 3; ctc 2 / 1.4486[24]
11975	9-Undecenoic acid, (E)- / trans-9-Undecenoic acid	$C_{11}H_{20}O_2$ / 184.28	37973-84-5 / 19	273	4-02-00-01616 / 0.9118[25]	eth 4 / 1.4519[20]
11976	9-Undecenoic acid, (Z)-	$C_{11}H_{20}O_2$ / 184.28	116836-30-7 / 2.5	156[4]; 119[0.8]	4-02-00-01616 / 0.9150[20]	1.4530[20]
11977	10-Undecenoic acid, ethyl ester	$C_{13}H_{24}O_2$ / 212.33	692-86-4 / -38	264.5	4-02-00-01614 / 0.8827[15]	H_2O 1; EtOH 3; eth 3; ctc 2 / 1.4449[25]
11978	10-Undecenoic acid, methyl ester	$C_{12}H_{22}O_2$ / 198.31	111-81-9 / -27.5	248	4-02-00-01613 / 0.889[15]	H_2O 1; EtOH 3; eth 3; ctc 2 / 1.4393[20]
11979	10-Undecen-1-ol	$C_{11}H_{22}O$ / 170.30	112-43-6 / -2	250	4-01-00-02194 / 0.8495[15]	H_2O 1; EtOH 3; eth 3; ctc 2 / 1.4500[20]
11980	2-Undecen-4-ol	$C_{11}H_{22}O$ / 170.30	22381-86-8	114-5[13]	4-01-00-02194 / 0.843[22]	1.4485[22]
11981	9-Undecen-1-ol	$C_{11}H_{22}O$ / 170.30	112-46-9 / 6	248.5	3-01-00-01957 / 0.8507[15]	1.4535[19]
11982	10-Undecenoyl chloride	$C_{11}H_{19}ClO$ / 202.72	38460-95-6	120-2[10]	4-02-00-01615 / 0.944[20]	1.454[20]
11983	1-Undecen-3-yne	$C_{11}H_{18}$ / 150.26	74744-28-8	74[9]	3-01-00-01054 / 0.7962[20]	1.4606[20]

No.	Name Synonym	Mol. Form. Mol. Wt.	CAS RN mp/°C	Merck No. bp/°C	Beil. Ref. den/g cm^{-3}	Solubility n_D
11984	1-Undecyne	C$_{11}$H$_{20}$ 152.28	2243-98-3 -25	196	4-01-00-01064 0.7728[20]	ace 4; bz 4; eth 4; EtOH 4 1.4306[20]
11985	2-Undecyne	C$_{11}$H$_{20}$ 152.28	60212-29-5 -30.1	204.2	0-01-00-00261 0.7827[20]	1.4391[20]
11986	4-Undecyne	C$_{11}$H$_{20}$ 152.28	60212-31-9 -74.7	198.5	3-01-00-01022 0.7752[20]	1.4369
11987	5-Undecyne	C$_{11}$H$_{20}$ 152.28	2294-72-6 -74.1	198	4-01-00-01064 0.7753[20]	1.4369[20]
11988	6-Undecynoic acid	C$_{11}$H$_{18}$O$_2$ 182.26	55182-83-7	124-5[0.17]	4-02-00-01739 0.9537[25]	1.4566[25]
11989	9-Undecynoic acid	C$_{11}$H$_{18}$O$_2$ 182.26	22202-65-9 61	170[15;] 132[0.1]	3-02-00-01471	H$_2$O 4; eth 4; EtOH 4
11990	9-Undecynoic acid, methyl ester Methyl 9-undecynoate	C$_{12}$H$_{20}$O$_2$ 196.29	18937-76-3	93-4[1]	3-02-00-01472 0.9245[17]	1.4732[19]
11991	10-Undecynoic acid, methyl ester Methyl 10-undecynoate	C$_{12}$H$_{20}$O$_2$ 196.29	2777-66-4 42	121[15]	4-02-00-01738 0.9177[20]	1.4421[21]
11992	Urea Carbamide	CH$_4$N$_2$O 60.06	57-13-6 132.7	9781 dec	4-03-00-00094 1.3230[20]	H$_2$O 4; EtOH 4; eth 1; bz 1 1.484
11993	Urea, (aminoiminomethyl)-	C$_2$H$_6$N$_4$O 102.10	141-83-3 105	160 dec	4-03-00-00155	H$_2$O 3; EtOH 2; eth 1; bz 1
11994	Urea, N,N'-bis(2-chloroethyl)-N-nitroso- Carmustine	C$_5$H$_9$Cl$_2$N$_3$O$_2$ 214.05	154-93-8 31	1852		H$_2$O 4; EtOH 4
11995	Urea, N,N'-bis(hydroxymethyl)- Oxymethurea	C$_3$H$_8$N$_2$O$_3$ 120.11	140-95-4 126	6921 149[25]	4-03-00-00107	H$_2$O 3; EtOH 3; eth 1; MeOH 3
11996	Urea, N,N-bis(2-methylpropyl)- 1,1-Diisobutylurea	C$_9$H$_{20}$N$_2$O 172.27	77464-06-3 73	180[25]	0-04-00-00170	eth 4; EtOH 4
11997	Urea, N,N'-bis(4-nitrophenyl)- 4,4'-Dinitrocarbanilide	C$_{13}$H$_{10}$N$_4$O$_5$ 302.25	587-90-6 312 dec	3271	4-12-00-01646	
11998	Urea, (4-bromophenyl)-	C$_7$H$_7$BrN$_2$O 215.05	1967-25-5 226	260	4-12-00-01511	H$_2$O 2; EtOH 4; eth 4; bz 4
11999	Urea, N'-(4-chlorophenyl)-N,N-dimethyl- Monuron	C$_9$H$_{11}$ClN$_2$O 198.65	150-68-5 170.5	6169	4-12-00-01191	H$_2$O 1; EtOH 2; ace 2
12000	Urea, N'-(3,4-dichlorophenyl)-N,N-dimethyl- Diuron	C$_9$H$_{10}$Cl$_2$N$_2$O 233.10	330-54-1 158	3388	4-12-00-01263	
12001	Urea, N,N'-diethyl-	C$_5$H$_{12}$N$_2$O 116.16	623-76-7 112.5	263	4-04-00-00370 1.0415[25]	H$_2$O 4; EtOH 4; eth 4 1.4616[40]
12002	Urea, N,N-diethyl-	C$_5$H$_{12}$N$_2$O 116.16	634-95-7 75	94-6[0.02]	4-04-00-00380	H$_2$O 4; EtOH 4; eth 3; bz 4
12003	Urea, N,N'-diethyl-N,N'-diphenyl- N,N'-Diethylcarbanilide	C$_{17}$H$_{20}$N$_2$O 268.36	85-98-3 79	3107	4-12-00-00844	H$_2$O 1; EtOH 4; chl 3
12004	Urea, N,N'-dimethyl-	C$_3$H$_8$N$_2$O 88.11	96-31-1 108	269	4-04-00-00207 1.142[25]	H$_2$O 4; EtOH 4; eth 1; chl 2
12005	Urea, N,N-dimethyl-	C$_3$H$_8$N$_2$O 88.11	598-94-7 183.5		4-04-00-00224 1.2555[25]	H$_2$O 3; EtOH 2; eth 1; tfa 2
12006	Urea, N,N'-dimethyl-N,N'-diphenyl-	C$_{15}$H$_{16}$N$_2$O 240.30	611-92-7 122	350	4-12-00-00839	H$_2$O 4; EtOH 4; eth 2; ace 4
12007	Urea, N,N-dimethyl-N'-[3-(trifluoromethyl)phenyl]- Fluometuron	C$_{10}$H$_{11}$F$_3$N$_2$O 232.21	2164-17-2 164	4080		ace 4; EtOH 4
12008	Urea, N,N'-di-1-naphthalenyl-	C$_{21}$H$_{16}$N$_2$O 312.37	607-56-7 296 sub		4-12-00-03078	py 4
12009	Urea, (2,5-dioxo-4-imidazolidinyl)- Allantoin	C$_4$H$_6$N$_4$O$_3$ 158.12	97-59-6 239	246	4-25-00-04071	H$_2$O 2; EtOH 3; eth 1; MeOH 1
12010	Urea, N,N'-diphenyl- Carbanilide	C$_{13}$H$_{12}$N$_2$O 212.25	102-07-8 236	1787 262	4-12-00-00741 1.239[25]	H$_2$O 2; EtOH 2; eth 3; ace 2
12011	Urea, N,N-diphenyl- 1,1-Diphenylurea	C$_{13}$H$_{12}$N$_2$O 212.25	603-54-3 189	dec	4-12-00-00852 1.276[25]	H$_2$O 2; EtOH 3; eth 3; chl 3
12012	Urea, N,N'-dipropyl-	C$_7$H$_{16}$N$_2$O 144.22	623-95-0 105	255	4-04-00-00482	eth 4; EtOH 4
12013	Urea, (4-ethoxyphenyl)- Dulcin	C$_9$H$_{12}$N$_2$O$_2$ 180.21	150-69-6 173.5	3448 dec	4-13-00-01154	H$_2$O 2; EtOH 3; AcOEt 4
12014	Urea, ethyl-	C$_3$H$_8$N$_2$O 88.11	625-52-5 92.5	dec	4-04-00-00369 1.2130[18]	H$_2$O 4; EtOH 4; eth 3; bz 4
12015	Urea, N-ethyl-N-nitroso- N-Ethyl-N-nitrosourea	C$_3$H$_7$N$_3$O$_2$ 117.11	759-73-9 98-100 dec	3788	3-04-00-00233	chl 3
12016	Urea, hydroxy- Hydroxyurea	CH$_4$N$_2$O$_2$ 76.06	127-07-1 141	4785 dec	4-03-00-00170	H$_2$O 4
12017	Urea, methyl-	C$_2$H$_6$N$_2$O 74.08	598-50-5 103	dec	4-04-00-00205 1.2040[0]	H$_2$O 4; EtOH 4; eth 1; bz 1
12018	Urea, N-methyl-N,N'-diphenyl-	C$_{14}$H$_{14}$N$_2$O 226.28	612-01-1 106	203	4-12-00-00839	H$_2$O 2; EtOH 4; eth 4; bz 3
12019	Urea, N-methyl-N-nitroso- Methylnitrosourea	C$_2$H$_5$N$_3$O$_2$ 103.08	684-93-5 123-4 dec		4-04-00-03385	H$_2$O 2; EtOH 2; eth 2
12020	Urea, monohydrochloride Urea hydrochloride	CH$_5$ClN$_2$O 96.52	506-89-8 145 dec	9782	4-03-00-00094	H$_2$O 3
12021	Urea, mononitrate Urea nitrate	CH$_5$N$_3$O$_4$ 123.07	124-47-0 152 dec	9784	4-03-00-00094 1.690[20]	EtOH 4

No.	Name Synonym	Mol. Form. Mol. Wt.	CAS RN mp/°C	Merck No. bp/°C	Beil. Ref. den/g cm^{-3}	Solubility n_D
12022	Urea, 1-naphthalenyl- 1-Naphthylurea	$C_{11}H_{10}N_2O$ 186.21	6950-84-1 219.5		4-12-00-03076	H_2O 1; EtOH 4; eth 3
12023	Urea, 2-naphthalenyl- 2-Naphthylurea	$C_{11}H_{10}N_2O$ 186.21	13114-62-0 219		4-12-00-03149	EtOH 4
12024	Urea, nitro- Nitrourea	$CH_3N_3O_3$ 105.05	556-89-8 158-9 dec	6573	4-03-00-00248	ace 4; EtOH 4
12025	Urea, (4-nitrophenyl)- 4-Nitrophenylurea	$C_7H_7N_3O_3$ 181.15	556-10-5 238	6547	4-12-00-01645	H_2O 4; EtOH 4
12026	Urea, phenyl- Phenylurea	$C_7H_8N_2O$ 136.15	64-10-8 147	7290 238	4-12-00-00734 1.302[25]	H_2O 2; EtOH 3; eth 2; AcOEt 3
12027	Urea, (phenylmethyl)- Benzylurea	$C_8H_{10}N_2O$ 150.18	538-32-9 148	1165 200 dec	4-12-00-02253	ace 4; EtOH 4
12028	Urea, 2-propenyl- Allylurea	$C_4H_8N_2O$ 100.12	557-11-9 85	296	4-04-00-01070	H_2O 5; EtOH 5; eth 2; chl 2
12029	Urea, tetraethyl- Urea, 1,1,3,3-tetraethyl-	$C_9H_{20}N_2O$ 172.27	1187-03-7 	 209	4-04-00-00380 0.919[20]	H_2O 1; alk 1; acid 1 1.4474[20]
12030	Urea, tetramethyl- Tetramethylurea	$C_5H_{12}N_2O$ 116.16	632-22-4 -1.2	9160 176.5	4-04-00-00225 0.9687[20]	EtOH 2; eth 2; ctc 2 1.4496[23]
12031	Urea, tetraphenyl- Tetraphenylurea	$C_{25}H_{20}N_2O$ 364.45	632-89-3 183		4-12-00-00853 1.222[25]	H_2O 1; EtOH 2
12032	Urea, trimethyl- Trimethylurea	$C_4H_{10}N_2O$ 102.14	632-14-4 75.5	 232.5	4-04-00-00224 1.1900[20]	H_2O 3; EtOH 3; eth 2; bz 2
12033	Uridine 1-β-D-Ribofuranosyluracil	$C_9H_{12}N_2O_6$ 244.20	58-96-8 165	9792	5-24-06-00132	H_2O 3; EtOH 3; py 3
12034	Uridine, 2'-deoxy-5-fluoro- Floxuridine	$C_9H_{11}FN_2O_5$ 246.20	50-91-9	4045	5-24-06-00291	
12035	5'-Uridylic acid Uridine 5'-phosphoric acid	$C_9H_{13}N_2O_9P$ 324.18	58-97-9 202 dec	9796	5-24-06-00173	H_2O 4; MeOH 3
12036	Urs-12-ene-27,28-dioic acid, 3-hydroxy-, (3β)- Quinovic acid	$C_{30}H_{46}O_5$ 486.69	465-74-7 298 dec	8104	4-10-00-02197	
12037	Urs-12-en-28-oic acid, 3-hydroxy-, (3β)- Ursolic acid	$C_{30}H_{48}O_3$ 456.71	77-52-1 284	9802	4-10-00-01160	ace 4; eth 4; chl 4
12038	Urs-12-en-3-ol, (3β)- α-Amyrin	$C_{30}H_{50}O$ 426.73	638-95-9 186	653 243[0.5]	4-06-00-04191	EtOH 3; eth 3; bz 3; chl 3
12039	Valeric acid, 3-chloro-, ethyl ester	$C_7H_{13}ClO_2$ 164.63	6513-13-9 	 189	3-02-00-00678 1.0330[20]	eth 4; EtOH 4 1.4278[20]
12040	DL-Valine	$C_5H_{11}NO_2$ 117.15	516-06-3 298 dec	9818 sub	3-04-00-01370 1.316[25]	H_2O 3; EtOH 2; eth 2; bz 2
12041	L-Valine 2-Aminoisovaleric acid	$C_5H_{11}NO_2$ 117.15	72-18-4 315	9818 sub	4-04-00-02659 1.23[25]	H_2O 4
12042	D-Valine, 3-[(2-amino-2-carboxyethyl)dithio]-, (R)- Penicillamine cysteine disulfide	$C_8H_{16}N_2O_4S_2$ 268.36	18840-45-4 195	7030	4-04-00-03231	
12043	D-Valine, 3-mercapto- Penicillamine	$C_5H_{11}NO_2S$ 149.21	52-67-5 198.5	7029	4-04-00-03228	
12044	Veratraman-3,23-diol, 14,15,16,17-tetradehydro-, (3β,23β)- Veratramine	$C_{27}H_{39}NO_2$ 409.61	60-70-8 206	9853	5-21-05-00297	EtOH 3; bz 3; chl 3; dil acid 3
12045	Veratraman-11-one, 17,23-epoxy-3-(β-D-glucopyranosyloxy)-, (3β,23β)- Pseudojervine	$C_{33}H_{49}NO_8$ 587.75	36069-05-3 304-5 dec	7932	4-27-00-03592	H_2O 1; EtOH 3; eth 1; bz 1
12046	Veratraman-11-one, 17,23-epoxy-3-hydroxy-, (3β,23β)- Jervine	$C_{27}H_{39}NO_3$ 425.61	469-59-0 243 dec	5145	4-27-00-03590	H_2O 1; EtOH 3; eth 2; ace 3
12047	Vernolate Carbamothioic acid, dipropyl-, S-propyl ester	$C_{10}H_{21}NOS$ 203.35	1929-77-7 150[30]	9866	0.952[20]	
12048	Vinclozolin 2,4-Oxazolidinedione, 3-(3,5-dichlorophenyl)-5-ethenyl-5-methyl-	$C_{12}H_9Cl_2NO_3$ 286.11	50471-44-8 108	9890 131[0.05]	1.51	
12049	9H-Xanthene 10H-9-Oxaanthracene	$C_{13}H_{10}O$ 182.22	92-83-1 100.5	 311	5-17-02-00252	H_2O 1; EtOH 2; eth 3; bz 3
12050	9H-Xanthen-9-one Xanthone	$C_{13}H_8O_2$ 196.21	90-47-1 174	9971 351; 146[3]	5-17-10-00430	H_2O 1; EtOH 3; eth 3; bz 3
12051	9H-Xanthen-9-one, 1,3-dihydroxy-	$C_{13}H_8O_4$ 228.20	3875-68-1 259	 sub	5-18-04-00004	EtOH 4
12052	9H-Xanthen-9-one, 2,7-dihydroxy- Xanthone, 2,7-dihydroxy	$C_{13}H_8O_4$ 228.20	64632-72-0 330	 sub	1-18-00-00357	eth 4; EtOH 4
12053	9H-Xanthen-9-one, 3,6-dihydroxy-	$C_{13}H_8O_4$ 228.20	1214-24-0 350 dec	 sub	5-18-04-00010	EtOH 4; HOAc 4
12054	9H-Xanthen-9-one, 1,7-dihydroxy-3-methoxy- Gentisin	$C_{14}H_{10}O_5$ 258.23	437-50-3 266.5	4291	5-18-04-00497	H_2O 1; EtOH 4; ace 1; py 3
12055	peri-Xanthenoxanthene Dinaphthalene dioxide	$C_{20}H_{10}O_2$ 282.30	191-28-6 243.5		5-19-02-00242 400[20-25]	bz 3; chl 3
12056	Xanthylium, 9-(2-carboxyphenyl)-3,6-bis(diethylamino)-, chloride Rhodamine, B	$C_{28}H_{32}ClN_2O_3$ 480.03	81-88-9 165	8182	5-19-08-00669	H_2O 3; EtOH 3; eth 3; bz 3
12057	o-Xylene, 4-propyl-	$C_{11}H_{16}$ 148.25	3982-66-9 -20	 208.9	3-05-00-01005 0.8715[20]	ace 4; bz 4; eth 4; EtOH 4 1.5000[20]

No.	Name Synonym	Mol. Form. Mol. Wt.	CAS RN mp/°C	Merck No. bp/°C	Beil. Ref. den/g cm^{-3}	Solubility n_D
12058	Xylitol Xylite	$C_5H_{12}O_5$ 152.15	87-99-0 93.5	9992 216	4-01-00-02832	H_2O 4; py 4; EtOH 4
12059	D-Xylose	$C_5H_{10}O_5$ 150.13	58-86-6 90.5	9995	4-01-00-04223 1.525[20]	H_2O 4; EtOH 3; eth 2
12060	Yohimban-16-carboxylic acid, 11,17-dimethoxy-18-[[1-oxo-3-(3,4,5-trimethoxyphenyl)-2-propenyl]oxy]-, methyl ester, (3β,16β,17α,18β,20α)- Rescinnamine	$C_{35}H_{42}N_2O_9$ 634.73	24815-24-5 238.5	8146	4-25-00-01323	H_2O 1; EtOH 2; ace 3; chl 3
12061	Yohimban-16-carboxylic acid, 11,17-dimethoxy-18-[(3,4,5-trimethoxybenzoyl)oxy]-, methyl ester, Reserpine	$C_{33}H_{40}N_2O_9$ 608.69	50-55-5 264.5	8149	4-25-00-01319	H_2O 2; EtOH 3; eth 2; ace 2
12062	Yohimban-16-carboxylic acid, 17-hydroxy-, methyl ester, (16α,17α)- Yohimbine	$C_{21}H_{26}N_2O_3$ 354.45	146-48-5 241	10011 sub 160	4-25-00-01237	H_2O 2; EtOH 3; eth 3; bz 2
12063	Yohimban-16-carboxylic acid, 18-hydroxy-11,17-dimethoxy-, (3β,16β,17α,18β,20 Reserpic acid	$C_{22}H_{28}N_2O_5$ 400.47	83-60-3 242	8147	4-25-00-01305	
12064	Yohimban-16-carboxylic acid, 17-hydroxy-18-[(3,4,5-trimethoxybenzoyl)oxy]-, methyl ester, (3β,16β Raunescine	$C_{31}H_{36}N_2O_8$ 564.64	117-73-7 165	8130	4-25-00-01281	H_2O 1; EtOH 3; chl 3; HOAc 3
12065	Yohimban-16-carboxylic acid, 17-methoxy-18-[(3,4,5-trimethoxybenzoyl)oxy]-, methyl ester, (3β,16β Deserpidine	$C_{32}H_{38}N_2O_8$ 578.66	131-01-1 230.5	2901	4-25-00-01282	H_2O 1; EtOH 3; chl 3
12066	Zinc, dimethyl Dimethyl zinc	C_2H_6Zn 95.46	544-97-8 -40	3252 46	1.386[10]	eth 3; peth 5
12067	Zinc, [[1,2-ethanediylbis[carbamodithioato]](2-)]- Zineb	$C_4H_6N_2S_4Zn$ 275.76	12122-67-7 157 dec	10071		
12068	Ziram Zinc, bis(dimethylcarbamodithioato-S,S')-, (T-4)-	$C_6H_{12}N_2S_4Zn$ 305.83	137-30-4			

STRUCTURAL FORMULAS OF ORGANIC COMPOUNDS

In numeric order as they occur in the Table of Physical Constants of Organic Compounds

47

48

49

50

51

52

53

54

55

56

57

58

59

60

61

62

63

64

65

66

67

68

69

70

71

72

73

74

75

76

77

78

79

80

81

82

83

84

85

STRUCTURAL FORMULAS OF ORGANIC COMPOUNDS (continued)

In numeric order as they occur in the Table of Physical Constants of Organic Compounds

86

87

88

89

90

91

92

93

95

96

97

98

99

100

101

102

103

104

105

107

108

109

111

112

113

114

115

116

117

118

119

120

121

123

124

125

126

127

128

129

130

131

STRUCTURAL FORMULAS OF ORGANIC COMPOUNDS (continued)

In numeric order as they occur in the Table of Physical Constants of Organic Compounds

132

133

134

135

136

137

138

139

140

141

142

143

144

145

146

147

148

149

150

151

152

153

154

157

158

159

160

161

162

163

164

165

166

167

168

169

170

171

172

173

174

175

176

177

STRUCTURAL FORMULAS OF ORGANIC COMPOUNDS (continued)

In numeric order as they occur in the Table of Physical Constants of Organic Compounds

178

179

180

181

182

183

184

185

186

187

188

189

190

191

192

193

194

195

196

197

198

199

200

201

202

203

204

205

206

207

208

209

211

212

213

214

215

216

217

218

219

220

221

222

223

224

225

226

236

246

255

263

227

238

247

256

264

228

239

248

257

265

229

240

249

258

266

230

241

250

259

267

231

242

252

260

268

232

243

253

261

269

233

244

254

262

270

235

245

271

STRUCTURAL FORMULAS OF ORGANIC COMPOUNDS (continued)

In numeric order as they occur in the Table of Physical Constants of Organic Compounds

272

273

274

277

278

279

280

281

282

283

284

285

286

287

288

289

290

291

292

293

294

295

296

297

298

299

300

301

302

303

304

306

312

313

314

315

316

317

318

319

320

321

322

323

324

325

326

327

329

331

332

333

334

335

336

337

338

339

340

341

342

343

344

345

346

347

348

349

350

351

352

353

354

355

356

357

358

359

360

STRUCTURAL FORMULAS OF ORGANIC COMPOUNDS (continued)

In numeric order as they occur in the Table of Physical Constants of Organic Compounds

362

364

365

366

367

368

369

370

371

372

373

374

375

376

377

379

380

381

382

383

384

385

386

387

388

389

390

391

392

393

394

395

396

397

398

399

400

401

402

403

404

405

406

407

408

409

410

411

412

413

414

415

416

417

418

419

420

421

422

423

424

425

426

427

428

429

430

431

432

433

434

435

436

437

438

439

440

441

442

443

444

445

446

447

448

449

450

451

452

453

454

455

456

457

458

459

460

461

462

463

464

465

466

467

468

470

471

472

473

474

475

476

477

478

479

480

481

482

483

484

485

486

487

488

489

490

491

492

493

494

495

496

497

498

499

500

501

502

503

504

505

506

507

509

510

511

512

513

514

515

516

517

518

519

520

521

522

523

524

525

526

527

528

529

530

531

532

533

535

536

537

538

539

540

541

542

543

544

545

546

547

548

549

550

551

552

553

554

555

556

557

558

559

560

561

562

563

564

565

566

567

568

569

570

571

572

573

574

575

In numeric order as they occur in the Table of Physical Constants of Organic Compounds

576

577

578

579

580

581

582

583

584

585

586

587

588

589

590

591

592

593

594

595

596

597

598

599

600

601

603

604

605

606

607

608

609

610

611

612

613

614

615

616

617

618

619

620

621

622

623

624

625

626

627

628

629

630

631

632

633

635

636

637

638

639

640

641

642

643

644

645

646

647

648

649

650

651

652

653

654

655

656

657

658

659

660

661

662

663

666

667

668

669

670

671

672

673

674

675

676

677

678

679

680

681

682

683

684

685

686

687

688

689

690

692

693

694

695

696

697

698

699

700

701

702

703

704

705

706

707

708

709

710

711

712

713

714

715

716

718

719

720

721

722

723

724

725

726

727

728

729

730

731

732

733

734

735

736

737

738

740

741

742

743

744

745

746

747

748

749

750

751

752

753

754

755

756

757

758

759

760

761

762

763

764

765

766

767

768

769

770

771

772

773

775

776

777

778

779

780

781

782

783

784

785

786

787

788

789

790

791

793

794

795

796

797

798

799

800

801

802

803

804

805

806

807

808

809

810

811

812

813

814

815

816

817

818

819

820

822

823

824

825

826

828

829

830

831

832

833

834

835

836

837

838

839

840

841

842

843

844

845

846

847

848

849

850

851

852

853

854

855

856

857

858

859

860

861

862

863

864

867

868

869

870

871

872

873

874

875

877

878

879

880

881

882

883

884

886

887

888

890

891

892

893

894

895

896

897

898

899

900

901

902

903

904

905

906

907

908

909

910

911

912

913

914

915

916

917

918

919

920

921

923

924

925

926

927

928

930

931

932

933

934

935

937

938

939

940

941

942

944

945

946

947

948

949

951

952

953

954

956

957

959

960

961

962

963

964

965

966

967

968

969

970

971

972

973

974

975

976

977

978

979

980

981

982

983

984

985

986

987

988

989

990

991

992

993

994

995

996

997

998

999

1000

1001

1002

1003

1004

1005

1006

1007

1008

1010

1011

1012

1013

1014

1015

1016

1017

1018

1019

1020

1021

1022

1023

1024

1025

1026

1027

1028

1029

1030

1031

1032

1033

1034

1035

1036

1037

1038

1039

1040

1041

1042

1043

1044

1045

1046

1047

1048

1049

1050

1051

1052

1053

1054

1055

1056

1057

1058

1059

1060

1061

1062

1063

1064

1065

1066

1067

1068

1069

1070

1071

1072

1073

1074

1075

1076

1077

1078

1079

1080

1081

1082

1083

1084

1085

1086

1087

1088

1089

STRUCTURAL FORMULAS OF ORGANIC COMPOUNDS (continued)

In numeric order as they occur in the Table of Physical Constants of Organic Compounds

1090

1091

1092

1093

1094

1095

1096

1097

1098

1099

1100

1101

1102

1103

1104

1105

1106

1108

1109

1110

1111

1112

1113

1114

1115

1116

1117

1118

1119

1120

1121

1122

1123

1124

1125

1126

1127

1128

1129

1130

1131

1132

1133

STRUCTURAL FORMULAS OF ORGANIC COMPOUNDS (continued)

In numeric order as they occur in the Table of Physical Constants of Organic Compounds

1134

1135

1136

1137

1138

1139

1140

1141

1142

1143

1144

1145

1146

1147

1148

1149

1150

1151

1153

1154

1155

1156

1157

1158

1159

1160

1161

1162

1163

1164

1166

1167

1168

1169

1170

1171

1172

1173

1175

1176

1177

1178

1179

1180

1181

1191

1200

1208

1216

1182

1192

1201

1209

1217

1183

1193

1202

1210

1218

1184

1194

1203

1211

1219

1185

1195

1204

1212

1220

1186

1196

1205

1213

1221

1187

1197

1206

1214

1222

1189

1198

1207

1215

1223

1190

1199

STRUCTURAL FORMULAS OF ORGANIC COMPOUNDS (continued)

In numeric order as they occur in the Table of Physical Constants of Organic Compounds

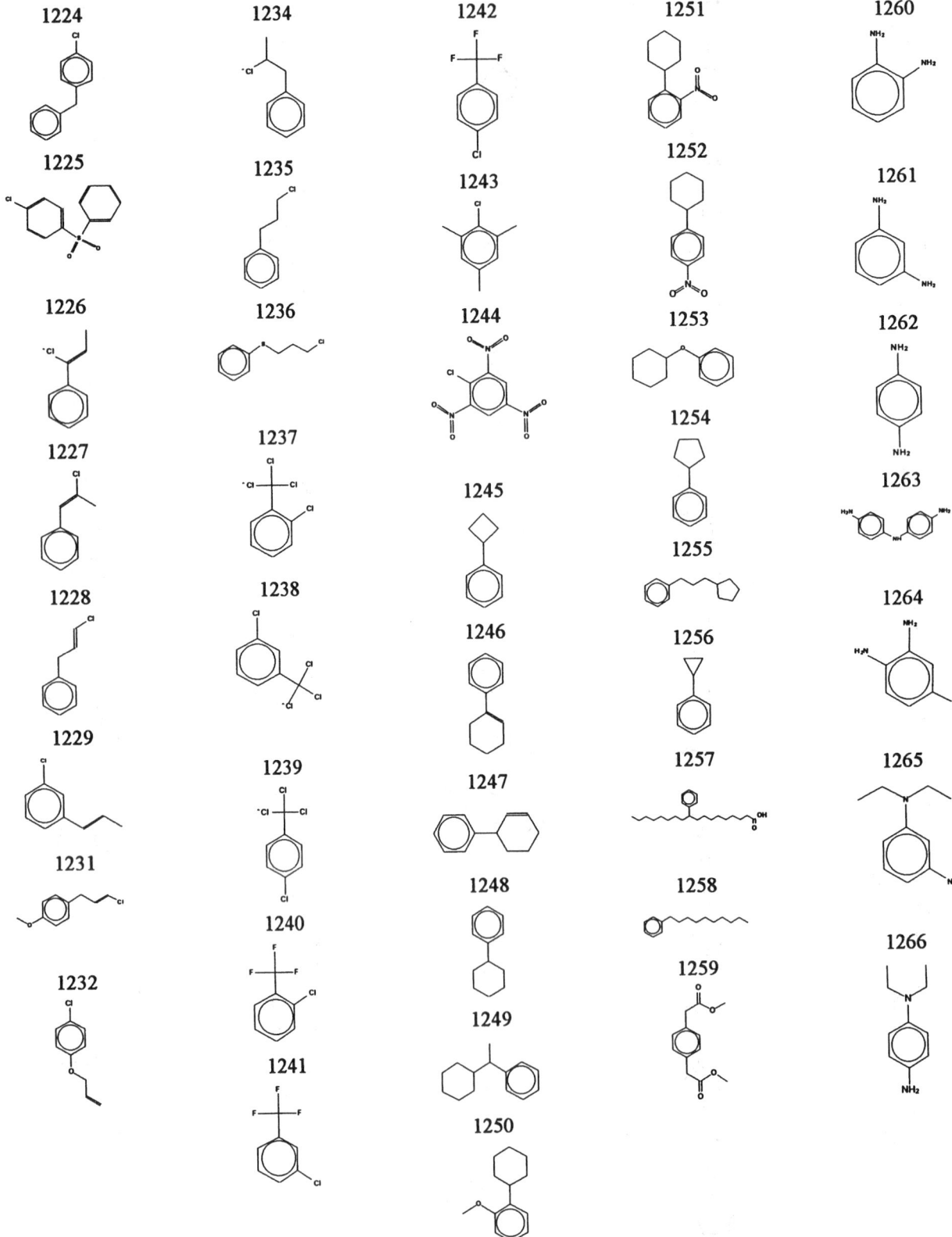

1224

1225

1226

1227

1228

1229

1231

1232

1234

1235

1236

1237

1238

1239

1240

1241

1242

1243

1244

1245

1246

1247

1248

1249

1250

1251

1252

1253

1254

1255

1256

1257

1258

1259

1260

1261

1262

1263

1264

1265

1266

1267

1275

1281

1288

1297

1268

1276

1289

1298

1282

1270

1277

1283

1290

1299

1284

1291

1300

1271

1278

1272

1285

1292

1301

1279

1273

1293

1302

1286

1294

1280

1295

1303

1274

1287

1304

1311

1319

1328

1335

1305

1312

1320

1329

1336

1306

1313

1321

1330

1337

1307

1314

1322

1331

1338

1308

1315

1323

1332

1339

1309

1316

1324

1333

1340

1310

1317

1325

1334

1341

1318

1327

1342

1343

1344

1345

1346

1347

1348

1349

1350

1351

1352

1353

1354

1355

1356

1357

1358

1359

1361

1362

1363

1364

1365

1366

1367

1368

1369

1370

1371

1372

1373

1374

1375

1376

1377

1378

1379

1380

1381

1382

1383

1384

1385

1386

1387

1388

1389

1390

1391

1392

1393

1394

1395

1396

1397

1398

1399

1400

1401

1402

1403

1404

1405

1406

1407

1408

1409

1410

1411

1412

1413

1414

1415

1416

1417

1418

1419

1420

1421

1422

1423

1424

1425

1426

1427

1428

1429

1430

1431

1432

1433

1434

1435

1436

1437

1438

1439

1440

1441

1442

1443

1444

1445

1447

1448

1449

1450

1451

1452

1453

1454

1455

1456

1457

1458

1459

1460

1461

1462

1463

1464

1465

1466

1467

1468

1469

1470

1471

1472

1473

1474

1475

1476

1477

1478

1488

1495

1503

1512

1479

1489

1496

1504

1513

1480

1490

1497

1505

1514

1481

1491

1498

1506

1515

1482

1492

1499

1507

1516

1483

1493

1500

1508

1517

1484

1494

1501

1509

1518

1486

1502

1510

1487

1511

1519

STRUCTURAL FORMULAS OF ORGANIC COMPOUNDS (continued)

In numeric order as they occur in the Table of Physical Constants of Organic Compounds

1520

1521

1522

1523

1524

1525

1526

1527

1528

1529

1530

1531

1532

1533

1534

1535

1536

1537

1538

1539

1540

1541

1542

1543

1544

1545

1546

1547

1548

1549

1550

1551

1552

1553

1554

1555

1556

1557

1558

1560

1561

1562

1563

1564

1565

1566

1567

1568

1569

1570

1571

1572

1573

1574

1575

1576

1577

1578

1579

1580

1581

1583

1584

1585

1586

1587

1588

1589

1590

1591

1592

1593

1594

1595

1596

1597

1604

1613

1621

1631

1598

1606

1614

1622

1632

1599

1607

1615

1623

1633

1600

1608

1616

1624

1634

1601

1609

1617

1626

1635

1602

1610

1618

1627

1636

1603

1619

1628

1637

1612

1620

1629

1638

1630

1639

1640

1641

1642

1643

1644

1645

1646

1647

1648

1649

1650

1651

1652

1653

1654

1655

1656

1657

1658

1659

1660

1661

1662

1663

1664

1665

1666

1667

1668

1669

1670

1671

1672

1673

1674

1675

1676

1677

1678

1679

1680

1681

1682

1683

1684

1693

1702

1710

1719

1703

1685

1694

1711

1720

1686

1695

1704

1712

1721

1687

1696

1705

1713

1722

1689

1697

1706

1714

1723

1690

1698

1707

1715

1724

1691

1699

1708

1716

1725

1700

1709

1717

1692

1701

1718

1726

1727

1728

1729

1730

1731

1732

1733

1734

1735

1736

1737

1738

1739

1740

1741

1742

1743

1744

1745

1746

1747

1748

1749

1750

1751

1752

1753

1754

1755

1756

1757

1758

1759

1760

1761

1762

1763

1764

1765

1766

1767

1768

1769

1770

1771

1772

1773

1774

1775

1776

1777

1778

1779

1780

1781

1782

1783

1784

1785

1786

1787

1788

1789

1790

1791

1792

1793

1794

1795

1796

1797

1798

1799

1800

1801

1802

1803

1804

1805

1806

1807

1808

1809

STRUCTURAL FORMULAS OF ORGANIC COMPOUNDS (continued)

In numeric order as they occur in the Table of Physical Constants of Organic Compounds

1810

1811

1812

1813

1814

1815

1816

1817

1818

1819

1820

1821

1822

1823

1826

1827

1828

1829

1830

1831

1832

1833

1834

1835

1836

1837

1838

1839

1840

1841

1842

1843

1844

1845

1846

1847

1848

1849

1850

1851

1852

1853

1854

1855

1856

1857

1858

1859

1860

1861

1862

1863

1864

1865

1866

1867

1868

1869

1870

1871

1872

1873

1874

1875

1876

1877

1878

1879

1880

1881

1882

1883

1884

1885

1886

1887

1895

1903

1911

1919

1888

1896

1904

1912

1920

1889

1897

1905

1913

1921

1890

1898

1906

1914

1923

1891

1899

1907

1915

1924

1892

1900

1908

1916

1925

1893

1901

1909

1917

1926

1894

1902

1910

1918

1927

1928

1929

1930

1931

1932

1933

1934

1935

1936

1937

1938

1939

1940

1941

1942

1943

1944

1945

1946

1947

1948

1949

1950

1951

1952

1953

1954

1955

1956

1957

1958

1959

1960

1961

1962

1963

1964

1965

1966

1967

1968

1969

1970

1971

1972

1973

1974

1975

1976

1977

1978

1979

1980

1981

1982

1983

1984

1985

1986

1987

1988

1989

1990

1991

1992

1993

1995

1996

1997

1998

1999

2000

2001

2002

2003

2004

2005

2006

2007

2009

2010

2011

In numeric order as they occur in the Table of Physical Constants of Organic Compounds

2012

2022

2031

2040

2049

2013

2023

2033

2041

2050

2014

2024

2034

2042

2051

2015

2025

2035

2044

2052

2016

2026

2036

2053

2017

2027

2037

2045

2018

2028

2038

2047

2054

2019

2029

2039

2048

2055

2021

2030

2056

2057

2058

2059

2060

2061

2062

2063

2064

2065

2066

2067

2068

2069

2070

2071

2072

2073

2074

2075

2076

2077

2078

2079

2080

2081

2083

2084

2085

2086

2087

2088

2089

2090

2091

2092

2093

2094

2095

2097

2098

2099

2100

2101

STRUCTURAL FORMULAS OF ORGANIC COMPOUNDS (continued)

In numeric order as they occur in the Table of Physical Constants of Organic Compounds

2102

2103

2104

2105

2106

2107

2108

2110

2111

2113

2114

2116

2117

2118

2119

2121

2122

2123

2124

2125

2126

2127

2129

2130

2131

2132

2133

2134

2137

2138

2139

2140

2141

2142

2143

2144

2145

2146

2147

2148

2149

2150

2151

2152

2153

2154

2155

2156

2157

2158

2159

2160

2161

2162

2163

2164

2165

2166

2167

2168

2169

2170

2171

2172

2173

2174

2175

2176

2177

2178

2179

2181

2182

2183

2184

2185

2186

2187

2188

2189

STRUCTURAL FORMULAS OF ORGANIC COMPOUNDS (continued)

In numeric order as they occur in the Table of Physical Constants of Organic Compounds

2190

2197

2205

2213

2223

2191

2198

2206

2214

2224

2199

2207

2215

2225

2192

2208

2216

2226

2193

2200

2217

2227

2210

2194

2202

2220

2228

2195

2203

2211

2221

2229

2204

2212

2196

2230

2231

2232

2233

2234

2235

2236

2237

2238

2239

2240

2242

2243

2244

2245

2246

2247

2248

2249

2250

2251

2252

2253

2254

2255

2256

2257

2258

2259

2260

2261

2262

2264

2265

2266

2267

2268

2269

2270

2281

2292

2300

2308

2271

2283

2293

2301

2309

2272

2284

2294

2302

2310

2276

2285

2295

2303

2311

2277

2286

2296

2304

2312

2278

2287

2297

2305

2313

2279

2289

2299

2306

2314

2280

2290

2291

2316

2317

2318

2319

2321

2322

2323

2324

2325

2326

2327

2328

2329

2330

2331

2332

2333

2334

2335

2336

2337

2339

2340

2341

2342

2343

2344

2345

2346

2347

2348

2352

2354

2355

2356

2357

2358

2359

2360

2361

2362

2363

2364

2365

2366

2367

2368

2369

2370

2371

2372

2373

2374

2375

2376

2377

2378

2379

2380

2381

2382

2383

2384

2385

2386

2387

2388

2389

2390

2391

2392

2393

2394

2395

2396

2397

2398

2399

In numeric order as they occur in the Table of Physical Constants of Organic Compounds

2400

2409

2418

2426

2436

2401

2419

2427

2437

2410

2402

2428

2438

2411

2420

2403

2429

2439

2421

2404

2430

2440

2412

2422

2405

2413

2431

2441

2423

2414

2432

2442

2406

2424

2415

2434

2443

2407

2425

2444

2416

2435

2408

2445

2417

STRUCTURAL FORMULAS OF ORGANIC COMPOUNDS (continued)

In numeric order as they occur in the Table of Physical Constants of Organic Compounds

2446

2447

2448

2449

2451

2452

2453

2454

2455

2456

2457

2458

2459

2460

2461

2462

2463

2464

2465

2466

2468

2469

2470

2471

2472

2474

2475

2476

2477

2478

2479

2480

2481

2482

2483

2484

2485

2486

2487

2488

2489

2490

2491

2492

2493

2494

2495

2496

2497

2498

2499

2500

2501

2502

2503

2504

2505

2506

2507

2508

2509

2510

2511

2512

2513

2514

2515

2516

2517

2518

2519

2520

2521

2522

2523

2524

2525

2526

2527

2528

2529

2530

2539

2547

2555

2562

2532

2540

2548

2556

2563

2533

2541

2549

2557

2564

2534

2542

2550

2558

2565

2535

2543

2551

2559

2566

2536

2544

2552

2560

2567

2537

2545

2553

2561

2568

2538

2546

2554

2569

2576

2583

2590

2598

2570

2577

2584

2591

2599

2600

2571

2578

2585

2592

2601

2572

2579

2586

2593

2602

2573

2580

2587

2594

2603

2574

2581

2588

2595

2596

2604

2582

2589

2597

2575

2605

2606

2607

2608

2609

2610

2611

2612

2613

2614

2615

2616

2617

2618

2619

2620

2621

2622

2624

2625

2626

2627

2628

2629

2630

2631

2632

2633

2634

2635

2636

2637

2638

2639

2640

2641

2642

2643

2644

2645

2646

2647

2648

2649

2650

2651

2652

2653

2654

2655

2656

2657

2658

2659

2660

2661

2662

2663

2664

2665

2666

2667

2668

2669

2670

2671

2672

2673

2674

2675

2676

2677

2678

2686

2693

2700

2707

2679

2687

2694

2701

2708

2680

2688

2695

2702

2709

2681

2689

2696

2703

2710

2682

2690

2697

2704

2711

2683

2691

2698

2705

2712

2684

2692

2699

2706

2713

2685

2714

2715

2716

2717

2718

2719

2720

2721

2722

2723

2724

2726

2727

2728

2729

2730

2731

2732

2733

2734

2735

2736

2737

2738

2739

2740

2741

2742

2744

2745

2746

2747

2748

2749

2750

2751

STRUCTURAL FORMULAS OF ORGANIC COMPOUNDS (continued)

In numeric order as they occur in the Table of Physical Constants of Organic Compounds

2752

2753

2754

2755

2756

2757

2758

2759

2760

2761

2762

2763

2764

2765

2766

2767

2768

2769

2770

2771

2772

2773

2774

2775

2776

2777

2778

2780

2781

2782

2783

2784

2785

2787

2788

2789

2790

2791

2792

2803

2811

2819

2830

2793

2804

2812

2820

2831

2795

2805

2813

2822

2832

2797

2806

2814

2825

2833

2798

2807

2815

2826

2834

2799

2808

2816

2827

2835

2800

2809

2817

2828

2836

2801

2810

2818

2829

2837

2802

2838

2839

2840

2841

2842

2843

2844

2845

2847

2849

2850

2851

2852

2853

2854

2855

2856

2857

2858

2859

2860

2861

2862

2863

2864

2865

2866

2867

2868

2869

2870

2871

2872

2873

2874

2875

2876

2877

2878

2879

2880

2881

2882

2883

2884

2885

2886

2887

2888

2889

2890

2891

2892

2893

2894

2895

2896

2897

2898

2899

2900

2901

2902

2903

2904

2905

2906

2907

2908

2909

2910

2911

2912

2913

2914

2915

2916

2918

2919

2928

2937

2951

2960

2920

2929

2939

2952

2961

2921

2930

2941

2953

2962

2922

2932

2944

2954

2963

2923

2933

2945

2955

2964

2925

2934

2946

2957

2966

2926

2935

2947

2958

2967

2927

2936

2948

2959

2968

2950

2969

2977

2986

2995

3003

2970

2978

2987

2996

3004

2971

2980

2988

2997

3005

2972

2981

2989

2998

3007

2973

2982

2990

2999

3008

2974

2983

2991

3000

3009

2975

2984

2992

3001

3010

2976

2985

2993

3002

3011

2994

3013

STRUCTURAL FORMULAS OF ORGANIC COMPOUNDS (continued)

In numeric order as they occur in the Table of Physical Constants of Organic Compounds

3014

3015

3016

3017

3018

3019

3020

3021

3022

3023

3024

3025

3026

3027

3028

3029

3030

3031

3032

3033

3034

3035

3037

3038

3039

3040

3041

3042

3043

3044

3045

3046

3050

3051

3052

3053

3054

3055

3056

3057

STRUCTURAL FORMULAS OF ORGANIC COMPOUNDS (continued)

In numeric order as they occur in the Table of Physical Constants of Organic Compounds

3058

3059

3060

3061

3062

3063

3064

3065

3066

3070

3071

3072

3073

3074

3075

3076

3077

3078

3079

3080

3081

3082

3083

3084

3085

3086

3087

3089

3090

3091

3092

3093

3094

3095

3096

3098

3099

3100

3101

3102

STRUCTURAL FORMULAS OF ORGANIC COMPOUNDS (continued)

In numeric order as they occur in the Table of Physical Constants of Organic Compounds

3103

3104

3105

3108

3109

3111

3112

3113

3114

3115

3116

3117

3118

3119

3120

3121

3122

3123

3124

3125

3126

3127

3128

3129

3130

3131

3132

3134

3135

3136

3137

3138

3139

3140

3141

STRUCTURAL FORMULAS OF ORGANIC COMPOUNDS (continued)

In numeric order as they occur in the Table of Physical Constants of Organic Compounds

3143

3153

3161

3170

3179

3144

3154

3162

3171

3180

3145

3155

3163

3172

3181

3146

3156

3164

3173

3182

3147

3157

3166

3174

3183

3148

3158

3167

3175

3184

3150

3168

3176

3185

3151

3159

3169

3177

3186

3152

3160

3178

3188

3189

3190

3191

3192

3194

3195

3196

3197

3198

3199

3200

3201

3202

3203

3205

3206

3207

3208

3209

3210

3211

3212

3213

3214

3215

3216

3217

3218

3220

3221

3222

3223

3224

3225

3226

3227

3228

3229

3230

3231

3232

3233

3234

3235

3236

3237

3238

3239

3240

3241

3242

3243

3244

3245

3246

3247

3248

3249

3250

3251

3252

3253

3254

3255

3256

3257

3258

3259

3262

3263

3264

3265

3266

3267

3268

3269

3270

3271

3272

3273

3274

3275

STRUCTURAL FORMULAS OF ORGANIC COMPOUNDS (continued)

In numeric order as they occur in the Table of Physical Constants of Organic Compounds

3276

3284

3292

3301

3311

3277

3285

3293

3302

3312

3278

3286

3294

3303

3313

3279

3287

3295

3304

3316

3280

3288

3296

3305

3281

3289

3297

3307

3317

3282

3290

3298

3308

3318

3283

3291

3299

3309

3319

3300

3320

3310

STRUCTURAL FORMULAS OF ORGANIC COMPOUNDS (continued)

In numeric order as they occur in the Table of Physical Constants of Organic Compounds

3321

3322

3323

3324

3325

3326

3327

3328

3329

3330

3331

3332

3333

3334

3336

3339

3340

3341

3342

3343

3344

3345

3346

3347

3348

3349

3350

3351

3352

3353

3354

3355

3356

3357

3359

3360

3361

3362

3363

3364

3373

3383

3395

3404

3405

3365

3374

3384

3396

3375

3385

3397

3406

3366

3376

3387

3398

3407

3367

3377

3388

3399

3368

3378

3390

3400

3408

3369

3379

3391

3401

3409

3370

3381

3392

3402

3410

3371

3382

3394

3403

3411

3372

3412

3414

3415

3416

3417

3418

3419

3420

3421

3422

3423

3426

3427

3428

3429

3430

3431

3433

3434

3435

3436

3437

3438

3439

3440

3441

3442

3443

3444

3445

3446

3447

3448

3449

3450

3451

3452

3453

3454

3455

3456

3457

3458

3459

3470

3479

3488

3498

3460

3471

3480

3489

3499

3461

3472

3481

3490

3500

3462

3473

3482

3492

3502

3464

3474

3483

3493

3503

3465

3475

3484

3494

3504

3467

3476

3485

3495

3505

3468

3477

3486

3497

3506

3469

3478

3487

3507

3508

3509

3510

3511

3512

3513

3514

3515

3516

3517

3518

3519

3520

3521

3522

3523

3524

3525

3526

3527

3528

3529

3530

3531

3532

3533

3534

3535

3537

3538

3539

3540

3541

3542

3543

3544

3545

3546

3547

3548

3549

3551

3552

STRUCTURAL FORMULAS OF ORGANIC COMPOUNDS (continued)

In numeric order as they occur in the Table of Physical Constants of Organic Compounds

3553

3563

3573

3582

3594

3555

3565

3574

3584

3595

3556

3566

3575

3585

3596

3557

3567

3576

3586

3597

3558

3568

3577

3587

3598

3560

3569

3578

3588

3599

3570

3590

3561

3571

3579

3601

3562

3572

3580

3592

3604

3581

3593

3605

3616

3624

3632

3641

3606

3617

3633

3642

3625

3608

3618

3634

3643

3626

3609

3619

3635

3644

3627

3636

3612

3620

3628

3645

3637

3613

3621

3629

3646

3638

3614

3622

3647

3630

3639

3615

3623

3640

3648

3631

STRUCTURAL FORMULAS OF ORGANIC COMPOUNDS (continued)

In numeric order as they occur in the Table of Physical Constants of Organic Compounds

3649

3658

3666

3675

3683

3650

3659

3667

3676

3684

3651

3660

3668

3677

3685

3653

3661

3669

3678

3688

3654

3662

3670

3679

3690

3655

3663

3671

3680

3691

3656

3664

3673

3681

3692

3657

3665

3682

3693

3695

3696

3697

3698

3699

3700

3702

3704

3705

3706

3707

3708

3709

3710

3711

3712

3713

3714

3715

3716

3717

3718

3720

3721

3722

3723

3724

3725

3726

3727

3728

3729

3730

3731

3732

3733

3734

3735

3736

3737

3738

3747

3755

3763

3773

3739

3748

3756

3764

3774

3740

3749

3757

3765

3775

3741

3750

3758

3766

3776

3742

3751

3759

3767

3777

3743

3752

3769

3778

3744

3760

3770

3779

3753

3745

3754

3761

3771

3781

3746

3762

3772

3782

STRUCTURAL FORMULAS OF ORGANIC COMPOUNDS (continued)

In numeric order as they occur in the Table of Physical Constants of Organic Compounds

3783

3784

3785

3786

3787

3788

3789

3790

3791

3792

3793

3794

3795

3796

3797

3798

3799

3800

3801

3802

3803

3804

3805

3806

3807

3808

3809

3810

3812

3813

3814

3815

3816

3817

3819

3820

3821

3822

3823

3824

3825

3826

3827

3828

3829

3830

3831

3832

3833

3834

3835

3836

3837

3838

3839

3840

3841

3842

3843

3844

3845

3846

3847

3848

3849

3850

3851

3852

3856

3857

3858

3859

3860

3861

3862

3864

3865

3866

3868

3869

3870

3871

3873

3874

STRUCTURAL FORMULAS OF ORGANIC COMPOUNDS (continued)

In numeric order as they occur in the Table of Physical Constants of Organic Compounds

3875

3887

3895

3905

3915

3877

3888

3896

3907

3916

3878

3889

3897

3908

3917

3879

3890

3898

3909

3918

3883

3891

3899

3910

3919

3884

3892

3900

3911

3920

3885

3893

3901

3912

3921

3886

3894

3902

3913

3922

3904

3914

STRUCTURAL FORMULAS OF ORGANIC COMPOUNDS (continued)

In numeric order as they occur in the Table of Physical Constants of Organic Compounds

3923

3925

3926

3927

3929

3930

3931

3932

3933

3934

3935

3936

3937

3938

3939

3940

3941

3942

3943

3944

3947

3948

3949

3950

3953

3954

3955

3956

3957

3959

3960

3961

3962

3963

3964

3966

3969

3970

3971

3972

3973

3976

3978

3979

3980

3982

3989

3990

3991

3992

3993

3994

3995

3997

3998

3999

4000

4002

4003

4005

4006

4007

4008

4009

4012

4013

4014

4015

4016

4017

4018

4019

4020

4021

4022

4023

4024

4025

4028

4029

4030

4031

4032

4033

STRUCTURAL FORMULAS OF ORGANIC COMPOUNDS (continued)

In numeric order as they occur in the Table of Physical Constants of Organic Compounds

4034

4035

4036

4037

4038

4039

4040

4043

4044

4046

4047

4048

4049

4050

4065

4066

4067

4068

4069

4071

4072

4075

4076

4077

4078

4079

4080

4081

4082

4083

4084

4085

4086

4087

4088

4089

4090

4091

4092

4093

4094

4095

4096

4097

4098

4099

4100

4101

4102

4103

4106

4107

4108

4110

4111

4113

4116

4118

4119

4120

4121

4122

4123

4124

4125

4126

4127

4128

4130

4133

4134

4137

4138

4139

4140

4141

4142

4143

4144

4146

4147

4150

4151

4152

4153

4154

4155

4156

4157

4159

4161

4162

4163

4164

4165

4166

4167

4168

4169

4170

4171

4172

4173

4174

4175

4176

4177

4178

4179

4180

STRUCTURAL FORMULAS OF ORGANIC COMPOUNDS (continued)

In numeric order as they occur in the Table of Physical Constants of Organic Compounds

4181

4182

4183

4184

4185

4186

4187

4188

4189

4190

4191

4192

4193

4194

4195

4196

4197

4198

4199

4200

4201

4202

4203

4204

4205

4206

4207

4208

4209

4210

4211

4212

4213

4214

4215

4216

4217

4218

4219

4220

4221

4222

4223

4224

4225

4226

4227

4228

4229

4230

4231

4233

4234

4235

4236

4237

4238

4239

4240

4241

4242

4243

4244

4245

4246

4247

4248

4249

4250

4251

4252

4253

4254

4255

4256

4257

4258

4259

4260

4261

4262

4263

4264

4265

4266

4267

4268

4269

4270

4271

In numeric order as they occur in the Table of Physical Constants of Organic Compounds

4273

4274

4275

4276

4277

4278

4279

4280

4281

4282

4283

4284

4285

4286

4287

4288

4289

4290

4291

4292

4293

4294

4295

4296

4297

4298

4299

4300

4301

4302

4303

4304

4305

4306

4307

4308

4309

4310

4311

4312

4313

4314

4315

4316

4317

4318

4319

4320

4322

4323

4324

4325

4326

4327

4328

4329

4330

4331

4332

4333

4334

4335

4336

4337

4339

4340

4341

4342

4343

4344

4345

4346

4347

4348

4349

4350

4351

4352

4353

4354

4356

4357

4358

4359

4360

4361

4362

4364

4365

4366

4367

4368

4369

4370

4371

4372

4373

4374

4375

4376

4377

4378

4379

4380

4381

4382

4383

4384

4385

4386

4387

4388

4389

4390

4391

4392

4393

4394

4395

4396

4397

4398

4399

4400

4401

4402

4403

4404

4405

4406

In numeric order as they occur in the Table of Physical Constants of Organic Compounds

4407

4408

4409

4410

4411

4413

4414

4415

4416

4417

4418

4419

4420

4421

4422

4423

4424

4425

4426

4427

4428

4429

4430

4431

4432

4433

4434

4435

4436

4437

4438

4439

4440

4441

4442

4443

4444

4445

4446

4447

4448

4449

4450

4451

4452

4453

4454

4455

4456

4457

4458

4459

4460

4461

4462

4463

4464

4465

4466

4467

4468

4469

4470

4471

4472

4473

4474

4475

4476

4477

4478

4479

4480

4481

4482

4483

4484

4485

4486

4487

4488

4489

4490

4491

4492

4493

4494

4495

4496

4497

4498

4499

4500

4501

4502

4503

4504

4505

4507

4508

4509

4510

4511

4513

4514

4515

4516

4517

4518

4519

4521

4522

4523

4525

4526

4527

4528

4529

4530

4531

4532

4533

4534

4535

4536

4537

4538

4539

4540

4541

4542

4543

4544

4545

4546

4547

4548

4549

4550

4551

4552

4553

STRUCTURAL FORMULAS OF ORGANIC COMPOUNDS (continued)

In numeric order as they occur in the Table of Physical Constants of Organic Compounds

4554

4563

4573

4582

4592

4555

4564

4574

4583

4593

4556

4565

4575

4585

4594

4557

4566

4576

4586

4595

4558

4568

4577

4587

4596

4559

4569

4578

4588

4597

4560

4570

4579

4589

4598

4562

4571

4580

4590

4599

4572

4581

4591

4600

4601

4610

4619

4628

4636

4602

4611

4620

4629

4637

4603

4612

4621

4630

4638

4604

4613

4622

4631

4639

4605

4614

4623

4632

4640

4606

4615

4624

4633

4642

4607

4616

4625

4634

4643

4608

4617

4626

4635

4644

4609

4618

4627

4645

STRUCTURAL FORMULAS OF ORGANIC COMPOUNDS (continued)

In numeric order as they occur in the Table of Physical Constants of Organic Compounds

4646

4647

4648

4649

4650

4651

4652

4653

4654

4655

4656

4657

4658

4659

4660

4661

4662

4663

4664

4665

4666

4667

4668

4669

4670

4671

4672

4673

4674

4675

4676

4677

4678

4679

4680

4681

4682

4683

4684

4685

4686

4687

4688

4689

4690

4691

4692

4694

4695

4696

4697

4698

4699

4700

4701

4702

4703

4704

4705

4706

4707

4708

4709

4710

4711

4712

4713

4714

4715

4716

4718

4719

4720

4721

4722

4723

4724

4725

4726

4727

4728

4729

4731

4732

4733

4734

4735

4736

4737

4738

4739

4740

4741

4742

4743

4744

4745

4746

4747

4748

4749

4750

4751

4752

4753

4754

4755

4756

4757

4758

4759

4760

4761

4762

4763

4764

4765

4766

4767

4768

4770

4771

4772

4773

4774

4775

4776

4777

4778

4779

4780

4781

4782

4783

4784

4785

4786

4789

4790

4791

4792

4793

4794

4796

4797

4798

4799

4800

4801

4802

4803

4804

4805

4806

4807

4808

4809

4810

4811

4812

4813

4814

4815

4816

4817

4818

4819

4820

4821

4822

4823

4824

4825

4826

4827

4828

4829

4830

4831

4832

4833

4834

4835

4836

4837

4838

4839

4840

4841

4842

4843

4844

4845

4846

4847

4848

4849

4850

4851

4852

4853

4854

4855

4856

4857

4858

4859

4860

4861

4862

4863

4864

4865

4866

4867

4868

4869

4870

4871

4872

4873

4874

4875

4876

4877

4878

4879

4880

4881

4882

4883

4884

4885

4886

4887

4888

4889

4890

4891

4892

4893

4894

STRUCTURAL FORMULAS OF ORGANIC COMPOUNDS (continued)

In numeric order as they occur in the Table of Physical Constants of Organic Compounds

4895

4896

4897

4898

4899

4900

4901

4902

4903

4904

4905

4906

4907

4908

4909

4910

4911

4912

4913

4914

4915

4916

4917

4918

4919

4920

4921

4922

4923

4924

4925

4926

4927

4928

4929

4930

4931

4932

4933

4934

4935

4936

4937

4938

4939

4940

4941

4942

4943

4944

4945

4946

4947

4948

4949

4950

4951

4952

4953

4954

4955

4956

4957

4958

4959

4960

4961

4962

4963

4964

4966

4967

4968

4969

4970

4971

4972

4973

4974

4975

4976

4977

4978

4979

4980

4981

4982

4983

4984

4985

4986

4987

4988

4989

4990

4991

4992

4993

4994

4995

4996

4997

4998

4999

5000

5001

5002

5003

5004

5005

5006

5008

5009

5010

5011

5012

5013

5014

5015

5016

5017

5018

5020

5021

5022

5023

5024

5026

5027

5028

5029

5030

5032

5033

5034

5035

5036

5037

5038

5039

5040

5041

5042

5043

5044

5045

5046

5047

5048

5049

5050

5051

5052

5053

5054

5055

5056

5057

5058

5059

5060

5061

5062

5063

5064

5065

5066

5067

5068

5069

5070

5071

5072

5073

5074

5075

5076

5077

5078

5080

5081

5082

5083

5084

5085

5086

5087

5088

5089

5090

5091

5092

5093

5094

5095

5097

5098

5099

5100

5101

5102

5103

5104

5105

5106

5109

5110

5111

5112

5113

5114

5115

5116

5118

5120

5121

5122

5123

5124

5125

5126

5127

5128

5129

5130

5131

5132

5133

5134

5135

5136

5137

5140

5141

5142

5143

5144

5145

5146

5147

5148

5149

5150

5151

5152

5153

5154

5155

5156

5157

5158

5159

5160

5161

5162

5163

5164

5165

5166

5167

5168

5169

STRUCTURAL FORMULAS OF ORGANIC COMPOUNDS (continued)

In numeric order as they occur in the Table of Physical Constants of Organic Compounds

5170

5171

5172

5173

5174

5175

5176

5177

5178

5179

5180

5181

5182

5183

5184

5186

5187

5188

5189

5190

5192

5193

5194

5195

5196

5197

5200

5201

5202

5203

5204

5205

5206

5207

5208

5209

5210

5211

In numeric order as they occur in the Table of Physical Constants of Organic Compounds

5212

5222

5230

5239

5247

5213

5224

5232

5240

5248

5214

5225

5233

5241

5249

5215

5226

5234

5242

5250

5217

5227

5235

5243

5251

5219

5228

5236

5244

5252

5220

5229

5237

5245

5253

5221

5238

5246

5254

5257

STRUCTURAL FORMULAS OF ORGANIC COMPOUNDS (continued)

In numeric order as they occur in the Table of Physical Constants of Organic Compounds

5258

5259

5260

5261

5262

5263

5264

5265

5266

5267

5268

5269

5270

5271

5272

5274

5275

5276

5277

5278

5279

5280

5281

5282

5283

5284

5285

5286

5287

5288

5289

5290

5291

5292

5293

5294

5295

5296

5297

5298

5299

5300

5301

5302

5303

5304

5305

5307

5308

5319

5329

5339

5348

5309

5320

5330

5340

5349

5310

5321

5331

5341

5350

5311

5322

5332

5342

5351

5312

5323

5333

5343

5352

5314

5324

5334

5344

5354

5315

5325

5335

5345

5355

5316

5326

5337

5346

5356

5317

5327

5338

5347

5357

5318

5328

5358

5359

5361

5362

5363

5364

5365

5366

5367

5368

5369

5370

5371

5372

5373

5374

5376

5379

5383

5384

5385

5386

5387

5391

5393

5394

5396

5397

5398

5399

5400

5403

5404

5405

5406

5409

5410

STRUCTURAL FORMULAS OF ORGANIC COMPOUNDS (continued)

In numeric order as they occur in the Table of Physical Constants of Organic Compounds

5411

5412

5413

5414

5415

5416

5417

5419

5420

5421

5422

5423

5424

5426

5427

5429

5430

5431

5432

5436

5441

5442

5445

5448

5449

5450

5451

5452

5453

5454

5455

5456

5457

5458

5459

5460

5461

5462

5463

5464

5465

3-454

5466

5467

5468

5469

5470

5471

5472

5473

5474

5475

5476

5477

5478

5479

5480

5481

5482

5483

5484

5485

5486

5487

5488

5489

5490

5491

5492

5493

5495

5496

5497

5498

5499

5500

5501

5502

5503

5505

5506

5507

5508

STRUCTURAL FORMULAS OF ORGANIC COMPOUNDS (continued)

In numeric order as they occur in the Table of Physical Constants of Organic Compounds

5509

5510

5511

5512

5513

5514

5515

5516

5517

5518

5519

5520

5521

5522

5523

5524

5525

5526

5527

5528

5529

5530

5531

5532

5533

5534

5535

5536

5537

5538

5539

5540

5541

5542

5543

5544

5545

5546

5547

5548

5549

5550

5551

5552

5553

5554

5555

5556

5557

5558

5559

5560

5562

5563

5564

5565

5567

5569

5570

5571

5572

5573

5574

5576

5577

5578

5579

5580

5582

5583

5584

5585

5586

5588

5589

5590

5591

5592

5593

5595

5596

5597

5598

5599

5600

5601

5602

5603

5604

5605

5606

5607

5608

5609

5610

5611

5612

5613

5614

5615

5616

5617

5618

5619

5620

5621

5622

5626

5627

5628

5629

5630

5631

5632

5633

5634

5635

5636

5637

5638

5639

5640

5641

5642

5643

5653

5660

5667

5678

5644

5654

5661

5668

5679

5645

5655

5662

5669

5680

5646

5656

5663

5672

5681

5647

5657

5664

5673

5682

5648

5658

5665

5674

5683

5650

5659

5666

5675

5684

5651

5676

5652

5677

5685

5686

5696

5704

5712

5723

5687

5697

5705

5713

5724

5688

5698

5706

5714

5726

5689

5699

5707

5715

5727

5690

5700

5708

5716

5728

5691

5701

5709

5717

5729

5692

5702

5710

5718

5730

5693

5703

5711

5719

5731

5694

5695

5720

5721

5722

5732

5733

5734

5735

5736

5737

5738

5739

5740

5741

5742

5743

5744

5745

5746

5747

5748

5749

5750

5751

5752

5753

5754

5755

5756

5757

5758

5759

5760

5761

5762

5763

5764

5765

5766

5767

5768

5769

5770

5771

5772

5773

5774

5775

5776

5777

5779

5780

5781

5782

5783

5784

5786

5787

5788

5789

5790

5791

5792

5793

5794

5795

5796

5797

5798

5800

5801

5802

5803

5805

5806

5807

5808

5809

5810

5811

5812

5813

5814

5815

5816

5817

5818

5819

5820

5821

5822

5823

5824

5826

5827

5828

5829

5830

5831

5832

5833

5834

5835

5836

5837

5838

5839

5840

5841

5842

5843

5844

5845

5846

5847

5848

5849

5850

5851

5852

5853

5855

5856

5857

5858

5859

5860

5861

5862

5863

5864

5865

5866

5867

5868

5869

5870

5871

5880

5888

5895

5902

5872

5881

5889

5896

5903

5873

5882

5890

5897

5904

5874

5883

5891

5898

5905

5875

5884

5892

5899

5906

5876

5885

5893

5900

5907

5877

5886

5894

5901

5908

5879

5887

5909

5910

5911

5912

5913

5914

5915

5916

5917

5918

5919

5920

5921

5922

5923

5924

5925

5926

5927

5929

5930

5931

5932

5933

5934

5935

5936

5937

5938

5939

5940

5941

5942

5943

5944

5945

5946

5947

5948

STRUCTURAL FORMULAS OF ORGANIC COMPOUNDS (continued)

In numeric order as they occur in the Table of Physical Constants of Organic Compounds

5950

5951

5952

5953

5954

5955

5956

5957

5960

5961

5962

5963

5964

5965

5967

5968

5970

5971

5973

5974

5975

5977

5978

5979

5980

5981

5982

5984

5985

5986

5987

5988

5989

5990

5991

5992

5993

5994

5995

5996

5997

5999

6000

6001

6002

6004

6005

6006

6007

6008

6009

6010

6011

6012

6013

6014

6015

6016

6017

6018

6019

6020

6021

6024

6025

6026

6027

6028

6029

6031

6032

6033

6034

6035

6036

6037

6039

6040

6041

6042

6043

6044

6045

6046

6047

6049

6050

6051

6052

6053

6055

6056

6057

6058

6059

6060

6061

6062

6063

6064

6066

6067

6068

6069

6070

6071

6072

6073

6074

6075

6076

6077

6081

6082

6083

6084

6085

6086

6087

6088

6089

6090

6091

6092

6093

6094

6095

6096

6097

6098

6099

6101

6102

6103

6104

6105

6106

6107

6108

6109

6110

6111

6112

6113

6114

6115

6116

6117

6118

6119

6120

6121

6122

6123

6124

6125

6126

6127

6128

6129

6130

6131

6132

6133

6134

6135

6137

6138

6139

6140

6141

6142

6143

6145

6146

6147

6148

6149

6150

6151

6152

6153

6154

6155

6156

6157

6158

6159

6160

6163

6164

6165

6166

6167

6168

6169

6170

6171

6172

6174

6175

6176

6180

6181

6183

6184

6185

6186

6187

6188

6190

6191

6192

6194

6206

6221

6240

6260

6197

6208

6241

6262

6222

6199

6212

6243

6263

6200

6214

6227

6249

6265

6201

6216

6228

6250

6266

6202

6217

6229

6251

6267

6203

6218

6252

6268

6238

6254

6204

6220

6272

6239

6255

6275

6276

6277

6278

6279

6280

6282

6284

6287

6288

6289

6291

6292

6294

6295

6296

6297

6298

6299

6300

6301

6302

6304

6305

6306

6307

6308

6309

6310

6311

6312

6313

6314

6315

6316

6317

6318

6319

6320

6321

6322

6323

6324

6325

6326

6327

6328

6329

6330

6331

6341

6352

6363

6372

6332

6342

6353

6364

6373

6333

6343

6354

6365

6374

6334

6344

6355

6366

6375

6335

6345

6356

6367

6376

6336

6346

6357

6377

6337

6347

6358

6368

6378

6338

6348

6359

6369

6379

6339

6349

6360

6370

6380

6340

6350

6361

6371

6381

6351

6362

6383

6392

6402

6411

6420

6384

6393

6403

6412

6421

6385

6394

6404

6413

6422

6386

6395

6405

6423

6387

6396

6406

6414

6424

6388

6397

6407

6415

6425

6389

6398

6408

6416

6426

6390

6399

6409

6417

6427

6391

6400

6410

6418

6428

6401

6419

6429

6430

6431

6432

6433

6434

6435

6436

6437

6438

6439

6440

6441

6442

6443

6444

6445

6446

6447

6448

6449

6450

6451

6452

6453

6454

6455

6456

6457

6459

6461

6462

6463

6464

6465

6466

6467

6468

6469

6470

6471

6472

6473

6474

6475

6476

6477

6478

6479

6480

6481

6482

6483

6484

6485

6486

6487

6488

6489

6490

6491

6492

6493

6494

6495

6496

6497

6498

6499

6500

6501

6502

6503

6504

6505

6506

6508

6509

6510

6511

6512

6513

6514

6515

6517

6518

6519

6520

6521

6522

6523

6524

6525

6526

6527

6528

6529

6530

6531

6532

6533

6534

6535

6536

6537

6538

6539

6540

6541

6542

6543

6544

6545

6546

6547

6548

6549

6550

6551

6552

6553

6554

6555

6556

6557

6558

6559

6560

6561

6562

6563

6564

6565

6566

6567

6568

6570

6572

6573

6574

6575

6578

6579

6588

6596

6605

6615

6580

6589

6597

6606

6616

6581

6590

6598

6607

6617

6582

6591

6599

6608

6618

6583

6592

6600

6609

6619

6584

6593

6601

6610

6620

6585

6594

6602

6611

6621

6586

6595

6603

6612

6622

6587

6604

6613

6623

6614

6625

6626

6627

6628

6629

6630

6631

6632

6633

6634

6635

6636

6637

6638

6639

6640

6641

6642

6643

6644

6647

6649

6650

6651

6652

6653

6654

6655

6656

6657

6658

6659

6660

6661

6662

6663

6664

6665

6666

6667

6668

6669

6670

6671

6672

6673

6674

6675

6676

6677

6678

6679

6680

6681

6682

6683

6684

6685

6686

6687

6688

6689

6690

6691

6692

6693

6694

6695

6696

6697

6698

6699

6700

6701

6702

6703

6705

6706

6707

6708

6709

6710

6711

6712

6713

6714

6715

6716

6717

6718

6719

6720

6721

6722

6723

6724

6725

6726

6727

6728

6729

6730

6731

6732

6733

6734

6735

6736

6737

6738

6739

6740

6741

6742

6743

6744

6745

6746

6747

6748

6749

6750

6751

6752

6754

6755

6756

6757

6758

6759

6760

6761

6762

6763

6764

6766

6767

6768

6769

6770

6771

6772

6773

6774

6775

STRUCTURAL FORMULAS OF ORGANIC COMPOUNDS (continued)

In numeric order as they occur in the Table of Physical Constants of Organic Compounds

6776

6777

6778

6779

6781

6782

6783

6784

6785

6786

6787

6788

6789

6790

6791

6792

6793

6794

6795

6796

6797

6798

6799

6800

6801

6802

6803

6804

6805

6806

6807

6808

6809

6810

6811

6812

6813

6814

6815

6816

6817

6818

6819

6820

6821

6822

STRUCTURAL FORMULAS OF ORGANIC COMPOUNDS (continued)

In numeric order as they occur in the Table of Physical Constants of Organic Compounds

6823

6824

6825

6826

6827

6828

6829

6830

6831

6832

6833

6834

6835

6836

6837

6838

6839

6840

6841

6842

6843

6844

6845

6846

6847

6848

6849

6850

6851

6852

6853

6854

6855

6856

6857

6858

6859

6860

6861

6862

6863

6864

6865

6866

6868

6869

6870

6872

6873

In numeric order as they occur in the Table of Physical Constants of Organic Compounds

6874

6883

6892

6904

6914

6875

6884

6893

6905

6915

6876

6885

6894

6906

6916

6877

6886

6895

6907

6917

6878

6887

6896

6908

6918

6879

6888

6897

6909

6919

6880

6889

6898

6910

6921

6881

6890

6899

6911

6922

6882

6891

6900

6912

6923

6901

6913

6924

6932

6941

6949

6957

6925

6933

6942

6950

6958

6926

6934

6943

6951

6959

6927

6935

6944

6952

6960

6928

6936

6945

6953

6961

6929

6937

6946

6954

6962

6930

6938

6947

6955

6963

6931

6939

6948

6956

6964

6940

STRUCTURAL FORMULAS OF ORGANIC COMPOUNDS (continued)

In numeric order as they occur in the Table of Physical Constants of Organic Compounds

6965

6966

6967

6968

6969

6970

6971

6972

6973

6974

6975

6976

6977

6978

6979

6980

6981

6982

6983

6984

6985

6986

6987

6988

6989

6990

6991

6992

6993

6994

6995

6996

6997

6998

6999

7000

7001

7002

7003

7004

7005

7007

STRUCTURAL FORMULAS OF ORGANIC COMPOUNDS (continued)

In numeric order as they occur in the Table of Physical Constants of Organic Compounds

7008

7009

7010

7011

7012

7013

7014

7015

7016

7017

7018

7019

7020

7021

7022

7023

7024

7025

7026

7027

7028

7029

7030

7031

7032

7033

7034

7035

7036

7037

7038

7039

7040

7041

7042

7043

7044

7045

7046

7047

7048

7049

7050

7051

7053

7054

7055

7056

7057

7058

STRUCTURAL FORMULAS OF ORGANIC COMPOUNDS (continued)

In numeric order as they occur in the Table of Physical Constants of Organic Compounds

7060

7061

7062

7063

7064

7065

7066

7067

7068

7069

7070

7071

7072

7073

7074

7075

7076

7077

7078

7079

7080

7081

7082

7083

7084

7085

7086

7087

7088

7089

7090

7091

7092

7093

7094

7095

7096

7097

7098

7099

7100

7101

7102

7103

7104

7105

7106

7107

7108

7109

7113

7114

7115

7116

7117

7118

7119

7120

7121

7122

7123

7124

7125

7126

7127

7128

7129

7130

7131

7132

7136

7137

7138

7140

7142

7144

7145

7146

7147

7148

7149

7150

7151

7152

7153

7154

7155

7156

7158

7159

7160

7161

7164

7165

7167

7168

7170

7172

7175

7176

7177

7178

7179

7180

7181

7182

7183

7184

7185

7186

7187

7188

7189

7190

7192

7193

7194

7195

7196

7197

7198

7199

7200

7201

7202

7203

7205

7206

7207

7208

7209

STRUCTURAL FORMULAS OF ORGANIC COMPOUNDS (continued)

In numeric order as they occur in the Table of Physical Constants of Organic Compounds

7210

7218

7231

7240

7249

7211

7219

7232

7241

7250

7212

7222

7233

7242

7251

7213

7224

7235

7243

7252

7214

7225

7236

7244

7253

7215

7226

7237

7245

7254

7216

7227

7238

7246

7255

7217

7228

7239

7247

7257

7230

7258

7259

STRUCTURAL FORMULAS OF ORGANIC COMPOUNDS (continued)

In numeric order as they occur in the Table of Physical Constants of Organic Compounds

7260

7262

7263

7264

7267

7268

7269

7271

7272

7273

7274

7276

7277

7278

7279

7281

7282

7284

7285

7286

7287

7289

7290

7291

7292

7293

7294

7295

7296

7297

7298

7299

7300

7302

7303

7304

7305

7307

7308

7309

STRUCTURAL FORMULAS OF ORGANIC COMPOUNDS (continued)

In numeric order as they occur in the Table of Physical Constants of Organic Compounds

7310

7319

7327

7337

7346

7312

7320

7328

7338

7347

7313

7321

7329

7339

7350

7314

7322

7330

7340

7351

7315

7323

7331

7342

7354

7316

7324

7335

7343

7355

7317

7325

7336

7344

7356

7326

7345

7357

7367

7375

7386

7403

7368

7387

7404

7359

7376

7388

7405

7369

7378

7361

7389

7406

7370

7380

7362

7390

7408

7371

7381

7363

7398

7409

7383

7364

7372

7399

7410

7384

7365

7374

7402

7411

7385

STRUCTURAL FORMULAS OF ORGANIC COMPOUNDS (continued)

In numeric order as they occur in the Table of Physical Constants of Organic Compounds

7412

7413

7415

7416

7417

7418

7419

7420

7421

7422

7423

7425

7426

7427

7430

7431

7432

7433

7434

7435

7436

7438

7439

7441

7442

7443

7444

7446

7447

7448

7449

7450

7451

7452

7454

7455

7459

7460

7461

7462

7463

7464

7465

7466

7467

7468

7469

7470

7471

7472

7473

7474

7475

7476

7477

7478

7479

7480

7481

7482

7483

7484

7485

7486

7487

7489

7490

7491

7492

7493

7494

7495

7496

7497

7498

7499

7500

7501

7502

7503

7504

7505

7506

7507

7508

7509

7510

7512

7513

7515

7516

7517

7518

7519

7520

7521

7522

7523

7524

7525

7526

7527

7528

7529

7530

7531

7532

7533

7534

7535

7536

7537

7538

7539

7540

7541

7542

7549

7556

7564

7573

7543

7550

7557

7565

7574

7544

7551

7558

7567

7575

7545

7552

7559

7568

7576

7546

7553

7560

7569

7577

7547

7554

7561

7570

7578

7548

7555

7562

7571

7579

7572

7580

7581

7593

7604

7618

7631

7582

7594

7605

7619

7632

7585

7595

7606

7621

7633

7586

7597

7607

7622

7635

7587

7598

7609

7623

7636

7588

7599

7612

7624

7639

7590

7600

7614

7625

7640

7591

7601

7615

7626

7641

7592

7602

7616

7629

7642

7617

7643

7644

7645

7646

7647

7648

7649

7650

7651

7653

7654

7655

7656

7657

7658

7659

7660

7661

7662

7663

7665

7666

7667

7669

7670

7671

7672

7674

7675

7678

7679

7680

7681

7682

7683

7684

7693

7702

7713

7723

7685

7694

7703

7714

7724

7686

7696

7704

7715

7725

7687

7697

7705

7716

7727

7689

7698

7706

7717

7728

7690

7699

7708

7718

7730

7691

7700

7709

7720

7731

7692

7701

7710

7721

7732

7712

7722

7733

7734

7743

7758

7775

7790

7735

7746

7759

7776

7791

7736

7747

7764

7778

7792

7737

7749

7765

7779

7793

7738

7750

7766

7780

7794

7739

7751

7770

7781

7795

7740

7752

7771

7782

7796

7742

7753

7772

7783

7797

7754

7773

7784

7798

7756

7785

7789

7799

7801

7802

7803

7804

7806

7807

7808

7809

7810

7811

7812

7813

7814

7815

7816

7817

7818

7819

7820

7821

7822

7823

7824

7825

7826

7827

7828

7829

7830

7831

7832

7833

7834

7835

7836

7838

7839

7842

7843

7846

7848

7850

7852

7853

7859

7860

7884

7898

7909

7924

7885

7899

7911

7925

7861

7886

7900

7913

7926

7862

7887

7901

7915

7863

7889

7902

7916

7927

7864

7890

7903

7917

7928

7891

7904

7865

7892

7918

7929

7906

7920

7879

7894

7931

7907

7921

7880

7895

7932

7881

7908

7922

7896

7933

7882

7923

7897

7934

7946

7958

7981

7993

7935

7948

7959

7982

7994

7949

7966

7983

7995

7936

7950

7984

7996

7937

7951

7975

7985

7997

7938

7976

7986

7998

7952

7987

7999

7939

7977

7988

8000

7941

7954

7989

8001

7942

7955

7978

7990

8002

7943

7956

7979

7991

8003

7944

7957

7980

7992

7945

8005

8017

8030

8039

8052

8006

8018

8031

8041

8054

8007

8020

8032

8043

8055

8008

8022

8033

8044

8056

8009

8024

8034

8045

8057

8010

8026

8035

8046

8059

8011

8027

8037

8047

8061

8012

8028

8038

8048

8062

8013

8050

8015

8063

STRUCTURAL FORMULAS OF ORGANIC COMPOUNDS (continued)

In numeric order as they occur in the Table of Physical Constants of Organic Compounds

8064

8065

8066

8067

8068

8069

8070

8071

8072

8073

8074

8075

8076

8077

8078

8079

8080

8081

8082

8083

8084

8085

8086

8087

8088

8089

8090

8091

8092

8093

8094

8095

8096

8097

8098

8099

8101

8103

8104

8105

8106

8107

8108

8109

8110

8111

8112

8113

8114

STRUCTURAL FORMULAS OF ORGANIC COMPOUNDS (continued)

In numeric order as they occur in the Table of Physical Constants of Organic Compounds

8115

8116

8117

8118

8119

8120

8121

8122

8123

8124

8125

8126

8127

8128

8129

8130

8131

8132

8133

8134

8135

8136

8137

8138

8139

8140

8141

8142

8143

8144

8145

8146

8147

8148

8149

8150

8151

8152

8154

8155

8156

8157

8158

8159

8160

8161

8162

8163

8164

8165

8166

8167

8168

8169

8170

8171

8172

8173

8174

8175

8176

8177

8178

8179

8180

8181

8182

8183

8184

8185

8186

8187

8188

8189

8190

8191

8192

8193

8194

8195

8196

8197

8198

8199

8200

8201

8202

8203

8204

8205

8206

8207

8208

8209

8212

8213

8214

8215

8216

8218

8219

8220

8223

8237

8260

8281

8293

8226

8238

8262

8282

8294

8227

8263

8240

8283

8228

8241

8264

8284

8297

8265

8229

8242

8266

8285

8298

8230

8243

8267

8287

8299

8231

8244

8268

8288

8232

8245

8289

8301

8269

8246

8302

8233

8271

8290

8247

8303

8234

8273

8291

8250

8235

8278

8304

8251

8279

8236

8292

8257

8280

8305

8306

8307

8308

8309

8310

8311

8312

8313

8314

8315

8316

8317

8318

8319

8320

8321

8322

8324

8325

8327

8328

8329

8330

8331

8332

8333

8334

8335

8336

8337

8338

8339

8340

8341

8342

8343

8344

8345

8346

8347

8348

8349

8350

8351

8352

8353

8354

8355

8356

8357

8358

8359

8360

8361

8363

8364

8365

8366

8367

8368

8369

8370

8371

8372

8373

8374

8375

8376

8377

8378

8379

8380

8381

8382

8383

8384

8385

8386

8387

8388

8389

8390

8391

8392

8393

8394

8395

8396

8398

8399

8400

8401

8402

8403

8404

8405

8406

8407

8408

8409

8410

8411

8412

8415

8416

8417

8418

8419

8420

8421

8422

8424

STRUCTURAL FORMULAS OF ORGANIC COMPOUNDS (continued)

In numeric order as they occur in the Table of Physical Constants of Organic Compounds

8425

8436

8447

8457

8468

8426

8437

8449

8458

8469

8427

8438

8450

8459

8470

8428

8439

8451

8460

8471

8429

8440

8452

8461

8472

8430

8441

8453

8462

8473

8431

8442

8454

8464

8474

8432

8443

8455

8465

8475

8433

8444

8456

8466

8476

8434

8445

8477

8435

8446

8478

8479

8480

8481

8482

8484

8485

8486

8487

8488

8489

8490

8491

8492

8493

8494

8495

8496

8497

8498

8499

8500

8502

8503

8504

8505

8507

8508

8509

8510

8511

8512

8513

8514

8515

8516

8517

8518

8519

8520

8521

8523

8524

8525

8526

8527

8528

8529

8530

8531

8532

8533

8534

8535

8536

8537

8538

8539

8540

8541

8542

8543

8544

8545

8547

8548

8552

8553

8554

8555

8556

8557

8558

8559

8560

8562

8563

8564

8565

8566

8568

8569

8570

8571

8574

8575

8576

8577

8578

8579

8580

8581

8582

8583

8584

8585

8586

8587

STRUCTURAL FORMULAS OF ORGANIC COMPOUNDS (continued)

In numeric order as they occur in the Table of Physical Constants of Organic Compounds

8588

8589

8590

8591

8592

8593

8594

8595

8596

8597

8598

8599

8600

8601

8602

8603

8604

8605

8607

8608

8609

8610

8611

8612

8613

8614

8615

8616

8617

8618

8619

8620

8621

8622

8623

8624

8625

8626

8627

8628

8629

8632

8633

8635

8636

8637

8638

8639

8640

8641

STRUCTURAL FORMULAS OF ORGANIC COMPOUNDS (continued)

In numeric order as they occur in the Table of Physical Constants of Organic Compounds

8642

8643

8644

8645

8646

8647

8648

8649

8650

8651

8652

8653

8654

8655

8656

8657

8658

8659

8660

8661

8662

8663

8664

8665

8666

8667

8668

8669

8670

8671

8672

8673

8674

8675

8676

8679

8680

8681

8682

8683

8684

8685

8686

8687

8688

8689

8690

8691

8692

8693

8694

8695

8696

8697

8698

8707

8716

8726

8734

8699

8708

8717

8735

8709

8727

8736

8700

8718

8710

8737

8701

8728

8711

8719

8729

8739

8702

8720

8712

8730

8740

8703

8721

8713

8731

8741

8704

8722

8714

8732

8742

8705

8723

8733

8715

8743

8706

STRUCTURAL FORMULAS OF ORGANIC COMPOUNDS (continued)

In numeric order as they occur in the Table of Physical Constants of Organic Compounds

8744

8745

8746

8747

8748

8749

8750

8751

8752

8753

8754

8755

8756

8757

8758

8760

8761

8762

8763

8764

8765

8766

8767

8768

8769

8770

8771

8772

8773

8774

8775

8776

8777

8778

8780

8781

8783

8784

8785

8786

8787

8788

8789

8790

8791

8792

8793

8794

8795

8796

8797

8798

8799

8801

8802

8803

8804

8805

8806

8807

8808

8809

8810

8811

8812

8813

8814

8815

8816

8817

8818

8819

8820

8821

8822

8823

8824

8825

8826

8827

8828

8829

8830

8831

8832

8833

8834

8835

8836

8837

8838

8839

8840

8841

8842

8843

8844

8845

8846

8847

8848

8849

STRUCTURAL FORMULAS OF ORGANIC COMPOUNDS (continued)

In numeric order as they occur in the Table of Physical Constants of Organic Compounds

8850

8851

8852

8853

8854

8855

8856

8857

8858

8859

8860

8861

8862

8865

8866

8867

8868

8869

8870

8871

8872

8873

8874

8875

8876

8878

8879

8881

8882

8884

8885

8886

8887

8888

8889

8891

8892

8893

8894

8895

8896

8897

8898

8900

8903

8904

8905

8906

8907

8908

8909

8910

8911

8912

8913

8914

8915

8916

8917

8918

8919

8920

8922

8923

8924

8925

8927

8928

8929

8930

8931

8932

8933

8934

8935

8936

8937

8938

8939

8940

8941

8942

8943

8944

8945

8946

8947

8950

8952

8953

8954

8955

8964

8972

8980

8989

8965

8956

8973

8981

8990

8966

8957

8974

8982

8991

8967

8958

8975

8983

8992

8959

8968

8976

8984

8995

8985

8996

8961

8969

8977

8997

8986

8998

8962

8970

8978

8987

8999

8963

8979

9000

8971

8988

9001

STRUCTURAL FORMULAS OF ORGANIC COMPOUNDS (continued)

In numeric order as they occur in the Table of Physical Constants of Organic Compounds

9002

9004

9005

9006

9007

9008

9009

9010

9011

9012

9013

9014

9015

9016

9017

9018

9019

9020

9021

9022

9023

9024

9025

9026

9027

9028

9029

9030

9031

9032

9033

9034

9035

9036

9037

9038

9039

9040

9041

9042

9043

9044

9045

9046

9047

9048

9049

9050

9051

9052

9053

9054

9055

9056

9057

STRUCTURAL FORMULAS OF ORGANIC COMPOUNDS (continued)

In numeric order as they occur in the Table of Physical Constants of Organic Compounds

9058

9059

9060

9061

9062

9065

9066

9067

9068

9069

9070

9071

9072

9073

9074

9075

9076

9077

9078

9079

9080

9081

9082

9083

9084

9085

9086

9087

9088

9089

9090

9091

9092

9093

9094

9095

9096

9097

9098

9099

9100

9101

9102

9103

9104

9105

9106

9107

9108

9117

9126

9135

9143

9109

9118

9128

9136

9144

9110

9119

9129

9137

9145

9111

9121

9130

9138

9146

9112

9122

9131

9139

9147

9113

9123

9132

9140

9149

9114

9124

9133

9141

9150

9115

9125

9134

9142

9151

9116

9153

9154

9156

9158

9159

9160

9161

9162

9164

9167

9168

9170

9171

9172

9173

9174

9179

9180

9181

9182

9183

9184

9185

9186

9187

9188

9189

9190

9191

9192

9193

9195

9198

9199

9201

9202

9204

9205

9206

9208

9210

9211

9212

9213

9216

9217

STRUCTURAL FORMULAS OF ORGANIC COMPOUNDS (continued)

In numeric order as they occur in the Table of Physical Constants of Organic Compounds

9218

9219

9220

9221

9223

9224

9225

9226

9227

9228

9230

9231

9232

9233

9249

9250

9251

9253

9254

9255

9256

9257

9260

9262

9263

9264

9265

9266

9267

9270

9271

9272

9273

9276

9277

9279

9280

9282

9283

9284

9285

9287

9288

9289

9290

9291

9292

9294

9295

9296

9297

9298

9300

9305

9306

9307

9308

9309

9310

9312

9313

9314

9315

9316

9317

9319

9320

9321

9322

9323

9324

9326

9327

9328

9330

9331

9333

9334

9335

9336

9344

9355

9364

9377

9337

9345

9356

9365

9378

9338

9348

9357

9370

9379

9340

9349

9358

9371

9381

9341

9350

9359

9372

9382

9342

9351

9360

9374

9385

9343

9352

9362

9375

9387

9353

9376

9388

9389

9391

9392

9394

9395

9396

9397

9401

9402

9403

9404

9405

9407

9409

9410

9411

9412

9413

9414

9416

9417

9418

9419

9420

9421

9423

9424

9425

9426

9427

9428

9430

9431

9432

9434

9435

9436

9438

9439

9440

9441

9442

9443

9445

9446

9447

9448

9451

9452

9454

9455

9456

9457

9458

9459

9461

9462

9463

9465

9466

9469

9471

9472

9473

9474

9475

9479

9482

9483

9484

9485

9486

9489

9492

9493

9494

9495

9496

9497

9499

9501

9503

9504

9505

9507

9510

9512

9513

9514

9515

9517

9518

9519

9520

9521

9522

9523

9524

9525

9527

9528

9529

9530

9532

9535

9536

9537

9539

9540

9541

9543

9545

9546

9548

9549

9550

9551

9553

9554

9555

9556

9557

9558

9559

9560

9562

9563

9564

9565

9566

9567

9568

9569

9570

9571

9572

9573

9574

9575

9576

9577

9578

9579

9580

9581

9582

9583

9584

9585

9586

9587

9588

9589

9590

9600

9609

9616

9625

9591

9601

9610

9617

9626

9592

9602

9611

9618

9627

9593

9603

9612

9619

9628

9594

9604

9613

9620

9629

9595

9605

9614

9621

9630

9596

9606

9615

9622

9631

9597

9608

9624

9632

9599

STRUCTURAL FORMULAS OF ORGANIC COMPOUNDS (continued)

In numeric order as they occur in the Table of Physical Constants of Organic Compounds

9634

9644

9656

9666

9675

9635

9645

9658

9667

9676

9636

9648

9659

9668

9678

9637

9649

9660

9669

9679

9638

9650

9661

9670

9680

9639

9651

9662

9671

9681

9642

9652

9663

9673

9682

9643

9655

9664

9674

9683

9684

9692

9701

9710

9718

9685

9693

9702

9711

9719

9686

9695

9703

9712

9720

9687

9696

9704

9713

9721

9688

9697

9705

9714

9722

9689

9698

9706

9715

9723

9690

9699

9707

9716

9724

9691

9700

9708

9717

9726

9709

STRUCTURAL FORMULAS OF ORGANIC COMPOUNDS (continued)

In numeric order as they occur in the Table of Physical Constants of Organic Compounds

9727

9728

9729

9730

9731

9732

9733

9734

9736

9738

9739

9740

9741

9743

9745

9746

9748

9749

9750

9751

9752

9753

9754

9756

9758

9759

9760

9761

9762

9763

9764

9765

9766

9767

9768

9769

9770

9771

9772

9773

9774

9775

9784

9798

9808

9817

9776

9785

9799

9809

9818

9786

9800

9810

9777

9801

9819

9787

9811

9820

9778

9788

9803

9812

9821

9779

9790

9804

9813

9822

9780

9791

9805

9814

9823

9781

9792

9806

9815

9824

9782

9793

9807

9816

9825

9783

9826

9827

9828

9829

9832

9834

9835

9836

9837

9838

9839

9840

9841

9842

9843

9844

9845

9846

9847

9848

9849

9850

9851

9852

9853

9854

9855

9856

9857

9858

9859

9860

9861

9862

9863

9864

9865

9866

9867

9868

9869

9870

9871

9872

9873

9874

9875

9876

9885

9894

9903

9913

9877

9886

9895

9904

9914

9878

9887

9896

9905

9915

9879

9888

9897

9906

9916

9880

9889

9898

9907

9917

9881

9890

9899

9908

9918

9882

9891

9900

9909

9919

9883

9892

9901

9910

9920

9884

9893

9902

9911

9912

STRUCTURAL FORMULAS OF ORGANIC COMPOUNDS (continued)

In numeric order as they occur in the Table of Physical Constants of Organic Compounds

9921

9922

9923

9924

9925

9926

9927

9928

9930

9931

9932

9933

9934

9935

9936

9937

9938

9939

9940

9941

9942

9943

9944

9945

9946

9947

9948

9949

9950

9951

9952

9953

9954

9955

9957

9958

9959

9960

9961

9962

9963

9964

9966

STRUCTURAL FORMULAS OF ORGANIC COMPOUNDS (continued)

In numeric order as they occur in the Table of Physical Constants of Organic Compounds

9967

9978

9989

9999

10007

9969

9980

9990

10000

10008

9970

9981

9991

10001

10009

9972

9982

9992

10002

10010

9973

9983

9994

10003

10011

9974

9985

9994

10004

10012

9975

9986

9995

10005

10013

9976

9987

9997

10006

10014

9977

9998

10016

STRUCTURAL FORMULAS OF ORGANIC COMPOUNDS (continued)

In numeric order as they occur in the Table of Physical Constants of Organic Compounds

10017

10018

10019

10020

10021

10022

10023

10025

10026

10027

10028

10029

10030

10031

10032

10033

10034

10035

10036

10037

10038

10039

10040

10041

10042

10043

10044

10045

10046

10047

10048

10049

10050

10051

10052

10053

10054

10055

10056

10057

10058

10059

10060

10061

10062

10063

10064

STRUCTURAL FORMULAS OF ORGANIC COMPOUNDS (continued)

In numeric order as they occur in the Table of Physical Constants of Organic Compounds

10065

10075

10084

10093

10102

10066

10076

10085

10094

10103

10067

10077

10086

10095

10105

10068

10078

10087

10096

10106

10069

10079

10088

10097

10107

10070

10080

10089

10098

10108

10071

10081

10090

10099

10109

10072

10082

10091

10100

10110

10073

10083

10092

10101

10111

10074

STRUCTURAL FORMULAS OF ORGANIC COMPOUNDS (continued)

In numeric order as they occur in the Table of Physical Constants of Organic Compounds

10112

10113

10114

10115

10116

10117

10118

10119

10120

10121

10122

10123

10124

10125

10126

10127

10128

10129

10130

10131

10132

10133

10134

10135

10136

10138

10139

10140

10141

10142

10143

10144

10145

10146

10147

10148

10149

10150

10151

10152

10153

10154

10155

STRUCTURAL FORMULAS OF ORGANIC COMPOUNDS (continued)

In numeric order as they occur in the Table of Physical Constants of Organic Compounds

10156

10157

10158

10159

10160

10161

10162

10163

10164

10166

10167

10168

10169

10170

10171

10172

10173

10174

10175

10176

10177

10178

10179

10180

10181

10182

10183

10184

10185

10186

10187

10188

10189

10190

10191

10192

10193

10194

10195

10196

10197

10198

10199

10200

10201

10202

10203

10204

10205

10207

10208

10209

10210

10211

10212

10213

10214

10215

10216

10217

10218

10219

10220

10221

10222

10223

10225

10226

10227

10228

10229

10230

10231

10232

10233

10234

10235

10236

10237

10238

10239

10240

10241

10242

10243

STRUCTURAL FORMULAS OF ORGANIC COMPOUNDS (continued)

In numeric order as they occur in the Table of Physical Constants of Organic Compounds

10244

10254

10263

10275

10283

10245

10255

10264

10276

10285

10246

10256

10265

10277

10286

10248

10258

10268

10278

10287

10249

10259

10269

10279

10288

10250

10260

10270

10280

10289

10251

10261

10271

10281

10290

10253

10262

10273

10274

10282

10291

10292

10300

10309

10318

10328

10293

10301

10310

10319

10329

10294

10302

10311

10320

10330

10295

10303

10312

10321

10331

10296

10304

10313

10322

10332

10297

10306

10314

10323

10333

10298

10307

10315

10324

10334

10299

10308

10316

10325

10335

10317

10326

10336

STRUCTURAL FORMULAS OF ORGANIC COMPOUNDS (continued)

In numeric order as they occur in the Table of Physical Constants of Organic Compounds

10337

10338

10339

10340

10341

10342

10343

10344

10345

10346

10347

10348

10349

10350

10351

10352

10353

10354

10355

10356

10357

10358

10359

10360

10361

10362

10363

10364

10365

10366

10367

10368

10369

10370

10371

10372

10373

10374

10375

10376

10378

10380

10381

10382

10383

STRUCTURAL FORMULAS OF ORGANIC COMPOUNDS (continued)

In numeric order as they occur in the Table of Physical Constants of Organic Compounds

10384

10385

10386

10387

10388

10389

10390

10391

10393

10394

10395

10396

10398

10399

10401

10402

10404

10405

10406

10407

10408

10409

10410

10411

10412

10413

10414

10415

10416

10417

10418

10419

10420

10421

10422

10423

10424

10425

10426

10427

10428

10429

10430

10431

10432

10433

10436

10437

10438

10439

10440

10441

10442

10443

10444

10445

10446

10447

10448

10449

10450

10451

10452

10454

10455

10457

10458

10460

10461

10462

10463

10464

10465

10466

10467

10468

10469

10470

10471

10472

10473

10474

10475

10476

10478

10486

10493

10501

10509

10479

10487

10494

10502

10510

10480

10488

10495

10503

10511

10481

10489

10496

10504

10512

10482

10490

10497

10505

10514

10483

10491

10498

10506

10515

10484

10492

10500

10507

10516

10485

10508

10517

10518

STRUCTURAL FORMULAS OF ORGANIC COMPOUNDS (continued)

In numeric order as they occur in the Table of Physical Constants of Organic Compounds

10519

10528

10537

10546

10554

10520

10529

10538

10547

10555

10521

10530

10539

10548

10556

10522

10531

10540

10549

10557

10523

10532

10541

10550

10558

10524

10533

10542

10551

10559

10525

10534

10543

10552

10560

10526

10535

10544

10553

10561

10527

10536

10545

10562

10571

10580

10589

10599

10600

10563

10572

10581

10590

10601

10564

10573

10582

10591

10602

10565

10583

10574

10592

10603

10566

10575

10584

10593

10604

10567

10576

10585

10594

10605

10568

10577

10586

10596

10569

10578

10587

10597

10606

10570

10579

10588

10598

10607

10618

10628

10638

10649

10608

10620

10629

10639

10650

10609

10621

10631

10640

10652

10611

10622

10632

10643

10653

10613

10624

10633

10644

10654

10614

10625

10634

10645

10655

10616

10626

10635

10647

10656

10617

10627

10636

10648

10657

10637

10658

10659

10668

10680

10692

10702

10660

10670

10681

10693

10703

10661

10671

10683

10694

10704

10662

10672

10685

10695

10706

10663

10673

10686

10696

10707

10664

10674

10687

10697

10708

10665

10677

10688

10698

10709

10666

10679

10689

10699

10710

10691

10700

10711

10721

10730

10739

10750

10713

10722

10731

10743

10751

10714

10723

10732

10744

10752

10715

10724

10733

10745

10753

10716

10725

10734

10746

10755

10717

10726

10735

10747

10718

10727

10736

10748

10756

10719

10728

10737

10749

10757

10720

10729

10738

10758

10768

10779

10789

10799

10769

10780

10790

10800

10759

10781

10791

10801

10770

10760

10782

10792

10802

10773

10761

10793

10803

10783

10774

10794

10763

10784

10805

10775

10764

10787

10795

10806

10777

10767

10796

10807

10788

10778

10798

10809

10810

10811

10813

10814

10815

10816

10818

10819

10820

10821

10822

10823

10824

10825

10826

10827

10828

10829

10830

10831

10832

10833

10834

10835

10836

10837

10838

10839

10840

10841

10842

10844

10845

10846

10847

10849

10851

10852

10853

10855

10864

10873

10883

10892

10856

10875

10884

10893

10865

10858

10876

10866

10886

10894

10859

10877

10868

10887

10895

10860

10878

10869

10888

10896

10861

10879

10870

10890

10897

10862

10880

10871

10891

10898

10863

10882

10872

STRUCTURAL FORMULAS OF ORGANIC COMPOUNDS (continued)

In numeric order as they occur in the Table of Physical Constants of Organic Compounds

10899

10901

10903

10905

10906

10907

10908

10909

10911

10914

10915

10916

10917

10918

10920

10921

10922

10923

10924

10925

10926

10927

10929

10932

10933

10935

10936

10937

10938

10939

10940

10941

10942

10943

10944

10945

10946

10947

10948

10949

10959

10968

10977

10988

10951

10960

10969

10980

10989

10952

10961

10970

10981

10990

10953

10962

10972

10982

10991

10954

10963

10973

10983

10992

10955

10965

10974

10984

10994

10956

10966

10975

10985

10995

10958

10967

10976

10987

10996

10997

10998

10999

11001

11002

11003

11004

11005

11006

11007

11008

11009

11010

11012

11013

11014

11016

11017

11018

11019

11020

11021

11023

11024

11025

11026

11027

11028

11029

11031

11032

11033

11034

11035

11036

11038

11039

11040

11041

11042

11043

11045

11049

11050

11051

11052

11054

11055

11056

11057

11058

11059

11062

11064

11066

11067

11068

11069

11071

11072

11075

11076

11078

11080

11081

11082

11083

11084

11085

11087

11088

11090

11091

11092

11093

11094

11095

11098

11099

11100

11101

11103

11104

11105

11106

11109

11110

11112

11113

11114

11115

11116

11117

11118

11119

11120

11121

11122

11124

11125

11126

11127

11128

11129

11130

11131

11132

11133

11134

11135

11136

11137

11138

11139

11141

11142

11143

11146

11147

11148

11149

11150

11151

11152

11153

11154

11155

11156

11157

11158

11160

11161

11163

11164

11165

11166

11167

11168

11169

11170

11171

11172

11173

11175

11176

11180

11181

11182

11183

11184

11185

11186

11187

11188

11189

11190

11191

11193

11194

11195

11208

11219

11231

11242

11199

•11209

11220

11232

11243

11201

11210

11222

11236

11244

11202

11223

11211

11237

11245

11203

11213

11224

11238

11246

11204

11214

11225

11239

11247

11206

11217

11227

11240

11248

11218

11230

11241

11250

11251

11262

11270

11280

11288

11254

11263

11271

11281

11289

11255

11264

11274

11282

11290

11257

11265

11275

11283

11291

11258

11266

11276

11284

11292

11259

11267

11277

11285

11296

11260

11268

11278

11286

11297

11261

11269

11279

11287

11298

11299

11307

11318

11330

11340

11300

11308

11319

11331

11341

11301

11311

11320

11333

11342

11302

11312

11321

11334

11343

11303

11313

11325

11335

11344

11304

11314

11327

11337

11345

11305

11315

11328

11338

11348

11306

11329

11339

11350

11317

11351

11361

11369

11377

11385

11362

11370

11378

11386

11353

11354

11363

11371

11380

11387

11388

11356

11364

11372

11381

11357

11365

11373

11382

11389

11390

11358

11366

11374

11383

11391

11375

11359

11367

11376

11384

11392

11360

11368

STRUCTURAL FORMULAS OF ORGANIC COMPOUNDS (continued)

In numeric order as they occur in the Table of Physical Constants of Organic Compounds

11393

11394

11395

11396

11397

11398

11399

11400

11401

11402

11403

11404

11405

11406

11407

11408

11409

11410

11411

11412

11413

11414

11415

11417

11418

11419

11420

11421

11422

11423

11424

11425

11426

11427

11428

11429

11430

11431

11432

11433

11434

11435

11436

11437

11438

11439

11440

11443

11444

11445

11446

11447

11448

11449

11450

11451

11452

11453

11454

11455

11456

11457

11458

11459

11460

11461

11462

11463

11464

11465

11466

11467

11468

11469

11470

11471

11472

11474

11475

11476

11477

11478

11479

11480

11481

11482

11483

11484

11485

11486

11487

11488

11489

11490

11491

11492

11493

11494

11495

11496

11497

11498

11499

11500

11501

11502

11503

11504

11505

11506

11507

11508

11509

11510

11511

11512

11513

11514

11517

11519

11520

11521

11527

11541

11551

11564

11575

11528

11542

11555

11567

11576

11529

11543

11556

11568

11577

11531

11544

11558

11569

11578

11532

11545

11559

11570

11579

11536

11546

11560

11571

11580

11537

11549

11562

11572

11581

11539

11550

11563

11573

11574

11582

11540

11583

11593

11603

11617

11631

11604

11618

11632

11584

11594

11605

11619

11633

11585

11595

11606

11621

11634

11607

11622

11635

11609

11586

11596

11624

11636

11610

11625

11598

11637

11611

11588

11626

11638

11599

11612

11627

11640

11613

11590

11628

11614

11641

11600

11615

11629

11642

11591

11601

11616

11630

11643

11592

11602

11644

11656

11666

11675

11683

11646

11657

11667

11676

11684

11647

11658

11668

11677

11685

11648

11659

11670

11678

11687

11649

11660

11671

11679

11688

11650

11661

11672

11680

11689

11651

11662

11673

11681

11690

11652

11664

11674

11682

11691

11654

11665

11692

11693

11694

11695

11696

11697

11698

11699

11700

11701

11703

11704

11705

11706

11707

11708

11709

11710

11712

11713

11714

11715

11716

11717

11719

11720

11721

11722

11723

11724

11725

11726

11727

11728

11729

11730

11731

11732

11733

11734

11735

11736

11737

11738

STRUCTURAL FORMULAS OF ORGANIC COMPOUNDS (continued)

In numeric order as they occur in the Table of Physical Constants of Organic Compounds

11739

11748

11756

11764

11773

11740

11749

11757

11765

11774

11741

11750

11758

11767

11775

11742

11751

11759

11768

11776

11743

11752

11760

11769

11777

11744

11753

11761

11770

11778

11746

11754

11762

11771

11779

11747

11755

11763

11772

11780

11781

11789

11797

11807

11819

11782

11790

11798

11808

11822

11783

11791

11799

11809

11823

11784

11792

11800

11810

11824

11785

11793

11801

11812

11826

11786

11794

11803

11813

11827

11787

11795

11804

11816

11828

11788

11796

11805

11817

11829

11806

11831

11842

11854

11865

11878

11866

11833

11845

11855

11879

11856

11867

11834

11846

11857

11880

11868

11835

11847

11858

11881

11869

11836

11848

11859

11870

11882

11860

11871

11883

11837

11849

11861

11872

11884

11852

11873

11885

11840

11862

11874

11886

11853

11875

11887

11841

11863

11877

11888

11889

11890

11901

11911

11921

11938

11891

11939

11902

11912

11922

11940

11892

11903

11926

11941

11893

11913

11894

11904

11914

11929

11942

11895

11943

11905

11915

11944

11897

11931

11945

11898

11906

11916

11934

11946

11899

11908

11917

11935

11947

11900

11909

11919

11936

11948

11910

11920

11937

11949

11950

11962

11974

11987

11999

11951

11963

11975

11988

12000

11953

11964

11977

11990

12001

11954

11965

11978

11991

12002

11955

11966

11979

11992

12004

11956

11967

11980

11993

12005

11957

11968

11981

11994

12006

11958

11969

11982

11995

11959

11970

11983

11996

12007

11960

11971

11984

11961

11972

11985

11973

11986

STRUCTURAL FORMULAS OF ORGANIC COMPOUNDS (continued)

In numeric order as they occur in the Table of Physical Constants of Organic Compounds

12009

12018

12032

12043

12058

12010

12019

12033

12044

12059

12012

12026

12034

12047

12061

12013

12027

12037

12048

12062

12014

12028

12038

12049

12065

12015

12029

12039

12050

12066

12016

12030

12040

12055

12067

12017

12031

12041

12056

12068

12057

3-585

Abadol	: 11662
Abamectin	: 477
7,13-Abietadien-18-oic acid	: 9167
Abietic acid	: 9167
L-Abrine	: 11923
Abscisic acid	: 8679
(+)-Abscisin II	: 8679
Abufene	: 323
Acacetin	: 2809
Acedapsone	: 91
Acemin	: 6871
Acenaphthalene	: 3
Acenaphthaquinone	: 8
1-Acenaphthenamine	: 2
Acenaphthene	: 6
Acenaphthene, 1-amino	: 2
Acenaphthene, 5-bromo-	: 4
Acenaphthene, 5-chloro-	: 5
Acenaphthenedione	: 8
Acenaphthene, 5-iodo-	: 7
Acenaphthenequinone	: 8
Acepramin	: 6873
Acepromazine	: 5807
Acetal	: 5541
Acetaldehyde ammonia	: 5674
Acetaldehyde, cyanohydrin	: 10105
Acetaldehyde methyl acetal	: 5548
Acetaldehyde, tribromo, hydrate	: 5581
Acetaldehyde, trichloro-, diethyl acetal	: 5657
Acetaldol	: 3184
Acetaldoxime	: 17
Acetamide, N-(butoxymethyl)-2-chloro-N-(2,6-diethylphenyl)-	: 3144
Acetamide, 2-chloro-N-(2,6-diethylphenyl)-N-(methoxymethyl)-	: 322
Acetamide, 2-chloro-N-(ethoxymethyl)-N-(2-ethyl-6-methylphenyl)-	: 270
Acetamide, 2-chloro-N-(2-ethyl-6-methylphenyl)-N-(2-methoxy-1-methylethyl)-	: 7657
Acetamide, 2-chloro-N-(1-methylethyl)-N-phenyl-	: 9771
Acetamide, N-[2,4-dimethyl-5-[[(trifluoromethyl)sulfonyl]amino]phenyl]-	: 7427
Acetamidine, hydrochloride	: 5671
2-Acetamidoacetic acid	: 6251
p-Acetamidoaniline	: 29
4-Acetamidobenzenesulfonamide	: 31
ε -Acetamidocaproic acid	: 6871
2-Acetamido-5-nitrothiazole	: 85
p-Acetamidophenol	: 67
Acetaminobenzene	: 86
Acetaminophen	: 67
Acetanilide	: 86
Acetanilide, 4'-hydroxy-	: 67
Acetanilide, N-isopropyl	: 74
Acetanilide, 2-nitro-	: 80
Acetanilide, N-propyl-	: 88
Acetanilide, 4'-sulfamoyl-	: 31
Acetanilide-p-sulfonyl chloride	: 2243
p-Acetanisidine	: 71
Acetazolamide	: 32
Acetic acid amide	: 22
Acetic acid, amide,(4(carboxy(propyl ester)methyl)-2-methoxyphenoxy)-N,N-diethyl	: 894
Acetic acid, amide,N,N-dimethyl	: 55
Acetic acid, amide,N(4-methoxyphenyl)	: 71
Acetic acid, amide, 1-naphthyl	: 7733
Acetic acid, [(benzoylamino)oxy]-	: 524
Acetic acid, bromide	: 291
Acetic acid, (4-bromophenyl)-hydroxy(DL)	: 889
Acetic acid, chlorophenyl-, ethyl ester, L-(+)-	: 891
Acetic acid, cyclohexyl	: 4306
Acetic acid, cyclopentyl ester	: 4865
Acetic acid, (1,1-dimethylbutyl) ester	: 8946
Acetic acid, diphenyl,2(diethylamino) ethyl ester,hydrochloride	: 938
Acetic acid, fluoride	: 301
Acetic acid, (p-fluorophenyl)-	: 901
Acetic acid, 2-methoxyethyl ester	: 5734
Acetic acid, (N-methyldodecanamido), sodium salt	: 6269
Acetic acid, [3,4-(methylenedioxy)phenyl]-, methyl ester	: 2416
Acetic acid, (6-methyl-2-heptyl) ester	: 6506
Acetic acid, (6-methyl-3-heptyl) ester	: 6507
Acetic acid, (2-methylpentyl) ester	: 8945
Acetic acid, (2-methyl-3-pentyl) ester	: 8948
Acetic acid, (3-methyl-3-pentyl) ester	: 8949
Acetic acid, 2-octyl ester(DL)	8447
Acetic acid, (2-pentyl) ester (l)	: 8918
Acetic acid, 3-phenylpropyl	: 2099
Acetic acid, sulfo, diethyl ester	: 184
Acetic acid, [(3,5,6-trichloro-2-pyridinyl)oxy]-	: 11854
Acetic acid, trifluoro,(1,2-dimethyl-1-ethylbutyl) ester	: 11888
Acetic anhydride	: 107
Acetimidic acid	: 22
Acetimidic acid, N-[(methylcarbamoyl)oxy]thio-, methyl ester	: 7650
Acetoacetanilide	: 3207
Acetoacetic acid	: 3568
Acetoacetic acid, 2-(2-hydroxyethyl)-, γ-lactone	: 6139
Acetoacetylaniline	: 3207
Acetocinnamone	: 3881
Acetoin	: 3661
1'-Acetonaphthone, 2-phenyl-	: 5859
Acetone	10411
Acetone, 1-bromo-3-chloro	10417
Acetonecarboxylic acid	: 3568
Acetone cyanohydrin	10108
Acetonedicarboxylic acid	: 8782
Acetone diethyl sulfone	: 9835
Acetone monoglycerol ketal	: 5207
Acetone, oxime	10476
Acetone, phenylhydrazone	10482
Acetonylacetone	: 6805
3-(α-Acetonylbenzyl)-4-hydroxycoumarin	: 2832
Acetophene,2-nitro	: 5863
Acetophenone	: 5864
Acetophenone, 3-amino	: 5772
Acetophenone, 4-amino	: 5773
Acetophenone, 3-bromo	: 5780
Acetophenone, 4-bromo	: 5781
Acetophenone, 2-bromo-	: 5782
Acetophenone, 2-bromo-4'-chloro-	: 5778
Acetophenone, 2-bromo-4'-methyl-	: 5779
Acetophenone, 3-chloro	: 5788
Acetophenone, 4-chloro	: 5789
Acetophenone, 2-chloro-	: 5790
Acetophenone, 2,4'-dibromo-	: 5777
Acetophenone, 3,5-dibromo	: 5797
Acetophenone, 2,5-dichloro	: 5800
Acetophenone, 3,4-dichloro	: 5801
Acetophenone, 2,2-dichloro-	: 5802
Acetophenone, 2,4-dihydroxy	: 5803
Acetophenone, 2-ethoxy-	: 5814
Acetophenone, 4-fluoro	: 5815
Acetophenone, 3-iodo	: 5836
Acetophenone, 4-iodo	: 5837
Acetophenone, 2-iodo-	: 5838
Acetoxime	10476
2-Acetoxyethanol	: 5577
Acetoxytributylstannane	11538
Acetoxytriphenylstannane	11539
Acetozone	: 9151
Aceturic acid	: 6251
α-Acetylacetanilide	: 3207
Acetyl acetate	: 107
Acetylacetone	: 8798
Acetylacetone, hexafluoro	: 8802
Acetylaminoacetic acid	: 6251
p-Acetylaminobenzenesulfonyl chloride	: 2243
2-(Acetylamino)fluorene	: 64
N-Acetylaniline	: 86
3-Acetylaniline	: 5772
4-Acetylaniline	: 5773
Acetylbenzene	: 5864
2-Acetylbenzofuran	: 5775
4-Acetylbiphenyl	: 5776
p-Acetylbromobenzene	: 5781
α-Acetylbutyrolactone	: 6139
2-Acetyl-5-chlorothiophene	: 5791
Acetyl choline bromide	: 5438
Acetyl choline chloride	: 5439
1-Acetylcyclohexene	: 5793
2-Acetylcyclopentanone	: 4879
Acetylcysteine	: 5015
Acetyl disulfide	: 5250
Acetylene	: 5936
Acetylene, bromo-	: 5937
Acetylene carbamide	: 7191
Acetylene, chloro-	: 5938
Acetylene, dibromo-	: 5939
Acetylene, dichloro-	: 5940
Acetylene, diiodo-	: 5941
Acetylene, fluoro-	: 5943
Acetylene, phenyl	: 1715
Acetylene tetrabromide	: 5632
Acetylene tetrachloride	: 5635
Acetyleneurea	: 7191
2-Acetyl-4-hydroxybutyric acid γ-lactone	: 6139
N-Acetylmethionine	: 7649
Acetyl methyl carbinol	: 3661
Acetyl nitrate	: 108
Acetyl peroxide	: 9156
Acetyl-p-phenylenediamine	: 29
Acetylpropionyl	: 8797
3-Acetylpyridine	: 5868
4-Acetylpyrogallol	: 5873
Acetylsalicylic acid	: 2480
Acetylsalicylic acid, methyl ester	: 2482
Acetylsalicylic acid, phenyl ester	: 2483
Acetylsulfanilamide	: 31
Acetylsulfanilyl chloride	: 2243
Acetylthiourea	: 33
Acetyl valeryl	: 6402
Acexamic acid	: 6871
Acifloctin	: 6779
Acocantherin	: 4040
Aconine	: 305
Aconitic acid, triethyl ester (trans)	: 10601
Aconitine	: 306
Acorn Sugar	: 4074
Acroleic acid	10809
Acrolein	10509
Acrolein, diethyl acetal	10560
Acrylaldehyde	10509
Acrylamide	10519
Acrylic acid	10609
Acrylic acid chloride	10714
Acrylic acid, chloro-, ethyl ester	10619
Acrylic acid, 3,3-dibromo-	10628
Acrylic acid, 3,3-dichloro-	10630
Acrylic acid, 3-(2-furyl), pentyl ester	10641
Acrylic acid, 3-(2-furyl), propyl ester	10642
Acrylic acid, (2-methylbutyl) ester	10652
Acrylic acid, 3-(1-naphtyl) (trans)	10875
Acrylic aldehyde	10509
Acrylic amide	10519
Acrylon	10583
Acrylonitrile	10583
Actinoquinol	11293
Active amyl alcohol	: 3628
Activol	: 9203
Actriol	: 5405
1-Adamantamine	11859
Adamantane	11860
Adamantane, 1(pyrid-3-ylmethylamino),dihydrochloride	: 7175
1-Adamantylamine	11859
Adenine	10751
Adenosine 5'-monophosphate	: 320
Adiphenine hydrochloride	: 938
Adipic acid	: 6779
Adipic acid bis(3-carboxy-2,4,6-triiodoanilide)	: 2594
Adipic acid, di-2-ethylbutyl ester	: 6780
Adipic acid, 3-methyl (DL)	: 6789
Adipic acid, 3-methyl, diethyl ester	: 6790
Adipic ketone	: 4878
Adipiodone	: 2594

Adiponitrile	: 6777	Aluminum lactate	: 340	2-Aminoglutaric acid	: 6241
Adol	: 4900	Amanozine	: 11830	Aminoglycol	: 9987
Adonitol	: 11343	Amantadine	: 11859	Aminoguanidine	: 7147
Adrenalone	: 5804	Amediol	: 9987	2-Aminohexanoic acid	: 8214
Adrenochrome	: 7301	Amethocaine monohydrochloride	: 2531	p-Aminohippuric acid	: 6252
Adrenosterone	: 353	Amidazophen	: 10851	1-Amino-2-hydroxybenzene	: 9201
Aflatoxin B2	: 4757	N-Amidinoglycine	: 6253	1-Amino-3-hydroxybenzene	: 9202
Aflatoxin G1	: 6184	Amidocyanogen	: 4155	4-Amino-1-hydroxybenzene	: 9203
Agaricic acid	: 8102	Amido-G-Acid	: 7870	(S)-2-Amino-4-hydroxybutanoic acid	: 7134
Agaricin	: 8102	Amikapron	: 4333	[R-(R*,S*)]-2-Amino-3-	: 11813
Agaritine	: 6247	Aminacrine	: 313	hydroxybutanoic acid	
Agifutol S	: 6264	Aminate base	: 7147	(S)-2-Amino-3-hydroxypropanoic	: 11357
Agroxone	: 141	Aminetrimethylboron	: 3120	acid	
Ahistan	: 9502	Aminic acid	: 6002	2-Amino-6-hydroxypurine	: 10768
Ajmaline	: 321	Aminitrozole	: 85	2-Aminohypoxanthine	: 10768
Akalbir	: 2842	2-(Aminoacetamido)acetic acid	: 6265	(RS)-1-Aminoindan	: 7243
Alanine, 3-[(carboxymethyl)thio]-, L-	: 5017	p-Aminoacetanilide	: 29	l-α-Aminoindole-3-propionic acid	: 11922
Alant camphor	: 8059	2-Aminoacetic acid	: 6250	2-Aminoisobutanol	: 10341
Alantolactone	: 8059	Aminoacetic acid	: 6250	α-Aminoisobutyric acid	: 331
Alar	: 3334	Aminoacetic acid, N,N-	: 6258	5-Aminoisoquinoline	: 7372
Albutoin	: 7229	(dicarboethoxy), ethyl ester		2-Aminoisovaleric acid	: 12041
Alcanfor	: 2961	Aminoacetic acid hydrazide	: 6266	(R)-2-Amino-3-mercaptopropanoic	: 5014
Aldocortene	: 9746	Aminoacetic acid, N-succinyl, ethyl	: 11145	acid	
Aldocortin	: 9746	ester		Aminomethanamidine	: 6278
Aldol	: 3184	Aminoacetonitrile	: 272	1-Amino-3-methylbutane	: 3220
Aldosterone	: 9746	4-Amino-2-anilino-1,3,5-triazine	: 11830	2-Amino-2-methylbutyric acid	: 7401
Aldrin	: 5177	4-Aminoanthranilic acid	: 2553	trans-4-(Aminomethyl)-1-	: 4333
Aldrin epoxide	: 5178	1-Aminoanthraquinone	: 380	cyclohexanecarboxylic acid	
Alizarin	: 391	2-Aminoanthraquinone	: 381	2-Amino-3-methylpentanoic acid	: 7370
Alizarine Blue	: 8062	p-Aminoazobenzene	: 837	2-Amino-4-methylpentanoic acid	: 7413
Alizarine Bordeaux B	: 413	1,2-Aminoazophenylene	: 2871	4-Amino-4-methyl-2-pentanone	: 8959
Alizarine Brown R	: 415	(S)-α-Aminobenzenepropanoic acid	: 9510	3-Amino-5-methylphenol	: 9212
Alizarine Maroon	: 382	3-(p-Aminobenzenesulfamido)-6-	: 2197	4-(Aminomethyl)piperidine	: 9673
Alizarine Orange	: 399	methoxypyridazine		2-Amino-2-methyl-1,3-propanediol	: 9987
Alizarine Yellow C	: 5873	p-Aminobenzenesulfonamide	: 2187	7-Amino-2-methylquinoline	: 11211
Alizarine Yellow R	: 2648	4-(p-Aminobenzenesulfonamido)-	: 2194	8-Amino-6-methylquinoline	: 11212
Alkannin	: 7854	2,6-dimethoxypyrimidine		2-Amino-4-methylthiazole	: 11665
Allacil	: 11060	2-(4-Aminobenzenesulfonamido)-	: 2196	2-Amino-4-methylthiobutanoic acid	: 7648
Allantoin	: 12009	4,6-dimethylpyrimidine		3-(Aminomethyl)-3,5,5-	: 4502
Allene	: 9772	2-(4-Aminobenzenesulfonamido)-4-	: 2199	trimethylcyclohexanol	
Allenolic acid	: 7953	methylpyrimidine		Aminometradine	: 11060
Allergosil	: 5587	2-(p-Aminobenzenesulfonamido)-4-	: 2200	2-Aminonaphthalene-6,8-disulfonic	: 7870
Allicin	: 10595	methylthiazole		acid	
Allidochlor	: 44	N'-(4-Aminobenzenesulfonyl)-N-	: 2192	3-Amino-2-naphthoic acid	: 7768
Alliin	: 333	butylurea		1-Amino-8-naphthol-3,6-disulfonic	: 7874
Allobarbital	: 11091	3-Aminobenzoic acid, 6-chloro	: 2490	acid	
Allocholesterol	: 4111	4-Aminobenzoic acid, isobutyl ester	: 2505	1-Amino-2-naphthol-4-sulfonic acid	: 7966
Allocyclogeranic acid	: 4307	2-Aminobenzothiazole	: 2859	6-Aminonicotinic acid	: 10897
Allomaleic acid	: 3764	6-Aminobenzothiazole	: 2860	2-Amino-5-nitrothiazole	: 11666
Allomethadione	: 8586	N-(p-Aminobenzoyl)-L-(+)-glutamic	: 6242	(S)-2-Aminopentanedioic acid	: 6241
Allopregnane	: 9740	acid		(S)-2-Aminopentanoic acid	: 8218
Allopurinol	: 10852	N-(p-Aminobenzoyl)glycine	: 6252	o-Aminophenol	: 9201
Alloxan	: 11073	p-Aminobiphenyl	: 3011	m-Aminophenol	: 9202
Alloxantin	: 3106	1-Amino-5-bromonaphthalene	: 7711	p-Aminophenol	: 9203
Allyl acetate	: 235	4-Amino-3-bromoquinoline	: 11207	p-Aminophenylacetic acid	: 886
Allylacetic acid	: 9073	1-Aminobutane	: 3209	3-Amino-1-phenylbutane	: 2119
Allyl alcohol	: 10697	γ-Aminobutyric acid	: 3464	4-Amino-3-phenylbutyric acid	: 2128
Allyl alcohol, 3-chloro	: 10701	ε-Aminocaproic acid	: 6873	5-(4-Aminophenylsulfonamido)-3,4-	: 2195
Allyl alcohol oxide	: 8618	(L)-2-Amino-3-	: 5017	dimethylisoxazole	
Allyl aldehyde	: 10509	(carboxymethylthio)propionic acid		2-Aminophenylsulfonic acid	: 2215
Allylamine	: 10522	5-Amino-2-chlorobenzoic acid	: 2490	3-Aminophenylsulfonic acid	: 2216
Allyl bromide	: 10531	2-Amino-5-chlorobenzoxazole	: 2876	4-Aminophenylsulfonic acid	: 2217
Allylcarbamide	: 12028	4-Amino-4'-chlorodiphenyl	: 3012	p-Aminophenylsulfonylthiourea	: 2189
Allylcatechol methylene ether	: 2429	6-Aminochrysene	: 4113	2-Amino-3-picoline	: 10866
Allyl chloride	: 10536	Aminocyclobutane	: 4169	2-Amino-4-picoline	: 10867
Allyl cinnamate	: 10692	Aminocyclooctane	: 4735	2-Amino-6-picoline	: 10868
Allyl crotonate	: 3856	1-Aminocyclopentanecarboxylic acid	: 4782	3-Amino-4-picoline	: 10870
Allyl cyanide	: 3803	1-Amino-1-deoxysorbitol	: 6207	1-Aminopropane-1,3-dicarboxylic	: 6241
5-Allyl-4,7-dimethoxy-1,3-	: 2423	[(4-Amino-3,5-dichloro-6-fluoro-2-	: 5982	acid	
benzodioxole		pyridyl)oxy]acetic acid		3-Aminopropanoic acid	: 323
Allyl ether	: 10592	2-Amino-4,6-dichlorophenol	: 9205	DL-2-Aminopropanoic acid	: 324
Allyl ethyl ether	: 10566	1-Amino-3-diethylamino-2-propanol	: 10339	(S)-2-Aminopropanoic acid	: 325
Allyl fluoride	: 10570	2-Amino-1,9-dihydro-9-β-D-	: 6289	DL-1-Amino-2-propanol	: 10338
Allyl formate	: 6017	ribofuranosyl-6H-purin-6-one		3-Aminopropionitrile	: 10092
Allyl glycidyl ether	: 8626	2-Amino-2,2-dimethylethanol	: 10341	β-Aminopropionitrile	: 10092
Allyl iodide	: 10573	2-Amino-4,5-dimethylpyrimidine	: 11044	4-Aminopropiophenone	: 10413
Allyl isocyanide	: 10574	p-Aminodiphenylamine	: 1283	p-Aminopropiophenone	: 10413
Allyl isothiocyanate	: 10575	2-Amino-6,8-disulfonaphthalene	: 7870	α-(α-Aminopropyl)benzyl alcohol	: 1840
4-Allyl-1,2-(methylenedioxy)benzene	: 2429	Aminoethanoic acid	: 6250	N-(3-Aminopropyl)diethanolamine	: 5679
m-Allylpyrocatechin methylene ether	: 2429	1-Aminoethanol	: 5674	Aminopropylin	: 9795
Allyl sulfide	: 10599	2-Aminoethanol	: 5675	4-(3-Aminopropyl)morpholine	: 7701
Allylurea	: 12028	N-(Aminoethyl)ethanolamine	: 5676	Aminopropylon	: 9795
Allyltrichlorosilane	: 11475	1-(2-Aminoethyl)piperazine	: 9615	6-Aminopurine	: 10751
Aloe-emodol	: 397	2-Amino-2-ethyl-1,3-propanediol	: 9985	4-Aminopyrazole	: 10812
Aloperidin	: 3651	2-Aminoethylsulfonic acid	: 5623	2-Aminopyridine	: 10857
Aloxidone	: 8586	Aminoformamidine	: 6278	α-Aminopyridine	: 10857
Alstonidine	: 10772	2-Aminoglutaramic acid	: 6249		

β-Aminopyridine : 10858
4-Aminopyridine : 10859
Aminopyrine : 10851
2-Aminoquinoline : 11202
4-Aminoquinoline : 11203
5-Aminoquinoline : 11204
6-Aminoquinoline : 11205
8-Aminoquinoline : 11206
6-Aminoquinoxaline : 11332
Aminorex : 8574
4-Amino-1-β-D-ribofuranosyl-2(1H)- : 5021
pyrimidinone
p-Aminosalicylic acid : 2498
5-Aminosalicylic acid : 2499
p-Aminosalicylic acid hydrazide : 2500
α-Aminosuccinamic acid : 468
L-Aminosuccinic acid : 471
2-Amino-1,3-thiazole : 11662
2-Aminothiazole : 11662
Aminoxafen : 8574
Aminozide : 3334
Amiphenazole : 11671
Amitrole : 11845
Amixetrine : 11169
Amizepin : 5144
Ammonium, acetate : 106
Ammonium, (3-carbamoyl-3,3- : 2120
diphenylpropyl) diisopropyl
methyl,iodide
Ammonium, diethyl(2- : 5445
hydroxyethyl)methyl-, bromide,
xanthene-9-carboxylate
Ammonium, dimethyl ethyl 3- : 865
hydroxyphenyl,chloride
Ammonium formate : 6003
Ammonium, pentamethylenebis : 8738
trimethyl, bromide
Ammonium, phenyl trimethyl,iodide : 866
Ammonium, tetraethyl,bromide : 5447
Ammonium, tetramethyl,bromide : 7456
Ammonium, tetramethyl,chloride : 7457
Ammonium, tetramethyl,iodide : 7458
Ammonium, tetrapropyl,iodide : 9833
Amobarbital : 11092
Amolanone : 2463
Amphecloral : 1583
Amphetamine : 1578
Ampyrone : 10845
Amudane : 11520
Amygdalin : 954
Amyl acetate : 227
sec-Amyl acetate (R) : 8918
Amylacetylene : 6609
tert-Amyl alcohol : 3628
Amyl alcohol : 8915
sec-Amyl alcohol : 8916
iso-Amylamine : 3223
Amylamine : 8700
Amylbenzene : 2107
d-Amyl bromide : 3238
tert-Amyl bromide : 3240
Amyl bromide : 8710
Amyl butyrate : 3579
Amyl caproate : 6907
tert-Amyl carbamate : 3634
Amyl chloride : 8719
6-n-Amyl-m-cresol : 9425
Amylene : 3801
α-Amylene : 9004
Amylene dichloride : 3286
2,4-Amylene glycol : 8789
Amyl ether : 8835
Amyl ethyl ketone : 8472
Amylethylmethylcarbinol : 8466
Amyl formate : 6015
Amyl heptanoate : 6470
Amyl hexanoate : 6907
Amyl hexyl ketone : 5328
Amyl iodide : 8818
Amylisopropylmethylcarbinol : 8451
tert-Amyl isovalerate : 3548
Amyl ketone : 11962
Amyl mercaptan : 8844
sec-Amylmercaptan : 8845
Amyl nitrite : 8090
: 8096
Amylol : 8915

4-Amylresorcinol : 1544
Amyl salicylate : 2649
Amyl stearate : 8270
Amyl sulfide : 8843
Amyltrichlorosilane : 11472
Amyl valerate : 8913
β-Amyrin : 8548
α-Amyrin : 12038
Anabasine : 11008
Anagyrine : 7573
Analeptin : 1887
Anastil : 9379
Anatabine : 3105
Anchovyxanthin, all-trans- : 4057
Androfluorone : 355
4-Androstene-3,17-dione : 352
Androsterone : 348
Anemonin : 5185
Anesthesin : 2497
Anethole : 1942
Angelic acid : 3845
Anhalamine : 7393
Anhalonidine : 7395
Anhydroecgonine : 491
Anhydron : 2851
Aniline : 633
Aniline, 2-bromo : 638
Aniline, 3-bromo : 639
Aniline, 4-bromo : 640
Aniline, 2-bromo-6-nitro- : 653
Aniline, 4-butyl : 655
Aniline, N-butyl- : 656
Aniline, 2-tert-butyl : 714
Aniline, 4-tert-butyl : 715
Aniline, N-tert-butyl- : 716
Aniline-4-sec-butyl : 813
Aniline, N-sec-butyl- : 816
Aniline-3-carboxylic acid : 2485
Aniline-4-carboxylic acid : 2486
Aniline, 2-chloro : 658
Aniline, 3-chloro : 659
Aniline, 4-chloro : 660
Aniline, 2-chloro, hydrochloride : 665
Aniline, 2,4-dibromo- : 681
Aniline, 2,6-dibromo- : 682
Aniline, 3,4-dibromo- : 683
Aniline, 2,3-dichloro- : 685
Aniline, 2,4-dichloro- : 686
Aniline, 2,5-dichloro- : 687
Aniline, 2,6-dichloro- : 688
Aniline, 3,4-dichloro- : 689
Aniline, 3,5-dichloro- : 690
Aniline, 2,6-dichloro-4-ethoxy : 691
Aniline, 3,4-diethoxy- : 693
Aniline, N,N-diethyl-4-nitroso : 699
Aniline, 2,4-difluoro- : 700
Aniline, 2,4-diiodo- : 701
Aniline, 2,6-diisopropyl- : 636
Aniline, 2,6-dimethoxy- : 703
Aniline, N,N-dimethyl-p-[(α,α,α- : 723
trifluoro-m-tolyl)azo]-
Aniline, 2,3-dinitro- : 724
Aniline, 2,6-dinitro- : 728
Aniline, 3,5-dinitro- : 727
Aniline, hydrochloride : 756
Aniline, 2-methylthio : 819
Aniline nitrate : 821
Aniline, 5-nitro-2-propoxy : 827
p-Anilinesulfonamide : 2187
2-Anilinesulfonic acid : 2215
3-Anilinesulfonic acid : 2216
Aniline-4-sulfonic acid : 2217
Aniline, 2,3,4,5-tetramethyl : 847
Aniline, 2,3,4-trichloro- : 852
p-Anilinobenzenesulfonic acid : 2234
o-Anilinobenzoic acid : 2715
Anilotic acid : 2647
p-Anisaldehyde : 585
Anise alcohol : 1891
p-Anisic acid : 2666
m-Anisic acid, 2-(methylamino)-, : 2672
methyl ester
o-Anisidine : 766
m-Anisidine : 767
p-Anisidine : 768
m-Anisidine, 4-nitro- : 776
Anisindione : 7269

Anisole : 1915
Anisole, 3,5-dinitro- : 1924
Anisole, 2,5-dinitro- : 1926
Anisole, 2,3-dinitro : 1922
Anisole, 4-nitro : 1937
Anisole, 2,3,4,6-tetrabromo : 2256
Anisole, 3,4,5-tribromo : 2310
Anisole, 2,4,5-tribromo : 2311
Anisole, 2,3,5-tribromo : 2312
o-(p-Anisoyl)benzoic acid : 2667
p-Anisoyl chloride : 2901
p-Anisyl alcohol : 1891
o-Anisyl benzoate : 9383
1-(p-Anisyl)ethanol : 1895
[6]Annulene : 867
[8]Annulene : 4742
Anthorine : 475
Anthracene, 9,10-dibenzyl- : 367
Anthracene, 9,3-dimethyl : 378
Anthragallol : 415
Anthralin : 430
Anthranil : 2390
o-Anthranilic acid : 2484
Anthranilic acid, N-methyl-, ethyl : 2680
ester
Anthranilic acid, 3-nitro- : 2506
Anthranilic acid, N-(α,α,α-trifluoro- : 2729
m-tolyl)-
Anthranol : 432
Anthraquinone : 379
Anthraquinone, 2-amino-1-hydroxy : 383
Anthraquinone, 2,7-dibromo- : 390
9,10-Anthraquinone, 1,7-dihydroxy : 394
Anthrarobin : 429
Anthrarufin : 393
1-Anthroic acid : 368
2-Anthroic acid : 369
9-Anthroic acid : 370
Anthrone : 433
Anthropodeoxycholic acid : 4097
Anticanitic vitamin : 2486
Antifebrin : 86
Antimonate(2-), bis[μ-[2,3- : 11588
dihydroxybutanedioato(4-)-
O1,O2:O3,O4]]di-, dipotassium,
trihydrate, stereoisomer
Antipyrine : 10847
Antipyrine, 4-bromo- : 10846
ANTU : 11807
Apacil : 2498
Apacizin : 2500
Aphylline : 7572
Apigenin : 2806
Apigenin 4'-methyl ether : 2809
Apiole : 2423
Apoatropine : 928
Apocodeine : 5160
Apocynin : 5824
Apomorphine : 5155
6aα-Aporphine-2,9-diol, 1,10- : 5156
dimethoxy-
6aα-Aporphine, 1,2,9,10- : 5157
tetramethoxy-
6aα-Aporphin-11-ol, 10-methoxy- : 2470
1,2-(methylenedioxy)-
6aα-Aporphin-11-ol, 1,2,10- : 5161
trimethoxy-
Aprobarbital : 11098
Arabinose (D)(β) : 436
Arachic alcohol : 5358
Arachidic acid : 5355
Arachidonic acid : 5361
Aramite : 11579
Arborine : 11200
Arbutin : 6224
Arecaidine : 10913
Arecoline : 10914
Argobyl : 10489
Armstrong's acid : 7867
Arsanilic acid : 456
Arsine, difluoromethyl- : 483
Arsine, oxybis[dimethyl- : 452
Ascaridole : 5184
Ascensil : 10867
Asculetine : 2804
Asmacoril : 2840
Aspartame : 9511

Compound	Number
Aspergillic acid	: 10811
Aspidospermine	: 473
Aspirin	: 2480
Atrolactic acid	: 914
Atropic acid	: 927
Atropine	: 902
Aureothin	: 10797
Aurin	: 4278
12-Azabenz[a]anthracene	: 630
Azacycloheptane	: 492
Azacyclohexane	: 9629
Azacyclopentane	: 11143
7-Azadibenz[a,j]anthracene	: 5143
8-Azaguanine	: 11851
7-Aza-7H-benzo[c]fluorene	: 2405
2-Azanaphthalene	: 7373
1-Azanaphthalene	: 11214
6-Azathymine	: 11831
6-Azauridine	: 11832
Azelaic acid	: 8130
Azelaic acid, 3-methyl	: 8135
1H-Azepine-1-carbothioic acid, hexahydro-, S-ethyl ester	: 7661
Aziminobenzene	: 2871
Azindole	: 2376
Azine	: 10872
trans-Azobenzene	: 5129
cis-Azobenzene	: 5130
Azobenzene, 3,3'-bis(dimethylamino)	: 361
3-Azobenzenecarboxylic acid, 4-hydroxy-4'-nitro	: 2648
Azobenzene, 2,2'-diethoxy-	: 5112
Azobenzene, 4-dimethylamino-3'-trifluoromethyl	: 723
Azobenzene, 4-methyl-	: 5139
Azobenzene, 3-methyl	: 5138
Azo, benzene 1'-naphthalene,2'-hydroxy	: 8037
2,2'-Azobis(isobutyronitrile)	: 10093
1,1'-Azobisnaphthalene	: 5127
Azobutane	: 5123
Azodicarbonamide	: 5124
Azodicarboxylic acid diamide	: 5124
Azoformic acid, diamide	: 5124
Azole	: 11112
Azomethane	: 5125
Azomycin	: 7211
1,1'-Azonaphthalene	: 5127
2,2'-Azonaphthalene	: 5128
Azopropane	: 5132
Azopyrin	: 2654
cis-Azoxybenzene	: 5131
Azoxybenzene, 4-bromo-	: 5120
Azoxybenzene, 2,2'-dimethyl (trans)	: 5116
Azoxybenzene, 3,3'-dimethyl (trans)	: 5118
Azoxybenzene, 2,2'-dimethyl(trans)	: 5117
Azoxybenzene, 3,3'-dimethyl (trans)	: 5119
Badische acid	: 7964
Baicalein	: 2844
Bamipine	: 9627
Barbital	: 11087
Barbituric acid	: 11076
Barbituric acid, 5-allyl-5-isobutyl-2-thio, sodium salt	: 11063
Barbituric acid, 5-sec-butyl-5-ethyl-2-thio-	: 11065
Basedol	: 11662
Batyl alcohol	: 10023
BCPE	: 1850
Bebeerine	: 11924
Behenic acid	: 5276
Bemegride	: 9656
Bendazol	: 2385
Benemid	: 2595
Benovocylin	: 5403
Benzalacetone	: 3881
β-Benzalbutyramide	: 3703
Benzal chloride	: 1382
Benzaldehyde, 3-nitro, diacetate	: 7511
Benzalhydrazone	: 563
Benzamide, N-[[(4-chlorophenyl)amino]carbonyl]-2,6-difluoro-	: 5176
Benzamide, 3,5-dichloro-N-(1,1-dimethyl-2-propynyl)-	: 10746
Benzamide, N-(3-tolyl)	: 622
Benzanilide	: 626
1,2-Benzanthracene	: 519
2,3-Benzanthracene	: 7703
1,2-Benzanthracene, 3-methyl	: 522
Benzanthrone	: 6292
Benzathine	: 5500
Benzazide	: 2884
2-Benzazine	: 7373
Benz[b]anthracene	: 7703
Benzcurine iodide	: 5441
Benzenamine, N-butyl-N-ethyl-2,6-dinitro-4-(trifluoromethyl)-	: 509
Benzenamine, N-(2-chloroethyl)-2,6-dinitro-N-propyl-4-(trifluoromethyl)-	: 5964
Benzenamine, N-(cyclopropylmethyl)-2,6-dinitro-N-propyl-4-(trifluoromethyl)-	: 9763
Benzenamine, 4-(1,1-dimethylethyl)-N-(1-methylpropyl)-2,6-dinitro-	: 3893
Benzenamine, 2,6-dinitro-N,N-dipropyl-4-(trifluoromethyl)-	: 11901
Benzenamine, N,N'-1,2-ethanediylbis-	: 5508
Benzenamine, N-ethyl-N-(2-methyl-2-propenyl)-2,6-dinitro-4-(trifluoromethyl)-	: 5409
Benzenamine, N-(1-ethylpropyl)-3,4-dimethyl-2,6-dinitro-	: 8635
Benzenamine, 4-(1-methylethyl)-2,6-dinitro-N,N-dipropyl-	: 7371
Benzenamine, 4-(methylsulfonyl)-2,6-dinitro-N,N-dipropyl-	: 8070
Benzeneacetamide, N,N-dimethyl-α-phenyl-	: 5219
Benzeneacetic acid, 4-bromo-α-(4-bromophenyl)-α-hydroxy-, 1-methylethyl ester	: 3140
Benzeneacetic acid, 4-chloro-α-(4-chlorophenyl)-α-hydroxy-, ethyl ester	: 4082
Benzeneacetic acid, 4-chloro-α-(1-methylethyl)-, cyano(3-phenoxyphenyl)methyl ester	: 5956
Benzeneamine, N-(1-methylpropyl)-2,6-dinitro	: 3894
Benzene, 3-aminobutyl	: 2119
Benzenearsonic acid	: 460
Benzene, 1,3-bis(diethylamino)	: 1285
Benzeneboronic acid	: 3135
Benzene, (1-bromopropyl) (DL)	: 1086
Benzenecarbonyl chloride	: 2886
Benzenecarboxaldehyde	: 525
Benzenecarboxylic acid	: 2475
Benzene, 1(3-chloroallyl)-4-methoxy	: 1231
Benzene, 1-chloro-4-[[(4-chlorophenyl)methyl]thio]-	: 4075
Benzene, 2-chloro-1-(3-ethoxy-4-nitrophenoxy)-4-(trifluoromethyl)-	: 8633
Benzene, 2-chloro-1-(4-nitrophenoxy)-4-(trifluoromethyl)-	: 8082
Benzene, 1-chlorophosphino-4-isopropyl	: 9555
Benzene, (1-decylundecyl)-	: 6297
1,3-Benzenediamine, N3,N3-diethyl-2,4-dinitro-6-(trifluoromethyl)-	: 5181
1,3-Benzenediamine, 2,4-dinitro-N3,N3-dipropyl-6-(trifluoromethyl)-	: 9761
Benzene, 2,3-dibromo-1-chloro	: 1292
Benzene, 3,4-dibromo-1-chloro	: 1293
Benzene, 2,6-dibromo-1-chloro	: 1294
Benzene, 2,5-dibromo-1-chloro	: 1295
Benzene, 1,4-di-tert-butyl	: 990
1,4-Benzenedicarboxylic acid, 2,3,5,6-tetrachloro-, dimethyl ester	: 5180
Benzene, 1,4-dichloro-2,5-dimethoxy-	: 4084
Benzene, 2,5-dichloro-1-ethyl	: 1377
Benzene, 2,4-dichloro-1-(4-nitrophenoxy)-	: 8081
Benzene, 1,3-diisopropyl-4-methyl-	: 1948
Benzene, 1,4-diisopropyl-2-methoxy	: 1926
1,3-Benzenedisulfonic acid, diamide,4,5-dichloro	: 1559
Benzene, (1-dodecyltridecyl)-	: 8638
Benzene, 1-ethyl-2-isopropyl	: 1688
Benzene, 1-ethyl-3-isopropyl	: 1689
Benzene, 1-ethyl-4-isopropyl	: 1690
Benzene, 4-ethyl-1,2,3-trimethyl	: 1708
Benzene, 6-ethyl-1,2,4-trimethyl	: 1709
Benzene, 5-ethyl-1,2,4-trimethyl	: 1710
Benzene, 3-ethyl-1,2,4-trimethyl	: 1711
Benzene hexachloride	: 4409
Benzene, 1-hydroxy-2-methoxy-4-propenyl,acetate(cis)	: 9394
Benzene, 1-isobutoxy-2-nitro	: 2034
Benzene, mercuriodi-	: 7436
Benzenenitrile	: 2740
Benzene, 2-nitro-1-(4-nitrophenoxy)-4-(trifluoromethyl)-	: 5980
Benzene, 1-nitro-3,4,5-tribromo	: 2315
Benzeneselenic acid	: 2174
Benzenestibonic acid	: 11548
Benzenesulfonamide, 4-(acetylamino)-	: 31
Benzenesulfonamide, 4-(dipropylamino)-3,5-dinitro-	: 8553
Benzenesulfonic acid, amide,4(4-aminophenylsulfonamido)	: 2188
Benzenesulfonic acid, amide,4(benzylamino)	: 2213
Benzenesulfonic acid, chloride	: 2219
Benzenesulfonic acid, p-chlorothio-, S-phenyl ester	: 2236
Benzenesulfonic acid, (2,4-dichlorophenyl) ester	: 9289
Benzenesulfonic acid, fluoride	: 2251
Benzenesulfonic acid, guanidine,4-amino	: 2191
Benzenesulfonic acid, thiolo, 4-chloro, S-phenyl ester	: 2236
Benzenesulfonic acid, thiolo,2-methyl,phenyl ester	: 11818
Benzene, 1,2,4,5-tetraisopropyl-	: 2276
Benzene, 1,2,3-triamino-	: 2305
Benzene, 1,3,5-tribenzoyl-	: 7587
Benzene, 1,3,5-trichloro-2,4,6-trihydroxy	: 2371
Benzene, 1,2,4-triisopropyl-	: 2372
Benzene, 1,2,3-tris(2(diethylamino) ethoxy),tri(ethyl iodide)	: 5441
Benzene, uneicosyl	: 1741
Benzestrol	: 9303
Benzhydramine	: 5422
Benzhydrol, α-(1-naphthyl)	: 7910
Benzhydrylamine	: 1826
Benzhydryl bromide	: 1058
2-(Benzhydryloxy)-N,N-dimethylethylamine	: 5422
p-Benzidine	: 3031
Benzidine	: 3031
Benzidine, 2,2'-dichloro-	: 3032
Benzil	: 5584
Benzilic acid	: 918
Benzimidazole, 1,5-dimethyl-	: 2378
Benzimidazole, 1-ethyl-2-methyl-	: 2381
1H-Benzimidazole, 2-(4-thiazolyl)-	: 11656
2-Benzimidazolethiol	: 2386
Benzo[a]chrysene	: 9601
1H-Benzo[a]fluorene, veratraman-3,23-diol deriv.	: 12044
2H,8H-Benzo[1,2-b:3,4-b']dipyran, butanoic acid deriv.	: 3540
Benzo[b]fluoranthene	: 631
Benzo[b]naphthacene	: 8836
Benzo[b]quinoline	: 314
3,4-Benzocarbazole	: 2405
Benzo[def]phenanthrene	: 10853
1,2-Benzodiazine	: 4136
2,3-Benzodiazine	: 9599
1,3-Benzodiazine	: 11195
1,4-Benzodiazine	: 11333
1,3-Benzodiazole	: 2376
Benzodihydrofuran	: 2442
1,3-Benzodioxole, 5-[[2-(2-butoxyethoxy)ethoxy]methyl]-6-propyl-	: 9727
1,3-Benzodioxol-4-ol, 2,2-dimethyl-, methylcarbamate	: 514
Benzo[d,e,f]chrysene	: 2395
Benzo[7,8]fluoreno[2,1-b]quinolizine, cevane-3,4,12,14,16,17,20-heptol deriv.	: 4067
Benzofuran, 2-hydroxymethyl	: 2455
Benzofuran, 7-isopropyl-4-methyl	: 2460

7-Benzofuranol, 2,3-dihydro-2,2-dimethyl-, methylcarbamate	: 3991
5-Benzofuranpropanol, 7-methoxy-2-[3,4-(methylenedioxy)phenyl]-	: 2464
6H-Benzofuro[3a,3,2-ef][2]benzazepin-6-ol, 4a,5,9,10,11,12-hexahydro-3-methoxy-11-methyl-, [4aS-(4aα,6β	: 6200
Benzoglyoxaline	: 2376
Benzoguanamine	: 11829
Benzohydrol	: 1910
Benzoic acid amide	: 600
Benzoic acid azide	: 2884
Benzoic acid, bromide	: 2885
Benzoic acid, chloride,2,4-dichloro	: 2893
Benzoic acid, chloride,3,4-dichloro	: 2894
Benzoic acid, chloride,4-methoxy	: 2901
Benzoic acid, 5-[2-chloro-4-(trifluoromethyl)phenoxy]-2-nitro-	: 304
Benzoic acid, 3,6-dichloro-2-methoxy-, methyl ester	: 7653
Benzoic acid, 5-(2,4-dichlorophenoxy)-2-nitro-, methyl ester	: 2997
Benzoic acid, o-ethoxy-, ethyl ester	: 2598
Benzoic acid, p-ethoxy-, ethyl ester	: 2599
Benzoic acid, 2-[[ethoxy[(1-methylethyl)amino]phosphinothioyl]oxy]-, 1-methylethyl ester	: 7359
Benzoic acid, fluoride	: 2909
Benzoic acid, 2-(2-hydroxyethenyl)-, δ-lactone	: 2791
Benzoic acid, 2-hydroxy,2-hydroxybenzoate	: 2616
Benzoic acid, 4-isopropyl, isobutyl ester	: 2695
Benzoic acid, 2-propyl	: 2723
Benzoic anhydride	: 2510
(±)-Benzoin	: 5822
1H,12H-Benzo[i]pyrano[3,4-g]indolizine, 16(15H)-oxaerythrinan-15-one deriv.	: 8567
Benzo[d]isothiazole	: 2387
Benzonitrile, o-ethoxy-	: 2754
3,4-Benzophenanthrene	: 2406
1,2-Benzophenanthrene	: 4114
9,10-Benzophenanthrene	: 11910
1,2-Benzophenazine	: 2394
Benzophenone 6	: 7595
Benzophenone	: 7616
Benzophenone, 2-bromo-	: 7600
Benzophenone, 2-carboxy-4'-methoxy-	: 2667
Benzophenone, 4,4'-diiodo-	: 7598
Benzophenone, 2,4-dimethoxy-	: 7612
Benzophenone, 3,4-dimethyl-	: 7615
Benzophenone, 2,4'-dimethyl-	: 7628
Benzophenone, 4-methyl-2-nitro	: 7627
Benzophenone, 2,2',4,4'-tetramethyl-	: 7594
Benzopinacol	: 5580
1,2-Benzopyran	: 2772
1H-Benzopyrazole	: 7235
2,3-Benzopyrene	: 2395
6,7-Benzopyrene	: 2395
3,4-Benzopyrene	: 2395
Benzo[d]pyridazine	: 9599
Benzopyridine	: 7373
1,2-Benzopyrone	: 2792
2,3-Benzopyrrole	: 7294
6,7-Benzoquinoline	: 2473
1,2-Benzoquinolizine-2-one, 9,10-dimethoxy-1,2,3,4,6,7,11b-heptahydro-3-isobutyl	: 2396
1,4-Benzoquinone	: 4257
p-Benzoquinone	: 4257
1,4-Benzoquinone, 2,5-dichloro-3,6-dihydroxy	: 4258
p-Benzoquinone, 2,5-dihydroxy-3-methoxy-6-methyl-	: 4260
1,4-Benzoquinone, 2,6-dimethoxy	: 4262
p-Benzoquinone imine, 2,6-dibromo-N-chloro-	: 4279
p-Benzoquinone, 2,3,5,6-tetrachloro-	: 4268
p-Benzoquinone, tetrahydroxy-	: 4269
p-Benzoquinone, tetramethyl-	: 4270
4-Benzoquinonimine, N,2,6-trichloro	: 4280
Benzoresorcinol	: 7609
Benzo[rst]pentacene	: 2850
Benzosol	: 9383
Benzosulfimide	: 2388
Benzosulfonazole	: 2861
1,2,4-Benzothiadiazine, 6-chloro-3(5-norbornen-2-yl)-7-sulfamyl-2,3,4-trihydro	: 2851
1,2,4-Benzothiadiazine-7-sulfonic acid,amide,6-chloro,1,1-dioxide	: 2852
1,2,4-Benzothiadiazine-7-sulfonic acid,amide,6-chloro-3-cyclopentylmethyl-2,3,4-trihydro,1,1-dioxide	: 2853
1,2,4-Benzothiadiazine-7-sulfonic acid,amide,6-chloro-3(dichloromethyl)-3,4-dihydro,1,1-dioxide	: 2854
1,2,4-Benzothiadiazine-7-sulfonic acid,amide,6-chloro-2,3,4-trihydro,1,1-dioxide	: 2855
1,2,4-Benzothiadiazine-7-sulfonic acid,amide,6-chloro-3-isobutyl-2,3,4-trihydro	: 2856
1,2,4-Benzothiadiazine-7-sulfonic acid,amide,6-chloro-2-methyl-3(2,2,2-trifluoroethylthiomethyl)	: 2857
1,2,4-Benzothiadiazine-7-sulfonic acid,amide,6(trifluoromethyl)-2,3,4-trihydro,1,1-dioxide	: 2858
1H-2,1,3-Benzothiadiazin-4(3H)-one, 3-(1-methylethyl)-, 2,2-dioxide	: 518
Benzothiazole, 6-amino-	: 2860
2-Benzothiazolesulfonic acid, amide,6-ethoxy	: 2865
2-Benzothiazolethiol	: 2866
Benzothiazole-2-thione	: 2866
2-Benzothiazolylamine	: 2859
Benzothiofuran	: 2399
2,3-Benzothiophene	: 2399
m-Benzotoluidide	: 622
Benzotrifluoride, 3-chloro	: 1241
2,3-Benzoxadiazole, 5,7-dinitro	: 2874
Benzoxale	: 5505
Benzoxazole, 2-hydroxy	: 2881
Benzoxiquine	: 11309
Benzoylacetic acid ethyl ester	: 2141
Benzoylacetonitrile	: 2124
Benzoyl tert-butyl peroxide	: 1128
Benzoylecgonine	: 481
Benzoylguaiacol	: 9383
Benzoyl-PAS	: 2514
Benzoyl peroxide	: 9157
1,12-Benzperylene	: 2471
Benzphetamine	: 1575
Benzpiperylon	: 10850
1,2-Benzpyrene	: 2436
Benzylacetaldehyde	: 2114
Benzyl acetate	: 233
Benzylacetic acid	: 2126
Benzyl alcohol	: 1836
Benzyl alcohol, α-butyl-α-methyl-	: 1847
Benzyl alcohol, α,α-dipropyl-	: 1867
Benzyl alcohol, α-ethyl-α-propyl-	: 1875
Benzyl alcohol, o-methyl-	: 1897
Benzyl alcohol, m-methyl-	: 1898
Benzyl alcohol, p-methyl-	: 1899
Benzyl alcohol, 3,4,5-trimethoxy-	: 1914
Benzyl alcohol, α-vinyl-	: 1868
Benzylamine	: 1795
Benzyl amine, 4-bromo	: 1799
Benzyl amine, 2-ethoxy	: 1808
Benzylaniline	: 1827
Benzyl benzenecarboxylate	: 2720
2-Benzylbenzimidazole	: 2385
Benzyl benzoate	: 2720
Benzyl bromide	: 1053
Benzyl bromide, 3,5-dibromo	: 1291
Benzyl-tert-butanol	: 2150
Benzyl sec-butyl ketone	: 8984
Benzylcarbonochloridate	: 4027
Benzyl chloride	: 1182
Benzyl chlorocarbonate	: 4027
Benzyl chloroformate	: 4027
Benzyl trans-cinnamate	: 10690
Benzyl cyanide	: 945
2-Benzylcyclopentanone	: 4893
2-[Benzyl(2-dimethylaminoethyl)amino]pyridine	: 5505
Benzyl disulfide	: 5248
N-Benzylethanolamine	: 5762
Benzyl ether	: 2079
Benzyl ethyl ether	: 1658
Benzyl formate	: 6016
Benzyl fumarate	: 3768
Benzylideneacetone	: 3881
Benzylideneaniline	: 840
Benzylidene chloride, 3,4-dichloro	: 1359
Benzylidene chloride, 2,3,4-trichloro	: 2326
Benzylidene chloride, 2,4,5-trichloro	: 2327
Benzylidene hydrazine	: 563
Benzylimidobis(p-methoxyphenyl)methane	: 1797
Benzyl isobutyl ether	: 2033
Benzyl levulinate	: 8911
Benzyl methyl ether	: 1930
2-Benzylnaphthalene	: 7950
Benzyl 1-naphthyl ketone	: 5859
Benzyl nicotinate	: 10911
Benzyl nitrite	: 8097
Benzyl oleate	: 8309
Benzyloxycarbonyl chloride	: 4027
Benzyl palmitate	: 6655
o-Benzylphenol	: 9454
p-Benzylphenol	: 9455
2-Benzylpiperidine	: 9699
4-Benzylpiperidine	: 9700
2-Benzylpropionic acid	: 2137
2-Benzylpyridine	: 11004
3-Benzylpyridine	: 11005
Benzyl salicylate	: 2651
N-Benzylsuccinimide	: 11163
N4-Benzylsulfanilamide	: 2213
Benzyl sulfide	: 2287
p-(Benzylsulfonamido)benzoic acid	: 2721
Benzylurea	: 12027
Berbamine	: 2911
Berberine	: 2467
13aβ-Berbin-3-ol, 2,9,10-trimethoxy-13α-methyl-	: 5151
Bergenin	: 10786
Beryllium acetylacetonate	: 2912
Beryllium bis(acetylacetonate)	: 2912
Besylic acid	: 2214
Betaine	: 7454
Bethanidine	: 6281
Betonicine	: 11178
Betulin	: 7416
Biacetyl	: 3359
Biallyl	: 6678
4,4'-Bianiline	: 3031
N,N'-Bianiline	: 7158
4,4'-Bi-o-anisidine	: 3033
Biarsine, tetraethyl	: 5109
Bibenzene	: 3008
α,α'-Bibenzhydrol	: 5580
2,2'-Bibenzoic acid	: 3038
2,2'-Bi-p-benzoquinone, 3,3'-dihydroxy-5,5'-dimethyl-	: 2979
Bicetyl	: 5348
Bicine	: 6259
Bicolorin	: 2822
Bicyclo[5.3.0]decapentaene	: 504
cis-Bicyclo[4.1.0]heptane	: 2915
Bicyclo[4.1.0]-heptane, 7-aza	: 478
Bicyclo[3.3.0]octane-2,6-dione	: 8685
Bicyclo[3.3.0]octane-2-one	: 8688
cis-Bicyclo[3.3.0]octan-3-one	: 8689
Bicyclopentyl, 2-hydroxy	: 2994
Bicylo[2.2.1]heptane, 1,3,3,4,6,6-hexamethyl-2,5,7-trioxa	: 11903
Biflorine	: 3109
Biformyl	: 5491
Biguanide	: 7233
[2,2'-Biindan]-1,1',3,3'-tetrone, 2,2'-dihydroxy-	: 2999
Biisobutyl	: 6774
Biisobutyryl	: 6807
Biisopropenyl	: 3159
Bikhaconitine	: 310
Bilirubin	: 3000
(Z,Z)-Bilirubin	: 3000
Biliverdine	: 3001
1,1'-Binaphthyl	: 3004

2,3,1',8'-Binaphthylene	: 2739
Bineopentyl	: 6849
Bineophyl	: 2281
Biochanin B	: 2830
Biotin	: 11689
Biphenyl	: 3008
p-Biphenylacetic acid	: 3009
p-Biphenylamine	: 3011
4-Biphenylamine, 4'-methyl-	: 3015
Biphenyl, 2-amino-6-methyl	: 3013
Biphenyl, 4-amino-4'-methyl	: 3015
2,4'-Biphenyldiamine	: 3029
2,2'-Biphenyldicarboxylic acid	: 3038
Biphenyl, 2,2'-dimethoxy-	: 3045
Biphenyl, 3,3'-dimethoxy-	: 3046
Biphenyl, 2,2'-dimethoxy-5,5'-dimethyl	: 3048
Biphenyl, 2,5'-dimethoxy-2',5-dimethyl	: 3049
Biphenyl, 2,3-dimethyl-	: 3052
Biphenyl, 2,4-dimethyl-	: 3054
Biphenyl, 3,4-dimethyl-	: 3059
Biphenyl, 3,5-dimethyl-	: 3060
4,4'-Biphenyldisulfonic acid	: 3070
o-Biphenylenemethane	: 5967
2,2'-Biphenylene oxide	: 5163
Biphenyl, 3-ethoxy	: 3071
2-Biphenylol	: 3088
4-Biphenylol	: 3090
2-Biphenylol, 3-chloro-	: 3092
2,3'-Bipiperidyl	: 3098
Bipropargyl	: 6716
Bipropionyl	: 6806
3,3'-Bipyridyl	: 3102
3,4'-Bipyridyl	: 3103
4,4'-Bipyridyl	: 3104
2,3'-Bipyridyl	: 3100
2,3'-Biquinolyl	: 3107
5,5'-Biresorcinol	: 3097
Birodan	: 2136
Bis(4-acetamidophenyl) sulfone	: 91
N,N'-Bis(acetoacetyl)-3,3'-dimethylbenzidine	: 3204
Bisacodyl	: 9465
Bis(β-amino-β-carboxyethyl) disulfide	: 5020
Bis(p-aminophenyl) ether	: 832
Bis(4-aminophenyl) sulfide	: 850
Bis(3-aminopropyl)amine	: 9874
1,4-Bis(3-aminopropyl)piperazine	: 9614
N,N'-Bis(3-aminopropyl)-1,4-tetramethylenediamine	: 3259
1,3-Bis(bromomethyl)tetramethyl-disiloxane	: 5229
1,2-Bis(chlorocarbonyl)ethane	: 3366
Bis(2-chloroethyl) carbonate	: 5697
Bis(2-chloroethyl) ether	: 5611
Bis(2-chloroethyl) sulfide	: 5645
Bis(chloromethyl) ether	: 7525
Bis(p-chlorophenoxy)methane	: 1965
1,1-Bis(4-chlorophenyl)ethane	: 1678
1,1-Bis(4-chlorophenyl)ethanol	: 1850
3,6-Bis(2-chlorophenyl)-1,2,4,5-tetrazine	: 4139
1,1-Bis(4-chlorophenyl)-2,2,2-trichloroethanol	: 1851
1,1-Bis(p-chlorophenyl)-2,2,2-trichloroethanol	: 1851
Bis(3-chloropropyl) sulfide	: 10153
Bis(cyclohexyl)carbodiimide	: 4291
Bis(cyclopentadienyl)iron	: 5960
1,3-Bis-(dichloromethyl)tetramethyl disiloxane	: 5231
Bis(2,2-diethoxyethyl) ether	: 5612
Bis-O,O-diethylphosphoric anhydride	: 5221
Bis(diethylthiocarbamoyl) sulfide	: 11714
Bis(dimethylamino)phosphonous anhydride	: 5220
2,3-Bis(ethoxycarbonyl)furan	: 6082
Bis-(2-ethoxyethyl) amine	: 5425
Bis(ethoxymethyl)ether	: 5614
Bis(2-ethylhexyl) adipate	: 6781
Bis(2-ethylhexyl) decanedioate	: 5043
Bis(2-ethylhexyl) phthalate	: 1334
Bis(2-ethylhexyl) sebacate	: 5043
2,2-Bis(ethylsulfonyl)propane	: 9835

Bis(2-furyl)glyoxime	: 5583
Bis-(3-hydroxy-2-butyl)amine	: 3624
Bishydroxycoumarin	: 2839
Bis(hydroxymethyl)acetylene	: 3911
Bis-(2-methallyl)carbonate	: 10703
Bis(methoxydimethylsilyl) oxide	: 5234
Bis(2-methoxyethyl) carbonate	: 5736
Bis(2-methylbutyl) sulfide	: 3435
Bis[methylmercapto]methane	: 7461
Bis(methylthio)methane	: 7461
1,2'-Bis-naphthyl ether	: 7928
Bis(m-nitrophenyl) disulfide	: 5247
Bisphenol A	: 9413
Bisphenol B	: 9433
1,2-Bis(phenylamino)ethane	: 5508
Bis-(1-phenylethyl)ether (DL)	: 2081
Bis(1,2,2,2-tetrachloroethyl) ether	: 5615
1,5-Bis(2-tetrahydrofuryl)-3-pentanol	: 6166
Bis(p-tolyl)mercury	: 7432
Bis(tributyltin) oxide	: 5241
Bithionol	: 9478
Bitoscanate	: 1432
Δ 1,1'-Biurea	: 5124
Biuret	: 7232
Bivinyl	: 3146
Bixin	: 5108
Boldenone	: 343
Boldine	: 5156
Boldine dimethyl ether	: 5157
Bolusanthin	: 2801
Borane, triisopentyl-	: 3122
Borane-trimethylamine (1:1)	: 3132
Borine, dimethoxy-	: 3117
Borine, triisopentyl-	: 3122
Bornane, 2-chloro-	: 2919
10-Bornanesulfonic acid, 2-oxo-, (1S,4R)-(+)-	: 2929
3-Bornanol, acetate, (1S,3R,4S)-(-)-	: 2947
(+)-2-Bornanone	: 2961
2-Bornanone, 3-acetyl-17O-	: 2950
Borneol, acetate	: 2948
Bornyl chloride	: 2920
Bornyl formate (d)	: 2949
Bornyl isovalerate (d)	: 3565
d-Bornyl isovalerate	: 3565
Boron fluoride-diethyl ether compound	: 3137
Boron trifluoride compd. with methyl ether	: 3138
Boron trifluoride etherate	: 3137
Boron trimethoxide	: 3126
Brassidic acid	: 5282
Brilliant green	: 5444
Bristurin	: 2858
Bromacetocarbamide	: 3195
ω-Bromacetophenone	: 5782
Bromal	: 20
Bromcarbamide	: 3196
Bromcholitin	: 5157
Bromcresol green	: 9367
Bromcresol purple	: 9364
Bromdiethylacetylurea	: 3195
Bromindione	: 7265
Bromine cyanide	: 4165
Bromisovalum	: 3196
5-Bromoacenaphthene	: 4
N-Bromoacetamide	: 36
p-Bromoacetanilide	: 37
Bromoacetic acid	: 111
Bromoacetic acid, propyl ester	: 120
Bromoacetone	: 10416
p-Bromoacetophenone	: 5781
Bromoacetyl bromide	: 292
Bromoacetylene	: 5937
5-Bromoacetylsalicylic acid	: 2481
o-Bromoaniline	: 638
m-Bromoaniline	: 639
p-Bromoaniline	: 640
5-Bromoanthranilic acid	: 2487
4-Bromoantipyrine	: 10846
Bromoaspirin	: 2481
Bromobenzene	: 998
p-Bromobenzenesulfonyl chloride	: 2244
p-Bromobenzoic acid	: 2520
2-Bromobenzophenone	: 7600
p-Bromobenzyl bromide	: 1000
p-Bromobenzyl chloride	: 1008

α-Bromobenzyl cyanide	: 947
4-Bromo-1,2-butadiene	: 3147
1-Bromo-1,3-butadiene	: 3148
2-Bromo-1-butene	: 3709
3-Bromo-1-butene	: 3710
(±)-2-Bromobutyric acid	: 3468
DL-α-Bromobutyric acid	: 3468
(RS)-α-Bromobutyric acid	: 3468
Bromochloroacetic acid	: 112
Bromochloroacetic acid, ethyl ester	: 113
1-Bromo-3-chloroacetone	: 10417
2-Bromo-1-chlorobutane	: 3234
Bromochlorodifluoromethane	: 7464
Bromochlorodinitromethane	: 7465
1-Bromo-1-chloroethane	: 5455
Bromochlorofluoromethane	: 7466
Bromochloromethane	: 7463
1-Bromo-4-chloronaphthalene	: 7741
4-Bromo-2-chlorophenol	: 9232
1-Bromo-2-chloropropane	: 9838
1-Bromo-3-chloropropane	: 9839
2-Bromo-1-chloropropane	: 9840
2-Bromo-2-chloropropane	: 9841
2-Bromo-5-chlorothiophene	: 11728
4-Bromo-3-chloro-3,4,4-trifluoro-1-butene	: 3715
1-Bromo-2-chloro-1,1,2-trifluoroethane	: 5457
6-Bromo-m-cresol	: 9237
Bromocycloheptane	: 4222
2-Bromo-1-cyclohexanol	: 4503
Bromocyclopentane	: 4777
2-Bromo-1-decene	: 5086
1-Bromo-2-decene	: 5087
2-Bromodibenzofuran	: 5164
3-Bromodibenzofuran	: 5165
Bromodichlorofluoromethane	: 7468
Bromodichloromethane	: 7467
2-Bromo-2,2-diethylacetamide	: 3198
1-Bromo-1,2-difluoroethylene	: 5883
Bromodifluoromethane	: 7469
2-Bromo-2,3-dimethylbutane	: 3236
1-Bromo-1,1-dimethylbutane	: 8717
4-Bromo-1,3-dimethylpyrazole	: 10816
1-Bromo-2,3-dinitrobenzene	: 1024
Bromodiphenylmethane	: 1058
2-Bromoethanesulfonyl chloride	: 5628
2-Bromoethanol	: 5685
2-Bromoethyl ethyl ether	: 5461
2-Bromo-2-ethylisovaleramide	: 3199
2-Bromofluorene	: 5968
Bromofluoroacetic acid	: 116
1-Bromo-4-fluorobutane	: 3237
1-Bromo-10-fluorodecane	: 5035
1-Bromo-12-fluorododecane	: 5294
1-Bromo-6-fluorohexane	: 6735
Bromofluoromethane	: 7470
1-Bromo-5-fluoropentane	: 8713
1-Bromo-11-fluoroundecane	: 11939
Bromoform	: 7548
2-Bromofumaric acid	: 3770
3-Bromofuran	: 6034
1-Bromoheptadecane	: 6307
2-Bromoheptane	: 6357
4-Bromoheptane	: 6358
2-Bromoheptanoic acid	: 6457
1-Bromo-1-heptyne	: 6612
1-Bromo-2-heptyne	: 6613
2-Bromohexane	: 6733
3-Bromohexane	: 6734
1-Bromo-3-iodobenzene	: 1047
m-Bromoiodobenzene	: 1047
1-Bromo-2-iodoethane	: 5463
Bromoiodomethane	: 7471
α-Bromoisobutyric acid	: 10209
4-Bromoisoquinoline	: 7374
β-Bromoisovaleric acid	: 3472
2-Bromoisovalerylurea	: 3196
Bromomethane	: 7462
Bromomethoxymethane	: 7472
2-Bromo-p-methylacetophenone	: 5779
2-Bromo-5-methylaniline	: 648
(+)-1-Bromo-2-methylbutane	: 3238
(S)-1-Bromo-2-methylbutane	: 3238
2-Bromo-3-methyl-1-butene	: 3716
(Bromomethyl)chlorodimethylsilane	: 11365
(Bromomethyl)dimethylchlorosilane	: 11365

2-Bromo-4-methyl-1-nitrobenzene : 1073
1-Bromo-2-methylpentane : 8714
1-Bromo-3-methylpentane : 8715
2-Bromo-2-methylpentane : 8717
3-Bromo-3-methylpentane : 8718
4-Bromo-3-methylpyrazole : 10817
2-Bromo-3-methylthiophene : 11729
Bromomethyltrimethylsilane : 11366
2-Bromonaphthalene : 7740
5-Bromo-2-naphthoic acid : 7769
3-Bromo-1,2-naphthoquinone : 7844
2-Bromo-4-nitrobenzoic acid : 2527
2-Bromo-5-nitrobenzoic acid : 2528
4-Bromo-3-nitrobenzoic acid : 2529
Bromonitromethane : 7473
2-Bromo-1-nitronaphthalene : 7745
4-Bromo-2-nitrophenol : 9244
6-Bromo-5-nitroquinoline : 11221
3-Bromo-2-nitrotoluene : 1068
4-Bromo-2-nitrotoluene : 1074
18-Bromooctadecanoic acid : 8250
Bromopentafluorobenzene : 1078
2-Bromopentane : 8711
3-Bromopentane : 8712
1-Bromo-1-pentene : 9007
2-Bromo-1-pentene : 9008
3-Bromo-1-pentene : 9009
2-Bromo-2-pentene : 9012
3-Bromo-2-pentene : 9013
5-Bromo-2-pentene : 9015
1-Bromo-1-pentyne : 9124
p-Bromophenacyl bromide : 5777
9-Bromophenanthrene : 9164
o-Bromophenylacetylene : 1040
m-Bromophenylacetylene : 1041
2-Bromo-3-phenyl-2-butene : 3719
(p-Bromophenyl)hydrazine : 7141
p-Bromophenyl isocyanate : 1049
Bromophenylmethane : 1053
p-Bromophenyl methyl ketone : 5781
4-Bromophenyl phenyl ether : 1079
3-Bromo-1-propene : 10531
β-Bromopropionic acid : 10205
3-Bromopropylene : 10531
3-Bromopyridine : 10876
4-Bromopyridine : 10877
3-Bromoquinoline : 11217
5-Bromoquinoline : 11218
6-Bromoquinoline : 11219
7-Bromoquinoline : 11220
5-Bromosalicylhydroxamic acid : 606
Bromosaligenin : 1843
trans-2-Bromostilbene : 1080
Bromosuccinic acid : 3300
N-Bromosuccinimide : 11153
2-Bromo-1,3-thiazole : 11668
2-Bromothiophene : 11728
o-Bromotoluene : 1054
m-Bromotoluene : 1055
p-Bromotoluene : 1056
Bromotrichloromethane : 7474
Bromotrifluoromethane : 7475
1-Bromo-2,3,5-trimethylbenzene : 1092
Bromotrimethylmethane : 9846
Bromotrinitromethane : 7476
Bromotriphenylmethane : 1061
11-Bromoundecanoic acid : 11944
2-Bromoundecanoic acid : 11945
5-Bromouracil : 11062
Bromoxynil : 2751
Bromphenol blue : 9366
Bromthymol blue : 9385
Broncholysin : 5015
Bronopol : 9991
p-Brosyl chloride : 2244
Broxyquinoline : 11313
Brucine : 11564
Buclosamide : 607
Bucolome : 11081
5β-Bufa-20,22-dienolide, 3β,14,16β- : 3141
trihydroxy-, 16-acetate
Bufexamac : 880
Bufotalin : 3141
Bufotenine : 7326
Bulbocapnine : 2470
1,3-Butadiene, hexafluoro : 3164

1,3-Butadiene, pentafluoro-2- : 3172
(trifluoromethyl)-
Butalbital : 11101
Butaldehyde : 3178
Butallylonal : 11080
Butamben : 2488
Butane, 2-amino (DL) : 3210
Butane, 2-benzyl-1-phenyl- : 1698
Butane, 1,4-bis(diethylamino) : 3262
Butane, 1-chloro-2-methyl (DL) : 3251
Butane, 1,4-diamino, : 3260
dihydrochloride
Butane, 1,4-dibromo-2-methyl (d) : 3273
1,4-Butanedicarboxylic acid : 6779
Butane, 1,4-dicyclohexyl- : 4323
Butane, 1,1-dicyclohexyl- : 4326
Butane, 1,2:3,4-diepoxy-, meso- : 3007
Butane, 2,2-di(2-furyl) : 6038
1,3-Butanediol, sulfite : 5198
Butane, 1,4-diphenyl- : 1099
Butane, 1,1-diphenyl- : 1120
Butane, 1,3-diphenyl (DL) : 2020
Butane, 2-iodo (DL) : 3382
Butane, 1-iodo-2-methyl (d) : 3383
Butane, 2-methyl-1,2,3-trichloro : 3453
Butane, 2-methyl-2,3,3-trichloro : 3454
Butane, 1,1,2,3,4-pentachloro : 3419
(liquid)
Butane, 1,2,3,4-tetrabromo (DL) : 3426
2-Butanethiol (DL) : 3438
Butanimidic acid : 3194
Butanoic acid, amide : 3194
Butanoic acid, amide,3-methyl : 3206
Butanoic acid, sec-butyl ester(d) : 3564
Butanoic acid, chloride : 3681
Butanoic acid, 4-(2,4- : 3925
dichlorophenoxy)-
Butanoic anhydride : 3467
3-Butanolal : 3184
2-Butanol, 3-bromo-, erythro-(±)- : 3601
2-Butanol, 3-chloro (erythro, DL) : 1605
1-Butanol, 4-cyclohexyl : 4324
2-Butanol, 4-cyclohexyl : 4469
1-Butanol, 3,5-dinitrobenzoate : 2592
1-Butanol, 2,4-diphenyl-, (-)- : 1108
2-Butanol, 3-methyl (DL) : 3629
1-Butanol, 2-methyl-4-phenyl (DL) : 1107
1-Butanol, 3-methyl-1-phenyl : 1905
1-Butanol, 3-methyl-2-phenyl : 1611
2-Butanol, 2-methyl-3-phenyl : 1616
2-Butanol, 2-methyl-4-phenyl : 2150
2-Butanol, 3-methyl-3-phenyl : 1617
2,3-Butanolone : 3661
2-Butanone, 1-(4-chlorophenoxy)- : 512
3,3-dimethyl-1-(1H-1,2,4-triazol-1-
yl)-
2-Butanone, 3,3-dimethyl-1- : 11717
(methylthio)-, O-
[(methylamino)carbonyl]oxime
Butazolamide : 3197
2-Butenal, 2-bromo, diethyl acetal : 4149
2-Butenal, diethyl acetal : 3758
2-Butenal, 3-ethoxy, diethyl acetal : 3811
trans-2-Butene : 3705
cis-2-Butene : 3706
1-Butene, 2-bromo-4-phenyl : 1001
2-Butene, 1-bromo-4-phenyl : 1002
2-Butene, 2,3-dichlorohexafluoro- : 3754
trans-Butenedioic acid : 3764
trans-2-Butene-1,4-diol : 3792
cis-2-Butene-1,4-diol : 3793
1-Butene, 1,3-diphenyl- : 2022
1-Butene, 1,2-diphenyl : 11560
2-Butene, 1,1,2,3,4,4-hexachloro : 3798
(liquid)
1-Butene, 2-isopropyl-3-methyl- : 8765
1-Butene, 3-methyl-2-isopropyl : 8765
1-Butene, 3-ol, 2-methyl : 3871
2-Butene-1-ol, 3-phenyl : 3872
1-Butene, 1,3,4,4-tetrachloro- : 3808
tetrafluoro
2-Butenoic acid, 4-chloro (trans) : 4151
2-Butenoic acid, 2-chloro, ethyl : 3824
ester (trans)
2-Butenoic acid, 2-chloro, ethyl : 3825
ester (cis)

2-Butenoic acid, 3- : 7659
[(dimethoxyphosphinyl)oxy]-, methyl
ester, (E)-
2-Butenoic acid, 3- : 4137
[(dimethoxyphosphinyl)oxy]-, 1-
phenylethyl ester, (E)-
2-Butenoic acid, 2-ethyl (trans) : 3837
2-Butenoic acid, 3- : 10719
[[(ethylamino)methoxyphosphinothio
yl]oxy]-, 1-methylethyl ester, (E)-
2-Butenoic acid, 2-(1-methylheptyl)- : 5182
4,6-dinitrophenyl ester
2-Butenoic acid, 3-methyl-, 2-(1- : 3003
methylpropyl)-4,6-dinitrophenyl ester
3-Buten-2-ol (DL) : 3858
3-Buten-2-ol, 2-phenyl- : 1869
3-Buten-1-ol, 3-phenyl- : 2157
trans-1-Butenyl bromide : 3707
1-Butenyl methyl ketone : 7086
Butethal : 11082
Butethamate : 898
Butethamine : 5748
Buthiazide : 2856
Butoform : 2488
2,4-D 2-Butoxyethyl ester : 170
Butyl acetate : 121
sec-Butyl acetate : 214
tert-Butyl acetate : 177
tert-Butylacetic acid : 3508
tert-Butylacetylene : 3907
Butylacetylene : 7114
Butylacrolein : 6530
Butyl acrylate : 10616
Butyl alcohol : 3593
sec-Butyl alcohol : 3594
tert-Butyl alcohol : 10389
Isobutyl alcohol, 1-phenyl : 1901
Butyl aldehyde : 3178
Butylallene : 6326
Butylamine : 3209
sec-Butylamine, (±)- : 3210
tert-Butylamine : 9818
Butyl 4-aminobenzoate : 2488
2-(Butylamino)ethanethiol : 5650
N-Butylaniline : 656
N-tert-Butylaniline : 716
N-sec-Butylaniline : 816
Butylated hydroxyanisole : 9329
Butylbenzene : 1118
tert-Butylbenzene : 1459
sec-Butylbenzene : 2035
Butyl benzoate : 2532
p-tert-Butylbenzoic acid : 2588
α-Butylbenzyl alcohol : 1846
Butyl bromide : 3231
(±)-sec-Butyl bromide : 3232
tert-Butyl bromide : 9846
2-tert-Butylbutane : 8847
Butyl butanoate : 3479
Butyl butyrate : 3479
Butyl carbamate : 3935
sec-Butylcarbinol : 3626
Butyl Carbitol : 5688
Butyl carbonate : 3998
Butyl Cellosolve : 5687
Butylchloral : 3191
Butyl chloride : 3241
sec-Butyl chloride (DL) : 3242
tert-Butyl chloride : 9863
tert-Butyl chloroacetate : 128
Butyl chloroformate : 4012
Butyl citrate : 10168
6-sec-Butyl-m-cresol : 9421
tert-Butyl cyanide : 10100
Butylcyclohexane : 4325
sec-Butylcyclohexane : 4448
trans-2-Butylcyclohexanol : 4504
2-Butylcyclohexanone : 4575
Butylcyclooctane : 4737
Butylcyclopentane : 4778
1-Butyl-1-cyclopentene : 4906
Butyl 2,3-dibromopropionate : 10239
Butyl dichloroacetate : 163
Butyl 3,5-dinitrobenzoate : 2592
2-tert-Butyl-4,6-dinitrophenol : 9328
5-n-Butyldocosane : 5269

11-Butyldocosane	: 5272
7-Butyldocosane	: 5270
9-Butyldocosane	: 5271
Butyl enanthate	: 6459
1-Butylene	: 3704
α-Butylene chlorohydrin	: 3605
β-Butylene chlorohydrin	: 3606
1,4-Butylenediamine	: 3258
α-Butylene dibromide	: 3265
1,3-Butylene glycol	: 3345
1,4-Butylene glycol	: 3346
Butylene glycol	: 3346
2,4-D Butyl ester	: 171
Butyl ethanoate	: 121
sec-Butylethene	: 9053
Butyl ether	: 3415
sec-Butyl ether, (±)-	: 3416
Butylethylamine	: 3218
sec-Butyl ethyl amine (DL)	: 3219
Butyl ethyl ether	: 3372
sec-Butyl ethyl ether	: 3373
(RS)-S-sec-Butyl O-ethyl 2-oxo-1,3-thiazolidin-3-ylphosphonothioate	: 6021
Butyl ethyl sulfide	: 3376
Butyl fluoride	: 3378
sec-Butyl fluoride	: 3379
Butyl formate	: 6004
2-tert-Butylfuran	: 6090
Butyl glycidyl ether	: 8602
Butyl heptanoate	: 6459
1-Butylheptyl alcohol	: 11953
2-Butyl-2-hexanol	: 8171
Butyl hexyl ketone	: 11961
tert-Butyl hydroperoxide	: 7178
Butyl 4-hydroxybenzoate	: 2615
Butyl 2-hydroxypropionate	: 10268
tert-Butyl hypochlorite	: 7183
Butylidene chloride	: 3276
2-Butylidenecyclohexanone	: 4576
Butyl iodide	: 3381
tert-Butyl iodide	: 10058
Butyl isobutyl ether	: 3401
tert-Butyl isocyanate	: 10062
Butyl isocyanide	: 3387
Butyl isopropyl ether	: 3399
Butyl isopropyl ketone	: 6522
tert-Butyl isopropyl ketone	: 8989
Butylisopropylmethylcarbinol	: 6480
tert-Butyl isothiocyanate	: 10066
sec-Butyl isothiocyanate (DL)	: 3389
Butyl isovalerate	: 3543
Butyl ketone	: 8175
n-Butylmalonic acid	: 9929
Butyl mercaptan	: 3437
tert-Butyl mercaptan	: 10159
Butyl methacrylate	: 10650
Butylmethylamine	: 3221
Butyl m-methylbenzoate	: 2685
Butyl methyl ether	: 3391
sec-Butyl methyl ether (d)	: 3392
tert-Butyl methyl ether	: 10072
sec-Butyl methyl ether (DL)	: 3393
tert-Butyl methyl ketone	: 3657
Butyl methyl ketone	: 6959
1-Butyl-3-(p-methylphenylsulfonyl)urea	: 2207
Butyl methyl sulfide	: 3402
tert-Butyl methyl sulfide	: 10083
2-Butylnaphthalene	: 7747
Butyl nitrate	: 8072
Butyl nitrite	: 8083
tert-Butyl nitrite	: 8085
5-Butylnonane	: 8122
Butyl nonanoate	: 8147
5-Butyl-4-nonene	: 8184
Butyl oxide	: 3415
Butylparaben	: 2615
Butyl parasept	: 2615
p-tert-Butylphenol	: 9323
o-sec-Butylphenol	: 9430
5-Butylpicolinic acid	: 10898
3-Butylpiperidine	: 9636
4-n-Butylpiperidine	: 9637
Butyl propionate	: 10214
Butyl propyl ether	: 3421
tert-Butyl propyl ketone	: 6965

Butyl propyl ketone	: 8473
2-Butylpyridine	: 10878
2-tert-Butylpyridine	: 10942
3-tert-Butylpyridine	: 10943
5-Butyl-2-pyridinecarboxylic acid	: 10898
2-Butylpyrrolidine	: 11147
2-Butylquinoline	: 11222
Butyl stearate	: 8251
1-Butyl-3-sulfanilylurea	: 2192
Butyl sulfide, 2,2'-dimethyl	: 3435
Butyl sulfite	: 11580
tert-Butyl tetrasulfide	: 11649
Butyl thiocyanate	: 11699
tert-Butyl thiocyanate	: 11701
Butylthiol	: 3437
2-Butylthiophene	: 11731
Butyl m-toluate	: 2685
p-tert-Butyltoluene	: 1466
1-Butyl-s-triazole	: 11847
Butyltrichlorosilane	: 11370
Butyl trifluoroacetate	: 265
2-Butyl-1,1,3-trimethylcyclohexane	: 4327
Butyltrimethylsilane	: 11371
Butyl vinyl ether	: 3369
Butyl vinyl ketone	: 6583
2-Butyne-1-al	: 3895
1-Butyne-3,4-diol	: 3912
Butyraldehyde	: 3178
Butyraldehyde, 3-chloro-, diethyl acetal	: 3243
Butyraldehyde, 3-chloro-, dimethyl acetal	: 3244
Butyraldehyde, 3-methyl, oxime	: 3187
Butyraldehyde, oxime	: 3188
Butyraldehyde, phenylhydrazone	: 3189
Butyramide	: 3194
p-Butyramidophenol	: 3205
Butyrhydrazide	: 3529
Butyric acid	: 3460
Butyric acid, 4-acetyl, ethyl ester	: 6906
Butyric acid, 3-bromo-3-methyl-	: 3472
Butyric acid chloride	: 3681
Butyric acid, sec-butyl ester (d)	: 3564
Butyric acid, 2-chloro-2-methyl-	: 3486
Butyric acid, 2-chloro-3-methyl, ethyl ester	: 3490
Butyric acid, 2-chloro-2-methyl, ethyl ester	: 3491
Butyric acid, 2-chloro-2-methyl-3-oxo, ethyl ester	: 3492
Butyric acid, 2-cyano-2-ethyl-, ethyl ester	: 3495
Butyric acid, 2,2-dichloro-	: 3500
Butyric acid, 3,4-dichloro, butyl ester	: 3501
Butyric acid, 2,3-dimethyl	: 3507
Butyric acid, heptafluoro-, isopropyl ester	: 3526
Butyric acid, 2-hydroxy, ethyl ester (DL)	: 3532
Butyric acid, 2-hydroxy, isobutyl ester	: 3536
Butyric acid, 3-methylamino-, ethyl ester	: 3541
Butyric acid, (2-methylbutyl) ester	: 3545
Butyric acid, 3-methyl, isopropyl ester	: 3553
Butyric acid (1-methylpentyl) ester (d)	: 3559
Butyric acid, 4-phenoxy-, ethyl ester	: 3581
Butyric acid, phenyl ester	: 3584
Butyric acid, 3-thioxo, ethyl ester	: 3589
Butyric acid, 2,2,3-trichloro-	: 3590
Butyric acid, 2,2,3-trichloro, ethyl ester	: 3591
Butyric aldehyde	: 3178
Butyric anhydride	: 3467
Butyroin	: 8474
γ-Butyrolactone	: 6142
1-Butyronaphthone	: 3674
Butyronitrile	: 3403
Butyrophenone, 4-[4-(p-chlorophenyl)-4-hydroxypiperidino]-4'-fluoro-	: 3651
Butyrophenone, 2-hydroxy-	: 3666
Butyrophenone, 4-methyl	: 3672
p-(Butyrylamino)phenol	: 3205
N-Butyryl-p-aminophenol	: 3205

Butyryl bromide	: 3680
Butyryl chloride	: 3681
Butyryl chloride, 3-chloro-	: 3684
Butyryl chloride, 2-chloro-3-methyl	: 3686
Butyryl chloride, 4-chloro-3-oxo	: 3687
Butyryl chloride, 2,3-dimethyl	: 3689
Butyryl chloride, 3-methyl	: 3693
Butyryl chloride, 2-methyl (DL)	: 3694
Butyryl oxide	: 3467
Butyryl triglyceride	: 3586
C.I. Direct blue 53	: 7877
C.I. Solvent Orange 7	: 8028
C.I. Solvent Orange 35	: 384
Cacodyl	: 5110
Cacodylic acid	: 451
Cacodyl oxide	: 452
Cacotheline	: 11352
Cadalene	: 7830
Cadalene, 3-methyl	: 7905
Cadalene, 8-methyl	: 8015
Cadalene, 5-methyl	: 8014
Cadaverine	: 8735
γ-Cadinene	: 7936
Caffeic acid	: 10632
Caffeine	: 10760
Calcyan	: 3930
Calusterone	: 357
Campesterol	: 5386
Camphane-3-carboxylic acid	: 3131
d-Camphocarboxylic acid	: 2917
Camphocean	: 4855
Campholic acid	: 4786
α-Campholytic acid	: 4913
(+)-Camphor	: 2961
Camphor, 3-acetyl	: 2950
d-Camphor, 10-bromo-	: 2951
Camphor, 3,3-dibromo(d)	: 2953
Camphoric acid, dimethyl ester, (+)-	: 4795
(-)-Camphor oxime	: 2962
D-Camphorsulfonic acid	: 2929
α-Camphylamine	: 2927
Canavanine	: 7135
Cannabidiol	: 1537
Cannabinol	: 5154
Canrenone	: 9736
Cantharidin	: 5371
Capraldehyde	: 5030
Capramide	: 5031
Capric acid	: 5059
Capric acid, 10-fluoro	: 5064
Capric alcohol	: 5070
Caprinitrile	: 5057
Caprinoyl chloride	: 5078
Caproaldehyde	: 6718
Caproaldehyde, 3-methyl	: 6720
Caproic acid	: 6870
Caproic acid, 5-bromo,ethyl ester	: 6878
Caproic acid, 6-cyclohexyl	: 4412
Caproic acid, 2,2-dimethyl-3-hydroxy, β-lactone	: 6892
Caproic acid, 6-fluoro	: 6888
Caproic acid, 2-methyl (DL)	: 6893
Caprolactam	: 493
Caprolactone	: 8591
Caproleic acid	: 5091
Capronitrile	: 6840
Caproyl alcohol	: 6911
Caproyl chloride	: 6986
Caproyl chloride, 3-methyl	: 6988
Caproyl chloride, 4-methyl	: 6989
Caprylene	: 8491
Caprylic acid	: 8422
Caprylic acid, 8-fluoro	: 8429
Caprylic acid, 3-methyl	: 8433
Caprylic acid, 2-methyl-3-oxo, ethyl ester	: 8436
Caprylic alcohol	: 8443
Caprylic aldehyde	: 8346
Caprylnitrile	: 8411
Capsaicin	: 8181
Capsanthin	: 4063
Captan	: 7368
Carbamamidine	: 6278
Carbamazepine	: 5144
Carbamic acid, 1H-benzimidazol-2-yl-, methyl ester	: 3990

Carbamic acid, [1-[(butylamino)carbonyl]-1H-benzimidazol-2-yl]-, methyl ester	: 515
Carbamic acid, sec-butyl, ethyl ester	: 3960
Carbamic acid, butyl-, 2-(hydroxymethyl)-2-methylpentyl ester, carbamate	: 3933
Carbamic acid, N-sec-butyl, methyl ester	: 3962
Carbamic acid, (3-chlorophenyl)-, 4-chloro-2-butynyl ester	: 510
Carbamic acid, N,N-diethyl	: 3941
Carbamic acid, dimethyl-, 2-(dimethylamino)-5,6-dimethyl-4-pyrimidinyl ester	: 9728
Carbamic acid, dimethyldithio-, manganese(2+) salt	: 7421
Carbamic acid, N-ethyl, butyl ester	: 3947
Carbamic acid, N-ethyl-N-nitro, butyl ester	: 3951
Carbamic acid, N-ethyl-N-nitro, ethyl ester	: 3950
Carbamic acid, isobutyl ester	: 3959
Carbamic acid, isopentyl-, ethyl ester	: 3952
Carbamic acid, isopropyl ester	: 3955
Carbamic acid, N-isopropyl-N-nitro, ethyl ester	: 3957
Carbamic acid, (3-methylphenyl)-, 3-[(methoxycarbonyl)amino]phenyl ester	: 9198
Carbamic acid, [3-[[(phenylamino)carbonyl]oxy]phenyl]-, ethyl ester	: 5104
Carbamic acid, [1,2-phenylenebis-(iminocarbonothioyl)]bis-, dimethyl ester	: 11723
Carbamic acid, 3-phenylpropyl ester	: 2149
Carbamic acid, 1-phenyl-2-propynyl ester	: 1877
Carbamic acid, N-propyl-N-nitro, ethyl ester	: 3964
Carbamide	: 11992
Carbamimidoselenoic acid	: 11354
Carbamodithioic acid, diethyl-, 2-chloro-2-propenyl ester	: 4065
Carbamodithioic acid, dimethyl-, sodium salt	: 11514
Carbamonitrile	: 4155
Carbamothioic acid, bis(2-methylpropyl)-, S-ethyl ester	: 11586
Carbamothioic acid, bis(1-methylethyl)-, S-(2,3,3-trichloro-2-propenyl) ester	: 11823
Carbamothioic acid, bis(1-methylethyl)-, S-(2,3-dichloro-2-propenyl) ester	: 5106
Carbamothioic acid, cyclohexyl-ethyl-, S-ethyl ester	: 4168
Carbamothioic acid, diethyl-, S-[(4-chlorophenyl)methyl] ester	: 11697
Carbamothioic acid, diethyl-, S-ethyl ester	: 5930
Carbamothioic acid, dipropyl-, S-ethyl ester	: 5372
Carbamothioic acid, dipropyl-, S-propyl ester	: 12047
Carbamoyl chloride, N,N-diethyl	: 3972
Carbamoylcholine chloride	: 5440
5-Carbamoyl-5H-dibenz[b,f]azepine	: 5144
Carbamyl chloride	: 3971
Carbanilaldehyde	: 5999
Carbanilide	: 12010
Carbarsone	: 455
Carbaryl	: 8032
Carbazide	: 4009
(N-Carbazolyl)acetic acid	: 3982
Carbazone, 1,5-diphenyl	: 5122
Carbic anhydride	: 7580
Carbimazole	: 7194
Carbitol	: 5720
Carbitol acetate	: 5721
Carbobenzoxy chloride	: 4027
Carbocysteine	: 5017
Carbodiimide, dicyclohexyl-	: 4291
Carbohydrazide	: 4009
Carbomethene	: 5927

o-Carbomethoxyaniline	: 2503
2-Carbomethoxyphenol	: 2634
Carbon bisulfide	: 3993
Carbon dibromide dichloride	: 7495
Carbon diselenide	: 4036
Carbonic acid, bis(2-chloroethyl) ester	: 5697
Carbonic acid, bis(2-ethoxyethyl) ester	: 5719
Carbonic acid, bis(2-methoxyethyl) ester	: 5736
Carbonic acid, bis(2-methoxyphenyl) ester	: 9384
Carbonic acid, bis(2-methylallyl) ester	: 10703
Carbonic acid, dithio, cyclic S,S-(6-methyl-2,3-quinoxalinediyl) ester	: 4072
Carbonic acid, monochloride,monoethyl ester	: 4018
Carbonic anhydride	: 3992
Carbon oxide	: 4011
Carbon oxysulfide	: 4035
Carbon suboxide	: 9773
Carbon tetrabromide	: 7539
Carbon tetrachloride	: 7540
Carbon tetrafluoride	: 7541
Carbon tetraiodide	: 7542
Carbonyl bromide	: 4006
Carbonyl chloride	: 4007
Carbonyl chloride fluoride	: 4005
Carbonyl cyanide	: 9915
N,N'-Carbonyldiimidazole	: 7193
Carbonyl fluoride	: 4008
Carbonyl sulfide	: 4035
Carbostyril	: 11325
3-Carboxyaniline	: 2485
4-Carboxyaniline	: 2486
9-Carboxyanthracene	: 370
4-Carboxybenzenesulfonamide	: 2508
3-Carboxycoumarin	: 2775
4-Carboxy-1,3-dimethylbenzene	: 2580
2-Carboxyethylamine	: 323
2-Carboxyfuran	: 6050
4-Carboxymethylbiphenyl	: 3009
S-(Carboxymethyl)-L-cysteine	: 5017
3-[(Carboxymethyl)thio]-L-alanine	: 5017
m-Carboxynitrobenzene	: 2704
p-Carboxynitrobenzene	: 2705
4-Carboxyphenol	: 2613
4'-Carboxyphenylmethane-sulfonanilide	: 2721
4-Carboxyphthalic acid	: 2319
3-Carboxypropylamine	: 3464
8-Carboxyquinoline	: 11229
2-Carboxythiophene	: 11737
Carbromal	: 3195
Carbromide	: 3198
Carbutamide	: 2192
Carbyloxime	: 6024
Cardilate	: 3432
Cardine	: 3540
Carditoxin	: 4043
Cardrase	: 2865
Δ4-Carene (l)	: 4050
Carfimate	: 1877
Carinamide	: 2721
Carminic acid	: 373
Carmustine	: 11994
Carnegine	: 7390
Carnosine	: 7133
γ-Carotene	: 4052
α-Carotene	: 4054
β-Carotene, all-trans-	: 4051
α-Carotene, 6',7'-didehydro-5,6-epoxy-4',5,5',6,7,8-hexahydro-3,3'	: 4055
Carvacrol	: 9415
Carvacrol, acetate	: 9419
Carvacryl acetate	: 9419
Carvenol	: 4698
Carvenone (DL)	: 4717
(+)-Carvone	: 4715
(-)-β-Caryophyllene	: 2996
β-Caryophyllene	: 2996
trans-Caryophyllene	: 2996
Caryophyllene	: 2996
Carzenide	: 2508
Casimiroin	: 5218

Catapyrin	: 11060
Catechol, 5-iodo	: 1523
Caulophylline	: 7578
CDT	: 4212
Cedrin	: 8058
Cedrol	: 7570
Celidoniol, deoxy-	: 8099
Cellocidin	: 3902
Cellosolve	: 5717
Cephaeline	: 5367
Cepharanthine	: 8630
Cerotic acid	: 6624
Cetane	: 6632
1-Cetene	: 6667
Cetyl alcohol	: 6659
Cetylamine	: 6631
Cetyl bromide	: 6633
Cetyl chloride	: 6634
α-Cetylcitric acid	: 8102
Cetyl iodide	: 6640
Cetyl lactate	: 10271
Cetyl mercaptan	: 6643
Cetyl palmitate	: 6651
Cetyl stearate	: 8258
Cevadic acid	: 3844
Cevadine	: 4069
CFC 11	: 7555
CFC 12	: 7500
CFC 13	: 7486
CFC 13B1	: 7475
CFC 21	: 7502
CFC 22	: 7478
CFC 113	: 5663
CFC 114	: 5538
CFC 115	: 5484
CFC 123	: 5540
CFC 123a	: 5539
C16 Guerbet fatty acid	: 5065
Chalcone	: 10708
Champacol	: 507
Chaulmoogric acid	: 4936
Chavicine	: 9633
Chavicol	: 9458
Cheirolin	: 10067
Chelerythrine	: 2433
Chelidonic acid	: 10776
Chenic acid	: 4097
Chenodeoxycholic acid	: 4097
Chenodiol	: 4097
Chimyl alcohol	: 10010
Chitosamine	: 6233
Chloral	: 21
Chloral alcoholate	: 5769
Chloral ammonia	: 5680
Chloral, diethyl acetal	: 5657
Chloral formamide	: 6000
Chloral hydrate	: 5582
Chloralide	: 5212
Chloramben	: 2492
Chlorambucil	: 1101
Chloraminophenamide	: 1558
Chloramphenicol	: 52
Chloramphenicol palmitate	: 6648
Chloranil	: 4268
Chloranilic acid	: 4258
Chlorocyanohydrin	: 10115
Chlordantoin	: 7220
Chlordimeform	: 7564
Chlorfenac	: 944
Chlorfenethol	: 1850
Chlorfenson	: 2221
Chlorindanol	: 7283
Chlorine cyanide	: 4166
Chlornaphazine	: 7710
5-Chloroacenaphthene	: 5
2-Chloroacetaldehyde	: 12
Chloroacetaldehyde, trimer	: 13
Chloroacetamide	: 41
Chloroacetamide, N-3-chloroallyl	: 43
Chloroacetic acid (β)	: 123
Chloroacetic acid, methyl ester	: 138
Chloroacetic acid, phenyl ester	: 147
Chloroacetic acid, propyl ester	: 149
Chloroacetic anhydride	: 124
Chloroacetone	: 10420
m-Chloroacetophenone	: 5788
p-Chloroacetophenone	: 5789

ω-Chloroacetophenone	: 5790
Chloroacetyl chloride	: 295
Chloroacetylene	: 5938
9-Chloroacridine	: 315
o-Chloroaniline	: 658
m-Chloroaniline	: 659
p-Chloroaniline	: 660
4-Chloro-o-anisidine	: 667
1-Chloroanthracene	: 374
Chlorobenzene	: 1131
p-Chlorobenzenesulfonic acid	: 2220
1-(p-Chlorobenzenesulfonyl)-3-propylurea	: 2208
2-Chloro-1,3,5-benzenetricarboxylic acid	: 2320
N-(p-Chlorobenzhydryl)-N'-methylhomopiperazine	: 5140
o-Chlorobenzoic acid	: 2534
m-Chlorobenzoic acid	: 2535
p-Chlorobenzoic acid	: 2536
2-Chlorobenzothiazole	: 2862
2-Chlorobenzoxazole	: 2878
m-Chlorobenzoyl chloride	: 2891
4-Chlorobenzyl chloride	: 1134
2-(2-Chlorobenzyl)-4,4-dimethyl-1,2-oxazolidin-3-one	: 4140
4'-Chloro-4-biphenylamine	: 3012
Chlorobromoacetic acid	: 112
Chlorobromodifluoromethane	: 7464
Chlorobromomethane	: 7463
2-Chloro-4-bromophenol	: 9232
2-Chloro-1-bromopropane	: 9838
1-Chloro-2-bromopropane	: 9840
2-Chloro-1,3-butadiene	: 3152
1-Chlorobutane	: 3241
1-Chloro-2-butanol	: 3605
3-Chloro-2-butanol	: 3606
Chlorobutanol	: 10407
1-Chloro-2-butanone	: 3648
1-Chloro-3-butanone	: 3648
trans-1-Chloro-1-butene	: 3720
cis-1-Chloro-1-butene	: 3721
2-Chloro-1-butene	: 3722
3-Chloro-1-butene	: 3723
3-Chloro-2-buten-1-ol	: 3864
4-Chloro-2-buten-1-ol	: 3865
1-Chloro-3-buten-2-ol	: 3866
3-Chloro-3-buten-2-ol	: 3868
Chloro-tert-butyl alcohol	: 10354
3-Chlorobutyryl chloride	: 3684
m-Chlorocarbanilic acid isopropyl ester	: 3939
3-Chloro-2-(chloromethyl)-1-butene	: 3729
4-Chloro-o-cresol	: 9266
4-Chloro-m-cresol	: 9267
6-Chloro-m-cresol	: 9262
α-Chlorocrotonaldehyde	: 3697
γ-Chlorocrotonic acid	: 3822
2-Chlorocrotonic acid	: 4150
3-Chlorocrotonic acid ethyl ester	: 3826
γ-Chlorocrotyl alcohol	: 3864
trans-2-Chlorocyclohexanol	: 4505
trans-4-Chlorocyclohexanol	: 4507
2-Chlorocyclohexanone	: 4577
4-Chlorocyclohexanone	: 4578
1-Chlorocyclohexene	: 4642
2-Chloro-4-cyclohexylphenol	: 9254
Chlorocyclooctatetraene	: 4745
Chlorocyclopentane	: 4789
Chlorocyclopropylmethane	: 4966
2-Chloro-N,N-diallylacetamide	: 44
Chlorodiazepoxide	: 2409
Chlorodibromomethane	: 7493
N-Chloro-2,6-dibromoquinoneimine	: 4279
Chloro(dichloromethyl)dimethyl-silane	: 11373
Chlorodifluoroacetic acid	: 127
1-Chloro-1,1-difluoroethane	: 5470
1-Chloro-2,2-difluoroethane	: 5471
2-Chloro-1,1-difluoroethylene	: 5885
Chlorodifluoromethane	: 7478
Chlorodiiodomethane	: 7479
1-Chloro-2,3-dimethylbutane	: 3245
1-Chloro-2,3-dimethyl-2-butene	: 3731
1-Chloro-2,3-dimethyl-2-butanol	: 3608
1-Chloro-3,3-dimethyl-2-butanol	: 3609
2-Chloro-2,5-dimethylhexane	: 6740
1-Chloro-3,3-dimethylpropane	: 3252
3-Chloro-1,5-dimethylpyrazole	: 10818
5-Chloro-1,3-dimethylpyrazole	: 10820
Chlorodimethyl(vinylethynyl)-methane	: 7107
1-Chloro-2,4-dinitrobenzene	: 1144
2-Chloro-1,3-dinitrobenzene	: 1145
Chlorodinitromethane	: 7480
4-Chloro-2,6-dinitrophenol	: 9258
3-Chloro-4,6-dinitrophenol	: 9259
Chlorodiphenylarsine	: 454
2-Chloroethanal	: 12
2-Chloroethanethiol	: 5651
2-(2-Chloroethoxy)-N-[[(4-methoxy-6-methyl-1,3,5-triaziN-2-yl)amino]carbonyl]benzene sulfonamide	: 11824
β-Chloroethyl acetate	: 5695
1-Chloroethyl 2-chloroethyl ether	: 5468
2-Chloroethyl chloroformate	: 4013
N-(2-Chloroethyl)dibenzylamine hydrochloride	: 1801
2-Chloroethyl isobutyrate	: 10292
6-Chloro-N-ethyl-N'-(1-methylethyl)-1,3,5-triazine-2,4-diamine	: 476
3-Chloro-3-ethyl-1-pentyne	: 9125
1-Chloroethyl propyl ether	: 9854
2-Chloroethyl vinyl ether	: 5887
Chlorofluoroacetic acid	: 131
Chlorofluoroacetic acid, butyl ester	: 132
Chlorofluoroacetic acid, methyl ester	: 134
Chlorofluoroacetic acid, propyl ester	: 135
Chlorofluoroacetyl chloride	: 296
1-Chloro-4-fluorobutane	: 3250
1-Chloro-10-fluorodecane	: 5037
1-Chloro-1-fluoroethane	: 5475
1-Chlorofluoroethane	: 5475
1-Chloro-6-fluorohexane	: 6742
Chlorofluoromethane	: 7481
1-Chloro-3-fluoropropane	: 9856
Chloroflurazole	: 2377
Chloroform	: 7554
Chloroformic acid, isopropenyl ester	: 4022
Chloroformic digitalin	: 4041
2-Chlorofuran	: 6073
3-Chlorofuran	: 6074
Chloroheptafluorocyclobutane	: 4177
2-Chloroheptane	: 6361
3-Chloroheptane	: 6362
4-Chloroheptane	: 6363
7-Chloro-1-heptanol	: 6477
1-Chloro-2-heptanol	: 6478
1-Chloro-2-heptanone	: 6513
1-Chloro-1-heptene	: 6538
1-Chloro-1-heptyne	: 6614
1-Chloro-5-heptyne	: 6615
1-Chloro-4-heptyne	: 6616
2-Chlorohexane	: 6738
3-Chlorohexane	: 6739
1-Chloro-2-hexanol	: 6917
1-Chloro-3-hexanol	: 6918
1-Chloro-1-hexene	: 7004
2-Chloro-1-hexene	: 7005
5-Chloro-1-hexene	: 7006
4-Chloro-2-hexene	: 7007
α-Chloro-N-hexylacetamide	: 45
α-Chlorohydrin	: 9994
1-Chloro-3-hydroxypentane	: 8925
3-Chloroindazole	: 7236
Chloroiodomethane	: 7482
4-Chloroisoprene	: 3154
2-Chloroisopropyl ether	: 10123
1-Chloroisoquinoline	: 7376
Chloromethane	: 7477
Chloromethoxymethane	: 7483
2-Chloro-N-[[(4-methoxy-6-methyl-1,3,5-triaziN-2-yl)amino]carbonyl]benzene-sulfonamide	: 4091
α-Chloro-N-(3-methoxypropyl)acetamide	: 46
2-(4-Chloro-6-methoxypyrimidin-2-ylcarbamoylsulfamoyl) benzoic acid, ethyl ester	: 4081
2-Chloro-2'-methylbiphenyl	: 3026
Chloromethylbutadiene	: 3153
3-Chloro-3-methyl-2-butanone	: 3649
1-Chloro-3-methyl-1-butene	: 3733
1-Chloro-2-methyl-1-butene	: 3735
1-Chloro-2-methyl-1-butene	: 3732
3-Chloro-2-methyl-1-butene	: 3734
2-Chloro-3-methyl-1-butene	: 3737
Chloromethyl chloroformate	: 4014
Chloromethyl cyanide	: 274
2-(Chloromethyl)-2,3-dihydrobenzofuran	: 2441
Chloromethyl ethyl ether	: 5478
Chloromethyl ethyl ketone	: 3646
2-Chloro-2-methylheptane	: 6366
2-Chloro-6-methylheptane	: 6367
3-Chloro-3-methylheptane	: 6369
4-Chloro-4-methylheptane	: 6370
2-Chloro-5-methylhexane	: 6745
Chloromethyl iodide	: 7482
Chloromethyl methyl dichloro silane	: 11390
Chloromethyl methyl ether	: 7483
1-Chloro-2-methylnaphthalene	: 7784
4-Chloro-4-methyloctane	: 8363
2-Chloro-2-methyloct-3-yne	: 8534
2-Chloro-4-methylpentane	: 8730
5-Chloro-methyl-2-pentene	: 9026
3-Chloro-3-methyl-1-pentyne	: 9126
4-(4-Chloro-2-methylphenoxy)butyric acid	: 3493
2-Chloromethylpropanol	: 10353
1-Chloro-2-methylpropene	: 10539
3-Chloro-2-methylpropene	: 10540
2-Chloro-8-methylquinoline	: 11244
Chloromethylsilane	: 11378
2-Chloro-5-methylthiophene	: 11741
cis,trans-3-Chloro-4-[4-methyl-2-(1H-1,2,4-triazol-1-ylmethyl)-1,3-dioxolan-2-yl]phenyl 4-chlorophenyl ether	: 5174
(Chloromethyl)trichlorosilane	: 11456
2-Chloronaphthalene	: 7781
5-Chloro-1-naphthalenecarboxylic acid	: 7762
5-Chloro-1,4-naphthoquinone	: 7847
o-Chloronitrobenzene	: 1214
m-Chloronitrobenzene	: 1215
p-Chloronitrobenzene	: 1216
2-Chloro-3-nitrobenzoic acid	: 2544
3-Chloro-2-nitrobenzoic acid	: 2546
1-Chloro-1-nitroethane	: 5482
2-Chloro-1-nitroethane	: 5483
Chloronitromethane	: 7485
1-Chloro-5-nitronaphthalene	: 7786
1-Chloro-8-nitronaphthalene	: 7787
2-Chloro-1-nitronaphthalene	: 7788
5-Chloro-2-nitrophenol	: 9269
1-Chloro-1-nitropropane	: 9866
3-Chloro-1-nitropropane	: 9868
1-Chloro-3-nitropropane	: 9868
2-Chloro-2-nitropropane	: 9870
3-Chloro-5-nitrotoluene	: 1207
2-Chlorononane	: 8124
5-Chlorononane	: 8125
9-Chlorononan-1-ol	: 8163
(Z)-4-Chloro-4-octene	: 8500
1-Chloro-1-octyne	: 8533
3-Chloro-1,3-pentadiene	: 8663
Chloropentafluoroacetone	: 10421
Chloropentafluorobenzene	: 1220
Chloropentafluoroethane	: 5484
3-Chloropentane	: 8721
3-Chloropentanoic acid	: 8864
1-Chloro-3-pentanol	: 8925
2-Chloro-1-pentene	: 9016
3-Chloro-1-pentene	: 9017
4-Chloro-1-pentene	: 9018
1-Chloro-2-pentene	: 9020
2-Chloro-2-pentene	: 9021
3-Chloro-2-pentene	: 9022
3-Chloroperoxybenzoic acid	: 1127
p-Chlorophenacyl bromide	: 5778
o-Chlorophenol	: 9251
m-Chlorophenol	: 9252
p-Chlorophenol	: 9253
(p-Chlorophenoxy)acetic acid	: 146
p'-Chloro-p-phenylaniline	: 3012

1-(p-Chloro-α-phenylbenzyl)-4-methylhomopiperazine : 5140
p-Chlorophenyl p-chlorobenzenesulphonate : 2221
(RS)-1-p-Chlorophenyl-4,4-dimethyl-3-(1H-1,2,4-triazol-1-ylmethyl)pentan-3-ol : 11589
m-Chlorophenyl ethyl ether : 1157
Chlorophenylmethane : 1182
4-Chlorophenyl phenyl ether : 1221
4-(4-Chlorophenyl)-2-phenyl-2-(1H-1,2,4-triazol-1-ylmethyl)butyronitrile : 5949
R-(-)-2-Chloro-1-phenylpropane : 1234
Chlorophenylsilane : 11384
1-(2-Chlorophenyl)thiourea : 11801
p-Chlorophenyl 2,4,5-trichlorophenyl sulfide : 11648
Chloropicrin : 7557
Chloropropamide : 2208
2-Chloro-1,3-propanediol : 9995
3-Chloro-1-propanol : 10348
trans-1-Chloropropene : 10533
cis-1-Chloropropene : 10534
3-Chloro-1-propene : 10536
Chloropropham : 3939
β-Chloropropionic acid : 10216
β-Chloropropionitrile : 10097
(S)-2-Chloropropionyl chloride : 10496
3-Chloropropylene : 10536
3-Chloropropyl methyl sulfide : 9864
2-Chloropropyl methyl sulfide : 9865
2-Chloro-2-propylpentane : 6370
6-Chloropurine : 10753
N1-(6-Chloro-3-pyridazinyl)sulfanilamide : 2193
2-Chloropyridine : 10915
3-Chloropyridine : 10916
4-Chloropyridine : 10917
2-Chloroquinoline : 11237
4-Chloroquinoline : 11238
5-Chloroquinoline : 11239
6-Chloroquinoline : 11240
7-Chloroquinoline : 11241
8-Chloroquinoline : 11242
6-Chloroquinoxaline : 11334
o-Chlorostyrene : 1149
N-Chlorosuccinimide : 11154
3-Chloro-6-sulfanilamidopyridazine : 2193
1-Chloro-1,1,2,2-tetrafluoroethane : 5485
1-Chloro-1,2,2,2-tetrafluoroethane : 5486
Chlorothalonil : 1317
Chlorothiazide : 2852
5-Chloro-2-thienyl chloride : 11748
2-Chlorothiophene : 11740
Chlorothymol : 9268
o-Chlorotoluene : 1183
m-Chlorotoluene : 1184
p-Chlorotoluene : 1185
3-Chloro-o-toluidine : 671
4-Chloro-o-toluidine : 673
5-Chloro-o-toluidine : 676
4-Chloro-m-toluidine : 674
6-Chloro-m-toluidine : 669
2-Chloro-p-toluidine : 668
3-Chloro-p-toluidine : 672
Chlorotrianisene : 1154
Chlorotribromomethane : 7549
1-Chloro-2-(trichloromethyl)benzene : 1237
1-Chloro-3-(trichloromethyl)benzene : 1238
1-Chloro-4-(trichloromethyl)benzene : 1239
2-Chloro-1,1,1-trifluoroethane : 5489
1-Chloro-1,1,2-trifluoroethane : 5487
1-Chloro-1,2,2-trifluoroethane : 5488
Chlorotrifluoroethylene : 5889
Chlorotrifluoromethane : 7486
4-Chloro-3-(trifluoromethyl)nitrobenzene : 1218
4-Chloro-2-(trifluoromethyl)nitrobenzene : 1219
5-[2-Chloro-4-(trifluoromethyl)phenoxy]-N-(methylsulfonyl)-2-nitrobenzamide : 5985
(E)-1-[1-[[4-Chloro-2-(trifluoromethyl)phenyl]imino]-2-propoxyethyl]-1H-imidazole : 11900

2-[[2-Chloro-4-(trifluoromethyl)phenyl]amino]-3-methylbutanoic acid, (±)-cyano(3-phenoxy phenyl)methyl ester : 5983
(RS)-2-[4-(3-Chloro-5-trifluoromethyl-2-pyridyloxy)phenoxy]propionic acid, ethyl ester : 6291
Chlorotriphenylstannane : 11541
4-Chlorovaleric acid : 8865
2-Chlorovinyl ethyl ether : 5886
Chloroxine : 11314
Chloroxylenol : 9257
Chlorphenesin : 10000
Chlorphenesin carbamate : 10001
Chlorpromazine : 9505
Chlortetracycline : 7704
Chlorzoxazone : 2883
Cholagon : 4101
5β-Cholan-24-amide, N-(carboxymethyl)-3α,7α,12α-trihydroxy- : 6273
Cholanic acid : 4094
Cholanthrene : 2391
Cholerythrin : 3000
5α-Cholestan-3β-ol : 4107
Δ 5-Cholesten-3β-ol : 4110
Δ 5,7-Cholesterol : 4103
Cholesterol : 4110
Cholesteryl alcohol : 4110
Cholic acid : 4100
Choline, o-acetyl,chloride : 5439
Choline, carbamate,chloride : 5440
Choline chloride : 5446
Cholografin : 2594
N-Cholylglycine : 6273
Chroman, 2-methyl- : 2778
Chroman, 6-methyl- : 2780
Chroman, 7-methyl- : 2781
Chroman, 3-methyl : 2779
6-Chromanol, 2,8-dimethyl-2-(4,8,12-trimethyltridecyl)- : 2785
6-Chromanol, 2,5,7,8-tetramethyl-2-(4,8,12-trimethyltridecyl)-, acetate, (+)- : 2788
6-Chromanol, 2,5,8-trimethyl-2-(4,8,12-trimethyltridecyl)- : 2789
6-Chromanol, 2,7,8-trimethyl-2-(4,8,12-trimethyltridecyl)- : 2790
4-Chromanone : 2799
3-Chromene : 2772
1,2-Chromene : 2772
Chromotropic acid : 7875
Chrysamminic acid : 401
Chrysanthemum monocarboxylic acid, methyl ester (DL, trans) : 4965
Chrysanthenone : 2977
Chrysatropic acid : 2829
Chrysin : 2811
Chrysoidine hydrochloride : 1284
Chrysophanic acid : 398
Chugaev's reagent : 3360
Cicrotoic acid : 3834
Cinchol : 11558
Cinchomeronic acid : 10930
Cinchonamine : 7306
Cinchonidine : 4133
Cinchoninaldehyde : 11225
(+)-Cinchonine : 4120
Cinchonine : 4120
Cinchophen : 11235
Cinchotin : 4122
Cinchotoxine : 10435
trans-Cinnamaldehyde : 10518
Cinnamedrine : 1903
trans-Cinnamic acid : 10682
Cinnamic acid, allyl ester : 10692
Cinnamic acid, benzyl ester (trans) : 10690
Cinnamic acid, β-bromo-, ethyl ester : 10614
Cinnamic acid, α-bromo-, methyl ester : 10615
Cinnamic acid, α-chloro-, methyl ester, (E)- : 10622
Cinnamic acid, β-chloro-, methyl ester : 10623
Cinnamic acid, α-fluoro- : 10637

Cinnamic acid, isopropyl ester (trans) : 10687
Cinnamic acid, 2-nitro, methyl ester (trans) : 10677
Cinnamic acid, 3-nitro, methyl ester (trans) : 10678
Cinnamic acid, phenyl ester (trans) : 10689
Cinnamic acid, propyl ester (trans) : 10693
Cinnamic anhydride : 10684
Cinnamoyl chloride : 10716
Cinnamyl cinnamate : 10691
Cirantin : 2794
Citraconic acid : 3786
β-Citraurin : 435
Citrazinic acid : 10900
Citric acid : 10167
Citric acid, trimethyl ester : 10170
Citrinin : 2774
(±)-Citronellol : 8514
(-)-α-Citronellol : 8516
Citropten : 2818
Citrulline : 8551
cis-Civetone : 4218
Civetone : 4218
1,7-Cleve's acid : 7967
Clitizina : 10904
Clofibrate : 10231
Clomethiazole : 11670
Clonitrate : 9997
Clopoxide : 2409
Clorindione : 7266
Clorophene : 9270
Clostebol : 354
Cloxyquin : 11311
Cluytyl alcohol : 8222
Cocaine : 482
Coclaurine : 7396
Codamine : 7392
Codeine : 7669
Codeine, phosphate : 7673
Colchiceine : 92
Colchicine : 93
2,3,4-Collidine : 11024
2,3,5-Collidine : 11025
2,3,6-Collidine : 11026
2,4,5-Collidine : 11027
2,4,6-Collidine : 11028
Collinomycin : 2773
Columbin : 5926
Con-5-enine, 3β-(dimethylamino)- : 4143
Conessine : 4143
Congo red : 7969
Conhydrine (DL) (lower m.p.) : 9677
γ-Coniceine : 11021
Coniferin : 6226
Coniferyl alcohol : 9370
Coniine : 9708
Conquinamine : 6177
Copaene : 11861
Copper, [μ-[carbonato(2-)-O:O']]dihydroxydi- : 511
Coproergostane : 5378
Coprostane : 4106
4:5-Coprosten-3-ol : 4111
Coprostenol : 4111
Coprosterol : 4109
Corallin : 4278
Corchorin : 4049
Cordianine : 12009
Cordycepic acid : 7422
Corticosterone : 9748
Cortisone : 9754
Corybulbine : 5151
Corycavamine : 3108
(+)-Corycavine : 3108
Corydaldine : 7398
Corydaline (d) : 5149
Corydine : 5162
Corynantheine : 4146
Cotarnine : 5216
Cotoin : 7608
Cotoran : 12007
Coumalic acid : 10773
Coumaran : 2442
p-Coumaric acid : 10644
Coumarilic acid : 2439
Coumarin : 2792

Coumarin-3-carboxylic acid : 2775
Coumarin, 3-chloro-7-hydroxy-4-methyl, O-ester with O,O-diethylphosphorothioate : 4147
Coumarin, 3-ethyl- : 2820
Coumarin, 7-hydroxy- : 2826
Coumarin, 7-hydroxy-6-methoxy- : 2829
cis-o-Coumarinic acid lactone : 2792
Coumarin, 3-methyl- : 2833
Coumarin, 5-methyl- : 2834
Coumarin, 7-methyl- : 2836
Coumarin, 8-methyl- : 2837
Coumarone : 2438
Coumarone-2-carboxylic acid : 2439
Coumestrol : 2466
Creatine : 6254
Creatinine : 7230
Creosol : 9385
o-Cresol : 9400
m-Cresol : 9401
p-Cresol : 9402
m-Cresol, 5-amino- : 9212
m-Cresol, 6-bromo- : 9237
m-Cresol, 6-sec-butyl- : 9421
m-Cresol, 2,4-dichloro : 9292
m-Cresol, 2,6-dichloro- : 9294
p-Cresol, 2,6-di-tert-pentyl- : 9225
p-Cresol, 2-iodo- : 9374
o-Cresol, 4-isobutyryl : 10448
o-Cresol, 3-isopropyl, methyl ether : 363
m-Cresol, 6-pentyl- : 9425
o-Cresol, α-phenyl- : 9454
p-Cresol, α-phenyl- : 9455
o-Cresolphthalein : 7353
Cresol red : 9368
o-Cresol, 4,5,6-trichloro- : 9487
p-Cresol, 2,3,6-trichloro- : 9488
m-Cresol, 2,4,6-trichloro- : 9489
o-Cresotic acid : 2629
m-Cresotic acid : 2630
p-Cresotic acid : 2631
o-Cresyl acetate : 211
m-Cresyl acetate : 212
Crimidine : 11043
Crithminic acid : 2678
Croceic acid : 7972
Crocetin (trans) : 5107
Croconic acid : 4939
Crodimyl : 2838
Cromaril : 2840
trans-Crotonaldehyde : 3695
Crotonaldehyde, 2-chloro- : 3697
Crotonaldehyde-2-chloro, dimethylacetal : 3730
Crotonaldehyde, diethyl acetal : 3758
Crotonchloral : 3191
trans-Crotonic acid : 3814
cis-Crotonic acid : 3815
Crotonic acid, allyl ester : 3856
Crotonic acid anhydride : 3818
Crotonic acid, 4-chloro- : 3822
Crotonic acid, 2-chloro-, ethyl ester : 3823
Crotonic acid, 3-chloro-, ethyl ester : 3826
Crotonic acid, 4-chloro-, ethyl ester : 3829
Crotonic acid, 4-chloro-, methyl ester : 3833
Crotonic acid, 2,3-dimethyl-, ethyl ester : 3836
Crotonic acid, 2-ethyl-, stereoisomer : 3837
trans-Crotyl alcohol : 3859
cis-Crotyl alcohol : 3860
Cryptocavine : 2435
Cryptopine : 2435
all-trans-β-Cryptoxanthin : 4061
Cryptoxanthin : 4061
Cryptoxanthol, all-trans- : 4061
Crystal violet : 7453
Cubebin : 6136
Cumene : 1975
Cumene hydroperoxide : 7181
3-p-Cumenyl-2-methylpropionaldehyde : 2117
Cumic acid : 2689
Cumic alcohol : 1902
Cumidine : 789
Cuminaldehyde : 590

Cuminic acid : 2689
Cuminol : 1902
Cumotocopherol : 2789
α-Cumyl alcohol : 1857
Cupferron : 759
Cupreine : 4119
Cupreol : 11558
Cupreol : 11558
ar-Curcumene : 1468
Curcumin : 6334
Curine : 11925
Curling factor : 11520
Cuscohygrine : 10415
Cusparine : 11216
Cyacetacide : 153
Cyamelide : 11907
Cyanidelonon 1522 : 2817
Cyanoacetamide : 49
Cyanoacetic acid : 150
Cyanoacetic acid, diethyl, ethyl ester : 3495
2-Cyanoacetophenone : 2124
Cyanoacetylene : 10737
N-Cyanoamine : 4155
Cyanobenzene : 2740
2-Cyanocaproic acid, ethyl ester : 6881
Cyanocyclobutane : 4173
1-Cyanocyclohexene : 4630
Cyanocyclopentane : 4779
1-Cyanocyclopentene : 4907
4-Cyano-2,6-dibromophenol : 2751
2-Cyanoethylamine : 10092
(2-Cyanoethyl)benzene : 2123
Cyanoethylene : 10583
Cyano(4-fluoro-3-phenoxyphenyl)methyl 3-(2,2-dichloroethenyl)-2,2-dimethylcyclopropanecarboxylate : 5008
1-Cyano-2-(2-furyl)ethene : 10585
Cyanogen : 5550
Cyanogenamide : 4155
Cyanogen iodide : 7337
Cyanogen nitride : 4155
Cyanoiodomethane : 283
Cyanomethane : 271
Cyanomethyl methyl sulfide : 286
Cyanomethyl phenyl ketone : 2124
Cyano(3-phenoxyphenyl)methyl-3-(2,2-dibromoethenyl)-2,2-dimethylcyclopropanecarboxylate : 5102
Cyano(3-phenoxyphenyl)methyl 2,2,3,3-tetramethylcyclo-propanecarboxylate : 5953
1-Cyano-2-phenylethane : 2123
1-Cyanopropane : 3403
2-Cyano-2-propanol : 10108
3-Cyano-1-propene : 3803
2-Cyanoquinoline : 11223
2-Cyanothiophene : 11732
2-Cyanotoluene : 2763
m-Cyanotoluene : 2764
p-Cyanotoluene : 2765
1-Cyanoundecane : 5303
Cyanuric acid, dihydrate : 11839
Cyanuric acid trichloride : 11837
Cyclamal : 2117
Cyclamic acid : 11568
Cyclobarbital : 11084
Cyclobutaneacetic acid, 3-carboxy-2,2-dimethyl-, diethyl ester : 4171
Cyclobutane-1-carboxylic acid : 4174
Cyclobutane-1,1-dicarboxylic acid, diethyl ester : 4180
Cyclobutane-1,2-dicarboxylic acid, dimethyl ester (trans) : 4181
Cyclobutane-1,3-dicarboxylic acid (cis), 2,2-dimethyl : 4183
Cyclobutane-1,3-dicarboxylic acid, 2,2-dimethyl, dimethyl ester : 4184
Cyclobutane, 1,2-divinyl-, trans- : 4187
Cyclobutylamine : 4169
Cyclobutylcarboxylic acid : 4174
Cyclobutyl hydroxide : 4196
Cyclobutyl methyl ketone : 5792
Cyclobutyrol : 4309
Cyclocapron : 4333
β-Cyclocitral : 4636

1,5,9-Cyclododecatriene, (cis, trans, trans) : 4213
cis,cis,cis-Cyclododeca-1,5,9-triene : 4214
Cyclodoecanone : 4211
Cyclofenchene : 11866
Cycloform : 2505
β-Cyclogeraniol : 4666
α-Cyclogeraniol : 4667
Cyclogeraniolane : 4486
α-Cyclogeraniolene : 4691
Cyclogeranyl alcohol : 4666
Cycloglycylglycine : 9612
9-Cycloheptadecene-1-one : 4217
Cycloheptyl bromide : 4222
Cycloheptylcarboxylic acid : 4223
1,3-Cyclohexadiene, 3,5-dimethyl : 4256
1,3-Cyclohexadiene, 5-methyl (DL) : 4272
Cyclohexanacetic acid, 2-methyl : 4311
Cyclohexanacetic acid, 3-methyl : 4312
Cyclohexane, 1,3-bis(aminomethyl) : 4362
Cyclohexane, 1,3-bis-(hydroxymethyl) : 4363
Cyclohexane, sec-butyl- : 4448
Cyclohexane, decafluoro-1,2-bis(trifluoromethyl)- : 4342
Cyclohexane-1,2-dicarboxylic acid, diethyl ester (trans, DL) : 4350
Cyclohexane-1,3-dicarboxylic acid, diethyl ester (cis) : 4351
Cyclohexane-1,3-dicarboxylic acid, dimethyl ester (cis) : 4352
Cyclohexane-1,3-dicarboxylic acid, dimethyl ester (trans) : 4353
Cyclohexane-1,4-dicarboxylic acid, diethyl ester (trans) : 4356
Cyclohexane-1,4-dicarboxylic acid, dimethyl ester (cis) : 4357
Cyclohexane-1,4-dicarboxylic acid, diethyl ester (cis) : 4355
Cyclohexane, 2,4-diisopropenyl-1-methyl-1-vinyl-, stereoisomer : 4389
Cyclohexane, 1,2-dimethyl, perfluoro : 4342
Cyclohexane, 1,2-epoxy-4-vinyl : 8555
Cyclohexane, 1,2,3,4,5,6-hexachloro-, (α, DL) : 4408
Cyclohexane hydroperoxide : 7177
Cyclohexane, 1-hydroxymethyl-2-methyl (cis) : 4422
Cyclohexane, 1-hydroxymethyl-2-methyl (trans) : 4423
Cyclohexane, 1-isopropenyl-3-methyl : 4435
Cyclohexane, isopropyl- : 4431
Cyclohexane, 1-methylamino-4-methyl : 4723
Cyclohexane, 1-methyl-2-ethyl (trans) : 4396
Cyclohexane, 2-methyl-1-ethylidene : 4393
Cyclohexane, 2-nitrophenyl : 1251
1-L-1,3,4/2,5-Cyclohexanepentol : 4074
Cyclohexanol, acetate : 155
Cyclohexanol, 2-allyl-, trans- : 4556
Cyclohexanol, 2-amino (trans, DL) : 4501
Cyclohexanol, 1,2-dimethyl-, stereoisomer : 4508
: 4509
Cyclohexanol, 2-ethyl (trans, DL) : 4523
Cyclohexanol, 2-isopropyl-, cis- : 4539
Cyclohexanol, 4-isopropyl-, cis- : 4540
Cyclohexanol, 4-isopropyl-, trans- : 4541
Cyclohexanol-4-isopropyl-2-methyl : 7428
Cyclohexanol, 2-methyl (cis, DL) : 4531
Cyclohexanol, 2-phenyl (cis, DL) : 4553
Cyclohexanol, 2-phenyl (trans, DL) : 4554
Cyclohexanone-4-carboxylic acid, 2-methyl : 4336
Cyclohexanone, 2,4-dimethyl (trans, d) : 4583
Cyclohexanone, 2-isopropyl- : 4599
Cyclohexanone, 4-isopropyl- : 4600
Cyclohexanone, 2-isopropyl-5-methyl (l) : 4606
1-Cyclohexene-1-carboxaldehyde, 4-isopropenyl- : 4633
1-Cyclohexene-1-carboxaldehyde, 4-isopropyl- : 4635

Cyclohexene-1-carboxylic acid	: 4637
1-Cyclohexene, 1-chloro	: 4642
Cyclohexene-1,2-dicarboxylic acid, diethyl ester	: 4644
Cyclohexene-1,4-dicarboxylic acid, dimethyl ester	: 4646
Cyclohexene, 2,4-dimethyl-	: 4651
Cyclohexene, 1-isopropenyl-5-methyl (DL)	: 4679
Cyclohexene, 1-isopropyl-4-methyl (l)	: 7429
Cyclohexene, 1,1'-(3,7,12,16-tetramethyl-1,3,5,7,9,11,13,15,17-octadecanonaene-1,18-diyl)bis[2,6,6-trimethyl-,	: 4051
Cyclohexen-3-ol, 1,3,5-trimethyl	: 4700
Cyclohexen-3-ol, 2,6,6-trimethyl	: 4701
Cyclohexen-3-ol, 1,5,5-trimethyl	: 4702
Cyclohexen-3-one, 1,6,6-trimethyl	: 4719
Cyclohexen-3-one, 1,4,6-trimethyl	: 4720
2,3-Cyclohexenopyridine	: 11297
4-(Cyclohex-3-enyl)pyridine	: 10922
Cycloheximide	: 9655
Cyclohexyl acetate	: 155
Cyclohexylacetic acid	: 4306
Cyclohexylacetic acid, methyl ester	: 4310
Cyclohexyl alcohol	: 4500
Cyclohexylamine	: 4283
Cyclohexylamine, 2-ethyl-	: 4287
Cyclohexylamine, 2-methyl-, trans-	: 4293
Cyclohexylbenzene	: 1248
Cyclohexyl bromide	: 4315
Cyclohexylcarbinol	: 4419
Cyclohexyl chloride	: 4339
Cyclohexylcyclohexane	: 2988
5-Cyclohexyl-5,7-diaza-2,3,4,5,6,7-hexahydroindene-4,6-dione	: 4900
Cyclohexyldimethylamine	: 4286
1-Cyclohexyleicosane	: 4384
Cyclohexyl ethyl ketone	: 10423
Cyclohexyl fluoride	: 4403
Cyclohexyl 2-furyl ether	: 6076
Cyclohexyl hydroperoxide	: 7177
Cyclohexylideneacetonitrile	: 276
Cyclohexyl iodide	: 4415
Cyclohexyl isocyanate	: 4416
Cyclohexyl mercaptan	: 4480
β-Cyclohexyl-β-methacrylic acid	: 3834
Cyclohexylmethyl alcohol	: 4419
Cyclohexyl methyl ketone	: 5794
3-Cyclohexyl-1-methylpropanol	: 4469
3-Cyclohexyl-5,6-trimethyleneuracil	: 4900
β-Cyclolavandulol	: 4665
Cycloleucine	: 4782
Cyclomethiazide	: 2853
Cyclomethone	: 4378
Cyclonite	: 11833
Cyclooctatetraene	: 4742
Cyclooctatetraene, chloro	: 4745
Cyclooctatetraene, propyl	: 4748
Cyclooctylamine	: 4735
1H-Cyclopenta[a]phenanthrene, hexadecahydro-10,13-dimethyl-17-(1-methylbutyl)-, [8R-[8α,9β,10α,1	: 4093
Cyclopentacycloheptene	: 504
Cyclopenta[de]naphthalene	: 3
Cyclopentadrin	: 4810
Cyclopentamine	: 4810
Cyclopentanecarboxylic acid, 2-isopropyl-5-methyl-	: 4784
Cyclopentanecarboxylic acid, 2-isopropyl-5-methyl, ethyl ester	: 4788
Cyclopentane-1,2-dicarboxylic acid, dimethyl ester (DL, trans)	: 4792
Cyclopentane-1,3-diol (trans)	: 4807
Cyclopentane, eicosyl	: 5351
Cyclopentaneethylamine, N,α-dimethyl-	: 4810
Cyclopentane, isobutyl-	: 4838
Cyclopentane, 1-isopropyl-2-methyl-	: 4829
Cyclopentane, 1,2,3-trimethyl-, trans-1,2,cis-1,3-	: 4857
Cyclopentane, 1,2,3-trimethyl-, cis-1,2,cis-1,3-	: 4858
Cyclopentane, 1,2,4-trimethyl-, cis-1,2,cis-1,4-	: 4861
Cyclopentanoic acid	: 4781

Cyclopentanone, 2-benzyl-	: 4893
Cyclopentanone, 2-isopentyl-3-methyl	: 4888
Cyclopentanone, 3-isopropyl-5-methyl	: 4884
Cyclopentanone, 2-isopropyl-2-methyl-	: 4889
Cyclopentanone, 2-isopropyl-4-methyl	: 4891
Cyclopentene-4-carboxylic acid, 2,3,3-trimethyl	: 4913
Cyclopentene-1,3-dicarboxylic acid, diethyl ester	: 4916
Cyclopentene, 2,3-dimethyl-4-isopropyl	: 4923
Cyclopentene, 1-isopropyl-3-methyl-	: 4931
Cyclopenten-3-one	: 4941
Cyclopenten-3-one, 4-methyl	: 4946
1,2-Cyclopentenophenanthrene	: 4756
Cyclopenthiazide	: 2853
Cyclopentobarbital	: 11085
Cyclopentyl acetate	: 4865
2-Cyclopentylacetic acid	: 4775
Cyclopentyl alcohol	: 4864
Cyclopentylamine	: 4771
Cyclopentyl bromide	: 4777
Cyclopentylcarboxylic acid	: 4781
Cyclopentyl chloride	: 4789
Cyclopentyl iodide	: 4824
Cyclopentyl mercaptan	: 4852
Cyclopropanecarboxylic acid, 3-(2,2-dichloroethenyl)-2,2-dimethyl-, cyano(3-phenoxy-phenyl)methyl ester	: 5012
Cyclopropanecarboxylic acid, 3-(2,2-dichloroethenyl)-2,2-dimethyl-, (3-phenoxyphenyl), methyl ester	: 9150
Cyclopropanecarboxylic acid, 2,2-dimethyl-3-(2-methyl-1-propenyl)-, [5-(phenylmethyl)- 3-furanyl] methyl ester	: 11340
Cyclopropanecarboxylic acid, hexadecyl ester	: 4952
Cyclopropane-1,2-dicarboxylic acid, diethyl ester (cis)	: 4968
Cyclopropane-1,2-dicarboxylic acid, dimethyl ester	: 4969
Cyclopropane-1,2-dicarboxylic acid (trans, DL)	: 4970
Cyclopropane, 1-methyl-1-ethyl	: 4975
Cyclopropylamine	: 4953
Cyclopropylbenzene	: 1256
Cyclopropyl cyanide	: 4956
Cyclopropylmethane	: 4981
Cyclopropyl methyl ether	: 4980
Cyclopropyl methyl ketone	: 5796
Cycloserine	: 7406
Cyclotetrasiloxane, 1,3,5,7-tetraethyl-1,3,5,7-tetraphenyl	: 4998
Cyclotetrasiloxane, 2,4,6,8-tetramethyl-2,4,6,8-tetraethyl	: 4997
Cyclothiazide	: 2851
Cymarose	: 7112
o-Cymene	: 1988
m-Cymene	: 1989
p-Cymene	: 1990
p-Cymene, 2-iodo	: 1781
p-Cymene, 3-iodo-	: 1782
p-Cymen-7-ol	: 1902
Cymonic acid	: 189
Cysteamine	: 5649
Cysteine disulfide	: 5020
Cythrin	: 5965
Cytidine 3'-monophosphate	: 5022
Cytofol	: 6243
Cytosine	: 11104
Cytosine, 5-methyl-	: 11106
Cytosine riboside	: 5021
2,4-D	: 169
2,4-D Isopropyl ester	: 173
2,4-D Methyl ester	: 172
Dagenan chloride	: 2243
Dahl's acid II	: 7873
Daidzein	: 2827
Daidzein 4'-methyl ether	: 2830
Daidzeol	: 2827
Dalapon	: 10244

Damascenine	: 2672
Daminozide	: 3334
Dansyl chloride	: 7977
Danthron	: 395
Daphnetin	: 2805
Dapsone	: 846
Daptazole	: 11671
Datiscetin	: 2842
DBMC	: 9222
DDT	: 2328
Deanol	: 5709
Decacil	: 2409
α,ω-Decadiene	: 5024
2,4-Decadienoic acid (trans-2, trans-4)	: 5025
2,4-Decadienoic acid, methyl ester (cis-2, trans-4)	: 5028
2,4-Decadienoic acid, methyl ester (cis-2, cis-4)	: 5029
Decafluorobutane	: 3255
cis-Decahydroquinoline	: 11246
Decalin	: 7789
cis-Decalin	: 7790
trans-Decalin	: 7791
Decalin, 2-chloro	: 7782
Decalin, 2-methylene (trans)	: 7792
1-Decalone (cis)	: 8053
Decamethylcyclopentasiloxane	: 4901
Decamethylene dibromide	: 5038
Decamethylene dichloride	: 5040
Decamethylene glycol	: 5047
Decamethyltetrasiloxane	: 11646
Decane, 1-cyclopentyl-	: 4790
Decane, 1-(1-naphthyl)-	: 7793
2-Decanol (DL)	: 5071
Decasiloxane, dicosamethyl	: 5079
Δ 2-Decenoic acid	: 5088
Decholin	: 4101
Decicaine	: 2531
Deconic acid telomer	: 5090
Decyl acetate	: 157
Decylacetylene	: 5343
Decylbenzene	: 1258
Decyl cyanide	: 11941
Decylenic alcohol	: 5092
Decyl fluoride	: 5052
Decyl isopropyl ketone	: 11892
Decyl mercaptan	: 5058
Decyl methyl ketone	: 5325
1-Decylnaphthalene	: 7793
Decyl nitrate	: 8073
Decyl nitrite	: 8084
Decyl vinyl ether	: 5051
DEET	: 612
Dehydroacetic acid	: 10779
Dehydrobilirubin	: 3001
7-Dehydrocholesterol	: 4103
Dehydrocholic acid	: 4101
11-Dehydrocorticosterone	: 9755
2,2'-Dehydrodimorphine	: 3002
Dehydroergosterol	: 5380
7-Dehydroestrone	: 5398
Dehydromorphine	: 3002
Dehydromucic acid	: 6081
7-Dehydrositosterol	: 11554
Delfinidol chloride	: 2849
Delphinic acid	: 3539
Delphinidin	: 2849
Delphinine	: 311
Deltathione	: 6264
N-Demethylanhalidine	: 7393
Deoxycholic acid	: 4095
Deoxycorticosterone	: 9751
Deoxyepinephrine	: 1533
2-Deoxy-D-glucose	: 7110
11-Deoxy-17-hydroxycorticosterone	: 9749
7-Deoxyhyocholic acid	: 4096
1-(2-Deoxy-β-D-ribofuranosyl)-5-fluorouracil	: 12034
Derris	: 2783
Deserpidine	: 12065
Desmethyldiazepam	: 2411
6-Desoxy-L-ascorbic acid	: 467
5-Desoxyquercetin	: 2813
Desthiobiotin	: 7223
Desthiobiotin, methyl ester	: 7224
Dexamethasone	: 9737

Dexpanthenol	: 3202
Dextroamphetamine sulfate	: 1582
Dextrotartaric acid	: 3309
4,4'-Diacetamidodiphenyl sulfone	: 91
Diacetic acid	: 3568
Diacetin	: 10189
1,3-Diacetin	: 10190
Diacetonamine	: 8959
Diacetone acrylamide	: 10521
Diacetone alcohol	: 8976
Diacetonyl	: 6805
1,4-Diacetoxy-2-butyne	: 3913
1,2-Diacetoxyethane	: 5571
Diacetyl	: 3359
Diacetyldioxime	: 3360
Diacetyl disulfide	: 5250
1,2-Diacetylethane	: 6805
Diacetylmethane	: 8798
Diacetylmorphine	: 7667
Diacetylperoxide	: 9156
Diadol	: 11091
Δ 2-Dialin	: 7815
Diallylamine	: 10526
5,5-Diallylbarbituric acid	: 11091
N,N-Diallyl-2-chloroacetamide	: 44
Diallylcyanamide	: 4159
Diallyl fumarate	: 3781
Diallyl maleate	: 3782
Diallyl oxalate	: 5564
Diallyl sulfide	: 10599
p-Diaminoazobenzene	: 635
2,3-Diaminobenzoic acid	: 2552
2,4-Diaminobenzoic acid	: 2553
3,5-Diaminobenzoic acid	: 2554
o,p'-Diaminobiphenyl	: 3029
4,4'-Diamino-1,1'-biphenyl	: 3031
1,4-Diaminobutane	: 3258
1,8-Diamino-3,6-diazaoctane	: 5499
β,β'-Diamino-β,β'-dicarboxydiethyl disulfide	: 5020
4,4'-Diamino-3,3'-dimethylbiphenyl	: 3034
p-Diaminodiphenyl	: 3031
4,4'-Diaminodiphenylamine	: 1263
p,p'-Diaminodiphenylmethane	: 784
3,3'-Diaminodipropylamine	: 9874
2,7-Diaminofluorene	: 5971
2,6-Diaminohexanoic acid	: 7418
	7419
2,6-Diamino-1-methylbenzene	: 1277
1,2-Diaminonaphthalene	: 7795
1,4-Diaminonaphthalene	: 7796
2,3-Diaminonaphthalene	: 7799
(S)-2,5-Diamino-5-oxopentanoic acid	: 6249
(S)-2,5-Diaminopentanoic acid	: 8550
2,3-Diaminophenazine	: 9194
2,4-Diaminophenol	: 9274
2,4-Diamino-5-phenylthiazole	: 11671
1,3-Diaminopropane	: 9873
2,3-Diaminopropionic acid	: 326
2,6-Diaminopurine	: 10754
2,5-Diaminopyridine	: 10924
1,3-Diamino-2-thiourea	: 4033
2,6-Diaminotoluene	: 1277
Diamylamine	: 8708
Diamyl ketone	: 11962
Dianat	: 2587
1,2,3,4-Dianhydroerythritol	: 3007
o,p'-Dianiline	: 3029
1,2-Dianilinoethane	: 5508
o-Dianisidine	: 3033
7,12-Diazabenz[a]anthracene	: 2394
1,2-Diazabenzene	: 10855
1,4-Diazabicyclooctane	: 5111
4,9-Diazadodecane-1,12-diamine	: 3259
2,3-Diazaindole	: 2871
Diazald	: 2212
1,5-Diazanaphthalene	: 8065
3,6-Diazaoctane-1,8-diamine	: 5499
Diazepam	: 2410
1,4-Diazine	: 10801
1,3-Diazine	: 11049
Diazinon	: 9582
Diazoacetic ester	: 158
Diazoaminobenzene	: 11826
1,3-Diazole	: 7192
1,2-Diazole	: 10815

Diazomethane	: 7491
Dibenzalacetone	: 8683
1,2:5,6-Dibenzanthracene	: 5142
2,3:6,7-Dibenzazepine-1-carboxylic acid,amide	: 5144
Dibenz[de,kl]anthracene	: 9161
Dibenzo[a,e]pyrene	: 8061
Dibenzo[a,i]phenanthrene	: 9601
Dibenzo[a,i]pyrene	: 2850
Dibenzo[a,jk]fluorene	: 2738
2,3:6,7-Dibenzoanthracene	: 8636
Dibenzo[b,e]pyridine	: 314
Dibenzo[b,h]pyrene	: 2850
Dibenzo[b,d]pyrrole	: 3981
1,2:6,7-Dibenzo-9-fluorenone	: 5152
Dibenzopyrazine	: 9193
1,2:7,8-Dibenzopyrene	: 2850
1,2:4,5-Dibenzopyrene	: 8061
Dibenzothiapyran	: 11811
Dibenzothiazine	: 9499
2,3,:5,6-Dibenzo-1,4-thiazine, 10(dimethyaminacetyl)	: 9502
Dibenzoylmethane	: 10027
2,3:6,7-Dibenzphenanthrene	: 8996
Dibenzyl	: 1587
Dibenzylamine	: 1828
9,10-Dibenzylanthracene	: 367
Dibenzyl disulfide	: 5248
Dibenzyl phosphite	: 9586
Diborane, methylthio	: 5168
Dibromantine	: 7217
Dibromoacetic acid	: 159
Dibromoacetic acid ethyl ester	: 160
Dibromoacetonitrile	: 277
β,β-Dibromoacrylic acid	: 10628
3,4-Dibromoaniline	: 683
9,10-Dibromoanthracene	: 375
2,7-Dibromoanthraquinone	: 390
o-Dibromobenzene	: 1288
m-Dibromobenzene	: 1289
p-Dibromobenzene	: 1290
2,4-Dibromobenzoic acid	: 2555
2,5-Dibromobenzoic acid	: 2556
2,6-Dibromobenzoic acid	: 2557
3,5-Dibromobenzyl bromide	: 1291
d-3,3-Dibromo-bornanone	: 2953
1,1-Dibromobutane	: 3264
1,2-Dibromobutane	: 3265
1,3-Dibromobutane	: 3266
1,4-Dibromobutane	: 3287
2,3-Dibromobutane	: 3268
2,3-Dibromo-1-butanol	: 3612
3,4-Dibromobutanol	: 3613
1,4-Dibromo-2-butanol	: 3614
3,4-Dibromo-2-butanone	: 3653
1,1-Dibromo-1-butene	: 3738
3,4-Dibromo-1-butene	: 3740
1,2-Dibromo-3-butene	: 3740
2,4-Dibromo-2-butene	: 3741
1,3-Dibromo-2-butene	: 3741
1,2-dibromo-1-butene	: 3739
1,4-Dibromo-2-butyne	: 3904
(+)-3,3-Dibromocamphor	: 2953
α,α'-Dibromo-d-camphor	: 2953
1,2-Dibromo-3-chlorobenzene	: 1292
1,2-Dibromo-4-chlorobenzene	: 1293
1,3-Dibromo-2-chlorobenzene	: 1294
1,4-Dibromo-2-chlorobenzene	: 1295
Dibromochlorofluoromethane	: 7494
Dibromochloromethane	: 7493
Dibromochloropropane	: 9885
1,2-Dibromo-1-chloro-1,2,2-trifluoroethane	: 5519
1,1-Dibromocyclobutane	: 4178
1,2-Dibromocyclobutane	: 4179
1,1-Dibromocyclohexane	: 4345
cis-1,2-Dibromocyclohexane	: 4346
trans-1,3-Dibromocyclohexane	: 4348
cis-1,4-Dibromocyclohexane	: 4349
1,10-Dibromodecane	: 5038
5,6-Dibromodecane	: 5039
Dibromodichloromethane	: 7495
1,2-Dibromo-2,4-dicyanobutane	: 8768
1,2-Dibromo-1,1-difluoroethane	: 5522
Dibromodifluoromethane	: 7496
1,4-Dibromo-2,3-dimethylbutane	: 3270
2,3-Dibromo-2,3-dimethylbutane	: 3271

1,2-Dibromo-2,3-dimethylbutane	: 3269
1,1-Dibromo-2,2-dimethylpropane	: 9886
Dibromodinitromethane	: 7497
1,1-Dibromoethane	: 5517
1,1-Dibromoethylene	: 5890
trans-1,2-Dibromoethylene	: 5891
cis-1,2-Dibromoethylene	: 5892
1,2-Dibromoethyl methyl ketone	: 3653
Dibromofluoroacetic acid, ethyl ester	: 161
Dibromofluoromethane	: 7498
2,3-Dibromofuran	: 6077
2,5-Dibromofuran	: 6078
Dibromogallic acid	: 2560
1,2-Dibromoheptane	: 6372
1,5-Dibromoheptane	: 6373
1,7-Dibromoheptane	: 6374
2,3-Dibromoheptane	: 6375
3,4-Dibromoheptane	: 6376
1,16-Dibromohexadecane	: 6635
1,2-Dibromohexane	: 6754
1,4-Dibromohexane	: 6755
1,5-Dibromohexane	: 6756
2,3-Dibromohexane	: 6758
1,2-Dibromoindan	: 7247
Dibromomethane	: 7492
1,2-Dibromo-2-methylbutane	: 3272
2,3-Dibromo-2-methylbutane	: 3274
1,3-Dibromo-2-methylpropene	: 10546
Dibromomonochloromethane	: 7493
1,3-Dibromonaphthalene	: 7801
1,4-Dibromonaphthalene	: 7802
1,5-Dibromonaphthalene	: 7803
3,4-Dibromonitrobenzene	: 1307
1,2-Dibromo-4-nitrobenzene	: 1307
2,6-Dibromonitrobenzene	: 1308
1,3-Dibromo-2-nitrobenzene	: 1308
1,2-Dibromononane	: 8127
1,4-Dibromooctafluorobutane	: 3275
1,2-Dibromooctane	: 8366
1,5-Dibromooctane	: 8368
1,8-Dibromooctane	: 8369
1,3-Dibromopentane	: 8739
1,4-Dibromopentane	: 8740
2,3-Dibromopentane	: 8742
2,4-Dibromopentane	: 8743
2,3-Dibromopentanoic acid	: 8870
2,4-Dibromophenol	: 9276
2,6-Dibromophenol	: 9277
3,5-Dibromophenol	: 9278
1,1-Dibromopropane	: 9880
1,2-Dibromopropane	: 9881
1,3-Dibromopropane	: 9882
2,2-Dibromopropane	: 9883
2,3-Dibromopropane	: 10545
2,3-Dibromopropylene	: 10545
3,5-Dibromopyridine	: 10926
2,6-Dibromoquinone-4-chlorimide	: 4279
3,5-Dibromosalicylaldehyde	: 534
3,5-Dibromosalicylic acid	: 2558
1,14-Dibromotetradecane	: 11616
1,2-Dibromotetrafluoroethane	: 5525
1,2-Dibromo-1,1,2,2-tetramethylethane	: 3271
2,3-Dibromothiophene	: 11742
2,4-Dibromothiophene	: 11743
2,5-Dibromothiophene	: 11744
2,3-Dibromotoluene	: 1302
3,4-Dibromotoluene	: 1303
2,6-Dibromotoluene	: 1304
2,5-Dibromotoluene	: 1305
1,2-Dibromo-1,1,2-trifluoroethane	: 5526
1,2-Dibromo-1,1,2-trimethylethane	: 3274
1,11-Dibromoundecane	: 11940
3,4-Dibromo-o-xylene	: 1297
Dibucaine hydrochloride	: 11226
Dibutyacetylene	: 5100
Di-tert-Butylacyloin	: 6975
Dibutylamine	: 3211
Dibutylbutadiyne	: 5287
Dibutylcarbinol	: 8161
Dibutyl carbonate	: 3998
Dibutyl ether	: 3415
Di-tert-Butyl ether	: 10124
N,N'-Dibutylhexamethylenediamine	: 6750
Di-sec-butyl ketone	: 6515
Dibutyl ketone	: 8175

Di-*tert*-butyl malonate	: 9925
Dibutylmercury	: 7433
Di-*tert*-butylmethane	: 8841
2,6-Di-*tert*-butyl-4-methylphenol	: 9224
Dibutylnitrosamine	: 3212
Dibutyl oxide	: 3415
2,6-Di-*sec*-butylphenol	: 9227
Dibutyl phthalate	: 1339
2,6-Di-*tert*-butylpyridine	: 10874
Dibutyl sebacate	: 5044
Di-*tert*-butyl succinate	: 3296
Di-*sec*-butyl succinate	: 3297
Dibutyl succinate	: 3301
Di-*tert*-butyl sulfide	: 10155
Dibutyl sulfide	: 3433
Dibutyl sulfite	: 11580
Di-*tert*-Butyl tetrasulfide	: 11649
Dicamba	: 2567
1,3-Dicarboxyacetone	: 8782
2,2'-Dicarboxybiphenyl	: 3038
2,2'-Dicarboxydiphenylamine	: 2656
2,3-Dicarboxypyrazine	: 10804
2,5-Dicarboxythiophene	: 11745
Dicentrine	: 2468
Dichlobenil	: 2752
Dichlofenthion	: 9581
Dichlone	: 7850
Dichloramine-T	: 2210
Dichloroacetic acid	: 162
Dichloroacetic acid butyl ester	: 163
1,1-Dichloroacetone	: 10425
1,3-Dichloroacetone	: 10426
Dichloroacetonitrile	: 278
Dichloroacetyl chloride	: 297
Dichloroacetylene	: 5940
β,β-Dichloroacrylic acid	: 10630
2,4-Dichloro-6-aminophenol	: 9205
2,3-Dichloroaniline	: 685
2,4-Dichloroaniline	: 686
2,5-Dichloroaniline	: 687
2,6-Dichloroaniline	: 688
3,4-Dichloroaniline	: 689
3,5-Dichloroaniline	: 690
3,6-Dichloro-*o*-anisic acid	: 2567
2,5-Dichlorobenzaldehyde	: 535
3,4-Dichlorobenzaldehyde	: 536
3,5-Dichlorobenzaldehyde	: 537
o-Dichlorobenzene	: 1354
m-Dichlorobenzene	: 1355
p-Dichlorobenzene	: 1356
3,4-Dichlorobenzenecarboxaldehyde	: 536
o-Dichlorobenzidine	: 3032
2,4-Dichlorobenzoic acid	: 2561
2,5-Dichlorobenzoic acid	: 2562
2,6-Dichlorobenzoic acid	: 2563
3,5-Dichlorobenzoic acid	: 2564
2,6-Dichlorobenzonitrile	: 2752
2,6-Dichloro-*p*-benzoquinone-4-chlorimide	: 4280
2,4-Dichlorobenzoyl chloride	: 2893
3,4-Dichlorobenzoyl chloride	: 2894
2,4-Dichlorobenzyl alcohol	: 1853
2,2-Dichloro-1,1-bis(4-chlorophenyl)ethylene	: 1374
1,1-Dichloro-2,2-bis(*p*-chlorophenyl)ethane	: 1378
Dichlorobromomethane	: 7467
1,1-Dichloro-1,3-butadiene	: 3156
1,2-Dichloro-1,3-butadiene	: 3157
1,1-Dichlorobutane	: 3276
3,4-Dichlorobutane	: 3277
1,2-Dichlorobutane	: 3277
1,3-Dichlorobutane	: 3278
1,4-Dichlorobutane	: 3279
2,2-Dichlorobutane	: 3280
1,4-Dichloro-2,3-butanediol	: 3349
1,3-Dichloro-2-butanone	: 3654
1,3-Dichloro-1-butene	: 3743
2,3-Dichloro-1-butene	: 3744
3,4-Dichloro-1-butene	: 3745
1,2-Dichloro-3-butene	: 3745
1,1-Dichloro-2-butene	: 3746
1,2-Dichloro-2-butene	: 3747
1,4-Dichloro-*trans*-2-butene	: 3750
3,4-Dichlorobutyric acid	: 3926
Dichloro(chloromethyl)methylsilane	: 11390
2,6-Dichloro-*m*-cresol	: 9294

4,4-Dichlorocrotonic acid	: 4152
2,6-Dichlorocyanobenzene	: 2752
Dichlorocyanuric acid	: 11838
1,1-Dichlorocyclohexane	: 4358
cis-1,2-Dichlorocyclohexane	: 4359
cis-1,4-Dichlorocyclohexane	: 4361
1,10-Dichlorodecane	: 5040
2,2'-Dichloro-4,4'-diaminobiphenyl	: 3032
Dichloro(dichloromethyl)methyl-silane	: 11391
2,3-Dichloro-5,6-dicyano-1,4-benzoquinone	: 4251
2,3-Dichloro-5,6-dicyanobenzoquinone	: 4251
Dichlorodicyano-*p*-benzoquinone	: 4251
Dichlorodicyanoquinone	: 4251
Dichlorodiethylsilane	: 11393
1,2-Dichloro-1,1-difluoroethane	: 5531
1,2-Dichloro-1,2-difluoroethane	: 5532
1,1-Dichloro-2,2-difluoroethylene	: 5897
1,2-Dichloro-1,2-difluoroethylene	: 5898
Dichlorodifluoromethane	: 7500
2,5-Dichloro-3,6-dihydroxy-*p*-benzoquinone	: 4258
2,2'-Dichlorodiisopropyl ether	: 10123
2,2-Dichloro-1,1-dimethylethanol	: 10364
2,5-Dichloro-2,5-dimethylhexane	: 6765
1,2-Dichloro-4,4-dimethylpentane	: 8752
1,5-Dichloro-3,3-dimethylpentane	: 8753
2,4-Dichloro-2,4-dimethylpentane	: 8754
Dichlorodimethylsilane	: 11394
Dichlorodinitromethane	: 7501
Dichlorodiphenylsilane	: 11395
2,2-Dichloroethanol	: 5704
2,5-Dichloroethylbenzene	: 1377
1,1-Dichloroethylene	: 5894
trans-1,2-Dichloroethylene	: 5895
cis-1,2-Dichloroethylene	: 5896
Dichloroethylmethylsilane	: 11397
2,7-Dichlorofluorene	: 5972
Dichlorofluoroacetonitrile	: 279
1,1-Dichloro-1-fluoroethane	: 5534
1,1-Dichloro-2-fluoroethylene	: 5900
Dichlorofluoromethane	: 7502
1,1-Dichloroheptane	: 6377
1,2-Dichloroheptane	: 6378
1,7-Dichloroheptane	: 6379
2,2-Dichloroheptane	: 6380
2,3-Dichlorohexafluoro-2-butene	: 3754
1,2-Dichlorohexafluoropropane	: 9898
1,2-Dichlorohexane	: 6761
1,6-Dichlorohexane	: 6762
2,2-Dichlorohexane	: 6763
2,3-Dichlorohexane	: 6764
(±)-2,5-Dichlorohexane	: 6766
Di(5-Chloro-2-hydroxyphenyl)methane	: 9408
1,2-Dichloroindan	: 7249
Dichloroiodomethane	: 7503
2,4-Dichloromesitylene	: 1402
Dichloromethane	: 7499
1,1-Dichloro-3-methylbutane	: 3282
1,2-Dichloro-2-methylbutane	: 3283
2,4-Dichloro-2-methylbutane	: 3284
2,3-Dichloro-2-methylbutane	: 3286
1,4-Dichloro-2-methylbutane	: 3285
3,3-Dichloro-2-methyl-1-butene	: 3755
1,3-Dichloro-2-methyl-2-butene	: 3756
1,4-Dichloro-2-methyl-2-butene	: 3757
Dichloromethyl cyanide	: 278
(Dichloromethyl)dimethylchloro-silane	: 11373
Dichloromethyl methyl dichlorosilane	: 11391
Dichloromethylphenylsilane	: 11401
1,1-Dichloro-2-methylpropane	: 9899
1,2-Dichloro-2-methylpropane	: 9900
1,3-Dichloro-2-methylpropane	: 9901
2,4-Dichloro-5-methylpyrimidine	: 11052
2,6-Dichloro-4-methylpyrimidine	: 11053
Dichloromethylsilane	: 11398
(Dichloromethyl)trichlorosilane	: 11459
1,2-Dichloronaphthalene	: 7807
1,3-Dichloronaphthalene	: 7808
1,4-Dichloronaphthalene	: 7809
1,5-Dichloronaphthalene	: 7810
1,7-Dichloronaphthalene	: 7811

1,8-Dichloronaphthalene	: 7812
2,6-Dichloronaphthalene	: 7813
2,3-Dichloro-1-naphthol	: 8025
2,6-Dichloro-4-nitroaniline	: 692
2,3-Dichloronitrobenzene	: 1391
2,4-Dichloronitrobenzene	: 1396
2,5-Dichloronitrobenzene	: 1395
2,6-Dichloronitrobenzene	: 1393
3,4-Dichloronitrobenzene	: 1392
3,5-Dichloronitrobenzene	: 1394
1,1-Dichloro-1-nitroethane	: 5536
1,1-Dichloro-1-nitropropane	: 9902
2,4-Dichloro-5-nitropyrimidine	: 11054
1,9-Dichlorononane	: 8128
2,4-Dichloro-6-(o-chloroanilino)-s-triazine	: 360
1,8-Dichlorooctane	: 8371
1,2-Dichloropentane	: 8744
1,3-Dichloropentane	: 8745
1,4-Dichloropentane	: 8746
1,5-Dichloropentane	: 8747
2,2-Dichloropentane	: 8748
2,3-Dichloropentane	: 8749
2,4-Dichloropentane	: 8750
3,3-Dichloropentane	: 8751
3,5-Dichloro-2-pentanone	: 8963
2,5-Dichloro-2-pentene	: 9028
1,4-Dichloro-3-pentene	: 9028
Dichlorophene	: 9408
2,3-Dichlorophenol	: 9282
2,4-Dichlorophenol	: 9283
2,5-Dichlorophenol	: 9284
2,6-Dichlorophenol	: 9285
3,4-Dichlorophenol	: 9286
3,5-Dichlorophenol	: 9287
2,4-Dichlorophenyl acetate	: 9288
Dichlorophenylarsine	: 462
N-(3,5-Dichlorophenyl)-1,2-dimethylcyclopropane-1,2-dicarboximide	: 9760
Dichlorophenylphosphine	: 9556
Dichlorophenylphosphine sulfide	: 9552
1-[[2-(2,4-Dichlorophenyl)-4-propyl-1,3-dioxolan-2-yl]methyl]-1H-1,2,4-triazole	: 10720
Dichlorophenylsilane	: 11403
(±)-2-(3,5-Dichlorophenyl)-2-(2,2,2-trichloroethyl)oxirane	: 11899
4,5-Dichlorophthalic anhydride	: 7344
1,1-Dichloropropane	: 9891
1,3-Dichloropropane	: 9893
1,3-Dichloro-2-propanol	: 10362
2,3-Dichloropropene	: 10552
3,3-Dichloropropionyl chloride	: 10502
N,N-Dichloropropylamine	: 9805
1,1-Dichloropropylene	: 10547
trans-1,2-Dichloropropylene	: 10548
cis-1,2-Dichloropropylene	: 10549
trans-1,3-Dichloropropylene	: 10550
cis-1,3-Dichloropropylene	: 10551
2,3-Dichloropropylene	: 10552
3,3-Dichloropropylene	: 10553
3,6-Dichloro-2-pyridinecarboxylic acid	: 4142
2,4-Dichloroquinoline	: 11248
2,7-Dichloroquinoline	: 11249
4,7-Dichloroquinoline	: 11251
3,7-Dichloroquinoline-8-carboxylic acid	: 11201
2,6-Dichloroquinone-4-chloroimide	: 4280
2,3-Dichlorostyrene	: 1367
2,6-Dichlorostyrene	: 1369
3,5-Dichlorostyrene	: 1370
2,5-Dichlorostyrene	: 1371
2,4-Dichlorostyrene	: 1373
p-Dichlorosulfamoylbenzoic Acid	: 2565
p-(N,N-Dichlorosulfamyl)benzoic acid	: 2565
2,5-Dichloroterephthalic acid	: 1349
1,2-Dichlorotetrafluorobenzene	: 1398
1,3-Dichlorotetrafluorobenzene	: 1399
1,4-Dichlorotetrafluorobenzene	: 1400
1,1-Dichlorotetrafluoroethane	: 5537
1,2-Dichloro-1,1,2,2-tetrafluoroethane	: 5538
2,3-Dichlorothiophene	: 11746
2,4-Dichlorothiophene	: 11747

2,5-Dichlorothiophene	: 11748
3,4-Dichlorothiophene	: 11749
2,3-Dichlorotoluene	: 1383
3,4-Dichlorotoluene	: 1384
2,6-Dichlorotoluene	: 1385
3,5-Dichlorotoluene	: 1386
2,5-Dichlorotoluene	: 1387
2,4-Dichlorotoluene	: 1388
N,N-Dichloro-p-toluenesulfonamide	: 2210
2,2-Dichloro-1,1,1-trifluoroethane	: 5540
Dichloroxylenol	: 9290
Dichlorprop	: 10250
Dicopur M	: 141
Dicryl	: 10520
Dictamnine	: 6180
Dicumarol	: 2839
Dicurone	: 6239
1,3-Dicyanobenzene	: 1315
Dicyanodiamide	: 6280
1,2-Dicyanoethane	: 3291
1,6-Dicyanohexane	: 8385
Dicyanomethane	: 9911
1,3-Dicyanopropane	: 8767
Dicyclohexylamine	: 4284
1,4-Dicyclohexylbutane	: 4323
Dicyclohexylcarbodiimide	: 4291
1,2-Dicyclohexylethane	: 4386
1,1-Dicyclohexylethane	: 4392
Dicyclohexylmethane	: 4429
1,3-Dicyclohexyl-2-methylpropane	: 4447
Dicyclohexyl oxalate	: 5559
Dicyclohexylphosphine	: 9516
1,2-Dicyclohexylpropane	: 4430
1,3-Dicyclohexylpropane	: 4466
1,1-Dicyclohexyltetradecane	: 4476
Dicyclopropyl ketone	: 7607
Dicysteine	: 5020
Dieldrin	: 5178
Dienestrol	: 9302
meso-Diepoxybutane	: 3007
Diethadione	: 8573
Diethanolamine	: 5730
Diethazine	: 9503
3,4-Diethoxyaniline	: 693
3,4-Diethoxybenzaldehyde	: 538
Diethoxydimethylsilane	: 11405
Diethoxydiphenylsilane	: 11406
1,1-Diethoxyethane	: 5541
Diethoxymethane	: 5603
3,3-Diethoxy-1-propene	: 10560
2,5-Diethoxytoluene	: 1410
Diethylacetaldehyde	: 3183
Diethylacetic acid	: 3514
Diethylacetylene	: 7116
O,O-Diethyl acetylphosphonate	: 9530
Diethylamine	: 5427
2-(Diethylamino)ethanol	: 5706
2-(Diethylamino)ethyl p-aminobenzoate	: 2493
S-[2-(Diethylamino)ethyl] diphenylthiocarbamate	: 3977
3-(Diethylamino)-2-hydroxypropylamine	: 10339
Diethyl aminomalonate	: 9924
2-Diethylamino-1-propanol	: 10366
3-(Diethylamino)propylamine	: 9876
N,N-Diethylaniline	: 695
Diethylarsine	: 441
o-Diethylbenzene	: 1413
m-Diethylbenzene	: 1414
p-Diethylbenzene	: 1415
N,N-Diethylbenzhydrylamine	: 1804
Diethylbromoacetamide	: 3198
Diethyl butanedioate	: 3303
Diethylcarbamazine	: 9603
N,N '-Diethylcarbanilide	: 12003
Diethyl carbinol	: 8917
Diethyl Carbitol	: 5612
Diethyl carbonate	: 3999
Diethyl chloromaleate	: 3771
Diethyl chlorophosphate	: 9574
Diethyl chlorothiophosphate	: 9576
Diethyl 1,1-cyclobutanedicarboxylate	: 4180
Diethyl 1-cyclohexene-1,3-dicarboxylate	: 4645
1,1-Diethylcyclopentane	: 4796
cis-1,2-Diethylcyclopentane	: 4797
trans-1,2-Diethylcyclopentane	: 4798
1,3-Diethylcyclopentane	: 4799
Diethyl 1-cyclopentene-1,2-dicarboxylate	: 4915
Diethyl dibenzylmalonate	: 9926
Diethyl 2,5-dibromovalerate	: 8873
Diethyl dibutylmalonate	: 9937
Diethyl dicarbonate	: 5169
Diethyl 2,2-diethylmalonate	: 9940
Diethyldifluorosilane	: 11410
Diethyl dimethylmalonate	: 9942
Diethyl dipropargylmalonate	: 9946
Diethyl disulfide	: 5253
Diethylenediamine	: 9602
Diethylenediamine dihydrochloride	: 9607
Diethylene glycol	: 5753
Diethylene glycol, diacetate	: 5754
Diethylene glycol, dibenzoate	: 5755
Diethylene glycol diethyl ether	: 5612
Diethylene glycol dimethyl ether	: 5613
Diethylene glycol, dodecanoate	: 5313
Diethylene glycol, monoacetate	: 5673
Diethylene glycol monobutyl ether	: 5688
Diethylene glycol monoethyl ether	: 5720
Diethylene glycol monoethyl ether acetate	: 5721
Diethylene glycol monolaurate	: 5313
Diethylenetriamine	: 5497
Diethyl ether	: 5608
Diethyl ether-trifluoroborane (1:1)	: 3137
N,N-Diethylethylenediamine	: 5502
Diethyl ethylphosphonate	: 9536
Diethyl 3,4-furandicarboxylate	: 6082
Diethyl glutaconate	: 9041
3,4-Diethylhexane	: 6768
3,4-Diethyl-3,4-hexanediol	: 6800
Diethylhexyl adipate	: 6781
1,1-Diethylhydrazine	: 7152
1,2-Diethylhydrazine	: 7153
Diethyl(2-hydroxyethyl)methylammonium bromide, xanthene-9-carboxylate	: 5445
Diethyl 3-hydroxyglutarate	: 8778
1,2-Diethylimidazole	: 7196
Diethyl isobutylmalonate	: 9969
Diethyl isophthalate	: 1346
Diethylisopropylcarbinol	: 8934
Diethyl isopropylidenemalonate	: 9968
N,N-Diethylisovaleramide	: 3201
Diethylketene	: 3877
Diethyl ketone	: 8958
Diethyl maleate	: 3776
Diethyl malonate	: 9941
Diethylmalonic acid	: 9939
Diethyl mercury	: 7434
Diethyl mesaconate	: 3787
Diethyl 3-methyladipate	: 6790
N,N-Diethyl-3-methylbenzamide	: 612
3,3-Diethyl-2-methylpentane	: 8757
3,4-Diethyl-2-methylpyrrole	: 11117
Diethyl oxalacetate	: 3340
Diethyl oxalate	: 5560
Diethyl oxydiformate	: 5169
3,3-Diethylpentane	: 8756
Diethylphosphinic acid	: 9525
Diethyl phthalate	: 1342
Diethyl pimelate	: 6394
N,N-Diethyl-p-phenylenediamine	: 1266
Diethylpropion	: 10428
2,4-Diethylpyridine	: 10933
Diethyl pyrocarbonate	: 5169
Diethylsilane	: 11409
Diethylstilbestrol	: 9300
Diethylstilbestrol dipropionate	: 9301
Diethyl succinate	: 3303
Diethyl sulfate	: 11573
Diethyl sulfide	: 5644
Diethyl sulfite	: 11581
Diethyl DL-tartrate	: 3307
Diethyl terephthalate	: 1350
2,2-Diethyltetrahydrofuran	: 6085
Diethylvinylethynylcarbinol	: 6608
Difluoroacetic acid	: 175
Difluoroacetic acid, ethyl ester	: 176
2,4-Difluoroaniline	: 700
o-Difluorobenzene	: 1422
m-Difluorobenzene	: 1423
p-Difluorobenzene	: 1424
Difluorochlorobromomethane	: 7464
1,1-Difluoro-2-chloroethylene	: 5885
Difluorochloromethane	: 7478
Difluorodibromomethane	: 7496
1,1-Difluoro-1,2-dichloroethane	: 5531
1,2-Difluoro-1,2-dichloroethane	: 5532
1,1-Difluoro-2,2-dichloroethene	: 5897
1,1-Difluoro-2,2-dichloroethylene	: 5897
Difluorodichloromethane	: 7500
4,4'-Difluorodiphenyl	: 3044
1,1-Difluoroethane	: 5544
1,2-Difluoroethane	: 5545
2,2-Difluoroethanol	: 5708
1,1-Difluoroethylene	: 5902
cis-1,2-Difluoroethylene	: 5903
Difluoroiodomethane	: 7505
Difluoromethane	: 7504
Difluoromethylarsine	: 463
1,1-Difluoroperchloroethane	: 5636
1,3-Difluoropropane	: 9905
2,2-Difluoropropane	: 9906
1,3-Difluoro-2-propanol	: 10369
1,1-Difluoro-1-propene	: 10561
2,2-Difluoropropyl chloride	: 9851
2,2-Difluoro-1,1,1,2-tetrachloroethane	: 5636
1,2-Difluoro-1,1,2,2-tetrachloroethane	: 5637
1,2-Difluorotetrachloroethane	: 5637
1,2-Difluoro-1,1,2,2-tetramethyldisilane	: 5227
1,2-Difluorotetramethyldisilane	: 5227
2,2-Difluoro-1,1,2-trichloroethane	: 5859
1,1-Difluoro-1,2,2-trichloroethane	: 5659
Difurfurylamine	: 6114
Digallic acid	: 2575
Digin	: 11530
Digitalose	: 6198
Digitoflavone	: 2814
Digitogenin	: 11535
Digitonin	: 6195
Digitoxigenin	: 4044
Digitoxin	: 4043
Digitoxose	: 7111
Diglycidyl resorcinol ether	: 8625
Diglycine	: 6265
Diglycol laurate	: 5313
Digoxigenin	: 4047
Digoxin	: 4041
Diheptylamine	: 6352
Dihexyl ketone	: 11891
1,2-Dihydroacenaphthylene	: 6
9,10-Dihydroanthracene	: 376
1,4-Dihydrobenzene	: 4249
2,3-Dihydrobenzofuran	: 2442
2,3-Dihydro-2-chloromethylbenzofuran	: 2441
Dihydrocholesterol	: 4107
(+)-Dihydrocinchonine	: 4122
Dihydrocinnamaldehyde	: 2114
Dihydrocodeine	: 7674
Dihydro-p-coumaric acid	: 2134
2,5-Dihydro-2,5-dimethoxyfuran	: 6087
2,3-Dihydro-2,2-dimethyl-7-benzofuranyl (dibutylamino)thio methylcarbamate	: 4038
2,3-Dihydro-5,6-dimethyl-1,4-dithiin, 1,1,4,4-tetraoxide	: 5179
α-Dihydrofucosterol	: 11558
Dihydrofumaric acid	: 3294
2,3-Dihydrofuran	: 6086
3,4-Dihydro-2H-pyran	: 10777
4,6-Dihydro-8-hydroxy-3,4,5-trimethyl-6-oxo-3H-2-benzopyran-7-carboxylic acid	: 2774
3,6-Dihydro-6-iminopurine	: 10751
1,2-Dihydro-2-iminopyridine	: 10857
Dihydroisophorone	: 4622
2,3-Dihydro-5-methylbenzofuran	: 2444
1,4-Dihydronaphthalene	: 7815
3,4-Dihydro-1-naphthoic acid	: 7764
2,3-Dihydro-3-oxobenzisosulfonazole	: 2388
9,10-Dihydrophenanthrene	: 9171
Dihydrosafrole	: 2431
Dihydrothebaine	: 7664

2,3-Dihydrothiophene	: 11750
2,5-Dihydrothiophene	: 11751
4,5-Dihydrotoluene	: 4271
Dihydroxyacetone	: 10429
1,3-Dihydroxy-2-amino-2-methylpropane	: 9987
2,5-Dihydroxybenzyl alcohol	: 1520
1,3-Dihydroxybutane	: 3345
1,4-Dihydroxybutane	: 3346
2,3-Dihydroxybutanedioic acid	: 3309
1,4-Dihydroxy-2-butyne	: 3911
3α,12α-Dihydroxycholanic acid	: 4095
3α,6α-Dihydroxy-5β-cholanic acid	: 4096
3α,7α-Dihydroxy-5β-cholan-24-oic acid	: 4097
3α,7β-Dihydroxy-5β-cholan-24-oic acid	: 4098
6,7-Dihydroxycoumarin	: 2804
6,7-Dihydroxycoumarin 6-glucoside	: 2822
trans-1,4-Dihydroxycyclohexane	: 4374
(+)-2,9-Dihydroxy-1,10-dimethoxyaporphine	: 5156
3,3'-Dihydroxy-5,5'-dimethylbiquinone	: 2979
2,3-Dihydroxy-2,3-dimethylbutane	: 3350
4,4'-Dihydroxydiphenylmethane-3,3'-dicarboxylic acid	: 2686
1,12-Dihydroxydodecane	: 5301
3,17-Dihydroxyestratriene	: 5400
Dihydroxyestrin	: 5399
1,2-Dihydroxyethylbenzene	: 5579
2,2'-Dihydroxyethyl ether monododecanoate	: 5313
4,4'-Dihydroxyfuchsone	: 4278
5',7-Dihydroxy-4'-glucosyloxyisoflavone	: 2824
2,5-Dihydroxyhexane	: 6799
1,3-Dihydroxy-5-iodobenzene	: 1524
4',7-Dihydroxyisoflavone	: 2827
Dihydroxymaleic acid	: 3777
5,7-Dihydroxy-4'-methoxyflavone	: 2809
5,7-Dihydroxy-6-methoxyflavone	: 2810
3,6-Dihydroxy-4-methoxy-2,5-toluquinone	: 4260
1,7-Dihydroxy-3-methoxyxanthone	: 12054
2,3-Dihydroxy-2-methylbutane	: 3355
1,4-Dihydroxy-1-methylbutane	: 8786
1,1-Di(hydroxymethyl)ethylamine	: 9987
2,5-Dihydroxy-3-methyl-6-methoxy-1,4-benzoquinone	: 4260
1,9-Dihydroxynonane	: 8136
9,10-Dihydroxyoctadecanoic acid	: 8254
2,4-Dihydroxypentane	: 8789
2,3-Dihydroxyphenol	: 2366
2,5-Dihydroxyphenol	: 2367
Dihydroxyphenylborane	: 3135
2,3-Dihydroxypropionic acid	: 10252
(±)-2,3-Dihydroxypropyl acetate	: 10187
3,6-Dihydroxyspiro[xanthene-9,3'-phthalide]	: 11522
9,10-Dihydroxystearic acid	: 8254
Dihydroxytartaric acid	: 3342
4,6-Dihydroxy-o-toluic acid	: 2573
Diiodoacetylene	: 5941
2,4-Diiodoaniline	: 701
4,4'-Diiodobenzophenone	: 7598
Diiodochloromethane	: 7479
1,1-Diiodoethane	: 5546
Diiodofluoromethane	: 7516
1,6-Diiodohexane	: 6769
Diiodomethane	: 7506
1,5-Diiodopentane	: 8758
3,5-Diiodo-α-phenylphloretic acid	: 2135
1,3-Diiodopropane	: 9907
2,2-Diiodopropane	: 9908
3,5-Diiodosalicylic acid	: 2617
Diiodotariric acid	: 8300
2,5-Diiodothiophene	: 11753
3,5-Diiodotyrosine	: 11928
Diisoamylamine	: 3226
Diisoamyl ether	: 3418
Diisoamyl phthalate	: 1335
Diisobutylamine	: 9823
Diisobutylcarbinol	: 6485
Diisobutylene	: 9064
Diisobutyl fumarate	: 3767

Diisobutyl ketone	: 6514
Diisobutyl oxalate	: 5557
Diisobutyl sulfide	: 10154
1,1-Diisobutylurea	: 11996
2,3-Diisonitrosobutane	: 3360
Diisopentyl ether	: 3418
Diisopentyl 2-hydroxyethylamine	: 5682
Diisopentyl phthalate	: 1335
Diisopentyl sulfide	: 3436
Diisopropanol	: 6799
Diisopropanolamine	: 10384
Diisopropenyl	: 3159
Diisopropenylacetylene	: 6714
Diisopropylamine	: 9821
N,N-Diisopropyl-2-aminoethanol	: 5683
Diisopropylbutadiyne	: 8345
Diisopropyl ether	: 10120
Diisopropyl fumarate	: 3766
Diisopropyl ketone	: 8967
Diisopropyl ketone, cyanohydrin	: 3408
Diisopropyl ketone oxime	: 8968
Diisopropyl sulfide	: 10152
Diketene	: 8598
Dilauroyl peroxide	: 9155
Diluran	: 32
Dimedazole	: 2380
Dimedone	: 4378
Dimefline	: 2819
Dimercaprol	: 10370
1,4-Dimercaptobutane	: 3388
1,2-Dimercaptoethane	: 5590
Dimethadione	: 8584
1,4:5,8-Dimethanonaphthalene, 1,2,3,4,10,10-hexachloro-6,7-epoxy-1,4,4a,5,6,7,8,8a-octahydro-, endo,exo-	: 5178
Dimethisoquin	: 5411
2,6-Dimethoxyaniline	: 703
2,4-Dimethoxybenzaldehyde	: 543
2,5-Dimethoxybenzaldehyde	: 544
3,5-Dimethoxybenzaldehyde	: 546
3,5-Dimethoxybenzoic acid	: 2577
2,6-Dimethoxybenzonitrile	: 2753
2,4-Dimethoxybenzophenone	: 7612
2,6-Dimethoxy-p-benzoquinone	: 4262
2,2'-Dimethoxybiphenyl	: 3045
3,3'-Dimethoxybiphenyl	: 3046
Dimethoxyborane	: 3117
5,7-Dimethoxycoumarin	: 2818
3,3'-Dimethoxy-4,4'-diaminodiphenyl	: 3033
2,5-Dimethoxy-2,5-dihydrofuran	: 6087
2,4-Dimethoxy-1-nitrobenzene	: 1448
1,3-Dimethoxy-4-nitrobenzene	: 1448
[[(Dimethoxyphosphinothioyl)thio]butanedioic acid diethyl ester	: 3316
5,6-Dimethoxyphthalaldehydic acid	: 2608
1,10-Dimethoxy-6α-aporphine-2,9-diol	: 5156
1,3-Dimethoxy-2-propanol	: 10372
2,6-Dimethoxypyridine	: 10935
2-[[[[[(4,6-Dimethoxypyrimidin-2-yl)amino]carbonyl]amino]sulfonyl]methyl]benzoic acid, methyl ester	: 518
N1-(2,6-Dimethoxy-4-pyrimidinyl)sulfanilamide	: 2194
2-(4,6-DimethoxypyrimidiN-2-ylcarbamoylsulfamoyl)-N,N-dimethylnicotinamide	: 8069
2,6-Dimethoxy-p-quinone	: 4262
Dimethyl acetal	: 5548
N,N-Dimethylacetamide	: 55
3,3-Dimethylacetylacetone	: 8799
Dimethylacetylene	: 3898
Dimethylacetylenecarbinol	: 3922
3,3-Dimethylacrolein	: 3700
trans-2,3-Dimethylacrylic acid	: 3844
1,3-Dimethylallene	: 8660
7,8-Dimethylalloxazine	: 2472
Dimethylamide acetate	: 55
Dimethylamine	: 7448
4-(Dimethylamino)azobenzene	: 721
2',3-Dimethyl-4-aminoazobenzene	: 798
p-(Dimethylamino)azobenzene-o-carboxylic acid	: 2587
p-(Dimethylamino)benzal-5-rhodanine	: 11684

p-(Dimethylamino)benzoic acid	: 2585
4-(Dimethylamino)benzophenone	: 7613
3-(Dimethylamino)benzophenone	: 2771
4,4'-Dimethylaminobenzophenonimide	: 657
2-(Dimethylamino)ethanol	: 5709
β-Dimethylaminoethanol diphenylmethyl ether	: 5422
β-Dimethylaminoethyl methacrylate	: 10654
2-[(2-Dimethylaminoethyl)(p-methoxybenzyl)amino]pyridine	: 5511
2-[(Dimethylamino)methyl]-cyclohexanone	: 4591
3-[(Dimethylamino)methyl]indole	: 7310
4-Dimethylaminophenazone	: 10851
2-Dimethylamino-1-propanol	: 10374
4-[2-(Dimethylamino)propionamido]-antipyrine	: 9795
3-(Dimethylamino)propionitrile	: 10101
N,N-Dimethylaniline	: 713
1,3-Dimethylanthracene	: 377
Dimethylarsine	: 442
2,4-Dimethylbenzaldehyde	: 547
2,5-Dimethylbenzaldehyde	: 548
9,10-Dimethyl-1,2-benzanthracene	: 520
2,4-Dimethylbenzenecarboxaldehyde	: 547
3,3'-Dimethylbenzidine	: 3034
1,5-Dimethylbenzimidazole	: 2378
2,5-Dimethylbenzofuran	: 2445
2,6-Dimethylbenzofuran	: 2446
3,5-Dimethylbenzofuran	: 2448
4,7-Dimethylbenzofuran	: 2451
5,6-Dimethylbenzofuran	: 2452
5,7-Dimethylbenzofuran	: 2453
2,4-Dimethylbenzoic acid	: 2580
2,6-Dimethylbenzoic acid	: 2582
3,5-Dimethylbenzoic acid	: 2583
3,4-Dimethylbenzophenone	: 7615
2,4'-Dimethylbenzophenone	: 7628
2,5-Dimethylbenzoxazole	: 2879
α,α-Dimethylbenzyl alcohol	: 1857
m-α-Dimethylbenzyl alcohol	: 1858
p-α-Dimethylbenzyl alcohol	: 1859
Dimethylbenzylamine	: 1807
3,5-Dimethylbenzyl bromide	: 1057
2,3-Dimethylbiphenyl	: 3052
2,4-Dimethylbiphenyl	: 3054
3,4-Dimethylbiphenyl	: 3059
3,5-Dimethylbiphenyl	: 3060
2,3-Dimethylbutadiene	: 3159
1,1-Dimethylbutadiene	: 8670
Dimethylbutadiyne	: 6717
2,2-Dimethylbutane	: 3289
2,3-Dimethylbutane	: 3290
2,2-Dimethyl-1-butanol	: 3615
3,3-Dimethylbutan-2-ol (DL-)	: 3619
2,3-Dimethyl-1-butene	: 3759
2,3-Dimethyl-2-butene	: 3761
2,2-Dimethyl-3-butenoic acid	: 3835
4,4-Dimethyl-2-butenoic acid	: 9078
1,1-Dimethylbutyl acetate	: 8946
1,3-Dimethylbutyl acetate	: 8947
3,4'-Dimethylbutyrophenone	: 3670
Dimethyl cadmium	: 3929
Dimethylcarbamoyl chloride	: 3973
Dimethylchloroarsine	: 453
Dimethyl chloromaleate	: 3773
5,6-Dimethylchrysene	: 4115
trans-1,2-Dimethylcyclobutane	: 4188
cis-1,3-Dimethylcyclobutane	: 4189
trans-1,3-Dimethylcyclobutane	: 4190
Dimethyl trans-1,2-cyclobutanedicarboxylate	: 4181
1,3-Dimethyl-1,3-cyclohexadiene	: 4252
1,4-Dimethyl-1,3-cyclohexadiene	: 4253
1,5-Dimethyl-1,3-cyclohexadiene	: 4254
2,5-Dimethyl-1,3-cyclohexadiene	: 4255
1,1-Dimethylcyclohexane	: 4364
gem-Dimethylcyclohexane	: 4364
cis-1,2-Dimethylcyclohexane	: 4365
trans-1,2-Dimethylcyclohexane	: 4366
cis-1,3-Dimethylcyclohexane	: 4367
cis-1,4-Dimethylcyclohexane	: 4368
trans-1,3-Dimethylcyclohexane	: 4370
trans-1,4-Dimethylcyclohexane	: 4371

Dimethyl cis-1,3-cyclohexanedicarboxylate	: 4352
cis-Dimethyl 1,4-cyclohexanedicarboxylate	: 4357
2,2-Dimethylcyclohexanol	: 4513
2,4-Dimethylcyclohexanol	: 4514
2,6-Dimethyl-1-cyclohexanol	: 4515
3,3-Dimethylcyclohexanol	: 4516
3,5-Dimethylcyclohexanol	: 4518
4,4-Dimethylcyclohexanol	: 4519
cis-1,3-Dimethylcyclohexanol	: 4510
cis-2,3-Dimethylcyclohexanone	: 4581
2,2-Dimethyl-1-cyclohexanone	: 4580
trans-2,3-Dimethylcyclohexanone	: 4582
trans-(±)-2,5-Dimethylcyclohexanone	: 4585
2,6-Dimethylcyclohexanone	: 4586
3,3-Dimethylcyclohexanone	: 4587
3,4-Dimethylcyclohexanone	: 4588
3,5-Dimethylcyclohexanone	: 4589
4,4-Dimethylcyclohexanone	: 4590
1,3-Dimethyl-1-cyclohexene	: 4649
1,3-Dimethylcyclohexene	: 4649
1,4-Dimethyl-1-cyclohexene	: 4650
3,3-Dimethylcyclohexene	: 4653
3,6-Dimethylcyclohexene	: 4655
3,5-Dimethylcyclohexene	: 4654
2,5-Dimethyl-2-cyclohexenone	: 4706
3,4-Dimethylcyclohexyl alcohol	: 4517
1,1-Dimethylcyclopentane	: 4800
gem-Dimethylcyclopentane	: 4800
cis-1,2-Dimethylcyclopentane	: 4801
trans-1,2-Dimethylcyclopentane	: 4802
cis-1,3-Dimethylcyclopentane	: 4803
trans-1,3-Dimethylcyclopentane	: 4804
2,5-Dimethylcyclopentanone	: 4882
1,4-Dimethylcyclopentene	: 4920
1,3-Dimethylcyclopentene	: 4919
1,1-Dimethylcyclopropane	: 4971
gem-Dimethylcyclopropane	: 4971
cis-1,2-Dimethylcyclopropane	: 4972
trans-1,2-Dimethylcyclopropane	: 4973
3,3-Dimethyl-4-decyne	: 5101
Dimethyldiacetylene	: 6717
1,6-Dimethyldiallyl	: 8320
2,7-Dimethyl-2,7-diazaoctane	: 3263
Dimethyldiazene	: 5125
O,O-Dimethyl S-(1,2-dicarbethoxyethyl) dithiophosphate	: 3316
2,5-Dimethyl-2,5-dichlorohexane	: 6765
3,3-Dimethyl-1,5-dichloropentane	: 8753
Dimethyldiethoxysilane	: 11405
Dimethyldifluoromethane	: 9906
S,S'-Dimethyl-2-difluoromethyl-4-isobutyl-6-trifluoromethylpyridine-3,5-dicarbothioate	: 5267
5,5-Dimethyldihydroresorcinol	: 4378
2,2-Dimethyl-1,3-dihydroxypentane	: 8792
Dimethyl diketone	: 3359
2,2-Dimethyl-1,3-dioxolane-4-methanol	: 5207
Dimethyldiphenoxysilane	: 11415
Dimethyldiphenylsilane	: 11416
Dimethyl disulfide	: 5254
7,8-Dimethylenebenz[a]anthracene	: 2391
3,4-Dimethylenedioxybenzaldehyde	: 2417
N,N-Dimethylethanamide	: 55
Dimethyl ether	: 7523
1,1-Dimethylethyl methyl ketone	: 3657
Dimethylethynylcarbinol	: 3922
Dimethylformaldehyde	: 10411
N,N-Dimethylformamide	: 5992
3,5-Dimethylfuran	: 6088
2,4-Dimethylfuran	: 6088
2,5-Dimethylfuran	: 6089
2,7-Dimethylfuran	: 2447
6,7-Dimethylfuran	: 2454
3,6-DimethylfuranBenzofuran	: 2449
Dimethyl 2,5-furandicarboxylate	: 6083
N,N-Dimethylglycine	: 6261
Dimethylglyoxal	: 3359
Dimethylglyoxime	: 3360
2,2-Dimethylheptane	: 6383
2,3-Dimethylheptane	: 6384
2,4-Dimethylheptane	: 6385
3,6-Dimethylheptane	: 6386
2,5-Dimethylheptane	: 6386
2,6-Dimethylheptane	: 6387
3,3-Dimethylheptane	: 6388
3,4-Dimethylheptane	: 6389
3,5-Dimethylheptane	: 6390
4,4-Dimethylheptane	: 6391
2,6-Dimethyl-2-heptanol	: 6479
2,3-Dimethyl-3-heptanol	: 6480
2,6-Dimethyl-3-heptanol	: 6481
3,6-Dimethylheptan-3-ol	: 6483
2,4-Dimethyl-4-heptanol	: 6484
2,6-Dimethyl-4-heptanol	: 6485
3,5-Dimethyl-4-heptanol	: 6486
3,5-Dimethyl-4-heptanone	: 6515
2,3-Dimethyl-2-heptene	: 6542
5,5-Dimethyl-3-heptyne	: 6618
2,5-Dimethyl-1,5-hexadien-3-yne	: 6714
2,2-Dimethylhexane	: 6771
2,3-Dimethylhexane	: 6772
2,5-Dimethylhexane	: 6774
3,3-Dimethylhexane	: 6775
3,4-Dimethylhexane	: 6776
2,4-Dimethylhexane	: 6773
2,5-Dimethyl-2,5-hexanediol	: 6801
2,5-Dimethyl-3,4-hexanedione	: 6807
2,3-Dimethyl-2-hexanol	: 6922
2,5-Dimethyl-2-hexanol	: 6923
2,2-Dimethyl-3-hexanol	: 6925
2,5-Dimethyl-3-hexanol	: 6927
3,3-Dimethyl-2-hexanone	: 6963
3,4-Dimethyl-2-hexanone	: 6964
2,2-Dimethyl-3-hexanone	: 6965
2,3-Dimethyl-1-hexene	: 7013
2,4-Dimethyl-1-hexene	: 7014
2,5-Dimethyl-1-hexene	: 7015
3,3-Dimethyl-1-hexene	: 7016
3,4-Dimethyl-1-hexene	: 7017
4,4-Dimethyl-1-hexene	: 7019
4,5-Dimethyl-1-hexene	: 7020
5,5-Dimethyl-1-hexene	: 7021
2,3-Dimethyl-2-hexene	: 7022
2,4-Dimethyl-2-hexene	: 7023
2,5-Dimethyl-2-hexene	: 7024
2,3-Dimethyl-4-hexene	: 7025
cis-5,5-Dimethyl-2-hexene	: 7027
2,4-Dimethyl-trans-3-hexene	: 7030
1,1-Dimethyl-5-hexen-1-ol	: 6582
2,5-Dimethyl-4-hexen-3-ol	: 7075
2,3-Dimethyl-2H-indazole	: 7238
1,1-Dimethylhydrazine	: 7154
1,2-Dimethylhydrazine	: 7155
1,1-Dimethyl-2-hydroxyethylamine	: 10341
2,2-Dimethyl-4-hydroxymethyldioxolane	: 5207
1,1-Dimethyl-3-hydroxypyrrolidinium bromide α-cyclopentylmandelate	: 11179
2,4-Dimethyl-6-hydroxyquinoline	: 11317
1,4-Dimethylimidazole	: 7201
1,1-Dimethylindan	: 7251
1,2-Dimethylindan	: 7252
4,7-Dimethylindan	: 7253
5,6-Dimethylindan	: 7254
1,5-Dimethylindazole	: 7237
2,5-Dimethylindole	: 7299
4,7-Dimethyl-2-indolinone	: 7330
Dimethyl isophthalate	: 1347
3,5-Dimethylisoxazole	: 7403
Dimethylketol	: 3661
Dimethyl ketone	: 10411
Dimethyl ketoxime	: 10476
Dimethyl mercury	: 7435
(E)-Dimethyl 1-methyl-3-(methylamino)-3-oxo-1-propenyl phosphate	: 7663
2,6-Dimethylmorpholine	: 7687
1,2-Dimethylnaphthalene	: 7818
1,3-Dimethylnaphthalene	: 7819
1,4-Dimethylnaphthalene	: 7820
1,5-Dimethylnaphthalene	: 7821
1,6-Dimethylnaphthalene	: 7822
1,7-Dimethylnaphthalene	: 7823
1,8-Dimethylnaphthalene	: 7824
2,3-Dimethylnaphthalene	: 7825
2,6-Dimethylnaphthalene	: 7826
2,7-Dimethylnaphthalene	: 7827
N,N-Dimethyl-1-naphthylamine	: 7715
2,8-Dimethylnonan-5-ol	: 8164
3,3-Dimethyl-4-nonyne	: 8201
3,7-Dimethylocta-2,4-diene	: 8322
2,7-Dimethyl-2,6-octadiene	: 8325
4,5-Dimethyl-2,6-octadiene	: 8327
cis-2,6-Dimethyl-2,6-octadiene	: 8324
3,6-Dimethyl-2,6-octadiene-4-yne	: 8340
2,2-Dimethyloctane	: 8373
2,3-Dimethyloctane	: 8374
2,4-Dimethyloctane	: 8375
2,5-Dimethyloctane	: 8376
2,6-Dimethyloctane	: 8377
3,3-Dimethyloctane	: 8379
3,4-Dimethyloctane	: 8380
3,5-Dimethyloctane	: 8381
3,6-Dimethyloctane	: 8382
4,4-Dimethyloctane	: 8383
4,5-Dimethyloctane	: 8384
2,6-Dimethyl-2-octanol	: 8450
2,7-Dimethyloctan-4-ol	: 8458
3,7-Dimethyl-1,3,6-octatriene	: 8485
2,6-Dimethyl-7-octene	: 8502
2,6-Dimethyl-2-octene	: 8503
2,2-Dimethyl-3-octyne	: 8535
Dimethylolpropionic acid	: 10272
Dimethyl oxalate	: 5563
2,4-Dimethyloxazole	: 8577
2,5-Dimethyloxazole	: 8578
3,3-Dimethyloxetane	: 8593
O,O-Dimethyl S-[(4-oxo-1,2,3-benzotriazin-3(4H)-yl)methyl]phosphorodithioate	: 497
2,2-Dimethylpentane	: 8760
2,3-Dimethylpentane	: 8761
2,4-Dimethylpentane	: 8762
3,3-Dimethylpentane	: 8763
3,3-Dimethyl-2,4-pentanedione	: 8799
2,4-Dimethyl-2-pentanol	: 8928
2,2-Dimethyl-3-pentanol	: 8929
2,3-Dimethyl-3-pentanol	: 8930
2,2-Dimethyl-3-pentanone	: 8966
2,4-Dimethyl-3-pentanone oxime	: 8968
2,3-Dimethyl-1-pentene	: 9029
2,4-Dimethyl-1-pentene	: 9030
3,3-Dimethyl-1-pentene	: 9031
3,4-Dimethyl-1-pentene	: 9032
2,3-Dimethyl-2-pentene	: 9034
2,4-Dimethyl-2-pentene	: 9035
3,4-Dimethyl-2-pentene	: 9036
2,2-Dimethylpent-4-enoic acid	: 9075
3,3-Dimethyl-1-pentyne	: 9128
4,4-Dimethyl-1-pentyne	: 9129
4,4-Dimethyl-2-pentynoyl chloride	: 9147
9,10-Dimethylphenanthrene	: 9172
Dimethylphenylcarbinol	: 1857
Dimethyl-p-phenylenediamine	: 1270
Dimethylphenylmethanol	: 1857
N-(2,6-Dimethylphenyl)-N-(methoxyacetyl)alanine, methyl ester	: 7439
N-(2,6-Dimethylphenyl)-2-methoxy-N-(2-oxo-3-oxazolidinyl)acetamide	: 8566
1,1-Dimethyl-3-phenylpropanol	: 2150
Dimethylphosphinic acid	: 9528
Dimethylphosphoramidothioic dichloride	: 9557
Dimethyl phthalate	: 1343
cis-2,5-Dimethylpiperazine	: 9609
trans-2,5-Dimethylpiperazine	: 9610
cis-2,6-Dimethylpiperazine	: 9611
2,5-Dimethylpiperidine	: 9649
3,3-Dimethylpiperidine	: 9651
3,5-Dimethylpiperidine	: 9652
1,1-Dimethylpropargyl alcohol	: 3922
Dimethylpropylamine	: 9807
1,1-Dimethylpropyl bromide	: 3240
1,1-Dimethyl-2-propyn-1-ol	: 3922
2,3-Dimethylpyrazine	: 10805
2,5-Dimethylpyrazine	: 10806
2,6-Dimethylpyrazine	: 10807
2,3-Dimethylpyridine	: 10936
2,5-Dimethylpyridine	: 10938
3,4-Dimethylpyridine	: 10940
3,5-Dimethylpyridine	: 10941
2,4-Dimethylpyrimidine	: 11055
2,5-Dimethylpyrimidine	: 11056
4,5-Dimethylpyrimidine	: 11057
4,6-Dimethylpyrimidine	: 11058

N1-(4,6-Dimethyl-2-pyrimidyl)sulfanilamide	: 2196
2,4-Dimethylpyrrole	: 11119
2,2-Dimethylpyrrolidine	: 11149
cis-2,5-Dimethylpyrrolidine	: 11151
3,3-Dimethyl-2-pyrrolidone	: 11182
4,6-Dimethylquinaldine	: 11302
2,3-Dimethylquinoline	: 11254
2,4-Dimethylquinoline	: 11255
2,7-Dimethylquinoline	: 11258
3,4-Dimethylquinoline	: 11260
3,8-Dimethylquinoline	: 11261
4,6-Dimethylquinoline	: 11262
5,8-Dimethylquinoline	: 11263
6,8-Dimethylquinoline	: 11264
Dimethylsilane	: 11414
Dimethylsiloxane cyclic trimer	: 5002
2,2'-Dimethylstilbene	: 1622
Dimethyl sulfate	: 11574
Dimethyl sulfide	: 7545
2,4-Dimethylsulfolane	: 11784
Dimethyl sulfone	: 7534
Dimethyl sulfoxide	: 7529
Dimethyl terephthalate	: 1351
1,3-Dimethyl-1,2,3,4-tetrahydronaphthalene	: 7982
1,1-Dimethyltetralin	: 7980
1,4-Dimethyltetralin	: 7983
2,3-Dimethyltetralin	: 7986
2,5-Dimethyltetralin	: 7987
2,7-Dimethyltetralin	: 7989
2,8-Dimethyltetralin	: 7990
5,6-Dimethyltetralin	: 7991
5,7-Dimethyltetralin	: 7992
2,5-Dimethyl-1,3,4-thiadiazole	: 11659
2,4-Dimethylthiazole	: 11674
4,5-Dimethylthiazole	: 11675
Dimethyl N,N'-[thiobis[[(methylimino)carbonyl]oxyl]]bis[ethanimidothioate]	: 11713
2,3-Dimethylthiophene	: 11754
2,4-Dimethylthiophene	: 11755
2,5-Dimethylthiophene	: 11756
3,4-Dimethylthiophene	: 11757
N,N'-Dimethylthiourea	: 11804
7,8-Dimethyltocol	: 2790
N,N-Dimethyl-o-toluidine	: 860
1,5-Dimethyl-1,2,3-triazole	: 11849
N,N-Dimethyltryptamine	: 7303
Dimethyl(vinylethyl)carbinol	: 7083
Dimethylvinylethynylchloromethane	: 7107
Dimethyl zinc	: 12066
Dimetridazole	: 7202
Dimorpholamine	: 7686
Dinaphthalene dioxide	: 12055
β,β'-Dinaphthylamine	: 7727
Dinaphthylene dioxide	: 12055
1,2'-Dinaphthyl ketone	: 7634
2,3-Dinitroaniline	: 724
2,4-Dinitroaniline	: 725
2,6-Dinitroaniline	: 726
3,5-Dinitroaniline	: 727
3,5-Dinitroanisole	: 1924
2,4-Dinitrobenzaldehyde	: 551
o-Dinitrobenzene	: 1491
m-Dinitrobenzene	: 1492
p-Dinitrobenzene	: 1493
2,4-Dinitrobenzenesulfenyl chloride	: 2178
3,4-Dinitrobenzoic acid	: 2590
3,5-Dinitrobenzoic acid	: 2591
3,5-Dinitrobenzoyl chloride	: 2896
1,4-Dinitrobutane	: 3293
4,4'-Dinitrocarbanilide	: 11997
Dinitrochloromethane	: 7480
2,6-Dinitro-4-chlorophenol	: 9258
Dinitrocresol	: 9405
4,6-Dinitro-o-cyclohexylphenol	: 9273
3,3'-Dinitrodiphenyl disulfide	: 5247
1,1-Dinitroethane	: 5551
1,2-Dinitroethane	: 5552
2,4-Dinitro-1-fluorobenzene	: 1718
1,6-Dinitrohexane	: 6778
Dinitromethane	: 7508
1,3-Dinitronaphthalene	: 7831
1,5-Dinitronaphthalene	: 7832
1,8-Dinitronaphthalene	: 7833
1,5-Dinitropentane	: 8769
2,3-Dinitrophenol	: 9341
2,4-Dinitrophenol	: 9342
2,5-Dinitrophenol	: 9343
2,6-Dinitrophenol	: 9344
3,5-Dinitrophenol	: 9346
2,4-Dinitrophenyl fluoride	: 1718
(2,4-Dinitrophenyl)hydrazine	: 7156
1,1-Dinitropropane	: 9918
1,3-Dinitropropane	: 9919
2,2-Dinitropropane	: 9920
2,4-Dinitroresorcinol	: 1517
2,5-Dinitrothiophene	: 11758
2,4-Dinitrotoluene	: 1956
2,6-Dinitrotoluene	: 1958
3,4-Dinitrotoluene	: 1960
17,18-Dinor-8βH,9βH,10α-labda-2,13(16),14-triene-19,20-dioic acid, 15,16-epoxy-1β,4,12	: 5926
Dinoroleic acid	: 6668
Dioctyl adipate	: 6781
Dioctyl phthalate	: 1344
Dioctyl sebacate	: 5043
Diosgenin	: 11536
Diosphenol	: 4709
1,3-Dioxacyclohexane	: 5186
1,4-Dioxacyclohexane	: 5187
1,3-Dioxacyclopentane	: 5204
2,15-Dioxa-12,25-diazatricyclo[22.2.2.211,14]triacontane, carpaine deriv.	: 4064
1,4-Dioxane, heptachloro	: 5192
m-Dioxane, 4-methyl-4-phenyl-	: 5194
1,2-Dioxocyclohexane	: 4375
1,4-Dioxocyclohexane	: 4377
2,5-Dioxohexane	: 6805
1,3-Dioxolane, 2,2-dimethyl-4-hydroxymethyl	: 5207
2,4-Dioxopentane	: 8798
Dioxybenzone	: 7619
3,4-Dioxymethylenebenzoic acid	: 2418
Dipentene	: 4677
Dipentylamine	: 8708
Dipentylamine, 2,2'-dimethyl,2,2'-dihydroxy	: 8936
2,5-Di-tert-pentylhydroquinone	: 1501
Dipentyl ketone	: 11962
Diphemethoxidine	: 9665
Diphenadione	: 7268
Diphenethyl ketone	: 8969
Diphenhydramine	: 5422
o,o'-Diphenic acid	: 3038
Diphenolic acid	: 1103
Diphenoxydimethylsilane	: 11415
Diphenyl	: 3008
Diphenylacetamide	: 57
Diphenylacetic acid	: 936
Diphenylacetylene	: 1714
Diphenylamine	: 836
Diphenylamine-2-carboxylic acid	: 2715
Diphenylamine-2,2'-dicarboxylic acid	: 2656
Diphenylamine-4-sulfonic acid	: 2234
Diphenylarsine	: 443
N,N'-Diphenylbenzidine	: 3035
1,4-Diphenylbutane	: 1099
1,1-Diphenylbutane	: 1120
2,2-Diphenyl-3-butanone	: 3658
1,3-Diphenyl-1-butene	: 2022
2,2'-Diphenylcarbazide	: 4010
sym-Diphenylcarbazide	: 4010
N,N'-Diphenylcarbazide	: 4010
Diphenylcarbazone	: 5122
Diphenylcarbinol	: 1910
1,5-Diphenylcarbohydrazide	: 4010
Diphenylcyanamide	: 4158
Diphenyldiethoxysilane	: 11406
Diphenyl(dimethyl)silane	: 11416
Diphenyl diselenide	: 5225
1,4-Diphenyl-2,3-dithiabutane	: 5248
1,1-Diphenylethane	: 1677
2,2-Diphenylethanol	: 1615
1,1-Diphenylethene	: 1636
Diphenyl ether	: 2071
N,N'-Diphenylethylenediamine	: 5508
Diphenyl glutarate	: 8776
1,3-Diphenylguanidine	: 6282
1,1-Diphenylheptane	: 1746
1,1-Diphenyl-1-heptene	: 1744
1,1-Diphenylhydrazine	: 7157
N,N'-Diphenylhydrazine	: 7158
1,3-Diphenylindene	: 7271
2,3-Diphenylindene	: 7272
Diphenyline	: 3029
Diphenylketene	: 5928
8,10-Diphenyllobelidiol	: 9646
8,10-Diphenyllobelidione	: 5854
Diphenylmercury	: 7436
Diphenylmethane	: 1961
Diphenylmethane, 2,2'-dibromo	: 1962
Diphenylmethane, 3,3'-dihydroxy	: 9406
2-(Diphenylmethoxy)-N,N-dimethylethylamine	: 5422
Diphenylmethyl alcohol	: 1910
Di(phenylmethyl)disulfide	: 5248
Diphenyl methylphosphonate	: 9541
3,5-Diphenyl-1,2,4-oxadiazole	: 8564
2,4-Diphenyloxazole	: 8579
4,5-Diphenyloxazole	: 8581
1,5-Diphenylpentane	: 2098
1,5-Diphenyl-3-pentanone	: 8969
N,N'-Diphenyl-p-phenylenediamine	: 1271
1,3-Diphenylpropane	: 2121
1,1-Diphenylpropane	: 2171
1,3-Diphenylpyrazole	: 10830
1,5-Diphenylpyrazole	: 10831
2,6-Diphenylpyridine	: 10947
1,1-Diphenylsilacyclohexane	: 11360
Diphenylsilane	: 11420
Diphenyl succinate	: 3320
Diphenyl sulfone	: 2237
1,1-Diphenyltetradecane	: 2267
2,4-Diphenylthiazole	: 11676
Diphenylthiocarbamic acid S-[2-(diethylamino)ethyl] ester	: 3977
Diphenylthiocarbazone	: 5121
sym-Diphenylthiourea	: 11805
1,1-Diphenylurea	: 12011
Diphosgene	: 4029
Dipicrylamine	: 864
Dipicryl sulfide	: 2288
Dipivaloylmethane	: 6404
Diploicin	: 5153
Diploicine	: 5153
Dipropargyl	: 6716
1,3-Dipropoxybenzene	: 1556
1,2-Dipropoxyethane	: 5586
Dipropylacetylene	: 8532
Dipropylamine	: 9826
Dipropylcarbinol	: 6475
Dipropyl carbonate	: 4002
Dipropylene glycol	: 10397
Dipropyl ether	: 10119
Dipropyl fumarate	: 3783
Dipropylglyoxime	: 8397
N,N-Dipropylhydroxylamine	: 9814
Dipropylisobutylamine, perfluoro	: 11912
Dipropyl ketone	: 6512
Dipropyl malonate	: 9945
Dipropylnitrosamine	: 9825
5,5-Dipropyl-2,4-oxazolidinedione	: 8585
2,6-Dipropylphenol	: 9347
p-(Dipropylsulfamoyl)benzoic acid	: 2595
Dipropyl sulfide	: 10151
Dipyridamole	: 5711
2,2'-Dipyridine	: 3099
Dipyrido[1,2-a:2',1'-c]pyrazinediium, 6,7-dihydro-, dibromide	: 5224
2,2'-Dipyridyl	: 3099
α,α'-Dipyridyl	: 3099
3,3'-Dipyridyl	: 3102
4,4'-Dipyridyl	: 3104
γ,γ'-Dipyridyl	: 3104
Dipyrine	: 10851
Diresorcinol	: 3097
Disalcid	: 2616
Disalicylic acid	: 2616
Disiloxane, 1-(cyclopentadienyl)-1,1,3,3-tetramethyl-3-phenyl-, Fe deriv.	: 5961
Disophenol	: 9304
Dispiro[5,1,5,1]tetradecane	: 5240
Distannoxane, hexakis(2-methyl-2-phenylpropyl)-	: 5950
Disulfan	: 2188
Disulfiram	: 11721

1,3-Dithiacyclopentane	: 5266
Dithianone	: 8056
1,3,5-Dithiazine, 4,5-dihydro-5-methyl	: 5284
Di-(2-thienyl) ketone	: 7617
3,3'-Dithiobis(2-aminopropanoic acid)	: 5020
2,4-Dithiobiuret	: 11718
4,4'-Dithiodimorpholine	: 7688
Dithioglycol	: 5590
Dithiooxamide	: 5588
Dithiosalicylic acid	: 1125
Dithizone	: 5121
1,2-Di-p-tolylethane	: 1590
2,4'-Ditolyl sulfide	: 1993
3,4'-Ditolyl sulfide	: 11570
Diundecyl ketone	: 11857
Diurea glyoxalate	: 7191
Diurite	: 2852
Diuron	: 12000
Diuroral	: 2854
Divinyl	: 3146
Divinylacetylene	: 6713
o-Divinylbenzene	: 1403
m-Divinylbenzene	: 1404
p-Divinylbenzene	: 1405
trans-1,2-Divinylcyclobutane	: 4187
Divinyl ether	: 5912
Djenkolic acid	: 5018
n-Docosane	: 5268
Docosane, 1-cyclohexyl-	: 4380
Docosane, 11-decyl	: 5273
Docosane, 1-phenyl-	: 1565
cis-13-Docosenol	: 5284
Docosylcyclopentane	: 4808
2,4-Dodecadienoic acid, 11-methoxy-3,7,11-trimethyl-, 1-methylethyl ester, (E,E)-	: 7651
2,4-Dodecadienoic acid, 3,7,11-trimethyl-, ethyl ester, (E,E)-	: 7182
Dodecahydrophenanthrene	: 9178
Dodecamethylcyclohexasiloxane	: 4624
Dodecamethylpentasiloxane	: 8997
Dodecanal dimethyl acetal	: 5297
1,12-Dodecanediol, dimethyl ether	: 5298
Dodecane, 1,1-di-(4-tolyl)	: 1567
Dodecane, 1-(1-naphthyl)-	: 7879
Dodecane, 1-(2-naphthyl)-	: 7880
N-Dodecanoyl-N-methylglycine sodium salt	: 6269
Dodecanoyl peroxide	: 9155
Dodecylmethylketone	: 11633
2-Dodecylnaphthalene	: 7880
Dodecyltrichlorosilane	: 11460
Donaxine	: 7310
Dormin	: 8679
2,6,8,10,12,14,16,18,20,22,24,26,30-Dotriacontatridecaene, 2,6,10,14,19,23,27,31-octamethyl-, (all-E)-	: 4053
Dotriacontylcyclohexane	: 4383
Dowicide 9	: 9255
Doxenitoin	: 7228
Draconic acid	: 2666
Dromostanolone propionate	: 351
Drosophilin A	: 9472
DTBP	: 9153
Dulcin	: 12013
Dulcitol	: 6194
Dulcose	: 6194
Duotal	: 9384
Duracaine	: 2493
Durene	: 2280
Durene aldehyde	: 598
Durohydroquinone	: 1552
Duroquinone	: 4270
1,2-Dyclodecanedione	: 4203
Dyphylline	: 10761
Dypnone	: 3876
Ecgonidine	: 491
Ecgonine	: 483
Echinochrome A	: 7857
Echinopsine	: 11330
Echitamine	: 4730
Edetic acid	: 6262
EEDQ	: 11231
Efloxate	: 225

Egonol	: 2464
Ehrlich's reagent	: 550
Eicosamethylnonasiloxane	: 8178
Eicosane, 1-cyclohexyl-	: 4384
Eicosane, 9-octyl	: 5354
Eicosane, 5-phenyl	: 1119
Eicosane, 3-phenyl-	: 1695
Eicosane, 2-phenyl-	: 2007
Eicosane, 9-phenyl-	: 2068
Eicosane, 4-phenyl-	: 2170
cis-5,8,11,14-Eicosatetraenoic acid	: 5361
Δ 11-Eicosenoic acid	: 5363
5,8,11,14-Eicosotetraenoic acid	: 5360
Eiver-pick acid	: 7868
Elaidic acid	: 8297
Elaidic acid	: 9746
Electrocortin	: 9746
β-Elemene	: 4389
Eleostearic acid	: 8246
Embelin	: 4261
Emetine	: 5368
Emodin	: 419
Enallylpropymal	: 11099
Enanthaldoxime	: 6346
Enanthylic acid	: 6455
Enheptin	: 11666
Enidin	: 2847
ENTPROL	: 10379
Eosin	: 11525
Ephdine	: 2849
(±)-Ephedrine	: 1900
Epiandrosterone	: 347
Epiborneol, acetate	: 2947
DL-α-Epichlorohydrin	: 8605
Epicholestanol	: 4108
Epicholesterol	: 4112
3,17-Epidihydroxyestratriene	: 5399
Epiestradiol	: 5400
16-Epiestriol	: 5405
Epihydrinaldehyde	: 8603
Epinephrine	: 1521
Epiquinidine	: 4125
EPN	: 9551
1,2-Epoxybutane	: 8615
2,3-Epoxypropanal	: 8603
1,2-Epoxypropane	: 8621
2,3-Epoxy-1-propanol	: 8618
2,3-Epoxypropyl ethyl ether	: 8614
Epsilan	: 2787
Epsilcapramin	: 6873
Equilenin	: 5397
Equilin	: 5398
Equol	: 2786
Ergobasinine	: 5373
Ergocornine	: 5387
Ergocorninine	: 5388
Ergocristine	: 5391
Ergocristinine	: 5392
Ergocryptine	: 5389
Ergocryptinine	: 5390
Ergoline-8α-carboxamide, 9,10-didehydro-N-((S)-2-hydroxy-1-methylethyl)-6-methyl-	: 5373
Ergometrinine	: 5373
Ergosine	: 5393
Ergostanol	: 5379
γ-Ergostenol	: 5384
α-Ergostenol	: 5385
Ergosterol	: 5383
Ergotamine	: 5394
Ergotaminine	: 5395
Ergothioneine	: 7204
Eriodictyol	: 2812
Eriodictyol 4'-monomethyl ether	: 2800
Erucic acid	: 5283
Erucyl alcohol	: 5284
Erysimupicrone	: 4049
Erythraric acid	: 3304
meso-Erythritol	: 3431
Erythritol	: 3431
Erythritol anhydride	: 3007
Erythrityl tetranitrate	: 3432
β-Erythroidine	: 8567
D-Erythrose	: 3192
Erythrosin	: 11524
L-Erythrulose	: 3679
Erytrin	: 31
Esculetin	: 2804

Esculin	: 2822
Esculoside	: 2822
d-3,17β-Estradiol	: 5399
Estradiol	: 5399
α-Estradiol	: 5400
17β-Estradiol 3-benzoate	: 5403
Estradiol benzoate	: 5403
Estragole	: 1943
Estra-1,3,5(10)-triene-3,17β-diol	: 5399
Estra-1,3,5(10)-triene-3,17α-diol	: 5400
Estra-1,3,5(10)-triene-3,16α,17β-triol	: 5404
Estratriol	: 5404
Estriol	: 5404
Estrone	: 5406
Ethacridine	: 317
Ethacrynic acid	: 168
Ethamivan	: 611
Ethanal	: 11
Ethanamide	: 22
Ethanamidine hydrochloride	: 5671
Ethanearsonic acid	: 457
Ethane, 2-bis(di-2-hydroxypropylamino)	: 10379
Ethane, 1-chloro-2-iodo-1,1,2-trifluoro	: 5490
1,2-Ethanedicarboxylic acid	: 3294
Ethane, 1,1-dichloro-2,2-bis(p-ethylphenyl)-	: 9160
Ethane, 1,2-dicyclohexyl-	: 4386
Ethane, 1,1-dicyclohexyl-	: 4392
Ethane, 1,2-dinitro-1,1,2,2-tetrafluoro	: 5642
Ethanedione	: 5491
1,2-Ethanediylbis(chloroformate)	: 3366
Ethane, 1-ethoxy-2-methylamino	: 5426
Ethane, 1-ethoxy-2-phenoxy	: 1652
Ethanenitrile	: 271
Ethane, perbromo	: 5594
Ethanephosphonic acid, diethyl ester	: 9536
Ethane, 1,1,1,2-tetraphenyl-	: 1618
Ethanethiol, 1-phenyl (1)	: 1834
Ethanimidothioic acid, 2-(dimethylamino)-N-[[(methylamino)carbonyl]oxy]-2-oxo-, methyl ester	: 8568
3,5-Ethano-3H-pyrrolo[2,3-d]carbazole, curan-17-ol deriv.	: 4154
Ethanoic acid	: 100
Ethanoic anhydride	: 107
Ethanolamine	: 5675
Ethanolamine, N,N-diisobutyl	: 5684
Ethanolamine, N-isobutyl	: 5747
Ethanolamine, N-(β-naphthyl)	: 5750
Ethanol, 2-(2,4-dichlorophenoxy)-, hydrogen sulfate, sodium salt	: 11358
Ethanol, 1(3,4-dihydroxyphenyl)-2(isopropylamino)	: 1522
Ethanol, 2-(diisobutylamino)-	: 5684
Ethanol, 2-(diisopentylamino)-	: 5682
Ethanol, 2,2-diphenyl-	: 1615
Ethanol, 2-(isobutylamino)-	: 5747
Ethanol, 2,2'-oxybis-, diethyl ether	: 5612
Ethanol, 1(3-tolyl)	: 1858
Ethanoylaminoethanoic acid	: 6251
Ethanoyl chloride	: 294
11aH-12,3,6α-Ethanylylidene-7,9-methanonaphth[2,3-b]azocine, aconitane-3,8,13,14,15-pentol deriv.	: 305
Ethenylmethylphenylcarbinol	: 1869
Ethenzamide	: 615
Ether, allyl p-chlorophenyl	: 1232
Ether, allyl ethenyl	: 10565
Ether, allyl ethyl	: 10566
Ether, allyl isopropyl	: 10579
Ether, allyl methyl	: 10576
Ether, allyl phenyl	: 2167
Ether, p-allylphenyl methyl	: 1943
Ether, allyl propyl	: 10594
Ether, allyl o-tolyl	: 2026
Ether, allyl m-tolyl	: 2027
Ether, allyl p-tolyl	: 2028
Ether, allyl vinyl	: 10565
Ether, benzyl butyl	: 1115
Ether, benzyl chloromethyl	: 1181
Ether, benzyl ethyl	: 1658
Ether, benzyl isobutyl	: 2033

Ether, benzyl isopentyl : 1951
Ether, benzyl methyl : 1930
Ether, benzyl (3-methylbutyl) : 1951
Ether, bis(2-bromoethyl) : 5609
Ether, bis(bromomethyl) : 7524
Ether, bis(p-bromophenyl) : 2072
Ether, bis(2-butoxyethyl) : 3417
Ether, bis(1-chloroethyl) : 5610
Ether, bis(chloroethyl) : 5611
Ether, bis(2-chloroethyl) : 5611
Ether, bis(chloromethyl) : 7525
Ether, bis(2-chloro-1-methylethyl) : 10123
Ether, bis(p-chlorophenyl) : 2073
Ether, bis(2-chloropropyl) : 10121
Ether, bis(3-chloropropyl) : 10122
Ether, bis(2-ethoxyethyl) : 5612
Ether, bis[2-(2-methoxyethoxy)ethyl] : 8995
Ether, bis(2-methoxyethyl) : 5613
Ether, bis(o-methoxyphenyl) : 2075
Ether, 2-bromethyl phenyl : 1033
Ether, 4-bromobutyl phenyl : 1003
Ether, 2-bromocyclohexyl methyl : 4317
Ether, 2-bromoethyl ethyl : 5461
Ether, 2-bromoethyl methyl : 5465
Ether, 6-bromohexyl methyl : 6736
Ether, bromomethyl ethyl : 5464
Ether, bromomethyl methyl : 7472
Ether, 4-bromophenyl hexyl : 1045
Ether, p-bromophenyl hexyl : 1045
Ether, p-bromophenyl phenyl : 1079
Ether, 3-bromopropyl phenyl : 1085
Ether, 1,3-butadienyl ethyl : 3160
Ether, 2,3-butadienyl methyl : 3167
Ether, 1,3-butadienyl methyl : 3168
Ether, 1-buten-3-ynyl methyl : 3891
Ether, butyl 2-chloroethyl : 3249
Ether, butyl ethyl : 3372
Ether, sec-butyl ethyl : 3373
Ether, tert-butyl ethyl : 10040
Ether, butyl ethynyl : 3377
Ether, butyl isobutyl : 3401
Ether, butyl isopropyl : 3399
Ether, tert-butyl isopropyl : 10081
Ether, butyl methyl : 3391
Ether, tert-butyl methyl : 10072
Ether, butyl phenyl : 1114
Ether, tert-butyl phenyl : 1458
Ether, sec-butyl phenyl : 2030
Ether, butyl propyl : 3421
Ether, tert-butyl propyl : 10087
Ether, butyl m-tolyl : 1116
Ether, butyl p-tolyl : 1117
Ether, butyl vinyl : 3369
Ether, sec-butyl vinyl : 3370
Ether, tert-butyl vinyl : 10036
Ether, 2-butynyl methyl : 3915
Ether chloratus : 5467
Ether, 1-chloroethyl 2-chloroethyl : 5468
Ether, 1-chloroethyl ethyl : 5473
Ether, 2-chloroethyl ethyl : 5474
Ether, 1-chloroethyl methyl : 5479
Ether, 2-chloroethyl methyl : 5480
Ether, 1-chloroethyl propyl : 9854
Ether, 2-chloroethyl vinyl : 5887
Ether, chloromethyl ethyl : 5478
Ether, 2-chloro-1-methylethyl methyl : 9859
Ether, chloromethyl methyl : 7483
Ether, chloromethyl propyl : 9858
Ether, p-chlorophenyl phenyl : 1221
Ether, 3-chloropropyl methyl : 9860
Ether, 2-chloropropyl methyl : 9861
Ether, 2-chloro-α,α,α-trifluoro-p-tolyl 3-ethoxy-4-nitrophenyl : 8633
Ether, 2-chloro-α,α,α-trifluoro-p-tolyl p-nitrophenyl : 8082
Ether, 2-chlorovinyl ethyl : 5886
Ether, cinnamyl ethyl : 1666
Ether cyanatus : 10090
Ether, cyclohexyl methyl : 4425
Ether, cyclohexyl phenyl : 1253
Ether, cyclopropyl methyl : 4980
Ether, decyl vinyl : 5051
Ether, diallyl : 10592
Ether, dibenzyl : 2079
Ether, 1,2-dibromoethyl ethyl : 5523
Ether, 2,3-dibromopropyl methyl : 9888
Ether, dibutyl : 3415

Ether, di-tert-butyl : 10124
Ether, 1,2-dichloroethyl ethyl : 5533
Ether, 2,4-dichlorophenyl p-nitrophenyl : 8081
Ether, 1,2-dichlorovinyl ethyl : 5899
Ether, dicyclohexyl : 4461
Ether, diethyl : 5808
Ether, diethyl,2-amino : 5423
Ether, diethyl,2-bromo : 5461
Ether, diethyl,1-chloro : 5473
Ether, diethyl,2-chloro : 5474
Ether, diethyl,2,2'-dibromo : 5609
Ether, diethyl,1,2,-dichloro : 5533
Ether, diethyl,2,2,'-dichloro : 5611
Ether, diethyl,2,2'-dicyano : 10114
Ether, diethyl,2,2'-diphenyl : 2074
Ether, diethyl,1,1'-diphenyl(DL) : 2081
Ether, difurfuryl : 6156
Ether, dihexdecyl : 6642
Ether, dihexyl : 6842
Ether, dihexyl,2,2'-diethyl : 6431
Ether, diisopropyl : 10120
Ether, diisopropyl,β,β'-dichloro : 10123
Ether, dimethyl : 7523
Ether, dimethyl,chloro : 7483
Ether, 2,2'-dinaphthyl : 7940
Ether, 2,4-dinitrophenyl phenyl : 1495
Ether, 2,4'-dinitro-4-trifluoromethyldiphenyl : 5980
Ether, dioctyl : 8416
Ether, dipentyl : 8835
Ether, diphenyl : 2071
Ether, diphenyl,4-bromo : 1079
Ether, diphenyl,4-chloro : 1221
Ether, diphenyl,4,4'-diamino : 832
Ether, diphenyl,4,4'-dibromo : 2072
Ether, diphenyl,4,4'-dichloro : 2073
Ether, diphenyl,2,2'-dimethoxy : 2075
Ether, diphenyl,2,2'-dimethyl : 2076
Ether, diphenyl,3,3'-dimethyl : 2077
Ether, diphenyl,4,4'-dimethyl : 2078
Ether, diphenyl,2-methoxy : 1938
Ether, diphenyl,3-methyl : 2013
Ether, diphenyl,2-nitro : 2053
Ether, diphenyl,4-nitro : 2054
Ether, dipropyl : 10119
Ether, dipropyl,2,2'-dichloro : 10121
Ether, dipropyl,3,3'-dichloro : 10122
Ether, divinyl : 5912
Ether, 2,3-epoxypropyl ethyl : 8614
Ether, 2,3-epoxypropyl phenyl : 8623
Ether, ethyl ethynyl : 5942
Ether, ethyl furfuryl : 6104
Ether, ethyl heptyl : 6406
Ether, ethyl hexyl : 6814
Ether, 2-ethylhexyl vinyl : 6405
Ether, ethyl isobutyl : 10039
Ether, ethyl isopentyl : 3374
Ether, ethyl isopropyl : 10038
Ether, ethyl methyl : 5602
Ether, ethyl 1-methyleneallyl : 3161
Ether, ethyl octyl : 8402
Ether, ethyl tert-pentyl : 3375
Ether, ethyl pentyl : 8808
Ether, ethyl propyl : 10037
Ether, ethyl 1-propynyl : 10732
Ether, ethyl 2-propynyl : 10733
Ether, ethyl styryl : 1651
Ether, ethyl vinyl : 5905
Ether, ethynyl methyl : 5944
Ether, ethynyl phenyl : 1716
Ether, ethynyl propyl : 10045
Ether, furfuryl methyl : 6128
Ether, 2-furyl phenyl : 6157
Ether, heptyl methyl : 6421
Ether, heptyl phenyl : 1747
Ether, hexadecyl phenyl : 1752
Ether, hexadecyl vinyl : 6638
Ether, hexyl phenyl : 1765
Ether hydrochloric : 5467
Ether, iodomethyl methyl : 7519
Ether, isobutyl methyl : 10071
Ether, isobutyl phenyl : 2031
Ether, isobutyl propyl : 10086
Ether, isobutyl vinyl : 10035
Ether, isopentyl methyl : 3395
Ether, isopentyl vinyl : 3371

Ether, isopropyl methyl : 10070
Ether, isopropyl phenyl : 1974
Ether, isopropyl propyl : 10076
Ether, isopropyl m-tolyl : 1987
Ether, isopropyl vinyl : 10034
Ether, m-methoxyphenyl o-methoxyphenyl : 1929
Ether, methyl 2-methylallyl : 10577
Ether, methyl 2-methylbutyl : 3394
Ether, methyl 1-methyleneallyl : 3169
Ether, methyl 2-octynyl : 8536
Ether, methyl tert-pentyl : 3396
Ether, methyl pentyl : 8822
Ether, methyl propyl : 10069
Ether, methyl 2-propynyl : 10736
Ether, methyl styryl : 1928
Ether, methyl vinyl : 5908
Ether muriatic : 5467
Ether, 1-naphthyl 2-naphthyl : 7928
Ether, 1-naphthyl phenyl : 7945
Ether, 2-naphthyl phenyl : 7946
Ether, o-nitrophenyl phenyl : 2053
Ether, p-nitrophenyl phenyl : 2054
Ether, p-nitrophenyl p-tolyl : 2006
Ether, p-nitrophenyl α,α,α-trifluoro-2-nitro-p-tolyl : 5980
Ether, octadecyl vinyl : 8240
Ether, octyl phenyl : 2069
Ether, octyl vinyl : 8401
Ether, phenyl m-tolyl : 2013
Ether, phenyl propyl : 2168
Ether, phenyl o-tolyl : 2012
Ether, phenyl vinyl : 1646
Ether, propyl vinyl : 10033
Ethide : 5536
17α-Ethinyl-17β-estradiol : 8216
Ethionamide : 10885
Ethionine : 991
Ethisterone : 9758
Ethohexadiol : 6802
Ethosuximide : 11158
α-Ethoxyacetaldehyde : 14
2-Ethoxyacetic acid : 180
Ethoxyacetic acid methyl ester : 182
2-Ethoxyacetophenone : 5814
o-Ethoxybenzonitrile : 2754
6-Ethoxy-2-benzothiazolesulfonamide : 2865
2-Ethoxybenzylamine : 1808
3-Ethoxybiphenyl : 3071
N-(Ethoxycarbonyl)aniline : 3965
Ethoxycarbonyl chloride : 4018
Ethoxydiethylphosphine oxide : 9527
2-Ethoxy-2,3-dihydro-3,3-dimethyl-5-benzofuranyl methanesulfonate : 5933
2-Ethoxy-1,3-dinitrobenzene : 1650
2-Ethoxyethyl acetate : 5718
(±)-2-Ethoxy-1-ethyl-2-oxoethyl 5-[2-chloro-4-(trifluoromethyl)phenoxy]-2-nitrobenzoate : 7410
5-Ethoxy-2-hydroxybenzaldehyde : 556
(±)-2-[1-(Ethoxyimino)butyl]-5-[2-(ethylthio)propyl]-3-hydroxy-2-cyclohexen-1-one : 11359
1-Ethoxy-2-methoxyethane : 5591
o-Ethoxy-β-methylstyrene : 1665
2-Ethoxynaphthalene : 7884
2-Ethoxy-1-propanol : 10380
1-Ethoxy-2-propanol : 10381
2-Ethoxypropionitrile : 10102
Ethoxyquin : 11265
8-Ethoxy-5-quinolinesulfonic acid : 11293
Ethoxytrifluorosilicon : 11429
Ethozxolamide : 2865
N-Ethylacetamide : 61
Ethyl acetate : 186
Ethylacetic acid : 3460
Ethyl acetoacetate : 3571
Ethyl acetylacetate : 3571
Ethylacetylene : 3897
Ethyl acrylate : 10635
Ethyl adipate : 6785
Ethyl alcohol : 5672
Ethylallene : 8658
3-Ethyl-1-allyl-6-aminouracil : 11060
1-Ethylallyl bromide : 9009
1-Ethylallyl chloride : 9017

γ-Ethylallyl chloride	: 9020
Ethyl allyl ether	: 10566
Ethylamine	: 5410
Ethyl aminoacetate	: 6263
Ethyl aminobenzoate	: 2497
Ethylammonium chloride	: 5433
N-Ethylamphetamine	: 1576
Ethylamylcarbinol	: 8445
Ethyl tert-amyl ether	: 3375
Ethyl amyl ether	: 8808
Ethyl amyl ketone	: 8472
2-Ethylaniline	: 740
p-Ethylaniline	: 742
N-Ethylaniline	: 743
Ethyl anisate	: 2671
9-Ethylanthracene	: 420
Ethyl anthranilate	: 2495
Ethyl arachate	: 5356
Ethyl arachidate	: 5356
Ethyl arsenate	: 439
Ethyl arsenite	: 440
Ethylarsine	: 444
Ethylarsonic acid	: 457
Ethyl behenate	: 5277
Ethylbenzene	: 1668
Ethyl benzeneacetate	: 899
Ethyl benzenesulfonate	: 2223
Ethyl benzoate	: 2602
Ethyl benzoin ether	: 5813
Ethyl benzoylacetate	: 2141
α-Ethylbenzyl alcohol	: 1870
Ethylbenzylamine	: 1810
Ethylbenzylaniline	: 1811
Ethyl benzyl ether	: 1658
Ethyl benzyl ketone	: 3677
Ethyl bromide	: 5454
Ethyl bromoacetate	: 115
2-Ethyl-2-bromobutyramide	: 3198
Ethyl bromochloroacetate	: 113
2-Ethyl-2-bromo-3-methylbutyramide	: 3199
Ethyl α-bromopropionate	: 10207
Ethyl butanoate	: 3515
2-Ethyl-1-butanol	: 3620
2-Ethyl-1-butene	: 8825
Ethyl 3-butenoate	: 3841
2-Ethylbutyl acetate	: 185
Ethylbutylamine	: 3218
Ethyl butyl ether	: 3372
Ethyl sec-butyl ether	: 3373
Ethyl tert-butyl ether	: 10040
Ethyl butyl ketone	: 6511
Ethyl sec-butyl ketone	: 6980
Ethyl tert-butyl ketone	: 8966
Ethyl butyl sulfide	: 3376
Ethyl butyrate	: 3515
2-Ethylbutyric acid	: 3514
α-Ethylbutyric acid	: 3514
Ethyl caprate	: 5063
Ethyl caproate	: 6884
Ethyl carbanilate	: 3965
Ethyl carbitol	: 5720
Ethyl carbonate	: 3999
Ethyl Cellosolve	: 5717
Ethyl cellosolve acetate	: 5718
Ethyl chloride	: 5467
Ethyl chloroacetate	: 130
Ethyl chlorocarbonate	: 4018
Ethyl β-chlorocrotonate	: 3823
Ethyl chloroethanoate	: 130
Ethyl chloroformate	: 4018
Ethyl 5-chloro-2-furancarboxylate	: 6053
Ethyl 2-chloro-3-methylbutanoate	: 3490
Ethyl chloromethyl ether	: 5478
Ethyl α-chloropropionate	: 10220
Ethyl β-chloropropionate	: 10221
Ethyl chlorosilicate	: 11462
Ethyl chlorosulfate	: 4085
Ethyl 4-chlorovalerate	: 8867
Ethyl 5-chlorovalerate	: 8868
Ethyl trans-cinnamate	: 10685
Ethyl cinnamylideneacetate	: 8680
Ethyl crotonate	: 3838
Ethyl trans-crotonate	: 3839
cis-Ethyl crotonate	: 3840
Ethyl cyanate	: 4163
Ethyl cyanide	: 10090

Ethyl cyanoacetate	: 152
Ethylcyclobutane	: 4191
Ethyl cyclobutanecarboxylate	: 4175
Ethyl cycloheptanecarboxylate	: 4224
Ethylcyclohexane	: 4390
1-Ethylcyclohexanol	: 4522
4-Ethylcyclohexanol	: 4525
4-Ethylcyclohexanone	: 4593
3-Ethyl-1-cyclohexene	: 4661
4-Ethyl-1-cyclohexene	: 4662
2-Ethylcyclohexylamine	: 4287
Ethyl cyclohexyl ketone	: 10423
Ethylcyclopentane	: 4812
4-Ethylcyclopentene	: 4926
Ethylcyclopropane	: 4974
Ethyl cyclopropanecarboxylate	: 4961
Ethyl cyclopropylcarboxylate	: 4961
Ethyl 2-diazopropionate	: 10236
Ethyl dibromoacetate	: 160
Ethyl dibromofluoroacetate	: 161
Ethyl 2,3-dibromopentanoate	: 8872
Ethyl dibunate	: 7968
Ethyl dichloroacetate	: 184
Ethyl dichlorophosphate	: 9577
Ethyl 3-(diethylamino)propionate	: 328
Ethyl diethylphosphinate	: 9527
Ethyl difluoroacetate	: 176
9-Ethyl-9,10-dihydroanthracene	: 421
Ethyl 2,3-dimethyl-2-butenoate	: 3836
Ethyl 4,4-dimethylcrotonate	: 9080
3-Ethyl-2,2-dimethylhexane	: 6816
3-Ethyl-2,3-dimethylhexane	: 6817
3-Ethyl-2,4-dimethylhexane	: 6818
3-Ethyl-2,5-dimethylhexane	: 6819
3-Ethyl-3,4-dimethylhexane	: 6820
4-Ethyl-2,3-dimethylhexane	: 6822
4-Ethyl-2,4-dimethylhexane	: 6823
4-Ethyl-3,3-dimethylhexane	: 6824
Ethyl 2,2-dimethylpropionate	: 10259
Ethyl 3,5-dinitrobenzoate	: 2593
Ethylene	: 5879
Ethylene, 1-bromo-2-chloro-	: 5881
Ethylene, 1-bromo-1,2-dichloro (dis)	: 5935
Ethylene, 1-bromo-1,2-difluoro-	: 5883
Ethylene bromohydrin	: 5685
Ethylenecarboxamide	: 10519
Ethylenecarboxylic acid	: 10609
Ethylene chlorohydrin	: 5693
Ethylenediamine	: 5496
Ethylenediamine, N-benzyl-N',N'-dimethyl-N-phenyl-	: 5506
Ethylenediamine, N,N-dimethyl-N'-ethyl	: 5510
Ethylene dibromide	: 5518
Ethylene, 1,1-dibromo-2-ethoxy	: 5893
Ethylene dichloride	: 5529
Ethylene dicyanide	: 3291
Ethylene diethanoate	: 5571
Ethylenedimercaptan	: 5590
Ethylenedisulfonic acid	: 5587
Ethylene fluorohydrin	: 5727
Ethylene glycol	: 5569
Ethylene glycol carbonate	: 5211
Ethylene glycol diacetate	: 5571
Ethylene glycol dibutyl ether	: 5527
Ethylene glycol didodecanoate	: 5311
Ethylene glycol diethyl ether	: 5542
Ethylene glycol diisobutyl ether	: 10032
Ethylene glycol dilaurate	: 5311
Ethylene glycol dimethyl ether	: 5549
Ethylene glycol dinitrate	: 5574
Ethylene glycol, dinitrite	: 8086
Ethylene glycol, methyl ethyl ether	: 5591
Ethylene glycol, monoacetate	: 5577
Ethylene glycol monoallyl ether	: 5763
Ethylene glycol monobenzyl ether	: 5761
Ethylene glycol monobutyl ether	: 5687
Ethylene glycol monoethyl ether	: 5717
Ethylene glycol monoethyl ether acetate	: 5718
Ethylene glycol monomethyl ether	: 5733
Ethylene glycol monopentyl ether	: 5757
Ethylene glycol monosalicylate	: 2622
Ethylene glycol, monovalerate	: 8885
Ethyleneimine	: 498
Ethylene iodohydrin	: 5731
1,8-Ethylenenaphthalene	: 6

Ethylene nitrite	: 8086
Ethylene oxide	: 8599
1,4-Ethylenepiperazine	: 5111
Ethylenethiourea	: 7225
Ethyleneurea	: 7227
Ethyl 2-ethoxybenzoate	: 2598
Ethyl 4-ethoxybenzoate	: 2599
Ethylethynylcarbinol	: 9140
17β-Ethyletiocholane	: 9741
Ethyl fluoride	: 5593
N-Ethylformamide	: 5994
Ethyl formate	: 6007
2-Ethylfuran	: 6105
Ethyl 2-furyl ether	: 6103
γ-Ethyl L-glutamate	: 6245
Ethyl glycidyl ether	: 8614
3-Ethylheptane	: 6407
4-Ethylheptane	: 6408
Ethyl heptyl ketone	: 5076
2-Ethylhexanal	: 6719
3-Ethylhexane	: 6815
Ethyl hexanoate	: 6884
2-Ethylhexanoic acid	: 6882
3-Ethylhexanoic acid	: 6883
2-Ethyl-2-hexanol	: 6500
2-Ethyl-1-hexanol	: 6930
2-Ethyl-2-hexene	: 6555
3-Ethyl-1-hexene	: 7033
Ethyl 2-hexenoate	: 7053
Ethyl 3-hexenoate	: 7054
2-Ethylhexyl acetate	: 187
2-Ethylhexyl acrylate	: 10636
2-Ethyl-1-hexylamine	: 6726
Ethyl hexyl ketone	: 8173
2-Ethylhexyl methacrylate	: 10659
Ethyl hydrogen peroxide	: 7179
Ethyl hydrogen phosphonate, aluminum salt	: 6020
Ethyl hydrogen sulfate	: 11576
Ethyl 4-hydroxybenzoate	: 2621
Ethyl 2-hydroxy-3-butenoate	: 3843
N-Ethylhydroxylamine	: 5434
Ethyl 4-hydroxypentanoate	: 8884
Ethyl hypochlorite	: 7184
Ethylidenecyclohexane	: 4391
2-Ethylidenecyclohexanone	: 4594
Ethylidenecyclopentane	: 4813
Ethylidene diacetate	: 5570
Ethylidene dibromide	: 5517
Ethylidene dichloride	: 5528
Ethylidene diethyl ether	: 5541
1-Ethylindan	: 7273
Ethyl iodide	: 5597
o-Ethyliodobenzene	: 1680
Ethyl isobutanoate	: 10298
Ethyl isobutyl sulfide	: 10044
Ethyl isocyanate	: 5598
Ethyl isopentyl ether	: 3374
1-Ethyl-2-isopropylacetylene	: 7122
Ethylisopropylamine	: 9810
Ethyl isopropyl ether	: 10038
Ethyl isopropyl ketone	: 8982
(RS)-5-Ethyl-2-(4-isopropyl-4-methyl-5-oxo-2-imidazolin-2-yl)nicotinic acid	: 7190
Ethyl isopropyl sulfide	: 10043
Ethyl isothiocyanate	: 5600
Ethyl isovalerate	: 3548
Ethyl laurate	: 5312
Ethyl levulinate	: 8907
Ethyl linoleate	: 8226
N-Ethylmaleimide	: 11122
Ethyl mercaptan	: 5648
Ethyl methacrylate	: 10658
Ethyl methanesulfonate	: 7531
Ethyl methoxyacetate	: 206
1-Ethyl-1-methylallene	: 8666
Ethyl methyl amine, hydrochloride	: 5435
Ethyl 2-(methylamino)benzoate	: 2680
1-Ethyl-2-methylbenzimidazole	: 2381
2-Ethyl-3-methylbut-1-ene	: 3797
Ethyl methyl carbonate	: 4003
cis-1-Ethyl-3-methylcyclohexane	: 4397
cis-1-Ethyl-4-methylcyclohexane	: 4399
trans-1-Ethyl-4-methylcyclohexane	: 4400
1-Ethyl-2-methylcyclohexanol	: 4526
1-Ethyl-3-methylcyclohexanol	: 4527

2-Ethyl-2-methylcyclohexanone	: 4595
1-Ethyl-4-methylcyclohexene	: 4663
1-Ethyl-1-methylcyclopentane	: 4814
cis-1-Ethyl-2-methylcyclopentane	: 4815
cis-1-Ethyl-3-methylcyclopentane	: 4817
trans-1-Ethyl-3-methylcyclopentane	: 4818
2-Ethyl-5-methylcyclopentanone	: 4883
cis-1-Ethyl-2-methylcyclopropane	: 4976
Ethyl methyl ether	: 5602
N-Ethyl-N'-(1-methylethyl)-6-(methylthio)1,3,5-triazine-2,4-diamine	: 341
2-Ethyl-5-methylfuran	: 6106
3-Ethyl-3-methylglutaric acid	: 8777
3-Ethyl-4-methylheptane	: 6411
3-Ethyl-5-methylheptane	: 6412
4-Ethyl-2-methylheptane	: 6413
4-Ethyl-3-methylheptane	: 6414
4-Ethyl-4-methylheptane	: 6415
5-Ethyl-5-methyl-3-heptyne	: 6619
3-Ethyl-2-methylhexane	: 6825
3-Ethyl-3-methylhexane	: 6826
4-Ethyl-3-methylhexane	: 6827
3-Ethyl-4-methylhexane	: 6827
4-Ethyl-2-methylhexane	: 6828
Ethyl 3-methylhexanoate	: 6899
Ethyl 4-methylhexanoate	: 6900
2-Ethyl-3-methylhexanoic acid	: 6886
3-Ethyl-3-methylhexanoic acid	: 6887
Ethyl 4-methyl-5-oxohexanoate	: 6902
3-Ethyl-2-methylpentane	: 8813
3-Ethyl-3-methylpentane	: 8814
3-Ethyl-3-methyl-1-pentene	: 9047
3-Ethyl-2-methyl-2-pentene	: 9049
Ethyl 4-methyl-2-pentenoate	: 9080
Ethyl 3-methyl-3-pentenoate	: 9081
Ethyl 4-methyl-3-pentenoate	: 9082
3-Ethyl-3-methyl-1-pentyne	: 9132
1-Ethyl-1-methylpropylbenzene	: 1691
1-Ethyl-3-methylpyrazole	: 10833
2-Ethyl-3-methylquinoline	: 11269
2-Ethyl-4-methylquinoline	: 11270
Ethyl methyl sulfide	: 5606
2-Ethyl-3-methylthiophene	: 11761
2-Ethyl-5-methylthiophene	: 11762
N-Ethylmorpholine	: 7694
1-Ethylnaphthalene	: 7885
2-Ethylnaphthalene	: 7886
Ethyl 1-naphthylacetate	: 7735
Ethyl nitrate	: 8074
Ethyl nitrite	: 8087
Ethyl nitroacetate	: 218
O-Ethyl O-p-nitrophenyl benzenethiophosphonate	: 9551
2-Ethyl-2-nitro-1,3-propanediol	: 10008
N-Ethyl-N-nitrosoethanamine	: 5432
N-Ethyl-N-nitrosourea	: 12015
Ethyl nonanoate	: 8150
N-Ethylnorephedrine	: 1871
3-Ethyloctane	: 8403
4-Ethyloctane	: 8404
Ethyl octanoate	: 8428
Ethyl octyl ketone	: 11959
Ethyl oenanthate	: 6462
Ethyl orthoformate	: 5604
Ethyl 3-oxo-3-phenylpropanoate	: 2141
Ethylparaben	: 2621
3-Ethylpentane	: 8809
2-Ethyl-2-pentanol	: 6946
2-Ethyl-1-pentene	: 9043
3-Ethyl-1-pentene	: 9044
3-Ethyl-2-pentene	: 9045
4-Ethyl-3-pentenoic acid	: 7080
Ethyl pentyl ether	: 8808
3-Ethyl-1-pentyne	: 9131
9-Ethylphenanthrene	: 9179
o-Ethylphenol	: 9353
m-Ethylphenol	: 9354
p-Ethylphenol	: 9355
Ethyl phenoxyacetate	: 229
Ethyl 4-phenoxybutyrate	: 3581
Ethyl 2-(4-phenoxyphenoxy)ethylcarbamate	: 5952
Ethyl phenylacetate	: 899
Ethyl N-phenylcarbamate	: 3965
1-Ethyl-1-phenylhydrazine	: 7159
1-Ethyl-2-phenylhydrazine	: 7160

Ethyl 5-phenyl-2,4-pentadienoate	: 8680
Ethyl N-phenylurethane	: 3965
Ethyl phosphate	: 9563
3-Ethyl-4-picoline	: 10964
4-Ethyl-2-picoline	: 10965
5-Ethyl-2-picoline	: 10966
N-Ethylpiperidine	: 9667
4-Ethylpiperidine	: 9670
1-Ethyl-3-piperidinol	: 9715
5-Ethyl-5-(1-piperidyl)barbituric acid	: 11096
Ethyl 2-propenyl ether	: 10566
Ethylpropiolic acid	: 9137
Ethyl propionate	: 10263
1-Ethyl-2-propylacetylene	: 6611
1-Ethyl-4-propylbenzene	: 1703
1-Ethylpropyl chloride	: 8721
Ethyl propyl ether	: 10037
Ethyl propyl ketone	: 6960
Ethyl propyl sulfide	: 10042
2-Ethylpyridine	: 10956
3-Ethylpyridine	: 10957
4-Ethylpyridine	: 10958
2-Ethylquinoline	: 11266
3-Ethylquinoline	: 11267
6-Ethylquinoline	: 11268
Ethyl salicylate	: 2619
Ethyl silicate	: 11506
Ethyl succinate	: 3303
Ethyl sulfate	: 11576
Ethyl sulfite	: 11581
Ethyl sulfocyanate	: 11703
Ethylsulfonal	: 3230
Ethylsulfonic acid	: 5622
2-(Ethylsulfonyl)ethanol	: 5725
Ethylsulfonylethyl alcohol	: 5725
Ethyl 2,3,3,3-tetrafluoropropionate	: 10328
2-Ethyltetrahydrofuran	: 6107
3-Ethyltetrahydrothiophene	: 11786
1-Ethyltetralin	: 7888
2-Ethyltetralin	: 7889
6-Ethyltetralin	: 7891
Ethyl tetraphosphate	: 11645
Ethyl thiocyanate	: 11703
2-(Ethylthio)ethanol	: 5726
2-Ethylthiophene	: 11759
3-Ethylthiophene	: 11760
3-Ethylthiophenol	: 2297
o-Ethyltoluene	: 1685
m-Ethyltoluene	: 1686
p-Ethyltoluene	: 1687
Ethyl p-toluenesulfonate	: 2231
N-Ethyl-4-toluidine	: 748
Ethyl m-tolyl sulfide	: 1706
Ethyl tosylate	: 2231
Ethyl tribromoacetate	: 242
Ethyl trichloroacetate	: 248
Ethyl(trichloromethyl)carbinol	: 3643
Ethyltrichlorosilane	: 11463
Ethyltrifluorosilane	: 11432
Ethyltrimethoxysilane	: 11433
1-Ethyl-2,3,5-trimethylbenzene	: 1709
1-Ethyl-2,4,5-trimethylbenzene	: 1710
5-Ethyl-1,2,3-trimethylbenzene	: 1713
Ethyltrimethylethylene	: 9034
3-Ethyl-2,2,3-trimethylpentane	: 8815
3-Ethyl-2,3,4-trimethylpentane	: 8816
Ethyl undecanoate	: 11946
Ethyl undecylate	: 11946
Ethyl urethane	: 3948
O-Ethylurethane	: 3948
Ethyl valerate	: 8878
Ethylvanillin	: 555
Ethylvinylacetylene	: 7103
Ethyl vinyl ether	: 5905
Ethyl vinyl ketone	: 9104
3-Ethyl-o-xylene	: 1670
Ethynylbenzene	: 1715
α-Ethynylbenzyl alcohol	: 1876
α-Ethynylbenzyl carbamate	: 1877
1-Ethynyl-1-cyclopentanol	: 4866
Ethynyldimethylcarbinol	: 3922
Ethynylestradiol	: 8216
Ethynyl methyl ketone	: 3923
Ethynylmethylpropylcarbinol	: 7128
Ethynylsilane	: 11435
Etiocholanic acid	: 346
Etofylline	: 10758

Eucalyptol	: 8560
Eucarvone	: 4220
Eucazulen	: 505
4αH-Eudesma-5,11(13)-dien-12-oic acid, 8β-hydroxy-, γ-lactone	: 8059
Eugenol	: 9389
Euonymit	: 6194
Euparin	: 5825
Euprax	: 7229
Evasprin	: 1913
Evodiamine	: 7333
Exaltol	: 4760
Exaltolide	: 8562
Fagarine	: 6179
(E,E)-α-Farnesene	: 5329
α-Farnesene	: 5329
β-Farnesene	: 5331
Felbinac	: 3009
Fenamic acid	: 2715
Fenamisal	: 2502
Fenchane(d)	: 2931
α-Fenchene (DL)	: 2923
l-α-Fenchyl alcohol	: 2941
Fenibut	: 2128
Fenicol	: 1870
Fenpiprane	: 9658
Fentanyl	: 9801
Fenticlor	: 9477
Fenyramidol	: 1913
Ferrocene, 1,1'-bis(3-phenyl-1,1,3,3-tetramethyldisiloxanyl)	: 5961
Ferron	: 11295
Fisetin	: 2813
Flacitran	: 2814
Flavan	: 2782
Flavanone, 4'-methoxy-3',5,7-trihydroxy	: 2801
Flavanone, 3',5,7-trihydroxy-4'-methoxy-, 7-(6-O-α-L-rhamnosyl-D-glucoside)	: 2794
Flavone	: 2840
Flexin	: 2876
Flopropione	: 10489
Floxuridine	: 12034
Flufenamic acid	: 2729
Fluimucetin	: 5015
Fluometuron	: 12007
Fluonilid	: 11802
Fluorakil 100	: 63
3',6'-Fluorandiol	: 11522
Fluorene	: 5967
Fluorene, 2-bromo-	: 5968
9H-Fluorene-9-carboxylic acid, 2-chloro-9-hydroxy-	: 4079
Fluorene, 9-(2-chlorobenzylidene)	: 5970
Fluorene, 9-methyl-	: 5974
N-2-Fluorenylacetamide	: 64
Fluorescein	: 11522
Fluorescein, 4',5'-diiodo	: 11523
Fluorescin	: 2572
Fluorine cyanide	: 4167
Fluoroacetic acid amide	: 63
Fluoroacetic acid, methyl ester	: 192
Fluoroacetylene	: 5943
o-Fluoroaniline	: 752
m-Fluoroaniline	: 753
p-Fluoroaniline	: 754
Fluorobenzene	: 1717
4-Fluorobenzeneacetic acid	: 901
p-Fluorobenzoic acid	: 2605
Fluorobromomethane	: 7470
2-Fluoro-1,3-butadiene	: 3162
1-Fluorobutane	: 3378
2-Fluorobutane	: 3379
4-Fluoro-1-butanol	: 3621
4-Fluorobutyric acid	: 3927
Fluorochlorobromomethane	: 7466
α-Fluorocinnamic acid	: 10637
Fluorocyclohexane	: 4403
1-Fluorodecane	: 5052
10-Fluorodecanoic acid	: 5064
10-Fluoro-1-decanol	: 5074
Fluorodichloroacetonitrile	: 279
1-Fluoro-1,1-dichloroethane	: 5534
Fluorodichloromethane	: 7502
Fluorodiiodomethane	: 7516
Fluoroethanoic acid	: 189

Fluoroform	: 7558
Fluoroformaldehyde	: 6019
1-Fluorohexadecane	: 6639
1-Fluorohexane	: 6829
2-Fluorohexane	: 6830
9α-Fluoro-11β-hydroxy-17-methyltestosterone	: 355
Fluoroiodomethane	: 7517
Fluoromethane	: 7515
1-Fluoronaphthalene	: 7893
α-Fluoronaphthalene	: 7893
2-Fluoronaphthalene	: 7894
β-Fluoronaphthalene	: 7894
o-Fluoronitrobenzene	: 1733
m-Fluoronitrobenzene	: 1734
p-Fluoronitrobenzene	: 1735
3-Fluoro-4-nitrotoluene	: 1730
8-Fluorooctanoic acid	: 8429
1-Fluoropentane	: 8817
5-Fluoropentanoic acid	: 8879
(p-Fluorophenyl)acetic acid	: 901
p-Fluorophenylamine	: 754
Fluoroplex	: 11067
Fluoroprene	: 3162
2-Fluoropropane	: 10047
3-Fluoro-1-propanol	: 10382
2-Fluoropropene	: 10569
3-Fluoropropene	: 10570
cis-1-Fluoropropene	: 10567
trans-1-Fluoropropene	: 10568
4-(Fluorosulfonyl)aniline	: 2252
o-Fluorotoluene	: 1726
m-Fluorotoluene	: 1727
p-Fluorotoluene	: 1728
Fluorotrichloromethane	: 7555
5-Fluorouracil	: 11067
5-Fluorouracil 2'-deoxyriboside	: 12034
Fluoxymesterone	: 355
Fluracil	: 11067
Folic acid	: 6243
Foligan	: 10852
Folinic acid	: 6244
Follestrol	: 5406
Follicyclin	: 5399
Formaldehyde, (2,2'-dichloroisopropyl) ethyl acetal	: 9896
Formaldehyde dimethyl mercaptal	: 7461
Formaldehyde, trimer	: 11905
Formalin	: 5987
Formamide, 1,1'-azobis-	: 5124
Formamide, N,N'-[1,4-piperazinediylbis(2,2,2-trichloroethylidene)]bis-	: 11902
Formamidine	: 7563
Formanilide	: 5999
Formic acid, isopropyl ester	: 6011
Formononetin	: 2830
N-Formylaniline	: 5999
Formylcyclooctane	: 4738
Formylformic acid	: 223
2-Formylfuran	: 6041
Formylpteroic acid	: 2494
4-Formylquinoline	: 11225
N5-Formyl-5,6,7,8-tetrahydrofolic acid	: 6244
5-Formyl-5,6,7,8-tetrahydrofolic acid	: 6244
3-Formylthiophene	: 11734
1-Formyl-2,6,6-trimethyl-1-cyclohexene	: 4636
Fraxetin	: 2808
Fraxin	: 2823
Freon 11	: 7555
Freon 12	: 7500
Freon 12B1	: 7464
Freon 12B2	: 7496
Freon 13	: 7486
Freon 13B1	: 7475
Freon 13I1	: 7559
Freon 14	: 7541
Freon 21	: 7502
Freon 22	: 7478
Freon 23	: 7558
Freon 31	: 7481
Freon 32	: 7504
Freon 41	: 7515
Freon 111	: 5618
Freon 112	: 5637

Freon 112a	: 5618
Freon 113	: 5663
Freon 114	: 5538
Freon 114a	: 5537
Freon 114B2	: 5525
Freon 115	: 5484
Freon 116	: 5596
Freon 121	: 5638
Freon 122	: 5659
Freon 123	: 5540
Freon 132	: 5532
Freon 133a	: 5489
Freon 152	: 5545
Freon 215	: 10179
Freon 218	: 10118
Freon C 318	: 4194
Freon 1110	: 5917
Freon 1111	: 5923
Frigen 11	: 7555
Frigen 11A	: 7555
Frigen 12	: 7500
Frigen 13	: 7486
Frigen 22	: 7478
Frigen 113	: 5663
Frigen 113A	: 5663
Frigen 114	: 5538
Frigen 114A	: 5537
Frigen 122	: 5659
Fuchsin, acid	: 2218
Fucoxanthin	: 4055
Fulvene	: 4767
Fulvene, 6-vinyl(trans)	: 4769
Fumaric acid	: 3764
Fumaric acid, bromo-	: 3770
Fumaric acid, diallyl ester	: 3781
Fumaric acid dichloride	: 3795
Fumaric acid, diisobutyl ester	: 3767
Fumaric acid, diisopropyl ester	: 3766
Fumaric acid, dipropyl ester	: 3783
Fumarine	: 3109
Fumaronitrile, chloro-	: 3763
Fumaryl chloride	: 3795
Fumigacin	: 8211
Fumigatin	: 4265
Furacrylic acid	: 10638
2-Furalacetic acid	: 10638
Furaldehyde	: 6041
2-Furanacrylic acid	: 10638
2-Furanacrylonitrile	: 10585
Furan, 2-tert-butyl-	: 6090
2-Furancarbinol	: 6118
Furan, 2(2-cyanovinyl)	: 10585
2-Furanmethanimine, 5-nitro-N-(2-oxo-3-oxazolidinyl)-	: 8589
2-Furanpropionic acid, tetrahydro-	: 6164
Furan, tetrahydro-2-ethyl	: 6107
2-Furanylmethanol	: 6118
Furazolidone	: 8589
Furfural	: 6041
Furfuran	: 6025
Furfuryl alcohol	: 6118
Furfuryl alcohol, tetrahydro-, stearate	: 8275
Furfurylamine	: 6113
Furfurylideneacetic acid	: 10638
Furfuryl phenyl ether	: 6158
Furfuryl valerate	: 8880
Furilazone	: 10904
α-Furildioxime	: 5583
2-Furoic acid	: 6050
β-Furoic acid, 5-bromo, ethyl ester	: 6190
2-Furoic acid, 5-chloro-, ethyl ester	: 6053
α-Furoic acid, hexyl ester	: 6058
3-Furoic acid, 5-methyl-	: 6065
α-Furoic acid, octyl ester	: 6067
2-Furoic acid, pentyl ester	: 6068
Furole	: 6041
Furonazide	: 10904
2-Furylacetic acid, butyl ester	: 6027
2-Furylacetonitrile	: 6031
3-(2-Furyl)acrylonitrile	: 10585
2-Furylcarbinol	: 6118
2-Furyl iodide	: 6110
2-Furyl isopropyl ether	: 6112
2-Furyl methyl ketone isonicotinoylhydrazone	: 10904
2-Furyl octyl ether	: 6135

2-Furyl phenyl ether	: 6157
2-Furyl phenyl sulfide	: 6160
Fusaric acid	: 10898
Fussol	: 63
Gainex	: 2384
Galegine	: 6283
Galipine	: 11253
Gallacetophenone	: 5873
Gallamine triethiodide	: 5441
Gallein	: 11526
Gallic acid	: 2731
Gallic acid, 3-gallate	: 2575
Gallodesoxycholic acid	: 4097
Gallomide blue	: 8062
m-Galloylgallic acid	: 2575
Gamarex	: 3464
Gardol	: 6269
Gastrodigenin	: 1884
Geissoschizoline	: 4154
Gelseminic acid	: 2829
Genetron 132b-B2	: 5522
Genetron 11	: 7555
Genetron 23	: 7558
Genetron 218	: 10118
Genistein	: 2807
Genistein, 4'-β-D-glucopyranoside	: 2824
Genisteol	: 2807
Genite	: 9289
Gentianic acid	: 12054
Gentisic acid	: 2570
Gentisin	: 12054
Gentisyl alcohol	: 1520
trans-Geraniol	: 8333
Geraniol, tetrahydro (DL)	: 8449
Geranyl alcohol	: 8333
trans-Geranyl bromide	: 8321
Germine	: 4068
Geronic acid	: 6460
Gibberellic acid	: 6204
Gibberellin A3	: 6204
Gibberellin X	: 6204
Gibbs' reagent	: 4280
Gingerone	: 3663
Gitogenin	: 11530
Gitoxigenin	: 4048
Gitoxin	: 4042
d-Glaucine	: 5157
Glucamine	: 6207
Gluconic acid	: 6209
D-Gluconic acid-1,5-lactone	: 6212
Gluconic acid, sodium salt	: 6213
δ-D-Gluconolactone	: 6212
1,5-Gluconolactone	: 6212
Glucopyranoside, genistein-4', β-D-	: 2824
D-Glucosaccharic acid	: 6205
D-Glucosamine	: 6233
L-Glucose, methylamino	: 6215
α-D-Glucose, pentamethyl ether	: 6228
Glucurone	: 6239
D-Glucuronolactone	: 6239
Glutaconic acid, diethyl ester	: 9041
α-Glutamic acid	: 6241
Glutamic acid 5-amide	: 6249
Glutamic acid, N-[p-[[(2-amino-5-formyl-5,6,7,8-tetrahydro-4-hydroxy-6-pteridinyl)methyl]amino]benzoyl]-, L-	: 6244
Glutamic acid, 5-[2-(α-hydroxy-p-tolyl)hydrazide], L-	: 6247
Glutamic acid, monosodium salt	: 6248
γ-L-Glutamyl-L-cysteinylglycine	: 6264
Glutaral	: 8734
Glutaraldehyde	: 8734
Glutaric acid	: 8770
Glutaric acid, 3(3-benzoxy-4-methoxyphenyl)	: 6892
Glutaric acid dialdehyde	: 8734
Glutaric acid dinitrile	: 8767
Glutaric acid, diphenyl ester	: 8776
Glutaric acid, 3-ethyl-3-methyl-	: 8777
Glutaric acid, 3-hydroxy-, diethyl ester	: 8778
Glutaric acid, 3-methyl	: 8779
Glutaronitrile	: 8767
Glutatione	: 6264
Glycamine	: 6207
Glyceric acid	: 10252

Glyceric acid, ethyl ester (DL)	: 10253
Glyceric acid, methyl ester, DL-	: 10254
Glyceric acid, 2-methyl, ethyl ester	: 10255
Glyceric acid, 2-methyl-, methyl ester	: 10256
Glycerin	: 10186
Glycerin tributyrate	: 3586
Glycerol	: 10186
Glycerol acetonide	: 5207
Glycerol β-chlorohydrin	: 9995
Glycerol diacetate	: 10189
Glycerol 1,3-dimethyl ether	: 10372
Glycerol, 1,3-diphenyl ether-2-acetate	: 10378
Glycerol guaiacolate	: 10016
Glycerol, 1-laurate (DL)	: 5310
Glycerol, 1-(2-methoxyphenyl)ether	: 10016
Glycerol, 2-methyl ether	: 10015
Glycerol, 1-monoacetate(DL)	: 10187
Glycerol, 1-monododecanoate(DL)	: 5310
Glycerol, 1-phenyl	: 10194
Glycerol tributanoate	: 3586
Glycerol trimethyl ether	: 10185
Glyceryl p-aminobenzoate	: 10188
Glycidaldehyde	: 8603
Glycidol	: 8618
Glycidyl allyl ether	: 8626
Glycine anhydride	: 9612
Glycine, N,N-bis(phosphonomethyl)-	: 6277
Glycine, N-(chloroacetyl)-N-(2,6-diethylphenyl)-, ethyl ester	: 5173
Glycine cholate	: 6273
Glycine, N-(phosphonomethyl)-	: 6276
Glycinol	: 5675
Glycinonitrile, N,N-dimethyl-	: 281
Glycocholic acid	: 6273
Glycocyamine	: 6253
Glycolaldehyde	: 15
Glycol diacetate	: 5571
Glycolic acid	: 196
Glycolic aldehyde	: 15
Glycol salicylate	: 2622
Glyconitrile	: 282
Glycopyrrolate	: 11179
Glycopyrronium bromide	: 11179
Glycosine	: 11200
Glycosthene	: 6250
Glycurone	: 6239
Glycyl alcohol	: 10186
α-Glycylglycine	: 6265
Glycyrrhetinic acid glycoside	: 6230
Glycyrrhizic acid	: 6230
Glycyrrhizin	: 6230
Glyoxal	: 5491
Glyoxaldiureine	: 7191
Glyoxaline-5-alanine	: 7132
Glyoxyaldehyde	: 5491
Glyoxylic acid	: 223
Glyoxylic diureide	: 12009
Gramine	: 7310
Grasex	: 21
Griseofulvin, (+)-	: 11520
Guaiacol	: 9379
Guaiacol benzoate	: 9383
Guaiacol carbonate	: 9384
Guaiacol glycerol ether	: 10016
Guaiacol phosphate	: 9387
Guaiacol valerate	: 8886
Guaiazulene	: 505
Guai-1(5)-en-11-ol	: 507
Guaifenesin	: 10016
Guaiol	: 507
Guajen	: 7825
Guanidine, dodecyl-, monoacetate	: 5347
Guanidinoacetic acid	: 6253
Guanine	: 10768
Guanyl glycine	: 6253
Guanyl hydrazine	: 7147
Gulose, D-	: 6290
Guvacine	: 10912
Gynecormone	: 5403
Halazone	: 2565
Halon 2302	: 5528
Haloperidol	: 3651
Halotestin	: 355
Halothane	: 5458
Harmaline	: 11038
Harman	: 11040
Harmine	: 11039
Hecogenin	: 11534
Hederagenin	: 8545
Helenin	: 8059
Helicin	: 562
Heliotropic acid	: 2418
Heliotropin	: 2417
Helvolic acid	: 8211
Hematein	: 628
Hematin	: 5959
Hematoidin	: 3000
Hematoporphyrin	: 9735
Hematoxylin	: 629
Hemimellitic acid	: 2318
Hemin	: 5958
Hemipyocyanine	: 9197
Hendecane	: 11938
Heneicosylbenzene	: 1741
Hepin	: 6873
Heptachlor	: 7576
1,1,1,2,2,3,3-Heptachloropropane	: 10048
1,1,1,2,3,3,3-Heptachloropropane	: 10049
Heptacosylbenzene	: 1742
Heptadecane, 1-cyclohexyl-	: 4404
Heptadecane, 9-phenethyl-	: 2070
Heptadecane, 1-phenyl-	: 1743
Heptadecyl alcohol	: 6317
Heptadecylbenzene	: 1743
Heptadecylcyclohexane	: 4404
1,6-Heptadiene-4-ol, 4-methyl	: 6338
1,6-Heptadiene-3-yne	: 6341
Heptaldehyde	: 6344
Heptamethylcyclotetrasiloxane	: 4992
Heptamethylene chlorohydrin	: 6477
Heptamethylene dibromide	: 6374
Heptane, 1-cyclopentyl-	: 4820
Heptane, 1,1-diphenyl-	: 1746
Heptane, 4-isopropyl-	: 6427
Heptane, 2-methyl-3-ethyl	: 6409
4-Heptanol, 3-methyl(DL)	: 6504
4-Heptanol-2-one, 3-methyl	: 6518
3-Heptanol-5-one, 3-methyl	: 6519
4-Heptanol, 4-phenyl-	: 1867
1-Heptanol, 1-phenyl-	: 1881
3-Heptanol, 7-phenyl	: 2102
cis-2-Heptene	: 6535
trans-3-Heptene	: 6536
cis-3-Heptene	: 6537
1-Heptene, 1,1-diphenyl-	: 1744
2-Heptene, 3-methyl (trans)	: 6549
2-Heptene, 3-methyl (cis)	: 6550
2-Heptene, 4-methyl (trans)	: 6551
2-Heptene, 4-methyl (cis)	: 6552
1-Heptene-4-ol	: 6572
1-Heptene-4-ol, 4-methyl	: 6573
cis-2-Hepten-1-ol	: 6568
1-Hepten-5-one	: 6586
3-Hepten-2-one (cis)	: 6588
2-Hepten-6-one, 4-ethyl-2-methyl	: 6593
6-Hepten-4-yne-3-ol, 3-ethyl	: 6608
Heptyl acetate	: 193
Heptylacetylene	: 8196
Heptyl alcohol	: 6472
Heptylamine	: 6349
Heptyl bromide	: 6356
2-Heptyl bromide	: 6357
4-Heptyl bromide	: 6358
Heptyl chloride	: 6360
Heptylcyclopentane	: 4820
Heptyl formate	: 6008
Heptyl heptanoate	: 6464
Heptylidene acetone	: 5093
Heptyl mercaptan	: 6436
Heptyl methyl ketone	: 8172
1-Heptylnaphthalene	: 7895
2-Heptylnaphthalene	: 7896
Heptyl nitrite	: 8088
Heptyl propyl ketone	: 11960
Heptyl thiocyanate	: 11704
2-Heptynyl bromide	: 6613
4-Heptynyl chloride	: 6616
Herbicide 634	: 4900
Hesperetin	: 2800
Hesperetin 7-rutinoside	: 2794
Hesperidin	: 2794
Hexabromoethane	: 5594
Hexachlorane	: 4409
Hexachloroacetone	: 10440
Hexachlorobenzene	: 1749
Hexachloro-1,3-butadiene	: 3163
γ-Hexachlorocyclohexane	: 4409
Hexachloroethane	: 5595
1,2,3,4,5,6-Hexachloro-3-hexene	: 7035
Hexachlorophene	: 9409
1,1,1,2,2,3-Hexachloropropane	: 10051
1,1,1,3,3,3-Hexachloropropane	: 10052
1,1,2,2,3,3-Hexachloropropane	: 10053
Hexachloropropylene	: 10571
Hexacosane, 1-phenyl-	: 1750
Hexacose	: 6707
Hexacosylcyclohexane	: 4410
Hexacosylcyclopentane	: 4821
Hexadecamethylcyclooctasiloxane	: 4741
Hexadecamethylene dibromide	: 6635
Hexadecamethylheptasiloxane	: 6528
Hexadecanal dimethyl acetal	: 6636
Hexadecane, 1-cyclohexyl-	: 4411
Hexadecane, 1-cyclopentyl-	: 4822
Hexadecylcyclopentane	: 4822
Hexadecyl 3-hydroxy-2-naphthoate	: 7774
Hexadecyl mercaptan	: 6643
Hexadecyl pyridinium chloride	: 11030
1,5-Hexadiene, 2-methyl-3-ethyl	: 6694
1,3-Hexadiene, 1-phenyl	: 1753
1,5-Hexadiene, 1-phenyl-	: 1754
1,5-Hexadiene, 6-phenyl	: 1755
1,5-Hexadiene, 3-phenyl	: 1625
1,5-Hexadiene-3-yne	: 6713
1,3-Hexadien-5-ol	: 6708
1,5-Hexadienylbenzene	: 1754
Hexaethylbenzene	: 1756
1,1,1,3,3,3-Hexaethyldisiloxane	: 5235
Hexafluorenium bromide	: 6753
Hexafluoro-2,4-pentanedione	: 8802
Hexafluoropropylene oxide	: 8629
1,2,3,4,5,6-Hexahydroanthracene	: 422
Hexahydroaplotaxene	: 6322
Hexahydroazepine	: 492
Hexahydrobenzene	: 4305
Hexahydrobenzoic acid	: 4332
Hexahydrobenzyl alcohol	: 4419
Hexahydroequilenin	: 5402
Hexahydroindan	: 7281
Hexahydronicotinic acid	: 9641
Hexahydrophthalic anhydride	: 7345
Hexahydropyrazine	: 9602
3,3',4',5,5',7-Hexahydroxyflavylium chloride	: 2849
3,3',4',5,5',7-Hexahydroxy-2-phenylbenzopyrylium chloride	: 2849
Hexamethylbenzene	: 1758
Hexamethylcyclohexasiloxane	: 4625
1,1,3,3,5,5-Hexamethylcyclotrisilazane	: 5001
Hexamethylcyclotrisiloxane	: 5002
Hexamethyldisilane	: 5228
Hexamethyldisilazane	: 11363
Hexamethyldisiloxane	: 5236
Hexamethylenediamine	: 6749
Hexamethylene diiodide	: 6769
Hexamethylene diisocyanate	: 6770
Hexamethylene glycol	: 6796
Hexamethylenetetramine, allyl iodide	: 11825
Hexamethylenimine	: 492
Hexamethylolmelamine	: 7584
Hexane, 1-cyclohexyl-	: 4414
Hexane, 1-cyclopentyl-	: 4823
Hexane, 2,5-dibromo(DL)	: 6759
3,4-Dibromohexane	: 6760
Hexane, 2,5-dimethyl-2,5-diphenyl-	: 2281
Hexane, 2,2-dimethyl-3-ethyl	: 6816
Hexane, 2,2-dimethyl-4-ethyl	: 6821
Hexane, 2,3-dimethyl-3-ethyl	: 6817
Hexane, 2,3-dimethyl-4-ethyl	: 6822
Hexane, 2,4-dimethyl-3-ethyl	: 6818
Hexane, 2,4-dimethyl-4-ethyl	: 6823
Hexane, 2,5-dimethyl-3-ethyl	: 6819
Hexane, 3,3-dimethyl-4-ethyl	: 6824
Hexane, 3,4-dimethyl-3-ethyl	: 6820
Hexane, 2-iodo (l)	: 6832
Hexane, 3-isopropyl-2-methyl-	: 6839
Hexane, 2-methyl-3-isopropyl	: 6839

Hexane, 3-phenyl-	: 1669
Hexane, 1-phenyl-	: 1763
Hexane, 2-phenyl-	: 2009
5-Hexanoic acid	: 7051
Hexanoic acid, 2-amino(DL)	: 8215
Hexanoic acid, 6-cyclohexyl	: 4412
Hexanoic acid, 2-methyl(DL)	: 6893
Hexanoic acid, 2,2,3,4,4,5,6,6-octafluoro-3,5,6-trichloro	: 6910
3-Hexanol (l)	: 6915
3-Hexanol, 3,5-dimethyl (DL)	: 6928
1-Hexanol, 2-isopropyl-5-methyl-	: 6948
1-Hexanol, 2-methyl (DL)	: 6937
1-Hexanol, 4-methyl (DL)	: 6940
3-Hexanol, 5-methyl (DL)	: 6947
2-Hexanol-3-one, 2-methyl	: 6973
2-Hexanol-4-one, 5-methyl	: 6974
2-Hexanol, 2-phenyl-	: 1847
3-Hexanol, 3-phenyl-	: 1875
2-Hexanol, 6-phenyl-	: 2103
1-Hexanol, 3-phenyl-	: 2159
Hexatriacontylcyclohexane	: 4413
Hexazole	: 11848
trans-2-Hexene	: 7000
cis-2-Hexene	: 7001
3-Hexene, 3-chloro (cis)	: 7009
1-Hexene, 1,2-dichloro (trans)	: 7011
1-Hexene 1,2-dichloro (cis)	: 7012
3-Hexene, 2,4-dimethyl (cis)	: 7031
3-Hexene, 3-(4-hydroxyphenyl)-4-(4-methoxyphenyl)	: 9358
5-Hexene-2-one,5-methyl	: 7099
2-Hexene, 1-phenyl	: 1760
2-Hexene, 2-phenyl	: 2008
2-Hexene-4-yne	: 7106
4-Hexenoic acid, 6-(4-hydroxy-6-methoxy-7-methyl-3-oxo-5-phthalanyl)-4-methyl-, (E)-	: 7052
2-Hexenoic acid, 3-phenyl-	: 4135
2-Hexen-4-ol, 5,5-dimethyl	: 7074
2-Hexen-4-ol, 2,4-dimethyl	: 7076
2-Hexen-5-one	: 7087
2-Hexen-5-one, 3,4-dimethyl	: 7092
3-Hexen-1-yl chloride	: 7008
Hex-5-en-1-yne	: 7105
Hexestrol	: 9298
Hexobarbital	: 11083
Hexocyclium methyl sulfate	: 9623
Hexylacetylene	: 8529
Hexylamine	: 6723
Hexylamine, 3-methyl-	: 6728
Hexylamine, 4-methyl-	: 6729
Hexyl bromide	: 6732
1-Hexyl-1,3-butadiene	: 5023
3-Hexylbutanol-1	: 8166
Hexyl chloride	: 6737
2-Hexyl chloride	: 6738
2-Hexyldecanoic acid	: 5065
7-Hexyldocosane	: 5274
Hexylene glycol	: 8793
Hexyl fluoride	: 6829
Hexyl formate	: 6009
9-Hexylheptadecane	: 6309
Hexyl heptanoate	: 6465
Hexyl iodide	: 6831
Hexyl mercaptan	: 6856
Hexyl methyl ether	: 6833
Hexyl methyl ketone	: 8471
1-Hexylnaphthalene	: 7900
Hexyl nitrite	: 8089
1-Hexylpiperidine	: 9672
Hexyl propyl ketone	: 5077
4-Hexylresorcinol	: 1519
Hexyltrichlorosilane	: 11464
Hexylvinylcarbinol	: 8189
Hexyl vinyl ether	: 6813
3-Hexyne, 2,5-dimethyl-2,5-dichloro	: 7117
3-Hexyne-2-one	: 7130
3-Hexynol	: 7126
Hippuric acid	: 6255
Hiragonic acid	: 6666
Histamine	: 7203
Histantin Richter	: 9502
Holocaine, hydrochloride	: 5670
Homatropine	: 911
Homochlorcyclizine	: 5140
Homochlorocyclizine	: 5140

Homogentisic acid	: 895
Homomesityl oxide	: 6600
o-Homosalicylic acid	: 2629
m-Homosalicylic acid	: 2630
p-Homosalicylic acid	: 2631
Homovanillic acid	: 913
Hordenine	: 9320
Humulene	: 5006
Humulon	: 4282
Hydantoin	: 7216
Hydantoin, 3-allyl-5-isobutyl-2-thio-	: 7229
Hydnocarpic acid	: 4940
Hydracrylic acid	: 10267
Hydracrylic acid, isopropyl ester	: 10277
Hydracrylonitrile	: 10106
Hydrangin	: 2826
Hydrastine	: 7357
Hydrastinine	: 5217
Hydrazinecarboxamide, N-amino-	: 4009
Hydrazinecarboxamide, 2-phenyl-N-(phenylimino)-	: 5122
Hydrazinecarboxylic acid, hydrazide	: 4009
Hydrazine, 1-isopentyl-1-phenyl	: 7162
Hydrazine, 1-methyl-2-(3-tolyl)	: 7163
Hydrazine, 1-phenyl-2-(3-tolyl)	: 7174
2-Hydrazinoethanol	: 5729
Hydrindantin	: 2999
Hydrocarbostyril	: 11327
Hydrochlorothiazide	: 2855
Hydrochol	: 4101
Hydrocinchonidine	: 4121
Hydrocinchonine	: 4122
Hydrocinnamaldehyde, p-isopropyl-α-methyl-	: 2117
Hydrocinnamaldehyde, o-methyl-	: 2115
Hydrocinnamic acid	: 2126
Hydrocinnamic acid, α-bromo-	: 2147
Hydrocinnamic acid, α,α-dimethyl-, methyl ester	: 2129
Hydrocinnamic acid, ester with 17β-hydroxyestr-4-en-3-one	: 5408
Hydrocinnamic acid, 4-(4-hydroxy-3-iodophenoxy)-3,5-diiodo-	: 2136
Hydrocinnamic aldehyde	: 2114
Hydrocinnamonitrile	: 2123
Hydrocinnamyl alcohol	: 2148
Hydrocodone	: 7678
Hydrocortisone	: 9753
Hydrocortisone acetate	: 9747
Hydrocotarnine	: 5214
Hydroflumethiazide	: 2858
Hydrofuramide	: 7488
Hydrogen carboxylic acid	: 6002
Hydrogen cyanate (HOCN)	: 4162
Hydrogen cyanide	: 7176
Hydrohydrastinine	: 5215
Hydroquinidine	: 4124
Hydroquinine	: 4123
p-Hydroquinone	: 1499
Hydroquinonecarboxylic acid	: 2570
Hydrosorbic acid	: 7050
Hydrosorbic acid, ethyl ester	: 7054
Hydrosorbic acid, 2-methyl-	: 7058
Hydrotrichlorothiazide	: 2854
2-Hydroxyacetaldehyde	: 15
Hydroxyacetic acid, propyl ester	: 199
4-Hydroxy-3-aminobutyric acid	: 3465
γ-Hydroxy-β-aminobutyric acid	: 3465
1-Hydroxy-3-aminomethyl-3,5,5-trimethylcyclohexane	: 4502
Hydroxyamphetamine	: 9215
2-Hydroxyaniline	: 9201
3-Hydroxyaniline	: 9202
4-Hydroxyaniline	: 9203
o-Hydroxyanisole	: 9379
p-Hydroxybenzaldehyde	: 566
Hydroxybenzene	: 9199
3-Hydroxybenzoic acid, 4-formyl	: 9598
3-Hydroxybenzoic acid, 6-iodo	: 2624
4-Hydroxybenzoic acid, 3-iodo	: 2623
2-Hydroxybenzoxazole	: 2881
o-Hydroxybenzyl alcohol	: 1882
p-Hydroxybenzyl alcohol	: 1884
2-Hydroxybiphenyl	: 3088
2-Hydroxy-3-butenoic acid	: 3842
Hydroxy-tert-butylamine	: 10341

N-(2-Hydroxybutyl)-N-(2-hydroxyethyl)amine	: 3623
3-Hydroxybutyraldehyde	: 3184
4'-Hydroxybutyranilide	: 3205
3-Hydroxycamphor	: 2957
3-Hydroxy-4-carboxyaniline	: 2498
3-Hydroxy-β-carotene	: 4061
2-Hydroxy-5-chloroaniline	: 9204
2-Hydroxy-3-chlorobiphenyl	: 3092
3α-Hydroxy-5α-cholestane	: 4108
3β-Hydroxy-5β-cholestanol	: 4109
3β-Hydroxycholest-5-ene	: 4110
3β-Hydroxycholest-4-ene	: 4111
DL-α-Hydroxycinnamic acid	: 2133
o-Hydroxycinnamic acid lactone	: 2792
Hydroxycodeinone	: 7675
7-Hydroxycoumarin 7-glucoside	: 2821
1-Hydroxycumene	: 1857
Hydroxycyclobutane	: 4196
Hydroxycyclopentane	: 4864
3-Hydroxy-2-p-cymenecarboxylic acid	: 2642
2-Hydroxy-1,4-dibromobutane	: 3614
12β-Hydroxydigitoxin	: 4041
2-Hydroxy-3,5-diiodobenzoic acid	: 2617
1-Hydroxy-2,11-dimethoxy-6aα-aporphine	: 5158
Hydroxydimethylphosphine oxide	: 9528
3-Hydroxy-1,1-dimethylpyrrolidinium bromide α-cyclopentylmandelate	: 11179
4-Hydroxydiphenyl	: 3090
3-Hydroxy-1,5-diphenylformazan	: 5122
16β-Hydroxyestradiol	: 5405
2-Hydroxyethanamine	: 5675
Hydroxyethanoic acid	: 196
2-Hydroxyethyl acetate	: 5577
2-Hydroxyethyl 2-hydroxybutyl amine	: 3623
1-Hydroxyethyl-4-methoxybenzene	: 1895
2-(2-Hydroxyethyl)piperidine	: 9663
4-(2-Hydroxyethyl)piperidine	: 9664
α-(2-Hydroxyethyl)styrene	: 2157
4-Hydroxy-1,6-heptadiene	: 6337
3-Hydroxy-1,5-hexadiene	: 6706
1-Hydroxy-2,4-hexadiene	: 6707
3-Hydroxy-2,5-hexanedione	: 6808
5-Hydroxy-2-hexanone	: 6969
5-Hydroxy-3-hexanone	: 6972
p-Hydroxyhydrocinnamic acid	: 2134
Hydroxyhydroquinone	: 2367
(Hydroxyimino)cyclohexane	: 4610
2-Hydroxyisobutyronitrile	: 10108
4-Hydroxyisophthalic acid	: 1352
β-Hydroxylactic acid	: 10252
Hydroxylamine, N,N-dipropyl-	: 9814
Hydroxylamine, N-ethyl-	: 5434
Hydroxylamine, N-methyl	: 7447
6α-Hydroxylithocholic acid	: 4096
Hydroxylupanine	: 7574
3-Hydroxy-5-methoxybenzaldehyde	: 573
1-Hydroxy-2-methoxybenzene	: 9379
4-Hydroxy-3-methoxybenzylacetone	: 3663
7-Hydroxy-4'-methoxyisoflavone	: 2830
3-Hydroxy-2-methoxy-5-methyl-p-benzoquinone	: 4265
(4-Hydroxy-3-methoxyphenyl)ethyl methyl ketone	: 3663
3-Hydroxy-4-methoxy-2,5-toluquinone	: 4265
p-Hydroxy-α-[(methylamino)-methyl]benzyl alcohol	: 1887
2-Hydroxymethylbenzofuran	: 2455
3-Hydroxy-3-methyl-1-butyne	: 3922
7-Hydroxy-4-methylcoumarin	: 2831
5-Hydroxymethylcytosine	: 11105
5-(Hydroxymethyl)-2-furaldehyde	: 6044
4-Hydroxy-4-methylheptane	: 6505
p-(Hydroxymethyl)nitrobenzene	: 1908
o-(Hydroxymethyl)phenol	: 1882
p-(Hydroxymethyl)phenol	: 1884
2'-Hydroxy-4'-methylpropiophenone	: 10445
3-Hydroxy-2-methylundecane	: 11956
1-Hydroxy-2-naphthoic acid	: 7771
3-Hydroxy-2-naphthoic acid	: 7772
2-Hydroxy-1-nitrobutane	: 3641
4-Hydroxynonanoic acid lactone	: 6149
2'-Hydroxypelargidenolon 1522	: 2816

4-Hydroxy-1-pentanol	: 8786	Indan, 4-methyl-	: 7257	Iodopyrrole	: 11140
1-Hydroxy-2-pentanone	: 8971	Indan, 5-methyl-	: 7258	Iodoquinol	: 11315
2-Hydroxy-3-pentanone	: 8975	2-Indanone	: 7287	4-Iodoquinoline	: 11272
3-Hydroxy-1-pentyne	: 9140	1-Indanone, 4-methyl-	: 7290	Iodostarin	: 8300
1-Hydroxypent-2-yn-4-ene	: 9118	1-Indanone, 2-phenyl-	: 7291	N-Iodosuccinimide	: 11159
3-Hydroxy-9,10-phenanthraquinone	: 9176	Indan, 2-phenyl-	: 7259	Iodotrifluoroethylene	: 5925
2-(β-Hydroxyphenethyl-amino)pyridine	: 1913	Indan, 1-phenyl-1,3,3-trimethyl	: 7263	Iodotrifluoromethane	: 7559
Hydroxyphenylamine	: 9200	Indan, 1,1,4,7-tetramethyl-	: 7260	3-Iodotyrosine	: 11932
o-Hydroxyphenylamine	: 9201	Indanthrene	: 434	3-Iodoveratrole	: 1767
m-Hydroxyphenylamine	: 9202	Indan, 1,1,5-trimethyl-	: 7261	4-Iodo-o-xylene	: 1774
N-(4-Hydroxyphenyl)glycine	: 6267	Indan, 1,4,7-trimethyl	: 7262	β-Ionol	: 3873
1-Hydroxy-3-phenylpropane	: 2148	1-Indenecarboxylic acid	: 7246	α-Ionol	: 3874
1-(m-Hydroxyphenyl)-1-propanol	: 1872	Indene, 1,3-diphenyl-	: 7271	Irene	: 7999
DL-2-Hydroxy-3-phenylpropionic acid	: 2133	Indene, 2,3-diphenyl-	: 7272	Iron dicyclopentadienyl	: 5960
4-Hydroxyphenylpropionic acid	: 2134	Indene, 4-methyl-	: 7278	β-Irone	: 3883
3-Hydroxy-3-phenylprop-1-yne	: 1876	Indene, 6-methyl-	: 7279	α-Irone	: 3884
4-Hydroxyphenylsulfonic acid	: 2226	Indene, 7-methyl-	: 7280	Iron pentacarbonyl	: 7339
17α-Hydroxyprogesterone	: 9750	Indigo	: 7331	Iron, tris(dimethylcarbamodithioato-S,S')-, (OC-6-11)-	: 5957
4-Hydroxy-L-proline	: 9767	Indigo Yellow	: 2843	Isatin	: 7300
3-Hydroxy-1-propene	: 10697	Indoleacetic acid	: 7295	Isoamylaldehyde	: 3186
2-Hydroxypropionitrile	: 10105	Indolebutyric acid	: 7296	Isoamylamine	: 3220
(3-Hydroxypropyl)benzene	: 2148	Indole, 7-methyl-	: 7317	Isoamyl benzoate	: 3632
1-Hydroxy-1-propylcyclohexane	: 4557	Indole, 1,2,3-trimethyl-	: 7322	Isoamyl bromide	: 3239
3-Hydroxypropylene oxide	: 8618	Indole, 2,3,5-trimethyl-	: 7323	Isoamyl chloride	: 3252
N-(2-Hydroxypropyl)ethylenediamine	: 10340	2-Indolinone, 1-ethyl-3-methyl-	: 7332	Isoamyl cyanide	: 8831
3-Hydroxypropyl fluoride	: 10382	1H-Indolo[3,2,1-de]pyrido[3,2,1-ij][1,5]naphthyridine, eburnamenine-14-carboxylic acid deriv.	: 5349	Isoamyl ether	: 3418
2-(2-Hydroxypropyl)pyridine	: 10951			Isoamyl iodide	: 3384
2-(3-Hydroxypropyl)pyridine	: 11012	Indolo[4,3-fg]quinoline, ergoline-8-carboxamide deriv.	: 5373	Isoamyl isovalerate	: 3551
(2-Hydroxypropyl)trimethyl ammonium chloride	: 9831	3-Indolylacetone	: 10442	Isoamyl 3-methylbutyrate	: 3551
4-Hydroxypyrazolo[3,4-d]pyrimidine	: 10852	Indonaphthene	: 7245	Isoamyl nitrate	: 3637
5-Hydroxyquinoline	: 11305	Indoxazene	: 2389	Isoamyl nitrite	: 8090
7-Hydroxyquinoline	: 11307	Inosine, 2-amino-	: 6289	Isoamyl salicylate	: 2633
8-Hydroxyquinoline	: 11308	Iodipamide	: 2594	Isoamyl sulfide	: 3436
8-Hydroxyquinoline sulfate	: 11324	5-Iodoacenaphthene	: 7	Isobutamben	: 2505
8-Hydroxy-5-quinolinesulfonic acid	: 11294	Iodoacetic acid	: 200	Isobutanal	: 9785
p-Hydroxysalicylic acid	: 2569	Iodoacetonitrile	: 283	Isobutane	: 10073
5-Hydroxysalicylic acid	: 2570	Iodoalphionic acid	: 2135	Isobutane, 1,1-dichloro	: 9899
2α-Hydroxytigogenin	: 11530	p-Iodoaniline	: 761	Isobutane, 1,2-dichloro	: 9900
2-Hydroxy-m-toluic acid	: 2629	o-Iodoanisole	: 1775	Isobutane, 1,1,1,2,3-pentachloro	: 10127
2-Hydroxy-p-toluic acid	: 2630	Iodobenzene	: 1766	Isobutane, 1,1,1,2-tetrachloro	: 10147
6-Hydroxy-m-toluic acid	: 2631	p-Iodobenzenesulfonyl chloride	: 2247	Isobutane, 1,1,2-trichloro	: 10177
2-Hydroxytridecane	: 11884	o-Iodobenzoic acid	: 2657	Isobutene	: 10578
11-Hydroxyundecanoic acid	: 11948	4-Iodo-1,2-butadiene	: 3166	2-Isobutoxyethanol	: 5746
4-Hydroxyundecanoic acid lactone	: 6152	4-Iodocatechol	: 1523	1-Isobutoxy-2-nitrobenzene	: 2034
Hydroxyurea	: 12016	Iodochlorhydroxyquin	: 11312	Isobutryic anhydride	: 10291
Hydroxyzine	: 5700	Iodocyanide (ICN)	: 7337	Isobutyl acetate	: 215
Hygrine	: 10471	Iodocyclohexane	: 4415	Isobutyl acrylate	: 10673
Hymecromone	: 2831	Iodocyclopentane	: 4824	Isobutyl alcohol	: 10388
Hyodeoxycholic acid	: 4096	2-Iodo-p-cymene	: 1781	Isobutyl alcohol, 2-chloro	: 10352
Hyoscyamine	: 917	1-Iododecane	: 5053	Isobutyl alcohol, 3-chloro	: 10353
Hypothiazide	: 2855	Iododifluoromethane	: 7505	Isobutylamine	: 9817
Hypoxanthine	: 10770	1-Iodo-3,5-dimethylbenzene	: 1771	Isobutyl p-aminobenzoate	: 2505
Hypoxanthine riboside	: 7335	4-Iododiphenyl	: 3079	Isobutylbenzene	: 2039
Ibogaine	: 7186	1-Iododecane	: 5302	Isobutyl bromide	: 9845
Ibrotamide	: 3199	Iodoethane	: 5597	Isobutyl butanoate	: 3563
Imecromone	: 2831	2-Iodoethanol	: 5731	Isobutyl carbamate	: 3959
1H-Imidazole, 1-[2-(2,4-dichlorophenyl)-2-(2-propenyloxy)ethyl]-	: 7187	Iodoethene	: 5907	Isobutyl chloride	: 9862
		Iodoethylene	: 5907	Isobutyl chlorocarbonate	: 4024
1H-Imidazole, 2-heptadecyl-4,5-dihydro-, monoacetate	: 6275	Iodoform	: 7560	Isobutyl chloroformate	: 4024
		2-Iodofuran	: 6110	Isobutyl cyanide	: 3410
2-Imidazolidine, 2-benzyl	: 7198	3-Iodofuran	: 6111	Isobutylcyclopentane	: 4838
1-Imidazolidinecarboxamide, 3-(3,5-dichlorophenyl)-N-(1-methylethyl)-2,4-dioxo-	: 7338	1-Iodoheptane	: 6419	Isobutyldimethylcarbinol	: 8928
		2-Iodoheptane	: 6420	Isobutylene chlorohydrin	: 10354
7H-Imidazo[4,5-d]pyrimidine	: 10752	1-Iodo-1-heptyne	: 6620	Isobutylene, 1,3-dibromo	: 10546
2-Iminobenzothiazoline	: 2859	1-Iodohexadecane	: 6640	Isobutylene glycol, 3-chloro	: 9998
Iminodiacetic acid	: 6260	1-Iodohexane	: 6831	Isobutylene (trimer)	: 10582
2,2'-Iminodibenzoic acid	: 2656	Iodomethane	: 7518	Isobutylethene	: 9054
Iminourea	: 6278	2-Iodo-4-methylaniline	: 762	Isobutyl formate	: 6013
Imperatorin	: 6189	4-Iodo-2-methylbutane	: 3384	Isobutylhydrochlorothiazide	: 2856
Indaconitine	: 307	(Iodomethyl) methyl ether	: 7519	Isobutyl iodide	: 10057
Indan	: 7250	1-Iodonaphthalene	: 7902	Isobutyl isobutanoate	: 10304
Indan, 1-amino (DL)	: 7243	1-Iodooctadecane	: 8242	Isobutyl isocyanate	: 10061
Indan, 1,2-dibromo-	: 7247	1-Iodooctane	: 8406	Isobutyl isovalerate	: 3555
Indan, 1,2-dichloro	: 7249	1-Iodopentane	: 8818	Isobutyl ketone	: 6514
Indan, 1,1-diemthyl	: 7251	3-Iodopentane	: 8820	Isobutyl mercaptan	: 10158
Indan, 1,2-dimethyl-	: 7252	1-Iodo-1-pentyne	: 9133	Isobutyl methacrylate	: 10666
Indan, 4,7-dimethyl-	: 7253	o-Iodophenol	: 9371	Isobutyl methyl ether	: 10071
Indan, 5,6-dimethyl-	: 7254	p-Iodophenol	: 9373	Isobutyl methyl ketone	: 8981
Indan, 1-ethyl-	: 7273	1-Iodopropane	: 10054	Isobutyl nitrate	: 8078
Indan, 5-hexyl-	: 7274	2-Iodopropane	: 10055	Isobutyl nitrite	: 8094
Indan, 1-methyl-	: 7255	3-Iodopropene	: 10573	Isobutyloxycarbonyl chloride	: 4024
Indan, 2-methyl-	: 7256	3-Iodopropylene	: 10573	Isobutyl propionate	: 10310
		1-Iodopropyne	: 10734	Isobutyl propyl ketone	: 6524
		3-Iodopropyne	: 10735	4-Isobutylresorcinol	: 1539
		2-Iodopyridine	: 10970	Isobutyl salicylate	: 2643
				Isobutyl stearate	: 8268

Isobutyl thiocyanate	: 11707
Isobutyl 2-tolyl ketone	: 3669
Isobutyl 4-tolyl ketone	: 3670
Isobutyl urethane	: 3961
Isobutyl vinyl ether	: 10035
Isobutyl vinyl ketone	: 7093
Isobutyramide, 2-bromo	: 9792
Isobutyric acid	: 10290
Isobutyric acid chloride	: 10504
Isobutyric acid, 2-chloroethyl ester	: 10292
Isobutyric acid, cyclohexyl ester	: 10293
Isobutyric acid, (2-nitropentyl) ester	: 10305
Isobutyric acid, 3-phenyl	: 2137
1-Isobutyronaphthone	: 10464
2-Isobutyronaphthone	: 10465
Isobutyronitrile	: 10111
Isobutyronitrile, 2-hydroxy-3-chloro	: 10098
Isobutyryl chloride, 3-chloro	: 10499
Isocaine	: 2505
Isocamphoric acid, (DL)	: 3931
Isochroman	: 2776
Isochrysene	: 11910
Isocinchomeronic acid	: 10929
Isocitrate lactone	: 6079
Isocorybulbine	: 5150
L-(+)-Isocorydine	: 5161
Isocorydine	: 5161
Isocoumarin	: 2791
Isodurene	: 2279
Isoestradiol	: 5401
8-Isoestrone	: 5407
Isoeugenol	: 9388
Isoeugenol, acetate (cis)	: 9394
Isoflavone	: 2841
Isoflavone, 4',7-dihydroxy-	: 2827
Isoflavone, 4',5-dihydroxy-7-methoxy-	: 2828
Isoflurophate	: 9580
Isogeronic acid	: 6461
Isoguanine	: 10767
Isohexane	: 8823
Isohexyl alcohol	: 8939
1H-Isoindole-1,3(2H)-dione, 2-[(trichloromethyl)thio]-	: 5984
Isolysergic acid	: 5375
β-Isomalic acid	: 9959
Isometheptene	: 6532
Isoniazid	: 10905
Isonicotinic acid	: 10895
Isonicotinic acid diethylamide	: 10892
Isonicotinic acid, (α-methylfurfurylidene)hydrazide	: 10904
Isonipecotic acid	: 9642
Isonipecotic acid, methyl ester	: 9644
Isonipecotic acid, 1-methyl-, methyl ester	: 9645
Isonitrosoacetone	: 9789
Isonitrosoacetophenone	: 876
β-Isonitrosopropane	: 10476
Isooctane	: 8848
Isopentadiene	: 3171
Isopentanal	: 3186
Isopentane	: 3397
Isopentanoic acid	: 3539
Isopentanoyl chloride	: 3693
Isopentyl acetate	: 3631
Isopentyl alcohol	: 3627
Isopentyl alcohol, orthoformate (3:1)	: 8552
Isopentyl alcohol, stearate	: 8265
Isopentylamine	: 3220
Isopentyl bromide	: 3239
Isopentyl 3-bromopropionate	: 10210
Isopentyl butanoate	: 3544
Isopentyl chloride	: 3252
N-Isopentylethanolamine	: 5739
Isopentyl ether	: 3418
Isopentyl formate	: 3636
Isopentyl iodide	: 3384
Isopentyl isovalerate	: 3551
Isopentyl mercaptan	: 3440
Isopentyl 3-methylbutanoate	: 3551
Isopentyl β-Naphthyl ether	: 7918
Isopentyl oleate	: 8306
Isopentyl propanoate	: 3639
Isopentyl salicylate	: 2633
2-Isopentylthiophene	: 11771
Isopentyl vinyl ether	: 3371

Isophorol	: 4702
Isophorone	: 4721
Isophthalamide, N,N,N',N'-tetraethyl	: 1326
Isophthalic acid	: 1328
Isophthalic acid, diethyl ester	: 1346
Isophytol	: 6871
Isopilosine	: 6144
Isoprene	: 3171
Isopropanesulfonic acid	: 10130
Isopropenyl acetate	: 10698
Isopropenylacetylene	: 3892
Isopropenyl carbinol	: 10702
Isopropenyl chloride	: 10535
Isopropenyl chloroformate	: 4022
4-Isopropenyl-1-cyclohexenecarboxaldehyde	: 4633
4-Isopropenyl-1-methylcyclohexene	: 4677
Isopropenyl methyl ketone	: 3879
9-Isopropenylphenanthrene	: 9182
2-Isopropoxyethanol	: 5740
1-Isopropoxy-2-propanol	: 10390
N-Isopropylacetanilide	: 74
Isopropyl acetate	: 209
Isopropylacetic acid	: 3539
Isopropyl acetoacetate	: 3574
Isopropyl alcohol	: 10335
Isopropylamine	: 9804
2-Isopropylaniline	: 788
N-Isopropylaniline	: 790
Isopropylbenzene hydroperoxide	: 7181
p-Isopropylbenzoic acid	: 2689
α-Isopropylbenzyl alcohol	: 1901
4-Isopropylbenzyl alcohol	: 1902
Isopropyl borate	: 3128
Isopropyl bromide	: 9837
α-Isopropyl-α-bromobutyramide	: 3199
Isopropyl-tert-butylcarbinol	: 8955
Isopropyl butyrate	: 3549
Isopropyl carbamate	: 3955
Isopropyl carbanilate	: 3966
Isopropyl chloride	: 9849
Isopropyl chloroacetate	: 139
Isopropyl 3-chlorocarbanilate	: 3939
Isopropyl chloroformate	: 4023
Isopropyl N-(3-chlorophenyl)carbamate	: 3939
Isopropyl 2-chloropropionate	: 10223
Isopropyl trans-cinnamate	: 10687
Isopropylcyclohexane	: 4431
cis-2-Isopropylcyclohexanol	: 4539
cis-4-Isopropylcyclohexanol	: 4540
trans-4-Isopropylcyclohexanol	: 4541
2-Isopropylcyclohexanone	: 4599
4-Isopropylcyclohexanone	: 4600
Isopropylcyclopentane	: 4828
2-Isopropylcyclopentanone	: 4887
7-Isopropyl-1,4-dimethylazulene	: 505
N-Isopropylethanolamine	: 5741
Isopropyl fluoride	: 10047
Isopropyl formate	: 6011
Isopropyl heptafluorobutyrate	: 3526
4-Isopropylheptane	: 6427
Isopropylidene glycerol	: 5207
Isopropyl iodide	: 10055
Isopropyl isobutyrate	: 10303
Isopropyl isocyanate	: 10060
Isopropyl isocyanide	: 10063
Isopropyl isonitrile	: 10063
Isopropyl lactate	: 10275
Isopropyl mercaptan	: 10157
Isopropylmethylamine	: 9819
4-Isopropyl-1-methylcyclopentene	: 4927
Isopropyl methyl ether	: 10070
2-Isopropyl-5-methyl-1-hexanol	: 6948
2-(4-Isopropyl-4-methyl-5-oxo-2-imidazolin-2-yl)nicotinic acid	: 7188
(RS)-2-(4-Isopropyl-4-methyl-5-oxo-2-imidazolin-2-yl)quinoline-3-carboxylic acid	: 7189
Isopropyl methyl sulfide	: 10089
Isopropyl myristate	: 11626
Isopropyl 1-naphthyl ketone	: 10464
Isopropyl nitrate	: 8076
Isopropyl nitrite	: 8092
2-Isopropyl-1-pentene	: 6836
Isopropyl phenylcarbamate	: 3966

Isopropylphenylcarbinol	: 1901
3-(p-Isopropylphenyl)-2-methylpropionaldehyde	: 2117
Isopropyl phenyl urethane	: 3966
Isopropyl propionate	: 10299
Isopropylpropylamine	: 9827
Isopropyl propyl sulfide	: 10077
3-Isopropylpyridine	: 10985
Isopropyl salicylate	: 2635
Isopropyl 2-thienyl ketone	: 10472
Isopropyl thiocyanate	: 11706
2-Isopropylthiophene	: 11772
3-Isopropylthiophene	: 11773
Isoquinaldine	: 7381
Isoquinoline, 5-amino-	: 7372
Isoquinoline, 3-benzyl	: 7387
Isoquinoline, 3-butyl-1-[2-(dimethylamino)ethyl]	: 5411
Isoquinoline, 6,7-diethoxy-1(3,4-diethoxybenzyl)	: 7377
Isoquinoline, 6,7-dimethoxy-1,2-dimethyl-1,2,3,4-tetrahydro	: 7391
Isoquinoline, 8-methoxy-2-methyl-6,7-methylenedioxy-1,2,3,4-tetrahydro	: 5214
Isorubijervine	: 11516
(+)-Isothebaine	: 5158
Isothebaine	: 5158
3(2H)-Isothiazolone, 2-octyl-	: 8528
Isothiocyanic acid, acetyl	: 303
Isothiocyanic acid, tert-butyl ester	: 10066
Isovaleraldehyde	: 3186
Isovaleramide	: 3206
Isovaleric acid	: 3539
Isovaleric acid chloride	: 3693
Isovaleric aldehyde	: 3186
Isovaleric amide	: 3206
Isovalerylamine	: 3220
Isovaleryl chloride	: 3693
Isovaleryl diethylamide	: 3201
Isoxylaldehyde	: 548
Itaconic acid	: 3328
Jasmone	: 4949
Javanicin	: 7855
Jervine	: 12046
Jilkon	: 6200
Juglone	: 7860
Julolidine	: 2737
Julolidine, 1,6-dioxo	: 2736
Kaempferol	: 2843
Kainic acid	: 11144
Kelthane	: 1851
Ketamine	: 4579
Ketene	: 5927
Ketene, diethyl-	: 3877
β-Ketobutyranilide	: 3207
α-Ketoglutaric acid	: 8781
3-Ketoglutaric acid	: 8782
Ketohydroxyestrin	: 5406
Ketone, 2-benzofuranyl methyl	: 5775
Ketone, 2-benzofuranyl phenyl	: 7588
Ketone, benzyl phenyl	: 5812
Ketone, 5-bromo-2-thienyl methyl	: 5783
Ketone, α-chlorobenzyl phenyl	: 5785
Ketone, 5-chloro-2-thienyl methyl	: 5791
Ketone, cyano phenyl	: 968
Ketone, cyclobutyl methyl	: 5792
Ketone, cyclobutyl phenyl	: 7605
Ketone, 1-cyclohexen-1-yl methyl	: 5793
Ketone, cyclohexyl methyl	: 5794
Ketone, cyclopentyl methyl	: 5795
Ketone, cyclopropyl methyl	: 5796
Ketone, dicyclohexyl	: 7606
Ketone, dicyclopropyl	: 7607
Ketone, 6,7-dimethoxy-1-isoquinolyl 3,4-dimethoxyphenyl	: 7611
Ketone, 10-[3-(dimethylamino)propyl]phenothiazin-2-yl methyl	: 5807
Ketone, 2-furyl methyl	: 5817
Ketone, 2-furyl phenyl	: 7618
Ketone, 1-hydroxycyclohexyl methyl	: 5821
Ketone, 6-hydroxy-2-isopropenyl-5-benzofuranyl methyl	: 5825
Ketone, 3-hydroxy-2-thienyl methyl	: 5835
Ketone, indol-3-yl methyl	: 5819
Ketone, methyl 1-methylcyclohexyl	: 5843
Ketone, methyl 3-methylcyclohexyl	: 5844

Ketone, methyl 4-methylcyclohexyl	: 5845
Ketone, methyl 1-methylcyclopentyl	: 5846
Ketone, methyl 2-methylcyclopentyl	: 5847
Ketone, methyl 6-methyl-3-pyridyl	: 5855
Ketone, methyl 1-methylpyrrol-2-yl	: 5849
Ketone, methyl 5-methyl-2-thienyl	: 5856
Ketone, methyl 2-pyridyl	: 5867
Ketone, methyl 3-pyridyl	: 5868
Ketone, methyl 4-pyridyl	: 5869
Ketone, methyl pyrrol-2-yl	: 5820
Ketone, methyl styryl	: 3881
Ketone, methyl 2-thienyl	: 5870
Ketone, 1-naphthyl o-tolyl	: 7629
Ketone, phenyl 2-pyridyl	: 7638
Ketone, phenyl 4-pyridyl	: 7639
Ketone, phenyl 2-thienyl	: 7641
Ketone, phenyl 3-thienyl	: 7642
2-Ketosuccinic acid	: 3339
Khellin	: 6185
Kojic acid	: 10795
KSCN	: 11711
Kyanmethin	: 11045
Kynurenic acid	: 11232
Kynurenine	: 1102
Lactic acid (DL)	: 10266
Lactic acid, butyl ester (d)	: 10268
Lactic acid, isopropyl ester	: 10275
Lactic acid, methyl ester (DL)	: 10273
Lactic acid, 3,3,3-trichloro-	: 10333
Lactobionic acid	: 6210
Lactonic acid	: 6210
Lactonitrile	: 10105
Lactonitrile, 3-chloro-2-methyl-	: 10098
Lactonitrile, 3,3,3-trichloro-	: 10115
p-Lactophenetide	: 9796
Lactyl chloride, acetate (DL)	: 10494
Lamoryl	: 11520
Lanex	: 12007
Lanosterol	: 7411
Lanoxin	: 4041
L-Lanthionine	: 5016
Lapachol	: 7862
Lappaconitine	: 309
Laricic acid	: 8102
Laudanidine	: 9399
Laudanine	: 9398
Laudanosine	: 7379
Lauraldehyde	: 5288
Lauraldehyde, dimethyl acetal	: 5297
Laureline	: 2469
Lauric acid	: 5305
Lauric acid, phenyl ester	: 5317
Lauric acid, propyl ester	: 5320
Laurolenic acid	: 4912
Laurone	: 11857
Lauronitrile	: 5303
Lauroyl peroxide	: 9155
Lauryl alcohol	: 5321
Lauryl bromide	: 5293
Lawsone	: 7859
Leaf aldehyde violet	: 8108
Ledol	: 4989
Lemonol	: 8333
Lenacil	: 4900
Lenthionine	: 8998
Lepidine	: 11279
Lepidine, 2-amino-	: 11208
Lepidine, 2-ethyl-	: 11270
Lergitin	: 5506
Leucethane	: 3948
Leucoline	: 7373
Levodopa	: 11929
Levopimaric acid	: 9169
Levulinic acid	: 8903
Levulinic acid, benzyl ester	: 8911
Levulinic acid, isopropyl ester	: 8909
Levulinic acid, 3-methyl-, methyl ester	: 8898
Levulinic acid, propyl ester	: 8912
β-Levulose	: 6023
Lichenic acid	: 3764
Lidocaine	: 54
Lignoceric acid	: 11607
Limettin	: 2818
α-Limonene	: 4677
Limonene	: 4677
Linalool formate	: 8337
Linalool, tetrahydro (DL)	: 8454
Linalyl formate	: 8337
Linamarin	: 10104
Linarigenin	: 2809
β-Lindane	: 4406
δ-Lindane	: 4407
Lindane	: 4409
Linderic acid	: 5338
Linoleic acid	: 8223
Linolenic acid	: 8287
Lithocholic acid	: 4099
Lobelanidine	: 9646
Lobelanine	: 5854
Lobeline	: 5834
Loflucarban	: 11802
Longifolene	: 7568
Luciculine	: 5007
Lumazine	: 10748
Lumichrome	: 2472
Luminol	: 9600
Lumisterol	: 5382
Lunacrine	: 6182
Lupeol	: 7417
3,5-Lupetidine	: 9652
Lupinine	: 11331
Lupulon	: 4281
Lupulon	: 4058
trans-Lutein	: 2814
Luteolin	: 2814
2,3-Lutidine	: 10936
2,4-Lutidine	: 10937
2,5-Lutidine	: 10938
2,6-Lutidine	: 10939
3,4-Lutidine	: 10940
3,5-Lutidine	: 10941
Lycomarasmine	: 472
trans-Lycopene	: 4053
Lycoremine	: 6200
Lycorine	: 6201
Lycoxanthin	: 4060
Lysergamide	: 5374
Lysergic acid	: 5376
D-Lysergic acid D-propanolamide	: 5373
Lysidine	: 7197
Lysilan	: 5140
Maclurin	: 7610
Magenta I	: 634
Magneson	: 1542
Malachite green	: 7455
Malathion	: 3316
Maleamic acid	: 3817
Maleanilic acid	: 3853
Maleic acid	: 3765
Maleic acid, chloro-, diethyl ester	: 3771
Maleic acid, diallyl ester	: 3782
Maleic acid, dihydroxy-	: 3777
Maleic anhydride	: 6091
Maleimide, N-phenyl-	: 11123
Malidone	: 8586
Malilum	: 11091
Malonic acid	: 9921
Malonic acid, amino-, diethyl ester	: 9924
Malonic acid, 2-cyclopentenyl, diethyl ester	: 9935
Malonic acid, cyclopentylidene, diethyl ester	: 9936
Malonic acid, dibenzyl-, diethyl ester	: 9926
Malonic acid, dibutyl-, diethyl ester	: 9937
Malonic acid, diethyl-, diethyl ester	: 9940
Malonic acid, dimethyl-, diethyl ester	: 9942
Malonic acid, dipropyl ester	: 9945
Malonic acid, di-2-propynyl-, diethyl ester	: 9946
Malonic acid, ethylisopropyl-, diethyl ester	: 9951
Malonic acid, (hydroxymethyl)-	: 9959
Malonic acid, isobutyl-, diethyl ester	: 9969
Malonic acid, isopropylidene-, diethyl ester	: 9966
Malonomononitrile, benzylidene,monoethyl ester	: 10645
Malononitrile	: 9911
Malononitrile, phenyl-	: 9916
Maltol	: 10796
Malvidin chloride	: 2847
Mandelic acid isoamyl ester	: 912
Mandelonitrile glucoside	: 955
Manganese, [[1,2-ethanediylbis-[carbamodithioato]](2-)]-	: 7420
Mannite	: 7422
Mannitol, hexanitrate	: 7424
Margaric acid	: 6314
Margaric aldehyde	: 6304
Matrine	: 7426
MCPA	: 141
Mechlorethamine	: 5413
Meclizine	: 9605
Meconic acid	: 10775
Meconin	: 7355
Mecrylate	: 10624
Medroxyprogesterone	: 9752
Melamine	: 11835
Melampyrin	: 6194
Melatonin	: 69
Meldrum's acid	: 5190
Melezitose	: 6221
Mellitene	: 1758
Mellitic acid	: 1748
Menadiol diacetate	: 7841
Menadione	: 7864
p-Mentha-1,8-dien-7-al	: 4633
p-Mentha-1(7),2-diene	: 4673
p-Menthane, 4-bromo	: 4321
trans-(-)-p-Menthan-3-one	: 4606
m-Menth-8-ene	: 4435
p-Menth-2-ene, 1,4-epidioxy-	: 5184
p-Menth-3-en-2-ol	: 4698
(±)-Menthol	: 4548
Menthol(DL)	: 4548
Menthone(d)	: 4805
l-Menthone	: 4606
(-)-Menthone	: 4606
Meparfynol	: 9146
Mephenesin	: 10019
Mephenoxalone	: 8588
Mephobarbital	: 11094
Mepivacaine	: 9639
Meprobamate	: 10021
β-Mercaptoalanine	: 5014
Mercaptobenzene	: 2289
Mercaptobenzimidazole	: 2386
2-Mercaptobenzoic acid	: 2663
2-Mercaptobenzothiazole	: 2866
1-Mercaptobutane	: 3437
Mercaptocyclopentane	: 4852
2-Mercaptoethanol	: 5732
3-Mercaptohexane	: 6858
Mercaptomethane	: 7546
6-Mercaptopurine	: 10763
Mercapturic acid	: 5015
Mercury, chloro(2-methoxyethyl)-	: 11822
Mesaconic acid	: 3785
Mesaconic acid, diethyl ester	: 3787
Mesalamine	: 2499
Mescaline	: 1586
Mesitylamine	: 859
Mesitylene	: 2359
Mesitylene, α-bromo-	: 1057
Mesitylene, 2,4-dibromo-	: 1313
Mesitylene, 2,4-dichloro-	: 1402
Mesitylenic acid	: 2583
Mesityl oxide	: 9108
Mesoxalylcarbamide	: 11073
Mesoxalylurea	: 11073
Mestanolone	: 350
Mestilbol	: 9358
Mesulphen	: 11661
Metacrolein	: 11906
Metaldehyde II (tetramer)	: 19
Metanilic acid	: 2216
Methacetin	: 71
Methacholine chloride	: 9830
Methacrolein	: 10515
Methacrolein diacetate	: 10564
Methacrylic acid	: 10648
Methacrylic acid anhydride	: 10649
Methacrylic acid, 3-bromo	: 10612
Methacrylic acid chloride	: 10715
Methacrylic acid, 2-ethylhexyl ester	: 10659
Methadone hydrochloride	: 6516
Methallatal	: 11064
Methallyl alcohol	: 10702
α-Methallyl bromide	: 3710
Methaminodiazepoxide	: 2409

Methamphetamine	: 1573
Methanal	: 5987
Methanamide	: 5989
Methandrostenolone	: 344
Methanearsonic acid	: 458
Methanecarbonitrile	: 271
Methanecarboxamide	: 22
Methanecarboxylic acid	: 100
Methane, dicyclohexyl-	: 4429
Methane, diphenyl-(3-tolyl)	: 7513
Methane, iodo-trichloro	: 7556
Methanesulfonic acid chloride	: 7535
Methanesulfonic acid fluoride	: 7538
Methane, tetraphenyl-	: 1831
Methane, tricyclohexyl-	: 4433
Methanimidamide, N,N-dimethyl-N'- [3-[[(methylamino)carbonyl]- oxy]phenyl]-, monohydrochloride	: 6001
1H-3a,7-Methanoazulene, cedrene deriv.	: 4066
5,9-Methano-2,4,3- benzodioxathiepin, 6,7,8,9,10,10- hexachloro-1,5,5a,6,9,9a- hexahydro-, 3-oxide	: 5369
7,9-Methano-7H-oxepino[3,4- a]pyrrolo[2,3-d]carbazole, 9,10- secostrychnidin-10-oic acid deriv.	: 11352
Methanoic acid	: 6002
Methanoic acid amidine	: 7563
4,7-Methano-1H-indene, 1,2,4,5,6,7,8,8-octachloro- 2,3,3a,4,7,7a-hexahydro-	: 4077
2,5-Methano-2H-indeno[1,2- b]oxirene, 2,3,4,5,6,7,7-heptachloro- 1a,1b,5,5a,6,6a-hexahydro-, (1aα,1bβ,2α,5α,5aβ,6β,6aα)-	: 6301
4,7-Methano-1H-isoindole-1,3(2H)- dione, 2-(2-ethylhexyl)-3a,4,7,7a- tetrahydro-	: 7660
Methanol, 2-tolyl	: 1897
Methanol, 3-tolyl	: 1898
Methanol, 4-tolyl	: 1899
Methantheline bromide	: 5445
Methapyrilene	: 5507
Methaqualone	: 11199
Metharbital	: 11088
Methenamine	: 11601
1,3,4-Metheno-2H- cyclobuta[cd]pentalen-2-one, 1,1a,3,3a,4,5,5,5a,5b,6-decachloro- octahydro-	: 7408
N,N'-Methenyl-o-phenylenediamine	: 2376
Methestrol	: 9299
Methimazole	: 7214
Methionamine	: 7649
Methionic acid	: 7514
Methone	: 4378
Methoxone	: 141
Methoxsalen	: 6187
Methoxyacetaldehyde	: 16
p-Methoxyacetanilide	: 71
l-Methoxyacetic acid	: 210
Methoxyacetic acid, ethyl ester	: 206
6-Methoxybaicalein	: 2810
4-Methoxybenzoic acid chloride	: 2901
4-Methoxybenzoyl chloride	: 2901
4-Methoxybenzyl alcohol	: 1891
3-Methoxy-1-butanol	: 3625
2-Methoxy-4-carboxyphenol	: 2625
Methoxychlor	: 2329
2-Methoxy-3,6-dichlorobenzoic acid	: 2567
2-(2-Methoxyethoxy)ethanol	: 5737
2-Methoxyethyl acetate	: 5735
2-Methoxyethyl trichloroacetate	: 250
2-Methoxyfuran	: 6127
7-Methoxyisoquinoline	: 7380
2-(Methoxymethyl)-5-nitrofuran	: 6129
2-Methoxynaphthalene	: 7914
3-Methoxy-4-nitroaniline	: 776
4-Methoxyphenol	: 9381
4-Methoxy-m-phenylenediamine	: 1274
1-(4-Methoxyphenyl)ethanol	: 1895
5-Methoxy-2-phenyloxazole	: 8582
2-Methoxy-1-propanol	: 10386
1-Methoxy-2-propanol	: 10387
2-Methoxypropionitrile	: 10109
3-Methoxypropiophenone	: 10456

N1-(6-Methoxy-3- pyridazinyl)sulfanilamide	: 2197
7-Methoxyquercetin	: 2815
4-Methoxyquinoline	: 11274
6-Methoxyquinoline	: 11275
8-Methoxyquinoline	: 11276
6-Methoxyquinoxaline	: 11336
6-Methoxy-α-tetralone	: 8048
5-Methoxytryptamine	: 7304
Methsuximide	: 11156
Methyl abietate	: 9168
N-Methylacetanilide	: 79
Methyl acetate	: 208
Methyl acetoacetate	: 3573
Methylacetylene	: 10727
Methyl acetylsalicylate	: 2482
Methyl acrylate	: 10655
Methylacrylonitrile	: 10586
Methylal	: 7507
Methyl alcohol	: 7581
Methylallene	: 3145
1-Methyl-1-allyl acetone	: 7097
Methylamine	: 7442
Methyl 2-aminobenzoate	: 2503
Methyl m-amino-p-hydroxybenzoate	: 2501
Methyl [(4-aminophenyl)sulfonyl]- carbamate	: 474
4-Methyl-2-aminopyridine	: 10867
4-Methyl-2-aminothiazole	: 11665
N-Methylaniline	: 781
2-(N-Methylanilino)ethanol	: 5743
5-Methyl-o-anisidine	: 770
1-Methylanthracene	: 423
2-Methylanthracene	: 424
9-Methylanthracene	: 425
Methyl anthranilate	: 2503
2-Methylanthraquinone	: 409
Methylarsine	: 446
3-Methylazelaic acid	: 8135
Methyl azide	: 7460
p-Methylazobenzene	: 5139
3-Methylbenz[a]anthracene	: 522
2-Methylbenzenemethanol	: 1897
4-Methylbenzenemethanol	: 1899
p-Methylbenzenesulfinic acid	: 2182
Methyl benzenesulfonate	: 2227
N-(4-Methylbenzenesulfonyl)-N'- butylurea	: 2207
Methyl benzoate	: 2687
3-Methylbenzo[b]furan	: 2457
5-Methylbenzo[b]furan	: 2458
7-Methylbenzo[b]furan	: 2459
4'-Methylbenzophenone-2-carboxylic acid	: 2684
2-Methylbenzothiazole	: 2863
2-Methylbenzoxazole	: 2880
α-Methylbenzyl alcohol	: 1896
o-Methylbenzyl alcohol	: 1897
m-Methylbenzyl alcohol	: 1898
p-Methylbenzyl alcohol	: 1899
1-Methylbicyclo[3,1,0]hexane	: 2980
2-Methylbicyclo[2.2.2]octane	: 2992
2-Methylbiphenyl-3-ylmethyl (Z)- (1RS,3RS)-3-(2-chloro-3,3,3- trifluoroprop-1-enyl)- 2,2- dimethylcyclopropanecarboxylate	: 2998
N-Methylbis(2,4- xylyliminomethyl)amine	: 342
Methyl borate	: 3126
Methyl bromide	: 7482
Methyl 2-bromoacetate	: 117
3-Methyl-1-bromobutane	: 3239
Methyl α-bromocinnamate	: 10615
Methyl p-bromophenyl ketone	: 5781
2-Methylbutadiene	: 3171
1-Methyl-2,3-butadiene	: 8658
Methylbutadiyne	: 8684
3-Methyl-1,2-butanediol	: 3352
2-Methyl-1,3-butanediol	: 3353
2-Methyl-2,3-butanediol	: 3355
2-Methylbutanenitrile	: 3409
(+)-2-Methyl-1-butanethiol	: 3439
2-Methyl-2-butanethiol	: 3441
1-Methylbutanethiol	: 8845
Methyl butanoate	: 3547
2-Methyl-1-butanol	: 3626
(±)-3-Methyl-2-butanol	: 3629

3-Methyl-2-butenal	: 3700
2-Methyl-1-butene	: 3799
3-Methyl-1-butene	: 3800
2-Methyl-2-butene	: 3801
trans-2-Methyl-2-butenoic acid	: 3844
Methyl 1-butenyl ketone	: 7086
1-Methyl-2-butylacetylene	: 6810
2-Methylbutyl alcohol	: 3626
3-Methylbutyl chloride	: 3252
3-Methylbutyl iodide	: 3384
Methyl tert-butyl ketone	: 3657
3-Methylbutyl 3-methylbutanoate	: 3551
3-Methylbutyramide	: 3206
(±)-2-Methylbutyric acid	: 3538
3-Methylbutyryl chloride	: 3693
Methyl caproate	: 6898
3-Methylcaproic acid, ethyl ester	: 6899
4-Methylcaproic acid, ethyl ester	: 6900
Methyl caprylate	: 8434
Methyl carbamate	: 3953
Methylcarbinol	: 5672
Methyl carbonate	: 4000
Methylcarbonyl fluoride	: 301
Methyl chemosept	: 2637
Methyl chloride	: 7477
Methyl 2-chloroacrylate	: 10621
Methyl chlorocarbonate	: 4020
Methyl γ-chlorocrotonate	: 3833
Methyl chlorofluoroacetate	: 134
Methylchloroform	: 5655
3-Methyl-3-chloroheptane	: 6369
Methyl 4-chloro-3-nitrobenzoate	: 2548
Methyl 5-chloro-2-nitrobenzoate	: 2549
2-Methyl-4-chlorophenoxyacetic acid	: 141
Methyl 2-chloropropionate	: 10222
2-Methyl-5-chlorothiophene	: 11741
3-Methylcholanthrene	: 2392
(24R)-Methylcholest-5-en-3β-ol	: 5386
2-Methylchroman	: 2778
3-Methylchromone	: 2838
1-Methylchrysene	: 4117
N-Methylconhydrine	: 4144
Methylconiine	: 9686
3-Methylcoumarin	: 2833
5-Methylcoumarin	: 2834
7-Methylcoumarin	: 2836
8-Methylcoumarin	: 2837
3-Methylcrotonaldehyde	: 3700
Methyl trans-crotonate	: 3847
Methyl cis-crotonate	: 3848
trans-2-Methylcrotonic acid	: 3844
Methyl cyanide	: 271
Methyl cyanoacetate	: 154
4-Methylcyanobenzene	: 2765
Methylcyclobutane	: 4192
Methyl cyclobutanecarboxylate	: 4176
1-Methylcyclobutene	: 4201
Methyl cyclobutyl ketone	: 5792
Methylcycloheptane	: 4226
1-Methyl-1-cycloheptanol	: 4230
2-Methylcycloheptanol	: 4231
2-Methylcycloheptanone	: 4235
1-Methylcycloheptene	: 4245
5-Methylcycloheptene	: 4246
2-Methyl-1,3-cyclohexadiene	: 4271
Methylcyclohexane	: 4426
Methyl cyclohexaneacetate	: 4310
β-Methylcyclohexaneacrylic acid	: 3834
cis-2-Methylcyclohexanemethanol	: 4422
trans-2-Methylcyclohexanemethanol	: 4423
β-Methylcyclohexanepropenoic acid	: 3834
1-Methyl-1-cyclohexanol	: 4530
1-Methylcyclohexanol	: 4530
(±)-cis-2-Methylcyclohexanol	: 4531
(±)-trans-2-Methylcyclohexanol	: 4532
cis-4-Methyl-1-cyclohexanol	: 4535
cis-4-Methylcyclohexanol	: 4535
trans-4-Methylcyclohexan-1-ol	: 4536
trans-4-Methylcyclohexanol	: 4536
cis-3-Methylcyclohexanol	: 4533
trans-3-Methylcyclohexanol	: 4534
(±)-2-Methylcyclohexanone	: 4596
(±)-3-Methylcyclohexanone	: 4597
4-Methyl-1-cyclohexanone	: 4598
4-Methylcyclohexanone	: 4598
(±)-3-Methylcyclohexene	: 4671
5-Methyl-2-cyclohexenone	: 4712

trans-2-Methylcyclohexylamine	: 4293
3-Methylcyclohexyl bromide	: 4320
Methyl cyclohexylideneacetate	: 156
Methyl cyclohexyl ketone	: 5794
2-Methylcyclopentadecanone	: 4762
Methylcyclopentane	: 4826
3-Methylcyclopentaneacetic acid	: 4776
1-Methylcyclopentanecarboxylic acid	: 4783
cis-2-Methylcyclopentanol	: 4868
trans-2-Methylcyclopentanol	: 4869
3-Methylcyclopentanol	: 4870
(±)-3-Methylcyclopentanone	: 4886
3-Methylcyclopentene	: 4929
3-Methyl-2-cyclopentenone	: 4945
Methylcyclopropane	: 4981
Methyl cyclopropanecarboxylate	: 4963
2-Methylcyclopropanecarboxylic acid	: 4964
Methyl cyclopropylcarboxylate	: 4963
Methylcysteine hydrochloride	: 5019
5-Methylcytosine	: 11106
2-Methyldecane	: 5054
3-Methyldecane	: 5055
4-Methyldecane	: 5056
Methyl decyl ketone	: 5325
Methyldiacetylene	: 8684
1-Methyldiallyl	: 6328
Methyldiazepinone	: 2410
3-Methyl-1,2-diazine	: 10856
2-Methyl-1,3-diazine	: 11070
4-Methyl-1,3-diazine	: 11071
5-Methyl-1,3-diazine	: 11072
Methyl 2-diazopropionate	: 10237
Methyl 3,5-dibromosalicylate	: 2559
Methyl dichloroacetate	: 166
5-Methyl-2,4-dichloropyrimidine	: 11052
N-Methyldiethanolamine	: 5742
3-Methyldigitoxose	: 7112
Methyl dihydrocinnamate	: 2139
α-Methyldihydrocinnamic acid	: 2137
Methyl 2,4-dimethylbenzoate	: 2589
Methyl α,α-dimethylhydrocinnamate	: 2129
Methyl N-(2,6-dimethylphenyl)-N-(phenylacetyl)-DL-alaninate	: 513
Methyl 2,2-dimethylpropionate	: 10260
Methyldiphenylamine	: 807
Methyl 11-dodecenoate	: 5341
Methyldopa	: 11931
Methyleneaminoacetonitrile	: 285
2,2'-Methylenebiphenyl	: 5967
4,4'-Methylenebis[2-chloroaniline]	: 785
3,3'-Methylenebis[6-hydroxybenzoic acid]	: 2686
3,3'-Methylenebis[4-hydroxycoumarin]	: 2839
Methylene blue	: 9506
Methylene bromide	: 7492
α-Methylene butyrolactone	: 6147
Methylene chloride	: 7499
exo-Methylenecyclobutane	: 4193
Methylenecyclohexane	: 4428
Methylenecyclopentane	: 4827
Methylene difluoride	: 7504
1,2-Methylenedioxy-9,10-dimethoxy-6aα-aporphine	: 2468
1,2-(Methylenedioxy)-4-propylbenzene	: 2431
5,5'-Methylenedisalicylic acid	: 2686
Methylenedisulfonic acid	: 7514
Methylene iodide	: 7506
3-Methylene-2-norbornanone	: 2958
Methylene oxide	: 5987
Methylenesuccinic acid	: 3328
(1R,2S)-N-Methylephedrine	: 1863
1-Methylethanesulfonic acid	: 10130
N-Methyl-2-ethanolamine	: 5738
Methyl ethoxyacetate	: 182
Methyl(2-ethoxyethyl)amine	: 5426
Methylethylallene	: 6879
Methylethylallylcarbinol	: 7084
cis-1-Methyl-3-ethylcyclohexane	: 4397
cis-1-Methyl-4-ethylcyclohexane	: 4399
trans-1-Methyl-4-ethylcyclohexane	: 4400
1-Methyl-1-ethylcyclopentane	: 4814
cis-1-Methyl-2-ethylcyclopentane	: 4815
1-Methyl-cis-3-ethylcyclopentane	: 4817
1-Methyl-trans-3-ethylcyclopentane	: 4818
cis-1-Methyl-2-ethylcyclopropane	: 4976
trans-1-Methyl-2-ethylcyclopropane	: 4977
Methylethyldichlorosilane	: 11397
2-Methyl-5-ethylfuran	: 6106
3-Methyl-3-ethylhexane	: 6826
Methyl ethyl ketone	: 3645
1-Methyl-5-ethylnaphthalene	: 7887
2-Methyl-3-ethylpentane	: 8813
3-Methyl-3-ethylpentane	: 8814
Methyl α-ethylstyryl ketone	: 8987
5-Methyl-3-ethylthiophene	: 11764
Methyl ethynyl ketone	: 3923
9-Methylfluorene	: 5974
Methyl fluoride	: 7515
Methyl fluoroacetate	: 192
Methylfluoroform	: 5665
N-Methylformamide	: 5995
Methyl formate	: 6010
2-Methylfuran	: 6130
3-Methylfuran	: 6131
5-Methyl-3-furancarboxylic acid	: 6065
N-Methylglucamine	: 6208
N-Methyl-L-glucosamidinostreptosidostreptidine	: 11562
N-Methyl-α-L-glucosamine	: 6215
α-Methylglucoside	: 6227
Methyl DL-glycerate	: 10254
1-Methylglycerol	: 3458
β-Methylglycerol α,γ-dichlorohydrin	: 10385
7-Methylguanine	: 10769
2-Methylheptadecane	: 6310
9-Methylheptadecane	: 6311
3-Methyl-1,5-heptadiene	: 6335
3-Methylheptan-2,4-diol	: 6401
2-Methylheptane	: 6422
4-Methylheptane	: 6424
Methyl heptanoate	: 6466
3-Methyl-1-heptanol	: 6492
4-Methyl-1-heptanol	: 6493
6-Methyl-1-heptanol	: 6494
2-Methyl-2-heptanol	: 6495
3-Methyl-2-heptanol	: 6496
5-Methyl-2-heptanol	: 6497
3-Methyl-3-heptanol	: 6500
4-Methyl-3-heptanol	: 6501
6-Methyl-3-heptanol	: 6503
2-Methyl-4-heptanol	: 6503
4-Methyl-4-heptanol	: 6505
3-Methyl-2-heptanone	: 6520
2-Methyl-3-heptanone	: 6522
2-Methyl-4-heptanone	: 6524
3-Methyl-4-heptanone	: 6525
2-Methyl-1-heptene	: 6543
3-Methyl-1-heptene	: 6544
4-Methyl-1-heptene	: 6545
5-Methyl-1-heptene	: 6546
6-Methyl-1-heptene	: 6547
2-Methyl-2-heptene	: 6548
3-Methyl-trans-2-heptene	: 6549
4-Methyl-trans-2-heptene	: 6551
2-Methyl-5-heptene	: 6553
2-Methyl-3-heptene	: 6554
3-Methyl-3-heptene	: 6555
4-Methyl-3-heptene	: 6556
2-Methyl-4-heptene	: 6558
6-Methyl-2-hepten-4-ol	: 6575
5-Methyl-2-hepten-4-one	: 6596
6-Methyl-4-hepten-2-one	: 6599
2-Methyl-3-buten-6-one	: 6599
5-Methylhept-4-en-3-one	: 6600
3-Methyl-5-hepten-2-one	: 6601
5-Methylhept-5-en-3-one	: 6605
Methyl hexacosanoate	: 6625
Methyl 2,4-hexadienoate	: 6702
2-Methylhexa-4,5-dione	: 6809
3-Methylhexanal	: 6720
2-Methylhexane	: 6834
(+)-3-Methylhexane	: 6835
Methylhexaneamine	: 6730
3-Methyl-2,5-hexanedione	: 6810
(R)-2-Methyl-1-hexanol	: 6938
5-Methyl-1-hexanol	: 6941
3-Methyl-2-hexanol	: 6943
5-Methyl-2-hexanol	: 6944
(±)-2-Methyl-3-hexanol	: 6945
3-Methyl-3-hexanol	: 6946
(±)-5-Methyl-3-hexanol	: 6947
3-Methyl-2-hexanone	: 6976
4-Methyl-2-hexanone	: 6977
2-Methyl-1-hexene	: 7036
3-Methyl-1-hexene	: 7037
4-Methyl-1-hexene	: 7038
5-Methyl-1-hexene	: 7039
2-Methyl-2-hexene	: 7040
Methyl 3-hexenoate	: 7062
4-Methyl-2-hexenoic acid	: 7056
5-Methyl-2-hexenoic acid	: 7057
5-Methyl-2-hexenol	: 7078
4-Methyl-5-hexen-3-ol	: 7085
4-Methyl-5-hexen-2-one	: 7098
5-Methyl-1-hexyne	: 7120
2-Methyl-3-hexyne	: 7122
2-Methyl-3-hexyn-2-ol	: 7129
2-Methyl-2H-indazole	: 7242
Methylhydrazine	: 7161
Methyl hydrocinnamate	: 2139
α-Methylhydrocinnamic acid	: 2137
Methyl hydrogen sulfate	: 11577
Methyl p-hydroxybenzoate	: 2637
Methyl 11-hydroxyhendecanoate	: 11949
Methyl 11-hydroxyundecanoate	: 11949
1-Methylindan	: 7255
2-Methylindan	: 7256
4-Methylindan	: 7257
5-Methylindan	: 7258
4-Methyl-1-indanone	: 7290
1-Methylindazole	: 7239
3-Methylindazole	: 7240
5-Methylindazole	: 7241
1-Methylindene	: 7275
2-Methylindene	: 7276
3-Methylindene	: 7277
4-Methylindene	: 7278
6-Methylindene	: 7279
7-Methylindene	: 7280
7-Methylindole	: 7317
Methyl iodide	: 7518
6-Methyl-α-ionone	: 3884
Methylisobutylcarbinol acetate	: 8947
Methyl isobutyrate	: 10301
Methyl isocrotonate	: 3848
Methyl isocyanate	: 7520
	: 7655
Methyl isonipecotate	: 9644
Methyl isopentyl ketone	: 6978
Methylisopropenylacetylene	: 9116
Methyl isopropenyl carbinol	: 3871
α-Methyl-p-isopropylhydrocinnamaldehyde	: 2117
Methyl isopropyl ketone	: 3668
6-Methyl-3-isopropylsalicylic acid	: 2642
1-Methylisoquinoline	: 7381
3-Methylisoquinoline	: 7382
4-Methylisoquinoline	: 7383
6-Methylisoquinoline	: 7384
8-Methylisoquinoline	: 7385
Methyl isothiocyanate	: 7521
Methyl isovalerate	: 3552
5-Methylisoxazole	: 7405
Methyl ketone	: 10411
2-Methyllactonitrile	: 10108
Methyl laurate	: 5315
6-Methyllepidine	: 11262
Methyl linoleate	: 8228
Methyl malonate	: 9943
Methyl mercaptan	: 7546
Methyl mercaptan, perfluoro	: 7528
Methyl methacrylate	: 10663
Methyl methanesulfonate	: 7532
3-Methyl-6-methoxy-2,5-dihydroxy-1,4-benzoquinone	: 4260
4-Methyl-4-methoxy-2-pentanone	: 8978
Methyl (2-methylbutyl) ether	: 3394
Methyl 2-methylbutyl ketone	: 6977
Methyl N-methylisonipectinate	: 9645
Methyl 4-methyl-5-oxohexanoate	: 6903
Methyl 3-methyl-4-oxopentanoate	: 8898
Methyl 3-methyl-3-pentenoate	: 9086
N-Methylmorpholine	: 7696
1-Methylnaphthalene	: 7915
α-Methylnaphthalene	: 7915
2-Methylnaphthalene	: 7916
β-Methylnaphthalene	: 7916

4-Methylnaphthalene-1-carboxaldehyde	: 7756	
Methyl nicotinate	: 10906	
Methyl nitrate	: 8075	
Methyl nitrite	: 8091	
4-Methyl-2-nitrobenzophenone	: 7627	
3-Methyl-1-nitrobutane	: 3400	
trans-Methyl o-nitrocinnamate	: 10677	
2-Methylnitrocyclohexane	: 4444	
1-Methyl-1-nitrocyclohexane	: 4443	
1-Methyl-4-nitrocyclohexane	: 4445	
1-Methyl-1-nitrocyclopentane	: 4830	
1-Methyl-2-nitrocyclopentane	: 4831	
2-Methyl-2'-nitrodiphenyl ether	: 2005	
4-Methyl-1-nitronaphthalene	: 7925	
2-Methyl-1-nitronaphthalene	: 7926	
N-Methyl-N'-nitro-N-nitrosoguanidine	: 6284	
N-Methyl-N-nitroso-p-toluenesulfonamide	: 2212	
Methylnitrosourea	: 12019	
1-Methyl-1-nonadecanol	: 5359	
2-Methylnonane	: 8140	
3-Methylnonane	: 8141	
4-Methylnonane	: 8142	
5-Methylnonane	: 8143	
3-Methyl-1-nonanol	: 8166	
4-Methyl-4-nonanol	: 8170	
5-Methyl-5-nonanol	: 8171	
8-Methyl-1-nonanol	: 8168	
2-Methyl-5-nonanone	: 8176	
2-Methylnon-1-ene	: 8185	
Methyl nonyl ketone	: 11958	
8-Methyl-4-nonyne	: 8202	
Methyl 9,12-octadecadienoate	: 8229	
9-Methyloctadecanoic acid	: 8?34	
4-Methyl-3,5-octadiene	: 8329	
2-Methyloctane	: 8408	
3-Methyloctane	: 8409	
4-Methyloctane	: 8410	
3-Methyloctanoic acid	: 8433	
3-Methyl-1-octanol	: 8461	
4-Methyl-1-octanol	: 8462	
2-Methyl-2-octanol	: 8464	
3-Methyl-3-octanol	: 8466	
6-Methyl-3-octanol	: 8468	
4-Methyl-4-octanol	: 8469	
3-Methyl-2-octanone	: 8475	
4-Methyl-3-octanone	: 8476	
7-Methyl-3-octanone	: 8477	
7-Methyl-4-octanone	: 8479	
2-Methyl-1-octene	: 8505	
2-Methyl-5-octen-4-one	: 8523	
Methyl octyl ketone	: 5075	
7-Methyl-3-octyne	: 8537	
Methyl (2-octyn-1-yl) ether	: 8536	
Methyl oleate	: 8307	
4-Methylolphenol	: 1884	
Methyl orange	: 2222	
4-Methyloxazole	: 8583	
Methyloxirane	: 8621	
Methyl β-oxoheptyl ketone	: 8137	
2-Methyl-4-oxo-3-(2-propenyl)-2-cyclopenten-1-yl 2,2-dimethyl-3-(2-methyl-1-propenyl) cyclopropane carboxylate	: 336	
Methylparaben	: 2637	
Methyl pentadecanoate	: 8646	
2-Methyl-1-pentadecanol	: 8649	
3-Methyl-1,2-pentadiene	: 8666	
cis-2-Methyl-1,3-pentadiene	: 8668	
4-Methyl-1,3-pentadiene	: 8670	
2-Methyl-1,4-pentadiene	: 8871	
3-Methylpentane	: 8824	
3-Methyl-2,4-pentanediol	: 8794	
2-Methyl-1-pentanol	: 8937	
2-Methyl-2-pentanol	: 8940	
3-Methyl-2-pentanol	: 8941	
4-Methyl-2-pentanol	: 8942	
2-Methyl-3-pentanol	: 8943	
2-Methyl-1-pentene	: 9052	
3-Methyl-1-pentene	: 9053	
4-Methyl-1-pentene	: 9054	
2-Methyl-2-pentene	: 9055	
(cis)-3-Methyl-2-pentene	: 9057	
4-Methyl-trans-2-pentene	: 9058	
cis-4-Methyl-2-pentene	: 9059	
4-Methyl-2-pentenoic acid	: 9078	
cis-3-Methyl-2-pentenoic acid methyl ester	: 9085	
4-Methyl-3-pentenol	: 9096	
2-Methyl-3-penten-2-ol	: 9097	
3-Methyl-3-penten-2-ol	: 9098	
Methyl 4-pentenyl ketone	: 6585	
2-Methyl-1-penten-3-yne	: 9116	
3-Methyl-3-penten-1-yne	: 9117	
Methylpentylacetylene	: 8530	
3-Methylpentylamine	: 8706	
(3-Methylpentyl)benzene	: 2010	
2-Methylpentyl bromide	: 8714	
3-Methylpentyl bromide	: 8715	
1-Methyl-2-pentylcyclohexane	: 4446	
Methyl pentyl ether	: 8822	
Methyl pentyl ketone	: 6510	
Methyl pentyl sulfide	: 8826	
2-Methylpentyl trifluoroacetate	: 268	
3-Methyl-1-pentyne	: 9134	
Methyl 2-pentynoate	: 9139	
3-Methylphenanthrene	: 9180	
4-Methylphenanthrene	: 9181	
2-Methylphenazine	: 9195	
Methyl phenoxyacetate	: 230	
2-Methyl-4-phenyl-2-butanol	: 2150	
1-(3-Methylphenyl)ethanol	: 1858	
1-(4-Methylphenyl)ethanol	: 1859	
1-(3-Methylphenyl)ethyl alcohol	: 1858	
3-Methyl-5-phenylhydantoin	: 7221	
1-Methyl-2-phenylhydrazine	: 7166	
Methyl phenyl ketone	: 5864	
Methyl β-phenylpropionate	: 2139	
2-Methyl-3-phenylpropionic acid	: 2137	
1-Methyl-3-phenylpropylamine	: 2119	
3-Methyl-2-phenylpyridine	: 10991	
2-Methyl-5-phenylthiophene	: 11774	
Methyl 2-phenylvinyl ketone	: 3881	
Methyl picrate	: 1946	
3-Methylpicric acid	: 9434	
2-Methylpiperazine	: 9619	
N-Methylpiperidine	: 9679	
1-Methyl-1,3-propanediol	: 3345	
Methyl propionate	: 10296	
2-Methyl-6-propionylphenol	: 10444	
1-Methyl-2-propylacetylene	: 7115	
Methylpropylcarbinol	: 8916	
2-Methylpropyl chloroformate	: 4024	
1-Methyl-1-propylcyclohexane	: 4449	
1-Methyl-2-propylcyclohexane	: 4450	
1-Methyl-3-propylcyclohexane	: 4451	
1-Methyl-4-propylcyclohexane	: 4452	
3-Methyl-1-propylcyclohexanol	: 4551	
2-Methyl-1-propylcyclohexanol	: 4550	
1-Methyl-1-propylcyclopentane	: 4833	
cis-1-Methyl-2-propylcyclopentane	: 4834	
trans-1-Methyl-2-propylcyclopentane	: 4835	
cis-1-Methyl-3-propylcyclopentane	: 4836	
trans-1-Methyl-3-propylcyclopentane	: 48?7	
Methyl propyl ether	: 10069	
1-Methyl-1-propylhexanol	: 8170	
Methyl propyl ketone	: 8957	
Methyl propyl ketone oxime	: 8985	
Methyl propyl sulfide	: 10088	
2-Methyl-2-propyltrimethylene butylcarbamate carbamate	: 3933	
2-Methylpyrazine	: 10810	
1-Methylpyrazole	: 10834	
4-Methylpyrazole	: 10836	
3-Methylpyridazine	: 10856	
Methyl 3-pyridyl ketone	: 5868	
2-Methylpyrimidine	: 11070	
4-Methylpyrimidine	: 11071	
5-Methylpyrimidine	: 11072	
N-(4-Methyl-2-pyrimidyl)sulfanilamide	: 2199	
3-Methylpyrrole	: 11132	
N-Methylpyrrolidine	: 11168	
N-Methyl-2-pyrrolidone	: 11185	
4-Methylquinaldine	: 11255	
3-Methylquinoline	: 11278	
5-Methylquinoline	: 11280	
6-Methylquinoline	: 11281	
7-Methylquinoline	: 11282	
8-Methylquinoline	: 11283	
6-Methyl-7-quinolinol	: 11320	
2-Methylquinoxaline	: 11337	
6-Methylquinoxaline	: 11338	
Methyl red	: 2587	
6-Methyl-β-resorcylic acid	: 2573	
Methyl salicylate	: 2634	
3-Methylsalicylic acid	: 2629	
4-Methylsalicylic acid	: 2630	
5-Methylsalicylic acid	: 2631	
Methylsilane	: 11437	
Methyl silicate	: 11508	
Methyl silyl ether	: 11436	
o-Methylstyrene	: 1640	
m-Methylstyrene	: 1641	
p-Methylstyrene	: 1642	
α-Methylstyrene	: 1972	
Methyl styryl ketone	: 3881	
5-Methyl-3-sulfanilamidoisoxazole	: 2198	
Methyl sulfate	: 11577	
Methyl sulfocyanate	: 11705	
3-Methylsulfolane	: 11789	
Methylsulfonic acid	: 7530	
N-Methyltaurine	: 5624	
17-Methyltestosterone	: 358	
2-Methyltetrahydrofuran	: 6171	
3-Methyltetrahydrofuran	: 6172	
2-Methylthiazole	: 11678	
4-Methylthiazole	: 11679	
4-Methyl-5-thiazolecarboxylic acid	: 11669	
4-Methyl-5-thiazoleethanol	: 11677	
N1-(4-Methyl-2-thiazolyl)sulfanilamide	: 2200	
(Methylthio)acetonitrile	: 286	
Methyl thiocyanate	: 11705	
2-(Methylthio)ethanol	: 5749	
2-Methylthiophene	: 11769	
3-Methylthiophene	: 11770	
3-(Methylthio)propionaldehyde	: 9786	
Methylthiouracil	: 11108	
8-Methyltocol	: 2785	
Methyl trichloroacetate	: 253	
Methyltrichlorosilane	: 11465	
2-Methyltricosane	: 11856	
7-Methyltridecane	: 11879	
Methyl tridecanoate	: 11882	
12-Methyl-1-tridecanol	: 11886	
Methyltrimethylene glycol	: 3345	
3-Methyl-2,4,6-trinitrophenol	: 9434	
Methyltriphenoxysilane	: 11443	
Methyltriphenylsilane	: 11444	
N-Methyltryptophan	: 11923	
β-Methylumbelliferone	: 2831	
2-Methyl-1-undecanol	: 11955	
Methyl undecyl ketone	: 11889	
Methyl 10-undecynoate	: 11991	
Methyl 9-undecynoate	: 11990	
6-Methyluracil	: 11069	
3-Methyluric acid	: 10765	
7-Methyluric acid	: 10766	
2-Methylvaleraldehyde	: 8693	
Methyl valerate	: 8891	
Methyl vincaminate	: 5349	
Methylvinylacetylene	: 9111	
Methyl vinyl ether	: 5908	
Methyl vinyl ketone	: 3875	
Methysticin	: 10790	
Metiazinic acid	: 9500	
Metoxon	: 3939	
Metronidazole	: 7206	
Metyrapone	: 10462	
Metyridine	: 10979	
Michler's Base	: 786	
Michler's ketone	: 7593	
Mimosine	: 11011	
Mirex	: 7644	
Molybdenum hexacarbonyl	: 7662	
Monobenzone	: 9453	
Monofluoroacetic acid	: 189	
Monomethylarsine	: 446	
Monophenylphosphine	: 9519	
Montanic acid	: 8221	
Montanyl alcohol	: 8222	
Monuron	: 11999	
Moquizone	: 11198	
Morin	: 2816	
Morphine	: 7666	
Morphine, N-oxide	: 7668	
Morpholine, 2,6-dimethyl-4-tridecyl-	: 11898	
Morpholine, N-(2-ethoxyethyl)	: 7693	

Mucic acid	: 6193
Murexide	: 11097
Muscone	: 4763
Mustard gas	: 5645
Mustard oil	: 10575
Myacide	: 1853
Mycophenolic acid	: 7052
β-Myrcene	: 8330
Myrcene	: 8330
Myricyl alcohol	: 11821
Myristaldehyde, dimethyl acetal	: 11617
Myristic acid	: 11622
Myristicin	: 2428
Myristonitrile	: 11619
Myristoyl chloride	: 11635
Nandrolone phenpropionate	: 5408
9,10-Naphthacenequinone	: 7707
1-Naphthaldehyde, 4-methyl-	: 7756
Naphthalene, 2-benzyl-	: 7950
Naphthalene, 2-amino-1,4-dimethyl	: 8064
Naphthalene, 1,2-diamino	: 7795
Naphthalene, 1,4-diamino	: 7796
Naphthalene, 1,8-diamino	: 7798
Naphthalene, 2,3-diamino	: 7799
2,5-Naphthalenedisulfonic acid	: 7868
Naphthalene-2,7-disulfonic acid	: 7869
Naphthalene, 1,2,3,4,9,10-hexahydro	: 7897
Naphthalene, 1-(1-hydroxyethyl) (DL)	: 7911
Naphthalene, 2-(isopentyloxy)-	: 7918
Naphthalene, 1-isopropyl-7-methyl	: 7924
Naphthalene, 5-isopropyl-1,3,8-trimethyl-	: 8015
Naphthalene, 8-isopropyl-1,2,5-trimethyl-	: 8014
1-Naphthalenesulfonic acid, 8-hydroxy, sulfone	: 7840
Naphthalene, 2-vinyl-	: 7882
2-Naphthalenol, 1,1'-methylenebis-, monosodium salt	: 11517
Naphthalic acid	: 7805
Naphthalic acid, diethyl ester	: 7806
Naphthalol	: 2645
Naphthaloyl chloride	: 7804
Naphthanil Blue B Base	: 3033
1H-Naphth[2',1':4,5]indeno[1,7a-c]pyrrole, con-5-enin-3-amine deriv.	: 4143
Naphtho[1,8-de]-1,3,2-dioxathiin, 2-oxide	: 7840
1-Naphthoic acid	: 7758
2-Naphthoic acid	: 7759
2-Naphthoic acid, 5-bromo-	: 7769
1-Naphthoic acid, 5-chloro-	: 7762
1-Naphthoic acid, 3,4-dihydro-	: 7764
2-Naphthoic acid, 1,2,3,4-tetrahydro-	: 7778
1-Naphthol	: 8017
2-Naphthol	: 8018
2-Naphthol, 1-benzyl	: 8040
1-Naphthol, 2,3-dichloro	: 8025
2-Naphthol, 1,4-dimethyl-	: 8027
2-Naphthol-6,8-disulfonic acid	: 7871
2-Naphthol-3,6-disulfonic acid	: 7876
1-Naphthol-4-sulfonic acid	: 7971
1-Naphthol-2-sulfonic acid	: 7973
2-Naphthol-6-sulfonic acid	: 7974
β-Naphthoquinoline	: 2437
1,2-Naphthoquinone	: 7842
1,4-Naphthoquinone	: 7843
1,4-Naphthoquinone, 3-acetonyl-5,8-dihydroxy-6-methoxy-2-methyl-	: 7855
1,2-Naphthoquinone, 3-bromo-	: 7844
1,4-Naphthoquinone, 5-chloro-	: 7847
1,4-Naphthoquinone, 5,6-dichloro	: 7851
1,4-Naphthoquinone, 5-hydroxy-2-methyl	: 7863
1,2-Naphthoquinone-4-sulfonic acid,sodium salt	: 7970
Naphthoresorcinol	: 7834
2-Naphthoxyacetic acid	: 217
1-Naphthylacetic acid	: 7734
1-Naphthylamine	: 7708
2-Naphthylamine	: 7709
1-Naphthylamine, 5-bromo-	: 7711
β-Naphthylamine-3-carboxylic acid	: 7768
1-Naphthylamine, 5,8-dihydro-	: 7714
1-Naphthylamine, N,N-diphenyl-	: 7717
2-Naphthylamine-6,8-disulfonic acid	: 7870
1-Naphthylamine-4,7-disulfonic acid	: 7872
1-Naphthylamine-4,6-disulfonic acid	: 7873
2-Naphthylamine-1-sulfonic acid	: 7961
1-Naphthylamine-4-sulfonic acid	: 7962
1-Naphthylamine-5-sulfonic acid	: 7963
1-Naphthylamine-8-sulfonic acid	: 7965
1-Naphthylamine, N-o-tolyl-	: 7723
2-Naphthylamine, N-o-tolyl-	: 7725
2-Naphthyl benzoate	: 8021
1-Naphthyl bromide	: 7739
1-Naphthyl chloride	: 7780
1,2-(1,8-Naphthylene)benzene	: 5966
Naphthyleneethylene	: 6
(±)-1-(α-Naphthyl)ethanol	: 7911
1-Naphthyl ether	: 7939
N-(1-Naphthyl)ethylenediamine	: 5514
1-Naphthyl isocyanate	: 7904
1-Naphthyl isothiocyanate	: 7906
1-Naphthyl mercaptan	: 8007
2-Naphthyl mercaptan	: 8008
1-Naphthyl pentyl ether	: 7943
2-Naphthyl pentyl ether	: 7944
2-Naphthyl 1-pentyl ether	: 7944
1-Naphthyl salicylate	: 2644
α-Naphthyl salicylate	: 2644
2-Naphthyl salicylate	: 2645
β-Naphthyl salicylate	: 2645
β-Naphthylsulfonic acid	: 7960
1-Naphthyl(2-tolyl)amine	: 7723
2-Naphthyl(2-tolyl)amine	: 7725
1-Naphthylurea	: 12022
2-Naphthylurea	: 12023
Naptalam	: 2702
Narceine	: 2586
Narcobarbital	: 11079
Naringenin	: 2802
Naringin	: 2795
Navidrex	: 2853
Nealbarbital	: 11090
Neocuproine	: 9192
Neohexane	: 3289
Neopentane	: 9910
Neopentane, 1,1-dibromo	: 9886
Neopentanoyl chloride	: 10503
Neopentyl alcohol	: 10373
Neopentyl glycol	: 10005
Neophyl chloride	: 1143
Neopine	: 7671
Nepetalactone	: 4758
Nephramid	: 32
Nephril	: 2857
Nerol	: 8334
Neurine	: 5878
Niacinamide	: 10890
Nialamide	: 10910
Nickel tetracarbonyl	: 8068
Nicotine	: 10992
Nicotinic acid	: 10894
Nicotinic acid benzyl ester	: 10911
Nicotinyl alcohol	: 10976
Nigelline	: 2672
Nikethamide	: 10891
Nimbecetin	: 2843
Ninhydrin	: 7267
Nioxime	: 4379
Nipagin A	: 2621
Nipecotic acid	: 9641
Nirvotin	: 1877
Nitarsone	: 459
Nitramine	: 818
Nitranilic acid	: 4259
Nitric acid, anhydride with acetic acid	: 108
Nitrilotriacetic acid	: 6256
5-Nitro-2-acetamidothiazole	: 85
α-Nitroacetanilide	: 80
Nitroacetic acid, ethyl ester	: 218
5-Nitro-2-aminothiazole	: 11666
o-Nitroaniline	: 822
m-Nitroaniline	: 823
p-Nitroaniline	: 824
5-Nitro-o-anisidine	: 775
2-Nitroanisole	: 1935
9-Nitroanthracene	: 426
5-Nitroanthracene	: 426
3-Nitroanthranilic acid	: 2506
5-Nitrobarbituric acid	: 11102
Nitrobenzene	: 2049
p-Nitrobenzeneazosalicylic acid	: 2648
m-Nitrobenzenecarboxylic acid	: 2704
p-Nitrobenzenecarboxylic acid	: 2705
o-Nitrobenzoic acid	: 2703
m-Nitrobenzoic acid	: 2704
p-Nitrobenzoic acid	: 2705
4-Nitrobenzoic acid chloride	: 2906
5-Nitrobenzotriazole	: 2873
p-Nitrobenzoyl chloride	: 2906
p-Nitrobenzyl alcohol	: 1908
4-Nitrobenzyl chloride	: 1208
p-Nitrobenzyl cyanide	: 967
o-Nitrobiphenyl	: 3085
p-Nitrobiphenyl	: 3087
p-Nitrobromobenzene	: 1077
3-Nitro-4-bromobenzoic acid	: 2529
2-Nitro-5-bromobenzoic acid	: 2530
Nitrobromomethane	: 7473
1-Nitrobutane	: 3412
2-Nitro-1-butanol	: 3640
3-Nitro-2-butanol	: 3642
Nitrocarbol	: 7522
2-Nitro-3-chlorobenzoic acid	: 2546
2-Nitro-5-chlorothiophenol	: 9269
Nitrocyclohexane	: 4454
Nitrocyclopentane	: 4839
3-Nitrodibenzofuran	: 5166
p-Nitrodiphenyl	: 3087
Nitroerythrol	: 3432
Nitroethane	: 5607
2-Nitroethanol	: 5752
2-Nitro-2-ethyl-1,3-propanediol	: 10008
Nitroform	: 7562
Nitrofurantoin	: 7222
Nitrofurazone	: 7144
3-[(5-Nitrofurfurylidene)amino]-2-oxazolidinone	: 8589
Nitroglycerin	: 10197
Nitroguanidine	: 6287
1-Nitroheptane	: 6429
1-Nitrohexane	: 6841
5-Nitroisoquinoline	: 7386
Nitromersol	: 7437
Nitromethane	: 7522
1-Nitro-3-methylbutane	: 3400
(Nitromethyl)cyclohexane	: 4455
(Nitromethyl)cyclopentane	: 4840
Nitromide	: 614
1-Nitronaphthalene	: 7929
2-Nitronaphthalene	: 7930
1-Nitro-2-naphthol	: 8034
1-Nitrooctane	: 8412
Nitropentaerythritol	: 9990
1-Nitropentane	: 8832
2-Nitropentane	: 8833
3-Nitropentane	: 8834
2-Nitro-1-pentanol	: 8950
3-Nitro-2-pentanol	: 8952
Nitrophenide	: 5247
o-Nitrophenol	: 9435
m-Nitrophenol	: 9436
p-Nitrophenol	: 9437
p-Nitrophenylacetic acid	: 933
1-Nitro-4-phenylbenzene	: 3087
4-Nitro-o-phenylenediamine	: 1281
(4-Nitrophenyl)hydrazine	: 7168
(4-Nitrophenyl)methanol	: 1908
o-Nitrophenylpropiolic acid	: 10740
2-Nitrophenyl 2-tolyl ether	: 2005
4-Nitrophenylurea	: 12025
1-Nitropropane	: 10116
2-Nitropropane	: 10117
1-Nitro-2-propanol	: 10396
5'-Nitro-2'-propoxyacetanilide	: 84
4-Nitropyridine N-oxide	: 10993
5-Nitroquinoline	: 11284
6-Nitroquinoline	: 11285
7-Nitroquinoline	: 11286
4-Nitroquinoline oxide	: 11287
3-Nitrosalicylic acid	: 2646
5-Nitrosalicylic acid	: 2647
Nitrosobenzene	: 2059
N-Nitrosodiethylamine	: 5432
N-Nitrosodimethylamine	: 7450

p-Nitroso-N,N-dimethylaniline	: 720
p-Nitrosodiphenylamine	: 829
N-Nitroso-N-diphenylamine	: 830
N-Nitrosomorpholine	: 7699
1-Nitroso-2-naphthol	: 8035
4-Nitrosophenol	: 9438
N-Nitrosopiperidine	: 9688
N-Nitrosopyrrolidine	: 11170
5-Nitrothiazolylamine-2	: 11666
2-Nitrothiophene	: 11775
3-Nitrothiophene	: 11776
o-Nitrotoluene	: 2000
m-Nitrotoluene	: 2001
p-Nitrotoluene	: 2002
5-Nitro-o-toluidine	: 801
2-Nitro-4-(trifluoromethyl)-1-chlorobenzene	: 1217
Nitrourea	: 12024
Nitrous acid, benzyl ester	: 8097
Nitrous acid, isopropyl ester	: 8092
Nitroxoline	: 11323
4-Nitro-o-xylene	: 1475
Nonacosanyl alcohol	: 8100
6,8-Nonadien-2-one, 5-isopropyl-8-methyl-, (E)-(S)-(+)-	: 8113
Nonaldehyde	: 8117
2-Nonen-1-al	: 8179
Nonyl acetate	: 220
Nonyl alcohol	: 8157
Nonyl mercaptan	: 8145
Nonyl methyl ketone	: 11958
19-Norandrostenolone phenylpropionate	: 5408
Norboral	: 2192
2-Norbornanecarboxaldehyde	: 2916
Norbornane, 7,7-dimethyl-2-methylene-, (±)-	: 2923
2-Norbornanone, 3-methylene-	: 2958
2-Norbornene, 2,3-dimethyl-	: 2968
Norcarane	: 2915
Norcholanic acid	: 8210
Nordazepam	: 2411
Nordihydroguaiaretic acid	: 1515
Norephedrine, N-ethyl	: 1871
Norepinephrine	: 1500
Norhyoscyamine	: 916
Normethadone	: 6968
Normorphine	: 7665
Normotiroide	: 11665
Normuscol	: 4760
Nornicotine	: 11017
Norpempidine	: 9711
cis-Norpinic acid	: 4183
Norpinic acid, dimethyl ester	: 4184
Norsulfazole	: 2206
Nortestosterone phenylpropionate	: 5408
Norvalamine	: 3209
Noscapine	: 7356
Novoldiamine	: 8736
Novonal	: 9003
Novoscabin	: 2720
Ocimene	: 8485
Octaacetyl sucrose	: 6229
Octabenzone	: 7621
Octachlorocyclopentene	: 4932
Octacosyl alcohol	: 8222
Octacosylbenzene	: 2065
Octacosylcyclohexane	: 4458
Octadecamethylcyclononasiloxane	: 4726
Octadecamethyloctasiloxane	: 8482
Octadecanal, dimethyl acetal	: 8237
Octadecane, 1-cyclohexyl-	: 4459
Octadecane, 1-cyclopentyl-	: 4842
1,18-Octadecanedicarboxylic acid	: 5352
Octadecane, 1-phenyl-	: 2066
3-Octadecanol (DL)	: 8279
Octadecenyl aldehyde	: 8288
Octadecylbenzene	: 2066
Octadecyl mercaptan	: 8244
Octadecyltrichlorosilane	: 11470
α,ω-Octadiene	: 8319
2,2,3,3,4,4,5,5-Octafluoro-1-pentanol	: 8953
1,2,3,4,4a,9,10,10-Octahydrophenanthrene	: 9184
1,2,3,4,5,6,7,8-Octahydrophenanthrene	: 9185

Octalene	: 5177
Octamethylcyclotetrasiloxane	: 4993
(all-E)-2,6,10,14,19,23,27,31-	: 4053
Octamethyl-2,6,8,10,12,14,16,18,20,22,24,26,30-dotriacontatridecaene	
Octamethylene dibromide	: 8369
Octamethylpyrophosphoramide	: 5220
1,1,1,3,5,7,7,7-Octamethyltetrasiloxane	: 11647
Octamethyltrisiloxane	: 11914
Octamylamine	: 6354
Octane, 2-bromo (DL)	: 8357
Octane, 1-cyclohexyl-	: 4460
Octane, 1-cyclopentyl-	: 4843
Octane, 2-phenyl-	: 1981
Octane, 1-phenyl-	: 2067
2-Octanethiol (DL)	: 8419
Octane, 2,2,4-trimethyl	: 8420
Octanoic acid, 3,5,7,8-tetrachloro-2,2,3,4,4,5,6,6,7,8,8-undecafluoro	: 8442
Octatetraene	: 8483
Octatrienal	: 8484
4-Octene, 4-chloro (cis)	: 8500
3-Octen-2-ol, 2-methyl (cis)	: 8519
3-Octen-5-ol, 4-methyl	: 8521
1-Octen-5-one, 7-methyl	: 8525
1-Octen-3-yne-5-ol, 5-methyl	: 8527
Octodecylcyclohexane	: 4459
Octylacetylene	: 5097
Octyl alcohol	: 8443
Octyl bromide	: 8356
Octyl chloride	: 8359
9-Octyldocosane	: 5275
Octyl enanthate	: 6468
Octyl fluoride	: 8405
Octyl heptanoate	: 6468
(±)-2-Octyl iodide	: 8407
Octyl mercaptan	: 8418
Octyl methyl ketone	: 5075
1-Octylnaphthalene	: 7937
Octyl phthalate	: 1344
Octyltrichlorosilane	: 11471
1-Octyne-3-ol, 3-methyl	: 8544
2-Octynol	: 8543
Oenanthylidene acetone	: 5093
Oestra-1,3,5(10)-triene-3,17β-diol	: 5399
Oleanolic acid	: 8547
Oleic acid	: 8294
Oleic acid, benzyl ester	: 8309
Oleic acid, isopentyl ester	: 8306
Oleyl alcohol	: 8312
Opianic acid	: 2608
Orcinol	: 1531
Orcinolcarboxylic acid	: 2573
Oretic	: 2855
Orexine	: 11196
Orotic acid	: 11050
Oroxylin A	: 2810
o-Orsellinic acid	: 2573
Orthanilic acid	: 2215
Orthesin	: 2497
Orthocaine	: 2501
Orthoform	: 2501
Orthoformic acid, triisobutyl ester	: 10080
Ouabain	: 4040
Ovahormon benzoate	: 5403
Ovex	: 2221
10H-9-Oxaanthracene	: 12049
1-Oxa-2-azacyclopentadiene	: 7402
1-Oxa-2-aza-1H-indene	: 2389
1-Oxa-3-azaindene	: 2877
7-Oxabicyclo[4.1.0]heptane, β,β-carotene-3,3'-diol deriv.	: 4056
7-Oxabicyclo[2.2.1]heptane-2,3-dicarboxylic acid, disodium salt	: 5370
Oxaceprol	: 9766
Oxacyclohexane	: 10800
Oxacyclopentadiene	: 6025
Oxacyclotetradecane-2,10-dione, 4-[[(2,6-dideoxy-3-C-methyl-3-O-methyl-α-L-ribo-hexopyranosyl)oxy]-14-eth	: 5396
1-Oxa-3,4-diazacyclopentadiene	: 8563
1,2,4-Oxadiazolidin-3,5-dione, 2-tert-butyl-4-(2,4-dichloro-5-isopropoxyphenyl)-	: 8565

1,2,4-Oxadiazolidine-3,5-dione, 2-(3,4-dichlorophenyl)-4-methyl-	: 7643
Oxalacetic acid	: 3339
Oxalacetic ester	: 3340
Oxalic acid	: 5553
Oxalic acid, diallyl ester	: 5564
Oxalic acid dichloride	: 5585
Oxalic acid, dicyclohexyl ester	: 5559
Oxalic acid dihydrate	: 5561
Oxalic acid, diisobutyl ester	: 5557
Oxalic acid, disodium salt	: 5566
Oxalic acid, lead salt	: 5568
Oxalyl chloride	: 5585
Oxamic acid	: 104
Oxamide	: 5493
Oxane	: 10800
1,4-OxathiiN-3-carboxamide, 5,6-dihydro-2-methyl-N-phenyl-	: 4039
1,3-Oxazole	: 8575
2,4-Oxazolidinedione, 3-allyl-5-methyl-	: 8586
2,4-Oxazolidinedione, 3-(3,5-dichlorophenyl)-5-ethenyl-5-methyl-	: 12048
2-Oxazoline, 2-amino-5-phenyl-	: 8574
Oxetane, gluconic acid deriv.	: 6211
3-Oximino-2-octanone	: 8399
3-Oximino-2-pentanone	: 8803
1-Oxindene	: 2438
Oxindole, 4,7-dimethyl	: 7330
Oxindole, 1-methyl-3-ethyl	: 7332
Oxiniacic acid	: 10909
2-Oxo-3-acetyltetrahydrofuran	: 6139
(+)-2-Oxo-10-bornanesulfonic acid	: 2929
2-Oxobutanedioic acid diethyl ester	: 3340
18-Oxocorticosterone	: 9746
Oxoethanoic acid	: 223
β-Oxoglutaric acid	: 8782
2-Oxo-2H-1-benzopyran-3-carboxylic acid	: 2775
Oxole	: 6025
2-Oxo-5-methylhex-3-ene	: 7094
Oxophenylarsine	: 447
2-Oxosuccinic acid	: 3339
Oxotremorine	: 11188
Oxyacanthine	: 8631
Oxybenzone	: 7620
Oxycholin	: 4101
Oxycinchophen	: 11233
Oxydiethylene acetate	: 5754
Oxydimorphine	: 3002
Oxymethurea	: 11995
Oxyphenonium bromide	: 5443
Oxytetracycline	: 7705
Palmitaldehyde, dimethyl acetal	: 6636
Palmitic acid	: 6644
Palmitic acid, benzyl ester	: 6655
Palmitoleic acid	: 6670
Palustric acid	: 9166
Pamacyl	: 2498
Pamoic acid	: 7777
Panthesin	: 8926
Pantocaine	: 2531
Pantocid	: 2565
Pantolactone	: 6143
Pantothenic acid	: 329
D(+)-Pantothenyl alcohol	: 3202
d-Pantothenyl alcohol	: 3202
Papaveraldine	: 7611
Papaverine	: 7378
Parabanic acid	: 7226
Paracetamin	: 29
Paraldehyde	: 11908
Paraldol	: 5191
Parasorbic acid	: 10791
Parasympatol	: 1887
Parathion	: 9583
Paraxenol	: 3090
Pargyline	: 1823
Paroxypropione	: 10450
Parsley camphor	: 2423
Patchouli alcohol	: 7586
Patulin	: 6183
Pebulate	: 3976
Pelargonic acid	: 8146
Pelargonic acid, butyl ester	: 8147
Pelargonic acid, 9-fluoro	: 8151
Pelargonic acid, heptyl ester	: 8152

Pelargonic acid, 3-methyl : 8153
Pelargonic acid 3-methyl, ethyl ester : 8155
Pelargonic acid, pentyl ester : 8156
Pelargonidin chloride : 2848
(-)-Pelletierine : 10485
Pellotine : 7394
Pempidine : 9691
Penicillamine : 12043
Penicillamine cysteine disulfide : 12042
Penicillic acid : 6701
Penicillin G, sodium salt : 11655
Pentabromoacetone : 10477
1,2,3,4,5-Pentabromobenzene : 2082
Pentabromobenzene : 2082
1,1,1,2,2-Pentabromoethane : 5616
Pentabromophenol : 9439
1,2,3,4,5-Pentachlorobenzene : 2084
Pentachlorobenzene : 2084
Pentachlorobenzoic acid : 2712
1,2,2,3,4-Pentachlorobutane : 3420
1,1,1,4,4-Pentachloro-2-butene : 3805
Pentachloroethane : 5617
Pentachlorophenol : 9440
1,1,1,2,3-Pentachloropropane : 10125
1,1,2,3,3-Pentachloropropane : 10126
Pentacosylbenzene : 2088
Pentacosylcyclohexane : 4462
1H-Pentadecafluoroheptane : 6432
Pentadecaldehyde : 8640
Pentadecane, 1-cyclopentyl- : 4844
Pentadecanoic acid, 15-hydroxy,ω- : 8562
lactone
sec-Pentadecyl alcohol : 8647
1-Pentadecylamine : 8641
3-Pentadecylcatechol : 1543
Pentadecylic acid : 8645
Pentadecyl methyl ketone : 6320
1,3-Pentadiene-1-carboxylic acid : 6698
2,4-Pentadienoic acid, 5(3,4- : 8677
metylenedioxyphenyl)
Pentaerythritol : 9989
Pentaerythrityl tetraacetate : 9988
Pentaerythrityl tetranitrate : 9990
Pentaethylbenzene : 2090
1,2,3,4,5-Pentafluorobenzene : 2091
Pentafluorobenzene : 2091
Pentafluorophenol : 9442
1,1,1,2,2-Pentafluoropropane : 10128
Pentafluoropropionic acid, methyl : 10317
ester
2,3,4,5,6-Pentafluorotoluene : 2094
3,5,7,2',4'-Pentahydroxyflavone : 2816
3,5,7,3',4'-Pentahydroxyflavone : 2817
Pentaiodobenzene : 2096
Pentalin : 5617
1,2,3,4,5-Pentamethylbenzene : 2097
Pentamethylbenzene : 2097
Pentamethylbenzoic acid : 2713
2,4,6,8,10- : 4902
Pentamethylcyclopentasiloxane
Pentamethylene : 4774
Pentamethylene acetate : 8791
Pentamethylene chloride : 8747
Pentamethylene diiodide : 8758
Pentamethylene glycol : 8787
Pentamethylene oxide : 10800
2,2,4,6,6-Pentamethylheptane : 6433
2,2,3,3,4-Pentamethylpentane : 8836
2,2,3,4,4-Pentamethylpentane : 8837
Pentamethylphenol : 9443
1,1,3,5,5-Pentamethyl-1,3,5- : 11915
triphenyltrisiloxane
Pentane, 1-amino-3-methyl : 8706
3-Pentanecarboxylic acid : 3514
Pentane, 2-chloro (d) : 8720
Pentane 3-chloro-2,2-dimethyl-3- : 8727
ethyl
Pentane, 1,5-dicyclohexyl- : 4464
1,2-Pentanediol (DL) : 8790
1,5-Pentanedione : 8734
2,4-Pentanedione, : 2912
bis,beryllium(II)complex
Pentane, 1,3-diphenyl- : 1697
Pentane, 1,5-diphenyl- : 2098
Pentane, 1,1-diphenyl- : 2108
Pentane, 2-iodo (DL) : 8819
Pentane, 3-isopropyl-2,4-dimethyl- : 8766

Pentane, 2-methyl-3,3-diethyl : 8757
Pentane, 3-methyl-3-phenyl : 1691
Pentane, 3-methyl-1-phenyl : 2010
Pentane, 2-methyl-2-phenyl : 1455
Pentane, perfluoro : 8807
Pentane, perfluoro,2-trifluoro methyl : 8854
Pentane, 3-phenyl : 1701
Pentane, 1-phenyl : 2107
Pentane, 2,2,3-trimethyl-3-ethyl : 8815
Pentane, 2,3,4-trimethyl-3-ethyl : 8816
1,4,5-Pentanetriol : 8851
1,2,3-Pentanetriol : 9119
Pentanoic acid, 4-[(17β)-androstan- : 4094
17-yl]-
Pentanoic acid, 5-chloro-4- : 6141
hydroxy,γ-lactone
3-Pentanol, 1,5-bis(tetrahydro-2- : 6166
furyl)-
1-Pentanol, 2,4-dimethyl (DL) : 8927
1-Pentanol, 3-methyl (DL) : 8938
1-Pentanol-2-one : 8971
1-Pentanol-4-one : 8974
2-Pentanol-3-one : 8975
3-Pentanol-2-one : 8972
1-Pentanol, 1-phenyl- : 1846
1-Pentanol, 5-phenyl- : 2101
2-Pentanol, 2-phenyl- : 1904
2-Pentanol, 5-phenyl- : 1106
3-Pentanol, 1-phenyl (d) : 2152
3-Pentanol, 3-phenyl- : 1855
2-Pentanone, 3-benzylidene- : 8987
2-Pentanone, 3-methyl (DL) : 8980
Pentasa : 2499
Pentatriacontylcyclopentane : 4845
Pent-4-en-1-al : 9001
trans-2-Pentene : 9005
cis-2-Pentene : 9006
trans-2-Pentenedioic acid diethyl : 9042
ester
1-Pentene, 2-ethyl-4-methyl- : 6837
1-Pentene, 2-isopropyl- : 6836
2-Pentene, 3-methyl (cis) : 9057
1-Pentene, 3-methyl-3-ethyl : 9047
1-Pentene, 4-methyl-3-ethyl : 9048
2-Pentene, 2-methyl-3-ethyl : 9049
2-Pentene, 4-methyl-3-ethyl (trans) : 9050
2-Pentene, 4-methyl-3-ethyl (cis) : 9051
trans-3-Pentenenitrile : 9061
1-Pentene-3-one : 9104
2-Pentene, 3-phenyl- : 1699
2-Pentene, 2-phenyl- : 1950
1-Pentene, 5-phenyl- : 2106
1-Pentene, 2-propyl- : 6426
2-Pentenoic acid, 2-methyl (trans) : 9076
2-Pentenoic acid, 3-methyl (cis) : 9077
3-Pentenyl bromide : 9015
trans-3-Penten-1-yne : 9113
cis-3-Penten-1-yne : 9114
Pentetic acid : 6257
Pentoxone : 8978
Pentyl acetate : 227
sec-Pentyl acetate (R) : 8918
sec-Pentyl alcohol : 8916
: 8917
tert-Pentyl bromide : 3240
2-Pentyl bromide : 8711
3-Pentyl bromide : 8712
Pentyl butanoate : 3579
Pentyl butyrate : 3579
Pentylcyclohexane : 4465
Pentylenetetrazole : 11654
Pentyl fluoride : 8817
Pentyl formate : 6015
Pentyl heptanoate : 6470
Pentyl isocyanide : 8821
Pentyl ketone : 11962
Pentyl mercaptan : 8844
3-Pentyl mercaptan : 8846
2-Pentylnaphthalene : 7942
Pentyl nitrite : 8096
Pentyl nonanoate : 8156
p-tert-Pentylphenol : 9339
Pentyl phenyl ether : 2109
Pentylpropiolic acid : 8538
4-(3-Pentyl)pyridine : 10968
1-Pentynyl bromide : 9124
Perbenzoic acid : 1126

Perchloroacetone : 10440
Perchlorobenzoic acid : 2712
Perchlorocyclopentadiene : 4766
Perchlorocyclopentene : 4932
Perchloromethyl mercaptan : 7527
Perchloronaphthalene : 7933
Perchloropropylene : 10571
Pereirine : 4154
Perfluoroacetone : 10441
Perfluorobenzene : 1757
Perfluorobutane : 3255
Perfluoro-2-butene : 3804
Perfluoro-2-butyltetrahydrofuran : 6109
Perfluorocyclobutane : 4194
Perfluorocyclohexane : 4381
Perfluorocyclohexene : 4643
Perfluorodecalin : 7934
Perfluorodecane : 5050
Perfluoro-2,3-dimethylbutane : 3414
Perfluoroethane : 5596
Perfluoroheptane : 6418
Perfluoro-1-heptene : 6560
Perfluorohexane : 6843
Perfluoro-1-hexene : 7032
Perfluoroisobutane : 10050
Perfluoro(methylcyclohexane) : 4499
Perfluoro(2-methylpentane) : 8854
Perfluoro-3-methylpentane : 8853
Perfluoronaphthalene : 7935
Perfluorononane : 8139
Perfluorooctane : 8415
Perfluoropentane : 8807
Perfluoropropane : 10118
Perfluoropropene : 10572
Perfluoropropylene oxide : 8629
Perfluorotoluene : 2095
Perillaldehyde : 4633
Perillyl aldehyde : 4633
Permethylbenzoic acid : 2713
Permethyldisilane : 5228
1,4-Peroxido-p-menthene-2 : 5184
Peroxyacetic acid : 5619
Perphenazine : 9617
Peucedanin : 6188
Phellandral : 4635
β-Phellandrene : 4673
Phenacemide : 879
Phenacetin : 60
Phenacyl acetate : 5771
Phenacylamine : 5774
Phenacyl bromide : 5782
Phenacyl cyanide : 2124
Phenallymal : 11103
Phenamizole : 11671
9,10-Phenanthraquinodimethane : 9172
9,10-Phenanthraquinone, 3-hydroxy : 9176
Phenanthrene, 9-isopropenyl- : 9182
Phenanthrenequinone : 9174
9-Phenanthroic acid : 9165
o-Phenanthroline : 9189
9-Phenanthryl bromide : 9164
Phenazine, 2,3-diamino- : 9194
Phenbenzamine : 5506
Phencarbamide : 3977
Phencyclidine : 9696
[v-Phenenyltris(oxyethylene)]- : 5441
tris[triethylammonium iodide]
Phenetharbital : 11089
Phenethyl alcohol : 1592
Phenethyl alcohol, 2-ethyl : 1603
Phenethyl alcohol, 4-ethyl : 1604
Phenethyl alcohol, α,α,β-trimethyl- : 1616
Phenethylamine, N-benzyl-N,α- : 1575
dimethyl-, (+)-
Phenethylamine, 4-bromo : 1570
Phenethylamine, 2-methoxy : 1577
2-Phenethyl propionate : 10324
o-Phenetidine : 736
p-Phenetidine : 737
o-Phenetidine, 5-nitro : 739
Phenetole : 1649
Phenetole, m-chloro- : 1157
Phenetole, 2,6-dinitro- : 1650
Phenetole, 2-propenyl : 1665
Phenetole, 2-propyl : 1687
Phenicarbazide : 7145
Phenicin : 2979

Phenigam	: 2128
Phenindione	: 7270
Pheniramine	: 11010
Phenmetrazine	: 7697
Phenobarbital	: 11095
Phenol, benzoate	: 2716
Phenol, 6-tert-butyl-3,4-dimethyl	: 9324
Phenol-2-carboxylic acid	: 2612
Phenol, 2,6-di-sec-butyl-	: 9227
Phenol, 3,5-dimethyl-4-(methylthio)-, methylcarbamate	: 7646
Phenoldisulfonic acid	: 1560
Phenol, 3-(1-hydroxypropyl)	: 1872
Phenol, o-methoxy-, phosphate (3:1)	: 9387
Phenol, 2-(1-methylethoxy)-, methylcarbamate	: 10722
Phenol, 3-methyl-5-(1-methylethyl)-, methylcarbamate	: 9769
Phenol, 2-methyl-4-(3-methyl pentyl)	: 4148
Phenol, 2-(1-methylpropyl)-4,6-dinitro-	: 5183
Phenolphthalein	: 7354
Phenolphthalein, 5',5"-diisopropyl-2',2"-dimethyl-	: 7352
Phenolphthalin	: 2517
Phenolphthalol	: 1841
o-Phenolsulfonic acid	: 2225
p-Phenolsulfonic acid	: 2226
Phenolsulfonphthalein	: 9363
Phenol, 2,4,5-trichloro-, O-phosphorodichloridothioate	: 9578
Phenopyrazone	: 10843
Phenothiazine, 2-chloro-10(3(4-methyl-1-piperazinyl)propyl)	: 9501
Phenoxazine	: 9507
Phenoxyacetic acid	: 228
Phenoxyacetic acid methyl ester	: 230
Phenoxyacetic acid, 2,4,5-trichloro	: 256
Phenoxyacetonitrile	: 287
2-Phenoxybutyric acid, phenyl ester	: 3583
2-Phenoxycarbonylphenol	: 2650
2-Phenoxyethanol	: 5758
Phenprobamate	: 2149
Phensuximide	: 11161
Phentermine	: 1572
Phentydrone	: 5978
Phenylacetaldehyde	: 868
α-Phenylacetamide	: 878
Phenyl acetate	: 231
Phenylacetic acid	: 883
Phenylacetic acid, α-chloro, ethyl ester (d)	: 891
Phenylacetic acid, 3,4-methylenedioxy, methyl ester	: 2416
N-Phenylacetoacetamide	: 3207
Phenylacetone	: 10481
Phenylacetyl chloride	: 971
3-(α-Phenyl-β-acetylethyl)-4-hydroxycoumarin	: 2832
Phenyl acetylsalicylate	: 2483
9-Phenylacridine	: 318
L-Phenylalanine, 4-hydroxy-	: 11926
1-Phenylallyl alcohol	: 1868
Phenylaminoacetic acid	: 6270
4-Phenyl-2-aminobutane	: 2119
Phenyl p-aminosalicylate	: 2502
p-Phenylaniline	: 3011
9-Phenylanthracene	: 428
N-Phenylanthranilic acid	: 2715
Phenylarsonic acid	: 460
(Phenylazo)thioformic acid 2-phenylhydrazide	: 5121
Phenylbenzene	: 3008
Phenyl benzoate	: 2716
2-Phenylbenzothiazole	: 2864
Phenyl biguanide	: 7234
Phenylboric acid	: 3135
N-Phenylbutanedioic acid monoamide	: 3575
1-Phenylbutane-1,3-diol	: 3357
2-Phenyl-1,2-butanediol	: 3356
4-Phenyl-3-butene-2-one	: 3881
4-Phenyl-3-butenoic acid	: 3855
Phenylbutylcarbinol	: 1846
α-Phenylbutyramide	: 881
Phenyl butyrate	: 3584
Phenyl carbonate	: 4001
Phenyl chloroacetate	: 147
Phenyl chloroformate	: 4026
2-Phenyl-6-chlorophenol	: 3092
Phenylchlorosilane	: 11384
2-Phenylchroman	: 2782
2-Phenyl-4-chromone	: 2840
3-Phenylchromone	: 2841
α-Phenylcinnamic acid	: 941
Phenyl cyanide	: 2740
1-Phenyl-2-cyanoethane	: 2123
2-Phenylcycloheptanone	: 4236
1-Phenylcyclohexan-1-ol	: 4552
trans-2-Phenylcyclopropanecarbonyl chloride	: 4957
Phenyldichlorosilane	: 11403
N-Phenyldiethanolamine	: 5760
Phenyldihydroxyborane	: 3135
Phenyl diselenide	: 5225
1-Phenyldodecane	: 1566
3-Phenyleicosane	: 1695
o-Phenylenediamine	: 1260
m-Phenylenediamine	: 1261
p-Phenylenediamine	: 1262
o-Phenylenediamine, N,N'-dimethyl	: 1269
o-Phenylenethiourea	: 2386
1-Phenyl-1,2-ethanediol	: 5579
1-Phenylethanol	: 1896
Phenylethanolamine	: 1839
N-Phenylethanolamine	: 5759
Phenylethylamine	: 1569
Phenyl ethyl carbinol	: 1870
Phenylethyl dimethyl carbinol	: 2150
1-Phenylethylene glycol	: 5579
Phenylethylene oxide	: 8624
9-(2-Phenylethyl)heptadecane	: 2070
1-Phenyl-1-ethylhydrazine	: 7159
1-Phenylethyl propyl ketone	: 6985
Phenylethynylcarbinol	: 1876
N-Phenylformamide	: 5999
N-Phenylformoguanamine	: 11830
2-Phenylfuran	: 6159
Phenylglyceryl ether	: 10026
Phenyl glycidyl ether	: 8623
α-Phenylglycine	: 885
N-Phenylglycine	: 6270
1-Phenyl-1-heptanol	: 1881
7-Phenylheptan-3-ol	: 2102
1-Phenyl-1,3-hexadiene	: 1753
6-Phenyl-1,3-hexadiene	: 1755
6-Phenyl-2-hexanol	: 2103
6-Phenyl-5-hexen-2-one	: 7102
Phenylhydrazine	: 7189
Phenylhydrazine-4-sulfonic acid	: 2224
Phenylhydroxylamine	: 757
2-Phenylindan	: 7259
2-Phenyl-1-indanone	: 7291
Phenyl isocyanate	: 1786
2-Phenylisopropanol	: 1857
Phenyl isothiocyanate	: 1792
3-Phenyl-3-ketopropionitrile	: 2124
(±)-3-Phenyllactic acid	: 2133
DL-β-Phenyllactic acid	: 2133
Phenyl laurate	: 5317
N-Phenylmaleimide	: 11123
Phenylmalononitrile	: 9916
Phenyl mercaptan	: 2289
Phenylmercuric acetate	: 7431
Phenylmethylbarbituric acid	: 11100
Phenyl methyl ketone	: 5864
2-(Phenylmethyl)phenol	: 9454
4-(Phenylmethyl)phenol	: 9455
1-Phenyl-2-methylpropyl alcohol	: 1901
1-Phenylnaphthalene	: 7947
N-Phenyl-β-naphthylamine	: 7728
2-Phenyloctane	: 1981
5-Phenyl-2,4-pentadienal	: 8657
5-Phenylpentanoic acid	: 2099
1-Phenyl-1-pentanol	: 1846
3-Phenyl-2-pentene	: 1699
2-Phenyl-2-pentene	: 1950
o-Phenylphenol	: 3088
m-Phenylphenol	: 3089
p-Phenylphenol	: 3090
Phenylphosphine	: 9519
Phenyl phthalate	: 1345
4-Phenylpiperidine	: 9695
3-Phenyl-1-propanal	: 2114
1-Phenyl-1,2,3-propanetriol	: 10194
3-Phenylpropanoic acid	: 2126
2-Phenyl-2-propanol	: 1857
1-Phenyl-1-propanol	: 1870
3-Phenyl-1-propanol	: 2148
Phenylpropanolamine hydrochloride	: 1838
1-Phenylpropargyl alcohol	: 1876
3-Phenylpropionaldehyde	: 2114
β-Phenylpropionic acid methyl ester	: 2139
3-Phenylpropionitrile	: 2123
3-Phenylpropyl alcohol	: 2148
Phenylpropylmethylamine	: 1574
1-Phenyl-2-propyn-1-ol	: 1876
1-Phenyl-3-pyrazolidinone	: 10844
1-Phenyl-2-pyrazoline	: 10823
2-Phenylpyridine	: 10997
3-Phenylpyridine	: 10998
4-Phenylpyridine	: 10999
2-Phenylpyrrole	: 11135
2-Phenylquinoline	: 11288
3-Phenylquinoline	: 11289
4-Phenylquinoline	: 11290
6-Phenylquinoline	: 11291
Phenyl salicylate	: 2650
Phenylseleninic acid	: 2174
4-Phenylsemicarbazide	: 7146
Phenyl stearate	: 8271
N-Phenylsuccinamic acid	: 3575
2-Phenyl-3-sulfanilamidopyrazole	: 2201
N-Phenylsulfanilic acid	: 2234
Phenyl sulfide	: 2283
Phenylsulfonyl chloride	: 2242
Phenylsulfonyl fluoride	: 2251
Phenyl thiocyanate	: 11709
2-(Phenylthio)furan	: 6160
Phenylthiourea	: 11808
Phenyltrichlorosilane	: 11473
7-Phenyltridecane	: 1764
1-Phenyltridecane	: 2342
Phenyl trimethyl ammonium iodide	: 866
Phenyltripropylsilane	: 11448
Phenylurea	: 12026
Phenylurethane	: 3965
5-Phenylvaleric acid	: 2099
δ-Phenylvaleric acid	: 2099
Phenylvinylcarbinol	: 1868
Phenyramidol	: 1913
Phenytoin	: 7219
Phenzidole	: 2384
Phloionic acid	: 8239
Phloionolic acid	: 8276
Phloretic acid	: 2134
Phloretin	: 10453
Phloroglucinol	: 2368
Phloropropiophenone	: 10489
Pholedrine	: 9403
Phorone	: 6340
Phosgene	: 4007
Phosphine, dichloro-(2,5-dimethylphenyl)	: 9553
Phosphonic acid, (2-chloroethyl)-	: 5929
Phosphonic acid, dibenzyl ester	: 9586
Phosphonic acid, isopropyl-	: 9542
Phosphonic acid, methyl diphenyl ester	: 9541
Phosphonic acid, thiono, methyl, diisopropyl ester	: 9549
Phosphonic acid, (2,2,2-trichloro-1-hydroxyethyl)-, dimethyl ester	: 11853
Phosphonodithioic acid, ethyl-, O-ethyl S-phenyl ester	: 5986
Phosphonous dichloride, 2,5-xylyl-	: 9553
Phosphoramidic acid, methyl-, 2-chloro-4-(1,1-dimethylethyl)phenyl methyl ester	: 4153
Phosphoramidic, (1-methylethyl)-, ethyl 3-methyl-4-(methylthio)phenyl ester	: 5947
Phosphoramidothioic acid, acetyl-, O,S-dimethyl ester	: 10
Phosphoramidothioic acid, O,S-dimethyl ester	: 7441
Phosphoric acid, amide,thiono,dichloride,N,N-dimethyl	: 9557
Phosphoric acid, 2-chloro-1-(2,4-dichlorophenyl)ethenyl diethyl ester	: 4078

Phosphoric acid, 2-chloro-1-(2,4,5-trichlorophenyl)ethenyl dimethyl ester : 6202
Phosphoric acid, 1,2-dibromo-2,2-dichloroethyl dimethyl ester : 7702
Phosphoric acid dichloride,thiono,2,4,5-trichlorophenyl ester : 9578
Phosphoric acid, 2,2-dichloroethenyl dimethyl ester : 5170
Phosphoric acid, 3-(dimethylamino)-1-methyl-3-oxo-1-propenyl dimethyl ester, (E)- : 5172
Phosphoric acid, triisobutyl ester : 9572
Phosphorodithioic acid, O,O-bis(1-methylethyl) S-[2-[(phenylsulfonyl)amino]ethyl] ester : 517
Phosphorodithioic acid, S-[[(4-chlorophenyl)thio]methyl] O,O-diethyl ester : 4037
Phosphorodithioic acid, S-[(6-chloro-2-oxo-3(2H)-benzoxazolyl)methyl]O,O-diethyl ester : 9514
Phosphorodithioic acid, S-[2-chloro-1-(1,3-dihydro-1,3-dioxo-2H-isoindol-2-yl)ethyl] : 5105
Phosphorodithioic acid, O,O-diethyl S-[2-(ethylthio)ethyl] ester : 5260
Phosphorodithioic acid, O,O-diethyl S-[(ethylthio)methyl] ester : 9513
Phosphorodithioic acid, S-[(1,3-dihydro-1,3-dioxo-2H-isoindol-2-yl)methyl] O,O-dimethyl ester : 9515
Phosphorodithioic acid, S-[[(1,1-dimethylethyl)thio]methyl] O,O-diethyl ester : 11592
Phosphorodithioic acid, O,O-dimethyl S-[2-(methylamino)-2-oxoethyl] ester : 5009
Phosphorodithioic acid, S,S'-1,4-dioxane-2,3-diyl O,O,O',O'-tetraethyl ester : 5201
Phosphorodithioic acid, O-ethyl S,S-dipropyl ester : 5934
Phosphorodithioic acid, S,S'-methylene O,O,O',O'-tetraethyl ester : 5931
Phosphorodithioic acid, S-[(5-methoxy-2-oxo-1,3,4-thiadiazol-3(2H)-yl)methyl], O,O-dimethyl ester : 7645
Phosphorothioic acid, O-(4-bromo-2-chlorophenyl) O-ethyl S-propyl ester : 9762
Phosphorothioic acid, O-[2-chloro-1-(l-methylethyl)-1H-imidazol-4-yl] O,O-diethyl ester : 7340
Phosphorothioic acid, O-[2,5-dichloro-4-(methylthio)phenyl] O,O-diethyl ester : 4092
Phosphorothioic acid, O,O-diethyl O-(3,5,6-trichloro-2-pyridinyl) ester : 4089
Phosphorothioic acid, O-[2-(diethylamino)-6-methyl-4-pyrimidinyl] O,O-dimethyl ester : 9730
Phosphorothioic acid, O-[2-(diethylamino)-6-methyl-4-pyrimidinyl] O,O-diethyl ester : 9729
Phosphorothioic acid, O,O-diethyl O-[4-(methylsulfinyl)phenyl] ester : 5954
Phosphorothioic acid, O,O-diethyl O-pyrazinyl ester : 11720
Phosphorothioic acid, O,O-dimethyl O-[3-methyl-4-(methylthio)phenyl] ester : 5955
Phosphorothioic acid, O,O-dimethyl O-(3,5,6-trichloro-2-pyridinyl) ester : 4090
Phosphorothioic acid, O-[4-[(dimethylamino)sulfonyl]phenyl] O,O-dimethyl ester : 5946
Phosphorothioic acid, O,O-dimethyl O-(3-methyl-4-nitrophenyl) ester : 5951
Phosphorothioic acid, O-(6-ethoxy-2-ethyl-4-pyrimidinyl) O,O-dimethyl ester : 5945
Phosphorothioic acid, S-[2-(ethylsulfinyl)ethyl] O,O-dimethyl ester : 8632

Phosphorothioic acid, S-[2-(ethylthio)ethyl] O,O-dimethyl ester : 5103
Phosphorothioic acid, O,O'-(thiodi-4,1-phenylene) O,O,O',O'-tetramethyl ester : 1
Phosphorotrithious acid, S,S,S-tributyl ester : 7438
Phthalamide : 1324
Phthalic acid : 1327
Phthalic acid, 3,5-dichloro- : 1340
Phthalic acid, dioctyl ester : 1344
Phthalic acid, monoamide,N(1-naphthyl) : 2702
Phthalic acid, tetrachloro-, ethyl ester : 1353
Phthalic acid, tetrachloro-, monoethyl ester : 1353
Phthalic anhydride : 7342
Phthalic anhydride, 3,4-dichloro : 7343
Phthalic anhydride, 4,5-dichloro- : 7344
Phthalimide : 7362
Phthaloyl chloride : 1318
Phthaloylmalonic acid, diethyl ester : 7248
2-(N4-Phthalylaminobenzenesulfonamide)thiazole : 2726
Phthalylsulfonazole : 2726
Phthalylsulphathiazole : 2726
Phthiocol : 7861
Phycitol : 3431
Phylloquinone : 7865
Physeteric acid : 11638
Physostigmine : 11191
Physoteric acid : 11638
Phytanic acid : 6658
Phytol : 6672
Phytomenadione : 7865
Piazthiole : 11658
Picein : 5818
Picloram : 10896
2-Picoline : 10980
3-Picoline : 10981
4-Picoline : 10982
γ-Picoline, 2-amino : 10887
3-Picoline, 2-phenyl- : 10991
Picolinic acid : 10893
4-Picolyl alcohol : 10977
Picramic acid : 9207
Picric acid : 9497
Picrolonic acid : 10848
Picropodophyllin : 6191
Picrotoxin : 7579
Picryl chloride : 1244
Picryl sulfide : 2288
Pilocarpine : 6151
Pimaric acid : 9170
Pimeclone : 4611
Pimelic acid : 6393
Pimelic acid, diethyl ester : 6394
Pimelic ketone : 4571
Pinacol : 3350
Pinacolone : 3657
Pinacone : 3350
Pinane (DL) : 2934
(-)-2-Pinene : 2971
(-)-α-Pinene : 2971
(1S,5S)-(-)-α-Pinene : 2971
(±)-α-Pinene : 2972
β-Pinene : 2925
2-Pinene : 2973
2-Pinen-7-one : 2977
2-Pinen-4-one, (1R,5R)-(+)- : 2978
Pinic acid diethyl ester : 4171
(±)-Pinol : 8561
Pinosylvin : 1545
Pipecolic acid : 9640
N,N'-Piperazinedicarboxylic acid, diethyl ester : 9606
Piperidic acid : 3464
Δ 3-Piperidine : 11020
Piperidine, 4-(aminomethyl)- : 9873
Piperidine, 4-benzoyl-N-methyl : 7409
Piperidine, 2-benzyl- : 9699
Piperidine, 4-benzyl- : 9700
Piperidine, N-tert-butyl : 9653
Piperidine, N-carboxyethyl, ethyl ester : 9704

Piperidine, 4(4-chlorophenyl)-1(3(4-fluorobenzoyl)propyl)-4-hydroxy : 3651
Piperidine, N,2-dimethyl (DL) : 9648
Piperidine, 2-ethyl (DL) : 9668
Piperidine, 3-ethyl (DL) : 9669
Piperidine, 2-(2-hydroxyethyl) : 9663
Piperidine, 4-(2-hydroxyethyl) : 9664
Piperidine, 3-methyl (DL) : 9681
Piperidine, N-(3-methyl-3-pentyl) : 9671
1-Piperidinepropionic acid, ethyl ester : 9704
Piperidine, 2(1-propionyl) : 10484
Piperidine, 3-vinyl- : 9666
Piperidinium, 1,1-dimethyl-, chloride : 7430
Piperidone : 9654
4-Piperidone, 1,2,3,6-tetramethyl : 9725
Piperine : 9632
Piperitenone : 4718
(±)-Piperitone : 4716
Piperonal : 2417
Piperonaldehyde : 2417
Piperonylic acid : 2418
trans-Piperylene : 8661
cis-Piperylene : 8682
Pipradrol : 9676
Pipsyl chloride : 2247
Pivalamide, N,N-diethyl : 9794
Pivaldehyde : 9783
Pivalic acid : 10258
Pivalic acid chloride : 10503
Pivalil : 6811
Pivaloin : 6975
Plasmocid : 9877
Plumericin : 11904
Podocarpa-7,13-dien-15-oic acid, 13-isopropyl- : 9167
Podophyllotoxin : 6192
Polythiazide : 2857
Populin : 6223
Pramoxine : 7683
Prednisolone : 9738
Prednisone : 9739
Pregnanediol : 9742
5β-Pregnane-3,20-dione : 9743
Pregnan-3α-ol-20-one : 9745
4-Pregnene-11β,17α,20β,21-tetrol-3-one : 9757
Pregnenolone : 9756
Prehnitenol : 9473
Prenoxdiazine hydrochloride : 9657
Primeverose : 6237
Primidone : 11066
Pristane : 8644
Probarbital : 11093
Probenecid : 2595
Procainamide hydrochloride : 602
Procaine : 2493
Proflavine : 316
Proformiphen : 2149
Progallin P : 2732
Promethazine : 9504
Prontosil : 2209
Propallylonal : 11078
Propanal, 2-methyl-2-(methylsulfonyl)-, O-[(methylamino)carbonyl]oxime : 335
Propanal, 2-methyl-2-(methylthio)-, O-[(methylamino)carbonyl]oxime : 334
Propanamide, N-(3,4-dichlorophenyl)- : 10200
Propanamide, N,N-diethyl-2-(1-naphthalenyloxy)- : 8067
1-Propanearsonic acid : 461
Propane, 1,3-bis(1-methyl-4-piperidyl) : 9702
Propane, 3-bromo-1,2-epoxy (DL) : 8600
1-Propanecarboxylic acid : 3460
Propane, 2-chloromethyl-1,1,2,3-tetrachloro : 10146
Propane, 2-chloro-1-phenyl (l) : 1234
Propane, 1-cyclohexyl-3-cyclopentyl- : 4341
Propane, 1-cyclohexyl- : 4984
Propane, 1-cyclopropyl- : 1698
Propane, 1,1-dibenzyl : 1698
Propane, 1,2-dichloro (DL) : 9892
Propane, 1,2-dicyclohexyl- : 4430
Propane, 2,2-dicyclohexyl : 4432
Propane, 1,3-dicyclohexyl- : 4466

Propane, 1,1-dicyclohexyl-	: 4472
Propane, 1,3-dicyclohexyl-2-ethyl	: 4402
Propane, 1,3-dicyclohexyl-2-methyl-	: 4447
1,2-Propanediol, 3-amino(DL)	: 9984
1,3-Propanediol, 2-methyl-2-propyl,N-butyl carbamate,carbamate	: 3933
Propane, 1,2-diphenoxy-	: 1970
Propane, 1,3-dipheny-1-ethyl	: 1697
Propane, 1,3-diphenyl-	: 2121
Propane, 1,1-diphenyl-	: 2171
Propanenitrile, 2-[[4-chloro-6-(cyclopropylamino)-1,3,5-triazin-2-yl]amino]-2-methyl-	: 9759
Propanenitrile, 2-[[4-chloro-6-(ethylamino)-1,3,5-triazin-2-yl]amino]-2-methyl-	: 4161
Propane, 1-nitro-3,3,3-trifluoro	: 10184
2-Propanephosphonic acid	: 9542
1,3-Propanesulfone	: 11694
Propane sultone	: 8570
Propane, 1,2,2,3-tetrachloro-tetrafluoro	: 10149
Propanoic acid, 3-amino-	: 323
Propanoic acid, 2-amino-3-mercapto-, (R)-	: 5014
Propanoic acid, chloride, 2-chloro(d)	: 10496
Propanoic acid, 2-(3-chlorophenoxy)-	: 4141
Propanoic acid, 2-[4-(2,4-dichlorophenoxy)phenoxy]-, methyl ester	: 5171
Propanoic acid, 2,2-dichloro-3,3,3-trifluoro, propyl ester	: 10721
Propanoic acid, 3,3'-dithiobis[2-amino-, [R-(R*,R*)]-	: 5020
Propanoic acid, methylene ester	: 7654
Propanoic acid, 2-(2,4,5-trichlorophenoxy)-	: 11512
Propanolamine	: 10337
Propanolamine, N-pantoyl-	: 3202
1-Propanol, 2-[2-(2-amino-2-methylpropyl)amino]-2-methyl	: 10342
2-Propanol, 1-chloro-3-isopropoxy-	: 10355
2-Propanol, 1-diethylamino (DL)	: 10368
2-Propanol, 1,1',1'',1'''-(ethylenedinitrilo)tetra-	: 10379
2-Propanol, 1-isopropoxy-	: 10390
2-Propanol, 1-nitro-3,3,3-trichloro	: 10408
2-Propanol, 1,1,1-trifluoro(DL)	: 10409
2-Propanone, acetyl-	: 8798
2-Propanone, (methoxyphenyl)-	: 10459
Propantheline bromide	: 9832
Proparagylacetic acid	: 9138
Propargyl alcohol	: 10743
Propargyl aldehyde	: 10723
Propargyl bromide	: 10728
Propargyl chloride	: 10729
Propargylic acid, (2-nitrophenyl)	: 10740
2-Propene-1,2-dicarboxylic acid	: 3328
Propene-3,3-diol, 2-methyl,diacetate	: 10564
Propene, 2-methyl(tetramer)	: 10580
2-Propene-1-ol, 1-phenyl	: 1868
Propenoic acid, 2-chloro,ethyl ester	: 10619
Propenoic acid, 2-methyl,butyl ester	: 10651
Propenoic acid, 3(1-naphthyl)(trans)	: 10675
2-Propenyl alcohol	: 10697
2-Propenyl bromide	: 10531
2-Propenyl chloride	: 10536
2-Propenyl isothiocyanate	: 10575
N-2-Propenylurea	: 12028
Propesin	: 2507
Propham	: 3966
Propiocine	: 5396
β-Propiolactone	: 8595
Propiolic acid	: 10738
Propiolic acid, propyl-	: 7123
Propionaldehyde	: 9774
Propionaldehyde, 3-(2-tolyl)	: 2115
Propionamide	: 9791
Propionamide, N-antipyrinyl-2-(dimethylamino)-	: 9795
Propionic acid	: 10201
Propionic acid bromide	: 10490
Propionic acid, 3-bromo, isopentyl ester	: 10210
Propionic acid, 2-chloro, isobutyl ester (d)	: 10224

Propionic acid, 2-diazo-, ethyl ester	: 10236
Propionic acid, 2-diazo-, methyl ester	: 10237
Propionic acid, 2,3-dibromo-, butyl ester	: 10239
Propionic acid, 2,3-dibromo-, propyl ester	: 10243
Propionic acid, 2,2-dichloro-3,3,3-trifluoro-, methyl ester	: 10251
Propionic acid, 3-(diethylamino), ethyl ester	: 328
Propionic acid, 2,3-dihydroxy-2-methyl-, methyl ester	: 10256
Propionic acid, 2,3-diphenyl (DL)	: 2143
Propionic acid, 3,3'-iminodi-N-methyl, diethyl ester	: 330
Propionic acid, 3,3'-(methylimino)-di-, diethyl ester	: 330
Propionic acid, 2,3,3,3-tetrafluoro-, ethyl ester	: 10328
Propionic acid, 3-(2-tetrahydrofuryl)	: 6164
Propionic acid, 3,3,3-trichloro-2-(2,2,2-trichloro-1-hydroxyethoxy)-, γ-lactone	: 5212
Propionic anhydride	: 10203
Propionitrile, 2-ethoxy-	: 10102
Propionitrile, 2-methoxy-	: 10109
Propionylacetone	: 6804
Propionyl bromide	: 10490
Propionyl chloride	: 10493
Propionyl chloride, 3,3-dichloro	: 10502
Propionyl fluoride	: 10506
Propiophenone	: 10480
Propiophenone, 2'-hydroxy-3'-methyl-	: 10444
Propiophenone, 2'-hydroxy-4'-methyl-	: 10445
Propiophenone, 3-methoxy	: 10456
Propiophenone, 2-methoxy-	: 10457
Propofol	: 9226
2-Propoxyethanol	: 5764
1-Propoxy-2-propanol	: 10401
N-Propylacetanilide	: 88
Propyl acetate	: 236
Propylacetylene	: 9122
Propyl alcohol	: 10334
Propylallene	: 6676
Propylamine	: 9803
Propylamine, N,N-dichloro	: 9805
Propyl p-aminobenzoate	: 2507
Propyl amyl ketone	: 8174
Propylbenzene	: 2169
Propyl benzenesulfonate	: 2235
Propyl benzoate	: 2724
Propyl bromide	: 9836
Propyl bromoacetate	: 120
Propyl butanoate	: 3588
Propyl-tert-butylcarbinol	: 6925
Propyl carbamate	: 3969
Propyl carbonate	: 4002
Propyl chloride	: 9848
Propyl chloroacetate	: 149
1-Propyl-3-(p-chlorobenzenesulfonyl)urea	: 2208
Propyl chlorocarbonate	: 4028
trans-Propyl cinnamate	: 10693
Propyl cyanide	: 3403
Propylcyclohexane	: 4471
Propyl cyclohexanecarboxylate	: 4337
trans-4-Propylcyclohexanol	: 4560
cis-2-Propylcyclohexanol	: 4558
trans-2-Propylcyclohexanol	: 4559
4-Propylcyclohexanone	: 4613
2-Propylcyclohexanone	: 4612
trans-2-Propylcyclopentanol	: 4874
Propylcyclopentane	: 4849
cis-2-Propylcyclopentanol	: 4873
2-Propylcyclopentanone	: 4894
3-Propylcyclopentanone	: 4895
1-Propyl-1-cyclopentene	: 4935
Propylcyclopropane	: 4984
Propyl 2,3-dibromopropionate	: 10243
Propyldimethylamine	: 9807
Propylene	: 10528
Propylene, 2-bromo-3-cyclohexyl	: 4322
Propylene, 3-bromo-3,3-difluoro	: 10532
Propylene chlorohydrin	: 10347

sec-Propylene chlorohydrin	: 10349
Propylenediamine	: 9872
Propylene dibromide	: 9881
Propylene, 1,3-dibromo (trans)	: 10543
Propylene, 1,3-dibromo (cis)	: 10544
Propylenedicarboxylic acid	: 3328
Propylene, 1,1-dichloro-2-fluoro	: 10554
Propylene, 1,2-dichloro-3,3,3-trifluoro	: 10558
1,2-Propylene glycol	: 9982
1,3-Propylene glycol	: 9983
Propylene glycol β-monoethyl ether	: 10380
1,2-Propylene glycol 1-propyl ether	: 10401
Propylene oxide	: 8621
1,2-Propylenimine	: 501
N-Propylethanolamine	: 5765
Propylethylene	: 9004
Propyl fluoride	: 10046
Propyl formate	: 6018
Propylformic acid	: 3460
2-Propylfuran	: 6167
Propyl gallate	: 2732
4-Propylheptane	: 6434
Propyl heptanoate	: 6471
4-Propyl-3-heptene	: 6559
1-Propylheptyl alcohol	: 5072
Propyl hexahydrobenzoate	: 4337
Propylhexedrine	: 4385
Propyl hexyl ketone	: 5077
Propyl p-hydroxybenzoate	: 2653
Propylidene chloride	: 9891
Propyl iodide	: 10054
Propyliodone	: 10873
Propyl isobutyrate	: 10308
Propyl isocyanate	: 10059
Propyl isopropyl sulfide	: 10077
Propyl isovalerate	: 3562
Propyl laurate	: 5320
Propyl levulinate	: 8912
Propyl mercaptan	: 10156
Propyl methacrylate	: 10672
Propyl nitrate	: 8080
Propyl nitrite	: 8098
5-Propyl-1,5-octadien-3-yne	: 8341
Propylparaben	: 2653
2-Propylpentanal	: 8697
2-Propyl-1-pentene	: 6426
3-Propyl-1-pentene	: 7033
Propyl phosphite	: 9594
4-Propylpiperidine	: 9709
Propyl propionate	: 10326
3-Propylpyrrole	: 11139
2-Propylquinoline	: 11292
Propyl sulfide, 3,3'-dichloro	: 10153
2-Propylthiophene	: 11778
3-Propylthiophene	: 11779
Propylthiouracil	: 11109
N-Propyl-p-toluidine	: 815
Propyltrichlorosilane	: 11476
Propyl trifluoroacetate	: 269
Propyl 3,4,5-trihydroxybenzoate	: 2732
Propyne, 1,3-dibromo-	: 10730
Protionamide	: 10886
Protoanemonin	: 6155
Protocatechualdehyde	: 542
Protocatechuic acid	: 2571
Protocatechuic acid methylene ether	: 2418
Protopine	: 3109
Protoverine	: 4070
Provitamin D3	: 4103
Prunetin	: 2828
Prunetol	: 2807
Pseudoaconitine	: 308
Pseudocinchonine	: 4122
Pseudocodeine	: 7672
Pseudoconhydrine	: 9718
Pseudocumene	: 2358
Pseudoionone	: 11965
Pseudojervine	: 12045
Pseudomorphine	: 3002
Pseudopelletierine	: 479
Pseudotropine	: 486
Pteroylglutamic acid	: 6243
Puleganic acid	: 4784
Pulegene	: 4931
Pulegone	: 4608

Purodigin	: 4043	L-Pyroglutamic acid	: 9768	Refrigerant 113	: 5663
Purostrophan	: 4040	Pyrolan	: 3945	Refrigerant 114	: 5538
Purpurin	: 416	Pyromellitic acid	: 2257	Refrigerant 121a	: 5639
Purpurogallin	: 2408	Pyromucic acid	: 6050	Refrigerant 122a	: 5660
Putrescine	: 3258	Pyropentylene	: 4764	Refrigerant 122b	: 5658
Pyocyanine	: 9196	Pyrrole, 2-benzoyl	: 7637	Refrigerant 124	: 5486
Pyran-4-carboxylic acid, tetrahydro-	: 10774	Pyrrole-3-propionic acid, 2-[[3-(2-	: 3001	Refrigerant 133	: 5488
Pyrazinamide	: 10802	carboxyethyl)-4-methyl-5-[(3-methyl-		Refrigerant 133b	: 5487
Pyrazinoic acid	: 10803	5-oxo-4-vinyl-3-pyrrolin-2-		Refrigerant 134	: 5641
Pyrazino[2,3-d]pyrimidine	: 10747	ylidene)meth		Refrigerant 142	: 5471
N1-2-Pyrazinylsulfanilamide	: 2202	2-Pyrrolidinecarboxylic acid	: 9784	Refrigerant 142b	: 5470
Pyrazole, 4-amino-3-methyl-1-	: 10813	2-Pyrrolidone	: 11180	Refrigerant 245	: 10128
phenyl-		2-Pyrrolidone, 3,3-dimethyl	: 11182	Refrigerant C317	: 4177
1H-Pyrazole, 3-chloro-5-methyl-4-	: 4083	3-Pyrroline	: 11118	Renazide	: 2851
nitro-		Pyruvaldehyde	: 9788	Renese	: 2857
2-Pyrazoline	: 10822	Pyruvic acid	: 10314	Resacetophenone	: 5803
2-Pyrazoline, 1-phenyl-	: 10823	Pyrylene	: 9111	Resazurin	: 9508
2-Pyrazoline, 3-phenyl	: 10824	Quercetin	: 2817	Rescinnamine	: 12060
1H-Pyrazolium, 1,2-dimethyl-3,5-	: 5175	Quercetin 7-methyl ether	: 2815	Reserpic acid	: 12063
diphenyl-, methyl sulfate		Quercetin-3-L-rhamnoside	: 2796	Reserpine	: 12061
3-Pyrenecarboxylic acid	: 10854	D-Quercitol	: 4074	Resorcinol	: 1498
Pyrethrin 1	: 4960	Quercitrin	: 2796	Resorcinol, 4-acetyl	: 8490
Pyrethrin 2	: 4962	Quillaic acid	: 8546	Resorcinol diglycidyl ether	: 8625
Pyribenzamine	: 5505	Quinaldic acid	: 11227	Resorcinol dipropyl ether	: 1556
3(2H)-Pyridazinone, 5-amino-4-	: 4080	Quinaldine	: 11277	Resorcinol, 5-iodo	: 1524
chloro-2-phenyl-		Quinaldinic acid	: 11227	Resorcinol, 4-isobutyl-	: 1539
3(2H)-Pyridazinone, 4-chloro-5-	: 8213	Quinaldonitrile	: 11223	Resorcinolphthalein	: 11522
(methylamino)-2-[3-		Quinalizarin	: 413	β-Resorcylaldehyde	: 541
(trifluoromethyl)phenyl]-		Quinamine	: 6178	β-Resorcylic acid	: 2569
Pyridine, 3-amino-2,6-dichloro	: 10862	4-Quinazolinone, 2-benzyl-1-methyl	: 11200	Retene	: 9183
Pyridine, 2-benzyl-	: 11004	Quinethazone	: 11197	Reychler's acid	: 2929
Pyridine, 3-benzyl-	: 11005	Quinic acid	: 4338	Rhamnetin	: 2815
Pyridine, 2-tert-butyl-	: 10942	Quinidine	: 4126	Rhamnol	: 11558
Pyridine, 3-tert-butyl-	: 10943	Quinine	: 4127	Rhamnol	: 11558
Pyridine-3-carbinol	: 10976	Quinine carbonate	: 4128	Rheadine	: 11342
4-Pyridinecarbinol	: 10977	Quinine, formate	: 4129	Rhein	: 371
Pyridine, 2-chloro-6-	: 8071	Quinine hydrochloride	: 4130	Rhizopterin	: 2494
(trichloromethyl)-		Quinine salicylate	: 4131	Rhodamine, B	: 12056
Pyridine, 2,6-di-tert-butyl-	: 10874	Quinine sulfate	: 4132	Rhodanine	: 11685
Pyridine, 2-[[2-	: 5507	Quininic acid	: 11234	Rhodanine, 5-[p-	: 11684
(dimethylamino)ethyl]-2-		Quininone	: 4134	(dimethylamino)benzylidene]-	
thenylamino]-		Quinizarin	: 392	(-)-Rhodinol	: 8516
Pyridine, 2(2-dimethylaminoethyl-4-	: 5511	Quinoline, 2-amino-	: 11202	Rhodizonic acid	: 4686
methoxybenzylamino)		Quinoline, 6-amino-	: 11205	Riboflavine	: 11344
Pyridine, 2-(2-hydroxypropyl)	: 10951	Quinoline, 4-amino-3-bromo	: 11207	Riboflavin lumichrome	: 2472
Pyridine, 2-(2-isopropoxyethyl)	: 10983	Quinoline, 2-amino-4-methyl	: 11208	β-D-Ribofuranoside, adenine-9	: 319
Pyridine, 3-isopropyl-	: 10985	Quinoline, 7-amino-2-methyl	: 11211	β-D-Ribofuranoside, guanine-9	: 6289
Pyridine N-oxide	: 10994	Quinoline, 8-amino-6-methyl-	: 11212	1-β-D-Ribofuranosyluracil	: 12033
Pyridine, 4-(3-pentyl)	: 10968	Quinoline, decahydro (trans, DL)	: 11247	Ricinine	: 10883
Pyridinium chloride	: 10969	Quinoline, 2,4-dimethyl-6-hydroxy	: 11317	Ricinoleic acid	: 8303
4(1H)-Pyridinone, 1-methyl-3-	: 5981	Quinoline, 2,4-dimethyl-8-hydroxy	: 11318	Risocaine	: 2507
phenyl-5-[3-(trifluoromethyl)phenyl]-		Quinoline, 5-hydroxy	: 11305	Ronnel	: 9584
1,5-Pyridopyridine	: 8065	Quinoline, 7-hydroxy	: 11307	Roquessine	: 4143
Pyridoxamine dihydrochloride	: 10978	Quinoline, 7-hydroxy-6-methyl	: 11320	p-Rosolic acid	: 4278
Pyridoxine hydrochloride	: 10934	Quinoline, 4-methoxy-2-[3,4-	: 11216	Rotenone	: 2783
α-Pyridylamine	: 10857	(methylenedioxy)phenethyl]-		Rubeanic acid	: 5588
3-Pyridylamine	: 10858	Quinolinic acid	: 10927	Rubijervine	: 11515
2-(2-Pyridyl)pyridine	: 3099	p-Quinone	: 4257	Rubixanthin	: 4062
4-(4-Pyridyl)pyridine	: 3104	Quinone	: 4257	Rufigallol	: 406
5-[p-(2-	: 2654	Quinone tetrachloride	: 4268	Rutecarpine	: 7334
Pyridylsulfamyl)phenylazo]salicylic		Quinovic acid	: 12036	Rutinose	: 6236
acid		Quinovose	: 6235	Sabadine	: 4067
N1-2-Pyridylsulfanilamide	: 2203	Quinoxaline, 6-amino-	: 11332	Sabinane (d)	: 11814
3-Pyridylsulfonic acid	: 11018	N1-(2-Quinoxalinyl)sulfanilamide	: 2205	Sabinene (DL)	: 2981
Pyrilamine	: 5511	Quintozene	: 2087	D-Saccharic acid	: 6205
Pyrimethamine	: 11051	Quinuclidine	: 480	Saccharin	: 2388
Pyrimidine, 2-amino-4,5-dimethyl-	: 11044	3-Quinuclidinol	: 485	Safranal	: 4250
2,4(1H,3H)-Pyrimidinedione, 5-	: 3139	Racemic tartaric acid	: 3311	Safrole	: 2429
bromo-6-methyl-3-(1-methylpropyl)-		Radicinin	: 10785	Salicin	: 6222
2,4(1H,3H)-Pyrimidinedione, 5-	: 11591	Raffinose	: 6219	Salicyl alcohol	: 1882
chloro-3-(1,1-dimethylethyl)-6-		Randox	: 44	Salicylaldehyde	: 564
methyl-		Raubasine	: 8571	Salicylaldoxime	: 579
5-Pyrimidinemethanol, α-(2-	: 5948	Raunescine	: 12064	Salicylamide	: 616
chlorophenyl)-α-(4-chlorophenyl)-		Reductic acid	: 4942	Salicylamide O-acetic acid	: 103
5-Pyrimidinemethanol, α-(2-	: 8219	Reductinic acid	: 4942	Salicylanilide	: 617
chlorophenyl)-α-(4-fluorophenyl)-		Refrigerant 10	: 7540	Salicylhydroxamic acid	: 613
4(1H)-Pyrimidinone, 5-butyl-2-	: 5932	Refrigerant 11	: 7555	Salicylic acid	: 2612
(ethylamino)-6-methyl-		Refrigerant 12	: 7500	p-Salicylic acid	: 2613
N1-2-Pyrimidinylsulfanilamide	: 2204	Refrigerant 13	: 7486	Salicylic acid, 4-benzamido-	: 2514
Pyrithyldione	: 10946	Refrigerant 13B1	: 7475	Salicylic acid, 3-chloro, ethyl ester	: 2541
Pyrocalciferol	: 5381	Refrigerant 14	: 7541	Salicylic acid, 3,5-dibromo-, methyl	: 2559
Pyrocarbonic acid diethyl ester	: 5169	Refrigerant 20	: 7554	ester	
Pyrocatechol	: 1497	Refrigerant 21	: 7502	Salicylic ether	: 2819
Pyrochol	: 4095	Refrigerant 22	: 7478	Salicylsulfonic acid	: 2655
Pyrocoll	: 5223	Refrigerant 31	: 7481	Saligenol	: 1882
Pyrogallic acid	: 2366	Refrigerant 32	: 7504	Salirepol	: 1520
Pyrogallol	: 2366	Refrigerant 50	: 7459	Salsalate	: 2616
				Salsoline	: 7397

Salvarsan	: 9275
Sanguinarine	: 2432
α-Santalol	: 9093
β-Santalol	: 9103
Santene	: 2968
β-Santenol	: 2938
α-Santonin	: 8057
Sarcosine	: 6268
Sarmentogenin	: 4046
Sarpagine	: 11347
Sarsasapogenin	: 11531
Scarlet red	: 8033
Schradan	: 5220
Scilliroside	: 3142
Scopolamine	: 908
Scopoletin	: 2829
Scopoline	: 7575
Sebacic acid	: 5042
Sebacic acid di(2-ethylhexyl) ester	: 5043
Sebacic acid, monomitrile, methyl ester	: 8148
Sebacil	: 4203
Sebacoin	: 4206
7,13a-Secoberbin-13a-one, 7-methyl-2,3:9,10-bis(methylenedioxy)-	: 3109
Seedrin	: 5177
Selanthrene	: 11353
Semicarbazide hydrochloride	: 7143
Seminose	: 7425
Semioxamazide	: 105
Senecialdehyde	: 3700
Senfoel	: 10575
Septochol	: 4095
Serenase	: 3651
Serpentine	: 8572
Silane, allyldiethoxymethyl-	: 11408
Silane, methyl tri-(4-tolyl)	: 11445
Silane, (oxy-p-phenylene)bis[trimethyl-	: 11497
Silane, phenyl-tributyl	: 11455
Silane, propyl trimethyl	: 11496
Silane, tetravinyl	: 11449
Silane, 4-trimethylsilylphenoxy-trimethyl	: 11497
Silane, tris(allyloxy)vinyl-	: 11427
Silanol, dimethyl ethyl	: 11501
3-Silapentane	: 11409
2-Silapropane	: 11414
Silicon tetraethoxide	: 11506
Silicon tetramethoxide	: 11508
Silin, 1,1-diphenyl perhydro	: 11360
Silmurin	: 3142
Siloxane, butylmethyl(cyclic tetramer)	: 4995
Sinomenine	: 7676
α1-Sitosterol	: 11556
β-Sitosterol	: 11558
γ-Sitosterol	: 11559
Skatole	: 7314
Skimmetin	: 2826
Skimmianine	: 6181
Skimmin	: 2821
Smilagenin	: 11533
Sodium cyanate	: 4164
Sodium N-lauroylsarcosinate	: 6269
Solanone	: 8113
Solasodine	: 11529
Solbrol A	: 2621
Solestro	: 5403
Sophoricoside	: 2824
Sorbic acid	: 6698
L-Sorbinose	: 11519
Sorbitol	: 6206
Spantol	: 2149
Sparteine	: 7571
Spasmocromona	: 2838
Spermine	: 3259
Spinulosin	: 4260
Spiro[3,5,8-ethanylylidene-1H-pyrano[3,4-c]pyridine-10,3'-[3H]indol]-2'(1'H)-one, 5-ethenyl	: 6203
Spiro[2.2]pentane	: 11527
Spirosal	: 2622
Spir[5.]undecane	: 11537
Squalane	: 11606

Squalene	: 11602
Stachydrine	: 11177
Stannane, hydroxytriphenyl-	: 11911
Stannane, tricyclohexylhydroxy-	: 5011
Stanolone	: 349
Staticin	: 2721
Stearic acid	: 8247
Stearic acid, 18-bromo	: 8250
Stearic acid, cyclohexyl ester	: 8253
Stearic acid, isopentyl ester	: 8265
Stearic acid, 14-methyl	: 8262
Stearic acid, 9-methyl	: 8264
Stearic acid, 12-oxo, ethyl ester	: 8269
Stearic acid, pentyl ester	: 8270
Stearic acid, phenyl ester	: 8271
Stearic acid, 9,10,12,13-tetrabromo, methyl ester	: 8274
Stearic acid, tetrahydrofurfuryl ester	: 8275
9-Stearolic acid	: 8316
Stearyl alcohol	: 8278
Stearyl mercaptan	: 8244
Stercorin	: 4109
Stigmastanol	: 11557
Stigmasterol	: 11555
trans-Stilbene	: 1619
cis-Stilbene	: 1620
Stilbene, 2-bromo (trans)	: 1080
Stilbene, 2-bromo (cis)	: 1081
4-Stilbenecarbonitrile (cis)	: 2768
Stilbene, 2,2'-dimethyl-	: 1622
Stilbene, 2,4-dinitro (trans)	: 1496
Stimulina	: 6249
Streptomycin, sulfate	: 11561
Strodival	: 4040
Strophanthidin	: 4049
Strychnine	: 11563
Strychnine nitrate	: 11565
Strychnine, sulfate	: 11566
Stylophorine	: 4071
Styphnic acid	: 1554
Styrallyl alcohol	: 1896
Styrene	: 1624
Styrene, α-chloro-2,4-dimethyl	: 1152
Styrene, β-chloro, α-4-dimethyl	: 1188
Styrene, 2,3-dichloro-	: 1367
Styrene, 2,6-dichloro-	: 1369
Styrene, 3,5-dichloro-	: 1370
Styrene, 2,5-dichloro-	: 1371
Styrene, 2,4-dichloro-	: 1373
Styrene epoxide	: 8624
Styrene glycol	: 5579
Styrylacetic acid	: 3855
Styryl methyl ketone	: 3881
Suberic acid	: 8386
Suberone	: 4234
Suberonitrile	: 8385
Succinamide	: 3257
Succinanil	: 11162
Succinanilic acid	: 3575
Succinanilic acid, 4'-(2-thiazolylsulfamoyl)-	: 3578
Succinic acid	: 3294
Succinic acid, amide,N-phenyl	: 3575
Succinic acid, sec-butyl-, diethyl ester	: 3332
Succinic acid, di-sec-butyl ester	: 3297
Succinic acid, dibutyl ester	: 3301
Succinic acid dichloride	: 3366
Succinic acid, 2,3-dichloro-, (±)-	: 3302
Succinic acid, diethyl ester	: 3303
Succinic acid N,N-dimethylhydrazide	: 3334
Succinic acid dinitrile	: 3291
Succinic acid, diphenyl ester	: 3320
Succinic acid, ethyl ester	: 3335
Succinic acid, mercapto-, diethyl ester, S-ester with O,O-dimethyl phosphorodithioate	: 3316
Succinic acid monoanilide	: 3575
Succinic acid, monopropyl ester	: 3338
Succinic acid, oxo,diethyl ester	: 3340
Succinic acid, pentyl ester	: 3337
Succinic acid, propyl ester	: 3338
Succinic anhydride	: 6093
Succinic anhydride, phenyl (DL)	: 6096
Succinimide	: 11152

Succinimide, N-benzyl-	: 11163
Succinonitrile	: 3291
Succinyl chloride	: 3366
Succinylsulfathiazole	: 3578
Sucrose	: 6218
Sucrose, octaacetate	: 6229
Sudan III	: 8038
Sulfabenzamide	: 603
Sulfabenzpyrazine	: 2205
Sulfacetamide	: 30
Sulfachlorpyridazine	: 2193
Sulfadiazine	: 2204
Sulfadigesin	: 3578
Sulfadimethoxine	: 2194
Sulfafurazole	: 2195
Sulfaguanidine	: 2191
Sulfamerazine	: 2199
Sulfamethazine	: 2196
Sulfamethoxazole	: 2198
Sulfamethoxypyridazine	: 2197
Sulfamethyldiazine	: 2199
Sulfamethylthiazole	: 2200
p-Sulfamidoaniline	: 2187
p-Sulfamoylbenzoic acid	: 2508
p-Sulfamylacetanilide	: 31
Sulfanilamide	: 2187
Sulfanilamide-N4-acetate	: 31
Sulfanilamide, N1-(5-methyl-3-isoxazolyl)-	: 2198
Sulfanilamide, N1-(thiocarbamoyl)-	: 2189
5-Sulfanilamido-1-phenylpyrazole	: 2201
2-Sulfanilamidopyrazine	: 2202
2-Sulfanilamidopyridine	: 2203
o-Sulfanilic acid	: 2215
m-Sulfanilic acid	: 2216
Sulfanilic acid	: 2217
2-(Sulfanilylamino)thiazole	: 2206
p-Sulfanilyl fluoride	: 2252
N4-Suifanilylsulfanilamide	: 2188
Sulfanilylurea	: 2190
Sulfaphenazole	: 2201
Sulfapyrazine	: 2202
Sulfapyridine	: 2203
Sulfapyrimidine	: 2204
Sulfaquinoxaline	: 2205
Sulfasalazine	: 2654
Sulfasuccidine	: 3578
Sulfathalidine	: 2726
Sulfathiazole	: 2206
Sulfathiourea	: 2189
Sulfazole	: 2200
o-Sulfhydrylbenzoic acid	: 2663
Sulfide, 3-chloropropyl methyl	: 9864
Sulfide, 2-chloropropyl methyl	: 9865
Sulfide, ethyl isobutyl	: 10044
Sulfide, o-tolyl p-tolyl	: 1993
4,4'-Sulfinyldianiline	: 845
Sulfiram	: 11714
Sulfisoxazole	: 2195
Sulfoacetic acid	: 238
o-Sulfobenzoic acid imide	: 2388
Sulfodiamine	: 91
2-Sulfoethylamine	: 5623
Sulfolane	: 11786
Sulfolane, 2,4-dimethyl	: 11784
3-Sulfolene	: 11752
Sulfonethylmethane	: 3230
Sulfonmethane	: 9835
Sulfonyldiacetic acid	: 239
p-Sulfophenol	: 2226
p-Sulfophenylhydrazine	: 2224
Sulfosalicylic acid	: 2655
Sulfotep	: 11716
Sulfovinic acid	: 11576
Sulfurous acid, 2-[4-(1,1-dimethylethyl)phenoxy]cyclohexyl 2-propynyl ester	: 10507
Sulphamidine	: 2196
Sulphenone	: 1225
Surinamine	: 11933
Sylvic acid	: 9167
Symclosene	: 11840
Synephrine	: 1887
Syringaldehyde	: 569
Syringidin chloride	: 2847
Syringin	: 6225
2,4,5-T	: 256

2,4,5-T Butoxyethyl ester	: 257
2,4,5-T Butyl ester	: 258
2,4,5-T *sec*-Butyl ester	: 260
2,4,5-T Isopropyl ester	: 259
Tabutrex	: 3301
Tachysterol	: 11349
Tagatose, *D*-	: 11587
L-Tartaric acid	: 3309
(+)-Tartaric acid	: 3309
DL-Tartaric acid	: 3311
(±)-Tartaric acid	: 3311
(*R,S*)-Tartaric acid	: 3304
meso-Tartaric acid	: 3304
(-)-*D*-Tartaric acid	: 3312
D-Tartaric acid	: 3312
(*S,S*)-Tartaric acid	: 3312
Tartaric acid, diethyl ester (*DL*)	: 3307
Tartaric acid, diisopropyl ester (*DL*)	: 3305
Tartaric acid, dipropyl ester, (+)-	: 3310
Tartaric acid, disodium salt	: 3314
Tartaric acid, ethyl ester	: 3315
Tartronic acid	: 9956
Taurine	: 5623
Taurocholic acid	: 5625
Tebanyl	: 2502
Tegosept P	: 2653
Tephrosin	: 3110
Terephthalic acid	: 1329
Terephthalic acid, 2,5-dichloro-	: 1349
Terephthalic acid, diethyl ester	: 1350
Terpenolic acid	: 6030
Terpenylic acid	: 6030
o-Terphenyl	: 11595
m-Terphenyl	: 11596
p-Terphenyl	: 11597
β-Terpineol	: 4542
Terpinolene	: 4684
α-Terthienyl	: 11600
Testosterone	: 356
Testosterone-17-propionate	: 359
1,4,7,10-Tetraazadecane	: 5499
1,5,10,14-Tetraazatetradecane	: 3259
Tetrabenzocyclooctatetraene	: 11644
2,3,4,6-Tetrabromoanisole	: 2256
1,2,3,5-Tetrabromobenzene	: 2253
2,3,5,6-Tetrabromobenzene	: 2254
1,2,4,5-Tetrabromobenzene	: 2254
3,4,5,6-Tetrabromo-*o*-cresol	: 9468
1,2,3,4-Tetrabromocyclobutane	: 4195
1,2,2,2-Tetrabromoethane	: 5631
Tetrabromoethylene	: 5914
2,3,4,6-Tetrabromophenol	: 9467
3',3'',5',5''-	: 7350
Tetrabromophenolphthalein	
1,1,2,2-Tetrabromopropane	: 10136
1,2,2,3-Tetrabromopropane	: 10138
2,3,4,5-Tetrabromothiophene	: 11781
Tetracaine hydrochloride	: 2531
1,1,3,3-Tetracarbethoxypropane	: 10140
1,2,4,5-Tetracarboxybenzene	: 2257
Tetracene	: 7703
1,2,3,4-Tetrachlorobenzene	: 2259
1,2,4,5-Tetrachlorobenzene	: 2260
s-Tetrachlorobenzene	: 2260
Tetrachloro-*p*-benzoquinone	: 4268
1,1,1,2-Tetrachlorobutane	: 3427
1,2,2,3-Tetrachlorobutane	: 3428
1,2,3,3-Tetrachlorobutane	: 3429
1,3,4,4-Tetrachloro-1-butene	: 3806
2,3,3,4-Tetrachloro-1-butene	: 3807
1,1,2,3-Tetrachloro-2-	: 10146
(chloromethyl)propane	
Tetrachloro-1,2-difluoroethane	: 5637
1,1,1,2-Tetrachloroethane	: 5634
Tetrachloroethylene	: 5917
1,1,1,2-Tetrachloro-2-fluoroethane	: 5639
1,1,1,2-Tetrachloro-2-methylpropane	: 10147
1,1,1,5-Tetrachloropentane	: 8838
2,3,4,6-Tetrachlorophenate	: 9470
2,3,4,5-Tetrachlorophenol	: 9469
2,3,5,6-Tetrachlorophenol	: 9471
Tetrachlorophthalic anhydride	: 7346
1,1,1,2-Tetrachloropropane	: 10141
1,1,1,3-Tetrachloropropane	: 10142
1,1,2,2-Tetrachloropropane	: 10143
1,1,2,3-Tetrachloropropane	: 10144
1,2,2,3-Tetrachloropropane	: 10145

1,1,3,3-Tetrachloro-2-propanol	: 10402
2,3,5,6-Tetrachloropyridine	: 11019
Tetrachloro-*p*-quinone	: 4268
3,3',4',5-Tetrachlorosalicylanilide	: 610
2,3,4,6-Tetrachlorotoluene	: 2262
2,3,5,6-Tetrachlorotoluene	: 2263
Tetracosamethylhendecasiloxane	: 11964
Tetracosamethylundecasiloxane	: 11964
Tetracosylbenzene	: 2265
Tetracosylcyclohexane	: 4474
Tetracyanoethylene	: 5915
Tetracycline	: 7706
Tetradecamethylcycloheptasiloxane	: 4239
Tetradecamethylene dibromide	: 11616
Tetradecamethylhexasiloxane	: 6990
Tetradecane, 1-cyclopentyl-	: 4850
Tetradecane, 1,1-dicyclohexyl	: 4476
Tetradecane, 1,1-diphenyl-	: 2267
Tetradecanoic acid, chloride	: 11635
Tetradecyl alcohol	: 11630
Tetradecylsulfonic acid	: 11620
Tetradifon	: 2325
Tetraethylammonium bromide	: 5447
1,2,3,4-Tetraethylbenzene	: 2268
1,2,4,5-Tetraethylbenzene	: 2269
N,N,N',*N'*-Tetraethyl-1,4-	: 3262
butanediamine	
Tetraethyldiarsenic	: 5109
Tetraethyl diphosphate	: 5221
Tetraethylene glycol	: 5758
Tetraethylene glycol dimethyl ether	: 8995
Tetraethylenepentamine	: 5498
Tetraethyl lead	: 9731
Tetraethylmethane	: 8756
N,N,N',N'-Tetraethyl-*m*-	: 1285
phenylenediamine	
N,N,N',N'-Tetraethylphthalamide	: 1325
Tetraethyl 1,1,2,3-	: 10139
propanetetracarboxylate	
Tetraethyl 1,1,3,3-	: 10140
propanetetracarboxylate	
Tetraethyl pyrophosphate	: 5221
Tetraethylsilane	: 11450
Tetraethyl silicate	: 11506
2,3,4,5-Tetrafluorobenzene	: 2270
1,2,3,4-Tetrafluorobenzene	: 2270
1,2,3,5-Tetrafluorobenzene	: 2271
2,3,5,6-Tetrafluorobenzene	: 2272
1,2,4,5-Tetrafluorobenzene	: 2272
1,1,1,2-Tetrafluoroethane	: 5640
1,1,2,2-Tetrafluoroethane	: 5641
Tetrafluoroethylene	: 5918
2,2,3,3-Tetrafluoro-1-propanol	: 10404
Tetrahelicene	: 2406
(1-4)-Tetrahydroacenaphthene	: 9
Tetrahydrobenzene	: 4626
2,3,6,7-Tetrahydro-1H,5H-	: 2737
benzo[ij]quinolizine	
Tetrahydrocortisone	: 9744
1,2,3,4-Tetrahydro-1,2-	: 7981
dimethylnaphthalene	
Tetrahydro-5,5-dimethyl-2-(1H)-	: 7136
pyrimidinone[3[4-(trifluoromethyl-	
styryl)-cinnamylidene- hydrazone	
Tetrahydroeucarvone	: 4238
Tetrahydrofuran	: 6168
Tetrahydro-2,5-furandimethanol	: 7109
Tetrahydrofurfuryl alcohol	: 6124
Tetrahydrofurfurylamine	: 6116
Tetrahydro-2H-pyran	: 10800
5,6,7,8-Tetrahydroisoquinoline	: 7389
Tetrahydro-3-methylfuran	: 6172
Tetrahydromyrcenol	: 8450
5,6,7,8-Tetrahydronaphthalene-2-	: 7978
sulfonyl chloride	
1,2,3,4-Tetrahydro-2-naphthoic acid	: 7778
Tetrahydro-1,4-oxazine	: 7681
1,2,3,4-Tetrahydrophenanthrene	: 9187
3,4,5,6-Tetrahydrophthalic anhydride	: 7347
1,2,3,6-Tetrahydropyridine	: 11020
5,6,7,8-Tetrahydroquinoline	: 11297
Tetrahydroquinone	: 4377
1,2,3,4-Tetrahydroquinoxaline	: 11339
D-Tetrahydroxyadipic acid	: 6205
Tetrahydroxy-1,4-benzoquinone	: 4269
3,3',5,5'-Tetrahydroxybiphenyl	: 3097

3,4',5,7-Tetrahydroxy-3',5'-	: 2847
dimethoxyflavylium chloride	
3,3',4',7-Tetrahydroxyflavone	: 2813
3',4',5,7-Tetrahydroxyflavone	: 2814
2',3,5,7-Tetrahydroxyflavone	: 2842
3,4',5,7-Tetrahydroxyflavone	: 2843
3,5,3',4'-Tetrahydroxy-7-	: 2815
methoxyflavone	
N,N,N',*N'*-Tetra(2-	: 10379
hydroxypropyl)ethylenediamine	
1,2,4,5-Tetraiodobenzene	: 2275
1,2,3,4-Tetraiodobenzene	: 2273
1,2,3,5-Tetraiodobenzene	: 2274
Tetraiodoethylene	: 5919
Tetraiodomethane	: 7542
1,2,4,5-Tetraisopropylbenzene	: 2276
Tetrakisdimethylaminophosphonous	: 5220
anhydride	
Tetrakis(2-hydroxypropyl)-	: 10379
ethylenediamine	
Tetralin	: 7979
Tetralin, 1,1-dimethyl	: 7980
Tetralin, 1,2-dimethyl	: 7981
Tetralin, 1,3-dimethyl	: 7982
Tetralin, 1,4-dimethyl	: 7983
Tetralin, 2,2-dimethyl	: 7985
Tetralin, 2,3-dimethyl	: 7986
Tetralin, 2,5-dimethyl	: 7987
Tetralin, 2,7-dimethyl	: 7989
Tetralin, 5,6-dimethyl	: 7990
Tetralin, 5,6-dimethyl	: 7991
Tetralin, 5,7-dimethyl	: 7992
Tetralin, 1-ethyl	: 7888
Tetralin, 2-ethyl	: 7889
Tetralin, 6-ethyl	: 7891
6-Tetralinsulfonyl chloride	: 7978
2-Tetralol	: 8043
1-Tetralone, 7-ethyl	: 8051
Tetramethylammonium bromide	: 7456
Tetramethylammonium chloride	: 7457
Tetramethylammonium iodide	: 7458
2,3,5,6-Tetramethylbenzaldehyde	: 598
2,2',4,4'-Tetramethylbenzophenone	: 7594
Tetramethyl-*p*-benzoquinone	: 4270
2,2,3,3-Tetramethylbutane	: 3430
1,1,4,4-Tetramethyl-1,4-butanediol	: 6801
1,1,3,4-Tetramethylcyclohexane	: 4477
cis-1,1,3,5-Tetramethylcyclohexane	: 4478
4-(2,5,6,6-Tetramethyl-2-	: 3884
cyclohexen-1-yl)-3-butene-2-one	
1,1,3,3-Tetramethylcyclopentane	: 4851
N,N,N',*N'*-Tetramethyl-1,4-	: 3263
diaminobutane	
N,N,N',N'-	: 5516
Tetramethyldiaminoethane	
Tetramethylene	: 4170
2,3-Tetramethyleneaziridine	: 478
1,4-Tetramethylenediamine	: 3258
Tetramethylenedithiol	: 3368
Tetramethylene glycol	: 3346
Tetramethylethylene glycol	: 3350
2,2,6,6-Tetramethyl-3,5-	: 6404
heptanedione	
2,2,3,3-Tetramethylhexane	: 6844
2,2,3,4-Tetramethylhexane	: 6845
2,2,3,5-Tetramethylhexane	: 6846
2,2,4,4-Tetramethylhexane	: 6847
2,2,4,5-Tetramethylhexane	: 6848
2,2,5,5-Tetramethylhexane	: 6849
2,3,3,4-Tetramethylhexane	: 6850
2,3,3,5-Tetramethylhexane	: 6851
2,3,4,4-Tetramethylhexane	: 6852
2,3,4,5-Tetramethylhexane	: 6853
3,3,4,4-Tetramethylhexane	: 6854
2,2,5,5-Tetramethyl-3,4-	: 6811
hexanedione	
1,1,4,7-Tetramethylindan	: 7260
Tetramethyl lead	: 9732
Tetramethylmethanediamine	: 7490
1,3,6,8-Tetramethylnaphthalene	: 8003
2,2,7,7-Tetramethyl-3,6-octanedione	: 8400
2,2,3,3-Tetramethylpentane	: 8839
2,2,3,4-Tetramethylpentane	: 8840
2,2,4,4-Tetramethylpentane	: 8841
2,3,3,4-Tetramethylpentane	: 8842
2,3,4,5-Tetramethylphenol	: 9473
Tetramethyl-*p*-phenylenediamine	: 1287

Compound	Number
2,2,4,6-Tetramethylpiperidine	: 9710
2,2,6,6-Tetramethylpiperidine	: 9711
1,2,5,6-Tetramethyl-4-piperidinone	: 9725
Tetramethylputrescine	: 3263
2,3,4,6-Tetramethylpyridine	: 11022
Tetramethylsilane	: 11451
Tetramethyl silicate	: 11508
Tetramethylstannane	: 11544
Tetramethylsuccinonitrile	: 3292
2,2,4,4-Tetramethyltetrahydrofuran	: 6174
N,N,N',N'-Tetramethyltetra-methylenediamine	: 3263
Tetramethylurea	: 12030
Tetranitrin	: 3432
Tetranitromethane	: 7544
Tetranitropentaerythritol	: 9990
2,3,4,6-Tetranitrophenol	: 9476
2,4,5,6-Tetraoxohexahydropyrimidine	: 11073
1,1,1,2-Tetraphenylethane	: 1618
1,1,2,2-Tetraphenyl-1,2-ethanediol	: 5580
1,1,2,2-Tetraphenylethylene glycol	: 5580
2,3,4,5-Tetraphenylfuran	: 6175
Tetraphenylfuran	: 6175
Tetraphenylmethane	: 1831
Tetraphenylsilane	: 11452
Tetraphenylurea	: 12031
Tetraphthene	: 9
Tetrapropylammonium iodide	: 9833
Tetratriacontylcyclohexane	: 4479
Tetravinylsilane	: 11449
sym-Tetrazine	: 11651
Tetrolaldehyde	: 3895
Tetrole	: 6025
Tetroquinone	: 4269
Thalidomide	: 7363
Thebaine	: 7679
Thebainone	: 7677
3-Thenoic acid	: 11738
Thenylpyramine	: 5507
Theobromine	: 10757
Theophylline	: 10756
THF glycol	: 7109
1-Thia-2-azaindene	: 2387
Thiabutazide	: 2856
Thiacetazone	: 34
Thiachromine-8-ethanol, 2,7-dimethyl-	: 11111
Thiacyclohexane	: 11795
Thiacyclopentane	: 11783
1-Thia-2,4-diazacyclopentadiene	: 11657
1,2,4-Thiadiazole, 5-ethoxy-3-(trichloromethyl)-	: 11599
1,3,4-Thiadiazole-2-sulfonic acid,amide,5-acetamido	: 32
1-Thiaindene	: 2399
Thiamine hydrochloride	: 11686
Thiamorpholine	: 11719
Thianaphthene	: 2399
Thiazolsulfone	: 11663
2-Thiazolylamine	: 11662
4'-(2-Thiazolylsulfamoyl)succinanilic acid	: 3578
2-Thienonitrile	: 11732
2-Thienyl bromide	: 11726
2-Thienyl chloride	: 11740
Thioacetaldehyde trimer	: 11920
Thioacetamide	: 5643
Thioacetic acid	: 5646
Thiobenzyl alcohol	: 1832
Thiobutyl alcohol	: 3437
Thiocarbamide	: 11799
Thiochrome	: 11111
Thiocyanatobenzene	: 11709
Thiocyanic acid, tert-butyl ester	: 11701
Thiocyanic acid, isopropyl ester	: 11706
1-Thiocyanobutane	: 11699
Thiocyclohexanone	: 4481
2,2'-Thiodiethanol	: 5766
Thiodiglycolic acid	: 240
Thiodiphenylamine	: 9499
3,3'-Thiodipropionic acid	: 10329
Thiofuran	: 11724
Thioglycerol	: 10013
Thioglycolic acid	: 202
Thioguanine	: 10762

Compound	Number
Thiolactic acid	: 10285
Thiomedon	: 7649
Thionalide	: 68
Thionaphthene-2-carboxylic acid	: 2400
Thiophene, 2-isobutyryl	: 10472
Thiophene, 2-isopentyl-	: 11771
Thiophene, 2-isopropyl-	: 11772
Thiophene, 3-isopropyl-	: 11773
Thiophenol	: 2289
Thiophenol, 3-ethyl	: 2297
Thiophosgene	: 4032
o-Thiosalicylic acid	: 2663
Thiosemicarbazide	: 7142
Thioserine	: 5014
Thiosinamine	: 11809
Thiotax	: 2866
3-Thiotolene	: 11770
2-Thiouracil	: 11110
Thioxanthone	: 11812
Thiram	: 11722
THP	: 10800
Threose (D-)	: 3193
4(10)-Thujene, (±)-	: 2981
4(10)-Thujene (d)	: 11815
4(10)-Thujene-3-ol	: 2983
Thujic acid	: 4241
Thujopsene	: 4954
Thymine	: 11068
Thymine 2-desoxyriboside	: 11816
Thymol	: 9418
Thymol, acetate	: 9420
Thymol blue	: 9369
2-Thymolcarboxylic acid	: 2642
Thymol, 2,6-dibromo	: 9281
Thymolphthalein	: 7352
o-Thymotic acid	: 2642
Thymyl iodide	: 1782
Thyroxine	: 11930
Tiglic acid	: 3844
Tigogenin	: 11532
Timonacic	: 11682
Tiocarlide	: 11800
δ-Tocopherol	: 2785
α-Tocopherol	: 2787
β-Tocopherol	: 2789
γ-Tocopherol	: 2790
α-Tocopheryl acetate	: 2788
Tokoferol acetate	: 2788
Tolazine	: 9195
Tolbutamide	: 2207
3,3'-Tolidine	: 3034
o-Tolidine	: 3034
o-Tolualdehyde	: 587
p-Tolualdehyde	: 589
α-Tolualdehyde	: 868
m-Tolualdehyde, 2-hydroxy	: 575
o-Toluamide	: 619
Toluene	: 1947
Toluene, 3-bromo-2-nitro-	: 1068
Toluene, 3-bromo-5-nitro-	: 1069
Toluene, 3-bromo-4-nitro-	: 1073
Toluene, 4-bromo-2-nitro-	: 1074
Toluene, 2-bromo-6-nitro	: 1070
o-Toluenecarbonitrile	: 2763
Toluene, 4-chloro-3-iodo-	: 1173
Toluene, 5-chloro-2-iodo	: 1174
Toluene, 2-chloro-α-methylimino	: 7652
Toluene, 3-chloro-5-nitro-	: 1207
Toluene-3,4-diamine	: 1276
Toluene-2,6-diamine	: 1277
Toluene-2,4-diamine	: 1278
Toluene, 2,3-dibromo	: 1302
Toluene, 3,4-dibromo	: 1303
Toluene, 2,6-dibromo	: 1304
Toluene, 2,5-dibromo-	: 1305
Toluene, 2,3-dichloro-	: 1383
Toluene, 2,5-dichloro-	: 1387
Toluene, 2,4-dichloro-α-hydroxy	: 1853
Toluene, 2,5-diethoxy	: 1410
Toluene-α,4-dihydroxy	: 1884
Toluene 2,4-diisocyanate	: 1431
Toluene-2,6-diisocyanate	: 1430
Toluene, 2,4-diisopropyl-	: 1948
Toluene-α,α-diol, m-nitro-, diacetate (ester)	: 7511
Toluene-3,4-dithiol	: 1564

Compound	Number
Toluene, 3-fluoro-4-nitro-	: 1730
Toluene, 4-fluoro-3-nitro	: 1729
Toluene, 3-(1-hydroxyethyl)	: 1858
Toluene, α-hydroxy-4-methoxy-α-methyl	: 1895
Toluene, α-hydroxy-4-nitro	: 1908
m-Toluenenitrile	: 2764
p-Toluenenitrile	: 2765
Toluene, p-(1-octyldecyl)-	: 8245
Toluene, α,α,2,3,4-pentachloro-	: 2326
Toluene, α,α,2,4,5-pentachloro-	: 2327
Toluene, 4-propoxy	: 2032
o-Toluenesulfinic acid	: 2181
p-Toluenesulfinic acid	: 2182
p-Toluenesulfonamide, N,N-dichloro-	: 2210
p-Toluenesulfonic acid	: 2229
p-Toluenesulfonic acid chloride	: 2250
p-Toluenesulfonic acid, ethyl ester	: 2231
o-Toluenesulfonyl chloride	: 2249
p-Toluenesulfonyl chloride	: 2250
Toluene, 2,3,4,6-tetrachloro-	: 2262
Toluene, 2,3,5,6-tetrachloro-	: 2263
α-Toluenethiol, α-methyl-, (S)-(-)-	: 1834
Toluene, 2,4,6-tribromo-	: 2314
Toluene, 2,3,4-trichloro-	: 2334
Toluene, 3,4,5-trichloro-	: 2335
Toluene, 2,4,5-trichloro-	: 2336
Toluene, 2,3,5-trichloro-	: 2337
Toluene, 2,4,6-triiodo-	: 2352
Toluene, 2,3,4-trinitro-	: 2046
Toluene, 2,4,5-trinitro-	: 2047
o-Toluic acid	: 2676
m-Toluic acid	: 2677
p-Toluic acid	: 2678
o-Toluidine	: 778
m-Toluidine	: 779
p-Toluidine	: 780
m-Toluidine, N-benzyl	: 1824
p-Toluidine, N-benzyl	: 1825
m-Toluidine, 6-bromo-	: 648
o-Toluidine hydrochloride	: 791
p-Toluidine, 2-iodo-	: 762
o-Toluidine, 5-iodo	: 763
o-Toluidine, 4-(phenylsulfonyl)-N-[(trifluoromethyl)sulfono]-	: 9149
p-Toluidine, N-propyl-	: 815
o-Tolunitrile	: 2763
m-Tolunitrile	: 2764
p-Tolunitrile	: 2765
m-Toluquinaldine	: 11258
m-Toluquinoline	: 11282
2-(p-Toluyl)benzoic acid	: 2684
o-Tolyl alcohol	: 1897
p-Tolyl alcohol	: 1899
2-o-Tolylaminoethanol	: 5744
2-p-Tolylaminoethanol	: 5745
p-Tolylcarbinol	: 1899
1-(4-Tolyl)ethanol	: 1859
m-Tolylmethanol	: 1898
p-Tolylsulfonylmethylnitrosamide	: 2212
o-Tolylthiourea	: 11806
Tomatidine	: 11528
Tomatine	: 6196
Topazone	: 8589
Torularhodin	: 4059
Tosic acid	: 2229
p-Tosyl chloride	: 2250
Tranexamic acid	: 4333
trans-1,3-Dimethylcyclohexanol	: 4511
Traumatic acid	: 5336
Trehalose	: 6220
Triacetamide	: 50
Triacetin	: 10195
N,N,N-Triacetylammonia	: 50
Triacontyl alcohol	: 11821
Triacontylcyclohexane	: 4482
Triacontylcyclopentane	: 4853
Triallylamine	: 10523
1,2,3-Triaminobenzene	: 2305
Triamterene	: 10750
1,2,3-Triaza-1H-indene	: 2871
1,2,4-Triazanaphthalene	: 2870
1,3,5-Triazine-2,4-diamine, N,N'-bis(1-methylethyl)-6-(methylthio)-	: 9770

1,3,5-Triazine-2,4-diamine, 6-chloro-N,N'-bis(1-methylethyl)-	: 10508
1,3,5-Triazine-2,4-diamine, 6-chloro-N-cyclopropyl-N'-(1-methylethyl)-	: 5013
1,3,5-Triazine-2,4-diamine, 6-chloro-N,N'-diethyl-	: 11513
1,3,5-Triazine-2,4-diamine, 6-chloro-N-(1,1-dimethylethyl)-N'-ethyl-	: 11593
1,3,5-Triazine-2,4-diamine, N-(1,1-dimethylethyl)-N'-ethyl-6-(methylthio)-	: 11594
1,3,5-Triazine-2,4-diamine, N-ethyl-6-methoxy-N'-(1-methylpropyl)-	: 3143
1,3,5-Triazine-2,4-diamine, 6-(ethylthio)-N,N'-bis(1-methylethyl)-	: 5222
1,3,5-Triazine-2,4(1H,3H)-dione, 3-cyclohexyl-6-(dimethylamino)-1-methyl-	: 6995
1,2,4-Triazin-5(4H)-one, 4-amino-6-(1,1-dimethylethyl)-3-(methylthio)-	: 7658
1,2,4-Triazolo[3,4-b]benzothiazole, 5-methyl-	: 11858
Tribenzobicyclo[2.2.2]octatriene	: 2375
1,3,5-Tribenzoylbenzene	: 7587
Tribromoacetamide	: 94
Tribromoacetic acid	: 241
Tribromoacetic acid, ethyl ester	: 242
2,4,6-Tribromoaniline	: 851
2,4,5-Tribromoanisole	: 2311
1,2,3-Tribromobenzene	: 2307
1,2,4-Tribromobenzene	: 2308
1,3,5-Tribromobenzene	: 2309
1,2,4-Tribromobutane	: 3445
1,2,2-Tribromobutane	: 3443
1,2,3-Tribromobutane	: 3444
1,3,3-Tribromobutane	: 3446
2,2,3-Tribromobutane	: 3447
Tribromo-tert-butyl alcohol	: 10405
Tribromochloromethane	: 7549
2,4,6-Tribromo-m-cresol	: 9480
1,1,2-Tribromocyclobutane	: 3442
1,1,2-Tribromoethane	: 5653
Tribromoethanol	: 5787
Tribromofluoromethane	: 7550
Tribromomethane	: 7548
3,4,5-Tribromonitrobenzene	: 2315
Tribromonitromethane	: 7551
2,4,6-Tribromophenol	: 9479
1,2,3-Tribromopropane	: 10163
2,3,5-Tribromothiophene	: 11790
2,4,6-Tribromotoluene	: 2314
Tribromsalan	: 609
Tributylamine	: 3213
Tributyl phosphate	: 9562
Tributylphosphine	: 9520
Tributylsilane	: 11454
Tributyltin acetate	: 11538
S,S,S-Tributyl trithiophosphate	: 11852
Tributyrin	: 3586
Tributyryl glyceride	: 3586
Tricarballylic acid	: 10165
Tricarbethoxymethane	: 7552
1,2,3-Tricarboxybenzene	: 2318
1,2,4-Tricarboxybenzene	: 2319
2,2,2-Trichloroacetaldehyde	: 21
Trichloroacetaldehyde hydrate	: 5582
Trichloroacetic acid	: 243
Trichloroacetic acid, ethyl ester	: 248
Trichloroacetic acid, (2-methoxyethyl) ester	: 250
Trichloroacetic acid, methyl ester	: 253
Trichloroacetic anhydride	: 244
Trichloroacetonitrile	: 288
Trichloroacetyl bromide	: 293
Trichloroacetyl chloride	: 300
2,3,4-Trichloroaniline	: 852
2,4,6-Trichloroanisole	: 2332
1,2,3-Trichlorobenzene	: 2321
1,2,4-Trichlorobenzene	: 2322
1,3,5-Trichlorobenzene	: 2323
2,3,6-Trichlorobenzoic acid	: 2727
2,4,5-Trichlorobenzoic acid	: 2728
2,2,2-Trichloro-1,1-bis(4-chlorophenyl)ethanol	: 1851
Trichlorobromomethane	: 7474
1,2,3-Trichloro-1,3-butadiene	: 3174
1,2,3-Trichlorobutane	: 3449

1,2,4-Trichlorobutane	: 3450
2,2,3-Trichlorobutane	: 3451
1,1,3-Trichlorobutane	: 3448
1,1,1-Trichloro-2-butanol	: 3643
2,3,4-Trichloro-1-butene	: 3809
1,2,4-Trichloro-2-butene	: 3810
Trichlorobutylsilane	: 11370
2,2,3-Trichlorobutyraldehyde	: 3191
α,α,β-Trichlorobutyraldehyde	: 3191
2,2,3-Trichlorobutyric acid	: 3590
Trichloro(chloromethyl)silane	: 11456
4,5,6-Trichloro-o-cresol	: 9487
2,3,6-Trichloro-p-cresol	: 9488
2,4,6-Trichloro-m-cresol	: 9489
Trichloro(dichloromethyl)silane	: 11459
1,1,1-Trichloro-2,2-difluoroethane	: 5658
1,2,2-Trichloro-1,2-difluoroethane	: 5660
1,1,1-Trichloro-2,2-dihydroxyethane	: 5582
Trichlorododecylsilane	: 11460
2,2,2-Trichloroethanal	: 21
1,1,1-Trichloroethane	: 5655
1,1,2-Trichloroethane	: 5656
2,2,2-Trichloro-1,1-ethanediol	: 5582
2,2,2-Trichloroethanol	: 5768
Trichloroethylene	: 5922
Trichlorofluoromethane	: 7555
Trichlorohexylsilane	: 11464
Trichloroiodomethane	: 7556
Trichloromethane	: 7554
Trichloromethiazide	: 2854
(Trichloromethyl)benzene	: 2333
1,1,2-Trichloro-2-methylpropane	: 10177
2-(Trichloromethyl)pyridine	: 11023
3,4,6-Trichloro-2-nitrophenol	: 9490
Trichloro perchlorate	: 9148
2,3,4-Trichlorophenol	: 9481
2,3,5-Trichlorophenol	: 9482
2,3,6-Trichlorophenol	: 9483
2,4,5-Trichlorophenol	: 9484
2,4,6-Trichlorophenol	: 9485
3,4,5-Trichlorophenol	: 9486
1,1,1-Trichloropropane	: 10171
1,1,2-Trichloropropane	: 10172
1,1,3-Trichloropropane	: 10173
1,2,2-Trichloropropane	: 10174
1,2,3-Trichloropropane	: 10175
1,1,1-Trichloro-2-propanol	: 10406
2,4,6-Trichloropyrimidine	: 11075
3',4',5-Trichlorosalicylanilide	: 608
2,3,5-Trichlorothiophene	: 11791
2,3,4-Trichlorotoluene	: 2334
2,4,5-Trichlorotoluene	: 2336
2,3,5-Trichlorotoluene	: 2337
2,2',2"-Trichlorotriethylamine	: 5412
1,3,5-Trichloro-2,4,6-trifluorobenzene	: 2340
1,1,1-Trichlorotrifluoroethane	: 5662
1,1,2-Trichlorotrifluoroethane	: 5663
Trichloro-1,3,5-trihydroxybenzene	: 2371
Trichlorovinylsilane	: 11461
Trichocid	: 85
Tricosylbenzene	: 2341
Tricosylcyclohexane	: 4483
Tricromyl	: 2838
Tricyclenic acid	: 11862
Tricyclenic acid, ethyl ester	: 11863
Tricyclenic acid, methyl ester	: 11864
Tridecamethylene dibromide	: 11874
Tridecane, 1-amino-	: 11870
Tridecane, 1-cyclohexyl-	: 4484
Tridecane, 1-cyclopentyl-	: 4854
Tridecane, 2-methyl-13-bromo-	: 11873
Tridecane, 7-phenyl-	: 1764
Tridecyl alcohol	: 11883
Tridecylcyclohexane	: 4484
Trientine	: 5499
Triethanolamine	: 5751
1,3,5-Triethoxybenzene	: 2343
Triethoxysilane	: 11477
Triethylamine	: 5417
Triethylantimony	: 11549
Triethyl arsenate	: 439
Triethyl arsenite	: 440
Triethylarsine	: 448
1,2,4-Triethylbenzene	: 2344
1,3,5-Triethylbenzene	: 2345
Triethylborane	: 3118

Triethylenediamine	: 5111
Triethylene glycol	: 5713
Triethylene glycol, (3-aminopropyl) ether	: 5678
Triethylene glycol, diacetate	: 5714
Triethylenemelamine	: 11844
Triethylenetetramine	: 5499
Triethyl methanetricarboxylate	: 7552
Triethyl phosphate	: 9563
Triethyl phosphine	: 9521
Triethyl phosphite	: 9591
Triethylphosphorus	: 9521
Triethylsilane	: 11483
Triethylstibine	: 11549
2,4,6-Triethyl-2,4,6-trimethylcyclotrisiloxane	: 5003
2,4,6-Triethyl-2,4,6-triphenylcyclotrisiloxane	: 5004
Trifluoroacetic acid	: 263
Trifluoroacetic acid, propyl ester	: 269
Trifluoroacetonitrile	: 289
1,3,5-Trifluorobenzene	: 2347
sym-Trifluorobenzene	: 2347
Trifluoroboron etherate	: 3137
Trifluorobromomethane	: 7475
Trifluorochloromethane	: 7486
1,1,2-Trifluoroethane	: 5666
2,2,2-Trifluoroethanol	: 5770
Trifluoroethylene	: 5924
Trifluoroethylsilane	: 11432
Trifluoroiodomethane	: 7559
Trifluoromethane	: 7558
o-(Trifluoromethyl)aniline	: 855
m-(Trifluoromethyl)aniline	: 856
(Trifluoromethyl)benzene	: 2348
3'-Trifluoromethyldiphenylamine-2-carboxylic acid	: 2729
Trifluoromethylhydrothiazide	: 2858
3-(Trifluoromethyl)-1-nitrobenzene	: 2061
N-(m-Trifluoromethylphenyl)-2-aminobenzoic acid	: 2729
N-(3-Trifluoromethylphenyl)-N',N'-dimethylurea	: 12007
(±)-2-[4-[[5-(Trifluoromethyl)-2-pyridinyl]oxy]phenoxy]propanoic acid, butyl ester	: 5962
(R)-2-[4-(5-Trifluoromethyl-2-pyridyloxy)phenoxy]propionic acid, butyl ester	: 5963
3,3,3-Trifluoropropene	: 10608
Triglycol diacetate	: 5714
Triglyme	: 11642
Trigonelline	: 11029
Trihexyphenidyl hydrochloride	: 9705
2',3',4'-Trihydroxyacetophenone	: 5873
1,3,5-Trihydroxybenzene	: 2368
3,4,5-Trihydroxybenzoic acid	: 2731
1,2,3-Trihydroxybutane	: 3458
Trihydroxyestrin	: 5404
4',5,7-Trihydroxyflavone	: 2806
1,2,6-Trihydroxyhexane	: 6868
4',5,7-Trihydroxyisoflavone	: 2807
3',5,7-Trihydroxy-4'-methoxyflavanone	: 2800
1,2,3-Trihydroxypropane	: 10186
2,4,6-Trihydroxypropiophenone	: 10489
1,2,3-Triiodobenzene	: 2349
1,2,4-Triiodobenzene	: 2350
1,3,5-Triiodobenzene	: 2351
Triiodomethane	: 7560
2,4,6-Triiodophenol	: 9492
3,3',5-Triiodothyropropionic acid	: 2136
2,4,6-Triiodotoluene	: 2352
Triisobutylamine	: 9820
Triisobutyl phosphate	: 9572
1,2,4-Triisopropylbenzene	: 2372
3,7,12-Triketocholanic acid	: 4101
Trimellitic acid	: 2319
Trimellitic anhydride	: 7341
Trimethadione	: 8587
3,4,5-Trimethoxybenzyl alcohol	: 1914
Trimethoxyborane	: 3126
Trimethoxyphosphine	: 9592
1,1,1-Trimethylacetone	: 3657
Trimethyl aluminum	: 339
Trimethylamine	: 7444
Trimethylamineborane	: 3132

Trimethylantimony	: 11550	2,2,3-Trimethylhexane	: 6859	Triphenylarsane	: 450
Trimethylarsine	: 449	2,2,4-Trimethylhexane	: 6860	Triphenylarsine	: 450
1,2,3-Trimethylbenzene	: 2357	2,2,5-Trimethylhexane	: 6861	Triphenylbismuth	: 3111
Trimethylborane	: 3119	2,3,3-Trimethylhexane	: 6862	Triphenylcarbinol	: 1866
Trimethyl borate	: 3126	2,3,4-Trimethylhexane	: 6863	Triphenylhydrazine	: 7173
2,2,3-Trimethylbutane	: 3457	2,3,5-Trimethylhexane	: 6864	Triphenylmethane	: 1982
Trimethylbutylsilane	: 11371	2,4,4-Trimethylhexane	: 6865	Triphenylmethane, 3-methyl	: 7513
Trimethylchlorosilane	: 11389	3,3,4-Trimethylhexane	: 6866	Triphenylmethanol	: 1866
Trimethyl citrate	: 10170	3,5,5-Trimethylhexan-3-ol	: 6958	Triphenylmethylsilane	: 11444
1,1,2-Trimethylcycloheptane	: 4227	1,1,5-Trimethylindan	: 7261	Triphenyl phosphate	: 9566
1,1,4-Trimethylcycloheptane	: 4228	1,4,7-Trimethylindan	: 7262	Triphenylphosphine	: 9523
3,3,7-Trimethylcycloheptanol	: 4233	1,2,3-Trimethylindole	: 7322	Triphenyltetrazolium chloride	: 11653
2,6,6-Trimethylcycloheptanone	: 4238	2,3,5-Trimethylindole	: 7323	Triphenyltin acetate	: 11539
1,4,4-Trimethyl-1-cycloheptene	: 4247	Trimethyl methanetricarboxylate	: 7553	Triphenyltin chloride	: 11541
1,1,2-Trimethylcyclohexane	: 4485	1,2,4-Trimethylnaphthalene	: 8009	Tripropylamine	: 9808
1,1,3-Trimethylcyclohexane	: 4486	1,2,5-Trimethylnaphthalene	: 8010	Tripropyl phosphate	: 9567
1,1,4-Trimethylcyclohexane	: 4487	1,2,6-Trimethylnaphthalene	: 8011	Tripropyl phosphite	: 9594
1,2,3-Trimethylcyclohexane	: 4488	2,4,6-Trimethylnaphthalene	: 8013	Tripropylsilane	: 11500
trans-1,3,5-Trimethylcyclohexane	: 4497	2,4,7-Trimethyloctane	: 8421	Triptycene	: 2375
1,1,3-Trimethyl-2-cyclohexanol	: 4566	Trimethylolmelamine	: 7583	2,4,6-Tripyridyl-s-triazine	: 11843
2,3,6-Trimethylcyclohexanol	: 4568	Trimethylolpropane	: 10006	Tris(allyloxy)vinylsilane	: 11427
cis-3,3,5-Trimethylcyclohexanol	: 4569	Trimethylolpropane phosphite	: 11909	Tris-BP	: 10358
trans-3,5,5-Trimethylcyclohexanol	: 4570	2,2,3-Trimethylpentane	: 8847	Tris-(2-chloroethyl) amine	: 5412
trans-3,3,5-Trimethylcyclohexanol	: 4570	2,3,3-Trimethylpentane	: 8849	Tris(chloromethyl)phosphine	: 9524
1,2,2-Trimethylcyclohexanol	: 4562	2,3,4-Trimethylpentane	: 8850	Tris(2,3-dibromopropyl) phosphate	: 10358
1,2,6-Trimethylcyclohexanol	: 4563	2,2,4-Trimethyl-1,3-pentanediol	: 8795	Tris(dimethylamino)phosphine oxide	: 9573
1,3,5-Trimethylcyclohexanol	: 4564	2,4,4-Trimethyl-2-pentanol	: 8954	Tris(hydroxymethyl)nitromethane	: 10012
2,2,5-Trimethylcyclohexanol	: 4565	2,2,4-Trimethyl-3-pentanol	: 8955	Tristearin	: 8272
2,2,3-Trimethylcyclohexanone	: 4614	2,4,4-Trimethyl-1-pentene	: 9064	Trithioacetaldehyde	: 11920
2,2,4-Trimethylcyclohexanone	: 4615	2,3,3-Trimethyl-1-pentene	: 9065	Trithiocarbonic acid	: 4034
2,3,6-Trimethylcyclohexanone	: 4618	2,3,4-Trimethyl-1-pentene	: 9066	Tri-o-tolyl phosphate	: 9569
2,4,4-Trimethylcyclohexanone	: 4619	2,2,4-Trimethyl-4-pentene	: 9067	Trityl alcohol	: 1866
2,4,5-Trimethylcyclohexanone	: 4620	3,4,4-Trimethyl-1-pentene	: 9069	Tromasedan	: 2385
3,3,5-Trimethylcyclohexanone	: 4622	2,3,4-Trimethyl-2-pentene	: 9070	Tromethamine	: 9986
3,4,4-Trimethylcyclohexanone	: 4623	2,4,4-Trimethyl-2-pentene	: 9071	Tropacocaine	: 488
2,2,5-Trimethylcyclohexanone	: 4616	3,4,4-Trimethyl-2-pentene	: 9072	Tropane	: 484
2,4,6-Trimethylcyclohexanone	: 4621	2,4,5-Trimethylphenyl acetate	: 9496	Tropic acid	: 915
1,4,4-Trimethylcyclohexene	: 4689	Trimethylphosphine	: 9522	Tropilidene	: 4240
1,5,5-Trimethylcyclohexene	: 4691	Trimethyl phosphite	: 9592	Tropine	: 489
1,2,3-Trimethylcyclohexene	: 4687	1,2,4-Trimethylpiperazine	: 9622	Tropone		
1,3,5-Trimethylcyclohexene	: 4688	2,3,6-Trimethylpiperidine	: 9713	Tryptamine	: 7302
1,4,5-Trimethylcyclohexene	: 4690	2,4,6-Trimethylpiperidine	: 9714	Tryptophol	: 7305
1,6,6-Trimethylcyclohexene	: 4692	1,1,2-Trimethylpropyl bromide	: 3236	Tsuduranine	: 5159
2,4,4-Trimethyl-2-cyclohexen-1-ol	: 4701	Trimethylpropylsilane	: 11496	Tsuzuic acid	: 11637
3,6,6-Trimethyl-2-cyclohexenone	: 4722	1,3,4-Trimethylpyrazole	: 10839	Tuaminoheptane	: 6350
4-(2,6,6-Trimethyl-1-cyclohexen-1-yl)-3-buten-2-ol	: 3873	1,3,5-Trimethylpyrazole	: 10840	Tubatoxin	: 2783
		1,4,5-Trimethylpyrazole	: 10841	Tuduic acid	: 11637
1,1,2-Trimethylcyclopentane	: 4855	2,3,4-Trimethylpyridine	: 11024	D-Turanose	: 6022
1,1,3-Trimethylcyclopentane	: 4856	2,3,5-Trimethylpyridine	: 11025	Tybamate	: 3933
1,2,2-Trimethylcyclopentanol	: 4875	2,3,6-Trimethylpyridine	: 11026	Tylenol	: 67
1,2,4-Trimethylcyclopentanol	: 4876	2,4,5-Trimethylpyridine	: 11027	Tyramine	: 9208
1,2,5-Trimethylcyclopentanol	: 4877	1,2,5-Trimethylpyrrole	: 11142	Tyrosine, 3,5-dibromo(L)	: 11927
2,3,5-Trimethylcyclopentanol	: 4898	2,2,4-Trimethylpyrrolidine	: 11175	Ugurol	: 4333
2,2,4-Trimethylcyclopentanone	: 4896	2,4,6-Trimethylquinoline	: 11302	Umbelliferone	: 2826
2,2,5-Trimethylcyclopentanone	: 4897	2,4,7-Trimethylquinoline	: 11303	Umbelliferone glucoside	: 2821
2,4,4-Trimethylcyclopentanone	: 4899	Trimethylsilane	: 11489	Undecafluorocyclohexane	: 4498
1,5,5-Trimethylcyclopentene	: 4938	(Trimethylsilyl)methyl bromide	: 11366	trans-9-Undecenoic acid	: 11975
1,1,2-Trimethylcyclopropane	: 4985	Trimethylstibine	: 11550	(10-Undecenyl)acetic acid	: 11895
trans-3,7,11-Trimethyl-1,3,6,10-dodecatetraene	: 5329	2,3,4-Trimethylthiophene	: 11792	Undecylenic acid	: 11974
		2,3,5-Trimethylthiophene	: 11793	Undecyl ketone	: 11857
trans-2,6,10-Trimethyl-2,6,9,11-dodecatetraene	: 5329	5,7,8-Trimethyltocol	: 2787	Undecyl mercaptan	: 11942
		1,1,1-Trimethyl-3,3,3-triphenyldisiloxane	: 5239	Uneicosane, 11-phenyl	: 6297
Trimethylene	: 4955			Unidigin	: 4043
Trimethylene chlorohydrin	: 10348	Trimethylurea	: 12032	Untriacontane	: 6299
Trimethylene dichloride	: 9893	Trimyristin	: 11628	Uracil	: 11059
Trimethylene diiodide	: 9907	Trinitroacetonitrile	: 290	Uracil mustard	: 11061
Trimethylene dimercaptan	: 10031	2,3,5-Trinitroanisole	: 364	Uramil	: 11077
Trimethylene glycol	: 9983	sym-Trinitrobenzene	: 2365	Urazole	: 11850
Trimethylene oxide	: 8592	2,4,6-Trinitrobenzoic acid	: 2735	Urea, N'-(4-bromo-3-chlorophenyl)-N-methoxy-N-methyl-	: 4076
Trimethylene sulfide	: 11691	Trinitro-m-cresol	: 9434		
Trimethylene sulfone	: 11692	2,4,6-Trinitro-3,5-dimethyl-tert-butylbenzene	: 1461	Urea, 1-(2-bromo-2-ethylbutanoyl)	: 3195
2,2,3-Trimethylheptane	: 6437			Urea, 1-(2-bromo-3-methylbutanoyl)	: 3196
2,2,4-Trimethylheptane	: 6438	2,4,7-Trinitrofluorenone	: 5979	Urea, 3-(p-bromophenyl)-1-methoxy-1-methyl-	: 7656
2,2,5-Trimethylheptane	: 6439	Trinitromethane	: 7562		
2,2,6-Trimethylheptane	: 6440	2,3,4-Trinitrotoluene	: 2046	Urea, N'-[4-(4-chlorophenoxy)phenyl]-N,N-dimethyl-	: 4088
2,3,3-Trimethylheptane	: 6441	2,4,5-Trinitrotoluene	: 2047		
2,3,4-Trimethylheptane	: 6442	2,4,6-Trinitrotoluene	: 2048		
2,3,5-Trimethylheptane	: 6443	28,29,30-Trinorlanostane	: 4105	Urea, N,N'-diamino-	: 4009
2,3,6-Trimethylheptane	: 6444	Triolein	: 8310	Urea, N'-(3,4-dichlorophenyl)-N-methoxy-N-methyl-	: 7415
2,4,4-Trimethylheptane	: 6445	Trional	: 3230		
2,4,5-Trimethylheptane	: 6446	Triopron	: 2136	Urea, 1,1-diisobutyl	: 11996
2,4,6-Trimethylheptane	: 6447	s-Trioxane, 2,4,6-trivinyl-	: 11906	Urea, N-[5-(1,1-dimethylethyl)-1,3,4-thiadiazol-2-yl]-N,N'-dimethyl-	: 11590
2,5,5-Trimethylheptane	: 6448	3,6,9-Trioxaundecane	: 5612		
3,3,4-Trimethylheptane	: 6449	3,7,12-Trioxocholanic acid	: 4101	Urea, N,N-dimethyl-N'-(octahydro-4,7-methano-1H-indeN-5-yl)-, (3aα,4α, 5α,7α,7aα)-	: 8212
3,3,5-Trimethylheptane	: 6450	Tripalmitin	: 6656		
3,4,4-Trimethylheptane	: 6451	Tripelennamine	: 5505		
3,4,5-Trimethylheptane	: 6452	Triphenylamine	: 728	Urea hydrochloride	: 12020
				Urea nitrate	: 12021

Urea, N-phenyl-N'-1,2,3-thiadiazol-5-yl-	: 11687
Urea, 1-sulfanilyl-2-thio-	: 2189
Urea, 1,1,3,3-tetraethyl-	: 12029
5-Ureidohydantoin	: 12009
Urethan	: 3948
Urethane, sec-butyl	: 3960
Urethane, isopentyl	: 3952
Uric acid	: 10764
Uric acid, 3-methyl-	: 10765
Uric acid, 7-methyl-	: 10766
Uridine 5'-phosphoric acid	: 12035
Urinex	: 2852
Uriprim	: 10852
Urochloralic acid	: 6231
Urofort	: 11830
Ursodeoxycholic acid	: 4098
Ursodiol	: 4098
Ursolic acid	: 12037
Uteroverdine	: 3001
Uvic acid	: 3311
Uzarin	: 4045
Vaccenic acid	: 8295
Valeraldehyde	: 8690
Valeraldehyde, 2-propyl-	: 8697
Valeric acid	: 8855
Valeric acid, 4-acetyl, ethyl ester	: 6902
Valeric acid, 4-acetyl, methyl ester	: 6903
Valeric acid, sec-butyl ester (d)	: 8899
Valeric acid, 4-chloro-	: 8865
Valeric acid, 3-chloro	: 8864
Valeric acid, 4-chloro-, ethyl ester	: 8867
Valeric acid, 5-chloro-4-hydroxy-, γ-lactone	: 6141
Valeric acid, 2,3-dibromo-	: 8870
Valeric acid, 2,5-dibromo-	: 8871
Valeric acid, 2,3-dibromo, ethyl ester	: 8872
Valeric acid, 2,5-dibromo-, ethyl ester	: 8873
Valeric acid, 4,4-dimethyl-5-acetyl	: 6461
Valeric acid, 2,2-dimethyl-5-acetyl	: 6460
Valeric acid, 2-ethyl (DL)	: 8877
Valeric acid, 5-fluoro-	: 8879
Valeric acid, furfuryl ester	: 8880
Valeric acid, 4-hydroxy-, ethyl ester	: 8884
Valeric acid, 2-hydroxyethyl ester	: 8885
Valeric acid, 3-methyl (DL)	: 8888
Valeric acid, 2-methylene	: 8890
Valeric acid, 3-methyl-3-phenyl	: 7175
Valeric acid, octyl ester	: 8901
Valeric acid, 2-oxo-, ethyl ester	: 8905
Valeric acid, 2-phenyl-, (±)-	: 943
Valeronitrile	: 8827
Valeronitrile, 3-imino-2-methyl	: 8830
Valeronitrile, 4-methyl-2-phenyl-	: 964
Valeroyl chloride	: 8990
Valerylacetone	: 8393
Valerylacetyl	: 6402
Valeryl chloride, 2-methyl (DL)	: 8993
Valeryl chloride, 3-methyl (DL)	: 8994
Valimyl	: 3201
Valium	: 2410
Vanillic acid	: 2625
Vanillin	: 574
Vanillylacetone	: 3663
Vanilmandelic acid	: 896
Vasicine	: 11192

Vellosimine	: 11346
Veratraldehyde	: 545
Veratramine	: 12044
Veratric acid	: 2576
Veratrole	: 1436
Veratrole, 3-iodo	: 1767
(R)-(+)-Verbenone	: 2978
d-Verbenone	: 2978
Vernine	: 6289
Vesipyrin	: 2483
Vibeline	: 3540
Vicianose	: 6234
Vicine	: 11107
(+)-cis-Vincamine	: 5349
Vincamine	: 5349
Vinyl acetate	: 179
Vinylacetonitrile	: 3803
Vinylacetylene	: 3889
Vinyl amide	: 10519
α-Vinylbenzyl alcohol	: 1868
Vinyl bromide	: 5880
Vinyl butyl ketone	: 6583
Vinyl butyrate	: 3513
Vinylcarbinol	: 10697
Vinyl chloride	: 5884
Vinyl cyanide	: 10583
4-Vinyl-1-cyclohexene dioxide	: 8557
Vinylethylacetylene	: 7103
Vinylethylene	: 3146
Vinyl fluoride	: 5906
Vinyl formate	: 6006
Vinylformic acid	: 10609
Vinylglycolic acid	: 3842
1-Vinylheptanol	: 8189
Vinylidene chloride	: 5894
Vinylidene fluoride	: 5902
Vinyl isobutyl ketone	: 7093
2-Vinylnaphthalene	: 7882
5-Vinyl-2-picoline	: 10955
Vinyl propionate	: 10261
Vinyltrichlorosilane	: 11481
Violaxanthin	: 4056
Violaxanthin, all-trans-	: 4056
Violet-leaf aldehyde	: 8108
Violuric acid	: 11074
Viquidil	: 10434
Visnadine	: 3540
Visnagin	: 6186
Vitamin A	: 11341
Vitamin B1, hydrochloride	: 11686
Vitamin B4	: 10751
Vitamin Bc	: 6243
Vitamin BX	: 2486
Vitamin C	: 466
Vitamin D2	: 11350
Vitamin D3	: 11348
Vitamin E	: 2787
Vitamin E acetate	: 2788
Vitamin K1	: 7865
Vitamin K6	: 7800
Vitamin L1	: 2484
Vomicine	: 11351
Warfarin	: 2832
Wintergreen oil	: 2634
Woodward's Reagent K	: 7407
Xanthaurine	: 2817
9-Xanthenecarboxylic acid, 2(N,N-diethylamino)ethyl ester,methylbromide	: 5445

Xanthine	: 10755
Xanthone	: 12050
Xanthone, 2,7-dihydroxy	: 12052
Xanthophyll 3	: 4057
all-trans-(+)-Xanthophyll	: 4058
Xanthophyll	: 4058
Xanthopterin	: 10749
Xanthoxyletin	: 2398
Xanthurenic acid	: 11230
Xanthyletin	: 2397
o-Xenol	: 3088
p-Xenylamine	: 3011
o-Xylene	: 1452
m-Xylene	: 1453
p-Xylene	: 1454
m-Xylene, 4-sec-butyl-	: 1473
m-Xylene, 5-tert-butyl-	: 1460
m-Xylenediamine	: 1433
o-Xylene, 3,4-dibromo-	: 1297
o-Xylene, 3,4-dichloro-	: 1360
p-Xylene, α,α'-dimethyl-	: 1415
p-Xylene, 2,5-dimethyl-	: 2280
p-Xylene, α,α'-diphenyl-	: 995
o-Xylene, 3-ethyl-	: 1670
o-Xylene, 4-ethyl-	: 1675
m-Xylene, 2-ethyl-	: 1673
m-Xylene, 4-ethyl-	: 1671
m-Xylene, 5-ethyl-	: 1672
p-Xylene, 2-ethyl-	: 1674
o-Xylene, 4-iodo-	: 1774
m-Xylene, 5-iodo-	: 1771
o-Xylene, 3-isopropyl-	: 1469
m-Xylene, 2-isopropyl-	: 1471
o-Xylene, 3-propyl-	: 1485
m-Xylene, 2-propyl-	: 1482
m-Xylene, 4-propyl-	: 1487
m-Xylene, 5-propyl-	: 1483
p-Xylene, 2-propyl-	: 1486
p-Xylene, 2,3,5-trinitro-	: 1489
2,3-Xylenol	: 9307
2,4-Xylenol	: 9308
2,5-Xylenol	: 9309
2,6-Xylenol	: 9310
3,4-Xylenol	: 9311
3,5-Xylenol	: 9312
3,4-Xylenol, 6-tert-butyl-	: 9324
2,3-Xylidine	: 704
2,4-Xylidine	: 705
2,5-Xylidine	: 706
2,6-Xylidine	: 707
3,4-Xylidine	: 708
3,5-Xylidine	: 709
Xylite	: 12058
p-Xylotocopherol	: 2789
o-Xylotocopherol	: 2790
1-(2,5-Xylylazo)-2-naphthol	: 8029
Yellow AB	: 7729
Yohimbine	: 12062
Zeaxanthin	: 4057
Zinc, bis(dimethylcarbamodithioato-S,S')-, (T-4)-	: 12068
Zineb	: 12067
Zingerone	: 3663
Zoxazolamine	: 2876
Zymostanol	: 4107

CBrClF$_2$
7464 - Methane, bromochlorodifluoro-
CBrClN$_2$O$_4$
7465 - Methane, bromochlorodinitro-
CBrCl$_2$F
7468 - Methane, bromodichlorofluoro-
CBrCl$_3$
7474 - Methane, bromotrichloro-
CBrF$_3$
7475 - Methane, bromotrifluoro-
CBrN
4165 - Cyanogen bromide
CBrN$_3$O$_6$
7476 - Methane, bromotrinitro-
CBr$_2$ClF
7494 - Methane, dibromochlorofluoro-
CBr$_2$Cl$_2$
7495 - Methane, dibromodichloro-
CBr$_2$F$_2$
7496 - Methane, dibromodifluoro-
CBr$_2$N$_2$O$_4$
7497 - Methane, dibromodinitro-
CBr$_2$O
4006 - Carbonic dibromide
CBr$_3$Cl
7549 - Methane, tribromochloro-
CBr$_3$F
7550 - Methane, tribromofluoro-
CBr$_3$NO$_2$
7551 - Methane, tribromonitro-
CBr$_4$
7539 - Methane, tetrabromo-
CClFO
4005 - Carbonic chloride fluoride
CClF$_3$
7486 - Methane, chlorotrifluoro-
CClF$_3$O$_2$S
7537 - Methanesulfonyl chloride, trifluoro-
CClN
4166 - Cyanogen chloride
CClNO$_3$S
11585 - Sulfuryl chloride isocyanate
CClN$_3$O$_6$
7487 - Methane, chlorotrinitro-
CCl$_2$F$_2$
7500 - Methane, dichlorodifluoro-
CCl$_2$N$_2$O$_4$
7501 - Methane, dichlorodinitro-
CCl$_2$O
4007 - Carbonic dichloride
CCl$_2$S
4032 - Carbonothioic dichloride
CCl$_3$F
7555 - Methane, trichlorofluoro-
CCl$_3$I
7556 - Methane, trichloroiodo-
CCl$_3$NO$_2$
7557 - Methane, trichloronitro-
CCl$_4$
7540 - Methane, tetrachloro-
CCl$_4$O$_2$S
7536 - Methanesulfonyl chloride, trichloro-
CCl$_4$O$_4$
9148 - Perchloric acid, trichloromethyl ester
CCl$_4$S
7527 - Methanesulfenyl chloride, trichloro-
CFN
4167 - Cyanogen fluoride
CF$_2$O
4008 - Carbonic difluoride
CF$_3$I
7559 - Methane, trifluoroiodo-
CF$_4$
7541 - Methane, tetrafluoro-
CF$_4$S
7528 - Methanesulfenyl fluoride, trifluoro-

CF$_8$S
11584 - Sulfur, pentafluoro(trifluoromethyl)-, (OC-6-21)-
CHBrClF
7466 - Methane, bromochlorofluoro-
CHBrCl$_2$
7467 - Methane, bromodichloro-
CHBrF$_2$
7469 - Methane, bromodifluoro-
CHBr$_2$Cl
7493 - Methane, dibromochloro-
CHBr$_2$F
7498 - Methane, dibromofluoro-
CHBr$_3$
7548 - Methane, tribromo-
CHClF$_2$
7478 - Methane, chlorodifluoro-
CHClI$_2$
7479 - Methane, chlorodiiodo-
CHClN$_2$O$_4$
7480 - Methane, chlorodinitro-
CHCl$_2$F
7502 - Methane, dichlorofluoro-
CHCl$_2$I
7503 - Methane, dichloroiodo-
CHCl$_3$
7554 - Methane, trichloro-
CHCl$_5$Si
11459 - Silane, trichloro(dichloromethyl)-
CHFI$_2$
7516 - Methane, fluorodiiodo-
CHFO
6019 - Formyl fluoride
CHF$_2$I
7505 - Methane, difluoroiodo-
CHF$_3$
7558 - Methane, trifluoro-
CHI$_3$
7560 - Methane, triiodo-
CHN
7176 - Hydrocyanic acid
CHNO
4162 - Cyanic acid
6024 - Fulminic acid
7360 - Isofulminic acid
CHNS
7399 - Isothiocyanic acid
CHN$_3$O$_6$
7562 - Methane, trinitro-
CHNaO$_3$
4004 - Carbonic acid monosodium salt
CH$_2$BrCl
7463 - Methane, bromochloro-
CH$_2$BrF
7470 - Methane, bromofluoro-
CH$_2$BrI
7471 - Methane, bromoiodo-
CH$_2$BrNO$_2$
7473 - Methane, bromonitro-
CH$_2$Br$_2$
7492 - Methane, dibromo-
CH$_2$ClF
7481 - Methane, chlorofluoro-
CH$_2$ClI
7482 - Methane, chloroiodo-
CH$_2$ClNO
3971 - Carbamic chloride
CH$_2$ClNO$_2$
7485 - Methane, chloronitro-
CH$_2$Cl$_2$
7499 - Methane, dichloro-
CH$_2$Cl$_4$Si
11456 - Silane, trichloro(chloromethyl)-
CH$_2$Cl$_6$Si$_2$
11440 - Silane, methylenebis[trichloro-
CH$_2$Cu$_2$O$_5$
511 - Basic copper carbonate

CH$_2$FI
7517 - Methane, fluoroiodo-
CH$_2$F$_2$
7504 - Methane, difluoro-
CH$_2$I$_2$
7506 - Methane, diiodo-
CH$_2$N$_2$
4155 - Cyanamide
7491 - Methane, diazo-
CH$_2$N$_2$O$_4$
7508 - Methane, dinitro-
CH$_2$N$_4$
11652 - 1H-Tetrazole
CH$_2$O
5987 - Formaldehyde
(CH$_2$O)$_x$
8634 - Paraformaldehyde
CH$_2$O$_2$
6002 - Formic acid
CH$_2$S$_3$
4034 - Carbonotrithioic acid
CH$_3$AsF$_2$
463 - Arsonous difluoride, methyl-
CH$_3$Br
7462 - Methane, bromo-
CH$_3$Br$_3$Si
11453 - Silane, tribromomethyl-
CH$_3$Cl
7477 - Methane, chloro-
CH$_3$ClO$_2$S
7535 - Methanesulfonyl chloride
CH$_3$ClO$_3$S
4086 - Chlorosulfuric acid, methyl ester
CH$_3$Cl$_3$Si
11465 - Silane, trichloromethyl-
CH$_3$F
7515 - Methane, fluoro-
CH$_3$FO$_2$S
7538 - Methanesulfonyl fluoride
CH$_3$I
7518 - Methane, iodo-
CH$_3$NO
5988 - Formaldehyde, oxime
5989 - Formamide
CH$_3$NO$_2$
7522 - Methane, nitro-
8091 - Nitrous acid, methyl ester
CH$_3$NO$_3$
8075 - Nitric acid, methyl ester
CH$_3$N$_3$
7460 - Methane, azido
CH$_3$N$_3$O$_3$
12024 - Urea, nitro-
CH$_4$
7459 - Methane
CH$_4$Cl$_2$Si
11398 - Silane, dichloromethyl-
CH$_4$N$_2$
7563 - Methanimidamide
CH$_4$N$_2$O
7566 - Methanimidamide, N-hydroxy-
11992 - Urea
CH$_4$N$_2$O$_2$
7451 - Methanamine, N-nitro-
12016 - Urea, hydroxy-
CH$_4$N$_2$S
11799 - Thiourea
CH$_4$N$_2$Se
11354 - Selenourea
CH$_4$N$_4$O$_2$
6287 - Guanidine, nitro-
CH$_4$O
7581 - Methanol
CH$_4$O$_2$
7180 - Hydroperoxide, methyl
CH$_4$O$_3$S
7530 - Methanesulfonic acid

CH_4O_4S
  11577 - Sulfuric acid, monomethyl ester
$CH_4O_6S_2$
  7514 - Methanedisulfonic acid
CH_4S
  7546 - Methanethiol
CH_5As
  446 - Arsine, methyl-
CH_5AsO_3
  458 - Arsonic acid, methyl-
CH_5ClN_2O
  12020 - Urea, monohydrochloride
CH_5ClSi
  11378 - Silane, chloromethyl
CH_5N
  7442 - Methanamine
CH_5NO
  7447 - Methanamine, *N*-hydroxy-
CH_5NO_2
  6003 - Formic acid, ammonium salt
CH_5N_3
  6278 - Guanidine
$CH_5N_3O_4$
  12021 - Urea, mononitrate
CH_5N_3S
  7142 - Hydrazinecarbothioamide
CH_5O_3P
  9538 - Phosphonic acid, methyl-
CH_6ClN
  7446 - Methanamine, hydrochloride
CH_6ClN_3
  6285 - Guanidine, monohydrochloride
CH_6ClN_3O
  7143 - Hydrazinecarboxamide, monohydrochloride
CH_6N_2
  7161 - Hydrazine, methyl-
CH_6N_4
  7147 - Hydrazinecarboximidamide
CH_6N_4O
  4009 - Carbonic dihydrazide
$CH_6N_4O_3$
  6286 - Guanidine, mononitrate
CH_6N_4S
  4033 - Carbonothioic dihydrazide
CH_6OSi
  11436 - Silane, methoxy
CH_6Si
  11437 - Silane, methyl-
CH_8B_2S
  5168 - Diborane(6), mu-(methylthio)-
CH_8Si_2
  11439 - Silane, methylenebis-
CIN
  7337 - Iodine cyanide
CI_4
  7542 - Methane, tetraiodo-
CKNS
  11711 - Thiocyanic acid, potassium salt
CNNaO
  4164 - Cyanic acid, sodium salt
CN_4O_8
  7544 - Methane, tetranitro-
CO
  4011 - Carbon monoxide
COS
  4035 - Carbon oxide sulfide
CO_2
  3992 - Carbon dioxide
CS_2
  3993 - Carbon disulfide
CSe_2
  4036 - Carbon selenide
C_2BrCl_3O
  293 - Acetyl bromide, trichloro-
C_2Br_2
  5939 - Ethyne, dibromo-

$C_2Br_2ClF_3$
  5519 - Ethane, 1,2-dibromo-1-chloro-1,2,2-trifluoro-
$C_2Br_2Cl_4$
  5524 - Ethane, 1,2-dibromo-1,1,2,2-tetrachloro-
$C_2Br_2F_4$
  5525 - Ethane, 1,2-dibromo-1,1,2,2-tetrafluoro-
C_2Br_4
  5914 - Ethene, tetrabromo-
C_2Br_6
  5594 - Ethane, hexabromo-
C_2CaN_2
  3930 - Calcium cyanide
C_2ClF_3
  5889 - Ethene, chlorotrifluoro-
C_2ClF_5
  5484 - Ethane, chloropentafluoro-
C_2Cl_2
  5940 - Ethyne, dichloro-
C_2Cl_2FN
  279 - Acetonitrile, dichlorofluoro-
$C_2Cl_2F_2$
  5897 - Ethene, 1,1-dichloro-2,2-difluoro-
  5898 - Ethene, 1,2-dichloro-1,2-difluoro-
$C_2Cl_2F_4$
  5537 - Ethane, 1,1-dichloro-1,2,2,2-tetrafluoro-
  5538 - Ethane, 1,2-dichloro-1,1,2,2-tetrafluoro-
$C_2Cl_2O_2$
  5585 - Ethanedioyl dichloride
C_2Cl_3F
  5923 - Ethene, trichlorofluoro-
$C_2Cl_3F_3$
  5662 - Ethane, 1,1,1-trichloro-2,2,2-trifluoro-
  5663 - Ethane, 1,1,2-trichloro-1,2,2-trifluoro-
C_2Cl_3N
  288 - Acetonitrile, trichloro-
C_2Cl_4
  5917 - Ethene, tetrachloro-
$C_2Cl_4F_2$
  5636 - Ethane, 1,1,1,2-tetrachloro-2,2-difluoro-
  5637 - Ethane, 1,1,2,2-tetrachloro-1,2-difluoro-
C_2Cl_4O
  300 - Acetyl chloride, trichloro-
$C_2Cl_4O_2$
  4029 - Carbonochloridic acid, trichloromethyl ester
C_2Cl_5F
  5618 - Ethane, pentachlorofluoro-
C_2Cl_6
  5595 - Ethane, hexachloro-
C_2F_3I
  5925 - Ethene, trifluoroiodo-
C_2F_3N
  289 - Acetonitrile, trifluoro-
C_2F_4
  5918 - Ethene, tetrafluoro-
$C_2F_4N_2O_4$
  5642 - Ethane, 1,1,2,2-tetrafluoro-1,2-dinitro-
C_2F_6
  5596 - Ethane, hexafluoro-
$C_2F_6S_2$
  5249 - Disulfide, bis(trifluoromethyl)
C_2HBr
  5937 - Ethyne, bromo-
$C_2HBrClF_3$
  5457 - Ethane, 1-bromo-2-chloro-1,1,2-trifluoro-
  5458 - Ethane, 2-bromo-2-chloro-1,1,1-trifluoro-
C_2HBrCl_2
  5882 - Ethene, 2-bromo-1,1-dichloro-
  5935 - Ethylene, 1-bromo-1,2-dichloro-, *trans*-
C_2HBrF_2
  5883 - Ethene, 1-bromo-1,2-difluoro-
$C_2HBr_2F_3$
  5526 - Ethane, 1,2-dibromo-1,1,2-trifluoro-
C_2HBr_2N
  277 - Acetonitrile, dibromo-
C_2HBr_3
  5921 - Ethene, tribromo-

C_2HBr_3O
  20 - Acetaldehyde, tribromo-
$C_2HBr_3O_2$
  241 - Acetic acid, tribromo-
C_2HBr_5
  5616 - Ethane, pentabromo-
C_2HCl
  5938 - Ethyne, chloro-
C_2HClF_2
  5885 - Ethene, 2-chloro-1,1-difluoro-
$C_2HClF_2O_2$
  127 - Acetic acid, chlorodifluoro-
C_2HClF_3I
  5490 - Ethane, 1-chloro-1,1,2-trifluoro-2-iodo-
C_2HClF_4
  5485 - Ethane, 1-chloro-1,1,2,2-tetrafluoro-
  5486 - Ethane, 1-chloro-1,2,2,2-tetrafluoro-
C_2HCl_2F
  5900 - Ethene, 1,1-dichloro-2-fluoro-
C_2HCl_2FO
  296 - Acetyl chloride, chlorofluoro-
$C_2HCl_2F_3$
  5539 - Ethane, 1,2-dichloro-1,1,2-trifluoro-
  5540 - Ethane, 2,2-dichloro-1,1,1-trifluoro-
C_2HCl_2N
  278 - Acetonitrile, dichloro-
C_2HCl_3
  5922 - Ethene, trichloro-
$C_2HCl_3F_2$
  5658 - Ethane, 1,1,1-trichloro-2,2-difluoro
  5659 - Ethane, 1,2,2-trichloro-1,1-difluoro
  5660 - Ethane, 1,2,2-trichloro-1,2-difluoro
C_2HCl_3O
  21 - Acetaldehyde, trichloro-
  297 - Acetyl chloride, dichloro-
$C_2HCl_3O_2$
  243 - Acetic acid, trichloro-
  4016 - Carbonochloridic acid, dichloromethyl ester
C_2HCl_4F
  5638 - Ethane, 1,1,2,2-tetrachloro-1-fluoro-
  5639 - Ethane, 1,1,1,2-tetrachloro-2-fluoro-
C_2HCl_5
  5617 - Ethane, pentachloro-
C_2HF
  5943 - Ethyne, fluoro-
C_2HF_3
  5924 - Ethene, trifluoro-
$C_2HF_3O_2$
  263 - Acetic acid, trifluoro-
C_2H_2
  5936 - Ethyne
C_2H_2BrCl
  5881 - Ethene, 1-bromo-2-chloro-
$C_2H_2BrClO_2$
  112 - Acetic acid, bromochloro-
$C_2H_2BrFO_2$
  116 - Acetic acid, bromofluoro-
$C_2H_2BrF_3$
  5466 - Ethane, 2-bromo-1,1,1-trifluoro-
$C_2H_2Br_2$
  5890 - Ethene, 1,1-dibromo-
  5891 - Ethene, 1,2-dibromo-, (*E*)-
  5892 - Ethene, 1,2-dibromo-, (*Z*)-
$C_2H_2Br_2Cl_2$
  5520 - Ethane, 1,2-dibromo-1,1-dichloro-
  5521 - Ethane, 1,2-dibromo-1,2-dichloro-
$C_2H_2Br_2F_2$
  5522 - Ethane, 1,2-dibromo-1,1-difluoro-
$C_2H_2Br_2O$
  292 - Acetyl bromide, bromo-
$C_2H_2Br_2O_2$
  159 - Acetic acid, dibromo-
$C_2H_2Br_3NO$
  94 - Acetamide, 2,2,2-tribromo-
$C_2H_2Br_4$
  5631 - Ethane, 1,1,1,2-tetrabromo-

5632 - Ethane, 1,1,2,2-tetrabromo-

$C_2H_2ClFO_2$
131 - Acetic acid, chlorofluoro-

$C_2H_2ClF_3$
5487 - Ethane, 1-chloro-1,1,2-trifluoro
5488 - Ethane, 1-chloro-1,2,2-trifluoro
5489 - Ethane, 2-chloro-1,1,1-trifluoro-

C_2H_2ClI
5888 - Ethene, 1-chloro-2-iodo-

C_2H_2ClN
274 - Acetonitrile, chloro-

C_2H_2ClNS
11700 - Thiocyanic acid, chloromethyl ester

$C_2H_2Cl_2$
5894 - Ethene, 1,1-dichloro-
5895 - Ethene, 1,2-dichloro-, (E)-
5896 - Ethene, 1,2-dichloro-, (Z)-

$C_2H_2Cl_2F_2$
5531 - Ethane, 1,2-dichloro-1,1-difluoro-
5532 - Ethane, 1,2-dichloro-1,2-difluoro-

$C_2H_2Cl_2O$
295 - Acetyl chloride, chloro-

$C_2H_2Cl_2O_2$
162 - Acetic acid, dichloro-
4014 - Carbonochloridic acid, chloromethyl ester

$C_2H_2Cl_3F$
5661 - Ethane, 1,1,2-trichloro-2-fluoro-

$C_2H_2Cl_3NO$
95 - Acetamide, 2,2,2-trichloro-

$C_2H_2Cl_4$
5634 - Ethane, 1,1,1,2-tetrachloro-
5635 - Ethane, 1,1,2,2-tetrachloro-

$C_2H_2F_2$
5902 - Ethene, 1,1-difluoro-
5903 - Ethene, 1,2-difluoro-, (Z)-

$C_2H_2F_2O_2$
175 - Acetic acid, difluoro-

$C_2H_2F_3I$
5667 - Ethane, 1,1,1-trifluoro-2-iodo-

$C_2H_2F_3NO_2$
5668 - Ethane, 1,1,1-trifluoro-2-nitro-

$C_2H_2F_4$
5640 - Ethane, 1,1,1,2-tetrafluoro-
5641 - Ethane, 1,1,2,2-tetrafluoro-

C_2H_2IN
283 - Acetonitrile, iodo-

$C_2H_2I_2$
5904 - Ethene, 1,2-diiodo-, (Z)-

$C_2H_2N_2O$
8563 - 1,3,4-Oxadiazole

$C_2H_2N_2S$
11657 - 1,2,4-Thiadiazole
11658 - 1,2,5-Thiadiazole

$C_2H_2N_4$
11651 - 1,2,4,5-Tetrazine

C_2H_2O
5927 - Ethenone

$C_2H_2O_2$
5491 - Ethanedial

$C_2H_2O_3$
223 - Acetic acid, oxo-

$C_2H_2O_4$
5553 - Ethanedioic acid

C_2H_3Br
5880 - Ethene, bromo-

C_2H_3BrO
291 - Acetyl bromide

$C_2H_3BrO_2$
111 - Acetic acid, bromo-

$C_2H_3Br_3$
5653 - Ethane, 1,1,2-tribromo-

$C_2H_3Br_3O$
5767 - Ethanol, 2,2,2-tribromo-

$C_2H_3Br_3O_2$
5581 - 1,1-Ethanediol, 2,2,2-tribromo-

C_2H_3Cl
5884 - Ethene, chloro-

$C_2H_3ClF_2$
5470 - Ethane, 1-chloro-1,1-difluoro
5471 - Ethane, 2-chloro-1,1-difluoro

C_2H_3ClO
12 - Acetaldehyde, chloro-
294 - Acetyl chloride

$C_2H_3ClO_2$
123 - Acetic acid, chloro-
4020 - Carbonochloridic acid, methyl ester

$C_2H_3Cl_2F$
5534 - Ethane, 1,1-dichloro-1-fluoro-
5535 - Ethane, 1,2-dichloro-1-fluoro-

$C_2H_3Cl_2NO$
51 - Acetamide, 2,2-dichloro-

$C_2H_3Cl_2NO_2$
5536 - Ethane, 1,1-dichloro-1-nitro-

$C_2H_3Cl_3$
5655 - Ethane, 1,1,1-trichloro-
5656 - Ethane, 1,1,2-trichloro-

$C_2H_3Cl_3O$
5768 - Ethanol, 2,2,2-trichloro-

$C_2H_3Cl_3O_2$
5582 - 1,1-Ethanediol, 2,2,2-trichloro-

$C_2H_3Cl_3Si$
11461 - Silane, trichloroethenyl-

C_2H_3F
5906 - Ethene, fluoro-

C_2H_3FO
301 - Acetyl fluoride

$C_2H_3FO_2$
189 - Acetic acid, fluoro-

$C_2H_3F_3$
5665 - Ethane, 1,1,1-trifluoro-
5666 - Ethane, 1,1,2-trifluoro-

$C_2H_3F_3O$
5770 - Ethanol, 2,2,2-trifluoro-

C_2H_3I
5907 - Ethene, iodo-

C_2H_3IO
302 - Acetyl iodide

$C_2H_3IO_2$
200 - Acetic acid, iodo-

$C_2H_3KO_2$
234 - Acetic acid, potassium salt

C_2H_3N
271 - Acetonitrile
7655 - Methylisonitrile

C_2H_3NO
282 - Acetonitrile, hydroxy-
7520 - Methane, isocyanato-

$C_2H_3NO_2$
5911 - Ethene, nitro-

$C_2H_3NO_3$
104 - Acetic acid, aminooxo-

$C_2H_3NO_4$
108 - Acetic acid, anhydride with nitric acid

C_2H_3NS
7521 - Methane, isothiocyanato-
11705 - Thiocyanic acid, methyl ester

$C_2H_3N_3$
11846 - 1H-1,2,3-Triazole

$C_2H_3N_3O_2$
109 - Acetic acid, azido-
11850 - 1,2,4-Triazolidine-3,5-dione

$C_2H_3NaO_2$
237 - Acetic acid, sodium salt

C_2H_4
5879 - Ethene

C_2H_4BrCl
5455 - Ethane, 1-bromo-1-chloro-
5456 - Ethane, 1-bromo-2-chloro-

$C_2H_4BrClO_2S$
5628 - Ethanesulfonyl chloride, 2-bromo-

C_2H_4BrF
5462 - Ethane, 1-bromo-2-fluoro-

C_2H_4BrI
5463 - Ethane, 1-bromo-2-iodo-

C_2H_4BrNO
36 - Acetamide, N-bromo-

$C_2H_4Br_2$
5517 - Ethane, 1,1-dibromo-
5518 - Ethane, 1,2-dibromo-

$C_2H_4Br_2O$
7524 - Methane, oxybis[bromo-

C_2H_4ClF
5475 - Ethane, 1-chloro-1-fluoro-
5476 - Ethane, 1-chloro-2-fluoro-

C_2H_4ClI
5477 - Ethane, 1-chloro-2-iodo-

C_2H_4ClNO
41 - Acetamide, 2-chloro-

$C_2H_4ClNO_2$
5482 - Ethane, 1-chloro-1-nitro-
5483 - Ethane, 1-chloro-2-nitro-

$C_2H_4Cl_2$
5528 - Ethane, 1,1-dichloro-
5529 - Ethane, 1,2-dichloro-

$C_2H_4Cl_2O$
5704 - Ethanol, 2,2-dichloro-
7525 - Methane, oxybis[chloro-

$C_2H_4Cl_2O_2S$
5629 - Ethanesulfonyl chloride, 2-chloro-

$C_2H_4Cl_3NO$
5680 - Ethanol, 1-amino-2,2,2-trichloro-

$C_2H_4Cl_4Si$
11391 - Silane, dichloro(dichloromethyl)methyl-

C_2H_4FNO
63 - Acetamide, 2-fluoro-

$C_2H_4F_2$
5544 - Ethane, 1,1-difluoro-
5545 - Ethane, 1,2-difluoro-

$C_2H_4F_2O$
5708 - Ethanol, 2,2-difluoro-

$C_2H_4I_2$
5546 - Ethane, 1,1-diiodo-
5547 - Ethane, 1,2-diiodo-

$C_2H_4N_2$
272 - Acetonitrile, amino-

$C_2H_4N_2O_2$
5492 - Ethanedial, dioxime
5493 - Ethanediamide

$C_2H_4N_2O_4$
5551 - Ethane, 1,1-dinitro-
5552 - Ethane, 1,2-dinitro-
8086 - Nitrous acid, 1,2-ethanediyl ester

$C_2H_4N_2O_6$
5574 - 1,2-Ethanediol, dinitrate

$C_2H_4N_2S_2$
5588 - Ethanedithioamide

$C_2H_4N_4$
6280 - Guanidine, cyano-
11845 - 1H-1,2,4-Triazol-3-amine

$C_2H_4N_4O_2$
5124 - Diazenedicarboxamide

C_2H_4O
8599 - Oxirane
11 - Acetaldehyde

$(C_2H_4O)_x$
7440 - Metaldehyde

C_2H_4OS
5646 - Ethanethioic acid

$C_2H_4O_2$
15 - Acetaldehyde, hydroxy-
100 - Acetic acid
6010 - Formic acid, methyl ester

$C_2H_4O_2S$
202 - Acetic acid, mercapto-

$C_2H_4O_3$
196 - Acetic acid, hydroxy-
5619 - Ethaneperoxoic acid

$C_2H_4O_3S$
5200 - 1,3,2-Dioxathiolane, 2-oxide

$C_2H_4O_4S$
5199 - 1,3,2-Dioxathiolane, 2,2-dioxide

$C_2H_4O_5S$
238 - Acetic acid, sulfo-
C_2H_4S
11695 - Thiirane
$C_2H_4S_2$
5589 - Ethane(dithioic) acid
$C_2H_4S_5$
8998 - 1,2,3,5,6-Pentathiepane
C_2H_4Si
11435 - Silane, ethynyl
$C_2H_5AsF_2$
445 - Arsine, ethyldifluoro-
C_2H_5Br
5454 - Ethane, bromo-
C_2H_5BrO
5685 - Ethanol, 2-bromo-
7472 - Methane, bromomethoxy-
C_2H_5Cl
5467 - Ethane, chloro-
C_2H_5ClO
5693 - Ethanol, 2-chloro-
7184 - Hypochlorous acid, ethyl ester
7483 - Methane, chloromethoxy-
$C_2H_5ClO_2S$
4087 - Chlorosulfurous acid, ethyl ester
5627 - Ethanesulfonyl chloride
$C_2H_5ClO_3S$
4085 - Chlorosulfuric acid, ethyl ester
C_2H_5ClS
5651 - Ethanethiol, 2-chloro-
7484 - Methane, chloro(methylthio)-
$C_2H_5Cl_2O_2P$
9577 - Phosphorodichloridic acid, ethyl ester
$C_2H_5Cl_3OSi$
11462 - Silane, trichloroethoxy-
$C_2H_5Cl_3Si$
11390 - Silane, dichloro(chloromethyl)methyl-
11463 - Silane, trichloroethyl-
C_2H_5F
5593 - Ethane, fluoro-
C_2H_5FO
5727 - Ethanol, 2-fluoro-
$C_2H_5F_3OSi$
11429 - Silane, ethoxytrifluoro-
$C_2H_5F_3Si$
11432 - Silane, ethyltrifluoro-
C_2H_5I
5597 - Ethane, iodo-
C_2H_5IO
5731 - Ethanol, 2-iodo-
7519 - Methane, iodomethoxy-
C_2H_5N
498 - Aziridine
C_2H_5NO
17 - Acetaldehyde, oxime
22 - Acetamide
5995 - Formamide, N-methyl-
$C_2H_5NO_2$
3953 - Carbamic acid, methyl ester
5607 - Ethane, nitro-
6250 - Glycine
8087 - Nitrous acid, ethyl ester
$C_2H_5NO_3$
5752 - Ethanol, 2-nitro-
8074 - Nitric acid, ethyl ester
C_2H_5NS
5643 - Ethanethioamide
$C_2H_5N_3O$
5681 - Ethanol, 2-azido-
$C_2H_5N_3O_2$
105 - Acetic acid, aminooxo-, hydrazide
7232 - Imidodicarbonic diamide
12019 - Urea, N-methyl-N-nitroso-
$C_2H_5N_3S_2$
11718 - Thioimidodicarbonic diamide
$C_2H_5N_5O_3$
6284 - Guanidine, N-methyl-N'-nitro-N-nitroso-

C_2H_6
5448 - Ethane
C_2H_6AsCl
453 - Arsinous chloride, dimethyl-
$C_2H_6BF_3O$
3138 - Boron, trifluoro[oxybis[methane]]-, (T-4)-
C_2H_6Cd
3929 - Cadmium, dimethyl-
$C_2H_6ClO_3P$
5929 - Ethephon
$C_2H_6Cl_2NPS$
9557 - Phosphoramidothioic dichloride, dimethyl-
$C_2H_6Cl_2Si$
11394 - Silane, dichlorodimethyl-
C_2H_6Hg
7435 - Mercury, dimethyl-
$C_2H_6N_2$
5125 - Diazene, dimethyl-
5126 - Diazene, dimethyl-, (E)-
$C_2H_6N_2O$
195 - Acetic acid, hydrazide
7450 - Methanamine, N-methyl-N-nitroso-
12017 - Urea, methyl-
$C_2H_6N_2O_2$
7149 - Hydrazinecarboxylic acid, methyl ester
7449 - Methanamine, N-methyl-N-nitro-
$C_2H_6N_4O$
11993 - Urea, (aminoiminomethyl)-
$C_2H_6N_4O_2$
5562 - Ethanedioic acid, dihydrazide
C_2H_6O
5672 - Ethanol
7523 - Methane, oxybis-
C_2H_6OS
5732 - Ethanol, 2-mercapto-
7529 - Methane, sulfinylbis-
$C_2H_6O_2$
5569 - 1,2-Ethanediol
7179 - Hydroperoxide, ethyl
9159 - Peroxide, dimethyl
$C_2H_6O_2S$
7534 - Methane, sulfonylbis-
$C_2H_6O_3S$
5622 - Ethanesulfonic acid
7532 - Methanesulfonic acid, methyl ester
11582 - Sulfurous acid, dimethyl ester
$C_2H_6O_4S$
11574 - Sulfuric acid, dimethyl ester
11576 - Sulfuric acid, monoethyl ester
$C_2H_6O_6$
5561 - Ethanedioic acid, dihydrate
$C_2H_6O_6S_2$
5587 - 1,2-Ethanedisulfonic acid
C_2H_6S
5648 - Ethanethiol
7545 - Methane, thiobis-
$C_2H_6S_2$
5254 - Disulfide, dimethyl
5590 - 1,2-Ethanedithiol
C_2H_6Se
7526 - Methane, selenobis-
C_2H_6Zn
12066 - Zinc, dimethyl
C_2H_7As
442 - Arsine, dimethyl-
444 - Arsine, ethyl-
$C_2H_7AsO_2$
451 - Arsinic acid, dimethyl-
$C_2H_7AsO_3$
457 - Arsonic acid, ethyl-
$C_2H_7BO_2$
3117 - Borane, dimethoxy-
3133 - Boronic acid, ethyl-
$C_2H_7ClN_2$
5671 - Ethanimidamide, monohydrochloride
C_2H_7N
5410 - Ethanamine

7448 - Methanamine, N-methyl-
C_2H_7NO
5434 - Ethanamine, N-hydroxy-
5674 - Ethanol, 1-amino-
5675 - Ethanol, 2-amino-
$C_2H_7NO_2$
106 - Acetic acid, ammonium salt
$C_2H_7NO_3S$
5623 - Ethanesulfonic acid, 2-amino-
C_2H_7NS
5649 - Ethanethiol, 2-amino-
$C_2H_7N_3O$
6266 - Glycine, hydrazide
$C_2H_7N_5$
7233 - Imidodicarbonimidic diamide
$C_2H_7O_2P$
9528 - Phosphinic acid, dimethyl-
$C_2H_7O_3P$
9588 - Phosphorous acid, dimethyl ester
$C_2H_7O_4P$
9559 - Phosphoric acid, dimethyl ester
9561 - Phosphoric acid, monoethyl ester
C_2H_8ClN
5433 - Ethanamine, hydrochloride
$C_2H_8NO_2PS$
7441 - Methamidophos
$C_2H_8N_2$
5496 - 1,2-Ethanediamine
7154 - Hydrazine, 1,1-dimethyl-
7155 - Hydrazine, 1,2-dimethyl-
$C_2H_8N_2O$
5729 - Ethanol, 2-hydrazino-
C_2H_8Si
11414 - Silane, dimethyl-
$C_2H_{10}N_2O$
5513 - 1,2-Ethanediamine, monohydrate
C_2I_2
5941 - Ethyne, diiodo-
C_2I_4
5919 - Ethene, tetraiodo-
C_2N_2
5550 - Ethanedinitrile
C_2N_2S
11572 - Sulfur cyanide
$C_2N_4O_6$
290 - Acetonitrile, trinitro-
$C_2Na_2O_4$
5566 - Ethanedioic acid, disodium salt
C_2O_4Pb
5568 - Ethanedioic acid, lead salt
C_3ClF_5
10541 - 1-Propene, 2-chloro-1,1,3,3,3-pentafluoro-
C_3ClF_5O
10421 - 2-Propanone, 1-chloro-1,1,3,3,3-pentafluoro-
$C_3Cl_2F_4$
10557 - 1-Propene, 1,2-dichloro-1,3,3,3-tetrafluoro-
$C_3Cl_2F_6$
9898 - Propane, 1,2-dichloro-1,1,2,3,3,3-hexafluoro-
$C_3Cl_2N_2$
9912 - Propanedinitrile, dichloro-
$C_3Cl_3F_3$
10607 - 1-Propene, 1,1,2-trichloro-3,3,3-trifluoro-
$C_3Cl_3F_5$
10179 - Propane, 1,1,1-trichloro-2,2,3,3,3-pentafluoro-
10180 - Propane, 1,2,2-trichloro-1,1,3,3,3-pentafluoro-
10181 - Propane, 1,2,3-trichloro-1,1,2,3,3-pentafluoro-
$C_3Cl_3N_3$
11837 - 1,3,5-Triazine, 2,4,6-trichloro-
$C_3Cl_3N_3O_3$
11840 - 1,3,5-Triazine-2,4,6(1H,3H,5H)-trione, 1,3,5-trichloro-
$C_3Cl_4F_4$
10149 - Propane, 1,2,2,3-tetrachloro-1,1,3,3-tetrafluoro-

C_3Cl_4O
10718 - 2-Propenoyl chloride, 2,3,3-trichloro-
C_3Cl_6
10571 - 1-Propene, 1,1,2,3,3,3-hexachloro-
C_3Cl_6O
10440 - 2-Propanone, 1,1,1,3,3,3-hexachloro-
$C_3Cl_6O_2$
262 - Acetic acid, trichloro-, trichloromethyl ester
$C_3Cl_6O_3$
7585 - Methanol, trichloro-, carbonate (2:1)
C_3F_6
10572 - 1-Propene, 1,1,2,3,3,3-hexafluoro-
C_3F_6O
8629 - Oxirane, trifluoro(trifluoromethyl)-
10441 - 2-Propanone, 1,1,1,3,3,3-hexafluoro-
C_3F_8
10118 - Propane, octafluoro-
C_3HBr_5O
10477 - 2-Propanone, 1,1,1,3,3-pentabromo-
$C_3HCl_2F_3$
10558 - Propene, 1,2-dichloro-3,3,3-trifluoro-
$C_3HCl_2N_3O_3$
11838 - 1,3,5-Triazine-2,4,6(1H,3H,5H)-trione, 1,3-dichloro-
$C_3HCl_3O_2$
10696 - 2-Propenoic acid, 2,3,3-trichloro-
C_3HCl_5
10593 - 1-Propene, 1,1,2,3,3-pentachloro-
C_3HCl_5O
10478 - 2-Propanone, 1,1,1,3,3-pentachloro-
C_3HCl_7
10048 - Propane, 1,1,1,2,2,3,3-heptachloro-
10049 - Propane, 1,1,1,2,3,3,3-heptachloro-
C_3HN
10737 - 2-Propynenitrile
C_3H_2BrNS
11668 - Thiazole, 2-bromo-
$C_3H_2Br_2$
10730 - 1-Propyne, 1,3-dibromo-
$C_3H_2Br_2O_2$
10627 - 2-Propenoic acid, 2,3-dibromo-, (Z)-
10628 - 2-Propenoic acid, 3,3-dibromo-
C_3H_2ClN
10584 - 2-Propenenitrile, 2-chloro-
$C_3H_2Cl_2O_2$
10029 - Propanedioyl dichloride
10630 - 2-Propenoic acid, 3,3-dichloro-
$C_3H_2Cl_3NO$
10115 - Propanenitrile, 3,3,3-trichloro-2-hydroxy-
$C_3H_2Cl_4$
10598 - 1-Propene, 1,2,3,3-tetrachloro-
$C_3H_2Cl_4O$
10486 - 2-Propanone, 1,1,1,3-tetrachloro-
10487 - 2-Propanone, 1,1,3,3-tetrachloro-
$C_3H_2Cl_6$
10051 - Propane, 1,1,1,2,2,3-hexachloro-
10052 - Propane, 1,1,1,3,3,3-hexachloro-
10053 - Propane, 1,1,2,2,3,3-hexachloro-
$C_3H_2N_2$
9911 - Propanedinitrile
$C_3H_2N_2O_3$
7226 - Imidazolidinetrione
C_3H_2O
10723 - 2-Propynal
$C_3H_2O_2$
10738 - 2-Propynoic acid
C_3H_3Br
10728 - 1-Propyne, 3-bromo-
$C_3H_3BrF_2$
10532 - 1-Propene, 3-bromo-3,3-difluoro-
$C_3H_3BrO_2$
10610 - 2-Propenoic acid, 2-bromo-
C_3H_3Cl
10729 - 1-Propyne, 3-chloro-
C_3H_3ClO
10511 - 2-Propenal, 2-chloro-
10714 - 2-Propenoyl chloride

$C_3H_3ClO_2$
10617 - 2-Propenoic acid, 3-chloro-, (E)-
10618 - 2-Propenoic acid, 3-chloro-, (Z)-
$C_3H_3ClO_3$
145 - Acetic acid, chlorooxo-, methyl ester
$C_3H_3Cl_2F$
10554 - 1-Propene, 1,1-dichloro-2-fluoro-
$C_3H_3Cl_2N$
10099 - Propanenitrile, 2,3-dichloro-
$C_3H_3Cl_3$
10602 - 1-Propene, 1,1,2-trichloro-
10603 - 1-Propene, 1,2,3-trichloro-
10604 - 1-Propene, 3,3,3-trichloro-
$C_3H_3Cl_3O$
8628 - Oxirane, (trichloromethyl)-
9790 - Propanal, 2,2,3-trichloro-
10488 - 2-Propanone, 1,1,1-trichloro-
10500 - Propanoyl chloride, 2,2-dichloro-
10501 - Propanoyl chloride, 2,3-dichloro-
10502 - Propanoyl chloride, 3,3-dichloro-
$C_3H_3Cl_3O_2$
253 - Acetic acid, trichloro-, methyl ester
10332 - Propanoic acid, 2,2,3-trichloro-
$C_3H_3Cl_3O_3$
10333 - Propanoic acid, 3,3,3-trichloro-2-hydroxy-
$C_3H_3Cl_5$
10125 - Propane, 1,1,1,2,3-pentachloro-
10126 - Propane, 1,1,2,3,3-pentachloro-
$C_3H_3F_3$
10608 - 1-Propene, 3,3,3-trifluoro-
$C_3H_3F_5$
10128 - Propane, 1,1,1,2,2-pentafluoro
C_3H_3I
10734 - 1-Propyne, 1-iodo-
10735 - 1-Propyne, 3-iodo-
C_3H_3N
10583 - 2-Propenenitrile
C_3H_3NO
7402 - Isoxazole
8575 - Oxazole
10113 - Propanenitrile, 2-oxo-
C_3H_3NOS
303 - Acetyl isothiocyanate
$C_3H_3NOS_2$
11685 - 4-Thiazolidinone, 2-thioxo-
$C_3H_3NO_2$
150 - Acetic acid, cyano-
$C_3H_3NO_2S$
11683 - 2,4-Thiazolidinedione
C_3H_3NS
11667 - Thiazole
$C_3H_3N_3$
11827 - 1,3,5-Triazine
$C_3H_3N_3O_2$
7211 - 1H-Imidazole, 2-nitro-
$C_3H_3N_3O_2S$
11666 - 2-Thiazolamine, 5-nitro-
$C_3H_3N_3O_3$
11907 - 1,3,5-Trioxane-2,4,6-triimine
C_3H_4
4990 - Cyclopropene
9772 - 1,2-Propadiene
10727 - 1-Propyne
C_3H_4BrClO
10417 - 2-Propanone, 1-bromo-3-chloro-
10495 - Propanoyl chloride, 2-bromo-
C_3H_4BrN
10094 - Propanenitrile, 2-bromo-
10095 - Propanenitrile, 3-bromo-
$C_3H_4Br_2$
10542 - 1-Propene, 1,1-dibromo-
10543 - 1-Propene, 1,3-dibromo-, (E)-
10544 - 1-Propene, 1,3-dibromo-, (Z)-
10545 - 1-Propene, 2,3-dibromo-
$C_3H_4Br_2O$
9780 - Propanal, 2,3-dibromo-
10424 - 2-Propanone, 1,3-dibromo-

$C_3H_4Br_2O_2$
10238 - Propanoic acid, 2,3-dibromo-
$C_3H_4Br_4$
10135 - Propane, 1,1,1,2-tetrabromo-
10136 - Propane, 1,1,2,2-tetrabromo-
10137 - Propane, 1,1,3,3-tetrabromo-
10138 - Propane, 1,2,2,3-tetrabromo-
$C_3H_4ClFO_2$
134 - Acetic acid, chlorofluoro-, methyl ester
$C_3H_4ClF_3$
9871 - Propane, 3-chloro-1,1,1-trifluoro-
C_3H_4ClN
10097 - Propanenitrile, 3-chloro-
$C_3H_4Cl_2$
10547 - 1-Propene, 1,1-dichloro-
10548 - 1-Propene, 1,2-dichloro-, (E)-
10549 - 1-Propene, 1,2-dichloro-, (Z)-
10550 - 1-Propene, 1,3-dichloro-, (E)-
10551 - 1-Propene, 1,3-dichloro-, (Z)-
10552 - 1-Propene, 2,3-dichloro-
10553 - 1-Propene, 3,3-dichloro-
$C_3H_4Cl_2O$
9781 - Propanal, 2,3-dichloro-
10425 - 2-Propanone, 1,1-dichloro-
10426 - 2-Propanone, 1,3-dichloro-
10496 - Propanoyl chloride, 2-chloro-, (S)-
10497 - Propanoyl chloride, 3-chloro-
$C_3H_4Cl_2O_2$
166 - Acetic acid, dichloro-, methyl ester
4013 - Carbonochloridic acid, 2-chloroethyl ester
10244 - Propanoic acid, 2,2-dichloro-
10245 - Propanoic acid, 2,3-dichloro-
$C_3H_4Cl_3NO_2$
6000 - Formamide, N-(2,2,2-trichloro-1-hydroxyethyl)-
$C_3H_4Cl_3NO_3$
10408 - 2-Propanol, 1,1,1-trichloro-3-nitro-
$C_3H_4Cl_4$
10141 - Propane, 1,1,1,2-tetrachloro-
10142 - Propane, 1,1,1,3-tetrachloro-
10143 - Propane, 1,1,2,2-tetrachloro-
10144 - Propane, 1,1,2,3-tetrachloro-
10145 - Propane, 1,2,2,3-tetrachloro-
$C_3H_4Cl_4O$
10402 - 2-Propanol, 1,1,3,3-tetrachloro-
10403 - 2-Propanol, 1,1,1,3-tetrachloro-
$C_3H_4F_2$
10561 - 1-Propene, 1,1-difluoro
$C_3H_4F_3NO_2$
10184 - Propane, 1,1,1-trifluoro-3-nitro-
$C_3H_4F_4O$
10404 - 1-Propanol, 2,2,3,3-tetrafluoro-
$C_3H_4N_2$
285 - Acetonitrile, (methyleneamino)-
7192 - 1H-Imidazole
10815 - 1H-Pyrazole
$C_3H_4N_2O$
49 - Acetamide, 2-cyano-
$C_3H_4N_2O_2$
7216 - 2,4-Imidazolidinedione
$C_3H_4N_2S$
11662 - 2-Thiazolamine
C_3H_4O
4988 - Cyclopropanone
5944 - Ethyne, methoxy-
10509 - 2-Propenal
10743 - 2-Propyn-1-ol
$C_3H_4O_2$
6006 - Formic acid, ethenyl ester
8595 - 2-Oxetanone
8596 - 3-Oxetanone
8603 - Oxiranecarboxaldehyde
9788 - Propanal, 2-oxo-
10609 - 2-Propenoic acid
$C_3H_4O_3$
5211 - 1,3-Dioxolan-2-one
10314 - Propanoic acid, 2-oxo-

$C_3H_4O_4$
  9921 - Propanedioic acid
$C_3H_4O_5$
  9956 - Propanedioic acid, hydroxy-
C_3H_5Br
  10529 - 1-Propene, 1-bromo-, (Z)-
  10530 - 1-Propene, 2-bromo-
  10531 - 1-Propene, 3-bromo-
C_3H_5BrO
  8600 - Oxirane, (bromomethyl)-, (±)-
  9775 - Propanal, 2-bromo-
  10416 - 2-Propanone, 1-bromo-
  10490 - Propanoyl bromide
  10699 - 2-Propen-1-ol, 2-bromo-
$C_3H_5BrO_2$
  117 - Acetic acid, bromo-, methyl ester
  10204 - Propanoic acid, 2-bromo-, (±)-
  10205 - Propanoic acid, 3-bromo-
$C_3H_5Br_2Cl$
  9885 - Propane, 1,2-dibromo-3-chloro-
$C_3H_5Br_3$
  10160 - Propane, 1,1,1-tribromo-
  10161 - Propane, 1,1,2-tribromo-
  10162 - Propane, 1,2,2-tribromo-
  10163 - Propane, 1,2,3-tribromo-
C_3H_5Cl
  10533 - 1-Propene, 1-chloro-, (E)-
  10534 - 1-Propene, 1-chloro-, (Z)-
  10535 - 1-Propene, 2-chloro-
  10536 - 1-Propene, 3-chloro-
$C_3H_5ClF_2$
  9851 - Propane, 1-chloro-2,2-difluoro-
$C_3H_5ClN_2O_6$
  9997 - 1,2-Propanediol, 3-chloro-, dinitrate
C_3H_5ClO
  8605 - Oxirane, (chloromethyl)-, (±)-
  9777 - Propanal, 2-chloro-
  9778 - Propanal, 3-chloro-
  10420 - 2-Propanone, 1-chloro-
  10493 - Propanoyl chloride
  10700 - 2-Propen-1-ol, 2-chloro-
  10701 - 2-Propen-1-ol, 3-chloro-
$C_3H_5ClO_2$
  138 - Acetic acid, chloro-, methyl ester
  299 - Acetyl chloride, methoxy-
  4018 - Carbonochloridic acid, ethyl ester
  7582 - Methanol, chloro-, acetate
  10215 - Propanoic acid, 2-chloro-
  10216 - Propanoic acid, 3-chloro-
$C_3H_5Cl_2F$
  9897 - Propane, 1,2-dichloro-2-fluoro-
$C_3H_5Cl_2NO_2$
  3940 - Carbamic acid, dichloro-, ethyl ester
  9902 - Propane, 1,1-dichloro-1-nitro-
$C_3H_5Cl_3$
  10171 - Propane, 1,1,1-trichloro-
  10172 - Propane, 1,1,2-trichloro-
  10173 - Propane, 1,1,3-trichloro-
  10174 - Propane, 1,2,2-trichloro-
  10175 - Propane, 1,2,3-trichloro-
$C_3H_5Cl_3O$
  10406 - 2-Propanol, 1,1,1-trichloro-
$C_3H_5Cl_3Si$
  11475 - Silane, trichloro-2-propenyl-
C_3H_5F
  10567 - 1-Propene, 1-fluoro- (Z)
  10568 - 1-Propene, 1-fluoro- (E)
  10569 - 1-Propene, 2-fluoro-
  10570 - 1-Propene, 3-fluoro-
C_3H_5FO
  8616 - Oxirane, (fluoromethyl)-
  10437 - 2-Propanone, 1-fluoro-
  10506 - Propanoyl fluoride
$C_3H_5FO_2$
  192 - Acetic acid, fluoro-, methyl ester
$C_3H_5F_3O$
  10409 - 2-Propanol, 1,1,1-trifluoro-, (±)-

$C_3H_5F_3O_3S$
  7533 - Methanesulfonic acid, trifluoro-, ethyl ester
C_3H_5I
  10573 - 1-Propene, 3-iodo-
C_3H_5IO
  8617 - Oxirane, (iodomethyl)-
  10454 - 2-Propanone, 1-iodo-
$C_3H_5IO_2$
  10282 - Propanoic acid, 2-iodo-, (±)-
C_3H_5N
  5599 - Ethane, isocyano-
  10090 - Propanenitrile
  10725 - 2-Propyn-1-amine
C_3H_5NO
  284 - Acetonitrile, methoxy-
  496 - 2-Azetidinone
  4163 - Cyanic acid, ethyl ester
  5598 - Ethane, isocyanato-
  10105 - Propanenitrile, 2-hydroxy-
  10106 - Propanenitrile, 3-hydroxy-
  10519 - 2-Propenamide
$C_3H_5NO_2$
  9789 - Propanal, 2-oxo-, 1-oxime
  10589 - 1-Propene, 1-nitro-
  10590 - 1-Propene, 2-nitro-
  10591 - 1-Propene, 3-nitro-
$C_3H_5NO_3$
  10475 - 2-Propanone, 1-nitro-
C_3H_5NS
  286 - Acetonitrile, (methylthio)-
  5600 - Ethane, isothiocyanato-
  11703 - Thiocyanic acid, ethyl ester
$C_3H_5N_3$
  10812 - 1H-Pyrazol-4-amine
$C_3H_5N_3O$
  153 - Acetic acid, cyano-, hydrazide
$C_3H_5N_3O_9$
  10197 - 1,2,3-Propanetriol, trinitrate
$C_3H_5NaO_2$
  10327 - Propanoic acid, sodium salt
C_3H_6
  4955 - Cyclopropane
  10528 - 1-Propene
C_3H_6BrCl
  9838 - Propane, 1-bromo-2-chloro-
  9839 - Propane, 1-bromo-3-chloro-
  9840 - Propane, 2-bromo-1-chloro-
  9841 - Propane, 2-bromo-2-chloro-
C_3H_6BrF
  9844 - Propane, 1-bromo-3-fluoro-
$C_3H_6BrNO_2$
  9847 - Propane, 2-bromo-2-nitro-
$C_3H_6BrNO_4$
  9991 - 1,3-Propanediol, 2-bromo-2-nitro-
$C_3H_6Br_2$
  9880 - Propane, 1,1-dibromo-
  9881 - Propane, 1,2-dibromo-
  9882 - Propane, 1,3-dibromo-
  9883 - Propane, 2,2-dibromo-
$C_3H_6Br_2O$
  10357 - 2-Propanol, 1,3-dibromo-
C_3H_6ClF
  9856 - Propane, 1-chloro-3-fluoro-
C_3H_6ClI
  9857 - Propane, 1-chloro-3-iodo-
C_3H_6ClNO
  3973 - Carbamic chloride, dimethyl-
$C_3H_6ClNO_2$
  9866 - Propane, 1-chloro-1-nitro-
  9867 - Propane, 1-chloro-2-nitro-
  9868 - Propane, 1-chloro-3-nitro-
  9869 - Propane, 2-chloro-1-nitro-
  9870 - Propane, 2-chloro-2-nitro-
C_3H_6ClNS
  3980 - Carbamothioic chloride, dimethyl-
$C_3H_6Cl_2$
  9891 - Propane, 1,1-dichloro-

  9892 - Propane, 1,2-dichloro-, (±)-
  9893 - Propane, 1,3-dichloro-
  9894 - Propane, 2,2-dichloro-
$C_3H_6Cl_2O$
  10359 - 1-Propanol, 2,3-dichloro-
  10360 - 1-Propanol, 3,3-dichloro-
  10361 - 2-Propanol, 1,1-dichloro-
  10362 - 2-Propanol, 1,3-dichloro-
$C_3H_6Cl_2Si$
  11396 - Silane, dichloroethenylmethyl-
$C_3H_6Cl_3P$
  9524 - Phosphine, tris(chloromethyl)-
$C_3H_6Cl_4Si$
  11458 - Silane, trichloro(3-chloropropyl)-
$C_3H_6F_2$
  9905 - Propane, 1,3-difluoro-
  9906 - Propane, 2,2-difluoro-
$C_3H_6F_2O$
  10369 - 2-Propanol, 1,3-difluoro-
$C_3H_6I_2$
  9907 - Propane, 1,3-diiodo-
  9908 - Propane, 2,2-diiodo-
$C_3H_6N_2$
  4157 - Cyanamide, dimethyl-
  10092 - Propanenitrile, 3-amino-
  10822 - 1H-Pyrazole, 4,5-dihydro-
$C_3H_6N_2O$
  7227 - 2-Imidazolidinone
$C_3H_6N_2OS$
  33 - Acetamide, N-(aminothioxomethyl)-
$C_3H_6N_2O_2$
  27 - Acetamide, N-(aminocarbonyl)-
  7406 - 3-Isoxazolidinone, 4-amino-, (R)-
  9784 - Propanal, 2-(hydroxyimino)-, oxime
$C_3H_6N_2O_4$
  3963 - Carbamic acid, nitro-, ethyl ester
  9918 - Propane, 1,1-dinitro-
  9919 - Propane, 1,3-dinitro-
  9920 - Propane, 2,2-dinitro-
$C_3H_6N_2O_7$
  10191 - 1,2,3-Propanetriol, 1,3-dinitrate
$C_3H_6N_2S$
  7225 - 2-Imidazolidinethione
  11664 - 2-Thiazolamine, 4,5-dihydro-
$C_3H_6N_6$
  11835 - 1,3,5-Triazine-2,4,6-triamine
$C_3H_6N_6O_6$
  11833 - 1,3,5-Triazine, hexahydro-1,3,5-trinitro-
C_3H_6O
  5908 - Ethene, methoxy-
  8592 - Oxetane
  8621 - Oxirane, methyl-
  9774 - Propanal
  10411 - 2-Propanone
  10697 - 2-Propen-1-ol
$C_3H_6O_2$
  16 - Acetaldehyde, methoxy-
  208 - Acetic acid, methyl ester
  5204 - 1,3-Dioxolane
  6007 - Formic acid, ethyl ester
  8618 - Oxiranemethanol
  10201 - Propanoic acid
  10443 - 2-Propanone, 1-hydroxy-
$C_3H_6O_2S$
  204 - Acetic acid, mercapto-, methyl ester
  216 - Acetic acid, (methylthio)-
  5909 - Ethene, (methylsulfonyl)-
  10285 - Propanoic acid, 2-mercapto-
  10286 - Propanoic acid, 3-mercapto-
  11692 - Thietane, 1,1-dioxide
$C_3H_6O_3$
  198 - Acetic acid, hydroxy-, methyl ester
  205 - Acetic acid, methoxy-
  4000 - Carbonic acid, dimethyl ester
  9782 - Propanal, 2,3-dihydroxy-, (±)-
$C_3H_6O_2$
  10266 - Propanoic acid, 2-hydroxy-, (±)-

$C_3H_6O_3$
10267 - Propanoic acid, 3-hydroxy-
10429 - 2-Propanone, 1,3-dihydroxy-
11905 - 1,3,5-Trioxane
$C_3H_6O_3S$
8570 - 1,2-Oxathiolane, 2,2-dioxide
11694 - Thietane, 1-oxide
$C_3H_6O_4$
10252 - Propanoic acid, 2,3-dihydroxy-
C_3H_6S
5910 - Ethene, (methylthio)-
10600 - 2-Propene-1-thiol
11691 - Thietane
11696 - Thiirane, methyl-
$C_3H_6S_2$
5266 - 1,3-Dithiolane
$C_3H_6S_3$
11919 - 1,3,5-Trithiane
C_3H_7Br
9836 - Propane, 1-bromo-
9837 - Propane, 2-bromo-
C_3H_7BrO
5464 - Ethane, (bromomethoxy)-
5465 - Ethane, 1-bromo-2-methoxy-
10344 - 1-Propanol, 3-bromo-
10345 - 2-Propanol, 1-bromo-
C_3H_7Cl
9848 - Propane, 1-chloro-
9849 - Propane, 2-chloro-
C_3H_7ClHgO
11822 - Triadimenol
C_3H_7ClO
5478 - Ethane, (chloromethoxy)-
5479 - Ethane, 1-chloro-1-methoxy-
5480 - Ethane, 1-chloro-2-methoxy-
10347 - 1-Propanol, 2-chloro-
10348 - 1-Propanol, 3-chloro-
10349 - 2-Propanol, 1-chloro-
$C_3H_7ClO_2$
9994 - 1,2-Propanediol, 3-chloro-
9995 - 1,3-Propanediol, 2-chloro-
$C_3H_7ClO_2S$
10133 - 1-Propanesulfonyl chloride
$C_3H_7ClO_3S$
5699 - Ethanol, 2-chloro-, methanesulfonate
C_3H_7ClS
5481 - Ethane, 1-chloro-2-(methylthio)-
$C_3H_7Cl_2N$
9805 - 1-Propanamine, N,N-dichloro-
$C_3H_7Cl_3Si$
11373 - Silane, chloro(dichloromethyl)dimethyl-
11466 - Silane, trichloro(1-methylethyl)-
11476 - Silane, trichloropropyl-
C_3H_7F
10046 - Propane, 1-fluoro-
10047 - Propane, 2-fluoro-
C_3H_7FO
10382 - 1-Propanol, 3-fluoro-
C_3H_7I
10054 - Propane, 1-iodo-
10055 - Propane, 2-iodo-
C_3H_7IO
10385 - 1-Propanol, 3-iodo-
C_3H_7N
494 - Azetidine
500 - Aziridine, 1-methyl-
501 - Aziridine, 2-methyl-
4953 - Cyclopropanamine
10522 - 2-Propen-1-amine
$C_3H_7NNaS_2$
11514 - Sodium dimethyldithiocarbamate
C_3H_7NO
72 - Acetamide, N-methyl-
5992 - Formamide, N,N-dimethyl-
5994 - Formamide, N-ethyl-
9787 - Propanal, oxime
9791 - Propanamide

10476 - 2-Propanone, oxime
C_3H_7NOS
3978 - Carbamothioic acid, O-ethyl ester
3979 - Carbamothioic acid, S-ethyl ester
$C_3H_7NO_2$
323 - β-Alanine
324 - DL-Alanine
325 - L-Alanine
3948 - Carbamic acid, ethyl ester
6268 - Glycine, N-methyl-
8092 - Nitrous acid, 1-methylethyl ester
8098 - Nitrous acid, propyl ester
9797 - Propanamide, 2-hydroxy-, (±)-
10116 - Propane, 1-nitro-
10117 - Propane, 2-nitro-
$C_3H_7NO_2S$
5014 - L-Cysteine
$C_3H_7NO_3$
8076 - Nitric acid, 1-methylethyl ester
8080 - Nitric acid, propyl ester
10394 - 1-Propanol, 2-nitro-
10395 - 1-Propanol, 3-nitro-
10396 - 2-Propanol, 1-nitro-
11355 - D-Serine
11356 - DL-Serine
11357 - L-Serine
$C_3H_7NO_4$
10022 - 1,2-Propanediol, 1-nitrate
$C_3H_7NO_5$
10192 - 1,2,3-Propanetriol, 1-nitrate
10193 - 1,2,3-Propanetriol, 2-nitrate
C_3H_7NS
11681 - Thiazolidine
$C_3H_7N_3O_2$
6253 - Glycine, N-(aminoiminomethyl)-
12015 - Urea, N-ethyl-N-nitroso-
$C_3H_7N_3O_5$
11839 - 1,3,5-Triazine-2,4,6(1H,3H,5H)-trione,
dihydrate
C_3H_8
9834 - Propane
$C_3H_8BrClSi$
11365 - Silane, (bromomethyl)chlorodimethyl-
$C_3H_8Cl_2Si$
11372 - Silane, chloro(chloromethyl)dimethyl-
11397 - Silane, dichloroethylmethyl-
$C_3H_8NO_5P$
6276 - Glyphosate
$C_3H_8N_2O$
12004 - Urea, N,N'-dimethyl-
12005 - Urea, N,N-dimethyl-
12014 - Urea, ethyl-
$C_3H_8N_2O_2$
326 - Alanine, 3-amino-
7148 - Hydrazinecarboxylic acid, ethyl ester
9824 - 1-Propanamine, N-nitro-
$C_3H_8N_2O_3$
11995 - Urea, N,N'-bis(hydroxymethyl)-
$C_3H_8N_2S$
11804 - Thiourea, N,N'-dimethyl-
C_3H_8O
5602 - Ethane, methoxy-
10334 - 1-Propanol
10335 - 2-Propanol
C_3H_8OS
5749 - Ethanol, 2-(methylthio)-
$C_3H_8OS_2$
10370 - 1-Propanol, 2,3-dimercapto-
10371 - 2-Propanol, 1,3-dimercapto-
$C_3H_8O_2$
5733 - Ethanol, 2-methoxy-
7507 - Methane, dimethoxy-
9982 - 1,2-Propanediol
9983 - 1,3-Propanediol
$C_3H_8O_2S$
10013 - 1,2-Propanediol, 3-mercapto-

$C_3H_8O_3$
10186 - 1,2,3-Propanetriol
$C_3H_8O_3S$
7531 - Methanesulfonic acid, ethyl ester
10130 - 2-Propanesulfonic acid
C_3H_8S
5606 - Ethane, (methylthio)-
10156 - 1-Propanethiol
10157 - 2-Propanethiol
$C_3H_8S_2$
7461 - Methane, bis(methylthio)-
10030 - 1,2-Propanedithiol
10031 - 1,3-Propanedithiol
C_3H_9Al
339 - Aluminum, trimethyl
C_3H_9As
449 - Arsine, trimethyl-
$C_3H_9AsO_3$
461 - Arsonic acid, propyl-
C_3H_9B
3119 - Borane, trimethyl-
$C_3H_9BO_2$
3136 - Boronic acid, propyl-
$C_3H_9BO_3$
3126 - Boric acid, trimethyl ester
C_3H_9BS
3129 - Borinic acid, dimethylthio-, methyl ester
$C_3H_9BS_3$
11698 - Thioboric acid, trimethyl ester
C_3H_9ClSi
11389 - Silane, chlorotrimethyl-
C_3H_9ClSn
11540 - Stannane, chlorotrimethyl-
C_3H_9N
7444 - Methanamine, N,N-dimethyl-
9803 - 1-Propanamine
9804 - 2-Propanamine
C_3H_9NO
5738 - Ethanol, 2-(methylamino)-
10336 - 1-Propanol, 2-amino-, (±)-
10337 - 1-Propanol, 3-amino-
10338 - 2-Propanol, 1-amino-, (±)-
$C_3H_9NO_2$
9984 - 1,2-Propanediol, 3-amino-, (±)-
$C_3H_9NO_3S$
5624 - Ethanesulfonic acid, 2-(methylamino)-
$C_3H_9O_2PS$
9550 - Phosphonothioic acid, methyl-, O,O-dimethyl
ester
$C_3H_9O_3P$
9540 - Phosphonic acid, methyl-, dimethyl ester
9542 - Phosphonic acid, (1-methylethyl)-
9543 - Phosphonic acid, methyl-, monoethyl ester
9547 - Phosphonic acid, propyl-
9592 - Phosphorous acid, trimethyl ester
$C_3H_9O_4P$
9564 - Phosphoric acid, trimethyl ester
C_3H_9P
9522 - Phosphine, trimethyl-
C_3H_9Sb
11550 - Stibine, trimethyl-
$C_3H_{10}ClN$
5435 - Ethanamine, N-methyl-, hydrochloride
7445 - Methanamine, N,N-dimethyl-, hydrochloride
$C_3H_{10}N_2$
5512 - 1,2-Ethanediamine, N-methyl-
9872 - 1,2-Propanediamine
9873 - 1,3-Propanediamine
$C_3H_{10}Si$
11489 - Silane, trimethyl-
$C_3H_{12}BN$
3120 - Borane, trimethyl-, monoammoniate
3132 - Boron, (N,N-dimethylmethanamine)trihydro-,
(T-4)-
C_3N_2O
9915 - Propanedinitrile, oxo-

C$_3$O$_2$
9773 - 1,2-Propadiene-1,3-dione
C$_4$Br$_2$F$_8$
3275 - Butane, 1,4-dibromo-1,1,2,2,3,3,4,4-
octafluoro-
C$_4$Br$_4$S
11781 - Thiophene, tetrabromo-
C$_4$ClF$_7$
4177 - Cyclobutane, 1-chloro-1,2,2,3,3,4,4-
heptafluoro-
C$_4$ClF$_7$O
3692 - Butanoyl chloride, heptafluoro-
C$_4$Cl$_2$F$_4$
4199 - Cyclobutene, 1,2-dichloro-3,3,4,4-tetrafluoro-
C$_4$Cl$_2$F$_6$
3754 - 2-Butene, 2,3-dichloro-1,1,1,4,4,4-hexafluoro-
C$_4$Cl$_2$F$_8$
3287 - Butane, 2,3-dichloro-1,1,1,2,3,4,4,4-
octafluoro-
C$_4$Cl$_3$F$_7$
3452 - Butane, 2,2,3-trichloro-1,1,1,3,4,4,4-
heptafluoro-
C$_4$Cl$_4$F$_4$
3808 - 1-Butene, 1,3,4,4-tetrachloro-1,2,3,4-
tetrafluoro-
C$_4$Cl$_4$S
11782 - Thiophene, tetrachloro-
C$_4$Cl$_6$
3163 - 1,3-Butadiene, 1,1,2,3,4,4-hexachloro-
C$_4$Cl$_6$O$_3$
244 - Acetic acid, trichloro-, anhydride
C$_4$F$_4$O$_3$
6099 - 2,5-Furandione, 3,3,4,4-tetrafluorodihydro-
C$_4$F$_6$
3164 - 1,3-Butadiene, 1,1,2,3,4,4-hexafluoro-
3914 - 2-Butyne, 1,1,1,4,4,4-hexafluoro-
4200 - Cyclobutene, hexafluoro-
C$_4$F$_6$O$_3$
264 - Acetic acid, trifluoro-, anhydride
C$_4$F$_8$
3804 - 2-Butene, 1,1,1,2,3,4,4,4-octafluoro-
4194 - Cyclobutane, octafluoro-
C$_4$F$_{10}$
3255 - Butane, decafluoro-
10050 - Propane, 1,1,1,2,3,3,3-heptafluoro-2-
(trifluoromethyl)-
C$_4$HBr$_3$S
11790 - Thiophene, 2,3,5-tribromo-
C$_4$HClN$_2$
3763 - 2-Butenedinitrile, 2-chloro-, (Z)-
C$_4$HClO$_3$
6092 - 2,5-Furandione, 3-chloro-
C$_4$HCl$_2$N$_3$O$_2$
11054 - Pyrimidine, 2,4-dichloro-5-nitro-
C$_4$HCl$_3$N$_2$
11075 - Pyrimidine, 2,4,6-trichloro-
C$_4$HCl$_3$O$_2$
3796 - 2-Butenedioyl dichloride, 2-chloro-, (Z)-
C$_4$HCl$_3$S
11791 - Thiophene, 2,3,5-trichloro-
C$_4$HCl$_7$O$_2$
5192 - 1,4-Dioxane, 2,2,3,3,5,5,6-heptachloro-
C$_4$HF$_7$O$_2$
3521 - Butanoic acid, heptafluoro-
C$_4$HI$_4$N
11140 - 1H-Pyrrole, 2,3,4,5-tetraiodo-
C$_4$H$_2$
3177 - 1,3-Butadiyne
C$_4$H$_2$BrClS
11728 - Thiophene, 2-bromo-5-chloro-
C$_4$H$_2$Br$_2$N$_2$O$_3$
11086 - 2,4,6(1H,3H,5H)-Pyrimidinetrione, 5,5-
dibromo-
C$_4$H$_2$Br$_2$O
6077 - Furan, 2,3-dibromo-
6078 - Furan, 2,5-dibromo-

C$_4$H$_2$Br$_2$S
11742 - Thiophene, 2,3-dibromo-
11743 - Thiophene, 2,4-dibromo-
11744 - Thiophene, 2,5-dibromo-
C$_4$H$_2$Cl$_2$O$_2$
3795 - 2-Butenedioyl dichloride, (E)-
C$_4$H$_2$Cl$_2$S
11746 - Thiophene, 2,3-dichloro-
11747 - Thiophene, 2,4-dichloro-
11748 - Thiophene, 2,5-dichloro-
11749 - Thiophene, 3,4-dichloro-
C$_4$H$_2$Cl$_4$
3173 - 1,3-Butadiene, 1,2,3,4-tetrachloro-
C$_4$H$_2$Cl$_6$
3798 - 2-Butene, 1,1,2,3,4,4-hexachloro-
C$_4$H$_2$Cl$_8$O
5615 - Ethane, 1,1'-oxybis[1,2,2,2-tetrachloro-
C$_4$H$_2$I$_2$S
11753 - Thiophene, 2,5-diiodo-
C$_4$H$_2$N$_2$
3762 - 2-Butenedinitrile, (E)-
C$_4$H$_2$N$_2$O$_4$
11073 - 2,4,5,6(1H,3H)-Pyrimidinetetrone
C$_4$H$_2$N$_2$O$_4$S
11758 - Thiophene, 2,5-dinitro-
C$_4$H$_2$O$_3$
6091 - 2,5-Furandione
C$_4$H$_3$BrClF$_3$
3715 - 1-Butene, 4-bromo-3-chloro-3,4,4-trifluoro-
C$_4$H$_3$BrN$_2$O$_2$
11062 - 2,4(1H,3H)-Pyrimidinedione, 5-bromo-
C$_4$H$_3$BrO
6033 - Furan, 2-bromo-
6034 - Furan, 3-bromo-
C$_4$H$_3$BrO$_4$
3769 - 2-Butenedioic acid, 2-bromo-, (E)-
3770 - 2-Butenedioic acid, 2-bromo-, (Z)-
C$_4$H$_3$BrS
11726 - Thiophene, 2-bromo-
11727 - Thiophene, 3-bromo-
C$_4$H$_3$Cl
3890 - 1-Buten-3-yne, 4-chloro-
C$_4$H$_3$ClO
6073 - Furan, 2-chloro-
6074 - Furan, 3-chloro-
C$_4$H$_3$ClO$_2$S$_2$
11780 - 2-Thiophenesulfonyl chloride
C$_4$H$_3$ClS
11740 - Thiophene, 2-chloro-
C$_4$H$_3$Cl$_2$F$_3$O$_2$
10251 - Propanoic acid, 2,2-dichloro-3,3,3-trifluoro-,
methyl ester
C$_4$H$_3$Cl$_3$
3174 - 1,3-Butadiene, 1,2,3-trichloro-
C$_4$H$_3$Cl$_5$
3805 - 2-Butene, 1,1,1,4,4-pentachloro-
C$_4$H$_3$Cl$_7$
3380 - Butane, 1,1,2,2,3,4,4-heptachloro-
C$_4$H$_3$FN$_2$O$_2$
11067 - 2,4(1H,3H)-Pyrimidinedione, 5-fluoro-
C$_4$H$_3$F$_5$O$_2$
10317 - Propanoic acid, pentafluoro-, methyl ester
C$_4$H$_3$F$_7$O
3622 - 1-Butanol, 2,2,3,3,4,4,4-heptafluoro-
C$_4$H$_3$IO
6110 - Furan, 2-iodo-
6111 - Furan, 3-iodo-
C$_4$H$_3$IS
11767 - Thiophene, 2-iodo-
C$_4$H$_3$NO$_2$
11121 - 1H-Pyrrole-2,5-dione
C$_4$H$_3$NO$_2$S
11775 - Thiophene, 2-nitro-
11776 - Thiophene, 3-nitro-
C$_4$H$_3$NO$_3$
6134 - Furan, 2-nitro-

C$_4$H$_3$N$_3$O$_4$
11074 - 2,4,5,6(1H,3H)-Pyrimidinetetrone, 5-oxime
C$_4$H$_3$N$_3$O$_5$
11102 - 2,4,6(1H,3H,5H)-Pyrimidinetrione, 5-nitro-
C$_4$H$_4$
3889 - 1-Buten-3-yne
C$_4$H$_4$BrNO$_2$
11153 - 2,5-Pyrrolidinedione, 1-bromo-
C$_4$H$_4$Br$_2$
3904 - 2-Butyne, 1,4-dibromo-
C$_4$H$_4$Br$_4$
4195 - Cyclobutane, 1,2,3,4-tetrabromo-
C$_4$H$_4$ClNO$_2$
11154 - 2,5-Pyrrolidinedione, 1-chloro-
C$_4$H$_4$ClNS
10538 - 1-Propene, 2-chloro-3-isothiocyanato-
C$_4$H$_4$ClN$_3$O$_2$
4083 - 5-Chloro-3-methyl-4-nitro-1H-pyrazole
C$_4$H$_4$Cl$_2$
3156 - 1,3-Butadiene, 1,1-dichloro-
3157 - 1,3-Butadiene, 1,2-dichloro-
3158 - 1,3-Butadiene, 2,3-dichloro-
3905 - 2-Butyne, 1,4-dichloro-
C$_4$H$_4$Cl$_2$O$_2$
3366 - Butanedioyl dichloride
3687 - Butanoyl chloride, 4-chloro-3-oxo-
4152 - Crotonic acid, 4,4-dichloro-
C$_4$H$_4$Cl$_2$O$_3$
124 - Acetic acid, chloro-, anhydride
C$_4$H$_4$Cl$_2$O$_4$
3302 - Butanedioic acid, 2,3-dichloro-, (±)-
C$_4$H$_4$Cl$_4$
3806 - 1-Butene, 1,3,4,4-tetrachloro-
3807 - 2-Butene, 2,3,3,4-tetrachloro-
C$_4$H$_4$Cl$_4$O$_2$
246 - Acetic acid, trichloro-, 2-chloroethyl ester
C$_4$H$_4$INO$_2$
11159 - 2,5-Pyrrolidinedione, 1-iodo-
C$_4$H$_4$I$_2$
3906 - 2-Butyne, 1,4-diiodo-
C$_4$H$_4$N$_2$
3291 - Butanedinitrile
10801 - Pyrazine
10855 - Pyridazine
11049 - Pyrimidine
C$_4$H$_4$N$_2$OS
11110 - 4(1H)-Pyrimidinone, 2,3-dihydro-2-thioxo-
C$_4$H$_4$N$_2$O$_2$
3902 - 2-Butynediamide
11059 - 2,4(1H,3H)-Pyrimidinedione
C$_4$H$_4$N$_2$O$_3$
11076 - 2,4,6(1H,3H,5H)-Pyrimidinetrione
C$_4$H$_4$N$_2$S$_2$
11702 - Thiocyanic acid, 1,2-ethanediyl ester
C$_4$H$_4$N$_6$O
11851 - 7H-1,2,3-Triazolo[4,5-d]pyrimidin-7-one, 5-
amino-1,4-dihydro-
C$_4$H$_4$Na$_2$O$_6$
3314 - Butanedioic acid, 2,3-dihydroxy- [R-(R*,R*)]-,
disodium salt
C$_4$H$_4$O
3895 - 2-Butynal
3923 - 3-Butyn-2-one
6025 - Furan
C$_4$H$_4$OS
11777 - Thiophene-2-ol
C$_4$H$_4$O$_2$
3917 - 2-Butynoic acid
5202 - 1,4-Dioxin
8598 - 2-Oxetanone, 4-methylene-
C$_4$H$_4$O$_3$
6093 - 2,5-Furandione, dihydro-
C$_4$H$_4$O$_4$
3764 - 2-Butenedioic acid (E)-
3765 - 2-Butenedioic acid (Z)-
C$_4$H$_4$O$_5$
3339 - Butanedioic acid, oxo-

$C_4H_4O_6$
3777 - 2-Butenedioic acid, 2,3-dihydroxy-, (Z)-
C_4H_4S
11724 - Thiophene
C_4H_5Br
3147 - 1,2-Butadiene, 4-bromo-
3148 - 1,3-Butadiene, 1-bromo-
3149 - 1,3-Butadiene, 2-bromo-
$C_4H_5BrN_2$
10817 - 1H-Pyrazole, 4-bromo-3-methyl-
C_4H_5BrO
3696 - 2-Butenal, 2-bromo-
$C_4H_5BrO_2$
10612 - 2-Propenoic acid, 3-bromo-2-methyl-
$C_4H_5BrO_4$
3299 - Butanedioic acid, bromo-, (±)-
3300 - Butanedioic acid, bromo-
$C_4H_5Br_2FO_2$
161 - Acetic acid, dibromofluoro-, ethyl ester
$C_4H_5Br_3O_2$
242 - Acetic acid, tribromo-, ethyl ester
C_4H_5Cl
3150 - 1,2-Butadiene, 4-chloro-
3151 - 1,3-Butadiene, 1-chloro-
3152 - 1,3-Butadiene, 2-chloro-
3899 - 1-Butyne, 3-chloro-
3900 - 2-Butyne, 1-chloro-
$C_4H_5ClN_2$
10821 - 1H-Pyrazole, 3-chloro-5-methyl-
C_4H_5ClO
3697 - 2-Butenal, 2-chloro-
3887 - 2-Butenoyl chloride
10715 - 2-Propenoyl chloride, 2-methyl-
$C_4H_5ClO_2$
3820 - 2-Butenoic acid, 3-chloro-, (E)-
3821 - 2-Butenoic acid, 3-chloro-, (Z)-
3822 - 2-Butenoic acid, 4-chloro-
4022 - Carbonochloridic acid, 1-methylethenyl ester
4150 - Crotonic acid, 2-chloro-
4151 - Crotonic acid, 4-chloro-, (E)-
10621 - 2-Propenoic acid, 2-chloro-, methyl ester
$C_4H_5ClO_3$
144 - Acetic acid, chlorooxo-, ethyl ester
$C_4H_5Cl_3$
3809 - 1-Butene, 2,3,4-trichloro-
3810 - 2-Butene, 1,2,4-trichloro-
10605 - 1-Propene, 1,1,3-trichloro-2-methyl-
10606 - 1-Propene, 3,3,3-trichloro-2-methyl-
$C_4H_5Cl_3O$
3191 - Butanal, 2,2,3-trichloro-
$C_4H_5Cl_3O_2$
248 - Acetic acid, trichloro-, ethyl ester
3590 - Butanoic acid, 2,2,3-trichloro-
$C_4H_5Cl_3O_3$
249 - Acetic acid, trichloro-, 2-hydroxyethyl ester
$C_4H_5Cl_5$
3419 - Butane, 1,1,2,3,4-pentachloro-
3420 - Butane, 1,2,2,3,4-pentachloro-
10127 - Propane, 1,1,1,2,3-pentachloro-2-methyl-
10146 - Propane, 1,1,2,3-tetrachloro-2-(chloromethyl)-
C_4H_5F
3162 - 1,3-Butadiene, 2-fluoro-
$C_4H_5F_3O_2$
267 - Acetic acid, trifluoro-, ethyl ester
C_4H_5I
3165 - 1,3-Butadiene, 2-iodo-
3166 - 1,2-Butadiene, 4-iodo-
C_4H_5N
3802 - 2-Butenenitrile, (E)-
3803 - 3-Butenenitrile
4956 - Cyclopropanecarbonitrile
10574 - 1-Propene, 3-isocyano-
10586 - 2-Propenenitrile, 2-methyl-
11112 - 1H-Pyrrole
C_4H_5NO
7405 - Isoxazole, 5-methyl-

8583 - Oxazole, 4-methyl-
$C_4H_5NO_2$
154 - Acetic acid, cyano-, methyl ester
4030 - Carbonocyanidic acid, ethyl ester
10233 - Propanoic acid, 2-cyano-
11152 - 2,5-Pyrrolidinedione
$C_4H_5NO_3$
3817 - 2-Butenoic acid, 4-amino-4-oxo-, (Z)-
C_4H_5NS
10575 - 1-Propene, 3-isothiocyanato-
11678 - Thiazole, 2-methyl-
11679 - Thiazole, 4-methyl-
11712 - Thiocyanic acid, 2-propenyl ester
$C_4H_5NS_2$
11680 - 2(3H)-Thiazolethione, 4-methyl-
$C_4H_5N_3$
11042 - 2-Pyrimidinamine
$C_4H_5N_3O$
11104 - 2(1H)-Pyrimidinone, 4-amino-
$C_4H_5N_3O_2$
11831 - 1,2,4-Triazine-3,5(2H,4H)-dione, 6-methyl-
$C_4H_5N_3O_3$
11077 - 2,4,6(1H,3H,5H)-Pyrimidinetrione, 5-amino-
C_4H_6
3145 - 1,2-Butadiene
3146 - 1,3-Butadiene
3897 - 1-Butyne
3898 - 2-Butyne
4198 - Cyclobutene
$C_4H_6BaO_4$
110 - Acetic acid, barium salt
C_4H_6BrClO
3682 - Butanoyl chloride, 2-bromo-
$C_4H_6BrClO_2$
113 - Acetic acid, bromochloro-, ethyl ester
C_4H_6BrN
3404 - Butanenitrile, 4-bromo-
10096 - Propanenitrile, 2-bromo-2-methyl-
$C_4H_6Br_2$
3738 - 1-Butene, 1,1-dibromo-
3739 - 1-Butene, 1,2-dibromo-
3740 - 1-Butene, 3,4-dibromo-
3741 - 2-Butene, 1,3-dibromo-
3742 - 2-Butene, 1,4-dibromo-, (E)-
4178 - Cyclobutane, 1,1-dibromo-
4179 - Cyclobutane, 1,2-dibromo-
10546 - 1-Propene, 1,3-dibromo-2-methyl-
$C_4H_6Br_2O$
3653 - 2-Butanone, 3,4-dibromo-
5893 - Ethene, 1,1-dibromo-2-ethoxy-
10491 - Propanoyl bromide, 2-bromo-2-methyl-
$C_4H_6Br_2O_2$
160 - Acetic acid, dibromo-, ethyl ester
3497 - Butanoic acid, 2,3-dibromo-
10241 - Propanoic acid, 2,3-dibromo-, methyl ester
$C_4H_6Br_4$
3424 - Butane, 1,1,4,4-tetrabromo-
3425 - Butane, 1,2,2,3-tetrabromo-
3426 - Butane, 1,2,3,4-tetrabromo-, DL-
$C_4H_6CaO_4$
122 - Acetic acid, calcium salt
$C_4H_6ClFO_2$
133 - Acetic acid, chlorofluoro-, ethyl ester
C_4H_6ClN
3405 - Butanenitrile, 4-chloro-
C_4H_6ClNO
10098 - Propanenitrile, 3-chloro-2-hydroxy-2-methyl-
$C_4H_6Cl_2$
3743 - 1-Butene, 1,3-dichloro-
3744 - 1-Butene, 2,3-dichloro-
3745 - 1-Butene, 3,4-dichloro-
3746 - 2-Butene, 1,1-dichloro-
3747 - 2-Butene, 1,2-dichloro-
3748 - 2-Butene, 1,3-dichloro-, (E)-
3749 - 2-Butene, 1,3-dichloro-, (Z)-
3750 - 2-Butene, 1,4-dichloro-, (E)-
3751 - 2-Butene, 1,4-dichloro-, (Z)-

3752 - 2-Butene, 2,3-dichloro-, (E)-
3753 - 2-Butene, 2,3-dichloro-, (Z)-
10537 - 1-Propene, 3-chloro-2-(chloromethyl)-
10555 - 1-Propene, 1,1-dichloro-2-methyl-
10556 - 1-Propene, 3,3-dichloro-2-methyl-
$C_4H_6Cl_2O$
3182 - Butanal, 2,3-dichloro-
3654 - 2-Butanone, 1,3-dichloro-
3655 - 2-Butanone, 3,4-dichloro-
3683 - Butanoyl chloride, 2-chloro-
3684 - Butanoyl chloride, 3-chloro-
3685 - Butanoyl chloride, 4-chloro-
5899 - Ethene, 1,2-dichloro-1-ethoxy-
10498 - Propanoyl chloride, 2-chloro-2-methyl-
10499 - Propanoyl chloride, 3-chloro-2-methyl-
$C_4H_6Cl_2O_2$
126 - Acetic acid, chloro-, 2-chloroethyl ester
164 - Acetic acid, dichloro-, ethyl ester
3500 - Butanoic acid, 2,2-dichloro-
3926 - Butyric acid, 3,4-dichloro-
4015 - Carbonochloridic acid, 3-chloropropyl ester
5188 - 1,4-Dioxane, 2,3-dichloro-
10248 - Propanoic acid, 2,3-dichloro-, methyl ester
$C_4H_6Cl_2O_3$
165 - Acetic acid, dichloro-, 2-hydroxyethyl ester
10247 - Propanoic acid, 3,3-dichloro-2-hydroxy-2-methyl-
$C_4H_6Cl_3NO$
97 - Acetamide, 2,2,2-trichloro-N,N-dimethyl-
$C_4H_6Cl_4$
3427 - Butane, 1,1,1,2-tetrachloro-
3428 - Butane, 1,2,2,3-tetrachloro-
3429 - Butane, 1,2,3,3-tetrachloro-
10147 - Propane, 1,1,1,2-tetrachloro-2-methyl-
10148 - Propane, 1,1,2,3-tetrachloro-2-methyl-
10176 - Propane, 1,2,3-trichloro-2-(chloromethyl)-
$C_4H_6F_2O_2$
176 - Acetic acid, difluoro-, ethyl ester
$C_4H_6F_3NO$
99 - Acetamide, 2,2,2-trifluoro-N,N-dimethyl-
$C_4H_6MnN_2S_4$
7420 - Maneb
$C_4H_6N_2$
7208 - 1H-Imidazole, 1-methyl-
7209 - 1H-Imidazole, 2-methyl-
7210 - 1H-Imidazole, 4-methyl-
10834 - 1H-Pyrazole, 1-methyl-
10835 - 1H-Pyrazole, 3-methyl-
10836 - 1H-Pyrazole, 4-methyl-
$C_4H_6N_2O$
6176 - Furazan, dimethyl-
$C_4H_6N_2O_2$
158 - Acetic acid, diazo-, ethyl ester
9612 - 2,5-Piperazinedione
10237 - Propanoic acid, 2-diazo-, methyl ester
$C_4H_6N_2S$
7214 - 2H-Imidazole-2-thione, 1,3-dihydro-1-methyl-
11659 - 1,3,4-Thiadiazole, 2,5-dimethyl-
11665 - 2-Thiazolamine, 4-methyl-
$C_4H_6N_2S_4Zn$
12067 - Zinc, [[1,2-ethanediylbis
[carbamodithioato](2-)]-
$C_4H_6N_4O_2$
7191 - Imidazo[4,5-d]imidazole-2,5(1H,3H)-dione, tetrahydro-
$C_4H_6N_4O_3$
12009 - Urea, (2,5-dioxo-4-imidazolidinyl)-
$C_4H_6N_4O_3S_2$
32 - Acetamide, N-[5-(aminosulfonyl)-1,3,4-thiadiazol-2-yl]-
$C_4H_6N_4O_{12}$
3432 - 1,2,3,4-Butanetetrol, tetranitrate, (R*,S*)-
C_4H_6O
3175 - 2,3-Butadien-1-ol
3695 - 2-Butenal, (E)-
3875 - 3-Buten-2-one
3919 - 2-Butyn-1-ol

3920 - 3-Butyn-1-ol
3921 - 3-Butyn-2-ol
4197 - Cyclobutanone
5912 - Ethene, 1,1'-oxybis-
5942 - Ethyne, ethoxy-
6086 - Furan, 2,3-dihydro-
8613 - Oxirane, ethenyl-
10515 - 2-Propenal, 2-methyl-
10736 - 1-Propyne, 3-methoxy-

C₄H₆OS
11794 - 2(3H)-Thiophenone, dihydro-

C₄H₆O₂
179 - Acetic acid ethenyl ester
3007 - 2,2'-Bioxirane, (R*,S*)-
3256 - Butanedial
3359 - 2,3-Butanedione
3813 - 3-Butenoic acid
3814 - 2-Butenoic acid, (E)-
3815 - 2-Butenoic acid, (Z)-
3911 - 2-Butyne-1,4-diol
3912 - 3-Butyne-1,2-diol
4958 - Cyclopropanecarboxylic acid
5203 - 1,4-Dioxin, 2,3-dihydro-
6017 - Formic acid, 2-propenyl ester
6142 - 2(3H)-Furanone, dihydro-
10648 - 2-Propenoic acid, 2-methyl-
10655 - 2-Propenoic acid, methyl ester

C₄H₆O₂S
5913 - Ethene, 1,1'-sulfonylbis-
11752 - Thiophene, 2,5-dihydro-, 1,1-dioxide

C₄H₆O₂S₂
5250 - Disulfide, diacetyl

C₄H₆O₃
107 - Acetic acid, anhydride
3567 - Butanoic acid, 2-oxo-
3568 - Butanoic acid, 3-oxo-
3842 - 3-Butenoic acid, 2-hydroxy-
5213 - 1,3-Dioxolan-2-one, 4-methyl-
10316 - Propanoic acid, 2-oxo-, methyl ester

C₄H₆O₄
101 - Acetic acid, (acetyloxy)-
3294 - Butanedioic acid
5563 - Ethanedioic acid, dimethyl ester
5573 - 1,2-Ethanediol, diformate
9156 - Peroxide, diacetyl
9960 - Propanedioic acid, methyl-

C₄H₆O₄S
240 - Acetic acid, 2,2'-thiobis-

C₄H₆O₅
226 - Acetic acid, 2,2'-oxybis-
9959 - Propanedioic acid, (hydroxymethyl)-

C₄H₆O₆
3304 - Butanedioic acid, 2,3-dihydroxy-, (R*,S*)-
3309 - Butanedioic acid, 2,3-dihydroxy- [R-(R*,R*)]-
3311 - Butanedioic acid, 2,3-dihydroxy-, (R*,R*)-(±)-
3312 - Butanedioic acid, 2,3-dihydroxy-, [S-(R*,R*)]-

C₄H₆O₆S
239 - Acetic acid, 2,2'-sulfonylbis-

C₄H₆O₈
3342 - Butanedioic acid, tetrahydroxy-

C₄H₆S
5920 - Ethene, 1,1'-thiobis-
11750 - Thiophene, 2,3-dihydro-
11751 - Thiophene, 2,5-dihydro-

C₄H₇Br
3707 - 1-Butene, 1-bromo-, (E)-
3708 - 1-Butene, 1-bromo-, (Z)-
3709 - 1-Butene, 2-bromo-
3710 - 1-Butene, 3-bromo-
3711 - 1-Butene, 4-bromo-
3712 - 2-Butene, 2-bromo-
3713 - 2-Butene, 2-bromo-, (E)-
3714 - 2-Butene, 2-bromo-, (Z)-

C₄H₇BrO
3179 - Butanal, 2-bromo-
3680 - Butanoyl bromide
9776 - Propanal, 2-bromo-2-methyl-

10492 - Propanoyl bromide, 2-methyl-

C₄H₇BrO₂
115 - Acetic acid, bromo-, ethyl ester
3468 - Butanoic acid, 2-bromo-, (±)-
5686 - Ethanol, 2-bromo-, acetate
10209 - Propanoic acid, 2-bromo-2-methyl-
10211 - Propanoic acid, 2-bromo-, methyl ester, (±)-
10212 - Propanoic acid, 3-bromo-, methyl ester

C₄H₇Br₂Cl₂O₄P
7702 - Naled

C₄H₇Br₃
3442 - Butane, 1,1,2-tribromo-
3443 - Butane, 1,2,2-tribromo-
3444 - Butane, 1,2,3-tribromo-
3445 - Butane, 1,2,4-tribromo-
3446 - Butane, 1,3,3-tribromo-
3447 - Butane, 2,2,3-tribromo-
10164 - Propane, 1,2,3-tribromo-2-methyl-

C₄H₇Br₃O
10405 - 2-Propanol, 1,1,1-tribromo-2-methyl-

C₄H₇Cl
3720 - 1-Butene, 1-chloro-, (E)-
3721 - 1-Butene, 1-chloro-, (Z)-
3722 - 1-Butene, 2-chloro-
3723 - 1-Butene, 3-chloro-
3724 - 1-Butene, 4-chloro-
3725 - 2-Butene, 1-chloro-, (E)-
3726 - 2-Butene, 1-chloro-, (Z)-
3727 - 2-Butene, 2-chloro-, (E)-
3728 - 2-Butene, 2-chloro-, (Z)-
4966 - Cyclopropane, (chloromethyl)-
10539 - 1-Propene, 1-chloro-2-methyl-
10540 - 1-Propene, 3-chloro-2-methyl-

C₄H₇ClO
3180 - Butanal, 2-chloro-
3181 - Butanal, 4-chloro-
3646 - 2-Butanone, 1-chloro-
3647 - 2-Butanone, 3-chloro-
3648 - 2-Butanone, 4-chloro-
3681 - Butanoyl chloride
3863 - 2-Buten-1-ol, 2-chloro-
3864 - 2-Buten-1-ol, 3-chloro-
3865 - 2-Buten-1-ol, 4-chloro-
3866 - 3-Buten-2-ol, 1-chloro-
3867 - 3-Buten-2-ol, 2-chloro-
3868 - 3-Buten-2-ol, 3-chloro-
5886 - Ethene, 1-chloro-2-ethoxy-
5887 - Ethene, (2-chloroethoxy)-
8606 - Oxirane, (chloromethyl)-2-methyl-
9779 - Propanal, 2-chloro-2-methyl-
10504 - Propanoyl chloride, 2-methyl-

C₄H₇ClO₂
130 - Acetic acid, chloro-, ethyl ester
298 - Acetyl chloride, ethoxy-
3480 - Butanoic acid, 2-chloro-
3481 - Butanoic acid, 3-chloro-, (±)-
3482 - Butanoic acid, 4-chloro-
4023 - Carbonochloridic acid, 1-methylethyl ester
4028 - Carbonochloridic acid, propyl ester
5694 - Ethanol, 1-chloro-, acetate
5695 - Ethanol, 2-chloro-, acetate
10222 - Propanoic acid, 2-chloro-, methyl ester
10225 - Propanoic acid, 3-chloro-, methyl ester
10228 - Propanoic acid, 2-chloro-2-methyl-
10230 - Propanoic acid, 3-chloro-2-methyl-

C₄H₇ClO₃
136 - Acetic acid, chloro-, 2-hydroxyethyl ester
4019 - Carbonochloridic acid, 2-methoxyethyl ester

C₄H₇Cl₂O₄P
5170 - Dichlorvos

C₄H₇Cl₃
3448 - Butane, 1,1,3-trichloro-
3449 - Butane, 1,2,3-trichloro-
3450 - Butane, 1,2,4-trichloro-
3451 - Butane, 1,3,3-trichloro-
10177 - Propane, 1,1,2-trichloro-2-methyl-
10178 - Propane, 1,2,3-trichloro-2-methyl-

C₄H₇Cl₃O
3643 - 2-Butanol, 1,1,1-trichloro-
10407 - 2-Propanol, 1,1,1-trichloro-2-methyl-

C₄H₇Cl₃O₂
5769 - Ethanol, 2,2,2-trichloro-1-ethoxy-

C₄H₇FO₂
191 - Acetic acid, fluoro-, ethyl ester
3927 - Butyric acid, 4-fluoro-

C₄H₇IO₂
201 - Acetic acid, iodo-, ethyl ester
10284 - Propanoic acid, 3-iodo-, methyl ester

C₄H₇N
3403 - Butanenitrile
10063 - Propane, 2-isocyano-
10111 - Propanenitrile, 2-methyl-
10726 - 2-Propyn-1-amine, N-methyl-
11118 - 1H-Pyrrole, 2,5-dihydro-

C₄H₇NO
3702 - 2-Butenamide, (E)-
8576 - Oxazole, 4,5-dihydro-2-methyl-
10059 - Propane, 1-isocyanato-
10060 - Propane, 2-isocyanato-
10108 - Propanenitrile, 2-hydroxy-2-methyl-
10109 - Propanenitrile, 2-methoxy-
10110 - Propanenitrile, 3-methoxy-
11180 - 2-Pyrrolidinone

C₄H₇NO₂
23 - Acetamide, N-acetyl-
495 - 2-Azetidinecarboxylic acid
3361 - 2,3-Butanedione, monooxime

C₄H₇NO₂S
11682 - 4-Thiazolidinecarboxylic acid

C₄H₇NO₃
6251 - Glycine, N-acetyl-

C₄H₇NO₄
218 - Acetic acid, nitro-, ethyl ester
470 - DL-Aspartic acid
471 - L-Aspartic acid
3466 - Butanoic acid, 4-amino-2-hydroxy-4-oxo-
6260 - Glycine, N-(carboxymethyl)-

C₄H₇NS
10064 - Propane, 1-isothiocyanato-
11672 - Thiazole, 4,5-dihydro-2-methyl-
11706 - Thiocyanic acid, 1-methylethyl ester

C₄H₇N₃
11849 - 1H-1,2,3-Triazole, 1,5-dimethyl-

C₄H₇N₃O
7230 - 4H-Imidazol-4-one, 2-amino-1,5-dihydro-1-
        methyl-

C₄H₈
3704 - 1-Butene
3705 - 2-Butene, (E)-
3706 - 2-Butene, (Z)-
4170 - Cyclobutane
4981 - Cyclopropane, methyl-
10578 - 1-Propene, 2-methyl-

C₄H₈BrCl
3233 - Butane, 1-bromo-4-chloro-
3234 - Butane, 2-bromo-1-chloro-
9842 - Propane, 1-bromo-3-chloro-2-methyl-

C₄H₈BrF
3237 - Butane, 1-bromo-4-fluoro-

C₄H₈BrNO
9792 - Propanamide, 2-bromo-2-methyl-

C₄H₈Br₂
3264 - Butane, 1,1-dibromo-
3265 - Butane, 1,2-dibromo-
3266 - Butane, 1,3-dibromo-
3267 - Butane, 1,4-dibromo-
3268 - Butane, 2,3-dibromo-
9889 - Propane, 1,2-dibromo-2-methyl-
9890 - Propane, 1,3-dibromo-2-methyl-

C₄H₈Br₂O
3612 - 1-Butanol, 2,3-dibromo-
3613 - 1-Butanol, 3,4-dibromo-
3614 - 2-Butanol, 1,4-dibromo-
5523 - Ethane, 1,2-dibromo-1-ethoxy-

5609 - Ethane, 1,1'-oxybis[2-bromo-
9888 - Propane, 1,2-dibromo-3-methoxy-

C₄H₈ClF
3250 - Butane, 1-chloro-4-fluoro-

C₄H₈Cl₂
3276 - Butane, 1,1-dichloro-
3277 - Butane, 1,2-dichloro-
3278 - Butane, 1,3-dichloro-
3279 - Butane, 1,4-dichloro-
3280 - Butane, 2,2-dichloro-
3281 - Butane, 2,3-dichloro-, (R^*,R^*)-(±)-
9899 - Propane, 1,1-dichloro-2-methyl-
9900 - Propane, 1,2-dichloro-2-methyl-
9901 - Propane, 1,3-dichloro-2-methyl-

C₄H₈Cl₂O
5468 - Ethane, 1-chloro-1-(2-chloroethoxy)-
5533 - Ethane, 1,2-dichloro-1-ethoxy-
5610 - Ethane, 1,1'-oxybis[1-chloro-
5611 - Ethane, 1,1'-oxybis[2-chloro-
10364 - 2-Propanol, 1,1-dichloro-2-methyl-
10365 - 2-Propanol, 1,3-dichloro-2-methyl-

C₄H₈Cl₂O₂
3349 - 2,3-Butanediol, 1,4-dichloro-

C₄H₈Cl₂S
5645 - Ethane, 1,1'-thiobis[2-chloro-

C₄H₈Cl₂Si
11402 - Silane, dichloromethyl-2-propenyl-

C₄H₈Cl₃O₄P
11853 - Trichlorfon

C₄H₈I₂
3288 - Butane, 1,4-diiodo-

C₄H₈N₂
281 - Acetonitrile, (dimethylamino)-
3903 - 2-Butyne-1,4-diamine
7197 - 1H-Imidazole, 4,5-dihydro-2-methyl-
10112 - Propanenitrile, 3-(methylamino)-

C₄H₈N₂O
11170 - Pyrrolidine, 1-nitroso-
12028 - Urea, 2-propenyl-

C₄H₈N₂O₂
73 - Acetamide, N-[(methylamino)carbonyl]-
3257 - Butanediamide
3360 - 2,3-Butanedione, dioxime
7699 - Morpholine, 4-nitroso-

C₄H₈N₂O₃
468 - L-Asparagine
3932 - Carbamic acid, (aminocarbonyl)-, ethyl ester
6265 - Glycine, N-glycyl-

C₄H₈N₂O₄
3293 - Butane, 1,4-dinitro-

C₄H₈N₂S
11809 - Thiourea, 2-propenyl-

C₄H₈O
3178 - Butanal
3645 - 2-Butanone
3858 - 3-Buten-2-ol, (±)-
3859 - 2-Buten-1-ol, (E)-
3860 - 2-Buten-1-ol, (Z)-
3861 - 3-Buten-1-ol
4196 - Cyclobutanol
4980 - Cyclopropane, methoxy-
5905 - Ethene, ethoxy-
6168 - Furan, tetrahydro-
8594 - Oxetane, 2-methyl-
8607 - Oxirane, 2,2-dimethyl-
8608 - Oxirane, 2,3-dimethyl-, cis-
8609 - Oxirane, 2,3-dimethyl-, trans-
9785 - Propanal, 2-methyl-
10576 - 1-Propene, 3-methoxy-
10702 - 2-Propen-1-ol, 2-methyl-

C₄H₈OS
5647 - Ethanethioic acid, S-ethyl ester
8569 - 1,4-Oxathiane
9786 - Propanal, 3-(methylthio)-

C₄H₈O₂
14 - Acetaldehyde, ethoxy-
186 - Acetic acid ethyl ester

3184 - Butanal, 3-hydroxy-
3460 - Butanoic acid
3660 - 2-Butanone, 1-hydroxy-
3661 - 2-Butanone, 3-hydroxy-
3662 - 2-Butanone, 4-hydroxy-
3792 - 2-Butene-1,4-diol, (E)-
3793 - 2-Butene-1,4-diol, (Z)-
3794 - 3-Butene-1,3-diol
5186 - 1,3-Dioxane
5187 - 1,4-Dioxane
5210 - 1,3-Dioxolane, 2-methyl-
5716 - Ethanol, 2-(ethenyloxy)-
6011 - Formic acid, 1-methylethyl ester
6018 - Formic acid, propyl ester
6137 - 3-Furanol, tetrahydro-
8615 - Oxirane, ethyl-
8620 - Oxirane, (methoxymethyl)-
10290 - Propanoic acid, 2-methyl-
10296 - Propanoic acid, methyl ester
10455 - 2-Propanone, 1-methoxy-

C₄H₈O₂S
188 - Acetic acid, (ethylthio)-
203 - Acetic acid, mercapto-, ethyl ester
3537 - Butanoic acid, 3-mercapto-
10287 - Propanoic acid, 3-mercapto-, methyl ester
10288 - Propanoic acid, 2-mercapto-2-methyl-
11786 - Thiophene, tetrahydro-, 1,1-dioxide

C₄H₈O₃
180 - Acetic acid, ethoxy-
197 - Acetic acid, hydroxy-, ethyl ester
207 - Acetic acid, methoxy-, methyl ester
3530 - Butanoic acid, 2-hydroxy-, (±)-
3531 - Butanoic acid, 3-hydroxy-, (±)-
4003 - Carbonic acid, ethyl methyl ester
5577 - 1,2-Ethanediol, monoacetate
10273 - Propanoic acid, 2-hydroxy-, methyl ester, (±)-
10276 - Propanoic acid, 3-hydroxy-, methyl ester
10278 - Propanoic acid, 2-hydroxy-2-methyl-

C₄H₈O₃S
5198 - 1,3,2-Dioxathiane, 4-methyl-, 2-oxide

C₄H₈O₄
3192 - Butanal, 2,3,4-trihydroxy-, [R-(R^*,R^*)]-
3193 - Butanal, 2,3,4-trihydroxy-, [S-(R^*,S^*)]-
3679 - 2-Butanone, 1,3,4-trihydroxy-, (S)-
10254 - Propanoic acid, 2,3-dihydroxy-, methyl ester, (±)-

C₄H₈S
10581 - 1-Propene, 3-(methylthio)-
11693 - Thietane, 2-methyl-
11783 - Thiophene, tetrahydro-

C₄H₈S₂
5261 - 1,2-Dithiane
5262 - 1,3-Dithiane
5263 - 1,4-Dithiane

C₄H₉Br
3231 - Butane, 1-bromo-
3232 - Butane, 2-bromo-, (±)-
9845 - Propane, 1-bromo-2-methyl-
9846 - Propane, 2-bromo-2-methyl-

C₄H₉BrO
3600 - 1-Butanol, 3-bromo-
3601 - 2-Butanol, 3-bromo-, (R^*,S^*)-(±)-
5461 - Ethane, 1-bromo-2-ethoxy-

C₄H₉BrO₂
5460 - Ethane, 2-bromo-1,1-dimethoxy-

C₄H₉Cl
3241 - Butane, 1-chloro-
3242 - Butane, 2-chloro-, (±)-
9862 - Propane, 1-chloro-2-methyl-
9863 - Propane, 2-chloro-2-methyl-

C₄H₉ClO
1605 - Benzeneethanol, α-ethyl-α-methyl-, (±)-
3602 - 1-Butanol, 2-chloro-
3604 - 1-Butanol, 4-chloro-
3605 - 2-Butanol, 1-chloro-
3606 - 2-Butanol, 3-chloro-
3607 - 2-Butanol, 4-chloro-

5473 - Ethane, 1-chloro-1-ethoxy-
5474 - Ethane, 1-chloro-2-ethoxy-
7183 - Hypochlorous acid, 1,1-dimethylethyl ester
9858 - Propane, 1-(chloromethoxy)-
9859 - Propane, 1-chloro-2-methoxy-
9860 - Propane, 1-chloro-3-methoxy-
9861 - Propane, 2-chloro-1-methoxy-
10352 - 1-Propanol, 2-chloro-2-methyl-
10353 - 1-Propanol, 3-chloro-2-methyl-
10354 - 2-Propanol, 1-chloro-2-methyl-

C₄H₉ClO₂
5472 - Ethane, 2-chloro-1,1-dimethoxy-
5698 - Ethanol, 2-(2-chloroethoxy)-
9998 - 1,2-Propanediol, 3-chloro-2-methyl-

C₄H₉ClO₂S
10134 - 1-Propanesulfonyl chloride, 2-methyl-

C₄H₉ClS
9864 - Propane, 1-chloro-3-(methylthio)-
9865 - Propane, 2-chloro-1-(methylthio)-

C₄H₉ClSi
11376 - Silane, chloroethenyldimethyl-

C₄H₉Cl₃Si
11370 - Silane, butyltrichloro-
11469 - Silane, trichloro(2-methylpropyl)-

C₄H₉F
3378 - Butane, 1-fluoro-
3379 - Butane, 2-fluoro-

C₄H₉FO
3621 - 1-Butanol, 4-fluoro-

C₄H₉I
3381 - Butane, 1-iodo-
3382 - Butane, 2-iodo-, (±)-
10057 - Propane, 1-iodo-2-methyl-
10058 - Propane, 2-iodo-2-methyl-

C₄H₉N
4169 - Cyclobutanamine
10524 - 2-Propen-1-amine, N-methyl-
11143 - Pyrrolidine

C₄H₉NO
55 - Acetamide, N,N-dimethyl-
61 - Acetamide, N-ethyl-
499 - 1-Aziridineethanol
3188 - Butanal, oxime
3194 - Butanamide
3675 - 2-Butanone, oxime
7681 - Morpholine
9798 - Propanamide, 2-methyl-
9799 - Propanamide, N-methyl-

C₄H₉NO₂
65 - Acetamide, N-(2-hydroxyethyl)-
331 - Alanine, 2-methyl-
3412 - Butane, 1-nitro-
3413 - Butane, 2-nitro-, (±)-
3463 - Butanoic acid, 2-amino-, (±)-
3464 - Butanoic acid, 4-amino-
3954 - Carbamic acid, methyl-, ethyl ester
3955 - Carbamic acid, 1-methylethyl ester
3969 - Carbamic acid, propyl ester
6261 - Glycine, N,N-dimethyl-
6263 - Glycine, ethyl ester
8083 - Nitrous acid, butyl ester
8085 - Nitrous acid, 1,1-dimethylethyl ester
8093 - Nitrous acid, 1-methylpropyl ester
8094 - Nitrous acid, 2-methylpropyl ester
10084 - Propane, 2-methyl-1-nitro-
10085 - Propane, 2-methyl-2-nitro-

C₄H₉NO₃
3465 - Butanoic acid, 3-amino-4-hydroxy-
3640 - 1-Butanol, 2-nitro-
3641 - 2-Butanol, 1-nitro-
3642 - 2-Butanol, 2-nitro-
7134 - Homoserine
8072 - Nitric acid, butyl ester
8077 - Nitric acid, 1-methylpropyl ester
8078 - Nitric acid, 2-methylpropyl ester
10391 - 1-Propanol, 2-methyl-2-nitro-
11813 - L-Threonine

C$_4$H$_9$NO$_4$
  10018 - 1,3-Propanediol, 2-methyl-2-nitro-
C$_4$H$_9$NO$_5$
  10012 - 1,3-Propanediol, 2-(hydroxymethyl)-2-nitro-
C$_4$H$_9$NS
  11719 - Thiomorpholine
C$_4$H$_9$NS$_2$
  5264 - 4H-1,3,5-Dithiazine, dihydro-5-methyl-
C$_4$H$_9$N$_3$O$_2$
  6254 - Glycine, N-(aminoiminomethyl)-N-methyl-
C$_4$H$_{10}$
  3229 - Butane
  10073 - Propane, 2-methyl-
C$_4$H$_{10}$BF$_3$O
  3137 - Boron, trifluoro[1,1'-oxybis[ethane]]-, (T-4)-
C$_4$H$_{10}$ClNO$_2$S
  5019 - L-Cysteine, methyl ester, hydrochloride
C$_4$H$_{10}$ClOP
  9529 - Phosphinic chloride, diethyl-
C$_4$H$_{10}$ClO$_2$PS
  9576 - Phosphorochloridothioic acid, O,O-diethyl
    ester
C$_4$H$_{10}$ClO$_3$P
  9574 - Phosphorochloridic acid, diethyl ester
C$_4$H$_{10}$Cl$_2$O$_2$Si
  11392 - Silane, dichlorodiethoxy-
C$_4$H$_{10}$Cl$_2$Si
  11393 - Silane, dichlorodiethyl-
  11399 - Silane, dichloromethyl(1-methylethyl)-
C$_4$H$_{10}$F$_2$Si
  11410 - Silane, diethyldifluoro-
C$_4$H$_{10}$Hg
  7434 - Mercury, diethyl-
C$_4$H$_{10}$NO$_3$PS
  10 - Acephate
C$_4$H$_{10}$N$_2$
  9602 - Piperazine
C$_4$H$_{10}$N$_2$O
  28 - Acetamide, N-(2-aminoethyl)-
  3529 - Butanoic acid, hydrazide
  5432 - Ethanamine, N-ethyl-N-nitroso-
  7680 - 4-Morpholinamine
  12032 - Urea, trimethyl-
C$_4$H$_{10}$N$_2$O$_2$
  5431 - Ethanamine, N-ethyl-N-nitro-
C$_4$H$_{10}$N$_2$O$_4$
  469 - DL-Asparagine, monohydrate
C$_4$H$_{10}$O
  3593 - 1-Butanol
  3594 - 2-Butanol
  5608 - Ethane, 1,1'-oxybis-
  10069 - Propane, 1-methoxy-
  10070 - Propane, 2-methoxy-
  10388 - 1-Propanol, 2-methyl-
  10389 - 2-Propanol, 2-methyl-
C$_4$H$_{10}$OS
  5621 - Ethane, 1,1'-sulfinylbis-
  5726 - Ethanol, 2-(ethylthio)-
C$_4$H$_{10}$O$_2$
  3344 - 1,2-Butanediol, (±)-
  3345 - 1,3-Butanediol
  3346 - 1,4-Butanediol
  3347 - 2,3-Butanediol, (R*,R*)-(±)-
  5548 - Ethane, 1,1-dimethoxy-
  5549 - Ethane, 1,2-dimethoxy-
  5717 - Ethanol, 2-ethoxy-
  7178 - Hydroperoxide, 1,1-dimethylethyl
  9158 - Peroxide, diethyl
  10017 - 1,2-Propanediol, 2-methyl-
  10386 - 1-Propanol, 2-methoxy-
  10387 - 2-Propanol, 1-methoxy-
C$_4$H$_{10}$O$_2$S
  5626 - Ethane, 1,1'-sulfonylbis-
  5766 - Ethanol, 2,2'-thiobis-
  11571 - Sulfoxylic acid, diethyl ester
C$_4$H$_{10}$O$_3$
  3458 - 1,2,3-Butanetriol

  3459 - 1,2,4-Butanetriol
  5753 - Ethanol, 2,2'-oxybis-
  7561 - Methane, trimethoxy-
  10014 - 1,2-Propanediol, 3-methoxy-
  10015 - 1,3-Propanediol, 2-methoxy-
C$_4$H$_{10}$O$_3$S
  5725 - Ethanol, 2-(ethylsulfonyl)-
  11581 - Sulfurous acid, diethyl ester
C$_4$H$_{10}$O$_4$
  3431 - 1,2,3,4-Butanetetrol, (R*,S*)-
C$_4$H$_{10}$O$_4$S
  11573 - Sulfuric acid, diethyl ester
C$_4$H$_{10}$S
  3437 - 1-Butanethiol
  3438 - 2-Butanethiol, (±)-
  5644 - Ethane, 1,1'-thiobis-
  10088 - Propane, 1-(methylthio)-
  10089 - Propane, 2-(methylthio)-
  10158 - 1-Propanethiol, 2-methyl-
  10159 - 2-Propanethiol, 2-methyl-
C$_4$H$_{10}$S$_2$
  3368 - 1,4-Butanedithiol
  5253 - Disulfide, diethyl
  5452 - Ethane, 1,2-bis(methylthio)-
C$_4$H$_{10}$S$_3$
  5652 - Ethanethiol, 2,2'-thiobis-
  11916 - Trisulfide, diethyl
C$_4$H$_{10}$Se
  5620 - Ethane, 1,1'-selenobis-
C$_4$H$_{10}$Te
  5630 - Ethane, 1,1'-tellurobis-
C$_4$H$_{11}$As
  441 - Arsine, diethyl-
C$_4$H$_{11}$BrSi
  11366 - Silane, (bromomethyl)trimethyl-
C$_4$H$_{11}$ClSi
  11377 - Silane, chloroethyldimethyl-
  11382 - Silane, (chloromethyl)trimethyl-
C$_4$H$_{11}$N
  3209 - 1-Butanamine
  3210 - 2-Butanamine, (±)-
  5421 - Ethanamine, N,N-dimethyl-
  5427 - Ethanamine, N-ethyl-
  9817 - 1-Propanamine, 2-methyl-
  9818 - 2-Propanamine, 2-methyl-
  9819 - 2-Propanamine, N-methyl-
C$_4$H$_{11}$NO
  3595 - 1-Butanol, 2-amino-, (±)-
  3596 - 1-Butanol, 3-amino-
  3597 - 1-Butanol, 4-amino-
  3598 - 2-Butanol, 3-amino-
  5423 - Ethanamine, 2-ethoxy-
  5429 - Ethanamine, N-ethyl-N-hydroxy-
  5709 - Ethanol, 2-(dimethylamino)-
  5722 - Ethanol, 2-(ethylamino)-
  9816 - 1-Propanamine, 3-methoxy-
  10341 - 1-Propanol, 2-amino-2-methyl-
C$_4$H$_{11}$NO$_2$
  5419 - Ethanamine, 2,2-dimethoxy-
  5730 - Ethanol, 2,2'-iminobis-
  9987 - 1,3-Propanediol, 2-amino-2-methyl-
C$_4$H$_{11}$NO$_3$
  9986 - 1,3-Propanediol, 2-amino-2-(hydroxymethyl)-
C$_4$H$_{11}$NO$_8$P$_2$
  6277 - Glyphosine
C$_4$H$_{11}$O$_2$P
  9525 - Phosphinic acid, diethyl-
C$_4$H$_{11}$O$_3$P
  9533 - Phosphonic acid, (1,1-dimethylethyl)-
  9537 - Phosphonic acid, ethyl-, dimethyl ester
  9544 - Phosphonic acid, (2-methylpropyl)-
  9587 - Phosphorous acid, diethyl ester
C$_4$H$_{11}$O$_4$P
  9558 - Phosphoric acid, diethyl ester
C$_4$H$_{12}$As$_2$
  5110 - Diarsine, tetramethyl-

C$_4$H$_{12}$As$_2$O
  452 - Arsinous acid, dimethyl-, anhydride
C$_4$H$_{12}$BFN$_2$
  3114 - Boranediamine, 1-fluoro-N,N,N',N'-
    tetramethyl-
C$_4$H$_{12}$BN
  3113 - Boranamine, N,N,1,1-tetramethyl-
C$_4$H$_{12}$BrN
  7456 - Methanaminium, N,N,N-trimethyl-, bromide
C$_4$H$_{12}$ClN
  5428 - Ethanamine, N-ethyl-, hydrochloride
  7457 - Methanaminium, N,N,N-trimethyl-, chloride
C$_4$H$_{12}$Cl$_2$N$_2$
  9607 - Piperazine, dihydrochloride
C$_4$H$_{12}$Cl$_2$OSi$_2$
  5232 - Disiloxane, 1,3-dichloro-1,1,3,3-tetramethyl-
C$_4$H$_{12}$Cl$_2$Si$_2$
  5226 - Disilane, 1,2-dichloro-1,1,2,2-tetramethyl-
C$_4$H$_{12}$F$_2$Si$_2$
  5227 - Disilane, 1,2-difluoro-1,1,2,2-tetramethyl-
C$_4$H$_{12}$IN
  7458 - Methanaminium, N,N,N-trimethyl-, iodide
C$_4$H$_{12}$N$_2$
  3258 - 1,4-Butanediamine
  3261 - 2,3-Butanediamine, (R*,R*)-(±)-
  5503 - 1,2-Ethanediamine, N,N'-dimethyl-
  5509 - 1,2-Ethanediamine, N-ethyl-
  7152 - Hydrazine, 1,1-diethyl-
  7153 - Hydrazine, 1,2-diethyl-
  9878 - 1,2-Propanediamine, 2-methyl-
C$_4$H$_{12}$N$_2$O
  5676 - Ethanol, 2-[(2-aminoethyl)amino]-
C$_4$H$_{12}$N$_2$OS
  11583 - Sulfurous diamide, tetramethyl-
C$_4$H$_{12}$N$_2$O$_2$S
  11569 - Sulfamide, tetramethyl-
C$_4$H$_{12}$OSi
  11501 - Silanol, ethyldimethyl-
C$_4$H$_{12}$O$_2$Si
  11412 - Silane, dimethoxydimethyl-
C$_4$H$_{12}$O$_3$Si
  11487 - Silane, trimethoxymethyl-
C$_4$H$_{12}$O$_4$
  19 - Acetaldehyde, tetramer
C$_4$H$_{12}$O$_4$Si
  11508 - Silicic acid (H$_4$SiO$_4$), tetramethyl ester
C$_4$H$_{12}$Pb
  9732 - Plumbane, tetramethyl
C$_4$H$_{12}$Si
  11409 - Silane, diethyl-
  11451 - Silane, tetramethyl-
C$_4$H$_{12}$Sn
  11544 - Stannane, tetramethyl-
C$_4$H$_{13}$N$_3$
  5497 - 1,2-Ethanediamine, N-(2-aminoethyl)-
C$_4$H$_{14}$Cl$_2$N$_2$
  3260 - 1,4-Butanediamine, dihydrochloride
C$_4$H$_{14}$OSi$_2$
  5237 - Disiloxane, 1,1,3,3-tetramethyl-
C$_4$H$_{16}$O$_4$Si$_4$
  4999 - Cyclotetrasiloxane, 2,4,6,8-tetramethyl-
C$_4$N$_2$
  3908 - 2-Butynedinitrile
C$_4$NiO$_4$
  8068 - Nickel carbonyl
C$_5$Cl$_2$F$_6$
  4917 - Cyclopentene, 1,2-dichloro-3,3,4,4,5,5-
    hexafluoro-
C$_5$Cl$_5$N
  10995 - Pyridine, pentachloro-
C$_5$Cl$_6$
  4766 - 1,3-Cyclopentadiene, 1,2,3,4,5,5-hexachloro-
C$_5$Cl$_8$
  4932 - Cyclopentene, octachloro-
C$_5$F$_8$
  3172 - 1,3-Butadiene, 1,1,2,4,4-pentafluoro-3-
    (trifluoromethyl)-

C_5F_{10}
9027 - 1-Pentene, 1,1,2,3,3,4,4,5,5,5-decafluoro-
C_5F_{12}
8807 - Pentane, dodecafluoro-
C_5FeO_5
7339 - Iron carbonyl
C_5HCl_4N
11019 - Pyridine, 2,3,5,6-tetrachloro-
$C_5H_2Cl_6O_3$
5212 - 1,3-Dioxolan-4-one, 2,5-bis(trichloromethyl)-
$C_5H_2F_6O_2$
8802 - 2,4-Pentanedione, 1,1,1,5,5,5-hexafluoro-
$C_5H_2O_5$
4939 - 4-Cyclopentene-1,2,3-trione, 4,5-dihydroxy-
$C_5H_3BrO_2$
6043 - 2-Furancarboxaldehyde, 5-bromo-
$C_5H_3Br_2N$
10926 - Pyridine, 3,5-dibromo-
$C_5H_3ClN_4$
10753 - 1H-Purine, 6-chloro-
$C_5H_3ClO_2$
6040 - 2-Furancarbonyl chloride
$C_5H_3F_7O_2$
3525 - Butanoic acid, heptafluoro-, methyl ester
C_5H_3NO
6039 - 2-Furancarbonitrile
$C_5H_3NO_4$
6046 - 2-Furancarboxaldehyde, 5-nitro-
$C_5H_3NO_5$
6066 - 2-Furancarboxylic acid, 5-nitro-
C_5H_3NS
11732 - 2-Thiophenecarbonitrile
C_5H_4
8684 - 1,3-Pentadiyne
C_5H_4BrN
10875 - Pyridine, 2-bromo-
10876 - Pyridine, 3-bromo-
10877 - Pyridine, 4-bromo-
C_5H_4ClN
10915 - Pyridine, 2-chloro-
10916 - Pyridine, 3-chloro-
10917 - Pyridine, 4-chloro-
C_5H_4ClOS
11735 - 2-Thiophenecarboxaldehyde, 5-chloro-
$C_5H_4Cl_2N_2$
10862 - 3-Pyridinamine, 2,6-dichloro-
11052 - Pyrimidine, 2,4-dichloro-5-methyl-
11053 - Pyrimidine, 2,4-dichloro-6-methyl-
$C_5H_4Cl_2O_2$
3367 - Butanedioyl dichloride, methylene-
$C_5H_4F_8O$
8953 - 1-Pentanol, 2,2,3,3,4,4,5,5-octafluoro-
C_5H_4IN
10970 - Pyridine, 2-iodo-
$C_5H_4N_2O_2$
10803 - Pyrazinecarboxylic acid
$C_5H_4N_2O_3$
10993 - Pyridine, 4-nitro-, 1-oxide
$C_5H_4N_2O_4$
7195 - 1H-Imidazole-4,5-dicarboxylic acid
11050 - 4-Pyrimidinecarboxylic acid, 1,2,3,6-
tetrahydro-2,6-dioxo-
$C_5H_4N_4$
10752 - 1H-Purine
$C_5H_4N_4O$
10770 - 6H-Purin-6-one, 1,7-dihydro-
10852 - 4H-Pyrazolo[3,4-d]pyrimidin-4-one, 1,5-
dihydro-
$C_5H_4N_4O_2$
10755 - 1H-Purine-2,6-dione, 3,7-dihydro-
$C_5H_4N_4O_3$
10764 - 1H-Purine-2,6,8(3H)-trione, 7,9-dihydro-
$C_5H_4N_4S$
10763 - 6H-Purine-6-thione, 1,7-dihydro-
C_5H_4OS
11733 - 2-Thiophenecarboxaldehyde
11734 - 3-Thiophenecarboxaldehyde

$C_5H_4O_2$
6041 - 2-Furancarboxaldehyde
6042 - 3-Furancarboxaldehyde
6155 - 2(5H)-Furanone, 5-methylene-
10788 - 2H-Pyran-2-one
10789 - 4H-Pyran-4-one
$C_5H_4O_2S$
11737 - 2-Thiophenecarboxylic acid
11738 - 3-Thiophenecarboxylic acid
$C_5H_4O_3$
6050 - 2-Furancarboxylic acid
6051 - 3-Furancarboxylic acid
6095 - 2,5-Furandione, dihydro-3-methylene-
6098 - 2,5-Furandione, 3-methyl-
10794 - 2H-Pyran-2-one, 3-hydroxy-
$C_5H_4O_4$
6054 - 3-Furancarboxylic acid, 4,5-dihydro-5-oxo-
C_5H_5BrO
6035 - Furan, 2-(bromomethyl)-
C_5H_5BrS
11729 - Thiophene, 2-bromo-3-methyl-
11730 - Thiophene, 2-bromo-5-methyl-
$C_5H_5ClN_2$
10861 - 2-Pyridinamine, 5-chloro-
C_5H_5ClO
6075 - Furan, 2-(chloromethyl)-
C_5H_5ClS
11741 - Thiophene, 2-chloro-5-methyl-
$C_5H_5Cl_3N_2OS$
11599 - Terrazole
C_5H_5N
8673 - 2,4-Pentadienenitrile
8674 - 2,4-Pentadienenitrile, (E)-
8675 - 2,4-Pentadienenitrile, (Z)-
10872 - Pyridine
C_5H_5NO
10994 - Pyridine, 1-oxide
11031 - 2-Pyridinol
11032 - 4-Pyridinol
11033 - 2(1H)-Pyridinone
11114 - 1H-Pyrrole-2-carboxaldehyde
$C_5H_5NO_2$
6047 - 2-Furancarboxaldehyde, oxime, (E)-
6048 - 2-Furancarboxaldehyde, oxime, (Z)-
10624 - 2-Propenoic acid, 2-cyano-, methyl ester
$C_5H_5NO_2S$
11669 - 5-Thiazolecarboxylic acid, 4-methyl-
$C_5H_5NO_3S$
11018 - 3-Pyridinesulfonic acid
$C_5H_5N_3O$
10802 - Pyrazinecarboxamide
$C_5H_5N_3O_3S$
85 - Acetamide, N-(5-nitro-2-thiazolyl)-
$C_5H_5N_5$
10751 - 1H-Purin-6-amine
$C_5H_5N_5O$
10767 - 2H-Purin-2-one, 6-amino-1,3-dihydro-
10768 - 6H-Purin-6-one, 2-amino-1,7-dihydro-
$C_5H_5N_5S$
10762 - 6H-Purine-6-thione, 2-amino-1,7-dihydro-
C_5H_6
3892 - 1-Buten-3-yne, 2-methyl-
4764 - 1,3-Cyclopentadiene
9111 - 1-Penten-3-yne
9112 - 1-Penten-4-yne
9113 - 3-Penten-1-yne, (E)-
9114 - 3-Penten-1-yne, (Z)-
$C_5H_6Br_2N_2O_2$
7217 - 2,4-Imidazolidinedione, 1,3-dibromo-5,5-
dimethyl-
C_5H_6ClN
10969 - Pyridine, hydrochloride
$C_5H_6ClNO_2$
11155 - 2,5-Pyrrolidinedione, 1-(chloromethyl)-
$C_5H_6Cl_2O_2$
8806 - Pentanedioyl dichloride

$C_5H_6F_4O_2$
10328 - Propanoic acid, 2,3,3,3-tetrafluoro-, ethyl
ester
$C_5H_6N_2$
8767 - Pentanedinitrile
9913 - Propanedinitrile, dimethyl-
10810 - Pyrazine, methyl-
10856 - Pyridazine, 3-methyl-
10857 - 2-Pyridinamine
10858 - 3-Pyridinamine
10859 - 4-Pyridinamine
11070 - Pyrimidine, 2-methyl-
11071 - Pyrimidine, 4-methyl-
11072 - Pyrimidine, 5-methyl-
$C_5H_6N_2OS$
11108 - 4(1H)-Pyrimidinone, 2,3-dihydro-6-methyl-2-
thioxo-
$C_5H_6N_2O_2$
11068 - 2,4(1H,3H)-Pyrimidinedione, 5-methyl-
11069 - 2,4(1H,3H)-Pyrimidinedione, 6-methyl-
$C_5H_6N_6$
10754 - 1H-Purine-2,6-diamine
C_5H_6O
3891 - 1-Buten-3-yne, 1-methoxy-
4941 - 2-Cyclopenten-1-one
6130 - Furan, 2-methyl-
6131 - Furan, 3-methyl-
9118 - 4-Penten-2-yn-1-ol
10771 - Pyran
C_5H_6OS
6117 - 2-Furanmethanethiol
11768 - 2-Thiophenemethanol
$C_5H_6O_2$
6118 - 2-Furanmethanol
6127 - Furan, 2-methoxy-
6147 - 2(3H)-Furanone, dihydro-3-methylene-
6148 - 2(3H)-Furanone, dihydro-5-methylene-
6153 - 2(3H)-Furanone, 5-methyl-
6154 - 2(5H)-Furanone, 5-methyl-
8676 - 2,4-Pentadienoic acid
9137 - 2-Pentynoic acid
9138 - 4-Pentynoic acid
10739 - 2-Propynoic acid, ethyl ester
10744 - 2-Propyn-1-ol, acetate
$C_5H_6O_3$
4942 - 2-Cyclopenten-1-one, 2,3-dihydroxy-
6094 - 2,5-Furandione, dihydro-3-methyl-
10780 - 2H-Pyran-2,6(3H)-dione, dihydro-
$C_5H_6O_4$
3328 - Butanedioic acid, methylene-
3785 - 2-Butenedioic acid, 2-methyl-, (E)-
3786 - 2-Butenedioic acid, 2-methyl-, (Z)-
4970 - 1,2-Cyclopropanedicarboxylic acid, trans-(±)-
8875 - Pentanoic acid, 2,4-dioxo-
$C_5H_6O_5$
8781 - Pentanedioic acid, 2-oxo-
8782 - Pentanedioic acid, 3-oxo-
C_5H_6S
11769 - Thiophene, 2-methyl-
11770 - Thiophene, 3-methyl-
C_5H_7Br
9124 - 1-Pentyne, 1-bromo-
$C_5H_7BrN_2$
10816 - 1H-Pyrazole, 4-bromo-1,3-dimethyl-
$C_5H_7BrO_2$
10611 - 2-Propenoic acid, 2-bromoethyl ester
C_5H_7Cl
3153 - 1,3-Butadiene, 1-chloro-2-methyl-
3154 - 1,3-Butadiene, 1-chloro-3-methyl-
3155 - 1,3-Butadiene, 2-chloro-3-methyl-
3901 - 1-Butyne, 3-chloro-3-methyl-
4914 - Cyclopentene, 3-chloro-
8663 - 1,3-Pentadiene, 3-chloro-
$C_5H_7ClN_2$
10818 - 1H-Pyrazole, 3-chloro-1,5-dimethyl-
10819 - 1H-Pyrazole, 4-chloro-3,5-dimethyl-
10820 - 1H-Pyrazole, 5-chloro-1,3-dimethyl-

C_5H_7ClO
3888 - 2-Butenoyl chloride, 3-methyl-
4880 - Cyclopentanone, 2-chloro-
$C_5H_7ClO_2$
3830 - 2-Butenoic acid, 2-chloro-, methyl ester, (E)-
3831 - 2-Butenoic acid, 3-chloro-, methyl ester, (E)-
3832 - 2-Butenoic acid, 3-chloro-, methyl ester, (Z)-
3833 - 2-Butenoic acid, 4-chloro-, methyl ester
6141 - 2(3H)-Furanone, 5-(chloromethyl)dihydro-
10619 - 2-Propenoic acid, chloro-, ethyl ester
10620 - 2-Propenoic acid, 2-chloroethyl ester
$C_5H_7ClO_3$
10494 - Propanoyl chloride, 2-(acetyloxy)-, (±)-
$C_5H_7Cl_2NO$
43 - Acetamide, 2-chloro-N-3-chloroallyl-
$C_5H_7Cl_3O_2$
254 - Acetic acid, trichloro-, 1-methylethyl ester
261 - Acetic acid, trichloro-, propyl ester
$C_5H_7Cl_3O_3$
250 - Acetic acid, trichloro-, 2-methoxyethyl ester
$C_5H_7F_3O_2$
269 - Acetic acid, trifluoro-, propyl ester
C_5H_7I
9133 - 1-Pentyne, 1-iodo-
C_5H_7N
4173 - Cyclobutanecarbonitrile
9061 - 3-Pentenenitrile, (E)
9062 - 4-Pentenenitrile
11130 - 1H-Pyrrole, 1-methyl-
11131 - 1H-Pyrrole, 2-methyl-
11132 - 1H-Pyrrole, 3-methyl-
C_5H_7NO
6113 - 2-Furanmethanamine
7403 - Isoxazole, 3,5-dimethyl-
8577 - Oxazole, 2,4-dimethyl-
8578 - Oxazole, 2,5-dimethyl-
9063 - 3-Pentenenitrile, 2-hydroxy-, (Z)-
$C_5H_7NO_2$
152 - Acetic acid, cyano-, ethyl ester
7404 - 5-Isoxazolemethanol, 3-methyl-
10091 - Propanenitrile, 2-(acetyloxy)-
10234 - Propanoic acid, 3-cyano-, methyl ester
11160 - 2,5-Pyrrolidinedione, 1-methyl-
$C_5H_7NO_3$
8584 - 2,4-Oxazolidinedione, 5,5-dimethyl-
9768 - L-Proline, 5-oxo-
C_5H_7NS
11674 - Thiazole, 2,4-dimethyl-
11675 - Thiazole, 4,5-dimethyl-
$C_5H_7N_3$
10923 - 2,3-Pyridinediamine
10924 - 2,5-Pyridinediamine
10925 - 2,6-Pyridinediamine
11034 - 2(1H)-Pyridinone, hydrazone
11046 - 2-Pyrimidinamine, 4-methyl-
11047 - 2-Pyrimidinamine, 5-methyl-
11048 - 4-Pyrimidinamine, 6-methyl-
$C_5H_7N_3O$
11106 - 2(1H)-Pyrimidinone, 4-amino-5-methyl-
$C_5H_7N_3O_2$
7202 - 1H-Imidazole, 1,2-dimethyl-5-nitro-
11105 - 2(1H)-Pyrimidinone, 4-amino-5-
(hydroxymethyl)-
C_5H_8
3170 - 1,2-Butadiene, 3-methyl-
3171 - 1,3-Butadiene, 2-methyl-
3916 - 1-Butyne, 3-methyl-
4193 - Cyclobutane, methylene-
4201 - Cyclobutene, 1-methyl-
4903 - Cyclopentene
8658 - 1,2-Pentadiene
8659 - 1,4-Pentadiene
8660 - 2,3-Pentadiene
8661 - 1,3-Pentadiene, (E)-
8662 - 1,3-Pentadiene, (Z)-
9122 - 1-Pentyne
9123 - 2-Pentyne

11527 - Spiropentane
C_5H_8BrN
8828 - Pentanenitrile, 5-bromo-
$C_5H_8Br_2O_2$
8870 - Pentanoic acid, 2,3-dibromo-
8871 - Pentanoic acid, 2,5-dibromo-
10240 - Propanoic acid, 2,3-dibromo-, ethyl ester
10242 - Propanoic acid, 2,3-dibromo-2-methyl-,
methyl ester
$C_5H_8Br_4$
9884 - Propane, 1,3-dibromo-2,2-bis(bromomethyl)-
$C_5H_8ClFO_2$
135 - Acetic acid, chlorofluoro-, propyl ester
$C_5H_8Cl_2$
3729 - 1-Butene, 3-chloro-2-(chloromethyl)-
3755 - 1-Butene, 3,3-dichloro-2-methyl-
3756 - 2-Butene, 1,3-dichloro-2-methyl-
3757 - 2-Butene, 1,4-dichloro-2-methyl-
9028 - 2-Pentene, 2,5-dichloro-
$C_5H_8Cl_2O$
3686 - Butanoyl chloride, 2-chloro-3-methyl-
8963 - 2-Pentanone, 3,5-dichloro-
8991 - Pentanoyl chloride, 2-chloro-
8992 - Pentanoyl chloride, 5-chloro-
$C_5H_8Cl_2O_2$
167 - Acetic acid, dichloro-, 1-methylethyl ester
3502 - Butanoic acid, 3,4-dichloro-, methyl ester
10246 - Propanoic acid, 2,3-dichloro-, ethyl ester
10363 - 2-Propanol, 1,3-dichloro-, acetate
$C_5H_8Cl_2O_3$
5697 - Ethanol, 2-chloro-, carbonate (2:1)
$C_5H_8Cl_4$
8838 - Pentane, 1,1,1,5-tetrachloro-
9895 - Propane, 1,3-dichloro-2,2-bis(chloromethyl)-
$C_5H_8NNaO_4$
6248 - L-Glutamic acid, monosodium salt
$C_5H_8N_2$
7200 - 1H-Imidazole, 1,2-dimethyl-
7201 - 1H-Imidazole, 1,4-dimethyl-
7207 - 1H-Imidazole, 1-ethyl-
10825 - 1H-Pyrazole, 1,3-dimethyl-
10826 - 1H-Pyrazole, 1,5-dimethyl-
10827 - 1H-Pyrazole, 3,4-dimethyl-
10828 - 1H-Pyrazole, 3,5-dimethyl-
10832 - 1H-Pyrazole, 1-ethyl-
$C_5H_8N_2O$
7205 - 1H-Imidazole-1-ethanol
$C_5H_8N_2O_2$
7218 - 2,4-Imidazolidinedione, 5,5-dimethyl-
10236 - Propanoic acid, 2-diazo-, ethyl ester
$C_5H_8N_2O_3$
40 - Acetamide, N,N'-carbonylbis-
$C_5H_8N_4O_{12}$
9990 - 1,3-Propanediol, 2,2-bis[(nitrooxy)methyl]-,
dinitrate (ester)
C_5H_8O
3167 - 1,2-Butadiene, 4-methoxy-
3168 - 1,3-Butadiene, 1-methoxy-
3169 - 1,3-Butadiene, 2-methoxy-
3698 - 2-Butenal, 2-methyl-
3699 - 2-Butenal, 2-methyl-, (E)-
3700 - 2-Butenal, 3-methyl-
3879 - 3-Buten-2-one, 3-methyl-
3915 - 2-Butyne, 1-methoxy-
3922 - 3-Butyn-2-ol, 2-methyl-
4878 - Cyclopentanone
5796 - Ethanone, 1-cyclopropyl-
8558 - 6-Oxabicyclo[3.1.0]hexane
8681 - 1,4-Pentadien-3-ol
9001 - 4-Pentenal
9104 - 1-Penten-3-one
9105 - 3-Penten-2-one, (E)-
9140 - 1-Pentyn-3-ol
9141 - 2-Pentyn-1-ol
9142 - 3-Pentyn-1-ol
9143 - 4-Pentyn-1-ol
10045 - Propane, 1-(ethynyloxy)-

10565 - 1-Propene, 3-(ethenyloxy)-
10732 - 1-Propyne, 1-ethoxy-
10733 - 1-Propyne, 3-ethoxy-
10777 - 2H-Pyran, 3,4-dihydro-
$C_5H_8O_2$
235 - Acetic acid, 2-propenyl ester
3844 - 2-Butenoic acid, 2-methyl-, (E)-
3845 - 2-Butenoic acid, 2-methyl-, (Z)-
3846 - 2-Butenoic acid, 3-methyl-
3847 - 2-Butenoic acid, methyl ester, (E)-
3848 - 2-Butenoic acid, methyl ester, (Z)-
4174 - Cyclobutanecarboxylic acid
4963 - Cyclopropanecarboxylic acid, methyl ester
4964 - Cyclopropanecarboxylic acid, 2-methyl-
6049 - 2-Furancarboxaldehyde, tetrahydro-
6145 - 2(3H)-Furanone, dihydro-3-methyl-
6146 - 2(3H)-Furanone, dihydro-5-methyl-, (±)-
8695 - Pentanal, 2-oxo-
8696 - Pentanal, 4-oxo-
8734 - Pentanedial
8797 - 2,3-Pentanedione
8798 - 2,4-Pentanedione
9073 - 4-Pentenoic acid
10261 - Propanoic acid, ethenyl ester
10635 - 2-Propenoic acid, ethyl ester
10663 - 2-Propenoic acid, 2-methyl-, methyl ester
10698 - 1-Propen-2-ol, acetate
10798 - 2H-Pyran-2-one, tetrahydro-
10799 - 4H-Pyran-4-one, tetrahydro-
$C_5H_8O_3$
3556 - Butanoic acid, 3-methyl-2-oxo-
3573 - Butanoic acid, 3-oxo-, methyl ester
6072 - 2-Furancarboxylic acid, tetrahydro-
8902 - Pentanoic acid, 2-oxo-
8903 - Pentanoic acid, 4-oxo-
10315 - Propanoic acid, 3-oxo-, ethyl ester
10412 - 2-Propanone, 1-(acetyloxy)-
$C_5H_8O_4$
3325 - Butanedioic acid, methyl-, (±)-
3336 - Butanedioic acid, monomethyl ester
5567 - Ethanedioic acid, ethyl methyl ester
8770 - Pentanedioic acid
9943 - Propanedioic acid, dimethyl ester
9948 - Propanedioic acid, ethyl-
10202 - Propanoic acid, 2-(acetyloxy)-, (±)-
C_5H_9Br
3716 - 1-Butene, 2-bromo-3-methyl-
3717 - 2-Butene, 1-bromo-3-methyl-
3718 - 2-Butene, 2-bromo-3-methyl-
4777 - Cyclopentane, bromo-
9007 - 1-Pentene, 1-bromo-
9008 - 1-Pentene, 2-bromo-
9009 - 1-Pentene, 3-bromo-
9010 - 1-Pentene, 5-bromo-
9011 - 2-Pentene, 1-bromo-
9012 - 2-Pentene, 2-bromo-
9013 - 2-Pentene, 3-bromo-
9014 - 2-Pentene, 4-bromo-
9015 - 2-Pentene, 5-bromo-
C_5H_9BrO
6036 - Furan, 2-(bromomethyl)tetrahydro-
$C_5H_9BrO_2$
120 - Acetic acid, bromo-, propyl ester
3471 - Butanoic acid, 2-bromo-3-methyl-, (±)-
3472 - Butanoic acid, 3-bromo-3-methyl-
3473 - Butanoic acid, 2-bromo-, methyl ester
3474 - Butanoic acid, 4-bromo-, methyl ester
8859 - Pentanoic acid, 2-bromo-, (±)-
10207 - Propanoic acid, 2-bromo-, ethyl ester
10208 - Propanoic acid, 3-bromo-, ethyl ester
C_5H_9Cl
3732 - 1-Butene, 1-chloro-2-methyl-
3733 - 1-Butene, 1-chloro-3-methyl-
3734 - 1-Butene, 3-chloro-3-methyl-
3735 - 2-Butene, 1-chloro-2-methyl-
3736 - 2-Butene, 1-chloro-3-methyl-
3737 - 2-Butene, 2-chloro-3-methyl-

4789 - Cyclopentane, chloro-
9016 - 1-Pentene, 2-chloro-
9017 - 1-Pentene, 3-chloro-
9018 - 1-Pentene, 4-chloro-
9019 - 1-Pentene, 5-chloro-
9020 - 2-Pentene, 1-chloro-
9021 - 2-Pentene, 2-chloro-
9022 - 2-Pentene, 3-chloro-
9023 - 2-Pentene, 4-chloro-
9024 - 2-Pentene, 5-chloro-

C_5H_9ClO
3603 - 1-Butanol, 3-chloro-
3649 - 2-Butanone, 3-chloro-3-methyl-
3693 - Butanoyl chloride, 3-methyl-
3694 - Butanoyl chloride, 2-methyl-, (±)-
8961 - 2-Pentanone, 1-chloro-
8962 - 3-Pentanone, 1-chloro-
8990 - Pentanoyl chloride
10503 - Propanoyl chloride, 2,2-dimethyl-

$C_5H_9ClO_2$
139 - Acetic acid, chloro-, 1-methylethyl ester
149 - Acetic acid, chloro-, propyl ester
3486 - Butanoic acid, 2-chloro-2-methyl-
3487 - Butanoic acid, 2-chloro-3-methyl-
3488 - Butanoic acid, 2-chloro-, methyl ester
3489 - Butanoic acid, 4-chloro-, methyl ester
4012 - Carbonochloridic acid, butyl ester
4024 - Carbonochloridic acid, 2-methylpropyl ester
8863 - Pentanoic acid, 2-chloro-, (±)-
8864 - Pentanoic acid, 3-chloro-
8865 - Pentanoic acid, 4-chloro-
8866 - Pentanoic acid, 5-chloro-
10219 - Propanoic acid, 3-chloro-2,2-dimethyl-
10220 - Propanoic acid, 2-chloro-, ethyl ester
10221 - Propanoic acid, 3-chloro-, ethyl ester
10229 - Propanoic acid, 2-chloro-2-methyl-, methyl ester
10350 - 1-Propanol, 3-chloro-, acetate
10351 - 2-Propanol, 1-chloro-, acetate

$C_5H_9ClO_3$
137 - Acetic acid, chloro-, 2-methoxyethyl ester
4017 - Carbonochloridic acid, 2-ethoxyethyl ester

$C_5H_9Cl_2N_3O_2$
11994 - Urea, N,N'-bis(2-chloroethyl)-N-nitroso-

$C_5H_9Cl_3$
3453 - Butane, 1,2,3-trichloro-2-methyl-
3454 - Butane, 2,2,3-trichloro-3-methyl-

$C_5H_9FO_2$
8879 - Pentanoic acid, 5-fluoro-

C_5H_9I
4824 - Cyclopentane, iodo-

$C_5H_9IO_2$
10283 - Propanoic acid, 3-iodo-, ethyl ester

C_5H_9N
3387 - Butane, 1-isocyano-
3409 - Butanenitrile, 2-methyl-
3410 - Butanenitrile, 3-methyl-
3896 - 3-Butyn-2-amine, 2-methyl-
8827 - Pentanenitrile
10100 - Propanenitrile, 2,2-dimethyl-
11020 - Pyridine, 1,2,3,6-tetrahydro-
11189 - 2-Pyrroline, 2-methyl-

C_5H_9NO
4892 - Cyclopentanone, oxime
8977 - 2-Pentanone, 4-imino-
9719 - 2-Piperidinone
10061 - Propane, 1-isocyanato-2-methyl-
10062 - Propane, 2-isocyanato-2-methyl-
10102 - Propanenitrile, 2-ethoxy-
10103 - Propanenitrile, 3-ethoxy-
11185 - 2-Pyrrolidinone, 1-methyl-
11186 - 2-Pyrrolidinone, 5-methyl-
11187 - 3-Pyrrolidinone, 1-methyl-

$C_5H_9NO_2$
25 - Acetamide, N-acetyl-N-methyl-
4839 - Cyclopentane, nitro-
7685 - 4-Morpholinecarboxaldehyde

8803 - 2,3-Pentanedione, 3-oxime
9764 - L-Proline
9765 - DL-Proline

$C_5H_9NO_2S_2$
10067 - Propane, 1-isothiocyanato-3-(methylsulfonyl)-

$C_5H_9NO_3$
9767 - L-Proline, 4-hydroxy-, trans-

$C_5H_9NO_3S$
5015 - L-Cysteine, N-acetyl-

$C_5H_9NO_4$
6240 - DL-Glutamic acid
6241 - L-Glutamic acid
10312 - Propanoic acid, 2-nitro-, ethyl ester

$C_5H_9NO_4S$
5017 - L-Cysteine, S-(carboxymethyl)-

$C_5H_9NO_5$
6246 - DL-Glutamic acid, 3-hydroxy-

C_5H_9NS
3388 - Butane, 1-isothiocyanato-
3389 - Butane, 2-isothiocyanato-, (±)-
10065 - Propane, 1-isothiocyanato-2-methyl-
10066 - Propane, 2-isothiocyanato-2-methyl-
11699 - Thiocyanic acid, butyl ester
11701 - Thiocyanic acid, 1,1-dimethylethyl ester
11707 - Thiocyanic acid, 2-methylpropyl ester

$C_5H_9N_3$
7203 - 1H-Imidazole-4-ethanamine

C_5H_{10}
3799 - 1-Butene, 2-methyl-
3800 - 1-Butene, 3-methyl-
3801 - 2-Butene, 2-methyl-
4192 - Cyclobutane, methyl-
4774 - Cyclopentane
4971 - Cyclopropane, 1,1-dimethyl-
4972 - Cyclopropane, 1,2-dimethyl-, cis-
4973 - Cyclopropane, 1,2-dimethyl-, trans-
4974 - Cyclopropane, ethyl-
9004 - 1-Pentene
9005 - 2-Pentene, (E)-
9006 - 2-Pentene, (Z)-

$C_5H_{10}BrF$
8713 - Pentane, 1-bromo-5-fluoro-

$C_5H_{10}Br_2$
3272 - Butane, 1,2-dibromo-2-methyl-
3273 - Butane, 1,4-dibromo-2-methyl-, (R)-
3274 - Butane, 2,3-dibromo-2-methyl-
8739 - Pentane, 1,3-dibromo-
8740 - Pentane, 1,4-dibromo-
8741 - Pentane, 1,5-dibromo-
8742 - Pentane, 2,3-dibromo-
8743 - Pentane, 2,4-dibromo-
9886 - Propane, 1,1-dibromo-2,2-dimethyl-
9887 - Propane, 1,3-dibromo-2,2-dimethyl-

$C_5H_{10}ClF$
8728 - Pentane, 1-chloro-5-fluoro-

$C_5H_{10}ClNO$
3972 - Carbamic chloride, diethyl-

$C_5H_{10}Cl_2$
3282 - Butane, 1,1-dichloro-3-methyl-
3283 - Butane, 1,2-dichloro-2-methyl-
3284 - Butane, 1,3-dichloro-3-methyl-
3285 - Butane, 1,4-dichloro-3-methyl-
3286 - Butane, 2,3-dichloro-2-methyl-
8744 - Pentane, 1,2-dichloro-
8745 - Pentane, 1,3-dichloro-
8746 - Pentane, 1,4-dichloro-
8747 - Pentane, 1,5-dichloro-
8748 - Pentane, 2,2-dichloro-
8749 - Pentane, 2,3-dichloro-
8750 - Pentane, 2,4-dichloro-
8751 - Pentane, 3,3-dichloro-

$C_5H_{10}I_2$
8758 - Pentane, 1,5-diiodo-
8759 - Pentane, 2,4-diiodo-

$C_5H_{10}N_2$
4156 - Cyanamide, diethyl-

10101 - Propanenitrile, 3-(dimethylamino)-

$C_5H_{10}N_2O$
9688 - Piperidine, 1-nitroso-

$C_5H_{10}N_2O_2$
8800 - 2,3-Pentanedione, dioxime
9687 - Piperidine, 1-nitro-

$C_5H_{10}N_2O_2S$
7650 - Methomyl

$C_5H_{10}N_2O_3$
6249 - L-Glutamine

$C_5H_{10}N_2O_4$
3950 - Carbamic acid, ethylnitro-, ethyl ester
8769 - Pentane, 1,5-dinitro-

$C_5H_{10}O$
3185 - Butanal, 2-methyl-, (±)-
3186 - Butanal, 3-methyl-
3668 - 2-Butanone, 3-methyl-
3869 - 2-Buten-1-ol, 3-methyl-
3870 - 3-Buten-2-ol, 2-methyl-
3871 - 3-Buten-2-ol, 3-methyl-
4864 - Cyclopentanol
4979 - Cyclopropanemethanol, α-methyl-
6171 - Furan, tetrahydro-2-methyl-
6172 - Furan, tetrahydro-3-methyl-
8593 - Oxetane, 3,3-dimethyl-
8690 - Pentanal
8957 - 2-Pentanone
8958 - 3-Pentanone
9087 - 1-Penten-3-ol
9088 - 4-Penten-1-ol
9089 - 4-Penten-2-ol
9090 - 2-Penten-1-ol, (E)-
9091 - 2-Penten-1-ol, (Z)-
9092 - 3-Penten-2-ol, (±)-
9783 - Propanal, 2,2-dimethyl-
10033 - Propane, 1-(ethenyloxy)-
10034 - Propane, 2-(ethenyloxy)-
10566 - 1-Propene, 3-ethoxy-
10577 - 1-Propene, 3-methoxy-2-methyl-
10800 - 2H-Pyran, tetrahydro-

$C_5H_{10}OS_2$
4031 - Carbonodithioic acid, S,S-diethyl ester

$C_5H_{10}O_2$
209 - Acetic acid, 1-methylethyl ester
236 - Acetic acid, propyl ester
3538 - Butanoic acid, 2-methyl-, (±)-
3539 - Butanoic acid, 3-methyl-
3547 - Butanoic acid, methyl ester
3664 - 2-Butanone, 4-hydroxy-3-methyl-
4806 - 1,3-Cyclopentanediol, cis-
4807 - 1,3-Cyclopentanediol, trans-
5193 - 1,3-Dioxane, 4-methyl-
5763 - Ethanol, 2-(2-propenyloxy)-
6004 - Formic acid, butyl ester
6012 - Formic acid, 1-methylpropyl ester
6013 - Formic acid, 2-methylpropyl ester
6124 - 2-Furanmethanol, tetrahydro-
8614 - Oxirane, (ethoxymethyl)-
8855 - Pentanoic acid
8971 - 2-Pentanone, 1-hydroxy-
8972 - 2-Pentanone, 3-hydroxy-
8973 - 2-Pentanone, 4-hydroxy-
8974 - 2-Pentanone, 5-hydroxy-
8975 - 3-Pentanone, 2-hydroxy-
10258 - Propanoic acid, 2,2-dimethyl-
10263 - Propanoic acid, ethyl ester
10301 - Propanoic acid, 2-methyl-, methyl ester
10562 - 1-Propene, 3,3-dimethoxy-
10787 - 2H-Pyran-4-ol, tetrahydro-

$C_5H_{10}O_2S$
10311 - Propanoic acid, 3-(methylthio)-, methyl ester
11789 - Thiophene, tetrahydro-3-methyl-, 1,1-dioxide

$C_5H_{10}O_3$
182 - Acetic acid, ethoxy-, methyl ester
199 - Acetic acid, hydroxy-, propyl ester
206 - Acetic acid, methoxy-, ethyl ester
3535 - Butanoic acid, 3-hydroxy-3-methyl-

3999 - Carbonic acid, diethyl ester
5197 - 1,3-Dioxan-5-ol, 2-methyl-
5208 - 1,3-Dioxolane-4-methanol, 2-methyl-
5734 - Ethanol, methoxy-, acetate
5735 - Ethanol, 2-methoxy-, acetate
8883 - Pentanoic acid, 2-hydroxy-
10269 - Propanoic acid, 2-hydroxy-, ethyl ester, (±)-
10270 - Propanoic acid, 3-hydroxy-, ethyl ester
10280 - Propanoic acid, 2-hydroxy-2-methyl-, methyl ester
10289 - Propanoic acid, 3-methoxy-, methyl ester

$C_5H_{10}O_4$
10187 - 1,2,3-Propanetriol, 1-acetate, (±)-
10253 - Propanoic acid, 2,3-dihydroxy-, ethyl ester, (±)-
10256 - Propanoic acid, 2,3-dihydroxy-2-methyl-, methyl ester
10272 - Propanoic acid, 3-hydroxy-2-(hydroxymethyl)-2-methyl-

$C_5H_{10}O_5$
436 - β-D-Arabinopyranose
11345 - D-Ribose
12059 - D-Xylose

$C_5H_{10}S$
4852 - Cyclopentanethiol
11787 - Thiophene, tetrahydro-2-methyl-
11788 - Thiophene, tetrahydro-3-methyl-
11795 - 2H-Thiopyran, tetrahydro-

$C_5H_{11}Br$
3238 - Butane, 1-bromo-2-methyl-, (S)-
3239 - Butane, 1-bromo-3-methyl-
3240 - Butane, 2-bromo-2-methyl-
8710 - Pentane, 1-bromo-
8711 - Pentane, 2-bromo-
8712 - Pentane, 3-bromo-
9843 - Propane, 1-bromo-2,2-dimethyl-

$C_5H_{11}Cl$
3251 - Butane, 1-chloro-2-methyl-, (±)-
3252 - Butane, 1-chloro-3-methyl-
3253 - Butane, 2-chloro-2-methyl-
8719 - Pentane, 1-chloro-
8720 - Pentane, 2-chloro-, (+)-
8721 - Pentane, 3-chloro-
9853 - Propane, 1-chloro-2,2-dimethyl-

$C_5H_{11}ClO$
3610 - 2-Butanol, 1-chloro-2-methyl-
3611 - 2-Butanol, 3-chloro-2-methyl-
8923 - 1-Pentanol, 5-chloro-
8924 - 2-Pentanol, 1-chloro-
8925 - 3-Pentanol, 1-chloro-
9854 - Propane, 1-(1-chloroethoxy)-

$C_5H_{11}ClO_2$
9852 - Propane, 3-chloro-1,1-dimethoxy-

$C_5H_{11}ClSi$
11375 - Silane, chlorodimethyl-2-propenyl-

$C_5H_{11}Cl_2N$
5413 - Ethanamine, 2-chloro-N-(2-chloroethyl)-N-methyl-

$C_5H_{11}Cl_3Si$
11472 - Silane, trichloropentyl-

$C_5H_{11}F$
8817 - Pentane, 1-fluoro-

$C_5H_{11}FO$
8935 - 1-Pentanol, 5-fluoro-

$C_5H_{11}I$
3383 - Butane, 1-iodo-2-methyl-, (S)-
3384 - Butane, 1-iodo-3-methyl-
3385 - Butane, 2-iodo-2-methyl-
8818 - Pentane, 1-iodo-
8819 - Pentane, 2-iodo-, (±)-
8820 - Pentane, 3-iodo-
10056 - Propane, 1-iodo-2,2-dimethyl-

$C_5H_{11}N$
4771 - Cyclopentanamine
9629 - Piperidine
11168 - Pyrrolidine, 1-methyl-

$C_5H_{11}NO$
3187 - Butanal, 3-methyl-, oxime
3206 - Butanamide, 3-methyl-
5991 - Formamide, N,N-diethyl-
6116 - 2-Furanmethanamine, tetrahydro-
7696 - Morpholine, 4-methyl-
8698 - Pentanamide
8985 - 2-Pentanone, oxime

$C_5H_{11}NO_2$
3400 - Butane, 3-methyl-1-nitro-
3935 - Carbamic acid, butyl ester
3941 - Carbamic acid, diethyl-
3949 - Carbamic acid, ethyl-, ethyl ester
3959 - Carbamic acid, 2-methylpropyl ester
7400 - DL-Isovaline
7401 - L-Isovaline
7454 - Methanaminium, 1-carboxy-N,N,N-trimethyl-, hydroxide, inner salt
8090 - Nitrous acid, 3-methylbutyl ester
8096 - Nitrous acid, pentyl ester
8217 - DL-Norvaline
8218 - L-Norvaline
8832 - Pentane, 1-nitro-
8833 - Pentane, 2-nitro-
8834 - Pentane, 3-nitro-
8857 - Pentanoic acid, 5-amino-
12040 - DL-Valine
12041 - L-Valine

$C_5H_{11}NO_2S$
7647 - DL-Methionine
7648 - L-Methionine
12043 - D-Valine, 3-mercapto-

$C_5H_{11}NO_3$
3637 - 1-Butanol, 3-methyl-, nitrate
3638 - 1-Butanol, 3-methyl-2-nitro-
8950 - 1-Pentanol, 2-nitro-
8951 - 2-Pentanol, 1-nitro-
8952 - 3-Pentanol, 1-nitro-

$C_5H_{11}NO_4$
10008 - 1,3-Propanediol, 2-ethyl-2-nitro-

C_5H_{12}
3397 - Butane, 2-methyl-
8709 - Pentane
9910 - Propane, 2,2-dimethyl-

$C_5H_{12}NO_3PS_2$
5009 - Cygon

$C_5H_{12}N_2$
9618 - Piperazine, 1-methyl-
9619 - Piperazine, 2-methyl-
9624 - 1-Piperidinamine

$C_5H_{12}N_2O$
12001 - Urea, N,N'-diethyl-
12002 - Urea, N,N-diethyl-
12030 - Urea, tetramethyl-

$C_5H_{12}N_2O_2$
8550 - L-Ornithine

$C_5H_{12}N_2S$
11803 - Thiourea, N,N'-diethyl-
11810 - Thiourea, tetramethyl-

$C_5H_{12}N_4O_3$
7135 - L-Homoserine, O-[(aminoiminomethyl)amino]-

$C_5H_{12}O$
3391 - Butane, 1-methoxy-
3392 - Butane, 2-methoxy-, (S)-
3393 - Butane, 2-methoxy-, (±)-
3626 - 1-Butanol, 2-methyl-
3627 - 1-Butanol, 3-methyl-
3628 - 2-Butanol, 2-methyl-
3629 - 2-Butanol, 3-methyl-, (±)-
8915 - 1-Pentanol
8916 - 2-Pentanol
8917 - 3-Pentanol
10037 - Propane, 1-ethoxy-
10038 - Propane, 2-ethoxy-
10071 - Propane, 1-methoxy-2-methyl-
10072 - Propane, 2-methoxy-2-methyl-
10373 - 1-Propanol, 2,2-dimethyl-

$C_5H_{12}O_2$
3352 - 1,2-Butanediol, 3-methyl-
3353 - 1,3-Butanediol, 2-methyl-
3354 - 1,3-Butanediol, 3-methyl-
3355 - 2,3-Butanediol, 2-methyl-
3625 - 1-Butanol, 3-methoxy-
5591 - Ethane, 1-ethoxy-2-methoxy-
5603 - Ethane, 1,1'-[methylenebis(oxy)]bis-
5740 - Ethanol, 2-(1-methylethoxy)-
5764 - Ethanol, 2-propoxy-
8785 - 1,3-Pentanediol
8786 - 1,4-Pentanediol
8787 - 1,5-Pentanediol
8788 - 2,3-Pentanediol
8789 - 2,4-Pentanediol
8790 - 1,2-Pentanediol, (±)-
9909 - Propane, 2,2-dimethoxy-
10005 - 1,3-Propanediol, 2,2-dimethyl-
10380 - 1-Propanol, 2-ethoxy-
10381 - 2-Propanol, 1-ethoxy-

$C_5H_{12}O_3$
5669 - Ethane, 1,1,1-trimethoxy-
5737 - Ethanol, 2-(2-methoxyethoxy)-
8851 - 1,2,5-Pentanetriol
8852 - 1,3,5-Pentanetriol
9119 - Pentitol, 1,2-dideoxy-
10011 - 1,3-Propanediol, 2-(hydroxymethyl)-2-methyl-
10372 - 2-Propanol, 1,3-dimethoxy-

$C_5H_{12}O_4$
7543 - Methane, tetramethoxy-
9989 - 1,3-Propanediol, 2,2-bis(hydroxymethyl)-

$C_5H_{12}O_5$
11343 - Ribitol
12058 - Xylitol

$C_5H_{12}S$
3402 - Butane, 1-(methylthio)-
3439 - 1-Butanethiol, 2-methyl-, (S)-
3440 - 1-Butanethiol, 3-methyl-
3441 - 2-Butanethiol, 3-methyl-
8844 - 1-Pentanethiol
8845 - 2-Pentanethiol
8846 - 3-Pentanethiol
10042 - Propane, 1-(ethylthio)-
10043 - Propane, 2-(ethylthio)-
10082 - Propane, 2-methyl-1-(methylthio)-
10083 - Propane, 2-methyl-2-(methylthio)-

$C_5H_{12}Si$
11425 - Silane, ethenyltrimethyl-

$C_5H_{13}Cl_2N$
5442 - Ethanaminium, 2-chloro-N,N,N-trimethyl-, chloride

$C_5H_{13}N$
3220 - 1-Butanamine, 3-methyl-
3221 - 1-Butanamine, N-methyl-
3222 - 2-Butanamine, 2-methyl-
3223 - 2-Butanamine, 3-methyl-
5430 - Ethanamine, N-ethyl-N-methyl-
8700 - 1-Pentanamine
8701 - 2-Pentanamine
8702 - 3-Pentanamine
9806 - 1-Propanamine, 2,2-dimethyl-
9807 - 1-Propanamine, N,N-dimethyl-
9809 - 1-Propanamine, N-ethyl-
9810 - 2-Propanamine, N-ethyl-

$C_5H_{13}NO$
3599 - 1-Butanol, 2-amino-3-methyl-, (±)-
5426 - Ethanamine, 2-ethoxy-N-methyl-
5741 - Ethanol, 2-[(1-methylethyl)amino]-
5765 - Ethanol, 2-(propylamino)-
5878 - Ethanaminium, N,N,N-trimethyl-, hydroxide
8920 - 1-Pentanol, 5-amino-
8921 - 3-Pentanol, 2-amino-
10374 - 1-Propanol, 2-(dimethylamino)-
10375 - 1-Propanol, 3-(dimethylamino)-
10376 - 2-Propanol, 1-(dimethylamino)-

C5H13NO2
5420 - Ethanamine, 2,2-dimethoxy-N-methyl-
5742 - Ethanol, 2,2'-(methylimino)bis-
9985 - 1,3-Propanediol, 2-amino-2-ethyl-
10383 - 2-Propanol, 1-[(2-hydroxyethyl)amino]-

C5H13O3P
9539 - Phosphonic acid, methyl-, diethyl ester

C5H14ClNO
5446 - Ethanaminium, 2-hydroxy-N,N,N-trimethyl-, chloride

C5H14N2
7490 - Methanediamine, N,N,N',N'-tetramethyl-
8735 - 1,5-Pentanediamine

C5H14N2O
10340 - 2-Propanol, 1-[(2-aminoethyl)amino]-

C5H14OSi
11430 - Silane, ethoxytrimethyl-

C5H14O3Si
11433 - Silane, ethyltrimethoxy-

C5H15NSi
11361 - Silanamine, pentamethyl-

C5H20O5Si5
4902 - Cyclopentasiloxane, 2,4,6,8,10-pentamethyl-

C6BrF5
1078 - Benzene, bromopentafluoro-

C6Br2F4
1312 - Benzene, 1,2-dibromo-3,4,5,6-tetrafluoro-

C6ClF5
1220 - Benzene, chloropentafluoro-

C6Cl2F4
1398 - Benzene, 1,2-dichloro-3,4,5,6-tetrafluoro
1399 - Benzene, 1,3-dichloro-2,4,5,6-tetrafluoro-
1400 - Benzene, 1,4-dichloro-2,3,5,6-tetrafluoro-

C6Cl2F8
4647 - Cyclohexene, 1,2-dichloro-3,3,4,4,5,5,6,6-octafluoro-

C6Cl3F3
2340 - Benzene, 1,3,5-trichloro-2,4,6-trifluoro-

C6Cl4O2
4268 - 2,5-Cyclohexadiene-1,4-dione, 2,3,5,6-tetrachloro-

C6Cl5NO2
2087 - Benzene, pentachloronitro-

C6Cl6
1749 - Benzene, hexachloro-

C6Cl10
6689 - 1,5-Hexadiene, 1,1,2,3,3,4,4,5,6,6-decachloro-

C6F5I
2092 - Benzene, pentafluoroiodo-

C6F6
1757 - Benzene, hexafluoro-

C6F8
4276 - 1,3-Cyclohexadiene, 1,2,3,4,5,5,6,6-octafluoro-
4277 - 1,4-Cyclohexadiene, 1,2,3,3,4,5,6,6-octafluoro-

C6F10
4643 - Cyclohexene, decafluoro-

C6F12
4381 - Cyclohexane, dodecafluoro-
7032 - 1-Hexene, 1,1,2,3,3,4,4,5,5,6,6-dodecafluoro-

C6F14
3414 - Butane, 1,1,1,2,3,4,4,4-octafluoro-2,3-bis(trifluoromethyl)-
6843 - Hexane, tetradecafluoro-
8853 - Pentane, 1,1,1,2,2,3,4,4,5,5,5-undecafluoro-3-(trifluoromethyl)
8854 - Pentane, 1,1,1,2,2,3,3,4,5,5,5-undecafluoro-4-(trifluoromethyl)-

C6F15N
5436 - Ethanamine, 1,1,2,2,2-pentafluoro-N,N-bis(pentafluoroethyl)-

C6HBr5
2082 - Benzene, pentabromo-

C6HBr5O
9439 - Phenol, pentabromo-

C6HCl3F8O2
6910 - Hexanoic acid, 3,5,6-trichloro-2,2,3,4,4,5,6,6-octafluoro-

C6HCl4NO2
2264 - Benzene, 1,2,4,5-tetrachloro-3-nitro-

C6HCl5
2084 - Benzene, pentachloro-

C6HCl5O
9440 - Phenol, pentachloro-

C6HF5
2091 - Benzene, pentafluoro-

C6HF5O
9442 - Phenol, pentafluoro-

C6HF5S
2304 - Benzenethiol, pentafluoro-

C6HF11
4498 - Cyclohexane, undecafluoro-

C6HI5
2096 - Benzene, pentaiodo-

C6H2Br2ClNO
4279 - 2,5-Cyclohexadien-1-one, 2,6-dibromo-4-(chloroimino)-

C6H2Br3NO2
2315 - Benzene, 1,2,3-tribromo-5-nitro-
2316 - Benzene, 1,3,5-tribromo-2-nitro-

C6H2Br4
2253 - Benzene, 1,2,3,5-tetrabromo-
2254 - Benzene, 1,2,4,5-tetrabromo-

C6H2Br4O
9467 - Phenol, 2,3,4,6-tetrabromo-

C6H2Br4O2
1551 - 1,4-Benzenediol, 2,3,5,6-tetrabromo-

C6H2ClN3O6
1244 - Benzene, 2-chloro-1,3,5-trinitro-

C6H2Cl2O4
4258 - 2,5-Cyclohexadiene-1,4-dione, 2,5-dichloro-3,6-dihydroxy-

C6H2Cl3NO
4280 - 2,5-Cyclohexadien-1-one, 2,6-dichloro-4-(chloroimino)-

C6H2Cl3NO2
2338 - Benzene, 1,2,3-trichloro-5-nitro-
2339 - Benzene, 1,2,4-trichloro-5-nitro-

C6H2Cl3NO3
9490 - Phenol, 3,4,6-trichloro-2-nitro-

C6H2Cl4
2259 - Benzene, 1,2,3,4-tetrachloro-
2260 - Benzene, 1,2,4,5-tetrachloro-

C6H2Cl4O
9469 - Phenol, 2,3,4,5-tetrachloro-
9470 - Phenol, 2,3,4,6-tetrachloro-
9471 - Phenol, 2,3,5,6-tetrachloro-

C6H2Cl5OPS
9578 - Phosphorodichloridothioic acid, O-(2,4,5-trichlorophenyl) ester

C6H2F4
2270 - Benzene, 1,2,3,4-tetrafluoro-
2271 - Benzene, 1,2,3,5-tetrafluoro-
2272 - Benzene, 1,2,4,5-tetrafluoro-

C6H2I3NO2
2353 - Benzene, 1,2,3-triiodo-5-nitro-

C6H2I4
2273 - Benzene, 1,2,3,4-tetraiodo-
2274 - Benzene, 1,2,3,5-tetraiodo-
2275 - Benzene, 1,2,4,5-tetraiodo-

C6H2N2O8
4259 - 2,5-Cyclohexadiene-1,4-dione, 2,5-dihydroxy-3,6-dinitro-

C6H2N4O5
2874 - 1,2,3-Benzoxadiazole, 5,7-dinitro-

C6H2N4O9
9476 - Phenol, 2,3,4,6-tetranitro-

C6H2O6
4686 - 5-Cyclohexene-1,2,3,4-tetrone, 5,6-dihydroxy-

C6H3BrCl2
1010 - Benzene, 1-bromo-2,3-dichloro-
1011 - Benzene, 1-bromo-2,4-dichloro-
1012 - Benzene, 1-bromo-3,5-dichloro-
1013 - Benzene, 2-bromo-1,3-dichloro-
1014 - Benzene, 2-bromo-1,4-dichloro-
1015 - Benzene, 4-bromo-1,2-dichloro-

C6H3BrCl2O
9233 - Phenol, 2-bromo-4,6-dichloro-

C6H3BrN2O4
1024 - Benzene, 1-bromo-2,3-dinitro-

C6H3BrN2O5
9234 - Phenol, 2-bromo-4,6-dinitro-
9235 - Phenol, 4-bromo-2,6-dinitro-

C6H3Br2Cl
1292 - Benzene, 1,2-dibromo-3-chloro-
1293 - Benzene, 1,2-dibromo-4-chloro-
1294 - Benzene, 1,3-dibromo-2-chloro-
1295 - Benzene, 1,4-dibromo-2-chloro-
1296 - Benzene, 2,4-dibromo-1-chloro-

C6H3Br2F
1300 - Benzene, 2,4-dibromo-1-fluoro-

C6H3Br2NO2
1307 - Benzene, 1,2-dibromo-4-nitro-
1308 - Benzene, 1,3-dibromo-2-nitro-
1309 - Benzene, 1,3-dibromo-5-nitro-
1310 - Benzene, 1,4-dibromo-2-nitro-
1311 - Benzene, 2,4-dibromo-1-nitro-

C6H3Br3
2307 - Benzene, 1,2,3-tribromo-
2308 - Benzene, 1,2,4-tribromo-
2309 - Benzene, 1,3,5-tribromo-

C6H3Br3O
9479 - Phenol, 2,4,6-tribromo-

C6H3ClNO3
9296 - Phenol, 2,6-dichloro-4-nitro-

C6H3ClN2O4
1144 - Benzene, 1-chloro-2,4-dinitro-
1145 - Benzene, 2-chloro-1,3-dinitro-

C6H3ClN2O4S
2178 - Benzenesulfenyl chloride, 2,4-dinitro-

C6H3ClN2O5
9258 - Phenol, 4-chloro-2,6-dinitro-
9259 - Phenol, 5-chloro-2,4-dinitro-

C6H3Cl2I
1380 - Benzene, 1,4-dichloro-2-iodo-

C6H3Cl2NO2
1391 - Benzene, 1,2-dichloro-3-nitro-
1392 - Benzene, 1,2-dichloro-4-nitro-
1393 - Benzene, 1,3-dichloro-2-nitro-
1394 - Benzene, 1,3-dichloro-5-nitro-
1395 - Benzene, 1,4-dichloro-2-nitro-
1396 - Benzene, 2,4-dichloro-1-nitro-
4142 - Clopyralid

C6H3Cl3
2321 - Benzene, 1,2,3-trichloro-
2322 - Benzene, 1,2,4-trichloro-
2323 - Benzene, 1,3,5-trichloro-

C6H3Cl3N2O2
10896 - 2-Pyridinecarboxylic acid, 4-amino-3,5,6-trichloro-

C6H3Cl3O
9481 - Phenol, 2,3,4-trichloro-
9482 - Phenol, 2,3,5-trichloro-
9483 - Phenol, 2,3,6-trichloro-
9484 - Phenol, 2,4,5-trichloro-
9485 - Phenol, 2,4,6-trichloro-
9486 - Phenol, 3,4,5-trichloro-

C6H3Cl3O3
2371 - 1,3,5-Benzenetriol, 2,4,6-trichloro-

C6H3Cl4N
8071 - Nitrapyrin

C6H3FN2O4
1718 - Benzene, 1-fluoro-2,4-dinitro-

C6H3F2NO2
1426 - Benzene, 2,4-difluoro-1-nitro-

$C_6H_3F_3$
2346 - Benzene, 1,2,4-trifluoro-
2347 - Benzene, 1,3,5-trifluoro-
$C_6H_3F_7O_2$
3523 - Butanoic acid, heptafluoro-, ethenyl ester
$C_6H_3I_2NO_3$
9304 - Phenol, 2,6-diiodo-4-nitro-
$C_6H_3I_3$
2349 - Benzene, 1,2,3-triiodo-
2350 - Benzene, 1,2,4-triiodo-
2351 - Benzene, 1,3,5-triiodo-
$C_6H_3I_3O$
9492 - Phenol, 2,4,6-triiodo-
$C_6H_3N_3O_6$
2365 - Benzene, 1,3,5-trinitro-
$C_6H_3N_3O_7$
9497 - Phenol, 2,4,6-trinitro-
$C_6H_3N_3O_8$
1554 - 1,3-Benzenediol, 2,4,6-trinitro-
C_6H_4BrCl
1004 - Benzene, 1-bromo-2-chloro-
1005 - Benzene, 1-bromo-3-chloro-
1006 - Benzene, 1-bromo-4-chloro-
C_6H_4BrClO
9231 - Phenol, 3-bromo-5-chloro-
9232 - Phenol, 4-bromo-2-chloro-
$C_6H_4BrClO_2S$
2244 - Benzenesulfonyl chloride, 4-bromo-
$C_6H_4BrCl_2N$
641 - Benzenamine, 2-bromo-4,6-dichloro-
C_6H_4BrF
1043 - Benzene, 1-bromo-3-fluoro-
1044 - Benzene, 1-bromo-4-fluoro-
C_6H_4BrI
1046 - Benzene, 1-bromo-2-iodo-
1047 - Benzene, 1-bromo-3-iodo-
1048 - Benzene, 1-bromo-4-iodo-
$C_6H_4BrNO_2$
1075 - Benzene, 1-bromo-2-nitro-
1076 - Benzene, 1-bromo-3-nitro-
1077 - Benzene, 1-bromo-4-nitro-
$C_6H_4BrNO_3$
9243 - Phenol, 2-bromo-3-nitro-
9244 - Phenol, 4-bromo-2-nitro-
$C_6H_4BrN_3$
977 - Benzene, 1-azido-4-bromo-
$C_6H_4BrN_3O_4$
646 - Benzenamine, 2-bromo-4,6-dinitro-
$C_6H_4Br_2$
1288 - Benzene, 1,2-dibromo-
1289 - Benzene, 1,3-dibromo-
1290 - Benzene, 1,4-dibromo-
$C_6H_4Br_2O$
9276 - Phenol, 2,4-dibromo-
9277 - Phenol, 2,6-dibromo-
9278 - Phenol, 3,5-dibromo-
$C_6H_4Br_3N$
851 - Benzenamine, 2,4,6-tribromo-
C_6H_4ClF
1166 - Benzene, 1-chloro-2-fluoro-
1167 - Benzene, 1-chloro-3-fluoro-
1168 - Benzene, 1-chloro-4-fluoro-
$C_6H_4ClFO_2S$
2246 - Benzenesulfonyl chloride, 4-fluoro-
C_6H_4ClI
1171 - Benzene, 1-chloro-2-iodo-
1172 - Benzene, 1-chloro-4-iodo-
$C_6H_4ClIO_2S$
2247 - Benzenesulfonyl chloride, 4-iodo-
$C_6H_4ClNO_2$
1214 - Benzene, 1-chloro-2-nitro-
1215 - Benzene, 1-chloro-3-nitro-
1216 - Benzene, 1-chloro-4-nitro-
$C_6H_4ClNO_2S$
2179 - Benzenesulfenyl chloride, 4-nitro-
$C_6H_4ClNO_3$
9269 - Phenol, 5-chloro-2-nitro-

$C_6H_4ClN_3$
978 - Benzene, 1-azido-4-chloro-
$C_6H_4ClO_2P$
2413 - 1,3,2-Benzodioxaphosphole, 2-chloro-
$C_6H_4Cl_2$
1354 - Benzene, 1,2-dichloro-
1355 - Benzene, 1,3-dichloro-
1356 - Benzene, 1,4-dichloro-
$C_6H_4Cl_2N_2O_2$
692 - Benzenamine, 2,6-dichloro-4-nitro-
$C_6H_4Cl_2O$
9282 - Phenol, 2,3-dichloro-
9283 - Phenol, 2,4-dichloro-
9284 - Phenol, 2,5-dichloro-
9285 - Phenol, 2,6-dichloro-
9286 - Phenol, 3,4-dichloro-
9287 - Phenol, 3,5-dichloro-
$C_6H_4Cl_2O_2$
1510 - 1,3-Benzenediol, 4,6-dichloro-
$C_6H_4Cl_2O_2S$
2245 - Benzenesulfonyl chloride, 4-chloro-
$C_6H_4Cl_3N$
852 - Benzenamine, 2,3,4-trichloro-
853 - Benzenamine, 2,4,5-trichloro-
854 - Benzenamine, 2,4,6-trichloro-
11023 - Pyridine, 2-(trichloromethyl)-
$C_6H_4Cl_4Si$
11457 - Silane, trichloro(4-chlorophenyl)-
C_6H_4FI
1719 - Benzene, 1-fluoro-2-iodo-
1720 - Benzene, 1-fluoro-4-iodo-
$C_6H_4FNO_2$
1733 - Benzene, 1-fluoro-2-nitro-
1734 - Benzene, 1-fluoro-3-nitro-
1735 - Benzene, 1-fluoro-4-nitro-
$C_6H_4F_2$
1422 - Benzene, 1,2-difluoro-
1423 - Benzene, 1,3-difluoro-
1424 - Benzene, 1,4-difluoro-
$C_6H_4INO_2$
1783 - Benzene, 1-iodo-2-nitro-
1784 - Benzene, 1-iodo-3-nitro-
1785 - Benzene, 1-iodo-4-nitro-
$C_6H_4IN_3$
979 - Benzene, 1-azido-2-iodo-
$C_6H_4I_2$
1427 - Benzene, 1,2-diiodo-
1428 - Benzene, 1,3-diiodo-
1429 - Benzene, 1,4-diiodo-
$C_6H_4N_2$
10881 - 2-Pyridinecarbonitrile
10882 - 3-Pyridinecarbonitrile
$C_6H_4N_2O_4$
1491 - Benzene, 1,2-dinitro-
1492 - Benzene, 1,3-dinitro-
1493 - Benzene, 1,4-dinitro-
10804 - 2,3-Pyrazinedicarboxylic acid
$C_6H_4N_2O_5$
9341 - Phenol, 2,3-dinitro-
9342 - Phenol, 2,4-dinitro-
9343 - Phenol, 2,5-dinitro-
9344 - Phenol, 2,6-dinitro-
9345 - Phenol, 3,4-dinitro-
9346 - Phenol, 3,5-dinitro-
$C_6H_4N_2O_6$
1517 - 1,3-Benzenediol, 2,4-dinitro-
$C_6H_4N_4$
10747 - Pteridine
$C_6H_4N_4O_2$
2873 - 1H-Benzotriazole, 5-nitro-
10748 - 2,4(1H,3H)-Pteridinedione
$C_6H_4N_4O_6$
863 - Benzenamine, 2,4,6-trinitro-
C_6H_4O
6108 - Furan, 2-ethynyl-
$C_6H_4O_2$
4257 - 2,5-Cyclohexadiene-1,4-dione

$C_6H_4O_4$
10773 - 2H-Pyran-5-carboxylic acid, 2-oxo-
$C_6H_4O_4S$
11745 - 2,5-Thiophenedicarboxylic acid
$C_6H_4O_5$
6079 - 2,3-Furandicarboxylic acid
6080 - 2,4-Furandicarboxylic acid
6081 - 2,5-Furandicarboxylic acid
$C_6H_4O_6$
4269 - 2,5-Cyclohexadiene-1,4-dione, 2,3,5,6-
    tetrahydroxy-
$C_6H_4S_2$
11688 - Thieno[2,3-b]thiophene
$C_6H_5AsCl_2$
462 - Arsonous dichloride, phenyl-
$C_6H_5AsI_2$
464 - Arsonous diiodide, phenyl-
C_6H_5AsO
447 - Arsine, oxophenyl-
$C_6H_5BF_2$
3116 - Borane, difluorophenyl-
$C_6H_5BO_2$
2412 - 1,3,2-Benzodioxaborole
C_6H_5Br
998 - Benzene, bromo-
$C_6H_5BrN_2O_2$
653 - Benzenamine, 2-bromo-6-nitro-
654 - Benzenamine, 4-bromo-2-nitro-
C_6H_5BrO
9228 - Phenol, 2-bromo-
9229 - Phenol, 3-bromo-
9230 - Phenol, 4-bromo-
C_6H_5BrOS
5783 - Ethanone, 1-(5-bromo-2-thienyl)-
$C_6H_5BrO_2$
1502 - 1,3-Benzenediol, 4-bromo-
1503 - 1,4-Benzenediol, 2-bromo-
C_6H_5BrS
2292 - Benzenethiol, 4-bromo-
$C_6H_5Br_2N$
681 - Benzenamine, 2,4-dibromo-
682 - Benzenamine, 2,6-dibromo-
683 - Benzenamine, 3,4-dibromo-
C_6H_5Cl
1131 - Benzene, chloro-
C_6H_5ClFN
664 - Benzenamine, 2-chloro-6-fluoro-
$C_6H_5ClN_2O_2$
677 - Benzenamine, 5-chloro-2-nitro-
C_6H_5ClO
9251 - Phenol, 2-chloro-
9252 - Phenol, 3-chloro-
9253 - Phenol, 4-chloro-
C_6H_5ClOS
2185 - Benzenesulfinyl chloride
5791 - Ethanone, 1-(5-chloro-2-thienyl)-
$C_6H_5ClO_2$
1504 - 1,2-Benzenediol, 3-chloro-
1505 - 1,2-Benzenediol, 4-chloro-
1506 - 1,3-Benzenediol, 5-chloro-
1507 - 1,4-Benzenediol, 2-chloro-
$C_6H_5ClO_2S$
2242 - Benzenesulfonyl chloride
$C_6H_5ClO_3S$
2220 - Benzenesulfonic acid, 4-chloro-
C_6H_5ClS
2293 - Benzenethiol, 2-chloro-
2294 - Benzenethiol, 3-chloro-
2295 - Benzenethiol, 4-chloro-
$C_6H_5Cl_2IS$
1563 - 1,3-Benzenedithiol, 4-chloro-
$C_6H_5Cl_2N$
685 - Benzenamine, 2,3-dichloro-
686 - Benzenamine, 2,4-dichloro-
687 - Benzenamine, 2,5-dichloro-
688 - Benzenamine, 2,6-dichloro-
689 - Benzenamine, 3,4-dichloro-

690 - Benzenamine, 3,5-dichloro-

$C_6H_5Cl_2NO$
9205 - Phenol, 2-amino-4,6-dichloro-
9206 - Phenol, 4-amino-2,6-dichloro-

$C_6H_5Cl_2OP$
9548 - Phosphonic dichloride, phenyl-

$C_6H_5Cl_2P$
9556 - Phosphonous dichloride, phenyl-

$C_6H_5Cl_2PS$
9552 - Phosphonothioic dichloride, phenyl-

$C_6H_5Cl_2Sb$
11547 - Stibine, dichlorophenyl-

$C_6H_5Cl_3Si$
11473 - Silane, trichlorophenyl-

C_6H_5F
1717 - Benzene, fluoro-

C_6H_5FO
9360 - Phenol, 2-fluoro-
9361 - Phenol, 3-fluoro-
9362 - Phenol, 4-fluoro-

$C_6H_5FO_2S$
2251 - Benzenesulfonyl fluoride

$C_6H_5F_2N$
700 - Benzenamine, 2,4-difluoro-

$C_6H_5F_3Si$
11486 - Silane, trifluorophenyl-

$C_6H_5F_7O_2$
3524 - Butanoic acid, heptafluoro-, ethyl ester

C_6H_5I
1766 - Benzene, iodo-

C_6H_5IO
9371 - Phenol, 2-iodo-
9372 - Phenol, 3-iodo-
9373 - Phenol, 4-iodo-

$C_6H_5IO_2$
1523 - 1,2-Benzenediol, 4-iodo-
1524 - 1,3-Benzenediol, 5-iodo-

$C_6H_5I_2N$
701 - Benzenamine, 2,4-diiodo-

C_6H_5NO
2059 - Benzene, nitroso-
6031 - 2-Furanacetonitrile
10887 - 2-Pyridinecarboxaldehyde
10888 - 3-Pyridinecarboxaldehyde
10889 - 4-Pyridinecarboxaldehyde

C_6H_5NOS
844 - Benzenamine, N-sulfinyl-

$C_6H_5NO_2$
2049 - Benzene, nitro-
9438 - Phenol, 4-nitroso-
10893 - 2-Pyridinecarboxylic acid
10894 - 3-Pyridinecarboxylic acid
10895 - 4-Pyridinecarboxylic acid

$C_6H_5NO_3$
9435 - Phenol, 2-nitro-
9436 - Phenol, 3-nitro-
9437 - Phenol, 4-nitro-
10899 - 3-Pyridinecarboxylic acid, 1,6-dihydro-6-oxo-
10909 - 3-Pyridinecarboxylic acid, 1-oxide

$C_6H_5NO_4$
1541 - 1,3-Benzenediol, 4-nitro-
10900 - 4-Pyridinecarboxylic acid, 1,2-dihydro-6-hydroxy-2-oxo-

C_6H_5NS
11725 - 2-Thiopheneacetonitrile

$C_6H_5N_3$
976 - Benzene, azido-
2871 - 1H-Benzotriazole

$C_6H_5N_3O_4$
724 - Benzenamine, 2,3-dinitro-
725 - Benzenamine, 2,4-dinitro-
726 - Benzenamine, 2,6-dinitro-
727 - Benzenamine, 3,5-dinitro-

$C_6H_5N_3O_5$
9207 - Phenol, 2-amino-4,6-dinitro-

$C_6H_5N_5O_2$
10749 - 4,6-Pteridinedione, 2-amino-1,5-dihydro-

C_6H_6
867 - Benzene
4767 - 1,3-Cyclopentadiene, 5-methylene-
6712 - 1,3-Hexadien-5-yne
6713 - 1,5-Hexadien-3-yne
6715 - 1,4-Hexadiyne
6716 - 1,5-Hexadiyne
6717 - 2,4-Hexadiyne

$C_6H_6AsNO_5$
459 - Arsonic acid, (4-nitrophenyl)-

C_6H_6BrN
638 - Benzenamine, 2-bromo-
639 - Benzenamine, 3-bromo-
640 - Benzenamine, 4-bromo-

$C_6H_6Br_2N_2$
8768 - Pentanedinitrile, 2-bromo-2-(bromomethyl)-

C_6H_6ClN
658 - Benzenamine, 2-chloro-
659 - Benzenamine, 3-chloro-
660 - Benzenamine, 4-chloro-
10919 - Pyridine, 2-chloro-4-methyl-
10920 - Pyridine, 2-chloro-6-methyl-

C_6H_6ClNO
9204 - Phenol, 2-amino-4-chloro-
10918 - Pyridine, 2-chloro-6-methoxy-

$C_6H_6ClN_2O_4$
1559 - 1,3-Benzenedisulfonamide, 4,5-dichloro-

$C_6H_6Cl_2Si$
11403 - Silane, dichlorophenyl-

$C_6H_6Cl_4$
6697 - 2,4-Hexadiene, 1,3,4,6-tetrachloro-

$C_6H_6Cl_6$
4406 - Cyclohexane, 1,2,3,4,5,6-hexachloro-, (1α,2β,3α,4β,5α,6β)-
4407 - Cyclohexane, 1,2,3,4,5,6-hexachloro-, (1α,2α,3α,4β,5α,6β)-
4408 - Cyclohexane, 1,2,3,4,5,6-hexachloro-, (1α,2α,3β,4α,5β,6β)-(±)-
4409 - Cyclohexane, 1,2,3,4,5,6-hexachloro-, (1α,2α,3β,4α,5α,6β)-
7035 - 3-Hexene, 1,2,3,4,5,6-hexachloro-

C_6H_6FN
752 - Benzenamine, 2-fluoro-
753 - Benzenamine, 3-fluoro-
754 - Benzenamine, 4-fluoro-

$C_6H_6FNO_2S$
2252 - Benzenesulfonyl fluoride, 4-amino-

C_6H_6IN
760 - Benzenamine, 3-iodo-
761 - Benzenamine, 4-iodo-

$C_6H_6N_2O$
828 - Benzenamine, 4-nitroso-
9914 - Propanedinitrile, (ethoxymethylene)-
10890 - 3-Pyridinecarboxamide

$C_6H_6N_2O_2$
822 - Benzenamine, 2-nitro-
823 - Benzenamine, 3-nitro-
824 - Benzenamine, 4-nitro-
825 - Benzenamine, N-nitro-
10897 - 3-Pyridinecarboxylic acid, 6-amino-

$C_6H_6N_2O_3$
9214 - Phenol, 2-amino-3-nitro-

$C_6H_6N_4O_3$
10765 - 1H-Purine-2,6,8(3H)-trione, 4,9-dihydro-3-methyl-
10766 - 1H-Purine-2,6,8(3H)-trione, 7,9-dihydro-7-methyl-

$C_6H_6N_4O_4$
7144 - Hydrazinecarboxamide, 2-[(5-nitro-2-furanyl)methylene]-
7156 - Hydrazine, (2,4-dinitrophenyl)-

C_6H_6O
6102 - Furan, 2-ethenyl-
9199 - Phenol

C_6H_6OS
5870 - Ethanone, 1-(2-thienyl)-
9376 - Phenol, 2-mercapto-

9377 - Phenol, 3-mercapto-
9378 - Phenol, 4-mercapto-
11736 - 2-Thiophenecarboxaldehyde, 5-methyl-

$C_6H_6O_2$
1497 - 1,2-Benzenediol
1498 - 1,3-Benzenediol
1499 - 1,4-Benzenediol
5817 - Ethanone, 1-(2-furanyl)-
6045 - 2-Furancarboxaldehyde, 5-methyl-

$C_6H_6O_2S$
2180 - Benzenesulfinic acid
5835 - Ethanone, 1-(3-hydroxy-2-thienyl)-

$C_6H_6O_2Se$
2174 - Benzeneseleninic acid

$C_6H_6O_3$
2366 - 1,2,3-Benzenetriol
2367 - 1,2,4-Benzenetriol
2368 - 1,3,5-Benzenetriol
6026 - 2-Furanacetic acid
6044 - 2-Furancarboxaldehyde, 5-(hydroxymethyl)-
6059 - 2-Furancarboxylic acid, methyl ester
6060 - 2-Furancarboxylic acid, 5-methyl-
6063 - 3-Furancarboxylic acid, methyl ester
6065 - 3-Furancarboxylic acid, 5-methyl-
6097 - 2,5-Furandione, 3,4-dimethyl-
10796 - 4H-Pyran-4-one, 3-hydroxy-2-methyl-

$C_6H_6O_3S$
2214 - Benzenesulfonic acid

$C_6H_6O_4$
3910 - 2-Butynedioic acid, dimethyl ester
10795 - 4H-Pyran-4-one, 5-hydroxy-2-(hydroxymethyl)-

$C_6H_6O_4S$
2225 - Benzenesulfonic acid, 2-hydroxy-
2226 - Benzenesulfonic acid, 4-hydroxy-

$C_6H_6O_7S_2$
1560 - 1,3-Benzenedisulfonic acid, 4-hydroxy-

C_6H_6S
2289 - Benzenethiol

$C_6H_6S_2$
1561 - 1,2-Benzenedithiol
1562 - 1,3-Benzenedithiol

C_6H_6Se
2177 - Benzeneselenol

$C_6H_7AsO_3$
460 - Arsonic acid, phenyl-

$C_6H_7BO_2$
3135 - Boronic acid, phenyl-

$C_6H_7BrN_2$
7141 - Hydrazine, (4-bromophenyl)-

$C_6H_7ClN_2$
1264 - 1,2-Benzenediamine, 4-chloro-

C_6H_7ClO
6711 - 2,4-Hexadienoyl chloride, (E,E)-

$C_6H_7ClO_4$
3773 - 2-Butenedioic acid, 2-chloro-, dimethyl ester, (E)-

C_6H_7ClSi
11384 - Silane, chlorophenyl-

$C_6H_7Cl_2F_3O_2$
10721 - Propionic acid, 2,2-dichloro-3,3,3-trifluoro-, propyl ester

$C_6H_7Cl_2N$
665 - Benzenamine, 2-chloro-, hydrochloride

$C_6H_7F_3O_3$
3592 - Butanoic acid, 4,4,4-trifluoro-3-oxo-, ethyl ester

$C_6H_7KO_2$
6704 - 2,4-Hexadienoic acid, potassium salt, (E,E)-

C_6H_7N
633 - Benzenamine
4907 - 1-Cyclopentenecarbonitrile
10980 - Pyridine, 2-methyl-
10981 - Pyridine, 3-methyl-
10982 - Pyridine, 4-methyl-

C_6H_7NO
757 - Benzenamine, N-hydroxy-

5820 - Ethanone, 1-(1H-pyrrol-2-yl)-
9200 - Phenol, amino-
9201 - Phenol, 2-amino-
9202 - Phenol, 3-amino-
9203 - Phenol, 4-amino-
10975 - 2-Pyridinemethanol
10976 - 3-Pyridinemethanol
10977 - 4-Pyridinemethanol
10988 - Pyridine, 3-methyl-, 1-oxide
11035 - 2(1H)-Pyridinone, 1-methyl-
11036 - 2(1H)-Pyridinone, 3-methyl-
11037 - 2(1H)-Pyridinone, 4-methyl-

$C_6H_7NO_2$
11122 - 1H-Pyrrole-2,5-dione, 1-ethyl-

$C_6H_7NO_3S$
2215 - Benzenesulfonic acid, 2-amino-
2216 - Benzenesulfonic acid, 3-amino-
2217 - Benzenesulfonic acid, 4-amino-

$C_6H_7NO_4$
6129 - Furan, 2-(methoxymethyl)-5-nitro-

C_6H_7NS
2290 - Benzenethiol, 2-amino-
2291 - Benzenethiol, 4-amino-

$C_6H_7N_3O$
10905 - 4-Pyridinecarboxylic acid, hydrazide

$C_6H_7N_3O_2$
1281 - 1,2-Benzenediamine, 4-nitro-
7168 - Hydrazine, (4-nitrophenyl)-

$C_6H_7N_5O$
10769 - 6H-Purin-6-one, 2-amino-1,7-dihydro-7-
      methyl

$C_6H_7O_3Sb$
11548 - Stibine oxide, dihydroxyphenyl-

C_6H_7P
9519 - Phosphine, phenyl-

C_6H_8
4172 - Cyclobutane, 1,2-bis(methylene)-
4248 - 1,3-Cyclohexadiene
4249 - 1,4-Cyclohexadiene
6992 - 1,3,5-Hexatriene, (E)-
6993 - 1,3,5-Hexatriene, (Z)-
7103 - 1-Hexen-3-yne
7104 - 1-Hexen-4-yne
7105 - 1-Hexen-5-yne
7106 - 2-Hexen-4-yne
9116 - 1-Penten-3-yne, 2-methyl-
9117 - 3-Penten-1-yne, 3-methyl-

$C_6H_8AsNO_3$
456 - Arsonic acid, (4-aminophenyl)-

C_6H_8ClN
756 - Benzenamine, hydrochloride

C_6H_8ClNS
11670 - Thiazole, 5-(2-chloroethyl)-4-methyl-

$C_6H_8ClN_3O_4S_2$
1558 - 1,3-Benzenedisulfonamide, 4-amino-6-chloro-

$C_6H_8Cl_2$
6690 - 2,4-Hexadiene, 1,3-dichloro-

$C_6H_8Cl_2O_3$
3503 - Butanoic acid, 2,2-dichloro-3-oxo-, ethyl ester

$C_6H_8Cl_2O_4$
129 - Acetic acid, chloro-, 1,2-ethanediyl ester
5554 - Ethanedioic acid, bis(2-chloroethyl) ester

$C_6H_8N_2$
1260 - 1,2-Benzenediamine
1261 - 1,3-Benzenediamine
1262 - 1,4-Benzenediamine
6777 - Hexanedinitrile
7169 - Hydrazine, phenyl-
10805 - Pyrazine, 2,3-dimethyl-
10806 - Pyrazine, 2,5-dimethyl-
10807 - Pyrazine, 2,6-dimethyl-
10866 - 2-Pyridinamine, 3-methyl-
10867 - 2-Pyridinamine, 4-methyl-
10868 - 2-Pyridinamine, 6-methyl-
10869 - 2-Pyridinamine, N-methyl-
10870 - 3-Pyridinamine, 4-methyl-
10971 - 2-Pyridinemethanamine

10972 - 3-Pyridinemethanamine
10973 - 4-Pyridinemethanamine
11055 - Pyrimidine, 2,4-dimethyl-
11056 - Pyrimidine, 2,5-dimethyl-
11057 - Pyrimidine, 4,5-dimethyl-
11058 - Pyrimidine, 4,6-dimethyl-

$C_6H_8N_2O$
9274 - Phenol, 2,4-diamino-
10114 - Propanenitrile, 3,3'-oxybis-
10808 - Pyrazineethanol
10865 - 3-Pyridinamine, 6-methoxy-

$C_6H_8N_2O_2S$
2187 - Benzenesulfonamide, 4-amino-

$C_6H_8N_2O_3$
821 - Benzenamine, nitrate

$C_6H_8N_2O_3S$
2224 - Benzenesulfonic acid, 4-hydrazino-

$C_6H_8N_6O_{18}$
7424 - D-Mannitol, hexanitrate

C_6H_8O
4703 - 2-Cyclohexen-1-one
4704 - 3-Cyclohexen-1-one
4908 - 1-Cyclopentene-1-carboxaldehyde
4944 - 2-Cyclopenten-1-one, 2-methyl-
4945 - 2-Cyclopenten-1-one, 3-methyl-
4946 - 2-Cyclopenten-1-one, 5-methyl-
6088 - Furan, 2,4-dimethyl-
6089 - Furan, 2,5-dimethyl-
6105 - Furan, 2-ethyl-
6675 - 2,4-Hexadienal, (E,E)-
7113 - 3-Hexynal
7130 - 3-Hexyn-2-one
7131 - 5-Hexyn-2-one

$C_6H_8O_2$
3176 - 1,3-Butadien-1-ol, acetate
3918 - 2-Butynoic acid, ethyl ester
4375 - 1,2-Cyclohexanedione
4376 - 1,3-Cyclohexanedione
4377 - 1,4-Cyclohexanedione
4909 - 1-Cyclopentene-1-carboxylic acid
6103 - Furan, 2-ethoxy-
6121 - 2-Furanmethanol, α-methyl-
6122 - 2-Furanmethanol, 5-methyl-
6128 - Furan, 2-(methoxymethyl)-
6698 - 2,4-Hexadienoic acid
7123 - 2-Hexynoic acid
9139 - 2-Pentynoic acid, methyl ester
10694 - 2-Propenoic acid, 2-propenyl ester
10791 - 2H-Pyran-2-one, 5,6-dihydro-6-methyl-, (S)-

$C_6H_8O_3$
6139 - 2(3H)-Furanone, 3-acetyldihydro-
10680 - 2-Propenoic acid, oxiranylmethyl ester

$C_6H_8O_4$
3778 - 2-Butenedioic acid (E)-, dimethyl ester
3779 - 2-Butenedioic acid (Z)-, dimethyl ester
3791 - 2-Butenedioic acid (E)-, monoethyl ester
4182 - 1,3-Cyclobutanedicarboxylic acid, cis-
4185 - 1,3-Cyclobutanedicarboxylic acid, trans-
5190 - 1,3-Dioxane-4,6-dione, 2,2-dimethyl-
9965 - Propanedioic acid, (1-methylethylidene)-
9979 - Propanedioic acid, 2-propenyl-

$C_6H_8O_5$
467 - L-Ascorbic acid, 6-deoxy-

$C_6H_8O_6$
465 - DL-Ascorbic acid
466 - L-Ascorbic acid
6239 - D-Glucuronic acid, γ-lactone
10165 - 1,2,3-Propanetricarboxylic acid

$C_6H_8O_7$
10167 - 1,2,3-Propanetricarboxylic acid, 2-hydroxy-

C_6H_8S
11754 - Thiophene, 2,3-dimethyl-
11755 - Thiophene, 2,4-dimethyl-
11756 - Thiophene, 2,5-dimethyl-
11757 - Thiophene, 3,4-dimethyl-
11759 - Thiophene, 2-ethyl-
11760 - Thiophene, 3-ethyl-

C_6H_8Si
11446 - Silane, phenyl-

C_6H_9Br
4627 - Cyclohexene, 1-bromo-
4628 - Cyclohexene, 3-bromo-

C_6H_9BrO
4574 - Cyclohexanone, 2-bromo-

$C_6H_9BrO_2$
3819 - 2-Butenoic acid, 4-bromo-, ethyl ester, (E)-

$C_6H_9BrO_3$
3477 - Butanoic acid, 2-bromo-3-oxo-, ethyl ester
3478 - Butanoic acid, 4-bromo-3-oxo-, ethyl ester

C_6H_9Cl
4642 - Cyclohexene, 1-chloro-
6688 - 1,3-Hexadiene, 3-chloro-
8664 - 1,2-Pentadiene, 1-chloro-3-methyl-
9126 - 1-Pentyne, 3-chloro-3-methyl-
9127 - 2-Pentyne, 4-chloro-4-methyl-

$C_6H_9ClN_2$
7172 - Hydrazine, phenyl-, monohydrochloride

C_6H_9ClO
4577 - Cyclohexanone, 2-chloro-
4578 - Cyclohexanone, 4-chloro-

$C_6H_9ClO_2$
3823 - 2-Butenoic acid, 2-chloro-, ethyl ester
3824 - 2-Butenoic acid, 2-chloro-, ethyl ester, (E)-
3825 - 2-Butenoic acid, 2-chloro-, ethyl ester, (Z)-
3826 - 2-Butenoic acid, 3-chloro-, ethyl ester
3827 - 2-Butenoic acid, 3-chloro-, ethyl ester, (E)-
3828 - 2-Butenoic acid, 3-chloro-, ethyl ester, (Z)-
3829 - 2-Butenoic acid, 4-chloro-, ethyl ester

$C_6H_9ClO_3$
3494 - Butanoic acid, 4-chloro-3-oxo-, ethyl ester
3570 - Butanoic acid, 3-oxo-, 2-chloroethyl ester

$C_6H_9Cl_3$
7048 - 1-Hexene, 1,1,2-trichloro-

$C_6H_9Cl_3O_2$
245 - Acetic acid, trichloro-, butyl ester
247 - Acetic acid, trichloro-, 1,1-dimethylethyl ester
251 - Acetic acid, trichloro-, 1-methylpropyl ester
252 - Acetic acid, trichloro-, 2-methylpropyl ester
3591 - Butanoic acid, 2,2,3-trichloro-, ethyl ester

$C_6H_9Cl_3O_3$
13 - Acetaldehyde, chloro-, trimer

$C_6H_9F_3O_2$
265 - Acetic acid, trifluoro-, butyl ester
266 - Acetic acid, trifluoro-, 1,1-dimethylethyl ester

C_6H_9N
4779 - Cyclopentanecarbonitrile
11119 - 1H-Pyrrole, 2,4-dimethyl-
11120 - 1H-Pyrrole, 2,5-dimethyl-
11124 - 1H-Pyrrole, 1-ethyl-
11125 - 1H-Pyrrole, 2-ethyl-

C_6H_9NO
6115 - 2-Furanmethanamine, N-methyl-
11183 - 2-Pyrrolidinone, 1-ethenyl-

C_6H_9NOS
11677 - 5-Thiazoleethanol, 4-methyl-

$C_6H_9NO_2$
10912 - 3-Pyridinecarboxylic acid, 1,2,5,6-tetrahydro-
11157 - 2,5-Pyrrolidinedione, 1-ethyl-

$C_6H_9NO_3$
50 - Acetamide, N,N-diacetyl-
8587 - 2,4-Oxazolidinedione, 3,5,5-trimethyl-

$C_6H_9NO_6$
6256 - Glycine, N,N-bis(carboxymethyl)-

$C_6H_9N_3$
2305 - 1,2,3-Benzenetriamine
2306 - 1,2,4-Benzenetriamine
11044 - 2-Pyrimidinamine, 4,5-dimethyl-
11045 - 4-Pyrimidinamine, 2,6-dimethyl-

$C_6H_9N_3O_2$
759 - Benzenamine, N-hydroxy-N-nitroso-,
     ammonium salt
7132 - L-Histidine

$C_6H_9N_3O_3$
7206 - 1H-Imidazole-1-ethanol, 2-methyl-5-nitro-

11841 - 1,3,5-Triazine-2,4,6(1H,3H,5H)-trione, 1,3,5-trimethyl-

C₆H₁₀

3159 - 1,3-Butadiene, 2,3-dimethyl-
3907 - 1-Butyne, 3,3-dimethyl-
4626 - Cyclohexene
4827 - Cyclopentane, methylene-
4928 - Cyclopentene, 1-methyl-
4929 - Cyclopentene, 3-methyl-
4930 - Cyclopentene, 4-methyl-
4982 - Cyclopropane, (1-methylethenyl)-
6676 - 1,2-Hexadiene
6677 - 1,4-Hexadiene
6678 - 1,5-Hexadiene
6679 - 2,3-Hexadiene
6680 - 2,4-Hexadiene
6681 - 1,3-Hexadiene, (E)-
6682 - 1,3-Hexadiene, (Z)-
6683 - 1,4-Hexadiene, (E)-
6684 - 1,4-Hexadiene, (Z)-
6685 - 2,4-Hexadiene, (E,E)-
6686 - 2,4-Hexadiene, (E,Z)-
6687 - 2,4-Hexadiene, (Z,Z)-
7114 - 1-Hexyne
7115 - 2-Hexyne
7116 - 3-Hexyne
8666 - 1,2-Pentadiene, 3-methyl-
8667 - 1,2-Pentadiene, 4-methyl-
8668 - 1,3-Pentadiene, 2-methyl-, (Z)-
8669 - 1,3-Pentadiene, 3-methyl-
8670 - 1,3-Pentadiene, 4-methyl-
8671 - 1,4-Pentadiene, 2-methyl-
8672 - 2,3-Pentadiene, 2-methyl-
9060 - 1-Pentene, 3-methylene-
9134 - 1-Pentyne, 3-methyl-
9135 - 1-Pentyne, 4-methyl-
9136 - 2-Pentyne, 4-methyl-

C₆H₁₀Br₂

4345 - Cyclohexane, 1,1-dibromo-
4346 - Cyclohexane, 1,2-dibromo-, cis-
4347 - Cyclohexane, 1,2-dibromo-, trans-(±)-
4348 - Cyclohexane, 1,3-dibromo-, trans-
4349 - Cyclohexane, 1,4-dibromo-, cis-

C₆H₁₀Br₂O₂

3498 - Butanoic acid, 2,3-dibromo-, ethyl ester
3499 - Butanoic acid, 2,4-dibromo-, ethyl ester
10243 - Propanoic acid, 2,3-dibromo-, propyl ester

C₆H₁₀ClFO₂

132 - Acetic acid, chlorofluoro-, butyl ester

C₆H₁₀Cl₂

4358 - Cyclohexane, 1,1-dichloro-
4359 - Cyclohexane, 1,2-dichloro-, cis-
4360 - Cyclohexane, 1,2-dichloro-, trans-(±)-
4361 - Cyclohexane, 1,4-dichloro-, cis-
7011 - 1-Hexene, 1,2-dichloro-, (E)-
7012 - 1-Hexene, 1,2-dichloro-, (Z)-

C₆H₁₀Cl₂O₂

163 - Acetic acid, dichloro-, butyl ester
10249 - Propanoic acid, 2,3-dichloro-, 1-methylethyl ester

C₆H₁₀Cl₃NO

96 - Acetamide, 2,2,2-trichloro-N,N-diethyl-

C₆H₁₀N₂

8830 - Pentanenitrile, 3-imino-2-methyl-
10833 - 1H-Pyrazole, 1-ethyl-3-methyl-
10839 - 1H-Pyrazole, 1,3,4-trimethyl-
10840 - 1H-Pyrazole, 1,3,5-trimethyl-
10841 - 1H-Pyrazole, 1,4,5-trimethyl-
10842 - 1H-Pyrazole, 3,4,5-trimethyl-

C₆H₁₀N₂O₂

4379 - 1,2-Cyclohexanedione, dioxime

C₆H₁₀N₄

11654 - 5H-Tetrazolo[1,5-a]azepine, 6,7,8,9-tetrahydro-

C₆H₁₀N₄O₃S₂

3197 - Butanamide, N-[5-(aminosulfonyl)-1,3,4-thiadiazol-2-yl]-

C₆H₁₀O

3160 - 1,3-Butadiene, 1-ethoxy-
3161 - 1,3-Butadiene, 2-ethoxy-
3377 - Butane, 1-(ethynyloxy)-
3877 - 1-Buten-1-one, 2-ethyl-
4571 - Cyclohexanone
4694 - 2-Cyclohexen-1-ol
4695 - 3-Cyclohexen-1-ol
4780 - Cyclopentanecarboxaldehyde
4885 - Cyclopentanone, 2-methyl-
4886 - Cyclopentanone, 3-methyl-, (±)-
5792 - Ethanone, 1-cyclobutyl-
6705 - 1,4-Hexadien-3-ol
6706 - 1,5-Hexadien-3-ol
6707 - 2,4-Hexadien-1-ol
6708 - 3,5-Hexadien-2-ol
6709 - 2,4-Hexadien-1-ol, (E,E)-
6996 - 2-Hexenal, (E)-
6997 - 3-Hexenal, (E)-
6998 - 3-Hexenal, (Z)-
7086 - 3-Hexen-2-one
7087 - 4-Hexen-2-one
7088 - 4-Hexen-3-one
7089 - 5-Hexen-2-one
7090 - 3-Hexen-2-one, (E)-
7124 - 1-Hexyn-3-ol
7125 - 2-Hexyn-1-ol
7126 - 3-Hexyn-1-ol
7127 - 5-Hexyn-2-ol
8554 - 7-Oxabicyclo[4.1.0]heptane
9002 - 2-Pentenal, 2-methyl-
9107 - 1-Penten-3-one, 2-methyl-
9108 - 3-Penten-2-one, 4-methyl-
9109 - 4-Penten-2-one, 4-methyl-
9146 - 1-Pentyn-3-ol, 3-methyl-
10592 - 1-Propene, 3,3'-oxybis-
10778 - 2H-Pyran, 3,6-dihydro-4-methyl-

C₆H₁₀OS

10596 - 1-Propene, 3,3'-sulfinylbis-

C₆H₁₀OS₂

10595 - 2-Propene-1-sulfinothioic acid, S-2-propenyl ester

C₆H₁₀O₂

3513 - Butanoic acid, ethenyl ester
3835 - 3-Butenoic acid, 2,2-dimethyl-
3837 - 2-Butenoic acid, 2-ethyl-, (E)-
3838 - 2-Butenoic acid, ethyl ester
3839 - 2-Butenoic acid, ethyl ester, (E)-
3840 - 2-Butenoic acid, ethyl ester, (Z)-
3841 - 3-Butenoic acid, ethyl ester
3851 - 2-Butenoic acid, 2-methyl-, methyl ester, (E)-
3852 - 2-Butenoic acid, 3-methyl-, methyl ester
3862 - 2-Buten-1-ol, acetate
4176 - Cyclobutanecarboxylic acid, methyl ester
4781 - Cyclopentanecarboxylic acid
4961 - Cyclopropanecarboxylic acid, ethyl ester
6150 - 2(3H)-Furanone, 5-ethyldihydro-
6693 - 1,5-Hexadiene-3,4-diol, (R*,R*)-(±)-
6748 - Hexanedial
6804 - 2,4-Hexanedione
6805 - 2,5-Hexanedione
6806 - 3,4-Hexanedione
7049 - 2-Hexenoic acid
7050 - 3-Hexenoic acid
7051 - 5-Hexenoic acid
7118 - 3-Hexyne-2,5-diol
8591 - 2-Oxepanone
8626 - Oxirane, [(2-propenyloxy)methyl]-
8890 - Pentanoic acid, 2-methylene-
9076 - 2-Pentenoic acid, 2-methyl-, (E)-
9077 - 2-Pentenoic acid, 3-methyl-, (Z)-
9078 - 2-Pentenoic acid, 4-methyl-
9079 - 3-Pentenoic acid, 2-methyl-
10297 - Propanoic acid, 2-methyl-, ethenyl ester
10325 - Propanoic acid, 2-propenyl ester
10658 - 2-Propenoic acid, 2-methyl-, ethyl ester

C₆H₁₀O₂S

3589 - Butanoic acid, 3-thioxo-, ethyl ester
10597 - 1-Propene, 3,3'-sulfonylbis-

C₆H₁₀O₂S₂

5453 - Ethanebis(thioic) acid, S,S-diethyl ester

C₆H₁₀O₃

3510 - Butanoic acid, 3,3-dimethyl-2-oxo-
3558 - Butanoic acid, 2-methyl-3-oxo-, methyl ester
3571 - Butanoic acid, 3-oxo-, ethyl ester
3572 - Butanoic acid, 4-oxo-, ethyl ester
3843 - 3-Butenoic acid, 2-hydroxy-, ethyl ester
5206 - 1,3-Dioxolane-4-carboxaldehyde, 2,2-dimethyl-
6087 - Furan, 2,5-dihydro-2,5-dimethoxy-
6143 - 2(3H)-Furanone, dihydro-3-hydroxy-4,4-dimethyl-, (R)-
6808 - 2,5-Hexanedione, 3-hydroxy-
6905 - Hexanoic acid, 5-oxo-
8895 - Pentanoic acid, 2-methyl-4-oxo-
8896 - Pentanoic acid, 3-methyl-4-oxo-
8908 - Pentanoic acid, 4-oxo-, methyl ester
10203 - Propanoic acid, anhydride
10281 - Propanoic acid, 2-hydroxy-, 2-propenyl ester
10661 - 2-Propenoic acid, 2-methyl-, 2-hydroxyethyl ester
10774 - 2H-Pyran-4-carboxylic acid, tetrahydro-

C₆H₁₀O₄

102 - Acetic acid, (acetyloxy)-, ethyl ester
3317 - Butanedioic acid, 2,3-dimethyl-
3318 - Butanedioic acid, dimethyl ester
3322 - Butanedioic acid, ethyl-, (R)-
3335 - Butanedioic acid, monoethyl ester
5560 - Ethanedioic acid, diethyl ester
5570 - 1,1-Ethanediol, diacetate
5571 - 1,2-Ethanediol, diacetate
6779 - Hexanedioic acid
7423 - D-Mannitol, 1,4:3,6-dianhydro-
8779 - Pentanedioic acid, 3-methyl-
8780 - Pentanedioic acid, monomethyl ester
9963 - Propanedioic acid, methyl-, dimethyl ester

C₆H₁₀O₄S

10329 - Propanoic acid, 3,3'-thiobis-

C₆H₁₀O₄S₂

5179 - Dimethipin

C₆H₁₀O₅

5169 - Dicarbonic acid, diethyl ester

(C₆H₁₀O₅)ₓ

6274 - Glycogen

C₆H₁₀O₆

3308 - Butanedioic acid, 2,3-dihydroxy-, dimethyl ester, (R*,R*)-(±)-
3315 - Butanedioic acid, 2,3-dihydroxy- [R-(R*,R*)]-, monoethyl ester
6211 - Gluconic acid, β-lactone
6212 - D-Gluconic acid, δ-lactone

C₆H₁₀O₇

6199 - D-Galacturonic acid
6238 - D-Glucuronic acid

C₆H₁₀O₈

6193 - Galactaric acid
6205 - D-Glucaric acid

C₆H₁₀S

4481 - Cyclohexanethione
10599 - 1-Propene, 3,3'-thiobis-

C₆H₁₀S₃

11917 - Trisulfide, di-2-propenyl

C₆H₁₁Br

4315 - Cyclohexane, bromo-

C₆H₁₁BrN₂O₂

3196 - Butanamide, N-(aminocarbonyl)-2-bromo-3-methyl-

C₆H₁₁BrO

4503 - Cyclohexanol, 2-bromo-
6961 - 2-Hexanone, 6-bromo-

C₆H₁₁BrO₂

114 - Acetic acid, bromo-, 1,1-dimethylethyl ester
118 - Acetic acid, bromo-, 2-methylpropyl ester

3469 - Butanoic acid, 2-bromo-, ethyl ester
3470 - Butanoic acid, 4-bromo-, ethyl ester
3476 - Butanoic acid, 2-bromo-3-methyl-, methyl ester
6875 - Hexanoic acid, 2-bromo-, (±)-
6876 - Hexanoic acid, 6-bromo-
10213 - Propanoic acid, 2-bromo-2-methyl-, ethyl ester

$C_6H_{11}Cl$
3731 - 2-Butene, 1-chloro-2,3-dimethyl-
4339 - Cyclohexane, chloro-
7004 - 1-Hexene, 1-chloro-
7005 - 1-Hexene, 2-chloro-
7006 - 1-Hexene, 5-chloro-
7007 - 2-Hexene, 4-chloro-
7008 - 3-Hexene, 1-chloro-
7009 - 3-Hexene, 3-chloro-, (Z)-
9025 - 1-Pentene, 3-chloro-2-methyl-
9026 - 2-Pentene, 5-chloro-2-methyl-

$C_6H_{11}ClO$
3688 - Butanoyl chloride, 2,2-dimethyl-
3689 - Butanoyl chloride, 2,3-dimethyl-
3690 - Butanoyl chloride, 3,3-dimethyl-
3691 - Butanoyl chloride, 2-ethyl-
4505 - Cyclohexanol, 2-chloro-, trans-
4506 - Cyclohexanol, 2-chloro-, cis-(±)-
4507 - Cyclohexanol, 4-chloro-, trans-
6986 - Hexanoyl chloride
8993 - Pentanoyl chloride, 2-methyl-, (±)-
8994 - Pentanoyl chloride, 3-methyl-, (±)-

$C_6H_{11}ClO_2$
125 - Acetic acid, chloro-, butyl ester
128 - Acetic acid, chloro-, 1,1-dimethylethyl ester
142 - Acetic acid, chloro-, 1-methylpropyl ester
143 - Acetic acid, chloro-, 2-methylpropyl ester
3483 - Butanoic acid, 2-chloro-, ethyl ester
3484 - Butanoic acid, 3-chloro-, ethyl ester
3485 - Butanoic acid, 4-chloro-, ethyl ester
3730 - 2-Butene, 2-chloro-1,1-dimethoxy-
4021 - Carbonochloridic acid, 3-methylbutyl ester
4025 - Carbonochloridic acid, pentyl ester
10223 - Propanoic acid, 2-chloro-, 1-methylethyl ester
10232 - Propanoic acid, 3-chloro-, propyl ester
10292 - Propanoic acid, 2-methyl-, 2-chloroethyl ester

$C_6H_{11}Cl_3O_2$
5657 - Ethane, 1,1,1-trichloro-2,2-diethoxy-

$C_6H_{11}F$
4403 - Cyclohexane, fluoro-

$C_6H_{11}FO_2$
190 - Acetic acid, fluoro-, 1,1-dimethylethyl ester
6888 - Hexanoic acid, 6-fluoro-

$C_6H_{11}I$
4415 - Cyclohexane, iodo-

$C_6H_{11}IO_2$
3928 - Butyric acid, 2-iodo-, ethyl ester

$C_6H_{11}N$
478 - 7-Azabicyclo[4.1.0]heptane
3407 - Butanenitrile, 2-ethyl-
6840 - Hexanenitrile
8821 - Pentane, 1-isocyano-
8831 - Pentanenitrile, 4-methyl-
10526 - 2-Propen-1-amine, N-2-propenyl-

$C_6H_{11}NO$
493 - 2H-Azepin-2-one, hexahydro-
3386 - Butane, 1-isocyanato-3-methyl-
4610 - Cyclohexanone, oxime
9638 - 1-Piperidinecarboxaldehyde
9721 - 2-Piperidinone, 1-methyl-
9722 - 2-Piperidinone, 3-methyl-
9723 - 3-Piperidinone, 1-methyl-
9724 - 4-Piperidinone, 1-methyl-
11181 - 2-Pyrrolidinone, 1,5-dimethyl-
11182 - 2-Pyrrolidinone, 3,3-dimethyl-

$C_6H_{11}NO_2$
24 - Acetamide, N-acetyl-N-ethyl-

3816 - 2-Butenoic acid, 3-amino-, ethyl ester, (E)-
4454 - Cyclohexane, nitro-
4782 - Cyclopentanecarboxylic acid, 1-amino-
4830 - Cyclopentane, 1-methyl-1-nitro-
4831 - Cyclopentane, 1-methyl-2-nitro-
4840 - Cyclopentane, (nitromethyl)-
7682 - Morpholine, 4-acetyl-
9640 - 2-Piperidinecarboxylic acid
9641 - 3-Piperidinecarboxylic acid
9642 - 4-Piperidinecarboxylic acid
11184 - 2-Pyrrolidinone, 1-(2-hydroxyethyl)-

$C_6H_{11}NO_3S$
333 - L-Alanine, 3-(2-propenylsulfinyl)-, (S)-

$C_6H_{11}NO_4$
7231 - Imidodicarbonic acid, diethyl ester

$C_6H_{11}NS$
3390 - Butane, 1-isothiocyanato-3-methyl-
11673 - Thiazole, 4,5-dihydro-2,4,4-trimethyl-

$C_6H_{11}N_2O_4PS_3$
7645 - Methidathion

$C_6H_{11}N_3$
11847 - 1H-1,2,4-Triazole, 1-butyl

$C_6H_{11}NaO_7$
6213 - D-Gluconic acid, monosodium salt

$C_6H_{11}O_3P$
11909 - 2,6,7-Trioxa-1-phosphabicyclo[2.2.2]octane, 4-ethyl-

C_6H_{12}
3759 - 1-Butene, 2,3-dimethyl-
3760 - 1-Butene, 3,3-dimethyl-
3761 - 2-Butene, 2,3-dimethyl-
4188 - Cyclobutane, 1,2-dimethyl-, trans-
4189 - Cyclobutane, 1,3-dimethyl-, cis-
4190 - Cyclobutane, 1,3-dimethyl-, trans-
4191 - Cyclobutane, ethyl-
4305 - Cyclohexane
4826 - Cyclopentane, methyl-
4975 - Cyclopropane, 1-ethyl-1-methyl-
4976 - Cyclopropane, 1-ethyl-2-methyl-, cis-
4977 - Cyclopropane, 1-ethyl-2-methyl-, trans-
4983 - Cyclopropane, (1-methylethyl)-
4984 - Cyclopropane, propyl-
4985 - Cyclopropane, 1,1,2-trimethyl-
4986 - Cyclopropane, 1,2,3-trimethyl-, (1α,2α,3α)-
4987 - Cyclopropane, 1,2,3-trimethyl-, (1α,2α,3β)-
6999 - 1-Hexene
7000 - 2-Hexene, (E)-
7001 - 2-Hexene, (Z)-
7002 - 3-Hexene, (E)-
7003 - 3-Hexene, (Z)-
8825 - Pentane, 3-methylene-
9052 - 1-Pentene, 2-methyl-
9053 - 1-Pentene, 3-methyl-
9054 - 1-Pentene, 4-methyl-
9055 - 2-Pentene, 2-methyl-
9056 - 2-Pentene, 3-methyl-, (E)-
9057 - 2-Pentene, 3-methyl-, (Z)-
9058 - 2-Pentene, 4-methyl-, (E)-
9059 - 2-Pentene, 4-methyl-, (Z)-

$C_6H_{12}BrF$
6735 - Hexane, 1-bromo-6-fluoro-

$C_6H_{12}BrNO$
3198 - Butanamide, 2-bromo-2-ethyl-

$C_6H_{12}Br_2$
3269 - Butane, 1,2-dibromo-2,3-dimethyl-
3270 - Butane, 1,4-dibromo-2,3-dimethyl-
3271 - Butane, 2,3-dibromo-2,3-dimethyl-
6754 - Hexane, 1,2-dibromo-
6755 - Hexane, 1,4-dibromo-
6756 - Hexane, 1,5-dibromo-
6757 - Hexane, 1,6-dibromo-
6758 - Hexane, 2,3-dibromo-
6759 - Hexane, 2,5-dibromo-, (R*,R*)-(±)-
6760 - Hexane, 3,4-dibromo-

$C_6H_{12}ClF$
6742 - Hexane, 1-chloro-6-fluoro-

$C_6H_{12}ClNO$
47 - Acetamide, 2-chloro-N-(1-methylpropyl)-

$C_6H_{12}ClNO_2$
46 - Acetamide, 2-chloro-N-(3-methoxypropyl)-

$C_6H_{12}Cl_2$
6761 - Hexane, 1,2-dichloro-
6762 - Hexane, 1,6-dichloro-
6763 - Hexane, 2,2-dichloro-
6764 - Hexane, 2,3-dichloro-
6766 - Hexane, 2,5-dichloro-, (R*,R*)-(±)-

$C_6H_{12}Cl_2O$
10121 - Propane, 1,1'-oxybis[2-chloro-
10122 - Propane, 1,1'-oxybis[3-chloro-
10123 - Propane, 2,2'-oxybis[1-chloro-

$C_6H_{12}Cl_2O_2$
5449 - Ethane, 1,2-bis(2-chloroethoxy)-
5530 - Ethane, 1,1-dichloro-2,2-diethoxy-
5592 - Ethane, 1,1'-[ethylidenebis(oxy)]bis[2-chloro-
9896 - Propane, 1,3-dichloro-2-(ethoxymethoxy)-

$C_6H_{12}Cl_2S$
10153 - Propane, 1,1'-thiobis[3-chloro-

$C_6H_{12}Cl_3N$
5412 - Ethanamine, 2-chloro-N,N-bis(2-chloroethyl)-

$C_6H_{12}Cl_3O_3P$
5702 - Ethanol, 2-chloro-, phosphite (3:1)

$C_6H_{12}Cl_3O_4P$
5701 - Ethanol, 2-chloro-, phosphate (3:1)

$C_6H_{12}I_2$
6769 - Hexane, 1,6-diiodo-

$C_6H_{12}MnN_2S_4$
7421 - Manganous dimethyldithiocarbamate

$C_6H_{12}N_2$
273 - Acetonitrile, (butylamino)-
280 - Acetonitrile, (diethylamino)-
5111 - 1,4-Diazabicyclo[2.2.2]octane
10463 - 2-Propanone, (1-methylethylidene)hydrazone

$C_6H_{12}N_2O_2$
5494 - Ethanediamide, N,N'-diethyl-

$C_6H_{12}N_2O_3$
3334 - Butanedioic acid, mono (2,2-dimethylhydrazide)

$C_6H_{12}N_2O_4$
3957 - Carbamic acid, (1-methylethyl)nitro-, ethyl ester
3964 - Carbamic acid, nitropropyl-, ethyl ester
6778 - Hexane, 1,6-dinitro-
7150 - 1,1-Hydrazinedicarboxylic acid, diethyl ester
7151 - 1,2-Hydrazinedicarboxylic acid, diethyl ester

$C_6H_{12}N_2O_4S$
5016 - L-Cysteine, S-(2-amino-2-carboxyethyl)-, (R)-

$C_6H_{12}N_2O_4S_2$
5020 - L-Cystine

$C_6H_{12}N_2S_3$
11715 - Thiodicarbonic diamide ([[(H2N)C(S)]2S), tetramethyl-

$C_6H_{12}N_2S_4$
11722 - Thioperoxydicarbonic diamide ([[(H2N)C(S)]2S2), tetramethyl-

$C_6H_{12}N_2S_4Zn$
12068 - Ziram

$C_6H_{12}N_4$
11601 - 1,3,5,7-Tetraazatricyclo[3.3.1.13,7]decane

$C_6H_{12}N_6O_3$
7583 - Methanol, (1,3,5-triazine-2,4,6-triyltriimino)tris-

$C_6H_{12}O$
3183 - Butanal, 2-ethyl-
3369 - Butane, 1-(ethenyloxy)-
3370 - Butane, 2-(ethenyloxy)-
3657 - 2-Butanone, 3,3-dimethyl-
4500 - Cyclohexanol
4825 - Cyclopentanemethanol
4867 - Cyclopentanol, 1-methyl-
4868 - Cyclopentanol, 2-methyl-, cis-
4869 - Cyclopentanol, 2-methyl-, trans-
4870 - Cyclopentanol, 3-methyl-
6107 - Furan, 2-ethyltetrahydro-

6718 - Hexanal
6959 - 2-Hexanone
6960 - 3-Hexanone
7066 - 1-Hexen-3-ol
7067 - 4-Hexen-2-ol
7068 - 5-Hexen-2-ol
7069 - 2-Hexen-1-ol, (E)-
7070 - 2-Hexen-1-ol, (Z)-
7071 - 3-Hexen-1-ol, (Z)-
7072 - 4-Hexen-1-ol, (E)-
8590 - Oxepane
8627 - Oxirane, tetramethyl-
8693 - Pentanal, 2-methyl-
8694 - Pentanal, 3-methyl-
8980 - 2-Pentanone, 3-methyl-, (±)-
8981 - 2-Pentanone, 4-methyl-
8982 - 3-Pentanone, 2-methyl-
9094 - 2-Penten-1-ol, 2-methyl-
9095 - 2-Penten-1-ol, 4-methyl-
9096 - 3-Penten-1-ol, 4-methyl-
9097 - 3-Penten-2-ol, 4-methyl-
9098 - 3-Penten-2-ol, 3-methyl-
9099 - 3-Penten-2-ol, 4-methyl-
9100 - 4-Penten-2-ol, 2-methyl-
9101 - 4-Penten-2-ol, 3-methyl-
9102 - 4-Penten-2-ol, 4-methyl-
10035 - Propane, 1-(ethenyloxy)-2-methyl-
10036 - Propane, 2-(ethenyloxy)-2-methyl-
10579 - 1-Propene, 3-(1-methylethoxy)-
10594 - 1-Propene, 3-propoxy-
$C_6H_{12}O_2$
121 - Acetic acid, butyl ester
177 - Acetic acid, 1,1-dimethylethyl ester
214 - Acetic acid, 1-methylpropyl ester
215 - Acetic acid, 2-methylpropyl ester
3506 - Butanoic acid, 2,2-dimethyl-
3507 - Butanoic acid, 2,3-dimethyl-
3508 - Butanoic acid, 3,3-dimethyl-
3514 - Butanoic acid, 2-ethyl-
3515 - Butanoic acid, ethyl ester
3552 - Butanoic acid, 3-methyl-, methyl ester
3636 - 1-Butanol, 3-methyl-, formate
4372 - 1,2-Cyclohexanediol, cis-
4373 - 1,2-Cyclohexanediol, trans-(±)-
4374 - 1,4-Cyclohexanediol, trans-
5189 - 1,3-Dioxane, 2,4-dimethyl-
5901 - Ethene, 1,1-diethoxy-
6015 - Formic acid, pentyl ester
6170 - Furan, tetrahydro-2-(methoxymethyl)-
6870 - Hexanoic acid
6969 - 2-Hexanone, 5-hydroxy-
6970 - 2-Hexanone, 6-hydroxy-
6971 - 3-Hexanone, 4-hydroxy-
6972 - 3-Hexanone, 5-hydroxy-
7177 - Hydroperoxide, cyclohexyl
8691 - Pentanal, 3-hydroxy-2-methyl-
8887 - Pentanoic acid, 2-methyl-, (±)-
8888 - Pentanoic acid, 3-methyl-, (±)-
8889 - Pentanoic acid, 4-methyl-
8891 - Pentanoic acid, methyl ester
8976 - 2-Pentanone, 4-hydroxy-4-methyl-
10260 - Propanoic acid, 2,2-dimethyl-, methyl ester
10298 - Propanoic acid, 2-methyl-, ethyl ester
10299 - Propanoic acid, 1-methylethyl ester
10326 - Propanoic acid, propyl ester
10784 - 2H-Pyran-2-methanol, tetrahydro-
$C_6H_{12}O_2S$
11784 - Thiophene, tetrahydro-2,4-dimethyl-, 1,1-dioxide
$C_6H_{12}O_3$
181 - Acetic acid, ethoxy-, ethyl ester
3532 - Butanoic acid, 2-hydroxy-, ethyl ester, (±)-
3533 - Butanoic acid, 3-hydroxy-, ethyl ester, (±)-
3534 - Butanoic acid, 3-hydroxy-, ethyl ester
5207 - 1,3-Dioxolane-4-methanol, 2,2-dimethyl-
5718 - Ethanol, 2-ethoxy-, acetate
6169 - Furan, tetrahydro-2,5-dimethoxy-

6891 - Hexanoic acid, 2-hydroxy-, (±)-
7109 - Hexitol, 2,5-anhydro-3,4-dideoxy-
10275 - Propanoic acid, 2-hydroxy-, 1-methylethyl ester
10277 - Propanoic acid, 3-hydroxy-, 1-methylethyl ester
10279 - Propanoic acid, 2-hydroxy-2-methyl-, ethyl ester
11908 - 1,3,5-Trioxane, 2,4,6-trimethyl-
$C_6H_{12}O_4$
5673 - Ethanol, 2-[2-(acetyloxy)ethoxy]-
7111 - D-ribo-Hexose, 2,6-dideoxy-
10255 - Propanoic acid, 2,3-dihydroxy-2-methyl-, ethyl ester
10257 - Propanoic acid, 2,3-dihydroxy-, propyl ester
$C_6H_{12}O_4Si$
11419 - Silanediol, dimethyl-, diacetate
$C_6H_{12}O_5$
4074 - D-chiro-Inositol, 2-deoxy-
6235 - D-Glucose, 6-deoxy-
7110 - D-arabino-Hexose, 2-deoxy-
$C_6H_{12}O_5S$
184 - Acetic acid, (ethoxysulfonyl)-, ethyl ester
$C_6H_{12}O_6$
337 - D-Allose
338 - D-Altrose
4073 - DL-chiro-Inositol
6023 - β-D-Fructose
6197 - D-Galactose
6232 - β-D-Glucose
6290 - D-Gulose
7336 - scyllo-Inositol
7425 - D-Mannose
11518 - DL-Sorbose
11519 - L-Sorbose
11587 - D-Tagatose
$C_6H_{12}O_7$
6209 - D-Gluconic acid
$C_6H_{12}S$
4480 - Cyclohexanethiol
11690 - Thiepane
11765 - Thiophene, 2-ethyltetrahydro-
11766 - Thiophene, 3-ethyltetrahydro-
11785 - Thiophene, tetrahydro-2,5-dimethyl-, cis-
11796 - 2H-Thiopyran, tetrahydro-2-methyl-
11797 - 2H-Thiopyran, tetrahydro-3-methyl-
11798 - 2H-Thiopyran, tetrahydro-4-methyl-
$C_6H_{12}S_3$
11920 - 1,3,5-Trithiane, 2,4,6-trimethyl-
11921 - 1,3,5-Trithiane, 2,4,6-trimethyl-, (2α,4α,6β)-
$C_6H_{13}Br$
3235 - Butane, 1-bromo-3,3-dimethyl-
3236 - Butane, 2-bromo-2,3-dimethyl-
6732 - Hexane, 1-bromo-
6733 - Hexane, 2-bromo-
6734 - Hexane, 3-bromo-
8714 - Pentane, 1-bromo-2-methyl-
8715 - Pentane, 1-bromo-3-methyl-
8716 - Pentane, 1-bromo-4-methyl-
8717 - Pentane, 2-bromo-2-methyl-
8718 - Pentane, 3-bromo-3-methyl-
$C_6H_{13}BrO_2$
5459 - Ethane, 2-bromo-1,1-diethoxy-
$C_6H_{13}Cl$
3245 - Butane, 1-chloro-2,3-dimethyl-
3246 - Butane, 1-chloro-3,3-dimethyl-
3247 - Butane, 2-chloro-2,3-dimethyl-
3248 - Butane, 3-chloro-2,2-dimethyl-
6737 - Hexane, 1-chloro-
6738 - Hexane, 2-chloro-
6739 - Hexane, 3-chloro-
8729 - Pentane, 2-chloro-2-methyl-
8730 - Pentane, 2-chloro-4-methyl-
8731 - Pentane, 3-(chloromethyl)-
8732 - Pentane, 3-chloro-3-methyl-
$C_6H_{13}ClO$
3249 - Butane, 1-(2-chloroethoxy)-

3608 - 2-Butanol, 1-chloro-2,3-dimethyl-
3609 - 2-Butanol, 1-chloro-3,3-dimethyl-
6916 - 1-Hexanol, 6-chloro-
6917 - 2-Hexanol, 1-chloro-
6918 - 3-Hexanol, 1-chloro-
6919 - 3-Hexanol, 5-chloro-
$C_6H_{13}ClO_2$
3244 - Butane, 3-chloro-1,1-dimethoxy-
5469 - Ethane, 2-chloro-1,1-diethoxy-
10355 - 2-Propanol, 1-chloro-3-(1-methylethoxy)-
10356 - 2-Propanol, 1-chloro-3-propoxy-
$C_6H_{13}ClS$
9855 - Propane, 2-[(2-chloroethyl)thio]-2-methyl-
$C_6H_{13}Cl_3Si$
11464 - Silane, trichlorohexyl-
$C_6H_{13}F$
6829 - Hexane, 1-fluoro-
6830 - Hexane, 2-fluoro-
$C_6H_{13}FO$
6936 - 1-Hexanol, 6-fluoro-
$C_6H_{13}I$
6831 - Hexane, 1-iodo-
6832 - Hexane, 2-iodo-, (R)-
$C_6H_{13}IO_2$
5543 - Ethane, 1,1-diethoxy-2-iodo-
$C_6H_{13}N$
492 - 1H-Azepine, hexahydro-
4283 - Cyclohexanamine
4772 - Cyclopentanamine, 1-methyl-
4773 - Cyclopentanamine, 2-methyl-
9679 - Piperidine, 1-methyl-
9680 - Piperidine, 2-methyl-, (±)-
9681 - Piperidine, 3-methyl-, (±)-
9682 - Piperidine, 4-methyl-
10525 - 2-Propen-1-amine, N-(1-methylethyl)-
10527 - 2-Propen-1-amine, N-propyl-
11149 - Pyrrolidine, 2,2-dimethyl-
11150 - Pyrrolidine, 2,4-dimethyl-
11151 - Pyrrolidine, 2,5-dimethyl-, cis-
11166 - Pyrrolidine, 1-ethyl-
$C_6H_{13}NO$
38 - Acetamide, N-butyl-
53 - Acetamide, N,N-diethyl-
3203 - Butanamide, N,N-dimethyl-
4501 - Cyclohexanol, 2-amino-, trans-(±)-
6721 - Hexanamide
7687 - Morpholine, 2,6-dimethyl-
7694 - Morpholine, 4-ethyl-
8959 - 2-Pentanone, 4-amino-4-methyl-
9674 - 2-Piperidinemethanol
9675 - 3-Piperidinemethanol
9716 - 3-Piperidinol, 1-methyl-
9717 - 4-Piperidinol, 1-methyl-
9802 - Propanamide, N-propyl-
11165 - 1-Pyrrolidineethanol
$C_6H_{13}NO_2$
332 - β-Alanine, N-methyl, ethyl ester
3633 - 1-Butanol, 3-methyl-, carbamate
3634 - 2-Butanol, 2-methyl-, carbamate
3937 - Carbamic acid, butyl-, methyl ester
3943 - Carbamic acid, (1,1-dimethylethyl)-, methyl ester
3956 - Carbamic acid, (1-methylethyl)-, ethyl ester
3962 - Carbamic acid, (1-methylpropyl)-, methyl ester
3970 - Carbamic acid, propyl-, ethyl ester
6120 - 2-Furanmethanol, 5-(aminomethyl)tetrahydro-
6841 - Hexane, 1-nitro-
6873 - Hexanoic acid, 6-amino-
7370 - L-Isoleucine
7412 - DL-Leucine
7413 - L-Leucine
7691 - 4-Morpholineethanol
8089 - Nitrous acid, hexyl ester
8214 - Norleucine
8215 - DL-Norleucine
$C_6H_{13}NO_3S$
11568 - Sulfamic acid, cyclohexyl-

$C_6H_{13}NO_4$
6259 - Glycine, N,N-bis(2-hydroxyethyl)-
$C_6H_{13}NO_5$
6233 - D-Glucose, 2-amino-2-deoxy-
$C_6H_{13}NS_2$
5265 - 4H-1,3,5-Dithiazine, dihydro-2,4,6-trimethyl-, (2α,4α,6α)-
$C_6H_{13}N_3$
6283 - Guanidine, (3-methyl-2-butenyl)-
$C_6H_{13}N_3O_3$
8551 - L-Ornithine, N5-(aminocarbonyl)-
$C_6H_{13}O_3P$
9535 - Phosphonic acid, ethenyl-, diethyl ester
$C_6H_{13}O_4P$
9530 - Phosphonic acid, acetyl-, diethyl ester
C_6H_{14}
3289 - Butane, 2,2-dimethyl-
3290 - Butane, 2,3-dimethyl-
6731 - Hexane
8823 - Pentane, 2-methyl-
8824 - Pentane, 3-methyl-
$C_6H_{14}Cl_4OSi_2$
5231 - Disiloxane, 1,3-bis(dichloromethyl)-1,1,3,3-tetramethyl-
$C_6H_{14}FO_3P$
9580 - Phosphorofluoridic acid, bis(1-methylethyl) ester
$C_6H_{14}N_2$
5132 - Diazene, dipropyl
5141 - 1H-1,4-Diazepine, hexahydro-1-methyl-
9608 - Piperazine, 1,4-dimethyl-
9609 - Piperazine, 2,5-dimethyl-, cis-
9610 - Piperazine, 2,5-dimethyl-, trans-
9611 - Piperazine, 2,6-dimethyl-, cis-
9673 - 4-Piperidinemethanamine
11164 - 1-Pyrrolidineethanamine
$C_6H_{14}N_2O$
7689 - 4-Morpholineethanamine
9616 - 1-Piperazineethanol
9822 - 2-Propanamine, N-(1-methylethyl)-N-nitroso-
9825 - 1-Propanamine, N-nitroso-N-propyl-
$C_6H_{14}N_2O_2$
7418 - DL-Lysine
7419 - L-Lysine
$C_6H_{14}N_4O_2$
437 - DL-Arginine
438 - L-Arginine
$C_6H_{14}O$
3372 - Butane, 1-ethoxy-
3373 - Butane, 2-ethoxy-
3394 - Butane, 1-methoxy-2-methyl-
3395 - Butane, 1-methoxy-3-methyl-
3396 - Butane, 2-methoxy-2-methyl-
3615 - 1-Butanol, 2,2-dimethyl-
3616 - 1-Butanol, 2,3-dimethyl-, (±)-
3617 - 1-Butanol, 3,3-dimethyl-
3618 - 2-Butanol, 2,3-dimethyl-
3619 - 2-Butanol, 3,3-dimethyl-, (±)-
3620 - 1-Butanol, 2-ethyl-
6911 - 1-Hexanol
6912 - 2-Hexanol, (±)-
6913 - 2-Hexanol, (R)-
6914 - 3-Hexanol, (±)-
6915 - 3-Hexanol, (R)-
8822 - Pentane, 1-methoxy-
8937 - 1-Pentanol, 2-methyl-
8938 - 1-Pentanol, 3-methyl-, (±)-
8939 - 1-Pentanol, 4-methyl-
8940 - 2-Pentanol, 2-methyl-
8941 - 2-Pentanol, 3-methyl-
8942 - 2-Pentanol, 4-methyl-
8943 - 3-Pentanol, 2-methyl-
8944 - 3-Pentanol, 3-methyl-
10039 - Propane, 1-ethoxy-2-methyl-
10040 - Propane, 2-ethoxy-2-methyl-
10076 - Propane, 1-(1-methylethoxy)-
10119 - Propane, 1,1'-oxybis-

10120 - Propane, 2,2'-oxybis-
$C_6H_{14}OS$
10129 - Propane, 1,1'-sulfinylbis-
$C_6H_{14}OSi$
11423 - Silane, ethenylethoxydimethyl-
$C_6H_{14}O_2$
3350 - 2,3-Butanediol, 2,3-dimethyl-
5541 - Ethane, 1,1-diethoxy-
5542 - Ethane, 1,2-diethoxy-
5687 - Ethanol, 2-butoxy-
5746 - Ethanol, 2-(2-methylpropoxy)-
6793 - 1,3-Hexanediol
6794 - 1,4-Hexanediol
6795 - 1,5-Hexanediol
6796 - 1,6-Hexanediol
6797 - 2,3-Hexanediol
6798 - 2,4-Hexanediol
6799 - 2,5-Hexanediol
8793 - 2,4-Pentanediol, 2-methyl-
8794 - 2,4-Pentanediol, 3-methyl-
10007 - 1,3-Propanediol, 2-ethyl-2-methyl-
10390 - 2-Propanol, 1-(1-methylethoxy)-
10401 - 2-Propanol, 1-propoxy-
$C_6H_{14}O_2S$
10131 - Propane, 1,1'-sulfonylbis-
$C_6H_{14}O_3$
5613 - Ethane, 1,1'-oxybis[2-methoxy-
5614 - Ethane, 1,1'-[oxybis(methyleneoxy)]bis-
5705 - Ethanol, 2,2-diethoxy-
5720 - Ethanol, 2-(2-ethoxyethoxy)-
6867 - 1,2,5-Hexanetriol
6868 - 1,2,6-Hexanetriol
6869 - 2,3,4-Hexanetriol
10006 - 1,3-Propanediol, 2-ethyl-2-(hydroxymethyl)-
10185 - Propane, 1,2,3-trimethoxy-
10397 - Propanol, oxybis-
$C_6H_{14}O_4$
5713 - Ethanol, 2,2'-[1,2-ethanediylbis(oxy)]bis-
$C_6H_{14}O_4S$
11575 - Sulfuric acid, dipropyl ester
$C_6H_{14}O_5$
10024 - Propanediol, oxybis-
10025 - 1,2-Propanediol, 3,3'-oxybis-
$C_6H_{14}O_6$
6194 - Galactitol
6206 - D-Glucitol
7422 - D-Mannitol
$C_6H_{14}S$
3376 - Butane, 1-(ethylthio)-
6856 - 1-Hexanethiol
6857 - 2-Hexanethiol
6858 - 3-Hexanethiol
8826 - Pentane, 1-(methylthio)-
10044 - Propane, 1-(ethylthio)-2-methyl-
10077 - Propane, 1-[(1-methylethyl)thio]-
10151 - Propane, 1,1'-thiobis-
10152 - Propane, 2,2'-thiobis-
$C_6H_{14}S_2$
5244 - Disulfide, bis(1-methylethyl)
5259 - Disulfide, dipropyl
5450 - Ethane, 1,1-bis(ethylthio)-
5451 - Ethane, 1,2-bis(ethylthio)-
6812 - 1,6-Hexanedithiol
$C_6H_{14}Si$
11495 - Silane, trimethyl-2-propenyl-
$C_6H_{15}As$
448 - Arsine, triethyl-
$C_6H_{15}AsO_3$
440 - Arsenous acid, triethyl ester
$C_6H_{15}AsO_4$
439 - Arsenic acid, triethyl ester
$C_6H_{15}B$
3118 - Borane, triethyl
$C_6H_{15}BO_3$
3125 - Boric acid, triethyl ester
$C_6H_{15}BrSi$
11368 - Silane, bromotriethyl-

$C_6H_{15}ClN_2O_2$
5440 - Ethanaminium, 2-[(aminocarbonyl)oxy]-N,N,N-trimethyl-, chloride
$C_6H_{15}ClO_3Si$
11387 - Silane, chlorotriethoxy-
$C_6H_{15}ClSi$
11369 - Silane, butylchlorodimethyl-
11386 - Silane, (3-chloropropyl)trimethyl-
11388 - Silane, chlorotriethyl-
$C_6H_{15}FSi$
11484 - Silane, triethylfluoro-
$C_6H_{15}N$
3215 - 1-Butanamine, N,N-dimethyl-
3216 - 2-Butanamine, 2,3-dimethyl-
3217 - 2-Butanamine, 3,3-dimethyl-
3218 - 1-Butanamine, N-ethyl-
3219 - 2-Butanamine, N-ethyl-, (±)-
5417 - Ethanamine, N,N-diethyl-
6723 - 1-Hexanamine
6724 - 2-Hexanamine, (±)-
8706 - 1-Pentanamine, 3-methyl-
8707 - 1-Pentanamine, 4-methyl-
9821 - 2-Propanamine, N-(1-methylethyl)-
9826 - 1-Propanamine, N-propyl-
9827 - 1-Propanamine, N-propyl
9829 - 1-Propanamine, N,N,2-trimethyl-
$C_6H_{15}NO$
5424 - Ethanamine, 2-ethoxy-N,N-dimethyl-
5691 - Ethanol, 2-(butylamino)-
5706 - Ethanol, 2-(diethylamino)-
5710 - Ethanol, 2-[(1,1-dimethylethyl)amino]-
5747 - Ethanol, 2-[(2-methylpropyl)amino]-
8922 - 1-Pentanol, 2-amino-4-methyl-, (±)-
9814 - 1-Propanamine, N-hydroxy-N-propyl-
$C_6H_{15}NO_2$
3623 - 2-Butanol, 1-[(2-hydroxyethyl)amino]-
5414 - Ethanamine, 2,2-diethoxy-
5723 - Ethanol, 2,2'-(ethylimino)bis-
10384 - 2-Propanol, 1,1'-iminobis-
$C_6H_{15}NO_3$
5751 - Ethanol, 2,2',2''-nitrilotris-
$C_6H_{15}NO_5$
6207 - D-Glucitol, 1-amino-1-deoxy-
$C_6H_{15}NS$
5650 - Ethanethiol, 2-(butylamino)-
$C_6H_{15}N_2$
5510 - 1,2-Ethanediamine, N'-ethyl-N,N-dimethyl-
$C_6H_{15}N_3$
9615 - 1-Piperazineethanamine
$C_6H_{15}OP$
9517 - Phosphine oxide, triethyl-
$C_6H_{15}O_2P$
9527 - Phosphinic acid, diethyl-, ethyl ester
$C_6H_{15}O_2PS_2$
9579 - Phosphorodithioic acid, O,O,S-triethyl ester
$C_6H_{15}O_3P$
9536 - Phosphonic acid, ethyl-, diethyl ester
9591 - Phosphorous acid, triethyl ester
$C_6H_{15}O_3PS$
9585 - Phosphorothioic acid, O,O,O-triethyl ester
$C_6H_{15}O_3PS_2$
5103 - Demeton S methyl
$C_6H_{15}O_4P$
9563 - Phosphoric acid, triethyl ester
$C_6H_{15}O_4PS_2$
8632 - Oxydemeton-methyl
$C_6H_{15}P$
9521 - Phosphine, triethyl-
$C_6H_{15}Sb$
11549 - Stibine, triethyl-
$C_6H_{16}Br_2OSi_2$
5229 - Disiloxane, 1,3-bis(bromomethyl)-1,1,3,3-tetramethyl-
$C_6H_{16}ClN$
5418 - Ethanamine, N,N-diethyl-, hydrochloride

C$_6$H$_{16}$ClNO
9831 - 1-Propanaminium, 2-hydroxy-*N,N,N*-trimethyl-, chloride
C$_6$H$_{16}$Cl$_2$OSi$_2$
5230 - Disiloxane, 1,3-bis(chloromethyl)-1,1,3,3-tetramethyl-
C$_6$H$_{16}$N$_2$
5501 - 1,2-Ethanediamine, *N,N*'-diethyl-
5502 - 1,2-Ethanediamine, *N,N*-diethyl-
5516 - 1,2-Ethanediamine, *N,N,N*',*N*'-tetramethyl-
6749 - 1,6-Hexanediamine
7137 - Hydrazine, 1,2-bis(1-methylethyl)-
C$_6$H$_{16}$N$_2$O
5677 - Ethanol, 2-[(2-amino-2-methylpropyl)amino]-
C$_6$H$_{16}$N$_2$O$_2$
5715 - Ethanol, 2,2'-(1,2-ethanediyldiimino)bis-
C$_6$H$_{16}$OSi
10410 - 1-Propanol, 3-(trimethylsilyl)-
11503 - Silanol, triethyl-
C$_6$H$_{16}$O$_2$Si
11405 - Silane, diethoxydimethyl-
C$_6$H$_{16}$O$_3$Si
11477 - Silane, triethoxy-
C$_6$H$_{16}$Si
11483 - Silane, triethyl-
11496 - Silane, trimethylpropyl-
C$_6$H$_{17}$N$_3$
9874 - 1,3-Propanediamine, *N*-(3-aminopropyl)-
C$_6$H$_{18}$AlO$_9$P$_3$
6020 - Fosetyl-Al
C$_6$H$_{18}$Cl$_2$O$_2$Si$_3$
11913 - Trisiloxane, 1,5-dichloro-1,1,3,3,5,5-hexamethyl-
C$_6$H$_{18}$N$_3$OP
9573 - Phosphoric triamide, hexamethyl-
C$_6$H$_{18}$N$_4$
5499 - 1,2-Ethanediamine, *N,N*'-bis(2-aminoethyl)-
C$_6$H$_{18}$OSi$_2$
5236 - Disiloxane, hexamethyl-
C$_6$H$_{18}$O$_3$Si$_2$
5234 - Disiloxane, 1,3-dimethoxy-1,1,3,3-tetramethyl-
C$_6$H$_{18}$O$_3$Si$_3$
5002 - Cyclotrisiloxane, hexamethyl-
C$_6$H$_{18}$Si$_2$
5228 - Disilane, hexamethyl-
C$_6$H$_{19}$NSi$_2$
11363 - Silanamine, 1,1,1-trimethyl-*N*-(trimethylsilyl)-
C$_6$H$_{21}$N$_3$Si$_3$
5001 - Cyclotrisilazane, 2,2,4,4,6,6-hexamethyl-
C$_6$H$_{24}$O$_6$Si$_6$
4625 - Cyclohexasiloxane, 2,4,6,8,10,12-hexamethyl-
C$_6$H$_{26}$O$_8$
3351 - 2,3-Butanediol, 2,3-dimethyl-, hexahydrate
C$_6$MoO$_6$
7662 - Molybdenum carbonyl
C$_6$N$_4$
5915 - Ethenetetracarbonitrile
C$_6$N$_6$
11836 - 1,3,5-Triazine-2,4,6-tricarbonitrile
C$_7$F$_5$N
2767 - Benzonitrile, pentafluoro-
C$_7$F$_8$
2095 - Benzene, pentafluoro(trifluoromethyl)-
C$_7$F$_{14}$
4499 - Cyclohexane, undecafluoro(trifluoromethyl)-
6560 - 1-Heptene, 1,1,2,3,3,4,4,5,5,6,6,7,7,7-tetradecafluoro-
C$_7$F$_{16}$
6418 - Heptane, hexadecafluoro-
C$_7$HCl$_5$O$_2$
2712 - Benzoic acid, pentachloro-
C$_7$HCl$_7$
2085 - Benzene, pentachloro(dichloromethyl)-
C$_7$HF$_5$O
596 - Benzaldehyde, pentafluoro-

C$_7$HF$_{15}$
6432 - Heptane, 1,1,1,2,2,3,3,4,4,5,5,6,6,7,7-pentadecafluoro-
C$_7$H$_3$Br$_2$NO
2751 - Benzonitrile, 3,5-dibromo-4-hydroxy-
C$_7$H$_3$Br$_5$
2083 - Benzene, pentabromomethyl-
C$_7$H$_3$ClF$_3$NO$_2$
1217 - Benzene, 1-chloro-2-nitro-4-(trifluoromethyl)-
1218 - Benzene, 1-chloro-4-nitro-2-(trifluoromethyl)-
1219 - Benzene, 4-chloro-1-nitro-2-(trifluoromethyl)-
C$_7$H$_3$ClN$_2$O$_5$
2896 - Benzoyl chloride, 3,5-dinitro-
C$_7$H$_3$Cl$_2$N
2752 - Benzonitrile, 2,6-dichloro-
C$_7$H$_3$Cl$_3$O
2893 - Benzoyl chloride, 2,4-dichloro-
2894 - Benzoyl chloride, 3,4-dichloro-
C$_7$H$_3$Cl$_3$O$_2$
2727 - Benzoic acid, 2,3,6-trichloro-
2728 - Benzoic acid, 2,4,5-trichloro-
C$_7$H$_3$Cl$_5$
1401 - Benzene, 1,2-dichloro-4-(trichloromethyl)-
2086 - Benzene, pentachloromethyl-
2326 - Benzene, 1,2,3-trichloro-4-(dichloromethyl)-
2327 - Benzene, 1,2,4-trichloro-5-(dichloromethyl)-
C$_7$H$_3$F$_5$
2094 - Benzene, pentafluoromethyl-
C$_7$H$_3$F$_5$O
2093 - Benzene, pentafluoromethoxy-
C$_7$H$_3$N$_3$O$_8$
2735 - Benzoic acid, 2,4,6-trinitro-
C$_7$H$_4$BrClO
2887 - Benzoyl chloride, 2-bromo-
2888 - Benzoyl chloride, 4-bromo-
C$_7$H$_4$BrF$_3$
1089 - Benzene, 1-bromo-2-(trifluoromethyl)-
1090 - Benzene, 1-bromo-3-(trifluoromethyl)-
1091 - Benzene, 1-bromo-4-(trifluoromethyl)-
C$_7$H$_4$BrN
2744 - Benzonitrile, 2-bromo-
2745 - Benzonitrile, 3-bromo-
2746 - Benzonitrile, 4-bromo-
C$_7$H$_4$BrNO
1049 - Benzene, 1-bromo-4-isocyanato-
C$_7$H$_4$BrNO$_4$
2527 - Benzoic acid, 2-bromo-4-nitro-
2528 - Benzoic acid, 2-bromo-5-nitro-
2529 - Benzoic acid, 4-bromo-3-nitro-
2530 - Benzoic acid, 5-bromo-2-nitro-
C$_7$H$_4$Br$_2$O$_2$
534 - Benzaldehyde, 3,5-dibromo-2-hydroxy-
2555 - Benzoic acid, 2,4-dibromo-
2556 - Benzoic acid, 2,5-dibromo-
2557 - Benzoic acid, 2,6-dibromo-
C$_7$H$_4$Br$_2$O$_3$
2558 - Benzoic acid, 3,5-dibromo-2-hydroxy-
C$_7$H$_4$Br$_2$O$_5$
2560 - Benzoic acid, 2,6-dibromo-3,4,5-trihydroxy-
C$_7$H$_4$Br$_4$O
2256 - Benzene, 1,2,3,5-tetrabromo-4-methoxy-
9468 - Phenol, 2,3,4,5-tetrabromo-6-methyl-
C$_7$H$_4$ClFO
2897 - Benzoyl chloride, 2-fluoro-
2898 - Benzoyl chloride, 3-fluoro-
2899 - Benzoyl chloride, 4-fluoro-
C$_7$H$_4$ClF$_3$
1240 - Benzene, 1-chloro-2-(trifluoromethyl)-
1241 - Benzene, 1-chloro-3-(trifluoromethyl)-
1242 - Benzene, 1-chloro-4-(trifluoromethyl)-
C$_7$H$_4$ClN
2747 - Benzonitrile, 2-chloro-
2748 - Benzonitrile, 3-chloro-
2749 - Benzonitrile, 4-chloro-
C$_7$H$_4$ClNO
1176 - Benzene, 1-chloro-2-isocyanato-
2878 - Benzoxazole, 2-chloro-

C$_7$H$_4$ClNO$_2$
2883 - 2(3H)-Benzoxazolone, 5-chloro-
C$_7$H$_4$ClNO$_3$
2905 - Benzoyl chloride, 3-nitro-
2906 - Benzoyl chloride, 4-nitro-
C$_7$H$_4$ClNO$_4$
2544 - Benzoic acid, 2-chloro-3-nitro-
2545 - Benzoic acid, 2-chloro-5-nitro-
2546 - Benzoic acid, 3-chloro-2-nitro-
2547 - Benzoic acid, 4-chloro-3-nitro-
C$_7$H$_4$ClNS
1177 - Benzene, 1-chloro-4-isothiocyanato-
2862 - Benzothiazole, 2-chloro-
C$_7$H$_4$Cl$_2$O
535 - Benzaldehyde, 2,5-dichloro-
536 - Benzaldehyde, 3,4-dichloro-
537 - Benzaldehyde, 3,5-dichloro-
2890 - Benzoyl chloride, 2-chloro-
2891 - Benzoyl chloride, 3-chloro-
2892 - Benzoyl chloride, 4-chloro-
C$_7$H$_4$Cl$_2$O$_2$
2561 - Benzoic acid, 2,4-dichloro-
2562 - Benzoic acid, 2,5-dichloro-
2563 - Benzoic acid, 2,6-dichloro-
2564 - Benzoic acid, 3,5-dichloro-
C$_7$H$_4$Cl$_2$O$_3$
2566 - Benzoic acid, 3,5-dichloro-2-hydroxy-
C$_7$H$_4$Cl$_3$F
1736 - Benzene, 1-fluoro-2-(trichloromethyl)-
C$_7$H$_4$Cl$_3$NO$_3$
11854 - Triclopyr
C$_7$H$_4$Cl$_4$
1237 - Benzene, 1-chloro-2-(trichloromethyl)-
1238 - Benzene, 1-chloro-3-(trichloromethyl)-
1239 - Benzene, 1-chloro-4-(trichloromethyl)-
1359 - Benzene, 1,2-dichloro-4-(trichloromethyl)-
2262 - Benzene, 1,2,3,5-tetrachloro-4-methyl-
2263 - Benzene, 1,2,4,5-tetrachloro-3-methyl-
2324 - Benzene, 1,2,4-trichloro-5-(chloromethyl)-
C$_7$H$_4$Cl$_4$O$_2$
9472 - Phenol, 2,3,5,6-tetrachloro-4-methoxy-
C$_7$H$_4$FN
2756 - Benzonitrile, 4-fluoro-
C$_7$H$_4$FNS
1721 - Benzene, 1-fluoro-3-isothiocyanato-
C$_7$H$_4$F$_3$NO$_2$
2060 - Benzene, 1-nitro-2-(trifluoromethyl)-
2061 - Benzene, 1-nitro-3-(trifluoromethyl)-
C$_7$H$_4$F$_4$
1737 - Benzene, 1-fluoro-2-(trifluoromethyl)-
1738 - Benzene, 1-fluoro-3-(trifluoromethyl)-
1739 - Benzene, 1-fluoro-4-(trifluoromethyl)-
C$_7$H$_4$I$_2$O$_3$
2617 - Benzoic acid, 2-hydroxy-3,5-diiodo-
2618 - Benzoic acid, 4-hydroxy-3,5-diiodo-
C$_7$H$_4$N$_2$O$_3$
1788 - Benzene, 1-isocyanato-2-nitro-
1789 - Benzene, 1-isocyanato-3-nitro-
1790 - Benzene, 1-isocyanato-4-nitro-
C$_7$H$_4$N$_2$O$_5$
551 - Benzaldehyde, 2,4-dinitro-
C$_7$H$_4$N$_2$O$_6$
2590 - Benzoic acid, 3,4-dinitro-
2591 - Benzoic acid, 3,5-dinitro-
C$_7$H$_4$O$_4$S
2875 - 3H-2,1-Benzoxathiol-3-one, 1,1-dioxide
C$_7$H$_4$O$_6$
10776 - 4H-Pyran-2,6-dicarboxylic acid, 4-oxo-
C$_7$H$_4$O$_7$
10775 - 4H-Pyran-2,6-dicarboxylic acid, 3-hydroxy-4-oxo-
C$_7$H$_5$BrO
526 - Benzaldehyde, 2-bromo-
527 - Benzaldehyde, 3-bromo-
528 - Benzaldehyde, 4-bromo-
2885 - Benzoyl bromide

$C_7H_5BrO_2$
2518 - Benzoic acid, 2-bromo-
2519 - Benzoic acid, 3-bromo-
2520 - Benzoic acid, 4-bromo-
$C_7H_5BrO_3$
2524 - Benzoic acid, 5-bromo-2-hydroxy-
$C_7H_5Br_3$
1291 - Benzene, 1,3-dibromo-5-(bromomethyl)-
2314 - Benzene, 1,3,5-tribromo-2-methyl-
$C_7H_5Br_3O$
2310 - Benzene, 1,2,3-tribromo-5-methoxy-
2311 - Benzene, 1,2,4-tribromo-5-methoxy-
2312 - Benzene, 1,2,5-tribromo-3-methoxy-
2313 - Benzene, 1,3,5-tribromo-2-methoxy-
9480 - Phenol, 2,4,6-tribromo-3-methyl-
$C_7H_5ClN_2$
7236 - 1H-Indazole, 3-chloro-
$C_7H_5ClN_2O$
2876 - 2-Benzoxazolamine, 5-chloro-
C_7H_5ClO
529 - Benzaldehyde, 2-chloro-
530 - Benzaldehyde, 3-chloro-
531 - Benzaldehyde, 4-chloro-
2886 - Benzoyl chloride
$C_7H_5ClO_2$
532 - Benzaldehyde, 3-chloro-4-hydroxy-
533 - Benzaldehyde, 5-chloro-2-hydroxy-
2534 - Benzoic acid, 2-chloro-
2535 - Benzoic acid, 3-chloro-
2536 - Benzoic acid, 4-chloro-
2900 - Benzoyl chloride, 2-hydroxy-
4026 - Carbonochloridic acid, phenyl ester
$C_7H_5ClO_3$
1127 - Benzenecarboperoxoic acid, 3-chloro-
2540 - Benzoic acid, 3-chloro-4-hydroxy-
$C_7H_5Cl_2F$
1379 - Benzene, (dichlorofluoromethyl)-
$C_7H_5Cl_2FN_2O_3$
5982 - Fluroxypyr
$C_7H_5Cl_2NO_2$
1135 - Benzene, 1-chloro-4-(chloromethyl)-2-nitro-
2491 - Benzoic acid, 2-amino-3,6-dichloro-
2492 - Benzoic acid, 3-amino-2,5-dichloro-
$C_7H_5Cl_2NO_4S$
2565 - Benzoic acid, 4-[(dichloroamino)sulfonyl]-
$C_7H_5Cl_3$
1357 - Benzene, 1,2-dichloro-4-(chloromethyl)-
1358 - Benzene, 1,3-dichloro-5-(chloromethyl)-
2333 - Benzene, (trichloromethyl)-
2334 - Benzene, 1,2,3-trichloro-4-methyl-
2335 - Benzene, 1,2,3-trichloro-5-methyl-
2336 - Benzene, 1,2,4-trichloro-5-methyl-
2337 - Benzene, 1,2,5-trichloro-3-methyl-
$C_7H_5Cl_3O$
2330 - Benzene, 1,2,4-trichloro-3-methoxy-
2331 - Benzene, 1,2,4-trichloro-5-methoxy-
2332 - Benzene, 1,3,5-trichloro-2-methoxy-
9487 - Phenol, 2,3,4-trichloro-6-methyl-
9488 - Phenol, 2,3,6-trichloro-4-methyl-
9489 - Phenol, 2,4,6-trichloro-3-methyl-
C_7H_5FO
559 - Benzaldehyde, 2-fluoro-
560 - Benzaldehyde, 3-fluoro-
561 - Benzaldehyde, 4-fluoro-
2909 - Benzoyl fluoride
$C_7H_5FO_2$
2603 - Benzoic acid, 2-fluoro-
2604 - Benzoic acid, 3-fluoro-
2605 - Benzoic acid, 4-fluoro-
$C_7H_5F_3$
2348 - Benzene, (trifluoromethyl)-
$C_7H_5F_3O$
9491 - Phenol, 3-(trifluoromethyl)-
$C_7H_5HgNO_3$
7437 - Mercury, [2-methyl-5-nitrophenolato(2-)-
C6,O1]-

C_7H_5IO
580 - Benzaldehyde, 2-iodo-
581 - Benzaldehyde, 3-iodo-
582 - Benzaldehyde, 4-iodo-
2910 - Benzoyl iodide
$C_7H_5IO_2$
2657 - Benzoic acid, 2-iodo-
2658 - Benzoic acid, 3-iodo-
2659 - Benzoic acid, 4-iodo-
$C_7H_5IO_3$
2623 - Benzoic acid, 4-hydroxy-3-iodo-
2624 - Benzoic acid, 5-hydroxy-2-iodo-
$C_7H_5I_3$
2352 - Benzene, 1,3,5-triiodo-2-methyl-
C_7H_5N
2740 - Benzonitrile
C_7H_5NO
1786 - Benzene, isocyanato-
2389 - 1,2-Benzisoxazole
2390 - 2,1-Benzisoxazole
2759 - Benzonitrile, 2-hydroxy-
2877 - Benzoxazole
10585 - 2-Propenenitrile, 3-(2-furanyl)-
C_7H_5NOS
2868 - 2(3H)-Benzothiazolone
$C_7H_5NO_2$
2881 - 2-Benzoxazolol
2882 - 2(3H)-Benzoxazolone
$C_7H_5NO_3$
591 - Benzaldehyde, 2-nitro-
592 - Benzaldehyde, 3-nitro-
593 - Benzaldehyde, 4-nitro-
$C_7H_5NO_3S$
2388 - 1,2-Benzisothiazol-3(2H)-one, 1,1-dioxide
$C_7H_5NO_4$
2703 - Benzoic acid, 2-nitro-
2704 - Benzoic acid, 3-nitro-
2705 - Benzoic acid, 4-nitro-
10927 - 2,3-Pyridinedicarboxylic acid
10928 - 2,4-Pyridinedicarboxylic acid
10929 - 2,5-Pyridinedicarboxylic acid
10930 - 3,4-Pyridinedicarboxylic acid
10931 - 3,5-Pyridinedicarboxylic acid
$C_7H_5NO_5$
2646 - Benzoic acid, 2-hydroxy-3-nitro-
2647 - Benzoic acid, 2-hydroxy-5-nitro-
C_7H_5NS
1792 - Benzene, isothiocyanato-
2387 - 1,2-Benzisothiazole
2861 - Benzothiazole
11709 - Thiocyanic acid, phenyl ester
$C_7H_5NS_2$
2866 - 2(3H)-Benzothiazolethione
$C_7H_5N_3$
2870 - 1,2,4-Benzotriazine
$C_7H_5N_3O$
2884 - Benzoyl azide
$C_7H_5N_3O_5$
614 - Benzamide, 3,5-dinitro-
$C_7H_5N_3O_6$
2046 - Benzene, 1-methyl-2,3,4-trinitro-
2047 - Benzene, 1-methyl-2,4,5-trinitro-
2048 - Benzene, 2-methyl-1,3,5-trinitro-
$C_7H_5N_3O_7$
364 - Anisole, 2,3,5-trinitro-
1946 - Benzene, 2-methoxy-1,3,5-trinitro-
9434 - Phenol, 3-methyl-2,4,6-trinitro-
$C_7H_5N_5O_8$
818 - Benzenamine, N-methyl-N,2,4,6-tetranitro-
$C_7H_5NaO_2$
2725 - Benzoic acid, sodium salt
C_7H_6BrCl
1007 - Benzene, 1-bromo-3-(chloromethyl)-
1008 - Benzene, 1-bromo-4-(chloromethyl)-
C_7H_6BrNO
604 - Benzamide, 2-bromo-
605 - Benzamide, 3-bromo-

$C_7H_6BrNO_2$
1066 - Benzene, 1-bromo-2-methyl-4-nitro-
1067 - Benzene, 1-(bromomethyl)-3-nitro-
1068 - Benzene, 1-bromo-3-methyl-2-nitro-
1069 - Benzene, 1-bromo-3-methyl-5-nitro-
1070 - Benzene, 1-bromo-4-methyl-3-nitro-
1071 - Benzene, 2-bromo-1-methyl-3-nitro-
1072 - Benzene, 2-bromo-1-methyl-4-nitro-
1073 - Benzene, 2-bromo-4-methyl-1-nitro-
1074 - Benzene, 4-bromo-1-methyl-2-nitro-
2487 - Benzoic acid, 2-amino-5-bromo-
$C_7H_6BrNO_3$
606 - Benzamide, 5-bromo-N,2-dihydroxy-
$C_7H_6Br_2$
999 - Benzene, 1-bromo-2-(bromomethyl)-
1000 - Benzene, 1-bromo-4-(bromomethyl)-
1301 - Benzene, (dibromomethyl)-
1302 - Benzene, 1,2-dibromo-3-methyl-
1303 - Benzene, 1,2-dibromo-4-methyl-
1304 - Benzene, 1,3-dibromo-2-methyl-
1305 - Benzene, 1,4-dibromo-2-methyl-
1306 - Benzene, 2,4-dibromo-1-methyl-
$C_7H_6Br_2O$
9279 - Phenol, 2,4-dibromo-6-methyl-
9280 - Phenol, 3,5-dibromo-2-methyl-
C_7H_6ClF
1169 - Benzene, 1-chloro-3-fluoro-2-methyl-
1170 - Benzene, 2-chloro-4-fluoro-1-methyl-
1194 - Benzene, 1-(chloromethyl)-2-fluoro-
1195 - Benzene, 1-(chloromethyl)-4-fluoro-
C_7H_6ClI
1173 - Benzene, 1-chloro-2-iodo-4-methyl-
1174 - Benzene, 4-chloro-1-iodo-2-methyl-
1175 - Benzene, 4-chloro-2-iodo-1-methyl-
11817 - Toluene, 3-chloro-2-iodo-
$C_7H_6ClNO_2$
1203 - Benzene, 1-(chloromethyl)-2-nitro-
1204 - Benzene, 1-chloro-2-methyl-3-nitro-
1205 - Benzene, 1-chloro-2-methyl-4-nitro-
1206 - Benzene, 1-(chloromethyl)-3-nitro-
1207 - Benzene, 1-chloro-3-methyl-5-nitro-
1208 - Benzene, 1-(chloromethyl)-4-nitro-
1209 - Benzene, 1-chloro-4-methyl-2-nitro-
1210 - Benzene, 2-chloro-1-methyl-4-nitro-
1211 - Benzene, 2-chloro-4-methyl-1-nitro-
1212 - Benzene, 4-chloro-1-methyl-2-nitro-
1213 - Benzene, 4-chloro-2-methyl-1-nitro-
2490 - Benzoic acid, 5-amino-2-chloro-
$C_7H_6ClN_3O_4S_2$
2852 - 2H-1,2,4-Benzothiadiazine-7-sulfonamide, 6-
chloro-, 1,1-dioxide
$C_7H_6Cl_2$
1132 - Benzene, 1-chloro-2-(chloromethyl)-
1133 - Benzene, 1-chloro-3-(chloromethyl)-
1134 - Benzene, 1-chloro-4-(chloromethyl)-
1382 - Benzene, (dichloromethyl)-
1383 - Benzene, 1,2-dichloro-3-methyl-
1384 - Benzene, 1,2-dichloro-4-methyl-
1385 - Benzene, 1,3-dichloro-2-methyl-
1386 - Benzene, 1,3-dichloro-5-methyl-
1387 - Benzene, 1,4-dichloro-2-methyl-
1388 - Benzene, 2,4-dichloro-1-methyl-
$C_7H_6Cl_2O$
1381 - Benzene, 2,4-dichloro-1-methoxy-
1853 - Benzenemethanol, 2,4-dichloro-
9292 - Phenol, 2,4-dichloro-3-methyl-
9293 - Phenol, 2,4-dichloro-5-methyl-
9294 - Phenol, 2,6-dichloro-3-methyl-
9295 - Phenol, 2,6-dichloro-4-methyl-
$C_7H_6Cl_2S$
1136 - Benzene, 1-chloro-4-[(chloromethyl)thio]-
$C_7H_6Cl_4O_2$
4770 - 1,3-Cyclopentadiene, 1,2,3,4-tetrachloro-5,5-
dimethoxy-
$C_7H_6FNO_2$
1729 - Benzene, 1-fluoro-4-methyl-2-nitro-
1730 - Benzene, 2-fluoro-4-methyl-1-nitro-

1731 - Benzene, 4-fluoro-1-methyl-2-nitro-
1732 - Benzene, 4-fluoro-2-methyl-1-nitro-
$C_7H_6F_2$
1425 - Benzene, (difluoromethyl)-
$C_7H_6F_3N$
855 - Benzenamine, 2-(trifluoromethyl)-
856 - Benzenamine, 3-(trifluoromethyl)-
857 - Benzenamine, 4-(trifluoromethyl)-
$C_7H_6N_2$
2376 - 1H-Benzimidazole
2741 - Benzonitrile, 2-amino-
2742 - Benzonitrile, 3-amino-
7235 - 1H-Indazole
$C_7H_6N_2O$
10884 - 3-Pyridinecarbonitrile, 1,2-dihydro-1-methyl-2-oxo-
$C_7H_6N_2O_3$
624 - Benzamide, 2-nitro-
625 - Benzamide, 3-nitro-
$C_7H_6N_2O_4$
1956 - Benzene, 1-methyl-2,4-dinitro-
1957 - Benzene, 1-methyl-3,5-dinitro-
1958 - Benzene, 2-methyl-1,3-dinitro-
1959 - Benzene, 2-methyl-1,4-dinitro-
1960 - Benzene, 4-methyl-1,2-dinitro-
2506 - Benzoic acid, 2-amino-3-nitro-
$C_7H_6N_2O_5$
1922 - Benzene, 1-methoxy-2,3-dinitro-
1923 - Benzene, 1-methoxy-2,4-dinitro-
1924 - Benzene, 1-methoxy-3,5-dinitro-
1925 - Benzene, 2-methoxy-1,3-dinitro-
1926 - Benzene, 2-methoxy-1,4-dinitro-
1927 - Benzene, 4-methoxy-1,2-dinitro-
9405 - Phenol, 2-methyl-4,6-dinitro-
$C_7H_6N_2S$
2386 - 2H-Benzimidazole-2-thione, 1,3-dihydro-
2859 - 2-Benzothiazolamine
2860 - 6-Benzothiazolamine
$C_7H_6N_4O$
7193 - 1H-Imidazole, 1,1'-carbonylbis-
C_7H_6O
525 - Benzaldehyde
4242 - 2,4,6-Cycloheptatrien-1-one
C_7H_6OS
1130 - Benzenecarbothioic acid
$C_7H_6OS_2$
1125 - Benzenecarbodithioic acid, 2-hydroxy-
$C_7H_6O_2$
564 - Benzaldehyde, 2-hydroxy-
565 - Benzaldehyde, 3-hydroxy-
566 - Benzaldehyde, 4-hydroxy-
2415 - 1,3-Benzodioxole
2475 - Benzoic acid
4243 - 2,4,6-Cycloheptatrien-1-one, 2-hydroxy-
4266 - 2,5-Cyclohexadiene-1,4-dione, 2-methyl-
10512 - 2-Propenal, 3-(2-furanyl)-
10513 - 2-Propenal, 3-(2-furanyl)-, (E)-
$C_7H_6O_2S$
2663 - Benzoic acid, 2-mercapto-
$C_7H_6O_3$
540 - Benzaldehyde, 2,3-dihydroxy-
541 - Benzaldehyde, 2,4-dihydroxy-
542 - Benzaldehyde, 3,4-dihydroxy-
1126 - Benzenecarboperoxoic acid
2612 - Benzoic acid, 2-hydroxy-
2613 - Benzoic acid, 4-hydroxy-
10638 - 2-Propenoic acid, 3-(2-furanyl)-
$C_7H_6O_4$
2568 - Benzoic acid, 2,3-dihydroxy-
2569 - Benzoic acid, 2,4-dihydroxy-
2570 - Benzoic acid, 2,5-dihydroxy-
2571 - Benzoic acid, 3,4-dihydroxy-
6183 - 4H-Furo[3,2-c]pyran-2(6H)-one, 4-hydroxy-
$C_7H_6O_5$
2730 - Benzoic acid, 2,3,4-trihydroxy-
2731 - Benzoic acid, 3,4,5-trihydroxy-

$C_7H_6O_6S$
2655 - Benzoic acid, 2-hydroxy-5-sulfo-
$C_7H_7BF_2$
3115 - Borane, difluoro(4-methylphenyl)-
C_7H_7Br
1053 - Benzene, (bromomethyl)-
1054 - Benzene, 1-bromo-2-methyl-
1055 - Benzene, 1-bromo-3-methyl-
1056 - Benzene, 1-bromo-4-methyl-
$C_7H_7BrN_2O$
11998 - Urea, (4-bromophenyl)-
C_7H_7BrO
1050 - Benzene, 1-bromo-2-methoxy-
1051 - Benzene, 1-bromo-3-methoxy-
1052 - Benzene, 1-bromo-4-methoxy-
9236 - Phenol, 2-bromo-4-methyl-
9237 - Phenol, 2-bromo-5-methyl-
9238 - Phenol, 3-bromo-2-methyl-
9239 - Phenol, 3-bromo-4-methyl-
9240 - Phenol, 3-bromo-5-methyl-
9241 - Phenol, 4-bromo-2-methyl-
9242 - Phenol, 4-bromo-3-methyl-
$C_7H_7BrO_2$
1843 - Benzenemethanol, 5-bromo-2-hydroxy-
$C_7H_7BrO_3$
6190 - 3-Furoic acid, 5-bromo-, ethyl ester
C_7H_7Cl
1182 - Benzene, (chloromethyl)-
1183 - Benzene, 1-chloro-2-methyl-
1184 - Benzene, 1-chloro-3-methyl-
1185 - Benzene, 1-chloro-4-methyl-
$C_7H_7ClN_2S$
11801 - Thiourea, (2-chlorophenyl)-
C_7H_7ClO
1178 - Benzene, 1-chloro-2-methoxy-
1179 - Benzene, 1-chloro-3-methoxy-
1180 - Benzene, 1-chloro-4-methoxy-
1848 - Benzenemethanol, 2-chloro-
1849 - Benzenemethanol, 4-chloro-
9260 - Phenol, 2-chloro-3-methyl-
9261 - Phenol, 2-chloro-4-methyl-
9262 - Phenol, 2-chloro-5-methyl-
9263 - Phenol, 2-chloro-6-methyl-
9264 - Phenol, 3-chloro-2-methyl-
9265 - Phenol, 3-chloro-4-methyl-
9266 - Phenol, 4-chloro-2-methyl-
9267 - Phenol, 4-chloro-3-methyl-
C_7H_7ClOS
2186 - Benzenesulfinyl chloride, 4-methyl-
$C_7H_7ClO_2S$
2249 - Benzenesulfonyl chloride, 2-methyl-
2250 - Benzenesulfonyl chloride, 4-methyl-
$C_7H_7ClO_3$
6053 - 2-Furancarboxylic acid, 5-chloro-, ethyl ester
$C_7H_7ClO_3S$
2248 - Benzenesulfonyl chloride, 2-methoxy-
C_7H_7ClS
1833 - Benzenemethanethiol, 4-chloro-
$C_7H_7Cl_2N$
1803 - Benzenemethanamine, 2,4-dichloro-
$C_7H_7Cl_2NO_2S$
2210 - Benzenesulfonamide, N,N-dichloro-4-methyl-
$C_7H_7Cl_3NO_3PS$
4090 - Chlorpyrifos-methyl
$C_7H_7Cl_3Si$
11467 - Silane, trichloro(2-methylphenyl)-
11468 - Silane, trichloro(3-methylphenyl)-
C_7H_7F
1725 - Benzene, (fluoromethyl)-
1726 - Benzene, 1-fluoro-2-methyl-
1727 - Benzene, 1-fluoro-3-methyl-
1728 - Benzene, 1-fluoro-4-methyl-
C_7H_7FO
1722 - Benzene, 1-fluoro-2-methoxy-
1723 - Benzene, 1-fluoro-3-methoxy-
1724 - Benzene, 1-fluoro-4-methoxy-
1879 - Benzenemethanol, 3-fluoro-

1880 - Benzenemethanol, 4-fluoro-
$C_7H_7F_7O_2$
3526 - Butanoic acid, heptafluoro-, 1-methylethyl ester
C_7H_7I
1777 - Benzene, (iodomethyl)-
1778 - Benzene, 1-iodo-2-methyl-
1779 - Benzene, 1-iodo-3-methyl-
1780 - Benzene, 1-iodo-4-methyl-
C_7H_7IO
1775 - Benzene, 1-iodo-2-methoxy-
1776 - Benzene, 1-iodo-4-methoxy-
1888 - Benzenemethanol, 2-iodo-
9374 - Phenol, 2-iodo-4-methyl-
9375 - Phenol, 2-iodo-6-methyl-
C_7H_7N
10952 - Pyridine, 2-ethenyl-
10953 - Pyridine, 3-ethenyl-
10954 - Pyridine, 4-ethenyl-
C_7H_7NO
594 - Benzaldehyde, oxime, (E)-
595 - Benzaldehyde, oxime, (Z)-
600 - Benzamide
5867 - Ethanone, 1-(2-pyridinyl)-
5868 - Ethanone, 1-(3-pyridinyl)-
5869 - Ethanone, 1-(4-pyridinyl)-
5999 - Formamide, N-phenyl-
$C_7H_7NO_2$
579 - Benzaldehyde, 2-hydroxy-, oxime
616 - Benzamide, 2-hydroxy-
2000 - Benzene, 1-methyl-2-nitro-
2001 - Benzene, 1-methyl-3-nitro-
2002 - Benzene, 1-methyl-4-nitro-
2052 - Benzene, (nitromethyl)-
2484 - Benzoic acid, 2-amino-
2485 - Benzoic acid, 3-amino-
2486 - Benzoic acid, 4-amino-
8097 - Nitrous acid, phenylmethyl ester
10906 - 3-Pyridinecarboxylic acid, methyl ester
10908 - 4-Pyridinecarboxylic acid, methyl ester
11029 - Pyridinium, 3-carboxy-1-methyl-, hydroxide, inner salt
$C_7H_7NO_2S$
2043 - Benzene, 1-(methylthio)-2-nitro-
2044 - Benzene, 1-(methylthio)-4-nitro-
$C_7H_7NO_3$
613 - Benzamide, N,2-dihydroxy-
1906 - Benzenemethanol, 2-nitro-
1907 - Benzenemethanol, 3-nitro-
1908 - Benzenemethanol, 4-nitro-
1935 - Benzene, 1-methoxy-2-nitro-
1936 - Benzene, 1-methoxy-3-nitro-
1937 - Benzene, 1-methoxy-4-nitro-
2498 - Benzoic acid, 4-amino-2-hydroxy-
2499 - Benzoic acid, 5-amino-2-hydroxy-
9422 - Phenol, 2-methyl-4-nitro-
9423 - Phenol, 2-methyl-6-nitro-
9424 - Phenol, 4-methyl-2-nitro-
$C_7H_7NO_4$
9386 - Phenol, 2-methoxy-3-nitro-
$C_7H_7NO_4S$
2508 - Benzoic acid, 4-(aminosulfonyl)-
$C_7H_7N_3$
980 - Benzene, (azidomethyl)-
981 - Benzene, 1-azido-2-methyl-
982 - Benzene, 1-azido-4-methyl-
2872 - 1H-Benzotriazole, 1-methyl-
$C_7H_7N_3O_3$
12025 - Urea, (4-nitrophenyl)-
C_7H_8
1947 - Benzene, methyl-
2913 - Bicyclo[2.2.1]hepta-2,5-diene
4240 - 1,3,5-Cycloheptatriene
4765 - 1,3-Cyclopentadiene, 5-ethylidene-
6341 - 1,6-Heptadien-3-yne
6342 - 1,5-Heptadiyne
6343 - 1,6-Heptadiyne

C_7H_8BrN
647 - Benzenamine, 2-bromo-4-methyl-
648 - Benzenamine, 2-bromo-5-methyl-
649 - Benzenamine, 3-bromo-4-methyl-
650 - Benzenamine, 4-bromo-2-methyl-
651 - Benzenamine, 4-bromo-3-methyl-
652 - Benzenamine, 5-bromo-2-methyl-
1799 - Benzenemethanamine, 4-bromo-

C_7H_8ClN
668 - Benzenamine, 2-chloro-4-methyl-
669 - Benzenamine, 2-chloro-5-methyl-
670 - Benzenamine, 2-chloro-N-methyl-
671 - Benzenamine, 3-chloro-2-methyl-
672 - Benzenamine, 3-chloro-4-methyl-
673 - Benzenamine, 4-chloro-2-methyl-
674 - Benzenamine, 4-chloro-3-methyl-
675 - Benzenamine, 4-chloro-N-methyl-
676 - Benzenamine, 5-chloro-2-methyl-

C_7H_8ClNO
666 - Benzenamine, 2-chloro-5-methoxy-
667 - Benzenamine, 4-chloro-2-methoxy-

$C_7H_8ClN_3O_4S_2$
2855 - 2H-1,2,4-Benzothiadiazine-7-sulfonamide, 6-chloro-3,4-dihydro-, 1,1-dioxide

$C_7H_8Cl_2Si$
11401 - Silane, dichloromethylphenyl-

C_7H_8FN
755 - Benzenamine, 4-fluoro-2-methyl-

C_7H_8IN
762 - Benzenamine, 2-iodo-4-methyl-
763 - Benzenamine, 5-iodo-2-methyl-

$C_7H_8N_2$
563 - Benzaldehyde, hydrazone
11137 - 1H-Pyrrole-1-propanenitrile

$C_7H_8N_2O$
89 - Acetamide, N-3-pyridinyl-
601 - Benzamide, 2-amino-
806 - Benzenamine, N-methyl-N-nitroso-
2611 - Benzoic acid, hydrazide
12026 - Urea, phenyl-

$C_7H_8N_2O_2$
799 - Benzenamine, 2-methyl-3-nitro-
800 - Benzenamine, 2-methyl-4-nitro-
801 - Benzenamine, 2-methyl-5-nitro-
802 - Benzenamine, 2-methyl-6-nitro-
803 - Benzenamine, 4-methyl-2-nitro-
804 - Benzenamine, N-methyl-2-nitro-
805 - Benzenamine, N-methyl-4-nitro-
2552 - Benzoic acid, 2,3-diamino-
2553 - Benzoic acid, 2,4-diamino-
2554 - Benzoic acid, 3,5-diamino-

$C_7H_8N_2O_3$
775 - Benzenamine, 2-methoxy-5-nitro-
776 - Benzenamine, 3-methoxy-4-nitro-
777 - Benzenamine, 5-methoxy-2-nitro-

$C_7H_8N_2S$
11808 - Thiourea, phenyl-

$C_7H_8N_4O_2$
10756 - 1H-Purine-2,6-dione, 3,7-dihydro-1,3-dimethyl-
10757 - 1H-Purine-2,6-dione, 3,7-dihydro-3,7-dimethyl-

C_7H_8O
1836 - Benzenemethanol
1915 - Benzene, methoxy-
9400 - Phenol, 2-methyl-
9401 - Phenol, 3-methyl-
9402 - Phenol, 4-methyl-

C_7H_8OS
2040 - Benzene, (methylsulfinyl)-
2299 - Benzenethiol, 3-methoxy-
2300 - Benzenethiol, 4-methoxy-
5856 - Ethanone, 1-(5-methyl-2-thienyl)-

$C_7H_8O_2$
1527 - 1,2-Benzenediol, 3-methyl-
1528 - 1,2-Benzenediol, 4-methyl-
1529 - 1,3-Benzenediol, 2-methyl-

1530 - 1,3-Benzenediol, 4-methyl-
1531 - 1,3-Benzenediol, 5-methyl-
1532 - 1,4-Benzenediol, 2-methyl-
1882 - Benzenemethanol, 2-hydroxy-
1883 - Benzenemethanol, 3-hydroxy-
1884 - Benzenemethanol, 4-hydroxy-
9379 - Phenol, 2-methoxy-
9380 - Phenol, 3-methoxy-
9381 - Phenol, 4-methoxy-
10438 - 1-Propanone, 1-(2-furanyl)-
10439 - 2-Propanone, 1-(2-furanyl)-
10792 - 2H-Pyran-2-one, 4,6-dimethyl-
10793 - 4H-Pyran-4-one, 2,6-dimethyl-

$C_7H_8O_2S$
2181 - Benzenesulfinic acid, 2-methyl-
2182 - Benzenesulfinic acid, 4-methyl-
11739 - 2-Thiophenecarboxylic acid, ethyl ester

$C_7H_8O_3$
1520 - 1,4-Benzenediol, 2-(hydroxymethyl)-
1525 - 1,2-Benzenediol, 3-methoxy-
1526 - 1,3-Benzenediol, 5-methoxy-
2369 - 1,2,3-Benzenetriol, 5-methyl-
6029 - 2-Furanacetic acid, methyl ester
6056 - 2-Furancarboxylic acid, ethyl ester
6119 - 2-Furanmethanol, acetate
6161 - 2-Furanpropanoic acid

$C_7H_8O_3S$
2227 - Benzenesulfonic acid, methyl ester
2228 - Benzenesulfonic acid, 2-methyl-
2229 - Benzenesulfonic acid, 4-methyl-

C_7H_8S
1832 - Benzenemethanethiol
2041 - Benzene, (methylthio)-
2301 - Benzenethiol, 2-methyl-
2302 - Benzenethiol, 3-methyl-
2303 - Benzenethiol, 4-methyl-

$C_7H_8S_2$
1564 - 1,2-Benzenedithiol, 4-methyl-

$C_7H_9AsN_2O_4$
455 - Arsonic acid, [4-[(aminocarbonyl)amino]phenyl]-

$C_7H_9BO_2$
3134 - Boronic acid, (2-methylphenyl)-

C_7H_9Cl
7107 - 1-Hexen-3-yne, 5-chloro-5-methyl-

C_7H_9ClO
9147 - 2-Pentynoyl chloride, 4,4-dimethyl-

C_7H_9ClSi
11381 - Silane, chloromethylphenyl-

C_7H_9N
778 - Benzenamine, 2-methyl-
779 - Benzenamine, 3-methyl-
780 - Benzenamine, 4-methyl-
781 - Benzenamine, N-methyl-
1795 - Benzenemethanamine
4629 - 3-Cyclohexene-1-carbonitrile
4630 - 1-Cyclohexenecarbonitrile
4905 - 1-Cyclopentene-1-acetonitrile
10936 - Pyridine, 2,3-dimethyl-
10937 - Pyridine, 2,4-dimethyl-
10938 - Pyridine, 2,5-dimethyl-
10939 - Pyridine, 2,6-dimethyl-
10940 - Pyridine, 3,4-dimethyl-
10941 - Pyridine, 3,5-dimethyl-
10956 - Pyridine, 2-ethyl-
10957 - Pyridine, 3-ethyl-
10958 - Pyridine, 4-ethyl-

C_7H_9NO
758 - Benzenamine, N-hydroxy-4-methyl-
766 - Benzenamine, 2-methoxy-
767 - Benzenamine, 3-methoxy-
768 - Benzenamine, 4-methoxy-
1837 - Benzenemethanol, 2-amino-
5849 - Ethanone, 1-(1-methyl-1H-pyrrol-2-yl)-
9209 - Phenol, 2-amino-3-methyl-
9210 - Phenol, 2-amino-4-methyl-
9211 - Phenol, 3-amino-4-methyl-

9212 - Phenol, 3-amino-5-methyl-
9213 - Phenol, 4-amino-2-methyl-
10945 - Pyridine, 2,6-dimethyl-, 1-oxide
10950 - 2-Pyridineethanol

$C_7H_9NO_2$
2509 - Benzoic acid, ammonium salt
10935 - Pyridine, 2,6-dimethoxy-

$C_7H_9NO_3$
8586 - 2,4-Oxazolidinedione, 5-methyl-3-(2-propenyl)-

C_7H_9NS
819 - Benzenamine, 2-(methylthio)-
820 - Benzenamine, 4-(methylthio)-

$C_7H_9N_3O$
7145 - Hydrazinecarboxamide, 2-phenyl-
7146 - Hydrazinecarboxamide, N-phenyl-

$C_7H_9N_3O_2$
2500 - Benzoic acid, 4-amino-2-hydroxy-, hydrazide

$C_7H_9N_3O_2S_2$
2189 - Benzenesulfonamide, 4-amino-N-(aminothioxomethyl)-

$C_7H_9N_3O_3S$
2190 - Benzenesulfonamide, 4-amino-N-(aminocarbonyl)-

C_7H_{10}
4219 - 1,3-Cycloheptadiene
4271 - 1,3-Cyclohexadiene, 2-methyl-
4272 - 1,3-Cyclohexadiene, 5-methyl-, (±)-
6529 - 1,3,5-Heptatriene
6606 - 1-Hepten-3-yne
9115 - 3-Penten-1-yne, 3-ethyl-

$C_7H_{10}ClN$
791 - Benzenamine, 2-methyl-, hydrochloride
792 - Benzenamine, N-methyl-, hydrochloride

$C_7H_{10}ClN_3$
11043 - 4-Pyrimidinamine, 2-chloro-N,N,6-trimethyl-

$C_7H_{10}N_2$
1275 - 1,2-Benzenediamine, 3-methyl-
1276 - 1,2-Benzenediamine, 4-methyl-
1277 - 1,3-Benzenediamine, 2-methyl-
1278 - 1,3-Benzenediamine, 4-methyl-
1279 - 1,4-Benzenediamine, 2-methyl-
1280 - 1,4-Benzenediamine, N-methyl-
1796 - Benzenemethanamine, 2-amino-
4159 - Cyanamide, di-2-propenyl-
6392 - Heptanedinitrile
7164 - Hydrazine, (3-methylphenyl)-
7165 - Hydrazine, 1-methyl-1-phenyl-
7166 - Hydrazine, 1-methyl-2-phenyl-
10863 - 2-Pyridinamine, N,N-dimethyl-
10864 - 4-Pyridinamine, 2,6-dimethyl-
10948 - 2-Pyridineethanamine
10949 - 4-Pyridineethanamine

$C_7H_{10}N_2O$
1273 - 1,2-Benzenediamine, 4-methoxy-
1274 - 1,3-Benzenediamine, 4-methoxy-

$C_7H_{10}N_2OS$
11109 - 4(1H)-Pyrimidinone, 2,3-dihydro-6-propyl-2-thioxo-

$C_7H_{10}N_2O_2S$
7194 - 1H-Imidazole-1-carboxylic acid, 2,3-dihydro-3-methyl-2-thioxo-, ethyl ester

$C_7H_{10}N_4O_2S$
2191 - Benzenesulfonamide, 4-amino-N-(aminoiminomethyl)-

$C_7H_{10}O$
4631 - 1-Cyclohexene-1-carboxaldehyde
4632 - 3-Cyclohexene-1-carboxaldehyde
4710 - 2-Cyclohexen-1-one, 2-methyl-
4711 - 2-Cyclohexen-1-one, 3-methyl-
4712 - 2-Cyclohexen-1-one, 5-methyl-
4713 - 3-Cyclohexen-1-one, 3-methyl-
4714 - 3-Cyclohexen-1-one, 4-methyl-
4866 - Cyclopentanol, 1-ethynyl-
4943 - 2-Cyclopenten-1-one, 2,4-dimethyl-
6106 - Furan, 2-ethyl-5-methyl-
6167 - Furan, 2-propyl-

6325 - 2,4-Heptadienal, (E,E)-
6339 - 3,5-Heptadien-2-one
7607 - Methanone, dicyclopropyl-

$C_7H_{10}O_2$
3856 - 2-Butenoic acid, 2-propenyl ester
3857 - 2-Butenoic acid, 2-propenyl ester, (E)-
4225 - 1,2-Cycloheptanedione
4637 - 1-Cyclohexene-1-carboxylic acid
4638 - 3-Cyclohexene-1-carboxylic acid
4879 - Cyclopentanone, 2-acetyl-
4904 - 3-Cyclopentene-1-acetic acid
6104 - Furan, 2-(ethoxymethyl)-
6112 - Furan, 2-isopropoxy-
6702 - 2,4-Hexadienoic acid, methyl ester
6703 - 2,4-Hexadienoic acid, methyl ester, (E,E)-
10671 - 2-Propenoic acid, 2-methyl-, 2-propenyl ester

$C_7H_{10}O_3$
3577 - Butanoic acid, 3-oxo-, 2-propenyl ester
6454 - 2,4,6-Heptanetrione
10668 - 2-Propenoic acid, 2-methyl-, oxiranylmethyl ester

$C_7H_{10}O_4$
3330 - Butanedioic acid, methylene-, dimethyl ester
3331 - Butanedioic acid, (1-methylethylidene)-
3789 - 2-Butenedioic acid, 2-methyl-, dimethyl ester, (E)-
3790 - 2-Butenedioic acid, 2-methyl-, dimethyl ester, (Z)-
4793 - 1,3-Cyclopentanedicarboxylic acid, cis-
4969 - 1,2-Cyclopropanedicarboxylic acid, dimethyl ester
8876 - Pentanoic acid, 2,4-dioxo-, ethyl ester
10563 - 2-Propene-1,1-diol, diacetate

$C_7H_{10}O_5$
8784 - Pentanedioic acid, 3-oxo-, dimethyl ester
9973 - Propanedioic acid, oxo-, diethyl ester

$C_7H_{10}O_6$
7553 - Methanetricarboxylic acid, trimethyl ester

$C_7H_{10}S$
11761 - Thiophene, 2-ethyl-3-methyl-
11762 - Thiophene, 2-ethyl-5-methyl-
11763 - Thiophene, 3-ethyl-4-methyl-
11764 - Thiophene, 4-ethyl-2-methyl-
11772 - Thiophene, 2-(1-methylethyl)-
11773 - Thiophene, 3-(1-methylethyl)-
11778 - Thiophene, 2-propyl-
11779 - Thiophene, 3-propyl-
11792 - Thiophene, 2,3,4-trimethyl-
11793 - Thiophene, 2,3,5-trimethyl-

$C_7H_{10}Si$
11442 - Silane, methylphenyl-

$C_7H_{11}Br$
6612 - 1-Heptyne, 1-bromo-
6613 - 2-Heptyne, 1-bromo-

$C_7H_{11}BrO_4$
9928 - Propanedioic acid, bromo-, diethyl ester

$C_7H_{11}Cl$
2918 - Bicyclo[2.2.1]heptane, 2-chloro-
6614 - 1-Heptyne, 1-chloro-
6615 - 2-Heptyne, 7-chloro-
6616 - 3-Heptyne, 7-chloro-
9125 - 1-Pentyne, 3-chloro-3-ethyl-

$C_7H_{11}ClO$
4330 - Cyclohexanecarbonyl chloride

$C_7H_{11}ClO_3$
3492 - Butanoic acid, 2-chloro-2-methyl-3-oxo-, ethyl ester

$C_7H_{11}ClO_4$
9931 - Propanedioic acid, chloro-, diethyl ester
9996 - 1,2-Propanediol, 3-chloro-, diacetate

$C_7H_{11}Cl_3O_2$
255 - Acetic acid, trichloro-, 3-methylbutyl ester

$C_7H_{11}I$
6620 - 1-Heptyne, 1-iodo-

$C_7H_{11}N$
4328 - Cyclohexanecarbonitrile
11128 - 1H-Pyrrole, 3-ethyl-4-methyl-

11129 - Pyrrole, 2-isopropyl-
11138 - 1H-Pyrrole, 1-propyl-
11139 - 1H-Pyrrole, 3-propyl-
11142 - 1H-Pyrrole, 1,2,5-trimethyl-

$C_7H_{11}NO$
4329 - Cyclohexanecarbonitrile, 1-hydroxy-
4416 - Cyclohexane, isocyanato-

$C_7H_{11}NO_2$
151 - Acetic acid, cyano-, butyl ester
9720 - 4-Piperidinone, 1-acetyl-
10913 - 3-Pyridinecarboxylic acid, 1,2,5,6-tetrahydro-1-methyl-
11158 - 2,5-Pyrrolidinedione, 3-ethyl-3-methyl-

$C_7H_{11}NO_3$
11567 - Succinimide, N-(ethoxymethyl)-

$C_7H_{11}NO_4$
9766 - L-Proline, 1-acetyl-4-hydroxy-, trans-

$C_7H_{11}NO_5$
9958 - Propanedioic acid, (hydroxyimino)-, diethyl ester

$C_7H_{11}NS$
4417 - Cyclohexane, isothiocyanato-

C_7H_{12}
2914 - Bicyclo[2.2.1]heptane
2915 - Bicyclo[4.1.0]heptane
2980 - Bicyclo[3,1,0]hexane, 1-methyl-
4244 - Cycloheptene
4428 - Cyclohexane, methylene-
4670 - Cyclohexene, 1-methyl-
4671 - Cyclohexene, 3-methyl-, (±)-
4672 - Cyclohexene, 4-methyl-
4813 - Cyclopentane, ethylidene-
4918 - Cyclopentene, 1,2-dimethyl-
4919 - Cyclopentene, 1,3-dimethyl-
4920 - Cyclopentene, 1,4-dimethyl-
4921 - Cyclopentene, 1,5-dimethyl-
4922 - Cyclopentene, 4,4-dimethyl-
4924 - Cyclopentene, 1-ethyl-
4925 - Cyclopentene, 3-ethyl-
4926 - Cyclopentene, 4-ethyl-
6326 - 1,2-Heptadiene
6327 - 1,4-Heptadiene
6328 - 1,5-Heptadiene
6329 - 2,4-Heptadiene
6609 - 1-Heptyne
6610 - 2-Heptyne
6611 - 3-Heptyne
6695 - 1,5-Hexadiene, 2-methyl-
6696 - 2,4-Hexadiene, 2-methyl-
7120 - 1-Hexyne, 5-methyl-
7121 - 2-Hexyne, 5-methyl-
7122 - 3-Hexyne, 2-methyl-
8665 - 1,3-Pentadiene, 2,4-dimethyl-
9128 - 1-Pentyne, 3,3-dimethyl-
9129 - 1-Pentyne, 4,4-dimethyl-
9130 - 1-Pentyne, 4,4-dimethyl-
9131 - 1-Pentyne, 3-ethyl-

$C_7H_{12}Br_2O_2$
8872 - Pentanoic acid, 2,3-dibromo-, ethyl ester
8873 - Pentanoic acid, 2,5-dibromo-, ethyl ester
10239 - Propanoic acid, 2,3-dibromo-, butyl ester

$C_7H_{12}ClN_5$
11513 - Simazine

$C_7H_{12}N_2$
7196 - 1H-Imidazole, 1,2-diethyl-
11190 - Pyrrolo[1,2-a]pyrimidine, 2,3,4,6,7,8-hexahydro-

$C_7H_{12}O$
4234 - Cycloheptanone
4331 - Cyclohexanecarboxaldehyde
4537 - Cyclohexanol, 2-methylene-
4596 - Cyclohexanone, 2-methyl-, (±)-
4597 - Cyclohexanone, 3-methyl-, (±)-
4598 - Cyclohexanone, 4-methyl-
4882 - Cyclopentanone, 2,5-dimethyl-
5795 - Ethanone, 1-cyclopentyl-
6336 - 1,5-Heptadien-4-ol

6337 - 1,6-Heptadien-4-ol
6530 - 2-Heptenal
6531 - 3-Heptenal
6583 - 1-Hepten-3-one
6584 - 4-Hepten-2-one
6585 - 6-Hepten-2-one
6586 - 6-Hepten-3-one
6587 - 3-Hepten-2-one, (E)-
6588 - 3-Hepten-2-one, (Z)-
6589 - 5-Hepten-2-one, (E)-
7093 - 1-Hexen-3-one, 5-methyl-
7094 - 1-Hexen-3-one, 5-methyl-
7095 - 4-Hexen-2-one, 5-methyl-
7096 - 4-Hexen-3-one, 2-methyl-
7097 - 5-Hexen-2-one, 3-methyl-
7098 - 5-Hexen-2-one, 5-methyl-
7099 - 5-Hexen-2-one, 5-methyl-
7128 - 1-Hexyn-3-ol, 3-methyl-
7129 - 3-Hexyn-2-ol, 2-methyl-
9144 - 1-Pentyn-3-ol, 3,4-dimethyl-
9145 - 1-Pentyn-3-ol, 3-ethyl-

$C_7H_{12}O_2$
3587 - Butanoic acid, 2-propenyl ester
3849 - 2-Butenoic acid, 2-methyl-, ethyl ester, (E)-
3850 - 2-Butenoic acid, 3-methyl-, ethyl ester
4175 - Cyclobutanecarboxylic acid, ethyl ester
4332 - Cyclohexanecarboxylic acid
4775 - Cyclopentaneacetic acid
4783 - Cyclopentanecarboxylic acid, 1-methyl-
4865 - Cyclopentanol, acetate
6005 - Formic acid, cyclohexyl ester
6402 - 2,3-Heptanedione
6403 - 2,6-Heptanedione
6561 - 2-Heptenoic acid
6562 - 4-Heptenoic acid
6563 - 5-Heptenoic acid
6564 - 6-Heptenoic acid
6809 - 2,3-Hexanedione, 5-methyl-
6810 - 2,5-Hexanedione, 3-methyl-
7055 - 2-Hexenoic acid, 2-methyl-, (E)-
7056 - 2-Hexenoic acid, 4-methyl-
7057 - 2-Hexenoic acid, 5-methyl-
7058 - 3-Hexenoic acid, 2-methyl-
7059 - 3-Hexenoic acid, 3-methyl-
7060 - 3-Hexenoic acid, 4-methyl-
7061 - 4-Hexenoic acid, 5-methyl-
7062 - 3-Hexenoic acid, methyl ester
8799 - 2,4-Pentanedione, 3,3-dimethyl-
8801 - 2,4-Pentanedione, 3-ethyl-
9075 - 4-Pentenoic acid, 2,2-dimethyl-
9083 - 2-Pentenoic acid, 2-methyl-, methyl ester
9084 - 2-Pentenoic acid, 3-methyl-, methyl ester, (E)-
9085 - 2-Pentenoic acid, 3-methyl-, methyl ester, (Z)-
9086 - 3-Pentenoic acid, 3-methyl-, methyl ester
10616 - 2-Propenoic acid, butyl ester
10664 - 2-Propenoic acid, 2-methyl-, 1-methylethyl ester
10672 - 2-Propenoic acid, 2-methyl-, propyl ester
10673 - 2-Propenoic acid, 2-methylpropyl ester
10731 - 1-Propyne, 3,3-diethoxy-
10783 - 2H-Pyran, 2-ethoxy-3,4-dihydro-

$C_7H_{12}O_3$
3519 - Butanoic acid, 2-ethyl-3-oxo-, methyl ester
3557 - Butanoic acid, 2-methyl-3-oxo-, ethyl ester
3574 - Butanoic acid, 3-oxo-, 1-methylethyl ester
6125 - 2-Furanmethanol, tetrahydro-, acetate
6164 - 2-Furanpropanoic acid, tetrahydro-
6469 - Heptanoic acid, 6-oxo-
6901 - Hexanoic acid, 4-methyl-5-oxo-
8874 - Pentanoic acid, 2,2-dimethyl-4-oxo-
8898 - Pentanoic acid, 3-methyl-4-oxo-, methyl ester
8905 - Pentanoic acid, 2-oxo-, ethyl ester
8906 - Pentanoic acid, 3-oxo-, ethyl ester
8907 - Pentanoic acid, 4-oxo-, ethyl ester
10634 - 2-Propenoic acid, 2-ethoxyethyl ester
10662 - 2-Propenoic acid, 2-methyl-, 2-hydroxypropyl ester

C$_7$H$_{12}$O$_4$
3323 - Butanedioic acid, ethyl methyl ester
3327 - Butanedioic acid, methyl-, dimethyl ester
3338 - Butanedioic acid, monopropyl ester
6393 - Heptanedioic acid
6788 - Hexanedioic acid, 2-methyl-
6789 - Hexanedioic acid, 3-methyl-, (±)-
6792 - Hexanedioic acid, monomethyl ester
7654 - Methylene bispropionate
8773 - Pentanedioic acid, 3,3-dimethyl-
8775 - Pentanedioic acid, dimethyl ester
9929 - Propanedioic acid, butyl-
9939 - Propanedioic acid, diethyl-
9941 - Propanedioic acid, diethyl ester
9968 - Propanedioic acid, (2-methylpropyl)-
10002 - 1,2-Propanediol, diacetate
10003 - 1,3-Propanediol, diacetate
11643 - 2,4,8,10-Tetraoxaspiro[5.5]undecane

C$_7$H$_{12}$O$_5$
9957 - Propanedioic acid, hydroxy-, diethyl ester
10189 - 1,2,3-Propanetriol, diacetate
10190 - 1,2,3-Propanetriol, 1,3-diacetate

C$_7$H$_{12}$O$_6$
4338 - Cyclohexanecarboxylic acid, 1,3,4,5-
  tetrahydroxy-, [1R-(1α,3α,4α,5β)]-

C$_7$H$_{12}$O$_6$Si
11499 - Silanetriol, methyl-, triacetate

C$_7$H$_{13}$Br
4222 - Cycloheptane, bromo-
4318 - Cyclohexane, (bromomethyl)-
4319 - Cyclohexane, 1-bromo-1-methyl-
4320 - Cyclohexane, 1-bromo-3-methyl-

C$_7$H$_{13}$BrN$_2$O$_2$
3195 - Butanamide, N-(aminocarbonyl)-2-bromo-2-
  ethyl-

C$_7$H$_{13}$BrO
4317 - Cyclohexane, 1-bromo-2-methoxy-

C$_7$H$_{13}$BrO$_2$
3475 - Butanoic acid, 2-bromo-3-methyl-, ethyl ester
6457 - Heptanoic acid, 2-bromo-
6458 - Heptanoic acid, 7-bromo-
8860 - Pentanoic acid, 2-bromo-, ethyl ester
8861 - Pentanoic acid, 5-bromo-, ethyl ester
10206 - Propanoic acid, 3-bromo-, butyl ester

C$_7$H$_{13}$Cl
6538 - 1-Heptene, 1-chloro-
6539 - 1-Heptene, 2-chloro-
6540 - 3-Heptene, 4-chloro-

C$_7$H$_{13}$ClO
6513 - 2-Heptanone, 1-chloro-
6527 - Heptanoyl chloride
6988 - Hexanoyl chloride, 3-methyl-
6989 - Hexanoyl chloride, 4-methyl-

C$_7$H$_{13}$ClO$_2$
3490 - Butanoic acid, 2-chloro-3-methyl-, ethyl ester
3491 - Butanoic acid, 2-chloro-2-methyl-, ethyl ester
8867 - Pentanoic acid, 4-chloro-, ethyl ester
8868 - Pentanoic acid, 5-chloro-, ethyl ester
8869 - Pentanoic acid, 2-chloroethyl ester
10217 - Propanoic acid, 2-chloro-, butyl ester
10218 - Propanoic acid, 3-chloro-, butyl ester
10224 - Propanoic acid, 2-chloro-, 2-methylpropyl
  ester, (R)-
10227 - Propanoic acid, 3-chloro-, 2-methylpropyl
  ester
12039 - Valeric acid, 3-chloro-, ethyl ester

C$_7$H$_{13}$FO$_2$
6463 - Heptanoic acid, 7-fluoro-

C$_7$H$_{13}$N
480 - 1-Azabicyclo[2.2.2]octane
6428 - Heptanenitrile
9121 - 1-Pentyn-3-amine, 3-ethyl-
9666 - Piperidine, 3-ethenyl-

C$_7$H$_{13}$NO
485 - 1-Azabicyclo[2.2.2]octan-3-ol
5990 - Formamide, N-cyclohexyl-
9630 - Piperidine, 1-acetyl-

C$_7$H$_{13}$NO$_2$
4443 - Cyclohexane, 1-methyl-1-nitro-
4444 - Cyclohexane, 1-methyl-2-nitro-
4445 - Cyclohexane, 1-methyl-4-nitro-
4455 - Cyclohexane, (nitromethyl)-
9644 - 4-Piperidinecarboxylic acid, methyl ester
10633 - 2-Propenoic acid, 2-(dimethylamino)ethyl
  ester
11177 - Pyrrolidinium, 2-carboxy-1,1-dimethyl-,
  hydroxide, inner salt, (S)-

C$_7$H$_{13}$NO$_3$
10107 - Propanenitrile, 3-[2-
  (2-hydroxyethoxy)ethoxy]-
11178 - Pyrrolidinium, 2-carboxy-4-hydroxy-1,1-
  dimethyl-, hydroxide, inner salt, (2S-trans)-

C$_7$H$_{13}$NO$_3$S
7649 - L-Methionine, N-acetyl-

C$_7$H$_{13}$NO$_4$
6245 - L-Glutamic acid, 5-ethyl ester
9924 - Propanedioic acid, amino-, diethyl ester

C$_7$H$_{13}$N$_3$O$_3$S
8568 - Oxamyl

C$_7$H$_{13}$O$_6$P
7659 - Mevinphos

C$_7$H$_{14}$
3797 - 1-Butene, 2-ethyl-3-methyl-
3812 - 1-Butene, 2,3,3-trimethyl-
4221 - Cycloheptane
4426 - Cyclohexane, methyl-
4800 - Cyclopentane, 1,1-dimethyl-
4801 - Cyclopentane, 1,2-dimethyl-, cis-
4802 - Cyclopentane, 1,2-dimethyl-, trans-
4803 - Cyclopentane, 1,3-dimethyl-, cis-
4804 - Cyclopentane, 1,3-dimethyl-, trans-
4812 - Cyclopentane, ethyl-
6533 - 1-Heptene
6534 - 2-Heptene, (E)-
6535 - 2-Heptene, (Z)-
6536 - 3-Heptene, (E)-
6537 - 3-Heptene, (Z)-
7036 - 1-Hexene, 2-methyl-
7037 - 1-Hexene, 3-methyl-
7038 - 1-Hexene, 4-methyl-
7039 - 1-Hexene, 5-methyl-
7040 - 2-Hexene, 2-methyl-
7041 - 2-Hexene, 3-methyl-, (Z)-
7042 - 2-Hexene, 4-methyl-, (E)-
7043 - 2-Hexene, 4-methyl-, (Z)-
7044 - 2-Hexene, 5-methyl-, (E)-
7045 - 2-Hexene, 5-methyl-, (Z)-
7046 - 3-Hexene, 2-methyl-, (E)-
7047 - 3-Hexene, 2-methyl-, (Z)-
9029 - 1-Pentene, 2,3-dimethyl-
9030 - 1-Pentene, 2,4-dimethyl-
9031 - 1-Pentene, 3,3-dimethyl-
9032 - 1-Pentene, 3,4-dimethyl-
9033 - 1-Pentene, 4,4-dimethyl-
9034 - 2-Pentene, 2,3-dimethyl-
9035 - 2-Pentene, 2,4-dimethyl-
9036 - 2-Pentene, 3,4-dimethyl-
9037 - 2-Pentene, 3,4-dimethyl-, (E)-
9038 - 2-Pentene, 3,4-dimethyl-, (Z)-
9039 - 2-Pentene, 4,4-dimethyl-, (E)-
9040 - 2-Pentene, 4,4-dimethyl-, (Z)-
9043 - 1-Pentene, 2-ethyl-
9044 - 1-Pentene, 3-ethyl-
9045 - 2-Pentene, 3-ethyl-

C$_7$H$_{14}$BrF
6359 - Heptane, 1-bromo-7-fluoro-

C$_7$H$_{14}$BrNO
3199 - Butanamide, 2-bromo-2-ethyl-3-methyl-

C$_7$H$_{14}$Br$_2$
6372 - Heptane, 1,2-dibromo-
6373 - Heptane, 1,5-dibromo-
6374 - Heptane, 1,7-dibromo-
6375 - Heptane, 2,3-dibromo-
6376 - Heptane, 3,4-dibromo-

C$_7$H$_{14}$ClF
6365 - Heptane, 1-chloro-7-fluoro-

C$_7$H$_{14}$Cl$_2$
6377 - Heptane, 1,1-dichloro-
6378 - Heptane, 1,2-dichloro-
6379 - Heptane, 1,7-dichloro-
6380 - Heptane, 2,2-dichloro-
6381 - Heptane, 4,4-dichloro-
8752 - Pentane, 1,2-dichloro-4,4-dimethyl-
8753 - Pentane, 1,5-dichloro-3,3-dimethyl-
8754 - Pentane, 2,4-dichloro-2,4-dimethyl-

C$_7$H$_{14}$NO$_5$P
7663 - Monocrotophos

C$_7$H$_{14}$N$_2$
9815 - 2-Propanamine, N,N′-methanetetraylbis-

C$_7$H$_{14}$N$_2$O$_2$
9604 - 1-Piperazinecarboxylic acid, ethyl ester

C$_7$H$_{14}$N$_2$O$_2$S
334 - Aldicarb

C$_7$H$_{14}$N$_2$O$_4$
3944 - Carbamic acid, (1,1-dimethylethyl)nitro-, ethyl
  ester
3951 - Carbamic acid, ethylnitro-, methyl ester

C$_7$H$_{14}$N$_2$O$_4$S
335 - Aldoxycarb

C$_7$H$_{14}$N$_2$O$_4$S$_2$
5018 - L-Cysteine, S,S′-methylenebis-

C$_7$H$_{14}$O
3371 - Butane, 1-(ethenyloxy)-3-methyl-
4229 - Cycloheptanol
4419 - Cyclohexanemethanol
4425 - Cyclohexane, methoxy-
4530 - Cyclohexanol, 1-methyl-
4531 - Cyclohexanol, 2-methyl-, cis-(±)-
4532 - Cyclohexanol, 2-methyl-, trans-(±)-
4533 - Cyclohexanol, 3-methyl-, cis-(±)-
4534 - Cyclohexanol, 3-methyl-, trans-(±)-
4535 - Cyclohexanol, 4-methyl-, cis-
4536 - Cyclohexanol, 4-methyl-, trans-
4811 - Cyclopentaneethanol
6173 - Furan, tetrahydro-2-propyl-
6344 - Heptanal
6510 - 2-Heptanone
6511 - 3-Heptanone
6512 - 4-Heptanone
6566 - 1-Hepten-4-ol
6567 - 2-Hepten-1-ol, (E)-
6568 - 2-Hepten-1-ol, (Z)-
6569 - 2-Hepten-4-ol, (±)-
6570 - 3-Hepten-2-ol
6571 - 6-Hepten-2-ol
6572 - 6-Hepten-3-ol
6720 - Hexanal, 3-methyl-
6976 - 2-Hexanone, 3-methyl-
6977 - 2-Hexanone, 4-methyl-
6978 - 2-Hexanone, 5-methyl-
6979 - 3-Hexanone, 2-methyl-
6980 - 3-Hexanone, 4-methyl-
6981 - 3-Hexanone, 5-methyl-
7077 - 1-Hexen-3-ol, 2-methyl-
7078 - 1-Hexen-1-ol, 5-methyl-
7079 - 3-Hexen-2-ol, 2-methyl-
7080 - 3-Hexen-2-ol, 3-methyl-
7081 - 4-Hexen-3-ol, 2-methyl-
7082 - 4-Hexen-3-ol, 3-methyl-
7083 - 5-Hexen-2-ol, 2-methyl-
7084 - 5-Hexen-3-ol, 3-methyl-
7085 - 5-Hexen-3-ol, 4-methyl-
8965 - 2-Pentanone, 4,4-dimethyl-
8966 - 3-Pentanone, 2,2-dimethyl-
8967 - 3-Pentanone, 2,4-dimethyl-

C$_7$H$_{14}$O$_2$
227 - Acetic acid, pentyl ester
3516 - Butanoic acid, 2-ethyl-2-methyl-
3517 - Butanoic acid, 2-ethyl-, methyl ester
3548 - Butanoic acid, 3-methyl-, ethyl ester
3549 - Butanoic acid, 1-methylethyl ester

3550 - Butanoic acid, 2-methyl-, ethyl ester, (S)-
3588 - Butanoic acid, propyl ester
3630 - 1-Butanol, 2-methyl-, acetate
3631 - 1-Butanol, 3-methyl-, acetate
6009 - Formic acid, hexyl ester
6455 - Heptanoic acid
6517 - 4-Heptanone, 2-hydroxy-
6893 - Hexanoic acid, 2-methyl-, (±)-
6894 - Hexanoic acid, 3-methyl-
6895 - Hexanoic acid, 4-methyl-, (±)-
6896 - Hexanoic acid, 5-methyl-
6898 - Hexanoic acid, methyl ester
6973 - 3-Hexanone, 2-hydroxy-2-methyl-
6974 - 3-Hexanone, 5-hydroxy-2-methyl-
8602 - Oxirane, (butoxymethyl)-
8610 - Oxirane, [(1,1-dimethylethoxy)methyl]-
8877 - Pentanoic acid, 2-ethyl-, (±)-
8878 - Pentanoic acid, ethyl ester
8918 - 2-Pentanol, acetate, (R)-
8919 - 3-Pentanol, acetate
8978 - 2-Pentanone, 4-methoxy-4-methyl-
10214 - Propanoic acid, butyl ester
10259 - Propanoic acid, 2,2-dimethyl-, ethyl ester
10303 - Propanoic acid, 2-methyl-, 1-methylethyl ester
10308 - Propanoic acid, 2-methyl-, propyl ester
10309 - Propanoic acid, 1-methylpropyl ester
10310 - Propanoic acid, 2-methylpropyl ester
10559 - 1-Propene, 1,1-diethoxy-
10560 - 1-Propene, 3,3-diethoxy-

$C_7H_{14}O_3$
3995 - Carbonic acid, bis(1-methylethyl) ester
4002 - Carbonic acid, dipropyl ester
8884 - Pentanoic acid, 4-hydroxy-, ethyl ester
8885 - Pentanoic acid, 2-hydroxyethyl ester
10262 - Propanoic acid, 3-ethoxy-, ethyl ester
10268 - Propanoic acid, 2-hydroxy-, butyl ester, (R)-

$C_7H_{14}O_4$
3505 - Butanoic acid, 2,3-dihydroxypropyl ester
7112 - ribo-Hexose, 2,6-dideoxy-3-O-methyl-

$C_7H_{14}O_5$
5736 - Ethanol, 2-methoxy-, carbonate (2:1)
6198 - Galactose, 6-deoxy-3-O-methyl-

$C_7H_{14}O_6$
6227 - α-D-Glucopyranoside, methyl

$C_7H_{15}Br$
6356 - Heptane, 1-bromo-
6357 - Heptane, 2-bromo-
6358 - Heptane, 4-bromo-

$C_7H_{15}BrO$
6736 - Hexane, 1-bromo-6-methoxy-

$C_7H_{15}Cl$
3254 - Butane, 2-chloro-2,3,3-trimethyl-
6360 - Heptane, 1-chloro-
6361 - Heptane, 2-chloro-
6362 - Heptane, 3-chloro-
6363 - Heptane, 4-chloro-
6743 - Hexane, 1-chloro-3-methyl-
6744 - Hexane, 2-chloro-2-methyl-
6745 - Hexane, 2-chloro-5-methyl-
6746 - Hexane, 3-chloro-3-methyl-
8722 - Pentane, 2-chloro-2,3-dimethyl-
8723 - Pentane, 2-chloro-2,4-dimethyl-
8724 - Pentane, 2-chloro-2,3-dimethyl-
8725 - Pentane, 4-chloro-2,2-dimethyl-
8726 - Pentane, 3-chloro-3-ethyl-

$C_7H_{15}ClO$
6477 - 1-Heptanol, 7-chloro-
6478 - 2-Heptanol, 1-chloro-

$C_7H_{15}ClO_2$
9850 - Propane, 3-chloro-1,1-diethoxy-

$C_7H_{15}F$
6417 - Heptane, 1-fluoro-

$C_7H_{15}FO$
6491 - 1-Heptanol, 7-fluoro-

$C_7H_{15}I$
6419 - Heptane, 1-iodo-

6420 - Heptane, 2-iodo-
$C_7H_{15}N$
502 - Azocine, octahydro-
4292 - Cyclohexanamine, 2-methyl-, cis-
4293 - Cyclohexanamine, 2-methyl-, trans-
4294 - Cyclohexanamine, 3-methyl-, cis-
4295 - Cyclohexanamine, 3-methyl-, trans-
4296 - Cyclohexanamine, 4-methyl-, cis-
4297 - Cyclohexanamine, 4-methyl-, trans-
4298 - Cyclohexanamine, N-methyl-
4418 - Cyclohexanemethanamine
9648 - Piperidine, 1,2-dimethyl-, (±)-
9649 - Piperidine, 2,5-dimethyl-
9650 - Piperidine, 2,6-dimethyl-
9651 - Piperidine, 3,3-dimethyl-
9652 - Piperidine, 3,5-dimethyl-
9667 - Piperidine, 1-ethyl-
9668 - Piperidine, 2-ethyl-, (±)-
9669 - Piperidine, 3-ethyl-, (±)-
9670 - Piperidine, 4-ethyl-
11173 - Pyrrolidine, 1-propyl-
11174 - Pyrrolidine, 3-propyl-
11175 - Pyrrolidine, 2,2,4-trimethyl-
11176 - Pyrrolidine, 2,2,5-trimethyl-

$C_7H_{15}NO$
6346 - Heptanal, oxime
6348 - Heptanamide
6982 - 2-Hexanone, 5-methyl-, oxime
8699 - Pentanamide, N,N-dimethyl-
8968 - 3-Pentanone, 2,4-dimethyl-, oxime
9662 - 1-Piperidineethanol
9663 - 2-Piperidineethanol
9664 - 4-Piperidineethanol
9678 - 3-Piperidinemethanol, 1-methyl-
9715 - 3-Piperidinol, 1-ethyl-
9793 - Propanamide, N,N-diethyl-
10427 - 2-Propanone, 1-(diethylamino)-

$C_7H_{15}NOS$
5930 - Ethiolate

$C_7H_{15}NO_2$
3541 - Butanoic acid, 3-(methylamino)-, ethyl ester
3936 - Carbamic acid, butyl-, ethyl ester
3942 - Carbamic acid, (1,1-dimethylethyl)-, ethyl ester
3947 - Carbamic acid, ethyl-, butyl ester
3960 - Carbamic acid, (1-methylpropyl)-, ethyl ester
3961 - Carbamic acid, (2-methylpropyl)-, ethyl ester
6429 - Heptane, 1-nitro-
7692 - 4-Morpholineethanol, α-methyl-
8088 - Nitrous acid, heptyl ester

$C_7H_{15}NO_5$
6215 - α-L-Glucopyranose, 2-(methylamino)-2-deoxy-

C_7H_{16}
3457 - Butane, 2,2,3-trimethyl-
6355 - Heptane
6834 - Hexane, 2-methyl-
6835 - Hexane, 3-methyl-, (S)-
6941 - 1-Hexanol, 5-methyl-
8760 - Pentane, 2,2-dimethyl-
8761 - Pentane, 2,3-dimethyl-
8762 - Pentane, 2,4-dimethyl-
8763 - Pentane, 3,3-dimethyl-
8809 - Pentane, 3-ethyl-

$C_7H_{16}BrNO_2$
5438 - Ethanaminium, 2-(acetyloxy)-N,N,N-trimethyl-, bromide

$C_7H_{16}NCl$
7430 - Mepiquat chloride

$C_7H_{16}ClNO_2$
5439 - Ethanaminium, 2-(acetyloxy)-N,N,N-trimethyl-, chloride

$C_7H_{16}N_2$
9622 - Piperazine, 1,2,4-trimethyl-
9625 - 1-Piperidinamine, 2,6-dimethyl-
9626 - 3-Piperidinamine, 1-ethyl-
11167 - 2-Pyrrolidinemethanamine, 1-ethyl-

$C_7H_{16}N_2O$
7701 - 4-Morpholinepropanamine
12012 - Urea, N,N'-dipropyl-

$C_7H_{16}O$
3374 - Butane, 1-ethoxy-3-methyl-
3375 - Butane, 2-ethoxy-2-methyl-
3399 - Butane, 1-(1-methylethoxy)-
3421 - Butane, 1-propoxy-
3644 - 2-Butanol, 2,3,3-trimethyl-
6472 - 1-Heptanol
6473 - 2-Heptanol, (±)-
6474 - 3-Heptanol, (S)-
6475 - 4-Heptanol
6833 - Hexane, 1-methoxy-
6937 - 1-Hexanol, 2-methyl-, (±)-
6938 - 1-Hexanol, 2-methyl-, (R)-
6939 - 1-Hexanol, 3-methyl-, (±)-
6940 - 1-Hexanol, 4-methyl-, (±)-
6942 - 2-Hexanol, 2-methyl-
6943 - 2-Hexanol, 3-methyl-
6944 - 2-Hexanol, 5-methyl-
6945 - 3-Hexanol, 2-methyl-, (±)-
6946 - 3-Hexanol, 3-methyl-
6947 - 3-Hexanol, 5-methyl-, (±)-
8808 - Pentane, 1-ethoxy-
8927 - 1-Pentanol, 2,4-dimethyl-, (±)-
8928 - 2-Pentanol, 2,4-dimethyl-
8929 - 3-Pentanol, 2,2-dimethyl-
8930 - 3-Pentanol, 2,3-dimethyl-
8931 - 3-Pentanol, 2,4-dimethyl-
8932 - 3-Pentanol, 3-ethyl-
10081 - Propane, 2-methyl-2-(1-methylethoxy)-
10086 - Propane, 2-methyl-1-propoxy-
10087 - Propane, 2-methyl-2-propoxy-

$C_7H_{16}O_2$
5757 - Ethanol, 2-(pentyloxy)-
6397 - 1,4-Heptanediol
6398 - 1,5-Heptanediol
6399 - 1,7-Heptanediol
6400 - 2,4-Hexanediol, 2-methyl-
6803 - 2,4-Hexanediol, 2-methyl-
8792 - 1,3-Pentanediol, 2,2-dimethyl-
9903 - Propane, 1,1-diethoxy-
9904 - Propane, 2,2-diethoxy-
10004 - 1,3-Propanediol, 2,2-diethyl-
10020 - 1,3-Propanediol, 2-methyl-2-propyl-
10074 - Propane, 1,1'-[methylenebis(oxy)]bis-
10346 - 2-Propanol, 1-butoxy-

$C_7H_{16}O_2Si$
11422 - Silane, ethenyldiethoxymethyl-

$C_7H_{16}O_3$
3455 - Butane, 1,1,3-trimethoxy-
3456 - Butane, 1,3,3-trimethoxy-
5604 - Ethane, 1,1,1''-[methylidynetris(oxy)]tris-
6453 - 1,4,7-Heptanetriol

$C_7H_{16}O_4$
10150 - Propane, 1,1,3,3-tetramethoxy-

$C_7H_{16}O_4S_2$
9835 - Propane, 2,2-bis(ethylsulfonyl)-

$C_7H_{16}S$
6436 - 1-Heptanethiol

$C_7H_{16}S_3$
5605 - Ethane, 1,1',1''-[methylidynetris(thio)]tris-

$C_7H_{17}N$
6349 - 1-Heptanamine
6350 - 2-Heptanamine
6351 - 4-Heptanamine
6728 - 1-Hexanamine, 3-methyl-
6729 - 1-Hexanamine, 4-methyl-
6730 - 2-Hexanamine, 4-methyl-
8703 - 1-Pentanamine, N,N-dimethyl-
8704 - 2-Pentanamine, 2,4-dimethyl-

$C_7H_{17}NO$
5739 - Ethanol, 2-[(3-methylbutyl)amino]-
10366 - 1-Propanol, 2-(diethylamino)-
10367 - 1-Propanol, 3-(diethylamino)-

10368 - 2-Propanol, 1-(diethylamino)-, (±)-

$C_7H_{17}NO_2$
7443 - Methanamine, 1,1-diethoxy-N,N-dimethyl-

$C_7H_{17}NO_5$
6208 - D-Glucitol, 1-deoxy-1-(methylamino)-

$C_7H_{17}O_2PS$
9549 - Phosphonothioic acid, methyl-, O,O-bis(1-methylethyl) ester

$C_7H_{17}O_2PS_3$
9513 - Phorate

$C_7H_{18}N_2$
6371 - 1,7-Heptanediamine
9876 - 1,3-Propanediamine, N,N-diethyl-
9879 - 1,3-Propanediamine, N,N,N′,N′-tetramethyl-

$C_7H_{18}N_2O$
10339 - 2-Propanol, 1-amino-3-(diethylamino)-
10343 - 2-Propanol, 1,3-bis(dimethylamino)-

$C_7H_{18}N_2O_2$
5679 - Ethanol, 2,2′-[(3-aminopropyl)imino]bis-

$C_7H_{18}O_3Si$
11479 - Silane, triethoxymethyl-

$C_7H_{18}Si$
11371 - Silane, butyltrimethyl-
11491 - Silane, trimethyl(2-methylpropyl)-

$C_7H_{19}N_3$
9875 - 1,3-Propanediamine, N-(3-aminopropyl)-N-methyl-

$C_7H_{22}O_4Si_4$
4992 - Cyclotetrasiloxane, heptamethyl-

$C_8Cl_2N_2O_2$
4251 - 1,4-Cyclohexadiene-1,2-dicarbonitrile, 4,5-dichloro-3,6-dioxo-

$C_8Cl_4N_2$
1317 - 1,3-Benzenedicarbonitrile, 2,4,5,6-tetrachloro-

$C_8Cl_4O_3$
7346 - 1,3-Isobenzofurandione, 4,5,6,7-tetrachloro-

$C_8F_{14}O_3$
3522 - Butanoic acid, heptafluoro-, anhydride

C_8F_{16}
4342 - Cyclohexane, 1,1,2,2,3,3,4,4,5,6-decafluoro-5,6-bis(trifluoromethyl)-
4343 - Cyclohexane, 1,1,2,2,3,4,4,5,5,6-decafluoro-3,6-bis(trifluoromethyl)-

$C_8F_{16}O$
6109 - Furan, 2,2,3,3,4,4,5-heptafluorotetrahydro-5-(nonafluorobutyl)-

C_8F_{18}
8415 - Octane, octadecafluoro-

$C_8HCl_4F_{11}O_2$
8442 - Octanoic acid, 3,5,7,8-tetrachloro-2,2,3,4,4,5,6,6,7,8,8-undecafluoro-

$C_8H_2Cl_2O_3$
7343 - 1,3-Isobenzofurandione, 4,5-dichloro-
7344 - 1,3-Isobenzofurandione, 5,6-dichloro-

$C_8H_3Cl_2F_3N_2$
2377 - 1H-Benzimidazole, 4,5-dichloro-2-(trifluoromethyl)-

$C_8H_4Cl_2O_2$
1318 - 1,2-Benzenedicarbonyl dichloride
1319 - 1,3-Benzenedicarbonyl dichloride
1320 - 1,4-Benzenedicarbonyl dichloride

$C_8H_4Cl_2O_4$
1340 - 1,2-Benzenedicarboxylic acid, 3,5-dichloro-
1349 - 1,4-Benzenedicarboxylic acid, 2,5-dichloro-

$C_8H_4Cl_6$
996 - Benzene, 1,4-bis(trichloromethyl)-

$C_8H_4F_2O_2$
1321 - 1,2-Benzenedicarbonyl difluoride

$C_8H_4F_3N$
2769 - Benzonitrile, 2-(trifluoromethyl)-
2770 - Benzonitrile, 3-(trifluoromethyl)-

$C_8H_4F_6$
997 - Benzene, 1,3-bis(trifluoromethyl)-

$C_8H_4N_2$
1315 - 1,3-Benzenedicarbonitrile
1316 - 1,4-Benzenedicarbonitrile

$C_8H_4N_2S_2$
1432 - Benzene, 1,4-diisothiocyanato-

$C_8H_4O_2S$
2401 - Benzo[b]thiophene-2,3-dione

$C_8H_4O_3$
7342 - 1,3-Isobenzofurandione

C_8H_5Br
1040 - Benzene, 1-bromo-2-ethynyl-
1041 - Benzene, 1-bromo-3-ethynyl-
1042 - Benzene, 1-bromo-4-ethynyl-

$C_8H_5ClN_2$
11334 - Quinoxaline, 6-chloro-

$C_8H_5Cl_3O$
5871 - Ethanone, 2,2,2-trichloro-1-phenyl-

$C_8H_5Cl_3O_2$
944 - Benzeneacetic acid, 2,3,6-trichloro-

$C_8H_5Cl_3O_3$
256 - Acetic acid, (2,4,5-trichlorophenoxy)-

$C_8H_5F_3O$
5872 - Ethanone, 2,2,2-trifluoro-1-phenyl-

$C_8H_5F_6N$
637 - Benzenamine, 3,5-bis(trifluoromethyl)-

C_8H_5NO
968 - Benzeneacetonitrile, α-oxo-
2757 - Benzonitrile, 3-formyl-
2758 - Benzonitrile, 4-formyl-

$C_8H_5NO_2$
2550 - Benzoic acid, 3-cyano-
7300 - 1H-Indole-2,3-dione
7362 - 1H-Isoindole-1,3(2H)-dione

C_8H_6
1715 - Benzene, ethynyl-

C_8H_6BrClO
5778 - Ethanone, 2-bromo-1-(4-chlorophenyl)-

C_8H_6BrN
947 - Benzeneacetonitrile, α-bromo-
948 - Benzeneacetonitrile, 2-bromo-

$C_8H_6Br_2O$
5777 - Ethanone, 2-bromo-1-(4-bromophenyl)-
5797 - Ethanone, 1-(3,5-dibromophenyl)-
5798 - Ethanone, 2,2-dibromo-1-phenyl-

$C_8H_6Br_2O_3$
2559 - Benzoic acid, 3,5-dibromo-2-hydroxy-, methyl ester

$C_8H_6Br_4$
2255 - Benzene, 1,2,3,4-tetrabromo-5,6-dimethyl-

C_8H_6ClN
949 - Benzeneacetonitrile, 2-chloro-
950 - Benzeneacetonitrile, 3-chloro-
2750 - Benzonitrile, 2-(chloromethyl)-

$C_8H_6ClNO_4$
2548 - Benzoic acid, 4-chloro-3-nitro-, methyl ester
2549 - Benzoic acid, 5-chloro-2-nitro-, methyl ester

$C_8H_6Cl_2$
1367 - Benzene, 1,2-dichloro-3-ethenyl-
1368 - Benzene, 1,2-dichloro-4-ethenyl-
1369 - Benzene, 1,3-dichloro-4-ethenyl-
1370 - Benzene, 1,3-dichloro-5-ethenyl-
1371 - Benzene, 1,4-dichloro-2-ethenyl-
1372 - Benzene, (2,2-dichloroethenyl)-
1373 - Benzene, 2,4-dichloro-1-ethenyl-

$C_8H_6Cl_2O$
973 - Benzeneacetyl chloride, α-chloro-
5784 - Ethanone, 2-chloro-1-(4-chlorophenyl)-
5799 - Ethanone, 1-(2,4-dichlorophenyl)-
5800 - Ethanone, 1-(2,5-dichlorophenyl)-
5801 - Ethanone, 1-(3,4-dichlorophenyl)-
5802 - Ethanone, 2,2-dichloro-1-phenyl-

$C_8H_6Cl_2O_2$
9288 - Phenol, 2,4-dichloro-, acetate

$C_8H_6Cl_2O_3$
169 - Acetic acid, (2,4-dichlorophenoxy)-
2567 - Benzoic acid, 3,6-dichloro-2-methoxy-

$C_8H_6Cl_3NO$
98 - Acetamide, 2,2,2-trichloro-N-phenyl-

$C_8H_6Cl_4$
989 - Benzene, 1,4-bis(dichloromethyl)-

2261 - Benzene, 1,2,3,5-tetrachloro-4,6-dimethyl-

C_8H_6FN
952 - Benzeneacetonitrile, 2-fluoro-
953 - Benzeneacetonitrile, 4-fluoro-

$C_8H_6N_2$
4136 - Cinnoline
8065 - 1,5-Naphthyridine
9599 - Phthalazine
11195 - Quinazoline
11333 - Quinoxaline

$C_8H_6N_2O_2$
965 - Benzeneacetonitrile, 2-nitro-
966 - Benzeneacetonitrile, 3-nitro-
967 - Benzeneacetonitrile, 4-nitro-
2766 - Benzonitrile, 2-methyl-5-nitro-

$C_8H_6N_4O_5$
7222 - 2,4-Imidazolidinedione, 1-[[(5-nitro-2-furanyl)methylene]amino]-

$C_8H_6N_4O_8$
3106 - [5,5′Bipyrimidine]2,2′,4,4′,6,6′(1H,1′H,3H,3′H,5H,5′H)-hexone-5,5′-dihydroxy

C_8H_6O
1716 - Benzene, (ethynyloxy)-
2438 - Benzofuran

C_8H_6OS
2403 - Benzo[b]thiophene-4-ol

$C_8H_6O_2$
1322 - 1,3-Benzenedicarboxaldehyde
1323 - 1,4-Benzenedicarboxaldehyde
2461 - 2(3H)-Benzofuranone
2462 - 3(2H)-Benzofuranone
7349 - 1(3H)-Isobenzofuranone

$C_8H_6O_3$
934 - Benzeneacetic acid, α-oxo-
2417 - 1,3-Benzodioxole-5-carboxaldehyde
2607 - Benzoic acid, 2-formyl-

$C_8H_6O_4$
1327 - 1,2-Benzenedicarboxylic acid
1328 - 1,3-Benzenedicarboxylic acid
1329 - 1,4-Benzenedicarboxylic acid
2418 - 1,3-Benzodioxole-5-carboxylic acid
2609 - Benzoic acid, 3-formyl-4-hydroxy-
9598 - p-Phthalaldehydic acid, 3-hydroxy-

$C_8H_6O_5$
1352 - 1,3-Benzenedicarboxylic acid, 4-hydroxy-

C_8H_6S
2399 - Benzo[b]thiophene

$C_8H_6S_2$
2434 - 1,4-Benzodithiin
3112 - 2,2′-Bithiophene

C_8H_7Br
1025 - Benzene, (1-bromoethenyl)-
1026 - Benzene, 1-bromo-2-ethenyl-
1027 - Benzene, 1-bromo-3-ethenyl-
1028 - Benzene, 1-bromo-4-ethenyl-
1029 - Benzene, (2-bromoethenyl)-, (E)-
1031 - Benzene, (2-bromoethenyl)-, (Z)-
4743 - 1,3,5,7-Cyclooctatetraene, 1-bromo-

C_8H_7BrO
5780 - Ethanone, 1-(3-bromophenyl)-
5781 - Ethanone, 1-(4-bromophenyl)-
5782 - Ethanone, 2-bromo-1-phenyl-

$C_8H_7BrO_2$
119 - Acetic acid, bromo-, phenyl ester
888 - Benzeneacetic acid, 4-bromo-
2525 - Benzoic acid, 3-bromo-, methyl ester
2526 - Benzoic acid, 4-bromo-, methyl ester

$C_8H_7BrO_3$
889 - Benzeneacetic acid, 4-bromo-α-hydroxy-, (±)-

C_8H_7Cl
1148 - Benzene, (1-chloroethenyl)-
1149 - Benzene, 1-chloro-2-ethenyl-
1150 - Benzene, 1-chloro-3-ethenyl-
1151 - Benzene, 1-chloro-4-ethenyl-
1153 - Benzene, (2-chloroethenyl)-, (E)-
1155 - Benzene, (2-chloroethenyl)-, (Z)-
4745 - 1,3,5,7-Cyclooctatetraene, 1-chloro-

C$_8$H$_7$ClO
  971 - Benzeneacetyl chloride
  2902 - Benzoyl chloride, 2-methyl-
  2903 - Benzoyl chloride, 3-methyl-
  2904 - Benzoyl chloride, 4-methyl-
  5788 - Ethanone, 1-(3-chlorophenyl)-
  5789 - Ethanone, 1-(4-chlorophenyl)-
  5790 - Ethanone, 2-chloro-1-phenyl-
C$_8$H$_7$ClO$_2$
  147 - Acetic acid, chloro-, phenyl ester
  2421 - 1,3-Benzodioxole, 5-(chloromethyl)-
  2543 - Benzoic acid, 4-chloro-, methyl ester
  2901 - Benzoyl chloride, 4-methoxy-
  4027 - Carbonochloridic acid, phenylmethyl ester
C$_8$H$_7$ClO$_3$
  146 - Acetic acid, (4-chlorophenoxy)-
  2542 - Benzoic acid, 5-chloro-2-hydroxy-, methyl
         ester
C$_8$H$_7$F
  1633 - Benzene, 1-ethenyl-2-fluoro-
  1634 - Benzene, 1-ethenyl-3-fluoro-
  1635 - Benzene, 1-ethenyl-4-fluoro-
C$_8$H$_7$FO
  5815 - Ethanone, 1-(4-fluorophenyl)-
  5816 - Ethanone, 2-fluoro-1-phenyl-
C$_8$H$_7$FO$_2$
  901 - Benzeneacetic acid, 4-fluoro-
C$_8$H$_7$F$_3$
  2045 - Benzene, 1-methyl-4-(trifluoromethyl)-
C$_8$H$_7$IO
  5836 - Ethanone, 1-(3-iodophenyl)-
  5837 - Ethanone, 1-(4-iodophenyl)-
  5838 - Ethanone, 2-iodo-1-phenyl-
C$_8$H$_7$IO$_2$
  2660 - Benzoic acid, 2-iodo-, methyl ester
  2661 - Benzoic acid, 3-iodo-, methyl ester
  2662 - Benzoic acid, 4-iodo-, methyl ester
C$_8$H$_7$N
  945 - Benzeneacetonitrile
  1791 - Benzene, (isocyanomethyl)-
  2763 - Benzonitrile, 2-methyl-
  2764 - Benzonitrile, 3-methyl-
  2765 - Benzonitrile, 4-methyl-
  7294 - 1H-Indole
  7324 - Indolizine
C$_8$H$_7$NO
  287 - Acetonitrile, phenoxy-
  956 - Benzeneacetonitrile, α-hydroxy-, (±)-
  1787 - Benzene, 1-isocyanato-2-methyl-
  2760 - Benzonitrile, 2-methoxy-
  2761 - Benzonitrile, 3-methoxy-
  2762 - Benzonitrile, 4-methoxy-
  2880 - Benzoxazole, 2-methyl-
  7327 - 2H-Indol-2-one, 1,3-dihydro-
  7369 - 1H-Isoindol-1-one, 2,3-dihydro-
C$_8$H$_7$NOS
  1793 - Benzene, 1-isothiocyanato-2-methoxy-
C$_8$H$_7$NO$_2$
  876 - Benzeneacetaldehyde, α-oxo-, aldoxime
  1644 - Benzene, 1-ethenyl-3-nitro-
  1645 - Benzene, 1-ethenyl-4-nitro-
  2050 - Benzene, (2-nitroethenyl)-, (E)-
  8414 - Octane, 3-nitro-
C$_8$H$_7$NO$_3$
  875 - Benzeneacetaldehyde, 2-nitro-
  2489 - Benzoic acid, 4-(aminocarbonyl)-
  5860 - Ethanone, 1-(3-nitrophenyl)-
  5861 - Ethanone, 1-(4-nitrophenyl)-
  5862 - Ethanone, 2-nitro-1-phenyl-
C$_8$H$_7$NO$_4$
  219 - Acetic acid, 2-nitrophenyl ester
  933 - Benzeneacetic acid, 4-nitro-
  1330 - 1,3-Benzenedicarboxylic acid, 2-amino-
  1331 - 1,3-Benzenedicarboxylic acid, 5-amino-
  2710 - Benzoic acid, 2-nitro-, methyl ester
  2711 - Benzoic acid, 3-nitro-, methyl ester

C$_8$H$_7$NS
  1794 - Benzene, (isothiocyanatomethyl)-
  2863 - Benzothiazole, 2-methyl-
  11710 - Thiocyanic acid, phenylmethyl ester
C$_8$H$_7$NS$_2$
  2867 - 2(3H)-Benzothiazolethione, 3-methyl-
C$_8$H$_7$N$_3$
  11332 - 6-Quinoxalinamine
C$_8$H$_7$N$_3$O
  5863 - Ethanone, 1-(2-nitrophenyl)-
C$_8$H$_7$N$_3$O$_2$
  9600 - 1,4-Phthalazinedione, 5-amino-2,3-dihydro-
C$_8$H$_7$N$_3$O$_5$
  8589 - 2-Oxazolidinone, 3-[[(5-nitro-
         2-furanyl)methylene]amino]-
C$_8$H$_7$N$_3$O$_6$
  1489 - Benzene, 1,4-dimethyl-2,3,5-trinitro-
  1490 - Benzene, 2,4-dimethyl-1,3,5-trinitro-
C$_8$H$_8$
  1624 - Benzene, ethenyl-
  4742 - 1,3,5,7-Cyclooctatetraene
  4769 - 1,3-Cyclopentadiene, 5-(2-propenylidene)-,
         (E)-
C$_8$H$_8$BrNO
  37 - Acetamide, N-(4-bromophenyl)-
C$_8$H$_8$Br$_2$
  983 - Benzene, 1,2-bis(bromomethyl)-
  984 - Benzene, 1,3-bis(bromomethyl)-
  985 - Benzene, 1,4-bis(bromomethyl)-
  1297 - Benzene, 1,2-dibromo-3,4-dimethyl-
  1298 - Benzene, 1,5-dibromo-2,4-dimethyl-
  1299 - Benzene, (1,2-dibromoethyl)-
C$_8$H$_8$ClN
  7652 - Methylamine, N-(o-chlorobenzylidene)-
C$_8$H$_8$ClNO
  48 - Acetamide, N-(4-chlorophenyl)-
  3974 - Carbamic chloride, methylphenyl-
C$_8$H$_8$ClNO$_3$S
  2243 - Benzenesulfonyl chloride, 4-(acetylamino)-
C$_8$H$_8$Cl$_2$
  986 - Benzene, 1,2-bis(chloromethyl)-
  987 - Benzene, 1,3-bis(chloromethyl)-
  988 - Benzene, 1,4-bis(chloromethyl)-
  1360 - Benzene, 1,2-dichloro-3,4-dimethyl-
  1361 - Benzene, 1,3-dichloro-2,5-dimethyl-
  1362 - Benzene, 1,4-dichloro-2,5-dimethyl-
  1363 - Benzene, 1,5-dichloro-2,4-dimethyl-
  1364 - Benzene, 2,3-dichloro-1,4-dimethyl-
  1376 - Benzene, (1,2-dichloroethyl)-
  1377 - Benzene, 1,4-dichloro-2-ethyl-
  1390 - Benzene, 1-(dichloromethyl)-4-methyl-
C$_8$H$_8$Cl$_2$NaO$_5$S
  11358 - Sesone
C$_8$H$_8$Cl$_2$O
  1375 - Benzene, 2,4-dichloro-1-ethoxy-
  1854 - Benzenemethanol, 2,4-dichloro-α-methyl-
  9290 - Phenol, 2,4-dichloro-3,5-dimethyl-
  9291 - Phenol, 2,6-dichloro-3,5-dimethyl-
C$_8$H$_8$Cl$_2$O$_2$
  4084 - Chloroneb
C$_8$H$_8$Cl$_3$N$_3$O$_4$S$_2$
  2854 - 2H-1,2,4-Benzothiadiazine-7-sulfonamide, 6-
         chloro-3-(dichloromethyl)-3,4-dihydro-, 1,1-
         dioxide
C$_8$H$_8$Cl$_3$O$_3$PS
  9584 - Phosphorothioic acid, O,O-dimethyl O-(2,4,5-
         trichlorophenyl) ester
C$_8$H$_8$F$_3$N$_3$O$_4$S$_2$
  2858 - 2H-1,2,4-Benzothiadiazine-7-sulfonamide,
         3,4-dihydro-6-(trifluoromethyl)-, 1,1-dioxide
C$_8$H$_8$HgO$_2$
  7431 - Mercury, (acetato-O)phenyl-
C$_8$H$_8$N$_2$
  946 - Benzeneacetonitrile, 4-amino-
  2382 - 1H-Benzimidazole, 1-methyl-
  2743 - Benzonitrile, 5-amino-2-methyl-
  4160 - Cyanamide, methylphenyl-

  7239 - 1H-Indazole, 1-methyl-
  7240 - 1H-Indazole, 3-methyl-
  7241 - 1H-Indazole, 5-methyl-
  7242 - 2H-Indazole, 2-methyl-
C$_8$H$_8$N$_2$O
  1324 - 1,2-Benzenedicarboxamide
  10883 - 3-Pyridinecarbonitrile, 1,2-dihydro-4-
          methoxy-1-methyl-2-oxo-
C$_8$H$_8$N$_2$O$_3$
  80 - Acetamide, 2-nitro-N-phenyl-
  81 - Acetamide, N-(2-nitrophenyl)-
  82 - Acetamide, N-(3-nitrophenyl)-
  83 - Acetamide, N-(4-nitrophenyl)-
C$_8$H$_8$N$_2$O$_5$
  1650 - Benzene, 2-ethoxy-1,3-dinitro-
C$_8$H$_8$O
  587 - Benzaldehyde, 2-methyl-
  588 - Benzaldehyde, 3-methyl-
  589 - Benzaldehyde, 4-methyl-
  868 - Benzeneacetaldehyde
  1646 - Benzene, (ethenyloxy)-
  2442 - Benzofuran, 2,3-dihydro-
  5864 - Ethanone, 1-phenyl-
  8624 - Oxirane, phenyl-
  9348 - Phenol, 2-ethenyl-
  9349 - Phenol, 3-ethenyl-
C$_8$H$_8$O$_2$
  231 - Acetic acid, phenyl ester
  575 - Benzaldehyde, 2-hydroxy-3-methyl-
  576 - Benzaldehyde, 2-hydroxy-4-methyl-
  577 - Benzaldehyde, 2-hydroxy-5-methyl-
  578 - Benzaldehyde, 2-hydroxy-6-methyl-
  583 - Benzaldehyde, 2-methoxy-
  584 - Benzaldehyde, 3-methoxy-
  585 - Benzaldehyde, 4-methoxy-
  883 - Benzeneacetic acid
  2414 - 1,4-Benzodioxin, 2,3-dihydro-
  2676 - Benzoic acid, 2-methyl-
  2677 - Benzoic acid, 3-methyl-
  2678 - Benzoic acid, 4-methyl-
  2687 - Benzoic acid, methyl ester
  3878 - 3-Buten-2-one, 4-(2-furanyl)-
  4263 - 2,5-Cyclohexadiene-1,4-dione, 2,3-dimethyl-
  4264 - 2,5-Cyclohexadiene-1,4-dione, 2,6-dimethyl-
  5830 - Ethanone, 1-(2-hydroxyphenyl)-
  5831 - Ethanone, 1-(3-hydroxyphenyl)-
  5832 - Ethanone, 1-(4-hydroxyphenyl)-
  5833 - Ethanone, 2-hydroxy-1-phenyl-
  6016 - Formic acid, phenylmethyl ester
C$_8$H$_8$O$_3$
  228 - Acetic acid, phenoxy-
  570 - Benzaldehyde, 2-hydroxy-3-methoxy-
  571 - Benzaldehyde, 2-hydroxy-5-methoxy-
  572 - Benzaldehyde, 3-hydroxy-4-methoxy-
  573 - Benzaldehyde, 3-hydroxy-5-methoxy-
  574 - Benzaldehyde, 4-hydroxy-3-methoxy-
  903 - Benzeneacetic acid, α-hydroxy-, (S)-
  904 - Benzeneacetic acid, 2-hydroxy-
  905 - Benzeneacetic acid, 3-hydroxy-
  906 - Benzeneacetic acid, 4-hydroxy-
  1540 - 1,2-Benzenediol, monoacetate
  2427 - 1,3-Benzodioxole-5-methanol
  2629 - Benzoic acid, 2-hydroxy-3-methyl-
  2630 - Benzoic acid, 2-hydroxy-4-methyl-
  2631 - Benzoic acid, 2-hydroxy-5-methyl-
  2632 - Benzoic acid, 3-hydroxy-5-methyl-
  2634 - Benzoic acid, 2-hydroxy-, methyl ester
  2636 - Benzoic acid, 3-hydroxy-, methyl ester
  2637 - Benzoic acid, 4-hydroxy-, methyl ester
  2664 - Benzoic acid, 2-methoxy-
  2665 - Benzoic acid, 3-methoxy-
  2666 - Benzoic acid, 4-methoxy-
  5803 - Ethanone, 1-(2,4-dihydroxyphenyl)-
  6070 - 2-Furancarboxylic acid, 2-propenyl ester
  7347 - 1,3-Isobenzofurandione, 4,5,6,7-tetrahydro-
  10640 - 2-Propenoic acid, 3-(2-furanyl)-, methyl ester

C$_8$H$_8$O$_4$
895 - Benzeneacetic acid, 2,5-dihydroxy-
2573 - Benzoic acid, 2,4-dihydroxy-6-methyl-
2625 - Benzoic acid, 4-hydroxy-3-methoxy-
4262 - 2,5-Cyclohexadiene-1,4-dione, 2,6-dimethoxy-
4265 - 2,5-Cyclohexadiene-1,4-dione, 3-hydroxy-2-
    methoxy-5-methyl-
5873 - Ethanone, 1-(2,3,4-trihydroxyphenyl)-
10779 - 2H-Pyran-2,4(3H)-dione, 3-acetyl-6-methyl-

C$_8$H$_8$O$_5$
4260 - 2,5-Cyclohexadiene-1,4-dione, 2,5-dihydroxy-
    3-methoxy-6-methyl-
6083 - 2,5-Furandicarboxylic acid, dimethyl ester

C$_8$H$_9$Br
1017 - Benzene, 1-bromo-2,3-dimethyl-
1018 - Benzene, 1-bromo-2,4-dimethyl-
1019 - Benzene, 1-bromo-3,5-dimethyl-
1020 - Benzene, 2-bromo-1,3-dimethyl-
1021 - Benzene, 2-bromo-1,4-dimethyl-
1022 - Benzene, 4-bromo-1,2-dimethyl-
1034 - Benzene, (1-bromoethyl)-, (±)-
1035 - Benzene, 1-bromo-2-ethyl-
1036 - Benzene, 1-bromo-3-ethyl-
1037 - Benzene, 1-bromo-4-ethyl-
1038 - Benzene, (2-bromoethyl)-
1039 - Benzene, (1-bromoethyl)-, (R)-
1062 - Benzene, 1-(bromomethyl)-2-methyl-
1063 - Benzene, 1-(bromomethyl)-3-methyl-
1064 - Benzene, 1-(bromomethyl)-4-methyl-

C$_8$H$_9$BrO
1032 - Benzene, 1-bromo-4-ethoxy-
1033 - Benzene, (2-bromoethoxy)-
1595 - Benzeneethanol, 4-bromo-
1844 - Benzenemethanol, α-(bromomethyl)-
1845 - Benzenemethanol, 2-bromo-α-methyl-

C$_8$H$_9$BrO$_2$
1016 - Benzene, 4-bromo-1,2-dimethoxy-

C$_8$H$_9$Cl
1137 - Benzene, 1-chloro-2,3-dimethyl-
1138 - Benzene, 1-chloro-2,4-dimethyl-
1139 - Benzene, 2-chloro-1,3-dimethyl-
1140 - Benzene, 2-chloro-1,4-dimethyl-
1141 - Benzene, 4-chloro-1,2-dimethyl-
1160 - Benzene, (1-chloroethyl)-, (±)-
1161 - Benzene, 1-chloro-2-ethyl-
1162 - Benzene, 1-chloro-3-ethyl-
1163 - Benzene, 1-chloro-4-ethyl-
1164 - Benzene, (2-chloroethyl)-
1198 - Benzene, 1-(chloromethyl)-2-methyl-
1199 - Benzene, 1-(chloromethyl)-3-methyl-
1200 - Benzene, 1-(chloromethyl)-4-methyl-

C$_8$H$_9$ClO
1156 - Benzene, 1-chloro-2-ethoxy-
1157 - Benzene, 1-chloro-3-ethoxy-
1158 - Benzene, 1-chloro-4-ethoxy-
1159 - Benzene, (2-chloroethoxy)-
1181 - Benzene, [(chloromethoxy)methyl]-
1197 - Benzene, 1-(chloromethyl)-4-methoxy-
1596 - Benzeneethanol, 3-chloro-
1597 - Benzeneethanol, 4-chloro-
1852 - Benzenemethanol, α-(chloromethyl)-
9256 - Phenol, 2-chloro-3,4-dimethyl-
9257 - Phenol, 4-chloro-3,5-dimethyl-

C$_8$H$_9$ClS
1165 - Benzene, [(2-chloroethyl)thio]-

C$_8$H$_9$Cl$_2$NO
691 - Benzenamine, 2,6-dichloro-4-ethoxy-

C$_8$H$_9$Cl$_2$P
9553 - Phosphonous dichloride,
    (2,5-dimethylphenyl)-
9554 - Phosphonous dichloride, (4-ethylphenyl)-

C$_8$H$_9$Cl$_3$Si
11474 - Silane, trichloro(2-phenylethyl)-

C$_8$H$_9$FO
1654 - Benzene, 1-ethoxy-2-fluoro-
1655 - Benzene, 1-ethoxy-3-fluoro-
1656 - Benzene, 1-ethoxy-4-fluoro-

C$_8$H$_9$I
1680 - Benzene, 1-ethyl-2-iodo-
1681 - Benzene, 1-ethyl-4-iodo-
1769 - Benzene, 1-iodo-2,3-dimethyl-
1770 - Benzene, 1-iodo-2,4-dimethyl-
1771 - Benzene, 1-iodo-3,5-dimethyl-
1772 - Benzene, 2-iodo-1,3-dimethyl-
1773 - Benzene, 2-iodo-1,4-dimethyl-
1774 - Benzene, 4-iodo-1,2-dimethyl-

C$_8$H$_9$IO
1657 - Benzene, 1-ethoxy-4-iodo-

C$_8$H$_9$IO$_2$
1767 - Benzene, 1-iodo-2,3-dimethoxy-
1768 - Benzene, 4-iodo-1,2-dimethoxy-

C$_8$H$_9$N
733 - Benzenamine, 2-ethenyl-
734 - Benzenamine, 3-ethenyl-
735 - Benzenamine, 4-ethenyl-
2966 - Bicyclo[2.2.1]hept-5-ene-2-carbonitrile
7297 - 1H-Indole, 2,3-dihydro-
7452 - Methanamine, N-(phenylmethylene)-
10955 - Pyridine, 5-ethenyl-2-methyl-
11014 - Pyridine, 2-(2-propenyl)-

C$_8$H$_9$NO
86 - Acetamide, N-phenyl-
619 - Benzamide, 2-methyl-
620 - Benzamide, N-methyl-
878 - Benzeneacetamide
5772 - Ethanone, 1-(3-aminophenyl)-
5773 - Ethanone, 1-(4-aminophenyl)-
5774 - Ethanone, 2-amino-1-phenyl-
5855 - Ethanone, 1-(6-methyl-3-pyridinyl)-
5866 - Ethanone, 1-phenyl-, oxime
5996 - Formamide, N-(2-methylphenyl)-
5997 - Formamide, N-(3-methylphenyl)-
5998 - Formamide, N-methyl-N-phenyl-

C$_8$H$_9$NO$_2$
67 - Acetamide, N-(4-hydroxyphenyl)-
618 - Benzamide, 4-methoxy-
885 - Benzeneacetic acid, α-amino-
886 - Benzeneacetic acid, 4-amino-
1474 - Benzene, 1,2-dimethyl-3-nitro-
1475 - Benzene, 1,2-dimethyl-4-nitro-
1476 - Benzene, 1,3-dimethyl-2-nitro-
1477 - Benzene, 1,3-dimethyl-5-nitro-
1478 - Benzene, 1,4-dimethyl-2-nitro-
1479 - Benzene, 2,4-dimethyl-1-nitro-
1692 - Benzene, 1-ethyl-2-nitro-
1693 - Benzene, 1-ethyl-3-nitro-
1694 - Benzene, 1-ethyl-4-nitro-
2004 - Benzene, 1-methyl-4-(nitromethyl)-
2051 - Benzene, (2-nitroethyl)-
2426 - 1,3-Benzodioxole-5-methanamine
2503 - Benzoic acid, 2-amino-, methyl ester
2504 - Benzoic acid, 3-amino-, methyl ester
2679 - Benzoic acid, 2-(methylamino)-
6270 - Glycine, N-phenyl-
10901 - 2-Pyridinecarboxylic acid, ethyl ester
10902 - 3-Pyridinecarboxylic acid, ethyl ester
10903 - 4-Pyridinecarboxylic acid, ethyl ester

C$_8$H$_9$NO$_3$
1612 - Benzeneethanol, 2-nitro-
1662 - Benzene, 1-ethoxy-2-nitro-
1663 - Benzene, 1-ethoxy-3-nitro-
1664 - Benzene, 1-ethoxy-4-nitro-
1934 - Benzene, 4-methoxy-1-methyl-2-nitro-
2501 - Benzoic acid, 3-amino-4-hydroxy-, methyl
    ester
6267 - Glycine, N-(4-hydroxyphenyl)-
9334 - Phenol, 3,6-dimethyl-2-nitro-

C$_8$H$_9$NO$_4$
1445 - Benzene, 1,2-dimethoxy-4-nitro-
1446 - Benzene, 1,3-dimethoxy-2-nitro-
1447 - Benzene, 1,4-dimethoxy-2-nitro-
1448 - Benzene, 2,4-dimethoxy-1-nitro-

C$_8$H$_{10}$
1452 - Benzene, 1,2-dimethyl-

1453 - Benzene, 1,3-dimethyl-
1454 - Benzene, 1,4-dimethyl-
1668 - Benzene, ethyl-
4749 - 1,3,5-Cyclooctatriene
4750 - 1,3,6-Cyclooctatriene
4768 - 1,3-Cyclopentadiene, 5-(1-methylethylidene)-
6714 - 1,5-Hexadien-3-yne, 2,5-dimethyl-
8342 - 1,7-Octadiyne
8343 - 2,6-Octadiyne
8344 - 3,5-Octadiyne
8483 - 1,3,5,7-Octatetraene

C$_8$H$_{10}$BrN
643 - Benzenamine, 2-bromo-N,N-dimethyl-
644 - Benzenamine, 3-bromo-N,N-dimethyl-
645 - Benzenamine, 4-bromo-N,N-dimethyl-
1570 - Benzeneethanamine, 4-bromo-

C$_8$H$_{10}$ClN
662 - Benzenamine, 2-chloro-N,N-dimethyl-
663 - Benzenamine, 4-chloro-N,N-dimethyl-
1802 - Benzenemethanamine, 3-chloro-N-methyl-

C$_8$H$_{10}$Cl$_2$Si
11400 - Silane, dichloromethyl(4-methylphenyl)-

C$_8$H$_{10}$K$_2$O$_{15}$Sb$_2$
11588 - Tartar emetic

C$_8$H$_{10}$N$_2$
18 - Acetaldehyde, phenylhydrazone
5134 - Diazene, ethylphenyl-
11339 - Quinoxaline, 1,2,3,4-tetrahydro-

C$_8$H$_{10}$N$_2$O
29 - Acetamide, N-(4-aminophenyl)-
720 - Benzenamine, N,N-dimethyl-4-nitroso-
750 - Benzenamine, N-ethyl-N-nitroso-
12027 - Urea, (phenylmethyl)-

C$_8$H$_{10}$N$_2$O$_2$
718 - Benzenamine, N,N-dimethyl-2-nitro-
719 - Benzenamine, N,N-dimethyl-3-nitro-

C$_8$H$_{10}$N$_2$O$_3$
739 - Benzenamine, 2-ethoxy-5-nitro-

C$_8$H$_{10}$N$_2$O$_3$S
30 - Acetamide, N-[(4-aminophenyl)sulfonyl]-
31 - Acetamide, N-[4-(aminosulfonyl)phenyl]-
2212 - Benzenesulfonamide, N,4-dimethyl-N-nitroso-

C$_8$H$_{10}$N$_2$O$_4$
11011 - 1(4H)-Pyridinepropanoic acid, α-amino-3-
    hydroxy-4-oxo-, (S)-

C$_8$H$_{10}$N$_2$O$_4$S
474 - Asulam

C$_8$H$_{10}$N$_2$S
10885 - 4-Pyridinecarbothioamide, 2-ethyl-
11806 - Thiourea, (2-methylphenyl)-

C$_8$H$_{10}$N$_4$O$_2$
10760 - 1H-Purine-2,6-dione, 3,7-dihydro-1,3,7-
    trimethyl-

C$_8$H$_{10}$N$_6$O$_7$
11097 - 2,4,6(1H,3H,5H)-Pyrimidinetrione, 5-
    [(hexahydro-2,4,6-trioxo-5-pyrimidinyl)imino]-,
    monoammonium salt

C$_8$H$_{10}$Na$_2$O$_5$
5370 - Endothall, disodium salt

C$_8$H$_{10}$O
1592 - Benzeneethanol
1649 - Benzene, ethoxy-
1896 - Benzenemethanol, α-methyl-
1897 - Benzenemethanol, 2-methyl-
1898 - Benzenemethanol, 3-methyl-
1899 - Benzenemethanol, 4-methyl-
1930 - Benzene, (methoxymethyl)-
1931 - Benzene, 1-methoxy-2-methyl-
1932 - Benzene, 1-methoxy-3-methyl-
1933 - Benzene, 1-methoxy-4-methyl-
2916 - Bicyclo[2.2.1]heptane-2-carboxaldehyde
2958 - Bicyclo[2.2.1]heptan-2-one, 3-methylene-
2967 - Bicyclo[2.2.1]hept-5-ene-2-carboxaldehyde
8484 - 2,4,6-Octatrienal
9307 - Phenol, 2,3-dimethyl-
9308 - Phenol, 2,4-dimethyl-
9309 - Phenol, 2,5-dimethyl-

9310 - Phenol, 2,6-dimethyl-
9311 - Phenol, 3,4-dimethyl-
9312 - Phenol, 3,5-dimethyl-
9353 - Phenol, 2-ethyl-
9354 - Phenol, 3-ethyl-
9355 - Phenol, 4-ethyl-

$C_8H_{10}OS$
2296 - Benzenethiol, 4-ethoxy-
10472 - 1-Propanone, 2-methyl-1-(2-thienyl)-

$C_8H_{10}O_2$
1434 - 1,3-Benzenedimethanol
1435 - 1,4-Benzenedimethanol
1436 - Benzene, 1,2-dimethoxy-
1437 - Benzene, 1,3-dimethoxy-
1438 - Benzene, 1,4-dimethoxy-
1511 - 1,2-Benzenediol, 4,5-dimethyl-
1512 - 1,3-Benzenediol, 2,5-dimethyl-
1513 - 1,3-Benzenediol, 4,5-dimethyl-
1514 - 1,3-Benzenediol, 4,6-dimethyl-
1518 - 1,3-Benzenediol, 4-ethyl-
1606 - Benzeneethanol, 3-hydroxy-
1889 - Benzenemethanol, 2-methoxy-
1890 - Benzenemethanol, 3-methoxy-
1891 - Benzenemethanol, 4-methoxy-
3659 - 2-Butanone, 4-(2-furanyl)-
5579 - 1,2-Ethanediol, 1-phenyl-
5758 - Ethanol, 2-phenoxy-
8685 - 1,5-Pentanedione, hexahydro-
9350 - Phenol, 2-ethoxy-
9351 - Phenol, 3-ethoxy-
9352 - Phenol, 4-ethoxy-
9385 - Phenol, 2-methoxy-4-methyl-

$C_8H_{10}O_2S$
1704 - Benzene, (ethylsulfonyl)-

$C_8H_{10}O_3$
1885 - Benzenemethanol, 2-hydroxy-3-methoxy-
1886 - Benzenemethanol, 4-hydroxy-3-methoxy-
3818 - 2-Butenoic acid, anhydride
6028 - 2-Furanacetic acid, ethyl ester
6061 - 2-Furancarboxylic acid, 1-methylethyl ester
6064 - 3-Furancarboxylic acid, 2-methyl-, ethyl ester
6071 - 2-Furancarboxylic acid, propyl ester
6123 - 2-Furanmethanol, propanoate
6162 - 2-Furanpropanoic acid, methyl ester
7345 - 1,3-Isobenzofurandione, hexahydro-
9305 - Phenol, 2,6-dimethoxy-
9306 - Phenol, 3,5-dimethoxy-
10649 - 2-Propenoic acid, 2-methyl-, anhydride

$C_8H_{10}O_3S$
2223 - Benzenesulfonic acid, ethyl ester
2232 - Benzenesulfonic acid, 4-methyl-, methyl ester

$C_8H_{10}O_4$
3909 - 2-Butynedioic acid, diethyl ester
3913 - 2-Butyne-1,4-diol, diacetate
5564 - Ethanedioic acid, di-2-propenyl ester
6701 - 2,5-Hexadienoic acid, 3-methoxy-5-methyl-4-oxo-

$C_8H_{10}S$
1705 - Benzene, (ethylthio)-
1834 - Benzenemethanethiol, α-methyl-, (S)-
2042 - Benzene, [(methylthio)methyl]-
2297 - Benzenethiol, 3-ethyl-
2298 - Benzenethiol, 4-ethyl-

$C_8H_{11}ClO_4$
3771 - 2-Butenedioic acid, 2-chloro-, diethyl ester, (E)-
3772 - 2-Butenedioic acid, 2-chloro-, diethyl ester, (Z)-

$C_8H_{11}ClSi$
11374 - Silane, chlorodimethylphenyl-

$C_8H_{11}Cl_2N_3O_2$
11061 - 2,4(1H,3H)-Pyrimidinedione, 5-[bis(2-chloroethyl)amino]-

$C_8H_{11}Cl_3O_7$
6231 - β-D-Glucopyranosiduronic acid, 2,2,2-trichloroethyl

$C_8H_{11}N$
276 - Acetonitrile, cyclohexylidene-
704 - Benzenamine, 2,3-dimethyl-
705 - Benzenamine, 2,4-dimethyl-
706 - Benzenamine, 2,5-dimethyl-
707 - Benzenamine, 2,6-dimethyl-
708 - Benzenamine, 3,4-dimethyl-
709 - Benzenamine, 3,5-dimethyl-
710 - Benzenamine, N,2-dimethyl-
711 - Benzenamine, N,3-dimethyl-
712 - Benzenamine, N,4-dimethyl-
713 - Benzenamine, N,N-dimethyl-
740 - Benzenamine, 2-ethyl-
741 - Benzenamine, 3-ethyl-
742 - Benzenamine, 4-ethyl-
743 - Benzenamine, N-ethyl-
1569 - Benzeneethanamine
1814 - Benzenemethanamine, α-methyl-, (±)-
1815 - Benzenemethanamine, 2-methyl-
1816 - Benzenemethanamine, 3-methyl-
1817 - Benzenemethanamine, 4-methyl-
1818 - Benzenemethanamine, N-methyl-
10962 - Pyridine, 2-ethyl-4-methyl-
10963 - Pyridine, 2-ethyl-6-methyl-
10964 - Pyridine, 3-ethyl-4-methyl-
10965 - Pyridine, 4-ethyl-2-methyl-
10966 - Pyridine, 5-ethyl-2-methyl-
10984 - Pyridine, 2-(1-methylethyl)-
10985 - Pyridine, 3-(1-methylethyl)-
10986 - Pyridine, 4-(1-methylethyl)-
11015 - Pyridine, 2-propyl-
11016 - Pyridine, 4-propyl-
11024 - Pyridine, 2,3,4-trimethyl-
11025 - Pyridine, 2,3,5-trimethyl-
11026 - Pyridine, 2,3,6-trimethyl-
11027 - Pyridine, 2,4,5-trimethyl-
11028 - Pyridine, 2,4,6-trimethyl-

$C_8H_{11}NO$
736 - Benzenamine, 2-ethoxy-
737 - Benzenamine, 4-ethoxy-
770 - Benzenamine, 2-methoxy-5-methyl-
771 - Benzenamine, 3-methoxy-4-methyl-
772 - Benzenamine, 4-methoxy-2-methyl-
773 - Benzenamine, 5-methoxy-2-methyl-
774 - Benzenamine, 5-methoxy-6-methyl-
1593 - Benzeneethanol, β-amino-
1594 - Benzeneethanol, 2-amino-
1812 - Benzenemethanamine, 2-methoxy-
1813 - Benzenemethanamine, 4-methoxy-
1839 - Benzenemethanol, α-(aminomethyl)-
5759 - Ethanol, 2-(phenylamino)-
9208 - Phenol, 4-(2-aminoethyl)-
9317 - Phenol, 2-(dimethylamino)-
9318 - Phenol, 3-(dimethylamino)-
9319 - Phenol, 4-(dimethylamino)-
9356 - Phenol, 3-(ethylamino)-
10951 - 2-Pyridineethanol, α-methyl-
10979 - Pyridine, 2-(2-methoxyethyl)-
11012 - 2-Pyridinepropanol
11013 - 3-Pyridinepropanol

$C_8H_{11}NO_2$
702 - Benzenamine, 2,5-dimethoxy-
703 - Benzenamine, 2,6-dimethoxy-

$C_8H_{11}NO_2S$
2211 - Benzenesulfonamide, N,4-dimethyl-

$C_8H_{11}NO_3$
1500 - 1,2-Benzenediol, 4-(2-amino-1-hydroxyethyl)-, (R)-

$C_8H_{11}NO_4$
11145 - 1-Pyrrolidineacetic acid, 2,5-dioxo-, ethyl ester

$C_8H_{11}N_3O_6$
11832 - 1,2,4-Triazine-3,5(2H,4H)-dione, 2-β-D-ribofuranosyl-

$C_8H_{11}N_5$
7234 - Imidodicarbonimidic diamide, N-phenyl-

$C_8H_{11}O_4$
5191 - 1,3-Dioxane-2-ethanol, 4-hydroxy-α,6-dimethyl-

C_8H_{12}
2993 - Bicyclo[4.2.0]oct-2-ene
4186 - Cyclobutane, 1,2-diethenyl-, cis-
4187 - Cyclobutane, 1,2-diethenyl-, trans-
4252 - 1,3-Cyclohexadiene, 1,3-dimethyl-
4253 - 1,3-Cyclohexadiene, 1,4-dimethyl-
4254 - 1,3-Cyclohexadiene, 1,5-dimethyl-
4255 - 1,3-Cyclohexadiene, 2,5-dimethyl-
4256 - 1,3-Cyclohexadiene, 2,6-dimethyl-
4314 - Cyclohexane, 1,2-bis(methylene)-
4657 - Cyclohexene, 1-ethenyl-
4658 - Cyclohexene, 4-ethenyl-
4732 - 1,4-Cyclooctadiene
4733 - 1,5-Cyclooctadiene
4734 - 1,5-Cyclooctadiene, (Z,Z)-
4755 - Cyclooctyne
6994 - 1,3,5-Hexatriene, 2,5-dimethyl-
8489 - 2,4,6-Octatriene, (E,E,E)-
8526 - 1-Octen-3-yne

$C_8H_{12}ClN$
717 - Benzenamine, N,N-dimethyl-, hydrochloride

$C_8H_{12}ClNO$
44 - Acetamide, 2-chloro-N,N-di-2-propenyl-

$C_8H_{12}ClNO_3$
10934 - 3,4-Pyridinedimethanol, 5-hydroxy-6-methyl-, hydrochloride

$C_8H_{12}Cl_2$
7117 - 3-Hexyne, 2,5-dichloro-2,5-dimethyl-

$C_8H_{12}N_2$
1267 - 1,2-Benzenediamine, N,N-dimethyl-
1268 - 1,3-Benzenediamine, N,N-dimethyl-
1269 - 1,2-Benzenediamine, N,N'-dimethyl-
1270 - 1,4-Benzenediamine, N,N-dimethyl-
1433 - 1,3-Benzenedimethanamine
3292 - Butanedinitrile, tetramethyl-
7159 - Hydrazine, 1-ethyl-1-phenyl-
7160 - Hydrazine, 1-ethyl-2-phenyl-
7163 - Hydrazine, 1-methyl-2-(3-methylphenyl)-
7170 - Hydrazine, (1-phenylethyl)-
8385 - Octanedinitrile
10809 - Pyrazine, 3-ethyl-2,5-dimethyl-

$C_8H_{12}N_2O$
1272 - 1,2-Benzenediamine, 4-ethoxy-

$C_8H_{12}N_2O_2$
6770 - Hexane, 1,6-diisocyanato-

$C_8H_{12}N_2O_3$
11087 - 2,4,6(1H,3H,5H)-Pyrimidinetrione, 5,5-diethyl-

$C_8H_{12}N_4$
10093 - Propanenitrile, 2,2'-azobis[2-methyl-

$C_8H_{12}O$
4529 - Cyclohexanol, 1-ethynyl-
4594 - Cyclohexanone, 2-ethylidene-
4705 - 2-Cyclohexen-1-one, 2,3-dimethyl-
4706 - 2-Cyclohexen-1-one, 2,5-dimethyl-
4707 - 2-Cyclohexen-1-one, 3,5-dimethyl-
4708 - 2-Cyclohexen-1-one, 3,6-dimethyl-
4754 - 3-Cyclooecten-1-one
4947 - 2-Cyclopenten-1-one, 3-(1-methylethyl)-
5793 - Ethanone, 1-(1-cyclohexen-1-yl)-
6090 - Furan, 2-(1,1-dimethylethyl)-
8509 - 1-Octen-4-ol
8555 - 7-Oxabicyclo[4.1.0]heptane, 3-ethenyl-
8688 - 2(1H)-Pentalenone, hexahydro-
8689 - 2(1H)-Pentalenone, hexahydro-, cis-

$C_8H_{12}O_2$
4378 - 1,3-Cyclohexanedione, 5,5-dimethyl-
4572 - Cyclohexanone, 2-acetyl-
4640 - 3-Cyclohexene-1-carboxylic acid, methyl ester
6699 - 2,4-Hexadienoic acid, ethyl ester
6700 - 2,4-Hexadienoic acid, ethyl ester, (E,E)-
8504 - 4-Octene-2,7-dione
8538 - 2-Octynoic acid
8539 - 3-Octynoic acid

8540 - 5-Octynoic acid
8541 - 7-Octynoic acid
8557 - 7-Oxabicyclo[4.1.0]heptane, 3-oxiranyl-
8597 - 2-Oxetanone, 3,3-dimethyl-4-
(1-methylethylidene)-
8805 - 2,4-Pentanedione, 3-(2-propenyl)-

$C_8H_{12}O_3$
4336 - Cyclohexanecarboxylic acid, 3-methyl-4-oxo-
4785 - Cyclopentanecarboxylic acid, 2-oxo-, ethyl
ester
10695 - 2-Propenoic acid, (tetrahydro-2-
furanyl)methyl ester

$C_8H_{12}O_4$
3462 - Butanoic acid, 2-acetyl-3-oxo-, ethyl ester
3775 - 2-Butenedioic acid (E)-, diethyl ester
3776 - 2-Butenedioic acid (Z)-, diethyl ester
4181 - 1,2-Cyclobutanedicarboxylic acid, dimethyl
ester, trans-
4183 - 1,3-Cyclobutanedicarboxylic acid, 2,2-
dimethyl-, cis-
4354 - 1,4-Cyclohexanedicarboxylic acid, trans-
6030 - 3-Furanacetic acid, tetrahydro-2,2-dimethyl-5-
oxo-
10564 - 2-Propene-1,1-diol, 2-methyl-, diacetate

$C_8H_{12}O_5$
3340 - Butanedioic acid, oxo-, diethyl ester

$C_8H_{12}O_6Si$
11498 - Silanetriol, ethenyl-, triacetate

$C_8H_{12}S$
11731 - Thiophene, 2-butyl-

$C_8H_{12}Si$
11418 - Silane, dimethylphenyl-
11449 - Silane, tetraethenyl-

$C_8H_{12}Sn$
11543 - Stannane, tetraethenyl-

$C_8H_{13}Cl$
8533 - 1-Octyne, 1-chloro-

$C_8H_{13}F_3O_2$
268 - Acetic acid, trifluoro-, 2-methylpentyl ester

$C_8H_{13}N$
4290 - Cyclohexanamine, 1-ethynyl-
4313 - Cyclohexaneacetonitrile
11113 - 1H-Pyrrole, 1-butyl-
11126 - 1H-Pyrrole, 3-ethyl-2,4-dimethyl-
11127 - 1H-Pyrrole, 4-ethyl-2,3-dimethyl-
11141 - 1H-Pyrrole, 2,3,4,5-tetramethyl-

$C_8H_{13}NO$
490 - 8-Azabicyclo[3.2.1]octan-3-one, 8-methyl-

$C_8H_{13}NO_2$
7575 - 2,5-Methano-2H-furo[3,2-b]pyrrol-6-ol,
hexahydro-4-methyl-, (2α,3aβ,5α,6β,6aβ
9656 - 2,6-Piperidinedione, 4-ethyl-4-methyl-
10914 - 3-Pyridinecarboxylic acid, 1,2,5,6-tetrahydro-
1-methyl-, methyl ester

$C_8H_{13}NO_3$
8573 - 2H-1,3-Oxazine-2,4(3H)-dione, 5,5-
diethyldihydro-

$C_8H_{13}NO_5$
9953 - Propanedioic acid, (formylamino)-, diethyl
ester

$C_8H_{13}N_2O_3PS$
11720 - Thionazin

C_8H_{14}
2991 - Bicyclo[4.2.0]octane
4245 - Cycloheptene, 1-methyl-
4246 - Cycloheptene, 5-methyl-
4388 - Cyclohexane, ethenyl-
4391 - Cyclohexane, ethylidene-
4434 - Cyclohexane, 1-methyl-4-methylene-
4648 - Cyclohexene, 1,2-dimethyl-
4649 - Cyclohexene, 1,3-dimethyl-
4650 - Cyclohexene, 1,4-dimethyl-
4651 - Cyclohexene, 1,5-dimethyl-
4652 - Cyclohexene, 1,6-dimethyl-
4653 - Cyclohexene, 3,3-dimethyl-
4654 - Cyclohexene, 3,5-dimethyl-
4655 - Cyclohexene, 3,6-dimethyl-

4656 - Cyclohexene, 4,4-dimethyl-
4660 - Cyclohexene, 1-ethyl-
4661 - Cyclohexene, 3-ethyl-
4662 - Cyclohexene, 4-ethyl-
4751 - Cyclooctene, (E)-
4752 - Cyclooctene, (Z)-
4848 - Cyclopentane, 2-propenyl-
4935 - Cyclopentene, 1-propyl-
4937 - Cyclopentene, 1,2,3-trimethyl-
4938 - Cyclopentene, 1,5,5-trimethyl-
6335 - 1,5-Heptadiene, 2-methyl-
6691 - 1,5-Hexadiene, 2,5-dimethyl-
6692 - 2,4-Hexadiene, 2,5-dimethyl-
8319 - 1,7-Octadiene
8320 - 2,6-Octadiene
8529 - 1-Octyne
8530 - 2-Octyne
8531 - 3-Octyne
8532 - 4-Octyne
8686 - Pentalene, octahydro-, cis-
8687 - Pentalene, octahydro-, trans-
9132 - 1-Pentyne, 3-ethyl-3-methyl-

$C_8H_{14}ClNS_2$
4065 - CDEC

$C_8H_{14}ClN_5$
476 - Atrazine

$C_8H_{14}Cl_2N_2O_2$
10978 - 3-Pyridinemethanol, 4-(aminomethyl)-5-
hydroxy-6-methyl-, dihydrochloride

$C_8H_{14}Cl_2O_2$
3501 - Butanoic acid, 3,4-dichloro-, butyl ester

$C_8H_{14}N_2$
275 - Acetonitrile, (cyclohexylamino)-
9703 - 1-Piperidinepropanenitrile

$C_8H_{14}N_4OS$
7658 - Metribuzin

$C_8H_{14}O$
4235 - Cycloheptanone, 2-methyl-
4580 - Cyclohexanone, 2,2-dimethyl-
4581 - Cyclohexanone, 2,3-dimethyl-, cis-
4582 - Cyclohexanone, 2,3-dimethyl-, trans-
4583 - Cyclohexanone, 2,4-dimethyl-, (2S-trans)-
4584 - Cyclohexanone, 2,4-dimethyl-, cis-
4585 - Cyclohexanone, 2,5-dimethyl-, trans-(±)-
4586 - Cyclohexanone, 2,6-dimethyl-
4587 - Cyclohexanone, 3,3-dimethyl-
4588 - Cyclohexanone, 3,4-dimethyl-
4589 - Cyclohexanone, 3,5-dimethyl-
4590 - Cyclohexanone, 4,4-dimethyl-
4593 - Cyclohexanone, 4-ethyl-
4740 - Cyclooctanone
4753 - 2-Cycloocten-1-ol
4883 - Cyclopentanone, 2-ethyl-5-methyl-
4887 - Cyclopentanone, 2-(1-methylethyl)-
4894 - Cyclopentanone, 2-propyl-
4895 - Cyclopentanone, 3-propyl-
4896 - Cyclopentanone, 2,2,4-trimethyl-
4897 - Cyclopentanone, 2,2,5-trimethyl-
4898 - Cyclopentanone, 2,3,5-trimethyl-
4899 - Cyclopentanone, 2,4,4-trimethyl-
5794 - Ethanone, 1-cyclohexyl-
5846 - Ethanone, 1-(1-methylcyclopentyl)-
5847 - Ethanone, 1-(2-methylcyclopentyl)-
6338 - 1,6-Heptadien-4-ol, 4-methyl-
6595 - 1-Hepten-3-one, 5-methyl-
6596 - 2-Hepten-4-one, 5-methyl-
6597 - 2-Hepten-4-one, 6-methyl-
6598 - 3-Hepten-2-one, 6-methyl-
6599 - 4-Hepten-3-one, 6-methyl-
6600 - 4-Hepten-3-one, 5-methyl-
6601 - 5-Hepten-2-one, 3-methyl-
6602 - 5-Hepten-2-one, 5-methyl-
6603 - 5-Hepten-2-one, 6-methyl-
6604 - 5-Hepten-3-one, 5-methyl-
6605 - 5-Hepten-3-one, 5-methyl-
7091 - 3-Hexen-2-one, 3,4-dimethyl-
7092 - 4-Hexen-2-one, 3,4-dimethyl-

8543 - 2-Octyn-1-ol

$C_8H_{14}O_2$
155 - Acetic acid, cyclohexyl ester
3836 - 2-Butenoic acid, 2,3-dimethyl-, ethyl ester
4223 - Cycloheptanecarboxylic acid
4306 - Cyclohexaneacetic acid
4335 - Cyclohexanecarboxylic acid, methyl ester
4776 - Cyclopentaneacetic acid, 3-methyl-
4847 - Cyclopentanepropanoic acid
5821 - Ethanone, 1-(1-hydroxycyclohexyl)-
6140 - 2(3H)-Furanone, 5-butyldihydro-
6565 - 2-Heptenoic acid, 6-methyl-
6807 - 3,4-Hexanedione, 2,5-dimethyl-
6892 - Hexanoic acid, 3-hydroxy-2,2-dimethyl-, β-
lactone
7053 - 2-Hexenoic acid, ethyl ester
7054 - 3-Hexenoic acid, ethyl ester
7119 - 3-Hexyne-2,5-diol, 2,5-dimethyl-
8364 - Octanedial
8393 - 2,4-Octanedione
8394 - 2,7-Octanedione
8395 - 3,6-Octanedione
8396 - 4,5-Octanedione
9080 - 2-Pentenoic acid, 4-methyl-, ethyl ester
9081 - 3-Pentenoic acid, 3-methyl-, ethyl ester
9082 - 3-Pentenoic acid, 4-methyl-, ethyl ester
10650 - 2-Propenoic acid, 2-methyl-, butyl ester
10651 - 2-Propenoic acid, methylbutyl ester
10652 - 2-Propenoic acid, 2-methylbutyl ester
10665 - 2-Propenoic acid, 2-methyl-, 1-methylpropyl
ester
10666 - 2-Propenoic acid, 2-methyl-, 2-methylpropyl
ester

$C_8H_{14}O_3$
3467 - Butanoic acid, anhydride
3511 - Butanoic acid, 2,2-dimethyl-3-oxo-, ethyl ester
3518 - Butanoic acid, 2-ethyl-3-oxo-, ethyl ester
3569 - Butanoic acid, 3-oxo-, butyl ester
6126 - 2-Furanmethanol, tetrahydro-, propanoate
6903 - Hexanoic acid, 4-methyl-5-oxo-, methyl ester
6906 - Hexanoic acid, 5-oxo-, ethyl ester
8897 - Pentanoic acid, 4-methyl-3-oxo-, ethyl ester
8909 - Pentanoic acid, 4-oxo-, 1-methylethyl ester
8912 - Pentanoic acid, 4-oxo-, propyl ester
10291 - Propanoic acid, 2-methyl-, anhydride

$C_8H_{14}O_4$
3303 - Butanedioic acid, diethyl ester
3333 - Butanedioic acid, monobutyl ester
3343 - Butanedioic acid, tetramethyl-
3348 - 1,4-Butanediol, diacetate
5556 - Ethanedioic acid, bis(1-methylethyl) ester
5565 - Ethanedioic acid, dipropyl ester
5576 - 1,2-Ethanediol, dipropanoate
6786 - Hexanedioic acid, dimethyl ester
6791 - Hexanedioic acid, monoethyl ester
8386 - Octanedioic acid
8777 - Pentanedioic acid, 3-ethyl-3-methyl-
9962 - Propanedioic acid, methyl-, diethyl ester

$C_8H_{14}O_4S$
10331 - Propanoic acid, 3,3′-thiobis-, dimethyl ester

$C_8H_{14}O_5$
5754 - Ethanol, 2,2′-oxybis-, diacetate

$C_8H_{14}O_6$
3307 - Butanedioic acid, 2,3-dihydroxy-, diethyl ester,
(R*,R*)-(±)-

$C_8H_{15}Br$
4316 - Cyclohexane, (2-bromoethyl)-

$C_8H_{15}BrO_2$
4149 - Crotonaldehyde, 2-bromo-, diethyl acetal
6877 - Hexanoic acid, 2-bromo-, ethyl ester, (±)-
6878 - Hexanoic acid, 5-bromo-, ethyl ester
6879 - Hexanoic acid, 6-bromo-, ethyl ester
8425 - Octanoic acid, 2-bromo-
8426 - Octanoic acid, 8-bromo-
10210 - Propanoic acid, 3-bromo-, 3-methylbutyl
ester

$C_8H_{15}Cl$
6541 - 2-Heptene, 6-chloro-2-methyl-
7010 - 3-Hexene, 1-chloro-4-ethyl-
8498 - 1-Octene, 2-chloro-
8499 - 2-Octene, 2-chloro-
8500 - 4-Octene, 4-chloro-, (Z)-
8501 - 4-Octene, 4-chloro-
$C_8H_{15}ClO$
6987 - Hexanoyl chloride, 2-ethyl-
8481 - Octanoyl chloride
$C_8H_{15}ClO_2$
10226 - Propanoic acid, 3-chloro-, 3-methylbutyl
ester
$C_8H_{15}FO_2$
8429 - Octanoic acid, 8-fluoro-
$C_8H_{15}N$
484 - 8-Azabicyclo[3.2.1]octane, 8-methyl-
7319 - 1H-Indole, octahydro-
7325 - Indolizine, octahydro-
8411 - Octanenitrile
9706 - Piperidine, 2-(1-propenyl)-, (±)-
11021 - Pyridine, 2,3,4,5-tetrahydro-6-propyl-
$C_8H_{15}NO$
486 - 8-Azabicyclo[3.2.1]octan-3-ol, 8-methyl-, exo-
489 - 8-Azabicyclo[3.2.1]octan-3-ol, 8-methyl-, endo-
3408 - Butanenitrile, 2-hydroxy-3-methyl-2-(1-methylethyl)-
8829 - Pentanenitrile, 2-hydroxy-2-propyl-
9631 - Piperidine, 1-acetyl-3-methyl-
10471 - 2-Propanone, 1-(1-methyl-2-pyrrolidinyl)-, (R)-
10484 - 1-Propanone, 1-(2-piperidinyl)-
10485 - 2-Propanone, 1-(2-piperidinyl)-, (R)-
$C_8H_{15}NO_2$
3406 - Butanenitrile, 4,4-diethoxy-
4333 - Cyclohexanecarboxylic acid, 4-(aminomethyl)-, trans-
8399 - 2,3-Octanedione, 3-oxime
9643 - 4-Piperidinecarboxylic acid, ethyl ester
9645 - 4-Piperidinecarboxylic acid, 1-methyl-, methyl ester
10654 - 2-Propenoic acid, 2-methyl-, 2-(dimethylamino)ethyl ester
$C_8H_{15}NO_3$
6871 - Hexanoic acid, 6-(acetylamino)-
$C_8H_{15}NS$
11704 - Thiocyanic acid, heptyl ester
C_8H_{16}
4226 - Cycloheptane, methyl-
4364 - Cyclohexane, 1,1-dimethyl-
4365 - Cyclohexane, 1,2-dimethyl-, cis-
4366 - Cyclohexane, 1,2-dimethyl-, trans-
4367 - Cyclohexane, 1,3-dimethyl-, cis-
4368 - Cyclohexane, 1,4-dimethyl-, cis-
4370 - Cyclohexane, 1,3-dimethyl-, trans-
4371 - Cyclohexane, 1,4-dimethyl-, trans-
4390 - Cyclohexane, ethyl-
4736 - Cyclooctane
4814 - Cyclopentane, 1-ethyl-1-methyl-
4815 - Cyclopentane, 1-ethyl-2-methyl-, cis-
4816 - Cyclopentane, 1-ethyl-2-methyl-, trans-
4817 - Cyclopentane, 1-ethyl-3-methyl-, cis-
4818 - Cyclopentane, 1-ethyl-3-methyl-, trans-
4828 - Cyclopentane, (1-methylethyl)-
4849 - Cyclopentane, propyl-
4855 - Cyclopentane, 1,1,2-trimethyl-
4856 - Cyclopentane, 1,1,3-trimethyl-
4857 - Cyclopentane, 1,2,3-trimethyl-, (1α,2β,3α)-
4858 - Cyclopentane, 1,2,3-trimethyl-, (1α,2α,3α)-
4859 - Cyclopentane, 1,2,3-trimethyl-, (1α,2α,3β)-
4860 - Cyclopentane, 1,2,4-trimethyl-, (1α,2β,4α)-
4861 - Cyclopentane, 1,2,4-trimethyl-, (1α,2α,4α)-
4862 - Cyclopentane, 1,2,4-trimethyl-, (1α,2α,4β)-
6425 - Heptane, 3-methylene-
6426 - Heptane, 4-methylene-
6543 - 1-Heptene, 2-methyl-
6544 - 1-Heptene, 3-methyl-
6545 - 1-Heptene, 4-methyl-
6546 - 1-Heptene, 5-methyl-
6547 - 1-Heptene, 6-methyl-
6548 - 2-Heptene, 2-methyl-
6549 - 2-Heptene, 3-methyl-, (E)-
6550 - 2-Heptene, 3-methyl-, (Z)-
6551 - 2-Heptene, 4-methyl-, (E)-
6552 - 2-Heptene, 4-methyl-, (Z)-
6553 - 2-Heptene, 6-methyl-
6554 - 3-Heptene, 2-methyl-
6555 - 3-Heptene, 3-methyl-
6556 - 3-Heptene, 4-methyl-
6557 - 3-Heptene, 5-methyl-
6558 - 3-Heptene, 6-methyl-
6836 - Hexane, 2-methyl-3-methylene-
6837 - Hexane, 2-methyl-4-methylene-
6838 - Hexane, 3-methyl-4-methylene-
7013 - 1-Hexene, 2,3-dimethyl-
7014 - 1-Hexene, 2,4-dimethyl-
7015 - 1-Hexene, 2,5-dimethyl-
7016 - 1-Hexene, 3,3-dimethyl-
7017 - 1-Hexene, 3,4-dimethyl-
7018 - 1-Hexene, 3,5-dimethyl-
7019 - 1-Hexene, 4,4-dimethyl-
7020 - 1-Hexene, 4,5-dimethyl-
7021 - 1-Hexene, 5,5-dimethyl-
7022 - 2-Hexene, 2,3-dimethyl-
7023 - 2-Hexene, 2,4-dimethyl-
7024 - 2-Hexene, 2,5-dimethyl-
7025 - 2-Hexene, 4,5-dimethyl-
7026 - 2-Hexene, 5,5-dimethyl-, (E)-
7027 - 2-Hexene, 5,5-dimethyl-, (Z)-
7028 - 3-Hexene, 2,2-dimethyl-, (E)-
7029 - 3-Hexene, 2,2-dimethyl-, (Z)-
7030 - 3-Hexene, 2,4-dimethyl-, (E)-
7031 - 3-Hexene, 2,4-dimethyl-, (Z)-
7033 - 1-Hexene, 3-ethyl-
7034 - 3-Hexene, 3-ethyl-
8491 - 1-Octene
8492 - 2-Octene, (E)-
8493 - 2-Octene, (Z)-
8494 - 3-Octene, (E)-
8495 - 3-Octene, (Z)-
8496 - 4-Octene, (E)-
8497 - 4-Octene, (Z)-
8764 - Pentane, 2,2-dimethyl-3-methylene-
8765 - Pentane, 2,4-dimethyl-3-methylene-
9046 - 1-Pentene, 3-ethyl-2-methyl-
9047 - 1-Pentene, 3-ethyl-3-methyl-
9048 - 1-Pentene, 3-ethyl-4-methyl-
9049 - 2-Pentene, 3-ethyl-2-methyl-
9050 - 2-Pentene, 3-ethyl-4-methyl-, (E)-
9051 - 2-Pentene, 3-ethyl-4-methyl-, (Z)-
9064 - Pentene, 2,4,4-trimethyl-
9065 - 1-Pentene, 2,3,3-trimethyl-
9066 - 1-Pentene, 2,3,4-trimethyl-
9067 - 1-Pentene, 2,4,4-trimethyl-
9068 - 1-Pentene, 3,3,4-trimethyl-
9069 - 1-Pentene, 3,4,4-trimethyl-
9070 - 2-Pentene, 2,3,4-trimethyl-
9071 - 2-Pentene, 2,4,4-trimethyl-
9072 - 2-Pentene, 3,4,4-trimethyl-
$C_8H_{16}BrF$
8358 - Octane, 1-bromo-8-fluoro-
$C_8H_{16}Br_2$
8366 - Octane, 1,2-dibromo-
8367 - Octane, 1,4-dibromo-
8368 - Octane, 1,5-dibromo-
8369 - Octane, 1,8-dibromo-
8370 - Octane, 4,5-dibromo-
$C_8H_{16}ClF$
8361 - Octane, 1-chloro-8-fluoro-
$C_8H_{16}ClNO$
42 - Acetamide, 2-chloro-N,N-bis(1-methylethyl)-
45 - Acetamide, 2-chloro-N-hexyl-
$C_8H_{16}Cl_2$
6765 - Hexane, 2,5-dichloro-2,5-dimethyl-
8371 - Octane, 1,8-dichloro-
8372 - Octane, 2,3-dichloro-
$C_8H_{16}NO_5P$
5172 - Dicrotophos
$C_8H_{16}N_2$
3673 - 2-Butanone, (1-methylpropylidene)hydrazone
$C_8H_{16}N_2O_2$
8397 - 4,5-Octanedione, dioxime
$C_8H_{16}N_2O_2S_2$
7688 - Morpholine, 4,4'-dithiobis-
$C_8H_{16}N_2O_3$
7414 - DL-Leucine, N-glycyl-
$C_8H_{16}N_2O_4S_2$
12042 - D-Valine, 3-[(2-amino-2-carboxyethyl)dithio]-, (R)-
$C_8H_{16}O$
4230 - Cycloheptanol, 1-methyl-
4231 - Cycloheptanol, 2-methyl-
4232 - Cycloheptanol, 4-methyl-
4387 - Cyclohexaneethanol
4420 - Cyclohexanemethanol, α-methyl-
4421 - Cyclohexanemethanol, α-methyl-, (S)-
4422 - Cyclohexanemethanol, 2-methyl-, cis-
4423 - Cyclohexanemethanol, 2-methyl-, trans-
4424 - Cyclohexanemethanol, 4-methyl-
4508 - Cyclohexanol, 1,2-dimethyl-, cis-
4509 - Cyclohexanol, 1,2-dimethyl-, trans-
4510 - Cyclohexanol, 1,3-dimethyl-, cis-
4511 - Cyclohexanol, 1,3-dimethyl-, trans-
4512 - Cyclohexanol, 1,4-dimethyl-, cis-
4513 - Cyclohexanol, 2,2-dimethyl-
4514 - Cyclohexanol, 2,4-dimethyl-
4515 - Cyclohexanol, 2,6-dimethyl-
4516 - Cyclohexanol, 3,3-dimethyl-
4517 - Cyclohexanol, 3,4-dimethyl-
4518 - Cyclohexanol, 3,5-dimethyl-
4519 - Cyclohexanol, 4,4-dimethyl-
4522 - Cyclohexanol, 1-ethyl-
4523 - Cyclohexanol, 2-ethyl-, trans-(±)-
4524 - Cyclohexanol, 2-ethyl-, cis-(±)-
4525 - Cyclohexanol, 4-ethyl-
4739 - Cyclooctanol
4872 - Cyclopentanol, 1-propyl-
4873 - Cyclopentanol, 2-propyl-, cis-
4874 - Cyclopentanol, 2-propyl-, trans-
4875 - Cyclopentanol, 1,2,2-trimethyl-
4876 - Cyclopentanol, 1,2,4-trimethyl-
4877 - Cyclopentanol, 1,2,5-trimethyl-
6085 - Furan, 2,2-diethyltetrahydro-
6174 - Furan, tetrahydro-2,2,4,4-tetramethyl-
6345 - Heptanal, 5-methyl-
6520 - 2-Heptanone, 3-methyl-
6521 - 2-Heptanone, 6-methyl-
6522 - 3-Heptanone, 2-methyl-
6523 - 3-Heptanone, 6-methyl-
6524 - 4-Heptanone, 2-methyl-
6525 - 4-Heptanone, 3-methyl-
6573 - 1-Hepten-4-ol, 4-methyl-
6574 - 2-Hepten-4-ol, 5-methyl-
6575 - 2-Hepten-4-ol, 6-methyl-
6576 - 3-Hepten-2-ol, 2-methyl-
6577 - 4-Hepten-2-ol, 2-methyl-
6578 - 4-Hepten-3-ol, 3-methyl-
6579 - 4-Hepten-3-ol, 4-methyl-
6580 - 5-Hepten-2-ol, 6-methyl-
6581 - 6-Hepten-1-ol, 3-methyl-
6582 - 6-Hepten-2-ol, 2-methyl-
6719 - Hexanal, 2-ethyl-
6813 - Hexane, 1-(ethenyloxy)-
6963 - 2-Hexanone, 3,3-dimethyl-
6964 - 2-Hexanone, 3,4-dimethyl-
6965 - 3-Hexanone, 2,2-dimethyl-
6966 - 3-Hexanone, 2,4-dimethyl-
6967 - 3-Hexanone, 4,4-dimethyl-
7073 - 1-Hexen-3-ol, 3,5-dimethyl-
7074 - 4-Hexen-3-ol, 2,2-dimethyl-
7075 - 4-Hexen-3-ol, 2,5-dimethyl-

7076 - 4-Hexen-3-ol, 3,5-dimethyl-
8346 - Octanal
8471 - 2-Octanone
8472 - 3-Octanone
8473 - 4-Octanone
8510 - 2-Octen-1-ol
8511 - 2-Octen-4-ol
8512 - 4-Octen-1-ol
8513 - 5-Octen-1-ol
8612 - Oxirane, 2-(2,2-dimethylpropyl)-2-methyl-
8697 - Pentanal, 2-propyl-
8970 - 2-Pentanone, 3-ethyl-4-methyl-
8989 - 3-Pentanone, 2,2,4-trimethyl-

$C_8H_{16}O_2$
185 - Acetic acid, 2-ethylbutyl ester
194 - Acetic acid, hexyl ester
3479 - Butanoic acid, butyl ester
3509 - Butanoic acid, 1,1-dimethylethyl ester
3553 - Butanoic acid, 3-methyl-, 1-methylethyl ester
3562 - Butanoic acid, 3-methyl-, propyl ester
3563 - Butanoic acid, 2-methylpropyl ester
3564 - Butanoic acid, 1-methylpropyl ester, (S)-
3639 - 1-Butanol, 3-methyl-, propanoate
3758 - 2-Butene, 1,1-diethoxy-
4363 - 1,3-Cyclohexanedimethanol
6008 - Formic acid, heptyl ester
6466 - Heptanoic acid, methyl ester
6518 - 2-Heptanone, 4-hydroxy-3-methyl-
6519 - 3-Heptanone, 5-hydroxy-5-methyl-
6882 - Hexanoic acid, 2-ethyl-
6883 - Hexanoic acid, 3-ethyl-
6884 - Hexanoic acid, ethyl ester
8422 - Octanoic acid
8474 - 4-Octanone, 5-hydroxy-
8692 - Pentanal, 3-hydroxy-2,2,4-trimethyl-
8892 - Pentanoic acid, 4-methyl-, ethyl ester
8893 - Pentanoic acid, 1-methylethyl ester
8914 - Pentanoic acid, propyl ester
8945 - 1-Pentanol, 2-methyl-, acetate
8946 - 2-Pentanol, 2-methyl-, acetate
8947 - 2-Pentanol, 4-methyl-, acetate
8948 - 3-Pentanol, 2-methyl-, acetate
8949 - 3-Pentanol, 3-methyl-, acetate
10294 - Propanoic acid, 2-methyl-, 1,1-dimethylethyl ester
10304 - Propanoic acid, 2-methyl-, 2-methylpropyl ester
10318 - Propanoic acid, pentyl ester

$C_8H_{16}O_3$
3536 - Butanoic acid, 2-hydroxy-, 2-methylpropyl ester
8432 - Octanoic acid, 2-hydroxy-
10274 - Propanoic acid, 2-hydroxy-, 3-methylbutyl ester

$C_8H_{16}O_4$
174 - Acetic acid, diethoxy-, ethyl ester
5721 - Ethanol, 2-(2-ethoxyethoxy)-, acetate

$C_8H_{16}Si$
11417 - Silane, dimethyldi-2-propenyl-

$C_8H_{17}Br$
8356 - Octane, 1-bromo-
8357 - Octane, 2-bromo-, (±)-

$C_8H_{17}Cl$
6366 - Heptane, 2-chloro-2-methyl-
6367 - Heptane, 2-chloro-6-methyl-
6368 - Heptane, 3-(chloromethyl)-
6369 - Heptane, 3-chloro-3-methyl-
6370 - Heptane, 4-chloro-4-methyl-
6740 - Hexane, 2-chloro-2,5-dimethyl-
6741 - Hexane, 3-chloro-2,3-dimethyl-
8359 - Octane, 1-chloro-
8360 - Octane, 2-chloro-
8733 - Pentane, 2-chloro-2,4,4-trimethyl-

$C_8H_{17}ClO$
8448 - 1-Octanol, 8-chloro-

$C_8H_{17}ClO_2$
3243 - Butane, 3-chloro-1,1-diethoxy-

$C_8H_{17}Cl_3Si$
11471 - Silane, trichlorooctyl-

$C_8H_{17}F$
8405 - Octane, 1-fluoro-

$C_8H_{17}FO$
8460 - 1-Octanol, 8-fluoro-

$C_8H_{17}I$
8406 - Octane, 1-iodo-
8407 - Octane, 2-iodo-, (±)-

$C_8H_{17}N$
503 - 1H-Azonine, octahydro-
4286 - Cyclohexanamine, N,N-dimethyl-
4287 - Cyclohexanamine, 2-ethyl-
4288 - Cyclohexanamine, N-ethyl-
4723 - Cyclohexylamine, N,4-dimethyl-
4735 - Cyclooctanamine
9684 - Piperidine, 1-(1-methylethyl)-
9707 - Piperidine, 1-propyl-
9708 - Piperidine, 2-propyl-, (S)-
9709 - Piperidine, 4-propyl-
9712 - Piperidine, 2,2,4-trimethyl-
9713 - Piperidine, 2,3,6-trimethyl-
9714 - Piperidine, 2,4,6-trimethyl-
11146 - Pyrrolidine, 1-butyl-
11147 - Pyrrolidine, 1-butyl-

$C_8H_{17}NO$
58 - Acetamide, N,N-dipropyl-
3200 - Butanamide, N,N-diethyl-
3656 - 2-Butanone, 4-(diethylamino)-
4520 - Cyclohexanol, 3-(dimethylamino)-
7684 - Morpholine, 4-butyl-
8348 - Octanal, oxime
8349 - Octanamide
9677 - 2-Piperidinemethanol, α-ethyl-, (R*,S*)-(±)-
9718 - 3-Piperidinol, 6-propyl-, (3S-trans)-

$C_8H_{17}NO_2$
3952 - Carbamic acid, (3-methylbutyl)-, ethyl ester
7693 - Morpholine, 4-(2-ethoxyethyl)-
8095 - Nitrous acid, octyl ester
8412 - Octane, 1-nitro-
8413 - Octane, 2-nitro-, (+)-
8423 - Octanoic acid, 2-amino-, (±)-

$C_8H_{17}NO_3$
8079 - Nitric acid, octyl ester

C_8H_{18}
3430 - Butane, 2,2,3,3-tetramethyl-
6422 - Heptane, 2-methyl-
6423 - Heptane, 3-methyl-, (S)-
6424 - Heptane, 4-methyl-
6771 - Hexane, 2,2-dimethyl-
6772 - Hexane, 2,3-dimethyl-
6773 - Hexane, 2,4-dimethyl-
6774 - Hexane, 2,5-dimethyl-
6775 - Hexane, 3,3-dimethyl-
6776 - Hexane, 3,4-dimethyl-
6815 - Hexane, 3-ethyl-
8355 - Octane
8813 - Pentane, 3-ethyl-2-methyl-
8814 - Pentane, 3-ethyl-3-methyl-
8847 - Pentane, 2,2,3-trimethyl-
8848 - Pentane, 2,2,4-trimethyl-
8849 - Pentane, 2,3,3-trimethyl-
8850 - Pentane, 2,3,4-trimethyl-

$C_8H_{18}ClNO_2$
9830 - 1-Propanaminium, 2-(acetyloxy)-N,N,N-trimethyl-, chloride

$C_8H_{18}Hg$
7433 - Mercury, dibutyl-

$C_8H_{18}N_2$
4362 - 1,3-Cyclohexanedimethanamine
5123 - Diazene, dibutyl

$C_8H_{18}N_2O$
3212 - 1-Butanamine, N-butyl-N-nitroso-

$C_8H_{18}O$
3401 - Butane, 1-(2-methylpropoxy)-
3415 - Butane, 1,1'-oxybis-
3416 - Butane, 2,2'-oxybis-, (±)-

$C_8H_{18}O$
6421 - Heptane, 1-methoxy-
6492 - 1-Heptanol, 3-methyl-
6493 - 1-Heptanol, 4-methyl-
6494 - 1-Heptanol, 6-methyl-
6495 - 2-Heptanol, 2-methyl-
6496 - 2-Heptanol, 3-methyl-
6497 - 2-Heptanol, 5-methyl-
6498 - 2-Heptanol, 6-methyl-
6499 - 3-Heptanol, 2-methyl-, (±)-
6500 - 3-Heptanol, 3-methyl-
6501 - 3-Heptanol, 4-methyl-
6502 - 3-Heptanol, 5-methyl-
6503 - 4-Heptanol, 2-methyl-
6504 - 4-Heptanol, 3-methyl-, (R*,S*)-(±)-
6505 - 4-Heptanol, 4-methyl-
6814 - Hexane, 1-ethoxy-
6921 - 1-Hexanol, 2,5-dimethyl-
6922 - 2-Hexanol, 2,3-dimethyl-
6923 - 2-Hexanol, 2,5-dimethyl-
6924 - 2-Hexanol, 3,4-dimethyl-
6925 - 3-Hexanol, 2,2-dimethyl-
6926 - 3-Hexanol, 2,3-dimethyl-
6927 - 3-Hexanol, 2,5-dimethyl-
6928 - 3-Hexanol, 3,5-dimethyl-, (±)-
6929 - 3-Hexanol, 4,4-dimethyl-
6930 - 1-Hexanol, 2-ethyl-
6931 - 3-Hexanol, 3-ethyl-
8443 - 1-Octanol
8444 - 2-Octanol, (±)-
8445 - 3-Octanol
8446 - 4-Octanol, (±)-
8933 - 2-Pentanol, 3-ethyl-2-methyl-
8934 - 3-Pentanol, 3-ethyl-2-methyl-
8954 - 2-Pentanol, 2,4,4-trimethyl-
8955 - 3-Pentanol, 2,2,4-trimethyl-
8956 - 3-Pentanol, 2,3,4-trimethyl-
10124 - Propane, 2,2'-oxybis[2-methyl-

$C_8H_{18}OS$
3422 - Butane, 1,1'-sulfinylbis-

$C_8H_{18}OSi_2$
5233 - Disiloxane, 1,3-diethenyl-1,1,3,3-tetramethyl-

$C_8H_{18}O_2$
5586 - Ethane, 1,1-diproxy-
5728 - Ethanol, 2-(hexyloxy)-
6401 - 2,4-Heptanediol, 3-methyl-
6801 - 2,5-Hexanediol, 2,5-dimethyl-
6802 - 1,3-Hexanediol, 2-ethyl-
8389 - 1,8-Octanediol
8390 - 2,4-Octanediol
8391 - 4,5-Octanediol, (±)-
8795 - 1,3-Pentanediol, 2,2,4-trimethyl-
8796 - 1,4-Pentanediol, 2,2,4-trimethyl-
9153 - Peroxide, bis(1,1-dimethylethyl)
9993 - 1,3-Propanediol, 2-butyl-2-methyl-

$C_8H_{18}O_2S$
3423 - Butane, 1,1'-sulfonylbis-
10132 - Propane, 1,1'-sulfonylbis[2-methyl-

$C_8H_{18}O_2Si$
11408 - Silane, diethoxymethyl-2-propenyl-

$C_8H_{18}O_3$
5612 - Ethane, 1,1'-oxybis[2-ethoxy-
5664 - Ethane, 1,1,1-triethoxy-
5688 - Ethanol, 2-(2-butoxyethoxy)-
6920 - 1-Hexanol, 3,5-dimethoxy-

$C_8H_{18}O_3S$
11578 - Sulfurous acid, bis(2-methylpropyl) ester
11580 - Sulfurous acid, dibutyl ester

$C_8H_{18}O_3Si$
11424 - Silane, ethenyltriethoxy-

$C_8H_{18}O_4$
11642 - 2,5,8,11-Tetraoxadodecane

$C_8H_{18}O_4S_2$
3230 - Butane, 2,2-bis(ethylsulfonyl)-

$C_8H_{18}O_5$
5756 - Ethanol, 2,2'-[oxybis(2,1-ethanediyloxy)]bis-

$C_8H_{18}S$
3433 - Butane, 1,1'-thiobis-

3434 - Butane, 2,2'-thiobis-
8418 - 1-Octanethiol
8419 - 2-Octanethiol, (±)-
10154 - Propane, 1,1'-thiobis[2-methyl-
10155 - Propane, 2,2'-thiobis[2-methyl-

$C_8H_{18}S_2$
5242 - Disulfide, bis(1,1-dimethylethyl)
5252 - Disulfide, dibutyl

$C_8H_{18}S_4$
11649 - Tetrasulfide, bis(1,1-dimethylethyl)

$C_8H_{18}Si_2$
11434 - Silane, 1,2-ethynediylbis[trimethyl-

$C_8H_{19}N$
3211 - 1-Butanamine, N-butyl-
3227 - 2-Butanamine, N-(1-methylpropyl)-
6353 - 2-Heptanamine, 6-methyl-, (±)-
6726 - 1-Hexanamine, 2-ethyl-
8350 - 1-Octanamine
8351 - 2-Octanamine, (±)-
9811 - 2-Propanamine, N-ethyl-N-(1-methylethyl)-
9812 - 1-Propanamine, N-ethyl-N-propyl-
9823 - 1-Propanamine, 2-methyl-N-(2-methylpropyl)-

$C_8H_{19}NO$
5683 - Ethanol, 2-[bis(1-methylethyl)amino]-
5712 - Ethanol, 2-(dipropylamino)-

$C_8H_{19}NO_2$
3214 - 1-Butanamine, 4,4-diethoxy-
3624 - 2-Butanol, 3,3'-iminobis-
5416 - Ethanamine, 2,2-diethoxy-N,N-dimethyl-
5425 - Ethanamine, 2-ethoxy-N-(2-ethoxyethyl)-
5692 - Ethanol, 2,2'-(butylimino)bis-
5707 - Ethanol, 2-[2-(diethylamino)ethoxy]-

$C_8H_{19}O_2PS_2$
5934 - Ethoprop

$C_8H_{19}O_2PS_3$
5260 - Disulfoton

$C_8H_{19}O_3P$
9531 - Phosphonic acid, bis(1,1-dimethylethyl) ester
9532 - Phosphonic acid, dibutyl ester

$C_8H_{20}As_2$
5109 - Diarsine, tetraethyl-

$C_8H_{20}BrN$
5447 - Ethanaminium, N,N,N-triethyl-, bromide

$C_8H_{20}N_2$
3263 - 1,4-Butanediamine, N,N,N',N'-tetramethyl-
6751 - 2,5-Hexanediamine, 2,5-dimethyl-
7140 - Hydrazine, 1,2-bis(2-methylpropyl)-
8365 - 1,8-Octanediamine

$C_8H_{20}N_2O$
10342 - 1-Propanol, 2-[(2-amino-
2-methylpropyl)amino]-2-methyl-

$C_8H_{20}OSi$
11428 - Silane, ethoxytriethyl-

$C_8H_{20}O_3P_2$
9526 - Phosphinic acid, diethyl-, anhydride

$C_8H_{20}O_3Si$
11478 - Silane, triethoxyethyl-

$C_8H_{20}O_4Si$
11506 - Silicic acid (H4SiO4), tetraethyl ester

$C_8H_{20}O_5P_2S_2$
11716 - Thiodiphosphoric acid ([(HO)2P(S)]2O),
tetraethyl ester

$C_8H_{20}O_6P_2$
7185 - Hypophosphoric acid, tetraethyl ester

$C_8H_{20}O_7P_2$
5221 - Diphosphoric acid, tetraethyl ester

$C_8H_{20}Pb$
9731 - Plumbane, tetraethyl

$C_8H_{20}Si$
11450 - Silane, tetraethyl-

$C_8H_{20}Si_2$
11421 - Silane, 1,2-ethenediylbis[trimethyl-, (E)-

$C_8H_{23}N_5$
5498 - 1,2-Ethanediamine, N-(2-aminoethyl)-N'-[2-
[(2-aminoethyl)amino]ethyl]-

$C_8H_{24}N_4O_3P_2$
5220 - Diphosphoramide, octamethyl-

$C_8H_{24}O_2Si_3$
11914 - Trisiloxane, octamethyl-

$C_8H_{24}O_4Si_4$
4993 - Cyclotetrasiloxane, octamethyl-

$C_8H_{26}O_3Si_4$
11647 - Tetrasiloxane, 1,1,1,3,5,7,7,7-octamethyl-

$C_8I_4O_3$
7348 - 1,3-Isobenzofurandione, 4,5,6,7-tetraiodo-

C_9F_{20}
8139 - Nonane, icosofluoro-

$C_9F_{21}N$
9813 - 1-Propanamine, 1,1,2,2,3,3,3-heptafluoro-
N,N-bis(heptafluoropropyl)-

$C_9H_3Cl_3O_3$
2317 - 1,3,5-Benzenetricarbonyl trichloride

$C_9H_3F_{15}O_2$
8438 - Octanoic acid, pentadecafluoro-, methyl ester

$C_9H_4Cl_3NO_2S$
5984 - Folpet

$C_9H_4O_5$
7341 - 5-Isobenzofurancarboxylic acid, 1,3-dihydro-
1,3-dioxo-

$C_9H_5BrN_2O_2$
11221 - Quinoline, 6-bromo-5-nitro-

$C_9H_5Br_2NO$
11313 - 8-Quinolinol, 5,7-dibromo-

C_9H_5ClINO
11312 - 8-Quinolinol, 5-chloro-7-iodo-

$C_9H_5ClO_6$
2320 - 1,3,5-Benzenetricarboxylic acid, 2-chloro-

$C_9H_5Cl_2N$
11248 - Quinoline, 2,4-dichloro-
11249 - Quinoline, 2,7-dichloro-
11250 - Quinoline, 4,5-dichloro-
11251 - Quinoline, 4,7-dichloro-
11252 - Quinoline, 5,8-dichloro-

$C_9H_5Cl_2NO$
11314 - 8-Quinolinol, 5,7-dichloro-

$C_9H_5Cl_3N_4$
360 - Anilazine

$C_9H_5I_2NO$
11315 - 8-Quinolinol, 5,7-diiodo-

$C_9H_5NO_4$
10740 - 2-Propynoic acid, 3-(2-nitrophenyl)-

C_9H_6BrN
7374 - Isoquinoline, 4-bromo-
11217 - Quinoline, 3-bromo-
11218 - Quinoline, 4-bromo-
11219 - Quinoline, 6-bromo-
11220 - Quinoline, 7-bromo-

C_9H_6BrNO
11310 - 8-Quinolinol, 5-bromo-
11326 - 2(1H)-Quinolinone, 4-bromo-

$C_9H_6Br_2O_2$
10629 - 2-Propenoic acid, 2,3-dibromo-3-phenyl-,
(Z)-

C_9H_6ClN
7376 - Isoquinoline, 1-chloro-
11237 - Quinoline, 2-chloro-
11238 - Quinoline, 4-chloro-
11239 - Quinoline, 5-chloro-
11240 - Quinoline, 6-chloro-
11241 - Quinoline, 7-chloro-
11242 - Quinoline, 8-chloro-

C_9H_6ClNO
11311 - 8-Quinolinol, 5-chloro-

$C_9H_6Cl_2N_2O_3$
7643 - Methazole

$C_9H_6Cl_6O_3S$
5369 - Endosulfan

C_9H_6IN
11272 - Quinoline, 4-iodo-
11273 - Quinoline, 6-iodo-

$C_9H_6INO_4S$
11295 - 5-Quinolinesulfonic acid, 8-hydroxy-7-iodo-

$C_9H_6N_2$
9916 - Propanedinitrile, phenyl-

$C_9H_6N_2O_2$
1430 - Benzene, 1,3-diisocyanato-2-methyl-
1431 - Benzene, 2,4-diisocyanato-1-methyl-
7386 - Isoquinoline, 5-nitro-
11284 - Quinoline, 5-nitro-
11285 - Quinoline, 6-nitro-
11286 - Quinoline, 7-nitro-

$C_9H_6N_2O_3$
11287 - Quinoline, 4-nitro-, 1-oxide
11323 - 8-Quinolinol, 5-nitro-

C_9H_6O
10724 - 2-Propynal, 3-phenyl-

$C_9H_6OS_2$
7617 - Methanone, di-2-thienyl-

$C_9H_6O_2$
2791 - 1H-2-Benzopyran-1-one
2792 - 2H-1-Benzopyran-2-one
2793 - 4H-1-Benzopyran-4-one
7264 - 1H-Indene-1,3(2H)-dione

$C_9H_6O_2S$
2400 - Benzo[b]thiophene-2-carboxylic acid

$C_9H_6O_3$
2439 - 2-Benzofurancarboxylic acid
2775 - 2H-1-Benzopyran-3-carboxylic acid, 2-oxo-
2825 - 2H-1-Benzopyran-2-one, 6-hydroxy-
2826 - 2H-1-Benzopyran-2-one, 7-hydroxy-

$C_9H_6O_4$
2804 - 2H-1-Benzopyran-2-one, 6,7-dihydroxy-
2805 - 2H-1-Benzopyran-2-one, 7,8-dihydroxy-
7267 - 1H-Indene-1,3(2H)-dione, 2,2-dihydroxy-

$C_9H_6O_6$
2318 - 1,2,3-Benzenetricarboxylic acid
2319 - 1,2,4-Benzenetricarboxylic acid

$C_9H_7BrN_2$
11207 - 4-Quinolinamine, 3-bromo-
11215 - Quinoline, 6-amino-3-bromo-

C_9H_7BrO
10510 - 2-Propenal, 3-bromo-3-phenyl-

$C_9H_7BrO_4$
2481 - Benzoic acid, 2-(acetyloxy)-5-bromo-

C_9H_7ClO
10716 - 2-Propenoyl chloride, 3-phenyl-
10717 - 2-Propenoyl chloride, 3-phenyl-, (E)-

$C_9H_7Cl_3O_3$
11512 - Silvex

$C_9H_7FO_2$
10637 - 2-Propenoic acid, 2-fluoro-3-phenyl-

C_9H_7N
7373 - Isoquinoline
10587 - 2-Propenenitrile, 3-phenyl-, (E)-
10588 - 2-Propenenitrile, 3-phenyl-, (Z)-
11214 - Quinoline

C_9H_7NO
2124 - Benzenepropanenitrile, β-oxo-
11305 - 5-Quinolinol
11306 - 6-Quinolinol
11307 - 7-Quinolinol
11308 - 8-Quinolinol
11325 - 2(1H)-Quinolinone

$C_9H_7NO_2$
7365 - 1H-Isoindole-1,3(2H)-dione, 2-methyl-

$C_9H_7NO_4$
10676 - 2-Propenoic acid, 3-(2-nitrophenyl)-, (E)-

$C_9H_7NO_4S$
11294 - 5-Quinolinesulfonic acid, 8-hydroxy-

$C_9H_7N_3S$
11858 - Tricyclazole

C_9H_8
2173 - Benzene, 1-propynyl-
7245 - 1H-Indene

$C_9H_8Br_2$
7247 - 1H-Indene, 1,2-dibromo-2,3-dihydro-

$C_9H_8Cl_2$
1397 - Benzene, (3,3-dichloro-2-propenyl)-
7249 - 1H-Indene, 1,2-dichloro-2,3-dihydro-

$C_9H_8Cl_2O_3$
172 - Acetic acid, (2,4-dichlorophenoxy)-, methyl ester
7653 - Methyl 3,6-dichloro-*o*-anisate
10250 - Propanoic acid, 2-(2,4-dichlorophenoxy)-

$C_9H_8Cl_3NO_2S$
7368 - 1H-Isoindole-1,3(2H)-dione, 3a,4,7,7a-tetrahydro-2-[(trichloromethyl)thio]-

$C_9H_8NO_3S$
11324 - 8-Quinolinol, sulfate (2:1) (salt)

$C_9H_8N_2$
7212 - 1H-Imidazole, 1-phenyl-
7213 - 1H-Imidazole, 2-phenyl-
7372 - 5-Isoquinolinamine
11202 - 2-Quinolinamine
11203 - 4-Quinolinamine
11204 - 5-Quinolinamine
11205 - 6-Quinolinamine
11206 - 8-Quinolinamine
11337 - Quinoxaline, 2-methyl-
11338 - Quinoxaline, 6-methyl-

$C_9H_8N_2O$
11336 - Quinoxaline, 6-methoxy-

$C_9H_8N_2O_6$
2593 - Benzoic acid, 3,5-dinitro-, ethyl ester

$C_9H_8N_4OS$
11687 - Thidiazuron

C_9H_8O
1876 - Benzenemethanol, α-ethynyl-
2456 - Benzofuran, 2-methyl-
2457 - Benzofuran, 3-methyl-
2458 - Benzofuran, 5-methyl-
2459 - Benzofuran, 7-methyl-
2772 - 2H-1-Benzopyran
7282 - 6H-Indeno[1,2-b]oxirene, 1a,6a-dihydro-
7286 - 1H-Inden-1-one, 2,3-dihydro-
7287 - 2H-Inden-2-one, 1,3-dihydro-
10518 - 2-Propenal, 3-phenyl-, (*E*)-
10745 - 2-Propyn-1-ol, 3-phenyl-

C_9H_8OS
2869 - 4H-1-Benzothiopyran-4-one, 2,3-dihydro-

$C_9H_8O_2$
927 - Benzeneacetic acid, α-methylene-
2425 - 1,3-Benzodioxole, 5-ethenyl-
2455 - 2-Benzofuranmethanol
2797 - 1H-2-Benzopyran-1-one, 3,4-dihydro-
2798 - 2H-1-Benzopyran-2-one, 3,4-dihydro-
2799 - 4H-1-Benzopyran-4-one, 2,3-dihydro-
6132 - Furan, 2,2′-methylenebis-
10028 - 1,2-Propanedione, 1-phenyl-
10682 - 2-Propenoic acid, 3-phenyl-, (*E*)-
10683 - 2-Propenoic acid, 3-phenyl-, (*Z*)-

$C_9H_8O_3$
2476 - Benzoic acid, 2-acetyl-
2477 - Benzoic acid, 3-acetyl-
2478 - Benzoic acid, 4-acetyl-
7580 - 4,7-Methanoisobenzofuran-1,3-dione, 3a,4,7,7a-tetrahydro-, (3aα,4α,7α,7aα)-
10644 - 2-Propenoic acid, 3-(4-hydroxyphenyl)-

$C_9H_8O_4$
2420 - 1,3-Benzodioxole-5-carboxylic acid, methyl ester
2480 - Benzoic acid, 2-(acetyloxy)-
9151 - Peroxide, acetyl benzoyl
10632 - 2-Propenoic acid, 3-(3,4-dihydroxyphenyl)-

C_9H_8S
2402 - Benzo[b]thiophene, 5-methyl-

C_9H_9Br
1082 - Benzene, 1-bromo-4-(1-propenyl)-
1083 - Benzene, 1-bromo-4-(2-propenyl)-
1084 - Benzene, (3-bromo-1-propenyl)-

C_9H_9BrO
5779 - Ethanone, 2-bromo-1-(4-methylphenyl)-
10418 - 1-Propanone, 1-(4-bromophenyl)-
10419 - 1-Propanone, 2-bromo-1-phenyl-

$C_9H_9BrO_2$
2147 - Benzenepropanoic-β-14C acid, α-bromo-

2521 - Benzoic acid, 2-bromo-, ethyl ester
2522 - Benzoic acid, 3-bromo-, ethyl ester
2523 - Benzoic acid, 4-bromo-, ethyl ester
5205 - 1,3-Dioxolane, 2-(3-bromophenyl)-

$C_9H_9Br_2NO_3$
11927 - *L*-Tyrosine, 3,5-dibromo-

C_9H_9Cl
1226 - Benzene, (1-chloro-1-propenyl)-
1227 - Benzene, (2-chloro-1-propenyl)-
1228 - Benzene, (3-chloro-1-propenyl)-
1229 - Benzene, 1-chloro-3-(1-propenyl)-, (*E*)-
1230 - Benzene, (3-chloro-1-propenyl)-, (*E*)-

C_9H_9ClO
1232 - Benzene, 1-chloro-4-(2-propenyloxy)-
2160 - Benzenepropanoyl chloride
2441 - Benzofuran, 2-(chloromethyl)-2,3-dihydro-
5787 - Ethanone, 2-chloro-1-(4-methylphenyl)-
7283 - 1H-Inden-4-ol, 7-chloro-2,3-dihydro-
9271 - Phenol, 4-chloro-2-(2-propenyl)-
10422 - 1-Propanone, 1-(4-chlorophenyl)-

$C_9H_9ClO_2$
140 - Acetic acid, chloro-, 4-methylphenyl ester
148 - Acetic acid, chloro-, phenylmethyl ester
2537 - Benzoic acid, 2-chloro-, ethyl ester
2538 - Benzoic acid, 3-chloro-, ethyl ester
2539 - Benzoic acid, 4-chloro-, ethyl ester
5696 - Ethanol, 2-chloro-, benzoate
10505 - Propanoyl chloride, 2-phenoxy-

$C_9H_9ClO_3$
141 - Acetic acid, (4-chloro-2-methylphenoxy)-
2541 - Benzoic acid, 3-chloro-2-hydroxy-, ethyl ester
4141 - Cloprop
5786 - Ethanone, 2-chloro-1-(4-hydroxy-3-methoxyphenyl)-

$C_9H_9Cl_2NO$
10200 - Propanil

$C_9H_9FO_2$
2606 - Benzoic acid, 4-fluoro-, ethyl ester

$C_9H_9I_2NO_3$
11928 - Tyrosine, 3,5-diiodo-

C_9H_9N
959 - Benzeneacetonitrile, α-methyl-
960 - Benzeneacetonitrile, 2-methyl-
961 - Benzeneacetonitrile, 3-methyl-
962 - Benzeneacetonitrile, 4-methyl-
2123 - Benzenepropanenitrile
7312 - 1H-Indole, 1-methyl-
7313 - 1H-Indole, 2-methyl-
7314 - 1H-Indole, 3-methyl-
7315 - 1H-Indole, 4-methyl-
7316 - 1H-Indole, 5-methyl-
7317 - 1H-Indole, 7-methyl-

C_9H_9NO
957 - Benzeneacetonitrile, 2-methoxy-
958 - Benzeneacetonitrile, 4-methoxy-
2754 - Benzonitrile, 2-ethoxy-
2755 - Benzonitrile, 4-ethoxy-
2879 - Benzoxazole, 2,5-dimethyl-
7311 - 1H-Indole, 7-methoxy-
11327 - 2(1H)-Quinolinone, 3,4-dihydro-

$C_9H_9NO_2$
2753 - Benzonitrile, 2,6-dimethoxy-
3363 - 1,3-Butanedione, 1-(2-pyridinyl)-
3364 - 1,3-Butanedione, 1-(3-pyridinyl)-
3365 - 1,3-Butanedione, 1-(4-pyridinyl)-

$C_9H_9NO_3$
6255 - Glycine, *N*-benzoyl-
7301 - 1H-Indole-5,6-dione, 2,3-dihydro-3-hydroxy-1-methyl-

$C_9H_9NO_4$
103 - Acetic acid, [2-(aminocarbonyl)phenoxy]-
524 - Benzadox
2707 - Benzoic acid, 2-nitro-, ethyl ester
2708 - Benzoic acid, 3-nitro-, ethyl ester
2709 - Benzoic acid, 4-nitro-, ethyl ester

$C_9H_9N_3O_2$
3990 - Carbendazim

$C_9H_9N_3O_2S_2$
2206 - Benzenesulfonamide, 4-amino-*N*-2-thiazolyl-
11663 - 2-Thiazolamine, 5-[(4-aminophenyl)sulfonyl]-

$C_9H_9N_3O_4S_2$
3975 - Carbamodithioic acid, dimethyl-, 2,4-dinitrophenyl ester

$C_9H_9N_3O_6$
2364 - Benzene, 1,3,5-trimethyl-2,4,6-trinitro-

$C_9H_9N_3S$
11671 - 2,4-Thiazolediamine, 5-phenyl-

$C_9H_9N_5$
11829 - 1,3,5-Triazine-2,4-diamine, 6-phenyl-
11830 - 1,3,5-Triazine-2,4-diamine, *N*-phenyl-

C_9H_{10}
1256 - Benzene, cyclopropyl-
1640 - Benzene, 1-ethenyl-2-methyl-
1641 - Benzene, 1-ethenyl-3-methyl-
1642 - Benzene, 1-ethenyl-4-methyl-
1972 - Benzene, (1-methylethenyl)-
2162 - Benzene, 1-propenyl-
2163 - Benzene, 1-propenyl-, (*E*)-
2164 - Benzene, 1-propenyl-, (*Z*)-
2165 - Benzene, 2-propenyl-
4747 - 1,3,5,7-Cyclooctatetraene, 1-methyl-
7250 - 1H-Indene, 2,3-dihydro-

$C_9H_{10}BrClN_2O_2$
4076 - Chlorbromuron

$C_9H_{10}Br_2$
1313 - Benzene, 2,4-dibromo-1,3,5-trimethyl-

$C_9H_{10}Cl_2$
1402 - Benzene, 2,4-dichloro-1,3,5-trimethyl-

$C_9H_{10}Cl_2N_2O$
12000 - Urea, *N*′-(3,4-dichlorophenyl)-*N*,*N*-dimethyl-

$C_9H_{10}Cl_2N_2O_2$
7415 - Linuron

$C_9H_{10}INO_3$
11932 - *L*-Tyrosine, 3-iodo-

$C_9H_{10}N_2$
2378 - 1H-Benzimidazole, 1,5-dimethyl-
2379 - 1H-Benzimidazole, 2,5-dimethyl-
2380 - 1H-Benzimidazole, 5,6-dimethyl-
7237 - 1H-Indazole, 1,5-dimethyl-
7238 - 2H-Indazole, 2,3-dimethyl-
10823 - 1H-Pyrazole, 4,5-dihydro-1-phenyl-
10824 - 1H-Pyrazole, 4,5-dihydro-3-phenyl-
11133 - 1H-Pyrrole, 2,2′-methylenebis-

$C_9H_{10}N_2O$
8574 - 2-Oxazolamine, 4,5-dihydro-5-phenyl-
10844 - 3-Pyrazolidinone, 1-phenyl-

$C_9H_{10}N_2O_2$
879 - Benzeneacetamide, *N*-(aminocarbonyl)-

$C_9H_{10}N_2O_3$
6252 - Glycine, *N*-(4-aminobenzoyl)-

$C_9H_{10}N_2O_3S_2$
2865 - 2-Benzothiazolesulfonamide, 6-ethoxy-

$C_9H_{10}N_2O_4$
2360 - Benzene, 1,3,5-trimethyl-2,4-dinitro-

$C_9H_{10}O$
547 - Benzaldehyde, 2,4-dimethyl-
548 - Benzaldehyde, 2,5-dimethyl-
549 - Benzaldehyde, 3,5-dimethyl-
558 - Benzaldehyde, 2-ethyl-
872 - Benzeneacetaldehyde, α-methyl-
873 - Benzeneacetaldehyde, 2-methyl-
874 - Benzeneacetaldehyde, 4-methyl-
1637 - Benzene, 1-ethenyl-2-methoxy-
1638 - Benzene, 1-ethenyl-3-methoxy-
1639 - Benzene, 1-ethenyl-4-methoxy-
1868 - Benzenemethanol, α-ethenyl-
1928 - Benzene, (2-methoxyethenyl)-
2114 - Benzenepropanal
2167 - Benzene, (2-propenyloxy)-
2443 - Benzofuran, 2,3-dihydro-2-methyl-
2444 - Benzofuran, 2,3-dihydro-5-methyl-
2776 - 1H-2-Benzopyran, 3,4-dihydro-
2777 - 2H-1-Benzopyran, 3,4-dihydro-
5851 - Ethanone, 1-(2-methylphenyl)-

5852 - Ethanone, 1-(3-methylphenyl)-
5853 - Ethanone, 1-(4-methylphenyl)-
7284 - 1H-Inden-1-ol, 2,3-dihydro-
7285 - 1H-Inden-5-ol, 2,3-dihydro-
8622 - Oxirane, 2-methyl-2-phenyl-
9456 - Phenol, 2-(1-propenyl)-
9457 - Phenol, 2-(2-propenyl)-
9458 - Phenol, 4-(2-propenyl)-
9459 - Phenol, 2-(1-propenyl)-, (E)-
10480 - 1-Propanone, 1-phenyl-
10481 - 2-Propanone, 1-phenyl-
10704 - 2-Propen-1-ol, 3-phenyl-, (E)-
10705 - 2-Propen-1-ol, 3-phenyl-, (Z)-

$C_9H_{10}O_2$

211 - Acetic acid, 2-methylphenyl ester
212 - Acetic acid, 3-methylphenyl ester
213 - Acetic acid, 4-methylphenyl ester
233 - Acetic acid, phenylmethyl ester
552 - Benzaldehyde, 2-ethoxy-
553 - Benzaldehyde, 3-ethoxy-
554 - Benzaldehyde, 4-ethoxy-
586 - Benzaldehyde, 4-methoxy-3-methyl-
870 - Benzeneacetaldehyde, 2-methoxy-
871 - Benzeneacetaldehyde, 4-methoxy-
924 - Benzeneacetic acid, α-methyl-, (±)-
925 - Benzeneacetic acid, 3-methyl-
926 - Benzeneacetic acid, 4-methyl-
929 - Benzeneacetic acid, methyl ester
1547 - 1,2-Benzenediol, 3-(2-propenyl)-
2126 - Benzenepropanoic acid
2580 - Benzoic acid, 2,4-dimethyl-
2581 - Benzoic acid, 2,5-dimethyl-
2582 - Benzoic acid, 2,6-dimethyl-
2583 - Benzoic acid, 3,5-dimethyl-
2600 - Benzoic acid, 2-ethyl-
2601 - Benzoic acid, 3-ethyl-
2602 - Benzoic acid, ethyl ester
2696 - Benzoic acid, 2-methyl-, methyl ester
2697 - Benzoic acid, 3-methyl-, methyl ester
2698 - Benzoic acid, 4-methyl-, methyl ester
5826 - Ethanone, 1-(2-hydroxy-4-methylphenyl)-
5827 - Ethanone, 1-(2-hydroxy-5-methylphenyl)-
5828 - Ethanone, 1-(4-hydroxy-2-methylphenyl)-
5840 - Ethanone, 1-(3-methoxyphenyl)-
5841 - Ethanone, 1-(4-methoxyphenyl)-
5842 - Ethanone, 2-methoxy-1-phenyl-
8619 - Oxiranemethanol, 3-phenyl-
8623 - Oxirane, (phenoxymethyl)-
10323 - Propanoic acid, phenyl ester
10449 - 1-Propanone, 1-(2-hydroxyphenyl)-
10450 - 1-Propanone, 1-(4-hydroxyphenyl)-
10451 - 1-Propanone, 2-hydroxy-1-phenyl-
10452 - 1-Propanone, 3-hydroxy-1-phenyl-
10479 - 2-Propanone, 1-phenoxy-

$C_9H_{10}O_3$

230 - Acetic acid, phenoxy-, methyl ester
543 - Benzaldehyde, 2,4-dimethoxy-
544 - Benzaldehyde, 2,5-dimethoxy-
545 - Benzaldehyde, 3,4-dimethoxy-
546 - Benzaldehyde, 3,5-dimethoxy-
555 - Benzaldehyde, 3-ethoxy-4-hydroxy-
556 - Benzaldehyde, 5-ethoxy-2-hydroxy-
910 - Benzeneacetic acid, α-hydroxy-, methyl ester, (±)-
914 - Benzeneacetic acid, α-hydroxy-α-methyl-
915 - Benzeneacetic acid, α-(hydroxymethyl)-
921 - Benzeneacetic acid, 2-methoxy-
922 - Benzeneacetic acid, 4-methoxy-
2133 - Benzenepropanoic acid, α-hydroxy-, (±)-
2134 - Benzenepropanoic acid, 4-hydroxy-
2596 - Benzoic acid, 2-ethoxy-
2597 - Benzoic acid, 3-ethoxy-
2619 - Benzoic acid, 2-hydroxy-, ethyl ester
2620 - Benzoic acid, 3-hydroxy-, ethyl ester
2621 - Benzoic acid, 4-hydroxy-, ethyl ester
2638 - Benzoic acid, 2-hydroxy-3-methyl-, methyl ester

2639 - Benzoic acid, 2-hydroxy-4-methyl-, methyl ester
2640 - Benzoic acid, 2-hydroxy-5-methyl-, methyl ester
2673 - Benzoic acid, 2-methoxy-, methyl ester
2674 - Benzoic acid, 3-methoxy-, methyl ester
2675 - Benzoic acid, 4-methoxy-, methyl ester
5578 - 1,2-Ethanediol, monobenzoate
5823 - Ethanone, 1-(2-hydroxy-4-methoxyphenyl)-
5824 - Ethanone, 1-(4-hydroxy-3-methoxyphenyl)-
9382 - Phenol, 2-methoxy-, acetate
10319 - Propanoic acid, 2-phenoxy-, (±)-
10320 - Propanoic acid, 3-phenoxy-
10430 - 1-Propanone, 1-(2,4-dihydroxyphenyl)-
10639 - 2-Propenoic acid, 3-(2-furanyl)-, ethyl ester

$C_9H_{10}O_4$

567 - Benzaldehyde, 2-hydroxy-4,6-dimethoxy-
568 - Benzaldehyde, 3-hydroxy-4,5-dimethoxy-
569 - Benzaldehyde, 4-hydroxy-3,5-dimethoxy-
913 - Benzeneacetic acid, 4-hydroxy-3-methoxy-
2576 - Benzoic acid, 3,4-dimethoxy-
2577 - Benzoic acid, 3,5-dimethoxy-
2622 - Benzoic acid, 2-hydroxy-, 2-hydroxyethyl ester
2627 - Benzoic acid, 4-hydroxy-3-methoxy-, methyl ester
6163 - 2-Furanpropanoic acid, β-oxo-, ethyl ester
10489 - 1-Propanone, 1-(2,4,6-trihydroxyphenyl)-

$C_9H_{10}O_5$

896 - Benzeneacetic acid, α,4-dihydroxy-3-methoxy-
7510 - Methanediol, 2-furanyl-, diacetate

$C_9H_{11}Br$

1057 - Benzene, 1-(bromomethyl)-3,5-dimethyl-
1059 - Benzene, 1-bromo-4-(1-methylethyl)-
1060 - Benzene, (2-bromo-1-methylethyl)-
1086 - Benzene, (1-bromopropyl)-, (±)-
1087 - Benzene, (2-bromopropyl)-
1088 - Benzene, (3-bromopropyl)-
1092 - Benzene, 1-bromo-2,3,5-trimethyl-
1093 - Benzene, 1-bromo-2,4,5-trimethyl-
1096 - Benzene, 2-bromo-1,3,5-trimethyl-

$C_9H_{11}BrN_2O_2$

7656 - Metobromuron

$C_9H_{11}BrO$

1085 - Benzene, (3-bromopropoxy)-

$C_9H_{11}Cl$

1186 - Benzene, 1-(chloromethyl)-2,4-dimethyl-
1189 - Benzene, (1-chloro-1-methylethyl)-
1190 - Benzene, 1-chloro-2-(1-methylethyl)-
1191 - Benzene, 1-(chloromethyl)-4-ethyl-
1192 - Benzene, 1-chloro-4-(1-methylethyl)-
1193 - Benzene, (2-chloro-1-methylethyl)-
1234 - Benzene, (2-chloropropyl)-, (R)-
1235 - Benzene, (3-chloropropyl)-
1243 - Benzene, 2-chloro-1,3,5-trimethyl-

$C_9H_{11}ClN_2O$

11999 - Urea, N'-(4-chlorophenyl)-N,N-dimethyl-

$C_9H_{11}ClO$

1233 - Benzene, (3-chloropropoxy)-

$C_9H_{11}ClO_3$

9999 - 1,2-Propanediol, 3-(2-chlorophenoxy)-
10000 - 1,2-Propanediol, 3-(4-chlorophenoxy)-

$C_9H_{11}ClS$

1236 - Benzene, [(3-chloropropyl)thio]-

$C_9H_{11}Cl_2P$

9555 - Phosphonous dichloride, [4-(1-methylethyl)phenyl]-

$C_9H_{11}Cl_3NO_3PS$

4089 - Chlorpyrifos

$C_9H_{11}F$

1740 - Benzene, 2-fluoro-1,3,5-trimethyl-

$C_9H_{11}FN_2O_5$

12034 - Uridine, 2'-deoxy-5-fluoro-

$C_9H_{11}N$

787 - Benzenamine, 2-(1-methylethenyl)-
841 - Benzenamine, N-2-propenyl-
5437 - Ethanamine, N-(phenylmethylene)-
7243 - 1H-Inden-1-amine, 2,3-dihydro-, (±)-

7244 - 1H-Inden-5-amine, 2,3-dihydro-
7388 - Isoquinoline, 1,2,3,4-tetrahydro-
7389 - Isoquinoline, 5,6,7,8-tetrahydro-
11296 - Quinoline, 1,2,3,4-tetrahydro-
11297 - Quinoline, 5,6,7,8-tetrahydro-

$C_9H_{11}NO$

76 - Acetamide, N-(2-methylphenyl)-
77 - Acetamide, N-(3-methylphenyl)-
78 - Acetamide, N-(4-methylphenyl)-
79 - Acetamide, N-methyl-N-phenyl-
87 - Acetamide, N-(phenylmethyl)-
550 - Benzaldehyde, 4-(dimethylamino)-
7567 - Methanimidic acid, N-phenyl-, ethyl ester
9800 - Propanamide, N-phenyl-
10413 - 1-Propanone, 1-(4-aminophenyl)-
10414 - 1-Propanone, 2-amino-1-phenyl-
10483 - 1-Propanone, 1-phenyl-, oxime

$C_9H_{11}NO_2$

70 - Acetamide, N-(2-methoxyphenyl)-
71 - Acetamide, N-(4-methoxyphenyl)-
615 - Benzamide, 2-ethoxy-
1977 - Benzene, 1-(1-methylethyl)-2-nitro-
1978 - Benzene, 1-(1-methylethyl)-4-nitro-
2003 - Benzene, (1-methyl-1-nitroethyl)-
2362 - Benzene, 1,2,4-trimethyl-5-nitro-
2363 - Benzene, 1,3,5-trimethyl-2-nitro-
2424 - 1,3-Benzodioxole-5-ethanamine
2495 - Benzoic acid, 2-amino-, ethyl ester
2496 - Benzoic acid, 3-amino-, ethyl ester
2497 - Benzoic acid, 4-amino-, ethyl ester
2584 - Benzoic acid, 2-(dimethylamino)-
2585 - Benzoic acid, 4-(dimethylamino)-
2681 - Benzoic acid, 2-(methylamino)-, methyl ester
3958 - Carbamic acid, methylphenyl-, methyl ester
3965 - Carbamic acid, phenyl-, ethyl ester
9509 - DL-Phenylalanine
9510 - L-Phenylalanine
10907 - 3-Pyridinecarboxylic acid, 1-methylethyl ester

$C_9H_{11}NO_3$

5804 - Ethanone, 1-(3,4-dihydroxyphenyl)-2-(methylamino)-
11926 - L-Tyrosine

$C_9H_{11}NO_4$

1443 - Benzene, 1-(dimethoxymethyl)-3-nitro-
1444 - Benzene, 1-(dimethoxymethyl)-4-nitro-
11929 - L-Tyrosine, 3-hydroxy-

C_9H_{12}

1685 - Benzene, 1-ethyl-2-methyl-
1686 - Benzene, 1-ethyl-3-methyl-
1687 - Benzene, 1-ethyl-4-methyl-
1975 - Benzene, (1-methylethyl)-
2169 - Benzene, propyl-
2357 - Benzene, 1,2,3-trimethyl-
2358 - Benzene, 1,2,4-trimethyl-
2359 - Benzene, 1,3,5-trimethyl-
8114 - 1,4-Nonadiyne
8115 - 1,8-Nonadiyne
8116 - 2,7-Nonadiyne

$C_9H_{12}NO_5PS$

5951 - Fenitrothion

$C_9H_{12}N_2$

10482 - 2-Propanone, phenylhydrazone
11017 - Pyridine, 3-(2-pyrrolidinyl)-, (S)-

$C_9H_{12}N_2O_2$

12013 - Urea, (4-ethoxyphenyl)-

$C_9H_{12}N_2O_3$

827 - Benzenamine, 5-nitro-2-propoxy-

$C_9H_{12}N_2O_6$

12033 - Uridine

$C_9H_{12}N_2S$

10886 - 4-Pyridinecarbothioamide, 2-propyl-

$C_9H_{12}N_4O_3$

10758 - 1H-Purine-2,6-dione, 3,7-dihydro-7-(2-hydroxyethyl)-1,3-dimethyl-
10759 - 1H-Purine-2,6-dione, 3,7-dihydro-8-methoxy-1,3,7-trimethyl-

$C_9H_{12}N_6$
11844 - 1,3,5-Triazine, 2,4,6-tris(1-aziridinyl)-
$C_9H_{12}O$
1607 - Benzeneethanol, α-methyl-
1608 - Benzeneethanol, β-methyl-
1609 - Benzeneethanol, 2-methyl-
1610 - Benzeneethanol, 4-methyl-
1658 - Benzene, (ethoxymethyl)-
1659 - Benzene, 1-ethoxy-2-methyl-
1660 - Benzene, 1-ethoxy-3-methyl-
1661 - Benzene, 1-ethoxy-4-methyl-
1682 - Benzene, 1-ethyl-2-methoxy-
1683 - Benzene, 1-ethyl-3-methoxy-
1684 - Benzene, 1-ethyl-4-methoxy-
1857 - Benzenemethanol, α,α-dimethyl-
1858 - Benzenemethanol, α,3-dimethyl-
1859 - Benzenemethanol, α,4-dimethyl-
1860 - Benzenemethanol, 2,4-dimethyl-
1861 - Benzenemethanol, 3,5-dimethyl-
1870 - Benzenemethanol, α-ethyl-
1916 - Benzene, 1-methoxy-2,3-dimethyl-
1917 - Benzene, 1-methoxy-2,4-dimethyl-
1918 - Benzene, 1-methoxy-3,5-dimethyl-
1919 - Benzene, 2-methoxy-1,3-dimethyl-
1920 - Benzene, 2-methoxy-1,4-dimethyl-
1921 - Benzene, 4-methoxy-1,2-dimethyl-
1974 - Benzene, (1-methylethoxy)-
2148 - Benzenepropanol
2168 - Benzene, propoxy-
9410 - Phenol, 2-(1-methylethyl)-
9411 - Phenol, 3-(1-methylethyl)-
9412 - Phenol, 4-(1-methylethyl)-
9461 - Phenol, 2-propyl-
9462 - Phenol, 3-propyl-
9463 - Phenol, 4-propyl-
9493 - Phenol, 2,3,4-trimethyl-
9494 - Phenol, 2,4,5-trimethyl-
9495 - Phenol, 2,4,6-trimethyl-

$C_9H_{12}O_2$
1439 - Benzene, 1,2-dimethoxy-3-methyl-
1440 - Benzene, 1,2-dimethoxy-4-methyl-
1441 - Benzene, 1,3-dimethoxy-5-methyl-
1535 - 1,2-Benzenediol, 4-(1-methylethyl)-
1548 - 1,2-Benzenediol, 4-propyl-
1549 - 1,3-Benzenediol, 4-propyl-
1550 - 1,3-Benzenediol, 5-propyl-
1553 - 1,3-Benzenediol, 2,4,6-trimethyl-
1872 - Benzenemethanol, α-ethyl-3-hydroxy-
1893 - Benzenemethanol, 2-methoxy-α-methyl-
1894 - Benzenemethanol, 3-methoxy-α-methyl-
1895 - Benzenemethanol, 4-methoxy-α-methyl-
5761 - Ethanol, 2-(phenylmethoxy)-
7181 - Hydroperoxide, 1-methyl-1-phenylethyl
9357 - Phenol, 4-ethyl-2-methoxy-
9460 - Phenol, 2-propoxy-
10398 - 1-Propanol, 2-phenoxy-
10399 - 2-Propanol, 1-phenoxy-

$C_9H_{12}O_3$
1856 - Benzenemethanol, 3,4-dimethoxy-
2354 - Benzene, 1,2,3-trimethoxy-
2355 - Benzene, 1,3,5-trimethoxy-
3520 - Butanoic acid, 2-furanylmethyl ester
6052 - 2-Furancarboxylic acid, butyl ester
6055 - 3-Furancarboxylic acid, 2,5-dimethyl-, ethyl ester
6062 - 2-Furancarboxylic acid, 2-methylpropyl ester
10026 - 1,2-Propanediol, 3-phenoxy-
10194 - 1,2,3-Propanetriol, 1-phenyl-
10300 - Propanoic acid, 2-methyl-, 2-furanylmethyl ester
11906 - 1,3,5-Trioxane, 2,4,6-triethenyl-

$C_9H_{12}O_3S$
2231 - Benzenesulfonic acid, 4-methyl-, ethyl ester
2235 - Benzenesulfonic acid, propyl ester

$C_9H_{12}S$
1706 - Benzene, 1-(ethylthio)-3-methyl-
1707 - Benzene, 1-(ethylthio)-4-methyl-
1980 - Benzene, [(1-methylethyl)thio]-
2125 - Benzenepropanethiol
2172 - Benzene, (propylthio)-

$C_9H_{13}BrN_2O_2$
3139 - Bromacil

$C_9H_{13}BrOSi$
11367 - Silane, (4-bromophenoxy)trimethyl-

$C_9H_{13}ClN_2O_2$
11591 - Terbacil

$C_9H_{13}ClN_6$
4161 - Cyanazine

$C_9H_{13}ClOSi$
11383 - Silane, (4-chlorophenoxy)trimethyl-

$C_9H_{13}ClSi$
11379 - Silane, (chloromethyl)dimethylphenyl-
11385 - Silane, (3-chlorophenyl)trimethyl-

$C_9H_{13}N$
745 - Benzenamine, 2-ethyl-6-methyl-
746 - Benzenamine, N-ethyl-2-methyl-
747 - Benzenamine, N-ethyl-3-methyl-
748 - Benzenamine, N-ethyl-4-methyl-
749 - Benzenamine, N-ethyl-N-methyl-
788 - Benzenamine, 2-(1-methylethyl)-
789 - Benzenamine, 4-(1-methylethyl)-
790 - Benzenamine, N-(1-methylethyl)-
842 - Benzenamine, 2-propyl-
843 - Benzenamine, N-propyl-
858 - Benzenamine, 2,4,5-trimethyl-
859 - Benzenamine, 2,4,6-trimethyl-
860 - Benzenamine, N,N,2-trimethyl-
861 - Benzenamine, N,N,3-trimethyl-
862 - Benzenamine, N,N,4-trimethyl-
1578 - Benzeneethanamine, α-methyl-, (±)-
1579 - Benzeneethanamine, β-methyl-
1580 - Benzeneethanamine, 4-methyl-
1581 - Benzeneethanamine, N-methyl-
1806 - Benzenemethanamine, α,α-dimethyl-
1807 - Benzenemethanamine, N,N-dimethyl-
1809 - Benzenemethanamine, α-ethyl-
1810 - Benzenemethanamine, N-ethyl-
10878 - Pyridine, 2-butyl-
10879 - Pyridine, 4-butyl-
10933 - Pyridine, 2,4-diethyl-
10942 - Pyridine, 2-(1,1-dimethylethyl)-
10943 - Pyridine, 3-(1,1-dimethylethyl)-
10944 - Pyridine, 4-(1,1-dimethylethyl)-
10959 - Pyridine, 2-ethyl-3,5-dimethyl-
10960 - Pyridine, 3-ethyl-2,6-dimethyl-
10961 - Pyridine, 4-ethyl-2,6-dimethyl-
11022 - Pyridine, 2,3,4,6-tetramethyl-

$C_9H_{13}NO$
769 - Benzenamine, 2-methoxy-N,N-dimethyl-
1577 - Benzeneethanamine, 2-methoxy-
1808 - Benzenemethanamine, 2-ethoxy-
1862 - Benzenemethanol, 4-(dimethylamino)-
4467 - Cyclohexanepropanenitrile, 2-oxo-
5743 - Ethanol, 2-(methylphenylamino)-
5744 - Ethanol, 2-[(2-methylphenyl)amino]-
5745 - Ethanol, 2-[(4-methylphenyl)amino]-
5762 - Ethanol, 2-[(phenylmethyl)amino]-
9215 - Phenol, 4-(2-aminopropyl)-, (±)-

$C_9H_{13}NO_2$
491 - 8-Azabicyclo[3.2.1]oct-2-ene-2-carboxylic acid, 8-methyl-, (1R)-
1533 - 1,2-Benzenediol, 4-[2-(methylamino)ethyl]-
1805 - Benzenemethanamine, 3,4-dimethoxy-
1887 - Benzenemethanol, 4-hydroxy-α-[(methylamino)methyl]-
10946 - 2,4(1H,3H)-Pyridinedione, 3,3-diethyl-
11115 - 1H-Pyrrole-3-carboxylic acid, 2,4-dimethyl-, ethyl ester
11116 - 1H-Pyrrole-3-carboxylic acid, 2,5-dimethyl-, ethyl ester

$C_9H_{13}NO_3$
1521 - 1,2-Benzenediol, 4-[1-hydroxy-2-(methylamino)ethyl]-, (R)-

$C_9H_{13}N_2O_9P$
12035 - 5'-Uridylic acid

$C_9H_{13}N_3O_2$
11060 - 2,4(1H,3H)-Pyrimidinedione, 6-amino-3-ethyl-1-(2-propenyl)-

$C_9H_{13}N_3O_5$
5021 - Cytidine

C_9H_{14}
2968 - Bicyclo[2.2.1]hept-2-ene, 2,3-dimethyl-
4473 - Cyclohexane, 2-propynyl-
4729 - Cyclononyne
6607 - 1-Hepten-4-yne, 6,6-dimethyl-
7108 - 3-Hexen-1-yne, 3-propyl-
8193 - 1-Nonen-3-yne
8194 - 1-Nonen-4-yne
8195 - 2-Nonen-4-yne
11865 - Tricyclo[2.2.1.02,6]heptane, 3,3-dimethyl-

$C_9H_{14}ClNO$
1838 - Benzenemethanol, α-(1-aminoethyl)-, hydrochloride, (R*,S*)-(±)-

$C_9H_{14}ClN_5$
5013 - Cyprazine

$C_9H_{14}Cl_2O_2$
8138 - Nonanedioyl dichloride

$C_9H_{14}IN$
866 - Benzenaminium, N,N,N-trimethyl-, iodide

$C_9H_{14}N_2$
8129 - Nonanedinitrile

$C_9H_{14}N_2O$
10860 - 3-Pyridinamine, 6-butoxy-

$C_9H_{14}N_2O_3$
11088 - 2,4,6(1H,3H,5H)-Pyrimidinetrione, 5,5-diethyl-1-methyl-
11093 - 2,4,6(1H,3H,5H)-Pyrimidinetrione, 5-ethyl-5-(1-methylethyl)-

$C_9H_{14}N_3O_8P$
5022 - 3'-Cytidylic acid

$C_9H_{14}N_4O_3$
7133 - L-Histidine, N-β-alanyl-

$C_9H_{14}O$
2954 - Bicyclo[2.2.1]heptan-2-one, 3,3-dimethyl-, (1S)-
2955 - Bicyclo[3.1.1]heptan-2-one, 6,6-dimethyl-, (1R)-
4719 - 2-Cyclohexen-1-one, 3,4,4-trimethyl-
4720 - 2-Cyclohexen-1-one, 3,4,6-trimethyl-
4721 - 2-Cyclohexen-1-one, 3,5,5-trimethyl-
4722 - 2-Cyclohexen-1-one, 3,6,6-trimethyl-
4948 - 2-Cyclopenten-1-one, 5-methyl-2-(1-methylethyl)-
6340 - 2,5-Heptadien-4-one, 2,6-dimethyl-
6608 - 6-Hepten-4-yn-3-ol, 3-ethyl-
7292 - 2H-Inden-2-one, octahydro-, cis-
7293 - 2H-Inden-2-one, octahydro-, trans-
8108 - 2,6-Nonadienal
8109 - 2,4-Nonadienal, (E,E)-
8527 - 7-Octen-5-yn-4-ol, 4-methyl-

$C_9H_{14}OSi$
11492 - Silane, trimethylphenoxy-

$C_9H_{14}O_2$
156 - Acetic acid, cyclohexylidene-, methyl ester
4639 - 3-Cyclohexene-1-carboxylic acid, ethyl ester
4910 - 1-Cyclopentene-1-carboxylic acid, 2,3,3-trimethyl-
4912 - 2-Cyclopentene-1-carboxylic acid, 1,2,3-trimethyl-
4913 - 3-Cyclopentene-1-carboxylic acid, 2,2,3-trimethyl-
6037 - Furan, 2-(butoxymethyl)-
8203 - 2-Nonynoic acid
8204 - 3-Nonynoic acid
8205 - 5-Nonynoic acid
8206 - 6-Nonynoic acid
8207 - 8-Nonynoic acid
8542 - 2-Octynoic acid, methyl ester
10626 - 2-Propenoic acid, cyclohexyl ester

$C_9H_{14}O_3$
6084 - Furan, 2-(diethoxymethyl)-
9074 - 4-Pentenoic acid, 2-acetyl-, ethyl ester
10674 - 2-Propenoic acid, 2-methyl-, (tetrahydro-2-furanyl)methyl ester
10703 - 2-Propen-1-ol, 2-methyl-, carbonate (2:1)

$C_9H_{14}O_3Si$
11488 - Silane, trimethoxyphenyl-

$C_9H_{14}O_4$
3329 - Butanedioic acid, methylene-, diethyl ester
3787 - 2-Butenedioic acid, 2-methyl-, diethyl ester, (E)-
3788 - 2-Butenedioic acid, 2-methyl-, diethyl ester, (Z)-
4792 - 1,2-Cyclopentanedicarboxylic acid, dimethyl ester, trans-(±)-
4967 - 1,1-Cyclopropanedicarboxylic acid, diethyl ester
4968 - 1,2-Cyclopropanedicarboxylic acid, diethyl ester, cis-
9041 - 2-Pentenedioic acid, diethyl ester
9042 - 2-Pentenedioic acid, diethyl ester, (E)-
9950 - Propanedioic acid, ethylidene-, diethyl ester

$C_9H_{14}O_5$
3324 - Butanedioic acid, formyl-, diethyl ester
8783 - Pentanedioic acid, 3-oxo-, diethyl ester
9923 - Propanedioic acid, acetyl-, diethyl ester

$C_9H_{14}O_6$
10195 - 1,2,3-Propanetriol, triacetate

$C_9H_{14}O_7$
10170 - 1,2,3-Propanetricarboxylic acid, 2-hydroxy-, trimethyl ester

$C_9H_{14}S$
11771 - Thiophene, 2-(3-methylbutyl)-

$C_9H_{14}Si$
11493 - Silane, trimethylphenyl-

$C_9H_{15}AlO_9$
340 - Aluminum, tris(2-hydroxypropanoato-$O1,O2$)-

$C_9H_{15}Br$
4322 - Cyclohexane, (2-bromo-2-propenyl)-

$C_9H_{15}Br_6O_4P$
10358 - 1-Propanol, 2,3-dibromo-, phosphate (3:1)

$C_9H_{15}Cl$
8200 - 1-Nonyne, 1-chloro-
8534 - 3-Octyne, 2-chloro-2-methyl-

$C_9H_{15}N$
10523 - 2-Propen-1-amine, N,N-di-2-propenyl-
11117 - 1H-Pyrrole, 3,4-diethyl-2-methyl-

$C_9H_{15}NO$
479 - 9-Azabicyclo[3.3.1]nonan-3-one, 9-methyl-

$C_9H_{15}NO_2$
3495 - Butanoic acid, 2-cyano-2-ethyl-, ethyl ester
6881 - Hexanoic acid, 2-cyano-, ethyl ester
9654 - 2,4-Piperidinedione, 3,3-diethyl-
10521 - 2-Propenamide, N-(1,1-dimethyl-3-oxobutyl)-

$C_9H_{15}NO_3$
483 - 8-Azabicyclo[3.2.1]octane-2-carboxylic acid, 3-hydroxy-8-methyl-, [1R-(exo,exo)]-
8585 - 2,4-Oxazolidinedione, 5,5-dipropyl-

$C_9H_{15}NO_5$
9922 - Propanedioic acid, (acetylamino)-, diethyl ester

$C_9H_{15}N_3O_2S$
7204 - 1H-Imidazole-4-ethanaminium, α-carboxy-2,3-dihydro-N,N,N-trimethyl-2-thioxo-, hydroxide, in

$C_9H_{15}N_3O_7$
472 - L-Aspartic acid, N-[2-[(2-amino-2-oxoethyl)amino]-2-carboxyethyl]-

C_9H_{16}
2990 - Bicyclo[3.3.1]nonane
2992 - Bicyclo[2.2.2]octane, 2-methyl-
4393 - Cyclohexane, 1-ethylidene-2-methyl-
4470 - Cyclohexane, 2-propenyl-
4663 - Cyclohexene, 1-ethyl-4-methyl-
4687 - Cyclohexene, 1,2,3-trimethyl-
4688 - Cyclohexene, 1,3,5-trimethyl-

4689 - Cyclohexene, 1,4,4-trimethyl-
4690 - Cyclohexene, 1,4,5-trimethyl-
4691 - Cyclohexene, 1,5,5-trimethyl-
4692 - Cyclohexene, 1,6,6-trimethyl-
4693 - Cyclohexene, 1,5,6-trimethyl-
4727 - Cyclononene, (E)-
4728 - Cyclononene, (Z)-
4906 - Cyclopentene, 1-butyl-
4927 - Cyclopentene, 4-isopropyl-1-methyl-
4931 - Cyclopentene, 3-methyl-1-(1-methylethyl)-
6330 - 1,3-Heptadiene, 2,6-dimethyl-
6331 - 1,5-Heptadiene, 2,6-dimethyl-
6332 - 2,4-Heptadiene, 2,4-dimethyl-
6333 - 2,4-Heptadiene, 3,5-dimethyl-
6617 - 3-Heptyne, 2,6-dimethyl-
6618 - 3-Heptyne, 5,5-dimethyl-
6694 - 1,5-Hexadiene, 3-ethyl-2-methyl-
7281 - 1H-Indene, octahydro-
8110 - 1,8-Nonadiene
8111 - 2,7-Nonadiene
8196 - 1-Nonyne
8197 - 2-Nonyne
8198 - 3-Nonyne
8199 - 4-Nonyne
8328 - 2,4-Octadiene, 7-methyl-
8329 - 3,5-Octadiene, 4-methyl-
8537 - 3-Octyne, 7-methyl-

$C_9H_{16}ClN_5$
10508 - Propazine
11593 - Terbuthylazine

$C_9H_{16}N_4OS$
11590 - Tebuthiuron

$C_9H_{16}O$
2938 - Bicyclo[2.2.1]heptan-2-ol, 1,7-dimethyl-, (exo,syn)-
4555 - Cyclohexanol, 1-(2-propenyl)-
4556 - Cyclohexanol, 2-(2-propenyl)-, trans-
4595 - Cyclohexanone, 2-ethyl-2-methyl-
4599 - Cyclohexanone, 2-(1-methylethyl)-
4600 - Cyclohexanone, 4-(1-methylethyl)-
4612 - Cyclohexanone, 2-propyl-
4613 - Cyclohexanone, 4-propyl-
4614 - Cyclohexanone, 2,2,3-trimethyl-
4615 - Cyclohexanone, 2,2,4-trimethyl-
4616 - Cyclohexanone, 2,2,5-trimethyl-
4617 - Cyclohexanone, 2,2,6-trimethyl-
4618 - Cyclohexanone, 2,3,6-trimethyl-
4619 - Cyclohexanone, 2,4,4-trimethyl-
4620 - Cyclohexanone, 2,4,5-trimethyl-
4621 - Cyclohexanone, 2,4,6-trimethyl-
4622 - Cyclohexanone, 3,3,5-trimethyl-
4623 - Cyclohexanone, 3,4,4-trimethyl-
4700 - 2-Cyclohexen-1-ol, 1,3,5-trimethyl-
4701 - 2-Cyclohexen-1-ol, 2,4,4-trimethyl-
4702 - 2-Cyclohexen-1-ol, 3,5,5-trimethyl-
4725 - Cyclononanone
4738 - Cyclooctanecarboxaldehyde
4884 - Cyclopentanone, 4-isopropyl-2-methyl-
4889 - Cyclopentanone, 2-methyl-2-(1-methylethyl)-
4890 - Cyclopentanone, 2-methyl-5-(1-methylethyl)-
4891 - Cyclopentanone, 3-methyl-2-(1-methylethyl)-
5843 - Ethanone, 1-(1-methylcyclohexyl)-
5844 - Ethanone, 1-(3-methylcyclohexyl)-
5845 - Ethanone, 1-(4-methylcyclohexyl)-
8112 - 2,6-Nonadien-1-ol
8179 - 2-Nonenal
8180 - 2-Nonenal, (E)-
8208 - 1-Nonyn-3-ol
8209 - 2-Nonyn-1-ol
8522 - 1-Octen-3-one, 5-methyl-
8523 - 5-Octen-4-one, 2-methyl-
8524 - 5-Octen-4-one, 7-methyl-
8525 - 7-Octen-4-one, 2-methyl-
8536 - 2-Octyne, 1-methoxy-
8544 - 1-Octyn-3-ol, 3-methyl-
10423 - 1-Propanone, 1-cyclohexyl-

$C_9H_{16}O_2$
4310 - Cyclohexaneacetic acid, methyl ester
4311 - Cyclohexaneacetic acid, 2-methyl-
4312 - Cyclohexaneacetic acid, 3-methyl-
4334 - Cyclohexanecarboxylic acid, ethyl ester
4468 - Cyclohexanepropanoic acid
6149 - 2(3H)-Furanone, dihydro-5-pentyl-
7063 - 2-Hexenoic acid, 2-methyl-, ethyl ester
7064 - 3-Hexenoic acid, 2-methyl-, ethyl ester
7065 - 3-Hexenoic acid, 3-methyl-, ethyl ester
8137 - 2,4-Nonanedione
8187 - 3-Nonenoic acid
8188 - 8-Nonenoic acid
10235 - Propanoic acid, cyclohexyl ester
10643 - 2-Propenoic acid, hexyl ester

$C_9H_{16}O_3$
3461 - Butanoic acid, 2-acetyl-3-methyl-, ethyl ester
6165 - 2-Furanpropanoic acid, tetrahydro-, ethyl ester
6460 - Heptanoic acid, 2,2-dimethyl-6-oxo-
6461 - Heptanoic acid, 4,4-dimethyl-6-oxo-
6902 - Hexanoic acid, 4-methyl-5-oxo-, ethyl ester
8856 - Pentanoic acid, 2-acetyl-, ethyl ester
8904 - Pentanoic acid, 4-oxo-, butyl ester
8910 - Pentanoic acid, 4-oxo-, 2-methylpropyl ester

$C_9H_{16}O_4$
3326 - Butanedioic acid, methyl-, diethyl ester
3337 - Butanedioic acid, monopentyl ester
6395 - Heptanedioic acid, dimethyl ester
6396 - Heptanedioic acid, monoethyl ester
8130 - Nonanedioic acid
8772 - Pentanedioic acid, diethyl ester
8774 - Pentanedioic acid, 3,3-dimethyl-, dimethyl ester
8791 - 1,5-Pentanediol, diacetate
9942 - Propanedioic acid, dimethyl-, diethyl ester
9945 - Propanedioic acid, dipropyl ester
9949 - Propanedioic acid, ethyl-, diethyl ester

$C_9H_{16}O_5$
8778 - Pentanedioic acid, 3-hydroxy-, diethyl ester

$C_9H_{17}ClN_3O_3PS$
7340 - Isazophos

$C_9H_{17}ClO$
8177 - Nonanoyl chloride

$C_9H_{17}FO_2$
8151 - Nonanoic acid, 9-fluoro-

$C_9H_{17}IN_4$
11825 - 3,5,7-Triaza-1-azoniatricyclo[3.3.1.13,7]decane, 1-(2-propenyl)-, iodide

$C_9H_{17}N$
4304 - Cyclohexanamine, N-2-propenyl-
8144 - Nonanenitrile
11246 - Quinoline, decahydro-, cis-
11247 - Quinoline, decahydro-, trans-(±)-

$C_9H_{17}NO$
4591 - Cyclohexanone, 2-[(dimethylamino)methyl]-
9003 - 4-Pentenamide, 2,2-diethyl-
9725 - 4-Piperidinone, 1,2,3,6-tetramethyl-
9726 - 4-Piperidinone, 2,2,6,6-tetramethyl-
10470 - 2-Propanone, 1-(1-methyl-2-piperidinyl)-, (±)-

$C_9H_{17}NOS$
7661 - Molinate

$C_9H_{17}NO_2$
10631 - 2-Propenoic acid, 2-(diethylamino)ethyl ester

$C_9H_{17}NO_4$
10305 - Propanoic acid, 2-methyl-, 2-nitropentyl ester

$C_9H_{17}NO_5$
329 - β-Alanine, N-(2,4-dihydroxy-3,3-dimethyl-1-oxobutyl)-, (R)-

$C_9H_{17}NS$
11708 - Thiocyanic acid, octyl ester

$C_9H_{17}N_5S$
341 - Ametryn

C_9H_{18}
4394 - Cyclohexane, 1-ethyl-1-methyl-
4395 - Cyclohexane, 1-ethyl-2-methyl-, cis-
4396 - Cyclohexane, 1-ethyl-2-methyl-, trans-

4397 - Cyclohexane, 1-ethyl-3-methyl-, *cis*-
4398 - Cyclohexane, 1-ethyl-3-methyl-, *trans*-
4399 - Cyclohexane, 1-ethyl-4-methyl-, *cis*-
4400 - Cyclohexane, 1-ethyl-4-methyl-, *trans*-
4431 - Cyclohexane, (1-methylethyl)-
4471 - Cyclohexane, propyl-
4485 - Cyclohexane, 1,1,2-trimethyl-
4486 - Cyclohexane, 1,1,3-trimethyl-
4487 - Cyclohexane, 1,1,4-trimethyl-
4488 - Cyclohexane, 1,2,3-trimethyl-
4489 - Cyclohexane, 1,2,3-trimethyl-, (1α,2β,3α)-
4490 - Cyclohexane, 1,2,3-trimethyl-, (1α,2α,3α)-
4491 - Cyclohexane, 1,2,3-trimethyl-, (1α,2α,3β)-
4492 - Cyclohexane, 1,2,4-trimethyl-, (1α,2α,4α)-
4493 - Cyclohexane, 1,2,4-trimethyl-, (1α,2β,4α)-
4494 - Cyclohexane, 1,2,4-trimethyl-, (1α,2α,4β)-
4495 - Cyclohexane, 1,2,4-trimethyl-, (1α,2β,4β)-
4496 - Cyclohexane, 1,3,5-trimethyl-, (1α,3α,5α)-
4497 - Cyclohexane, 1,3,5-trimethyl-, (1α,3α,5β)-
4724 - Cyclononane
4778 - Cyclopentane, butyl-
4796 - Cyclopentane, 1,1-diethyl-
4797 - Cyclopentane, 1,2-diethyl-, *cis*-
4798 - Cyclopentane, 1,2-diethyl-, *trans*-
4799 - Cyclopentane, 1,3-diethyl-
4805 - Cyclopentane, (1,1-dimethylethyl)-
4829 - Cyclopentane, 1-methyl-2-(1-methylethyl)-
4832 - Cyclopentane, (1-methylpropyl)-
4833 - Cyclopentane, 1-methyl-1-propyl-
4834 - Cyclopentane, 1-methyl-2-propyl-, *cis*-
4835 - Cyclopentane, 1-methyl-2-propyl-, *trans*-
4836 - Cyclopentane, 1-methyl-3-propyl-, *cis*-
4837 - Cyclopentane, 1-methyl-3-propyl-, *trans*-
4838 - Cyclopentane, (2-methylpropyl)-
4851 - Cyclopentane, 1,1,3,3-tetramethyl-
6542 - 2-Heptene, 2,3-dimethyl-
8182 - 1-Nonene
8183 - 3-Nonene, (*E*)-
8505 - 1-Octene, 2-methyl-
8506 - 2-Octene, 3-methyl-
8507 - 3-Octene, 7-methyl-
8508 - 4-Octene, 2-methyl-

$C_9H_{18}Br_2$
8127 - Nonane, 1,2-dibromo-
$C_9H_{18}ClF$
8126 - Nonane, 1-chloro-9-fluoro-
$C_9H_{18}Cl_2$
6382 - Heptane, 2,6-dichloro-2,6-dimethyl-
8128 - Nonane, 1,9-dichloro-
$C_9H_{18}FeN_3S_6$
5957 - Ferbam
$C_9H_{18}NO_3PS_2$
6021 - Fosthiazate
$C_9H_{18}N_2O_2S$
11717 - Thiofanox
$C_9H_{18}N_2O_4$
3938 - Carbamic acid, *N*-butyl-*N*-nitro,butyl ester
10021 - 1,3-Propanediol, 2-methyl-2-propyl-,
    dicarbamate
$C_9H_{18}N_6O_6$
7584 - Methanol, (1,3,5-triazine-
    2,4,6-triyltrinitrilo)hexakis-
$C_9H_{18}O$
4526 - Cyclohexanol, 1-ethyl-2-methyl-
4527 - Cyclohexanol, 1-ethyl-3-methyl-
4538 - Cyclohexanol, 1-(1-methylethyl)-
4539 - Cyclohexanol, 2-(1-methylethyl)-, *cis*-
4540 - Cyclohexanol, 4-(1-methylethyl)-, *cis*-
4541 - Cyclohexanol, 4-(1-methylethyl)-, *trans*-
4557 - Cyclohexanol, 1-propyl-
4558 - Cyclohexanol, 2-propyl-, *cis*-
4559 - Cyclohexanol, 2-propyl-, *trans*-
4560 - Cyclohexanol, 4-propyl-, *trans*-
4561 - Cyclohexanol, 4-propyl-, *cis*-
4562 - Cyclohexanol, 1,2,2-trimethyl-
4563 - Cyclohexanol, 1,2,6-trimethyl-
4564 - Cyclohexanol, 1,3,5-trimethyl-

4565 - Cyclohexanol, 2,2,5-trimethyl-
4566 - Cyclohexanol, 2,2,6-trimethyl-
4567 - Cyclohexanol, 2,3,3-trimethyl-
4568 - Cyclohexanol, 2,3,6-trimethyl-
4569 - Cyclohexanol, 3,3,5-trimethyl-, *cis*-
4570 - Cyclohexanol, 3,3,5-trimethyl-, *trans*-
6514 - 4-Heptanone, 2,6-dimethyl-
6515 - 4-Heptanone, 3,5-dimethyl-
8117 - Nonanal
8172 - 2-Nonanone
8173 - 3-Nonanone
8174 - 4-Nonanone
8175 - 5-Nonanone
8189 - 1-Nonen-3-ol
8190 - 8-Nonen-1-ol
8475 - 2-Octanone, 3-methyl-
8476 - 3-Octanone, 4-methyl-
8477 - 3-Octanone, 7-methyl-
8478 - 4-Octanone, 3-methyl-
8479 - 4-Octanone, 7-methyl-
8517 - 1-Octen-4-ol, 4-methyl-
8518 - 2-Octen-4-ol, 7-methyl-
8519 - 3-Octen-2-ol, 2-methyl-, (*Z*)-
8520 - 4-Octen-3-ol, 3-methyl-
8521 - 5-Octen-4-ol, 5-methyl-
8988 - 3-Pentanone, 2,2,4,4-tetramethyl-
$C_9H_{18}O_2$
193 - Acetic acid, heptyl ester
3543 - Butanoic acid, 3-methyl-, butyl ester
3544 - Butanoic acid, 3-methylbutyl ester
3545 - Butanoic acid, 2-methyl-, butyl ester
3555 - Butanoic acid, 3-methyl-, 2-methylpropyl ester
3579 - Butanoic acid, pentyl ester
6014 - Formic acid, octyl ester
6462 - Heptanoic acid, ethyl ester
6476 - 4-Heptanol, acetate
6886 - Hexanoic acid, 2-ethyl-3-methyl-
6887 - Hexanoic acid, 3-ethyl-3-methyl-
6899 - Hexanoic acid, 3-methyl-, ethyl ester
6900 - Hexanoic acid, 4-methyl-, ethyl ester
6909 - Hexanoic acid, propyl ester
8146 - Nonanoic acid
8433 - Octanoic acid, 3-methyl-
8434 - Octanoic acid, methyl ester
8862 - Pentanoic acid, butyl ester
8899 - Pentanoic acid, 1-methylpropyl ester, (+)-
8900 - Pentanoic acid, 2-methylpropyl ester
10265 - Propanoic acid, hexyl ester
10302 - Propanoic acid, 2-methyl-, 3-methylbutyl
    ester
$C_9H_{18}O_3$
3994 - Carbonic acid, bis(1,1-dimethylethyl) ester
3997 - Carbonic acid, bis(2-methylpropyl) ester
3998 - Carbonic acid, dibutyl ester
$C_9H_{18}O_5$
5719 - Ethanol, 2-ethoxy-, carbonate (2:1)
$C_9H_{19}Br$
8121 - Nonane, 1-bromo-
$C_9H_{19}Cl$
6364 - Heptane, 3-chloro-3-ethyl-
6747 - Hexane, 3-chloro-2,2,3-trimethyl-
8123 - Nonane, 1-chloro-
8124 - Nonane, 2-chloro-
8125 - Nonane, 5-chloro-
8362 - Octane, 3-chloro-3-methyl-
8363 - Octane, 4-chloro-4-methyl-
8727 - Pentane, 3-chloro-3-ethyl-2,2-dimethyl-
$C_9H_{19}ClO$
8163 - 1-Nonanol, 9-chloro-
$C_9H_{19}FO$
8165 - 1-Nonanol, 9-fluoro-
$C_9H_{19}N$
4303 - Cyclohexanamine, *N*-(1-methylethyl)-
4810 - Cyclopentaneethanamine, *N*,α-dimethyl-
6532 - 5-Hepten-2-amine, *N*,6-dimethyl-
9635 - Piperidine, 1-butyl-
9636 - Piperidine, 3-butyl-

9637 - Piperidine, 4-butyl-
9647 - Piperidine, 2,5-diethyl-
9653 - Piperidine, 1-(1,1-dimethylethyl)-
9685 - Piperidine, 1-(2-methylpropyl)-
9686 - Piperidine, 1-methyl-2-propyl-, (*S*)-
9710 - Piperidine, 2,2,4,6-tetramethyl-
9711 - Piperidine, 2,2,6,6-tetramethyl-
$C_9H_{19}NO$
3201 - Butanamide, *N*,*N*-diethyl-3-methyl-
4144 - Conhydrine, *N*-methyl-
8118 - Nonanamide
8964 - 2-Pentanone, 5-(diethylamino)-
9794 - Propanamide, *N*,*N*-diethyl-2,2-dimethyl-
$C_9H_{19}NOS$
5372 - EPTC
$C_9H_{19}NO_2$
328 - β-Alanine, *N*,*N*-diethyl-, ethyl ester
3934 - Carbamic acid, butyl-, butyl ester
$C_9H_{19}NO_4$
3202 - Butanamide, 2,4-dihydroxy-*N*-
    (3-hydroxypropyl)-3,3-dimethyl-, (*R*)-
C_9H_{20}
6383 - Heptane, 2,2-dimethyl-
6384 - Heptane, 2,3-dimethyl-
6385 - Heptane, 2,4-dimethyl-
6386 - Heptane, 2,5-dimethyl-
6387 - Heptane, 2,6-dimethyl-
6388 - Heptane, 3,3-dimethyl-
6389 - Heptane, 3,4-dimethyl-
6390 - Heptane, 3,5-dimethyl-
6391 - Heptane, 4,4-dimethyl-
6407 - Heptane, 3-ethyl-
6408 - Heptane, 4-ethyl-
6825 - Hexane, 3-ethyl-2-methyl-
6826 - Hexane, 3-ethyl-3-methyl-
6827 - Hexane, 3-ethyl-4-methyl-
6828 - Hexane, 4-ethyl-2-methyl-
6859 - Hexane, 2,2,3-trimethyl-
6860 - Hexane, 2,2,4-trimethyl-
6861 - Hexane, 2,2,5-trimethyl-
6862 - Hexane, 2,3,3-trimethyl-
6863 - Hexane, 2,3,4-trimethyl-
6864 - Hexane, 2,3,5-trimethyl-
6865 - Hexane, 2,4,4-trimethyl-
6866 - Hexane, 3,3,4-trimethyl-
8120 - Nonane
8408 - Octane, 2-methyl-
8409 - Octane, 3-methyl-
8410 - Octane, 4-methyl-
8756 - Pentane, 3,3-diethyl-
8810 - Pentane, 3-ethyl-2,2-dimethyl-
8811 - Pentane, 3-ethyl-2,3-dimethyl-
8812 - Pentane, 3-ethyl-2,4-dimethyl-
8839 - Pentane, 2,2,3,3-tetramethyl-
8840 - Pentane, 2,2,3,4-tetramethyl-
8841 - Pentane, 2,2,4,4-tetramethyl-
8842 - Pentane, 2,3,3,4-tetramethyl-
$C_9H_{20}N_2$
9628 - 4-Piperidinamine, 2,2,6,6-tetramethyl-
$C_9H_{20}N_2O$
11996 - Urea, *N*,*N*-bis(2-methylpropyl)-
12029 - Urea, tetraethyl-
$C_9H_{20}O$
6406 - Heptane, 1-ethoxy-
6479 - 2-Heptanol, 2,6-dimethyl-
6480 - 3-Heptanol, 2,3-dimethyl-
6481 - 3-Heptanol, 2,6-dimethyl-
6482 - 3-Heptanol, 3,5-dimethyl-
6483 - 3-Heptanol, 3,6-dimethyl-
6484 - 4-Heptanol, 2,4-dimethyl-
6485 - 4-Heptanol, 2,6-dimethyl-
6486 - 4-Heptanol, 3,5-dimethyl-
6487 - 4-Heptanol, 4-ethyl-
6932 - 1-Hexanol, 2-ethyl-3-methyl-
6933 - 2-Hexanol, 3-ethyl-2-methyl-
6934 - 3-Hexanol, 3-ethyl-2-methyl-
6935 - 3-Hexanol, 3-ethyl-5-methyl-

6951 - 1-Hexanol, 3,5,5-trimethyl-
6952 - 2-Hexanol, 2,3,4-trimethyl-
6953 - 3-Hexanol, 2,2,3-trimethyl-
6954 - 3-Hexanol, 2,3,4-trimethyl-
6955 - 3-Hexanol, 2,4,4-trimethyl-
6956 - 3-Hexanol, 2,5,5-trimethyl-
6957 - 3-Hexanol, 3,4,4-trimethyl-
6958 - 3-Hexanol, 3,5,5-trimethyl-
8157 - 1-Nonanol
8158 - 2-Nonanol, (±)-
8159 - 3-Nonanol, (±)-
8160 - 4-Nonanol, (S)-
8161 - 5-Nonanol
8461 - 1-Octanol, 3-methyl-
8462 - 1-Octanol, 4-methyl-
8463 - 1-Octanol, 7-methyl-
8464 - 2-Octanol, 2-methyl-
8465 - 2-Octanol, 3-methyl-
8466 - 3-Octanol, 3-methyl-
8467 - 3-Octanol, 4-methyl-
8468 - 3-Octanol, 6-methyl-
8469 - 4-Octanol, 4-methyl-
8470 - 4-Octanol, 5-methyl-

$C_9H_{20}O_2$
3398 - Butane, 1,1'-[methylenebis(oxy)]bis-
8136 - 1,9-Nonanediol
8755 - Pentane, 1,1-diethoxy-
9992 - 1,3-Propanediol, 2-butyl-2-ethyl-
10075 - Propane, 1,1'-[methylenebis(oxy)]bis[2-methyl-

$C_9H_{20}O_3$
10182 - Propane, 1,1,1-triethoxy-
10183 - Propane, 1,1,3-triethoxy-

$C_9H_{20}O_3Si$
11482 - Silane, triethoxy-2-propenyl-

$C_9H_{20}O_4$
5601 - Ethane, 1,1',1''1'''-[methanetetrayl
tetrakis(oxy)]tetrakis-

$C_9H_{20}S$
8145 - 1-Nonanethiol

$C_9H_{21}B$
3121 - Borane, tripropyl-

$C_9H_{21}BO_3$
3127 - Boric acid, tripropyl ester
3128 - Boric acid, tris(1-methylethyl) ester

$C_9H_{21}ClSn$
11542 - Stannane, chlorotripropyl-

$C_9H_{21}N$
8119 - 1-Nonanamine
9808 - 1-Propanamine, N,N-dipropyl-

$C_9H_{21}NO_3$
10393 - 2-Propanol, 1,1',1''-nitrilotris-

$C_9H_{21}NO_4$
5678 - Ethanol, 2-[2-[2-(3-aminopropoxy)
ethoxy]ethoxy]-

$C_9H_{21}O_2PS_3$
11592 - Terbufos

$C_9H_{21}O_3P$
9594 - Phosphorous acid, tripropyl ester
9595 - Phosphorous acid, tris(1-methylethyl) ester

$C_9H_{21}O_4P$
9567 - Phosphoric acid, tripropyl ester
9568 - Phosphoric acid, tris(1-methylethyl) ester

$C_9H_{22}N_2$
7489 - Methanediamine, N,N,N',N'-tetraethyl-
8736 - 1,4-Pentanediamine, N1,N1-diethyl-

$C_9H_{22}O_4P_2S_4$
5931 - Ethion

$C_9H_{22}Si$
11500 - Silane, tripropyl-

$C_9H_{23}NO_3Si$
9828 - 1-Propanamine, 3-(triethoxysilyl)-

$C_9H_{24}O_3Si_3$
5003 - Cyclotrisiloxane, 2,4,6-triethyl-2,4,6-trimethyl-

$C_9H_{24}O_4Si_4$
4991 - Cyclotetrasiloxane, ethenylheptamethyl-

$C_{10}Cl_8$
7933 - Naphthalene, octachloro-

$C_{10}Cl_{10}O$
7408 - Kepone

$C_{10}Cl_{12}$
7644 - 1,3,4-Metheno-1H-cyclobuta[cd]pentalene,
1,1a,2,2,3,3a,4,5,5,5a,5b,
6-dodecachlorooctahydro-

$C_{10}F_8$
7935 - Naphthalene, octafluoro-

$C_{10}F_{18}$
7934 - Naphthalene, octadecafluorodecahydro-

$C_{10}F_{22}$
5050 - Decane, docosafluoro-

$C_{10}F_{23}N$
11912 - Tripropylamine, eicosafluoro-
2-(trifluoromethyl)-

$C_{10}H_4Cl_2O_2$
7849 - 1,2-Naphthalenedione, 3,4-dichloro-
7850 - 1,4-Naphthalenedione, 2,3-dichloro-
7851 - 1,4-Naphthalenedione, 5,6-dichloro-

$C_{10}H_4N_4O_8$
8004 - Naphthalene, 1,3,6,8-tetranitro-

$C_{10}H_5BrO_2$
7844 - 1,2-Naphthalenedione, 3-bromo-

$C_{10}H_5BrO_3$
7845 - 1,4-Naphthalenedione, 2-bromo-3-hydroxy-

$C_{10}H_5ClO_2$
7847 - 1,4-Naphthalenedione, 5-chloro-

$C_{10}H_5ClO_3$
7848 - 1,4-Naphthalenedione, 2-chloro-3-hydroxy-

$C_{10}H_5Cl_2NO_2$
11201 - Quinclorac

$C_{10}H_5Cl_7$
7576 - 4,7-Methano-1H-indene, 1,4,5,6,7,8,8-
heptachloro-3a,4,7,7a-tetrahydro-

$C_{10}H_5Cl_7O$
6301 - Heptachlor epoxide

$C_{10}H_5N_3O_6$
8016 - Naphthalene, 1,3,5-trinitro-

$C_{10}H_5NaO_5S$
7970 - 1-Naphthalenesulfonic acid, 3,4-dihydro-3,4-
dioxo-, sodium salt

$C_{10}H_6BrCl$
7741 - Naphthalene, 1-bromo-4-chloro-

$C_{10}H_6BrNO_2$
7744 - Naphthalene, 1-bromo-3-nitro-
7745 - Naphthalene, 2-bromo-1-nitro-

$C_{10}H_6Br_2$
7801 - Naphthalene, 1,3-dibromo-
7802 - Naphthalene, 1,4-dibromo-
7803 - Naphthalene, 1,5-dibromo-

$C_{10}H_6ClNO_2$
7786 - Naphthalene, 1-chloro-5-nitro-
7787 - Naphthalene, 1-chloro-8-nitro-
7788 - Naphthalene, 2-chloro-1-nitro-

$C_{10}H_6Cl_2$
7807 - Naphthalene, 1,2-dichloro-
7808 - Naphthalene, 1,3-dichloro-
7809 - Naphthalene, 1,4-dichloro-
7810 - Naphthalene, 1,5-dichloro-
7811 - Naphthalene, 1,7-dichloro-
7812 - Naphthalene, 1,8-dichloro-
7813 - Naphthalene, 2,6-dichloro-

$C_{10}H_6Cl_2O$
8025 - 1-Naphthalenol, 2,3-dichloro-
8026 - 1-Naphthalenol, 2,4-dichloro-

$C_{10}H_6Cl_4O_4$
1353 - 1,2-Benzenedicarboxylic acid, 3,4,5,6-
tetrachloro-, monoethyl ester
5180 - Dimethyl tetrachloroterephthalate

$C_{10}H_6Cl_8$
4077 - Chlordane

$C_{10}H_6N_2$
7375 - 5-Isoquinolinecarbonitrile
11223 - 2-Quinolinecarbonitrile
11224 - 4-Quinolinecarbonitrile

$C_{10}H_6N_2OS_2$
4072 - Chinomethionat

$C_{10}H_6N_2O_2$
5223 - 5H,10H-Dipyrrolo[1,2-a:1',2'-d]pyrazine-5,10-
dione

$C_{10}H_6N_2O_4$
7831 - Naphthalene, 1,3-dinitro-
7832 - Naphthalene, 1,5-dinitro-
7833 - Naphthalene, 1,8-dinitro-

$C_{10}H_6O_2$
7842 - 1,2-Naphthalenedione
7843 - 1,4-Naphthalenedione

$C_{10}H_6O_3$
7859 - 1,4-Naphthalenedione, 2-hydroxy-
7860 - 1,4-Naphthalenedione, 5-hydroxy-

$C_{10}H_6O_3S$
7840 - 1,8-Naphthalenediol, cyclic sulfite

$C_{10}H_6O_4$
7852 - 1,4-Naphthalenedione, 2,3-dihydroxy-
7853 - 1,4-Naphthalenedione, 5,8-dihydroxy-

$C_{10}H_6O_8$
2257 - 1,2,4,5-Benzenetetracarboxylic acid

$C_{10}H_7Br$
7739 - Naphthalene, 1-bromo-
7740 - Naphthalene, 2-bromo-

$C_{10}H_7BrO$
8022 - 2-Naphthalenol, 1-bromo-

$C_{10}H_7Cl$
7780 - Naphthalene, 1-chloro-
7781 - Naphthalene, 2-chloro-

$C_{10}H_7ClO$
8023 - 2-Naphthalenol, 6-chloro-
8063 - 2-Naphthol, 8-chloro-

$C_{10}H_7ClO_2S$
7975 - 1-Naphthalenesulfonyl chloride
7976 - 2-Naphthalenesulfonyl chloride

$C_{10}H_7Cl_5O$
11899 - Tridiphane

$C_{10}H_7F$
7893 - Naphthalene, 1-fluoro-
7894 - Naphthalene, 2-fluoro-

$C_{10}H_7I$
7902 - Naphthalene, 1-iodo-
7903 - Naphthalene, 2-iodo-

$C_{10}H_7NO$
11225 - 4-Quinolinecarboxaldehyde

$C_{10}H_7NO_2$
7929 - Naphthalene, 1-nitro-
7930 - Naphthalene, 2-nitro-
8035 - 2-Naphthalenol, 1-nitroso-
11123 - 1H-Pyrrole-2,5-dione, 1-phenyl-
11227 - 2-Quinolinecarboxylic acid
11228 - 7-Quinolinecarboxylic acid
11229 - 8-Quinolinecarboxylic acid

$C_{10}H_7NO_3$
8034 - 2-Naphthalenol, 1-nitro-
11232 - 2-Quinolinecarboxylic acid, 4-hydroxy-

$C_{10}H_7NO_4$
11230 - 2-Quinolinecarboxylic acid, 4,8-dihydroxy-

$C_{10}H_7N_3$
7738 - Naphthalene, 1-azido-

$C_{10}H_7N_3S$
11656 - Thiabendazole

$C_{10}H_8$
504 - Azulene
7732 - Naphthalene

$C_{10}H_8BrN$
7711 - 1-Naphthalenamine, 5-bromo-

$C_{10}H_8ClN$
11243 - Quinoline, 2-chloro-4-methyl-
11244 - Quinoline, 2-chloro-8-methyl-
11245 - Quinoline, 4-chloro-2-methyl-

$C_{10}H_8ClN_3O$
4080 - Chloridazon

$C_{10}H_8N_2$
3099 - 2,2'-Bipyridine
3100 - 2,3'-Bipyridine

$C_{10}H_8N_2O_4$
3101 - 2,4′-Bipyridine
3102 - 3,3′-Bipyridine
3103 - 3,4′-Bipyridine
3104 - 4,4′-Bipyridine
9917 - Propanedinitrile, (phenylmethyl)-

$C_{10}H_8N_2O_4$
5583 - Ethanedione, di-2-furanyl-, dioxime

$C_{10}H_8N_4O_5$
10848 - 3H-Pyrazol-3-one, 2,4-dihydro-5-methyl-4-nitro-2-(4-nitrophenyl)-

$C_{10}H_8O$
3924 - 3-Butyn-2-one, 4-phenyl-
6159 - Furan, 2-phenyl-
8017 - 1-Naphthalenol
8018 - 2-Naphthalenol

$C_{10}H_8OS$
6160 - Furan, 2-(phenylthio)-

$C_{10}H_8O_2$
2833 - 2H-1-Benzopyran-2-one, 3-methyl-
2834 - 2H-1-Benzopyran-2-one, 5-methyl-
2835 - 2H-1-Benzopyran-2-one, 6-methyl-
2836 - 2H-1-Benzopyran-2-one, 7-methyl-
2837 - 2H-1-Benzopyran-2-one, 8-methyl-
2838 - 4H-1-Benzopyran-4-one, 6-methyl-
5775 - Ethanone, 1-(2-benzofuranyl)-
6157 - Furan, 2-phenoxy-
7246 - 1H-Indene-1-carboxylic acid
7834 - 1,3-Naphthalenediol
7835 - 1,5-Naphthalenediol
7836 - 1,6-Naphthalenediol
7837 - 1,7-Naphthalenediol
7838 - 2,6-Naphthalenediol
7839 - 2,7-Naphthalenediol
10742 - 2-Propynoic acid, 3-phenyl-, methyl ester

$C_{10}H_8O_3$
2831 - 2H-1-Benzopyran-2-one, 7-hydroxy-4-methyl-
6096 - 2,5-Furandione, dihydro-3-phenyl-, (±)-

$C_{10}H_8O_3S$
7959 - 1-Naphthalenesulfonic acid
7960 - 2-Naphthalenesulfonic acid

$C_{10}H_8O_4$
2829 - 2H-1-Benzopyran-2-one, 7-hydroxy-6-methoxy-
5185 - 1,7-Dioxadispiro[4.0.4.2]dodeca-3,9-diene-2,8-dione, trans-
6057 - 2-Furancarboxylic acid, 2-furanylmethyl ester

$C_{10}H_8O_4S$
7971 - 1-Naphthalenesulfonic acid, 4-hydroxy-
7972 - 1-Naphthalenesulfonic acid, 7-hydroxy-
7973 - 2-Naphthalenesulfonic acid, 1-hydroxy-
7974 - 2-Naphthalenesulfonic acid, 6-hydroxy-

$C_{10}H_8O_5$
2808 - 2H-1-Benzopyran-2-one, 7,8-dihydroxy-6-methoxy-

$C_{10}H_8O_6S_2$
7867 - 1,5-Naphthalenedisulfonic acid
7868 - 1,6-Naphthalenedisulfonic acid
7869 - 2,7-Naphthalenedisulfonic acid

$C_{10}H_8O_7S_2$
7871 - 1,3-Naphthalenedisulfonic acid, 7-hydroxy-
7876 - 2,7-Naphthalenedisulfonic acid, 3-hydroxy-

$C_{10}H_8O_8S_2$
7875 - 2,7-Naphthalenedisulfonic acid, 4,5-dihydroxy-

$C_{10}H_8S$
8007 - 1-Naphthalenethiol
8008 - 2-Naphthalenethiol

$C_{10}H_9BrO_2$
10615 - 2-Propenoic acid, 2-bromo-3-phenyl-, methyl ester

$C_{10}H_9ClN_4O_2S$
2193 - Benzenesulfonamide, 4-amino-N-(6-chloro-3-pyridazinyl)-

$C_{10}H_9ClO$
4957 - Cyclopropanecarbonyl chloride, 2-phenyl-, trans-

$C_{10}H_9ClO_2$
10622 - 2-Propenoic acid, 2-chloro-3-phenyl-, methyl ester, (E)-
10623 - 2-Propenoic acid, 3-chloro-3-phenyl-, methyl ester

$C_{10}H_9ClO_3$
972 - Benzeneacetyl chloride, α-(acetyloxy)-

$C_{10}H_9Cl_2NO$
10520 - 2-Propenamide, N-(3,4-dichlorophenyl)-2-methyl-

$C_{10}H_9Cl_4O_4P$
6202 - Gardona

$C_{10}H_9N$
7381 - Isoquinoline, 1-methyl-
7382 - Isoquinoline, 3-methyl-
7383 - Isoquinoline, 4-methyl-
7384 - Isoquinoline, 6-methyl-
7385 - Isoquinoline, 8-methyl-
7708 - 1-Naphthalenamine
7709 - 2-Naphthalenamine
11134 - 1H-Pyrrole, 1-phenyl-
11135 - 1H-Pyrrole, 2-phenyl-
11277 - Quinoline, 2-methyl-
11278 - Quinoline, 3-methyl-
11279 - Quinoline, 4-methyl-
11280 - Quinoline, 5-methyl-
11281 - Quinoline, 6-methyl-
11282 - Quinoline, 7-methyl-
11283 - Quinoline, 8-methyl-

$C_{10}H_9NO$
5819 - Ethanone, 1-(1H-indol-3-yl)-
7380 - Isoquinoline, 7-methoxy-
8019 - 2-Naphthalenol, 6-amino-
8020 - 2-Naphthalenol, 8-amino-
11274 - Quinoline, 4-methoxy-
11275 - Quinoline, 6-methoxy-
11276 - Quinoline, 8-methoxy-
11319 - 6-Quinolinol, 2-methyl-
11320 - 7-Quinolinol, 6-methyl-
11321 - 8-Quinolinol, 2-methyl-
11322 - 8-Quinolinol, 6-methyl-
11328 - 2(1H)-Quinolinone, I-methyl-
11329 - 2(1H)-Quinolinone, 6-methyl-
11330 - 4(1H)-Quinolinone, 1-methyl-

$C_{10}H_9NO_2$
1877 - Benzenemethanol, α-ethynyl-, carbamate
7295 - 1H-Indole-3-acetic acid
7364 - 1H-Isoindole-1,3(2H)-dione, N-ethyl-
8582 - Oxazole, 5-methoxy-2-phenyl-
11162 - 2,5-Pyrrolidinedione, 1-phenyl-

$C_{10}H_9NO_3$
3853 - 2-Butenoic acid, 4-oxo-4-(phenylamino)-, (Z)-

$C_{10}H_9NO_3S$
7961 - 1-Naphthalenesulfonic acid, 2-amino-
7962 - 1-Naphthalenesulfonic acid, 4-amino-
7963 - 1-Naphthalenesulfonic acid, 5-amino-
7964 - 1-Naphthalenesulfonic acid, 7-amino-
7965 - 1-Naphthalenesulfonic acid, 8-amino-
7967 - 2-Naphthalenesulfonic acid, 8-amino-

$C_{10}H_9NO_4$
10677 - 2-Propenoic acid, 3-(2-nitrophenyl)-, methyl ester, (E)-
10678 - 2-Propenoic acid, 3-(3-nitrophenyl)-, methyl ester, (E)-
10679 - 2-Propenoic acid, 3-(4-nitrophenyl)-, methyl ester, (E)-

$C_{10}H_9NO_4S$
7966 - 1-Naphthalenesulfonic acid, 4-amino-3-hydroxy-

$C_{10}H_9NO_6S_2$
7870 - 1,3-Naphthalenedisulfonic acid, 7-amino-
7872 - 1,6-Naphthalenedisulfonic acid, 4-amino-
7873 - 1,7-Naphthalenedisulfonic acid, 4-amino-

$C_{10}H_9NO_7S_2$
7874 - 2,7-Naphthalenedisulfonic acid, 4-amino-5-hydroxy-

$C_{10}H_9N_3$
10871 - 2-Pyridinamine, N-2-pyridinyl-

$C_{10}H_{10}$
1097 - Benzene, 1,3-butadienyl-, (E)-
1098 - Benzene, 1,3-butadienyl-, (Z)-
1124 - Benzene, 3-butynyl-
1403 - Benzene, 1,2-diethenyl-
1404 - Benzene, 1,3-diethenyl-
1405 - Benzene, 1,4-diethenyl-
1967 - Benzene, (1-methylene-2-propenyl)-
7275 - 1H-Indene, 1-methyl-
7276 - 1H-Indene, 2-methyl-
7277 - 1H-Indene, 3-methyl-
7278 - 1H-Indene, 4-methyl-
7279 - 1H-Indene, 6-methyl-
7280 - 1H-Indene, 7-methyl-
7814 - Naphthalene, 1,2-dihydro-
7815 - Naphthalene, 1,4-dihydro-

$C_{10}H_{10}Cl_2O_3$
3925 - Butyrac 118

$C_{10}H_{10}Fe$
5960 - Ferrocene

$C_{10}H_{10}N_2$
7167 - Hydrazine, 1-naphthalenyl-
7795 - 1,2-Naphthalenediamine
7796 - 1,4-Naphthalenediamine
7797 - 1,5-Naphthalenediamine
7798 - 1,8-Naphthalenediamine
7799 - 2,3-Naphthalenediamine
10837 - 1H-Pyrazole, 1-methyl-3-phenyl-
10838 - 1H-Pyrazole, 3-methyl-1-phenyl-
10987 - Pyridine, 3-(1-methyl-1H-pyrrol-2-yl)-
11208 - 2-Quinolinamine, 4-methyl-
11209 - 3-Quinolinamine, 2-methyl-
11210 - 4-Quinolinamine, 2-methyl-
11211 - 7-Quinolinamine, 2-methyl-
11212 - 8-Quinolinamine, 6-methyl-
11335 - Quinoxaline, 2,6-dimethyl-

$C_{10}H_{10}N_2O$
10849 - 3H-Pyrazol-3-one, 1,2-dihydro-5-methyl-2-phenyl-

$C_{10}H_{10}N_2O_2$
7221 - 2,4-Imidazolidinedione, 3-methyl-5-phenyl-

$C_{10}H_{10}N_4O_2S$
2202 - Benzenesulfonamide, 4-amino-N-pyrazinyl-
2204 - Benzenesulfonamide, 4-amino-N-2-pyrimidinyl-

$C_{10}H_{10}O$
1878 - Benzenemethanol, α-ethynyl-α-methyl-
2445 - Benzofuran, 2,5-dimethyl-
2446 - Benzofuran, 2,6-dimethyl-
2447 - Benzofuran, 2,7-dimethyl-
2448 - Benzofuran, 3,5-dimethyl-
2449 - Benzofuran, 3,6-dimethyl-
2450 - Benzofuran, 4,6-dimethyl-
2451 - Benzofuran, 4,7-dimethyl-
2452 - Benzofuran, 5,6-dimethyl-
2453 - Benzofuran, 5,7-dimethyl-
2454 - Benzofuran, 6,7-dimethyl-
3701 - 2-Butenal, 3-phenyl-
3880 - 2-Buten-1-one, 1-phenyl-
3881 - 3-Buten-2-one, 4-phenyl-
3882 - 3-Buten-2-one, 4-phenyl-, (E)-
7289 - 1H-Inden-1-one, 2,3-dihydro-2-methyl-
7290 - 1H-Inden-1-one, 2,3-dihydro-4-methyl-
8046 - 1(2H)-Naphthalenone, 3,4-dihydro-
8047 - 2(1H)-Naphthalenone, 3,4-dihydro-
10516 - 2-Propenal, 2-methyl-3-phenyl-
10517 - 2-Propenal, 3-(4-methylphenyl)-

$C_{10}H_{10}O_2$
2118 - Benzenepropanal, α-methyl-β-oxo-
2429 - 1,3-Benzodioxole, 5-(2-propenyl)-
2430 - 1,3-Benzodioxole, 5-(1-propenyl)-, (E)-
2722 - Benzoic acid, 2-propenyl ester
2803 - 4H-1-Benzopyran-4-one, 2,3-dihydro-6-methyl-
3362 - 1,3-Butanedione, 1-phenyl-

3854 - 2-Butenoic acid, 3-phenyl-, (Z)-
3855 - 3-Butenoic acid, 4-phenyl-
5865 - Ethanone, 1,1'-(1,3-phenylene)bis-
10670 - 2-Propenoic acid, 2-methyl-3-phenyl-, (Z)-
10686 - 2-Propenoic acid, 3-phenyl-, methyl ester, (E)-
10688 - 2-Propenoic acid, phenylmethyl ester
10707 - 2-Propen-1-ol, 3-phenyl-, formate

$C_{10}H_{10}O_2S_2$
6101 - Furan, 2,2'-[dithiobis(methylene)]bis-

$C_{10}H_{10}O_3$
935 - Benzeneacetic acid, α-oxo-, ethyl ester
2142 - Benzenepropanoic acid, β-oxo-, methyl ester
2479 - Benzoic acid, 4-acetyl-, methyl ester
5771 - Ethanone, 2-(acetyloxy)-1-phenyl-
6156 - Furan, 2,2'-[oxybis(methylene)]bis-
10514 - 2-Propenal, 3-(4-hydroxy-3-methoxyphenyl)-

$C_{10}H_{10}O_4$
1343 - 1,2-Benzenedicarboxylic acid, dimethyl ester
1347 - 1,3-Benzenedicarboxylic acid, dimethyl ester
1348 - 1,3-Benzenedicarboxylic acid, 4,6-dimethyl-
1351 - 1,4-Benzenedicarboxylic acid, dimethyl ester
1508 - 1,2-Benzenediol, diacetate
1509 - 1,4-Benzenediol, diacetate
2416 - 1,3-Benzodioxole-5-acetic acid, methyl ester
2419 - 1,3-Benzodioxole-5-carboxylic acid, ethyl ester
2482 - Benzoic acid, 2-(acetyloxy)-, methyl ester
3341 - Butanedioic acid, phenyl-, (±)-
7355 - 1(3H)-Isobenzofuranone, 6,7-dimethoxy-

$C_{10}H_{10}O_5$
2608 - Benzoic acid, 6-formyl-2,3-dimethoxy-

$C_{10}H_{11}Br$
1001 - Benzene, (3-bromo-3-butenyl)-
1002 - Benzene, (4-bromo-2-butenyl)-
3719 - 2-Butene, 2-bromo-3-phenyl-

$C_{10}H_{11}Cl$
1152 - Benzene, 1-(1-chloroethenyl)-2,4-dimethyl-
1188 - Benzene, 1-(2-chloro-1-methylethenyl)-4-methyl-

$C_{10}H_{11}ClO$
974 - Benzeneacetyl chloride, α-ethyl-
1231 - Benzene, 1-(3-chloro-2-propenyl)-4-methoxy-
3650 - 1-Butanone, 4-chloro-1-phenyl-

$C_{10}H_{11}ClO_2$
891 - Benzeneacetic acid, α-chloro-, ethyl ester, (S)-

$C_{10}H_{11}ClO_2S$
7978 - 2-Naphthalenesulfonyl chloride, 5,6,7,8-tetrahydro-

$C_{10}H_{11}F_3N_2O$
12007 - Urea, N,N-dimethyl-N'-[3-(trifluoromethyl)phenyl]-

$C_{10}H_{11}F_7O_2$
8398 - 3,5-Octanedione, 6,6,7,7,8,8,8-heptafluoro-2,2-dimethyl-

$C_{10}H_{11}I_2NO_3$
10873 - 1(4H)-Pyridineacetic acid, 3,5-diiodo-4-oxo-, propyl ester

$C_{10}H_{11}N$
951 - Benzeneacetonitrile, α-ethyl-
7298 - 1H-Indole, 1,3-dimethyl-
7299 - 1H-Indole, 2,5-dimethyl-
7307 - 1H-Indole, 1-ethyl-
7308 - 1H-Indole, 3-ethyl-
7714 - 1-Naphthalenamine, 5,8-dihydro-

$C_{10}H_{11}NO$
3411 - Butanenitrile, 4-phenoxy-
7305 - 1H-Indole-3-ethanol
7328 - 2H-Indol-2-one, 1,3-dihydro-1,3-dimethyl-
7329 - 2H-Indol-2-one, 1,3-dihydro-3,3-dimethyl-
7330 - 2H-Indol-2-one, 1,3-dihydro-4,7-dimethyl-

$C_{10}H_{11}NO_2$
26 - Acetamide, N-acetyl-N-phenyl-
3207 - Butanamide, 3-oxo-N-phenyl-
6114 - 2-Furanmethanamine, N-(2-furanylmethyl)-

$C_{10}H_{11}NO_3$
224 - Acetic acid, oxo(phenylamino)-, ethyl ester

327 - DL-Alanine, N-benzoyl-
3575 - Butanoic acid, 4-oxo-4-(phenylamino)-

$C_{10}H_{11}N_3$
10813 - 1H-Pyrazol-4-amine, 3-methyl-1-phenyl-
10814 - 1H-Pyrazol-5-amine, 3-methyl-1-phenyl-

$C_{10}H_{11}N_3O_2S_2$
2200 - Benzenesulfonamide, 4-amino-N-(4-methyl-2-thiazolyl)-

$C_{10}H_{11}N_3O_3S$
2198 - Benzenesulfonamide, 4-amino-N-(5-methyl-3-isoxazolyl)-

$C_{10}H_{12}$
1109 - Benzene, 1-butenyl-, (E)-
1110 - Benzene, 1-butenyl-, (Z)-
1111 - Benzene, 2-butenyl-
1112 - Benzene, 3-butenyl-
1245 - Benzene, cyclobutyl-
1626 - Benzene, 1-ethenyl-2,4-dimethyl-
1627 - Benzene, 1-ethenyl-3,5-dimethyl-
1628 - Benzene, 2-ethenyl-1,4-dimethyl-
1629 - Benzene, 4-ethenyl-1,2-dimethyl-
1630 - Benzene, 1-ethenyl-2-ethyl-
1631 - Benzene, 1-ethenyl-3-ethyl-
1632 - Benzene, 1-ethenyl-4-ethyl-
1955 - Benzene, (1-methylcyclopropyl)-
1968 - Benzene, (1-methylenepropyl)-
1984 - Benzene, 1-methyl-2-(1-methylethenyl)-
1985 - Benzene, 1-methyl-3-(1-methylethenyl)-
1986 - Benzene, 1-methyl-4-(1-methylethenyl)-
2023 - Benzene, 1-methyl-2-(2-propenyl)-
2024 - Benzene, (2-methyl-1-propenyl)-
2025 - Benzene, (1-methyl-1-propenyl)-, (E)-
2029 - Benzene, (1-methyl-1-propenyl)-, (Z)-
4746 - 1,3,5,7-Cyclooctatetraene, 1-ethyl-
7255 - 1H-Indene, 2,3-dihydro-1-methyl-
7256 - 1H-Indene, 2,3-dihydro-2-methyl-
7257 - 1H-Indene, 2,3-dihydro-4-methyl-
7258 - 1H-Indene, 2,3-dihydro-5-methyl-
7577 - 4,7-Methano-1H-indene, 3a,4,7,7a-tetrahydro-, (3aα,4α,7α,7aα)-
7979 - Naphthalene, 1,2,3,4-tetrahydro-

$C_{10}H_{12}Br_2O$
9281 - Phenol, 2,4-dibromo-3-methyl-6-(1-methylethyl)-

$C_{10}H_{12}ClNO$
3939 - Carbamic acid, (3-chlorophenyl)-, 1-methylethyl ester

$C_{10}H_{12}ClNO_4$
10001 - 1,2-Propanediol, 3-(4-chlorophenoxy)-, 1-carbamate

$C_{10}H_{12}ClN_3O_3S$
11197 - 6-Quinazolinesulfonamide, 7-chloro-2-ethyl-1,2,3,4-tetrahydro-4-oxo-

$C_{10}H_{12}N_2$
2381 - 1H-Benzimidazole, 1-ethyl-2-methyl-
3105 - 2,3'-Bipyridine, 1,2,3,6-tetrahydro-, (S)-
7198 - 1H-Imidazole, 4,5-dihydro-2-(phenylmethyl)-
7302 - 1H-Indole-3-ethanamine

$C_{10}H_{12}N_2O_3$
1102 - Benzenebutanoic acid, α,2-diamino-γ-oxo-
11091 - 2,4,6(1H,3H,5H)-Pyrimidinetrione, 5,5-di-2-propenyl-

$C_{10}H_{12}N_2O_3S$
518 - Bentazon

$C_{10}H_{12}N_2O_4$
2282 - Benzene, 1,2,4,5-tetramethyl-3,6-dinitro-

$C_{10}H_{12}N_2O_5$
5183 - Dinoseb
9328 - Phenol, 2-(1,1-dimethylethyl)-4,6-dinitro-

$C_{10}H_{12}N_3O_3PS_2$
497 - Azinphos-methyl

$C_{10}H_{12}N_4OS$
34 - Acetamide, N-[4-[[(aminothioxomethyl)hydrazono]methyl]phenyl]-

$C_{10}H_{12}N_4O_5$
7335 - Inosine

$C_{10}H_{12}O$
590 - Benzaldehyde, 4-(1-methylethyl)-
599 - Benzaldehyde, 2,4,6-trimethyl-
1651 - Benzene, (2-ethoxyethenyl)-
1869 - Benzenemethanol, α-ethenyl-α-methyl-
1911 - Benzenemethanol, α-2-propenyl-
1941 - Benzene, 1-methoxy-2-(1-propenyl)-
1942 - Benzene, 1-methoxy-4-(1-propenyl)-
1943 - Benzene, 1-methoxy-4-(2-propenyl)-
1944 - Benzene, 1-methoxy-4-(1-propenyl)-, (E)-
2026 - Benzene, 1-methyl-2-(2-propenyloxy)-
2027 - Benzene, 1-methyl-3-(2-propenyloxy)-
2028 - Benzene, 1-methyl-4-(2-propenyloxy)-
2115 - Benzenepropanal, 2-methyl-
2116 - Benzenepropanal, 4-methyl-
2157 - Benzenepropanol, γ-methylene-
2778 - 2H-1-Benzopyran, 3,4-dihydro-2-methyl-
2779 - 2H-1-Benzopyran, 3,4-dihydro-3-methyl-
2780 - 2H-1-Benzopyran, 3,4-dihydro-6-methyl-
2781 - 2H-1-Benzopyran, 3,4-dihydro-7-methyl-
3676 - 1-Butanone, 1-phenyl-
3677 - 2-Butanone, 1-phenyl-
3678 - 2-Butanone, 4-phenyl-
3872 - 3-Buten-1-ol, 3-phenyl-
5809 - Ethanone, 1-(2,4-dimethylphenyl)-
5810 - Ethanone, 1-(2,5-dimethylphenyl)-
5811 - Ethanone, 1-(3,4-dimethylphenyl)-
8041 - 1-Naphthalenol, 1,2,3,4-tetrahydro-
8042 - 1-Naphthalenol, 5,6,7,8-tetrahydro-
8043 - 2-Naphthalenol, 1,2,3,4-tetrahydro-
8044 - 2-Naphthalenol, 5,6,7,8-tetrahydro-
9427 - Phenol, 2-methyl-6-(2-propenyl)-
9428 - Phenol, 4-methyl-2-(1-propenyl)-
9429 - Phenol, 4-methyl-2-(2-propenyl)-
10466 - 1-Propanone, 1-(2-methylphenyl)-
10467 - 1-Propanone, 1-(3-methylphenyl)-
10468 - 1-Propanone, 1-(4-methylphenyl)-
10469 - 1-Propanone, 2-methyl-1-phenyl-

$C_{10}H_{12}O_2$
232 - Acetic acid, 2-phenylethyl ester
897 - Benzeneacetic acid, α-ethyl-
899 - Benzeneacetic acid, ethyl ester
1100 - Benzenebutanoic acid
2137 - Benzenepropanoic acid, α-methyl-
2138 - Benzenepropanoic acid, β-methyl-, (±)-
2139 - Benzenepropanoic acid, methyl ester
2431 - 1,3-Benzodioxole, 5-propyl-
2589 - Benzoic acid, 2,4-dimethyl-, methyl ester
2688 - Benzoic acid, 2-(1-methylethyl)-
2689 - Benzoic acid, 4-(1-methylethyl)-
2690 - Benzoic acid, 2-methyl-, ethyl ester
2691 - Benzoic acid, 3-methyl-, ethyl ester
2692 - Benzoic acid, 4-methyl-, ethyl ester
2693 - Benzoic acid, 1-methylethyl ester
2723 - Benzoic acid, o-propyl-
2724 - Benzoic acid, propyl ester
3584 - Butanoic acid, phenyl ester
3665 - 1-Butanone, 1-(2-hydroxyphenyl)-
3666 - 1-Butanone, 2-hydroxy-1-phenyl-
3667 - 1-Butanone, 4-hydroxy-1-phenyl-
4241 - 1,3,6-Cycloheptatriene-1-carboxylic acid, 5,5-dimethyl-
4267 - 2,5-Cyclohexadiene-1,4-dione, 2-methyl-5-(1-methylethyl)-
4270 - 2,5-Cyclohexadiene-1,4-dione, 2,3,5,6-tetramethyl-
5195 - 1,3-Dioxane, 2-phenyl-
5196 - 1,3-Dioxane, 4-phenyl-
5814 - Ethanone, 2-ethoxy-1-phenyl-
9313 - Phenol, 2,4-dimethyl-, acetate
9314 - Phenol, 2,5-dimethyl-, acetate
9315 - Phenol, 3,4-dimethyl-, acetate
9316 - Phenol, 3,5-dimethyl-, acetate
9388 - Phenol, 2-methoxy-4-(1-propenyl)-
9389 - Phenol, 2-methoxy-4-(2-propenyl)-
9390 - Phenol, 2-methoxy-5-(2-propenyl)-
9391 - Phenol, 2-methoxy-6-(1-propenyl)-

9392 - Phenol, 2-methoxy-6-(2-propenyl)-
9396 - Phenol, 2-methoxy-4-(1-propenyl)-, (E)-
9397 - Phenol, 2-methoxy-4-(1-propenyl)-, (Z)-
10444 - 1-Propanone, 1-(2-hydroxy-3-methylphenyl)-
10445 - 1-Propanone, 1-(2-hydroxy-4-methylphenyl)-
10446 - 1-Propanone, 1-(2-hydroxy-5-methylphenyl)-
10456 - 1-Propanone, 1-(3-methoxyphenyl)-
10457 - 1-Propanone, 2-methoxy-1-phenyl-
10458 - 1-Propanone, 3-methoxy-1-phenyl-
10459 - 2-Propanone, 1-(methoxyphenyl)-
10460 - 2-Propanone, 1-(3-methoxyphenyl)-
10461 - 2-Propanone, 1-(4-methoxyphenyl)-

$C_{10}H_{12}O_3$
229 - Acetic acid, phenoxy-, ethyl ester
557 - Benzaldehyde, 4-ethoxy-3-methoxy-
869 - Benzeneacetaldehyde, 3,4-dimethoxy-
907 - Benzeneacetic acid, α-hydroxy-, ethyl ester, (±)-
909 - Benzeneacetic acid, α-hydroxy-, ethyl ester
1892 - Benzenemethanol, 4-methoxy-, acetate
2635 - Benzoic acid, 2-hydroxy-, 1-methylethyl ester
2652 - Benzoic acid, 2-hydroxy-, propyl ester
2653 - Benzoic acid, 4-hydroxy-, propyl ester
2669 - Benzoic acid, 2-methoxy-, ethyl ester
2670 - Benzoic acid, 3-methoxy-, ethyl ester
2671 - Benzoic acid, 4-methoxy-, ethyl ester
3580 - Butanoic acid, 2-phenoxy-
5209 - 1,3-Dioxolane-4-methanol, 2-phenyl-
5805 - Ethanone, 1-(3,4-dimethoxyphenyl)-
9370 - Phenol, 4-(3-hydroxy-1-propenyl)-2-methoxy-
10642 - 2-Propenoic acid, 3-(2-furanyl)-, propyl ester

$C_{10}H_{12}O_4$
2574 - Benzoic acid, 2,4-dihydroxy-6-methyl-, ethyl ester
2579 - Benzoic acid, 3,4-dimethoxy-, methyl ester
2626 - Benzoic acid, 4-hydroxy-3-methoxy-, ethyl ester
3781 - 2-Butenedioic acid (E)-, di-2-propenyl ester
3782 - 2-Butenedioic acid (Z)-, di-2-propenyl ester
5371 - 4,7-Epoxyisobenzofuran-1,3-dione, hexahydro-3a,7a-dimethyl-, (3aα,4β,7β,7aα)-

$C_{10}H_{12}O_5$
2732 - Benzoic acid, 3,4,5-trihydroxy-, propyl ester
2733 - Benzoic acid, 2,4,5-trimethoxy-
2734 - Benzoic acid, 3,4,5-trimethoxy-
6082 - 3,4-Furandicarboxylic acid, diethyl ester

$C_{10}H_{13}Br$
1023 - Benzene, 1-bromo-4-(1,1-dimethylethyl)-
1065 - Benzene, 2-bromo-1-methyl-4-(1-methylethyl)-

$C_{10}H_{13}BrN_2O_3$
11078 - 2,4,6(1H,3H,5H)-Pyrimidinetrione, 5-(2-bromo-2-propenyl)-5-(1-methylethyl)-

$C_{10}H_{13}BrO$
1003 - Benzene, (4-bromobutoxy)-

$C_{10}H_{13}Cl$
1142 - Benzene, 1-chloro-4-(1,1-dimethylethyl)-
1143 - Benzene, (2-chloro-1,1-dimethylethyl)-
1201 - Benzene, 2-chloro-1-methyl-4-(1-methylethyl)-
1202 - Benzene, 2-chloro-4-methyl-1-(1-methylethyl)-

$C_{10}H_{13}ClN_2$
7564 - Methanimidamide, N'-(4-chloro-2-methylphenyl)-N,N-dimethyl-

$C_{10}H_{13}ClN_2O_3S$
2208 - Benzenesulfonamide, 4-chloro-N-[(propylamino)carbonyl]-

$C_{10}H_{13}ClN_6$
9759 - Procyazine

$C_{10}H_{13}ClO$
1598 - Benzeneethanol, 4-chloro-α,α-dimethyl-
9268 - Phenol, 4-chloro-5-methyl-2-(1-methylethyl)-

$C_{10}H_{13}Cl_2O_3PS$
9581 - Phosphorothioic acid, O-(2,4-dichlorophenyl) O,O-diethyl ester

$C_{10}H_{13}I$
1781 - Benzene, 2-iodo-1-methyl-4-(1-methylethyl)-
1782 - Benzene, 2-iodo-4-methyl-1-(1-methylethyl)-

$C_{10}H_{13}N$
7731 - 1-Naphthalenamine, 5,6,7,8-tetrahydro-
10996 - Pyridine, 2-(1-pentenyl)-
11171 - Pyrrolidine, 1-phenyl-
11299 - Quinoline, 1,2,3,4-tetrahydro-1-methyl-

$C_{10}H_{13}NO$
56 - Acetamide, N-(2,4-dimethylphenyl)-
62 - Acetamide, N-ethyl-N-phenyl-
75 - Acetamide, N-methyl-N-(2-methylphenyl)-
627 - Benzamide, N,N,4-trimethyl-
881 - Benzeneacetamide, α-ethyl-
3208 - Butanamide, N-phenyl-
5806 - Ethanone, 1-[3-(dimethylamino)phenyl]-
11298 - Quinoline, 1,2,3,4-tetrahydro-6-methoxy-

$C_{10}H_{13}NO_2$
59 - Acetamide, N-(2-ethoxyphenyl)-
60 - Acetamide, N-(4-ethoxyphenyl)-
1467 - Benzene, 1-(1,1-dimethylethyl)-3-nitro-
1991 - Benzene, 1-methyl-4-(1-methylethyl)-2-nitro-
2128 - Benzenepropanoic acid, β-(aminomethyl)-
2149 - Benzenepropanol, carbamate
2507 - Benzoic acid, 4-amino-, propyl ester
2680 - Benzoic acid, 2-(methylamino)-, ethyl ester
3205 - Butanamide, N-(4-hydroxyphenyl)-
3966 - Carbamic acid, phenyl-, 1-methylethyl ester
3967 - Carbamic acid, (phenylmethyl)-, ethyl ester
6271 - Glycine, N-phenyl-, ethyl ester
10898 - 2-Pyridinecarboxylic acid, 5-butyl-

$C_{10}H_{13}NO_3$
2034 - Benzene, 1-(2-methylpropoxy)-2-nitro-
2672 - Benzoic acid, 3-methoxy-2-(methylamino)-, methyl ester
11933 - L-Tyrosine, N-methyl-

$C_{10}H_{13}NO_4$
10188 - 1,2,3-Propanetriol, 1-(4-aminobenzoate)
11931 - L-Tyrosine, 3-hydroxy-α-methyl-

$C_{10}H_{13}N_3O_4$
3894 - N-sec-Butyl-2,6-dinitrobenzene

$C_{10}H_{13}N_5O_4$
319 - Adenosine

$C_{10}H_{13}N_5O_5$
6289 - Guanosine

$C_{10}H_{14}$
1118 - Benzene, butyl-
1413 - Benzene, 1,2-diethyl-
1414 - Benzene, 1,3-diethyl-
1415 - Benzene, 1,4-diethyl-
1459 - Benzene, (1,1-dimethylethyl)-
1670 - Benzene, 1-ethyl-2,3-dimethyl-
1671 - Benzene, 1-ethyl-2,4-dimethyl-
1672 - Benzene, 1-ethyl-3,5-dimethyl-
1673 - Benzene, 2-ethyl-1,3-dimethyl-
1674 - Benzene, 2-ethyl-1,4-dimethyl-
1675 - Benzene, 4-ethyl-1,2-dimethyl-
1988 - Benzene, 1-methyl-2-(1-methylethyl)-
1989 - Benzene, 1-methyl-3-(1-methylethyl)-
1990 - Benzene, 1-methyl-4-(1-methylethyl)-
2035 - Benzene, 1-methylpropyl
2036 - Benzene, 1-methyl-2-propyl-
2037 - Benzene, 1-methyl-3-propyl-
2038 - Benzene, 1-methyl-4-propyl-
2039 - Benzene, (2-methylpropyl)-
2278 - Benzene, 1,2,3,4-tetramethyl-
2279 - Benzene, 1,2,3,5-tetramethyl-
2280 - Benzene, 1,2,4,5-tetramethyl-
7897 - Naphthalene, 1,2,3,4,4a,8a-hexahydro-
8340 - 2,6-Octadien-4-yne, 3,6-dimethyl-
8345 - 3,5-Octadiyne, 2,7-dimethyl-

$C_{10}H_{14}BeO_4$
2912 - Beryllium, bis(2,4-pentanedionato-O,O')-, (T-4)-

$C_{10}H_{14}BrN$
642 - Benzenamine, 4-bromo-N,N-diethyl-

$C_{10}H_{14}Br_2O$
2953 - Bicyclo[2.2.1]heptan-2-one, 3,3-dibromo-1,7,7-trimethyl-, (1R)-

$C_{10}H_{14}ClN$
661 - Benzenamine, 4-chloro-N,N-diethyl-

$C_{10}H_{14}Cl_6N_4O_2$
11902 - Triforine

$C_{10}H_{14}NO_5PS$
9583 - Phosphorothioic acid, O,O-diethyl O-(4-nitrophenyl) ester

$C_{10}H_{14}N_2$
3189 - Butanal, phenylhydrazone
9620 - Piperazine, 1-phenyl-
10992 - Pyridine, 3-(1-methyl-2-pyrrolidinyl)-, (S)-
11008 - Pyridine, 3-(2-piperidinyl)-, (S)-
11009 - Pyridine, 4-(4-piperidinyl)-

$C_{10}H_{14}N_2O$
699 - Benzenamine, N,N-diethyl-4-nitroso-
10891 - 3-Pyridinecarboxamide, N,N-diethyl-
10892 - 4-Pyridinecarboxamide, N,N-diethyl-

$C_{10}H_{14}N_2O_2$
698 - Benzenamine, N,N-diethyl-4-nitro-

$C_{10}H_{14}N_2O_2S$
11064 - 4,6(1H,5H)-Pyrimidinedione, 5-ethyldihydro-5-(2-methyl-2-propenyl)-2-thioxo-

$C_{10}H_{14}N_2O_3$
11098 - 2,4,6(1H,3H,5H)-Pyrimidinetrione, 5-(1-methylethyl)-5-(2-propenyl)-

$C_{10}H_{14}N_2O_5$
11816 - Thymidine

$C_{10}H_{14}N_4O_4$
10761 - 1H-Purine-2,6-dione, 7-(2,3-dihydroxypropyl)-3,7-dihydro-1,3-dimethyl-

$C_{10}H_{14}N_5O_7P$
320 - 5'-Adenylic acid

$C_{10}H_{14}O$
1114 - Benzene, butoxy-
1458 - Benzene, (1,1-dimethylethoxy)-
1600 - Benzeneethanol, α,α-dimethyl-
1602 - Benzeneethanol, β-ethyl-
1603 - Benzeneethanol, 2-ethyl-
1604 - Benzeneethanol, 4-ethyl-
1653 - Benzene, 1-ethoxy-4-ethyl-
1874 - Benzenemethanol, α-ethyl-α-methyl-, (R)-
1901 - Benzenemethanol, α-(1-methylethyl)-
1902 - Benzenemethanol, 4-(1-methylethyl)-
1912 - Benzenemethanol, α-propyl-, (R)-
1945 - Benzene, 1-methoxy-4-propyl-
1987 - Benzene, 1-methyl-3-(1-methylethoxy)-
2030 - Benzene, (1-methylpropoxy)-
2031 - Benzene, (2-methylpropoxy)-
2032 - Benzene, 1-methyl-4-propoxy-
2155 - Benzenepropanol, α-methyl-
2156 - Benzenepropanol, γ-methyl-
2465 - Benzofuran, 4,5,6,7-tetrahydro-3,6-dimethyl-
2956 - Bicyclo[3.1.1]heptan-3-one, 6,6-dimethyl-2-methylene-, (1S)-
2977 - Bicyclo[3.1.1]hept-2-en-6-one, 2,7,7-trimethyl-
2978 - Bicyclo[3.1.1]hept-3-en-2-one, 4,6,6-trimethyl-, (1R)-
2987 - Bicyclo[3.1.0]hex-3-en-2-one, 4-methyl-1-(1-methylethyl)-
4220 - 2,4-Cycloheptadien-1-one, 2,6,6-trimethyl-
4250 - 1,3-Cyclohexadiene-1-carboxaldehyde, 2,6,6-trimethyl-
4633 - 1-Cyclohexene-1-carboxaldehyde, 4-(1-methylethenyl)-
4634 - 1-Cyclohexene-1-carboxaldehyde, 4-(1-methylethenyl)-, (R)-
4715 - 2-Cyclohexen-1-one, 2-methyl-5-(1-methylethenyl)-, (S)-
4718 - 2-Cyclohexen-1-one, 3-methyl-6-(1-methylethylidene)-
4881 - Cyclopentanone, 2-cyclopentylidene-
6133 - Furan, 3-(4-methyl-3-pentenyl)-
9246 - Phenol, 2-butyl-
9247 - Phenol, 3-butyl-
9248 - Phenol, 4-butyl-
9321 - Phenol, 2-(1,1-dimethylethyl)-

9322 - Phenol, 3-(1,1-dimethylethyl)-
9323 - Phenol, 4-(1,1-dimethylethyl)-
9414 - Phenol, 2-methyl-4-(1-methylethyl)-
9415 - Phenol, 2-methyl-5-(1-methylethyl)-
9416 - Phenol, 2-methyl-6-(1-methylethyl)-
9417 - Phenol, 4-methyl-2-(1-methylethyl)-
9418 - Phenol, 5-methyl-2-(1-methylethyl)-
9430 - Phenol, 2-(1-methylpropyl)-
9431 - Phenol, 4-(1-methylpropyl)-
9432 - Phenol, 4-(2-methylpropyl)-
9473 - Phenol, 2,3,4,5-tetramethyl-
9474 - Phenol, 2,3,4,6-tetramethyl-
9475 - Phenol, 2,3,5,6-tetramethyl-

$C_{10}H_{14}O_2$
1406 - Benzene, 1,2-diethoxy-
1407 - Benzene, 1,3-diethoxy-
1408 - Benzene, 1,4-diethoxy-
1516 - 1,2-Benzenediol, 4-(1,1-dimethylethyl)-
1536 - 1,2-Benzenediol, 3-methyl-6-(1-methylethyl)-
1538 - 1,4-Benzenediol, 2-methyl-5-(1-methylethyl)-
1539 - 1,3-Benzenediol, 4-(2-methylpropyl)-
1552 - 1,4-Benzenediol, 2,3,5,6-tetramethyl-
1652 - Benzene, (2-ethoxyethoxy)-
1873 - Benzenemethanol, α-ethyl-4-methoxy-
2154 - Benzenepropanol, 4-methoxy-
2926 - Bicyclo[2.2.1]heptane-2,3-dione, 1,7,7-trimethyl-, (1R)-
3356 - 1,2-Butanediol, 2-phenyl-
3357 - 1,3-Butanediol, 1-phenyl-
3358 - 1,3-Butanediol, 3-phenyl-
4758 - Cyclopenta[c]pyran-1(4aH)-one, 5,6,7,7a-tetrahydro-4,7-dimethyl-
6076 - Furan, 2-(cyclohexyloxy)-
9245 - Phenol, 2-butoxy-
10400 - 1-Propanol, 3-(phenylmethoxy)-
11862 - Tricyclo[2.2.1.02,6]heptane-1-carboxylic acid, 7,7-dimethyl-

$C_{10}H_{14}O_3$
1599 - Benzeneethanol, 2,5-dimethoxy-
2153 - Benzenepropanol, 4-hydroxy-3-methoxy-
6027 - 2-Furanacetic acid, butyl ester
6068 - 2-Furancarboxylic acid, pentyl ester
8559 - 3-Oxabicyclo[3.2.1]octane-2,4-dione, 1,8,8-trimethyl-, (1R)-
8880 - Pentanoic acid, 2-furanylmethyl ester
10019 - 1,2-Propanediol, 3-(2-methylphenoxy)-

$C_{10}H_{14}O_3S$
2219 - Benzenesulfonic acid, sec-butyl ester
2233 - Benzenesulfonic acid, 4-methyl-, propyl ester

$C_{10}H_{14}O_4$
1914 - Benzenemethanol, 3,4,5-trimethoxy-
4646 - 1-Cyclohexene-1,4-dicarboxylic acid, dimethyl ester
10016 - 1,2-Propanediol, 3-(2-methoxyphenoxy)-
10656 - 2-Propenoic acid, 2-methyl-, 1,2-ethanediyl ester

$C_{10}H_{15}BrO$
2951 - Bicyclo[2.2.1]heptan-2-one, 1-(bromomethyl)-7,7-dimethyl-, (1S)-
2952 - Bicyclo[2.2.1]heptan-2-one, 3-bromo-1,7,7-trimethyl-, (1R-endo)-

$C_{10}H_{15}N$
655 - Benzenamine, 4-butyl-
656 - Benzenamine, N-butyl-
694 - Benzenamine, 2,6-diethyl-
695 - Benzenamine, N,N-diethyl-
714 - Benzenamine, 2-(1,1-dimethylethyl)-
715 - Benzenamine, 4-(1,1-dimethylethyl)-
716 - Benzenamine, N-(1,1-dimethylethyl)-
793 - Benzenamine, 2-methyl-5-(1-methylethyl)-
794 - Benzenamine, 4-methyl-N-(1-methylethyl)-
812 - Benzenamine, 2-(1-methylpropyl)-
813 - Benzenamine, 4-(1-methylpropyl)-
814 - Benzenamine, 4-(2-methylpropyl)-
815 - Benzenamine, 4-methyl-N-propyl-
816 - Benzenamine, N-(1-methylpropyl)-
817 - Benzenamine, N-(2-methylpropyl)-

847 - Benzenamine, 2,3,4,5-tetramethyl-
848 - Benzenamine, 2,3,4,6-tetramethyl-
849 - Benzenamine, N,N,2,6-tetramethyl-
1572 - Benzeneethanamine, α,α-dimethyl-
1573 - Benzeneethanamine, N,α-dimethyl-, (S)-
1574 - Benzeneethanamine, N,β-dimethyl-
1819 - Benzenemethanamine, α-(1-methylethyl)-
1820 - Benzenemethanamine, N-(1-methylethyl)-
1830 - Benzenemethanamine, α-propyl-
2119 - Benzenepropanamine, α-methyl-
10967 - Pyridine, 2-(1-ethylpropyl)-
10968 - Pyridine, 4-(1-ethylpropyl)-

$C_{10}H_{15}NO$
1840 - Benzenemethanol, α-(1-aminopropyl)-
1900 - Benzenemethanol, α-[1-(methylamino)ethyl]-, (R*,S*)-(±)-
9297 - Phenol, 3-(diethylamino)-
9320 - Phenol, 4-[2-(dimethylamino)ethyl]-
9403 - Phenol, 4-[2-(methylamino)propyl]-
10983 - Pyridine, 2-[2-(1-methylethoxy)ethyl]-

$C_{10}H_{15}NO_2$
693 - Benzenamine, 3,4-diethoxy-
1571 - Benzeneethanamine, 3,4-dimethoxy-
5760 - Ethanol, 2,2'-(phenylimino)bis-

$C_{10}H_{15}NO_4$
9933 - Propanedioic acid, (2-cyanoethyl)-, diethyl ester
11144 - 3-Pyrrolidineacetic acid, 2-carboxy-4-(1-methylethenyl)-, [2S-(2α,3β,4β)]-

$C_{10}H_{15}N_3$
6281 - Guanidine, N,N'-dimethyl-N''-(phenylmethyl)-

$C_{10}H_{15}OPS_2$
5986 - Fonofos

$C_{10}H_{15}O_3PS_2$
5955 - Fenthion

$C_{10}H_{16}$
2921 - Bicyclo[2.2.1]heptane, 2,2-dimethyl-3-methylene-, (±)-
2922 - Bicyclo[2.2.1]heptane, 2,2-dimethyl-5-methylene-
2923 - Bicyclo[2.2.1]heptane, 7,7-dimethyl-2-methylene-, (±)-
2924 - Bicyclo[3.1.1]heptane, 6,6-dimethyl-2-methylene-, (1R)-
2925 - Bicyclo[3.1.1]heptane, 6,6-dimethyl-2-methylene-
2970 - Bicyclo[2.2.1]hept-2-ene, 1,7,7-trimethyl-
2971 - Bicyclo[3.1.1]hept-2-ene, 2,6,6-trimethyl-, (1S)-
2972 - Bicyclo[3.1.1]hept-2-ene, 2,6,6-trimethyl-, (±)-
2973 - Bicyclo[3.1.1]hept-2-ene, 2,6,6-trimethyl
2974 - Bicyclo[4.1.0]hept-3-ene, 3,7,7-trimethyl-, (1S)-
2981 - Bicyclo[3.1.0]hexane, 4-methylene-1-(1-methylethyl)-, (±)-
2986 - Bicyclo[3.1.0]hex-2-ene, 2-methyl-5-(1-methylethyl)-
4050 - 4-Carene, (1S,3R,6R)-(-)-
4209 - Cyclodecyne
4273 - 1,3-Cyclohexadiene, 1-methyl-4-(1-methylethyl)-
4274 - 1,3-Cyclohexadiene, 2-methyl-5-(1-methylethyl)-, (±)-
4275 - 1,4-Cyclohexadiene, 1-methyl-4-(1-methylethyl)-
4673 - Cyclohexene, 3-methylene-6-(1-methylethyl)-
4674 - Cyclohexene, 3-methylene-6-(1-methylethyl)-, (+)-
4675 - Cyclohexene, 4-methylene-1-(1-methylethyl)-
4676 - Cyclohexene, 1-methyl-3-(1-methylethenyl)-
4677 - Cyclohexene, 1-methyl-4-(1-methylethenyl)-
4678 - Cyclohexene, 1-methyl-5-(1-methylethenyl)-
4679 - Cyclohexene, 5-methyl-1-(1-methylethenyl)-, (±)-
4684 - Cyclohexene, 1-methyl-4-(1-methylethylidene)-
5094 - 1-Decen-3-yne

5095 - 1-Decen-4-yne
5096 - 2-Decen-4-yne
8330 - 1,6-Octadiene, 7-methyl-3-methylene-
8485 - 1,3,6-Octatriene, 3,7-dimethyl-
8486 - 1,3,7-Octatriene, 3,7-dimethyl-
8487 - 2,4,6-Octatriene, 2,6-dimethyl-, (E,E)-
8488 - 2,4,6-Octatriene, 2,6-dimethyl-, (E,Z)-
11815 - 4(10)-Thujene, (+)-
11860 - Tricyclo[3.3.1.13,7]decane
11866 - Tricyclo[2.2.1.02,6]heptane, 1,3,3-trimethyl-
11867 - Tricyclo[2.2.1.02,6]heptane, 1,7,7-trimethyl-

$C_{10}H_{16}ClNO$
865 - Benzenaminium, N-ethyl-3-hydroxy-N,N-dimethyl-, chloride

$C_{10}H_{16}Cl_2O_2$
5049 - Decanedioyl dichloride

$C_{10}H_{16}Cl_3NOS$
11823 - Triallate

$C_{10}H_{16}NO_5PS_2$
5946 - Famphur

$C_{10}H_{16}N_2$
1265 - 1,3-Benzenediamine, N,N-diethyl-
1266 - 1,4-Benzenediamine, N,N-diethyl-
1286 - 1,2-Benzenediamine, N,N,N',N'-tetramethyl-
1287 - 1,4-Benzenediamine, N,N,N',N'-tetramethyl-
5041 - Decanedinitrile

$C_{10}H_{16}N_2OS$
7229 - 4-Imidazolidinone, 5-(2-methylpropyl)-3-(2-propenyl)-2-thioxo-

$C_{10}H_{16}N_2O_2S$
11065 - 4,6(1H,5H)-Pyrimidinedione, 5-ethyldihydro-5-(1-methylpropyl)-2-thioxo-

$C_{10}H_{16}N_2O_3$
11082 - 2,4,6(1H,3H,5H)-Pyrimidinetrione, 5-butyl-5-ethyl-

$C_{10}H_{16}N_2O_3S$
11689 - 1H-Thieno[3,4-d]imidazole-4-pentanoic acid, hexahydro-2-oxo-, [3aS-(3aα,4β,6aα)]-

$C_{10}H_{16}N_2O_8$
6262 - Glycine, N,N'-1,2-ethanediylbis[N-(carboxymethyl)-

$C_{10}H_{16}N_4O_7$
11107 - 4(1H)-Pyrimidinone, 2,6-diamino-5-(β-D-glucopyranosyloxy)-

$C_{10}H_{16}O$
2960 - Bicyclo[2.2.1]heptan-2-one, 1,3,3-trimethyl-, (±)-
2961 - Bicyclo[2.2.1]heptan-2-one, 1,7,7-trimethyl-, (1R)-
2963 - Bicyclo[2.2.1]heptan-2-one, 4,7,7-trimethyl-, (1S)-
2964 - Bicyclo[3.1.1]heptan-2-one, 4,6,6-trimethyl-, [1R-(1α,4α,5α)]-
2965 - Bicyclo[3.1.1]heptan-3-one, 2,6,6-trimethyl-, [1S-(1α,2β,5α)]-
2975 - Bicyclo[3.1.1]hept-3-en-2-ol, 4,6,6-trimethyl-, (1α,2β,5α)-
2976 - Bicyclo[3.1.1]hept-3-en-2-ol, 4,6,6-trimethyl-, (1α,2α,5α)-
2983 - Bicyclo[3.1.0]hexan-3-ol, 4-methylene-1-(1-methylethyl)-
2984 - Bicyclo[3.1.0]hexan-3-ol, 4-methylene-1-(1-methylethyl)-, [1S-(1α,3β,5α)]-
2995 - [1,1'-Bicyclopentyl]-2-one
4576 - Cyclohexanone, 2-butylidene-
4601 - Cyclohexanone, 2-methyl-5-(1-methylethenyl)-, (2S-trans)-
4602 - Cyclohexanone, 5-methyl-2-(1-methylethenyl)-
4603 - Cyclohexanone, 5-methyl-2-(1-methylethenyl)-, trans-
4607 - Cyclohexanone, 5-methyl-2-(1-methylethylidene)-
4608 - Cyclohexanone, 5-methyl-2-(1-methylethylidene)-, (R)-
4635 - 1-Cyclohexene-1-carboxaldehyde, 4-(1-methylethyl)-

4636 - 1-Cyclohexene-1-carboxaldehyde, 2,6,6-trimethyl-
4664 - 1-Cyclohexene-1-methanol, 4-(1-methylethenyl)-
4696 - 2-Cyclohex-1-ol, 2-methyl-5-(1-methylethenyl)-
4716 - 2-Cyclohexen-1-one, 3-methyl-6-(1-methylethyl)-, (±)-
4717 - 2-Cyclohexen-1-one, 6-methyl-3-(1-methylethyl)-, (±)-
8053 - 1(2H)-Naphthalenone, octahydro-, *cis*-
8054 - 1(2H)-Naphthalenone, octahydro-, *trans*-
8055 - 2(1H)-Naphthalenone, octahydro-
8317 - 2,6-Octadienal, 3,7-dimethyl-, (*E*)-
8318 - 2,6-Octadienal, 3,7-dimethyl-, (*Z*)-
8561 - 6-Oxabicyclo[3.2.1]oct-3-ene, 4,7,7-trimethyl-, (±)-

C₁₀H₁₆O₂

2957 - Bicyclo[2.2.1]heptan-2-one, 3-hydroxy-1,7,7-trimethyl-
3834 - 2-Butenoic acid, 3-cyclohexyl-
4203 - 1,2-Cyclodecanedione
4307 - Δ1,α-Cyclohexaneacetic acid, 3,3-dimethyl-
4709 - 2-Cyclohexen-1-one, 2-hydroxy-3-methyl-6-(1-methylethyl)-
4911 - 1-Cyclopentene-1-carboxylic acid, 2,3,3-trimethyl-, methyl ester
4959 - Cyclopropanecarboxylic acid, 2,2-dimethyl-3-(2-methyl-1-propenyl)-, (1*R-cis*)-
5025 - 2,4-Decadienoic acid, (*E,E*)-
5184 - 2,3-Dioxabicyclo[2.2.2]oct-5-ene, 1-methyl-4-(1-methylethyl)-
8331 - 2,4-Octadienoic acid, 3,7-dimethyl-
10653 - 2-Propenoic acid, 2-methyl-, cyclohexyl ester

C₁₀H₁₆O₄

3766 - 2-Butenedioic acid (*E*)-, bis(1-methylethyl) ester
3783 - 2-Butenedioic acid (*E*)-, dipropyl ester
3784 - 2-Butenedioic acid (*Z*)-, dipropyl ester
3931 - Camphoric acid, *trans*-(±)-
4180 - 1,1-Cyclobutanedicarboxylic acid, diethyl ester
4184 - 1,3-Cyclobutanedicarboxylic acid, 2,2-dimethyl-, dimethyl ester
4352 - 1,3-Cyclohexanedicarboxylic acid, dimethyl ester, *cis*-
4353 - 1,3-Cyclohexanedicarboxylic acid, dimethyl ester, *trans*-
4357 - 1,4-Cyclohexanedicarboxylic acid, dimethyl ester, *trans*-
4794 - 1,3-Cyclopentanedicarboxylic acid, 1,2,2-trimethyl-, (1*R-cis*)-
9966 - Propanedioic acid, (1-methylethylidene)-, diethyl ester
9980 - Propanedioic acid, 2-propenyl-, diethyl ester

C₁₀H₁₆O₄S

2929 - Bicyclo[2.2.1]heptane-1-methanesulfonic acid, 7,7-dimethyl-2-oxo-, (1*S*)-

C₁₀H₁₆O₅

3295 - Butanedioic acid, acetyl-, diethyl ester
9947 - Propanedioic acid, (ethoxymethylene)-, diethyl ester

C₁₀H₁₆O₆

7552 - Methanetricarboxylic acid, triethyl ester

C₁₀H₁₆Si

11490 - Silane, trimethyl(4-methylphenyl)-
11494 - Silane, trimethyl(phenylmethyl)-

C₁₀H₁₇Br

8321 - 2,6-Octadiene, 1-bromo-3,7-dimethyl-, (*E*)-

C₁₀H₁₇Cl

2919 - Bicyclo[2.2.1]heptane, 2-chloro-1,7,7-trimethyl-
2920 - Bicyclo[2.2.1]heptane, 2-chloro-1,7,7-trimethyl-, *endo*-
7782 - Naphthalene, 2-chlorodecahydro-

C₁₀H₁₇ClO₄

9932 - Propanedioic acid, (3-chloropropyl)-, diethyl ester

C₁₀H₁₇Cl₂NOS

5106 - Diallate

C₁₀H₁₇N

11859 - Tricyclo[3.3.1.13,7]decan-1-amine

C₁₀H₁₇NO

2962 - Bicyclo[2.2.1]heptan-2-one, 1,7,7-trimethyl-, oxime, (1*S*)-

C₁₀H₁₇NO₆

6258 - Glycine, *N,N*-bis(ethoxycarbonyl)-, ethyl ester
10104 - Propanenitrile, 2-((β-*D*-glucopyranosyloxy)-2-methyl-

C₁₀H₁₇N₂O₄PS

5945 - Etrimfos

C₁₀H₁₇N₃

11848 - 4H-1,2,4-Triazole, 4-cyclohexyl-3-ethyl-

C₁₀H₁₇N₃O₆S

6264 - Glycine, *N*-(*N-L*-γ-glutamyl-*L*-cysteinyl)-

C₁₀H₁₈

2931 - Bicyclo[2.2.1]heptane, 1,3,3-trimethyl-, (+)-
2932 - Bicyclo[2.2.1]heptane, 1,7,7-trimethyl-
2933 - Bicyclo[2.2.1]heptane, 2,2,3-trimethyl-
2934 - Bicyclo[3.1.1]heptane, 2,6,6-trimethyl-, (1α,2α,5α)-(±)-
2935 - Bicyclo[3.1.1]heptane, 2,6,6-trimethyl-, [1*R*-(1α,2β,5α)]-
2936 - Bicyclo[4.1.0]heptane, 3,7,7-trimethyl-, [1*S*-(1α,3β,6α)]-
2937 - Bicyclo[4.1.0]heptane, 3,7,7-trimethyl-, [1*S*-(1α,3α,6α)]-
2982 - Bicyclo[3.1.0]hexane, 4-methyl-1-(1-methylethyl)-
4207 - Cyclodecene, (*E*)-
4208 - Cyclodecene, (*Z*)-
4247 - Cycloheptene, 1,4,4-trimethyl-
4435 - Cyclohexane, 1-methyl-3-(1-methylethenyl)-
4441 - Cyclohexane, 1-methyl-3-(1-methylethylidene)-
4442 - Cyclohexane, 1-methyl-4-(1-methylethylidene)-
4680 - Cyclohexene, 1-methyl-4-(1-methylethyl)-
4681 - Cyclohexene, 1-methyl-4-(1-methylethyl)-, (*R*)-
4682 - Cyclohexene, 3-methyl-6-(1-methylethyl)-, (3*R-trans*)-
4683 - Cyclohexene, 4-methyl-1-(1-methylethyl)-, (*R*)-
4923 - Cyclopentene, 1,5-dimethyl-4-(1-methylethyl)-
4934 - Cyclopentene, 1-pentyl-
5023 - 1,3-Decadiene
5024 - 1,9-Decadiene
5097 - 1-Decyne
5098 - 3-Decyne
5099 - 4-Decyne
5100 - 5-Decyne
6619 - 3-Heptyne, 5-ethyl-5-methyl-
7429 - *p*-Menth-3-ene, (*S*)-(-)-
7789 - Naphthalene, decahydro-
7790 - Naphthalene, decahydro-, *cis*-
7791 - Naphthalene, decahydro-, *trans*-
8202 - 4-Nonyne, 8-methyl-
8322 - 2,4-Octadiene, 3,7-dimethyl-
8323 - 2,5-Octadiene, 2,6-dimethyl-
8324 - 2,6-Octadiene, 2,6-dimethyl-, (*Z*)-
8325 - 2,6-Octadiene, 2,7-dimethyl-
8326 - 2,6-Octadiene, 3,6-dimethyl-
8327 - 2,6-Octadiene, 4,5-dimethyl-
8535 - 3-Octyne, 2,2-dimethyl-
11814 - Thujane, (1*S*,4*R*,5*S*)-(+)-

C₁₀H₁₈N₂O₃

7223 - 4-Imidazolidinehexanoic acid, 5-methyl-2-oxo-, (4*R-cis*)-

C₁₀H₁₈N₂O₄

9606 - 1,4-Piperazinedicarboxylic acid, diethyl ester

C₁₀H₁₈N₄O₄S₃

11713 - Thiodicarb

C₁₀H₁₈O

2930 - Bicyclo[3.1.1]heptane-2-methanol, 6,6-dimethyl-, [1*S*-(1α,2β,5α)]-
2940 - Bicyclo[2.2.1]heptan-2-ol, 1,3,3-trimethyl-, *endo*-(±)-
2941 - Bicyclo[2.2.1]heptan-2-ol, 1,3,3-trimethyl-, (1*S-endo*)-
2942 - Bicyclo[2.2.1]heptan-2-ol, 1,7,7-trimethyl-, *endo*-(±)-
2943 - Bicyclo[2.2.1]heptan-2-ol, 1,7,7-trimethyl-, *exo*-(±)-
2944 - Bicyclo[3.1.1]heptan-2-ol, 4,6,6-trimethyl-, [1*R*-(1α,2α,4α,5α)]-
2985 - Bicyclo[3.1.0]hexan-3-ol, 4-methyl-1-(1-methylethyl)-
2994 - [1,1'-Bicyclopentyl]-2-ol
3652 - 1-Butanone, 1-cyclohexyl-
4205 - Cyclodecanone
4237 - Cycloheptanone, 2,2,6-trimethyl-
4238 - Cycloheptanone, 2,6,6-trimethyl-
4542 - Cyclohexanol, 1-methyl-4-(1-methylethenyl)-
4543 - Cyclohexanol, 2-methyl-5-(1-methylethenyl)-, [1*S*-(1α,2β,5α)]-
4544 - Cyclohexanol, 5-methyl-2-(1-methylethenyl)-, [1*R*-(1α,2β,5α)]-
4575 - Cyclohexanol, 2-butyl-
4604 - Cyclohexanone, 2-methyl-5-(1-methylethyl)-, (2*S-trans*)-
4605 - Cyclohexanone, 5-methyl-2-(1-methylethyl)-, (2*R-cis*)-
4606 - Cyclohexanone, 5-methyl-2-(1-methylethyl)-, (2*S-trans*)-
4609 - Cyclohexanone, 4-(2-methylpropyl)-
4665 - 1-Cyclohexene-1-methanol, 2,4,4-trimethyl-
4666 - 1-Cyclohexene-1-methanol, 2,6,6-trimethyl-
4667 - 2-Cyclohexene-1-methanol, 2,6,6-trimethyl-
4668 - 3-Cyclohexene-1-methanol, α,α,4-trimethyl-, (±)-
4697 - 2-Cyclohexen-1-ol, 3-methyl-6-(1-methylethyl)-
4698 - 2-Cyclohexen-1-ol, 6-methyl-3-(1-methylethyl)-
4699 - 3-Cyclohexen-1-ol, 4-methyl-1-(1-methylethyl)-
4978 - Cyclopropanemethanol, 2,2-dimethyl-3-(2-methyl-1-propenyl)-
5080 - 2-Decenal
5081 - 3-Decenal
5093 - 3-Decen-2-one
6590 - 3-Hepten-2-one, 3-ethyl-4-methyl-
6591 - 4-Hepten-2-one, 3-ethyl-4-methyl-
6592 - 4-Hepten-3-one, 5-ethyl-4-methyl-
6593 - 5-Hepten-2-one, 4-ethyl-6-methyl-
6594 - 5-Hepten-3-one, 5-ethyl-4-methyl-
8024 - 2-Naphthalenol, decahydro-
8191 - 5-Nonen-4-one, 6-methyl-
8192 - 6-Nonen-4-one, 6-methyl-
8332 - 1,6-Octadien-3-ol, 3,7-dimethyl-, (±)-
8333 - 2,6-Octadien-1-ol, 3,7-dimethyl-, (*E*)-
8334 - 2,6-Octadien-1-ol, 3,7-dimethyl-, (*Z*)-
8490 - 6-Octenal, 3,7-dimethyl-
8556 - 7-Oxabicyclo[2.2.1]heptane, 1-methyl-4-(1-methylethyl)-
8560 - 2-Oxabicyclo[2.2.2]octane, 1,3,3-trimethyl-

C₁₀H₁₈O₂

3496 - Butanoic acid, cyclohexyl ester
4206 - Cyclodecanone, 2-hydroxy-
4224 - Cycloheptanecarboxylic acid, ethyl ester
4308 - Cyclohexaneacetic acid, ethyl ester
4337 - Cyclohexanecarboxylic acid, propyl ester
4784 - Cyclopentanecarboxylic acid, 2-methyl-5-(1-methylethyl)-
4786 - Cyclopentanecarboxylic acid, 1,2,2,3-tetramethyl-
5088 - 2-Decenoic acid
5089 - 3-Decenoic acid

5090 - 4-Decenoic acid
5091 - 9-Decenoic acid
6811 - 3,4-Hexanedione, 2,2,5,5-tetramethyl-
10293 - Propanoic acid, 2-methyl-, cyclohexyl ester
10660 - 2-Propenoic acid, 2-methyl-, hexyl ester

$C_{10}H_{18}O_3$
3504 - Butanoic acid, 2,2-diethyl-3-oxo-, ethyl ester
3542 - Butanoic acid, 3-methyl-, anhydride
4309 - Cyclohexaneacetic acid, α-ethyl-1-hydroxy-
6872 - Hexanoic acid, 2-acetyl-, ethyl ester
8858 - Pentanoic acid, anhydride
11903 - 2,5,7-Trioxabicyclo[2.2.1]heptane, 1,3,3,4,6,6-hexamethyl-

$C_{10}H_{18}O_4$
3321 - Butanedioic acid, dipropyl ester
3512 - Butanedioic acid, 1,2-ethanediyl ester
5042 - Decanedioic acid
5557 - Ethanedioic acid, bis(2-methylpropyl) ester
5558 - Ethanedioic acid, dibutyl ester
6785 - Hexanedioic acid, diethyl ester
8135 - Nonanedioic acid, 3-methyl-
8388 - Octanedioic acid, dimethyl ester
8601 - Oxirane, 2,2'-[1,4-butanediylbis (oxymethylene)]bis-
9964 - Propanedioic acid, (1-methylethyl)-, diethyl ester
9981 - Propanedioic acid, propyl-, diethyl ester

$C_{10}H_{18}O_4S$
10330 - Propanoic acid, 3,3'-thiobis-, diethyl ester

$C_{10}H_{18}O_6$
3305 - Butanedioic acid, 2,3-dihydroxy-, bis(1-methylethyl) ester, (R*,R*)-(±)-
3310 - Butanedioic acid, 2,3-dihydroxy- [R-(R*,R*)]-, dipropyl ester
5714 - Ethanol, 2,2'-[1,2-ethanediylbis(oxy)]bis-, diacetate

$C_{10}H_{19}Br$
4321 - Cyclohexane, 1-bromo-4-methyl-1-(1-methylethyl)-
5086 - 1-Decene, 2-bromo-
5087 - 2-Decene, 1-bromo-

$C_{10}H_{19}BrO_2$
5061 - Decanoic acid, 2-bromo-

$C_{10}H_{19}ClO$
5078 - Decanoyl chloride

$C_{10}H_{19}FO_2$
5064 - Decanoic acid, 10-fluoro-

$C_{10}H_{19}N$
2927 - Bicyclo[2.2.1]heptane-1-methanamine, 7,7-dimethyl-
2928 - Bicyclo[3.1.1]heptane-2-methanamine, 6,6-dimethyl-, [1S-(1α,2β,5α)]-
5057 - Decanenitrile
7712 - 1-Naphthalenamine, decahydro-

$C_{10}H_{19}NO$
11331 - 2H-Quinolizine-1-methanol, octahydro-, (1R-trans)-

$C_{10}H_{19}NO_2$
9704 - 1-Piperidinepropanoic acid, ethyl ester

$C_{10}H_{19}N_5O$
3143 - sec-Bumeton

$C_{10}H_{19}N_5S$
9770 - Prometryn
11594 - Terbutryn

$C_{10}H_{19}O_6PS_2$
3316 - Butanedioic acid, [(dimethoxyphosphinothioyl)thio]-, diethyl ester

$C_{10}H_{20}$
4202 - Cyclodecane
4227 - Cycloheptane, 1,1,2-trimethyl-
4228 - Cycloheptane, 1,1,4-trimethyl-
4325 - Cyclohexane, butyl-
4369 - Cyclohexane, (1,1-dimethylethyl)-
4436 - Cyclohexane, 1-methyl-2-(1-methylethyl)-
4437 - Cyclohexane, 1-methyl-3-(1-methylethyl)-
4438 - Cyclohexane, 1-methyl-4-(1-methylethyl)-

4439 - Cyclohexane, 1-methyl-4-(1-methylethyl)-, cis-
4440 - Cyclohexane, 1-methyl-4-(1-methylethyl)-, trans-
4448 - Cyclohexane, (1-methylpropyl)-
4449 - Cyclohexane, 1-methyl-1-propyl-
4450 - Cyclohexane, 1-methyl-2-propyl-
4451 - Cyclohexane, 1-methyl-3-propyl-
4452 - Cyclohexane, 1-methyl-4-propyl-
4453 - Cyclohexane, (2-methylpropyl)-
4477 - Cyclohexane, 1,1,3,4-tetramethyl-
4478 - Cyclohexane, 1,1,3,5-tetramethyl-, cis-
4846 - Cyclopentane, pentyl-
5082 - 1-Decene
5083 - 4-Decene
5084 - 5-Decene, (E)-
5085 - 5-Decene, (Z)-
6559 - 3-Heptene, 4-propyl-
8185 - 1-Nonene, 2-methyl-
8186 - 3-Nonene, 2-methyl-
8502 - 1-Octene, 3,7-dimethyl-
8503 - 2-Octene, 2,6-dimethyl-

$C_{10}H_{20}BrF$
5035 - Decane, 1-bromo-10-fluoro-

$C_{10}H_{20}Br_2$
5038 - Decane, 1,10-dibromo-
5039 - Decane, 5,6-dibromo-

$C_{10}H_{20}ClF$
5037 - Decane, 1-chloro-10-fluoro-

$C_{10}H_{20}Cl_2$
5040 - Decane, 1,10-dichloro-

$C_{10}H_{20}NO_4PS$
10719 - Propetamphos

$C_{10}H_{20}N_2$
3098 - 2,3'-Bipiperidine

$C_{10}H_{20}N_2O_2$
7690 - Morpholine, 4,4'-(1,2-ethanediyl)bis-

$C_{10}H_{20}N_2S_2$
9659 - Piperidine, 1,1'-dithiobis-

$C_{10}H_{20}N_2S_3$
11714 - Thiodicarbonic diamide ([(H2N)C(S)]2S), tetraethyl-

$C_{10}H_{20}N_2S_4$
11721 - Thioperoxydicarbonic diamide ([(H2N)C(S)]2S2), tetraethyl-

$C_{10}H_{20}O$
4204 - Cyclodecanol
4233 - Cycloheptanol, 2,6,6-trimethyl-
4324 - Cyclohexanebutanol
4469 - Cyclohexanepropanol, α-methyl-
4504 - Cyclohexanol, 2-butyl-, trans-
4521 - Cyclohexanol, 2-(1,1-dimethylethyl)-
4546 - Cyclohexanol, 1-methyl-4-(1-methylethyl)-
4547 - Cyclohexanol, 2-methyl-5-(1-methylethyl)-, [1R-(1α,2α,5β)]-
4548 - Cyclohexanol, 5-methyl-2-(1-methylethyl)-, (1α,2β,5α)-(±)-
4550 - Cyclohexanol, 2-methyl-1-propyl-
4551 - Cyclohexanol, 3-methyl-1-propyl-
5030 - Decanal
5075 - 2-Decanone
5076 - 3-Decanone
5077 - 4-Decanone
5092 - 9-Decen-1-ol
6405 - Heptane, 3-[(ethenyloxy)methyl]-
7428 - m-Menthan-6-ol
8176 - 5-Nonanone, 2-methyl-
8401 - Octane, 1-(ethenyloxy)-
8514 - 6-Octen-1-ol, 3,7-dimethyl-, (±)-
8515 - 6-Octen-3-ol, 3,7-dimethyl-
8516 - 7-Octen-1-ol, 3,7-dimethyl-, (S)-

$C_{10}H_{20}O_2$
187 - Acetic acid, 2-ethylhexyl ester
222 - Acetic acid, octyl ester
3528 - Butanoic acid, hexyl ester
3546 - Butanoic acid, 3-methyl-, 1,1-dimethylpropyl ester
3551 - Butanoic acid, 3-methyl-, 3-methylbutyl ester

3559 - Butanoic acid, 1-methylpentyl ester, (S)-
5059 - Decanoic acid
6471 - Decanoic acid, propyl ester
6506 - 2-Heptanol, 6-methyl-, acetate
6507 - 3-Heptanol, 6-methyl-, acetate
6880 - Hexanoic acid, butyl ester
6885 - Hexanoic acid, 2-ethyl-, ethyl ester
6975 - 3-Hexanone, 4-hydroxy-2,2,5,5-tetramethyl-
8153 - Nonanoic acid, 3-methyl-
8154 - Nonanoic acid, methyl ester
8347 - Octanal, 7-hydroxy-3,7-dimethyl-
8428 - Octanoic acid, ethyl ester
8447 - 2-Octanol, acetate, (±)-
8894 - Pentanoic acid, 4-methyl-, 2-methylpropyl ester
8913 - Pentanoic acid, pentyl ester
10264 - Propanoic acid, heptyl ester

$C_{10}H_{20}O_3$
3811 - 2-Butene, 1,1,3-triethoxy-

$C_{10}H_{20}O_4$
5689 - Ethanol, 2-(2-butoxyethoxy)-, acetate

$C_{10}H_{21}Br$
5034 - Decane, 1-bromo-

$C_{10}H_{21}Cl$
5036 - Decane, 1-chloro-

$C_{10}H_{21}ClO$
5073 - 1-Decanol, 10-chloro-

$C_{10}H_{21}F$
5052 - Decane, 1-fluoro-

$C_{10}H_{21}FO$
5074 - 1-Decanol, 10-fluoro-

$C_{10}H_{21}I$
5053 - Decane, 1-iodo-

$C_{10}H_{21}N$
4285 - Cyclohexanamine, N,N-diethyl-
4385 - Cyclohexaneethanamine, N,α-dimethyl-
9691 - Piperidine, 1,2,2,6,6-pentamethyl-
9692 - Piperidine, 1-pentyl-

$C_{10}H_{21}NO$
4502 - Cyclohexanol, 3-(aminomethyl)-3,5,5-trimethyl-
5031 - Decanamide

$C_{10}H_{21}NOS$
3976 - Carbamothioic acid, butylethyl-, S-propyl ester
12047 - Vernolate

$C_{10}H_{21}NO_2$
8084 - Nitrous acid, decyl ester

$C_{10}H_{21}NO_3$
8073 - Nitric acid, decyl ester

$C_{10}H_{21}N_3O$
9603 - 1-Piperazinecarboxamide, N,N-diethyl-4-methyl-

$C_{10}H_{22}$
5033 - Decane
6409 - Heptane, 3-ethyl-2-methyl-
6410 - Heptane, 3-ethyl-3-methyl-
6411 - Heptane, 3-ethyl-4-methyl-
6412 - Heptane, 3-ethyl-5-methyl-
6413 - Heptane, 4-ethyl-2-methyl-
6414 - Heptane, 4-ethyl-3-methyl-
6415 - Heptane, 4-ethyl-4-methyl-
6416 - Heptane, 5-ethyl-2-methyl-
6427 - Heptane, 4-(1-methylethyl)-
6434 - Heptane, 4-propyl-
6437 - Heptane, 2,2,3-trimethyl-
6438 - Heptane, 2,2,4-trimethyl-
6439 - Heptane, 2,2,5-trimethyl-
6440 - Heptane, 2,2,6-trimethyl-
6441 - Heptane, 2,3,3-trimethyl-
6442 - Heptane, 2,3,4-trimethyl-
6443 - Heptane, 2,3,5-trimethyl-
6444 - Heptane, 2,3,6-trimethyl-
6445 - Heptane, 2,4,4-trimethyl-
6446 - Heptane, 2,4,5-trimethyl-
6447 - Heptane, 2,4,6-trimethyl-
6448 - Heptane, 2,5,5-trimethyl-
6449 - Heptane, 3,3,4-trimethyl-

6450 - Heptane, 3,3,5-trimethyl-
6451 - Heptane, 3,4,4-trimethyl-
6452 - Heptane, 3,4,5-trimethyl-
6767 - Hexane, 3,3-diethyl-
6768 - Hexane, 3,4-diethyl-
6816 - Hexane, 3-ethyl-2,2-dimethyl-
6817 - Hexane, 3-ethyl-2,3-dimethyl-
6818 - Hexane, 3-ethyl-2,4-dimethyl-
6819 - Hexane, 3-ethyl-2,5-dimethyl-
6820 - Hexane, 3-ethyl-3,4-dimethyl-
6821 - Hexane, 4-ethyl-2,2-dimethyl-
6822 - Hexane, 4-ethyl-2,3-dimethyl-
6823 - Hexane, 4-ethyl-2,4-dimethyl-
6824 - Hexane, 4-ethyl-3,3-dimethyl-
6839 - Hexane, 3-methyl-3-(1-methylethyl)-
6844 - Hexane, 2,2,3,3-tetramethyl-
6845 - Hexane, 2,2,3,4-tetramethyl-
6846 - Hexane, 2,2,3,5-tetramethyl-
6847 - Hexane, 2,2,4,4-tetramethyl-
6848 - Hexane, 2,2,4,5-tetramethyl-
6849 - Hexane, 2,2,5,5-tetramethyl-
6850 - Hexane, 2,3,3,4-tetramethyl-
6851 - Hexane, 2,3,3,5-tetramethyl-
6852 - Hexane, 2,3,4,4-tetramethyl-
6853 - Hexane, 2,3,4,5-tetramethyl-
6854 - Hexane, 3,3,4,4-tetramethyl-
8140 - Nonane, 2-methyl-
8141 - Nonane, 3-methyl-
8142 - Nonane, 4-methyl-
8143 - Nonane, 5-methyl-
8373 - Octane, 2,2-dimethyl-
8374 - Octane, 2,3-dimethyl-
8375 - Octane, 2,4-dimethyl-
8376 - Octane, 2,5-dimethyl-
8377 - Octane, 2,6-dimethyl-
8378 - Octane, 2,7-dimethyl-
8379 - Octane, 3,3-dimethyl-
8380 - Octane, 3,4-dimethyl-
8381 - Octane, 3,5-dimethyl-
8382 - Octane, 3,6-dimethyl-
8383 - Octane, 4,4-dimethyl-
8384 - Octane, 4,5-dimethyl-
8403 - Octane, 3-ethyl-
8404 - Octane, 4-ethyl-
8757 - Pentane, 3,3-diethyl-2-methyl-
8766 - Pentane, 2,4-dimethyl-3-(1-methylethyl)-
8815 - Pentane, 3-ethyl-2,2,3-trimethyl-
8816 - Pentane, 3-ethyl-2,3,4-trimethyl-
8836 - Pentane, 2,2,3,3,4-pentamethyl-
8837 - Pentane, 2,2,3,4,4-pentamethyl-

$C_{10}H_{22}O$

3418 - Butane, 1,1'-oxybis[3-methyl-
5070 - 1-Decanol
5071 - 2-Decanol, (±)-
5072 - 4-Decanol
6488 - 3-Heptanol, 3-ethyl-2-methyl-
6489 - 3-Heptanol, 3-ethyl-6-methyl-
6490 - 3-Heptanol, 5-ethyl-4-methyl-
6508 - 4-Heptanol, 4-propyl-
6509 - 4-Heptanol, 2,4,6-trimethyl-
6948 - 1-Hexanol, 5-methyl-2-(1-methylethyl)-
6950 - 3-Hexanol, 2,2,5,5-tetramethyl-
8166 - 1-Nonanol, 3-methyl-
8167 - 1-Nonanol, 4-methyl-
8168 - 1-Nonanol, 8-methyl
8169 - 3-Nonanol, 2-methyl-
8170 - 4-Nonanol, 4-methyl-
8171 - 5-Nonanol, 5-methyl-
8402 - Octane, 1-ethoxy-
8449 - 1-Octanol, 3,7-dimethyl-, (±)-
8450 - 2-Octanol, 2,6-dimethyl-
8451 - 3-Octanol, 2,3-dimethyl-
8452 - 3-Octanol, 2,7-dimethyl-
8453 - 3-Octanol, 3,6-dimethyl-
8454 - 3-Octanol, 3,6-dimethyl-(+-)-
8455 - 4-Octanol, 2,4-dimethyl-
8456 - 4-Octanol, 2,5-dimethyl-

8457 - 4-Octanol, 2,6-dimethyl-
8458 - 4-Octanol, 2,7-dimethyl-
8459 - 3-Octanol, 3-ethyl-
8835 - Pentane, 1,1'-oxybis-

$C_{10}H_{22}O_2$

5047 - 1,10-Decanediol
5048 - 2,9-Decanediol
5527 - Ethane, 1,2-dibutoxy
6800 - 3,4-Hexanediol, 3,4-diethyl-
8392 - 1,7-Octanediol, 3,7-dimethyl-
9154 - Peroxide, bis(1,1-dimethylpropyl)
10032 - Propane, 1,1'-[1,2-ethanediylbis(oxy)]bis[2-methyl-
10041 - Propane, 1,1'-[ethylidenebis(oxy)]bis[2-methyl-

$C_{10}H_{22}O_3$

10078 - Propane, 1,1',1''-[methylidynetris(oxy)]tris-
10079 - Propane, 2,2',2''-[methylidynetris(oxy)]tris-

$C_{10}H_{22}O_4$

5690 - Ethanol, 2-[2-(2-butoxyethoxy)ethoxy]-

$C_{10}H_{22}O_5$

8995 - 2,5,8,11,14-Pentaoxapentadecane

$C_{10}H_{22}S$

3435 - Butane, 1,1'-thiobis[2-methyl-
3436 - Butane, 1,1'-thiobis[3-methyl-
5058 - 1-Decanethiol
8843 - Pentane, 1,1'-thiobis-

$C_{10}H_{22}S_2$

5243 - Disulfide, bis(3-methylbutyl)
5257 - Disulfide, dipentyl

$C_{10}H_{23}N$

3226 - 1-Butanamine, 3-methyl-N-(3-methylbutyl)-
5032 - 1-Decanamine
8708 - 1-Pentanamine, N-pentyl-

$C_{10}H_{23}NO$

5684 - Ethanol, 2-[bis(2-methylpropyl)amino]-
5703 - Ethanol, 2-(dibutylamino)-

$C_{10}H_{23}NO_2$

5415 - Ethanamine, 2,2-diethoxy-N,N-diethyl-

$C_{10}H_{24}N_2$

5515 - 1,2-Ethanediamine, N,N,N',N'-tetraethyl-
6752 - 1,6-Hexanediamine, N,N,N',N'-tetramethyl-

$C_{10}H_{24}N_4$

9614 - 1,4-Piperazinedipropanamine

$C_{10}H_{26}N_4$

3259 - 1,4-Butanediamine, N,N'-bis(3-aminopropyl)-

$C_{10}H_{30}O_3Si_4$

11646 - Tetrasiloxane, decamethyl-

$C_{10}H_{30}O_5Si_5$

4901 - Cyclopentasiloxane, decamethyl-

$C_{11}H_7BrO_2$

7761 - 1-Naphthalenecarboxylic acid, 5-bromo-
7769 - 2-Naphthalenecarboxylic acid, 5-bromo-
7846 - 1,4-Naphthalenedione, 2-bromo-3-methyl-

$C_{11}H_7ClO$

7750 - 2-Naphthalenecarbonyl chloride

$C_{11}H_7ClO_2$

7762 - 1-Naphthalenecarboxylic acid, 5-chloro-
7763 - 1-Naphthalenecarboxylic acid, 8-chloro-

$C_{11}H_7N$

7748 - 1-Naphthalenecarbonitrile
7749 - 2-Naphthalenecarbonitrile

$C_{11}H_7NO$

7904 - Naphthalene, 1-isocyanato-

$C_{11}H_7NO_4$

7767 - 1-Naphthalenecarboxylic acid, 5-nitro-

$C_{11}H_7NS$

7906 - Naphthalene, 1-isothiocyanato-

$C_{11}H_8O$

7751 - 1-Naphthalenecarboxaldehyde
7752 - 2-Naphthalenecarboxaldehyde

$C_{11}H_8OS$

7641 - Methanone, phenyl-2-thienyl-
7642 - Methanone, phenyl-3-thienyl-

$C_{11}H_8O_2$

7618 - Methanone, 2-furanylphenyl-
7754 - 1-Naphthalenecarboxaldehyde, 2-hydroxy-

7758 - 1-Naphthalenecarboxylic acid
7759 - 2-Naphthalenecarboxylic acid
7864 - 1,4-Naphthalenedione, 2-methyl-

$C_{11}H_8O_3$

7771 - 2-Naphthalenecarboxylic acid, 1-hydroxy-
7772 - 2-Naphthalenecarboxylic acid, 3-hydroxy-
7861 - 1,4-Naphthalenedione, 2-hydroxy-3-methyl-
7863 - 1,4-Naphthalenedione, 5-hydroxy-2-methyl-

$C_{11}H_8O_4$

7856 - 1,4-Naphthalenedione, 3,5-dihydroxy-2-methyl-

$C_{11}H_8O_5$

2408 - 5H-Benzocyclohepten-5-one, 2,3,4,6-tetrahydroxy-

$C_{11}H_9Br$

7742 - Naphthalene, 1-(bromomethyl)-
7743 - Naphthalene, 2-(bromomethyl)-

$C_{11}H_9Cl$

7783 - Naphthalene, 1-(chloromethyl)-
7784 - Naphthalene, 1-chloro-2-methyl-
7785 - Naphthalene, 2-(chloromethyl)-

$C_{11}H_9Cl_2NO_2$

510 - Barban

$C_{11}H_9N$

10997 - Pyridine, 2-phenyl-
10998 - Pyridine, 3-phenyl-
10999 - Pyridine, 4-phenyl-

$C_{11}H_9NO$

7637 - Methanone, phenyl-1H-pyrrol-2-yl-
7757 - 1-Naphthalenecarboxamide

$C_{11}H_9NO_2$

7760 - 1-Naphthalenecarboxylic acid, 5-amino-
7768 - 2-Naphthalenecarboxylic acid, 3-amino-
7925 - Naphthalene, 1-methyl-4-nitro-
7926 - Naphthalene, 2-methyl-1-nitro-

$C_{11}H_9NO_3$

11234 - 4-Quinolinecarboxylic acid, 6-methoxy-

$C_{11}H_{10}$

7915 - Naphthalene, 1-methyl-
7916 - Naphthalene, 2-methyl-

$C_{11}H_{10}NO_3$

7695 - Morpholine, 4-(2-hydroxybenzoyl)-

$C_{11}H_{10}N_2O$

90 - Acetamide, N-4-quinolinyl-
12022 - Urea, 1-naphthalenyl-
12023 - Urea, 2-naphthalenyl-

$C_{11}H_{10}N_2O_3$

11100 - 2,4,6(1H,3H,5H)-Pyrimidinetrione, 5-methyl-5-phenyl-

$C_{11}H_{10}N_2S$

11807 - Thiourea, 1-naphthalenyl-

$C_{11}H_{10}O$

4951 - 2-Cyclopenten-1-one, 3-phenyl-
7908 - 1-Naphthalenemethanol
7909 - 2-Naphthalenemethanol
7913 - Naphthalene, 1-methoxy-
7914 - Naphthalene, 2-methoxy-
8030 - 1-Naphthalenol, 4-methyl-
8031 - 2-Naphthalenol, 1-methyl-
8657 - 2,4-Pentadienal, 5-phenyl-

$C_{11}H_{10}O_2$

2820 - 2H-1-Benzopyran-2-one, 3-ethyl-
6158 - Furan, 2-(phenoxymethyl)-
7764 - 1-Naphthalenecarboxylic acid, 3,4-dihydro-
10741 - 2-Propynoic acid, 3-phenyl-, ethyl ester

$C_{11}H_{10}O_3$

2440 - 2-Benzofurancarboxylic acid, ethyl ester
10781 - 2H-Pyran-2,6(3H)-dione, dihydro-3-phenyl-
10782 - 2H-Pyran-2,6(3H)-dione, dihydro-4-phenyl-

$C_{11}H_{10}O_4$

2818 - 2H-1-Benzopyran-2-one, 5,7-dimethoxy-

$C_{11}H_{10}S$

11774 - Thiophene, 2-methyl-5-phenyl-

$C_{11}H_{11}BrN_2O$

10846 - 3H-Pyrazol-3-one, 4-bromo-1,2-dihydro-1,5-dimethyl-2-phenyl-

$C_{11}H_{11}BrO_2$
10613 - 2-Propenoic acid, 2-bromo-3-phenyl-, ethyl ester
10614 - 2-Propenoic acid, 3-bromo-3-phenyl-, ethyl ester, (Z)-

$C_{11}H_{11}Cl_3O_3$
259 - Acetic acid, (2,4,5-trichlorophenoxy)-, 1-methylethyl ester

$C_{11}H_{11}N$
7720 - 1-Naphthalenamine, 4-methyl-
7721 - 1-Naphthalenamine, N-methyl-
7722 - 2-Naphthalenamine, N-methyl-
7907 - 2-Naphthalenemethanamine
11136 - 1H-Pyrrole, 1-(phenylmethyl)-
11254 - Quinoline, 2,3-dimethyl-
11255 - Quinoline, 2,4-dimethyl-
11256 - Quinoline, 2,5-dimethyl-
11257 - Quinoline, 2,6-dimethyl-
11258 - Quinoline, 2,7-dimethyl-
11259 - Quinoline, 2,8-dimethyl-
11260 - Quinoline, 3,4-dimethyl-
11261 - Quinoline, 3,8-dimethyl-
11262 - Quinoline, 4,6-dimethyl-
11263 - Quinoline, 5,8-dimethyl-
11264 - Quinoline, 6,8-dimethyl-
11266 - Quinoline, 2-ethyl-
11267 - Quinoline, 3-ethyl-
11268 - Quinoline, 6-ethyl-

$C_{11}H_{11}NO$
10442 - 2-Propanone, 1-(1H-indol-3-yl)-
11316 - 4-Quinolinol, 2,8-dimethyl-
11317 - 6-Quinolinol, 2,4-dimethyl-
11318 - 8-Quinolinol, 2,4-dimethyl-

$C_{11}H_{11}NO_2$
893 - Benzeneacetic acid, α-cyano-, ethyl ester
11161 - 2,5-Pyrrolidinedione, 1-methyl-3-phenyl-
11163 - 2,5-Pyrrolidinedione, 1-(phenylmethyl)-

$C_{11}H_{11}NO_4S$
7407 - Isoxazolium, 2-ethyl-5-(3-sulfophenyl)-, hydroxide, inner salt
11293 - 5-Quinolinesulfonic acid, 8-ethoxy-

$C_{11}H_{11}NO_6$
7511 - Methanediol, (3-nitrophenyl)-, diacetate (ester)

$C_{11}H_{11}N_3O_2S$
2203 - Benzenesulfonamide, 4-amino-N-2-pyridinyl-

$C_{11}H_{12}$
7816 - Naphthalene, 1,2-dihydro-3-methyl-
7817 - Naphthalene, 1,2-dihydro-4-methyl-

$C_{11}H_{12}Cl_2N_2O_5$
52 - Acetamide, 2,2-dichloro-N-[2-hydroxy-1-(hydroxymethyl)-2-(4-nitrophenyl)ethyl]-, [R-(R*,R*)]-

$C_{11}H_{12}Cl_2O_3$
173 - Acetic acid, (2,4-dichlorophenoxy)-, 1-methylethyl ester

$C_{11}H_{12}Cl_3N$
1583 - Benzeneethanamine, α-methyl-N-(2,2,2-trichloroethylidene)-

$C_{11}H_{12}NO_4PS_2$
9515 - Phosmet

$C_{11}H_{12}N_2$
7800 - 1,4-Naphthalenediamine, 2-methyl-
10829 - 1H-Pyrazole, 3,5-dimethyl-1-phenyl-

$C_{11}H_{12}N_2O$
10847 - 3H-Pyrazol-3-one, 1,2-dihydro-1,5-dimethyl-2-phenyl-
11192 - Pyrrolo[2,1-b]quinazolin-3-ol, 1,2,3,9-tetrahydro-, (R)-

$C_{11}H_{12}N_2O_2$
11922 - L-Tryptophan

$C_{11}H_{12}N_2O_6$
2592 - Benzoic acid, 3,5-dinitro-, butyl ester

$C_{11}H_{12}N_4O_2S$
2199 - Benzenesulfonamide, 4-amino-N-(4-methyl-2-pyrimidinyl)-

$C_{11}H_{12}N_4O_3S$
2197 - Benzenesulfonamide, 4-amino-N-(6-methoxy-3-pyridazinyl)-

$C_{11}H_{12}O$
2407 - 5H-Benzocyclohepten-5-one, 6,7,8,9-tetrahydro-
3190 - Butanal, 2-(phenylmethylene)-
7288 - 1H-Inden-1-one, 2,3-dihydro-3,3-dimethyl-
7605 - Methanone, cyclobutylphenyl-
8049 - 1(2H)-Naphthalenone, 3,4-dihydro-2-methyl-
8050 - 1(2H)-Naphthalenone, 3,4-dihydro-4-methyl-
9110 - 1-Penten-3-one, 1-phenyl-

$C_{11}H_{12}O_2$
7778 - 2-Naphthalenecarboxylic acid, 1,2,3,4-tetrahydro-
7779 - 2-Naphthalenecarboxylic acid, 5,6,7,8-tetrahydro-
8048 - 1(2H)-Naphthalenone, 3,4-dihydro-6-methoxy-
8804 - 1,4-Pentanedione, 1-phenyl-
10685 - 2-Propenoic acid, 3-phenyl-, ethyl ester, (E)-
10706 - 2-Propen-1-ol, 3-phenyl-, acetate, (E)-

$C_{11}H_{12}O_3$
2141 - Benzenepropanoic acid, β-oxo-, ethyl ester
2428 - 1,3-Benzodioxole, 4-methoxy-6-(2-propenyl)-
6138 - 2-Furanol, tetrahydro-, benzoate
10647 - 2-Propenoic acid, 3-(2-methoxyphenyl)-, methyl ester
10681 - 2-Propenoic acid, 2-phenoxyethyl ester

$C_{11}H_{12}O_4$
7512 - Methanediol, phenyl-, diacetate
9976 - Propanedioic acid, phenyl-, dimethyl ester

$C_{11}H_{13}ClF_3N_3O_4S_3$
2857 - 2H-1,2,4-Benzothiadiazine-7-sulfonamide, 6-chloro-3,4-dihydro-2-methyl-3-[[(2,2,2-trifluoroethyl)thio]methyl]-

$C_{11}H_{13}ClO$
2889 - Benzoyl chloride, 4-butyl-
2895 - Benzoyl chloride, 4-(1,1-dimethylethyl)-
9255 - Phenol, 4-chloro-2-cyclopentyl-

$C_{11}H_{13}ClO_3$
3493 - Butanoic acid, 4-(4-chloro-2-methylphenoxy)-

$C_{11}H_{13}F_3N_2O_3S$
7427 - Mefluidide

$C_{11}H_{13}F_3N_4O_4$
5181 - Dinitramine

$C_{11}H_{13}N$
963 - Benzeneacetonitrile, α-(1-methylethyl)-
970 - Benzeneacetonitrile, α-propyl-
1823 - Benzenemethanamine, N-methyl-N-2-propynyl-
7309 - 1H-Indole, 3-ethyl-2-methyl-
7322 - 1H-Indole, 1,2,3-trimethyl-
7323 - 1H-Indole, 2,3,5-trimethyl-
10922 - Pyridine, 4-(3-cyclohexen-1-yl)-

$C_{11}H_{13}NO$
3703 - 3-Butenamide, 3-methyl-4-phenyl-
7332 - 2H-Indol-2-one, 1-ethyl-1,3-dihydro-3-methyl-

$C_{11}H_{13}NO_2$
5215 - 1,3-Dioxolo[4,5-g]isoquinoline, 5,6,7,8-tetrahydro-6-methyl-

$C_{11}H_{13}NO_3$
5217 - 1,3-Dioxolo[4,5-g]isoquinolin-5-ol, 5,6,7,8-tetrahydro-6-methyl-
7398 - 1(2H)-Isoquinolinone, 3,4-dihydro-6,7-dimethoxy-

$C_{11}H_{13}NO_4$
514 - Bendiocarb
2706 - Benzoic acid, 4-nitro-, butyl ester
8588 - 2-Oxazolidinone, 5-[(2-methoxyphenoxy)methyl]-

$C_{11}H_{13}N_3O$
10845 - 3H-Pyrazol-3-one, 4-amino-1,2-dihydro-1,5-dimethyl-2-phenyl-

$C_{11}H_{13}N_3O_3S$
2195 - Benzenesulfonamide, 4-amino-N-(3,4-dimethyl-5-isoxazolyl)-

$C_{11}H_{14}$
1254 - Benzene, cyclopentyl-
1643 - Benzene, 1-ethenyl-4-(1-methylethyl)-
1647 - Benzene, 2-ethenyl-1,3,5-trimethyl-
1699 - Benzene, (1-ethyl-1-propenyl)-
1700 - Benzene, (1-ethyl-2-propenyl)-
1950 - Benzene, 1-methyl-1-butenyl)-
2104 - Benzene, 1-pentenyl-
2105 - Benzene, 3-pentenyl-
2106 - Benzene, 4-pentenyl-
4748 - 1,3,5,7-Cyclooctatetraene, propyl-
7251 - 1H-Indene, 2,3-dihydro-1,1-dimethyl-
7252 - 1H-Indene, 2,3-dihydro-1,2-dimethyl-
7253 - 1H-Indene, 2,3-dihydro-4,7-dimethyl-
7254 - 1H-Indene, 2,3-dihydro-5,6-dimethyl-
7273 - 1H-Indene, 1-ethyl-2,3-dihydro-
7995 - Naphthalene, 1,2,3,4-tetrahydro-1-methyl-
7996 - Naphthalene, 1,2,3,4-tetrahydro-2-methyl-
7997 - Naphthalene, 1,2,3,4-tetrahydro-5-methyl-
7998 - Naphthalene, 1,2,3,4-tetrahydro-6-methyl-

$C_{11}H_{14}ClNO$
9771 - Propachlor

$C_{11}H_{14}ClNO_2$
607 - Benzamide, N-butyl-4-chloro-2-hydroxy-

$C_{11}H_{14}N_2$
7310 - 1H-Indole-3-methanamine, N,N-dimethyl-

$C_{11}H_{14}N_2O$
7304 - 1H-Indole-3-ethanamine, 5-methoxy-

$C_{11}H_{14}N_2O_4$
84 - Acetamide, N-(5-nitro-2-propoxyphenyl)-

$C_{11}H_{14}O$
598 - Benzaldehyde, 2,3,5,6-tetramethyl-
1665 - Benzene, 1-ethoxy-2-(1-propenyl)-
1666 - Benzene, (3-ethoxy-1-propenyl)-
3671 - 1-Butanone, 3-methyl-1-phenyl-
3672 - 1-Butanone, 1-(4-methylphenyl)-
4871 - Cyclopentanol, 1-phenyl-
5848 - Ethanone, 1-[4-(1-methylethyl)phenyl]-
5875 - Ethanone, 1-(2,4,5-trimethylphenyl)-
5876 - Ethanone, 1-(2,4,6-trimethylphenyl)-
5877 - Ethanone, 1-(3,4,5-trimethylphenyl)-
8986 - 1-Pentanone, 1-phenyl-
10431 - 1-Propanone, 2,2-dimethyl-1-phenyl-

$C_{11}H_{14}O_2$
900 - Benzeneacetic acid, α-ethyl-, methyl ester
930 - Benzeneacetic acid, α-(1-methylethyl)-
943 - Benzeneacetic acid, α-propyl-, (±)-
1104 - Benzenebutanoic acid, γ-methyl-
1449 - Benzene, 1,2-dimethoxy-4-(1-propenyl)-
1450 - Benzene, 1,2-dimethoxy-4-(2-propenyl)-
1451 - Benzene, 1,2-dimethoxy-4-(1-propenyl)-, (Z)-
2099 - Benzenepentanoic acid
2130 - Benzenepropanoic acid, β,β-dimethyl-
2131 - Benzenepropanoic acid, ethyl ester
2532 - Benzoic acid, butyl ester
2588 - Benzoic acid, 4-(1,1-dimethylethyl)-
2694 - Benzoic acid, 4-(1-methylethyl)-, methyl ester
2701 - Benzoic acid, 2-methylpropyl ester
3585 - Butanoic acid, phenylmethyl ester
5194 - 1,3-Dioxane, 4-methyl-4-phenyl-
9496 - Phenol, 2,4,5-trimethyl-, acetate
10307 - Propanoic acid, 2-methyl-, phenylmethyl ester
10324 - Propanoic acid, 2-phenylethyl ester
10436 - 1-Propanone, 3-ethoxy-1-phenyl-
10447 - 1-Propanone, 1-(2-hydroxy-5-methylphenyl)-2-methyl-
10448 - 1-Propanone, 1-(4-hydroxy-3-methylphenyl)-2-methyl-

$C_{11}H_{14}O_3$
538 - Benzaldehyde, 3,4-diethoxy-
923 - Benzeneacetic acid, 4-methoxy-, ethyl ester
1128 - Benzenecarboperoxoic acid, 1,1-dimethylethyl ester
2598 - Benzoic acid, 2-ethoxy-, ethyl ester
2599 - Benzoic acid, 4-ethoxy-, ethyl ester
2614 - Benzoic acid, 2-hydroxy-, butyl ester

2615 - Benzoic acid, 4-hydroxy-, butyl ester
2641 - Benzoic acid, 2-hydroxy-3-methyl-6-(1-methylethyl)-
2642 - Benzoic acid, 2-hydroxy-6-methyl-3-(1-methylethyl)-
2643 - Benzoic acid, 2-hydroxy-, 2-methylpropyl ester
3663 - 2-Butanone, 4-(4-hydroxy-3-methoxyphenyl)-
10321 - Propanoic acid, 2-phenoxy-, ethyl ester
10322 - Propanoic acid, 3-phenoxy-, ethyl ester

$C_{11}H_{14}O_4$
2578 - Benzoic acid, 3,4-dimethoxy-, ethyl ester
5874 - Ethanone, 1-(2,3,4-trimethoxyphenyl)-

$C_{11}H_{15}BrClO_3PS$
9762 - Profenofos

$C_{11}H_{15}BrN_2O_3$
11079 - 2,4,6(1H,3H,5H)-Pyrimidinetrione, 5-(2-bromo-2-propenyl)-1-methyl-5-(1-methylethyl)-
11080 - 2,4,6(1H,3H,5H)-Pyrimidinetrione, 5-(2-bromo-2-propenyl)-5-(1-methylpropyl)-

$C_{11}H_{15}Cl_2O_3PS_2$
4092 - Chlorthiophos

$C_{11}H_{15}N$
9693 - Piperidine, 1-phenyl-
9694 - Piperidine, 3-phenyl-
9695 - Piperidine, 4-phenyl-
11172 - Pyrrolidine, 1-(phenylmethyl)-

$C_{11}H_{15}NO$
74 - Acetamide, N-(1-methylethyl)-N-phenyl-
88 - Acetamide, N-phenyl-N-propyl-
539 - Benzaldehyde, 4-(diethylamino)-
882 - Benzeneacetamide, α-(1-methylethyl)-
7697 - Morpholine, 3-methyl-2-phenyl-
7698 - Morpholine, 4-(4-methylphenyl)-
7700 - Morpholine, 4-(phenylmethyl)-
8960 - 1-Pentanone, 1-(4-aminophenyl)-

$C_{11}H_{15}NO_2$
2488 - Benzoic acid, 4-amino-, butyl ester
2505 - Benzoic acid, 4-amino-, 2-methylpropyl ester
3968 - Carbamic acid, phenyl-, 2-methylpropyl ester
6272 - Glycine, N-(phenylmethyl)-, ethyl ester
7397 - 6-Isoquinolinol, 1,2,3,4-tetrahydro-7-methoxy-1-methyl-, (S)-
9512 - L-Phenylalanine, ethyl ester

$C_{11}H_{15}NO_2S$
7646 - Methiocarb

$C_{11}H_{15}NO_3$
7393 - 8-Isoquinolinol, 1,2,3,4-tetrahydro-6,7-dimethoxy-
9796 - Propanamide, N-(4-ethoxyphenyl)-2-hydroxy-
10722 - Propoxur

$C_{11}H_{15}N_2NaO_2S$
11063 - 4,6(1H,5H)-Pyrimidinedione, dihydro-5-(2-methylpropyl)-5-(2-propenyl)-2-thioxo-, monosodium salt

$C_{11}H_{16}$
1121 - Benzene, 1-butyl-2-methyl-
1122 - Benzene, 1-butyl-3-methyl-
1123 - Benzene, 1-butyl-4-methyl-
1416 - Benzene, 1,2-diethyl-3-methyl-
1417 - Benzene, 1,2-diethyl-4-methyl-
1418 - Benzene, 1,3-diethyl-2-methyl-
1419 - Benzene, 1,3-diethyl-5-methyl-
1420 - Benzene, 1,4-diethyl-2-methyl-
1421 - Benzene, 2,4-diethyl-1-methyl-
1464 - Benzene, 1-(1,1-dimethylethyl)-2-methyl-
1465 - Benzene, 1-(1,1-dimethylethyl)-3-methyl-
1466 - Benzene, 1-(1,1-dimethylethyl)-4-methyl-
1469 - Benzene, 1,2-dimethyl-3-(1-methylethyl)-
1470 - Benzene, 1,2-dimethyl-4-(1-methylethyl)-
1471 - Benzene, 1,3-dimethyl-2-(1-methylethyl)-
1472 - Benzene, 2,4-dimethyl-1-(1-methylethyl)-
1480 - Benzene, (1,1-dimethylpropyl)-
1481 - Benzene, (1,2-dimethylpropyl)-
1482 - Benzene, 1,3-dimethyl-2-propyl-
1483 - Benzene, 1,3-dimethyl-5-propyl-
1484 - Benzene, (2,2-dimethylpropyl)-
1485 - Benzene, 1,2-dimethyl-3-propyl-

1486 - Benzene, 1,4-dimethyl-2-propyl-
1487 - Benzene, 2,4-dimethyl-1-propyl-
1688 - Benzene, 1-ethyl-2-(1-methylethyl)-
1689 - Benzene, 1-ethyl-3-(1-methylethyl)-
1690 - Benzene, 1-ethyl-4-(1-methylethyl)-
1701 - Benzene, (1-ethylpropyl)-
1702 - Benzene, 1-ethyl-2-propyl-
1703 - Benzene, 1-ethyl-4-propyl-
1708 - Benzene, 1-ethyl-2,3,4-trimethyl-
1709 - Benzene, 1-ethyl-2,3,5-trimethyl-
1710 - Benzene, 1-ethyl-2,4,5-trimethyl-
1711 - Benzene, 2-ethyl-1,3,4-trimethyl-
1712 - Benzene, 2-ethyl-1,3,5-trimethyl-
1713 - Benzene, 5-ethyl-1,2,3-trimethyl-
1952 - Benzene, (1-methylbutyl)-
1953 - Benzene, (2-methylbutyl)-
1954 - Benzene, (3-methylbutyl)-
1994 - Benzene, 1-methyl-2-(1-methylpropyl)-
1995 - Benzene, 1-methyl-2-(2-methylpropyl)-
1996 - Benzene, 1-methyl-3-(1-methylpropyl)-
1997 - Benzene, 1-methyl-3-(2-methylpropyl)-
1998 - Benzene, 1-methyl-4-(1-methylpropyl)-
1999 - Benzene, 1-methyl-4-(2-methylpropyl)-
2097 - Benzene, pentamethyl-
2107 - Benzene, pentyl-
8341 - 1,5-Octadien-3-yne, 5-propyl-
11934 - 1,10-Undecadiyne
12057 - o-Xylene, 4-propyl-

$C_{11}H_{16}ClN_3O_2$
6001 - Formetanate hydrochloride

$C_{11}H_{16}ClN_3O_4S_2$
2856 - 2H-1,2,4-Benzothiadiazine-7-sulfonamide, 6-chloro-3,4-dihydro-3-(2-methylpropyl)-, 1,1-dioxide

$C_{11}H_{16}ClO_2PS_3$
4037 - Carbophenothion

$C_{11}H_{16}N_2$
9621 - Piperazine, 1-(phenylmethyl)-

$C_{11}H_{16}N_2O_2$
6151 - 2(3H)-Furanone, 3-ethyldihydro-4-[(1-methyl-1H-imidazol-5-yl)methyl]-, (3S-cis)-

$C_{11}H_{16}N_2O_3$
11099 - 2,4,6(1H,3H,5H)-Pyrimidinetrione, 1-methyl-5-(1-methylethyl)-5-(2-propenyl)-
11101 - 2,4,6(1H,3H,5H)-Pyrimidinetrione, 5-(2-methylpropyl)-5-(2-propenyl)-

$C_{11}H_{16}O$
363 - Anisole, 3-isopropyl-2-methyl-
1106 - Benzenebutanol, α-methyl-
1107 - Benzenebutanol, β-methyl-, (±)-
1115 - Benzene, (butoxymethyl)-
1116 - Benzene, 1-butoxy-3-methyl-
1117 - Benzene, 1-butoxy-4-methyl-
1462 - Benzene, 1-(1,1-dimethylethyl)-4-methoxy-
1611 - Benzeneethanol, β-(1-methylethyl)-
1616 - Benzeneethanol, α,α,β-trimethyl-
1617 - Benzeneethanol, α,β,β-trimethyl-
1667 - Benzene, 1-ethoxy-2-propyl-
1846 - Benzenemethanol, α-butyl-
1855 - Benzenemethanol, α,α-diethyl-
1864 - Benzenemethanol, α-(1,1-dimethylethyl)-
1865 - Benzenemethanol, 4-(1,1-dimethylethyl)-
1904 - Benzenemethanol, α-methyl-α-propyl-
1905 - Benzenemethanol, α-(2-methylpropyl)-
2033 - Benzene, [(2-methylpropoxy)methyl]-
2101 - Benzenepentanol
2109 - Benzene, (pentyloxy)-
2150 - Benzenepropanol, α,α-dimethyl-
2151 - Benzenepropanol, β,β-dimethyl-
2152 - Benzenepropanol, α-ethyl-, (S)-
4949 - 2-Cyclopenten-1-one, 3-methyl-2-(2-pentenyl)-, (Z)-
9250 - Phenol, 2-butyl-4-methyl-
9330 - Phenol, 2-(1,1-dimethylethyl)-4-methyl-
9331 - Phenol, 2-(1,1-dimethylethyl)-5-methyl-
9332 - Phenol, 2-(1,1-dimethylethyl)-6-methyl-
9333 - Phenol, 4-(1,1-dimethylethyl)-2-methyl-

9339 - Phenol, 4-(1,1-dimethylpropyl)-
9404 - Phenol, 4-(3-methylbutyl)-
9421 - Phenol, 5-methyl-2-(1-methylpropyl)-
9443 - Phenol, pentamethyl-
9444 - Phenol, 4-pentyl-

$C_{11}H_{16}O_2$
1105 - Benzenebutanol, 4-methoxy-
1410 - Benzene, 1,4-diethoxy-2-methyl-
1442 - Benzene, 1,4-dimethoxy-2-(1-methylethyl)-
1534 - 1,3-Benzenediol, 4-(3-methylbutyl)-
1544 - 1,3-Benzenediol, 4-pentyl-
9329 - Phenol, (1,1-dimethylethyl)-4-methoxy-
11864 - Tricyclo[2.2.1.0.2,6]heptane-1-carboxylic acid, 7,7-dimethyl-, methyl ester

$C_{11}H_{16}O_3$
1409 - Benzene, (diethoxymethyl)-
2917 - Bicyclo[2.2.1]heptane-2-carboxylic acid, 4,7,7-trimethyl-3-oxo-
4641 - 2-Cyclohexene-1-carboxylic acid, 2,6-dimethyl-4-oxo-, ethyl ester
6058 - 2-Furancarboxylic acid, hexyl ester

$C_{11}H_{16}O_3S$
2230 - Benzenesulfonic acid, 4-methyl-, butyl ester

$C_{11}H_{16}O_4$
4915 - 1-Cyclopentene-1,2-dicarboxylic acid, diethyl ester
4916 - 1-Cyclopentene-1,3-dicarboxylic acid, diethyl ester

$C_{11}H_{17}Cl_3N_2O_2S$
7220 - 2,4-Imidazolidinedione, 5-(1-ethylpentyl)-3-[(trichloromethyl)thio]-

$C_{11}H_{17}N$
362 - Aniline, p-isopentyl-
696 - Benzenamine, N,N-diethyl-2-methyl-
697 - Benzenamine, N,N-diethyl-4-methyl-
783 - Benzenamine, N-(3-methylbutyl)-
833 - Benzenamine, 2,3,4,5,6-pentamethyl-
1576 - Benzeneethanamine, N-methyl-α-methyl-

$C_{11}H_{17}NO$
1863 - Benzenemethanol, α-[1-(dimethylamino)ethyl]-, [R-(R*,S*)]-
1871 - Benzenemethanol, α-[1-(ethylamino)ethyl]-, (R*,S*)-
5724 - Ethanol, 2-[ethyl(3-methylphenyl)amino]-
10392 - 1-Propanol, 3-[methyl(phenylmethyl)amino]-

$C_{11}H_{17}NO_3$
1522 - 1,2-Benzenediol, 4-[1-hydroxy-2-[(1-methylethyl)amino]ethyl]-
1586 - Benzeneethanamine, 3,4,5-trimethoxy-

$C_{11}H_{17}N_3O_3$
11096 - 2,4,6(1H,3H,5H)-Pyrimidinetrione, 5-ethyl-5-(1-piperidinyl)-

$C_{11}H_{17}N_3O_3S$
2192 - Benzenesulfonamide, 4-amino-N-[(butylamino)carbonyl]-

$C_{11}H_{17}O_3P$
9545 - Phosphonic acid, (phenylmethyl)-, diethyl ester

$C_{11}H_{17}O_4PS_2$
5954 - Fensulfothion

$C_{11}H_{18}$
7792 - Naphthalene, decahydro-2-methylene-, trans-
11983 - 1-Undecen-3-yne

$C_{11}H_{18}N_2$
5504 - 1,2-Ethanediamine, N,N-dimethyl-N'-(phenylmethyl)-
7162 - Hydrazine, 1-(3-methylbutyl)-1-phenyl-

$C_{11}H_{18}N_2O_3$
11092 - 2,4,6(1H,3H,5H)-Pyrimidinetrione, 5-ethyl-5-(3-methylbutyl)-

$C_{11}H_{18}N_4O_2$
9728 - Pirimicarb

$C_{11}H_{18}O$
2959 - Bicyclo[2.2.1]heptan-2-one, 1,4,7,7-tetramethyl-
2969 - Bicyclo[3.1.1]hept-2-ene-2-ethanol, 6,6-dimethyl-

4950 - 2-Cyclopenten-1-one, 3-methyl-2-pentyl-

$C_{11}H_{18}O_2$

2949 - Bicyclo[2.2.1]heptan-2-ol, 1,7,7-trimethyl-, formate, (1R-endo)-

3131 - 3-Bornanecarboxylic acid

4965 - Cyclopropanecarboxylic acid, 2,2-dimethyl-3-(2-methyl-1-propenyl)-, methyl ester, trans-(±)-

5026 - 2,4-Decadienoic acid, methyl ester, (E,E)-

5027 - 2,4-Decadienoic acid, methyl ester, (E,Z)-

5028 - 2,4-Decadienoic acid, methyl ester, (Z,E)-

5029 - 2,4-Decadienoic acid, methyl ester, (Z,Z)-

8337 - 1,6-Octadien-3-ol, 3,7-dimethyl-

8338 - 2,6-Octadien-1-ol, 3,7-dimethyl-, formate, (E)-

11988 - 6-Undecynoic acid

11989 - 9-Undecynoic acid

$C_{11}H_{18}O_2Si$

11407 - Silane, diethoxymethylphenyl-

$C_{11}H_{18}O_3Si$

11427 - Silane, ethenyltris(2-propenyloxy)-

$C_{11}H_{18}O_4$

4791 - 1,1-Cyclopentanedicarboxylic acid, diethyl ester

$C_{11}H_{18}O_5$

8771 - Pentanedioic acid, 2-acetyl-, diethyl ester

9972 - Propanedioic acid, (1-oxobutyl)-, diethyl ester

$C_{11}H_{18}O_6$

5654 - 1,1,2-Ethanetricarboxylic acid, triethyl ester

$C_{11}H_{19}ClO$

11982 - 10-Undecenoyl chloride

$C_{11}H_{19}NOS$

8528 - Octhilinone

$C_{11}H_{19}NO_2$

8148 - Nonanoic acid, 9-cyano-, methyl ester

$C_{11}H_{19}N_3O$

5932 - Ethirimol

$C_{11}H_{20}$

4340 - Cyclohexane, cyclopentyl-

8201 - 4-Nonyne, 3,3-dimethyl-

11537 - Spiro[5.5]undecane

11984 - 1-Undecyne

11985 - 2-Undecyne

11986 - 4-Undecyne

11987 - 5-Undecyne

$C_{11}H_{20}ClN_5$

11828 - 1,3,5-Triazine-2,4-diamine, 6-chloro-N,N,N',N'-tetraethyl-

$C_{11}H_{20}N_2O_3$

7224 - 4-Imidazolidinehexanoic acid, 5-methyl-2-oxo-, methyl ester, (4R-cis)-

$C_{11}H_{20}N_3O_3PS$

9730 - Pirimiphos-methyl

$C_{11}H_{20}O$

4592 - Cyclohexanone, 4-(1,1-dimethylpropyl)-

4888 - Cyclopentanone, 3-methyl-2-(3-methylbutyl)-

$C_{11}H_{20}O_2$

6152 - 2(3H)-Furanone, 5-heptyldihydro-

6404 - 3,5-Heptanedione, 2,2,6,6-tetramethyl-

8149 - Nonanoic acid, ethenyl ester

10636 - 2-Propenoic acid, 2-ethylhexyl ester

11974 - 10-Undecenoic acid

11975 - 9-Undecenoic acid, (E)-

11976 - 9-Undecenoic acid, (Z)-

$C_{11}H_{20}O_3$

8436 - Octanoic acid, 2-methyl-3-oxo-, ethyl ester

$C_{11}H_{20}O_4$

6394 - Heptanedioic acid, diethyl ester

6790 - Hexanedioic acid, 3-methyl-, diethyl ester

8134 - Nonanedioic acid, dimethyl ester

9925 - Propanedioic acid, bis(1,1-dimethylethyl) ester

9930 - Propanedioic acid, butyl-, diethyl ester

9938 - Propanedioic acid, dibutyl ester

9940 - Propanedioic acid, diethyl-, diethyl ester

9967 - Propanedioic acid, (1-methylpropyl)-, diethyl ester

9969 - Propanedioic acid, (2-methylpropyl)-, diethyl ester

9971 - Propanedioic acid, octyl-

$C_{11}H_{20}O_{10}$

6234 - D-Glucose, 6-O-α-L-arabinopyranosyl-

6237 - D-Glucose, 6-O-β-D-xylopyranosyl-

$C_{11}H_{21}BrO_2$

11944 - Undecanoic acid, 11-bromo-

11945 - Undecanoic acid, 2-bromo-

$C_{11}H_{21}FO_2$

11947 - Undecanoic acid, 11-fluoro-

$C_{11}H_{21}N$

11941 - Undecanenitrile

$C_{11}H_{21}NOS$

4168 - Cycloate

$C_{11}H_{21}NO_4$

330 - β-Alanine, N-(3-ethoxy-3-oxopropyl)-N-methyl-, ethyl ester

$C_{11}H_{21}N_5S$

5222 - Dipropetryn

$C_{11}H_{22}$

4427 - Cyclohexane, (3-methylbutyl)-

4465 - Cyclohexane, pentyl-

4823 - Cyclopentane, hexyl-

11966 - 1-Undecene

11967 - 3-Undecene

11968 - 2-Undecene, (E)-

11969 - 2-Undecene, (Z)-

11970 - 4-Undecene, (E)-

11971 - 4-Undecene, (Z)-

11972 - 5-Undecene, (E)-

11973 - 5-Undecene, (Z)-

$C_{11}H_{22}BrF$

11939 - Undecane, 1-bromo-11-fluoro-

$C_{11}H_{22}Br_2$

11940 - Undecane, 1,11-dibromo-

$C_{11}H_{22}N_2$

9683 - Piperidine, 1,1'-methylenebis-

$C_{11}H_{22}O$

11935 - Undecanal

11958 - 2-Undecanone

11959 - 3-Undecanone

11960 - 4-Undecanone

11961 - 5-Undecanone

11962 - 6-Undecanone

11979 - 10-Undecen-1-ol

11980 - 2-Undecen-4-ol

11981 - 1-Undecen-1-ol

$C_{11}H_{22}O_2$

220 - Acetic acid, nonyl ester

3527 - Butanoic acid, heptyl ester

5066 - Decanoic acid, methyl ester

6459 - Heptanoic acid, butyl ester

6467 - Heptanoic acid, 2-methylpropyl ester

6897 - Hexanoic acid, 3-methylbutyl ester

6907 - Hexanoic acid, pentyl ester

8150 - Nonanoic acid, ethyl ester

8435 - Octanoic acid, 1-methylethyl ester

8441 - Octanoic acid, propyl ester

8882 - Pentanoic acid, hexyl ester

10313 - Propanoic acid, octyl ester

11943 - Undecanoic acid

$C_{11}H_{22}O_3$

3635 - 1-Butanol, 3-methyl-, carbonate (2:1)

11948 - Undecanoic acid, 11-hydroxy-

$C_{11}H_{22}O_6$

6228 - α-D-Glucopyranoside, methyl 2,3,4,6-tetra-O-methyl-

$C_{11}H_{23}N$

9671 - Piperidine, 1-(1-ethyl-1-methylpropyl)-

9672 - Piperidine, 1-hexyl-

$C_{11}H_{23}NOS$

11586 - Sutan

$C_{11}H_{24}$

5054 - Decane, 2-methyl-

5055 - Decane, 3-methyl-

5056 - Decane, 4-methyl-

8420 - Octane, 2,2,4-trimethyl-

8421 - Octane, 2,4,7-trimethyl

11938 - Undecane

$C_{11}H_{24}O$

8164 - 5-Nonanol, 2,8-dimethyl-

11950 - 1-Undecanol

11951 - 2-Undecanol, (±)-

11952 - 3-Undecanol, (R)-

11953 - 5-Undecanol

11954 - 6-Undecanol

$C_{11}H_{24}O_3Si$

11426 - Silane, ethenyltris(1-methylethoxy)-

$C_{11}H_{24}S$

11942 - 1-Undecanethiol

$C_{11}H_{25}N$

11937 - 1-Undecanamine

$C_{11}H_{26}O_3Si$

11480 - Silane, triethoxypentyl-

$C_{11}H_{26}O_6P_2$

9546 - Phosphonic acid, 1,3-propanediylbis-, tetraethyl ester

$C_{11}H_{28}Br_2N_2$

8738 - 1,5-Pentanediaminium, N,N,N,N',N',N'-hexamethyl-, dibromide

$C_{12}F_{10}$

3027 - 1,1'-Biphenyl, 2,2',3,3',4,4',5,5',6,6'-decafluoro-

$C_{12}F_{27}N$

3228 - 1-Butanamine, 1,1,2,2,3,3,4,4,4-nonafluoro-N,N-bis(nonafluorobutyl)-

$C_{12}H_4N_6O_{12}S$

2288 - Benzene, 1,1'-thiobis[2,4,6-trinitro-

$C_{12}H_5N_7O_{12}$

864 - Benzenamine, 2,4,6-trinitro-N-(2,4,6-trinitrophenyl)-

$C_{12}H_6Cl_2O_2$

7804 - 1,8-Naphthalenedicarbonyl dichloride

$C_{12}H_6Cl_4O_2S$

2325 - Benzene, 1,2,4-trichloro-5-[(4-chlorophenyl)sulfonyl]-

9478 - Phenol, 2,2'-thiobis[4,6-dichloro-

$C_{12}H_6Cl_4S$

11648 - Tetrasul

$C_{12}H_6O_2$

8 - 1,2-Acenaphthylenedione

$C_{12}H_6O_{12}$

1748 - Benzenehexacarboxylic acid

$C_{12}H_7BrO$

5164 - Dibenzofuran, 2-bromo-

5165 - Dibenzofuran, 3-bromo-

$C_{12}H_7Cl_2NO_3$

8081 - Nitrofen

$C_{12}H_7NO_3$

5166 - Dibenzofuran, 3-nitro-

$C_{12}H_7NO_4$

9508 - 3H-Phenoxazin-3-one, 7-hydroxy-, 10-oxide

$C_{12}H_8$

3 - Acenaphthylene

7892 - Naphthalene, 1-ethynyl-

$C_{12}H_8Br_2$

3037 - 1,1'-Biphenyl, 4,4'-dibromo-

$C_{12}H_8Br_2O$

2072 - Benzene, 1,1'-oxybis[4-bromo-

$C_{12}H_8Br_2O_2S$

2238 - Benzene, 1,1'-sulfonylbis[4-bromo-

$C_{12}H_8Br_2S$

2284 - Benzene, 1,1'-thiobis[4-bromo-

$C_{12}H_8Cl_2$

3040 - 1,1'-Biphenyl, 3,3'-dichloro-

3041 - 1,1'-Biphenyl, 4,4'-dichloro-

$C_{12}H_8Cl_2O$

2073 - Benzene, 1,1'-oxybis[4-chloro-

$C_{12}H_8Cl_2O_2S$

2239 - Benzene, 1,1'-sulfonylbis[4-chloro-

9477 - Phenol, 2,2'-thiobis[4-chloro-

$C_{12}H_8Cl_2O_3S$

2221 - Benzenesulfonic acid, 4-chloro-, 4-chlorophenyl ester

9289 - Phenol, 2,4-dichloro-, benzenesulfonate

$C_{12}H_8Cl_6$
5177 - 1,4 5,8-Dimethanonaphthalene,1,2,3,4,10,10-hexachloro-1,4,4a,5,8,8a-hexahydro-,

$C_{12}H_8Cl_6O$
5178 - 2,7 3,6-Dimethanonaphth[2,3-b]oxirene, 3,4,5,6,9,9-hexachloro-1a,2,2a,3,6,6a,7,7a-octahydro-,(1aα,2β,2aα,3β,6β,6aα,7β,7aα)-

$C_{12}H_8F_2$
3042 - 1,1′-Biphenyl, 2,2′-difluoro-
3043 - 1,1′-Biphenyl, 3,3′-difluoro-
3044 - 1,1′-Biphenyl, 4,4′-difluoro-

$C_{12}H_8N_2$
9189 - 1,10-Phenanthroline
9190 - 1,7-Phenanthroline
9191 - 4,7-Phenanthroline
9193 - Phenazine

$C_{12}H_8N_2O$
9197 - 1-Phenazinol

$C_{12}H_8N_2O_4$
3062 - 1,1′-Biphenyl, 2,2′-dinitro-
3063 - 1,1′-Biphenyl, 2,4′-dinitro-

$C_{12}H_8N_2O_4S_2$
5247 - Disulfide, bis(3-nitrophenyl)

$C_{12}H_8N_2O_5$
1495 - Benzene, 2,4-dinitro-1-phenoxy-

$C_{12}H_8O$
5163 - Dibenzofuran

$C_{12}H_8O_3$
8060 - Naphtho[1,2-c]furan-1,3-dione, 4,5-dihydro-

$C_{12}H_8O_4$
6187 - 7H-Furo[3,2-g][1]benzopyran-7-one, 9-methoxy-
7805 - 1,8-Naphthalenedicarboxylic acid

$C_{12}H_8S$
5167 - Dibenzothiophene

$C_{12}H_8S_2$
11660 - Thianthrene

$C_{12}H_8S_3$
11600 - 2,2′:5′,2″-Terthiophene

$C_{12}H_8Se_2$
11353 - Selenanthrene

$C_{12}H_9Br$
4 - Acenaphthylene, 5-bromo-1,2-dihydro-
3016 - 1,1′-Biphenyl, 2-bromo-
3017 - 1,1′-Biphenyl, 3-bromo-
3018 - 1,1′-Biphenyl, 4-bromo-

$C_{12}H_9BrN_2O$
5120 - Diazene, (4-bromophenyl)phenyl-, 1-oxide

$C_{12}H_9BrO$
1079 - Benzene, 1-bromo-4-phenoxy-

$C_{12}H_9Cl$
5 - Acenaphthylene, 5-chloro-1,2-dihydro-
3023 - 1,1′-Biphenyl, 2-chloro-
3024 - 1,1′-Biphenyl, 3-chloro-
3025 - 1,1′-Biphenyl, 4-chloro-

$C_{12}H_9ClF_3N_3O$
8213 - Norflurazon

$C_{12}H_9ClO$
1221 - Benzene, 1-chloro-4-phenoxy-
3092 - [1,1′-Biphenyl]-2-ol, 3-chloro-

$C_{12}H_9ClO_2S$
1225 - Benzene, 1-chloro-4-(phenylsulfonyl)-

$C_{12}H_9ClO_2S_2$
2236 - Benzenesulfonothioic acid, 4-chloro-, S-phenyl ester

$C_{12}H_9Cl_2NO_3$
12048 - Vinclozolin

$C_{12}H_9Cl_3Si$
11364 - Silane, [1,1′-biphenyl]-4-yltrichloro-

$C_{12}H_9F$
3074 - 1,1′-Biphenyl, 2-fluoro-
3075 - 1,1′-Biphenyl, 4-fluoro-

$C_{12}H_9I$
7 - Acenaphthylene, 1,2-dihydro-5-iodo-
3077 - 1,1′-Biphenyl, 2-iodo-
3078 - 1,1′-Biphenyl, 3-iodo-
3079 - 1,1′-Biphenyl, 4-iodo-

$C_{12}H_9N$
3981 - 9H-Carbazole
7736 - 1-Naphthaleneacetonitrile
7737 - 2-Naphthaleneacetonitrile

$C_{12}H_9NO$
7638 - Methanone, phenyl-2-pyridinyl-
7639 - Methanone, phenyl-4-pyridinyl-
9507 - 10H-Phenoxazine

$C_{12}H_9NO_2$
3085 - 1,1′-Biphenyl, 2-nitro-
3086 - 1,1′-Biphenyl, 3-nitro-
3087 - 1,1′-Biphenyl, 4-nitro-
6180 - Furo[2,3-b]quinoline, 4-methoxy-

$C_{12}H_9NO_2S$
2057 - Benzene, 1-nitro-2-(phenylthio)-
2058 - Benzene, 1-nitro-4-(phenylthio)-

$C_{12}H_9NO_3$
2053 - Benzene, 1-nitro-2-phenoxy-
2054 - Benzene, 1-nitro-4-phenoxy-

$C_{12}H_9NS$
9499 - 10H-Phenothiazine

$C_{12}H_9N_3O_4$
1542 - 1,3-Benzenediol, 4-[(4-nitrophenyl)azo]-

$C_{12}H_{10}$
6 - Acenaphthylene, 1,2-dihydro-
3008 - 1,1′-Biphenyl
7881 - Naphthalene, 1-ethenyl-
7882 - Naphthalene, 2-ethenyl-

$C_{12}H_{10}AsCl$
454 - Arsinous chloride, diphenyl-

$C_{12}H_{10}ClN$
678 - Benzenamine, 3-chloro-N-phenyl-
679 - Benzenamine, 4-chloro-N-phenyl-
3012 - [1,1′-Biphenyl]-4-amine, 4′-chloro-
10921 - Pyridine, 2-[(4-chlorophenyl)methyl]-

$C_{12}H_{10}ClO_3P$
9575 - Phosphorochloridic acid, diphenyl ester

$C_{12}H_{10}Cl_2N_2$
3032 - [1,1′-Biphenyl]-4,4′-diamine, 2,2′-dichloro-

$C_{12}H_{10}Cl_2Si$
11395 - Silane, dichlorodiphenyl-

$C_{12}H_{10}F_2Si$
11411 - Silane, difluorodiphenyl-

$C_{12}H_{10}Hg$
7436 - Mercury, diphenyl-

$C_{12}H_{10}N_2$
5129 - Diazene, diphenyl-, (E)-
5130 - Diazene, diphenyl-, (Z)-
11040 - 9H-Pyrido[3,4-b]indole, 1-methyl-

$C_{12}H_{10}N_2O$
829 - Benzenamine, 4-nitroso-N-phenyl-
830 - Benzenamine, N-nitroso-N-phenyl-
5131 - Diazene, diphenyl-, 1-oxide, (Z)-
9448 - Phenol, 4-(phenylazo)-

$C_{12}H_{10}N_2O_2$
826 - Benzenamine, 4-nitro-N-phenyl-
9216 - Phenol, 2,2′-azobis-

$C_{12}H_{10}N_4$
9194 - 2,3-Phenazinediamine

$C_{12}H_{10}N_4O_2$
2472 - Benzo[g]pteridine-2,4(1H,3H)-dione, 7,8-dimethyl-

$C_{12}H_{10}O$
2071 - Benzene, 1,1′-oxybis-
3088 - [1,1′-Biphenyl]-2-ol
3089 - [1,1′-Biphenyl]-3-ol
3090 - [1,1′-Biphenyl]-4-ol
5857 - Ethanone, 1-(1-naphthalenyl)-
5858 - Ethanone, 1-(2-naphthalenyl)-
7756 - 1-Naphthalenecarboxaldehyde, 4-methyl-

$C_{12}H_{10}OS$
2183 - Benzene, 1,1′-sulfinylbis-

$C_{12}H_{10}O_2$
3064 - [1,1′-Biphenyl]-2,2′-diol
3065 - [1,1′-Biphenyl]-2,4′-diol
3066 - [1,1′-Biphenyl]-3,3′-diol
5829 - Ethanone, 1-(1-hydroxy-2-naphthalenyl)-

$C_{12}H_{10}O_2S$
2237 - Benzene, 1,1′-sulfonylbis-

$C_{12}H_{10}O_3$
217 - Acetic acid, (2-naphthalenyloxy)
6069 - 2-Furancarboxylic acid, phenylmethyl ester
7775 - 2-Naphthalenecarboxylic acid, 3-hydroxy-, methyl ester

$C_{12}H_{10}O_4$
3097 - [1,1′-Biphenyl]-3,3′,5,5′-tetrol
8677 - 2,4-Pentadienoic acid, 5-(1,3-benzodioxol-5-yl)-
8678 - 2,4-Pentadienoic acid, 5-(1,3-benzodioxol-5-yl)-, (E,E)-

$C_{12}H_{10}O_4S$
9466 - Phenol, 4,4′-sulfonylbis-

$C_{12}H_{10}O_6S_2$
3070 - [1,1′-Biphenyl]-4,4′-disulfonic acid

$C_{12}H_{10}O_7$
7857 - 1,4-Naphthalenedione, 2-ethyl-3,5,6,7,8-pentahydroxy-
7858 - 1,4-Naphthalenedione, 6-ethyl-2,3,5,7,8-pentahydroxy-

$C_{12}H_{10}S$
2283 - Benzene, 1,1′-thiobis-

$C_{12}H_{10}S_2$
5258 - Disulfide, diphenyl

$C_{12}H_{10}Se$
2175 - Benzene, 1,1′-selenobis-

$C_{12}H_{10}Se_2$
5225 - Diselenide, diphenyl

$C_{12}H_{11}As$
443 - Arsine, diphenyl-

$C_{12}H_{11}BO$
3130 - Borinic acid, diphenyl-

$C_{12}H_{11}Cl_2NO$
10746 - Propyzamide

$C_{12}H_{11}N$
2 - 1-Acenaphthylenamine, 1,2-dihydro-
836 - Benzenamine, N-phenyl-
3010 - [1,1′-Biphenyl]-2-amine
3011 - [1,1′-Biphenyl]-4-amine
10989 - Pyridine, 2-(4-methylphenyl)-
10990 - Pyridine, 2-methyl-5-phenyl-
10991 - Pyridine, 3-methyl-2-phenyl-
11004 - Pyridine, 2-(phenylmethyl)-
11005 - Pyridine, 3-(phenylmethyl)-
11006 - Pyridine, 4-(phenylmethyl)-

$C_{12}H_{11}NO$
834 - Benzenamine, 2-phenoxy-
835 - Benzenamine, 3-phenoxy-
7733 - 1-Naphthaleneacetamide
9445 - Phenol, 2-(phenylamino)-
9446 - Phenol, 3-(phenylamino)-
9447 - Phenol, 4-(phenylamino)-

$C_{12}H_{11}NOS$
68 - Acetamide, 2-mercapto-N-2-naphthalenyl-

$C_{12}H_{11}NO_2$
66 - Acetamide, N-(2-hydroxy-1-naphthalenyl)-
2736 - 1H,5H-Benzo[ij]quinolizine-1,6(7H)-dione, 2,3-dihydro-
8032 - 1-Naphthalenol, methylcarbamate
10625 - 2-Propenoic acid, 2-cyano-3-phenyl-, ethyl ester
10645 - 2-Propenoic acid, 2-isocyano-3-phenyl-, ethyl ester

$C_{12}H_{11}NO_3S$
2234 - Benzenesulfonic acid, 4-(phenylamino)-

$C_{12}H_{11}NO_4$
5218 - 1,3-Dioxolo[4,5-h]quinolin-8(9H)-one, 6-methoxy-9-methyl-
7361 - 2H-Isoindole-2-acetic acid, 1,3-dihydro-1,3-dioxo-, ethyl ester

C₁₂H₁₁N₃ — $C_{12}H_{11}N_3$
837 - Benzenamine, 4-(phenylazo)-
11826 - 1-Triazene, 1,3-diphenyl-

$C_{12}H_{11}N_3O_2$
10904 - 4-Pyridinecarboxylic acid, [1-(2-furanyl)ethylidene]hydrazide

$C_{12}H_{11}N_7$
10750 - 2,4,7-Pteridinetriamine, 6-phenyl-

$C_{12}H_{11}O_3P$
9534 - Phosphonic acid, diphenyl ester

$C_{12}H_{12}$
7818 - Naphthalene, 1,2-dimethyl-
7819 - Naphthalene, 1,3-dimethyl-
7820 - Naphthalene, 1,4-dimethyl-
7821 - Naphthalene, 1,5-dimethyl-
7822 - Naphthalene, 1,6-dimethyl-
7823 - Naphthalene, 1,7-dimethyl-
7824 - Naphthalene, 1,8-dimethyl-
7825 - Naphthalene, 2,3-dimethyl-
7826 - Naphthalene, 2,6-dimethyl-
7827 - Naphthalene, 2,7-dimethyl-
7885 - Naphthalene, 1-ethyl-
7886 - Naphthalene, 2-ethyl-

$C_{12}H_{12}N_2Br$
5224 - Diquat dibromide

$C_{12}H_{12}ClNO_2S$
7977 - 1-Naphthalenesulfonyl chloride, 5-(dimethylamino)-

$C_{12}H_{12}ClN_5O_4S$
4091 - Chlorsulfuron

$C_{12}H_{12}N_2$
1282 - 1,2-Benzenediamine, N-phenyl-
1283 - 1,4-Benzenediamine, N-phenyl-
3028 - [1,1'-Biphenyl]-2,2'-diamine
3029 - [1,1'-Biphenyl]-2,4'-diamine
3030 - [1,1'-Biphenyl]-3,3'-diamine
3031 - [1,1'-Biphenyl]-4,4'-diamine
7157 - Hydrazine, 1,1-diphenyl-
7158 - Hydrazine, 1,2-diphenyl-

$C_{12}H_{12}N_2O$
832 - Benzenamine, 4,4'-oxybis-

$C_{12}H_{12}N_2OS$
845 - Benzenamine, 4,4'-sulfinylbis-

$C_{12}H_{12}N_2OS_2$
11684 - 4-Thiazolidinone, 5-[[4-(dimethylamino)phenyl]methylene]-2-thioxo-

$C_{12}H_{12}N_2O_2S$
846 - Benzenamine, 4,4'-sulfonylbis-

$C_{12}H_{12}N_2O_3$
8066 - 1,8-Naphthyridine-3-carboxylic acid, 1-ethyl-1,4-dihydro-7-methyl-4-oxo-
11095 - 2,4,6(1H,3H,5H)-Pyrimidinetrione, 5-ethyl-5-phenyl-

$C_{12}H_{12}N_2S$
850 - Benzenamine, 4,4'-thiobis-

$C_{12}H_{12}N_4$
635 - Benzenamine, 4,4'-azobis-

$C_{12}H_{12}O$
6710 - 3,5-Hexadien-2-one, 6-phenyl-
7883 - Naphthalene, 1-ethoxy-
7884 - Naphthalene, 2-ethoxy-
7911 - 1-Naphthalenemethanol, α-methyl-, (±)-
8027 - 2-Naphthalenol, 1,4-dimethyl-

$C_{12}H_{12}O_2$
8045 - 1(2H)-Naphthalenone, 2-acetyl-3,4-dihydro-
10692 - 2-Propenoic acid, 3-phenyl-, 2-propenyl ester

$C_{12}H_{12}O_5$
2845 - 2H-1-Benzopyran-2-one, 6,7,8-trimethoxy-
10785 - 4H,5H-Pyrano[4,3-b]pyran-4,5-dione, 2,3-dihydro-3-hydroxy-2-methyl-7-(1-propenyl)-, [2S-[2α,3β,7(E)]]-

$C_{12}H_{12}O_6$
2370 - 1,2,4-Benzenetriol, triacetate

$C_{12}H_{12}Si$
11420 - Silane, diphenyl-

$C_{12}H_{13}ClF_3N_3O_4$
5964 - Fluchloralin

$C_{12}H_{13}ClN_4$
1284 - 1,3-Benzenediamine, 4-(phenylazo)-, monohydrochloride
11051 - 2,4-Pyrimidinediamine, 5-(4-chlorophenyl)-6-ethyl-

$C_{12}H_{13}Cl_3O_3$
258 - Acetic acid, (2,4,5-trichlorophenoxy)-, butyl ester
260 - Acetic acid, (2,4,5-trichlorophenoxy)-, 1-methylpropyl ester

$C_{12}H_{13}N$
3989 - 1H-Carbazole, 2,3,4,9-tetrahydro-
7715 - 1-Naphthalenamine, N,N-dimethyl-
7716 - 2-Naphthalenamine, N,N-dimethyl-
7718 - 1-Naphthalenamine, N-ethyl-
7719 - 2-Naphthalenamine, N-ethyl-
8064 - 2-Naphthylamine, 1,4-dimethyl-
11269 - Quinoline, 2-ethyl-3-methyl-
11270 - Quinoline, 2-ethyl-4-methyl-
11271 - Quinoline, 2-ethyl-6-methyl-
11292 - Quinoline, 2-propyl-
11300 - Quinoline, 2,3,6-trimethyl-
11301 - Quinoline, 2,3,8-trimethyl-
11302 - Quinoline, 2,4,6-trimethyl-
11303 - Quinoline, 2,4,7-trimethyl-
11304 - Quinoline, 2,5,7-trimethyl-

$C_{12}H_{13}NO$
5750 - Ethanol, 2-(2-naphthalenylamino)-

$C_{12}H_{13}NO_2$
7296 - 1H-Indole-3-butanoic acid
11156 - 2,5-Pyrrolidinedione, 1,3-dimethyl-3-phenyl-

$C_{12}H_{13}NO_2S$
4039 - Carboxin

$C_{12}H_{13}N_3$
1263 - 1,4-Benzenediamine, N-(4-aminophenyl)-
10974 - 2-Pyridinemethanamine, N-(2-pyridinylmethyl)-

$C_{12}H_{13}N_3O_4S_2$
2188 - Benzenesulfonamide, 4-amino-N-[4-(aminosulfonyl)phenyl]-

$C_{12}H_{14}$
9 - Acenaphthylene, 1,2,2a,3,4,5-hexahydro-
1246 - Benzene, 1-cyclohexen-1-yl-
1247 - Benzene, 2-cyclohexen-1-yl-
1625 - Benzene, 1-ethenyl-3-butenyl-
1753 - Benzene, 1,3-hexadienyl-
1754 - Benzene, 1,5-hexadienyl-
1755 - Benzene, 3,5-hexadienyl-

$C_{12}H_{14}ClNO_2$
4140 - Clomazone

$C_{12}H_{14}ClN_5O_2S$
2209 - Benzenesulfonamide, 4-[(2,4-diaminophenyl)azo]-

$C_{12}H_{14}Cl_2O_3$
171 - Acetic acid, (2,4-dichlorophenoxy)-, butyl ester

$C_{12}H_{14}Cl_3O_4P$
4078 - Chlorfenvinphos

$C_{12}H_{14}N_2$
5514 - 1,2-Ethanediamine, N-1-naphthalenyl-
11213 - 6-Quinolinamine, N,N,2-trimethyl-

$C_{12}H_{14}N_2O_2$
11066 - 4,6(1H,5H)-Pyrimidinedione, 5-ethyldihydro-5-phenyl-
11923 - L-Tryptophan, N-methyl-

$C_{12}H_{14}N_2O_3$
11085 - 2,4,6(1H,3H,5H)-Pyrimidinetrione, 5-(2-cyclopenten-1-yl)-5-(2-propenyl)-

$C_{12}H_{14}N_2O_5$
6242 - L-Glutamic acid, N-(4-aminobenzoyl)-
9273 - Phenol, 2-cyclohexyl-4,6-dinitro-

$C_{12}H_{14}N_4OS$
11111 - 5H-Pyrimido[4,5-d]thiazolo[3,2-a]pyrimidine-8-ethanol, 2,7-dimethyl-

$C_{12}H_{14}N_4O_2S$
2196 - Benzenesulfonamide, 4-amino-N-(4,6-dimethyl-2-pyrimidinyl)-

$C_{12}H_{14}N_4O_4S$
2194 - Benzenesulfonamide, 4-amino-N-(2,6-dimethoxy-4-pyrimidinyl)-

$C_{12}H_{14}N_4O_4S_2$
11723 - Thiophanate-methyl

$C_{12}H_{14}O$
2460 - Benzofuran, 4-methyl-7-(1-methylethyl)-
4893 - Cyclopentanone, 2-(phenylmethyl)-
7101 - 5-Hexen-2-one, 4-phenyl-
7102 - 5-Hexen-2-one, 6-phenyl-
8051 - 1(2H)-Naphthalenone, 7-ethyl-3,4-dihydro-
8987 - 2-Pentanone, 3-(phenylmethylene)-

$C_{12}H_{14}O_2$
4135 - Cinnamic acid, β-propyl-
6038 - Furan, 2,2'-sec-butylidenedi-
10687 - 2-Propenoic acid, 3-phenyl-, 1-methylethyl ester, (E)-
10693 - 2-Propenoic acid, 3-phenyl-, propyl ester, (E)-

$C_{12}H_{14}O_3$
884 - Benzeneacetic acid, α-acetyl-, ethyl ester
2100 - Benzenepentanoic acid, δ-oxo-, methyl ester
8604 - Oxiranecarboxylic acid, 3-methyl-3-phenyl-, ethyl ester
8911 - Pentanoic acid, 4-oxo-, phenylmethyl ester
9393 - Phenol, 2-methoxy-4-(1-propenyl)-, acetate, (E)-
9394 - Phenol, 2-methoxy-4-(1-propenyl)-, acetate, (Z)-
9395 - Phenol, 2-methoxy-4-(2-propenyl)-, acetate

$C_{12}H_{14}O_4$
1259 - 1,4-Benzenediacetic acid, dimethyl ester
1342 - 1,2-Benzenedicarboxylic acid, diethyl ester
1346 - 1,3-Benzenedicarboxylic acid, diethyl ester
1350 - 1,4-Benzenedicarboxylic acid, diethyl ester
2422 - 1,3-Benzodioxole, 4,7-dimethoxy-5-(1-propenyl)-, (E)-
2423 - 1,3-Benzodioxole, 4,7-dimethoxy-5-(2-propenyl)-
8625 - Oxirane, 2,2'-[1,3-phenylenebis(oxymethylene)]bis-

$C_{12}H_{15}Br$
1009 - Benzene, 1-bromo-4-cyclohexyl-

$C_{12}H_{15}ClNO_4PS_2$
9514 - Phosalone

$C_{12}H_{15}ClO$
2907 - Benzoyl chloride, 4-pentyl-
9254 - Phenol, 2-chloro-4-cyclohexyl-

$C_{12}H_{15}ClO_2$
2908 - Benzoyl chloride, 4-(pentyloxy)-

$C_{12}H_{15}ClO_3$
10231 - Propanoic acid, 2-(4-chlorophenoxy)-2-methyl-, ethyl ester

$C_{12}H_{15}N$
964 - Benzeneacetonitrile, α-(2-methylpropyl)-
1800 - Benzenemethanamine, N-3-butynyl-N-methyl-
2737 - 1H,5H-Benzo[ij]quinolizine, 2,3,6,7-tetrahydro-

$C_{12}H_{15}NO$
9634 - Piperidine, 1-benzoyl-

$C_{12}H_{15}NO_2$
1251 - Benzene, 1-cyclohexyl-2-nitro-
1252 - Benzene, 1-cyclohexyl-4-nitro-

$C_{12}H_{15}NO_3$
3991 - Carbofuran
5214 - 1,3-Dioxolo[4,5-g]isoquinoline, 5,6,7,8-tetrahydro-4-methoxy-6-methyl-

$C_{12}H_{15}NO_4$
5216 - 1,3-Dioxolo[4,5-g]isoquinolin-5-ol, 5,6,7,8-tetrahydro-4-methoxy-6-methyl-

$C_{12}H_{15}N_3O_3$
11842 - 1,3,5-Triazine-2,4,6(1H,3H,5H)-trione, 1,3,5-tri-2-propenyl-

C₁₂H₁₅N₃O₆
1461 - Benzene, 1-(1,1-dimethylethyl)-3,5-dimethyl-
    2,4,6-trinitro-
C₁₂H₁₆
1248 - Benzene, cyclohexyl-
1760 - Benzene, 2-hexenyl-
1761 - Benzene, 3-hexenyl-
1762 - Benzene, 5-hexenyl-
1973 - Benzene, 1-(1-methylethenyl)-4-(1-
    methylethyl)-
2008 - Benzene, (1-methyl-1-pentenyl)-
4744 - 1,3,5,7-Cyclooctatetraene, 1-butyl-
7261 - 1H-Indene, 2,3-dihydro-1,1,5-trimethyl-
7262 - 1H-Indene, 2,3-dihydro-1,4,7-trimethyl-
7888 - Naphthalene, 1-ethyl-1,2,3,4-tetrahydro-
7889 - Naphthalene, 2-ethyl-1,2,3,4-tetrahydro-
7890 - Naphthalene, 5-ethyl-1,2,3,4-tetrahydro-
7891 - Naphthalene, 6-ethyl-1,2,3,4-tetrahydro-
7980 - Naphthalene, 1,2,3,4-tetrahydro-1,1-dimethyl-
7981 - Naphthalene, 1,2,3,4-tetrahydro-1,2-dimethyl-
7982 - Naphthalene, 1,2,3,4-tetrahydro-1,3-dimethyl-
7983 - Naphthalene, 1,2,3,4-tetrahydro-1,4-dimethyl-
7984 - Naphthalene, 1,2,3,4-tetrahydro-1,5-dimethyl-
7985 - Naphthalene, 1,2,3,4-tetrahydro-2,2-dimethyl-
7986 - Naphthalene, 1,2,3,4-tetrahydro-2,3-dimethyl-
7987 - Naphthalene, 1,2,3,4-tetrahydro-2,5-dimethyl-
7988 - Naphthalene, 1,2,3,4-tetrahydro-2,6-dimethyl-
7989 - Naphthalene, 1,2,3,4-tetrahydro-2,7-dimethyl-
7990 - Naphthalene, 1,2,3,4-tetrahydro-2,8-dimethyl-
7991 - Naphthalene, 1,2,3,4-tetrahydro-5,6-dimethyl-
7992 - Naphthalene, 1,2,3,4-tetrahydro-5,7-dimethyl-
7993 - Naphthalene, 1,2,3,4-tetrahydro-5,8-dimethyl-
7994 - Naphthalene, 1,2,3,4-tetrahydro-6,7-dimethyl-
C₁₂H₁₆ClNOS
11697 - Thiobencarb
C₁₂H₁₆N₂
7303 - 1H-Indole-3-ethanamine, N,N-dimethyl-
C₁₂H₁₆N₂O
7326 - 1H-Indol-5-ol, 3-[2-(dimethylamino)ethyl]-
7578 - 1,5-Methano-8H-pyrido[1,2-a][1,5]diazocin-8-
    one, 1,2,3,4,5,6-hexahydro-3-methyl-, (1R)-
C₁₂H₁₆N₂O₃
11083 - 2,4,6(1H,3H,5H)-Pyrimidinetrione, 5-(1-
    cyclohexen-1-yl)-1,5-dimethyl-
11084 - 2,4,6(1H,3H,5H)-Pyrimidinetrione, 5-(1-
    cyclohexen-1-yl)-5-ethyl-
C₁₂H₁₆N₂O₅
1463 - Benzene, 1-(1,1-dimethylethyl)-2-methoxy-4-
    methyl-3,5-dinitro-
C₁₂H₁₆O
1253 - Benzene, (cyclohexyloxy)-
3669 - 1-Butanone, 3-methyl-1-(2-methylphenyl)-
3670 - 1-Butanone, 3-methyl-1-(4-methylphenyl)-
4552 - Cyclohexanol, 1-phenyl-
4553 - Cyclohexanol, 2-phenyl-, cis-(±)-
4554 - Cyclohexanol, 2-phenyl-, trans-(±)-
5808 - Ethanone, 1-[4-(1,1-dimethylethyl)phenyl]-
5850 - Ethanone, 1-[2-methyl-5-
    (1-methylethyl)phenyl]-
6983 - 1-Hexanone, 1-phenyl-
6984 - 2-Hexanone, 3-phenyl-
6985 - 3-Hexanone, 2-phenyl-
8983 - 1-Pentanone, 4-methyl-1-phenyl-
8984 - 2-Pentanone, 3-methyl-1-phenyl-
9272 - Phenol, 4-cyclohexyl-
C₁₂H₁₆O₂
890 - Benzeneacetic acid, α-butyl-
932 - Benzeneacetic acid, 2-methylpropyl ester
1759 - Benzenehexanoic acid
2129 - Benzenepropanoic acid, α,α-dimethyl-, methyl
    ester
2140 - Benzenepropanoic acid, 1-methylethyl ester
2146 - Benzenepropanoic acid, β-propyl-
2685 - Benzoic acid, 3-methyl-, butyl ester
2713 - Benzoic acid, pentamethyl-
3560 - Butanoic acid, 3-methyl-, phenylmethyl ester
3632 - 1-Butanol, 3-methyl-, benzoate

7175 - Hydrocinnamic acid, β-ethyl-β-methyl-
8979 - 1-Pentanone, 1-(3-methoxyphenyl)-
9419 - Phenol, 2-methyl-5-(1-methylethyl)-, acetate
9420 - Phenol, 5-methyl-2-(1-methylethyl)-, acetate
10306 - Propanoic acid, 2-methyl-, 2-phenylethyl
    ester
C₁₂H₁₆O₃
2356 - Benzene, 1,2,4-trimethoxy-5-(1-propenyl)-
2633 - Benzoic acid, 2-hydroxy-, 3-methylbutyl ester
2649 - Benzoic acid, 2-hydroxy-, pentyl ester
2668 - Benzoic acid, 4-methoxy-, butyl ester
3581 - Butanoic acid, 4-phenoxy-, ethyl ester
3582 - Butanoic acid, 2-phenoxyethyl ester
6962 - 1-Hexanone, 1-(2,4-dihydroxyphenyl)-
8886 - Pentanoic acid, 2-methoxyphenyl ester
10641 - 2-Propenoic acid, 3-(2-furanyl)-, pentyl ester
C₁₂H₁₆O₄
2628 - Benzoic acid, 4-hydroxy-3-methoxy-, 2-
    methylpropyl ester
C₁₂H₁₆O₇
6224 - β-D-Glucopyranoside, 4-hydroxyphenyl
C₁₂H₁₇BrO
1045 - Benzene, 1-bromo-4-(hexyloxy)-
C₁₂H₁₇N
680 - Benzenamine, N-cyclohexyl-
9698 - Piperidine, 1-(phenylmethyl)-
9699 - Piperidine, 2-(phenylmethyl)-
9700 - Piperidine, 4-(phenylmethyl)-
C₁₂H₁₇NO
39 - Acetamide, N-butyl-N-phenyl-
612 - Benzamide, N,N-diethyl-3-methyl-
6722 - Hexanamide, N-phenyl-
C₁₂H₁₇NO₂
9769 - Promecarb
C₁₂H₁₇NO₃
611 - Benzamide, N,N-diethyl-4-hydroxy-3-methoxy-
880 - Benzeneacetamide, 4-butoxy-N-hydroxy-
7395 - 8-Isoquinolinol, 1,2,3,4-tetrahydro-6,7-
    dimethoxy-1-methyl-, (S)-
C₁₂H₁₇N₃O₄
6247 - L-Glutamic acid, 5-[2-
    [4-(hydroxymethyl)phenyl]hydrazide]
C₁₂H₁₈
992 - Benzene, 1,2-bis(1-methylethyl)-
993 - Benzene, 1,3-bis(1-methylethyl)-
994 - Benzene, 1,4-bis(1-methylethyl)-
1455 - Benzene, (1,1-dimethylbutyl)-
1460 - Benzene, 1-(1,1-dimethylethyl)-3,5-dimethyl-
1473 - Benzene, 2,4-dimethyl-1-(1-methylpropyl)-
1557 - Benzene, 1,4-dipropyl-
1669 - Benzene, 1-ethylbutyl-
1691 - Benzene, (1-ethyl-1-methylpropyl)-
1758 - Benzene, hexamethyl-
1763 - Benzene, hexyl-
2009 - Benzene, (1-methylpentyl)-
2010 - Benzene, (3-methylpentyl)-
2011 - Benzene, (4-methylpentyl)-
2344 - Benzene, 1,2,4-triethyl-
2345 - Benzene, 1,3,5-triethyl-
2361 - Benzene, 1,2,4-trimethyl-5-(1-methylethyl)-
4212 - 1,5,9-Cyclododecatriene
4213 - 1,5,9-Cyclododecatriene, (Z,E,E)-
4214 - 1,5,9-Cyclododecatriene, (Z,Z,Z)-
5286 - 3,9-Dodecadiyne
5287 - 5,7-Dodecadiyne
8339 - 2,6-Octadien-4-yne, 3,6-diethyl-
C₁₂H₁₈Cl₂N₄OS
11686 - Thiazolium, 3-[(4-amino-2-methyl-5-
    pyrimidinyl)methyl]-5-(2-hydroxyethyl)-4-
    methyl- chloride, monohydrochlorid
C₁₂H₁₈N₂O
11188 - 2-Pyrrolidinone, 1-[4-(1-pyrrolidinyl)-2-
    butynyl]-
C₁₂H₁₈N₂O₃
11090 - 2,4,6(1H,3H,5H)-Pyrimidinetrione, 5-(2,2-
    dimethylpropyl)-5-(2-propenyl)-

C₁₂H₁₈N₂O₃S
2207 - Benzenesulfonamide, N-[(butylamino)
    carbonyl]-4-methyl-
C₁₂H₁₈N₄
7171 - Hydrazine, phenyl-, hemihydrate
C₁₂H₁₈N₄O₆S
8553 - Oryzalin
C₁₂H₁₈O
1488 - Benzene, 1-(1,1-dimethylpropyl)-4-methoxy-
1765 - Benzene, (hexyloxy)-
1847 - Benzenemethanol, α-butyl-α-methyl-
1875 - Benzenemethanol, α-ethyl-α-propyl-
1909 - Benzenemethanol, α-pentyl-
1951 - Benzene, [(3-methylbutoxy)methyl]-
2103 - Benzenepentanol, α-methyl-
2159 - Benzenepropanol, γ-propyl-
6949 - 3-Hexanol, 1-phenyl-
9226 - Phenol, 2,6-bis(1-methylethyl)-
9324 - Phenol, 2-(1,1-dimethylethyl)-4,5-dimethyl-
9325 - Phenol, 2-(1,1-dimethylethyl)-4,6-dimethyl-
9326 - Phenol, 4-(1,1-dimethylethyl)-2,5-dimethyl-
9327 - Phenol, 4-(1,1-dimethylethyl)-2,6-dimethyl-
9340 - Phenol, 4-(1,1-dimethylpropyl)-2-methyl-
9347 - Phenol, 2,6-dipropyl-
9425 - Phenol, 5-methyl-2-pentyl-
C₁₂H₁₈O₂
1519 - 1,3-Benzenediol, 4-hexyl-
1555 - Benzene, 1,2-dipropoxy-
1556 - Benzene, 1,3-dipropoxy-
2950 - Bicyclo[2.2.1]heptan-2-one, 3-(acetyl-17O)-
    1,7,7-trimethyl-
11863 - Tricyclo[2.2.1.02,6]heptane-1-carboxylic
    acid, 7,7-dimethyl-, ethyl ester
C₁₂H₁₈O₃
2343 - Benzene, 1,3,5-triethoxy-
C₁₂H₁₈O₄
4644 - 1-Cyclohexene-1,2-dicarboxylic acid, diethyl
    ester
4645 - 1-Cyclohexene-1,3-dicarboxylic acid, diethyl
    ester
9935 - Propanedioic acid, 2-cyclopenten-1-yl-, diethyl
    ester
9936 - Propanedioic acid, cyclopentylidene-, diethyl
    ester
C₁₂H₁₈O₅
10669 - 2-Propenoic acid, 2-methyl-, oxydi-2,1-
    ethanediyl ester
C₁₂H₁₈O₆
10601 - 1-Propene-1,2,3-tricarboxylic acid, triethyl
    ester, (E)-
C₁₂H₁₉ClNO₃P
4153 - Crufomate
C₁₂H₁₉N
636 - Benzenamine, 2,6-bis(1-methylethyl)-
730 - Benzenamine, N,N-dipropyl-
C₁₂H₁₉NO
738 - Benzenamine, 3-ethoxy-N,N-diethyl-
C₁₂H₂₀
5342 - 1-Dodecen-3-yne
C₁₂H₂₀As₂Cl₂O₅
9275 - Phenol, 4,4'-(1,2-diarsenediyl)bis[2-amino-,
    dihydrochloride
C₁₂H₂₀N₂O₂
10811 - 2(1H)-Pyrazinone, 1-hydroxy-6-(1-
    methylpropyl)-3-(2-methylpropyl)-, (+)-
C₁₂H₂₀N₄O₂
6995 - Hexazinone
C₁₂H₂₀O
2989 - [1,1'-Bicyclohexyl]-2-one
6032 - Furan, 2,5-bis(1,1-dimethylethyl)-
C₁₂H₂₀O₂
2945 - Bicyclo[2.2.1]heptan-2-ol, 1,7,7-trimethyl-,
    acetate, endo-(±)-
2946 - Bicyclo[2.2.1]heptan-2-ol, 1,7,7-trimethyl-,
    acetate, exo-(±)-
2947 - Bicyclo[2.2.1]heptan-2-ol, 4,7,7-trimethyl-,
    acetate, (1R-endo)-

4545 - Cyclohexanol, 5-methyl-2-(1-methylethenyl)-, acetate, [1R-(1α,2β,5α)]-
4669 - 3-Cyclohexene-1-methanol, α,α,4-trimethyl-, acetate
6135 - Furan, 2-(octyloxy)-
8335 - 1,6-Octadien-3-ol, 3,7-dimethyl-, acetate, (R)-
8336 - 2,6-Octadien-1-ol, 3,7-dimethyl-, acetate, (Z)-
11990 - 9-Undecynoic acid, methyl ester
11991 - 10-Undecynoic acid, methyl ester

$C_{12}H_{20}O_3Si$
11481 - Silane, triethoxyphenyl-

$C_{12}H_{20}O_4$
3767 - 2-Butenedioic acid (E)-, bis(2-methylpropyl) ester
3774 - 2-Butenedioic acid (E)-, dibutyl ester
4350 - 1,2-Cyclohexanedicarboxylic acid, diethyl ester, trans-(±)-
4351 - 1,3-Cyclohexanedicarboxylic acid, diethyl ester, cis-
4355 - 1,4-Cyclohexanedicarboxylic acid, diethyl ester, cis-
4356 - 1,4-Cyclohexanedicarboxylic acid, diethyl ester, trans-
4795 - 1,3-Cyclopentanedicarboxylic acid, 1,2,2-trimethyl-, dimethyl ester, (1R-cis)-
5336 - 2-Dodecenedioic acid, (E)-

$C_{12}H_{20}O_4Si$
11511 - Silicic acid (H_4SiO_4), triethyl phenyl ester

$C_{12}H_{20}O_6$
10198 - 1,2,3-Propanetriol, tripropanoate

$C_{12}H_{20}O_7$
10169 - 1,2,3-Propanetricarboxylic acid, 2-hydroxy-, triethyl ester

$C_{12}H_{20}Si$
11485 - Silane, triethylphenyl-

$C_{12}H_{21}NO$
4611 - Cyclohexanone, 2-(1-piperidinylmethyl)-

$C_{12}H_{21}N_2O_3PS$
9582 - Phosphorothioic acid, O,O-diethyl O-[6-methyl-2-(1-methylethyl)-4-pyrimidinyl] ester

$C_{12}H_{22}$
2988 - 1,1'-Bicyclohexyl
4215 - Cyclododecene, (E)-
4216 - Cyclododecene, (Z)-
5101 - 4-Decyne, 3,3-dimethyl-
5285 - .1,11-Dodecadiene
5343 - 1-Dodecyne
5344 - 2-Dodecyne
5345 - 3-Dodecyne
5346 - 6-Dodecyne

$C_{12}H_{22}O$
4211 - Cyclododecanone
4461 - Cyclohexane, 1,1'-oxybis-

$C_{12}H_{22}OSi_2$
11497 - Silane, trimethyl[4-[(trimethylsilyl)oxy]phenyl]-

$C_{12}H_{22}O_2$
2948 - Bicyclo[2.2.1]heptan-2-ol, 1,7,7-trimethyl-, acetate, endo-
4412 - Cyclohexanehexanoic acid
4549 - Cyclohexanol, 5-methyl-2-(1-methylethyl)-, acetate, [1R-(1α,2β,5α)]-
4787 - Cyclopentanecarboxylic acid, 3-isopropyl-1-methyl-
4788 - Cyclopentanecarboxylic acid, 2-methyl-5-(1-methylethyl)-, ethyl ester
5337 - 2-Dodecenoic acid
5338 - 4-Dodecenoic acid
5339 - 5-Dodecenoic acid
5340 - 11-Dodecenoic acid
8400 - 3,6-Octanedione, 2,2,7,7-tetramethyl-
10659 - 2-Propenoic acid, 2-methyl-, 2-ethylhexyl ester
11978 - 10-Undecenoic acid, methyl ester

$C_{12}H_{22}O_2Si_2$
11447 - Silane, [1,3-phenylenebis(oxy)]bis[trimethyl-

$C_{12}H_{22}O_3$
210 - Acetic acid, [[5-methyl-2-(1-methylethyl)cyclohexyl]oxy]-, [1R-(1α,2β,5α)]-
6874 - Hexanoic acid, anhydride

$C_{12}H_{22}O_4$
3296 - Butanedioic acid, bis(1,1-dimethylethyl) ester
3297 - Butanedioic acid, bis(1-methylpropyl) ester
3301 - Butanedioic acid, dibutyl ester
3332 - Butanedioic acid, (1-methylpropyl)-, diethyl ester
5046 - Decanedioic acid, dimethyl ester
5299 - Dodecanedioic acid
5555 - Ethanedioic acid, bis(3-methylbutyl) ester
6782 - Hexanedioic acid, bis(1-methylethyl) ester
6787 - Hexanedioic acid, dipropyl ester
8387 - Octanedioic acid, diethyl ester
9951 - Propanedioic acid, ethyl(1-methylethyl)-, diethyl ester
9961 - Propanedioic acid, (3-methylbutyl)-, diethyl ester
9974 - Propanedioic acid, pentyl-, diethyl ester

$C_{12}H_{22}O_6$
3313 - Butanedioic acid, 2,3-dihydroxy- [R-(R^*,R^*)]-, dibutyl ester

$C_{12}H_{22}O_{10}$
6236 - D-Glucose, 6-O-(6-deoxy-α-L-mannopyranosyl)-

$C_{12}H_{22}O_{11}$
6022 - D-Fructopyranose, 3-O-α-D-glucopyranosyl-
6214 - β-D-Glucopyranose, 4-O-β-D-galactopyranosyl-
6218 - α-D-Glucopyranoside, β-D-fructofuranosyl
6220 - α-D-Glucopyranoside, α-D-glucopyranosyl

$C_{12}H_{22}O_{12}$
6210 - D-Gluconic acid, 4-O-β-D-galactopyranosyl-

$C_{12}H_{23}BrO_2$
5307 - Dodecanoic acid, 2-bromo-

$C_{12}H_{23}ClO$
5328 - Dodecanoyl chloride

$C_{12}H_{23}N$
4284 - Cyclohexanamine, N-cyclohexyl-
5303 - Dodecanenitrile

$C_{12}H_{23}NSi_2$
11362 - Silanamine, 1,1,1-trimethyl-N-phenyl-N-(trimethylsilyl)-

$C_{12}H_{23}P$
9516 - Phosphine, dicyclohexyl-

$C_{12}H_{24}$
4210 - Cyclododecane
4414 - Cyclohexane, hexyl-
4446 - Cyclohexane, 1-methyl-2-pentyl-
4737 - Cyclooctane, butyl-
4820 - Cyclopentane, heptyl-
5335 - 1-Dodecene
10582 - 1-Propene, 2-methyl-, trimer

$C_{12}H_{24}BrF$
5294 - Dodecane, 1-bromo-12-fluoro-

$C_{12}H_{24}Br_2$
5296 - Dodecane, 1,12-dibromo-

$C_{12}H_{24}N_2$
9661 - Piperidine, 1,1'-(1,2-ethanediyl)bis-

$C_{12}H_{24}O$
5051 - Decane, 1-(ethenyloxy)-
5288 - Dodecanal
5325 - 2-Dodecanone
5326 - 6-Dodecanone
11936 - Undecanal, 2-methyl-

$C_{12}H_{24}O_2$
157 - Acetic acid, decyl ester
3566 - Butanoic acid, octyl ester
5063 - Decanoic acid, ethyl ester
5305 - Dodecanoic acid
6470 - Heptanoic acid, pentyl ester
6890 - Hexanoic acid, hexyl ester
8155 - Nonanoic acid, 3-methyl-, ethyl ester
8427 - Octanoic acid, butyl ester

8881 - Pentanoic acid, heptyl ester

$C_{12}H_{24}O_3$
11949 - Undecanoic acid, 11-hydroxy-, methyl ester

$C_{12}H_{24}O_4Si_4$
4996 - Cyclotetrasiloxane, 2,4,6,8-tetraethenyl-2,4,6,8-tetramethyl-

$C_{12}H_{25}Br$
5293 - Dodecane, 1-bromo-

$C_{12}H_{25}Cl$
5295 - Dodecane, 1-chloro-

$C_{12}H_{25}Cl_3Si$
11460 - Silane, trichlorododecyl-

$C_{12}H_{25}I$
5302 - Dodecane, 1-iodo-

$C_{12}H_{25}NO$
5289 - Dodecanamide

$C_{12}H_{26}$
5292 - Dodecane
6433 - Heptane, 2,2,4,6,6-pentamethyl-

$C_{12}H_{26}O$
5321 - 1-Dodecanol
5322 - 2-Dodecanol
5323 - 3-Dodecanol
5324 - 6-Dodecanol
6842 - Hexane, 1,1'-oxybis-
11955 - 1-Undecanol, 2-methyl-
11956 - 3-Undecanol, 2-methyl-
11957 - 5-Undecanol, 2-methyl-

$C_{12}H_{26}O_2$
5301 - 1,12-Dodecanediol

$C_{12}H_{26}O_3$
3417 - Butane, 1,1'-[oxybis(2,1-ethanediyloxy)]bis-

$C_{12}H_{26}O_6P_2S_4$
5201 - Dioxathion

$C_{12}H_{26}S$
5304 - 1-Dodecanethiol
6855 - Hexane, 1,1'-thiobis-

$C_{12}H_{27}B$
3123 - Borane, tris(2-methylpropyl)-

$C_{12}H_{27}BO_3$
3124 - Boric acid, tributyl ester

$C_{12}H_{27}N$
3213 - 1-Butanamine, N,N-dibutyl-
5291 - 1-Dodecanamine
6727 - 1-Hexanamine, N-hexyl-
9820 - 1-Propanamine, 2-methyl-N,N-bis(2-methylpropyl)-

$C_{12}H_{27}NO$
5682 - Ethanol, 2-[bis(3-methylbutyl)amino]-

$C_{12}H_{27}NO_2$
8936 - 2-Pentanol, 1,1'-iminobis[2-methyl-

$C_{12}H_{27}OPS_3$
11852 - S,S,S-Tributyl phosphorotrithioate

$C_{12}H_{27}O_3P$
9590 - Phosphorous acid, tributyl ester

$C_{12}H_{27}O_4P$
9562 - Phosphoric acid tributyl ester
9572 - Phosphoric acid, tris(2-methylpropyl) ester

$C_{12}H_{27}P$
9520 - Phosphine, tributyl-

$C_{12}H_{27}PS_3$
7438 - Merphos

$C_{12}H_{28}IN$
9833 - 1-Propanaminium, N,N,N-tripropyl-, iodide

$C_{12}H_{28}N_2$
3262 - 1,4-Butanediamine, N,N,N',N'-tetraethyl-

$C_{12}H_{28}O_4Si$
11510 - Silicic acid (H_4SiO_4), tetrapropyl ester

$C_{12}H_{28}Si$
11454 - Silane, tributyl-

$C_{12}H_{28}Sn$
11546 - Stannane, tetrapropyl-

$C_{12}H_{30}OSi_2$
5235 - Disiloxane, hexaethyl-

$C_{12}H_{30}O_{13}P_4$
11645 - Tetraphosphoric acid, hexaethyl ester

$C_{12}H_{32}O_4Si_4$
4997 - Cyclotetrasiloxane, 2,4,6,8-tetraethyl-2,4,6,8-
tetramethyl-

$C_{12}H_{36}O_4Si_5$
8997 - Pentasiloxane, dodecamethyl-

$C_{12}H_{36}O_6Si_6$
4624 - Cyclohexasiloxane, dodecamethyl-

$C_{13}H_5N_3O_7$
5979 - 9H-Fluoren-9-one, 2,4,7-trinitro-

$C_{13}H_6Cl_6O_2$
9409 - Phenol, 2,2'-methylenebis[3,4,6-trichloro-

$C_{13}H_7ClF_3NO_3$
8082 - Nitrofluorfen

$C_{13}H_7ClO$
5976 - 9H-Fluoren-9-one, 2-chloro-

$C_{13}H_7Cl_4NO_2$
610 - Benzamide, 3,5-dichloro-
N-(3,4-dichlorophenyl)-2-hydroxy-

$C_{13}H_7F_3N_2O_5$
5980 - Fluorodifen

$C_{13}H_7NO_3$
5977 - 9H-Fluoren-9-one, 2-nitro-

$C_{13}H_8Br_2O$
7590 - Methanone, bis(4-bromophenyl)-

$C_{13}H_8Br_3NO_2$
609 - Benzamide, 3,5-dibromo-N-(4-bromophenyl)-2-
hydroxy-

$C_{13}H_8ClN$
315 - Acridine, 9-chloro-

$C_{13}H_8Cl_2$
5972 - 9H-Fluorene, 2,7-dichloro-

$C_{13}H_8Cl_2O$
7591 - Methanone, bis(3-chlorophenyl)-
7592 - Methanone, bis(4-chlorophenyl)-
7603 - Methanone, (2-chlorophenyl)(4-chlorophenyl)-

$C_{13}H_8Cl_3NO_2$
608 - Benzamide, 5-chloro-N-(3,4-dichlorophenyl)-2-
hydroxy-

$C_{13}H_8I_2O$
7598 - Methanone, bis(4-iodophenyl)-

$C_{13}H_8O$
5975 - 9H-Fluoren-9-one

$C_{13}H_8OS$
11812 - 9H-Thioxanthen-9-one

$C_{13}H_8O_2$
12050 - 9H-Xanthen-9-one

$C_{13}H_8O_4$
12051 - 9H-Xanthen-9-one, 1,3-dihydroxy-
12052 - 9H-Xanthen-9-one, 2,7-dihydroxy-
12053 - 9H-Xanthen-9-one, 3,6-dihydroxy-

$C_{13}H_9Br$
5968 - 9H-Fluorene, 2-bromo-

$C_{13}H_9BrO$
7600 - Methanone, (2-bromophenyl)phenyl-
7601 - Methanone, (3-bromophenyl)phenyl-
7602 - Methanone, (4-bromophenyl)phenyl-

$C_{13}H_9ClO$
7604 - Methanone, (4-chlorophenyl)phenyl-

$C_{13}H_9Cl_2FN_2S$
11802 - Thiourea, N-(3,5-dichlorophenyl)-N'-(4-
fluorophenyl)-

$C_{13}H_9N$
314 - Acridine
2437 - Benzo[f]quinoline
2473 - Benzo[g]quinoline
2474 - Benzo[h]quinoline
3020 - [1,1'-Biphenyl]-2-carbonitrile
9188 - Phenanthridine

$C_{13}H_9NOS$
9217 - Phenol, 2-(2-benzothiazolyl)-

$C_{13}H_9NO_2$
9218 - Phenol, 2-(2-benzoxazolyl)-

$C_{13}H_9NO_3$
7635 - Methanone, (3-nitrophenyl)phenyl-
7636 - Methanone, (4-nitrophenyl)phenyl-

$C_{13}H_9NS$
2864 - Benzothiazole, 2-phenyl-

$C_{13}H_9N_3O_5$
2648 - Benzoic acid, 2-hydroxy-5-
[(4-nitrophenyl)azo]-

$C_{13}H_{10}$
5967 - 9H-Fluorene

$C_{13}H_{10}Br_2$
1962 - Benzene, 1,1'-methylenebis[2-bromo-

$C_{13}H_{10}Cl_2$
1389 - Benzene, 1,1'-(dichloromethylene)bis-
1963 - Benzene, 1,1'-methylenebis[4-chloro-

$C_{13}H_{10}Cl_2O_2$
1965 - Benzene, 1,1'-[methylenebis(oxy)]bis[4-
chloro-
9408 - Phenol, 2,2'-methylenebis[4-chloro-

$C_{13}H_{10}Cl_2S$
4075 - Chlorbenside

$C_{13}H_{10}N_2$
312 - 4-Acridinamine
313 - 9-Acridinamine
764 - Benzenamine, N,N'-methanetetraylbis-
2383 - 1H-Benzimidazole, 1-phenyl-
2384 - 1H-Benzimidazole, 2-phenyl-
4158 - Cyanamide, diphenyl-
9195 - Phenazine, 2-methyl-

$C_{13}H_{10}N_2O$
9196 - Phenazinium, 1-hydroxy-5-methyl-, hydroxide,
inner salt

$C_{13}H_{10}N_2O_4$
7363 - 1H-Isoindole-1,3(2H)-dione, 2-(2,6-dioxo-3-
piperidinyl)-

$C_{13}H_{10}N_4O_5$
11997 - Urea, N,N'-bis(4-nitrophenyl)-

$C_{13}H_{10}O$
7616 - Methanone, diphenyl-
12049 - 9H-Xanthene

$C_{13}H_{10}O_2$
597 - Benzaldehyde, 3-phenoxy-
2716 - Benzoic acid, phenyl ester
3021 - [1,1'-Biphenyl]-2-carboxylic acid
3022 - [1,1'-Biphenyl]-4-carboxylic acid
7622 - Methanone, (2-hydroxyphenyl)phenyl-
10675 - 2-Propenoic acid, 3-(1-naphthalenyl)-, (E)-
10710 - 2-Propen-1-one, 3-(2-furanyl)-1-phenyl-

$C_{13}H_{10}O_3$
2650 - Benzoic acid, 2-hydroxy-, phenyl ester
2714 - Benzoic acid, 2-phenoxy-
4001 - Carbonic acid, diphenyl ester
7596 - Methanone, bis(2-hydroxyphenyl)-
7597 - Methanone, bis(4-hydroxyphenyl)-
7609 - Methanone, (2,4-dihydroxyphenyl)phenyl-
8682 - 1,4-Pentadien-3-one, 1,5-di-2-furanyl-

$C_{13}H_{10}O_4$
6186 - 5H-Furo[3,2-g][1]benzopyran-5-one, 4-
methoxy-7-methyl-

$C_{13}H_{10}O_6$
7610 - Methanone, (3,4-dihydroxyphenyl)(2,4,6-
trihydroxyphenyl)-

$C_{13}H_{10}S$
7547 - Methanethione, diphenyl-
11811 - 9H-Thioxanthene

$C_{13}H_{11}Br$
1058 - Benzene, 1,1'-(bromomethylene)bis-

$C_{13}H_{11}Cl$
1187 - Benzene, 1,1'-(chloromethylene)bis-
1224 - Benzene, 1-chloro-4-(phenylmethyl)-
3026 - Biphenyl, 2-chloro-2'-methyl-

$C_{13}H_{11}ClO$
9270 - Phenol, 4-chloro-2-(phenylmethyl)-

$C_{13}H_{11}Cl_2NO_2$
9760 - Procymidone

$C_{13}H_{11}N$
840 - Benzenamine, N-(phenylmethylene)-
1835 - Benzenemethanimine, α-phenyl-
3986 - 9H-Carbazole, 3-methyl-
3987 - 9H-Carbazole, 9-methyl-
11000 - Pyridine, 2-(2-phenylethenyl)-, (E)-
11001 - Pyridine, 4-(2-phenylethenyl)-, (E)-

11002 - Pyridine, 2-(2-phenylethenyl)-, (Z)-

$C_{13}H_{11}NO$
626 - Benzamide, N-phenyl-
5993 - Formamide, N,N-diphenyl-
9450 - Phenol, 2-[(phenylimino)methyl]-

$C_{13}H_{11}NO_2$
617 - Benzamide, 2-hydroxy-N-phenyl-
2715 - Benzoic acid, 2-(phenylamino)-
10911 - 3-Pyridinecarboxylic acid, phenylmethyl ester

$C_{13}H_{11}NO_3$
2005 - Benzene, 1-methyl-2-(2-nitrophenoxy)-
2006 - Benzene, 1-methyl-4-(4-nitrophenoxy)-
2502 - Benzoic acid, 4-amino-2-hydroxy-, phenyl
ester
6179 - Furo[2,3-b]quinoline, 4,8-dimethoxy-

$C_{13}H_{11}NS$
1129 - Benzenecarbothioamide, N-phenyl-

$C_{13}H_{11}N_3$
316 - 3,6-Acridinediamine

$C_{13}H_{12}$
632 - 1H-Benz[e]indene, 2,3-dihydro-
1961 - Benzene, 1,1'-methylenebis-
3082 - 1,1'-Biphenyl, 2-methyl-
3083 - 1,1'-Biphenyl, 3-methyl-
3084 - 1,1'-Biphenyl, 4-methyl-
7954 - Naphthalene, 1-(2-propenyl)-

$C_{13}H_{12}Cl_2N_2$
785 - Benzenamine, 4,4'-methylenebis[2-chloro-

$C_{13}H_{12}Cl_2O_4$
168 - Acetic acid, [2,3-dichloro-4-(2-methylene-1-
oxobutyl)phenoxy]-

$C_{13}H_{12}N_2$
5138 - Diazene, (3-methylphenyl)phenyl-
5139 - Diazene, (4-methylphenyl)phenyl-
5971 - 9H-Fluorene-2,7-diamine
7565 - Methanimidamide, N,N'-diphenyl-

$C_{13}H_{12}N_2O$
2717 - Benzoic acid, 2-phenylhydrazide
5135 - Diazene, (2-methoxyphenyl)phenyl-
5136 - Diazene, (3-methoxyphenyl)phenyl-
5137 - Diazene, (4-methoxyphenyl)phenyl-
7589 - Methanone, (3-aminophenyl)phenyl-
9426 - Phenol, 4-methyl-2-(phenylazo)-
11039 - 9H-Pyrido[3,4-b]indole, 7-methoxy-1-methyl-
12010 - Urea, N,N'-diphenyl-
12011 - Urea, N,N-diphenyl-

$C_{13}H_{12}N_2O_3$
11103 - 2,4,6(1H,3H,5H)-Pyrimidinetrione, 5-phenyl-
5-(2-propenyl)-

$C_{13}H_{12}N_2O_3S$
603 - Benzamide, N-[(4-aminophenyl)sulfonyl]-

$C_{13}H_{12}N_2S$
11805 - Thiourea, N,N'-diphenyl-

$C_{13}H_{12}N_4O$
5122 - Diazenecarboxylic acid, phenyl-, 2-
phenylhydrazide

$C_{13}H_{12}N_4S$
5121 - Diazenecarbothioic acid, phenyl-, 2-
phenylhydrazide

$C_{13}H_{12}O$
1910 - Benzenemethanol, α-phenyl-
2012 - Benzene, 1-methyl-2-phenoxy-
2013 - Benzene, 1-methyl-3-phenoxy-
3080 - 1,1'-Biphenyl, 2-methoxy-
3081 - 1,1'-Biphenyl, 4-methoxy-
5978 - 9H-Fluoren-9-one, 1,2,3,4-tetrahydro-
9454 - Phenol, 2-(phenylmethyl)-
9455 - Phenol, 4-(phenylmethyl)-
10473 - 1-Propanone, 1-(1-naphthalenyl)-
10474 - 1-Propanone, 1-(2-naphthalenyl)-

$C_{13}H_{12}O_2$
1546 - 1,4-Benzenediol, 2-(phenylmethyl)-
1938 - Benzene, 1-methoxy-2-phenoxy-
7753 - 1-Naphthalenecarboxaldehyde, 2-ethoxy-
7765 - 1-Naphthalenecarboxylic acid, ethyl ester
7770 - 2-Naphthalenecarboxylic acid, ethyl ester
7952 - 1-Naphthalenepropanoic acid

9406 - Phenol, 3,3'-methylenebis-
9407 - Phenol, 4,4'-methylenebis-
9451 - Phenol, 2-(phenylmethoxy)-
9452 - Phenol, 3-(phenylmethoxy)-
9453 - Phenol, 4-(phenylmethoxy)-

$C_{13}H_{12}O_2S$
2112 - Benzene, [(phenylmethyl)sulfonyl]-

$C_{13}H_{12}O_2S_2$
11818 - o-Toluenesulfonic acid, thio-, S-phenyl ester

$C_{13}H_{12}O_3$
5825 - Ethanone, 1-[6-hydroxy-2-(1-methylethenyl)-5-benzofuranyl]-
7773 - 2-Naphthalenecarboxylic acid, 3-hydroxy-, ethyl ester
7953 - 2-Naphthalenepropanoic acid, 6-hydroxy-

$C_{13}H_{12}S$
2017 - Benzene, 1-methyl-2-(phenylthio)-
2018 - Benzene, 1-methyl-3-(phenylthio)-
2019 - Benzene, 1-methyl-4-(phenylthio)-
2113 - Benzene, [(phenylmethyl)thio]-

$C_{13}H_{13}ClSi$
11380 - Silane, chloromethyldiphenyl-

$C_{13}H_{13}Cl_2N_3O_3$
7338 - Iprodione

$C_{13}H_{13}N$
807 - Benzenamine, N-methyl-N-phenyl-
808 - Benzenamine, 3-methyl-N-phenyl-
839 - Benzenamine, 4-(phenylmethyl)-
1826 - Benzenemethanamine, α-phenyl-
1827 - Benzenemethanamine, N-phenyl-
3013 - [1,1'-Biphenyl]-2-amine, 6-methyl-
3014 - [1,1'-Biphenyl]-4-amine, 3-methyl-
3015 - [1,1'-Biphenyl]-4-amine, 4'-methyl-
11003 - Pyridine, 2-(2-phenylethyl)-

$C_{13}H_{13}N_3$
6282 - Guanidine, N,N'-diphenyl-

$C_{13}H_{13}N_3O_5S_2$
3578 - Butanoic acid, 4-oxo-4-[[4-[(2-thiazolylamino)sulfonyl]phenyl]amino]-

$C_{13}H_{13}O_3P$
9541 - Phosphonic acid, methyl-, diphenyl ester

$C_{13}H_{14}$
7887 - Naphthalene, 1-ethyl-5-methyl-
7920 - Naphthalene, 1-(1-methylethyl)-
7921 - Naphthalene, 2-(1-methylethyl)-
7957 - Naphthalene, 1-propyl-
7958 - Naphthalene, 2-propyl-
8009 - Naphthalene, 1,2,4-trimethyl-
8010 - Naphthalene, 1,2,5-trimethyl-
8011 - Naphthalene, 1,2,6-trimethyl-
8012 - Naphthalene, 1,2,7-trimethyl-
8013 - Naphthalene, 1,3,7-trimethyl-

$C_{13}H_{14}F_3N_3O_4$
5409 - Ethalfluralin

$C_{13}H_{14}N_2$
784 - Benzenamine, 4,4'-methylenebis-
7174 - Hydrazobenzene, 3-methyl-

$C_{13}H_{14}N_2O$
1913 - Benzenemethanol, α-[(2-pyridinylamino)methyl]-

$C_{13}H_{14}N_2O_2S$
2213 - Benzenesulfonamide, 4-[(phenylmethyl)amino]-

$C_{13}H_{14}N_2O_3$
11094 - 2,4,6(1H,3H,5H)-Pyrimidinetrione, 5-ethyl-1-methyl-5-phenyl-

$C_{13}H_{14}N_2O_{13}$
11038 - 3H-Pyrido[3,4-b]indole, 4,9-dihydro-7-methoxy-1-methyl-

$C_{13}H_{14}N_4O$
4010 - Carbonic dihydrazide, 2,2'-diphenyl-

$C_{13}H_{14}O$
7955 - Naphthalene, 1-propoxy-
7956 - Naphthalene, 2-propoxy-

$C_{13}H_{14}OSi$
11502 - Silanol, methyldiphenyl-

$C_{13}H_{14}O_2$
8680 - 2,4-Pentadienoic acid, 5-phenyl-, ethyl ester

$C_{13}H_{14}O_3$
3576 - Butanoic acid, 3-oxo-2-(phenylmethylene)-, ethyl ester

$C_{13}H_{14}O_5$
2774 - 3H-2-Benzopyran-7-carboxylic acid, 4,6-dihydro-8-hydroxy-3,4,5-trimethyl-6-oxo-, (3R-trans)-

$C_{13}H_{14}Si$
11438 - Silane, methyldiphenyl-

$C_{13}H_{15}N$
11222 - Quinoline, 2-butyl-

$C_{13}H_{15}NO_4$
11148 - 1,2-Pyrrolidinedicarboxylic acid, 1-(phenylmethyl) ester, (S)-

$C_{13}H_{15}N_3O_2$
3945 - Carbamic acid, dimethyl-, 3-methyl-1-phenyl-1H-pyrazol-5-yl ester

$C_{13}H_{15}N_3O_3$
7188 - Imazapyr

$C_{13}H_{16}ClNO$
4579 - Cyclohexanone, 2-(2-chlorophenyl)-2-(methylamino)-, (±)-

$C_{13}H_{16}F_3N_3O_4$
509 - Balan
11901 - Trifluralin

$C_{13}H_{16}N_2O_2$
69 - Acetamide, N-[2-(5-methoxy-1H-indol-3-yl)ethyl]-

$C_{13}H_{16}O$
4236 - Cycloheptanone, 2-phenyl-
7100 - 1-Hexen-3-one, 5-methyl-1-phenyl-
9106 - 1-Penten-3-one, 4,4-dimethyl-1-phenyl-

$C_{13}H_{16}O_2$
1412 - Benzene, (3,3-diethoxy-1-propynyl)-
2551 - Benzoic acid, cyclohexyl ester

$C_{13}H_{16}O_3$
2132 - Benzenepropanoic acid, α-ethyl-β-oxo-, ethyl ester

$C_{13}H_{16}O_4$
9946 - Propanedioic acid, di-2-propynyl-, diethyl ester
9975 - Propanedioic acid, phenyl-, diethyl ester

$C_{13}H_{16}O_7$
562 - Benzaldehyde, 2-(β-D-glucopyranosyloxy)-

$C_{13}H_{17}F_3N_4O_4$
9761 - Prodiamine

$C_{13}H_{17}NO$
7409 - Ketone, 1-methyl-4-piperidyl phenyl

$C_{13}H_{17}NO_4$
10932 - 3,5-Pyridinedicarboxylic acid, 2,6-dimethyl-, diethyl ester

$C_{13}H_{17}N_3O$
10851 - 3H-Pyrazol-3-one, 4-(dimethylamino)-1,2-dihydro-1,5-dimethyl-2-phenyl-

$C_{13}H_{18}$
7260 - 1H-Indene, 2,3-dihydro-1,1,4,7-tetramethyl-
8000 - Naphthalene, 1,2,3,4-tetrahydro-1,1,6-trimethyl-

$C_{13}H_{18}ClN_3O_4S_2$
2853 - 2H-1,2,4-Benzothiadiazine-7-sulfonamide, 6-chloro-3-(cyclopentylmethyl)-3,4-dihydro-, 1,1-dioxide

$C_{13}H_{18}N_2O_2$
4900 - 1H-Cyclopentapyrimidine-2,4(3H,5H)-dione, 3-cyclohexyl-6,7-dihydro-

$C_{13}H_{18}O$
1250 - Benzene, 1-cyclohexyl-2-methoxy-
2117 - Benzenepropanal, α-methyl-4-(1-methylethyl)-
6526 - 1-Heptanone, 1-phenyl-

$C_{13}H_{18}O_2$
2610 - Benzoic acid, hexyl ester
8611 - Oxirane, [[4-(1,1-dimethylethyl)phenoxy]methyl]-

$C_{13}H_{18}O_3$
912 - Benzeneacetic acid, α-hydroxy-, 3-methylbutyl ester

$C_{13}H_{18}O_4$
2277 - Benzene, 1,2,3,4-tetramethoxy-5-(2-propenyl)-

$C_{13}H_{18}O_5S$
5933 - Ethofumesate

$C_{13}H_{18}O_7$
6222 - β-D-Glucopyranoside, 2-(hydroxymethyl)phenyl

$C_{13}H_{19}N$
9697 - Piperidine, 1-(1-phenylethyl)-

$C_{13}H_{19}NO$
10428 - 1-Propanone, 2-(diethylamino)-1-phenyl-

$C_{13}H_{19}NO_2$
7390 - Isoquinoline, 1,2,3,4-tetrahydro-6,7-dimethoxy-1,2-dimethyl-, (±)-
7391 - Isoquinoline, 1,2,3,4-tetrahydro-6,7-dimethoxy-1,2-dimethyl-

$C_{13}H_{19}NO_3$
7394 - 8-Isoquinolinol, 1,2,3,4-tetrahydro-6,7-dimethoxy-1,2-dimethyl-, (S)-

$C_{13}H_{19}NO_4S$
2595 - Benzoic acid, 4-[(dipropylamino)sulfonyl]-

$C_{13}H_{19}N_3O_4$
8635 - Pendimethalin

$C_{13}H_{19}N_3O_6S$
8070 - Nitralin

$C_{13}H_{20}$
1745 - Benzene, heptyl-
1948 - Benzene, 1-methyl-2,4-bis(1-methylethyl)-
1949 - Benzene, 1-methyl-3,5-bis(1-methylethyl)-
1992 - Benzene, [2-methyl-1-(1-methylethyl)propyl]-
11868 - 1,12-Tridecadiyne

$C_{13}H_{20}N_2O_2$
2493 - Benzoic acid, 4-amino-, 2-(diethylamino)ethyl ester
5748 - Ethanol, 2-[(2-methylpropyl)amino]-, 4-aminobenzoate (ester)

$C_{13}H_{20}O$
1747 - Benzene, (heptyloxy)-
1842 - Benzenemethanol, α,α-bis(1-methylethyl)-
1867 - Benzenemethanol, α,α-dipropyl-
1881 - Benzenemethanol, α-hexyl-
2102 - Benzenepentanol, α-ethyl-
3885 - 3-Buten-2-one, 4-(2,6,6-trimethyl-1-cyclohexen-1-yl)-, (E)-
3886 - 3-Buten-2-one, 4-(2,6,6-trimethyl-2-cyclohexen-1-yl)-, (E)-(±)-
4148 - o-Cresol, 4-(3-methylpentyl)-
11965 - 3,5,9-Undecatrien-2-one, 6,10-dimethyl-

$C_{13}H_{20}O_3$
6067 - 2-Furancarboxylic acid, octyl ester

$C_{13}H_{20}O_4$
9944 - Propanedioic acid, di-2-propenyl-, diethyl ester

$C_{13}H_{20}O_8$
9988 - 1,3-Propanediol, 2,2-bis[(acetyloxy)methyl]-, diacetate

$C_{13}H_{21}N$
10874 - Pyridine, 2,6-bis(1,1-dimethylethyl)-

$C_{13}H_{22}$
5973 - 1H-Fluorene, dodecahydro-

$C_{13}H_{22}ClN_3O$
602 - Benzamide, 4-amino-N-[2-(diethylamino)ethyl]-, monohydrochloride

$C_{13}H_{22}NO_3PS$
5947 - Fenamiphos

$C_{13}H_{22}N_2$
4291 - Cyclohexanamine, N,N'-methanetetraylbis-

$C_{13}H_{22}N_2O$
8212 - Norea

$C_{13}H_{22}O$
3873 - 3-Buten-2-ol, 4-(2,6,6-trimethyl-1-cyclohexen-1-yl)-
3874 - 3-Buten-2-ol, 4-(2,6,6-trimethyl-2-cyclohexen-1-yl)-
7606 - Methanone, dicyclohexyl-

8113 - 6,8-Nonadien-2-one, 8-methyl-5-(1-
  methylethyl)-, [S-(E)]-

$C_{13}H_{22}O_4$
  4171 - Cyclobutaneacetic acid, 3-(ethoxycarbonyl)-
    2,2-dimethyl-, ethyl ester
  9934 - Propanedioic acid, cyclohexyl-, diethyl ester

$C_{13}H_{24}$
  4429 - Cyclohexane, 1,1'-methylenebis-
  11897 - 1-Tridecyne

$C_{13}H_{24}N_2O$
  10415 - 2-Propanone, 1,3-bis(1-methyl-2-
    pyrrolidinyl)-, (R*,S*)-

$C_{13}H_{24}N_3O_3PS$
  9729 - Pirimiphos-ethyl

$C_{13}H_{24}O_2$
  5341 - 11-Dodecenoic acid, methyl ester
  10646 - 2-Propenoic acid, isodecyl ester
  11894 - 2-Tridecenoic acid
  11895 - 12-Tridecenoic acid
  11977 - 10-Undecenoic acid, ethyl ester

$C_{13}H_{24}O_3$
  6166 - 2-Furanpropanol, tetrahydro-α-[2-(tetrahydro-
    2-furanyl)ethyl]-

$C_{13}H_{24}O_4$
  8133 - Nonanedioic acid, diethyl ester
  9955 - Propanedioic acid, hexyl-, diethyl ester

$C_{13}H_{25}BrO_2$
  5309 - Dodecanoic acid, 2-bromo-, methyl ester

$C_{13}H_{25}N$
  11880 - Tridecanenitrile

$C_{13}H_{26}$
  4327 - Cyclohexane, 2-butyl-1,1,3-trimethyl-
  4405 - Cyclohexane, heptyl-
  4843 - Cyclopentane, octyl-
  8184 - 4-Nonene, 5-butyl-
  11893 - 1-Tridecene

$C_{13}H_{26}Br_2$
  11874 - Tridecane, 1,13-dibromo-

$C_{13}H_{26}N_2$
  4299 - Cyclohexanamine, 4,4'-methylenebis-
  4300 - Cyclohexanamine, 4,4'-methylenebis-,
    [cis(cis)]-
  4301 - Cyclohexanamine, 4,4'-methylenebis-,
    [trans(cis)]-
  4302 - Cyclohexanamine, 4,4'-methylenebis-,
    [trans(trans)]-
  9701 - Piperidine, 4,4'-(1,3-propanediyl)bis-

$C_{13}H_{26}N_2O_4$
  3933 - Carbamic acid, butyl-,2-[[(aminocarbonyl)oxy]
    methyl]-2-methylpentyl ester

$C_{13}H_{26}O$
  11869 - Tridecanal
  11889 - 2-Tridecanone
  11890 - 3-Tridecanone
  11891 - 7-Tridecanone

$C_{13}H_{26}O_2$
  5067 - Decanoic acid, 1-methylethyl ester
  5069 - Decanoic acid, propyl ester
  5314 - Dodecanoic acid, 2-methyl-
  5315 - Dodecanoic acid, methyl ester
  6465 - Heptanoic acid, hexyl ester
  6889 - Hexanoic acid, heptyl ester
  8147 - Nonanoic acid, butyl ester
  8439 - Octanoic acid, pentyl ester
  8901 - Pentanoic acid, octyl ester
  11881 - Tridecanoic acid
  11946 - Undecanoic acid, ethyl ester

$C_{13}H_{27}Br$
  11872 - Tridecane, 1-bromo-

$C_{13}H_{27}N$
  9690 - Piperidine, 1-octyl-

$C_{13}H_{28}$
  8122 - Nonane, 5-butyl-
  11871 - Tridecane

$C_{13}H_{28}O$
  8162 - 5-Nonanol, 5-butyl-
  11883 - 1-Tridecanol

11884 - 2-Tridecanol
11885 - 3-Tridecanol

$C_{13}H_{28}O_2$
  11875 - 1,12-Tridecanediol
  11876 - 1,13-Tridecanediol

$C_{13}H_{28}O_3$
  10080 - Propane, 1,1',1''-[methylidynetris(oxy)]tris[2-
    methyl-

$C_{13}H_{28}O_4$
  10068 - Propane, 1,1',1'',1'''-[methanetetrayltetrakis
    (oxy)]tetrakis-

$C_{13}H_{29}N$
  6354 - 2-Heptanamine, 6-methyl-N-(3-methylbutyl)-
  11870 - 1-Tridecanamine

$C_{14}H_4N_2O_2S_2$
  8056 - Naphtho[2,3-b]-1,4-dithiin-2,3-dicarbonitrile,
    5,10-dihydro-5,10-dioxo-

$C_{14}H_4N_4O_{12}$
  401 - 9,10-Anthracenedione, 1,8-dihydroxy-2,4,5,7-
    tetranitro-

$C_{14}H_4O_6$
  2784 - [2]Benzopyrano[6,5,4-def][2]benzopyran-
    1,3,6,8-tetrone

$C_{14}H_6Br_2O_2$
  390 - 9,10-Anthracenedione, 2,7-dibromo-

$C_{14}H_6Cl_4O_4$
  9152 - Peroxide, bis(2,4-dichlorobenzoyl)

$C_{14}H_6N_2O_6$
  405 - 9,10-Anthracenedione, 1,5-dinitro-

$C_{14}H_7BrO_2$
  385 - 9,10-Anthracenedione, 1-bromo-
  386 - 9,10-Anthracenedione, 2-bromo-

$C_{14}H_7ClF_3NO_5$
  304 - Acifluorfen

$C_{14}H_7ClO_2$
  387 - 9,10-Anthracenedione, 1-chloro-
  388 - 9,10-Anthracenedione, 2-chloro-

$C_{14}H_7NO_4$
  410 - 9,10-Anthracenedione, 1-nitro-
  411 - 9,10-Anthracenedione, 2-nitro-

$C_{14}H_7NO_6$
  399 - 9,10-Anthracenedione, 1,2-dihydroxy-3-nitro-
  400 - 9,10-Anthracenedione, 1,2-dihydroxy-4-nitro-

$C_{14}H_8Br_2$
  375 - Anthracene, 9,10-dibromo-

$C_{14}H_8Cl_2N_4$
  4139 - Clofentezine

$C_{14}H_8Cl_4$
  1374 - Benzene, 1,1'-(dichloroethenylidene)bis[4-
    chloro-

$C_{14}H_8O_2$
  379 - 9,10-Anthracenedione
  9174 - 9,10-Phenanthrenedione

$C_{14}H_8O_3$
  407 - 9,10-Anthracenedione, 1-hydroxy-
  408 - 9,10-Anthracenedione, 2-hydroxy-
  5145 - Dibenz[c,e]oxepin-5,7-dione
  5969 - 9H-Fluorene-2-carboxylic acid, 9-oxo-
  9175 - 9,10-Phenanthrenedione, 2-hydroxy-
  9176 - 9,10-Phenanthrenedione, 3-hydroxy-

$C_{14}H_8O_4$
  391 - 9,10-Anthracenedione, 1,2-dihydroxy-
  392 - 9,10-Anthracenedione, 1,4-dihydroxy-
  393 - 9,10-Anthracenedione, 1,5-dihydroxy-
  394 - 9,10-Anthracenedione, 1,7-dihydroxy-
  395 - 9,10-Anthracenedione, 1,8-dihydroxy-
  396 - 9,10-Anthracenedione, 2,7-dihydroxy-

$C_{14}H_8O_5$
  415 - 9,10-Anthracenedione, 1,2,3-trihydroxy-
  416 - 9,10-Anthracenedione, 1,2,4-trihydroxy-
  417 - 9,10-Anthracenedione, 1,2,6-trihydroxy-
  418 - 9,10-Anthracenedione, 1,2,7-trihydroxy-

$C_{14}H_8O_6$
  412 - 9,10-Anthracenedione, 1,2,5,6-tetrahydroxy-
  413 - 9,10-Anthracenedione, 1,2,5,8-tetrahydroxy-
  414 - 9,10-Anthracenedione, 1,3,5,7-tetrahydroxy-

$C_{14}H_8O_8$
  406 - 9,10-Anthracenedione, 1,2,3,5,6,7-
    hexahydroxy-

$C_{14}H_9Br$
  9164 - Phenanthrene, 9-bromo-

$C_{14}H_9Cl$
  374 - Anthracene, 1-chloro-

$C_{14}H_9ClF_2N_2O_2$
  5176 - Diflubenzuron

$C_{14}H_9ClO_3$
  4079 - Chlorflurecol

$C_{14}H_9Cl_2NO_5$
  2997 - Bifenox

$C_{14}H_9Cl_5$
  2328 - Benzene, 1,1'-(2,2,2-trichloroethylidene)bis[4-
    chloro-

$C_{14}H_9Cl_5O$
  1851 - Benzenemethanol, 4-chloro-α-(4-
    chlorophenyl)-α-(trichloromethyl)-

$C_{14}H_9NO_2$
  380 - 9,10-Anthracenedione, 1-amino-
  381 - 9,10-Anthracenedione, 2-amino-
  426 - Anthracene, 9-nitro-
  7366 - 1H-Isoindole-1,3(2H)-dione, 2-phenyl-

$C_{14}H_9NO_3$
  383 - 9,10-Anthracenedione, 2-amino-1-hydroxy-

$C_{14}H_9NO_4$
  382 - 9,10-Anthracenedione, 3-amino-1,2-dihydroxy-

$C_{14}H_{10}$
  366 - Anthracene
  1714 - Benzene, 1,1'-(1,2-ethynediyl)bis-
  9163 - Phenanthrene

$C_{14}H_{10}Cl_2$
  1365 - Benzene, 1,1'-(1,2-dichloro-1,2-ethenediyl)
    bis-, (E)-
  1366 - Benzene, 1,1'-(1,2-dichloro-1,2-ethenediyl)
    bis-, (Z)-

$C_{14}H_{10}Cl_4$
  1378 - Benzene, 1,1'-(2,2-dichloroethylidene)bis[4-
    chloro-

$C_{14}H_{10}F_3NO_2$
  2729 - Benzoic acid, 2-[[3-(trifluoromethyl)
    phenyl]amino]-

$C_{14}H_{10}N_2O$
  8564 - 1,2,4-Oxadiazole, 3,5-diphenyl-

$C_{14}H_{10}N_2O_2$
  389 - 9,10-Anthracenedione, 1,5-diamino-

$C_{14}H_{10}N_2O_4$
  1494 - Benzene, 1,1'-(1,2-dinitro-1,2-ethenediyl)bis-,
    (Z)-
  1496 - Benzene, 2,4-dinitro-1-(2-phenylethenyl)-, (E)-

$C_{14}H_{10}O$
  431 - 1-Anthracenol
  432 - 9-Anthracenol
  433 - 9(10H)-Anthracenone
  5928 - Ethenone, diphenyl-

$C_{14}H_{10}O_2$
  5584 - Ethanedione, diphenyl-
  9173 - 3,4-Phenanthrenediol

$C_{14}H_{10}O_2S_2$
  5251 - Disulfide, dibenzoyl

$C_{14}H_{10}O_3$
  429 - 1,2,10-Anthracenetriol
  430 - 1,8,9-Anthracenetriol
  2510 - Benzoic acid, anhydride
  2511 - Benzoic acid, 3-benzoyl-
  2512 - Benzoic acid, 4-benzoyl-

$C_{14}H_{10}O_4$
  3038 - [1,1'-Biphenyl]-2,2'-dicarboxylic acid
  9157 - Peroxide, dibenzoyl

$C_{14}H_{10}O_5$
  2616 - Benzoic acid, 2-hydroxy-, 2-carboxyphenyl
    ester
  12054 - 9H-Xanthen-9-one, 1,7-dihydroxy-3-
    methoxy-

$C_{14}H_{10}O_6$
2979 - [Bi-1,4-cyclohexadien-1-yl]-3,3',6,6'-tetrone, 2,2'-dihydroxy-4,4'-dimethyl-

$C_{14}H_{10}O_9$
2575 - Benzoic acid, 3,4-dihydroxy-5-[(3,4,5-trihydroxybenzoyl)oxy]-

$C_{14}H_{11}Br$
1030 - Benzene, 1,1'-(bromoethenylidene)bis-
1080 - Benzene, 1-bromo-2-(2-phenylethenyl)-, (E)-
1081 - Benzene, 1-bromo-2-(2-phenylethenyl)-, (Z)-

$C_{14}H_{11}Cl$
1146 - Benzene, 1,1'-(1-chloro-1,2-ethenediyl)bis-, (E)-
1147 - Benzene, 1,1'-(1-chloro-1,2-ethenediyl)bis-, (Z)-
1222 - Benzene, 1-chloro-2-(2-phenylethenyl)-, (E)-
1223 - Benzene, 1-chloro-3-(2-phenylethenyl)-, (E)-

$C_{14}H_{11}ClO$
975 - Benzeneacetyl chloride, α-phenyl-
5785 - Ethanone, 2-chloro-1,2-diphenyl-

$C_{14}H_{11}N$
365 - 2-Anthracenamine
969 - Benzeneacetonitrile, α-phenyl-
7320 - 1H-Indole, 2-phenyl-
9162 - 9-Phenanthrenamine

$C_{14}H_{11}NO$
3983 - 9H-Carbazole, 9-acetyl-

$C_{14}H_{11}NO_2$
2055 - Benzene, 1-nitro-2-(2-phenylethenyl)-, (E)-
2056 - Benzene, 1-nitro-2-(2-phenylethenyl)-, (Z)-
3982 - 9H-Carbazole-9-acetic acid

$C_{14}H_{11}NO_3$
2513 - Benzoic acid, 3-(benzoylamino)-
7627 - Methanone, (4-methyl-2-nitrophenyl)phenyl-
7630 - Methanone, (4-methylphenyl)(4-nitrophenyl)-

$C_{14}H_{11}NO_4$
2514 - Benzoic acid, 4-(benzoylamino)-2-hydroxy-
2656 - Benzoic acid, 2,2'-iminobis-

$C_{14}H_{12}$
376 - Anthracene, 9,10-dihydro-
1619 - Benzene, 1,1'-(1,2-ethenediyl)bis-, (E)-
1620 - Benzene, 1,1'-(1,2-ethenediyl)bis-, (Z)-
1636 - Benzene, 1,1'-ethenylidenebis-
5974 - 9H-Fluorene, 9-methyl-
9171 - Phenanthrene, 9,10-dihydro-

$C_{14}H_{12}Cl_2$
1678 - Benzene, 1,1'-ethylidenebis[4-chloro-

$C_{14}H_{12}Cl_2O$
1850 - Benzenemethanol, 4-chloro-α-(4-chlorophenyl)-α-methyl-

$C_{14}H_{12}F_3NO_4S_2$
9149 - Perfluidone

$C_{14}H_{12}N_2$
2385 - 1H-Benzimidazole, 2-(phenylmethyl)-
9192 - 1,10-Phenanthroline, 2,9-dimethyl-
11196 - Quinazoline, 3,4-dihydro-3-phenyl-

$C_{14}H_{12}N_2O_2$
5495 - Ethanediamide, N,N'-diphenyl-
9449 - Phenol, 4-(phenylazo)-, acetate (ester)

$C_{14}H_{12}N_2S$
782 - Benzenamine, 4-(6-methyl-2-benzothiazolyl)-

$C_{14}H_{12}N_4O_2S$
2205 - Benzenesulfonamide, 4-amino-N-2-quinoxalinyl-

$C_{14}H_{12}O$
877 - Benzeneacetaldehyde, α-phenyl-
5776 - Ethanone, 1-[1,1'-biphenyl]-4-yl-
5812 - Ethanone, 1,2-diphenyl-
7631 - Methanone, (2-methylphenyl)phenyl-
7632 - Methanone, (3-methylphenyl)phenyl-
7633 - Methanone, (4-methylphenyl)phenyl-

$C_{14}H_{12}O_2$
936 - Benzeneacetic acid, α-phenyl-
1545 - 1,3-Benzenediol, 5-(2-phenylethenyl)-, (E)-
2699 - Benzoic acid, 2-methylphenyl ester
2700 - Benzoic acid, 4-methylphenyl ester
2718 - Benzoic acid, 2-(phenylmethyl)-

2719 - Benzoic acid, 4-(phenylmethyl)-
2720 - Benzoic acid, phenylmethyl ester
3009 - [1,1'-Biphenyl]-4-acetic acid
3091 - [1,1'-Biphenyl]-4-ol, acetate
5822 - Ethanone, 2-hydroxy-1,2-diphenyl-, (±)-
7623 - Methanone, (2-methoxyphenyl)phenyl-
7624 - Methanone, (3-methoxyphenyl)phenyl-
7625 - Methanone, (4-methoxyphenyl)phenyl-

$C_{14}H_{12}O_3$
918 - Benzeneacetic acid, α-hydroxy-α-phenyl-
2397 - 2H,8H-Benzo[1,2-b:5,4-b']dipyran-2-one, 8,8-dimethyl-
2651 - Benzoic acid, 2-hydroxy-, phenylmethyl ester
7620 - Methanone, (2-hydroxy-4-methoxyphenyl)phenyl-
9383 - Phenol, 2-methoxy-, benzoate

$C_{14}H_{12}O_4$
7608 - Methanone, (2,6-dihydroxy-4-methoxyphenyl)phenyl-
7619 - Methanone, (2-hydroxy-4-methoxyphenyl)(2-hydroxyphenyl)-

$C_{14}H_{12}O_5$
6185 - 5H-Furo[3,2-g][1]benzopyran-5-one, 4,9-dimethoxy-7-methyl-

$C_{14}H_{12}S_2$
11661 - Thianthrene, 2,7-dimethyl-

$C_{14}H_{13}N$
809 - Benzenamine, 2-methyl-N-(phenylmethylene)-
810 - Benzenamine, 3-methyl-N-(phenylmethylene)-
811 - Benzenamine, 4-methyl-N-(phenylmethylene)-
3985 - 9H-Carbazole, 9-ethyl-

$C_{14}H_{13}NO$
35 - Acetamide, N-[1,1'-biphenyl]-2-yl-
57 - Acetamide, N,N-diphenyl-
621 - Benzamide, N-(2-methylphenyl)-
622 - Benzamide, N-(3-methylphenyl)-
623 - Benzamide, N-(4-methylphenyl)-
7626 - Methanone, [2-(methylamino)phenyl]phenyl-

$C_{14}H_{13}NO_4$
6181 - Furo[2,3-b]quinoline, 4,7,8-trimethoxy-

$C_{14}H_{13}NO_4S$
2721 - Benzoic acid, 4-[[(phenylmethyl)sulfonyl]amino]-

$C_{14}H_{14}$
1587 - Benzene, 1,1'-(1,2-ethanediyl)bis-
1677 - Benzene, 1,1'-ethylidenebis-
2014 - Benzene, 1-methyl-2-(phenylmethyl)-
2015 - Benzene, 1-methyl-3-(phenylmethyl)-
2016 - Benzene, 1-methyl-4-(phenylmethyl)-
3050 - 1,1'-Biphenyl, 2,2'-dimethyl-
3051 - 1,1'-Biphenyl, 2,3'-dimethyl-
3052 - 1,1'-Biphenyl, 2,3-dimethyl-
3053 - 1,1'-Biphenyl, 2,4'-dimethyl-
3054 - 1,1'-Biphenyl, 2,4-dimethyl-
3055 - 1,1'-Biphenyl, 2,5-dimethyl-
3056 - 1,1'-Biphenyl, 2,6-dimethyl-
3057 - 1,1'-Biphenyl, 3,3'-dimethyl-
3058 - 1,1'-Biphenyl, 3,4'-dimethyl-
3059 - 1,1'-Biphenyl, 3,4-dimethyl-
3060 - 1,1'-Biphenyl, 3,5-dimethyl-
3061 - 1,1'-Biphenyl, 4,4'-dimethyl-
3072 - 1,1'-Biphenyl, 2-ethyl-
3073 - 1,1'-Biphenyl, 3-ethyl-
9187 - Phenanthrene, 1,2,3,4-tetrahydro-

$C_{14}H_{14}Cl_2N_2O$
7187 - Imazalil

$C_{14}H_{14}Hg$
7432 - Mercury, bis(4-methylphenyl)-

$C_{14}H_{14}NO_4PS$
9551 - Phosphonothioic acid, phenyl-, O-ethyl O-(4-nitrophenyl) ester

$C_{14}H_{14}N_2$
732 - Benzenamine, 4,4'-(1,2-ethenediyl)bis-, (E)-
5114 - Diazene, bis(2-methylphenyl)-
5115 - Diazene, bis(3-methylphenyl)-, (E)-

$C_{14}H_{14}N_2O$
5116 - Diazene, bis(2-methylphenyl)-, 1-oxide, (Z)-

5117 - Diazene, bis(2-methylphenyl)-, 1-oxide, (E)-
5118 - Diazene, bis(3-methylphenyl)-, 1-oxide, (Z)-
5119 - Diazene, bis(3-methylphenyl)-, 1-oxide, (E)-
5133 - Diazene, (4-ethoxyphenyl)phenyl-
10462 - 1-Propanone, 2-methyl-1,2-di-3-pyridinyl-
12018 - Urea, N-methyl-N,N'-diphenyl-

$C_{14}H_{14}N_3NaO_3S$
2222 - Benzenesulfonic acid, 4-[[4-(dimethylamino)phenyl]azo]-, sodium salt

$C_{14}H_{14}O$
1613 - Benzeneethanol, α-phenyl-
1614 - Benzeneethanol, α-phenyl-, (S)-
1615 - Benzeneethanol, β-phenyl-
2076 - Benzene, 1,1'-oxybis[2-methyl-
2077 - Benzene, 1,1'-oxybis[3-methyl-
2078 - Benzene, 1,1'-oxybis[4-methyl-
2079 - Benzene, 1,1'-[oxybis(methylene)]bis-
3071 - 1,1'-Biphenyl, 3-ethoxy-
3674 - 1-Butanone, 1-(1-naphthalenyl)-
10464 - 1-Propanone, 2-methyl-1-(1-naphthalenyl)-
10465 - 1-Propanone, 2-methyl-1-(2-naphthalenyl)-

$C_{14}H_{14}OS$
2184 - Benzene, 1,1'-[sulfinylbis(methylene)]bis-

$C_{14}H_{14}O_2$
1588 - Benzene, 1,1'-[1,2-ethanediylbis(oxy)]bis-
3045 - 1,1'-Biphenyl, 2,2'-dimethoxy-
3046 - 1,1'-Biphenyl, 3,3'-dimethoxy-
3047 - 1,1'-Biphenyl, 4,4'-dimethoxy-
3067 - [1,1'-Biphenyl]-2,2'-diol, 3,3'-dimethyl-
3068 - [1,1'-Biphenyl]-2,2'-diol, 5,5'-dimethyl-
5575 - 1,2-Ethanediol, 1,2-diphenyl-, (R*,R*)-(±)-
7735 - 1-Naphthaleneacetic acid, ethyl ester

$C_{14}H_{14}O_2S$
2240 - Benzene, 1,1'-sulfonylbis[4-methyl-
2241 - Benzene, 1,1'-[sulfonylbis(methylene)]bis-

$C_{14}H_{14}O_3$
1929 - Benzene, 1-methoxy-2-(3-methoxyphenoxy)-
2075 - Benzene, 1,1'-oxybis[2-methoxy-
2786 - 2H-1-Benzopyran-7-ol, 3,4-dihydro-3-(4-hydroxyphenyl)-, (S)-

$C_{14}H_{14}O_8$
2258 - 1,2,4,5-Benzenetetracarboxylic acid, tetramethyl ester

$C_{14}H_{14}S$
1993 - Benzene, 1-methyl-2-[(4-methylphenyl)thio]-
2285 - Benzene, 1,1'-thiobis[2-methyl-
2286 - Benzene, 1,1'-thiobis[4-methyl-
2287 - Benzene, 1,1'-[thiobis(methylene)]bis-
11570 - Sulfide, m-tolyl p-tolyl

$C_{14}H_{14}S_2$
5245 - Disulfide, bis(3-methylphenyl)
5246 - Disulfide, bis(4-methylphenyl)
5248 - Disulfide, bis(phenylmethyl)

$C_{14}H_{14}Se$
2176 - Benzene, 1,1'-selenobis[4-methyl-

$C_{14}H_{15}Cl_2N$
7710 - 2-Naphthalenamine, N,N-bis(2-chloroethyl)-

$C_{14}H_{15}N$
751 - Benzenamine, N-ethyl-N-phenyl-
795 - Benzenamine, 2-methyl-N-(2-methylphenyl)-
796 - Benzenamine, 3-methyl-N-(3-methylphenyl)-
797 - Benzenamine, 4-methyl-N-(4-methylphenyl)-
838 - Benzenamine, 2-(2-phenylethyl)-
1584 - Benzeneethanamine, α-phenyl-
1821 - Benzenemethanamine, N-(2-methylphenyl)-
1824 - Benzenemethanamine, N-(3-methylphenyl)-
1825 - Benzenemethanamine, N-(4-methylphenyl)-
1828 - Benzenemethanamine, N-(phenylmethyl)-
11007 - Pyridine, 4-(3-phenylpropyl)-

$C_{14}H_{15}N_3$
721 - Benzenamine, N,N-dimethyl-4-(phenylazo)-
798 - Benzenamine, 2-methyl-4-[(2-methylphenyl)azo]-

$C_{14}H_{15}O_3P$
9586 - Phosphorous acid, bis(phenylmethyl) ester

$C_{14}H_{16}$
422 - Anthracene, 1,2,3,4,5,6-hexahydro-

7746 - Naphthalene, 1-butyl-
7828 - Naphthalene, 1-(1,1-dimethylethyl)-
7829 - Naphthalene, 2-(1,1-dimethylethyl)-
7922 - Naphthalene, 1-methyl-4-(1-methylethyl)-
7923 - Naphthalene, 1-methyl-7-(1-methylethyl)-
7924 - Naphthalene, 7-methyl-1-(1-methylethyl)-
7927 - Naphthalene, 1-(1-methylpropyl)-
8001 - Naphthalene, 1,2,4,7-tetramethyl-
8002 - Naphthalene, 1,2,5,6-tetramethyl-
8003 - Naphthalene, 1,3,6,8-tetramethyl-

$C_{14}H_{16}ClN_3O_2$
512 - Bayleton

$C_{14}H_{16}ClN_3O_4S_2$
2851 - 2H-1,2,4-Benzothiadiazine-7-sulfonamide, 3-
bicyclo[2.2.1]hept-5-en-2-yl-6-chloro-3,4-
dihydro-, 1,1-dioxide

$C_{14}H_{16}ClN_5O_5S$
11824 - Triasulfuron

$C_{14}H_{16}ClO_5PS$
4147 - Coumaphos

$C_{14}H_{16}F_3N_3O_4$
9763 - Profluralin

$C_{14}H_{16}N_2$
731 - Benzenamine, 4,4'-(1,2-ethanediyl)bis-
3034 - [1,1'-Biphenyl]-4,4'-diamine, 3,3'-dimethyl-
5508 - 1,2-Ethanediamine, N,N'-diphenyl-
7138 - Hydrazine, 1,2-bis(3-methylphenyl)-
7139 - Hydrazine, 1,2-bis(4-methylphenyl)-

$C_{14}H_{16}N_2O_2$
3033 - [1,1'-Biphenyl]-4,4'-diamine, 3,3'-dimethoxy-

$C_{14}H_{16}N_2O_3$
11089 - 2,4,6(1H,3H,5H)-Pyrimidinetrione, 5,5-
diethyl-1-phenyl-

$C_{14}H_{16}O_2Si$
11413 - Silane, dimethoxydiphenyl-
11415 - Silane, dimethyldiphenoxy-

$C_{14}H_{16}O_4$
9978 - Propanedioic acid, (phenylmethylene)-, diethyl
ester

$C_{14}H_{16}O_9$
10786 - Pyrano[3,2-c][2]benzopyran-6(2H)-one,
3,4,4a,10b-tetrahydro-3,4,8,10-tetrahydroxy-2-
(hydroxymethyl)-9-methoxy-
[2R-(2α,3β,4α,4aα,10bβ)]-

$C_{14}H_{16}Si$
11416 - Silane, dimethyldiphenyl-

$C_{14}H_{17}ClNO_4PS_2$
5105 - Dialifor

$C_{14}H_{17}Cl_3O_4$
257 - Acetic acid, (2,4,5-trichlorophenoxy)-, 2-
butoxyethyl ester

$C_{14}H_{17}N$
7713 - 1-Naphthalenamine, N,N-diethyl-

$C_{14}H_{17}NO_3$
11231 - 1(2H)-Quinolinecarboxylic acid, 2-ethoxy-,
ethyl ester

$C_{14}H_{17}NO_6$
955 - Benzeneacetonitrile,
α-(β-D-glucopyranosyloxy)-

$C_{14}H_{18}$
427 - Anthracene, 1,2,3,4,5,6,7,8-octahydro-
7747 - Naphthalene, 2-butyl-
9184 - Phenanthrene, 1,2,3,4,4a,9,10,10a-octahydro-
9185 - Phenanthrene, 1,2,3,4,5,6,7,8-octahydro-

$C_{14}H_{18}Cl_2O_4$
170 - Acetic acid, (2,4-dichlorophenoxy)-, 2-
butoxyethyl ester

$C_{14}H_{18}N_2O_4$
8566 - Oxadixyl

$C_{14}H_{18}N_2O_5$
9511 - L-Phenylalanine, N-L-α-aspartyl-, 1-methyl
ester

$C_{14}H_{18}N_4O_3$
515 - Benomyl

$C_{14}H_{18}O$
6347 - Heptanal, 2-(phenylmethylene)-

$C_{14}H_{18}O_4$
9977 - Propanedioic acid, (phenylmethyl)-, diethyl
ester

$C_{14}H_{18}O_7$
5818 - Ethanone, 1-[4-(β-D-glucopyranosyloxy)
phenyl]-

$C_{14}H_{19}Cl_2NO_2$
1101 - Benzenebutanoic acid, 4-[bis
(2-chloroethyl)amino]-

$C_{14}H_{19}NO$
11265 - Quinoline, 6-ethoxy-1,2-dihydro-2,2,4-
trimethyl-

$C_{14}H_{19}N_3S$
5507 - 1,2-Ethanediamine, N,N-dimethyl-N'-2-
pyridinyl-N'-(2-thienylmethyl)-

$C_{14}H_{19}O_6P$
4137 - Ciodrin

$C_{14}H_{20}$
1249 - Benzene, (1-cyclohexylethyl)-
1255 - Benzene, (3-cyclopentylpropyl)-
7999 - Naphthalene, 1,2,3,4-tetrahydro-1,1,2,6-
tetramethyl-

$C_{14}H_{20}ClNO_2$
270 - Acetochlor
322 - Alachlor

$C_{14}H_{20}O$
8480 - 1-Octanone, 1-phenyl-

$C_{14}H_{20}O_2$
2695 - Benzoic acid, 4-(1-methylethyl)-, 2-
methylpropyl ester

$C_{14}H_{20}O_8$
5916 - Ethenetetracarboxylic acid, tetraethyl ester

$C_{14}H_{21}N_3O_4$
3893 - Butralin

$C_{14}H_{22}$
990 - Benzene, 1,4-bis(1,1-dimethylethyl)-
1981 - Benzene, (1-methylheptyl)-
2067 - Benzene, octyl-
2268 - Benzene, 1,2,3,4-tetraethyl-
2269 - Benzene, 1,2,4,5-tetraethyl-
9178 - Phenanthrene, dodecahydro-
11610 - 6,8-Tetradecadiyne

$C_{14}H_{22}N_2O$
54 - Acetamide, 2-(diethylamino)-N-(2,6-
dimethylphenyl)-

$C_{14}H_{22}N_2O_3$
11081 - 2,4,6(1H,3H,5H)-Pyrimidinetrione, 5-butyl-1-
cyclohexyl-

$C_{14}H_{22}O$
2069 - Benzene, (octyloxy)-
3883 - 3-Buten-2-one, 4-(2,5,6,6-tetramethyl-1-
cyclohexen-1-yl)-
3884 - 3-Buten-2-one, 4-(2,5,6,6-tetramethyl-2-
cyclohexen-1-yl)-
9219 - Phenol, 2,4-bis(1,1-dimethylethyl)-
9220 - Phenol, 2,6-bis(1,1-dimethylethyl)-
9227 - Phenol, 2,6-bis(1-methylpropyl)-

$C_{14}H_{22}O_2$
1314 - Benzene, 1,4-dibutoxy-
1411 - Benzene, 1-(diethoxymethyl)-4-(1-
methylethyl)-
4528 - Cyclohexanol, 1,1'-(1,2-ethynediyl)bis-

$C_{14}H_{22}O_4$
5559 - Ethanedioic acid, dicyclohexyl ester

$C_{14}H_{22}O_6$
10657 - 2-Propenoic acid, 2-methyl-, 1,2-
ethanediylbis(oxy-2,1-ethanediyl) ester

$C_{14}H_{22}O_8$
5633 - 1,1,2,2-Ethanetetracarboxylic acid, tetraethyl
ester
10166 - 1,2,3-Propanetricarboxylic acid, 2-
(acetyloxy)-, triethyl ester

$C_{14}H_{23}N$
684 - Benzenamine, N,N-dibutyl-
831 - Benzenamine, 4-octyl-
10880 - Pyridine, 4-(1-butylpentyl)-

$C_{14}H_{23}N_3O_{10}$
6257 - Glycine, N,N-bis[2-[bis(carboxymethyl)
amino]ethyl]-

$C_{14}H_{24}$
5240 - Dispiro[5.1.5.1]tetradecane
9186 - Phenanthrene, tetradecahydro-

$C_{14}H_{244}NO_4PS_3$
517 - Bensulide

$C_{14}H_{24}N_2$
1285 - 1,3-Benzenediamine, N,N,N',N'-tetraethyl-

$C_{14}H_{24}O_2$
10295 - Propanoic acid, 2-methyl-, 3,7-dimethyl-2,6-
octadienyl ester, (E)-

$C_{14}H_{26}$
4341 - Cyclohexane, (3-cyclopentylpropyl)-
4386 - Cyclohexane, 1,1'-(1,2-ethanediyl)bis-
4392 - Cyclohexane, 1,1'-ethylidenebis-
11640 - 2-Tetradecyne
11641 - 7-Tetradecyne

$C_{14}H_{26}O$
11637 - 4-Tetradecenoic acid
11638 - 5-Tetradecenoic acid
11639 - 9-Tetradecenoic acid
11896 - 12-Tridecenoic acid, methyl ester

$C_{14}H_{26}O_3$
183 - Acetic acid, ethoxy-, 5-methyl-2-(1-
methylethyl)cyclohexyl ester, (1α,2β,5α)-
6456 - Heptanoic acid, anhydride

$C_{14}H_{26}O_4$
3319 - Butanedioic acid, dipentyl ester
5045 - Decanedioic acid, diethyl ester
5300 - Dodecanedioic acid, dimethyl ester
6783 - Hexanedioic acid, bis(2-methylpropyl) ester
6784 - Hexanedioic acid, dibutyl ester

$C_{14}H_{27}BrO_2$
5308 - Dodecanoic acid, 2-bromoethyl ester

$C_{14}H_{27}ClO$
11635 - Tetradecanoyl chloride

$C_{14}H_{27}N$
11619 - Tetradecanenitrile

$C_{14}H_{28}$
4460 - Cyclohexane, octyl-
4841 - Cyclopentane, nonyl-
11636 - 1-Tetradecene

$C_{14}H_{28}Br_2$
11616 - Tetradecane, 1,14-dibromo-

$C_{14}H_{28}O$
11633 - 2-Tetradecanone
11634 - 3-Tetradecanone
11892 - 3-Tridecanone, 2-methyl-
11963 - 4-Undecanone, 7-ethyl-2-methyl-

$C_{14}H_{28}O_2$
178 - Acetic acid, dodecyl ester
5312 - Dodecanoic acid, ethyl ester
6464 - Heptanoic acid, heptyl ester
6904 - Hexanoic acid, octyl ester
8156 - Nonanoic acid, pentyl ester
8431 - Octanoic acid, hexyl ester
11622 - Tetradecanoic acid
11882 - Tridecanoic acid, methyl ester

$C_{14}H_{29}Br$
11614 - Tetradecane, 1-bromo-
11873 - Tridecane, 1-bromo-12-methyl-

$C_{14}H_{29}Cl$
11615 - Tetradecane, 1-chloro-

$C_{14}H_{29}N$
4289 - Cyclohexanamine, N-(2-ethylhexyl)-
9689 - Piperidine, 1-nonyl-

$C_{14}H_{29}NO$
11611 - Tetradecanamide

$C_{14}H_{30}$
11613 - Tetradecane
11879 - Tridecane, 7-methyl-

$C_{14}H_{30}O$
6430 - Heptane, 1,1'-oxybis-
11630 - 1-Tetradecanol
11631 - 2-Tetradecanol

11632 - 3-Tetradecanol
11886 - 1-Tridecanol, 12-methyl-
11887 - 3-Tridecanol, 2-methyl-
11888 - 4-Tridecanol, 2-methyl

$C_{14}H_{30}O_2$
5297 - Dodecane, 1,1-dimethoxy-
5298 - Dodecane, 1,12-dimethoxy-
11618 - 1,14-Tetradecanediol

$C_{14}H_{30}O_2Sn$
11538 - Stannane, (acetyloxy)tributyl-

$C_{14}H_{30}O_3S$
11620 - 1-Tetradecanesulfonic acid

$C_{14}H_{30}S$
6435 - Heptane, 1,1'-thiobis-
11621 - 1-Tetradecanethiol

$C_{14}H_{31}N$
6352 - 1-Heptanamine, N-heptyl-
11612 - 1-Tetradecanamine

$C_{14}H_{32}N_2$
6750 - 1,6-Hexanediamine, N,N'-dibutyl-

$C_{14}H_{32}N_2O_4$
10379 - 2-Propanol, 1,1',1'',1'''-
(1,2-ethanediyldinitrilo)tetrakis-

$C_{14}H_{42}O_5Si_6$
6990 - Hexasiloxane, tetradecamethyl-

$C_{14}H_{42}O_7Si_7$
4239 - Cycloheptasiloxane, tetradecamethyl-

$C_{15}H_8O_4$
372 - 2-Anthracenecarboxylic acid, 9,10-dihydro-
9,10-dioxo-

$C_{15}H_8O_5$
2466 - 6H-Benzofuro[3,2-c][1]benzopyran-6-one, 3,9-
dihydroxy-

$C_{15}H_8O_6$
371 - 2-Anthracenecarboxylic acid, 9,10-dihydro-4,5-
dihydroxy-9,10-dioxo-

$C_{15}H_9BrO_2$
7265 - 1H-Indene-1,3(2H)-dione, 2-(4-bromophenyl)-

$C_{15}H_9ClO_2$
7266 - 1H-Indene-1,3(2H)-dione, 2-(4-chlorophenyl)-

$C_{15}H_{10}ClF_3N_2O_6S$
5985 - Fomesafen

$C_{15}H_{10}O_2$
368 - 1-Anthracenecarboxylic acid
369 - 2-Anthracenecarboxylic acid
370 - 9-Anthracenecarboxylic acid
409 - 9,10-Anthracenedione, 2-methyl-
2840 - 4H-1-Benzopyran-4-one, 2-phenyl-
2841 - 4H-1-Benzopyran-4-one, 3-phenyl-
7270 - 1H-Indene-1,3(2H)-dione, 2-phenyl-
7588 - Methanone, 2-benzofuranylphenyl-
9165 - 9-Phenanthrenecarboxylic acid

$C_{15}H_{10}O_3$
10199 - Propanetrione, diphenyl-

$C_{15}H_{10}O_4$
398 - 9,10-Anthracenedione, 1,8-dihydroxy-3-methyl-
2811 - 4H-1-Benzopyran-4-one, 5,7-dihydroxy-2-
phenyl-
2827 - 4H-1-Benzopyran-4-one, 7-hydroxy-3-(4-
hydroxyphenyl)-

$C_{15}H_{10}O_5$
397 - 9,10-Anthracenedione, 1,8-dihydroxy-3-
(hydroxymethyl)-
419 - 9,10-Anthracenedione, 1,3,8-trihydroxy-6-
methyl-
1332 - 1,2-Benzenedicarboxylic acid, 3-benzoyl-
2533 - Benzoic acid, 4,4'-carbonylbis-
2806 - 4H-1-Benzopyran-4-one, 5,7-dihydroxy-2-(4-
hydroxyphenyl)-
2807 - 4H-1-Benzopyran-4-one, 5,7-dihydroxy-3-(4-
hydroxyphenyl)-
2844 - 4H-1-Benzopyran-4-one, 5,6,7-trihydroxy-2-
phenyl-

$C_{15}H_{10}O_6$
2813 - 4H-1-Benzopyran-4-one, 2-
(3,4-dihydroxyphenyl)-3,7-dihydroxy-

2814 - 4H-1-Benzopyran-4-one,
2-(3,4-dihydroxyphenyl)-5,7-dihydroxy-
2842 - 4H-1-Benzopyran-4-one, 3,5,7-trihydroxy-2-
(2-hydroxyphenyl)-
2843 - 4H-1-Benzopyran-4-one, 3,5,7-trihydroxy-2-
(4-hydroxyphenyl)-

$C_{15}H_{10}O_7$
2816 - 4H-1-Benzopyran-4-one,
2-(2,4-dihydroxyphenyl)-3,5,7-trihydroxy-
2817 - 4H-1-Benzopyran-4-one,
2-(3,4-dihydroxyphenyl)-3,5,7-trihydroxy-

$C_{15}H_{11}ClF_3NO_4$
8633 - Oxyfluorfen

$C_{15}H_{11}ClN_2O$
2411 - 2H-1,4-Benzodiazepin-2-one, 7-chloro-1,3-
dihydro-5-phenyl-

$C_{15}H_{11}ClO_5$
2848 - 1-Benzopyrylium, 3,5,7-trihydroxy-2-(4-
hydroxyphenyl)-, chloride

$C_{15}H_{11}ClO_7$
2849 - 1-Benzopyrylium, 3,5,7-trihydroxy-2-(3,4,5-
trihydroxyphenyl)-, chloride

$C_{15}H_{11}I_3O_4$
2136 - Benzenepropanoic acid, 4-(4-hydroxy-3-
iodophenoxy)-3,5-diiodo-

$C_{15}H_{11}I_4NO_4$
11930 - L-Tyrosine, O-(4-hydroxy-3,5-diiodophenyl)-
3,5-diiodo-

$C_{15}H_{11}N$
2768 - Benzonitrile, 4-(2-phenylethenyl)-, (Z)-
11288 - Quinoline, 2-phenyl-
11289 - Quinoline, 3-phenyl-
11290 - Quinoline, 4-phenyl-
11291 - Quinoline, 6-phenyl-

$C_{15}H_{11}NO$
8579 - Oxazole, 2,4-diphenyl-
8580 - Oxazole, 2,5-diphenyl-
8581 - Oxazole, 4,5-diphenyl-

$C_{15}H_{11}NO_2$
384 - 9,10-Anthracenedione, 1-amino-2-methyl-
7367 - 1H-Isoindole-1,3(2H)-dione, 2-(phenylmethyl)-

$C_{15}H_{11}NS$
11676 - Thiazole, 2,4-diphenyl-

$C_{15}H_{12}$
423 - Anthracene, 1-methyl-
424 - Anthracene, 2-methyl-
425 - Anthracene, 9-methyl-
9180 - Phenanthrene, 3-methyl-
9181 - Phenanthrene, 4-methyl-

$C_{15}H_{12}I_2O_3$
2135 - Benzenepropanoic acid, 4-hydroxy-3,5-diiodo-
α-phenyl-

$C_{15}H_{12}N_2$
10830 - 1H-Pyrazole, 1,3-diphenyl-
10831 - 1H-Pyrazole, 1,5-diphenyl-

$C_{15}H_{12}N_2O$
5144 - 5H-Dibenz[b,f]azepine-5-carboxamide

$C_{15}H_{12}N_2O_2$
7219 - 2,4-Imidazolidinedione, 5,5-diphenyl-
10843 - 3,5-Pyrazolidinedione, 1,4-diphenyl-

$C_{15}H_{12}N_2O_3$
7488 - Methanediamine, 1-(2-furanyl)-N,N'-bis(2-
furanylmethylene)-

$C_{15}H_{12}N_6O_4$
2494 - Benzoic acid, 4-[[(2-amino-1,4-dihydro-4-oxo-
6-pteridinyl)methyl]formylamino]-

$C_{15}H_{12}O$
5146 - 10H-Dibenzo[a,d]cyclohepten-10-one, 5,11-
dihydro-
5147 - 5H-Dibenzo[a,d]cyclohepten-5-one, 10,11-
dihydro-
7291 - 1H-Inden-1-one, 2,3-dihydro-2-phenyl-
10708 - 2-Propen-1-one, 1,3-diphenyl-
10709 - 2-Propen-1-one, 1,3-diphenyl-, (E)-

$C_{15}H_{12}O_2$
941 - Benzeneacetic acid, α-(phenylmethylene)-
10027 - 1,3-Propanedione, 1,3-diphenyl-

10689 - 2-Propenoic acid, 3-phenyl-, phenyl ester,
(E)-
10711 - 2-Propen-1-one, 3-hydroxy-1,3-diphenyl-

$C_{15}H_{12}O_3$
2516 - Benzoic acid, 2-benzoyl-, methyl ester
2684 - Benzoic acid, 2-(4-methylbenzoyl)-

$C_{15}H_{12}O_4$
2483 - Benzoic acid, 2-(acetyloxy)-, phenyl ester
2667 - Benzoic acid, 2-(4-methoxybenzoyl)-
7509 - Methanediol, dibenzoate

$C_{15}H_{12}O_5$
2802 - 4H-1-Benzopyran-4-one, 2,3-dihydro-5,7-
dihydroxy-2-(4-hydroxyphenyl)-, (S)-

$C_{15}H_{12}O_6$
2686 - Benzoic acid, 3,3'-methylenebis[6-hydroxy-
2812 - 4H-1-Benzopyran-4-one,
2-(3,4-dihydroxyphenyl)-2,3-dihydro-5,7-
dihydroxy-, (S)-

$C_{15}H_{13}N$
7318 - 1H-Indole, 3-methyl-2-phenyl-
7321 - 1H-Indole, 1-(phenylmethyl)-

$C_{15}H_{13}NO$
64 - Acetamide, N-9H-fluoren-2-yl-

$C_{15}H_{13}NO_2S$
9500 - 10H-Phenothiazine-2-acetic acid, 10-methyl-

$C_{15}H_{14}$
1966 - Benzene, 1,1'-(1-methyleneethyl)-
1,2-ethanediyl)bis-
1971 - Benzene, 1,1'-(1-methyl-1,2-ethenediyl)bis-
2161 - Benzene, 1,1'-(1-propene-1,3-diyl)bis-
2166 - Benzene, 1,1'-(1-propenylidene)bis-
7259 - 1H-Indene, 2,3-dihydro-2-phenyl-

$C_{15}H_{14}F_3N_3$
723 - Benzenamine, N,N-dimethyl-4-[[3-
(trifluoromethyl)phenyl]azo]-

$C_{15}H_{14}N_2$
765 - Benzenamine, N,N'-methanetetraylbis[4-
methyl-

$C_{15}H_{14}N_2O$
7228 - 4-Imidazolidinone, 5,5-diphenyl-

$C_{15}H_{14}N_4O_2S$
2201 - Benzenesulfonamide, 4-amino-N-(1-phenyl-
1H-pyrazol-5-yl)-

$C_{15}H_{14}O$
1939 - Benzene, 1-methoxy-4-(2-phenylethenyl)-,
(E)-
2782 - 2H-1-Benzopyran, 3,4-dihydro-2-phenyl-
7599 - Methanone, bis(4-methylphenyl)-
7614 - Methanone, (2,4-dimethylphenyl)phenyl-
7615 - Methanone, (3,4-dimethylphenyl)phenyl-
7628 - Methanone, (2-methylphenyl)
(4-methylphenyl)-
10432 - 2-Propanone, 1,1-diphenyl-
10433 - 2-Propanone, 1,3-diphenyl-

$C_{15}H_{14}O_2$
931 - Benzeneacetic acid, α-methyl-α-phenyl-
2143 - Benzenepropanoic acid, α-phenyl-, (±)-
5839 - Ethanone, 2-methoxy-1,2-diphenyl-

$C_{15}H_{14}O_3$
920 - Benzeneacetic acid, α-hydroxy-α-phenyl-,
methyl ester
3996 - Carbonic acid, bis(2-methylphenyl) ester
7612 - Methanone, (2,4-dimethoxyphenyl)phenyl-
7862 - 1,4-Naphthalenedione, 2-hydroxy-3-(3-methyl-
2-butenyl)-

$C_{15}H_{14}O_4$
2398 - 2H,8H-Benzo[1,2-b:5,4-b']dipyran-2-one, 5-
methoxy-8,8-dimethyl-
6188 - 7H-Furo[3,2-g][1]benzopyran-7-one, 3-
methoxy-2-(1-methylethyl)-
7841 - 1,4-Naphthalenediol, 2-methyl-, diacetate

$C_{15}H_{14}O_5$
7595 - Methanone, bis(2-hydroxy-4-methoxyphenyl)-
9384 - Phenol, 2-methoxy-, carbonate (2:1)
10453 - 1-Propanone, 3-(4-hydroxyphenyl)-1-(2,4,6-
trihydroxyphenyl)-

10790 - 2H-Pyran-2-one, 6-[2-(1,3-benzodioxol-5-yl)ethenyl]-5,6-dihydro-4-methoxy-, [R-(E)]-

C₁₅H₁₄O₆

2846 - 2H-1-Benzopyran-3,5,7-triol, 2-(3,4-dihydroxyphenyl)-3,4-dihydro-, (2S-cis)-

7248 - 2H-Indene-2,2-dicarboxylic acid, 1,3-dihydro-1,3-dioxo-, diethyl ester

7855 - 1,4-Naphthalenedione, 5,8-dihydroxy-6-methoxy-2-methyl-3-(2-oxopropyl)-

11904 - 2H,4aH-1,4,5-Trioxadicyclopent[a,hi]indene-7-carboxylic acid, 3-ethylidene-3,3a,7a,9b-tetrahydro-2-oxo-, methyl ester, [3aS-(3E,3aα,4aβ,7aβ,9aR*,9bβ)]-

C₁₅H₁₅ClF₃N₃O

11900 - Triflumizole

C₁₅H₁₅ClN₂O₂

4088 - Chloroxuron

C₁₅H₁₅ClN₄O₆S

4081 - Chlorimuron

C₁₅H₁₅NO

2771 - Benzophenone, 3-(dimethylamino)-

7613 - Methanone, [4-(dimethylamino)phenyl]phenyl-

C₁₅H₁₅NO₂

3946 - Carbamic acid, diphenyl-, ethyl ester

C₁₅H₁₅N₃O

317 - 3,9-Acridinediamine, 7-ethoxy-

C₁₅H₁₅N₃O₂

2587 - Benzoic acid, 2-[[4-(dimethylamino)phenyl]azo]-

C₁₅H₁₆

1696 - Benzene, 1-ethyl-4-(phenylmethyl)-

1964 - Benzene, 1,1'-methylenebis[4-methyl-

1969 - Benzene, 1,1'-(1-methyl-1,2-ethanediyl)bis-

1976 - Benzene, 1,1'-(1-methylethylidene)bis-

2121 - Benzene, 1,1'-(1,3-propanediyl)bis-

2171 - Benzene, 1,1'-propylidenebis-

C₁₅H₁₆F₅NO₂S₂

5267 - Dithiopyr

C₁₅H₁₆N₂O

12006 - Urea, N,N'-dimethyl-N,N'-diphenyl-

C₁₅H₁₆O

1940 - Benzene, 1-methoxy-4-(1-phenylethyl)-

2158 - Benzenepropanol, β-phenyl-

C₁₅H₁₆O₂

1970 - Benzene, 1,1'-[(1-methyl-1,2-ethanediyl)bis(oxy)]bis-

2122 - Benzene, 1,1'-[1,3-propanediylbis(oxy)]bis-

9413 - Phenol, 4,4'-(1-methylethylidene)bis-

9464 - Phenol, 4,4'-propylidenebis-

C₁₅H₁₆O₃

10377 - 2-Propanol, 1,3-diphenoxy-

C₁₅H₁₆O₈

2821 - 2H-1-Benzopyran-2-one, 7-(β-D-glucopyranosyloxy)-

C₁₅H₁₆O₉

2822 - 2H-1-Benzopyran-2-one, 6-(β-D-glucopyranosyloxy)-7-hydroxy-

C₁₅H₁₇Cl₂N₃O₂

10720 - Propiconazole

C₁₅H₁₇N

1811 - Benzenemethanamine, N-ethyl-N-phenyl-

C₁₅H₁₇N₃

6279 - Guanidine, N,N'-bis(2-methylphenyl)-

C₁₅H₁₈

505 - Azulene, 1,4-dimethyl-7-(1-methylethyl)-

506 - Azulene, 4,8-dimethyl-2-(1-methylethyl)-

7830 - Naphthalene, 1,6-dimethyl-4-(1-methylethyl)-

7941 - Naphthalene, 1-pentyl-

7942 - Naphthalene, 2-pentyl-

C₁₅H₁₈Cl₂N₂O₃

8565 - Oxadiazon

C₁₅H₁₈N₂O₆

3003 - Binapacryl

C₁₅H₁₈N₆O₆S

8069 - Nicosulfuron

C₁₅H₁₈O

7917 - Naphthalene, 1-(3-methylbutoxy)-

7918 - Naphthalene, 2-(3-methylbutoxy)-

7943 - Naphthalene, 1-(pentyloxy)-

7944 - Naphthalene, 2-(pentyloxy)-

C₁₅H₁₈O₂

2939 - Bicyclo[2.2.1]heptan-2-ol, 1-phenyl-, acetate

C₁₅H₁₈O₃

8057 - Naphtho[1,2-b]furan-2,8(3H,4H)-dione, 3a,5,5a,9b-tetrahydro-3,5a,9-trimethyl-,

C₁₅H₁₈O₆

8058 - Naphtho[1,2-b]furan-2,8(3H,4H)-dione, 3a,5,5a,9b-tetrahydro-4,5-dihydroxy-9-methoxy-3,5a-dimethyl-

C₁₅H₁₉NO₂

487 - 8-Azabicyclo[3.2.1]octan-3-ol, 8-methyl-, benzoate (ester), endo-

488 - 8-Azabicyclo[3.2.1]octan-3-ol, 8-methyl-, benzoate (ester), exo-

C₁₅H₁₉N₃O₃

7190 - Imazethapyr

C₁₅H₂₀N₂O

7573 - 7,14-Methano-4H,6H-dipyrido[1,2-a1',2'-e][1,5]diazocin-4-one, 7,7a,8,9,10,11,13,14-octahydro-,

C₁₅H₂₀O₂

6293 - Helenine

8059 - Naphtho[2,3-b]furan-2(3H)-one, 3a,5,6,7,8,8a,9,9a-octahydro-5,8a-dimethyl-3-methylene-,

C₁₅H₂₀O₄

8679 - 2,4-Pentadienoic acid, 5-(1-hydroxy-2,6,6-trimethyl-4-oxo-2-cyclohexen-1-yl)-3-methyl-, [S-(Z,E)]-

9952 - Propanedioic acid, ethylphenyl-, diethyl ester

C₁₅H₂₁NO₄

7439 - Metalaxyl

C₁₅H₂₁N₃O₂

11191 - Pyrrolo[2,3-b]indol-5-ol, 1,2,3,3a,8,8a-hexahydro-1,3a,8-trimethyl-, methylcarbamate (ester), (3aS-cis)-

C₁₅H₂₂

1468 - Benzene, 1-(1,5-dimethyl-4-hexenyl)-4-methyl-

7274 - 1H-Indene, 5-hexyl-2,3-dihydro-

C₁₅H₂₂ClNO₂

7657 - Metolachlor

C₁₅H₂₂N₂O

9639 - 2-Piperidinecarboxamide, N-(2,6-dimethylphenyl)-1-methyl-

C₁₅H₂₂O

8052 - 2(3H)-Naphthalenone, 4,4a,5,6,7,8-hexahydro-4,4a-dimethyl-6-(1-methylethylidene)-, (4R-cis)-

11521 - Spiro[4.5]dec-6-en-8-one, 6,10-dimethyl-2-(1-methylethylidene)-, (5R-cis)-

C₁₅H₂₃ClO₄S

11579 - Sulfurous acid, 2-chloroethyl 2-[4-(1,1-dimethylethyl)phenoxy]-1-methylethyl ester

C₁₅H₂₃NO₄

9655 - 2,6-Piperidinedione, 4-[2-(3,5-dimethyl-2-oxocyclohexyl)-2-hydroxyethyl]-,

C₁₅H₂₃N₃O₄

7371 - Isopropalin

C₁₅H₂₄

2064 - Benzene, nonyl-

2372 - Benzene, 1,2,4-tris(1-methylethyl)-

2373 - Benzene, 1,3,5-tris(1-methylethyl)-

2996 - Bicyclo[7.2.0]undec-4-ene, 4,11,11-trimethyl-8-methylene-, [1R-(1R*,4E,9S*)]-

4066 - Cedrene

4389 - Cyclohexane, 1-ethenyl-1-methyl-2,4-bis(1-methylethenyl)-, (1α,2β,4β)-

4659 - Cyclohexene, 6-ethenyl-6-methyl-1-(1-methylethyl)-3-(1-methylethylidene)-, (S)-

4685 - Cyclohexene, 1-methyl-4-(5-methyl-1-methylene-4-hexenyl)-, (S)-

4954 - Cyclopropa[d]naphthalene, 1,1a,4,4a,5,6,7,8-octahydro-2,4a,8,8-tetramethyl-, [1aS-(1aα,4aβ,8aR*)]-

5006 - 2,6,8-Cycloundecatriene, 2,6,6,9-tetramethyl-, (E,E,E)-

5329 - 1,3,6,10-Dodecatetraene, 3,7,11-trimethyl-, (E,E)-

5331 - 1,6,10-Dodecatriene, 7,11-dimethyl-3-methylene-, (E)-

7568 - 1,4-Methanoazulene, decahydro-4,8,8-trimethyl-9-methylene-, [1S-(1-(1α,3aβ,4α,8aβ)]-

7898 - Naphthalene, 1,2,4a,5,8,8a-hexahydro-4,7-dimethyl-1-(1-methylethyl)-, [1S-(1α,4aβ,8aα)]-

7899 - Naphthalene, 1,2,3,5,6,8a-hexahydro-4,7-dimethyl-1-(1-methylethyl)-, (1S-cis)-

7936 - Naphthalene, 1,2,3,4,4a,5,6,8a-octahydro-7-methyl-4-methylene-1-(1-methylethyl)-, (1α,4aβ,8aα)-

11861 - Tricyclo[4.4.0.02,7]dec-3-ene, 1,3-dimethyl-8-(1-methylethyl)-, stereoisomer

C₁₅H₂₄NO₄PS

7359 - Isofenphos

C₁₅H₂₄N₂O

7426 - Matridin-15-one

7572 - 7,14-Methano-2H,6H-dipyrido[1,2-a1',2'-e][1,5]diazocin-6-one, dodecahydro-, [7R-(7α,7aβ,14α,14aalp

C₁₅H₂₄N₂O₂

7574 - 7,14-Methano-4H,6H-dipyrido[1,2-a1',2'-e][1,5]diazocin-4-one, dodecahydro-9-hydroxy-,

C₁₅H₂₄O

5330 - 2,6,10-Dodecatrienal, 3,7,11-trimethyl-

9093 - 2-Penten-1-ol, 5-(2,3-dimethyltricyclo[2.2.1.02,6]hept-3-yl)-2-methyl-, stereoisomer

9103 - 2-Penten-1-ol, 2-methyl-5-(2-methyl-3-methylenebicyclo[2.2.1]hept-2-yl)-, [1S-[1α,2α(Z),4α]

9222 - Phenol, 2,4-bis(1,1-dimethylethyl)-5-methyl-

9223 - Phenol, 2,4-bis(1,1-dimethylethyl)-6-methyl-

9224 - Phenol, 2,6-bis(1,1-dimethylethyl)-4-methyl-

C₁₅H₂₄O₈

10139 - 1,1,2,3-Propanetetracarboxylic acid, tetraethyl ester

10140 - 1,1,3,3-Propanetetracarboxylic acid, tetraethyl ester

C₁₅H₂₅ClN₂O₂

2531 - Benzoic acid, 4-(butylamino)-, 2-(dimethylamino)ethyl ester, monohydrochloride

C₁₅H₂₆N₂

7571 - 7,14-Methano-2H,6H-dipyrido[1,2-a1',2'-e][1,5]diazocine, dodecahydro-,

C₁₅H₂₆O

507 - 5-Azulenemethanol, 1,2,3,4,5,6,7,8-octahydro-α,α,3,8-tetramethyl-,

4989 - 1H-Cycloprop[e]azulen-4-ol, decahydro-1,1,4,7tetramethyl,(1aR,4R,4aS,7R,7aS,7bS)-

5332 - 1,6,10-Dodecatrien-3-ol, 3,7,11-trimethyl-, [S-(Z)]-

5333 - 2,6,10-Dodecatrien-1-ol, 3,7,11-trimethyl-, (E,E)-

5334 - 2,6,10-Dodecatrien-1-ol, 3,7,11-trimethyl-, (Z,E)-

7569 - 1,4-Methanoazulen-9-ol, decahydro-1,5,5,8a-tetramethyl-, [1R-(1α,3aβ,4α,8aβ,9S*)]

7570 - 1H-3a,7-Methanoazulen-6-ol, octahydro-3,6,8,8-tetramethyl-,

7586 - 1,6-Methanonaphthalen-1(2H)-ol, octahydro-4a,8,9,9-tetramethyl-,

C₁₅H₂₆O₂

3565 - Butanoic acid, 3-methyl-, 1,7,7-trimethylbicyclo[2.2.1]hept-2-yl ester, (1R-endo)-

C₁₅H₂₆O₆

3586 - Butanoic acid, 1,2,3-propanetriyl ester

$C_{15}H_{26}Si$
11448 - Silane, phenyltripropyl-

$C_{15}H_{28}$
4430 - Cyclohexane, 1,1'-(1-methyl-1,2-ethanediyl)bis-
4432 - Cyclohexane, 1,1'-(1-methylethylidene)bis-
4466 - Cyclohexane, 1,1'-(1,3-propanediyl)bis-
4472 - Cyclohexane, 1,1'-propylidenebis-
8656 - 1-Pentadecyne

$C_{15}H_{28}NNaO_3$
6269 - Glycine, N-methyl-N-(1-oxododecyl)-, sodium salt

$C_{15}H_{28}O$
4761 - Cyclopentadecanone

$C_{15}H_{28}O_2$
3554 - Butanoic acid, 3-methyl-, 5-methyl-2-(1-methylethyl)cyclohexyl ester
8562 - Oxacyclohexadecan-2-one

$C_{15}H_{28}O_4$
9937 - Propanedioic acid, dibutyl-, diethyl ester

$C_{15}H_{30}$
4401 - Cyclohexane, 1-ethyl-1-methyl-2,4-bis(1-methylethyl)-, [1R-(1α,2β,4β)]-
4457 - Cyclohexane, nonyl-
4759 - Cyclopentadecane
4790 - Cyclopentane, decyl-
8653 - 1-Pentadecene
8654 - 7-Pentadecene

$C_{15}H_{30}N_2$
9702 - Piperidine, 4,4'-(1,3-propanediyl)bis[1-methyl-

$C_{15}H_{30}O$
4760 - Cyclopentadecanol
8640 - Pentadecanal
8651 - 2-Pentadecanone
8652 - 8-Pentadecanone
8655 - 1-Pentadecen-3-ol

$C_{15}H_{30}O_2$
5316 - Dodecanoic acid, 1-methylethyl ester
5320 - Dodecanoic acid, propyl ester
6468 - Heptanoic acid, octyl ester
8430 - Octanoic acid, heptyl ester
8645 - Pentadecanoic acid
11625 - Tetradecanoic acid, methyl ester

$C_{15}H_{30}O_4$
5310 - Dodecanoic acid, 2,3-dihydroxypropyl ester, (±)-

$C_{15}H_{31}Br$
8643 - Pentadecane, 1-bromo-

$C_{15}H_{32}$
8642 - Pentadecane

$C_{15}H_{32}O$
8647 - 2-Pentadecanol
8648 - 3-Pentadecanol

$C_{15}H_{33}B$
3122 - Borane, tris(3-methylbutyl)-

$C_{15}H_{33}N$
3224 - 1-Butanamine, 2-methyl-N,N-bis(2-methylbutyl)-
3225 - 1-Butanamine, 3-methyl-N,N-bis(3-methylbutyl)-
8641 - 1-Pentadecanamine
8705 - 1-Pentanamine, N,N-dipentyl-

$C_{13}H_{29}N_3 \cdot C_2H_4O_2$
5347 - Dodine

$C_{15}H_{33}O_4P$
9565 - Phosphoric acid, tripentyl ester

$C_{16}H_8O_2S_2$
2404 - Benzo[b]thiophen-3(2H)-one, 2-(3-oxobenzo[b]thien-2(3H)-ylidene)-

$C_{16}H_9N_3Na_2O_{10}S_2$
7878 - 2,7-Naphthalenedisulfonic acid, 4,5-dihydroxy-3-[(4-nitrophenyl)azo]-, disodium salt

$C_{16}H_{10}$
5966 - Fluoranthene
10853 - Pyrene

$C_{16}H_{10}Cl_4O_5$
5153 - 11H-Dibenzo[b,e][1,4]dioxepin-11-one, 2,4,7,9-tetrachloro-3-hydroxy-8-methoxy-1,6-dimethyl-

$C_{16}H_{10}N_2$
2394 - Benzo[a]phenazine

$C_{16}H_{10}N_2O_2$
7331 - 3H-Indol-3-one, 2-(1,3-dihydro-3-oxo-2H-indol-2-ylidene)-1,2-dihydro-

$C_{16}H_{11}N$
2405 - 7H-Benzo[c]carbazole

$C_{16}H_{11}NO$
7640 - Methanone, phenyl-4-quinolinyl-

$C_{16}H_{11}NO_2$
7866 - 1,4-Naphthalenedione, 2-(phenylamino)-
11235 - 4-Quinolinecarboxylic acid, 2-phenyl-
11309 - 8-Quinolinol, benzoate (ester)

$C_{16}H_{11}NO_3$
11233 - 4-Quinolinecarboxylic acid, 3-hydroxy-2-phenyl-

$C_{16}H_{12}$
7947 - Naphthalene, 1-phenyl-
7948 - Naphthalene, 2-phenyl-

$C_{16}H_{12}N_2O$
8036 - 1-Naphthalenol, 2-(phenylazo)-
8037 - 2-Naphthalenol, 1-(phenylazo)-

$C_{16}H_{12}O$
6100 - Furan, 2,5-diphenyl-
7945 - Naphthalene, 1-phenoxy-
7946 - Naphthalene, 2-phenoxy-

$C_{16}H_{12}O_2$
402 - 9,10-Anthracenedione, 1,4-dimethyl-
403 - 9,10-Anthracenedione, 2,3-dimethyl-
404 - 9,10-Anthracenedione, 2,6-dimethyl-

$C_{16}H_{12}O_3$
7269 - 1H-Indene-1,3(2H)-dione, 2-(4-methoxyphenyl)-

$C_{16}H_{12}O_4$
2830 - 4H-1-Benzopyran-4-one, 7-hydroxy-3-(4-methoxyphenyl)-
3780 - 2-Butenedioic acid (Z)-, diphenyl ester

$C_{16}H_{12}O_5$
2809 - 4H-1-Benzopyran-4-one, 5,7-dihydroxy-2-(4-methoxyphenyl)-
2810 - 4H-1-Benzopyran-4-one, 5,7-dihydroxy-6-methoxy-2-phenyl-
2828 - 4H-1-Benzopyran-4-one, 5-hydroxy-3-(4-hydroxyphenyl)-7-methoxy-

$C_{16}H_{12}O_6$
628 - Benz[b]indeno[1,2-d]pyran-9(6H)-one, 6a,7-dihydro-3,4,6a,10-tetrahydroxy-

$C_{16}H_{12}O_7$
2815 - 4H-1-Benzopyran-4-one, 2-(3,4-dihydroxyphenyl)-3,5-dihydroxy-7-methoxy-

$C_{16}H_{12}S$
7951 - Naphthalene, 1-(phenylthio)-

$C_{16}H_{13}ClN_2O$
2410 - 2H-1,4-Benzodiazepin-2-one, 7-chloro-1,3-dihydro-1-methyl-5-phenyl-

$C_{16}H_{13}N$
7387 - Isoquinoline, 3-(phenylmethyl)-
7728 - 2-Naphthalenamine, N-phenyl-

$C_{16}H_{13}N_3$
7729 - 2-Naphthalenamine, 1-(phenylazo)-

$C_{16}H_{14}$
377 - Anthracene, 1,3-dimethyl-
378 - Anthracene, 2,10-dimethyl-
420 - Anthracene, 9-ethyl-
1095 - Benzene, 1,1'-(1,3-butadiene-1,4-diyl)bis-, (E,E)-
1096 - Benzene, 1,1'-(1,3-butadiene-1,4-diyl)bis-, (Z,Z)-
9172 - Phenanthrene, 9,10-dimethyl-
9179 - Phenanthrene, 9-ethyl-

$C_{16}H_{14}ClN_3O$
2409 - 3H-1,4-Benzodiazepin-2-amine, 7-chloro-N-methyl-5-phenyl-, 4-oxide

$C_{16}H_{14}Cl_2O_3$
4082 - Chlorobenzilate

$C_{16}H_{14}Cl_2O_4$
5171 - Diclofop-methyl

$C_{16}H_{14}N_2O$
11199 - 4(3H)-Quinazolinone, 2-methyl-3-(2-methylphenyl)-
11200 - 4(1H)-Quinazolinone, 1-methyl-2-(phenylmethyl)-

$C_{16}H_{14}O$
3876 - 2-Buten-1-one, 1,3-diphenyl-

$C_{16}H_{14}O_2$
10690 - 2-Propenoic acid, 3-phenyl-, phenylmethyl ester, (E)-
10712 - 2-Propen-1-one, 3-(3-methoxyphenyl)-1-phenyl-
10713 - 2-Propen-1-one, 3-(4-methoxyphenyl)-1-phenyl-

$C_{16}H_{14}O_3$
887 - Benzeneacetic acid, anhydride
2515 - Benzoic acid, 2-benzoyl-, ethyl ester
2682 - Benzoic acid, 2-methyl-, anhydride
2683 - Benzoic acid, 3-methyl-, anhydride

$C_{16}H_{14}O_4$
3039 - [1,1'-Biphenyl]-2,2'-dicarboxylic acid, dimethyl ester
3320 - Butanedioic acid, diphenyl ester
5572 - 1,2-Ethanediol, dibenzoate
6189 - 7H-Furo[3,2-g][1]benzopyran-7-one, 9-[(3-methyl-2-butenyl)oxy]-

$C_{16}H_{14}O_6$
629 - Benz[b]indeno[1,2-d]pyran-3,4,6a,9,10(6H)-pentol, 7,11b-dihydro-, cis-(+)-
2800 - 4H-1-Benzopyran-4-one, 2,3-dihydro-5,7-dihydroxy-2-(3-hydroxy-4-methoxyphenyl)-, (S)-
2801 - 4H-1-Benzopyran-4-one, 2,3-dihydro-5,7-dihydroxy-3-(3-hydroxy-4-methoxyphenyl)-

$C_{16}H_{15}ClO_2$
892 - Benzeneacetic acid, α-chloro-α-phenyl-, ethyl ester

$C_{16}H_{15}Cl_3O_2$
2329 - Benzene, 1,1'-(2,2,2-trichloroethylidene)bis[4-methoxy-

$C_{16}H_{16}$
421 - Anthracene, 9-ethyl-9,10-dihydro-
1113 - Benzene, 1,1'-(1-butenylidene)bis-
1456 - Benzene, 1,1'-(1,2-dimethyl-1,2-ethenediyl)bis-, (E)-
1457 - Benzene, 1,1'-(1,2-dimethyl-1,2-ethenediyl)bis-, (Z)-
1622 - Benzene, 1,1'-(1,2-ethenediyl)bis[2-methyl-
2022 - Benzene, 1,1'-(3-methyl-1-propene-1,3-diyl)bis-
11560 - Stilbene, α-ethyl-

$C_{16}H_{16}N_2OS$
9502 - 10H-Phenothiazine, 10-[(dimethylamino)acetyl]-

$C_{16}H_{16}N_2O_2$
5375 - Ergoline-8-carboxylic acid, 9,10-didehydro-6-methyl-, (8α)-
5376 - Ergoline-8-carboxylic acid, 9,10-didehydro-6-methyl-, (8β)-

$C_{16}H_{16}N_2O_4$
5104 - Desmedipham
9198 - Phenmedipham

$C_{16}H_{16}N_2O_4S$
91 - Acetamide, N,N'-(sulfonyldi-4,1-phenylene)bis-

$C_{16}H_{16}O$
3658 - 2-Butanone, 3,3-diphenyl-

$C_{16}H_{16}O_2$
939 - Benzeneacetic acid, α-phenyl-, ethyl ester
940 - Benzeneacetic acid, 2-phenylethyl ester
1621 - Benzene, 1,1'-(1,2-ethenediyl)bis[4-methoxy-

2144 - Benzenepropanoic acid, α-(phenylmethyl)-
2145 - Benzenepropanoic acid, phenylmethyl ester
3019 - [1,1'-Biphenyl]-4-butanoic acid
5813 - Ethanone, 2-ethoxy-1,2-diphenyl-

$C_{16}H_{16}O_3$
919 - Benzeneacetic acid, α-hydroxy-α-phenyl-, ethyl ester
3583 - Butanoic acid, 2-phenoxy-, phenyl ester

$C_{16}H_{16}O_4$
7806 - 1,8-Naphthalenedicarboxylic acid, diethyl ester

$C_{16}H_{16}O_5$
7854 - 1,4-Naphthalenedione, 5,8-dihydroxy-6-(1-hydroxy-4-methyl-3-pentenyl)-, (S)-

$C_{16}H_{16}Si$
11404 - Silane, diethenyldiphenyl-

$C_{16}H_{17}N$
722 - Benzenamine, N,N-dimethyl-4-(2-phenylethenyl)-, (E)-
3984 - 9H-Carbazole, 9-butyl-

$C_{16}H_{17}NO$
5219 - Diphenamid

$C_{16}H_{17}NO_3$
7665 - Morphinan-3,6-diol, 7,8-didehydro-4,5-epoxy-, (5α,6α)-

$C_{16}H_{17}NO_4$
6201 - Galanthan-1,2-diol, 3,12-didehydro-9,10-[methylenebis(oxy)]-, (1α,2β)-

$C_{16}H_{17}N_2NaO_4S$
11655 - 4-Thia-1-azabicyclo[3.2.0]heptane-2-carboxylic acid, 3,3-dimethyl-7-oxo-6-[(phenylacetyl)amino]-

$C_{16}H_{17}N_3O$
5374 - Ergoline-8-carboxamide, 9,10-didehydro-6-methyl-, (8β)-

$C_{16}H_{18}$
1099 - Benzene, 1,1'-(1,4-butanediyl)bis-
1120 - Benzene, 1,1'-butylidenebis-
1589 - Benzene, 1,1'-(1,2-ethanediyl)bis[3-methyl-
1590 - Benzene, 1,1'-(1,2-ethanediyl)bis[4-methyl-
1676 - Benzene, 1,1'-(1-ethyl-1,2-ethanediyl)bis-
1979 - Benzene, 1-(1-methylethyl)-4-(phenylmethyl)-
2020 - Benzene, 1,1'-(1-methyl-1,3-propanediyl)bis-, (±)-
2021 - Benzene, 1,1'-(2-methyl-1,3-propanediyl)bis-
2111 - Benzene, 1-(phenylmethyl)-4-propyl-
3095 - 1,1'-Biphenyl, 2,2',4,4'-tetramethyl-
3096 - 1,1'-Biphenyl, 2,2',5,5'-tetramethyl-

$C_{16}H_{18}ClN_3S$
9506 - Phenothiazin-5-ium, 3,7-bis(dimethylamino)-, chloride

$C_{16}H_{18}N_2$
9613 - Piperazine, 1,4-diphenyl-

$C_{16}H_{18}N_2O_2$
5112 - Diazene, bis(2-ethoxyphenyl)-
5113 - Diazene, bis(4-ethoxyphenyl)-

$C_{16}H_{18}N_2O_3$
6144 - 2(3H)-Furanone, dihydro-3-(hydroxyphenylmethyl)-4-[(1-methyl-1H-imidazol-5-yl)methyl]-,

$C_{16}H_{18}N_4O_2$
10910 - 4-Pyridinecarboxylic acid, 2-[3-oxo-3-[(phenylmethyl)amino]propyl]hydrazide

$C_{16}H_{18}N_4O_7S$
516 - Bensulfuron-methyl

$C_{16}H_{18}O$
1108 - Benzenebutanol, β-phenyl-, (R)-
2074 - Benzene, 1,1'-oxybis[2-ethyl-
2080 - Benzene, 1,1'-(oxydi-2,1-ethanediyl)bis-
2081 - Benzene, 1,1'-(oxydiethylidene)bis-, (R*,R*)-(±)-

$C_{16}H_{18}O_2$
3048 - 1,1'-Biphenyl, 2,2'-dimethoxy-5,5'-dimethyl-
3049 - 1,1'-Biphenyl, 2,5'-dimethoxy-2'-5-dimethyl-
3069 - [1,1'-Biphenyl]-4,4'-diol, 3,3',5,5'-tetramethyl-
9249 - Phenol, 4,4'-butylidenebis-
9433 - Phenol, 4,4'-(1-methylpropylidene)bis-

$C_{16}H_{18}O_{10}$
2823 - 2H-1-Benzopyran-2-one, 8-(β-D-glucopyranosyloxy)-7-hydroxy-6-methoxy-

$C_{16}H_{19}Cl_2N$
1801 - Benzenemethanamine, N-(2-chloroethyl)-N-(phenylmethyl)-, hydrochloride

$C_{16}H_{19}N$
744 - Benzenamine, 2-ethyl-N-(2-ethylphenyl)-
1585 - Benzeneethanamine, N-(2-phenylethyl)-
1822 - Benzenemethanamine, α-methyl-N-(1-phenylethyl)-

$C_{16}H_{19}NO_3$
6182 - Furo[2,3-b]quinolin-4(2H)-one, 3,9-dihydro-8-methoxy-9-methyl-2-(1-methylethyl)-, (R)-
8567 - 16(15H)-Oxaerythrinan-15-one, 1,2,6,7-tetradehydro-14,17-dihydro-3-methoxy-, (3β)-

$C_{16}H_{19}NO_4$
481 - 8-Azabicyclo[3.2.1]octane-2-carboxylic acid, 3-(benzoyloxy)-8-methyl-, [1R-(exo,exo)]-

$C_{16}H_{20}$
7900 - Naphthalene, 1-hexyl-
7901 - Naphthalene, 2-hexyl-
7905 - Naphthalene, 1-isopropyl-2,4,7-trimethyl-
8014 - Naphthalene, 1,2,5-trimethyl-8-(1-methylethyl)-
8015 - Naphthalene, 1,3,8-trimethyl-5-(1-methylethyl)-

$C_{16}H_{20}N_2$
3036 - [1,1'-Biphenyl]-4,4'-diamine, tetramethyl-
5500 - 1,2-Ethanediamine, N,N'-bis(phenylmethyl)-
11010 - 2-Pyridinepropanamine, N,N-dimethyl-γ-phenyl-

$C_{16}H_{20}N_4$
361 - Aniline, 3,3'-azobis[N,N-dimethyl-

$C_{16}H_{20}O_2Si$
11406 - Silane, diethoxydiphenyl-

$C_{16}H_{20}O_6P_2S_3$
1 - Abate

$C_{16}H_{21}NO_3$
911 - Benzeneacetic acid, α-hydroxy-, 8-methyl-8-azabicyclo[3.2.1]oct-3-yl ester, endo-(±)-
916 - Benzeneacetic acid, α-(hydroxymethyl)-, 8-azabicyclo[3.2.1]oct-3-yl ester, [3(S)-endo]-

$C_{16}H_{21}N_3$
5505 - 1,2-Ethanediamine, N,N-dimethyl-N'-(phenylmethyl)-N'-2-pyridinyl-

$C_{16}H_{22}ClNO_3$
5173 - Diethatyl-ethyl

$C_{16}H_{22}N_4O_2$
9795 - Propanamide, N-(2,3-dihydro-1,5-dimethyl-3-oxo-2-phenyl-1H-pyrazol-4-yl)-2-(dimethylamino)-

$C_{16}H_{22}OSi_2$
5238 - Disiloxane, 1,1,3,3-tetramethyl-1,3-diphenyl-

$C_{16}H_{22}O_4$
1336 - 1,2-Benzenedicarboxylic acid, bis(2-methylpropyl) ester
1338 - 1,4-Benzenedicarboxylic acid, bis(2-methylpropyl) ester
1339 - 1,2-Benzenedicarboxylic acid, dibutyl ester

$C_{16}H_{22}O_6$
1333 - 1,2-Benzenedicarboxylic acid, bis(2-ethoxyethyl) ester

$C_{16}H_{22}O_8$
6226 - β-D-Glucopyranoside, 4-(3-hydroxy-1-propenyl)-2-methoxyphenyl

$C_{16}H_{22}O_{11}$
6216 - α-D-Glucopyranose, pentaacetate
6217 - β-D-Glucopyranose, pentaacetate

$C_{16}H_{23}ClN_3O$
11589 - Tebuconazole

$C_{16}H_{24}N_2O_2$
1325 - 1,2-Benzenedicarboxamide, N,N,N',N'-tetraethyl-
1326 - 1,3-Benzenedicarboxamide, N,N,N',N'-tetraethyl-

$C_{16}H_{25}NO_2$
898 - Benzeneacetic acid, α-ethyl-, 2-(diethylamino)ethyl ester

$C_{16}H_{26}$
1258 - Benzene, decyl-
2090 - Benzene, pentaethyl-
6627 - 6,10-Hexadecadiyne
11609 - 5,9-Tetradecadien-7-yne, 6,9-dimethyl-

$C_{16}H_{26}O$
9221 - Phenol, 2,6-bis(1,1-dimethylethyl)-4-ethyl-

$C_{16}H_{26}O_2$
1501 - 1,4-Benzenediol, 2,5-bis(1,1-dimethylpropyl)-
6666 - 6,10,14-Hexadecatrienoic acid

$C_{16}H_{28}O_2$
4940 - 2-Cyclopentene-1-undecanoic acid, (R)-

$C_{16}H_{30}$
4323 - Cyclohexane, 1,1'-(1,4-butanediyl)bis-
4326 - Cyclohexane, 1,1'-butylidenebis-
4447 - Cyclohexane, 1,1'-(2-methyl-1,3-propanediyl)bis-
6673 - 1-Hexadecyne
6674 - 2-Hexadecyne

$C_{16}H_{30}O$
4762 - Cyclopentadecanone, 2-methyl-
4763 - Cyclopentadecanone, 3-methyl-

$C_{16}H_{30}O_2$
6668 - 7-Hexadecenoic acid
6669 - 9-Hexadecenoic acid
6670 - 9-Hexadecenoic acid, (Z)-

$C_{16}H_{30}O_3$
8424 - Octanoic acid, anhydride

$C_{16}H_{31}ClO$
6665 - Hexadecanoyl chloride

$C_{16}H_{31}N$
6641 - Hexadecanenitrile

$C_{16}H_{32}$
4344 - Cyclohexane, decyl-
4863 - Cyclopentane, undecyl-
6667 - 1-Hexadecene
10580 - 1-Propene, 2-methyl-, tetramer

$C_{16}H_{32}Br_2$
6635 - Hexadecane, 1,16-dibromo-

$C_{16}H_{32}O$
6628 - Hexadecanal
6663 - 3-Hexadecanone

$C_{16}H_{32}O_2$
5065 - Decanoic acid, 2-hexyl-
6644 - Hexadecanoic acid
8152 - Nonanoic acid, heptyl ester
8437 - Octanoic acid, octyl ester
8646 - Pentadecanoic acid, methyl ester
11624 - Tetradecanoic acid, ethyl ester

$C_{16}H_{32}O_4$
5313 - Dodecanoic acid, 2-(2-hydroxyethoxy)ethyl ester

$C_{16}H_{33}Br$
6633 - Hexadecane, 1-bromo-

$C_{16}H_{33}Cl$
6634 - Hexadecane, 1-chloro-

$C_{16}H_{33}F$
6639 - Hexadecane, 1-fluoro-

$C_{16}H_{33}I$
6640 - Hexadecane, 1-iodo-

$C_{16}H_{33}NO$
5290 - Dodecanamide, N,N-diethyl-
6629 - Hexadecanamide

$C_{16}H_{34}$
6632 - Hexadecane

$C_{16}H_{34}O$
6431 - Heptane, 3,3'-[oxybis(methylene)]bis-
6659 - 1-Hexadecanol
6660 - 2-Hexadecanol
6661 - 3-Hexadecanol
8416 - Octane, 1,1'-oxybis-
8649 - 1-Pentadecanol, 2-methyl-
8650 - 6-Pentadecanol, 6-methyl-

$C_{16}H_{34}O_2$
11617 - Tetradecane, 1,1-dimethoxy-
$C_{16}H_{34}O_3$
8552 - Orthoformic acid, triisopentyl ester
$C_{16}H_{34}S$
6643 - 1-Hexadecanethiol
8417 - Octane, 1,1'-thiobis-
$C_{16}H_{35}N$
6631 - 1-Hexadecanamine
8352 - 2-Octanamine, N-(1-methylheptyl)-
8354 - 1-Octanamine, N-octyl-
$C_{16}H_{36}O_4Si$
11505 - Silicic acid (H_4SiO_4), tetrabutyl ester
$C_{16}H_{48}O_6Si_7$
6528 - Heptasiloxane, hexadecamethyl-
$C_{16}H_{48}O_8Si_8$
4741 - Cyclooctasiloxane, hexadecamethyl-
$C_{17}H_9NO_4$
8062 - Naphtho[2,3-f]quinoline-7,12-dione, 5,6-dihydroxy-
$C_{17}H_{10}O$
6292 - 7H-Benz[de]anthracen-7-one
$C_{17}H_{10}O_2$
10854 - 1-Pyrenecarboxylic acid
$C_{17}H_{11}N$
630 - Benz[c]acridine
$C_{17}H_{12}$
2393 - 11H-Benzo[a]fluorene
$C_{17}H_{12}ClFN_2O$
8219 - Nuarimol
$C_{17}H_{12}Cl_2N_2O$
5948 - Fenarimol
$C_{17}H_{12}O_2$
8021 - 2-Naphthalenol, benzoate
$C_{17}H_{12}O_3$
2644 - Benzoic acid, 2-hydroxy-, 1-naphthalenyl ester
2645 - Benzoic acid, 2-hydroxy-, 2-naphthalenyl ester
$C_{17}H_{12}O_7$
6184 - 1H,12H-Furo[3',2':4,5]furo[2,3-h]pyrano[3,4-c][1]benzopyran-1,12-dione, 3,4,7a,10a-tetrahydro-5-meth
$C_{17}H_{13}N$
10947 - Pyridine, 2,6-diphenyl-
$C_{17}H_{13}N_3O_5S_2$
2726 - Benzoic acid, 2-[[[4-[(2-thiazolylamino)sulfonyl]phenyl]amino]carbonyl]-
$C_{17}H_{14}$
4756 - 15H-Cyclopenta[a]phenanthrene, 16,17-dihydro-
7949 - Naphthalene, 1-(phenylmethyl)-
7950 - Naphthalene, 2-(phenylmethyl)-
9182 - Phenanthrene, 9-(1-methylethenyl)-
$C_{17}H_{14}O$
7912 - 1-Naphthalenemethanol, α-phenyl-
8039 - 1-Naphthalenol, 2-(phenylmethyl)-
8040 - 2-Naphthalenol, 1-(phenylmethyl)-
8683 - 1,4-Pentadien-3-one, 1,5-diphenyl-
$C_{17}H_{14}O_6$
4757 - Cyclopenta[c]furo[3',2':4,5]furo[2,3-h][1]benzopyran-1,11-dione, 2,3,6a,8,9,9a-hexahydro-4-methoxy-, (6aR-
$C_{17}H_{15}ClF_3NO_4$
6291 - Haloxyfop-(2-ethoxyethyl)
$C_{17}H_{15}ClO_7$
2847 - 1-Benzopyrylium, 3,5,7-trihydroxy-2-(4-hydroxy-3,5-dimethoxyphenyl)-, chloride
$C_{17}H_{15}N$
7723 - 1-Naphthalenamine, N-(2-methylphenyl)-
7724 - 1-Naphthalenamine, N-(4-methylphenyl)-
7725 - 2-Naphthalenamine, N-(2-methylphenyl)-
7730 - 2-Naphthalenamine, N-(phenylmethyl)-
$C_{17}H_{16}Br_2O_3$
3140 - Bromopropylate

$C_{17}H_{16}O_2$
942 - Benzeneacetic acid, α-(phenylmethylene)-, ethyl ester
$C_{17}H_{16}O_4$
8776 - Pentanedioic acid, diphenyl ester
9927 - Propanedioic acid, bis(phenylmethyl) ester
$C_{17}H_{17}ClO_6$
11520 - Spiro[benzofuran-2(3H),1'-[2]cyclohexene]-3,4'-dione, 7-chloro-2',4,6-trimethoxy-
$C_{17}H_{17}NO_2$
5155 - 4H-Dibenzo[de,g]quinoline-10,11-diol, 5,6,6a,7-tetrahydro-6-methyl-, (R)-
$C_{17}H_{17}N_3O_3$
7189 - Imazaquin
$C_{17}H_{18}O$
7594 - Methanone, bis(2,4-dimethylphenyl)-
8969 - 3-Pentanone, 1,5-diphenyl-
$C_{17}H_{18}O_4$
1103 - Benzenebutanoic acid, 4-hydroxy-γ-(4-hydroxyphenyl)-γ-methyl-
9970 - Propanedioic acid, 1-naphthalenyl-, diethyl ester
10378 - 2-Propanol, 1,3-diphenoxy-, acetate
$C_{17}H_{19}ClN_2S$
9505 - 10H-Phenothiazine-10-propanamine, 2-chloro-N,N-dimethyl-
$C_{17}H_{19}NO_3$
7396 - 7-Isoquinolinol, 1,2,3,4-tetrahydro-1-[(4-hydroxyphenyl)methyl]-6-methoxy-, (S)-
7666 - Morphinan-3,6-diol, 7,8-didehydro-4,5-epoxy-17-methyl- (5α,6α)-
9632 - Piperidine, 1-[5-(1,3-benzodioxol-5-yl)-1-oxo-2,4-pentadienyl]-, (E,E)-
9633 - Piperidine, 1-[5-(1,3-benzodioxol-5-yl)-1-oxo-2,4-pentadienyl]-, (Z,Z)-
$C_{17}H_{19}NO_4$
5952 - Fenoxycarb
7668 - Morphinan-3,6-diol, 7,8-didehydro-4,5-epoxy-17-methyl- (5α,6α)-, 17-oxide
$C_{17}H_{20}$
1697 - Benzene, 1,1'-(1-ethyl-1,3-propanediyl)bis-
1698 - Benzene, 1,1'-(2-ethyl-1,3-propanediyl)bis-
2098 - Benzene, 1,1'-(1,5-pentanediyl)bis-
2108 - Benzene, 1,1'-pentylidenebis-
$C_{17}H_{20}N_2O$
7593 - Methanone, bis[4-(dimethylamino)phenyl]-
12003 - Urea, N,N'-diethyl-N,N'-diphenyl-
$C_{17}H_{20}N_2S$
9504 - 10H-Phenothiazine-10-ethanamine, N,N,α-trimethyl-
$C_{17}H_{20}N_4O_6$
11344 - Riboflavin
$C_{17}H_{20}O_6$
7052 - 4-Hexenoic acid, 6-(1,3-dihydro-4-hydroxy-6-methoxy-7-methyl-3-oxo-5-isobenzofuranyl)-4-methyl-, (E)-
$C_{17}H_{20}Si$
11360 - Silacyclohexane, 1,1-diphenyl-
$C_{17}H_{21}N$
1575 - Benzeneethanamine, N,α-dimethyl-N-(phenylmethyl)-, (+)-
1804 - Benzenemethanamine, N,N-diethyl-α-phenyl-
$C_{17}H_{21}NO$
5422 - Ethanamine, 2-(diphenylmethoxy)-N,N-dimethyl-
$C_{17}H_{21}NO_2$
928 - Benzeneacetic acid, α-methylene-, 8-methyl-8-azabicyclo[3.2.1]oct-3-yl ester, endo-
8067 - Napropamide
$C_{17}H_{21}NO_3$
6200 - Galanthamine
$C_{17}H_{21}NO_4$
482 - 8-Azabicyclo[3.2.1]octane-2-carboxylic acid, 3-(benzoyloxy)-8-methyl-, methyl ester, [1R-(exo,exo)]-

908 - Benzeneacetic acid, α-(hydroxymethyl)-, 9-methyl-3-oxa-9-azatricyclo[3.3.1.02,4]non-7-yl ester, [(7S)-(1α,2β,4β,5α,7β)]-
$C_{17}H_{21}N_3$
657 - Benzenamine, 4,4'-carbonimidoylbis[N,N-dimethyl-
$C_{17}H_{22}$
7895 - Naphthalene, 1-heptyl-
7896 - Naphthalene, 2-heptyl-
$C_{17}H_{22}N_2$
786 - Benzenamine, 4,4'-methylenebis[N,N-dimethyl-
5506 - 1,2-Ethanediamine, N,N-dimethyl-N'-phenyl-N'-(phenylmethyl)-
$C_{17}H_{23}NO_3$
902 - Benzeneacetic acid, α-(hydroxymethyl)- 8-methyl-8-azabicyclo[3.2.1]oct-3-yl ester endo-(±)-
917 - Benzeneacetic acid, α-(hydroxymethyl)-, 8-methyl-8-azabicyclo[3.2.1]oct-3-yl ester, [3(S)-endo]-
$C_{17}H_{23}N_3O$
5511 - 1,2-Ethanediamine, N-[(4-methoxyphenyl)methyl]-N',N'-dimethyl-N-2-pyridinyl-
$C_{17}H_{24}N_2O$
5411 - Ethanamine, 2-[(3-butyl-1-isoquinolinyl)oxy]-N,N-dimethyl-
$C_{17}H_{24}O_9$
6225 - β-D-Glucopyranoside, 4-(3-hydroxy-1-propenyl)-2,6-dimethoxyphenyl
$C_{17}H_{25}N$
9696 - Piperidine, 1-(1-phenylcyclohexyl)-
$C_{17}H_{25}NO_2$
7660 - MGK 264
$C_{17}H_{25}N_3O$
9877 - 1,3-Propanediamine, N,N-diethyl-N'-(6-methoxy-8-quinolinyl)-
$C_{17}H_{26}ClNO_2$
3144 - Butachlor
$C_{17}H_{26}ClNO_3S$
4138 - Clethodim
$C_{17}H_{26}O_4$
4261 - 2,5-Cyclohexadiene-1,4-dione, 2,5-dihydroxy-3-undecyl-
$C_{17}H_{27}NO$
11169 - Pyrrolidine, 1-[2-(3-methylbutoxy)-2-phenylethyl]-
$C_{17}H_{27}NO_3$
7683 - Morpholine, 4-[3-(4-butoxyphenoxy)propyl]-
$C_{17}H_{28}$
2374 - Benzene, undecyl-
$C_{17}H_{28}O$
9225 - Phenol, 2,6-bis(1,1-dimethylpropyl)-4-methyl-
$C_{17}H_{29}NO_3S$
11359 - Sethoxydim
$C_{17}H_{30}O$
4217 - 9-Cycloheptadecen-1-one
4218 - 9-Cycloheptadecen-1-one, (Z)-
$C_{17}H_{30}O_3$
7182 - Hydroprene
$C_{17}H_{32}$
4402 - Cyclohexane, 1,1'-(2-ethyl-1,3-propanediyl)bis-
4464 - Cyclohexane, 1,1'-(1,5-pentanediyl)bis-
$C_{17}H_{32}O_2$
6324 - 9-Heptadecenoic acid
$C_{17}H_{33}N$
6312 - Heptadecanenitrile
$C_{17}H_{34}$
4809 - Cyclopentane, dodecyl-
6322 - 1-Heptadecene
$C_{17}H_{34}Br_2$
6308 - Heptadecane, 1,17-dibromo-
$C_{17}H_{34}O$
6304 - Heptadecanal
6320 - 2-Heptadecanone
6321 - 9-Heptadecanone

$C_{17}H_{34}O_2$
6314 - Heptadecanoic acid
6653 - Hexadecanoic acid, methyl ester
11626 - Tetradecanoic acid, 1-methylethyl ester
11629 - Tetradecanoic acid, propyl ester

$C_{17}H_{35}Br$
6307 - Heptadecane, 1-bromo-

$C_{17}H_{35}N$
9660 - Piperidine, 1-dodecyl-

$C_{17}H_{36}$
6306 - Heptadecane

$C_{17}H_{36}O$
6317 - 1-Heptadecanol
6318 - 2-Heptadecanol
6319 - 9-Heptadecanol

$C_{17}H_{37}N$
6305 - 1-Heptadecanamine
8353 - 1-Octanamine, N-methyl-N-octyl-

$C_{18}H_{10}O_2$
4116 - 5,6-Chrysenedione
7707 - 1,2-Naphthacenedione

$C_{18}H_{10}O_6$
2999 - [2,2'-Bi-1H-indene]-1,1',3,3'(2H,2'H)-tetrone,
2,2'-dihydroxy-

$C_{18}H_{12}$
519 - Benz[a]anthracene
2406 - Benzo[c]phenanthrene
4114 - Chrysene
7703 - Naphthacene
11910 - Triphenylene

$C_{18}H_{12}N_2$
3107 - 2,3'-Biquinoline

$C_{18}H_{12}N_6$
11843 - 1,3,5-Triazine, 2,4,6-tri-2-pyridinyl-

$C_{18}H_{13}N$
4113 - 6-Chrysenamine

$C_{18}H_{13}NO_3$
2702 - Benzoic acid,
2-[(1-naphthalenylamino)carbonyl]-

$C_{18}H_{13}N_3O$
7334 - Indolo[2',3':3,4]pyrido[2,1-b]quinazolin-5(7H)-
one, 8,13-dihydro-

$C_{18}H_{14}$
11595 - 1,1':2',1''-Terphenyl
11596 - 1,1':3',1''-Terphenyl
11597 - 1,1':4',1''-Terphenyl

$C_{18}H_{14}N_4O_5S$
2654 - Benzoic acid, 2-hydroxy-5-[[4-[(2-
pyridinylamino)sulfonyl]phenyl]azo]-

$C_{18}H_{14}O$
5859 - Ethanone, 1-(1-naphthalenyl)-2-phenyl-
7629 - Methanone, (2-methylphenyl)-1-naphthalenyl-

$C_{18}H_{14}O_3$
10684 - 2-Propenoic acid, 3-phenyl-, anhydride

$C_{18}H_{15}As$
450 - Arsine, triphenyl-

$C_{18}H_{15}Bi$
3111 - Bismuthine, triphenyl-

$C_{18}H_{15}ClSn$
11541 - Stannane, chlorotriphenyl-

$C_{18}H_{15}N$
728 - Benzenamine, N,N-diphenyl-

$C_{18}H_{15}OP$
9518 - Phosphine oxide, triphenyl-

$C_{18}H_{15}O_3P$
9593 - Phosphorous acid, triphenyl ester

$C_{18}H_{15}O_4P$
9566 - Phosphoric acid, triphenyl ester

$C_{18}H_{15}P$
9523 - Phosphine, triphenyl-

$C_{18}H_{15}Sb$
11551 - Stibine, triphenyl-

$C_{18}H_{16}N_2$
1271 - 1,4-Benzenediamine, N,N'-diphenyl-
7173 - Hydrazine, triphenyl-

$C_{18}H_{16}N_2O$
8028 - 2-Naphthalenol, 1-[(2,4-dimethylphenyl)azo]-

8029 - 2-Naphthalenol, 1-[(2,5-dimethylphenyl)azo]-

$C_{18}H_{16}OSi$
11504 - Silanol, triphenyl-

$C_{18}H_{16}OSn$
11911 - Triphenyltin hydroxide

$C_{18}H_{16}O_2$
9177 - 9,10-Phenanthrenedione, 1-methyl-7-(1-
methylethyl)-
10691 - 2-Propenoic acid, 3-phenyl-, 3-phenyl-2-
propenyl ester

$C_{18}H_{16}O_4$
3768 - 2-Butenedioic acid (E)-, bis(phenylmethyl)
ester

$C_{18}H_{18}$
9183 - Phenanthrene, 1-methyl-7-(1-methylethyl)-

$C_{18}H_{18}O_2$
5397 - Estra-1,3,5,7,9-pentaen-17-one, 3-hydroxy-
9302 - Phenol, 4,4'-(1,2-diethylidene-1,2-
ethanediyl)bis-

$C_{18}H_{18}O_4$
3298 - Butanedioic acid, bis(phenylmethyl) ester

$C_{18}H_{18}O_5$
5755 - Ethanol, 2,2'-oxybis-, dibenzoate

$C_{18}H_{18}O_6$
3306 - Butanedioic acid, 2,3-dihydroxy- [R-(R*,R*)]-,
bis(phenylmethyl) ester

$C_{18}H_{19}NO_2$
5160 - 4H-Dibenzo[de,g]quinolin-11-ol, 5,6,6a,7-
tetrahydro-10-methoxy-6-methyl-, (R)-

$C_{18}H_{19}NO_3$
5159 - 4H-Dibenzo[de,g]quinolin-10-ol, 5,6,6a,7-
tetrahydro-1,2-dimethoxy-, (R)-

$C_{18}H_{19}NO_4$
7675 - Morphinan-6-one, 7,8-didehydro-4,5-epoxy-
14-hydroxy-3-methoxy-17-methyl-, (5α)-

$C_{18}H_{20}$
7263 - 1H-Indene, 2,3-dihydro-1,1,3-trimethyl-3-
phenyl-

$C_{18}H_{20}Cl_2$
9160 - Perthane

$C_{17}H_{17}N_2 \cdot CH_3O_4S$
5175 - Difenzoquat methyl sulfate

$C_{18}H_{20}O_2$
5398 - Estra-1,3,5(10),7-tetraen-17-one, 3-hydroxy-
9300 - Phenol, 4,4'-(1,2-diethyl-1,2-ethenediyl)bis-,
(E)-

$C_{18}H_{21}NO$
9676 - 2-Piperidinemethanol, α,α-diphenyl-

$C_{18}H_{21}NO_3$
7669 - Morphinan-6-ol, 7,8-didehydro-4,5-epoxy-3-
methoxy-17-methyl-, (5α,6α)-
7670 - Morphinan-6-ol, 7,8-didehydro-4,5-epoxy-3-
methoxy-17-methyl-, (5α,6β)-
7671 - Morphinan-6-ol, 8,14-didehydro-4,5-epoxy-3-
methoxy-17-methyl-, (5α,6α)-
7672 - Morphinan-8-ol, 6,7-didehydro-4,5-epoxy-3-
methoxy-17-methyl-, (5α,8β)-
7677 - Morphinan-6-one, 7,8-didehydro-4-hydroxy-3-
methoxy-17-methyl-
7678 - Morphinan-6-one, 4,5-epoxy-3-methoxy-17-
methyl-, (5α)-

$C_{18}H_{22}$
3076 - 1,1'-Biphenyl, 2,2',4,4',6,6'-hexamethyl-

$C_{18}H_{22}N_2S$
9503 - 10H-Phenothiazine-10-ethanamine, N,N-
diethyl-

$C_{18}H_{22}O_2$
5406 - Estra-1,3,5(10)-trien-17-one, 3-hydroxy-
5407 - Estra-1,3,5(10)-trien-17-one, 3-hydroxy-, (8α)-
9298 - Phenol, 4,4'-(1,2-diethyl-1,2-ethanediyl)bis-,
(R*,S*)-

$C_{18}H_{22}O_4$
1515 - 1,2-Benzenediol, 4,4'-(2,3-dimethyl-1,4-
butanediyl)bis-

$C_{18}H_{23}ClN_2O_2$
5670 - Ethanimidamide, N,N'-bis(4-ethoxyphenyl)-,
monohydrochloride

$C_{18}H_{23}NO_3$
7674 - Morphinan-6-ol, 4,5-epoxy-3-methoxy-17-
methyl-, (5α,6α)-

$C_{18}H_{24}$
7937 - Naphthalene, 1-octyl-
7938 - Naphthalene, 2-octyl-

$C_{18}H_{24}NO_7P$
7673 - Morphinan-6-ol, 7,8-didehydro-4,5-epoxy-3-
methoxy-17-methyl-, (5α,6α)-, phosphate (1
1) (salt)

$C_{18}H_{24}N_2O_6$
5182 - Dinocap

$C_{18}H_{24}O_2$
5399 - Estra-1,3,5(10)-triene-3,17-diol (17β)-
5400 - Estra-1,3,5(10)-triene-3,17-diol, (17α)-
5401 - Estra-1,3,5(10)-triene-3,17-diol, (8α,17β)-
5402 - Estra-5,7,9-triene-3,17-diol, (3β,17β)-

$C_{18}H_{24}O_3$
5404 - Estra-1,3,5(10)-triene-3,16,17-triol, (16α,17β)-
5405 - Estra-1,3,5(10)-triene-3,16,17-triol, (16β,17β)-

$C_{18}H_{26}O_4$
1335 - 1,2-Benzenedicarboxylic acid, bis(3-
methylbutyl) ester

$C_{18}H_{27}NO_3$
8181 - 6-Nonenamide, N-[(4-hydroxy-
3-methoxyphenyl)methyl]-8-methyl-, (E)-

$C_{18}H_{27}NO_5$
894 - Benzeneacetic acid, 4-[2-(diethylamino)-2-
oxoethoxy]-3-methoxy-, propyl ester

$C_{18}H_{27}N_3O$
8737 - 1,5-Pentanediamine, N-(6-methoxy-8-
quinolinyl)-N'-(1-methylethyl)-

$C_{18}H_{28}N_2O_4S$
1582 - Benzeneethanamine, α-methyl-, (S)-, sulfate
(2:1)

$C_{18}H_{28}O$
5327 - 1-Dodecanone, 1-phenyl-

$C_{18}H_{28}O_2$
5317 - Dodecanoic acid, phenyl ester

$C_{18}H_{30}$
1566 - Benzene, dodecyl-
1756 - Benzene, hexaethyl-
2276 - Benzene, 1,2,4,5-tetrakis(1-methylethyl)-
4118 - Chrysene, octadecahydro-
8230 - 7,11-Octadecadiyne

$C_{18}H_{30}O$
9498 - Phenol, 2,4,6-tris(1,1-dimethylethyl)-

$C_{18}H_{30}O_2$
8246 - 9,11,13-Octadecatrienoic acid (Z,Z,Z)
8282 - 9,11,13-Octadecatrienoic acid, (E,E,E)-
8284 - 9,11,13-Octadecatrienoic acid, (E,Z,E)-
8286 - 9,11,13-Octadecatrienoic acid, (Z,Z,E)-
8287 - 9,12,15-Octadecatrienoic acid, (Z,Z,Z)-

$C_{18}H_{30}O_4$
4933 - 1-Cyclopentene-1-pentanoic acid, δ-hydroxy-
3,5-bis(1-methylpropyl)-β-oxo-

$C_{18}H_{32}$
4731 - 1,3-Cyclooctadecadiene

$C_{18}H_{32}I_2$
8225 - 9,12-Octadecadienoic acid

$C_{18}H_{32}I_2O_2$
8300 - 6-Octadecenoic acid, 6,7-diiodo-

$C_{18}H_{32}N_2O_5S$
8926 - 1-Pentanol, 2-(diethylamino)-4-methyl-,
4-aminobenzoate (ester),
monomethanesulfonate (salt)

$C_{18}H_{32}O_2$
4936 - 2-Cyclopentene-1-tridecanoic acid, (S)-
8223 - 9,12-Octadecadienoic acid (Z,Z)-
8224 - 10,12-Octadecadienoic acid, (E,E)-
8316 - 9-Octadecynoic acid

$C_{18}H_{32}O_5$
9120 - Pentonic acid, 5-C-[3,5-bis(1-methylpropyl)-1-
cyclopenten-1-yl]-4-deoxy-

$C_{18}H_{32}O_6$
3561 - Butanoic acid, 3-methyl-, 1,2,3-propanetriyl
ester

$C_{18}H_{32}O_7$
10168 - 1,2,3-Propanetricarboxylic acid, 2-hydroxy-, tributyl ester

$C_{18}H_{32}O_{16}$
6219 - α-D-Glucopyranoside, β-D-fructofuranosyl O-α-D-galactopyranosyl-(1
6221 - α-D-Glucopyranoside, O-α-D-glucopyranosyl-(13)-β-D-fructofuranosyl

$C_{18}H_{32}Si$
11455 - Silane, tributylphenyl-

$C_{18}H_{33}N$
8293 - 9-Octadecenenitrile, (Z)-

$C_{18}H_{34}$
8313 - 1-Octadecyne
8314 - 2-Octadecyne
8315 - 9-Octadecyne

$C_{18}H_{34}O$
8288 - 9-Octadecenal
8289 - 9-Octadecenal, (Z)-

$C_{18}H_{34}OSn$
5011 - Cyhexatin

$C_{18}H_{34}O_2$
8294 - 9-Octadecenoic acid (Z)-
8295 - 11-Octadecenoic acid, (E)-
8296 - 2-Octadecenoic acid, (E)-
8297 - 9-Octadecenoic acid, (E)-

$C_{18}H_{34}O_3$
8303 - 9-Octadecenoic acid, 12-hydroxy-, [R-(Z)-

$C_{18}H_{34}O_4$
5044 - Decanedioic acid, dibutyl ester
6780 - Hexanedioic acid, bis(2-ethylbutyl) ester

$C_{18}H_{34}O_6$
8239 - Octadecanedioic acid, 9,10-dihydroxy-, (R*,R*)-(±)-

$C_{18}H_{35}BrO_2$
8250 - Octadecanoic acid, 18-bromo-

$C_{18}H_{35}ClO$
8281 - Octadecanoyl chloride

$C_{18}H_{35}N$
8243 - Octadecanenitrile

$C_{18}H_{36}$
4382 - Cyclohexane, dodecyl-
4854 - Cyclopentane, tridecyl-
8291 - 1-Octadecene
8292 - 9-Octadecene

$C_{18}H_{36}O$
6638 - Hexadecane, 1-(ethenyloxy)-
8280 - 3-Octadecanone
8311 - 9-Octadecen-1-ol, (E)-
8312 - 9-Octadecen-1-ol, (Z)-

$C_{18}H_{36}O_2$
5068 - Decanoic acid, 2-octyl-
6316 - Heptadecanoic acid, methyl ester
6650 - Hexadecanoic acid, ethyl ester
6662 - 1-Hexadecanol, acetate
8247 - Octadecanoic acid

$C_{18}H_{36}O_3$
6652 - Hexadecanoic acid, 2-hydroxyethyl ester

$C_{18}H_{36}O_4$
8254 - Octadecanoic acid, 9,10-dihydroxy-

$C_{18}H_{36}O_5$
8276 - Octadecanoic acid, 9,10,18-trihydroxy-, (R*,R*)-

$C_{18}H_{37}Br$
8235 - Octadecane, 1-bromo-

$C_{18}H_{37}Cl$
8236 - Octadecane, 1-chloro-

$C_{18}H_{37}Cl_3Si$
11470 - Silane, trichlorooctadecyl-

$C_{18}H_{37}I$
8242 - Octadecane, 1-iodo-

$C_{18}H_{37}NO$
8231 - Octadecanamide

$C_{18}H_{38}$
6310 - Heptadecane, 2-methyl-
6311 - Heptadecane, 9-methyl-
8234 - Octadecane

$C_{18}H_{38}O$
8278 - 1-Octadecanol
8279 - 3-Octadecanol, (±)-

$C_{18}H_{38}O_2$
6636 - Hexadecane, 1,1-dimethoxy-

$C_{18}H_{38}S$
8244 - 1-Octadecanethiol

$C_{18}H_{39}N$
6725 - 1-Hexanamine, N,N-dihexyl-
8233 - 1-Octadecanamine

$C_{18}H_{54}O_7Si_8$
8482 - Octasiloxane, octadecamethyl-

$C_{18}H_{54}O_9Si_9$
4726 - Cyclononasiloxane, octadecamethyl-

$C_{19}H_{10}Br_4O_5S$
9366 - Phenol, 4,4'-(3H-2,1-benzoxathiol-3-ylidene)bis[2,6-dibromo-, S,S-dioxide

$C_{19}H_{12}O_6$
2839 - 2H-1-Benzopyran-2-one, 3,3'-methylenebis[4-hydroxy-

$C_{19}H_{13}N$
318 - Acridine, 9-phenyl-

$C_{19}H_{14}$
521 - Benz[a]anthracene, 11-methyl-
522 - Benz[a]anthracene, 3-methyl-
523 - Benz[a]anthracene, 8-methyl-
4117 - Chrysene, 1-methyl-

$C_{19}H_{14}F_3NO$
5981 - Fluridone

$C_{19}H_{14}O_3$
4278 - 2,5-Cyclohexadien-1-one, 4-[bis(4-hydroxyphenyl)methylene]-

$C_{19}H_{14}O_5S$
9363 - Phenol, 4,4'-(3H-2,1-benzoxathiol-3-ylidene)biS-, S,S-dioxide

$C_{19}H_{15}Br$
1061 - Benzene, 1,1',1''-(bromomethylidyne)tris-

$C_{19}H_{15}Cl$
1196 - Benzene, 1,1',1''-(chloromethylidyne)tris-

$C_{19}H_{15}ClF_3NO_7$
7410 - Lactofen

$C_{19}H_{15}ClN_4$
11653 - 2H-Tetrazolium, 2,3,5-triphenyl-, chloride

$C_{19}H_{15}N$
3988 - 9H-Carbazole, 9-(phenylmethyl)-

$C_{19}H_{15}NO_2$
11236 - 4-Quinolinecarboxylic acid, 2-phenyl-, 2-propenyl ester

$C_{19}H_{16}$
1982 - Benzene, 1,1',1''-methylidynetris-
3093 - 1,1'-Biphenyl, 2-(phenylmethyl)-
3094 - 1,1'-Biphenyl, 4-(phenylmethyl)-

$C_{19}H_{16}N_2$
11041 - 9H-Pyrido[3,4-b]indole, 1-[(2-methylphenyl)methyl]-

$C_{19}H_{16}O$
1866 - Benzenemethanol, α,α-diphenyl-

$C_{19}H_{16}O_3$
1983 - Benzene, 1,1',1''-[methylidynetris(oxy)]tris-

$C_{19}H_{16}O_4$
2832 - 2H-1-Benzopyran-2-one, 4-hydroxy-3-(3-oxo-1-phenylbutyl)-

$C_{19}H_{16}O_5$
225 - Acetic acid, [(4-oxo-2-phenyl-4H-1-benzopyran-7-yl)oxy]-, ethyl ester

$C_{19}H_{17}Cl_2N_3O_3$
5174 - Difenoconazole

$C_{19}H_{17}N$
729 - Benzenamine, 4-(diphenylmethyl)-

$C_{19}H_{17}NO_3$
11216 - Quinoline, 2-[2-(1,3-benzodioxol-5-yl)ethyl]-4-methoxy-

$C_{19}H_{17}N_3$
6288 - Guanidine, N,N',N''-triphenyl-

$C_{19}H_{17}N_3O$
7333 - Indolo[2',3':3,4]pyrido[2,1-b]quinazolin-5(7H)-one, 8,13,13b,14-tetrahydro-14-methyl-, (S)-

$C_{19}H_{18}O_3Si$
11443 - Silane, methyltriphenoxy-

$C_{19}H_{18}O_5$
2464 - 5-Benzofuranpropanol, 2-(1,3-benzodioxol-5-yl)-7-methoxy-

$C_{19}H_{18}Si$
11444 - Silane, methyltriphenyl-

$C_{19}H_{19}NO_3$
2469 - 5H-Benzo[g]-1,3-benzodioxolo[6,5,4-de]quinoline, 6,7,7a,8-tetrahydro-11-methoxy-7-methyl-, (R)-

$C_{19}H_{19}NO_4$
2470 - 5H-Benzo[g]-1,3-benzodioxolo[6,5,4-de]quinolin-12-ol, 6,7,7a,8-tetrahydro-11-methoxy-7-methyl-, (S)-

$C_{19}H_{19}N_7O_6$
6243 - L-Glutamic acid, N-[4-[[(2-amino-1,4-dihydro-4-oxo-6-pteridinyl)methyl]amino]benzoyl]-

$C_{19}H_{20}F_3NO_4$
5962 - Fluazipop-butyl
5963 - Fluazipop-P-butyl

$C_{19}H_{20}N_2O$
11346 - Sarpagan-17-al

$C_{19}H_{21}NO_3$
5158 - 4H-Dibenzo[de,g]quinolin-1-ol, 5,6,6a,7-tetrahydro-2,11-dimethoxy-6-methyl-, (S)-
7679 - Morphinan, 6,7,8,14-tetradehydro-4,5-epoxy-3,6-dimethoxy-17-methyl-, (5α)-

$C_{19}H_{21}NO_4$
5156 - 4H-Dibenzo[de,g]quinoline-2,9-diol, 5,6,6a,7-tetrahydro-1,10-dimethoxy-6-methyl-, (S)-

$C_{19}H_{22}$
1744 - Benzene, 1,1'-(1-heptenylidene)bis-

$C_{19}H_{22}N_2O$
4120 - Cinchonan-9-ol, (9S)-
4133 - Cinchonan-9-ol, (8α,9R)-
10435 - 1-Propanone, 3-(3-ethenyl-4-piperidinyl)-1-(4-quinolinyl)-, (3R-cis)-

$C_{19}H_{22}N_2OS$
5807 - Ethanone, 1-[10-[3-(dimethylamino)propyl]-10H-phenothiazin-2-yl]-

$C_{19}H_{22}N_2O_2$
4119 - Cinchonan-6',9-diol, (8α,9R)-
11347 - Sarpagan-10,17-diol

$C_{19}H_{22}O_2$
9358 - Phenol, 4-[1-ethyl-2-(4-methoxyphenyl)-1-butenyl]-
9359 - Phenol, 4-[1-ethyl-2-(4-methoxyphenyl)-1-butenyl]-, (E)-

$C_{19}H_{22}O_6$
6204 - Gibb-3-ene-1,10-dicarboxylic acid, 2,4a,7-trihydroxy-1-methyl-8-methylene-, 1,4a-lactone,

$C_{19}H_{23}ClN_2$
5140 - 1H-1,4-Diazepine, 1-[(4chlorophenyl)phenylmethyl]hexahydro-4-methyl-

$C_{19}H_{23}NO$
1903 - Benzenemethanol, α-[1-[methyl(3-phenyl-2-propenyl)amino]ethyl]-

$C_{19}H_{23}NO_3$
7664 - Morphinan, 6,7-didehydro-4,5-epoxy-3,6-dimethoxy-17-methyl-, (5α)-

$C_{19}H_{23}NO_4$
7676 - Morphinan-6-one, 7,8-didehydro-4-hydroxy-3,7-dimethoxy-17-methyl-, (9α,13α,14α)-

$C_{19}H_{23}N_3$
342 - Amitraz

$C_{19}H_{23}N_3O_2$
5373 - Ergoline-8-carboxamide, 9,10-didehydro-N-(2-hydroxy-1-methylethyl)-6-methyl-, [8α(S)]-

$C_{19}H_{24}$
1746 - Benzene, 1,1'-heptylidenebis-

C$_{19}$H$_{24}$N$_2$
9627 - 4-Piperidinamine, 1-methyl-*N*-phenyl-*N*-
(phenylmethyl)-
C$_{19}$H$_{24}$N$_2$O
4121 - Cinchonan-9-ol, 10,11-dihydro-, (8α,9*R*)-
4122 - Cinchonan-9-ol, 10,11-dihydro-, (9*S*)-
7306 - 1H-Indole-3-ethanol, 2-(5-ethenyl-1-
azabicyclo[2.2.2]oct-2-yl)-
C$_{19}$H$_{24}$N$_2$OS
3977 - Carbamothioic acid, diphenyl-, *S*-[2-
(diethylamino)ethyl] ester
C$_{19}$H$_{24}$N$_2$O$_2$
6177 - 3aH-Furo[2,3-b]indol-3a-ol, 8a-(5-ethenyl-1-
azabicyclo[2.2.2]oct-2-yl)-2,3,8,8a-tetrahydro-
6178 - 3aH-Furo[2,3-b]indol-3a-ol, 8a-(5-ethenyl-1-
azabicyclo[2.2.2]oct-2-yl)-2,3,8,8a-tetrahydro-
C$_{19}$H$_{24}$O$_3$
353 - Androst-4-ene-3,11,17-trione
C$_{19}$H$_{26}$
7931 - Naphthalene, 1-nonyl-
7932 - Naphthalene, 2-nonyl-
C$_{19}$H$_{26}$N$_2$O
4154 - Curan-17-ol, (16α)-
C$_{19}$H$_{26}$O$_2$
343 - Androsta-1,4-dien-3-one, 17-hydroxy-, (17β)-
352 - Androst-4-ene-3,17-dione
C$_{19}$H$_{26}$O$_3$
336 - Allethrin
C$_{19}$H$_{26}$O$_4$S
10507 - Propargite
C$_{19}$H$_{27}$ClO$_2$
354 - Androst-4-en-3-one, 4-chloro-17-hydroxy-,
(17β)-
C$_{19}$H$_{27}$NO$_3$
2396 - 2H-Benzo[a]quinolizin-2-one, 1,3,4,6,7,11b-
hexahydro-9,10-dimethoxy-3-(2-methylpropyl)-
C$_{19}$H$_{28}$BrNO$_3$
11179 - Pyrrolidinium,
3-[(cyclopentylhydroxyphenylacetyl)oxy]-1,1-
dimethyl-, bromide
C$_{19}$H$_{28}$O$_2$
356 - Androst-4-en-3-one, 17-hydroxy-, (17β)-
C$_{19}$H$_{30}$O$_2$
347 - Androstan-17-one, 3-hydroxy-, (3β,5α)-
348 - Androstan-17-one, 3-hydroxy-, (3α,5α)-
349 - Androstan-3-one, 17-hydroxy-, (5α,17β)-
5318 - Dodecanoic acid, phenylmethyl ester
C$_{19}$H$_{30}$O$_5$
9727 - Piperonyl butoxide
C$_{19}$H$_{32}$
345 - Androstane
C$_{20}$H$_{27}$O$_4$P
9560 - Phosphoric acid, 2-ethylhexyl diphenyl ester
C$_{20}$H$_{28}$
7793 - Naphthalene, 1-decyl-
7794 - Naphthalene, 2-decyl-
C$_{20}$H$_{28}$O$_2$
344 - Androsta-1,4-dien-3-one, 17-hydroxy-17-
methyl-, (17β)-
C$_{20}$H$_{28}$O$_3$S
7968 - 1-Naphthalenesulfonic acid, 3,6-bis(1,1-
dimethylethyl)-, ethyl ester
C$_{20}$H$_{29}$FO$_3$
355 - Androst-4-en-3-one, 9-fluoro-11,17-dihydroxy-
17-methyl-, (11β,17β)-
C$_{20}$H$_{30}$ClN$_3$O$_2$
11226 - 4-Quinolinecarboxamide, 2-butoxy-*N*-[2-
(diethylamino)ethyl]-, monohydrochloride
C$_{20}$H$_{30}$O
11341 - Retinol
C$_{20}$H$_{30}$O$_2$
358 - Androst-4-en-3-one, 17-hydroxy-17-methyl-,
(17β)-
9166 - 1-Phenanthrenecarboxylic acid,
1,2,3,4,4a,5,6,9,10,10a-decahydro-1,4a-
dimethyl-7-(1-methylethyl)-,

9167 - 1-Phenanthrenecarboxylic acid,
1,2,3,4,4a,4b,5,6,10,10a-decahydro-1,4a-
dimethyl-7-(1-methylethyl)-
9169 - 1-Phenanthrenecarboxylic acid,
1,2,3,4,4a,4b,5,9,10,10a-decahydro-1,4a-
dimethyl-7-(1-methylethyl)-,
9170 - 1-Phenanthrenecarboxylic acid, 7-ethenyl-
1,2,3,4,4a,4b,5,6,7,9,10,10a-dodecahydro-
1,4a,7-trimethyl-, [1R-(1α,4aβ,4bα,7β,10aα)]-
C$_{20}$H$_{32}$ClNO
9705 - 1-Piperidinepropanol, α-cyclohexyl-α-phenyl-,
hydrochloride
C$_{20}$H$_{32}$N$_2$O$_3$S
4038 - Carbosulfan
C$_{20}$H$_{32}$O$_2$
346 - Androstane-17-carboxylic acid, (5β,17β)-
350 - Androstan-3-one, 17-hydroxy-17-methyl-,
(5α,17β)-
5360 - 5,8,11,14-Eicosatetraenoic acid
5361 - 5,8,11,14-Eicosatetraenoic acid, (all-Z)-
C$_{20}$H$_{34}$
2266 - Benzene, tetradecyl-
C$_{20}$H$_{34}$O$_2$
8283 - 9,12,15-Octadecatrienoic acid, ethyl ester,
(Z,Z,Z)-
C$_{20}$H$_{36}$O$_2$
8226 - 9,12-Octadecadienoic acid (Z,Z)-, ethyl ester
8227 - 9,12-Octadecadienoic acid, ethyl ester
C$_{20}$H$_{38}$
5366 - 1-Eicosyne
C$_{20}$H$_{38}$N$_4$O$_4$
7686 - 4-Morpholinecarboxamide, N,N'-1,2-
ethanediylbis[N-butyl-
C$_{20}$H$_{38}$O$_2$
4952 - Cycloprate
5363 - 11-Eicosenoic acid
5364 - 9-Eicosenoic acid
5365 - 9-Eicosenoic acid, (Z)-
8301 - 9-Octadecenoic acid (Z)-, ethyl ester
8302 - 9-Octadecenoic acid, ethyl ester, (E)-
C$_{20}$H$_{38}$O$_3$
5060 - Decanoic acid, anhydride
8269 - Octadecanoic acid, 12-oxo-, ethyl ester
8305 - 9-Octadecenoic acid, 12-hydroxy-, ethyl ester,
[R-(Z)]-
C$_{20}$H$_{38}$O$_4$
5352 - Eicosanedioic acid
C$_{20}$H$_{40}$
4475 - Cyclohexane, tetradecyl-
4844 - Cyclopentane, pentadecyl-
5362 - 1-Eicosene
C$_{20}$H$_{40}$O
6671 - 1-Hexadecen-3-ol, 3,7,11,15-tetramethyl-
6672 - 2-Hexadecen-1-ol, 3,7,11,15-tetramethyl-, [R-
[R*,R*-(E)]]-
8240 - Octadecane, 1-(ethenyloxy)-
C$_{20}$H$_{40}$O$_2$
221 - Acetic acid, octadecyl ester
5062 - Decanoic acid, decyl ester
5355 - Eicosanoic acid
6647 - Hexadecanoic acid, butyl ester
6658 - Hexadecanoic acid, 3,7,11,15-tetramethyl-
8257 - Octadecanoic acid, ethyl ester
C$_{20}$H$_{40}$O$_3$
8260 - Octadecanoic acid, 2-hydroxyethyl ester
C$_{20}$H$_{42}$
5350 - Eicosane
C$_{20}$H$_{42}$O
5358 - 1-Eicosanol
5359 - 2-Eicosanol
C$_{20}$H$_{42}$O$_2$
8237 - Octadecane, 1,1-dimethoxy-
C$_{20}$H$_{48}$O$_4$Si$_4$
4995 - Cyclotetrasiloxane, 2,4,6,8-tetrabutyl-2,4,6,8-
tetramethyl-
C$_{20}$H$_{60}$O$_8$Si$_9$
8178 - Nonasiloxane, eicosamethyl-

C$_{21}$H$_{12}$O
5152 - 13H-Dibenzo[a,h]fluoren-13-one
C$_{21}$H$_{13}$N
5143 - Dibenz[a,j]acridine
C$_{21}$H$_{14}$
5148 - 13H-Dibenzo[a,g]fluorene
C$_{21}$H$_{14}$Br$_4$O$_5$S
9367 - Phenol, 4,4'-(3H-2,1-benzoxathiol-3-
ylidene)bis[2,6-dibromo-3-methyl-, *S*,*S*-dioxide
C$_{21}$H$_{14}$O
7634 - Methanone, 1-naphthalenyl-2-naphthalenyl-
C$_{21}$H$_{16}$
2392 - Benz[j]aceanthrylene, 1,2-dihydro-3-methyl-
7271 - 1H-Indene, 1,3-diphenyl-
7272 - 1H-Indene, 2,3-diphenyl-
7919 - Naphthalene, 1,1'-methylenebis-
C$_{21}$H$_{16}$Br$_2$O$_5$S
9364 - Phenol, 4,4'-(3H-2,1-benzoxathiol-3-
ylidene)bis[2-bromo-6-methyl-, *S*,*S*-dioxide
C$_{21}$H$_{16}$N$_2$
7215 - 1H-Imidazole, 2,4,5-triphenyl-
C$_{21}$H$_{16}$N$_2$O
12008 - Urea, N,N'-di-1-naphthalenyl-
C$_{21}$H$_{16}$NaO$_2$
11517 - Sonar
C$_{21}$H$_{18}$N$_2$
7199 - 1H-Imidazole, 4,5-dihydro-2,4,5-triphenyl-, *cis*-
C$_{21}$H$_{18}$O$_5$S
9368 - Phenol, 4,4'-(3H-2,1-benzoxathiol-3-
ylidene)bis[2-methyl-, *S*,*S*-dioxide
C$_{21}$H$_{19}$NO$_5$
2433 - [1,3]Benzodioxolo[5,6-c]phenanthridinium,
1,2-dimethoxy-12-methyl-
C$_{21}$H$_{20}$
2110 - Benzene, 1,1'-(phenylmethylene)bis[4-methyl-
C$_{21}$H$_{20}$Cl$_2$O$_3$
9150 - Permethrin
C$_{21}$H$_{20}$N$_2$O$_3$
8572 - Oxayohimbanium, 3,4,5,6,16,17
hexahydro-16-(methoxycarbonyl)-19-
methyl-, (19α)-
C$_{21}$H$_{20}$O$_6$
6334 - 1,6-Heptadiene-3,5-dione, 1,7-bis(4-hydroxy-
3-methoxyphenyl)-, (E,E)-
C$_{21}$H$_{20}$O$_{10}$
2824 - 4H-1-Benzopyran-4-one, 3-[4-(β-*D*-
glucopyranosyloxy)phenyl]-5,7-dihydroxy-
C$_{21}$H$_{20}$O$_{11}$
2796 - 4H-1-Benzopyran-4-one, 3-[(6-deoxy-α-*L*-
mannopyranosyl)oxy]-2-(3,4-dihydroxyphenyl)-
5,7-dihydroxy-
C$_{21}$H$_{21}$N
1798 - Benzenemethanamine,
N,N-bis(phenylmethyl)-
C$_{21}$H$_{21}$NO$_5$
3108 - Bis[1,3]benzodioxolo[4,5-c:5',6'-g]azecin-
13(5H)-one, 4,6,7,14-tetrahydro-5,14-
dimethyl-, (*R*)-
C$_{21}$H$_{21}$NO$_6$
7357 - 1(3H)-Isobenzofuranone, 6,7-dimethoxy-3-
(5,6,7,8-tetrahydro-6-methyl-1,3-dioxolo[4,5-
g]isoquinolin-5-yl)-, [*S*-
11342 - Rheadan, 8-methoxy-16-methyl-2,3
10,11-bis[methylenebis(oxy)]-, (8β)-
C$_{21}$H$_{21}$N$_3$
11834 - 1,3,5-Triazine, hexahydro-1,3,5-triphenyl-
C$_{21}$H$_{21}$N$_3$O$_7$
11352 - 9,10-Secostrychnidin-10-oic acid, 2,3-
dihydro-4-nitro-2,3-dioxo-
C$_{21}$H$_{21}$O$_3$P
9596 - Phosphorous acid, tris(2-methylphenyl) ester
9597 - Phosphorous acid, tris(4-methylphenyl) ester
C$_{21}$H$_{21}$O$_4$P
9569 - Phosphoric acid, tris(2-methylphenyl) ester
9570 - Phosphoric acid, tris(3-methylphenyl) ester
9571 - Phosphoric acid, tris(4-methylphenyl) ester

$C_{21}H_{21}O_7P$
9387 - Phenol, 2-methoxy-, phosphate (3:1)
$C_{21}H_{21}Sb$
11552 - Stibine, tris(3-methylphenyl)-
11553 - Stibine, tris(4-methylphenyl)-
$C_{21}H_{22}N_2O_2$
11563 - Strychnidin-10-one
$C_{21}H_{23}ClFNO_2$
3651 - 1-Butanone, 4-[4-(4-chlorophenyl)-4-hydroxy-
1-piperidinyl]-1-(4-fluorophenyl)-
$C_{21}H_{23}NO_5$
2435 - Benzo[e]-1,3-dioxolo[4,5-l][2]benzazecin-
12(5H)-one, 4,6,7,13-tetrahydro-9,10-
dimethoxy-5-methyl-
7667 - Morphinan-3,6-diol, 7,8-didehydro-4,5-epoxy-
17-methyl- (5α,6α)-, diacetate (ester)
$C_{21}H_{23}NO_6$
92 - Acetamide, N-(5,6,7,9-tetrahydro-10-hydroxy-
1,2,3-trimethoxy-9-oxobenzo[a]heptalen-7-yl)-,
(S)-
$C_{21}H_{23}N_3O_5$
11565 - Strychnidin-10-one, mononitrate
$C_{21}H_{24}N_2O_3$
8571 - Oxayohimban-16-carboxylic acid, 16,17-
didehydro-19-methyl-, methyl ester, (19α)-
$C_{21}H_{24}OSi_2$
5239 - Disiloxane, 1,1,1-trimethyl-3,3,3-triphenyl-
$C_{21}H_{24}O_3Si_3$
5005 - Cyclotrisiloxane, 2,4,6-trimethyl-2,4,6-
triphenyl-
$C_{21}H_{24}O_4$
9926 - Propanedioic acid, bis(phenylmethyl)-, diethyl
ester
$C_{21}H_{24}O_7$
3540 - Butanoic acid, 2-methyl-, 10-(acetyloxy)-9,10-
dihydro-8,8-dimethyl-2-oxo-2H,8H-benzo[1,2-b
3,4-b']dipyran-9-yl
$C_{21}H_{25}NO_4$
5150 - 6H-Dibenzo[a,g]quinolizin-2-ol, 5,8,13,13a-
tetrahydro-3,9,10-trimethoxy-13-methyl-, (13S-
trans)-
5151 - 6H-Dibenzo[a,g]quinolizin-3-ol, 5,8,13,13a-
tetrahydro-2,9,10-trimethoxy-13-methyl-, (13S-
trans)-
5157 - 4H-Dibenzo[de,g]quinoline, 5,6,6a,7-
tetrahydro-1,2,9,10-tetramethoxy-6-methyl-,
(S)-
$C_{21}H_{26}BrNO_3$
5445 - Ethanaminium, N,N-diethyl-N-methyl-2-[(9H-
xanthen-9-ylcarbonyl)oxy]-, bromide
$C_{21}H_{26}ClN_3OS$
9617 - 1-Piperazineethanol, 4-[3-(2-chloro-10H-
phenothiazin-10-yl)propyl]-
$C_{21}H_{26}N_2O_3$
5349 - Eburnamenine-14-carboxylic acid, 14,15-
dihydro-14-hydroxy-, methyl ester,
(3α,14β,16α)-
12062 - Yohimban-16-carboxylic acid, 17-hydroxy-,
methyl ester, (16α,17α)-
$C_{21}H_{26}N_2O_4$
4129 - Cinchonan-9-ol, 6'-methoxy-, (8α,9R)-,
monoformate (salt)
$C_{21}H_{26}O_2$
5154 - 6H-Dibenzo[b,d]pyran-1-ol, 6,6,9-trimethyl-3-
pentyl-
$C_{21}H_{26}O_3$
7621 - Methanone, [2-hydroxy-4-(octyloxy)
phenyl]phenyl-
$C_{21}H_{26}O_5$
9739 - Pregna-1,4-diene-3,11,20-trione, 17,21-
dihydroxy-
$C_{21}H_{27}ClN_2O_2$
5700 - Ethanol, 2-[2-[4-[(4-chlorophenyl)
phenylmethyl]-1-piperazinyl]ethoxy]-

$C_{21}H_{27}NO_4$
7379 - Isoquinoline, 1-[(3,4-dimethoxyphenyl)methyl]-
1,2,3,4-tetrahydro-6,7-dimethoxy-2-methyl-,
(S)-
$C_{21}H_{28}ClNO$
6516 - 3-Heptanone, 6-(dimethylamino)-4,4-
diphenyl-, hydrochloride
$C_{21}H_{28}O_2$
9758 - Pregn-4-en-20-yn-3-one, 17-hydroxy-, (17α)-
$C_{21}H_{28}O_3$
4960 - Cyclopropanecarboxylic acid, 2,2-dimethyl-3-
(2-methyl-1-propenyl)-, 2-methyl-4-oxo-3-(2,4-
pentadienyl)-2-cyclopenten-1-yl ester, [1R-
[1α[S*(Z)],3β]]-
$C_{21}H_{28}O_4$
9755 - Pregn-4-ene-3,11,20-trione, 21-hydroxy-
$C_{21}H_{28}O_5$
9738 - Pregna-1,4-diene-3,20-dione, 11,17,21-
trihydroxy-, (11β)-
9746 - Pregn-4-en-18-al, 11,21-dihydroxy-3,20-
dioxo-, (11β)-
9754 - Pregn-4-ene-3,11,20-trione, 17,21-dihydroxy-
$C_{21}H_{30}O_2$
1537 - 1,3-Benzenediol, 2-[3-methyl-6-(1-
methylethenyl)-2-cyclohexen-1-yl]-5-pentyl-,
(1R-trans)-
$C_{21}H_{30}O_3$
9750 - Pregn-4-ene-3,20-dione, 17-hydroxy-
9751 - Pregn-4-ene-3,20-dione, 21-hydroxy-
$C_{21}H_{30}O_4$
9748 - Pregn-4-ene-3,20-dione, 11,21-dihydroxy-,
(11β)-
9749 - Pregn-4-ene-3,20-dione, 17,21-dihydroxy-
$C_{21}H_{30}O_5$
4282 - 2,4-Cyclohexadien-1-one, 3,5,6-trihydroxy-
4,6-bis(3-methyl-2-butenyl)-2-(3-methyl-1-
oxobutyl)-, (R)-
9753 - Pregn-4-ene-3,20-dione, 11,17,21-trihydroxy-,
(11β)-
$C_{21}H_{32}O_2$
9168 - 1-Phenanthrenecarboxylic acid,
1,2,3,4,4a,4b,5,6,10,10a-decahydro-1,4a-
dimethyl-7-(1-methylethyl)-, methyl ester, [1R-
(1α,4aβ,4bα,10aα)]-
9743 - Pregnane-3,20-dione, (5β)-
9756 - Pregn-5-en-20-one, 3-hydroxy-, (3β)-
$C_{21}H_{32}O_5$
9744 - Pregnane-11,20-dione, 3,17,21-trihydroxy-,
(3α,5β)-
9757 - Pregn-4-en-3-one, 11,17,20,21-tetrahydroxy-,
(11β,20R)-
$C_{21}H_{34}BrNO_3$
5443 - Ethanaminium,
2-[(cyclohexylhydroxyphenylacetyl)
oxy]-N,N-diethyl-N-methyl-, bromide
$C_{21}H_{34}O_2$
9745 - Pregnan-20-one, 3-hydroxy-, (3α,5β)-
11627 - Tetradecanoic acid, phenylmethyl ester
$C_{21}H_{36}$
2089 - Benzene, pentadecyl-
9740 - Pregnane, (5α)-
9741 - Pregnane, (5β)-
$C_{21}H_{36}N_2O_5S$
9623 - Piperazinium, 4-(2-cyclohexyl-2-hydroxy-2-
phenylethyl)-1,1-dimethyl-, methyl sulfate
(salt)
$C_{21}H_{36}O$
9441 - Phenol, 3-pentadecyl-
$C_{21}H_{36}O_2$
1543 - 1,2-Benzenediol, 3-pentadecyl-
9742 - Pregnane-3,20-diol, (3α,5β,20S)-
$C_{21}H_{38}ClN$
11030 - Pyridinium, 1-hexadecyl-, chloride
$C_{21}H_{38}O_6$
6908 - Hexanoic acid, 1,2,3-propanetriyl ester
$C_{21}H_{39}N_7O_{12}$
11562 - Streptomycin

$C_{21}H_{40}O_4$
8131 - Nonanedioic acid, bis(2-ethylbutyl) ester
8299 - 9-Octadecenoic acid (Z)-, 2,3-dihydroxypropyl
ester
$C_{21}H_{41}O_2$
5357 - Eicosanoic acid, methyl ester
$C_{21}H_{42}$
4463 - Cyclohexane, pentadecyl-
4822 - Cyclopentane, hexadecyl-
$C_{21}H_{42}O_2$
8267 - Octadecanoic acid, 1-methylethyl ester
8273 - Octadecanoic acid, propyl ester
$C_{21}H_{42}O_4$
8255 - Octadecanoic acid, 2,3-dihydroxypropyl ester,
(±)-
$C_{21}H_{44}$
6294 - Heneicosane
$C_{21}H_{44}O_3$
10023 - 1,2-Propanediol, 3-(octadecyloxy)-
$C_{22}H_{12}$
2471 - Benzo[ghi]perylene
$C_{22}H_{14}$
5142 - Dibenz[a,h]anthracene
8636 - Pentacene
8996 - Pentaphene
9601 - Picene
$C_{22}H_{16}N_4O$
8038 - 2-Naphthalenol, 1-[[4-(phenylazo)phenyl]azo]-
$C_{22}H_{17}N$
7717 - 1-Naphthalenamine, N,N-diphenyl-
$C_{22}H_{18}Cl_2FNO_3$
5008 - Cyfluthrin
$C_{22}H_{18}O_4$
1337 - 1,2-Benzenedicarboxylic acid,
bis(phenylmethyl) ester
7353 - 1(3H)-Isobenzofuranone, 3,3-bis(4-hydroxy-3-
methylphenyl)-
$C_{22}H_{19}Br_2NO_3$
5102 - Deltamethrin
$C_{22}H_{19}Cl_2NO_3$
5012 - Cypermethrin
$C_{22}H_{19}NO_4$
9465 - Phenol, 4,4'-(2-pyridinylmethylene)bis-,
diacetate (ester)
$C_{22}H_{20}O_{13}$
373 - 2-Anthracenecarboxylic acid, 7-β-D-
glucopyranosyl-9,10-dihydro-3,5,6,8-
tetrahydroxy-1-methyl-9,10-dioxo-
$C_{22}H_{21}NO_2$
1797 - Benzenemethanamine, N-[bis
(4-methoxyphenyl)methylene]-
$C_{22}H_{22}O_8$
6191 - Furo[3',4':6,7]naphtho[2,3-d]-1,3-dioxol-
6(5aH)-one, 5,8,8a,9-tetrahydro-9-hydroxy-5-
(3,4,5-trimethoxyphenyl)-
6192 - Furo[3',4':6,7]naphtho[2,3-d]-1,3-dioxol-
6(5aH)-one, 5,8,8a,9-tetrahydro-9-hydroxy-5-
(3,4,5-trimetho
$C_{22}H_{23}ClN_2O_8$
7704 - 2-Naphthacenecarboxamide, 7-chloro-4-
(dimethylamino)-1,4,4a,5,5a,6,11,12a-
octahydro-3,6,10,12,12a-pentahydroxy-
6-methyl-1,11-dioxo, [4S
(4α,4aα,5aα,6β,12aα)]-
$C_{22}H_{23}NO_3$
5953 - Fenpropathrin
$C_{22}H_{23}NO_6$
10797 - 4H-Pyran-4-one, 2-methoxy-3,5-dimethyl-6-
[tetrahydro-4-[2-methyl-3-(4-nitrophenyl)-2-
propenylidene]-2-furanyl]
$C_{22}H_{23}NO_7$
7356 - 1(3H)-Isobenzofuranone, 6,7-dimethoxy-3-
(5,6,7,8-tetrahydro-4-methoxy-6-methyl-1,3-
dioxolo[4,5-g]isoquinolin-5
$C_{22}H_{24}N_2O_4$
3204 - Butanamide, N,N'-(3,3'-dimethyl[1,1'-
biphenyl]-4,4'-diyl)bis[3-oxo-

10772 - 2H-Pyran-5-carboxylic acid, 3,4-dihydro-3-
(hydroxymethyl)-2-methyl-4-[(9-methyl-9H-
pyrido[3,4-b]indol-1-yl)methyl]-, methyl ester,
[2S-(2α,3β,4β)]-

11351 - 16,19-Secostrychnidine-10,16-dione, 4-
hydroxy-19-methyl-

$C_{22}H_{24}N_2O_8$
7706 - 2-Naphthacenecarboxamide,
4-(dimethylamino)-1,4,4a,5,5a,6,11,12a-
octahydro-3,6,10,12,12a-pentahydroxy-6-
methyl-1,11-dioxo-,
[4S-(4α,4aα,5aα,6β,12aα)]-

$C_{22}H_{24}N_2O_9$
7705 - 2-Naphthacenecarboxamide,
4-(dimethylamino)-1,4,4a,5,5a,6,11,12a-
octahydro-3,5,6,10,12,12a-hexahydroxy-
6-methyl-1,11-dioxo-,[4S-
(4α,4aα,5aα,6β,12aα)]-

$C_{22}H_{24}Si$
11445 - Silane, methyltri-p-tolyl-

$C_{22}H_{25}NO_2$
5854 - Ethanone, 2,2'-(1-methyl-
2,6-piperidinediyl)bis[1-phenyl-, cis-

$C_{22}H_{25}NO_6$
93 - Acetamide, N-(5,6,7,9-tetrahydro-1,2,3,10-
tetramethoxy-9-oxobenzo[a]heptalen-7-yl)-,
(S)-

$C_{22}H_{25}N_3O$
10850 - 3H-Pyrazol-3-one, 1,2-dihydro-2-(1-methyl-4-
piperidinyl)-5-phenyl-4-(phenylmethyl)-

$C_{22}H_{26}N_2O_3$
4146 - Corynan-16-carboxylic acid, 16,17,18,19-
tetradehydro-17-methoxy-, methyl ester,
(16E)-

$C_{22}H_{26}O_3$
11340 - (-)-cis-Resmethrin

$C_{22}H_{27}NO_2$
5834 - Ethanone, 2-[6-(2-hydroxy-2-phenylethyl)-1-
methyl-2-piperidinyl]-1-phenyl-, [2R-
[2α,6α(S*)]]-

$C_{22}H_{27}NO_4$
5149 - 6H-Dibenzo[a,g]quinolizine, 5,8,13,13a-
tetrahydro-2,3,9,10-tetramethoxy-13-methyl-,
(13S-trans)-

$C_{22}H_{28}N_2O$
9801 - Propanamide, N-phenyl-N-[1-(2-phenylethyl)-
4-piperidinyl]-

$C_{22}H_{28}N_2O_5$
12063 - Yohimban-16-carboxylic acid, 18-hydroxy-
11,17-dimethoxy-, (3β,16β,17α,18β,20

$C_{22}H_{28}O_3$
9736 - Pregna-4,6-diene-21-carboxylic acid, 17-
hydroxy-3-oxo-, γ-lactone, (17α)-

$C_{22}H_{28}O_5$
4962 - Cyclopropanecarboxylic acid, 3-(3-methoxy-2-
methyl-3-oxo-1-propenyl)-2,2-dimethyl-, 2-
methyl-4-oxo-3-(2,4-pentadienyl)-2-
cyclopenten-1-yl ester [1R-[1α[S*(Z)],3β(E)]]-

$C_{22}H_{29}FO_5$
9737 - Pregna-1,4-diene-3,20-dione, 9-fluoro-
11,17,21-trihydroxy-16-methyl-, (11β,16α)-

$C_{22}H_{29}NO_2$
9646 - 2,6-Piperidinediethanol, 1-methyl-α,α'-
diphenyl-, [2α(R*),6α(S*)]-

$C_{22}H_{30}N_2O_2$
473 - Aspidospermidine, 1-acetyl-17-methoxy-

$C_{22}H_{30}N_2O_5$
4730 - 2,4(1H)-Cyclo-3,4-secoakuammilanium, 3,17-
dihydroxy-16-(methoxycarbonyl)-4-methyl-,
(3β,16R)-

$C_{22}H_{32}$
7879 - Naphthalene, 1-dodecyl-
7880 - Naphthalene, 2-dodecyl-

$C_{22}H_{32}O_3$
359 - Androst-4-en-3-one, 17-(1-oxopropoxy)-, (17β)-
9752 - Pregn-4-ene-3,20-dione, 17-hydroxy-6-
methyl-, (6α)-

$C_{22}H_{33}NO_2$
475 - Atisine

$C_{22}H_{35}NO_3$
5007 - 7,20-Cycloveatchane-1,12,15-triol, 21-ethyl-4-
methyl-16-methylene-, (1α,12α,15β)-

$C_{22}H_{36}O$
6664 - 1-Hexadecanone, 1-phenyl-

$C_{22}H_{37}NO$
6630 - Hexadecanamide, N-phenyl-

$C_{22}H_{38}$
1751 - Benzene, hexadecyl-

$C_{22}H_{38}O$
1752 - Benzene, (hexadecyloxy)-

$C_{22}H_{40}O_7$
8102 - 1,2,3-Nonadecanetricarboxylic acid, 2-
hydroxy-

$C_{22}H_{42}O_2$
5281 - 13-Docosenoic acid
5282 - 13-Docosenoic acid, (E)-
5283 - 13-Docosenoic acid, (Z)-
8298 - 9-Octadecenoic acid (Z)-, butyl ester
10667 - 2-Propenoic acid, 2-methyl-, octadecyl ester

$C_{22}H_{42}O_3$
8304 - 9-Octadecenoic acid, 12-hydroxy-, butyl ester,
[R-(Z)]-

$C_{22}H_{42}O_4$
6781 - Hexanedioic acid, bis(2-ethylhexyl) ester
8238 - Octadecanedioic acid, diethyl ester

$C_{22}H_{44}$
4411 - Cyclohexane, hexadecyl-
5280 - 1-Docosene

$C_{20}H_{40}N_2 \cdot C_2H_4O_2$
6275 - Glyodin

$C_{22}H_{44}O$
5284 - 13-Docosen-1-ol, (Z)-

$C_{22}H_{44}O_2$
5276 - Docosanoic acid
5356 - Eicosanoic acid, ethyl ester
8251 - Octadecanoic acid, butyl ester
8268 - Octadecanoic acid, 2-methylpropyl ester

$C_{22}H_{46}$
5268 - Docosane

$C_{22}H_{46}O$
5279 - 1-Docosanol

$C_{22}H_{66}O_9Si_{10}$
5079 - Decasiloxane, docosamethyl-

$C_{23}H_{16}O_3$
7268 - 1H-Indene-1,3(2H)-dione, 2-(diphenylacetyl)-

$C_{23}H_{16}O_5$
7777 - 2-Naphthalenecarboxylic acid, 4,4'-
methylenebis[3-hydroxy-

$C_{23}H_{18}O$
7910 - 1-Naphthalenemethanol, α,α-diphenyl-

$C_{23}H_{19}ClF_3NO_3$
5010 - Cyhalothrin

$C_{23}H_{21}ClO_3$
1154 - Benzene, 1,1',1''-(1-chloro-1-ethenyl-2-
ylidene)tris[4-methoxy-

$C_{23}H_{22}ClF_3O_2$
2998 - Bifenthrin

$C_{23}H_{22}O_6$
2783 - [1]Benzopyrano[3,4-b]furo
[2,3-h][1]benzopyran-6(6aH)-one, 1,2,12,12a-
tetrahydro-8,9-dimethoxy-2-(1-methylethen

$C_{23}H_{22}O_7$
3110 - 3H-Bis[1]benzopyrano[3,4-b:6',5'-e]pyran-
7(7aH)-one, 13,13a-dihydro-7a-hydroxy-9,10-
dimethoxy-3,3-dimethyl

$C_{23}H_{23}O_6$
9747 - Pregn-4-ene-3,20-dione, 21-(acetyloxy)-
11,17-dihydroxy-, (11β)-

$C_{23}H_{25}ClN_2$
7455 - Methanaminium, N-[4-
[[4-(dimethylamino)phenyl]phenylmethylene]-
2,5-cyclohexadien-1-ylidene]-N-methyl-,
chlorid

$C_{23}H_{26}N_2O_4$
11564 - Strychnidin-10-one, 2,3-dimethoxy-

$C_{23}H_{27}NO_8$
2586 - Benzoic acid, 6-[[6-[2-(dimethylamino)ethyl]-4-
methoxy-1,3-benzodioxol-5-yl]acetyl]-2,3-
dimethoxy-

$C_{23}H_{28}ClN_3O$
9657 - Piperidine, 1-[2-[3-(2,2-diphenylethyl)-1,2,4-
oxadiazol-5-yl]ethyl]-, monohydrochloride

$C_{23}H_{30}BrNO_3$
9832 - 2-Propanaminium, N-methyl-N-(1-
methylethyl)-N-[2-[(9H-xanthen-9-
ylcarbonyl)oxy]ethyl]-, bromide

$C_{23}H_{30}O_2Si_3$
11915 - Trisiloxane, 1,1,3,5,5-pentamethyl-1,3,5-
triphenyl-

$C_{23}H_{32}N_2O_2S$
11800 - Thiourea, N,N'-bis[4-
(3-methylbutoxy)phenyl]-

$C_{23}H_{32}O_6$
4049 - Card-20(22)-enolide, 3,5,14-trihydroxy-19-
oxo-, (3β,5β)-

$C_{23}H_{33}IN_2O$
2120 - Benzenepropanaminium, γ-(aminocarbonyl)-
N-methyl-N,N-bis(1-methylethyl)-γ-phenyl-, iod

$C_{23}H_{34}O_4$
4044 - Card-20(22)-enolide, 3,14-dihydroxy-, (3β,5β)-

$C_{23}H_{34}O_5$
4046 - Card-20(22)-enolide, 3,11,14-trihydroxy-,
(3β,5β,11α)-
4047 - Card-20(22)-enolide, 3,12,14-trihydroxy-,
(3β,5β,12β)-
4048 - Card-20(22)-enolide, 3,14,16-trihydroxy-,
(3β,5β,16β)-

$C_{23}H_{36}O_3$
351 - Androstan-3-one, 2-methyl-17-(1-oxopropoxy)-,
(2α,5α,17β)-

$C_{23}H_{38}O_2$
6655 - Hexadecanoic acid, phenylmethyl ester
8210 - 24-Norcholan-23-oic acid, (5β)-

$C_{23}H_{40}$
1743 - Benzene, heptadecyl-

$C_{23}H_{44}O_2$
8306 - 9-Octadecenoic acid (Z)-, 3-methylbutyl ester

$C_{23}H_{44}O_3$
8275 - Octadecanoic acid, (tetrahydro-2-
furanyl)methyl ester

$C_{23}H_{44}O_4$
9954 - Propanedioic acid, hexadecyl-, diethyl ester

$C_{23}H_{46}$
4404 - Cyclohexane, heptadecyl-
4842 - Cyclopentane, octadecyl-

$C_{23}H_{46}O$
11857 - 12-Tricosanone

$C_{23}H_{46}O_2$
5278 - Docosanoic acid, methyl ester
8265 - Octadecanoic acid, 3-methylbutyl ester
8270 - Octadecanoic acid, pentyl ester

$C_{23}H_{48}$
6309 - Heptadecane, 9-hexyl-
11855 - Tricosane

$C_{24}H_{12}$
4145 - Coronene

$C_{24}H_{14}$
2850 - Benzo[rst]pentaphene
8061 - Naphtho[1,2,3,4-def]chrysene

$C_{24}H_{16}$
11644 - Tetraphenylene

$C_{24}H_{18}$
11193 - 1,1':2',1'':2'',1'''-Quaterphenyl
11194 - 1,1':4',1'':4'',1'''-Quaterphenyl
11598 - 1,1':3',1''-Terphenyl, 5'-phenyl-

$C_{24}H_{20}N_2$
3035 - [1,1'-Biphenyl]-4,4'-diamine, N,N'-diphenyl-

$C_{24}H_{20}N_4O$
8033 - 2-Naphthalenol, 1-[[2-methyl-4-[(2-
methylphenyl)azo]phenyl]azo]-

C$_{24}$H$_{20}$O$_4$Si
11509 - Silicic acid (H4SiO4), tetraphenyl ester
C$_{24}$H$_{20}$O$_6$
10196 - 1,2,3-Propanetriol, tribenzoate
C$_{24}$H$_{20}$Pb
9733 - Plumbane, tetraphenyl-
C$_{24}$H$_{20}$Si
11452 - Silane, tetraphenyl-
C$_{24}$H$_{20}$Sn
11545 - Stannane, tetraphenyl-
C$_{24}$H$_{27}$O$_4$P
9335 - Phenol, 2,4-dimethyl-, phosphate (3:1)
9336 - Phenol, 2,5-dimethyl-, phosphate (3:1)
9337 - Phenol, 2,6-dimethyl-, phosphate (3:1)
9338 - Phenol, 3,4-dimethyl-, phosphate (3:1)
C$_{24}$H$_{28}$O$_4$
9301 - Phenol, 4,4'-(1,2-diethyl-1,2-ethenediyl)bis-, dipropanoate, (E)-
C$_{24}$H$_{29}$NO$_4$
7377 - Isoquinoline, 1-[(3,4-diethoxyphenyl)methyl]-6,7-diethoxy-
C$_{24}$H$_{30}$O$_3$Si$_3$
5004 - Cyclotrisiloxane, 2,4,6-triethyl-2,4,6-triphenyl-
C$_{24}$H$_{34}$O$_5$
4101 - Cholan-24-oic acid, 3,7,12-trioxo-, (5β)-
C$_{24}$H$_{38}$O$_4$
1334 - 1,2-Benzenedicarboxylic acid, bis(2-ethylhexyl) ester
1344 - 1,2-Benzenedicarboxylic acid, dioctyl ester
C$_{24}$H$_{39}$NO
8290 - 9-Octadecenamide, N-phenyl-, (Z)-
C$_{24}$H$_{40}$N$_2$
4143 - Con-5-enin-3-amine, N,N-dimethyl-, (3β)-
C$_{24}$H$_{40}$N$_8$O$_4$
5711 - Ethanol, 2,2',2'',2'''-[(4,8-di-1-piperidinylpyrimido[5,4-d]pyrimidine-
C$_{24}$H$_{40}$O$_2$
1257 - Benzenedecanoic acid, ι-octyl-
2063 - Benzenenonanoic acid, θ-nonyl-
4094 - Cholan-24-oic acid
8271 - Octadecanoic acid, phenyl ester
C$_{24}$H$_{40}$O$_3$
4099 - Cholan-24-oic acid, 3-hydroxy-, (3α,5β)-
C$_{24}$H$_{40}$O$_4$
4095 - Cholan-24-oic acid, 3,12-dihydroxy-, (3α,5β,12α)-
4096 - Cholan-24-oic acid, 3,6-dihydroxy-, (3α,5β,6α)-
4097 - Cholan-24-oic acid, 3,7-dihydroxy-, (3α,5β,7α)-
4098 - Cholan-24-oic acid, 3,7-dihydroxy-, (3α,5β,7β)-
C$_{24}$H$_{40}$O$_5$
4100 - Cholan-24-oic acid, 3,7,12-trihydroxy-, (3α,5β,7α,12α)-
C$_{24}$H$_{41}$NO$_2$
8232 - Octadecanamide, N-(4-hydroxyphenyl)-
C$_{24}$H$_{42}$
2066 - Benzene, octadecyl-
4093 - Cholane
C$_{24}$H$_{46}$O$_2$
8253 - Octadecanoic acid, cyclohexyl ester
11608 - 15-Tetracosenoic acid
C$_{24}$H$_{46}$O$_3$
5306 - Dodecanoic acid, anhydride
C$_{24}$H$_{46}$O$_4$
5353 - Eicosanedioic acid, diethyl ester
9155 - Peroxide, bis(1-oxododecyl)
C$_{24}$H$_{48}$
4459 - Cyclohexane, octadecyl-
C$_{24}$H$_{48}$O$_2$
5277 - Docosanoic acid, ethyl ester
11607 - Tetracosanoic acid
C$_{24}$H$_{50}$
11603 - Tetracosane
11856 - Tricosane, 2-methyl-

C$_{24}$H$_{52}$O$_4$Si
11507 - Silicic acid (H4SiO4), tetrakis(2-ethylbutyl) ester
C$_{24}$H$_{54}$OSn$_2$
5241 - Distannoxane, hexabutyl-
C$_{24}$H$_{72}$O$_{10}$Si$_{11}$
11964 - Undecasiloxane, tetracosamethyl-
C$_{25}$H$_{20}$
1831 - Benzene, 1,1',1'',1'''-methanetetrayltetrakis-
C$_{25}$H$_{20}$N$_2$O
12031 - Urea, tetraphenyl-
C$_{25}$H$_{22}$ClNO$_3$
5956 - Fenvalerate
C$_{25}$H$_{22}$OSi
11441 - Silane, (2-methylphenoxy)triphenyl-
C$_{25}$H$_{24}$F$_6$N$_4$
7136 - Hydramethylnon
C$_{25}$H$_{27}$ClN$_2$
9605 - Piperazine, 1-[(4-chlorophenyl)phenylmethyl]-4-[(3-methylphenyl)methyl]-
C$_{25}$H$_{28}$O$_3$
5403 - Estra-1,3,5(10)-triene-3,17-diol (17β)-, 3-benzoate
C$_{25}$H$_{30}$ClN$_3$
7453 - Methanaminium, N-[4-[bis[4-(dimethylamino)phenyl]methylene]-2,5-cyclohexadien-1-ylidene]-N-methyl, chloride
C$_{25}$H$_{30}$O$_4$
5108 - 6,6'-Diapo-ψ,ψ-carotenedioic acid, monomethyl ester, 9-cis-
C$_{25}$H$_{40}$O$_2$
8309 - 9-Octadecenoic acid (Z)-, phenylmethyl ester
C$_{25}$H$_{41}$NO$_9$
305 - Aconitane-3,8,13,14,15-pentol, 20-ethyl-1,6,16-trimethoxy-4-(methoxymethyl)-,
C$_{25}$H$_{44}$
2062 - Benzene, nonadecyl-
2070 - Benzene, (3-octylundecyl)-
8245 - Octadecane, 9-p-tolyl-
C$_{25}$H$_{48}$O$_4$
8132 - Nonanedioic acid, bis(2-ethylhexyl) ester
C$_{25}$H$_{50}$
4456 - Cyclohexane, nonadecyl-
5351 - Eicosane, 1-cyclopentyl-
6323 - 8-Heptadecene, 9-octyl-
C$_{25}$H$_{52}$
6313 - Heptadecane, 9-octyl-
8637 - Pentacosane
C$_{26}$H$_{16}$
6621 - Hexacene
C$_{26}$H$_{20}$
1623 - Benzene, 1,1',1'',1'''-(1,2-ethenediylidene)tetrakis-
C$_{26}$H$_{22}$
1591 - Benzene, 1,1',1'',1'''-(1,2-ethenediylidene)tetrakis-
1618 - Benzene, 1,1',1'',1'''-(1-ethanyl-2-ylidyne)tetrakis-
C$_{26}$H$_{22}$ClF$_3$N$_2$O$_3$
5983 - Fluvalinate
C$_{26}$H$_{22}$O$_2$
5580 - 1,2-Ethanediol, 1,1,2,2-tetraphenyl-
C$_{26}$H$_{23}$F$_2$NO$_4$
5965 - Flucythrinate
C$_{26}$H$_{36}$O$_6$
3141 - Bufa-20,22-dienolide, 16-(acetyloxy)-3,14-dihydroxy-, (3β,5β,16β)-
C$_{26}$H$_{38}$
1567 - Benzene, 1,1'-dodecylidenebis[4-methyl-
2267 - Benzene, 1,1'-tetradecylidenebis-
C$_{26}$H$_{38}$O$_4$
4281 - 2,4-Cyclohexadien-1-one, 3,5-dihydroxy-2,6,6-tris(3-methyl-2-butenyl)-4-(3-methyl-1-oxobutyl)-
C$_{26}$H$_{43}$NO$_6$
6273 - Glycine, N-[(3α,5β,7α,12α)-3,7,12-trihydroxy-24-oxocholan-24-yl]-

C$_{26}$H$_{45}$NO$_7$S
5625 - Ethanesulfonic acid, 2-[[(3α,5β,7α,12α)-3,7,12-trihydroxy-24-oxocholan-24-yl]amino]-
C$_{26}$H$_{46}$
1119 - Benzene, (1-butylhexadecyl)-
1568 - Benzene, eicosyl-
1695 - Benzene, (1-ethyloctadecyl)-
2007 - Benzene, (1-methylnonadecyl)-
2068 - Benzene, (1-octyldodecyl)-
2170 - Benzene, (1-propylheptadecyl)-
C$_{26}$H$_{50}$
4476 - Cyclohexane, 1,1'-tetradecylidenebis-
C$_{26}$H$_{50}$O$_4$
5043 - Decanedioic acid, bis(2-ethylhexyl) ester
5311 - Dodecanoic acid, 1,2-ethanediyl ester
C$_{26}$H$_{52}$
4384 - Cyclohexane, eicosyl-
4819 - Cyclopentane, heneicosyl-
C$_{26}$H$_{52}$O$_2$
6624 - Hexacosanoic acid
C$_{26}$H$_{52}$O$_6$
8259 - Octadecanoic acid, 2-[2-[2-(2-hydroxyethoxy)ethoxy]ethoxy]ethyl ester
C$_{26}$H$_{54}$
5269 - Docosane, 5-butyl-
5270 - Docosane, 7-butyl-
5271 - Docosane, 9-butyl-
5272 - Docosane, 11-butyl-
6296 - Heneicosane, 11-(2,2-dimethylpropyl)-
6622 - Hexacosane
6637 - Hexadecane, 6,11-dipentyl-
8241 - Octadecane, 3-ethyl-5-(2-ethylbutyl)-
11605 - Tetracosane, 3-ethyl-
C$_{26}$H$_{54}$O
6626 - 1-Hexacosanol
C$_{27}$H$_{18}$O$_3$
7587 - Methanone, 1,3,5-benzenetriyltris[phenyl-
C$_{27}$H$_{20}$O$_{12}$
2773 - 1H-2-Benzopyran-3-carboxylic acid, 6-[2-(4,9-dihydro-8-hydroxy-5,7-dimethoxy-4,9-dioxonaphtho[2,3-b]furan-2-yl
C$_{27}$H$_{28}$Br$_2$O$_5$S
9365 - Phenol, 4,4'-(3H-2,1-benzoxathiol-3-ylidene)bis[2-bromo-3-methyl-6-(1-methylethyl)-, S,S-dioxide
C$_{27}$H$_{30}$O$_5$S
9369 - Phenol, 4,4'-(3H-2,1-benzoxathiol-3-ylidene)bis[5-methyl-2-(1-methylethyl)-, S,S-dioxide
C$_{27}$H$_{32}$N$_2$O$_6$
4131 - Cinchonan-9-ol, 6'-methoxy-, (8α,9R)-, mono(2-hydroxybenzoate) (salt)
C$_{27}$H$_{32}$O$_{14}$
2795 - 4H-1-Benzopyran-4-one, 7-[[2-O-(6-deoxy-α-L-mannopyranosyl)-β-D-glucopyranosyl]oxy]-2,3-dihydro-5-hy
C$_{27}$H$_{34}$N$_2$O$_4$S
5444 - Ethanaminium, N-[4-[[4-(diethylamino)phenyl]phenylmethylene]-2,5-cyclohexadien-1-ylidene]-N-ethyl-, sulfate (1
C$_{27}$H$_{34}$O$_3$
5408 - Estr-4-en-3-one, 17-(1-oxo-3-phenylpropoxy)-, (17β)-
C$_{27}$H$_{39}$NO$_2$
12044 - Veratraman-3,23-diol, 14,15,16,17-tetradehydro-, (3β,23β)-
C$_{27}$H$_{39}$NO$_3$
12046 - Veratraman-11-one, 17,23-epoxy-3-hydroxy-, (3β,23β)-
C$_{27}$H$_{40}$O$_3$
7774 - 2-Naphthalenecarboxylic acid, 3-hydroxy-, hexadecyl ester

$C_{27}H_{42}Cl_2N_2O_6$
6648 - Hexadecanoic acid, 2-[(dichloroacetyl)amino]-3-hydroxy-3-(4-nitrophenyl)propyl ester, [R-(R^*,R^*)]-

$C_{27}H_{42}O_3$
11536 - Spirost-5-en-3-ol, (3β,25R)-

$C_{27}H_{42}O_4$
11534 - Spirostan-12-one, 3-hydroxy-, (3β,5α,25R)-

$C_{27}H_{43}NO_2$
11515 - Solanid-5-ene-3,12-diol, (3β,12α)-
11516 - Solanid-5-ene-3,18-diol, (3β)-
11529 - Spirosol-5-en-3-ol, (3β,22α,25R)-

$C_{27}H_{43}NO_8$
4068 - Cevane-3,4,7,14,15,16,20-heptol, 4,9-epoxy-, (3β,4α,7α,15α,16β)-

$C_{27}H_{43}NO_9$
4070 - Cevane-3,4,6,7,14,15,16,20-octol, 4,9-epoxy-, (3β,4α,6α,7α,15α,16β

$C_{27}H_{44}$
4102 - Cholesta-3,5-diene

$C_{27}H_{44}O$
4103 - Cholesta-5,7-dien-3-ol, (3β)-
4104 - Cholesta-8,24-dien-3-ol, (3β,5α)-
11348 - 9,10-Secocholesta-5,7,10(19)-trien-3-ol, (3β,5Z,7E)-

$C_{27}H_{44}O_3$
11531 - Spirostan-3-ol, (3β,5β,25S)-
11532 - Spirostan-3-ol, (3β,5α,25R)-
11533 - Spirostan-3-ol, (3β,5β,25R)-

$C_{27}H_{44}O_4$
11530 - Spirostan-2,3-diol, (2α,3β,5α,25R)-

$C_{27}H_{44}O_5$
11535 - Spirostan-2,3,15-triol, (2α,3β,5α,15β,25R)-

$C_{27}H_{45}NO_2$
11528 - Spirosolan-3-ol, (3β,5α,22β,25S)-

$C_{27}H_{46}$
6298 - 10-Heneicosene, 11-phenyl-

$C_{27}H_{46}O$
4110 - Cholest-5-en-3-ol (3β)-
4111 - Cholest-4-en-3-ol, (3β)-
4112 - Cholest-5-en-3-ol, (3α)-

$C_{27}H_{46}O_2$
2785 - 2H-1-Benzopyran-6-ol, 3,4-dihydro-2,8-dimethyl-2-(4,8,12-trimethyltridecyl)-, [2R-[2R^*(4R^*,8R^*)]]-

$C_{27}H_{48}$
1741 - Benzene, heneicosyl-
4105 - Cholestane, (5α)-
4106 - Cholestane, (5β)-
6297 - Heneicosane, 11-phenyl-

$C_{27}H_{48}O$
4107 - Cholestanol
4108 - Cholestan-3-ol, (3α,5α)-
4109 - Cholestan-3-ol, (3β,5β)-

$C_{27}H_{50}O_6$
8440 - Octanoic acid, 1,2,3-propanetriyl ester

$C_{27}H_{54}$
4808 - Cyclopentane, docosyl-

$C_{27}H_{54}O$
6303 - 14-Heptacosanone

$C_{27}H_{54}O_2$
6625 - Hexacosanoic acid, methyl ester

$C_{27}H_{56}$
6302 - Heptacosane

$C_{28}H_{14}N_2O_4$
434 - 5,9,14,18-Anthrazinetetrone, 6,15-dihydro-

$C_{28}H_{20}O$
6175 - Furan, tetraphenyl-

$C_{28}H_{22}$
367 - Anthracene, 9,10-bis(phenylmethyl)-

$C_{28}H_{22}O_3$
937 - Benzeneacetic acid, α-phenyl-, anhydride

$C_{28}H_{24}N_2O_7$
8549 - Orcein

$C_{28}H_{30}O_4$
7352 - 1(3H)-Isobenzofuranone, 3,3-bis[4-hydroxy-2-methyl-5-(1-methylethyl)phenyl]-

$C_{28}H_{32}ClN_2O_3$
12056 - Xanthylium, 9-(2-carboxyphenyl)-3,6-bis(diethylamino)-, chloride

$C_{28}H_{32}O_4Si_4$
5000 - Cyclotetrasiloxane, 2,4,6,8-tetramethyl-2,4,6,8-tetraphenyl-

$C_{28}H_{34}O_{15}$
2794 - 4H-1-Benzopyran-4-one, 7-[[6-O-(6-deoxy-α-L-mannopyranosyl)-β-D-glucopyranosyl]oxy]-2,3-dih

$C_{28}H_{38}N_2O_4$
5367 - Emetan-6′-ol, 7′,10,11-trimethoxy-

$C_{28}H_{38}O_{19}$
6229 - α-D-Glucopyranoside, 1,3,4,6-tetra-O-acetyl-β-D-fructofuranosyl, tetraacetate

$C_{28}H_{42}O$
5380 - Ergosta-5,7,9(11),22-tetraen-3-ol, (3β,22E)-

$C_{28}H_{44}O$
5381 - Ergosta-5,7,22-trien-3-ol, (3β,10α,22E)-
5382 - Ergosta-5,7,22-trien-3-ol, (3β,9β,10α,22E)-
5383 - Ergosta-5,7,22-trien-3-ol, (3β,22E)-
11349 - 9,10-Secoergosta-5(10),6,8,22-tetraen-3-ol, (3β,6E,22E)-
11350 - 9,10-Secoergosta-5,7,10(19),22-tetraen-3-ol, (3β,5Z,7E,22E)-

$C_{28}H_{48}O$
357 - Androst-4-en-3-one, 17-hydroxy-7,17-dimethyl-, (7β,17β)-
5384 - Ergost-7-en-3-ol, (3β,5α)-
5385 - Ergost-8(14)-en-3-ol, (3β,5α)-
5386 - Ergost-5-en-3-ol, (3β,24R)-

$C_{28}H_{48}O_2$
2789 - 2H-1-Benzopyran-6-ol, 3,4-dihydro-2,5,8-trimethyl-2-(4,8,12-trimethyltridecyl)-
2790 - 2H-1-Benzopyran-6-ol, 3,4-dihydro-2,7,8-trimethyl-2-(4,8,12-trimethyltridecyl)-

$C_{28}H_{50}$
1565 - Benzene, docosyl-
5377 - Ergostane, (5α)-
5378 - Ergostane, (5β)-

$C_{28}H_{50}N_2O_4$
4064 - Carpaine

$C_{28}H_{50}O$
5379 - Ergostan-3-ol, (3β,5α)-

$C_{28}H_{54}O_3$
11623 - Tetradecanoic acid, anhydride

$C_{28}H_{56}$
4380 - Cyclohexane, docosyl-

$C_{28}H_{56}O_2$
8221 - Octacosanoic acid

$C_{28}H_{58}$
5274 - Docosane, 7-hexyl-
5354 - Eicosane, 9-octyl-
8220 - Octacosane
11877 - Tridecane, 2,2,4,10,12,12-hexamethyl-7-(3,5,5-trimethylhexyl)-

$C_{28}H_{58}O$
8222 - 1-Octacosanol

$C_{29}H_{40}N_2O_4$
5368 - Emetan, 6′,7′,10,11-tetramethoxy-

$C_{29}H_{44}O_{12}$
4040 - Card-20(22)-enolide, 3-[(6-deoxy-α-L-mannopyranosyl)oxy]-1,5,11,14,19-pentahydroxy-, (1β,3β

$C_{29}H_{47}NO_8$
4067 - Cevane-3,4,12,14,16,17,20-heptol, 3-acetate, (3β,4α,16β)-

$C_{29}H_{48}O$
11554 - Stigmasta-5,7-dien-3-ol, (3β)
11555 - Stigmasta-5,22-dien-3-ol, (3β,22E)-

$C_{29}H_{50}O$
11558 - Stigmast-5-en-3-ol, (3β)-
11559 - Stigmast-5-en-3-ol, (3β,24S)-

$C_{29}H_{50}O_2$
2787 - 2H-1-Benzopyran-6-ol, 3,4-dihydro-2,5,7,8-tetramethyl-2-(4,8,12-trimethyltridecyl)-, [2R-[2R^*(4R^*,8R^*)]]-

$C_{29}H_{52}$
2341 - Benzene, tricosyl-

$C_{29}H_{52}O$
11557 - Stigmastan-3-ol, (3β,5α)-

$C_{29}H_{58}$
4483 - Cyclohexane, tricosyl-

$C_{29}H_{60}$
8099 - Nonacosane

$C_{29}H_{60}O$
8100 - 1-Nonacosanol

$C_{30}H_{34}O_{13}$
7579 - 3,6-Methano-8H-1,5,7-trioxacyclopenta[ij]cycloprop[a]azulene-4,8(3H)-dione, hexahydro-2a-hydroxy

$C_{30}H_{37}N_5O_5$
5393 - Ergotaman-3′,6′,18-trione, 12′-hydroxy-2′-methyl-5′-(2-methylpropyl)-,

$C_{30}H_{40}O_2$
435 - 8′-Apo-β,ψ-carotenal, 3-hydroxy-, (3R)-

$C_{30}H_{42}FeO_2Si_4$
5961 - Ferrocene, 1,1′-bis(1,1,3,3-tetramethyl-3-phenyldisiloxanyl)-

$C_{30}H_{46}O_5$
8546 - Olean-12-en-28-oic acid, 3,16-dihydroxy-23-oxo-, (3β,4α,16α)-
12036 - Urs-12-ene-27,28-dioic acid, 3-hydroxy-, (3β)-

$C_{30}H_{48}O_3$
8547 - Olean-12-en-28-oic acid, 3-hydroxy-, (3β)-
12037 - Urs-12-en-28-oic acid, 3-hydroxy-, (3β)-

$C_{30}H_{48}O_4$
8545 - Olean-12-en-28-oic acid, 3,23-dihydroxy-, (3β,4α)-

$C_{30}H_{50}$
11602 - 2,6,10,14,18,22-Tetracosahexaene, 2,6,10,15,19,23-hexamethyl-, (all-E)-

$C_{30}H_{50}O$
7411 - Lanosta-8,24-dien-3-ol, (3β)-
7417 - Lup-20(29)-en-3-ol, (3β)-
8548 - Olean-12-en-3-ol, (3β)-
11556 - Stigmasta-7,24(28)-dien-3-ol, 4-methyl-, (3β,4α,5α,24Z)-
12038 - Urs-12-en-3-ol, (3β)-

$C_{30}H_{50}O_2$
7416 - Lup-20(29)-ene-3,28-diol, (3β)-

$C_{30}H_{54}$
2265 - Benzene, tetracosyl-

$C_{30}H_{60}$
4474 - Cyclohexane, tetracosyl-

$C_{30}H_{60}I_3N_3O_3$
5441 - Ethanaminium, 2,2′,2′′-[1,2,3-benzenetriyltris(oxy)]tris[N,N,N-triethyl-, tr

$C_{30}H_{60}O_2$
11820 - Triacontanoic acid

$C_{30}H_{62}$
5275 - Docosane, 9-octyl-
11606 - Tetracosane, 2,6,10,15,19,23-hexamethyl-
11819 - Triacontane

$C_{30}H_{62}O$
11821 - 1-Triacontanol

$C_{31}H_{38}N_2O_9$
12064 - Yohimban-16-carboxylic acid, 17-hydroxy-18-[(3,4,5-trimethoxybenzoyl)oxy]-, methyl ester, (3β,16β

$C_{31}H_{39}N_5O_5$
5387 - Ergotaman-3′,6′,18-trione, 12′-hydroxy-2′,5′-bis(1-methylethyl)-,
5388 - Ergotaman-3′,6′,18-trione, 12′-hydroxy-2′,5′-bis(1-methylethyl)-,

$C_{31}H_{46}O_2$
7865 - 1,4-Naphthalenedione, 2-methyl-3-(3,7,11,15-tetramethyl-2-hexadecenyl)-, [R-[R^*,R^*-(E)]]-

$C_{31}H_{52}O_3$
2788 - 2H-1-Benzopyran-6-ol, 3,4-dihydro-2,5,7,8-tetramethyl-2-(4,8,12-trimethyltridecyl)-, acetate, [2R-[2R^*(4R^*,8R^*

$C_{31}H_{56}$
2088 - Benzene, pentacosyl-
8638 - Pentacosane, 13-phenyl-

$C_{31}H_{62}$
4462 - Cyclohexane, pentacosyl-
4821 - Cyclopentane, hexacosyl-

$C_{31}H_{62}O$
6300 - 16-Hentriacontanone

$C_{31}H_{64}$
6295 - Heneicosane, 11-decyl-
6299 - Hentriacontane

$C_{32}H_{22}N_6Na_2O_6S_2$
7969 - 1-Naphthalenesulfonic acid, 3,3'-[[1,1'-biphenyl]-4,4'-diylbis(azo)]bis[4-amino-, disodium salt

$C_{32}H_{38}N_2O_8$
12065 - Yohimban-16-carboxylic acid, 17-methoxy-18-[(3,4,5-trimethoxybenzoyl)oxy]-, methyl ester, (3β,16β

$C_{32}H_{40}O_4Si_4$
4998 - Cyclotetrasiloxane, 2,4,6,8-tetraethyl-2,4,6,8-tetraphenyl-

$C_{32}H_{41}N_5O_5$
5389 - Ergotaman-3',6',18-trione, 12'-hydroxy-2'-(1-methylethyl)-5'-(2-methylpropyl)-

$C_{32}H_{44}N_2O_8$
309 - Aconitane-4,8,9-triol, 20-ethyl-1,14,16-trimethoxy-, 4-[2-(acetylamino)benzoate], (1α,14α,16β)-

$C_{32}H_{44}O_{12}$
3142 - Bufa-4,20,22-trienolide, 6-(acetyloxy)-3-(β-D-glucopyranosyloxy)-8,14-dihydroxy-, (3β,6β)-

$C_{32}H_{49}NO_9$
4069 - Cevane-3,4,12,14,16,17,20-heptol, 4,9-epoxy-, 3-(2-methyl-2-butenoate), [3β(Z),4α,16β]-

$C_{32}H_{58}$
1750 - Benzene, hexacosyl-

$C_{32}H_{62}O_3$
6646 - Hexadecanoic acid, anhydride

$C_{32}H_{64}$
4410 - Cyclohexane, hexacosyl-

$C_{32}H_{64}O_2$
6651 - Hexadecanoic acid, hexadecyl ester

$C_{32}H_{66}$
5273 - Docosane, decyl-
5348 - Dotriacontane

$C_{32}H_{66}O$
6642 - Hexadecane, 1,1'-oxybis-

$C_{33}H_{34}N_4O_6$
3001 - 21H-Biline-8,12-dipropanoic acid, 3,18-diethenyl-1,19,22,24-tetrahydro-2,7,13,17-tetramethyl-1,19-dioxo-

$C_{33}H_{35}N_5O_5$
5395 - Ergotaman-3',6',18-trione, 12'-hydroxy-2'-methyl-5'-(phenylmethyl)-,

$C_{33}H_{36}N_4O_6$
3000 - 21H-Biline-8,12-dipropanoic acid, 2,17-diethenyl-1,10,19,22,23,24-hexahydro-3,7,13,18-tetramethyl-1,19-dioxo-

$C_{33}H_{40}N_2O_9$
12061 - Yohimban-16-carboxylic acid, 11,17-dimethoxy-18-[(3,4,5-trimethoxybenzoyl)oxy]-, methyl ester,

$C_{33}H_{41}N_5O_5$
5390 - Ergotaman-3',6',18-trione, 12'-hydroxy-2'-(1-methylethyl)-5'-(2-methylpropyl)-, (5'α,8α)-

$C_{33}H_{44}O_8$
8211 - 29-Nordammara-1,17(20),24-trien-21-oic acid, 6,16-bis(acetyloxy)-3,7-dioxo-,

$C_{33}H_{45}NO_9$
311 - Aconitane-8,13,14-triol, 1,6,16-trimethoxy-4-(methoxymethyl)-20-methyl-, 8-acetate 14-benzoate,

$C_{33}H_{49}NO_8$
12045 - Veratraman-11-one, 17,23-epoxy-3-(β-D-glucopyranosyloxy)-, (3β,23β)-

$C_{33}H_{55}N_5O_5$
5394 - Ergotaman-3',6',18-trione, 12'-hydroxy-2'-methyl-5'-(phenylmethyl)-, (5'α)-

$C_{33}H_{60}$
1742 - Benzene, heptacosyl-

$C_{34}H_{24}N_6Na_4O_{14}S_4$
7877 - 1,3-Naphthalenedisulfonic acid, 6,6'-[(3,3'-dimethyl[1,1'-biphenyl]-4,4'-diyl)bis(azo)]bis[4-amino-5-hydroxy-,

$C_{34}H_{32}ClFeN_4O_4$
5958 - Ferrate(2-), chloro[7,12-diethenyl-3,8,13,17-tetramethyl-21H,23H-porphine-2,18-dipropanoato(4-)-N21,N22,N23,

$C_{34}H_{33}FeN_4O_5$
5959 - Ferrate(2-), [7,12-diethenyl-3,8,13,17-tetramethyl-21H,23H-porphine-2,18-dipropanoato(4-)-N21,N22,N23,N24

$C_{34}H_{36}N_2O_6$
3002 - [2,2'-Bimorphinan]-3,3',6,6'-tetrol, 7,7',8,8'-tetradehydro-4,5:4',5'-diepoxy-17,17'dimethyl-

$C_{34}H_{38}N_4O_6$
9735 - 21H,23H-Porphine-2,18-dipropanoic acid, 7,12-bis(1-hydroxyethyl)-3,8,13,17-tetramethyl-

$C_{34}H_{47}NO_{10}$
307 - Aconitane-3,8,13,14-tetrol, 20-ethyl-1,6,16-trimethoxy-4-(methoxymethyl)-, 8-acetate 14-benzoate, (1α,3alp

$C_{34}H_{47}NO_{11}$
306 - Aconitane-3,8,13,14,15-pentol, 20-ethyl-1,6,16-trimethoxy-4-(methoxymethyl)-, 8-acetate 14-benzoate,

$C_{34}H_{62}$
2065 - Benzene, octacosyl-

$C_{34}H_{66}O_4$
6649 - Hexadecanoic acid, 1,2-ethanediyl ester

$C_{34}H_{68}$
4458 - Cyclohexane, octacosyl-

$C_{34}H_{68}O_2$
8258 - Octadecanoic acid, hexadecyl ester

$C_{34}H_{70}$
11604 - Tetracosane, 11-decyl-
11650 - Tetratriacontane

$C_{35}H_{39}N_5O_5$
5391 - Ergotaman-3',6',18-trione, 12'-hydroxy-2'-(1-methylethyl)-5'-(phenylmethyl)-
5392 - Ergotaman-3',6',18-trione, 12'-hydroxy-2'-(1-methylethyl)-5'-(phenylmethyl)-

$C_{35}H_{42}N_2O_9$
12060 - Yohimban-16-carboxylic acid, 11,17-dimethoxy-18-[[1-oxo-3-(3,4,5-trimethoxyphenyl)-2-propenyl]oxy]-, methyl ester, (3β,16β,17α,18β,20α)-

$C_{35}H_{54}O_{14}$
4045 - Card-20(22)-enolide, 3-[(6-O-β-D-glucopyranosyl-β-D-glucopyranosyl)oxy]-14-hydroxy-,

$C_{35}H_{70}$
4853 - Cyclopentane, triacontyl-

$C_{35}H_{70}O$
9000 - 18-Pentatriacontanone

$C_{35}H_{72}$
8999 - Pentatriacontane

$C_{36}H_{38}N_2O_6$
11924 - Tubocuraran-7',12'-diol, 6,6'-dimethoxy-2,2'-dimethyl-, (1'α)-
11925 - Tubocuraran-7',12'-diol, 6,6'-dimethoxy-2,2'-dimethyl-, (1β)-

$C_{36}H_{42}Br_2N_2$
6753 - 1,6-Hexanediaminium, N,N'-di-9H-fluoren-9-yl-N,N,N',N'-tetramethyl-dibromide

$C_{36}H_{51}NO_{11}$
310 - Aconitane-8,13,14-triol, 20-ethyl-1,6,16-trimethoxy-4-(methoxymethyl)-, 8-acetate 14-(3,4-dimethoxybenzoate),

$C_{36}H_{51}NO_{12}$
308 - Aconitane-3,8,13,14-tetrol, 20-ethyl-1,6,16-trimethoxy-4-(methoxymethyl)-, 8-acetate 14-(3,4-dimethoxybenzoate),(1α,3α,6α,14α,16β)-

$C_{36}H_{70}CaO_4$
8252 - Octadecanoic acid, calcium salt

$C_{36}H_{70}O_3$
8249 - Octadecanoic acid, anhydride

$C_{36}H_{70}O_4Pb$
8261 - Octadecanoic acid, lead salt

$C_{36}H_{70}O_4Zn$
8277 - Octadecanoic acid, zinc salt

$C_{36}H_{72}$
4482 - Cyclohexane, triacontyl-

$C_{36}H_{74}$
6991 - Hexatriacontane
8639 - Pentacosane, 13-undecyl-

$C_{37}H_{38}N_2O_6$
8630 - Oxyacanthan, 6',12'-dimethoxy-2,2'-dimethyl-6,7-[methylenebis(oxy)]-

$C_{37}H_{40}N_2O_6$
2911 - Berbaman-12-ol, 6,6',7-trimethoxy-2,2'-dimethyl-
8631 - Oxyacanthan-12'-ol, 6,6',7-trimethoxy-2,2'-dimethyl-

$C_{37}H_{67}NO_{13}$
5396 - Erythromycin

$C_{38}H_{74}O_4$
8256 - Octadecanoic acid, 1,2-ethanediyl ester

$C_{38}H_{76}$
4383 - Cyclohexane, dotriacontyl-

$C_{38}H_{78}$
6623 - Hexacosane, 13-dodecyl-

$C_{39}H_{74}O_6$
5319 - Dodecanoic acid, 1,2,3-propanetriyl ester

$C_{40}H_{36}$
4054 - β,ε-Carotene, (6'R)-

$C_{40}H_{50}N_4O_8S$
4132 - Cinchonan-9-ol, 6'-methoxy-, (8α,9R)-, sulfate (2:1) (salt)

$C_{40}H_{52}O_2$
4059 - β,ψ-Caroten-16'-oic acid, 3',4'-didehydro-

$C_{40}H_{56}$
4051 - β,β-Carotene
4052 - β,ψ-Carotene
4053 - ψ,ψ-Carotene

$C_{40}H_{56}O$
4060 - ψ,ψ-Caroten-16-ol
4061 - β,β-Caroten-3-ol, (3R)-
4062 - β,ψ-Caroten-3-ol, (3R)-

$C_{40}H_{56}O_2$
4057 - β,β-Carotene-3,3'-diol, (3R,3'R)-
4058 - β,ε-Carotene-3,3'-diol, (3R,3'R,6'R)-

$C_{40}H_{56}O_3$
4063 - β,κ-Caroten-6'-one, 3,3'-dihydroxy-, (3R,3'S,5'R)-

$C_{40}H_{56}O_4$
4056 - β,β-Carotene-3,3'-diol, 5,6:5',6'-diepoxy-5,5',6,6'-tetrahydro-

$C_{40}H_{60}O_6$
4055 - β,β-Carotene, 3'-(acetyloxy)-6',7'-didehydro-5,6-epoxy-5,5',6,6',7,8-hexahydro-3,5'-dihydroxy-8-oxo-,

$C_{40}H_{80}$
4479 - Cyclohexane, tetratriacontyl-
4845 - Cyclopentane, pentatriacontyl-

$C_{41}H_{46}N_4O_5$
4128 - Cinchonan-9-ol, 6'-methoxy-, carbonate (2:1) (ester), (8α,9R)-

$C_{41}H_{64}O_{13}$
4043 - Card-20(22)-enolide, 3-[(O-2,6-dideoxy-β-D-ribo-hexopyranosyl-(1-4)-O-2,6-dideoxy-β-D-ribo-hexopyranosyl-(1-4)-2,6-dideoxy-β-D, 14-dihydroxy

$C_{41}H_{64}O_{14}$

4041 - Card-20(22)-enolide, 3-[(O-2,6-dideoxy-β-D-ribo-hexopyranosyl-(1-4)-O-2,6-dideoxy-β-D-ribo-hexopyranosyl-(1-4)-2,6-dideoxy-β-D, 12,14-dihydroxy

4042 - Card-20(22)-enolide, 3-[(O-2,6-dideoxy-β-D-ribo-hexopyranosyl-(1-4)-O-2,6-dideoxy-β-D-ribo-hexopyranosyl-(1-4)-2,6-dideoxy-β-D, 14,16-dihydroxy

$C_{42}H_{46}N_4O_8S$

11566 - Strychnidin-10-one, sulfate (2:1)

$C_{42}H_{62}O_{16}$

6230 - α-D-Glucopyranosiduronic acid, (3β,20β)-20-carboxy-11-oxo-30-norolean-12-en-3-yl 2-O-

$C_{42}H_{84}$

4413 - Cyclohexane, hexatriacontyl-

$C_{42}H_{84}N_{14}O_{36}S_3$

11561 - D-Streptamine, O-2-deoxy-2-(methylamino)-α-L-glucopyranosyl-(12)-O-5-deoxy-3-C-formyl-α-L-lyxofuranosyl-(1-4)-N,N,N' -55-bis(aminoiminomethyl)-, sulfate (2:3)(salt)

$C_{43}H_{88}$

11918 - Tritetracontane

$C_{45}H_{86}O_6$

11628 - Tetradecanoic acid, 1,2,3-propanetriyl ester

$C_{48}H_{40}O_4Si_4$

4994 - Cyclotetrasiloxane, octaphenyl-

$C_{48}H_{72}O_{14}$

477 - Avermectin B1a

$C_{48}H_{93}AlO_6$

6645 - Hexadecanoic acid, aluminum salt

$C_{50}H_{83}NO_{21}$

6196 - β-D-Galactopyranoside, (3β,5α,22β,25S)-spirosolan-3-yl O-β-

$C_{51}H_{98}O_6$

6656 - Hexadecanoic acid, 1,2,3-propanetriyl ester

$C_{54}H_{105}AlO_6$

8248 - Octadecanoic acid, aluminum salt

$C_{56}H_{92}O_{29}$

6195 - β-D-Galactopyranoside, (2α,3β,5α,15β,25R)-2,15-dihydroxyspirostan-3-yl

$C_{57}H_{104}O_6$

8310 - 9-Octadecenoic acid (Z)-, 1,2,3-propanetriyl ester

$C_{57}H_{110}O_6$

8272 - Octadecanoic acid, 1,2,3-propanetriyl ester

$C_{60}H_{78}OSn_2$

5950 - Fenbutatin oxide

$C_{66}H_{103}N_{17}O_{16}S$

508 - Bacitracin

CAS	No.	CAS	No.	CAS	No.	CAS	No.	CAS	No.
50-00-0	: 5987	54-95-5	: 11654	58-72-0	: 1648	63-74-1	: 2187	71-30-7	: 11104
50-01-1	: 6285	55-10-7	: 896	58-73-1	: 5422	63-75-2	: 10914	71-36-3	: 3593
50-02-2	: 9737	55-18-5	: 5432	58-74-2	: 7378	63-91-2	: 9510	71-41-0	: 8915
50-03-3	: 9747	55-21-0	: 600	58-85-5	: 11689	63-98-9	: 879	71-43-2	: 867
50-06-6	: 11095	55-22-1	: 10895	58-86-6	: 12059	64-04-0	: 1569	71-44-3	: 3259
50-10-2	: 5443	55-38-9	: 5955	58-89-9	: 4409	64-10-8	: 12026	71-55-6	: 5655
50-11-3	: 11088	55-43-6	: 1801	58-90-2	: 9470	64-17-5	: 5672	71-63-6	: 4043
50-14-6	: 11350	55-63-0	: 10197	58-93-5	: 2855	64-18-6	: 6002	71-81-8	: 2120
50-22-6	: 9748	55-73-2	: 6281	58-94-6	: 2852	64-19-7	: 100	71-91-0	: 5447
50-23-7	: 9753	55-91-4	: 9580	58-95-7	: 2788	64-20-0	: 7456	72-14-0	: 2206
50-24-8	: 9738	56-03-1	: 7233	58-96-8	: 12033	64-65-3	: 9656	72-18-4	: 12041
50-27-1	: 5404	56-04-2	: 11108	58-97-9	: 12035	64-67-5	: 11573	72-19-5	: 11813
50-28-2	: 5399	56-12-2	: 3464	59-00-7	: 11230	64-69-7	: 200	72-23-1	: 9755
50-29-3	: 2328	56-18-8	: 9874	59-02-9	: 2787	64-77-7	: 2207	72-43-5	: 2329
50-30-6	: 2563	56-23-5	: 7540	59-23-4	: 6197	64-85-7	: 9751	72-44-6	: 11199
50-31-7	: 2727	56-25-7	: 5371	59-26-7	: 10891	64-86-8	: 93	72-48-0	: 391
50-32-8	: 2395	56-29-1	: 11083	59-30-3	: 6243	65-29-2	: 5441	72-54-8	: 1378
50-34-0	: 9832	56-33-7	: 5238	59-31-4	: 11325	65-45-2	: 616	72-55-9	: 1374
50-35-1	: 7363	56-35-9	: 5241	59-40-5	: 2205	65-46-3	: 5021	72-56-0	: 9160
50-36-2	: 482	56-36-0	: 11538	59-46-1	: 2493	65-49-6	: 2498	72-63-9	: 344
50-42-0	: 938	56-38-2	: 9583	59-47-2	: 10019	65-64-5	: 7170	73-22-3	: 11922
50-44-2	: 10763	56-40-6	: 6250	59-48-3	: 7327	65-71-4	: 11068	73-24-5	: 10751
50-50-0	: 5403	56-41-7	: 325	59-49-4	: 2882	65-82-7	: 7649	73-31-4	: 69
50-53-3	: 9505	56-45-1	: 11357	59-50-7	: 9267	65-85-0	: 2475	73-32-5	: 7370
50-55-5	: 12061	56-49-5	: 2392	59-51-8	: 7647	65-86-1	: 11050	73-40-5	: 10768
50-69-1	: 11345	56-53-1	: 9300	59-52-9	: 10370	66-02-4	: 11928	73-49-4	: 11197
50-70-4	: 6206	56-54-2	: 4126	59-66-5	: 32	66-22-8	: 11059	74-11-3	: 2536
50-71-5	: 11073	56-55-3	: 519	59-67-6	: 10894	66-23-9	: 5438	74-31-7	: 1271
50-78-2	: 2480	56-57-5	: 11287	59-87-0	: 7144	66-25-1	: 6718	74-39-5	: 1542
50-79-3	: 2562	56-72-4	: 4147	59-88-1	: 7172	66-27-3	: 7532	74-79-3	: 438
50-81-7	: 466	56-75-7	: 52	59-89-2	: 7699	66-28-4	: 4049	74-82-8	: 7459
50-82-8	: 2728	56-81-5	: 10186	59-92-7	: 11929	66-32-0	: 11565	74-83-9	: 7462
50-84-0	: 2561	56-82-6	: 9782	59-98-3	: 7198	66-56-8	: 9341	74-84-0	: 5448
50-85-1	: 2630	56-84-8	: 471	60-00-4	: 6262	66-71-7	: 9189	74-85-1	: 5879
50-89-5	: 11816	56-85-9	: 6249	60-01-5	: 3586	66-75-1	: 11061	74-86-2	: 5936
50-91-9	: 12034	56-86-0	: 6241	60-09-3	: 837	66-76-2	: 2839	74-87-3	: 7477
51-03-6	: 9727	56-87-1	: 7419	60-10-6	: 5121	66-77-3	: 7751	74-88-4	: 7518
51-12-7	: 10910	56-89-3	: 5020	60-11-7	: 721	66-81-9	: 9655	74-89-5	: 7442
51-17-2	: 2376	57-00-1	: 6254	60-12-8	: 1592	66-99-9	: 7752	74-90-8	: 7176
51-18-3	: 11844	57-06-7	: 10575	60-18-4	: 11926	67-03-8	: 11686	74-93-1	: 7546
51-20-7	: 11062	57-08-9	: 6871	60-23-1	: 5649	67-20-9	: 7222	74-95-3	: 7492
51-21-8	: 11067	57-10-3	: 6644	60-24-2	: 5732	67-43-6	: 6257	74-96-4	: 5454
51-26-3	: 2136	57-11-4	: 8247	60-27-5	: 7230	67-45-8	: 8589	74-97-5	: 7463
51-28-5	: 9342	57-13-6	: 11992	60-29-7	: 5608	67-47-0	: 6044	74-98-6	: 9834
51-34-3	: 908	57-14-7	: 7154	60-31-1	: 5439	67-48-1	: 5446	74-99-7	: 10727
51-35-4	: 9767	57-15-8	: 10407	60-32-2	: 6873	67-51-6	: 10828	75-00-3	: 5467
51-36-5	: 2564	57-24-9	: 11563	60-33-3	: 8223	67-52-7	: 11076	75-01-4	: 5884
51-41-2	: 1500	57-27-2	: 7686	60-34-4	: 7161	67-56-1	: 7581	75-02-5	: 5906
51-43-4	: 1521	57-41-0	: 7219	60-35-5	: 22	67-63-0	: 10335	75-03-6	: 5597
51-45-6	: 7203	57-43-2	: 11092	60-41-3	: 11586	67-64-1	: 10411	75-04-7	: 5410
51-48-9	: 11930	57-44-3	: 11087	60-51-5	: 5009	67-66-3	: 7554	75-05-8	: 271
51-52-5	: 11109	57-47-6	: 11191	60-54-8	: 7706	67-68-5	: 7529	75-07-0	: 11
51-55-8	: 902	57-50-1	: 6218	60-56-0	: 7214	67-71-0	: 7534	75-08-1	: 5648
51-63-8	: 1582	57-53-4	: 10021	60-57-1	: 5178	67-72-1	: 5595	75-09-2	: 7499
51-66-1	: 71	57-55-6	: 9982	60-70-8	: 12044	67-97-0	: 11348	75-10-5	: 7504
51-67-2	: 9208	57-57-8	: 8595	60-80-0	: 10847	68-11-1	: 202	75-11-6	: 7506
51-75-2	: 5413	57-62-5	: 7704	60-82-2	: 10453	68-12-2	: 5992	75-12-7	: 5989
51-79-6	: 3948	57-63-6	: 8216	60-87-7	: 9504	68-26-8	: 11341	75-15-0	: 3993
51-80-9	: 7490	57-66-9	: 2595	60-91-3	: 9503	68-35-9	: 2204	75-17-2	: 5988
51-83-2	: 5440	57-67-0	: 2191	61-00-7	: 5807	68-36-0	: 996	75-18-3	: 7545
52-28-8	: 7673	57-68-1	: 2196	61-12-1	: 11226	68-41-7	: 7406	75-19-4	: 4955
52-31-3	: 11084	57-71-6	: 3361	61-19-8	: 320	68-88-2	: 5700	75-21-8	: 8599
52-39-1	: 9746	57-74-9	: 4077	61-50-7	: 7303	68-94-0	: 10770	75-22-9	: 3132
52-43-7	: 11091	57-85-2	: 359	61-54-1	: 7302	68-96-2	: 9750	75-24-1	: 339
52-49-3	: 9705	57-87-4	: 5383	61-73-4	: 9506	69-24-9	: 10435	75-25-2	: 7548
52-51-7	: 9991	57-88-5	: 4110	61-78-9	: 6252	69-57-8	: 11655	75-26-3	: 9837
52-52-8	: 4782	57-91-0	: 5400	61-80-3	: 2876	69-65-8	: 7422	75-27-4	: 7487
52-67-5	: 12043	57-92-1	: 11562	61-82-5	: 11845	69-72-7	: 2612	75-28-5	: 10073
52-68-6	: 11853	57-97-6	: 520	61-90-5	: 7413	69-89-6	: 10755	75-29-6	: 9849
52-85-7	: 5946	58-00-4	: 5155	62-23-7	: 2705	69-91-0	: 885	75-30-7	: 10055
52-86-8	: 3651	58-05-9	: 6244	62-38-4	: 7431	69-93-2	: 10764	75-31-0	: 9804
52-90-4	: 5014	58-08-2	: 10760	62-44-2	: 60	70-07-5	: 8588	75-33-2	: 10157
53-03-2	: 9739	58-14-0	: 11051	62-50-0	: 7531	70-11-1	: 5782	75-34-3	: 5528
53-05-4	: 9744	58-15-1	: 10851	62-51-1	: 9830	70-18-8	: 6264	75-35-4	: 5894
53-06-5	: 9754	58-18-4	: 358	62-53-3	: 633	70-22-4	: 11188	75-36-5	: 294
53-16-7	: 5406	58-22-0	: 356	62-54-4	: 122	70-25-7	: 6284	75-37-6	: 5544
53-41-8	: 348	58-25-3	: 2409	62-55-5	: 5643	70-26-8	: 8550	75-38-7	: 5902
53-46-3	: 5445	58-27-5	: 7864	62-56-6	: 11799	70-29-1	: 5621	75-39-8	: 5874
53-70-3	: 5142	58-32-2	: 5711	62-57-7	: 331	70-30-4	: 9409	75-43-4	: 7502
53-89-4	: 10850	58-38-8	: 9501	62-59-9	: 4069	70-34-8	: 1718	75-44-5	: 4007
53-96-3	: 64	58-39-9	: 9617	62-73-7	: 5170	70-47-3	: 468	75-45-6	: 7478
54-04-6	: 1586	58-46-8	: 2396	62-75-9	: 7450	70-54-2	: 7418	75-46-7	: 7558
54-06-8	: 7301	58-54-8	: 168	62-76-0	: 5566	70-69-9	: 10413	75-47-8	: 7560
54-11-5	: 10992	58-55-9	: 10756	62-90-8	: 5408	70-70-2	: 10450	75-50-3	: 7444
54-25-1	: 11832	58-56-0	: 10934	63-05-8	: 352	70-78-0	: 11932	75-52-5	: 7522
54-36-4	: 10462	58-61-7	: 319	63-25-2	: 8032	71-00-1	: 7132	75-54-7	: 11398
54-85-3	: 10905	58-63-9	: 7335	63-68-3	: 7648	71-23-8	: 10334	75-55-8	: 501

CAS	Page	CAS	Page	CAS	Page	CAS	Page	CAS	Page
75-56-9	8621	77-09-8	7354	79-10-7	10609	81-92-5	1841	85-46-1	7975
75-57-0	7457	77-10-1	9696	79-11-8	123	82-02-0	6185	85-47-2	7959
75-58-1	7458	77-16-7	11904	79-14-1	196	82-05-3	6292	85-55-2	2684
75-60-5	451	77-25-8	9940	79-15-2	36	82-12-2	406	85-64-3	9398
75-61-6	7496	77-26-9	11101	79-16-3	72	82-28-0	384	85-66-5	9196
75-62-7	7474	77-28-1	11082	79-17-4	7147	82-29-1	417	85-73-4	2726
75-63-8	7475	77-40-7	9433	79-19-6	7142	82-34-8	410	85-74-5	7873
75-64-9	9818	77-41-8	11156	79-20-9	208	82-35-9	405	85-75-6	7872
75-65-0	10389	77-42-9	9103	79-21-0	5619	82-40-6	6182	85-82-5	8029
75-66-1	10159	77-46-3	91	79-22-1	4020	82-44-0	387	85-83-6	8033
75-68-3	5470	77-47-4	4766	79-24-3	5607	82-45-1	380	85-84-7	7729
75-69-4	7555	77-48-5	7217	79-27-6	5632	82-54-2	5216	85-86-9	8038
75-71-8	7500	77-49-6	10018	79-28-7	5914	82-57-5	6186	85-90-5	2838
75-72-9	7486	77-52-1	12037	79-29-8	3290	82-58-6	5376	85-91-6	2681
75-73-0	7541	77-53-2	7570	79-30-1	10504	82-62-2	9490	85-95-0	9303
75-74-1	9732	77-59-8	11528	79-31-2	10290	82-66-6	7288	85-97-2	3092
75-75-2	7530	77-60-1	11532	79-34-5	5635	82-68-8	2087	85-98-3	12003
75-76-3	11451	77-63-4	5000	79-35-6	5897	82-71-3	1554	86-00-0	3085
75-77-4	11389	77-65-6	3195	79-36-7	297	82-75-7	7965	86-21-5	11010
75-78-5	11394	77-67-8	11158	79-37-8	5585	82-86-0	8	86-26-0	3080
75-79-6	11465	77-71-4	7218	79-38-9	5889	83-07-8	10845	86-28-2	3985
75-80-9	5767	77-74-7	8944	79-40-3	5588	83-12-5	7270	86-29-3	969
75-81-0	5520	77-75-8	9146	79-41-4	10648	83-13-6	9975	86-30-6	830
75-82-1	5522	77-76-9	9909	79-42-5	10285	83-14-7	7394	86-34-0	11161
75-83-2	3289	77-77-0	5913	79-43-6	162	83-25-0	11162	86-48-6	7771
75-84-3	10373	77-78-1	11574	79-44-7	3973	83-27-2	9967	86-50-0	497
75-85-4	3628	77-79-2	11752	79-46-9	10117	83-32-9	6	86-52-2	7783
75-86-5	10108	77-83-8	8604	79-49-2	10477	83-33-0	7286	86-53-3	7748
75-87-6	21	77-84-9	10007	79-53-8	10421	83-34-1	7314	86-55-5	7758
75-88-7	5489	77-85-0	10011	79-54-9	9169	83-40-9	2629	86-56-6	7715
75-89-8	5770	77-86-1	9986	79-55-0	9691	83-41-0	1474	86-57-7	7929
75-91-2	7178	77-89-4	10166	79-57-2	7705	83-42-1	1204	86-59-9	11229
75-93-4	11577	77-92-9	10167	79-58-3	11515	83-44-3	4095	86-60-2	7964
75-94-5	11461	77-93-0	10169	79-63-0	7411	83-45-4	11557	86-65-7	7870
75-95-6	5616	77-94-1	10168	79-69-6	3884	83-46-5	11558	86-68-8	11234
75-96-7	241	77-95-2	4338	79-70-9	3883	83-47-6	11559	86-73-7	5967
75-97-8	3657	77-99-6	10006	79-74-3	1501	83-48-7	11555	86-74-8	3981
75-98-9	10258	78-00-2	9731	79-77-6	3885	83-49-8	4096	86-75-9	11309
75-99-0	10244	78-07-9	11478	79-83-4	329	83-53-4	7802	86-76-0	5164
76-00-6	10406	78-08-0	11424	80-00-2	1225	83-54-5	11330	86-78-2	8737
76-01-7	5617	78-09-1	5801	80-05-7	9413	83-56-7	7835	86-80-6	5411
76-02-8	300	78-10-4	11506	80-06-8	1850	83-60-3	12063	86-84-0	7904
76-03-9	243	78-11-5	9990	80-07-9	2239	83-66-9	1463	86-86-2	7733
76-04-0	127	78-13-7	11507	80-08-0	846	83-67-0	10757	86-87-3	7734
76-05-1	263	78-26-2	10020	80-09-1	9466	83-68-1	7800	86-88-4	11807
76-06-2	7557	78-27-3	4529	80-10-4	11395	83-72-7	7859	86-89-5	7941
76-08-4	10405	78-30-8	9569	80-11-5	2212	83-73-8	11315	86-98-6	11251
76-09-5	3350	78-32-0	9571	80-13-7	2565	83-74-9	7186	87-00-3	911
76-11-9	5636	78-34-2	5201	80-14-8	11915	83-79-4	2783	87-10-5	609
76-12-0	5637	78-38-6	9536	80-15-9	7181	83-81-8	1325	87-13-8	9947
76-13-1	5663	78-39-7	5664	80-18-2	2227	83-88-5	11344	87-17-2	617
76-14-2	5538	78-40-0	9563	80-26-2	4669	83-95-4	6181	87-19-4	2643
76-15-3	5484	78-48-8	11852	80-32-0	2193	84-11-7	9174	87-20-7	2633
76-16-4	5596	78-54-6	4528	80-33-1	2221	84-15-1	11595	87-24-1	2690
76-17-5	10181	78-59-1	4721	80-35-3	2197	84-16-2	9298	87-25-2	2495
76-19-7	10118	78-62-6	11405	80-40-0	2231	84-17-3	9302	87-28-5	2622
76-20-0	3230	78-67-1	10093	80-42-2	2235	84-26-4	7334	87-31-0	2874
76-24-4	3106	78-74-0	5653	80-46-6	9339	84-31-1	4134	87-39-8	11074
76-28-8	4046	78-75-1	9881	80-48-8	2232	84-52-6	5022	87-40-1	2332
76-30-2	3342	78-77-3	9845	80-55-7	10279	84-54-8	409	87-41-2	7349
76-36-8	3191	78-78-4	3397	80-56-8	2973	84-55-9	10434	87-42-3	10753
76-37-9	10404	78-79-5	3171	80-59-1	3844	84-58-2	4251	87-44-5	2996
76-39-1	10391	78-80-8	3892	80-62-6	10663	84-61-7	1341	87-47-8	3945
76-43-7	355	78-81-9	9817	80-63-7	10621	84-62-8	1345	87-51-4	7295
76-44-8	7576	78-82-0	10111	80-69-3	9956	84-65-1	379	87-52-5	7310
76-45-9	4070	78-83-1	10388	80-72-8	4942	84-66-2	1342	87-58-1	11140
76-49-3	2948	78-84-2	9785	80-92-2	9742	84-68-4	3032	87-59-2	704
76-57-3	7669	78-85-3	10515	80-97-7	4107	84-69-5	1336	87-60-5	671
76-59-5	9365	78-88-6	10552	81-04-9	7867	84-74-2	1339	87-61-6	2321
76-60-8	9387	78-89-7	10347	81-07-2	2388	84-79-7	7862	87-62-7	707
76-61-9	9369	78-90-0	9872	81-08-3	2875	84-80-0	7865	87-64-9	9263
76-62-0	7350	78-92-2	3594	81-13-0	3202	84-86-6	7962	87-65-0	9285
76-65-3	2463	78-93-3	3845	81-15-2	1461	84-87-7	7971	87-66-1	2366
76-67-5	9952	78-94-4	3875	81-16-3	7961	84-88-8	11294	87-68-3	3163
76-68-6	11085	78-95-5	10420	81-20-9	1476	84-89-9	7963	87-69-4	3309
76-76-6	11093	78-97-7	10105	81-23-2	4101	84-95-7	7713	87-73-0	6205
76-80-2	3110	78-98-8	9788	81-24-3	5625	84-99-1	2398	87-79-6	11519
76-83-5	1196	78-99-9	9891	81-25-4	4100	85-00-7	5224	87-81-0	11587
76-84-6	1866	79-00-5	5656	81-30-1	2784	85-01-8	9163	87-83-2	2083
76-87-9	11911	79-01-6	5922	81-38-9	2489	85-02-9	2437	87-85-4	1758
76-89-1	920	79-03-8	10493	81-54-9	416	85-23-4	4260	87-86-5	9440
76-93-7	918	79-04-9	295	81-61-8	413	85-29-0	7603	87-88-7	4258
76-94-8	11100	79-05-0	9791	81-64-1	392	85-34-7	944	87-90-1	11840
77-02-1	11098	79-06-1	10519	81-77-6	434	85-38-1	2646	87-92-3	3313
77-03-2	9654	79-07-2	41	81-81-2	2832	85-41-6	7362	87-99-0	12058
77-04-3	10946	79-08-3	111	81-88-9	12056	85-42-7	7345	88-04-0	9257
77-06-5	6204	79-09-4	10201	81-90-3	2517	85-44-9	7342	88-05-1	859

CAS	Page	CAS	Page	CAS	Page	CAS	Page	CAS	Page
88-06-2	9485	90-80-2	6212	93-15-2	1450	95-50-1	1354	97-30-3	6227
88-09-5	3514	90-81-3	1900	93-16-3	1449	95-51-2	658	97-39-2	6279
88-10-8	3972	90-84-6	10428	93-18-5	7884	95-52-3	1726	97-53-0	9389
88-12-0	11183	90-86-8	1903	93-22-1	7942	95-53-4	778	97-54-1	9388
88-13-1	11738	90-89-1	9603	93-25-4	921	95-54-5	1260	97-56-3	798
88-14-2	6050	90-90-4	7602	93-26-5	70	95-55-6	9201	97-59-6	12009
88-15-3	5870	90-94-8	7593	93-28-7	9395	95-56-7	9228	97-62-1	10298
88-16-4	1240	90-98-2	7592	93-35-6	2826	95-57-8	9251	97-63-2	10658
88-17-5	855	90-99-3	1187	93-37-8	11258	95-59-0	5188	97-65-4	3328
88-18-6	9321	91-00-9	1826	93-39-0	2821	95-63-6	2358	97-72-3	10291
88-20-0	2228	91-01-0	1910	93-42-5	68	95-64-7	708	97-74-5	11715
88-21-1	2215	91-02-1	7638	93-44-7	8021	95-65-8	9311	97-77-8	11721
88-60-8	9331	91-08-7	1430	93-50-5	667	95-66-9	1138	97-85-8	10304
88-65-3	2518	91-10-1	9305	93-51-6	9385	95-68-1	705	97-86-9	10666
88-67-5	2657	91-13-4	983	93-52-7	1299	95-69-2	673	97-88-1	10650
88-68-6	601	91-14-5	1403	93-53-8	872	95-70-5	1279	97-90-5	10656
88-69-7	9410	91-16-7	1436	93-54-9	1870	95-71-6	1532	97-94-9	3118
88-72-2	2000	91-17-8	7789	93-55-0	10480	95-72-7	1140	97-95-0	3620
88-73-3	1214	91-18-9	10747	93-56-1	5579	95-73-8	1388	97-96-1	3183
88-74-4	822	91-19-0	11333	93-58-3	2687	95-74-9	672	97-97-2	5472
88-75-5	9435	91-20-3	7732	93-59-4	1126	95-75-0	1384	97-99-4	6124
88-85-7	5183	91-21-4	7388	93-60-7	10906	95-76-1	689	98-00-0	6118
88-87-9	9258	91-22-5	11214	93-61-8	5998	95-77-2	9286	98-01-1	6041
88-88-0	1244	91-23-6	1935	93-71-0	44	95-78-3	706	98-02-2	6117
88-89-1	9497	91-40-7	2715	93-72-1	11512	95-79-4	676	98-03-3	11733
88-95-9	1318	91-49-6	39	93-76-5	256	95-80-7	1278	98-04-4	866
88-96-0	1324	91-53-2	11265	93-78-7	259	95-81-8	669	98-05-5	460
88-99-3	1327	91-56-5	7300	93-79-8	258	95-82-9	687	98-06-6	1459
89-00-9	10927	91-57-6	7916	93-88-9	1574	95-83-0	1264	98-07-7	2333
89-01-0	10804	91-58-7	7781	93-89-0	2602	95-84-1	9210	98-08-8	2348
89-05-4	2257	91-59-8	7709	93-90-3	5743	95-85-2	9204	98-09-9	2242
89-31-6	7397	91-60-1	8008	93-91-4	3382	95-86-3	9274	98-11-3	2214
89-37-2	3975	91-62-3	11281	93-97-0	2510	95-87-4	9309	98-13-5	11473
89-39-4	1447	91-63-4	11277	93-98-1	628	95-92-1	5560	98-15-7	1241
89-54-3	2490	91-64-5	2792	93-99-2	2716	95-93-2	2280	98-16-8	856
89-55-4	2524	91-65-6	4285	94-02-0	2141	95-94-3	2260	98-17-9	9491
89-56-5	2631	91-66-7	695	94-07-5	1887	95-95-4	9484	98-19-1	1460
89-57-6	2499	91-68-9	9297	94-08-6	2692	96-00-4	4206	98-27-1	9333
89-58-7	1478	91-73-6	1829	94-09-7	2497	96-01-5	4203	98-29-3	1516
89-59-8	1212	91-76-9	11829	94-11-1	173	96-02-6	6092	98-46-4	2061
89-60-1	1209	91-78-1	11834	94-12-2	2507	96-04-8	6402	98-50-0	456
89-61-2	1395	91-80-5	5507	94-13-3	2653	96-05-9	10671	98-51-1	1466
89-62-3	803	91-81-6	5505	94-14-4	2505	96-09-3	8624	98-54-4	9323
89-68-9	9268	91-84-9	5511	94-20-2	2208	96-11-7	10163	98-56-6	1242
89-69-0	2339	91-88-3	5724	94-25-7	2488	96-12-8	9885	98-58-8	2244
89-71-4	2696	91-96-3	3204	94-26-8	2615	96-14-0	8824	98-59-9	2250
89-72-5	9430	92-06-9	11596	94-30-4	2871	96-18-4	10175	98-60-2	2245
89-73-6	613	92-13-7	6151	94-33-7	5578	96-19-5	10603	98-61-3	2247
89-74-7	5809	92-24-0	7703	94-36-0	9157	96-21-9	10357	98-62-4	2252
89-75-8	2893	92-35-3	11111	94-41-7	10708	96-22-0	8958	98-66-8	2220
89-79-2	4544	92-36-4	782	94-44-0	10911	96-23-1	10362	98-67-9	2226
89-82-7	4608	92-41-1	7869	94-46-2	3632	96-24-2	9994	98-71-5	2224
89-83-8	9418	92-43-3	10844	94-49-5	5572	96-26-4	10429	98-72-6	459
89-84-9	5803	92-48-8	2835	94-53-1	2418	96-27-5	10013	98-73-7	2588
89-86-1	2569	92-51-3	2988	94-58-6	2431	96-29-7	3675	98-79-3	9768
89-87-2	1479	92-52-4	3008	94-59-7	2429	96-31-1	12004	98-80-6	3135
89-92-9	1062	92-54-6	9620	94-62-2	9632	96-32-2	117	98-81-7	1025
89-93-0	1815	92-59-1	1811	94-65-5	4612	96-33-3	10655	98-82-8	1975
89-95-2	1897	92-61-5	2829	94-67-7	579	96-34-4	138	98-83-9	1972
89-96-3	1161	92-62-6	316	94-68-8	746	96-35-5	198	98-85-1	1896
89-98-5	529	92-66-0	3018	94-69-9	5996	96-37-7	4826	98-86-2	5864
90-00-6	9353	92-67-1	3011	94-70-2	736	96-40-2	4914	98-87-3	1382
90-01-7	1882	92-69-3	3090	94-71-3	9350	96-41-3	4864	98-88-4	2886
90-02-8	564	92-70-6	7772	94-74-6	141	96-43-5	11740	98-89-5	4332
90-04-0	766	92-71-7	8580	94-75-7	169	96-45-7	7225	98-91-9	1130
90-05-1	9379	92-82-0	9193	94-80-4	171	96-47-9	6171	98-92-0	10890
90-11-9	7739	92-83-1	12049	94-81-5	3493	96-48-0	6142	98-94-2	4286
90-12-0	7915	92-84-2	9499	94-82-6	3925	96-49-1	5211	98-95-3	2049
90-13-1	7780	92-85-3	11660	94-96-2	6802	96-50-4	11662	98-96-4	10802
90-14-2	7902	92-86-4	3037	95-00-1	1803	96-54-8	11130	98-97-5	10803
90-15-3	8017	92-87-5	3031	95-01-2	541	96-76-4	9219	98-98-6	10893
90-19-7	2815	92-91-1	5776	95-05-6	11714	96-77-5	1560	99-02-5	5788
90-20-0	7874	92-92-2	3022	95-06-7	4065	96-80-0	5683	99-03-6	5772
90-26-6	881	92-93-3	3087	95-11-4	2966	96-82-2	6210	99-04-7	2677
90-27-7	897	92-94-4	11597	95-13-6	7245	96-88-8	9639	99-05-8	2485
90-33-5	2831	92-99-9	11213	95-14-7	2871	96-91-3	9207	99-07-0	9318
90-39-1	7571	93-01-6	7974	95-15-6	2399	96-97-9	2647	99-08-1	2001
90-41-5	3010	93-02-7	544	95-16-9	2861	96-99-1	2547	99-09-2	823
90-42-6	2989	93-03-8	1856	95-20-5	7313	97-00-7	1144	99-11-6	10900
90-43-7	3088	93-04-9	7914	95-21-6	2880	97-02-9	725	99-12-7	1477
90-44-8	433	93-05-0	1266	95-25-0	2883	97-05-2	2655	99-14-9	10165
90-45-9	313	93-07-2	2576	95-43-2	3193	97-16-5	9289	99-17-2	6223
90-47-1	12050	93-08-3	5858	95-45-4	3360	97-17-6	9581	99-20-7	6220
90-59-5	534	93-09-4	7759	95-46-5	1054	97-18-7	9478	99-30-9	692
90-65-3	8701	93-10-7	11227	95-47-6	1452	97-23-4	9408	99-31-0	1331
90-69-7	5834	93-11-8	7976	95-48-7	9400	97-24-5	9477	99-32-1	10776
90-74-4	6236	93-14-1	10016	95-49-8	1183	97-25-6	6231	99-33-2	2896

CAS	No.	CAS	No.	CAS	No.	CAS	No.	CAS	No.	CAS	No.
99-34-3	2591	100-71-0	10956	103-19-5	5246	105-07-7	2758	106-96-7	10728		
99-35-4	2365	100-72-1	10784	103-23-1	6781	105-13-5	1891	106-97-8	3229		
99-36-5	2697	100-74-3	7694	103-24-2	8132	105-14-6	8964	106-98-9	3704		
99-45-6	5804	100-75-4	9688	103-25-3	2139	105-30-6	8937	106-99-0	3146		
99-48-9	4696	100-76-5	480	103-28-6	10307	105-31-7	7124	107-00-6	3897		
99-50-3	2571	100-79-8	5207	103-29-7	1587	105-34-0	154	107-02-8	10509		
99-51-4	1475	100-80-1	1641	103-30-0	1619	105-36-2	115	107-03-9	10156		
99-52-5	800	100-81-2	1816	103-32-2	1827	105-37-3	10263	107-04-0	5456		
99-53-6	9422	100-83-4	565	103-34-4	7688	105-38-4	10261	107-05-1	10536		
99-54-7	1392	100-84-5	1932	103-37-7	3585	105-39-5	130	107-06-2	5529		
99-55-8	801	100-86-7	1600	103-38-8	3560	105-40-8	3954	107-07-3	5693		
99-56-9	1281	100-88-9	11568	103-43-5	3298	105-41-9	6730	107-08-4	10054		
99-59-2	775	100-97-0	11601	103-44-6	6405	105-42-0	6977	107-10-8	9803		
99-61-6	592	101-01-9	6288	103-45-7	232	105-45-3	3573	107-11-9	10522		
99-62-7	993	101-02-0	9593	103-48-0	10306	105-46-4	214	107-12-0	10090		
99-63-8	1319	101-05-3	360	103-49-1	1828	105-48-6	139	107-13-1	10583		
99-65-0	1492	101-10-0	4141	103-50-4	2079	105-50-0	8783	107-14-2	274		
99-71-8	9431	101-14-4	785	103-55-9	5504	105-53-3	9941	107-15-3	5496		
99-73-0	5777	101-17-7	678	103-63-9	1038	105-54-4	3515	107-16-4	282		
99-75-2	2698	101-18-8	9446	103-65-1	2169	105-55-5	11803	107-18-6	10697		
99-76-3	2637	101-21-3	3939	103-67-3	1818	105-56-6	152	107-19-7	10743		
99-77-4	2709	101-27-9	510	103-69-5	743	105-57-7	5541	107-20-0	12		
99-82-1	4438	101-31-5	917	103-70-8	5999	105-58-8	3999	107-21-1	5569		
99-84-3	4675	101-38-2	4280	103-71-9	1786	105-59-9	5742	107-22-2	5491		
99-85-4	4275	101-39-3	10516	103-72-0	1792	105-60-2	493	107-25-5	5908		
99-86-5	4273	101-40-6	4385	103-73-1	1649	105-66-8	3588	107-29-9	17		
99-87-6	1990	101-41-7	929	103-74-2	10950	105-67-9	9308	107-30-2	7483		
99-88-7	789	101-43-9	10653	103-75-3	10783	105-68-0	3639	107-31-3	6010		
99-89-8	9412	101-53-1	9455	103-76-4	9816	105-70-4	10190	107-34-6	481		
99-90-1	5781	101-54-2	1283	103-79-7	10481	105-72-6	3512	107-35-7	5623		
99-91-2	5789	101-55-3	1079	103-80-0	971	105-74-8	9155	107-37-9	11475		
99-92-3	5773	101-57-5	2234	103-81-1	878	105-75-9	3774	107-39-1	9067		
99-93-4	5832	101-60-0	9734	103-82-2	883	105-83-9	9875	107-40-4	9071		
99-94-5	2678	101-61-1	786	103-83-3	1807	105-86-2	8338	107-41-5	8793		
99-96-7	2613	101-76-8	1963	103-84-4	86	105-99-7	6784	107-43-7	7454		
99-97-8	862	101-77-9	784	103-85-5	11808	106-02-5	8562	107-46-0	5236		
99-98-9	1270	101-80-4	832	103-88-8	37	106-07-0	8259	107-47-1	10155		
99-99-0	2002	101-81-5	1961	103-89-9	78	106-19-4	6787	107-48-2	8612		
100-00-5	1216	101-82-6	11004	103-90-2	67	106-23-0	8490	107-49-3	5221		
100-01-6	824	101-83-7	4284	103-95-7	2117	106-24-1	8333	107-50-6	4239		
100-02-7	9437	101-84-8	2071	103-99-1	8232	106-25-2	8334	107-51-7	11914		
100-06-1	5841	101-90-6	8625	104-01-8	922	106-26-3	8318	107-52-8	6990		
100-07-2	2901	101-91-7	3205	104-03-0	933	106-27-4	3544	107-53-9	11964		
100-09-4	2666	101-93-3	899	104-04-1	83	106-28-5	5333	107-59-5	128		
100-10-7	550	101-99-5	3965	104-06-3	34	106-30-9	6462	107-68-6	5624		
100-12-9	1694	102-01-2	3207	104-07-4	8253	106-31-0	3467	107-70-0	8978		
100-13-0	1645	102-02-3	7234	104-09-6	874	106-32-1	8428	107-72-2	11472		
100-14-1	1208	102-04-5	10433	104-13-2	655	106-33-2	5312	107-74-4	8392		
100-15-2	805	102-06-7	6282	104-15-4	2229	106-35-4	6511	107-75-5	8347		
100-16-3	7168	102-07-8	12010	104-21-2	1892	106-36-5	10326	107-80-2	3266		
100-17-4	1937	102-08-9	11805	104-22-3	2213	106-37-6	1290	107-81-3	8711		
100-18-5	994	102-09-0	4001	104-29-0	10000	106-38-7	1056	107-82-4	3239		
100-19-6	5861	102-13-6	932	104-36-9	1314	106-39-8	1006	107-83-5	8823		
100-20-9	1320	102-14-7	3575	104-45-0	1945	106-40-1	640	107-84-6	3252		
100-21-0	1329	102-20-5	940	104-46-1	1942	106-41-2	9230	107-85-7	3220		
100-22-1	1287	102-25-0	2345	104-47-2	958	106-42-3	1454	107-86-8	3700		
100-25-4	1493	102-27-2	747	104-50-7	6140	106-43-4	1185	107-87-9	8957		
100-26-5	10929	102-45-4	4810	104-51-8	1118	106-44-5	9402	107-88-0	3345		
100-28-7	1790	102-47-6	1357	104-52-9	1235	106-45-6	2303	107-89-1	3184		
100-29-8	1664	102-50-1	772	104-53-0	2114	106-46-7	1356	107-91-5	49		
100-36-7	5502	102-51-2	1273	104-57-4	6016	106-47-8	660	107-92-6	3460		
100-37-8	5706	102-52-3	10150	104-61-0	6149	106-48-9	9253	107-93-7	3814		
100-39-0	1053	102-53-4	7489	104-63-2	5762	106-49-0	780	107-94-8	10216		
100-40-3	4658	102-54-5	5960	104-65-4	10707	106-50-3	1262	107-95-9	323		
100-41-4	1668	102-56-7	702	104-66-5	1588	106-51-4	4257	107-96-0	10286		
100-42-5	1624	102-60-3	10379	104-67-6	6152	106-52-5	9717	107-97-1	6268		
100-43-6	10954	102-69-2	9808	104-72-3	1258	106-53-6	2292	107-98-2	10387		
100-44-7	1182	102-70-5	10523	104-75-6	6726	106-54-7	2295	108-01-0	5709		
100-45-8	4629	102-71-6	5751	104-76-7	6930	106-57-0	9612	108-03-2	10116		
100-46-9	1795	102-76-1	10195	104-78-9	9876	106-58-1	9608	108-05-4	179		
100-47-0	2740	102-79-4	5692	104-80-3	7109	106-63-8	10673	108-08-7	8762		
100-49-2	4419	102-81-8	5703	104-81-4	1064	106-65-0	3318	108-10-1	8981		
100-50-5	4632	102-82-9	3213	104-82-5	1200	106-68-3	8472	108-11-2	8942		
100-51-6	1836	102-85-2	9590	104-83-6	1134	106-69-4	6868	108-12-3	3693		
100-52-7	525	102-86-3	6725	104-84-7	1817	106-70-7	6898	108-16-7	10376		
100-53-8	1832	102-92-1	10716	104-85-8	2765	106-73-0	6466	108-18-9	9821		
100-54-9	10882	102-94-3	10683	104-87-0	589	106-74-1	10634	108-19-0	7232		
100-55-0	10976	102-97-6	1820	104-88-1	531	106-79-6	5046	108-20-3	10120		
100-60-7	4298	103-01-5	6270	104-90-5	10966	106-86-5	8555	108-21-4	209		
100-61-8	781	103-02-6	10482	104-91-6	9438	106-87-6	8557	108-22-5	10698		
100-63-0	7169	103-03-7	7145	104-92-7	1052	106-88-7	8615	108-23-6	4023		
100-64-1	4610	103-05-9	2150	104-93-8	1933	106-90-1	10680	108-24-7	107		
100-65-2	757	103-09-3	187	104-94-9	768	106-91-2	10668	108-27-0	11186		
100-66-3	1915	103-11-7	10636	104-96-1	820	106-92-3	8626	108-28-1	6155		
100-68-5	2041	103-12-8	2209	105-03-3	8131	106-93-4	5518	108-30-5	6093		
100-69-6	10952	103-16-2	9453	105-05-5	1415	106-94-5	9836	108-31-6	6091		
100-70-9	10881	103-17-3	4075	105-06-6	1405	106-95-6	10531	108-32-7	5213		

108-36-1	: 1289	109-90-0	: 5598	111-17-1	: 10329	112-51-6	: 5257	117-37-3	: 7269
108-37-2	: 1005	109-92-2	: 5905	111-18-2	: 6752	112-52-7	: 5295	117-39-5	: 2817
108-38-3	: 1453	109-93-3	: 5912	111-19-3	: 5049	112-53-8	: 5321	117-73-7	: 12064
108-39-4	: 9401	109-94-4	: 6007	111-20-6	: 5042	112-54-9	: 5288	117-78-2	: 372
108-40-7	: 2302	109-95-5	: 8087	111-21-7	: 5714	112-55-0	: 5304	117-79-3	: 381
108-41-8	: 1184	109-96-6	: 11118	111-24-0	: 8741	112-57-2	: 5498	117-80-6	: 7850
108-42-9	: 659	109-97-7	: 11112	111-25-1	: 6732	112-58-3	: 6842	117-81-7	: 1334
108-43-0	: 9252	109-98-8	: 10822	111-26-2	: 6723	112-60-7	: 5756	117-84-0	: 1344
108-44-1	: 779	109-99-9	: 6168	111-27-3	: 6911	112-61-8	: 8266	117-93-1	: 66
108-45-2	: 1261	110-00-9	: 6025	111-28-4	: 6707	112-62-9	: 8307	117-99-7	: 7622
108-46-3	: 1498	110-01-0	: 11783	111-29-5	: 8787	112-63-0	: 8228	118-00-3	: 6289
108-47-4	: 10937	110-02-1	: 11724	111-30-8	: 8734	112-64-1	: 11635	118-08-1	: 7357
108-48-5	: 10939	110-03-2	: 6801	111-31-9	: 6856	112-66-3	: 178	118-10-5	: 4120
108-50-9	: 10807	110-04-3	: 5587	111-34-2	: 3369	112-67-4	: 6665	118-32-1	: 7871
108-52-1	: 11046	110-05-4	: 9153	111-40-0	: 5497	112-70-9	: 11883	118-34-3	: 6225
108-55-4	: 10780	110-06-5	: 5242	111-41-1	: 5676	112-71-0	: 11614	118-41-2	: 2734
108-56-5	: 3340	110-12-3	: 6978	111-42-2	: 5730	112-72-1	: 11630	118-44-5	: 7718
108-57-6	: 1404	110-13-4	: 6805	111-43-3	: 10119	112-73-2	: 3417	118-46-7	: 8020
108-59-8	: 9943	110-14-5	: 3257	111-44-4	: 5611	112-76-5	: 8281	118-55-8	: 2650
108-60-1	: 10123	110-15-6	: 3294	111-45-5	: 5763	112-79-8	: 8297	118-58-1	: 2651
108-64-5	: 3548	110-16-7	: 3765	111-46-6	: 5753	112-80-1	: 8294	118-61-6	: 2619
108-67-8	: 2359	110-17-8	: 3764	111-47-7	: 10151	112-82-3	: 6633	118-69-4	: 1385
108-68-9	: 9312	110-18-9	: 5516	111-48-8	: 5766	112-85-6	: 5276	118-71-8	: 10796
108-69-0	: 709	110-19-0	: 215	111-49-9	: 492	112-86-7	: 5283	118-74-1	: 1749
108-70-3	: 2323	110-22-5	: 9156	111-51-3	: 3263	112-88-9	: 8291	118-75-2	: 4268
108-73-6	: 2368	110-27-0	: 11626	111-55-7	: 5571	112-89-0	: 8235	118-76-3	: 4686
108-75-8	: 11028	110-38-3	: 5063	111-56-8	: 5307	112-91-4	: 8293	118-78-5	: 11077
108-77-0	: 11837	110-39-4	: 3566	111-60-4	: 8260	112-92-5	: 8278	118-79-6	: 9479
108-78-1	: 11835	110-40-7	: 5045	111-61-5	: 8257	112-95-8	: 5350	118-83-2	: 1219
108-82-7	: 6485	110-41-8	: 11936	111-62-6	: 8301	113-00-8	: 6278	118-90-1	: 2676
108-83-8	: 6514	110-42-9	: 5066	111-64-8	: 8481	113-15-5	: 5394	118-91-2	: 2534
108-84-9	: 8947	110-43-0	: 6510	111-65-9	: 8355	113-48-4	: 7660	118-92-3	: 2484
108-85-0	: 4315	110-44-1	: 6698	111-66-0	: 8491	114-07-8	: 5396	118-93-4	: 5830
108-86-1	: 998	110-45-2	: 3636	111-68-2	: 6349	114-25-0	: 3001	118-96-7	: 2048
108-87-2	: 4426	110-46-3	: 8090	111-69-3	: 6777	114-26-1	: 10722	119-13-1	: 2785
108-88-3	: 1947	110-49-6	: 5735	111-70-6	: 6472	114-91-0	: 10979	119-20-0	: 2494
108-89-4	: 10982	110-52-1	: 3267	111-71-7	: 6344	115-07-1	: 10528	119-26-6	: 7156
108-90-7	: 1131	110-53-2	: 8710	111-74-0	: 5501	115-10-6	: 7523	119-27-7	: 1923
108-91-8	: 4283	110-54-3	: 6731	111-75-1	: 5691	115-11-7	: 10578	119-28-8	: 7967
108-93-0	: 4500	110-56-5	: 3279	111-76-2	: 5687	115-17-3	: 20	119-33-5	: 9424
108-94-1	: 4571	110-57-6	: 3750	111-77-3	: 5737	115-18-4	: 3870	119-36-8	: 2634
108-95-2	: 9199	110-58-7	: 8700	111-78-4	: 4733	115-19-5	: 3922	119-41-5	: 225
108-97-4	: 10789	110-59-8	: 8827	111-81-9	: 11978	115-20-8	: 5768	119-44-8	: 10749
108-98-5	: 2289	110-60-1	: 3258	111-82-0	: 5315	115-21-9	: 11463	119-48-2	: 7686
108-99-6	: 10981	110-61-2	: 3291	111-83-1	: 8356	115-24-2	: 9835	119-59-5	: 845
109-01-3	: 9618	110-62-3	: 8690	111-84-2	: 8120	115-25-3	: 4194	119-60-8	: 7606
109-02-4	: 7696	110-63-4	: 3346	111-85-3	: 8359	115-29-7	: 5369	119-61-9	: 7616
109-04-6	: 10875	110-65-6	: 3911	111-86-4	: 8350	115-32-2	: 1851	119-64-2	: 7979
109-06-8	: 10980	110-66-7	: 8844	111-87-5	: 8443	115-37-7	: 7679	119-65-3	: 7373
109-07-9	: 9619	110-67-8	: 10110	111-88-6	: 8418	115-38-8	: 11094	119-67-5	: 2607
109-08-0	: 10810	110-68-9	: 3221	111-90-0	: 5720	115-39-9	: 9366	119-68-6	: 2679
109-09-1	: 10915	110-69-0	: 3188	111-92-2	: 3211	115-40-2	: 9364	119-84-6	: 2798
109-12-6	: 11042	110-70-3	: 5503	111-96-6	: 5613	115-43-5	: 11103	119-90-4	: 3033
109-16-0	: 10657	110-71-4	: 5549	112-04-9	: 11470	115-53-7	: 7676	119-93-7	: 3034
109-19-3	: 3543	110-72-5	: 5509	112-05-0	: 8146	115-56-0	: 11064	120-07-0	: 5760
109-21-7	: 3479	110-73-6	: 5722	112-06-1	: 193	115-61-7	: 11349	120-12-7	: 366
109-43-3	: 5044	110-74-7	: 6018	112-10-7	: 8287	115-63-9	: 9623	120-14-9	: 545
109-49-9	: 7089	110-75-8	: 5887	112-12-9	: 11958	115-69-5	: 9987	120-15-0	: 11298
109-52-4	: 8855	110-76-9	: 5423	112-13-0	: 5078	115-70-8	: 9985	120-18-3	: 7960
109-53-5	: 10035	110-77-0	: 5726	112-14-1	: 222	115-71-9	: 9093	120-20-7	: 1571
109-56-8	: 5741	110-78-1	: 10059	112-15-2	: 5721	115-76-4	: 10004	120-21-8	: 539
109-57-9	: 11809	110-80-5	: 5717	112-16-3	: 5328	115-77-5	: 9989	120-22-9	: 699
109-59-1	: 5740	110-81-6	: 5253	112-17-4	: 157	115-80-0	: 10182	120-23-0	: 217
109-60-4	: 236	110-82-7	: 4305	112-20-9	: 8119	115-84-4	: 9992	120-25-2	: 557
109-61-5	: 4028	110-83-8	: 4626	112-23-2	: 6008	115-86-6	: 9566	120-29-6	: 489
109-63-7	: 3137	110-85-0	: 9602	112-24-3	: 5499	115-90-2	: 5954	120-32-1	: 9270
109-64-8	: 9882	110-86-1	: 10872	112-25-4	: 5728	115-96-8	: 5701	120-33-2	: 2691
109-65-9	: 3231	110-87-2	: 10777	112-26-5	: 5449	115-99-1	: 8337	120-36-5	: 10250
109-66-0	: 8709	110-88-3	: 11905	112-27-6	: 5713	116-06-3	: 334	120-43-4	: 9604
109-67-1	: 9004	110-89-4	: 9629	112-29-8	: 5034	116-09-6	: 10443	120-46-7	: 10027
109-69-3	: 3241	110-91-8	: 7681	112-30-1	: 5070	116-14-3	: 5918	120-47-8	: 2621
109-70-6	: 9839	110-93-0	: 6603	112-31-2	: 5030	116-15-4	: 10572	120-48-9	: 2706
109-73-9	: 3209	110-94-1	: 8770	112-32-3	: 6014	116-16-5	: 10440	120-50-3	: 2701
109-74-0	: 3403	110-95-2	: 9879	112-34-5	: 5688	116-17-6	: 9595	120-51-4	: 2720
109-75-1	: 3803	110-96-3	: 9823	112-36-7	: 5612	116-26-7	: 4250	120-55-8	: 5755
109-76-2	: 9873	110-97-4	: 10384	112-37-8	: 11943	116-29-0	: 2325	120-57-0	: 2417
109-77-3	: 9911	110-99-6	: 228	112-38-9	: 11974	116-38-1	: 865	120-61-6	: 1351
109-78-4	: 10106	111-01-3	: 11606	112-39-0	: 6653	116-43-8	: 3578	120-66-1	: 76
109-79-5	: 3437	111-02-4	: 11602	112-40-3	: 5292	116-44-9	: 2202	120-71-8	: 770
109-80-8	: 10031	111-03-5	: 8299	112-41-4	: 5335	116-54-1	: 166	120-72-9	: 7294
109-81-9	: 5512	111-06-8	: 6647	112-42-5	: 11950	116-58-5	: 9757	120-73-0	: 10752
109-82-0	: 285	111-11-5	: 8434	112-43-6	: 11979	116-63-2	: 7966	120-75-2	: 2863
109-83-1	: 5738	111-12-6	: 8542	112-44-7	: 11935	117-08-8	: 7346	120-80-9	: 1497
109-84-2	: 5729	111-13-7	: 8471	112-46-9	: 11981	117-10-2	: 395	120-82-1	: 2322
109-86-4	: 5733	111-14-8	: 6455	112-47-0	: 5047	117-12-4	: 393	120-83-2	: 9283
109-87-5	: 7507	111-15-9	: 5718	112-48-1	: 5527	117-18-0	: 2264	120-85-4	: 9622
109-89-7	: 5427	111-16-0	: 6393	112-49-2	: 11642	117-34-0	: 936	120-87-6	: 8254

CAS	Page	CAS	Page	CAS	Page	CAS	Page	CAS	Page
120-89-8	: 7226	123-11-5	: 585	126-68-1	: 9585	133-59-5	: 2249	140-80-7	: 8736
120-92-3	: 4878	123-15-9	: 8693	126-71-6	: 9572	133-67-5	: 2854	140-82-9	: 5707
120-93-4	: 7227	123-19-3	: 6512	126-72-7	: 10358	133-90-4	: 2492	140-87-4	: 153
120-94-5	: 11168	123-20-6	: 3513	126-73-8	: 9562	133-91-5	: 2617	140-88-5	: 10635
120-97-8	: 1559	123-22-8	: 9587	126-81-8	: 4378	134-04-3	: 2848	140-95-4	: 11995
121-06-2	: 9337	123-25-1	: 3303	126-84-1	: 9904	134-11-2	: 2596	141-03-7	: 3301
121-14-2	: 1956	123-29-5	: 8150	126-98-7	: 10586	134-20-3	: 2503	141-04-8	: 6783
121-17-5	: 1217	123-30-8	: 9203	126-99-8	: 3152	134-31-6	: 11324	141-05-9	: 3776
121-21-1	: 4960	123-31-9	: 1499	127-00-4	: 10349	134-32-7	: 7708	141-06-0	: 8914
121-29-9	: 4962	123-32-0	: 10806	127-06-0	: 10476	134-49-6	: 7697	141-10-6	: 11965
121-30-2	: 1558	123-35-3	: 8330	127-07-1	: 12016	134-55-4	: 2483	141-12-8	: 8336
121-32-4	: 555	123-38-6	: 9774	127-08-2	: 234	134-58-7	: 11851	141-20-8	: 5313
121-33-5	: 574	123-39-7	: 5995	127-09-3	: 237	134-62-3	: 612	141-22-0	: 8303
121-34-6	: 2625	123-42-2	: 8976	127-17-3	: 10314	134-81-6	: 5584	141-27-5	: 8317
121-43-7	: 3128	123-43-3	: 238	127-18-4	: 5917	134-84-9	: 7633	141-28-6	: 6785
121-44-8	: 5417	123-45-5	: 239	127-19-5	: 55	134-85-0	: 7604	141-32-2	: 10616
121-45-9	: 9592	123-51-3	: 3627	127-25-3	: 9168	134-96-3	: 569	141-43-5	: 5675
121-46-0	: 2913	123-54-6	: 8798	127-27-5	: 9170	135-00-2	: 7641	141-46-8	: 15
121-47-1	: 2216	123-56-8	: 11152	127-29-7	: 308	135-01-3	: 1413	141-57-1	: 11476
121-57-3	: 2217	123-62-6	: 10203	127-40-2	: 4058	135-02-4	: 583	141-62-8	: 11646
121-59-5	: 455	123-63-7	: 11908	127-48-0	: 8587	135-09-1	: 2858	141-63-9	: 8997
121-60-8	: 2243	123-66-0	: 6884	127-63-9	: 2237	135-19-3	: 8018	141-66-2	: 5172
121-61-9	: 31	123-72-8	: 3178	127-66-2	: 1878	135-20-6	: 759	141-75-3	: 3681
121-66-4	: 11666	123-73-9	: 3695	127-69-5	: 2195	135-44-4	: 8926	141-78-6	: 186
121-69-7	: 713	123-75-1	: 11143	127-71-9	: 603	135-48-8	: 8636	141-79-7	: 9108
121-71-1	: 5831	123-76-2	: 8903	127-79-7	: 2199	135-58-0	: 11661	141-82-2	: 9921
121-72-2	: 861	123-77-3	: 5124	127-91-3	: 2925	135-67-1	: 9507	141-83-3	: 11993
121-73-3	: 1215	123-80-8	: 5576	128-04-1	: 11514	135-68-2	: 3012	141-84-4	: 11685
121-75-5	: 3316	123-82-0	: 6350	128-08-5	: 11153	135-70-6	: 11194	141-86-6	: 10925
121-79-9	: 2732	123-83-1	: 5510	128-09-6	: 11154	135-88-6	: 7728	141-90-2	: 11110
121-81-3	: 614	123-84-2	: 10340	128-13-2	: 4098	135-97-7	: 486	141-91-3	: 7687
121-82-4	: 11833	123-86-4	: 121	128-20-1	: 9745	135-98-8	: 2035	141-93-5	: 1414
121-86-8	: 1210	123-88-6	: 11822	128-23-4	: 9743	136-35-6	: 11826	141-97-9	: 3571
121-89-1	: 5860	123-90-0	: 11719	128-27-8	: 5381	136-44-7	: 10188	142-04-1	: 756
121-90-4	: 2905	123-91-1	: 5187	128-33-6	: 4104	136-47-0	: 2531	142-08-5	: 11033
121-91-5	: 1328	123-92-2	: 3631	128-37-0	: 9224	136-60-7	: 2532	142-09-6	: 10660
121-92-6	: 2704	123-93-3	: 240	128-39-2	: 9220	136-72-1	: 8678	142-26-7	: 65
121-98-2	: 2675	123-95-5	: 8251	128-50-7	: 2969	136-77-6	: 1519	142-28-9	: 9893
122-00-9	: 5853	123-98-8	: 8138	128-53-0	: 11122	136-78-7	: 11358	142-29-0	: 4903
122-01-0	: 2892	123-99-9	: 8130	128-62-1	: 7356	136-79-8	: 739	142-30-3	: 7119
122-03-2	: 590	124-02-7	: 10528	128-68-7	: 2979	136-80-1	: 5744	142-47-2	: 6248
122-04-3	: 2906	124-04-9	: 6779	129-00-0	: 10853	136-95-8	: 2859	142-50-7	: 5332
122-07-6	: 5420	124-06-1	: 11624	129-43-1	: 407	137-00-8	: 11677	142-60-9	: 10313
122-09-8	: 1572	124-07-2	: 8422	129-44-2	: 389	137-04-2	: 665	142-61-0	: 6986
122-11-2	: 2194	124-09-4	: 6749	129-64-6	: 7580	137-05-3	: 10624	142-62-1	: 6870
122-14-5	: 5951	124-10-7	: 11625	129-66-8	: 2735	137-06-4	: 2301	142-64-3	: 9607
122-20-3	: 10393	124-11-8	: 8182	129-79-3	: 5979	137-07-5	: 2290	142-68-7	: 10800
122-25-8	: 2686	124-12-9	: 8411	130-15-4	: 7843	137-16-6	: 6269	142-73-4	: 6260
122-28-1	: 82	124-13-0	: 8346	130-16-5	: 11311	137-17-7	: 858	142-77-8	: 8298
122-32-7	: 8310	124-17-4	: 5689	130-26-7	: 11312	137-19-9	: 1510	142-82-5	: 6355
122-34-9	: 11513	124-18-5	: 5033	130-73-4	: 9299	137-26-8	: 11722	142-83-6	: 6675
122-35-0	: 10505	124-19-6	: 8117	130-80-3	: 9301	137-30-4	: 12068	142-84-7	: 9826
122-37-2	: 9447	124-21-0	: 5425	130-85-8	: 7777	137-32-6	: 3626	142-91-6	: 6654
122-39-4	: 836	124-22-1	: 5291	130-86-9	: 3109	137-40-6	: 10327	142-92-7	: 194
122-40-7	: 6347	124-26-5	: 8231	130-89-2	: 4130	137-43-9	: 4777	142-96-1	: 3415
122-42-9	: 3966	124-30-1	: 8233	130-90-5	: 4129	137-58-6	: 54	143-07-7	: 5305
122-48-3	: 212	124-38-9	: 3992	130-95-0	: 4127	138-41-0	: 2508	143-08-8	: 8157
122-48-5	: 3663	124-40-3	: 7448	131-01-1	: 12065	138-52-3	: 6222	143-10-2	: 5058
122-51-0	: 5804	124-42-5	: 5671	131-09-9	: 388	138-53-4	: 955	143-13-5	: 220
122-52-1	: 9591	124-47-0	: 12021	131-11-3	: 1343	138-86-3	: 4677	143-15-7	: 5293
122-57-6	: 3881	124-48-1	: 7493	131-28-2	: 2586	138-87-4	: 4542	143-16-8	: 6727
122-59-8	: 228	124-58-3	: 458	131-53-3	: 7619	138-89-6	: 720	143-22-6	: 5690
122-60-1	: 8623	124-63-0	: 7535	131-54-4	: 7595	139-13-9	: 6256	143-24-8	: 8995
122-62-3	: 5043	124-68-5	: 10341	131-56-6	: 7609	139-40-2	: 10508	143-27-1	: 6631
122-66-7	: 7158	124-70-9	: 11396	131-57-7	: 7620	139-45-7	: 10198	143-28-2	: 8312
122-69-0	: 10691	124-73-2	: 5525	131-58-8	: 7631	139-65-1	: 850	143-50-0	: 7408
122-70-3	: 10324	124-80-1	: 4067	131-73-7	: 864	139-66-2	: 2283	143-62-4	: 4044
122-73-6	: 1951	124-83-4	: 4794	131-89-5	: 9273	139-85-5	: 542	143-74-8	: 9363
122-78-1	: 868	124-87-8	: 7579	131-91-9	: 8035	139-87-7	: 5723	144-19-4	: 8795
122-79-2	: 231	125-15-5	: 11351	132-57-0	: 7972	139-93-5	: 9275	144-49-0	: 189
122-80-5	: 29	125-20-2	: 7352	132-60-5	: 11235	140-09-3	: 5702	144-55-8	: 4004
122-84-9	: 10461	125-24-6	: 3002	132-64-9	: 5163	140-10-3	: 10682	144-62-7	: 5553
122-87-2	: 6267	125-28-0	: 7674	132-65-0	: 5167	140-11-4	: 233	144-68-3	: 4057
122-88-3	: 146	125-29-1	: 7878	132-66-1	: 2702	140-18-1	: 148	144-79-6	: 11380
122-95-2	: 1408	125-33-7	: 11066	132-75-2	: 7736	140-22-7	: 4010	144-80-9	: 30
122-97-4	: 2148	125-55-3	: 11079	132-86-5	: 7834	140-25-0	: 5318	144-83-2	: 2203
122-98-5	: 5759	126-00-1	: 1103	133-06-2	: 7368	140-28-3	: 5500	145-13-1	: 9756
122-99-6	: 5758	126-07-8	: 11520	133-07-3	: 5984	140-29-4	: 945	145-73-4	: 5370
123-00-2	: 7701	126-11-4	: 10012	133-08-4	: 9930	140-31-8	: 9615	145-94-8	: 7283
123-01-3	: 1566	126-14-7	: 6229	133-11-9	: 2502	140-39-6	: 213	146-06-5	: 4128
123-02-4	: 2342	126-17-0	: 11529	133-13-1	: 9949	140-40-9	: 85	146-48-5	: 12062
123-03-5	: 11030	126-18-1	: 11533	133-14-2	: 9152	140-55-6	: 9718	147-71-7	: 3312
123-04-6	: 6368	126-19-2	: 11531	133-26-6	: 6188	140-57-0	: 11579	147-73-9	: 3304
123-05-7	: 6719	126-29-4	: 4056	133-32-4	: 7296	140-65-8	: 7683	147-82-0	: 851
123-06-8	: 9914	126-30-7	: 10005	133-37-9	: 3311	140-67-0	: 1943	147-85-3	: 9764
123-07-9	: 9355	126-33-0	: 11786	133-53-9	: 9290	140-76-1	: 10955	147-93-3	: 2663
123-08-0	: 566	126-54-5	: 11643	133-58-4	: 7437	140-77-2	: 4847	148-03-8	: 2789

148-24-3	: 11308	258-31-1	: 6621	307-34-6	: 8415	353-50-4	: 4008	373-05-7	: 6888		
148-25-4	: 7875	260-36-6	: 2473	307-45-9	: 5050	353-54-8	: 7550	373-14-8	: 6829		
148-53-8	: 570	260-94-6	: 314	308-24-7	: 4498	353-55-9	: 7494	373-28-4	: 6735		
148-75-4	: 7876	261-31-4	: 11811	309-00-2	: 5177	353-58-2	: 7468	373-32-0	: 6936		
148-79-8	: 11656	262-30-6	: 11353	311-89-7	: 3228	353-59-3	: 7464	373-44-4	: 8365		
148-86-7	: 3091	271-44-3	: 7235	312-40-3	: 11411	353-82-2	: 279	373-49-9	: 6670		
149-29-1	: 6183	271-58-9	: 2390	312-84-5	: 11355	353-83-3	: 5667	373-52-4	: 7470		
149-30-4	: 2866	271-89-6	: 2438	313-72-4	: 7935	353-85-5	: 289	373-53-5	: 7517		
149-32-6	: 3431	271-95-4	: 2389	314-13-6	: 7877	353-89-9	: 11432	373-80-8	: 11584		
149-57-5	: 6882	272-16-2	: 2387	314-40-9	: 3139	354-04-1	: 5526	374-07-2	: 5537		
149-73-5	: 7561	273-53-0	: 2877	315-30-0	: 10852	354-06-3	: 5457	374-25-4	: 3715		
149-74-6	: 11401	274-07-7	: 2412	317-52-2	: 6753	354-11-0	: 5639	374-77-6	: 4343		
149-91-7	: 2731	274-09-9	: 2415	319-85-7	: 4406	354-12-1	: 5658	375-01-9	: 3622		
150-13-0	: 2486	274-40-8	: 7324	319-86-8	: 4407	354-14-3	: 5638	375-16-6	: 3692		
150-19-6	: 9380	275-51-4	: 504	319-88-0	: 2340	354-15-4	: 5660	375-22-4	: 3521		
150-25-4	: 6259	278-30-8	: 2991	319-89-1	: 4269	354-21-2	: 5659	375-83-7	: 6432		
150-30-1	: 9509	279-23-2	: 2914	320-72-9	: 2566	354-23-4	: 5539	375-96-2	: 8139		
150-46-9	: 3125	280-57-9	: 5111	321-28-8	: 1722	354-25-6	: 5485	376-27-2	: 8438		
150-50-5	: 7438	280-65-9	: 2990	321-38-0	: 7893	354-26-7	: 5490	376-87-4	: 9027		
150-60-7	: 5248	281-23-2	: 11860	321-60-8	: 3074	354-51-8	: 5519	377-41-3	: 4177		
150-61-8	: 5508	285-67-6	: 8558	323-09-1	: 7894	354-56-3	: 5618	377-70-8	: 4276		
150-68-5	: 11999	286-08-8	: 2915	324-74-3	: 3075	354-58-5	: 5662	377-93-5	: 4199		
150-69-6	: 12013	286-18-0	: 478	326-56-7	: 2420	354-92-7	: 10050	378-68-7	: 10251		
150-76-5	: 9381	286-20-4	: 8554	326-59-0	: 2416	354-96-1	: 3414	378-75-6	: 10317		
150-77-6	: 5515	287-23-0	: 4170	326-62-5	: 952	355-02-2	: 4499	381-61-3	: 3162		
150-78-7	: 1438	287-27-4	: 11691	327-54-8	: 2272	355-04-4	: 8854	381-73-7	: 175		
150-86-7	: 6672	287-92-3	: 4774	327-57-1	: 8214	355-20-4	: 3287	382-45-6	: 353		
151-10-0	: 1437	288-13-1	: 10815	328-39-2	: 7412	355-25-9	: 3255	383-63-1	: 267		
151-13-3	: 8304	288-14-2	: 7402	328-42-7	: 3339	355-63-5	: 6560	383-66-4	: 269		
151-18-8	: 10092	288-32-4	: 7192	328-50-7	: 8781	355-68-0	: 4381	383-90-4	: 3114		
151-19-9	: 8453	288-36-8	: 11846	328-74-5	: 637	355-75-9	: 4643	384-04-3	: 3172		
151-56-4	: 498	288-39-1	: 11658	328-96-1	: 723	355-80-6	: 8953	384-22-5	: 2060		
151-67-7	: 5458	288-42-6	: 8575	329-71-5	: 9343	356-16-1	: 5642	386-17-4	: 7351		
152-16-9	: 5220	288-47-1	: 11667	330-54-1	: 12000	356-24-1	: 3525	388-82-9	: 3042		
152-58-9	: 9749	288-92-6	: 11657	330-55-2	: 7415	356-27-4	: 3524	389-08-2	: 8066		
152-93-2	: 11107	288-94-8	: 11652	331-39-5	: 10632	356-28-5	: 3523	389-40-2	: 2093		
152-95-4	: 2824	288-99-3	: 8563	332-77-4	: 6087	357-20-0	: 3808	392-56-3	: 1757		
154-17-6	: 7110	289-80-5	: 10855	333-20-0	: 11711	357-49-3	: 10721	392-69-8	: 1740		
154-41-6	: 1838	289-95-2	: 11049	333-41-5	: 9582	357-57-3	: 11564	392-83-6	: 1089		
154-42-7	: 10762	290-37-9	: 10801	333-93-7	: 3260	357-67-5	: 11089	392-85-8	: 1737		
154-93-8	: 11994	290-67-5	: 5202	334-28-1	: 6463	357-70-0	: 6200	393-52-2	: 2897		
156-08-1	: 1575	290-87-9	: 11827	334-42-9	: 6359	358-06-5	: 11410	394-46-7	: 1633		
156-10-5	: 829	290-96-0	: 11651	334-43-0	: 6365	358-43-0	: 11484	396-01-0	: 10750		
156-38-7	: 906	291-21-4	: 11919	334-48-5	: 5059	358-99-6	: 5883	396-64-5	: 3043		
156-43-4	: 737	291-64-5	: 4221	334-56-5	: 5052	359-01-3	: 3379	398-23-2	: 3044		
156-59-2	: 5896	292-46-6	: 8998	334-59-8	: 5064	359-02-4	: 5900	399-92-8	: 10328		
156-60-5	: 5895	292-64-8	: 4736	334-61-2	: 5035	359-10-4	: 5885	400-52-2	: 266		
156-87-6	: 10337	293-55-0	: 4724	334-62-3	: 5037	359-11-5	: 5924	401-56-9	: 133		
157-40-4	: 11527	293-96-9	: 4202	334-64-5	: 5074	359-13-7	: 5708	401-78-5	: 1090		
180-43-8	: 11537	294-62-2	: 4210	334-88-3	: 7491	359-25-1	: 116	401-80-9	: 1738		
184-97-4	: 5240	295-48-7	: 4759	335-36-4	: 8109	359-28-4	: 5661	402-31-3	: 997		
189-55-9	: 2850	297-97-2	: 11720	335-44-4	: 3452	359-29-5	: 5923	402-43-7	: 1091		
191-07-1	: 4145	298-02-2	: 9513	335-48-8	: 3275	359-32-0	: 296	402-44-8	: 1739		
191-24-2	: 2471	298-04-4	: 5260	335-57-9	: 6418	359-35-3	: 5641	402-67-5	: 1734		
191-28-6	: 12055	298-12-4	: 223	336-19-6	: 4647	359-37-5	: 5925	403-42-9	: 5815		
192-65-4	: 8061	298-45-3	: 2470	336-59-4	: 3522	359-70-6	: 5436	403-43-0	: 2899		
192-97-2	: 2436	298-46-4	: 5144	338-65-8	: 5471	360-68-9	: 4109	404-72-8	: 1721		
195-19-7	: 2406	298-81-7	: 6187	338-83-0	: 9813	360-89-4	: 3804	404-86-4	: 8181		
198-55-0	: 9161	298-96-4	: 11653	339-43-5	: 2192	363-51-9	: 664	405-50-5	: 901		
205-25-4	: 2405	299-84-3	: 9584	343-65-7	: 1102	363-72-4	: 2091	405-99-2	: 1635		
205-82-3	: 2738	299-86-5	: 4153	344-04-7	: 1078	366-18-7	: 3099	406-74-6	: 190		
205-99-2	: 631	300-38-9	: 11927	344-07-0	: 1220	367-11-3	: 1422	407-25-0	: 264		
206-44-0	: 5966	300-57-2	: 2165	345-35-7	: 1194	367-12-4	: 9360	407-75-0	: 8879		
207-08-9	: 2739	300-62-9	: 1578	346-18-9	: 2857	367-23-7	: 2346	407-97-6	: 8713		
207-83-0	: 5148	300-76-5	: 7702	348-51-6	: 1166	367-25-9	: 700	407-98-7	: 8728		
208-96-8	: 3	300-87-8	: 7403	348-52-7	: 1719	367-64-6	: 265	408-16-2	: 6491		
212-74-8	: 11644	301-00-8	: 8285	348-54-9	: 752	368-34-3	: 132	408-27-5	: 8460		
213-46-7	: 9601	301-12-2	: 8632	348-70-9	: 135	368-43-4	: 2251	408-38-8	: 6639		
217-59-4	: 11910	301-21-3	: 9399	349-88-2	: 2246	368-47-8	: 11486	420-04-2	: 4155		
218-01-9	: 4114	302-17-0	: 5582	350-03-8	: 5868	368-77-4	: 2770	420-05-3	: 4162		
222-93-5	: 8996	302-27-2	: 306	350-46-9	: 1735	368-98-9	: 3116	420-12-2	: 11695		
224-42-0	: 5143	302-72-7	: 324	350-50-5	: 1725	370-14-9	: 9403	420-24-6	: 463		
225-51-4	: 630	302-84-1	: 11356	350-51-6	: 1634	371-40-4	: 754	420-26-8	: 10047		
225-61-6	: 2394	303-04-8	: 3754	350-90-3	: 10637	371-41-5	: 9362	420-45-1	: 9906		
229-87-8	: 9188	303-38-8	: 2568	352-11-4	: 1195	371-62-0	: 5727	420-46-2	: 5665		
230-07-9	: 9191	304-21-2	: 11038	352-32-9	: 1728	372-09-8	: 150	420-90-6	: 10532		
230-27-3	: 2474	304-84-7	: 611	352-33-0	: 1168	372-18-9	: 1423	420-97-3	: 9897		
230-46-6	: 9190	305-01-1	: 2804	352-34-1	: 1720	372-19-0	: 753	420-99-5	: 9851		
238-84-6	: 2393	305-03-3	: 1101	352-70-5	: 1727	372-20-3	: 9361	421-04-5	: 5487		
250-84-0	: 11688	305-84-0	: 7133	352-91-0	: 9844	372-31-6	: 3592	421-06-7	: 5466		
253-52-1	: 9599	305-85-1	: 9304	352-93-2	: 5644	372-38-3	: 2347	421-83-0	: 7537		
253-66-7	: 4136	306-08-1	: 913	352-97-6	: 6253	372-46-3	: 4403	425-23-0	: 3526		
253-82-7	: 11195	306-44-5	: 9789	353-25-3	: 8429	372-54-3	: 6830	425-75-2	: 7533		
254-04-6	: 2772	306-83-2	: 5540	353-29-7	: 5294	372-64-5	: 5249	428-59-1	: 8629		
254-79-5	: 8065	306-94-5	: 7934	353-36-6	: 5593	372-75-8	: 8551	430-40-0	: 445		
254-87-5	: 2870	306-98-9	: 4342	353-42-4	: 3138	372-93-0	: 3621	430-51-3	: 10437		
255-50-5	: 2434	307-11-9	: 11912	353-49-1	: 4005			430-57-9	: 5535		

430-63-7 : 10561	461-98-3 : 11045	474-69-1 : 5382	486-25-9 : 5975	495-45-4 : 3876
430-66-0 : 5666	462-02-2 : 11907	474-77-1 : 4112	486-34-0 : 8012	495-61-4 : 4685
430-71-7 : 10506	462-06-6 : 1717	474-86-2 : 5398	486-35-1 : 2805	495-69-2 : 6255
430-95-5 : 10554	462-08-8 : 10858	475-03-6 : 8000	486-39-5 : 7396	495-76-1 : 2427
431-03-8 : 3359	462-18-0 : 11891	475-20-7 : 7568	486-47-5 : 7377	495-85-2 : 10790
431-06-1 : 5532	462-23-7 : 3927	475-25-2 : 628	486-66-8 : 2827	495-91-0 : 9633
431-07-2 : 5488	462-38-4 : 9856	475-31-0 : 6273	486-70-4 : 11331	496-10-6 : 7281
431-27-6 : 10558	462-39-5 : 9905	475-38-7 : 7853	486-84-0 : 11040	496-11-7 : 7250
431-52-7 : 10607	462-43-1 : 10382	475-67-2 : 5161	486-86-2 : 7578	496-15-1 : 7297
431-53-8 : 10557	462-72-6 : 3237	475-81-0 : 5157	486-89-5 : 7573	496-16-2 : 2442
432-25-7 : 4636	462-73-7 : 3250	476-28-8 : 6201	487-06-9 : 2818	496-41-3 : 2439
433-52-3 : 134	462-94-2 : 8735	476-32-4 : 4071	487-10-5 : 5127	496-46-8 : 7191
434-03-7 : 9758	462-95-3 : 5603	476-45-9 : 7855	487-19-4 : 10987	496-49-1 : 10471
434-13-9 : 4099	463-04-7 : 8096	476-69-7 : 5162	487-21-8 : 10748	496-64-0 : 10794
434-16-2 : 4103	463-11-6 : 8405	476-70-0 : 5156	487-27-4 : 7575	496-67-3 : 3196
434-45-7 : 5872	463-16-1 : 8151	477-27-0 : 92	487-68-3 : 599	496-72-0 : 1276
434-64-0 : 2095	463-17-2 : 11947	477-32-7 : 3540	487-79-6 : 11144	496-73-1 : 1530
434-90-2 : 3027	463-23-0 : 8128	477-47-4 : 6191	487-93-4 : 7326	496-74-2 : 1564
436-05-5 : 11925	463-24-1 : 8165	477-60-1 : 11924	488-10-8 : 4949	496-77-5 : 8474
437-38-7 : 9801	463-33-2 : 11939	477-75-8 : 2375	488-17-5 : 1527	496-78-6 : 9494
437-50-3 : 12054	463-40-1 : 8287	477-89-4 : 5218	488-23-3 : 2278	497-03-0 : 3699
438-08-4 : 346	463-49-0 : 9772	477-90-7 : 10786	488-43-7 : 6207	497-04-1 : 9995
439-14-5 : 2410	463-51-4 : 5927	478-40-0 : 7856	488-59-5 : 7336	497-06-3 : 3794
442-16-0 : 317	463-52-5 : 7563	478-43-3 : 371	488-70-0 : 9473	497-18-7 : 4009
442-51-3 : 11039	463-58-1 : 4035	478-61-5 : 2911	488-71-1 : 848	497-20-1 : 4767
443-48-1 : 7206	463-71-8 : 4032	478-71-7 : 9173	488-73-3 : 4074	497-26-7 : 5210
443-83-4 : 1169	463-72-9 : 3971	478-94-4 : 5374	488-81-3 : 11343	497-30-3 : 7204
444-27-9 : 11682	463-82-1 : 9910	478-95-5 : 5375	488-86-8 : 4939	497-32-5 : 2922
445-29-4 : 2603	463-88-7 : 5878	479-00-5 : 5373	488-87-9 : 1512	497-39-2 : 9222
445-69-2 : 1321	464-06-2 : 3457	479-13-0 : 2466	488-92-6 : 11128	497-59-6 : 10775
446-10-6 : 1731	464-10-8 : 7551	479-18-5 : 10761	488-93-7 : 6051	497-76-7 : 6224
446-11-7 : 1729	464-15-3 : 2932	479-21-0 : 7608	488-97-1 : 11866	498-00-0 : 1886
446-33-3 : 1732	464-17-5 : 2970	479-22-1 : 4259	488-98-2 : 1736	498-02-2 : 5824
446-34-4 : 1730	464-41-5 : 2920	479-23-2 : 2391	489-84-9 : 505	498-15-7 : 2974
446-35-5 : 1426	464-49-3 : 2961	479-27-6 : 7798	489-86-1 : 507	498-23-7 : 3786
446-52-6 : 559	464-72-2 : 5580	479-45-8 : 818	489-98-5 : 863	498-24-8 : 3785
446-72-0 : 2807	464-85-7 : 6178	479-59-4 : 2737	490-02-8 : 10811	498-59-9 : 5018
447-31-4 : 5785	464-86-8 : 6177	480-11-5 : 2810	490-03-9 : 4709	498-60-2 : 6042
447-53-0 : 7814	464-88-0 : 4786	480-15-9 : 2842	490-06-2 : 1536	498-62-4 : 11734
447-60-9 : 2769	465-24-7 : 7569	480-16-0 : 2816	490-10-8 : 4758	498-67-9 : 1379
450-95-3 : 5816	465-42-9 : 4063	480-40-0 : 2811	490-11-9 : 10930	498-94-2 : 9642
451-13-8 : 895	465-74-7 : 12036	480-41-1 : 2802	490-53-9 : 7390	498-95-3 : 9641
451-40-1 : 5812	465-99-6 : 8545	480-44-4 : 2809	490-55-1 : 11671	498-96-4 : 10912
451-46-7 : 2606	466-14-8 : 3199	480-64-8 : 2573	490-64-2 : 2733	499-04-7 : 10913
451-80-9 : 1654	466-43-3 : 475	480-68-2 : 11102	490-65-3 : 7923	499-06-9 : 2583
452-35-7 : 2865	466-49-9 : 473	480-72-8 : 9	490-79-9 : 2570	499-75-2 : 9415
452-58-4 : 10923	466-81-9 : 8567	480-91-1 : 7369	490-91-5 : 4267	499-80-9 : 10928
452-71-1 : 755	466-96-6 : 7672	481-06-1 : 8057	491-04-3 : 4697	499-81-0 : 10931
452-73-3 : 1170	466-97-7 : 7665	481-20-9 : 4106	491-09-8 : 4718	499-89-8 : 4241
452-86-8 : 1528	467-14-1 : 7671	481-21-0 : 4105	491-14-5 : 9120	500-05-0 : 10773
453-13-4 : 10369	467-55-0 : 11534	481-26-5 : 9741	491-18-9 : 11127	500-22-1 : 10888
453-18-9 : 192	467-60-7 : 9676	481-29-8 : 347	491-31-6 : 2791	500-38-9 : 1515
453-20-3 : 6137	467-85-6 : 6968	481-37-8 : 483	491-34-9 : 11299	500-44-7 : 11011
454-14-8 : 10415	467-98-1 : 7677	481-39-0 : 7860	491-35-0 : 11279	500-49-2 : 1550
454-31-9 : 176	468-28-0 : 4281	481-42-5 : 7863	491-37-2 : 2799	500-55-0 : 928
455-14-1 : 857	468-45-1 : 11516	481-49-2 : 8630	491-38-3 : 2793	500-99-2 : 9306
455-31-2 : 1425	469-59-0 : 12046	481-72-1 : 397	491-67-8 : 2844	501-15-5 : 1533
455-32-3 : 2909	470-40-6 : 4954	481-74-3 : 398	491-70-3 : 2814	501-19-9 : 9390
455-38-9 : 2604	470-49-5 : 8874	482-05-3 : 3038	491-88-3 : 6144	501-24-6 : 9441
456-22-4 : 2605	470-67-7 : 8556	482-28-0 : 7306	492-17-1 : 3029	501-30-4 : 10795
456-47-3 : 1879	470-82-6 : 8560	482-44-0 : 6189	492-22-8 : 11812	501-52-0 : 2126
456-48-4 : 560	470-90-6 : 4078	482-66-6 : 4756	492-27-3 : 11232	501-53-1 : 4027
456-49-5 : 1723	470-99-5 : 4702	482-68-8 : 11347	492-38-6 : 927	501-65-5 : 1714
457-87-4 : 1576	471-04-5 : 6461	482-74-6 : 2435	492-80-8 : 657	501-92-8 : 9458
458-03-7 : 1655	471-12-5 : 2982	482-89-3 : 7331	492-89-7 : 1543	501-97-3 : 2134
458-35-5 : 9370	471-16-9 : 2984	483-04-5 : 8571	492-97-7 : 3112	502-41-0 : 4229
458-36-6 : 10514	471-25-0 : 10738	483-17-0 : 5367	492-99-9 : 4379	502-42-1 : 4234
458-37-7 : 6334	471-35-2 : 5110	483-18-1 : 5368	493-01-6 : 7790	502-44-3 : 8591
458-88-8 : 9708	471-44-3 : 131	483-55-6 : 7861	493-02-7 : 7791	502-49-8 : 4740
459-22-3 : 953	471-46-5 : 5493	483-64-7 : 2672	493-05-0 : 2776	502-56-7 : 8175
459-26-7 : 1656	471-47-6 : 104	483-65-8 : 9183	493-08-3 : 2777	502-59-0 : 6354
459-56-3 : 1880	471-87-4 : 11177	483-76-1 : 7899	493-09-4 : 2414	502-61-4 : 5329
459-57-4 : 561	471-95-4 : 3141	483-78-3 : 7830	493-49-2 : 7398	502-65-8 : 4053
459-60-9 : 1724	472-20-8 : 4666	484-11-7 : 9192	493-52-7 : 2587	502-72-7 : 4761
459-67-6 : 4940	472-70-8 : 4061	484-29-7 : 6180	493-57-2 : 2401	502-73-8 : 6300
459-72-3 : 191	472-93-5 : 4052	484-47-9 : 7215	494-03-1 : 7710	502-99-8 : 8486
459-73-4 : 6263	473-02-9 : 11865	484-67-3 : 9472	494-12-2 : 2782	503-01-5 : 6532
460-00-4 : 1044	473-06-3 : 2977	484-73-1 : 5223	494-40-6 : 2356	503-09-3 : 8616
460-12-8 : 3177	473-19-8 : 2933	484-89-9 : 4265	494-47-3 : 7488	503-17-3 : 3898
460-13-9 : 10046	473-30-3 : 11663	484-93-5 : 491	494-52-0 : 11008	503-28-6 : 5125
460-19-5 : 5550	473-34-7 : 2210	485-31-4 : 3003	494-55-3 : 5215	503-29-7 : 494
460-35-5 : 9871	473-81-4 : 10252	485-47-2 : 7267	494-90-6 : 2465	503-30-0 : 8592
460-55-9 : 11429	473-91-6 : 4937	485-64-3 : 4121	494-97-3 : 11017	503-38-8 : 4029
461-11-0 : 6460	473-98-3 : 7416	485-65-4 : 4122	494-99-9 : 1440	503-40-2 : 7514
461-35-8 : 10184	474-25-9 : 4097	485-71-2 : 4133	495-08-9 : 1520	503-46-8 : 4691
461-58-5 : 6280	474-40-8 : 11556	485-72-3 : 2830	495-40-9 : 3676	503-60-6 : 3736
461-72-3 : 7216	474-62-4 : 5386	485-89-2 : 11233	495-41-0 : 3880	503-64-0 : 3815

503-66-2	: 10267	512-04-9	: 11536	521-04-0	: 11554	529-86-2	: 432	537-29-1	: 916
503-74-2	: 3539	512-12-9	: 8585	521-11-9	: 350	529-92-0	: 11216	537-45-1	: 4279
503-76-4	: 9869	512-13-0	: 2941	521-12-0	: 351	530-14-3	: 5818	537-46-2	: 1573
503-80-0	: 452	512-16-3	: 4309	521-18-6	: 349	530-22-3	: 2464	537-47-3	: 7146
503-93-5	: 4220	512-48-1	: 9003	521-24-4	: 7970	530-40-5	: 10892	537-49-5	: 11933
504-02-9	: 4376	512-56-1	: 9564	521-31-3	: 9600	530-43-8	: 6648	537-64-4	: 7432
504-03-0	: 9650	512-60-7	: 11862	521-35-7	: 5154	530-44-9	: 7613	537-65-5	: 1263
504-15-4	: 1531	512-69-6	: 6219	521-74-4	: 11313	530-48-3	: 1636	537-91-7	: 5247
504-20-1	: 6340	512-85-6	: 5184	521-85-7	: 3108	530-50-7	: 7157	537-92-8	: 77
504-24-5	: 10859	513-02-0	: 9568	522-12-3	: 2796	530-55-2	: 4262	538-08-9	: 4159
504-29-0	: 10857	513-08-6	: 9587	522-27-0	: 5583	530-62-1	: 7193	538-23-8	: 8440
504-31-4	: 10788	513-12-2	: 5725	522-57-6	: 7811	530-78-9	: 2729	538-24-9	: 5319
504-53-0	: 9000	513-23-5	: 2985	522-66-7	: 4123	530-85-8	: 10740	538-32-9	: 12027
504-57-4	: 8105	513-31-5	: 10545	522-75-8	: 2404	530-91-6	: 8043	538-39-6	: 1590
504-61-0	: 3859	513-35-9	: 3801	523-27-3	: 375	530-93-8	: 8047	538-41-0	: 635
504-63-2	: 9983	513-36-0	: 9862	523-31-9	: 1337	531-02-2	: 3097	538-43-2	: 10026
504-64-3	: 9773	513-37-1	: 10539	523-47-7	: 7898	531-18-0	: 7584	538-44-3	: 9106
504-78-9	: 11681	513-38-2	: 10057	523-80-8	: 2423	531-29-3	: 6226	538-49-8	: 11000
505-20-4	: 5261	513-42-8	: 10702	524-15-2	: 6179	531-37-3	: 9383	538-51-2	: 840
505-22-6	: 5186	513-44-0	: 10158	524-30-1	: 2823	531-39-5	: 8886	538-56-7	: 10684
505-23-7	: 5262	513-81-5	: 3159	524-34-5	: 11236	531-75-9	: 2822	538-58-9	: 8683
505-29-3	: 5263	513-86-0	: 3661	524-36-7	: 10978	531-81-7	: 2775	538-62-5	: 5122
505-32-8	: 6671	513-88-2	: 10425	524-40-3	: 10883	531-84-0	: 7774	538-64-7	: 3768
505-34-0	: 10067	513-92-8	: 5919	524-42-5	: 7842	531-91-9	: 3035	538-68-1	: 2107
505-48-6	: 8386	513-96-2	: 10115	524-63-0	: 4119	531-95-3	: 2786	538-74-9	: 2287
505-60-2	: 5645	514-10-3	: 9187	524-80-1	: 3982	532-18-3	: 7727	538-75-0	: 4291
505-84-0	: 10074	514-12-5	: 2953	524-96-9	: 1797	532-24-1	: 490	538-81-8	: 1095
505-92-0	: 5338	514-92-1	: 4059	525-06-4	: 5928	532-27-4	: 5790	538-86-3	: 1930
506-03-6	: 10010	515-12-8	: 4401	525-15-5	: 11041	532-32-1	: 2725	538-93-2	: 2039
506-12-7	: 6314	515-25-3	: 11178	525-37-1	: 7868	532-48-9	: 5825	539-03-7	: 48
506-23-0	: 8284	515-30-0	: 914	525-64-4	: 5971	532-54-7	: 876	539-08-2	: 9796
506-24-1	: 8316	515-40-2	: 1143	525-68-8	: 11253	532-62-1	: 1284	539-15-1	: 9320
506-30-9	: 5355	515-46-8	: 2223	525-82-6	: 2840	532-96-7	: 2717	539-23-1	: 10860
506-31-0	: 5364	515-49-1	: 2189	526-08-9	: 2201	533-18-6	: 211	539-30-0	: 1658
506-32-1	: 5361	515-59-3	: 2200	526-31-8	: 11923	533-24-4	: 1544	539-47-9	: 10638
506-33-2	: 5282	515-82-2	: 6000	526-35-2	: 8586	533-30-2	: 2860	539-52-6	: 6133
506-42-3	: 8311	515-83-3	: 5769	526-55-6	: 7305	533-32-4	: 3201	539-74-2	: 10208
506-46-7	: 6624	515-84-4	: 248	526-73-8	: 2357	533-45-9	: 11670	539-80-0	: 4242
506-48-9	: 8221	515-88-8	: 2944	526-75-0	: 9307	533-48-2	: 7223	539-82-2	: 8878
506-50-3	: 11820	515-90-2	: 2964	526-84-1	: 3777	533-50-6	: 3679	539-86-6	: 10595
506-52-5	: 6626	515-94-6	: 326	526-85-2	: 9493	533-58-4	: 9371	539-88-8	: 8907
506-68-3	: 4165	515-96-8	: 105	526-86-3	: 4263	533-68-6	: 3469	539-89-9	: 3961
506-77-4	: 4166	516-05-2	: 9960	526-95-4	: 6209	533-70-0	: 701	539-90-2	: 3563
506-78-5	: 7337	516-06-3	: 12040	526-99-8	: 6193	533-73-3	: 2367	539-92-4	: 3997
506-80-9	: 4036	516-12-1	: 11159	527-07-1	: 6213	533-75-5	: 4243	540-07-8	: 6907
506-82-1	: 3929	516-78-9	: 5384	527-17-3	: 4270	533-86-8	: 8300	540-08-9	: 6321
506-85-4	: 6024	516-85-8	: 5380	527-18-4	: 1552	533-98-2	: 3265	540-09-0	: 11857
506-89-8	: 12020	516-95-0	: 4108	527-35-5	: 9475	534-00-9	: 3238	540-10-3	: 6651
506-93-4	: 6286	517-04-4	: 5401	527-52-6	: 7111	534-07-6	: 10426	540-18-1	: 3579
506-96-7	: 291	517-06-6	: 5407	527-53-7	: 2279	534-13-4	: 11804	540-36-3	: 1424
507-02-8	: 302	517-07-7	: 5402	527-55-9	: 1513	534-15-6	: 5548	540-37-4	: 761
507-09-5	: 5646	517-09-9	: 5397	527-60-6	: 9495	534-22-5	: 6130	540-38-5	: 9373
507-19-7	: 9846	517-10-2	: 4111	527-61-7	: 4264	534-26-9	: 7197	540-42-1	: 10310
507-20-0	: 9863	517-22-6	: 11126	527-62-8	: 9205	534-52-1	: 9405	540-47-6	: 4980
507-25-5	: 7542	517-23-7	: 6139	527-69-5	: 6040	534-59-8	: 9929	540-51-2	: 5685
507-32-4	: 457	517-25-9	: 7582	527-72-0	: 11737	534-84-9	: 4611	540-54-5	: 9848
507-36-8	: 3240	517-28-2	: 629	527-73-1	: 7211	534-85-0	: 1282	540-61-4	: 272
507-40-4	: 7183	517-60-2	: 1748	527-84-4	: 1988	535-11-5	: 10207	540-63-6	: 5590
507-42-6	: 5581	517-66-8	: 2468	527-85-5	: 619	535-13-7	: 10220	540-67-0	: 5602
507-45-9	: 3286	517-82-8	: 7857	527-89-9	: 1125	535-15-9	: 164	540-69-2	: 6003
507-47-1	: 5680	517-92-0	: 401	527-93-5	: 5153	535-46-6	: 11548	540-73-8	: 7155
507-60-8	: 3142	517-97-5	: 5159	528-21-2	: 5873	535-75-1	: 9640	540-80-7	: 8085
508-02-1	: 8547	518-05-8	: 7805	528-23-4	: 4784	535-77-3	: 1989	540-82-9	: 11576
508-32-7	: 11867	518-17-2	: 7333	528-29-0	: 1491	535-80-8	: 2535	540-84-1	: 8848
508-44-1	: 5185	518-28-5	: 6192	528-44-9	: 2319	535-83-1	: 11029	540-88-5	: 177
508-54-3	: 7675	518-44-5	: 2572	528-45-0	: 2590	535-87-5	: 2554	540-97-6	: 4824
508-65-6	: 4068	518-61-6	: 9502	528-48-3	: 2813	535-89-7	: 11043	541-01-5	: 6528
509-12-6	: 2938	518-69-4	: 5149	528-53-0	: 2849	536-08-3	: 2575	541-02-6	: 4901
509-14-8	: 7544	518-75-2	: 2774	528-71-2	: 9197	536-17-4	: 11684	541-05-9	: 5002
509-15-9	: 6203	518-77-4	: 5151	528-75-6	: 551	536-25-4	: 2501	541-16-2	: 9925
509-20-6	: 305	518-82-1	: 419	528-76-7	: 2178	536-33-4	: 10885	541-20-8	: 8738
509-64-8	: 7670	519-02-8	: 7426	528-79-0	: 9420	536-38-9	: 5778	541-28-6	: 3384
509-87-5	: 11096	519-05-1	: 2608	528-81-4	: 467	536-50-5	: 1859	541-31-1	: 3440
510-15-6	: 4082	519-09-5	: 481	529-00-0	: 4602	536-57-2	: 2182	541-33-3	: 3276
510-20-3	: 9939	519-34-6	: 7610	529-08-8	: 506	536-59-4	: 4664	541-35-5	: 3194
510-90-7	: 11063	519-37-9	: 10758	529-16-8	: 2968	536-60-7	: 1902	541-41-3	: 4018
511-07-9	: 5392	519-44-8	: 1517	529-17-9	: 484	536-66-3	: 2689	541-42-4	: 8092
511-08-0	: 5391	519-72-2	: 1804	529-19-1	: 2763	536-69-6	: 10898	541-46-8	: 3206
511-09-1	: 5389	519-73-3	: 1982	529-20-4	: 587	536-74-3	: 1715	541-47-9	: 3846
511-10-4	: 5390	519-87-9	: 57	529-21-5	: 10964	536-75-4	: 10958	541-50-4	: 3588
511-18-2	: 8210	520-03-6	: 7366	529-28-2	: 1775	536-78-7	: 10957	541-53-7	: 11718
511-20-6	: 5377	520-18-3	: 2843	529-33-9	: 8041	536-88-9	: 10965	541-58-2	: 11674
511-21-7	: 5378	520-26-3	: 2794	529-34-0	: 8046	536-89-0	: 7164	541-59-3	: 11121
511-34-2	: 11535	520-33-2	: 2800	529-35-1	: 8042	536-90-3	: 787	541-73-1	: 1355
511-67-1	: 11086	520-36-5	: 2806	529-36-2	: 8007	536-95-8	: 2721	541-88-8	: 124
511-70-6	: 3198	520-45-6	: 10779	529-64-6	: 915	537-17-7	: 11830	541-91-3	: 4763
511-96-6	: 11530	520-85-4	: 9752	529-65-7	: 62	537-26-8	: 488	542-05-2	: 8782

542-08-5 : 3574	551-09-7 : 5514	557-89-1 : 453	573-33-1 : 7199	583-53-9 : 1288	
542-10-9 : 5570	551-62-2 : 2270	557-91-5 : 5517	573-56-8 : 9344	583-55-1 : 1046	
542-15-4 : 821	551-76-8 : 9489	557-93-7 : 10530	573-58-0 : 7969	583-58-4 : 10940	
542-18-7 : 4339	551-77-9 : 9488	557-98-2 : 10535	573-97-7 : 8022	583-61-9 : 10936	
542-28-9 : 10798	551-78-0 : 9487	557-99-3 : 301	573-98-8 : 7818	583-68-6 : 647	
542-37-0 : 3546	551-88-2 : 8834	558-13-4 : 7539	574-09-4 : 5813	583-69-7 : 1503	
542-46-1 : 4218	551-92-8 : 7202	558-17-8 : 10058	574-12-9 : 2841	583-70-0 : 1018	
542-50-7 : 6303	552-16-9 : 2703	558-25-8 : 7538	574-39-0 : 3983	583-71-1 : 1022	
542-52-9 : 3998	552-30-7 : 7341	558-30-5 : 8607	574-84-5 : 2808	583-75-5 : 650	
542-54-1 : 8831	552-32-9 : 81	558-37-2 : 3760	575-37-1 : 7823	583-78-8 : 9284	
542-55-2 : 6013	552-41-0 : 5823	558-38-3 : 10352	575-38-2 : 7837	583-86-8 : 8276	
542-56-3 : 8094	552-45-4 : 1198	558-42-9 : 10354	575-41-7 : 7819	584-02-1 : 8917	
542-58-5 : 5695	552-58-9 : 2812	558-43-0 : 10017	575-43-9 : 7822	584-04-3 : 10371	
542-59-6 : 5577	552-59-0 : 2828	559-70-6 : 8548	575-44-0 : 7836	584-12-3 : 6033	
542-69-8 : 3381	552-70-5 : 479	560-08-7 : 3931	575-74-6 : 607	584-27-0 : 3331	
542-76-7 : 10097	552-72-7 : 9646	560-21-4 : 8849	576-22-7 : 1020	584-70-3 : 621	
542-81-4 : 5481	552-79-4 : 1863	560-22-5 : 9068	576-23-8 : 1017	584-79-2 : 336	
542-85-8 : 5800	552-82-9 : 807	560-23-6 : 9065	576-24-9 : 9282	584-84-9 : 1431	
542-88-1 : 7525	552-89-6 : 591	560-95-2 : 7476	576-26-1 : 9310	584-87-2 : 2609	
542-90-5 : 11703	552-94-3 : 2616	561-07-9 : 311	576-55-6 : 9488	584-90-7 : 5114	
542-91-6 : 11409	553-03-7 : 11327	561-20-6 : 11352	576-83-0 : 1094	584-92-9 : 8399	
542-92-7 : 4764	553-17-3 : 9384	561-25-1 : 7664	577-16-2 : 5851	584-94-1 : 6772	
543-20-4 : 3366	553-19-5 : 2397	561-27-3 : 7667	577-19-5 : 1075	584-98-5 : 3299	
543-21-5 : 3902	553-20-8 : 84	561-83-1 : 11090	577-27-5 : 4989	584-99-6 : 3769	
543-24-8 : 6251	553-26-4 : 3104	561-94-4 : 5393	577-33-3 : 429	585-32-0 : 1806	
543-27-1 : 4024	553-39-9 : 7953	562-49-2 : 8763	577-37-7 : 7572	585-34-2 : 9322	
543-28-2 : 3959	553-60-6 : 10907	562-74-3 : 4699	577-55-9 : 992	585-48-8 : 10874	
543-29-3 : 8078	553-69-5 : 1913	563-03-1 : 9387	577-56-0 : 2476	585-68-2 : 6054	
543-38-4 : 7135	553-79-7 : 827	563-04-2 : 9570	577-59-3 : 5863	585-74-0 : 5852	
543-59-9 : 8719	553-82-2 : 1381	563-12-2 : 5931	577-71-9 : 9345	585-76-2 : 2519	
543-67-9 : 8098	553-86-6 : 2461	563-16-6 : 6775	577-91-3 : 2135	585-79-5 : 1076	
543-75-9 : 5203	553-90-2 : 5563	563-41-7 : 7143	578-07-4 : 312	585-81-9 : 2632	
543-80-6 : 110	553-94-6 : 1021	563-45-1 : 3800	578-54-1 : 740	586-11-8 : 9346	
543-83-9 : 6283	553-97-9 : 4266	563-46-2 : 3799	578-57-4 : 1050	586-27-6 : 4698	
543-86-2 : 3633	554-00-7 : 686	563-47-3 : 10540	578-58-5 : 1931	586-37-8 : 5840	
543-87-3 : 3637	554-01-8 : 11106	563-52-0 : 3723	578-66-5 : 11206	586-38-9 : 2665	
544-00-3 : 3226	554-12-1 : 10296	563-57-5 : 10553	578-67-6 : 11305	586-39-0 : 1644	
544-01-4 : 3418	554-14-3 : 11769	563-58-6 : 10547	578-68-7 : 11203	586-42-5 : 2477	
544-02-5 : 3436	554-21-2 : 5212	563-70-2 : 7473	578-76-7 : 10769	586-61-8 : 1059	
544-10-5 : 6737	554-35-8 : 10104	563-78-0 : 3759	579-04-4 : 7112	586-62-9 : 4684	
544-13-8 : 8767	554-68-7 : 5418	563-79-1 : 3761	579-07-7 : 10028	586-75-4 : 2888	
544-16-1 : 8083	554-70-1 : 9521	563-80-4 : 3668	579-10-2 : 79	586-76-5 : 2520	
544-25-2 : 4240	554-84-7 : 9436	563-83-7 : 9798	579-18-0 : 2511	586-77-6 : 645	
544-35-4 : 8226	555-03-3 : 1936	563-84-8 : 3606	579-21-5 : 5854	586-78-7 : 1077	
544-40-1 : 3433	555-10-2 : 4673	564-00-1 : 3007	579-44-2 : 5822	586-84-5 : 6129	
544-62-7 : 10023	555-16-8 : 593	564-02-3 : 8847	579-60-2 : 9392	586-89-0 : 2478	
544-63-8 : 11622	555-21-5 : 967	564-03-4 : 9069	579-66-8 : 694	586-95-8 : 10977	
544-65-0 : 11637	555-30-6 : 11931	564-04-5 : 8966	579-75-9 : 2664	586-96-9 : 2059	
544-66-1 : 11638	555-35-1 : 6645	564-36-3 : 5387	579-92-0 : 2656	586-98-1 : 10975	
544-72-9 : 8286	555-43-1 : 8272	564-37-4 : 5388	579-94-2 : 183	587-02-0 : 741	
544-73-0 : 8282	555-44-2 : 6656	565-00-4 : 2921	580-02-9 : 2482	587-03-1 : 1898	
544-76-3 : 6632	555-45-3 : 11628	565-53-7 : 161	580-13-2 : 7740	587-04-2 : 530	
544-77-4 : 6640	555-57-7 : 1823	565-59-3 : 8761	580-15-4 : 11205	587-54-2 : 2513	
544-85-4 : 5348	555-59-9 : 3853	565-60-6 : 8941	580-16-5 : 11306	587-61-1 : 10873	
544-97-8 : 12066	555-77-1 : 5412	565-63-9 : 3845	580-20-1 : 11307	587-85-9 : 7436	
545-06-2 : 288	555-89-5 : 1965	565-64-0 : 10245	580-22-3 : 11202	587-90-6 : 11997	
545-26-6 : 4048	556-10-5 : 12025	565-67-3 : 8943	580-48-3 : 11828	588-01-2 : 11906	
545-47-1 : 7417	556-22-9 : 6275	565-69-5 : 8982	580-51-8 : 3089	588-46-5 : 87	
545-93-7 : 11078	556-24-1 : 3552	565-75-3 : 8850	581-08-8 : 59	588-47-6 : 817	
546-06-5 : 4143	556-27-4 : 333	565-76-4 : 9066	581-40-8 : 7825	588-52-3 : 5113	
546-33-8 : 5004	556-50-3 : 6265	565-77-5 : 9070	581-42-0 : 7826	588-63-6 : 1085	
546-43-0 : 8059	556-52-5 : 8618	565-80-0 : 8967	581-43-1 : 7838	588-67-0 : 1115	
546-45-2 : 5005	556-56-9 : 10573	567-18-0 : 7973	581-45-3 : 11009	588-72-7 : 1029	
546-56-5 : 4994	556-61-6 : 7521	568-02-5 : 8062	581-46-4 : 3102	588-73-8 : 1031	
546-97-4 : 5926	556-64-9 : 11705	568-21-8 : 5158	581-47-5 : 3101	588-96-5 : 1032	
547-25-1 : 6022	556-67-2 : 4993	568-93-4 : 399	581-49-7 : 3105	589-08-2 : 1581	
547-44-4 : 2190	556-68-3 : 4741	568-99-0 : 383	581-50-0 : 3100	589-09-3 : 841	
547-52-4 : 2188	556-69-4 : 8482	569-08-4 : 394	581-55-5 : 7512	589-10-6 : 1033	
547-58-0 : 2222	556-70-7 : 5079	569-31-3 : 7355	581-89-5 : 7930	589-14-0 : 6120	
547-63-7 : 10301	556-71-8 : 4726	569-34-6 : 10759	582-08-1 : 5128	589-15-1 : 1000	
547-65-9 : 6147	556-82-1 : 3869	569-41-5 : 7824	582-16-2 : 7827	589-16-2 : 742	
547-81-9 : 5405	556-88-7 : 6287	569-51-7 : 2318	582-17-2 : 7839	589-17-3 : 1008	
547-91-1 : 11295	556-89-8 : 12024	569-57-3 : 1154	582-22-9 : 1579	589-18-4 : 1899	
548-40-3 : 8631	556-96-7 : 1019	569-64-2 : 7455	582-24-1 : 5833	589-21-9 : 7141	
548-51-6 : 2842	557-00-6 : 3562	569-65-3 : 9605	582-33-2 : 2496	589-29-7 : 1435	
548-62-9 : 7453	557-05-1 : 8277	569-77-7 : 2408	582-60-5 : 2380	589-33-3 : 11113	
548-80-1 : 7878	557-11-9 : 12028	570-09-0 : 9923	582-61-6 : 2884	589-38-8 : 6960	
548-98-1 : 4093	557-17-5 : 10069	570-22-9 : 7195	582-82-7 : 3671	589-40-2 : 6012	
550-10-7 : 5214	557-24-4 : 3817	570-24-1 : 802	582-77-4 : 622	589-43-5 : 6773	
550-24-3 : 4281	557-25-5 : 3505	571-58-4 : 7820	582-78-5 : 623	589-44-6 : 3465	
550-44-7 : 7365	557-30-2 : 5492	571-61-9 : 7821	583-03-9 : 1846	589-53-7 : 6424	
550-60-7 : 8034	557-31-3 : 10566	572-59-8 : 4125	583-04-0 : 2722	589-55-9 : 6475	
550-74-3 : 10848	557-40-4 : 10592	572-83-8 : 386	583-05-1 : 8804	589-59-3 : 3555	
550-82-3 : 9508	557-59-5 : 11607	572-93-0 : 396	583-33-5 : 6052	589-63-9 : 8473	
550-97-0 : 2644	557-61-9 : 8222	573-17-1 : 9164	583-39-1 : 2386	589-75-3 : 8427	
551-01-9 : 9877	557-66-4 : 5433	573-20-6 : 7841	583-48-2 : 6776	589-87-7 : 1048	
551-06-4 : 7906	557-68-6 : 7471	573-26-2 : 75	583-50-6 : 3192	589-92-4 : 4598	

589-93-5 : 10938	593-59-9 : 444	597-90-0 : 6487	603-39-4 : 2110	608-93-5 : 2084	
589-98-0 : 8445	593-60-2 : 5880	597-93-3 : 8162	603-45-2 : 4278	608-96-8 : 2096	
590-01-2 : 10214	593-61-3 : 5937	597-96-6 : 6946	603-50-9 : 9465	608-98-0 : 1553	
590-02-3 : 125	593-63-5 : 5938	598-01-6 : 6505	603-52-1 : 3946	609-02-9 : 9963	
590-11-4 : 5892	593-66-8 : 5907	598-02-7 : 9558	603-54-3 : 12011	609-08-5 : 9962	
590-12-5 : 5891	593-70-4 : 7481	598-03-8 : 10131	603-69-0 : 3462	609-09-6 : 9973	
590-13-6 : 10529	593-71-5 : 7482	598-04-9 : 3423	603-71-4 : 2363	609-11-0 : 3498	
590-16-9 : 5463	593-74-8 : 7435	598-05-0 : 11575	603-76-9 : 7312	609-12-1 : 3475	
590-18-1 : 3706	593-75-9 : 7655	598-09-4 : 8806	603-81-6 : 2552	609-13-2 : 3477	
590-19-2 : 3145	593-77-1 : 7447	598-16-3 : 5921	603-83-8 : 799	609-14-3 : 3557	
590-26-1 : 5904	593-79-3 : 7526	598-17-4 : 9880	603-85-0 : 9214	609-19-8 : 9486	
590-35-2 : 8760	593-81-7 : 7445	598-19-6 : 10699	604-53-5 : 3004	609-22-3 : 9279	
590-36-3 : 8940	593-88-4 : 449	598-21-0 : 292	604-61-5 : 2515	609-23-4 : 9492	
590-42-1 : 10066	593-90-8 : 3119	598-22-1 : 10490	604-68-2 : 6216	609-25-6 : 2369	
590-50-1 : 8965	593-92-0 : 5890	598-23-2 : 3916	604-69-3 : 6217	609-26-7 : 8813	
590-60-3 : 3634	593-95-3 : 4006	598-25-4 : 3170	604-83-1 : 9172	609-29-0 : 8803	
590-66-9 : 4364	593-96-4 : 5455	598-31-2 : 10416	604-88-6 : 1756	609-31-4 : 3640	
590-67-0 : 4530	593-98-6 : 7466	598-38-9 : 5704	605-01-6 : 2090	609-36-9 : 9765	
590-73-8 : 6771	594-02-5 : 5546	598-45-8 : 10063	605-02-7 : 7947	609-39-2 : 6134	
590-86-3 : 3186	594-03-6 : 5589	598-50-5 : 12017	605-03-8 : 11290	609-40-5 : 11775	
590-90-9 : 3662	594-04-7 : 7503	598-53-8 : 10070	605-27-6 : 411	609-46-1 : 2225	
590-92-1 : 10205	594-08-1 : 4034	598-55-0 : 3953	605-32-3 : 408	609-65-4 : 2890	
590-93-2 : 3917	594-09-2 : 9522	598-56-1 : 5421	605-37-8 : 7852	609-72-3 : 860	
591-07-1 : 27	594-10-5 : 11550	598-57-2 : 7451	605-39-0 : 3050	609-73-4 : 1783	
591-08-2 : 33	594-11-6 : 4981	598-58-3 : 8075	605-42-5 : 390	610-02-6 : 2730	
591-09-3 : 108	594-15-0 : 7549	598-61-8 : 4192	605-50-5 : 1335	610-15-1 : 624	
591-11-7 : 6154	594-16-1 : 9883	598-74-3 : 3223	605-54-9 : 1333	610-16-2 : 2584	
591-12-8 : 6153	594-18-3 : 7495	598-76-5 : 9899	605-63-0 : 7786	610-17-3 : 718	
591-17-3 : 1055	594-20-7 : 9894	598-77-6 : 10172	605-65-2 : 7977	610-25-3 : 2047	
591-18-4 : 1047	594-22-9 : 7556	598-78-7 : 10215	605-71-0 : 7832	610-34-4 : 2707	
591-19-5 : 639	594-27-4 : 11544	598-82-3 : 10266	605-79-8 : 7634	610-39-9 : 1960	
591-20-8 : 9229	594-34-3 : 9889	598-88-9 : 5898	605-81-2 : 6228	610-46-8 : 377	
591-22-0 : 10941	594-36-5 : 3253	598-92-5 : 5482	605-82-3 : 421	610-48-0 : 423	
591-27-5 : 9202	594-37-6 : 9900	598-94-7 : 12005	605-83-4 : 420	610-50-4 : 431	
591-31-1 : 584	594-38-7 : 3385	598-96-9 : 9072	605-85-6 : 5859	610-66-2 : 965	
591-34-4 : 10309	594-39-8 : 3222	598-98-1 : 10260	605-99-2 : 10765	610-67-3 : 1662	
591-35-5 : 9287	594-42-3 : 7527	598-99-2 : 253	606-17-7 : 2594	610-69-5 : 219	
591-47-9 : 4672	594-44-5 : 5627	599-01-9 : 10333	606-18-8 : 2506	610-71-9 : 2556	
591-49-1 : 4670	594-45-6 : 5622	599-04-2 : 6143	606-20-2 : 1958	610-72-0 : 2581	
591-50-4 : 1766	594-47-8 : 94	599-66-6 : 2240	606-21-3 : 1145	610-89-9 : 9074	
591-60-6 : 3569	594-51-4 : 3274	599-70-2 : 1704	606-22-4 : 726	610-91-3 : 2362	
591-62-8 : 3936	594-52-5 : 3236	599-79-1 : 2654	606-23-5 : 7264	610-97-9 : 2660	
591-68-4 : 8862	594-56-9 : 3812	599-91-7 : 2233	606-27-9 : 2710	610-99-1 : 10449	
591-76-4 : 6834	594-57-0 : 3247	599-97-3 : 5657	606-28-0 : 2516	611-00-7 : 2555	
591-78-6 : 6959	594-58-1 : 10228	599-99-5 : 242	606-35-9 : 1946	611-01-8 : 2580	
591-80-0 : 9073	594-60-5 : 3618	600-00-0 : 10213	606-37-1 : 7831	611-03-0 : 2553	
591-82-2 : 10065	594-61-6 : 10278	600-05-5 : 10238	606-43-9 : 11328	611-06-3 : 1396	
591-84-4 : 11707	594-65-0 : 95	600-06-6 : 3245	606-45-1 : 2673	611-07-4 : 9269	
591-87-7 : 235	594-70-7 : 10085	600-07-7 : 3538	606-46-2 : 696	611-10-9 : 4785	
591-93-5 : 8659	594-71-8 : 9870	600-11-3 : 8749	606-81-5 : 3063	611-12-1 : 10366	
591-95-7 : 8658	594-72-9 : 5536	600-13-5 : 4150	606-88-2 : 7173	611-13-2 : 6059	
591-96-8 : 8660	594-73-0 : 5594	600-14-6 : 8797	606-97-3 : 3093	611-14-3 : 1685	
592-01-8 : 3930	594-81-0 : 3271	600-15-7 : 3530	606-99-5 : 751	611-15-4 : 1640	
592-13-2 : 6774	594-82-1 : 3430	600-17-9 : 3842	607-00-1 : 5993	611-19-8 : 1132	
592-20-1 : 10412	594-83-2 : 3644	600-18-0 : 3567	607-31-8 : 11274	611-20-1 : 2759	
592-22-3 : 5250	594-89-8 : 10048	600-22-6 : 10316	607-32-9 : 7386	611-21-2 : 710	
592-27-8 : 6422	595-30-2 : 8559	600-25-9 : 9866	607-34-1 : 11284	611-32-5 : 11283	
592-34-7 : 4012	595-37-9 : 3506	600-30-6 : 3497	607-42-1 : 368	611-33-6 : 11242	
592-35-8 : 3935	595-38-0 : 8724	600-36-2 : 8931	607-50-1 : 7919	611-34-7 : 11204	
592-41-6 : 6999	595-39-1 : 7400	600-40-8 : 5551	607-52-3 : 7939	611-35-8 : 11238	
592-42-7 : 6678	595-40-4 : 7401	601-75-2 : 9948	607-53-4 : 8005	611-43-8 : 3051	
592-44-9 : 6676	595-41-5 : 8930	601-76-3 : 9918	607-56-7 : 12008	611-45-0 : 7949	
592-45-0 : 6677	595-44-8 : 9902	601-77-4 : 9822	607-81-8 : 9977	611-49-4 : 7928	
592-46-1 : 6680	595-49-3 : 9920	601-84-3 : 2557	607-85-2 : 2635	611-61-0 : 3053	
592-49-4 : 6679	595-89-1 : 9733	601-88-7 : 1393	607-86-3 : 2682	611-62-1 : 3065	
592-50-7 : 8817	595-90-4 : 11545	602-03-9 : 724	607-90-9 : 2652	611-69-8 : 1901	
592-51-8 : 9062	596-27-0 : 7353	602-21-1 : 1353	607-91-0 : 2428	611-70-1 : 10469	
592-55-2 : 5461	596-29-2 : 7358	602-27-7 : 1489	607-97-6 : 3518	611-73-4 : 934	
592-57-4 : 4248	596-43-0 : 1061	602-29-9 : 2046	607-99-8 : 2313	611-79-0 : 7589	
592-65-4 : 10154	596-51-0 : 11179	602-37-9 : 7787	608-07-1 : 7304	611-92-7 : 12006	
592-76-2 : 6533	596-75-8 : 9937	602-38-0 : 7833	608-21-9 : 2307	611-94-9 : 7625	
592-80-3 : 8935	597-04-6 : 3511	602-55-1 : 428	608-23-1 : 1137	611-95-0 : 2512	
592-82-5 : 3388	597-05-7 : 8934	602-56-2 : 318	608-25-3 : 1529	611-97-2 : 7599	
592-84-7 : 6004	597-09-1 : 10008	602-60-8 : 426	608-26-4 : 9260	611-99-4 : 7597	
592-88-1 : 10599	597-12-6 : 6221	602-64-2 : 415	608-27-5 : 685	612-00-0 : 1677	
592-90-5 : 8590	597-32-0 : 10365	602-65-3 : 418	608-28-6 : 1772	612-01-1 : 12018	
593-03-3 : 6661	597-33-1 : 9998	602-84-2 : 1332	608-29-7 : 2349	612-08-8 : 5109	
593-08-8 : 11889	597-35-3 : 5626	602-92-6 : 2560	608-30-0 : 682	612-12-4 : 986	
593-12-4 : 8358	597-49-9 : 8932	602-96-0 : 2364	608-31-1 : 688	612-13-5 : 2750	
593-14-6 : 8361	597-50-2 : 9517	602-99-3 : 9434	608-32-2 : 2305	612-15-7 : 1637	
593-45-3 : 8234	597-52-4 : 11503	603-26-9 : 7513	608-33-3 : 9277	612-16-8 : 1889	
593-49-7 : 6302	597-55-7 : 9926	603-32-7 : 450	608-50-4 : 2360	612-17-9 : 7815	
593-50-0 : 11821	597-67-1 : 11428	603-33-8 : 3111	608-66-2 : 6194	612-19-1 : 2600	
593-51-1 : 7446	597-71-7 : 9988	603-34-9 : 728	608-69-5 : 3308	612-22-8 : 1692	
593-52-2 : 446	597-72-8 : 10068	603-35-0 : 9523	608-71-9 : 9439	612-23-7 : 1203	
593-53-3 : 7515	597-76-2 : 8931	603-36-1 : 11551	608-89-9 : 3315	612-25-9 : 1906	
593-57-7 : 442	597-77-3 : 6935	603-38-3 : 729	608-90-2 : 2082	612-28-2 : 804	

CAS	No.	CAS	No.	CAS	No.	CAS	No.	CAS	No.	CAS	No.
612-35-1	2718	615-57-6	681	618-85-9	1957	622-04-8	10377	623-96-1	4002		
612-37-3	10766	615-58-7	9276	618-87-1	727	622-08-2	5761	623-97-2	5697		
612-55-5	7903	615-59-8	1305	618-89-3	2525	622-09-3	7693	624-03-3	6649		
612-57-7	11240	615-60-1	1141	618-91-7	2661	622-15-1	7565	624-04-4	5311		
612-58-8	11278	615-62-3	9265	618-95-1	2711	622-16-2	764	624-08-8	6319		
612-60-2	11282	615-65-6	668	618-98-4	2708	622-24-2	1164	624-09-9	6464		
612-61-3	11241	615-67-8	1507	619-02-3	4601	622-26-4	9664	624-13-5	8441		
612-62-4	11237	615-68-9	2350	619-12-5	9598	622-29-7	7452	624-16-8	5077		
612-64-6	750	615-71-4	2306	619-15-8	1959	622-31-1	594	624-17-9	8133		
612-71-5	11598	615-74-7	9262	619-20-5	2601	622-32-2	595	624-20-4	6754		
612-75-9	3057	615-79-2	8876	619-23-8	1206	622-36-6	7166	624-24-8	8891		
612-76-0	3066	615-81-6	5556	619-25-0	1907	622-37-7	976	624-29-3	4368		
612-78-2	3005	615-83-8	8860	619-31-8	719	622-38-8	1705	624-31-7	1780		
612-81-7	3107	615-84-9	5494	619-33-0	5530	622-39-9	11015	624-38-4	1429		
612-94-2	7948	615-85-0	5453	619-39-6	5068	622-40-2	7691	624-41-9	3630		
612-95-3	11291	615-87-2	1298	619-41-0	5779	622-42-4	2052	624-42-0	6523		
612-96-4	11288	615-89-4	1514	619-42-1	2526	622-45-7	155	624-43-1	10192		
612-97-5	11196	615-98-5	5565	619-44-3	2662	622-47-9	926	624-44-2	6982		
613-03-6	2370	615-99-6	5564	619-52-3	4683	622-57-1	748	624-45-3	8908		
613-08-1	369	616-02-4	6098	619-58-9	2659	622-60-6	1661	624-48-6	3779		
613-12-7	424	616-06-8	8215	619-60-3	9319	622-61-7	1158	624-49-7	3778		
613-13-8	365	616-12-6	9056	619-73-8	1908	622-62-8	9352	624-54-4	10318		
613-21-8	11319	616-19-3	9901	619-82-9	4354	622-63-9	1707	624-57-7	8617		
613-29-6	684	616-20-6	8721	619-84-1	2585	622-78-6	1794	624-60-2	5435		
613-31-0	376	616-21-7	3277	619-99-8	6815	622-79-7	980	624-61-3	5939		
613-33-2	3061	616-23-9	10359	620-02-0	6045	622-80-0	843	624-64-6	3705		
613-37-6	3081	616-24-0	8702	620-03-1	6047	622-82-2	7160	624-65-7	10729		
613-39-8	9613	616-25-1	9087	620-05-3	1777	622-85-5	2168	624-66-8	10734		
613-42-3	3094	616-27-3	3646	620-11-1	8919	622-86-6	1159	624-67-9	10723		
613-43-4	5112	616-28-4	3912	620-12-2	10193	622-93-5	10367	624-70-4	5477		
613-45-6	543	616-31-9	8846	620-13-3	1063	622-97-9	1642	624-72-6	5545		
613-46-7	7749	616-38-6	4000	620-14-4	1686	622-98-0	1163	624-73-7	5547		
613-48-9	697	616-39-7	5430	620-16-6	1162	623-00-7	2746	624-74-8	5941		
613-50-3	11285	616-40-0	7152	620-17-7	9354	623-03-0	2749	624-75-9	283		
613-51-1	11286	616-42-2	11582	620-18-8	9349	623-05-2	1884	624-76-0	5731		
613-59-2	7950	616-43-3	11132	620-19-9	1199	623-08-5	712	624-79-3	5599		
613-69-4	552	616-44-4	11770	620-20-2	1133	623-09-6	1280	624-81-7	5434		
613-70-7	9382	616-45-5	11180	620-22-4	2764	623-10-9	758	624-82-8	7566		
613-75-2	7510	616-47-7	7208	620-23-5	588	623-12-1	1180	624-83-9	7520		
613-77-4	11249	616-55-7	9223	620-24-6	1883	623-15-4	3878	624-85-1	7184		
613-78-5	2645	616-91-1	5015	620-32-6	2241	623-17-6	6119	624-89-5	5606		
613-80-9	7940	617-00-5	795	620-40-6	1798	623-18-7	10640	624-90-8	7460		
613-81-0	8006	617-02-7	2699	620-42-8	9597	623-19-8	6123	624-91-9	8091		
613-84-3	577	617-05-0	2626	620-43-9	3224	623-20-1	10639	624-92-0	5254		
613-88-7	956	617-09-4	3996	620-47-3	2015	623-21-2	3520	624-95-3	3617		
613-89-8	5774	617-30-1	6797	620-63-3	3561	623-22-3	10642	624-96-4	3284		
613-90-1	968	617-31-2	8883	620-71-3	9800	623-24-5	985	624-97-5	9100		
613-91-2	5866	617-33-4	160	620-72-4	119	623-25-6	988	625-01-4	4085		
613-93-4	620	617-35-6	10315	620-73-5	147	623-26-7	1316	625-04-7	8959		
613-94-5	2611	617-45-8	470	620-80-4	3578	623-27-8	1323	625-06-9	8928		
613-97-8	749	617-50-5	10303	620-81-5	5495	623-30-3	10512	625-08-1	3535		
614-00-6	806	617-51-6	10275	620-82-6	5559	623-34-7	3285	625-23-0	6942		
614-16-4	2124	617-52-7	3330	620-83-7	2016	623-36-9	9002	625-25-2	6495		
614-18-6	10902	617-53-8	3789	620-85-9	1696	623-39-2	10014	625-27-4	9055		
614-21-1	5862	617-54-9	3790	620-86-0	2719	623-40-5	8985	625-28-5	3410		
614-27-7	2142	617-60-7	5309	620-88-2	2054	623-42-7	3547	625-30-9	8701		
614-29-9	1613	617-65-2	6240	620-92-8	9407	623-43-8	3847	625-31-0	9089		
614-33-5	10196	617-73-2	8432	620-93-9	797	623-46-1	5533	625-36-5	10497		
614-34-6	2700	617-75-4	448	620-94-0	2286	623-47-2	10739	625-37-6	3702		
614-39-1	602	617-78-7	8809	620-95-1	11005	623-48-3	201	625-38-7	3813		
614-45-9	1128	617-80-1	3407	620-99-5	5670	623-49-4	4030	625-44-5	10071		
614-47-1	10709	617-83-4	4156	621-08-9	2184	623-50-7	197	625-45-6	205		
614-54-0	1881	617-84-5	5991	621-14-7	3320	623-51-8	203	625-46-7	10591		
614-68-6	1787	617-85-6	11549	621-15-8	6722	623-53-0	4003	625-47-8	5483		
614-71-1	1659	617-86-7	11483	621-23-8	2355	623-56-3	6981	625-48-9	5752		
614-72-2	1156	617-88-9	6075	621-25-0	7174	623-59-6	73	625-50-3	61		
614-75-5	904	617-89-0	6113	621-26-1	7138	623-60-9	10351	625-52-5	12014		
614-78-8	11806	617-90-3	6039	621-27-2	9462	623-65-4	6646	625-54-7	10038		
614-96-0	7316	617-92-5	11124	621-31-8	9356	623-66-5	8424	625-55-8	6011		
614-99-3	6056	617-94-7	1857	621-32-9	1660	623-68-7	3818	625-56-9	7582		
615-05-4	1274	618-31-5	1301	621-34-1	9351	623-69-8	10372	625-57-0	3978		
615-09-8	6163	618-32-6	2885	621-36-3	925	623-70-1	3839	625-58-1	8074		
615-10-1	6071	618-34-8	1148	621-37-4	905	623-71-2	10221	625-59-2	11706		
615-11-2	6057	618-36-0	1814	621-50-1	966	623-72-3	10270	625-60-5	5647		
615-13-4	7287	618-38-2	2910	621-51-2	2597	623-73-4	158	625-65-0	9035		
615-18-9	2878	618-40-6	7165	621-52-3	1683	623-76-7	12001	625-66-1	3282		
615-20-3	2862	618-41-7	2180	621-59-0	572	623-78-9	3949	625-67-2	8750		
615-30-5	8691	618-42-8	9630	621-62-5	5469	623-80-3	4031	625-68-3	3481		
615-36-1	638	618-45-1	9411	621-63-6	5705	623-81-4	11581	625-69-4	8789		
615-37-2	1778	618-46-2	2891	621-64-7	9825	623-84-7	10002	625-71-8	3531		
615-38-3	4531	618-51-9	2658	621-70-5	6908	623-85-8	3970	625-74-1	10084		
615-39-4	4532	618-62-2	1394	621-72-7	2385	623-86-9	102	625-76-3	7508		
615-41-8	1171	618-65-5	562	621-76-1	10078	623-87-0	10191	625-77-4	23		
615-42-9	1427	618-68-8	2144	621-77-2	8705	623-91-6	3775	625-80-9	10152		
615-52-1	5567	618-71-3	2593	621-87-4	10479	623-93-8	8161	625-82-1	11119		
615-54-3	2308	618-76-8	2618	621-95-4	731	623-95-0	12012	625-84-3	11120		
615-55-4	683	618-80-4	9296	622-00-4	3306			625-86-5	6089		

CAS	Page	CAS	Page	CAS	Page	CAS	Page	CAS	Page
625-88-7	: 11753	627-48-5	: 4163	629-39-0	: 8079	634-67-3	: 852	641-36-1	: 5160
625-92-3	: 10926	627-51-0	: 5920	629-40-3	: 8385	634-68-4	: 2273	641-74-7	: 7423
625-95-6	: 1779	627-52-1	: 11572	629-41-4	: 8389	634-89-9	: 2253	641-85-0	: 9740
625-96-7	: 4597	627-53-2	: 5620	629-43-6	: 8088	634-92-4	: 2274	641-91-8	: 8010
625-98-9	: 1167	627-54-3	: 5630	629-45-8	: 5252	634-93-5	: 854	641-96-3	: 11193
626-00-6	: 1428	627-58-7	: 6691	629-46-9	: 8095	634-95-7	: 12002	642-28-4	: 7912
626-01-7	: 760	627-59-8	: 6944	629-49-2	: 5344	635-03-0	: 10139	642-32-0	: 2268
626-02-8	: 9372	627-63-4	: 3795	629-50-5	: 11871	635-10-9	: 2258	642-44-4	: 11060
626-04-0	: 1562	627-67-8	: 3400	629-54-9	: 6629	635-11-0	: 2276	642-84-2	: 608
626-10-8	: 3771	627-70-3	: 10463	629-59-4	: 11613	635-27-8	: 11239	643-28-7	: 788
626-13-1	: 796	627-73-6	: 3323	629-60-7	: 11880	635-46-1	: 11296	643-58-3	: 3082
626-15-3	: 984	627-82-7	: 10025	629-62-9	: 8642	635-65-4	: 3000	643-60-7	: 7393
626-16-4	: 987	627-83-8	: 8256	629-63-0	: 11619	635-67-6	: 1508	643-65-2	: 7632
626-17-5	: 1315	627-88-3	: 8265	629-64-1	: 6430	635-81-4	: 2269	643-75-4	: 10199
626-18-6	: 1434	627-89-4	: 8306	629-65-2	: 6435	635-88-1	: 7918	643-84-5	: 2847
626-19-7	: 1322	627-90-7	: 11946	629-66-3	: 8106	635-90-5	: 11134	643-93-6	: 3083
626-20-0	: 1638	627-91-8	: 6792	629-70-9	: 6662	635-93-8	: 533	644-08-6	: 3084
626-21-1	: 10933	627-93-0	: 6786	629-72-1	: 8643	636-04-4	: 1129	644-15-5	: 7725
626-23-3	: 3227	627-96-3	: 6756	629-73-2	: 6667	636-09-9	: 1350	644-21-3	: 7159
626-26-6	: 3434	627-97-4	: 6548	629-74-3	: 6673	636-10-2	: 7162	644-30-4	: 1468
626-27-7	: 6456	627-98-5	: 6941	629-75-5	: 6674	636-21-5	: 791	644-31-5	: 9151
626-29-9	: 11623	628-02-4	: 6721	629-78-7	: 6306	636-28-2	: 2254	644-32-6	: 5251
626-31-3	: 3297	628-03-5	: 3390	629-79-8	: 6641	636-30-6	: 853	644-35-9	: 9461
626-33-5	: 8524	628-04-6	: 3374	629-80-1	: 6628	636-31-7	: 2275	644-49-5	: 10308
626-35-7	: 218	628-05-7	: 8832	629-82-3	: 8416	636-36-2	: 6891	644-71-3	: 9445
626-36-8	: 3932	628-08-0	: 3862	629-89-0	: 8313	636-41-9	: 11131	644-80-4	: 3770
626-37-9	: 3963	628-09-1	: 10350	629-90-3	: 6304	636-46-4	: 1352	644-90-6	: 8423
626-39-1	: 2309	628-11-5	: 4015	629-92-5	: 8101	636-53-3	: 1346	644-97-3	: 9556
626-41-5	: 9278	628-12-6	: 4019	629-93-6	: 8242	636-60-2	: 3325	644-98-4	: 10984
626-43-7	: 690	628-13-7	: 10969	629-94-7	: 6294	636-72-6	: 11768	645-00-1	: 1784
626-44-8	: 2351	628-16-0	: 6716	629-96-9	: 5358	636-73-7	: 11018	645-09-0	: 625
626-48-2	: 11069	628-17-1	: 8818	629-97-0	: 5268	636-82-8	: 4637	645-12-5	: 6066
626-51-7	: 8779	628-20-6	: 3405	629-98-1	: 5284	636-98-6	: 1785	645-13-6	: 5848
626-53-9	: 6454	628-21-7	: 3288	629-99-2	: 8637	637-03-6	: 447	645-36-3	: 5414
626-55-1	: 10876	628-28-4	: 3391	630-01-3	: 6622	637-07-0	: 10231	645-41-0	: 3225
626-58-4	: 9682	628-29-5	: 3402	630-02-4	: 8220	637-12-7	: 8248	645-45-4	: 2160
626-60-8	: 10916	628-30-8	: 10064	630-03-5	: 8099	637-27-4	: 10323	645-49-8	: 1620
626-61-9	: 10917	628-32-0	: 10037	630-04-6	: 6299	637-47-8	: 7139	645-55-6	: 825
626-62-0	: 4415	628-33-1	: 10733	630-06-8	: 6991	637-50-3	: 2162	645-56-7	: 9483
626-64-2	: 11032	628-34-2	: 5474	630-07-9	: 8999	637-55-8	: 8271	645-59-0	: 2123
626-67-5	: 9679	628-37-5	: 9158	630-08-0	: 4011	637-57-0	: 10679	645-66-9	: 5306
626-70-0	: 6788	628-41-1	: 4249	630-10-4	: 11354	637-59-2	: 1088	645-67-0	: 8912
626-77-7	: 6909	628-44-4	: 8464	630-13-7	: 9908	637-64-9	: 6125	645-69-2	: 3319
626-82-4	: 6880	628-46-6	: 6896	630-16-0	: 5631	637-65-0	: 6126	645-96-5	: 2177
626-84-6	: 5736	628-50-2	: 4021	630-17-1	: 9843	637-69-4	: 1639	646-04-8	: 9005
626-85-7	: 11580	628-61-5	: 8360	630-18-2	: 10100	637-78-5	: 10299	646-05-9	: 9111
626-86-8	: 6791	628-62-6	: 6348	630-19-3	: 9783	637-87-6	: 1172	646-06-0	: 5204
626-87-9	: 8740	628-63-7	: 227	630-20-6	: 5634	637-88-7	: 4377	646-07-1	: 8889
626-88-0	: 8716	628-64-8	: 4017	630-25-1	: 5524	637-89-8	: 9378	646-13-9	: 8268
626-89-1	: 8939	628-66-0	: 10003	630-51-3	: 3343	637-92-3	: 10040	646-14-0	: 6841
626-90-4	: 3187	628-67-1	: 3348	630-60-4	: 4040	637-98-9	: 3979	646-19-5	: 6371
626-91-5	: 3395	628-68-2	: 5754	630-72-8	: 290	638-00-6	: 11755	646-20-8	: 6392
626-92-6	: 8746	628-71-7	: 6609	630-76-2	: 1831	638-02-8	: 11756	646-30-0	: 8103
626-94-8	: 7068	628-72-8	: 6329	630-95-5	: 7910	638-04-0	: 4367	646-31-1	: 11603
626-95-9	: 8786	628-73-9	: 6840	631-01-6	: 8546	638-07-3	: 3494	650-69-1	: 435
626-96-0	: 8696	628-76-2	: 8747	631-28-7	: 10164	638-08-4	: 8249	653-37-2	: 596
626-97-1	: 8698	628-77-3	: 8758	631-36-7	: 11450	638-10-8	: 3850	655-48-1	: 5575
626-99-3	: 8676	628-80-8	: 8822	631-40-3	: 9833	638-11-9	: 3549	655-86-7	: 9194
627-00-9	: 3482	628-81-9	: 3372	631-57-2	: 10113	638-17-5	: 5265	659-04-1	: 10678
627-02-1	: 10039	628-83-1	: 11699	631-61-8	: 106	638-20-0	: 40	659-49-4	: 828
627-03-2	: 180	628-89-7	: 5698	631-64-1	: 159	638-21-1	: 9519	659-70-1	: 3551
627-04-3	: 188	628-92-2	: 4244	632-05-3	: 3444	638-23-3	: 5017	659-86-9	: 10735
627-05-4	: 3412	628-96-6	: 5574	632-07-5	: 10233	638-25-5	: 8439	660-68-4	: 5428
627-07-6	: 9824	628-97-7	: 6650	632-14-4	: 12032	638-28-8	: 6738	660-88-8	: 8857
627-08-7	: 10076	629-01-6	: 8349	632-15-5	: 11757	638-29-9	: 8990	661-11-0	: 6417
627-09-8	: 10744	629-03-8	: 6757	632-16-6	: 11754	638-37-9	: 3256	661-19-8	: 5279
627-11-2	: 4013	629-04-9	: 6356	632-21-3	: 10487	638-41-5	: 4025	661-68-7	: 5227
627-12-3	: 3969	629-05-0	: 8529	632-22-4	: 12030	638-45-9	: 6831	661-97-2	: 9898
627-13-4	: 8080	629-06-1	: 6360	632-32-6	: 5385	638-46-0	: 3376	666-99-9	: 8102
627-16-7	: 10730	629-08-3	: 6428	632-46-2	: 2582	638-49-3	: 6015	670-54-2	: 5915
627-18-9	: 10344	629-09-4	: 6769	632-50-8	: 1591	638-51-7	: 8089	670-96-2	: 7213
627-19-0	: 9122	629-11-8	: 6796	632-51-9	: 1623	638-53-9	: 11881	671-56-7	: 3866
627-20-3	: 9006	629-14-1	: 5542	632-56-4	: 5633	638-54-0	: 8364	672-13-9	: 571
627-21-4	: 9123	629-15-2	: 5573	632-77-9	: 412	638-58-4	: 11611	672-15-1	: 7134
627-22-5	: 3151	629-16-3	: 8086	632-80-4	: 7348	638-60-8	: 11640	673-31-4	: 2149
627-26-9	: 3802	629-17-4	: 11702	632-82-6	: 414	638-65-3	: 8243	673-32-5	: 2173
627-27-0	: 3861	629-19-6	: 5259	632-83-7	: 385	638-67-5	: 11855	673-79-0	: 10330
627-30-5	: 10348	629-20-9	: 4742	632-89-3	: 12031	638-68-6	: 11819	674-76-0	: 9058
627-31-6	: 9907	629-23-2	: 11634	632-92-8	: 1490	638-73-3	: 7479	674-82-8	: 8598
627-32-7	: 10385	629-27-6	: 8408	632-99-5	: 634	638-95-9	: 12038	675-09-2	: 10792
627-37-2	: 10524	629-30-1	: 6399	633-03-4	: 5444	639-46-3	: 7668	675-20-7	: 9719
627-39-4	: 9787	629-31-2	: 6346	634-19-5	: 5978	639-58-7	: 11541	677-21-4	: 10608
627-40-7	: 10576	629-32-3	: 6421	634-36-6	: 2354	639-81-6	: 5395	677-68-9	: 10149
627-41-8	: 10736	629-33-4	: 6009	634-41-3	: 7723	640-19-7	: 63	678-26-2	: 8807
627-42-9	: 5480	629-35-6	: 7433	634-43-5	: 7724	640-61-9	: 2211	679-37-8	: 7185
627-44-1	: 7434	629-36-7	: 10122	634-47-9	: 11243	641-06-5	: 50	680-31-9	: 9573
627-45-2	: 5994	629-37-8	: 8412	634-66-2	: 2259	641-16-7	: 9476	681-06-1	: 9550

681-84-5	: 11508	699-08-1	: 1657	764-96-5	: 11973	779-84-0	: 9450	827-52-1	: 1248
682-01-9	: 11510	699-09-2	: 2296	764-97-6	: 11972	780-69-8	: 11481	827-54-3	: 7882
682-30-4	: 9535	699-17-2	: 3659	765-03-7	: 5343	781-35-1	: 10432	828-01-3	: 2133
683-08-9	: 9539	699-30-9	: 6099	765-05-9	: 5051	782-05-8	: 1457	829-83-4	: 443
683-50-1	: 9777	700-12-9	: 2097	765-09-3	: 11872	782-06-9	: 1456	829-84-5	: 9516
683-51-2	: 10511	700-75-4	: 769	765-19-3	: 8656	784-50-9	: 5969	830-13-7	: 4211
683-68-1	: 5521	700-88-9	: 1254	765-27-5	: 5366	786-19-6	: 4037	830-89-7	: 7229
683-72-7	: 51	701-57-5	: 2044	765-30-0	: 4953	790-69-2	: 11802	834-12-8	: 341
684-16-2	: 10441	701-97-3	: 4468	765-34-4	: 8603	791-28-6	: 9518	835-11-0	: 7596
684-84-4	: 3353	702-28-3	: 4969	765-42-4	: 4979	791-29-7	: 11444	835-64-3	: 9218
684-93-5	: 12019	702-54-5	: 8573	765-43-5	: 5796	791-31-1	: 11504	836-30-6	: 826
685-63-2	: 3164	703-32-2	: 11252	765-47-9	: 4918	793-23-7	: 995	837-45-6	: 9165
685-73-4	: 6199	703-61-7	: 11248	765-58-2	: 11730	799-53-1	: 5239	838-41-5	: 8579
685-87-0	: 9928	703-80-0	: 5819	765-85-5	: 4176	804-63-7	: 4132	838-45-9	: 1697
685-91-6	: 53	704-01-8	: 1286	765-87-7	: 4375	811-93-8	: 9878	838-95-9	: 722
688-14-2	: 7414	704-38-1	: 7617	766-00-7	: 4811	811-97-2	: 5640	841-73-6	: 11081
688-52-8	: 7108	704-79-0	: 3854	766-05-2	: 4328	812-01-1	: 4086	842-07-9	: 8037
688-71-1	: 3127	706-78-5	: 4932	766-07-4	: 7177	812-03-3	: 10141	846-48-0	: 343
688-74-4	: 3124	706-79-6	: 4917	766-08-5	: 11442	813-76-3	: 9525	848-53-3	: 5140
688-84-6	: 10659	708-06-5	: 7754	766-09-6	: 9667	813-78-5	: 9559	865-71-4	: 8853
689-89-4	: 6703	708-76-9	: 567	766-20-1	: 5189	814-49-3	: 9574	868-18-8	: 3314
689-97-4	: 3889	708-82-7	: 10629	766-39-2	: 6097	814-67-5	: 10030	868-77-9	: 10661
690-02-8	: 9159	709-09-1	: 1445	766-46-1	: 1040	814-68-6	: 10714	868-85-9	: 9588
690-08-4	: 9039	709-98-8	: 10200	766-51-8	: 1178	814-78-8	: 3879	869-29-4	: 10563
690-37-9	: 8954	710-43-0	: 4968	766-53-0	: 2992	815-17-8	: 3510	869-50-1	: 9996
690-92-6	: 7029	711-79-5	: 5829	766-77-8	: 11418	815-24-7	: 8988	870-23-5	: 10600
690-93-7	: 7028	712-48-1	: 454	766-79-0	: 9684	815-58-7	: 10718	870-63-3	: 3717
691-37-2	: 9054	713-36-0	: 2014	766-81-4	: 1041	815-66-7	: 6975	870-74-6	: 8977
691-38-3	: 9059	716-79-0	: 2384	766-84-7	: 2748	816-11-5	: 3517	871-28-3	: 9112
691-83-8	: 3788	717-21-5	: 10710	766-90-5	: 2164	816-39-7	: 10424	871-83-0	: 8140
692-24-0	: 7046	717-74-8	: 2373	766-92-7	: 2042	816-79-5	: 9045	871-84-1	: 8342
692-35-3	: 9854	719-79-9	: 1120	766-93-8	: 5990	817-91-4	: 6493	872-05-9	: 5082
692-42-2	: 441	723-46-6	: 2198	766-94-9	: 1646	817-95-8	: 181	872-10-6	: 8843
692-45-5	: 6006	723-62-6	: 370	766-96-1	: 1042	818-23-5	: 8652	872-31-1	: 11727
692-50-2	: 3914	726-42-1	: 765	767-03-3	: 4904	818-38-2	: 8772	872-50-4	: 11185
692-86-4	: 11977	726-44-3	: 2122	767-10-2	: 11146	818-92-8	: 10570	872-53-7	: 4780
693-02-7	: 7114	732-11-6	: 9515	767-12-4	: 4516	819-07-8	: 5668	872-55-9	: 11759
693-05-0	: 10112	732-26-3	: 9498	767-54-4	: 4570	819-44-3	: 7107	872-85-5	: 10889
693-13-0	: 9815	737-89-3	: 7726	767-58-8	: 7255	819-93-2	: 3502	872-93-5	: 11789
693-23-2	: 5299	741-58-2	: 517	767-59-9	: 7275	821-06-7	: 3742	873-32-5	: 2747
693-39-0	: 6429	742-20-1	: 2853	767-60-2	: 7277	821-07-8	: 6992	873-49-4	: 1256
693-54-9	: 5075	747-90-0	: 4102	767-88-4	: 10884	821-08-9	: 6713	873-66-5	: 2163
693-58-3	: 8121	750-90-3	: 4131	767-99-7	: 2029	821-09-0	: 9088	873-76-7	: 1849
693-61-8	: 11968	753-89-9	: 9853	768-00-3	: 2025	821-10-3	: 3905	873-94-9	: 4622
693-62-9	: 11970	754-05-2	: 11425	768-22-9	: 7282	821-11-4	: 3792	874-23-7	: 4572
693-65-2	: 8835	755-25-9	: 7032	768-32-1	: 11493	821-25-0	: 6377	874-35-1	: 7258
693-72-1	: 8295	756-79-6	: 9540	768-33-2	: 11374	821-55-6	: 8172	874-41-9	: 1671
693-89-0	: 4928	757-58-4	: 11645	768-39-8	: 3115	821-67-0	: 5132	874-60-2	: 2904
693-93-6	: 8583	759-05-7	: 3556	768-49-0	: 2024	821-76-1	: 6379	874-63-5	: 1918
693-95-8	: 11879	759-36-4	: 9964	768-52-5	: 790	821-95-4	: 11966	874-79-3	: 2172
693-98-1	: 7209	759-73-9	: 12015	768-56-9	: 1112	821-96-5	: 11969	874-90-8	: 2762
694-05-3	: 11020	759-94-4	: 5372	768-66-1	: 9711	821-98-7	: 11971	875-30-9	: 7298
694-28-0	: 4880	760-20-3	: 9053	768-94-5	: 11859	821-99-8	: 128	875-40-1	: 2262
694-31-5	: 10826	760-21-4	: 8825	769-06-2	: 849	822-06-0	: 6770	875-51-4	: 654
694-35-9	: 4925	760-23-6	: 3745	769-25-5	: 1647	822-23-1	: 221	875-59-2	: 5828
694-48-4	: 10825	760-40-7	: 10408	769-68-6	: 951	822-28-6	: 6638	876-05-1	: 4793
694-53-1	: 11446	760-55-4	: 9971	769-92-6	: 715	822-35-5	: 4198	877-09-8	: 2261
694-59-7	: 10994	760-67-8	: 6987	770-09-2	: 11494	822-36-6	: 7210	877-11-2	: 2086
694-80-4	: 1004	760-78-1	: 8217	770-35-4	: 10399	822-50-4	: 4802	877-43-0	: 11257
694-81-5	: 11057	760-79-2	: 3203	771-56-2	: 2094	822-66-2	: 4695	877-44-1	: 2344
694-85-9	: 11035	760-93-0	: 10649	771-61-9	: 9442	822-67-3	: 4694	877-48-5	: 9314
695-06-7	: 6150	761-06-8	: 10015	771-62-0	: 2304	822-84-4	: 11776	877-53-2	: 9313
695-12-5	: 4388	762-21-0	: 3909	771-97-1	: 7799	822-85-5	: 4574	877-65-6	: 1865
695-28-3	: 4653	762-42-5	: 3910	771-98-2	: 1246	822-87-7	: 4577	877-82-7	: 9316
695-34-1	: 10867	762-49-2	: 5462	771-99-3	: 9695	823-17-6	: 4654	878-93-3	: 8064
695-53-4	: 8584	762-50-5	: 5476	772-00-9	: 5196	823-40-5	: 1277	879-97-0	: 9327
695-84-1	: 9348	762-62-9	: 9033	772-01-0	: 5195	823-76-7	: 5794	880-09-1	: 9683
695-98-7	: 11025	762-63-0	: 9040	772-17-8	: 2138	824-11-3	: 11909	880-93-3	: 7925
696-28-6	: 462	762-72-1	: 11495	773-76-2	: 11314	824-22-6	: 7257	881-03-8	: 7926
696-29-7	: 4431	763-29-1	: 9052	773-82-0	: 2767	824-42-0	: 575	881-67-4	: 1444
696-30-0	: 10986	763-30-4	: 8671	774-40-3	: 909	824-47-5	: 1193	882-25-7	: 4148
696-41-3	: 581	763-69-9	: 10282	775-12-2	: 11420	824-55-5	: 1186	882-33-7	: 5258
696-44-6	: 711	763-89-3	: 9096	775-51-9	: 4277	824-63-5	: 7256	883-93-2	: 2864
696-59-3	: 6169	763-93-9	: 7086	775-56-4	: 11407	824-72-6	: 9548	883-99-8	: 7775
696-62-8	: 1776	764-01-2	: 3919	776-35-2	: 9171	824-94-2	: 1197	886-50-0	: 11594
696-63-9	: 2300	764-13-6	: 6692	776-74-9	: 1058	825-25-2	: 4881	886-58-8	: 1979
696-71-9	: 4739	764-33-0	: 7123	776-75-0	: 9634	825-51-4	: 8024	886-74-8	: 10001
697-11-0	: 4200	764-35-2	: 7115	776-76-1	: 11438	825-94-5	: 11457	886-77-1	: 8682
697-86-9	: 641	764-42-1	: 3762	777-22-0	: 1981	826-18-6	: 2104	888-71-1	: 8564
698-00-0	: 643	764-47-6	: 10033	777-37-7	: 1218	826-36-8	: 9726	897-78-9	: 4573
698-01-1	: 662	764-48-7	: 5716	778-22-3	: 1976	826-73-3	: 2407	900-95-8	: 11539
698-27-1	: 576	764-49-8	: 11712	778-24-5	: 11416	826-74-4	: 7881	910-86-1	: 11800
698-63-5	: 6046	764-56-7	: 6342	778-25-6	: 11502	826-77-7	: 11282	917-61-3	: 4164
698-69-1	: 663	764-60-3	: 7125	778-28-9	: 2230	826-81-3	: 11321		
698-87-3	: 1607	764-73-8	: 8343	778-66-5	: 2166	827-08-7	: 1312		
698-88-4	: 1372	764-85-2	: 8177	779-02-2	: 425	827-15-6	: 2092		
699-02-5	: 1610	764-93-2	: 5097	779-51-1	: 1971	827-16-7	: 11841		

CAS	No.	CAS	No.	CAS	No.	CAS	No.	CAS	No.
917-92-0	3907	932-16-1	5849	982-43-4	9657	1070-87-7	8841	1119-49-9	38
917-93-1	9779	932-32-1	670	992-94-9	11437	1071-26-7	6383	1119-51-3	9010
918-00-3	10488	932-39-8	4797	993-00-0	11378	1071-73-4	8974	1119-60-4	6564
918-07-0	3254	932-40-1	4798	993-07-7	11489	1071-81-4	6849	1119-65-9	6610
918-79-6	8692	932-43-4	4834	993-13-5	9538	1071-83-6	6276	1120-07-6	8118
918-82-1	9128	932-44-5	4835	994-05-8	3396	1071-94-9	6589	1120-16-7	5289
918-84-3	8732	932-53-6	11831	994-25-2	8726	1072-05-5	6387	1120-21-4	11938
919-12-0	9132	932-56-9	4235	994-30-9	11388	1072-16-8	8378	1120-28-1	5357
919-19-7	9530	932-66-1	5793	994-49-0	5235	1072-21-5	6748	1120-29-2	5287
919-30-2	9828	932-77-4	1007	996-98-5	5562	1072-36-2	8224	1120-36-1	11636
919-86-8	5103	932-96-7	675	997-49-9	11698	1072-39-5	5281	1120-48-5	8354
919-94-8	3375	933-05-1	4865	998-29-8	11500	1072-43-1	11696	1120-56-5	4193
920-37-6	10584	933-21-1	4728	998-30-1	11477	1072-44-2	500	1120-59-8	11750
920-46-7	10715	933-36-8	4616	998-40-3	9520	1072-52-2	499	1120-62-3	4929
921-08-4	3487	933-48-2	4569	998-41-4	11454	1072-53-3	5199	1120-64-5	8576
921-13-1	7480	933-67-5	7317	998-93-6	6358	1072-83-9	5820	1120-71-4	8570
921-47-1	6863	933-75-5	9483	998-95-8	6363	1072-91-9	10840	1120-72-5	4885
922-28-1	6389	933-78-8	9482	999-21-3	3782	1072-98-6	10861	1120-73-6	4944
922-55-4	5016	933-88-0	2902	999-52-0	6362	1073-06-9	1043	1120-87-2	10877
922-59-8	9134	933-98-2	1670	999-55-3	10694	1073-07-0	4732	1120-97-4	5193
922-62-3	9057	934-00-9	1525	999-78-0	9130	1073-23-0	10945	1121-07-9	11160
922-65-6	8681	934-34-9	2868	999-79-1	9127	1073-67-2	1151	1121-18-2	4710
923-06-8	3300	934-53-2	1189	999-81-5	5442	1074-11-9	1376	1121-24-0	9376
923-26-2	10682	934-74-7	1672	999-97-3	11363	1074-17-5	2036	1121-37-5	7607
923-99-9	9594	934-80-5	1875	1000-49-3	11371	1074-43-7	2037	1121-55-7	10953
924-16-3	3212	935-05-7	8097	1000-50-6	11369	1074-55-1	2038	1121-60-4	10887
924-41-4	6706	935-07-9	18	1000-86-8	8665	1074-58-4	275	1121-92-2	502
924-43-6	8093	935-08-0	5134	1001-53-2	28	1074-92-6	1464	1121-95-5	4875
924-50-5	3852	935-12-6	6164	1001-89-4	6381	1075-22-5	7254	1122-20-9	4705
924-52-7	8077	935-13-7	6161	1002-28-4	7126	1075-38-3	1465	1122-25-4	4594
925-15-5	3321	935-31-9	4208	1002-69-3	5036	1075-74-7	2106	1122-39-0	11027
925-78-0	8173	935-95-5	9471	1002-84-2	8645	1076-26-2	8031	1122-42-5	1773
926-02-3	10036	936-48-1	10824	1003-03-8	4771	1076-55-7	9391	1122-54-9	5869
926-26-1	3296	936-53-8	10823	1003-09-4	11726	1076-61-5	7994	1122-60-7	4454
926-55-6	9116	936-58-3	1911	1003-10-7	11794	1076-66-0	1625	1122-62-9	5867
926-56-7	8670	936-72-1	2026	1003-29-8	11114	1077-16-3	1763	1122-69-6	10963
926-63-6	9807	937-14-4	1127	1003-30-1	6107	1078-04-2	7260	1122-81-2	11016
926-64-7	281	937-20-2	5784	1003-31-2	11732	1078-19-9	8048	1122-82-3	4417
926-65-8	10034	937-32-6	2179	1003-56-1	11036	1078-21-3	2128	1122-83-4	844
926-82-9	6390	937-33-7	716	1003-64-1	4391	1078-30-4	11228	1122-91-4	528
927-49-1	11962	938-16-9	10431	1003-73-2	10988	1078-71-3	1745	1123-00-8	4775
927-54-8	7006	938-25-0	7795	1003-78-7	11784	1079-71-6	427	1123-09-7	4707
927-62-8	3215	938-33-0	11276	1003-90-3	11141	1080-16-6	5130	1123-27-9	5821
927-68-4	5686	938-39-6	11326	1004-36-0	10793	1080-32-6	9545	1123-34-8	4555
927-73-1	3724	938-45-4	10446	1004-66-6	1919	1081-34-1	11600	1123-84-8	1371
927-74-2	3920	938-73-8	615	1004-77-9	4599	1081-75-0	2121	1123-85-9	1608
927-80-0	5942	939-23-1	10999	1005-64-7	1109	1081-77-2	2064	1123-86-0	10423
928-40-5	6795	939-26-4	7743	1005-67-0	7684	1083-56-3	1099	1123-95-1	11105
928-45-0	8072	939-27-5	7886	1006-06-0	1874	1086-80-2	2472	1123-96-2	10959
928-49-4	7116	939-48-0	2693	1006-31-1	2263	1088-11-5	2411	1124-05-6	1362
928-50-7	9019	939-52-6	3650	1007-26-7	1484	1093-58-9	354	1124-07-8	9293
928-51-8	3604	939-55-9	5696	1007-32-5	3677	1095-90-5	6516	1124-20-5	1985
928-56-3	5886	939-83-3	2766	1008-88-4	10998	1111-74-6	11414	1124-27-2	4442
928-65-4	11464	939-87-7	4957	1008-89-5	10997	1111-97-3	3901	1124-33-0	10993
928-68-7	6521	940-41-0	11474	1009-14-9	8986	1112-37-4	9529	1125-60-6	7372
928-80-3	5076	940-49-8	2033	1009-87-2	2137	1112-39-6	11412	1125-74-2	1547
928-92-7	7072	940-54-5	3189	1009-81-0	1754	1112-48-7	11368	1125-78-6	8044
928-94-9	7070	941-69-5	11123	1009-93-4	5001	1112-55-6	11449	1125-80-0	7382
928-95-0	7069	941-98-0	5857	1010-48-6	2130	1112-56-7	11543	1126-09-6	9643
928-96-1	7071	942-01-8	3989	1012-72-2	990	1113-12-8	11417	1126-18-7	4575
929-55-5	8348	942-06-3	7344	1012-84-6	2712	1113-30-0	3113	1126-46-1	2543
929-61-3	8402	942-92-7	6983	1013-08-7	9187	1113-68-4	25	1126-75-6	2031
929-62-4	8401	943-14-6	2528	1013-83-8	7769	1113-74-2	8968	1126-78-9	656
929-77-1	5278	943-15-7	1991	1013-88-3	1835	1114-09-6	3302	1126-79-0	1114
930-02-9	8240	943-27-1	5808	1013-96-3	10676	1114-51-8	9793	1127-76-0	7885
930-18-7	4972	944-22-9	5986	1016-77-9	7601	1114-71-2	3976	1128-08-1	4950
930-21-2	496	945-51-7	2183	1016-94-0	9195	1114-76-7	3200	1128-54-7	10838
930-22-3	8813	947-73-9	9162	1017-56-7	7583	1115-11-3	3698	1128-76-3	2538
930-27-8	6131	947-84-2	3021	1024-57-3	6301	1115-12-4	9915	1129-47-1	10293
930-28-9	4789	947-91-1	877	1025-15-6	11842	1115-30-6	3295	1129-50-6	3208
930-30-3	4941	948-32-3	2372	1048-08-4	11452	1116-24-1	58	1129-89-1	4216
930-36-9	10834	948-65-2	7320	1056-77-5	6175	1116-39-8	3123	1131-16-4	10829
930-37-0	8620	948-97-0	1966	1066-27-9	11435	1116-40-1	9820	1131-18-6	10814
930-55-2	11170	948-98-1	1146	1066-45-1	11540	1116-81-6	3121	1131-60-8	9272
930-66-5	4642	948-99-2	1147	1067-08-9	8814	1117-19-7	9945	1131-62-0	5805
930-68-7	4703	949-87-1	5139	1067-20-5	8756	1117-55-1	8431	1131-63-1	7779
930-87-0	11142	950-37-8	7645	1068-19-5	6391	1117-59-5	8882	1132-21-4	2577
930-89-2	4815	951-86-0	1365	1068-57-1	195	1118-09-8	11491	1132-39-4	2175
930-90-5	4816	952-47-6	9426	1068-87-7	8812	1118-68-9	6261	1132-66-7	1765
931-10-2	4293	952-97-6	2058	1068-90-2	9922	1118-71-4	6404	1133-80-8	5968
931-20-4	9721	955-83-9	6100	1069-53-0	6864	1118-84-9	3577	1134-23-2	4168
931-56-6	4425	957-51-7	5219	1070-14-0	6705	1119-06-8	6465	1134-62-9	7747
931-77-1	4319	959-33-1	10713	1070-32-2	6492	1119-19-3	3895	1135-12-2	839
931-87-3	4752	961-11-5	6202	1070-34-4	3335	1119-33-1	6245	1140-14-3	7614
931-89-5	4751	961-71-7	5506	1070-71-9	10737	1119-40-0	8775	1140-16-5	7628
931-97-5	4329	964-68-1	2533	1070-78-6	10142	1119-46-6	8866	1142-70-7	11080
932-01-4	4519	976-71-6	9736	1070-83-3	3508			1142-97-8	2236

1143-38-0 : 430	1214-24-0 : 12053	1501-27-5 : 8780	1574-40-9 : 9114	1639-09-4 : 6436	
1144-74-7 : 7636	1241-94-7 : 9560	1501-60-6 : 8668	1574-41-0 : 8662	1640-89-7 : 4812	
1146-98-1 : 7265	1260-17-9 : 373	1502-05-2 : 4204	1575-61-7 : 8992	1641-40-3 : 2413	
1146-99-2 : 7266	1300-94-3 : 9425	1502-06-3 : 4205	1576-13-2 : 9464	1643-49-8 : 5970	
1148-11-4 : 11148	1322-67-4 : 9178	1503-53-3 : 2481	1576-95-0 : 9091	1646-26-0 : 5775	
1149-24-2 : 10932	1330-61-6 : 10646	1504-54-7 : 3872	1576-96-1 : 9090	1648-86-4 : 335	
1151-15-1 : 2667	1400-62-0 : 8549	1504-58-1 : 10745	1577-22-6 : 7051	1647-08-1 : 7019	
1154-59-2 : 610	1405-86-3 : 6230	1504-70-7 : 10814	1578-21-8 : 10628	1647-16-1 : 5024	
1165-39-5 : 6184	1405-87-4 : 508	1504-75-2 : 10517	1579-40-4 : 2078	1647-23-0 : 3235	
1165-48-6 : 2819	1407-14-3 : 6293	1511-62-2 : 7469	1582-09-8 : 11901	1647-26-3 : 4316	
1174-72-7 : 11509	1420-07-1 : 9328	1515-76-0 : 3176	1585-06-4 : 1653	1649-08-7 : 5531	
1184-60-7 : 10569	1421-14-3 : 894	1515-80-6 : 6702	1585-07-5 : 1037	1653-19-6 : 3158	
1185-33-7 : 3615	1423-60-5 : 3923	1515-95-3 : 1684	1587-04-8 : 2023	1653-31-2 : 11884	
1185-55-3 : 11487	1435-50-3 : 1014	1516-80-9 : 11431	1587-20-8 : 10170	1653-32-3 : 11632	
1186-31-8 : 6573	1435-53-6 : 1300	1517-05-1 : 5681	1587-26-4 : 3753	1653-34-5 : 8647	
1186-53-4 : 8840	1435-55-8 : 4124	1518-75-8 : 11779	1587-29-7 : 3752	1653-40-3 : 6494	
1186-73-8 : 7553	1436-43-7 : 11223	1518-86-1 : 9215	1587-41-3 : 7501	1653-57-2 : 8084	
1187-03-7 : 12029	1441-87-8 : 2900	1519-36-4 : 402	1588-44-9 : 1782	1654-86-0 : 5062	
1187-13-9 : 3837	1445-23-4 : 9324	1519-59-1 : 11002	1589-47-5 : 10386	1655-70-5 : 2075	
1187-58-2 : 9799	1445-45-0 : 5669	1520-21-4 : 735	1589-60-2 : 4552	1655-71-6 : 1929	
1188-33-6 : 7443	1445-73-4 : 9724	1520-45-2 : 1698	1590-08-5 : 8049	1656-48-0 : 10114	
1189-71-5 : 11585	1446-34-0 : 3006	1520-46-3 : 2021	1591-31-7 : 3079	1657-52-9 : 1222	
1189-99-7 : 6448	1447-26-3 : 6600	1521-51-3 : 4628	1592-23-0 : 8252	1663-67-8 : 10029	
1190-22-3 : 3278	1450-14-2 : 5228	1522-22-1 : 8802	1592-38-7 : 7909	1666-13-3 : 5225	
1190-34-7 : 6605	1450-31-3 : 7547	1522-46-9 : 3461	1595-04-6 : 1122	1666-96-2 : 11289	
1190-39-2 : 9938	1450-58-4 : 6048	1526-73-4 : 7848	1595-05-7 : 1123	1667-01-2 : 5876	
1190-63-2 : 8258	1450-72-2 : 5827	1527-89-5 : 2761	1595-06-8 : 1994	1667-04-5 : 1243	
1190-83-6 : 6440	1450-81-3 : 8871	1528-30-9 : 4827	1595-11-5 : 1121	1670-46-8 : 4879	
1191-04-4 : 7049	1453-17-4 : 4254	1529-17-5 : 11492	1595-16-0 : 1998	1671-75-6 : 6526	
1191-08-8 : 3368	1453-24-3 : 4660	1529-41-5 : 950	1596-84-5 : 3334	1672-46-4 : 4047	
1191-41-9 : 8283	1453-25-4 : 4245	1530-03-6 : 2171	1599-41-3 : 10180	1674-10-8 : 4648	
1191-43-1 : 6812	1453-58-3 : 10835	1530-05-8 : 1746	1599-67-3 : 5280	1674-30-2 : 1852	
1191-75-9 : 6668	1454-80-4 : 3028	1530-37-9 : 1744	1600-37-9 : 10593	1674-33-5 : 8744	
1191-95-3 : 4197	1454-84-8 : 8104	1532-97-4 : 7374	1601-00-9 : 5847	1674-37-9 : 8480	
1191-96-4 : 4974	1454-85-9 : 6317	1534-26-5 : 11890	1603-40-3 : 10866	1674-38-0 : 5327	
1191-99-7 : 6086	1455-20-5 : 11731	1538-87-0 : 5479	1603-79-8 : 935	1674-56-2 : 10338	
1192-18-3 : 4801	1455-21-6 : 8145	1539-42-0 : 10974	1603-91-4 : 11665	1675-69-0 : 8129	
1192-28-5 : 4892	1457-85-8 : 224	1540-28-9 : 8856	1604-01-9 : 11021	1678-80-4 : 4492	
1192-30-9 : 6036	1458-99-7 : 9023	1540-29-0 : 6872	1604-02-0 : 4872	1678-81-5 : 4489	
1192-37-6 : 4428	1459-00-3 : 1060	1540-34-7 : 8801	1604-11-1 : 3327	1678-82-6 : 4440	
1192-62-7 : 5817	1459-09-2 : 1751	1541-23-7 : 6328	1606-08-2 : 4340	1678-91-7 : 4390	
1192-88-7 : 4631	1459-10-5 : 2266	1547-87-1 : 99	1609-47-8 : 5169	1678-92-8 : 4471	
1193-02-8 : 2291	1459-14-9 : 1039	1550-09-0 : 6742	1609-86-5 : 10062	1678-93-9 : 4325	
1193-12-0 : 9651	1459-93-4 : 1347	1551-06-0 : 11125	1609-93-4 : 10618	1678-97-3 : 4488	
1193-16-4 : 4294	1460-16-8 : 4223	1551-09-3 : 11139	1610-24-8 : 4433	1678-98-4 : 4453	
1193-17-5 : 4295	1460-88-4 : 4911	1551-21-9 : 10089	1610-29-3 : 9094	1679-06-7 : 6857	
1193-18-6 : 4711	1460-98-6 : 1948	1551-27-5 : 11778	1611-50-3 : 3942	1679-07-8 : 4852	
1193-35-7 : 6085	1462-03-9 : 4887	1551-32-2 : 11765	1611-52-5 : 3952	1679-09-0 : 3441	
1193-46-0 : 4513	1462-12-0 : 9950	1551-44-6 : 3496	1611-65-0 : 3386	1679-36-3 : 7130	
1193-47-1 : 4580	1462-27-7 : 3652	1551-88-8 : 4581	1613-32-7 : 11292	1679-47-6 : 6145	
1193-70-0 : 4894	1462-34-6 : 5468	1551-89-9 : 4582	1613-34-9 : 11266	1680-51-9 : 7998	
1193-72-2 : 1011	1462-84-6 : 11026	1552-12-1 : 4734	1613-45-2 : 10044	1680-53-1 : 7922	
1193-74-4 : 11044	1463-17-8 : 11259	1552-67-6 : 7053	1613-51-0 : 11795	1680-58-6 : 7927	
1193-81-3 : 4420	1467-05-6 : 1191	1552-95-0 : 8680	1615-14-1 : 7205	1681-22-7 : 7999	
1193-82-4 : 2040	1487-79-4 : 4157	1555-80-2 : 887	1615-70-9 : 8673	1686-41-5 : 4237	
1194-02-1 : 2756	1468-39-9 : 3542	1556-18-9 : 4824	1615-75-4 : 5475	1689-82-3 : 9448	
1194-65-6 : 2752	1470-94-6 : 7285	1558-17-4 : 11058	1615-80-1 : 7153	1689-84-5 : 2751	
1195-31-9 : 4681	1471-03-0 : 10594	1558-24-3 : 11459	1617-18-1 : 3841	1690-75-1 : 9645	
1195-32-0 : 1986	1471-96-1 : 7858	1558-25-4 : 11456	1617-90-9 : 5349	1694-19-5 : 1939	
1195-42-2 : 4303	1472-90-8 : 8238	1558-31-2 : 11391	1618-26-4 : 7461	1695-04-1 : 1938	
1196-31-2 : 4605	1475-13-4 : 1854	1558-33-4 : 11390	1619-28-9 : 4947	1696-20-4 : 7682	
1196-39-0 : 7383	1476-07-9 : 1666	1559-02-0 : 4967	1619-34-7 : 485	1698-60-8 : 4080	
1196-58-3 : 1701	1476-11-5 : 3751	1559-81-5 : 7995	1619-56-3 : 3495	1702-17-6 : 4142	
1196-67-4 : 3701	1477-55-0 : 1433	1560-06-1 : 1111	1619-57-4 : 3504	1703-46-4 : 1862	
1196-79-8 : 7299	1482-15-1 : 9144	1560-09-4 : 1110	1619-62-1 : 9942	1703-52-2 : 6106	
1197-18-8 : 4333	1482-91-3 : 8483	1560-11-8 : 4909	1620-14-0 : 10427	1704-15-0 : 10711	
1197-37-1 : 1272	1483-60-9 : 1473	1560-89-0 : 6310	1622-10-2 : 6373	1706-11-2 : 1920	
1197-40-6 : 6132	1484-08-8 : 3984	1561-10-0 : 6900	1622-32-8 : 5629	1706-73-6 : 1546	
1197-55-3 : 886	1484-12-4 : 3987	1561-20-2 : 10630	1623-14-9 : 9561	1708-32-3 : 11751	
1198-14-7 : 11310	1484-19-1 : 7308	1563-66-2 : 3991	1626-09-1 : 8394	1708-39-0 : 5209	
1198-37-4 : 11255	1484-84-0 : 9663	1563-83-3 : 24	1629-58-9 : 9104	1709-03-1 : 46	
1198-59-0 : 1398	1484-85-1 : 2424	1563-87-7 : 26	1630-77-9 : 5903	1710-98-1 : 2895	
1198-61-4 : 1399	1486-75-5 : 4215	1565-71-5 : 1855	1630-94-0 : 4971	1711-06-4 : 2903	
1198-62-5 : 1400	1487-18-9 : 6102	1565-86-2 : 1905	1631-82-9 : 11381	1711-07-5 : 2898	
1200-03-9 : 1003	1489-47-0 : 8167	1569-01-3 : 10401	1631-84-1 : 11403	1712-64-7 : 8076	
1200-73-3 : 5194	1489-57-2 : 4271	1569-02-4 : 10381	1632-16-2 : 6425	1717-00-6 : 5534	
1201-26-9 : 10442	1489-60-7 : 4201	1569-43-3 : 7076	1632-76-4 : 10856	1718-50-9 : 2098	
1202-08-0 : 5850	1490-05-7 : 7428	1569-44-4 : 7084	1632-83-3 : 2382	1719-53-5 : 11393	
1202-34-2 : 10871	1493-01-2 : 7516	1569-59-1 : 9101	1633-90-5 : 6858	1719-57-9 : 11372	
1203-39-0 : 7845	1493-02-3 : 6019	1569-60-4 : 6580	1634-04-4 : 10072	1719-58-0 : 11376	
1204-78-0 : 3015	1493-03-4 : 7505	1569-69-3 : 4480	1634-09-9 : 7746	1721-89-7 : 11254	
1205-02-3 : 327	1493-27-2 : 1733	1570-45-2 : 10903	1635-02-5 : 7091	1721-93-3 : 7381	
1205-64-7 : 808	1495-50-7 : 4167	1570-64-5 : 9266	1635-61-6 : 677	1723-94-0 : 7690	
1205-71-6 : 679	1498-51-7 : 9577	1573-17-7 : 3913	1637-31-6 : 3173	1725-74-2 : 7035	
1205-91-0 : 1509	1498-65-3 : 9557	1573-28-0 : 6483	1638-22-8 : 9248	1726-12-1 : 2108	
1207-69-8 : 315	1501-04-8 : 2100	1573-58-6 : 3174	1638-26-2 : 4800	1726-14-3 : 1113	
1210-35-1 : 5147	1501-06-0 : 8771	1574-33-0 : 9117	1638-63-7 : 972	1729-67-5 : 10241	

1731-79-9	: 5300	1830-54-2	: 8784	2004-67-3	: 9102	2051-53-8	: 793	2163-00-0	: 6762
1731-84-6	: 8154	1830-95-1	: 3120	2004-69-5	: 9113	2051-60-7	: 3023	2163-44-2	: 9934
1731-88-0	: 11882	1832-53-7	: 9543	2004-70-8	: 8661	2051-61-8	: 3024	2163-48-6	: 9981
1731-92-6	: 6316	1833-51-8	: 11379	2006-14-6	: 10675	2051-62-9	: 3025	2164-08-1	: 4900
1732-08-7	: 6395	1836-75-5	: 8081	2008-41-5	: 11586	2051-78-7	: 3587	2164-09-2	: 10520
1732-09-8	: 8388	1838-59-1	: 6017	2009-74-7	: 6598	2051-84-5	: 783	2164-17-2	: 12007
1732-10-1	: 8134	1838-77-3	: 7085	2009-83-8	: 6916	2051-90-3	: 1389	2164-19-4	: 4292
1733-12-6	: 9368	1839-88-9	: 4490	2016-42-4	: 11612	2051-97-0	: 11136	2167-39-7	: 8594
1738-25-6	: 10101	1843-05-6	: 7621	2016-57-1	: 5032	2051-98-1	: 4	2168-93-6	: 3422
1738-36-9	: 284	1845-30-3	: 2975	2018-90-8	: 7907	2052-01-9	: 10209	2170-03-8	: 6095
1739-84-0	: 7200	1846-70-4	: 8203	2021-28-5	: 2131	2052-06-4	: 642	2171-96-2	: 11436
1740-97-2	: 9414	1850-14-0	: 7543	2025-40-3	: 10625	2052-07-5	: 3016	2173-56-0	: 8913
1741-83-9	: 8826	1855-63-6	: 4630	2027-17-0	: 7921	2052-14-4	: 2614	2174-64-3	: 1526
1745-16-0	: 2105	1859-76-3	: 7626	2027-19-2	: 7958	2052-15-5	: 8904	2175-91-9	: 4768
1745-81-9	: 9457	1860-21-5	: 247	2028-63-9	: 3921	2055-40-5	: 1643	2176-62-7	: 10995
1746-11-8	: 2443	1860-27-1	: 3399	2029-94-9	: 538	2057-43-4	: 10996	2176-98-9	: 11546
1746-13-0	: 2167	1861-21-8	: 11099	2031-67-6	: 11479	2057-49-0	: 11007	2177-32-4	: 7093
1746-77-6	: 3955	1861-32-1	: 5180	2032-35-1	: 5459	2065-23-8	: 230	2177-47-1	: 7276
1754-62-7	: 10686	1861-40-1	: 509	2032-65-7	: 7646	2065-70-5	: 7813	2180-43-0	: 6949
1755-01-7	: 7577	1863-63-4	: 2509	2033-24-1	: 5190	2077-13-6	: 1190	2180-68-9	: 8674
1755-05-1	: 8686	1864-92-2	: 738	2035-94-1	: 1602	2078-54-8	: 9226	2180-69-0	: 8675
1758-10-7	: 2027	1866-31-5	: 10692	2036-41-1	: 11072	2079-95-0	: 11621	2182-66-3	: 11419
1758-33-4	: 8608	1867-37-4	: 9917	2037-31-2	: 2294	2081-44-9	: 10787	2189-60-8	: 2067
1758-85-6	: 1421	1868-53-7	: 7498	2038-03-1	: 7689	2082-59-9	: 8858	2196-23-8	: 6529
1758-88-9	: 1674	1871-52-9	: 4749	2039-82-9	: 1028	2082-76-0	: 5060	2197-37-7	: 8225
1759-53-1	: 4958	1871-57-4	: 10537	2039-83-0	: 1368	2083-91-2	: 11361	2198-20-1	: 4207
1759-58-6	: 4804	1871-58-5	: 10178	2039-85-2	: 1150	2084-19-7	: 8845	2198-61-0	: 6897
1759-64-4	: 4652	1871-76-7	: 975	2039-86-3	: 1027	2084-45-9	: 1363	2198-72-3	: 6508
1759-81-5	: 4930	1871-96-1	: 5041	2039-87-4	: 1149	2085-88-3	: 8622	2198-75-6	: 7808
1759-88-2	: 11439	1873-25-2	: 3605	2039-88-5	: 1026	2086-83-1	: 2467	2199-51-1	: 11115
1760-46-9	: 937	1873-29-6	: 10061	2039-89-6	: 1628	2090-14-4	: 4118	2199-52-2	: 11116
1761-71-3	: 4299	1873-54-7	: 11267	2039-93-2	: 1968	2090-89-3	: 5748	2203-34-1	: 3607
1768-31-6	: 10478	1873-92-3	: 11402	2040-07-5	: 5875	2091-29-4	: 6669	2203-35-2	: 3603
1772-10-7	: 1996	1875-88-3	: 1597	2040-14-4	: 10466	2093-20-1	: 5673	2203-80-7	: 7120
1776-37-0	: 7241	1877-72-1	: 2550	2040-95-1	: 4778	2095-57-0	: 11065	2206-38-4	: 1253
1777-82-8	: 1853	1878-68-8	: 888	2040-96-2	: 4849	2100-17-6	: 9001	2206-48-6	: 1250
1779-81-3	: 11664	1879-06-7	: 5845	2042-37-7	: 2744	2101-86-2	: 982	2206-89-5	: 10620
1780-31-0	: 11052	1879-09-0	: 9325	2043-38-1	: 2856	2101-88-4	: 977	2207-01-4	: 4365
1781-78-8	: 4755	1883-32-5	: 1615	2043-61-0	: 4331	2103-64-2	: 11526	2207-03-6	: 4370
1790-22-3	: 3450	1885-14-9	: 4026	2044-08-8	: 4627	2104-64-5	: 9551	2207-04-7	: 4371
1792-41-2	: 2379	1885-29-6	: 2741	2045-79-6	: 1577	2106-54-9	: 6910	2207-27-4	: 4770
1792-81-0	: 4372	1885-38-7	: 10587	2047-21-4	: 5877	2108-92-1	: 4358	2207-50-3	: 8574
1794-84-9	: 7485	1888-57-9	: 6966	2049-66-3	: 9951	2109-66-2	: 7692	2210-28-8	: 10672
1795-01-3	: 11760	1888-71-7	: 10571	2049-67-4	: 9041	2110-78-3	: 10280	2211-67-8	: 3281
1795-04-6	: 11792	1888-85-3	: 3370	2049-73-2	: 1407	2111-75-3	: 4633	2211-68-9	: 3727
1795-05-7	: 11793	1896-62-4	: 3882	2049-74-3	: 5719	2113-47-5	: 35	2211-69-0	: 3728
1795-09-1	: 11787	1897-45-6	: 1317	2049-80-1	: 9980	2113-51-1	: 3077	2211-70-3	: 3722
1795-15-9	: 4460	1899-24-7	: 6043	2049-94-7	: 1954	2113-57-7	: 3017	2212-67-1	: 7661
1795-16-0	: 4344	1904-98-9	: 10754	2049-95-8	: 1480	2113-58-8	: 3086	2213-08-3	: 332
1795-17-1	: 4382	1912-21-6	: 10319	2050-01-3	: 10302	2114-00-3	: 10419	2213-23-2	: 6385
1795-18-2	: 4475	1912-24-9	: 476	2050-03-5	: 1488	2114-39-8	: 1087	2213-32-3	: 9030
1795-20-6	: 4843	1917-15-3	: 6060	2050-04-6	: 2109	2114-42-3	: 4470	2213-43-6	: 9624
1795-21-7	: 4790	1918-00-9	: 2567	2050-07-9	: 8983	2116-62-3	: 11003	2214-14-4	: 1955
1795-22-8	: 4850	1918-02-1	: 10896	2050-08-0	: 2649	2116-65-6	: 11006	2216-12-8	: 2053
1795-26-2	: 4497	1918-16-7	: 9771	2050-14-8	: 9216	2121-66-6	: 10140	2216-15-1	: 698
1795-27-3	: 4496	1921-70-6	: 8644	2050-20-6	: 6394	2122-70-5	: 7735	2216-30-0	: 6386
1795-48-8	: 10060	1928-30-9	: 11856	2050-23-9	: 8387	2123-27-5	: 1373	2216-32-2	: 6408
1796-05-0	: 1494	1928-38-7	: 172	2050-24-0	: 1419	2123-28-6	: 1367	2216-33-3	: 8409
1804-15-5	: 9784	1929-73-3	: 170	2050-32-0	: 4256	2130-23-6	: 8504	2216-34-4	: 8410
1805-61-4	: 9404	1929-77-7	: 12047	2050-33-1	: 4255	2131-18-2	: 2089	2216-36-6	: 8124
1806-29-7	: 3064	1929-82-4	: 8071	2050-43-3	: 56	2131-38-6	: 8013	2216-67-3	: 7722
1807-36-9	: 2837	1931-60-8	: 3367	2050-46-6	: 1406	2131-43-3	: 8002	2216-68-4	: 7721
1809-02-5	: 3155	1932-04-3	: 9661	2050-47-7	: 2072	2131-55-7	: 1177	2216-69-5	: 7913
1809-05-8	: 8820	1937-54-8	: 8113	2050-48-8	: 2238	2132-80-1	: 3047	2216-70-8	: 8328
1809-10-5	: 8712	1937-62-8	: 8308	2050-60-4	: 5558	2136-81-4	: 1238	2216-81-1	: 10264
1809-19-4	: 9532	1938-17-6	: 1257	2050-61-5	: 5557	2136-89-2	: 1237	2216-87-7	: 11959
1809-26-3	: 9014	1938-22-3	: 2063	2050-67-1	: 3040	2136-95-0	: 2085	2216-92-4	: 6271
1812-51-7	: 3072	1940-18-7	: 4522	2050-68-2	: 3041	2138-22-9	: 1505	2216-94-6	: 10741
1814-88-6	: 10128	1942-45-6	: 8532	2050-69-3	: 7807	2138-43-4	: 1535	2217-06-3	: 2288
1817-47-6	: 1978	1942-46-7	: 5100	2050-73-9	: 7811	2141-62-0	: 10103	2217-07-4	: 730
1817-57-8	: 3924	1943-16-4	: 7487	2050-74-0	: 7812	2142-01-0	: 7367	2217-14-3	: 3310
1817-73-8	: 646	1943-54-0	: 9291	2050-76-2	: 8026	2142-06-5	: 11163	2217-41-6	: 7731
1818-07-1	: 2069	1945-53-5	: 9166	2050-77-3	: 5053	2142-63-4	: 5780	2217-45-0	: 847
1820-09-3	: 2976	1951-11-7	: 3684	2050-78-4	: 8073	2142-73-6	: 5810	2217-60-9	: 1538
1821-02-9	: 8902	1967-25-5	: 11998	2050-87-5	: 11917	2146-37-4	: 4813	2219-66-1	: 3904
1821-12-1	: 1100	1969-43-3	: 6406	2050-89-7	: 3030	2146-38-5	: 4924	2219-82-1	: 9332
1821-36-9	: 680	1969-73-9	: 875	2050-92-2	: 8708	2150-18-7	: 10962	2224-37-5	: 8951
1821-39-2	: 842	1971-46-6	: 7322	2050-95-5	: 3835	2150-38-1	: 2579	2227-13-6	: 11648
1822-74-8	: 5910	1973-22-4	: 1035	2051-00-5	: 5555	2155-30-8	: 10273	2230-70-8	: 4619
1822-86-2	: 3149	1974-04-5	: 6357	2051-04-9	: 5243	2155-42-2	: 1370	2231-57-4	: 4033
1823-91-2	: 959	1975-44-6	: 7787	2051-10-7	: 4116	2157-31-5	: 8792	2233-00-3	: 10604
1824-81-3	: 10868	1975-78-6	: 5057	2051-25-4	: 5023	2157-50-8	: 10483	2233-29-6	: 11024
1825-30-5	: 7810	1982-47-4	: 4088	2051-30-1	: 8377	2158-55-6	: 4228	2234-13-1	: 7933
1825-31-6	: 7809	1985-57-5	: 1455	2051-31-2	: 5072	2159-75-3	: 5123	2234-16-4	: 5799
1825-62-3	: 11430	1985-59-7	: 7980	2051-32-3	: 8459	2162-92-7	: 6761	2234-20-0	: 1626
1825-82-7	: 11462	1985-97-3	: 1691	2051-33-4	: 6948	2162-98-3	: 5040	2237-30-1	: 2742
1826-14-8	: 11676	1990-29-0	: 338	2051-49-2	: 6874	2162-99-4	: 8371	2239-78-3	: 6657

CAS	: Page	CAS	: Page	CAS	: Page	CAS	: Page	CAS	: Page
2243-27-8	: 8144	2385-70-8	: 3468	2444-89-5	: 2073	2612-46-6	: 6993	2807-30-9	: 5764
2243-30-3	: 833	2385-85-5	: 7644	2445-76-3	: 10265	2613-61-8	: 6447	2807-54-7	: 3781
2243-32-5	: 2713	2388-14-9	: 1973	2445-82-1	: 2833	2613-65-2	: 4818	2808-71-1	: 4661
2243-35-8	: 5771	2393-23-9	: 1813	2445-83-2	: 2836	2613-66-3	: 4817	2808-76-6	: 4649
2243-42-7	: 2714	2396-53-4	: 2080	2447-54-3	: 2432	2613-69-6	: 4858	2808-77-7	: 4651
2243-43-8	: 3411	2396-60-3	: 5137	2450-71-7	: 10725	2613-72-1	: 4861	2808-79-9	: 4650
2243-53-0	: 3855	2396-63-6	: 6343	2455-24-5	: 10674	2614-88-2	: 6711	2808-80-2	: 4434
2243-55-2	: 7167	2396-65-8	: 8115	2459-05-4	: 3791	2620-50-0	: 2426	2809-64-5	: 7997
2243-61-0	: 7796	2396-78-3	: 7062	2459-09-8	: 10908	2621-46-7	: 1192	2809-67-8	: 8530
2243-62-1	: 7797	2396-83-0	: 7054	2459-24-7	: 7766	2621-78-5	: 3967	2809-69-0	: 6717
2243-71-2	: 400	2396-84-1	: 6700	2459-25-8	: 7776	2622-08-4	: 9596	2810-04-0	: 11739
2243-76-7	: 2648	2398-09-6	: 4189	2462-85-3	: 8229	2622-21-1	: 4657	2810-69-7	: 2255
2243-80-3	: 7635	2398-10-9	: 4190	2462-94-4	: 5363	2622-60-8	: 2383	2815-34-1	: 9610
2243-81-4	: 7757	2398-16-5	: 4182	2463-53-8	: 8179	2622-89-1	: 3130	2816-57-1	: 4586
2243-83-6	: 7750	2398-37-0	: 1051	2463-63-0	: 6530	2623-23-6	: 4549	2817-71-2	: 10832
2243-89-2	: 11302	2398-65-4	: 2068	2464-37-1	: 4079	2623-50-9	: 11263	2819-48-9	: 4314
2243-94-9	: 8016	2398-66-5	: 2007	2467-09-6	: 3098	2623-54-3	: 2923	2819-86-5	: 9443
2243-98-3	: 11984	2398-68-7	: 1568	2467-10-9	: 8838	2623-82-7	: 8425	2820-37-3	: 10827
2244-07-7	: 11941	2398-81-4	: 10909	2470-68-0	: 6352	2623-84-9	: 11945	2825-00-5	: 10797
2244-16-8	: 4715	2399-48-6	: 10695	2473-01-0	: 8123	2623-95-2	: 5061	2825-79-8	: 8296
2254-94-6	: 2887	2400-01-3	: 1764	2476-31-1	: 5800	2624-01-3	: 6457	2834-05-1	: 11944
2257-35-4	: 10696	2400-02-4	: 1695	2486-07-9	: 1495	2625-30-1	: 4455	2835-21-4	: 10574
2258-58-4	: 9086	2400-03-5	: 2170	2489-86-3	: 7954	2625-31-2	: 4840	2835-33-8	: 11014
2259-96-3	: 2851	2400-04-6	: 1119	2492-22-0	: 8324	2627-95-4	: 5233	2835-81-6	: 3463
2270-20-4	: 2099	2401-22-1	: 1173	2493-02-9	: 1049	2631-37-0	: 9769	2835-96-3	: 9213
2272-45-9	: 811	2401-24-3	: 666	2495-35-4	: 10688	2639-63-6	: 3528	2835-97-4	: 9209
2279-76-7	: 11542	2401-73-2	: 5232	2497-21-4	: 7088	2641-89-6	: 1551	2836-00-2	: 9211
2288-18-8	: 1967	2402-06-4	: 4973	2498-75-1	: 522	2642-63-9	: 5801	2836-03-5	: 1267
2291-80-7	: 3968	2402-79-1	: 11019	2499-95-8	: 10643	2642-98-0	: 4113	2836-04-6	: 1268
2294-43-1	: 1083	2404-35-5	: 4222	2505-06-8	: 4246	2648-61-5	: 5802	2837-89-0	: 5486
2294-71-5	: 900	2408-20-0	: 10325	2506-41-4	: 7785	2652-13-3	: 8178	2845-89-8	: 1179
2294-72-6	: 11987	2408-37-9	: 4617	2508-29-4	: 8920	2655-83-6	: 1157	2847-72-5	: 5056
2294-94-2	: 1618	2409-52-1	: 3329	2512-81-4	: 9648	2657-65-0	: 3426	2848-25-1	: 1540
2295-31-0	: 11683	2409-55-4	: 9330	2514-91-2	: 11815	2675-77-6	: 4084	2855-08-5	: 3246
2295-58-1	: 10489	2412-73-9	: 2551	2516-34-9	: 4169	2676-33-7	: 10269	2856-63-5	: 949
2303-16-4	: 5106	2415-72-7	: 4984	2516-96-3	: 2545	2678-54-8	: 5901	2858-66-4	: 10485
2303-17-5	: 11823	2417-04-1	: 3069	2517-04-6	: 495	2679-87-0	: 3373	2859-67-8	: 11013
2305-13-7	: 2153	2417-90-5	: 10095	2517-43-3	: 3625	2681-83-6	: 6875	2859-68-9	: 11012
2306-88-9	: 8437	2418-31-7	: 3787	2523-37-7	: 5974	2681-92-7	: 8859	2859-78-1	: 1016
2308-38-5	: 3509	2419-73-0	: 3349	2523-55-9	: 4297	2687-25-4	: 1275	2867-05-2	: 2986
2310-17-0	: 9514	2420-16-8	: 532	2523-56-0	: 4296	2688-77-9	: 7379	2867-47-2	: 10654
2310-98-7	: 9841	2423-01-0	: 4906	2524-04-1	: 9576	2688-84-8	: 834	2867-59-6	: 3596
2311-59-3	: 5067	2423-10-1	: 8289	2524-09-6	: 9579	2689-59-0	: 7618	2867-63-2	: 4893
2312-35-8	: 10507	2424-92-2	: 5352	2524-37-0	: 2574	2689-88-5	: 9946	2868-37-3	: 4963
2313-65-7	: 6943	2424-98-8	: 10297	2524-52-9	: 10901	2690-08-6	: 8417	2869-34-3	: 11870
2314-78-5	: 11157	2425-28-7	: 1844	2524-64-3	: 9575	2706-56-1	: 10948	2870-04-4	: 1673
2314-97-8	: 7559	2425-54-9	: 11615	2525-02-2	: 1548	2713-09-9	: 5943	2873-18-9	: 11728
2315-68-6	: 2724	2425-66-3	: 9867	2528-38-3	: 9565	2717-42-2	: 8009	2873-74-7	: 8806
2316-50-9	: 9234	2425-79-8	: 8601	2528-61-2	: 6527	2717-44-4	: 7816	2873-89-4	: 661
2316-64-5	: 1843	2426-02-0	: 7347	2531-80-8	: 10312	2718-25-4	: 11342	2873-97-4	: 10521
2319-29-1	: 5031	2426-08-6	: 8602	2532-58-3	: 4803	2719-27-9	: 4330	2874-74-0	: 5314
2319-61-1	: 4392	2426-54-2	: 10631	2536-38-1	: 8250	2719-52-0	: 1952	2876-35-9	: 7829
2320-30-1	: 4589	2430-00-4	: 96	2545-59-7	: 257	2720-41-4	: 4481	2876-44-0	: 7938
2321-07-5	: 11522	2430-22-0	: 8463	2547-81-7	: 7536	2721-38-2	: 4796	2876-45-1	: 7896
2323-81-1	: 8868	2431-24-5	: 6540	2550-04-1	: 11482	2722-36-3	: 2156	2876-46-2	: 7901
2328-24-7	: 924	2431-30-3	: 7117	2550-06-3	: 11458	2724-59-6	: 8263	2876-51-9	: 7937
2338-12-7	: 2873	2431-50-7	: 3809	2550-21-2	: 6976	2725-82-8	: 1036	2876-52-0	: 7895
2344-70-9	: 2155	2431-52-9	: 3420	2550-22-3	: 7097	2734-70-5	: 703	2876-53-1	: 7900
2344-71-0	: 1106	2431-54-1	: 3810	2550-26-7	: 3678	2736-37-0	: 10492	2876-62-2	: 3674
2344-80-1	: 11382	2432-12-4	: 9295	2550-28-9	: 7131	2736-40-5	: 3691	2876-63-3	: 10473
2344-83-4	: 11386	2432-63-5	: 3784	2550-36-9	: 4318	2738-19-4	: 7040	2882-98-6	: 4841
2345-26-8	: 10295	2436-84-2	: 6330	2553-04-0	: 7623	2739-12-0	: 792	2883-02-5	: 4457
2345-27-9	: 11633	2436-85-3	: 7716	2553-19-7	: 11406	2741-16-4	: 1974	2883-07-0	: 4341
2345-28-0	: 8651	2436-92-2	: 11260	2554-06-5	: 4996	2745-25-7	: 6031	2883-08-1	: 4447
2345-61-1	: 10617	2436-93-3	: 11264	2555-49-9	: 229	2745-26-8	: 6026	2883-12-7	: 1255
2346-00-1	: 11672	2436-96-6	: 3062	2562-38-1	: 4839	2747-53-7	: 9098	2883-45-6	: 6337
2346-81-8	: 6739	2437-03-8	: 7719	2563-97-5	: 98	2757-90-6	: 6247	2885-00-9	: 8244
2350-19-8	: 6367	2437-25-4	: 5303	2565-39-1	: 6586	2758-18-1	: 4945	2886-52-4	: 10445
2351-13-5	: 5229	2437-56-1	: 11893	2568-33-4	: 3354	2759-28-6	: 9621	2887-61-8	: 3665
2351-97-5	: 9080	2437-73-2	: 11300	2568-90-3	: 3398	2761-24-2	: 11480	2890-62-2	: 5843
2358-84-1	: 10669	2437-76-5	: 1065	2568-91-4	: 10075	2761-84-4	: 5339	2892-18-4	: 7100
2362-10-9	: 5230	2437-88-9	: 2343	2570-10-7	: 4747	2765-04-0	: 11920	2896-60-8	: 1518
2362-12-1	: 9241	2437-95-8	: 2972	2570-26-5	: 8641	2765-11-9	: 8640	2899-90-3	: 8950
2364-59-2	: 3581	2437-98-1	: 88	2571-39-3	: 7615	2765-18-6	: 7957	2902-69-4	: 5871
2365-48-2	: 204	2438-01-3	: 2723	2575-20-4	: 3658	2768-41-4	: 3915	2902-96-7	: 10394
2366-52-1	: 3378	2438-04-2	: 2688	2579-20-6	: 4362	2769-64-4	: 3387	2905-56-8	: 9698
2367-82-0	: 2271	2438-12-2	: 4668	2579-22-8	: 10724	2777-66-4	: 11991	2912-62-1	: 973
2370-88-9	: 4999	2438-19-9	: 10641	2583-25-7	: 9979	2778-68-9	: 2936	2917-26-2	: 6643
2371-19-9	: 6520	2438-72-4	: 880	2586-89-2	: 6611	2781-85-3	: 4990	2917-47-7	: 10410
2373-51-5	: 7484	2439-01-2	: 4072	2591-86-8	: 9638	2782-57-2	: 11838	2918-13-0	: 6583
2381-31-9	: 523	2439-10-3	: 5347	2593-15-9	: 11599	2782-91-4	: 11810	2919-05-3	: 9118
2382-43-6	: 9831	2439-35-2	: 10633	2594-20-9	: 3956	2785-74-2	: 1511	2919-23-5	: 4196
2384-73-8	: 6606	2439-99-8	: 6277	2594-21-0	: 3937	2785-89-9	: 9357	2921-88-2	: 4089
2384-75-0	: 6538	2443-03-0	: 4837	2595-97-3	: 337	2790-09-2	: 1348	2922-51-2	: 6320
2384-85-2	: 5098	2443-04-1	: 4836	2601-89-0	: 10099	2791-29-9	: 6636	2923-68-4	: 8442
2384-86-3	: 5099	2443-91-6	: 9886	2605-18-7	: 1940	2798-73-4	: 3891	2933-74-6	: 5745
2384-90-9	: 6326	2444-37-3	: 216	2612-33-1	: 9997	2804-50-4	: 10541	2935-44-6	: 6799

CAS	Ref	CAS	Ref	CAS	Ref	CAS	Ref	CAS	Ref
2935-90-2	: 10287	3087-62-5	: 6211	3280-08-8	: 1616	3463-92-1	: 4064	3728-43-6	: 11490
2941-19-7	: 1830	3088-41-3	: 9703	3280-51-1	: 6708	3468-99-3	: 939	3730-60-7	: 6923
2941-20-0	: 1809	3096-47-7	: 5976	3282-30-2	: 10503	3476-89-9	: 11339	3731-51-9	: 10971
2941-55-1	: 5930	3096-52-4	: 5977	3283-12-3	: 9528	3478-88-4	: 7594	3731-52-0	: 10972
2943-70-6	: 5231	3101-60-8	: 8611	3288-04-8	: 1793	3497-00-5	: 9552	3731-53-1	: 10973
2944-49-2	: 1916	3102-33-8	: 9105	3289-28-9	: 4334	3508-00-7	: 6307	3735-90-8	: 3977
2947-60-6	: 961	3112-88-7	: 2112	3290-06-0	: 1358	3508-78-9	: 8805	3741-00-2	: 4846
2947-61-7	: 962	3113-98-2	: 4421	3296-05-7	: 978	3508-94-9	: 930	3741-38-6	: 5200
2955-65-9	: 8395	3118-97-6	: 8028	3299-38-5	: 3656	3510-70-1	: 11496	3742-38-9	: 4926
2955-88-6	: 11165	3121-83-3	: 9993	3319-15-1	: 1895	3512-80-9	: 10864	3742-42-5	: 4662
2959-96-8	: 10781	3128-06-1	: 6905	3320-83-0	: 1176	3521-91-3	: 6566	3743-22-4	: 9317
2961-47-9	: 10880	3128-07-2	: 6469	3320-86-3	: 1788	3522-94-9	: 6861	3744-02-3	: 9109
2969-81-5	: 3470	3129-39-3	: 7846	3320-87-4	: 1789	3524-62-7	: 5839	3757-32-2	: 8910
2971-79-1	: 9644	3129-90-6	: 7399	3321-50-4	: 4386	3524-73-0	: 7039	3761-94-2	: 4230
2973-00-4	: 7497	3132-99-8	: 527	3329-48-4	: 7073	3524-75-2	: 4848	3763-39-1	: 4991
2973-27-5	: 11224	3140-92-9	: 11743	3329-56-4	: 3377	3528-17-4	: 2869	3763-55-1	: 4062
2977-31-3	: 1617	3140-93-0	: 11742	3333-23-1	: 7764	3535-87-9	: 1925	3764-01-0	: 11075
2978-58-7	: 3896	3141-12-6	: 440	3333-44-6	: 6138	3536-29-6	: 2158	3768-43-2	: 9722
2979-19-3	: 4587	3141-24-0	: 11790	3333-52-6	: 3292	3538-65-6	: 3529	3768-60-3	: 11583
2980-71-4	: 8185	3141-27-3	: 11744	3337-71-1	: 474	3540-95-2	: 9658	3768-63-6	: 11569
2983-26-8	: 5523	3142-58-3	: 8799	3338-15-6	: 10444	3544-25-0	: 946	3769-23-1	: 7038
2983-37-1	: 6885	3142-66-3	: 8972	3344-70-5	: 5296	3547-04-4	: 1678	3769-41-3	: 9452
2984-40-9	: 3806	3142-84-5	: 6608	3347-22-6	: 8056	3552-01-0	: 4538	3773-93-1	: 5208
2987-53-3	: 819	3144-16-9	: 2929	3350-30-9	: 4725	3554-74-3	: 9716	3774-03-6	: 5197
2987-77-1	: 11485	3144-54-5	: 6962	3350-78-5	: 3888	3555-84-8	: 7612	3779-29-1	: 4180
2996-92-1	: 11488	3147-55-5	: 2558	3351-28-8	: 4117	3558-69-8	: 10947	3780-58-3	: 6894
2998-18-7	: 10665	3152-68-9	: 9110	3351-86-8	: 4055	3567-38-2	: 1877	3788-32-7	: 4838
3000-79-1	: 9680	3153-36-4	: 3485	3352-87-2	: 5290	3570-55-6	: 5652	3790-71-4	: 5334
3001-72-7	: 11190	3153-37-5	: 3489	3354-58-3	: 9427	3570-58-9	: 5699	3810-26-2	: 4951
3002-24-2	: 6804	3156-70-5	: 10589	3355-17-7	: 3900	3574-40-1	: 3157	3810-74-0	: 11561
3007-91-8	: 7770	3156-73-8	: 10396	3358-28-9	: 6174	3574-43-4	: 5682	3835-64-1	: 1864
3007-97-4	: 7765	3156-74-9	: 3641	3368-16-9	: 941	3581-87-1	: 11678	3837-38-5	: 404
3008-39-7	: 4225	3163-07-3	: 1541	3373-53-3	: 10767	3581-91-7	: 11675	3839-50-7	: 4765
3010-02-4	: 280	3163-27-7	: 7742	3375-23-3	: 8036	3582-71-6	: 11913	3840-31-1	: 1914
3010-04-6	: 273	3167-49-5	: 10897	3377-71-7	: 7321	3586-12-7	: 835	3848-12-2	: 126
3012-37-1	: 11710	3172-52-9	: 11748	3377-86-4	: 6733	3586-14-9	: 2013	3849-33-0	: 10049
3016-19-1	: 8487	3173-53-3	: 4416	3377-87-5	: 6734	3587-57-3	: 9858	3850-30-4	: 3217
3017-67-2	: 5086	3173-79-3	: 8597	3383-96-8	: 1	3587-60-8	: 1181	3852-09-3	: 10289
3017-68-3	: 3714	3174-67-2	: 8785	3384-04-1	: 1233	3588-30-5	: 3169	3853-80-3	: 10202
3017-70-7	: 3718	3174-83-2	: 2157	3386-33-2	: 8236	3593-00-8	: 6563	3856-25-5	: 11861
3017-71-8	: 3713	3175-23-3	: 10174	3393-64-4	: 3664	3594-37-4	: 3364	3857-25-8	: 6122
3017-95-6	: 9840	3176-62-3	: 7240	3393-78-0	: 2284	3598-14-9	: 287	3862-11-1	: 9338
3017-96-7	: 9838	3178-22-1	: 4369	3395-79-7	: 1443	3600-24-6	: 11916	3862-12-2	: 9335
3018-09-5	: 5881	3178-23-2	: 4429	3395-91-3	: 10212	3607-78-1	: 10052	3864-18-4	: 3052
3018-12-0	: 278	3178-24-3	: 4466	3399-21-1	: 4357	3609-53-8	: 2479	3867-18-3	: 733
3019-04-3	: 10454	3178-29-8	: 6434	3400-45-1	: 4781	3610-02-4	: 2403	3875-51-2	: 4828
3019-20-3	: 1980	3179-83-3	: 10375	3402-74-2	: 2006	3613-42-1	: 367	3875-52-3	: 4805
3019-25-8	: 5792	3180-09-4	: 9246	3402-76-4	: 7945	3615-21-2	: 2377	3875-68-1	: 12051
3022-41-1	: 4209	3187-94-8	: 6073	3404-57-7	: 6558	3616-56-6	: 5416	3877-15-4	: 10088
3024-72-4	: 2894	3188-13-4	: 5478	3404-58-8	: 7033	3616-57-7	: 5415	3877-19-8	: 7996
3027-13-2	: 10460	3189-22-8	: 7311	3404-61-3	: 7037	3634-92-2	: 8273	3878-55-5	: 3336
3031-05-8	: 8011	3194-15-8	: 10438	3404-63-5	: 9060	3637-01-2	: 5811	3884-88-6	: 8246
3031-66-1	: 7118	3195-24-2	: 9944	3404-67-9	: 6838	3637-61-4	: 4825	3910-35-8	: 7263
3031-73-0	: 7180	3196-15-4	: 3473	3404-71-5	: 9043	3638-35-5	: 4983	3913-71-1	: 5080
3031-74-1	: 7179	3199-61-9	: 2440	3404-72-6	: 9029	3638-64-0	: 5911	3913-85-7	: 5088
3032-32-4	: 2491	3202-84-4	: 7695	3404-73-7	: 9031	3643-64-9	: 4688	3916-64-1	: 6339
3034-53-5	: 11668	3208-16-0	: 6105	3404-77-1	: 7016	3657-07-6	: 245	3917-15-5	: 10565
3035-75-4	: 11877	3208-22-8	: 6173	3404-78-2	: 7024	3658-79-5	: 8755	3920-53-4	: 6453
3036-66-6	: 3168	3209-22-1	: 1391	3404-80-6	: 6837	3669-80-5	: 11948	3937-45-9	: 4422
3040-44-6	: 9662	3211-48-1	: 4183	3411-95-8	: 9217	3673-79-8	: 10242	3937-46-0	: 4423
3041-40-5	: 9916	3212-75-7	: 4753	3416-24-8	: 6233	3674-09-7	: 10248	3937-56-2	: 8136
3042-22-6	: 11135	3213-79-4	: 1269	3419-71-4	: 4689	3674-10-0	: 10363	3938-95-2	: 10259
3042-50-0	: 1486	3217-86-5	: 9802	3424-39-9	: 618	3674-13-3	: 10240	3938-96-3	: 206
3043-33-2	: 8672	3218-02-8	: 4418	3426-01-5	: 10843	3674-75-7	: 9179	3943-74-6	: 2627
3047-38-9	: 4907	3221-61-2	: 8408	3430-27-1	: 10870	3675-68-1	: 3442	3943-77-9	: 2578
3051-09-0	: 11097	3228-04-4	: 9416	3433-37-2	: 9674	3675-69-2	: 3443	3944-36-3	: 10390
3054-92-0	: 8956	3228-99-7	: 9895	3433-80-5	: 999	3678-62-4	: 10919	3955-26-8	: 2324
3054-95-3	: 10560	3229-00-3	: 9884	3435-28-7	: 11048	3680-02-2	: 5909	3958-03-0	: 11781
3055-14-9	: 1949	3230-23-7	: 9670	3437-89-6	: 8987	3681-78-5	: 5320	3958-38-1	: 4727
3058-47-7	: 2043	3232-84-6	: 11850	3437-95-4	: 11767	3682-35-7	: 11843	3958-57-4	: 1067
3060-89-7	: 7856	3234-64-8	: 9121	3438-46-8	: 11071	3683-19-0	: 7043	3959-07-7	: 1799
3062-81-5	: 3122	3238-38-8	: 9474	3439-97-2	: 11443	3683-22-5	: 7042	3962-77-4	: 1926
3066-71-5	: 10626	3238-40-2	: 6081	3440-02-6	: 11415	3689-24-5	: 11716	3963-78-8	: 382
3068-00-6	: 3459	3238-62-8	: 9677	3445-11-2	: 11184	3690-04-8	: 9795	3964-56-5	: 9232
3073-66-3	: 4486	3238-75-3	: 5712	3446-98-8	: 11147	3697-27-6	: 4115	3964-58-7	: 2540
3073-92-5	: 3421	3240-09-3	: 7099	3452-07-1	: 5362	3699-01-2	: 2019	3964-61-2	: 9254
3074-61-1	: 4935	3243-42-3	: 7952	3452-09-3	: 8196	3710-30-3	: 8319	3968-85-2	: 1953
3074-64-4	: 6542	3244-88-0	: 2218	3452-97-9	: 6951	3710-43-8	: 6088	3970-35-2	: 2544
3074-71-3	: 6384	3252-43-5	: 277	3454-07-7	: 1632	3710-84-7	: 5429	3970-62-5	: 8929
3074-75-7	: 6828	3254-93-1	: 7228	3457-58-7	: 2003	3711-34-0	: 7137	3971-28-6	: 4363
3074-76-8	: 6826	3256-88-0	: 10990	3458-28-4	: 7425	3711-37-3	: 7140	3972-56-3	: 1142
3074-77-9	: 6827	3260-87-5	: 9264	3460-18-2	: 1310	3721-95-7	: 4174	3972-65-4	: 1023
3075-84-1	: 3096	3261-62-9	: 1580	3460-20-6	: 2315	3722-73-4	: 2781	3973-62-4	: 9694
3077-16-5	: 7698	3268-49-3	: 9786	3460-24-0	: 1295	3722-74-5	: 2780	3974-99-0	: 254
3081-24-1	: 9512	3268-79-9	: 11700	3460-67-1	: 10904	3725-05-1	: 6714	3976-34-9	: 3056
3083-23-6	: 8628	3277-26-7	: 5237	3463-26-1	: 10837	3725-07-3	: 8340	3976-36-1	: 3095
3085-53-8	: 5997	3278-46-4	: 10332	3463-40-9	: 2316	3725-30-2	: 4750	3978-81-2	: 10944

3982-64-7 : 1483	4200-95-7 : 6305	4388-88-9 : 6811	4562-36-1 : 4042	4806-58-0 : 4986
3982-66-9 : 12057	4205-23-6 : 6290	4389-53-1 : 2841	4573-05-1 : 4252	4806-59-1 : 4987
3982-87-0 : 1712	4206-75-1 : 11384	4390-75-4 : 3408	4573-09-5 : 4897	4806-61-5 : 4191
3988-03-2 : 7590	4208-49-5 : 6070	4392-24-9 : 1084	4588-18-5 : 8505	4810-09-7 : 6544
3991-61-5 : 2012	4208-64-4 : 6121	4392-30-7 : 1245	4593-16-2 : 9631	4815-57-0 : 1557
4001-73-4 : 604	4209-24-9 : 5787	4393-06-0 : 1868	4594-78-9 : 4467	4823-47-6 : 10611
4008-48-4 : 11323	4209-90-9 : 6925	4394-11-0 : 3103	4597-87-9 : 10869	4829-04-3 : 5266
4016-11-9 : 8614	4214-28-2 : 1770	4394-85-8 : 7685	4604-28-8 : 1155	4830-99-3 : 7273
4018-65-9 : 1504	4215-86-5 : 2011	4395-73-7 : 1819	4606-07-9 : 4961	4831-43-0 : 11182
4021-75-4 : 3612	4218-48-8 : 1690	4395-79-3 : 1201	4606-65-9 : 9675	4832-17-1 : 8055
4028-23-3 : 11375	4219-24-3 : 7050	4395-80-6 : 1202	4609-89-6 : 8833	4835-11-4 : 6750
4032-86-4 : 6388	4219-49-2 : 6652	4395-81-7 : 1782	4609-92-1 : 8414	4838-00-0 : 7242
4032-92-2 : 6445	4224-70-8 : 6876	4396-98-9 : 1847	4619-74-3 : 9480	4839-46-7 : 8773
4032-93-3 : 6444	4225-85-8 : 11244	4397-05-1 : 1842	4620-70-6 : 5710	4850-28-6 : 4862
4032-94-4 : 8375	4228-00-6 : 5317	4398-65-6 : 6744	4625-24-5 : 2980	4850-32-2 : 4832
4038-04-4 : 9044	4229-91-8 : 6167	4403-69-4 : 1796	4628-21-1 : 3726	4860-03-1 : 6634
4041-09-2 : 4882	4253-34-3 : 11499	4405-42-9 : 11385	4630-20-0 : 3986	4877-77-4 : 1081
4043-71-4 : 2430	4253-89-8 : 5244	4407-36-7 : 10704	4630-82-4 : 4335	4877-93-4 : 3045
4044-65-9 : 1432	4253-91-2 : 10129	4412-16-2 : 5337	4635-59-0 : 3685	4884-24-6 : 2995
4048-30-0 : 6578	4254-02-8 : 4779	4413-16-5 : 1249	4636-16-2 : 5838	4884-25-7 : 2994
4048-31-1 : 6574	4255-62-3 : 4590	4422-95-1 : 2317	4645-15-2 : 4461	4887-30-3 : 6904
4048-32-2 : 6581	4258-93-9 : 4449	4426-11-3 : 4173	4654-39-1 : 1595	4891-38-7 : 10742
4049-81-4 : 6695	4259-00-1 : 4855	4426-51-1 : 5198	4655-34-9 : 10664	4894-61-5 : 3725
4050-45-7 : 7000	4259-43-2 : 10179	4427-56-9 : 9417	4662-96-8 : 1589	4894-62-6 : 6335
4053-34-3 : 11329	4264-29-3 : 2055	4428-13-1 : 1801	4663-22-3 : 4982	4897-84-1 : 3474
4054-38-0 : 4219	4265-25-2 : 2456	4433-10-7 : 3054	4667-38-3 : 11392	4900-30-5 : 8110
4057-42-5 : 8503	4265-97-8 : 8430	4433-11-8 : 3059	4667-99-6 : 11387	4901-51-3 : 9469
4065-80-9 : 4537	4268-36-4 : 3933	4433-63-0 : 3133	4669-01-6 : 4844	4904-61-4 : 4212
4068-78-4 : 2542	4271-30-1 : 6242	4435-14-7 : 4313	4670-56-8 : 2639	4911-55-1 : 8684
4074-24-2 : 3322	4279-22-5 : 3280	4435-18-1 : 276	4671-03-8 : 11848	4911-65-3 : 1236
4074-43-5 : 9247	4279-70-3 : 1993	4435-50-1 : 3458	4672-38-2 : 9547	4912-92-9 : 7251
4079-52-1 : 5842	4279-76-9 : 1716	4436-59-3 : 4238	4675-18-7 : 8581	4914-91-4 : 9038
4088-60-2 : 3860	4280-28-8 : 1927	4436-96-8 : 1867	4676-51-1 : 3326	4914-92-5 : 9037
4088-81-7 : 9175	4282-24-0 : 6079	4437-18-7 : 6035	4685-47-6 : 1921	4915-21-3 : 6028
4091-39-8 : 3647	4282-28-4 : 6080	4437-20-1 : 6101	4694-12-6 : 4899	4915-22-4 : 6029
4095-09-4 : 11453	4282-31-9 : 11745	4437-22-3 : 6156	4695-31-2 : 10288	4915-23-5 : 6027
4095-22-1 : 11772	4282-32-0 : 6083	4437-23-4 : 6158	4701-36-4 : 1699	4916-63-6 : 6994
4096-20-2 : 9693	4282-40-0 : 6419	4437-50-7 : 6810	4702-34-5 : 2797	4920-84-7 : 1448
4096-21-3 : 11171	4283-80-1 : 8717	4437-51-8 : 6806	4705-34-4 : 1621	4920-99-4 : 1689
4096-34-8 : 4704	4285-42-1 : 3974	4439-20-7 : 5715	4705-94-6 : 8027	4923-77-7 : 4395
4100-80-5 : 6094	4286-47-9 : 6778	4439-24-1 : 5746	4706-81-4 : 11631	4923-78-8 : 4396
4101-68-2 : 5038	4286-49-1 : 3293	4441-57-0 : 4324	4706-89-2 : 1472	4923-84-6 : 9533
4104-01-2 : 9025	4288-84-0 : 10355	4441-90-1 : 9965	4712-55-4 : 9534	4926-76-5 : 4398
4107-62-4 : 10234	4291-79-6 : 4450	4442-79-9 : 4387	4721-34-0 : 9544	4926-78-7 : 4399
4110-44-5 : 8379	4291-80-9 : 4451	4443-55-4 : 4384	4721-37-3 : 9542	4926-90-3 : 4394
4110-50-3 : 10042	4291-81-0 : 4452	4444-12-6 : 6666	4726-14-1 : 8070	4930-98-7 : 11034
4110-77-4 : 1153	4291-98-9 : 4934	4445-06-1 : 4459	4727-17-7 : 4760	4944-94-9 : 632
4113-12-6 : 6642	4292-19-7 : 5302	4445-07-2 : 2066	4730-22-7 : 6498	4945-28-2 : 11301
4114-28-7 : 7151	4292-75-5 : 4414	4445-08-3 : 8245	4731-34-4 : 2076	4945-47-5 : 9627
4114-31-2 : 7148	4292-92-6 : 4465	4447-60-3 : 10079	4731-84-4 : 9249	4945-48-6 : 9635
4117-15-1 : 11934	4295-06-1 : 11245	4455-26-9 : 8353	4734-90-1 : 4754	4946-13-8 : 2298
4124-31-6 : 244	4301-39-7 : 6053	4457-00-5 : 4823	4736-48-5 : 4214	4949-44-4 : 8906
4124-88-3 : 8187	4303-44-0 : 9855	4463-33-6 : 1439	4737-41-1 : 8731	4957-14-6 : 1964
4126-78-7 : 4226	4313-03-5 : 6325	4467-06-5 : 10989	4740-00-5 : 11788	4964-71-0 : 11218
4127-45-1 : 4985	4318-37-0 : 5141	4467-88-3 : 7271	4744-08-5 : 9903	4965-38-0 : 11220
4128-31-8 : 8444	4318-76-7 : 10924	4468-42-2 : 1669	4747-05-1 : 3161	4972-29-6 : 2185
4130-08-9 : 11498	4319-49-7 : 7680	4471-05-0 : 1909	4747-07-3 : 6833	4974-21-4 : 246
4130-42-1 : 9221	4325-48-8 : 8729	4481-08-7 : 6198	4747-15-3 : 1928	4974-27-0 : 8320
4131-74-2 : 10331	4325-49-9 : 6366	4481-30-5 : 1481	4747-21-1 : 9819	4975-21-7 : 6176
4132-71-2 : 1442	4325-82-0 : 9099	4482-03-5 : 3076	4749-27-3 : 10606	4984-01-4 : 8502
4132-77-8 : 1470	4325-97-7 : 5651	4484-72-4 : 11460	4749-28-4 : 10590	4984-85-4 : 6971
4142-85-2 : 11440	4328-94-3 : 8852	4484-80-4 : 251	4749-31-9 : 10127	4985-70-0 : 374
4143-41-3 : 5126	4339-05-3 : 7128	4485-09-0 : 8174	4753-75-7 : 6115	4985-85-7 : 5679
4145-94-2 : 11673	4340-76-5 : 5359	4485-13-6 : 6559	4753-80-4 : 11690	4996-48-9 : 6068
4147-51-7 : 5222	4341-76-8 : 3918	4485-16-9 : 6556	4755-36-6 : 4247	5006-66-6 : 10899
4147-89-1 : 11362	4342-61-4 : 5226	4489-20-3 : 1082	4755-77-5 : 144	5008-52-6 : 5007
4156-16-5 : 7407	4346-18-3 : 3584	4491-19-4 : 307	4766-33-0 : 7711	5008-69-5 : 10082
4160-52-5 : 3672	4350-41-8 : 10921	4492-01-7 : 10830	4766-57-8 : 11505	5008-73-1 : 10077
4160-80-9 : 10782	4351-54-6 : 6005	4493-42-9 : 5027	4767-03-7 : 10272	5009-11-0 : 7104
4161-60-8 : 8973	4354-56-7 : 4412	4497-29-4 : 7524	4771-47-5 : 2546	5009-27-8 : 4988
4164-28-7 : 7449	4358-59-2 : 3848	4510-34-3 : 10705	4771-80-6 : 4638	5020-21-3 : 7246
4166-46-5 : 6926	4358-75-2 : 3216	4516-69-2 : 4856	4773-53-9 : 10364	5022-29-7 : 7364
4167-74-2 : 9432	4358-87-6 : 910	4518-10-9 : 2504	4775-09-1 : 9527	5026-76-6 : 6547
4167-77-5 : 4791	4358-88-7 : 907	4520-29-0 : 11447	4775-86-4 : 8800	5029-66-3 : 10284
4169-04-4 : 10398	4360-12-7 : 321	4523-45-9 : 7720	4780-79-4 : 7908	5029-67-4 : 10970
4170-24-5 : 3480	4360-96-7 : 9959	4524-77-0 : 9233	4784-77-4 : 3712	5035-52-9 : 11547
4170-46-1 : 11466	4361-06-2 : 9968	4525-36-4 : 6165	4789-40-6 : 6032	5053-43-0 : 11070
4171-83-9 : 2057	4362-01-0 : 4135	4525-44-4 : 11397	4795-29-3 : 6116	5058-19-5 : 10878
4173-44-8 : 6710	4363-04-6 : 8019	4534-74-1 : 4525	4795-88-2 : 2935	5069-26-1 : 11774
4175-54-6 : 7983	4363-93-3 : 11225	4535-66-4 : 5684	4798-44-1 : 7066	5075-33-2 : 7129
4179-19-5 : 1441	4373-13-1 : 7817	4537-00-2 : 7763	4798-46-3 : 7077	5075-92-3 : 11181
4180-23-8 : 1944	4375-14-8 : 7319	4537-05-7 : 2285	4798-60-1 : 7081	5076-20-0 : 8627
4185-62-0 : 7745	4376-23-2 : 7090	4542-61-4 : 3117	4798-61-2 : 8511	5077-67-8 : 3660
4185-63-1 : 7788	4377-37-1 : 11023	4549-31-9 : 6374	4798-62-3 : 6575	5090-41-5 : 8288
4187-86-4 : 9140	4383-05-5 : 1885	4549-32-0 : 8369	4798-64-5 : 8518	5097-93-8 : 11001
4187-87-5 : 1876	4383-18-0 : 1904	4549-74-0 : 8669	4799-68-2 : 10400	5100-98-1 : 11817
4192-77-2 : 10685	4388-87-8 : 6807	4553-07-5 : 893		5103-42-4 : 2999

CAS	No.	CAS	No.	CAS	No.	CAS	No.	CAS	No.	CAS	No.
5112-21-0	9999	5362-55-0	9095	5535-49-9	1165	5842-00-2	5650	6094-02-6	7036		
5113-93-9	4682	5363-64-4	6813	5556-07-0	5835	5846-43-5	4559	6102-15-4	4641		
5131-66-8	10346	5367-28-2	1213	5557-31-3	8292	5856-77-9	3688	6108-61-8	6687		
5132-75-2	6468	5368-81-0	2674	5558-29-2	963	5856-82-6	3680	6109-22-4	422		
5137-45-1	5591	5370-25-2	5783	5558-31-6	964	5857-36-3	8989	6111-61-1	3643		
5145-64-2	11138	5371-48-2	9119	5558-66-7	931	5857-86-3	4558	6111-78-0	521		
5145-99-3	10043	5379-20-4	1627	5558-78-1	970	5859-45-0	7784	6111-88-2	8733		
5150-80-1	6341	5380-40-5	6069	5560-69-0	7968	5867-45-8	89	6111-99-5	10236		
5150-93-6	3333	5390-04-5	9143	5563-78-0	4547	5870-61-1	5882	6114-18-7	8302		
5153-67-3	2050	5390-71-6	9861	5581-35-1	1583	5870-93-9	3527	6117-80-2	3793		
5159-41-1	1888	5390-72-7	9859	5581-75-9	1759	5876-87-9	5285	6118-14-5	3858		
5160-99-6	1997	5392-86-9	1375	5582-82-1	6500	5877-55-4	809	6119-92-2	5182		
5161-04-6	1999	5395-43-7	11553	5586-15-2	5256	5877-58-7	810	6120-13-4	2919		
5161-13-7	11785	5395-75-5	5451	5587-63-3	4910	5878-19-3	10455	6125-21-9	9919		
5161-16-0	11796	5396-24-7	149	5587-79-1	4948	5881-17-4	8403	6125-24-2	2051		
5161-17-1	11798	5396-38-3	1462	5588-20-5	7220	5882-44-0	717	6127-92-0	3827		
5162-44-7	3711	5396-58-7	3355	5589-96-8	112	5891-21-4	8961	6127-93-1	3828		
5162-48-1	8955	5396-91-8	8969	5597-27-3	2958	5897-76-7	1840	6130-75-2	2331		
5166-35-8	3734	5398-08-3	9961	5597-81-9	7782	5902-51-2	11591	6131-24-4	6835		
5166-53-0	7094	5398-10-7	9955	5597-89-7	8687	5906-73-0	11501	6131-25-5	6423		
5171-84-6	6854	5398-25-4	8742	5598-13-0	4090	5908-87-2	5277	6134-66-3	3503		
5182-44-5	1596	5398-75-4	9177	5609-09-6	6587	5910-87-2	8109	6135-29-1	262		
5183-77-7	4347	5399-02-0	6312	5614-32-4	3160	5910-89-4	10805	6136-67-0	7624		
5183-79-9	4360	5405-13-0	1821	5617-41-4	4405	5911-04-6	8141	6137-08-2	8475		
5194-50-3	6686	5405-15-2	1825	5617-42-5	4820	5911-08-0	4966	6137-15-1	8476		
5194-51-4	6685	5405-17-4	1824	5617-92-5	4978	5912-58-3	5694	6138-90-5	8321		
5195-37-9	7982	5405-41-4	3534	5630-56-8	7598	5912-86-7	9397	6139-84-0	3181		
5195-40-4	7981	5405-79-8	6965	5634-22-0	1565	5912-87-8	9393	6140-17-6	2045		
5208-49-1	4050	5406-12-2	2116	5634-30-0	4809	5913-85-9	9706	6140-65-4	4908		
5209-18-7	2161	5406-18-8	2154	5636-65-7	7061	5915-41-3	11593	6140-80-3	10579		
5209-33-6	5	5407-98-7	7605	5637-96-7	2070	5917-47-5	9660	6142-95-6	1412		
5216-25-1	1239	5408-57-1	8477	5638-26-6	6336	5921-54-0	3673	6149-41-3	10276		
5216-32-0	1366	5408-86-6	3268	5648-29-3	5614	5921-73-3	8209	6149-46-8	8884		
5217-05-0	4783	5410-97-9	5166	5650-40-8	10451	5921-84-6	6476	6152-31-4	7171		
5221-17-0	9780	5411-50-7	1307	5650-41-9	10452	5930-28-9	9206	6153-16-8	4674		
5223-59-6	1676	5411-56-3	1845	5650-75-9	8890	5932-68-3	9396	6153-56-6	5561		
5234-68-4	4039	5412-92-0	8352	5661-71-2	503	5943-35-1	11649	6158-45-8	7920		
5251-93-4	524	5413-05-8	884	5663-96-7	8538	5944-41-2	10942	6159-55-3	11192		
5258-50-4	11797	5414-19-7	5609	5666-21-7	10527	5950-19-6	3649	6161-50-8	3046		
5259-65-4	4714	5414-21-1	8828	5668-93-9	3073	5951-67-7	4659	6162-79-4	3951		
5259-98-3	8923	5416-94-4	11704	5669-09-0	10041	5959-52-4	7768	6163-64-0	10083		
5263-87-6	11275	5419-55-6	3128	5675-22-9	6327	5963-77-9	9137	6163-66-2	10124		
5265-18-9	10414	5421-04-5	912	5675-51-4	5301	5965-66-2	6214	6163-75-3	9537		
5271-38-5	5749	5424-21-5	11053	5676-58-4	2879	5976-47-6	10700	6164-98-3	7564		
5271-39-6	1679	5426-65-3	10846	5683-33-0	10863	5984-58-7	6353	6165-44-2	4323		
5272-02-6	6369	5429-85-6	8290	5683-44-3	8794	5985-23-9	6246	6166-86-5	4902		
5274-50-0	9554	5432-61-1	4289	5685-06-3	11680	5986-55-0	7586	6166-87-6	4625		
5281-18-5	563	5432-85-9	4600	5687-92-3	11692	6001-87-2	10225	6168-72-5	10336		
5283-66-9	11471	5434-27-5	9887	5689-04-3	7292	6004-60-0	5795	6174-95-4	5916		
5285-18-7	8677	5441-51-0	4593	5697-85-8	838	6006-06-0	11895	6175-23-1	8137		
5285-87-0	11709	5441-52-1	4518	5699-58-1	8875	6006-33-3	4484	6175-49-1	5325		
5292-21-7	4306	5445-24-9	4557	5699-74-1	8829	6006-34-4	4854	6195-92-2	4886		
5292-43-3	114	5447-97-2	9847	5703-21-9	869	6006-90-2	8638	6196-58-3	5757		
5292-53-5	9978	5447-99-4	8952	5703-26-4	871	6006-95-7	4463	6196-60-7	9047		
5307-19-7	10951	5448-43-1	11334	5703-52-6	2146	6012-97-1	11782	6202-04-6	11839		
5311-96-6	7150	5451-80-9	8881	5704-20-1	8975	6020-51-5	7224	6213-90-7	3821		
5314-37-4	3070	5451-85-4	8901	5715-23-1	4517	6031-02-3	2009	6214-25-1	3832		
5314-55-6	11433	5452-37-9	4735	5728-52-9	3009	6032-29-7	8916	6214-28-4	3820		
5324-00-5	7272	5452-75-5	4308	5736-03-8	5206	6035-49-0	2845	6222-35-1	10235		
5325-97-3	9185	5453-44-1	3857	5743-97-5	9186	6040-62-6	8058	6223-78-5	6765		
5326-81-8	45	5453-80-5	2967	5744-03-6	5973	6044-68-4	10562	6225-06-5	8699		
5327-44-6	1924	5454-28-4	6459	5750-00-5	3248	6050-13-1	5145	6231-18-1	10935		
5328-01-8	7883	5454-79-5	4533	5756-43-4	6814	6051-43-0	2489	6236-88-0	4400		
5332-06-9	3404	5455-24-3	8396	5763-61-1	1805	6051-52-1	1869	6258-66-8	1833		
5332-24-1	11217	5458-48-0	1252	5765-44-6	7405	6057-60-9	3019	6261-22-9	9141		
5332-25-2	11219	5458-59-3	8435	5773-56-8	1614	6061-06-9	3156	6265-26-5	6166		
5332-52-5	11942	5459-58-5	151	5775-82-6	10816	6061-10-5	8433	6267-24-9	5605		
5332-73-0	9816	5459-93-8	4288	5779-94-2	548	6064-27-3	5326	6268-37-7	1228		
5335-75-1	10879	5460-32-2	1768	5779-95-3	549	6065-59-4	9974	6269-92-7	8366		
5337-03-1	10774	5461-51-8	74	5781-53-3	145	6065-82-3	174	6270-16-2	3642		
5337-70-2	8552	5465-09-8	4588	5787-31-5	3232	6065-90-3	8752	6270-34-4	6061		
5337-72-4	4515	5465-13-4	2282	5792-36-9	10430	6065-93-6	10555	6270-55-9	10300		
5337-93-9	10468	5468-97-3	1598	5794-88-7	2487	6068-69-5	816	6270-56-0	6104		
5339-85-5	1594	5469-19-2	1093	5798-75-4	2523	6069-71-2	2954	6272-38-4	9451		
5340-36-3	8466	5470-02-0	9707	5798-79-8	947	6069-97-2	2937	6272-40-8	7588		
5340-41-0	6953	5470-28-0	9606	5798-88-9	3472	6069-98-3	4439	6274-16-4	3950		
5342-31-4	7509	5470-47-3	882	5798-94-7	606	6078-26-8	310	6279-86-3	7552		
5344-20-7	8707	5470-91-7	10712	5802-82-4	6625	6080-79-1	9125	6280-03-1	8149		
5344-55-8	3590	5472-38-8	3324	5805-78-5	2381	6086-22-2	11847	6280-96-2	9460		
5344-82-1	11801	5500-21-0	4956	5807-64-7	3039	6088-94-4	8114	6280-98-4	1555		
5344-90-1	1837	5502-88-5	4680	5807-76-1	1096	6089-09-4	9138	6284-40-8	6208		
5345-01-7	8777	5503-12-8	4634	5809-07-4	6103	6089-12-9	3600	6284-84-0	9609		
5349-18-8	2032	5507-44-8	11422	5813-64-9	9806	6091-52-7	4716	6285-05-8	10422		
5349-55-3	10281	5510-99-6	9227	5814-42-6	10392	6091-58-3	3351	6285-06-9	9145		
5349-60-0	1873	5515-83-3	328	5814-85-7	1969	6091-64-1	2521	6287-38-3	536		
5350-47-0	7630	5519-42-6	10842	5836-66-8	9888	6093-68-1	2825	6290-24-0	2598		
5356-83-2	11423	5519-50-6	9723	5837-78-5	3849			6290-49-9	207		

CAS	No.	CAS	No.	CAS	No.	CAS	No.	CAS	No.
6292-82-6	9710	6630-33-7	526	6943-58-4	10356	7225-64-1	6313	7567-22-8	4743
6294-31-1	6855	6639-30-1	2336	6946-29-8	2500	7225-66-3	11878	7568-92-5	1593
6294-89-9	7149	6639-82-3	11336	6946-35-6	2668	7235-40-7	4051	7568-93-6	1839
6298-37-9	11332	6640-27-3	9261	6946-88-9	3338	7236-47-7	3703	7570-26-5	5552
6302-94-9	5264	6640-77-3	2685	6950-43-2	2530	7239-24-9	9829	7570-98-1	7951
6304-18-3	10985	6641-83-4	8895	6950-84-1	12022	7242-17-3	3780	7572-29-4	5940
6306-30-5	6984	6669-13-2	1458	6951-08-2	2419	7250-95-5	7712	7583-53-1	9678
6308-98-1	1585	6682-71-9	7253	6952-59-6	2745	7251-61-8	11337	7611-43-0	472
6309-50-8	5308	6688-11-5	4738	6959-06-4	3624	7252-83-7	5460	7611-86-1	3721
6310-09-4	5791	6693-29-4	4302	6963-44-6	8791	7283-69-4	3767	7611-87-2	3720
6310-21-0	714	6693-30-7	4301	6964-04-1	6401	7283-70-7	3766	7614-93-9	2022
6311-60-0	1309	6693-31-8	4300	6967-43-7	3623	7283-96-7	11735	7615-57-8	11836
6314-28-9	2400	6697-12-7	6664	6969-16-0	11894	7287-19-6	9770	7616-22-0	2790
6315-03-3	2695	6703-78-2	6298	6971-51-3	1890	7287-81-2	1858	7619-08-1	8227
6315-60-2	330	6703-80-6	6297	6973-79-1	10206	7291-33-0	97	7623-01-0	4997
6315-96-4	10474	6703-82-8	4819	6974-10-3	7361	7297-25-8	3432	7623-10-1	10499
6319-21-7	5135	6704-31-0	8596	6974-77-2	9842	7300-03-0	6555	7623-11-2	3683
6319-40-0	2529	6705-31-3	10808	6975-60-6	10439	7307-55-3	11937	7623-13-4	10501
6320-03-2	2293	6709-22-4	4913	6975-92-4	7015	7311-34-4	546	7642-04-8	8493
6320-40-7	2314	6709-39-3	6331	6975-98-0	5054	7314-06-9	732	7642-09-3	7003
6322-01-6	10339	6728-26-3	6996	6975-99-1	5346	7314-37-6	11620	7642-10-6	6537
6322-49-2	3648	6734-98-1	7007	6976-00-7	11963	7315-32-4	2425	7642-15-1	8497
6326-44-9	9953	6738-23-4	1917	6976-72-3	6889	7318-87-4	6884	7658-08-4	6235
6338-45-0	7201	6739-34-0	4337	6982-25-8	3347	7319-00-8	6683	7661-39-4	11222
6341-60-2	9970	6740-88-1	4579	6983-79-5	5108	7321-48-4	8500	7661-55-4	11280
6341-97-5	9288	6742-54-7	2374	6986-48-7	5610	7321-55-3	9913	7665-72-7	8610
6344-28-1	5786	6748-95-4	436	6995-30-8	7905	7328-33-8	5026	7667-55-2	4491
6344-72-5	11338	6753-98-6	5006	6995-79-5	4374	7332-93-6	8895	7667-58-5	4494
6346-09-4	3214	6765-39-5	6322	6996-92-5	2174	7335-01-5	9672	7667-59-6	4493
6351-10-6	7284	6776-19-8	3840	6998-82-9	4352	7335-02-6	9690	7667-60-9	4495
6361-23-5	535	6780-13-8	5513	7005-72-3	1221	7335-06-0	11166	7681-84-7	6049
6372-01-6	3831	6780-49-0	7567	7012-16-0	5972	7335-07-1	11173	7683-59-2	1522
6374-70-5	5146	6781-42-6	5865	7013-11-8	3744	7335-25-3	2537	7688-21-3	7001
6378-11-6	4087	6781-98-2	1139	7021-04-7	889	7335-26-4	2669	7693-52-9	9244
6378-65-0	6890	6784-18-5	4890	7035-03-2	957	7335-27-5	2539	7696-51-7	11129
6380-21-8	9456	6785-23-5	4863	7036-98-8	8191	7344-34-5	7278	7697-46-3	7637
6380-24-1	1451	6789-80-6	6998	7040-43-9	6090	7351-74-8	7803	7700-17-6	4137
6380-28-5	9419	6789-88-4	2610	7044-96-4	7654	7357-93-9	3797	7712-60-9	9026
6380-34-3	464	6789-94-2	9626	7045-89-8	11903	7359-72-0	2334	7712-66-5	11814
6382-13-4	8270	6790-27-8	5345	7058-01-7	4448	7367-38-6	8184	7731-28-4	4535
6402-36-4	5336	6790-42-7	4307	7065-46-5	3690	7369-50-8	1693	7731-29-5	4536
6414-69-3	10283	6795-75-1	5935	7069-38-7	10548	7371-67-7	4181	7735-33-3	8869
6423-02-5	6758	6802-75-1	9966	7081-78-9	5473	7372-85-2	3055	7740-33-2	3589
6423-29-6	7804	6812-07-9	4411	7086-07-9	10147	7372-92-1	7280	7745-91-7	649
6436-90-4	6272	6812-39-1	4822	7087-68-5	9811	7379-12-6	6979	7745-93-9	1072
6443-91-0	5944	6812-78-8	8516	7094-26-0	4485	7383-20-2	8208	7756-94-7	10582
6443-92-1	6535	6814-58-0	4083	7094-27-1	4487	7383-90-6	3058	7766-23-6	9238
6453-99-2	7642	6818-07-1	6901	7094-34-0	7591	7385-78-6	9032	7771-44-0	5360
6482-24-2	5465	6829-40-9	9924	7098-07-9	7207	7385-82-2	7044	7773-60-6	9358
6482-34-4	3995	6829-41-0	9958	7098-21-7	11918	7391-39-1	6881	7778-87-2	6471
6493-77-2	4640	6831-89-6	10831	7101-96-4	11215	7398-82-5	989	7779-80-8	6467
6493-83-0	10457	6832-98-0	6630	7116-86-1	7021	7399-49-7	1984	7781-98-8	2620
6496-48-6	11176	6836-38-0	5324	7119-92-8	5431	7399-50-0	10967	7782-21-0	943
6498-47-1	3868	6843-66-9	11413	7119-94-0	9687	7400-08-0	10644	7785-26-4	2971
6513-13-9	12039	6846-11-3	7221	7132-64-1	8846	7403-66-9	42	7786-34-7	7659
6526-72-3	1977	6848-84-6	8769	7133-68-8	3358	7415-31-8	3748	7786-44-9	8112
6531-35-7	403	6849-18-9	9082	7137-56-6	1251	7417-19-8	1599	7787-72-6	7792
6538-02-9	5379	6850-36-8	4287	7144-05-0	9673	7417-48-3	8666	7789-89-1	10171
6553-48-6	4187	6850-57-3	1812	7145-20-2	7022	7423-69-0	7018	7789-90-4	9790
6556-12-3	6238	6852-54-6	5437	7146-60-3	8374	7425-45-8	3483	7789-92-6	10183
6568-58-7	4731	6861-64-9	7	7148-74-5	10495	7425-47-0	3928	7789-99-3	8945
6573-52-0	4729	6871-44-9	4730	7149-70-4	1066	7425-48-1	3484	9005-79-2	6274
6579-55-1	10383	6873-15-0	11200	7149-75-9	674	7428-48-0	8261	10008-73-8	6148
6592-85-4	5217	6874-98-2	11346	7152-15-0	8897	7433-56-9	5084	10021-92-8	4353
6597-78-0	7653	6876-23-9	4366	7152-88-7	2828	7433-78-5	5085	10022-60-3	6780
6600-40-4	8218	6886-16-4	6921	7153-14-2	4576	7439-12-0	4185	10024-74-5	1822
6607-66-5	3456	6890-03-5	4520	7154-66-7	2887	7443-55-2	4534	10031-82-0	554
6609-56-9	2760	6891-45-8	9250	7154-73-6	11164	7446-43-7	9814	10031-87-5	185
6609-57-0	2754	6894-69-5	4912	7154-75-8	9135	7459-46-3	5654	10042-29-2	268
6617-04-5	4311	6897-76-3	8014	7154-79-2	8839	7462-74-0	9792	10048-32-5	10791
6622-76-0	3851	6897-98-7	8015	7154-80-5	6450	7466-38-8	5133	10061-01-5	10551
6624-76-6	8100	6912-05-6	4923	7163-25-9	7773	7474-10-4	4720	10061-02-6	10550
6626-15-9	1502	6917-76-6	11377	7164-98-9	7212	7474-83-1	7844	10071-60-0	9020
6626-32-0	9347	6921-35-3	8593	7168-85-2	3048	7488-99-5	4054	10075-38-4	3749
6627-55-0	9236	6921-40-0	7738	7169-34-8	2462	7493-82-5	6470	10088-95-6	10785
6627-72-1	2942	6921-64-8	5826	7170-38-9	10320	7495-97-8	9526	10124-73-9	7005
6627-74-3	4667	6923-20-2	10549	7187-01-1	10585	7498-57-9	7737	10130-87-7	2248
6628-00-8	4304	6923-22-4	7663	7193-78-4	5844	7499-05-2	7744	10130-91-3	4566
6628-04-2	11210	6931-71-1	6800	7200-25-1	437	7521-80-4	11370	10136-52-4	6324
6628-06-4	9429	6933-10-4	651	7205-90-5	1136	7523-44-6	3865	10138-10-0	3572
6628-18-8	5452	6938-94-9	6782	7208-05-1	8577	7524-63-2	7988	10138-89-3	3455
6628-21-3	10246	6939-75-9	8911	7208-92-6	5554	7525-62-4	1631	10140-89-3	9781
6628-77-9	10865	6940-09-6	8275	7209-38-3	9614	7526-26-3	9541	10143-54-1	10107
6628-79-1	8896	6940-76-7	9857	7214-50-8	4712	7529-63-7	9697	10143-60-9	6431
6628-80-4	4505	6940-78-9	3233	7216-56-0	8488	7539-25-5	364	10147-36-1	10133
6628-97-3	7866	6941-69-1	129	7220-26-0	6818	7554-65-6	10836	10151-95-8	80
6629-96-5	5192	6942-99-0	1313	7220-81-7	4757	7564-83-8	1630	10152-76-8	10581

10160-87-9	: 10731	10574-36-4	: 7041	13402-36-3	: 4748	14077-58-8	: 298	15068-90-3	: 2950
10160-99-3	: 6627	10574-37-5	: 9034	13402-37-4	: 4744	14090-87-0	: 8393	15104-61-7	: 10126
10166-08-2	: 873	10575-87-8	: 6378	13403-37-7	: 8605	14092-20-7	: 7106	15111-56-5	: 4639
10193-50-7	: 9406	10577-44-3	: 1941	13417-43-1	: 3735	14108-88-4	: 7993	15116-17-3	: 1534
10203-08-4	: 537	10588-10-0	: 8900	13423-15-9	: 6172	14116-69-9	: 6234	15121-84-3	: 1612
10203-28-8	: 5322	10602-34-3	: 3758	13436-48-1	: 7239	14122-00-0	: 9280	15150-39-7	: 140
10203-30-2	: 5323	10604-59-8	: 7307	13444-24-1	: 9715	14159-48-9	: 10130	15164-44-0	: 582
10203-58-4	: 9969	10605-21-7	: 3990	13483-39-3	: 8068	14167-59-0	: 11650	15192-80-0	: 8489
10210-64-7	: 2912	10606-73-2	: 891	13483-40-6	: 7339	14179-94-3	: 9126	15220-85-6	: 10580
10212-74-5	: 3960	11024-24-1	: 6195	13468-40-5	: 8657	14180-63-3	: 10596	15232-76-5	: 8531
10220-20-9	: 9659	11028-42-5	: 4066	13468-41-6	: 11037	14181-05-6	: 11145	15232-96-9	: 1247
10222-95-4	: 2361	12069-69-1	: 511	13471-24-6	: 6915	14188-79-5	: 7794	15240-91-2	: 1800
10226-29-6	: 6961	12122-67-7	: 12067	13475-76-8	: 5272	14214-31-4	: 10538	15250-29-0	: 10670
10229-10-4	: 9142	12321-04-9	: 5961	13475-77-9	: 5354	14222-60-7	: 10886	15268-49-2	: 10086
10230-68-9	: 10475	12427-38-2	: 7420	13475-78-0	: 6416	14248-22-7	: 2129	15299-99-7	: 8067
10233-13-3	: 5316	13012-59-4	: 9864	13475-79-1	: 8766	14252-42-7	: 5450	15301-40-3	: 11293
10236-47-2	: 2795	13014-24-9	: 1401	13475-81-5	: 6844	14255-23-3	: 7023	15320-72-6	: 3833
10240-08-1	: 8030	13019-20-0	: 6522	13475-82-6	: 6433	14272-54-9	: 9115	15329-10-9	: 4708
10250-47-2	: 8949	13019-22-2	: 5092	13481-70-4	: 9294	14273-06-4	: 10732	15333-22-9	: 3829
10257-92-8	: 7318	13023-00-2	: 3500	13482-62-7	: 9524	14282-76-9	: 11729	15339-36-3	: 7421
10259-22-0	: 2670	13024-80-1	: 1750	13491-79-7	: 4521	14287-61-7	: 3507	15356-70-4	: 4548
10265-92-6	: 7441	13027-43-5	: 1650	13505-34-5	: 6403	14296-80-1	: 4172	15358-48-2	: 7574
10273-90-2	: 10991	13030-26-7	: 2778	13511-38-1	: 10219	14303-70-9	: 11629	15361-99-6	: 2277
10276-09-2	: 3835	13038-21-6	: 8190	13513-82-1	: 1893	14304-30-4	: 5087	15409-60-6	: 4591
10283-15-5	: 9256	13045-99-3	: 3798	13529-27-6	: 6084	14315-14-1	: 2402	15411-43-5	: 734
10289-68-6	: 11885	13047-06-8	: 7600	13532-18-8	: 10311	14321-27-8	: 1810	15430-98-5	: 8654
10292-98-5	: 2963	13054-87-0	: 3595	13545-04-5	: 3317	14331-54-5	: 11055	15457-05-3	: 5980
10293-06-8	: 2952	13057-17-5	: 7472	13552-31-3	: 9984	14352-61-5	: 4310	15466-93-0	: 4511
10297-09-3	: 8541	13057-19-7	: 7519	13556-55-3	: 7985	14371-10-9	: 10518	15466-94-1	: 4510
10297-38-8	: 11571	13063-43-9	: 9912	13556-58-6	: 7888	14379-80-1	: 6266	15469-77-9	: 5089
10299-30-6	: 6867	13065-07-1	: 7989	13602-13-6	: 3747	14400-94-3	: 9487	15489-90-4	: 5959
10302-94-0	: 3772	13071-79-9	: 11592	13603-04-8	: 11150	14401-73-1	: 5797	15501-33-4	: 10056
10307-61-6	: 3550	13073-29-5	: 9423	13603-21-9	: 9636	14411-75-7	: 1471	15519-73-0	: 3068
10309-50-9	: 2959	13073-35-3	: 991	13618-93-4	: 7325	14436-32-9	: 5091	15521-18-3	: 10374
10311-74-7	: 1622	13086-84-5	: 9531	13632-93-4	: 1416	14444-77-0	: 1409	15543-70-1	: 4965
10311-84-9	: 5105	13099-50-8	: 1227	13632-94-5	: 1420	14446-69-6	: 9653	15570-12-4	: 2299
10312-83-1	: 16	13102-31-3	: 9449	13632-95-6	: 1418	14452-30-3	: 5836	15600-01-8	: 10053
10315-89-6	: 9685	13105-52-7	: 3934	13643-05-5	: 8667	14459-29-1	: 9735	15606-95-8	: 439
10316-00-4	: 7700	13114-62-0	: 12023	13665-04-8	: 5798	14472-14-1	: 9242	15644-80-1	: 11316
10317-10-9	: 10353	13116-53-5	: 10145	13672-18-9	: 7730	14476-30-3	: 10162	15647-08-2	: 9589
10321-71-8	: 9078	13116-60-4	: 10143	13678-54-1	: 11764	14476-37-0	: 11960	15647-11-7	: 4502
10323-40-7	: 3471	13121-70-5	: 5011	13679-46-4	: 6128	14484-64-1	: 5957	15657-96-2	: 10091
10324-58-0	: 9692	13129-23-2	: 6063	13679-70-4	: 11736	14490-79-0	: 11608	15679-02-4	: 4188
10327-08-9	: 10204	13138-51-7	: 3429	13679-74-8	: 5856	14548-31-3	: 7105	15706-73-7	: 3545
10334-26-6	: 2926	13147-16-4	: 11639	13684-56-5	: 5104	14548-46-0	: 7639	15721-05-8	: 4992
10339-67-0	: 6602	13151-04-7	: 6546	13684-63-4	: 9198	14558-14-6	: 8003	15726-15-5	: 6525
10340-91-7	: 1791	13151-05-8	: 6545	13688-75-0	: 11468	14575-93-0	: 2965	15727-65-8	: 7892
10342-83-3	: 10418	13151-17-2	: 7045	13698-16-3	: 3940	14595-35-8	: 3783	15764-04-2	: 8052
10343-99-4	: 11246	13151-34-3	: 5055	13698-87-8	: 1326	14596-92-0	: 6682	15764-16-6	: 547
10373-81-6	: 2957	13153-11-2	: 11694	13706-86-0	: 6809	14602-62-1	: 10161	15797-21-4	: 4795
10394-35-1	: 2378	13163-73-0	: 4604	13721-54-5	: 7103	14620-52-1	: 5297	15802-97-8	: 10841
10403-60-8	: 3451	13165-77-0	: 2181	13732-80-4	: 1417	14620-53-2	: 11617	15802-99-0	: 10839
10410-35-2	: 2448	13172-91-3	: 6557	13794-14-4	: 3580	14620-55-4	: 8237	15825-70-4	: 7424
10420-90-3	: 6712	13176-46-0	: 3478	13799-90-1	: 1349	14630-40-1	: 11434	15826-80-9	: 2981
10420-91-4	: 6715	13194-48-4	: 5934	13808-64-5	: 10817	14631-45-9	: 10102	15840-60-5	: 7047
10424-29-0	: 3341	13195-80-7	: 10542	13811-01-3	: 4274	14660-52-7	: 8861	15843-48-8	: 5568
10428-64-5	: 3272	13206-46-7	: 9776	13828-34-7	: 4441	14676-29-0	: 6409	15854-58-7	: 10647
10435-81-1	: 8419	13222-26-9	: 10498	13831-30-6	: 101	14685-29-1	: 4995	15869-80-4	: 6407
10436-75-6	: 794	13249-58-6	: 1030	13835-81-9	: 11467	14686-13-6	: 6534	15869-85-9	: 8143
10439-23-3	: 2186	13250-46-9	: 303	13855-77-1	: 3964	14686-14-7	: 6536	15869-86-0	: 8404
10443-65-9	: 10610	13254-34-7	: 6479	13862-07-2	: 9665	14689-97-5	: 5592	15869-87-1	: 8373
10453-86-8	: 11340	13258-63-4	: 10949	13865-48-0	: 2018	14697-46-2	: 8851	15869-89-3	: 8376
10463-48-6	: 3740	13269-52-8	: 7002	13877-91-3	: 8485	14716-89-3	: 7404	15869-92-8	: 8380
10473-14-0	: 3871	13279-87-3	: 3448	13898-58-3	: 2514	14719-83-6	: 2548	15869-93-9	: 8381
10476-95-6	: 10564	13290-74-9	: 1205	13898-73-2	: 4678	14720-74-2	: 6438	15869-94-0	: 8382
10478-02-1	: 2592	13313-91-2	: 261	13905-48-1	: 4320	14721-66-5	: 6658	15869-95-1	: 8383
10478-12-3	: 9083	13325-10-5	: 3597	13909-73-4	: 5874	14737-94-1	: 10622	15869-96-2	: 8384
10486-19-8	: 11869	13327-31-6	: 11273	13937-08-1	: 9957	14752-60-4	: 9124	15870-10-7	: 6543
10487-71-5	: 3887	13329-40-3	: 5837	13939-06-5	: 7662	14752-61-5	: 9133	15874-03-0	: 6637
10487-96-4	: 4871	13347-42-7	: 9255	13956-29-1	: 1537	14752-75-1	: 1743	15877-57-3	: 8694
10488-25-2	: 4539	13351-61-6	: 2151	13963-35-4	: 2017	14804-32-1	: 1682	15890-36-5	: 4541
10488-94-5	: 8436	13351-73-0	: 2872	13979-20-9	: 3836	14804-59-2	: 10510	15890-40-1	: 4859
10489-97-1	: 4345	13356-08-6	: 5950	13984-57-1	: 6906	14813-68-4	: 8345	15901-49-2	: 5003
10495-45-1	: 10132	13360-45-7	: 4076	13993-65-2	: 9500	14845-35-3	: 4706	15909-76-9	: 10254
10498-35-8	: 4359	13360-61-7	: 8653	13997-70-1	: 1232	14845-55-7	: 4887	15918-08-8	: 6426
10500-57-9	: 11297	13360-63-9	: 3218	13997-73-4	: 9271	14847-51-9	: 9237	15922-53-9	: 11849
10503-96-5	: 3249	13360-65-1	: 10809	14007-64-8	: 898	14850-22-7	: 8495	15932-80-6	: 4607
10508-09-5	: 9154	13361-35-8	: 143	14010-23-2	: 6315	14850-23-8	: 8496	15950-66-0	: 9481
10519-06-9	: 1117	13361-36-9	: 137	14021-92-2	: 5048	14852-31-4	: 6660	15953-45-4	: 10821
10519-33-2	: 5093	13361-63-2	: 9129	14045-26-2	: 9671	14869-39-7	: 5814	15953-73-8	: 10819
10521-91-2	: 2101	13362-52-2	: 11876	14062-18-1	: 923	14919-01-8	: 8494	15971-29-6	: 7755
10522-26-6	: 11955	13367-29-8	: 2993	14062-78-3	: 627	14924-53-9	: 4175	15972-60-8	: 322
10524-08-0	: 9018	13388-93-7	: 5846	14064-10-9	: 9931	14938-35-3	: 9444	15980-15-1	: 8569
10528-67-3	: 4469	13389-42-9	: 8492	14064-43-8	: 1223	14963-40-7	: 4946	16003-14-8	: 6160
10544-63-5	: 3838	13398-94-2	: 1606	14064-68-7	: 2768	14979-39-6	: 6501	16009-13-5	: 5958
10568-38-4	: 1683	13402-32-9	: 1308	14072-86-7	: 4856	14996-78-2	: 4236	16020-17-0	: 8001
10574-17-1	: 2030	13402-35-2	: 4746	14073-97-3	: 4606	15014-25-2	: 9927	16021-20-8	: 1702

16066-09-4	: 11647	16789-46-1	: 6825	17745-45-8	: 3136	18913-31-0	: 3175	19780-66-6	: 9046
16096-32-5	: 7315	16789-51-8	: 7034	17773-65-8	: 3737	18917-91-4	: 340	19780-87-7	: 9049
16106-59-5	: 7020	16790-49-1	: 3197	17788-00-0	: 9292	18932-14-4	: 8420	19780-90-6	: 6798
16109-68-5	: 5120	16805-78-0	: 9578	17789-14-9	: 5205	18936-17-9	: 3409	19780-96-2	: 8164
16133-49-6	: 777	16813-18-6	: 6318	17804-35-2	: 515	18937-76-3	: 11990	19781-10-3	: 8451
16136-84-8	: 10534	16837-43-7	: 2118	17837-41-1	: 11693	18963-00-3	: 10176	19781-11-4	: 8458
16136-85-9	: 10533	16841-48-8	: 10597	17849-38-6	: 1848	18963-01-4	: 10148	19781-68-1	: 4976
16147-07-2	: 1446	16874-33-2	: 6072	17851-27-3	: 1710	18970-44-0	: 1688	19781-69-2	: 4977
16177-46-1	: 4186	16883-48-0	: 4860	17878-44-3	: 11367	18971-59-0	: 8821	19781-73-8	: 4404
16179-36-5	: 8885	16889-72-8	: 10294	17880-86-3	: 10502	18979-60-7	: 1549	19781-77-2	: 6572
16183-46-3	: 3666	16898-52-5	: 9701	17912-17-3	: 8368	18979-62-9	: 1539	19812-64-7	: 11618
16188-57-1	: 9374	16932-49-3	: 2753	17937-88-7	: 11404	18992-39-7	: 10402	19814-71-2	: 2077
16197-90-3	: 3822	16939-57-4	: 1097	17952-11-3	: 8808	18992-80-8	: 5806	19819-98-8	: 1609
16245-79-7	: 831	16957-70-3	: 9076	17972-08-6	: 7652	18999-28-5	: 6561	19821-84-2	: 6612
16246-07-4	: 3947	16980-61-3	: 4512	17988-31-7	: 11427	19037-72-0	: 4922	19832-98-5	: 8050
16292-88-9	: 776	16995-35-0	: 10486	18002-79-4	: 11360	19044-88-3	: 8553	19860-56-1	: 10256
16302-35-5	: 10778	17005-59-3	: 11383	18023-33-1	: 11426	19074-59-0	: 9336	19866-51-4	: 9077
16306-39-1	: 9184	17015-11-1	: 6914	18023-36-4	: 11511	19089-47-5	: 10380	19879-11-9	: 4508
16308-92-2	: 1860	17021-26-0	: 357	18030-61-0	: 11364	19113-78-1	: 6928	19879-12-0	: 4509
16315-07-4	: 1922	17042-16-9	: 6980	18036-81-2	: 11497	19145-60-9	: 487	19889-37-3	: 3516
16326-97-9	: 4806	17057-82-8	: 7252	18169-57-8	: 11469	19145-96-1	: 4356	19890-00-7	: 2956
16326-98-0	: 4807	17057-88-4	: 3060	18171-59-0	: 11373	19155-24-9	: 7329	19891-74-8	: 4060
16357-59-8	: 11231	17057-92-0	: 7887	18178-59-1	: 11421	19163-05-4	: 3129	19902-08-0	: 2924
16369-05-4	: 3599	17059-44-8	: 1485	18181-80-1	: 3140	19177-04-9	: 1987	19915-11-8	: 8689
16369-17-8	: 8922	17059-45-9	: 1482	18187-24-1	: 5234	19184-10-2	: 10567	19942-78-0	: 11708
16369-21-4	: 5765	17059-52-8	: 2459	18231-53-3	: 8764	19184-67-9	: 8774	19947-22-9	: 1700
16386-93-9	: 9075	17071-47-5	: 3401	18236-57-2	: 11400	19221-28-4	: 3165	19961-27-4	: 9810
16387-70-5	: 8344	17082-09-6	: 10717	18236-89-0	: 11399	19230-27-4	: 1294	19967-57-8	: 9775
16387-72-7	: 11610	17082-12-1	: 5129	18240-50-1	: 6114	19246-38-9	: 4346	19987-13-4	: 9865
16409-46-4	: 3554	17085-91-5	: 7828	18243-41-9	: 11366	19246-39-0	: 3601	20009-26-1	: 7955
16419-60-6	: 3134	17091-40-6	: 5747	18261-92-2	: 8280	19269-28-4	: 6720	20024-90-2	: 1703
16420-13-6	: 3980	17094-21-2	: 3558	18262-85-6	: 1709	19317-11-4	: 5330	20027-77-4	: 7991
16423-68-0	: 11524	17096-37-6	: 3796	18281-05-5	: 5356	19329-89-6	: 10274	20030-32-4	: 4690
16426-64-5	: 2527	17102-64-6	: 6709	18282-40-1	: 1680	19354-27-9	: 6170	20063-92-7	: 8183
16432-53-4	: 6794	17113-33-6	: 6159	18282-59-2	: 1015	19374-46-0	: 4857	20089-07-0	: 3439
16433-43-5	: 1104	17176-77-1	: 9586	18295-59-5	: 8697	19393-45-4	: 3773	20098-48-0	: 2338
16435-50-0	: 9024	17199-29-0	: 903	18309-32-5	: 2978	19393-92-1	: 1013	20177-02-0	: 9028
16452-01-0	: 771	17206-52-9	: 4595	18338-40-4	: 3449	19395-58-5	: 11198	20184-89-8	: 8198
16484-17-6	: 7293	17216-08-9	: 8045	18362-36-2	: 578	19396-83-9	: 2916	20184-91-2	: 8199
16491-15-9	: 4921	17216-62-5	: 9933	18362-97-5	: 8893	19398-47-1	: 3614	20185-55-1	: 2694
16503-53-0	: 2147	17226-28-7	: 3416	18368-63-3	: 10920	19398-53-9	: 8743	20193-20-8	: 9809
16509-46-9	: 8335	17226-34-5	: 7009	18381-45-8	: 3406	19398-61-9	: 1387	20213-30-3	: 7917
16518-62-0	: 644	17228-64-7	: 10918	18388-45-9	: 11408	19398-75-5	: 4799	20231-81-6	: 4045
16520-62-0	: 1124	17236-58-7	: 11777	18397-07-4	: 4154	19398-77-7	: 6768	20232-11-5	: 7279
16532-02-8	: 11365	17249-75-1	: 11747	18398-36-2	: 7849	19402-87-0	: 3988	20237-34-7	: 6681
16538-93-5	: 4737	17249-76-2	: 11749	18423-69-3	: 6136	19404-07-0	: 5191	20244-71-7	: 5888
16545-78-1	: 9061	17249-77-3	: 11791	18441-43-5	: 2458	19420-29-2	: 7946	20247-89-6	: 2008
16560-43-3	: 11272	17249-79-5	: 11746	18444-79-6	: 11521	19434-65-2	: 9778	20266-65-3	: 10022
16580-23-7	: 4438	17249-82-0	: 11741	18476-57-8	: 8327	19438-10-9	: 2636	20278-84-6	: 6446
16580-24-8	: 4437	17283-45-3	: 2946	18479-51-1	: 8515	19447-29-1	: 8197	20278-85-7	: 6443
16582-38-0	: 1207	17297-04-0	: 1108	18479-57-7	: 8450	19472-74-3	: 948	20278-87-9	: 6449
16587-71-6	: 4592	17299-41-1	: 4719	18492-37-0	: 2960	19481-82-4	: 10094	20278-88-0	: 6451
16605-36-0	: 4435	17301-94-9	: 8142	18495-23-3	: 8536	19482-57-6	: 8535	20278-89-1	: 6452
16619-12-8	: 7291	17302-01-1	: 6410	18495-26-6	: 6613	19489-10-2	: 4397	20279-53-2	: 6062
16629-19-9	: 11780	17302-02-2	: 6767	18495-30-2	: 10144	19493-31-3	: 3026	20281-83-8	: 8938
16631-63-3	: 4833	17302-04-4	: 6415	18510-29-7	: 11455	19493-44-8	: 7376	20281-85-0	: 3616
16650-52-5	: 7762	17312-63-9	: 8122	18521-07-8	: 8519	19523-57-0	: 7753	20281-86-1	: 6912
16661-99-7	: 4349	17348-59-3	: 10081	18523-48-3	: 109	19549-71-4	: 6480	20281-91-8	: 3619
16672-87-0	: 5929	17356-19-3	: 4866	18530-30-8	: 2917	19549-73-6	: 6481	20291-91-2	: 6816
16674-04-7	: 10230	17372-87-1	: 11525	18530-56-8	: 8212	19549-74-7	: 6482	20291-95-6	: 6439
16694-52-3	: 9868	17406-45-0	: 6196	18589-29-2	: 6420	19549-77-0	: 6484	20327-65-5	: 10568
16696-65-4	: 11940	17432-37-0	: 598	18598-63-5	: 5019	19549-79-2	: 6486	20333-41-9	: 5245
16714-77-5	: 3654	17447-00-6	: 9085	18649-84-4	: 6108	19549-84-9	: 6515	20354-26-1	: 7643
16726-67-3	: 7761	17447-01-7	: 9084	18668-68-3	: 3147	19549-97-4	: 6617	20357-79-3	: 7247
16736-42-8	: 8325	17478-66-9	: 5138	18685-70-6	: 4073	19550-03-9	: 6922	20359-55-1	: 8979
16737-44-3	: 1983	17484-36-5	: 1934	18699-48-4	: 1338	19550-05-1	: 6924	20395-25-9	: 10173
16744-89-1	: 7083	17496-14-9	: 7289	18719-43-2	: 9932	19550-07-3	: 6927	20442-79-9	: 3078
16745-94-1	: 7017	17534-15-5	: 1561	18720-65-5	: 6502	19550-09-5	: 6929	20474-93-5	: 3856
16746-02-4	: 6836	17556-48-8	: 10409	18729-48-1	: 4870	19550-10-8	: 6964	20485-41-0	: 11669
16746-86-4	: 7013	17577-93-4	: 8523	18747-42-7	: 10470	19550-14-2	: 6987	20589-85-9	: 10598
16746-87-5	: 7014	17587-22-3	: 8398	18748-27-1	: 11578	19550-40-4	: 4655	20599-21-7	: 8534
16747-25-4	: 6859	17609-31-3	: 8484	18752-92-6	: 11445	19550-48-2	: 4920	20599-27-3	: 9011
16747-26-5	: 6860	17618-76-7	: 6554	18758-34-4	: 4998	19564-40-0	: 2115	20634-92-8	: 9812
16747-28-7	: 6862	17627-77-9	: 7395	18773-77-8	: 4160	19594-02-6	: 1652	20637-49-4	: 10185
16747-30-1	: 6865	17639-93-9	: 10222	18786-24-8	: 8572	19613-78-4	: 1024	20668-29-5	: 11261
16747-31-2	: 6866	17640-26-5	: 182	18787-64-9	: 6663	19617-44-8	: 7231	20699-48-3	: 3261
16747-32-3	: 8810	17648-05-4	: 2281	18791-49-6	: 10282	19653-33-9	: 9704	20731-93-5	: 1875
16747-33-4	: 8811	17655-74-2	: 1651	18794-84-8	: 5331	19655-60-8	: 11268	20734-71-8	: 9386
16747-38-9	: 8842	17658-32-1	: 8924	18812-82-9	: 7079	19666-30-9	: 8565	20739-58-6	: 8543
16747-42-5	: 6848	17672-88-7	: 2422	18829-56-6	: 8180	19686-73-8	: 10345	20748-86-1	: 6400
16747-44-7	: 8836	17679-92-4	: 8526	18839-90-2	: 9359	19689-18-0	: 5083	20754-04-5	: 8478
16747-45-8	: 8837	17696-11-6	: 8426	18840-45-4	: 12042	19694-02-1	: 10854	20769-85-1	: 10491
16747-50-5	: 4814	17696-37-6	: 9326	18854-56-3	: 5586	19700-97-1	: 6693	20809-46-5	: 8479
16749-11-4	: 4361	17696-64-9	: 142	18858-85-6	: 11441	19718-45-7	: 7956	20820-82-0	: 11022
16751-59-0	: 6351	17704-30-2	: 165	18869-72-2	: 6311	19735-89-8	: 10849	20830-75-5	: 4041
16752-77-5	: 7650	17715-00-3	: 4473	18904-54-6	: 4146	19752-55-7	: 1012	20850-43-5	: 2421
16754-49-7	: 10080	17742-04-0	: 7528	18905-91-4	: 6973	19780-63-3	: 8933	20893-30-5	: 11725

20972-43-4 : 5131	22401-25-8 : 9546	24342-04-9 : 9139	25871-69-6 : 7587	29091-05-2 : 5181	
20984-33-2 : 11322	22409-91-2 : 10322	24346-56-3 : 2132	26073-26-7 : 10500	29091-21-2 : 9761	
21020-24-6 : 3899	22418-49-1 : 10577	24356-00-1 : 9017	26106-95-6 : 3602	29106-32-9 : 4936	
21020-26-8 : 9131	22421-97-2 : 10229	24393-70-2 : 2943	26116-12-1 : 11167	29110-74-5 : 7236	
21020-27-9 : 9136	22445-41-6 : 1771	24395-10-6 : 6571	26118-38-7 : 6963	29113-63-1 : 6055	
21040-45-9 : 10706	22483-09-6 : 5419	24398-88-7 : 2522	26120-52-5 : 4253	29136-19-4 : 2062	
21040-59-5 : 7392	22520-40-7 : 8391	24410-50-2 : 2449	26153-88-8 : 8703	29172-20-1 : 6111	
21087-29-6 : 1230	22524-32-9 : 113	24410-51-3 : 2446	26171-83-5 : 3344	29204-02-2 : 5365	
21087-64-9 : 7658	22531-06-2 : 8051	24410-52-4 : 2452	26186-01-6 : 8107	29232-93-7 : 9730	
21102-13-6 : 6952	22531-20-0 : 7891	24425-40-9 : 7244	26186-02-7 : 11897	29276-40-2 : 7707	
21129-27-1 : 4546	22539-65-7 : 1469	24425-97-6 : 10051	26190-82-9 : 11256	29289-13-2 : 762	
21179-48-6 : 10239	22545-12-6 : 1603	24438-17-3 : 7328	26198-63-0 : 9892	29342-44-7 : 6740	
21211-65-4 : 11133	22545-13-7 : 1604	24446-63-7 : 942	26225-79-6 : 5933	29342-53-8 : 2918	
21232-53-1 : 361	22564-83-6 : 7429	24544-04-5 : 636	26259-45-0 : 3143	29394-58-9 : 3383	
21266-88-6 : 6375	22564-99-4 : 8332	24545-81-1 : 2987	26260-02-6 : 580	29400-42-8 : 8211	
21266-90-0 : 6376	22567-21-1 : 4543	24549-06-2 : 745	26303-90-2 : 5090	29472-95-5 : 9421	
21293-29-8 : 8679	22610-63-5 : 8255	24557-10-6 : 10627	26311-17-1 : 5424	29488-27-5 : 11773	
21296-92-4 : 7323	22618-23-1 : 9315	24579-70-2 : 3941	26311-33-1 : 7063	29538-77-0 : 4507	
21299-26-3 : 4578	22621-74-5 : 2947	24580-48-1 : 4550	26330-51-8 : 3476	29554-49-2 : 4745	
21326-62-5 : 3611	22646-79-3 : 4646	24580-52-7 : 4551	26340-58-9 : 4151	29555-02-0 : 4964	
21352-22-7 : 11209	22663-64-5 : 8166	24602-86-6 : 11898	26370-28-5 : 8108	29559-27-1 : 2004	
21370-71-8 : 8054	22672-74-8 : 5150	24612-75-7 : 4477	26399-36-0 : 9763	29559-52-2 : 10177	
21391-98-0 : 4635	22692-70-2 : 11560	24618-31-3 : 4317	26438-26-6 : 7931	29560-84-7 : 10701	
21400-25-9 : 10602	22717-57-3 : 2640	24622-72-7 : 11169	26438-27-7 : 7793	29604-75-9 : 1296	
21436-44-2 : 2683	22726-00-7 : 605	24634-61-5 : 6704	26438-28-8 : 7879	29606-79-9 : 4603	
21472-86-6 : 2335	22734-04-9 : 4905	24644-78-8 : 7290	26464-32-4 : 3488	29624-17-7 : 4348	
21490-63-1 : 8609	22767-95-9 : 2140	24677-78-9 : 540	26485-81-6 : 7288	29653-38-1 : 3926	
21504-43-8 : 10559	22767-96-0 : 2145	24716-09-4 : 890	26472-41-3 : 4282	29682-41-5 : 1380	
21531-91-9 : 6793	22768-19-0 : 6550	24724-07-0 : 11949	26473-49-4 : 3537	29780-00-5 : 11896	
21535-97-7 : 2457	22768-20-3 : 6549	24734-68-7 : 2125	26489-01-0 : 8514	29823-16-3 : 8873	
21564-91-0 : 7984	22781-23-3 : 514	24764-97-4 : 3179	26513-20-2 : 1265	29843-58-1 : 3757	
21564-92-1 : 7986	22839-47-0 : 9511	24796-87-0 : 4503	26530-20-1 : 8528	29865-90-5 : 568	
21570-35-4 : 6503	22868-76-4 : 11056	24815-24-5 : 12060	26531-85-1 : 6237	29882-57-3 : 8720	
21571-91-5 : 8751	22900-08-9 : 4540	24840-05-9 : 10588	26533-33-5 : 8169	29883-15-6 : 954	
21617-18-5 : 11250	22922-67-4 : 7004	24844-28-8 : 4556	26549-24-6 : 6913	29897-82-3 : 11172	
21633-79-4 : 9972	22924-15-8 : 553	24885-45-8 : 9375	26549-25-7 : 6474	29916-45-8 : 3269	
21655-48-1 : 9611	22927-13-5 : 558	24887-75-0 : 345	26608-06-0 : 5165	29921-50-4 : 8063	
21693-54-9 : 7992	22936-86-3 : 5013	24910-63-2 : 9036	26644-46-2 : 11902	29943-42-8 : 10799	
21700-31-2 : 10125	23010-00-8 : 3733	24932-49-8 : 1297	26730-14-3 : 11879	29949-19-7 : 7511	
21702-79-4 : 2559	23010-04-0 : 3283	24948-66-1 : 5095	26754-48-3 : 6030	29972-79-0 : 5341	
21725-46-2 : 4161	23012-11-5 : 8578	24964-64-5 : 2757	26771-11-9 : 4959	30090-17-6 : 814	
21777-84-4 : 1992	23048-13-7 : 4943	24965-84-2 : 4596	26963-33-7 : 11771	30122-12-4 : 8745	
21849-97-8 : 7840	23063-36-7 : 1390	24973-49-7 : 3020	27063-27-0 : 11208	30168-50-4 : 4830	
21858-89-3 : 6970	23074-36-4 : 3709	25006-60-4 : 167	27129-86-8 : 1057	30273-11-1 : 813	
21884-26-4 : 8909	23079-65-4 : 1116	25013-16-5 : 9329	27129-87-9 : 1861	30361-33-2 : 5025	
21889-88-3 : 6585	23097-98-5 : 6618	25044-01-3 : 9107	27287-69-2 : 2773	30389-18-5 : 4290	
21890-35-7 : 3148	23103-98-2 : 9728	25057-89-0 : 518	27314-13-2 : 8213	30433-57-9 : 10833	
21915-53-7 : 8619	23132-52-7 : 1467	25109-28-8 : 1009	27356-52-1 : 11269	30515-28-7 : 6458	
21923-23-9 : 4092	23135-22-0 : 8568	25109-57-3 : 3653	27387-79-7 : 10218	30525-89-4 : 8634	
21948-47-0 : 8372	23144-52-7 : 8448	25117-74-2 : 2755	27464-82-0 : 11659	30538-80-8 : 9689	
21964-23-8 : 9013	23184-86-9 : 3144	25118-28-9 : 6755	27532-76-9 : 378	30544-34-4 : 6077	
21964-44-3 : 8189	23204-80-0 : 1229	25144-04-1 : 4869	27598-85-2 : 9200	30560-19-1 : 10	
21968-17-2 : 9827	23287-26-5 : 2638	25144-05-2 : 4868	27610-88-4 : 8400	30600-19-2 : 10619	
21974-48-1 : 9714	23308-82-9 : 1894	25148-87-2 : 3756	27644-49-1 : 8465	30614-77-8 : 6082	
21984-93-0 : 6065	23378-11-2 : 3732	25167-70-8 : 9064	27655-41-0 : 7375	30673-60-0 : 5069	
21987-21-3 : 11886	23399-70-4 : 1174	25172-42-3 : 4873	27701-01-5 : 3943	30685-95-1 : 3886	
21994-77-4 : 7065	23418-37-3 : 8469	25172-43-4 : 4874	27779-01-7 : 4158	30689-41-9 : 4916	
21994-78-5 : 7064	23418-38-4 : 8170	25182-84-7 : 10395	27831-13-6 : 1629	30752-19-3 : 1045	
22029-76-1 : 3873	23422-53-9 : 6001	25186-47-4 : 1386	27876-94-4 : 5107	30926-60-4 : 1188	
22037-28-1 : 6034	23431-48-3 : 2028	25201-40-5 : 6338	27993-04-2 : 8984	30964-01-3 : 8207	
22037-73-6 : 3710	23438-77-9 : 4722	25229-42-9 : 3834	28056-54-4 : 4896	31053-99-3 : 1092	
22038-57-9 : 3830	23444-65-7 : 7854	25234-83-7 : 8898	28123-70-8 : 8125	31059-19-5 : 7944	
22056-82-2 : 14	23470-12-4 : 7127	25245-33-4 : 1767	28148-04-1 : 9890	31116-84-4 : 10459	
22077-55-0 : 2176	23505-41-1 : 9729	25285-71-8 : 10397	28249-77-6 : 11697	31161-71-4 : 11627	
22104-78-5 : 8510	23511-70-8 : 3582	25309-64-2 : 1681	28261-97-4 : 5273	31202-12-7 : 363	
22118-12-3 : 3682	23564-05-8 : 11723	25311-74-5 : 7359	28300-74-5 : 11588	31218-83-4 : 10719	
22128-62-7 : 4014	23578-35-0 : 6751	25312-34-9 : 3874	28320-88-9 : 8362	31252-42-3 : 9700	
22128-63-8 : 4016	23580-51-0 : 8544	25312-65-6 : 4094	28320-89-0 : 6364	31283-43-9 : 8498	
22139-77-1 : 1545	23580-52-1 : 10960	25346-31-0 : 8718	28332-44-7 : 7095	31340-44-0 : 10247	
22144-60-1 : 1912	23617-71-2 : 2589	25346-32-1 : 8730	28466-26-4 : 10812	31366-00-4 : 1713	
22160-12-9 : 6893	23619-59-2 : 9459	25346-33-2 : 8714	28467-92-7 : 3190	31367-46-1 : 6496	
22160-39-0 : 8887	23676-09-7 : 2599	25354-90-5 : 5065	28469-92-3 : 1369	31508-08-4 : 6607	
22160-40-3 : 8888	23708-56-7 : 11954	25354-98-7 : 8649	28664-03-1 : 4171	31508-12-0 : 8194	
22160-41-4 : 6895	23733-68-8 : 4717	25394-75-6 : 10772	28685-60-1 : 3958	31543-75-6 : 1306	
22202-65-9 : 11989	23769-39-3 : 11921	25395-22-6 : 103	28715-26-6 : 2451	31599-60-7 : 1789	
22224-92-6 : 5947	23843-52-9 : 8239	25395-31-7 : 10189	28788-62-7 : 2889	31599-61-8 : 1774	
22227-75-4 : 10556	23947-60-6 : 5932	25414-22-6 : 6127	28823-41-8 : 6696	31642-67-8 : 8188	
22232-54-8 : 7194	23950-58-5 : 10746	25415-67-2 : 8892	28832-55-5 : 3180	31656-92-5 : 981	
22253-11-8 : 7259	24127-62-6 : 10615	25415-70-7 : 8894	28905-12-6 : 6232	31699-35-1 : 8116	
22287-02-1 : 8176	24152-39-4 : 9637	25419-36-7 : 7990	28921-35-9 : 6064	31702-33-7 : 10605	
22319-28-4 : 6592	24247-53-8 : 3696	25419-37-8 : 7987	28994-41-4 : 9454	31720-69-1 : 4562	
22331-38-0 : 5351	24264-08-2 : 3877	25542-62-5 : 6879	28995-89-3 : 8004	31766-07-1 : 2771	
22349-03-7 : 4456	24280-93-1 : 7052	25611-78-3 : 1584	29003-73-4 : 163	31772-05-1 : 11874	
22364-88-1 : 960	24306-10-9 : 6323	25641-86-9 : 1340	29022-31-9 : 8201	31805-48-8 : 11761	
22374-89-6 : 2119	24307-26-4 : 7430	25659-22-7 : 7087	29030-84-0 : 6689	31844-95-8 : 9008	
22381-86-8 : 11980	24331-72-4 : 9794	25695-77-6 : 10689	29040-46-8 : 2445	31844-96-9 : 3716	
22398-09-0 : 9709	24332-22-7 : 6584	25790-55-0 : 3150	29072-93-3 : 10087	31849-78-2 : 3708	

31883-98-4	: 4713	34840-79-4	: 1364	37493-31-5	: 6162	40575-42-6	: 8509	50710-40-2	: 3244
31915-94-3	: 1098	34859-98-8	: 8946	37493-70-2	: 11953	40589-38-6	: 3890	50840-23-8	: 11047
32018-88-5	: 7760	34874-30-1	: 11117	37549-83-0	: 7056	40596-69-8	: 7651	50868-72-9	: 773
32064-78-1	: 8524	34885-03-5	: 4424	37549-89-6	: 7031	40604-49-7	: 8023	50868-73-0	: 774
32116-65-7	: 4889	34887-14-4	: 8748	37674-63-8	: 9079	40605-42-3	: 3864	50876-32-9	: 4478
32121-06-5	: 10544	34893-50-0	: 9649	37674-67-2	: 7074	40614-52-6	: 3363	50876-33-0	: 4851
32121-07-6	: 10543	34973-41-6	: 3380	37688-96-3	: 11616	40646-07-9	: 6397	50902-82-4	: 6887
32161-06-1	: 9720	35000-22-7	: 10525	37710-49-9	: 8663	40649-36-3	: 4613	51034-46-9	: 3154
32166-40-8	: 8053	35008-86-7	: 1761	37746-78-4	: 3819	40650-41-7	: 7281	51060-05-0	: 7360
32315-10-9	: 7585	35018-15-6	: 11149	37806-29-4	: 1808	40734-75-6	: 1002	51115-02-7	: 4931
32322-73-9	: 47	35018-28-1	: 11175	37845-14-0	: 8060	40738-26-9	: 5310	51116-73-5	: 8715
32328-03-3	: 8778	35021-70-6	: 1752	37866-06-1	: 3731	40745-44-6	: 2	51152-12-6	: 2930
32337-74-9	: 8664	35099-89-9	: 4227	37924-13-3	: 9149	40775-09-5	: 1741	51218-45-2	: 7657
32360-05-7	: 10667	35120-10-6	: 286	37935-39-0	: 3492	40960-69-8	: 4554	51225-20-8	: 691
32367-54-7	: 7889	35161-71-8	: 10726	37951-49-8	: 10456	40960-73-4	: 4553	51235-04-2	: 6995
32395-96-3	: 1747	35182-51-5	: 10968	37973-84-5	: 11975	41055-92-9	: 6513	51282-49-6	: 2549
32433-28-6	: 90	35194-37-7	: 6562	37985-18-5	: 11701	41065-91-2	: 6883	51284-68-5	: 5116
32449-92-8	: 6239	35205-71-1	: 7059	38031-78-6	: 10943	41085-43-2	: 1071	51299-55-9	: 4620
32460-00-7	: 6078	35205-79-9	: 8153	38204-89-6	: 1361	41096-46-2	: 7182	51308-99-7	: 8163
32460-20-1	: 6190	35216-11-6	: 11641	38237-74-0	: 8960	41198-08-7	: 9782	51309-10-5	: 5073
32541-33-6	: 8925	35223-80-4	: 120	38256-94-9	: 5426	41223-14-7	: 4773	51333-70-1	: 8111
32607-54-8	: 8885	35242-05-8	: 4504	38260-54-7	: 5945	41284-12-2	: 5615	51338-27-3	: 5171
32620-08-9	: 3707	35246-18-5	: 7309	38300-67-3	: 3445	41295-64-1	: 3687	51437-67-3	: 5115
32665-23-9	: 3553	35274-05-6	: 10271	38460-95-6	: 11982	41433-81-2	: 9954	51500-32-4	: 10818
32666-56-1	: 7714	35280-53-6	: 136	38487-94-4	: 2103	41589-42-8	: 9936	51525-97-4	: 10137
32718-56-2	: 5734	35305-13-6	: 9686	38511-07-8	: 4645	41635-77-2	: 1753	51556-10-6	: 6614
32723-67-4	: 586	35323-91-2	: 2846	38514-02-2	: 8461	41653-96-7	: 7057	51575-85-0	: 6616
32750-14-4	: 3067	35355-36-3	: 2454	38514-03-3	: 8462	41658-69-9	: 10096	51630-58-1	: 5956
32764-34-4	: 6507	35364-90-0	: 1411	38552-72-6	: 6601	41692-47-1	: 6899	51637-47-9	: 8864
32788-54-0	: 1383	35365-59-4	: 8315	38577-97-8	: 11523	41755-60-6	: 6655	51686-60-3	: 4883
32777-26-7	: 4224	35387-38-5	: 5176	38624-62-3	: 3826	41781-12-0	: 6766	51686-78-3	: 1311
32809-16-8	: 9760	35383-51-8	: 3153	38628-39-6	: 11868	41814-78-2	: 11858	51707-55-2	: 11687
32830-97-0	: 8962	35383-59-6	: 3686	38651-65-9	: 2955	41851-34-7	: 4430	51750-65-3	: 6847
32838-55-4	: 9699	35413-38-8	: 4615	38661-81-3	: 1034	41867-20-3	: 3816	51756-08-2	: 3519
32854-75-4	: 309	35432-36-1	: 10134	38661-82-4	: 1160	41901-72-8	: 1410	51760-90-8	: 3689
32889-48-8	: 9759	35449-34-4	: 250	38727-55-8	: 5173	42027-23-6	: 8788	51772-30-6	: 10467
33018-91-6	: 6396	35472-56-1	: 2680	38738-60-2	: 4676	42061-72-3	: 4609	51786-70-0	: 3243
33083-83-9	: 11961	35502-06-8	: 9852	38827-66-6	: 11517	42067-48-1	: 9051	51806-20-3	: 5543
33089-61-1	: 342	35554-44-0	: 7187	38862-78-1	: 1397	42067-49-2	: 9050	51807-53-5	: 7196
33092-82-9	: 4501	35569-54-1	: 11552	38870-89-2	: 299	42101-38-2	: 3755	51952-42-2	: 9015
33184-48-4	: 1175	35573-93-4	: 9850	38939-88-7	: 1211	42131-85-1	: 9016	52033-97-3	: 4888
33245-39-5	: 5964	35608-64-1	: 3533	39029-41-9	: 7936	42131-89-5	: 6763	52152-71-3	: 3382
33357-44-7	: 11270	35673-03-1	: 1226	39052-12-5	: 693	42185-47-7	: 4618	52152-72-4	: 8819
33401-05-7	: 10098	35691-65-7	: 8768	39052-57-8	: 2736	42235-39-2	: 5353	52182-15-7	: 919
33429-70-8	: 2327	35794-11-7	: 9652	39075-90-6	: 9245	42245-37-4	: 8706	52208-82-5	: 2056
33429-72-0	: 8725	35897-16-6	: 8948	39076-02-3	: 3962	42286-84-0	: 2834	52244-70-9	: 1105
33467-76-4	: 6567	35911-17-2	: 10546	39118-35-9	: 8527	42345-82-4	: 5899	52262-38-1	: 9281
33515-82-1	: 4233	35951-33-8	: 8723	39121-37-4	: 6519	42398-73-2	: 7384	52315-07-8	: 5012
33553-93-4	: 8754	35951-36-1	: 6382	39135-39-2	: 9625	42412-84-0	: 10321	52324-03-5	: 9147
33560-15-5	: 252	36045-56-4	: 6903	39148-24-8	: 6020	42474-20-4	: 8739	52358-73-3	: 7801
33560-17-7	: 249	36049-78-2	: 8407	39178-11-5	: 5255	42474-21-5	: 6372	52387-50-5	: 7067
33567-59-8	: 870	36065-15-3	: 2962	39191-07-6	: 1802	42551-55-3	: 3598	52390-72-4	: 6473
33622-26-3	: 5094	36069-05-3	: 12045	39196-18-4	: 11717	42569-16-4	: 9092	52414-97-8	: 1068
33629-47-9	: 3893	36076-25-2	: 1259	39228-29-0	: 10677	42576-02-3	: 2997	52418-81-2	: 6918
33662-96-3	: 10292	36190-77-9	: 5750	39251-86-0	: 6058	42597-26-2	: 2127	52460-86-3	: 892
33683-44-2	: 6972	36215-07-3	: 9860	39251-88-2	: 6067	42775-75-7	: 7890	52488-28-5	: 1069
33695-59-9	: 10109	36301-29-8	: 1995	39257-08-4	: 4614	42842-08-0	: 10045	52497-07-1	: 3743
33738-48-6	: 7756	36357-38-7	: 5855	39273-81-9	: 6599	42852-95-9	: 6215	52562-19-3	: 787
33742-81-3	: 4178	36386-49-9	: 2940	39478-78-9	: 652	42874-01-1	: 8082	52645-53-1	: 9150
33820-53-0	: 7371	36386-52-4	: 2945	39511-08-5	: 10513	42874-03-9	: 8633	52708-03-9	: 8160
33829-48-0	: 7465	36402-31-0	: 3609	39513-75-2	: 2803	43036-02-2	: 11137	52744-85-1	: 10645
33840-38-9	: 4877	36441-31-3	: 8040	39515-51-0	: 597	43121-43-3	: 512	52780-23-1	: 1506
33843-55-9	: 4700	36441-32-4	: 8039	39622-79-2	: 1330	43197-78-0	: 6746	52802-07-0	: 6917
33877-11-1	: 1834	36448-60-9	: 10472	39669-95-9	: 1755	43222-48-6	: 5175	52896-87-4	: 6427
33880-83-0	: 4389	36556-06-6	: 7389	39713-71-8	: 11151	44829-76-7	: 6595	52896-88-5	: 6413
33933-78-7	: 8171	36566-80-0	: 7122	39755-03-8	: 3667	44855-57-4	: 8351	52896-89-6	: 6414
33933-79-8	: 8455	36653-82-4	: 6659	39761-61-0	: 7027	44914-03-6	: 10652	52896-90-9	: 6412
33941-07-0	: 10771	36678-06-5	: 3167	39782-38-2	: 3371	45223-18-5	: 6635	52896-91-0	: 6411
33966-50-6	: 3210	36691-77-7	: 11888	39782-43-9	: 7026	47172-89-4	: 8776	52896-92-1	: 6437
33978-71-1	: 11957	36701-01-6	: 8880	39928-72-8	: 6141	48145-04-6	: 10681	52896-93-2	: 6441
33996-33-7	: 9766	36734-19-7	: 7338	39966-95-5	: 3427	49542-66-7	: 5678	52896-95-4	: 6442
34008-71-4	: 10436	36768-62-4	: 9628	39989-39-4	: 7380	49763-65-7	: 2907	52896-99-8	: 6821
34014-18-1	: 11590	36823-84-4	: 2908	40058-87-5	: 10223	49845-33-2	: 11054	52897-00-4	: 6817
34069-94-8	: 293	36825-36-2	: 11207	40203-74-5	: 156	49852-35-9	: 6597	52897-01-5	: 6822
34103-97-4	: 4876	36847-51-5	: 3499	40225-75-0	: 8468	50375-10-5	: 2330	52897-03-7	: 6823
34238-52-3	: 9022	36854-57-6	: 974	40239-35-8	: 11892	50402-72-7	: 9713	52897-04-8	: 6819
34238-81-8	: 5136	36895-62-2	: 11825	40242-15-7	: 7847	50407-18-6	: 7238	52897-05-9	: 6824
34240-76-1	: 5739	36903-89-6	: 8363	40248-63-3	: 210	50461-74-0	: 8905	52897-06-0	: 6820
34256-82-1	: 270	36903-92-1	: 8521	40248-84-8	: 9377	50468-22-9	: 3352	52897-08-2	: 6845
34283-68-8	: 6729	36903-95-4	: 8329	40323-88-4	: 11762	50471-44-8	: 12048	52897-09-3	: 6846
34314-08-2	: 3036	36917-36-9	: 10961	40385-54-4	: 1073	50552-30-2	: 7101	52897-10-6	: 6850
34316-15-9	: 2433	36994-79-3	: 2352	40441-35-8	: 4623	50592-87-5	: 6736	52897-11-7	: 6851
34451-18-8	: 10268	37025-57-3	: 1871	40466-95-3	: 9235	50594-66-6	: 304	52897-12-8	: 6852
34570-59-7	: 10136	37050-06-9	: 8537	40487-42-1	: 8835	50599-73-0	: 6989	52897-15-1	: 6853
34619-03-9	: 3994	37273-91-9	: 7440	40514-70-3	: 4393	50623-57-9	: 8147	52897-16-2	: 8757
34757-14-7	: 10237	37434-59-6	: 9976	40527-16-0	: 8582	50670-64-9	: 2743	52897-17-3	: 8815
34786-24-8	: 1706	37470-46-5	: 2623	40571-45-7	: 4772	50689-17-3	: 6074	52897-19-5	: 8816

CAS Number	Page	CAS Number	Page	CAS Number	Page	CAS Number	Page	CAS Number	Page	CAS Number	Page
52914-66-6	: 4762	55130-16-0	: 8309	57968-71-5	: 3307	61828-19-1	: 4479	66235-62-9	: 3608
52918-63-5	: 5102	55156-16-6	: 8980	58138-08-2	: 11899	61828-21-5	: 4413	66295-45-2	: 9549
53022-14-3	: 3565	55182-83-7	: 11988	58138-81-1	: 3669	61828-25-9	: 1742	66395-23-1	: 4585
53045-71-9	: 9009	55268-62-7	: 1567	58167-01-4	: 3305	61828-26-0	: 2065	66398-52-5	: 3466
53068-88-5	: 6902	55268-63-8	: 2267	58190-94-6	: 3491	61838-78-6	: 10464	66577-03-5	: 11763
53078-85-6	: 648	55282-10-5	: 6296	58210-03-0	: 4568	61847-78-7	: 7030	66577-17-1	: 7924
53109-18-5	: 4933	55282-12-7	: 8241	58371-98-5	: 8963	61847-80-1	: 9048	66610-39-7	: 3392
53152-98-0	: 9681	55282-14-9	: 5271	58474-80-9	: 5081	61847-97-0	: 8314	66719-29-7	: 8759
53172-84-2	: 1950	55282-15-0	: 5270	58588-28-6	: 6919	61886-67-7	: 7932	66719-37-7	: 6488
53175-28-3	: 3697	55282-16-1	: 5269	58593-78-5	: 1563	61892-85-1	: 6808	66719-40-2	: 6489
53178-20-4	: 3242	55282-17-2	: 11605	58616-95-8	: 4970	61949-83-5	: 7243	66719-53-7	: 8456
53188-23-1	: 6023	55283-68-6	: 5409	58618-94-9	: 7806	61971-91-3	: 3670	66719-54-8	: 8457
53220-82-9	: 7741	55285-14-8	: 4038	58625-77-7	: 3655	62016-13-1	: 6839	66719-55-9	: 8452
53252-08-7	: 6591	55289-35-5	: 1070	58766-17-9	: 6745	62016-38-0	: 8421	66731-94-0	: 6490
53252-19-0	: 7096	55290-64-7	: 5179	58873-44-2	: 9182	62016-48-2	: 3394	66793-72-4	: 6956
53252-20-3	: 8192	55319-83-0	: 5275	59020-74-5	: 2447	62016-53-9	: 4842	66793-74-6	: 6957
53252-21-4	: 7092	55320-06-4	: 6295	59022-71-8	: 7330	62016-55-1	: 4808	66793-80-4	: 8467
53346-71-7	: 8648	55334-06-2	: 4476	59113-36-9	: 10024	62016-58-4	: 4821	66793-92-8	: 6955
53381-90-1	: 9725	55335-06-3	: 11854	59204-02-3	: 8449	62059-56-7	: 4351	66794-02-3	: 6933
53440-12-3	: 7778	55373-86-9	: 5274	59255-95-7	: 853	62087-44-1	: 4527	66794-03-4	: 6934
53491-32-0	: 2441	55429-84-0	: 11604	59357-07-2	: 6974	62080-86-8	: 8274	66794-04-5	: 6932
53535-68-5	: 10417	55449-46-2	: 1234	59368-15-9	: 4443	62088-36-2	: 4663	66810-87-5	: 6958
53566-37-3	: 7121	55454-22-3	: 6568	59376-54-4	: 184	62108-06-9	: 8517	66898-39-3	: 2820
53584-29-5	: 10009	55505-26-5	: 8168	59433-08-8	: 9689	62108-66-1	: 10232	66997-45-3	: 10146
53588-92-4	: 5464	55517-73-2	: 6623	59434-05-8	: 11758	62108-68-3	: 10227	67191-37-1	: 1665
53608-85-8	: 4322	55517-89-0	: 8639	59654-13-6	: 6832	62108-70-7	: 10226	67329-04-8	: 7340
53608-93-8	: 9935	55563-72-9	: 10458	59669-26-0	: 11713	62127-46-2	: 3446	67451-76-7	: 4312
53660-21-2	: 6478	55665-79-7	: 7060	59697-51-7	: 7011	62127-47-3	: 3447	67485-29-4	: 7136
53663-23-3	: 2353	55696-02-1	: 3944	59697-55-1	: 7012	62127-49-5	: 10135	67535-07-3	: 10255
53663-26-6	: 8148	55702-49-3	: 3894	59734-23-5	: 8470	62127-61-1	: 10160	67590-77-6	: 1760
53776-69-5	: 2081	55751-54-7	: 812	59756-60-4	: 5981	62154-77-2	: 2297	67632-66-0	: 9148
53778-43-1	: 4975	55775-41-2	: 3182	59777-92-3	: 4231	62155-41-3	: 2111	67700-22-5	: 2159
53780-34-0	: 7427	55789-02-7	: 1872	59889-45-1	: 8722	62168-25-6	: 3264	67700-26-9	: 8512
53859-78-2	: 4213	55877-01-5	: 6790	59905-55-4	: 2983	62184-67-2	: 11766	67747-70-0	: 9021
53878-96-9	: 3903	55882-21-8	: 10153	59956-48-8	: 118	62184-82-1	: 4919	67885-08-9	: 3166
53889-94-4	: 9428	55944-70-2	: 6477	60012-63-7	: 2102	62184-83-2	: 4938	67952-57-2	: 6506
53894-19-2	: 10361	56072-27-6	: 6518	60096-09-5	: 6398	62261-05-6	: 3957	68057-83-0	: 3532
53966-53-3	: 8186	56103-67-4	: 9225	60111-14-0	: 7082	62353-69-9	: 8685	68077-28-1	: 10601
53977-99-4	: 7048	56106-37-7	: 1556	60168-88-9	: 5948	62476-56-6	: 10862	68107-05-1	: 6724
54105-76-9	: 4427	56180-61-1	: 8688	60207-90-1	: 10720	62496-53-1	: 8753	68165-06-0	: 11187
54106-40-0	: 2005	56216-12-7	: 3823	60212-29-5	: 11985	62521-69-1	: 3453	68266-67-1	: 1360
54131-44-1	: 2927	56245-02-4	: 2034	60212-31-9	: 11986	62624-30-0	: 465	68359-37-5	: 5008
54244-90-5	: 6590	56456-42-9	: 1496	60251-57-2	: 8357	62690-62-4	: 7897	68384-70-3	: 3541
54268-02-9	: 10138	56523-26-3	: 8322	60291-32-9	: 4408	62692-40-4	: 1001	68420-93-9	: 11212
54298-97-4	: 8525	56630-31-0	: 8206	60468-23-7	: 9007	62706-16-5	: 7008	68694-11-1	: 11900
54305-87-2	: 6764	56630-33-2	: 8204	60512-85-8	: 10687	62882-00-2	: 7385	68723-25-1	: 7629
54340-87-3	: 7262	56630-34-3	: 8205	60669-40-1	: 11967	63019-98-7	: 3014	68750-16-3	: 5677
54383-22-1	: 4373	56688-75-6	: 4871	60671-36-5	: 11956	63157-81-3	: 10194	69112-21-6	: 6997
54410-69-4	: 2010	56694-97-4	: 11864	60698-31-9	: 6157	63284-71-9	: 8219	69190-65-4	: 3811
54411-01-7	: 4446	56694-98-5	: 11863	60703-31-3	: 7075	63370-69-4	: 4776	69248-35-7	: 3332
54429-56-0	: 5628	56739-95-8	: 1781	60710-39-6	: 9239	63468-05-3	: 9097	69295-21-2	: 3729
54454-10-3	: 10820	56745-61-0	: 6969	60761-04-8	: 8561	63790-14-7	: 1086	69305-42-6	: 9496
54460-46-7	: 4952	56800-09-0	: 3746	60814-29-1	: 11335	63927-44-6	: 6877	69321-60-4	: 1304
54460-96-7	: 10121	56908-88-4	: 1291	60836-07-9	: 6509	64057-79-0	: 10703	69371-59-1	: 7102
54462-68-9	: 6759	56911-77-4	: 2074	60899-39-0	: 7880	64161-50-8	: 2951	69377-81-7	: 5982
54462-70-3	: 3270	56920-82-2	: 6037	60956-23-2	: 1303	64165-34-0	: 11317	69498-28-8	: 3273
54467-95-7	: 979	56961-23-0	: 2371	60956-24-3	: 1293	64168-11-2	: 9702	69542-91-2	: 4514
54484-63-8	: 1377	56961-24-1	: 2320	60956-26-5	: 1074	64207-03-0	: 10447	69564-68-7	: 2881
54527-68-3	: 3570	56961-32-1	: 2541	61050-68-8	: 8397	64257-84-7	: 5953	69668-88-8	: 6588
54553-14-9	: 11155	56961-77-4	: 1010	61228-10-2	: 6619	64287-26-9	: 1285	69704-44-5	: 3262
54573-13-6	: 6620	56961-82-1	: 2326	61231-45-6	: 7717	64334-96-9	: 11211	69745-49-9	: 4692
54630-50-1	: 6497	56961-84-3	: 1359	61531-45-1	: 8156	64339-43-1	: 1524	69770-98-5	: 4891
54638-10-7	: 8918	56961-86-5	: 2337	61539-75-1	: 8370	64346-45-8	: 3807	69806-50-4	: 5962
54644-23-4	: 1231	56961-93-4	: 7851	61551-49-3	: 7978	64379-30-2	: 8704	69813-63-4	: 1970
54676-39-0	: 4327	56962-04-0	: 9231	61563-25-5	: 1302	64501-77-5	: 8508	69833-18-7	: 469
54715-57-0	: 9259	56962-07-3	: 7343	61589-68-2	: 8991	64502-89-2	: 8971	69947-01-9	: 9805
54737-45-0	: 1080	57074-96-1	: 8539	61592-89-0	: 1962	64531-26-6	: 8533	70110-24-6	: 10496
54774-99-1	: 10249	57129-69-8	: 6146	61675-14-7	: 7098	64632-72-0	: 12052	70116-68-6	: 3629
54819-86-2	: 10217	57150-86-0	: 9553	61764-94-1	: 6370	64853-59-4	: 744	70124-77-5	: 5965
54829-48-0	: 6110	57179-35-8	: 573	61792-07-2	: 260	64902-72-3	: 4091	70127-93-4	: 7237
54833-31-7	: 4464	57223-18-4	: 8193	61827-85-8	: 1487	64930-16-1	: 3741	70202-92-5	: 4915
54833-34-0	: 4402	57296-00-1	: 10613	61827-86-9	: 1708	64965-91-9	: 2453	70396-13-3	: 6615
54837-90-0	: 815	57323-93-0	: 6988	61827-87-0	: 1711	65144-02-7	: 9797	70401-56-8	: 2779
54852-73-2	: 3071	57386-83-1	: 1402	61827-89-2	: 5286	65248-43-3	: 6985	70644-46-1	: 10922
54862-92-9	: 6517	57392-55-9	: 255	61827-93-8	: 4853	65423-25-8	: 5340	70690-24-3	: 8367
54885-00-6	: 7640	57403-35-7	: 1135	61827-98-3	: 4845	65438-35-9	: 13	70715-19-4	: 10983
54889-55-3	: 7274	57404-54-3	: 9176	61828-04-4	: 2341	65469-88-7	: 3357	70732-43-3	: 8202
54890-00-5	: 4326	57456-98-1	: 3185	61828-05-5	: 2285	65530-93-0	: 6728	70732-44-4	: 8341
54922-86-0	: 10032	57502-49-5	: 9666	61828-06-6	: 2088	65732-90-3	: 11518	70732-45-5	: 5101
54934-90-6	: 4432	57526-28-0	: 3694	61828-07-7	: 4380	65786-09-6	: 6380	70786-82-2	: 8887
54934-91-7	: 4472	57576-00-7	: 4545	61828-08-8	: 4483	65786-11-0	: 6539	71159-31-4	: 10378
55030-56-3	: 3739	57605-95-5	: 7911	61828-09-9	: 4474	65927-80-8	: 6954	71172-57-1	: 8970
55038-01-2	: 2455	57706-88-4	: 8454	61828-10-2	: 4462	66050-98-4	: 6938	71173-15-4	: 2939
55057-45-9	: 10494	57772-57-3	: 2824	61828-11-3	: 4410	66056-06-2	: 19	71200-79-8	: 3763
55066-53-0	: 8305	57837-19-1	: 7439	61828-13-5	: 4458	66225-16-9	: 6552	71284-96-3	: 8025
55073-86-4	: 6950	57885-43-5	: 10211	61828-15-7	: 4482	66225-17-0	: 6551	71297-97-7	: 5118
55124-79-3	: 6309	57933-83-2	: 4022	61828-17-9	: 4383	66225-35-2	: 6803	71605-85-1	: 8152

CAS	No.	CAS	No.	CAS	No.	CAS	No.	CAS	No.
71608-10-1	9334	77753-21-0	3805	89896-73-1	6531	99129-21-2	4138	112456-48-9	2934
71626-11-4	513	77753-24-3	3419	90112-26-8	4898	99172-57-3	6076	112489-85-7	6096
71633-43-7	11303	77825-53-7	3824	90112-75-7	6565	99365-26-1	2801	113426-22-5	8506
71745-63-6	9340	77825-54-8	3825	90112-82-6	6892	99706-68-0	11875	113666-64-1	11951
71747-37-0	2152	77866-58-1	4560	90162-24-6	8390	99814-65-0	8655	114180-21-1	3251
71751-41-2	477	77928-86-0	5039	90200-61-6	4232	100121-73-1	6038	114369-43-6	5949
71783-56-7	7391	77958-21-5	6604	90200-83-2	8513	100130-25-4	43	114489-96-2	10224
72178-02-0	5985	78277-23-3	10690	90202-08-7	6878	100224-75-7	7627	115113-98-9	6569
72312-48-2	4687	78738-36-0	10368	90202-21-4	8499	100295-82-7	6945	115310-98-0	11318
72622-74-3	10342	78738-37-1	9668	90210-56-3	7387	100295-83-8	6947	116435-30-4	2931
72762-00-6	11031	78938-11-1	11448	90226-23-6	4723	100296-26-2	6499	116435-95-1	10257
72935-12-7	3049	79127-80-3	5952	90357-58-7	199	100314-90-7	6135	116529-73-8	3906
73120-52-2	5298	79241-46-6	5963	90499-41-5	1611	100319-48-0	8339	116530-63-3	10305
73178-43-5	9042	79504-01-1	3234	90638-48-5	10651	100367-45-1	6697	116530-99-5	4788
73206-57-2	10448	79630-70-9	3428	90644-60-3	8195	100791-95-5	11189	116632-47-4	11174
73210-25-0	4565	79762-76-8	10243	90645-54-8	4621	100983-12-8	10210	116668-34-9	2450
73383-24-1	3738	79912-55-3	8872	90645-61-7	4884	101144-90-5	1687	116668-40-7	5096
73398-15-9	6694	79912-57-5	8870	90676-55-4	8520	101257-63-0	6743	116668-44-1	6577
73454-83-8	6690	80204-19-9	9012	90722-14-8	8200	101257-71-0	9712	116668-45-2	6576
73513-50-5	7058	80325-37-7	6541	90760-75-1	4564	101257-79-8	4195	116668-48-5	8323
73522-42-6	2928	80656-12-8	4792	90769-70-3	4927	101427-54-7	11609	116668-49-6	8326
73548-71-7	7025	80832-54-8	556	90841-14-8	3719	101587-36-6	4336	116668-50-9	8501
73548-72-8	6553	80864-10-4	8796	90925-48-7	3356	101654-30-2	6741	116697-35-9	4524
73557-60-5	2310	81177-02-8	6789	90952-10-6	6920	101715-60-0	8331	116723-89-8	5117
73642-91-8	8127	81280-12-8	6579	90982-32-4	4081	101870-06-8	6688	116723-90-1	5119
73741-61-4	4701	81334-34-1	7188	91049-43-3	8790	101935-28-8	6333	116723-93-4	3863
73758-54-0	3486	81335-37-7	7189	91057-79-3	4184	101935-40-6	9243	116723-94-5	3501
73918-56-6	1570	81335-77-5	7190	91465-08-6	5010	102153-88-8	4831	116723-95-6	3536
73931-44-9	2312	81726-82-1	6308	91572-15-5	5168	102370-18-3	4526	116723-96-7	3559
74051-80-2	11359	81777-89-1	4140	91840-99-2	3438	102249-95-6	6747	116723-97-8	3591
74112-36-0	8447	81925-81-7	6596	91890-87-6	3843	102851-06-9	5983	116724-10-8	3219
74115-24-5	4139	82097-50-5	11824	91913-99-4	3490	102871-67-0	11304	116724-11-9	3389
74204-00-5	9240	82322-93-8	4895	91965-23-0	3131	103095-51-8	10813	116724-15-3	4355
74219-20-8	2949	82507-04-8	7010	92040-00-1	7409	103275-78-1	2219	116724-17-5	4567
74221-86-6	4444	82584-73-4	8600	92687-41-7	4644	103697-14-9	8230	116724-18-6	4693
74244-64-7	4217	82657-04-3	2998	92737-91-2	6504	103985-40-6	4685	116779-77-2	3424
74283-48-0	3610	82906-78-3	9555	93710-40-7	10187	104035-66-7	6593	116779-78-3	3425
74346-30-8	1152	83055-99-6	516	93921-42-7	4583	104068-42-0	4787	116781-85-2	3413
74421-05-9	6332	83682-72-8	10360	93972-93-1	6869	104177-72-2	362	116781-86-3	4272
74513-58-9	10693	83863-33-6	763	94347-45-2	8863	104514-49-0	1292	116783-11-0	1107
74581-94-5	6886	84087-01-4	11201	94434-64-7	8262	105065-24-5	4679	116783-12-1	1605
74683-66-2	8158	84098-45-3	10277	94826-08-1	4144	105401-59-0	7175	116783-21-2	2020
74742-08-8	8159	84258-49-1	7332	94942-89-9	2143	105688-74-2	11271	116783-23-4	3393
74742-10-2	5071	84583-53-9	11320	95642-49-2	8830	105728-23-2	11247	116783-28-9	4506
74744-28-8	11983	85733-97-7	8936	95835-77-1	2460	107494-37-1	11952	116783-29-0	4584
74744-36-8	5342	86051-37-8	8155	95970-07-3	2256	107534-96-3	11589	116836-06-7	7163
74764-56-0	6594	86073-38-3	8264	95970-10-8	2311	107574-57-2	10465	116836-10-3	4321
74778-22-6	8446	86623-80-5	9081	96034-00-3	3435	107770-92-3	11570	116836-12-5	8413
74925-48-7	7249	86661-53-2	8727	96097-89-7	11818	108365-83-9	3730	116836-13-6	8522
75055-73-1	3365	86668-33-9	8507	96244-13-2	4563	108438-40-0	7943	116836-16-9	8921
75455-41-3	3867	87018-30-2	3613	96836-97-4	4350	108836-86-8	8650	116836-19-2	3583
75579-88-3	6345	87237-48-7	6291	97073-23-9	10484	108965-84-0	5028	116836-20-5	7248
76078-85-8	8135	87541-87-5	10623	97412-23-2	9394	108965-86-2	5029	116836-21-6	9647
76149-14-9	1523	88472-61-1	8269	97479-78-2	3337	110318-09-7	6699	116836-30-7	11976
76429-68-0	2444	88947-16-4	10403	97886-45-8	5287	110383-31-8	7080	116836-32-9	8899
76469-08-4	8540	89033-70-5	4179	97961-66-5	7055	110795-27-2	3938	116836-55-6	3564
76472-83-8	3013	89123-63-7	10612	98070-91-8	3454	111768-02-6	8927	116862-65-8	4769
76619-89-1	9212	89223-57-4	4829	98203-04-4	11221	111768-04-8	6937	116908-78-2	9063
77053-92-0	7078	89300-69-6	10253	98272-34-5	6112	111768-05-9	6940	116908-83-9	8877
77295-79-5	5893	89533-67-5	7113	98431-97-1	11567	111768-08-2	6939	116908-84-0	8993
77392-54-2	3638	89583-12-0	6760	98551-14-5	4149	111772-90-8	11873	116908-85-1	8994
77437-98-0	6582	89583-61-9	9896	98790-22-8	4561	111823-35-9	8765	119446-68-3	5174
77464-06-3	11996	89796-76-9	6381	98886-44-3	6021	111897-18-8	8279	127665-32-1	6258
77501-63-4	7410	89886-25-9	4523	98930-89-3	11887	111991-09-4	8069		
77732-09-3	8566	89895-45-4	4445	98991-54-9	6570				

MELTING POINT INDEX OF ORGANIC COMPOUNDS

In this index the temperature in °C is given in boldface. The numbers immediately below a temperature entry are identification numbers (given in the first column of the table *Physical Constants of Organic Compounds*) for compounds whose melting point values begin with the same integer as the temperature entry. Thus, the numbers that appear between temperature entries of **3** and **4**, for example, indicate the compounds that have melting points equal to or greater than 3.00°C and less than 4.00°C. The entries following **-8** are compounds melting between -8.00°C and -8.99°C, inclusive. The entries for **0** temperature are compounds with melting points between -0.99 and +0.99°C.

-205	10046	4814	6536	5470	3241	5544	3231	5900	5640
4011	10073	5593	6537	5898	4835	7444	3232	5923	5644
			7504	6611	6501	8668	3239	6168	6829
-196	**-158**	**-142**	9033	7040	7546	9849	3993	6542	7116
5430	5889	4191	9772	9862	8760	10042	4857	10072	7483
	7500	4826		10157			4859	10116	8531
-187		4856	**-135**	10541	**-122**	**-116**	4983		8940
9834	**-157**	6019	4903	11392	4188	4858	5085	**-107**	
	3759		7502		5894	5608	8599	4497	**-102**
-185	7478	**-141**	9040	**-129**	5908	5897	8847	5540	4982
3704		7001	9052	3255	7505	6384	9046	6554	6387
10528	**-156**	7038	9055	3804	8812	6862	10035	8409	6681
	10572	7046	11378	4813	9848	7122	10036	8848	7036
-183	11437	7515	11462	8709	10043			9773	8842
7541	11483	7523	11489	10697	11429	**-115**	**-111**	11527	9679
		10984				3391	3403		10151
-182	**-155**		**-134**	**-128**	**-121**	3437	3714	**-106**	10727
5448	7558	**-140**	3378	3290	3379	3760	4008	4428	
7459		4972	4193	6695	6424	4671	4390	4815	**-101**
	-153	5421	4804		6771	4672	4828	6494	1631
-181	5884	6678	8763	**-127**	8840	4838	5597	9071	3410
7486	8823	8662	9029	4007		5905	5665	9871	5168
	9053	9005	9031	4955	**-120**	7002		11973	5253
-177	9054	9058	9057	6864	4670	7022	**-110**		5531
4981		9059	9059	9035	6423	7466	279	**-105**	5532
	-151	10034	10534	9044	6860	8853	4219	3222	6866
-172	5927	10578	10536	10037	8817	9039	4848	3705	8491
7475	9006	**-139**	11409				5525	3727	8532
		5880		**-126**	**-119**	**-114**	7496	4816	10054
-169	**-150**	6999	**-133**	4426	4980	3713	7555	4917	10089
5879	8615		3709	4928	6835	5417	8494	5489	10388
8845	11414	**-138**	3801	5938	8762	5672	8846	5606	11672
		3229	4803	6775	9845	6407	8990	6440	
-168	**-149**	3706	7000	8495	10531	8665	9136	6498	**-100**
3800	4973	4035	7481	8712		8813	9836	6861	3707
	4974	4812	10421	10530	**-118**	10119		7507	4935
-165		4985			4797	10607		8721	5541
3438	**-148**	5467	**-132**	**-125**	4921		**-109**	10154	5596
9004	8659	5885	3164	3897	4924	**-113**	3812	11432	5726
		6683	3413	7042	5454	3170	4660	11463	5912
-162	**-147**	9056	4861	7120	6533	3892	4971		6405
8824	5648		4862	8660	6834	5548	6422	**-104**	7179
	6857	**-137**		10441	7041	5602	6534	3219	7693
-161	10118	3799	**-131**		8497	6383	8850	3252	8493
3119		7003	3242	**-124**	8809	6865	9123	3382	8849
4192	**-145**	7029	5918	3372	9034	8410	10558	4834	9159
	3171	8658	7114	7044		9038	11546	6830	9892
-160	7491	8720	8825	9030	**-117**	9070	11695	9135	10646
4930		10535		9036	3627	10088		9906	11787
5906	**-144**	10557	**-130**	9037	3728	10156	**-108**		
	5902		3117	10334	3914	10529	3146	**-103**	**-99**
-159		**-136**	3152		4802		4658	3381	2169
3397	**-143**	442	4860	**-123**	4849	**-112**	4778	4626	3548
7464	446	3145	4975	11	5485	294	4902	5534	5233

5484	8074	10504	5881	**-81**	10729	6417	9107	8965	5413
6010	8707	10636	9817	1738	11410	7550	11770	10141	5535
8142	8826	10958		6503		8916		10616	5992
8719	10040	11465	**-85**	6609	**-77**	9100	**-68**	10937	6908
8810	10411		1123	6712	498	9823	1640		8177
9893	10493	**-89**	1893	6856	4492	10263	3287	**-63**	8673
10533		3681	3547	8182	5056	10318	4894	743	9982
10573	**-93**	3916	3577	8766	8505	11691	5613	993	10715
11451	179	4248	4490	11788	8749		5688	1149	11764
	236	4431	4491		8915	**-72**	5796	1989	11797
	3118	4439	4764	**-80**	9945	7180	6406	2038	
-98	3636	5607	5259	187		8111	6909	3920	**-59**
208	3668	5641	5733	194	**-76**	8747	11762	5517	1060
214	4774	7115	6025	3281	3159	9109		6471	1483
215	5466	9837	6343	3512	3695	10181	**-67**	7554	4650
3289	7442	10214	6499	4018	8320	10571	1035	8481	4751
3515	7462	10259	8609	4156	8957	11916	1990	8482	5709
7545	8496	11760	8818	4400	10582		3754	8882	6990
10580	9067	11785	9522	5238	10957	**-71**	5693	9826	9108
11549	9808		10120	5410	11646	1988	6433	10956	10173
	11398	**-88**	10599	5752		4443	6459	11769	11707
-97		3593	11703	5896	**-75**	6898	6781	11970	
3402	**-92**	3594		5936	1630	10111	6884		**-58**
5204	3113	4366	**-84**	6840	2107	10310	8453	**-62**	3398
7477	3369	4851	301	6895	3467	10635		1163	4886
7499	5055	4906	1672	7124	3586	11740	**-66**	1626	9839
7581	5538	5882	3125	8408	4367		1675	1671	11796
8592	5987	6713	3129	8608	4885	**-70**	2345	1687	
10074	6018	6843	4648	8997	6691	227	4489	1726	**-57**
11971	7121	7687	5254	9774	8844	494	5082	3211	21
	7448	8710	5666	10179	9129	1112	5236	4750	781
-96	8862	9521	5922	10304	10091	3213	5603	6009	992
291	10090	10081	8141	10575	10326	3216	5940	6089	1098
1975	10698	10298	8981	11283		4086	6462	8603	1459
3178		10522	9803	11914	**-74**	4647	7518	11550	3527
5528	**-91**		10301		107	4734	8841	11756	4029
6686	3479	**-87**		**-79**	1190	4765	9818	11967	4316
8827	3564	449	**-83**	1685	1419	5634	10980		4465
11393	4494	1118	186	3433	3280	5717	11969	**-61**	6335
11783	6004	1727	1414	4240	3436	5729		1763	7154
	6130	3803	4495	4625	3761	5735	**-65**	2164	7467
-95	6502	4368	4749	6007	4325	5887	264	2925	7684
1686	6774	4438	4846	6680	4425	6331	2095	3745	8359
3376	8690	6986	4934	8419	4656	6474	3265	3901	10265
3415	8878	7463	5935	8529	5542	6694	4486	5522	11143
3588	9897	8143	6500	10031	5687	6930	4999	5718	
4453	10117	8492	6527	10992	6696	7891	6658	6497	**-56**
4677	11696	8661	6732		6831	8610	9785	8530	1241
4796		10296	8075	**-78**	8140	8914	10584	9821	1728
4798	**-90**	10335	9938	121	8325	9158		9943	3121
4805	4170	10509	10583	3528	11986		**-64**	10673	3992
5639	4370	11405		3631	11987	**-69**	741	11972	4244
6013	4918	11466	**-82**	3907		1423	2971		4315
6731	4966	11584	1161	5126	**-73**	4733	2973	**-60**	4342
8711	5230		1472	5414	209	4800	6428	696	5537
9804	6355	**-86**	2037	5419	1995	5244	6880	860	6356
11461	6492	1641	4167	5539	3253	5527	7557	1627	6466
	6543	2236	4771	6528	3579	5549	7696	1949	6718
-94	8814	3525	5638	7498	4597	6360	8185	2036	8355
1668	8942	3645	6505	8132	4823	8835	8327	3137	8742
1947	9122	4162	9130	8913	5084	8917	8405	3417	8858
4471	10055	4440	11506	10152	5100	8967	8612	3617	9563
6102	10499	4493		10195	6015	9065	8675	4200	9851
6737									

11130	4878	5213	3114	1719	8684	**-34**	5571	5617	1175
	5611	6419	3226	1739	8837	752	6146	5708	1452
-55	6418	6465	4499	4369	10058	1628	6392	6577	1478
1038	7250	11020	4843	5051	11606	1635	8444	6904	2283
1162	8241	11132	5097	5590	11724	1642		7061	2357
1334	8699	11373	5193	5642	11977	1931	**-30**	7522	4899
1669	8821	11968	5565	6475		2602	998	11453	5260
2972	8831		6685	6514	**-37**	4454	1624	11798	5721
3566	8843	**-47**	6759	6874	1647	5078	1815	11878	8988
5516	9556	1184	7890	7525	1693	6889	1832		9863
5911	10097	1422	8976	8888	1915	7274	1973	**-27**	11467
6890	10743	1453	10420	10872	2093	7516	2016	976	11938
6959	11387	1932	10943	10944	3279	7524	2042	1054	11984
6960		2091	11585		4872	8144	2898	1592	
8356	**-50**	3640		**-40**	5232	8393	4405	1655	**-24**
8372	193	4358	**-43**	1342	7051	8439	5558	1720	1355
8700	1464	6907	271	1629	10130	8740	5655	1779	1458
9148	2924	6911	740	2165	10563	9882	7791	2162	1696
9154	3223	9909	1037	3203	11579	11747	7915	2168	1992
9881	4197	10634	1109	4194	11879		8292	2788	2064
11492	5266	10663	1166	4225		**-33**	8353	3073	2304
11544	5425		2358	4598	**-36**	745	8431	3574	2885
	5678	**-46**	3522	4896	849	1242	8687	4571	3268
-54	5714	2271	3999	5560	1716	1747	9553	4992	3457
3879	5723	8441	4036	5872	1740	1974	9638	5550	4341
4801	6470	8881	4199	6323	2067	2021	9732	8115	5939
6844	6495	9564	4339	6802	3116	3973	9799	8156	6394
8378	8196	10106	4414	7998	3177	4364	9942	8689	8772
9928	8793	10148	4419	8245	4371	6464	9973	9983	9681
10986	9941	10290	4996	8434	4461	6472	10140	10052	11573
	10264	11028	4997	8446	5656	6512	10604	10466	
-53	10315		5770	8600	5663	8595	11985	10876	**-23**
1674	11002	**-45**	6142	9153	8150	8756		11574	945
2935	11658	636	6344	10105	8836	8855	**-29**	11978	1148
3368		1131	6436	10656	9915	8889	463		1424
4703	**-49**	1182	7789	10875	10416	9894	899	**-26**	1972
4732	1632	1724	7996	11246		10942	982	220	2025
4820	1670	2172	8428	11389	**-35**	11657	1649	525	2051
5892	3209	3571	8674	11645	192	11667	2092	1168	2279
7281	3739	5186	10555	11748	1139		2094	1178	2764
8120	4249	5612	10917	11767	1183	**-32**	2163	1379	2903
10291	4365	5728	12066		1339	138	3126	1828	4472
11913	4496	5815		**-39**	1723	684	3188	2070	4951
	5360	7520	**-42**	1055	1725	1138	3301	3895	7997
-52	5361	7655	288	1722	3569	1933	3539	4034	8798
1026	5427	8131	1415	3592	3813	2989	3675	4239	8944
1404	5895	8285	1717	5497	4769	3898	4345	5058	11112
1437	8418	8340	3847	6014	4953	4908	4485	5520	11185
1466	9141	8406	3864	6511	5052	4979	4841	5521	11415
1930	9855	8411	5635	8123	5335	5689	5033	5727	11542
3657	10800	8445	7790	8741	5529	6784	5034	5913	
7125	11368	8966	7889	8958	6453	8310	5653	8298	**-22**
7161	11754	10203	8427		6510		8121	8605	152
7492	11966	10643	8775	**-38**	6620	**-31**	8767	8733	154
10409		10985	8815	222	7979	232	10810	10269	158
	-48	11642	8901	445	8158	779	11592	11468	295
-51	1745		9227	695	8228	1078		11743	1059
166	2024	**-44**	9918	1981	8229	1256	**-28**	11746	2532
233	3428	559	10149	4901	8487	1413	1764	11792	3261
2039	3751	1025	10313	5888	8624	1766	2015		3936
2724	4735	1643		6041	10586	3514	2348	**-25**	4430
3186	5043	1715	**-41**	6766	11131	4617	2909	25	4790
3802	5054	2359	1465	8147		5036	3870	452	5574

MELTING POINT INDEX OF ORGANIC COMPOUNDS (continued)

5917	1765	8569	9874	**-10**	**-7**	2918	11055	1113	3316
6468	2913	10443	10175	561	564	3321	11705	1140	3750
6723	3007	11742	10537	659	700	3460	11935	1171	3766
7083	3444	11934	11214	1173	807	4533		1238	3922
7540	3779			1303	1031	4809	**-1**	2886	4179
7941	3798	**-16**	**-13**	1644	1289	5600	175	2956	4348
9073	4716	12	1388	2000	2373	6805	658	3016	4632
11281	5343	202	1710	3247	4242	7474	767	3082	4768
	6309	793	1874	3319	4781	8157	1053	3248	5076
-21	6561	1654	2710	4326	5713	9687	1094	3447	5270
130	6724	1673	2740	4457	6176	11871	1217	3507	5273
1005	6785	3267	2763	4863	6455		1364	3775	5988
1174	7746	5053	2995	5001	7556	**-4**	2123	3909	6565
1792	8068	5456	3678	5044	7886	2333	2640	4000	6777
2442	8117	5585	3774	5312	8172	2890	2790	4214	7892
3163	8307	6637	4464	5753	8416	3929	2904	4284	7896
3303	10108	6675	4596	5766	9357	5030	2955	4344	8053
4711		7822	5686	6806	9389	7479	3359	4447	8148
4713	**-18**	7920	5931	7917	10518	7681	3425	4534	8315
4855	55	8091	6313	7934	10950	8187	3595	5271	8400
5245	1460	8209	6727	8430	10973	9354	3919	5302	8556
5742	2523	8471	7176	8474		10467	3921	5338	9281
6296	2960	9846	7823	8807	**-6**	10588	4174	5339	9349
6395	3445	9910	7885	11023	633	11070	4359	5354	9415
6812	3937	10938	8955		1141	11639	4624	5619	9548
9820	4213	11394	9630	**-9**	1240	11849	5003	6669	10446
9824	4429	11791	9631	547	1733		5049	6670	10478
9870	4784		9697	1047	2018	**-3**	5071	6782	10901
9919	5066	**-15**	11893	1306	2278	4792	5811	7691	11015
10972	5810	157		1977	3072	5995	5945	7818	11128
	6150	263	**-12**	3628	3835	6870	6040	8134	11610
-20	6349	1384	796	4535	4166	8175	6757	8560	11649
718	6662	1712	1004	4766	4360	8205	6811	9491	11697
1986	7900	1836	1192	5077	4606	9186	7245	9608	11749
2302	8133	3414	1692	5292	5110		7253	9661	11789
3217	8137	3941	1694	5293	5304	**-2**	7901	10159	11961
4392	8437	4723	2108	5295	5756	742	7937	10338	11988
4460	8543	5477	2687	5344	5891	859	7958	10469	
4580	8597	6780	4752	5555	6564	1450	8001	10877	**2**
5063	8787	6787	5569	6994	6716	2061	8119	11277	713
5201	10981	7701	6456	7893	7129	2100	8293	11690	1032
7177	11421	8152	6849	8839	7208	2589	8385	11720	1097
7339	11486	8863	6992	8897	7819	3056	8388	12030	1387
7493		10404	6993	8986	7992	3173	8424		1639
7865	**-17**	10481	7293	10155	8598	3468	8591	**1**	2754
8145	253	11946	8223		9321	4347	8653	671	2897
8488	443		8932	**-8**	9456	4532	8983	694	3083
8954	994	**-14**	10463	188	9457	4742	9708	737	3452
9246	1018	656	10798	1656	9703	5242	10440	948	3793
9853	1044	705	11419	2098	9925	5412	11003	1050	5045
9907	1132	778	11636	2634	10177	7543	11508	1090	5061
10168	1354	1258		3494	10592	7747	11635	1105	5198
10201	1382	2289	**-11**	3776	10939	7780	11784	1301	5677
10354	1677	3138	1246	4221	10941	7829	11868	1467	5830
10396	1681	3506	4572	4721	10976	7942	11942	1609	5877
11242	3521	3618	4601	6748	11744	7986	11953	1612	5963
11728	4212	3960	5072	7155		7993	11979	1662	5989
12057	4283	4003	5200	7814	**-5**	8225		2079	6053
	4466	5904	5652	7895	1567	8929	**0**	2175	6667
-19	4864	6118	7828	7957	1620	9586	178	2296	6875
462	5057	7469	8287	8173	2347	9680	363	2767	7315
503	5328	8443	9629	8203	2374	9957	754	2861	7487
1114	5622	9416	10940	10855	2614	10180	1022	2910	7632

7902	2430	5842	6396	5073	**14**	9458	5337	3318	**21**
8188	3092	6002	6463	5088	250	9614	5808	3598	478
8538	3508	7548	6636	5499	597	9730	5855	3650	526
8791	4305	7931	6786	5800	599	10100	6190	4904	552
11263	4912	8046	7292	6324	755	10806	6634	5274	1062
11698	5070	8540	7381	6665	806	10869	7476	5322	1136
11897	7178	9390	7551	7050	1746	11136	7529	5768	1158
11960	7506	9414	7994	7738	2219	11950	8204	6643	1735
11976	7739	10364	8440	7932	2242	11958	8291	8207	1942
	8312	10788	8656	7938	2770		9628	8541	1944
3	8902	10902	9889	8013	2863	**16**	9662	8643	2012
1566	9583		10066	8146	3058	100	10132	8799	2361
1914	9706	**9**	10138	8402	4357	109	11937	9250	2541
2272	9783	549	10552	8649	4511	563	11952	11056	2720
2787	11120	709	10685	9077	4736	680		11280	3487
5184	11640	989	11604	9401	5075	1187	**18**	11599	3942
10982	11688	1021	11701	9534	5621	1912	275	11619	4822
11057	11872	1046	11886	10337	5662	2161	639	11625	5325
	11882	1360	11945	10609	5758	2266	962	11795	5826
4	11981	1426		11004	5998	2322	1079	11975	6072
571		1479	**11**	11006	6692	2892	1430		6693
2227	**7**	2880	644	11624	7682	2893	1449	**20**	7685
2776	668	2899	707	11908	7921	3024	2419	596	8015
2777	1028	2978	950	11951	8011	3481	2769	598	9454
2988	1029	3057	1346	11959	8206	3482	3190	831	
3924	1185	4346	1772		9578	4468	3307	977	**22**
5303	1209	4850	2004	**13**	9694	4519	3896	978	538
8116	1224	5062	2268	162	9926	5869	4361	1431	875
9693	1248	5518	2429	216	11869	6526	4484	1799	956
11615	1288	6078	2887	554	11962	6638	4787	1896	1218
	1297	6769	4168	1052		6647	4958	2402	1380
5	1604	6792	4386	1104	**15**	7057	5089	2421	1436
173	2671	7465	4516	1454	339	8422	5138	2596	1860
856	2878	8639	4518	1979	464	9185	5570	2994	1876
867	2946	8642	4724	2171	706	9360	5754	3346	2089
1304	3274	8735	5187	2495	1151	9377	5961	3366	2111
1305	3347	8758	5298	2966	1318	9430	6153	3880	3060
1343	4229	9251	5496	3609	1361	9700	6632	3908	3313
1757	4531	10738	6322	4403	1474	9975	6633	5250	4236
2049	5041	11008	6477	4509	1476	10163	6639	5340	5074
2643	5780	11279	9529	4775	2001	10266	7322	5751	5314
3288	5955	11777	9569	5228	2019	10648	8047	5774	6306
3619	6098	11874	9596	5645	2301	10908	8151	5789	6399
4854	7652	11880	10049	5868	2975	11127	8539	5802	8159
5315	7820		10436	5962	3338	11278	8646	5864	8275
6310	9461	**10**	11171	6623	3815	11362	8866	5920	8313
7497	10468	1085		6654	4299		8876	6297	8480
7883	11621	1935	**12**	6905	4432	**17**	9226	6657	9248
8161		2249	9	7212	5040	530	9353	6674	9315
8302	**8**	2269	97	7544	5491	575	9525	7169	9325
8387	1229	2342	212	7712	5874	715	10186	7530	9463
9228	1230	3665	529	8012	6673	1934	10286	7532	10129
9376	1286'	4202	1233	8294	6703	2068	10480	7794	10444
9385	1371	4541	1312	8308	7561	2267	11637	7988	10587
10737	1636	5240	1407	8969	7562	3026		8162	10922
11613	2069	5326	1817	9361	7793	3337	**19**	10323	11049
11614	3043	5429	2412	9702	8043	3644	1023	11184	11075
	3333	5513	3348	10197	8049	4638	1462	11296	11262
	3335	5675	3672	10314	8369	4844	1825	11627	
6	4513	5870	3676	10368	8412	4993	1833	11638	**23**
766	5275	5875	4323	11183	9340	5032	2681	11842	735
1383	5318	6101	4382	11217	9375	5220	2900	11887	848
2014	5697	6295	4714		9410	5313	3050		1759
2121									

1880	1385	**27**	6800	10881	4529	9252	2231	6876	1073
4508	1401	96	7333	11054	4741	9379	2756	7564	1204
4577	1718	666	8163	11790	4806	9699	3023	8299	1357
5321	1891	1120	8234	11823	5059	9911	3305	8354	1827
5364	1961	1296	8235		5300	9978	3816	8395	2063
6529	2277	1427	8236	**30**	6044	10789	4291	8429	2417
8281	2798	1562	8267	1	6352	11071	4412	8647	2622
8754	2894	1731	8268	46	6458	11885	4522	9078	3085
9444	3792	1732	8273	673	6641		4725	9374	4082
10596	4500	1751	9294	808	8454	**33**	5281	9541	4307
10615	4530	2639	9711	1119	8638	652	5283	10233	4476
10903	4739	3258	10352	1176	8896	760	5838	10258	4569
11846	5323	3573	10438	1302	9332	838	5857	10433	4875
11884	7201	3943	11333	1475	9382	868	6056	10945	5194
	7689	4510	11570	1890	9400	942	6094	11475	6778
	7815	4866	11780	1907	9913	2698	6151		7244
24	8265	5453	11943	2073	10228	2992	6312	**36**	7308
39	9570	5856		2246	11035	3112	6949	595	7376
685	9593	6057	**29**	2413	11236	3293	7755	972	7714
949	9673	6983	502	2440	11241	3567	7916	1280	7748
1130	10204	7123	669	2707	11567	4306	8041	1325	7884
1197	10341	7702	744	3667	11583	4411	8242	1358	8107
1211	10389	7879	772	4842	11632	5048	8883	1663	8262
1410	10891	8251	947	5146	11694	5136	8968	1780	8641
1440	11058	8343	1237	5147	11883	5755	9150	1857	9151
1444	11180	8655	1564	5188	11890	5799	9171	2137	9306
1600	11954	9256	1585	5324	11994	5817	9334	2317	9442
1777		9333	1637	6134		6092	9402	2799	9459
2150		9347	1645	6316	**32**	6403	9496	2905	9985
2503	**26**	11259	1657	6653	119	6668	9763	3046	10422
2760	647	11786	1695	6707	140	6879	10417	3296	10717
2862	649	11817	1916	6709	506	7751	10704	3568	10838
2901	672	11828	1976	7150	578	7756	10705	4301	10840
2976	676	11870	2007	7571	638	8010	11005	4506	11856
3236	940	11949	2066	7943	968	8054	11012	4867	
4475	1011		2423	8240	1007	8257	11220	5211	**38**
4512	1386	**28**	2638	8244	1214	8864	11238	5350	642
5060	1729	72	2765	8270	1375	8903	11631	5366	804
5317	2071	258	2869	8314	1743	9187	11634	5798	857
5363	2154	1056	2948	8391	1814	9229		6258	1212
5365	2290	1068	3040	8504	1970	9396	**35**	6304	1595
5637	2455	1135	3059	9146	2040	10378	189	6655	1784
6640	2606	1159	3080	9156	2060	10622	594	7049	1897
6650	3078	1250	3259	9395	2399	10866	663	7409	1936
7171	4538	1381	4463	10865	2525	10951	811	7521	2152
7282	5091	1547	4505	11072	2884	11633	855	7807	2185
7944	5500	1561	4740	11605	3422	11891	907	8311	2387
8640	7373	1902	5816	11782	3612		1014	9417	2682
8890	7880	1964	6307	11889	4160	**34**	1064	9424	3330
9076	7918	2232	6791		4217	221	1082	9551	4136
9308	8619	2672	7117	**31**	4218	262	1293	9726	4206
9425	8680	4205	7552	112	4515	517	1586	10640	4404
10322	9411	4567	7699	505	4542	839	1768	10686	4540
10452	9462	5038	9348	774	5261	919	1811	10716	4548
10639	10051	5291	9350	810	5307	987	2706	10835	4637
11219	10085	5362	9378	999	5865	1080	2803	11894	5046
11974	10191	5463	9404	1134	7247	1084	3071	11947	5068
25	10424	5730	9717	1405	7345	1269	3490		5280
127	10742	5853	9814	1563	7577	1333	4148	**37**	5351
247	11052	5863	10048	2170	7736	1396	4329	26	5841
1015	11157	6315	10131	2877	7770	1598	5284	580	6308
1213	11275	6340	10439	3556	7783	1826	6046	583	7138
1344	11434	6371	10623	4332	8101	2151	6628	835	7451

7495	4203	10216	5305	1058	7148	2941	116	5779	11156
7731	4204	10219	5559	1251	7303	3187	345	5805	11218
8269	4375	10271	5806	1314	7441	3583	687	5835	11289
8398	4668	10517	5814	1767	7512	3977	1008	6652	
8426	4807	10868	6382	1864	7553	4384	1203	7698	**53**
8648	4913	11274	6663	2330	7946	4433	2461	7720	770
8920	4932	11753	6799	2335	8587	6298	2542	7750	836
9110	5171	11881	7100	2337	8830	6320	3199	8052	1144
9182	5183		8652	2619	9324	7621	3306	8279	1146
9237	5615	**42**	8977	3351	9331	7785	3363	8280	1270
9276	5636	1070	9106	3411	9489	8246	3911	8795	1356
10585	6294	1126	9371	3423	9514	8316	3945	9223	1730
11237	6469	1205	9992	3529	9696	9270	4243	9266	1776
11490	7622	1321	9993	3837	10282	9271	5277	9329	2420
11540	7887	1525	11186	3845	10342	9362	5356	9330	3011
11630	8259	1704	11240	4155	10414	9517	5519	9432	3742
11895	8264	2510	11710	4267	10432	9603	5781	9976	3935
	9199	2888		4456	11034	9822	5782	10625	3965
39	9254	3052	**44**	4502	11053	9953	5827	10645	5581
260	9787	3980	147	4521	11114	10209	6278	10792	5946
688	10007	4089	454	4808	11775	10418	6461	10801	5952
1033	11499	4819	536	5241		11195	6661	10882	6317
1222		5745	570	5554	**47**	11247	7157	11225	6321
1844	**41**	5964	1215	5578	19	11643	7187	11774	7510
2005	290	7286	1337	7321	531		7231		7547
2504	487	7600	1350	7947	773	**49**	8300	**52**	9181
2680	539	7638	1406	8136	825	88	8483	4	9441
2745	627	8585	1568	8297	897	159	9137	43	9920
2940	1003	8625	2308	8685	1074	488	9232	544	10977
3878	1030	8657	2363	9262	2022	827	9566	667	11551
4590	1071	9253	2626	9283	2114	1125	10245	690	11623
4646	1734	9322	2768	9289	2259	2287	10406	795	11909
5118	1752	10482	2939	9310	2601	2675	10462	1099	
5119	1788	10683	3471	9317	2708	3084	10475	1100	**54**
5552	2441	10906	3530	9355	2771	3298	10513	1587	210
5990	2748	11245	3903	9946	2791	3419	11616	1878	479
7205	2954	11298	3958	10393	5246	3531	11906	1959	840
7645	3020	11865	5268	10408	5327	3926		1969	1516
8266	3095	11899	5288	10426	5953	3948	**51**	2147	1783
8651	3658		5710	10824	6357	3967	28	2166	1929
9220	3882	**43**	6660	11122	6869	5010	556	2179	1937
9295	3939	490	7309	11239	7615	5025	708	2474	2097
9719	3978	545	7624	11665	7616	5064	1108	2516	2341
10690	4459	591	8135	11864	9606	5771	1259	4165	2355
10988	4553	780	8253		10067	5839	1273	4539	2661
11282	4959	892	8295	**46**	10288	5947	1287	5009	2742
	5135	1075	8394	124	10710	6305	1565	5750	3029
40	5195	1319	8581	546	10807	6454	1789	5791	3906
74	5296	1392	8922	648	10818	6659	1871	5823	3953
219	5306	2113	9221	691	11855	6689	2002	6091	4380
322	6374	2303	9435	946	11892	7380	2078	6950	5115
762	6749	2334	10384	1177		7708	2158	7263	5262
874	6796	2343	10445	1206	**48**	8284	2245	7294	5278
1057	7374	2463	10691	1616	42	9155	2321	7716	5353
1257	7623	2543	11521	2138	693	9418	2381	8233	5357
1295	7686	2578		2291	763	9450	2482	8271	5563
1428	7951	3013	**45**	2747	1390	9634	2741	8365	6096
2062	7956	3014	17	2782	1504	11276	2921	8489	6318
2737	8243	3350	47	4700	1536	11509	3010	8645	6651
3133	8290	3842	129	5265	1741	11901	3096	8768	7284
3663	9269	4090	259	5974	1966		4349	9597	7572
3912	9584	4356	661	5999	2354	**50**	4566	10373	7903
3994	9658	4666	834	6631	2549	111	5239	10823	8238

MELTING POINT INDEX OF ORGANIC COMPOUNDS (continued)

8637	7242	11400	4211	10434	2933	1042	11620	8263	8103
8776	7586	11560	4474	10474	3086	1279	11659	9267	8247
10193	7641	11717	5434	11257	3103	1508	11819	9701	9445
10483	7742	11944	6302	11875	3365	1615	11948	10026	9452
10512	7743		6664	11905	4381	1742		11145	9484
11335	8045	**58**	7287		4462	2043	**66**	11248	9485
11603	8224	475	7305	**61**	5258	2056	150	11867	9663
11622	8484	535	7596	87	5813	2285	509		9674
	8635	592	7633	450	5996	2505	640	**68**	9789
55	9160	765	7766	1391	6283	2627	830	54	11444
62	9219	910	7806	1796	6644	2688	934	422	11547
75	9236	943	7907	1885	7237	2872	1210	858	
645	9239	1252	7949	1899	7513	3844	1261	1006	**70**
986	9240	1307	8399	2054	7752	3963	1283	1363	5
1338	9260	1593	9284	2281	7801	4302	1341	1519	567
1963	9277	1804	10435	2295	7808	4410	1958	1527	660
2744	9305	1960	10708	2332	8286	5002	2065	2252	1096
3093	10780	2427	10709	2493	8792	5378	2382	2406	1298
4263	10837	2488	11243	2750	9179	6646	3049	2600	1451
4570	11099	2723	11654	2755	9222	6788	4280	2615	1850
5114	11231	2944	11818	2762	9356	7629	4282	3008	1965
5192	11595	3329		2953	9412	7908	5215	3098	2314
5620	11628	3336	**60**	3101	9482	8220	5859	3102	3003
6642		3498	420	3432	9619	8463	6656	3985	3791
7328	**57**	3984	576	3820	10020	9164	7741	4458	5012
7549	579	5916	719	3821	10205	9241	7749	4501	6435
7663	581	6161	939	4300	11134	9659	7798	4936	7711
7710	768	6271	1010	4759	11804	10009	7882	5105	7977
7740	1172	7035	1548	5282		10010	7911	5785	8042
7945	1434	7285	1650	6314	**63**	10612	9230	5797	8081
8009	1540	7449	1664	6319	65	10858	9352	6026	8432
9265	1790	7781	1821	6891	123	11158	9488	6710	9231
9722	2099	7906	2050	7174	686	11334	9914	7382	9386
10831	2127	7950	2072	7313	930	11752	10238	7730	9423
	2186	7978	2130	7625	966		11182	7975	9470
56	2265	8296	2212	7894	1000	**65**		8082	9916
577	2286	8803	2481	7929	1275	77	**67**	9233	10019
771	2339	9279	2579	8104	4552	224	101	9285	10023
975	2709	9282	3576	9431	4761	537	195	9286	10027
1076	3130	9292	3590	9675	5225	1013	528	9287	10199
1395	3291	9483	3763	9970	5310	1101	761	11254	10815
1678	4127	10006	3893	10004	5359	1313	1274	11302	11206
1839	4483	10827	3969	10115	6625	1394	1347		11331
2058	4615	11155	3996	10192	7642	1528	1366	**69**	11857
2110	4892	11336	4153	10476	7811	2153	1457	493	
2248	5409		4210	11244	8071	2214	1534	847	**71**
2422	5582	**59**	4940	11256	8099	2834	1537	957	822
2628	5760	243	5008	11258	8274	4853	1613	1300	1208
3356	5787	650	5116	11290	8389	5264	1614	1829	1272
3362	6346	1067	5812	11608	9242	5267	2059	1862	1362
4554	6962	1207	5866	11989	9268	5358	2220	1910	1495
4821	7606	1397	6291		9312	7620	2228	1946	1927
5137	8044	1402	6797	**62**	9344	7824	2356	2006	1956
5214	8106	1438	7163	568	10618	8105	3027	2176	2183
5311	8258	1543	7239	925	11303	8181	3198	2998	2250
5616	8278	1853	7316	928	11341	9180	3959	3846	2362
5769	8870	2088	7640	1311	11572	9311	5129	3983	2683
5790	9138	2515	8260	1539	11900	10332	5579	4266	2700
5858	9381	2793	8348	1714		10486	6299	5348	2716
5914	9383	3088	8682	1750	**64**	10712	6717	5970	2792
6622	10008	3408	8998	2122	7	10994	6765	6095	3048
6635	10857	3633	9504	2323	489	11318	7603	7224	3814
7210	11269	3768	9695	2592	557	11431	7809	7626	4482

MELTING POINT INDEX OF ORGANIC COMPOUNDS (continued)

5130	3074	1903	7776	9336	11758	1308	3207	11345	786
5139	3120	2507	9319	10170		1892	3634	11821	1798
5248	3357	2870	9487	10477	**82**	2180	3968		2350
5933	3780	2906	9571	10711	200	2326	5163	**89**	2833
7439	4158	4075	9771	10713	512	5247	5837	1276	3018
7925	4520	7324	10630	11009	620	5975	6229	1322	5255
8282	5572	7349	10999	11810	702	8003	7570	1369	6048
9224	5626	7690	11853	12003	1216	8022	8796	1374	6100
9326	5680	7775			1549	8100	9264	1446	6630
9999	5793	8064	**78**	**80**	1550	8633	9318	1601	7318
10610	5993	8065	427	22	1971	9455	9405	3091	7588
10932	7149	8067	854	683	2057	9480	10617	3987	7960
11160	7914	8250	1066	723	2312	9740	10628	7192	9387
11721	8039	8999	1072	1526	2336	10514	11288	7379	9917
	9293	9003	1535	2048	4507	10519	11300	7812	11000
72	9447	9797	2211	2124	7602	11081		9000	11273
543	9495	11308	2711	2181	7821	11348	**87**	9384	
551	9547	11340	2951	4102	9327	11538	13	9449	**92**
689	10127	11692	3025	4105	10247		70	10391	720
1294	10677	12002	3089	4413	10278	**85**	682	11680	784
1393	11133	12032	3917	4760	10947	238	699	11848	799
1408	11212		4377	6219		423	922		1048
1447	11260	**76**	4479	6266	**83**	1310	995	**90**	1127
1544	11321	703	4544	6347	374	1463	1290	717	1523
1648	11569	883	6293	6626	377	1590	1511	1492	1524
2044	11996	1264	7650	7276	641	2247	1863	1505	1888
2338		1448	7863	7732	674	2360	1882	1797	2497
2584	**74**	1900	8048	9393	757	2997	2307	2144	4579
2820	550	1983	9190	9474	1012	3094	2329	2580	6143
3070	653	2244	9235	9523	1069	3364	2860	2867	7283
3099	679	2757	9274	11030	1244	3540	2914	3081	7539
3946	969	2835	9297	11827	1320	3834	3143	3131	7613
4106	1906	2952	10000		1622	4545	3497	3308	9244
4264	2620	3230	10016	**81**	1837	5133	3855	3315	9259
4599	2896	3361	11115	425	1913	5377	5131	3510	9490
5279	3039	5174	11148	574	2210	6188	7447	3966	9528
6649	3075	5355	11776	651	2548	7271	7612	4093	10200
7511	3472	5775	11803	2526	2644	7317	7912	4610	11445
7639	4383	5801	11809	2644	3822	7384	7935	5191	11676
7774	4845	6991		3028	4001	7773	8014	5820	12014
8249	5047	8059	**79**	4309	4151	7804	8777	5833	
9307	5368	8584	23	5276	4279	8030	8779	6030	**93**
9338	5508	8874	50	5301	4709	9473	9291	6648	6
9494	5929	10196	59	5767	5547	10062	9769	7617	926
9515	6594	10696	196	5860	6043	10627	9801	8221	1957
11161	7246	10817	513	5861	6300	10830	10722	8254	2263
11552	7347	11876	681	5941	6701	11618	11043	8565	2310
11650	8255		797	5978	7203	11664	11596	9599	3030
	8580	**77**	990	6975	7737	11918	11607	9728	3063
73	8960	555	1282	7563	7754	12028		10001	3786
496	9309	582	1840	7601	7802		**88**	10429	3955
497	9370	698	1938	7909	8222	**86**	10	10871	4147
852	9542	900	2075	7926	9290	342	341	10946	4281
887	10689	984	2280	7928	9481	534	1145	11019	4336
1093	11284	1851	2403	7976	9741	881	2313	11123	5120
1223	11328	2331	7238	8008	9988	901	2743	11702	7415
1292		3111	7364	9258	11612	953	3974	12059	9373
1345	**75**	3140	7585	9278	11675	1326	6624		10794
1848	510	5154	7637	9391		1877	6801	**91**	11251
1883	920	5633	7724	9446	**84**	2084	7423	3	11662
2055	1299	6303	7930	9493	378	2182	7735	507	11820
2292	1477	7291	8256	9791	729	2221	10774	607	12058
2636	1849	7604	8676	10377	965	2465	10813		
				10812					

MELTING POINT INDEX OF ORGANIC COMPOUNDS (continued)

94	2483	**99**	9183	4373	**108**	10924	106	9472	**119**
80	3322	18	9486	4376	45	11204	153	9971	172
81	3762	51	11213	4989	63	11291	572	10442	889
1225	4076	334	11920	5222	312		2469	10814	1922
1353	4592	457		7307	540	**111**	2473	10848	2085
1923	5778	504	**102**	7649	565	68	2476	11350	2118
1982	5831	911	84	7825	850	376	2662		2753
2262	6097	1278	798	7940	917	654	2746	**117**	3019
2437	6348	2178	1129	9157	1494	882	3021	69	3988
2593	7599	2253	1260	9800	1507	1503	3087	566	4097
3044	9422	2264	2149	9979	2328	3104	3194	721	6065
3132	9436	2370	2462	10021	2912	4083	3342	967	6093
3189	9922	3015	4109	10782	4134	4270	3852	1323	6198
4039	11196	5000	5786	11173	5031	5777	5967	1435	7193
4135		5102	6189	11191	7141	6183	7566	2621	9234
5190	**97**	5167	7355	11708	7534	6206	8040	2689	9544
5381	15	5199	9406	11993	7575	7394	9609	4573	9655
5584	1926	5649	11589	12012	7728	7786	9611	7167	9770
5783	2306	5854			8349	11141	9948	7241	10862
5980	2383	6995	**103**	**106**	9343		10847	7488	11836
7723	2652	7509	36	927	9426	**112**	11205	9189	11911
10523	2653	7788	57	1277	9538	2046		9320	
10670	3208	8118	79	1309	12048	2315	**115**	9358	**120**
11121	3400	8567	338	2766		4409	1154	9359	141
11226	5674	8875	1502	4077	**109**	4528	1886	10250	359
	5715	9163	2305	4786	486	5263	2284	10819	1529
	6789	10430	2676	4998	996	6083	2567	11116	2655
95	7314	11339	3076	5369	1378	7111	2611	11781	3031
611	7778	11651	3778	5773	1498	7361	2864		3989
701	7827		6023	5862	2459	7569	3325	**118**	5104
983	7840	**100**	7729	6393	2677	7674	4118	361	5157
1679	8351	533	7948	8130	2819	7826	4257	521	6000
2378	8573	988	8579	8698	3064	8031	5367	775	7358
2479	8770	1520	8773	9152	3979	9968	5643	902	7587
2645	9617	2758	11163	9602	5015	10926	5824	915	7646
2749	9895	2840	11224	9665	5433	10933	7299	1284	7796
3701	10320	2871	11541	9718	5832	12001	7726	1491	9433
5840	10407	3430		10870	6060		7753	1506	10775
6722	10878	3493	**104**	11215	7272	**113**	8781	1925	10923
7119	10898	4152	1497	12004	7919	569	9257	2276	11199
7290	10961	4372	2047	12018	8035	823	9342	2352	11249
7576	11252	5005	2229		8231	938	9471	2464	
7635		7377	2845	**107**	8564	1196	9627	2718	**121**
7656	**98**	7438	5177	364	8568	1510	10319	2962	35
7725	223	9280	6871	593	9434	1843	10833	3042	41
7787	228	10194	7387	1456	10779	2256	10919	3195	49
8017	482	10629	7745	1531	10845	2714	11079	3426	94
9205	937	10867	8034	1872	11715	2970	11253	3925	315
9238	1445	12049	8566	2126		3067	11822	5382	2401
9339	1518		9301	2665	**110**	3079		5696	3320
9479	1588	**101**	9654	3136	76	4125	**116**	5948	3431
10781	2133	1024	9929	6629	326	5968	803	7302	4064
11223	2607	2664	11343	7197	601	6216	888	8002	4793
11322	2759	2938	11594	7424	1461	7240	914	8248	5165
	5181	5561	11611	7810	2837	7709	2349	9195	5776
96	5772	5618	12017	7864	4104	7800	2582	9372	5825
98	7514	5784		8021	5164	7841	4265	9475	7304
739	7795	5829	**105**	8026	5289	8683	5212	9610	7323
758	9323	6721	801	9188	5966	9437	5217	9796	7343
802	10811	7112	1530	10828	7386	9467	6003	10821	7560
853	11272	8025	1546	11031	7939		7367	11193	8018
1291	11826	8063	1924	11033	8005	**114**	8023	11250	9677
1908	12015	8276	2311		9987	86	9469		10925

11539	7764	7372	9991	**135**	7202	9751	**145**	2274	2631
	7868	7843	11001	2	7265	10638	447	2646	3651
122	7974	9443	11080	541	7310	11036	621	2841	3991
168	8261	10520	11202	826	7636	11098	2225	3575	4088
520	9215	10849		2480	7836	12016	2467	4110	4142
792	9346	11082	**132**	4071	8036		2733	6187	7369
851	9757	11683	241	4154	8782	**142**	2736	7186	7677
955	9835	11811	305	4756	9162	95	2990	7235	8006
1541	10333		630	4910	10842	616	7266	7831	10910
2309	10678	**129**	905	5219	11101	625	8032	9792	11862
2365	10968	56	1432	7139	11738	2394	9299	11211	
2475	10995	240	2410	7151		4261	9782	12027	**152**
5575		285	2574	7734	**139**	6179	10104		78
6051	**126**	876	2581	8027	532	6231	12020	**149**	432
9453	728	1756	2859	9921	1489	7173		2243	477
9497	1514	2369	2919	9960	2260	7298	**146**	2559	805
10448	2572	2875	2920		2868	7565	426	2818	833
10846	2603	3193	3317	**136**	3205	7717	2112	3041	906
12006	3062	4115	4111	373	5256	8102	2325	3108	1265
	3349	9798	5216	617	6235	9959	2484	4129	2241
123	6200	10005	7357	657	7595		2667	4378	3975
225	7627	10329	7425	1513	7862	**143**	2684	5007	6779
1509	7658	11135	7744	1939	9212	829	5380	5162	7846
4909	8060	11737	9464	2273	10747	913	5384	7213	12021
5160	8219		9664	2371	11085	1365	5411	7236	
6177	8239	**130**	9766	2641	11548	2119	5438	7854	**153**
7194	9656	514	10793	3254		2501	5803	9745	256
7643	10844	573	11286	5149	**140**	4182	6288	10450	542
7834	11152	2134	11517	5347	169	4374	7110	11032	646
9202	11995	2650		7233	402	6892	7214	11854	895
9218		2732	**133**	7338	524	7234	7268		1061
9743	**127**	3765	89	7634	629	7319	7326	**150**	2502
11661	600	5042	615	7910	916	9149	7337	52	3068
12019	677	5834	800	9210	1332	9173	7389	304	4052
	837	7342	2366	9512	2367	9198	7842	722	6211
124	1077	7608	2398	10684	2530	11060	10969	1271	6221
921	2642	8277	2729	10886	2534		11083	1553	6753
1619	2917	10890	3703	10893	2553	**144**	11800	2498	7431
1884	6207	11037	6050	11198	5019	474	11801	2518	8029
2174	6270	11221	6180		7411	985		2557	9413
2604	7392	11311	8232	**137**	7536	1496	**147**	2560	10167
2727	7671		9196	60	7813	1618	435	2608	10795
3066	9296	**131**	9272	731	7959	2087	619	4103	11285
7296	9939	71	9465	2188	8550	2192	824	4939	
7591	11042	777	10682	2597	9204	2258	936	5439	**154**
7630	11553	818	11208	3033	9476	2563	1517	6232	40
7688		896	11992	3206	9990	2752	2190	6282	880
8016	**128**	1429		5374	10413	5370	2239	7270	933
9761	337	2397	**134**	5822	10884	5379	2531	8070	1095
11310	351	2637	461	7375	11558	6186	2616	8630	2562
	724	3034	903	7578		7209	2703	9169	3196
125	2207	3854	2184	7584	**141**	7609	3304	9209	3334
622	2237	5112	2405	8574	726	8386	3466	9646	4009
1532	2396	5385	3012	9249	1262	8588	5441	9736	6185
2316	2836	6136	4084	9337	1351	9172	7378	10018	7329
3061	5299	6209	5251	10790	2704	9341	7592	10770	9802
4143	5435	7158	6217	10861	2881	9438	9243	10853	11203
5352	5741	7227	6698		3769	9500	12026	11088	11287
5625	5828	7264	7199	**138**	4107	9502		11140	11504
5949	5972	7803	7365	518	4112	11554	**148**	11154	11805
5976	6208	9217	9345	2586	4407	11557	226		11808
6205	7146	9498	10851	2882	7052	11834	904	**151**	11837
7301	7327	9632	11587	5950	8553		1538	918	

MELTING POINT INDEX OF ORGANIC COMPOUNDS (continued)

155	2719	**162**	5179	**169**	7368	6175	6279	7142	5150
82	2874	1315	5336	764	7727	7143	7593	7228	5919
356	3139	3065	6238	776	8069	7398	8252	9755	6144
433	5185	4338	9613	2524	8800	8057	11314	11045	7393
436	7168	5113	9831	3142	9735	8572		12031	9203
605	8926	7341	10012	3309	9830	10489	**180**		9499
2491	9197	7664	11519	7574	11091	10753	73	**184**	11229
2519	10758	7676	12033	9207	11723		725	318	11858
3045	10797	9407	12056	10405		**176**	1501	411	
3777	10859	9768	12064	11065	**173**	90	2392	614	**188**
5180	11832	10679		11210	476	624	2528	1490	1533
5981	12000	11518	**166**	11568	828	931	2679	1887	2486
7406	12024	11806	29		2485	2573	5451	2351	2490
7779			344	**170**	4049	2783	5957	2876	2564
7860	**159**	**163**	618	67	5587	4063	6223	3770	2658
9448	2240	358	2527	1522	5873	4091	7837	7427	7344
10746	2612	387	2545	2376	6242	5979	7953	7705	9478
11722	4130	626	2590	2796	6603	7356	11102	7849	
11902	4408	727	2728	3292	7190	9193	11323	8213	**189**
	5111	759	2738	4010	7589	9213	11337	9399	354
156	7330	1512	3090	4033	7667	10759	11859	10856	832
93	9192	2511	3134	4072	7833	11094			2393
146	9492	2657	5971	4259	7861	11591	**181**	**185**	2786
217	9832	4094	6199	5221	9167	11598	349	348	2865
428	11332	5156	6273	5383	9216	11841	603	1515	4128
522	11660	5711	8028	5588	11084		2395	2577	4794
523	11845	7847	9409	6197	11711	**177**	2436	2605	5553
1545		8634	9746	6292	11713	409	2576	2666	6194
7269	**160**	9884	9760	7706	12013	2630	2866	2702	6236
7952	360	10885	10165	7971		4095	5166	4108	10772
9507	458	11327	11556	9300	**174**	5004	5401	5161	12011
9518	519	11590	11940	9759	455	5671	7672	6178	
11090	604	11671		9965	1493	5984	7851	6218	**190**
11103	2082	11802	**167**	11555	1785	6181	7856	6249	821
11162	3004		602	11685	2391	6239	9748	6261	2120
11227	3009	**164**	2353	11999	2555	7590	9795	7759	2751
	3331	310	5013		2623	8210	10850	11533	2775
157	4061	1340	5583	**171**	4126	8212	11078		5127
357	4062	1866	9398	1103	5148	9191	11353	**186**	5144
878	5932	2135	11502	2213	5445	9408	11512	385	7136
2556	6241	2561		2540	7721	11324	11718	2963	7232
5122	6301	2588	**168**	2583	9201			4081	7320
5175	7206	3037	37	3271	9440	**178**	**182**	4099	7797
5386	7395	6054	465	4185	9477	92	2189	5595	9744
7223	8679	7221	467	7188	11095	347	2197	6066	10272
7669	10993	7580	613	7542	11153	2774	2254	6226	11087
8857	11046	9166	631	7670	12050	2961	2547	7306	
9211	11064	9208	2663	7763		5399	4139	8397	**191**
9304	11209	9303	2717	9986	**175**	5492	5387	9298	466
9784		10998	2959	10905	562	6203	5580	9657	692
9956	**161**	12007	3341		845	7218	6285	10873	1102
10637	610		4055	**172**	846	7844	10863	11816	2191
11092	944	**165**	4060	941	1554	10147	12005	11824	2195
11652	2832	33	4124	1499	3047	11089		12038	2883
12067	3299	327	4161	2096	3107	11312	**183**		5396
	3339	333	5402	2238	3302	11564	30	**187**	5443
158	3702	343	6022	2477	3328	11593	2544	239	5670
431	7758	1758	6227	2734	4053	11799	2587	516	6245
460	9403	2187	7295	3312	4183		2715	2193	6255
480	10761	2629	7422	3817	4970	**179**	4051	2385	
623	10796	3032	7704	4123	5178	48	4933	3005	**192**
1263	11200	4146	9206	5960	5391	430	6192	3294	350
2535		5121		7145	5494	2678	6334	4054	664

2203	11313	**202**	11233	3204	3902	**222**	5388	**235**	5376
2439	11682	307	12044	4942	5145	353	5393	66	7867
3853		4119		5124	6286	1324	5428	319	9466
4379	**197**	4260	**207**	5389	11041	4047	6191	468	
5819	7217	5524	6222	6268		7610	9501	665	**241**
6225	7933	7845	6247	7760	**218**	7772	9733	2617	313
9533	8548	10904	8004	8211	27	8551	10927	4132	5624
9756	9177	11529	10934		2368	9754	11011	5804	7267
10802	11048	11666		**213**	5371		11111	6516	7767
10864	11910	12035	**208**	2377	5375	**223**	11357	9738	9364
11179			429	4069	7335	397	11545	11086	12062
	198	**203**	473	5394	9170	508		11177	
193	756	2194	2478	10508	9367	2261	**229**	11830	**242**
64	2196	2379	2506	11319	10896	2435	2418	11930	404
395	2624	2529	2942	11597	11338	3141	2647		2585
407	2911	3464	3109	11928		7353	2721	**236**	2705
863	3036	4098	3986		**219**	7430	4121	2199	2846
3578	3110	5158	4056	**214**	31	9747	8066	2409	6256
6259	4096	6220	5128	308	2319		9439	4074	7414
7679	4100	7134	5269	1591	2487	**224**	11355	7162	11515
7839	5108	7225	7855	1621	2550	866		7719	11516
7866	6252	7300	9468	2613	3135	954	**230**	10894	12063
8019	7678	7579		2857	7832	1316	1327	11040	
10804	11807	11074	**209**	4251	11109	2086	2517	11452	**243**
11047	12043	11093	424	5152	11663	5977	2826	12010	2536
		11207	2712	9752	12022	7418	2979		2686
194	**199**		6246	11235	12023	7419	5021	**237**	7172
782	311	**204**	9174	11317			6276	2200	9742
1838	1281	306	11812		**220**	**225**	7456	2618	11653
2509	2470	2829		**215**	2566	34	12065	4101	12046
2831	2512	3785	**210**	366	2654	1623		5151	12055
7156	2570	5159	104	791	2821	2569	**231**	5392	
7700	2926	10011	403	879	4068	10632	410	6195	**244**
	5915	11932	2632	3982	5985	10803	732	11329	399
195	6224		2713	4057	6230		2808		438
320	6240	**205**	4113	6265	7216	**226**	2810	**238**	864
451	7799	384	4133	7453	7396	317	5349	365	1551
481	7869	386	5440	8677	7838	375		2364	2609
2500	11325	483	6234	8678	7857	6079	**232**	2546	5562
2565		2380	6237	9749	7858	10978	456	2853	7226
2595	**200**	2568	7229	11044	8056	11305	606	7144	10676
2929	323	2591	7366	11096	9753	11513	5153	7362	11320
3931	392	2823	7597		11100	11829	5982	7598	
4131	511	2957	7611	**216**	11520	11998	7853	10760	**245**
5155	634	4080	11315	83	11919		10913	10786	3360
5373	886	6873	11528	2411		**227**	11689	12025	5390
5818	1542	7385	11843	3465	**221**	472		12060	6184
6264	2318	7757		5143	103	609	**233**		6244
7771	2492	9765	**211**	7417	105	2800	1552	**239**	6262
7850	2571	10773	388	7768	485	2801	3038	2828	6267
7859	3343	11532	1521	7848	2730	7446	5022	4116	7432
8038	4040	11536	2282	8631	2856	9302	8061	5176	7762
11306	4994	11833	2625	9214	3069		9739	5398	11358
12042	11159		4059	10752	4070	**228**	10843	5442	11927
	11531	**206**	6280		5400	346	11051	6287	
196	11566	321	10644	**217**	7189	491	11644	6289	**246**
398		2234	10675	309	7397	1559		11107	367
3932	**201**	3311	11192	370	8559	1749	**234**	11307	3981
4058	1841	4730	11687	437	9369	2388	2851	12009	7440
5011	2552	5017	11831	495	9764	2554	4048		9511
5403	9365	6251		1500	10785	2558	6204	**240**	11356
6281	10703	8020	**212**	2739	11924	2735	9598	2400	11840
9120	10883	11061	2943	2873	11925	3022		5014	

247
5495, 6214, 7346, 10909, 10929

248
4041, 4686, 11076, 11686

249
390, 2209, 10928, 11228, 11850

250
628, 635, 1317, 2128, 2825, 2999, 6243, 7416, 11038, 12068

251
368, 2202, 2802

252
2513, 5395, 11178

253
380, 2648, 2731, 3106, 4044, 4073, 7352

254
1558, 2275, 2520, 5407

255
2204, 4043, 4114, 6193, 7666, 8589

256
2830, 4117, 4262, 10930, 11073, 11813

257
419, 8996, 9165, 11342

258
383, 4067, 5397, 8571, 9705, 9989

259
416, 7334, 12051

260
32, 2514, 5020, 5406, 5418, 5566, 7133, 7805, 11295

261
8, 7761, 9641

262
2255, 2794, 2805, 7354, 7391, 7835, 9737, 10776, 11316

263
2809, 7222, 10453

264
85, 2844, 9194, 9640, 12061

265
4120

266
1348, 2432, 6080, 8058, 11326, 11534, 12054

267
2812

268
2575, 3257, 4122, 5223, 11860

269
4045, 5142, 8062

270
355, 471, 2659, 2854, 2858, 6196

271
11530

272
9758

273
991, 2726, 7665, 10756, 11039

274
2408, 2824, 2855, 7668, 7870, 9767, 10854, 11201

275
7215, 7675

276
2257, 2804

277
470, 2842, 2843, 7445, 9161, 11929

279
9366

280
393, 3260, 4046, 6201, 9833, 11344, 11669

281
369, 2773, 2850, 7648, 11066, 11535

282
1831, 3002, 6253, 7647, 7852, 11232, 11351

283
2499, 4258, 9510, 11067

284
7175, 7370, 9175, 9509, 12037

285
316, 2320, 2811, 4042, 5447, 11234

286
379, 1748, 2224, 5107, 7219, 11293

287
3764, 4757, 7132, 7211, 7970, 11563

288
2083

289
391, 400, 11230, 11922

290
91, 2508, 2839, 4268, 4900, 5405, 6250, 7195, 7204

291
372

292
885

293
394, 2384, 5016, 7412, 7413, 7454, 11933

294
7707, 8546

295
2815, 10912, 11525, 11565, 11923

296
2656, 7350, 12008

297
325

298
2386

2217, 5404 (**281** group adjacent: 12036, 12040)

300
324, 2489, 3006, 3990, 5568, 7191, 7230, 11851, 11931

301
2807, 8214, 8738

302
10754

303
2816, 8217

304
381, 3463, 12045

305
5446, 11346

306
408, 1349

307
2594, 8218

308
4278, 7351

309
234

310
1352, 3097, 8547, 10899

11062

312
4354, 9612, 10897, 11997

313
415, 10763

315
7777, 10895, 11522, 12041

316
2817, 7400, 11068

319
389

320
11194, 11347

321
371

322
11104, 11294

323
2827

324
10931

327
7348, 8215

328
237, 5623

329
2814

330
417

2813, 4782, 9176, 9600, 11108, 12052

333
8545

335
331

336
9642

337
5224

338
11059

340
412, 1330

341
5969

342
6081

343
11926

345
11050, 11835

347
1328, 2806

348
10748

350
2852, 5493, 7408, 10852, 12053

353
396, 7336

357
7703, 10757, 11018

359
2404, 11745

360
414, 1331, 9734, 10751, 10768

365
2533

368
9601

370
10769

374
418

380
6621

385
405, 2466

390
7331

420
7457

437
4145

450
2784

485
7644

550
4164

BOILING POINT INDEX OF ORGANIC COMPOUNDS

In this index the temperature in °C is given in boldface. The numbers immediately below a temperature entry are identification numbers (given in the first column of the table *Physical Constants of Organic Compounds*) for compounds whose normal boiling point values begin with the same integer as the temperature entry. Thus, the numbers that appear between temperature entries of **3** and **4**, for example, indicate the compounds that have normal boiling points equal to or greater than 3.00°C and less than 4.00°C. The entries following **-8** are compounds boiling between -8.00°C and -8.99°C, inclusive. The entries for **0** temperature are compounds with normal boiling points between -0.99 and +0.99°C.

-191	**-47**	**-21**	**-2**	**9**	**25**	**35**	7518	**52**	5528
4011	5665	5550	10046	9910	3117	4974	8661	449	7032
	10528				3379	5471	9112	4381	7526
-161	**-46**	**-20**	**-1**	**10**	7496	5648	10532	4643	8854
7459	4167	3119	3255	3145		5905	10737	5457	9134
		11414		3177	**26**	9849		8607	10529
-128	**-42**	11584	**0**	8599	3898		**44**	9113	11409
7541	9834		3229		3916	**36**	4903	10157	
		-19	3705	**12**	5466	442	4980	10509	**58**
-103	**-40**	5641	4981	3162	7176	444	8658		3722
5879	7478	5987	9906	4170	8659	4192	8662	**53**	4188
				5467	11451	5421	9114	5168	4976
-88	**-37**	**-18**	**1**	5487		7466	9818	7517	4983
5448	5484	5885	3804		**27**	8709	10506	9054	5642
	5593		5126	**13**	500	9005	10576	10522	8853
-85		**-17**		4166	3397	9006		10558	9058
5902	**-36**	10128	**2**		5540		**45**		10071
	10118	10608	446	**14**		**37**	9871	**54**	10729
-84			4198	9159	**28**	3907	10536	4985	
4008	**-34**	**-13**	7444		4973	4972		5667	**59**
5936	9772	5884		**15**	5539	5900	**46**	6007	3152
			3	5880	5912	7184	3993	6086	3393
-82	**-12**	3706	11439		7545	5531	6099	3414	
7558	**-32**	5486	5538	**16**	**29**	9522	5532	8802	4977
	4955	8091	7462	5410	8807	10533	5662	9053	4987
-81				5475	9004		9848	10038	5476
7486	**-30**	**-11**	**4**			**38**	12066		6678
	5938	10073	5537	**17**	**30**	3692		**55**	7483
-78			5937	5488	5545	3801	**47**	3759	8594
3992	**-29**	**-10**			5925	4201	5525	5427	9111
5596	7500	5485	**5**	**18**	9027	5454	5663	8684	9837
7515	10572		3889	8087	10070		8592	9851	10317
		-9	4200			**39**	9803	10034	
-75	**-27**	5470	5666	**19**	**31**	264	10557	10072	**60**
5918	5889	7481	5908	5883	3799	3172		11425	3392
	10441		7546	5897	5894	5233	**48**		4189
-72		**-7**		7470	6010	7520	5895	**56**	5598
5906	**-26**	11429	**6**		6025	10069	8098	208	5896
	5640		3164	**20**	9804	11527	8660	498	8608
-68	6019	**-6**	5489				9774	725	8823
289		3704	7448	**11**	**32**	**40**	10530	3890	10723
	-24	7442	9773	301	3378	3170		4975	11389
-57	3914	10578	10541	3800	3892	7499	**49**	5907	
7475	5544		11489	4971	5534	8092	3289	6843	**61**
11437	7477	**-5**			10534	9122	4774	8609	267
	7523	4194	**7**	**21**				8671	4165
-51	10569		5602	5898	**33**	**41**	**50**	9059	7554
5924		**-4**	11378	7505	279	3760	294	9123	9135
7504	**-23**	3146			5940	4764	4953	10411	11410
	7491	11432	**8**	**22**		9905	5458		
-50	10727		3897	10535	**34**	11398	5491	**57**	**62**
4035		**-3**	4007		3171		5942	3290	3210
	-22	7464	7502	**23**	5249	**42**	9819	4034	3727
-49	7559	10570	10421	4162	5608	4193	9863	4190	4498
5927	11435			7555	9898	7507		4277	8817

9052

63
494
3287
3721
4276
5585
6999
7154
8085
8615
8824
10037
10736

64
3723
4006
4172
4929
5548
6130
7498
7581
8825
9785
10159
10524

65
3113
4930
6131
6168
6677
6683
9158
9915
10033
10600
11465

66
4986
5430
5606
6684
7003
9117
9807
10565

67
501
3126
4199
7000
7002
8094
9055
9057
9817

10156
10566
10577

68
3151
3159
3242
3720
3754
3899
6011
6731
7001
7463
8093
8613
9862
10120
10515
10539

69
296
4984
5910
8575
9810
9894

70
3391
3728
4020
4191
4982
5436
7105
8666
8667
9056
9128
10531
11394

71
3129
4826
5237
5462
5479
5659
5923
7114
9836
10540

72
179
5597
5660
8672
9033

9788
10040
10180
11696

73
263
3761
4032
5535
5658
6681
6682
9136
9846
10181

74
3178
5655
5770
10060
10179

75
2347
3169
3724
4827
4928
8668
9060
10036
10045
10714

76
291
463
3901
3908
4499
5202
5526
6676
7540
8670
9039
9129
10547
11430

77
186
3209
3222
3812
7183
8669
9031
9783
10437
10548

10583

78
3241
5204
5672
6073
6171
6679
6992
6993
8083
10409
10554
11544

79
3523
3645
3896
5599
8760
10296

80
867
1757
3457
3525
4248
4305
6018
6074
6680
7494
8593
8762
9032
9040
9829
9856
10493
10655
10771
11550
11720

81
271
3370
3373
3875
5210
6560
6715
7116
7155
9030
9809
10039
11168

82
269
1423
4169
4626
5490
6418
6685
9806
10335
10389
10733
11412

83
266
5455
5478
5529
6017
6686
6712
7037
7490
9035
9130
9821
9909
10035
10059
10076
10579
10725
10726
11376

84
1717
2271
3726
3923
5881
5920
6325
7115
9029
9044
9131
9853
10089
10553
10732
11724

85
12
288
2091
3223
3253
3438
3713
3725

4249
5549
6687
6713
7039
7046
7103
7153
8616
11495

86
3396
3708
3710
6172
6716
6830
7038
7043
7047
7180
8763
9777
10777
11143
11361

87
3733
4800
5922
7042
7104
7161
7472
8074
10063
10081

88
209
3150
3167
3359
3709
3877
4922
4966
5603
6695
7044
7106
8583
9891
9897
10158
10562
10584
10607
10800

89
1424
2913
3797
5417
7045
8702
8761
9036
9038
10055
10728
11377
11496

90
2272
2346
3394
4000
4803
4917
6834
7467
8627
8701
9779
10119
10586
11118

91
3168
3221
3232
3395
4802
4804
5305
5636
6829
9022
9037
9845
10261
10594
11461

92
16
3185
3186
3372
4919
5480
5522
5637
5644
5890
6105
6835
7036
7120

10113
10301
10499
10504
10581
11396

93
2980
3148
3155
3714
4920
5473
5519
5646
6089
6327
6533
7435
8665
8809
9017
9574
10549

94
445
1422
2270
3161
3369
3668
3707
3734
5203
5227
6328
9043
10552
10592
10698
11658

95
177
3118
3215
3524
3622
4018
5078
5708
6088
6536
6537
7040
7041
7122
7179
7402
9841
10088

11691

96
3220
3732
3737
5668
6432
9016
9021
9045
9115
9827
9892
10525
10585
10715
11189

97
21
3275
3858
3870
4925
6012
7492
8720
8721
9018
9034
9912
10090
10334
10697

98
3116
3158
3219
3252
3437
3452
3711
3879
4926
5911
6013
6355
6534
6535
8086
10574
11382

99
3251
3441
3594
3715
3915
4197
4801

4921
5236
6102
6609
7152
8090
8822
8848
9001
9861
10083
10263
10635
11423

100
176
196
683
3154
4343
4426
7028
7516
8076
8663
10058
10087
10663
11463

101
236
1738
3231
4272
4342
4403
5187
6002
7522
8958
9064
9067
9132
9844
10260
11397
11462
11486

102
265
2348
3217
3375
3547
3628
3681
3695
3752
4022
4428

BOILING POINT INDEX OF ORGANIC COMPOUNDS (continued)

4672	6108	4771	7025	11769	6558	6426	9007	10057	3887
5541	6326	5517	7033		6772	6553	9100	10694	4196
5591	7029	5620	8080	**113**	7533	6724	10109	10744	4368
5661	7474	5669	8764	3276	8319	6776	10387	11390	4370
6109	7484	5887	8847	3861	8681	7009	10420	11399	4654
7121	8415	5891	9012	4647	8813	8578	11175	11466	5482
8558	8846	6329	9732	5228	9049	8693	11402	11508	5733
8957	9068	7550	10298	5651	9087	8711	11446	11657	6861
9126	9899	8577	10496	5656	10735	8808	11492	11783	8077
10054	10086	8719	10734	5935	10801	8980			8320
10490		8797	11484	6529	10872	9124	**120**	**122**	9109
11487	**106**	9020		6546	11371	9805	191	127	10174
	295	9024	**111**	6547	11372	9816	3125	4434	10376
103	3114	9065	3248	6836	11770	9840	3239	6548	11433
235	3430	9066	5796	7005		10492	3382	6549	11908
1739	3522	10171	6335	8620	**116**	10658	3387	6550	
3716	3526	10328	6544	8730	134	10778	3416	6688	**125**
3900	3657	10555	6696	8850	215	11475	3440	6738	298
4812	3895	10556	6775	8982	997		3619	7405	1893
5727	3921	11020	7014	9776	1728	**118**	3802	7485	3137
6033	4924	11491	7015	9883	2915	2236	4219	7519	3409
6034	5186		8576	10373	3513	3421	4367	8495	3743
7339	6004	**109**	9153	10743	3552	3618	5503	8496	3753
8690	6771	453	9592		3694	3848	5886	8497	4036
9023	7027	3736	9860	**114**	3745	4221	6425	8606	4655
9104	7525	4770	10526	1737	4861	4851	7006	8704	4848
9859	9629	4860	11375	3250	4862	5513	7493	8706	5821
10111	9843	4938		3693	5189	6106	9146	8841	6103
10573	9900	5254	**112**	3871	5610	6423	9893	8944	6527
	10061	6107	214	4789	5638	6815	10117	9105	6607
104	10100	6773	278	4855	5639	8605	10152	10326	7077
192	11149	6774	299	4963	5647	8712	10303	11751	7137
2095	11151	7020	3165	5132	7034	8814	10425		8139
3157	11693	7031	3247	5193	7582	9014	10739	**123**	8355
3216		7482	3371	5607	7696	9107	11501	3402	8492
3280	**107**	8990	3374	6167	8732	9680		3636	8493
3377	3153	9046	3629	6551	8917	9838	**121**	3860	8707
3712	3180	9775	3731	6552	8981	10042	3238	4366	8966
3922	3218	9826	3744	6691	9013	10259	3515	4773	8967
4671	4003	9858	4187	7543	9051	10500	3521	4858	9010
4856	4014	10299	5892	8849	9070	10602	3847	4979	9101
7016	4271	10404	6343	9050	9089	11456	3859	5536	9140
7018	5423	10511	6545	9854	10455	11571	4772	5694	9681
7021	5456	11483	6554	9904	11150	11667	4814	6714	10542
7026	5474		6557	10604	11366		4816	6739	10664
7561	5882	**110**	6610	10702		**119**	4817	6960	11363
8096	6611	19	6838	11405	**117**	145	4818	7007	
8700	7019	1947	7017	11827	100	284	4937	7538	**126**
8765	7030	3147	7024	11905	300	3281	5424	8078	121
9019	9008	3160	7557		1220	3439	5516	8494	274
9069	9048	3735	8603	**115**	2094	3718	5917	8703	3746
9071	9679	4016	8695	17	3403	3803	6174	9011	3999
10297	10043	4670	8845	1726	3593	3841	6555	9878	4038
10551	10124	4857	9047	1727	3649	4364	6556	9903	4651
11166	10388	5464	9072	3246	3698	4371	6998	10325	4828
	10503	5465	9097	3647	3699	4653	7004	10560	6833
	11585	5619	10082	4028	4240	5542	7022	11049	6860
105		6127	10149	4030	4656	5888	7501	11476	7544
441		6341	10527	4244	4859	6090	8491		8529
2914	**108**	6606	10550	4813	5496	6543	8694	**124**	8598
4023	297	6837	11130	5426	6417	7521	8940	3245	8731
4363	302	7013	11176	5512	6422	8590	9025	3277	8844
4632	3240	7023	11750	5769	6424	8916	9092	3385	8943
4918	3399								

8965	9102	3748	4999	8673	9090	6384	6734	6362	3488
9811	9521	3862	5002	8812	9091	6389	6857	6363	3509
10294	10980	4470	5510	9002	9145	6826	7083	6402	3888
10498	11112	4852	5709	9901	9618	6827	7124	6512	4492
11582	11124	5205	6692	10102	9812	6866	7556	6807	4494
		6128	6980	10476	10416	6912	7934	6977	4692
127	**130**	6383	7066	11756	10537	7085	8110	6978	4833
1167	339	6385	8526	11760	10599	7086	8405	8409	4835
3410	1168	6723	8941		11070	7090	8410	8505	4882
4255	3166	6730	9099	**137**	11865	8612	8507	8811	4908
4314	3282	7503	10044	144		8819	8717	9652	5501
4649	3381	8687	11434	281	**139**	8839	8918	9879	5590
5461	3549	8751	11759	1078	107	8953	8932	10211	5635
5472	3683	8919	11787	1166	1453	9062	9061	10273	5704
6959	3690	10077		1241	3606	9098	9668	10497	6085
7087	3747	10172	**135**	3386	3851	10066	9678	10822	6965
8891	3749	10222	3237	3646	4885	10074	9866	10939	6994
9648	4878	10495	3427	4660	5234	10084	10151	10981	6996
9650	5634	10673	3548	4777	6351	10380	10289	11388	7073
10085	6015	10783	3689	5630	6539	11442	10603		7129
10349	6617	11173	4176	6388	6540	11755	11071	**145**	7687
10369	6806	11453	4252	6849	6863		11132	1624	8182
10382	6865	11705	4487	6862	**141**	11479	3407	8756	
10591	8022		4956	6913	3605		11785	3866	8878
10834	8718	**133**	5618	7067	3611	**143**		4485	9602
	9108	99	5717	7097	3831	293	4657	10214	
128	9667	175	6112	7098	4688	1715		4690	10345
3115	9682	3236	6387	7128	5716	2304	4732	11674	
3175	9778	4425	6391	8530	6330	3588	4749	11703	
3626	9813	4662	6737	8600	6408	3617	5533		
4024	10386	4780	6746	8915	6614	4751	5664	**147**	
4029	11449	5429	6914	9649	8714	4829	5695	135	
4388		6111	6915	9651	8827	5604	5740	3384	
4650	**131**	6706	6979	10280	8842	5735	6042	3620	
4658	207	6828	6981	10310	8993	6331	6113	3697	
4815	1131	6859	8342	10390	8994	6407	6169	3758	
5693	3627	8072	8929	10810	9088	6720	6542	3995	
5899	3644	8531	8989	10825	9881	6733	6585	4298	
7681	4256	8810	10184	11417	10201	6813	6694	4798	
8614	4390	8840	10229		10410	6814	8506	6039	
8723	4652	8928	10308	**138**	10545	6942	8716	6447	
8725	4849	9026	10316	448	10609	6945	8750	6511	
8748	4935	9144	10700	492	10672	6946	8820	6821	
10354	5518	9880	11392	1242	11754	6976	8826	6848	
10597	5600	10309		1454	11795	7403	9442	6947	
11678	6718	10347	**136**	2093		8408	9714	6963	
11740	6864	10463	1668	2993	**142**	8726	9902	6966	
	8532	10559	2991	3007	166	8728	10173	7527	
129	8554	10606	3517	3235	206	9839	10443	7528	
133	8942	11422	3615	3839	3553	11469	10616	8183	
138	9608	11477	3838	4496	3587		10982	8489	
2045	9787	11679	3840	4648	3631	**144**	11138	8569	
3286	10116		3852	4752	3832	130	11421	8947	
3680	10381	**134**	4245	4931	4012	1452	11459	8954	
3920	11365	3227	4253	5232	4339	3376	11672	8972	
4254		3278	4391	5792	4488	3451	11757	9624	
4365	**132**	3283	4486	6332	4493	3616	11866	9691	
5509	117	3700	4867	6809	4495	4489		9815	
6717	190	4086	4941	6825	4642	4805	**146**	10220	
7089	303	4226	6173	7088	4927	4886	182	10361	
7443	3550	4283	6386	7694	6049	5050	2970	10505	
8089	3592	4661	6390	8798	6350	5502	3234	10665	
8710	3688	4961	7108	8931	6333	6586	3284	10701	

10917	10826	11386	9619	6450	4473	11675	8382	6818	11368
11131	11373		9666	6538	4866	11778	8697	6820	11764
11673	11391	**152**	9847	6732	4906	11797	8978	6822	11779
11912		1170	10064	8088	4944	11798	9519	6824	
	150	1240	10143	8373	4945		9611	7075	**164**
148	128	1975	11072	8821	5469	**159**	9685	7126	1641
292	139	2922	11706	9626	6117	1183	10065	7535	2359
3383	452	3564	11772	9842	6434	1723	10650	7537	2924
3637	478	3675	11796	10315	6437	2169	10651	8115	2934
3686	1043	3751	11914	10523	6474	2992	10652	8384	2981
3919	3739	4085		10666	6845	3121	10839	8595	3191
4836	3756	4297	**154**	10806	6911	3211	10963	8713	4228
4837	4293	4394	1169	10807	6927	3389	11424	8746	4288
4838	4398	4399	1722	11408	7069	3795	11543	9609	4451
4868	4687	4478	2909		7070	3863	11762	9610	4694
4870	4733	4834	3188	**156**	7660	4175		10178	4695
5226	4796	4897	3601	998	8198	4331	**161**	10221	4883
6438	4799	5768	4021	2165	8383	5601	149	10598	5921
6440	4869	6344	4431	2972	8766	5617	1184	10648	6078
6846	5685	6448	4832	2973	8818	6416	1686	10660	6503
6967	6439	6566	5141	3563	9096	6473	2932	10953	6520
7096	7099	6928	5253	3849	9591	6573	3187	11384	6523
7461	7495	7095	6524	4395	10142	7072	3268	11748	6612
8715	8120	7127	6587	4397	10175	7091	3279	11792	6616
8744	8196	7450	6819	4471	10192	7434	3828		6850
8949	8563	8610	6923	4778	10193	8378	3830	**163**	6937
9712	8955	8971	7092	4875	10820	8381	3864	142	6938
9799	9889	8975	7524	5718	11765	8837	6041	167	8351
10185	10279	8988	8199	6176	11773	8924	6087	1786	8374
10304	10374	9669	8326	6336	11814	8927	6358	3400	10258
10916	10401	10030	8502	6349		9095	6442	3460	10350
10956	11055	10141	8626	6444	**158**	10338	6449	3483	11125
11370	11726	10364	8834	6446	181	10368	6451	3608	11681
11540		10575	8970	6475	2921	10952	6505	3918	11793
	151	10833	9141	6495	3258	10984	6604	4157	11815
149	1044	11867	9142	6816	3264	11058	6823	4825	
198	1090		9143	6853	3994	11174	6852	4923	**165**
227	2931	**153**	9670	6925	4319	11727	7607	5414	55
4173	2986	253	9814	7071	4393		8186	5463	132
4400	3401	1915	10269	8375	4480	**160**	8197	5563	440
4663	3610	3412	10290	8831	4755	197	8379	5706	1685
4736	3682	3448	10699	8933	4876	1091	8617	6409	1972
5460	4295	3456	11146	8956	4896	3598	10936	6410	3434
5764	4396	3850	11228	8991	4898	3639	11549	6500	3435
6104	4490	4292	11450	9710	5732	3660	11712	6599	3905
6115	4491	4294	11503	9711	5738	4178	11761	6719	4531
6263	4734	4296	11741	9808	5763	4418		6764	4596
6728	6337	4797		10225	5795	4477	**162**	6768	4627
6898	6421	5872	**155**	10605	6338	4500	131	6840	4943
7131	6445	5992	164	10611	6412	4899	185	6988	4992
7473	6510	6116	1646	10805	6413	5746	1185	8344	5110
7548	6531	6525	2923	10938	6427	6063	1687	8380	6411
8328	6743	6569	2971	11387	6522	6360	2767	8404	6414
8329	6856	6589	3562	11418	6926	6415	2918	8473	6839
8628	6943	6729	3729		6964	6441	3572	8503	6924
8833	6944	6847	3750	**157**	7549	6443	4286	8892	6957
8937	8111	6851	3836	203	8376	6559	5613	8945	7489
8948	8939	6986	4013	1398	8438	6804	5686	10110	8114
9622	9028	8327	4530	1724	8934	6844	6005	10177	8142
9684	9084	8501	4571	2982	10718	6922	6121	10213	8143
10155	9360	8646	4768	3455	10937	6929	6452	10375	8322
10351	9707	8938	6009	3609	11478	6931	6504	10406	8500
10488	10223	8946	6353	3625	11581	8377	6515	10491	9625

10075	6989	5570	**171**	8360	5033	4450	5611	4839	10355
10145	7281	5722	194	8471	5097	4515	6008	4846	10538
10341	7429	6514	1654	8509	5573	4538	6124	5698	11440
10348	8140	6726	1740	8757	5627	5197	6356	5794	
10957	8141	6811	1931	8832	6466	5432	6567	6703	**183**
11419	8330	6941	2974	8894	6485	5710	6568	6900	116
11763	8472	7078	3573	9867	6498	6472	7117	8026	125
11766	8499	8498	3643	9869	6675	8446	8464	8116	312
	8914	8555	3673	10031	7053	8560	8479	8469	1055
166	8976	8602	4358	10091	8095	9635	8517	10232	1118
484	9094	8815	4369	11129	8478	9709	8742	10563	1162
2092	9882	8816	4436	11472	8544	9966	8801	10582	1415
2925	10079	9913	4453	11500	8899	10017	9361	10809	1672
2933	10275	10372	4673		9080	10224	9491	10887	1720
3265	10407	11493	4674	**173**	9251	10362	10360		2024
3479	11015		4810	155	9686	10841	10966	**181**	2038
3654	11747	**170**	5084	161	9963	12030	10986	561	2173
4315		143	5085	560	10336		11730	1054	2177
4437	**168**	199	5675	1256	10365	**177**		1414	3390
4522	115	280	6118	1355	10634	487	**179**	1807	3955
4824	189	700	6476	1656	10962	1112	102	3322	4230
6357	499	956	6484	3546	11021	1990	201	4320	4275
6406	3263	1930	6596	3685	11147	3469	525	4524	4335
6496	3285	3556	7140	4442		3476	1182	4564	4419
6530	3388	3954	8346	4535	**175**	3538	1379	6059	4616
6600	3473	4287	8445	4675	254	3558	3127	6137	4811
6615	3555	4435	8520	4681	559	3824	3398	6381	4891
6767	4002	4438	8921	4872	752	3953	3490	6509	4894
6817	4533	4440	9857	5200	1932	4015	3544	6831	4958
6971	4728	4598	10032	5483	1933	4513	3545	8968	5530
7054	5687	4703	10041	5567	1984	4514	4448	9199	6428
7680	6742	4704	10103	5741	1989	4584	4583	9540	8359
8325	8185	4714	10154	6040	2162	5098	4830	9943	8523
8403	8324	5082	10182	6479	3233	5100	4934	10343	8561
8836	8396	5083	10346	6480	3355	5134	5546	11403	8803
9653	8420	5416	10412	6483	3603	5244	6119	11458	10105
9708	8421	5459	10941	6761	4274	5991	6598	11731	10150
10262	8985	5847	11026	6940	4452	6433	6921	11767	10246
10366	9706	6036	11120	7010	4586	6436	8350		10291
10799	9876	6497	11142	8556	4676	6532	8444	**182**	10486
11164	10302	6501	11668	8799	4678	6766	8735	754	10487
11903	10318	6597		8897	4887	6899	8747	1919	10626
	10958	6854	**172**	8925	4993	6934	8900	1968	10920
167	11119	6939	1194	9908	5266	7250	8902	2023	
248	11506	6950	1642	10101	5817	8519	10125	2037	**184**
596		6953	1655	10426	6481	8973	10144	2877	633
1089	**169**	6955	2039	10876	6580	9890	10208	3347	1056
2164	277	7340	2415	11690	8511	10985	10276	3428	1138
3449	1459	7536	3132	11746	8752	11057	10417	3454	1163
3532	1640	8339	3272		10097		10439	3511	1413
3973	1649	8340	3351	**174**	11707	**178**	10935	3519	1659
4247	2289	8597	3418	20	11719	1161	10940	3662	3453
4532	2358	9588	3763	1356	11729	1988	10965	4526	3485
4534	2935	10203	4179	2438		2163	11426	4589	3814
4682	2936	10292	4416	3266	**176**	3228		4640	3827
4683	2937	10840	4439	3350	120	3595	**180**	4890	4679
5024	2990	10915	4449	3489	1111	4234	1354	5452	5450
5705	3813	10942	4580	3956	1974	4581	1818	5765	5595
6077	3815	11028	4585	4273	2357	4617	3571	6487	5726
6499	4050	11113	5893	4441	3491	4677	3826	7245	6507
6521	4597	11647	6368	4529	3539	4710	4325	7506	6751
6884	4902	11906	6502	4536	3823	4724	4330	8477	6930
6973	5023		6935	4680	3949	4729	4587	8905	9887

9918	3574	1671	3551	3405	1918	8457	1472	1021	**201**
10217	3579	1719	3928	3470	1920	8481	1569	1035	1053
10359	3664	1766	3997	3811	1996	8689	1616	1148	1200
11025	3762	2442	4174	4285	1997	8887	1617	1153	1386
11913	3972	2756	4555	4423	2025	9261	1637	1471	1388
	4519	3123	4618	5515	2029	10485	1683	1734	1879
185	4621	3173	4735	5867	2030	10967	1699	1834	2105
53	4684	3226	4895	8160	3492	11374	1998	1916	2106
283	5052	4156	5556	8434	3514	11938	2031	1952	2878
1458	5109	4422	5571	8739	3960		2443	2687	3461
1484	5843	4565	5683	9997	4375	**196**	2456	2940	4310
1787	6486	5001	6037	10519	4639	231	2457	4551	4558
1795	6492	5055	6590	10878	5415	243	2886	5003	4711
1985	8401	5577	7255	10919	5425	251	3267	5263	4713
1986	8402	5612	8124	11024	5752	781	3846	5691	5629
2036	8419	5615	8173	11441	6429	849	4031	5747	5793
3361	8855	5653	8835	11490	6805	1005	4567	5995	6043
3433	11699	6493	8885	11732	6951	1006	4612	7201	7792
3446		6494	9647	11837	8393	1690	5569	7275	8736
3533	**187**	8175	10002	11966	8512	1806	5657	8418	9310
3534	261	8357	10134	11968	8624	1994	5844	8459	9402
3835	856	8487	10148	11971	9228	1995	6340	8789	9962
3845	1140	9073	10162	11972	9539	1999	7897	8805	11473
3948	1481	9256	10265	11973	9637	2280	8793	8922	
4229	1630	9628	10312		9822	3214	8888	9079	**202**
4235	1682	10121	10787	**193**	10078	3318	9564	9317	1034
4516	1701	10284	11014	163	10943	3327	9942	10126	1036
4712	1814	10960	11464	193	11646	3536	11733	10209	1469
5198	3557	11027	11494	1179		3577		11126	1483
5264	4518			1688	**195**	3794	**198**		1857
5551	4566	**189**	**191**	2041	25	3969	1109	**200**	1921
5560	4588	123	262	2954	868	4334	1178	154	1950
5757	5056	1110	1190	2960	1199	4427	1192	587	2061
6133	5208	1465	1700	3366	1817	4723	1198	595	3354
7319	6045	1675	2167	3778	1953	4740	1312	778	3779
7452	6377	2168	2740	4593	1954	4892	1385	780	3936
8474	6462	2770	3507	4620	1987	5712	1684	844	4202
8862	6506	2898	4231	4701	2301	5720	2279	1176	4569
8869	6519	3510	4237	5259	2302	5815	2340	1387	5860
8963	6909	4360	4614	5737	2303	5845	2458	1464	5864
9362	7449	4420	4619	6125	2927	6056	3518	1470	6091
9888	7791	4517	4884	6488	3368	6092	3570	1820	7258
9920	8174	4570	4964	6841	3820	6758	3844	2664	7532
10092	8788	4622	6084	6932	4208	6810	4207	2880	8119
10194	9587	4706	6508	8158	4568	7197	5574	3353	8171
10215	9982	4948	7251	8161	4595	8454	5994	3467	8453
10593	10123	5054	8117	8189	5437	8867	6014	4313	8458
10784	10244	7529	8860	8452	5520	8908	6061	4754	8708
10918	10270	7539	8906	10235	5521	9262	7277	5547	8786
11016	10337	8451	9400	10875	5592	9636	8332	5754	9401
11700	10815	9263	9793	11967	6123	9885	9536	6145	9662
11749	10933	10367	9820	11970	6346	10423	9692	8169	9663
11834	10961	10668	10227		6699	10863	10959	8356	10274
	11165	12039	10564		6700	10944	10964	8889	10283
186				**194**	6982	11771	11791	8898	11117
1139	**188**	**190**	**192**	160	7208	11969	11986	9673	11185
1658	118	1124	1151	162	7256	11984	11987	9717	11659
1674	252	1466	1480	713	7790			9941	
2709	753	1631	1632	860	8159	**197**	**199**	10161	**203**
3203	1060	1673	1660	1141	8172	24	174	10869	124
3474	1137	2459	1689	1670	8191	564	187	11127	126
3475	1149	3344	1917	1810	8443	1164	423		205
3506	1661	3508	2938	1832	8447	1180	588		743

BOILING POINT INDEX OF ORGANIC COMPOUNDS (continued)

779	72	4359	3330	12029	5565	7684	6893	2765	**219**
790	152	4610	3528		5576	8154	7893	3303	841
872	662	5077	4238	**210**	6798	8157	8177	3326	1038
993	1018	5431	4387	220	6948	8792	8184	4446	1088
1020	1122	6146	4546	222	7815	9236	8591	4547	1235
1039	1243	7200	4560	712	8794	9410	9563	4719	1254
1417	1382	7252	4575	769	9284	9791	9802	5451	1406
1578	1419	7693	4707	994	9309	10234	9964	5564	1706
1702	1421	7775	4722	1114	10176	10948	10127	6053	1859
1816	1455	7814	5030	1156	10323			6895	1861
2251	1572	8372	5285	1485	10818	**214**	**216**	7061	1870
3659	1639	8463	5728	1579		705	695	7732	2450
3742	1691	9594	6055	1653	**212**	706	746	8122	2903
3789	1703	9794	6467	1880	212	741	929	8542	4501
3917	1705	9825	6471	2032	213	1017	1050	8806	4783
4465	1735	10052	6796	2042	861	1022	1133	9078	4969
4708	1836	10241	6877	2049	1028	1580	1201	9350	5729
4823	1896	10836	6880	2444	1063	2028	1202	9355	7196
4911	2107	10882	7050	2606	1438	3495	1711	9376	9568
5076	2278	11246	7207	2757	1573	3775	1762	9585	9672
5527	2293	11385	7276	3462	1712	4600	1904	9704	9688
6016	2763	11401	8412	3487	2027	4947	1949	9798	10466
7051	2769		8428	3790	2033	5684	2447	10480	10798
7585	3597	**207**	9074	4220	2414	5851	2451	10968	10903
8176	4385	710	9949	4542	2602	6801	2693	11181	11053
8913	4454	1123	10005	4791	2965	7567	3194		
8964	5739	1157	10226	4901	3520	8775	3213	**218**	**220**
10204	6126	1383	10855	4909	3591	8876	3442	547	548
10242	6749	1420	10868	5072	4316	8890	3504	670	668
10440	6797	1426	10908	5075	4781	9076	3968	1059	709
10971	6870	1460	11537	5346	5869	9252	4233	1159	794
10979	7055	1482	11573	5827	7273	9671	4548	1267	1064
11022	7119	1574	12057	6032	7894	9983	5292	1289	1440
12018	7257	2433		6071	8449	10240	5645	1290	1707
	7324	2919	**209**	8407	10278	10856	6896	1592	1708
204	7689	2920	58	8796	10789	11170	6961	2321	1973
245	8123	2961	558	9283	11075	11383	7049	2344	1992
589	8411	3345	696	9308			8167	2454	2010
749	8868	3516	740	9973	**213**	**215**	9307	2879	2172
992	8907	3959	1026	10245	233	112	9435	2885	2387
1004	9379	3998	1647	10264	530	707	10051	2944	2445
1019	9726	4606	1669	10653	531	716	10122	3952	2448
1037	10363	5497	1681	11578	1142	1052	10481	4609	2698
1439		6070	2955	11743	1158	1186	10996	5721	2987
1486	**206**	7279	2985	11744	1377	1286	11642	5830	3444
1643	180	7979	3837		1667	1600	12058	6799	3633
3641	711	10788	4699	**211**	1709	1713		7293	4362
4209	968	11768	6154	529	1710	1898	**217**	7389	4604
4605	1416		6752	862	1731	1943	246	7560	4668
5230	1436	**208**	7278	1051	1779	2345	255	8912	4698
6142	1487	111	7280	1228	2322	2390	577	9354	5032
6419	1581	211	7935	1778	2764	2696	742	9720	5820
6762	1809	575	8790	1780	2902	2777	748	9911	5852
9353	1815	658	8877	1874	2959	3163	1062	10053	5868
10293	2026	1121	8909	1928	2963	3445	1132	10828	5989
10835	2294	1384	8974	1945	3443	3542	1437	11114	6175
10906	2295	1418	8977	2297	3496	4337	1732	11407	7284
11247	3027	1980	9237	2298	4613	4340	1878	11698	7510
11846	3200	2009	10003	2724	4720	4667	1899	11739	7512
11985	3352	2323	10571	3201	5335	4721	2011	11742	7701
	3404	2983	10879	3436	5870	4967	2060	11777	7995
	3447	2984	10914	4308	6098	5343	2104		8145
205	3503	3323	11583	5071	6894	6150	2446		8335
61									

9253	6455	8856	9630	11951	11580	9418	**235**	11157	**239**
9285	8741	10068	9677	11954		9427	540	11485	287
9456	8863	10881	10385	11962	**231**	9463	590	11511	676
9457	9638	10902	10922		151	9494	770		986
9461	9800	11775	10972	**229**	584	9573	815	**237**	2155
9495	9931		11467	38	663	9875	1011	814	2652
9615	9980	**225**	11468	638	745	9923	1215	1015	3585
10014	10028	41	11510	669	959	10015	1247	1375	4766
10431	10163	242		697	1023	10639	1407	2539	5676
10469	10243	789	**227**	1116	1032	11471	1555	2870	6094
	10696	816	490	1117	1360	11481	1605	3590	6565
221	10866	1094	899	1769	1665	11709	1849	4776	7685
549	10926	1288	1172	1772	1666	11710	2148	6795	7984
704	10969	1372	1321	1842	1774	11893	2354	6812	8349
747	11297	2745	1721	2109	2336	12032	2638	7159	8422
788		2891	1819	3348	2861		2969	7312	8787
873	**223**	3527	2330	3787	4520	**233**	2994	7989	9631
874	23	3584	2690	4184	4715	489	3179	8047	
1792	576	4181	2978	5557	5070	945	3225	8904	**240**
1830	1115	4543	6970	6161	5688	1376	3346	9041	95
2333	1134	5036	7691	7297	5736	2127	3793	9042	537
2361	1143	5324	7712	7998	6050	3678	3958	9294	565
2399	1363	5762	8150	8317	6881	5229	4717	9314	647
2452	1855	6897	8557	9945	6886	5235	5453	9316	650
2697	1901	7292	8858	10472	7166	6052	6190	9323	651
2948	2008	7699	9311	10479	7239	8440	6984	9330	675
4417	2116	8334	9460	10612	9295	9232	7889	9333	904
4601	2119	8365	9664	11333	10342	9287	7981	9415	1174
4785	2749	8406	10146	11952	11958	9335	8795	9429	1248
5707	2945	8698	10640	11953		10399	9241	9432	1303
6872	3776	9264	11959		**232**	10842	9246	9604	1474
6906	4635	9416	11961	**230**	147	11457	9267	10239	1478
7980	5915	9556		526	232	11636	9315	11172	1745
7996	6097	10007	**228**	556	535	11782	10253	11182	1765
8121	6563	11379	708	578	660		10467	11214	3512
8920	7253	11569	900	659	736	**234**	11052		3624
9312	7433	11688	953	1057	817	51	11704	**238**	4212
9385	9266	11776	1812	1232	859	527	11871	599	5034
9721	9321		1911	1364	952	714		813	5058
9957	10080	**226**	2300	1652	1012	819	**236**	990	5840
9981	10339	842	2431	1771	1218	858	439	1082	6875
10106		1014	3329	1848	1381	1065	1008	1092	7988
10819	**224**	1049	3676	2292	1692	1093	1305	1204	8000
11454	766	1476	3677	2337	1860	1171	1577	1462	8427
	1083	1680	3788	3419	1912	1942	1813	1488	9322
222	1651	1763	4696	3967	2001	1944	1865	1561	9474
22	1897	2052	4775	5080	2097	2290	1905	1776	9520
843	1948	2904	5809	6800	2589	2429	1951	2002	10004
1217	2114	3206	6372	6855	2692	2691	2746	2139	10320
1361	2299	5252	6882	7985	2747	2780	3671	2296	11447
1362	2425	5853	7165	8333	2995	3124	5092	2373	
1761	2778	5914	8880	8866	3224	3635	5677	2635	**241**
1941	3773	6459	8910	9260	4267	4638	6727	2863	486
2000	4180	6561	9250	9332	4332	4951	7289	2890	674
2449	4414	6564	9265	9412	4716	5913	7987	2988	715
2453	4608	6842	9411	9414	5449	7982	7990	3911	793
2634	4820	6907	9417	9459	5789	7997	8772	4634	951
2892	4996	7057	9430	9532	5810	9293	9229	7534	1357
2964	5066	7231	9462	9605	5856	9951	9292	9230	1729
4549	6371	7983	10307	9683	6910	10020	9357	9276	1775
6062	7138	8441	10313	9940	7388	11134	9493	9458	2332
6165	8144	8882	10688	10397	8997	11160	9718	9930	2460
6403	8514	9313	11546	10973	9245	11897	10468	12026	4637

4913	9517	9698	**248**	10306	9389	2762	9693	3882	645
5063	9944	9996	552	10384	9390	3008	10821	5787	964
5558	10321	10483	585	11019	9553	3050	11281	6093	1237
5697	10901	10628	772	11035	11613	3297		6465	1292
5876		10792	961	11979	11941	3417	**259**	6725	1663
6985	**244**	11266	1410			4392	644	6889	1793
7160	157	11274	1527	**251**	**254**	5194	847	7242	2067
7888	673	11810	2674	639	140	5761	1295	8777	2989
7916	960	11950	2862	687	737	5871	1597	8901	4460
8366	1441		3074	767	783	6869	1891	9305	4795
8371	1610	**246**	3669	771	907	7383	2276	10686	7822
8791	2260	479	5211	949	1016	7781	2600	11891	9326
8896	2334	963	5626	987	1450	9226	2676		9572
9431	2372	1304	5682	1431	1467	9231	3670	**262**	11256
9961	2640	1322	7244	1475	1510	9347	4223	854	11258
10254	2675	1323	7381	2598	2259	9719	5291	1197	11977
11195	3566	1408	7575	3582	2331	11239	7780	2685	
11818	4405	2335	9247	3672	2521	11505	10189	2901	**265**
	4664	2673	9248	5774	3044		10195	3056	178
242	5788	2888	9327	6321	4786	**257**	10456	4066	570
730	6802	5325	10516	9554	5848	362	10460	4841	582
962	7891	5811	11186	9938	6348	636	11278	6399	883
1013	7915	5875	11861	10447	6394	1046		7826	926
1175	9288	6890	11978	10451	7993	1260	**260**	9339	1076
1212	10398	9239	11981	10793	8146	1359	62	9992	1249
1216	10580	9257		11180		1391	75	9993	1276
1397	11337	9351	**249**	11296	**255**	3565	667	11238	1294
1693		9352	9		462	4787	1210	11279	1411
2701	**245**	9419	554	**252**	649	5781	1287	11280	1529
2894	230	9421	1173	661	848	7192	2328	11567	1961
4461	553	9485	1177	685	871	8830	2670	11836	2362
5213	671	9616	1205	1047	970	9227	2754	12010	4502
6397	686	10518	1603	1048	1238	9694	3112		6983
7553	870	10864	1889	1246	1275	9695	3426	**263**	7109
8705	985	11277	1902	1884	1298	10704	5321	682	7384
11224	1010	11299	2461	1890	1392	10705	5973	924	7821
11474	1214	11411	2475	2268	2050	10716	5990	1270	7827
11937	1239	11920	5802	2744	2355	10717	7162	1507	8392
	1497		5850	2750	2363	11282	7237	1758	8886
243	1562	**247**	5960	4182	2681	11942	7605	2012	9224
583	1694	229	7382	5322	2760		7700	2040	9489
656	3560	536	9325	7307	3082	**258**	8128	2417	9661
672	4306	571	9722	7991	3295	1075	8439	2523	10132
694	4624	932	9972	7994	4910	1280	9340	2741	10319
768	4997	1269	10049	9567	5696	1296	10656	4352	10674
1402	5689	1433	10305	10707	6721	1320	11012	5053	10706
1609	5719	1479	10588	10858	8041	1396	11790	5295	10741
1794	5753	2131	11338	11640	8983	1528	11998	5808	10880
2537	5758	2401		11963	9277	1936		6164	11255
2538	5826	3818	**250**		9404	2071	**261**	6374	
2639	5842	4210	1233	**253**	10838	2755	655	7210	**266**
2887	5866	5196	1604	586	10870	4949	690	7819	1934
3190	6470	5243	1657	773	11259	5841	950	7823	2680
4843	6757	5723	1799	1380	11849	6430	1209	9219	2895
5057	6785	5742	2051	1556	12012	6790	1253	10587	3072
5632	6990	5790	2269	2430		7298	1266	11240	3291
5814	7890	9475	2532	3075	**256**	7385	1593	11254	5499
5998	8881	9484	3321	5195	79	7568	2123	11862	7314
7169	8986	10048	4429	7285	935	7885	2522	11889	7317
7286	9212	10419	7715	7294	1074	7886	2643	12001	7818
7373	9420	10992	8783	7552	1077	9268	2669		7954
8623	9496	11136	9392	7992	1293	9324	3362	**264**	8042
9381	9687	11283	9444	9286	2503	9548	3632	74	8601

9318	7904	1677	5692	680	4457	3096	5996	3025	4863
9388	7913	2013	6293	2015	5766	3774	7809	4989	5831
9473	9223	2137	6904	2126	7374	5713	8008	5238	5874
10976	9555	2610	8044	4790	7626	5744	8017	7274	7807
11237	9955	2798	8059	5759	7884	6791	10670	7309	8644
11257	10687	3054	8995	7308	7923	7364	11006	7747	11243
11263	11245	3083	9486	7731	7924	7618	11242	7751	11560
11267	11269	4386	11217	7828	8387	7690		11009	11615
	12004	5212	11870		8424	7713	**289**	11872	
267		7313	11974	**280**	9222	7811	1966		**297**
29	**270**	9221		3	9393	7813	2742	**293**	1493
1262	493	9697	**276**	224	11222	8018	3058	807	1696
1395	544	10829	1319	943		8480	5772	2020	2621
1536	642	10998	1498	**283**		9280	6639	4326	2708
1612	702	11135	1514	1343		10170	7746	5129	3016
1662	853	11270	2262	1401		10693	9562	6783	3076
1747	1268		2428	1532		10925	11003	7955	3103
3084	1449	**273**	2440	1664		11300	11036	11880	5824
5315	1451	641	2666	2166		11339			5857
5555	1530	1279	2905	2327		**286**		**294**	7323
5928	1892	1837	3026	2660		738	**290**	427	8354
7209	1906	2324	4468	3057		958	543	632	
7315	2117	2723	5293	3308		1964	600	1444	**298**
7316	2649	3099	5332	3505		2016	1100	2423	1258
9391	2752	5579	7212	3766		2382	1538	2496	1589
9443	2872	6271	7829	3974		2627	1783	5773	1756
9699	2912	7328	9297	6458		2791	2314	7241	1856
11241	3051	7927	11008	7240		3088	4741	7721	1910
11308	4289	7958	11789	7883		4432	6468	7830	2079
11321	4341	8604		8442		5714	7349	8651	2313
11892	5771	10823	**277**	10837		6152	7776	9272	5302
	5784	10859	833	10995		6632	7801	10169	6435
268	6528	11262	1265	11033		7238	8430		
88	6891	11975	1297	11218	**284**	7365	8564	**295**	**299**
1250	7824		1636	11302	822	8767	10186	618	1224
2495	8559	**274**	1935	11303	1221	9479	10637	751	2820
2580	8642	684	2034	11943	1513	9620	11220	1272	3010
4084	9186	1477	2084		1587	10638	11758	1342	4344
5059	9700	1553	2090	**281**	2077	11413		2578	7748
5730	10779	1937	2102	39	3024		**291**	3061	8647
7820	11017	2582	2661	545	4430	**287**	620	3100	
7825	11261	3023	3053	1318	6667	1120	1492	4447	**300**
7920	11331	3060	5303	2171	6673	1427	1676	5967	839
7921		3080	5304	2636	11013	1531	3102	6777	851
8388	**271**	3301	6460	3101	11271	3095	3423	7342	1714
8653	897	6905	6464	3307	11298	5163	4466	9185	2121
8656	1535	7376	8431	7739	11631	5745	4995	10613	2370
9329	2076	7898	11004	7740		5805	7773	11184	2378
10459	2309	7914	11416	9395	**285**	11005	7783	11380	2733
10461	2614	7957		9516	555	11244	7808	11624	3017
11264	2672	9278	**278**	10987	574	11304	8133	11625	6322
11335	5312	11841	1512	10999	1261	11630	8652		7329
	5999	11883	2326	11219	1313	11786	11115	**296**	7361
269	6352	11887	2416	11248	1428		11116	76	7641
1069	8369		2633	11318	1429	**288**	11260	1113	7708
1073	10685	**275**	5654	11784	1499	1351	11612	1307	9977
1796	10997	691	5690		1516	1785		1336	10682
2541	11774	1881	9498	**282**	1958	1811	**292**	1718	
2671		2308	11209	719	1971	1938	852	1822	**301**
3098	**272**	2599		1347	2069	2339	1278	2283	2046
6431	689	2707	**279**	1835	2078	3411	2232	2356	2086
6456	820	2710	6	1976	2285	4351	2626	2381	2175
7235	1255	3305	538	3059	2551	4408	2833	3576	2562

BOILING POINT INDEX OF ORGANIC COMPOUNDS (continued)

2792	4402	11628	**318**	**326**	7599	**342**	**352**	728	**388**
9093	7709	12049	811	89	7947	3065	1095	3986	8648
10932	7749		1491	5776		3876	2437	7629	
	7764	**312**	2074	6634	**335**		8236	11660	**390**
302	8437	625	2080	7617	4	**343**	8969		9183
836	10473	946	7917	7803	679	3021		**366**	
1346	11275	1824		11781	2867	3987	**353**	7931	**391**
1350		2439	**319**		2882	4217	4118		4842
1821	**307**	4809	5	**327**	5751	4218	7592	**367**	11603
2161	78	5139	796	1929	7944	5350		5280	
3011	1619	6659	1825	2098	7946	6962	**354**		**393**
3855	2017	9454	10847	3989	8278	7624	1283	**368**	11163
5858	2311		11213	4854		8251	3981	4822	
6138	3045	**313**		8107	**336**			5268	**395**
6306	3586	1282	**320**		2305	**344**	**355**		7590
7902	7941	2073	1024	**328**	6305	314	35	**371**	7728
11406	8063	5828	1678	1566	6633	5044	2714	2864	
	8178	7344	2054	3046		5822	3013	9499	**397**
303	8641	8313	2651	5756	**337**	6100	4844		1743
77	10432	10465	3040	8259	258	7291	7625	**372**	4404
2021	10464	10474	3064		454	11431		7932	10947
2310	10871		3079	**329**	2143		**356**		
2422	11614	**314**	3085	2253	5993	**345**	5358	**373**	**398**
3310		809	3313	4763	6779	1333	6294	2089	784
5215	**308**	2716	4299	7343	7594	7600	7937	4463	
7737	834	2717	5812	8101		7948		11812	**400**
7741	2108	6660	6320	9701	**338**	8104	**357**		5172
	2312	7930	9525		7369		3037	**374**	11162
304	7263	10813	9634	**330**	9218	**346**		2255	
70	7631		11208	797	10831	2342	**359**		**401**
86	7770	**315**		2075		4484	1982	**375**	3031
1697	7903	810	**321**	3320	**339**	8233	2266	9583	8637
1698	11037	835	421	3431	366				
1826		1144	7614	5701	678	**347**	**360**	**376**	**402**
2264	**309**	1145		9447	2072	5584	450	11597	7725
2835	2018	2365	**322**		2122		1585		
7908	2366	6656	1146	**331**	2474	**348**	1591	**377**	**404**
10647	2699	7639	6631	764	5133	9188	2510	379	318
11211		8027	7900	4382	8579		2868	9528	10853
11319	**310**	9606	7943	10433		**349**	3946		
	831		8643		**340**	5239	4475	**379**	**405**
305	840	**316**	9455	**332**	1339	6307	7588	2237	2393
376	1196	795	11007	824	2256	6312	7912	7793	7730
799	1979	808	11964	2782	2306	7945	8294		
3062	2158	2374		3561	3087		8580	**380**	**406**
3071	2497	2700	**323**	5167	5137	**350**	9190	1866	2240
3090	2753	7719	2720	7604	5366	1317	9593	11855	4476
3104	3018	8234	4409	11595	7213	2379	11306		
4323	3658	9599			7895	7602		**383**	**408**
5045	5258		**324**	**333**	9163	7949	**362**	8242	2066
7616	7765	**317**	6317	48	9446	7950	8235		
7637	7802	624	7901	1746		9180	8243	**384**	**409**
7716	7942	1099		6641	**341**	9195		1334	4459
7718	11204	2019	**325**	7596	4850	11606	**363**	5966	
7750	11621	3041	1749	8064	5362	12006	3029		**410**
7956		3674	2145	8311	5498		11288	**385**	9569
9884	**311**	7628	4433	10814	5975	**351**	11596	1751	
11395	1584	7632	4464	11210	7615	2516		1798	**412**
	2022	7638	7571		7896	6644		4411	6622
306	2111	7722	11000	**334**	11811	12050			
1827	2436	10710	11328	1519			**365**	**387**	**413**
4001	6310			2085				7794	5351

414	420	430	438	443	452	11598	476	498	525
7880	1623	4384	519	8266	3005		4462	4853	4145
	4456					467			
415	4819	431	439	448	458	4474	480	499	527
7879	11193	1831	1741	1565	6299	5348	2739	4458	4383
	11545	8220		2405					
417			440	4114	459	468	484	514	531
428	425	433	8099		417	4821	4410	4482	4845
6653	11910	4808		449	4483				
11509			442	4380		471	490	519	540
	429	434	6302	11819	462	7727	8999	9601	4479
419	1568	782			418				
2062									551
									4413

DIAMAGNETIC SUSCEPTIBILITY OF SELECTED ORGANIC COMPOUNDS

When a material is placed in a magnetic field H, a magnetization M is induced in the material which is related to H by $M = \kappa H$, where κ is called the volume susceptibility. Since H and M have the same dimensions, κ is dimensionless. A more useful parameter is the molar susceptibility χ_m, defined by

$$\chi_m = \kappa V_m = \kappa\, M/\rho$$

where V_m is the molar volume of the substance, M the molar mass, and ρ the mass density. When the cgs system is used, the customary unit for χ_m is $cm^3\ mol^{-1}$; the corresponding SI unit is $m^3\ mol^{-1}$. Substances with no unpaired electrons are called diamagnetic; they have negative values of χ_m.

This table gives values of the diamagnetic susceptibility for about 400 common organic compounds. All values refer to room temperature and atmospheric pressure and to the physical form that is stable under these conditions. Substances are arranged by molecular formula in Hill order. A more extensive table may be found in Reference 1.

In keeping with customary practice, the molar susceptibility is given here in units appropriate to the cgs system. These values should be multiplied by 4π to obtain values for use in SI equations (where the magnetic field strength H has units of A m^{-1}).

REFERENCES

1. *Landolt-Börnstein, Numerical Data and Functional Relationships in Science and Technology, New Series,* II/16, *Diamagnetic Susceptibility,* Gupta, R. R., Ed., Springer-Verlag, Heidelberg, 1986.
2. Barter, C., Meisenheimer, R. G., and Stevenson, D. P., *J. Phys. Chem.* 64, 1312, 1960.
3. Broersma, S., *J. Chem. Phys.* 17, 873, 1949.

Molecular Formula	Compound	$-\chi_m/10^{-6}\ cm^3\ mol^{-1}$	Molecular Formula	Compound	$-\chi_m/10^{-6}\ cm^3\ mol^{-1}$
CBrCl$_3$	Bromotrichloromethane	73.2	C$_2$HCl$_3$O	Trichloroacetaldehyde	73.0
CBr$_4$	Tetrabromomethane	93.7	C$_2$HCl$_3$O	Dichloroacetyl chloride	69.0
CClF$_3$	Chlorotrifluoromethane	45.3	C$_2$HCl$_3$O$_2$	Trichloroacetic acid	73.0
CClN	Cyanogen chloride	32.4	C$_2$HCl$_5$	Pentachloroethane	99.1
CCl$_2$F$_2$	Dichlorodifluoromethane	52.2	C$_2$HF$_3$O$_2$	Trifluoroacetic acid	43.3
CCl$_2$O	Carbonyl chloride	47.9	C$_2$H$_2$	Acetylene	20.8
CCl$_3$F	Trichlorofluoromethane	58.7	C$_2$H$_2$Br$_4$	1,1,2,2-Tetrabromoethane	123.4
CCl$_3$NO$_2$	Trichloronitromethane	75.3	C$_2$H$_2$Cl$_2$	1,1-Dichloroethylene	49.2
CCl$_4$	Tetrachloromethane	66.8	C$_2$H$_2$Cl$_2$	cis-1,2-Dichloroethylene	51.0
CHBrCl$_2$	Bromodichloromethane	66.3	C$_2$H$_2$Cl$_2$	trans-1,2-Dichloroethylene	48.9
CHBr$_3$	Tribromomethane	82.6	C$_2$H$_2$Cl$_4$	1,1,2,2-Tetrachloroethane	89.8
CHCl$_3$	Trichloromethane	58.9	C$_2$H$_3$Cl	Chloroethylene	35.9
CHI$_3$	Triiodomethane	117.1	C$_2$H$_3$ClO	Acetyl chloride	39.3
CH$_2$BrCl	Bromochloromethane	55.1	C$_2$H$_3$N	Acetonitrile	27.8
CH$_2$Br$_2$	Dibromomethane	65.1	C$_2$H$_4$	Ethylene	18.8
CH$_2$Cl$_2$	Dichloromethane	46.6	C$_2$H$_4$Br$_2$	1,2-Dibromoethane	78.9
CH$_2$I$_2$	Diiodomethane	93.1	C$_2$H$_4$Cl$_2$	1,1-Dichloroethane	57.4
CH$_2$N$_2$	Cyanamide	24.8	C$_2$H$_4$Cl$_2$	1,2-Dichloroethane	59.6
CH$_2$O	Formaldehyde	18.6	C$_2$H$_4$O	Acetaldehyde	22.2
CH$_2$O$_2$	Formic acid	19.9	C$_2$H$_4$O	Ethylene oxide	30.5
CH$_3$Br	Bromomethane	42.8	C$_2$H$_4$O$_2$	Acetic acid	31.8
CH$_3$Cl	Chloromethane	32.0	C$_2$H$_4$O$_2$	Methyl formate	31.1
CH$_3$F	Fluoromethane	17.8	C$_2$H$_5$Br	Bromoethane	78.8
CH$_3$I	Iodomethane	57.2	C$_2$H$_5$Cl	Chloroethane	69.9
CH$_3$NO	Formamide	23.0	C$_2$H$_5$I	Iodoethane	69.1
CH$_3$NO$_2$	Nitromethane	21.0	C$_2$H$_5$NO	Acetamide	33.9
CH$_4$	Methane	17.4	C$_2$H$_5$NO$_2$	Nitroethane	35.4
CH$_4$N$_2$O	Urea	33.5	C$_2$H$_5$NO$_2$	Glycine	39.6
CH$_4$O	Methanol	21.4	C$_2$H$_6$	Ethane	26.8
CH$_5$N	Methylamine	27.0	C$_2$H$_6$O	Ethanol	33.7
CI$_4$	Tetraiodomethane	136	C$_2$H$_6$O	Dimethyl ether	26.3
CN$_4$O$_8$	Tetranitromethane	43.0	C$_2$H$_6$O$_2$	Ethylene glycol	38.9
C$_2$ClF$_3$	Chlorotrifluoroethylene	49.1	C$_2$H$_6$S	Ethanethiol	47.0
C$_2$Cl$_4$	Tetrachloroethylene	81.6	C$_2$H$_6$S	Dimethyl sulfide	44.9
C$_2$Cl$_6$	Hexachloroethane	112.8	C$_2$H$_8$N$_2$	1,2-Ethanediamine	46.5
C$_2$HCl$_3$	Trichloroethylene	65.8	C$_2$N$_2$	Cyanogen	21.6

Molecular Formula	Compound	$-\chi_m/10^{-6}$ cm^3 mol^{-1}	Molecular Formula	Compound	$-\chi_m/10^{-6}$ cm^3 mol^{-1}
C_3H_4	Allene	25.3	$C_4H_8O_2$	1,4-Dioxane	52.2
$C_3H_4O_2$	Vinyl formate	34.7	C_4H_9Br	1-Bromobutane	77.1
C_3H_5Br	3-Bromopropene	58.6	C_4H_9Br	1-Bromo-2-methylpropane	79.9
C_3H_5Cl	2-Chloropropene	47.8	C_4H_9Cl	1-Chlorobutane	67.1
C_3H_5Cl	3-Chloropropene	47.8	C_4H_9Cl	2-Chlorobutane	67.4
C_3H_5N	Propanenitrile	38.6	C_4H_9I	1-Iodobutane	93.6
C_3H_6	Propene	30.7	C_4H_9N	Pyrrolidine	54.8
C_3H_6	Cyclopropane	39.2	C_4H_9NO	Morpholine	55.0
C_3H_6O	Allyl alcohol	36.7	C_4H_{10}	Butane	50.3
C_3H_6O	Propanal	34.2	C_4H_{10}	Isobutane	50.5
C_3H_6O	Acetone	33.8	$C_4H_{10}O$	1-Butanol	56.4
C_3H_6O	Methyloxirane	42.5	$C_4H_{10}O$	2-Butanol	57.6
$C_3H_6O_2$	Propanoic acid	43.2	$C_4H_{10}O$	2-Methyl-1-propanol	57.6
$C_3H_6O_2$	Ethyl formate	42.4	$C_4H_{10}O$	2-Methyl-2-propanol	56.6
C_3H_7Br	1-Bromopropane	65.6	$C_4H_{10}O$	Diethyl ether	55.5
C_3H_7Br	2-Bromopropane	65.1	$C_4H_{10}O_2$	1,3-Butanediol	61.8
C_3H_7Cl	1-Chloropropane	56.0	$C_4H_{10}O_2$	1,4-Butanediol	61.8
C_3H_7I	1-Iodopropane	84.3	$C_4H_{10}S$	1-Butanethiol	70.2
C_3H_7N	Allylamine	40.1	$C_4H_{11}N$	Butylamine	58.9
$C_3H_7NO_2$	1-Nitropropane	45.0	$C_4H_{11}N$	Isobutylamine	59.8
$C_3H_7NO_2$	2-Nitropropane	45.4	$C_4H_{11}N$	Diethylamine	56.8
$C_3H_7NO_2$	Ethyl carbamate	57.0	$C_5H_4O_2$	Furfural	47.2
C_3H_8	Propane	38.6	C_5H_5N	Pyridine	48.7
C_3H_8O	1-Propanol	44.8	$C_5H_6O_2$	Furfuryl alcohol	61.0
C_3H_8O	2-Propanol	45.7	$C_5H_7NO_2$	Ethyl cyanoacetate	67.3
$C_3H_8O_2$	1,3-Propylene glycol	50.2	C_5H_8	2-Methyl-1,3-butadiene	46.0
$C_3H_8O_2$	Dimethoxymethane	47.3	C_5H_8O	Cyclopentanone	51.6
$C_3H_8O_3$	Glycerol	57.1	$C_5H_8O_2$	Methyl methacrylate	57.3
$C_4H_2O_3$	Maleic anhydride	35.8	$C_5H_8O_2$	2,4-Pentanedione	54.9
$C_4H_4N_2$	Pyrazine	37.8	C_5H_{10}	1-Pentene	54.6
$C_4H_4N_2$	Pyrimidine	43.1	C_5H_{10}	2-Methyl-2-butene	54.7
C_4H_4O	Furan	43.1	C_5H_{10}	Cyclopentane	56.2
$C_4H_4O_3$	Succinic anhydride	47.5	$C_5H_{10}O$	Cyclopentanol	64.0
$C_4H_4O_4$	Maleic acid	49.6	$C_5H_{10}O$	Pentanal	57.5
$C_4H_4O_4$	Fumaric acid	49.1	$C_5H_{10}O$	2-Pentanone	57.5
C_4H_4S	Thiophene	57.3	$C_5H_{10}O$	3-Pentanone	57.7
C_4H_5N	Pyrrole	48.6	$C_5H_{10}O_2$	Pentanoic acid	66.5
C_4H_6	1,2-Butadiene	35.6	$C_5H_{10}O_2$	3-Methylbutanoic acid	67.7
C_4H_6	1,3-Butadiene	32.1	$C_5H_{10}O_2$	Butyl formate	65.8
$C_4H_6O_2$	Vinyl acetate	46.4	$C_5H_{10}O_2$	Isobutyl formate	66.8
$C_4H_6O_3$	Acetic anhydride	52.8	$C_5H_{10}O_2$	Propyl acetate	65.9
$C_4H_6O_4$	Succinic acid	58.0	$C_5H_{10}O_2$	Isopropyl acetate	67.0
$C_4H_6O_4$	Dimethyl oxalate	55.7	$C_5H_{10}O_2$	Ethyl propanoate	66.3
C_4H_7N	Butanenitrile	50.4	$C_5H_{10}O_2$	Tetrahydrofurfuryl alcohol	69.4
C_4H_8	1-Butene	41.0	$C_5H_{10}O_3$	Diethyl carbonate	75.4
C_4H_8	cis-2-Butene	42.6	$C_5H_{11}N$	Piperidine	64.2
C_4H_8	trans-2-Butene	43.3	C_5H_{12}	Pentane	61.5
C_4H_8	Isobutene	40.8	C_5H_{12}	Isopentane	63.0
C_4H_8	Cyclobutane	40.0	C_5H_{12}	Neopentane	63.0
C_4H_8O	Ethyl vinyl ether	47.9	$C_5H_{12}O$	1-Pentanol	67.0
C_4H_8O	1,2-Epoxybutane	54.8	$C_5H_{12}O$	2-Pentanol	69.1
C_4H_8O	Butanal	45.9	$C_5H_{12}O_2$	1,5-Pentanediol	73.5
C_4H_8O	2-Butanone	45.6	$C_5H_{13}N$	Pentylamine	69.3
$C_4H_8O_2$	Butanoic acid	55.2	C_6Cl_6	Hexachlorobenzene	147.0
$C_4H_8O_2$	2-Methylpropanoic acid	56.1	$C_6H_4ClNO_2$	1-Chloro-2-nitrobenzene	75.5
$C_4H_8O_2$	Propyl formate	55.0	$C_6H_4ClNO_2$	1-Chloro-3-nitrobenzene	77.2
$C_4H_8O_2$	Ethyl acetate	54.1	$C_6H_4ClNO_2$	1-Chloro-4-nitrobenzene	74.7
$C_4H_8O_2$	Methyl propanoate	54.5	$C_6H_4Cl_2$	o-Dichlorobenzene	84.4

DIAMAGNETIC SUSCEPTIBILITY OF SELECTED ORGANIC COMPOUNDS (continued)

Molecular Formula	Compound	$-\chi_m/10^{-6}$ cm^3 mol^{-1}	Molecular Formula	Compound	$-\chi_m/10^{-6}$ cm^3 mol^{-1}
$C_6H_4Cl_2$	m-Dichlorobenzene	84.1	$C_6H_{14}O$	Dipropyl ether	79.4
$C_6H_4Cl_2$	p-Dichlorobenzene	81.7	$C_6H_{14}O_2$	1,6-Hexanediol	84.3
$C_6H_4O_2$	p-Benzoquinone	36	$C_6H_{14}O_2$	1,1-Diethoxyethane	81.4
C_6H_5Br	Bromobenzene	78.4	$C_6H_{14}O_6$	D-Glucitol	107.8
C_6H_5Cl	Chlorobenzene	69.5	$C_6H_{15}N$	Triethylamine	83.3
C_6H_5ClO	o-Chlorophenol	77.3	C_7H_5N	Benzonitrile	65.2
C_6H_5ClO	m-Chlorophenol	77.6	C_7H_6O	Benzaldehyde	60.7
C_6H_5ClO	p-Chlorophenol	77.7	$C_7H_6O_2$	Salicylaldehyde	66.8
C_6H_5F	Fluorobenzene	58.4	$C_7H_6O_3$	Salicylic acid	75
C_6H_5I	Iodobenzene	92.0	C_7H_7Br	p-Bromotoluene	88.7
$C_6H_5NO_2$	Nitrobenzene	61.9	C_7H_7Cl	o-Chlorotoluene	82.4
$C_6H_5NO_3$	o-Nitrophenol	68.9	C_7H_7Cl	m-Chlorotoluene	79.7
$C_6H_5NO_3$	m-Nitrophenol	65.9	C_7H_7Cl	p-Chlorotoluene	80.3
$C_6H_5NO_3$	p-Nitrophenol	66.9	C_7H_7Cl	(Chloromethyl)benzene	81.6
C_6H_6	Benzene	54.8	C_7H_7NO	Benzamide	72.0
C_6H_6ClN	o-Chloroaniline	79.5	$C_7H_7NO_2$	o-Nitrotoluene	72.2
C_6H_6ClN	m-Chloroaniline	76.6	$C_7H_7NO_2$	m-Nitrotoluene	72.7
C_6H_6ClN	p-Chloroaniline	76.7	$C_7H_7NO_2$	p-Nitrotoluene	73.3
$C_6H_6N_2O_2$	o-Nitroaniline	67.4	C_7H_8	Toluene	65.6
$C_6H_6N_2O_2$	m-Nitroaniline	69.7	C_7H_8O	o-Cresol	73.3
$C_6H_6N_2O_2$	p-Nitroaniline	68.0	C_7H_8O	m-Cresol	72.2
C_6H_6O	Phenol	60.6	C_7H_8O	p-Cresol	72.4
$C_6H_6O_2$	p-Hydroquinone	64.7	C_7H_8O	Benzyl alcohol	71.8
$C_6H_6O_2$	Pyrocatechol	68.2	C_7H_8O	Anisole	72.2
$C_6H_6O_2$	Resorcinol	67.2	C_7H_9N	o-Methylaniline	74.9
C_6H_7N	Aniline	62.4	C_7H_9N	m-Methylaniline	74.6
C_6H_7N	4-Methylpyridine	59.8	C_7H_9N	p-Methylaniline	72.5
C_6H_8	1,4-Cyclohexadiene	48.7	C_7H_9N	N-Methylaniline	74.1
$C_6H_8N_2$	o-Phenylenediamine	72.5	C_7H_9N	2,4-Dimethylpyridine	71.3
$C_6H_8N_2$	m-Phenylenediamine	70.4	C_7H_9N	2,6-Dimethylpyridine	72.5
$C_6H_8N_2$	p-Phenylenediamine	70.7	C_7H_9NO	o-Methoxyaniline [o-Anisidine]	79.1
C_6H_{10}	1,5-Hexadiene	55.1	$C_7H_{12}O_4$	Diethyl malonate	92.6
C_6H_{10}	1-Hexyne	64.5	C_7H_{14}	1-Heptene	77.8
C_6H_{10}	Cyclohexene	58.0	C_7H_{14}	Cycloheptane	73.9
$C_6H_{10}O$	Cyclohexanone	62.0	C_7H_{14}	Methylcyclohexane	78.9
$C_6H_{10}O_3$	Ethyl acetoacetate	71.7	$C_7H_{14}O$	1-Heptanal	81.0
$C_6H_{10}O_4$	Diethyl oxalate	81.7	$C_7H_{14}O$	2-Heptanone	80.5
C_6H_{12}	1-Hexene	66.4	$C_7H_{14}O$	3-Heptanone	80.7
C_6H_{12}	2,3-Dimethyl-2-butene	65.9	$C_7H_{14}O$	4-Heptanone	80.5
C_6H_{12}	Cyclohexane	68	$C_7H_{14}O$	2,4-Dimethyl-3-pentanone	81.1
C_6H_{12}	Methylcyclopentane	70.2	$C_7H_{14}O_2$	Heptanoic acid	89.0
$C_6H_{12}O$	Hexanal	69.4	$C_7H_{14}O_2$	Pentyl acetate	88.9
$C_6H_{12}O$	2-Hexanone	69.2	$C_7H_{14}O_2$	Isopentyl acetate	89.4
$C_6H_{12}O$	3-Hexanone	69.0	$C_7H_{14}O_2$	Butyl propanoate	89.1
$C_6H_{12}O$	4-Methyl-2-pentanone	69.7	$C_7H_{14}O_2$	Ethyl 3-methylbutanoate	91.1
$C_6H_{12}O$	Cyclohexanol	73.4	C_7H_{16}	Heptane	85.2
$C_6H_{12}O_2$	Hexanoic acid	78.1	C_7H_{16}	3-Ethylpentane	86.2
$C_6H_{12}O_2$	Isopentyl formate	78.4	C_7H_{16}	2,2-Dimethylpentane	87.0
$C_6H_{12}O_2$	Isobutyl acetate	78.7	C_7H_{16}	2,3-Dimethylpentane	87.5
$C_6H_{12}O_2$	Propyl propanoate	77.7	C_7H_{16}	2,4-Dimethylpentane	87.5
$C_6H_{12}O_3$	Paraldehyde	86.1	C_7H_{16}	3,3-Dimethylpentane	89.5
C_6H_{14}	Hexane	74.1	$C_7H_{16}O$	1-Heptanol	91.7
C_6H_{14}	2-Methylpentane	75.3	$C_7H_{16}O$	4-Heptanol	92.1
C_6H_{14}	3-Methylpentane	75.5	$C_8H_4O_3$	Phthalic anhydride	66.7
C_6H_{14}	2,2-Dimethylbutane	76.2	$C_8H_6O_4$	Phthalic acid	83.6
C_6H_{14}	2,3-Dimethylbutane	76.2	$C_8H_6O_4$	Isophthalic acid	84.6
$C_6H_{14}O$	1-Hexanol	79.5	$C_8H_6O_4$	Terephthalic acid	83.5
$C_6H_{14}O$	4-Methyl-2-pentanol	80.4	C_8H_7N	Benzeneacetonitrile	76.9

DIAMAGNETIC SUSCEPTIBILITY OF SELECTED ORGANIC COMPOUNDS (continued)

Molecular Formula	Compound	$-\chi_m/10^{-6}$ cm^3 mol^{-1}	Molecular Formula	Compound	$-\chi_m/10^{-6}$ cm^3 mol^{-1}
C$_8$H$_7$N	Indole	85.0	C$_{10}$H$_{10}$O$_2$	Safrole	97.5
C$_8$H$_8$	Styrene	68.2	C$_{10}$H$_{10}$O$_4$	Dimethyl terephthalate	101.6
C$_8$H$_8$O	Acetophenone	72.5	C$_{10}$H$_{14}$	Butylbenzene	100.7
C$_8$H$_8$O$_2$	o-Toluic acid	84.3	C$_{10}$H$_{14}$	tert-Butylbenzene	101.8
C$_8$H$_8$O$_2$	m-Toluic acid	83.0	C$_{10}$H$_{14}$	Isobutylbenzene	101.7
C$_8$H$_8$O$_2$	p-Toluic acid	82.4	C$_{10}$H$_{14}$	p-Cymene	102.8
C$_8$H$_8$O$_2$	Benzeneacetic acid	82.4	C$_{10}$H$_{14}$	1,2,4,5-Tetramethylbenzene	101.2
C$_8$H$_8$O$_2$	Methyl benzoate	81.6	C$_{10}$H$_{14}$O	p-tert-Butylphenol	108.0
C$_8$H$_8$O$_3$	Methyl salicylate	86.6	C$_{10}$H$_{15}$N	N,N-Diethylaniline	107.9
C$_8$H$_{10}$	Ethylbenzene	77.3	C$_{10}$H$_{16}$	d-Limonene	98.0
C$_8$H$_{10}$	o-Xylene	77.7	C$_{10}$H$_{16}$	α-Pinene	100.7
C$_8$H$_{10}$	m-Xylene	76.4	C$_{10}$H$_{16}$	β-Pinene	101.9
C$_8$H$_{10}$	p-Xylene	77.0	C$_{10}$H$_{16}$O	Camphor, (+)	103.0
C$_8$H$_{10}$O	Phenetole	84.5	C$_{10}$H$_{18}$	cis-Decahydronaphthalene	107.0
C$_8$H$_{11}$N	N-Ethylaniline	85.6	C$_{10}$H$_{18}$	trans-Decahydronaphthalene	107.6
C$_8$H$_{11}$N	N,N-Dimethylaniline	85.1	C$_{10}$H$_{22}$	Decane	119.5
C$_8$H$_{11}$N	2,4,6-Trimethylpyridine	83.1	C$_{11}$H$_{10}$	1-Methylnaphthalene	102.9
C$_8$H$_{14}$O$_4$	Ethyl succinate	105.0	C$_{11}$H$_{10}$	2-Methylnaphthalene	102.7
C$_8$H$_{16}$	1-Octene	88.8	C$_{11}$H$_{24}$	Undecane	131.8
C$_8$H$_{16}$	Cyclooctane	85.3	C$_{12}$H$_8$	Acenaphthylene	111.6
C$_8$H$_{16}$O$_2$	Octanoic acid	99.5	C$_{12}$H$_9$N	Carbazole	119.9
C$_8$H$_{16}$O$_2$	Hexyl acetate	100.9	C$_{12}$H$_{10}$	Acenaphthene	109.9
C$_8$H$_{17}$Cl	1-Chlorooctane	114.9	C$_{12}$H$_{10}$	Biphenyl	103.3
C$_8$H$_{18}$	Octane	96.6	C$_{12}$H$_{10}$N$_2$	Azobenzene	106.8
C$_8$H$_{18}$	4-Methylheptane	97.3	C$_{12}$H$_{11}$N	Diphenylamine	108.4
C$_8$H$_{18}$	3-Ethylhexane	97.8	C$_{12}$H$_{14}$O$_4$	Diethyl phthalate	127.5
C$_8$H$_{18}$	3,4-Dimethylhexane	99.1	C$_{12}$H$_{18}$	Hexamethylbenzene	122.5
C$_8$H$_{18}$	2,2,4-Trimethylpentane	99.1	C$_{12}$H$_{24}$O$_2$	Dodecanoic acid	113.0
C$_8$H$_{18}$	2,3,4-Trimethylpentane	99.8	C$_{13}$H$_9$N	Acridine	118.8
C$_8$H$_{18}$O	1-Octanol	101.6	C$_{13}$H$_{10}$O	Benzophenone	109.6
C$_8$H$_{19}$N	Dibutylamine	103.7	C$_{13}$H$_{12}$	Diphenylmethane	116.0
C$_9$H$_7$N	Quinoline	86.1	C$_{13}$H$_{28}$	Tridecane	153.7
C$_9$H$_7$N	Isoquinoline	83.9	C$_{14}$H$_8$O$_2$	9,10-Anthracenedione	113.0
C$_9$H$_8$	Indene	83	C$_{14}$H$_{10}$	Anthracene	129.8
C$_9$H$_{10}$	Isopropenylbenzene	80.0	C$_{14}$H$_{10}$	Phenanthrene	127.6
C$_9$H$_{10}$O$_2$	Ethyl benzoate	93.8	C$_{14}$H$_{10}$	Diphenylacetylene	116
C$_9$H$_{10}$O$_2$	Benzyl acetate	93.2	C$_{14}$H$_{10}$O$_2$	Benzil	106.8
C$_9$H$_{12}$	Propylbenzene	89.1	C$_{14}$H$_{12}$O$_2$	Benzyl benzoate	132.2
C$_9$H$_{12}$	Isopropylbenzene [Cumene]	89.5	C$_{14}$H$_{14}$	1,2-Diphenylethane	127.8
C$_9$H$_{12}$	1,3,5-Trimethylbenzene [Mesitylene]	92.3	C$_{14}$H$_{28}$O$_2$	Tetradecanoic acid [Myristic acid]	176.0
C$_9$H$_{18}$	1-Nonene	100.1	C$_{14}$H$_{30}$	Tetradecane	166.2
C$_9$H$_{18}$O	2,6-Dimethyl-4-heptanone	104.3	C$_{16}$H$_{10}$	Pyrene	147
C$_9$H$_{20}$	Nonane	108.1	C$_{16}$H$_{32}$O$_2$	Hexadecanoic acid [Palmitic acid]	198.6
C$_{10}$H$_7$Br	1-Bromonaphthalene	123.6	C$_{16}$H$_{34}$	Hexadecane	187.6
C$_{10}$H$_7$Cl	1-Chloronaphthalene	107.6	C$_{16}$H$_{34}$O	1-Hexadecanol	183.5
C$_{10}$H$_8$	Naphthalene	91.6	C$_{18}$H$_{12}$	Chrysene	148.0
C$_{10}$H$_8$	Azulene	123.7	C$_{18}$H$_{14}$	o-Terphenyl	150.4
C$_{10}$H$_8$O	1-Naphthol	96.2	C$_{18}$H$_{14}$	m-Terphenyl	155.5
C$_{10}$H$_8$O	2-Naphthol	96.8	C$_{18}$H$_{14}$	p-Terphenyl	156.0
C$_{10}$H$_9$N	1-Naphthalenamine	92.5	C$_{18}$H$_{34}$O$_2$	cis-9-Octadecenoic acid [Oleic acid]	208.5
C$_{10}$H$_9$N	2-Naphthalenamine	98.0	C$_{18}$H$_{36}$O$_2$	Octadecanoic acid [Stearic acid]	220.8
			C$_{20}$H$_{12}$	Perylene	167.5

Section 4
Properties of the Elements and Inorganic Compounds

The Elements 4-1
Physical Constants of Inorganic Compounds 4-37
Synonym Index of Inorganic Compounds 4-101
CAS Registry Number Index of Inorganic Compounds 4-107
Physical Properties of the Rare Earth Metals 4-114
Melting, Boiling, and Critical Temperatures of the Elements 4-124
Heat Capacity of the Elements at 25°C 4-125
Vapor Pressure of the Metallic Elements 4-126
Density of Molten Elements and Representative Salts 4-129
Magnetic Susceptibility of the Elements and Inorganic Compounds 4-133
Index of Refraction of Inorganic Liquids 4-139
Physical and Optical Properties of Minerals 4-141
Crystallographic Data on Minerals 4-148

THE ELEMENTS
C. R. Hammond

One of the most striking facts about the elements is their unequal distribution and occurrence in nature. Present knowledge of the chemical composition of the universe, obtained from the study of the spectra of stars and nebulae, indicates that hydrogen is by far the most abundant element and may account for more than 90% of the atoms or about 75% of the mass of the universe. Helium atoms make up most of the remainder. All of the other elements together contribute only slightly to the total mass.

The chemical composition of the universe is undergoing continuous change. Hydrogen is being converted into helium, and helium is being changed into heavier elements. As time goes on, the ratio of heavier elements increases relative to hydrogen. Presumably, the process is not reversible.

Burbidge, Burbidge, Fowler, and Hoyle, and more recently, Peebles, Penzias, and others have studied the synthesis of elements in stars. To explain all of the features of the nuclear abundance curve — obtained by studies of the composition of the earth, meteorites, stars, etc. — it is necessary to postulate that the elements were originally formed by at least eight different processes: (1) hydrogen burning, (2) helium burning, (3) χ process, (4) e process, (5) s process, (6) r process, (7) p process, and (8) the X process. The X process is thought to account for the existence of light nuclei such as D, Li, Be, and B. Common metals such as Fe, Cr, Ni, Cu, Ti, Zn, etc. were likely produced early in the history of our galaxy. It is also probable that most of the heavy elements on earth and elsewhere in the universe were originally formed in supernovae, or in the hot interior of stars.

Studies of the solar spectrum have led to the identification of 67 elements in the sun's atmosphere; however, all elements cannot be identified with the same degree of certainty. Other elements may be present in the sun, although they have not yet been detected spectroscopically. The element helium was discovered on the sun before it was found on earth. Some elements such as scandium are relatively more plentiful in the sun and stars than here on earth.

Minerals in lunar rocks brought back from the moon on the Apollo missions consist predominantly of *plagioclase* {(Ca,Na)(Al,Si)O_4O_8} and *pyroxene* {(Ca,Mg,Fe)$_2$Si$_2O_6$} — two minerals common in terrestrial volcanic rock. No new elements have been found on the moon that cannot be accounted for on earth; however, three minerals, *armalcolite* {(Fe,Mg)Ti$_2O_5$}, *pyroxferroite* {CaFe$_6$(SiO$_3$)$_7$}, and *tranquillityite* {Fe$_8$(Zr,Y)Ti$_3$Si$_3O_2$}, are new. The oldest known terrestrial rocks are about 4 billion years old. One rock, known as the "Genesis Rock," brought back from the Apollo 15 Mission, is about 4.15 billion years old. This is only about one-half billion years younger than the supposed age of the moon and solar system. Lunar rocks appear to be relatively enriched in refractory elements such as chromium, titanium, zirconium, and the rare earths, and impoverished in volatile elements such as the alkali metals, in chlorine, and in noble metals such as nickel, platinum, and gold.

Even older than the "Genesis Rock" are *carbonaceous chondrites,* a type of meteorite that has fallen to earth and has been studied. These are some of the most primitive objects of the solar system yet found. The grains making up these objects probably condensed directly out the gaseous nebula from which the sun and planets were born. Most of the condensation of the grains probably was completed within 50,000 years of the time the disk of the nebula was first formed — about 4.6 billion years ago. It is now thought that this type of meteorite may contain a small percentage of presolar dust grains. The relative abundances of the elements of these meteorites are about the same as the abundances found in the solar chromosphere.

The X-ray fluorescent spectrometer sent with the Viking I spacecraft to Mars shows that the Martian soil contains about 12 to 16% iron, 14 to 15% silicon, 3 to 8% calcium, 2 to 7% aluminum, and one-half to 2% titanium. The gas chromatograph — mass spectrometer on Viking II found no trace of organic compounds that should be present if life ever existed there.

F. W. Clarke and others have carefully studied the composition of rocks making up the crust of the earth. Oxygen accounts for about 47% of the crust, by weight, while silicon comprises about 28% and aluminum about 8%. These elements, plus iron, calcium, sodium, potassium, and magnesium, account for about 99% of the composition of the crust.

Many elements such as tin, copper, zinc, lead, mercury, silver, platinum, antimony, arsenic, and gold, which are so essential to our needs and civilization, are among some of the rarest elements in the earth's crust. These are made available to us only by the processes of concentration in ore bodies. Some of the so-called *rare-earth* elements have been found to be much more plentiful than originally thought and are about as abundant as uranium, mercury, lead, or bismuth. The least abundant rare-earth or *lanthanide* element, thulium, is now believed to be more plentiful on earth than silver, cadmium, gold, or iodine, for example. Rubidium, the 16th most abundant element, is more plentiful than chlorine while its compounds are little known in chemistry and commerce.

It is now thought that at least 24 elements are essential to living matter. The four most abundant in the human body are hydrogen, oxygen, carbon, and nitrogen. The seven next most common, in order of abundance, are calcium, phosphorus, chlorine, potassium, sulfur, sodium, and magnesium. Iron, copper, zinc, silicon, iodine, cobalt, manganese, molybdenum, fluorine, tin, chromium, selenium, and vanadium are needed and play a role in living matter. Boron is also thought essential for some plants, and it is possible that aluminum, nickel, and germanium may turn out to be necessary.

Ninety-one elements occur naturally on earth. Minute traces of plutonium-244 have been discovered in rocks mined in Southern California. This discovery supports the theory that heavy elements were produced during creation of the solar system. While technetium and promethium have not yet been found naturally on earth, they have been found to be present in stars. Technetium has been identified in the spectra of certain "late" type stars, and promethium lines have been identified in the spectra of a faintly visible star HR465 in Andromeda. Promethium must have been made very recently near the star's surface for no known isotope of this element has a half-life longer than 17.7 years.

It has been suggested that californium is present in certain stellar explosions known as supernovae; however, this has not been proved. At present no elements are found elsewhere in the universe that cannot be accounted for here on earth.

All atomic mass numbers from 1 to 238 are found naturally on earth except for masses 5 and 8. About 285 relatively stable and 67 naturally radioactive isotopes occur on earth totaling 352. In addition, the neutron, technetium, promethium, and the transuranic elements (lying beyond uranium) have now been produced artificially. Scientists at the Lawrence Berkeley National Laboratory, in June 1999, reported they had found evidence of an isotope of Element 118 and its immediate decay products, Element Nos. 116 and 114. The sequence of events tend to reinforce the theories that have predicted, since the 1970s, an "island of stability" for nuclei with approximately 114 protons and 184 neutrons. This "island" refers to nuclei in which the decay lasts for a period of time instead of a decay that occurs instantly.

Laboratory processes have now extended the radioactive element mass numbers beyond 238 to 293. Each element from atomic numbers 1 to 110 is known to have at least one radioactive isotope. As of December 1999, 3288 isotopes and isomers were thought to be known and recognized. Many stable and radioactive isotopes are now produced and distributed by the Oak Ridge National Laboratory, Oak Ridge, Tenn., U.S.A., to customers licensed by the U.S. Department of Energy.

THE ELEMENTS (continued)

The nucleus of an atom is characterized by the number of protons it contains, denoted by Z, and by the number of neutrons, N. Isotopes of an element have the same value of Z, but different values of N. The *mass number A*, is the sum of Z and N. For example, Uranium-238 has a mass number of 238, and contains 92 protons and 146 neutrons.

There is evidence that the definition of chemical elements must be broadened to include the electron. Several compounds known as *electrides,* have recently been made of alkaline metal elements and electrons. A relatively stable combination of a positron and electron, known as *positronium,* has also been studied.

In addition to the proton, neutron, and electron, there are considerably more than 100 other fundamental particles which have been discovered or hypothesized. The majority of these fall into one of two classes, *leptons* or *hadrons*. The leptons comprise just four known particles, the *electron*, the *muon* (μ meson), and two kinds of *neutrinos*. The muon is essentially similar to the electron and has a charge of –1, but it is 200 times heavier. The neutrino is either of two stable particles of small (probably zero) rest mass, carrying no charge. Also there are four *antileptons,* identical to the corresponding leptons in some respects, such as mass, but they have properties exactly opposite those of the leptons. The *positron,* for example, is an antilepton, with a charge of +1. Leptons cannot be broken into smaller units and are considered to be elementary. On the other hand, hadrons are complex and thought to have internal structure. Protons and neutrons, which make up atomic nuclei, are hadrons.

Elementary particle physics is not yet clearly understood, but groupings and arrangements of these particles have been made resembling the periodic table of chemical elements. This has led to the speculation that hadrons are composed of simpler components called *quarks*. Quarks are presumed to be elementary particles. Quarks come in a variety of "flavors". Two kinds of quarks — up and down — combine to form the protons and neutrons of ordinary matter. Three other quarks, known as {strange", "charm", and "bottom", are found in particle accelerators, and one of these may exist in the center of very dense, massive stars. Theorists have thought that a sixth "top" quark, much more massive than the others, should exist to complete the quark family of sub-atomic particles. In 1994, scientists using Fermilab's Tevatron accelerator may have "glimpsed" the top quark. There is presently no evidence that quarks exist in isolation. Many physicists now hold that all the matter and energy in the universe is controlled by four fundamental natural forces: the electromagnetic force, gravity, a weak nuclear force, and a strong nuclear force. Each of these natural forces is passed back and forth among the basic particles of matter by unique force-carrying particles. The electromagnetic force is carried by the *photon,* the weak nuclear force by the intermediate vector *boson,* and gravity by the *gravitron.* There is now evidence of the existence of a particle, known as the *gluon,* that binds quarks together by carrying the strong nuclear force. *Hadrons* are thought to be particles composed of quarks and gluons.

The available evidence leads to the conclusion that elements 89 (actinium) through 103 (lawrencium) are chemically similar to the rare-earth or lanthanide elements (elements 57 to 71, inclusive). These elements therefore have been named *actinides* after the first member of this series. Those elements beyond uranium that have been produced artificially have the following names and symbols: neptunium, 93 (Np); plutonium, 94 (Pu); americium, 95 (Am); curium, 96 (Cm); berkelium, 97 (Bk); californium, 98 (Cf); einsteinium, 99 (Es); fermium, 100 (Fm); mendelevium, 101 (Md); nobelium, 102 (No); and lawrencium, 103 (Lr). It is now claimed that Elements 104 through 112 have been produced and identified. More recently, Elements 118, 116, and 114 are now thought to be found. In August 1997, the International Union of Pure and Applied Chemistry (IUPAC) gave final approval to the following names for Elements 104 to 109: Element 104 — rutherfordium (Rf); Element 105 — dubnium (Db); Element 106 — seaborgium (Sg); Element 107 — bohrium (Bh); Element 108 — hassium (Hs); and Element 109 — meitnerium (Mt). The recently discovered elements 110, 111, 112, 114, 116, and 118 have not yet been named, but may carry temporary names as designated by the International Union of Pure and Applied Chemistry. IUPAC recommends that until the existence of a new element is proven to their satisfaction, the elements are to have names and symbols derived according to these precise and simple rules: The name is based on the digits in the element's atomic number. Each digit is replaced with these expressions, with the end using the usual –ium suffix as follows: **0 nil, 1 un, 2 bi, 3 tri, 4 quad, 5 pent, 6 hex, 7 sept, 8 oct, 9 enn**. Double letter i's are not used, as for example Ununbiium, but would be Ununbium. The symbol used would be the first letter of the three main syllables. For example, Element 126 would be Unbihexium, with the symbol Ubh. It is thought there is a good possibility of producing elements beyond Element 118, and discovering, in the near future, Elements 117, 115, and 113 by altering the beams of ions and the targets from those now being used.

There are many claims in the literature of the existence of various allotropic modifications of the elements, some of which are based on doubtful or incomplete evidence. Also, the physical properties of an element may change drastically by the presence of small amounts of impurities. With new methods of purification, which are now able to produce elements with 99.9999% purity, it has been necessary to restudy the properties of the elements. For example, the melting point of thorium changes by several hundred degrees by the presence of a small percentage of ThO_2 as an impurity. Ordinary commercial tungsten is brittle and can be worked only with difficulty. Pure tungsten, however, can be cut with a hacksaw, forged, spun, drawn, or extruded. In general, the value of a physical property given here applies to the pure element, when it is known.

Many of the chemical elements and their compounds are toxic and should be handled with due respect and care. In recent years there has been a greatly increased knowledge and awareness of the health hazards associated with chemicals, radioactive materials, and other agents. Anyone working with the elements and certain of their compounds should become thoroughly familiar with the proper safeguards to be taken. Information on specific hazards and recommended exposure limits may also be found in Section 16. Reference should also be made to publications such as the following:

1. *Code of Federal Regulations, Title 29, Labor.* With additions found in issues of the *Federal Register.*
2. *Code of Federal Regulations, Title 10, Energy.* With additions found in issues of the *Federal Register.* (Published by the U.S. Government Printing Office. Supt. of Documents.)
3. *Occupational Safety and Health Reporter* (latest edition with amendments and corrections), Bureau of National Affairs, Washington, D.C.
4. *Atomic Energy Law Reporter,* Commerce Clearing House, Chicago, IL.
5. *Nuclear Regulation Reporter,* Commerce Clearing House, Chicago, IL.
6. *TLVs® Threshold Limit Values for Chemical Substances and Physical Agents* is issued annually be the American Conference of Governmental Industrial Hygienists, Cincinnati, Ohio.
7. *The Sigma Aldrich Library of Regulatory and Safety Data. Vol. 3,* Robert E. Lenga and Kristine L. Volonpal, Sigma Chemical Co. and Aldrich Chemical Co., Inc. 1993.
8. *Hazardous Chemicals Desk Reference,* Richard J. Lewis, Sr., 4th ed., John Wiley & Sons, New York, Dec. 1997.

THE ELEMENTS (continued)

The prices of elements as indicated in this article are intended to be only a rough guide. Prices may vary, over time, widely with supplier, quantity, and purity.

Actinium — (Gr. *aktis, aktinos,* beam or ray), Ac; at. wt. (227); at. no. 89; m.p. 1051°C, b.p. 3200 ± 300°C (est.); sp. gr. 10.07 (calc.). Discovered by Andre Debierne in 1899 and independently by F. Giesel in 1902. Occurs naturally in association with uranium minerals. Thirty four isotopes and isomers are now recognized. All are radioactive. Actinium-227, a decay product of uranium-235, is an alpha and beta emitter with a 21.77-year half-life. Its principal decay products are thorium-227 (18.72-day half-life), radium-223 (11.4-day half-life), and a number of short-lived products including radon, bismuth, polonium, and lead isotopes. In equilibrium with its decay products, it is a powerful source of alpha rays. Actinium metal has been prepared by the reduction of actinium fluoride with lithium vapor at about 1100 to 1300°C. The chemical behavior of actinium is similar to that of the rare earths, particularly lanthanum. Purified actinium comes into equilibrium with its decay products at the end of 185 days, and then decays according to its 21.77-year half-life. It is about 150 times as active as radium, making it of value in the production of neutrons. Actinium-225, with a purity of 99%, is available from the Oak Ridge National Laboratory to holders of a permit for about $500/millicurie, plus packing charges.

Aluminum — (L. *alumen, alum*), Al; at. wt. 26.981539(5); at. no. 13; f.p. 660.323°C; b.p. 2519°C; sp. gr. 2.6989 (20°C); valence 3. The ancient Greeks and Romans used *alum* in medicine as an astringent, and as a mordant in dyeing. In 1761 de Morveau proposed the name *alumine* for the base in alum, and Lavoisier, in 1787, thought this to be the oxide of a still undiscovered metal. Wohler is generally credited with having isolated the metal in 1827, although an impure form was prepared by Oersted two years earlier. In 1807, Davy proposed the name *alumium* for the metal, undiscovered at that time, and later agreed to change it to *aluminum*. Shortly thereafter, the name *aluminium* was adopted to conform with the "ium" ending of most elements, and this spelling is now in use elsewhere in the world. *Aluminium* was also the accepted spelling in the U.S. until 1925, at which time the American Chemical Society officially decided to use the name *aluminum* thereafter in their publications. The method of obtaining aluminum metal by the electrolysis of alumina dissolved in *cryolite* was discovered in 1886 by Hall in the U.S. and at about the same time by Heroult in France. Cryolite, a natural ore found in Greenland, is no longer widely used in commercial production, but has been replaced by an artificial mixture of sodium, aluminum, and calcium fluorides. *Bauxite,* an impure hydrated oxide ore, is found in large deposits in Jamaica, Australia, Suriname, Guyana, Russia, Arkansas, and elsewhere. The Bayer process is most commonly used today to refine bauxite so it can be accommodated in the Hall-Heroult refining process, used to make most aluminum. Aluminum can now be produced from clay, but the process is not economically feasible at present. Aluminum is the most abundant metal to be found in the earth's crust (8.1%), but is never found free in nature. In addition to the minerals mentioned above, it is found in feldspars, granite, and in many other common minerals. Twenty-two isotopes and isomers are known. Natural aluminum is made of one isotope, ^{27}Al. Pure aluminum, a silvery-white metal, possesses many desirable characteristics. It is light, nontoxic, has a pleasing appearance, can easily be formed, machined, or cast, has a high thermal conductivity, and has excellent corrosion resistance. It is nonmagnetic and nonsparking, stands second among metals in the scale of malleability, and sixth in ductility. It is extensively used for kitchen utensils, outside building decoration, and in thousands of industrial applications where a strong, light, easily constructed material is needed. Although its electrical conductivity is only about 60% that of copper, it is used in electrical transmission lines because of its light weight. Pure aluminum is soft and lacks strength, but it can be alloyed with small amounts of copper, magnesium, silicon, manganese, and other elements to impart a variety of useful properties. These alloys are of vital importance in the construction of modern aircraft and rockets. Aluminum, evaporated in a vacuum, forms a highly reflective coating for both visible light and radiant heat. These coatings soon form a thin layer of the protective oxide and do not deteriorate as do silver coatings. They have found application in coatings for telescope mirrors, in making decorative paper, packages, toys, and in many other uses. The compounds of greatest importance are aluminum oxide, the sulfate, and the soluble sulfate with potassium (alum). The oxide, alumina, occurs naturally as ruby, sapphire, corundum, and emery, and is used in glassmaking and refractories. Synthetic ruby and sapphire have found application in the construction of lasers for producing coherent light. In 1852, the price of aluminum was about $1200/kg, and just before Hall's discovery in 1886, about $25/kg. The price rapidly dropped to 60¢ and has been as low as 33¢/kg. The price in December 1999 was about $1.50/kg.

Americium — (the Americas), Am; at. wt. 243; at. no. 95; m.p. 1176°C; b.p. 2011°C; sp. gr. 13.67 (20°C); valence 2, 3, 4, 5, or 6. Americium was the fourth transuranium element to be discovered; the isotope ^{241}Am was identified by Seaborg, James, Morgan, and Ghiorso late in 1944 at the wartime Metallurgical Laboratory of the University of Chicago as the result of successive neutron capture reactions by plutonium isotopes in a nuclear reactor:

$$^{239}Pu(n,\gamma) \rightarrow {}^{240}Pu(n,\gamma) \rightarrow {}^{241}Pu \xrightarrow{\beta} {}^{241}Am$$

Since the isotope ^{241}Am can be prepared in relatively pure form by extraction as a decay product over a period of years from strongly neutron-bombarded plutonium, ^{241}Pu, this isotope is used for much of the chemical investigation of this element. Better suited is the isotope ^{243}Am due to its longer half-life (7.37×10^3 years as compared to 432.2 years for ^{241}Am). A mixture of the isotopes ^{241}Am, ^{242}Am, and ^{243}Am can be prepared by intense neutron irradiation of ^{241}Am according to the reactions ^{241}Am $(n, \gamma) \rightarrow {}^{242}$Am $(n, \gamma) \rightarrow {}^{243}$Am. Nearly isotopically pure ^{243}Am can be prepared by a sequence of neutron bombardments and chemical separations as follows: neutron bombardment of ^{241}Am yields ^{242}Pu by the reactions ^{241}Am (n, γ) $\rightarrow {}^{242}$Am $\rightarrow {}^{242}$Pu, after chemical separation the ^{242}Pu can be transformed to ^{243}Am via the reactions ^{242}Pu $(n, \gamma) \rightarrow {}^{243}$Pu $\rightarrow {}^{243}$Am, and the ^{243}Am can be chemically separated. Fairly pure ^{242}Pu can be prepared more simply by very intense neutron irradiation of ^{239}Pu as the result of successive neutron-capture reactions. Seventeen radioactive isotopes and isomers are now recognized. Americium metal has been prepared by reducing the trifluoride with barium vapor at 1000 to 1200°C or the dioxide by lanthanum metal. The luster of freshly prepared americium metal is white and more silvery than plutonium or neptunium prepared in the same manner. It appears to be more malleable than uranium or neptunium and tarnishes slowly in dry air at room temperature. Americium is thought to exist in two forms: an alpha form which has a double hexagonal close-packed structure and a loose-packed cubic beta form. Americium must be handled with great care to avoid personal contamination. As little as 0.03 μCi of ^{241}Am is the maximum permissible total body burden. The alpha activity from ^{241}Am is about three times that of radium. When gram quantities of ^{241}Am are handled, the intense gamma activity makes exposure a serious problem. Americium dioxide, AmO_2, is the most important oxide. AmF_3, AmF_4, $AmCl_3$, $AmBr_3$, AmI_3, and other compounds have been prepared. The isotope ^{241}Am has been used as a portable source for gamma radiography. It has also been used

THE ELEMENTS (continued)

as a radioactive glass thickness gage for the flat glass industry, and as a source of ionization for smoke detectors. Americium-243 (99%) is available from the Oak Ridge National Laboratory at a cost of about $750/g plus packing charges.

Antimony — (Gr. *anti* plus *monos* — a metal not found alone), Sb; at. wt. 121.760(1); at. no. 51; m.p. 630.63°C; b.p. 1587°C; sp. gr. 6.691 (20°C); valence 0, –3, +3, or +5. Antimony was recognized in compounds by the ancients and was known as a metal at the beginning of the 17th century and possibly much earlier. It is not abundant, but is found in over 100 mineral species. It is sometimes found native, but more frequently as the sulfide, *stibnite* (Sb_2S_3); it is also found as antimonides of the heavy metals, and as oxides. It is extracted from the sulfide by roasting to the oxide, which is reduced by salt and scrap iron; from its oxides it is also prepared by reduction with carbon. Two allotropic forms of antimony exist: the normal stable, metallic form, and the amorphous gray form. The so-called explosive antimony is an ill-defined material always containing an appreciable amount of halogen; therefore, it no longer warrants consideration as a separate allotrope. The yellow form, obtained by oxidation of *stibine*, SbH_3, is probably impure, and is not a distinct form. Natural antimony is made of two stable isotopes, ^{121}Sb and ^{123}Sb. Forty five other radioactive isotopes and isomers are now recognized. Metallic antimony is an extremely brittle metal of a flaky, crystalline texture. It is bluish white and has a metallic luster. It is not acted on by air at room temperature, but burns brilliantly when heated with the formation of white fumes of Sb_2O_3. It is a poor conductor of heat and electricity, and has a hardness of 3 to 3.5. Antimony, available commercially with a purity of 99.999 + %, is finding use in semiconductor technology for making infrared detectors, diodes, and Hall-effect devices. Commercial-grade antimony is widely used in alloys with percentages ranging from 1 to 20. It greatly increases the hardness and mechanical strength of lead. Batteries, antifriction alloys, type metal, small arms and tracer bullets, cable sheathing, and minor products use about half the metal produced. Compounds taking up the other half are oxides, sulfides, sodium antimonate, and antimony trichloride. These are used in manufacturing flame-proofing compounds, paints, ceramic enamels, glass, and pottery. Tartar emetic (hydrated potassium antimonyl tartate) has been used in medicine. Antimony and many of its compounds are toxic. Antimony costs about $140/kg or about $12/g (99.999%).

Argon — (Gr. *argos,* inactive), Ar; at. wt. 39.948(1); at. no. 18; m.p. –189.35°C; b.p. –185.85°C; t_c -122.28; density 1.7837 g/l. Its presence in air was suspected by Cavendish in 1785, discovered by Lord Rayleigh and Sir William Ramsay in 1894. The gas is prepared by fractionation of liquid air, the atmosphere containing 0.94% argon. The atmosphere of Mars contains 1.6% of ^{40}Ar and 5 p.p.m. of ^{36}Ar. Argon is two and one half times as soluble in water as nitrogen, having about the same solubility as oxygen. It is recognized by the characteristic lines in the red end of the spectrum. It is used in electric light bulbs and in fluorescent tubes at a pressure of about 400 Pa, and in filling photo tubes, glow tubes, etc. Argon is also used as an inert gas shield for arc welding and cutting, as a blanket for the production of titanium and other reactive elements, and as a protective atmosphere for growing silicon and germanium crystals. Argon is colorless and odorless, both as a gas and liquid. It is available in high-purity form. Commercial argon is available at a cost of about 3¢ per cubic foot. Argon is considered to be a very inert gas and is not known to form true chemical compounds, as do krypton, xenon, and radon. However, it does form a hydrate having a dissociation pressure of 105 atm at 0°C. Ion molecules such as $(ArKr)^+$, $(ArXe)^+$, $(NeAr)^+$ have been observed spectroscopically. Argon also forms a clathrate with β-hydroquinone. This clathrate is stable and can be stored for a considerable time, but a true chemical bond does not exist. Van der Waals' forces act to hold the argon. Naturally occurring argon is a mixture of three isotopes. Seventeen other radioactive isotopes are now known to exist. Commercial argon is priced at about $70/300 cu. ft. or 8.5 cu. meters.

Arsenic — (L. *arsenicum,* Gr. *arsenikon,* yellow orpiment, identified with *arsenikos,* male, from the belief that metals were different sexes; Arabic, *Az-zernikh,* the orpiment from Persian *zerni-zar*, gold), As; at. wt. 74.92160(2); at. no. 33; valence –3, 0, +3 or +5. Elemental arsenic occurs in two solid modifications: yellow, and gray or metallic, with specific gravities of 1.97, and 5.73, respectively. Gray arsenic, the ordinary stable form, has a triple point of 817°C and sublimes at 614°C and has a critical temperature of 1400°C. Several other allotropic forms of arsenic are reported in the literature. It is believed that Albertus Magnus obtained the element in 1250 A.D. In 1649 Schroeder published two methods of preparing the element. It is found native, in the sulfides *realgar* and *orpiment,* as arsenides and sulfarsenides of heavy metals, as the oxide, and as arsenates. *Mispickel,* arsenopyrite, (FeSAs) is the most common mineral, from which on heating the arsenic sublimes leaving ferrous sulfide. The element is a steel gray, very brittle, crystalline, semimetallic solid; it tarnishes in air, and when heated is rapidly oxidized to arsenous oxide (As_2O_3) with the odor of garlic. Arsenic and its compounds are poisonous. Exposure to arsenic and its compounds should not exceed 0.2 mg/m³ as elemental As during an 8-h work day. These values, however, are being studied, and may be lowered. Arsenic is also used in bronzing, pyrotechny, and for hardening and improving the sphericity of shot. The most important compounds are white arsenic (As_2O_3), the sulfide, Paris green $3Cu(AsO_2)_2 \cdot Cu(C_2H_3O_2)_2$, calcium arsenate, and lead arsenate; the last three have been used as agricultural insecticides and poisons. Marsh's test makes use of the formation and ready decomposition of arsine (AsH_3). Arsenic is available in high-purity form. It is finding increasing uses as a doping agent in solid-state devices such as transistors. Gallium arsenide is used as a laser material to convert electricity directly into coherent light. Natural arsenic is made of one isotope ^{75}As. Thirty other radioactive isotopes and isomers are known. Arsenic (99%) costs about $175/kg. Purified arsenic (99.9995%) costs about $3.50/g.

Astatine — (Gr. *astatos,* unstable), At; at. wt. (210); at. no. 85; m.p. 300°C (est.); valence probably 1, 3, 5, or 7. Synthesized in 1940 by D. R. Corson, K. R. MacKenzie, and E. Segre at the University of California by bombarding bismuth with alpha particles. The longest-lived isotope, ^{210}At, has a half-life of only 8.1 hours. Thirty-six other isotopes and isomers are now known. Minute quantities of ^{215}At, ^{218}At, and ^{219}At exist in nature with naturally occurring uranium and thorium isotopes, and traces of ^{217}At are equilibrium with ^{233}U and ^{239}Np resulting from interaction of thorium and uranium with naturally produced neutrons. The total amount of astatine present in the earth's crust, however, is probably less than 1 oz. Astatine can be produced by bombarding bismuth with energetic alpha particles to obtain the relatively long-lived $^{209-211}At$, which can be distilled from the target by heating it in air. Only about 0.05 µg of astatine has been prepared to date. The "time of flight" mass spectrometer has been used to confirm that this highly radioactive halogen behaves chemically very much like other halogens, particularly iodine. The interhalogen compounds AtI, AtBr, and AtCl are known to form, but it is not yet known if astatine forms diatomic astatine molecules. HAt and CH_3At (methyl astatide) have been detected. Astatine is said to be more metallic that iodine, and, like iodine, it probably accumulates in the thyroid gland. Workers at the Brookhaven National Laboratory have recently used reactive scattering in crossed molecular beams to identify and measure elementary reactions involving astatine.

Barium — (Gr. *barys,* heavy), Ba; at. wt. 137.327(7), at. no. 56; m.p. 727°C; b.p. 1897°C; sp. gr. 3.5 (20°C); valence 2. Baryta was distinguished from lime by Scheele in 1774; the element was discovered by Sir Humphrey Davy in 1808. It is found only in combination with other elements, chiefly in *barite* or *heavy spar* (sulfate) and *witherite* (carbonate) and is prepared by electrolysis of the chloride. Large deposits of barite are found in China, Germany, India, Morocco, and in the U.S. Barium is a metallic element, soft, and when pure is silvery white like lead; it belongs to the alkaline earth group, resembling calcium chemically. The metal oxidizes very easily and should be kept under petroleum or other suitable oxygen-free liquids to exclude air. It is decomposed by water or alcohol. The metal is used as a "getter" in vacuum tubes. The most important compounds are the peroxide (BaO_2), chloride, sulfate, carbonate, nitrate, and chlorate. Lithopone, a pigment containing barium sulfate and zinc sulfide, has good covering power, and does not darken in the presence of sulfides. The sulfate, as permanent white or *blanc fixe,* is also used in paint, in X-ray diagnostic work, and in glassmaking. *Barite* is extensively used as a weighting agent in oilwell drilling fluids, and also in making rubber. The carbonate has been used as a rat poison, while the nitrate and chlorate give green colors in pyrotechny. The impure sulfide phosphoresces after exposure to the light. The compounds and the metal are not expensive. Barium metal (99.2 + % pure) costs about $3/g. All barium compounds that are water or acid soluble are poisonous. Naturally occurring barium is a mixture of seven stable isotopes. Thirty six other radioactive isotopes and isomers are known to exist.

Berkelium — (*Berkeley,* home of the University of California), Bk; at. wt. (247); at. no. 97; m.p. 1050°C; valence 3 or 4; sp. gr. 14 (est.). Berkelium, the eighth member of the actinide transition series, was discovered in December 1949 by Thompson, Ghiorso, and Seaborg, and was the fifth transuranium element synthesized. It was produced by cyclotron bombardment of milligram amounts of ^{241}Am with helium ions at Berkeley, California. The first isotope produced had a mass number of 243 and decayed with a half-life of 4.5 hours. Thirteen isotopes are now known and have been synthesized. The existence of ^{249}Bk, with a half-life of 320 days, makes it feasible to isolate berkelium in weighable amounts so that its properties can be investigated with macroscopic quantities. One of the first visible amounts of a pure berkelium compound, berkelium chloride, was produced in 1962. It weighed 3 billionth of a gram. Berkelium probably has not yet been prepared in elemental form, but it is expected to be a silvery metal, easily soluble in dilute mineral acids, and readily oxidized by air or oxygen at elevated temperatures to form the oxide. X-ray diffraction methods have been used to identify the following compounds: BkO_2, BkO_3, BkF_3, $BkCl$, and $BkOCl$. As with other actinide elements, berkelium tends to accumulate in the skeletal system. The maximum permissible body burden of ^{249}Bk in the human skeleton is about 0.0004 µg. Because of its rarity, berkelium presently has no commercial or technological use. Berkelium most likely resembles terbium with respect to chemical properties. Berkelium-249 is available from O.R.N.L. at a cost of $185/µg plus packing charges.

Beryllium — (Gr. *beryllos, beryl;* also called Glucinium or Glucinum, Gr. *glykys,* sweet), Be; at. wt. 9.012182(3); at. no. 4; m.p. 1287°C; b.p. 2471°C; sp. gr. 1.848 (20°C); valence 2. Discovered as the oxide by Vauquelin in beryl and in emeralds in 1798. The metal was isolated in 1828 by Wohler and by Bussy independently by the action of potassium on beryllium chloride. Beryllium is found in some 30 mineral species, the most important of which are *bertrandite, beryl, chrysoberyl,* and *phenacite. Aquamarine* and *emerald* are precious forms of *beryl.* Beryllium minerals are found in the U.S., Brazil, Russia, Kazakhstan, and elsewhere. Colombia is known for its emeralds. *Beryl* ($3BeO \cdot Al_2O_3 \cdot 6SiO_2$) and *bertrandite* ($4BeO \cdot 2SiO_2 \cdot H_2O$) are the most important commercial sources of the element and its compounds. Most of the metal is now prepared by reducing beryllium fluoride with magnesium metal. Beryllium metal did not become readily available to industry until 1957. The metal, steel gray in color, has many desirable properties. It is one of the lightest of all metals, and has one of the highest melting points of the light metals. Its modulus of elasticity is about one third greater than that of steel. It resists attack by concentrated nitric acid, has excellent thermal conductivity, and is nonmagnetic. It has a high permeability to X-rays, and when bombarded by alpha particles, as from radium or polonium, neutrons are produced in the ratio of about 30 neutrons/ million alpha particles. At ordinary temperatures beryllium resists oxidation in air, although its ability to scratch glass is probably due to the formation of a thin layer of the oxide. Beryllium is used as an alloying agent in producing beryllium copper which is extensively used for springs, electrical contacts, spot-welding electrodes, and nonsparking tools. It has found application as a structural material for high-speed aircraft, missiles, spacecraft, and communication satellites. It is being used in the windshield frame, brake discs, support beams, and other structural components of the space shuttle. Because beryllium is relatively transparent to X-rays, ultra-thin Be-foil is finding use in X-ray lithography for reproduction of microminiature integrated circuits. Natural beryllium is made of ^9Be and is stable. Eight other radioactive isotopes are known.

Beryllium is used in nuclear reactors as a reflector or moderator for it has a low thermal neutron absorption cross section. It is used in gyroscopes, computer parts, and instruments where lightness, stiffness, and dimensional stability are required. The oxide has a very high melting point and is also used in nuclear work and ceramic applications. Beryllium and its salts are toxic and should be handled with the greatest of care. Beryllium and its compounds should not be tasted to verify the sweetish nature of beryllium (as did early experimenters). The metal, its alloys, and its salts can be handled safely if certain work codes are observed, but no attempt should be made to work with beryllium before becoming familiar with proper safeguards. Beryllium metal is available at a cost of about $4.50/g (99.5% pure).

Bismuth — (Ger. *Weisse Masse,* white mass; later *Wisuth* and *Bisemutum*), Bi; at. wt. 208.98038(2); at. no. 83; m.p. 271.4°C; b.p. 1564°C; sp. gr. 9.747 (20°C); valence 3 or 5. In early times bismuth was confused with tin and lead. Claude Geoffroy the Younger showed it to be distinct from lead in 1753. It is a white crystalline, brittle metal with a pinkish tinge. It occurs native. The most important ores are *bismuthinite* or bismuth glance (Bi_2S_3) and *bismite* (Bi_2O_3). Peru, Japan, Mexico, Bolivia, and Canada are major bismuth producers. Much of the bismuth produced in the U.S. is obtained as a by-product in refining lead, copper, tin, silver, and gold ores. Bismuth is the most diamagnetic of all metals, and the thermal conductivity is lower than any metal, except mercury. It has a high electrical resistance, and has the highest Hall effect of any metal (i.e., greatest increase in electrical resistance when placed in a magnetic field). "Bismanol" is a permanent magnet of high coercive force, made of MnBi, by the U.S. Naval Surface Weapons Center. Bismuth expands 3.32% on solidification. This property makes bismuth alloys particularly suited to the making of sharp castings of objects subject to damage by high temperatures. With other metals such as tin, cadmium, etc., bismuth forms low-melting alloys which are extensively used for safety devices in fire detection and extinguishing systems. Bismuth is used in producing malleable irons and is finding use as a catalyst for making acrylic fibers. When bismuth is heated in air it burns with a blue flame, forming yellow fumes of the oxide. The metal is also used as a thermocouple material, and has found application as a carrier for U^{235} or U^{233} fuel in atomic reactors. Its soluble salts are characterized by forming insoluble basic salts on the addition of water, a property sometimes used in detection work. Bismuth oxychloride is used extensively in cosmetics. Bismuth subnitrate and subcarbonate are used in medicine. Natural bismuth contains only one isotope ^{209}Bi. Forty-four isotopes and isomers of bismuth are known. Bismuth metal (99.5%) costs about $250/kg.

Bohrium — (Named after Niels Bohr [1885-1962], Danish atomic and nuclear physicist.) Bh; at.wt. [262]. at.no. 107. Bohrium is expected to have chemical properties similar to rhenium. This element was synthesized and unambiguously identified in 1981 using the Universal Linear Accelerator (UNILAC) at the Gesellschaft für Schwerionenforschung (G.S.I.) in Darmstadt, Germany. The discovery team was led by Armbruster and Münzenberg. The reaction producing the element was proposed and applied earlier by a Dubna Group led by Oganessian in 1976. A target of ^{209}Bi was bombarded by a beam of ^{54}Cr ions. In 1983 experiments at Dubna using the 157-inch cyclotron, produced $^{262}107$ by the reaction $^{209}Bi + ^{54}Cr$. The alpha decay of ^{246}Cf, the sixth member in the decay chain of $^{262}107$, served to establish a 1-neutron reaction channel. The IUPAC adopted the name *Bohrium* with the symbol Bh for Element 107 in August 1997. Five isotopes of bohrium are now recognized. One isotope of bohrium appears to have a relatively long life of 15 seconds. Work on this relatively long-lived isotope has been performed with the 88-inch cyclotron at the Lawrence-Berkeley National Laboratory.

Boron — (Ar. *Buraq,* Pers. *Burah*), B; at. wt. 10.811(7); at. no. 5; m.p. 2075°C; b.p. 4000°C; sp. gr. of crystals 2.34, of amorphous variety 2.37; valence 3. Boron compounds have been known for thousands of years, but the element was not discovered until 1808 by Sir Humphry Davy and by Gay-Lussac and Thenard. The element is not found free in nature, but occurs as orthoboric acid usually in certain volcanic spring waters and as borates in *borax* and *colemanite. Ulexite,* another boron mineral, is interesting as it is nature's own version of "fiber optics." Important sources of boron are the ores *rasorite (kernite)* and *tincal (borax ore).* Both of these ores are found in the Mojave Desert. *Tincal* is the most important source of boron from the Mojave. Extensive *borax* deposits are also found in Turkey. Boron exists naturally as 19.9% ^{10}B isotope and 80.1% ^{11}B isotope. Ten other isotopes of boron are known. High-purity crystalline boron may be prepared by the vapor phase reduction of boron trichloride or tribromide with hydrogen on electrically heated filaments. The impure, or amorphous, boron, a brownish-black powder, can be obtained by heating the trioxide with magnesium powder. Boron of 99.9999% purity has been produced and is available commercially. Elemental boron has an energy band gap of 1.50 to 1.56 eV, which is higher than that of either silicon or germanium. It has interesting optical characteristics, transmitting portions of the infrared, and is a poor conductor of electricity at room temperature, but a good conductor at high temperature. Amorphous boron is used in pyrotechnic flares to provide a distinctive green color, and in rockets as an igniter. By far the most commercially important boron compound in terms of dollar sales is $Na_2B_4O_7 \cdot 5H_2O$. This pentahydrate is used in very large quantities in the manufacture of insulation fiberglass and sodium perborate bleach. Boric acid is also an important boron compound with major markets in textile fiberglass and in cellulose insulation as a flame retardant. Next in order of importance is borax ($Na_2B_4O_7 \cdot 10H_2O$) which is used principally in laundry products. Use of borax as a mild antiseptic is minor in terms of dollars and tons. Boron compounds are also extensively used in the manufacture of borosilicate glasses. The isotope boron-10 is used as a control for nuclear reactors, as a shield for nuclear radiation, and in instruments used for detecting neutrons. Boron nitride has remarkable properties and can be used to make a material as hard as diamond. The nitride also behaves like an electrical insulator but conducts heat like a metal. It also has lubricating properties similar to graphite. The hydrides are easily oxidized with considerable energy liberation, and have been studied for use as rocket fuels. Demand is increasing for boron filaments, a high-strength, lightweight material chiefly employed for advanced aerospace structures. Boron is similar to carbon in that it has a capacity to form stable covalently bonded molecular networks. Carboranes, metalloboranes, phosphacarboranes, and other families comprise thousands of compounds. Crystalline boron (99.5%) costs about $30/g. Amorphous boron (94–96%) costs about $1.50/g. Elemental boron and the borates are not considered to be toxic, and they do not require special care in handling. However, some of the more exotic boron hydrogen compounds are definitely toxic and do require care.

Bromine — (Gr. *bromos,* stench), Br; at. wt. 79.904(1); at. no. 35; m.p. –7.2°C; b.p. 58.8°C; t_c 315°C; density of gas 7.59 g/l, liquid 3.12 (20°C); valence 1, 3, 5, or 7. Discovered by Balard in 1826, but not prepared in quantity until 1860. A member of the halogen group of elements, it is obtained from natural brines from wells in Michigan and Arkansas. Little bromine is extracted today from seawater, which contains only about 85 ppm. Bromine is the only liquid nonmetallic element. It is a heavy, mobile, reddish-brown liquid, volatilizing readily at room temperature to a red vapor with a strong disagreeable odor, resembling chlorine, and having a very irritating effect on the eyes and throat; it is readily soluble in water or carbon disulfide, forming a red solution, is less active than chlorine but more so than iodine; it unites readily with many elements and has a bleaching action; when spilled on the skin it produces painful sores. It presents a serious health hazard, and maximum safety precautions should be taken when handling it. Much of the bromine output in the U.S. was used in the production of ethylene dibromide, a lead scavenger used in making gasoline antiknock compounds. Lead in gasoline, however, has been drastically reduced, due to environmental considerations. This will greatly affect future production of bromine. Bromine is also used in making fumigants, flameproofing agents, water purification compounds, dyes, medicinals, sanitizers, inorganic bromides for photography, etc. Organic bromides are also important. Natural bromine is made of two isotopes, ^{79}Br and ^{81}Br. Thirty-four isotopes and isomers are known. Bromine costs about $100/kg.

Cadmium — (L. *cadmia;* Gr. *kadmeia*—ancient name for calamine, zinc carbonate), Cd; at. wt. 112.411(8); at. no. 48; m.p. 321.07°C; b.p. 767°C; sp. gr. 8.65 (20°C); valence 2. Discovered by Stromeyer in 1817 from an impurity in zinc carbonate. Cadmium most often occurs in small quantities associated with zinc ores, such as *sphalerite* (ZnS). *Greenockite* (CdS) is the only mineral of any consequence bearing cadmium. Almost all cadmium is obtained as a by-product in the treatment of zinc, copper, and lead ores. It is a soft, bluish-white metal which is easily cut with a knife. It is similar in many respects to zinc. It is a component of some of the lowest melting alloys; it is used in bearing alloys with low coefficients of friction and great resistance to fatigue; it is used extensively in electroplating, which accounts for about 60% of its use. It is also used in many types of solder, for standard E.M.F. cells, for Ni-Cd batteries, and as a barrier to control atomic fission. The market for Ni-Cd batteries is expected to grow significantly in the next few years. Cadmium compounds are used in black and white television phosphors and in blue and green phosphors for color TV tubes. It forms a number of salts, of which the sulfate is most common; the sulfide is used as a yellow pigment. Cadmium and solutions of its compounds are toxic. Failure to appreciate the toxic properties of cadmium may cause workers to be unwittingly exposed to dangerous fumes. Some silver solders, for example, contain cadmium and should be handled with care. Serious toxicity problems have been found from long-term exposure and work with cadmium plating baths. Cadmium is present in certain phosphate rocks. This has raised concerns that the long-term use of certain phosphate fertilizers might pose a health hazard from levels of cadmium that might enter the food chain. In 1927 the International Conference on Weights and Measures redefined the meter in terms of the wavelength of the red cadmium spectral line (i.e. 1 m = 1,553,164.13 wavelengths). This definition has been changed (see under Krypton). The current price of cadmium is about $2/g (99.5%). It is available in high purity form for about $300/kg. Natural cadmium is made of eight isotopes. Thirty four other isotopes and isomers are now known and recognized.

Calcium — (L. *calx,* lime), Ca; at. wt. 40.078(4); at. no. 20; m.p. 842°C; b.p. 1484°C; sp. gr. 1.55 (20°C); valence 2. Though lime was prepared by the Romans in the first century under the name calx, the metal was not discovered until 1808. After learning that Berzelius and Pontin prepared calcium amalgam by electrolyzing lime in mercury, Davy was able to isolate the impure metal. Calcium is a metallic element, fifth in abundance in the earth's crust, of which it forms more than 3%. It is an essential constituent of leaves, bones, teeth, and shells. Never found in nature uncombined, it occurs abundantly as *limestone* ($CaCO_3$), *gypsum* ($CaSO_4 \cdot 2H_2O$), and *fluorite* (CaF_2); *apatite* is the fluorophosphate or chlorophosphate of calcium. The metal has a silvery color, is rather hard, and is prepared by electrolysis of the fused chloride to which calcium fluoride is added to lower the melting point. Chemically it is one of the alkaline earth elements; it readily forms a white coating of oxide in air, reacts with water, burns with a yellow-red flame, forming largely the oxide. The metal is used as a reducing agent in preparing other metals such as thorium, uranium, zirconium, etc., and is used as a deoxidizer, desulfurizer, and inclusion modifier for various ferrous and nonferrous alloys. It is also used as an alloying agent for aluminum, beryllium, copper, lead, and magnesium alloys, and serves as a "getter" for residual gases in vacuum tubes, etc. Its natural and prepared compounds are widely used. Quicklime (CaO), made by heating limestone and changed into slaked lime by the careful addition of water, is the great cheap base of chemical industry with countless uses. Mixed with sand it hardens as mortar and plaster by taking up carbon dioxide from the air. Calcium from limestone is an important element in Portland cement. The solubility of the carbonate in water containing carbon dioxide causes the formation of caves with stalactites and stalagmites and is responsible for hardness in water. Other important compounds are the carbide (CaC_2), chloride ($CaCl_2$), cyanamide ($CaCN_2$), hypochlorite ($Ca(OCl)_2$), nitrate ($Ca(NO_3)_2$), and sulfide (CaS). Calcium sulfide is phosphorescent after being exposed to light. Natural calcium contains six isotopes. Sixteen other radioactive isotopes are known. Metallic calcium (99.5%) costs about $225/kg.

Californium — (State and University of California), Cf; at. wt. (251); m.p. 900°C; at. no. 98. Californium, the sixth transuranium element to be discovered, was produced by Thompson, Street, Ghiorso, and Seaborg in 1950 by bombarding microgram quantities of ^{242}Cm with 35 MeV helium ions in the Berkeley 60-inch cyclotron. Californium (III) is the only ion stable in aqueous solutions, all attempts to reduce or oxidize californium (III) having failed. The isotope ^{249}Cf results from the beta decay of ^{249}Bk while the heavier isotopes are produced by intense neutron irradiation by the reactions:

$$^{249}Bk(n,\gamma) \rightarrow {}^{250}Bk \xrightarrow{\beta} {}^{250}Cf \text{ and } {}^{249}Cf(n,\gamma) \rightarrow {}^{250}Cf$$

followed by

$$^{250}Cf(n,\gamma) \rightarrow {}^{251}Cf(n,\gamma) \rightarrow {}^{252}Cf$$

The existence of the isotopes ^{249}Cf, ^{250}Cf, ^{251}Cf, and ^{252}Cf makes it feasible to isolate californium in weighable amounts so that its properties can be investigated with macroscopic quantities. Californium-252 is a very strong neutron emitter. One microgram releases 170 million neutrons per minute, which presents biological hazards. Proper safeguards should be used in handling californium. Twenty isotopes of californium are now recognized. ^{249}Cf and ^{252}Cf have half-lives of 351 years and 900 years, respectively. In 1960 a few tenths of a microgram of californium trichloride, $CfCl_3$, californium oxychloride, CfOCl, and californium oxide, Cf_2O_3, were first prepared. Reduction of californium to its metallic state has not yet been accomplished. Because californium is a very efficient source of neutrons, many new uses are expected for it. It has already found use in neutron moisture gages and in well-logging (the determination of water and oil-bearing layers). It is also being used as a portable neutron source for discovery of metals such as gold or silver by on-the-spot activation analysis. ^{252}Cf is now being offered for sale by the Oak Ridge National Laboratory (O.R.N.L.) at a cost of $60/μg and ^{249}Cf at a cost of $185/μg plus packing charges. It has been suggested that californium may be produced in certain stellar explosions, called *supernovae*, for the radioactive decay of ^{254}Cf (55-day half-life) agrees with the characteristics of the light curves of such explosions observed through telescopes. This suggestion, however, is questioned. Californium is expected to have chemical properties similar to dysprosium.

Carbon — (L. *carbo*, charcoal), C; at. wt. 12.0107(8); at. no. 6; sublimes at 3642°C; triple point (graphite-liquid-gas), 4492°C at a pressure of 101.325 kPa; sp. gr. amorphous 1.8 to 2.1, graphite 1.9 to 2.3, diamond 3.15 to 3.53 (depending on variety); gem diamond 3.513 (25°C); valence 2, 3, or 4. Carbon, an element of prehistoric discovery, is very widely distributed in nature. It is found in abundance in the sun, stars, comets, and atmospheres of most planets. Carbon in the form of microscopic diamonds is found in some meteorites. Natural diamonds are found in *kimberlite* or *lamporite* of ancient formations called "pipes," such as found in South Africa, Arkansas, and elsewhere. Diamonds are now also being recovered from the ocean floor off the Cape of Good Hope. About 30% of all industrial diamonds used in the U.S. are now made synthetically. The energy of the sun and stars can be attributed at least in part to the well-known carbon-nitrogen cycle. Carbon is found free in nature in three allotropic forms: amorphous, graphite, and diamond. A fourth form, known as "white" carbon, is now thought to exist. Graphite is one of the softest known materials while diamond is one of the hardest. Graphite exists in two forms: alpha and beta. These have identical physical properties, except for their crystal structure. Naturally occurring graphites are reported to contain as much as 30% of the rhombohedral (beta) form, whereas synthetic materials contain only the alpha form. The hexagonal alpha type can be converted to the beta by mechanical treatment, and the beta form reverts to the alpha on heating it above 1000°C. In 1969 a new allotropic form of carbon was produced during the sublimation of pyrolytic graphite at low pressures. Under free-vaporization conditions above ~2550 K, "white" carbon forms as small transparent crystals on the edges of the basal planes of graphite. The interplanar spacings of "white" carbon are identical to those of carbon form noted in the graphitic gneiss from the Ries (meteoritic) Crater of Germany. "White" carbon is a transparent birefringent material. Little information is presently available about this allotrope. Of recent interest is the discovery of all-carbon molecules, known as "buckyballs" or fullerenes, which have a number of unusual properties. These interesting molecules, consisting of 60 or 70 carbon atoms linked together, seem capable of withstanding great pressure and trapping foreign atoms inside their network of carbon. They are said to be capable of magnetism and superconductivity and have potential as a nonlinear optical material. Buckyball films are reported to remain superconductive at temperatures as high as 45 K. In combination, carbon is found as carbon dioxide in the atmosphere of the earth and dissolved in all natural waters. It is a component of great rock masses in the form of carbonates of calcium (limestone), magnesium, and iron. Coal, petroleum, and natural gas are chiefly hydrocarbons. Carbon is unique among the elements in the vast number and variety of compounds it can form. With hydrogen, oxygen, nitrogen, and other elements, it forms a very large number of compounds, carbon atom often being linked to carbon atom. There are close to ten million known carbon compounds, many thousands of which are vital to organic and life processes. Without carbon, the basis for life would be impossible. While it has been thought that silicon might take the place of carbon in forming a host of similar compounds, it is now not possible to form stable compounds with very long chains of silicon atoms. The atmosphere of Mars contains 96.2% CO_2. Some of the most important compounds of carbon are carbon dioxide (CO_2), carbon monoxide (CO), carbon disulfide (CS_2), chloroform ($CHCl_3$), carbon tetrachloride (CCl_4), methane (CH_4), ethylene (C_2H_4), acetylene (C_2H_2), benzene (C_6H_6), ethyl alcohol (C_2H_5OH), acetic acid (CH_3COOH), and their derivatives. Carbon has fifteen isotopes. Natural carbon consists of 98.89% ^{12}C and 1.11% ^{13}C. In 1961 the International Union of Pure and Applied Chemistry adopted the isotope carbon-12 as the basis for atomic weights. Carbon-14, an isotope with a half-life of 5715 years, has been widely used to date such materials as wood, archeological specimens, etc. A new brittle form of carbon, known as "glassy carbon", has been developed. It can be obtained with high purity. It has a high resistance to corrosion, has good thermal stability, and is structurally impermeable to both gases and liquids. It has a randomized structure, making it useful in ultra-high technology applications, such as crystal growing, crucibles for high-temperature use, etc. Both Types 1 and 2 glassy carbon are available at a cost of about $35/10gms.

THE ELEMENTS (continued)

Cerium — (named for the asteroid *Ceres,* which was discovered in 1801 only 2 years before the element), Ce; at. wt. 140.115(4); at. no. 58; m.p. 798°C; b.p. 3424°C; sp. gr. 6.770 (25°C); valence 3 or 4. Discovered in 1803 by Klaproth and by Berzelius and Hisinger; metal prepared by Hillebrand and Norton in 1875. Cerium is the most abundant of the metals of the so-called rare earths. It is found in a number of minerals including *allanite* (also known as *orthite*), *monazite, bastnasite, cerite,* and *samarskite.* Monazite and bastnasite are presently the two most important sources of cerium. Large deposits of monazite found on the beaches of Travancore, India, in river sands in Brazil, and deposits of *allanite* in the western United States, and *bastnasite* in Southern California will supply cerium, thorium, and the other rare-earth metals for many years to come. Metallic cerium is prepared by metallothermic reduction techniques, such as by reducing cerous fluoride with calcium, or by electrolysis of molten cerous chloride or other cerous halides. The metallothermic technique is used to produce high-purity cerium. Cerium is especially interesting because of its variable electronic structure. The energy of the inner 4f level is nearly the same as that of the outer or valence electrons, and only small amounts of energy are required to change the relative occupancy of these electronic levels. This gives rise to dual valency states. For example, a volume change of about 10% occurs when cerium is subjected to high pressures or low temperatures. It appears that the valence changes from about 3 to 4 when it is cooled or compressed. The low temperature behavior of cerium is complex. Four allotropic modifications are thought to exist: cerium at room temperature and at atmospheric pressure is known as γ cerium. Upon cooling to −16°C, γ cerium changes to β cerium. The remaining γ cerium starts to change to α cerium when cooled to −172°C, and the transformation is complete at −269°C. α Cerium has a density of 8.16; δ cerium exists above 726°C. At atmospheric pressure, liquid cerium is more dense than its solid form at the melting point. Cerium is an iron-gray lustrous metal. It is malleable, and oxidizes very readily at room temperature, especially in moist air. Except for europium, cerium is the most reactive of the "rare-earth" metals. It slowly decomposes in cold water, and rapidly in hot water. Alkali solutions and dilute and concentrated acids attack the metal rapidly. The pure metal is likely to ignite if scratched with a knife. Ceric salts are orange red or yellowish; cerous salts are usually white. Cerium is a component of misch metal, which is extensively used in the manufacture of pyrophoric alloys for cigarette lighters, etc. Natural cerium is stable and contains four isotopes. Thirty-two other radioactive isotopes and isomers are known. While cerium is not radioactive, the impure commercial grade may contain traces of thorium, which is radioactive. The oxide is an important constituent of incandescent gas mantles and it is emerging as a hydrocarbon catalyst in "self-cleaning" ovens. In this application it can be incorporated into oven walls to prevent the collection of cooking residues. As ceric sulfate it finds extensive use as a volumetric oxidizing agent in quantitative analysis. Cerium compounds are used in the manufacture of glass, both as a component and as a decolorizer. The oxide is finding increased use as a glass polishing agent instead of rouge, for it is much faster than rouge in polishing glass surfaces. Cerium compounds are finding use in automobile exhaust catalysts. Cerium is also finding use in making permanent magnets. Cerium, with other rare earths, is used in carbon-arc lighting, especially in the motion picture industry. It is also finding use as an important catalyst in petroleum refining and in metallurgical and nuclear applications. In small lots, cerium costs about $5/g (99.9%).

Cesium — (L. *caesius,* sky blue), Cs; at. wt. 132.90545(2); at. no. 55; m.p. 28.44°C; b.p. 671°C; sp. gr. 1.873 (20°C); valence 1. Cesium was discovered spectroscopically by Bunsen and Kirchhoff in 1860 in mineral water from Durkheim. Cesium, an alkali metal, occurs in *lepidolite, pollucite* (a hydrated silicate of aluminum and cesium), and in other sources. One of the world's richest sources of cesium is located at Bernic Lake, Manitoba. The deposits are estimated to contain 300,000 tons of pollucite, averaging 20% cesium. It can be isolated by electrolysis of the fused cyanide and by a number of other methods. Very pure, gas-free cesium can be prepared by thermal decomposition of cesium azide. The metal is characterized by a spectrum containing two bright lines in the blue along with several others in the red, yellow, and green. It is silvery white, soft, and ductile. It is the most electropositive and most alkaline element. Cesium, gallium, and mercury are the only three metals that are liquid at room temperature. Cesium reacts explosively with cold water, and reacts with ice at temperatures above −116°C. Cesium hydroxide, the strongest base known, attacks glass. Because of its great affinity for oxygen the metal is used as a "getter" in electron tubes. It is also used in photoelectric cells, as well as a catalyst in the hydrogenation of certain organic compounds. The metal has recently found application in ion propulsion systems. Cesium is used in atomic clocks, which are accurate to 5 s in 300 years. A second of time is now defined as being the duration of 9,192,631,770 periods of the radiation corresponding to the transition between the two hyper-fine levels of the ground state of the cesium-133 atom. Its chief compounds are the chloride and the nitrate. Cesium has 52 isotopes and isomers with masses ranging from 112 to 148. The present price of cesium is about $65/g (99.98%) sealed in a glass ampoule.

Chlorine — (Gr. *chloros,* greenish yellow), Cl; at. wt. 35.4527(9); at. no. 17; m.p. −101.5°C; b.p. −34.04°C; t_c 143.8°C; density 3.214 g/l; sp. gr. 1.56 (−33.6°C); valence 1, 3, 5, or 7. Discovered in 1774 by Scheele, who thought it contained oxygen; named in 1810 by Davy, who insisted it was an element. In nature it is found in the combined state only, chiefly with sodium as common salt (NaCl), *carnallite* ($KMgCl_3 \cdot 6H_2O$), and *sylvite* (KCl). It is a member of the halogen (salt-forming) group of elements and is obtained from chlorides by the action of oxidizing agents and more often by electrolysis; it is a greenish-yellow gas, combining directly with nearly all elements. At 10°C one volume of water dissolves 3.10 volumes of chlorine, at 30°C only 1.77 volumes. Chlorine is widely used in making many everyday products. It is used for producing safe drinking water the world over. Even the smallest water supplies are now usually chlorinated. It is also extensively used in the production of paper products, dyestuffs, textiles, petroleum products, medicines, antiseptics, insecticides, foodstuffs, solvents, paints, plastics, and many other consumer products. Most of the chlorine produced is used in the manufacture of chlorinated compounds for sanitation, pulp bleaching, disinfectants, and textile processing. Further use is in the manufacture of chlorates, chloroform, carbon tetrachloride, and in the extraction of bromine. Organic chemistry demands much from chlorine, both as an oxidizing agent and in substitution, since it often brings desired properties in an organic compound when substituted for hydrogen, as in one form of synthetic rubber. Chlorine is a respiratory irritant. The gas irritates the mucous membranes and the liquid burns the skin. As little as 3.5 ppm can be detected as an odor, and 1000 ppm is likely to be fatal after a few deep breaths. It was used as a war gas in 1915. Natural chlorine contains two isotopes. Twenty other isotopes and isomers are known.

Chromium — (Gr. *chroma,* color), Cr; at. wt. 51.9961(6); at. no. 24; m.p. 1907°C; b.p. 2671°C; sp. gr. 7.18 to 7.20 (20°C); valence chiefly 2, 3, or 6. Discovered in 1797 by Vauquelin, who prepared the metal the next year, chromium is a steel-gray, lustrous, hard metal that takes a high polish. The principal ore is *chromite* ($FeCr_2O_4$), which is found in Zimbabwe, Russia, South Africa, Turkey, Iran, Albania, Finland, Democratic Republic of Madagascar, the Philippines, and elsewhere. The U.S. has no appreciable chromite ore reserves. The metal is usually produced by reducing the oxide with aluminum. Chromium is used to harden steel, to manufacture stainless steel, and to form many useful alloys. Much is used in plating to produce a hard, beautiful surface and to prevent corrosion. Chromium is used to give glass an emerald green color. It finds wide use as a catalyst. All compounds of chromium are colored; the most important are the chromates of sodium and potassium (K_2CrO_4) and the dichromates ($K_2Cr_2O_7$) and the potassium and ammonium chrome alums, as $KCr(SO_4)_2 \cdot 12H_2O$. The dichromates are used as oxidizing agents in quantitative analysis, also in tanning leather.

Other compounds are of industrial value; lead chromate is chrome yellow, a valued pigment. Chromium compounds are used in the textile industry as mordants, and by the aircraft and other industries for anodizing aluminum. The refractory industry has found chromite useful for forming bricks and shapes, as it has a high melting point, moderate thermal expansion, and stability of crystalline structure. Chromium is an essential trace element for human health. Many chromium compounds, however, are acutely toxic, chronically toxic, and may be carcinogenic. They should be handled with proper safeguards. Natural chromium contains four isotopes. Twenty other isotopes are known. Chromium metal (99.95%) costs about $600/kg. Commercial grade chromium (99%) costs about $75/kg.

Cobalt — (*Kobald,* from the German, goblin or evil spirit, *cobalos,* Greek, mine), Co; at. wt. 58.93320(1); at. no. 27; m.p. 1495°C; b.p. 2927°C; sp. gr. 8.9 (20°C); valence 2 or 3. Discovered by Brandt about 1735. Cobalt occurs in the mineral *cobaltite, smaltite,* and *erythrite,* and is often associated with nickel, silver, lead, copper, and iron ores, from which it is most frequently obtained as a by-product. It is also present in meteorites. Important ore deposits are found in Congo-Kinshasa, Australia, Zambia, Russia, Canada, and elsewhere. The U.S. Geological Survey has announced that the bottom of the north central Pacific Ocean may have cobalt-rich deposits at relatively shallow depths in waters close to the Hawaiian Islands and other U.S. Pacific territories. Cobalt is a brittle, hard metal, closely resembling iron and nickel in appearance. It has a magnetic permeability of about two thirds that of iron. Cobalt tends to exist as a mixture of two allotropes over a wide temperature range; the β-form predominates below 400°C, and the α above that temperature. The transformation is sluggish and accounts in part for the wide variation in reported data on physical properties of cobalt. It is alloyed with iron, nickel and other metals to make Alnico, an alloy of unusual magnetic strength with many important uses. Stellite alloys, containing cobalt, chromium, and tungsten, are used for high-speed, heavy-duty, high temperature cutting tools, and for dies. Cobalt is also used in other magnet steels and stainless steels, and in alloys used in jet turbines and gas turbine generators. The metal is used in electroplating because of its appearance, hardness, and resistance to oxidation. The salts have been used for centuries for the production of brilliant and permanent blue colors in porcelain, glass, pottery, tiles, and enamels. It is the principal ingredient in Sevre's and Thenard's blue. A solution of the chloride ($CoCl_2 \cdot 6H_2O$) is used as sympathetic ink. The cobalt ammines are of interest; the oxide and the nitrate are important. Cobalt carefully used in the form of the chloride, sulfate, acetate, or nitrate has been found effective in correcting a certain mineral deficiency disease in animals. Soils should contain 0.13 to 0.30 ppm of cobalt for proper animal nutrition. Cobalt is found in Vitamin B-12, which is essential for human nutrition. Cobalt of 99.9+% purity is priced at about $500/kg. Cobalt-60, an artificial isotope, is an important gamma ray source, and is extensively used as a tracer and a radiotherapeutic agent. Single compact sources of Cobalt-60 vary from about $1 to $10/curie, depending on quantity and specific activity. Thirty isotopes and isomers of cobalt are known.

Columbium — See Niobium.

Copper — (L. *cuprum,* from the island of Cyprus), Cu; at. wt. 63.546(3); at. no. 29; f.p. 1084.62 °C; b.p. 2562°C; sp. gr. 8.96 (20°C); valence 1 or 2. The discovery of copper dates from prehistoric times. It is said to have been mined for more than 5000 years. It is one of man's most important metals. Copper is reddish colored, takes on a bright metallic luster, and is malleable, ductile, and a good conductor of heat and electricity (second only to silver in electrical conductivity). The electrical industry is one of the greatest users of copper. Copper occasionally occurs native, and is found in many minerals such as *cuprite, malachite, azurite, chalcopyrite,* and *bornite.* Large copper ore deposits are found in the U.S., Chile, Zambia, Zaire, Peru, and Canada. The most important copper ores are the sulfides, oxides, and carbonates. From these, copper is obtained by smelting, leaching, and by electrolysis. Its alloys, brass and bronze, long used, are still very important; all American coins are now copper alloys; monel and gun metals also contain copper. The most important compounds are the oxide and the sulfate, blue vitriol; the latter has wide use as an agricultural poison and as an algicide in water purification. Copper compounds such as Fehling's solution are widely used in analytical chemistry in tests for sugar. High-purity copper (99.999 + %) is readily available commercially. The price of commercial copper has fluctuated widely. The price of copper in December 1999 was about $1.75/kg. Natural copper contains two isotopes. Twenty-six other radioactive isotopes and isomers are known.

Curium — (Pierre and Marie Curie), Cm; at. wt. (247); at. no. 96; m.p. 1345°C; sp. gr. 13.51 (calc.); valence 3 and 4. Although curium follows americium in the periodic system, it was actually known before americium and was the third transuranium element to be discovered. It was identified by Seaborg, James, and Ghiorso in 1944 at the wartime Metallurgical Laboratory in Chicago as a result of helium-ion bombardment of ^{239}Pu in the Berkeley, California, 60-inch cyclotron. Visible amounts (30 μg) of ^{242}Cm, in the form of the hydroxide, were first isolated by Werner and Perlman of the University of California in 1947. In 1950, Crane, Wallmann, and Cunningham found that the magnetic susceptibility of microgram samples of CmF_3 was of the same magnitude as that of GdF_3. This provided direct experimental evidence for assigning an electronic configuration to Cm^{+3}. In 1951, the same workers prepared curium in its elemental form for the first time. Sixteen isotopes of curium are now known. The most stable, ^{247}Cm, with a half-life of 16 million years, is so short compared to the earth's age that any primordial curium must have disappeared long ago from the natural scene. Minute amounts of curium probably exist in natural deposits of uranium, as a result of a sequence of neutron captures and β decays sustained by the very low flux of neutrons naturally present in uranium ores. The presence of natural curium, however, has never been detected. ^{242}Cm and ^{244}Cm are available in multigram quantities. ^{248}Cm has been produced only in milligram amounts. Curium is similar in some regards to gadolinium, its rare-earth homolog, but it has a more complex crystal structure. Curium is silver in color, is chemically reactive, and is more electropositive than aluminum. CmO_2, Cm_2O_3, CmF_3, CmF_4, $CmCl_3$, $CmBr_3$, and CmI_3 have been prepared. Most compounds of trivalent curium are faintly yellow in color. ^{242}Cm generates about three watts of thermal energy per gram. This compares to one-half watt per gram of ^{238}Pu. This suggests use for curium as a power source. ^{244}Cm is now offered for sale by the O.R.N.L. at $185/mg plus packing charges. ^{248}Cm is available at a cost of $160/μg, plus packing charges, from the O.R.N.L. Curium absorbed into the body accumulates in the bones, and is therefore very toxic as its radiation destroys the red-cell forming mechanism. The maximum permissible total body burden of ^{244}Cm (soluble) in a human being is 0.3 μCi (microcurie).

Deuterium, an isotope of hydrogen — see Hydrogen.

Dubnium — (named after the Joint Institute of Nuclear Research in Dubna, Russia). Db; at.wt. [262]; at.no. 105. In 1967 G. N. Flerov reported that a Soviet team working at the Joint Institute for Nuclear Research at Dubna may have produced a few atoms of ^{260}105 and ^{261}105 by bombarding ^{243}Am with ^{22}Ne. Their evidence was based on time-coincidence measurements of alpha energies. More recently, it was reported that early in 1970 Dubna scientists synthesized Element 105 and that by the end of April 1970 "had investigated all the types of decay of the new element and had determined its chemical properties." In late April 1970, it was announced that Ghiorso, Nurmia, Harris, K. A. Y. Eskola, and P. L. Eskola, working at the University of California at Berkeley, had positively identified Element 105. The discovery was made by bombarding a target of ^{249}Cf with a beam of 84 MeV nitrogen nuclei in the Heavy Ion Linear Accelerator (HILAC). When a ^{15}N nuclear is absorbed by a ^{249}Cf nucleus, four neutrons are emitted and a new atom of ^{260}105 with a half-life of 1.6 s is formed. While the first atoms of Element 105 are said to have been detected conclusively

THE ELEMENTS (continued)

on March 5, 1970, there is evidence that Element 105 had been formed in Berkeley experiments a year earlier by the method described. Ghiorso and his associates have attempted to confirm Soviet findings by more sophisticated methods without success.

In October 1971, it was announced that two new isotopes of Element 105 were synthesized with the heavy ion linear accelerator by A. Ghiorso and co-workers at Berkeley. Element ^{261}105 was produced both by bombarding ^{250}Cf with ^{15}N and by bombarding ^{249}Bk with ^{16}O. The isotope emits 8.93-MeV α particles and decays to ^{257}Lr with a half-life of about 1.8 s. Element ^{262}105 was produced by bombarding ^{249}Bk with ^{18}O. It emits 8.45 MeV α particles and decays to ^{258}Lr with a half-life of about 40 s. Nine isotopes of Dubnium are now recognized. Soon after the discovery the names *Hahnium* and *Joliotium,* named after Otto Hahn and Jean-Frederic Joliot and Mme. Joliot-Curie, were suggested as names for Element 105. The IUPAC in August 1997 finally resolved the issue, naming Element 105 **Dubnium** with the symbol Db. Dubnium is thought to have properties similar to tantalum.

Dysprosium — (Gr. *dysprositos,* hard to get at), Dy; at. wt. 162.50(3); at. no. 66; m.p. 1412°C; b.p. 2567°C; sp. gr. 8.551 (25°C); valence 3. Dysprosium was discovered in 1886 by Lecoq de Boisbaudran, but not isolated. Neither the oxide nor the metal was available in relatively pure form until the development of ion-exchange separation and metallographic reduction techniques by Spedding and associates about 1950. Dysprosium occurs along with other so-called rare-earth or lanthanide elements in a variety of minerals such as *xenotime, fergusonite, gadolinite, euxenite, polycrase,* and *blomstrandine.* The most important sources, however, are from *monazite* and *bastnasite.* Dysprosium can be prepared by reduction of the trifluoride with calcium. The element has a metallic, bright silver luster. It is relatively stable in air at room temperature, and is readily attacked and dissolved, with the evolution of hydrogen, by dilute and concentrated mineral acids. The metal is soft enough to be cut with a knife and can be machined without sparking if overheating is avoided. Small amounts of impurities can greatly affect its physical properties. While dysprosium has not yet found many applications, its thermal neutron absorption cross-section and high melting point suggest metallurgical uses in nuclear control applications and for alloying with special stainless steels. A dysprosium oxide-nickel cermet has found use in cooling nuclear reactor rods. This cermet absorbs neutrons readily without swelling or contracting under prolonged neutron bombardment. In combination with vanadium and other rare earths, dysprosium has been used in making laser materials. Dysprosium-cadmium chalcogenides, as sources of infrared radiation, have been used for studying chemical reactions. The cost of dysprosium metal has dropped in recent years since the development of ion-exchange and solvent extraction techniques, and the discovery of large ore bodies. Thirty two isotopes and isomers are now known. The metal costs about $5/g (99.9% purity).

Einsteinium — (Albert Einstein [1879–1955]), Es; at. wt. (252); m.p. 860°C (est.); at. no. 99. Einsteinium, the seventh transuranic element of the actinide series to be discovered, was identified by Ghiorso and co-workers at Berkeley in December 1952 in debris from the first large thermonuclear explosion, which took place in the Pacific in November 1952. The isotope produced was the 20-day ^{253}Es isotope. In 1961, a sufficient amount of einsteinium was produced to permit separation of a macroscopic amount of ^{253}Es. This sample weighed about 0.01 µg. A special magnetic-type balance was used in making this determination. ^{253}Es so produced was used to produce mendelevium. About 3 µg of einsteinium has been produced at Oak Ridge National Laboratories by irradiating for several years kilogram quantities of ^{239}Pu in a reactor to produce ^{242}Pu. This was then fabricated into pellets of plutonium oxide and aluminum powder, and loaded into target rods for an initial 1-year irradiation at the Savannah River Plant, followed by irradiation in a HFIR (High Flux Isotopic Reactor). After 4 months in the HFIR the targets were removed for chemical separation of the einsteinium from californium. Nineteen isotopes and isomers of einsteinium are now recognized. ^{254}Es has the longest half-life (276 days). Tracer studies using ^{253}Es show that einsteinium has chemical properties typical of a heavy trivalent, actinide element. Einsteinium is extremely radioactive. Great care must be taken when handling it.

Element 93 — See Neptunium.
Element 94 — See Plutonium.
Element 95 — See Americium.
Element 96 — See Curium.
Element 97 — See Berkelium.
Element 98 — See Californium.
Element 99 — See Einsteinium.
Element 100 — See Fermium (unnilnilium).
Element 101 — See Mendelevium (unnilunium).
Element 102 — See Nobelium (unnilbium).
Element 103 — See Lawrencium (unniltrium).
Element 104 — See Rutherfordium (unnilquadium).
Element 105 — See Dubnium (unnilpentium).
Element 106 — See Seaborgium (unnilhexium).
Element 107 — See Bohrium (unnilseptium).
Element 108 — See Hassium (unniloctium).
Element 109 — See Meitnerium (unnilennium).

Element 110 — In 1987 Oganessian, et al., at Dubna, claimed discovery of this element. Their experiments indicated the spontaneous fissioning nuclide ^{272}110 with a half-life of 10 ms. More recently a group led by Armbruster at G.S.I. in Darmstadt, Germany, reported evidence of ^{269}110, which was produced by bombarding lead for many days with more than 10^{18} nickel atoms. A detector searched each collision for Element 110's distinct decay sequence. On November 9, 1994, evidence of 110 was detected. Berkeley scientists, in 1991, performed similar experiments and reported evidence of 110, but this was not confirmed. Workers at Dubna have experiments underway to produce ^{273}110 by bombarding plutonium with sulfur atoms. Other experiments at G.S.I. and elsewhere are now searching for heavier isotopes. Five isotopes of Element 110 are now recognized. Several years ago the IUPAC suggested the use of the temporary name *ununnilium,* with the symbol Uun, for Element 110 when it was found.

Element 111 — On December 20, 1994, scientists at GSI Darmstadt, Germany announced they had detected three atoms of a new element with 111 protons and 161 neutrons. This element was made by bombarding ^{83}Bi with ^{28}Ni. Signals of Element 111 appeared for less than 0.002 sec, then decayed into lighter elements including Element ^{268}109 and Element ^{264}107. These isotopes had not previously been observed. A name for Element 111 has not been suggested although IUPAC has suggested a temporary name of Unununium, with the symbol Uuu. Element 111 is expected to have properties similar to gold.

THE ELEMENTS (continued)

Element 112 — In late February 1996, Siguard Hofmann and his collaborators at GSI Darmstadt announced their discovery of Element 112, having 112 protons and 165 neutrons, with an atomic mass of 277. This element was made by bombarding a lead target with high-energy zinc ions. A single nucleus of Element 112 was detected, which decayed after less than 0.001 sec by emitting an α particle, consisting of two protons and two neutrons. This created Element 110_{273}, which in turn decayed by emitting an α particle to form a new isotope of Element 108 and so on. Evidence indicates that nuclei with 162 neutrons are held together more strongly than nuclei with a smaller or larger number of neutrons. This suggests a narrow "peninsula" of relatively stable isotopes around Element 114. GSI scientists are experimenting to bombard targets with ions heavier than zinc to produce Elements 113 and 114. A name has not yet been suggested for Element 112, although the IUPAC suggested the temporary name of ununbium, with the symbol of Uub, when the element was discovered. Element 112 is expected to have properties similar to mercury.

Element 113 — (ununtrium) As of December 1999 this element remains undiscovered.

Element 114 — (ununquadium) Symbol Uuq. Element 114 is the first new element to be discovered since 1996. This element was found by a Russian–American team, including Livermore researchers, by bombarding a sheet of plutonium with a rare form of calcium hoping to make the atoms stick together in a new element. Radiation showed that the new element broke into smaller pieces. Data of radiation collected at the Russian Joint Institute for Nuclear Research in November and December 1998, were analyzed in January 1999. It was found that some of the heavy atoms created when 114 decayed lived up to 30 seconds, which was longr than ever seen before, for such a heavy element. This isotope decayed into a previously unknown isotope of Element 112, which itself lasted 15 minutes. That isotope, in turn, decayed to a previously undiscovered isotope of Element 108, which survived 17 minutes. Isotopes of these and those with longer life-times have been predicted for some time by theorists. It appears that these isotopes are on the edge of the "island of stability", and that some of the isotopes in this region might last long enough for studies of their nuclear behavior and for a chemical evaluation to be made. No name has yet been suggested for Element 114; however, the temporary name of ununquadium with symbol Uuq may be used.

In June 1999 it was announced that Elements 118 and 116 had been discovered. These disintegrated into Element 114 and other elements. See under Element 118.

Element 115 — (ununpentium) As of December 1999, this element remains undiscovered.

Element 116 — (ununhexium) Symbol Uuh. This element was discovered along with Element 118 in June 1999. See under Element 118.

Element 117 — (ununseptium) As of December 1999, this element remains undiscovered.

Element 118 — (ununoctium) Symbol Uuo. Victor Nonov of the Lawrence Berkeley National Laboratory led a team of university and government scientists in experiments that led to the discovery of Elements 118 and 116 in June 1999. A lead target was bombarded for more than 10 days with roughly 1 quintillion krypton ions. The team made three atoms of Element 118, which quickly decayed into Elements 116, 114, and other elements. The isotopes of Element 118 lasted only about 200 microseconds, while the isotope of Element 116 lasted only 1.2 milliseconds. Investigations seem to indicate that the investigations are converging on a cluster of very stable super-heavy isotopes, some of which may survive for very long times. Researchers at GSI, in Russia at Dubna, or at the Berkeley Laboratory may soon fire krypton at bismuth to make Element 119, which would decay into the yet-undiscovered Elements 117, 115, and 113.

Erbium — (*Ytterby*, a town in Sweden), Er; at. wt. 167.26(3); at. no. 68; m.p. 1529°C; b.p. 2868°C; sp. gr. 9.066 (25°C); valence 3, Erbium, one of the so-called rare-earth elements of the lanthanide series, is found in the minerals mentioned under dysprosium above. In 1842 Mosander separated "yttria," found in the mineral *gadolinite,* into three fractions which he called *yttria, erbia,* and *terbia.* The names *erbia* and *terbia* became confused in this early period. After 1860, Mosander's *terbia* was known as *erbia,* and after 1877, the earlier known *erbia* became *terbia.* The *erbia* of this period was later shown to consist of five oxides, now known as *erbia, scandia, holmia, thulia* and *ytterbia.* By 1905 Urbain and James independently succeeded in isolating fairly pure Er_2O_3. Klemm and Bommer first produced reasonably pure erbium metal in 1934 by reducing the anhydrous chloride with potassium vapor. The pure metal is soft and malleable and has a bright, silvery, metallic luster. As with other rare-earth metals, its properties depend to a certain extent on the impurities present. The metal is fairly stable in air and does not oxidize as rapidly as some of the other rare-earth metals. Naturally occurring erbium is a mixture of six isotopes, all of which are stable. Twenty-seven radioactive isotopes of erbium are also recognized. Recent production techniques, using ion-exchange reactions, have resulted in much lower prices of the rare-earth metals and their compounds in recent years. The cost of 99.9% erbium metal is about $10/g. Erbium is finding nuclear and metallurgical uses. Added to vanadium, for example, erbium lowers the hardness and improves workability. Most of the rare-earth oxides have sharp absorption bands in the visible, ultraviolet, and near infrared. This property, associated with the electronic structure, gives beautiful pastel colors to many of the rare-earth salts. Erbium oxide gives a pink color and has been used as a colorant in glasses and porcelain enamel glazes.

Europium — (Europe), Eu; at. wt. 151.964(1); at. no. 63; m.p. 822°C; b.p. 1596°C; sp. gr. 5.244 (25°C); valence 2 or 3. In 1890 Boisbaudran obtained basic fractions from samarium-gadolinium concentrates which had spark spectral lines not accounted for by samarium or gadolinium. These lines subsequently have been shown to belong to europium. The discovery of europium is generally credited to Demarcay, who separated the rare earth in reasonably pure form in 1901. The pure metal was not isolated until recent years. Europium is now prepared by mixing Eu_2O_3 with a 10%-excess of lanthanum metal and heating the mixture in a tantalum crucible under high vacuum. The element is collected as a silvery-white metallic deposit on the walls of the crucible. As with other rare-earth metals, except for lanthanum, europium ignites in air at about 150 to 180°C. Europium is about as hard as lead and is quite ductile. It is the most reactive of the rare-earth metals, quickly oxidizing in air. It resembles calcium in its reaction with water. *Bastnasite* and *monazite* are the principal ores containing europium. Europium has been identified spectroscopically in the sun and certain stars. Europium isotopes are good neutron absorbers and are being studied for use in nuclear control applications. Europium oxide is now widely used as a phosphor activator and europium-activated yttrium vanadate is in commercial use as the red phosphor in color TV tubes. Europium-doped plastic has been used as a laser material. With the development of ion-exchange techniques and special processes, the cost of the metal has been greatly reduced in recent years. Natural europium contains two stable isotopes. Thirty five other radioactive isotopes and isomers are known. Europium is one of the rarest and most costly of the rare-earth metals. It is priced at about $50/g (99.9% pure).

Fermium — (Enrico Fermi [1901–1954], nuclear physicist), Fm; at. wt. [257]; at. no. 100; m.p. 1527°C. Fermium, the eighth transuranium element of the actinide series to be discovered, was identified by Ghiorso and co-workers in 1952 in the debris from a thermonuclear explosion in the Pacific in work involving the University of California Radiation Laboratory, the Argonne National Laboratory, and the Los Alamos Scientific Laboratory. The isotope produced was the 20-hour ^{255}Fm. During 1953 and early 1954, while discovery of elements 99 and 100 was withheld from publication for security reasons, a group from the Nobel Institute of Physics in Stockholm bombarded ^{238}U with ^{16}O ions, and isolated a 30-min α-emitter, which

THE ELEMENTS (continued)

they ascribed to $^{250}100$, without claiming discovery of the element. This isotope has since been identified positively, and the 30-min half-life confirmed. The chemical properties of fermium have been studied solely with tracer amounts, and in normal aqueous media only the (III) oxidation state appears to exist. The isotope ^{254}Fm and heavier isotopes can be produced by intense neutron irradiation of lower elements such as plutonium by a process of successive neutron capture interspersed with beta decays until these mass numbers and atomic numbers are reached. Twenty isotopes and isomers of fermium are known to exist. ^{257}Fm, with a half-life of about 100.5 days, is the longest lived. ^{250}Fm, with a half-life of 30 min, has been shown to be a product of decay of Element $^{254}102$. It was by chemical identification of ^{250}Fm that production of Element 102 (nobelium) was confirmed. Fermium would probably have chemical properties resembling erbium.

Fluorine — (L. and F. *fluere,* flow, or flux), F; at. wt. 18.9984032(5); at. no. 9; m.p. –219.62°C (1 atm); b.p. –188.12°C (1 atm); t_c -129.02°C; density 1.696 g/L (0°C, 1 atm); liq. den. at b.p. 1.50 g/cm^3; valence 1. In 1529, Georigius Agricola described the use of fluorspar as a flux, and as early as 1670 Schwandhard found that glass was etched when exposed to fluorspar treated with acid. Scheele and many later investigators, including Davy, Gay-Lussac, Lavoisier, and Thenard, experimented with hydrofluoric acid, some experiments ending in tragedy. The element was finally isolated in 1886 by Moisson after nearly 74 years of continuous effort. Fluorine occurs chiefly in *fluorspar* (CaF$_2$) and *cryolite* (Na$_2$AlF$_6$), but is rather widely distributed in other minerals. It is a member of the halogen family of elements, and is obtained by electrolyzing a solution of potassium hydrogen fluoride in anhydrous hydrogen fluoride in a vessel of metal or transparent fluorspar. Modern commercial production methods are essentially variations on the procedures first used by Moisson. Fluorine is the most electronegative and reactive of all elements. It is a pale yellow, corrosive gas, which reacts with practically all organic and inorganic substances. Finely divided metals, glass, ceramics, carbon, and even water burn in fluorine with a bright flame. Until World War II, there was no commercial production of elemental fluorine. The atom bomb project and nuclear energy applications, however, made it necessary to produce large quantities. Safe handling techniques have now been developed and it is possible at present to transport liquid fluorine by the ton. Fluorine and its compounds are used in producing uranium (from the hexafluoride) and more than 100 commercial fluorochemicals, including many well-known high-temperature plastics. Hydrofluoric acid is extensively used for etching the glass of light bulbs, etc. Fluorochloro hydrocarbons have been extensively used in air conditioning and refrigeration. However, in recent years the U.S. and other countries have been phasing out ozone-depleting substances, such as the fluorochloro hydrocarbons that have been used in these applications. It has been suggested that fluorine might be substituted for hydrogen wherever it occurs in organic compounds, which could lead to an astronomical number of new fluorine compounds. The presence of fluorine as a soluble fluoride in drinking water to the extent of 2 ppm may cause mottled enamel in teeth, when used by children acquiring permanent teeth; in smaller amounts, however, fluorides are said to be beneficial and used in water supplies to prevent dental cavities. Elemental fluorine has been studied as a rocket propellant as it has an exceptionally high specific impulse value. Compounds of fluorine with rare gases have now been confirmed. Fluorides of xenon, radon, and krypton are among those known. Elemental fluorine and the fluoride ion are highly toxic. The free element has a characteristic pungent odor, detectable in concentrations as low as 20 ppb, which is below the safe working level. The recommended maximum allowable concentration for a daily 8-hour time-weighted exposure is 1 ppm. Fluorine is known to have fourteen isotopes.

Francium — (France), Fr; at. no. 87; at. wt. [223]; m.p. 27°C; valence 1. Discovered in 1939 by Mlle. Marguerite Perey of the Curie Institute, Paris. Francium, the heaviest known member of the alkali metal series, occurs as a result of an alpha disintegration of actinium. It can also be made artificially by bombarding thorium with protons. While it occurs naturally in uranium minerals, there is probably less than an ounce of francium at any time in the total crust of the earth. It has the highest equivalent weight of any element, and is the most unstable of the first 101 elements of the periodic system. Thirty-six isotopes and isomers of francium are recognized. The longest lived ^{223}Fr(Ac, K), a daughter of ^{227}Ac, has a half-life of 21.8 min. This is the only isotope of francium occurring in nature. Because all known isotopes of francium are highly unstable, knowledge of the chemical properties of this element comes from radiochemical techniques. No weighable quantity of the element has been prepared or isolated. The chemical properties of francium most closely resemble cesium. In 1996, researchers Orozco, Sprouse, and co-workers at the State University of New York, Stony Brook, reported that they had produced francium atoms by bombarding ^{18}O atoms at a gold target heated almost to its melting point. Collisions between gold and oxygen nuclei created atoms of francium-210 which had 87 protons and123 neutrons. This team reported they had generated about 1 million francium-210 ions per second and held 1000 or more atoms at a time for about 20 secs. in a magnetic trap they had devised before the atoms decayed or escaped. Enough francium was trapped so that a video camera could capture the light given off by the atoms as they fluoresced. A cluster of about 10,000 francium atoms appeared as a glowing sphere about 1 mm in diameter. It is thought that the francium atoms could serve as miniature laboratoires for probing interactions between electrons and quarks.

Gadolinium — (*gadolinite,* a mineral named for Gadolin, a Finnish chemist), Gd; at. wt. 157.25(3); at. no. 64; m.p. 1313°C; b.p. 3273°C; sp. gr. 7.901 (25°C); valence 3. Gadolinia, the oxide of gadolinium, was separated by Marignac in 1880 and Lecoq de Boisbaudran independently isolated the element from Mosander's "yttria" in 1886. The element was named for the mineral *gadolinite* from which this rare earth was originally obtained. Gadolinium is found in several other minerals, including *monazite* and *bastnasite,* which are of commercial importance. The element has been isolated only in recent years. With the development of ion-exchange and solvent extraction techniques, the availability and price of gadolinium and the other rare-earth metals have greatly improved. Thirty-one isotopes and isomers of gadolinium are now recognized; seven are stable and occur naturally. The metal can be prepared by the reduction of the anhydrous fluoride with metallic calcium. As with other related rare-earth metals, it is silvery white, has a metallic luster, and is malleable and ductile. At room temperature, gadolinium crystallizes in the hexagonal, close-packed α form. Upon heating to 1235°C, α gadolinium transforms into the β form, which has a body-centered cubic structure. The metal is relatively stable in dry air, but in moist air it tarnishes with the formation of a loosely adhering oxide film which spalls off and exposes more surface to oxidation. The metal reacts slowly with water and is soluble in dilute acid. Gadolinium has the highest thermal neutron capture cross-section of any known element (49,000 barns). Natural gadolinium is a mixture of seven isotopes. Two of these, ^{155}Gd and ^{157}Gd, have excellent capture characteristics, but they are present naturally in low concentrations. As a result, gadolinium has a very fast burnout rate and has limited use as a nuclear control rod material. It has been used in making gadolinium yttrium garnets, which have microwave applications. Compounds of gadolinium are used in making phosphors for color TV tubes. The metal has unusual superconductive properties. As little as 1% gadolinium has been found to improve the workability and resistance of iron, chromium, and related alloys to high temperatures and oxidation. Gadolinium ethyl sulfate has extremely low noise characteristics and may find use in duplicating the performance of amplifiers, such as the maser. The metal is ferromagnetic. Gadolinium is unique for its high magnetic moment and for its special Curie temperature (above which ferromagnetism vanishes) lying just at room temperature. This suggests uses as a magnetic component that senses hot and cold. The price of the metal is about $12/g (99.9% purity).

Gallium — (L. *Gallia,* France; also from Latin, *gallus,* a translation of Lecoq, a cock), Ga; at. wt. 69.723(1); at. no. 31; m.p. 29.76°C; b.p. 2204°C; sp. gr. 5.904 (29.6°C) solid; sp. gr. 6.095 (29.6°C) liquid; valence 2 or 3. Predicted and described by Mendeleev as ekaaluminum, and discovered spectroscopically by Lecoq de Boisbaudran in 1875, who in the same year obtained the free metal by electrolysis of a solution of the hydroxide in KOH. Gallium is often found as a trace element in *diaspore, sphalerite, germanite, bauxite,* and *coal.* Some flue dusts from burning coal have been shown to contain as much as 1.5% gallium. It is the only metal, except for mercury, cesium, and rubidium, which can be liquid near room temperatures; this makes possible its use in high-temperature thermometers. It has one of the longest liquid ranges of any metal and has a low vapor pressure even at high temperatures. There is a strong tendency for gallium to supercool below its freezing point. Therefore, seeding may be necessary to initiate solidification. Ultra-pure gallium has a beautiful, silvery appearance, and the solid metal exhibits a conchoidal fracture similar to glass. The metal expands 3.1% on solidifying; therefore, it should not be stored in glass or metal containers, as they may break as the metal solidifies. Gallium wets glass or porcelain, and forms a brilliant mirror when it is painted on glass. It is widely used in doping semiconductors and producing solid-state devices such as transistors. High-purity gallium is attacked only slowly by mineral acids. Magnesium gallate containing divalent impurities such as Mn^{+2} is finding use in commercial ultraviolet activated powder phosphors. Gallium nitride has been used to produce blue light-emitting diodes. Blue LED's used in compact disc applications can be used to store a 2-h movie, for example, on one 5-in. diameter disc. Extensive use of gallium has found recent application in the **Gallex Detector Experiment** located in the Gran Sasso Underground Laboratory in Italy. This underground facility has been built by the Italian Istituto Nazionale di Fisica Nucleare in the middle of a highway tunnel through the Abruzzese mountains, about 150 km east of Rome. The experiment is shielded by a 3300-m water-equivalent of rock. In this experiment, 30.3 tons of gallium in the form of 110 tons of $GaCl_3$-HCl solution are being used to detect solar neutrinos. The production of ^{71}Ge from gallium is being measured.Gallium arsenide is capable of converting electricity directly into coherent light. Gallium readily alloys with most metals, and has been used as a component in low-melting alloys. Its toxicity appears to be of a low order, but should be handled with care until more data are forthcoming. Natural gallium contains two stable isotopes. Twenty-six other isotopes, one of which is an isomer, are known. The metal can be supplied in ultrapure form (99.99999+%). The cost is about $5/g (99.999%).

Germanium — (L. *Germania,* Germany), Ge; at. wt. 72.61(2); at. no. 32; m.p. 938.25°C; b.p. 2833°C; sp. gr. 5.323 (25°C); valence 2 and 4. Predicted by Mendeleev in 1871 as ekasilicon, and discovered by Winkler in 1886. The metal is found in *argyrodite,* a sulfide of germanium and silver; in *germanite,* which contains 8% of the element; in zinc ores; in coal; and in other minerals. The element is frequently obtained commercially from flue dusts of smelters processing zinc ores, and has been recovered from the by-products of combustion of certain coals. Its presence in coal insures a large reserve of the element in the years to come. Germanium can be separated from other metals by fractional distillation of its volatile tetrachloride. The tetrachloride may then be hydrolyzed to give GeO_2; the dioxide can be reduced with hydrogen to give the metal. Recently developed zone-refining techniques permit the production of germanium of ultra-high purity. The element is a gray-white metalloid, and in its pure state is crystalline and brittle, retaining its luster in air at room temperature. It is a very important semiconductor material. Zone-refining techniques have led to production of crystalline germanium for semiconductor use with an impurity of only one part in 10^{10}. Doped with arsenic, gallium, or other elements, it is used as a transistor element in thousands of electronic applications. Its application in fiber optics and infra-red optical systems now provides the largest use for germanium. Germanium is also finding many other applications including use as an alloying agent, as a phosphor in fluorescent lamps, and as a catalyst. Germanium and germanium oxide are transparent to the infrared and are used in infrared spectroscopes and other optical equipment, including extremely sensitive infrared detectors. Germanium oxide's high index of refraction and dispersion has made it useful as a component of glasses used in wide-angle camera lenses and microscope objectives. The field of organogermanium chemistry is becoming increasingly important. Certain germanium compounds have a low mammalian toxicity, but a marked activity against certain bacteria, which makes them of interest as chemotherapeutic agents. The cost of germanium is about $10/g (99.999% purity). Thirty isotopes and isomers are known, five of which occur naturally.

Gold — (Sanskrit *Jval;* Anglo-Saxon *gold*), Au (L. *aurum,* gold); at. wt. 196.96654(3); at. no. 79; m.p. 1064.18°C; b.p. 2856°C; sp. gr. ~19.3 (20°C); valence 1 or 3. Known and highly valued from earliest times, gold is found in nature as the free metal and in tellurides; it is very widely distributed and is almost always associated with quartz or pyrite. It occurs in veins and alluvial deposits, and is often separated from rocks and other minerals by sluicing and panning operations. About 25% of the world's gold output comes from South Africa, and about two thirds of the total U.S. production now comes from South Dakota and Nevada. The metal is recovered from its ores by cyaniding, amalgamating, and smelting processes. Refining is also frequently done by electrolysis. Gold occurs in sea water to the extent of 0.1 to 2 mg/ton, depending on the location where the sample is taken. As yet, no method has been found for recovering gold from sea water profitably. It is estimated that all the gold in the world, so far refined, could be placed in a single cube 60 ft on a side. Of all the elements, gold in its pure state is undoubtedly the most beautiful. It is metallic, having a yellow color when in a mass, but when finely divided it may be black, ruby, or purple. The Purple of Cassius is a delicate test for auric gold. It is the most malleable and ductile metal; 1 oz. of gold can be beaten out to 300 ft^2. It is a soft metal and is usually alloyed to give it more strength. It is a good conductor of heat and electricity, and is unaffected by air and most reagents. It is used in coinage and is a standard for monetary systems in many countries. It is also extensively used for jewelry, decoration, dental work, and for plating. It is used for coating certain space satellites, as it is a good reflector of infrared and is inert. Gold, like other precious metals, is measured in troy weight; when alloyed with other metals, the term *carat* is used to express the amount of gold present, 24 carats being pure gold. For many years the value of gold was set by the U.S. at $20.67/troy ounce; in 1934 this value was fixed by law at $35.00/troy ounce, 9/10th fine. On March 17, 1968, because of a gold crisis, a two-tiered pricing system was established whereby gold was still used to settle international accounts at the old $35.00/troy ounce price while the price of gold on the private market would be allowed to fluctuate. Since this time, the price of gold on the free market has fluctuated widely. The price of gold on the free market reached a price of $620/troy oz. in January 1980. More recently, the U.K. and other nations, including the I.M.F. have sold or threatened to sell a sizeable portion of their gold reserves. This has caused wide fluctuations in the price of gold. Because this has damaged the economy of some countires, a moretorium for a few years has been declared. This has tended to stabilize temporarily the price of gold. The most common gold compounds are auric chloride ($AuCl_3$) and chlorauric acid ($HAuCl_4$), the latter being used in photography for toning the silver image. Gold has forty-eight recognized isotopes and isomers; ^{198}Au, with a half-life of 2.7 days, is used for treating cancer and other diseases. Disodium aurothiomalate is administered intramuscularly as a treatment for arthritis. A mixture of one part nitric acid with three of hydrochloric acid is called *aqua regia* (because it dissolved gold, the King of Metals). Gold is available commercially with a purity of 99.999+%. For many years the temperature assigned to the freezing point of gold has been 1063.0°C; this has served as a calibration point for the International Temperature Scales (ITS-27 and ITS-48) and the International Practical Temperature Scale (IPTS-

48). In 1968, a new International Practical Temperature Scale (IPTS-68) was adopted, which demanded that the freezing point of gold be changed to 1064.43°C. In 1990 a new International Temperature Scale (ITS-90) was adopted bringing the t.p. (triple point) of H_2O (t_{90} (°C)) to 0.01°C and the freezing point of gold to 1064.18°C. The specific gravity of gold has been found to vary considerably depending on temperature, how the metal is precipitated, and cold-worked. As of December 1999, gold was priced at about $300/troy oz. ($9.50/g).

Hafnium — (*Hafnia,* Latin name for Copenhagen), Hf; at. wt. 178.49(2); at. no. 72; m.p. 2233°C; b.p. 4603°C; sp. gr. 13.31 (20°C); valence 4. Hafnium was thought to be present in various minerals and concentrations many years prior to its discovery, in 1923, credited to D. Coster and G. von Hevesey. On the basis of the Bohr theory, the new element was expected to be associated with zirconium. It was finally identified in *zircon* from Norway, by means of X-ray spectroscopic analysis. It was named in honor of the city in which the discovery was made. Most zirconium minerals contain 1 to 5% hafnium. It was originally separated from zirconium by repeated recrystallization of the double ammonium or potassium fluorides by von Hevesey and Jantzen. Metallic hafnium was first prepared by van Arkel and deBoer by passing the vapor of the tetraiodide over a heated tungsten filament. Almost all hafnium metal now produced is made by reducing the tetrachloride with magnesium or with sodium (Kroll Process). Hafnium is a ductile metal with a brilliant silver luster. Its properties are considerably influenced by the impurities of zirconium present. Of all the elements, zirconium and hafnium are two of the most difficult to separate. Their chemistry is almost identical, however, the density of zirconium is about half that of hafnium. Very pure hafnium has been produced, with zirconium being the major impurity. Natural hafnium contains six isotopes, one of which is slightly radioactive. Hafnium has a total of 41 recognized isotopes and isomers. Because hafnium has a good absorption cross section for thermal neutrons (almost 600 times that of zirconium), has excellent mechanical properties, and is extremely corrosion resistant, it is used for reactor control rods. Such rods are used in nuclear submarines. Hafnium has been successfully alloyed with iron, titanium, niobium, tantalum, and other metals. Hafnium carbide is the most refractory binary composition known, and the nitride is the most refractory of all known metal nitrides (m.p. 3310°C). Hafnium is used in gas-filled and incandescent lamps, and is an efficient "getter" for scavenging oxygen and nitrogen. Finely divided hafnium is pyrophoric and can ignite spontaneously in air. Care should be taken when machining the metal or when handling hot sponge hafnium. At 700°C hafnium rapidly absorbs hydrogen to form the composition $HfH_{1.86}$. Hafnium is resistant to concentrated alkalis, but at elevated temperatures reacts with oxygen, nitrogen, carbon, boron, sulfur, and silicon. Halogens react directly to form tetrahalides. The price of the metal is in the broad range of $1/g to $5/g, depending on purity and quantity. The yearly demand for hafnium in the U.S. is now in excess of 50,000 kg.

Hahnium — A name previously used for Element 105, now named *dubnium.*

Hassium — (named for the German state, Hesse) Hs, at.wt. [265]; at.no. 108. This element was first synthesized and identified in 1964 by the same G.S.I. Darmstadt Group who first identified *Bohrium* and *Meitnerium.* Presumably this element has chemical properties similar to osmium. Isotope [265]108 was produced using a beam of [58]Fe projectiles, produced by the Universal Linear Accelerator (UNILAC) to bombard a [208]Pb target. Discovery of *Bohrium* and *Meitnerium* was made using detection of isotopes with odd proton and neutron numbers. Elements having even atomic numbers have been thought to be less stable against spontaneous fusion than odd elements. The production of [265]108 in the same reaction as was used at G.S.I. was confirmed at Dubna with detection of the seventh member of the decay chain [253]Es. Isotopes of *Hassium* are believed to decay by spontaneous fission, explaining why 109 was produced before 108. Isotope [265]108 and [266]108 are thought to decay to [261]106, which in turn decay to [257]104 and [253]102. The IUPAC adopted the name *Hassium* after the German state of Hesse in September 1997.

Helium — (Gr. *helios,* the sun), He; at. wt. 4.002602(2); at. no. 2; m.p. below — 272.2°C (26 atm); b.p. — 268.93°C; t_c -267.96°C; density 0.1785 g/l (0°C, 1 atm); liquid density 7.62 lb/ft³ at. b.p.; valence usually 0. Evidence of the existence of helium was first obtained by Janssen during the solar eclipse of 1868 when he detected a new line in the solar spectrum; Lockyer and Frankland suggested the name *helium* for the new element; in 1895, Ramsay discovered helium in the uranium mineral *clevite,* and it was independently discovered in cleveite by the Swedish chemists Cleve and Langlet about the same time. Rutherford and Royds in 1907 demonstrated that α particles are helium nuclei. Except for hydrogen, helium is the most abundant element found throughout the universe. Helium is extracted from natural gas; all natural gas contains at least trace quantities of helium. It has been detected spectroscopically in great abundance, especially in the hotter stars, and it is an important component in both the proton-proton reaction and the carbon cycle, which account for the energy of the sun and stars. The fusion of hydrogen into helium provides the energy of the hydrogen bomb. The helium content of the atmosphere is about 1 part in 200,000. It is present in various radioactive minerals as a decay product. Much of the world's supply of helium is obtained from wells in Texas, Colorado, and Kansas. The only other known helium extraction plants, outside the United States, in 1999 were in Poland, Russia, China, Algeria, and India. The cost of helium has fallen from $2500/ft³ in 1915 to about 2.5¢/cu.ft. (.028 cu meters) in 1999. Helium has the lowest melting point of any element and has found wide use in cryogenic research, as its boiling point is close to absolute zero. Its use in the study of superconductivity is vital. Using liquid helium, Kurti and co-workers, and others, have succeeded in obtaining temperatures of a few microkelvins by the adiabatic demagnetization of copper nuclei, starting from about 0.01 K. Liquid helium (He⁴) exists in two forms: He⁴I and He⁴II, with a sharp transition point at 2.174 K (3.83 cm Hg). He⁴I (above this temperature) is a normal liquid, but He⁴II (below it) is unlike any other known substance. It expands on cooling; its conductivity for heat is enormous; and neither its heat conduction nor viscosity obeys normal rules. It has other peculiar properties. Helium is the only liquid that cannot be solidified by lowering the temperature. It remains liquid down to absolute zero at ordinary pressures, but it can readily be solidified by increasing the pressure. Solid ³He and ⁴He are unusual in that both can readily be changed in volume by more than 30% by application of pressure. The specific heat of helium gas is unusually high. The density of helium vapor at the normal boiling point is also very high, with the vapor expanding greatly when heated to room temperature. Containers filled with helium gas at 5 to 10 K should be treated as though they contained liquid helium due to the large increase in pressure resulting from warming the gas to room temperature. While helium normally has a 0 valence, it seems to have a weak tendency to combine with certain other elements. Means of preparing helium diflouride have been studied, and species such as HeNe and the molecular ions He⁺ and He⁺⁺ have been investigated. Helium is widely used as an inert gas shield for arc welding; as a protective gas in growing silicon and germanium crystals, and in titanium and zirconium production; as a cooling medium for nuclear reactors, and as a gas for supersonic wind tunnels. A mixture of helium and oxygen is used as an artificial atmosphere for divers and others working under pressure. Different ratios of He/O_2 are used for different depths at which the diver is operating. Helium is extensively used for filling balloons as it is a much safer gas than hydrogen. One of the recent largest uses for helium has been for pressuring liquid fuel rockets. A Saturn booster such as used on the Apollo lunar missions required about 13 million ft³ of helium for a firing, plus more for checkouts. Liquid helium's use in magnetic resonance imaging (MRI) continues to increase as the medical profession accepts and develops new uses for the equipment. This equipment is providing accurate diagnoses of problems where exploratory surgery has previously been required to determine problems. Another medical application that is being developed uses MRI to determine by blood analysis whether a patient has any form of cancer. Lifting gas applications are increasing. Various companies in addition to Goodyear, are now using "blimps" for advertising. The Navy and the Air Force are investigating the use of airships to provide early warning systems to detect low-flying cruise missiles. The Drug Enforcement Agency has used radar-equipped blimps to detect drug smugglers along the southern border of the U.S. In addition, NASA is currently using helium-filled balloons to sample the atmosphere in Antarctica to determine what is depleting the ozone layer that protects Earth from harmful U.V. radiation. Research on and development of materials which become superconductive at temperatures well above the boiling point of helium could have a major impact on the demand for helium. Less costly refrigerants having boiling points considerably higher could replace the present need to cool such superconductive materials to the boiling point of helium. Natural helium contains two stable isotopes ^{3}He and ^{4}He. ^{3}He is present in very small quantities. Six other isotopes of helium are now recognized.

Holmium — (L. *Holmia* , for Stockholm), Ho; at. wt. 164.93032(2); at. no 67; m.p. 1474°C; b.p. 2700°C; sp. gr. 8.795 (25°C); valence + 3. The spectral absorption bands of holmium were noticed in 1878 by the Swiss chemists Delafontaine and Soret, who announced the existence of an "Element X". Cleve, of Sweden, later independently discovered the element while working on erbia earth. The element is named after Cleve's native city. Pure holmia, the yellow oxide, was prepared by Homberg in 1911. Holmium occurs in *gadolinite, monazite,* and in other rare-earth minerals. It is commercially obtained from monazite, occurring in that mineral to the extent of about 0.05%. It has been isolated by the reduction of its anhydrous chloride or fluoride with calcium metal. Pure holmium has a metallic to bright silver luster. It is relatively soft and malleable, and is stable in dry air at room temperature, but rapidly oxidizes in moist air and at elevated temperatures. The metal has unusual magnetic properties. Few uses have yet been found for the element. The element, as with other rare earths, seems to have a low acute toxic rating. Natural holmium consists of one isotope ^{165}Ho, which is not radioactive. Holmium has 49 other isotopes known, all of which are radioactive. The price of 99.9% holmium metal is about $35/g.

Hydrogen — (Gr. *hydro*, water, and *genes*, forming), H; at. wt. 1.00794(7); at. no. 1; m.p. –259.34°C; b.p. –252.87°C; t_c -240.18; density 0.08988 g/l; density (liquid) 70.8 g/l (–253°C); density (solid) 70.6 g/l (–262°C); valence 1. Hydrogen was prepared many years before it was recognized as a distinct substance by Cavendish in 1766. It was named by Lavoisier. Hydrogen is the most abundant of all elements in the universe, and it is thought that the heavier elements were, and still are, being built from hydrogen and helium. It has been estimated that hydrogen makes up more than 90% of all the atoms or three quarters of the mass of the universe. It is found in the sun and most stars, and plays an important part in the proton-proton reaction and carbon-nitrogen cycle, which accounts for the energy of the sun and stars. It is thought that hydrogen is a major component of the planet Jupiter and that at some depth in the planet's interior the pressure is so great that solid molecular hydrogen is converted into solid metallic hydrogen. In 1973, it was reported that a group of Russian experimenters may have produced metallic hydrogen at a pressure of 2.8 Mbar. At the transition the density changed from 1.08 to 1.3 g/cm^3. Earlier, in 1972, a Livermore (California) group also reported on a similar experiment in which they observed a pressure-volume point centered at 2 Mbar. It has been predicted that metallic hydrogen may be metastable; others have predicted it would be a superconductor at room temperature. On earth, hydrogen occurs chiefly in combination with oxygen in water, but it is also present in organic matter such as living plants, petroleum, coal, etc. It is present as the free element in the atmosphere, but only to the extent of less than 1 ppm by volume. It is the lightest of all gases, and combines with other elements, sometimes explosively, to form compounds. Great quantities of hydrogen are required commercially for the fixation of nitrogen from the air in the Haber ammonia process and for the hydrogenation of fats and oils. It is also used in large quantities in methanol production, in hydrodealkylation, hydrocracking, and hydrodesulfurization. It is also used as a rocket fuel, for welding, for production of hydrochloric acid, for the reduction of metallic ores, and for filling balloons. The lifting power of 1 ft^3 of hydrogen gas is about 0.076 lb at 0°C, 760 mm pressure. Production of hydrogen in the U.S. alone now amounts to about 3 billion cubic feet per year. It is prepared by the action of steam on heated carbon, by decomposition of certain hydrocarbons with heat, by the electrolysis of water, or by the displacement from acids by certain metals. It is also produced by the action of sodium or potassium hydroxide on aluminum. Liquid hydrogen is important in cryogenics and in the study of superconductivity, as its melting point is only a 20°C above absolute zero. Hydrogen consists of three isotopes, most of which is ^{1}H. The ordinary isotope of hydrogen, H, is known as *protium*. In 1932, Urey announced the discovery of a stable isotope, deuterium (^{2}H or D) with an atomic weight of 2. Deuterium is present in natural hydrogen to the extent of 0.015%. Two years later an unstable isotope, tritium (^{3}H), with an atomic weight of 3 was discovered. Tritium has a half-life of about 12.32 years. Tritium atoms are also present in natural hydrogen but in much smaller proportion. Tritium is readily produced in nuclear reactors and is used in the production of the hydrogen bomb. It is also used as a radioactive agent in making luminous paints, and as a tracer. Deuterium gas is readily available, without permit, at about $1/l. Heavy water, deuterium oxide (D$_2$O), which is used as a moderator to slow down neutrons, is available without permit at a cost of 6c to $1/g, depending on quantity and purity. About 1000 tons (4,400,000 kg) of deuterium oxide (heavy water) are now in use at the Sudbury (Ontario) Neutrino Observatory. This observatory is taking data to provide new revolutionary insight into the properties of neutrinos and into the core of the sun. The heavy water is on loan from Atomic Energy of Canada, Ltd. (AECL). The observatory and detectors are located 6800 ft (2072 m) deep in the Creighton mine of the International Nickel Co., near Sudbury. The heavy water is contained in an acrylic vessel, 12 m in diameter. Neutrinos react with the heavy water to produce Cherenkov radiation. This light is then detected with 9600 photomultiplier tubes surrounding the vessel. The detector laboratory is immensely clean to reduce background radiation, which otherwise hide the very weak signals from neutrinos. Quite apart from isotopes, it has been shown that hydrogen gas under ordinary conditions is a mixture of two kinds of molecules, known as *ortho-* and *para*-hydrogen, which differ from one another by the spins of their electrons and nuclei. Normal hydrogen at room temperature contains 25% of the *para* form and 75% of the *ortho* form. The *ortho* form cannot be prepared in the pure state. Since the two forms differ in energy, the physical properties also differ. The melting and boiling points of *para*hydrogen are about 0.1°C lower than those of normal hydrogen. Consideration is being given to an entire economy based on solar- and nuclear-generated hydrogen. Located in remote regions, power plants would electrolyze sea water; the hydrogen produced would travel to distant cities by pipelines. Pollution-free hydrogen could replace natural gas, gasoline, etc., and could serve as a reducing agent in metallurgy, chemical processing, refining, etc. It could also be used to convert trash into methane and ethylene. Public acceptance, high capital investment, and the high present cost of hydrogen with respect to present fuels are but a few of the problems facing establishment of such an economy. Hydrogen is being investigated as a substitute for deep-sea diving applications below 300 m. Hydrogen is readily available from air product suppliers.

Indium — (from the brilliant indigo line in its spectrum),In; at. wt. 114.818(3); at. no. 49; m.p. 156.60°C; b.p. 2072°C; sp. gr. 7.31 (20°C); valence 1, 2, or 3. Discovered by Reich and Richter, who later isolated the metal. Indium is most frequently associated with zinc materials, and it is from these that most commercial indium is now obtained; however, it is also found in iron, lead, and copper ores. Until 1924, a gram or so constituted the world's supply of this element in isolated form. It is probably about as abundant as silver. About 4 million troy ounces of indium are now produced annually in the Free World. Canada is presently producing more than 1,000,000 troy ounces annually. The present cost of indium is about $2 to $10/g, depending on quantity and purity. It is available in ultrapure form. Indium is a very soft, silvery-white metal with a brilliant luster. The pure metal gives a high-pitched "cry" when bent. It wets glass, as does gallium. It has found application in making low-melting alloys; an alloy of 24% indium-76% gallium is liquid at room temperature. Indium is used in making bearing alloys, germanium transistors, rectifiers, thermistors, liquid crystal displays, high definition television, batteries, and photoconductors. It can be plated onto metal and evaporated onto glass, forming a mirror as good as that made with silver but with more resistance to atmospheric corrosion. There is evidence that indium has a low order of toxicity; however, care should be taken until further information is available. Seventy isotopes and isomers are now recognized (more than any other element). Natural indium contains two isotopes. One is stable. The other, ^{115}In, comprising 95.71% of natural indium is slightly radioactive with a very long half-life.

Iodine — (Gr. *iodes,* violet), I; at. wt. 126.90447(3); at. no. 53; m.p. 113.7°C; b.p. 184.4°C; t_c 546°C; density of the gas 11.27 g/l; sp. gr. solid 4.93 (20°C); valence 1, 3, 5, or 7. Discovered by Courtois in 1811. Iodine, a halogen, occurs sparingly in the form of iodides in sea water from which it is assimilated by seaweeds, in Chilean saltpeter and nitrate-bearing earth, known as *caliche* in brines from old sea deposits, and in brackish waters from oil and salt wells. Ultrapure iodine can be obtained from the reaction of potassium iodide with copper sulfate. Several other methods of isolating the element are known. Iodine is a bluish-black, lustrous solid, volatilizing at ordinary temperatures into a blue-violet gas with an irritating odor; it forms compounds with many elements, but is less active than the other halogens, which displace it from iodides. Iodine exhibits some metallic-like properties. It dissolves readily in chloroform, carbon tetrachloride, or carbon disulfide to form beautiful purple solutions. It is only slightly soluble in water. Iodine compounds are important in organic chemistry and very useful in medicine. Forty two isotopes and isomers are recognized. Only one stable isotope, ^{127}I is found in nature. The artificial radioisotope ^{131}I, with a half-life of 8 days, has been used in treating the thyroid gland. The most common compounds are the iodides of sodium and potassium (KI) and the iodates (KIO_3). Lack of iodine is the cause of goiter. Iodides, and thyroxin which contains iodine, are used internally in medicine, and a solution of KI and iodine in alcohol is used for external wounds. Potassium iodide finds use in photography. The deep blue color with starch solution is characteristic of the free element. Care should be taken in handling and using iodine, as contact with the skin can cause lesions; iodine vapor is intensely irritating to the eyes and mucous membranes. Elemental iodine costs about 25 to 75¢/g depending on purity and quantity.

Iridium — (L. *iris*, rainbow), Ir; at. wt. 192.217(3); at. no. 77; m.p. 2446°C; b.p. 4428°C; sp. gr. 22.42 (17°C); valence 3 or 4. Discovered in 1803 by Tennant in the residue left when crude platinum is dissolved by aqua regia. The name iridium is appropriate, for its salts are highly colored. Iridium, a metal of the platinum family, is white, similar to platinum, but with a slight yellowish cast. It is very hard and brittle, making it very hard to machine, form, or work. It is the most corrosion-resistant metal known, and was used in making the standard meter bar of Paris, which is a 90% platinum-10% iridium alloy. This meter bar was replaced in 1960 as a fundamental unit of length (see under Krypton). Iridium is not attacked by any of the acids nor by aqua regia, but is attacked by molten salts, such as NaCl and NaCN. Iridium occurs uncombined in nature with platinum and other metals of this family in alluvial deposits. It is recovered as a by-product from the nickel mining industry. The largest reserves and production of the platinum group of metals, which includes iridium, is in South Africa, followed by Russia and Canada. The U.S. has only one active mine, located at Nye, MT. The presence of iridium has recently been used in examining the Cretaceous-Tertiary (K-T) boundary. Meteorites contain small amounts of iridium. Because iridium is found widely distributed at the K-T boundary, it has been suggested that a large meteorite or asteroid collided with the earth, killing the dinosaurs, and creating a large dust cloud and crater. Searches for such a crater point to one in the Yucatan, known as Chicxulub. Iridium has found use in making crucibles and apparatus for use at high temperatures. It is also used for electrical contacts. Its principal use is as a hardening agent for platinum. With osmium, it forms an alloy which is used for tipping pens and compass bearings. The specific gravity of iridium is only very slightly lower than that of osmium, which has been generally credited as being the heaviest known element. Calculations of the densities of iridium and osmium from the space lattices gives values of 22.65 and 22.61 g/cm^3, respectively. These values may be more reliable than actual physical measurements. At present, therefore, we know that either iridium or osmium is the densest known element, but the data do not yet allow selection between the two. Natural iridium contains two stable isotopes. Forty-five other isotopes, all radioactive, are now recognized. Iridium (99.9%) costs about $100/g.

Iron — (Anglo-Saxon, *iron*), Fe (L. *ferrum*); at. wt. 55.845(2); at. no. 26; m.p. 1538°C; b.p. 2861°C; sp. gr. 7.874 (20°C); valence 2, 3, 4, or 6. The use of iron is prehistoric. Genesis mentions that Tubal-Cain, seven generations from Adam, was "an instructor of every artificer in brass and iron." A remarkable iron pillar, dating to about A.D. 400, remains standing today in Delhi, India. This solid shaft of wrought iron is about $7^1/_4$ m high by 40 cm in diameter. Corrosion to the pillar has been minimal although it has been exposed to the weather since its erection. Iron is a relatively abundant element in the universe. It is found in the sun and many types of stars in considerable quantity. Its nuclei are very stable. It has been suggested that the iron we have here on earth may have originated in a super-nova. Iron is a very difficult element to produce in ordinary nuclear reactions, such as would take place in the sun. Iron is found native as a principal component of a class of iron-nickel meteorites known as *siderites*, and is a minor constituent of the other two classes of meteorites. The core of the earth, 2150 miles in radius, is thought to be largely composed of iron with about 10% occluded hydrogen. The metal is the fourth most abundant element, by weight, making up the crust of the earth. The most common ore is *hematite* (Fe_2O_3). Magnetite (Fe_3O_4) is frequently seen as *black sands* along beaches and banks of streams. *Lodestone* is another form of magnetite. *Taconite* is becoming increasingly important as a commercial ore. Iron is a vital constituent of plant and animal life, and appears in hemoglobin. The pure metal is not often encountered in commerce, but is usually alloyed with carbon or other metals. The pure metal is very reactive chemically, and rapidly corrodes, especially in moist air or at elevated temperatures. It has four allotropic forms, or ferrites, known as α, β, γ, and δ, with transition points at 700, 928, and 1530°C. The α form is magnetic, but when transformed into the β form, the magnetism disappears although the lattice remains unchanged. The relations of these forms are peculiar. Pig iron is an alloy containing about 3% carbon with varying amounts of S, Si, Mn, and P. It is hard, brittle, fairly fusible, and is used to produce other alloys, including steel. Wrought iron contains only a few tenths of a percent of carbon, is tough, malleable, less fusible, and has usually a "fibrous" structure. Carbon steel is an alloy of iron with carbon, with small amounts of Mn, S, P, and Si. Alloy steels are carbon steels with other additives such as nickel, chromium, vanadium, etc. Iron is the cheapest and most abundant, useful, and important of all metals. Natural iron contains four isotopes and isomers. Twenty-six other isotopes and isomers, all radioactive, are now recognized.

Krypton — (Gr. *kryptos*, hidden), Kr; at. wt. 83.80(1); at. no. 36; m.p. –157.36°C; b.p. –153.22 ± 0.10°C; t_c –63.74°C; density 3.733 g/l (0°C); valence usually 0. Discovered in 1898 by Ramsay and Travers in the residue left after liquid air had nearly boiled away. Krypton is present in the air to the extent of about 1 ppm. The atmosphere of Mars has been found to contain 0.3 ppm of krypton. It is one of the "noble" gases. It is characterized by its brilliant green and orange spectral lines. Naturally occurring krypton contains six stable isotopes. Thirty other unstable isotopes and isomers are now recognized. The spectral lines of krypton are easily produced and some are very sharp. In 1960 it was internationally agreed that the fundamental unit of length, the meter, should be defined in terms of the orange-red spectral line of ^{86}Kr. This replaced the standard meter of Paris, which was defined in terms of a bar made of a platinum-iridium alloy. In October 1983 the meter, which originally was defined as being one ten millionth of a quadrant of the earth's polar circumference, was again redefined by the International Bureau of Weights and Measures as being the length of path traveled by light in a vacuum during a time interval of 1/299,792,458 of a second. Solid krypton is a white crystalline substance with a face-centered cubic structure which is common to all the "rare gases". While krypton is generally thought of as a rare gas that normally does not combine with other elements to form compounds, it now appears that the existence of some krypton compounds is established. Krypton difluoride has been prepared in gram quantities and can be made by several methods. A higher fluoride of krypton and a salt of an oxyacid of krypton also have been reported. Molecule-ions of $ArKr^+$ and KrH^+ have been identified and investigated, and evidence is provided for the formation of KrXe or $KrXe^+$. Krypton clathrates have been prepared with hydroquinone and phenol. ^{85}Kr has found recent application in chemical analysis. By imbedding the isotope in various solids, *kryptonates* are formed. The activity of these kryptonates is sensitive to chemical reactions at the surface. Estimates of the concentration of reactants are therefore made possible. Krypton is used in certain photographic flash lamps for high-speed photography. Uses thus far have been limited because of its high cost. Krypton gas presently costs about $690/100 l.

Kurchatovium — See Rutherfordium.

Lanthanum — (Gr. *lanthanein*, to lie hidden), La; at. wt. 138.9055(2); at. no. 57; m.p. 918°C; b.p. 3464°C; sp. gr. 6.145 (25°C); valence 3. Mosander in 1839 extracted a new earth *lanthana*, from impure cerium nitrate, and recognized the new element. Lanthanum is found in rare-earth minerals such as *cerite, monazite, allanite,* and *bastnasite*. Monazite and bastnasite are principal ores in which lanthanum occurs in percentages up to 25 and 38%, respectively. Misch metal, used in making lighter flints, contains about 25% lanthanum. Lanthanum was isolated in relatively pure form in 1923. Ion-exchange and solvent extraction techniques have led to much easier isolation of the so-called "rare-earth" elements. The availability of lanthanum and other rare earths has improved greatly in recent years. The metal can be produced by reducing the anhydrous fluoride with calcium. Lanthanum is silvery white, malleable, ductile, and soft enough to be cut with a knife. It is one of the most reactive of the rare-earth metals. It oxidizes rapidly when exposed to air. Cold water attacks lanthanum slowly, and hot water attacks it much more rapidly. The metal reacts directly with elemental carbon, nitrogen, boron, selenium, silicon, phosphorus, sulfur, and with halogens. At 310°C, lanthanum changes from a hexagonal to a face-centered cubic structure, and at 865°C it again transforms into a body-centered cubic structure. Natural lanthanum is mixture of two isotopes, one of which is stable and one of which is radioactive with a very long half-life. Thirty other radioactive isotopes are recognized. Rare-earth compounds containing lanthanum are extensively used in carbon lighting applications, especially by the motion picture industry for studio lighting and projection. This application consumes about 25% of the rare-earth compounds produced. La_2O_3 improves the alkali resistance of glass, and is used in making special optical glasses. Small amounts of lanthanum, as an additive, can be used to produce nodular cast iron. There is current interest in hydrogen sponge alloys containing lanthanum. These alloys take up to 400 times their own volume of hydrogen gas, and the process is reversible. Heat energy is released every time they do so; therefore these alloys have possibilities in energy conservation systems. Lanthanum and its compounds have a low to moderate acute toxicity rating; therefore, care should be taken in handling them. The metal costs about $2/g (99.9%).

Lawrencium — (Ernest O. Lawrence [1901–1958], inventor of the cyclotron), Lr; at. no. 103; at. mass no. [262]; valence + 3(?). This member of the 5f transition elements (actinide series) was discovered in March 1961 by A. Ghiorso, T. Sikkeland, A. E. Larsh, and R. M. Latimer. A 3-μg californium target, consisting of a mixture of isotopes of mass number 249, 250, 251, and 252, was bombarded with either ^{10}B or ^{11}B. The electrically charged transmutation nuclei recoiled with an atmosphere of helium and were collected on a thin copper conveyor tape which was then moved to place collected atoms in front of a series of solid-state detectors. The isotope of element 103 produced in this way decayed by emitting an 8.6-MeV alpha particle with a half-life of 8 s. In 1967, Flerov and associates of the Dubna Laboratory reported their inability to detect an alpha emitter with a half-life of 8 s which was assigned by the Berkeley group to $^{257}103$. This assignment has been changed to ^{258}Lr or ^{259}Lr. In 1965, the Dubna workers found a longer-lived lawrencium isotope, ^{256}Lr, with a half-life of 35 s. In 1968, Ghiorso and associates at Berkeley were able to use a few atoms of this isotope to study the oxidation behavior of lawrencium. Using solvent extraction techniques and working very rapidly, they extracted lawrencium ions from a buffered aqueous solution into an organic solvent, completing each extraction in about 30 s. It was found that lawrencium behaves differently from dipositive nobelium and more like the tripositive elements earlier in the actinide series. Ten isotopes of lawrencium are now recognized.

Lead — (Anglo-Saxon *lead*), Pb (L. *plumbum*); at. wt. 207.2(1); at. no. 82; m.p. 327.46°C; b.p. 1749°C; sp. gr. 11.35 (20°C); valence 2 or 4. Long known, mentioned in Exodus. The alchemists believed lead to be the oldest metal and associated it with the planet Saturn. Native lead occurs in nature, but it is rare. Lead is obtained chiefly from *galena* (PbS) by a roasting process. *Anglesite* ($PbSO_4$), *cerussite* ($PbCO_3$), and *minim* (Pb_3O_4) are other common lead minerals. Lead is a bluish-white metal of bright luster, is very soft, highly malleable, ductile, and a poor conductor of electricity. It is very resistant to corrosion; lead pipes bearing the insignia of Roman emperors, used as drains from the baths, are still in service. It is used in containers for corrosive liquids (such as sulfuric acid) and may be toughened by the addition of a small percentage of antimony or other metals. Natural lead is a mixture of four stable isotopes: ^{204}Pb (1.4%), ^{206}Pb (24.1%), ^{207}Pb (22.1%), and ^{208}Pb (52.4%). Lead isotopes are the end products of each of the three series of naturally occurring radioactive elements: ^{206}Pb for the uranium series, ^{207}Pb for the actinium series, and ^{208}Pb for the thorium series. Forty-three other isotopes of lead, all of which are radioactive, are recognized. Its alloys include solder, type metal, and various antifriction metals. Great quantities of lead, both as the metal and as the dioxide, are used in storage batteries. Lead is also used for cable covering, plumbing, and ammunition. The metal is very effective as a sound absorber, is used as a radiation shield around X-ray equipment and nuclear reactors, and is used to absorb vibration. Lead, alloyed with tin, is used in making organ pipes. White lead, the basic carbonate, sublimed white lead ($PbSO_4$), chrome yellow ($PbCrO_4$), red lead (Pb_3O_4), and other lead compounds are used extensively in paints, although in recent years the use of lead in paints has been drastically curtailed to eliminate or reduce health hazards. Lead oxide is used in producing fine "crystal glass" and "flint glass" of a high index of refraction for achromatic lenses. The nitrate and the acetate are soluble salts. Lead salts such as lead arsenate have been used as insecticides, but their use in recent years has been practically eliminated in favor of less harmful organic compounds. Care must be used in handling lead as it is a cumulative poison. Environmental concern with lead poisoning has resulted in a national program to eliminate the lead tetraethyl in gasoline. The U.S. Occupational Safety and Health Administration (OSHA) has recommended that industries limit airborne lead to 50 μgms/cu. meter. Lead is priced at about 90¢/kg (99.9%).

Lithium — (Gr. *lithos*, stone), Li; at. wt. 6.941(2); at. no. 3; m.p. 180.5°C; b.p. 1342°C; sp. gr. 0.534 (20°C); valence 1. Discovered by Arfvedson in 1817. Lithium is the lightest of all metals, with a density only about half that of water. It does not occur free in nature; combined it is found in small amounts in nearly all igneous rocks and in the waters of many mineral springs. *Lepidolite, spodumene, petalite,* and *amblygonite* are the more important minerals containing it. Lithium is presently being recovered from brines of Searles Lake, in California, and from Nevada, Chile, and Argentina. Large deposits of spodumene are found in North Carolina. The metal is produced electrolytically from the fused chloride. Lithium is silvery in appearance,

THE ELEMENTS (continued)

much like Na and K, other members of the alkali metal series. It reacts with water, but not as vigorously as sodium. Lithium imparts a beautiful crimson color to a flame, but when the metal burns strongly the flame is a dazzling white. Since World War II, the production of lithium metal and its compounds has increased greatly. Because the metal has the highest specific heat of any solid element, it has found use in heat transfer applications; however, it is corrosive and requires special handling. The metal has been used as an alloying agent, is of interest in synthesis of organic compounds, and has nuclear applications. It ranks as a leading contender as a battery anode material as it has a high electrochemical potential. Lithium is used in special glasses and ceramics. The glass for the 200-inch telescope at Mt. Palomar contains lithium as a minor ingredient. Lithium chloride is one of the most hygroscopic materials known, and it, as well as lithium bromide, is used in air conditioning and industrial drying systems. Lithium stearate is used as an all-purpose and high-temperature lubricant. Other lithium compounds are used in dry cells and storage batteries. Seven isotopes of lithium are recognized. Natural lithium contains two isotopes. The metal is priced at about $1.50/g (99.9%).

Lutetium — (Lutetia, ancient name for Paris, sometimes called *cassiopeium* by the Germans), Lu; at. wt. 174.967(1); at. no. 71; m.p. 1663°C; b.p. 3402°C; sp. gr. 9.841 (25°C); valence 3. In 1907, Urbain described a process by which Marignac's ytterbium (1879) could be separated into the two elements, ytterbium (neoytterbium) and lutetium. These elements were identical with "aldebaranium" and "cassiopeium," independently discovered by von Welsbach about the same time. Charles James of the University of New Hampshire also independently prepared the very pure oxide, *lutecia*, at this time. The spelling of the element was changed from *lutecium* to *lutetium* in 1949. Lutetium occurs in very small amounts in nearly all minerals containing yttrium, and is present in *monazite* to the extent of about 0.003%, which is a commercial source. The pure metal has been isolated only in recent years and is one of the most difficult to prepare. It can be prepared by the reduction of anhydrous $LuCl_3$ or LuF_3 by an alkali or alkaline earth metal. The metal is silvery white and relatively stable in air. While new techniques, including ion-exchange reactions, have been developed to separate the various rare-earth elements, lutetium is still the most costly of all rare earths. It is priced at about $100/g (99.9%). ^{176}Lu occurs naturally (97.41%) with ^{175}Lu (2.59%), which is radioactive with a very long half-life. It is radioactive with a half-life of about 4×10^{10} years. Lutetium has 50 isotopes and isomers that are now recognized. Stable lutetium nuclides, which emit pure beta radiation after thermal neutron activation, can be used as catalysts in cracking, alkylation, hydrogenation, and polymerization. Virtually no other commercial uses have been found yet for lutetium. While lutetium, like other rare-earth metals, is thought to have a low toxicity rating, it should be handled with care until more information is available.

Magnesium — (*Magnesia,* district in Thessaly) Mg; at. wt. 24.3050(6); at. no. 12; m.p. 650°C; b.p. 1090°C; sp. gr. 1.738 (20°C); valence 2. Compounds of magnesium have long been known. Black recognized magnesium as an element in 1755. It was isolated by Davy in 1808, and prepared in coherent form by Bussy in 1831. Magnesium is the eighth most abundant element in the earth's crust. It does not occur uncombined, but is found in large deposits in the form of *magnesite, dolomite,* and other minerals. The metal is now principally obtained in the U.S. by electrolysis of fused magnesium chloride derived from brines, wells, and sea water. Magnesium is a light, silvery-white, and fairly tough metal. It tarnishes slightly in air, and finely divided magnesium readily ignites upon heating in air and burns with a dazzling white flame. It is used in flashlight photography, flares, and pyrotechnics, including incendiary bombs. It is one third lighter than aluminium, and in alloys is essential for airplane and missile contruction. The metal improves the mechanical, fabrication, and welding characteristics of aluminum when used as an alloying agent. Magnesium is used in producing nodular graphite in cast iron, and is used as an additive to conventional propellants. It is also used as a reducing agent in the production of pure uranium and other metals from their salts. The hydroxide (*milk of magnesia*), chloride, sulfate (*Epsom salts*), and citrate are used in medicine. Dead-burned magnesite is employed for refractory purposes such as brick and liners in furnaces and converters. Calcined magnesia is also used for water treatment and in the manufacture of rubber, paper, etc. Organic magnesium compounds (Grignard's reagents) are important. Magnesium is an important element in both plant and animal life. Chlorophylls are magnesium-centered porphyrins. The adult daily requirement of magnesium is about 300 mg/day, but this is affected by various factors. Great care should be taken in handling magnesium metal, especially in the finely divided state, as serious fires can occur. Water should not be used on burning magnesium or on magnesium fires. Natural magnesium contains three isotopes. Twelve other isotopes are recognized. Magnesium metal costs about $110/kg (99.8%).

Manganese — (L. *magnes*, magnet, from magnetic properties of pyrolusite; It. *manganese*, corrupt form of *magnesia*), Mn; at. wt. 54.938049(9); at. no. 25; m.p. 1246°C; b.p. 2061°C; sp. gr. 7.21 to 7.44, depending on allotropic form; valence 1, 2, 3, 4, 6, or 7. Recognized by Scheele, Bergman, and others as an element and isolated by Gahn in 1774 by reduction of the dioxide with carbon. Manganese minerals are widely distributed; oxides, silicates, and carbonates are the most common. The discovery of large quantities of manganese nodules on the floor of the oceans holds promise as a source of manganese. These nodules contain about 24% manganese together with many other elements in lesser abundance. Most manganese today is obtained from ores found in the Ukraine, Brazil, Australia, Republic of So. Africa, Gabon, China, and India. *Pyrolusite* (MnO_2) and *rhodochrosite* ($MnCO_3$) are among the most common manganese minerals. The metal is obtained by reduction of the oxide with sodium, magnesium, aluminum, or by electrolysis. It is gray-white, resembling iron, but is harder and very brittle. The metal is reactive chemically, and decomposes cold water slowly. Manganese is used to form many important alloys. In steel, manganese improves the rolling and forging qualities, strength, toughness, stiffness, wear resistance, hardness, and hardenability. With aluminum and antimony, especially with small amounts of copper, it forms highly ferromagnetic alloys. Manganese metal is ferromagnetic only after special treatment. The pure metal exists in four allotropic forms. The alpha form is stable at ordinary temperature; gamma manganese, which changes to alpha at ordinary temperatures, is said to be flexible, soft, easily cut, and capable of being bent. The dioxide (pyrolusite) is used as a depolarizer in dry cells, and is used to "decolorize" glass that is colored green by impurities of iron. Manganese by itself colors glass an amethyst color, and is responsible for the color of true amethyst. The dioxide is also used in the preparation of oxygen and chlorine, and in drying black paints. The permanganate is a powerful oxidizing agent and is used in quantitative analysis and in medicine. Manganese is widely distributed throughout the animal kingdom. It is an important trace element and may be essential for utilization of vitamin B_1. Twenty-seven isotopes and isomers are known. Manganese metal (99.95%) is priced at about $400/kg. Metal of 99.6% purity is priced at about $60/kg.

Meitnerium — (named for Lise Meitner [1878–1968], Austrian-Swedish physicist and mathematician), Mt; at. wt [266]; at. no. 109. On August 29, 1992, Element 109 was made and identified by physicists at the Heavy Ion Research Laboratory (G.S.I.), Darmstadt, Germany, by bombarding a target of ^{209}Bi with accelerated nuclei of ^{58}Fe. The production of Element 109 has been extremely small. It took a week of target bombardment (10^{11} nuclear encounters) to produce a single atom of 109. Oganessian and his team at Dubna in 1994 repeated the Darmstadt experiment using a tenfold irradiation dose. One fission event from seven alpha decays of 109 was observed, thus indirectly confirming the existence of isotope ^{266}109. In August 1997, the IUPAC adopted the name *meitnerium* for this element, honoring L. Meitner. Four isotopes of *meitnerium* are now recognized.

Mendelevium — (Dmitri Mendeleev [1834–1907]), Md; at. wt. (258); at. no. 101; m.p. 827°C; valence +2, +3. Mendelevium, the ninth transuranium element of the actinide series to be discovered, was first identified by Ghiorso, Harvey, Choppin, Thompson, and Seaborg early in 1955 as a result of the bombardment of the isotope ^{253}Es with helium ions in the Berkeley 60-inch cyclotron. The isotope produced was ^{256}Md, which has a half-life of 78 min. This first identification was notable in that ^{256}Md was synthesized on a one-atom-at-a-time basis. Nineteen isotopes and isomers are now recognized. ^{258}Md has a half-life of 51.5 days. This isotope has been produced by the bombardment of an isotope of einsteinium with ions of helium. It now appears possible that eventually enough ^{258}Md can be made so that some of its physical properties can be determined. ^{256}Md has been used to elucidate some of the chemical properties of mendelevium in aqueous solution. Experiments seem to show that the element possesses a moderately stable dipositive (II) oxidation state in addition to the tripositive (III) oxidation state, which is characteristic of actinide elements.

Mercury — (Planet *Mercury*), Hg (*hydrargyrum*, liquid silver); at. wt. 200.59(2); at. no. 80; t.p. –38.83°C; b.p. 356.73°C; t_c 1447°C; sp. gr. 13.546 (20°C); valence 1 or 2. Known to ancient Chinese and Hindus; found in Egyptian tombs of 1500 B.C. Mercury is the only common metal liquid at ordinary temperatures. It only rarely occurs free in nature. The chief ore is *cinnabar* (HgS). Spain and China produce about 75% of the world's supply of the metal. The commercial unit for handling mercury is the "flask," which weighs 76 lb (34.46 kg). The metal is obtained by heating cinnabar in a current of air and by condensing the vapor. It is a heavy, silvery-white metal; a rather poor conductor of heat, as compared with other metals, and a fair conductor of electricity. It easily forms alloys with many metals, such as gold, silver, and tin, which are called *amalgams*. Its ease in amalgamating with gold is made use of in the recovery of gold from its ores. The metal is widely used in laboratory work for making thermometers, barometers, diffusion pumps, and many other instruments. It is used in making mercury-vapor lamps and advertising signs, etc. and is used in mercury switches and other electrical apparatus. Other uses are in making pesticides, mercury cells for caustic soda and chlorine production, dental preparations, antifouling paint, batteries, and catalysts. The most important salts are mercuric chloride $HgCl_2$ (corrosive sublimate — a violent poison), mercurous chloride Hg_2Cl_2 (calomel, occasionally still used in medicine), mercury fulminate $(Hg(ONC)_2)$, a detonator widely used in explosives, and mercuric sulfide (HgS, vermillion, a high-grade paint pigment). Organic mercury compounds are important. It has been found that an electrical discharge causes mercury vapor to combine with neon, argon, krypton, and xenon. These products, held together with van der Waals' forces, correspond to HgNe, HgAr, HgKr, and HgXe. Mercury is a virulent poison and is readily absorbed through the respiratory tract, the gastrointestinal tract, or through unbroken skin. It acts as a cumulative poison and dangerous levels are readily attained in air. Air saturated with mercury vapor at 20°C contains a concentration that exceeds the toxic limit many times. The danger increases at higher temperatures. *It is therefore important that mercury be handled with care.* Containers of mercury should be securely covered and spillage should be avoided. If it is necessary to heat mercury or mercury compounds, it should be done in a well-ventilated hood. Methyl mercury is a dangerous pollutant and is now widely found in water and streams. The triple point of mercury, – 38.8344°C, is a fixed point on the International Temperature Scale (ITS-90). Mercury (99.98%) is priced at about $110/kg. Native mercury contains seven isotopes. Thirty-six other isotopes and isomers are known.

Molybdenum — (Gr. *molybdos*, lead), Mo; at. wt. 95.94(1); at. no. 42; m.p. 2623°C; b.p. 4639°C; sp. gr. 10.22 (20°C); valence 2, 3, 4?, 5?, or 6. Before Scheele recognized molybdenite as a distinct ore of a new element in 1778, it was confused with graphite and lead ore. The metal was prepared in an impure form in 1782 by Hjelm. Molybdenum does not occur native, but is obtained principally from *molybdenite* (MoS_2). *Wulfenite* $(PbMoO_4)$ and *powellite* $(Ca(MoW)O_4)$ are also minor commercial ores. Molybdenum is also recovered as a by-product of copper and tungsten mining operations. The U.S., Canada, Chile, and China produce most of the world's molybdenum ores. The metal is prepared from the powder made by the hydrogen reduction of purified molybdic trioxide or ammonium molybdate. The metal is silvery white, very hard, but is softer and more ductile than tungsten. It has a high elastic modulus, and only tungsten and tantalum, of the more readily available metals, have higher melting points. It is a valuable alloying agent, as it contributes to the hardenability and toughness of quenched and tempered steels. It also improves the strength of steel at high temperatures. It is used in certain nickel-based alloys, such as the "Hastelloys®" which are heat-resistant and corrosion-resistant to chemical solutions. Molybdenum oxidizes at elevated temperatures. The metal has found recent application as electrodes for electrically heated glass furnaces and forehearths. The metal is also used in nuclear energy applications and for missile and aircraft parts. Molybdenum is valuable as a catalyst in the refining of petroleum. It has found application as a filament material in electronic and electrical applications. Molybdenum is an essential trace element in plant nutrition. Some lands are barren for lack of this element in the soil. Molybdenum sulfide is useful as a lubricant, especially at high temperatures where oils would decompose. Almost all ultra-high strength steels with minimum yield points up to 300,000 psi(lb/in.²) contain molybdenum in amounts from 0.25 to 8%. Natural molybdenum contains seven isotopes. Thirty other isotopes and isomers are known, all of which are radioactive. Molybdenum metal costs about $1/g (99.999% purity). Commercial molybdenum metal (99.9%) costs about $300/kg.

Neodymium — (Gr. *neos*, new, and *didymos*, twin), Nd; at. wt. 144.24(3); at. no. 60; m.p. 1021°C; b.p. 3074°C; sp. gr. 7.008 (25°C); valence 3. In 1841, Mosander, extracted from *cerite* a new rose-colored oxide, which he believed contained a new element. He named the element *didymium*, as it was *an inseparable twin brother of lanthanum*. In 1885 von Welsbach separated didymium into two new elemental components, *neodymia* and *praseodymia*, by repeated fractionation of ammonium didymium nitrate. While the free metal is in *misch metal*, long known and used as a pyrophoric alloy for light flints, the element was not isolated in relatively pure form until 1925. Neodymium is present in misch metal to the extent of about 18%. It is present in the minerals *monazite* and *bastnasite*, which are principal sources of rare-earth metals. The element may be obtained by separating neodymium salts from other rare earths by ion-exchange or solvent extraction techniques, and by reducing anhydrous halides such as NdF_3 with calcium metal. Other separation techniques are possible. The metal has a bright silvery metallic luster. Neodymium is one of the more reactive rare-earth metals and quickly tarnishes in air, forming an oxide that spalls off and exposes metal to oxidation. The metal, therefore, should be kept under light mineral oil or sealed in a plastic material. Neodymium exists in two allotropic forms, with a transformation from a double hexagonal to a body-centered cubic structure taking place at 863°C. Natural neodymium is a mixture of seven isotopes, one of which has a very long half-life. Twenty seven other radioactive isotopes and isomers are recognized. Didymium, of which neodymium is a component, is used for coloring glass to make welder's goggles. By itself, neodymium colors glass delicate shades ranging from pure violet through wine-red and warm gray. Light transmitted through such glass shows unusually sharp absorption bands. The glass has been used in astronomical work to produce sharp bands by which spectral lines may be calibrated. Glass containing neodymium can be used as a laser material to produce coherent light. Neodymium salts are also used as a colorant for enamels. The element is also being used with iron and boron to produce extremely strong magnets having energy densities as high as 27 to 35 million gauss oersteds. These are the most compact magnets commercially available. The price of the metal is about $4/g. Neodymium has a low-to-moderate acute toxic rating. As with other rare earths, neodymium should be handled with care.

Neon — (Gr. *neos*, new), Ne; at. wt. 20.1797(6); at. no. 10; t.p. –248.59°C; b.p. –246.08°C; t_c -228.7°C (1 atm); density of gas 0.89990 g/l (1 atm, 0°C); density of liquid at b.p. 1.207 g/cm³; valence 0. Discovered by Ramsay and Travers in 1898. Neon is a rare gaseous element present in the atmosphere to the extent of 1 part in 65,000 of air. It is obtained by liquefaction of air and separated from the other gases by fractional distillation. Natural neon is a mixture of three isotopes. Fourteen other unstable isotopes are known. It is very inert element; however, it is said to form a compound with fluorine. It is still questionable if true compounds of neon exist, but evidence is mounting in favor of their existence. The following ions are known from optical and mass spectrometric studies: Ne^+, $(NeAr)^+$, $(NeH)^+$, and $(HeNe^+)$. Neon also forms an unstable hydrate. In a vacuum discharge tube, neon glows reddish orange. Of all the rare gases, the discharge of neon is the most intense at ordinary voltages and currents. Neon is used in making the common neon advertising signs, which accounts for its largest use. It is also used to make high-voltage indicators, lightning arrestors, wave meter tubes, and TV tubes. Neon and helium are used in making gas lasers. Liquid neon is now commercially available and is finding important application as an economical cryogenic refrigerant. It has over 40 times more refrigerating capacity per unit volume than liquid helium and more than three times that of liquid hydrogen. It is compact, inert, and is less expensive than helium when it meets refrigeration requirements. Neon costs about $800/80 cu. ft. (2265 l).

Neptunium — (Planet *Neptune*), Np; at. wt. (237); at. no. 93; m.p. 644°C; sp. gr. 20.25 (20°C); valence 3, 4, 5, and 6. Neptunium was the first synthetic transuranium element of the actinide series discovered; the isotope ^{239}Np was produced by McMillan and Abelson in 1940 at Berkeley, California, as the result of bombarding uranium with cyclotron-produced neutrons. The isotope ^{237}Np (half-life of 2.14×10^6 years) is currently obtained in gram quantities as a by-product from nuclear reactors in the production of plutonium. Twenty-three isotopes and isomers of neptunium are now recognized. Trace quantities of the element are actually found in nature due to transmutation reactions in uranium ores produced by the neutrons which are present. Neptunium is prepared by the reduction of NpF_3 with barium or lithium vapor at about 1200°C. Neptunium metal has a silvery appearance, is chemically reactive, and exists in at least three structural modifications: α-neptunium, orthorhombic, density 20.25 g/cm³, β-neptunium (above 280°C), tetragonal, density (313°C) 19.36 g/cm³; γ-neptunium (above 577°C), cubic, density (600°C) 18.0 g/cm³. Neptunium has four ionic oxidation states in solution: Np^{+3} (pale purple), analogous to the rare earth ion Pm^{+3}, Np^{+4} (yellow green); NpO^+ (green blue); and NpO^{++} (pale pink). These latter oxygenated species are in contrast to the rare earths which exhibit only simple ions of the (II), (III), and (IV) oxidation states in aqueous solution. The element forms tri- and tetrahalides such as NpF_3, NpF_4, $NpCl_4$, $NpBr_3$, NpI_3, and oxides of various compositions such as are found in the uranium-oxygen system, including Np_3O_8 and NpO_2.

Nickel — (Ger. *Nickel*, Satan or Old Nick's and from *kupfernickel*, Old Nick's copper), Ni; at. wt. 58.6934(2); at. no. 28; m.p. 1455°C; b.p. 2913°C; sp. gr. 8.902 (25°C); valence 0, 1, 2, 3. Discovered by Cronstedt in 1751 in kupfernickel (*niccolite*). Nickel is found as a constituent in most meteorites and often serves as one of the criteria for distinguishing a meteorite from other minerals. Iron meteorites, or *siderites*, may contain iron alloyed with from 5 to nearly 20% nickel. Nickel is obtained commercially from *pentlandite* and *pyrrhotite* of the Sudbury region of Ontario, a district that produces much of the world's nickel. It is now thought that the Sudbury deposit is the result of an ancient meteorite impact. Large deposits of nickel, cobalt, and copper have recently been developed at Voisey's Bay, Laborador. Other deposits of nickel are found in Russia, New Caledonia, Australia, Cuba, Indonesia, and elsewhere. Nickel is silvery white and takes on a high polish. It is hard, malleable, ductile, somewhat ferromagnetic, and a fair conductor of heat and electricity. It belongs to the iron-cobalt group of metals and is chiefly valuable for the alloys it forms. It is extensively used for making stainless steel and other corrosion-resistant alloys such as Invar®, Monel®, Inconel®, and the Hastelloys®. Tubing made of a copper-nickel alloy is extensively used in making desalination plants for converting sea water into fresh water. Nickel is also now used extensively in coinage and in making nickel steel for armor plate and burglar-proof vaults, and is a component in Nichrome®, Permalloy®, and constantan. Nickel added to glass gives a green color. Nickel plating is often used to provide a protective coating for other metals, and finely divided nickel is a catalyst for hydrogenating vegetable oils. It is also used in ceramics, in the manufacture of Alnico magnets, and in the Edison® storage battery. The sulfate and the oxides are important compounds. Natural nickel is a mixture of five stable isotopes; twenty-five other unstable isotopes are known. Nickel sulfide fume and dust is recognized as having carcinogenic potential. Nickel metal (99.9%) is priced at about $1.25/g or less in larger quantities.

Nielsbohrium — See Bohrium.

Niobium — (*Niobe*, daughter of Tantalus), Nb; or Columbium (*Columbia*, name for America); at. wt. 92.90638(2); at. no. 41; m.p. 2477°C; b.p. 4744°C, sp. gr. 8.57 (20°C); valence 2, 3, 4?, 5. Discovered in 1801 by Hatchett in an ore sent to England more that a century before by John Winthrop the Younger, first governor of Connecticut. The metal was first prepared in 1864 by Blomstrand, who reduced the chloride by heating it in a hydrogen atmosphere. The name *niobium* was adopted by the International Union of Pure and Applied Chemistry in 1950 after 100 years of controversy. Many leading chemical societies and government organizations refer to it by this name. Most metallurgists, leading metal societies, and all but one of the leading U.S. commercial producers, however, still refer to the metal as "columbium". The element is found in *niobite*(or *columbite*), *niobite-tantalite*, *pyrochlore*, and *euxenite*. Large deposits of niobium have been found associated with *carbonatites* (carbon-silicate rocks), as a constituent of *pyrochlore*. Extensive ore reserves are found in Canada, Brazil, Congo-Kinshasa, Rwanda, and Australia. The metal can be isolated from tantalum, and prepared in several ways. It is a shiny, white, soft, and ductile metal, and takes on a bluish cast when exposed to air at room temperatures for a long time. The metal starts to oxidize in air at 200°C, and when processed at even moderate temperatures must be placed in a protective atmosphere. It is used in arc-welding rods for stabilized grades of stainless steel. Thousands of pounds of niobium have been used in advance air frame systems such as were used in the Gemini space program. It has also found use in super-alloys for applications such as jet engine components, rocket subassemblies, and heat-resisting equipment. The element has superconductive properties; superconductive magnets have been made with Nb-Zr wire, which retains its superconductivity in strong magnetic fields. This type of application offers hope of direct large-scale generation of electric power. Natural niobium is composed of only one isotope, ^{93}Nb. Forty-seven other isotopes and isomers of niobium are now recognized. Niobium metal (99.9% pure) is priced at about 50¢/g.

Nitrogen — (L. *nitrum*, Gr. *nitron*, native soda; genes, *forming*, N; at. wt. 14.00674(7); at. no. 7; m.p. –210.00°C; b.p. –198.79°C; t_c -146.94°C; density 1.2506 g/l; sp. gr. liquid 0.808 (–195.8°C), solid 1.026 (–252°C); valence 3 or 5. Discovered by Daniel Rutherford in 1772, but Scheele, Cavendish,Priestley, and others about the same time studied "burnt or dephlogisticated air," as air without oxygen was then called. Nitrogen makes up 78% of the air, by volume. The atmosphere of Mars, by comparison, is 2.6% nitrogen. The estimated amount of this element in our atmosphere is more than 4000 trillion tons. From this inexhaustible source it can be obtained by liquefaction and fractional distillation. Nitrogen molecules give the orange-red, blue-green, blue-violet, and deep violet shades to the aurora.The element is so inert that Lavoisier named it *azote*, meaning without life, yet its compounds are so active as to be most important in foods, poisons, fertilizers, and explosives. Nitrogen can be also easily prepared by heating a water solution of ammonium nitrite. Nitrogen, as a gas, is colorless, odorless, and a generally inert element. As a liquid it is also colorless and odorless, and is similar in appearance to water. Two allotropic forms of solid nitrogen exist, with the transition from the α to the β form taking place at –237°C. When nitrogen is heated, it combines directly with magnesium, lithium, or calcium; when mixed with oxygen and subjected to electric sparks, it forms first nitric oxide (NO) and then the dioxide (NO_2); when heated under pressure with a catalyst with hydrogen, ammonia is formed (Haber process). The ammonia thus formed is of the utmost importance as it is used in fertilizers, and it can be oxidized to nitric acid (Ostwald process). The ammonia industry is the largest consumer of nitrogen. Large amounts of gas are also used by the electronics industry, which uses the gas as a blanketing medium during production of such components as transistors, diodes, etc. Large quantities of nitrogen are used in annealing stainless steel and other steel mill products. The drug industry also uses large quantities. Nitrogen is used as a refrigerant both for the immersion freezing of food products and for transportation of foods. Liquid nitrogen is also used in missile work as a purge for components, insulators for space chambers, etc., and by the oil industry to build up great pressures in wells to force crude oil upward. Sodium and potassium nitrates are formed by the decomposition of organic matter with compounds of the metals present. In certain dry areas of the world these saltpeters are found in quantity. Ammonia, nitric acid, the nitrates, the five oxides (N_2O, NO, N_2O_3, NO_2, and N_2O_5), TNT, the cyanides, etc. are but a few of the important compounds. Nitrogen gas prices vary from 2¢ to $2.75 per 100 ft^3 (2.83 cu. meters), depending on purity, etc. Production of elemental nitrogen in the U.S. is more than 9 million short tons per year. Natural nitrogen contains two isotopes, ^{14}N and ^{15}N. Ten other isotopes are known.

Nobelium — (Alfred Nobel, discoverer of dynamite), No; at. wt. [259]; at. no. 102; valence +2, +3. Nobelium was unambiguously discovered and identified in April 1958 at Berkeley by A. Ghiorso, T. Sikkeland, J. R. Walton, and G. T. Seaborg, who used a new double-recoil technique. A heavy-ion linear accelerator (HILAC) was used to bombard a thin target of curium (95% ^{244}Cm and 4.5% ^{246}Cm) with ^{12}C ions to produce 102^{254} according to the ^{246}Cm (^{12}C, $4n$) reaction. Earlier in 1957 workers of the U.S., Britain, and Sweden announced the discovery of an isotope of Element 102 with a 10-min half-life at 8.5 MeV, as a result of bombarding ^{244}Cm with ^{13}C nuclei. On the basis of this experiment the name *nobelium* was assigned and accepted by the Commission on Atomic Weights of the International Union of Pure and Applied Chemistry. The acceptance of the name was premature, for both Russian and American efforts now completely rule out the possibility of any isotope of Element 102 having a half-life of 10 min in the vicinity of 8.5 MeV. Early work in 1957 on the search for this element, in Russia at the Kurchatov Institute, was marred by the assignment of 8.9 ± 0.4 MeV alpha radiation with a half-life of 2 to 40 sec, which was too indefinite to support claim to discovery. Confirmatory experiments at Berkeley in 1966 have shown the existence of $^{254}102$ with a 55-s half-life, $^{252}102$ with a 2.3-s half-life, and $^{257}102$ with a 25-s half-life. Twelve isotopes are now recognized, one of which — $^{255}102$ has a half-life of 3.1 min. In view of the discover's traditional right to name an element, the Berkeley group, in 1967, suggested that the hastily given name *nobelium*, along with the symbol No, be retained.

Osmium — (Gr. *osme*, a smell), Os; at. wt. 190.23(3); at. no. 76; m.p. 3033°C; b.p. 5012°C; sp. gr. 22.57; valence 0 to +8, more usually +3, +4, +6, and +8. Discovered in 1803 by Tennant in the residue left when crude platinum is dissolved by *aqua regia*. Osmium occurs in *iridosmine* and in platinum-bearing river sands of the Urals, North America, and South America. It is also found in the nickel-bearing ores of Sudbury, Ontario, region along with other platinum metals. While the quantity of platinum metals in these ores is very small, the large tonnages of nickel ores processed make commercial recovery possible. The metal is lustrous, bluish white, extremely hard, and brittle even at high temperatures. It has the highest melting point and the lowest vapor pressure of the platinum group. The metal is very difficult to fabricate, but the powder can be sintered in a hydrogen atmosphere at a temperature of 2000°C. The solid metal is not affected by air at room temperature, but the powdered or spongy metal slowly gives off osmium tetroxide, which is a powerful oxidizing agent and has a strong smell. The tetroxide is highly toxic, and boils at 130°C (760 mm). Concentrations in air as low as 10^{-7} g/m^3 can cause lung congestion, skin damage, or eye damage. The tetroxide has been used to detect fingerprints and to stain fatty tissue for microscope slides. The metal is almost entirely used to produce very hard alloys, with other metals of the platinum group, for fountain pen tips, instrument pivots, phonograph needles, and electrical contacts. The price of 99.9% pure osmium powder — the form usually supplied commercially — is about $100/g, depending on quantity and supplier. Natural osmium contains seven isotopes, one of which, ^{186}Os, is radioactive with a very long half-life. Thirty four other isotopes and isomers are known, all of which are radioactive. The measured densities of iridium and osmium seem to indicate that osmium is slightly more dense than iridium, so osmium has generally been credited with being the heaviest known element. Calculations of the density from the space lattice, which may be more reliable for these elements than actual measurements, however, give a density of 22.65 for iridium compared to 226.61 for osmium. At present, therefore, we know either iridium or osmium is the heaviest element, but the data do not allow selection between the two.

Oxygen — (Gr. *oxys*, sharp, acid, and *genes*, forming; acid former), O; at. wt. 15.9994(3); at. no. 8; t.p. –218.79°C; t_c -118.56°C; valence 2. For many centuries, workers occasionally realized air was composed of more than one component. The behavior of oxygen and nitrogen as components of air led to the advancement of the phlogiston theory of combustion, which captured the minds of chemists for a century. Oxygen was prepared by several workers, including Bayen and Borch, but they did not know how to collect it, did not study its properties, and did not recognize it as an elementary substance. Priestley is generally credited with its discovery, although Scheele also discovered it independently. Oxygen is the third most abundant element found in the sun, and it plays a part in the carbon-nitrogen cycle, one process thought to give the sun and stars their energy. Oxygen under excited conditions is responsible for the bright red and yellow-green colors of the aurora. Oxygen, as a gaseous element, forms 21% of the atmosphere by volume from which it can be obtained by liquefaction and fractional distillation. The atmosphere of Mars contains about 0.15% oxygen. The element and its compounds make up 49.2%, by weight, of the earth's crust. About two thirds of the human body and nine tenths of water is oxygen. In the laboratory it can be prepared by the electrolysis of water or by heating potassium chlorate with manganese dioxide as a catalyst. The gas is colorless, odorless, and tasteless. The liquid and solid forms are a pale blue color and are strongly paramagnetic. Ozone (O_3), a highly active compound, is formed by the action of an electrical discharge or ultraviolet light on oxygen. Ozone's presence in the atmosphere (amounting to the equivalent of a layer 3 mm thick at ordinary pressures and temperatures) is of vital importance in preventing harmful ultraviolet rays of the sun from reaching the earth's surface. There has been recent concern that pollutants in the atmosphere may have a detrimental effect on this ozone layer. Ozone is toxic and exposure should not exceed 0.2 mg/m^3 (8-hour time-weighted average — 40-hour work week). Undiluted ozone has a bluish color. Liquid ozone is bluish black, and solid ozone is violet-black. Oxygen is very reactive and capable of combining with most elements. It is a component of hundreds of thousands of organic compounds. It is essential for respiration of all plants and animals and for practically all combustion. In hospitals it is frequently used to aid respiration of patients. Its atomic weight was used as a standard of comparison for each of the other elements until 1961 when the International Union of Pure and Applied Chemistry adopted carbon 12 as the new basis. Oxygen has thirteen recognized isotopes. Natural oxygen is a mixture of three isotopes. Oxygen 18 occurs naturally, is stable, and is available commercially. Water (H_2O with 1.5% ^{18}O) is also available. Commercial oxygen consumption in the U.S. is estimated to be 20 million short tons per year and the demand is expected to increase substantially in the next few years. Oxygen enrichment of steel blast furnaces accounts for the greatest use of the gas. Large quantities are also used in making synthesis gas for ammonia and methanol, ethylene oxide, and for oxy-acetylene welding. Air separation plants produce about 99% of the gas, electrolysis plants about 1%. The gas costs 5¢/ft^3 ($1.75/cu. meters) in small quantities.

Palladium — (named after the asteroid *Pallas*, discovered about the same time; Gr. *Pallas*, goddess of wisdom), Pd. at. wt. 106.42(1) at. no. 46; m.p. 1554.9°C; b.p. 2963°C; sp. gr. 12.02 (20°C); valence 2, 3, or 4. Discovered in 1803 by Wollaston. Palladium is found along with platinum and other metals of the platinum group in deposits of Russia, South Africa, Canada (Ontario), and elsewhere. Natural palladium contains six stable isotopes. Twenty-nine other isotopes are recognized, all of which are radioactive. It is frequently found associated with the nickel-copper deposits such as those found in Ontario. Its separation from the platinum metals depends upon the type of ore in which it is found. It is a steel-white metal, does not tarnish in air, and is the least dense and lowest melting of the platinum group of metals. When annealed, it is soft and ductile; cold working greatly increases its strength and hardness. Palladium is attacked by nitric and sulfuric acid. At room temperatures the metal has the unusual property of absorbing up to 900 times its own volume of hydrogen, possibly forming Pd_2H. It is not yet clear if this a true compound. Hydrogen readily diffuses through heated palladium and this provides a means of purifying the gas. Finely divided palladium is a good catalyst and is used for hydrogenation and dehydrogenation reactions. It is alloyed and used in jewelry trades. White gold is an alloy of gold decolorized by the addition of palladium. Like gold, palladium can be beaten into leaf as thin as 1/250,000 in. The metal is used in dentistry, watchmaking, and in making surgical instruments and electrical contacts. Palladium recently has been substituted for higher priced platinum in catalytic converters by some automobile companies. This has caused a large increase in the cost of palladium. The price of the two metals are now, in 1999, about the same. Palladium, however, is less resistant to poisoning by sulfur and lead, than platinum, but it may prove useful in controlling emissions from diesel vehicles. The metal sells for about $425/tr. oz. ($13/g).

Phosphorus — (Gr. *phosphoros*, light bearing; ancient name for the planet Venus when appearing before sunrise), P; at. wt. 30.973762(4); at. no. 15; m.p. (white) 44.15°C; b.p. 280.5°C; sp. gr. (white) 1.82 (red) 2.20, (black) 2.25 to 2.69; valence 3 or 5. Discovered in 1669 by Brand, who prepared it from urine. Phosphorus exists in four or more allotropic forms: white (or yellow), red, and black (or violet). White phosphorus has two modifications: α and β with a transition temperature at −3.8°C. Never found free in nature, it is widely distributed in combination with minerals. Twenty-one isotopes of phosphorus are recognized. *Phosphate* rock, which contains the mineral *apatite*, an impure tri-calcium phosphate, is an important source of the element. Large deposits are found in the Russia,China, Morocco, and in Florida, Tennessee, Utah, Idaho, and elsewhere. Phosphorus in an essential ingredient of all cell protoplasm, nervous tissue, and bones. Ordinary phosphorus is a waxy white solid; when pure it is colorless and transparent. It is insoluble in water, but soluble in carbon disulfide. It takes fire spontaneously in air, burning to the pentoxide. It is very poisonous, 50 mg constituting an approximate fatal dose. Exposure to white phosphorus should not exceed 0.1 mg/m^3 (8-hour time-weighted average — 40-hour work week). White phosphorus should be kept under water, as it is dangerously reactive in air, and it should be handled with forceps, as contact with the skin may cause severe burns. When exposed to sunlight or when heated in its own vapor to 250°C, it is converted to the red variety, which does not phosphoresce in air as does the white variety. This form does not ignite spontaneously and it is not as dangerous as white phosphorus. It should, however, be handled with care as it does convert to the white form at some temperatures and it emits highly toxic fumes of the oxides of phosphorus when heated. The red modification is fairly stable, sublimes with a vapor pressure of 1 atm at 417°C,and is used in the manufacture of safety matches, pyrotechnics, pesticides, incendiary shells, smoke bombs, tracer bullets, etc. White phosphorus may be made by several methods. By one process, tri-calcium phosphate, the essential ingredient of phosphate rock, is heated in the presence of carbon and silica in an electric furnace or fuel-fired furnace. Elementary phosphorus is liberated as vapor and may be collected under water. If desired, the phosphorus vapor and carbon monoxide produced by the reaction can be oxidized at once in the presence of moisture to produce phosphoric acid, an important compound in making super-phosphate fertilizers. In recent years, concentrated phosphoric acids, which may contain as much as 70 to 75% P_2O_5 content, have become of great importance to agriculture and farm production. World-wide demand for fertilizers has caused record phosphate production. Phosphates are used in the production of special glasses, such as those used for sodium lamps. Bone-ash, calcium phosphate, is also used to produce fine chinaware and to produce mono-calcium phosphate used in baking powder. Phosphorus is also important in the production of steels, phosphor bronze, and many other products. Trisodium phosphate is important as a cleaning agent, as a water softener, and for preventing boiler scale and corrosion of pipes and boiler tubes. Organic compounds of phosphorus are important. Amorphous (red) phosphorus costs about $80/kg (99.5%).

Platinum — (It. *platina*, silver), Pt; at. wt. 195.078(2); at. no. 78; m.p. 1768.4°C; b.p. 3825°C; sp. gr. 21.45 (20°C); valence 1?, 2, 3, or 4. Discovered in South America by Ulloa in 1735 and by Wood in 1741. The metal was used by pre-Columbian Indians. Platinum occurs native, accompanied by small quantities of iridium, osmium, palladium, ruthenium, and rhodium, all belonging to the same group of metals. These are found in the alluvial deposits of the Ural mountains and in Columbia. *Sperrylite* ($PtAs_2$), occurring with the nickel-bearing deposits of Sudbury, Ontario, is a source of a considerable amount of metal. The large production of nickel offsets there being only one part of the platinum metals in two million parts of ore. The largest supplier of the platinum group of metals is now South Africa, followed by Russia and Canada. Platinum is a beautiful silvery-white metal, when pure, and is malleable and ductile. It has a coefficient of expansion almost equal to that of soda-lime-silica glass, and is therefore used to make sealed electrodes in glass systems. The metal does not oxidize in air at any temperature, but is corroded by halogens, cyanides, sulfur, and caustic alkalis. It is insoluble in hydrochloric and nitric acid, but dissolves when they are mixed as *aqua regia*, forming chloroplatinic acid (H_2PtCl_6), an important compound. Natural platinum contains six isotopes, one of which, ^{190}Pt, is radioactive with a long half-life. Thirty-seven other radioactive isotopes and isomers are recognized. The metal is extensively used in jewelry, wire, and vessels for laboratory use, and in many valuable instruments including thermocouple elements. It is also used for electrical contacts, corrosion-resistant apparatus, and in dentistry. Platinum-cobalt alloys have magnetic properties. One such alloy made of 76.7% Pt and 23.3% Co, by weight, is an extremely powerful magnet that offers a B-H (max) almost twice that of Alnico V. Platinum resistance wires are used for constructing high-temperature electric furnaces. The metal is used for coating missile nose cones, jet engine fuel nozzles, etc., which must perform reliably for long periods of time at high temperatures. The metal, like palladium, absorbs large volumes, of hydrogen, retaining it at ordinary temperatures but giving it up at red heat. In the finely divided state platinum is an excellent catalyst, having long been used in the contact process for producing sulfuric acid. It is also used as a catalyst in cracking petroleum products. There is also much current interest in the use of platinum as a catalyst in fuel cells and in its use as antipollution devices for automobiles. Platinum anodes are extensively used in cathodic protection systems for large ships and ocean-going vessels, pipelines, steel piers, etc. Pure platinum wire will glow red hot when placed in the vapor of methyl alcohol. It acts here as a catalyst, converting the alcohol to formaldehyde. This phenomenon has been used commercially to produce cigarette lighters and hand warmers. Hydrogen and oxygen explode in the presence of platinum. The price of platinum has varied widely; more than a century ago it was used to adulterate gold. It was nearly eight times as valuable as gold in 1920. The price in December 1999 was about $460/troy oz. ($15/g), somewhat higher than the price of gold.

Plutonium — (Planet *pluto*), Pu; at. wt. (244); at. no. 94; sp. gr. (α modification) 19.84 (25°C); m.p. 640°C; b.p. 3228°C; valence 3, 4, 5, or 6. Plutonium was the second transuranium element of the actinide series to be discovered. The isotope ^{238}Pu was produced in 1940 by Seaborg, McMillan,

Kennedy, and Wahl by deuteron bombardment of uranium in the 60-inch cyclotron at Berkeley, California. Plutonium also exists in trace quantities in naturally occurring uranium ores. It is formed in much the same manner as neptunium, by irradiation of natural uranium with the neutrons which are present. By far of greatest importance is the isotope Pu^{239}, with a half-life of 24,100 years, produced in extensive quantities in nuclear reactors from natural uranium:

$$^{238}U(n,\gamma) \longrightarrow ^{239}U \xrightarrow{\beta} ^{239}Np \xrightarrow{\beta} ^{239}Pu$$

Nineteen isotopes of plutonium are now known. Plutonium has assumed the position of dominant importance among the transuranium elements because of its successful use as an explosive ingredient in nuclear weapons and the place which it holds as a key material in the development of industrial use of nuclear power. One kilogram is equivalent to about 22 million kilowatt hours of heat energy. The complete detonation of a kilogram of plutonium produces an explosion equal to about 20,000 tons of chemical explosive. Its importance depends on the nuclear property of being readily fissionable with neutrons and its availability in quantity. The world's nuclear-power reactors are now producing about 20,000 kg of plutonium/yr. By 1982 it was estimated that about 300,000 kg had accumulated. The various nuclear applications of plutonium are well known. ^{238}Pu has been used in the Apollo lunar missions to power seismic and other equipment on the lunar surface. As with neptunium and uranium, plutonium metal can be prepared by reduction of the trifluoride with alkaline-earth metals. The metal has a silvery appearance and takes on a yellow tarnish when slightly oxidized. It is chemically reactive. A relatively large piece of plutonium is warm to the touch because of the energy given off in alpha decay. Larger pieces will produce enough heat to boil water. The metal readily dissolves in concentrated hydrochloric acid, hydroiodic acid, or perchloric acid with formation of the Pu^{+3} ion. The metal exhibits six allotropic modifications having various crystalline structures. The densities of these vary from 16.00 to 19.86 g/cm^3. Plutonium also exhibits four ionic valence states in aqueous solutions: Pu^{+3}(blue lavender), Pu^{+4} (yellow brown), PuO$^+$ (pink?), and PuO^{+2} (pink orange). The ion PuO$^+$ is unstable in aqueous solutions, disproportionating into Pu^{+4} and PuO^{+2}. The Pu^{+4} thus formed, however, oxidizes the PuO$^+$ into PuO^{+2}, itself being reduced to Pu^{+3}, giving finally Pu^{+3} and PuO^{+2}. Plutonium forms binary compounds with oxygen: PuO, PuO$_2$, and intermediate oxides of variable composition; with the halides: PuF$_3$, PuF$_4$, PuCl$_3$, PuBr$_3$, PuI$_3$; with carbon, nitrogen, and silicon: PuC, PuN, PuSi$_2$. Oxyhalides are also well known: PuOCl, PuOBr, PuOI. Because of the high rate of emission of alpha particles and the element being specifically absorbed by bone marrow, plutonium, as well as all of the other transuranium elements except neptunium, are radiological poisons and must be handled with very special equipment and precautions. Plutonium is a very dangerous radiological hazard. Precautions must also be taken to prevent the unintentional formation of a critical mass. Plutonium in liquid solution is more likely to become critical than solid plutonium. The shape of the mass must also be considered where criticality is concerned. Plutonium-239 is available to authorized users from the O.R.N.L. at a cost of about $4.80/mg (99.9%) plus packing costs.

Polonium — (Poland, native country of Mme. Curie [1867–1934]), Po; at. wt. (209); at. no. 84; m.p. 254°C; b.p. 962°C; sp. gr. (alpha modification) 9.32; valence –2, 0, +2, +3(?), +4, and +6. Polonium was the first element discovered by Mme. Curie in 1898, while seeking the cause of radioactivity of pitchblende from Joachimsthal, Bohemia. The electroscope showed it separating with bismuth. Polonium is also called Radium F. Polonium is a very rare natural element. Uranium ores contain only about 100 µg of the element per ton. Its abundance is only about 0.2% of that of radium. In 1934, it was found that when natural bismuth (^{209}Bi) was bombarded by neutrons, ^{210}Bi, the parent of polonium, was obtained. Milligram amounts of polonium may now be prepared this way, by using the high neutron fluxes of nuclear reactors. Polonium-210 is a low-melting, fairly volatile metal, 50% of which is vaporized in air in 45 hours at 55°C. It is an alpha emitter with a half-life of 138.39 days. A milligram emits as many alpha particles as 5 g of radium. The energy released by its decay is so large (140 W/g) that a capsule containing about half a gram reaches a temperature above 500°C. The capsule also presents a contact gamma-ray dose rate of 0.012 Gy/h. A few curies (1 curie = 3.7 × 10^{10} Bq) of polonium exhibit a blue glow, caused by excitation of the surrounding gas. Because almost all alpha radiation is stopped within the solid source and its container, giving up its energy, polonium has attracted attention for uses as a lightweight heat source for thermoelectric power in space satellites. Thirty-eight isotopes and isomers of polonium are known, with atomic masses ranging from 192 to 218. All are radioactive. Polonium-210 is the most readily available. Isotopes of mass 209 (half-life 102 years) and mass 208 (half-life 2.9 years) can be prepared by alpha, proton, or deuteron bombardment of lead or bismuth in a cyclotron, but these are expensive to produce. Metallic polonium has been prepared from polonium hydroxide and some other polonium compounds in the presence of concentrated aqueous or anhydrous liquid ammonia. Two allotropic modifications are known to exist. Polonium is readily dissolved in dilute acids, but is only slightly soluble in alkalis. Polonium salts of organic acids char rapidly; halide amines are reduced to the metal. Polonium can be mixed or alloyed with beryllium to provide a source of neutrons. It has been used in devices for eliminating static charges in textile mills, etc.; however, beta sources are more commonly used and are less dangerous. It is also used on brushes for removing dust from photographic films. The polonium for these is carefully sealed and controlled, minimizing hazards to the user. Polonium-210 is very dangerous to handle in even milligram or microgram amounts, and special equipment and strict control is necessary. Damage arises from the complete absorption of the energy of the alpha particle into tissue. The maximum permissible body burden for ingested polonium is only 0.03 µCi, which represents a particle weighing only 6.8 × 10^{-12} g. Weight for weight it is about 2.5 × 10^{11} times as toxic as hydrocyanic acid. The maximum allowable concentration for soluble polonium compounds in air is about 2 × 10^{11} µCi/cm^3. Polonium-209 is available on special order from the Oak Ridge National Laboratory at a cost of $3600/µCi plus packing costs..

Potassium — (English, *potash* — pot ashes; L. *kalium*, Arab. *qali*, alkali), K; at. wt. 39.0983(1); at. no. 19; m.p. 63.38°C; b.p. 759°C; sp. gr. 0.862 (20°C); valence 1. Discovered in 1807 by Davy, who obtained it from caustic potash (KOH); this was the first metal isolated by electrolysis. The metal is the seventh most abundant and makes up about 2.4% by weight of the earth's crust. Most potassium minerals are insoluble and the metal is obtained from them only with great difficulty. Certain minerals, however, such as *sylvite, carnallite, langbeinite,* and *polyhalite* are found in ancient lake and sea beds and form rather extensive deposits from which potassium and its salts can readily be obtained. Potash is mined in Germany, New Mexico, California, Utah, and elsewhere. Large deposits of potash, found at a depth of some 1000 m in Saskatchewan, promise to be important in coming years. Potassium is also found in the ocean, but is present only in relatively small amounts, compared to sodium. The greatest demand for potash has been in its use for fertilizers. Potassium is an essential constituent for plant growth and it is found in most soils. Potassium is never found free in nature, but is obtained by electrolysis of the hydroxide, much in the same manner as prepared by Davy. Thermal methods also are commonly used to produce potassium (such as by reduction of potassium compounds with CaC$_2$, C, Si, or Na). It is one of the most reactive and electropositive of metals. Except for lithium, it is the lightest known metal. It is soft, easily cut with a knife, and is silvery in appearance immediately after a fresh surface is exposed. It rapidly oxidizes in air and should be preserved in a mineral oil. As with other metals of the alkali group, it decomposes in water with the evolution of hydrogen. It catches fire spontaneously on water. Potassium and its salts impart a violet color to flames. Twenty one isotopes, one of which is an isomer, of potassium are known. Ordinary potassium is composed of three isotopes, one of which is ^{40}K (0.0117%), a radioactive isotope with a half-life of 1.26×10^9 years. The radioactivity presents no appreciable hazard. An alloy of sodium and potassium (NaK) is used as a heat-transfer medium. Many potassium salts are of utmost importance, including the hydroxide, nitrate, carbonate, chloride, chlorate, bromide, iodide, cyanide, sulfate, chromate, and dichromate. Metallic potassium is available commercially for about $1200/kg (98% purity) or $55/g (99.95% purity).

Praseodymium — (Gr. *prasios,* green, and *didymos,* twin), Pr; at. wt. 140.90765(2); at. no. 59; m.p. 931°C; b.p. 3520°C; sp. gr. 6.773; valence 3. In 1841 Mosander extracted the rare earth *didymia* from *lanthana*; in 1879, Lecoq de Boisbaudran isolated a new earth, *samaria,* from didymia obtained from the mineral *samarskite.* Six years later, in 1885, von Welsbach separated didymia into two others, *praseodymia* and *neodymia,* which gave salts of different colors. As with other rare earths, compounds of these elements in solution have distinctive sharp spectral absorption bands or lines, some of which are only a few Angstroms wide. The element occurs along with other rare-earth elements in a variety of minerals. *Monazite* and *bastnasite* are the two principal commercial sources of the rare-earth metals. Ion-exchange and solvent extraction techniques have led to much easier isolation of the rare earths and the cost has dropped greatly in the past few years. Thirty-seven isotopes and isomers are now recognized. Praseodymium can be prepared by several methods, such as by calcium reduction of the anhydrous chloride of fluoride. Misch metal, used in making cigarette lighters, contains about 5% praseodymium metal. Praseodymium is soft, silvery, malleable, and ductile. It was prepared in relatively pure form in 1931. It is somewhat more resistant to corrosion in air than europium, lanthanum, cerium, or neodymium, but it does develop a green oxide coating that spalls off when exposed to air. As with other rare-earth metals it should be kept under a light mineral oil or sealed in plastic. The rare-earth oxides, including Pr_2O_3, are among the most refractory substances known. Along with other rare earths, it is widely used as a core material for carbon arcs used by the motion picture industry for studio lighting and projection. Salts of praseodymium are used to color glasses and enamels; when mixed with certain other materials, praseodymium produces an intense and unusually clean yellow color in glass. Didymium glass, of which praseodymium is a component, is a colorant for welder's goggles. The metal (99.9% pure) is priced at about $5/g.

Promethium — (*Prometheus,* who, according to mythology, stole fire from heaven), Pm; at. no. 61; at. wt. (145); m.p. 1042°C; b.p. 3000°C (est.); sp. gr. 7.264 (25°C); valence 3. In 1902 Branner predicted the existence of an element between neodymium and samarium, and this was confirmed by Moseley in 1914. Unsuccessful searches were made for this predicted element over two decades, and various investigators proposed the names "illinium", "florentium", and "cyclonium" for this element. In 1941, workers at Ohio State University irradiated neodymium and praseodymium with neutrons, deuterons, and alpha particles, resp., and produced several new radioactivities, which most likely were those of element 61. Wu and Segre, and Bethe, in 1942, confirmed the formation; however, chemical proof of the production of element 61 was lacking because of the difficulty in separating the rare earths from each other at that time. In 1945, Marinsky, Glendenin, and Coryell made the first chemical identification by use of ion-exchange chromatography. Their work was done by fission of uranium and by neutron bombardment of neodymium. These investigators named the newly discovered element. Searches for the element on earth have been fruitless, and it now appears that promethium is completely missing from the earth's crust. Promethium, however, has been reported to be in the spectrum of the star HR^{465} in Andromeda. This element is being formed recently near the star's surface, for no known isotope of promethium has a half-life longer than 17.7 years. Thirty five isotopes and isomers of promethium, with atomic masses from 130 to 158 are now known. Promethium-145, with a half-life of 17.7 years, is the most useful. Promethium-145 has a specific activity of 940 Ci/g. It is a soft beta emitter; although no gamma rays are emitted, X-radiation can be generated when beta particles impinge on elements of a high atomic number, and great care must be taken in handling it. Promethium salts luminesce in the dark with a pale blue or greenish glow, due to their high radioactivity. Ion-exchange methods led to the preparation of about 10 g of promethium from atomic reactor fuel processing wastes in early 1963. Little is yet generally known about the properties of metallic promethium. Two allotropic modifications exist. The element has applications as a beta source for thickness gages, and it can be absorbed by a phosphor to produce light. Light produced in this manner can be used for signs or signals that require dependable operation; it can be used as a nuclear-powered battery by capturing light in photocells which convert it into electric current. Such a battery, using ^{147}Pm, would have a useful life of about 5 years. It is being used for fluorescent lighting starters and coatings for self-luminous watch dials. Promethium shows promise as a portable X-ray source, and it may become useful as a heat source to provide auxiliary power for space probes and satellites. More than 30 promethium compounds have been prepared. Most are colored. Promethium-147 is available upon special order from the Idaho National Engineering Laboratory, Idaho Falls, ID, or from the Westinghouse Hanford Co., Richland, WA.

Protactinium — (Gr. *protos,* first), Pa; at. wt. 231.03588(2); at. no. 91; m.p. 1572°C; sp. gr. 15.37 (calc.); valence 4 or 5. The first isotope of element 91 to be discovered was ^{234}Pa, also known as UX_2, a short-lived member of the naturally occurring ^{238}U decay series. It was identified by K. Fajans and O. H. Gohring in 1913 and they named the new element *brevium.* When the longer-lived isotope ^{231}Pa was identified by Hahn and Meitner in 1918, the name protoactinium was adopted as being more consistent with the characteristics of the most abundant isotope. Soddy, Cranson, and Fleck were also active in this work. The name *protoactinium* was shortened to *protactinium* in 1949. In 1927, Grosse prepared 2 mg of a white powder, which was shown to be Pa_2O_5. Later, in 1934, from 0.1 g of pure Pa_2O_5 he isolated the element by two methods, one of which was by converting the oxide to an iodide and "cracking" it in a high vacuum by an electrically heated filament by the reaction

$$2PaI_5 \rightarrow 2Pa + 5I_2$$

Protactinium has a bright metallic luster which it retains for some time in air. The element occurs in *pitchblende* to the extent of about 1 part ^{231}Pa to 10 million of ore. Ores from Congo-Kinshasa have about 3 ppm. Protactinium has twenty-eight isotopes and isomers, the most common of which is ^{231}Pr with a half-life of 32,500 years. A number of protactinium compounds are known, some of which are colored. The element is superconductive below 1.4 K. The element is a dangerous toxic material and requires precautions similar to those used when handling plutonium. In 1959 and 1961, it was announced that the Great Britain Atomic Energy Authority extracted by a 12-stage process 125 g of 99.9% protactinium, the world's only stock of the metal for many years to come. The extraction was made from 60 tons of waste material at a cost of about $500,000. Protactinium is one of the rarest and most expensive naturally occurring elements.

Radium — (L. *radius,* ray), Ra; at. wt. (226); at. no. 88; m.p. 700°C; sp. gr. 5; valence 2. Radium was discovered in 1898 by M. and Mme. Curie in the *pitchblende* or *uraninite* of North Bohemia (Czech Republic), where it occurs. There is about 1 g of radium in 7 tons of pitchblende. The element was isolated in 1911 by Mme. Curie and Debierne by the electrolysis of a solution of pure radium chloride, employing a mercury cathode; on distillation in an atmosphere of hydrogen this amalgam yielded the pure metal. Originally, radium was obtained from the rich pitchblende ore found at Joachimsthal, Bohemia. The *carnotite* sands of Colorado furnish some radium, but richer ores are found in the Republic of Congo-Kinshasa and the

Great Bear Lake region of Canada. Radium is present in all uranium minerals, and could be extracted, if desired, from the extensive wastes of uranium processing. Large uranium deposits are located in Ontario, New Mexico, Utah, Australia, and elsewhere. Radium is obtained commercially as the bromide or chloride; it is doubtful if any appreciable stock of the isolated element now exists. The pure metal is brilliant white when freshly prepared, but blackens on exposure to air, probably due to formation of the nitride. It exhibits luminescence, as do its salts; it decomposes in water and is somewhat more volatile than barium. It is a member of the alkaline-earth group of metals. Radium imparts a carmine red color to a flame. Radium emits alpha, beta, and gamma rays and when mixed with beryllium produce neutrons. One gram of ^{226}Ra undergoes 3.7×10^{10} disintegrations per s. The *curie (Ci)* is defined as that amount of radioactivity which has the same disintegration rate as 1 g of ^{226}Ra. Thirty-six isotopes are now known; radium 226, the common isotope, has a half-life of 1599 years. One gram of radium produces about 0.0001 ml (stp) of emanation, or radon gas, per day. This is pumped from the radium and sealed in minute tubes, which are used in the treatment of cancer and other diseases. One gram of radium yields about 4186 kJ per year. Radium is used in producing self-luminous paints, neutron sources, and in medicine for the treatment of disease. Some of the more recently discovered radioisotopes, such as ^{60}Co, are now being used in place of radium. Some of these sources are much more powerful, and others are safer to use. Radium loses about 1% of its activity in 25 years, being transformed into elements of lower atomic weight. Lead is a final product of disintegration. Stored radium should be ventilated to prevent build-up of radon. Inhalation, injection, or body exposure to radium can cause cancer and other body disorders. The maximum permissible burden in the total body for ^{226}Ra is 7400 becquerel.

Radon — (from *radium*; called *niton* at first, L. *nitens*, shining), Rn; at. wt. (222); at. no. 86; m.p. –71°C; b.p. –61.7°C; t_c 104°C; density of gas 9.73 g/l; sp. gr. liquid 4.4 at –62°C, solid 4; valence usually 0. The element was discovered in 1900 by Dorn, who called it *radium emanation*. In 1908 Ramsay and Gray, who named it *niton*, isolated the element and determined its density, finding it to be the heaviest known gas. It is essentially inert and occupies the last place in the zero group of gases in the Periodic Table. Since 1923, it has been called radon. Thirty-seven isotopes and isomers are known. Radon-222, coming from radium, has a half-life of 3.823 days and is an alpha emitter; Radon-220, emanating naturally from thorium and called *thoron*, has a half-life of 55.6 s and is also an alpha emitter. Radon-219 emanates from actinium and is called *actinon*. It has a half-life of 3.96 s and is also on alpha emitter. It is estimated that every square mile of soil to a depth of 6 inch contains about 1 g of radium, which releases radon in tiny amounts to the atmosphere. Radon is present in some spring waters, such as those at Hot Springs, Arkansas. On the average, one part of radon is present to 1×10^{21} part of air. At ordinary temperatures radon is a colorless gas; when cooled below the freezing point, radon exhibits a brilliant phosphorescence which becomes yellow as the temperature is lowered and orange-red at the temperature of liquid air. It has been reported that fluorine reacts with radon, forming radon fluoride. Radon clathrates have also been reported. Radon is still produced for therapeutic use by a few hospitals by pumping it from a radium source and sealing it in minute tubes, called seeds or needles, for application to patients. This practice has now been largely discontinued as hospitals can order the seeds directly from suppliers, who make up the seeds with the desired activity for the day of use. Care must be taken in handling radon, as with other radioactive materials. The main hazard is from inhalation of the element and its solid daughters, which are collected on dust in the air. Good ventilation should be provided where radium, thorium, or actinium is stored to prevent build-up of this element. Radon build-up is a health consideration in uranium mines. Recently radon build-up in homes has been a concern. Many deaths from lung cancer are caused by radon exposure. In the U.S. it is recommended that remedial action be taken if the air from radon in homes exceeds 4 pCi/l.

Rhenium — (L. *Rhenus,* Rhine), Re; at. wt. 186.207(1); at. no. 75; m.p. 3186°C; b.p. 5596°C; sp. gr. 21.02 (20°C); valence –1, +1, 2, 3, 4, 5, 6, 7. Discovery of rhenium is generally attributed to Noddack, Tacke, and Berg, who announced in 1925 they had detected the element in platinum ores and *columbite*. They also found the element in *gadolinite* and *molybdenite*. By working up 660 kg of molybdenite they were able in 1928 to extract 1 g of rhenium. The price in 1928 was $10,000/g. Rhenium does not occur free in nature or as a compound in a distinct mineral species. It is, however, widely spread throughout the earth's crust to the extent of about 0.001 ppm. Commercial rhenium in the U.S. today is obtained from molybdenite roaster-flue dusts obtained from copper-sulfide ores mined in the vicinity of Miami, Arizona, and elsewhere in Arizona and Utah. Some molybdenites contain from 0.002 to 0.2% rhenium. It is estimated that in 1999 about 16,000 kg of rhenium was being produced. The total estimated world reserves of rhenium is 11,000,000 kg. The total estimated Free World reserve of rhenium metal is 3500 tons. Natural rhenium is a mixture of two isotopes, one of which has a very long half-life. Thirty nine other unstable isotopes are recognized. Rhenium metal is prepared by reducing ammonium perrhenate with hydrogen at elevated temperatures. The element is silvery white with a metallic luster; its density is exceeded only by that of platinum, iridium, and osmium, and its melting point is exceeded only by that of tungsten and carbon. It has other useful properties. The usual commercial form of the element is a powder, but it can be consolidated by pressing and resistance-sintering in a vacuum or hydrogen atmosphere. This produces a compact shape in excess of 90% of the density of the metal. Annealed rhenium is very ductile, and can be bent, coiled, or rolled. Rhenium is used as an additive to tungsten and molybdenum-based alloys to impart useful properties. It is widely used for filaments for mass spectrographs and ion gages. Rhenium-molybdenum alloys are superconductive at 10 K. Rhenium is also used as an electrical contact material as it has good wear resistance and withstands arc corrosion. Thermocouples made of Re-W are used for measuring temperatures up to 2200°C, and rhenium wire has been used in photoflash lamps for photography. Rhenium catalysts are exceptionally resistant to poisoning from nitrogen, sulfur, and phosphorus, and are used for hydrogenation of fine chemicals, hydrocracking, reforming, and disproportionation of olefins. Rhenium has recently become especially important as a catalyst for petroleum refining and in making super-alloys for jet engines. Rhenium costs about $16/g (99.99% pure). Little is known of its toxicity; therefore, it should be handled with care until more data are available.

Rhodium — (Gr. *rhodon*, rose), Rh; at. wt. 102.90550(3); at. no. 45; m.p. 1964°C; b.p. 3695°C; sp. gr. 12.41 (20°C); valence 2, 3, 4, 5, and 6. Wollaston discovered rhodium in 1803-4 in crude platinum ore he presumably obtained from South America. Rhodium occurs native with other platinum metals in river sands of the Urals and in North and South America. It is also found with other platinum metals in the copper-nickel sulfide ores of the Sudbury, Ontario region. Although the quantity occurring here is very small, the large tonnages of nickel processed make the recovery commercially feasible. The annual world production of rhodium in 1999 was only about 9000 kg. The metal is silvery white and at red heat slowly changes in air to the sesquioxide. At higher temperatures it converts back to the element. Rhodium has a higher melting point and lower density than platinum. Its major use is as an alloying agent to harden platinum and palladium. Such alloys are used for furnace windings, thermocouple elements, bushings for glass fiber production, electrodes for aircraft spark plugs, and laboratory crucibles. It is useful as an electrical contact material as it has a low electrical resistance, a low and stable contact resistance, and is highly resistant to corrosion. Plated rhodium, produced by electroplating or evaporation, is exceptionally hard and is used for optical instruments. It has a high reflectance and is hard and durable. Rhodium is also used for jewelry, for decoration, and as a catalyst. Fifty-two isotopes and isomers are now known. Soluble salts should not exceed 0.01 mg/m^3. Rhodium metal (powder) costs about $180/g (99.9%).

THE ELEMENTS (continued)

Rubidium — (L. *rubidus*, deepest red), Rb; at. wt. 85.4678(3); at. no. 37; m.p. 39.31°C; b.p. 688°C; sp. gr. (solid) 1.532 (20°C), (liquid) 1.475 (39°C); valence 1, 2, 3, 4. Discovered in 1861 by Bunsen and Kirchoff in the mineral *lepidolite* by use of the spectroscope. The element is much more abundant than was thought several years ago. It is now considered to be the 16th most abundant element in the earth's crust. Rubidium occurs in *pollucite, carnallite, leucite,* and *zinnwaldite*, which contains traces up to 1%, in the form of the oxide. It is found in lepidolite to the extent of about 1.5%, and is recovered commercially from this source. Potassium minerals, such as those found at Searles Lake, California, and potassium chloride recovered from brines in Michigan also contain the element and are commercial sources. It is also found along with cesium in the extensive deposits of *pollucite* at Bernic Lake, Manitoba. Rubidium can be liquid at room temperature. It is a soft, silvery-white metallic element of the alkali group and is the second most electropositive and alkaline element. It ignites spontaneously in air and reacts violently in water, setting fire to the liberated hydrogen. As with other alkali metals, it forms amalgams with mercury and it alloys with gold, cesium, sodium, and potassium. It colors a flame yellowish violet. Rubidium metal can be prepared by reducing rubidium chloride with calcium, and by a number of other methods. It must be kept under a dry mineral oil or in a vacuum or inert atmosphere. Thirty five isotopes and isomers of rubidium are known. Naturally occurring rubidium is made of two isotopes, ^{85}Rb and ^{87}Rb. Rubidium-87 is present to the extent of 27.83% in natural rubidium and is a beta emitter with a half-life of 4.9×10^{10} years. Ordinary rubidium is sufficiently radioactive to expose a photographic film in about 30 to 60 days. Rubidium forms four oxides: Rb_2O, Rb_2O_2, Rb_2O_3, Rb_2O_4. Because rubidium can be easily ionized, it has been considered for use in "ion engines" for space vehicles; however, cesium is somewhat more efficient for this purpose. It is also proposed for use as a working fluid for vapor turbines and for use in a thermoelectric generator using the magnetohydro-dynamic principle where rubidium ions are formed by heat at high temperature and passed through a magnetic field. These conduct electricity and act like an armature of a generator thereby generating an electric current. Rubidium is used as a getter in vacuum tubes and as a photocell component. It has been used in making special glasses. $RbAg_4I_5$ is important, as it has the highest room conductivity of any known ionic crystal. At 20°C its conductivity is about the same as dilute sulfuric acid. This suggests use in thin film batteries and other applications. The present cost in small quantities is about $80/g (99.8% pure).

Ruthenium — (L. *Ruthenia*, Russia), Ru; at. wt. 101.07(2); at. no. 44, m.p. 2334°C; b.p. 4150°C; sp. gr. 12.41 (20°C); valence 0, 1, 2, 3, 4, 5, 6, 7, 8. Berzelius and Osann in 1827 examined the residues left after dissolving crude platinum from the Ural mountains in *aqua regia*. While Berzelius found no unusual metals, Osann thought he found three new metals, one of which he named ruthenium. In 1844 Klaus, generally recognized as the discoverer, showed that Osann's ruthenium oxide was very impure and that it contained a new metal. Klaus obtained 6 g of ruthenium from the portion of crude platinum that is insoluble in *aqua regia*. A member of the platinum group, ruthenium occurs native with other members of the group in ores found in the Ural mountains and in North and South America. It is also found along with other platinum metals in small but commercial quantities in *pentlandite* of the Sudbury, Ontario, nickel-mining region, and in *pyroxinite* deposits of South Africa. Natural ruthenium contains seven isotopes. Twenty-eight other isotopes and isomers are known, all of which are radioactive. The metal is isolated commercially by a complex chemical process, the final stage of which is the hydrogen reduction of ammonium ruthenium chloride, which yields a powder. The powder is consolidated by powder metallurgy techniques or by argon-arc welding. Ruthenium is a hard, white metal and has four crystal modifications. It does not tarnish at room temperatures, but oxidizes in air at about 800°C. The metal is not attacked by hot or cold acids or *aqua regia*, but when potassium chlorate is added to the solution, it oxidizes explosively. It is attacked by halogens, hydroxides, etc. Ruthenium can be plated by electrodeposition or by thermal decomposition methods. The metal is one of the most effective hardeners for platinum and palladium, and is alloyed with these metals to make electrical contacts for severe wear resistance. A ruthenium-molybdenum alloy is said to be superconductive at 10.6 K. The corrosion resistance of titanium is improved a hundredfold by addition of 0.1% ruthenium. It is a versatile catalyst. Hydrogen sulfide can be split catalytically by light using an aqueous suspension of CdS particles loaded with ruthenium dioxide. It is thought this may have application to removal of H_2S from oil refining and other industrial processes. Compounds in at least eight oxidation states have been found, but of these, the +2, +3, and +4 states are the most common. Ruthenium tetroxide, like osmium tetroxide, is highly toxic. In addition, it may explode. Ruthenium compounds show a marked resemblance to those of osmium. The metal is priced at about $30/g (99.95% pure).

Rutherfordium — (named for Ernest Rutherford [1871–1937], New Zealand, Canadian, and British physicist); Rf; at. wt. [261]; at. no. 104. In 1964, workers of the Joint Nuclear Research Institute at Dubna (Russia) bombarded plutonium with accelerated 113 to 115 MeV neon ions. By measuring fission tracks in a special glass with a microscope, they detected an isotope that decays by spontaneous fission. They suggested that this isotope, which has a half-life of 0.3 ± 0.1 s, might be ^{260}104, produced by the following reaction:

$$^{242}_{94}Pu + ^{22}_{10}Ne \rightarrow ^{260}104 + 4n$$

Element 104, the first *transactinide* element, is expected to have chemical properties similar to those of hafnium. It would, for example, form a relatively volatile compound with chlorine (a tetrachloride). The Soviet scientists have performed experiments aimed at chemical identification, and have attempted to show that the 0.3-s activity is more volatile than that of the relatively nonvolatile actinide trichlorides. This experiment does not fulfill the test of chemically separating the new element from all others, but it provides important evidence for evaluation. New data, reportedly issued by Soviet scientists, have reduced the half-life of the isotope they worked with from 0.3 to 0.15 s. The Dubna scientists suggest the name *kurchatovium* and symbol *Ku* for Element 104, in honor of Igor Vasilevich Kurchatov (1903—1960), late Head of Soviet Nuclear Research. The Dubna Group also has proposed the name *dubnium* for Element 104. In 1969, Ghiorso, Nurmia, Harris, K. A. Y. Eskola, and P. L. Eskola of the University of California at Berkeley reported they had positively identified two, and possibly three, isotopes of Element 104. The group also indicated that after repeated attempts so far they have been unable to produce isotope ^{260}104 reported by the Dubna groups in 1964. The discoveries at Berkeley were made by bombarding a target of ^{249}Cf with ^{12}C nuclei of 71 MeV, and ^{13}C nuclei of 69 MeV. The combination of ^{12}C with ^{249}Cf followed by instant emission of four neutrons produced Element ^{257}104. This isotope has a half-life of 4 to 5 s, decaying by emitting an alpha particle into ^{253}No, with a half-life of 105 s. The same reaction, except with the emission of three neutrons, was thought to have produced ^{258}104 with a half-life of about 1/100 s. Element ^{259}104 is formed by the merging of a ^{13}C nuclei with ^{249}Cf, followed by emission of three neutrons. This isotope has a half-life of 3 to 4 s, and decays by emitting an alpha particle into ^{255}No, which has a half-life of 185 s. Thousands of atoms of ^{257}104 and ^{259}104 have been detected. The Berkeley group believe their identification of ^{258}104 was correct. Eleven isotopes of Element 104 have now been identified. The Berkeley group proposed for the new element the name *rutherfordium* (symbol Rf), in honor of Ernest Rutherford. This name was formally adapted by IUPAC in August 1997.

THE ELEMENTS (continued)

Samarium — (*Samarskite* a mineral), Sm; at. wt. 150.36(3); at. no. 62; m.p. 1074°C; b.p. 1794°C; sp. gr (α) 7.520 (25°C); valence 2 or 3. Discovered spectroscopically by its sharp absorption lines in 1879 by Lecoq de Boisbaudran in the mineral *samarskite*, named in honor of a Russian mine official, Col. Samarski. Samarium is found along with other members of the rare-earth elements in many minerals, including *monazite* and *bastnasite*, which are commercial sources. The largest producer of rare earth minerals is now China, followed by the U.S., India, and Russia. It occurs in monazite to the extent of 2.8%. While *misch metal* containing about 1% of samarium metal, has long been used, samarium has not been isolated in relatively pure form until recent years. Ion-exchange and solvent extraction techniques have recently simplified separation of the rare earths from one another; more recently, electrochemical deposition, using an electrolytic solution of lithium citrate and a mercury electrode, is said to be a simple, fast, and highly specific way to separate the rare earths. Samarium metal can be produced by reducing the oxide with barium or lanthanum. Samarium has a bright silver luster and is reasonably stable in air. Three crystal modifications of the metal exist, with transformations at 734 and 922°C. The metal ignites in air at about 150°C. Thirty-three isotopes and isomers of samarium are now recognized. Natural samarium is a mixture of seven isotopes, three of which are unstable but have long half-lives. Samarium, along with other rare earths, is used for carbon-arc lighting for the motion picture industry. The sulfide has excellent high-temperature stability and good thermoelectric efficiencies up to 1100°C. $SmCo_5$ has been used in making a new permanent magnet material with the highest resistance to demagnetization of any known material. It is said to have an intrinsic coercive force as high as 2200 kA/m. Samarium oxide has been used in optical glass to absorb the infrared. Samarium is used to dope calcium fluoride crystals for use in optical masers or lasers. Compounds of the metal act as sensitizers for phosphors excited in the infrared; the oxide exhibits catalytic properties in the dehydration and dehydrogenation of ethyl alcohol. It is used in infrared absorbing glass and as a neutron absorber in nuclear reactors. The metal is priced at about \$3.50/g (99.9%). Little is known of the toxicity of samarium; therefore, it should be handled carefully.

Scandium — (L. *Scandia*, Scandinavia), Sc; at. wt. 44.955910(8); at. no. 21; m.p. 1541°C; b.p. 2836°C; sp. gr. 2.989 (25°C); valence 3. On the basis of the Periodic System, Mendeleev predicted the existence of *ekaboron*, which would have an atomic weight between 40 of calcium and 48 of titanium. The element was discovered by Nilson in 1878 in the minerals *euxenite* and *gadolinite*, which had not yet been found anywhere except in Scandinavia. By processing 10 kg of euxenite and other residues of rare-earth minerals, Nilson was able to prepare about 2 g of scandium oxide of high purity. Cleve later pointed out that Nilson's scandium was identical with Mendeleev's ekaboron. Scandium is apparently a much more abundant element in the sun and certain stars than here on earth. It is about the 23rd most abundant element in the sun, compared to the 50th most abundant on earth. It is widely distributed on earth, occurring in very minute quantities in over 800 mineral species. The blue color of beryl (aquamarine variety) is said to be due to scandium. It occurs as a principal component in the rare mineral *thortveitite*, found in Scandinavia and Malagasy. It is also found in the residues remaining after the extraction of tungsten from Zinnwald *wolframite*, and in *wiikite* and *bazzite*. Most scandium is presently being recovered from *thortveitite* or is extracted as a by-product from uranium mill tailings. Metallic scandium was first prepared in 1937 by Fischer, Brunger, and Grieneisen, who electrolyzed a eutectic melt of potassium, lithium, and scandium chlorides at 700 to 800°C. Tungsten wire and a pool of molten zinc served as the electrodes in a graphite crucible. Pure scandium is now produced by reducing scandium fluoride with calcium metal. The production of the first pound of 99% pure scandium metal was announced in 1960. Scandium is a silver-white metal which develops a slightly yellowish or pinkish cast upon exposure to air. It is relatively soft, and resembles yttrium and the rare-earth metals more than it resembles aluminum or titanium. It is a very light metal and has a much higher melting point than aluminum, making it of interest to designers of spacecraft. Scandium is not attacked by a 1:1 mixture of conc. HNO_3 and 48% HF. Scandium reacts rapidly with many acids. Twenty-three isotopes and isomers of scandium are recognized. The metal is expensive, costing about \$200/g with a purity of about 99.9%. About 20 kg of scandium (as Sc_2O_3) are now being used yearly in the U.S. to produce high-intensity lights, and the radioactive isotope ^{46}Sc is used as a tracing agent in refinery crackers for crude oil, etc. Scandium iodide added to mercury vapor lamps produces a highly efficient light source resembling sunlight, which is important for indoor or night-time color TV. Little is yet known about the toxicity of scandium; therefore, it should be handled with care.

Seaborgium — (named for Glenn T. Seaborg [1912–1999], American chemist and nuclear physicist). Sg; at. wt. [263]; at no. 106. The discovery of *Seaborgium*, Element 106, took place in 1974 almost simultaneously at the Lawrence-Berkeley Laboratory and at the Joint Institute for Nuclear Research at Dubna, Russia. The Berkeley Group, under direction of Ghiorso, used the Super-Heavy Ion Linear Accelerator (Super HILAC) as a source of heavy ^{18}O ions to bombard a 259-µg target of ^{249}Cf. This resulted in the production and positive identification of $^{263}106$, which decayed with a half-life of 0.9 ± 0.2 s by the emission of alpha particles as follows:

$$^{263}106 \xrightarrow{\alpha} {}^{259}104 \xrightarrow{\alpha} {}^{255}No \xrightarrow{\alpha}.$$

The Dubna Team, directed by Flerov and Organessian, produced heavy ions of ^{54}Cr with their 310-cm heavy-ion cyclotron to bombard ^{207}Pb and ^{208}Pb and found a product that decayed with a half-life of 7 ms. They assigned $^{259}106$ to this isotope. It is now thought seven isotopes of *Seaborgium* have been identified. Two of the isotopes are believed to have half-lives of about 30 s. *Seaborgium* most likely would have properties resembling tungsten. The IUPAC adopted the name *Seaborgium* in August 1997. Normally the naming of an element is not given until after the death of the person for which the element is named; however, in this case, it was named while Dr. Seaborg was still alive.

Selenium — (Gr. *Selene*, moon), Se; at. wt. 78.96(3); at. no. 34; m.p. (gray) 221°C; b.p. (gray) 685°C; sp. gr. (gray) 4.79, (vitreous) 4.28; valence –2, +4, or +6. Discovered by Berzelius in 1817, who found it associated with tellurium, named for the earth. Selenium is found in a few rare minerals, such as *crooksite* and *clausthalite*. In years past it has been obtained from flue dusts remaining from processing copper sulfide ores, but the anode muds from electrolytic copper refineries now provide the source of most of the world's selenium. Selenium is recovered by roasting the muds with soda or sulfuric acid, or by smelting them with soda and niter. Selenium exists in several allotropic forms. Three are generally recognized, but as many as six have been claimed. Selenium can be prepared with either an amorphous or crystalline structure. The color of amorphous selenium is either red, in powder form, or black, in vitreous form. Crystalline monoclinic selenium is a deep red; crystalline hexagonal selenium, the most stable variety, is a metallic gray. Natural selenium contains six stable isotopes. Twenty-nine other isotopes and isomers have been characterized. The element is a member of the sulfur family and resembles sulfur both in its various forms and in its compounds. Selenium exhibits both photovoltaic action, where light is converted directly into electricity, and photoconductive action, where the electrical resistance decreases with increased illumination. These properties make selenium useful in the production of photocells and exposure meters for photographic use, as well as solar cells. Selenium is also able to convert a.c. electricity to d.c., and is extensively used in rectifiers. Below its melting point selenium is a p-type semiconductor and is finding many uses in electronic and solid-state applications. It is used in Xerography for reproducing and copying documents, letters, etc., but recently its use in this application has been decreasing in favor of certain organic compounds. It is used by the glass industry to decolorize glass and to make ruby-colored glasses and enamels. It is also used as a photographic toner, and as an additive to stainless steel. Elemental selenium has been said to be practically nontoxic and is considered to be an essential trace element; however, hydrogen selenide and other selenium compounds are extremely toxic, and resemble arsenic in their physiological reactions. Hydrogen selenide in a concentration of 1.5 ppm is intolerable to man. Selenium occurs in some soils in amounts sufficient to produce serious effects on animals feeding on plants, such as locoweed, grown in such soils. Selenium (99.5%) is priced at about $250/kg. It is also available in high-purity form at a cost of about $750/kg (99.999%).

Silicon — (L. *silex, silicis*, flint), Si; at. wt. 28.0855(3); at. no. 14; m.p. 1414°C; b.p. 3265°C; sp. gr. 2.33 (25°C); valence 4. Davy in 1800 thought silica to be a compound and not an element; later in 1811, Gay Lussac and Thenard probably prepared impure amorphous silicon by heating potassium with silicon tetrafluoride. Berzelius, generally credited with the discovery, in 1824 succeeded in preparing amorphous silicon by the same general method as used earlier, but he purified the product by removing the fluosilicates by repeated washings. Deville in 1854 first prepared crystalline silicon, the second allotropic form of the element. Silicon is present in the sun and stars and is a principal component of a class of meteorites known as "aerolites". It is also a component of *tektites*, a natural glass of uncertain origin. Natural silicon contains three isotopes. Twenty-four other radioactive isotopes are recognized. Silicon makes up 25.7% of the earth's crust, by weight, and is the second most abundant element, being exceeded only by oxygen. Silicon is not found free in nature, but occurs chiefly as the oxide and as silicates. *Sand, quartz, rock crystal, amethyst, agate, flint, jasper,* and *opal* are some of the forms in which the oxide appears. *Granite, hornblende, asbestos, feldspar, clay mica,* etc. are but a few of the numerous silicate minerals. Silicon is prepared commercially by heating silica and carbon in an electric furnace, using carbon electrodes. Several other methods can be used for preparing the element. Amorphous silicon can be prepared as a brown powder, which can be easily melted or vaporized. Crystalline silicon has a metallic luster and grayish color. The Czochralski process is commonly used to produce single crystals of silicon used for solid-state or semiconductor devices. Hyperpure silicon can be prepared by the thermal decomposition of ultra-pure trichlorosilane in a hydrogen atmosphere, and by a vacuum float zone process. This product can be doped with boron, gallium, phosphorus, or arsenic to produce silicon for use in transistors, solar cells, rectifiers, and other solid-state devices which are used extensively in the electronics and space-age industries. Hydrogenated amorphous silicon has shown promise in producing economical cells for converting solar energy into electricity. Silicon is a relatively inert element, but it is attacked by halogens and dilute alkali. Most acids except hydrofluoric, do not affect it. Silicones are important products of silicon. They may be prepared by hydrolyzing a silicon organic chloride, such as dimethyl silicon chloride. Hydrolysis and condensation of various substituted chlorosilanes can be used to produce a very great number of polymeric products, or silicones, ranging from liquids to hard, glasslike solids with many useful properties. Elemental silicon transmits more than 95% of all wavelengths of infrared, from 1.3 to 6.7 μm. Silicon is one of man's most useful elements. In the form of sand and clay it is used to make concrete and brick; it is a useful refractory material for high-temperature work, and in the form of silicates it is used in making enamels, pottery, etc. Silica, as sand, is a principal ingredient of glass, one of the most inexpensive of materials with excellent mechanical, optical, thermal, and electrical properties. Glass can be made in a very great variety of shapes, and is used as containers, window glass, insulators, and thousands of other uses. Silicon tetrachloride can be used to iridize glass. Silicon is important in plant and animal life. Diatoms in both fresh and salt water extract silica from the water to build up their cell walls. Silica is present in ashes of plants and in the human skeleton. Silicon is an important ingredient in steel; silicon carbide is one of the most important abrasives and has been used in lasers to produce coherent light of 4560 Å. A remarkable material, first discovered in 1930, is *Aerogel,* developed and now used by NASA in their *Stardust* mission, which is expected to encounter Comet Wild 2 in 2004, returning cometary and interplanet dust to Earth in 2006. *Aerogel* is a highly insulative material that has the lowest density of any known solid. One form of *Aerogel* is 99.9% air and 0.1% SiO_2, by volume. It is 1000 times less dense than glass. It has been called "blue smoke" or "solid smoke". A block of *Aerogel* as large as a person may weigh less than a pound and yet support the weight of 1000 lbs (455 kg). This material is expected to trap cometary particles traveling at speeds of 32 km/sec. *Aerogel* is said to be non-toxic and non-inflammable. It has high thermal insulating qualities that could be used in home insulation. Its light weight may have aircraft applications. Regular grade silicon (99.5%) costs about $160/kg. Silicon (99.9999%) pure costs about $250/kg; hyperpure silicon is available at a higher cost. Miners, stonecutters, and other engaged in work where siliceous dust is breathed in large quantities often develop a serious lung disease known as *silicosis.*

Silver — (Anglo-Saxon, *Seolfor siolfur*), Ag (L. argentum), at. wt. 107.8682(2); at. no. 47; m.p. 961.78°C; b.p. 2162°C; sp. gr. 10.50 (20°C); valence 1, 2. Silver has been known since ancient times. It is mentioned in Genesis. Slag dumps in Asia Minor and on islands in the Aegean Sea indicate that man learned to separate silver from lead as early as 3000 B.C. Silver occurs native and in ores such as *argentite* (Ag_2S) and *horn silver* (AgCl); lead, lead-zinc, copper, gold, and copper-nickel ores are principal sources. Mexico, Canada, Peru, and the U.S. are the principal silver producers in the western hemisphere. Silver is also recovered during electrolytic refining of copper. Commercial fine silver contains at least 99.9% silver. Purities of 99.999+% are available commercially. Pure silver has a brilliant white metallic luster. It is a little harder than gold and is very ductile and malleable, being exceeded only by gold and perhaps palladium. Pure silver has the highest electrical and thermal conductivity of all metals, and possesses the lowest contact resistance. It is stable in pure air and water, but tarnishes when exposed to ozone, hydrogen sulfide, or air containing sulfur. The alloys of silver are important. Sterling silver is used for jewelry, silverware, etc. where appearance is paramount. This alloy contains 92.5% silver, the remainder being copper or some other metal. Silver is of utmost importance in photography, about 30% of the U.S. industrial consumption going into this application. It is used for dental alloys. Silver is used in making solder and brazing alloys, electrical contacts, and high capacity silver-zinc and silver-cadmium batteries. Silver paints are used for making printed circuits. It is used in mirror production and may be deposited on glass or metals by chemical deposition, electrodeposition, or by evaporation. When freshly deposited, it is the best reflector of visible light known, but is rapidly tarnishes and loses much of its reflectance. It is a poor reflector of ultraviolet. Silver fulminate ($Ag_2C_2N_2O_2$), a powerful explosive, is sometimes formed during the silvering process. Silver iodide is used in seeding clouds to produce rain. Silver chloride has interesting optical properties as it can be made transparent; it also is a cement for glass. Silver nitrate, or *lunar caustic*, the most important silver compound, is used extensively in photography. While silver itself is not considered to be toxic, most of its salts are poisonous. Natural silver contains two stable isotopes. Fifty-six other radioactive isotopes and isomers are known. Silver compounds can be absorbed in the circulatory system and reduced silver deposited in the various tissues of the body. A condition, known as *argyria*, results with a greyish pigmentation of the skin and mucous membranes. Silver has germicidal effects and kills many lower organisms effectively without harm to higher animals. Silver for centuries has been used traditionally for coinage by many countries of the world. In recent times, however, consumption of silver has at times greatly exceeded the output. In 1939, the price of silver was fixed by the U.S. Treasury at 71¢/troy oz., and at 90.5¢/troy oz. in 1946. In November 1961 the U.S. Treasury suspended sales of nonmonetized silver, and the price stabilized for a time at about $1.29, the melt-down value of silver U.S. coins. The Coinage Act of 1965 authorized a change in the metallic composition of the three U.S. subsidiary denominations to clad or composite type coins. This was the first change in U.S. coinage since the monetary system was established in 1792. Clad dimes and quarters are made of an outer layer of 75% Cu and 25% Ni bonded to a central core of pure Cu. The composition of the one- and five-cent pieces remains unchanged. One-cent coins are 95% Cu and 5% Zn. Five-cent coins are 75% Cu and 25% Ni. Old silver dollars are 90% Ag and 10% Cu. Earlier subsidiary coins of 90% Ag and 10% Cu officially were to circulate alongside the clad coins; however, in practice they have largely disappeared (Gresham's Law), as the value of the silver is now greater than their exchange value. Silver coins of other countries have largely been replaced with coins made of other metals. On June 24, 1968, the U.S. Government ceased to redeem U.S. Silver Certificates with silver. Since that time, the price of silver has fluctuated widely. As of December 1999, the price of silver was about $5.20/troy oz. (17¢/g); however the price has fluctuated considerably due to market instability. The price of silver in 1999 was only about five times the cost of the metal about 150 years ago. This has largely been caused by Central Banks disposing of some of their silver reserves and the development of more productive mines with better refining methods. Also, silver has been displaced by other metals or processes, such as digital photography.

Sodium — (English, *soda*; Medieval Latin, *sodanum,* headache remedy), Na (L. *natrium*); at. wt. 22.989770(2); at. no. 11; m.p. 97.80°C; b.p. 883°C; sp. gr. 0.971 (20°C); valence 1. Long recognized in compounds, sodium was first isolated by Davy in 1807 by electrolysis of caustic soda. Sodium is present in fair abundance in the sun and stars. The D lines of sodium are among the most prominent in the solar spectrum. Sodium is the sixth most abundant element on earth, comprising about 2.6% of the earth's crust; it is the most abundant of the alkali group of metals of which it is a member. The most common compound is sodium chloride, but it occurs in many other minerals, such as *soda niter, cryolite, amphibole, zeolite, sodalite,*etc. It is a very reactive element and is never found free in nature. It is now obtained commercially by the electrolysis of absolutely dry fused sodium chloride. This method is much cheaper than that of electrolyzing sodium hydroxide, as was used several years ago. Sodium is a soft, bright, silvery metal which floats on water, decomposing it with the evolution of hydrogen and the formation of the hydroxide. It may or may not ignite spontaneously on water, depending on the amount of oxide and metal exposed to the water. It normally does not ignite in air at temperatures below 115°C. Sodium should be handled with respect, as it can be dangerous when improperly handled. Metallic sodium is vital in the manufacture of sodamide and esters, and in the preparation of organic compounds. The metal may be used to improve the structure of certain alloys, to descale metal, to purify molten metals, and as a heat transfer agent. An alloy of sodium with potassium, NaK, is also an important heat transfer agent. Sodium compounds are important to the paper, glass, soap, textile, petroleum, chemical, and metal industries. Soap is generally a sodium salt of certain fatty acids. The importance of common salt to animal nutrition has been recognized since prehistoric times. Among the many compounds that are of the greatest industrial importance are common salt (NaCl), soda ash (Na_2CO_3), baking soda ($NaHCO_3$), caustic soda (NaOH), Chile saltpeter ($NaNO_3$), di- and tri-sodium phosphates, sodium thiosulfate (hypo, $Na_2S_2O_3 \cdot 5H_2O$), and borax ($Na_2B_4O_7 \cdot 10H_2O$). Seventeen isotopes of sodium are recognized. Metallic sodium is priced at about $130/kg (99.95%). On a volume basis, it is the cheapest of all metals. Sodium metal should be handled with great care. It should be kept in an inert atmosphere and contact with water and other substances with which sodium reacts should be avoided.

Strontium — (*Strontian*, town in Scotland), Sr; at. wt. 87.62(1); at. no. 38; m.p. 777°C; b.p. 1382°C; sp. gr. 2.54; valence 2. Isolated by Davey by electrolysis in 1808; however, Adair Crawford in 1790 recognized a new mineral (strontianite) as differing from other barium minerals (baryta). Strontium is found chiefly as *celestite* ($SrSO_4$) and *strontianite* ($SrCO_3$). *Celestite* is found in Mexico, Turkey, Iran, Spain, Algeria, and in the U.K. The U.S. has no active *celestite* mines. The metal can be prepared by electrolysis of the fused chloride mixed with potassium chloride, or is made by reducing strontium oxide with aluminum in a vacuum at a temperature at which strontium distills off. Three allotropic forms of the metal exist, with transition points at 235 and 540°C. Strontium is softer than calcium and decomposes water more vigorously. It does not absorb nitrogen below 380°C. It should be kept under mineral oil to prevent oxidation. Freshly cut strontium has a silvery appearance, but rapidly turns a yellowish color with the formation of the oxide. The finely divided metal ignites spontaneously in air. Volatile strontium salts impart a beautiful crimson color to flames, and these salts are used in pyrotechnics and in the production of flares. Natural strontium is a mixture of four stable isotopes. Thirty-two other unstable isotopes and isomers are known to exist. Of greatest importance is ^{90}Sr with a half-life of 29 years. It is a product of nuclear fallout and presents a health problem. This isotope is one of the best long-lived high-energy beta emitters known, and is used in SNAP (Systems for Nuclear Auxiliary Power) devices. These devices hold promise for use in space vehicles, remote weather stations, navigational buoys, etc., where a lightweight, long-lived, nuclear-electric power source is needed. The major use for strontium at present is in producing glass for color television picture tubes. All color TV and cathode ray tubes sold in the U.S. are required by law to contain strontium in the face plate glass to block X-ray emission. Strontium also improves the brilliance of the glass and the quality of the picture. It has also found use in producing ferrite magnets and in refining zinc. Strontium titanate is an interesting optical material as it has an extremely high refractive index and an optical dispersion greater than that of diamond. It has been used as a gemstone, but it is very soft. It does not occur naturally. Strontium metal (99% pure) costs about $220/kg.

Sulfur — (Sanskrit, *sulvere*; L. *sulphurium*), S; at. wt. 32.066(6); at. no. 16; m.p. 115.21°C; b.p. 444.60°C; t_c 1041°C; sp. gr. (rhombic) 2.07, (monoclinic) 1.957 (20°C); valence 2, 4, or 6. Known to the ancients; referred to in Genesis as *brimstone*. Sulfur is found in meteorites. A dark area near the crater Aristarchus on the moon has been studied by R. W. Wood with ultraviolet light. This study suggests strongly that it is a sulfur deposit. Sulfur occurs native in the vicinity of volcanoes and hot springs. It is widely distributed in nature as *iron pyrites, galena, sphalerite, cinnabar, stibnite, gypsum, Epsom salts, celestite, barite,*etc. Sulfur is commercially recovered from wells sunk into the salt domes along the Gulf Coast of the U.S. It is obtained from these wells by the Frasch process, which forces heated water into the wells to melt the sulfur, which is then brought to the surface. Sulfur also occurs in natural gas and petroleum crudes and must be removed from these products. Formerly this was done chemically, which wasted the sulfur. New processes now permit recovery, and these sources promise to be very important. Large amounts of sulfur are being recovered from Alberta gas fields. Sulfur is a pale yellow, odorless, brittle solid, which is insoluble in water but soluble in carbon disulfide. In every state, whether gas, liquid or solid, elemental sulfur occurs in more than one allotropic form or modification; these present a confusing multitude of forms whose relations are not yet fully understood. Amorphous or "plastic" sulfur is obtained by fast cooling of the crystalline form. X-ray studies indicate that amorphous sulfur may have a helical structure with eight atoms per spiral. Crystalline sulfur seems to be made of rings, each containing eight sulfur atoms, which fit together to give a normal X-ray pattern. Twenty-one isotopes of sulfur are now recognized. Four occur in natural sulfur, none of which is radioactive. A finely divided form of sulfur, known as *flowers of sulfur*, is obtained by sublimation. Sulfur readily forms sulfides with many elements. Sulfur is a component of black gunpowder, and is used in the vulcanization of natural rubber and a fungicide. It is also used extensively is making phosphatic fertilizers. A tremendous tonnage is used to produce sulfuric acid, the most important manufactured chemical. It is used in making sulfite paper and other papers, as a fumigant, and in the bleaching of dried fruits. The element is a good electrical insulator. Organic compounds containing sulfur are very important. Calcium sulfate, ammonium sulfate, carbon disulfide, sulfur dioxide, and hydrogen sulfide are but a few of the many other important compounds of sulfur. Sulfur is essential to life. It is a minor constituent of fats, body fluids, and skeletal minerals. Carbon disulfide, hydrogen sulfide, and sulfur dioxide should be handled carefully. Hydrogen sulfide in small concentrations can be metabolized, but in higher concentrations it quickly can cause death by respiratory paralysis. It is insidious in that it quickly deadens the sense of smell. Sulfur dioxide is a dangerous component in atmospheric air pollution. In 1975, University of Pennsylvania scientists reported synthesis of polymeric sulfur nitride, which has the properties of a metal, although it contains no metal atoms. The material has unusual optical and electrical properties.

Tantalum — (Gr. *Tantalos*, mythological character, father of *Niobe*), Ta; at. wt. 180.9479(1); at. no. 73; m.p. 3017°C; b.p. 5458°C; sp. gr. 16.654; valence 2?, 3, 4?, or 5. Discovered in 1802 by Ekeberg, but many chemists thought niobium and tantalum were identical elements until Rose, in 1844, and Marignac, in 1866, showed that niobic and tantalic acids were two different acids. The early investigators only isolated the impure metal. The first relatively pure ductile tantalum was produced by von Bolton in 1903. Tantalum occurs principally in the mineral *columbite-tantalite* (Fe, Mn)(Nb, Ta)$_2$O$_6$. Tantalum ores are found in Australia, Brazil, Rwanda, Zimbabwe, Congo-Kinshasa, Nigeria, and Canada. Separation of tantalum from niobium requires several complicated steps. Several methods are used to commercially produce the element, including electrolysis of molten potassium fluorotantalate, reduction of potassium fluorotantalate with sodium, or reacting tantalum carbide with tantalum oxide. Thirty four isotopes and isomers of tantalum are known to exist. Natural tantalum contains two isotopes, one of which is radioactive with a very long half-life. Tantalum is a gray, heavy, and very hard metal. When pure, it is ductile and can be drawn into fine wire, which is used as a filament for evaporating metals such as aluminum. Tantalum is almost completely immune to chemical attack at temperatures below 150°C, and is attacked only by hydrofluoric acid, acidic solutions containing the fluoride ion, and free sulfur trioxide. Alkalis attack it only slowly. At high temperatures, tantalum becomes much more reactive. The element has a melting point exceeded only by tungsten and rhenium. Tantalum is used to make a variety of alloys with desirable properties such as high melting point, high strength, good ductility, etc. Scientists at Los Alamos have produced a tantalum carbide graphite composite material, which is said to be one of the hardest materials ever made. The compound has a melting point of 3738°C. Tantalum has good "gettering" ability at high temperatures, and tantalum oxide films are stable and have good rectifying and dielectric properties. Tantalum is used to make electrolytic capacitors and vacuum furnace parts, which account for about 60% of its use. The metal is also widely used to fabricate chemical process equipment, nuclear reactors, and aircraft and missile parts. Tantalum is completely immune to body liquids and is a nonirritating metal. It has, therefore, found wide use in making surgical appliances. Tantalum oxide is used to make special glass with high index of refraction for camera lenses. The metal has many other uses. The price of (99.9%) tantalum is about $2/g.

Technetium — (Gr. *technetos*, artificial), Tc; at. wt. (98); at. no. 43; m.p. 2157°C; b.p. 4265°C; sp. gr. 11.50 (calc.); valence 0, +2, +4, +5, +6, and +7. Element 43 was predicted on the basis of the periodic table, and was erroneously reported as having been discovered in 1925, at which time it was named *masurium*. The element was actually discovered by Perrier and Segre in Italy in 1937. It was found in a sample of molybdenum, which was bombarded by deuterons in the Berkeley cyclotron, and which E. Lawrence sent to these investigators. Technetium was the first element to be produced artificially. Since its discovery, searches for the element in terrestrial materials have been made without success. If it does exist, the concentration must be very small. Technetium has been found in the spectrum of S-, M-, and N-type stars, and its presence in stellar matter is leading to new theories of the production of heavy elements in the stars. Forty-three isotopes and isomers of technetium, with atomic masses ranging from 86 to 113, are known. ^{97}Tc has a half-life of 2.6×10^6 years. ^{98}Tc has a half-life of 4.2×10^6 years. The isomeric isotope ^{95m}Tc, with a half-life of 61 days, is useful for tracer work, as it produces energetic gamma rays. Technetium metal has been produced in kilogram quantities. The metal was first prepared by passing hydrogen gas at 1100°C over Tc$_2$S$_7$. It is now conveniently prepared by the reduction of ammonium pertechnetate with hydrogen. Technetium is a silvery-gray metal that tarnishes slowly in moist air. Until 1960, technetium was available only in small amounts and the price was as high as $2800/g. ^{99}Tc is now commercially available to holders of O.R.N.L. permits at a price of $83/g plus packing charges. ^{99}Tc is available at a cost of $1.56/μCi. The chemistry of technetium is said to be similar to that of rhenium. Technetium dissolves in nitric acid, aqua regia, and conc. sulfuric acid, but is not soluble in hydrochloric acid of any strength. The element is a remarkable corrosion inhibitor for steel. It is reported that mild carbon steels may be effectively protected by as little as 55 ppm of KTcO$_4$ in aerated distilled water at temperatures up to 250°C. This corrosion protection is limited to closed systems, since technetium is radioactive and must be confined. ^{99}Tc has a specific activity of 6.2×10^8 Bq/g. Activity of this level must not be allowed to spread. ^{99}Tc is a contamination hazard and should be handled in a glove box. The metal is an excellent superconductor at 11°K and below.

Tellurium — (L. *tellus*, earth), Te; at. wt. 127.60(3); at. no. 52; m.p. 449.51°C; b.p. 988°C; sp. gr. 6.24 (20°C); valence 2, 4, or 6. Discovered by Muller von Reichenstein in 1782; named by Klaproth, who isolated it in 1798. Tellurium is occasionally found native, but is more often found as the telluride of gold (*calaverite*), and combined with other metals. It is recovered commercially from the anode muds produced during the electrolytic refining of blister copper. The U.S., Canada, Peru, and Japan are the largest Free World producers of the element. Crystalline tellurium has a silvery-white appearance, and when pure exhibits a metallic luster. It is brittle and easily pulverized. Amorphous tellurium is formed by precipitating tellurium from a solution of telluric or tellurous acid. Whether this form is truly amorphous, or made of minute crystals, is open to question. Tellurium is a p-type semiconductor, and shows greater conductivity in certain directions, depending on alignment of the atoms. Its conductivity increases slightly with exposure to light. It can be doped with silver, copper, gold, tin, or other elements. In air, tellurium burns with a greenish-blue flame, forming the dioxide. Molten tellurium corrodes iron, copper, and stainless steel. Tellurium and its compounds are probably toxic and should be handled with care. Workmen exposed to as little as 0.01 mg/m^3 of air, or less, develop "tellurium breath," which has a garlic-like odor. Forty two isotopes and isomers of tellurium are known, with atomic masses ranging from 106 to 138. Natural tellurium consists of eight isotopes, two of which are radioactive with very long half-lives. Tellurium improves the machinability of copper and stainless steel, and its addition to lead decreases the corrosive action of sulfuric acid on lead and improves its strength and hardness. Tellurium catalysts are used in the oxidation of organic compounds and are used in hydrogenation and halogenation reactions. Tellurium is also used in electronic and semi-conductor devices. It is also used as a basic ingredient in blasting caps, and is added to cast iron for chill control. Tellurium is used in ceramics. Bismuth telluride has been used in thermoelectric devices. Tellurium costs about 20¢/g, with a purity of about 99.5%. The metal with a purity of 99.9999% costs about $6/g.

Terbium — (*Ytterby*, village in Sweden), Tb; at. wt. 158.92534(2); at. no. 65; m.p. 1356°C; b.p. 3230°C; sp. gr. 8.230; valence 3, 4. Discovered by Mosander in 1843. Terbium is a member of the lanthanide or "rare earth" group of elements. It is found in *cerite, gadolinite*, and other minerals along with other rare earths. It is recovered commercially from *monazite* in which it is present to the extent of 0.03%, from *xenotime*, and from *euxenite*,

THE ELEMENTS (continued)

a complex oxide containing 1% of more of terbia. Terbium has been isolated only in recent years with the development of ion-exchange techniques for separating the rare-earth elements. As with other rare earths, it can be produced by reducing the anhydrous chloride or fluoride with calcium metal in a tantalum crucible. Calcium and tantalum impurities can be removed by vacuum remelting. Other methods of isolation are possible. Terbium is reasonably stable in air. It is a silver-gray metal, and is malleable, ductile, and soft enough to be cut with a knife. Two crystal modifications exist, with a transformation temperature of 1289°C. Forty-two isotopes and isomers are recognized. The oxide is a chocolate or dark maroon color. Sodium terbium borate is used as a laser material and emits coherent light at 0.546 μm. Terbium is used to dope calcium fluoride, calcium tungstate, and strontium molybdate, used in solid-state devices. The oxide has potential application as an activator for green phosphors used in color TV tubes. It can be used with ZrO_2 as a crystal stabilizer of fuel cells which operate at elevated temperature. Few other uses have been found. The element is priced at about $40/g (99.9%). Little is known of the toxicity of terbium. It should be handled with care as with other lanthanide elements.

Thallium — (Gr. *thallos*, a green shoot or twig), Tl; at. wt. 204.3833(2); at. no. 81; m.p. 304°C; b.p. 1473°C; sp. gr. 11.85 (20°C); valence 1, or 3. Thallium was discovered spectroscopically in 1861 by Crookes. The element was named after the beautiful green spectral line, which identified the element. The metal was isolated both by Crookes and Lamy in 1862 about the same time. Thallium occurs in *crooksite, lorandite,* and *hutchinsonite.* It is also present in *pyrites* and is recovered from the roasting of this ore in connection with the production of sulfuric acid. It is also obtained from the smelting of lead and zinc ores. Extraction is somewhat complex and depends on the source of the thallium. Manganese nodules, found on the ocean floor, contain thallium. When freshly exposed to air, thallium exhibits a metallic luster, but soon develops a bluish-gray tinge, resembling lead in appearance. A heavy oxide builds up on thallium if left in air, and in the presence of water the hydroxide is formed. The metal is very soft and malleable. It can be cut with a knife. Forty-seven isotopes of thallium, with atomic masses ranging from 179 to 210 are recognized. Natural thallium is a mixture of two isotopes. The element and its compounds are toxic and should be handled carefully. Contact of the metal with skin is dangerous, and when melting the metal adequate ventilation should be provided. Thallium is suspected of carcinogenic potential for man. Thallium sulfate has been widely employed as a rodenticide and ant killer. It is odorless and tasteless, giving no warning of its presence. Its use, however, has been prohibited in the U.S. since 1975 as a household insecticide and rodenticide. The electrical conductivity of thallium sulfide changes with exposure to infrared light, and this compound is used in photocells. Thallium bromide-iodide crystals have been used as infrared optical materials. Thallium has been used, with sulfur or selenium and arsenic, to produce low melting glasses which become fluid between 125 and 150°C. These glasses have properties at room temperatures similar to ordinary glasses and are said to be durable and insoluble in water. Thallium oxide has been used to produce glasses with a high index of refraction. Thallium has been used in treating ringworm and other skin infections; however, its use has been limited because of the narrow margin between toxicity and therapeutic benefits. A mercury-thallium alloy, which forms a eutectic at 8.5% thallium, is reported to freeze at –60°C, some 20° below the freezing point of mercury. Thallium metal (99.999%) costs about $2/g.

Thorium — (*Thor,* Scandinavian god of war), Th; at. wt. 232.0381(1); at. no. 90; m.p. 1750°C; b.p. 4788°C; sp. gr. 11.72; valence +2(?), +3(?), +4. Discovered by Berzelius in 1828. Thorium occurs in *thorite* ($ThSiO_4$) and in *thorianite* ($ThO_2 + UO_2$). Large deposits of thorium minerals have been reported in New England and elsewhere, but these have not yet been exploited. Thorium is now thought to be about three times as abundant as uranium and about as abundant as lead or molybdenum. The metal is a source of nuclear power. There is probably more energy available for use from thorium in the minerals of the earth's crust than from both uranium and fossil fuels. Any sizable demand for thorium as a nuclear fuel is still several years in the future. Work has been done in developing thorium cycle converter-reactor systems. Several prototypes, including the HTGR (high-temperature gas-cooled reactor) and MSRE (molten salt converter reactor experiment), have operated. While the HTGR reactors are efficient, they are not expected to become important commercially for many years because of certain operating difficulties. Thorium is recovered commercially from the mineral *monazite*, which contains from 3 to 9% ThO_2 along with rare-earth minerals. Much of the internal heat the earth produces has been attributed to thorium and uranium. Several methods are available for producing thorium metal: it can be obtained by reducing thorium oxide with calcium, by electrolysis of anhydrous thorium chloride in a fused mixture of sodium and potassium chlorides, by calcium reduction of thorium tetrachloride mixed with anhydrous zinc chloride, and by reduction of thorium tetrachloride with an alkali metal. Thorium was originally assigned a position in Group IV of the periodic table. Because of its atomic weight, valence, etc., it is now considered to be the second member of the *actinide* series of elements. When pure, thorium is a silvery-white metal which is air-stable and retains its luster for several months. When contaminated with the oxide, thorium slowly tarnishes in air, becoming gray and finally black. The physical properties of thorium are greatly influenced by the degree of contamination with the oxide. The purest specimens often contain several tenths of a percent of the oxide. High-purity thorium has been made. Pure thorium is soft, very ductile, and can be cold-rolled, swaged, and drawn. Thorium is dimorphic, changing at 1400°C from a cubic to a body-centered cubic structure. Thorium oxide has a melting point of 3300°C, which is the highest of all oxides. Only a few elements, such as tungsten, and a few compounds, such as tantalum carbide, have higher melting points. Thorium is slowly attacked by water, but does not dissolve readily in most common acids, except hydrochloric. Powdered thorium metal is often pyrophoric and should be carefully handled. When heated in air, thorium turnings ignite and burn brilliantly with a white light. The principal use of thorium has been in the preparation of the Welsbach mantle, used for portable gas lights. These mantles, consisting of thorium oxide with about 1% cerium oxide and other ingredients, glow with a dazzling light when heated in a gas flame. Thorium is an important alloying element in magnesium, imparting high strength and creep resistance at elevated temperatures. Because thorium has a low work-function and high electron emission, it is used to coat tungsten wire used in electronic equipment. The oxide is also used to control the grain size of tungsten used for electric lamps; it is also used for high-temperature laboratory crucibles. Glasses containing thorium oxide have a high refractive index and low dispersion. Consequently, they find application in high quality lenses for cameras and scientific instruments. Thorium oxide has also found use as a catalyst in the conversion of ammonia to nitric acid, in petroleum cracking, and in producing sulfuric acid. Thorium has not found many uses due to its radioactive nature and its handling and disposal problems. Thirty isotopes of thorium are known with atomic masses ranging from 210 to 237. All are unstable. ^{232}Th occurs naturally and has a half-life of 1.4×10^{10} years. It is an alpha emitter. ^{232}Th goes through six alpha and four beta decay steps before becoming the stable isotope ^{208}Pb. ^{232}Th is sufficiently radioactive to expose a photographic plate in a few hours. Thorium disintegrates with the production of "thoron" (^{220}Rn), which is an alpha emitter and presents a radiation hazard. Good ventilation of areas where thorium is stored or handled is therefore essential. Thorium metal (99.8%) costs about $25/g.

Thulium — (*Thule,* the earliest name for Scandinavia), Tm; at. wt. 168.93421(3); at. no. 69; m.p. 1545°C; b.p. 1950°C; sp. gr. 9.321 (25°C); valence 3. Discovered in 1879 by Cleve. Thulium occurs in small quantities along with other rare earths in a number of minerals. It is obtained commercially from *monazite*, which contains about 0.007% of the element. Thulium is the least abundant of the rare earth elements, but with new sources recently discovered, it is now considered to be about as rare as silver, gold, or cadmium. Ion-exchange and solvent extraction techniques have recently permitted much easier separation of the rare earths, with much lower costs. Only a few years ago, thulium metal was not obtainable at any cost; in 1996 the oxide cost $20/g. Thulium metal powder now costs $70/g (99.9%). Thulium can be isolated by reduction of the oxide with lanthanum metal or by calcium reduction of the anhydrous fluoride. The pure metal has a bright, silvery luster. It is reasonably stable in air, but the metal should be protected from moisture in a closed container. The element is silver-gray, soft, malleable, and ductile, and can be cut with a knife. Forty-one isotopes and isomers are known, with atomic masses ranging from 146 to 176. Natural thulium, which is 100% ^{169}Tm, is stable. Because of the relatively high price of the metal, thulium has not yet found many practical applications. ^{169}Tm bombarded in a nuclear reactor can be used as a radiation source in portable X-ray equipment. ^{171}Tm is potentially useful as an energy source. Natural thulium also has possible use in *ferrites* (ceramic magnetic materials) used in microwave equipment. As with other lanthanides, thulium has a low-to-moderate acute toxic rating. It should be handled with care.

Tin — (anglo-Saxon, *tin*), Sn (L. *stannum*); at. wt. 118.710(7); at. no. 50; m.p. 231.93°C; b.p. 2602°C; sp. gr. (gray) 5.75, (white) 7.31; valence 2, 4. Known to the ancients. Tin is found chiefly in *cassiterite* (SnO_2). Most of the world's supply comes from China, Indonesia, Peru, Brazil, and Bolivia. The U.S. produces almost none, although occurrences have been found in Alaska and Colorado. Tin is obtained by reducing the ore with coal in a reverberatory furnace. Ordinary tin is composed of ten stable isotopes; thirty-six unstable isotopes and isomers are also known. Ordinary tin is a silver-white metal, is malleable, somewhat ductile, and has a highly crystalline structure. Due to the breaking of these crystals, a "tin cry" is heard when a bar is bent. The element has two allotropic forms at normal pressure. On warming, gray, or α tin, with a cubic structure, changes at 13.2°C into white, or β tin, the ordinary form of the metal. White tin has a tetragonal structure. When tin is cooled below 13.2°C, it changes slowly from white to gray. This change is affected by impurities such as aluminum and zinc, and can be prevented by small additions of antimony or bismuth. This change from the α to β form is called the tin pest. Tin-lead alloys are used to make organ pipes. There are few if any uses for gray tin. Tin takes a high polish and is used to coat other metals to prevent corrosion or other chemical action. Such tin plate over steel is used in the so-called tin can for preserving food. Alloys of tin are very important. Soft solder, type metal, fusible metal, pewter, bronze, bell metal, Babbitt metal, White metal, die casting alloy, and phosphor bronze are some of the important alloys using tin. Tin resists distilled sea and soft tap water, but is attacked by strong acids, alkalis, and acid salts. Oxygen in solution accelerates the attack. When heated in air, tin forms SnO_2, which is feebly acid, forming stannate salts with basic oxides. The most important salt is the chloride ($SnCl_2 \cdot H_2O$), which is used as a reducing agent and as a mordant in calico printing. Tin salts sprayed onto glass are used to produce electrically conductive coatings. These have been used for panel lighting and for frost-free windshields. Most window glass is now made by floating molten glass on molten tin (float glass) to produce a flat surface (Pilkington process). Of recent interest is a crystalline tin-niobium alloy that is superconductive at very low temperatures. This promises to be important in the construction of superconductive magnets that generate enormous field strengths but use practically no power. Such magnets, made of tin-niobium wire, weigh but a few pounds and produce magnetic fields that, when started with a small battery, are comparable to that of a 100 ton electromagnet operated continuously with a large power supply. The small amount of tin found in canned foods is quite harmless. The agreed limit of tin content in U.S. foods is 300 mg/kg. The trialkyl and triaryl tin compounds are used as biocides and must be handled carefully. Over the past 25 years the price of commercial tin has varied from 50¢/lb ($1.10/kg) to its present price of about $9/kg as of December 1999.

Titanium — (L. *Titans*, the first sons of the Earth, myth.), Ti; at. wt. 47.867(1); at. no. 22; m.p. 1668°C; b.p. 3287°C; sp. gr. 4.54; valence 2, 3, or 4. Discovered by Gregor in 1791; named by Klaproth in 1795. Impure titanium was prepared by Nilson and Pettersson in 1887; however, the pure metal (99.9%) was not made until 1910 by Hunter by heating $TiCl_4$ with sodium in a steel bomb. Titanium is present in meteorites and in the sun. Rocks obtained during the Apollo 17 lunar mission showed presence of 12.1% TiO_2. Analyses of rocks obtained during earlier Apollo missions show lower percentages. Titanium oxide bands are prominent in the spectra of M-type stars. The element is the ninth most abundant in the crust of the earth. Titanium is almost always present in igneous rocks and in the sediments derived from them. It occurs in the minerals *rutile, ilmenite,* and *sphene,* and is present in titanates and in many iron ores. Deposits of ilmenite and rutile are found in Florida, California, Tennessee, and New York. Australia, Norway, Malaysia, India, and China are also large suppliers of titanium minerals. Titanium is present in the ash of coal, in plants, and in the human body. The metal was a laboratory curiosity until Kroll, in 1946, showed that titanium could be produced commercially by reducing titanium tetrachloride with magnesium. This method is largely used for producing the metal today. The metal can be purified by decomposing the iodide. Titanium, when pure, is a lustrous, white metal. It has a low density, good strength, is easily fabricated, and has excellent corrosion resistance. It is ductile only when it is free of oxygen. The metal burns in air and is the only element that burns in nitrogen. Titanium is resistant to dilute sulfuric and hydrochloric acid, most organic acids, moist chlorine gas, and chloride solutions. Natural titanium consists of five isotopes with atomic masses from 46 to 50. All are stable. Eighteen other unstable isotopes are known. The metal is dimorphic. The hexagonal α form changes to the cubic β form very slowly at about 880°C. The metal combines with oxygen at red heat, and with chlorine at 550°C. Titanium is important as an alloying agent with aluminum, molybdenum, manganese, iron, and other metals. Alloys of titanium are principally used for aircraft and missiles where lightweight strength and ability to withstand extremes of temperature are important. Titanium is as strong as steel, but 45% lighter. It is 60% heavier than aluminum, but twice as strong. Titanium has potential use in desalination plants for converting sea water into fresh water. The metal has excellent resistance to sea water and is used for propeller shafts, rigging, and other parts of ships exposed to salt water. A titanium anode coated with platinum has been used to provide cathodic protection from corrosion by salt water. Titanium metal is considered to be physiologically inert; however, titanium powder may be a carcinogenic hazard. When pure, titanium dioxide is relatively clear and has an extremely high index of refraction with an optical dispersion higher than diamond. It is produced artificially for use as a gemstone, but it is relatively soft. Star sapphires and rubies exhibit their asterism as a result of the presence of TiO_2. Titanium dioxide is extensively used for both house paint and artist's paint, as it is permanent and has good covering power. Titanium oxide pigment accounts for the largest use of the element. Titanium paint is an excellent reflector of infrared, and is extensively used in solar observatories where heat causes poor seeing conditions. Titanium tetrachloride is used to iridize glass. This compound fumes strongly in air and has been used to produce smoke screens. The price of titanium metal (99.8%) is about $800/kg.

Tungsten — (Swedish, *tung sten,* heavy stone); also known as *wolfram* (from *wolframite,* said to be named from *wolf rahm* or *spumi lupi,* because the ore interfered with the smelting of tin and was supposed to devour the tin), W; at. wt. 183.84(1); at. no. 74; m.p. 3422°C; b.p. 5555°C; sp. gr. 19.3 (20°C); valence 2, 3, 4, 5, or 6. In 1779 Peter Woulfe examined the mineral now known as *wolframite* and concluded it must contain a new substance. Scheele, in 1781, found that a new acid could be made from *tung sten* (a name first applied about 1758 to a mineral now known as *scheelite*). Scheele and Berman suggested the possibility of obtaining a new metal by reducing this acid. The de Elhuyar brothers found an acid in *wolframite* in 1783 that was identical to the acid of *tungsten* (tungstic acid) of Scheele, and in that year they succeeded in obtaining the element by reduction of this acid with charcoal. Tungsten occurs in *wolframite,* (Fe, Mn)WO_4; *scheelite,* $CaWO_4$; *huebnerite,* $MnWO_4$; and *ferberite,* $FeWO_4$. Important deposits of tungsten occur in California, Colorado, Bolivia, Russia, and Portugal. China is reported to have about 75% of the world's tungsten resources. Natural tungsten contains five stable isotopes. Thirty two other unstable isotopes and isomers are recognized. The metal is obtained commercially by reducing tungsten oxide with hydrogen or carbon. Pure tungsten is a steel-gray to tin-white metal. Very pure tungsten can be cut with a hacksaw, and can be forged, spun, drawn, and extruded. The impure metal is brittle and can be worked only with difficulty. Tungsten has the highest melting point of all metals, and at temperatures over 1650°C has the highest tensile strength. The metal oxidizes in air and must be protected at elevated temperatures. It has excellent corrosion resistance and is attacked only slightly by most mineral acids. The thermal expansion is about the same as borosilicate glass, which makes the metal useful for glass-to-metal seals. Tungsten and its alloys are used extensively for filaments for electric lamps, electron and television tubes, and for metal evaporation work; for electrical contact points for automobile distributors; X-ray targets; windings and heating elements for electrical furnaces; and for numerous spacecraft and high-temperature applications. High-speed tool steels, Hastelloy®, Stellite®, and many other alloys contain tungsten. Tungsten carbide is of great importance to the metal-working, mining, and petroleum industries.Calcium and magnesium tungstates are widely used in fluorescent lighting; other salts of tungsten are used in the chemical and tanning industries. Tungsten disulfide is a dry, high-temperature lubricant, stable to 500°C. Tungsten bronzes and other tungsten compounds are used in paints. Zirconium tungstate has found recent applications (see under Zirconium). Tungsten powder (99.999%) costs about $2900/kg.

Uranium — (Planet *Uranus*), U; at. wt. 238.0289(1); at. no. 92; m.p. 1135°C; b.p. 4131°C; sp. gr. ~18.95; valence 2, 3, 4, 5, or 6. Yellow-colored glass, containing more than 1% uranium oxide and dating back to 79 A.D., has been found near Naples, Italy. Klaproth recognized an unknown element in *pitchblende* and attempted to isolate the metal in 1789. The metal apparently was first isolated in 1841 by Peligot, who reduced the anhydrous chloride with potassium. Uranium is not as rare as it was once thought. It is now considered to be more plentiful than mercury, antimony, silver, or cadmium, and is about as abundant as molybdenum or arsenic. It occurs in numerous minerals such as *pitchblende, uraninite, carnotite, autunite, uranophane, davidite,* and *tobernite.* It is also found in *phosphate rock, lignite, monazite sands,* and can be recovered commercially from these sources. Large deposits of uranium ore occur in Utah, Colorado, New Mexico, Canada, and elsewhere. Uranium can be made by reducing uranium halides with alkali or alkaline earth metals or by reducing uranium oxides by calcium, aluminum, or carbon at high temperatures. The metal can also be produced by electrolysis of KUF_5 or UF_4, dissolved in a molten mixture of $CaCl_2$ and NaCl. High-purity uranium can be prepared by the thermal decomposition of uranium halides on a hot filament. Uranium exhibits three crystallographic modifications as follows:

$$\alpha \xrightarrow{688°C} \beta \xrightarrow{776°C} \gamma$$

Uranium is a heavy, silvery-white metal which is pyrophoric when finely divided. It is a little softer than steel, and is attacked by cold water in a finely divided state. It is malleable, ductile, and slightly paramagnetic. In air, the metal becomes coated with a layer of oxide. Acids dissolve the metal, but it is unaffected by alkalis. Uranium has twenty three isotopes, one of which is an isomer and all of which are radioactive. Naturally occurring uranium contains 99.2745% by weight ^{238}U, 0.720% ^{235}U, and 0.0055% ^{234}U. Studies show that the percentage weight of ^{235}U in natural uranium varies by as much as 0.1%, depending on the source. The U.S.D.O.E. has adopted the value of 0.711 as being their "official" percentage of ^{235}U in natural uranium. Natural uranium is sufficiently radioactive to expose a photographic plate in an hour or so. Much of the internal heat of the earth is thought to be attributable to the presence of uranium and thorium. ^{238}U with a half-life of 4.46×10^9 years, has been used to estimate the age of igneous rocks. The origin of uranium, the highest member of the naturally occurring elements — except perhaps for traces of neptunium or plutonium — is not clearly understood, although it has been thought that uranium might be a decay product of elements of higher atomic weight, which may have once been present on earth or elsewhere in the universe. These original elements may have been formed as a result of a primordial "creation," known as "the big bang," in a supernova, or in some other stellar processes. The fact that recent studies show that most trans-uranic elements are extremely rare with very short half-lives indicates that it may be necessary to find some alternative explanation for the very large quantities of radioactive uranium we find on earth. Studies of meteorites from other parts of the solar system show a relatively low radioactive content, compared to terrestrial rocks. Uranium is of great importance as a nuclear fuel. ^{238}U can be converted into fissionable plutonium by the following reactions:

$$^{238}U(n,\gamma) \to ^{239}U \xrightarrow{\beta} ^{239}Np \xrightarrow{\beta} ^{239}Pu$$

This nuclear conversion can be brought about in "breeder" reactors where it is possible to produce more new fissionable material than the fissionable material used in maintaining the chain reaction. ^{235}U is of even greater importance, for it is the key to the utilization of uranium. ^{235}U, while occurring in natural uranium to the extent of only 0.72%, is so fissionable with slow neutrons that a self-sustaining fission chain reaction can be made to occur in a reactor constructed from natural uranium and a suitable moderator, such as heavy water or graphite, alone. ^{235}U can be concentrated by gaseous diffusion and other physical processes, if desired, and used directly as a nuclear fuel, instead of natural uranium, or used as an explosive. Natural uranium, slightly enriched with ^{235}U by a small percentage, is used to fuel nuclear power reactors for the generation of electricity. Natural thorium can be irradiated with neutrons as follows to produce the important isotope ^{233}U.

$$^{232}Th(n,\gamma) \to ^{233}Th \xrightarrow{\beta} ^{233}Pa \xrightarrow{\beta} ^{233}U$$

While thorium itself is not fissionable, ^{233}U is, and in this way may be used as a nuclear fuel. One pound of completely fissioned uranium has the fuel value of over 1500 tons of coal. The uses of nuclear fuels to generate electrical power, to make isotopes for peaceful purposes, and to make explosives are well known. The estimated world-wide production of the 437 nuclear power reactors in operation in 1998 amounted to about 352,000 Megawatt hours. In 1998 the U.S. had about 107 commercial reactors with an output of about 100,000 Megawatt-hours. Some nuclear-powered electric generating plants have recently been closed because of safety concerns. There are also serious problems with nuclear waste disposal that have not been completely resolved. Uranium in the U.S.A. is controlled by the U.S. Nuclear Regulatory Commission, under the Department of Energy. Uses are being found for the large quantities of "depleted" uranium, now available, where uranium-235 has been lowered to about 0.2%. Depleted uranium has been used for inertial guidance devices, gyrocompasses, counterweights for aircraft control surfaces, ballast for missile reentry vehicles, and as a shielding material for tanks, etc. Concerns, however, have been raised over its low radioactive properties. Uranium metal is used for X-ray targets for production of high-energy X-rays. The nitrate has been used as photographic toner, and the acetate is used in analytical chemistry. Crystals of uranium nitrate are triboluminescent. Uranium salts have also been used for producing yellow "vaseline" glass and glazes. Uranium and its compounds are highly toxic, both from a chemical and radiological standpoint. Finely divided uranium metal, being pyrophoric, presents a fire hazard. The maximum permissible total body burden of natural uranium (based on radiotoxicity) is $0.2\,\mu$Ci for soluble compounds. Recently, the natural presence of uranium and thorium in many soils has become of concern to homeowners because of the generation of radon and its daughters (see under Radon). Uranium metal is available commercially at a cost of about \$6/g (99.7%) in air-tight glass under argon.

Unilnilium, Unnilbium, etc. — See under the opening paragraphs of this article and also under Elements 110 to 118.

Vanadium — (Scandinavian goddess, *Vanadis*), V; at. wt. 50.9415(1); at. no. 23; m.p. 1910°C; b.p. 3407°C; sp. gr. 6.11 (18.7°C); valence 2, 3, 4, or 5. Vanadium was first discovered by del Rio in 1801. Unfortunately, a French chemist incorrectly declared del Rio's new element was only impure chromium; del Rio thought himself to be mistaken and accepted the French chemist's statement. The element was rediscovered in 1830 by Sefstrom, who named the element in honor of the Scandinavian goddess *Vanadis* because of its beautiful multicolored compounds. It was isolated in nearly pure form by Roscoe, in 1867, who reduced the chloride with hydrogen. Vanadium of 99.3 to 99.8% purity was not produced until 1927. Vanadium is found in about 65 different minerals among which are *carnotite, roscoelite, vanadinite,* and *patronite* important sources of the metal. Vanadium is also found in phosphate rock and certain iron ores, and is present in some crude oils in the form of organic complexes. It is also found in small percentages in meteorites. Commercial production from petroleum ash holds promise as an important source of the element. China, South Africa, and Russia supply much of the world's vanadium ores. High-purity ductile vanadium can be obtained by reduction of vanadium trichloride with magnesium or with magnesium-sodium mixtures. Much of the vanadium metal being produced is now made by calcium reduction of V_2O_5 in a pressure vessel, an adaption of a process developed by McKechnie and Seybolt. Natural vanadium is a mixture of two isotopes, ^{50}V (0.25%) and ^{51}V (99.75%). ^{50}V is slightly radioactive, having a long half-life. Twenty other unstable isotopes are recognized. Pure vanadium is a bright white metal, and is soft and ductile. It has good corrosion resistance to alkalis, sulfuric and hydrochloric acid, and salt water, but the metal oxidizes readily above 660°C. The metal has good structural strength and a low fission neutron cross section, making it useful in nuclear applications. Vanadium is used in producing rust resistant, spring, and highspeed tool steels. It is an important carbide stabilizer in making steels. About 80% of the vanadium now produced is used as ferrovanadium or as a steel additive. Vanadium foil is used as a bonding agent in cladding titanium to steel. Vanadium pentoxide is used in ceramics and as a catalyst. It is also used in producing a superconductive magnet with a field of 175,000 gauss. Vanadium and its compounds are toxic and should be handled with care. Ductile vanadium is commercially available. Vanadium metal (99.7%) costs about \$3/g.

Wolfram — see Tungsten.

Xenon — (Gr. *xenon*, stranger), Xe; at. wt. 131.29(2); at. no. 54; m.p. −111.79°C; b.p. −108.12°C; t_c 16.62°C; density (gas) 5.887 ± 0.009 g/l, sp. gr (liquid) 3.52 (−109°C); valence usually 0. Discovered by Ramsay and Travers in 1898 in the residue left after evaporating liquid air components. Xenon is a member of the so-called noble or "inert" gases. It is present in the atmosphere to the extent of about one part in twenty million. Xenon is present in the Martian atmosphere to the extent of 0.08 ppm. The element is found in the gases evolved from certain mineral springs, and is commercially obtained by extraction from liquid air. Natural xenon is composed of nine stable isotopes. In addition to these, thirty five unstable isotopes and isomers have been characterized. Before 1962, it had generally been assumed that xenon and other noble gases were unable to form compounds. Evidence has been mounting in the past few years that xenon, as well as other members of the zero valence elements, do form compounds. Among the "compounds" of xenon now reported are xenon hydrate, sodium perxenate, xenon deuterate, difluoride, tetrafluoride, hexafluoride, and $XePtF_6$ and $XeRhF_6$. Xenon trioxide, which is highly explosive, has been prepared. More than 80 xenon compounds have been made with xenon chemically bonded to fluorine and oxygen. Some xenon compounds are colored. Metallic xenon has been produced, using several hundred kilobars of pressure. Xenon in a vacuum tube produces a beautiful blue glow when excited by an electrical discharge. The gas is used in making electron tubes, stroboscopic lamps, bactericidal lamps, and lamps used to excite ruby lasers for generating coherent light. Xenon is used in the atomic energy field in bubble chambers, probes, and other applications where its high molecular weight is of value. The perxenates are used in analytical chemistry as oxidizing agents. ^{133}Xe and ^{135}Xe are produced by neutron irradiation in air cooled nuclear reactors. ^{133}Xe has useful applications as a radioisotope. The element is available in sealed glass containers for about \$20/l of gas at standard pressure. Xenon is not toxic, but its compounds are highly toxic because of their strong oxidizing characteristics.

Ytterbium — (Ytterby, village in Sweden), Yb; at. wt. 173.04(3); at. no. 70; m.p. 819°C; b.p. 1196°C; sp. gr (α) 6.903 (β) 6.966; valence 2, 3. Marignac in 1878 discovered a new component, which he called *ytterbia*, in the earth then known as *erbia*. In 1907, Urbain separated ytterbia into two components, which he called *neoytterbia* and *lutecia*. The elements in these earths are now known as *ytterbium* and *lutetium*, respectively. These elements are identical with *aldebaranium* and *cassiopeium*, discovered independently and at about the same time by von Welsbach. Ytterbium occurs along with other rare earths in a number of rare minerals. It is commercially recovered principally from *monazite sand*, which contains about 0.03%. Ion-exchange and solvent extraction techniques developed in recent years have greatly simplified the separation of the rare earths from one another. The element was first prepared by Klemm and Bonner in 1937 by reducing ytterbium trichloride with potassium. Their metal was mixed, however, with KCl. Daane, Dennison, and Spedding prepared a much purer form in 1953 from which the chemical and physical properties of the element could be determined. Ytterbium has a bright silvery luster, is soft, malleable, and quite ductile. While the element is fairly stable, it should be kept in closed containers to protect it from air and moisture. Ytterbium is readily attacked and dissolved by dilute and concentrated mineral acids and reacts slowly with water. Ytterbium has three allotropic forms with transformation points at −13° and 795°C. The beta form is a room-temperature, face-centered, cubic modification, while the high-temperature gamma form is a body-centered cubic form. Another body-centered cubic phase has recently been found to be stable at high pressures at room temperatures. The beta form ordinarily has metallic-type conductivity, but becomes a semiconductor when the pressure is increased above 16,000 atm. The electrical resistance increases tenfold as the pressure is increased to 39,000 atm and drops to about 80% of its standard temperature-pressure resistivity at a pressure of 40,000 atm. Natural ytterbium is a mixture of seven stable isotopes. Twenty six other unstable isotopes and isomers are known. Ytterbium metal has possible use in improving the grain refinement, strength, and other mechanical properties of stainless steel. One isotope is reported to have been used as a radiation source as a substitute for a portable X-ray machine where electricity is unavailable. Few other uses have been found. Ytterbium metal is available with a purity of about 99.99% for about \$20/g. Ytterbium has a low acute toxic rating, but may present a carcinogenic hazard.

Yttrium — (*Ytterby*, village in Sweden near Vauxholm), Y; at. wt. 88.90585(2); at. no. 39; m.p. 1522°C; b.p. 3345°C; sp. gr. 4.469 (25°C); valence 3. *Yttria*, which is an earth containing yttrium, was discovered by Gadolin in 1794. *Ytterby* is the site of a quarry which yielded many unusually minerals containing rare earths and other elements. This small town, near Stockholm, bears the honor of giving names to *erbium, terbium*, and *ytterbium* as well as *yttrium*. In 1843 Mosander showed that yttria could be resolved into the oxides (or earths) of three elements. The name yttria was reserved for the most basic one; the others were named *erbia* and *terbia*. Yttrium occurs in nearly all of the rare-earth minerals. Analysis of lunar rock samples obtained during the Apollo missions show a relatively high yttrium content. It is recovered commercially from *monazite sand*, which contains about 3%, and from *bastnasite*, which contains about 0.2%. Wohler obtained the impure element in 1828 by reduction of the anhydrous chloride with potassium. The metal is now produced commercially by reduction of the fluoride with calcium metal. It can also be prepared by other techniques. Yttrium has a silver-metallic luster and is relatively stable in air. Turnings of the metal, however, ignite in air if their temperature exceeds 400°C, and finely divided yttrium is very unstable in air. Yttrium oxide is one of the most important compounds of yttrium and accounts for the largest use. It is widely used in making YVO_4 europium, and Y_2O_3 europium phosphors to give the red color in color television tubes. Many hundreds of thousands of pounds are now used in this application. Yttrium oxide also is used to produce yttrium-iron-garnets, which are very effective microwave filters. Yttrium iron, aluminum, and gadolinium garnets, with formulas such as $Y_3Fe_5O_{12}$ and $Y_3Al_5O_{12}$, have interesting magnetic properties. Yttrium iron garnet is also exceptionally efficient as both a transmitter and transducer of acoustic energy. Yttrium aluminum garnet, with a hardness of 8.5, is also finding use as a gemstone (simulated diamond). Small amounts of yttrium (0.1 to 0.2%) can be used to reduce the grain size in chromium, molybdenum, zirconium, and titanium, and to increase strength of aluminum and magnesium alloys. Alloys with other useful properties can be obtained by using yttrium as an additive. The metal can be used as a deoxidizer for vanadium and other nonferrous metals. The metal has a low cross section for nuclear capture. ^{90}Y, one of the isotopes of yttrium, exists in equilibrium with its parent ^{90}Sr, a product of atomic explosions. Yttrium has been considered for use as a nodulizer for producing nodular cast iron, in which the graphite forms compact nodules instead of the usual flakes. Such iron has increased ductility. Yttrium is also finding application in laser systems and as a catalyst for ethylene polymerization. It has also potential use in ceramic and glass formulas, as the oxide has a high melting point and imparts shock resistance and low expansion characteristics to glass. Natural yttrium contains but one isotope, ^{89}Y. Forty-three other unstable isotopes and isomers have been characterized. Yttrium metal of 99.9% purity is commercially available at a cost of about $3/g.

Zinc — (Ger. *Zink*, of obscure origin), Zn; at. wt. 65.39(2); at. no. 30; m.p. 419.53°C; b.p. 907°C; sp. gr. 7.133 (25°C); valence 2. Centuries before zinc was recognized as a distinct element, zinc ores were used for making brass. Tubal-Cain, seven generations from Adam, is mentioned as being an "instructor in every artificer in brass and iron." An alloy containing 87% zinc has been found in prehistoric ruins in Transylvania. Metallic zinc was produced in the 13th century A.D. in India by reducing calamine with organic substances such as wool. The metal was rediscovered in Europe by Marggraf in 1746, who showed that it could be obtained by reducing *calamine* with charcoal. The principal ores of zinc are *sphalerite* or *blende* (sulfide), *smithsonite* (carbonate), *calamine* (silicate), and *franklinite* (zinc, manganese, iron oxide). Canada, Japan, Belgium, Germany, and The Netherlands are suppliers of zinc ores. Zinc is also mined in Alaska, Tennessee, Missouri, and elsewhere in the U.S. Zinc can be obtained by roasting its ores to form the oxide and by reduction of the oxide with coal or carbon, with subsequent distillation of the metal. Other methods of extraction are possible. Naturally occurring zinc contains five stable isotopes. Twenty-five other unstable isotopes and isomers are recognized. Zinc is a bluish-white, lustrous metal. It is brittle at ordinary temperatures but malleable at 100 to 150°C. It is a fair conductor of electricity, and burns in air at high red heat with evolution of white clouds of the oxide. The metal is employed to form numerous alloys with other metals. Brass, nickel silver, typewriter metal, commercial bronze, spring brass, German silver, soft solder, and aluminum solder are some of the more important alloys. Large quantities of zinc are used to produce die castings, used extensively by the automotive, electrical, and hardware industries. An alloy called *Prestal*®, consisting of 78% zinc and 22% aluminum is reported to be almost as strong as steel but as easy to mold as plastic. It is said to be so plastic that it can be molded into form by relatively inexpensive die casts made of ceramics and cement. It exhibits superplasticity. Zinc is also extensively used to galvanize other metals such as iron to prevent corrosion. Neither zinc nor zirconium is ferromagnetic; but $ZrZn_2$ exhibits ferromagnetism at temperatures below 35 K. Zinc oxide is a unique and very useful material to modern civilization. It is widely used in the manufacture of paints, rubber products, cosmetics, pharmaceuticals, floor coverings, plastics, printing inks, soap, storage batteries, textiles, electrical equipment, and other products. It has unusual electrical, thermal, optical, and solid-state properties that have not yet been fully investigated. Lithopone, a mixture of zinc sulfide and barium sulfate, is an important pigment. Zinc sulfide is used in making luminous dials, X-ray and TV screens, and fluorescent lights. The chloride and chromate are also important compounds. Zinc is an essential element in the growth of human beings and animals. Tests show that zinc-deficient animals require 50% more food to gain the same weight as an animal supplied with sufficient zinc. Zinc is not considered to be toxic, but when freshly formed ZnO is inhaled a disorder known as the *oxide shakes* or *zinc chills* sometimes occurs. It is recommended that where zinc oxide is encountered good ventilation be provided. The commercial price of zinc was roughly 60¢/lb ($1.30 kg). Zinc metal with a purity of 99.9999% is priced at about $6/g.

Zirconium — (Syriac, *zargun*, color of gold), Zr; at. wt. 91.224(2); at. no. 40; m.p. 1855°C; b.p. 4409°C; sp. gr. 6.506 (20°C); valence +2, +3, and +4. The name *zircon* may have originated from the Syriac word *zargono*, which describes the color of certain gemstones now known as *zircon, jargon, hyacinth, jacinth,* or *ligure.*This mineral, or its variations, is mentioned in biblical writings. These minerals were not known to contain this element until Klaproth, in 1789, analyzed a *jargon* from Sri Lanka and found a new earth, which Werner named zircon (*silex circonius*), and Klaproth called *Zirkonerde (zirconia)*. The impure metal was first isolated by Berzelius in 1824 by heating a mixture of potassium and potassium zirconium fluoride in a small iron tube. Pure zirconium was first prepared in 1914. Very pure zirconium was first produced in 1925 by van Arkel and de Boer by an iodide decomposition process they developed. Zirconium is found in abundance in S-type stars, and has been identified in the sun and meteorites. Analyses of lunar rock samples obtained during the various Apollo missions to the moon show a surprisingly high zirconium oxide content, compared with terrestial rocks. Naturally occurring zirconium contains five isotopes. Thirty-one other radioactive isotopes and isomers are known to exist. *Zircon*, $ZrSiO_4$, the principal ore, is found in deposits in Florida, South Carolina, Australia, South Africa, and elsewhere. *Baddeleyite*, found in Brazil, is an important zirconium mineral. It is principally pure ZrO_2in crystalline form having a hafnium content of about 1%. Zirconium also occurs in some 30 other recognized mineral species. Zirconium is produced commercially by reduction of the chloride with magnesium (the Kroll Process), and by other methods. It is a grayish-white lustrous metal. When finely divided, the metal may ignite spontaneously in air, especially at elevated temperatures. The solid metal is much more difficult to ignite. The inherent toxicity of zirconium compounds is low. Hafnium is invariably found in zirconium ores, and the separation is difficult. Commercial-grade zirconium contains from 1 to 3% hafnium. Zirconium has a low absorption cross section for neutrons, and is therefore used for nuclear energy applications, such as for cladding fuel elements. Commercial nuclear power generation now takes more than 90% of zirconium metal production. Reactors of the size now being made may use as much as a half-million lineal feet of zirconium alloy tubing. Reactor-grade zirconium is essentially free of hafnium. *Zircaloy*® is an important alloy developed specifically for nuclear applications. Zirconium is exceptionally resistant to corrosion by many common acids and alkalis, by sea water, and by other agents. It is used extensively by the chemical industry where corrosive agents are employed. Zirconium is used as a getter in vacuum tubes, as an alloying agent in steel, in surgical appliances, photoflash bulbs, explosive primers, rayon spinnerets, lamp filaments, etc. It is used in poison ivy lotions in the form of the carbonate as it combines with *urushiol*. With niobium, zirconium is superconductive at low temperatures and is used to make superconductive magnets, which offer hope of direct large-scale generation of electric power. Alloyed with zinc, zirconium becomes magnetic at temperatures below 35 K. Zirconium oxide (zircon) has a high index of refraction and is used as a gem material. The impure oxide, zirconia, is used for laboratory crucibles that will withstand heat shock, for linings of metallurgical furnaces, and by the glass and ceramic industries as a refractory material. Its use as a refractory material accounts for a large share of all zirconium consumed. Zirconium tungstate is an unusual material that shrinks, rather than expands, when heated. While this compound has been known for more than 30 years, it is only now that it is being studied to determine the nature of this unusual behavior. A few other compounds are known to possess this property, but they tend to shrink in one direction, while they stretch out in others in order to maintain an overall volume. Zirconium tungstate shrinks in all directions over a wide temperature range of from near absolute zero to +777°C. This material is being considered for use in very small computer chips, which are subject to severe temperature changes. It is also being considered for use in composite materials where thermal expansion may be a problem. Zirconium of about 99.5% purity is available at a cost of about $2100/kg or about $3/g.

PHYSICAL CONSTANTS OF INORGANIC COMPOUNDS

The compounds in this table were selected on the basis of their laboratory and industrial importance, as well as their value in illustrating trends in the variation of physical properties with position in the periodic table. An effort has been made to include the most frequently encountered inorganic substances; organometallics are not covered, with the exception of metal salts of organic acids. Many, if not most, of the compounds that are solids at ambient temperature can exist in more than one crystalline modification. The information given here applies to the most stable or common crystalline form. In cases where two or more forms are of practical importance, separate entries will be found in the table.

Compounds are arranged primarily in alphabetical order by the most commonly used name. However, adjustments are made in many instances in order to bring closely related compounds together. For example, hydrides of elements such as boron, silicon, and germanium are grouped together immediately following the entry for the parent element, since they would otherwise be scattered throughout the table. Likewise, the oxoacids of an element are given in one group when a strict alphabetical order would separate them (e.g., sulfuric acid and fluorosulfuric acid). Whenever deviations from alphabetical order occur, the names are included in the Synonym Index to provide an easy way to locate these substances.

The data fields in the table are described below:

- **No.:** An identification number used in the indexes.
- **Name:** Common name for the substance. The valence state of a metallic element is indicated by a Roman numeral , e.g., copper in the +2 state is written as copper(II) rather than cupric (but names beginning with cupric, ferrous, etc. are included in the synonym index).
- **Formula:** The simplest descriptive formula is given, but this does not necessarily specify the actual structure of the compound. For example, aluminum chloride is designated as $AlCl_3$, even though a more accurate representation of the structure in the solid phase (and, under some conditions, in the gas phase) is Al_2Cl_6. A few exceptions are made, such as the use of Hg_2^{+2} for the mercury(I) ion.
- **CAS RN:** Chemical Abstracts Service Registry Number. An asterisk * following the CAS RN for a hydrate indicates that the number refers to the anhydrous compound. In most cases the generic CAS RN for the compound is given rather than the number for a specific crystalline form or mineral.
- **Mol. Wt.:** Molecular weight (relative molar mass) as calculated with the 1993 IUPAC Recommended Atomic Weights. The number of decimal places corresponds to the number of places in the atomic weight of the least accurately known element (e.g., one place for lead compounds, two places for compounds of selenium, germanium, etc.); a maximum of three places is given. For compounds of radioactive elements for which IUPAC makes no recommendation, the mass number of the isotope with the longest half-life is used, and the result is rounded to the nearest integer.
- **Physical form:** The crystal system is given ,when available, for compounds that are solid at room temperature, together with color and other descriptive features. Abbreviations are listed below.
- **mp:** Normal melting point in °C. A "tp" following the number indicates a triple point, where solid, liquid, and gas are in equilibrium at a pressure greater than one atmosphere (i.e., the normal melting point does not exist). When available, the triple point pressure is listed.
- **bp:** Normal boiling point in °C (referred to 101.325 kPa or 760 mmHg pressure). An "sp" following the number indicates a sublimation point, where the pressure of the vapor in equilibrium with the solid reaches 101.325 kPa. See Reference 8, p. 23, for further discussion of sublimation points and triple points. Note: The abbreviation "subl" indicates that there is a perceptible sublimation pressure above the solid, but does not imply any quantitative information on the pressure-temperature behavior.
- **den:** Density. Values for solids and liquids are always in grams per cubic centimeter and can be assumed to refer to temperatures near room temperature unless otherwise stated. Values for gases are the calculated ideal gas densities in grams per liter at 25°C and 101.325 kPa; the unit is always specified with a gas value.
- **Solubility:** Qualitative information on the solubility in various solvents is given. The abbreviations are listed below.
- **Other Data:** The codes refer to other tables in this handbook which include an entry for the compound in question. These codes are:

    a. Data on enthalpy of formation, entropy, and heat capacity in the table "Standard Thermodynamic Properties of Chemical Substances" (Section 5).
    b Critical temperature, pressure, and volume in the table "Critical Constants, Boiling Points, and Melting Points of Selected Compounds" (Section 6).
    c. Table of "Enthalpy of Fusion" (Section 6).
    d. Table of "Enthalpy of Vaporization" (Section 6).
    e. Vapor pressure data in one or more of the vapor pressure tables in Section 6.
    f. Optical constants and hardness in the table "Physical and Optical Properties of Minerals" (Section 4).

A Synonym Index and a CAS Registry Number Index follow the table.

LIST OF ABBREVIATIONS

Ac	acetyl	gl	glass, glassy	rhom	rhombohedral
ace	acetone	grn	green	s	soluble in
acid	acid solutions	hc	hydrocarbon solvents	silv	silvery
alk	alkaline solutions	hex	hexagonal	sl	slightly soluble in
amorp	amorphous	hp	heptane	soln	solution
anh	anhydrous	hx	hexane	sp	sublimation point
aq	aqueous	hyd	hydrate	stab	stable
blk	black	hyg	hygroscopic	subl	sublimes
brn	brown	i	insoluble in	temp	temperature
bz	benzene	liq	liquid	tetr	tetragonal
chl	chloroform	MeOH	methanol	thf	tetrahydrofuran
col	colorless	mono	monoclinic	tol	toluene
conc	concentrated	octahed	octahedral	tp	triple point
cry	crystals, crystalline	oran	orange	trans	transition, transformation
cub	cubic	orth	orthorhombic	tricl	triclinic
cyhex	cyclohexane	os	organic solvents	trig	trigonal
dec	decomposes	peth	petroleum ether	unstab	unstable
dil	dilute	pow	powder	viol	violet
diox	dioxane	prec	precipitate	visc	viscous
eth	ethyl ether	pur	purple	vs	very soluble in
EtOH	ethanol	py	pyridine	wh	white
exp	explodes, explosive	reac	reacts with	xyl	xylene
flam	flammable	refrac	refractory	yel	yellow

REFERENCES

1. Phillips, S.L., and Perry, D.L., *Handbook of Inorganic Compounds*, CRC Press, Boca Raton, FL, 1995.
2. Trotman-Dickenson, A.F., Executive Editor, *Comprehensive Inorganic Chemistry*, Vol. 1-5, Pergamon Press, Oxford, 1973.
3. Greenwood, N.N., and Earnshaw, A., *Chemistry of the Elements*, Pergamon Press, Oxford, 1984.
4. Budavari, S., Editor, *The Merck Index, Eleventh Edition*, Merck & Co., Rahway, NJ, 1989.
5. *GMELIN Handbook of Inorganic and Organometallic Chemistry*, Springer-Verlag, Heidelberg.
6. Chase, M.W., Davies, C.A., Downey, J.R., Frurip, D.J., McDonald, R.A., and Syverud, A.N.; *JANAF Thermochemical Tables, Third Edition, J. Phys. Chem. Ref. Data*, Vol. 14, Suppl. 1, 1985.
7. Donnay, J.D.H., and Ondik, H.M., *Crystal Data Determinative Tables, Third Edition, Volumes 2 and 4, Inorganic Compounds*, Joint Committee on Powder Diffraction Standards, Swarthmore, PA, 1973.
8. Lide, D.R., and Kehiaian, H.V., *CRC Handbook of Thermophysical and Thermochemical Data*, CRC Press, Boca Raton, FL, 1994.
9. *Kirk-Othmer Concise Encyclopedia of Chemical Technology*, Wiley-Interscience, New York, 1985.
10. *Dictionary of Inorganic Compounds*, Chapman & Hall, New York, 1992.
11. Massalski, T.B., Editor, *Binary Alloy Phase Diagrams*, American Society for Metals, Metals Park, Ohio, 1986.
12. *Landolt-Börnstein, Numerical Data and Functional Relationships in Science and Technology, Sixth Edition, II/4, Caloric Quantities of State*, Springer-Verlag, Heidelberg, 1961.
13. Deer, W.A., Howie, R.A., and Zussman, J., *An Introduction to the Rock-Forming Minerals, 2nd Edition*, Longman Scientific & Technical, Harlow, Essex, 1992.
14. Carmichael, R.S., *Practical Handbook of Physical Properties of Rocks and Minerals*, CRC Press, Boca Raton, FL, 1989.
15. Dinsdale, A.T., "SGTE Data for Pure Elements", *CALPHAD*, 15, 317-425, 1991.
16. Madelung, O., *Semiconductors: Group IV Elements and III-IV Compounds*, Springer-Verlag, Heidelberg, 1991.
17. Daubert, T.E., Danner, R.P., Sibul, H.M., and Stebbins, C.C., *Physical and Thermodynamic Properties of Pure Compounds: Data Compilation*, extant 1994 (core with 4 supplements), Taylor & Francis, Bristol, PA.

No.	Name Formula	CAS RN Mol. Wt.	Physical Form	mp/°C den/g cm⁻³	bp/°C Other Data	Solubility
1	Actinium Ac	7440-34-8 227	silv metal; cub	1051 10	3198 a	
2	Actinium bromide AcBr$_3$	33689-81-5 467	wh hex cry	5.85		s H$_2$O
3	Actinium chloride AcCl$_3$	22986-54-5 333	wh hex cry	4.81		
4	Actinium fluoride AcF$_3$	33689-80-4 284	wh hex cry	7.88		i H$_2$O
5	Actinium iodide AcI$_3$	33689-82-6 608	wh cry			s H$_2$O
6	Actinium oxide Ac$_2$O$_3$	12002-61-8 502	wh hex cry	1977 9.19		i H$_2$O
7	Aluminum Al	7429-90-5 26.982	silv-wh metal; cub cry	660.32 2.70	2519 a,c,d,e	i H$_2$O; s acid, alk
8	Aluminum ammonium sulfate	7784-25-0 237.148	wh powder			sl H$_2$O; i EtOH
9	Aluminum ammonium sulfate dodecahydrate AlNH$_4$(SO$_4$)$_2$•12H$_2$O	7784-26-1 453.331	col cry or powder	94.5 1.65	dec >280	s H$_2$O; i EtOH
10	Aluminum antimonide AlSb	25152-52-7 148.742	cub cry	1065 4.26		
11	Aluminum arsenide AlAs	22831-42-1 101.903	oran cub cry; hyg	1740 3.76		
12	Aluminum borate 2Al$_2$O$_3$•B$_2$O$_3$	11121-16-7 273.543	needles	≈1050		i H$_2$O
13	Aluminum boride AlB$_2$	12041-50-8 48.604	powder	dec >920 3.19		s dil HCl
14	Aluminum borohydride Al(BH$_4$)$_3$	16962-07-5 71.510	flam liq	-64.5	44.5 a,d	reac H$_2$O
15	Aluminum bromate nonahydrate Al(BrO$_3$)$_3$•9H$_2$O	11126-81-1* 572.826	wh hyg cry	62	dec >100	s H$_2$O
16	Aluminum bromide AlBr$_3$	7727-15-3 266.694	wh-yel monocl cry; hyg	97.5 3.2	255 a,b,c,d,e	reac H$_2$O; s bz, tol
17	Aluminum bromide hexahydrate AlBr$_3$•6H$_2$O	7784-11-4 374.785	col-yel hyg cry	93 2.54		s H$_2$O, EtOH, CS$_2$
18	Aluminum carbide Al$_4$C$_3$	1299-86-1 143.959	yel hex cry	2100 2.36	dec >2200	reac H$_2$O
19	Aluminum chlorate nonahydrate Al(ClO$_3$)$_3$•9H$_2$O	15477-33-5 439.472	hyg cry			vs H$_2$O; s EtOH
20	Aluminum chloride AlCl$_3$	7446-70-0 133.340	wh hex cry or powder; hyg	192.6 2.48	a,b,c,e	reac H$_2$O; s bz, ctc, chl
21	Aluminum chloride hexahydrate AlCl$_3$•6H$_2$O	7784-13-6 241.431	col hyg cry	dec 100 2.398		vs H$_2$O; s EtOH, eth
22	Aluminum diacetate Al(OH)(C$_2$H$_3$O$_2$)$_2$	142-03-0 162.079	wh amorp powder			i H$_2$O
23	Aluminum ethoxide Al(C$_2$H$_5$O)$_3$	555-75-9 162.165	liq, condenses to wh solid	140		reac H$_2$O; sl xyl
24	Aluminum fluoride AlF$_3$	7784-18-1 83.977	wh hex cry	≈2250 tp (220 MPa) 3.10	1276 sp a,c,e	sl H$_2$O
25	Aluminum fluoride monohydrate AlF$_3$•H$_2$O	32287-65-3 101.992	orth cry	2.17		sl H$_2$O
26	Aluminum fluoride trihydrate AlF$_3$•3H$_2$O	15098-87-0 138.023	wh hyg cry	1.914		sl H$_2$O
27	Aluminum hexafluorosilicate nonahydrate Al$_2$(SiF$_6$)$_3$•9H$_2$O	17099-70-6 642.329	hex prisms	dec >500		s H$_2$O
28	Aluminum hydride AlH$_3$	7784-21-6 30.006	col hex cry	dec >150	a	reac H$_2$O
29	Aluminum hydroxide Al(OH)$_3$	21645-51-2 78.004	wh amorp powder	2.42	f	i H$_2$O; s alk, acid
30	Aluminum hydroxychloride Al$_2$(OH)$_5$Cl•2H$_2$O	1327-41-9 210.483	gl solid			s H$_2$O
31	Aluminum hypophosphite Al(H$_2$PO$_2$)$_3$	7784-22-7 221.948	cry powder	dec 220		i H$_2$O; s alk, acid
32	Aluminum iodide AlI$_3$	7784-23-8 407.695	wh leaflets	188.32 3.98	382 a,b,c,d,e	reac H$_2$O
33	Aluminum iodide hexahydrate AlI$_3$•6H$_2$O	10090-53-6 515.786	yel hyg cry powder			vs H$_2$O; s EtOH, eth
34	Aluminum lactate Al(C$_3$H$_5$O$_3$)$_3$	18917-91-4 294.195	powder			vs H$_2$O
35	Aluminum nitrate nonahydrate Al(NO$_3$)$_3$•9H$_2$O	7784-27-2 375.134	wh hyg cry; monocl	73 1.72	dec 135	vs H$_2$O, EtOH; i pyr
36	Aluminum nitride AlN	24304-00-5 40.989	blue-wh hex cry	3000 3.255	a	reac H$_2$O
37	Aluminum oleate Al(C$_{18}$H$_{33}$O$_2$)$_3$	688-37-9 871.358	yel solid			i H$_2$O; s EtOH, bz
38	Aluminum phosphate AlPO$_4$	7784-30-7 121.953	wh rhomb plates	>1460 2.56	a	i H$_2$O; sl acid

No.	Name Formula	CAS RN Mol. Wt.	Physical Form	mp/°C den/g cm^{-3}	bp/°C Other Data	Solubility
39	Aluminum metaphosphate $Al(PO_3)_3$	32823-06-6 263.898	col powder; tetr	≈1525 2.78		i H_2O
40	Aluminum oxide Al_2O_3	1344-28-1 101.961	wh powder; hex	2054 3.97	≈3000 a,c,e,f	i H_2O, os; sl alk
41	Aluminum oxyhydroxide AlO(OH)	14457-84-2 59.989	ortho cry	 3.44	f	i H_2O; s acid, alk
42	Aluminum palmitate $Al(C_{15}H_{31}COO)_3$	555-35-1 793.244	wh-yel powder			i H_2O, EtOH; s peth
43	Aluminum perchlorate nonahydrate $Al(ClO_4)_3 \cdot 9H_2O$	14452-39-2 487.470	wh hyg cry	82 dec 2.0		
44	Aluminum phosphide AlP	20859-73-8 57.956	grn or yel cub cry	2550 2.40	a	reac H_2O
45	Aluminum selenide Al_2Se_3	1302-82-5 290.84	yel-brown powder	960 3.437		reac H_2O
46	Aluminum silicate Al_2SiO_5	12183-80-1 162.046	gray-grn cry	 3.145	f	
47	Aluminum silicate dihydrate $Al_2O_3 \cdot 2SiO_2 \cdot 2H_2O$	1332-58-7 258.161	wh-yel powder; tricl	 2.59		i H_2O, acid, alk
48	Aluminum stearate $Al(C_{18}H_{35}O_2)_3$	637-12-7 877.406	wh powder	115 1.070		i H_2O, EtOH, eth; s alk
49	Aluminum sulfate $Al_2(SO_4)_3$	10043-01-3 342.154	wh cry	dec 1040		s H_2O; i EtOH
50	Aluminum sulfate octadecahydrate $Al_2(SO_4)_3 \cdot 18H_2O$	7784-31-8 666.429	col monocl cry	dec 86 1.69	f	s H_2O
51	Aluminum sulfide Al_2S_3	1302-81-4 150.161	yel-gray powder	1100 2.02	a	
52	Aluminum telluride Al_2Te_3	12043-29-7 436.76	gray-blk hex cry	≈895 4.5		
53	Aluminum thiocyanate $Al(CNS)_3$	538-17-0 201.233	yel powder			s H_2O; i EtOH, eth
54	Americium Am	7440-35-9 243	silv metal; hex or cub	1176 12	2011 a,c	s acid
55	Americium(III) oxide Am_2O_3	12254-64-7 534	tan hex cry	 11.77		s acid
56	Americium(III) bromide $AmBr_3$	14933-38-1 483	wh orth cry	 6.85		s H_2O
57	Americium(III) chloride $AmCl_3$	13464-46-5 349	pink hex cry	500 5.87		
58	Americium(III) fluoride AmF_3	13708-80-0 300	pink hex cry	1393 9.53		
59	Americium(III) iodide AmI_3	13813-47-3 624	yel ortho cry	≈950 6.9		
60	Americium(IV) fluoride AmF_4	15947-41-8 319	tan monocl cry	 7.23		
61	Americium(IV) oxide AmO_2	12005-67-3 275	blk cub cry	dec >1000 11.68		s acid
62	Ammonia NH_3	7664-41-7 17.031	col gas	-77.74 0.696 g/L	-33.33 a,b,c,d,e	vs H_2O; s EtOH, eth
63	Ammonium acetate $NH_4C_2H_3O_2$	631-61-8 77.084	wh hyg cry	114 1.073		vs H_2O
64	Ammonium azide NH_4N_3	12164-94-2 60.059	ortho cry; flam	160 1.346	exp a,e	s H_2O
65	Ammonium benzoate $NH_4C_7H_5O_2$	1863-63-4 139.154	wh cry or powder	198 1.26		s H_2O; sl EtOH
66	Ammonium bimalate $NH_4OOCCH_2CH(OH)COOH$	5972-71-4 151.119	ortho cry	160 1.15		s H_2O; sl EtOH
67	Ammonium borate tetrahydrate $(NH_4)_2B_4O_7 \cdot 4H_2O$	12228-87-4 263.377	tetr cry			s H_2O; i EtOH
68	Ammonium bromide NH_4Br	12124-97-9 97.943	wh hyg tetr cry	542 dec 2.429	396 sp a,e	s H_2O, EtOH, ace; sl eth
69	Ammonium caprylate $NH_4C_8H_{15}O_2$	5972-76-9 161.245	hyg monocl cry	≈75		reac H_2O; s EtOH; i chl, bz
70	Ammonium carbamate NH_2COONH_4	1111-78-0 78.071	cry powder			vs H_2O; s EtOH
71	Ammonium carbonate $(NH_4)_2CO_3$	506-87-6 96.086	col cry powder	dec 58		
72	Ammonium cerium(III) sulfate tetrahydrate $NH_4Ce(SO_4)_2 \cdot 4H_2O$	21995-38-0* 422.342	monocl cry			s H_2O
73	Ammonium cerium(IV) nitrate $(NH_4)_2Ce(NO_3)_6$	16774-21-3 548.222	red-oran cry			vs H_2O
74	Ammonium chlorate NH_4ClO_3	10192-29-7 101.490	wh cry	102 exp 1.80		s H_2O
75	Ammonium chloride NH_4Cl	12125-02-9 53.492	col cub cry	520 tp (dec) 1.519	338 sp a,b,e	s H_2O
76	Ammonium chromate $(NH_4)_2CrO_4$	7788-98-9 152.071	yel cry	dec 185 1.90		s H_2O; sl ace, MeOH; i EtOH

No.	Name Formula	CAS RN Mol. Wt.	Physical Form	mp/°C den/g cm^{-3}	bp/°C Other Data	Solubility
77	Ammonium chromic sulfate dodecahydrate $NH_4Cr(SO_4)_2 \cdot 12H_2O$	10022-47-6 478.345	blue-viol cry	94 dec 1.72		s H_2O; sl EtOH
78	Ammonium cobalt(II) phosphate $CoNH_4PO_4$	14590-13-7 171.943	red-viol powder; hyg			i H_2O; s acid
79	Ammonium cobalt(II) sulfate hexahydrate $(NH_4)_2Co(SO_4)_2 \cdot 6H_2O$	13586-38-4 395.229	red monocl prisms	1.90		s H_2O; i EtOH
80	Ammonium copper(II) chloride dihydrate $2NH_4Cl \cdot CuCl_2 \cdot 2H_2O$	10060-13-6 277.464	blue-grn tetr cry	≈115 dec 1.99		s H_2O, EtOH
81	Ammonium cyanide NH_4CN	12211-52-8 44.056	col tetr cry	dec 1.10	a,e	vs H_2O
82	Ammonium dichromate $(NH_4)_2Cr_2O_7$	7789-09-5 252.065	oran-red monocl cry; hyg	dec 180 2.155		vs H_2O
83	Ammonium dihydrogen arsenate $NH_4H_2AsO_4$	13462-93-6 158.975	tetr cry	300 dec 2.311		vs H_2O
84	Ammonium dihydrogen phosphate $NH_4H_2PO_4$	7722-76-1 115.026	wh tetr cry	190 1.80		vs H_2O; sl EtOH; i ace
85	Ammonium dithiocarbamate NH_4NH_2CSS	513-74-6 110.204	yel ortho cry	99 dec 1.45		s H_2O
86	Ammonium ferric chromate $NH_4Fe(CrO_4)_2$	7789-08-4 305.871	red powder			i H_2O
87	Ammonium ferric oxalate trihydrate $(NH_4)_3Fe(C_2O_4)_3 \cdot 3H_2O$	13268-42-3 428.065	grn monocl cry; hyg	≈160 dec 1.780		vs H_2O; i EtOH
88	Ammonium ferric sulfate dodecahydrate $NH_4Fe(SO_4)_2 \cdot 12H_2O$	10138-04-2 482.194	col to viol cry	≈37 1.71		vs H_2O; i EtOH
89	Ammonium ferricyanide trihydrate $(NH_4)_3Fe(CN)_6 \cdot 3H_2O$	14221-48-8* 320.113	red cry			s H_2O; i EtOH
90	Ammonium ferrocyanide trihydrate $(NH_4)_4Fe(CN)_6 \cdot 3H_2O$	14481-29-9* 338.151	yel cry	dec		s H_2O; i EtOH
91	Ammonium ferrous sulfate hexahydrate $(NH_4)_2Fe(SO_4)_2 \cdot 6H_2O$	10045-89-3 392.141	blue-grn monocl cry	dec ≈100 1.86		s H_2O; i EtOH
92	Ammonium fluoride NH_4F	12125-01-8 37.037	wh hex cry; hyg	dec 1.015	a	vs H_2O; sl EtOH
93	Ammonium fluoroborate NH_4BF_4	13826-83-0 104.844	wh powder; orth	487 dec 1.871		sl H_2O
94	Ammonium fluorosulfonate NH_4SO_3F	13446-08-7 117.101	col needles	245		s H_2O, EtOH, MeOH
95	Ammonium formate NH_4CHO_2	540-69-2 63.057	hyg cry	116 1.27		vs H_2O; s EtOH
96	Ammonium hexachloroiridate(IV) $(NH_4)_2IrCl_6$	16940-92-4 441.010	blk cry powder	dec 2.856		sl H_2O
97	Ammonium hexachloroosmiate(IV) $(NH_4)_2OsCl_6$	12125-08-5 439.02	red cry or powder	2.93	subl	s H_2O, EtOH
98	Ammonium hexachloropalladate(IV) $(NH_4)_2PdCl_6$	19168-23-1 355.21	red-brn hyg cry	dec 2.418		
99	Ammonium hexachloroplatinate(IV) $(NH_4)_2PtCl_6$	16919-58-7 443.87	red-oran cub cry	dec 380 3.065		sl H_2O; i EtOH
100	Ammonium hexafluoroaluminate $(NH_4)_3AlF_6$	7784-19-2 195.087	cub cry	1.78		s H_2O
101	Ammonium hexafluorogallate $(NH_4)_3GaF_6$	14639-94-2 237.828	col cub cry	dec >200 2.10		
102	Ammonium hexafluorogermanate $(NH_4)_2GeF_6$	16962-47-3 222.68	wh cry	380 2.564	subl	s H_2O; i EtOH
103	Ammonium hexafluorophosphate NH_4PF_6	16941-11-0 163.003	wh cub cry	dec 58 2.180		vs H_2O; s ace, EtOH, MeOH
104	Ammonium hexafluorosilicate $(NH_4)_2SiF_6$	16919-19-0 178.153	wh cub or trig cry	dec 2.011	a	s H_2O; i EtOH, ace
105	Ammonium hydrogen arsenate $(NH_4)_2HAsO_4$	7784-44-3 176.004	wh powder	1.99		s H_2O
106	Ammonium hydrogen borate trihydrate $NH_4HB_4O_7 \cdot 3H_2O$	10135-84-9 228.332	col cry	≈2.5		s H_2O
107	Ammonium hydrogen carbonate NH_4HCO_3	1066-33-7 79.056	col or wh prisms	107 dec 1.586	a	s H_2O; i EtOH, bz
108	Ammonium hydrogen citrate $(NH_4)_2HC_6H_5O_7$	3012-65-5 226.186	col cry	1.48		vs H_2O; sl EtOH
109	Ammonium hydrogen fluoride NH_4HF_2	1341-49-7 57.044	wh orth cry	125 1.50	240 dec	s H_2O
110	Ammonium hydrogen oxalate monohydrate $NH_4HC_2O_4 \cdot H_2O$	5972-72-5* 125.082	col rhomb cry	dec 1.56		sl H_2O, EtOH
111	Ammonium hydrogen phosphate $(NH_4)_2HPO_4$	7783-28-0 132.055	wh cry	155 dec 1.619	a	s H_2O; i EtOH, ace
112	Ammonium hydrogen sulfate NH_4HSO_4	7803-63-6 115.111	wh hyg cry	147 1.78	a	vs H_2O; i EtOH, ace, py
113	Ammonium hydrogen sulfide NH_4HS	12124-99-1 51.113	wh tetr or orth cry	dec 1.17	e	vs H_2O; sl ace; i bz, eth
114	Ammonium hydrogen sulfite NH_4HSO_3	10192-30-0 99.111	col cry	dec 2.03	a	vs H_2O

No.	Name Formula	CAS RN Mol. Wt.	Physical Form	mp/°C den/g cm^{-3}	bp/°C Other Data	Solubility
115	Ammonium hydrogen tartrate $NH_4OOCCH(OH)CH(OH)COOH$	3095-65-6 167.118	wh cry	 1.68		sl H_2O; s alk; i EtOH
116	Ammonium hydroxide NH_4OH	1336-21-6 35.046	exists only in soln		a	
117	Ammonium hypophosphite $NH_4H_2PO_2$	7803-65-8 83.028	wh hyg cry	dec		vs H_2O; sl EtOH; i ace
118	Ammonium iodate NH_4IO_3	13446-09-8 192.941	wh powder	150 3.3		
119	Ammonium iodide NH_4I	12027-06-4 144.943	wh tetr cry; hyg	551 dec 2.514	405 sp a,c,e	vs H_2O; sl EtOH, MeOH
120	Ammonium lactate $NH_4C_3H_5O_3$	52003-58-4 107.109	col cry	92		s H_2O, EtOH; sl MeOH; i ace, eth
121	Ammonium metatungstate hexahydrate $(NH_4)_6W_7O_{24} \cdot 6H_2O$	12028-48-7 1887.19	wh cry			s H_2O; i EtOH
122	Ammonium metavanadate $NH_4VO_3 \cdot$	7803-55-6 116.979	wh-yel cry	dec 2.326		sl H_2O
123	Ammonium molybdate(IV) tetrahydrate $(NH_4)_6Mo_7O_{24} \cdot 4H_2O$	12054-85-2 1235.86	col or grn-yel cry	dec 2.498		s H_2O; i EtOH
124	Ammonium nickel chloride hexahydrate $NH_4NiCl_3 \cdot 6H_2O$	16122-03-5* 291.181	grn hyg cry	 1.65		s H_2O
125	Ammonium nickel sulfate hexahydrate $(NH_4)_2Ni(SO_4)_2 \cdot 6H_2O$	7785-20-8 394.989	blue-grn cry	dec 1.923		sl H_2O; i EtOH
126	Ammonium nitrate NH_4NO_3	6484-52-2 80.043	wh hyg cry; orth	dec 210; exp 1.72	 a,c	vs H_2O; sl MeOH
127	Ammonium nitroferricyanide $(NH_4)_2Fe(CN)_5NO$	14402-70-1 252.017	red-brn cry			s H_2O, EtOH
128	Ammonium oleate $NH_4C_{18}H_{33}O_2$	544-60-5 299.498	yel-brn paste	21		s H_2O; sl ace
129	Ammonium oxalate monohydrate $(NH_4)_2C_2O_4 \cdot H_2O$	6009-70-7 142.111	wh orth cry	dec 1.50		sl H_2O, EtOH
130	Ammonium palmitate $NH_4C_{15}H_{31}COO$	593-26-0 273.460	yel-wh powder	22		s H_2O; sl bz, xyl; i ace, EtOH, ctc
131	Ammonium pentachlorozincate $(NH_4)_3ZnCl_5$	14639-98-6 296.77	hyg orth cry	 1.81		vs H_2O
132	Ammonium perchlorate NH_4ClO_4	7790-98-9 117.490	wh orth cry	dec, exp 1.95	 a	vs H_2O; s MeOH; sl EtOH, ace; i eth
133	Ammonium permanganate NH_4MnO_4	13446-10-1 136.975	purp rhomb cry	dec 70 2.22		sl H_2O
134	Ammonium peroxydisulfate $(NH_4)_2S_2O_8$	7727-54-0 228.204	monocl cry or wh powder	dec 1.982		vs H_2O
135	Ammonium perrhenate NH_4ReO_4	13598-65-7 268.244	col powder	 3.97		sl H_2O
136	Ammonium phosphite, dibasic, monohydrate $(NH_4)_2HPO_3 \cdot H_2O$	51503-61-8 134.071	hyg cry			s H_2O
137	Ammonium phosphomolybdate monohydrate $(NH_4)_3PO_4 \cdot 12MoO_3 \cdot H_2O$	54723-94-3 1894.36	yel cry or powder	dec		sl H_2O
138	Ammonium phosphotungstate dihydrate $(NH_4)_3PO_4 \cdot 12WO_3 \cdot 2H_2O$	1311-90-6 2967.18	cry powder			sl H_2O
139	Ammonium picrate $NH_4C_6H_2N_3O_7$	131-74-8 246.137	yel orth cry	exp 1.72		sl H_2O
140	Ammonium salicylate $NH_4C_7H_5O_3$	528-94-9 155.153	wh cry powder			vs H_2O; s EtOH
141	Ammonium selenate $(NH_4)_2SeO_4$	7783-21-3 179.04	wh monocl cry	dec 2.194		vs H_2O; i EtOH, ace
142	Ammonium selenite $(NH_4)_2SeO_3$	7783-19-9 163.04	wh or red hyg cry	dec		s H_2O
143	Ammonium stearate $NH_4C_{18}H_{35}O_2$	1002-89-7 301.514	yel-wh powder	22 0.89		sl H_2O, bz; s EtOH, MeOH; i ace
144	Ammonium sulfamate $NH_4NH_2SO_3$	7773-06-0 114.125	wh hyg cry	131	dec 160	vs H_2O; sl EtOH
145	Ammonium sulfate $(NH_4)_2SO_4$	7783-20-2 132.141	wh or brn orth cry	dec 280 1.77	 a,f	vs H_2O; i EtOH, ace
146	Ammonium sulfide $(NH_4)_2S$	12135-76-1 68.143	yel-oran cry	dec ≈0		s H_2O, EtOH, alk
147	Ammonium sulfite monohydrate $(NH_4)_2SO_3 \cdot H_2O$	7783-11-1 134.156	col cry	dec 1.41		s H_2O; i EtOH, ace
148	Ammonium tartrate $(NH_4)_2C_4H_4O_6$	3164-29-2 184.148	wh cry	dec 1.601		s H_2O
149	Ammonium tellurate $(NH_4)_2TeO_4$	13453-06-0 227.68	wh powder	dec 3.024		
150	Ammonium tetrachloroaluminate NH_4AlCl_4	7784-14-7 186.832	wh hyg solid	304		s H_2O, eth
151	Ammonium tetrachloroplatinate(II) $(NH_4)_2PtCl_4$	13820-41-2 372.97	red cry	dec 2.936		s H_2O; i EtOH
152	Ammonium tetrachlorozincate $(NH_4)_2ZnCl_4$	14639-97-5 243.28	wh orth plates; hyg	150 dec 1.879		vs H_2O

No.	Name Formula	CAS RN Mol. Wt.	Physical Form	mp/°C den/g cm^{-3}	bp/°C Other Data	Solubility
153	Ammonium tetrathiotungstate $(NH_4)_2WS_4$	13862-78-7 348.18	oran cry	dec 2.71		s H_2O
154	Ammonium thiocyanate NH_4SCN	1762-95-4 76.122	col hyg cry	≈149 1.30	dec	vs H_2O, EtOH; s ace; i chl
155	Ammonium thiosulfate $(NH_4)_2S_2O_3$	7783-18-8 148.207	wh cry	dec 150 1.678		vs H_2O; i EtOH, eth
156	Ammonium titanium oxalate monohydrate $(NH_4)_2TiO(C_2O_4)_2 \cdot H_2O$	10580-03-7 293.997	hyg cry			vs H_2O
157	Ammonium uranate(VI) $(NH_4)_2U_2O_7$	7783-22-4 624.131	red-yel amorp powder			i H_2O, akl; s acid
158	Ammonium uranium fluoride $UO_2(NH_4)_3F_5$	18433-40-4 419.135	grn-yel monocl cry			s H_2O; i EtOH
159	Ammonium valerate $NH_4C_4H_9COO$	42739-38-8 119.164	hyg cry	108		vs H_2O, EtOH; s eth
160	Ammonium zirconyl carbonate dihydrate $(NH_4)_3ZrOH(CO_3)_3 \cdot 2H_2O$	12616-24-9* 362.405	prisms; unstable			s H_2O
161	Antimony Sb	7440-36-0 121.760	silv metal; hex	630.63 6.68	1587 a,c,e,f	i dil acid
162	Stibine SbH_3	7803-52-3 124.784	col gas; flam	-88 5.100 g/L	-17 a,d	sl H_2O; s EtOH
163	Antimony arsenide SbAs	12322-34-8 196.682	hex cry	≈680 6.0		
164	Antimony(III) bromide $SbBr_3$	7789-61-9 361.472	yel orth cry; hyg	96.6 4.35	280 a,b,d,e	reac H_2O; s ace, bz, chl
165	Antimony(III) chloride $SbCl_3$	10025-91-9 228.118	col orth cry; hyg	73.4 3.14	220.3 a,b,c,d,e	vs H_2O; s acid, EtOH, bz, ace
166	Antimony(III) fluoride SbF_3	7783-56-4 178.755	wh orth cry; hyg	292 4.38	≈345 a	vs H_2O
167	Antimony(III) iodide SbI_3	7790-44-5 502.473	red rhomb cry	168 4.92	401 a,b,d,e	reac H_2O; s EtOH, ace; i ctc
168	Antimony(III) oxide (senarmontite) Sb_2O_3	1309-64-4 291.518	col cub cry	570 trans 5.58	1425 e,f	sl H_2O; i os
169	Antimony(III) oxide (valentinite) Sb_2O_3	1309-64-4 291.518	wh orth cry	655 5.7	1425 e,f	sl H_2O; i os
170	Antimony(III) oxychloride SbOCl	7791-08-4 173.212	wh momo cry	170 dec		reac H_2O; i EtOH, eth
171	Antimony(III) selenide Sb_2Se_3	1315-05-5 480.40	grn orth cry	611 5.81		sl H_2O
172	Antimony(III) sulfate $Sb_2(SO_4)_3$	7446-32-4 531.711	wh cry powder; hyg	dec 3.62		sl H_2O
173	Antimony(III) sulfide Sb_2S_3	1345-04-6 339.718	gray-blk orth cry	550 4.562	f	i H_2O; s conc HCl
174	Antimony(III) telluride Sb_2Te_3	1327-50-0 626.32	gray cry	620 6.5		
175	Antimony(III,V) oxide Sb_2O_4	1332-81-6 307.518	yel orth cry	 6.64	f	
176	Antimony(V) chloride $SbCl_5$	7647-18-9 299.024	col or yel liq	4 2.34	140 dec e	reac H_2O; s chl, ctc
177	Antimony(V) fluoride SbF_5	7783-70-2 216.752	hyg visc liq	8.3 3.10	141	reac H_2O
178	Antimony(V) dichlorotrifluoride $SbCl_2F_3$	7791-16-4 249.660	visc liq			reac H_2O
179	Antimony(V) oxide Sb_2O_5	1314-60-9 323.517	yel powder; cub	dec 3.78	a	sl H_2O
180	Antimony(V) sulfide Sb_2S_5	1315-04-4 403.850	oran-yel powder	75 dec 4.120		i H_2O; s acid, alk
181	Argon Ar	7440-37-1 39.948	col gas	-189.35 1.633 g/L	-185.85 a,b,c,d,e	sl H_2O
182	Arsenic As	7440-38-2 74.922	gray metal; rhomb	817 tp 5.75	614 sp a,b,c,e,f	i H_2O
183	Arsine AsH_3	7784-42-1 77.946	col gas	-116 3.186 g/L	-62.5 a,b,d,e	sl H_2O
184	Diarsine As_2H_4	15942-63-9 153.875	unstable liq		≈100	
185	Arsenic acid H_3AsO_4	7778-39-4 141.944	exists only in soln		a	
186	Arsenic acid hemihydrate $H_3AsO_4 \cdot 0.5H_2O$	7778-39-4* 150.951	wh hyg cry	35.5 ≈2		vs H_2O, EtOH
187	Arsenious acid $H_3AsO_3 \cdot$	13464-58-9 125.944	exists only in soln			
188	Arsenic diiodide As_2I_4	13770-56-4 657.461	red cry	137		reac H_2O; s os
189	Arsenic hemiselenide As_2Se	1303-35-1 228.80	blk cry			i H_2O, os; dec acid, alk
190	Arsenic sulfide As_4S_4	12279-90-2 427.950	red monocl cry	320 3.5	565 f	i H_2O; sl bz; s alk

No.	Name Formula	CAS RN Mol. Wt.	Physical Form	mp/°C den/g cm^{-3}	bp/°C Other Data	Solubility
191	Arsenic(III) bromide $AsBr_3$	7784-33-0 314.634	yel orth cry; hyg	31.1 3.40	221 a,c,d,e	reac H_2O; s hc, ctc; vs eth, bz
192	Arsenic(III) chloride $AsCl_3$	7784-34-1 181.280	col liq	-16 2.150	130 a,b,c,d,e	reac H_2O; vs chl, ctc, eth
193	Arsenic(III) fluoride AsF_3	7784-35-2 131.917	col liq	-5.9 2.7	57.8 a,c,d,e	reac H_2O; s EtOH, eth, bz
194	Arsenic(III) iodide AsI_3	7784-45-4 455.635	red hex cry	140.9 4.73	424 a,d	sl H_2O, EtOH, eth; s bz. tol
195	Arsenic(III) oxide (arsenolite) As_2O_3	1327-53-3 197.841	wh cub cry	274 3.86	460 e,f	sl H_2O
196	Arsenic(III) oxide (claudetite) As_2O_3	1327-53-3 197.841	wh monocl cry	313 3.74	460 e,f	sl H_2O; s dil acid, alk; i EtOH
197	Arsenic(III) selenide As_2Se_3	1303-36-2 386.72	brn-blk solid	260 4.75		i H_2O; s alk
198	Arsenic(III) sulfide As_2S_3	1303-33-9 246.041	yel-oran monocl cry	310 3.46	707 a,f	i H_2O; s alk
199	Arsenic(III) telluride As_2Te_3	12044-54-1 532.64	blk monocl cry	621 6.50		
200	Arsenic(V) chloride $AsCl_5$	22441-45-8 252.186	stable at low temp	≈-50 dec		
201	Arsenic(V) fluoride AsF_5	7784-36-3 169.914	col gas	-79.8 6.945 g/L	-52.8 d,e	reac H_2O; s EtOH, bz, eth
202	Arsenic(V) oxide As_2O_5	1303-28-2 229.840	wh amorp powder	315 4.32	a	vs H_2O, EtOH
203	Arsenic(V) selenide As_2Se_5	1303-37-3 544.64	blk solid	dec		i H_2O, EtOH, eth; s alk
204	Arsenic(V) sulfide As_2S_5	1303-34-0 310.173	brn-yel amorp solid	dec		i H_2O; s alk
205	Astatine At	7440-68-8 210	cry	302	a	s HNO_3, os
206	Barium Ba	7440-39-3 137.327	silv-yel metal; cub	727 3.62	1897 a,c,d,e	reac H_2O; sl EtOH
207	Barium acetate $Ba(C_2H_3O_2)_2$	543-80-6 255.417	wh powder	 2.47		vs H_2O
208	Barium acetate monohydrate $Ba(C_2H_3O_2)_2 \cdot H_2O$	5908-64-5 273.432	wh cry	dec 110 2.19		vs H_2O; sl EtOH
209	Barium aluminate $BaAl_2O_4$	12004-04-5 255.288	hex cry	1827		
210	Barium azide $Ba(N_3)_2$	18810-58-7 221.367	monocl cry; exp	dec ≈120 2.936		s H_2O; sl EtOH; i eth
211	Barium bromate monohydrate $Ba(BrO_3)_2 \cdot H_2O$	10326-26-8 411.147	wh monocl cry	dec 260 3.99		sl H_2O; i EtOH
212	Barium bromide $BaBr_2$	10553-31-8 297.135	wh orth cry	857 4.781	1835 a,c	vs H_2O
213	Barium bromide dihydrate $BaBr_2 \cdot 2H_2O$	7791-28-8 333.166	wh cry	dec 75 3.7		vs H_2O; s MeOH; i EtOH, ace, diox
214	Barium carbide BaC_2	50813-65-5 161.349	gray tetr cry	dec 3.74		reac H_2O
215	Barium carbonate $BaCO_3$	513-77-9 197.336	wh orth cry	1555 4.2865	a,f	i H_2O; s acid
216	Barium chlorate monohydrate $Ba(ClO_3)_2 \cdot H_2O$	10294-38-9 322.244	wh monocl cry	414 3.179		s H_2O, acid; sl EtOH, ace
217	Barium chloride $BaCl_2$	10361-37-2 208.232	wh orth cry; hyg	962 3.9	1560 a,c	s H_2O
218	Barium chloride dihydrate $BaCl_2 \cdot 2H_2O$	10326-27-9 244.263	wh monocl cry	dec ≈120 3.097		s H_2O; i EtOH
219	Barium chromate(V) $Ba_3(CrO_4)_2$	12345-14-1 643.968	grn-blk hex cry	 5.25		s H_2O
220	Barium chromate(VI) $BaCrO_4$	10294-40-3 253.321	yel orth cry	dec 4.50		i H_2O; reac acid
221	Barium citrate monohydrate $Ba_3(C_6H_5O_7)_2 \cdot H_2O$	512-25-4* 808.199	gray-wh cry			s H_2O, acid
222	Barium cyanide $Ba(CN)_2$	542-62-1 189.362	wh cry powder			vs H_2O; s EtOH
223	Barium dichromate dihydrate $BaCr_2O_7 \cdot 2H_2O$	10031-16-0 389.346	brn-red needles	dec		reac H_2O
224	Barium dithionate dihydrate $BaS_2O_6 \cdot 6H_2O$	13845-17-5 333.486	cry	 4.54		s H_2O; sl EtOH
225	Barium ferrocyanide hexahydrate $Ba_2Fe(CN)_6 \cdot H_2O$	13821-06-2* 594.696	yel monocl cry	dec 80		i H_2O, EtOH
226	Barium fluoride BaF_2	7787-32-8 175.324	wh cub cry	1368 4.893	2260 a,c	sl H_2O
227	Barium formate $Ba(CHO_2)_2$	541-43-5 227.363	cry	 3.21		s H_2O; i EtOH
228	Barium hexaboride BaB_6	12046-08-1 202.193	blk cub cry	2070 4.36		i H_2O; s acid; i EtOH
229	Barium hexafluorosilicate $BaSiF_6$	17125-80-3 279.403	wh orth needles	dec 300 4.29		i H_2O, EtOH; sl acid

No.	Name Formula	CAS RN Mol. Wt.	Physical Form	mp/°C den/g cm^{-3}	bp/°C Other Data	Solubility
230	Barium hydride BaH$_2$	13477-09-3 139.343	gray orth cry	1200 4.16	a	reac H$_2$O
231	Barium Hydrogen phosphate BaHPO$_4$	10048-98-3 233.306	wh cry powder	dec 400 4.16		i H$_2$O; s dil acid
232	Barium hydrosulfide Ba(HS)$_2$	25417-81-6 203.475	yel hyg cry			s H$_2$O
233	Barium hydrosulfide tetrahydrate Ba(HS)$_2$·4H$_2$O	12230-74-9 275.536	yel rhomb cry	dec 50		s H$_2$O
234	Barium hydroxide Ba(OH)$_2$	17194-00-2 171.342	wh powder	408	a,c	sl H$_2$O
235	Barium hydroxide monohydrate Ba(OH)$_2$·H$_2$O	22326-55-2 189.357	wh powder	3.743		sl H$_2$O; s acid
236	Barium hydroxide octahydrate Ba(OH)$_2$·8H$_2$O	12230-71-6 315.464	wh monocl cry	dec 78 2.18		sl H$_2$O
237	Barium hypophosphite monohydrate Ba(H$_2$PO$_2$)$_2$·H$_2$O	14871-79-5* 285.320	monocl plates	2.90		s H$_2$O; i EtOH
238	Barium iodate Ba(IO$_3$)$_2$	10567-69-8 487.132	wh cry powder	dec 476 5.23		sl H$_2$O
239	Barium iodate monohydrate Ba(IO$_3$)$_2$·H$_2$O	7787-34-0 505.148	cry	dec 130 5.00		sl H$_2$O; s acid; i EtOH
240	Barium iodide BaI$_2$	13718-50-8 391.136	wh orth cry	711 5.15	a,c	vs H$_2$O
241	Barium iodide dihydrate BaI$_2$·2H$_2$O	7787-33-9 427.167	col cry	740 dec 5.0		vs H$_2$O; s EtOH, ace
242	Barium manganate(VI) BaMnO$_4$	7787-35-1 256.263	grn-gray hyg cry	4.85		reac H$_2$O
243	Barium metaborate monohydrate Ba(BO$_2$)$_2$·H$_2$O	26124-86-7 240.962	wh powder	>900 3.3		sl H$_2$O
244	Barium molybdate BaMoO$_4$	7787-37-3 297.27	wh powder	1450 4.975		i H$_2$O
245	Barium niobate Ba(NbO$_3$)$_2$	12009-14-2 419.136	yel orth cry	1455 5.44		i H$_2$O
246	Barium nitrate Ba(NO$_3$)$_2$	10022-31-8 261.336	wh cub cry	590 3.24	a	s H$_2$O; sl EtOH, ace
247	Barium nitride Ba$_3$N$_2$	12047-79-9 439.994	yel-brn cry	4.78		reac H$_2$O
248	Barium nitrite Ba(NO$_2$)$_2$	13465-94-6 229.338	col hex cry	267 3.234	a	
249	Barium nitrite monohydrate Ba(NO$_2$)$_2$·H$_2$O	7787-38-4 247.353	yel-wh hex cry	217 dec 3.18		vs H$_2$O; i EtOH
250	Barium oxalate BaC$_2$O$_4$	516-02-9 225.347	wh powder	400 dec 2.658		i H$_2$O
251	Barium oxalate monohydrate BaC$_2$O$_4$·H$_2$O	13463-22-4 243.362	wh cry powder	2.66		i H$_2$O; s acid
252	Barium oxide BaO	1304-28-5 153.326	wh-yel powder; cub and hex	1973 5.72(cub)	a,c	sl H$_2$O; s dil acid, EtOH; i ace
253	Barium perchlorate Ba(ClO$_4$)$_2$	13465-95-7 336.227	col hex cry	505 3.20		vs H$_2$O, EtOH
254	Barium perchlorate trihydrate Ba(ClO$_4$)$_2$·3H$_2$O	10294-39-0 390.273	col cry	2.74		s H$_2$O, MeOH; sl EtOH, ace; i eth
255	Barium permanganate Ba(MnO$_4$)$_2$	7787-36-2 375.198	brn-viol cry	dec 200 3.77		sl H$_2$O; reac EtOH
256	Barium peroxide BaO$_2$	1304-29-6 169.326	gray-wh tetr cry	450 dec 4.96		i H$_2$O; reac dil acid
257	Barium metaphosphate Ba(PO$_3$)$_2$	13466-20-1 295.271	wh powder	1560		i H$_2$O; sl acid
258	Barium potassium chromate BaK$_2$(CrO$_4$)$_2$	27133-66-0 447.511	yel hex cry	3.63		vs H$_2$O
259	Barium pyrophosphate Ba$_2$P$_2$O$_7$	13466-21-2 448.597	wh powder	3.9		sl H$_2$O; s acid
260	Barium selenate BaSeO$_4$	7787-41-9 280.29	wh rhomb cry	dec 4.75		i H$_2$O
261	Barium selenide BaSe	1304-39-8 216.29	cub cry powder	1780 5.02		reac H$_2$O
262	Barium selenite BaSeO$_3$	13718-59-7 264.29	solid			i H$_2$O
263	Barium disilicate BaSi$_2$O$_5$	12650-28-1 273.495	wh orth cry	1420 3.70		
264	Barium metasilicate BaSiO$_3$	13255-26-0 213.411	col rhomb powder	1605 4.40		i H$_2$O; s acid
265	Barium silicide BaSi$_2$	1304-40-1 193.498	gray lumps	1180		reac H$_2$O
266	Barium sodium niobate Ba$_2$Na(NbO$_3$)$_5$	12323-03-4 1002.167	wh orth cry	1437 5.40		i H$_2$O
267	Barium stannate BaSnO$_3$	12009-18-6 304.035	cub cry	7.24		sl H$_2$O
268	Barium stannate trihydrate BaSnO$_3$·3H$_2$O	12009-18-6* 358.081	wh cry powder			sl H$_2$O; s acid

No.	Name Formula	CAS RN Mol. Wt.	Physical Form	mp/°C den/g cm⁻³	bp/°C Other Data	Solubility
269	Barium stearate $Ba(C_{18}H_{35}O_2)_2$	6865-35-6 704.277	wh powder	160 1.145		i H_2O, alc
270	Barium sulfate $BaSO_4$	7727-43-7 233.391	wh orth cry	1580 4.49	a,c,f	i H_2O, EtOH
271	Barium sulfide BaS	21109-95-5 169.393	col cub cry or gray powder	2230 4.3	a	sl H_2O
272	Barium sulfite $BaSO_3$	7787-39-5 217.391	wh monocl cry	dec 4.44		i H_2O, EtOH
273	Barium tartrate $BaC_4H_4O_6$	5908-81-6 285.399	wh cry	2.98		s H_2O; i EtOH
274	Barium tetracyanoplatinate(II) tetrahydrate $BaPt(CN)_4{\cdot}4H_2O$	13755-32-3 508.54	yel powder or cry	2.076		sl H_2O; i EtOH
275	Barium tetraiodomercurate(II) $BaHgI_4$	10048-99-4 845.54	yel-red hyg cry			vs H_2O, EtOH
276	Barium thiocyanate $Ba(SCN)_2$	2092-17-3 253.494	hyg cry			vs H_2O; s ace, MeOH, EtOH
277	Barium thiocyanate dihydrate $Ba(SCN)_2{\cdot}2H_2O$	2092-17-3* 289.525	hyg wh cry			s H_2O, EtOH
278	Barium thiocyanate trihydrate $Ba(SCN)_2{\cdot}3H_2O$	68016-36-4 307.540	wh needles; hyg	2.286		s H_2O, EtOH
279	Barium thiosulfate BaS_2O_3	35112-53-9 249.457	wh cry powder	dec 220		sl H_2O; i EtOH
280	Barium thiosulfate monohydrate $BaS_2O_3{\cdot}H_2O$	7787-40-8 267.473	wh cry powder	dec 3.5		sl H_2O; i EtOH
281	Barium titanate $BaTiO_3$	12047-27-7 233.192	wh tetr cry	1625 6.02		i H_2O
282	Barium tungstate $BaWO_4$	7787-42-0 385.17	wh tetr cry	5.04		i H_2O
283	Barium uranium oxide BaU_2O_7	10380-31-1 725.381	oran-yel powder			i H_2O; s acid
284	Barium orthovanadate $Ba_3(VO_4)_2$	39416-30-3 641.859	hex cry	707 5.14		
285	Barium zirconate $BaZrO_3$	12009-21-1 276.549	gray-wh cub cry	2500 5.52		i H_2O, alk; sl acid
286	Berkelium (α form) Bk	7440-40-6 247	hex	1050 14.78	a	
287	Berkelium (β form) Bk	7440-40-6 247	cub cry	986 13.25	a	
288	Beryllium Be	7440-41-7 9.012	hex	1287 1.85	2471 a,c,e	s acid, alk
289	Beryllium acetate $Be(C_2H_3O_2)_2$	543-81-7 127.102	wh cry	dec 60		i H_2O, EtOH
290	Beryllium acetylacetonate $Be(C_5H_7O_2)_2$	10210-64-7 207.231	monocl cry powder	108 1.168	270	i H_2O, vs EtOH, eth
291	Beryllium aluminate $BeAl_2O_4$	12004-06-7 126.973	orth cry	3.65	f	
292	Beryllium aluminum metasilicate $Be_3Al_2(SiO_3)_6$	1302-52-9 537.502	col or grn-yel cry; hex	2.64	f	
293	Beryllium basic acetate $Be_4O(C_2H_3O_2)_6$	1332-52-1 406.316	wh cry	285 1.25	330	i H_2O; s eth, os
294	Beryllium boride BeB_2	12228-40-9 30.634	refrac solid	>1970		
295	Beryllium borohydride $Be(BH_4)_2$	17440-85-6 36.682	solid	dec 125; flam	subl e	reac H_2O
296	Beryllium bromide $BeBr_2$	7787-46-4 168.820	orth cry; hyg	508 3.465	520 a,c,e	vs H_2O; s EtOH, pyr
297	Beryllium carbide Be_2C	506-66-1 30.035	red cub cry	dec >2100 1.90		reac H_2O
298	Beryllium carbonate tetrahydrate $BeCO_3{\cdot}4H_2O$	60883-64-9 93.085	unstable solid			
299	Beryllium chloride $BeCl_2$	7787-47-5 79.917	wh-yel orth cry; hyg	415 1.90	482 a,c,d,e	vs H_2O; s EtOH, eth, py; i bz, tol
300	Beryllium fluoride BeF_2	7787-49-7 47.009	tetr cry or gl; hyg	552 2.1	1169 a,c	vs H_2O; sl EtOH
301	Beryllium formate $Be(CHO_2)_2$	1111-71-3 99.048	powder	dec >250		reac H_2O; i os
302	Beryllium hydride BeH_2	7787-52-2 11.028	wh amorp solid	dec 250 0.65		reac H_2O; i eth, tol
303	Beryllium hydroxide $Be(OH)_2$	13327-32-7 43.027	amorp powder or cry	1.92	a	sl H_2O, alk; s acid
304	Beryllium iodide BeI_2	7787-53-3 262.821	hyg needles	470 4.32	487 a,c,d,e	reac H_2O; s EtOH
305	Beryllium nitrate trihydrate $Be(NO_3)_2{\cdot}3H_2O$	13597-99-4 187.068	yel-wh hyg cry	≈30	dec	vs H_2O; s EtOH
306	Beryllium nitride Be_3N_2	1304-54-7 55.050	gray refrac cry; cub	2200 2.71		reac acid, alk
307	Beryllium oxide BeO	1304-56-9 25.011	wh amorp powder	2578 3.01	a,c	i H_2O; sl acid, alk

No.	Name Formula	CAS RN Mol. Wt.	Physical Form	mp/°C den/g cm^{-3}	bp/°C Other Data	Solubility
308	Beryllium perchlorate tetrahydrate $Be(ClO_4)_2 \cdot 4H_2O$	7787-48-6 279.974	hyg cry			vs H_2O
309	Beryllium selenate tetrahydrate $BeSeO_4 \cdot 4H_2O$	10039-31-3 224.03	orth cry	dec 100 2.03		vs H_2O
310	Beryllium sulfate $BeSO_4$	13510-49-1 105.076	col tetr cry; hyg	1127 2.5	a	s H_2O
311	Beryllium sulfate tetrahydrate $BeSO_4 \cdot 4H_2O$	7787-56-6 177.137	col tetr cry	dec ≈100 1.71		vs H_2O; i EtOH
312	Beryllium sulfide BeS	13598-22-6 41.078	col cub cry	dec 2.36	a	reac hot H_2O
313	Bismuth Bi	7440-69-9 208.980	gray-wh soft metal	271.40 9.79	1564 a,c,d,e,f	s acid
314	Bismuth basic carbonate $(BiO)_2CO_3$	5892-10-4 509.969	wh powder	6.86		i H_2O; s acid
315	Bismuth bromide $BiBr_3$	7787-58-8 448.692	yel cub cry	218 5.72	453 b,d,e	reac H_2O; s dil acid, ace; i EtOH
316	Bismuth chloride $BiCl_3$	7787-60-2 315.338	yel-wh cub cry; hyg	230 4.75	447 a,b,c,d,e	reac H_2O; s acid, EtOH, ace
317	Bismuth citrate $BiC_6H_5O_7$	813-93-4 398.082	wh powder	3.458		i H_2O; sl EtOH
318	Bismuth fluoride BiF_3	7787-61-3 265.975	wh-gray cub cry	725 8.3	900 a	i H_2O
319	Bismuth pentafluoride BiF_5	7787-62-4 303.972	wh tetr needles; hyg	154 5.55	230	reac H_2O
320	Bismuth hydride BiH_3	18288-22-7 212.004	col gas; unstable	-67 8.665 g/L	≈17	
321	Bismuth hydroxide $Bi(OH)_3$	10361-43-0 260.002	wh-yel amorp powder	4.962	a	i H_2O; s acid
322	Bismuth iodide BiI_3	7787-64-6 589.693	blk hex cry	408.6 5.778	542 a	i H_2O; s EtOH
323	Bismuth molybdate $Bi_2(MoO_4)_3$	51898-99-8 897.77	monocl cry	5.95		
324	Bismuth nitrate pentahydrate $Bi(NO_3)_3 \cdot 5H_2O$	10035-06-0 485.071	col tricl cry; hyg	≈75 dec 2.83		reac H_2O; s ace; i EtOH
325	Bismuth oleate $Bi(C_{18}H_{33}O_2)_3$	52951-38-9 1053.356	soft yel-brn solid			i H_2O; s eth; sl bz
326	Bismuth orthovanadate $BiVO_4$	14059-33-7 323.920	oran-yel orth cry	trans 500 6.25		
327	Bismuth oxalate $Bi_2(C_2O_4)_3$	6591-55-5 682.020	wh powder			i H_2O, EtOH; s dil acid
328	Bismuth oxide Bi_2O_3	1304-76-3 465.959	yel monocl cry or powder	817 8.9	1890 a	i H_2O, s acid
329	Bismuth oxybromide $BiOBr$	7787-57-7 304.883	col tetr cry	8.08		i H_2O, EtOH; s acid
330	Bismuth oxychloride $BiOCl$	7787-59-9 260.432	wh tetr cry	7.72	a	i H_2O
331	Bismuth oxyiodide $BiOI$	7787-63-5 351.883	red tetr cry	>300 dec 7.92		i H_2O, EtOH, chl; s HCl
332	Bismuth oxynitrate $BiONO_3$	10361-46-3 286.985	wh powder	260 dec 4.93		i H_2O, EtOH; s acid
333	Bismuth phosphate $BiPO_4$	10049-01-1 303.951	monocl cry	6.32		sl H_2O, dil acid; i EtOH
334	Bismuth selenide Bi_2Se_3	12068-69-8 654.84	blk hex cry	710 dec 7.5		i H_2O
335	Bismuth stannate pentahydrate $Bi_2(SnO_3)_3 \cdot 5H_2O$	12777-45-6 1008.162	wh cry			i H_2O
336	Bismuth sulfate $Bi_2(SO_4)_3$	7787-68-0 706.152	wh needles or powder	405 dec 5.08	a	reac H_2O, EtOH
337	Bismuth sulfide Bi_2S_3	1345-07-9 514.159	blk-brn orth cry	850 6.78	a,f	i H_2O; s acid
338	Bismuth telluride Bi_2Te_3	1304-82-1 800.76	gray hex plates	580 7.74	f	i H_2O; s EtOH
339	Bismuth tetroxide Bi_2O_4	12048-50-9 481.959	red-oran powder	305 5.6		reac H_2O
340	Bismuth titanate $Bi_4(TiO_4)_3$	12048-51-0 1171.515	wh orth cry	7.85		
341	Boron B	7440-42-8 10.811	blk rhomb cry	2075 2.34	4000 a,c,d	i H_2O
342	Diborane(6) B_2H_6	19287-45-7 27.670	col gas; flam	-165.5 1.131 g/L	-92.4 a,b,d,e	reac H_2O
343	Tetraborane(10) B_4H_{10}	18283-93-7 53.323	col gas	-121 2.180 g/L	18 a,d,e	reac H_2O
344	Pentaborane(9) B_5H_9	19624-22-7 63.126	flam liq	-46.6 0.60	60 a,e	reac hot H_2O
345	Pentaborane(11) B_5H_{11}	18433-84-6 65.142	col liq; unstable	-123 0.67	63 a,d,e	reac H_2O
346	Hexaborane(10) B_6H_{10}	23777-80-2 74.945	col liq	-62.3 0.67	108 dec a	reac hot H_2O

No.	Name Formula	CAS RN Mol. Wt.	Physical Form	mp/°C den/g cm^{-3}	bp/°C Other Data	Solubility
347	Hexaborane(12) B_6H_{12}	12008-19-4 76.961	col liq; unstable	-82	≈80	reac H_2O
348	Nonaborane(15) B_9H_{15}	19465-30-6 112.418	col liq	2.6		
349	Decaborane(14) $B_{10}H_{14}$	17702-41-9 122.221	wh orth cry	99.6 0.94	≈213 e	sl H_2O; s EtOH, bz, CS_2, ctc
350	Borane carbonyl BH_3CO	13205-44-2 41.845	col gas	-137 1.710g/L	-64 e	reac H_2O
351	Borazine $B_3N_3H_6$	6569-51-3 80.501	col liq	-58 0.824	53 a	reac H_2O
352	Boric acid (orthoboric acid) $B(OH)_3$	10043-35-3 61.833	col tricl cry	170.9 1.5	a	sl H_2O, EtOH
353	Metaboric acid (α form) HBO_2	13460-50-9 43.818	col orth cry; hyg	176 1.784	a	s H_2O
354	Metaboric acid (β form) HBO_2	13460-50-9 43.818	col monocl cry; hyg	201 2.045	a	s H_2O
355	Metaboric acid (γ form) HBO_2	13460-50-9 43.818	col cub cry	236 2.487	a	s H_2O
356	Fluoroboric acid HBF_4	16872-11-0 87.813	col liq	≈1.8	130 dec	vs H_2O, EtOH
357	Boron arsenide BAs	12005-69-5 85.733	cub cry	dec 920 5.22		
358	Boron bromide BBr_3	10294-33-4 250.523	col liq; hyg	-45 2.6	91 a,b,d,e	reac H_2O, EtOH
359	Boron carbide B_4C	12069-32-8 55.255	hard blk cry	2350 2.50	>3500	i H_2O, acid
360	Boron chloride BCl_3	10294-34-5 117.169	col liq or gas	-107 4.789g/L	12.65 a,b,c,d,e	reac H_2O, EtOH
361	Boron chloride B_2Cl_4	13701-67-2 163.433	col liq; flam	-92.6	65 a	reac H_2O
362	Boron fluoride BF_3	7637-07-2 67.806	col gas	-126.8 2.771 g/L	-101 a,b,c,d,e	s H_2O
363	Boron fluoride B_2F_4	13965-73-6 97.616	col gas; flam	-56 3.990 g/L	-34 a,d	reac H_2O
364	Boron iodide BI_3	13517-10-7 391.524	needles	43 3.96	210 a,b,d	
365	Boron nitride BN	10043-11-5 24.818	wh powder; hex or cub cry	2967 2.18	a	i H_2O, acid
366	Boron oxide B_2O_3	1303-86-2 69.620	col gl or hex cry; hyg	450 2.55	a,c	sl H_2O; s EtOH
367	Boron phosphide BP	20205-91-8 41.785	red cub cry or powder	dec 1125		reac H_2O, acid
368	Boron sulfide B_2S_3	12007-33-9 117.820	yel amorp solid	softens ≈320 ≈1.7	a	
369	Bromine Br_2	7726-95-6 159.808	red liq	-7.2 3.1028	58.8 a,b,c,d,e	sl H_2O
370	Bromic acid $HBrO_3$	7789-31-3 128.910	stable only in aq soln			s H_2O
371	Bromine oxide Br_2O	21308-80-5 175.807	brn solid	-17.5 dec		
372	Bromine dioxide BrO_2	21255-83-4 111.903	unstable yel cry	dec ≈0	a	
373	Bromine azide BrN_3	13973-87-0 121.924	red cry; exp	≈45	exp	
374	Bromine chloride BrCl	13863-41-7 115.357	unstable red-brn gas	≈-66 4.715 g/L	≈5 dec a	reac H_2O; s eth, CS_2
375	Bromine fluoride BrF	13863-59-7 98.902	unstable red-brn gas	≈-33 4.043 g/L	≈20 dec a,d	
376	Bromine trifluoride BrF_3	7787-71-5 136.899	col hyg liq	8.77 2.803	125.8 a,d	reac H_2O
377	Bromine pentafluoride BrF_5	7789-30-2 174.896	col liq	-60.5 2.460	40.76 a,c,d,e	reac H_2O (exp)
378	Cadmium Cd	7440-43-9 112.411	silv-wh metal	321.07 8.69	767 a,c,d,e	i H_2O; reac acid
379	Cadmium acetate $Cd(C_2H_3O_2)_2$	543-90-8 230.501	col cry	255 2.34		s H_2O, EtOH
380	Cadmium acetate dihydrate $Cd(C_2H_3O_2)_2 \cdot 2H_2O$	5743-04-4 266.530	wh cry	dec 130 2.01		vs H_2O; s EtOH
381	Cadmium antimonide CdSb	12014-29-8 234.171	orth cry	456 6.92		
382	Cadmium arsenide Cd_3As_2	12006-15-4 487.076	gray tetr cry	721 6.25		
383	Cadmium azide $Cd(N_3)_2$	14215-29-3 196.451	yel-wh orth cry; exp	exp 3.24		
384	Cadmium bromide $CdBr_2$	7789-42-6 272.219	wh hex powder or flakes; hyg	568 5.19	844 a,c,d	s H_2O; sl ace, eth
385	Cadmium bromide tetrahydrate $CdBr_2 \cdot 4H_2O$	13464-92-1 344.281	wh-yel cry			s H_2O, ace, EtOH

No.	Name Formula	CAS RN Mol. Wt.	Physical Form	mp/°C den/g cm^{-3}	bp/°C Other Data	Solubility
386	Cadmium carbonate $CdCO_3$	513-78-0 172.420	wh hex cry	dec 500 4.258	a	i H_2O; s acid
387	Cadmium chlorate dihydrate $Cd(ClO_3)_2 \cdot 2H_2O$	22750-54-5* 315.343	col hyg cry	80 dec 2.28		vs H_2O
388	Cadmium chloride $CdCl_2$	10108-64-2 183.316	rhomb cry; hyg	564 4.08	960 a,c,d,e	vs H_2O; s ace; sl EtOH; i eth
389	Cadmium chloride hemipentahydrate $CdCl_2 \cdot 2.5H_2O$	7790-78-5 228.354	wh rhomb leaflets 3.327			vs H_2O; s ace
390	Cadmium chloride monohydrate $CdCl_2 \cdot H_2O$	34330-64-8 201.331	wh cry			vs H_2O
391	Cadmium chromate $CdCrO_4$	14312-00-6 228.405	yel orth cry	4.5		i H_2O
392	Cadmium cyanide $Cd(CN)_2$	542-83-6 164.446	wh cub cry	2.23		sl H_2O
393	Cadmium fluoride CdF_2	7790-79-6 150.408	cub cry	1110 6.33	1748 a,c,d,e	sl H_2O; s acid; i EtOH
394	Cadmium hydroxide $Cd(OH)_2$	21041-95-2 146.426	wh trig or hex cry	dec 130 4.79	a	i H_2O, s dil acid
395	Cadmium iodate $Cd(IO_3)_2$	7790-81-0 462.216	wh powder	6.48		i H_2O; s HNO_3
396	Cadmium iodide CdI_2	7790-80-9 366.220	hex flakes	387 5.64	742 a,c,d,e	s H_2O, EtOH, eth, ace
397	Cadmium metasilicate $CdSiO_3$	13477-19-5 188.495	grn monocl cry	1252 5.10		
398	Cadmium molybdate $CdMoO_4$	13972-68-4 272.35	col tetr cry	≈900 dec 5.4		i H_2O; s acid
399	Cadmium niobate $Cd_2Nb_2O_7$	12187-14-3 522.631	cub cry	≈1410 6.28		i H_2O
400	Cadmium nitrate $Cd(NO_3)_2$	10325-94-7 236.420	wh cub cry; hyg	350 3.6		vs H_2O; s EtOH
401	Cadmium nitrate tetrahydrate $Cd(NO_3)_2 \cdot 4H_2O$	10022-68-1 308.482	col orth cry; hyg	59.5 2.45		vs H_2O; s EtOH, ace
402	Cadmium oxalate CdC_2O_4	814-88-0 200.431	wh solid	3.32		i H_2O
403	Cadmium oxalate trihydrate $CdC_2O_4 \cdot 3H_2O$	20712-42-9 254.477	wh amorp powder	340 dec		i H_2O, EtOH; s dil acid
404	Cadmium oxide CdO	1306-19-0 128.410	brn cub cry	8.15	1559 sp a,e	i H_2O; s dil acid
405	Cadmium perchlorate hexahydrate $Cd(ClO_4)_2 \cdot 6H_2O$	10326-28-0 419.403	wh hex cry	2.37		vs H_2O
406	Cadmium phosphide Cd_3P_2	12014-28-7 399.181	gr tetr needles	700 5.96		s dil HCl
407	Cadmium selenate dihydrate $CdSeO_4 \cdot 2H_2O$	10060-09-0 291.40	orth cry	dec 100 3.62		vs H_2O
408	Cadmium selenide $CdSe$	1306-24-7 191.37	wh cub cry	1240 5.81		i H_2O
409	Cadmium sulfate $CdSO_4$	10124-36-4 208.475	col orth cry	1000 4.69	a	vs H_2O; i EtOH
410	Cadmium sulfate monohydrate $CdSO_4 \cdot H_2O$	7790-84-3 226.490	monocl cry	105 3.79		
411	Cadmium sulfate octahydrate $CdSO_4 \cdot 8H_2O$	15244-35-6 352.597	col monocl cry	dec 40 3.08		vs H_2O
412	Cadmium sulfide CdS	1306-23-6 144.477	yel-oran cub cry	1750 4.83	a,f	i H_2O; s acid
413	Cadmium telluride $CdTe$	1306-25-8 240.01	brn-blk cub cry	1042 6.2	a	i H_2O, dil acid
414	Cadmium titanate $CdTiO_3$	12014-14-1 208.276	orth cry	6.5		
415	Cadmium tungstate $CdWO_4$	7790-85-4 360.25	wh monocl cry	8.0		i H_2O, acid; s NH_4OH
416	Calcium Ca	7440-70-2 40.078	silv-wh metal	842 1.54	1484 a,c,e	reac H_2O; i bz
417	Calcium acetate $Ca(C_2H_3O_2)_2$	62-54-4 158.168	wh hyg cry	dec 160 1.50		s H_2O; sl EtOH
418	Calcium acetate monohydrate $Ca(C_2H_3O_2)_2 \cdot H_2O$	5743-26-0 176.183	wh needles or powder	dec ≈150		s H_2O; sl EtOH
419	Calcium aluminate $CaAl_2O_4$	12042-68-1 158.039	wh monocl cry	1605 2.98		reac H_2O
420	Calcium aluminate (β form) $Ca_3Al_2O_6$	12042-78-3 270.193	wh cub cry; refr	1535 3.04		i H_2O
421	Calcium arsenate $Ca_3(AsO_4)_2$	7778-44-1 398.072	wh powder	dec 3.6		sl H_2O; s dil acid
422	Calcium boride CaB_6	12007-99-7 104.944	refrac solid	2235 2.49		
423	Calcium bromide $CaBr_2$	7789-41-5 199.886	rhomb cry; hyg	742 3.38	1815 a,c	vs H_2O; s EtOH, ace
424	Calcium bromide hexahydrate $CaBr_2 \cdot 6H_2O$	13477-28-6 307.977	wh hyg powder	38 dec 2.29		vs H_2O

No.	Name Formula	CAS RN Mol. Wt.	Physical Form	mp/°C den/g cm^{-3}	bp/°C Other Data	Solubility
425	Calcium carbide CaC$_2$	75-20-7 64.100	gray-blk orth cry	2300 2.22	a	reac H$_2$O
426	Calcium carbonate (aragonite) CaCO$_3$	471-34-1 100.087	wh orth cry or powder	825 dec 2.83	a,c,f	i H$_2$O; s dil acid
427	Calcium carbonate (calcite) CaCO$_3$	471-34-1 100.087	wh hex cry or powder	1330 2.71	a,c,f	i H$_2$O; s dil acid
428	Calcium chlorate Ca(ClO$_3$)$_2$	10137-74-3 206.979	wh cry	340		vs H$_2$O
429	Calcium chlorate dihydrate Ca(ClO$_3$)$_2$·2H$_2$O	10035-05-9 243.010	wh monocl cry; hyg	100 dec 2.711		vs H$_2$O; s EtOH
430	Calcium chloride CaCl$_2$	10043-52-4 110.983	wh cub cry or powder; hyg	775 2.15	1935.5 a,c	vs H$_2$O, EtOH
431	Calcium chloride dihydrate CaCl$_2$·2H$_2$O	10035-04-8 147.014	hyg flakes or powder	175 dec 1.85		vs H$_2$O, EtOH
432	Calcium chloride hexahydrate CaCl$_2$·6H$_2$O	7774-34-7 219.074	wh hex cry; hyg	30 dec 1.71		vs H$_2$O
433	Calcium chloride monohydrate CaCl$_2$·H$_2$O	13477-29-7 128.998	wh hyg cry	260 dec 2.24		s H$_2$O, EtOH
434	Calcium chromate dihydrate CaCrO$_4$·2H$_2$O	13765-19-0 192.102	yel orth cry	2.50		s H$_2$O
435	Calcium cyanamide CaCN$_2$	156-62-7 80.102	col hex cry	≈1340 2.29	subl	reac H$_2$O
436	Calcium cyanide Ca(CN)$_2$	592-01-8 92.113	wh rhomb cry; hyg		a	s H$_2$O, EtOH
437	Calcium dichromate trihydrate CaCr$_2$O$_7$·3H$_2$O	14307-33-6* 310.112	red-oran cry	dec 100 2.37		vs H$_2$O; reac EtOH; i eth, ctc
438	Calcium dihydrogen phosphate monohydrate Ca(H$_2$PO$_4$)$_2$·H$_2$O	10031-30-8 252.068	col tricl plates	dec ≈100 2.22		sl H$_2$O; s dil acid
439	Calcium fluoride CaF$_2$	7789-75-5 78.075	wh cub cry or powder	1418 3.18	2533.4 a,c,f	i H$_2$O; sl acid
440	Calcium formate Ca(CHO$_2$)$_2$	544-17-2 130.114	orth cry	300 dec 2.02		s H$_2$O; i EtOH
441	Calcium hexafluorosilicate dihydrate CaSiF$_6$·2H$_2$O	16925-39-6 218.185	col tetr cry	2.25		i H$_2$O, ace; reac hot H$_2$O
442	Calcium hydride CaH$_2$	7789-78-8 42.094	gray orth cry or powder	1000 1.7	a	reac H$_2$O, EtOH
443	Calcium hydrogen phosphate CaHPO$_4$	7757-93-9 136.057	wh tricl cry	dec 2.92	f	i H$_2$O, EtOH
444	Calcium hydrogen phosphate dihydrate CaHPO$_4$·2H$_2$O	7789-77-7 172.088	monocl cry	dec ≈100 2.31		i H$_2$O, EtOH; s dil acid
445	Calcium hydroxide Ca(OH)$_2$	1305-62-0 74.093	soft hex cry	≈2.2	a	sl H$_2$O; s acid
446	Calcium hypophosphite Ca(H$_2$PO$_2$)$_2$	7789-79-9 170.055	wh monocl cry	dec 300		s H$_2$O; i EtOH
447	Calcium iodate Ca(IO$_3$)$_2$	7789-80-2 389.883	wh monocl cry	4.52		sl H$_2$O; s HNO$_3$; i EtOH
448	Calcium iodide CaI$_2$	10102-68-8 293.887	hyg hex cry	783 3.96	a,c	vs H$_2$O; s MeOH, EtOH, ace; i eth
449	Calcium iodide hexahydrate CaI$_2$·6H$_2$O	71626-98-7 401.978	wh hex needles or powder	42 dec 2.55		vs H$_2$O, EtOH
450	Calcium molybdate CaMoO$_4$	7789-82-4 200.02	wh tetr cry	965 dec 4.35	f	i H$_2$O, EtOH; s conc acid
451	Calcium nitrate Ca(NO$_3$)$_2$	10124-37-5 164.087	wh cub cry; hyg	561 2.5	a	vs H$_2$O; s EtOH, MeOH, ace
452	Calcium nitrate tetrahydrate Ca(NO$_3$)$_2$·4H$_2$O	13477-34-4 236.149	wh cry	dec ≈40 1.82		vs H$_2$O; s EtOH, ace
453	Calcium nitride Ca$_3$N$_2$	12013-82-0 148.247	red-brn cub cry	1195 2.67		s H$_2$O, acid; i EtOH
454	Calcium nitrite Ca(NO$_2$)$_2$	13780-06-8 132.089	wh-yel hex cry; hyg	2.23		vs H$_2$O; sl EtOH
455	Calcium oxalate CaC$_2$O$_4$	563-72-4 128.098	wh cry powder	2.2	a	i H$_2$O
456	Calcium oxalate monohydrate CaC$_2$O$_4$·H$_2$O	5794-28-5 146.113	cub cry	dec 200 2.2	f	i H$_2$O; s dil acid
457	Calcium oxide CaO	1305-78-8 56.077	gray-wh cub cry	2899 3.34	a,c	reac H$_2$O; s acid
458	Calcium perchlorate Ca(ClO$_4$)$_2$	13477-36-6 238.978	wh cry	dec 270 2.65		vs H$_2$O; s EtOH
459	Calcium permanganate Ca(MnO$_4$)$_2$	10118-76-0 277.949	purp hyg cry	2.4		vs H$_2$O; reac EtOH
460	Calcium peroxide CaO$_2$	1305-79-9 72.077	wh-yel tetr cry; hyg	dec ≈200 2.9		sl H$_2$O; s acid
461	Calcium phosphate Ca$_3$(PO$_4$)$_2$	7758-87-4 310.177	wh amorp powder	1670 3.14	a	i H$_2$O, EtOH; s dil acid
462	Calcium dihydrogen phosphate monohydrate Ca(H$_2$PO$_4$)$_2$·H$_2$O	7758-23-8* 252.068	col hyg cry	dec ≈100 2.22		s H$_2$O, acid

No.	Name Formula	CAS RN Mol. Wt.	Physical Form	mp/°C den/g cm^{-3}	bp/°C Other Data	Solubility
463	Calcium phosphide Ca_3P_2	1305-99-3 182.182	red-brn hyg cry	≈1600 2.51		reac H_2O; i EtOH, eth
464	Calcium pyrophosphate $Ca_2P_2O_7$	7790-76-3 254.099	wh powder	1353 3.09		i H_2O; s dil acid
465	Calcium selenate dihydrate $CaSeO_4·2H_2O$	7790-74-1 219.07	wh monocl cry	 2.75		s H_2O
466	Calcium selenide CaSe	1305-84-6 119.04	wh-brn cub cry	 3.8		reac H_2O
467	Calcium metasilicate $CaSiO_3$	1344-95-2 116.162	wh monocl cry	 2.92	f	i H_2O
468	Calcium silicide $CaSi_2$	12013-56-8 96.249	gray hex cry	1040 2.50		i cold H_2O; reac hot H_2O; s acid
469	Calcium silicide CaSi	12013-55-7 68.164	orth cry	1324 2.39		
470	Calcium sulfate $CaSO_4$	7778-18-9 136.142	orth cry	1460 2.96	a,c,f	sl H_2O
471	Calcium sulfate dihydrate $CaSO_4·2H_2O$	10101-41-4 172.172	monocl cry or powder	dec 150 2.32	f	s H_2O; i os
472	Calcium sulfate hemihydrate $CaSO_4·0.5H_2O$	10034-76-1 145.149	wh powder			s H_2O
473	Calcium sulfide CaS	20548-54-3 72.144	wh-yel cub cry; hyg	2525 2.59	a,f	sl H_2O; i EtOH
474	Calcium sulfite dihydrate $CaSO_3·2H_2O$	10257-55-3 156.173	wh powder			sl H_2O, EtOH; s acid
475	Calcium telluride CaTe	12013-57-9 167.68	cub cry	 4.87		
476	Calcium tetrahydroaluminate $Ca(AlH_4)_2$	16941-10-9 102.105	gray powder; flam			reac H_2O; s thf; i eth, bz
477	Calcium thiocyanate tetrahydrate $Ca(SCN)_2·4H_2O$	2092-16-2 228.307	hygr cry	dec 160		vs H_2O; s EtOH, ace
478	Calcium thiosulfate hexahydrate $CaS_2O_3·6H_2O$	10124-41-1 260.300	tricl cry	dec 45 1.87		s H_2O; i EtOH
479	Calcium titanate $CaTiO_3$	12049-50-2 135.943	cub cry	1980 3.98	f	
480	Calcium tungstate $CaWO_4$	7790-75-2 287.92	wh tetr cry	1620 6.06	f	i H_2O; s hot acid
481	Californium Cf	7440-71-3 251	hex or cub metal	900 15.1	a	
482	Carbon (diamond) C	7782-40-3 12.011	col cub cry	4440 (12.4 GPa) 3.513	e,f	i H_2O
483	Carbon (graphite) C	7440-44-0 12.011	soft blk hex cry	4492 tp (10.3 MPa) 2.2	3825 sp a,c,f	i H_2O
484	Carbon (fullerene-C_{60}) C_{60}	99685-96-8 720.660	yel needles or plates	>280		s os
485	Carbon (fullerene-C_{70}) C_{70}	115383-22-7 840.770	red-brn solid	>280		s bz, tol
486	Fullerene fluoride $C_{60}F_{60}$	134929-59-2 1860.564	col plates	287		vs ace; s thf; i chl
487	Carbon monoxide CO	630-08-0 28.010	col gas	-205 1.145 g/L	-191.5 a,b,c,d,e	sl H_2O; s chl, EtOH
488	Carbon dioxide CO_2	124-38-9 44.010	col gas	-56.57 tp 1.799 g/L	-78.4 sp a,b,c,e	s H_2O
489	Carbon diselenide CSe_2	506-80-9 169.93	yel liq	-45.5 2.6626	125.5 a,e	i H_2O; vs ctc, tol
490	Carbon disulfide CS_2	75-15-0 76.143	col or yel liq	-111.5 1.2555	46 a,b,c,d,e	i H_2O; vs EtOH, bz, os
491	Carbon oxyselenide COSe	1603-84-5 106.97	col gas; unstable	-122 4.372 g/L	-21.5 e	reac H_2O
492	Carbon oxysulfide COS	463-58-1 60.076	col gas	-138.8 2.456 g/L	-50 a,b,e	s H_2O, EtOH
493	Carbon suboxide C_3O_2	504-64-3 68.032	col gas	-111.3 2.781 g/L	6.8 e	reac H_2O
494	Carbon sulfide telluride CSTe	10340-06-4 171.68	red-yel liq; unstable			reac H_2O
495	Carbonyl bromide $COBr_2$	593-95-3 187.818	col liq	 2.5	64.5 a	reac H_2O
496	Carbonyl chloride $COCl_2$	75-44-5 98.915	col gas	-127.9 4.043 g/L	8 a,b,c,e	sl H_2O; s bz, tol
497	Carbonyl fluoride COF_2	353-50-4 66.007	col gas	-111.26 2.698 g/L	-84.57 a	reac H_2O
498	Cyanogen C_2N_2	460-19-5 52.035	col gas	-27.9 2.127 g/L	-21.1 a,b,d,e	sl H_2O, eth; s EtOH
499	Cyanogen bromide BrCN	506-68-3 105.922	wh hyg needles	52 2.005	61.5 a	s H_2O, EtOH, eth
500	Cyanogen chloride ClCN	506-77-4 61.471	col gas	-6.55 2.513 g/L	13 a	s H_2O, EtOH, eth

No.	Name Formula	CAS RN Mol. Wt.	Physical Form	mp/°C den/g cm^{-3}	bp/°C Other Data	Solubility
501	Cyanogen fluoride FCN	1495-50-7 45.016	col gas	-82 1.840 g/L	-46 a	
502	Cyanogen iodide ICN	506-78-5 152.922	col needles	146.7 1.84	a	s H_2O, EtOH, eth
503	Cerium Ce	7440-45-1 140.115	silv metal; cub or hex	798 6.770	3443 a,c	s dil acid
504	Cerium boride CeB$_6$	12008-02-5 204.981	blue refrac solid; hex	2550 4.87		i H_2O, HCl
505	Cerium carbide CeC$_2$	12012-32-7 164.137	red hex cry	2250 5.47		reac H_2O
506	Cerium nitride CeN	25764-08-3 154.122	refrac cub cry	2557 7.89		
507	Cerium silicide CeSi$_2$	12014-85-6 196.286	tetr cry	1620 5.31		i H_2O
508	Cerium(II) hydride CeH$_2$	13569-50-1 142.131	cub cry	5.45		reac H_2O
509	Cerium(II) iodide CeI$_2$	19139-47-0 393.924	bronze cry	808		
510	Cerium(II) sulfide CeS	12014-82-3 172.181	yel cub cry	2445 5.9	a	
511	Cerium(III) bromide CeBr$_3$	14457-87-5 379.827	wh hex cry; hyg	733	1457	s H_2O
512	Cerium(III) bromide heptahydrate CeBr$_3$·7H$_2$O	14457-87-5* 505.934	col hyg needles	732		s H_2O, EtOH
513	Cerium(III) carbide Ce$_2$C$_3$	12115-63-8 316.263	yel-brn cub cry	1505 6.9		
514	Cerium(III) carbonate hydrate Ce$_2$(CO$_3$)$_3$·5H$_2$O	72520-94-6 550.334	wh powder			i H_2O; s dil acid
515	Cerium(III) chloride CeCl$_3$	7790-86-5 246.473	wh hex cry	817 3.97	a	s H_2O, EtOH
516	Cerium(III) chloride heptahydrate CeCl$_3$·7H$_2$O	18618-55-8 372.580	yel orth cry; hyg	dec 90		vs H_2O, EtOH
517	Cerium(III) fluoride CeF$_3$	7758-88-5 197.110	wh hex cry; hyg	1430 6.157		i H_2O
518	Cerium(III) iodide CeI$_3$	7790-87-6 520.828	yel orth cry; hyg	766		s H_2O
519	Cerium(III) iodide nonahydrate CeI$_3$·9H$_2$O	7790-87-6* 682.966	wh-red cry			vs H_2O; s EtOH
520	Cerium(III) oxide Ce$_2$O$_3$	1345-13-7 328.228	yel-grn cub cry	2230 6.2	a	i H_2O; s acid
521	Cerium(III) sulfate octahydrate Ce$_2$(SO$_4$)$_3$·8H$_2$O	13454-94-9 712.543	wh orth cry	dec ≈250 2.87		s H_2O
522	Cerium(III) sulfide Ce$_2$S$_3$	12014-93-6 376.428	red cub cry	2450 5.02		i H_2O
523	Cerium(IV) fluoride CeF$_4$	10060-10-3 216.109	wh hyg powder	dec ≈600 4.77		i H_2O
524	Cerium(IV) oxide CeO$_2$	1306-38-3 172.114	wh-yel powder; cub	2400 7.65	a	i H_2O, dil acid
525	Cerium(IV) sulfate tetrahydrate Ce(SO$_4$)$_2$·4H$_2$O	10294-42-5 404.304	yel-oran orth cry	dec 180 3.91		reac H_2O
526	Cesium Cs	7440-46-2 132.905	silv-wh metal	28.44 1.93	671 a,c,e	reac H_2O
527	Cesium amide CsNH$_2$	22205-57-8 148.928	wh tetr cry	3.70	a	
528	Cesium azide CsN$_3$	22750-57-8 174.925	hyg tetr cry; exp	326 ≈3.5		
529	Cesium bromate CsBrO$_3$	13454-75-6 260.807	col hex cry	4.11		sl H_2O
530	Cesium bromide CsBr	7787-69-1 212.809	wh cub cry; hyg	636 4.43	≈1300 a,e	vs H_2O; s EtOH; i ace
531	Cesium carbonate Cs$_2$CO$_3$	534-17-8 325.820	wh monocl cry; hyg	792 4.24	a	s H_2O, EtOH, eth
532	Cesium chlorate CsClO$_3$	13763-67-2 216.356	col hex cry	3.57		sl H_2O
533	Cesium chloride CsCl	7647-17-8 168.358	wh cub cry; hyg	645 3.988	1297 a,c,e	vs H_2O; s EtOH
534	Cesium cyanide CsCN	21159-32-0 158.923	wh cub cry; hyg	350 3.34		vs H_2O
535	Cesium fluoride CsF	13400-13-0 151.903	wh cub cry; hyg	703 4.64	a,c,e	vs H_2O; s MeOH; i diox, py
536	Cesium hydride CsH	58724-12-2 133.913	wh cub cry; flam	dec ≈170 3.42	a	reac H_2O
537	Cesium hydrogen fluoride CsHF$_2$	12280-52-3 171.910	tetr cry	170 3.86	a	
538	Cesium hydroxide CsOH	21351-79-1 149.912	wh-yel hyg cry	342.3 3.68	a	vs H_2O; s EtOH
539	Cesium iodide CsI	7789-17-5 259.809	col cub cry; hyg	621 4.51	≈1280 a,e	vs H_2O; s EtOH, MeOH, ace

No.	Name Formula	CAS RN Mol. Wt.	Physical Form	mp/°C den/g cm^{-3}	bp/°C Other Data	Solubility
540	Cesium metaborate $CsBO_2$	92141-86-1 175.715	cub cry	732 ≈3.7	 a	
541	Cesium nitrate $CsNO_3$	7789-18-6 194.910	wh hex or cub cry	414 3.66	 a	s H_2O, ace; sl EtOH
542	Cesium oxide Cs_2O	20281-00-9 281.810	yel-oran hex cry	490 4.65	 a	vs H_2O
543	Cesium perchlorate $CsClO_4$	13454-84-7 232.356	wh orth cry; hyg	250 3.327	 a	sl H_2O
544	Cesium superoxide CsO_2	12018-61-0 164.904	yel tetr cry	432 3.77	 a	reac H_2O
545	Cesium sulfate Cs_2SO_4	10294-54-9 361.875	wh orth cry or hex prisms; hyg	1005 4.24	 a,c	vs H_2O; i EtOH, ace, py
546	Chlorine Cl_2	7782-50-5 70.905	grn-yel gas	-101.5 2.898g/L	-34.04 a,b,c,d,e	sl H_2O
547	Hypochlorous acid HOCl .	7790-92-3 52.460	grn-yel; stable only in aq soln	 	 a	s H_2O
548	Perchloric acid $HClO_4$	7601-90-3 100.459	col hyg liq	-112 1.77	dec ≈90 a	s H_2O
549	Chlorine oxide Cl_2O	7791-21-1 86.904	yel-brn gas	-120.6 3.552 g/L	2.2 a,d,e	vs H_2O
550	Chlorine dioxide ClO_2	10049-04-4 67.452	oran-grn gas	-59 2.757 g/L	11 a,d,e	sl H_2O
551	Chlorine trioxide Cl_2O_3	17496-59-2 118.903	dark brn solid	exp <25 		
552	Chlorine hexoxide Cl_2O_6	12442-63-6 166.901	red liq	3.5 	≈200 e	reac H_2O
553	Chlorine heptoxide Cl_2O_7	10294-48-1 182.901	col oily liq; exp	-91.5 1.9	82 e	reac H_2O
554	Chlorine fluoride ClF	7790-89-8 54.451	col gas	-155.6 2.226 g/L	-101.1 a,d,e	reac H_2O
555	Chlorine trifluoride ClF_3	7790-91-2 92.448	gas	-76.34 3.779g/L	11.75 a,d,e	reac H_2O
556	Chlorine pentafluoride ClF_5	13637-63-3 130.445	col gas	-103 5.332 g/L	-13.1 b	
557	Chlorine trioxide fluoride (perchloryl fluoride) ClO_3F	7616-94-6 102.449	col gas	-147 4.187 g/L	-46.75 a,b,d	
558	Chlorine perchlorate $ClOClO_3$	27218-16-2 134.903	unstable liq	-117 1.81(0°C)	dec ≈25 	
559	Chromium Cr	7440-47-3 51.996	blue-wh metal; cub	1907 7.15	2671 a,c,e	reac dil acid
560	Chromium acetylacetonate $Cr(CH_3COCHCOCH_3)_3$	21679-31-2 349.324	red monocl cry	208 1.34	345 	i H_2O; s bz
561	Chromium antimonide CrSb	12053-12-2 173.756	hex cry	1110 7.11		
562	Chromium arsenide Cr_2As	12254-85-2 178.914	tetr cry	 7.04		
563	Chromium boride CrB	12006-79-0 62.807	refrac orth cry	2100 6.1		
564	Chromium boride CrB_2	12007-16-8 73.618	refrac solid; hex	2200 5.22		
565	Chromium boride Cr_5B_3	12007-38-4 292.414	tetr cry	1900 6.10		
566	Chromium carbide Cr_3C_2	12012-35-0 180.010	gray orth cry	1895 6.68		
567	Chromium carbonyl $Cr(CO)_6$	13007-92-6 220.058	col orth cry	dec 130 1.77	subl e	i H_2O, EtOH; s eth, chl
568	Chromium nitride Cr_2N	12053-27-9 117.999	hex cry	1650 6.8		
569	Chromium nitride CrN	24094-93-7 66.003	gray cub cry	dec 1080 5.9		
570	Chromium phosphide CrP	26342-61-0 82.970	orth cry	 5.25		
571	Chromium selenide CrSe	12053-13-3 130.96	hex cry	≈1500 6.1		
572	Chromium silicide Cr_3Si	12018-36-9 184.074	cub cry	1770 6.4		
573	Chromium silicide $CrSi_2$	12018-09-6 108.167	gray hex cry	1490 4.91		
574	Chromium(II) acetate monohydrate $Cr(C_2H_3O_2)_2·H_2O$	628-52-4* 188.101	red monocl cry	 1.79		sl H_2O
575	Chromium(II) bromide $CrBr_2$	10049-25-9 211.804	wh monocl cry; aq soln blue	842 4.236	 a	s H_2O, EtOH
576	Chromium(II) chloride $CrCl_2$	10049-05-5 122.901	hyg needles; aq soln blue	814 2.88	1300 a,c,d	s H_2O
577	Chromium(II) chloride tetrahydrate $Cr(H_2O)_4Cl_2·4H_2O$	13931-94-7 267.023	blue hyg cry	dec 51 		s H_2O

No.	Name Formula	CAS RN Mol. Wt.	Physical Form	mp/°C den/g cm^{-3}	bp/°C Other Data	Solubility
578	Chromium(II) fluoride CrF_2	10049-10-2 89.993	blue-grn monocl cry	894 3.79	a	sl H_2O; i EtOH
579	Chromium(II) iodide CrI_2	13478-28-9 305.805	red-brn cry; hyg	868 5.1	a	
580	Chromium(II) oxalate monohydrate $CrC_2O_4 \cdot H_2O$	814-90-4* 158.031	yel-grn powder	2.468		sl H_2O
581	Chromium(II) sulfate pentahydrate $CrSO_4 \cdot 5H_2O$	13825-86-0 238.136	blue cry			s H_2O, dil acid; sl EtOH; i ace
582	Chromium(II,III) oxide Cr_3O_4	12018-34-7 219.986	cub cry	6.1	a	
583	Chromium(III) acetate hexahydrate $Cr(C_2H_3O_2)_3 \cdot 6H_2O$	1066-30-4* 285.226	blue needles			s H_2O
584	Chromium(III) bromide $CrBr_3$	10031-25-1 291.708	dark grn hex cry	1130 4.68		s hot H_2O
585	Chromium(III) bromide hexahydrate $Cr(H_2O)_6Br_3$	10031-25-1* 399.799	viol hyg cry			s H_2O; i EtOH, eth
586	Chromium(III) bromide hexahydrate $CrBr_3(H_2O)_4 \cdot 2H_2O$	18721-05-6 399.799	grn hyg cry			s H_2O, EtOH
587	Chromium(III) chloride $CrCl_3$	10025-73-7 158.354	purp hex plates	1152 2.87	dec 1300 a	sl H_2O
588	Chromium(III) chloride hexahydrate $[CrCl_2(H_2O)_4]Cl \cdot 2H_2O$	10060-12-5 266.445	grn monocl cry; hyg			s H_2O, EtOH; sl ace; i eth
589	Chromium(III) fluoride CrF_3	7788-97-8 108.991	grn needles	1400 3.8	a	i H_2O; EtOH
590	Chromium(III) fluoride trihydrate $CrF_3 \cdot 3H_2O$	16671-27-5 163.037	grn hex cry	2.2		sl H_2O
591	Chromium(III) hydroxide trihydrate $Cr(OH)_3 \cdot 3H_2O$	1308-14-1 157.063	blue-grn powder			i H_2O; s acid
592	Chromium(III) iodide CrI_3	13569-75-0 432.709	dark grn hex cry	dec 500 5.32	a	sl H_2O
593	Chromium(III) nitrate $Cr(NO_3)_3$	13548-38-4 238.011	grn hyg powder	dec >60		vs H_2O
594	Chromium(III) nitrate nonahydrate $Cr(NO_3)_3 \cdot 9H_2O$	7789-02-8 400.148	grn-blk monocl cry	66.3 1.80	dec >100	vs H_2O
595	Chromium(III) oxide Cr_2O_3	1308-38-9 151.990	grn hex cry	2330 5.22	≈3000 a,c	i H_2O, EtOH; sl acid, alk
596	Chromium(III) phosphate $CrPO_4$	7789-04-0 146.967	blue orth cry	>1800 4.6		i H_2O, acid, aqua regia
597	Chromium(III) phosphate hemiheptahydrate $CrPO_4 \cdot 3.5H_2O$	84359-31-9 210.021	blue-grn powder	2.15		i H_2O; s acid
598	Chromium(III) phosphate hexahydrate $CrPO_4 \cdot 6H_2O$	84359-31-9 255.059	viol cry	dec >500 2.121		i H_2O; s acid, alk
599	Chromium(III) potassium sulfate dodecahydrate $CrK(SO_4)_2 \cdot 12H_2O$	7788-99-0 499.405	viol-blk cub cry	89 dec 1.83		s H_2O; i EtOH
600	Chromium(III) sulfate $Cr_2(SO_4)_3$	10101-53-8 392.183	red-brn hex cry	3.1		i H_2O, acid
601	Chromium(III) sulfide Cr_2S_3	12018-22-3 200.190	brn-blk hex cry	3.8		
602	Chromium(III) telluride Cr_2Te_3	12053-39-3 486.79	hex cry	≈1300 7.0		
603	Chromium(IV) chloride $CrCl_4$	15597-88-3 193.807	gas, stable at high temp	8.504 g/L	dec >600	
604	Chromium(IV) fluoride CrF_4	10049-11-3 127.990	grn cry	277		
605	Chromium(IV) oxide CrO_2	12018-01-8 83.995	brn-blk tetr powder	dec ~400 4.89	a	i H_2O; s acid
606	Chromium(V) fluoride CrF_5	14884-42-5 146.988	red orth cry	34	117	
607	Chromium(VI) fluoride CrF_6	13843-28-2 165.986	yel solid; stable at low temp	dec -100		
608	Chromium(VI) oxide CrO_3	1333-82-0 99.994	red orth cry	197 2.7	dec ≈250	vs H_2O
609	Chromyl chloride CrO_2Cl_2	14977-61-8 154.900	red liq	-96.5 1.91	117 a,d,e	reac H_2O; s ctc, chl, bz
610	Cobalt Co	7440-48-4 58.933	gray metal; hex or cub	1495 8.86	2927 a,c	s dil acid
611	Cobalt antimonide CoSb	12052-42-5 180.69	hex cry	1202 8.8		
612	Cobalt arsenic sulfide CoAsS	12254-82-9 165.92	silv-wh solid	≈6.1	f	
613	Cobalt arsenide CoAs	27016-73-5 133.855	orth cry	1180 8.22		
614	Cobalt arsenide $CoAs_2$	12044-42-7 208.776	monocl cry	7.2		
615	Cobalt arsenide $CoAs_3$	12256-04-1 283.698	cub cry	942 6.84		

No.	Name Formula	CAS RN Mol. Wt.	Physical Form	mp/°C den/g cm^{-3}	bp/°C Other Data	Solubility
616	Cobalt boride Co_2B	12045-01-1 128.677	refrac solid	1280 8.1		
617	Cobalt boride CoB	12006-77-8 69.744	refrac solid	1460 7.25		reac H_2O, HNO_3
618	Cobalt carbonyl $Co_2(CO)_8$	10210-68-1 341.949	oran cry	51 dec 1.78		i H_2O; s EtOH, eth, CS_2
619	Cobalt phosphide Co_2P	12134-02-0 148.840	gray needles	1386 6.4		i H_2O; s HNO_3
620	Cobalt silicide $CoSi_2$	12017-12-8 115.10	gray cub cry	1326 4.9		s hot HCl
621	Cobalt disulfide CoS_2	12013-10-4 123.07	cub cry	 4.3		
622	Cobalt dodecacarbonyl $Co_4(CO)_{12}$	17786-31-1 571.858	blk cry	dec 60 2.09		
623	Cobalt(II) acetate $Co(C_2H_3O_2)_2$	71-48-7 177.023	pink cry			vs H_2O; s EtOH
624	Cobalt(II) acetate tetrahydrate $Co(C_2H_3O_2)_2 \cdot 4H_2O$	6147-53-1 249.083	red monocl cry	 1.705		s H_2O, EtOH, dil acid
625	Cobalt(II) aluminate $CoAl_2O_4$	13820-62-7 176.894	blue cub cry	 4.37		i H_2O
626	Cobalt(II) arsenate octahydrate $Co_3(AsO_4)_2 \cdot 8H_2O$	24719-19-5 598.760	red monocl needles	dec 1000 3.0		i H_2O; s dil acid
627	Cobalt(II) bromate hexahydrate $Co(BrO_3)_2 \cdot 6H_2O$	13476-01-2 422.829	viol cry	 ≈2.5		vs H_2O
628	Cobalt(II) bromide $CoBr_2$	7789-43-7 218.741	grn hex cry; hyg	678 4.91	a	vs H_2O; s MeOH, EtOH, ace
629	Cobalt(II) bromide hexahydrate $CoBr_2 \cdot 6H_2O$	13762-12-4 326.832	red hyg cry	47 dec 2.46	dec 100	s H_2O
630	Cobalt(II) carbonate $CoCO_3$	513-79-1 118.942	pink rhomb cry	 4.2		i H_2O, EtOH
631	Cobalt(II) chloride $CoCl_2$	7646-79-9 129.838	blue hyg leaflets	740 3.36	1049 a,c,e	s H_2O, EtOH, eth, ace, py
632	Cobalt(II) chloride dihydrate $CoCl_2 \cdot 2H_2O$	16544-92-6 165.869	viol-blue cry	 2.477		
633	Cobalt(II) chloride hexahydrate $CoCl_2 \cdot 6H_2O$	7791-13-1 237.929	pink-red monocl cry	87 dec 1.924		vs H_2O; s EtOH, ace, eth
634	Cobalt(II) chromate $CoCrO_4$	24613-38-5 174.927	yel-brn orth cry	 ≈4.0		i H_2O; s acid
635	Cobalt(II) chromite $CoCr_2O_4$	13455-25-9 226.923	blue-grn cub cry	 5.14		i H_2O, conc acid
636	Cobalt(II) cyanide $Co(CN)_2$	542-84-7 110.968	blue hyg cry	 1.872		i H_2O
637	Cobalt(II) cyanide dihydrate $Co(CN)_2 \cdot 2H_2O$	20427-11-6 146.999	pink-brn needles			i H_2O, acid
638	Cobalt(II) ferricyanide $Co_3[Fe(CN)_6]_2$	14049-81-1 600.703	red needles			i H_2O, HCl; s NH_4OH
639	Cobalt(II) fluoride CoF_2	10026-17-2 96.930	red tetr cry	1127 4.46	≈1400 a,c	sl H_2O; s acid
640	Cobalt(II) fluoride tetrahydrate $CoF_2 \cdot 4H_2O$	13817-37-3 168.992	red orth cry	dec 2.22		s H_2O
641	Cobalt(II) formate dihydrate $Co(CHO_2)_2 \cdot 2H_2O$	6424-20-0 184.999	red cry powder	dec 140 2.13		s H_2O; i EtOH
642	Cobalt(II) hexafluorosilicate hexahydrate $CoSiF_6 \cdot 6H_2O$	12021-68-0 309.100	pale red cry	 2.087		vs H_2O
643	Cobalt(II) hydroxide $Co(OH)_2$	21041-93-0 92.948	blue-grn cry	dec ≈160 3.60	a	sl H_2O; s acid
644	Cobalt(II) iodate $Co(IO_3)_2$	13455-28-2 408.738	blk-viol needles	dec 200 5.09		sl H_2O
645	Cobalt(II) iodide CoI_2	15238-00-3 312.742	blk hex cry; hyg	520 5.60	a	vs H_2O
646	Cobalt(II) iodide hexahydrate $CoI_2 \cdot 6H_2O$	15238-00-3* 420.833	red hex prisms	dec 130 2.90		s H_2O, EtOH, eth, ace
647	Cobalt(II) titanate $CoTiO_3$	12017-01-5 154.798	grn rhomb cry	 5.0		
648	Cobalt(II) molybdate $CoMoO_4$	13762-14-6 218.87	blk monocl cry	1040 4.7		
649	Cobalt(II) nitrate $Co(NO_3)_2$	10141-05-6 182.942	pale red powder	dec 100 2.49	a	vs H_2O
650	Cobalt(II) nitrate hexahydrate $Co(NO_3)_2 \cdot 6H_2O$	10026-22-9 291.034	red monocl cry; hyg	≈55 1.88		vs H_2O; s EtOH
651	Cobalt(II) oxalate CoC_2O_4	814-89-1 146.953	pink powder	250 dec 3.02		i H_2O; s acid, NH_4OH
652	Cobalt(II) oxalate dihydrate $CoC_2O_4 \cdot 2H_2O$	814-89-1* 166.984	pink needles	dec		i H_2O; sl acid; s NH_4OH
653	Cobalt(II) oxide CoO	1307-96-6 74.932	gray cub cry	1830 6.44	a	i H_2O; s acid
654	Cobalt(II) perchlorate $Co(ClO_4)_2$	13455-31-7 257.833	red needles	 3.33		vs H_2O; i EtOH, ace

No.	Name Formula	CAS RN Mol. Wt.	Physical Form	mp/°C den/g cm⁻³	bp/°C Other Data	Solubility
655	Cobalt(II) phosphate octahydrate $Co_3(PO_4)_2 \cdot 8H_2O$	10294-50-5 510.865	pink amorp powder	 2.77		i H_2O; s acid
656	Cobalt(II) potassium sulfate hexahydrate $CoK_2(SO_4)_2 \cdot 6H_2O$	10026-20-7 437.349	red monocl cry	dec 75 2.22		vs H_2O
657	Cobalt(II) selenate pentahydrate $CoSeO_4 \cdot 5H_2O$	14590-19-3 291.97	red tricl cry	dec 2.51		vs H_2O
658	Cobalt(II) selenide CoSe	1307-99-9 137.89	yel hex cry	1055 7.65		i H_2O, alk; s aqua regia
659	Cobalt(II) selenite dihydrate $CoSeO_3 \cdot 2H_2O$	19034-13-0 221.92	blue-red powder			i H_2O
660	Cobalt(II) orthosilicate Co_2SiO_4	12017-08-2 209.950	red-viol orth cry	1345 4.63		i H_2O; s dil HCl
661	Cobalt(II) stannate Co_2SnO_4	12139-93-4 300.574	grn-blue cub cry	 6.30		i H_2O; s alk
662	Cobalt(II) sulfate $CoSO_4$	10124-43-3 154.997	red orth cry	>700 3.71	a	s H_2O
663	Cobalt(II) sulfate heptahydrate $CoSO_4 \cdot 7H_2O$	10026-24-1 281.103	pink monocl cry	41 dec 2.03		vs H_2O; sl EtOH, MeOH
664	Cobalt(II) sulfate monohydrate $CoSO_4 \cdot H_2O$	13455-34-0 173.012	red monocl cry	 3.08		sl H_2O
665	Cobalt(II) sulfide CoS	1317-42-6 91.00	blk amorp powder	1182 5.45	a	i H_2O; s acid
666	Cobalt(II) telluride CoTe	12017-13-9 186.53	hex cry	≈8.8		
667	Cobalt(II) thiocyanate trihydrate $Co(SCN)_2 \cdot 3H_2O$	97126-35-7 229.146	viol rhomb cry			s H_2O, EtOH, eth, ace
668	Cobalt(II) tungstate $CoWO_4$	12640-47-0 306.77	blue monocl cry	≈7.8		i H_2O; s hot conc acid
669	Cobalt(II,III) oxide Co_3O_4	1308-06-1 240.798	blk cub cry	dec 900 6.11	a	i H_2O; s acid, alk
670	Cobalt(III) acetate $Co(C_2H_3O_2)_3$	917-69-1 236.066	grn hyg cry	dec 100		s H_2O, EtOH
671	Cobalt(III) ammonium tetranitrodiammine $NH_4[Co(NH_3)_2(NO_2)_4]$	13600-89-0 295.054	red-brn orth cry	 1.97		s H_2O
672	Cobalt(III) fluoride CoF_3	10026-18-3 115.928	brn hex cry	927 3.88		reac H_2O
673	Cobalt(III) hexammine chloride $Co(NH_3)_6Cl_3$	10534-89-1 267.474	red monocl cry	 1.71		s H_2O; i EtOH
674	Cobalt(III) hydroxide $Co(OH)_3$	1307-86-4 109.955	brn powder	dec ≈4		i H_2O; s acid
675	Cobalt(III) nitrate $Co(NO_3)_3$	15520-84-0 244.948	grn cub cry; hyg	≈3.0		s H_2O; reac os
676	Cobalt(III) oxide Co_2O_3	1308-04-9 165.864	gray-blk powder	dec 895 5.18		i H_2O; s conc acid
677	Cobalt(III) oxide monohydrate $Co_2O_3 \cdot H_2O$	12016-80-7 183.880	brn-blk hex cry	dec 150		i H_2O; s acid
678	Cobalt(III) potassium nitrite sesquihydrate $CoK_3(NO_2)_6 \cdot 1.5H_2O$	13782-01-9* 479.284	yel cub cry	 2.6		sl H_2O; reac acid; i EtOH
679	Cobalt(III) sulfide Co_2S_3	1332-71-4 214.064	blk cub cry	 4.8	a	reac acid
680	Cobalt(III) titanate Co_2TiO_4	12017-38-8 229.731	grn-blk cub cry	 5.1		s conc HCl
681	Copper Cu	7440-50-8 63.546	red metal; cub	1084.62 8.96	2562 a,c,e,f	sl dil acid
682	Copper acetylacetonate $Cu(CH_3COCHCOCH_3)_2$	13395-16-9 261.765	blue powder	284 dec	subl	sl H_2O; s chl
683	Copper nitride Cu_3N	1308-80-1 204.645	cub cry	300 dec 5.84		
684	Copper phosphide CuP_2	12019-11-3 125.494	monocl cry	≈900 4.20		
685	Copper(I) acetate $CuC_2H_3O_2$	598-54-9 122.591	col cry	dec	subl	reac H_2O
686	Copper(I) acetylide Cu_2C_2	1117-94-8 151.114	red amorp powder; exp			
687	Copper(I) azide CuN_3	14336-80-2 105.566	tetr cry; exp			
688	Copper(I) bromide CuBr	7787-70-4 143.450	wh cub cry; hyg	497 4.98	1345 a,e	sl H_2O; i ace
689	Copper(I) chloride CuCl	7758-89-6 98.999	wh cub cry	430 4.14	≈1400 a,c,e,f	sl H_2O; i EtOH, ace
690	Copper(I) cyanide CuCN	544-92-3 89.564	wh powder or grn orth cry	474 2.9	dec a	i H_2O, EtOH; s KCN soln
691	Copper(I) fluoride CuF	13478-41-6 82.544	cub cry	 7.1		
692	Copper(I) hydride CuH	13517-00-5 64.554	red-brn solid	dec 60		

No.	Name Formula	CAS RN Mol. Wt.	Physical Form	mp/°C den/g cm^{-3}	bp/°C Other Data	Solubility
693	Copper(I) iodide CuI	7681-65-4 190.450	wh cub cry	606 5.67	≈1290 a,f	i H_2O, dil acid
694	Copper(I) mercury iodide Cu_2HgI_4	13876-85-2 835.30	red cry powder	trans ≈60 (brn)		i H_2O, EtOH
695	Copper(I) oxide Cu_2O	1317-39-1 143.091	red-brn cub cry	1235 6.0	dec 1800 a,f	i H_2O
696	Copper(I) selenide Cu_2Se	20405-64-5 206.05	blue-blk tetr cry	1113 6.84		i H_2O; s acid
697	Copper(I) sulfide Cu_2S	22205-45-4 159.158	blue-blk orth cry	≈1100 5.6	a,f	i H_2O; sl acid
698	Copper(I) sulfite hemihydrate $Cu_2SO_3 \cdot 0.5H_2O$	13982-53-1* 216.164	wh-yel hex cry			sl H_2O; s acid, alk; i EtOH, eth
699	Copper(I) telluride Cu_2Te	12019-52-2 254.69	blue hex cry	1127 4.6		
700	Copper(I) thiocyanate CuSCN	1111-67-7 121.630	wh-yel amorp powder	1084 2.85		i H_2O, dil acid, EtOH, ace; s eth
701	Copper(I,II) sulfite dihydrate $Cu_2SO_3 \cdot CuSO_3 \cdot 2H_2O$	13814-81-8 386.797	red prisms or powder			i H_2O, EtOH; s HCl
702	Copper(II) acetate $Cu(C_2H_3O_2)_2$	142-71-2 181.636	blue-grn hyg powder			
703	Copper(II) acetate metaarsenite $Cu(C_2H_3O_2)_2 \cdot 3Cu(AsO_2)_2$	12002-03-8 1013.796	grn cry powder			i H_2O; reac acid
704	Copper(II) acetate monohydrate $Cu(C_2H_3O_2)_2 \cdot H_2O$	6046-93-1 199.651	grn monocl cry	115 1.88	dec 240	s H_2O, EtOH; sl eth
705	Copper(II) acetylide CuC_2	12540-13-5 87.568	brn-blk solid; exp	exp 100		
706	Copper(II) arsenate $Cu_3(AsO_4)_2$	10103-61-4 468.476	blue-grn cry			i H_2O, EtOH; s dil acid
707	Copper(II) arsenite $CuHAsO_3$	10290-12-7 187.474	yel-grn powder			i H_2O, EtOH; s acid
708	Copper(II) azide $Cu(N_3)_2$	14215-30-6 147.586	brn orth cry; exp	≈2.6		
709	Copper(II) basic acetate $Cu(C_2H_3O_2)_2 \cdot CuO \cdot 6H_2O$	52503-64-7 369.272	blue-grn cry or powder			sl H_2O, EtOH; s dil acid, NH_4OH
710	Copper(II) borate $Cu(BO_2)_2$	39290-85-2 149.166	blue-grn powder	3.859		i H_2O; s acid
711	Copper(II) bromide $CuBr_2$	7789-45-9 223.354	blk monocl cry; hyg	498 4.710	900 a	vs H_2O; s EtOH, ace; i bz, eth
712	Copper(II) butanoate monohydrate $Cu(C_4H_7O_2)_2 \cdot H_2O$	540-16-9 255.758	grn monocl plates			s H_2O, diox, bz; sl EtOH
713	Copper(II) carbonate hydroxide $CuCO_3 \cdot Cu(OH)_2$	12069-69-1 221.116	grn monocl cry	200 dec 4.0		i H_2O, EtOH; s dil acid
714	Copper(II) chlorate hexahydrate $Cu(ClO_3)_2 \cdot 6H_2O$	14721-21-2 338.539	blue-grn hyg cry	65	dec 100	vs H_2O, EtOH
715	Copper(II) chloride $CuCl_2$	7447-39-4 134.451	yel-brn monocl cry; hyg	630 dec 3.4	a,c	s H_2O, EtOH, ace
716	Copper(II) chloride dihydrate $CuCl_2 \cdot 2H_2O$	10125-13-0 170.482	grn-blue orth cry; hyg	100 dec 2.51		vs H_2O, EtOH, MeOH; s ace; i eth
717	Copper(II) chromate $CuCrO_4$	13548-42-0 179.540	red-brn cry			i H_2O; s EtOH
718	Copper(II) chromite $CuCr_2O_4$	12018-10-9 231.536	gray-blk tetr cry	5.4		i H_2O, dil acid
719	Copper(II) citrate hemipentahydrate $Cu_2C_6H_4O_7 \cdot 2.5H_2O$	10402-15-0 360.223	blue-grn cry	dec 100		sl H_2O; s dil acid
720	Copper(II) cyanide $Cu(CN)_2$	14763-77-0 115.581	grn powder			i H_2O; s acid, alk
721	Copper(II) dichromate dihydrate $CuCr_2O_7 \cdot 2H_2O$	13675-47-3 315.565	red-brn tricl cry	2.286		vs H_2O
722	Copper(II) ferrocyanide $Cu_2Fe(CN)_6$	13601-13-3 339.043	red-br cub cry or powder	2.2		i H_2O, acid, os
723	Copper(II) ferrous sulfide $CuFeS_2$	1308-56-1 183.523	yel tetr cry	950 4.2	f	i H_2O, HCl; s HNO_3
724	Copper(II) fluoride CuF_2	7789-19-7 101.543	wh monocl cry	836 4.23	1676 a,c	sl H_2O
725	Copper(II) fluoride dihydrate $CuF_2 \cdot 2H_2O$	13454-88-1 137.574	blue monocl cry	dec 130 2.934		sl H_2O
726	Copper(II) formate $Cu(CHO_2)_2$	544-19-4 153.582	blue cry			s H_2O; i os
727	Copper(II) formate tetrahydrate $Cu(CHO_2)_2 \cdot 4H_2O$	5893-61-8 225.642	blue monocl cry			s H_2O; sl EtOH; i os
728	Copper(II) hexafluorosilicate tetrahydrate $CuSiF_6 \cdot 4H_2O$	12062-24-7 277.684	blue monocl cry	dec 2.56		vs H_2O; sl EtOH
729	Copper(II) hydroxide $Cu(OH)_2$	20427-59-2 97.561	blue-grn powder	3.37	a	i H_2O; s acid, conc alk
730	Copper(II) molybdate $CuMoO_4$	13767-34-5 223.48	grn cry	820 dec 3.4		i H_2O
731	Copper(II) nitrate $Cu(NO_3)_2$	3251-23-8 187.555	blue-grn orth cry; hyg	255	subl a	s H_2O, diox; reac eth

No.	Name Formula	CAS RN Mol. Wt.	Physical Form	mp/°C den/g cm^{-3}	bp/°C Other Data	Solubility
732	Copper(II) nitrate hexahydrate $Cu(NO_3)_2 \cdot 6H_2O$	13478-38-1 295.647	blue rhomb cry; hyg	 2.07		vs H_2O; s EtOH
733	Copper(II) nitrate trihydrate $Cu(NO_3)_2 \cdot 3H_2O$	10031-43-3 241.602	blue rhomb cry	114 2.32	170 dec	vs H_2O, EtOH
734	Copper(II) oleate $Cu(C_{18}H_{33}O_2)_2$	1120-44-1 626.464	blue-grn solid			i H_2O; sl EtOH; s eth
735	Copper(II) oxalate CuC_2O_4	814-91-5 151.566	blue-wh powder	dec 310		i H_2O, EtOH, eth; s NH_4OH
736	Copper(II) oxide CuO	1317-38-0 79.545	blk powder or monocl cry	1446 6.31	a,c	i H_2O, EtOH; s dil acid
737	Copper(II) perchlorate $Cu(ClO_4)_2$	13770-18-8 262.446	grn hyg cry	dec 130		s H_2O, eth, diox; i bz, ctc
738	Copper(II) perchlorate hexahydrate $Cu(ClO_4)_2 \cdot 6H_2O$	10294-46-9 370.538	blue monocl cry; hyg	82 2.22	dec 120	vs H_2O, EtOH, HOAc, ace; sl eth
739	Copper(II) phosphate trihydrate $Cu_3(PO_4)_2 \cdot 3H_2O$	10031-48-8 434.627	blue-grn orth cry			i H_2O; s acid, NH_4OH
740	Copper(II) selenate pentahydrate $CuSeO_4 \cdot 5H_2O$	10031-45-5 296.58	blue tricl cry	dec 80 2.56		s H_2O, acid, NH_4OH; sl ace; i EtOH
741	Copper(II) selenide $CuSe$	1317-41-5 142.51	blue-blk needles or plates	550 dec 5.99	a	reac acid
742	Copper(II) selenite dihydrate $CuSeO_3 \cdot 2H_2O$	15168-20-4 226.54	blue orth cry	 3.31		i H_2O; s acid, NH_4OH
743	Copper(II) stearate $Cu(C_{18}H_{35}O_2)_2$	660-60-6 630.496	blue-grn amorp powder	≈250		i H_2O, EtOH, eth; s py
744	Copper(II) sulfate $CuSO_4$	7758-98-7 159.610	wh-grn amorp powder or rhomb cry	560 dec 3.60	a	s H_2O, i EtOH
745	Copper(II) sulfate pentahydrate $CuSO_4 \cdot 5H_2O$	7758-99-8 249.686	blue tricl cry	dec 110 2.286	f	vs H_2O; s MeOH; sl EtOH
746	Copper(II) sulfate, basic $Cu_3(OH)_4SO_4$	1332-14-5 354.732	grn rhomb cry	 3.88		i H_2O
747	Copper(II) sulfide CuS	1317-40-4 95.612	blk hex cry	trans 507 4.76	a,f	i H_2O, EtOH, dil acid, alk
748	Copper(II) tartrate trihydrate $CuC_4H_4O_6 \cdot 3H_2O$	815-82-7 265.664	blue-grn powder			sl H_2O; s acid, alk
749	Copper(II) telluride $CuTe$	12019-23-7 191.15	yel orth cry	trans ≈400 7.09		
750	Copper(II) tungstate $CuWO_4$	13587-35-4 311.38	yel-brn powder	 7.5	a	
751	Copper(II) tungstate dihydrate $CuWO_4 \cdot 2H_2O$	13587-35-4* 347.41	grn powder			i H_2O; sl HOAc; reac conc acid
752	Curium Cm	7440-51-9 247	silv metal; hex or cub	1345 13.51	≈3100 a	
753	Dysprosium Dy	7429-91-6 162.50	silv metal; hex	1412 8.55	2567 a,c	s dil acid
754	Dysprosium boride DyB_4	12310-43-9 205.74	tetr cry	2500 6.98		
755	Dysprosium nitride DyN	12019-88-4 176.51	cub cry	 9.93		
756	Dysprosium silicide $DySi_2$	12133-07-2 218.67	orth cry	 5.2		
757	Dysprosium(II) chloride $DyCl_2$	13767-31-2 233.41	blk cry	721 dec		reac H_2O
758	Dysprosium(II) iodide DyI_2	36377-94-3 416.31	purp cry	659		reac H_2O
759	Dysprosium(III) bromide $DyBr_3$	14456-48-5 402.21	wh hyg cry	879		s H_2O
760	Dysprosium(III) chloride $DyCl_3$	10025-74-8 268.86	yel cry	680 3.67	a	s H_2O
761	Dysprosium(III) fluoride DyF_3	13569-80-7 219.50	grn cry	1154		
762	Dysprosium(III) hydride DyH_3	13537-09-2 165.52	hex cry	 7.1		
763	Dysprosium(III) iodide DyI_3	15474-63-2 543.21	grn cry	978		
764	Dysprosium(III) oxide Dy_2O_3	1308-87-8 373.00	wh cub cry	2408 7.81	a	s acid
765	Dysprosium(III) sulfide Dy_2S_3	12133-10-7 421.20	red-brn monocl cry	 6.08		
766	Einsteinium Es	7429-92-7 252	metal; cub	860	a	
767	Erbium Er	7440-52-0 167.26	silv metal; hex	1529 9.07	2868 a,c	i H_2O; s acid
768	Erbium boride ErB_4	12310-44-0 210.50	tetr cry	2450 7.0		
769	Erbium bromide $ErBr_3$	13536-73-7 406.97	viol hyg cry	923		s H_2O
770	Erbium chloride $ErCl_3$	10138-41-7 273.62	viol monocl cry; hyg	776 4.1	a	s H_2O

No.	Name Formula	CAS RN Mol. Wt.	Physical Form	mp/°C den/g cm^{-3}	bp/°C Other Data	Solubility
771	Erbium chloride hexahydrate $ErCl_3 \cdot 6H_2O$	10025-75-9 381.71	pink hyg cry	dec		s H_2O; sl EtOH
772	Erbium fluoride ErF_3	13760-83-3 224.26	pink orth cry	1147 7.8	a	i H_2O
773	Erbium hydride ErH_3	13550-53-3 170.28	hex cry	≈7.6		
774	Erbium iodide ErI_3	13813-42-8 547.97	viol hex cry; hyg	1014 ≈5.5		s H_2O
775	Erbium nitride ErN	12020-21-2 181.27	cub cry	 10.6		
776	Erbium oxide Er_2O_3	12061-16-4 382.52	pink powder	2418 8.64	a	i H_2O; s acid
777	Erbium silicide $ErSi_2$	12020-28-9 223.43	orth cry	 7.26		
778	Erbium sulfate $Er_2(SO_4)_3$	13478-49-4 622.71	hyg powder	dec 3.68		s H_2O
779	Erbium sulfate octahydrate $Er_2(SO_4)_3 \cdot 8H_2O$	10031-52-4 766.83	pink monocl cry	dec 3.20		s H_2O
780	Erbium sulfide Er_2S_3	12159-66-9 430.72	red-brn monocl cry	1730 6.07		
781	Erbium telluride Er_2Te_3	12020-39-2 717.32	orth cry	1213 7.11		
782	Europium Eu	7440-53-1 151.965	soft silv metal; cub	822 5.24	1529 a,c	reac H_2O
783	Europium boride EuB_6	12008-05-8 216.831	cub cry	≈2600 4.91		
784	Europium nitride EuN	12020-58-5 165.972	cub cry	 8.7		
785	Europium silicide $EuSi_2$	12434-24-1 208.136	tetr cry	1500 5.46		
786	Europium(II) bromide $EuBr_2$	13780-48-8 311.773	wh cry	683		s H_2O
787	Europium(II) chloride $EuCl_2$	13769-20-5 222.870	wh orth cry	731 4.9		s H_2O
788	Europium(II) fluoride EuF_2	14077-39-5 189.962	grn-yel cub cry	≈1380 6.5		
789	Europium(II) iodide EuI_2	22015-35-6 405.774	grn cry	580		s H_2O
790	Europium(II) selenide EuSe	12020-66-5 230.93	brn cub cry	 6.45		
791	Europium(II) sulfide EuS	12020-65-4 184.031	cub cry	 5.7		
792	Europium(II) telluride EuTe	12020-69-8 279.57	blk cub cry	1526 6.48		
793	Europium(III) bromide $EuBr_3$	13759-88-1 391.677	gray cry	dec		s H_2O
794	Europium(III) chloride $EuCl_3$	10025-76-0 258.323	grn-yel needles	623 4.89	a	
795	Europium(III) chloride hexahydrate $EuCl_3 \cdot 6H_2O$	13759-92-7 366.414	wh-yel hyg cry	850 4.89		s H_2O
796	Europium(III) fluoride EuF_3	13765-25-8 208.960	wh hyg cry	1276		i H_2O
797	Europium(III) nitrate hexahydrate $Eu(NO_3)_3 \cdot 6H_2O$	10031-53-5 446.071	wh-pink hyg cry	dec 85		s H_2O
798	Europium(III) oxide Eu_2O_3	1308-96-9 351.928	pink powder	2350 7.42	a	i H_2O; s acid
799	Europium(III) sulfate $Eu_2(SO_4)_3$	13537-15-0 592.121	pale pink cry	 4.99		sl H_2O
800	Europium(III) sulfate octahydrate $Eu_2(SO_4)_3 \cdot 8H_2O$	10031-52-4 736.243	pink cry	dec 375		sl H_2O
801	Fermium Fm	7440-72-4 257	metal	1527	a	
802	Fluorine F_2	7782-41-4 37.997	pale yel gas	-219.66 1.553 g/L	-188.12 a,b,c,d,e	reac H_2O
803	Fluorine monoxide F_2O	7783-41-7 53.996	col gas	-223.8 2.207 g/L	-144.75 a,b,d,e	sl H_2O
804	Fluorine dioxide F_2O_2	7783-44-0 69.996	gas, stable only at low temp	-154 2.861 g/L	-57 a,d	
805	Fluorine nitrate FNO_3	7789-26-6 81.003	col gas	-175 3.311 g/L	-46	reac H_2O, EtOH, eth; s ace
806	Fluorine perchlorate $FOClO_3$	10049-03-3 118.449	col gas; exp	-167.3 4.841 g/L	-16	reac H_2O
807	Francium Fr	7440-73-5 223	short-lived alkali metal	27	a	
808	Gadolinium Gd	7440-54-2 157.25	silv metal; hex	1313 7.90	3273 a,c	s dil acid
809	Gadolinium boride GdB_6	12008-06-9 222.12	blk-brn cub cry	2510 5.31		

No.	Name Formula	CAS RN Mol. Wt.	Physical Form	mp/°C den/g cm^{-3}	bp/°C Other Data	Solubility
810	Gadolinium nitride GdN	25764-15-2 171.26	cub cry	9.10		
811	Gadolinium silicide GdSi$_2$	12134-75-7 213.42	orth cry	5.9		
812	Gadolinium(II) iodide GdI$_2$	13814-72-7 411.06	bronze cry	831		
813	Gadolinium(II) selenide GdSe	12024-81-6 236.21	cub cry	2170 8.1		
814	Gadolinium(III) bromide GdBr$_3$	13818-75-2 396.96	wh monocl cry; hyg	770 4.56		
815	Gadolinium(III) chloride GdCl$_3$	10138-52-0 263.61	wh monocl cry; hyg	609 4.52	a	s H$_2$O
816	Gadolinium(III) chloride hexahydrate GdCl$_3$·6H$_2$O	19423-81-5 371.70	col hyg cry	2.424		s H$_2$O
817	Gadolinium(III) fluoride GdF$_3$	13765-26-9 214.25	wh cry	1231	a	
818	Gadolinium(III) iodide GdI$_3$	13572-98-0 537.96	yel cry	925		
819	Gadolinium(III) nitrate hexahydrate Gd(NO$_3$)$_3$·6H$_2$O	19598-90-4 451.36	hyg tricl cry	91 dec 2.33		s H$_2$O, EtOH
820	Gadolinium(III) nitrate pentahydrate Gd(NO$_3$)$_3$·5H$_2$O	52788-53-1 433.34	wh cry	92 dec 2.41		i H$_2$O
821	Gadolinium(III) oxide Gd$_2$O$_3$	12064-62-9 362.50	wh hyg powder	2420 7.07	a	i H$_2$O; s acid
822	Gadolinium(III) sulfate octahydrate Gd$_2$(SO$_4$)$_3$·8H$_2$O	13450-87-8 746.81	col monocl cry	dec 400 4.14		s H$_2$O
823	Gadolinium(III) sulfide Gd$_2$S$_3$	12134-77-9 410.70	yel cub cry	6.1		
824	Gadolinium(III) telluride Gd$_2$Te$_3$	12160-99-5 697.30	orth cry	1255 7.7		
825	Gallium Ga	7440-55-3 69.723	silv liq or gray orth cry	29.76 5.91	2204 a,c,d,e	reac alk
826	Gallium antimonide GaSb	12064-03-8 191.483	cub cry	712 5.6137	a,c	
827	Gallium arsenide GaAs	1303-00-0 144.645	gray cub cry	1238 5.3176	a	
828	Gallium nitride GaN	25617-97-4 83.730	gray hex cry	600 6.1	dec >600 a	
829	Gallium phosphide GaP	12063-98-8 100.697	yel cub cry	1457 4.138	a	
830	Gallium suboxide Ga$_2$O	12024-20-3 155.445	brn powder	>660 4.77	dec >800 a	
831	Gallium(II) chloride GaCl$_2$	24597-12-4 140.628	wh orth cry	172.4 2.74	535	
832	Gallium(II) selenide GaSe	12024-11-2 148.68	hex cry	960 5.03		
833	Gallium(II) sulfide GaS	12024-10-1 101.789	hex cry	965 3.86		
834	Gallium(II) telluride GaTe	12024-14-5 197.32	monocl cry	824 5.44		
835	Gallium(III) acetylacetonate Ga(CH$_3$COCHCOCH$_3$)$_3$	14405-43-7 367.051	wh powder	193 1.42	subl	
836	Gallium(III) bromide GaBr$_3$	13450-88-9 309.435	wh orth cry	121.5 3.69	279 a,b,c,d	
837	Gallium(III) chloride GaCl$_3$	13450-90-3 176.081	col needles or gl solid	77.9 2.47	201 a,b,c,d,e	
838	Gallium(III) fluoride GaF$_3$	7783-51-9 126.718	wh powder or col needles	>1000 4.47	a	i H$_2$O
839	Gallium(III) fluoride trihydrate GaF$_3$·3H$_2$O	22886-66-4 180.764	wh cry	dec >140		sl H$_2$O
840	Gallium(III) hydride GaH$_3$	13572-93-5 72.747	visc liq	-15	dec ≈0	
841	Gallium(III) hydroxide Ga(OH)$_3$	12023-99-3 120.745	unstable prec		a	
842	Gallium(III) iodide GaI$_3$	13450-91-4 450.436	monocl cry	212 4.5	340 a,b,c,d	
843	Gallium(III) nitrate Ga(NO$_3$)$_3$	13494-90-1 255.738	wh cry powder			s H$_2$O, EtOH, eth
844	Gallium(III) oxide Ga$_2$O$_3$	12024-21-4 187.444	wh cry	1807 ≈6.0	a	s hot acid
845	Gallium(III) oxide hydroxide GaOOH	20665-52-5 102.730	orth cry	5.23		
846	Gallium(III) selenide Ga$_2$Se$_3$	12024-24-7 376.33	cub cry	937 4.92		
847	Gallium(III) sulfate Ga$_2$(SO$_4$)$_3$	13494-91-2 427.637	hex cry			
848	Gallium(III) sulfate octadecahydrate Ga$_2$(SO$_4$)$_3$·18H$_2$O	13780-42-2 751.912	octahed cry			s H$_2$O, EtOH

No.	Name Formula	CAS RN Mol. Wt.	Physical Form	mp/°C den/g cm⁻³	bp/°C Other Data	Solubility
849	Gallium(III) sulfide Ga$_2$S$_3$	12024-22-5 235.644	monocl cry	1090 3.7		
850	Gallium(III) telluride Ga$_2$Te$_3$	12024-27-0 522.25	cub cry	790 5.57		
851	Germanium Ge	7440-56-4 72.61	gray-wh cub cry	938.25 5.3234	2833 a,c,d	i H$_2$O, dil acid, alk
852	Germanium nitride Ge$_3$N$_4$	12065-36-0 273.86	orth cry	dec 900		i H$_2$O, acid, aqua regia
853	Germane GeH$_4$	7782-65-2 76.64	col gas; flam	-165 3.133 g/L	-88.1 a,b,d,e	i H$_2$O
854	Bromogermane GeH$_3$Br	13569-43-2 155.54	col liq	-32 2.34	52 a	reac H$_2$O
855	Chlorogermane GeH$_3$Cl	13637-65-5 111.09	col liq	-52 1.75	28 a	reac H$_2$O
856	Chlorotrifluorogermane GeF$_3$Cl	14188-40-0 165.06	gas	-66.2 6.747 g/L	-20.3	
857	Dibromogermane GeH$_2$Br$_2$	13769-36-3 234.43	col liq	-15 2.80	89	reac H$_2$O
858	Dichlorogermane GeH$_2$Cl$_2$	15230-48-5 145.53	col liq	-68 1.90	69.5	reac H$_2$O
859	Dichlorodifluorogermane GeF$_2$Cl$_2$	24422-21-7 181.51	col gas	-51.8 7.419 g/L	-2.8	
860	Fluorogermane GeH$_3$F	13537-30-9 94.63	col gas	3.868 g/L	a	reac H$_2$O
861	Iodogermane GeH$_3$I	13573-02-9 202.54	liq	-15	≈90 a	reac H$_2$O
862	Tribromogermane GeHBr$_3$	14779-70-5 313.33	col liq	-25	dec	reac H$_2$O
863	Trichlorogermane GeHCl$_3$	1184-65-2 179.98	liq	-71 1.93	75.3 e	reac H$_2$O
864	Trichlorofluorogermane GeCl$_3$F	24422-20-6 197.97	liq	-49.8	37.5	
865	Methylgermane GeH$_3$CH$_3$	1449-65-6 90.67	col gas	-158 3.706 g/L	-23	
866	Digermane Ge$_2$H$_6$	13818-89-8 151.27	col liq; flam	-109	30.8 a,d,e	
867	Trigermane Ge$_3$H$_8$	14691-44-2 225.89	col liq	-105.6 2.20(-105°C)	110.5 a,d,e	i H$_2$O
868	Germanium(II) bromide GeBr$_2$	24415-00-7 232.42	yel monocl cry	122	dec 150	reac H$_2$O
869	Germanium(II) chloride GeCl$_2$	10060-11-4 143.51	wh-yel hyg powder	dec		reac H$_2$O; s eth, bz
870	Germanium(II) fluoride GeF$_2$	13940-63-1 110.61	wh orth cry; hyg	110 3.64	dec 130	reac H$_2$O
871	Germanium(II) iodide GeI$_2$	13573-08-5 326.42	oran-yel hex cry	dec 550 5.4		reac H$_2$O
872	Germanium(II) oxide GeO	20619-16-3 88.61	blk solid	dec 700	a	
873	Germanium(II) selenide GeSe	12065-10-0 151.57	gray orth cry or brn powder	667 5.6		
874	Germanium(II) sulfide GeS	12025-32-0 104.68	gray orth cry	615 4.1	a	
875	Germanium(II) telluride GeTe	12025-39-7 200.21	cub cry	725 6.16		i H$_2$O; s conc HNO$_3$
876	Germanium(IV) bromide GeBr$_4$	13450-92-5 392.23	wh cry	26.1 3.132	186.35 a,b,d,e	reac H$_2$O
877	Germanium(IV) chloride GeCl$_4$	10038-98-9 214.42	col liq	-49.5 1.88	86.55 a,b,d,e	reac H$_2$O; s bz, eth, EtOH, ctc
878	Germanium(IV) fluoride GeF$_4$	7783-58-6 148.60	col gas	-15 tp 6.074 g/L	-36.5 sp a	reac H$_2$O
879	Germanium(IV) iodide GeI$_4$	13450-95-8 580.23	red-oran cub cry	144 4.4	377 a,b	reac H$_2$O
880	Germanium(IV) oxide GeO$_2$	1310-53-8 104.61	wh hex cry	1115 4.25	a	i H$_2$O
881	Germanium(IV) selenide GeSe$_2$	12065-11-1 230.53	yel-oran orth ccry	707 dec 4.56		
882	Germanium(IV) sulfide GeS$_2$	12025-34-2 136.74	blk orth cry	530 3.01		
883	Gold Au	7440-57-5 196.967	soft yel metal	1064.18 19.3	2856 a,c,d,e,f	s aqua regia
884	Bromoauric acid pentahydrate HAuBr$_4$·5H$_2$O	17083-68-0 607.667	red-brn hyg cry	27		s H$_2$O, EtOH
885	Chloroauric acid tetrahydrate HAuCl$_4$·4H$_2$O	16903-35-8 411.847	yel monocl cry; hyg	≈3.9		vs H$_2$O, EtOH; s eth
886	Gold(I) bromide AuBr	10294-27-6 276.871	yel-gray tetr cry	dec 165 8.20	a	i H$_2$O
887	Gold(I) chloride AuCl	10294-29-8 232.420	yel orth cry	dec 289 7.6	a	i H$_2$O

No.	Name Formula	CAS RN Mol. Wt.	Physical Form	mp/°C den/g cm^{-3}	bp/°C Other Data	Solubility
888	Gold(I) cyanide AuCN	506-65-0 222.985	yel hex cry	dec 7.2		i H_2O, EtOH, eth, dil acid
889	Gold(I) iodide AuI	10294-31-2 323.871	yel-grn powder; tetr	dec 120 8.25	a	i H_2O; s CN soln
890	Gold(I) sulfide Au_2S	1303-60-2 425.999	brn-blk cub cry; unstable	240 dec ≈11		i H_2O, acid; s aqua regia
891	Gold(III) bromide $AuBr_3$	10294-28-7 436.679	red-br monocl cry	dec ≈160 	a	s H_2O, EtOH
892	Gold(III) chloride $AuCl_3$	13453-07-1 303.325	red monocl cry	dec >160 4.7	a	
893	Gold(III) cyanide trihydrate $Au(CN)_3 \cdot 3H_2O$	535-37-5* 329.066	wh hyg cry	dec 50		vs H_2O; sl EtOH
894	Gold(III) fluoride AuF_3	14720-21-9 253.962	oran-yel hex cry	>300 6.75	subl a	
895	Gold(III) hydroxide $Au(OH)_3$	1303-52-2 247.989	brn powder	dec ≈100		i H_2O; s acid
896	Gold(III) iodide AuI_3	31032-13-0 577.680	unstable grn powder			
897	Gold(III) oxide Au_2O_3	1303-58-8 441.931	brn powder	dec ≈150		i H_2O; s acid
898	Gold(III) selenate $Au_2(SeO_4)_3$	10294-32-3 822.81	yel cry			i H_2O; s acid
899	Gold(III) selenide Au_2Se_3	1303-62-4 630.81	blk amorp solid	dec 4.65		s aqua regia
900	Gold(III) sulfide Au_2S_3	1303-61-3 490.131	unstable blk powder	dec 200		
901	Hafnium Hf	7440-58-6 178.49	gray metal; hex	2233 13.3	4603 a,c	s HF
902	Hafnium boride HfB_2	12007-23-7 200.11	gray hex cry	3100 10.5		
903	Hafnium bromide $HfBr_4$	13777-22-5 498.11	wh cub cry	424 tp 4.90	323 sp b	
904	Hafnium carbide HfC	12069-85-1 190.50	refrac cub cry	≈3000 12.2		
905	Hafnium chloride $HfCl_4$	13499-05-3 320.30	wh monocl cry	432 tp 320.30	317 sp a,b	reac H_2O
906	Hafnium fluoride HfF_4	13709-52-9 254.48	wh monocl cry	>970 7.1	970 sp a	
907	Hafnium hydride HfH_2	12770-26-2 180.51	refrac tetr cry	11.4		
908	Hafnium iodide HfI_4	13777-23-6 686.11	yel-oran cub cry	449 tp 5.6	394 sp b	
909	Hafnium nitride HfN	25817-87-2 192.50	yel-brn cub cry	3305 13.8		
910	Hafnium oxide HfO_2	12055-23-1 210.49	wh cub cry	2774 9.68	a	i H_2O
911	Hafnium oxychloride octahydrate $HfOCl_2 \cdot 8H_2O$	14456-34-9 409.52	wh tetr cry	dec		s H_2O
912	Hafnium phosphide HfP	12325-59-6 209.46	hex cry	9.78		
913	Hafnium selenide $HfSe_2$	12162-21-9 336.41	brn hex cry	7.46		
914	Hafnium orthosilicate $HfSiO_4$	13870-13-8 270.57	tetr cry	7.0		
915	Hafnium silicide $HfSi_2$	12401-56-8 234.66	gray orth cry	≈1700 7.6		
916	Hafnium sulfate $Hf(SO_4)_2$	15823-43-5 370.62	wh cry	dec >500		
917	Hafnium sulfide HfS_2	18855-94-2 242.62	purp-brn hex cry	6.03		
918	Helium He	7440-59-7 4.003	col gas	0.164 g/L	-268.93 a,b,d,e	sl H_2O; l EtOH
919	Holmium Ho	7440-60-0 164.930	silv metal; hex	1474 8.80	2700 a,c	s dil acid
920	Holmium bromide $HoBr_3$	13825-76-8 404.642	yel hyg cry	919	1470	
921	Holmium chloride $HoCl_3$	10138-62-2 271.288	yel monocl cry; hyg	718 3.7	1500 a	s H_2O
922	Holmium fluoride HoF_3	13760-78-6 221.925	pink-yel orth cry; hyg	1143 7.664	>2200 a	s H_2O
923	Holmium iodide HoI_3	13813-41-7 545.643	yel hex cry	994 5.4		
924	Holmium nitride HoN	12029-81-1 178.937	cub cry	10.6		
925	Holmium oxide Ho_2O_3	12055-62-8 377.859	yel cub cry	2415 8.41	a	s acid
926	Holmium silicide $HoSi_2$	12136-24-2 221.10	hex cry	7.1		

No.	Name Formula	CAS RN Mol. Wt.	Physical Form	mp/°C den/g cm⁻³	bp/°C Other Data	Solubility
927	Holmium sulfide Ho_2S_3	12162-59-3 426.059	yel-oran monocl cry	 5.92		
928	Hydrazine N_2H_4	302-01-2 32.045	col oily liq	1.4 1.0036	113.55 a,b,c,d	vs H_2O, EtOH, MeOH
929	Hydrazine hydrate $N_2H_4 \cdot H_2O$	7803-57-8 50.060	fuming liq	-51.7 1.030	119	vs H_2O, EtOH; i chl, eth
930	Hydrazine hydrobromide $N_2H_4 \cdot HBr$	13775-80-9 112.957	wh monocl cry flakes	84 2.3	dec ≈190	s H_2O, EtOH
931	Hydrazine hydrochloride $N_2H_4 \cdot HCl$	2644-70-4 68.506	wh orth cry	89 1.5	dec 240	s H_2O; i os
932	Hydrazine dihydrochloride $N_2H_4 \cdot 2HCl$	5341-61-7 104.966	wh orth cry	198 dec 1.42		s H_2O; sl EtOH
933	Hydrazine hydroiodide $N_2H_4 \cdot HI$	10039-55-1 159.957	hyg cry	125		s H_2O
934	Hydrazine nitrate $N_2H_4 \cdot HNO_3$	13464-97-6 95.058	monocl cry; exp	70		vs H_2O
935	Hydrazine sulfate $N_2H_4 \cdot H_2SO_4$	10034-93-2 130.125	col orth cry	254 1.378		sl H_2O; i EtOH
936	Hydrazoic acid HN_3	7782-79-8 43.028	col liq; exp	-80	35.7 a,d	s H_2O
937	Hydroxylamine H_2NOH	7803-49-8 33.030	wh orth flakes or needles	33.1 1.21	58 a	vs H_2O, MeOH
938	Hydroxylamine sulfate $(H_2NOH)_2 \cdot H_2SO_4$	10039-54-0 164.139	cry	170		vs H_2O
939	Hydrogen H_2	1333-74-0 2.016	col gas; flam	-259.34 0.082 g/L	-252.87 a,b,c,d,e	sl H_2O
940	Hydrogen bromide HBr	10035-10-6 80.912	col gas	-86.81 3.307 g/L	-66.38 a,b,c,d,e	vs H_2O; s EtOH
941	Hydrogen chloride HCl	7647-01-0 36.461	col gas	-114.18 1.490 g/L	-85 a,b,c,d,e	vs H_2O
942	Hydrogen cyanide HCN	74-90-8 27.026	col liq	-13.4 0.684	26 a,b,c,e	vs H_2O, EtOH; sl eth
943	Hydrogen fluoride HF	7664-39-3 20.006	col gas	-83.36 0.818 g/L	20 a,b,c,e	vs H_2O, EtOH; sl eth
944	Hydrogen iodide HI	10034-85-2 127.912	col or yel gas	-50.77 5.228 g/L	-35.55 a,b,c,d,e	vs H_2O; s os
945	Hydrogen peroxide H_2O_2	7722-84-1 34.015	col liq	-0.43 1.44	150.2 a,c,d,e	vs H_2O
946	Hydrogen selenide H_2Se	7783-07-5 80.98	col gas; flam	-65.73 3.310 g/L	-41.25 a,b,d,e	s H_2O
947	Hydrogen sulfide H_2S	7783-06-4 34.082	col gas; flam	-85.5 1.393 g/L	-59.55 a,b,c,d,e	s H_2O
948	Hydrogen disulfide H_2S_2	13465-07-1 66.148	col liq	 1.334	70.7 a,d,e	
949	Hydrogen telluride H_2Te	7783-09-7 129.62	col gas	-49 5.298 g/L	-2 a,d,e	s H_2O, EtOH, alk
950	Indium In	7440-74-6 114.818	soft wh metal	156.60 7.31	2072 a,c	s acid
951	Indium antimonide InSb	1312-41-0 236.578	blk cub cry	525 5.7747	 a,c	
952	Indium arsenide InAs	1303-11-3 189.740	gray cub cry	942 5.67	 a	i acids
953	Indium nitride InN	25617-98-5 128.825	hex cry	1100 6.88		
954	Indium phosphide InP	22398-80-7 145.792	blk cub cry	1062 4.81	 a	sl acid
955	Indium(I) bromide InBr	14280-53-6 194.722	oran-red orth cry	290 4.96	656 a,d	reac H_2O
956	Indium(I) chloride InCl	13465-10-6 150.271	yel cub cry	211 4.19	608 a,c	reac H_2O
957	Indium(I) iodide InI	13966-94-4 241.722	orth cry	364.4 5.32	712 a,d	
958	Indium(II) bromide $InBr_2$	21264-43-7 274.626	orth cry			reac H_2O
959	Indium(II) chloride $InCl_2$	13465-11-7 185.723	col orth cry	235 3.64		reac H_2O
960	Indium(II) sulfide InS	12030-14-7 146.884	red-brn orth cry	692 5.2	 a	
961	Indium(III) bromide $InBr_3$	13465-09-3 354.530	hyg yel-wh monocl cry	420 4.74	 a	
962	Indium(III) chloride $InCl_3$	10025-82-8 221.176	yel monocl cry; hyg	583 4.0	 a	vs H_2O; s EtOH
963	Indium(III) fluoride InF_3	7783-52-0 171.813	wh hex cry; hyg	1170 4.39	>1200	sl H_2O; s dil acid
964	Indium(III) fluoride trihydrate $InF_3 \cdot 3H_2O$	14166-78-0 225.859	wh cry	dec 100		s H_2O
965	Indium(III) hydroxide $In(OH)_3$	20661-21-6 165.840	cub cry	 4.4		

No.	Name Formula	CAS RN Mol. Wt.	Physical Form	mp/°C den/g cm^{-3}	bp/°C Other Data	Solubility
966	Indium(III) iodide InI_3	13510-35-5 495.531	yel-red monocl cry; hyg	207 4.69	a	
967	Indium(III) oxide In_2O_3	1312-43-2 277.634	yel cub cry	1913 7.18	a	i H_2O; s hot acid
968	Indium(III) perchlorate octahydrate $In(ClO_4)_3 \cdot 8H_2O$	13465-15-1 557.291	wh cry	≈80	dec 200	
969	Indium(III) phosphate $InPO_4$	14693-82-4 209.789	wh orth cry	4.9		i H_2O
970	Indium(III) selenide In_2Se_3	1312-42-1 466.52	blk hex cry	660 5.8		
971	Indium(III) sulfate $In_2(SO_4)_3$	13464-82-9 517.827	hyg wh powder	3.44		s H_2O
972	Indium(III) sulfide In_2S_3	12030-24-9 325.834	oran cub cry	1050 4.45	a	
973	Indium(III) telluride In_2Te_3	1312-45-4 612.44	blk cub cry	667 5.75		
974	Iodine I_2	7553-56-2 253.809	blue-blk plates	113.7 4.933	184.4 a,b,c,d,e	sl H_2Os; s bz, EtOH, eth, ctc, chl
975	Iodic acid HIO_3	7782-68-5 175.910	col orth cry	110 dec 4.63	a	vs H_2O; i EtOH, eth
976	Periodic acid dihydrate $HIO_4 \cdot 2H_2O$	10450-60-9 227.940	monocl hyg cry	122 dec		s H_2O, EtOH; sl eth
977	Iodine tetroxide I_2O_4	12399-08-5 317.807	yel cry	dec 85 4.2		sl H_2O
978	Iodine pentoxide I_2O_5	12029-98-0 333.806	hyg wh cry	≈300 dec 4.98		s H_2O; i EtOH, eth, CS_2
979	Iodine nonaoxide I_4O_9	73560-00-6 651.613	hyg yel powder	dec 75		
980	Iodine bromide IBr	7789-33-5 206.808	blk orth cry	40 4.3	116 dec a,b	s H_2O, EtOH, eth
981	Iodine chloride ICl	7790-99-0 162.357	red cry or oily liq	27.38 3.24	100 dec a,c	reac H_2O; s EtOH
982	Iodine trichloride ICl_3	865-44-1 233.262	yel tricl cry; hyg	101 tp (16 atm) 3.2	64 sp (dec)	reac H_2O; s EtOH, bz
983	Iodine fluoride IF	13873-84-2 145.902	disproportionates at room temp		a	
984	Iodine trifluoride IF_3	22520-96-3 183.899	yel solid, stable at low temp	dec -28		
985	Iodine pentafluoride IF_5	7783-66-6 221.896	yel liq	9.43 3.19	100.5 a,d,e	reac H_2O
986	Iodine heptafluoride IF_7	16921-96-3 259.893	col liq	6.5 tp 2.67	4.8 sp e	s H_2O
987	Iridium Ir	7439-88-5 192.217	silv-wh metal; cub	2446 22.5	4428 a,c	s aqua regia
988	Iridium(III) sulfide Ir_2S_3	12136-42-4 480.632	orth cry	10.2	a	
989	Iridium(III) bromide $IrBr_3$	10049-24-8 431.929	red-brn monocl cry	6.82		i H_2O, acid, alk
990	Iridium(III) bromide tetrahydrate $IrBr_3 \cdot 4H_2O$	10049-24-8* 503.991	grn-brn cry			s H_2O; i EtOH
991	Iridium(III) chloride $IrCl_3$	10025-83-9 298.575	brn monocl cry	dec 763 5.30	a	i H_2O, acid, alk
992	Iridium(III) fluoride IrF_3	23370-59-4 249.212	blk hex cry	dec 250 ≈8.0		i H_2O, dil acid
993	Iridium(III) iodide IrI_3	7790-41-2 572.930	dark brn monocl cry	≈7.4		i H_2O, acid, bz, chl; s alk
994	Iridium(III) oxide Ir_2O_3	1312-46-5 432.432	blue-blk cry	dec 1000		i H_2O; sl hot HCl
995	Iridium(IV) oxide IrO_2	12030-49-8 224.216	brn tetr cry	1100 dec 11.7	a	
996	Iridium(IV) sulfide IrS_2	12030-51-2 256.349	orth cry	9.3	a	
997	Iridium(VI) fluoride IrF_6	7783-75-7 306.207	yel cub cry; hyg	44 4.8	53 a,c,d	reac H_2O
998	Iron Fe	7439-89-6 55.845	silv-wh or gray met	1538 7.87	2861 a,c,e	s dil acid
999	Iron arsenide $FeAs$	12044-16-5 130.767	gray orth cry	1030 7.85		
1000	Iron boride Fe_2B	12006-86-9 122.501	refr solid; tetr	1389 7.3		
1001	Iron boride FeB	12006-84-7 66.656	refr solid; orth	1650 ≈7		
1002	Iron pentacarbonyl $Fe(CO)_5$	13463-40-6 195.897	yel oily liq; flam	-20 1.490	103 a,e	i H_2O; s eth, bz, ace
1003	Iron nonacarbonyl $Fe_2(CO)_9$	15321-51-4 363.784	oran-yel cry	100 dec 2.85		
1004	Iron dodecacarbonyl $Fe_3(CO)_{12}$	12088-65-2 503.660	blk cry	140 2.00		

No.	Name Formula	CAS RN Mol. Wt.	Physical Form	mp/°C den/g cm⁻³	bp/°C Other Data	Solubility
1005	Iron hydrocarbonyl $H_2Fe(CO)_4$	17440-90-3 169.903	col liq; unstable	-70	dec	s alk
1006	Iron phosphide Fe_2P	1310-43-6 142.664	gray hex needles	1370 6.8		i H_2O,dil acid, alk
1007	Iron disulfide FeS_2	1317-66-4 119.977	blk cub cry	dec >600 5.02	a,f	i H_2O
1008	Iron silicide FeSi	12022-95-6 83.931	gray cub cry	1410 6.1		
1009	Iron silicide $FeSi_2$	12022-99-0 112.016	gray tetr cry	1220 4.74		
1010	Iron(II) aluminate $Fe(AlO_2)_2$	12068-49-4 173.806	blk cub cry	4.3		
1011	Iron(II) bromide $FeBr_2$	7789-46-0 215.653	yel-brn hex cry; hyg	691 4.636	dec a,c	vs H_2O, EtOH
1012	Iron(II) bromide hexahydrate $FeBr_2 \cdot 6H_2O$	13463-12-2 323.744	grn hyg cry	27 dec 4.64		s H_2O, EtOH
1013	Iron(II) carbonate $FeCO_3$	563-71-3 115.854	gray-brn hex cry	3.9	f	
1014	Iron(II) chloride $FeCl_2$	7758-94-3 126.750	wh hex cry; hyg	677 3.16	1023 a,c,e	vs H_2O, EtOH, ace; sl bz
1015	Iron(II) chloride dihydrate $FeCl_2 \cdot 2H_2O$	16399-77-2 162.781	wh-grn monocl cry	dec 120 2.39		s H_2O
1016	Iron(II) chloride tetrahydrate $FeCl_2 \cdot 4H_2O$	13478-10-9 198.812	grn monocl cry	dec 105 1.93		s H_2O, EtOH
1017	Iron(II) chromite $FeCr_2O_4$	1308-31-2 223.835	blk cub cry	5.0	f	
1018	Iron(II) fluoride FeF_2	7789-28-8 93.842	wh tetr cry	1100 4.09	a,c	sl H_2O; s dil HF; i EtOH, eth
1019	Iron(II) fluoride tetrahydrate $FeF_2 \cdot 4H_2O$	13940-89-1 165.904	col hex cry	2.20		
1020	Iron(II) hydroxide $Fe(OH)_2$	18624-44-7 89.860	wh-grn hex cry	3.4		i H_2O
1021	Iron(II) iodide FeI_2	7783-86-0 309.654	red-viol hex cry; hyg	587 5.3	a,c	s H_2O, EtOH, eth
1022	Iron(II) iodide tetrahydrate $FeI_2 \cdot 4H_2O$	7783-86-0* 381.716	blk hyg leaflets	dec 90 2.87		s H_2O, EtOH
1023	Iron(II) molybdate $FeMoO_4$	13718-70-2 215.78	brn-yel monocl cry	1115 5.6	a	i H_2O
1024	Iron(II) nitrate $Fe(NO_3)_2$	14013-86-6 179.854	grn solid			s H_2O
1025	Iron(II) nitrate hexahydrate $Fe(NO_3)_2 \cdot 6H_2O$	13476-08-9 287.946	grn solid	60 dec		vs H_2O
1026	Iron(II) oxalate dihydrate $FeC_2O_4 \cdot 2H_2O$	6047-25-2 179.895	yel cry	dec 150 2.28		sl H_2O; s acid
1027	Iron(II) oxide FeO	1345-25-1 71.844	blk cub cry	1377 6.0	a,c	i H_2O, alk; s acid
1028	Iron(II) phosphate octahydrate $Fe_3(PO_4)_2 \cdot 8H_2O$	14940-41-1 501.600	gray-blue monocl cry; hyg	2.58	f	i H_2O; s acid
1029	Iron(II) selenide FeSe	1310-32-3 134.81	blk hex cry	6.7		i H_2O
1030	Iron(II) orthosilicate Fe_2SiO_4	10179-73-4 203.774	brn orth cry	4.30	a,f	
1031	Iron(II) sulfate $FeSO_4$	7720-78-7 151.909	wh orth cry; hyg	3.65	a	s H_2O
1032	Iron(II) sulfate monohydrate $FeSO_4 \cdot H_2O$	17375-41-6 169.924	wh-yel monocl cry	dec 300 3.0		s H_2O
1033	Iron(II) sulfate heptahydrate $FeSO_4 \cdot 7H_2O$	7782-63-0 278.015	blue-grn monocl cry	dec ≈60 1.895	f	vs H_2O; i EtOH
1034	Iron(II) sulfide FeS	1317-37-9 87.911	col hex or tetr cry; hyg	1188 4.7	dec a,c	i H_2O; reac acid
1035	Iron(II) telluride FeTe	12125-63-2 183.45	tetr cry	914 6.8		
1036	Iron(II) thiocyanate trihydrate $Fe(SCN)_2 \cdot 3H_2O$	6010-09-9 226.058	grn monocl cry			s H_2O, EtOH, eth
1037	Iron(II) titanate $FeTiO_3$	12168-52-4 151.710	blk rhomb cry	≈1470 4.72	f	
1038	Iron(II) tungstate $FeWO_4$	13870-24-1 303.68	monocl cry	7.51	a,f	
1039	Iron(II,III) oxide Fe_3O_4	1317-61-9 231.533	blk cub cry or amorp powder	1597 5.17	a,c,f	i H_2O; s acid
1040	Iron(III) acetate, basic $FeOH(C_2H_3O_2)_2$	10450-55-2 190.942	brn-red amorp powder			i H_2O; s EtOH, acid
1041	Iron(III) acetylacetonate $Fe(OC(CH_3)=CHC(O)CH_3)_3$	14024-18-1 353.173	red-oran cry	179 5.24		sl H_2O; s os
1042	Iron(III) arsenate dihydrate $FeAsO_4 \cdot 2H_2O$	10102-49-5 230.795	grn-brn powder	dec 3.18		i H_2O; s dil acid
1043	Iron(III) bromide $FeBr_3$	10031-26-2 295.557	dark red hex cry; hyg	dec 4.5	a	s H_2O, EtOH, eth

No.	Name Formula	CAS RN Mol. Wt.	Physical Form	mp/°C den/g cm^{-3}	bp/°C Other Data	Solubility
1044	Iron(III) chloride FeCl$_3$	7705-08-0 162.203	grn hex cry; hyg	304 2.90	≈316 a,c,e	s H$_2$O, EtOH, eth, ace
1045	Iron(III) chloride hexahydrate FeCl$_3$·6H$_2$O	10025-77-1 270.294	yel-oran monocl cry; hyg	dec 37 1.82		s H$_2$O, EtOH, eth, ace
1046	Iron(III) chromate Fe$_2$(CrO$_4$)$_3$	10294-52-7 459.671	yel powder			i H$_2$O, EtOH; s acid
1047	Iron(III) citrate pentahydrate FeC$_6$H$_5$O$_7$·5H$_2$O	3522-50-7 335.023	red-brn cry			s H$_2$O; i EtOH
1048	Iron(III) dichromate Fe$_2$(Cr$_2$O$_7$)$_3$	10294-53-8 759.654	red-brn solid			s H$_2$O, acid
1049	Iron(III) ferrocyanide Fe$_4$[Fe(CN)$_6$]$_3$	14038-43-8 859.234	dark blue powder	1.80		i H$_2$O, dil acid, os
1050	Iron(III) fluoride FeF$_3$	7783-50-8 112.840	grn hex cry	>1000 3.87		sl H$_2$O; i EtOH, eth, bz
1051	Iron(III) fluoride trihydrate FeF$_3$·3H$_2$O	15469-38-2 166.886	yel-brn tetr cry	2.3		
1052	Iron(III) hydroxide Fe(OH)$_3$	1309-33-7 106.867	yel monocl cry	3.12		
1053	Iron(III) hydroxide oxide FeO(OH)	20344-49-4 88.852	red-brn orth cry	4.26	f	i H$_2$O; s acid
1054	Iron(III) nitrate nonahydrate Fe(NO$_3$)$_3$·9H$_2$O	7782-61-8 403.997	viol-gray hyg cry	47 dec 1.68		vs H$_2$O, EtOH, ace
1055	Iron(III) oxalate Fe$_2$(C$_2$O$_4$)$_3$	19469-07-9 375.749	yel amorp powder	dec 100		s H$_2$O, acid; i alk
1056	Iron(III) oxide Fe$_2$O$_3$	1309-37-1 159.688	red-brn hex cry	1565 5.25	a,f	i H$_2$O; s acid
1057	Iron(III) phosphate dihydrate FePO$_4$·2H$_2$O	10045-86-0 186.847	gray-wh orth cry	2.87	f	i H$_2$O; s HCl
1058	Iron(III) pyrophosphate nonahydrate Fe$_4$(P$_2$O$_7$)$_3$·9H$_2$O	10058-44-3 907.348	yel powder			i H$_2$O; s acid
1059	Iron(III) hypophosphite Fe(H$_2$PO$_2$)$_3$	7783-84-8 250.811	wh-gray powder			i H$_2$O
1060	Iron(III) sulfate Fe$_2$(SO$_4$)$_3$	10028-22-5 399.881	gray-wh rhomb cry; hyg	3.10		sl H$_2$O
1061	Iron(III) sulfate nonahydrate Fe$_2$(SO$_4$)$_3$·9H$_2$O	13520-56-4 562.018	yel hex cry	dec 400 2.1	f	vs H$_2$O
1062	Iron(III) thiocyanate monohydrate Fe(SCN)$_3$·H$_2$O	4119-52-2 248.111	red hyg cry	dec		s H$_2$O, EtOH, ace; i tol, chl
1063	Iron(III) metavanadate Fe(VO$_3$)$_3$	65842-03-7 352.665	gray-brn powder			i H$_2$O, EtOH; s acid
1064	Krypton Kr	7439-90-9 83.80	col gas	-157.38 tp 3.425g/L	-153.22 a,b,c,d,e	sl H$_2$O
1065	Krypton difluoride KrF$_2$	13773-81-4 121.80	col tetr cry	dec ≈25 3.24		reac H$_2$O
1066	Lanthanum La	7439-91-0 138.906	silv metal; hex	918 6.15	3464 a,c	s dil acid
1067	Lanthanum boride LaB$_6$	12008-21-8 203.772	blk cub cry; refrac	2715 4.76		
1068	Lanthanum bromide LaBr$_3$	13536-79-3 378.618	wh hex cry; hyg	788 5.1		s H$_2$O
1069	Lanthanum carbide LaC$_2$	12071-15-7 162.928	tetr cry	2360 5.29		
1070	Lanthanum carbonate octahydrate La$_2$(CO$_3$)$_3$·8H$_2$O	6487-39-4 601.961	wh cry powder	2.6		i H$_2$O; s dil acid
1071	Lanthanum chloride LaCl$_3$	10099-58-8 245.264	wh hex cry; hyg	859 3.84	a	s H$_2$O
1072	Lanthanum chloride heptahydrate LaCl$_3$·7H$_2$O	20211-76-1 371.371	wh tricl cry; hyg	dec 91		s H$_2$O, EtOH
1073	Lanthanum fluoride LaF$_3$	13709-38-1 195.901	wh hex cry; hyg	1493 5.9		i H$_2$O, acid
1074	Lanthanum hydride LaH$_3$	13864-01-2 141.930	blk cub cry	5.36		
1075	Lanthanum hydroxide La(OH)$_3$	14507-19-8 189.928	wh amorp solid	dec		i H$_2$O
1076	Lanthanum iodide LaI$_3$	13813-22-4 519.619	wh orth cry; hyg	778 5.6		s H$_2$O
1077	Lanthanum nitrate hexahydrate La(NO$_3$)$_3$·6H$_2$O	10277-43-7 433.012	wh hyg cry	dec ≈40		vs H$_2$O, EtOH
1078	Lanthanum nitride LaN	25764-10-7 152.913	cub cry	6.73		
1079	Lanthanum oxide La$_2$O$_3$	1312-81-8 325.809	wh amorp powder	2305 6.51	4200 a	i H$_2$O; s dil acid
1080	Lanthanum silicide LaSi$_2$	12056-90-5 195.077	gray tetr cry	5.0		
1081	Lanthanum sulfate nonahydrate La$_2$(SO$_4$)$_3$·9H$_2$O	10294-62-9 728.139	hex cry	2.82		sl H$_2$O; i EtOH
1082	Lanthanum sulfide La$_2$S$_3$	12031-49-1 374.009	red cub cry	2110 4.9		

No.	Name Formula	CAS RN Mol. Wt.	Physical Form	mp/°C den/g cm⁻³	bp/°C Other Data	Solubility
1083	Lanthanum sulfide LaS	12031-30-0 170.972	yel cub cry	2300 5.61	a	
1084	Lawrencium Lr	22537-19-5 262	metal	1627	a	
1085	Lead Pb	7439-92-1 207.2	soft silv-gray metal; cub	327.46 11.3	1749 a,c,d,e	s conc acid
1086	Lead(II) acetate Pb(C₂H₃O₂)₂	301-04-2 325.3	wh cry	280 3.25	dec	vs H₂O
1087	Lead(II) acetate trihydrate Pb(C₂H₃O₂)₂·3H₂O	6080-56-4 427.3	col cry	75 dec 2.55		vs H₂O; sl EtOH
1088	Lead(II) acetate, basic Pb(C₂H₃O₂)₂·2PbO·2H₂O	1335-32-6 807.7	wh powder	dec		
1089	Lead(II) antimonate Pb₃(SbO₄)₂	13510-89-9 993.1	oran-yel powder	6.58		i H₂O, dil acid
1090	Lead(II) arsenate Pb₃(AsO₄)₂	3687-31-8 899.4	wh cry	1042 dec 5.8		i H₂O; s HNO₃
1091	Lead(II) arsenite Pb(AsO₂)₂	10031-13-7 421.0	wh powder	5.85		i H₂O; s dil HNO₃
1092	Lead(II) azide Pb(N₃)₂	13424-46-9 291.2	col orth needles; exp	exp ≈350 4.7		i H₂O; vs HOAc
1093	Lead(II) borate monohydrate Pb(BO₂)₂·H₂O	10214-39-8 310.8	wh powder	500 dec 5.6		i H₂O; s dil HNO₃
1094	Lead(II) bromate monohydrate Pb(BrO₃)₂·H₂O	10031-21-7 481.0	col cry	≈180 dec 5.53		sl H₂O
1095	Lead(II) bromide PbBr₂	10031-22-8 367.0	wh orth cry	371 6.69	892 a,c,d,e	sl H₂O; i EtOH
1096	Lead(II) butanoate Pb(C₄H₇O₂)₂	819-73-8 381.4	col solid	≈90		i H₂O; s dil HNO₃
1097	Lead(II) carbonate PbCO₃	598-63-0 267.2	col orth cry	dec ≈315 6.6	a,f	i H₂O
1098	Lead(II) carbonate, basic Pb(OH)₂·2PbCO₃	1319-46-6 775.6	wh hex cry	400 dec ≈6.5		i H₂O, EtOH; s acid
1099	Lead(II) chlorate Pb(ClO₃)₂	10294-47-0 374.1	col hyg cry	dec 230 3.9		vs H₂O, EtOH
1100	Lead(II) chloride PbCl₂	7758-95-4 278.1	wh orth needles or powder	501 5.98	951 a,c,d,e,f	sl H₂O; s alk
1101	Lead(II) chloride fluoride PbClF	13847-57-9 261.7	tetr cry	7.05		i H₂O
1102	Lead(II) chromate PbCrO₄	7758-97-6 323.2	yel-oran monocl cry	844 6.12	a,f	i H₂O; s akl, dil acid
1103	Lead(II) chromate(VI) oxide PbCrO₄·PbO	18454-12-1 546.4	red powder			i H₂O
1104	Lead(II) citrate trihydrate Pb₃(C₆H₅O₇)₂·3H₂O	512-26-5 1053.8	wh cry powder			s H₂O; sl EtOH
1105	Lead(II) cyanide Pb(CN)₂	592-05-2 259.2	wh-yel powder			sl H₂O; reac acid
1106	Lead(II) fluoride PbF₂	7783-46-2 245.2	wh orth cry	830 8.44	1293 a,c,d,e	sl H₂O
1107	Lead(II) fluoroborate Pb(BF₄)₂	13814-96-5 380.8	stable only in aq soln			s H₂O
1108	Lead(II) formate Pb(CHO₂)₂	811-54-1 297.2	wh prisms or needles	dec 190 4.63		sl H₂O; i EtOH
1109	Lead(II) hydrogen arsenate PbHAsO₄	7784-40-9 347.1	wh monocl cry	dec 280 5.943		i H₂O; s HNO₃, alk
1110	Lead(II) hydrogen phosphate PbHPO₄	15845-52-0 303.2	wh monocl cry	dec 5.66		
1111	Lead(II) hydroxide Pb(OH)₂	19783-14-3 241.2	wh powder	145 dec 7.59		i H₂O; s acid
1112	Lead(II) iodate Pb(IO₃)₂	25659-31-8 557.0	wh orth cry	6.50		i H₂O
1113	Lead(II) iodide PbI₂	10101-63-0 461.0	yel hex cry or powder	410 6.16	872 dec a,c,d,e	sl H₂O; i EtOH
1114	Lead(II) lactate Pb(C₃H₅O₃)₂	18917-82-3 385.3	wh cry powder			s H₂O, hot EtOH
1115	Lead(II) molybdate PbMoO₄	10190-55-3 367.1	yel tetr cry	≈1060 6.7	a,f	i H₂O; HNO₃, NaOH
1116	Lead(II) niobate Pb(NbO₃)₂	12034-88-7 489.0	rhomb or tetr cry	1343 6.6		i H₂O
1117	Lead(II) nitrate Pb(NO₃)₂	10099-74-8 331.2	col cub cry	470 4.53	a	vs H₂O; sl EtOH
1118	Lead(II) oleate Pb(C₁₈H₃₃O₂)₂	1120-46-3 770.1	wax-like solid			i H₂O; s EtOH, bz, eth
1119	Lead(II) oxalate PbC₂O₄	814-93-7 295.2	wh powder	dec 300 5.28	a	i H₂O; s dil HNO₃
1120	Lead(II) oxide (litharge) PbO	1317-36-8 223.2	red tetr cry	888 9.35	a,e,f	i H₂O, EtOH; s dil HNO₃
1121	Lead(II) oxide (massicot) PbO	1317-36-8 223.2	yel orth cry	stab >489 9.64	a,e,f	i H₂O, EtOH; s dil HNO₃

No.	Name Formula	CAS RN Mol. Wt.	Physical Form	mp/°C den/g cm⁻³	bp/°C Other Data	Solubility
1122	Lead(II) oxide hydrate $3PbO \cdot H_2O$	1311-11-1 687.6	wh powder	7.41		i H_2O; s dil acid
1123	Lead(II) perchlorate $Pb(ClO_4)_2$	13453-62-8 406.1	wh cry			vs H_2O
1124	Lead(II) perchlorate trihydrate $Pb(ClO_4)_2 \cdot 3H_2O$	13637-76-8 460.1	wh cry	100 dec 2.6		vs cold H_2O; s EtOH
1125	Lead(II) phosphate $Pb_3(PO_4)_2$	7446-27-7 811.5	wh hex cry	1014 7.01		i H_2O, EtOH
1126	Lead(II) hypophosphite $Pb(H_2PO_2)_2$	10294-58-3 337.2	hyg cry powder	dec		sl H_2O; i EtOH
1127	Lead(II) metasilicate $PbSiO_3$	10099-76-0 283.3	wh monocl cry powder	764 6.49	a	i H_2O, os
1128	Lead(II) orthosilicate Pb_2SiO_4	13566-17-1 506.5	monocl cry	743 7.60	a	
1129	Lead(II) hexafluorosilicate dihydrate $PbSiF_6 \cdot 2H_2O$	1310-03-8 385.3	col cry	dec		vs H_2O
1130	Lead(II) selenate $PbSeO_4$	7446-15-3 350.2	orth cry	6.37	a	i H_2O; s conc acid
1131	Lead(II) selenide $PbSe$	12069-00-0 286.2	gray cub cry	1078 8.1	a	i H_2O; s HNO_3
1132	Lead(II) selenite $PbSeO_3$	7488-51-9 334.2	wh monocl cry	≈500 7.0		i H_2O
1133	Lead(II) sodium thiosulfate $Na_4Pb(S_2O_3)_3$	10101-94-7 635.6	wh cry			sl H_2O
1134	Lead(II) stearate $Pb(C_{18}H_{35}O_2)_2$	1072-35-1 774.2	wh powder	≈100 1.4		i H_2O, s hot EtOH
1135	Lead(II) sulfate $PbSO_4$	7446-14-2 303.3	orth cry	1087 6.29	a,f	i H_2O, acid; sl alk
1136	Lead(II) sulfide PbS	1314-87-0 239.3	blk powder or silv cub cry	1118 7.60	a,c,e,f	i H_2O; s acid
1137	Lead(II) sulfite $PbSO_3$	7446-10-8 287.3	wh powder	dec	a	i H_2O; s HNO_3
1138	Lead(II) tantalate $PbTa_2O_6$	12065-68-8 665.1	orth cry	7.9		i H_2O
1139	Lead(II) telluride $PbTe$	1314-91-6 334.8	gray cub cry	924 8.164	a,f	i H_2O, acid
1140	Lead(II) thiocyanate $Pb(SCN)_2$	592-87-0 323.4	wh-yel powder	3.82		sl H_2O
1141	Lead(II) thiosulfate PbS_2O_3	13478-50-7 319.3	wh cry	dec 5.18		i H_2O; s acid
1142	Lead(II) titanate $PbTiO_3$	12060-00-3 303.1	yel tetr cry	7.9		i H_2O; reac HCl
1143	Lead(II) tungstate (stolzite) $PbWO_4$	7759-01-5 455.0	yel tetr cry	1130 8.24	f	i H_2O; s alk
1144	Lead(II) tungstate (raspite) $PbWO_4$	7759-01-5 455.0	monocl cry	trans 400 8.46	f	i H_2O; s alk
1145	Lead(II) metavanadate $Pb(VO_3)_2$	10099-79-3 405.1	yel powder			i H_2O; reac HNO_3
1146	Lead(II) zirconate $PbZrO_3$	12060-01-4 346.4	col orth cry	≈8		i H_2O, alk; s acid
1147	Lead(II,IV) oxide Pb_2O_3	1314-27-8 462.4	blk monocl cry or red amorp powder	dec 530 10.05		i H_2O; s alk; reac conc HCl
1148	Lead(II,II,IV) oxide Pb_3O_4	1314-41-6 685.6	red tetr cry	830 8.92	a,f	i H_2O, EtOH; s hot HCl
1149	Lead(IV) acetate $Pb(C_2H_3O_2)_4$	546-67-8 443.4	col monocl cry	≈175 2.23		reac H_2O, EtOH; s bz, chl
1150	Lead(IV) bromide $PbBr_4$	13701-91-2 526.8	unstable liq			
1151	Lead(IV) chloride $PbCl_4$	13463-30-4 349.0	yel oily liq	-15	dec ≈50 a	
1152	Lead(IV) fluoride PbF_4	7783-59-7 283.2	wh tetr cry; hyg	≈600 6.7		
1153	Lead(IV) oxide PbO_2	1309-60-0 239.2	red tetr cry or brn powder	290 dec 9.64	a	
1154	Lithium Li	7439-93-2 6.941	soft silv-wh metal	180.5 0.534	1342 a,c,e	reac H_2O
1155	Lithium acetate dihydrate $LiC_2H_3O_2 \cdot 2H_2O$	6108-17-4 102.017	wh rhomb cry	58 dec 1.3		vs H_2O; s EtOH
1156	Lithium aluminum hydride $LiAlH_4$	16853-85-3 37.955	gray-wh monocl cry	dec >125 0.917	a	reac H_2O, EtOH; s eth, thf
1157	Lithium amide $LiNH_2$	7782-89-0 22.964	tetr cry	380 1.18	a	reac H_2O
1158	Lithium arsenate Li_3AsO_4	13478-14-3 159.743	col orth cry	3.07		sl H_2O; s HOAc
1159	Lithium azide LiN_3	19597-69-4 48.961	hyg monocl cry; exp	1.83		vs H_2O
1160	Lithium borate $LiBO_2$	13453-69-5 49.751	wh monocl cry; hyg	849 2.18	a	vs H_2O; s EtOH

No.	Name / Formula	CAS RN / Mol. Wt.	Physical Form	mp/°C / den/g cm^{-3}	bp/°C / Other Data	Solubility
1161	Lithium borohydride / $LiBH_4$	16949-15-8 / 21.784	wh-gray orth cry or powder	268 / 0.66	dec 380 / a	s alk, eth, thf
1162	Lithium bromide / $LiBr$	7550-35-8 / 86.845	wh cub cry; hyg	552 / 3.464	≈1300 / a,c,e	s H_2O, EtOH, eth
1163	Lithium carbonate / Li_2CO_3	554-13-2 / 73.891	wh monocl cry	723 / 2.11	dec 1300 / a,c	sl H_2O; s acid; i EtOH
1164	Lithium chloride / $LiCl$	7447-41-8 / 42.394	wh cub cry or powder; hyg	610 / 2.07	1383 / a,c,e	s H_2O, EtOH, ace, py
1165	Lithium chromate dihydrate / $Li_2CrO_4 \cdot 2H_2O$	7789-01-7 / 165.906	yel orth cry; hyg	dec 75 / 2.15		vs H_2O; s EtOH
1166	Lithium dichromate dihydrate / $Li_2Cr_2O_7 \cdot 2H_2O$	10022-48-7 / 265.901	yel-red hyg cry	130 dec / 2.34		vs H_2O
1167	Lithium fluoride / LiF	7789-24-4 / 25.939	wh cub cry or powder	848.2 / 2.640	1673 / a,c,d,e	sl H_2O; s acid
1168	Lithium formate monohydrate / $Li(CHO_2) \cdot H_2O$	6108-23-2 / 69.974	col-wh cry	/ 1.46		s H_2O
1169	Lithium hydride / LiH	7580-67-8 / 7.949	gray cub cry or powder; hyg	688.7 / 0.78	/ a,c	reac H_2O, EtOH
1170	Lithium hydroxide / $LiOH$	1310-65-2 / 23.948	col tetr cry	471.2 / 1.45	1626 / a,c,d	s H_2O; sl EtOH
1171	Lithium hydroxide monohydrate / $LiOH \cdot H_2O$	1310-66-3 / 41.964	wh monocl cry or powder	/ 1.51		s H_2O; sl EtOH
1172	Lithium iodide / LiI	10377-51-2 / 133.845	wh cub cry; hyg	469 / 4.06	1171 / a,c,e	vs H_2O
1173	Lithium iodide trihydrate / $LiI \cdot 3H_2O$	7790-22-9 / 187.891	wh hyg cry	73 / 3.48		vs H_2O, EtOH, ace
1174	Lithium niobate / $LiNbO_3$	12031-63-9 / 147.845	wh hex cry	≈1240 / 4.30		
1175	Lithium nitrate / $LiNO_3$	7790-69-4 / 68.946	col hex cry; hyg	253 / 2.38	/ a,c	vs H_2O; s EtOH
1176	Lithium nitride / Li_3N	26134-62-3 / 34.830	red hex cry	813 / 1.27		reac H_2O
1177	Lithium phosphate / Li_3PO_4	10377-52-3 / 115.794	wh orth cry	1205 / 2.46	/ a	i H_2O
1178	Lithium oxide / Li_2O	12057-24-8 / 29.881	wh cub cry	1570 / 2.013	/ a	
1179	Lithium perchlorate / $LiClO_4$	7791-03-9 / 106.392	wh orth cry or powder	236 / 2.428	dec 430 / a,c	s H_2O, EtOH, ace, eth
1180	Lithium peroxide / Li_2O_2	12031-80-0 / 45.881	wh hex cry	/ 2.31	/ a	s H_2O; i EtOH
1181	Lithium selenate monohydrate / $Li_2SeO_4 \cdot H_2O$	7790-71-8 / 174.86	monocl cry	/ 2.56		vs H_2O
1182	Lithium metasilicate / Li_2SiO_3	10102-24-6 / 89.966	wh orth needles	1201 / 2.52	/ a,c	i cold H_2O; reac dil acid
1183	Lithium sulfate / Li_2SO_4	10377-48-7 / 109.946	wh monocl cry; hyg	859 / 2.21	/ a,c	s H_2O
1184	Lithium sulfate monohydrate / $Li_2SO_4 \cdot H_2O$	10102-25-7 / 127.961	col cry	dec 130 / 2.06		vs H_2O; sl EtOH
1185	Lithium sulfide / Li_2S	12136-58-2 / 45.948	wh cub cry; hyg	1372 / 1.64	/ a	
1186	Lutetium / Lu	7439-94-3 / 174.967	silv metal; hex	1663 / 9.84	3402 / a,c	s dil acid
1187	Lutetium boride / LuB_4	12688-52-7 / 218.211	tetr cry	2600 / ≈7.0		
1188	Lutetium bromide / $LuBr_3$	14456-53-2 / 414.679	wh hyg cry	1025 /		vs H_2O
1189	Lutetium chloride / $LuCl_3$	10099-66-8 / 281.325	wh monocl cry; hyg	925 / 3.98	/ a	s H_2O
1190	Lutetium fluoride / LuF_3	13760-81-1 / 231.962	orth cry	1182 / 8.3	2200	i H_2O
1191	Lutetium iodide / LuI_3	13813-45-1 / 555.680	brn hex cry; hyg	1050 / ≈5.6	/ a	vs H_2O
1192	Lutetium nitride / LuN	12125-25-6 / 188.974	cub cry	/ 11.6		
1193	Lutetium oxide / Lu_2O_3	12032-20-1 / 397.932	wh cub cry or powder	2490 / 9.41	/ a	
1194	Lutetium sulfate octahydrate / $Lu_2(SO_4)_3 \cdot 8H_2O$	13473-77-3 / 782.247	wh cry			vs H_2O
1195	Lutetium sulfide / Lu_2S_3	12163-20-1 / 446.132	gray rhomb cry	1750 dec / 6.26		
1196	Lutetium telluride / Lu_2Te_3	12163-22-3 / 732.73	orth cry	/ 7.8		
1197	Magnesium / Mg	7439-95-4 / 24.305	silv-wh metal	650 / 1.74	1090 / a,c,e	s dil acid
1198	Magnesium acetate tetrahydrate / $Mg(C_2H_3O_2)_2 \cdot 4H_2O$	16674-78-5 / 214.455	col monocl cry; hyg	80 dec / 1.45		vs H_2O, EtOH
1199	Magnesium amide / $Mg(NH_2)_2$	7803-54-5 / 56.350	wh powder; flam	dec / 1.39		reac H_2O

No.	Name Formula	CAS RN Mol. Wt.	Physical Form	mp/°C den/g cm^{-3}	bp/°C Other Data	Solubility
1200	Magnesium antimonide Mg_3Sb_2	12057-75-9 316.435	hex cry	1245 3.99		
1201	Magnesium boride MgB_2	12007-25-9 45.927	hex cry	dec 800 2.57		
1202	Magnesium bromate hexahydrate $Mg(BrO_3)_2 \cdot 6H_2O$	7789-36-8 388.201	col cub cry	dec 200 2.29		vs H_2O
1203	Magnesium bromide $MgBr_2$	7789-48-2 184.113	wh hex cry; hyg	711 3.72	a,c	vs H_2O
1204	Magnesium bromide hexahydrate $MgBr_2 \cdot 6H_2O$	13446-53-2 292.204	col monocl cry	dec 165 2.0		vs H_2O; s EtOH
1205	Magnesium carbonate $MgCO_3$	546-93-0 84.314	wh hex cry	990 3.05	f	i H_2O, EtOH; s acid
1206	Magnesium chlorate hexahydrate $Mg(ClO_3)_2 \cdot 6H_2O$	10326-21-3* 299.298	wh hyg cry	dec ≈35 1.80		vs H_2O; sl EtOH
1207	Magnesium chloride $MgCl_2$	7786-30-3 95.210	wh hex leaflets; hyg	714 2.325	1412 a,c,e	vs H_2O
1208	Magnesium chloride hexahydrate $MgCl_2 \cdot 6H_2O$	7791-18-6 203.301	wh hyg cry	dec ≈100 1.56		vs H_2O; s EtOH
1209	Magnesium fluoride MgF_2	7783-40-6 62.302	wh tetr cry	1263 3.148	2227 a,c,f	sl H_2O
1210	Magnesium formate dihydrate $Mg(CHO_2)_2 \cdot 2H_2O$	6150-82-9 150.371	wh cry	dec		s H_2O; i EtOH
1211	Magnesium germanide Mg_2Ge	1310-52-7 121.22	cub cry	1117 3.09		
1212	Magnesium hydride MgH_2	60616-74-2 26.321	wh tetr cry	327 1.45		reac H_2O
1213	Magnesium hydrogen phosphate trihydrate $MgHPO_4 \cdot 3H_2O$	7757-86-0 174.331	wh powder	dec 550 2.13	f	sl H_2O; s dil acid
1214	Magnesium hydroxide $Mg(OH)_2$	1309-42-8 58.320	wh hex cry	350 2.37	a,f	i H_2O; s dil acid
1215	Magnesium iodide MgI_2	10377-58-9 278.114	wh hex cry; hyg	634 4.43	a,c	vs H_2O
1216	Magnesium iodide octahydrate $MgI_2 \cdot 8H_2O$	7790-31-0 422.236	wh orth cry; hyg	41 dec 2.10		vs H_2O; s EtOH
1217	Magnesium nitrate $Mg(NO_3)_2$	10377-60-3 148.314	wh cub cry	≈2.3	a	vs H_2O
1218	Magnesium nitrate dihydrate $Mg(NO_3)_2 \cdot 2H_2O$	15750-45-5 184.345	wh cry	≈100 dec 1.45		s H_2O, EtOH
1219	Magnesium nitrate hexahydrate $Mg(NO_3)_2 \cdot 6H_2O$	13446-18-9 256.406	col monocl cry; hyg	dec ≈95 1.46		vs H_2O; s EtOH
1220	Magnesium nitride Mg_3N_2	12057-71-5 100.928	yel cub cry	dec ~1500 2.71		
1221	Magnesium oxalate dihydrate $MgC_2O_4 \cdot 2H_2O$	6150-88-5 148.355	wh powder			i H_2O, EtOH; s dil acid
1222	Magnesium oxide MgO	1309-48-4 40.304	wh cub cry	2826 3.6	3600 a,c,f	sl H_2O; i EtOH
1223	Magnesium perchlorate $Mg(ClO_4)_2$	10034-81-8 223.205	wh hyg powder	dec 250 2.2		vs H_2O
1224	Magnesium perchlorate hexahydrate $Mg(ClO_4)_2 \cdot 6H_2O$	13446-19-0 331.297	wh hyg cry	dec 190 1.98		vs H_2O; s EtOH
1225	Magnesium permanganate hexahydrate $Mg(MnO_4)_2 \cdot 6H_2O$	10377-62-5 370.268	blue-blk cry	dec 2.18		s H_2O
1226	Magnesium peroxide MgO_2	1335-26-8 56.304	wh cub cry	dec 100 ≈3.0		i H_2O; s dil acid
1227	Magnesium phosphate octahydrate $Mg_3(PO_4)_2 \cdot 8H_2O$	13446-23-6 406.980	wh monocl cry	2.17		i H_2O; s acid
1228	Magnesium pyrophosphate trihydrate $Mg_2P_2O_7 \cdot 3H_2O$	10102-34-8 276.600	wh powder	dec 100 2.56		i H_2O; s acid
1229	Magnesium phosphide Mg_3P_2	12057-74-8 134.863	yel cub cry	2.06		reac H_2O
1230	Magnesium selenate hexahydrate $MgSeO_4 \cdot 6H_2O$	13446-28-1 275.35	wh monocl cry	1.928		s H_2O
1231	Magnesium selenide $MgSe$	1313-04-8 103.27	brn cub cry	4.2		reac H_2O
1232	Magnesium selenite hexahydrate $MgSeO_3 \cdot 6H_2O$	15593-61-0 259.36	col hex cry	2.09		i H_2O; s dil acid
1233	Magnesium metasilicate $MgSiO_3$	13776-74-4 100.389	wh monocl cry	dec ≈1550 3.19	f	i H_2O; sl HF
1234	Magnesium orthosilicate Mg_2SiO_4	26686-77-1 140.694	wh orth cry	1898 3.21	a,c,f	i H_2O
1235	Magnesium hexafluorosilicate hexahydrate $MgSiF_6 \cdot 6H_2O$	60950-56-3 274.472	wh cry	120 dec 1.79		s H_2O; i EtOH
1236	Magnesium silicide Mg_2Si	22831-39-6 76.696	gray cub cry	1102 1.99		reac H_2O
1237	Magnesium stannide Mg_2Sn	1313-08-2 167.320	blue cub cry	771 3.60		s H_2O, dil HCl

No.	Name Formula	CAS RN Mol. Wt.	Physical Form	mp/°C den/g cm^{-3}	bp/°C Other Data	Solubility
1238	Magnesium sulfate $MgSO_4$	7487-88-9 120.369	col orth cry	1127 2.66	a,c	s H_2O
1239	Magnesium sulfate heptahydrate $MgSO_4 \cdot 7H_2O$	10034-99-8 246.475	col orth cry	dec 150 1.67	f	vs H_2O; sl EtOH
1240	Magnesium sulfate monohydrate $MgSO_4 \cdot H_2O$	14168-73-1 138.384	col monocl cry	dec 150 2.57	f	vs H_2O
1241	Magnesium sulfide MgS	12032-36-9 56.371	red-brn cub cry	2227 2.68	a	reac H_2O
1242	Magnesium sulfite hexahydrate $MgSO_3 \cdot 6H_2O$	13446-29-2 212.461	wh hex cry	dec 200 1.72		sl H_2O; i EtOH
1243	Magnesium sulfite trihydrate $MgSO_3 \cdot 3H_2O$	19086-20-5 158.415	col orth cry	 2.12		sl H_2O
1244	Magnesium thiosulfate hexahydrate $MgS_2O_3 \cdot 6H_2O$	13446-30-5 244.527	col cry	dec 170 1.82		vs H_2O; i EtOH
1245	Magnesium titanate $MgTiO_3$	1312-99-8 120.170	col hex cry	1565 3.85	f	
1246	Magnesium tungstate $MgWO_4$	13573-11-0 272.14	wh monocl cry	 6.89		i H_2O, EtOH
1247	Manganese Mn	7439-96-5 54.938	hard gray metal	1246 7.3	2061 a,c,e	s dil acids
1248	Manganèse antimonide $MnSb$	12032-82-5 176.698	hex cry	840 6.9		
1249	Manganese antimonide Mn_2Sb	12032-97-2 231.636	tetr cry	948 7.0		
1250	Manganese boride MnB	12045-15-7 65.749	orth cry	1890 6.45		
1251	Manganese boride MnB_2	12228-50-1 76.560	hex cry	1827 5.3		
1252	Manganese boride Mn_2B	12045-16-8 120.687	red-brn tetr cry	1580 7.20		
1253	Manganese carbide Mn_3C	12266-65-8 176.825	refrac solid	1520 6.89		
1254	Manganese carbonyl $Mn_2(CO)_{10}$	10170-69-1 389.980	yel monocl cry	154 1.75		i H_2O; s os
1255	Manganese phosphide Mn_2P	12333-54-9 140.850	hex cry	1327 6.0		
1256	Manganese phosphide MnP	12032-78-9 85.912	orth cry	1147 5.49		
1257	Manganese(II) acetate tetrahydrate $Mn(C_2H_3O_2)_2 \cdot 4H_2O$	15243-27-3 245.088	red monocl cry	80 1.59		s H_2O, EtOH
1258	Manganese(II) tetraborate octahydrate $MnB_4O_7 \cdot 8H_2O$	12228-91-0 354.300	red solid			i H_2O, EtOH; s dil acid
1259	Manganese(II) bromide $MnBr_2$	13446-03-2 214.746	pink hex cry	698 4.385	a	vs H_2O
1260	Manganese(II) bromide tetrahydrate $MnBr_2 \cdot 4H_2O$	10031-20-6 286.808	red hyg cry	64 dec		vs H_2O
1261	Manganese(II) carbonate $MnCO_3$	598-62-9 114.947	pink hex cry	dec >200 3.70	f	sl H_2O; s dil acid
1262	Manganese(II) chloride $MnCl_2$	7773-01-5 125.843	pink trig cry; hyg	650 2.977	1190 a,c,e	s H_2O, py, EtOH; i eth
1263	Manganese(II) chloride tetrahydrate $MnCl_2 \cdot 4H_2O$	13446-34-9 197.905	red monocl cry; hyg	87.5 1.913		vs H_2O; s EtOH; i eth
1264	Manganese(II) dihydrogen phosphate dihydrate $Mn(H_2PO_4)_2 \cdot 2H_2O$	18718-07-5 284.944	col hyg cry			s H_2O; i EtOH
1265	Manganese(II) fluoride MnF_2	7782-64-1 92.935	red tetr cry	930 3.98		sl H_2O; i EtOH
1266	Manganese(II) hydroxide $Mn(OH)_2$	18933-05-6 88.953	pink hex cry	dec 3.26	f	i H_2O
1267	Manganese(II) iodide MnI_2	7790-33-2 308.747	wh hex cry; hyg	638 5.04		s H_2O, EtOH
1268	Manganese(II) iodide tetrahydrate $MnI_2 \cdot 4H_2O$	7790-33-2* 380.809	red cry			vs H_2O; s EtOH
1269	Manganese(II) molybdate $MnMoO_4$	14013-15-1 214.88	yel monocl cry	 4.05		
1270	Manganese(II) nitrate $Mn(NO_3)_2$	10377-93-2 178.948	col orth cry; hyg	 2.2		vs H_2O; s diox, thf
1271	Manganese(II) nitrate hexahydrate $Mn(NO_3)_2 \cdot 6H_2O$	10377-66-9 287.040	rose monocl cry	28 dec 1.8		vs H_2O, EtOH
1272	Manganese(II) nitrate tetrahydrate $Mn(NO_3)_2 \cdot 4H_2O$	20694-39-7 251.010	pink hyg cry	37.1 dec 2.13		vs H_2O; s EtOH
1273	Manganese(II) oxalate dihydrate $MnC_2O_4 \cdot 2H_2O$	6556-16-7 178.988	wh cry powder	dec 150 2.45		sl H_2O; s acid
1274	Manganese(II) oxide MnO	1344-43-0 70.937	gr cub cry or powder	1840 5.37	a,c,f	i H_2O; s acid
1275	Manganese(II) perchlorate hexahydrate $Mn(ClO_4)_2 \cdot 6H_2O$	15364-94-0 361.930	pink hex cry	 2.10		
1276	Manganese(II) pyrophosphate $Mn_2P_2O_7$	53731-35-4 283.819	wh monocl cry	1196 3.71		i H_2O

No.	Name Formula	CAS RN Mol. Wt.	Physical Form	mp/°C den/g cm^{-3}	bp/°C Other Data	Solubility
1277	Manganese(II) metasilicate $MnSiO_3$	7759-00-4 131.022	red orth cry	1291 3.48	a,f	i H_2O
1278	Manganese(II) orthosilicate Mn_2SiO_4	13568-32-6 201.960	orth cry	 4.11	a	i H_2O
1279	Manganese(II) selenide MnSe	1313-22-0 133.90	gray cub cry	1460 5.45	a	i H_2O
1280	Manganese(II) sulfate $MnSO_4$	7785-87-7 151.002	wh orth cry	700 3.25	dec 850	vs H_2O
1281	Manganese(II) sulfate monohydrate $MnSO_4 \cdot H_2O$	10034-96-5 169.017	red monocl cry	 2.95		vs H_2O; i EtOH
1282	Manganese(II) sulfate tetrahydrate $MnSO_4 \cdot 4H_2O$	10101-68-5 223.063	red monocl cry	38 dec 2.26		s H_2O; i EtOH
1283	Manganese(II) sulfide (α) MnS	18820-29-6 87.004	grn cub cry	1610 4.0	a,f	i H_2O; s dil acid
1284	Manganese(II) sulfide (β) MnS	18820-29-6 87.004	red cub cry	 3.3	a,f	i H_2O; s dil acid
1285	Manganese(II) sulfide (γ) MnS	18820-29-6 87.004	red hex cry	≈3.3	a,f	i H_2O; s dil acid
1286	Manganese(II) telluride MnTe	12032-88-1 182.54	hex cry	≈1150 6.0		
1287	Manganese(II) titanate $MnTiO_3$	12032-74-5 150.803	red hex cry	1360 4.55		
1288	Manganese(II) tungstate $MnWO_4$	13918-22-4 302.78	wh monocl cry	 7.2	f	
1289	Manganese(II,III) oxide Mn_3O_4	1317-35-7 228.812	brn tetr cry	1567 4.84	a,f	i H_2O; s HCl
1290	Manganese(III) fluoride MnF_3	7783-53-1 111.933	red monocl cry; hyg	dec >600 3.54		reac H_2O
1291	Manganese(III) hydroxide MnO(OH)	1332-63-4 87.945	blk monocl cry	dec 250 ≈4.3	f	i H_2O
1292	Manganese(III) oxide Mn_2O_3	1317-34-6 157.874	blk cub cry	dec 1080 ≈5.0	a	i H_2O
1293	Manganese(IV) oxide MnO_2	1313-13-9 86.937	blk tetr cry	535 dec 5.08	a,f	i H_2O, HNO_3
1294	Manganese(VII) oxide Mn_2O_7	12057-92-0 221.872	grn oil; exp	5.9 2.40	exp 95	vs H_2O
1295	Mendelevium Md	7440-11-1 258	Metal	827	a	
1296	Mercury Hg	7439-97-6 200.59	heavy silv liq	-38.83 13.5336	356.73 a,b,c,d,e	i H_2O
1297	Mercury(I) acetate $Hg_2(C_2H_3O_2)_2$	631-60-7 519.27	col scales	dec		sl H_2O; i EtOH, eth
1298	Mercury(I) bromide Hg_2Br_2	15385-58-7 560.99	wh tetr cry or powder	407 7.307	a	i H_2O, EtOH, eth
1299	Mercury(I) chlorate $Hg_2(ClO_3)_2$	10294-44-7 568.08	wh cry	dec ≈250 6.41		sl H_2O; s EtOH
1300	Mercury(I) chloride Hg_2Cl_2	10112-91-1 472.09	wh tetr cry	525 tp 7.16	383 sp a,f	i H_2O, EtOH, eth
1301	Mercury(I) fluoride Hg_2F_2	13967-25-4 439.18	yel cub cry	570 dec 8.73	subl	reac H_2O
1302	Mercury(I) iodide Hg_2I_2	15385-57-6 654.99	yel amorp powder	290 7.70	a,c	i H_2O, EtOH, eth
1303	Mercury(I) nitrate dihydrate $Hg_2(NO_3)_2 \cdot 2H_2O$	7782-86-7 561.22	col cry	70 dec 4.8		reac H_2O
1304	Mercury(I) oxide Hg_2O	15829-53-5 417.18	prob mixture of HgO+Hg	100 dec 9.8		i H_2O; s HNO_3
1305	Mercury(I) sulfate Hg_2SO_4	7783-36-0 497.24	wh-yel cry powder	 7.56	a	i H_2O; s dil HNO_3
1306	Mercury(II) acetate $Hg(C_2H_3O_2)_2$	1600-27-7 318.68	wh-yel cry or powder	179 dec 3.28		s H_2O, EtOH
1307	Mercury(II) amide chloride $Hg(NH_2)Cl$	10124-48-8 252.07	wh solid	 5.38	subl	i H_2O, EtOH; s warm acid
1308	Mercury(II) bromide $HgBr_2$	7789-47-1 360.40	wh rhomb cry or powder	236 6.05	322 a,b,c,d,e	sl H_2O, chl; s EtOH, MeOH
1309	Mercury(II) chloride $HgCl_2$	7487-94-7 271.50	wh orth cry	276 5.6	304 a,b,c,d,e	sl H_2O, bz; s EtOH, MeOH, ace, eth
1310	Mercury(II) chromate $HgCrO_4$	13444-75-2* 316.58	red monocl cry	 6.06		sl H_2O
1311	Mercury(II) cyanide $Hg(CN)_2$	592-04-1 252.63	col tetr cry	dec 320 4.00		s H_2O, EtOH; sl eth
1312	Mercury(II) dichromate $HgCr_2O_7$	7789-10-8 416.58	red cry powder			i H_2O; s acid
1313	Mercury(II) fluoride HgF_2	7783-39-3 238.59	wh cub cry; hyg	dec 645 8.95		reac H_2O
1314	Mercury(II) fulminate $Hg(CNO)_2$	628-86-4 284.62	gray cry	exp 4.42		sl H_2O; s EtOH, NH_4OH
1315	Mercury(II) hydrogen arsenate $HgHAsO_4$	7784-37-4 340.52	yel powder			i H_2O; s acid

No.	Name Formula	CAS RN Mol. Wt.	Physical Form	mp/°C den/g cm^{-3}	bp/°C Other Data	Solubility
1316	Mercury(II) iodate $Hg(IO_3)_2$	7783-32-6 550.40	wh powder	dec 175		i H_2O
1317	Mercury(II) iodide HgI_2	7774-29-0 454.40	red tetr cry or powder	259 6.28	354 a,b,c,d,e	i H_2O; sl EtOH, ace, eth
1318	Mercury(II) nitrate $Hg(NO_3)_2$	10045-94-0 324.60	col hyg cry	79 4.3		s H_2O; i EtOH
1319	Mercury(II) nitrate dihydrate $Hg(NO_3)_2 \cdot 2H_2O$	10045-94-0* 360.63	monocl cry	4.78		s H_2O
1320	Mercury(II) nitrate monohydrate $Hg(NO_3)_2 \cdot H_2O$	7783-34-8 342.62	wh-yel hyg cry	4.3		s H_2O, dil acid
1321	Mercury(II) oxide HgO	21908-53-2 216.59	red or yel orth cry	dec 500 11.14	a,f	i H_2O, EtOH; s dil acid
1322	Mercury(II) oxycyanide $Hg(CN)_2 \cdot HgO$	1335-31-5 469.21	wh orth cry	exp 4.44		s H_2O
1323	Mercury(II) phosphate $Hg_3(PO_4)_2$	7782-66-3 791.71	wh-yel powder			i H_2O, EtOH; s acid
1324	Mercury(II) selenide $HgSe$	20601-83-6 279.55	gray cub cry	8.21	subl	i H_2O
1325	Mercury(II) sulfate $HgSO_4$	7783-35-9 296.65	wh monocl cry	6.47	a	reac H_2O
1326	Mercury(II) sulfide (black) HgS	1344-48-5 232.66	blk cub cry or powder	583 7.70	a,f	i H_2O; s acid, EtOH
1327	Mercury(II) sulfide (red) HgS	1344-48-5 232.66	red hex cry	trans to blk HgO 386 8.17	a,f	i H_2O, acid; s aqua regia
1328	Mercury(II) telluride $HgTe$	12068-90-5 328.19	gray cub cry	673 8.63	a	
1329	Mercury(II) thiocyanate $Hg(SCN)_2$	592-85-8 316.76	monocl cry	dec ≈165 3.71		i H_2O; s dil HCl
1330	Molybdenum Mo	7439-98-7 95.94	gray-blk metal; cub	2623 10.2	4639 a,c,e	i H_2O, dil acid, alk
1331	Molybdenum boride Mo_2B	12006-99-4 202.69	refrac tetr cry	2000 9.2		
1332	Molybdenum boride Mo_2B_5	12007-97-5 245.94	refrac hex cry	1600 ≈7.2		
1333	Molybdenum carbide MoC	12011-97-1 107.95	refrac solid; cub	2577		
1334	Molybdenum carbide Mo_2C	12069-89-5 203.89	gray orth cry	2687 9.18		
1335	Molybdenum carbonyl $Mo(CO)_6$	13939-06-5 264.00	wh orth cry	dec 150 1.96	subl a,d	i H_2O; s bz; sl eth
1336	Molybdenum nitride MoN	12033-19-1 109.95	hex cry	1750 9.20		
1337	Molybdenum nitride Mo_2N	12033-31-7 205.89	gray cub cry	790 dec 9.46		
1338	Molybdenum phosphide MoP	12163-69-8 126.91	blk hex cry	7.34		
1339	Molybdenum silicide $MoSi_2$	12136-78-6 152.11	gray tetr cry	≈1900 6.2		i H_2O; s HF
1340	Molybdenum(II) bromide $MoBr_2$	13446-56-5 255.75	yel-red cry	dec 900		
1341	Molybdenum(II) chloride $MoCl_2$	13478-17-6 166.85	yel cry	dec 530		
1342	Molybdenum(II) iodide MoI_2	14055-74-4 349.75	blk hyg cry	5.278		
1343	Molybdenum(III) bromide $MoBr_3$	13446-57-6 335.65	grn hex cry	977 4.89		i H_2O
1344	Molybdenum(III) chloride $MoCl_3$	13478-18-7 202.30	dark red monocl cry	1027 3.74		i H_2O
1345	Molybdenum(III) fluoride MoF_3	20193-58-2 152.94	brn hex cry	>600 4.64		i H_2O
1346	Molybdenum(III) iodide MoI_3	14055-75-5 476.65	blk solid	927		i H_2O
1347	Molybdenum(III) oxide Mo_2O_3	1313-29-7 239.88	gray-blk powder			i H_2O; sl acid
1348	Molybdenum(IV) bromide $MoBr_4$	13520-59-7 415.56	blk cry	dec		reac H_2O
1349	Molybdenum(IV) chloride $MoCl_4$	13320-71-3 237.75	blk cry	dec >170		reac H_2O
1350	Molybdenum(IV) fluoride MoF_4	23412-45-5 171.93	grn cry	dec		reac H_2O
1351	Molybdenum(IV) oxide MoO_2	18868-43-4 127.94	brn-viol tetr cry	dec ≈1100 6.47	a	sl H_2O
1352	Molybdenum(IV) selenide $MoSe_2$	12058-18-3 253.86	gray hex cry	>1200 6.90		
1353	Molybdenum(IV) sulfide MoS_2	1317-33-5 160.07	blk powder or hex cry	2375 5.06	a,f	i H_2O; s conc acid

No.	Name Formula	CAS RN Mol. Wt.	Physical Form	mp/°C den/g cm^{-3}	bp/°C Other Data	Solubility
1354	Molybdenum(IV) telluride MoTe$_2$	12058-20-7 351.14	gray hex cry	 7.7		
1355	Molybdenum(V) chloride MoCl$_5$	10241-05-1 273.20	gr-blk monocl cry; hyg	194 2.93	268 b,c,d	s EtOH, eth
1356	Molybdenum(V) fluoride MoF$_5$	13819-84-6 190.93	yel monocl cry	67 3.5	213	
1357	Molybdenum(V) oxytrichloride MoOCl$_3$	13814-74-9 218.30	blk monocl cry	297	subl	reac H$_2$O
1358	Molybdenum(VI) acid monohydrate H$_2$MoO$_4$·H$_2$O	7782-91-4 179.97	wh powder	 3.1		sl H$_2$O; s alk
1359	Molybdenum(VI) fluoride MoF$_6$	7783-77-9 209.93	wh cub cry or col liq; hyg	17.5 2.54	34 a,b,c,d,e	reac H$_2$O
1360	Molybdenum(VI) oxytetrafluoride MoOF$_4$	14459-59-7 187.93	volatile solid	98		
1361	Molybdenum(VI) oxytetrachloride MoOCl$_4$	13814-75-0 253.75	grn hyg powder	101		
1362	Molybdenum(VI) dioxydichloride MoO$_2$Cl$_2$	13637-68-8 198.84	yel-oran solid	≈175 3.31		reac H$_2$O
1363	Molybdenum(VI) oxide MoO$_3$	1313-27-5 143.94	wh-yel rhomb cry	801 4.70	1155 a,c,d,e	sl H$_2$O; s conc acid
1364	Molybdenum(VI) metaphosphate Mo(PO$_3$)$_6$	133863-98-6 569.77	yel powder	 3.28		i H$_2$O, acid
1365	Neodymium Nd	7440-00-8 144.24	silv metal; hex	1021 7.01	3074 a,c	
1366	Neodymium boride NdB$_6$	12008-23-0 209.11	blk cub cry	2610 4.93		
1367	Neodymium bromide NdBr$_3$	13536-80-6 383.95	viol orth cry; hyg	682 5.3	1540	s H$_2$O
1368	Neodymium chloride NdCl$_3$	10024-93-8 250.60	viol hex cry	758 4.13	1600 a	vs H$_2$O, EtOH; i eth, chl
1369	Neodymium chloride hexahydrate NdCl$_3$·6H$_2$O	13477-89-9 358.69	purp cry	dec 124 2.3		vs H$_2$O; s EtOH
1370	Neodymium fluoride NdF$_3$	13709-42-7 201.24	viol hex cry; hyg	1377 6.51	2300 a	i H$_2$O
1371	Neodymium iodide NdI$_3$	13813-24-6 524.95	grn orth cry; hyg	784 5.85		s H$_2$O
1372	Neodymium nitrate hexahydrate Nd(NO$_3$)$_3$·6H$_2$O	14517-29-4 438.35	purp hyg cry			vs H$_2$O; s EtOH, ace
1373	Neodymium nitride NdN	25764-11-8 158.25	blk cub cry	 7.69		
1374	Neodymium oxide Nd$_2$O$_3$	1313-97-9 336.48	blue hex cry; hyg	2320 7.24	 a	i H$_2$O; s dil acid
1375	Neodymium sulfate Nd$_2$(SO$_4$)$_3$	13477-91-3 576.67	pink needles	dec ≈700		s H$_2$O
1376	Neodymium sulfide Nd$_2$S$_3$	12035-32-4 384.68	orth cry	2207 5.46		
1377	Neodymium telluride Nd$_2$Te$_3$	12035-35-7 671.28	gray orth cry	1377 7.0		
1378	Neon Ne	7440-01-9 20.180	col gas	-248.59 0.825 g/L	-246.08 a,b,c,d,e	sl H$_2$O
1379	Neptunium Np	7439-99-8 237	silv metal	644 20.2	 c	s HCl
1380	Neptunium(IV) oxide NpO$_2$	12035-79-9 269	grn cub cry	2547 11.1		
1381	Nickel Ni	7440-02-0 58.693	wh metal; cub	1455 8.90	2913 a,c,e	i H$_2$O; sl dil acid
1382	Nickel antimonide NiSb	12035-52-8 180.453	hex cry	1147 8.74	 f	
1383	Nickel arsenide NiAs	27016-75-7 133.615	hex cry	967 7.77	 f	
1384	Nickel boride Ni$_3$B	12007-02-2 186.891	refrac solid	1156 8.17		
1385	Nickel boride NiB	12007-00-0 69.504	grn refrac solid	1035 7.13		
1386	Nickel boride Ni$_2$B	12007-01-1 128.198	refrac solid	1125 7.90		
1387	Nickel carbonyl Ni(CO)$_4$	13463-39-3 170.735	col liq	-19.3 1.31	43 (exp ≈60) a,e	i H$_2$O; s EtOH, bz, ace, ctc
1388	Nickel phosphide Ni$_2$P	12035-64-2 148.361	hex cry	1100 7.33		
1389	Nickel silicide Ni$_2$Si	12059-14-2 145.473	orth cry	1255 7.40		
1390	Nickel silicide NiSi$_2$	12201-89-7 114.864	cub cry	993 4.83		
1391	Nickel(II) arsenate octahydrate Ni$_3$(AsO$_4$)$_2$·8H$_2$O	7784-48-7 598.040	yel-grn powder	dec 4.98		i H$_2$O; s acid
1392	Nickel(II) bromide NiBr$_2$	13462-88-9 218.501	yel hex cry; hyg	963 5.10	subl a	vs H$_2$O

No.	Name Formula	CAS RN Mol. Wt.	Physical Form	mp/°C den/g cm⁻³	bp/°C Other Data	Solubility
1393	Nickel(II) bromide trihydrate $NiBr_2 \cdot 3H_2O$	13462-88-9* 272.547	yel-grn hyg cry	dec 200		vs H_2O; s EtOH, eth
1394	Nickel(II) carbonate $NiCO_3$	3333-67-3 118.702	grn rhomb cry	4.39		i H_2O; s dil acid
1395	Nickel(II) chloride $NiCl_2$	7718-54-9 129.598	yel hex cry; hyg	1009 3.51	subl a,c,e	s H_2O, EtOH
1396	Nickel(II) chloride hexahydrate $NiCl_2 \cdot 6H_2O$	7791-20-0 237.689	grn monocl cry			vs H_2O; s EtOH
1397	Nickel(II) cyanide tetrahydrate $Ni(CN)_2 \cdot 4H_2O$	13477-95-7 182.790	grn plates	dec 200		i H_2O; sl dil acid; s NH_4OH
1398	Nickel(II) fluoride NiF_2	10028-18-9 96.690	yel tetr cry	1474 4.7	a	sl H_2O; i EtOH, eth
1399	Nickel(II) hydroxide $Ni(OH)_2$	12054-48-7 92.708	grn hex cry	dec 230 4.1	a	i H_2O
1400	Nickel(II) hydroxide monohydrate $Ni(OH)_2 \cdot H_2O$	36897-37-7 110.723	grn powder			i H_2O; s dil acid
1401	Nickel(II) iodide NiI_2	13462-90-3 312.502	blk hex cry; hyg	780 5.22	subl a	vs H_2O
1402	Nickel(II) iodide hexahydrate $NiI_2 \cdot 6H_2O$	7790-34-3 420.593	grn monocl cry; hyg			vs H_2O, EtOH
1403	Nickel(II) nitrate hexahydrate $Ni(NO_3)_2 \cdot 6H_2O$	13478-00-7 290.794	grn monocl cry; hyg	dec 56 2.05		vs H_2O; s EtOH
1404	Nickel(II) oxide NiO	1313-99-1 74.692	grn cub cry	1955 6.72	f	i H_2O; s acid
1405	Nickel(II) selenide $NiSe$	1314-05-2 137.65	yel-grn hex cry	980 7.2		
1406	Nickel(II) sulfate $NiSO_4$	7786-81-4 154.757	grn-yel orth cry	dec 840 4.01	a	vs H_2O
1407	Nickel(II) sulfate heptahydrate $NiSO_4 \cdot 7H_2O$	10101-98-1 280.863	grn orth cry	1.98		s H_2O, EtOH
1408	Nickel(II) sulfate hexahydrate $NiSO_4 \cdot 6H_2O$	10101-97-0 262.848	blue-grn tetr cry	dec ≈100 2.07		vs H_2O; sl EtOH
1409	Nickel(II) sulfide NiS	16812-54-7 90.759	yel hex cry	976 5.5	a,c,f	i H_2O
1410	Nickel(II) titanate $NiTiO_3$	12035-39-1 154.558	brn hex cry	5.0		
1411	Nickel(II,III) sulfide Ni_3S_4	12137-12-1 304.344	cub cry	995 4.77		
1412	Nickel(III) oxide Ni_2O_3	1314-06-3 165.385	gray-blk cub cry	dec ≈600	a	i H_2O; s hot acid
1413	Nickel(III) sulfide Ni_3S_2	12035-72-2 240.212	hex cry	787 5.87		
1414	Niobium Nb	7440-03-1 92.906	gray metal; cub	2477 8.57	4744 a,c	i acid
1415	Niobium boride NbB	12045-19-1 103.717	gray orth cry	2270 7.5		
1416	Niobium boride NbB_2	12007-29-3 114.528	gray hex cry	2900 6.8		
1417	Niobium carbide NbC	12069-94-2 104.917	gray cub cry	3608 7.82	4300	i H_2O, acid
1418	Niobium carbide Nb_2C	12011-99-3 197.824	refrac hex cry	3080 7.8		i H_2O
1419	Niobium nitride NbN	24621-21-4 106.913	gray cry; cub	2300 8.47		i HCl, acid
1420	Niobium phosphide NbP	12034-66-1 123.880	tetr cry	6.5		
1421	Niobium silicide $NbSi_2$	12034-80-9 149.077	gray hex cry	1950 5.7		
1422	Niobium(II) oxide NbO	12034-57-0 108.905	gray cub cry	1937 7.30	a,c	
1423	Niobium(III) bromide $NbBr_3$	15752-41-7 332.618	dark brn solid		subl	
1424	Niobium(III) chloride $NbCl_3$	13569-59-0 199.264	blk solid			
1425	Niobium(III) fluoride NbF_3	15195-53-6 149.901	blue cub cry	4.2		
1426	Niobium(IV) chloride $NbCl_4$	13569-70-5 234.717	viol-blk monocl cry	3.2		
1427	Niobium(IV) fluoride NbF_4	13842-88-1 168.900	blk tetr cry; hyg	dec >350 4.01		
1428	Niobium(IV) iodide NbI_4	13870-21-8 600.524	gray orth cry	503 5.6		
1429	Niobium(IV) oxide NbO_2	12034-59-2 124.905	wh tetr cry or powder	1902 5.9	a,c	
1430	Niobium(IV) selenide $NbSe_2$	12034-77-4 250.83	gray hex cry	>1300 6.3		
1431	Niobium(IV) sulfide NbS_2	12136-97-9 157.038	blk rhomb cry	4.4		

No.	Name Formula	CAS RN Mol. Wt.	Physical Form	mp/°C den/g cm^{-3}	bp/°C Other Data	Solubility
1432	Niobium(IV) telluride NbTe$_2$	12034-83-2 348.11	hex cry	7.6		
1433	Niobium(V) bromide NbBr$_5$	13478-45-0 492.426	oran orth cry	254 4.36	360	s H$_2$O, EtOH
1434	Niobium(V) chloride NbCl$_5$	10026-12-7 270.170	yel monocl cry; hyg	204.7 2.78	254.0 a,b,c,d	reac H$_2$O; s HCl, ctc
1435	Niobium(V) fluoride NbF$_5$	7783-68-8 187.898	col monocl cry; hyg	80 2.70	229 a,b,c,d,e	reac H$_2$O; sl CS$_2$, chl
1436	Niobium(V) iodide NbI$_5$	13779-92-5 727.428	yel-blk monocl cry	dec ≈200 5.32		
1437	Niobium(V) oxide Nb$_2$O$_5$	1313-96-8 265.810	wh orth cry	1512 4.6	a,c	i H$_2$O; s HF
1438	Niobium(V) oxybromide NbOBr$_3$	14459-75-7 348.617	yel-brn cry	dec ≈320	subl	
1439	Niobium(V) oxychloride NbOCl$_3$	13597-20-1 215.263	wh tetr cry	3.72	subl	
1440	Niobium(V) dioxyfluoride NbO$_2$F	15195-33-2 143.903	wh cub cry	4.0		
1441	Nitrogen N$_2$	7727-37-9 28.013	col gas	-210.00 1.145 g/L	-195.79 a,b,c,d,e	sl H$_2$O; i EtOH
1442	Nitramide NO$_2$NH$_2$	7782-94-7 62.028	unstable wh cry	72 dec	a	s H$_2$O, EtOH, ace, eth; i chl
1443	Nitric acid HNO$_3$	7697-37-2 63.013	col liq; hyg	-41.6 1.55	83 a,c,d	vs H$_2$O
1444	Nitrous acid HNO$_2$	7782-77-6 47.014	stable only in soln		a	
1445	Nitrous oxide N$_2$O	10024-97-2 44.012	col gas	-90.8 1.799 g/L	-88.48 a,b,c,d,e	sl H$_2$O; s EtOH, eth
1446	Nitric oxide NO	10102-43-9 30.006	col gas	-163.6 1.226 g/L	-151.74 a,b,c,d,e	sl H$_2$O
1447	Nitrogen dioxide NO$_2$	10102-44-0 46.006	brn gas; equil with N$_2$O$_4$	1.880 g/L	see N$_2$O$_4$ a	reac H$_2$O
1448	Nitrogen trioxide N$_2$O$_3$	10544-73-7 76.011	blue solid or liq (low temp)	-101.1 1.4(2°C)	≈3 dec a	reac H$_2$O
1449	Nitrogen tetroxide N$_2$O$_4$	10544-72-6 92.011	col liq; equil with NO$_2$	-9.3 1.45(20°C)	21.15 a,b,c,d,e	reac H$_2$O
1450	Nitrogen pentoxide N$_2$O$_5$	10102-03-1 108.010	col hex cry	30 2.0	47 a,e	s chl; sl ctc
1451	Nitrogen tribromide NBr$_3$	15162-90-0 253.719	unstable solid	exp -100		
1452	Nitrogen trichloride NCl$_3$	10025-85-1 120.365	yel oily liq; exp	-40 1.653	71 a	i H$_2$O; s CS$_2$, bz, ctc
1453	Nitrogen trifluoride NF$_3$	7783-54-2 71.002	col gas	-206.79 2.902 g/L	-128.75 a,b,d,e	i H$_2$O
1454	Nitrogen triiodide NI$_3$	13444-85-4 394.720	unstable blk cry; exp			
1455	Difluoramine NHF$_2$	10405-27-3 53.012	col gas	2.167 g/L	-23 a,b	
1456	cis-Difluorodiazine N$_2$F$_2$	13812-43-6 66.010	col gas	<-195 2.698 g/L	-105.75 a,b	
1457	trans-Difluorodiazine N$_2$F$_2$	13776-62-0 66.010	col gas	-172 2.698 g/L	-111.45 a,b	
1458	Tetrafluorohydrazine N$_2$F$_4$	10036-47-2 104.007	col gas	-164.5 4.251 g/L	-74 a,b,d	
1459	Nitrosyl bromide NOBr	13444-87-6 109.910	red gas	-56 4.492 g/L	≈0 a	reac H$_2$O
1460	Nitrosyl chloride NOCl	2696-92-6 65.459	yel gas	-59.6 2.676 g/L	-5.5 a,b,d,e	reac H$_2$O
1461	Nitrosyl fluoride NOF	7789-25-5 49.004	col gas	-132.5 2.003 g/L	-59.9 a,d,e	
1462	Trifluoramine oxide NOF$_3$	13847-65-9 87.001	col gas	-161 3.556 g/L	-87.5 b	
1463	Nitryl chloride NO$_2$Cl	13444-90-1 81.459	col gas	-145 3.330 g/L	-15 a,d	
1464	Nitryl fluoride NO$_2$F	10022-50-1 65.004	col gas	-166 2.657 g/L	-72.4 a,b,d,e	reac H$_2$O
1465	Nitrogen selenide N$_4$Se$_4$	12033-88-4 371.87	red monocl cry; hyg	exp 4.2		i H$_2$O, eth, EtOH; sl bz, CS$_2$
1466	Nobelium No	10028-14-5 259	metal	827	a	
1467	Osmium Os	7440-04-2 190.23	blue-wh metal; hex	3033 22.59	5012 a,c	s aqua regia
1468	Osmium carbonyl Os$_3$(CO)$_{12}$	15696-40-9 906.82	yel cry	3.48		
1469	Osmium(III) bromide OsBr$_3$	59201-51-3 429.94	dark gray cry	dec 340		
1470	Osmium(III) chloride OsCl$_3$	13444-93-4 296.59	gray cub cry	dec >450	a	i H$_2$O; s HNO$_3$

No.	Name Formula	CAS RN Mol. Wt.	Physical Form	mp/°C den/g cm⁻³	bp/°C Other Data	Solubility
1471	Osmium(IV) chloride $OsCl_4$	10026-01-4 332.04	red-blk orth cry	4.38	450 sp	reac H_2O
1472	Osmium(IV) fluoride OsF_4	54120-05-7 266.22	yel cry	230		
1473	Osmium(IV) oxide OsO_2	12036-02-1 222.23	yel-brn tetr cry	11.4		i H_2O, acid
1474	Osmium(V) fluoride OsF_5	31576-40-6 285.22	blue cry	70		reac H_2O
1475	Osmium(VI) fluoride OsF_6	13768-38-2 304.22	yel cub cry	33.2 4.1	46 a	reac H_2O
1476	Osmium(VIII) oxide OsO_4	20816-12-0 254.23	yel monocl cry	41 5.1	135 a,b,c,e	sl H_2O
1477	Oxygen O_2	7782-44-7 31.999	col gas	-218.79 1.308 g/L	-182.95 a,b,c,d,e	sl H_2O, EtOH, os
1478	Ozone O_3	10028-15-6 47.998	blue gas	-193 1.962 g/L	-111.35 a,b,e	sl H_2O
1479	Palladium Pd	7440-05-3 106.42	silv-wh metal; cub	1554.9 12.0	2963 a,c	s aqua regia
1480	Palladium(II) sulfide PdS	12125-22-3 138.49	gray tetr cry	6.7	a	
1481	Palladium(II) bromide $PdBr_2$	13444-94-5 266.23	red-blk monocl cry; hyg	250 dec ≈5.2		i H_2O
1482	Palladium(II) chloride $PdCl_2$	7647-10-1 177.33	red rhomb cry; hyg	679 4.0		s H_2O, EtOH, ace
1483	Palladium(II) fluoride PdF_2	13444-96-7 144.42	viol tetr cry; hyg	952 5.76		reac H_2O
1484	Palladium(II) iodide PdI_2	7790-38-7 360.23	blk cry	360 dec 6.0		i H_2O, EtOH, eth
1485	Palladium(II) nitrate $Pd(NO_3)_2$	10102-05-3 230.43	brn hyg cry	dec		sl H_2O; s dil HNO_3
1486	Palladium(II) oxide PdO	1314-08-5 122.42	grn-blk tetr cry	750 dec 8.3	a	i H_2O, acid; sl aqua regia
1487	Phosphorus (white) P	7723-14-0 30.974	col waxlike cub cry	44.15 1.823	280.5 a,b,c,d,e	i H_2O; sl bz, EtOH, chl; s CS_2
1488	Phosphorus (red) P	7723-14-0 30.974	red-viol amorp powder	590tp 2.16	431sp a,b,c,d,e	i H_2O, os
1489	Phosphorus (black) P	7723-14-0 30.974	blk orth cry or amorp solid	610 2.69	a,b,c,d,e	i os
1490	Phosphine PH_3	7803-51-2 33.998	col gas; flam	-133 1.390 g/L	-87.75 a,b,d,e	i H_2O; sl EtOH, eth
1491	Diphosphine P_2H_4	13445-50-6 65.980	col liq	-99	63.5 dec a,d	reac H_2O
1492	Phosphoric acid (orthophosphoric acid) H_3PO_4	7664-38-2 97.995	col visc liq	42.4	407 a,c	s H_2O, EtOH
1493	Metaphosphoric acid HPO_3	37267-86-0 79.980	gl solid; hyg			sl H_2O; s EtOH
1494	Pyrophosphoric acid $H_4P_2O_7$	2466-09-3 177.975	wh cry	71.5	a	s H_2O
1495	Hypophosphoric acid $H_4P_2O_6$	7803-60-3 161.976	col orth cry	73 dec		vs H_2O
1496	Difluorophosphoric acid HPO_2F_2	13779-41-4 101.978	col liq	≈-94 1.583	dec 110	reac H_2O
1497	Phosphorous acid H_3PO_3	13598-36-2 81.996	wh hyg cry	74.4 1.65	200 a,c	vs H_2O, EtOH
1498	Hypophosphorous acid HPH_2O_2	6303-21-5 65.997	hyg cry or col oily liq	26.5 1.49	130 a,c	vs H_2O, EtOH, eth
1499	Phosphonium iodide PH_4I	12125-09-6 161.910	col tetr cry	18.5 2.86	62.5 e	reac H_2O, EtOH
1500	Phosphorus heptasulfide P_4S_7	12037-82-0 348.357	pale yel monocl cry	308 2.19	523	sl CS_2
1501	Phosphorus nitride P_3N_5	12136-91-3 162.955	wh amorp solid	dec 800		i H_2O; s os
1502	Phosphonitrilic chloride trimer $(PNCl_2)_3$	940-71-6 347.657	wh hyg cry	128.8 1.98		reac H_2O
1503	Phosphorus(III) bromide PBr_3	7789-60-8 270.686	col liq	-40 2.8	172.95 a,b,d,e	reac H_2O, EtOH; s ace, CS_2
1504	Phosphorus(III) chloride PCl_3	7719-12-2 137.332	col liq	-112 1.574	75.95 a,b,c,d,e	reac H_2O, EtOH; s bz, chl, eth
1505	Phosphorus(III) chloride difluoride $PClF_2$	14335-40-1 104.424	col gas	4.268 g/L	-47.25 b,d	
1506	Phosphorus(III) dichloride fluoride PCl_2F	15597-63-4 120.877	col gas	4.941 g/L	14 b,d	
1507	Phosphorus(III) fluoride PF_3	7783-55-3 87.969	col gas	-151.5 3.596 g/L	-101.5 a,b,d,e	reac H_2O
1508	Phosphorus(III) iodide PI_3	13455-01-1 411.687	red-oran hex cry; hyg	61.5 4.18	227 dec a,d	reac H_2O; s EtOH
1509	Phosphorus(III) oxide P_2O_3	1314-24-5 109.946	col monocl cry or liq	23.8 2.13	173 e	reac H_2O

No.	Name Formula	CAS RN Mol. Wt.	Physical Form	mp/°C den/g cm^{-3}	bp/°C Other Data	Solubility
1510	Phosphorus(III) selenide P_2Se_3	1314-86-9 298.83	oran-red cry	245 1.31	≈380	reac H_2O; s bz, ctc, CS_2, ace
1511	Phosphorus(III) sulfide P_2S_3	12165-69-4 158.146	yel solid	290	490	reac H_2O; s EtOH, eth, CS_2
1512	Phosphorus(V) bromide PBr_5	7789-69-7 430.494	yel orth cry, Donnay	dec >100 3.61	a	reac H_2O, EtOH; s CS_2, ctc
1513	Phosphorus(V) chloride PCl_5	10026-13-8 208.238	wh-yel tetr cry; hyg	167 tp 2.1	160 sp a,b,e	reac H_2O; s CS_2, ctc
1514	Phosphorus(V) fluoride PF_5	7647-19-0 125.966	col gas	-93.8 5.149 g/L	-84.6 a,d	reac H_2O
1515	Phosphorus(V) oxide P_2O_5	1314-56-3 141.945	wh orth cry; hyg	562 2.30	605 c,e	reac H_2O, EtOH
1516	Phosphorus(V) oxybromide $POBr_3$	7789-59-5 286.685	faint oran plates	56 2.822	191.7 a,d	reac H_2O; s bz, eth, chl
1517	Phosphorus(V) oxychloride $POCl_3$	10025-87-3 153.331	col liq	1 1.645	105.5 a,c,d,e	reac H_2O, EtOH
1518	Phosphorus(V) oxyfluoride POF_3	13478-20-1 103.968	col gas	-39.1 tp 4.250 g/L	-39.7 sp a	reac H_2O
1519	Phosphorus(V) selenide P_2Se_5	1314-82-5 456.75	blk-purp amorp solid			reac hot H_2O, ctc; i CS_2
1520	Phosphorus(V) sulfide P_2S_5	1314-80-3 222.278	grn-yel hyg cry	285 2.03	515	reac H_2O; s CS_2
1521	Phosphorus(V) sulfide trifluoride PSF_3	2404-52-6 120.035	gas	-148.8 4.906 g/L	-52.25 b,d	
1522	Phosphorus(V) sulfide trichloride $PSCl_3$	3982-91-0 169.398	fuming liq	-36.2 1.635	125 e	reac H_2O, s bz, ctc, chl, CS_2
1523	Phosphorus(V) sulfide chloride difluoride $PSClF_2$	2524-02-9 136.490	col gas	-155 5.579 g/L	6.3 b	
1524	Platinum Pt	7440-06-4 195.08	silv-gray metal; cub	1768.4 21.5	3825 a,c,e,f	i acid; s aqua regia
1525	Hexachloroplatinic acid hexahydrate $H_2PtCl_6 \cdot 6H_2O$	16941-12-1 517.90	brn-yel hyg cry	60 2.43		vs H_2O, EtOH
1526	Platinum silicide PtSi	12137-83-6 223.17	orth cry	1229 12.4		
1527	Platinum(II) bromide $PtBr_2$	13455-12-4 354.89	red-brn powder	dec 250 6.65	a	i H_2O
1528	Platinum(II) chloride $PtCl_2$	10025-65-7 265.99	grn hex cry	dec 581 6.0	a	i H_2O, EtOH, eth; s HCl
1529	Platinum(II) iodide PtI_2	7790-39-8 448.89	blk powder	dec 325 6.4		i H_2O
1530	Platinum(II) oxide PtO	12035-82-4 211.08	blk tetr cry	dec 325 14.1		i H_2O, EtOH; s aqua regia
1531	Platinum(II) sulfide PtS	12038-20-9 227.15	tetr cry	10.25	a,f	
1532	Platinum(III) bromide $PtBr_3$	25985-07-3 434.79	grn-blk cry	dec 200	a	
1533	Platinum(III) chloride $PtCl_3$	25909-39-1 301.44	grn-blk cry	435 dec 5.26	a	
1534	Platinum(IV) bromide $PtBr_4$ ·	68938-92-1 514.70	brn-blk cry	dec 180	a	i H_2O; sl EtOH, eth
1535	Platinum(IV) chloride $PtCl_4$	37773-49-2 336.89	red-brn cub cry	327 dec 4.30		sl H_2O
1536	Platinum(IV) chloride pentahydrate $PtCl_4 \cdot 5H_2O$	13454-96-1 426.97	red cry	2.43		s H_2O, EtOH
1537	Platinum(IV) fluoride PtF_4	13455-15-7 271.07	red cry	600		
1538	Platinum(IV) iodide PtI_4	7790-46-7 702.70	brn-blk powder	dec 130	a	s H_2O
1539	Platinum(IV) oxide PtO_2	1314-15-4 227.08	blk hex cry	450 11.8		i H_2O; s conc acid, dil alk
1540	Platinum(IV) sulfide PtS_2	12038-21-0 259.21	hex cry	7.85	a	
1541	Platinum(VI) fluoride PtF_6	13693-05-5 309.07	red cub cry	61.3 ≈4.0	69.1 a	
1542	Plutonium Pu	7440-07-5 244	silv-wh metal; monocl	640 19.7	3228 a,c	
1543	Plutonium nitride PuN	12033-54-4 258	gray cub cry	2550 14.4		
1544	Plutonium(II) oxide PuO	12035-83-5 260	cub cry	14.0		
1545	Plutonium(III) bromide $PuBr_3$	15752-46-2 484	grn orth cry	681 6.75		s H_2O
1546	Plutonium(III) chloride $PuCl_3$	13569-62-5 350	grn hex cry	760 5.71		s H_2O
1547	Plutonium(III) fluoride PuF_3	13842-83-6 301	purp hex cry	1425 9.33		i H_2O; sl acid
1548	Plutonium(III) iodide PuI_3	13813-46-2 625	grn orth cry; hyg	777 6.92		s H_2O

No.	Name Formula	CAS RN Mol. Wt.	Physical Form	mp/°C den/g cm⁻³	bp/°C Other Data	Solubility
1549	Plutonium(III) oxide Pu_2O_3	12036-34-9 536	blk cub cry	10.5		
1550	Plutonium(IV) fluoride PuF_4	13709-56-3 320	red-brn monocl cry	1027 7.1		
1551	Plutonium(IV) oxide PuO_2	12059-95-9 276	yel-brn cub cry	2400 11.5		
1552	Plutonium(VI) fluoride PuF_6	13693-06-6 358	red-brn orth cry	52 5.08		
1553	Polonium Po	7440-08-6 209	silv metal; cub	254 9.20	962 a	
1554	Polonium(IV) chloride $PoCl_4$	10026-02-5 351	yel hyg cry	≈300	390	s H_2O, EtOH, ace
1555	Polonium(IV) oxide PoO_2	7446-06-2 241	yel cub cry	dec 500 8.9		
1556	Potassium K	7440-09-7 39.098	soft silv-wh metal; cub	63.38 0.89	759 a,c,e	reac H_2O
1557	Potassium acetate $KC_2H_3O_2$	127-08-2 98.143	wh hyg cry	292 1.57	a	vs H_2O; s EtOH; i eth
1558	Potassium aluminate trihydrate $K_2Al_2O_4 \cdot 3H_2O$	12003-63-3* 250.204	wh orth cry	2.13		vs H_2O; i EtOH
1559	Potassium aluminum silicate $KAlSi_3O_8$	1327-44-2 278.332	col monocl cry	2.56		i H_2O
1560	Potassium aluminum sulfate $KAl(SO_4)_2$	10043-67-1 258.207	wh hyg powder			sl H_2O
1561	Potassium aluminum sulfate dodecahydrate $KAl(SO_4)_2 \cdot 12H_2O$	7784-24-9 474.391	col cry	dec ≈100 1.72		sl H_2O
1562	Potassium arsenate K_3AsO_4	7784-41-0 256.215	col cry	2.8		s H_2O
1563	Potassium azide KN_3	20762-60-1 81.118	tetr cry; exp	2.04		vs H_2O
1564	Potassium borohydride KBH_4	13762-51-1 53.941	wh cub cry	dec ≈500 1.11	a	s H_2O
1565	Potassium bromate $KBrO_3$	7758-01-2 167.000	wh hex cry	434 dec 3.27	a	sl H_2O; i EtOH
1566	Potassium bromide KBr	7758-02-3 119.002	col cub cry; hyg	734 2.74	1435 a,c,e	vs H_2O; sl EtOH
1567	Potassium carbonate K_2CO_3	584-08-7 138.206	wh monocl cry; hyg	898 2.29	dec a,c	s H_2O; i EtOH
1568	Potassium carbonate sesquihydrate $K_2CO_3 \cdot 1.5H_2O$	6381-79-9 165.229	granular cry			vs H_2O
1569	Potassium chlorate $KClO_3$	3811-04-9 122.549	wh monocl cry	368 2.32	dec a	s H_2O
1570	Potassium chloride KCl	7447-40-7 74.551	wh cub cry	771 1.988	a,c,e,f	s H_2O; i eth, ace
1571	Potassium chromate K_2CrO_4	7789-00-6 194.191	yel orth cry	975 2.73		vs H_2O
1572	Potassium cyanate KCNO	590-28-3 81.115	wh tetr cry	dec ≈700 2.05		s H_2O; sl EtOH
1573	Potassium cyanide KCN	151-50-8 65.116	wh cub cry; hyg	634 1.55	a	vs H_2O; sl EtOH
1574	Potassium dichromate $K_2Cr_2O_7$	7778-50-9 294.185	oran-red tricl cry	398 2.68	dec ≈500	s H_2O
1575	Potassium dihydrogen arsenate KH_2AsO_4	7784-41-0 180.034	col cry	288 2.87		s H_2O; i EtOH
1576	Potassium dihydrogen phosphate KH_2PO_4	7778-77-0 136.085	wh tetr cry	253 2.34	a	vs H_2O; sl EtOH
1577	Potassium ferricyanide $K_3Fe(CN)_6$	13746-66-2 329.246	red cry	dec 1.89		vs H_2O
1578	Potassium ferrocyanide trihydrate $K_4Fe(CN)_6 \cdot 3H_2O$	14459-95-1 422.390	yel monocl cry	dec 60 1.85		s H_2O; i EtOH, eth
1579	Potassium fluoride KF	7789-23-3 58.096	wh cub cry	858 2.48	1502 a,c,e	s H_2O
1580	Potassium fluoride dihydrate $KF \cdot 2H_2O$	13455-21-5 94.127	monocl cry	41 dec 2.5		vs H_2O
1581	Potassium fluoroborate KBF_4	14075-53-7 125.903	col orth cry	530 2.505		sl H_2O, EtOH
1582	Potassium formate $KCHO_2$	590-29-4 84.116	col hyg cry	167 1.91	a	vs H_2O
1583	Potassium hexachloroosmate(IV) K_2OsCl_6	16871-60-6 481.14	red cub cry			vs H_2O; sl EtOH
1584	Potassium hexachloroplatinate K_2PtCl_6	16921-30-5 485.99	yel-oran cub cry	dec 250 3.50		sl H_2O; i EtOH
1585	Potassium hexacyanocobalt $K_3Co(CN)_6$	13963-58-1 332.334	yel monocl cry	dec 1.91		vs H_2O; i EtOH
1586	Potassium hexafluoromanganate(IV) K_2MnF_6	16962-31-5 247.125	yel hex cry			reac H_2O
1587	Potassium hexafluorosilicate K_2SiF_6	16871-90-2 220.273	wh cry	dec 2.27	a	sl H_2O; i EtOH

No.	Name Formula	CAS RN Mol. Wt.	Physical Form	mp/°C den/g cm⁻³	bp/°C Other Data	Solubility
1588	Potassium hydride KH	7693-26-7 40.106	cub cry	1.43	a	reac H_2O
1589	Potassium hydrogen arsenite $KAsO_2 \cdot HAsO_2$	10124-50-2 253.947	wh hyg powder			s H_2O
1590	Potassium hydrogen carbonate $KHCO_3$	298-14-6 100.115	col monocl cry	dec ≈100 2.17	a	vs H_2O; i EtOH
1591	Potassium hydrogen fluoride KHF_2	7789-29-9 78.103	col tetr cry	238.9 2.37	a,c	s H_2O; i EtOH
1592	Potassium hydrogen phosphate K_2HPO_4	7758-11-4 174.176	wh hyg cry	dec		vs H_2O; sl EtOH
1593	Potassium hydrogen phosphite K_2HPO_3	13492-26-7 158.177	wh hyg powder	dec		vs H_2O; i EtOH
1594	Potassium hydrogen selenite $KHSeO_3$	7782-70-9 167.06	hyg orth cry	dec >100		s H_2O; sl EtOH
1595	Potassium hydrogen sulfate $KHSO_4$	7646-93-7 136.170	wh monocl cry; hyg	≈200 2.32	a	vs H_2O
1596	Potassium hydrogen sulfide KHS	1310-61-8 72.172	wh hex cry; hyg	≈450 1.69		s H_2O, EtOH
1597	Potassium hydrogen sulfide hemihydrate $KHS \cdot 0.5H_2O$	1310-61-8* 81.179	wh-yel hyg cry	≈175 1.7		vs H_2O, EtOH
1598	Potassium hydrogen sulfite $KHSO_3$	7773-03-7 120.170	wh cry powder	dec 190		s H_2O; i EtOH
1599	Potassium hydrogen tartrate $KHC_4H_4O_6$	868-14-4 188.178	wh cry	1.98		sl H_2O; s acid, alk; i EtOH
1600	Potassium hydroxide KOH	1310-58-3 56.105	wh rhomb cry; hyg	406 2.044	1327 a,c,e	vs H_2O; s EtOH; s MeOH
1601	Potassium hypophosphite KH_2PO_2	7782-87-8 104.087	wh hyg cry	dec		vs H_2O; s EtOH
1602	Potassium iodate KIO_3	7758-05-6 214.001	wh monocl cry	560 dec 3.89	a	sl H_2O
1603	Potassium iodide KI	7681-11-0 166.003	col cub cry	681 3.12	1323 a,c,e	vs H_2O; sl EtOH
1604	Potassium manganate K_2MnO_4	10294-64-1 197.133	grn cry	190 dec		s H_2O; reac HCl
1605	Potassium metabisulfite $K_2S_2O_5$	16731-55-8 222.326	wh powder	dec ≈150 2.3		vs H_2O; reac acid; i EtOH
1606	Potassium metaborate KBO_2	13709-94-9 81.908	wh hex cry	≈2.3	a	
1607	Potassium molybdate K_2MoO_4	13446-49-6 238.14	wh hyg cry	919 2.3		vs H_2O; i EtOH
1608	Potassium niobate $KNbO_3$	12030-85-2 180.002	wh rhomb cry	≈1100 4.64		i H_2O
1609	Potassium nitrate KNO_3	7757-79-1 101.103	col rhomb cry or powder	337 2.11	dec 400 a,c	s H_2O; i EtOH
1610	Potassium nitrite KNO_2	7758-09-0 85.104	wh hyg cry	441 1.915	exp 537 a	vs H_2O; sl EtOH
1611	Potassium oxalate monohydrate $K_2C_2O_4 \cdot H_2O$	6487-48-5 184.232	col cry	dec 160 2.13		vs H_2O
1612	Potassium oxide K_2O	12136-45-7 94.196	gray cub cry	350 dec 2.35	a	s H_2O, EtOH, eth
1613	Potassium perchlorate $KClO_4$	7778-74-7 138.549	col orth cry; hyg	525 2.52	a	sl H_2O
1614	Potassium periodate KIO_4	7790-21-8 230.001	col tetr cry	582 3.618	exp a	sl H_2O
1615	Potassium permanganate $KMnO_4$	7722-64-7 158.034	purp orth cry	dec 2.7	a	s H_2O; reac EtOH
1616	Potassium peroxide K_2O_2	17014-71-0 110.196	yel amorp solid	490	a	reac H_2O
1617	Potassium persulfate $K_2S_2O_8$	7727-21-1 270.324	col cry	dec ≈100 2.48		sl H_2O
1618	Potassium phosphate K_3PO_4	7778-53-2 212.266	wh orth cry; hyg	1340 2.564	a	vs H_2O; i EtOH
1619	Potassium pyrophosphate trihydrate $K_4P_2O_7 \cdot 3H_2O$	7320-34-5* 384.383	col hyg cry	1090 2.33		vs H_2O; i EtOH
1620	Potassium pyrosulfate $K_2S_2O_7$	7790-62-7 254.325	col needles	≈325 2.28		s H_2O
1621	Potassium selenate K_2SeO_4	7790-59-2 221.16	wh powder	3.07		vs H_2O
1622	Potassium selenide K_2Se	1312-74-9 157.16	red cub cry; hyg	800 2.29		s H_2O
1623	Potassium stannate trihydrate $K_2SnO_3 \cdot 3H_2O$	12142-33-5* 298.951	col cry	3.20		vs H_2O; i EtOH
1624	Potassium sulfate K_2SO_4	7778-80-5 174.261	wh orth cry	1069 2.66	a,c,f	s H_2O; i EtOH
1625	Potassium sulfide K_2S	1312-73-8 110.263	red-yel cub cry; hyg	948 1.74	a,c	s H_2O, EtOH; i eth
1626	Potassium sulfide pentahydrate $K_2S \cdot 5H_2O$	1312-73-8* 200.339	col rhomb cry	60		vs H_2O, EtOH; i eth

No.	Name Formula	CAS RN Mol. Wt.	Physical Form	mp/°C den/g cm^{-3}	bp/°C Other Data	Solubility
1627	Potassium sulfite dihydrate $K_2SO_3 \cdot 2H_2O$	7790-56-9 194.292	wh monocl cry	dec		s H_2O; sl EtOH; dec dil acid
1628	Potassium superoxide KO_2	12030-88-5 71.097	yel tetr cry; hyg	380 2.16	a	reac H_2O
1629	Potassium tellurate(VI) trihydrate $K_2TeO_4 \cdot 3H_2O$	15571-91-2* 323.84	wh cry powder			s H_2O
1630	Potassium tellurite K_2TeO_3	7790-58-1 253.80	wh hyg cry	dec ≈460		vs H_2O
1631	Potassium tetraborate pentahydrate $K_2B_4O_7 \cdot 5H_2O$	1332-77-0 323.513	wh cry powder			s H_2O; sl EtOH
1632	Potassium tetrachloroaurate dihydrate $KAuCl_4 \cdot 2H_2O$	13682-61-6 413.907	yel monocl cry			s H_2O, EtOH, eth
1633	Potassium tetrachloroplatinate K_2PtCl_4	10025-99-7 415.09	pink-red tetr cry	dec 500 3.38		s H_2O; i EtOH
1634	Potassium tetracyanoplatinate(II) trihydrate $K_2Pt(CN)_4 \cdot 3H_2O$	562-76-5* 431.39	col rhomb prisms			s H_2O
1635	Potassium tetraiodomercurate(II) K_2HgI_4	7783-33-7 786.40	yel hyg cry	4.29		vs H_2O; s EtOH, eth, ace
1636	Potassium thiocyanate KSCN	333-20-0 97.182	col tetr cry; hyg	173 1.88	dec 500 a	vs H_2O; s EtOH
1637	Potassium thiosulfate $K_2S_2O_3$	10294-66-3 190.327	col hyg cry			vs H_2O; i EtOH
1638	Potassium titanate K_2TiO_3	12030-97-6 174.062	wh orth cry	1515 3.1		reac H_2O
1639	Potassium triiodide monohydrate $KI_3 \cdot H_2O$	7790-42-3 437.827	brn monocl cry; hyg	dec 225 3.5		s H_2O; reac EtOH, eth
1640	Potassium thiocarbonate K_2CS_3	26750-66-3 186.406	yel-red hyg cry			vs H_2O
1641	Potassium tungstate K_2WO_4	7790-60-5 326.04	hyg cry	921 3.12		vs H_2O; i EtOH
1642	Potassium uranate $K_2U_2O_7$	7790-63-8 666.251	oran cub cry	6.12		i H_2O; s acid
1643	Praseodymium Pr	7440-10-0 140.908	silv metal; hex	931 6.77	3520 a,c	
1644	Praseodymium boride PrB_6	12008-27-4 205.774	blk cub cry	2610 4.84		
1645	Praseodymium bromide $PrBr_3$	13536-53-3 380.620	grn hex cry; hyg	693 5.28		s H_2O
1646	Praseodymium chloride $PrCl_3$	10361-79-2 247.266	grn hex needles; hyg	786 4.0	a	s H_2O, EtOH
1647	Praseodymium chloride heptahydrate $PrCl_3 \cdot 7H_2O$	10025-90-8 373.373	grn cry	dec 110		s H_2O, EtOH
1648	Praseodymium fluoride PrF_3	13709-46-1 197.903	grn hex cry	1395 6.3		
1649	Praseodymium iodide PrI_3	13813-23-5 521.621	orth hyg cry	737 ≈5.8		s H_2O
1650	Praseodymium nitrate hexahydrate $Pr(NO_3)_3 \cdot 6H_2O$	14483-17-1 435.014	ligtt grn cry			vs H_2O
1651	Praseodymium nitride PrN	25764-09-4 154.915	cub cry	7.46		
1652	Praseodymium oxide Pr_2O_3	12036-32-7 329.813	wh hex cry	2300 6.9	a	
1653	Praseodymium silicide $PrSi_2$	12066-83-0 197.079	tetr cry	1712 5.46		
1654	Praseodymium sulfide Pr_2S_3	12038-13-0 378.013	cub cry	1765 5.1		
1655	Praseodymium telluride Pr_2Te_3	12038-12-9 664.62	cub cry	1500 ≈7.0		
1656	Promethium Pm	7440-12-2 145	silv metal; hex	1042 7.26	3000 a	
1657	Protactinium Pa	7440-13-3 231.036	shiny metal; tetr or cub	1572 15.4	a,c	
1658	Protactinium(V) chloride $PaCl_5$	13760-41-3 408.300	yel monocl cry	306 3.74	a	
1659	Radium Ra	7440-14-4 226	wh metal; cub	700 5	a	
1660	Radium bromide $RaBr_2$	10031-23-9 386	wh orth cry	728 5.79		s H_2O, EtOH
1661	Radium chloride $RaCl_2$	10025-66-8 297	wh orth cry	1000 4.9		s H_2O, EtOH
1662	Radium fluoride RaF_2	20610-49-5 264	wh cub cry	6.7		
1663	Radium sulfate $RaSO_4$	7446-16-4 322	wh cry		a	i H_2O, acid
1664	Radon Rn	10043-92-2 222	col gas	-71 9.074 g/L	-61.7 a,b,e	sl H_2O
1665	Rhenium Re	7440-15-5 186.207	silv-gray metal	3186 20.8	5596 a,c	i HCl

No.	Name Formula	CAS RN Mol. Wt.	Physical Form	mp/°C den/g cm^{-3}	bp/°C Other Data	Solubility
1666	Perrhenic acid $HReO_4$	13768-11-1 251.213	exists only in soln		a	vs H_2O, os
1667	Rhenium carbonyl $Re_2(CO)_{10}$	14285-68-8 652.518	yel-wh cry	170 dec 2.87		s os
1668	Rhenium(III) bromide $ReBr_3$	13569-49-8 425.919	red-brn monocl cry	6.10	subl 500 a	s ace, MeOH, EtOH
1669	Rhenium(III) chloride $ReCl_3$	13569-63-6 292.565	red-blk hyg cry	dec 500 4.81	a	s H_2O
1670	Rhenium(III) iodide ReI_3	15622-42-1 566.920	blk solid	dec		
1671	Rhenium(IV) chloride $ReCl_4$	13569-71-6 328.018	purp-blk cry; hyg	dec 300 4.9		
1672	Rhenium(IV) fluoride ReF_4	15192-42-4 262.201	blue tetr cry	7.49	subl >300	
1673	Rhenium(IV) oxide ReO_2	12036-09-8 218.206	gray orth cry	dec 900 11.4		
1674	Rhenium(IV) sulfide ReS_2	12038-63-0 250.339	tricl cry	7.6		
1675	Rhenium(IV) telluride $ReTe_2$	12067-00-4 441.41	orth cry	8.50		
1676	Rhenium(V) bromide $ReBr_5$	30937-53-2 585.727	brn solid	110 dec		
1677	Rhenium(V) chloride $ReCl_5$	39368-69-9 363.471	brn-blk solid	220 4.9		reac H_2O
1678	Rhenium(V) fluoride ReF_5	30937-52-1 281.199	yel-grn solid	48	≈220	
1679	Rhenium(V) oxide Re_2O_5	12165-05-8 452.411	blue-blk tetr cry	≈7		
1680	Rhenium(VI) chloride $ReCl_6$	31234-26-1 398.923	red-grn solid	29		
1681	Rhenium(VI) fluoride ReF_6	10049-17-9 300.197	yel liq or cub cry	18.5 4.06(cry)	33.7	s HNO_3
1682	Rhenium(VI) oxide ReO_3	1314-28-9 234.205	redcub cry	400 dec 6.9		i H_2O, acid, alk
1683	Rhenium(VI) oxytetrachloride $ReOCl_4$	13814-76-1 344.017	brn cry	29.3	223 b	reac H_2O
1684	Rhenium(VI) oxytetrafluoride $ReOF_4$	17026-29-8 278.200	blue solid	108	171	
1685	Rhenium(VII) fluoride ReF_7	17029-21-9 319.196	yel cub cry	48.3 4.32	73.7	
1686	Rhenium(VII) oxide Re_2O_7	1314-68-7 484.410	yel hyg cry	297 6.10	360 a,b,c,e	s H_2O, EtOH, eth, diox, py
1687	Rhenium(VII) trioxychloride ReO_3Cl	7791-09-5 269.658	col liq	4.5 3.87	128	reac H_2O
1688	Rhenium(VII) trioxyfluoride ReO_3F	42246-24-2 253.203	yel solid	147	164	
1689	Rhenium(VII) dioxytrifluoride ReO_2F_3	57246-89-6 275.201	yel solid	90	185	reac H_2O
1690	Rhenium(VII) oxypentafluoride $ReOF_5$	23377-53-9 297.198	cream solid	43.8	73.0	
1691	Rhenium(VII) sulfide Re_2S_7	12038-67-4 596.876	brn-blk tetr cry	4.87		i H_2O
1692	Rhodium Rh	7440-16-6 102.906	silv-wh metal; cub	1964 12.4	3695 a,c	i acid, sl aqua regia
1693	Rhodium carbonyl chloride $[Rh(CO)_2Cl]_2$	14523-22-9 388.758	red-oran cry	124		s os
1694	Rhodium dodecacarbonyl $Rh_4(CO)_{12}$	19584-30-6 747.747	red hyg cry	2.52		reac H_2O
1695	Rhodium(III) chloride $RhCl_3$	10049-07-7 209.264	red monocl cry	5.38	717 a	i H_2O; s alk
1696	Rhodium(III) fluoride RhF_3	60804-25-3 159.901	red hex cry	5.4		
1697	Rhodium(III) iodide RhI_3	15492-38-3 483.619	blk monocl cry; hyg	6.4		
1698	Rhodium(III) oxide Rh_2O_3	12036-35-0 253.809	gray hex cry	1100 dec 8.2	a	
1699	Rhodium(III) sulfate $Rh_2(SO_4)_3$	10489-46-0 494.002	red-yel solid	dec >500		
1700	Rhodium(IV) oxide RhO_2	12137-27-8 134.905	blk tetr cry	7.2		
1701	Rhodium(VI) fluoride RhF_6	13693-07-7 216.896	blk cub cry	≈70 3.1		
1702	Rubidium Rb	7440-17-7 85.468	soft silv metal; cub	39.31 1.53	688 a,c,e	reac H_2O
1703	Rubidium aluminum sulfate $RbAl(SO_4)_2$	13530-57-9 304.577	hex cry	≈3.1		sl H_2O; i EtOH

No.	Name / Formula	CAS RN / Mol. Wt.	Physical Form	mp/°C / den/g cm^{-3}	bp/°C / Other Data	Solubility
1704	Rubidium aluminum sulfate dodecahydrate RbAl(SO$_4$)$_2$•12H$_2$O	7784-29-4 520.761	col cub cry	≈100 dec ≈1.9		s H$_2$O; i EtOH
1705	Rubidium azide RbN$_3$	22756-36-1 127.488	tetr cry; exp	317 2.79		
1706	Rubidium bromide RbBr	7789-39-1 165.372	wh cub cry; hyg	682 3.35	1340 a,c,e	vs H$_2$O
1707	Rubidium carbonate Rb$_2$CO$_3$	584-09-8 230.945	col monocl cry; hyg	837	a	vs H$_2$O
1708	Rubidium chloride RbCl	7791-11-9 120.921	wh cub cry; hyg	715 2.76	1390 a,c,e	s H$_2$O; sl EtOH
1709	Rubidium cyanide RbCN	19073-56-4 111.486	wh cub cry	2.3		s H$_2$O; i EtOH, eth
1710	Rubidium fluoride RbF	13446-74-7 104.466	wh cub cry; hyg	833 3.2	1410 a,c,e	vs H$_2$O; i EtOH
1711	Rubidium fluoride RbHF$_2$	12280-64-7 124.473	tetr cry	188 3.3	a	
1712	Rubidium hydride RbH	13446-75-8 86.476	wh cub cry; flam	dec ≈170 2.60	a	reac H$_2$O
1713	Rubidium hydrogen carbonate RbHCO$_3$	19088-74-5 146.485	wh rhomb cry	dec 175		vs H$_2$O
1714	Rubidium hydrogen sulfate RbHSO$_4$	15587-72-1 182.540	col monocl cry	208 2.9	a	s H$_2$O
1715	Rubidium hydroxide RbOH	1310-82-3 102.475	gray-wh orth cry; hyg	382 3.2	a	vs H$_2$O; s EtOH
1716	Rubidium iodide RbI	7790-29-6 212.372	wh cub cry	642 3.55	1300 a,c,e	s H$_2$O, EtOH
1717	Rubidium nitrate RbNO$_3$	13126-12-0 147.473	wh hex cry; hyg	305 3.11	a,c	vs H$_2$O
1718	Rubidium oxide Rb$_2$O	18088-11-4 186.935	yel-brn cub cry; hyg	dec 400 4.0	a	reac H$_2$O
1719	Rubidium perchlorate RbClO$_4$	13510-42-4 184.919	wh hyg cry	281 2.8	dec 600 a	sl H$_2$O
1720	Rubidium peroxide Rb$_2$O$_2$	23611-30-5 202.935	wh orth cry	3.8	a	reac H$_2$O
1721	Rubidium selenide Rb$_2$Se	31052-43-4 249.90	wh cub cry	733 3.22		reac H$_2$O
1722	Rubidium sulfate Rb$_2$SO$_4$	7488-54-2 267.000	wh orth cry	1050 3.6	a	vs H$_2$O
1723	Rubidium sulfide Rb$_2$S	31083-74-6 203.002	wh cub cry	425 2.91		s H$_2$O
1724	Rubidium superoxide RbO$_2$	12137-25-6 117.467	tetr cry	412 ≈3.0	a	
1725	Ruthenium Ru	7440-18-8 101.07	silv-wh metal; hex	2334 12.1	4150 a,c	i acid, aqua regia
1726	Ruthenium dodecacarbonyl Ru$_3$(CO)$_{12}$	15243-33-1 639.34	oran cry	dec 150		
1727	Ruthenium(III) acetylacetonate Ru(CH$_3$COCHCOCH$_3$)$_3$	14284-93-6 398.40	red-brn cry	230		
1728	Ruthenium(III) bromide RuBr$_3$	14014-88-1 340.78	brn hex cry	dec >400 5.3	a	
1729	Ruthenium(III) chloride RuCl$_3$	10049-08-8 207.43	brn hex cry	dec >500 3.1	a	i H$_2$O; s EtOH
1730	Ruthenium(III) fluoride RuF$_3$	51621-05-7 158.07	brn rhomb cry	dec >600 5.36		
1731	Ruthenium(III) iodide RuI$_3$	13896-65-6 481.78	blk hex cry	6.0	a	
1732	Ruthenium(IV) fluoride RuF$_4$	71500-16-8 177.06	yel cry			reac H$_2$O
1733	Ruthenium(IV) oxide RuO$_2$	12036-10-1 133.07	gray-blk tetr cry	7.05	a	i H$_2$O, acid
1734	Ruthenium(V) fluoride RuF$_5$	14521-18-7 196.06	grn monocl cry	86.5 3.90	227	
1735	Ruthenium(VI) fluoride RuF$_6$	13693-08-8 215.06	dark brn orth cry	54 3.54		reac H$_2$O
1736	Ruthenium(VIII) oxide RuO$_4$	20427-56-9 165.07	yel monocl prisms	25.4 3.29	40 a	sl H$_2$O; vs ctc; reac EtOH
1737	Samarium Sm	7440-19-9 150.36	silv metal; rhomb	1074 7.52	1794 a,c	
1738	Samarium boride SmB$_6$	12008-29-6 215.23	refrac solid	2580 5.07		
1739	Samarium silicide SmSi$_2$	12300-22-0 206.53	orth cry	5.14		
1740	Samarium(II) bromide SmBr$_2$	50801-97-3 310.17	brn cry	669		reac H$_2$O
1741	Samarium(II) chloride SmCl$_2$	13874-75-4 221.27	brn cry	855 3.69		reac H$_2$O
1742	Samarium(II) fluoride SmF$_2$	15192-17-3 188.36	purp cry			reac H$_2$O

No.	Name / Formula	CAS RN / Mol. Wt.	Physical Form	mp/°C / den/g cm⁻³	bp/°C / Other Data	Solubility
1743	Samarium(II) iodide SmI_2	32248-43-4 404.17	grn cry	520		reac H_2O
1744	Samarium(III) bromide $SmBr_3$	13759-87-0 390.07	yel cry	640		reac H_2O
1745	Samarium(III) chloride $SmCl_3$	10361-82-7 256.72	yel cry	682 4.46	a	reac H_2O
1746	Samarium(III) chloride hexahydrate $SmCl_3 \cdot 6H_2O$	13465-55-9 364.81	yel cry	dec 2.38		reac H_2O
1747	Samarium(III) fluoride SmF_3	13765-24-7 207.36	wh cry	1306	a	reac H_2O
1748	Samarium(III) iodide SmI_3	13813-25-7 531.07	oran cry	850		reac H_2O
1749	Samarium(III) oxide Sm_2O_3	12060-58-1 348.72	yel-wh cub cry	2335 7.6	a	
1750	Samarium(III) sulfate octahydrate $Sm_2(SO_4)_3 \cdot 8H_2O$	13465-58-2 733.03	yel cry	2.93		sl H_2O
1751	Samarium(III) sulfide Sm_2S_3	12067-22-0 396.92	gray-brn cub cry	1720 5.87		
1752	Samarium(III) telluride Sm_2Te_3	12040-00-5 683.52	orth cry	7.31		
1753	Scandium Sc	7440-20-2 44.956	silv metal; hex	1541 2.99	2836 a,c	
1754	Scandium boride ScB_2	12007-34-0 66.578	refrac solid	2250 3.17		
1755	Scandium bromide $ScBr_3$	13465-59-3 284.668	wh hyg cry	969 9.33	a	s H_2O
1756	Scandium chloride $ScCl_3$	10361-84-9 151.314	wh hyg cry	967 2.4	a	s H_2O; i EtOH
1757	Scandium fluoride ScF_3	13709-47-2 101.951	wh powder	1515	a	sl H_2O
1758	Scandium oxide Sc_2O_3	12060-08-1 137.910	wh cub cry	2485 3.864	a	s conc acid
1759	Scandium sulfide Sc_2S_3	12166-29-9 186.110	yel orth cry	1775 2.91		
1760	Scandium telluride Sc_2Te_3	12166-44-8 472.71	blk hex cry	5.29		
1761	Selenium (α form) Se	7782-49-2 78.96	red monocl cry	221 4.39	685 a,b,c,d,e	i H_2O, EtOH; sl eth
1762	Selenium (gray) Se	7782-49-2 78.96	gray metallic cry; hex	220.5 4.81	685 a,b,c,d,e	i H_2O, CS_2
1763	Selenium (vitreous) Se	7782-49-2 78.96	blk amorp solid	trans to gray Se 180 4.28	685 a,b,c,d,e	i H_2O; sl CS_2
1764	Selenic acid H_2SeO_4	7783-08-6 144.97	wh hyg solid	58 2.95	260 dec a	vs H_2O; reac EtOH
1765	Selenous acid H_2SeO_3	7783-00-8 128.97	wh hyg cry	dec 70 3.0		vs H_2O; s EtOH
1766	Selenium dioxide SeO_2	7446-08-4 110.96	wh tetr needles or powder	340 tp 3.95	315 sp a,e	s H_2O, EtOH, MeOH; sl ace
1767	Selenium trioxide SeO_3	13768-86-0 126.96	wh tetr cry; hyg	118 3.44	subl	s H_2O, os
1768	Selenium bromide Se_2Br_2	7789-52-8 317.73	red liq	3.60	225 dec	reac H_2O; s CS_2, chl
1769	Selenium chloride Se_2Cl_2	10025-68-0 228.83	yel-brn oily liq	-85 2.774	130 dec	reac H_2O; s CS_2, bz, ctc, chl
1770	Selenium tetrabromide $SeBr_4$	7789-65-3 398.58	oran-red cry	123		reac H_2O; s CS_2, chl
1771	Selenium tetrachloride $SeCl_4$	10026-03-6 220.77	wh-yel cry	305 2.6	subl e	reac H_2O
1772	Selenium tetrafluoride SeF_4	13465-66-2 154.95	col liq	-10 2.75	106 d	reac H_2O, vs EtOH, eth
1773	Selenium hexafluoride SeF_6	7783-79-1 192.95	col gas	-34.6 tp 7.887 g/L	-46.6 sp a,b,e	i H_2O
1774	Selenium oxybromide $SeOBr_2$	7789-51-7 254.77	red-yel solid	41.6 3.38	220 dec	reac H_2O; s CS_2, bz, ctc
1775	Selenium oxychloride $SeOCl_2$	7791-23-3 165.86	col or yel liq	8.5 2.44	177 b,e	reac H_2O; s ctc, chl, bz, tol
1776	Selenium oxyfluoride $SeOF_2$	7783-43-9 132.96	col liq	15 2.8	125	reac H_2O
1777	Selenium dioxydifluoride SeO_2F_2	14984-81-7 148.96	col gas	-99.5 6.088 g/L	-8.4	reac H_2O
1778	Selenium sulfide Se_2S_6	75926-26-0 350.32	oran needles	121.5 2.44		s CS_2; sl bz
1779	Selenium sulfide Se_4S_4	75926-28-2 444.10	red cry	113 dec 3.29		s bz; sl CS_2
1780	Silicon Si	7440-21-3 28.086	gray cry or brn amorp solid	1414 2.3290	3265 a,c,e	i H_2O, acid; s alk
1781	Silane SiH_4	7803-62-5 32.118	col gas; flam	-185 1.313 g/L	-111.9 a,d,e	reac H_2O; i EtOH, bz

No.	Name Formula	CAS RN Mol. Wt.	Physical Form	mp/°C den/g cm⁻³	bp/°C Other Data	Solubility
1782	Bromosilane SiH_3Br	13465-73-1 111.014	col gas	-94 4.538 g/L	1.9 a,d,e	
1783	Bromotrichlorosilane $SiCl_3Br$	13465-74-2 214.348	col liq	-62 1.826	80.3 a	reac H_2O
1784	Chlorosilane SiH_3Cl	13465-78-6 66.563	col gas	-118 2.721 g/L	-30.4 a,d,e	
1785	Chlorotrifluorosilane SiF_3Cl	14049-36-6 120.534	col gas	-138 4.927 g/L	-70.0 b,d	reac H_2O
1786	Dibromosilane SiH_2Br_2	13768-94-0 189.910	liq	-70.1 	66 a,d,e	
1787	Dichlorosilane SiH_2Cl_2	4109-96-0 101.007	col gas; flam	-122 4.129 g/L	8.3 a,d	reac H_2O
1788	Dichlorodifluorosilane $SiCl_2F_2$	18356-71-3 136.988	col gas	-44 5.599 g/L	-32 b,d,e	reac H_2O
1789	Difluorosilane SiH_2F_2	13824-36-7 68.099	col gas	-122 2.783 g/L	-77.8 d,e	
1790	Diiodosilane SiH_2I_2	13760-02-6 283.911	col liq	-1 	150 e	
1791	Fluorosilane SiH_3F	13537-33-2 50.108	col gas	 2.048 g/L	-98.6 a,d,e	
1792	Iodosilane SiH_3I	13598-42-0 158.014	col liq	-57 	45.6 a,e	
1793	Tetrabromosilane $SiBr_4$	7789-66-4 347.702	col fuming liq	5.2 2.8	154 a,b,d	reac H_2O
1794	Tetrachlorosilane $SiCl_4$	10026-04-7 169.897	col fuming liq	-68.85 1.5	57.65 a,b,c,d,e	reac H_2O
1795	Tetrafluorosilane SiF_4	7783-61-1 104.080	col gas	-90.2 4.254 g/L	-86 a,b,e	reac H_2O
1796	Tetraiodosilane SiI_4	13465-84-4 535.704	wh powder	120.5 4.1	287.35 a,b,c,d	
1797	Tribromosilane $SiHBr_3$	7789-57-3 268.806	flam liq	-73 2.7	109 a,b,d,e	reac H_2O
1798	Tribromochlorosilane $SiBr_3Cl$	13465-76-4 303.251	col liq	-20.8 2.497	127 a	reac H_2O
1799	Trichlorosilane $SiHCl_3$	10025-78-2 135.452	fuming liq	-128.2 1.331	33 a,b,d,e	reac H_2O
1800	Trichlorofluorosilane $SiCl_3F$	14965-52-7 153.442	col gas	 6.272 g/L	12.25 b,e	
1801	Trifluorosilane $SiHF_3$	13465-71-9 86.089	col gas	-131 3.519 g/L	-95 a,d,e	
1802	Triiodosilane $SiHI_3$	13465-72-0 409.807	liq	8 	220 dec	
1803	Disilane Si_2H_6	1590-87-0 62.219	col gas; flam	-132.5 2.543 g/L	-14.3 a,d,e	reac H_2O, ctc, chl; s EtOH, bz
1804	Trisilane Si_3H_8	7783-26-8 92.321	flam liq	-117.4 0.725(-117°C)	52.9 a,d,e	reac H_2O
1805	Tetrasilane Si_4H_{10}	7783-29-1 122.421	col liq	-84.3 0.79(-84°C)	107.4 e	reac H_2O
1806	Disiloxane $(SiH_3)_2O$	13597-73-4 78.218	gas	-144 3.197 g/L	-15.2 e	
1807	Metasilicic acid H_2SiO_3	7699-41-4 78.100	wh amorp powder	 	 a,f	i H_2O; s HF
1808	Orthosilicic acid H_4SiO_4	10193-36-9 96.116	exists only in soln	 	 a	
1809	Fluorosilicic acid H_2SiF_6	16961-83-4 144.092	stable only in aq soln	 		s H_2O
1810	Silicon carbide SiC	409-21-2 40.097	hard grn-black hex cry	2830 3.16	 a,f	i H_2O, EtOH
1811	Silicon nitride Si_3N_4	12033-89-5 140.284	gray refrac solid; hex	1900 3.17	 a	
1812	Silicon monoxide SiO	10097-28-6 44.085	blk cub cry, stable >1200	 2.18	 a	
1813	Silicon dioxide (α quartz) SiO_2	14808-60-7 60.085	col hex cry	573 trans to quartz 2.648	2950 a,c,f	i H_2O, acid; s HF
1814	Silicon dioxide (β quartz) SiO_2	14808-60-7 60.085	col hex cry	867 trans to tridymite 2.533(600°C)	2950 a,c,f	i H_2O, acid; s HF
1815	Silicon dioxide (tridymite) SiO_2	15468-32-3 60.085	col hex cry	1470 trans to cristobalite 2.265	2950	i H_2O, acid; s HF
1816	Silicon dioxide (cristobalite) SiO_2	14464-46-1 60.085	col hex cry	1713 2.334	2950 e	i H_2O, acid; s HF
1817	Silicon dioxide (vitreous) SiO_2	60676-86-0 60.085	col amorp solid	1713 2.196	2950	i H_2O, acid; s HF
1818	Silicon monosulfide SiS	50927-81-6 60.152	yel-red hyg powder	≈900 1.85	940	reac H_2O
1819	Silicon disulfide SiS_2	13759-10-9 92.218	wh rhomb cry	1090 2.04	subl	reac H_2O, EtOH; i bz

No.	Name Formula	CAS RN Mol. Wt.	Physical Form	mp/°C den/g cm^{-3}	bp/°C Other Data	Solubility
1820	Silver Ag	7440-22-4 107.868	silv metal; cub	961.78 10.5	2162 a,c,e,f	
1821	Silver azide AgN$_3$	13863-88-2 149.888	orth cry; exp	exp ≈250 4.9		
1822	Silver subfluoride Ag$_2$F	1302-01-8 234.734	yel hex cry	dec 100 8.6		reac H$_2$O
1823	Silver(I) acetate AgC$_2$H$_3$O$_2$	563-63-3 166.913	wh needles or powder	dec 3.26		s H$_2$O
1824	Silver(I) acetylide Ag$_2$C$_2$	7659-31-6 239.758	wh powder; exp			
1825	Silver(I) acetylide AgC$_2$H	13092-75-6 132.898	wh powder; exp			
1826	Silver(I) bromate AgBrO$_3$	7783-89-3 235.770	wh tetr cry	dec 5.2	a	sl H$_2$O
1827	Silver(I) bromide AgBr	7785-23-1 187.772	yel cub cry	432 6.47	1502 a,c,d,f	i H$_2$O, acid
1828	Silver(I) carbonate Ag$_2$CO$_3$	534-16-7 275.745	yel monocl cry	218 6.077		sl H$_2$O; s acid
1829	Silver(I) chlorate AgClO$_3$	7783-92-8 191.319	wh tetr cry	230 4.430	dec 270 a	s H$_2$O; sl EtOH
1830	Silver(I) chloride AgCl	7783-90-6 143.321	wh cub cry	455 5.56	1547 a,c,d,e,f	i H$_2$O
1831	Silver(I) chromate Ag$_2$CrO$_4$	7784-01-2 331.730	brn-red monocl cry	 5.625	a	i H$_2$O
1832	Silver(I) citrate Ag$_3$C$_6$H$_5$O$_7$	126-45-4 512.707	wh cry powder			i H$_2$O; s HNO$_3$
1833	Silver(I) cyanide AgCN	506-64-9 133.886	wh-gray hex cry	320 dec 3.95	a	i H$_2$O, EtOH, dil acid
1834	Silver(I) dichromate Ag$_2$Cr$_2$O$_7$	7784-02-3 431.724	red cry	 4.770		sl H$_2$O
1835	Silver(I) fluoride AgF	7775-41-9 126.866	yel-brn cub cry; hyg	435 5.852	1159 a	s H$_2$O
1836	Silver(I) hydrogen fluoride AgHF$_2$	12249-52-4 146.873	hyg cry	dec		
1837	Silver(I) iodate AgIO$_3$	7783-97-3 282.770	wh orth cry	>200 5.53	a	i H$_2$O
1838	Silver(I) iodide AgI	7783-96-2 234.772	yel powder; hex	558 5.68	1506 a,c,d,e,f	i H$_2$O, acid
1839	Silver(I) lactate monohydrate AgC$_3$H$_5$O$_3$·H$_2$O	128-00-7 214.955	gray cry powder			sl H$_2$O, EtOH
1840	Silver(I) molybdate Ag$_2$MoO$_4$	13765-74-7 375.67	yel cub cry	483 6.18		sl H$_2$O
1841	Silver(I) nitrate AgNO$_3$	7761-88-8 169.873	col rhomb cry	212 4.35	dec 440 a,c	vs H$_2$O; sl EtOH, ace
1842	Silver(I) nitrite AgNO$_2$	7783-99-5 153.874	yel needles	dec 140 4.453		sl H$_2$O; i EtOH; reac acid
1843	Silver(I) oxalate Ag$_2$C$_2$O$_4$	533-51-7 303.756	wh cry powder	exp 140 5.03		i H$_2$O
1844	Silver(I) oxide Ag$_2$O	20667-12-3 231.735	brn-blk cub cry	dec ≈200 7.2	a	i H$_2$O, EtOH; s acid, alk
1845	Silver(I) perchlorate AgClO$_4$	7783-93-9 207.319	col cub cry; hyg	dec 486 2.806	a	vs H$_2$O; s bz, py, os
1846	Silver(I) perchlorate monohydrate AgClO$_4$·H$_2$O	14242-05-8 225.334	hyg wh cry	dec 43		
1847	Silver(I) permanganate AgMnO$_4$	7783-98-4 226.804	viol monocl cry	dec 4.49		sl H$_2$O; reac EtOH
1848	Silver(I) phosphate Ag$_3$PO$_4$	7784-09-0 418.576	yel powder	849 6.37		i H$_2$O; sl dil acid
1849	Silver(I) picrate monohydrate AgC$_6$H$_2$N$_3$O$_7$·H$_2$O	146-84-9 353.981	yel cry			sl H$_2$O, EtOH; i chl, eth
1850	Silver(I) selenate Ag$_2$SeO$_4$	7784-07-8 358.69	orth cry	 5.72		sl H$_2$O
1851	Silver(I) selenide Ag$_2$Se	1302-09-6 294.70	gray hex needles	880 8.216		i H$_2$O
1852	Silver(I) selenite Ag$_2$SeO$_3$	7784-05-6 342.69	needles	530 5.930	dec >550	sl H$_2$O; s acid
1853	Silver(I) sulfate Ag$_2$SO$_4$	10294-26-5 311.800	col cry or powder	652 5.45	a	sl H$_2$O
1854	Silver(I) sulfide Ag$_2$S	21548-73-2 247.802	gray-blk orth powder	825 7.23	a,c,f	i H$_2$O; s acid
1855	Silver(I) telluride Ag$_2$Te	12002-99-2 343.34	blk orth cry	955 8.4	f	
1856	Silver(I) tetraiodomercurate(II) Ag$_2$HgI$_4$	7784-03-4 923.94	yel tetr cry	trans to red cub ≈40 6.1		i H$_2$O, dil acid
1857	Silver(I) thiocyanate AgSCN	14104-20-2 165.952	wh powder	dec		i H$_2$O

No.	Name / Formula	CAS RN / Mol. Wt.	Physical Form	mp/°C / den/g cm^{-3}	bp/°C / Other Data	Solubility
1858	Silver(II) oxide / AgO	1301-96-8 / 123.867	gray powder; monocl or cub	dec >100 / 7.5		i H$_2$O; s alk; reac acid
1859	Silver(II) fluoride / AgF$_2$	7783-95-1 / 145.865	wh or gray hyg cry	690 / 4.58	a	reac H$_2$O
1860	Sodium / Na	7440-23-5 / 22.990	soft silv met; cub	97.80 / 0.97	883 / a,c,e	reac H$_2$O
1861	Sodium acetate / NaC$_2$H$_3$O$_2$	127-09-3 / 82.035	col cry	324 / 1.528	a	vs H$_2$O
1862	Sodium acetate trihydrate / NaC$_2$H$_3$O$_2$•3H$_2$O	6131-90-4 / 136.080	col cry	58 dec / 1.45		vs H$_2$O; sl EtOH
1863	Sodium aluminate / NaAlO$_2$	1302-42-7 / 81.971	wh orth cry; hyg	1650 / 4.63		vs H$_2$O; i EtOH
1864	Sodium aluminum sulfate dodecahydrate / NaAl(SO$_4$)$_2$•12H$_2$O	10102-71-3 / 458.283	col cry	≈60 / 1.61		vs H$_2$O; i EtOH
1865	Sodium amide / NaNH$_2$	7782-92-5 / 39.013	wh-grn orth cry	210 / 1.39	dec 500 / a	reac H$_2$O
1866	Sodium ammonium phosphate tetrahydrate / NaNH$_4$HPO$_4$•4H$_2$O	13011-54-6 / 209.069	monocl cry	≈80 dec / 1.54		s H$_2$O; i EtOH
1867	Sodium arsenite / NaAsO$_2$	7784-46-5 / 129.911	wh-gray hyg powder	/ 1.87		vs H$_2$O; i EtOH
1868	Sodium azide / NaN$_3$	26628-22-8 / 65.010	col hex cry	dec 300; exp / 1.846	a	vs H$_2$O; sl EtOH; i eth
1869	Sodium borohydride / NaBH$_4$	16940-66-2 / 37.833	wh cub cry; hyg	dec ≈400 / 1.07	a	vs H$_2$O; reac EtOH
1870	Sodium bromate / NaBrO$_3$	7789-38-0 / 150.892	col cub cry	381 / 3.34	a,c	s H$_2$O; i EtOH
1871	Sodium bromide / NaBr	7647-15-6 / 102.894	wh cub cry	747 / 3.200	1390 / a,c,e	vs H$_2$O; s EtOH
1872	Sodium bromide dihydrate / NaBr•2H$_2$O	13466-08-5 / 138.925	wh cry	36 dec / 2.18		s H$_2$O; sl EtOH
1873	Sodium carbonate / Na$_2$CO$_3$	497-19-8 / 105.989	wh hyg powder	858.1 / 2.54	a,c	s H$_2$O; i EtOH
1874	Sodium carbonate decahydrate / Na$_2$CO$_3$•10H$_2$O	6132-02-1 / 286.142	col cry	34 dec / 1.46		vs H$_2$O; i EtOH
1875	Sodium carbonate monohydrate / Na$_2$CO$_3$•H$_2$O	5968-11-6 / 124.005	col orth cry	dec 100 / 2.25	f	vs H$_2$O; i EtOH
1876	Sodium chlorate / NaClO$_3$	7775-09-9 / 106.441	col cub cry	248 / 2.5	dec >300 / a,c	vs H$_2$O; sl EtOH
1877	Sodium chloride / NaCl	7647-14-5 / 58.443	col cub cry	800.7 / 2.17	1465 / a,c,e,f	s H$_2$O; sl EtOH
1878	Sodium chlorite / NaClO$_2$	7758-19-2 / 90.442	wh hyg cry	dec ≈180 /	a	vs H$_2$O
1879	Sodium chromate / Na$_2$CrO$_4$	7775-11-3 / 161.974	yel orth cry	792 / 2.72		vs H$_2$O; sl EtOH
1880	Sodium chromate tetrahydrate / Na$_2$CrO$_4$•4H$_2$O	10034-82-9 / 234.035	yel hyg cry	dec /		vs H$_2$O; sl EtOH
1881	Sodium citrate dihydrate / Na$_3$C$_6$H$_5$O$_7$•2H$_2$O	6132-04-3 / 294.101	wh cry	dec 150 /		vs H$_2$O; i EtOH, eth
1882	Sodium cyanate / NaOCN	917-61-3 / 65.007	col needles	550 / 1.89	a	s H$_2$O; sl EtOH; i eth
1883	Sodium cyanide / NaCN	143-33-9 / 49.008	wh cub cry; hyg	563 / 1.6	a,e	vs H$_2$O; sl EtOH
1884	Sodium cyanoborohydride / NaBH$_3$(CN)	25895-60-7 / 62.843	wh hyg powder	240 dec / 1.12		vs H$_2$O; s thf; sl EtOH; i bz, eth
1885	Sodium dichromate / Na$_2$Cr$_2$O$_7$	10588-01-9 / 261.968	red hyg cry	357 /	dec 400	vs H$_2$O
1886	Sodium dihydrogen phosphate monohydrate / NaH$_2$PO$_4$•H$_2$O	10049-21-5 / 137.993	wh hyg cry	dec 100 /		vs H$_2$O; i EtOH
1887	Sodium dihydrogen phosphate dihydrate / NaH$_2$PO$_4$•2H$_2$O	13472-35-0 / 156.008	col orth cry	60 dec / 1.91		vs H$_2$O; i EtOH
1888	Sodium dihydrogen pyrophosphate / Na$_2$H$_2$P$_2$O$_7$	7758-16-9 / 221.939	wh powder	dec 220 / ≈1.9		s H$_2$O
1889	Sodium dithionate dihydrate / Na$_2$S$_2$O$_6$•2H$_2$O	7631-94-9* / 242.139	col orth cry	dec 110 / 2.19		s H$_2$O; i EtOH
1890	Sodium ferricyanide monohydrate / Na$_3$Fe(CN)$_6$•H$_2$O	14217-21-1* / 298.935	red hyg cry			s H$_2$O; i EtOH
1891	Sodium ferrocyanide decahydrate / Na$_4$Fe(CN)$_6$•10H$_2$O	13601-19-9 / 484.063	yel monocl cry	dec ≈50 / 1.46		s H$_2$O; i os
1892	Sodium fluoride / NaF	7681-49-4 / 41.988	col cub or tetr cry	996 / 2.78	1704 / a,c,e,f	sl H$_2$O; i EtOH
1893	Sodium fluoroborate / NaBF$_4$	13755-29-8 / 109.795	wh orth prisms	384 / 2.47	a	vs H$_2$O; sl EtOH
1894	Sodium formate / NaCHO$_2$	141-53-7 / 68.008	wh hyg cry	253 / 1.92	dec / a	vs H$_2$O; sl EtOH
1895	Sodium hexafluoroaluminate / Na$_3$AlF$_6$	13775-53-6 / 209.941	col monocl cry; trans cub 560	1009 / 2.97	f	i H$_2$O

No.	Name Formula	CAS RN Mol. Wt.	Physical Form	mp/°C den/g cm^{-3}	bp/°C Other Data	Solubility
1896	Sodium hexafluorosilicate Na_2SiF_6	16893-85-9 188.056	wh hex cry	dec 2.7	a	sl H_2O; i EtOH
1897	Sodium hydride NaH	7646-69-7 23.998	silv cub cry; flam	425 dec 1.39	a	reac H_2O, EtOH
1898	Sodium hydrogen arsenate heptahydrate $Na_2HAsO_4 \cdot 7H_2O$	10048-95-0 312.014	wh monocl cry	dec ≈50 1.87		vs H_2O; sl EtOH
1899	Sodium hydrogen carbonate $NaHCO_3$	144-55-8 84.007	wh monocl cry	dec ≈50 2.20	a	s H_2O; i EtOH
1900	Sodium hydrogen fluoride $NaHF_2$	1333-83-1 61.995	wh hex cry	dec >160 2.08	a	s H_2O
1901	Sodium hydrogen phosphate Na_2HPO_4	7558-79-4 141.959	wh hyg powder	1.7	a	s H_2O
1902	Sodium hydrogen phosphate dodecahydrate $Na_2HPO_4 \cdot 12H_2O$	10039-32-4 358.143	col cry	dec ≈35 ≈1.5		s H_2O; i EtOH
1903	Sodium hydrogen phosphate heptahydrate $Na_2HPO_4 \cdot 7H_2O$	7782-85-6 268.066	col cry	≈1.7		s H_2O; i EtOH
1904	Sodium hydrogen sulfate $NaHSO_4$	7681-38-1 120.062	wh hyg cry	≈315 2.43	a	vs H_2O
1905	Sodium hydrogen sulfate monohydrate $NaHSO_4 \cdot H_2O$	10034-88-5 138.077	wh monocl cry	2.10		vs H_2O; reac EtOH
1906	Sodium hydrogen sulfide NaHS	16721-80-5 56.064	col rhomb cry	350 1.79		s H_2O, EtOh, eth
1907	Sodium hydrogen sulfide dihydrate $NaHS \cdot 2H_2O$	16721-80-5 92.095	yel hyg needles	dec 55		vs H_2O, EtOH, eth
1908	Sodium hydrogen sulfite $NaHSO_3$	7631-90-5 104.062	wh cry	1.48		s H_2O; sl EtOH
1909	Sodium hydroxide NaOH	1310-73-2 39.997	wh orth cry; hyg	323 2.13	1388 a,c,d,e	vs H_2O; s EtOH, MeOH
1910	Sodium hypochlorite NaClO	7681-52-9 74.442	stable in aq soln	anh form exp		vs H_2O
1911	Sodium hypochlorite pentahydrate $NaOCl \cdot 5H_2O$	7681-52-9* 164.518	pale grn orth cry	18 1.6		s H_2O
1912	Sodium iodate $NaIO_3$	7681-55-2 197.892	wh orth cry	dec 4.28	a	s H_2O; i EtOH
1913	Sodium iodide NaI	7681-82-5 149.894	wh cub cry; hyg	660 3.67	1304 a,c,e	vs H_2O; s EtOH, ace
1914	Sodium bismuthate $NaBiO_3$	12232-99-4 279.968	yel-brn hyg cry			i cold H_2O, reac acid
1915	Sodium metabisulfite $Na_2S_2O_5$	7681-57-4 190.109	wh cry			vs H_2O; sl EtOH
1916	Sodium metaborate $NaBO_2$	7775-19-1 65.800	wh hex cry	966 2.46	1434 a,c	s H_2O
1917	Sodium metasilicate Na_2SiO_3	6834-92-0 122.064	wh amorp solid; hyg	1089 2.61	a,c	s cold H_2O; reac hot H_2O
1918	Sodium molybdate Na_2MoO_4	7631-95-0 205.92	col cub cry	687 ≈3.5	a	vs H_2O
1919	Sodium molybdate dihydrate $Na_2MoO_4 \cdot 2H_2O$	10102-40-6 241.95	cry powder	dec 100 ≈3.5		vs H_2O
1920	Sodium niobate $NaNbO_3$	12034-09-2 163.894	rhomb cry	1422 4.55		i H_2O
1921	Sodium nitrate $NaNO_3$	7631-99-4 84.995	col hex cry; hyg	307 2.26	a,c	vs H_2O; sl EtOH, MeOH
1922	Sodium nitrite $NaNO_2$	7632-00-0 68.996	wh orth cry; hyg	271 2.17	dec >320 a	vs H_2O; sl EtOH; reac acid
1923	Sodium oxalate $Na_2C_2O_4$	62-76-0 134.000	wh powder	dec ≈250 2.34	a	sl H_2O; i EtOH
1924	Sodium oxide Na_2O	1313-59-3 61.979	wh amorp powder	1132 dec 2.27	a,c	reac H_2O
1925	Sodium perchlorate $NaClO_4$	7601-89-0 122.441	wh orth cry; hyg	480 dec 2.52	a	vs H_2O
1926	Sodium perchlorate monohydrate $NaClO_4 \cdot H_2O$	7791-07-3 140.456	wh hyg cry	dec ≈130 2.02		vs H_2O
1927	Sodium periodate $NaIO_4$	7790-28-5 213.892	wh tetr cry	dec ≈300 3.86	a	s H_2O, acid
1928	Sodium periodate trihydrate $NaIO_4 \cdot 3H_2O$	13472-31-6 267.938	wh hex cry	dec 175 3.22		s H_2O
1929	Sodium permanganate trihydrate $NaMnO_4 \cdot 3H_2O$	10101-50-5* 195.972	red-blk hyg cry	dec 170 2.47		vs H_2O; reac EtOH
1930	Sodium peroxide Na_2O_2	1313-60-6 77.979	yel hyg powder	675 2.805	a	reac H_2O
1931	Sodium persulfate $Na_2S_2O_8$	7775-27-1 238.107	wh hyg cry			vs H_2O; reac EtOH
1932	Sodium phosphate dodecahydrate $Na_3PO_4 \cdot 12H_2O$	7601-54-9 380.124	col hex cry	≈75 1.62		s H_2O; i EtOH
1933	Sodium potassium tartrate tetrahydrate $NaK(C_4H_4O_6) \cdot 4H_2O$	304-59-6 282.221	wh cry	dec ≈70 1.79	anh at 130	vs H_2O; i EtOH

No.	Name / Formula	CAS RN / Mol. Wt.	Physical Form	mp/°C / den/g cm^{-3}	bp/°C / Other Data	Solubility
1934	Sodium pyrophosphate $Na_4P_2O_7$	7722-88-5 265.902	col cry	988 2.53		sl H_2O
1935	Sodium selenate decahydrate $Na_2SeO_4 \cdot 10H_2O$	10102-23-5 369.09	wh cry	1.61		vs H_2O
1936	Sodium selenide Na_2Se	1313-85-5 124.94	amorp solid	>875 2.62		reac H_2O
1937	Sodium selenite Na_2SeO_3	10102-18-8 172.94	wh tetr cry			vs H_2O; i EtOH
1938	Sodium sulfate Na_2SO_4	7757-82-6 142.044	wh orth cry or powder	884 2.7	a,c,f	s H_2O; i EtOH
1939	Sodium sulfate decahydrate $Na_2SO_4 \cdot 10H_2O$	7727-73-3 322.197	col monocl cry	32 dec 1.46	f	s H_2O; i EtOH
1940	Sodium sulfide Na_2S	1313-82-2 78.046	wh cub cry; hyg	1172 1.856	a,c	s H_2O; sl EtOH; i eth
1941	Sodium sulfide nonahydrate $Na_2S \cdot 9H_2O$	1313-84-4 240.184	wh-yel hyg cry	dec ≈50 1.43		vs H_2O; sl EtOH; i eth
1942	Sodium sulfide pentahydrate $Na_2S \cdot 5H_2O$	1313-83-3 168.122	col orth cry	120 dec 1.58		vs H_2O; s EtOH; i eth
1943	Sodium sulfite Na_2SO_3	7757-83-7 126.044	wh hex cry	dec 2.63	a	s H_2O; i EtOH
1944	Sodium sulfite heptahydrate $Na_2SO_3 \cdot 7H_2O$	10102-15-5 252.151	wh monocl cry; unstable	1.56		vs H_2O; sl EtOH
1945	Sodium superoxide NaO_2	12034-12-7 54.989	yel cub cry	552 2.2	a	reac H_2O
1946	Sodium tellurate Na_2TeO_4	10102-20-2 237.58	wh powder			sl H_2O
1947	Sodium tetraborate $Na_2B_4O_7$	1330-43-4 201.220	col gl solid; hyg	743 2.4	1575 a	sl H_2O, MeOH
1948	Sodium tetraborate decahydrate $Na_2B_4O_7 \cdot 10H_2O$	1303-96-4 381.373	wh monocl cry	75 dec 1.73	f	sl H_2O; i EtOH
1949	Sodium tetraborate pentahydrate $Na_2B_4O_7 \cdot 5H_2O$	12045-88-4 291.296	hex cry	dec 1.88		sl H_2O
1950	Sodium tetraborate tetrahydrate $Na_2B_4O_7 \cdot 4H_2O$	12045-87-3 273.281	wh monocl cry	1.95	f	sl H_2O
1951	Sodium tetrachloroaluminate $NaAlCl_4$	7784-16-9 191.783	orth cry	2.01		s H_2O
1952	Sodium tetrafluoroberyllate Na_2BeF_4	13871-27-7 130.986	orth cry	575 2.47		sl H_2O
1953	Sodium thiocyanate NaSCN	540-72-7 81.074	col hyg cry	287		vs H_2O
1954	Sodium thiosulfate pentahydrate $Na_2S_2O_3 \cdot 5H_2O$	10102-17-7 248.186	col cry	dec ≈50 1.69		vs H_2O; i EtOH
1955	Sodium tungstate dihydrate $Na_2WO_4 \cdot 2H_2O$	10213-10-2 329.85	wh orth cry	dec 100 3.25		vs H_2O; i EtOH
1956	Strontium Sr	7440-24-6 87.62	silv-wh metal; cub	777 2.64	1382 a,c,e	reac H_2O; s EtOH
1957	Strontium bromate monohydrate $Sr(BrO_3)_2 \cdot H_2O$	14519-18-7 361.44	wh-yel hyg cry	dec 240 3.77		s H_2O
1958	Strontium bromide $SrBr_2$	10476-81-0 247.43	wh tetr cry	657 4.216	a,c	vs H_2O
1959	Strontium bromide hexahydrate $SrBr_2 \cdot 6H_2O$	7789-53-9 355.52	col hyg cry	88 dec		vs H_2O; s EtOH; i eth
1960	Strontium carbide SrC_2	12071-29-3 111.64	tetr cry	3.19		
1961	Strontium carbonate $SrCO_3$	1633-05-2 147.63	wh orth cry; hyg	1494 3.5	a,f	i H_2O; s dil acid
1962	Strontium chlorate $Sr(ClO_3)_2$	7791-10-8 254.52	col cry	120 dec 3.15		vs H_2O; sl EtOH
1963	Strontium chloride $SrCl_2$	10476-85-4 158.53	wh cub cry; hyg	874 3.052	1250 a,c	vs H_2O
1964	Strontium chloride hexahydrate $SrCl_2 \cdot 6H_2O$	10025-70-4 266.62	col hyg cry	dec 100 1.96		vs H_2O; s EtOH
1965	Strontium chromate $SrCrO_4$	7789-06-2 203.61	yel monocl cry	dec 3.9		sl H_2O; s dil acid
1966	Strontium fluoride SrF_2	7783-48-4 125.62	wh cub cry or powder	1477 4.24	2460 a,c	s H_2O, dil acid
1967	Strontium hydride SrH_2	13598-33-9 89.64	orth cry	1050 3.26	a	reac H_2O
1968	Strontium hydroxide $Sr(OH)_2$	18480-07-4 121.64	col orth cry; hyg	535 3.625	dec 710 a,c	s H_2O
1969	Strontium iodide SrI_2	10476-86-5 341.43	wh hyg cry	538 4.55	1773 dec a,c	vs H_2O
1970	Strontium iodide hexahydrate $SrI_2 \cdot 6H_2O$	73796-25-5 449.52	wh-yel hex cry; hyg	dec 120 4.4		vs H_2O; s EtOH
1971	Strontium niobate $SrNb_2O_6$	12034-89-8 369.43	monocl cry	1225 5.11		i H_2O
1972	Strontium nitrate $Sr(NO_3)_2$	10042-76-9 211.63	wh cub cry	570 2.99	645 a	vs H_2O; sl EtOH, ace

PHYSICAL CONSTANTS OF INORGANIC COMPOUNDS (continued)

No.	Name / Formula	CAS RN / Mol. Wt.	Physical Form	mp/°C / den/g cm⁻³	bp/°C / Other Data	Solubility
1973	Strontium nitrite / Sr(NO₂)₂	13470-06-9 / 179.63	wh-yel hyg needles	dec 240 / 2.8	a	vs H_2O
1974	Strontium oxide / SrO	1314-11-0 / 103.62	col cub cry	2532 / 5.1	a,c,e	reac H_2O
1975	Strontium peroxide / SrO₂	1314-18-7 / 119.62	wh tetr cry; unstable	215 dec / 4.78		reac H_2O
1976	Strontium phosphate / Sr₃(PO₄)₂	7446-28-8 / 452.80	wh powder			i H_2O; s acid
1977	Strontium selenate / SrSeO₄	7446-21-1 / 230.58	orth cry	/ 4.25		i H_2O; s hot HCl
1978	Strontium selenide / SrSe	1315-07-7 / 166.58	wh cub cry	1600 / 4.54	a	
1979	Strontium orthosilicate / Sr₂SiO₄	13597-55-2 / 267.32	orth cry	/ 4.5	a	
1980	Strontium silicide / SrSi₂	12138-28-2 / 143.79	silv-gray cub cry	1100 / 3.35		
1981	Strontium sulfate / SrSO₄	7759-02-6 / 183.68	wh orth cry	1607 / 3.96	a,f	i H_2O, EtOH; sl acid
1982	Strontium sulfide / SrS	1314-96-1 / 119.69	gray cub cry	2227 / 3.70	a	sl H_2O; s acid
1983	Strontium titanate / SrTiO₃	12060-59-2 / 183.49	wh cub cry	2080 / 5.1		i H_2O
1984	Sulfur (α form) / S	7704-34-9 / 32.066	yel orth cry	trans to β form 95.3 / 2.07	444.60 / a,b,c,d,e,f	i H_2O; sl EtOH, bz, eth; s CS_2
1985	Sulfur (β form) / S	7704-34-9 / 32.066	yel monocl needles, stable 94.5-115	115.21 / 2.07	444.60 / a,b,c,d,e,f	i H_2O; sl EtOH, bz, eth; s CS_2
1986	Sulfuric acid / H₂SO₄	7664-93-9 / 98.080	col oily liq	10.31 / 1.8	337 / a,c,e	vs H_2O
1987	Peroxysulfuric acid / H₂SO₅	7722-86-3 / 114.079	wh cry; unstable	dec 45		vs H_2O
1988	Chlorosulfuric acid / SO₂(OH)Cl	7790-94-5 / 116.525	col-yel liq	-80 / 1.75	152	reac H_2O; s py
1989	Fluorosulfonic acid / SO₂(OH)F	7789-21-1 / 100.070	col liq	-89 / 1.726	163	reac H_2O
1990	Nitrosylsulfuric acid / HNOSO₄	7782-78-7 / 127.078	prisms	dec 73		reac H_2O; s H_2SO_4
1991	Sulfurous acid / H₂SO₃	7782-99-2 / 82.080	exists only in soln			soln of SO_2 in H_2O
1992	Sulfur dioxide / SO₂	7446-09-5 / 64.065	col gas	-75.5 / 2.619 g/L	-10.05 / a,b,d,e	s H_2O, EtOh, eth, chl
1993	Sulfur trioxide / SO₃	7446-11-9 / 80.064	col liq	16.8 / 1.92	45 / a,b,d,e	reac H_2O
1994	Sulfur bromide / SSBr₂	13172-31-1 / 223.940	red oily liq	-46 / 2.63	dec >25 / a	reac H_2O
1995	Sulfur chloride / SSCl₂	10025-67-9 / 135.037	yel-red oily liq	-77 / 1.69	137 / a,e	reac H_2O; s EtOH, bz, eth, ctc
1996	Sulfur fluoride / SSF₂	16860-99-4 / 102.129	col gas	-164.6 / 4.174 g/L	-10.6	reac H_2O
1997	Sulfur fluoride / FSSF	13709-35-8 / 102.129	col gas	-133 / 4.174 g/L	15	reac H_2O
1998	Sulfur dichloride / SCl₂	10545-99-0 / 102.971	red visc liq	-122 / 1.62	59.6 / a	reac H_2O
1999	Sulfur tetrafluoride / SF₄	7783-60-0 / 108.060	col gas	-125 / 4.417 g/L	-40.45 / a,b,d	reac H_2O
2000	Sulfur hexafluoride / SF₆	2551-62-4 / 146.056	col gas	-50.7 tp / 5.970 g/L	-63.8 sp / a,b,c,d,e	sl H_2O; s EtOH
2001	Sulfur bromide pentafluoride / SF₅Br	15607-89-3 / 206.962	col gas	-79 / 8.459 g/L	3.1	
2002	Sulfur chloride pentafluoride / SF₅Cl	13780-57-9 / 162.511	col gas	-64 / 6.642 g/L	-19.05 / a,b	
2003	Sulfur decafluoride / S₂F₁₀	5714-22-7 / 254.116	liq	-52.7 / 2.08	30; dec 150	i H_2O
2004	Sulfuryl amide / SO₂(NH₂)₂	7803-58-9 / 96.110	orth plates	93	dec 250	s H_2O, hot EtOH, ace
2005	Sulfuryl chloride / SO₂Cl₂	7791-25-5 / 134.970	col liq	-51 / 1.680	69.4 / a,d,e	reac H_2O; s bz, tol, eth
2006	Sulfuryl fluoride / SO₂F₂	2699-79-8 / 102.062	col gas	-135.8 / 4.172 g/L	-55.4 / a	sl H_2O, EtOH; s tol, ctc
2007	Thionyl bromide / SOBr₂	507-16-4 / 207.873	yel liq	-50	140 / e	reac H_2O
2008	Thionyl chloride / SOCl₂	7719-09-7 / 118.970	yel fuming liq	-101 / 1.631	75.6 / a,d,e	reac H_2O; s bz, ctc, chl
2009	Thionyl fluoride / SOF₂	7783-42-8 / 86.062	col gas	-129.5 / 3.518 g/L	-43.8 / a,d	reac H_2O; s bz, eth
2010	Sulfur fluoride hypofluorite / F₅SOF	15179-32-5 / 162.055	col gas	-86 / 6.624 g/L	-35.1	
2011	Tantalum / Ta	7440-25-7 / 180.948	gray metal; cub	3017 / 16.4	5458 / a,c	reac HF

No.	Name Formula	CAS RN Mol. Wt.	Physical Form	mp/°C den/g cm^{-3}	bp/°C Other Data	Solubility
2012	Tantalum aluminide $TaAl_3$	12004-76-1 261.893	gray refrac powder	≈1400 7.02		i H_2O, acid, alk
2013	Tantalum boride TaB	12007-07-7 191.759	refrac orth cry	2040 14.2		
2014	Tantalum boride TaB_2	12007-35-1 202.570	blk hex cry	3140 11.2		i H_2O, acid, alk
2015	Tantalum carbide TaC	12070-06-3 192.959	gold-brown powder; cub	3880 14.3	4780	s HF-HNO_3 mixture
2016	Tantalum carbide Ta_2C	12070-07-4 373.907	refrac hex cry	3327 15.1		
2017	Tantalum nitride TaN	12033-62-4 194.955	blk hex cry	3090 13.7		i H_2O; sl aqua regia; reac alk
2018	Tantalum silicide $TaSi_2$	12039-79-1 237.119	gray powder	2200 9.14		
2019	Tantalum(IV) oxide TaO_2	12036-14-5 212.947	tetr cry	 10.0		
2020	Tantalum(IV) selenide $TaSe_2$	12039-55-3 338.87	hex cry	 6.7		
2021	Tantalum(IV) sulfide TaS_2	12143-72-5 245.080	blk hex cry	>3000 6.86		i H_2O
2022	Tantalum(IV) telluride $TaTe_2$	12067-66-2 436.15	monocl cry	 9.4		
2023	Tantalum(V) bromide $TaBr_5$	13451-11-1 580.468	yel cry powder	265 4.99	349 a,b,c,d	
2024	Tantalum(V) chloride $TaCl_5$	7721-01-9 358.212	yel monocl cry; hyg	216 3.68	239.35 a,b,c,d	reac H_2O; s EtOH
2025	Tantalum(V) fluoride TaF_5	7783-71-3 275.940	wh monocl cry; hyg	95.1 5.0	229.2 a,d,e	s H_2O, eth; sl CS_2, ctc
2026	Tantalum(V) iodide TaI_5	14693-81-3 815.470	blk hex cry; hyg	496 5.80	543	
2027	Tantalum(V) oxide Ta_2O_5	1314-61-0 441.893	wh rhomb cry or powder	1785 8.2	 a,c	i H_2O, EtOH, acid; s HF
2028	Technetium Tc	7440-26-8 98	hex cry	2157 11	4265 a,c	
2029	Technetium(V) fluoride TcF_5	31052-14-9 193	yel solid	50 dec		
2030	Technetium(VI) fluoride TcF_6	13842-93-8 212	yel cub cry	37.4 3.0	55.3	
2031	Tellurium Te	13494-80-9 127.60	gray-wh rhomb cry	449.51 6.24	988 a,c,d,e	i H_2O, bz, CS_2
2032	Telluric(VI) acid H_6TeO_6	7803-68-1 229.64	wh monocl cry	136 3.07		s H_2O
2033	Tellurous acid H_2TeO_3	10049-23-7 177.61	wh cry	dec 40 3.0		sl H_2O; s dil acid, alk
2034	Tellurium dioxide TeO_2	7446-07-3 159.60	wh orth cry	733 5.9	1245 a	i H_2O; s alk, acid
2035	Tellurium trioxide TeO_3	13451-18-8 175.60	yel-oran cry	430 5.07		i H_2O
2036	Tellurium dibromide $TeBr_2$	7789-54-0 287.41	grn-brn hyg cry	210	339	reac H_2O; s eth; sl chl
2037	Tellurium dichloride $TeCl_2$	10025-71-5 198.51	blk amorp solid; hyg	208 6.9	328	reac H_2O; i ctc
2038	Tellurium tetrabromide $TeBr_4$	10031-27-3 447.22	yel-oran monocl cry	388 4.3	≈420 dec a	reac H_2O; s eth
2039	Tellurium tetrachloride $TeCl_4$	10026-07-0 269.41	wh monocl cry; hyg	224 3.0	387 a,b,d,e	reac H_2O; s EtOH, tol
2040	Tellurium tetrafluoride TeF_4	15192-26-4 203.59	col cry	129	dec 195	reac H_2O
2041	Tellurium tetraiodide TeI_4	7790-48-9 635.22	blk orth cry	280 5.05		reac H_2O; sl ace
2042	Tellurium hexafluoride TeF_6	7783-80-4 241.59	col gas	-37.6 tp 9.875 g/L	-38.9 sp a,b,e	reac H_2O
2043	Terbium Tb	7440-27-9 158.925	silv metal; hex	1356 8.23	3230 a,c	
2044	Terbium chloride $TbCl_3$	10042-88-3 265.283	wh orth cry; hyg	588 4.35	 a	s H_2O
2045	Terbium chloride hexahydrate $TbCl_3 \cdot 6H_2O$	13798-24-8 373.374	hyg cry	 4.35		vs H_2O
2046	Terbium iodide TbI_3	13813-40-6 539.638	hex cry; hyg	957 ≈5.2		s H_2O
2047	Terbium nitride TbN	12033-64-6 172.932	cub cry	 9.55		
2048	Terbium oxide Tb_2O_3	12036-41-8 365.849	wh cub cry	2410 7.91	 a	
2049	Terbium silicide $TbSi_2$	12039-80-4 215.096	orth cry	 6.66		
2050	Terbium sulfide Tb_2S_3	12138-11-3 414.049	cub cry	 6.35		

PHYSICAL CONSTANTS OF INORGANIC COMPOUNDS (continued)

No.	Name / Formula	CAS RN / Mol. Wt.	Physical Form	mp/°C / den/g cm⁻³	bp/°C / Other Data	Solubility
2051	Thallium / Tl	7440-28-0 / 204.383	soft blue-wh metal	304 / 11.8	1473 / a,c,e	i H₂O; reac acid
2052	Thallium(I) acetate / TlC₂H₃O₂	563-68-8 / 263.428	hyg wh cry	131 / 3.68		s H₂O, EtOH
2053	Thallium(I) bromide / TlBr	7789-40-4 / 284.287	yel cub cry	460 / 7.5	819 / a,c,d,e	i H₂O
2054	Thallium(I) carbonate / Tl₂CO₃	6533-73-9 / 468.776	wh monocl cry	272 / 7.11	a,c	sl H₂O; i EtOH
2055	Thallium(I) chlorate / TlClO₃	13453-30-0 / 287.834	col hex cry	5.5		sl H₂O
2056	Thallium(I) chloride / TlCl	7791-12-0 / 239.836	wh cub cry	430 / 7.0	720 / a,c,d,e	sl H₂O; i EtOH
2057	Thallium(I) cyanide / TlCN	13453-34-4 / 230.401	wh hex plates	6.523		s H₂O, acid, EtOH
2058	Thallium(I) ethoxide / TlC₂H₅O	20398-06-5 / 249.444	cloudy liq	-3 / 3.49	dec 130	reac H₂O
2059	Thallium(I) fluoride / TlF	7789-27-7 / 223.381	wh orth cry	326 / 8.36	826 / a,c	s H₂O
2060	Thallium(I) formate / TlCHO₂	992-98-3 / 249.401	hyg col needles	101 / 4.97		vs H₂O; s MeOH
2061	Thallium(I) hexafluorophosphate / TlPF₆	60969-19-9 / 349.347	wh cub cry	4.6		
2062	Thallium(I) hydroxide / TlOH	12026-06-1 / 221.390	yel needles	139 dec / 7.44	a	vs H₂O
2063	Thallium(I) iodide / TlI	7790-30-9 / 331.287	yel cry powder	441.8 / 7.1	824 / a,c,d,e	i H₂O, EtOH
2064	Thallium(I) molybdate / Tl₂MoO₄	34128-09-1 / 568.71	yel-wh cub cry			i H₂O
2065	Thallium(I) nitrate / TlNO₃	10102-45-1 / 266.388	wh cry	206 / 5.55	450 dec / a,c	s H₂O; i EtOH
2066	Thallium(I) nitrite / TlNO₂	13826-63-6 / 250.389	cub cry	5.7		vs H₂O
2067	Thallium(I) oxalate / Tl₂C₂O₄	30737-24-7 / 496.787	wh powder	6.31		
2068	Thallium(I) oxide / Tl₂O	1314-12-1 / 424.766	blk rhomb cry; hyg	579 / 9.52	≈1080 / a	s H₂O, EtOH
2069	Thallium(I) perchlorate / TlClO₄	13453-40-2 / 303.834	col orth cry	4.8		s H₂O
2070	Thallium(I) selenate / Tl₂SeO₄	7446-22-2 / 551.73	orth cry	>400 / 6.875		sl H₂O; i EtOH, eth
2071	Thallium(I) selenide / Tl₂Se	15572-25-5 / 487.73	gray plates	340	a	i H₂O, acid
2072	Thallium(I) sulfate / Tl₂SO₄	7446-18-6 / 504.831	wh rhomb prisms	632 / 6.77	a,c	sl H₂O
2073	Thallium(I) sulfide / Tl₂S	1314-97-2 / 440.833	blue-blk cry	448 / 8.39	1367 / a,c,d	sl H₂O, alk; s acid
2074	Thallium(III) bromide tetrahydrate / TlBr₃·4H₂O	13701-90-1 / 516.157	yel orth cry	3.65		s H₂O, EtOH
2075	Thallium(III) chloride / TlCl₃	13453-32-2 / 310.741	monocl cry	155 / 4.7	a	vs H₂O, EtOH, eth
2076	Thallium(III) chloride tetrahydrate / TlCl₃·4H₂O	13453-32-2* / 382.803	orth cry	3.00		s H₂O
2077	Thallium(III) fluoride / TlF₃	7783-57-5 / 261.378	wh orth cry; hyg	550 dec / 8.65		reac H₂O
2078	Thallium(III) nitrate / Tl(NO₃)₃	13746-98-0 / 390.398	col cry			reac H₂O
2079	Thallium(III) oxide / Tl₂O₃	1314-32-5 / 456.765	brn cub cry	834 / 10.2		i H₂O; reac acid
2080	Thorium / Th	7440-29-1 / 232.038	soft gray-wh metal; cub	1750 / 11.7	4788 / a,c	s acids
2081	Thorium hydride / ThH₂	16689-88-6 / 234.054	tetr cry	9.5	a	
2082	Thorium boride / ThB₆	12229-63-9 / 296.904	refrac solid	2450 / 6.99		
2083	Thorium bromide / ThBr₄	13453-49-1 / 551.654	wh hyg cry	679		reac H₂O
2084	Thorium carbide / ThC	12012-16-7 / 244.049	cub cry	2500 / 10.6		reac H₂O
2085	Thorium dicarbide / ThC₂	12071-31-7 / 256.060	yel monocl cry	≈2650 / 9.0		reac H₂O
2086	Thorium chloride / ThCl₄	10026-08-1 / 373.849	gray-wh tetr needles; hyg	770 / 4.59	921 / a,c,d	s H₂O, EtOH
2087	Thorium fluoride / ThF₄	13709-59-6 / 308.032	wh monocl cry; hyg	1110 / 6.1	1680 / a,d	
2088	Thorium iodide / ThI₄	7790-49-0 / 739.656	wh-yel monocl cry	570	837	
2089	Thorium nitrate tetrahydrate / Th(NO₃)₄·4H₂O	33088-16-3 / 552.119	wh hyg cry	dec 500		vs H₂O; s EtOH

No.	Name Formula	CAS RN Mol. Wt.	Physical Form	mp/°C den/g cm^{-3}	bp/°C Other Data	Solubility
2090	Thorium nitride ThN	12033-65-7 246.045	refrac cub cry	2820 11.6		reac H_2O
2091	Thorium oxide ThO_2	1314-20-1 264.037	wh cub cry	3390 10.0	4400 a,f	i H_2O, alk; sl acid
2092	Thorium selenide $ThSe_2$	60763-24-8 389.96	orth cry	 8.5		
2093	Thorium orthosilicate $ThSiO_4$	14553-44-7 324.122	brn tetr cry	 6.7	f	
2094	Thorium silicide $ThSi_2$	12067-54-8 288.209	tetr cry	1850 7.9		
2095	Thorium sulfate nonahydrate $Th(SO_4)_2 \cdot 9H_2O$	10381-37-0 586.303	wh monocl cry	dec 2.8		s H_2O
2096	Thorium sulfide ThS_2	12138-07-7 296.170	dark brn cry	1905 7.30		i H_2O; s acid
2097	Thulium Tm	7440-30-4 168.934	silv metal; hex	1545 9.32	1950 a,c	s dil acid
2098	Thulium bromide $TmBr_3$	14456-51-0 408.646	wh hyg cry	954		s H_2O
2099	Thulium chloride $TmCl_3$	13537-18-3 275.292	yel hyg cry	824	 a	s H_2O
2100	Thulium chloride heptahydrate $TmCl_3 \cdot 7H_2O$	13778-39-7 401.399	hyg cry			s H_2O, EtOH
2101	Thulium fluoride TmF_3	13760-79-7 225.929	wh cry	1158		s H_2O
2102	Thulium iodide TmI_3	13813-43-9 549.647	yel hyg cry	1021		
2103	Thulium oxide Tm_2O_3	12036-44-1 385.866	grn-wh cub cry	2425 8.6	 a	sl acid
2104	Tin (gray) Sn	7440-31-5 118.710	cub cry	trans to wh Sn 13.2 5.769	2602 a,c,e	
2105	Tin (white) Sn	7440-31-5 118.710	silv tetr cry	231.93 7.265	2602 a,c,e	
2106	Stannane SnH_4	2406-52-2 122.742	unstable col gas	-146 5.017 g/L	-51.8 a,d,e	
2107	Methylstannane SnH_3CH_3	1631-78-3 136.769	col gas	 5.590 g/L	0	reac H_2O
2108	Tin(II) acetate $Sn(C_2H_3O_2)_2$	638-39-1 236.800	wh orth cry	183 2.31	subl	s dil HCl
2109	Tin(II) bromide $SnBr_2$	10031-24-0 278.518	yel powder	215 5.12	639 a,d	s H_2O, EtOH, eth, ace
2110	Tin(II) chloride $SnCl_2$	7772-99-8 189.615	wh orth cry	247 3.90	623 a,c,d,e	s H_2O, EtOH, ace, eth; i xyl
2111	Tin(II) chloride dihydrate $SnCl_2 \cdot 2H_2O$	10025-69-1 225.646	wh monocl cry	37 dec 2.71		s H_2O, EtOH, NaOH; vs HCl
2112	Tin(II) fluoride SnF_2	7783-47-3 156.707	wh monocl cry; hyg	213 4.57	850	s H_2O; i EtOH, eth, chl
2113	Tin(II) hexafluorozirconate $SnZrF_6$	12419-43-1 323.924	cry	 4.21		s H_2O
2114	Tin(II) hydroxide $Sn(OH)_2$	12026-24-3 152.725	wh amorp solid	 a		
2115	Tin(II) iodide SnI_2	10294-70-9 372.519	red-oran powder	320 5.28	714 a,d	sl H_2O; s bz, chl, CS_2
2116	Tin(II) oxalate SnC_2O_4	814-94-8 206.730	wh powder	280 dec 3.56		i H_2O; s dil HCl
2117	Tin(II) oxide SnO	21651-19-4 134.709	blue-blk tetr cry	1080 dec 6.45	 a	i H_2O, EtOH; s acid
2118	Tin(II) pyrophosphate $Sn_2P_2O_7$	15578-26-4 411.363	wh amorp powder	dec 400 4.009		i H_2O; s conc acid
2119	Tin(II) selenide SnSe	1315-06-6 197.67	gray orth cry	861 6.18		i H_2O; s aqua regia
2120	Tin(II) sulfate $SnSO_4$	7488-55-3 214.774	wh orth cry	dec 378 4.15		reac H_2O
2121	Tin(II) sulfide SnS	1314-95-0 150.776	gray orth cry	880 5.08	1210 a	i H_2O; s conc acid
2122	Tin(II) tartrate $SnC_4H_4O_6$	815-85-0 266.782	wh cry powder			s H_2), dil HCl
2123	Tin(II) telluride SnTe	12040-02-7 246.31	gray cub cry	790 6.5		
2124	Tin(IV) bromide $SnBr_4$	7789-67-5 438.326	wh cry	31 3.34	205 a,b,c,d,e	vs H_2O; s EtOH
2125	Tin(IV) chloride $SnCl_4$	7646-78-8 260.521	col fuming liq	-33 2.234	114.15 a,b,c,d,e	reac H_2O; s EtOH, ctc, bz, ace
2126	Tin(IV) chloride pentahydrate $SnCl_4 \cdot 5H_2O$	10026-06-9 350.597	wh-yel cry	dec 56 2.04		vs H_2O; s EtOH
2127	Tin(IV) chromate $Sn(CrO_4)_2$	38455-77-5 350.697	brn-yel cry powder	dec		s H_2O
2128	Tin(IV) fluoride SnF_4	7783-62-2 194.704	wh tetr cry	 4.78	subl 705	reac H_2O

No.	Name / Formula	CAS RN / Mol. Wt.	Physical Form	mp/°C / den/g cm⁻³	bp/°C / Other Data	Solubility
2129	Tin(IV) iodide / SnI₄	7790-47-8 / 626.328	yel-brn cub cry	143 / 4.46	364.35 / a,b,d,e	reac H₂O; s EtOH, bz, chl, eth
2130	Tin(IV) oxide / SnO₂	18282-10-5 / 150.709	gray tetr cry	1630 / 6.85	a,f	i H₂O, EtOH; s hot conc alk
2131	Tin(IV) selenide / SnSe₂	20770-09-6 / 276.63	red-brn cry	650 / ≈5.0		i H₂O; s alk, conc acid
2132	Tin(IV) selenite / Sn(SeO₃)₂	7446-25-5 / 372.63	cry powder			i H₂O; s hot HCl
2133	Tin(IV) sulfide / SnS₂	1315-01-1 / 182.842	gold-yel hex cry	dec 600 / 4.5		i H₂O; s alk, aqua regia
2134	Titanium / Ti	7440-32-6 / 47.867	gray metal; hex	1668 / 4.506	3287 / a,c	
2135	Titanium hydride / TiH₂	7704-98-5 / 49.883	gray-blk powder	dec ≈450 / 3.75		i H₂O
2136	Titanium boride / TiB₂	12045-63-5 / 69.489	gray refrac solid; hex	3225 / 4.38		
2137	Titanium carbide / TiC	12070-08-5 / 59.878	cub cry	3067 / 4.93		i H₂O; s HNO₃
2138	Titanium nitride / TiN	25583-20-4 / 61.874	yel-brn cub cry	2950 / 5.21²⁵		i H₂O; s aqua regia
2139	Titanium phosphide / TiP	12037-65-9 / 78.841	gray hex cry	1990 / 4.08		
2140	Titanium silicide / TiSi₂	12039-83-7 / 104.038	blk orth cry	1500 / 4.0		i H₂O, acid, alk; s HF
2141	Titanium(II) bromide / TiBr₂	13783-04-5 / 207.675	blk powder	/ 4.0	a	reac H₂O
2142	Titanium(II) chloride / TiCl₂	10049-06-6 / 118.772	blk hex cryc	1035 / 3.13	1500 / a,d	reac H₂O; s EtOH; i chl, eth
2143	Titanium(II) iodide / TiI₂	13783-07-8 / 301.676	blk hex cry	/ 5.02		reac H₂O
2144	Titanium(II) oxide / TiO	12137-20-1 / 63.866	cub cry	1750 / 4.95	a	
2145	Titanium(II) sulfide / TiS	12039-07-5 / 79.933	brn hex cry	1780 / 3.85		s conc acid
2146	Titanium(III) bromide / TiBr₃	13135-31-4 / 287.579	blue-blk hex cry		a	s H₂O
2147	Titanium(III) chloride / TiCl₃	7705-07-9 / 154.225	red-viol hex cry; hyg	425 dec / 2.64	a,d	reac H₂O
2148	Titanium(III) fluoride / TiF₃	13470-08-1 / 104.862	viol hex cry	1200 / 2.98	1400	i H₂O, dil acid, alk
2149	Titanium(III) oxide / Ti₂O₃	1344-54-3 / 143.732	viol hex cry	1842 / 4.486	a	s hot HF
2150	Titanium(III) sulfate / Ti₂(SO₄)₃	10343-61-0 / 383.925	grn cry			i H₂O, EtOH; s dil HCl
2151	Titanium(III) sulfide / Ti₂S₃	12039-16-6 / 191.932	blk hex cry	/ 3.56		
2152	Titanium(III,IV) oxide / Ti₃O₅	12065-65-5 / 223.598	blk monocl cry	1777 / 4.24	a	
2153	Titanium(IV) bromide / TiBr₄	7789-68-6 / 367.483	yel-oran cub cry; hyg	39 / 3.37	230 / a,b,c,d	reac H₂O
2154	Titanium(IV) chloride / TiCl₄	7550-45-0 / 189.678	col or yel liq	-25 / 1.73	136.45 / a,b,c,d,e	reac H₂O; s EtOH
2155	Titanium(IV) fluoride / TiF₄	7783-63-3 / 123.861	wh hyg powder	284 / 2.798	subl	reac H₂O; s EtOH, py
2156	Titanium(IV) iodide / TiI₄	7720-83-4 / 555.485	red hyg powder	150 / 4.3	377 / a,b,c,d	reac H₂O
2157	Titanium(IV) oxide / TiO₂	13463-67-7 / 79.866	wh tetr cry	1843 / 4.23	a,f	i H₂O, dil acid; s conc acid
2158	Titanium(IV) oxysulfate monohydrate / TiOSO₄·H₂O	13825-74-6* / 177.945	col orth cry	/ 2.71		reac H₂O
2159	Titanium(IV) sulfide / TiS₂	12039-13-3 / 111.999	yel-brn hex cry; hyg	/ 3.37		s H₂SO₄
2160	Tungsten / W	7440-33-7 / 183.84	gray-wh metal; cub	3422 / 19.3	5555 / a,c,e	
2161	Tungstic acid / H₂WO₄	7783-03-1 / 249.85	yel amorp powder	dec 100 / 5.5		i H₂O, acid; s alk
2162	Tungsten boride / W₂B	12007-10-2 / 378.49	refrac blk powder	2670 / 16.0		i H₂O
2163	Tungsten boride / WB	12007-09-9 / 194.65	blk refrac powder	2665 / 15.2		i H₂O
2164	Tungsten boride / W₂B₅	12007-98-6 / 421.74	refrac solid	2365 / 11.0		i H₂O
2165	Tungsten carbide / W₂C	12070-13-2 / 379.69	refrac hex cry	≈2800 / 14.8		i H₂O
2166	Tungsten carbide / WC	12070-12-1 / 195.85	gray hex cry	2785 / 15.6		i H₂O; s HNO₃/HF
2167	Tungsten carbonyl / W(CO)₆	14040-11-0 / 351.90	wh cry	dec 170 / 2.65	subl	i H₂O; s os

No.	Name Formula	CAS RN Mol. Wt.	Physical Form	mp/°C den/g cm^{-3}	bp/°C Other Data	Solubility
2168	Tungsten nitride WN$_2$	60922-26-1 211.85	hex cry	dec 600 7.7		
2169	Tungsten nitride W$_2$N	12033-72-6 381.69	gray cub cry	dec 17.8		
2170	Tungsten silicide WSi$_2$	12039-88-2 240.01	blue-gray tetr cry	2160 9.3		i H$_2$O
2171	Tungsten silicide W$_5$Si$_3$	12039-95-1 1003.46	blue-gray refrac solid	2320 14.4		
2172	Tungsten(II) bromide WBr$_2$	13470-10-5 343.65	yel powder	dec 400		
2173	Tungsten(II) chloride WCl$_2$	13470-12-7 254.75	yel solid	dec >500		s H$_2$O
2174	Tungsten(II) iodide WI$_2$	13470-17-2 437.65	oran cry	6.79		
2175	Tungsten(III) bromide WBr$_3$	15163-24-3 423.55	blk hex cry	dec >80		i H$_2$O
2176	Tungsten(III) chloride WCl$_3$	20193-56-0 290.20	red solid	550 dec	subl	reac H$_2$O
2177	Tungsten(IV) bromide WBr$_4$	14055-81-3 503.46	blk orth cry			reac H$_2$O
2178	Tungsten(IV) chloride WCl$_4$	13470-13-8 325.65	blk hyg powder	4.62		reac H$_2$O
2179	Tungsten(IV) fluoride WF$_4$	13766-47-7 259.83	red-brn cry	dec >800		
2180	Tungsten(IV) iodide WI$_4$	14055-84-6 691.46	blk powder	dec		reac H$_2$O; s EtOH; i eth chl
2181	Tungsten(IV) oxide WO$_2$	12036-22-5 215.84	blue monocl cry	dec 1500-1700 10.8	a	i H$_2$O, os
2182	Tungsten(IV) selenide WSe$_2$	12067-46-8 341.76	gray hex cry	9.2		
2183	Tungsten(IV) sulfide WS$_2$	12138-09-9 247.97	gray hex cry	1250 dec 7.6		i H$_2$O, HCl, alk
2184	Tungsten(IV) telluride WTe$_2$	12067-76-4 439.04	gray orth cry	1020 9.43		
2185	Tungsten(V) bromide WBr$_5$	13470-11-6 583.36	brn-blk hyg solid	286	333	
2186	Tungsten(V) chloride WCl$_5$	13470-14-9 361.10	blk hyg cry	242	286	reac H$_2$O
2187	Tungsten(V) fluoride WF$_5$	19357-83-6 278.83	yel solid	dec >80		
2188	Tungsten(V) oxytribromide WOBr$_3$	20213-56-3 439.55	dark brn tetr cry	≈5.9		
2189	Tungsten(V) oxytrichloride WOCl$_3$	14249-98-0 306.20	grn tetr cry	≈4.6		
2190	Tungsten(VI) bromide WBr$_6$	13701-86-5 663.26	blue-blk cry	309	a	
2191	Tungsten(VI) chloride WCl$_6$	13283-01-7 396.56	purp hex cry; hyg	275 3.52	346.75 a,b,c,d	s EtOH, os
2192	Tungsten(VI) dioxydibromide WO$_2$Br$_2$	13520-75-7 375.65	red cry			
2193	Tungsten(VI) dioxydichloride WO$_2$Cl$_2$	13520-76-8 286.74	yel orth cry	265 4.67		i H$_2$O
2194	Tungsten(VI) dioxydiiodide WO$_2$I$_2$	14447-89-3 469.65	monocl cry	6.39		
2195	Tungsten(VI) fluoride WF$_6$	7783-82-6 297.83	col gas	2.3 12.173 g/L	17 a,b,c,d,e	reac H$_2$O
2196	Tungsten(VI) oxide WO$_3$	1314-35-8 231.84	yel powder	1472 7.2	a,c	i H$_2$O; sl acid; s alk
2197	Tungsten(VI) oxytetrabromide WOBr$_4$	13520-77-9 519.46	red tetr cry	277 ≈5.5	327	reac H$_2$O
2198	Tungsten(VI) oxytetrachloride WOCl$_4$	13520-78-0 341.65	red hyg cry	211 11.92	227.55 b,c,d	reac H$_2$O; s bz, CS$_2$
2199	Tungsten(VI) oxytetrafluoride WOF$_4$	13520-79-1 275.83	wh monocl cry	106 5.07	186	reac H$_2$O
2200	Tungsten(VI) sulfide WS$_3$	12125-19-8 280.04	brn powder			sl H$_2$O; s alk
2201	Uranium U	7440-61-1 238.029	silv-wh orth cry	1135 19.1	4131 a,c	
2202	Uranium boride UB$_2$	12007-36-2 259.651	refrac solid	2430 12.7		
2203	Uranium boride UB$_4$	12007-84-0 281.273	refrac solid	2530 9.32		i H$_2$O
2204	Uranium carbide UC	12070-09-6 250.040	gray cub cry	2790		
2205	Uranium carbide UC$_2$	12071-33-9 262.051	gray tetr cry	2350 11.3	4370	reac H$_2$O; sl EtOH
2206	Uranium carbide U$_2$C$_3$	12076-62-9 512.091	gray cub cry	dec ≈1700 12.7		

No.	Name Formula	CAS RN Mol. Wt.	Physical Form	mp/°C den/g cm^{-3}	bp/°C Other Data	Solubility
2207	Uranium nitride UN	25658-43-9 252.036	gray cub cry	2805 14.3		i H_2O
2208	Uranium nitride U_2N_3	12033-83-9 518.078	cub cry	dec 11.3		
2209	Uranium(III) bromide UBr_3	13470-19-4 477.741	red hyg cry	727		s H_2O
2210	Uranium(III) chloride UCl_3	10025-93-1 344.387	grn hyg cry	837 5.51	a	vs H_2O; i bz, ctc
2211	Uranium(III) fluoride UF_3	13775-06-9 295.024	blk hex cry	dec 8.9	a	i H_2O; s acid
2212	Uranium(III) hydride UH_3	13598-56-6 241.053	gray-blk cub cry	11.1	a	
2213	Uranium(III) iodide UI_3	13775-18-3 618.742	blk hyg cry	766		s H_2O
2214	Uranium(IV) bromide UBr_4	13470-20-7 557.645	brn hyg cry	519		s H_2O, EtOH
2215	Uranium(IV) chloride UCl_4	10026-10-5 379.840	grn octahed cry	590 4.72	791 a,c	reac H_2O; s EtOH
2216	Uranium(IV) fluoride UF_4	10049-14-6 314.023	grn monocl cry	1036 6.7	1417 a	i H_2O; s conc acid, alk
2217	Uranium(IV) iodide UI_4	13470-22-9 745.647	blk hyg cry	506		s H_2O, EtOH
2218	Uranium(IV) oxide UO_2	1344-57-6 270.028	brn cub cry	2827 10.97	a,f	i H_2O, dil acid; s conc acid
2219	Uranium(IV,V) oxide U_4O_9	12037-15-9 1096.111	cub cry	11.2	a	
2220	Uranium(V) bromide UBr_5	13775-16-1 637.549	brn hyg cry			reac H_2O
2221	Uranium(V) chloride UCl_5	13470-21-8 415.293	brn hyg cry	287		reac H_2O
2222	Uranium(V) fluoride UF_5	13775-07-0 333.021	pale blue tetr cry; hyg	348 5.81		s H_2O
2223	Uranium(V,VI) oxide U_3O_8	1344-59-8 842.082	grn-blk orth cry	1300 dec 8.38	a	
2224	Uranium(VI) chloride UCl_6	13763-23-0 450.745	green hex cry	177 3.6	a	
2225	Uranium(VI) fluoride UF_6	7783-81-5 352.019	wh monocl solid	64.0 tp 5.09	56.5 sp a,b,c,e	reac H_2O; s ctc, chl
2226	Uranium(VI) oxide UO_3	1344-58-7 286.027	oran-yel cry	≈7.3	a	i H_2O; s acid
2227	Uranium(VI) oxide monohydrate $UO_3 \cdot H_2O$	12326-21-5 304.043	yel orth cry	dec 570 7.05		
2228	Uranyl chloride UO_2Cl_2	7791-26-6 340.933	yel orth cry; hyg	577	a	vs H_2O; s EtOH, ace; i bz
2229	Uranyl nitrate hexahydrate $UO_2(NO_3)_2 \cdot 6H_2O$	13520-83-7 502.129	yel orth cry; hyg	60 2.81	dec 118	vs H_2O; s EtOH, eth
2230	Uranyl sulfate trihydrate $UO_2SO_4 \cdot 3H_2O$	20910-28-5 420.138	yel cry	3.28		s H_2O, sl EtOH
2231	Vanadium V	7440-62-2 50.942	gray-wh metal; cub	1910 6.0	3407 a,c	i H_2O; s acid
2232	Vanadium boride VB	12045-27-1 61.753	refrac solid	2250		i H_2O
2233	Vanadium boride VB_2	12007-37-3 72.564	refrac solid	2450		
2234	Vanadium carbide VC	12070-10-9 62.953	refrac blk cry; cub	2810 5.77		i H_2O
2235	Vanadium carbide V_2C	12012-17-8 113.894	hex cry	2167		
2236	Vanadium carbonyl $V(CO)_6$	20644-87-5 219.004	blue-grn cry; flam	dec 60	subl	
2237	Vanadium nitride VN	24646-85-3 64.949	blk powder; cub	2050 6.13		i H_2O; s aqua regia
2238	Vanadium silicide VSi_2	12039-87-1 107.113	metallic prisms	4.42		s HF
2239	Vanadium silicide V_3Si	12039-76-8 180.911	cub cry	1935 5.70		
2240	Vanadium(II) bromide VBr_2	14890-41-6 210.750	oran-brn hex cry	4.58	subl 800	reac H_2O
2241	Vanadium(II) chloride VCl_2	10580-52-6 121.847	grn hex plates	3.23	subl 910	reac H_2O; s EtOH, eth
2242	Vanadium(II) fluoride VF_2	13842-80-3 88.939	blue hyg cry			reac H_2O
2243	Vanadium(II) iodide VI_2	15513-84-5 304.751	red-viol hex cry	5.44		reac H_2O
2244	Vanadium(II) oxide VO	12035-98-2 66.941	grn cry	1790 5.758	a,c	s acid
2245	Vanadium(II) sulfate heptahydrate $VSO_4 \cdot 7H_2O$	36907-42-3 273.112	viol cry			

No.	Name Formula	CAS RN Mol. Wt.	Physical Form	mp/°C den/g cm^{-3}	bp/°C Other Data	Solubility
2246	Vanadium(III) acetylacetonate V(CHCOCHCOCH$_3$)$_3$	13476-99-8 348.270	brn cry	≈185 ≈1.0	subl	s MeOH, ace, bz chl
2247	Vanadium(III) bromide VBr$_3$	13470-26-3 290.654	gray-brn hyg cry	 4.00		reac H$_2$O
2248	Vanadium(III) chloride VCl$_3$	7718-98-1 157.300	red-viol hex cry; hyg	 3.00	 a	reac H$_2$O; s EtOH, eth
2249	Vanadium(III) fluoride VF$_3$	10049-12-4 107.937	yel-grn hex cry	≈1400 3.363	subl	i H$_2$O, EtOH
2250	Vanadium(III) fluoride trihydrate VF$_3$·3H$_2$O	10049-12-4* 161.983	grn rhomb cry	dec ≈100		sl H$_2$O
2251	Vanadium(III) iodide VI$_3$	15513-94-7 431.655	brn-blk rhomb cry; hyg	 5.21		reac H$_2$O
2252	Vanadium(III) oxide V$_2$O$_3$	1314-34-7 149.881	blk powder	2067 4.87	 a	i H$_2$O
2253	Vanadium(III) sulfate V$_2$(SO$_4$)$_3$	13701-70-7 390.074	yel powder	dec ≈400		sl H$_2$O
2254	Vanadium(III) sulfide V$_2$S$_3$	1315-03-3 198.081	grn-blk powder	dec 4.7		i H$_2$O, s hot HCl
2255	Vanadium(IV) bromide VBr$_4$	13595-30-7 370.558	unstable magenta cry	dec -23	 a	
2256	Vanadium(IV) chloride VCl$_4$	7632-51-1 192.753	unstable red liq	-25.7 1.816	148 a,c,d	reac H$_2$O; s EtOH, eth
2257	Vanadium(IV) fluoride VF$_4$	10049-16-8 126.936	grn hyg powder	dec 325 3.15	subl a	vs H$_2$O
2258	Vanadium(IV) oxide VO$_2$	12036-21-4 82.941	blue-blk powder	1967 4.339		i H$_2$O; s acid, alk
2259	Vanadium(V) fluoride VF$_5$	7783-72-4 145.934	col liq	19.5 2.50	48.3 a,c,d	reac H$_2$O
2260	Vanadium(V) oxide V$_2$O$_5$	1314-62-1 181.880	yel-brn orth cry	670 3.35	1800 dec a,c	sl H$_2$O; s conc acid, alk; i EtOH
2261	Vanadyl bromide VOBr	13520-88-2 146.845	viol cry	dec 480		
2262	Vanadyl chloride VOCl	13520-87-1 102.394	brn orth cry	dec >700 1.72	 a,b	
2263	Vanadyl dibromide VOBr$_2$	13520-89-3 226.749	yel-brn cry	dec 180		
2264	Vanadyl dichloride VOCl$_2$	10213-09-9 137.846	grn hyg cry	dec 380 2.88		reac H$_2$O; s EtOH
2265	Vanadyl difluoride VOF$_2$	13814-83-0 104.938	yel cry			
2266	Vanadyl selenite hydrate VOSeO$_3$·H$_2$O	133578-89-9 211.92	grn tricl plates	 3.506		
2267	Vanadyl sulfate dihydrate VOSO$_4$·2H$_2$O	27774-13-6 199.036	blue cry powder			s H$_2$O
2268	Vanadyl tribromide VOBr$_3$	13520-90-6 306.653	deep red liq		dec 180	reac H$_2$O
2269	Vanadyl trichloride VOCl$_3$	7727-18-6 173.299	fuming red liq	-79 1.829	127 a,d,e	reac H$_2$O; s MeOH, eth, ace
2270	Vanadyl trifluoride VOF$_3$	13709-31-4 123.936	yel hyg powder	300 2.459	480	reac H$_2$O
2271	Water H$_2$O	7732-18-5 18.015	col liq	0.00 0.9970	100.0 a,b,c,d,e	s EtOH, MeOH, ace
2272	Xenon Xe	7440-63-3 131.29	col gas	-111.79 tp 5.366 g/L	-108.12 a,b,c,d,e	sl H$_2$O
2273	Xenon trioxide XeO$_3$	13776-58-4 179.29	col orth cry	exp ≈25 4.55		s H$_2$O
2274	Xenon tetroxide XeO$_4$	12340-14-6 195.29	yel solid; exp	-35.9	dec ≈0	
2275	Xenon difluoride XeF$_2$	13709-36-9 169.29	col tetr cry	129.03 tp 4.32	114.35 sp b	sl H$_2$O
2276	Xenon tetrafluoride XeF$_4$	13709-61-0 207.28	col monocl cry	117.10 tp 4.04	115.75 sp a,b	reac H$_2$O
2277	Xenon hexafluoride XeF$_6$	13693-09-9 245.28	col monocl cry	49.5 3.56	75.6	reac H$_2$O
2278	Xenon dioxydifluoride XeO$_2$F$_2$	13875-06-4 201.29	col orth cry	30.8 exp 4.10		
2279	Xenon oxytetrafluoride XeOF$_4$	13774-85-1 223.28	col liq	-46.2 3.17(0°C)		reac H$_2$O
2280	Xenon fluoride hexafluororuthenate XeFRuF$_6$	22527-13-5 365.35	yel-grn monocl cry	110 3.78		
2281	Xenon fluoride undecafluoroantimonate XeFSb$_2$F$_{11}$	15364-10-0 602.79	yel monocl cry	63 3.69		
2282	Xenon fluoride hexafluoroarsenate Xe$_2$F$_3$AsF$_6$	50432-32-1 508.49	yel-grn monocl cry	99 3.62		reac H$_2$O
2283	Xenon fluoride hexafluoroantimonate XeF$_3$SbF$_6$	39797-63-2 424.04	yel-grn monocl cry	≈110 3.92		
2284	Xenon trifluoride undecafluoroantimonate XeF$_3$Sb$_2$F$_{11}$	35718-37-7 640.79	yel-grn tricl cry	82 3.98		

No.	Name Formula	CAS RN Mol. Wt.	Physical Form	mp/°C den/g cm^{-3}	bp/°C Other Data	Solubility
2285	Xenon pentafluoride hexafluoroarsenate XeF_5AsF_6	20328-94-3 415.19	wh monocl cry	130.5 3.51		
2286	Xenon pentafluoride hexafluororuthenate XeF_5RuF_6	39796-98-0 441.34	grn orth cry	152 3.79		
2287	Ytterbium Yb	7440-64-4 173.04	silv metal; cub	819 6.90	1196 a,c	s dil acid
2288	Ytterbium silicide $YbSi_2$	12039-89-3 229.21	hex cry	7.54		
2289	Ytterbium(II) bromide $YbBr_2$	25502-05-0 332.85	yel cry	673		reac H_2O
2290	Ytterbium(II) chloride $YbCl_2$	13874-77-6 243.95	grn cry	721 5.27		reac H_2O
2291	Ytterbium(II) iodide YbI_2	19357-86-9 426.85	blk cry	772		reac H_2O
2292	Ytterbium(III) chloride $YbCl_3$	10361-91-8 279.40	wh hyg powder	875	a	s H_2O
2293	Ytterbium(III) chloride hexahydrate $YbCl_3 \cdot 6H_2O$	19423-87-1 387.49	grn hyg cry	dec 150 2.57		vs H_2O
2294	Ytterbium(III) fluoride YbF_3	13760-80-0 230.04	wh cry	1157 8.2		i H_2O
2295	Ytterbium(III) oxide Yb_2O_3	1314-37-0 394.08	col cub cry	2435 9.2	a	s dil acid
2296	Ytterbium(III) sulfate octahydrate $Yb_2(SO_4)_3 \cdot 8H_2O$	10034-98-7 778.39	col cry	3.3		s H_2O
2297	Yttrium Y	7440-65-5 88.906	silv metal; hex	1522 4.47	3345 a,c	reac H_2O; s dil acid
2298	Yttrium aluminum oxide $Y_3Al_5O_{12}$	12005-21-9 593.619	grn cub cry	≈4.5		
2299	Yttrium antimonide YSb	12186-97-9 210.666	cub cry	2310 5.97		
2300	Yttrium arsenide YAs	12255-48-0 163.828	cub cry	5.59		
2301	Yttrium boride YB_6	12008-32-1 153.772	refrac solid	2600 3.72		
2302	Yttrium bromide YBr_3	13469-98-2 328.618	col hyg cry	904		vs H_2O
2303	Yttrium carbide YC_2	12071-35-1 112.928	refrac solid	≈2400 4.13		
2304	Yttrium carbonate trihydrate $Y_2(CO_3)_3 \cdot 3H_2O$	5970-44-5 411.886	red-brn powder			i H_2O; s dil acid
2305	Yttrium chloride YCl_3	10361-92-9 195.264	wh monocl cry; hyg	721 2.61	a	vs H_2O
2306	Yttrium fluoride YF_3	13709-49-4 145.901	wh hyg powder	≈1150 4.0	a	i H_2O
2307	Yttrium nitrate hexahydrate $Y(NO_3)_3 \cdot 6H_2O$	13494-98-9 383.012	hyg cry			vs H_2O
2308	Yttrium oxide Y_2O_3	1314-36-9 225.810	wh cry; cub	2439 5.03	a,c	s dil acid
2309	Yttrium phosphide YP	12294-01-8 119.880	cub cry	≈4.4		
2310	Yttrium sulfate octahydrate $Y_2(SO_4)_3 \cdot 8H_2O$	7446-33-5 610.125	red monocl cry	2.6		s H_2O
2311	Yttrium sulfide Y_2S_3	12039-19-9 274.010	yel cub cry	1925 3.87		
2312	Zinc Zn	7440-66-6 65.39	blue-wh metal; hex	419.53 7.14	907 a,c,e	s acid, alk
2313	Zinc acetate dihydrate $Zn(C_2H_3O_2)_2 \cdot 2H_2O$	5970-45-6 219.51	wh powder	237 dec 1.735		vs H_2O; s EtOH
2314	Zinc antimonide ZnSb	12039-35-9 187.15	silv-wh orth cry	565 6.33		reac H_2O
2315	Zinc arsenate $Zn_3(AsO_4)_2$	13464-44-3 474.01	wh powder			i H_2O; s acid, alk
2316	Zinc arsenate octahydrate $Zn_3(AsO_4)_2 \cdot 8H_2O$	13464-45-4 618.13	wh monocl cry	3.33		i H_2O; s acid, alk
2317	Zinc arsenite $Zn(AsO_2)_2$	10326-24-6 279.23	col powder			i H_2O; s acid
2318	Zinc borate $3ZnO \cdot 2B_2O_3$	27043-84-1 383.41	wh amorp powder	3.64		sl H_2O; s dil acid
2319	Zinc borate hemiheptahydrate $2ZnO \cdot 3B_2O_3 \cdot 3.5H_2O$	12513-27-8 434.69	wh cry	980 4.22		i H_2O
2320	Zinc bromate hexahydrate $Zn(BrO_3)_2 \cdot 6H_2O$	13517-27-6 429.29	wh hyg solid	100 2.57		vs H_2O
2321	Zinc bromide $ZnBr_2$	7699-45-8 225.20	wh hex cry; hyg	394 4.5	697 a,c,d	vs H_2O, EtOH; s eth
2322	Zinc caprylate $Zn(C_8H_{15}O_2)_2$	557-09-5 351.80	wh hyg cry	136		sl H_2O
2323	Zinc carbonate $ZnCO_3$	3486-35-9 125.40	wh rhomb cry	dec 4.4	a,f	i H_2O; s dil acid, alk

No.	Name Formula	CAS RN Mol. Wt.	Physical Form	mp/°C den/g cm^{-3}	bp/°C Other Data	Solubility
2324	Zinc chlorate $Zn(ClO_3)_2$	10361-95-2 232.29	yel hyg cry	dec 60 2.15		vs H_2O
2325	Zinc chloride $ZnCl_2$	7646-85-7 136.29	wh hyg cry	290 2.907	732 a,d,e	vs H_2O; s EtOH, ace
2326	Zinc chromite $ZnCr_2O_4$	12018-19-8 233.38	grn cub cry	5.29		
2327	Zinc citrate dihydrate $Zn_3(C_6H_5O_7)_2 \cdot 2H_2O$	546-46-3 610.40	col powder			sl H_2O; s dil acid, alk
2328	Zinc cyanide $Zn(CN)_2 \cdot$	557-21-1 117.43	wh powder	1.852		i H_2O; reac acid
2329	Zinc fluoride ZnF_2	7783-49-5 103.39	wh tetr needles; hyg	872 4.9	1500 a,d,e	sl H_2O
2330	Zinc fluoride tetrahydrate $ZnF_2 \cdot 4H_2O$	13986-18-0 175.45	wh orth cry	2.30		sl H_2O
2331	Zinc fluoroborate hexahydrate $Zn(BF_4)_2 \cdot 6H_2O$	27860-83-9 347.09	hex cry	2.12		vs H_2O; s EtOH
2332	Zinc formate dihydrate $Zn(CHO_2)_2 \cdot 2H_2O$	5970-62-7 191.46	wh cry	2.207		s H_2O, i EtOH
2333	Zinc hexafluorosilicate hexahydrate $ZnSiF_6 \cdot 6H_2O$	16871-71-9 315.56	wh cry			s H_2O
2334	Zinc hydroxide $Zn(OH)_2$	20427-58-1 99.41	col orth cry	125 dec 3.05	a	sl H_2O
2335	Zinc iodate $Zn(IO_3)_2$	7790-37-6 415.20	wh cry powder			sl H_2O
2336	Zinc iodide ZnI_2	10139-47-6 319.20	wh hyg cry	446 4.74	625 a	vs H_2O; s EtOH, eth
2337	Zinc laurate $Zn(C_{12}H_{23}O_2)_2$	2452-01-9 464.02	wh powder	128		sl H_2O
2338	Zinc molybdate $ZnMoO_4$	13767-32-3 225.33	wh tetr cry	>700 4.3		i H_2O
2339	Zinc nitrate $Zn(NO_3)_2$	7779-88-6 189.40	wh powder		a	vs H_2O
2340	Zinc nitrate hexahydrate $Zn(NO_3)_2 \cdot 6H_2O$	10196-18-6 297.49	col orth cry	36 dec 2.067		vs H_2O, EtOH
2341	Zinc nitride Zn_3N_2	1313-49-1 224.18	blue-gray cub cry			
2342	Zinc nitrite $Zn(NO_2)_2$	10102-02-0 157.40	hyg solid			reac H_2O
2343	Zinc oleate $Zn(C_{18}H_{33}O_2)_2$	557-07-3 628.31	wh powder	70 dec		i H_2O; s EtOH, eth, bz
2344	Zinc oxalate dihydrate $ZnC_2O_4 \cdot 2H_2O$	547-68-2 189.44	wh powder	100 dec 2.56		i H_2O; s dil acid
2345	Zinc oxide ZnO	1314-13-2 81.39	wh powder; hex	1975 5.6	a,c,f	i H_2O; s dil acid
2346	Zinc perchlorate hexahydrate $Zn(ClO_4)_2 \cdot 6H_2O$	10025-64-6 372.38	wh cub cry; hyg	106 dec 2.2		vs H_2O; s EtOH
2347	Zinc permanganate hexahydrate $Zn(MnO_4)_2 \cdot 6H_2O$	23414-72-4 411.35	blk orth cry; hyg	2.45		s H_2O; reac EtOH
2348	Zinc peroxide ZnO_2	1314-22-3 97.39	yel-wh powder	dec >150 1.57	exp 212	i H_2O; reac acid, EtOH, ace
2349	Zinc phosphate $Zn_3(PO_4)_2$	7779-90-0 386.11	wh monocl cry	900 4.0		i H_2O
2350	Zinc phosphate tetrahydrate $Zn_3(PO_4)_2 \cdot 4H_2O$	7543-51-3 458.17	col orth cry	3.04	f	i H_2O, EtOH; s dil acid, alk
2351	Zinc phosphide Zn_3P_2	1314-84-7 258.12	gray tetr cry	1160 4.55		i H_2O, EtOH; reac acid; s bz
2352	Zinc pyrophosphate $Zn_2P_2O_7$	7446-26-6 304.72	wh cry powder	3.75		i H_2O; s dil acid
2353	Zinc selenate pentahydrate $ZnSeO_4 \cdot 5H_2O$	13597-54-1 298.42	tricl cry	dec 50 2.59		s H_2O
2354	Zinc selenide $ZnSe$	1315-09-9 144.35	yel-red cub cry	>1100 5.65	subl a	i H_2O; s dil acid
2355	Zinc orthosilicate Zn_2SiO_4	13597-65-4 222.86	wh hex cry	1509 4.1	a,f	i H_2O, dil acid
2356	Zinc stearate $Zn(C_{18}H_{35}O_2)_2$	557-05-1 632.34	wh powder	130 1.095		i H_2O, EtOH, eth; s bz
2357	Zinc sulfate $ZnSO_4$	7733-02-0 161.45	col orth cry	680 dec 3.8	a	vs H_2O
2358	Zinc sulfate monohydrate $ZnSO_4 \cdot H_2O$	7446-19-7 179.47	wh monocl cry	238 dec 3.20		vs H_2O; i EtOH
2359	Zinc sulfate heptahydrate $ZnSO_4 \cdot 7H_2O$	7446-20-0 287.56	col orth cry	100 dec 1.97	f	vs H_2O; i EtOH
2360	Zinc sulfide (sphalerite) ZnS	1314-98-3 97.46	gray-wh cub cry	1700 4.04	a,f	i H_2O, EtOH; s dil acid
2361	Zinc sulfide (wurtzite) ZnS	1314-98-3 97.46	wh hex cry	1700 4.09	a,f	i H_2O; s dil acid
2362	Zinc sulfite dihydrate $ZnSO_3 \cdot 2H_2O$	123714-57-8 181.49	wh powder	200 dec		i H_2O, EtOH; reac hot H_2O

No.	Name Formula	CAS RN Mol. Wt.	Physical Form	mp/°C den/g cm^{-3}	bp/°C Other Data	Solubility
2363	Zinc telluride ZnTe	1315-11-3 192.99	red cub cry	1239 5.9		i H_2O
2364	Zinc thiocyanate Zn(SCN)$_2$	557-42-6 181.56	wh hyg cry			sl H_2O; s EtOH
2365	Zirconium Zr	7440-67-7 91.224	gray-wh metal; hex	1855 6.52	4409 a,c	s hot conc acid
2366	Zirconium boride ZrB$_2$	12045-64-6 112.846	gray refrac solid; hex	3245 6.17		
2367	Zirconium bromide ZrBr$_4$	13777-25-8 410.840	wh cub cry	450 tp 3.98	360 sp a,b,e	
2368	Zirconium carbide ZrC	12020-14-3 103.235	gray refrac solid; cub	3532 6.73		s HF
2369	Zirconium chloride ZrCl$_4$	10026-11-6 233.035	wh monocl cry; hyg	437 tp 2.80	331 sp a,b,c,e	reac H_2O; s EtOH, eth
2370	Zirconium fluoride ZrF$_4$	7783-64-4 167.218	wh monocl cry	932 tp 4.43	912 sp a,c	sl H_2O
2371	Zirconium hydride ZrH$_2$	7704-99-6 93.240	gray tetr cry	5.6	a	i H_2O
2372	Zirconium hydroxide Zr(OH)$_4$	14475-63-9 159.254	wh amorp powder	dec 3.25		i H_2O; s acid
2373	Zirconium iodide ZrI$_4$	13986-26-0 598.842	oran cub cry; hyg	499 tp 4.85	431 sp a,b,e	vs H_2O
2374	Zirconium nitrate pentahydrate Zr(NO$_3$)$_4$·5H$_2$O	13746-89-9 429.320	wh hyg cry	dec 100		vs H_2O; s EtOH
2375	Zirconium nitride ZrN	25658-42-8 105.231	yel cub cry	2960 7.09		s conc HF; sl dil acid
2376	Zirconium oxide ZrO$_2$	1314-23-4 123.223	wh amorp powder	2710 5.68	a,c,f	i H_2O; sl acid
2377	Zirconium phosphide ZrP$_2$	12037-80-8 153.172	orth cry	≈5.1		
2378	Zirconium orthosilicate ZrSiO$_4$	10101-52-7 183.308	wh tetr cry	dec 1540 4.6	a,f	i H_2O, acid
2379	Zirconium silicide ZrSi$_2$	12039-90-6 147.395	gray powder	1620 4.88		i H_2O, aqua regia; s HF
2380	Zirconium sulfate tetrahydrate Zr(SO$_4$)$_2$·4H$_2$O	7446-31-3 355.413	wh tetr cry	dec 100 2.80		vs H_2O
2381	Zirconium sulfide ZrS$_2$	12039-15-5 155.356	red-brn hex cry	1480 3.82		i H_2O
2382	Zirconyl chloride octahydrate ZrOCl$_2$·8H$_2$O	13520-92-8 322.251	tetr cry	dec 400 1.91		vs H_2O, EtOH

Acanthite	1854	Ammonium sulfocarbamate	85	Basic antimony chloride	170
Acetylenogen	425	Ammonium tetraborate tetrahydrate	67	Basic lead acetate	1088
Actinium tribromide	2	Ammonium tungstophosphate	138	Bayerite	29
Actinium trichloride	3	Ammonium vanadate	122	Berlinite	38
Actinium trifluoride	4	Andalusite	46	Beryl	292
Actinium triiodide	5	Anglesite	1135	Beryllia	307
Adams' catalyst	1539	Anhydrite	470	Beryllium sodium fluoride	1952
ADP	84	Antarcticite	432	Berzelianite	696
Alabandite	1283	Antimonous sulfate	172	Bieberite	663
Alamoisite	1127	Antimony pentachloride	176	Bischofite	1208
Allemontite	163	Antimony pentafluoride	177	Bismite	328
Altaite	1139	Antimony pentasulfide	180	Bismuth bromine oxide	329
Aluctyl	34	Antimony pentoxide	179	Bismuth iodide oxide	331
Alum	1561	Antimony tetroxide	175	Bismuth pentafluoride	319
Alumina	40	Antimony tribromide	164	Bismuth peroxide	339
Aluminum ammonium chloride	150	Antimony trichloride	165	Bismuth subcarbonate	314
Aluminum calcium hydride	476	Antimony trifluoride	166	Bismuth titanium oxide	340
Aluminum chlorohydroxide	30	Antimony triiodide	167	Bismuth tribromide	315
Aluminum ethylate	23	Antimony trioxide	169	Bismuth trichloride	316
Aluminum hexadecanoate	42	Antimony triselenide	171	Bismuth trifluoride	318
Aluminum iron oxide	1010	Antimony trisulfide	173	Bismuth triiodide	322
Aluminum metaphosphate	39	Antlerite	746	Bismuthine	320
Aluminum orthophosphate	38	Aragonite	426	Bismuthinite	337
Aluminum potassium sulfate	1560	Arcanite	1624	Black phosphorus	1489
Aluminum sodium oxide	1863	Argentite	1854	Blue verdigris	709
Aluminum subacetate	22	Arnaudon's green	597	Blue vitriol	745
Aluminum tetrahydroborate	14	Arsenic antimonide	163	Bobierrite	1227
Aluminum tribromide	16	Arsenic hydride	183	Boehmite	41
Aluminum trichloride	20	Arsenic pentachloride	200	Borax	1948
Aluminum trichloride hexahydrate	21	Arsenic pentafluoride	201	Borazole	351
Aluminum trifluoride	24	Arsenic pentaselenide	203	Boric oxide	366
Aluminum triiodide	32	Arsenic pentasulfide	204	Boron tetrachloride	361
Aluminum triiodide hexahydrate	33	Arsenic pentoxide	202	Boron tetrafluoride	363
Aluminum trioleate	37	Arsenic tribromide	191	Boron tribromide	358
Aluminum tripalmitate	42	Arsenic trichloride	192	Boron trichloride	360
Aluminum tristearate	48	Arsenic trifluoride	193	Boron trifluoride	362
Aluminum yttrium aluminate	2298	Arsenic triiodide	194	Boron triiodide	364
Alunogenite	49	Arsenic trioxide	195,196	Braggite	1531
Americium tetrafluoride	60	Arsenic triselenide	197	Breithauptite	1382
Americium tribromide	56	Arsenic trisulfide	198	Bremen green	713
Americium trichloride	57	Arsenolite	195	Bromocyanide	499
Americium trifluoride	58	Auric bromide	891	Bromogermane	854
Americium triiodide	59	Auric chloride	892	Bromophosgene	495
Aminomercuric chloride	1307	Auric cyanide trihydrate	893	Bromosilane	1782
Ammonia solution	116	Auric fluoride	894	Bromotrichlorosilane	1783
Ammonium alum	8	Auric hydroxide	895	Bromyrite	1827
Ammonium aluminum fluoride	100	Auric iodide	896	Brucite	1214
Ammonium bicarbonate	107	Auric oxide	897	Brushite	444
Ammonium bichromate	82	Auric selenate	898	Buckminsterfullerene	484,485
Ammonium bifluoride	109	Auric selenide	899	Bunsenite	1404
Ammonium binoxalate monohydrate	110	Auric sulfide	900	Burnt alum	1560
Ammonium bisulfate	112	Aurous bromide	886	Cadmium dibromide	384
Ammonium bisulfide	113	Aurous chloride	887	Cadmium dichloride	388
Ammonium bisulfite	114	Aurous cyanide	888	Cadmium dichloride hemipentahydrate	389
Ammonium bitartrate	115	Aurous iodide	889	Cadmium difluoride	393
Ammonium carbazoate	139	Aurous sulfide	890	Cadmium diiodide	396
Ammonium diuranate	157	Avogadrite	1581	Cadmium metasilicate	397
Ammonium heptamolybdate tetrahydrate	123	Baddeleyite	2376	Cadmoselite	408
Ammonium hexacyanoferrate(II) trihydrate	90	Barite	270	Calcite	427
Ammonium hexacyanoferrate(III) trihydrate	89	Barium boride	228	Calcium aluminum hydride	476
Ammonium hexadecanoate	130	Barium diphosphate	259	Calcium carbimide	435
Ammonium hydrogen malate	66	Barium disilicate	263	Calcium dioxide	460
Ammonium hyposulfite	155	Barium dioxide	460	Calcium diphosphate	464
Ammonium molybdophosphate	137	Barium hyposulfate dihydrate	224	Calcium metasilicate	467
Ammonium nitroprusside	127	Barium hyposulfite	279	Calcium phosphate, dibasic	443
Ammonium octadecanoate	143	Barium hyposulfite monohydrate	280	Calcium phosphate, monobasic,	
Ammonium octanoate	69	Barium mercuric iodide	275	monohydrate	438
Ammonium osmium chloride	97	Barium metaphosphate	257	Calomel	1300
Ammonium persulfate	134	Barium metasilicate	264	Carbon oxychloride	496
Ammonium phosphate monobasic	84	Barium metatitanate	281	Carbon sulfotelluride	494
Ammonium phosphate, dibasic	111	Barium monoxide	252	Carbonic acid anhydride	488
Ammonium platinic chloride	99	Barium phosphate, dibasic	231	Carbonyl selenide	491
Ammonium platinous chloride	151	Barium platinous cyanide tetrahydrate	274	Carbonyl sulfide	492
Ammonium rhodanide	154	Barium silicofluoride	229	Caro's acid	1987
Ammonium soap	128	Barium tungsten oxide	282	Cassiopeium	1186
Ammonium sodium phosphate tetrahydrate	1866	Barium white	282	Cassiterite	2130
		Barium zirconium oxide	285		

Caustic baryta	234	Chrysoberyl	291	Cupric acetoarsenite	703
Caustic soda	1909	Cianurina	1311	Cupric acetylide	705
Celestite	1981	Cinnabar	1327	Cupric ammonium chloride	80
Celphos	44	Claudetite	196	Cupric arsenate	706
Cerargyrite	1830	Clausthalite	1131	Cupric azide	708
Ceria	524	Clinoenstatite	1233	Cupric borate	710
Ceric ammonium nitrate	73	Cobalt black	676	Cupric bromide	711
Ceric fluoride	523	Cobalt hydroxide oxide	677	Cupric butyrate monohydrate	712
Ceric oxide	524	Cobalt yellow	678	Cupric chlorate hexahydrate	714
Ceric sulfate tetrahydrate	525	Cobaltic acetate	670	Cupric chloride	715
Cerium hexaboride	504	Cobaltic ammonium phosphate	78	Cupric chloride dihydrate	716
Cerous ammonium sulfate tetrahydrate	72	Cobaltic ammonium sulfate hexahydrate	79	Cupric chromate	717
Cerous bromide	511	Cobaltic fluoride	672	Cupric chromite	718
Cerous bromide heptahydrate	512	Cobaltic hexafluorosilicate hexahydrate	642	Cupric cyanide	720
Cerous carbide	513	Cobaltic hydroxide	674	Cupric dichromate dihydrate	721
Cerous carbonate hydrate	514	Cobaltic nitrate	675	Cupric ferrocyanide	722
Cerous chloride	515	Cobaltic orthotitanate	680	Cupric fluoride	724
Cerous chloride heptahydrate	516	Cobaltic oxide	676	Cupric fluoride dihydrate	725
Cerous fluoride	517	Cobaltic sulfide	679	Cupric formate	726
Cerous iodide	518	Cobaltic-cobaltous oxide	669	Cupric formate tetrahydrate	727
Cerous iodide nonahydrate	519	Cobaltite	612	Cupric hexafluorosilicate tetrahydrate	728
Cerous nitride	506	Cobaltous acetate	623	Cupric hydroxide	729
Cerous oxide	520	Cobaltous acetate tetrahydrate	624	Cupric molybdate	730
Cerous sulfate octahydrate	521	Cobaltous arsenate octahydrate	626	Cupric nitrate	731
Cerous sulfide	522	Cobaltous bromate hexahydrate	627	Cupric nitrate hexahydrate	732
Cerussite	1097	Cobaltous bromide	628	Cupric nitrate trihydrate	733
Cervantite	175	Cobaltous bromide hexahydrate	629	Cupric oleate	734
Chalcanthite	745	Cobaltous chloride	631	Cupric oxalate	735
Chalcocite	697	Cobaltous chloride dihydrate	632	Cupric oxide	736
Chalcocyanite	744	Cobaltous chloride hexahydrate	633	Cupric perchlorate	737
Chalcopyrite	723	Cobaltous chromate	634	Cupric perchlorate hexahydrate	738
Chereul's salt	701	Cobaltous chromate(III)	635	Cupric phosphate trihydrate	739
Chile saltpeter	1921	Cobaltous cyanide	636	Cupric selenate pentahydrate	740
China clay	47	Cobaltous cyanide dihydrate	637	Cupric selenide	741
Chlorine monofluoride	554	Cobaltous dipotassium sulfate hexahydrate	656	Cupric selenite dihydrate	742
Chlorine tetroxyfluoride	806	Cobaltous ferricyanide	638	Cupric stearate	743
Chlorocyanide	500	Cobaltous fluoride	639	Cupric sulfate	744
Chlorogermane	855	Cobaltous fluoride tetrahydrate	640	Cupric sulfate pentahydrate	745
Chlorosilane	1784	Cobaltous formate dihydrate	641	Cupric sulfide	747
Chlorosulfuric acid	1988	Cobaltous hydroxide	643	Cupric tartrate trihydrate	748
Chlorotrifluorogermane	856	Cobaltous iodate	644	Cupric tungstate	750
Chrome alum	599	Cobaltous iodide	645	Cupric tungstate dihydrate	751
Chrome alum ammonium	77	Cobaltous iodide hexahydrate	646	Cuprite	695
Chrome yellow	1102	Cobaltous metatitanate	647	Cuprous acetate	685
Chromic acetate hexahydrate	583	Cobaltous molybdate	648	Cuprous acetylide	686
Chromic acid	608	Cobaltous nitrate	649	Cuprous azide	687
Chromic anhydride	608	Cobaltous nitrate hexahydrate	650	Cuprous bromide	688
Chromic bromide	584	Cobaltous orthosilicate	660	Cuprous chloride	689
Chromic bromide hexahydrate	585,586	Cobaltous oxalate	651	Cuprous cyanide	690
Chromic chloride	587	Cobaltous oxalate dihydrate	652	Cuprous fluoride	691
Chromic chloride hexahydrate	588	Cobaltous oxide	653	Cuprous hydride	692
Chromic fluoride	589	Cobaltous perchlorate	654	Cuprous iodide	693
Chromic fluoride trihydrate	590	Cobaltous phosphate octahydrate	655	Cuprous oxide	695
Chromic hydroxide trihydrate	591	Cobaltous selenate pentahydrate	657	Cuprous selenide	696
Chromic iodide	592	Cobaltous selenide	658	Cuprous sulfide	697
Chromic nitrate	593	Cobaltous selenite dihydrate	659	Cuprous sulfite hemihydrate	698
Chromic nitrate nonahydrate	594	Cobaltous stannate	661	Cuprous telluride	699
Chromic oxide	595	Cobaltous sulfate	662	Cuprous tetraiodomercurate	694
Chromic phosphate	596	Cobaltous sulfate monohydrate	664	Cuprous thiocyanate	700
Chromic phosphate hemiheptahydrate	597	Cobaltous sulfide	665	Cyanogas	436
Chromic phosphate hexahydrate	598	Cobaltous telluride	666	Cyanogran	1883
Chromic sulfate	600	Cobaltous thiocyanate trihydrate	667	Decaborane	349
Chromic sulfide	601	Cobaltous tungstate	668	Dehydrite	1223
Chromic telluride	602	Coccinite	1317	Detia	44
Chromite	1017	Coloradoite	1328	Diammonium citrate	108
Chromium diboride	601	Copper hydroxide sulfate	746	Diammonium hydrogen phosphite	
Chromium dioxide	605	Coquimbite	1061	monohydrate	136
Chromium trioxide	608	Corundum	40	Diamond	482
Chromium(VI) oxychloride	609	Cotunnite	1100	Diarsane	184
Chromous acetate monohydrate	574	Covellite	747	Diaspore	41
Chromous bromide	575	Cristobalite	1816	Diborane	342
Chromous chloride	576	Crocoite	1102	Diboron tetrachloride	361
Chromous chloride tetrahydrate	577	Cryolite	1895	Dibromogermane	857
Chromous fluoride	578	Cryptohalite	104	Dibromosilane	1786
Chromous oxalate monohydrate	580	Cupric acetate	702	Dichlorine hexoxide	552
Chromous sulfate pentahydrate	581	Cupric acetate monohydrate	704	Dichlorine trioxide	551

Dichlorodifluorogermane	859	Ferric oxide	1056	Goslarite	2359
Dichlorodifluorosilane	1788	Ferric oxide hydroxide	1053	Graphite	483
Dichlorogermane	858	Ferric phosphate dihydrate	1057	Greenockite	412
Dichlorosilane	1787	Ferric pyrophosphate nonahydrate	1058	Guanajuatite	334
Dicobalt octacarbonyl	618	Ferric sulfate	1060	Gummite	2227
Difluoroazane	1455	Ferric sulfate nonahydrate	1061	Gunningite	2358
Difluorodiazine	1457	Ferric thiocyanate sesquihydrate	1062	Gypsum	471
Difluorophosphoric acid	1496	Ferric vanadate	1063	Hafnia	910
Difluorosilane	1789	Ferrous aluminate	1010	Hafnium orthosilicate	914
Digermane	866	Ferrous ammonium sulfate	91	Hafnium tetrabromide	903
Dihydrogen disulfide	948	Ferrous bromide	1011	Hafnium tetrachloride	905
Diiodine pentoxide	978	Ferrous bromide hexahydrate	1012	Hafnium tetrafluoride	906
Diiodine tetroxide	977	Ferrous carbonate	1013	Hafnium tetraiodide	908
Diiodosilane	1790	Ferrous chloride	1014	Halite	1877
Diniobium carbide	1418	Ferrous chloride dihydrate	1015	Hatchett's brown	722
Dinitrogen pentoxide	1450	Ferrous chloride tetrahydrate	1016	Hausmannite	1289
Dinitrogen tetraoxide	1449	Ferrous chromite	1017	Heazlewoodite	1413
Dinitrogen trioxide	1448	Ferrous fluoride	1018	Hematite	1056
Dioxygen difluoride	804	Ferrous fluoride tetrahydrate	1019	Hercynite	1010
Diphosphine	1491	Ferrous hydroxide	1020	Hessite	1855
Diphosphoric acid	1494	Ferrous iodide	1021	Hexaaminecobalt trichloride	673
Diselenium dibromide	1768	Ferrous iodide tetrahydrate	1022	Hieratite	1587
Diselenium dichloride	1769	Ferrous molybdate	1023	Holmia	925
Disilane	1803	Ferrous nitrate	1024	Hopeite	2350
Disiloxane	1806	Ferrous nitrate hexahydrate	1025	Huebnerite	1288
Disulfur decafluoride	2003	Ferrous oxalate dihydrate	1026	Hydrazine	928
Disulfur dibromide	1994	Ferrous oxide	1027	Hydrazoic acid	936
Disulfur dichloride	1995	Ferrous phosphate octahydrate	1028	Hydrobromic acid	940
Disulfur difluoride	1996, 1997	Ferrous phosphide	1006	Hydrocerussite	1098
		Ferrous selenide	1029	Hydrochloric acid	941
Dysprosia	764	Ferrous sulfate	1031	Hydrocyanic acid	942
Enneaborane(15)	348	Ferrous sulfate heptahydrate	1033	Hydrofluoric acid	943
Enstatite	1233	Ferrous sulfate monohydrate	1032	Hydrogen azide	936
Epsomite	1239	Ferrous sulfide	1034	Hydrogen hexafluorosilicate	1809
Erbia	776	Ferrous telluride	1035	Hydrogen tetrabromoaurate pentahydrate	884
Erdmann's salt	671	Ferrous thiocyanate trihydrate	1036	Hydrogen tetracarbonylferrate(II)	1005
Eremeyevite	12	Ferrous titanate	1037	Hydrogen tetrachloroaurate tetrahydrate	885
Eriochaleite	716	Fischer's yellow	678	Hydroiodic acid	944
Eskolaite	595	Fluellite	25	Hydrophilite	430
Etard's salt	698	Fluorine oxide	803	Hydrous gold bromide	884
Europia	798	Fluorite	439	Hydroxylamine	937
Europic bromide	793	Fluoroboric acid	356	Hypo	1954
Europic chloride	794	Fluorocyanide	501	Hypochlorous acid	547
Europic chloride hexahydrate	795	Fluorogermane	860	Ilmenite	1037
Europic fluoride	796	Fluorosilane	1791	Indium dibromide	958
Europic nitrate hexahydrate	797	Fluorosulfuric acid	1989	Indium dichloride	959
Europic sulfate	799	Fluorspar	439	Indium sesquioxide	967
Europic sulfate octahydrate	800	Forsterite	1234	Indium tribromide	961
Europous bromide	786	Fullerene	484,485	Indium trichloride	962
Europous chloride	787	Gadolinia	821	Indium trifluoride	963
Europous fluoride	788	Galena	1136	Indium trifluoride trihydrate	964
Europous iodide	789	Gallane	840	Indium triiodide	966
Fayalite	1030	Gallium dichloride	831	Iodine monobromide	980
Ferberite	1038	Gallium tribromide	836	Iodine monochloride	981
Ferric acetylacetonate	1041	Gallium trichloride	837	Iodine monofluoride	983
Ferric alum	88	Gallium trifluoride	838	Iodocyanide	502
Ferric ammonium chromate	86	Gallium trifluoride trihydrate	839	Iodogermane	861
Ferric ammonium oxalate trihydrate	87	Gallium triiodide	842	Iodosilane	1792
Ferric arsenate dihydrate	1042	Geikielite	1245	Iodyrite	1838
Ferric basic acetate	1040	Gerhardite	733	Iridium hexafluoride	997
Ferric bromide	1043	Germanium dibromide	868	Iridium tribromide	989
Ferric chloride	1044	Germanium dichloride	869	Iridium tribromide tetrahydrate	990
Ferric chloride hexahydrate	1045	Germanium difluoride	870	Iridium trichloride	991
Ferric chromate	1046	Germanium diiodide	871	Iridium trifluoride	992
Ferric citrate pentahydrate	1047	Germanium dioxide	880	Iridium triiodide	993
Ferric dichromate	1048	Germanium monoxide	872	Iridium trioxide	994
Ferric ferrocyanide	1049	Germanium tetrabromide	876	Iron orthosilicate	1030
Ferric ferrous oxide	1039	Germanium tetrachloride	877	Jeremejevite	12
Ferric fluoride	1050	Germanium tetrafluoride	878	Kalinite	1561
Ferric fluoride trihydrate	1051	Germanium tetrahydride	853	Kalium	1556
Ferric hexacyanoferrate(II)	1049	Germanium tetraiodide	879	Kaolin	47
Ferric hydroxide	1052	Gibbsite	29	Kernite	1950
Ferric hypophosphite	1059	Glauber's salt	1939	Kieserite	1240
Ferric metavanadate	1063	Glucinium	288	Koettigite	2316
Ferric nitrate nonahydrate	1054	Goethite	1053	Lautarite	447
Ferric oxalate	1055	Gold trichloride acid	885	Lead butyrate	1096

Lead diacetate	1086	Mercuric thiocyanate	1329	Niobium trichloride	1424
Lead dioxide	1153	Mercurous acetate	1297	Niobium trifluoride	1425
Lead hydroxide	1122	Mercurous bromide	1298	Niter	1921
Lead metasilicate	1127	Mercurous chlorate	1299	Nitrobarite	246
Lead metatantalate	1138	Mercurous chloride	1300	Nitromagnesite	1219
Lead metatitanate	1142	Mercurous fluoride	1301	Nitrosylsulfuric acid	1990
Lead monetite	1110	Mercurous iodide	1302	Nitroxy fluoride	805
Lead orthosilicate	1128	Mercurous nitrate dihydrate	1303	Norbide	359
Lead sesquioxide	1147	Mercurous oxide	1304	Oil of vitriol	1986
Lead tetraacetate	1149	Mercurous sulfate	1305	Oldhamite	473
Lead tetrabromide	1150	Metaboric acid	353,354,	Opal	1807
Lead tetrachloride	1151		355	Orpiment	198
Lead tetrafluoride	1152	Metacinnabar	1326	Orthoarsenic acid	185
Lead tetroxide	1148	Metaphosphoric acid	1493	Orthoboric acid	352
Lead vanadate	1145	Metasilicic acid	1807	Orthoclase	1559
Lepidocrocite	1053	Microcline	1559	Orthophosphoric acid	1492
Lime	457	Microcosmic salt	1866	Orthophosphorous acid	1497
Litharge	1120	Millerite	1409	Orthosilicic acid	1808
Lithia	1178	Minium	1148	Orthotelluric acid	2032
Lithiophosphate	1177	Mirabilite	1939	Osmic acid	1476
Lithium metaborate	1160	Misenite	1595	Osmium dioxide	1473
Lithium metasilicate	1182	Mohr's salt	91	Osmium hexafluoride	1475
Lutetium triiodide	1191	Moissanite	1810	Osmium pentafluoride	1474
Macquer's salt	1575	Molybdenite	1353	Osmium tetrachloride	1471
Maekinenite	1405	Molybdenum dibromide	1340	Osmium tetrafluoride	1472
Maghemite	1056	Molybdenum dichloride	1341	Osmium tetroxide	1476
Magnesia	1222	Molybdenum diiodide	1342	Osmium tribromide	1469
Magnesite	1205	Molybdenum dioxide	1351	Osmium trichloride	1470
Magnesium dioxide	1226	Molybdenum diselenide	1352	Otavite	386
Magnesium fluorosilicate hexahydrate	1235	Molybdenum disilicide	1339	Oxygen difluoride	803
Magnesium hyposulfite hexahydrate	1244	Molybdenum disulfide	1353	Palladium monoxide	1486
Magnesium metasilicate	1233	Molybdenum ditelluride	1354	Palladous bromide	1481
Magnesium orthosilicate	1234	Molybdenum hexacarbonyl	1335	Palladous chloride	1482
Magnetite	1039	Molybdenum hexafluoride	1359	Palladous fluoride	1483
Magnogene	1207	Molybdenum metaphosphate	1364	Palladous iodide	1484
Malacite	713	Molybdenum oxychloride	1362	Palladous nitrate	1485
Malladrite	1896	Molybdenum pentaboride	1332	Palladous sulfide	1480
Mallebrin	19	Molybdenum pentachloride	1355	Parawollastonite	467
Manganese dibromide	1259	Molybdenum pentafluoride	1356	Paris green	703
Manganese dichloride	1261	Molybdenum sesquioxide	1347	Pearl ash	1567
Manganese difluoride	1265	Molybdenum tetrabromide	1348	Pentaborane(11)	345
Manganese diiodide	1267	Molybdenum tetrachloride	1349	Pentaborane(9)	344
Manganese dioxide	1293	Molybdenum tetrafluoride	1350	Perchloric acid	548
Manganese green	242	Molybdenum tribromide	1343	Perchloric anhydride	553
Manganese heptoxide	1294	Molybdenum trichloride	1344	Perchloryl fluoride	557
Manganese trioxide	1292	Molybdenum trichloride oxide	1357	Periclase	1222
Manganite	1291	Molybdenum trifluoride	1345	Perovskite	479
Manganjustite	1277	Molybdenum triiodide	1346	Perrhenic acid	1666
Manganosite	1274	Molybdenum trioxide	1363	Phosgene	496
Marshite	693	Molybdic acid monohydrate	1358	Phosphorus hydride	1490
Mascagnite	145	Molybdic anhydride	1363	Phosphorus pentabromide	1512
Massicot	1121	Molysite	1044	Phosphorus pentachloride	1513
Matlockite	1101	Monetite	443	Phosphorus pentafluoride	1514
Melanterite	1033	Monosilver acetylide	1825	Phosphorus pentaselenide	1519
Mercallite	1595	Montroydite	1321	Phosphorus pentasulfide	1520
Mercuric acetate	1306	Nantokite	689	Phosphorus pentoxide	1515
Mercuric arsenate	1315	Naples yellow	1089	Phosphorus sulfochloride	1522
Mercuric bromide	1308	Natrium	1860	Phosphorus tribromide	1503
Mercuric chloride	1309	Neodymia	1374	Phosphorus trichloride	1504
Mercuric chromate	1310	Neptunium dioxide	1380	Phosphorus trifluoride	1507
Mercuric cyanide	1311	Newberyite	1213	Phosphorus triiodide	1508
Mercuric dichromate	1312	Niccolite	1383	Phosphorus trioxide	1509
Mercuric fluoride	1313	Nickel ammonium chloride	124	Phosphorus trisulfide	1511
Mercuric fulminate	1314	Nickel ammonium sulfate	125	Phosphoryl bromide	1516
Mercuric iodate	1316	Nickel sesquioxide	1412	Phosphoryl chloride	1517
Mercuric iodide	1317	Niobium dioxide	1429	Phosphoryl fluoride	1518
Mercuric nitrate	1318	Niobium fluorodioxide	1440	Phostoxin	44
Mercuric nitrate dihydrate	1319	Niobium pentabromide	1433	Photophor	463
Mercuric nitrate monohydrate	1320	Niobium pentachloride	1434	Pigment E	258
Mercuric oxide	1321	Niobium pentafluoride	1435	Plaster of Paris	472
Mercuric oxycyanide	1322	Niobium pentaiodide	1436	Platinic acid hexahydrate	1525
Mercuric phosphate	1323	Niobium pentoxide	1437	Platinic oxide	1539
Mercuric potassium iodide	1635	Niobium tetrachloride	1426	Platinum dibromide	1527
Mercuric silver iodide	1856	Niobium tetrafluoride	1427	Platinum dichloride	1528
Mercuric sulfate	1325	Niobium tetraiodide	1428	Platinum diiodide	1529
Mercuric sulfide	1327	Niobium tribromide	1423	Platinum hexafluoride	1541

Platinum monoxide	1530
Platinum tetrabromide	1534
Platinum tetrachloride	1535
Platinum tetrafluoride	1537
Platinum tetraiodide	1538
Platinum tribromide	1532
Platinum trichloride	1533
Plattnerite	1153
Plessy's green	597
Plutonium dioxide	1551
Plutonium hexafluoride	1552
Plutonium monoxide	1544
Plutonium sesquioxide	1549
Plutonium tetrafluoride	1550
Plutonium tribromide	1545
Plutonium trichloride	1546
Plutonium trifluoride	1547
Plutonium triiodide	1548
Polonium dioxide	1555
Polonium tetrachloride	1554
Polydymite	1411
Portlandite	445
Potassium bicarbonate	1590
Potassium bichromate	1574
Potassium bifluoride	1591
Potassium biselenite	1594
Potassium bisulfate	1595
Potassium bisulfide	1596
Potassium bisulfite	1598
Potassium bitartrate	1599
Potassium chloroplatinate	1633
Potassium cobalticyanide	1585
Potassium dioxide	1628
Potassium hexacyanoferrate(II) trihydrate	1578
Potassium hexacyanoferrate(III)	1577
Potassium hyposulfite	1637
Potassium monoxide	1612
Potassium orthodiuranate	1642
Potassium peroxydisulfate	1617
Potassium phosphate, dibasic	1592
Potassium pyrosulfite	1605
Potassium sodium tartrate	1933
Potassium trithiocarbonate	1640
Potcrate	1569
Powellite	450
Prussic acid	942
Pucherite	326
Pyrite	1007
Pyrochroite	1266
Pyrolusite	1293
Pyrophenite	1287
Pyrophosphoric acid	1494
Quartz	1813-1816
Quartz	1815
Quicklime	457
Quicksilver	1296
Quilonorm	1155
Raspite	1144
Realgar	190
Red lead	1148
Red lead chromate	1103
Red phosphorus	1488
Retgersite	1408
Rhenium dioxide	1673
Rhenium disulfide	1674
Rhenium ditelluride	1675
Rhenium heptafluoride	1685
Rhenium heptasulfide	1691
Rhenium heptoxide	1686
Rhenium hexachloride	1680
Rhenium hexafluoride	1681
Rhenium pentabromide	1676
Rhenium pentachloride	1677
Rhenium pentafluoride	1678
Rhenium pentoxide	1679

Rhenium tetrachloride	1671
Rhenium tetrafluoride	1672
Rhenium tribromide	1668
Rhenium trichloride	1669
Rhenium triiodide	1670
Rhenium trioxide	1682
Rhodium dioxide	1700
Rhodium hexafluoride	1701
Rhodium trichloride	1695
Rhodium trifluoride	1696
Rhodium triiodide	1697
Rhodium trioxide	1698
Rhodochrosite	1261
Rhodonite	1277
Rochelle salt	1933
Rock salt	1877
Rohrbach's solution	275
Rubidium alum	1704
Rubidium bicarbonate	1713
Rubidium bifluoride	1711
Rubidium bisulfate	1714
Rubidium dioxide	1724
Ruthenium dioxide	1733
Ruthenium hexafluoride	1735
Ruthenium monoxide	1718
Ruthenium pentafluoride	1734
Ruthenium tetrafluoride	1732
Ruthenium tetroxide	1736
Ruthenium tribromide	1728
Ruthenium trichloride	1729
Ruthenium trifluoride	1730
Ruthenium triiodide	1731
Rutile	2157
Sacchite	1262
Sal ammoniac	75
Sal soda	1874
Salt of tartar	1567
Saltpeter	1609
Samaria	1749
Sanbornite	263
Sassolite	352
Scandia	1758
Scheele's green	707
Scheelite	480
Schultenite	1109
Selenium hexasulfide	1778
Selenium tetrasulfide	1779
Selenium(IV) bromide	1770
Selenium(IV) oxide	1766
Sellaite	1209
Senarmontite	168
Siderite	1013
Silane	1781
Silica	1813
Silicon chloride trifluoride	1785
Silicon decahydride	1805
Silicon octahydride	1804
Silicon tetrabromide	1793
Silicon tetrachloride	1794
Silicon tetrafluoride	1795
Silicon tetraiodide	1796
Silicon(II) oxide	1812
Silver difluoride	1859
Sinjarite	431
Skutterudite	615
Smaltite	614
Smite	1868
Smithsonite	2323
Soda ash	1873
Soda lye	1909
Sodamide	1865
Sodium acid pyrophosphate	1888
Sodium alum	1864
Sodium bicarbonate	1899
Sodium bifluoride	1900
Sodium bisulfate	1904

Sodium bisulfate monohydrate	1905
Sodium bisulfide	1906
Sodium bisulfide dihydrate	1907
Sodium bisulfite	1908
Sodium borate	1947
Sodium borofluoride	1893
Sodium dioxide	1945
Sodium metaarsenite	1867
Sodium metabismuthate	1914
Sodium metaperiodate	1928
Sodium metasilicate	1917
Sodium monobasic phosphate	1886
Sodium monoxide	1924
Sodium peroxydisulfate	1931
Sodium phosphate, dibasic	1901
Sodium prussiate yellow	1891
Sodium pyrosulfite	1915
Sodium tetrafluoroborate	1893
Sphalerite	2360
Spherocobaltite	630
Stannane	2106
Stannic bromide	2124
Stannic chloride	2125
Stannic chloride pentahydrate	2126
Stannic chromate	2127
Stannic fluoride	2128
Stannic iodide	2129
Stannic oxide	2130
Stannic selenide	2131
Stannic selenite	2132
Stannic sulfide	2133
Stannous acetate	2108
Stannous bromide	2109
Stannous chloride	2110
Stannous chloride dihydrate	2111
Stannous fluoride	2112
Stannous hexafluorozirconate	2113
Stannous hydroxide	2114
Stannous iodide	2115
Stannous monoxide	2117
Stannous oxalate	2116
Stannous pyrophosphate	2118
Stannous selenide	2119
Stannous sulfate	2120
Stannous sulfide	2121
Stannous tartrate	2122
Stannous telluride	2123
Stibine	162
Stibium	161
Stibnite	173
Stilleite	2354
Stolzite	1143
Strengite	1057
Strontia	1974
Strontianite	1961
Strontium orthosilicate	1979
Sulfamide	2004
Sulfinyl dibromide	2007
Sulfinyl dichloride	2008
Sulfinyl difluoride	2009
Sulfur chloride	1998
Sulfur decafluoride	2003
Sulfur monobromide	1994
Sulfur monochloride	1995
Sulfur monofluoride	1996, 1997
Sycoporite	665
Sylvite	1570
Szmikite	1281
Szomolnokite	1032
Tantalum boride	2014
Tantalum dioxide	2019
Tantalum diselenide	2020
Tantalum disulfide	2021
Tantalum ditelluride	2022
Tantalum pentabromide	2023

Tantalum pentachloride	2024	Titanium dibromide	2141	Uranium pentachloride	2221
Tantalum pentafluoride	2025	Titanium dichloride	2142	Uranium pentafluoride	2222
Tantalum pentaiodide	2026	Titanium diiodide	2143	Uranium tetrabromide	2214
Tantalum pentoxide	2027	Titanium dioxide	2157	Uranium tetrachloride	2215
Tarapacaite	1571	Titanium disilicide	2140	Uranium tetrafluoride	2216
Technetium hexafluoride	2030	Titanium disulfide	2159	Uranium tetraiodide	2217
Technetium pentafluoride	2029	Titanium monosulfide	2145	Uranium tribromide	2209
Tellurite	2034	Titanium monoxide	2144	Uranium tricarbide	2206
Tellurous bromide	2036	Titanium pentoxide	2152	Uranium trichloride	2210
Tellurous chloride	2037	Titanium tetrabromide	2153	Uranium trifluoride	2211
Tenorite	736	Titanium tetrachloride	2154	Uranium trihydride	2212
Tephoroite	1277	Titanium tetrafluoride	2155	Uranium triiodide	2213
Tephroite	1278	Titanium tetraiodide	2156	Uranium trioxide	2226
Terbia	2048	Titanium tribromide	2146	Uranyl ammonium fluoride	158
Tetrabromosilane	1793	Titanium trichloride	2147	Vanadium dibromide	2240
Tetrachlorodiborane(4)	361	Titanium trifluoride	2148	Vanadium dichloride	2241
Tetrachlorosilane	1794	Titanium trioxide	2149	Vanadium difluoride	2242
Tetracobalt dodecacarbonyl	622	Titanium trisulfide	2151	Vanadium diiodide	2243
Tetradymite	338	Titanyl sulfate monohydrate	2158	Vanadium dioxide	2258
Tetrafluorosilane	1795	Triazaborane	351	Vanadium hexacarbonyl	2236
Tetraiodine nonaoxide	979	Tribromochlorosilane	1798	Vanadium monoxide	2244
Tetraiodosilane	1796	Tribromogermane	862	Vanadium oxide bromide	2261
Tetrasilane	1805	Tribromosilane	1797	Vanadium oxide chloride	2262
Thallic bromide tetrahydrate	2074	Trichlorofluorogermane	864	Vanadium pentafluoride	2259
Thallic chloride	2075	Trichlorofluorosilane	1800	Vanadium pentoxide	2260
Thallic chloride tetrahydrate	2076	Trichlorogermane	863	Vanadium tetrabromide	2255
Thallic fluoride	2077	Trichlorosilane	1799	Vanadium tetrachloride	2256
Thallic nitrate	2078	Tridymite	1815	Vanadium tetrafluoride	2257
Thallic oxide	2079	Trifluoroamine oxide	1462	Vanadium tribromide	2247
Thallium sesquioxide	2079	Trifluorosilane	1801	Vanadium trichloride	2248
Thallous acetate	2052	Trigermane	867	Vanadium trifluoride	2249
Thallous bromide	2053	Triiodosilane	1802	Vanadium trifluoride trihydrate	2250
Thallous carbonate	2054	Trisilane	1804	Vanadium triiodide	2251
Thallous chlorate	2055	Troilite	1034	Vanadium trioxide	2252
Thallous chloride	2056	Tschernigite	9	Vanadium trisulfate	2253
Thallous cyanide	2057	Tungsten dibromide	2172	Vanadium trisulfide	2254
Thallous ethoxide	2058	Tungsten dichloride	2173	Verdigris	704
Thallous fluoride	2059	Tungsten diiodide	2174	Villiaumite	1892
Thallous formate	2060	Tungsten dinitride	2168	Vitreous silica	1817
Thallous hexafluorophosphate	2061	Tungsten dioxide	2181	Vivianite	1028
Thallous hydroxide	2062	Tungsten dioxydibromide	2192	Vulcanite	749
Thallous iodide	2063	Tungsten diselenide	2182	Washing soda	1874
Thallous molybdate	2064	Tungsten disilicide	2170	Whewellite	456
Thallous nitrate	2065	Tungsten disulfide	2183	White lead	1098
Thallous nitrite	2066	Tungsten ditelluride	2184	White phosphorus	1487
Thallous oxalate	2067	Tungsten hexabromide	2190	Whitlockite	461
Thallous oxide	2068	Tungsten hexacarbonyl	2167	Willemite	2355
Thallous perchlorate	2069	Tungsten hexachloride	2191	Witherite	215
Thallous selenate	2070	Tungsten hexafluoride	2195	Wolfram	2160
Thallous selenide	2071	Tungsten oxybromide	2197	Wulfenite	1115
Thallous sulfate	2072	Tungsten oxychloride	2198	Wulframite	415
Thallous sulfide	2073	Tungsten oxyfluoride	2199	Wurtzite	2361
Thenard's blue	625	Tungsten oxyiodide	2194	Wustite	1027
Thenardite	1938	Tungsten oxytribromide	2188	YAG	2298
Thermonatrite	1875	Tungsten oxytrichloride	2189	Yellow phosphorus	1487
Thiophosphoryl chloride difluoride	1523	Tungsten pentaboride	2164	Ytterbia	2295
Thiophosphoryl trifluoride	1521	Tungsten pentabromide	2185	Yttria	2308
Thiothionyl bromide	1994	Tungsten pentachloride	2186	Zinc ammonium chloride	131,152
Thiothionyl chloride	1995	Tungsten pentafluoride	2187	Zinc dodecanoate	2337
Thiothionyl fluoride	1996	Tungsten tetrabromide	2177	Zinc metaarsenite	2317
Thoria	2091	Tungsten tetrachloride	2178	Zinc octanoate	2322
Thorianite	2091	Tungsten tetrafluoride	2179	Zinc orthoarsenate	2315
Thorite	2093	Tungsten tetraiodide	2180	Zinc orthosilicate	2355
Thorium hexaboride	2082	Tungsten tribromide	2175	Zincite	2345
Thorium orthosilicate	2093	Tungsten trichloride	2176	Zinkosite	2357
Thorium tetrabromide	2083	Tungsten trioxide	2196	Zircon	2378
Thorium tetrachloride	2086	Tungsten trisulfide	2200	Zirconia	2376
Thorium tetrafluoride	2087	Tungstenite	2183	Zirconium ammonium carbonate	160
Thorium tetraiodide	2088	Tysonite	1073	Zirconium orthosilicate	2378
Thulia	2103	Uraninite	2218	Zirconium tetrabromide	2367
Tiemannite	1324	Uranium dioxide	2218	Zirconium tetrachloride	2369
Tin hydride	2106	Uranium hexachloride	2224	Zirconium tetrafluoride	2370
Tincal	1948	Uranium hexafluoride	2225	Zirconium tetraiodide	2373
Tincalconite	1949	Uranium oxychloride	2228		
Titanium ammonium oxalate monohydrate	156	Uranium pentabromide	2220		

CAS		CAS		CAS		CAS		CAS	
62-54-4	417	555-75-9	23	1303-35-1	189	1313-29-7	1347	1332-52-1	293
62-76-0	1923	557-05-1	2356	1303-36-2	197	1313-49-1	2341	1332-58-7	47
71-48-7	623	557-07-3	2343	1303-37-3	203	1313-59-3	1924	1332-63-4	1291
74-90-8	942	557-09-5	2322	1303-52-2	895	1313-60-6	1930	1332-71-4	679
75-15-0	490	557-21-1	2328	1303-58-8	897	1313-82-2	1940	1332-77-0	1631
75-20-7	425	557-42-6	2364	1303-60-2	890	1313-83-3	1942	1332-81-6	175
75-44-5	496	562-76-5*	1634	1303-61-3	900	1313-84-4	1941	1333-74-0	939
124-38-9	488	563-63-3	1823	1303-62-4	899	1313-85-5	1936	1333-82-0	608
126-45-4	1832	563-68-8	2052	1303-86-2	366	1313-96-8	1437	1333-83-1	1900
127-08-2	1557	563-71-3	1013	1303-96-4	1948	1313-97-9	1374	1335-26-8	1226
127-09-3	1861	563-72-4	455	1304-28-5	252	1313-99-1	1404	1335-31-5	1322
128-00-7	1839	584-08-7	1567	1304-29-6	256	1314-05-2	1405	1335-32-6	1088
131-74-8	139	584-09-8	1707	1304-39-8	261	1314-06-3	1412	1336-21-6	116
141-53-7	1894	590-28-3	1572	1304-40-1	265	1314-08-5	1486	1341-49-7	109
142-03-0	22	590-29-4	1582	1304-54-7	306	1314-11-0	1974	1344-28-1	40
142-71-2	702	592-01-8	436	1304-56-9	307	1314-12-1	2068	1344-43-0	1274
143-33-9	1883	592-04-1	1311	1304-76-3	328	1314-13-2	2345	1344-48-5	1327
144-55-8	1899	592-05-2	1105	1304-82-1	338	1314-15-4	1539	1344-48-5	1326
146-84-9	1849	592-85-8	1329	1305-62-0	445	1314-18-7	1975	1344-54-3	2149
151-50-8	1573	592-87-0	1140	1305-78-8	457	1314-20-1	2091	1344-57-6	2218
156-62-7	435	593-26-0	130	1305-79-9	460	1314-22-3	2348	1344-58-7	2226
298-14-6	1590	593-95-3	495	1305-84-6	466	1314-23-4	2376	1344-59-8	2223
301-04-2	1086	598-54-9	685	1305-99-3	463	1314-24-5	1509	1344-95-2	467
302-01-2	928	598-62-9	1261	1306-19-0	404	1314-27-8	1147	1345-04-6	173
304-59-6	1933	598-63-0	1097	1306-23-6	412	1314-28-9	1682	1345-07-9	337
333-20-0	1636	628-52-4*	574	1306-24-7	408	1314-32-5	2079	1345-13-7	520
353-50-4	497	628-86-4	1314	1306-25-8	413	1314-34-7	2252	1345-25-1	1027
409-21-2	1810	630-08-0	487	1306-38-3	524	1314-35-8	2196	1449-65-6	865
460-19-5	498	631-60-7	1297	1307-86-4	674	1314-36-9	2308	1495-50-7	501
463-58-1	492	631-61-8	63	1307-96-6	653	1314-37-0	2295	1590-87-0	1803
471-34-1	426	637-12-7	48	1307-99-9	658	1314-41-6	1148	1600-27-7	1306
471-34-1	427	638-39-1	2108	1308-04-9	676	1314-56-3	1515	1603-84-5	491
497-19-8	1873	660-60-6	743	1308-06-1	669	1314-60-9	179	1631-78-3	2107
504-64-3	493	688-37-9	37	1308-14-1	591	1314-61-0	2027	1633-05-2	1961
506-64-9	1833	811-54-1	1108	1308-31-2	1017	1314-62-1	2260	1762-95-4	154
506-65-0	888	813-93-4	317	1308-38-9	595	1314-68-7	1686	1863-63-4	65
506-66-1	297	814-88-0	402	1308-56-1	723	1314-80-3	1520	2092-16-2	477
506-68-3	499	814-89-1	651	1308-80-1	683	1314-82-5	1519	2092-17-3	276
506-77-4	500	814-89-1*	652	1308-87-8	764	1314-84-7	2351	2092-17-3*	277
506-78-5	502	814-90-4*	580	1308-96-9	798	1314-86-9	1510	2404-52-6	1521
506-80-9	489	814-91-5	735	1309-33-7	1052	1314-87-0	1136	2406-52-2	2106
506-87-6	71	814-93-7	1119	1309-37-1	1056	1314-91-6	1139	2452-01-9	2337
507-16-4	2007	814-94-8	2116	1309-42-8	1214	1314-95-0	2121	2466-09-3	1494
512-25-4*	221	815-82-7	748	1309-48-4	1222	1314-96-1	1982	2524-02-9	1523
512-26-5	1104	815-85-0	2122	1309-60-0	1153	1314-97-2	2073	2551-62-4	2000
513-74-6	85	819-73-8	1096	1309-64-4	169	1314-98-3	2361	2644-70-4	931
513-77-9	215	865-44-1	982	1309-64-4	168	1314-98-3	2360	2696-92-6	1460
513-78-0	386	868-14-4	1599	1310-03-8	1129	1315-01-1	2133	2699-79-8	2006
513-79-1	630	917-61-3	1882	1310-32-3	1029	1315-03-3	2254	3012-65-5	108
516-02-9	250	917-69-1	670	1310-43-6	1006	1315-04-4	180	3095-65-6	115
528-94-9	140	940-71-6	1502	1310-52-7	1211	1315-05-5	171	3164-29-2	148
533-51-7	1843	992-98-3	2060	1310-53-8	880	1315-06-6	2119	3251-23-8	731
534-16-7	1828	1002-89-7	143	1310-58-3	1600	1315-07-7	1978	3333-67-3	1394
534-17-8	531	1066-30-4*	583	1310-61-8	1596	1315-09-9	2354	3486-35-9	2323
535-37-5*	893	1066-33-7	107	1310-61-8*	1597	1315-11-3	2363	3522-50-7	1047
538-17-0	53	1072-35-1	1134	1310-65-2	1170	1317-33-5	1353	3687-31-8	1090
540-16-9	712	1111-67-7	700	1310-66-3	1171	1317-34-6	1292	3811-04-9	1569
540-69-2	95	1111-71-3	301	1310-73-2	1909	1317-35-7	1289	3982-91-0	1522
540-72-7	1953	1111-78-0	70	1310-82-3	1715	1317-36-8	1120	4109-96-0	1787
541-43-5	227	1117-94-8	686	1311-11-1	1122	1317-36-8	1121	4119-52-2	1062
542-62-1	222	1120-44-1	734	1311-90-6	138	1317-37-9	1034	5341-61-7	932
542-83-6	392	1120-46-3	1118	1312-41-0	951	1317-38-0	736	5714-22-7	2003
542-84-7	636	1184-65-2	863	1312-42-1	970	1317-39-1	695	5743-04-4	380
543-80-6	207	1299-86-1	18	1312-43-2	967	1317-40-4	747	5743-26-0	418
543-81-7	289	1301-96-8	1858	1312-45-4	973	1317-41-5	741	5794-28-5	456
543-90-8	379	1302-01-8	1822	1312-46-5	994	1317-42-6	665	5892-10-4	314
544-17-2	440	1302-09-6	1851	1312-73-8	1625	1317-61-9	1039	5893-61-8	727
544-19-4	726	1302-42-7	1863	1312-73-8*	1626	1317-66-4	1007	5908-64-5	208
544-60-5	128	1302-52-9	292	1312-74-9	1622	1319-46-6	1098	5908-81-6	273
544-92-3	690	1302-81-4	51	1312-81-8	1079	1327-41-9	30	5968-11-6	1875
546-46-3	2327	1302-82-5	45	1312-99-8	1245	1327-44-2	1559	5970-44-5	2304
546-67-8	1149	1303-00-0	827	1313-04-8	1231	1327-50-0	174	5970-45-6	2313
546-93-0	1205	1303-11-3	952	1313-08-2	1237	1327-53-3	196	5970-62-7	2332
547-68-2	2344	1303-28-2	202	1313-13-9	1293	1327-53-3	195	5972-71-4	66
554-13-2	1163	1303-33-9	198	1313-22-0	1279	1330-43-4	1947	5972-72-5*	110
555-35-1	42	1303-34-0	204	1313-27-5	1363	1332-14-5	746	5972-76-9	69

CAS No.	No.	CAS No.	No.	CAS No.	No.	CAS No.	No.	CAS No.	No.
6009-70-7	129	7440-34-8	1	7580-67-8	1169	7757-86-0	1213	7783-11-1	147
6010-09-9	1036	7440-35-9	54	7601-54-9	1932	7757-93-9	443	7783-18-8	155
6046-93-1	704	7440-36-0	161	7601-89-0	1925	7758-01-2	1565	7783-19-9	142
6047-25-2	1026	7440-37-1	181	7601-90-3	548	7758-02-3	1566	7783-20-2	145
6080-56-4	1087	7440-38-2	182	7616-94-6	557	7758-05-6	1602	7783-21-3	141
6108-17-4	1155	7440-39-3	206	7631-90-5	1908	7758-09-0	1610	7783-22-4	157
6108-23-2	1168	7440-40-6	286	7631-94-9*	1889	7758-11-4	1592	7783-26-8	1804
6131-90-4	1862	7440-40-6	287	7631-95-0	1918	7758-16-9	1888	7783-28-0	111
6132-02-1	1874	7440-41-7	288	7631-99-4	1921	7758-19-2	1878	7783-29-1	1805
6132-04-3	1881	7440-42-8	341	7632-00-0	1922	7758-23-8*	462	7783-32-6	1316
6147-53-1	624	7440-43-9	378	7632-51-1	2256	7758-87-4	461	7783-33-7	1635
6150-82-9	1210	7440-44-0	483	7637-07-2	362	7758-88-5	517	7783-34-8	1320
6150-88-5	1221	7440-45-1	503	7646-69-7	1897	7758-89-6	689	7783-35-9	1325
6303-21-5	1498	7440-46-2	526	7646-78-8	2125	7758-94-3	1014	7783-36-0	1305
6381-79-9	1568	7440-47-3	559	7646-79-9	631	7758-95-4	1100	7783-39-3	1313
6424-20-0	641	7440-48-4	610	7646-85-7	2325	7758-97-6	1102	7783-40-6	1209
6484-52-2	126	7440-50-8	681	7646-93-7	1595	7758-98-7	744	7783-41-7	803
6487-39-4	1070	7440-51-9	752	7647-01-0	941	7758-99-8	745	7783-42-8	2009
6487-48-5	1611	7440-52-0	767	7647-10-1	1482	7759-00-4	1277	7783-43-9	1776
6533-73-9	2054	7440-53-1	782	7647-14-5	1877	7759-01-5	1143	7783-44-0	804
6556-16-7	1273	7440-54-2	808	7647-15-6	1871	7759-01-5	1144	7783-46-2	1106
6569-51-3	351	7440-55-3	825	7647-17-8	533	7759-02-6	1981	7783-47-3	2112
6591-55-5	327	7440-56-4	851	7647-18-9	176	7761-88-8	1841	7783-48-4	1966
6834-92-0	1917	7440-57-5	883	7647-19-0	1514	7772-99-8	2110	7783-49-5	2329
6865-35-6	269	7440-58-6	901	7659-31-6	1824	7773-01-5	1262	7783-50-8	1050
7320-34-5*	1619	7440-59-7	918	7664-38-2	1492	7773-03-7	1598	7783-51-9	838
7429-90-5	7	7440-60-0	919	7664-39-3	943	7773-06-0	144	7783-52-0	963
7429-91-6	753	7440-61-1	2201	7664-41-7	62	7774-29-0	1317	7783-53-1	1290
7429-92-7	766	7440-62-2	2231	7664-93-9	1986	7774-34-7	432	7783-54-2	1453
7439-88-5	987	7440-63-3	2272	7681-11-0	1603	7775-09-9	1876	7783-55-3	1507
7439-89-6	998	7440-64-4	2287	7681-38-1	1904	7775-11-3	1879	7783-56-4	166
7439-90-9	1064	7440-65-5	2297	7681-49-4	1892	7775-19-1	1916	7783-57-5	2077
7439-91-0	1066	7440-66-6	2312	7681-52-9	1910	7775-27-1	1931	7783-58-6	878
7439-92-1	1085	7440-67-7	2365	7681-52-9*	1911	7775-41-9	1835	7783-59-7	1152
7439-93-2	1154	7440-68-8	205	7681-55-2	1912	7778-18-9	470	7783-60-0	1999
7439-94-3	1186	7440-69-9	313	7681-57-4	1915	7778-39-4*	186	7783-61-1	1795
7439-95-4	1197	7440-70-0	416	7681-65-4	693	7778-39-4	185	7783-62-2	2128
7439-96-5	1247	7440-71-3	481	7681-82-5	1913	7778-44-1	421	7783-63-3	2155
7439-97-6	1296	7440-72-4	801	7693-26-7	1588	7778-50-9	1574	7783-64-4	2370
7439-98-7	1330	7440-73-5	807	7697-37-2	1443	7778-53-2	1618	7783-66-6	985
7439-99-8	1379	7440-74-6	950	7699-41-4	1807	7778-74-7	1613	7783-68-8	1435
7440-00-8	1365	7446-06-2	1555	7699-45-8	2321	7778-77-0	1576	7783-70-2	177
7440-01-9	1378	7446-07-3	2034	7704-34-9	1985	7778-80-5	1624	7783-71-3	2025
7440-02-0	1381	7446-08-4	1766	7704-34-9	1984	7779-88-6	2339	7783-72-4	2259
7440-03-1	1414	7446-09-5	1992	7704-98-5	2135	7779-90-0	2349	7783-75-7	997
7440-04-2	1467	7446-10-8	1137	7704-99-6	2371	7782-40-3	482	7783-77-9	1359
7440-05-3	1479	7446-11-9	1993	7705-07-9	2147	7782-41-4	802	7783-79-1	1773
7440-06-4	1524	7446-14-2	1135	7705-08-0	1044	7782-44-7	1477	7783-80-4	2042
7440-07-5	1542	7446-15-3	1130	7718-54-9	1395	7782-49-2	1761	7783-81-5	2225
7440-08-6	1553	7446-16-4	1663	7718-98-1	2248	7782-49-2	1762	7783-82-6	2195
7440-09-7	1556	7446-18-6	2072	7719-09-7	2008	7782-49-2	1763	7783-84-8	1059
7440-10-0	1643	7446-19-7	2358	7719-12-2	1504	7782-50-5	546	7783-86-0	1021
7440-11-1	1295	7446-20-0	2359	7720-78-7	1031	7782-61-8	1054	7783-86-0*	1022
7440-12-2	1656	7446-21-1	1977	7720-83-4	2156	7782-63-0	1033	7783-89-3	1826
7440-13-3	1657	7446-22-2	2070	7721-01-9	2024	7782-64-1	1265	7783-90-6	1830
7440-14-4	1659	7446-25-5	2132	7722-64-7	1615	7782-65-2	853	7783-92-8	1829
7440-15-5	1665	7446-26-6	2352	7722-76-1	84	7782-66-3	1323	7783-93-9	1845
7440-16-6	1692	7446-27-7	1125	7722-84-1	945	7782-68-5	975	7783-95-1	1859
7440-17-7	1702	7446-28-8	1976	7722-86-3	1987	7782-70-9	1594	7783-96-2	1838
7440-18-8	1725	7446-31-3	2380	7722-88-5	1934	7782-77-6	1444	7783-97-3	1837
7440-19-9	1737	7446-32-4	172	7723-14-0	1489	7782-78-7	1990	7783-98-4	1847
7440-20-2	1753	7446-33-5	2310	7723-14-0	1487	7782-79-8	936	7783-99-5	1842
7440-21-3	1780	7446-70-0	20	7723-14-0	1488	7782-85-6	1903	7784-01-2	1831
7440-22-4	1820	7447-39-4	715	7726-95-6	369	7782-86-7	1303	7784-02-3	1834
7440-23-5	1860	7447-40-7	1570	7727-15-3	16	7782-87-8	1601	7784-03-4	1856
7440-24-6	1956	7447-41-8	1164	7727-18-6	2269	7782-89-0	1157	7784-05-6	1852
7440-25-7	2011	7487-88-9	1238	7727-21-1	1617	7782-91-4	1358	7784-07-8	1850
7440-26-8	2028	7487-94-7	1309	7727-37-9	1441	7782-92-5	1865	7784-09-0	1848
7440-27-9	2043	7488-51-9	1132	7727-43-7	270	7782-94-7	1442	7784-11-4	17
7440-28-0	2051	7488-54-2	1722	7727-54-0	134	7782-99-2	1991	7784-13-6	21
7440-29-1	2080	7488-55-3	2120	7727-73-3	1939	7783-00-8	1765	7784-14-7	150
7440-30-4	2097	7543-51-3	2350	7732-18-5	2271	7783-03-1	2161	7784-16-9	1951
7440-31-5	2104	7550-35-8	1162	7733-02-0	2357	7783-06-4	947	7784-18-1	24
7440-31-5	2105	7550-45-0	2154	7757-79-1	1609	7783-07-5	946	7784-19-2	100
7440-32-6	2134	7553-56-2	974	7757-82-6	1938	7783-08-6	1764	7784-21-6	28
7440-33-7	2160	7558-79-4	1901	7757-83-7	1943	7783-09-7	949	7784-22-7	31

7784-23-8	32	7789-28-8	1018	7790-92-3	547
7784-24-9	1561	7789-29-9	1591	7790-94-5	1988
7784-25-0	8	7789-30-2	377	7790-98-9	132
7784-26-1	9	7789-31-3	370	7790-99-0	981
7784-27-2	35	7789-33-5	980	7791-03-9	1179
7784-29-4	1704	7789-36-8	1202	7791-07-3	1926
7784-30-7	38	7789-38-0	1870	7791-08-4	170
7784-31-8	50	7789-39-1	1706	7791-09-5	1687
7784-33-0	191	7789-40-4	2053	7791-10-8	1962
7784-34-1	192	7789-41-5	423	7791-11-9	1708
7784-35-2	193	7789-42-6	384	7791-12-0	2056
7784-36-3	201	7789-43-7	628	7791-13-1	633
7784-37-4	1315	7789-45-9	711	7791-16-4	178
7784-40-9	1109	7789-46-0	1011	7791-18-6	1208
7784-41-0	1562	7789-47-1	1308	7791-20-0	1396
7784-41-0	1575	7789-48-2	1203	7791-21-1	549
7784-42-1	183	7789-51-7	1774	7791-23-3	1775
7784-44-3	105	7789-52-8	1768	7791-25-5	2005
7784-45-4	194	7789-53-9	1959	7791-26-6	2228
7784-46-5	1867	7789-54-0	2036	7791-28-8	213
7784-48-7	1391	7789-57-3	1797	7803-49-8	937
7785-20-8	125	7789-59-5	1516	7803-51-2	1490
7785-23-1	1827	7789-60-8	1503	7803-52-3	162
7785-87-7	1280	7789-61-9	164	7803-54-5	1199
7786-30-3	1207	7789-65-3	1770	7803-55-6	122
7786-81-4	1406	7789-66-4	1793	7803-57-8	929
7787-32-8	226	7789-67-5	2124	7803-58-9	2004
7787-33-9	241	7789-68-6	2153	7803-60-3	1495
7787-34-0	239	7789-69-7	1512	7803-62-5	1781
7787-35-1	242	7789-75-5	439	7803-63-6	112
7787-36-2	255	7789-77-7	444	7803-65-8	117
7787-37-3	244	7789-78-8	442	7803-68-1	2032
7787-38-4	249	7789-79-9	446	10022-31-8	246
7787-39-5	272	7789-80-2	447	10022-47-6	77
7787-40-8	280	7789-82-4	450	10022-48-7	1166
7787-41-9	260	7790-11-8	1614	10022-50-1	1464
7787-42-0	282	7790-22-9	1173	10022-68-1	401
7787-46-4	296	7790-28-5	1927	10024-93-8	1368
7787-47-5	299	7790-29-6	1716	10024-97-2	1445
7787-48-6	308	7790-30-9	2063	10025-64-6	2346
7787-49-7	300	7790-31-0	1216	10025-65-7	1528
7787-52-2	302	7790-33-2	1267	10025-66-8	1661
7787-53-3	304	7790-33-2*	1268	10025-67-9	1995
7787-56-6	311	7790-34-3	1402	10025-68-0	1769
7787-57-7	329	7790-37-6	2335	10025-69-1	2111
7787-58-8	315	7790-38-7	1484	10025-70-4	1964
7787-59-9	330	7790-39-8	1529	10025-71-5	2037
7787-60-2	316	7790-41-2	993	10025-73-7	587
7787-61-3	318	7790-42-3	1639	10025-74-8	760
7787-62-4	319	7790-44-5	167	10025-75-9	771
7787-63-5	331	7790-46-7	1538	10025-76-0	794
7787-64-6	322	7790-47-8	2129	10025-77-1	1045
7787-68-0	336	7790-48-9	2041	10025-78-2	1799
7787-69-1	530	7790-49-0	2088	10025-82-8	962
7787-70-4	688	7790-56-9	1627	10025-83-9	991
7787-71-5	376	7790-58-1	1630	10025-85-1	1452
7788-97-8	589	7790-59-2	1621	10025-87-3	1517
7788-98-9	76	7790-60-5	1641	10025-90-8	1647
7788-99-0	599	7790-62-7	1620	10025-91-9	165
7789-00-6	1571	7790-63-8	1642	10025-93-1	2210
7789-01-7	1165	7790-69-4	1175	10025-99-7	1633
7789-02-8	594	7790-71-8	1181	10026-01-4	1471
7789-04-0	596	7790-74-1	465	10026-02-5	1554
7789-06-2	1965	7790-75-2	480	10026-03-6	1771
7789-08-4	86	7790-76-3	464	10026-04-7	1794
7789-09-5	82	7790-78-5	389	10026-06-9	2126
7789-10-8	1312	7790-79-6	393	10026-07-0	2039
7789-17-5	539	7790-80-9	396	10026-08-1	2086
7789-18-6	541	7790-81-0	395	10026-10-5	2215
7789-19-7	724	7790-84-3	410	10026-11-6	2369
7789-21-1	1989	7790-85-4	415	10026-12-7	1434
7789-23-3	1579	7790-86-5	515	10026-13-8	1513
7789-24-4	1167	7790-87-6	518	10026-17-2	639
7789-25-5	1461	7790-87-6*	519	10026-18-3	672
7789-26-6	805	7790-89-8	554	10026-20-7	656
7789-27-7	2059	7790-91-2	555	10026-22-9	650

10026-24-1	663	10058-44-3	1058
10028-14-5	1466	10060-09-0	407
10028-15-6	1478	10060-10-3	523
10028-18-9	1398	10060-11-4	869
10028-22-5	1060	10060-12-5	588
10031-13-7	1091	10060-13-6	80
10031-16-0	223	10090-53-6	33
10031-20-6	1260	10097-28-6	1812
10031-21-7	1094	10099-58-8	1071
10031-22-8	1095	10099-66-8	1189
10031-23-9	1660	10099-74-8	1117
10031-24-0	2109	10099-76-0	1127
10031-25-1	584	10099-79-3	1145
10031-25-1*	585	10101-41-4	471
10031-26-2	1043	10101-50-5*	1929
10031-27-3	2038	10101-52-7	2378
10031-30-8	438	10101-53-8	600
10031-43-3	733	10101-63-0	1113
10031-45-5	740	10101-68-5	1282
10031-48-8	739	10101-94-7	1133
10031-52-4	779	10101-97-0	1408
10031-52-4	800	10101-98-1	1407
10031-53-5	797	10102-02-0	2342
10034-76-1	472	10102-03-1	1450
10034-81-8	1223	10102-05-3	1485
10034-82-9	1880	10102-15-5	1944
10034-85-2	944	10102-17-7	1954
10034-88-5	1905	10102-18-8	1937
10034-93-2	935	10102-20-2	1946
10034-96-5	1281	10102-23-5	1935
10034-98-7	2296	10102-24-6	1182
10034-99-8	1239	10102-25-7	1184
10035-04-8	431	10102-34-8	1228
10035-05-9	429	10102-40-6	1919
10035-06-0	324	10102-43-9	1446
10035-10-6	940	10102-44-0	1447
10036-47-2	1458	10102-45-1	2065
10038-98-9	877	10102-49-5	1042
10039-31-3	309	10102-68-8	448
10039-32-4	1902	10102-71-3	1864
10039-54-0	938	10103-61-4	706
10039-55-1	933	10108-64-2	388
10042-76-9	1972	10112-91-1	1300
10042-88-3	2044	10118-76-0	459
10043-01-3	49	10124-36-4	409
10043-11-5	365	10124-37-5	451
10043-35-3	352	10124-41-1	478
10043-52-4	430	10124-43-3	662
10043-67-1	1560	10124-48-8	1307
10043-92-2	1664	10124-50-2	1589
10045-86-0	1057	10125-13-0	716
10045-89-3	91	10135-84-9	106
10045-94-0	1318	10137-74-3	428
10045-94-0*	1319	10138-04-2	88
10048-95-0	1898	10138-41-7	770
10048-98-3	231	10138-52-0	815
10048-99-4	275	10138-62-2	921
10049-01-1	333	10139-47-6	2336
10049-03-3	806	10141-05-6	649
10049-04-4	550	10170-69-1	1254
10049-05-5	576	10179-73-4	1030
10049-06-6	2142	10190-55-3	1115
10049-07-7	1695	10192-29-7	74
10049-08-8	1729	10192-30-0	114
10049-10-2	578	10193-36-9	1808
10049-11-3	604	10196-18-6	2340
10049-12-4	2249	10210-64-7	290
10049-12-4*	2250	10210-68-1	618
10049-14-6	2216	10213-09-9	2264
10049-16-8	2257	10213-10-2	1955
10049-17-9	1681	10214-39-8	1093
10049-21-5	1886	10241-05-1	1355
10049-23-7	2033	10257-55-3	474
10049-24-8*	990	10277-43-7	1077
10049-24-8	989	10290-12-7	707
10049-25-9	575	10294-26-5	1853

CAS	No.	CAS	No.	CAS	No.	CAS	No.	CAS	No.
10294-27-6	886	12004-76-1	2012	12019-52-2	699	12035-79-9	1380	12057-71-5	1220
10294-28-7	891	12005-21-9	2298	12019-88-4	755	12035-82-4	1530	12057-74-8	1229
10294-29-8	887	12005-67-3	61	12020-14-3	2368	12035-83-5	1544	12057-75-9	1200
10294-31-2	889	12005-69-5	357	12020-21-2	775	12035-98-2	2244	12057-92-0	1294
10294-32-3	898	12006-15-4	382	12020-28-9	777	12036-02-1	1473	12058-18-3	1352
10294-33-4	358	12006-77-8	617	12020-39-2	781	12036-09-8	1673	12058-20-7	1354
10294-34-5	360	12006-79-0	563	12020-58-5	784	12036-10-1	1733	12059-14-2	1389
10294-38-9	216	12006-84-7	1001	12020-65-4	791	12036-14-5	2019	12059-95-9	1551
10294-39-0	254	12006-86-9	1000	12020-66-5	790	12036-21-4	2258	12060-00-3	1142
10294-40-3	220	12006-99-4	1331	12020-69-8	792	12036-22-5	2181	12060-01-4	1146
10294-42-5	525	12007-00-0	1385	12021-68-0	642	12036-32-7	1652	12060-08-1	1758
10294-44-7	1299	12007-01-1	1386	12022-95-6	1008	12036-34-9	1549	12060-58-1	1749
10294-46-9	738	12007-02-2	1384	12022-99-0	1009	12036-35-0	1698	12060-59-2	1983
10294-47-0	1099	12007-07-7	2013	12023-99-3	841	12036-41-8	2048	12061-16-4	776
10294-48-1	553	12007-09-9	2163	12024-10-1	833	12036-44-1	2103	12062-24-7	728
10294-50-5	655	12007-10-2	2162	12024-11-2	832	12037-15-9	2219	12063-98-8	829
10294-52-7	1046	12007-16-8	564	12024-14-5	834	12037-65-9	2139	12064-03-8	826
10294-53-8	1048	12007-23-7	902	12024-20-3	830	12037-80-8	2377	12064-62-9	821
10294-54-9	545	12007-25-9	1201	12024-21-4	844	12037-82-0	1500	12065-10-0	873
10294-58-3	1126	12007-29-3	1416	12024-22-5	849	12038-12-9	1655	12065-11-1	881
10294-62-9	1081	12007-33-9	368	12024-24-7	846	12038-13-0	1654	12065-36-0	852
10294-64-1	1604	12007-34-0	1754	12024-27-0	850	12038-20-9	1531	12065-65-5	2152
10294-66-3	1637	12007-35-1	2014	12024-81-6	813	12038-21-0	1540	12065-68-8	1138
10294-70-9	2115	12007-36-2	2202	12025-32-0	874	12038-63-0	1674	12066-83-0	1653
10325-94-7	400	12007-37-3	2233	12025-34-2	882	12038-67-4	1691	12067-00-4	1675
10326-21-3*	1206	12007-38-4	565	12025-39-7	875	12039-07-5	2145	12067-22-0	1751
10326-24-6	2317	12007-84-0	2203	12026-06-1	2062	12039-13-3	2159	12067-46-8	2182
10326-26-8	211	12007-97-5	1332	12026-24-3	2114	12039-15-5	2381	12067-54-8	2094
10326-27-9	218	12007-98-6	2164	12027-06-4	119	12039-16-6	2151	12067-66-2	2022
10326-28-0	405	12007-99-7	422	12028-48-7	121	12039-19-9	2311	12067-76-4	2184
10340-06-4	494	12008-02-5	504	12029-81-1	924	12039-35-9	2314	12068-49-4	1010
10343-61-0	2150	12008-05-8	783	12029-98-0	978	12039-55-3	2020	12068-69-8	334
10361-37-2	217	12008-06-9	809	12030-14-7	960	12039-76-8	2239	12068-90-5	1328
10361-43-0	321	12008-19-4	347	12030-24-9	972	12039-79-1	2018	12069-00-0	1131
10361-46-3	332	12008-21-8	1067	12030-49-8	995	12039-80-4	2049	12069-32-8	359
10361-79-2	1646	12008-23-0	1366	12030-51-2	996	12039-83-7	2140	12069-69-1	713
10361-82-7	1745	12008-27-4	1644	12030-85-2	1608	12039-87-1	2238	12069-85-1	904
10361-84-9	1756	12008-29-6	1738	12030-88-5	1628	12039-88-2	2170	12069-89-5	1334
10361-91-8	2292	12008-32-1	2301	12030-97-6	1638	12039-89-3	2288	12069-94-2	1417
10361-92-9	2305	12009-14-2	245	12031-30-0	1083	12039-90-6	2379	12070-06-3	2015
10361-95-2	2324	12009-18-6*	268	12031-49-1	1082	12039-91-5	2171	12070-07-4	2016
10377-48-7	1183	12009-18-6	267	12031-63-9	1174	12040-00-5	1752	12070-08-5	2137
10377-51-2	1172	12009-21-1	285	12031-80-0	1180	12040-02-7	2123	12070-09-6	2204
10377-52-3	1177	12011-97-1	1333	12032-20-1	1193	12041-50-8	13	12070-10-9	2234
10377-58-9	1215	12011-99-3	1418	12032-36-9	1241	12042-68-1	419	12070-12-1	2166
10377-60-3	1217	12012-16-7	2084	12032-74-5	1287	12042-78-3	420	12070-13-2	2165
10377-62-5	1225	12012-17-8	2235	12032-78-9	1256	12043-29-7	52	12071-15-7	1069
10377-66-9	1271	12012-32-7	505	12032-82-5	1248	12044-16-5	999	12071-29-3	1960
10377-93-2	1270	12012-35-0	566	12032-88-1	1286	12044-42-7	614	12071-31-7	2085
10380-31-1	283	12013-10-4	621	12032-97-2	1249	12044-54-1	199	12071-33-9	2205
10381-37-0	2095	12013-55-7	469	12033-19-1	1336	12045-01-1	616	12071-35-1	2303
10402-15-0	719	12013-56-8	468	12033-31-7	1337	12045-15-7	1250	12076-62-9	2206
10405-27-3	1455	12013-57-9	475	12033-54-4	1543	12045-16-8	1252	12088-65-2	1004
10450-55-2	1040	12013-82-0	453	12033-62-4	2017	12045-19-1	1415	12115-63-8	513
10450-60-9	976	12014-14-1	414	12033-64-6	2047	12045-27-1	2232	12124-97-9	68
10476-81-0	1958	12014-28-7	406	12033-65-7	2090	12045-63-5	2136	12124-99-1	113
10476-85-4	1963	12014-29-8	381	12033-72-6	2169	12045-64-6	2366	12125-01-8	92
10476-86-5	1969	12014-82-3	510	12033-83-9	2208	12045-87-3	1950	12125-02-9	75
10489-46-0	1699	12014-85-6	507	12033-88-4	1465	12045-88-4	1949	12125-08-5	97
10534-89-1	673	12014-93-6	522	12033-89-5	1811	12046-08-1	228	12125-09-6	1499
10544-72-6	1449	12016-80-7	677	12034-09-2	1920	12047-27-7	281	12125-19-8	2200
10544-73-7	1448	12017-01-5	647	12034-12-7	1945	12047-79-9	247	12125-22-3	1480
10545-99-0	1998	12017-08-2	660	12034-57-0	1422	12048-50-9	339	12125-25-6	1192
10553-31-8	212	12017-12-8	620	12034-59-2	1429	12048-51-0	340	12125-63-2	1035
10567-69-8	238	12017-13-9	666	12034-66-1	1420	12049-50-2	479	12133-07-2	756
10580-03-7	156	12017-38-8	680	12034-77-4	1430	12052-42-5	611	12133-10-7	765
10580-52-6	2241	12018-01-8	605	12034-80-9	1421	12053-12-2	561	12134-02-0	619
10588-01-9	1885	12018-09-6	573	12034-83-2	1432	12053-13-3	571	12134-75-7	811
11121-16-7	12	12018-10-9	718	12034-88-7	1116	12053-27-9	568	12134-77-9	823
11126-81-1*	15	12018-19-8	2326	12034-89-8	1971	12053-39-3	602	12135-76-1	146
12002-03-8	703	12018-22-3	601	12035-32-4	1376	12054-48-7	1399	12136-24-2	926
12002-61-8	6	12018-34-7	582	12035-35-7	1377	12054-85-2	123	12136-42-4	988
12002-99-2	1855	12018-36-9	572	12035-39-1	1410	12055-23-1	910	12136-45-7	1612
12003-63-3*	1558	12018-61-0	544	12035-52-8	1382	12055-62-8	925	12136-58-2	1185
12004-04-5	209	12019-11-3	684	12035-64-2	1388	12056-90-5	1080	12136-78-6	1339
12004-06-7	291	12019-23-7	749	12035-72-2	1413	12057-24-8	1178	12136-91-3	1501

CAS	No.	CAS	No.	CAS	No.	CAS	No.	CAS	No.
12136-97-9	1431	13126-12-0	1717	13463-22-4	251	13494-90-1	843	13637-76-8	1124
12137-12-1	1411	13135-31-4	2146	13463-30-4	1151	13494-91-2	847	13675-47-3	721
12137-20-1	2144	13172-31-1	1994	13463-39-3	1387	13494-98-9	2307	13682-61-6	1632
12137-25-6	1724	13205-44-2	350	13463-40-6	1002	13499-05-3	905	13693-05-5	1541
12137-27-8	1700	13255-26-0	264	13463-67-7	2157	13510-35-5	966	13693-06-6	1552
12137-83-6	1526	13268-12-3	87	13464-44-3	2315	13510-42-4	1719	13693-07-7	1701
12138-07-7	2096	13283-01-7	2191	13464-45-4	2316	13510-49-1	310	13693-08-8	1735
12138-09-9	2183	13320-71-3	1349	13464-46-5	57	13510-89-9	1089	13693-09-9	2277
12138-11-3	2050	13327-32-7	303	13464-58-9	187	13517-00-5	692	13701-67-2	361
12138-28-2	1980	13395-16-9	682	13464-82-9	971	13517-10-7	364	13701-70-7	2253
12139-93-4	661	13400-13-0	535	13464-92-1	385	13517-27-6	2320	13701-86-5	2190
12142-33-5*	1623	13424-46-9	1092	13464-97-6	934	13520-56-4	1061	13701-90-1	2074
12143-72-5	2021	13444-75-2*	1310	13465-07-1	948	13520-59-7	1348	13701-91-2	1150
12159-66-9	780	13444-85-4	1454	13465-09-3	961	13520-75-7	2192	13708-80-0	58
12160-99-5	824	13444-87-6	1459	13465-10-6	956	13520-76-8	2193	13709-31-4	2270
12162-21-9	913	13444-90-1	1463	13465-11-7	959	13520-77-9	2197	13709-35-8	1997
12162-59-3	927	13444-93-4	1470	13465-15-1	968	13520-78-0	2198	13709-36-9	2275
12163-20-1	1195	13444-94-5	1481	13465-55-9	1746	13520-79-1	2199	13709-38-1	1073
12163-22-3	1196	13444-96-7	1483	13465-58-2	1750	13520-83-7	2229	13709-42-7	1370
12163-69-8	1338	13445-50-6	1491	13465-59-3	1755	13520-87-1	2262	13709-46-1	1648
12164-94-2	64	13446-03-2	1259	13465-66-2	1772	13520-88-2	2261	13709-47-2	1757
12165-05-8	1679	13446-08-7	94	13465-71-9	1801	13520-89-3	2263	13709-49-4	2306
12165-69-4	1511	13446-09-8	118	13465-72-0	1802	13520-90-6	2268	13709-52-9	906
12166-29-9	1759	13446-10-1	133	13465-73-1	1782	13520-92-8	2382	13709-56-3	1550
12166-44-8	1760	13446-18-9	1219	13465-74-2	1783	13530-57-9	1703	13709-59-6	2087
12168-52-4	1037	13446-19-0	1224	13465-76-4	1798	13536-53-3	1645	13709-61-0	2276
12183-80-1	46	13446-23-6	1227	13465-78-6	1784	13536-73-7	769	13709-94-9	1606
12186-97-9	2299	13446-28-1	1230	13465-84-4	1796	13536-79-3	1068	13718-50-8	240
12187-14-3	399	13446-29-2	1242	13465-94-6	248	13536-80-6	1367	13718-59-7	262
12201-89-7	1390	13446-30-5	1244	13465-95-7	253	13537-09-2	762	13718-70-2	1023
12211-52-8	81	13446-34-9	1263	13466-08-5	1872	13537-15-0	799	13746-66-2	1577
12228-40-9	294	13446-49-6	1607	13466-20-1	257	13537-18-3	2099	13746-89-9	2374
12228-50-1	1251	13446-53-2	1204	13466-21-2	259	13537-30-9	860	13746-98-0	2078
12228-87-4	67	13446-56-5	1340	13469-98-2	2302	13537-33-2	1791	13755-29-8	1893
12228-91-0	1258	13446-57-6	1343	13470-06-9	1973	13548-38-4	593	13755-32-3	274
12229-63-9	2082	13446-74-7	1710	13470-08-1	2148	13548-42-0	717	13759-10-9	1819
12230-71-6	236	13446-75-8	1712	13470-10-5	2172	13550-53-3	773	13759-87-0	1744
12230-74-9	233	13450-87-8	822	13470-11-6	2185	13566-17-1	1128	13759-88-1	793
12232-99-4	1914	13450-88-9	836	13470-12-7	2173	13568-32-6	1278	13759-92-7	795
12249-52-4	1836	13450-90-3	837	13470-13-8	2178	13569-43-2	854	13760-02-6	1790
12254-64-7	55	13450-91-4	842	13470-14-9	2186	13569-49-8	1668	13760-41-3	1658
12254-82-9	612	13450-92-5	876	13470-17-2	2174	13569-50-1	508	13760-78-6	922
12254-85-2	562	13450-95-8	879	13470-19-4	2209	13569-59-0	1424	13760-79-7	2101
12255-48-0	2300	13451-11-1	2023	13470-20-7	2214	13569-62-5	1546	13760-80-0	2294
12256-04-1	615	13451-18-8	2035	13470-21-8	2221	13569-63-6	1669	13760-81-1	1190
12266-65-8	1253	13453-06-0	149	13470-22-9	2217	13569-70-5	1426	13760-83-3	772
12279-90-2	190	13453-07-1	892	13470-26-3	2247	13569-71-6	1671	13762-12-4	629
12280-52-3	537	13453-30-0	2055	13472-31-6	1928	13569-75-0	592	13762-14-6	648
12280-64-7	1711	13453-32-2	2075	13472-35-0	1887	13569-80-7	761	13762-51-1	1564
12294-01-8	2309	13453-32-2*	2076	13473-77-3	1194	13572-93-5	840	13763-23-0	2224
12300-22-0	1739	13453-34-4	2057	13476-01-2	627	13572-98-0	818	13763-67-2	532
12310-43-9	754	13453-40-2	2069	13476-08-9	1025	13573-02-9	861	13765-19-0	434
12310-44-0	768	13453-49-1	2083	13476-99-8	2246	13573-08-5	871	13765-24-7	1747
12322-34-8	163	13453-62-8	1123	13477-09-3	230	13573-11-0	1246	13765-25-8	796
12323-03-4	266	13453-69-5	1160	13477-19-5	397	13586-38-4	79	13765-26-9	817
12325-59-6	912	13454-75-6	529	13477-28-6	424	13587-35-4	750	13765-74-7	1840
12326-21-5	2227	13454-84-7	543	13477-29-7	433	13587-35-4*	751	13766-47-7	2179
12333-54-9	1255	13454-88-1	725	13477-34-4	452	13595-30-7	2255	13767-31-2	757
12340-14-6	2274	13454-94-9	521	13477-36-6	458	13597-20-1	1439	13767-32-3	2338
12345-14-1	219	13454-96-1	1536	13477-89-9	1369	13597-54-1	2353	13767-34-5	730
12399-08-5	977	13455-01-1	1508	13477-91-3	1375	13597-55-2	1979	13768-11-1	1666
12401-56-8	915	13455-12-4	1527	13477-95-7	1397	13597-65-4	2355	13768-38-2	1475
12419-43-1	2113	13455-15-7	1537	13478-00-7	1403	13597-73-4	1806	13768-86-0	1767
12434-24-1	785	13455-21-5	1580	13478-10-9	1016	13597-99-4	305	13768-94-0	1786
12442-63-6	552	13455-25-9	635	13478-14-3	1158	13598-22-6	312	13769-20-5	787
12513-27-8	2319	13455-28-2	644	13478-17-6	1341	13598-33-9	1967	13769-36-3	857
12540-13-5	705	13455-31-7	654	13478-18-7	1344	13598-36-2	1497	13770-18-8	737
12616-24-9*	160	13455-34-0	664	13478-20-1	1518	13598-42-0	1792	13770-56-4	188
12640-47-0	668	13460-50-9	353	13478-28-9	579	13598-56-6	2212	13773-81-4	1065
12650-28-1	263	13460-50-9	354	13478-38-1	732	13598-65-7	135	13774-85-1	2279
12688-52-7	1187	13460-50-9	355	13478-41-6	691	13600-89-0	671	13775-06-9	2211
12770-26-2	907	13462-88-9	1392	13478-45-0	1433	13601-13-3	722	13775-07-0	2222
12777-45-6	335	13462-88-9*	1393	13478-49-4	778	13601-19-9	1891	13775-16-1	2220
13007-92-6	567	13462-90-3	1401	13478-50-7	1141	13637-63-3	556	13775-18-3	2213
13011-54-6	1866	13462-93-6	83	13492-26-7	1593	13637-65-5	855	13775-53-6	1895
13092-75-6	1825	13463-12-2	1012	13494-80-9	2031	13637-68-8	1362	13775-80-9	930

CAS	No.	CAS	No.	CAS	No.	CAS	No.	CAS	No.
13776-58-4	2273	13940-89-1	1019	14763-77-0	720	16893-85-9	1896	20328-94-3	2285
13776-62-0	1457	13963-58-1	1585	14779-70-5	862	16903-35-8	885	20344-49-4	1053
13776-74-4	1233	13965-73-6	363	14808-60-7	1813	16919-19-0	104	20398-06-5	2058
13777-22-5	903	13966-94-4	957	14808-60-7	1814	16919-58-7	99	20405-64-5	696
13777-23-6	908	13967-25-4	1301	14871-79-5*	237	16921-30-5	1584	20427-11-6	637
13777-25-8	2367	13972-68-4	398	14884-42-5	606	16921-96-3	986	20427-56-9	1736
13778-39-7	2100	13973-87-0	373	14890-41-6	2240	16925-39-6	441	20427-58-1	2334
13779-41-4	1496	13982-53-1*	698	14933-38-1	56	16940-66-2	1869	20427-59-2	729
13779-92-5	1436	13986-18-0	2330	14940-41-1	1028	16940-92-4	96	20548-54-3	473
13780-06-8	454	13986-26-0	2373	14965-52-7	1800	16941-10-9	476	20601-83-6	1324
13780-42-2	848	14013-15-1	1269	14977-61-8	609	16941-11-0	103	20610-49-5	1662
13780-48-8	786	14013-86-6	1024	14984-81-7	1777	16941-12-1	1525	20619-16-3	872
13780-57-9	2002	14014-88-1	1728	15098-87-0	26	16949-15-8	1161	20644-87-5	2236
13782-01-9*	678	14024-18-1	1041	15162-90-0	1451	16961-83-4	1809	20661-21-6	965
13783-04-5	2141	14038-43-8	1049	15163-24-3	2175	16962-07-5	14	20665-52-5	845
13783-07-8	2143	14040-11-0	2167	15168-20-4	742	16962-31-5	1586	20667-12-3	1844
13798-24-8	2045	14049-36-6	1785	15179-32-5	2010	16962-47-3	102	20694-39-7	1272
13812-43-6	1456	14049-81-1	638	15192-17-3	1742	17014-71-0	1616	20712-42-9	403
13813-22-4	1076	14055-74-4	1342	15192-26-4	2040	17026-29-8	1684	20762-60-1	1563
13813-23-5	1649	14055-75-5	1346	15192-42-4	1672	17029-21-9	1685	20770-09-6	2131
13813-24-6	1371	14055-81-3	2177	15195-33-2	1440	17083-68-0	884	20816-12-0	1476
13813-25-7	1748	14055-84-6	2180	15195-53-6	1425	17099-70-6	27	20859-73-8	44
13813-40-6	2046	14059-33-7	326	15238-00-3	645	17125-80-3	229	20910-28-5	2230
13813-41-7	923	14075-53-7	1581	15238-00-3*	646	17194-00-2	234	21041-93-0	643
13813-42-8	774	14077-39-5	788	15243-27-3	1257	17375-41-6	1032	21041-95-2	394
13813-43-9	2102	14104-20-2	1857	15243-33-1	1726	17440-85-6	295	21109-95-5	271
13813-45-1	1191	14166-78-0	964	15244-35-6	411	17440-90-3	1005	21159-32-0	534
13813-46-2	1548	14168-73-1	1240	15321-51-4	1003	17496-59-2	551	21255-83-4	372
13813-47-3	59	14188-40-0	856	15364-10-0	2281	17702-41-9	349	21264-43-7	958
13814-72-7	812	14215-29-3	383	15364-94-0	1275	17786-31-1	622	21308-80-5	371
13814-74-9	1357	14215-30-6	708	15385-57-6	1302	18088-11-4	1718	21351-79-1	538
13814-75-0	1361	14217-21-1*	1890	15385-58-7	1298	18282-10-5	2130	21548-73-2	1854
13814-76-1	1683	14221-48-8*	89	15468-32-3	1815	18283-93-7	343	21645-51-2	29
13814-81-8	701	14242-05-8	1846	15469-38-2	1051	18288-22-7	320	21651-19-4	2117
13814-83-0	2265	14249-98-0	2189	15474-63-2	763	18356-71-3	1788	21679-31-2	560
13814-96-5	1107	14280-53-6	955	15477-33-5	19	18433-40-4	158	21908-53-2	1321
13817-37-3	640	14284-93-6	1727	15492-38-3	1697	18433-84-6	345	21995-38-0*	72
13818-75-2	814	14285-68-8	1667	15513-84-5	2243	18454-12-1	1103	22015-35-6	789
13818-89-8	866	14307-33-6*	437	15513-94-7	2251	18480-07-4	1968	22205-45-4	697
13819-84-6	1356	14312-00-6	391	15520-84-0	675	18618-55-8	516	22205-57-8	527
13820-41-2	151	14335-40-1	1505	15571-91-2*	1629	18624-44-7	1020	22326-55-2	235
13820-62-7	625	14336-80-2	687	15572-25-5	2071	18718-07-5	1264	22398-80-7	954
13821-06-2*	225	14402-70-1	127	15578-26-4	2118	18721-05-6	586	22441-45-8	200
13824-36-7	1789	14405-43-7	835	15587-72-1	1714	18810-58-7	210	22520-96-3	984
13825-74-6*	2158	14447-89-3	2194	15593-61-0	1232	18820-29-6	1285	22527-13-5	2280
13825-76-8	920	14452-39-2	43	15597-63-4	1506	18820-29-6	1283	22537-19-5	1084
13825-86-0	581	14456-34-9	911	15597-88-3	603	18820-29-6	1284	22750-54-5*	387
13826-63-6	2066	14456-48-5	759	15607-89-3	2001	18855-94-2	917	22750-57-8	528
13826-83-0	93	14456-51-0	2098	15622-42-1	1670	18868-43-4	1351	22756-36-1	1705
13842-80-3	2242	14456-53-2	1188	15696-40-9	1468	18917-82-3	1114	22831-39-6	1236
13842-83-6	1547	14457-84-2	41	15750-04-5	1218	18917-91-4	34	22831-42-1	11
13842-88-1	1427	14457-87-5	511	15752-41-7	1423	18933-05-6	1266	22886-66-4	839
13842-93-8	2030	14457-87-5*	512	15752-46-2	1545	19034-13-0	659	22986-54-5	3
13843-28-2	607	14459-59-7	1360	15823-43-5	916	19073-56-4	1709	23370-59-4	992
13845-17-5	224	14459-75-7	1438	15829-53-5	1304	19086-20-5	1243	23377-53-9	1690
13847-57-9	1101	14459-95-1	1578	15845-52-0	1110	19088-74-5	1713	23412-45-5	1350
13847-65-9	1462	14464-46-1	1816	15942-63-9	184	19139-47-0	509	23414-72-4	2347
13862-78-7	153	14475-63-9	2372	15947-41-8	60	19168-23-1	98	23611-30-5	1720
13863-41-7	374	14481-29-9*	90	16122-03-5*	124	19287-45-7	342	23777-80-2	346
13863-59-7	375	14483-17-1	1650	16399-77-2	1015	19357-83-6	2187	24094-93-7	569
13863-88-2	1821	14507-19-8	1075	16544-92-6	632	19357-86-9	2291	24304-00-5	36
13864-01-2	1074	14517-29-4	1372	16671-27-5	590	19423-81-5	816	24415-00-7	868
13870-13-8	914	14519-18-7	1957	16674-78-5	1198	19423-87-1	2293	24422-20-6	864
13870-21-8	1428	14521-18-7	1734	16689-88-6	2081	19465-30-6	348	24422-21-7	859
13870-24-1	1038	14523-22-9	1693	16721-80-5	1906	19467-79-9	1055	24597-12-4	831
13871-27-7	1952	14553-44-7	2093	16721-80-5	1907	19584-30-6	1694	24613-38-5	634
13873-84-2	983	14590-13-7	78	16731-55-8	1605	19597-69-4	1159	24621-21-4	1419
13874-75-4	1741	14590-19-3	657	16774-21-3	73	19598-90-4	819	24646-85-3	2237
13874-77-6	2290	14639-94-2	101	16812-54-7	1409	19624-22-7	344	24719-19-5	626
13875-06-4	2278	14639-97-5	152	16853-85-3	1156	19783-14-3	1111	25152-52-7	10
13876-85-2	694	14639-98-6	131	16860-99-4	1996	20193-56-0	2176	25417-81-6	232
13896-65-6	1731	14691-44-2	867	16871-60-6	1583	20193-58-2	1345	25502-05-0	2289
13918-22-4	1288	14693-15-3	2026	16871-71-9	2333	20205-91-8	367	25583-20-4	2138
13931-94-7	577	14693-82-4	969	16871-90-2	1587	20211-76-1	1072	25617-97-4	828
13939-06-5	1335	14720-21-9	894	16872-11-0	356	20213-56-3	2188	25617-98-5	953
13940-63-1	870	14721-21-2	714			20281-00-9	542	25658-42-8	2375

25658-43-9	2207	27218-16-2	558	35112-53-9	279	51621-05-7	1730	65842-03-7	1063
25659-31-8	1112	27774-13-6	2267	35718-37-7	2284	51898-99-8	323	68016-36-4	278
25764-08-3	506	27860-83-9	2331	36377-94-3	758	52003-58-4	120	68938-92-1	1534
25764-09-4	1651	30737-24-7	2067	36897-37-7	1400	52503-64-7	709	71500-16-8	1732
25764-10-7	1078	30937-52-1	1678	36907-42-3	2245	52788-53-1	820	71626-98-7	449
25764-11-8	1373	30937-53-2	1676	37267-86-0	1493	52951-38-9	325	72520-94-6	514
25764-15-2	810	31032-13-0	896	37773-49-2	1535	53731-35-4	1276	73560-00-6	979
25817-87-2	909	31052-14-9	2029	38455-77-5	2127	54120-05-7	1472	73796-25-5	1970
25895-60-7	1884	31052-43-4	1721	39290-85-2	710	54723-94-3	137	75926-26-0	1778
25909-39-1	1533	31083-74-6	1723	39368-69-9	1677	57246-89-6	1689	75926-28-2	1779
25985-07-3	1532	31234-26-1	1680	39416-30-3	284	58724-12-2	536	84359-31-9	597
26124-86-7	243	31576-40-6	1474	39796-98-0	2286	59201-51-3	1469	84359-31-9	598
26134-62-3	1176	32248-43-4	1743	39797-63-2	2283	60616-74-2	1212	92141-86-1	540
26342-61-0	570	32287-65-3	25	42246-24-2	1688	60676-86-0	1817	97126-35-7	667
26628-22-8	1868	32823-06-6	39	42739-38-8	159	60763-24-8	2092	99685-96-8	484
26686-77-1	1234	33088-16-3	2089	50432-32-1	2282	60804-25-3	1696	115383-22-7	485
26750-66-3	1640	33689-80-4	4	50801-97-3	1740	60883-64-9	298	123714-57-8	2362
27016-73-5	613	33689-81-5	2	50813-65-5	214	60922-26-1	2168	133578-89-9	2266
27016-75-7	1383	33689-82-6	5	50927-81-6	1818	60950-56-3	1235	133863-98-6	1364
27043-84-1	2318	34128-09-1	2064	51503-61-8	136	60969-19-9	2061	134929-59-2	486
27133-66-0	258	34330-64-8	390						

* Registry number refers to the anhydrous compound.

PHYSICAL PROPERTIES OF THE RARE EARTH METALS

K.A. Gschneidner, Jr.

Table 1
Data for the Trivalent Ions of the Rare Earth Elements

Rare earth	Symbol	Atomic no.	Atomic wt.[a]	Electronic configuration for R^{3+}				Spectroscopic ground state symbol
				No. 4f electrons	S	L	J	
Scandium	Sc	21	44.955910	0	—	—	—	—
Yttrium	Y	39	88.90585	0	—	—	—	—
Lanthanum	La	57	138.9055	0	—	—	—	—
Cerium	Ce	58	140.115	1	1/2	3	5/2	$^2F_{5/2}$
Praseodymium	Pr	59	140.90765	2	1	5	4	$^3H_4$
Neodymium	Nd	60	144.24	3	3/2	6	9/2	$^4I_{9/2}$
Promethium	Pm	61	(145)	4	2	6	4	$^5I_4$
Samarium	Sm	62	150.36	5	5/2	5	5/2	$^6H_{5/2}$
Europium	Eu	63	151.965	6	3	3	0	$^7F_0$
Gadolinium	Gd	64	157.25	7	7/2	0	7/2	$^8S_{7/2}$
Terbium	Tb	65	158.92534	8	3	3	6	$^7F_6$
Dysprosium	Dy	66	162.50	9	5/2	5	15/2	$^6H_{15/2}$
Holmium	Ho	67	164.93032	10	2	6	8	$^5I_8$
Erbium	Er	68	167.26	11	3/2	6	15/2	$^4I_{15/2}$
Thulium	Tm	69	168.93421	12	1	5	6	$^3H_6$
Ytterbium	Yb	70	173.04	13	1/2	3	7/2	$^2F_{7/2}$
Lutetium	Lu	71	174.967	14	—	—	—	—

Note: For additional information, see Goldschmidt, Z.B., in *Handbook on the Physics and Chemistry of Rare Earths,* Vol. 1, Gschneidner, K.A., Jr. and Eyring, L., Eds., North-Holland Physics, Amsterdam, 1978; DeLaeter, J.R., and Heumann, K.G., *J. Phys. Chem. Ref. Data,* 20, 1313, 1991; *Pure Appl. Chem.,* 66, 2423, 1994.

[a] 1993 standard atomic weights.

Table 2
Crystallographic Data for the Rare Earth Metals at 24°C (297 K) or Below

Rare earth metal	Crystal structure[a]	Lattice constants(Å)			Metallic radius CN = 12 (Å)	Atomic volume (cm³/mol)	Density (g/cm³)
		a_o	b_o	c_o			
αSc	hcp	3.3088	—	5.2680	1.6406	15.039	2.989
αY	hcp	3.6482	—	5.7318	1.8012	19.893	4.469
αLa	dhcp	3.7740	—	12.171	1.8791	22.602	6.146
αCe[b]	fcc	4.85[b]	—	—	1.72[b]	17.2[b]	8.16[b]
βCe	dhcp	3.6810	—	11.857	1.8321	20.947	6.689
γCe[c]	fcc	5.1610	—	—	1.8247	20.696	6.770
αPr	dhcp	3.6721	—	11.8326	1.8279	20.803	6.773
αNd	dhcp	3.6582	—	11.7966	1.8214	20.583	7.008
αPm	dhcp	3.65	—	11.65	1.811	20.24	7.264
αSm	rhomb[d]	3.6290[d]	—	26.207	1.8041	20.000	7.520
Eu	bcc	4.5827	—	—	2.0418	28.979	5.244
αGd	hcp	3.6336	—	5.7810	1.8013	19.903	7.901
α′Tb[e]	ortho	3.605[e]	6.244[e]	5.706[e]	1.784[e]	19.34[e]	8.219[e]
αTb	hcp	3.6055	—	5.6966	1.7833	19.310	8.230
α′Dy[f]	ortho	3.595[f]	6.184[f]	5.678[f]	1.774[f]	19.00[f]	8.551[f]
αDy	hcp	3.5915	—	5.6501	1.7740	19.004	8.551
Ho	hcp	3.5778	—	5.6178	1.7661	18.752	8.795

Table 2
Crystallographic Data for the Rare Earth Metals at 24°C (297 K) or Below (continued)

Rare earth metal	Crystal structure[a]	Lattice constants(Å)			Metallic radius CN = 12 (Å)	Atomic volume (cm³/mol)	Density (g/cm³)
		a_o	b_o	c_o			
Er	hcp	3.5592	—	5.5850	1.7566	18.449	9.066
Tm	hcp	3.5375	—	5.5540	1.7462	18.124	9.321
αYb[g]	hcp	3.8799[g]	—	6.3859[g]	1.9451[g]	25.067[g]	6.903[g]
βYb	fcc	5.4848	—	—	1.9392	24.841	6.966
Lu	hcp	3.5052	—	5.5494	1.7349	17.779	9.841

Note: For additional information, see Gschneidner, K.A., Jr. and Calderwood, F.W., in *Handbook on the Physics and Chemistry of Rare Earths,* Vol. 8, Gschneidner, K.A., Jr. and Eyring, L., Eds., North-Holland Physics, Amsterdam, 1986; Gschneidner, K.A., Jr., Pecharsky, V.K., Cho, Jaephil and Martin, S.W., *Scripta Mater.,* 1996, to be published.

[a] hcp = hexagonal close-packed; P6₃/mmc, hP2, A3, Mg-type; dhcp = double-c hexagonal close-packed; P6₃/mmc, hP4, A3′, αLa-type; fcc = face-centered cubic; Fm3̄m, cF4, A1, Cu-type; rhomb = rhombohedral; R3̄m, hR3, αSm-type; bcc = body-centered cubic; Im3̄m, cI2, A2, W-type; ortho = orthorhombic; Cmcm, oC4, α′ Dy-type.

[b] At 77 K (–196°C).

[c] Equilibrium room temperature (standard state) phase.

[d] Rhombohedral is the primitive cell. Lattice parameters given are for the nonprimitive hexagonal cell.

[d] At 220 K (–53°C).

[f] At 86 K (–187°C).

[g] At 23°C.

Table 3
Crystallographic Data for Rare Earth Metals at High Temperature

Rare earth metal	Structure	Lattice parameter (Å)	Temp. (°C)	Metallic radius		Atomic volume (cm³/mol)	Density (g/cm³)
				CN = 8 (Å)	CN = 12 (Å)		
βSc	bcc	3.73 (est.)	1337	1.62	1.66	15.6	2.88
βY	bcc	4.10[a]	1478	1.78	1.83	20.8	4.28
βLa	fcc	5.303	325	—	1.875	22.45	6.187
γLa	bcc	4.26	887	1.84	1.90	23.3	5.97
δCe	bcc	4.12	757	1.78	1.84	21.1	6.65
βPr	bcc	4.13	821	1.79	1.84	21.2	6.64
βNd	bcc	4.13	883	1.79	1.84	21.2	6.80
βPm	bcc	4.10 (est.)	890	1.78	1.83	20.8	6.99
βSm	hcp	$a = 3.6630$ $c = 5.8448$	450[b]	—	1.8176	20.450	7.353
γSm	bcc	4.10 (est.)	922	1.77	1.82	20.8	7.25
βGd	bcc	4.06	1265	1.76	1.81	20.2	7.80
βTb	bcc	4.07[a]	1289	1.76	1.81	20.3	7.82
βDy	bcc	4.03[a]	1381	1.75	1.80	19.7	8.23
γYb	bcc	4.44	763[c]	1.92	1.98	26.4	6.57

Note: The rare earths Eu, Ho, Er, Tm, and Lu are monomorphic. For additional information, see Gschneidner, K.A., Jr. and Calderwood, F.W., in *Handbook on the Physics and Chemistry of Rare Earths,* Vol. 8, Gschneidner, K.A., Jr. and Eyring, L., Eds., North-Holland Physics, Amsterdam, 1986, 1.

[a] Determined by extrapolation to 0% solute of *a* vs. composition data for R-Mg alloys at 24°C and corrected for thermal expansion to temperature given.

[b] The hcp phase was stabilized by impurities and the temperature of measurement was below the equilibrium transition temperature (see Table 4).

[c] The bcc phase was stabilized by impurities and the temperature of measurement was below the equilibrium transition temperature (see Table 4).

Table 4
High Temperature Transition Temperatures and Melting Point of Rare Earth Metals

Rare earth metal	Transition I $(\alpha - \beta)^a$		Transition II $(\beta - \gamma)^a$		Melting point (C°)
	Temp. (°C)	Phases	Temp. (C°)	Phases	
Sc	1337	hcp ⇌ bcc	—	—	1541
Y	1478	hcp ⇌ bcc	—	—	1522
La[b]	310	dhcp → fcc	865	fcc ⇌ bcc	918
Ce[c,d]	139	dhcp → fcc (β - γ)	726	fcc ⇌ bcc (γ - δ)	798
Pr	795	dhcp ⇌ bcc	—	—	931
Nd	863	dhcp ⇌ bcc	—	—	1021
Pm	890	dhcp ⇌ bcc	—	—	1042
Sm[e]	734	rhom → hcp	922	hcp ⇌ bcc	1074
Eu	—	—	—	—	822
Gd	1235	hcp ⇌ bcc	—	—	1313
Tb	1289	hcp ⇌ bcc	—	—	1356
Dy	1381	hcp ⇌ bcc	—	—	1412
Ho	—	—	—	—	1474
Er	—	—	—	—	1529
Tm	—	—	—	—	1545
Yb	795	fcc ⇌ bcc (β - γ)	—	—	819
Lu	—	—	—	—	1663

Note: For additional information, see Gschneidner, K.A., Jr. and Calderwood, F.W., in *Handbook on the Physics and Chemistry of Rare Earths*, Vol. 8, Gschneidner, K.A., Jr. and Eyring, L., Eds., North-Holland Physics, Amsterdam, 1986; Gschneidner, K.A., Jr., Pecharsky, V.K., Cho, Jaephil and Martin, S.W., *Scripta Mater.*, 34, 1717, 1996.

[a] For all the transformations listed, unless otherwise noted.
[b] On cooling, fcc → dhcp ($\beta \rightarrow \alpha$), 260°C.
[c] The $\beta \rightleftharpoons \gamma$ equilibrium transition temperature is 10 ± 5°C.
[d] On cooling, fcc → dhcp ($\gamma \rightarrow \beta$), –16°C.
[e] On cooling, hcp → rhomb ($\beta \rightarrow \alpha$), 727°C.

Table 5
Low Temperature Transition Temperatures of the Rare Earth Metals

Rare earth metal	Cooling			Rare earth metal	Heating		
	Transformation	°C	K		Transformation	°C	K
Ce	$\gamma \rightarrow \beta^a$	–16	257	Ce	$\alpha \rightarrow \beta$	–148	125
	$\gamma \rightarrow \alpha$	–172	101		$\alpha \rightarrow \beta + \gamma$	–104	169
	$\beta \rightarrow \alpha$	–228	45		$\beta \rightarrow \gamma^a$	139	412
Tb	$\alpha \rightarrow \alpha'$	–53	220	Yb	$\alpha \rightarrow \beta$	7	280
Dy	$\alpha \rightarrow \alpha'$	–187	86				
Yb	$\beta \rightarrow \alpha$	–13	260				

Note: For additional information, see Beaudry, B.J. and Gschneidner, K.A., Jr., in *Handbook on the Physics and Chemistry of Rare Earths*, Vol. 1, Gschneidner, K.A., Jr. and Eyring, L., Eds., North-Holland Physics, Amsterdam, 1978, 173. Koskenmaki, D.C. and Gschneidner, K.A., Jr., 1978, in *Handbook on the Physics and Chemistry of Rare Earths*, Vol. 1, Gschneidner, K.A., Jr. and Eyring, L., Eds., North-Holland Physics, Amsterdam, 1978, 337. Gschneidner, K.A., Jr., Pecharsky, V.K., Cho, Jaephil and Martin, S.W., *Scripta Mater.*, 34, 1717, 1996.

[a] The $\beta \rightleftharpoons \gamma$ equilibrium transition temperature is 10 ± 5°C (283 ± 5K).

Table 6
Heat Capacity, Standard Entropy, Heats of Transformation, and Fusion of the Rare Earth Metals

Rare earth metal	Heat capacity at 298 K (J/mol K)	Standard entropy $S°_{298}$ (J/mol K)	Heat of transformation (kJ/mol)				Heat of fusion (kJ/mol)
			trans. 1	$\Delta H_{tr}{}^1$	trans. 2	$\Delta H_{tr}{}^2$	
Sc	25.5	34.6	$\alpha \rightleftharpoons \beta$	4.00	—	—	14.1
Y	26.5	44.4	$\alpha \rightleftharpoons \beta$	4.99	—	—	11.4
La	27.1	56.9	$\alpha \rightleftharpoons \beta$	0.36	$\beta \rightleftharpoons \gamma$	3.12	6.20
Ce	26.9	72.0	$\beta \rightleftharpoons \gamma$	0.05	$\gamma \rightleftharpoons \delta$	2.99	5.46
Pr	27.2	73.2	$\alpha \rightleftharpoons \beta$	3.17	—	—	6.89
Nd	27.5	71.5	$\alpha \rightleftharpoons \beta$	3.03	—	—	7.14
Pm	27.3[a]	71.6[a]	$\alpha \rightleftharpoons \beta$	3.0[a]	—	—	7.7[a]
Sm	29.5	69.6	$\alpha \rightleftharpoons \beta$	0.2[a]	$\beta \rightleftharpoons \gamma$	3.11	8.62
Eu	27.7	77.8	—	—	—	—	9.21
Gd	37.0	68.1	$\alpha \rightleftharpoons \beta$	3.91	—	—	10.0
Tb	28.9	73.2	$\alpha \rightleftharpoons \beta$	5.02	—	—	10.79
Dy	27.7	75.6	$\alpha \rightleftharpoons \beta$	4.16	—	—	11.06
Ho	27.2	75.3	—	—	—	—	17.0[a]
Er	28.1	73.2	—	—	—	—	19.9
Tm	27.0	74.0	—	—	—	—	16.8
Yb	26.7	59.9	$\beta \rightleftharpoons \gamma$	1.75	—	—	7.66
Lu	26.9	51.0	—	—	—	—	22[a]

Note: For additional information, see Hultgren, R., Desai, P.D., Hawkins, D.T., Gleiser, M., Kelley, K.K., and Wagman, D.D., *Selected Values of the Thermodynamic Properties of the Elements,* ASM International, Metals Park, Ohio, 1973; Wagman, D.D., Evans, W.H., Parker, V.B., Schumm, R.H., Halow, I., Bailey, S.M., Churney, K.L., and Nuttall, R.L., *The NBS Tables of Chemical Thermodynamic Properties, J. Phys. Chem. Ref. Data,* Vol. 11, Suppl 2, 1982; Amitin, E.B., Bessergenev, W.G., Kovalevskaya, Yu. A., and Paukov, I.E., *J. Chem. Thermodyn.*, 15, 181, 1983; Amitin, E.B., Bessergenev, W.G., Kovalevskaya, Yu. A., and Paukov, I.E., *J. Chem. Thermodyn.*, 15, 181, 1983.

[a] Estimated.

Table 7
Vapor Pressures, Boiling Points, and Heats of Sublimation of Rare Earth Metals

Rare earth metal	Temperature in °C[a] for a vapor pressure of				Boiling point[a] (°C)	Heat of sublimation at 25°C (kJ/mol)
	10^{-8} atm (0.001 Pa)	10^{-6} atm (0.101 Pa)	10^{-4} atm (10.1Pa)	10^{-2} atm (1013 Pa)		
Sc	1036	1243	1533	1999	2836	377.8
Y	1222	1460	1812	2360	3345	424.7
La	1301	1566	1938	2506	3464	431.0
Ce	1290	1554	1926	2487	3443	422.6
Pr	1083	1333	1701	2305	3520	355.6
Nd	955	1175	1500	2029	3074	327.6
Pm	—	—	—	—	3000[b]	348[b]
Sm	508	642	835	1150	1794	206.7
Eu	399	515	685	964	1529	175.3
Gd	1167	1408	1760	2306	3273	397.5
Tb	1124	1354	1698	2237	3230	388.7
Dy	804	988	1252	1685	2567	290.4
Ho	845	1036	1313	1771	2700	300.8
Er	908	1113	1405	1896	2868	317.1
Tm	599	748	964	1300	1950	232.2
Yb	301	400	541	776	1196	152.1
Lu	1241	1483	1832	2387	3402	427.6

Note: For additional information, see Hultgren, R., Desai, P.D., Hawkins, D.T., Gleiser, M., Kelley, K.K., and Wagman, D.D., *Selected Values of the Thermodynamic Properties of the Elements,* ASM International, Metals Park, Ohio, 1973 and Beaudry, B.J. and Gschneidner, K.A., Jr., in *Handbook on the Physics and Chemistry of Rare Earths,* Vol. 1, Gschneidner, K.A., Jr. and Eyring, L., Eds., North-Holland Physics, Amsterdam, 1978, 173.

[a] International Temperature Scale of 1990 (ITS-90) values.

[b] Estimated.

Table 8
Magnetic Properties of the Rare Earth Metals

Rare earth metal	$\chi_A \times 10^6$ at 298 K (emu/mol)	Effective magnetic moment — Paramagnetic at ~298 K Theory[a]	Obs.	Ferromagnetic at ~0 K Theory[b]	Obs.	Easy axis	Néel temp. T_N (K) Hex sites	Cubic sites	Curie temp. T_C (K)	θ_P (K) ∥c	⊥c	Polycryst. or avg.
αSc	295.2	—	—	—	—	—	—	—	—	—	—	—
αY	187.7	—	—	—	—	—	—	—	—	—	—	—
αLa	95.9	—	—	—	—	—	—	—	—	—	—	—
βLa	105	—	—	—	—	—	—	—	—	—	—	−50
γCe	2,270	2.54	2.52	2.14	—	—	13.7	14.4	—	—	—	−41
βCe	2,500	2.54	2.61	2.14	—	—	—	12.5	—	—	—	—
αPr	5,530	3.58	3.56	3.20	2.7[c]	a	0.03	—	—	0	5	0
αNd	5,930	3.62	3.45	3.27	2.2[c]	b	19.9	7.5	—	—	—	3.3
αPm	—	2.68	—	2.40	—	—	—	—	—	—	—	—
αSm	1,278[d]	0.85	1.74	0.71	0.5[c]	a	109	14.0	—	—	—	—
Eu	30,900	7.94	8.48	7.0	5.9	<110>	—	90.4	—	—	—	100
αGd	185,000[e]	7.94	7.98	7.0	7.63	30° to c	—	—	293.4	317	317	317
αTb	170,000	9.72	9.77	—	—	—	230.0	—	—	195	239	224
α'Tb	—	—	—	9.0	9.34	b	—	—	219.5	—	—	—
αDy	98,000	10.64	10.83	—	—	—	180.2	—	—	121	169	153
α'Dy	—	—	—	10.0	10.33	a	—	—	90.5[g]	—	—	—
Ho	72,900	10.60	11.2	10.0	10.34	b	132	—	19.5	73.0	88.0	83.0
Er	48,000	9.58	9.9	9.0	9.1	30° to c	85	—	18.7	61.7	32.5	42.2
Tm	24,700	7.56	7.61	7.0	7.14	c	58	—	32.0	41.0	−17.0	2.3
βYb	67[d]	—	—	—	—	—	—	—	—	—	—	—
Lu	182.9	—	—	—	—	—	—	—	—	—	—	—

Note: For additional information, see McEwen, K.A., in *Handbook on the Physics and Chemistry of Rare Earths*, Vol. 1, Gschneidner, K.A., Jr. and Eyring, L., Eds., North-Holland Physics, Amsterdam, 1978, 411 and Legvold, S., in *Ferromagnetic Materials*, Vol. 1, Wohlfarth, E.P., Ed. North-Holland Physics, Amsterdam, 1980, 183; Pecharsky, V.K., Gschneidner, K.A., Jr. and Fort, D., *Phys. Rev. B*, 47, 5063, 1993; Pecharsky, V.K., Gschneidner, K.A., Jr. and Fort, D., 1996, to be published; Steward, A.M. and Collocott, S.J., *J. Phys.: Condens. Matter*, 1, 677, 1988.

a $g[J(J + 1)]^{1/2}$
b gJ.
c At 38 T and 4.2 K.
d At 290 K.
e At 350 K.
g On cooling $T_C = 89.6$ K and on warming $T_C = 91.5$ K.

Table 9
Room Temperature Coefficient of Thermal Expansion, Thermal Conductivity, Electrical Resistance, and Hall Coefficient

Rare earth metal	Expansion $\alpha_i \times 10^6$ ($°C^{-1}$)			Thermal conductivity (W/cm-K)	Electrical resistance ($\mu\Omega \cdot cm$)			Hall coefficient ($R_i \times 10^{12}$) (V·cm/A·Oe)		
	α_a	α_c	α_{poly}		ρ_a	ρ_c	ρ_{poly}	R_a	R_c	R_{poly}
αSc	7.6	15.3	10.2	0.158	70.9	26.9	56.2[a]	—	—	-0.13
αY	6.0	19.7	10.6	0.172	72.5	35.5	59.6	-0.27	-1.6	—
aLa	4.5	27.2	12.1	0.134	—	—	61.5	—	—	-0.35
bCe	—	—	—	—	—	—	82.8	—	—	—
γCe	6.3	—	6.3	0.113	—	—	74.4	—	—	+1.81
αPr	4.5	11.2	6.7	0.125	—	—	70.0	—	—	+0.709
αNd	7.6	13.5	9.6	0.165	—	—	64.3	—	—	+0.971
αPm	9[b]	16[b]	11[b]	0.15[b]	—	—	75[b]	—	—	—
αSm	9.6	19.0	12.7	0.133	—	—	94.0	—	—	-0.21
Eu	35.0	—	35.0	0.139[b]	—	—	90.0	—	—	+24.4
αGd	9.1[c]	10.0[c]	9.4[c]	0.105	135.1	121.7	131.0	-10	-54	-4.48[d]
αTb	9.3	12.4	10.3	0.111	123.5	101.5	115.0	-1.0	-3.7	—
αDy	7.1	15.6	9.9	0.107	111.0	76.6	92.6	-0.3	-3.7	—
Ho	7.0	19.5	11.2	0.162	101.5	60.5	81.4	+0.2	-3.2	—
Er	7.9	20.9	12.2	0.145	94.5	60.3	86.0	+0.3	-3.6	—
Tm	8.8	22.2	13.3	0.169	88.0	47.2	67.6	—	—	-1.8
βYb	26.3	—	26.3	0.385	—	—	25.0	—	—	+3.77
Lu	4.8	20.0	9.9	0.164	76.6	34.7	58.2	+0.45	-2.6	-0.535

Note: For additional information, see Beaudry, B. J. and Gschneidner, K.A., Jr., in *Handbook on the Physics and Chemistry of Rare Earths*, Vol. 1, Gschneidner, K.A., Jr. and Eyring, L., Eds., North-Holland Physics, Amsterdam, 1978, 173 and McEwen, K.A., in *Handbook on the Physics and Chemistry of Rare Earths*, Vol. 1, Gschneidner, K.A., Jr. and Eyring, L., Eds., North-Holland Physics, Amsterdam, 1978, 411.

[a] Calculated from single crystal values.
[b] Estimated.
[c] At 100°C.
[d] At 77°C.

Table 10
Electronic Specific Heat Constant (γ), Electron-Electron (Coulomb) Coupling Constant (μ*), Electron-Phonon Coupling Constant (λ), Debye Temperature at 0 K(θ$_D$), and Superconducting Transition Temperature

Rare earth metal	γ (mJ/mol·K^2)	μ*	λ	θ$_D$ (K) from Heat capacity	θ$_D$ (K) from Elastic constants	Superconducting temperature (K)
αSc	10.334	0.16	0.30	345.3	—	0.050[a]
αY	7.878	0.15	0.30	244.4	258	1.3[b]
αLa	9.45	0.08	0.76	150	154	5.10
βLa	11.5	—	—	140	—	6.00
αCe	12.8	—	—	179	—	0.022[c]
αPr	20	—	1.07[d]	155[e]	153	—
αNd	f	—	0.86[d]	157[e]	163	—
αPm	—	—	—	159[e]	—	—
αSm	8.1 ± 1.5[g]	—	0.81[d]	162[e,f]	169	—
Eu	f	—	—	f	118	—
αGd	4.48	—	0.30	169	182	—
α'Tb	3.71	—	0.34[d]	169.6	177	—
α'Dy	4.9	—	0.32[d]	192	183	—
Ho	2.1	—	0.30[d]	175[e]	190	—
Er	8.7	—	0.33[d]	176.9	188	—
Tm	f	—	0.36[d]	179[e]	200	—
αYb	3.30	—	—	117.6	118	—
βYb	8.36	—	—	109	—	—
Lu	8.194	0.14	0.31	183.2	185	0.022[h]

Note: For additional information, see Sundström, L.J., in *Handbook on the Physics and Chemistry of Rare Earths,* Vol. 1, Gschneidner, K.A., Jr., and Eyring, L., Eds., North-Holland Physics, Amsterdam, 1978, 379, Scott, T., in *Handbook on the Physics and Chemistry of Rare Earths,* Vol. 1, Gschneidner, K.A., Jr. and Eyring, L., Eds., North-Holland Physics, Amsterdam, 1978, 591, Probst, C. and Wittig, J., in *Handbook on the Physics and Chemistry of Rare Earths,* Vol. 1, Gschneidner, K.A., Jr. and Eyring, L., Eds., North-Holland Physics, Amsterdam, 1978, 749, and Tsang, T.-W.E., Gschneidner, K.A., Jr., Schmidt, F.A., and Thome, D.K., *Phys. Rev.,* B, 31, 235, 1985. Collocott, S.J., Hill, R.W. and Stewart, A.M., *J. Phys. F,* 18, L223, 1988. Hill, R.W. and Gschneidner, K.A., Jr., *J. Phys. F,* 18, 2545, 1988. Skriver, H.L. and Mertig, I., *Phys. Rev. B,* 41, 6553, 1990. Collocott, S.J. and Stewart, A.M., *J. Phys.: Condens. Matter,* 4, 6743, 1992. Pecharsky, V.K., Gschneidner, K.A., Jr. and Fort, D., *Phys. Rev. B,* 47, 5063, 1993.

[a] At 18.6 GPa.

[b] At 11 GPa.

[c] At 2.2 GPa.

[d] Calculated value.

[e] Estimated.

[f] Heat capacity results have been reported, but the resultant γ and θ$_D$ values are unreliable because of the presence of impurities and/or there was no reliable procedure or model to correct for the magnetic contribution to the heat capacity.

[g] Based on the values reported for the purer Sm sample (IV).

[h] At 4.5 GPa.

Table 11
Room Temperature Elastic Moduli and Mechanical Properties

| Rare earth metal | Elastic moduli (GPa) | | | | Mechanical properties (MPa) | | | | Recryst. temp. (°C) |
	Young's (elastic) modulus	Shear modulus	Bulk modulus	Poisson's ratio	Yield strength 0.2% offset	Ultimate tensile strength	Uniform elongation (%)	Reduction in area (%)	
Sc	74.4	29.1	56.6	0.279	173[a]	255[a]	5.0[a]	8.0[a]	550
Y	63.5	25.6	41.2	0.243	42	129	34.0	—	550
αLa	36.6	14.3	27.9	0.280	126[a]	130	7.9[a]	—	300
βCe	—	—	—	—	86	138	—	24.0	—
γCe	33.6	13.5	21.5	0.24	28	117	22.0	30.0	325
αPr	37.3	14.8	28.8	0.281	73	147	15.4	67.0	400
αNd	41.4	16.3	31.8	0.281	71	164	25.0	72.0	400
αPm	46[b]	18[b]	33[b]	0.28[b]	—	—	—	—	400[b]
αSm	49.7	19.5	37.8	0.274	68	156	17.0	29.5	440
Eu	18.2	7.9	8.3	0.152	—	—	—	—	300
αGd	54.8	21.8	37.9	0.259	15	118	37.0	56.0	500
αTb	55.7	22.1	38.7	0.261	—	—	—	—	500
αDy	61.4	24.7	40.5	0.247	43	139	30.0	30.0	550
Ho	64.8	26.3	40.2	0.231	—	—	—	—	520
Er	69.9	28.3	44.4	0.237	60	136	11.5	11.9	520
Tm	74.0	30.5	44.5	0.213	—	—	—	—	600
βYb	23.9	9.9	30.5	0.207	7	58	43.0	92.0	300
Lu	68.6	27.2	47.6	0.261	—	—	—	—	600

Note: For additional information, see Scott, T., in *Handbook on the Physics and Chemistry of Rare Earths,* Vol. 1, Gschneidner, K.A., Jr. and Eyring, L., Eds., North-Holland Physics, Amsterdam, 1978, 591.

[a] Value is questionable.
[b] Estimated.

Table 12
Liquid Metal Properties Near the Melting Point

Rare earth metal	Density (g/cm³)	Surface tension (N/m)	Viscosity (centipoise)	Heat capacity (J/mol K)	Thermal conductivity (W/cm K)	Magnetic susceptibility $\chi \times 10^4$ (emu/mol)	Electrical resistivity (μΩ·cm)	$\Delta V_{L \to s}$[a] (%)	Spectral emittance at λ = 645 nm ε (%)	Temp. range (°C)
Sc	2.80	0.954	—	44.2[b]	—	—	—	—	36.8	1522–1647
Y	4.24	0.871	—	43.1	—	—	—	—	25.4	920–1287
La	5.96	0.718	2.65	34.3	0.238	1.20	133	−0.6	32.2	877–1547
Ce	6.68	0.706	3.20	37.7	0.210	9.37	130	+1.1	28.4	931–1537
Pr	6.59	0.707	2.85	43.0	0.251	17.3	139	−0.02	39.4	1021–1567
Nd	6.72	0.687	—	48.8	0.195	18.7	151	−0.9	—	—
Pm	6.9[b]	0.680[b]	—	50[b]	—	—	160[b]	—	43.7	1075
Sm	7.16	0.431	—	50.2[b]	—	18.3	182	−3.6	—	—
Eu	4.87	0.264	—	38.1	—	97	242	−4.8	34.2	1313–1600
Gd	7.4	0.664	—	37.2	0.149	67	195	−2.0	—	—
Tb	7.65	0.669	—	46.5	—	82	193	−3.1	29.7	1412–1437
Dy	8.2	0.648	—	49.9	0.187	95	210	−4.5	—	—
Ho	8.34	0.650	—	43.9	—	88	221	−7.4	37.2	1529–1587
Er	8.6	0.637	—	38.7	—	69	226	−9.0	—	—
Tm	9.0[b]	—	—	41.4	—	41	235[b]	−6.9	—	—
Yb	6.21	0.320	2.67	36.8	—	—	113	−5.1	—	—
Lu	9.3	0.940	—	47.9[b]	—	—	224	−3.6	—	—

Note: For additional information, see Van Zytveld, J., in *Handbook on the Physics and Chemistry of Rare Earths*, Vol. 12, Gschneidner, K.A., Jr. and Eyring, L., Eds., North-Holland Physics, Amsterdam, 1989, 357. Stretz, L.A. and Bautista, R.G., in *Temperature, Its Measurement and Control in Science and Industry*, Vol. 4, part I, H.H. Plumb, Ed., Instrument Society of America, Pittsburgh, 1972, 489. King, T.S., Baria, D.N., and Bautista, R.G., *Met. Trans. B*, 7, 411, 1976. Baria, D.N., King, T.S., and Bautista, R.G., *Met. Trans. B*, 7, 577, 1976.

[a] Volume change on freezing.
[b] Estimated.

PHYSICAL PROPERTIES OF THE RARE EARTH METALS (continued)

Table 13
Ionization Potentials (Electronvolts)

Rare earth	I Neutral atom	II Singly ionized	III Doubly ionized	IV Triply ionized	V Quadruply ionized
Sc	6.56144	12.79967	24.75666	73.4894	91.65
Y	6.217	12.24	20.52	60.597	77.0
La	5.5770	11.060	19.1773	49.95	61.6
Ce	5.5387	10.85	20.198	36.758	65.55
Pr	5.464	10.55	21.624	38.98	57.53
Nd	5.5250	10.73	22.1	40.41	—
Pm	5.554	10.90	22.3	41.1	—
Sm	5.6437	11.07	23.4	41.4	—
Eu	5.6704	11.241	24.92	42.7	—
Gd	6.1500	12.09	20.63	44.0	—
Tb	5.8639	11.52	21.91	39.79	—
Dy	5.9389	11.67	22.8	41.47	—
Ho	6.0216	11.80	22.84	42.5	—
Er	6.1078	11.93	22.74	42.7	—
Tm	6.18431	12.05	23.68	42.7	—
Yb	6.25416	12.1761	25.05	43.56	—
Lu	5.42585	13.9	20.9594	45.25	66.8

Note: For references, see the table "Ionization Potentials of Atoms and Atomic Ions" in Section 10.

Table 14
Effective Ionic Radii (Å)[a]

Rare earth ion	R^{2+} CN = 6	R^{2+} CN = 8	R^{3+} CN = 6	R^{3+} CN = 8	R^{3+} CN = 12	R^{4+} CN = 6	R^{4+} CN = 8
Sc	—	—	0.745	0.87	1.116	—	—
Y	—	—	0.900	1.015	1.220	—	—
La	—	—	1.045	1.18	1.320	—	—
Ce	—	—	1.010	1.14	1.290	0.80	0.97
Pr	—	—	0.997	1.14	1.286	0.78	0.96
Nd	—	—	0.983	1.12	1.276	—	—
Pm	—	—	0.97	1.10	1.267	—	—
Sm	1.19	1.27	0.958	1.09	1.260	—	—
Eu	1.17	1.25	0.947	1.07	1.252	—	—
Gd	—	—	0.938	1.06	1.246	—	—
Tb	—	—	0.923	1.04	1.236	0.76	0.88
Dy	—	—	0.912	1.03	1.228	—	—
Ho	—	—	0.901	1.02	1.221	—	—
Er	—	—	0.890	1.00	1.214	—	—
Tm	—	—	0.880	0.99	1.207	—	—
Yb	1.00	1.07	0.868	0.98	1.199	—	—
Lu	—	—	0.861	0.97	1.194	—	—

Note: For additional information, see Shannon, R.D. and Prewitt, C.T., *Acta Cryst.,* 25, 925, 1969 and Shannon, R.D. and Prewitt, C.T., *Acta Cryst.,* 26, 1046, 1970.

[a] Radius of O^{2-} is 1.40 Å for a coordination number (CN) of 6.

MELTING, BOILING, AND CRITICAL TEMPERATURES OF THE ELEMENTS

This table summarizes the melting point t_m, normal boiling point t_b, and critical temperature t_c (on the ITS-90 scale) for the elements for which data are available. A "tp" after a value indicates a solid-liquid-gas triple point, and "sp" indicates a sublimation point, where the vapor pressure of the solid phase reaches 101.325 kPa (1 atm). Transition temperatures between allotropic forms are included for several elements. References may be found in the tables *Physical Constants of Inorganic Compounds* and *Critical Constants*.

Name	t_m/°C	t_b/°C	t_c/°C
Actinium	1051	3198	
Aluminum	660.32	2519	
Americium	1176	2011	
Antimony	630.63	1587	
Argon	-189.35	-185.85	-122.28
Arsenic (gray)	817 tp (3.7 MPa)	603 sp	1400
Astatine	302		
Barium	727	1897	
Berkelium (α form)	1050		
Berkelium (β form)	986		
Beryllium	1287	2471	
Bismuth	271.40	1564	
Boron	2075	4000	
Bromine	-7.2	58.8	315
Cadmium	321.07	767	
Calcium	842	1484	
Californium	900		
Carbon (graphite)	4492 tp (10.3 MPa)	3825 sp	
Carbon (diamond)	4440 (12.4 GPa)		
Cerium	798	3443	
Cesium	28.5	671	1665
Chlorine	-101.5	-34.04	143.8
Chromium	1907	2671	
Cobalt	1495	2927	
Copper	1084.62	2562	
Curium	1345	3100	
Dysprosium	1412	2567	
Einsteinium	860		
Erbium	1529	2868	
Europium	822	1529	
Fermium	1527		
Fluorine	-219.67 tp	-188.12	-129.02
Francium	27		
Gadolinium	1313	3273	
Gallium	29.76	2204	
Germanium	938.25	2833	
Gold	1064.18	2856	
Hafnium	2233	4603	
Helium		-268.93	-267.96
Holmium	1474	2700	
Hydrogen	-259.34	-252.87	-240.18
Indium	156.60	2072	
Iodine	113.7	184.4	546
Iridium	2446	4428	
Iron	1538	2861	
Krypton	-157.38 tp (73.2 kPa)	-153.22	-63.74
Lanthanum	918	3464	
Lawrencium	1627		
Lead	327.46	1749	
Lithium	180.50	1342	2950
Lutetium	1663	3402	
Magnesium	650	1090	
Manganese	1246	2061	
Mendelevium	827		
Mercury	-38.83	356.73	1477

Name	t_m/°C	t_b/°C	t_c/°C
Molybdenum	2623	4639	
Neodymium	1021	3074	
Neon	-248.59	-246.08	-228.7
Neptunium	644		
Nickel	1455	2913	
Niobium	2477	4744	
Nitrogen	-210.00	-195.79	-146.94
Nobelium	827		
Osmium	3033	5012	
Oxygen	-218.79	-182.95	-118.56
Palladium	1554.9	2963	
Phosphorus (white)	44.15	280.5	721
Phosphorus (red)	590 tp	431 sp	721
Phosphorus (black)	610		
Platinum	1768.4	3825	
Plutonium	640	3228	
Polonium	254	962	
Potassium	63.5	759	1950
Praseodymium	931	3520	
Promethium	1042	3000	
Protactinium	1572		
Radium	700		
Radon	-71	-61.7	104
Rhenium	3186	5596	
Rhodium	1964	3695	
Rubidium	39.30	688	1820
Ruthenium	2334	4150	
Samarium	1074	1794	
Scandium	1541	2836	
Selenium(vitreous)	180 (trans to gray)	685	
Selenium (gray)	220.5	685	1493
Silicon	1414	3265	
Silver	961.78	2162	
Sodium	97.80	883	2300
Strontium	777	1382	
Sulfur (rhombic)	95.3 (trans to mono)	444.60	1041
Sulfur (monoclinic)	119.6	444.60	1041
Tantalum	3017	5458	
Technetium	2157	4265	
Tellurium	449.51	988	
Terbium	1356	3230	
Thallium	304	1473	
Thorium	1750	4788	
Thulium	1545	1950	
Tin (gray)	13.2 (trans to white)	2602	
Tin (white)	231.93	2602	
Titanium	1668	3287	
Tungsten	3422	5555	
Uranium	1135	4131	
Vanadium	1910	3407	
Xenon	-111.79 tp (81.6 kPa)	-108.12	16.62
Ytterbium	819	1196	
Yttrium	1522	3345	
Zinc	419.53	907	
Zirconium	1855	4409	

HEAT CAPACITY OF THE ELEMENTS AT 25°C

This table gives the specific heat capacity (c_p) in J/g K and the molar heat capacity (C_p) in J/mol K at a temperature of 25°C and a pressure of 100 kPa (1 bar or 0.987 standard atmospheres) for all the elements for which reliable data are available.

Name	c_p J/g K	C_p J/mol K	Name	c_p J/g K	C_p J/mol K
Actinium	0.120	27.2	Molybdenum	0.251	24.06
Aluminum	0.897	24.200	Neodymium	0.190	27.45
Antimony	0.207	25.23	Neon	1.030	20.786
Argon	0.520	20.786	Nickel	0.444	26.07
Arsenic	0.329	24.64	Niobium	0.265	24.60
Barium	0.204	28.07	Nitrogen (N_2)	1.040	29.124
Beryllium	1.825	16.443	Osmium	0.130	24.7
Bismuth	0.122	25.52	Oxygen (O_2)	0.918	29.378
Boron	1.026	11.087	Palladium	0.246	25.98
Bromine (Br_2)	0.226	36.057	Phosphorous (white)	0.769	23.824
Cadmium	0.232	26.020	Platinum	0.133	25.86
Calcium	0.647	25.929	Potassium	0.757	29.600
Carbon (graphite)	0.709	8.517	Praseodymium	0.193	27.20
Cerium	0.192	26.94	Radon	0.094	20.786
Cesium	0.242	32.210	Rhenium	0.137	25.48
Chlorine (Cl_2)	0.479	33.949	Rhodium	0.243	24.98
Chromium	0.449	23.35	Rubidium	0.363	31.060
Cobalt	0.421	24.81	Ruthenium	0.238	24.06
Copper	0.385	24.440	Samarium	0.197	29.54
Dysprosium	0.170	27.7	Scandium	0.568	25.52
Erbium	0.168	28.12	Selenium	0.321	25.363
Europium	0.182	27.66	Silicon	0.705	19.789
Fluorine (F_2)	0.824	31.304	Silver	0.235	25.350
Gadolinium	0.236	37.03	Sodium	1.228	28.230
Gallium	0.371	25.86	Strontium	0.301	26.4
Germanium	0.320	23.222	Sulfur (rhombic)	0.710	22.75
Gold	0.129	25.418	Tantalum	0.140	25.36
Hafnium	0.144	25.73	Tellurium	0.202	25.73
Helium	5.193	20.786	Terbium	0.182	28.91
Holmium	0.165	27.15	Thallium	0.129	26.32
Hydrogen (H_2)	14.304	28.836	Thorium	0.113	26.230
Indium	0.233	26.74	Thulium	0.160	27.03
Iodine (I_2)	0.145	36.888	Tin (white)	0.228	27.112
Iridium	0.131	25.10	Titanium	0.523	25.060
Iron	0.449	25.10	Tungsten	0.132	24.27
Krypton	0.248	20.786	Uranium	0.116	27.665
Lanthanum	0.195	27.11	Vanadium	0.489	24.89
Lead	0.129	26.650	Xenon	0.158	20.786
Lithium	3.582	24.860	Ytterbium	0.155	26.74
Lutetium	0.154	26.86	Yttrium	0.298	26.53
Magnesium	1.023	24.869	Zinc	0.388	25.390
Manganese	0.479	26.32	Zirconium	0.278	25.36
Mercury	0.140	27.983			

VAPOR PRESSURE OF THE METALLIC ELEMENTS

C. B. Alcock

This table gives coefficients in an equation for the vapor pressure of 65 metallic elements in both the solid and liquid state. Vapor pressures in the range 10^{-10} to 10^2 Pa (10^{-15} to 10^{-3} atm) are covered. The equation is:

for p in pascals: $\log(p/\text{Pa}) = 5.006 + A + BT^{-1} + C\log T + DT^{-3}$

for p in atmospheres: $\log(p/\text{atm}) = A + BT^{-1} + C\log T + DT^{-3}$, where T is the temperature in K

This equation reproduces the observed vapor pressures to an accuracy of ±5% or better. Reprinted with permission of the publisher, Pergamon Press.

REFERENCE

Alcock, C. B., Itkin, V. P., and Horrigan, M. K., *Canadian Metallurgical Quarterly,* 23, 309, 1984.

Element, state	A	B	C	D	Temperature range
Li sol	5.667	-8310			298-m.p.
Li liq	5.055	-8023			m.p.-1000
Na sol	5.298	-5603			298-m.p.
Na liq	4.704	-5377			m.p.-700
K sol	4.961	-4646			298-m.p.
K liq	4.402	-4453			m.p.-600
Rb sol	4.857	-4215			298-m.p.
Rb liq	4.312	-4040			m.p.-550
Cs sol	4.711	-3999			298-m.p.
Cs liq	4.165	-3830			m.p.-550
Be sol	8.042	-17020	-0.4440		298-m.p.
Be liq	5.786	-15731			m.p.-1800
Mg sol	8.489	-7813	-0.8253		298-m.p.
Ca sol	10.127	-9517	-1.4030		298-m.p.
Sr sol	9.226	-8572	-1.1926		298-m.p.
Ba sol	12.405	-9690	-2.2890		298-m.p.
Ba liq	4.007	-8163			m.p.-1200
Al sol	9.459	-17342	-0.7927		298-m.p.
Al liq	5.911	-16211			m.p.-1800
Ga sol	6.657	-14208			298-m.p.
Ga liq	6.754	-13984	-0.3413		m.p.-1600
In sol	5.991	-12548			298-m.p.
In liq	5.374	-12276			m.p.-1500
Tl sol	5.971	-9447			298-m.p.
Tl liq	5.259	-9037			m.p.-1100
Sn sol	6.036	-15710			298-m.p.
Sn liq	5.262	-15332			m.p.-1850
Pb sol	5.643	-10143			298-m.p.
Pb liq	4.911	-9701			m.p.-1200
Sc sol	6.650	-19721	0.2885	-0.3663	298-m.p.
Sc liq	5.795	-17681			m.p.-2000
Y sol	9.735	-22306	-0.8705		298-m.p.
Y liq	5.795	-20341			m.p.-2300
La sol	7.463	-22551	-0.3142		298-m.p.
La liq	5.911	-21855			m.p.-2450
Ti sol	11.925	-24991	-1.3376		298-m.p.
Ti liq	6.358	-22747			m.p.-2400
Zr sol	10.008	-31512	-0.7890		298-m.p
Zr liq	6.806	-30295			m.p.-2500
Hf sol	9.445	-32482	-0.6735		298-m.p.
V sol	9.744	-27132	-0.5501		298-m.p.

Element, state	A	B	C	D	Temperature range
V liq	6.929	-25011			m.p.-2500
Nb sol	8.822	-37818	-0.2575		298-2500
Ta sol	16.807	-41346	-3.2152	0.7437	248-2500
Cr sol	6.800	-20733	0.4391	-0.4094	298-2000
Mo sol	11.529	-34626	-1.1331		298-2500
W sol	2.945	-44094	1.3677		298-2350
W sol	-54.527	-57687	-12.2231		2200-2500
Mn sol	12.805	-15097	-1.7896		298-m.p.
Re sol	11.543	-40726	-1.1629		298-2500
Fe sol	7.100	-21723	0.4536	-0.5846	298-m.p.
Fe liq	6.347	-19574			m.p.-2100
Ru sol	9.755	-34154	-0.4723		298-m.p.
Os sol	9.419	-41198	-0.3896		298-2500
Co sol	10.976	-22576	-1.0280		298-m.p.
Co liq	6.488	-20578			m.p.-2150
Rh sol	10.168	-29010	-0.7068		298-m.p.
Rh liq	6.802	-26792			m.p.-2500
Ir sol	10.506	-35099	-0.7500		298-2500
Ni sol	10.557	-22606	-0.8717		298-m.p.
Ni liq	6.666	-20765			m.p.-2150
Pd sol	9.502	-19813	-0.9258		298-m.p.
Pd liq	5.426	-17899			m.p.-2100
Pt sol	4.882	-29387	1.1039	-0.4527	298-m.p.
Pt liq	6.386	-26856			m.p.-2500
Cu sol	9.123	-17748	-0.7317		298-m.p.
Cu liq	5.849	-16415			m.p.-1850
Ag sol	9.127	-14999	-0.7845		298-m.p.
Ag liq	5.752	-13827			m.p.-1600
Au sol	9.152	-19343	-0.7479		298-m.p.
Au liq	5.832	-18024			m.p.-2050
Zn sol	6.102	-6776			298-m.p.
Zn liq	5.378	-6286			m.p.-750
Cd sol	5.939	-5799			298-m.p.
Cd liq	5.242	-5392			m.p.-650
Hg liq	5.116	-3190			298-400
Ce sol	6.139	-21752			298-m.p.
Ce liq	5.611	-21200			m.p.-2450
Pr sol	8.859	-18720	-0.9512		298-m.p.
Pr liq	4.772	-17315			m.p.-2200
Nd sol	8.996	-17264	-0.9519		298-m.p.
Nd liq	4.912	-15824			m.p.-2000
Sm sol	9.988	-11034	-1.3287		298-m.p.
Eu sol	9.240	-9459	-1.1661		298-m.p.
Gd sol	8.344	-20861	-0.5775		298-m.p.
Gd liq	5.557	-19389			m.p.-2250
Tb sol	9.510	-20457	-0.9247		298-m.p.
Tb liq	5.411	-18639			m.p.-2200
Dy sol	9.579	-15336	-1.1114		298-m.p.
Ho sol	9.785	-15899	-1.1753		298-m.p.
Er sol	9.916	-16642	-1.2154		298-m.p.
Er liq	4.668	-14380			m.p.-1900
Tm sol	8.882	-12270	-0.9564		298-1400
Yb sol	9.111	-8111	-1.0849		298-900
Lu sol	8.793	-22423	-0.6200		298-m.p.
Lu liq	5.648	-20302			m.p.-2350
Th sol	8.668	-31483	-0.5288		298-m.p.
Th liq	-18.453	-24569	6.6473		m.p.-2500
Pa sol	10.552	-34869	-1.0075		298-m.p.

VAPOR PRESSURE OF THE METALLIC ELEMENTS (continued)

Element, state	A	B	C	D	Temperature range
Pa liq	6.177	-32874			m.p.-2500
U sol	0.770	-27729	2.6982	-1.5471	298-m.p.
U liq	20.735	-28776	-4.0962		m.p.-2500
Np sol	19.643	-24886	-3.9991		298-m.p.
Np liq	10.076	-23378	-1.3250		m.p.-2500
Pu sol	26.160	-19162	-6.6675		298-600
Pu sol	18.858	-18460	-4.4720		500-m.p.
Pu liq	3.666	-16658			m.p.-2450
Am sol	11.311	-15059	-1.3449		298-m.p.
Cm sol	8.369	-20364	-0.5770		298-m.p.
Cm liq	5.223	-18292			m.p.-2200

DENSITY OF MOLTEN ELEMENTS AND REPRESENTATIVE SALTS

This table lists the liquid density at the melting point, ρ_m, for elements that are solid at room temperature, as well as for some representative salts of these elements. Densities at higher temperatures (up to the t_{max} given in the last column) may be estimated from the equation

$$\rho(t) = \rho_m - k(t-t_m)$$

where t_m is the melting point and k is given in the fifth column of the table. If a value of t_{max} is not given, the equation should not be used to extrapolate more than about 20°C beyond the melting point.

Data for the elements were selected from the primary literature; the assistance of Gernot Lang in compiling these data is gratefully acknowledged. The molten salt data were derived from Reference 1.

REFERENCE

1. Janz, G. J., Thermodynamic and Transport Properties of Molten Salts: Correlation Equations for Critically Evaluated Density, Surface Tension, Electrical Conductance, and Viscosity Data, *J. Phys. Chem. Ref. Data*, 17, Suppl. 2, 1988.
2. Nasch, P. M., and Steinemann, S. G., *Phys. Chem. Liq.*, 29, 43, 1995.

Formula	Name	$t_m/°C$	$\rho_m/g\ cm^{-3}$	$k/g\ cm^{-3}\ °C^{-1}$	t_{max}
Ag	Silver	961.78	9.320	0.0009	1500
AgBr	Silver(I) bromide	432	5.577	0.001035	667
AgCl	Silver(I) chloride	455	4.83	0.00094	627
AgI	Silver(I) iodide	558	5.58	0.00101	802
AgNO$_3$	Silver(I) nitrate	212	3.970	0.001098	360
Ag$_2$SO$_4$	Silver(I) sulfate	652	4.84	0.001089	770
Al	Aluminum	660.32	2.375	0.000233	1340
AlBr$_3$	Aluminum bromide	97.5	2.647	0.002435	267
AlCl$_3$	Aluminum chloride	192.6	1.302	0.002711	296
AlI$_3$	Aluminum iodide	188.32	3.223	0.0025	240
As	Arsenic	817	5.22	0.000544	
Au	Gold	1064.18	17.31	0.001343	1200
B	Boron	2075	2.08		
Ba	Barium	727	3.338	0.000299	1550
BaBr$_2$	Barium bromide	857	3.991	0.000924	900
BaCl$_2$	Barium chloride	962	3.174	0.000681	1081
BaF$_2$	Barium fluoride	1368	4.14	0.000999	1727
BaI$_2$	Barium iodide	711	4.26	0.000977	975
Be	Beryllium	1287	1.690	0.00011	
BeCl$_2$	Beryllium chloride	415	1.54	0.0011	473
BeF$_2$	Beryllium fluoride	552	1.96	0.000015	850
Bi	Bismuth	271.40	10.05	0.00135	800
BiBr$_3$	Bismuth bromide	218	4.76	0.002637	927
BiCl$_3$	Bismuth chloride	230	3.916	0.0023	350
Ca	Calcium	842	1.378	0.000230	1484
CaBr$_2$	Calcium bromide	742	3.111	0.0005	791
CaCl$_2$	Calcium chloride	775	2.085	0.000422	950
CaF$_2$	Calcium fluoride	1418	2.52	0.000391	2027
CaI$_2$	Calcium iodide	783	3.443	0.000751	1028
Cd	Cadmium	321.07	7.996	0.001218	500
CdBr$_2$	Cadmium bromide	568	4.075	0.00108	720
CdCl$_2$	Cadmium chloride	564	3.392	0.00082	807
CdI$_2$	Cadmium iodide	387	4.396	0.001117	700
Ce	Cerium	799	6.55	0.000710	1460
CeCl$_3$	Cerium(III) chloride	817	3.25	0.00092	950
CeF$_3$	Cerium(III) fluoride	1430	4.659	0.000936	1927
Co	Cobalt	1495	7.75	0.00165	1580
Cr	Chromium	1907	6.3	0.0011	2100
Cs	Cesium	28.44	1.843	0.000556	510
CsBr	Cesium bromide	636	3.133	0.001223	860
CsCl	Cesium chloride	645	2.79	0.001065	906
CsF	Cesium fluoride	703	3.649	0.001282	912
CsI	Cesium iodide	621	3.197	0.001183	907

DENSITY OF MOLTEN ELEMENTS AND REPRESENTATIVE SALTS (continued)

Formula	Name	$t_m/°C$	$\rho_m/g\ cm^{-3}$	$k/g\ cm^{-3}\ °C^{-1}$	t_{max}
$CsNO_3$	Cesium nitrate	414	2.820	0.001166	491
Cs_2SO_4	Cesium sulfate	1005	3.1	0.00095	1530
Cu	Copper	1084.62	8.02	0.000609	1630
CuCl	Copper(I) chloride	430	3.692	0.00076	585
Dy	Dysprosium	1411	8.37	0.00143	1540
$DyCl_3$	Dysprosium(III) chloride	680	3.62	0.00068	987
Er	Erbium	1529	8.86	0.00157	1700
Eu	Europium	822	5.13	0.0028	980
Fe	Iron	1538	6.98	0.000572	1680
$FeCl_2$	Iron(II) chloride	677	2.348	0.000555	877
Ga	Gallium	29.76	6.08	0.00062	400
$GaBr_3$	Gallium(III) bromide	121.5	3.116	0.00246	135
$GaCl_3$	Gallium(III) chloride	77.9	2.053	0.002083	141
GaI_3	Gallium(III) iodide	212	3.630	0.002377	252
Gd	Gadolinium	1314	7.4		
$GdCl_3$	Gadolinium(III) chloride	609	3.56	0.000671	1007
GdI_3	Gadolinium(III) iodide	925	4.12	0.000908	1032
Ge	Germanium	938.25	5.60	0.00055	1600
Hf	Hafnium	2233	12		
$HgBr_2$	Mercury(II) bromide	236	5.126	0.003233	319
$HgCl_2$	Mercury(II) chloride	276	4.368	0.002862	304
HgI_2	Mercury(II) iodide	259	5.222	0.003235	354
Ho	Holmium	1472	8.34		
In	Indium	156.60	7.02	0.000836	500
$InBr_3$	Indium(III) bromide	420	3.121	0.0015	528
$InCl_3$	Indium(III) chloride	583	2.140	0.0021	666
InI_3	Indium(III) iodide	207	3.820	0.0015	360
Ir	Iridium	2446	19		
K	Potassium	63.38	0.828	0.000232	500
KBr	Potassium bromide	734	2.127	0.000825	930
KCl	Potassium chloride	771	1.527	0.000583	939
KF	Potassium fluoride	858	1.910	0.000651	1037
KI	Potassium iodide	681	2.448	0.000956	904
KNO_3	Potassium nitrate	337	1.865	0.000723	457
La	Lanthanum	920	5.94	0.00061	1600
$LaBr_3$	Lanthanum bromide	788	4.933	0.000096	912
$LaCl_3$	Lanthanum chloride	859	3.209	0.000777	973
LaF_3	Lanthanum fluoride	1493	4.589	0.000682	2177
LaI_3	Lanthanum iodide	778	4.29	0.001110	907
Li	Lithium	180.5	0.512	0.00052	285
LiBr	Lithium bromide	552	2.528	0.000652	739
LiCl	Lithium chloride	610	1.502	0.000432	781
LiF	Lithium fluoride	848.2	1.81	0.000490	1047
LiI	Lithium iodide	469	3.109	0.000917	667
$LiNO_3$	Lithium nitrate	253	1.781	0.000546	441
Li_2SO_4	Lithium sulfate	859	2.003	0.000407	1214
Lu	Lutetium	1663	9.3		
Mg	Magnesium	650	1.584	0.000234	900
$MgBr_2$	Magnesium bromide	711	2.62	0.000478	935
$MgCl_2$	Magnesium chloride	714	1.68	0.000271	826
MgI_2	Magnesium iodide	634	3.05	0.000651	888
Mn	Manganese	1246	5.95	0.00105	1590
$MnCl_2$	Manganese(II) chloride	650	2.353	0.000437	850
Mo	Molybdenum	2623	9.33		
Na	Sodium	97.80	0.927	0.00023	600
NaBr	Sodium bromide	747	2.342	0.000816	945
Na_2CO_3	Sodium carbonate	858.1	1.972	0.000448	1004
NaCl	Sodium chloride	800.7	1.556	0.000543	1027
NaF	Sodium fluoride	996	1.948	0.000636	1097
NaI	Sodium iodide	660	2.742	0.000949	912

DENSITY OF MOLTEN ELEMENTS AND REPRESENTATIVE SALTS (continued)

Formula	Name	$t_m/°C$	ρ_m/g cm^{-3}	k/g cm^{-3} °C^{-1}	t_{max}
$NaNO_3$	Sodium nitrate	307	1.90	0.000715	370
Na_2SO_4	Sodium sulfate	884	2.069	0.000483	1077
Nd	Neodymium	1016	6.89	0.00076	1350
Ni	Nickel	1455	7.81	0.000726	1700
$NiCl_2$	Nickel(II) chloride	1009	2.653	0.00066	1057
Os	Osmium	3033	20		
Pb	Lead	327.46	10.66	0.00122	700
$PbBr_2$	Lead(II) bromide	371	5.73	0.00165	600
$PbCl_2$	Lead(II) chloride	501	4.951	0.0015	710
PbI_2	Lead(II) iodide	410	5.691	0.001594	697
Pd	Palladium	1554.9	10.38	0.001169	1700
Pr	Praseodymium	931	6.50	0.00093	1460
$PrCl_3$	Praseodymium chloride	786	3.23	0.00074	977
Pt	Platinum	1768.4	19.77	0.0024	2200
Pu	Plutonium	640	16.63	0.001419	950
Rb	Rubidium	39.31	1.46	0.000451	800
RbBr	Rubidium bromide	682	2.715	0.001072	907
Rb_2CO_3	Rubidium carbonate	837	2.84	0.000640	1007
RbCl	Rubidium chloride	715	2.248	0.000883	923
RbF	Rubidium fluoride	833	2.87	0.00102	1067
RbI	Rubidium iodide	642	2.904	0.001143	902
$RbNO_3$	Rubidium nitrate	305	2.519	0.001068	417
Rb_2SO_4	Rubidium sulfate	1050	2.56	0.000665	1545
Re	Rhenium	3186	18.9		
Rh	Rhodium	1964	10.7	0.000895	2200
Ru	Ruthenium	2334	10.65		
S	Sulfur	115.21	1.819	0.00080	160
Sb	Antimony	630.63	6.53	0.00067	745
$SbCl_3$	Antimony(III) chloride	73.4	2.681	0.002293	77
$SbCl_5$	Antimony(V) chloride	4	2.37	0.001869	77
SbI_3	Antimony(III) iodide	168	4.171	0.002483	322
Sc	Scandium	1541	2.80		
Se	Selenium	221	3.99		
Si	Silicon	1414	2.57	0.000936	1500
Sm	Samarium	1072	7.16		
Sn	Tin	231.93	6.99	0.000601	1200
$SnCl_2$	Tin(II) chloride	247	3.36	0.001253	480
$SnCl_4$	Tin(IV) chloride	-33	2.37	0.002687	138
Sr	Strontium	777	6.980		
$SrBr_2$	Strontium bromide	657	3.70	0.000745	1004
$SrCl_2$	Strontium chloride	874	2.727	0.000578	1037
SrF_2	Strontium fluoride	1477	3.470	0.000751	1927
SrI_2	Strontium iodide	538	4.085	0.000885	1026
Ta	Tantalum	3017	15		
$TaCl_5$	Tantalum(V) chloride	216	2.700	0.004316	457
Tb	Terbium	1359	7.65		
Te	Tellurium	449.51	5.70	0.00035	600
$ThCl_4$	Thorium chloride	770	3.363	0.0014	847
ThF_4	Thorium fluoride	1110	6.058	0.000759	1378
Ti	Titanium	1668	4.11		
$TiCl_4$	Titanium(IV) chloride	-25	1.807	0.001735	137
Tl	Thallium	304	11.22	0.00144	600
TlBr	Thallium(I) bromide	460	5.98	0.001755	647
TlCl	Thallium(I) chloride	430	5.628	0.0018	642
TlI	Thallium(I) iodide	441.8	6.15	0.001761	737
$TlNO_3$	Thallium(I) nitrate	206	4.91	0.001873	279
Tl_2SO_4	Thallium(I) sulfate	632	5.62	0.00130	927
Tm	Thulium	1545	8.56	0.00050	1675
U	Uranium	1135	17.3		
UCl_3	Uranium(III) chloride	837	4.84	0.007943	1057

DENSITY OF MOLTEN ELEMENTS AND REPRESENTATIVE SALTS (continued)

Formula	Name	$t_m/°C$	$\rho_m/\text{g cm}^{-3}$	$k/\text{g cm}^{-3}\ °C^{-1}$	t_{max}
UCl_4	Uranium(IV) chloride	590	3.572	0.001945	667
UF_4	Uranium(IV) fluoride	1036	6.485	0.000992	1341
V	Vanadium	1910	5.5		
W	Tungsten	3422	17.6		
Y	Yttrium	1526	4.24		
YCl_3	Yttrium chloride	721	2.510	0.0005	845
Yb	Ytterbium	824	6.21		
Zn	Zinc	419.53	6.57	0.0011	700
$ZnBr_2$	Zinc bromide	394	3.47	0.000959	602
$ZnCl_2$	Zinc chloride	290	2.54	0.00053	557
ZnI_2	Zinc iodide	446	3.878	0.00136	588
$ZnSO_4$	Zinc sulfate	680	3.14	0.00047	987
Zr	Zirconium	1855	5.8		
$ZrCl_4$	Zirconium chloride	437	1.643	0.007464	492

MAGNETIC SUSCEPTIBILITY OF THE ELEMENTS AND INORGANIC COMPOUNDS

When a material is placed in a magnetic field H, a magnetization (magnetic moment per unit volume) M is induced in the material which is related to H by $M = \kappa H$, where κ is called the volume susceptibility. Since H and M have the same dimensions, κ is dimensionless. A more useful parameter is the molar susceptibility χ_m, defined by

$$\chi_m = \kappa V_m = \kappa\, M/\rho$$

where V_m is the molar volume of the substance, M the molar mass, and ρ the mass density. When the cgs system is used, the customary units for χ_m are $cm^3\ mol^{-1}$; the corresponding SI units are $m^3\ mol^{-1}$.

Substances that have no unpaired electron orbital or spin angular momentum generally have negative values of χ_m and are called diamagnetic. Their molar susceptibility varies only slightly with temperature. Substances with unpaired electrons, which are termed paramagnetic, have positive χ_m and show a much stronger temperature dependence, varying roughly as $1/T$. The net susceptibility of a paramagnetic substance is the sum of the paramagnetic and diamagnetic contributions, but the former almost always dominates.

This table gives values of χ_m for the elements and selected inorganic compounds. All values refer to nominal room temperature (285 to 300 K) unless otherwise indicated. When the physical state (s = solid, l = liquid, g = gas, aq = aqueous solution) is not given, the most common crystalline form is understood. An entry of "Ferro." indicates a ferromagnetic substance.

Substances are arranged in alphabetical order by the most common name, except that compounds such as hydrides, oxides, and acids are grouped with the parent element (the same ordering used in the table "Physical Constants of Inorganic Compounds").

In keeping with customary practice, the molar susceptibility is given here in units appropriate to the cgs system. These values should be multiplied by 4π to obtain values for use in SI equations (where the magnetic field strength H has units of $A\ m^{-1}$).

REFERENCES

1. *Landolt-Börnstein, Numerical Data and Functional Relationships in Science and Technology, New Series,* II/16, *Diamagnetic Susceptibility,* Springer-Verlag, Heidelberg, 1986.
2. *Landolt-Börnstein, Numerical Data and Functional Relationships in Science and Technology, New Series,* III/19, Subvolumes a to i2, *Magnetic Properties of Metals,* Springer-Verlag, Heidelberg, 1986-1992.
3. *Landolt-Börnstein, Numerical Data and Functional Relationships in Science and Technology, New Series,* II/2, II/8, II/10, II/11, and II/12a, *Coordination and Organometallic Transition Metal Compounds,* Springer-Verlag, Heidelberg, 1966-1984.
4. *Tables de Constantes et Données Numérique, Volume 7, Relaxation Paramagnetique,* Masson, Paris, 1957.

Name	Formula	$\chi_m/10^{-6}\ cm^3\ mol^{-1}$	Name	Formula	$\chi_m/10^{-6}\ cm^3\ mol^{-1}$
Aluminum	Al	+16.5	Arsenic (yellow)	As	-23.2
Aluminum trifluoride	AlF_3	-13.9	Arsine (g)	AsH_3	-35.2
Aluminum oxide	Al_2O_3	-37	Arsenic(III) bromide	$AsBr_3$	-106
Aluminum sulfate	$Al_2(SO_4)_3$	-93	Arsenic(III) chloride	$AsCl_3$	-72.5
Ammonia (g)	NH_3	-16.3	Arsenic(III) iodide	AsI_3	-142.2
Ammonia (aq)	NH_3	-18.3	Arsenic(III) oxide	As_2O_3	-30.34
Ammonium acetate	$NH_4C_2H_3O_2$	-41.1	Arsenic(III) sulfide	As_2S_3	-70
Ammonium bromide	NH_4Br	-47	Barium	Ba	+20.6
Ammonium carbonate	$(NH_4)_2CO_3$	-42.5	Barium bromide	$BaBr_2$	-92
Ammonium chlorate	NH_4ClO_3	-42.1	Barium bromide dihydrate	$BaBr_2 \cdot 2H_2O$	-119.3
Ammonium chloride	NH_4Cl	-36.7	Barium carbonate	$BaCO_3$	-58.9
Ammonium fluoride	NH_4F	-23	Barium chloride	$BaCl_2$	-72.6
Ammonium iodate	NH_4IO_3	-62.3	Barium chloride dihydrate	$BaCl_2 \cdot 2H_2O$	-100
Ammonium iodide	NH_4I	-66	Barium fluoride	BaF_2	-51
Ammonium nitrate	NH_4NO_3	-33	Barium hydroxide	$Ba(OH)_2$	-53.2
Ammonium sulfate	$(NH_4)_2SO_4$	-67	Barium iodate	$Ba(IO_3)_2$	-122.5
Ammonium thiocyanate	NH_4SCN	-48.1	Barium iodide	BaI_2	-124.4
Antimony	Sb	-99	Barium iodide dihydrate	$BaI_2 \cdot 2H_2O$	-163
Stibine (g)	SbH_3	-34.6	Barium nitrate	$Ba(NO_3)_2$	-66.5
Antimony(III) bromide	$SbBr_3$	-111.4	Barium oxide	BaO	-29.1
Antimony(III) chloride	$SbCl_3$	-86.7	Barium peroxide	BaO_2	-40.6
Antimony(III) fluoride	SbF_3	-46	Barium sulfate	$BaSO_4$	-65.8
Antimony(III) iodide	SbI_3	-147.2	Beryllium	Be	-9.0
Antimony(III) oxide	Sb_2O_3	-69.4	Beryllium chloride	$BeCl_2$	-26.5
Antimony(III) sulfide	Sb_2S_3	-86	Beryllium hydroxide	$Be(OH)_2$	-23.1
Antimony(V) chloride	$SbCl_5$	-120.5	Beryllium oxide	BeO	-11.9
Argon (g)	Ar	-19.32	Beryllium sulfate	$BeSO_4$	-37
Arsenic (gray)	As	-5.6	Bismuth	Bi	-280.1

Name	Formula	$\chi_m/10^{-6}$ cm^3 mol^{-1}	Name	Formula	$\chi_m/10^{-6}$ cm^3 mol^{-1}
Bismuth tribromide	$BiBr_3$	-147	Cesium bromate	$CsBrO_3$	-75.1
Bismuth trichloride	$BiCl_3$	-26.5	Cesium bromide	$CsBr$	-67.2
Bismuth fluoride	BiF_3	-61.2	Cesium carbonate	Cs_2CO_3	-103.6
Bismuth hydroxide	$Bi(OH)_3$	-65.8	Cesium chlorate	$CsClO_3$	-65
Bismuth triiodide	BiI_3	-200.5	Cesium chloride	$CsCl$	-56.7
Bismuth nitrate pentahydrate	$Bi(NO_3)_3 \cdot 5H_2O$	-159	Cesium fluoride	CsF	-44.5
Bismuth oxide	Bi_2O_3	-83	Cesium iodide	CsI	-82.6
Bismuth phosphate	$BiPO_4$	-77	Cesium superoxide	CsO_2	+1534
Bismuth sulfate	$Bi_2(SO_4)_3$	-199	Cesium sulfate	Cs_2SO_4	-116
Bismuth sulfide	Bi_2S_3	-123	Chlorine (l)	Cl_2	-40.4
Boron	B	-6.7	Chlorine trifluoride (g)	ClF_3	-26.5
Diborane (g)	B_2H_6	-21.0	Chromium	Cr	+167
Boric acid (orthoboric acid)	H_3BO_3	-34.1	Chromium(II) chloride	$CrCl_2$	+7230
Boron trichloride	BCl_3	-59.9	Chromium(III) chloride	$CrCl_3$	+6350
Boron oxide	B_2O_3	-38.7	Chromium(III) fluoride	CrF_3	+4370
Bromine (l)	Br_2	-56.4	Chromium(III) oxide	Cr_2O_3	+1960
Bromine (g)	Br_2	-73.5	Chromium(III) sulfate	$Cr_2(SO_4)_3$	+11800
Bromine trifluoride	BrF_3	-33.9	Chromium(VI) oxide	CrO_3	+40
Bromine pentafluoride	BrF_5	-45.1	Cobalt	Co	Ferro.
Cadmium	Cd	-19.7	Cobalt(II) bromide	$CoBr_2$	+13000
Cadmium bromide	$CdBr_2$	-87.3	Cobalt(II) chloride	$CoCl_2$	+12660
Cadmium bromide tetrahydrate	$CdBr_2 \cdot 4H_2O$	-131.5	Cobalt(II) chloride hexahydrate	$CoCl_2 \cdot 6H_2O$	+9710
Cadmium carbonate	$CdCO_3$	-46.7	Cobalt(II) cyanide	$Co(CN)_2$	+3825
Cadmium chloride	$CdCl_2$	-68.7	Cobalt(II) fluoride	CoF_2	+9490
Cadmium chromate	$CdCrO_4$	-16.8	Cobalt(II) iodide	CoI_2	+10760
Cadmium cyanide	$Cd(CN)_2$	-54	Cobalt(II) sulfate	$CoSO_4$	+10000
Cadmium fluoride	CdF_2	-40.6	Cobalt(II) sulfide	CoS	+225
Cadmium hydroxide	$Cd(OH)_2$	-41	Cobalt(II,III) oxide	Co_3O_4	+7380
Cadmium iodate	$Cd(IO_3)_2$	-108.4	Cobalt(III) fluoride	CoF_3	+1900
Cadmium iodide	CdI_2	-117.2	Cobalt(III) oxide	Co_2O_3	+4560
Cadmium nitrate	$Cd(NO_3)_2$	-55.1	Copper	Cu	-5.46
Cadmium nitrate tetrahydrate	$Cd(NO_3)_2 \cdot 4H_2O$	-140	Copper(I) bromide	$CuBr$	-49
Cadmium oxide	CdO	-30	Copper(I) chloride	$CuCl$	-40
Cadmium sulfate	$CdSO_4$	-59.2	Copper(I) cyanide	$CuCN$	-24
Cadmium sulfide	CdS	-50	Copper(I) iodide	CuI	-63
Calcium	Ca	+40	Copper(I) oxide	Cu_2O	-20
Calcium bromide	$CaBr_2$	-73.8	Copper(II) bromide	$CuBr_2$	+685
Calcium carbonate	$CaCO_3$	-38.2	Copper(II) chloride	$CuCl_2$	+1080
Calcium chloride	$CaCl_2$	-54.7	Copper(II) chloride dihydrate	$CuCl_2 \cdot 2H_2O$	+1420
Calcium fluoride	CaF_2	-28	Copper(II) fluoride	CuF_2	+1050
Calcium hydroxide	$Ca(OH)_2$	-22	Copper(II) fluoride dihydrate	$CuF_2 \cdot 2H_2O$	+1600
Calcium iodate	$Ca(IO_3)_2$	-101.4	Copper(II) hydroxide	$Cu(OH)_2$	+1170
Calcium iodide	CaI_2	-109	Copper(II) nitrate trihydrate	$Cu(NO_3)_2 \cdot 3H_2O$	+1570
Calcium oxide	CaO	-15.0	Copper(II) nitrate hexahydrate	$Cu(NO_3)_2 \cdot 6H_2O$	+1625
Calcium sulfate	$CaSO_4$	-49.7	Copper(II) oxide	CuO	+238
Calcium sulfate dihydrate	$CaSO_4 \cdot 2H_2O$	-74	Copper(II) sulfate	$CuSO_4$	+1330
Carbon (diamond)	C	-5.9	Copper(II) sulfate pentahydrate	$CuSO_4 \cdot 5H_2O$	+1460
Carbon (graphite)	C	-6.0	Copper(II) sulfide	CuS	-2.0
Carbon monoxide (g)	CO	-11.8	Dysprosium (α)	Dy	+98000
Carbon dioxide (g)	CO_2	-21.0	Dysprosium(III) oxide	Dy_2O_3	+89600
Cerium (β)	Ce	+2500	Dysprosium(III) sulfide	Dy_2S_3	+95200
Cerium(II) sulfide	CeS	+2110	Erbium	Er	+48000
Cerium(III) chloride	$CeCl_3$	+2490	Erbium oxide	Er_2O_3	+73920
Cerium(III) fluoride	CeF_3	+2190	Erbium sulfate octahydrate	$Er_2(SO_4)_3 \cdot 8H_2O$	+74600
Cerium(III) sulfide	Ce_2S_3	+5080	Erbium sulfide	Er_2S_3	+77200
Cerium(IV) oxide	CeO_2	+26	Europium	Eu	+30900
Cerium(IV) sulfate tetrahydrate	$Ce(SO_4)_2 \cdot 4H_2O$	-97	Europium(II) bromide	$EuBr_2$	+26800
Cesium	Cs	+29			

Name	Formula	$\chi_m/10^{-6}$ cm^3 mol^{-1}	Name	Formula	$\chi_m/10^{-6}$ cm^3 mol^{-1}
Europium(II) chloride	$EuCl_2$	+26500	Iodine pentoxide	I_2O_5	-79.4
Europium(II) fluoride	EuF_2	+23750	Iodine chloride	ICl	-54.6
Europium(II) iodide	EuI_2	+26000	Iodine trichloride	ICl_3	-90.2
Europium(II) sulfide	EuS	+23800	Iodine pentafluoride	IF_5	-58.1
Europium(III) oxide	Eu_2O_3	+10100	Iridium	Ir	+25
Europium(III) sulfate	$Eu_2(SO_4)_3$	+10400	Iridium(III) chloride	$IrCl_3$	-14.4
Fluorine	F_2	-9.63	Iridium(IV) oxide	IrO_2	+224
Gadolinium (350 K)	Gd	+185000	Iron	Fe	Ferro.
Gadolinium(III) chloride	$GdCl_3$	+27930	Iron(II) bromide	$FeBr_2$	+13600
Gadolinium(III) oxide	Gd_2O_3	+53200	Iron(II) carbonate	$FeCO_3$	+11300
Gadolinium(III) sulfate octahydrate	$Gd_2(SO_4)_3 \cdot 8H_2O$	+53280	Iron(II) chloride	$FeCl_2$	+14750
			Iron(II) chloride tetrahydrate	$FeCl_2 \cdot 4H_2O$	+12900
Gadolinium(III) sulfide	Gd_2S_3	+55500	Iron(II) fluoride	FeF_2	+9500
Gallium	Ga	-21.6	Iron(II) iodide	FeI_2	+13600
Gallium suboxide	Ga_2O	-34	Iron(II) oxide	FeO	+7200
Gallium(II) sulfide	GaS	-23	Iron(II) sulfate	$FeSO_4$	+12400
Gallium(III) chloride	$GaCl_3$	-63	Iron(II) sulfate monohydrate	$FeSO_4 \cdot H_2O$	+10500
Gallium(III) sulfide	Ga_2S_3	-80	Iron(II) sulfate heptahydrate	$FeSO_4 \cdot 7H_2O$	+11200
Germanium	Ge	-11.6	Iron(II) sulfide	FeS	+1074
Germane (g)	GeH_4	-29.7	Iron(III) chloride	$FeCl_3$	+13450
Germanium(II) oxide	GeO	-28.8	Iron(III) chloride hexahydrate	$FeCl_3 \cdot 6H_2O$	+15250
Germanium(II) sulfide	GeS	-40.9	Iron(III) fluoride	FeF_3	+13760
Germanium(IV) chloride	$GeCl_4$	-72	Iron(III) fluoride trihydrate	$FeF_3 \cdot 3H_2O$	+7870
Germanium(IV) fluoride	GeF_4	-50	Iron(III) nitrate nonahydrate	$Fe(NO_3)_3 \cdot 9H_2O$	+15200
Germanium(IV) iodide	GeI_4	-171	Krypton (g)	Kr	-29.0
Germanium(IV) oxide	GeO_2	-34.3	Lanthanum (α)	La	+95.9
Germanium(IV) sulfide	GeS_2	-53.9	Lanthanum oxide	La_2O_3	-78
Gold	Au	-28	Lanthanum sulfate nonahydrate	$La_2(SO_4)_3 \cdot 9H_2O$	-262
Gold(I) bromide	$AuBr$	-61			
Gold(I) chloride	$AuCl$	-67	Lanthanum sulfide	La_2S_3	-37
Gold(I) iodide	AuI	-91	Lead	Pb	-23
Gold(III) chloride	$AuCl_3$	-112	Lead(II) acetate	$Pb(C_2H_3O_2)_2$	-89.1
Hafnium	Hf	+71	Lead(II) bromide	$PbBr_2$	-90.6
Hafnium oxide	HfO_2	-23	Lead(II) carbonate	$PbCO_3$	-61.2
Helium (g)	He	-2.02	Lead(II) chloride	$PbCl_2$	-73.8
Holmium	Ho	+72900	Lead(II) chromate	$PbCrO_4$	-18
Holmium oxide	Ho_2O_3	+88100	Lead(II) fluoride	PbF_2	-58.1
Hydrazine (l)	N_2H_4	-201	Lead(II) iodate	$Pb(IO_3)_2$	-131
Hydrogen (l, 20.3 K)	H_2	-5.44	Lead(II) iodide	PbI_2	-126.5
Hydrogen (g)	H_2	-3.99	Lead(II) nitrate	$Pb(NO_3)_2$	-74
Hydrogen chloride (l)	HCl	-22.6	Lead(II) oxide	PbO	-42
Hydrogen chloride (aq)	HCl	-22	Lead(II) phosphate	$Pb_3(PO_4)_2$	-182
Hydrogen fluoride (l)	HF	-8.6	Lead(II) sulfate	$PbSO_4$	-69.7
Hydrogen fluoride (aq)	HF	-9.3	Lead(II) sulfide	PbS	-83.6
Hydrogen iodide (s, 195 K)	HI	-47.3	Lithium	Li	+14.2
Hydrogen iodide (l, 233 K)	HI	-48.3	Lithium bromide	$LiBr$	-34.3
Hydrogen iodide (aq)	HI	-50.2	Lithium carbonate	Li_2CO_3	-27
Hydrogen peroxide (l)	H_2O_2	-17.3	Lithium chloride	$LiCl$	-24.3
Hydrogen sulfide (g)	H_2S	-25.5	Lithium fluoride	LiF	-10.1
Indium	In	-10.2	Lithium hydride	LiH	-4.6
Indium(I) chloride	$InCl$	-30	Lithium hydroxide (aq)	$LiOH$	-12.3
Indium(II) chloride	$InCl_2$	-56	Lithium iodide	LiI	-50
Indium(II) sulfide	InS	-28	Lithium sulfate	Li_2SO_4	-41.6
Indium(III) bromide	$InBr_3$	-107	Lutetium	Lu	+182.9
Indium(III) chloride	$InCl_3$	-86	Magnesium	Mg	+13.1
Indium(III) oxide	In_2O_3	-56	Magnesium bromide	$MgBr_2$	-72
Indium(III) sulfide	In_2S_3	-98	Magnesium carbonate	$MgCO_3$	-32.4
Iodine	I_2	-90	Magnesium chloride	$MgCl_2$	-47.4
Iodic acid	HIO_3	-48	Magnesium fluoride	MgF_2	-22.7

Name	Formula	$\chi_m/10^{-6}$ cm^3 mol^{-1}	Name	Formula	$\chi_m/10^{-6}$ cm^3 mol^{-1}
Magnesium hydroxide	$Mg(OH)_2$	-22.1	Molybdenum(VI) oxide	MoO_3	+3
Magnesium iodide	MgI_2	-111	Neodymium (α)	Nd	+5930
Magnesium oxide	MgO	-10.2	Neodymium fluoride	NdF_3	+4980
Magnesium sulfate	$MgSO_4$	-42	Neodymium oxide	Nd_2O_3	+10200
Magnesium sulfate monohydrate	$MgSO_4 \cdot H_2O$	-61	Neodymium sulfate	$Nd_2(SO_4)_3$	+9990
			Neodymium sulfide	Nd_2S_3	+5550
Magnesium sulfate heptahydrate	$MgSO_4 \cdot 7H_2O$	-135.7	Neon (g)	Ne	-6.96
			Neptunium	Np	+575
Manganese	Mn	+511	Nickel	Ni	Ferro.
Manganese(II) bromide	$MnBr_2$	+13900	Nickel(II) bromide	$NiBr_2$	+5600
Manganese(II) carbonate	$MnCO_3$	+11400	Nickel(II) chloride	$NiCl_2$	+6145
Manganese(II) chloride	$MnCl_2$	+14350	Nickel(II) chloride hexahydrate	$NiCl_2 \cdot 6H_2O$	+4240
Manganese(II) chloride tetrahydrate	$MnCl_2 \cdot 4H_2O$	+14600	Nickel(II) fluoride	NiF_2	+2410
Manganese(II) fluoride	MnF_2	+10700	Nickel(II) hydroxide	$Ni(OH)_2$	+4500
Manganese(II) hydroxide	$Mn(OH)_2$	+13500	Nickel(II) iodide	NiI_2	+3875
Manganese(II) iodide	MnI_2	+14400	Nickel(II) nitrate hexahydrate	$Ni(NO_3)_2 \cdot 6H_2O$	+4300
Manganese(II) oxide	MnO	+4850	Nickel(II) oxide	NiO	+660
Manganese(II) sulfate	$MnSO_4$	+13660	Nickel(II) sulfate	$NiSO_4$	+4005
Manganese(II) sulfate monohydrate	$MnSO_4 \cdot H_2O$	+14200	Nickel(II) sulfide	NiS	+190
			Nickel(III) sulfide	Ni_3S_2	+1030
Manganese(II) sulfate tetrahydrate	$MnSO_4 \cdot 4H_2O$	+14600	Niobium	Nb	+208
			Niobium(V) oxide	Nb_2O_5	-10
Manganese(II) sulfide (α form)	MnS	+5630	Nitrogen (g)	N_2	-12.0
			Nitric acid (l)	HNO_3	-19.9
Manganese(II) sulfide (β form)	MnS	+3850	Nitrous oxide (g)	N_2O	-18.9
			Nitric oxide (s, 90 K)	NO	+19.8
Manganese(II,III) oxide	Mn_3O_4	+12400	Nitric oxide (l, 118 K)	NO	+114.2
Manganese(III) fluoride	MnF_3	+10500	Nitric oxide (g)	NO	+1461
Manganese(III) oxide	Mn_2O_3	+14100	Nitrogen dioxide (g, 408 K)	NO_2	+150
Manganese(IV) oxide	MnO_2	+2280	Nitrogen trioxide (g)	N_2O_3	-16
Mercury (s, 234 K)	Hg	-24.1	Nitrogen tetroxide (g)	N_2O_4	-23.0
Mercury (l)	Hg	-33.5	Osmium	Os	+11
Mercury(I) bromide	Hg_2Br_2	-105	Oxygen (s, 54 K)	O_2	+10200
Mercury(I) chloride	Hg_2Cl_2	-120	Oxygen (l, 90 K)	O_2	+7699
Mercury(I) fluoride	Hg_2F_2	-106	Oxygen (g)	O_2	+3449
Mercury(I) iodide	Hg_2I_2	-166	Ozone (l)	O_3	+6.7
Mercury(I) nitrate	$Hg_2(NO_3)_2$	-121	Palladium	Pd	+540
Mercury(I) oxide	Hg_2O	-76.3	Palladium(II) chloride	$PdCl_2$	-38
Mercury(I) sulfate	Hg_2SO_4	-123	Phosphorus (white)	P	-26.66
Mercury(II) bromide	$HgBr_2$	-94.2	Phosphorus (red)	P	-20.77
Mercury(II) chloride	$HgCl_2$	-82	Phosphine (g)	PH_3	-26.2
Mercury(II) cyanide	$Hg(CN)_2$	-67	Phosphoric acid (aq)	H_3PO_4	-43.8
Mercury(II) fluoride	HgF_2	-57.3	Phosphorous acid (aq)	H_3PO_3	-42.5
Mercury(II) iodide	HgI_2	-165	Phosphorus(III) chloride (l)	PCl_3	-63.4
Mercury(II) nitrate	$Hg(NO_3)_2$	-74	Platinum	Pt	+193
Mercury(II) oxide	HgO	-46	Platinum(II) chloride	$PtCl_2$	-54
Mercury(II) sulfate	$HgSO_4$	-78.1	Platinum(III) chloride	$PtCl_3$	-66.7
Mercury(II) sulfide	HgS	-55.4	Platinum(IV) chloride	$PtCl_4$	-93
Mercury(II) thiocyanate	$Hg(SCN)_2$	-96.5	Platinum(IV) fluoride	PtF_4	+445
Molybdenum	Mo	+72	Plutonium	Pu	+525
Molybdenum(III) bromide	$MoBr_3$	+525	Plutonium(IV) fluoride	PuF_4	+1760
Molybdenum(III) chloride	$MoCl_3$	+43	Plutonium(IV) oxide	PuO_2	+730
Molybdenum(III) oxide	Mo_2O_3	-42.0	Plutonium(VI) fluoride	PuF_6	+173
Molybdenum(IV) bromide	$MoBr_4$	+520	Potassium	K	+20.8
Molybdenum(IV) chloride	$MoCl_4$	+1750	Potassium bromate	$KBrO_3$	-52.6
Molybdenum(IV) oxide	MoO_2	+41	Potassium bromide	KBr	-49.1
Molybdenum(V) chloride	$MoCl_5$	+990	Potassium carbonate	K_2CO_3	-59
Molybdenum(VI) fluoride	MoF_6	-26.0	Potassium chlorate	$KClO_3$	-42.8

Name	Formula	$\chi_m/10^{-6}$ cm^3 mol^{-1}	Name	Formula	$\chi_m/10^{-6}$ cm^3 mol^{-1}
Potassium chloride	KCl	-38.8	Disilane (g)	Si_2H_6	-37.3
Potassium chromate	K_2CrO_4	-3.9	Tetramethylsilane (l)	$(CH_3)_4Si$	-74.80
Potassium cyanide	KCN	-37	Tetraethylsilane (l)	$(C_2H_5)_4Si$	-120.2
Potassium ferricyanide	$K_3Fe(CN)_6$	+2290	Tetrabromosilane (l)	$SiBr_4$	-126
Potassium ferrocyanide trihydrate	$K_4Fe(CN)_6 \cdot 3H_2O$	-172.3	Tetrachlorosilane (l)	$SiCl_4$	-87.5
			Silicon carbide	SiC	-12.8
Potassium fluoride	KF	-23.6	Silicon dioxide	SiO_2	-29.6
Potassium hydrogen sulfate	$KHSO_4$	-49.8	Silver	Ag	-19.5
Potassium hydroxide (aq)	KOH	-22	Silver(I) bromide	AgBr	-61
Potassium iodate	KIO_3	-63.1	Silver(I) carbonate	Ag_2CO_3	-80.90
Potassium iodide	KI	-63.8	Silver(I) chloride	AgCl	-49
Potassium nitrate	KNO_3	-33.7	Silver(I) chromate	Ag_2CrO_4	-40
Potassium nitrite	KNO_2	-23.3	Silver(I) cyanide	AgCN	-43.2
Potassium permanganate	$KMnO_4$	+20	Silver(I) fluoride	AgF	-36.5
Potassium sulfate	K_2SO_4	-67	Silver(I) iodide	AgI	-80
Potassium sulfide	K_2S	-60	Silver(I) nitrate	$AgNO_3$	-45.7
Potassium superoxide	KO_2	+3230	Silver(I) nitrite	$AgNO_2$	-42
Potassium thiocyanate	KSCN	-48	Silver(I) oxide	Ag_2O	-134
Praseodymium (α)	Pr	+5530	Silver(I) phosphate	Ag_3PO_4	-120
Praseodymium chloride	$PrCl_3$	+44.5	Silver(I) sulfate	Ag_2SO_4	-92.90
Praseodymium oxide	Pr_2O_3	+8994	Silver(I) thiocyanate	AgSCN	-61.8
Praseodymium sulfide	Pr_2S_3	+10770	Silver(II) oxide	AgO	-19.6
Protactinium	Pa	+277	Sodium	Na	+16
Rhenium	Re	+67	Sodium acetate	$NaC_2H_3O_2$	-37.6
Rhenium(IV) oxide	ReO_2	+44	Sodium bromate	$NaBrO_3$	-44.2
Rhenium(IV) sulfide	ReS_2	+38	Sodium bromide	NaBr	-41
Rhenium(V) chloride	$ReCl_5$	+1225	Sodium carbonate	Na_2CO_3	-41
Rhenium(VI) oxide	ReO_3	+16	Sodium chlorate	$NaClO_3$	-34.7
Rhenium(VII) oxide	Re_2O_7	-16	Sodium chloride	NaCl	-30.2
Rhodium	Rh	+102	Sodium dichromate	$Na_2Cr_2O_7$	+55
Rhodium(III) chloride	$RhCl_3$	-7.5	Sodium fluoride	NaF	-15.6
Rhodium(III) oxide	Rh_2O_3	+104	Sodium hydrogen phosphate	Na_2HPO_4	-56.6
Rubidium	Rb	+17	Sodium hydroxide (aq)	NaOH	-15.8
Rubidium bromide	RbBr	-56.4	Sodium iodate	$NaIO_3$	-53
Rubidium carbonate	Rb_2CO_3	-75.4	Sodium iodide	NaI	-57
Rubidium chloride	RbCl	-46	Sodium nitrate	$NaNO_3$	-25.6
Rubidium fluoride	RbF	-31.9	Sodium nitrite	$NaNO_2$	-14.5
Rubidium iodide	RbI	-72.2	Sodium oxide	Na_2O	-19.8
Rubidium nitrate	$RbNO_3$	-41	Sodium peroxide	Na_2O_2	-28.10
Rubidium sulfate	Rb_2SO_4	-88.4	Sodium sulfate	Na_2SO_4	-52
Rubidium superoxide	RbO_2	+1527	Sodium sulfate decahydrate	$Na_2SO_4 \cdot 10H_2O$	-184
Ruthenium	Ru	+39	Sodium sulfide	Na_2S	-39
Ruthenium(III) chloride	$RuCl_3$	+1998	Sodium tetraborate	$Na_2B_4O_7$	-85
Ruthenium(IV) oxide	RuO_2	+162	Strontium	Sr	+92
Samarium (α)	Sm	+1278	Strontium bromide	$SrBr_2$	-86.6
Samarium(II) bromide	$SmBr_2$	+5337	Strontium bromide hexahydrate	$SrBr_2 \cdot 6H_2O$	-160
Samarium(III) bromide	$SmBr_3$	+972			
Samarium(III) oxide	Sm_2O_3	+1988	Strontium carbonate	$SrCO_3$	-47
Samarium(III) sulfate octahydrate	$Sm_2(SO_4)_3 \cdot 8H_2O$	+1710	Strontium chlorate	$Sr(ClO_3)_2$	-73
			Strontium chloride	$SrCl_2$	-61.5
Samarium(III) sulfide	Sm_2S_3	+3300	Strontium chloride hexahydrate	$SrCl_2 \cdot 6H_2O$	-145
Scandium (α)	Sc	+295.2			
Selenium	Se	-25	Strontium chromate	$SrCrO_4$	-5.1
Selenium dioxide	SeO_2	-27.2	Strontium fluoride	SrF_2	-37.2
Selenium bromide	Se_2Br_2	-113	Strontium hydroxide	$Sr(OH)_2$	-40
Selenium chloride (l)	Se_2Cl_2	-94.8	Strontium iodate	$Sr(IO_3)_2$	-108
Selenium hexafluoride (g)	SeF_6	-51	Strontium iodide	SrI_2	-112
Silicon	Si	-3.12	Strontium nitrate	$Sr(NO_3)_2$	-57.2
Silane (g)	SiH_4	-20.4	Strontium oxide	SrO	-35

Name	Formula	$\chi_m/10^{-6}$ cm^3 mol^{-1}	Name	Formula	$\chi_m/10^{-6}$ cm^3 mol^{-1}
Strontium peroxide	SrO_2	-32.3	Tungsten carbide	WC	+10
Strontium sulfate	$SrSO_4$	-57.9	Tungsten(II) chloride	WCl_2	-25
Sulfur (rhombic)	S	-15.5	Tungsten(IV) oxide	WO_2	+57
Sulfur (monoclinic)	S	-14.9	Tungsten(IV) sulfide	WS_2	+5850
Sulfuric acid (l)	H_2SO_4	-39	Tungsten(V) bromide	WBr_5	+270
Sulfur dioxide (g)	SO_2	-18.2	Tungsten(V) chloride	WCl_5	+387
Sulfur trioxide (l)	SO_3	-28.54	Tungsten(VI) chloride	WCl_6	-71
Sulfur chloride (l)	$SSCl_2$	-62.2	Tungsten(VI) fluoride (g)	WF_6	-53
Sulfur dichloride (l)	SCl_2	-49.4	Tungsten(VI) oxide	WO_3	-15.8
Sulfur hexafluoride (g)	SF_6	-44	Uranium	U	+409
Thionyl chloride (l)	$SOCl_2$	-44.3	Uranium(III) bromide	UBr_3	+4740
Tantalum	Ta	+154	Uranium(III) chloride	UCl_3	+3460
Tantalum(V) chloride	$TaCl_5$	+140	Uranium(III) hydride	UH_3	+6244
Tantalum(V) oxide	Ta_2O_5	-32	Uranium(III) iodide	UI_3	+4460
Technetium	Tc	+115	Uranium(IV) bromide	UBr_4	+3530
Tellurium	Te	-38	Uranium(IV) chloride	UCl_4	+3680
Tellurium dibromide	$TeBr_2$	-106	Uranium(IV) fluoride	UF_4	+3530
Tellurium dichloride	$TeCl_2$	-94	Uranium(IV) oxide	UO_2	+2360
Tellurium hexafluoride (g)	TeF_6	-66	Uranium(VI) fluoride	UF_6	+43
Terbium (α)	Tb	+170000	Uranium(VI) oxide	UO_3	+128
Terbium oxide	Tb_2O_3	+78340	Vanadium	V	+285
Thallium	Tl	-50	Vanadium(II) bromide	VBr_2	+3230
Thallium(I) bromate	$TlBrO_3$	-75.9	Vanadium(II) chloride	VCl_2	+2410
Thallium(I) bromide	$TlBr$	-63.9	Vanadium(III) bromide	VBr_3	+2910
Thallium(I) carbonate	Tl_2CO_3	-101.6	Vanadium(III) chloride	VCl_3	+3030
Thallium(I) chlorate	$TlClO_3$	-65.5	Vanadium(III) fluoride	VF_3	+2757
Thallium(I) chloride	$TlCl$	-57.8	Vanadium(III) oxide	V_2O_3	+1976
Thallium(I) chromate	Tl_2CrO_4	-39.3	Vanadium(III) sulfide	V_2S_3	+1560
Thallium(I) cyanide	$TlCN$	-49	Vanadium(IV) chloride	VCl_4	+1215
Thallium(I) fluoride	TlF	-44.4	Vanadium(IV) oxide	VO_2	+99
Thallium(I) iodate	$TlIO_3$	-86.8	Vanadium(V) oxide	V_2O_5	+128
Thallium(I) iodide	TlI	-82.2	Water (s, 273 K)	H_2O	-12.63
Thallium(I) nitrate	$TlNO_3$	-56.5	Water (l, 293 K)	H_2O	-12.96
Thallium(I) nitrite	$TlNO_2$	-50.8	Water (l, 373 K)	H_2O	-13.09
Thallium(I) sulfate	Tl_2SO_4	-112.6	Water (g, 373 K))	H_2O	-13.1
Thallium(I) sulfide	Tl_2S	-88.8	Xenon (g)	Xe	-45.5
Thorium	Th	+97	Ytterbium (β)	Yb	+67
Thorium(IV) oxide	ThO_2	-16	Yttrium (α)	Y	+187.7
Thulium	Tm	+24700	Yttrium oxide	Y_2O_3	+44.4
Thulium oxide	Tm_2O_3	+51444	Yttrium sulfide	Y_2S_3	+100
Tin (gray)	Sn	-37.4	Zinc	Zn	-9.15
Tin(II) chloride	$SnCl_2$	-69	Zinc carbonate	$ZnCO_3$	-34
Tin(II) chloride dihydrate	$SnCl_2 \cdot 2H_2O$	-91.4	Zinc chloride	$ZnCl_2$	-55.33
Tin(II) oxide	SnO	-19	Zinc cyanide	$Zn(CN)_2$	-46
Tin(IV) bromide	$SnBr_4$	-149	Zinc fluoride	ZnF_2	-34.3
Tin(IV) chloride (l)	$SnCl_4$	-115	Zinc hydroxide	$Zn(OH)_2$	-67
Tin(IV) oxide	SnO_2	-41	Zinc iodide	ZnI_2	-108
Titanium	Ti	+151	Zinc oxide	ZnO	-27.2
Titanium(II) bromide	$TiBr_2$	+720	Zinc phosphate	$Zn_3(PO_4)_2$	-141
Titanium(II) chloride	$TiCl_2$	+484	Zinc sulfate	$ZnSO_4$	-47.8
Titanium(II) iodide	TiI_2	+1790	Zinc sulfate monohydrate	$ZnSO_4 \cdot H_2O$	-63
Titanium(II) sulfide	TiS	+432	Zinc sulfate heptahydrate	$ZnSO_4 \cdot 7H_2O$	-138
Titanium(III) bromide	$TiBr_3$	+660	Zinc sulfide	ZnS	-25
Titanium(III) chloride	$TiCl_3$	+1110	Zirconium	Zr	+120
Titanium(III) fluoride	TiF_3	+1300	Zirconium carbide	ZrC	-26
Titanium(III) oxide	Ti_2O_3	+132	Zirconium nitrate pentahydrate	$Zr(NO_3)_4 \cdot 5H_2O$	-77
Titanium(IV) chloride	$TiCl_4$	-54			
Titanium(IV) oxide	TiO_2	+5.9	Zirconium(IV) oxide	ZrO_2	-13.8
Tungsten	W	+53			

INDEX OF REFRACTION OF INORGANIC LIQUIDS

This table gives the index of refraction n of several inorganic substances in the liquid state at specified temperatures. The measurements refer to ambient atmospheric pressure except for substances whose normal boiling points are greater than the indicated temperature; in this case the pressure is the saturated vapor pressure of the substance. All values refer to a wavelength of 589 nm unless otherwise indicated. Entries are arranged in alphabetical order by chemical formula as normally written.

Data on the index of refraction at other temperatures and wavelengths may be found in Reference 1.

REFERENCES

1. Wohlfarth, C., and Wohlfarth, B., *Landolt-Börnstein, Numerical Data and Functional Relationships in Science and Technology, New Series*, III/38A, Martienssen, W., Editor, Springer-Verlag, Heidelberg, 1996.
2. Francis, A.W., *J. Chem. Eng. Data*, 5, 534, 1960.

Formula	Name	$t/°C$	n
Ar	Argon	-188	1.2312
$AsCl_3$	Arsenic(III) chloride	16	1.604
BBr_3	Boron tribromide	16	1.312
BrF_3	Bromine trifluoride	25	1.4536
BrF_5	Bromine pentafluoride	25	1.3529
Br_2	Bromine	15	1.659
COS	Carbon oxysulfide	25	1.3506
CO_2	Carbon dioxide	24	1.6630
CS_2	Carbon disulfide	20	1.62774
C_3O_2	Carbon suboxide	0	1.453
Cl_2	Chlorine	20	1.3834
CrO_2Cl_2	Chromyl chloride	23	1.524
$Fe(CO)_5$	Iron pentacarbonyl	14	1.523
$GeBr_4$	Germanium(IV) bromide	26	1.6269
$GeCl_4$	Germanium(IV) chloride	25	1.4614
HBr	Hydrogen bromide	10	1.325
HCN	Hydrogen cyanide	20	1.26136
HCl	Hydrogen chloride	18	1.3287 [a]
$HClO_4$	Perchloric acid	50	1.3819
HF	Hydrogen fluoride	25	1.1574
HI	Hydrogen iodide	16	1.466
HNO_3	Nitric acid	25	1.393
H_2	Hydrogen	-253	1.1096
H_2O	Water	20	1.33336
H_2O_2	Hydrogen peroxide	28	1.4061
H_2S	Hydrogen sulfide	-80	1.460
		20	1.3682
H_2SO_4	Sulfuric acid	20	1.4183
H_2S_2	Hydrogen disulfide	20	1.630
He	Helium	-269	1.02451 [c]
Kr	Krypton	-157	1.3032 [c]
NH_3	Ammonia	-77	1.3944 [b]
		20	1.3327
NO	Nitric oxide	-90	1.330
N_2	Nitrogen	-196	1.19876 [b]
N_2H_4	Hydrazine	22	1.470
N_2O	Nitrous oxide	25	1.238
O_2	Oxygen	-183	1.2243 [c]
PBr_3	Phosphorus(III) bromide	25	1.687
PCl_3	Phosphorus(III) chloride	21	1.5122
PH_3	Phosphine	17	1.317
P_2O_3	Phosphorus(III) oxide	27	1.540
S	Sulfur	125	1.9170
SCl_2	Sulfur dichloride	14	1.557
SF_6	Sulfur hexafluoride	25	1.167
$SOCl_2$	Thionyl chloride	10	1.527
SO_2	Sulfur dioxide	25	1.3396
SO_2Cl_2	Sulfuryl chloride	12	1.444

Formula	Name	$t/^\circ C$	n
SO_3	Sulfur trioxide	20	1.40965
$SSCl_2$	Sulfur chloride	20	1.671
$SbCl_5$	Antimony(V) chloride	22	1.5925
$SiBr_4$	Tetrabromosilane	31	1.5685
$SiCl_4$	Tetrachlorosilane	25	1.41156
$SnBr_4$	Tin(IV) bromide	31	1.6628
$SnCl_4$	Tin(IV) chloride	25	1.5086
$TiCl_4$	Titanium(IV) chloride	18	1.6076
Xe	Xenon	-112	1.3918 [c]

[a]   At 581 nm
[b]   At 578 nm
[c]   At 546 nm

PHYSICAL AND OPTICAL PROPERTIES OF MINERALS

The chemical formula, crystal system, density, hardness, and index of refraction of some common minerals are given in this table. Entries are arranged alphabetically by mineral name. The columns are:

- **Formula:** Chemical formula for a typical sample of the mineral. Composition often varies considerably with the origin of the sample.
- **Crystal system:** tricl = triclinic; monocl = monoclinic; orth = orthorhombic; tetr = tetragonal; hex = hexagonal; rhomb = rhombohedral; cub = cubic.
- **Density:** Typical density in g/cm^3. Individual samples may vary by a few percent.
- **Hardness:** On the Mohs' scale (range of 1 to 10, with talc = 1 and diamond = 10).
- **Index of refraction:** Values are given for the three coordinate axes in the order of least, intermediate, and greatest index. For cubic crystals there is only a single value. See Reference 1 for details on the axis systems. Variations of several percent, depending on the origin and exact composition of the sample, are common.

REFERENCES

1. Deer, W.A., Howie, R.A., and Zussman, J., *An Introduction to the Rock-Forming Minerals*, 2nd Edition, Longman Scientific & Technical, Harlow, Essex, 1992.
2. Carmichael, R.S., *Practical Handbook of Physical Properties of Rocks and Minerals*, CRC Press, Boca Raton, FL, 1989.
3. Donnay, J.D.H., and Ondik, H.M., *Crystal Data Determinative Tables, Third Edition, Volume 2, Inorganic Compounds*, Joint Committee on Powder Diffraction Standards, Swarthmore, PA, 1973.

Name	Formula	Crystal system	Density g/cm^3	Hard-ness	Index of refraction n_α	n_β	n_γ
Acanthite	Ag_2S	orth	7.2	2.3			
Actinolite	$Ca_2(Mg,Fe)_5Si_8O_{22}(OH,F)_2$	monocl	3.23	5.5	1.624	1.655	1.664
Aegirine	$NaFe(SiO_3)_2$	monocl	3.58	6	1.763	1.800	1.815
Akermanite	$Ca_2MgSi_2O_7$	tetr	2.94	5.5	1.632	1.640	
Alabandite	MnS	cub	4.0	3.8			
Albite	$NaAlSi_3O_8$	tricl	2.63	6.3	1.527	1.531	1.538
Allanite	$(Ca,Mn,Ce,La,Y,Th)_2(Fe,Ti)(Al,Fe)$ $O \cdot OH(Si_2O_7)(SiO_4)$	monocl	3.8	5.8	1.75	1.78	1.80
Allemontite	$SbAs$	hex	6.0	3.5			
Almandine	$Fe_3Al_2Si_3O_{12}$	cub	4.32	6.8	1.830		
Altaite	$PbTe$	cub	8.16	3			
Aluminite	$Al_2(SO_4)(OH)_4 \cdot 7H_2O$	monocl	1.74	1.5	1.459	1.464	1.470
Alunite	$(K,Na)Al_3(SO_4)_2(OH)_6$	rhomb	2.8	3.8	1.572	1.592	
Alunogen	$Al_2(SO_4)_3 \cdot 18H_2O$	monocl	1.69	1.8	1.467	1.47	1.478
Amblygonite	$(Li,Na)Al(PO_4)(F,OH)$	tricl	3.1	5.8	1.591	1.604	1.613
Analcite	$NaAlSi_2O_6 \cdot H_2O$	cub	2.27	5.5	1.486		
Anatase	TiO_2	tetr	4.23	5.8	2.488	2.561	
Andalusite	Al_2OSiO_4	orth	3.15	7.5	1.635	1.639	1.644
Andesine	$([NaSi]_{0.7-0.5}[CaAl]_{0.3-0.5}AlSi_2O_8$	tricl	2.67	6.3	1.550	1.553	1.557
Andorite	$PbAgSb_3S_6$	rhomb	5.35	3.3			
Andradite	$Ca_3(Fe,Ti)_2Si_3O_{12}$	cub	3.86	6.8	1.887		
Anglesite	$PbSO_4$	orth	6.29	2.8	1.877	1.883	1.894
Anhydrite	$CaSO_4$	orth	2.96	3.5	1.570	1.575	1.614
Ankerite	$Ca(Fe,Mg,Mn)(CO_3)_2$	rhomb	3.0	3.8	1.529	1.720	
Anorthite	$CaAl_2Si_2O_8$	tricl	2.76	6.3	1.577	1.585	1.590
Anorthoclase	$(Na,K)AlSi_3O_8$	tricl	2.58	6	1.523	1.528	1.529
Anthophyllite	$(Mg,Fe)_7Si_8O_{22}(OH,F)_2$	rhomb	3.21	5.8	1.645	1.658	1.668
Apatite	$Ca_5(PO_4)_3(OH,F,Cl)$	hex	3.2	5	1.645	1.648	
Apophyllite	$KFCa_4Si_8O_{20} \cdot 8H_2O$	tetr	2.35	4.8	1.535	1.536	
Aragonite	$CaCO_3$	orth	2.83	3.5	1.531	1.680	1.686
Arcanite	K_2SO_4	orth	2.66		1.494	1.494	1.497
Argentite	Ag_2S	orth	7.2	2.3			
Arsenolite	As_2O_3	cub	3.86	1.5	1.755		
Arsenopyrite	$FeAsS$	monocl	6.1	5.8			
Atacamite	$Cu_2(OH)_3Cl$	rhomb	3.76	3.3	1.831	1.861	1.880
Augelite	$Al_2(PO_4)(OH)_3$	monocl	2.70	4.8	1.574	1.576	1.588
Augite	$(Ca,Mg,Fe,Ti,Al)_2(Si,Al)_2O_6$	monocl	3.38	6	1.703	1.707	1.738
Autunite	$Ca(UO_{22})(PO_4)_2 \cdot 10H2O$	tetr	3.2	2.3	1.553	1.577	
Axinite	$(Ca,Mn,Fe)_3Al_2BO_3Si_4O_{12}(OH)$	tricl	3.31	6.8	1.684	1.691	1.694

Name	Formula	Crystal system	Density g/cm^3	Hardness	Index of refraction		
					n_α	n_β	n_γ
Azurite	$Cu_3(OH)_2(CO_3)_2$	monocl	3.77	3.8	1.730	1.758	1.838
Baddeleyite	ZrO_2	monocl	5.7	6.5	2.13	2.19	2.20
Barite	$BaSO_4$	orth	4.49	3.3	1.636	1.637	1.648
Benitoite	$BaTi(SiO_3)_3$	rhomb	3.65	6.3	1.757	1.804	
Bertrandite	$Be_4Si_2O_7(OH)_2$	rhomb	2.6	6	1.589	1.602	1.613
Beryl	$Be_3Al_2(SiO_3)_6$	hex	2.64	7.8	1.582	1.589	
Beryllonite	$NaBe(PO)_4$	monocl	2.81	5.8	1.552	1.558	1.561
Biotite	$K(Mg,Fe)_3AlSi_3O_{10}(OH,F)_2$	monocl	3.0	2.8	1.595	1.651	1.651
Bismuthinite	Bi_2S_3	orth	6.78	2			
Bixbyite	$(Mn,Fe)_2O_3$	cub	4.95	6.3			
Bloedite	$Na_2Mg(SO_4)_2 \cdot 4H_2O$	monocl	2.25	2.8	1.483	1.486	1.487
Boehmite	$AlO(OH)$	orth	3.44	3.8	1.64	1.65	1.66
Boracite	$Mg_3B_7O_{13}Cl$	rhomb	2.94	7.3	1.66	1.66	1.67
Borax	$Na_2B_4O_7 \cdot 10H_2O$	monocl	1.73	2.3	1.447	1.469	1.472
Bornite	Cu_5FeS_4	cub	5.07	3			
Boulangerite	$Pb_5Sb_4S_{11}$	monocl	6.1	2.8			
Bournonite	$PbCuSbS_3$	rhomb	5.83	2.8			
Braggite	PtS	tetr	10.2				
Braunite	$(Mn,Si)_2O_3$	tetr	4.78	6.3			
Bravoite	$(Ni,Fe)S_2$	cub	4.62	5.8			
Breithauptite	$NiSb$	hex	≈ 8.7	5.5			
Brochantite	$Cu_4(SO_4)(OH)_6$	monocl	3.79	3.8	1.728	1.771	1.800
Bromyrite	$AgBr$	cub	6.47	2.5	2.253		
Brookite	TiO_2	tetr	4.23	5.8	2.583	2.584	2.700
Brucite	$Mg(OH)_2$	hex	2.37	2.5	1.575	1.59	
Bunsenite	NiO	cub	6.72	5.5			
Cacoxenite	$Fe_4(PO_4)_3(OH)_3 \cdot 12H_2O$	hex	2.3	3.5	1.580	1.646	
Calcite	$CaCO_3$	hex	2.71	3	1.486	1.658	
Caledonite	$Cu_2Pb_5(SO_4)_3(CO_3)(OH)_6$	rhomb	5.76	2.8	1.818	1.866	1.909
Calomel	Hg_2Cl_2	tetr	7.16	1.5	1.973	2.656	
Cancrinite	$(Na,Ca,K)_7[Al_6Si_6O_{24}]$ $(CO_3,SO_4,Cl,OH)_2 \cdot H_2O$	hex	2.42	5.5	1.495	1.509	
Carnalite	$KMgCl_3 \cdot 6H_2O$	rhomb	1.60	2.5	1.466	1.475	1.494
Carnotite	$K_2(UO_2)_2(VO_4)_2 \cdot 3H_2O$	rhomb		1.5	1.75	1.92	1.95
Cassiterite	SnO_2	tetr	6.85	6.5	2.006	2.097	
Celestite	$SrSO_4$	orth	3.96	3.3	1.622	1.624	1.631
Celsian	$BaAl_2Si_2O_8$	monocl	3.25	6.3	1.583	1.588	1.594
Cerargyrite	$AgCl$	cub	5.56	2.5	2.071		
Cerussite	$PbCO_3$	orth	6.6	3.3	1.804	2.076	2.079
Cervantite	Sb_2O_4	orth	6.64	4.5			
Chabazite	$Ca[Al_2Si_4O_{12}] \cdot 6H_2O$	trig	2.08	4.5	1.482		
Chalcanthite	$CuSO_4 \cdot 5H_2O$	tricl	2.29	2.5	1.514	1.537	1.543
Chalcocite	Cu_2S	orth	5.6	2.8			
Chalcopyrite	$CuFeS_2$	tetr	4.2	3.8			
Chiolite	$Na_5Al_3F_{14}$	tetr	3.00	3.8	1.342	1.349	
Chlorite	$(Mg,Al,Fe)_{12}(Si,Al)_8O_{20}(OH)_{16}$	monocl	3.0	2.5	1.61	1.62	1.62
Chloritoid	$FeAl_4O_2(SiO_4)_2(OH)_4$	monocl	3.66	6.5	1.717	1.721	1.726
Chondrodite	$Mg(OH,F)_2 \cdot 2Mg_2SiO_4$	monocl	3.21	6.5	1.604	1.615	1.634
Chromite	$FeCr_2O_4$	cub	5.0	5.5	2.16		
Chrysoberyl	$BeAl_2O_4$	orth	3.65	8.5	1.746	1.748	1.756
Chrysocolla	$CuSiO_3 \cdot 2H_2O$	rhomb	2.4	2	1.575	1.597	1.598
Cinnabar	HgS	hex	8.17	2.3	2.814	3.143	
Claudetite	As_2O_3	monocl	3.74	2.5	1.87	1.92	2.01
Clinohumite	$Mg(OH,F)_2 \cdot 4Mg_2SiO_4$	monocl	3.21	6	1.633	1.647	1.668
Clinozoisite	$Ca_2Al_3Si_3O_{12}(OH)$	monocl	3.30	6.5	1.693	1.700	1.712
Cobaltite	$CoAsS$	cub	≈ 6.1	5.5			
Colemanite	$Ca_2B_6O_{11} \cdot 5H_2O$	monocl	2.42	4.5	1.586	1.592	1.614
Columbite	$(Fe,Mn)(Nb,Ta)_2O_6$	rhomb	5.20	6			
Connellite	$Cu_{19}(SO_4)Cl_4(OH)_{32} \cdot 3H_2O$	hex	3.36	3	1.731	1.752	

Name	Formula	Crystal system	Density g/cm^3	Hardness	Index of refraction n_α	n_β	n_γ
Copiapite	$(Fe,Mg)Fe_4(SO_4)_6(OH)_2 \cdot 20H_2O$	tricl	2.13	2.8	1.52	1.54	1.59
Coquimbite	$Fe_2(SO_4)_3 \cdot 9H_2O$	hex	2.1	2.5	1.54	1.56	
Cordierite	$Al_3(Mg,Fe)_2Si_5AlO_{18}$	rhomb	2.66	7	1.540	1.549	1.553
Corundum	Al_2O_3	hex	3.97	9	1.761	1.769	
Cotunnite	$PbCl_2$	orth	5.98	2.5	2.199	2.217	2.260
Covellite	CuS	hex	4.8	1.8			
Cristobalite	SiO_2	hex	2.33	6.5	1.484	1.487	
Crocoite	$PbCrO_4$	monocl	6.12	2.8	2.29	2.36	2.66
Cryolite	Na_3AlF_6	monocl	2.97	2.5	1.338	1.338	1.339
Cryolithionite	$Na_3Li_3Al_2F_{12}$	cub	2.77	2.8	1.340		
Cubanite	$CuFe_2S_3$	rhomb	4.11	3.5			
Cummingtonite	$(Mg,Fe)_7Si_8O_{22}(OH)_2$	monocl	3.4	5.5	1.650	1.660	1.676
Cuprite	Cu_2O	cub	6.0	3.8			
Danburite	$CaSi_2B_2O_8$	rhomb	3.0	7	1.63	1.63	1.63
Datolite	$CaBSiO_4(OH)$	monocl	2.98	5.3	1.624	1.652	1.668
Daubreelite	Cr_2FeS_4	cub	3.81				
Derbylite	$Fe_6Ti_6Sb_2O_{23}$	rhomb	4.53	5	2.45	2.45	2.51
Diamond	C	cub	3.51	10	2.418		
Diaspore	$AlO(OH)$	orth	3.4	6.8	1.694	1.715	1.741
Digenite	$Cu_{2-x}S$	cub	5.55	2.8			
Diopside	$CaMgSi_2O_6$	monocl	3.30	6	1.680	1.687	1.708
Dioptase	$CuSiO_2(OH)_2$	rhomb	3.5	5	1.65	1.70	
Dolomite	$CaMg(CO_3)_2$	rhomb	2.86	3.5	1.500	1.679	
Douglasite	$K_2FeCl_4 \cdot 2H_2O$	orth	2.16		1.488	1.500	
Dyscrasite	Ag_3Sb	rhomb	9.74	3.8			
Eddingtonite	$BaAl_2Si_3O_{10} \cdot 4H_2O$	rhomb	2.8		1.541	1.553	1.557
Eglestonite	Hg_4OCl_2	cub	8.4	2.5	2.49		
Emplectite	$CuBiS_2$	rhomb	6.38	2			
Enargite	Cu_3AsS_4	rhomb	4.5	3			
Enstatite	$MgSiO_3$	monocl	3.19	5.5	1.656	1.662	1.669
Epidote	$Ca_2Al_2(Al,Fe)OH(SiO_4)_3$	monocl	3.44	6	1.733	1.755	1.765
Epsomite	$MgSO_4 \cdot 7H_2O$	orth	1.67	2.3	1.433	1.455	1.461
Erythrite	$(Co,Ni)_3(AsO_4)_2 \cdot 8H_2O$	monocl	3.06	2	1.626	1.661	1.699
Eucairite	$CuAgSe$	orth	7.7	2.5			
Euclasite	$BeAlSiO_4(OH)$	monocl	3.1	7.5	1.651	1.655	1.671
Eudialite	$(Na,Ca,Ce)_5(Fe,Mn)(Zr,Ti)(Si_3O_9)_2$ (OH,Cl)	hex	3.0	5.5	1.623	1.600	1.615
Eulytite	$Bi_4Si_3O_{12}$	cub	6.6	4.5	2.05		
Euxenite	$(Y,Ca,Ce,U,Th)(Nb,Ta,Ti)_2O_6$	rhomb	5.5	6	2.2		
Fayalite	Fe_2SiO_4	orth	4.30	6.5	1.827	1.869	1.879
Ferberite	$FeWO_4$	monocl	7.51	4.3			
Fergussonite	$(Y,Er,Ce,Fe)(Nb,Ta,Ti)O_4$	tetr	5.7	6	2.1		
Fluorite	CaF_2	cub	3.18	4	1.434		
Forsterite	Mg_2SiO_4	orth	3.21	7	1.635	1.651	1.670
Franklinite	$ZnFe_2O_4$	cub	5.21	6	2.36		
Gahnite	$ZnAl_2O_4$	cub	4.62	7.8	1.805		
Galaxite	$MnAl_2O_4$	cub	4.04	7.8	1.92		
Galena	PbS	cub	7.60	2.5	3.91		
Galenabismuthite	$PbBi_2S_4$	rhomb	7.04	3			
Ganomalite	$(Ca,Pb)_{10}(OH,Cl)_2(Si_2O_7)_3$	hex	5.6	3.5	1.910	1.945	
Gaylussite	$Na_2Ca(CO_3)_2 \cdot 5H_2O$	monocl	1.99	2.8	1.444	1.516	1.523
Gehlenite	$Ca_2Al_2SiO_7$	tetr	3.04	5.5	1.658	1.669	
Geikielite	$MgTiO_3$	hex	3.85	5.5	1.95	2.31	
Gibbsite	$Al(OH)_3$	monocl	2.42	3	1.57	1.57	1.59
Glauberite	$Na_2Ca(SO_4)_2$	monocl	2.80	2.8	1.515	1.535	1.536
Glauconite	$(K,Na,Ca)_{1.6}(Fe,Al,Mg)_{4.0}Si_{7.3}Al_{0.7}$ $O_{20}(OH)_4$	monocl	2.7	2	1.60	1.63	1.63
Glaucophane	$Na_2Mg_3Al_2Si_8O_{22}(OH)_2$	monocl	3.19	6	1.634	1.645	1.648
Gmelinite	$(Ca,Na_2)[Al_2Si_4O_{12}] \cdot 6H_2O$	hex	2.10	4.5	1.477	1.485	

Name	Formula	Crystal system	Density g/cm^3	Hard-ness	Index of refraction n_α	n_β	n_γ
Goethite	FeO(OH)	orth	4.3	5.3	2.268	2.401	2.457
Goslarite	ZnSO$_4$·7H$_2$O	orth	1.97	2.3	1.457	1.480	1.484
Greenockite	CdS	cub	4.8	3.3	2.506	2.529	
Grossularite	Ca$_3$Al$_2$Si$_3$O$_{12}$	cub	3.59	6.8	1.734		
Gummite	UO$_3$·H$_2$O	orth	7.05	3.8			
Gypsum	CaSO$_4$·2H$_2$O	monocl	2.32	2	1.520	1.525	1.530
Halite	NaCl	cub	2.17	2	1.544		
Hambergite	Be$_2$(OH)(BO$_3$)	rhomb	2.36	7.5	1.56	1.59	1.63
Hanksite	Na$_{22}$K(SO$_4$)$_9$(CO$_3$)$_2$Cl	hex	2.56	3.3	1.461	1.481	
Harmotome	Ba[Al$_2$Si$_6$O$_{16}$]·6H$_2$O	monocl	2.44	4.5	1.506	1.507	1.511
Hausmannite	Mn$_3$O$_4$	tetr	4.84	5.5	2.15	2.46	
Haüyne	(Na,Ca)$_{4-8}$Al$_6$Si$_6$O$_{24}$(SO$_4$,S)$_{1-2}$	cub	2.47	5.8	1.502		
Hedenbergite	CaFeSi$_2$O$_6$	monocl	3.53	6	1.721	1.727	1.746
Helvite	Mn$_4$Be$_3$Si$_3$O$_{12}$S	cub	3.32	6	1.739		
Hematite	Fe$_2$O$_3$	hex	5.25	6	2.91	3.19	
Hemimorphite	Zn$_4$Si$_2$O$_7$(OH)$_2$·H$_2$O	rhomb	3.45	5	1.614	1.617	1.636
Hercynite	Fe(AlO$_2$)$_2$	cub	4.3	7.8	1.835		
Herderite	CaBe(PO$_4$)(Fe,OH)	monocl	2.98	5.3	1.592	1.612	1.621
Hessite	Ag$_2$Te	orth	8.4	2.5			
Heulandite	(Ca,Na$_2$,K$_2$)[Al$_2$Si$_7$O$_{18}$]·6H$_2$O	monocl	2.2	3.8	1.498	1.498	1.506
Hopeite	Zn$_3$(PO$_4$)$_2$·4H$_2$O	orth	3.0	3.2	1.58	1.59	1.59
Hornblende	Ca$_2$(Mg,Fe)$_4$Al(Si$_7$AlO$_{22}$)(OH)$_2$	monocl	3.24	5.5	1.67	1.67	1.69
Huebnerite	MnWO$_4$	monocl	7.2	4.3	2.17	2.22	2.32
Humite	Mg(OH,F)$_2$·3Mg$_2$SiO$_4$	orth	3.3	6	1.625	1.636	1.657
Huntite	Mg$_3$Ca(CO$_3$)$_4$	trig	2.70				
Hydrogrossularite	Ca$_3$Al$_2$Si$_2$O$_8$(SiO$_4$)$_{1-m}$(OH)$_{4m}$	cub	3.4	6.8	1.70		
Hydromagnesite	3MgCO$_3$·Mg(OH)$_2$·3H$_2$O	monocl	2.24	3.5	1.523	1.527	1.545
Illite	KAl$_4$[Si$_7$AlO$_{20}$](OH)$_4$	monocl	2.8	1.5	1.56	1.59	1.59
Ilmenite	FeTiO$_3$	rhomb	4.72	5.5			
Iodyrite	AgI	hex	5.68	1.5	2.21	2.22	
Jacobsite	MnFe$_2$O$_4$	cub	4.87	7.8	2.3		
Jadeite	NaAlSi$_2$O$_6$	monocl	3.34	6	1.649	1.654	1.663
Jamesonite	Pb$_4$FeSb$_6$S$_{14}$	monocl	5.63	2.5			
Jarosite	KFe$_3$(SO$_4$)$_2$(OH)$_6$	rhomb	3.09	3	1.715	1.820	
Kainite	KMg(SO$_4$)Cl·3H$_2$O	monocl	2.15	2.8	1.494	1.505	1.516
Kaliophylite	KAlSiO$_4$	hex	2.61	6	1.532	1.537	
Kaolinite	Al$_4$Si$_4$O$_{10}$(OH)$_8$	tricl	2.65	2.3	1.549	1.564	1.565
Kernite	Na$_2$B$_4$O$_7$·4H$_2$O	monocl	1.95	2.5	1.454	1.472	1.488
Kieserite	MgSO$_4$·H$_2$O	monocl	2.57	3.5	1.520	1.533	1.584
Kyanite	Al$_2$OSiO$_4$	tricl	3.59	6.3	1.715	1.722	1.731
Lanarkite	Pb$_2$(SO$_4$)O	monocl	6.92	2.3	1.928	2.007	2.036
Lanthanite	(La,Ce)$_2$(CO$_3$)$_3$·8H$_2$O	rhomb	2.72	2.8	1.52	1.587	1.613
Laumontite	Ca$_4$[Al$_8$Si$_{16}$O$_{48}$]·16H$_2$O	monocl	2.3	3.3	1.508	1.517	1.519
Laurionite	Pb(OH)Cl	rhomb	6.24	3.3	2.08	2.12	2.16
Lawsonite	CaAl$_2$(OH)$_2$Si$_2$O$_7$·H$_2$O	rhomb	3.08	6	1.655	1.675	1.685
Lazulite	(Mg,Fe)Al$_2$(PO$_4$)$_2$(OH)$_2$	monocl	3.23	5.8	1.615	1.64	1.650
Lazurite	Na$_4$SSi$_3$Al$_3$O$_{12}$	cub	2.42	5.3	1.500		
Leadhillite	Pb$_4$(SO$_4$)(CO$_3$)$_2$(OH)$_2$	monocl	6.55	2.8	1.87	2.00	2.01
Lepidocrocite	FeO(OH)	orth	4.26	5	1.94	2.20	2.51
Lepidolite	K$_2$(Li,Al)$_{5-6}$[Si$_{6-7}$Al$_{2-1}$O$_{20}$](OH,F)$_4$	monocl	2.85	3.3	1.536	1.565	1.566
Leucite	KAlSi$_2$O$_6$	tetr	2.49	5.8	1.510		
Levyne	(Ca,Na$_2$)Al$_2$Si$_4$O$_{12}$·6H$_2$O	rhomb	2.10	4.5	1.496	1.501	
Litharge	PbO	tetr	9.35	2	2.535	2.665	
Loellingite	FeAs$_2$	rhomb	7.40	5.3			
Maghemite	Fe$_2$O$_3$	cub	4.88	7.8	2.63		
Magnesite	MgCO$_3$	hex	3.05	4	1.536	1.741	
Magnetite	Fe$_3$O$_4$	cub	5.17	6	2.42		
Malachite	Cu$_2$(OH)$_2$(CO$_3$)	monocl	4.05	3.8	1.655	1.875	1.909
Manganite	MnO(OH)	monocl	≈4.3	4	2.25	2.25	2.53

Name	Formula	Crystal system	Density g/cm³	Hardness	Index of refraction		
					n_α	n_β	n_γ
Manganosite	MnO	cub	5.37	5.5			
Marcasite	FeS_2	cub	5.02	6.3			
Marialite	$Na_4Al_3Si_9O_{24}Cl$	tetr	2.56	5.5	1.541	1.548	
Marshite	CuI	cub	5.67	2.5	2.346		
Mascagnite	$(NH_4)_2SO_4$	orth	1.77	2.3	1.520	1.523	1.533
Matlockite	PbClF	tetr	7.05	2.8	2.006	2.145	
Meionite	$Ca_4Al_6Si_6O_{24}CO_3$	tetr	2.78	5.5	1.559	1.595	
Melanterite	$FeSO_4 \cdot 7H_2O$	monocl	1.89	2	1.47	1.48	1.49
Melilite	$(Ca,Na)_2(Mg,Fe,Al,Si)_3O_7$	tetr	3.00	5.5	1.639	1.645	
Mellite	$Al_2C_{12}O_{12} \cdot 18H_2O$	tetr	1.64	2.3	1.511	1.539	
Mendipite	$Pb_3O_2Cl_2$	rhomb	7.24	2.5	2.24	2.27	2.31
Mesolite	$Na_2Ca_2(Al_2Si_3O_{10})_3 \cdot 8H_2O$	orth	2.26	5	1.506		
Metacinnabar	HgS	cub	7.70	3			
Microcline	$KAlSi_3O_8$	monocl	2.56	6.3	1.522	1.526	1.530
Miersite	AgI	hex	5.68	2.5	2.20		
Millerite	NiS	hex	5.5	3.3			
Mimetite	$Pb_5(AsO_4,PO_4)_3Cl$	hex	7.24	3.8	2.128	2.147	
Minium	Pb_3O_4	tetr	8.9	2.5			
Mirabilite	$Na_2SO_4 \cdot 10H_2O$	monocl	1.46	1.8	1.394	1.396	1.398
Moissanite	SiC	hex	3.16	9.5	2.648	2.691	
Molybdenite	MoS_2	hex	5.06	1.3			
Monazite	$(Ce,La,Th)PO_4$	monocl	5.2	5	1.787	1.789	1.840
Monetite	$CaHPO_4$	tricl	2.92	3.5	1.587	1.61	1.640
Monticellite	$Ca(Mg,Fe)SiO_4$	orth	3.18	5.5	1.647	1.655	1.664
Montmorillonite	$(0.5Ca,Na)_{0.7}(Al,Mg,Fe)_4$ $[(Si,Al)_8O_{20}](OH)_4 \cdot nH_2O$	monocl	2.5	1.5	1.55	1.57	1.57
Montroydite	HgO	orth	11.14	2.5	2.37	2.50	2.65
Mordenite	$(Na,K,Ca)[Al_2Si_{10}O_{24}] \cdot 7H_2O$	orth	2.13	3.5	1.478	1.480	1.482
Muscovite	$KAl_2Si_3AlO_{10}(OH,F)_2$	monocl	2.83	2.8	1.563	1.596	1.602
Nantokite	CuCl	cub	4.14	2.5	1.930		
Natrolite	$Na_2Al_2Si_3O_{10} \cdot 2H_2O$	orth	2.23	5	1.478	1.481	1.491
Nepheline	$Na_3KAl_4Si_4O_{16}$	hex	2.61	5.8	1.534	1.538	
Newberyite	$MgHPO_4 \cdot 3H_2O$	orth	2.13	3.3	1.514	1.517	1.533
Niccolite	NiAs	hex	7.77	5.3			
Norbergite	$Mg(OH,F)_2 \cdot Mg_2SiO_4$	orth	3.21	6.5	1.565	1.573	1.592
Nosean	$Na_8Al_6Si_6O_{24}SO_4$	cub	2.35	5.5	1.495		
Oldhamite	CaS	cub	2.59	4	2.137		
Oligoclase	$([NaSi]_{0.9-0.7}[CaAl]_{0.1-0.3})AlSi_2O_8$	tricl	2.64	6.3	1.539	1.543	1.547
Olivenite	$Cu_2(AsO_4)(OH)$	rhomb	4.2	3	1.77	1.80	1.85
Olivine	$(Mg,Fe)SiO_4$	rhomb	3.81	6.8	1.73	1.76	1.78
Opal	$SiO_2 \cdot nH_2O$	amorp	1.9	5	1.44		
Orpiment	As_2S_3	monocl	3.46	1.8	2.40	2.81	3.02
Orthoclase	$KAlSi_3O_8$	monocl	2.56	6	1.523	1.527	1.531
Orthopyroxene	$(Mg,Fe)SiO_3$	rhomb	3.6	5.5	1.709	1.712	1.723
Paragonite	$NaAl_2Si_3AlO_{10}(OH)_2$	monocl	2.85	2.5	1.572	1.602	1.605
Parisite	$(Ce,La,Na)FCO_3 \cdot CaCO_3$	hex	4.42	4.5	1.672	1.771	
Pectolite	$Ca_2NaH(SiO_3)_3$	tricl	2.88	4.8	1.603	1.610	1.639
Penfieldite	$Pb_4Cl_6(OH)_2$	hex	6.6		2.13	2.21	
Pentlandite	$(Fe,Ni)_9S_8$	cub	4.8	3.8			
Percylite	$PbCuCl_2(OH)_2$	cub		2.5	2.05		
Periclase	MgO	cub	3.6	5.5	1.735		
Perovskite	$CaTiO_3$	cub	3.98	5.5	2.34		
Petalite	$LiAlSi_4O_{10}$	monocl	2.42	6.5	1.506	1.511	1.519
Pharmacosiderite	$Fe_3(AsO_4)_2(OH)_3 \cdot 5H_2O$	cub	2.80	2.5	1.690		
Phenakite	Be_2SiO_4	rhomb	2.98	7.5	1.654	1.670	
Phillipsite	$K(Ca_{0.5},Na)_2[Al_3Si_5O_{16}] \cdot 6H_2O$	monocl	2.2	4.3	1.494	1.497	1.505
Phlogopite	$KMg_3AlSi_3O_{10}(OH,F)_2$	monocl	2.83	2.3	1.560	1.597	1.598
Phosgenite	$Pb_2(CO_3)Cl_2$	tetr	6.13	2.5	2.118	2.145	
Piemontite	$Ca_2(Mn,Fe,Al)_3O(Si_2O_7)(SiO_4)(OH)$	monocl	3.49	6	1.762	1.773	1.796

Name	Formula	Crystal system	Density g/cm^3	Hard-ness	Index of refraction		
					n_α	n_β	n_γ
Pigeonite	$(Mg,Fe,Ca)(Mg,Fe)Si_2O_6$	monocl	3.38	6	1.702	1.703	1.728
Pollucite	$CsAlSi_2O_6$	tetr	2.9	6.5	1.517		
Polybasite	$(Ag,Cu)_{16}Sb_2S_{11}$	monocl	6.1	2.5			
Powellite	$Ca(Mo,W)O_4$	tetr	4.35	3.8	1.971	1.980	
Prehnite	$Ca_2Al_2Si_3O_{10}(OH)_2$	rhomb	2.93	6.3	1.622	1.628	1.648
Proustite	Ag_3AsS_3	rhomb	5.57	2.3	2.792	3.088	
Pseudobrookite	Fe_2TiO_5	rhomb	4.36	6	2.38	2.39	2.42
Psilomelane	$BaMn_9O_{16}(OH)_4$	rhomb	4.71	5.5			
Pumpellyite	$Ca_2Al_2(Al,Fe,Mg)[Si_2(O,OH)_7]$ $(SiO_4)(OH,O)_3$	monocl	3.21	5.5	1.688	1.695	1.705
Pyrargyrite	Ag_3SbS_3	rhomb	5.85	2.5	2.88	3.08	
Pyrite	FeS_2	cub	5.02	6.3			
Pyrochlore	$NaCaNb_2O_6F$	cub	5.3	5.3			
Pyrochroite	$Mn(OH)_2$	hex	3.26	2.5	1.68	1.72	
Pyrolusite	MnO_2	tetr	5.08	6.3			
Pyromorphite	$Pb_5(PO_4,AsO_4)_3Cl$	hex	7.04	3.8	2.048	2.058	
Pyrope	$Mg_3Al_2Si_3O_{12}$	cub	3.58	6.8	1.714		
Pyrophyllite	$Al_2Si_4O_{10}(OH)_2$	monocl	2.78	1.5	1.545	1.579	1.599
Pyrrhotite	Fe_7S_8	hex	4.62	4			
Quartz	SiO_2	hex	2.65	7	1.544	1.553	
Rammelsbergite	$NiAs_2$	orth	7.1	5.8			
Raspite	$PbWO_4$	monocl	8.46	2.8	1.27	1.27	1.30
Realgar	As_4S_4	monocl	3.5	1.8	2.538	2.684	2.704
Rhodochrosite	$MnCO_3$	hex	3.70	3.8	1.597	1.816	
Rhodonite	$(Mn,Fe,Ca)SiO_3$	orth	3.48	6	1.725	1.729	1.737
Riebeckite	$Na_2Fe_5(Si_8O_{22})(OH)_2$	monocl	3.3	5	1.675	1.683	1.694
Rutile	TiO_2	tetr	4.23	6.2	2.609	2.900	
Safflorite	$(Co,Fe)As_2$	rhomb	7.3	4.8			
Samarskite	(Y,Er,Ce,U,Ca,Fe,Pb,Th) $(Nb,Ta,Ti,Sn)_2O_6$	rhomb	5.69	5.5	2.200		
Sapphirine	$(Mg,Fe)_2Al_4O_6SiO_4$	monocl	3.49	7.5	1.709	1.712	1.715
Scapolite	$(Na,Ca)_4Al_3(Al,Si)_3Si_6O_{24}$ (Cl,F,OH,CO_3,SO_4)	tetr	2.64	5.5	1.551	1.573	
Scheelite	$CaWO_4$	tetr	6.06	4.8	1.920	1.936	
Scolecite	$CaAl_2Si_3O_{10}\cdot 3H_2O$	monocl	2.27	5	1.510	1.518	1.519
Scorodite	$Fe(AsO_4)\cdot 2H_2O$	rhomb	3.28	3.8	1.784	1.795	1.814
Sellaite	MgF_2	tetr	3.15	5	1.378	1.390	
Senarmontite	Sb_2O_3	cub	5.58	2.3	2.087		
Serpentine	$Mg_3Si_2O_5(OH)_4$	monocl	2.55	3	1.55	1.56	1.56
Siderite	$FeCO_3$	hex	3.9	4.3	1.635	1.875	
Sillimanite	Al_2OSiO_4	rhomb	3.25	7	1.658	1.660	1.660
Skutterudite	$(Co,Ni)As_3$	cub	6.8	5.8			
Smithsonite	$ZnCO_3$	rhomb	4.4	4.3	1.621	1.848	
Sodalite	$Na_8Al_6Si_6O_{24}Cl_2$	cub	2.30	5.8	1.485		
Sperrylite	$PtAs_2$	cub	10.58	6.5			
Spessartite	$Mn_3Al_2Si_3O_{12}$	cub	4.19	6.8	1.800		
Sphalerite	ZnS	cub	4.0	3.8	2.369		
Sphene	$CaTiSiO_4(O,OH,F)$	monocl	3.50	5	1.90	1.95	2.03
Spinel	$MgAl_2O_4$	cub	3.55	7.8	1.719		
Spodumene	$LiAlSi_2O_6$	monocl	3.13	6.8	1.656	1.662	1.671
Stannite	Cu_2FeSn_4	tetr	4.4	4			
Staurolite	$(Fe,Mg,Zn)_2(Al,Fe,Ti)_9O_6$ $[(Si,Al)O_4]_4(O,OH)_2$	monocl	3.79	7.5	1.743	1.747	1.755
Stercorite	$Na(NH_4)H(PO_4)\cdot 4H_2O$	tricl	1.62	2	1.439	1.442	1.469
Stibiotantalite	$Sb(Ta,Nb)O_4$	rhomb	6.6	5.5	2.38	2.41	2.46
Stibnite	Sb_2S_3	orth	4.56	2			
Stilbite	$NaCa_2[Al_5Si_{13}O_{36}]\cdot 14H_2O$	monocl	2.2	3.8	1.492	1.499	1.503

PHYSICAL AND OPTICAL PROPERTIES OF MINERALS (continued)

Name	Formula	Crystal system	Density g/cm³	Hardness	n_α	n_β	n_γ
Stilpnomelane	$(K,Na,Ca)_{0.6}(Fe,Mg)_6Si_8Al$ $(O,OH)_{27} \cdot 2H_2O$	monocl	2.8	3.5	1.585	1.665	1.665
Stolzite	$PbWO_4$	tetr	8.2	2.8	2.19	2.27	
Strengite	$FePO_4 \cdot 2H_2O$	orth	2.87	4	1.707	1.719	1.741
Strontianite	$SrCO_3$	orth	3.5	3.5	1.518	1.666	1.668
Struvite	$Mg(NH_4)(PO_4) \cdot 6H_2O$	rhomb	1.71	2	1.495	1.496	1.504
Sulfur	S	orth	2.07	2	1.958	2.038	2.245
Sylvanite	$(Ag,Au)Te_2$	monocl	8.16	1.8			
Sylvite	KCl	cub	1.99	2	1.490		
Talc	$Mg_3Si_4O_{10}(OH)_2$	monocl	2.71	1	1.545	1.592	1.595
Tantalite	$(Fe,Mn)(Ta,Nb)_2O_6$	rhomb	7.95	6.5	2.26	2.32	2.43
Tapiolite	$FeTa_2O_6$	tetr	7.9	6.3	2.27	2.42	
Tellurobismuthite	Bi_2Te_3	hex	7.74	1.8			
Terlinguaite	Hg_2OCl	monocl	8.73	2.5	2.35	2.64	2.66
Tetrahedrite	$(Cu,Fe)_{12}Sb_4S_{13}$	cub	4.9	3.8			
Thenardite	Na_2SO_4	orth	2.7	2.8	1.468	1.475	1.483
Thermonatrite	$Na_2CO_3 \cdot H_2O$	orth	2.25	1.3	1.420	1.506	1.524
Thomsenolite	$NaCaAlF_6 \cdot H_2O$	monocl	2.98	2	1.407	1.414	1.415
Thorianite	ThO_2	cub	10.0	6.5	2.200		
Thorite	$ThSiO_4$	tetr	6.7	4.8	1.8		
Topaz	$Al_2SiO_3(OH,F)_2$	rhomb	3.53	8	1.618	1.620	1.627
Torbernite	$Cu(UO_2)_2(PO_4)_2 \cdot 8H_2O$	tetr	3.22	2.3	1.582	1.592	
Tourmaline	$Na(Mg,Fe,Mn,Li,Al)_3Al_6Si_6O_{18}$ $(BO_3)_3$	rhomb	3.14	7	1.62	1.65	
Tremolite	$Ca_2Mg_5Si_8O_{22}(OH,F)_2$	monocl	3.0	5.5	1.599	1.612	1.622
Trevorite	$NiFe_2O_4$	cub	5.33	7.8	2.3		
Tridymite	SiO_2	hex	2.27	7	1.475	1.476	1.479
Triphyllite-Lithiophyllite	$Li(Fe,Mn)PO_4$	rhomb	3.46	4.5	1.68	1.68	1.69
Troegerite	$(UO_2)_3(AsO_4)_2 \cdot 12H_2O$	tetr		2.5	1.59	1.630	
Troilite	FeS	hex	4.7	4			
Trona	$Na_3H(CO_3)_2 \cdot 2H_2O$	monocl	2.14	2.8	1.412	1.492	1.540
Turquois	$Cu(Al,Fe)_6(PO_4)_4(OH)_8 \cdot 4H_2O$	tricl	2.9	5.3	1.70	1.73	1.75
Ullmannite	$NiSbS$	cub	6.65	5.3			
Uraninite	UO_2	cub	11.0	5.5			
Uvarovite	$Ca_3Cr_2Si_3O_{12}$	cub	3.83	6.8	1.865		
Valentinite	Sb_2O_3	orth	5.7	2.8	2.18	2.35	2.35
Vanadinite	$Pb_5(VO_4)_3Cl$	hex	6.8	2.9	2.350	2.416	
Variseite-Strengite	$(Al,Fe)(PO_4) \cdot 2H_2O$	rhomb	2.72	4	1.635	1.654	1.668
Vaterite	$CaCO_3$	hex	2.71		1.550	1.645	
Vermiculite	$(Mg,Ca)_{0.7}(Mg,Fe,Al)_6[(Al,Si)_8O_{20}]$ $(OH)_4 \cdot 8H_2O$	monocl	2.3	1.5	1.542	1.556	1.556
Vesuvianite	$Ca_{10}(Mg,Fe)_2Al_4(Si_2O_7)_2(SiO_4)_5$ $(OH,F)_4$	tetr	3.33	6.5	1.72	1.73	
Villiaumite	NaF	cub	2.78	2.3	1.327		
Vivianite	$Fe_3(PO_4)_2 \cdot 8H_2O$	monocl	2.58	1.8	1.598	1.629	1.652
Wagnerite	$Mg_2(PO_4)F$	monocl	3.15	5.3	1.568	1.572	1.582
Wavellite	$Al_3(OH)_3(PO_4)_2 \cdot 5H_2O$	rhomb	2.36	3.6	1.527	1.535	1.553
Whewellite	$CaC_2O_4 \cdot H_2O$	cub	2.2	2.8	1.491	1.554	1.650
Willemite	Zn_2SiO_4	hex	4.1	5.5	1.691	1.719	
Witherite	$BaCO_3$	orth	4.29	3.5	1.529	1.676	1.677
Wolframite	$(Fe,Mn)WO_4$	monocl	7.3	4.3	2.26	2.32	2.42
Wollastonite	$CaSiO_3$	monocl	2.92	4.8	1.628	1.639	1.642
Wulfenite	$PbMoO_4$	tetr	6.7	2.9	2.283	2.403	
Wurtzite	ZnS	hex	4.09	3.8	2.356	2.378	
Xenotime	YPO_4	tetr	4.8	4.5	1.721	1.816	
Zeunerite	$Cu(UO_2)_2(AsO_4)_2 \cdot 10H_2O$	tetr			1.606		
Zincite	ZnO	hex	5.6	4	2.013	2.029	
Zircon	$ZrSiO_4$	tetr	4.6	7.5	1.94	1.99	
Zoisite	$Ca_2Al_3Si_3O_{12}(OH)$	rhomb	3.26	6	1.695	1.699	1.711

CRYSTALLOGRAPHIC DATA ON MINERALS

This table contains x-ray crystallographic data on about 400 common minerals, as well as selected crystalline elements. Entries are arranged alphabetically by mineral name. The columns are:

Name: Common name of the mineral.
Formula: Chemical formula for a typical sample of the mineral. Composition often varies considerably with the origin of the sample.
Crystal system: tricl = triclinic; monocl = monoclinic; orth = orthorhombic; tetr = tetragonal; hex = hexagonal; rhomb = rhombohedral; cubic = cubic.
Structure type: Prototype for the structural arrangement of the crystallographic cell.
Z: Number of formula units per the unit cell.
a, b, c: Lengths of the cell edges in Å ($1 Å = 10^{-8}$ cm).
α, β, γ : Angles between cell axes.

REFERENCES

1. Robie, R.A., Bethke, P.M., and Beardsley, K.M., *U. S. Geological Survey Bulletin 1248*, U. S. Government Printing Office, Washington, D.C.
2. Donnay, J.D.H., and Ondik, H.M., *Crystal Data Determinative Tables, Third Edition, Volume 2, Inorganic Compounds*, Joint Committee on Powder Diffraction Standards, Swarthmore, PA, 1973.
3. Deer, W.A., Howie, R.A., and Zussman, J., *An Introduction to the Rock-Forming Minerals, 2nd Edition*, Longman Scientific & Technical, Harlow, Essex, 1992.

CRYSTALLOGRAPHIC DATA ON MINERALS (continued)

Name	Formula	Crystal system	Structure type	Z	a/Å	b/Å	c/Å	α	β	γ
Acanthite	Ag_2S	monocl		4	4.228	6.928	7.862		99.58°	
Acmite (Aegirine)	$NaFe(SiO_3)_2$	monocl	diopside	4	9.658	8.795	5.294		107.42°	
Akermanite	$Ca_2MgSi_2O_7$	tetr	melilite	2	7.8435		5.010			
Alabandite	MnS	cubic	rock salt	4	5.223					
Almandine (Almandite)	$Fe_3Al_2Si_3O_{12}$	cubic	garnet	8	11.526					
Altaite	$PbTe$	cubic	rock salt	4	6.4606					
Aluminum	Al	cubic	copper	4	4.049					
Alunite	$KAl_3(SO_4)_2(OH)_6$	rhomb		3	6.982		17.32			
Analcite	$NaAlSi_2O_6 \cdot H_2O$	cubic		16	13.733					
Anatase	TiO_2	tetr		4	3.785		9.514			
Andalusite	Al_2OSiO_4	orth		4	7.7959	7.8983	5.5583			
Andradite	$Ca_3Fe_2Si_3O_{12}$	cubic	garnet	8	12.048					
Anglesite	$PbSO_4$	orth	barite	4	8.480	5.398	6.958			
Anhydrite	$CaSO_4$	orth	anhydrite	4	6.991	6.996	6.238			
Annite	$KFe_3[AlSi_3O_{10}](OH)_2$	monocl	1M mica	2	10.29	9.33	5.39		105.1°	
Anorthite	$CaAl_2Si_2O_8$	tricl	primitive cell	8	8.177	12.877	14.169	93.17°	115.85°	91.22°
Anthophyllite	$Mg_7Si_8O_{22}(OH)_2$	orth		4	18.61	18.01	5.24			
Antimony	Sb	rhomb	arsenic	6	4.2996		11.2516			
Aragonite	$CaCO_3$	orth	aragonite	4	5.741	7.968	4.959			
Arcanite	K_2SO_4	orth	arcanite	4	5.772	10.072	7.483			
Argentite	Ag_2S	cubic		2	4.870					
Argentopyrite	$AgFe_2S_3$	orth		4	6.64	11.47	6.45			
Arsenic	As	rhomb	arsenic	6	3.760		10.555			
Arsenolite	As_2O_3	cubic	diamond	16	11.074					
Arsenopyrite	$FeAsS$	tricl		4	5.760	5.690	5.785	90.00°	112.23°	90.00°
Azurite	$Cu_3(OH)_2(CO_3)_2$	monocl		2	5.008	5.844	10.336		92.45°	
Baddeleyite	ZrO_2	monocl	baddeleyite	4	5.1454	5.2075	5.3107		99.23°	
Banalsite	$BaNa_2Al_4Si_4O_{16}$	orth		4	8.50	9.97	16.72			
Barite	$BaSO_4$	orth	barite	4	8.878	5.450	7.152			
Berlinite	$AlPO_4$	hex	α-quartz	3	4.942		10.97			
Beryl	$Be_3Al_2(SiO_3)_6$	hex	beryl	2	9.215		9.192			
Berzelianite	Cu_2Se	cubic		4	5.85					
Bismite	Bi_2O_3	monocl	pseudo-orth	4	7.48	8.14	5.83		112.9°	
Bismuth	Bi	rhomb	arsenic	6	4.5367		11.8383			
Bismuthinite	Bi_2S_3	orth	stibnite	4	11.150	11.300	3.981			
Bixbyite	Mn_2O_3	cubic	thallium trioxide	16	9.411					
Boehmite	$AlO(OH)$	orth	lepidocrocite	4	2.868	12.227	3.700			
Borax	$Na_2B_4O_7 \cdot 10H_2O$	monocl		4	11.858	10.674	12.197		106.68°	
Bornite (metastable)	Cu_5FeS_4	cubic		8	10.94					
Breithauptite	$NiSb$	hex	niccolite	2	3.942		5.155			
Brochantite	$Cu_4SO_4(OH)_6$	monocl		4	13.066	9.85	6.022		103.27°	
Bromargyrite	$AgBr$	cubic	rock salt	4	5.7745					
Bromellite	BeO	hex	zincite	2	2.6979		4.3772			
Brookite	TiO_2	orth		8	5.456	9.182	5.143			
Brucite	$Mg(OH)_2$	hex	cadmium iodide	1	3.147		4.769			

Name	Formula	Crystal system	Structure type	Z	a/Å	b/Å	c/Å	α	β	γ
Bunsenite	NiO	cubic	rock salt	4	4.177					
Bustamite	$CaMn(SiO_3)_2$	tricl		6	7.736	7.157	13.824	90.52°	94.58°	103.87°
Cadmium telluride	$CdTe$	cubic	sphalerite	4	6.4805					
Cadmoselite	$CdSe$	hex	zincite	2	4.2977		7.0021			
Calcite	$CaCO_3$	rhomb	calcite	6	4.9899		17.064			
Calomel	Hg_2Cl_2	tetr		4	4.478		10.910			
Carbonate-apatite	$Ca_{10}(PO_4)_6CO_3 \cdot H_2O$	hex	apatite	1	9.436		6.883			
Cassiterite	SnO_2	tetr	rutile	2	4.738		3.188			
Cattierite	CoS_2	cubic	pyrite	4	5.5345					
Celestite	$SrSO_4$	orth	barite	4	8.359	5.352	6.866			
Celsian	$BaAl_2Si_2O_8$	monocl		8	8.627	13.045	14.408		115.20°	
Cerianite	CeO_2	cubic	fluorite	4	5.4110					
Cerussite	$PbCO_3$	orth	aragonite	4	6.152	8.436	5.195			
Cervantite	Sb_2O_4	orth		4	5.424	11.76	4.804			
Chalcanthite	$CuSO_4 \cdot 5H_2O$	tricl		2	6.1045	10.72	5.949	97.57°	107.28°	77.43°
Chalcocite	Cu_2S	orth		96	11.881	27.323	13.491			
Chalcopyrite	$CuFeS_2$	tetr		4	5.2988		10.434			
Chlorapatite	$Ca_5(PO_4)_3Cl$	hex	apatite	2	9.629		6.777			
Chlorargyrite	$AgCl$	cubic	rock salt	4	5.5491					
Chloritoid	$FeAl_4O_2(SiO_4)_2(OH)_4$	monocl		8	9.48	5.48	18.18		101.77°	
Chloromagnesite	$MgCl_2$	rhomb		3	3.632		17.795			
Chondrodite	$2Mg_2SiO_4 \cdot MgF_2$	monocl		2	7.89	4.743	10.29		109.03°	
Chrysoberyl	$BeAl_2O_4$	orth	olivine	4	5.4756	9.4041	4.4267			
Cinnabar	HgS	hex	cinnabar	3	4.149		9.495			
Claudetite	As_2O_3	monocl		4	5.339	12.984	4.5405		94.27°	
Clausthalite	$PbSe$	cubic	rock salt	4	6.1255					
Clinoenstatite	$MgSiO_3$	monocl		8	9.620	8.825	5.188		108.33°	
Clinoferrosilite	$FeSiO_3$	monocl		8	9.7085	9.0872	5.2284		108.43°	
Clinohumite	$4Mg_2SiO_4 \cdot MgF_2$	monocl		2	13.68	4.75	10.27		100.83°	
Clinozoisite	$Ca_2Al_3(SiO_4)_3OH$	monocl		2	8.887	5.581	10.14		115.93°	
Cobalt olivine	Co_2SiO_4	orth	olivine	4	4.782	10.301	6.003			
Cobalt oxide	CoO	cubic	rock salt	4	4.260					
Cobalt sulfide	CoS	cubic	sphalerite	4	5.339					
Cobalt titanate	$CoTiO_3$	rhomb	ilmenite	6	5.066		13.918			
Cobalticalcite	$CoCO_3$	rhomb	calcite	6	4.6581		14.958			
Cobaltite	$CoAsS$	cubic	NiSbS	4	5.60					
Coesite	SiO_2	monocl		16	7.152	12.379	7.152		120.00°	
Coffinite	$USiO_4$	tetr	zircon	4	6.995		6.263			
Colemanite	$Ca_2B_6O_{11} \cdot 5H_2O$	monocl		4	8.743	11.264	6.102		110.12°	
Coloradoite	$HgTe$	cubic	sphalerite	4	6.4600					
Cooperite	PtS	tetr		2	3.4699		6.1098			
Copper	Cu	cubic	face-centered cubic	4	3.6150					
Corundum	Al_2O_3	rhomb	corundum	6	4.7591		12.9894			
Cotunnite	$PbCl_2$	orth		4	4.535	7.62	9.05			
Covellite	CuS	hex		6	3.792		16.34			

CRYSTALLOGRAPHIC DATA ON MINERALS (continued)

Name	Formula	Crystal system	Structure type	Z	a/Å	b/Å	c/Å	α	β	γ
Cristobalite (α)	SiO_2	tetr		4	4.971		6.918			
Cristobalite (β)	SiO_2	cubic		8	7.1382					
Cryolite	Na_3AlF_6	monocl		2	5.40	5.60	7.776		90.18°	
Cubanite	$CuFe_2S_3$	orth		4	6.46	11.12	6.23			
Cummingtonite	$(Mg,Fe,Mn)_7(Si_4O_{11})_2(OH)_2$	monocl	tremolite	2	9.522	18.223	5.332		101.92°	
Cuprite	Cu_2O	cubic		2	4.2696					
Danburite	$CaB_2Si_2O_8$	orth		4	8.04	8.77	7.74			
Datolite	$CaBSiO_4(OH)$	monocl		4	9.62	7.60	4.84		90.15°	
Daubreeite	$FeCr_2S_4$	cubic	spinel	8	9.966					
Diamond	C	cubic	diamond	8	3.5670					
Diaspore	$AlO(OH)$	orth		4	4.401	9.421	2.845			
Dickite	$Al_2Si_2O_5(OH)_4$	monocl		4	5.150	8.940	14.736		103.58°	
Digenite	$Cu_{1.79}S$	cubic	deformed fluorite	4	5.5695					
Diopside	$CaMg(SiO_3)_2$	monocl	diopside	4	9.743	8.923	5.251		105.93°	
Dioptase	$CuSiO_2(OH)_2$	rhomb	phenacite	18	14.61		7.80			
Dolerophanite	$Cu_2O(SO_4)$	monocl		4	8.334	6.312	7.628		108.4°	
Dolomite	$CaMg(CO_3)_2$	rhomb	calcite	3	4.8079		16.010			
Dravite	$NaMg_3Al_6B_3Si_6O_{27}(OH)_4$	rhomb	tourmaline	3	15.942		7.224			
Elbaite	$NaLiAl_{7.67}B_3Si_6O_{27}(OH)_4$	rhomb	tourmaline	3	15.842		7.009			
Enargite	Cu_3AsS_4	orth		2	6.426	7.422	6.144			
Enstatite	$MgSiO_3$	orth		16	8.829	18.22	5.192			
Epidote	$Ca_2Al_2(Al,Fe)OH(SiO_4)_3$	monocl		2	8.89	5.63	10.19		115.40°	
Epsomite	$MgSO_4\cdot7H_2O$	orth		4	11.86	11.99	6.858			
Eskolaite	Cr_2O_3	rhomb	corundum	6	4.9607		13.599			
Eucairite	$AgCuSe$	orth		10	4.105	20.35	6.31			
Euclase	$AlBeSiO_4(OH)$	monocl		4	4.763	14.29	4.618		100.25°	
Famatinite	Cu_3SbS_4	tetr		2	5.384		10.770			
Fayalite	Fe_2SiO_4	orth	olivine	4	4.817	10.477	6.105			
Fe-Cordierite	$Fe_2Al_3(AlSi_5O_{18})$	orth	cordierite	4	9.726	17.065	9.287			
Fe-Gehlenite	$Ca_2Fe_2SiO_7$	tetr	melilite	2	7.54		4.855			
Fe-Indialite	$Fe_2Al_3(AlSi_5O_{18})$	hex	beryl	2	9.860		9.285			
Fe-Leucite	$KFeSi_2O_6$	tetr		16	13.205		13.970			
Fe-Microcline	$KFeSi_3O_8$	tricl		4	8.68	13.10	7.340	90.75°	116.05°	86.23°
Fe-Sanidine	$KFeSi_3O_8$	monocl		4	8.689	13.12	7.319		116.10°	
Fe-Skutterudite	$FeAs_{2.95}$	cubic		8	8.1814					
Ferberite	$FeWO_4$	monocl	wolframite	2	4.732	5.708	4.965		90.00°	
Ferriannite	$KFe_3[FeSi_3O_{10}](OH)_2$	monocl		2	5.430	9.404	10.341		100.07°	
Ferroselite	$FeSe_2$	orth	marcasite	2	4.801	5.778	3.587			
Ferrotremolite	$Ca_2Fe_5[Si_8O_{22}](OH)_2$	monocl	tremolite	2	9.97	18.34	5.30		104.50°	
Fluor-edenite	$NaCa_2Mg_5[AlSi_7O_{22}]F_2$	monocl	tremolite	2	9.847	18.00	5.282		104.83°	
Fluor-humite	$3Mg_2SiO_4\cdot MgF_2$	orth		4	10.243	20.72	4.735			
Fluor-norbergite	$Mg_2SiO_4\cdot MgF_2$	orth		4	8.727	10.271	4.709			
Fluor-phlogopite	$KMg_3[AlSi_3O_{10}]F_2$	monocl	1M mica	2	5.299	9.188	10.135		99.92°	
Fluor-richterite	$Na_2CaMg_5[Si_8O_{22}]F_2$	monocl	tremolite	2	9.823	17.96	5.268		104.33°	
Fluor-tremolite	$Ca_2Mg_5[Si_8O_{22}]F_2$	monocl	tremolite	2	9.781	18.01	5.267		104.52°	

CRYSTALLOGRAPHIC DATA ON MINERALS (continued)

Name	Formula	Crystal system	Structure type	Z	a/Å	b/Å	c/Å	α	β	γ
Fluorapatite	$Ca_5(PO_4)_3F$	hex	apatite	2	9.3684		6.8841			
Fluorite	CaF_2	cubic	fluorite	4	5.4638					
Forsterite	Mg_2SiO_4	orth	olivine	4	4.758	10.214	5.984			
Frohbergite	$FeTe_2$	orth	marcasite	2	5.265	6.265	3.869			
Gahnite	$ZnAl_2O_4$	cubic	spinel	8	8.0848					
Galaxite	$MnAl_2O_4$	cubic	spinel	8	8.258					
Galena	PbS	cubic	rock salt	4	5.9360					
Gallium oxide	Ga_2O_3	rhomb	corundum	6	4.9793		13.429			
Gehlenite	$Ca_2Al_2SiO_7$	tet	melilite	2	7.690		5.0675			
Geikielite	$MgTiO_3$	rhomb	ilmenite	6	5.054		13.898			
Gerhardite	$Cu_2(NO_3)(OH)_3$	orth		4	6.075	13.812	5.592			
Gersdorffite	$NiAsS$	cubic		4	5.693					
Gibbsite	$Al(OH)_3$	monocl		8	9.719	5.0705	8.6412		94.57°	
Glauchroite	$CaMnSiO_4$	orth	olivine	4	4.944	11.19	6.529			
Glaucodot	$(Co,Fe)AsS$	orth		24	6.64	28.39	5.64			
Glaucophane I	$Na_2Mg_3Al_2[Si_8O_{22}](OH)_2$	monocl	tremolite	2	9.748	17.915	5.273		102.78°	
Glaucophane II	$Na_2Mg_3Al_2[Si_8O_{22}](OH)_2$	monocl	tremolite	2	9.663	17.696	5.277		103.67°	
Goethite	$FeO(OH)$	orth		4	4.596	9.957	3.021			
Gold	Au	cubic	face-centered cubic	4	4.0786					
Goldmanite	$Ca_3V_2Si_3O_{12}$	cubic	garnet	8	12.070					
Goslarite	$ZnSO_4 \cdot 7H_2O$	orth	epsomite	4	11.779	12.050	6.822			
Graphite	C	hex	graphite	4	2.4612		6.7079			
Greenockite	CdS	hex	zincite	2	4.1354		6.7120			
Greigite	Fe_3S_4	cubic	spinel	8	9.876					
Grossularite	$Ca_3Al_2Si_3O_{12}$	cubic	garnet	8	11.851					
Grunerite	$Fe_7[Si_8O_{22}](OH)_2$	monocl	tremolite	2	9.572	18.44	5.342		101.77°	
Gudmundite	$FeSbS$	monocl		8	10.00	5.93	6.73		90.00°	
Gypsum	$CaSO_4 \cdot 2H_2O$	monocl		4	5.68	15.18	6.29		113.83°	
Hafnia	HfO_2	monocl	baddeleyite	4	5.1156	5.1722	5.2948		99.18°	
Halite	$NaCl$	cubic	rock salt	4	5.6402					
Hambergite	$Be_2(OH,F)BO_3$	orth		8	9.755	12.201	4.426			
Hardystonite	$Ca_2ZnSi_2O_7$	tet	melilite	2	7.87		5.01			
Hauerite	MnS_2	cubic	pyrite	4	6.1014					
Hausmannite	Mn_3O_4	tet		8	8.136		9.422			
Hawleyite	CdS	cubic	sphalerite	4	5.833					
Heazelwoodite	Ni_3S_2	rhomb		3	5.746		7.134			
Hedenbergite	$CaFe(SiO_3)_2$	monocl	diopside	4	9.854	9.024	5.263		104.23°	
Hematite	Fe_2O_3	rhomb	corundum	6	5.0329				13.749	
Hemimorphite	$Zn_4(OH)_2Si_2O_7 \cdot H_2O$	orth		2	8.370	10.719	5.120			
Hercynite	$Fe(AlO_2)_2$	cubic	spinel	8	8.150					
Herzenbergite	SnS	orth	germanium sulfide	4	4.328	11.190	3.978			
Hessite	Ag_2Te	monocl		4	8.13	4.48	8.09		111.9°	
Hexahydrite	$MgSO_4 \cdot 6H_2O$	monocl		8	10.110	7.212	24.41		98.30°	
High albite (Analbite)	$NaAlSi_3O_8$	tricl		4	8.160	12.870	7.106	93.54°	116.36°	90.19°
High argentite	Ag_2S	cubic		4	6.269					

CRYSTALLOGRAPHIC DATA ON MINERALS (continued)

Name	Formula	Crystal system	Structure type	Z	a/Å	b/Å	c/Å	α	β	γ
High bornite	Cu_5FeS_4	cubic		1	5.50					
High carnegeite	$NaAlSiO_4$	cubic		4	7.325					
High chalcocite	Cu_2S	hex		2	3.961		6.722			
High clinoenstatite	$MgSiO_3$	tricl		8	10.000	8.934	5.170	88.27°	70.03°	91.01°
High digenite	Cu_2S	cubic		4	5.725					
High germania	GeO_2	hex	α–quartz	3	4.987		5.652			
High leucite	$KAlSi_2O_6$	cubic		16	13.43					
High naumanite	Ag_2Se	cubic		2	4.993					
High sanidine	$KAlSi_3O_8$	monocl		4	8.615	13.031	7.177		115.98°	
Huebnerite	$MnWO_4$	monocl	wolframite	2	4.834	5.758	4.999		91.18°	
Huntite	$Mg_3Ca(CO_3)_4$	rhomb	calcite	3	9.498		7.816			
Hydroxylapatite	$Ca_5(PO_4)_3OH$	hex	apatite	2	9.418		6.883			
Ice	H_2O	hex		4	4.5212		7.3666			
Ilmenite	$FeTiO_3$	rhomb	ilmenite	6	5.093		14.055			
Indialite (Cordierite)	$Mg_2Al_3(AlSi_5O_{18})$	hex	beryl	2	9.7698		9.3517			
Iodargyrite	AgI	hex	zincite	2	4.5955		7.5005			
Iron (α)	Fe	cubic	body-centered cubic	2	2.8664					
Jacobsite	$MnFe_2O_4$	cubic	spinel	8	8.499					
Jadeite	$NaAl(SiO_3)_2$	monocl	diopside	4	9.409	8.564	5.220		107.50°	
Jalpaite	$Ag_{1.55}Cu_{0.45}S$	tetr		16	8.673		11.756			
Johannsenite	$CaMn(SiO_3)_2$	monocl	diopside	4	9.83	9.04	5.27		105.00°	
Kaliophilite	$KAlSiO_4$	hex		54	26.930		8.522			
Kalsilite	$KAlSiO_4$	hex		2	5.1597		8.7032			
Kaolinite	$Al_2Si_2O_5(OH)_4$	tricl		2	5.155	8.959	7.407	91.68°	104.87°	89.93°
Karelianite	V_2O_3	rhomb	corundum	6	4.952		14.002			
Keatite	SiO_2	tetr		12	7.456		8.604			
Kernite	$Na_2B_4O_7·4H_2O$	monocl		4	7.022	9.151	15.676		108.83°	
Kerschsteinite	$CaFeSiO_4$	orth	olivine	4	4.886	11.146	6.434			
Klockmannite	$CuSe$	hex	deformed covellite	78	14.206		17.25			
Knebelite	$MnFeSiO_4$	orth	olivine	4	4.854	10.602	6.162			
Kyanite	Al_2OSiO_4	tricl		4	7.123	7.848	5.564	89.92°	101.25°	105.97°
Larnite	Ca_2SiO_4	monocl		4	5.48	6.76	9.28		94.55°	
Laurite	RuS_2	cubic	pyrite	4	5.60					
Lawrencite	$FeCl_2$	rhomb		3	3.593		17.58			
Lawsonite	$CaAl_2Si_2O_7(OH)_2·H_2O$	orth		4	8.787	5.836	13.123			
Lead	Pb	cubic	face-centered cubic	4	4.9505					
Leonhardite	$MgSO_4·4H_2O$	monocl		4	5.922	13.604	7.905		90.85°	
Lepidocrocite	$FeO(OH)$	orth		4	3.868	12.525	3.066			
Lepidolite	$K_2Al_3Li_2AlSi_7O_{20}(OH)_4$	monocl	2M2 mica	2	9.2	5.3	20.0		98.00°	
Leucite	$KAlSi_2O_6$	tetr		16	13.074		13.738			
Lime	CaO	cubic	rock salt	4	4.8108					
Lime olivine	Ca_2SiO_4	orth	olivine	4	5.091	11.371	6.782			
Linnaeite	Co_3S_4	cubic	spinel	8	9.401					
Litharge	PbO	tetr		2	3.9759		5.023			
Loellingite	$FeAs_2$	orth	marcasite	2	5.300	5.981	2.882			

CRYSTALLOGRAPHIC DATA ON MINERALS (continued)

Name	Formula	Crystal system	Structure type	Z	a/Å	b/Å	c/Å	α	β	γ
Low albite	$NaAlSi_3O_8$	tricl		4	8.139	12.788	7.160	94.27°	116.57°	87.68°
Low bornite	Cu_5FeS_4	tetr		16	10.94		21.88			
Low cordierite	$Mg_2Al_3(AlSi_5O_{18})$	orth		4	9.721	17.062	9.339			
Low germania	GeO_2	tetr	rutile	2	4.3963		2.8626			
Low nepheline	$NaAlSiO_4$	hex		8	9.986		8.330			
Luzonite	Cu_3AsS_4	tetr		2	5.289		10.440			
Mackinawite	FeS	tetr		2	3.675		5.030			
Magnesioriebeckite	$Na_2Mg_3Fe_2[Si_8O_{22}](OH)_2$	monocl	tremolite	2	9.733	17.946	5.299		103.30°	
Magnesite	$MgCO_3$	rhomb	calcite	6	4.6330		15.016			
Magnetite	Fe_3O_4	cubic	spinel	8	8.3940					
Malachite	$Cu_2(OH)_2CO_3$	monocl		4	9.502	11.974	3.240		98.75°	
Maldonite	Au_2Bi	cubic		8	7.958					
Manganese sulfide (γ)	MnS	hex	zincite	2	3.976		6.432			
Manganese sulfide (β)	MnS	cubic	sphalerite	4	5.611					
Manganosite	MnO	cubic	rock salt	4	4.4448					
Marcasite	FeS_2	orth	marcasite	2	4.443	5.423	3.3876		95.00°	
Margarite	$CaAl_2[AlSi_2O_{10}](OH)_2$	monocl	2M mica	4	5.13	8.92	19.50			
Marialite	$Na_4Al_3Si_9O_{24}Cl$	tetr		2	12.064		7.514			
Marshite	CuI	cubic	sphalerite	4	6.0507					
Mascagnite	$(NH_4)_2SO_4$	orth	arcanite	4	7.782	5.993	10.636			
Massicot	PbO	orth		4	5.489	4.755	5.891			
Matlockite	$PbClF$	tetr		2	4.106		7.23			
Maucherite	$Ni_{11}As_8$	tetr		4	6.870		21.81			
Meionite	$Ca_4Al_6Si_6O_{24}CO_3$	tetr		2	12.174		7.652			
Melanophlogite	SiO_2	cubic	clathrate type	46	13.402					
Melanterite	$FeSO_4 \cdot 7H_2O$	monocl		4	14.072	6.503	11.041		105.57°	
Melonite	$NiTe_2$	hex	cadmium iodide	1	3.869		5.308			
Metacinnabar	HgS	cubic	sphalerite	4	5.8517					
Miargyrite	$AgSbS_2$	monocl		8	12.862	4.111	13.220	90.62°	98.63°	
Microcline	$KAlSi_3O_8$	tricl		4	8.582	12.964	7.222		115.92°	87.68°
Miersite	AgI	cubic	sphalerite	4	6.4963					
Millerite	NiS	rhomb		9	9.616		3.152			
Minium	Pb_3O_4	tetr		4	8.815		6.565			
Minnesotaite	$Fe_3Si_4O_{10}(OH)_2$	monocl		4	5.4	9.42	19.4		100.00°	
Mirabilite	$Na_2SO_4 \cdot 10H_2O$	monocl		4	11.51	10.38	12.83		107.75°	
Mn-Indialite	$Mn_2Al_3(AlSi_5O_{18})$	hex	beryl	2	9.925		9.297			
Molybdenite	MoS_2	hex	molybdenite	2	3.1604		12.295			
Molybdenum	Mo	cubic		2	3.1653					
Molybdite	MoO_3	orth		4	3.962	13.858	3.697			
Monteponite	CdO	cubic	rock salt	4	4.6953					
Monticellite	$CaMgSiO_4$	orth	olivine	4	4.827	11.084	6.376			
Montroydite	HgO	orth		4	6.608	5.518	3.519			
Mullite (2:1)	$2Al_2O_3 \cdot SiO_2$	orth		6	7.5788	7.6909	2.8883			
Mullite (3:2)	$3Al_2O_3 \cdot 2SiO_2$	orth		3	7.557	7.6876	2.8842			
Muscovite	$KAl_2AlSi_3O_{10}(OH)_2$	monocl	2M2 mica	4	5.203	8.995	20.030		94.47°	

Name	Formula	Crystal system	Structure type	Z	a/Å	b/Å	c/Å	α	β	γ
Nacrite	$Al_2Si_2O_5(OH)_4$	monocl		4	8.909	5.146	15.697		113.70°	
Nantokite	CuCl	cubic	sphalerite	4	5.416					
Natroalunite	$NaAl_3(SO_4)_2(OH)_6$	rhomb		3	6.974		16.69			
Natrolite	$Na_2Al_2Si_3O_{10} \cdot 2H_2O$	orth		8	18.30	18.63	6.60			
Neighborite	$NaMgF_3$	orth	perovskite	4	5.363	7.676	5.503			
Ni-Skutterudite	$NiAs_{2.95}$	cubic		8	8.3300					
Niccolite	NiAs	hex	niccolite	2	3.618		5.034			
Nickel	Ni	cubic	face-centered cubic	4	3.5238					
Nickel carbonate	$NiCO_3$	rhomb	calcite	6	4.5975		14.723			
Nickel olivine	Ni_2SiO_4	orth	olivine	4	4.727	10.121	5.915			
Nickel selenide	$NiSe_2$	cubic	pyrite	4	5.9604					
Niter	KNO_3	orth	aragonite	4	6.431	9.164	5.414			
Norsethite	$BaMg(CO_3)_2$	rhomb	calcite	3	5.020		16.75			
Oldhamite	CaS	cubic	rock salt	4	5.689					
Orpiment	As_2S_3	monocl		4	11.49	9.59	4.25		90.45°	
Orthoclase	$KAlSi_3O_8$	monocl		4	8.562	12.996	7.193		116.02°	
Orthoferrosilite	$FeSiO_3$	orth	enstatite	16	9.080	18.431	5.238			
Otavite	$CdCO_3$	rhomb	calcite	6	4.9204		16.298			
Paracelsian	$BaAl_2Si_2O_8$	monocl		4	8.58	9.583	9.08		90.00°	
Paragonite	$NaAl_2AlSi_3O_{10}(OH)_2$	monocl	2M1 mica	4	5.13	8.89	19.32		95.17°	
Pararammelsbergite	$NiAs_2$	orth		8	5.75	5.82	11.428			
Paratellurite	TeO_2	tetr		4	4.810		7.613			
Parawollastonite	$CaSiO_3$	monocl		12	15.417	7.321	7.066		95.40°	
Pectolite	$Ca_2NaH(SiO_3)_3$	tricl		2	7.99	7.04	7.02	90.05°	95.27°	102.47°
Pentlandite	$Fe_{5.25}Ni_{3.75}S_8$	cubic		4	10.196					
Pentlandite	$Fe_{4.75}Ni_{5.25}S_8$	cubic		4	10.095					
Periclase	MgO	cubic	rock salt	4	4.2117					
Perovskite	$CaTiO_3$	orth	perovskite	4	5.3670	7.6438	5.4439			
Petalite	$LiAlSi_4O_{10}$	monocl		2	11.32	5.14	7.62		105.90°	
Petzite	Ag_3AuTe_2	cubic		8	10.38					
Phenacite	Be_2SiO_4	rhomb	phenacite	18	12.472		8.252			
Phlogopite	$KMg_3AlSi_3O_{10}(OH)_2$	monocl	1M mica	2	5.326	9.210	10.311		100.17°	
Picrochromite	$MgCr_2O_4$	cubic	spinel	8	8.333					
Piemontite	$Ca_2Al_{1.5}Mn_{1.5}(SiO_4)_3OH$	monocl		2	8.95	5.70	9.41		115.70°	
Platinum	Pt	cubic	face-centered cubic	4	3.9231					
Polymidite	Ni_3S_4	cubic	spinel	8	9.480					
Portlandite	$Ca(OH)_2$	hex	cadmium iodide	1	3.5933		4.9086			
Powellite	$CaMoO_4$	tetr	scheelite	4	5.226		11.43			
Protoenstatite	$MgSiO_3$	orth		8	9.25	8.74	5.32			
Proustite	Ag_3AsS_3	rhomb		6	10.816		8.6948			
Pseudowollastonite	$CaSiO_3$	tricl		24	6.90	11.78	19.65	90.00°	90.80°	90.00°
Pyrargyrite	Ag_3SbS_3	rhomb		6	11.052		8.7177			
Pyrite	FeS_2	cubic	pyrite	4	5.4175					
Pyrolusite	MnO_2	tetr	rutile	2	4.388		2.865			
Pyrope	$Mg_3Al_2Si_3O_{12}$	cubic	garnet	8	11.459					

Name	Formula	Crystal system	Structure type	Z	a/Å	b/Å	c/Å	α	β	γ
Pyrophanite	$MnTiO_3$	rhomb	ilmenite	6	5.155		14.18			
Pyrophyllite	$Al_2Si_4O_{10}(OH)_2$	monocl	2M1 mica	4	5.14	8.90	18.55		99.92°	
Pyroxmangite	$MnFe(SiO_3)_2$	tricl		7	7.56	17.45	6.67	84.00°	94.30°	113.70°
Pyrrhotite	$Fe_{0.980}S$	hex	defect niccolite	2	3.446		5.848			
Pyrrhotite	$Fe_{0.885}S$	hex	defect niccolite	2	3.440		5.709			
Quartz (α)	SiO_2	hex		3	4.9136		5.4051			
Quartz (β)	SiO_2	hex		3	4.999		5.4592			
Rammelsbergite	$NiAs_2$	orth	marcasite	2	4.757	5.797	3.542			
Realgar	AsS	monocl		16	9.29	13.53	6.57		106.55°	
Retgersite	$NiSO_4 \cdot 4H_2O$	tetr		4	6.782		18.28			
Rhodochrosite	$MnCO_3$	rhomb	calcite	6	4.7771		15.664			
Rhodonite	$MnSiO_3$	tricl		10	7.682	11.818	6.707	92.36°	93.95°	105.66°
Riebeckite	$Na_2Fe_5FSi_8O_{22}(OH)_2$	monocl	tremolite	2	9.729	18.065	5.334		103.31°	
Rutile	TiO_2	tetr		2	4.5937		2.9618			
Safflorite	$Co_{0.5}Fe_{0.5}As_2$	orth	marcasite	2	5.231	5.953	2.962			
Sanmartinite	$ZnWO_4$	monocl	wolframite	2	4.691	5.720	4.925		89.36°	
Sapphirine	$Mg_2Al_4O_6SiO_4$	monocl		8	9.96	28.60	9.85		110.5°	
Scacchite	$MnCl_2$	rhomb		3	3.711		17.59			
Scheelite	$CaWO_4$	tetr	scheelite	4	5.242		11.372			
Schorl	$NaFe_3Al_6B_3Si_6O_{27}(OH)_4$	rhomb	tourmaline	3	16.032		7.149			
Selenium	Se	hex		3	4.3642		4.9588			
Selenolite	SeO_2	tetr		8	8.35		5.05			
Sellaite	MgF_2	tetr	rutile	2	4.621		3.050			
Senarmontite	Sb_2O_3	cubic	arsenic trioxide	16	11.152					
Shandite	$Ni_3Pb_2S_2$	rhomb		3	5.576		13.658			
Shortite	$Na_2Ca_2(CO_3)_3$	orth		2	4.961	11.03	7.12			
Siderite	$FeCO_3$	rhomb	calcite	6	4.6887		15.373			
Silicon	Si	cubic	diamond	8	5.4305					
Sillimanite	Al_2OSiO_4	orth		4	7.4843	7.6730	5.7711			
Silver	Ag	cubic	face-centered cubic	4	4.0862					
Silver telluride I	Ag_2Te	cubic		2	5.29					
Silver telluride II	Ag_2Te	cubic		4	6.585					
Smithsonite	$ZnCO_3$	rhomb	calcite	6	4.6528		15.025			
Soda niter	$NaNO_3$	rhomb	calcite	6	5.0696		16.829			
Sodium melilite	$NaCaAlSi_2O_7$	tetr	melilite	2	8.511		4.809			
Sperrylite	$PtAs_2$	cubic	pyrite	4	5.968					
Spessartite	$Mn_3Al_2Si_3O_{12}$	cubic	garnet	8	11.621					
Sphalerite	ZnS	cubic	sphalerite	4	5.4093					
Sphene	$CaTiSiO_5$	monocl		4	7.07	8.72	6.56		113.95°	
Spinel	$MgAl_2O_4$	cubic	spinel	8	8.080					
Spodumene	$LiAl(SiO_3)_2$	monocl	diopside	4	9.451	8.387	5.208		110.07°	
Spodumene (β)	$LiAl(SiO_3)_2$	tetr		4	7.5332		9.1540			
Staurolite	$Fe_2Al_9Si_4O_{22}(OH)_2$	monocl		2	7.90	16.65	5.63		90.00°	
Sternbergite	$AgFe_2S_3$	orth		8	11.60	12.675	6.63			
Stibnite	Sb_2S_3	orth	stibnite	4	11.229	11.310	3.8389			

CRYSTALLOGRAPHIC DATA ON MINERALS (continued)

Name	Formula	Crystal system	Structure type	Z	a/Å	b/Å	c/Å	α	β	γ
Stilleite	ZnSe	cubic	sphalerite	4	5.6685					
Stishovite	SiO_2	tetr	rutile	2	4.1790		2.6649			
Stolzite	$PbWO_4$	tetr	scheelite	4	5.4616		12.046			
Stromeyerite	$Ag_{0.93}Cu_{1.07}S$	orth		4	4.066	6.628	7.972			
Strontianite	$SrCO_3$	orth	aragonite	4	6.029	8.414	5.107			
Sulfur (monoclinic)	S	monocl	S8 ring molecules	48	11.04	10.98	10.92		96.73°	
Sulfur (orthorhombic)	S	orth	S8 ring molecules	128	10.4646	12.8660	24.4860			
Sulfur (rhombohedral)	S	rhomb	S6 ring molecules	18	10.818		4.280			
Sylvite	KCl	cubic	rock salt	4	6.2931					
Syngenite	$K_2Ca(SO_4)_2 \cdot H_2O$	monocl		2	9.775	7.156	6.251		104.00°	
Synthetic anorthite	$CaAl_2Si_2O_8$	hex		2	5.10		14.72			
Synthetic anorthite	$CaAl_2Si_2O_8$	orth		2	8.22	8.60	4.83			
Talc	$Mg_3Si_4O_{10}(OH)_2$	monocl	2M1 mica	4	5.287	9.158	18.95		99.50°	
Tantalum	Ta	cubic	tungsten	2	3.3058					
Teallite	$PbSnS_2$	orth	germanium sulfide	2	4.266	11.419	4.090			
Tellurite	TeO_2	orth	tellurite	8	5.607	12.034	5.463			
Tellurium	Te	hex	selenium	3	4.4570		5.9290			
Tellurobismuthite	Bi_2Te_3	rhomb	selenium	3	4.3835		30.487			
Tennantite	$Cu_{12}As_4S_{13}$	cubic	tetrahedrite	2	10.190					
Tenorite	CuO	monocl		4	4.684	3.425	5.129		99.47°	
Tephroite	Mn_2SiO_4	orth	olivine	4	4.871	10.636	6.232			
Tetrahedrite	$Cu_{12}Sb_4S_{13}$	cubic	tetrahedrite	2	10.327					
Thenardite	Na_2SO_4	orth	thenardite	8	5.863	12.304	9.821			
Thorianite	ThO_2	cubic	fluorite	4	5.5952					
Thorite	$ThSiO_4$	tetr	zircon	4	7.143		6.327			
Tiemannite	HgSe	cubic	sphalerite	4	6.0853					
Tin	Sn	tetr		4	5.8315		3.1813			
Titanium	Ti	hex		2	2.953		4.729			
Titanium(III) oxide	Ti_2O_3	rhomb	corundum	6	5.149		13.642			
Topaz	$Al_2(SiO_4)(OH)$	orth		4	8.394	8.792	4.649			
Tremolite	$Ca_2Mg_5Si_8O_{22}(OH)_2$	monocl	tremolite	2	9.840	18.052	5.275		104.70°	
Trevorite	$NiFe_2O_4$	cubic	spinel	8	8.339					
Tridymite (β)	SiO_2	hex		4	5.0463		8.2563			
Trogtalite	$CoSe_2$	cubic	pyrite	4	5.8588					
Troilite	FeS	hex	niccolite	2	3.446		5.877			
Tschermakite	$CaAl_2SiO_6$	monocl	diopside	4	9.615	8.661	5.272		106.12°	
Tungsten	W	cubic		2	3.1653					
Tungstenite	WS_2	hex	molybdenite	2	3.154		12.362			
Turquois	$CuAl_6(PO_4)_4(OH)_8 \cdot 4H_2O$	tricl		1	7.424	7.629	9.910	68.61°	69.71°	65.08°
Umangite	Cu_3Se_2	tetr		2	6.402		4.276			
Uraninite	UO_2	cubic	fluorite	4	5.4682					
Ureyite	$NaCr(SiO_3)_2$	monocl	diopside	4	9.550	8.712	5.273		107.44°	
Uvarovite	$Ca_3Cr_2Si_3O_{12}$	cubic	garnet	8	11.999					
Uvite	$CaMg_4Al_5B_3Si_6O_{27}(OH)_4$	rhomb	tourmaline	3	15.86		7.19			
Vaesite	NiS_2	cubic	pyrite	4	5.6873					

Name	Formula	Crystal system	Structure type	Z	a/Å	b/Å	c/Å	α	β	γ
Valentinite	Sb_2O_3	orth	antimony trioxide	4	4.914	12.468	5.421			
Vanthoffite	$MgSO_4 \cdot 3Na_2SO_4$	monocl		2	9.797	9.217	8.199		113.50°	
Vaterite	$CaCO_3$	hex		6	7.135		8.524			
Villiaumite	NaF	cubic	rock salt	4	4.6342					
Violarite	$FeNi_2S_4$	cubic	spinel	8	9.464					
Willemite	Zn_2SiO_4	rhomb	phenacite	18	13.94		9.309			
Witherite	$BaCO_3$	orth	aragonite	4	6.430	8.904	5.314			
Wolframite	$Fe_{0.5}Mn_{0.5}WO_4$	monocl	wolframite	2	4.782	5.731	4.982		90.57°	
Wollastonite	$CaSiO_3$	tricl		6	7.94	7.32	7.07	90.03°	95.37°	103.43°
Wulfenite	$PbMoO_4$	tetr	scheelite	4	5.435		12.110			
Wurtzite	ZnS	hex	zincite	2	3.8230		6.2565			
Wustite	$Fe_{0.953}O$	cubic	defect rock salt	4	4.3088					
Xenotime	YPO_4	tetr	zircon	4	6.885		5.982			
Zinc	Zn	hex	hexagonal close pack	2	2.665		4.947			
Zinc telluride	ZnTe	cubic	sphalerite	4	6.1020					
Zincite	ZnO	hex	zincite	2	3.2495		5.2069			
Zinkosite	$ZnSO_4$	orth	barite	4	8.588	6.740	4.770			
Zircon	$ZrSiO_4$	tetr	zircon	4	6.604		5.979			
Zoisite	$Ca_2Al_3(SiO_4)_3OH$	orth		4	16.15	5.581	10.06			

Section 5
Thermochemistry, Electrochemistry, and Kinetics

CODATA Key Values for Thermodynamics 5-1
Standard Thermodynamic Properties of Chemical Substances 5-4
Thermodynamic Properties as a Function of Temperature 5-61
Thermodynamic Properties of Aqueous Systems 5-85
Heat of Combustion 5-89
Electrical Conductivity of Water 5-90
Standard KCl Solutions for Calibrating Conductivity Cells 5-91
Molar Conductivity of Aqueous HF, HCl, HBr, and HI 5-92
Equivalent Conductivity of Electrolytes in Aqueous Solution 5-94
Ionic Conductivity and Diffusion at Infinite Dilution 5-95
Activity Coefficients of Acids, Bases, and Salts 5-97
Mean Activity Coefficients of Electrolytes as a Function of Concentration 5-99
Enthalpy of Dilution of Acids 5-104
Enthalpy of Solution of Electrolytes 5-105
Chemical Kinetic Data for Stratospheric Modeling 5-106

CODATA KEY VALUES FOR THERMODYNAMICS

The Committee on Data for Science and Technology (CODATA) has conducted a project to establish internationally agreed values for the thermodynamic properties of key chemical substances. This table presents the final results of the project. Use of these recommended, internally consistent values is encouraged in the analysis of thermodynamic measurements, data reduction, and preparation of other thermodynamic tables.

The table includes the standard enthalpy of formation at 298.15 K, the entropy at 298.15 K, and the quantity $H°$ (298.15 K)–$H°$ (0). A value of 0 in the $\Delta_f H°$ column for an element indicates the reference state for that element. The standard state pressure is 100000 Pa (1 bar). See the reference for information on the dependence of gas-phase entropy on the choice of standard state pressure.

Substances are listed in alphabetical order of their chemical formulas when written in the most common form.

The table is reprinted with permission of CODATA.

REFERENCE

Cox, J. D., Wagman, D. D., and Medvedev, V. A., *CODATA Key Values for Thermodynamics,* Hemisphere Publishing Corp., New York, 1989.

Substance	State	$\Delta_f H°$ (298.15 K) kJ·mol^{-1}	$S°$ (298.15 K) J·K^{-1}·mol^{-1}	$H°$ (298.15 K)–$H°$ (0) kJ·mol^{-1}
Ag	cr	0	42.55 ± 0.20	5.745 ± 0.020
Ag	g	284.9 ± 0.8	172.997 ± 0.004	6.197 ± 0.001
Ag$^+$	aq	105.79 ± 0.08	73.45 ± 0.40	
AgCl	cr	-127.01 ± 0.05	96.25 ± 0.20	12.033 ± 0.020
Al	cr	0	28.30 ± 0.10	4.540 ± 0.020
Al	g	330.0 ± 4.0	164.554 ± 0.004	6.919 ± 0.001
Al^{+3}	aq	-538.4 ± 1.5	-325 ± 10	
AlF$_3$	cr	-1510.4 ± 1.3	66.5 ± 0.5	11.62 ± 0.04
Al$_2$O$_3$	cr, corundum	-1675.7 ± 1.3	50.92 ± 0.10	10.016 ± 0.020
Ar	g	0	154.846 ± 0.003	6.197 ± 0.001
B	cr, rhombic	0	5.90 ± 0.08	1.222 ± 0.008
B	g	565 ± 5	153.436 ± 0.015	6.316 ± 0.002
BF$_3$	g	-1136.0 ± 0.8	254.42 ± 0.20	11.650 ± 0.020
B$_2$O$_3$	cr	-1273.5 ± 1.4	53.97 ± 0.30	9.301 ± 0.040
Be	cr	0	9.50 ± 0.08	1.950 ± 0.020
Be	g	324 ± 5	136.275 ± 0.003	6.197 ± 0.001
BeO	cr	-609.4 ± 2.5	13.77 ± 0.04	2.837 ± 0.008
Br	g	111.87 ± 0.12	175.018 ± 0.004	6.197 ± 0.001
Br$^-$	aq	-121.41 ± 0.15	82.55 ± 0.20	
Br$_2$	l	0	152.21 ± 0.30	24.52 ± 0.01
Br$_2$	g	30.91 ± 0.11	245.468 ± 0.005	9.725 ± 0.001
C	cr, graphite	0	5.74 ± 0.10	1.050 ± 0.020
C	g	716.68 ± 0.45	158.100 ± 0.003	6.536 ± 0.001
CO	g	-110.53 ± 0.17	197.660 ± 0.004	8.671 ± 0.001
CO$_2$	g	-393.51 ± 0.13	213.785 ± 0.010	9.365 ± 0.003
CO$_2$	aq, undissoc.	-413.26 ± 0.20	119.36 ± 0.60	
CO$_3^{-2}$	aq	-675.23 ± 0.25	-50.0 ± 1.0	
Ca	cr	0	41.59 ± 0.40	5.736 ± 0.040
Ca	g	177.8 ± 0.8	154.887 ± 0.004	6.197 ± 0.001
Ca^{+2}	aq	-543.0 ± 1.0	-56.2 ± 1.0	
CaO	cr	-634.92 ± 0.90	38.1 ± 0.4	6.75 ± 0.06
Cd	cr	0	51.80 ± 0.15	6.247 ± 0.015
Cd	g	111.80 ± 0.20	167.749 ± 0.004	6.197 ± 0.001
Cd^{+2}	aq	-75.92 ± 0.60	-72.8 ± 1.5	
CdO	cr	-258.35 ± 0.40	54.8 ± 1.5	8.41 ± 0.08
CdSO$_4$·8/3H$_2$O	cr	-1729.30 ± 0.80	229.65 ± 0.40	35.56 ± 0.04
Cl	g	121.301 ± 0.008	165.190 ± 0.004	6.272 ± 0.001
Cl$^-$	aq	-167.080 ± 0.10	56.60 ± 0.20	
ClO$_4^-$	aq	-128.10 ± 0.40	184.0 ± 1.5	
Cl$_2$	g	0	223.081 ± 0.010	9.181 ± 0.001
Cs	cr	0	85.23 ± 0.40	7.711 ± 0.020
Cs	g	76.5 ± 1.0	175.601 ± 0.003	6.197 ± 0.001
Cs$^+$	aq	-258.00 ± 0.50	132.1 ± 0.5	

Substance	State	$\Delta_f H°$ (298.15 K) $\overline{\text{kJ·mol}^{-1}}$	$S°$ (298.15 K) $\overline{\text{J·K}^{-1}\text{·mol}^{-1}}$	$H°$ (298.15 K)–$H°$ (0) $\overline{\text{kJ·mol}^{-1}}$
Cu	cr	0	33.15 ± 0.08	5.004 ± 0.008
Cu	g	337.4 ± 1.2	166.398 ± 0.004	6.197 ± 0.001
Cu^{+2}	aq	64.9 ± 1.0	-98 ± 4	
$CuSO_4$	cr	-771.4 ± 1.2	109.2 ± 0.4	16.86 ± 0.08
F	g	79.38 ± 0.30	158.751 ± 0.004	6.518 ± 0.001
F^-	aq	-335.35 ± 0.65	-13.8 ± 0.8	
F_2	g	0	202.791 ± 0.005	8.825 ± 0.001
Ge	cr	0	31.09 ± 0.15	4.636 ± 0.020
Ge	g	372 ± 3	167.904 ± 0.005	7.398 ± 0.001
GeF_4	g	-1190.20 ± 0.50	301.9 ± 1.0	17.29 ± 0.10
GeO_2	cr, tetragonal	-580.0 ± 1.0	39.71 ± 0.15	7.230 ± 0.020
H	g	217.998 ± 0.006	114.717 ± 0.002	6.197 ± 0.001
H^+	aq	0	0	
HBr	g	-36.29 ± 0.16	198.700 ± 0.004	8.648 ± 0.001
HCO_3^-	aq	-689.93 ± 0.20	98.4 ± 0.5	
HCl	g	-92.31 ± 0.10	186.902 ± 0.005	8.640 ± 0.001
HF	g	-273.30 ± 0.70	173.779 ± 0.003	8.599 ± 0.001
HI	g	26.50 ± 0.10	206.590 ± 0.004	8.657 ± 0.001
HPO_4^{-2}	aq	-1299.0 ± 1.5	-33.5 ± 1.5	
HS^-	aq	-16.3 ± 1.5	67 ± 5	
HSO_4^-	aq	-886.9 ± 1.0	131.7 ± 3.0	
H_2	g	0	130.680 ± 0.003	8.468 ± 0.001
H_2O	l	-285.830 ± 0.040	69.95 ± 0.03	13.273 ± 0.020
H_2O	g	-241.826 ± 0.040	188.835 ± 0.010	9.905 ± 0.005
$H_2PO_4^-$	aq	-1302.6 ± 1.5	92.5 ± 1.5	
H_2S	g	-20.6 ± 0.5	205.81 ± 0.05	9.957 ± 0.010
H_2S	aq, undissoc.	-38.6 ± 1.5	126 ± 5	
H_3BO_3	cr	-1094.8 ± 0.8	89.95 ± 0.60	13.52 ± 0.04
H_3BO_3	aq, undissoc.	-1072.8 ± 0.8	162.4 ± 0.6	
He	g	0	126.153 ± 0.002	6.197 ± 0.001
Hg	l	0	75.90 ± 0.12	9.342 ± 0.008
Hg	g	61.38 ± 0.04	174.971 ± 0.005	6.197 ± 0.001
Hg^{+2}	aq	170.21 ± 0.20	-36.19 ± 0.80	
HgO	cr, red	-90.79 ± 0.12	70.25 ± 0.30	9.117 ± 0.025
Hg_2^{+2}	aq	166.87 ± 0.50	65.74 ± 0.80	
Hg_2Cl_2	cr	-265.37 ± 0.40	191.6 ± 0.8	23.35 ± 0.20
Hg_2SO_4	cr	-743.09 ± 0.40	200.70 ± 0.20	26.070 ± 0.030
I	g	106.76 ± 0.04	180.787 ± 0.004	6.197 ± 0.001
I^-	aq	-56.78 ± 0.05	106.45 ± 0.30	
I_2	cr	0	116.14 ± 0.30	13.196 ± 0.040
I_2	g	62.42 ± 0.08	260.687 ± 0.005	10.116 ± 0.001
K	cr	0	64.68 ± 0.20	7.088 ± 0.020
K	g	89.0 ± 0.8	160.341 ± 0.003	6.197 ± 0.001
K^+	aq	-252.14 ± 0.08	101.20 ± 0.20	
Kr	g	0	164.085 ± 0.003	6.197 ± 0.001
Li	cr	0	29.12 ± 0.20	4.632 ± 0.040
Li	g	159.3 ± 1.0	138.782 ± 0.010	6.197 ± 0.001
Li^+	aq	-278.47 ± 0.08	12.24 ± 0.15	
Mg	cr	0	32.67 ± 0.10	4.998 ± 0.030
Mg	g	147.1 ± 0.8	148.648 ± 0.003	6.197 ± 0.001
Mg^{+2}	aq	-467.0 ± 0.6	-137 ± 4	
MgF_2	cr	-1124.2 ± 1.2	57.2 ± 0.5	9.91 ± 0.06
MgO	cr	-601.60 ± 0.30	26.95 ± 0.15	5.160 ± 0.020
N	g	472.68 ± 0.40	153.301 ± 0.003	6.197 ± 0.001
NH_3	g	-45.94 ± 0.35	192.77 ± 0.05	10.043 ± 0.010
NH_4^+	aq	-133.26 ± 0.25	111.17 ± 0.40	
NO_3^-	aq	-206.85 ± 0.40	146.70 ± 0.40	

Substance	State	$\Delta_f H°$ (298.15 K) kJ·mol^{-1}	$S°$ (298.15 K) J·K^{-1}·mol^{-1}	$H°$ (298.15 K)–$H°$ (0) kJ·mol^{-1}
N_2	g	0	191.609 ± 0.004	8.670 ± 0.001
Na	cr	0	51.30 ± 0.20	6.460 ± 0.020
Na	g	107.5 ± 0.7	153.718 ± 0.003	6.197 ± 0.001
Na$^+$	aq	-240.34 ± 0.06	58.45 ± 0.15	
Ne	g	0	146.328 ± 0.003	6.197 ± 0.001
O	g	249.18 ± 0.10	161.059 ± 0.003	6.725 ± 0.001
OH$^-$	aq	-230.015 ± 0.040	-10.90 ± 0.20	
O_2	g	0	205.152 ± 0.005	8.680 ± 0.002
P	cr, white	0	41.09 ± 0.25	5.360 ± 0.015
P	g	316.5 ± 1.0	163.199 ± 0.003	6.197 ± 0.001
P_2	g	144.0 ± 2.0	218.123 ± 0.004	8.904 ± 0.001
P_4	g	58.9 ± 0.3	280.01 ± 0.50	14.10 ± 0.20
Pb	cr	0	64.80 ± 0.30	6.870 ± 0.030
Pb	g	195.2 ± 0.8	175.375 ± 0.005	6.197 ± 0.001
Pb^{+2}	aq	0.92 ± 0.25	18.5 ± 1.0	
PbSO$_4$	cr	-919.97 ± 0.40	148.50 ± 0.60	20.050 ± 0.040
Rb	cr	0	76.78 ± 0.30	7.489 ± 0.020
Rb	g	80.9 ± 0.8	170.094 ± 0.003	6.197 ± 0.001
Rb$^+$	aq	-251.12 ± 0.10	121.75 ± 0.25	
S	cr, rhombic	0	32.054 ± 0.050	4.412 ± 0.006
S	g	277.17 ± 0.15	167.829 ± 0.006	6.657 ± 0.001
SO$_2$	g	-296.81 ± 0.20	248.223 ± 0.050	10.549 ± 0.010
SO$_4^{-2}$	aq	-909.34 ± 0.40	18.50 ± 0.40	
S_2	g	128.60 ± 0.30	228.167 ± 0.010	9.132 ± 0.002
Si	cr	0	18.81 ± 0.08	3.217 ± 0.008
Si	g	450 ± 8	167.981 ± 0.004	7.550 ± 0.001
SiF$_4$	g	-1615.0 ± 0.8	282.76 ± 0.50	15.36 ± 0.05
SiO$_2$	cr, alpha quartz	-910.7 ± 1.0	41.46 ± 0.20	6.916 ± 0.020
Sn	cr, white	0	51.18 ± 0.08	6.323 ± 0.008
Sn	g	301.2 ± 1.5	168.492 ± 0.004	6.215 ± 0.001
Sn^{+2}	aq	-8.9 ± 1.0	-16.7 ± 4.0	
SnO	cr, tetragonal	-280.71 ± 0.20	57.17 ± 0.30	8.736 ± 0.020
SnO$_2$	cr, tetragonal	-577.63 ± 0.20	49.04 ± 0.10	8.384 ± 0.020
Th	cr	0	51.8 ± 0.5	6.35 ± 0.05
Th	g	602 ± 6	190.17 ± 0.05	6.197 ± 0.003
ThO$_2$	cr	-1226.4 ± 3.5	65.23 ± 0.20	10.560 ± 0.020
Ti	cr	0	30.72 ± 0.10	4.824 ± 0.015
Ti	g	473 ± 3	180.298 ± 0.010	7.539 ± 0.002
TiCl$_4$	g	-763.2 ± 3.0	353.2 ± 4.0	21.5 ± 0.5
TiO$_2$	cr, rutile	-944.0 ± 0.8	50.62 ± 0.30	8.68 ± 0.05
U	cr	0	50.20 ± 0.20	6.364 ± 0.020
U	g	533 ± 8	199.79 ± 0.10	6.499 ± 0.020
UO$_2$	cr	-1085.0 ± 1.0	77.03 ± 0.20	11.280 ± 0.020
UO$_2^{+2}$	aq	-1019.0 ± 1.5	-98.2 ± 3.0	
UO$_3$	cr, gamma	-1223.8 ± 1.2	96.11 ± 0.40	14.585 ± 0.050
U$_3$O$_8$	cr	-3574.8 ± 2.5	282.55 ± 0.50	42.74 ± 0.10
Xe	g	0	169.685 ± 0.003	6.197 ± 0.001
Zn	cr	0	41.63 ± 0.15	5.657 ± 0.020
Zn	g	130.40 ± 0.40	160.990 ± 0.004	6.197 ± 0.001
Zn^{+2}	aq	-153.39 ± 0.20	-109.8 ± 0.5	
ZnO	cr	-350.46 ± 0.27	43.65 ± 0.40	6.933 ± 0.040

STANDARD THERMODYNAMIC PROPERTIES OF CHEMICAL SUBSTANCES

This table gives the standard state chemical thermodynamic properties of about 2400 individual substances in the crystalline, liquid, and gaseous states. Substances are listed by molecular formula in a modified Hill order; all compounds not containing carbon appear first, followed by those that contain carbon. The properties tabulated are:

$\Delta_f H°$ Standard molar enthalpy (heat) of formation at 298.15 K in kJ/mol
$\Delta_f G°$ Standard molar Gibbs energy of formation at 298.15 K in kJ/mol
$S°$ Standard molar entropy at 298.15 K in J/mol K
C_p Molar heat capacity at constant pressure at 298.15 K in J/mol K

The standard state pressure is 100 kPa (1 bar). The standard states are defined for different phases by:

- The standard state of a pure gaseous substance is that of the substance as a (hypothetical) ideal gas at the standard state pressure.
- The standard state of a pure liquid substance is that of the liquid under the standard state pressure.
- The standard state of a pure crystalline substance is that of the crystalline substance under the standard state pressure.

An entry of 0.0 for $\Delta_f H°$ for an element indicates the reference state of that element. See References 1 and 2 for further information on reference states. A blank means no value is available.

We are indebted to M.V. Korobov for providing data on fullerene compounds.

REFERENCES

1. Cox, J.D., Wagman, D.D., and Medvedev, V.A., *CODATA Key Values for Thermodynamics*, Hemisphere Publishing Corp., New York, 1989.
2. Wagman, D.D., Evans, W.H., Parker, V.B., Schumm, R.H., Halow, I., Bailey, S.M., Churney, K.L., and Nuttall, R.L., *The NBS Tables of Chemical of Chemical Thermodynamic Properties, J. Phys. Chem. Ref. Data*, Vol. 11, Suppl. 2, 1982.
3. Chase M.W., Davies, C.A., Downey, J.R., Frurip, D.J., McDonald, R.A., and Syverud, A.N., *JANAF Thermochemical Tables, Third Edition, J. Phys. Chem. Ref. Data*, Vol. 14, Suppl. 1, 1985.
4. Daubert, T.E., Danner, R.P., Sibul, H.M., and Stebbins, C.C., *Physical and Thermodynamic Properties of Pure Compounds: Data Compilation*, extant 1994 (core with 4 supplements), Taylor & Francis, Bristol, PA.
5. Pedley, J.B., Naylor, R.D., and Kirby, S.P., *Thermochemical Data of Organic Compounds, Second Edition*, Chapman & Hall, London, 1986.
6. Pedley, J.B., *Thermochemical Data and Structures of Organic Compounds*, Thermodynamic Research Center, Texas A & M University, College Station, TX, 1994.
7. Dolmalski, E.S., Evans, W.H., Hearing E.D., *Heat Capacities and Entropies of Organic Compounds in the Condensed Phase, J. Phys. Chem. Ref. Data*, Vol. 13, Suppl. 1, 1984; Vol. 19, No. 4, 881—1047, 1990.
8. Gurvich, L.V., Veyts, I.V., and Alcock, C.B., *Thermodynamic Properties of Individual Substances, Fourth Edition*, Vol. 1, Hermisphere Publishing Corp., New York, 1989.
9. Gurvich, L.V., Veyts, I.V., and Alcock, C.B., *Thermodynamic Properties of Individual Substances, Fourth Edition*, Vol. 3, CRC Press, Boca Raton, FL, 1994.

STANDARD THERMODYNAMIC PROPERTIES OF CHEMICAL SUBSTANCES (continued)

Molecular formula	Name	$\Delta_f H°$/kJ mol⁻¹			$\Delta_f G°$/kJ mol⁻¹			$S°$/J mol⁻¹ K⁻¹			C_p/J mol⁻¹ K⁻¹		
		Crys.	Liq.	Gas	Crys.	Liq.	Gas	Crys.	Liq.	Gas	Crys.	Liq.	Gas
Ac	Actinium	0.0		406.0			366.0	56.5		188.1	27.2		20.8
Ag	Silver	0.0		284.9			246.0	42.6		173.0	25.4		20.8
AgBr	Silver(I) bromide	-100.4			-96.9			107.1			52.4		
AgBrO₃	Silver(I) bromate	-10.5			71.3			151.9					
AgCl	Silver(I) chloride	-127.0			-109.8			96.3			50.8		
AgClO₃	Silver(I) chlorate	-30.3			64.5			142.0					
AgClO₄	Silver(I) perchlorate	-31.1											
AgF	Silver(I) fluoride	-204.6											
AgF₂	Silver(II) fluoride	-360.0											
AgI	Silver(I) iodide	-61.8			-66.2			115.5			56.8		
AgIO₃	Silver(I) iodate	-171.1			-93.7			149.4			102.9		
AgNO₃	Silver(I) nitrate	-124.4			-33.4			140.9			93.1		
Ag₂	Disilver			410.0			358.8			257.1			37.0
Ag₂CrO₄	Silver(I) chromate	-731.7			-641.8			217.6			142.3		
Ag₂O	Silver(I) oxide	-31.1			-11.2			121.3			65.9		
Ag₂O₂	Silver(II) oxide	-24.3			27.6			117.0			88.0		
Ag₂O₃	Silver(III) oxide	33.9			121.4			100.0					
Ag₂O₄S	Silver(I) sulfate	-715.9			-618.4			200.4			131.4		
Ag₂S	Silver(I) sulfide (argentite)	-32.6			-40.7			144.0			76.5		
Al	Aluminum	0.0		330.0			289.4	28.3		164.6	24.4		21.4
AlB₃H₁₂	Aluminum borohydride		-16.3	13.0		145.0	147.0		289.1	379.2		194.6	
AlBr	Aluminum monobromide			-4.0			-42.0			239.5			35.6
AlBr₃	Aluminum tribromide	-527.2		-425.1				180.2			100.6		
AlCl	Aluminum monochloride			-47.7			-74.1			228.1			35.0
AlCl₂	Aluminum dichloride			-331.0									
AlCl₃	Aluminum trichloride	-704.2		-583.2	-628.8			109.3			91.1		
AlF	Aluminum monofluoride			-258.2			-283.7			215.0			31.9
AlF₃	Aluminum trifluoride	-1510.4		-1204.6	-1431.1		-1188.2	66.5		277.1	75.1		62.6
AlF₄Na	Sodium tetrafluoroaluminate			-1869.0			-1827.5			345.7			105.9
AlH	Aluminum hydride			259.2			231.2			187.9			29.4
AlH₃	Aluminum hydride	-46.0						30.0			40.2		
AlH₄K	Potassium aluminum hydride	-183.7											
AlH₄Li	Lithium aluminum hydride	-116.3			-44.7			78.7			83.2		
AlH₄Na	Sodium aluminum hydride	-15.5											
AlI	Aluminum monoiodide			65.5									36.0
AlI₃	Aluminum triiodide	-313.8		-207.5	-300.8			159.0			98.7		
AlN	Aluminum nitride	-318.0			-287.0			20.2			30.1		
AlO	Aluminum monoxide			91.2			65.3			218.4			30.9
AlO₄P	Aluminum phosphate	-1733.8			-1617.9			90.8			93.2		
AlP	Aluminum phosphide	-166.5											
AlS	Aluminum monosulfide			200.9			150.1			230.6			33.4
Al₂	Dialuminum			485.9			433.3			233.2			36.4
Al₂Br₆	Aluminum hexabromide			-970.7									
Al₂Cl₆	Aluminum hexachloride			-1290.8			-1220.4			490.0			

STANDARD THERMODYNAMIC PROPERTIES OF CHEMICAL SUBSTANCES (continued)

Molecular formula	Name	$\Delta_f H°$/kJ mol⁻¹ Crys.	Liq.	Gas	$\Delta_f G°$/kJ mol⁻¹ Crys.	Liq.	Gas	$S°$/J mol⁻¹ K⁻¹ Crys.	Liq.	Gas	C_p/J mol⁻¹ K⁻¹ Crys.	Liq.	Gas
Al_2F_6	Aluminum hexafluoride			-2628.0									
Al_2I_6	Aluminum hexaiodide			-516.7									
Al_2O	Aluminum oxide (Al_2O)			-130.0			-159.0			259.4			45.7
Al_2O_3	Aluminum oxide (corundum)	-1675.7			-1582.3			50.9			79.0		
Al_2S_3	Aluminum sulfide	-724.0						116.9			105.1		
Am	Americium	0.0											
Ar	Argon			0.0						154.8			20.8
As	Arsenic (gray)	0.0		302.5			261.0	35.1		174.2	24.6		20.8
As	Arsenic (yellow)	14.6											
$AsBr_3$	Arsenic(III) bromide	-197.5		-130.0			-159.0			363.9			79.2
$AsCl_3$	Arsenic(III) chloride		-305.0	-261.5		-259.4	-248.9		216.3	327.2		126.6	75.7
AsF_3	Arsenic(III) fluoride		-821.3	-785.8		-774.2	-770.8		181.2	289.1			65.6
AsGa	Gallium arsenide	-71.0			-67.8			64.2			46.2		
AsH_3	Arsine			66.4			68.9			222.8			38.1
AsH_3O_4	Arsenic acid	-906.3											
AsI_3	Arsenic(III) iodide	-58.2			-59.4			213.1		388.3	105.8		80.6
AsIn	Indium arsenide	-58.6			-53.6			75.7			47.8		
AsO	Arsenic monoxide			70.0									
As_2	Diarsenic			222.2			171.9			239.4			35.0
As_2O_5	Arsenic(V) oxide	-924.9			-782.3			105.4			116.5		
As_2S_3	Arsenic(III) sulfide	-169.0			-168.6			163.6			116.3		
At	Astatine	0.0											
Au	Gold	0.0		366.1			326.3	47.4		180.5	25.4		20.8
AuBr	Gold(I) bromide	-14.0											
$AuBr_3$	Gold(II) bromide	-53.3											
AuCl	Gold(I) chloride	-34.7											
$AuCl_3$	Gold(III) chloride	-117.6											
AuF_3	Gold(III) fluoride	-363.6											
AuH	Gold hydride			295.0			265.7			211.2			29.2
AuI	Gold(I) iodide	0.0											
Au_2	Digold			515.1									36.9
B	Boron (rhombic)	0.0		565.0			521.0	5.9		153.4	11.1		20.8
BBr	Bromoborane			238.1			195.4			225.0			32.9
BBr_3	Boron tribromide		-239.7	-205.6		-238.5	-232.5		229.7	324.2			67.8
BCl	Chloroborane			149.5			120.9			213.2			31.7
BClO	Chloroxyborane			-314.0									
BCl_3	Boron trichloride		-427.2	-403.8		-387.4	-388.7		206.3	290.1		106.7	62.7
$BCsO_2$	Cesium metaborate	-972.0			-915.0			104.4			80.6		
BF	Fluoroborane			-122.2			-149.8			200.5			29.6
BFO	Fluorooxyborane			-607.0									
BF_3	Boron trifluoride			-1136.0			-1119.4			254.4			
BF_3H_3N	Aminetrifluoroboron	-1353.9											
BF_3H_3P	Trihydro(phosphorus trifluoride)boron			-854.0									
BF_4Na	Sodium tetrafluoroborate	-1844.7			-1750.1			145.3			120.3		

STANDARD THERMODYNAMIC PROPERTIES OF CHEMICAL SUBSTANCES (continued)

Molecular formula	Name	ΔfH°/kJ mol⁻¹ Crys.	Liq.	Gas	ΔfG°/kJ mol⁻¹ Crys.	Liq.	Gas	S°/J mol⁻¹ K⁻¹ Crys.	Liq.	Gas	Cp/J mol⁻¹ K⁻¹ Crys.	Liq.	Gas
BH	Boron monohydride			442.7			419.6			171.9			29.2
BHO₂	Metaboric acid (monoclinic)	-794.3		-561.9	-723.4		-551.0	38.0		240.1			42.2
BH₃	Borane (3)			89.2						188.2			36.0
BH₃O₃	Boric acid (orthoboric acid)	-1094.3		-994.1	-968.9			90.0			86.1		
BH₄K	Potassium borohydride	-227.4			-160.3			106.3			96.1		
BH₄Li	Lithium borohydride	-190.8			-125.0			75.9			82.6		
BH₄Na	Sodium borohydride	-188.6			-123.9			101.3			86.8		
BI₃	Boron triiodide			71.1			20.7			349.2			70.8
BKO₂	Potassium metaborate	-981.6			-923.4			80.0			66.7		
BLiO₂	Lithium metaborate	-1032.2			-976.1			51.5			59.8		
BN	Boron nitride	-254.4		647.5	-228.4		614.5	14.8		212.3	19.7		29.5
BNaO₂	Sodium metaborate	-977.0			-920.7			73.5			65.9		
BO	Boron monoxide			25.0			-4.0			203.5			29.2
BO₂	Boron dioxide			-300.4			-305.9			229.6			43.0
BO₂Rb	Rubidium metaborate	-971.0			-913.0			94.3			74.1		
BS	Boron monosulfide			342.0			288.8			216.2			30.0
B₂	Diboron			830.5			774.0			201.9			30.5
B₂Cl₄	Tetrachlorodiborane		-523.0	-490.4		-464.8	-460.6		262.3	357.4		137.7	95.4
B₂F₄	Tetrafluorodiborane			-1440.1			-1410.4			317.3			79.1
B₂H₆	Diborane			36.4			86.7			232.1			56.7
B₂O₂	Diboron dioxide			-454.8			-462.3			242.5			57.3
B₂O₃	Boron oxide	-1273.5		-843.8	-1194.3		-832.0	54.0		279.8	62.8		66.9
B₂S₃	Boron sulfide	-240.6	-541.0	67.0				100.0			111.7		
B₃H₆N₃	Borazine		-541.0			-392.7			199.6	280.3			93.2
B₄H₁₀	Tetraborane (10)			66.1									
B₄Na₂O₇	Sodium tetraborate	-3291.1			-3096.0			189.5			186.8		
B₅H₉	Pentaborane(9)		42.7	73.2		171.8	175.0		184.2	280.6		151.1	99.6
B₅H₁₁	Pentaborane (11)		73.2	103.3						321.0			130.3
B₆H₁₀	Hexaborane (10)		56.3	94.6						296.8			125.7
Ba	Barium	0.0		180.0			146.0	62.5		170.2	28.1		20.8
BaBr₂	Barium bromide	-757.3			-736.8			146.0					
BaCl₂	Barium chloride	-855.0			-806.7			123.7			75.1		
BaCl₂H₄O₂	Barium chloride dihydrate	-1456.9			-1293.2			203.0					
BaF₂	Barium fluoride	-1207.1			-1156.8			96.4			71.2		
BaH₂	Barium hydride	-177.0			-138.2			63.0			46.0		
BaH₂O₂	Barium hydroxide	-944.7											
BaI₂	Barium iodide	-602.1											
BaN₂O₄	Barium nitrite	-768.2											
BaN₂O₆	Barium nitrate	-988.0			-792.6			214.0			151.4		
BaO	Barium oxide	-548.0			-520.3			72.1			47.3		
BaO₄S	Barium sulfate	-1473.2			-1362.2			132.2			101.8		
BaS	Barium sulfide	-460.0			-456.0			78.2			49.4		
Be	Beryllium	0.0		324.0			286.6	9.5		136.3	16.4		20.8
BeBr₂	Beryllium bromide	-353.5						108.0			69.4		
BeCl₂	Beryllium chloride	-490.4			-445.6			75.8			62.4		

Molecular formula	Name	$\Delta_f H°$/kJ mol⁻¹ Crys.	Liq.	Gas	$\Delta_f G°$/kJ mol⁻¹ Crys.	Liq.	Gas	$S°$/J mol⁻¹ K⁻¹ Crys.	Liq.	Gas	C_p/J mol⁻¹ K⁻¹ Crys.	Liq.	Gas
BeF₂	Beryllium fluoride	-1026.8			-979.4			53.4			51.8		
BeH₂O₂	Beryllium hydroxide	-902.5			-815.0			45.5			62.1		
BeI₂	Beryllium iodide	-192.5						121.0			71.1		
BeO	Beryllium oxide	-609.4			-580.1			13.8			25.6		
BeO₄S	Beryllium sulfate	-1205.2			-1093.8			77.9			85.7		
BeS	Beryllium sulfide	-234.3						34.0			34.0		
Bi	Bismuth	0.0		207.1			168.2	56.7		187.0	25.5		20.8
BiClO	Bismuth oxychloride	-366.9			-322.1			120.5					
BiCl₃	Bismuth trichloride	-379.1		-265.7	-315.0		-256.0	177.0		358.9	105.0		79.7
BiH₃O₃	Bismuth hydroxide	-711.3											
BiI₃	Bismuth triiodide				-175.3								
Bi₂	Dibismuth			219.7									
Bi₂O₃	Bismuth oxide	-573.9			-493.7			151.5			113.5		
Bi₂O₁₂S₃	Bismuth sulfate	-2544.3											
Bi₂S₃	Bismuth sulfide	-143.1			-140.6			200.4			122.2		
Bk	Berkelium	0.0											
Br	Bromine (atomic)			111.9			82.4			175.0			20.8
BrCl	Bromine chloride			14.6			-1.0			240.1			35.0
BrCl₃Si	Bromotrichlorosilane									350.1			90.9
BrCs	Cesium bromide	-405.8			-391.4			113.1			52.9		
BrCu	Copper(I) bromide	-104.6			-100.8			96.1			54.7		
BrF	Bromine fluoride			-93.8			-109.2			229.0			33.0
BrF₃	Bromine trifluoride		-300.8	-255.6		-240.5	-229.4		178.2	292.5		124.6	66.6
BrF₅	Bromine pentafluoride		-458.6	-428.9		-351.8	-350.6		225.1	320.2			99.6
BrGe	Germanium monobromide			235.6						274.8			37.1
BrGeH₃	Bromogermane												56.4
BrH	Hydrogen bromide			-36.3			-53.4			198.7			29.1
BrHSi	Bromosilylene												
BrH₃Si	Bromosilane			-464.4						262.4			52.8
BrH₄N	Ammonium bromide	-270.8			-175.2			113.0			96.0		
BrI	Iodine bromide			40.8			3.7			258.8			36.4
BrIn	Indium(I) bromide	-175.3		-56.9	-169.0		-94.3	113.0		259.5			36.7
BrK	Potassium bromide	-393.8			-380.7			95.9			52.3		
BrKO₃	Potassium bromate	-360.2			-271.2			149.2			105.2		
BrKO₄	Potassium perbromate	-287.9			-174.4			170.1			120.2		
BrLi	Lithium bromide	-351.2			-342.0			74.3			51.4		
BrNO	Nitrosyl bromide			82.2			82.4			273.7			45.5
BrNa	Sodium bromide	-361.1		-143.1	-349.0		-177.1	86.8		241.2			36.3
BrNaO₃	Sodium bromate	-334.1			-242.6			128.9					
BrO	Bromine monoxide			125.8			108.2			237.6			32.1
BrO₂	Bromine dioxide	48.5											
BrRb	Rubidium bromide	-394.6			-381.8			110.0			52.8		
BrSi	Bromosilylidyne			209.0									38.6
BrTl	Thallium(I) bromide	-173.2		-37.7	-167.4			120.5					
Br₂	Bromine		0.0	30.9			3.1		152.2	245.5		75.7	36.0

Molecular formula	Name	$\Delta_f H°$/kJ mol⁻¹			$\Delta_f G°$/kJ mol⁻¹			$S°$/J mol⁻¹ K⁻¹			C_p/J mol⁻¹ K⁻¹		
		Crys.	Liq.	Gas	Crys.	Liq.	Gas	Crys.	Liq.	Gas	Crys.	Liq.	Gas
Br₂Ca	Calcium bromide	-682.8			-663.6			130.0					
Br₂Cd	Cadmium bromide	-316.2			-296.3			137.2			76.7		
Br₂Co	Cobalt(II) bromide	-220.9									79.5		
Br₂Cr	Chromium(II) bromide	-302.1											
Br₂Cu	Copper(II) bromide	-141.8											
Br₂Fe	Iron(II) bromide	-249.8			-238.1			140.6					
Br₂H₂Si	Dibromosilane									309.7			65.5
Br₂Hg	Mercury(II) bromide	-170.7			-153.1			172.0					
Br₂Hg₂	Mercury(I) bromide	-206.9			-181.1			218.0					
Br₂Mg	Magnesium bromide	-524.3			-503.8			117.2					
Br₂Mn	Manganese(II) bromide	-384.9											
Br₂Ni	Nickel(II) bromide	-212.1											
Br₂Pb	Lead(II) bromide	-278.7			-261.9			161.5			80.1		
Br₂Pt	Platinum(II) bromide	-82.0											
Br₂S₂	Sulfur bromide		-13.0										
Br₂Se	Selenium bromide			-21.0									
Br₂Sn	Tin(II) bromide	-243.5											
Br₂Sr	Strontium bromide	-717.6			-697.1			135.1			75.3		
Br₂Ti	Titanium(II) bromide	-402.0											
Br₂Zn	Zinc bromide	-328.7			-312.1			138.5					
Br₃ClSi	Tribromochlorosilane									377.1			95.3
Br₃Fe	Iron(III) bromide	-268.2											
Br₃Ga	Gallium(III) bromide	-386.6			-359.8			180.0					
Br₃HSi	Tribromosilane		-355.6	-317.6		-336.4	-328.5		248.1	348.6			80.8
Br₃In	Indium(III) bromide	-428.9											
Br₃OP	Phosphorus(V) oxybromide	-458.6		-282.0						359.8			89.9
Br₃P	Phosphorus(III) bromide		-184.5	-139.3		-175.7	-162.8		240.2	348.1			76.0
Br₃Pt	Platinum(III) bromide	-120.9											
Br₃Re	Rhenium(III) bromide	-167.0											
Br₃Ru	Ruthenium(III) bromide	-138.0											
Br₃Sb	Antimony(III) bromide	-259.4		-194.6	-239.3		-223.9	207.1		372.9			80.2
Br₃Sc	Scandium bromide	-743.1											
Br₃Ti	Titanium(III) bromide	-548.5			-523.8			176.6			101.7		
Br₄Ge	Germanium(IV) bromide		-347.7	-300.0		-331.4	-318.0		280.7	396.2			101.8
Br₄Pa	Protactinium(IV) bromide	-824.0			-787.8			234.0					
Br₄Pt	Platinum(IV) bromide	-156.5											
Br₄Si	Tetrabromosilane		-457.3	-415.5		-443.9	-431.8		277.8	377.9			97.1
Br₄Sn	Tin(IV) bromide	-377.4		-314.6	-350.2		-331.4	264.4		411.9			103.4
Br₄Te	Tellurium tetrabromide	-190.4											
Br₄Ti	Titanium(IV) bromide	-616.7		-549.4	-589.5		-568.2	243.5		398.4	131.5		100.8
Br₄V	Vanadium(IV) bromide			-336.8									
Br₄Zr	Zirconium(IV) bromide	-760.7											
Br₅P	Phosphorus(V) bromide	-269.9											
Br₅Ta	Tantalum(V) bromide	-598.3											
Br₆W	Tungsten(VI) bromide	-348.5											

STANDARD THERMODYNAMIC PROPERTIES OF CHEMICAL SUBSTANCES (continued)

Molecular formula	Name	$\Delta_f H°$/kJ mol⁻¹ Crys.	Liq.	Gas	$\Delta_f G°$/kJ mol⁻¹ Crys.	Liq.	Gas	$S°$/J mol⁻¹ K⁻¹ Crys.	Liq.	Gas	C_p/J mol⁻¹ K⁻¹ Crys.	Liq.	Gas
Ca	Calcium	0.0		177.8			144.0	41.6		154.9	25.9		20.8
$CaCl_2$	Calcium chloride	-795.4			-748.8			108.4			72.9		
CaF_2	Calcium fluoride	-1228.0			-1175.6			68.5			67.0		
CaH_2	Calcium hydride	-181.5			-142.5			41.4			41.0		
CaH_2O_2	Calcium hydroxide	-985.2			-897.5			83.4			87.5		
CaI_2	Calcium iodide	-533.5			-528.9			142.0					
CaN_2O_6	Calcium nitrate	-938.2			-742.8			193.2			149.4		
CaO	Calcium oxide	-634.9			-603.3			38.1			42.0		
CaO_4S	Calcium sulfate	-1434.5			-1322.0			106.5			99.7		
CaS	Calcium sulfide	-482.4			-477.4			56.5			47.4		
$Ca_3O_8P_2$	Calcium phosphate	-4120.8			-3884.7			236.0			227.8		
Cd	Cadmium	0.0		111.8				51.8		167.7	26.0		20.8
$CdCl_2$	Cadmium chloride	-391.5			-343.9			115.3			74.7		
CdF_2	Cadmium fluoride	-700.4			-647.7			77.4					
CdH_2O_2	Cadmium hydroxide	-560.7			-473.6			96.0					
CdI_2	Cadmium iodide	-203.3			-201.4			161.1			80.0		
CdO	Cadmium oxide	-258.4			-228.7			54.8			43.4		
CdO_4S	Cadmium sulfate	-933.3			-822.7			123.0			99.6		
CdS	Cadmium sulfide	-161.9			-156.5			64.9					
CdTe	Cadmium telluride	-92.5			-92.0			100.0					
Ce	Cerium (γ, fcc)	0.0		423.0			385.0	72.0		191.8	26.9		23.1
$CeCl_3$	Cerium(III) chloride	-1060.5			-984.8			151.0			87.4		
CeI_3	Cerium(III) iodide	-669.3											
CeO_2	Cerium(IV) oxide	-1088.7			-1024.6			62.3			61.6		
CeS	Cerium(II) sulfide	-459.4			-451.5			78.2			50.0		
Ce_2O_3	Cerium(III) oxide	-1796.2			-1706.2			150.6			114.6		
Cf	Californium	0.0											
Cl	Chlorine (atomic)			121.3			105.3			165.2			21.8
ClCs	Cesium chloride	-443.0			-414.5			101.2			52.5		
$ClCsO_4$	Cesium perchlorate	-443.1			-314.3			175.1			108.3		
ClCu	Copper(I) chloride	-137.2			-119.9			86.2			48.5		
ClF	Chlorine fluoride			-50.3			-51.8			217.9			32.1
$ClFO_3$	Perchloryl fluoride			-23.8			48.2			279.0			64.9
ClF_3	Chlorine trifluoride		-189.5	-163.2			-123.0			281.6			63.9
ClF_5S	Sulfur chloride pentafluoride		-1065.7										
ClGe	Germanium monochloride			155.2			124.2			247.0			36.9
$ClGeH_3$	Chlorogermane									263.7			54.7
ClH	Hydrogen chloride			-92.3			-95.3			186.9			29.1
ClHO	Hypochlorous acid			-78.7			-66.1			236.7			37.2
$ClHO_4$	Perchloric acid		-40.6										
ClH_3Si	Chlorosilane									250.7			51.0
ClH_4N	Ammonium chloride	-314.4			-202.9			94.6			84.1		
ClH_4NO_4	Ammonium perchlorate	-295.3			-88.8			186.2					
ClH_4P	Phosphonium chloride	-145.2											
ClI	Iodine chloride		-23.9	17.8		-13.6	-5.5		135.1	247.6			35.6

STANDARD THERMODYNAMIC PROPERTIES OF CHEMICAL SUBSTANCES (continued)

Molecular formula	Name	$\Delta_f H°$/kJ mol⁻¹ Crys.	Liq.	Gas	$\Delta_f G°$/kJ mol⁻¹ Crys.	Liq.	Gas	$S°$/J mol⁻¹ K⁻¹ Crys.	Liq.	Gas	C_p/J mol⁻¹ K⁻¹ Crys.	Liq.	Gas
ClIn	Indium(I) chloride	-186.2		-75.0			-233.3			239.1			36.5
ClK	Potassium chloride	-436.5		-214.6	-408.5			82.6			51.3		
ClKO₃	Potassium chlorate	-397.7			-296.3			143.1			100.3		
ClKO₄	Potassium perchlorate	-432.8			-303.1			151.0			112.4		
ClLi	Lithium chloride	-408.6			-384.4			59.3			48.0		
ClLiO₄	Lithium perchlorate	-381.0											
ClNO	Nitrosyl chloride			51.7			66.1			261.7			44.7
ClNO₂	Nitryl chloride			12.6			54.4			272.2			53.2
ClNa	Sodium chloride	-411.2			-384.1			72.1			50.5		
ClNaO₂	Sodium chlorite	-307.0											
ClNaO₃	Sodium chlorate	-365.8			-262.3			123.4					
ClNaO₄	Sodium perchlorate	-383.3			-254.9			142.3					
ClO	Chlorine monoxide			101.8			98.1			226.6			31.5
ClOV	Vanadyl chloride	-607.0			-556.0			75.0					
ClO₂	Chlorine dioxide			102.5			120.5			256.8			42.0
ClO₂	Chlorine superoxide (ClOO)			89.1			105.0			263.7			46.0
ClO₄Rb	Rubidium perchlorate	-437.2			-306.9			161.1			52.4		
ClRb	Rubidium chloride	-435.4			-407.8			95.9					
ClSi	Chlorosilylidyne			189.9									36.9
ClTl	Thallium(I) chloride	-204.1		-67.8	-184.9			111.3			50.9		
Cl₂	Chlorine			0.0						223.1			33.9
Cl₂Co	Cobalt(II) chloride	-312.5			-269.8			109.2			78.5		
Cl₂Cr	Chromium(II) chloride	-395.4			-356.0			115.3			71.2		
Cl₂CrO₂	Chromyl chloride		-579.5	-538.1		-510.8	-501.6		221.8	329.8			84.5
Cl₂Cu	Copper(II) chloride	-220.1			-175.7			108.1			71.9		
Cl₂Fe	Iron(II) chloride	-341.8			-302.3			118.0			76.7		
Cl₂H₂Si	Dichlorosilane									285.7			60.5
Cl₂Hg	Mercury(II) chloride	-224.3			-178.6			146.0					
Cl₂Hg₂	Mercury(I) chloride	-265.4			-210.7			191.6					
Cl₂Mg	Magnesium chloride	-641.3			-591.8			89.6			71.4		
Cl₂Mn	Manganese(II) chloride	-481.3			-440.5			118.2			72.9		
Cl₂Ni	Nickel(II) chloride	-305.3			-259.0			97.7			71.7		
Cl₂O	Chlorine oxide			80.3			97.9			266.2			45.4
Cl₂OS	Thionyl chloride		-245.6	-212.5			-198.3			309.8		121.0	66.5
Cl₂O₂S	Sulfuryl chloride		-394.1	-364.0			-320.0			311.9		134.0	77.0
Cl₂O₂U	Uranyl chloride	-1243.9			-1146.4			150.5			107.9		
Cl₂Pb	Lead(II) chloride	-359.4			-314.1			136.0					
Cl₂Pt	Platinum(II) chloride	-123.4											
Cl₂S	Sulfur dichloride		-50.0										
Cl₂S₂	Sulfur chloride		-59.4										
Cl₂Sn	Tin(II) chloride	-325.1											
Cl₂Sr	Strontium chloride	-828.9			-781.1			114.9			75.6		
Cl₂Ti	Titanium(II) chloride	-513.8			-464.4			87.4			69.8		
Cl₂Zn	Zinc chloride	-415.1		-266.1	-369.4			111.5			71.3		
Cl₂Zr	Zirconium(II) chloride	-502.0											

STANDARD THERMODYNAMIC PROPERTIES OF CHEMICAL SUBSTANCES (continued)

Molecular formula	Name	$\Delta_f H°$/kJ mol⁻¹ Crys.	Liq.	Gas	$\Delta_f G°$/kJ mol⁻¹ Crys.	Liq.	Gas	$S°$/J mol⁻¹ K⁻¹ Crys.	Liq.	Gas	C_p/J mol⁻¹ K⁻¹ Crys.	Liq.	Gas
Cl₃Cr	Chromium(III) chloride	-556.5			-486.1			123.0			91.8		
Cl₃Dy	Dysprosium(III) chloride	-1000.0									100.0		
Cl₃Er	Erbium chloride	-998.7											
Cl₃Eu	Europium(III) chloride	-936.0									96.7		
Cl₃Fe	Iron(II) chloride	-399.5			-334.0			142.3					
Cl₃Ga	Gallium(III) chloride	-524.7			-454.8			142.0					
Cl₃Gd	Gadolinium(III) chloride	-1008.0									88.0		
Cl₃HSi	Trichlorosilane		-539.3	-513.0		-482.5	-482.0		227.6	313.9			75.8
Cl₃Ho	Holmium chloride	-1005.4									88.0		
Cl₃In	Indium(III) chloride	-537.2		-374.0									
Cl₃Ir	Iridium(III) chloride	-245.6											
Cl₃La	Lanthanum chloride	-1072.2									108.8		
Cl₃Lu	Lutetium chloride	-945.6		-649.0									
Cl₃N	Nitrogen trichloride		230.0										
Cl₃Nd	Neodymium chloride	-1041.0									113.0		
Cl₃OP	Phosphorus(V) oxychloride		-597.1	-558.5		-520.8	-512.9		222.5	325.5		138.8	84.9
Cl₃OV	Vanadyl trichloride		-734.7	-695.6		-668.5	-659.3		244.3	344.3			89.9
Cl₃Os	Osmium(III) chloride	-190.4											
Cl₃P	Phosphorus(III) chloride		-319.7	-287.0		-272.3	-267.8		217.1	311.8			71.8
Cl₃Pr	Praseodymium chloride	-1056.9									100.0		
Cl₃Pt	Platinum(III) chloride	-182.0			-188.0			123.8			92.4		
Cl₃Re	Rhenium(III) chloride	-264.0											
Cl₃Rh	Rhodium(III) chloride	-299.2											
Cl₃Ru	Ruthenium(III) chloride	-205.0											
Cl₃Sb	Antimony(III) chloride	-382.2			-323.7			184.1			107.9		
Cl₃Sc	Scandium chloride	-925.1											
Cl₃Sm	Samarium(III) chloride	-1025.9											
Cl₃Tb	Terbium chloride	-997.0			-653.5			139.7			97.2		
Cl₃Ti	Titanium(III) chloride	-720.9											
Cl₃Tl	Thallium(III) chloride	-315.1											
Cl₃Tm	Thulium chloride	-986.6											
Cl₃U	Uranium(III) chloride	-866.5			-799.1			159.0			102.5		
Cl₃V	Vanadium(III) chloride	-580.7			-511.2			131.0			93.2		
Cl₃Y	Yttrium chloride	-1000.0		-750.2									75.0
Cl₃Yb	Ytterbium(III) chloride	-959.8											
Cl₄Ge	Germanium(IV) chloride		-531.8	-495.8		-462.7	-457.3		245.6	347.7			96.1
Cl₄Hf	Hafnium(IV) chloride	-990.4		-884.5	-901.3			190.8			120.5		
Cl₄Pa	Protactinium(IV) chloride	-1043.0			-953.0			192.0					
Cl₄Pb	Lead(IV) chloride		-329.3										
Cl₄Pt	Platinum(IV) chloride	-231.8											
Cl₄Si	Tetrachlorosilane		-687.0	-657.0		-619.8	-617.0		239.7	330.7		145.3	90.3
Cl₄Sn	Tin(IV) chloride		-511.3	-471.5		-440.1	-432.2		258.6	365.8		165.3	98.3
Cl₄Te	Tellurium tetrachloride	-326.4									138.5		
Cl₄Th	Thorium(IV) chloride	-1186.2		-964.4	-1094.1		-932.0	190.4		390.7	120.3		107.5
Cl₄Ti	Titanium(IV) chloride		-804.2	-763.2		-737.2	-726.3		252.3	353.2		145.2	95.4

STANDARD THERMODYNAMIC PROPERTIES OF CHEMICAL SUBSTANCES (continued)

Molecular formula	Name	$\Delta_f H°$/kJ mol⁻¹			$\Delta_f G°$/kJ mol⁻¹			S°/J mol⁻¹ K⁻¹			C_p/J mol⁻¹ K⁻¹		
		Crys.	Liq.	Gas	Crys.	Liq.	Gas	Crys.	Liq.	Gas	Crys.	Liq.	Gas
Cl₄U	Uranium(IV) chloride	-1019.2		-809.6	-930.0		-786.6	197.1		419.0	122.0		
Cl₄V	Vanadium(IV) chloride		-569.4	-525.5		-503.7	-492.0		255.0	362.4			96.2
Cl₄Zr	Zirconium(IV) chloride	-980.5			-889.9			181.6			119.8		
Cl₅Nb	Niobium(V) chloride	-797.5		-703.7	-683.2		-646.0	210.5		400.6	148.1		120.8
Cl₅P	Phosphorus(V) chloride	-443.5		-374.9			-305.0			364.6			112.8
Cl₅Pa	Protactinium(V) chloride	-1145.0			-1034.0			238.0					
Cl₅Ta	Tantalum(V) chloride	-859.0											
Cl₆U	Uranium(VI) chloride	-1092.0		-1013.0	-962.0		-928.0	285.8		431.0	175.7		
Cl₆W	Tungsten(VI) chloride	-602.5		-513.8									
Cm	Curium	0.0											
Co	Cobalt	0.0		424.7			380.3	30.0		179.5	24.8		23.0
CoF₂	Cobalt(II) fluoride	-692.0			-647.2			82.0			68.8		
CoH₂O₂	Cobalt(II) hydroxide	-539.7			-454.3			79.0					
CoI₂	Cobalt(II) iodide	-88.7											
CoN₂O₆	Cobalt(II) nitrate	-420.5											
CoO	Cobalt(II) oxide	-237.9			-214.2			53.0			55.2		
CoO₄S	Cobalt(II) sulfate	-888.3			-782.3			118.0					
CoS	Cobalt(II) sulfide	-82.8											
Co₂S₃	Cobalt(III) sulfide	-147.3											
Co₃O₄	Cobalt(II,III) oxide	-891.0			-774.0			102.5			123.4		
Cr	Chromium	0.0		396.6			351.8	23.8		174.5	23.4		20.8
CrF₂	Chromium(II) fluoride	-778.0											
CrF₃	Chromium(III) fluoride	-1159.0			-1088.0			93.9			78.7		
CrI₂	Chromium(II) iodide	-156.9											
CrI₃	Chromium(III) iodide	-205.0											
CrO₂	Chromium(IV) oxide	-598.0											
CrO₄Pb	Lead(II) chromate	-930.9											
Cr₂FeO₄	Chromium iron oxide	-1444.7			-1343.8			146.0			133.6		
Cr₂O₃	Chromium(III) oxide	-1139.7			-1058.1			81.2			118.7		
Cr₃O₄	Chromium(II,III) oxide	-1531.0											
Cs	Cesium	0.0		76.5			49.6	85.2		175.6	32.2		20.8
CsF	Cesium fluoride	-553.5			-525.5			92.8			51.1		
CsF₂H	Cesium hydrogen fluoride	-923.8			-858.9			135.2			87.3		
CsH	Cesium hydride	-54.2											
CsHO	Cesium hydroxide	-417.2											
CsHO₄S	Cesium hydrogen sulfate	-1158.1											
CsH₂N	Cesium amide	-118.4											
CsI	Cesium iodide	-346.6			-340.6			123.1			52.8		
CsNO₃	Cesium nitrate	-506.0			-406.5			155.2					
CsO₂	Cesium superoxide	-286.2											
Cs₂O	Cesium oxide	-345.8			-308.1			146.9			76.0		
Cs₂O₃S	Cesium sulfite	-1134.7											
Cs₂O₄S	Cesium sulfate	-1443.0			-1323.6			211.9			134.9		
Cs₂S	Cesium sulfide	-359.8											
Cu	Copper	0.0		337.4			297.7	33.2		166.4	24.4		20.8

Molecular formula	Name	$\Delta_f H°$/kJ mol⁻¹			$\Delta_f G°$/kJ mol⁻¹			$S°$/J mol⁻¹ K⁻¹			C_p/J mol⁻¹ K⁻¹		
		Crys.	Liq.	Gas	Crys.	Liq.	Gas	Crys.	Liq.	Gas	Crys.	Liq.	Gas
CuF₂	Copper(II) fluoride	-542.7											
CuH₂O₂	Copper(II) hydroxide	-449.8						96.7			54.1		
CuI	Copper(I) iodide	-67.8			-69.5								
CuN₂O₆	Copper(II) nitrate	-302.9											
CuO	Copper(II) oxide	-157.3			-129.7			42.6			42.3		
CuO₄S	Copper(II) sulfate	-771.4			-662.2			109.2					
CuO₄W	Copper(II) tungstate	-1105.0											
CuS	Copper(II) sulfide	-53.1			-53.6			66.5			47.8		
CuSe	Copper(II) selenide	-39.5											
Cu₂	Dicopper			484.2			431.9			241.6			36.6
Cu₂O	Copper(I) oxide	-168.6			-146.0			93.1			63.6		
Cu₂S	Copper(I) sulfide	-79.5			-86.2			120.9			76.3		
Dy	Dysprosium	0.0		290.4			254.4	75.6		196.6	27.7		20.8
Dy₂O₃	Dysprosium(III) oxide	-1863.1			-1771.5			149.8			116.3		
Er	Erbium	0.0		317.1			280.7	73.2		195.6	28.1		20.8
ErF₃	Erbium fluoride	-1711.0											
Er₂O₃	Erbium oxide	-1897.9			-1808.7			155.6			108.5		
Es	Einsteinium	0.0											
Eu	Europium	0.0		175.3			142.2	77.8		188.8	27.7		
Eu₂O₃	Europium(III) oxide	-1651.4			-1556.8			146.0			122.2		
Eu₃O₄	Europium(II,III) oxide	-2272.0			-2142.0			205.0					
F	Fluorine (atomic)			79.4			62.3			158.8			22.7
FGa	Gallium monofluoride			-251.9									33.3
FGe	Germanium monofluoride			-33.4									34.7
FGeH₃	Fluorogermane									252.8			51.6
FH	Hydrogen fluoride		-299.8	-273.3			-275.4			173.8			
FH₃Si	Fluorosilane									238.4			47.4
FH₄N	Ammonium fluoride	-464.0			-348.7			72.0			65.3		
FI	Iodine fluoride			-95.7			-118.5			236.2			33.4
FIn	Indium monofluoride			-203.4									
FK	Potassium fluoride	-567.3			-537.8			66.6			49.0		
FLi	Lithium fluoride	-616.0			-587.7			35.7			41.6		
FNO	Nitrosyl fluoride			-66.5			-51.0			248.1			41.3
FNO₂	Nitryl fluoride									260.4			49.8
FNS	Thionitrosyl fluoride (NSF)									259.8			44.1
FNa	Sodium fluoride	-576.6			-546.3			51.1			46.9		
FO	Fluorine monoxide			109.0			105.0			216.8			30.5
FRb	Rubidium fluoride	-557.7											
FSi	Fluorosilylidyne			7.1			-24.3			225.8			32.6
FTl	Thallium(I) fluoride	-324.7		-182.4									
F₂	Fluorine			0.0						202.8			31.3
F₂Fe	Iron(II) fluoride	-711.3			-668.6			87.0			68.1		
F₂HK	Potassium hydrogen fluoride	-927.7			-859.7			104.3			76.9		
F₂HN	Difluoramine									252.8			43.4
F₂HNa	Sodium hydrogen fluoride	-920.3			-852.2			90.9			75.0		

Molecular formula	Name	$\Delta_f H°$/kJ mol⁻¹ Crys.	Liq.	Gas	$\Delta_f G°$/kJ mol⁻¹ Crys.	Liq.	Gas	$S°$/J mol⁻¹ K⁻¹ Crys.	Liq.	Gas	C_p/J mol⁻¹ K⁻¹ Crys.	Liq.	Gas
F_2HRb	Rubidium hydrogen fluoride	-922.6			-855.6			120.1			79.4		
F_2Mg	Magnesium fluoride	-1124.2			-1071.1			57.2			61.6		
F_2N	Difluoroamidogen			43.1			57.8			249.9			41.0
F_2N_2	cis-Difluorodiazine			69.5									
F_2N_2	trans-Difluorodiazine			82.0									
F_2Ni	Nickel(II) fluoride	-651.4			-604.1			73.6			64.1		
F_2O	Fluorine monoxide			24.7			41.9			247.4			43.3
F_2OS	Thionyl fluoride									278.7			56.8
F_2O_2	Fluorine dioxide			18.0									
F_2O_2S	Sulfuryl fluoride									284.0			66.0
F_2O_2U	Uranyl fluoride	-1653.5			-1557.4			135.6			103.2		
F_2Pb	Lead(II) fluoride	-664.0			-617.1			110.5					
F_2Si	Difluorosilylene			-619.0			-628.0			252.7			43.9
F_2Sr	Strontium fluoride	-1216.3			-1164.8			82.1			70.0		
F_2Zn	Zinc fluoride	-764.4			-713.3			73.7			65.7		
F_3Ga	Gallium(III) fluoride	-1163.0			-1085.3			84.0					
F_3Gd	Gadolinium(III) fluoride			-1297.0									
F_3HSi	Trifluorosilane									271.9			60.5
F_3Ho	Holmium fluoride	-1707.0											
F_3N	Nitrogen trifluoride			-132.1			-90.6			260.8			53.4
F_3Nd	Neodymium fluoride	-1657.0											
F_3OP	Phosphorus(V) oxyfluoride			-1254.3			-1205.8			285.4			68.8
F_3P	Phosphorus(III) fluoride			-958.4			-936.9			273.1			58.7
F_3Sb	Antimony(III) fluoride	-915.5											
F_3Sc	Scandium fluoride	-1629.2		-1247.0	-1555.6		-1234.0	92.0		300.5			67.8
F_3Sm	Samarium(III) fluoride	-1778.0											
F_3Th	Thorium(III) fluoride			-1166.1			-1160.6			339.2			73.3
F_3U	Uranium(III) fluoride	-1502.1		-1058.5	-1433.4		-1051.9	123.4		331.9	95.1		74.3
F_3Y	Yttrium fluoride	-1718.8		-1288.7	-1644.7		-1277.8	100.0		311.8			70.3
F_4Ge	Germanium(IV) fluoride			-1190.2			-1150.0			301.9			
F_4Hf	Hafnium fluoride	-1930.5		-1669.8	-1830.4			113.0					
F_4N_2	Tetrafluorohydrazine			-8.4			79.9			301.2			79.2
F_4Pb	Lead(IV) fluoride	-941.8											
F_4S	Sulfur tetrafluoride			-763.2			-722.0			299.6			77.6
F_4Si	Tetrafluorosilane			-1615.0			-1572.8			282.8			73.6
F_4Th	Thorium(IV) fluoride	-2097.8		-1759.0	-2003.4		-1724.0	142.0		341.7	110.7		93.0
F_4U	Uranium(IV) fluoride	-1914.2		-1598.7	-1823.3		-1572.7	151.7		368.0	116.0		91.2
F_4V	Vanadium(IV) fluoride	-1403.3											
F_4Xe	Xenon tetrafluoride	-261.5											
F_4Zr	Zirconium(IV) fluoride	-1911.3			-1809.9			104.6			103.7		
F_5I	Iodine pentafluoride		-864.8	-822.5			-751.7			327.7			99.2
F_5Nb	Niobium(V) fluoride	-1813.8		-1739.7	-1699.0		-1673.6	160.2		321.9	134.7		97.1
F_5P	Phosphorus(V) fluoride			-1594.4			-1520.7			300.8			84.8
F_5Ta	Tantalum(V) fluoride	-1903.6											
F_5V	Vanadium(V) fluoride		-1480.3	-1433.9		-1373.1	-1369.8		175.7	320.9			98.6

Molecular formula	Name	ΔfH°/kJ mol⁻¹ Crys.	Liq.	Gas	ΔfG°/kJ mol⁻¹ Crys.	Liq.	Gas	S°/J mol⁻¹ K⁻¹ Crys.	Liq.	Gas	Cp/J mol⁻¹ K⁻¹ Crys.	Liq.	Gas
$F_6H_8N_2Si$	Ammonium hexafluorosilicate	-2681.7			-2365.3			280.2			228.1		
F_6Ir	Iridium(VI) fluoride	-579.7		-544.0	-461.6		-460.0	247.7		357.8			121.1
F_6K_2Si	Potassium hexafluorosilicate	-2956.0			-2798.6			226.0					
F_6Mo	Molybdenum(VI) fluoride		-1585.5	-1557.7		-1473.0	-1472.2		259.7	350.5		169.8	120.6
F_6Na_2Si	Sodium hexafluorosilicate	-2909.6			-2754.2			207.1			187.1		
F_6Os	Osmium(VI) fluoride							246.0		358.1			120.8
F_6Pt	Platinum(VI) fluoride							235.6		348.3			122.8
F_6S	Sulfur hexafluoride			-1220.5			-1116.5			291.5			97.0
F_6Se	Selenium hexafluoride			-1117.0			-1017.0			313.9			110.5
F_6Te	Tellurium hexafluoride			-1318.0									
F_6U	Uranium(VI) fluoride	-2197.0	-1747.7	-2147.4	-2068.5		-2063.7	227.6		377.9	166.8		129.6
F_6W	Tungsten(VI) fluoride			-1721.7		-1631.4	-1632.1		251.5	341.1			119.0
Fe	Iron	0.0		416.3			370.7	27.3		180.5	25.1		25.7
FeI_2	Iron(II) iodide	-113.0											
FeI_3	Iron(III) iodide			71.0									
$FeMoO_4$	Iron(II) molybdate	-1075.0			-975.0			129.3			118.5		
FeO	Iron(II) oxide	-272.0											
FeO_4S	Iron(II) sulfate	-928.4			-820.8			107.5			100.6		
FeO_4W	Iron(II) tungstate	-1155.0			-1054.0			131.8			114.6		
FeS	Iron(II) sulfide	-100.0			-100.4			60.3			50.5		
FeS_2	Iron disulfide	-178.2			-166.9			52.9			62.2		
Fe_2O_3	Iron(III) oxide	-824.2			-742.2			87.4			103.9		
Fe_2O_4Si	Iron(II) orthosilicate	-1479.9			-1379.0			145.2			132.9		
Fe_3O_4	Iron(II,III) oxide	-1118.4			-1015.4			146.4			143.4		
Fm	Fermium	0.0											
Fr	Francium	0.0						95.4					
Ga	Gallium	0.0	5.6	272.0	0.0		233.7	40.8		169.0	26.1		25.3
GaH_3O_3	Gallium(III) hydroxide	-964.4			-831.3			100.0			100.0		
GaI_3	Gallium(III) iodide	-238.9						205.0					
GaN	Gallium nitride	-110.5											
GaO	Gallium monoxide			279.5			253.5			231.1			32.1
GaP	Gallium phosphide	-88.0											
$GaSb$	Gallium antimonide	-41.8			-38.9			76.1			48.5		
Ga_2	Digallium			438.5									
Ga_2O	Gallium suboxide	-356.0											
Ga_2O_3	Gallium(III) oxide	-1089.1			-998.3			85.0			92.1		
Gd	Gadolinium	0.0		397.5			359.8	68.1		194.3	37.0		27.5
Gd_2O_3	Gadolinium(III) oxide	-1819.6									106.7		
Ge	Germanium	0.0		372.0			331.2	31.1		167.9	23.3		30.7
GeH_3I	Iodogermane									283.2			57.5
GeH_4	Germane			90.8			113.4			217.1			45.0
GeI_4	Germanium(IV) iodide	-141.8		-56.9	-144.3		-106.3	271.1		428.9			104.1
GeO	Germanium(II) oxide	-261.9		-46.2	-237.2		-73.2	50.0		224.3			30.9
GeO_2	Germanium(IV) oxide	-580.0			-521.4			39.7			52.1		

STANDARD THERMODYNAMIC PROPERTIES OF CHEMICAL SUBSTANCES (continued)

Molecular formula	Name	$\Delta_f H°$/kJ mol⁻¹			$\Delta_f G°$/kJ mol⁻¹			$S°$/J mol⁻¹ K⁻¹			C_p/J mol⁻¹ K⁻¹		
		Crys.	Liq.	Gas	Crys.	Liq.	Gas	Crys.	Liq.	Gas	Crys.	Liq.	Gas
GeP	Germanium phosphide	-21.0			-17.0			63.0					
GeS	Germanium(II) sulfide	-69.0		92.0	-71.5		42.0	71.0		234.0			33.7
GeTe	Germanium(II) telluride	20.0											
Ge₂	Digermanium			473.1			416.3			252.8			35.6
Ge₂H₆	Digermane		137.3	162.3									
Ge₃H₈	Trigermane		193.7	226.8									
H	Hydrogen (atomic)			218.0			203.3			114.7			20.8
HI	Hydrogen iodide			26.5			1.7			206.6			29.2
HIO₃	Iodic acid	-230.1											
HK	Potassium hydride	-57.7											
HKO	Potassium hydroxide	-424.6		-228.0	-378.7		-229.7	78.9		238.3	64.9		49.2
HKO₄S	Potassium hydrogen sulfate	-1160.6			-1031.3			138.1					
HLi	Lithium hydride	-90.5			-68.3			20.0			27.9		
HLiO	Lithium hydroxide	-484.9			-439.0			42.8			49.7		
HN	Imidogen			351.5			345.6			181.2			29.2
HNO₂	Nitrous acid			-79.5			-46.0			254.1			45.6
HNO₃	Nitric acid		-174.1	-133.9		-80.7	-73.5		155.6	266.9		109.9	54.1
HN₃	Hydrazoic acid		264.0	294.1		327.3	328.1		140.6	239.0			43.7
HNa	Sodium hydride	-56.3			-33.5			40.0			36.4		
HNaO	Sodium hydroxide	-425.6			-379.5			64.5			59.5		
HNaO₄S	Sodium hydrogen sulfate	-1125.5			-992.8			113.0					
HNa₂O₄P	Sodium hydrogen phosphate	-1748.1			-1608.2			150.5			135.3		
HO	Hydroxyl			39.0			34.2			183.7			29.9
HORb	Rubidium hydroxide	-418.2			-195.8			88.0					
HOTl	Thallium(I) hydroxide	-238.9											
HO₂	Hydroperoxy			10.5			22.6			229.0			34.9
HO₃P	Metaphosphoric acid	-948.5											
HO₄RbS	Rubidium hydrogen sulfate	-1159.0											
HO₄Re	Perrhenic acid	-762.3			-656.4			158.2					
HRb	Rubidium hydride	-52.3											
HS	Mercapto			142.7			113.3			195.7			32.3
HSi	Silylidyne			361.0									
HTa₂	Tantalum hydride	-32.6			-69.0			79.1			90.8		
H₂	Hydrogen			0.0						130.7			28.8
H₂KN	Potassium amide	-128.9											
H₂KO₄P	Potassium dihydrogen phosphate	-1568.3			-1415.9			134.9			116.6		
H₂LiN	Lithium amide	-179.5											
H₂Mg	Magnesium hydride	-75.3			-35.9			31.1			35.4		
H₂MgO₂	Magnesium hydroxide	-924.5			-833.5			63.2			77.0		
H₂N	Amidogen			184.9			194.6			195.0			33.9
H₂NNa	Sodium amide	-123.8			-64.0			76.9			66.2		
H₂NRb	Rubidium amide	-113.0											
H₂N₂O₂	Nitramide	-89.5											
H₂NiO₂	Nickel(II) hydroxide	-529.7			-447.2			88.0					

STANDARD THERMODYNAMIC PROPERTIES OF CHEMICAL SUBSTANCES (continued)

Molecular formula	Name	$\Delta_f H°$/kJ mol⁻¹ Crys.	Liq.	Gas	$\Delta_f G°$/kJ mol⁻¹ Crys.	Liq.	Gas	$S°$/J mol⁻¹ K⁻¹ Crys.	Liq.	Gas	C_p/J mol⁻¹ K⁻¹ Crys.	Liq.	Gas
H_2O	Water		-285.8	-241.8		-237.1	-228.6		70.0	188.8		75.3	33.6
H_2O_2	Hydrogen peroxide		-187.8	-136.3		-120.4	-105.6		109.6	232.7		89.1	43.1
H_2O_2Sn	Tin(II) hydroxide	-561.1			-491.6			155.0					
H_2O_2Sr	Strontium hydroxide	-959.0											
H_2O_2Zn	Zinc hydroxide	-641.9			-553.5			81.2					
H_2O_3Si	Metasilicic acid	-1188.7			-1092.4			134.0					
H_2O_4S	Sulfuric acid		-814.0			-690.0			156.9			138.9	
H_2O_4Se	Selenic acid	-530.1											
H_2S	Hydrogen sulfide			-20.6			-33.4			205.8			34.2
H_2S_2	Hydrogen disulfide		-18.1	15.5								84.1	51.5
H_2Se	Hydrogen selenide			29.7			15.9			219.0			34.7
H_2Sr	Strontium hydride	-180.3											
H_2Te	Hydrogen telluride			99.6									
H_2Th	Thorium hydride	-139.7			-100.0			50.7			36.7		
H_2Zr	Zirconium(II) hydride	-169.0			-128.8			35.0			31.0		
H_3ISi	Iodosilane												
H_3N	Ammonia			-45.9			-16.4			192.8			35.1
H_3NO	Hydroxylamine	-114.2											
H_3O_2P	Hypophosphorous acid	-604.6	-595.4										
H_3O_3P	Phosphorous acid	-964.4											
H_3O_4P	Phosphoric acid (orthophosphoric acid)	-1284.4	-1271.7		-1124.3	-1123.6		110.5	150.8		106.1	145.0	
H_3P	Phosphine			5.4			13.4			210.2			37.1
H_3Sb	Stibine			145.1			147.8			232.8			41.1
H_3U	Uranium(III) hydride	-127.2			-72.8			63.7			49.3		
H_4IN	Ammonium iodide	-201.4			-112.5			117.0					
H_4N_2	Hydrazine		50.6	95.4		149.3	159.4		121.2	238.5		98.9	48.4
$H_4N_2O_2$	Ammonium nitrite	-256.5											
$H_4N_2O_3$	Ammonium nitrate	-365.6			-183.9			151.1			139.3		
H_4N_4	Ammonium azide	115.5			274.2			112.5					
H_4O_4Si	Orthosilicic acid	-1481.1			-1332.9			192.0					
$H_4O_7P_2$	Pyrophosphoric acid	-2241.0	-2231.7										
H_4P_2	Diphosphine		-5.0	20.9									
H_4Si	Silane			34.3			56.9			204.6			42.8
H_4Sn	Stannane			162.8			188.3			227.7			49.0
H_5NO	Ammonium hydroxide		-361.2			-254.0			165.6			154.9	
H_5NO_3S	Ammonium hydrogen sulfite	-768.6											
H_5NO_4S	Ammonium hydrogen sulfate	-1027.0											
H_6Si_2	Disilane			80.3			127.3			272.7			80.8
$H_8N_2O_4S$	Ammonium sulfate	-1180.9			-901.7			220.1			187.5		
H_8Si_3	Trisilane		92.5	120.9									
$H_9N_2O_4P$	Diammonium hydrogen phosphate	-1566.9											
$H_{12}N_3O_4P$	Ammonium phosphate	-1671.9									188.0		
He	Helium			0.0						126.2			20.8

STANDARD THERMODYNAMIC PROPERTIES OF CHEMICAL SUBSTANCES (continued)

Molecular formula	Name	$\Delta_f H°$/kJ mol⁻¹ Crys.	Liq.	Gas	$\Delta_f G°$/kJ mol⁻¹ Crys.	Liq.	Gas	S°/J mol⁻¹ K⁻¹ Crys.	Liq.	Gas	C_p/J mol⁻¹ K⁻¹ Crys.	Liq.	Gas
Hf	Hafnium	0.0		619.2			576.5	43.6		186.9	25.7		20.8
HfO_2	Hafnium oxide	-1144.7			-1088.2			59.3			60.3		
Hg	Mercury		0.0	61.4			31.8		75.9	175.0		28.0	20.8
HgI_2	Mercury(II) iodide	-105.4			-101.7			180.0					
HgO	Mercury(II) oxide	-90.8			-58.5			70.3			44.1		
HgO_4S	Mercury(II) sulfate	-707.5											
HgS	Mercury(II) sulfide	-58.2			-50.6			82.4			48.4		
HgTe	Mercury(II) telluride	-42.0											
Hg_2	Dimercury			108.8			68.2			288.1			37.4
Hg_2I_2	Mercury(I) iodide	-121.3			-111.0			233.5			132.0		
Hg_2O_4S	Mercury(I) sulfate	-743.1			-625.8			200.7					
Ho	Holmium	0.0		300.8			264.8	75.3		195.6	27.2		20.8
Ho_2O_3	Holmium oxide	-1880.7			-1791.1			158.2			115.0		
I	Iodine (atomic)			106.8			70.2			180.8			20.8
IIn	Indium(I) iodide	-116.3			-120.5			130.0					
IK	Potassium iodide	-327.9		7.5	-324.9		-37.7	106.3		267.3	52.9		36.8
IKO_3	Potassium iodate	-501.4			-418.4			151.5			106.5		
IKO_4	Potassium periodate	-467.2			-361.4			175.7					
ILi	Lithium iodide	-270.4			-270.3			86.8			51.0		
INa	Sodium iodide	-287.8			-286.1			98.5			52.1		
$INaO_3$	Sodium iodate	-481.8			-323.0						92.0		
$INaO_4$	Sodium periodate	-429.3						163.0					
IO	Iodine monoxide			175.1			149.8			245.5			32.9
IRb	Rubidium iodide	-333.8			-328.9			118.4			53.2		
ITl	Thallium(I) iodide	-123.8			-125.4			127.6					
I_2	Iodine (rhombic)	0.0		62.4			19.3	116.1		260.7	54.4		36.9
I_2Mg	Magnesium iodide	-364.0			-358.2			129.7					
I_2Ni	Nickel(II) iodide	-78.2											
I_2Pb	Lead(II) iodide	-175.5			-173.6			174.9			77.4		
I_2Sn	Tin(II) iodide	-143.5											
I_2Sr	Strontium iodide	-558.1											
I_2Zn	Zinc iodide	-208.0			-209.0			161.1			81.6		
I_3In	Indium(III) iodide	-238.0		-120.5									
I_3La	Lanthanum iodide	-668.9											
I_3Lu	Lutetium iodide	-548.0											
I_3P	Phosphorus(III) iodide	-45.6											
I_3Ru	Ruthenium(III) iodide	-65.7								374.4			78.4
I_3Sb	Antimony(III) iodide	-100.4											
I_4Pt	Platinum(IV) iodide	-72.8											
I_4Si	Tetraiodosilane	-189.5											
I_4Sn	Tin(IV) iodide												
I_4Ti	Titanium(IV) iodide	-375.7		-277.8	-371.5			249.4		446.1	84.9		105.4
I_4V	Vanadium(IV) iodide			-122.6							125.7		
I_4Zr	Zirconium(IV) iodide	-481.6											
In	Indium	0.0		243.3			208.7	57.8		173.8	26.7		20.8

Molecular formula	Name	$\Delta_f H°$/kJ mol⁻¹			$\Delta_f G°$/kJ mol⁻¹			S°/J mol⁻¹ K⁻¹			C_p/J mol⁻¹ K⁻¹		
		Crys.	Liq.	Gas	Crys.	Liq.	Gas	Crys.	Liq.	Gas	Crys.	Liq.	Gas
InO	Indium monoxide			387.0			364.4			236.5			32.6
InP	Indium phosphide	-88.7			-77.0			59.8			45.4		
InS	Indium(II) sulfide	-138.1		238.0	-131.8			67.0			49.5		
InSb	Indium antimonide	-30.5		344.3	-25.5			86.2					
In₂	Diindium			380.9									
In₂O₃	Indium(III) oxide	-925.8			-830.7			104.2			92.0		
In₂S₃	Indium(III) sulfide	-427.0			-412.5			163.6			118.0		
Ir	Iridium	0.0		665.3			617.9	35.5		193.6	25.1		20.8
IrO₂	Iridium(IV) oxide	-274.1									57.3		
IrS₂	Iridium(IV) sulfide	-138.0											
Ir₂S₃	Iridium(III) sulfide	-234.0											
K	Potassium	0.0		89.0			60.5	64.7		160.3	29.6		20.8
KMnO₄	Potassium permanganate	-837.2			-737.6			171.7			117.6		
KNO₂	Potassium nitrite	-369.8			-306.6			152.1			107.4		
KNO₃	Potassium nitrate	-494.6			-394.9			133.1			96.4		
KNa	Potassium sodium		6.3										
KO₂	Potassium superoxide	-284.9			-239.4			116.7			77.5		
K₂	Dipotassium			123.7			87.5			249.7			37.9
K₂O	Potassium oxide	-361.5											
K₂O₂	Potassium peroxide	-494.1			-425.1			102.1					
K₂O₄S	Potassium sulfate	-1437.8			-1321.4			175.6			131.5		
K₂S	Potassium sulfide	-380.7			-364.0			105.0					
K₃O₄P	Potassium phosphate	-1950.2											
Kr	Krypton			0.0						164.1			20.8
La	Lanthanum	0.0		431.0			393.6	56.9		182.4	27.1		22.8
LaS	Lanthanum sulfide	-456.0			-451.5			73.2			59.0		
La₂O₃	Lanthanum oxide	-1793.7			-1705.8			127.3			108.8		
Li	Lithium	0.0		159.3			126.6	29.1		138.8	24.8		20.8
LiNO₂	Lithium nitrite	-372.4						96.0					
LiNO₃	Lithium nitrate	-483.1			-381.1			90.0					
Li₂	Dilithium			215.9			174.4			197.0			36.1
Li₂O	Lithium oxide	-597.9			-561.2			37.6			54.1		
Li₂O₂	Lithium peroxide	-634.3											
Li₂O₃Si	Lithium metasilicate	-1648.1			-1557.2			79.8			99.1		
Li₂O₄S	Lithium sulfate	-1436.5			-1321.7			115.1			117.6		
Li₂S	Lithium sulfide	-441.4											
Li₃O₄P	Lithium phosphate	-2095.8											
Lr	Lawrencium	0.0											
Lu	Lutetium	0.0		427.6			387.8	51.0		184.8	26.9		20.9
Lu₂O₃	Lutetium oxide	-1878.2			-1789.0			110.0			101.8		
Md	Mendelevium	0.0											
Mg	Magnesium	0.0		147.1			112.5	32.7		148.6	24.9		20.8
MgN₂O₆	Magnesium nitrate	-790.7			-589.4			164.0			141.9		
MgO	Magnesium oxide	-601.6			-569.3			27.0			37.2		
MgO₄S	Magnesium sulfate	-1284.9			-1170.6			91.6			96.5		

STANDARD THERMODYNAMIC PROPERTIES OF CHEMICAL SUBSTANCES (continued)

Molecular formula	Name	$\Delta_f H°$/kJ mol⁻¹ Crys.	Liq.	Gas	$\Delta_f G°$/kJ mol⁻¹ Crys.	Liq.	Gas	$S°$/J mol⁻¹ K⁻¹ Crys.	Liq.	Gas	C_p/J mol⁻¹ K⁻¹ Crys.	Liq.	Gas
MgO₄Se	Magnesium selenate	-968.5											
MgS	Magnesium sulfide	-346.0			-341.8			50.3			45.6		
Mg₂	Dimagnesium			287.7									20.8
Mg₂O₄Si	Magnesium orthosilicate	-2174.0			-2055.1			95.1			118.5		
Mn	Manganese	0.0		280.7			238.5	32.0		173.7	26.3		20.8
MnN₂O₆	Manganese(II) nitrate	-576.3											
MnNa₂O₄	Sodium permanganate	-1156.0											
MnO	Manganese(II) oxide	-385.2			-362.9			59.7			45.4		
MnO₂	Manganese(IV) oxide	-520.0			-465.1			53.1			54.1		
MnO₃Si	Manganese(II) metasilicate	-1320.9			-1240.5			89.1			86.4		
MnS	Manganese(II) sulfide	-214.2			-218.4			78.2			50.0		
MnSe	Manganese(II) selenide	-106.7			-111.7			90.8			51.0		
Mn₂O₃	Manganese(III) oxide	-959.0			-881.1			110.5			107.7		
Mn₂O₄Si	Manganese(II) orthosilicate	-1730.5			-1632.1			163.2			129.9		
Mn₃O₄	Manganese(II,III) oxide	-1387.8			-1283.2			155.6			139.7		
Mo	Molybdenum	0.0		658.1			612.5	28.7		182.0	24.1		20.8
MoNa₂O₄	Sodium molybdate	-1468.1			-1354.3			159.7			141.7		
MoO₂	Molybdenum(IV) oxide	-588.9			-533.0			46.3			56.0		
MoO₃	Molybdenum(VI) oxide	-745.1			-668.0			77.7			75.0		
MoO₄Pb	Lead(II) molybdate	-1051.9			-951.4			166.1			119.7		
MoS₂	Molybdenum(IV) sulfide	-235.1			-225.9			62.6			63.6		
N	Nitrogen (atomic)			472.7			455.5			153.3			20.8
NNaO₂	Sodium nitrite	-358.7			-284.6			103.8					
NNaO₃	Sodium nitrate	-467.9			-367.0			116.5			92.9		
NO	Nitric oxide			91.3			87.6			210.8			29.9
NO₂	Nitrogen dioxide			33.2			51.3			240.1			37.2
NO₂Rb	Rubidium nitrite	-367.4			-306.2			172.0					
NO₃Rb	Rubidium nitrate	-495.1			-395.8			147.3			102.1		
NO₃Tl	Thallium(I) nitrate	-243.9			-152.4			160.7			99.5		
NP	Phosphorus nitride	-63.0											
N₂	Nitrogen			0.0						191.6			29.1
N₂O	Nitrous oxide			81.6			103.7			220.0			38.6
N₂O₃	Nitrogen trioxide		50.3	86.6			142.4			314.7			72.7
N₂O₄	Nitrogen tetroxide		-19.5	11.1		97.5	99.8		209.2	304.4		142.7	79.2
N₂O₄Sr	Strontium nitrite	-762.3											
N₂O₅	Nitrogen pentoxide	-43.1		13.3	113.9		117.1	178.2		355.7	143.1		
N₂O₆Pb	Lead(II) nitrate	-451.9											
N₂O₆Ra	Radium nitrate	-992.0			-796.1			222.0					
N₂O₆Sr	Strontium nitrate	-978.2			-780.0			194.6			149.9		
N₂O₆Zn	Zinc nitrate	-483.7											
N₃Na	Sodium azide	21.7			93.8			96.9			76.6		
N₄Si₃	Silicon nitride	-743.5			-642.6			101.3					
Na	Sodium	0.0		107.5			77.0	51.3		153.7	28.2		20.8
NaO₂	Sodium superoxide	-260.2			-218.4			115.9			72.1		
Na₂	Disodium			142.1			103.9			230.2			37.6

Molecular formula	Name	$\Delta_f H°$/kJ mol⁻¹			$\Delta_f G°$/kJ mol⁻¹			$S°$/J mol⁻¹ K⁻¹			C_p/J mol⁻¹ K⁻¹		
		Crys.	Liq.	Gas	Crys.	Liq.	Gas	Crys.	Liq.	Gas	Crys.	Liq.	Gas
Na₂O	Sodium oxide	-414.2			-375.5			75.1			69.1		
Na₂O₂	Sodium peroxide	-510.9			-447.7			95.0			89.2		
Na₂O₃S	Sodium sulfite	-1100.8			-1012.5			145.9			120.3		
Na₂O₃Si	Sodium metasilicate	-1554.9			-1462.8			113.9			128.2		
Na₂O₄S	Sodium sulfate	-1387.1			-1270.2			149.6					
Na₂S	Sodium sulfide	-364.8			-349.8			83.7					
Nb	Niobium	0.0		725.9			681.1	36.4		186.3	24.6		30.2
NbO	Niobium(II) oxide	-405.8			-378.6			48.1			41.3		
NbO₂	Niobium(IV) oxide	-796.2			-740.5			54.5			57.5		
Nb₂O₅	Niobium(V) oxide	-1899.5			-1766.0			137.2			132.1		
Nd	Neodymium	0.0		327.6			292.4	71.5		189.4	27.5		22.1
Nd₂O₃	Neodymium oxide	-1807.9			-1720.8			158.6			111.3		
Ne	Neon			0.0						146.3			20.8
Ni	Nickel	0.0		429.7			384.5	29.9		182.2	26.1		23.4
NiO₄S	Nickel(II) sulfate	-872.9			-759.7			92.0			138.0		
NiS	Nickel(II) sulfide	-82.0			-79.5			53.0			47.1		
Ni₂O₃	Nickel(III) oxide	-489.5											
No	Nobelium	0.0											
O	Oxygen (atomic)			249.2			231.7			161.1			21.9
OP	Phosphorus monoxide			-28.5			-51.9			222.8			31.8
OPb	Lead(II) oxide (massicot)	-217.3			-187.9			68.7			45.8		
OPb	Lead(II) oxide (litharge)	-219.0			-188.9			66.5			45.8		
OPd	Palladium(II) oxide	-85.4		348.9			325.9			218.0	31.4		
ORa	Radium oxide	-523.0											
ORb₂	Rubidium oxide	-339.0											
ORh	Rhodium monoxide			385.0			-19.9			222.0			30.2
OS	Sulfur monoxide			6.3			26.8			234.0			31.3
OSe	Selenium monoxide			53.4						211.6			29.9
OSi	Silicon monoxide			-99.6			-126.4						
OSn	Tin(II) oxide	-280.7		15.1	-251.9		-8.4	57.2		232.1	44.3		31.6
OSr	Strontium oxide	-592.0			-561.9			54.4			45.0		
OTi	Titanium(II) oxide	-519.7			-495.0			50.0			40.0		
OTl₂	Thallium(I) oxide	-178.7			-147.3			126.0					
OU	Uranium(II) oxide			21.0									
OV	Vanadium(II) oxide	-431.8			-404.2			38.9			45.4		
OZn	Zinc oxide	-350.5			-320.5			43.7			40.3		
O₂	Oxygen			0.0						205.2			29.4
O₂P	Phosphorus dioxide			-279.9			-281.6			252.1			39.5
O₂Pb	Lead(IV) oxide	-277.4			-217.3			68.6			64.6		
O₂Rb	Rubidium superoxide	-278.7											
O₂Rb₂	Rubidium peroxide	-472.0											
O₂Ru	Ruthenium(IV) oxide	-305.0											
O₂S	Sulfur dioxide		-320.5	-296.8			-300.1			248.2			39.9
O₂Se	Selenium dioxide	-225.4											
O₂Si	Silicon dioxide (α-quartz)	-910.7		-322.0	-856.3			41.5			44.4		

STANDARD THERMODYNAMIC PROPERTIES OF CHEMICAL SUBSTANCES (continued)

Molecular formula	Name	$\Delta_f H°$/kJ mol⁻¹ Crys.	Liq.	Gas	$\Delta_f G°$/kJ mol⁻¹ Crys.	Liq.	Gas	$S°$/J mol⁻¹ K⁻¹ Crys.	Liq.	Gas	C_p/J mol⁻¹ K⁻¹ Crys.	Liq.	Gas
O₂Sn	Tin(IV) oxide	-577.6			-515.8			49.0			52.6		
O₂Te	Tellurium dioxide	-322.6			-270.3			79.5					
O₂Th	Thorium(IV) oxide	-1226.4			-1169.2			65.2			61.8		
O₂Ti	Titanium(IV) oxide	-944.0			-888.8			50.6			55.0		
O₂U	Uranium(IV) oxide	-1085.0		-465.7	-1031.8		-471.5	77.0		274.6	63.6		51.4
O₂W	Tungsten(IV) oxide	-589.7			-533.9			50.5			56.1		
O₂Zr	Zirconium(IV) oxide	-1100.6			-1042.8			50.4			56.2		
O₃	Ozone			142.7			163.2			238.9			39.2
O₃PbS	Lead(II) sulfite	-669.9						109.6					
O₃PbSi	Lead(II) metasilicate	-1145.7			-1062.1						90.0		
O₃Pr₂	Praseodymium oxide	-1809.6									117.4		
O₃Rh₂	Rhodium(III) oxide	-343.0									103.8		
O₃S	Sulfur trioxide	-454.5	-441.0	-395.7	-374.2	-373.8	-371.1	70.7	113.8	256.8			50.7
O₃Sc₂	Scandium oxide	-1908.8			-1819.4			77.0			94.2		
O₃SiSr	Strontium metasilicate	-1633.9			-1549.7			96.7			88.5		
O₃Sm₂	Samarium(III) oxide	-1823.0			-1734.6			151.0			114.5		
O₃Tb₂	Terbium oxide	-1865.2									115.9		
O₃Ti₂	Titanium(III) oxide	-1520.9			-1434.2			78.8			97.4		
O₃Tm₂	Thulium oxide	-1888.7			-1794.5			139.7			116.7		
O₃U	Uranium(VI) oxide	-1223.8			-1145.7			96.1			81.7		
O₃V₂	Vanadium(III) oxide	-1218.8			-1139.3			98.3			103.2		
O₃W	Tungsten(VI) oxide	-842.9			-764.0			75.9			73.8		
O₃Y₂	Yttrium oxide	-1905.3			-1816.6			99.1			102.5		
O₃Yb₂	Ytterbium(III) oxide	-1814.6			-1726.7			133.1			115.4		
O₄Os	Osmium(VIII) oxide	-394.1		-337.2	-304.9		-292.8	143.9		293.8	103.2		74.1
O₄PbS	Lead(II) sulfate	-920.0			-813.0			148.5			103.2		
O₄PbSe	Lead(II) selenate	-609.2			-504.9			167.8					
O₄Pb₂Si	Lead(II) orthosilicate	-1363.1			-1252.6			186.6			137.2		
O₄Pb₃	Lead(II,II,IV) oxide	-718.4			-601.2			211.3			146.9		
O₄RaS	Radium sulfate	-1471.1			-1365.6			138.0					
O₄Rb₂S	Rubidium sulfate	-1435.6			-1316.9			197.4			134.1		
O₄Ru	Ruthenium(VIII) oxide	-239.3			-152.2			146.4					
O₄SSr	Strontium sulfate	-1453.1			-1340.9			117.0					
O₄STl₂	Thallium(I) sulfate	-931.8			-830.4			230.5					
O₄SZn	Zinc sulfate	-982.8			-871.5			110.5			99.2		
O₄SiSr₂	Strontium orthosilicate	-2304.5			-2191.1			153.1			134.3		
O₄SiZn₂	Zinc orthosilicate	-1636.7			-1523.2			131.4			123.3		
O₄SiZr	Zirconium(IV) orthosilicate	-2033.4			-1919.1			84.1			98.7		
O₅Sb₂	Antimony(V) oxide	-971.9			-829.2			125.1					
O₅Ta₂	Tantalum(V) oxide	-2046.0			-1911.2			143.1			135.1		
O₅Ti₃	Titanium(III,IV) oxide	-2459.4			-2317.4			129.3			154.8		
O₅V₂	Vanadium(V) oxide	-1550.6			-1419.5			131.0			127.7		
O₅V₃	Vanadium(III,IV) oxide	-1933.0			-1803.0			163.0					
O₇Re₂	Rhenium(VII) oxide	-1240.1		-1100.0	-1066.0		-994.0	207.1		452.0	166.1		
O₇U₃	Uranium(IV,VI) oxide	-3427.1			-3242.9			250.5			215.5		

STANDARD THERMODYNAMIC PROPERTIES OF CHEMICAL SUBSTANCES (continued)

Molecular formula	Name	$\Delta_f H^\circ$/kJ mol^{-1}			$\Delta_f G^\circ$/kJ mol^{-1}			S°/J mol^{-1} K^{-1}			C_p/J mol^{-1} K^{-1}		
		Crys.	Liq.	Gas	Crys.	Liq.	Gas	Crys.	Liq.	Gas	Crys.	Liq.	Gas
O$_8$S$_2$Zr	Zirconium(IV) sulfate	-2217.1									172.0		
O$_8$U$_3$	Uranium(V,VI) oxide	-3574.8			-3369.5			282.6			238.4		
O$_9$U$_4$	Uranium(IV,V) oxide	-4510.4			-4275.1			334.1			293.3		
Os	Osmium	0.0		791.0			745.0	32.6		192.6	24.7		20.8
P	Phosphorus (white)	0.0		316.5			280.1	41.1		163.2	23.8		20.8
P	Phosphorus (red)	-17.6						22.8			21.2		
P	Phosphorus (black)	-39.3											
P$_2$	Diphosphorus			144.0			103.5			218.1			32.1
P$_4$	Tetraphosphorus			58.9			24.4			280.0			67.2
Pa	Protactinium	0.0		607.0			563.0	51.9		198.1			22.9
Pb	Lead	0.0		195.2			162.2	64.8		175.4	26.4		20.8
PbS	Lead(II) sulfide	-100.4			-98.7			91.2			49.5		
PbSe	Lead(II) selenide	-102.9			-101.7			102.5			50.2		
PbTe	Lead(II) telluride	-70.7			-69.5			110.0			50.5		20.8
Pd	Palladium	0.0		378.2			339.7	37.6		167.1	26.0		
PdS	Palladium(II) sulfide	-75.0			-67.0			46.0		187.1			24.3
Pm	Promethium	0.0											
Po	Polonium	0.0											
Pr	Praseodymium	0.0		355.6			320.9	73.2		189.8	27.2		21.4
Pt	Platinum	0.0		565.3			520.5	41.6		192.4	25.9		25.5
PtS	Platinum(II) sulfide	-81.6			-76.1			55.1			43.4		
PtS$_2$	Platinum(IV) sulfide	-108.8			-99.6			74.7			65.9		
Pu	Plutonium	0.0		0.0									20.8
Ra	Radium	0.0		159.0			130.0	71.0		176.5	31.1		20.8
Rb	Rubidium	0.0		80.9			53.1	76.8		170.1	25.5		20.8
Re	Rhenium	0.0		769.9			724.6	36.9		188.9	25.0		21.0
Rh	Rhodium	0.0		556.9			510.8	31.5		185.8			20.8
Rn	Radon			0.0						176.2			20.8
Ru	Ruthenium	0.0		642.7			595.8	28.5		186.5	24.1		21.5
S	Sulfur (rhombic)	0.0		277.2			236.7	32.1		167.8	22.6		23.7
S	Sulfur (monoclinic)	0.3											
SSi	Silicon monosulfide			112.5			60.9			223.7			32.3
SSn	Tin(II) sulfide	-100.0			-98.3			77.0			49.3		
SSr	Strontium sulfide	-472.4			-467.8			68.2			48.7		
STl$_2$	Thallium(I) sulfide	-97.1			-93.7			151.0					
SZn	Zinc sulfide (wurtzite)	-192.6											
SZn	Zinc sulfide (sphalerite)	-206.0			-201.3			57.7			46.0		
S$_2$	Disulfur			128.6			79.7			228.2			32.5
Sb	Antimony	0.0		262.3	0.0		222.1	45.7		180.3	25.2		20.8
Sb$_2$	Diantimony			235.6			187.0			254.9			36.4
Sc	Scandium	0.0		377.8			336.0	34.6		174.8	25.5		22.1
Se	Selenium	0.0		227.1			187.0	42.4		176.7	25.4		20.8
SeSr	Strontium selenide	-385.8						172.0					
SeTl$_2$	Thallium(I) selenide	-59.0			-59.0								
SeZn	Zinc selenide	-163.0			-163.0			84.0					

STANDARD THERMODYNAMIC PROPERTIES OF CHEMICAL SUBSTANCES (continued)

Molecular formula	Name	$\Delta_f H°$/kJ mol⁻¹ Crys.	Liq.	Gas	$\Delta_f G°$/kJ mol⁻¹ Crys.	Liq.	Gas	$S°$/J mol⁻¹ K⁻¹ Crys.	Liq.	Gas	C_p/J mol⁻¹ K⁻¹ Crys.	Liq.	Gas
Se₂	Diselenium			146.0			96.2			252.0			35.4
Si	Silicon	0.0		450.0			405.5	18.8		168.0	20.0		22.3
Si₂	Disilicon			594.0			536.0			229.9			34.4
Sm	Samarium	0.0		206.7			172.8	69.6		183.0	29.5		30.4
Sn	Tin (white)	0.0		301.2			266.2	51.2		168.5	27.0		21.3
Sn	Tin (gray)	-2.1			0.1			44.1			25.8		
Sr	Strontium	0.0		164.4			130.9	55.0		164.6	26.8		20.8
Ta	Tantalum	0.0		782.0			739.3	41.5		185.2	25.4		20.9
Tb	Terbium	0.0		388.7			349.7	73.2		203.6	28.9		24.6
Tc	Technetium	0.0		678.0						181.1			20.8
Te	Tellurium	0.0		196.7			157.1	49.7		182.7	25.7		20.8
Te₂	Ditellurium			168.2			118.0			268.1			36.7
Th	Thorium	0.0		602.0			560.7	51.8		190.2	27.3		20.8
Ti	Titanium	0.0		473.0			428.4	30.7		180.3	25.0		24.4
Tl	Thallium	0.0		182.2			147.4	64.2		181.0	26.3		20.8
Tm	Thulium	0.0		232.2			197.5	74.0		190.1	27.0		20.8
U	Uranium	0.0		533.0			488.4	50.2		199.8	27.7		23.7
V	Vanadium	0.0		514.2			754.4	28.9		182.3	24.9		26.0
W	Tungsten	0.0		849.4			807.1	32.6		174.0	24.3		21.3
Xe	Xenon			0.0						169.7			20.8
Y	Yttrium	0.0		421.3			381.1	44.4		179.5	26.5		25.9
Yb	Ytterbium	0.0		152.3			118.4	59.9		173.1	26.7		20.8
Zn	Zinc	0.0		130.4			94.8	41.6		161.0	25.4		20.8
Zr	Zirconium	0.0		608.8			566.5	39.0		181.4	25.4		26.7
C	Carbon (graphite)	0.0		716.7			671.3	5.7		158.1	8.5		20.8
C	Carbon (diamond)	1.9			2.9			2.4			6.1		
CAgN	Silver(I) cyanide	146.0			156.9			107.2			66.7		
CAg₂O₃	Silver(I) carbonate	-505.8			-436.8			167.4			112.3		
CBaO₃	Barium carbonate	-1213.0			-1134.4			112.1			86.0		
CBeO₃	Beryllium carbonate	-1025.0						52.0			65.0		
CBrClF₂	Bromochlorodifluoro-methane									318.5			74.6
CBrCl₂F	Bromodichlorofluoro-methane									330.6			80.0
CBrCl₃	Bromotrichloromethane			-41.1									85.3
CBrF₃	Bromotrifluoromethane			-648.3									69.3
CBrN	Cyanogen bromide	140.5		186.2			165.3			248.3			46.9
CBrN₃O₆	Bromotrinitromethane		32.5	80.3									
CBr₂ClF	Dibromochlorofluoro-methane									342.8			82.4
CBr₂Cl₂	Dibromodichloromethane									347.8			87.1
CBr₂F₂	Dibromodifluoromethane									325.3			77.0
CBr₂O	Carbonyl bromide		-127.2	-96.2			-110.9			309.1			61.8
CBr₃Cl	Tribromochloromethane									357.8			89.4
CBr₃F	Tribromofluoromethane									345.9			84.4

STANDARD THERMODYNAMIC PROPERTIES OF CHEMICAL SUBSTANCES (continued)

Molecular formula	Name	$\Delta_f H^\circ$/kJ mol⁻¹			$\Delta_f G^\circ$/kJ mol⁻¹			S°/J mol⁻¹ K⁻¹			C_p/J mol⁻¹ K⁻¹		
		Crys.	Liq.	Gas	Crys.	Liq.	Gas	Crys.	Liq.	Gas	Crys.	Liq.	Gas
CBr₄	Tetrabromomethane	29.4		83.9	47.7		67.0	212.5		358.1	144.3		91.2
CCaO₃	Calcium carbonate (calcite)	-1207.6			-1129.1			91.7			83.5		
CCaO₃	Calcium carbonate (aragonite)	-1207.8			-1128.2			88.0			82.3		
CCdO₃	Cadmium carbonate	-750.6			-669.4			92.5					
CClFO	Carbonyl chloride fluoride									276.7			52.4
CClF₃	Chlorotrifluoromethane			-706.3									66.9
CClN	Cyanogen chloride		112.1	138.0			131.0			236.2			45.0
CClN₃O₆	Chlorotrinitromethane		-27.1	18.4									
CCl₂F₂	Dichlorodifluoromethane			-477.4			-439.4			300.8			72.3
CCl₂O	Carbonyl chloride			-219.1			-204.9			283.5			57.7
CCl₃	Trichloromethyl			59.0									
CCl₃F	Trichlorofluoromethane		-301.3	-268.3		-236.8			225.4			121.6	78.1
CCl₄	Tetrachloromethane		-128.2	-95.7								130.7	83.3
CCoO₃	Cobalt(II) carbonate	-713.0											
CCs₂O₃	Cesium carbonate	-1139.7			-1054.3			204.5			123.9		
CCuN	Copper(I) cyanide	96.2			111.3			84.5					
CFN	Cyanogen fluoride									224.7			41.8
CF₂O	Carbonyl fluoride			-639.8									46.8
CF₃	Trifluoromethyl			-477.0			-464.0			264.5			49.6
CF₃I	Trifluoroiodomethane			-587.8						307.4			70.9
CF₄	Tetrafluoromethane			-933.6						261.6			61.1
CFeO₃	Iron(II) carbonate	-740.6			-666.7			92.9			82.1		
CFe₃	Iron carbide	25.1			20.1			104.6			105.9		
CH	Methylidyne			595.8									
CHBrClF	Bromochlorofluoromethane									304.3			63.2
CHBrCl₂	Bromodichloromethane									316.4			67.4
CHBrF₂	Bromodifluoromethane			-424.9						295.1			58.7
CHBr₂Cl	Chlorodibromomethane									327.7			69.2
CHBr₂F	Dibromofluoromethane									316.8			65.1
CHBr₃	Tribromomethane		-22.3	23.8		-5.0	8.0		220.9	330.9		130.7	71.2
CHClF₂	Chlorodifluoromethane			-482.6						280.9			55.9
CHCl₂F	Dichlorofluoromethane									293.1			60.9
CHCl₃	Trichloromethane		-134.1	-102.7		-73.7	6.0		201.7	295.7		114.2	65.7
CHCsO₃	Cesium hydrogen carbonate	-966.1											
CHFO	Formyl fluoride									246.6			39.9
CHF₃	Trifluoromethane			-695.4						259.7			51.0
CHI₃	Triiodomethane	-181.1		251.0						356.2			75.0
CHKO₂	Potassium formate	-679.7						115.5					
CHKO₃	Potassium hydrogen carbonate	-963.2			-863.5								
CHN	Hydrogen cyanide		108.9	135.1		125.0	124.7		112.8	201.8		70.6	35.9
CHNO	Isocyanic acid (HNCO)			127.6			113.0			238.0			44.9
CHNS	Isothiocyanic acid			-0.2						247.8			46.9
CHN₃O₆	Trinitromethane		-32.8										
CHNaO₂	Sodium formate	-666.5			-599.9			103.8			82.7		

Molecular formula	Name	$\Delta_f H^\circ$/kJ mol^{-1} Crys.	Liq.	Gas	$\Delta_f G^\circ$/kJ mol^{-1} Crys.	Liq.	Gas	S°/J mol^{-1} K^{-1} Crys.	Liq.	Gas	C_p/J mol^{-1} K^{-1} Crys.	Liq.	Gas
CHNaO₃	Sodium hydrogen carbonate	-950.8			-851.0			101.7			87.6		
CHO	Oxomethyl (HCO)			43.1			28.0			224.7			34.6
CH₂	Methylene			390.4			372.9			194.9			33.8
CH₂BrCl	Bromochloromethane									287.6			52.7
CH₂BrF	Bromofluoromethane									276.3			49.2
CH₂Br₂	Dibromomethane									293.2			54.7
CH₂ClF	Chlorofluoromethane									264.4			47.0
CH₂Cl₂	Dichloromethane		-124.2	-95.4					177.8	270.2		101.2	51.0
CH₂F₂	Difluoromethane			-452.3						246.7			42.9
CH₂I₂	Diiodomethane		68.5	119.5		90.4	95.8		174.1	309.7		134.0	57.7
CH₂N₂	Diazomethane									242.9			52.5
CH₂N₂	Cyanamide	58.8											
CH₂N₂O₄	Dinitromethane		-104.9	-58.9									
CH₂O	Formaldehyde			-108.6			-102.5			218.8			35.4
(CH₂O)ₓ	Paraformaldehyde	-177.6											
CH₂O₂	Formic acid		-425.0	-378.7		-361.4			129.0			99.0	
CH₂S₃	Trithiocarbonic acid		24.0										
CH₃	Methyl			145.7			147.9			194.2			38.7
CH₃BO	Borane carbonyl			-111.2			-92.9			249.4			59.5
CH₃Br	Bromomethane		-59.8	-35.4			-26.3			246.4			42.4
CH₃Cl	Chloromethane			-81.9						234.6			40.8
CH₃F	Fluoromethane									222.9			37.5
CH₃I	Iodomethane		-13.6	14.4						254.1			44.1
CH₃NO	Formamide		-254.0	-193.9					163.2	254.1		126.0	44.1
CH₃NO₂	Nitromethane		-112.6	-74.3		-14.4	-6.8		171.8	275.0		106.6	57.3
CH₃NO₂	Methyl nitrite			-66.1									
CH₃NO₃	Methyl nitrate		-156.3	-122.4		-43.4	-39.2		217.1	318.5		157.3	
CH₄	Methane			-74.6			-50.5			186.3			35.7
CH₄N₂	Ammonium cyanide	0.4									134.0		
CH₄N₂O	Urea	-333.1		-245.8									
CH₄N₂S	Thiourea	-89.1		22.9									
CH₄N₄O₂	Nitroguanidine	-92.4											
CH₄O	Methanol		-239.2	-201.0		-166.6	-162.3		126.8	239.9		81.1	44.1
CH₄S	Methanethiol		-46.7	-22.9		-7.7	-9.3		169.2	255.2		90.5	50.3
CH₅N	Methylamine		-47.3	-22.5		35.7	32.7		150.2	242.9		102.1	50.1
CH₅NO₃	Ammonium hydrogen carbonate	-849.4			-665.9			120.9					
CH₅N₃	Guanidine	-56.0											
CH₅N₃S	Hydrazinecarbothioamide	24.7											
CH₅N₅O₂	3-Amino-1-nitroguanidine	22.1											
CH₆ClN	Methylamine hydrochloride	-298.1											
CH₆N₂	Methylhydrazine		54.2	94.7		180.0	187.0		165.9	278.8		134.9	71.1
CH₆Si	Methylsilane									256.5			65.9
CHg₂O₃	Mercury(I) carbonate	-553.5			-468.1			180.0					
CIN	Cyanogen iodide	166.2		225.5	185.0		196.6	96.2		256.8			48.3

STANDARD THERMODYNAMIC PROPERTIES OF CHEMICAL SUBSTANCES (continued)

Molecular formula	Name	$\Delta_fH°$/kJ mol⁻¹ Crys.	Liq.	Gas	$\Delta_fG°$/kJ mol⁻¹ Crys.	Liq.	Gas	S°/J mol⁻¹ K⁻¹ Crys.	Liq.	Gas	C_p/J mol⁻¹ K⁻¹ Crys.	Liq.	Gas
CI₄	Tetraiodomethane	-392.9		474.0						391.9			95.9
CKN	Potassium cyanide	-113.0			-101.9			128.5			66.3		
CKNS	Potassium thiocyanate	-200.2			-178.3			124.3			88.5		
CK₂O₃	Potassium carbonate	-1151.0			-1063.5			155.5			114.4		
CLi₂O₃	Lithium carbonate	-1215.9			-1132.1			90.4			99.1		
CMgO₃	Magnesium carbonate	-1095.8			-1012.1			65.7			75.5		
CMnO₃	Manganese(II) carbonate	-894.1			-816.7			85.8			81.5		
CN	Cyanide			437.6			407.5			202.6			29.2
CNNa	Sodium cyanide	-87.5			-76.4			115.6			70.4		
CNNaO	Sodium cyanate	-405.4			-358.1			96.7			86.6		
CN₄O₈	Tetranitromethane		38.4	88.3									
CNa₂O₃	Sodium carbonate	-1130.7			-1044.4			135.0			112.3		
CO	Carbon monoxide			-110.5			-137.2			197.7			29.1
COS	Carbon oxysulfide			-142.0			-169.2			231.6			41.5
CO₂	Carbon dioxide			-393.5			-394.4			213.8			37.1
CO₃Pb	Lead(II) carbonate	-699.1			-625.5			131.0			87.4		
CO₃Rb₂	Rubidium carbonate	-1136.0			-1051.0			181.3			117.6		
CO₃Sr	Strontium carbonate	-1220.1			-1140.1			97.1			81.4		
CO₃Tl₂	Thallium(I) carbonate	-700.0			-614.6			155.2					
CO₃Zn	Zinc carbonate	-812.8			-731.5			82.4			79.7		
CS	Carbon sulfide			234.0			184.0			210.6			29.8
CS₂	Carbon disulfide		89.0	116.7		64.6	67.1		151.3	237.8		76.4	45.4
CSe₂	Carbon diselenide		164.8										
CSi	Silicon carbide (cubic)	-65.3			-62.8			16.6			26.9		
CSi	Silicon carbide (hexagonal)	-62.8			-60.2			16.5			26.7		
C₂	Dicarbon			831.9			775.9			199.4			43.2
C₂BrF₅	Bromopentafluoroethane			-1064.4									
C₂Br₂ClF₃	1,2-Dibromo-1-chloro-1,2,2-trifluoroethane		-691.7	-656.6									
C₂Br₂F₄	1,2-Dibromotetrafluoroethane		-817.7	-789.1									
C₂Br₄	Tetrabromoethylene									387.1			102.7
C₂Br₆	Hexabromoethane									441.9			139.3
C₂Ca	Calcium carbide	-59.8			-64.9			70.0			62.7		
C₂CaN₂	Calcium cyanide	-184.5											
C₂CaO₄	Calcium oxalate	-1360.6											
C₂ClF₃	Chlorotrifluoroethylene		-522.7	-505.5			-523.8			322.1			83.9
C₂ClF₅	Chloropentafluoroethane			-1118.8									184.2
C₂Cl₂F₄	1,2-Dichlorotetrafluoroethane												
C₂Cl₂O₂	Ethanedioyl dichloride		-960.2	-937.0								111.7	
C₂Cl₃F₃	1,1,2-Trichlorotrifluoroethane		-367.6	-335.8								170.1	
C₂Cl₃N	Trichloroacetonitrile		-745.0	-716.8						336.6			96.1
C₂Cl₄	Tetrachloroethylene	-50.6		-10.9		3.0			266.9			143.4	

Molecular formula	Name	$\Delta_f H°$/kJ mol⁻¹ Crys.	Liq.	Gas	$\Delta_f G°$/kJ mol⁻¹ Crys.	Liq.	Gas	S°/J mol⁻¹ K⁻¹ Crys.	Liq.	Gas	C_p/J mol⁻¹ K⁻¹ Crys.	Liq.	Gas
$C_2Cl_4F_2$	1,1,1,2-Tetrachloro-2,2-difluoroethane			-489.9			-407.0			382.9			123.4
$C_2Cl_4F_2$	1,1,2,2-Tetrachloro-1,2-difluoroethane											173.6	
C_2Cl_4O	Trichloroacetyl chloride	-202.8	-280.8	-239.8									
C_2Cl_6	Hexachloroethane			-143.6				237.3			198.2		
C_2F_3N	Trifluoroacetonitrile			-497.9						298.1			77.9
C_2F_4	Tetrafluoroethylene	-820.5		-658.9						300.1			80.5
C_2F_6	Hexafluoroethane			-1344.2						332.3			106.7
C_2HBr	Bromoacetylene									253.7			55.7
$C_2HBrClF_3$	1-Bromo-2-chloro-1,1,2-trifluoroethane		-675.3	-644.8									
$C_2HBrClF_3$	2-Bromo-2-chloro-1,1,1-trifluoroethane		-720.0	-690.4									
C_2HCl	Chloroacetylene									242.0			54.3
C_2HClF_2	1-Chloro-2,2-difluoroethylene			-315.5			-289.1			303.0			72.1
C_2HCl_2F	1,1-Dichloro-2-fluoroethylene									313.9			76.5
$C_2HCl_2F_3$	2,2-Dichloro-1,1,1-trifluoroethane									352.8			102.5
C_2HCl_3	Trichloroethylene		-43.6	-9.0						324.8		124.4	80.3
C_2HCl_3O	Trichloroacetaldehyde		-234.5	-196.6					228.4			151.0	
C_2HCl_3O	Dichloroacetyl chloride		-280.4	-241.0									
$C_2HCl_3O_2$	Trichloroacetic acid	-503.3											
C_2HCl_5	Pentachloroethane		-187.6	-142.0								173.8	
C_2HF	Fluoroacetylene									231.7			52.4
C_2HF_3	Trifluoroethylene			-490.5									
$C_2HF_3O_2$	Trifluoroacetic acid		-1069.9	-1031.4									
C_2HF_5	Pentafluoroethane			-1100.4									
C_2H_2	Acetylene			227.4			209.9			200.9			44.0
$C_2H_2BrF_3$	2-Bromo-1,1,1-trifluoroethane			-694.5									
$C_2H_2Br_2$	cis-1,2-Dibromoethylene									311.3			68.8
$C_2H_2Br_2$	trans-1,2-Dibromoethylene									313.5			70.3
$C_2H_2Br_2Cl_2$	1,2-Dibromo-1,2-dichloroethane			-36.9									
$C_2H_2Br_4$	1,1,2,2-Tetrabromoethane											165.7	
$C_2H_2ClF_3$	2-Chloro-1,1,1-trifluoroethane									326.5			89.1
$C_2H_2Cl_2$	1,1-Dichloroethylene		-23.9	2.8		24.1	25.4		201.5	289.0		111.3	67.1
$C_2H_2Cl_2$	cis-1,2-Dichloroethylene		-26.4	4.6					198.4	289.6		116.4	65.1
$C_2H_2Cl_2$	trans-1,2-Dichloroethylene		-24.3	5.0		27.3	28.6		195.9	290.0		116.8	66.7
$C_2H_2Cl_2O$	Chloroacetyl chloride		-283.7	-244.8									
$C_2H_2Cl_2O_2$	Dichloroacetic acid		-496.3										

Molecular formula	Name	ΔfH°/kJ mol⁻¹ Crys.	Liq.	Gas	ΔfG°/kJ mol⁻¹ Crys.	Liq.	Gas	S°/J mol⁻¹ K⁻¹ Crys.	Liq.	Gas	C_p/J mol⁻¹ K⁻¹ Crys.	Liq.	Gas
$C_2H_2Cl_3NO$	2,2,2-Trichloroacetamide	-358.0											
$C_2H_2Cl_4$	1,1,1,2-Tetrachloroethane			-149.2						356.0			102.7
$C_2H_2Cl_4$	1,1,2,2-Tetrachloroethane		-195.0	-335.0					246.9	362.8		162.3	100.8
$C_2H_2F_2$	1,1-Difluoroethylene									266.2			60.1
$C_2H_2F_2$	cis-1,2-Difluoroethylene									268.3			58.2
$C_2H_2F_3I$	1,1,1-Trifluoro-2-iodoethane			-644.5									
$C_2H_2I_2$	cis-1,2-Diiodoethylene			-207.4									
C_2H_2O	Ketene		-67.9	-47.5			-48.3			247.6			51.8
$C_2H_2O_2$	Glyoxal			-212.0									
$C_2H_2O_4$	Oxalic acid	-821.7		-723.7				109.8					
$C_2H_2O_4Sr$	Strontium formate	-1393.3									91.0		
$C_2H_2S_2$	Thiirene			300.0			275.8			255.3			54.7
C_2H_3Br	Bromoethylene			79.2			81.8			275.8			55.5
C_2H_3BrO	Acetyl bromide		-223.5	-190.4						264.0			
C_2H_3Cl	Chloroethylene	-94.1	14.6	37.2			53.6			264.0	59.4		53.7
$C_2H_3ClF_2$	1-Chloro-1,1-difluoroethane									307.2			82.5
C_2H_3ClO	Acetyl chloride		-272.9	-242.8		-208.0	-205.8		200.8	295.1		117.0	67.8
$C_2H_3ClO_2$	Chloroacetic acid	-510.5		-435.2									
$C_2H_3Cl_2F$	1,1-Dichloro-1-fluoroethane									320.2			88.7
$C_2H_3Cl_3$	1,1,1-Trichloroethane		-177.4	-144.4					227.4	323.1		144.3	93.3
$C_2H_3Cl_3$	1,1,2-Trichloroethane		-190.8	-151.3					232.6	337.2		150.9	89.0
C_2H_3F	Fluoroethylene			-138.8									
C_2H_3FO	Acetyl fluoride		-467.2	-442.1						279.9			78.2
$C_2H_3F_3$	1,1,1-Trifluoroethane			-744.6									
$C_2H_3F_3$	1,1,2-Trifluoroethane			-730.7									
$C_2H_3F_3O$	2,2,2-Trifluoroethanol		-932.4	-888.4						285.0			57.9
C_2H_3I	Iodoethylene			-126.4									
C_2H_3IO	Acetyl iodide		-163.5										
$C_2H_3KO_2$	Potassium acetate	-723.0											
C_2H_3N	Acetonitrile		40.6	74.0		86.5	91.9		149.6	243.4		91.5	52.2
C_2H_3N	Isocyanomethane		130.8	163.5		159.5	165.7		159.0	246.9			52.9
C_2H_3NO	Methylisocyanate		-92.0										
$C_2H_3NO_3$	Oxamic acid	-661.2		-552.3									
C_2H_3NS	Methyl isothiocyanate	79.4											
$C_2H_3NaO_2$	Sodium acetate	-708.8			-607.2			123.0			79.9		
C_2H_4	Ethylene			52.4			68.4			219.3			42.9
C_2H_4BrCl	1-Bromo-2-chloroethane		-66.2									130.1	
$C_2H_4Br_2$	1,1-Dibromoethane		-79.2	-37.5					223.3	327.7		136.0	80.8
$C_2H_4Br_2$	1,2-Dibromoethane												
C_2H_4ClF	1-Chloro-1-fluoroethane			-313.4									
$C_2H_4Cl_2$	1,1-Dichloroethane		-158.4	-127.7		-73.8			211.8	305.1		126.3	76.2
$C_2H_4Cl_2$	1,2-Dichloroethane		-166.8	-126.4			-70.8			308.4		128.4	78.7

Molecular formula	Name	$\Delta_f H^\circ/\text{kJ mol}^{-1}$			$\Delta_f G^\circ/\text{kJ mol}^{-1}$			$S^\circ/\text{J mol}^{-1}\,\text{K}^{-1}$			$C_p/\text{J mol}^{-1}\,\text{K}^{-1}$		
		Crys.	Liq.	Gas	Crys.	Liq.	Gas	Crys.	Liq.	Gas	Crys.	Liq.	Gas
$C_2H_4F_2$	1,1-Difluoroethane			-497.0						282.5			67.8
$C_2H_4I_2$	1,2-Diiodoethane	9.3		75.0									
$C_2H_4N_2O_2$	Oxamide	-504.4		-387.1									
$C_2H_4N_2O_2$	Ethanedial, dioxime	-90.5											
$C_2H_4N_2O_4$	1,1-Dinitroethane		-148.2										
$C_2H_4N_2O_4$	1,2-Dinitroethane		-165.2										
$C_2H_4N_2S_2$	Ethanedithioamide	-20.8											
$C_2H_4N_4$	1H-1,2,4-Triazol-3-amine	76.8		83.0									
C_2H_4O	Acetaldehyde		-192.2	-166.2		-127.6	-133.0		160.2	263.8		89.0	55.3
C_2H_4O	Ethylene oxide		-78.0	-52.6		-11.8	-13.0		153.9	242.5		88.0	47.9
C_2H_4OS	Thioacetic acid		-216.9	-175.1									
$C_2H_4O_2$	Acetic acid		-484.3	-432.2		-389.9	-374.2		159.8	283.5		123.3	63.4
$C_2H_4O_2$	Methyl formate		-386.1	-357.4						285.3		119.1	64.4
C_2H_4S	Thiirane		51.6	82.0			96.8			255.2			53.3
C_2H_4Si	Ethynylsilane									269.4			72.6
C_2H_5Br	Bromoethane		-90.5	-61.9		-25.8	-23.9		198.7	286.7		100.8	64.5
C_2H_5Cl	Chloroethane		-136.8	-112.1		-59.3	-60.4		190.8	276.0		104.3	62.8
C_2H_5ClO	Ethylene chlorohydrin		-295.4										
C_2H_5F	Fluoroethane									264.5			58.6
C_2H_5I	Iodoethane		-40.0	-8.1					211.7	306.0		115.1	66.9
C_2H_5N	Ethyleneimine		91.9	126.5		14.7	19.2						
C_2H_5NO	Acetamide	-317.0		-238.3				115.0			91.3		
C_2H_5NO	N-Methylformamide											123.8	
$C_2H_5NO_2$	Nitroethane		-143.9	-102.3								134.4	
$C_2H_5NO_2$	Glycine	-528.5		-392.1									
$C_2H_5NO_3$	2-Nitroethanol		-350.7										
$C_2H_5NO_3$	Ethyl nitrate		-190.4	-154.1									
C_2H_5NS	Thioacetamide	-71.7		11.4									
C_2H_6	Ethane			-84.0			-32.0			229.2			52.5
C_2H_6Cd	Dimethyl cadmium		63.6	101.6		139.0	146.9		201.9	303.0		132.0	
C_2H_6Hg	Dimethyl mercury		59.8	94.4		140.3	146.1		209.0	306.0			83.3
$C_2H_6N_2O$	N-Methylurea	-332.8											
$C_2H_6N_4O_2$	1,2-Hydrazinedicarboxamide	-498.7			-302.4								
$C_2H_6N_4O_2$	Ethanedioic acid, dihydrazide	-295.2						142.0					
C_2H_6O	Ethanol		-277.6	-234.8		-174.8	-167.9		160.7	281.6		112.3	65.6
C_2H_6O	Dimethyl ether		-203.3	-184.1			-112.6			266.4			64.4
C_2H_6OS	Dimethyl sulfoxide		-204.2	-151.3		-99.9			188.3			153.0	
$C_2H_6O_2$	Ethylene glycol		-460.0	-392.2			-272.7		163.2	303.8		148.6	82.7
$C_2H_6O_2S$	Dimethyl sulfone	-450.1		-373.1						310.6			100.0
$C_2H_6O_3S$	Dimethyl sulfite		-523.6	-483.4									
$C_2H_6O_4S$	Dimethyl sulfate		-735.5	-687.0									
C_2H_6S	Ethanethiol		-73.6	-46.1		-5.5	-4.8		207.0	296.2		117.9	72.7
C_2H_6S	Dimethyl sulfide		-65.3	-37.4					196.4	286.0		118.1	74.1
$C_2H_6S_2$	1,2-Ethanedithiol		-54.3	-9.7									
$C_2H_6S_2$	Dimethyl disulfide		-62.6	-24.7					235.4			146.1	

Molecular formula	Name	$\Delta_f H°$/kJ mol⁻¹ Crys.	Liq.	Gas	$\Delta_f G°$/kJ mol⁻¹ Crys.	Liq.	Gas	$S°$/J mol⁻¹ K⁻¹ Crys.	Liq.	Gas	C_p/J mol⁻¹ K⁻¹ Crys.	Liq.	Gas
C_2H_6Zn	Dimethyl zinc		23.4	53.0					201.6			129.2	
C_2H_7N	Ethylamine		-74.1	-47.5			36.3			283.8		130.0	71.5
C_2H_7N	Dimethylamine		-43.9	-18.8		70.0	68.5		182.3	273.1		137.7	70.7
C_2H_7NO	Ethanolamine											195.5	
C_2H_8ClN	Dimethylamine hydrochloride	-289.3											
$C_2H_8N_2$	1,2-Ethanediamine		-63.0	-18.0									
$C_2H_8N_2$	1,1-Dimethylhydrazine		48.9	84.1		206.4			198.0			172.6	
$C_2H_8N_2$	1,2-Dimethylhydrazine		52.7	92.2								164.1	
$C_2H_8N_2O_4$	Ammonium oxalate	-1123.0											
C_2HgO_4	Mercury(II) oxalate	-678.2									226.0		
C_2I_2	Diiodoacetylene	305.0								313.1			70.3
C_2I_4	Tetraiodoethylene												
$C_2K_2O_4$	Potassium oxalate	-1346.0											
C_2MgO_4	Magnesium oxalate	-1269.0											
C_2N_2	Cyanogen		285.9	306.7						241.9			56.8
$C_2N_4O_6$	Trinitroacetonitrile		183.7										
$C_2Na_2O_4$	Sodium oxalate			-1318.0									
C_2O_4Pb	Lead(II) oxalate	-851.4			-750.1			146.0			105.4		
C_3F_8	Perfluoropropane			-1783.2									
$C_3H_2N_2$	Malononitrile	186.4		265.5									
$C_3H_2O_2$	2-Propynoic acid		-193.2										
$C_3H_2O_3$	1,3-Dioxol-2-one		-459.9	-418.6									
$C_3H_3Cl_3$	1,2,3-Trichloropropene		-101.8										
$C_3H_3F_3$	3,3,3-Trifluoropropene			-614.2									
C_3H_3N	2-Propenenitrile		147.1	180.6									
C_3H_3NO	Oxazole		-48.0	-15.5									
C_3H_3NO	Isoxazole		42.1	78.6									
C_3H_4	Allene			190.5									
C_3H_4	Propyne			184.9									
C_3H_4	Cyclopropene			277.1									
$C_3H_4Cl_2$	2,3-Dichloropropene		-73.3										
$C_3H_4Cl_4$	1,1,1,3-Tetrachloropropane		-208.7										
$C_3H_4Cl_4$	1,2,2,3-Tetrachloropropane		-251.8										
$C_3H_4F_4O$	2,2,3,3-Tetrafluoro-1-propanol		-1114.9	-1061.3									
$C_3H_4N_2$	1H-Pyrazole	49.8	105.4	179.4									
$C_3H_4N_2$	Imidazole			132.9									
$C_3H_4O_2$	1,2-Propanedione		-309.1	-271.0									
$C_3H_4O_2$	Propenoic acid		-383.8									145.7	
$C_3H_4O_2$	2-Oxetanone		-329.9	-282.9					175.3			122.1	
$C_3H_4O_3$	Ethylene carbonate		-571.5	-508.4								133.9	
C_3H_5Br	cis-1-Bromopropene		7.9	40.8									
C_3H_5Br	3-Bromopropene		12.2	45.2									
C_3H_5BrO	Bromoacetone			-181.0									

STANDARD THERMODYNAMIC PROPERTIES OF CHEMICAL SUBSTANCES (continued)

Molecular formula	Name	$\Delta_f H°$/kJ mol⁻¹			$\Delta_f G°$/kJ mol⁻¹			$S°$/J mol⁻¹ K⁻¹			C_p/J mol⁻¹ K⁻¹		
		Crys.	Liq.	Gas	Crys.	Liq.	Gas	Crys.	Liq.	Gas	Crys.	Liq.	Gas
C_3H_5Cl	2-Chloropropene			-21.0								125.1	
C_3H_5Cl	3-Chloropropene											131.6	
C_3H_5ClO	Epichlorohydrin		-148.4	-107.8									
$C_3H_5ClO_2$	2-Chloropropanoic acid		-522.5										
$C_3H_5ClO_2$	3-Chloropropanoic acid	-549.3		-475.8									
$C_3H_5ClO_2$	Ethyl chloroformate		-505.3	-462.9								183.6	
$C_3H_5Cl_3$	1,2,3-Trichloropropane		-230.6	-182.9									
C_3H_5I	3-Iodopropene		53.7	91.5									
C_3H_5IO	Iodoacetone			-130.5									
$C_3H_5IO_2$	3-Iodopropanoic acid	-460.0											
C_3H_5N	Propanenitrile		15.5	51.7								119.3	
C_3H_5N	2-Propyn-1-amine		205.7										
C_3H_5N	Ethyl isocyanide		108.6	141.7								149.1	
$C_3H_5NO_3$	Nitroacetone		-278.6										
$C_3H_5NO_4$	Methyl nitroacetate		-464.0										
$C_3H_5N_3O_9$	Trinitroglycerol		-370.9	-270.9									
C_3H_6	Propene		4.0	20.0									
C_3H_6	Cyclopropane		35.2	53.3			104.5			237.5			55.6
$C_3H_6Br_2$	1,2-Dibromopropane		-113.6	-71.6									
$C_3H_6Cl_2$	1,2-Dichloropropane		-198.8	-162.8									
$C_3H_6Cl_2$	1,3-Dichloropropane		-199.9	-159.2									
$C_3H_6Cl_2$	2,2-Dichloropropane		-205.8	-173.2									
$C_3H_6Cl_2O$	2,3-Dichloro-1-propanol		-381.5	-316.3									
$C_3H_6Cl_2O$	2,3-Dichloro-2-propanol		-385.3	-318.4									
$C_3H_6I_2$	1,2-Diiodopropane			35.6									
$C_3H_6I_2$	1,3-Diiodopropane		-9.0										
$C_3H_6N_2O_2$	Propanediamide	-546.1											
$C_3H_6N_2O_2$	N-(Aminocarbonyl)acetamide	-544.2		-441.2									
$C_3H_6N_2O_4$	1,1-Dinitropropane		-163.2	-100.7									
$C_3H_6N_2O_4$	1,3-Dinitropropane		-207.1										
$C_3H_6N_2O_4$	2,2-Dinitropropane		-181.2										
C_3H_6O	Allyl alcohol		-171.8	-124.5								138.9	
C_3H_6O	Propanal		-215.6	-185.6						304.5			80.7
C_3H_6O	Acetone		-248.4	-217.1			-152.7		199.8	295.3		126.3	74.5
C_3H_6O	Methyloxirane		-123.0	-94.7					196.5	286.9		120.4	72.6
C_3H_6O	Oxetane		-110.8	-80.5									
$C_3H_6O_2$	Propanoic acid		-510.7	-455.7					191.0			152.8	
$C_3H_6O_2$	Ethyl formate											149.3	
$C_3H_6O_2$	Methyl acetate		-445.9	-413.3						324.4		141.9	86.0
$C_3H_6O_2$	1,3-Dioxolane		-333.5	-298.0								118.0	
$C_3H_6O_2S$	Thiolactic acid		-468.4										
$C_3H_6O_3$	Trioxane	-522.5		-465.9				133.0			111.4		
C_3H_6S	Thiacyclobutane		24.7	60.6			107.1		184.9				
C_3H_6S	Methylthiirane		11.3	45.8						285.0			68.3

Molecular formula	Name	$\Delta_f H°$/kJ mol⁻¹ Crys.	Liq.	Gas	$\Delta_f G°$/kJ mol⁻¹ Crys.	Liq.	Gas	$S°$/J mol⁻¹ K⁻¹ Crys.	Liq.	Gas	C_p/J mol⁻¹ K⁻¹ Crys.	Liq.	Gas
$C_3H_6S_2$	1,2-Dithiolane			0.0						313.5			86.5
$C_3H_6S_2$	1,3-Dithiolane			10.0						323.3			84.7
C_3H_7Br	1-Bromopropane		-121.9	-87.0									
C_3H_7Br	2-Bromopropane		-130.5	-99.4									
C_3H_7Cl	1-Chloropropane		-160.5	-131.9									
C_3H_7Cl	2-Chloropropane		-172.3	-144.9									
$C_3H_7ClO_2$	3-Chloro-1,2-propanediol		-525.3										
$C_3H_7ClO_2$	2-Chloro-1,3-propanediol		-517.5										
C_3H_7F	1-Fluoropropane			-285.9									
C_3H_7F	2-Fluoropropane			-293.5									
C_3H_7I	1-Iodopropane		-66.0	-30.0									
C_3H_7I	2-Iodopropane		-74.8	-40.3									
C_3H_7N	Allylamine		-10.0										
C_3H_7N	Cyclopropylamine		45.8	77.0									
C_3H_7NO	N,N-Dimethylformamide		-239.3	-192.4					187.7			147.1	
C_3H_7NO	Propanamide	-338.2		-259.0								150.6	
$C_3H_7NO_2$	1-Nitropropane		-167.2	-123.8								170.3	
$C_3H_7NO_2$	2-Nitropropane		-180.3	-138.9									
$C_3H_7NO_2$	DL-Alanine	-563.6											
$C_3H_7NO_2$	D-Alanine	-561.2											
$C_3H_7NO_2$	L-Alanine	-604.0		-465.9									
$C_3H_7NO_2$	β-Alanine	-558.0		-424.0									
$C_3H_7NO_2$	N-Methylglycine	-513.3		-367.3									
$C_3H_7NO_2S$	L-Cysteine	-534.1											
$C_3H_7NO_3$	Propyl nitrate		-214.5	-173.9									
$C_3H_7NO_3$	Isopropyl nitrate		-229.7	-191.0									
$C_3H_7NO_3$	DL-Serine	-739.0											
$C_3H_7NO_3$	L-Serine	-732.7											
C_3H_8	Propane		-120.9	-103.8			-23.4			270.3			73.6
$C_3H_8N_2O$	N-Ethylurea	-357.8											
$C_3H_8N_2O$	N,N-Dimethylurea	-319.1											
$C_3H_8N_2O$	N,N'-Dimethylurea	-312.1											
$C_3H_8N_2O_3$	Oxymethurea	-717.0											
C_3H_8O	1-Propanol		-302.6	-255.1					193.6	322.6		143.9	85.6
C_3H_8O	2-Propanol		-318.1	-272.6					181.1	309.2		156.5	89.3
C_3H_8O	Ethyl methyl ether			-216.4						309.2			93.3
$C_3H_8O_2$	1,2-Propylene glycol		-501.0	-429.8								190.8	
$C_3H_8O_2$	1,3-Propylene glycol		-480.8	-408.0									
$C_3H_8O_2$	Ethylene glycol monomethyl ether		-377.8	-348.5					244.0			171.1	
$C_3H_8O_2$	Dimethoxymethane											162.0	
$C_3H_8O_3$	Glycerol		-669.6	-577.9					206.3			218.9	
C_3H_8S	1-Propanethiol		-99.9	-67.8					242.5			144.6	
C_3H_8S	2-Propanethiol		-105.9	-76.2					233.5			145.3	
C_3H_8S	Ethyl methyl sulfide		-91.6	-59.6					239.1			144.6	

Molecular formula	Name	$\Delta_f H°$/kJ mol⁻¹ Crys.	Liq.	Gas	$\Delta_f G°$/kJ mol⁻¹ Crys.	Liq.	Gas	S°/J mol⁻¹ K⁻¹ Crys.	Liq.	Gas	C_p/J mol⁻¹ K⁻¹ Crys.	Liq.	Gas
$C_3H_8S_2$	1,3-Propanedithiol		-79.4	-29.8									
C_3H_9Al	Trimethyl aluminum		-136.4	-74.1		-9.9			209.4			155.6	
C_3H_9B	Trimethylborane		-143.1	-124.3		-32.1	-35.9		238.9	314.7			88.5
C_3H_9ClSi	Trimethylchlorosilane		-382.8	-352.8		-246.4	-243.5		278.2	369.1			
C_3H_9N	Propylamine		-101.5	-70.1			39.9			325.4		164.1	91.2
C_3H_9N	Isopropylamine		-112.3	-83.7			32.2		218.3	312.2		163.8	97.5
C_3H_9N	Trimethylamine		-45.7	-23.6					208.5	287.1		137.9	91.8
$C_3H_{10}ClN$	Propylamine hydrochloride	-354.7											
$C_3H_{10}ClN$	Trimethylamine hydrochloride	-282.9											
$C_3H_{10}N_2$	1,2-Propanediamine, (±)-		-97.8	-53.6									
$C_3H_{10}Si$	Trimethylsilane									331.0			117.9
$C_3H_{12}BN$	Trimethylamine borane	-142.5			70.7			187.0					
$C_3H_{12}BN$	Aminetrimethylboron	-284.1			-79.3			218.0					
C_4F_8	Perfluorocyclobutane			-1542.6									
C_4F_{10}	Perfluorobutane											127.2	
$C_4H_2N_2$	trans-2-Butenedinitrile	268.2		340.2									
$C_4H_2O_3$	Maleic anhydride	-469.8		-398.3									
$C_4H_2O_4$	2-Butynedioic acid	-577.3											
$C_4H_3NO_3$	2-Nitrofuran	-104.1		-28.8									
$C_4H_4BrNO_2$	N-Bromosuccinimide	-335.9											
$C_4H_4ClNO_2$	N-Chlorosuccinimide	-357.9											
$C_4H_4N_2$	Succinonitrile		139.7	209.7					191.6			145.6	
$C_4H_4N_2$	Pyrazine	139.8		196.1									
$C_4H_4N_2$	Pyrimidine		145.9	195.7									
$C_4H_4N_2$	Pyridazine		224.9	278.3									
$C_4H_4N_2O_2$	Uracil	-429.4		-302.9							120.5		
$C_4H_4N_2O_3$	2,4,6-Trihydroxypyrimidine	-634.7											
C_4H_4O	Furan		-62.3	-34.8					177.0	267.2		114.8	65.4
$C_4H_4O_2$	Diketene		-233.1	-190.3									
$C_4H_4O_3$	Succinic anhydride	-608.6		-527.9									
$C_4H_4O_4$	Maleic acid	-789.4		-679.4				160.8			137.0		
$C_4H_4O_4$	Fumaric acid	-811.7		-675.8				168.0			142.0		
C_4H_4S	Thiophene		80.2	114.9			126.1		181.2	278.8		123.8	72.8
C_4H_5N	trans-2-Butenenitrile		95.1	134.3									
C_4H_5N	3-Butenenitrile		117.8	159.7									
C_4H_5N	Pyrrole		63.1	108.2					156.4			127.7	
C_4H_5N	Cyclopropanecarbonitrile		140.8	182.8									
$C_4H_5NO_2$	Succinimide	-459.0		-375.4									
C_4H_5NS	4-Methylthiazole		67.9	111.8									
$C_4H_5N_3O$	Cytosine	-221.3									132.6		
C_4H_6	1,2-Butadiene		138.6	162.3									
C_4H_6	1,3-Butadiene		88.5	110.0					199.0			123.6	
C_4H_6	1-Butyne		141.4	165.2									
C_4H_6	2-Butyne		119.1	145.7									

STANDARD THERMODYNAMIC PROPERTIES OF CHEMICAL SUBSTANCES (continued)

Molecular formula	Name	ΔfH°/kJ mol⁻¹ Crys.	ΔfH°/kJ mol⁻¹ Liq.	ΔfH°/kJ mol⁻¹ Gas	ΔfG°/kJ mol⁻¹ Crys.	ΔfG°/kJ mol⁻¹ Liq.	ΔfG°/kJ mol⁻¹ Gas	S°/J mol⁻¹ K⁻¹ Crys.	S°/J mol⁻¹ K⁻¹ Liq.	S°/J mol⁻¹ K⁻¹ Gas	Cp/J mol⁻¹ K⁻¹ Crys.	Cp/J mol⁻¹ K⁻¹ Liq.	Cp/J mol⁻¹ K⁻¹ Gas
C_4H_6	Cyclobutene			156.7									
$C_4H_6N_2O_2$	2,5-Piperazinedione	-446.5											
C_4H_6O	Divinyl ether		-39.8	-13.6									
C_4H_6O	trans-2-Butenal		-138.7	-100.6									
$C_4H_6O_2$	trans-Crotonic acid											161.1	
$C_4H_6O_2$	Methacrylic acid		-349.2	-314.4									
$C_4H_6O_2$	Vinyl acetate		-362.2	-333.0								158.8	
$C_4H_6O_2$	Methyl acrylate			-366.5					239.5			141.4	
$C_4H_6O_2$	γ-Butyrolactone		-420.9										
$C_4H_6O_3$	Acetic anhydride		-624.4	-572.5								218.6	
$C_4H_6O_3$	Propylene carbonate		-613.2	-582.5									
$C_4H_6O_4$	Succinic acid	-940.5		-823.0				167.3			153.1		
$C_4H_6O_4$	Dimethyl oxalate	-756.3		-708.9									
C_4H_6S	2,3-Dihydrothiophene		52.9	90.7			133.5			303.5			79.8
C_4H_6S	2,5-Dihydrothiophene		47.0	86.9			131.6			297.1			83.3
C_4H_7ClO	2-Chloroethyl vinyl ether		-208.1	-170.1									
$C_4H_7ClO_2$	2-Chlorobutanoic acid		-575.5										
$C_4H_7ClO_2$	3-Chlorobutanoic acid		-556.3										
$C_4H_7ClO_2$	4-Chlorobutanoic acid		-566.3										
$C_4H_7ClO_2$	Propyl chlorocarbonate		-533.4	-492.7									
C_4H_7N	Butanenitrile		-5.8	33.6									
C_4H_7N	2-Methylpropanenitrile		-13.8	23.4									
C_4H_7NO	2-Pyrrolidone		-286.2										
C_4H_7NO	2-Methyl-2-oxazoline		-169.5	-130.5									
$C_4H_7NO_4$	Iminodiacetic acid	-932.6											
$C_4H_7NO_4$	Ethyl nitroacetate		-487.1										
$C_4H_7NO_4$	L-Aspartic acid	-973.3											
$C_4H_7N_3O$	4H-Imidazol-4-one, 2-amino-1,5-dihydro-1-methyl-	-238.5											
C_4H_8	1-Butene		-20.8	0.1					227.0			118.0	
C_4H_8	cis-2-Butene		-29.8	-7.1					219.9			127.0	
C_4H_8	trans-2-Butene		-33.3	-11.4									
C_4H_8	Isobutene		-37.5	-16.9									
C_4H_8	Cyclobutane		3.7	27.7									
C_4H_8	Methylcyclopropane		1.7										
$C_4H_8Br_2$	1,2-Dibromobutane		-142.1	-91.6									
$C_4H_8Br_2$	1,3-Dibromobutane		-148.0										
$C_4H_8Br_2$	1,4-Dibromobutane		-140.3	-87.8									
$C_4H_8Br_2$	2,3-Dibromobutane		-139.6	-102.0									
$C_4H_8Br_2$	1,2-Dibromo-2-methylpropane		-156.6	-113.3									
$C_4H_8Cl_2$	1,3-Dichlorobutane		-237.3	-195.0									
$C_4H_8Cl_2$	1,4-Dichlorobutane		-229.8	-183.4									
$C_4H_8Cl_2O$	Bis(2-chloroethyl) ether											220.9	

Molecular formula	Name	$\Delta_f H°$/kJ mol^{-1}			$\Delta_f G°$/kJ mol^{-1}			$S°$/J mol^{-1} K^{-1}			C_p/J mol^{-1} K^{-1}		
		Crys.	Liq.	Gas	Crys.	Liq.	Gas	Crys.	Liq.	Gas	Crys.	Liq.	Gas
C$_4$H$_8$I$_2$	1,4-Diiodobenzene		-30.0										
C$_4$H$_8$N$_2$O$_2$	Succinamide	-581.2											
C$_4$H$_8$N$_2$O$_2$	Dimethylglyoxime	-199.7											
C$_4$H$_8$N$_2$O$_3$	L-Asparagine	-789.4											
C$_4$H$_8$N$_2$O$_3$	N-Glycylglycine	-747.7											
C$_4$H$_8$N$_2$O$_4$	1,4-Dinitrobutane		-237.5										
C$_4$H$_8$O	Ethyl vinyl ether		-167.4	-140.8									
C$_4$H$_8$O	1,2-Epoxybutane		-168.9						230.9				
C$_4$H$_8$O	Butanal		-239.2	-204.8					246.6	343.7		147.0	103.4
C$_4$H$_8$O	Isobutanal		-247.3	-215.7								163.7	
C$_4$H$_8$O	2-Butanone		-273.3	-238.5					239.1	339.9		158.7	101.7
C$_4$H$_8$O	Tetrahydrofuran		-216.2	-184.1					204.3	302.4		124.0	76.3
C$_4$H$_8$OS	S-Ethyl thioacetate		-268.2	-228.1									
C$_4$H$_8$O$_2$	Butanoic acid		-533.8	-475.9					222.2			178.6	
C$_4$H$_8$O$_2$	2-Methylpropanoic acid											173.0	
C$_4$H$_8$O$_2$	Propyl formate		-500.3	-462.7					257.7			170.7	
C$_4$H$_8$O$_2$	Ethyl acetate		-479.3	-443.6								143.9	
C$_4$H$_8$O$_2$	1,3-Dioxane		-379.7	-340.6									
C$_4$H$_8$O$_2$	1,4-Dioxane		-353.9	-315.3					270.2			152.1	
C$_4$H$_8$O$_2$	2-Methyl-1,3-dioxolane		-386.9	-352.0									
C$_4$H$_8$O$_2$S	Sulfolane											180.0	
C$_4$H$_8$S	Tetrahydrothiophene		-72.9	-34.1			45.8			309.6			92.5
C$_4$H$_8$S$_2$	1,3-Dithiane			-10.0			72.4			333.5			110.4
C$_4$H$_9$Br	1-Bromobutane		-143.8	-107.1									
C$_4$H$_9$Br	2-Bromobutane		-154.9	-120.3									
C$_4$H$_9$Br	2-Bromo-2-methylpropane		-164.4	-132.4									
C$_4$H$_9$Cl	1-Chlorobutane		-188.1	-154.4									
C$_4$H$_9$Cl	2-Chlorobutane		-192.8	-161.1									
C$_4$H$_9$Cl	1-Chloro-2-methylpropane		-191.1	-159.3									
C$_4$H$_9$Cl	2-Chloro-2-methylpropane		-211.3	-182.2									
C$_4$H$_9$ClO	2-Chloroethyl ethyl ether		-335.6	-301.3								162.3	
C$_4$H$_9$I	1-Iodo-2-methylpropane		-107.5	-72.1									
C$_4$H$_9$I	2-Iodo-2-methylpropane		5.6	41.2									
C$_4$H$_9$N	Cyclobutanamine		-41.1	-3.6									
C$_4$H$_9$N	Pyrrolidine								204.1			156.6	
C$_4$H$_9$NO	Butanamide		-346.9	-282.0									
C$_4$H$_9$NO	N-Methylpropanamide			-282.6								179.0	
C$_4$H$_9$NO	2-Methylpropanamide	-368.6											
C$_4$H$_9$NO	N,N-Dimethylacetamide		-278.3	-228.0								175.6	
C$_4$H$_9$NO	Morpholine											164.8	
C$_4$H$_9$NO$_2$	1-Nitrobutane		-192.5	-143.9									
C$_4$H$_9$NO$_2$	2-Nitroisobutane		-217.2	-177.1									
C$_4$H$_9$NO$_2$	Propyl carbamate	-552.6											
C$_4$H$_9$NO$_2$	4-Aminobutanoic acid	-581.0		-471.4									
C$_4$H$_9$NO$_3$	3-Nitro-2-butanol		-390.0	-441.0									

Molecular formula	Name	$\Delta_f H°$/kJ mol⁻¹ Crys.	Liq.	Gas	$\Delta_f G°$/kJ mol⁻¹ Crys.	Liq.	Gas	$S°$/J mol⁻¹ K⁻¹ Crys.	Liq.	Gas	C_p/J mol⁻¹ K⁻¹ Crys.	Liq.	Gas
C₄H₉NO₃	2-Methyl-2-nitro-1-propanol	-410.1											
C₄H₉NO₃	DL-Threonine	-758.8											
C₄H₉NO₃	L-Threonine	-807.2											
C₄H₉N₃O₂	Creatine	-537.2											
C₄H₁₀	Butane		-147.3	-125.7								140.9	
C₄H₁₀	Isobutane		-154.2	-134.2								182.8	
C₄H₁₀Hg	Diethyl mercury		30.1	75.3									
C₄H₁₀N₂	Piperazine	-45.6											
C₄H₁₀N₂O	Trimethylurea	-330.5											
C₄H₁₀N₂O₂	N-Nitrodiethylamine		-106.2	-53.0									
C₄H₁₀N₂O₄	L-Asparagine, monohydrate	-1086.6											
C₄H₁₀O	1-Butanol		-327.3	-274.9					225.8			177.2	
C₄H₁₀O	2-Butanol		-342.6	-292.8					214.9	359.5		196.9	112.7
C₄H₁₀O	2-Methyl-1-propanol		-334.7	-283.8					214.7			181.5	
C₄H₁₀O	2-Methyl-2-propanol		-359.2	-312.5					193.3	326.7		218.6	113.6
C₄H₁₀O	Diethyl ether		-279.5	-252.1					172.4	342.7		175.6	119.5
C₄H₁₀O	Methyl propyl ether		-266.0	-238.1					262.9			165.4	
C₄H₁₀O	Isopropyl methyl ether		-278.8	-252.0					253.8			161.9	
C₄H₁₀OS	Diethyl sulfoxide		-268.0	-205.6									
C₄H₁₀O₂	1,2-Butanediol, (±)-		-523.6										
C₄H₁₀O₂	1,3-Butanediol		-501.0	-433.2								200.1	
C₄H₁₀O₂	1,4-Butanediol		-505.3	-428.7					223.4			213.0	
C₄H₁₀O₂	2,3-Butanediol		-541.5	-482.3									
C₄H₁₀O₂	2-Methyl-1,2-propanediol		-539.7										
C₄H₁₀O₂	Ethylene glycol monoethyl ether		-376.6									210.8	
C₄H₁₀O₂	Ethylene glycol dimethyl ether		-420.6	-389.7								193.3	
C₄H₁₀O₂	Dimethylacetal		-293.6	-245.9									
C₄H₁₀O₂	tert-Butyl hydroperoxide												
C₄H₁₀O₃	Diethylene glycol		-628.5	-571.2									
C₄H₁₀O₃S	Diethyl sulfite		-600.7									244.8	
C₄H₁₀O₄S	Sulfuric acid, diethyl ester		-813.2	-756.3									
C₄H₁₀S	1-Butanethiol		-124.7	-88.0								171.2	
C₄H₁₀S	2-Butanethiol		-131.0	-96.9									
C₄H₁₀S	2-Methyl-1-propanethiol		-132.0	-97.3									
C₄H₁₀S	2-Methyl-2-propanethiol		-140.5	-109.6									
C₄H₁₀S	Diethyl sulfide		-119.4	-83.5					269.3	368.1		171.4	117.0
C₄H₁₀S	Methyl propyl sulfide		-118.5	-82.2					272.5			171.6	
C₄H₁₀S	Isopropyl methyl sulfide		-124.7	-90.5					263.1			172.4	
C₄H₁₀S₂	1,4-Butanedithiol		-105.7	-50.6								171.4	
C₄H₁₀S₂	Diethyl disulfide		-120.1	-79.4					269.3			179.2	
C₄H₁₁N	Butylamine		-127.6	-91.9									
C₄H₁₁N	sec-Butylamine		-137.5	-104.6								192.1	
C₄H₁₁N	tert-Butylamine		-150.6	-121.0									
C₄H₁₁N	Isobutylamine		-132.6	-98.7								183.2	
C₄H₁₁N	Diethylamine		-103.7	-72.2								169.2	

Molecular formula	Name	ΔfH°/kJ mol⁻¹			ΔfG°/kJ mol⁻¹			S°/J mol⁻¹ K⁻¹			Cp/J mol⁻¹ K⁻¹		
		Crys.	Liq.	Gas	Crys.	Liq.	Gas	Crys.	Liq.	Gas	Crys.	Liq.	Gas
C₄H₁₁NO	N,N-Dimethylethanolamine			-203.6									
C₄H₁₁NO₂	Diethanolamine		-253.7	-397.1							233.5		
C₄H₁₁NO₃	2-Amino-2-(hydroxymethyl)-1,3-propanediol	-717.8											
C₄H₁₂BrN	Tetramethylammonium bromide	-251.0											
C₄H₁₂ClN	Diethylamine hydrochloride	-358.6											
C₄H₁₂ClN	Tetramethylammonium chloride	-276.4											
C₄H₁₂IN	Tetramethylammonium iodide	-203.9											
C₄H₁₂N₂	2-Methyl-1,2-propanediamine		-133.9	-90.3									
C₄H₁₂Pb	Tetramethyl lead		97.9	135.9									
C₄H₁₂Si	Tetramethylsilane		-264.0	-239.1		-100.0	-99.9		277.3	359.0		204.1	143.9
C₄H₁₂Sn	Tetramethylstannane		-52.3	-18.8								254.0	
C₄H₁₃N₃	Diethylenetriamine												
C₄N₂	2-Butynedinitrile		500.4	529.2									
C₄NiO₄	Nickel carbonyl		-633.0	-602.9		-588.2	-587.2		313.4	410.6		204.6	145.2
C₅FeO₅	Iron pentacarbonyl		-774.0			-705.3			338.1			240.6	
C₅H₂F₆O₂	Hexafluoroacetylacetone	-2286.7											
C₅H₃NO₅	5-Nitro-2-furancarboxylic acid	-516.8											
C₅H₄N₄	1H-Purine	169.4											
C₅H₄N₄O	Hypoxanthine	-110.8						145.6			134.5		
C₅H₄N₄O₂	Xanthine	-379.6						161.1			151.3		
C₅H₄N₄O₃	Uric acid	-618.8						173.2			166.1		
C₅H₄O₂	Furfural		-201.6	-151.0									
C₅H₄O₃	2-Furancarboxylic acid	-498.4		-390.0									
C₅H₄O₃	3-Methyl-2,5-furandione		-504.5	-447.2									
C₅H₅F₃O₂	1,1,1-Trifluoro-2,4-pentanedione		-1040.2	-993.3									
C₅H₅N	Pyridine		100.2	140.4								163.2	
C₅H₅NO	1H-Pyrrole-2-carboxaldehyde	-106.4											
C₅H₅N₅	Adenine	96.9		205.7							147.0		
C₅H₅N₅O	Guanine	-183.9											
C₅H₆	cis-3-Penten-1-yne		226.5										
C₅H₆	trans-3-Penten-1-yne		228.2										
C₅H₆	1,3-Cyclopentadiene		105.9	134.3								132.7	
C₅H₆N₂O₂	Thymine	-462.8		-328.7							150.8		
C₅H₆O₂	Furfuryl alcohol		-276.2									204.0	
C₅H₆O₄	trans-1-Propene-1,2-dicarboxylic acid	-824.4		-211.8									
C₅H₆S	2-Methylthiophene		44.6	83.5					218.5				
C₅H₆S	3-Methylthiophene		43.1	82.5								149.8	
C₅H₇N	trans-3-Pentenenitrile		80.9	125.7									
C₅H₇N	Cyclobutanecarbonitrile		103.0	147.4									

Molecular formula	Name	$\Delta_f H°$/kJ mol^{-1}			$\Delta_f G°$/kJ mol^{-1}			$S°$/J mol^{-1} K^{-1}			C_p/J mol^{-1} K^{-1}		
		Crys.	Liq.	Gas	Crys.	Liq.	Gas	Crys.	Liq.	Gas	Crys.	Liq.	Gas
C_5H_7N	1-Methylpyrrole		62.4	103.1								220.2	
C_5H_7N	2-Methylpyrrole		23.3	74.0									
C_5H_7N	3-Methylpyrrole		20.5	70.2									
$C_5H_7NO_2$	Ethyl cyanoacetate												
C_5H_8	1,2-Pentadiene			140.7									
C_5H_8	cis-1,3-Pentadiene			81.4									
C_5H_8	trans-1,3-Pentadiene			76.1									
C_5H_8	1,4-Pentadiene			105.7									
C_5H_8	2,3-Pentadiene			133.1									
C_5H_8	3-Methyl-1,2-butadiene		101.2										
C_5H_8	2-Methyl-1,3-butadiene		48.2	75.5					229.3			152.6	
C_5H_8	Cyclopentene		4.3	34.0					201.2			122.4	
C_5H_8	Methylenecyclobutane		93.8	121.6									
C_5H_8	Spiropentane		157.5	185.2					193.7			134.5	
$C_5H_8N_4O_{12}$	Pentaerythritol tetranitrate	-538.6											
C_5H_8O	Cyclopentanone		-235.9	-192.1									
$C_5H_8O_2$	4-Pentenoic acid	-430.6	-382.9	-341.9									
$C_5H_8O_2$	Methyl trans-2-butenoate		-423.8	-382.0								191.2	
$C_5H_8O_2$	Methyl methacrylate											184.1	
$C_5H_8O_2$	Allyl acetate												
$C_5H_8O_2$	2,4-Pentanedione												
$C_5H_8O_2$	4-Methyl-gamma-butyrolactone		-461.3	-406.5									
$C_5H_8O_2$	Tetrahydro-2H-pyran-2-one		-436.7	-379.6									
$C_5H_8O_4$	Glutaric acid	-960.0											
$C_5H_9ClO_2$	Propyl chloroacetate		-515.5	-467.0									
C_5H_9N	Pentanenitrile		-33.1	10.5					232.0			179.4	
C_5H_9N	2,2-Dimethylpropanenitrile		-39.8	-2.3									
C_5H_9N	1,2,5,6-Tetrahydropyridine		33.5										
C_5H_9NO	2-Piperidinone	-306.6											
C_5H_9NO	N-Methyl-2-pyrrolidone		-262.2	-366.2								307.8	
$C_5H_9NO_2$	L-Proline	-515.2											
$C_5H_9NO_4$	D-Glutamic acid	-1005.3											
$C_5H_9NO_4$	L-Glutamic acid	-1009.7											
C_5H_{10}	1-Pentene		-46.9	-21.1					262.6			154.0	
C_5H_{10}	cis-2-Pentene		-53.7	-27.6					258.6			151.7	
C_5H_{10}	trans-2-Pentene		-58.2	-31.9					256.5			157.0	
C_5H_{10}	2-Methyl-1-butene		-61.1	-35.2					254.0			157.2	
C_5H_{10}	3-Methyl-1-butene		-51.5	-27.5					253.3			156.1	
C_5H_{10}	2-Methyl-2-butene		-68.6	-41.7					251.0			152.8	
C_5H_{10}	Cyclopentane		-105.1	-76.4					204.5			128.8	
C_5H_{10}	Methylcyclobutane		-44.5										
C_5H_{10}	Ethylcyclopropane		-24.8										
C_5H_{10}	1,1-Dimethylcyclopropane		-33.3	-8.2									
C_5H_{10}	cis-1,2-Dimethylcyclopropane		-26.3										

STANDARD THERMODYNAMIC PROPERTIES OF CHEMICAL SUBSTANCES (continued)

Molecular formula	Name	$\Delta_f H°$/kJ mol⁻¹			$\Delta_f G°$/kJ mol⁻¹			S°/J mol⁻¹ K⁻¹			C_p/J mol⁻¹ K⁻¹		
		Crys.	Liq.	Gas	Crys.	Liq.	Gas	Crys.	Liq.	Gas	Crys.	Liq.	Gas
C_5H_{10}	trans-1,2-Dimethyl-cyclopropane		-30.7										
$C_5H_{10}Br_2$	1,3-Dibromo-2-methylpropane			-137.6									
$C_5H_{10}N_2O$	N-Nitrosopiperidine		-31.1	16.6									
$C_5H_{10}N_2O_2$	N-Nitropiperidine		-93.0	-44.5									
$C_5H_{10}N_2O_3$	L-Glutamine	-826.4											
$C_5H_{10}O$	Cyclopentanol		-300.1	-242.5					206.3			184.1	
$C_5H_{10}O$	Pentanal		-267.2	-228.4								184.1	
$C_5H_{10}O$	2-Pentanone		-297.3	-258.8					266.0			190.9	
$C_5H_{10}O$	3-Pentanone		-296.5	-257.9					268.5			179.9	
$C_5H_{10}O$	3-Methyl-2-butanone		-299.5	-262.6									
$C_5H_{10}O$	3,3-Dimethyloxetane		-182.2	-148.2									
$C_5H_{10}O$	Tetrahydropyran		-258.3	-223.4									
$C_5H_{10}OS$	S-Propyl thioacetate		-294.5	-250.4					259.8			210.3	
$C_5H_{10}O_2$	Pentanoic acid		-559.4	-491.9									
$C_5H_{10}O_2$	2-Methylbutanoic acid		-554.5	-510.0									
$C_5H_{10}O_2$	3-Methylbutanoic acid		-561.6	-491.3									
$C_5H_{10}O_2$	2,2-Dimethylpropanoic acid	-564.5											
$C_5H_{10}O_2$	Butyl formate		-518.9	-481.6								200.2	
$C_5H_{10}O_2$	Propyl acetate		-502.7	-463.4								196.2	
$C_5H_{10}O_2$	Isopropyl acetate											199.4	
$C_5H_{10}O_2$	Ethyl propanoate												
$C_5H_{10}O_2$	cis-1,2-Cyclopentanediol	-485.0											
$C_5H_{10}O_2$	trans-1,2-Cyclopentanediol	-490.1											
$C_5H_{10}O_2$	4-Methyl-1,3-dioxane	-416.1		-376.9									
$C_5H_{10}O_2$	(Ethoxymethyl)oxirane		-296.5										
$C_5H_{10}O_2$	Tetrahydrofurfuryl alcohol		-435.7	-369.1									
$C_5H_{10}O_3$	Diethyl carbonate		-681.5	-637.9									
$C_5H_{10}O_3$	Ethylene glycol momomethyl ether acetate												
$C_5H_{10}O_3$	Ethyl lactate		-909.2									310.0	
$C_5H_{10}O_4$	1,2,3-Propanetriol, 1-acetate, (±)-											254.0	
$C_5H_{10}O_5$	D-Ribose	-1047.2											
$C_5H_{10}O_5$	D-Xylose	-1057.8											
$C_5H_{10}O_5$	α-D-Arabinopyranose	-1057.9											
$C_5H_{10}S$	Thiacyclohexane		-106.3	-63.5			53.1		218.2	323.0		163.3	109.7
$C_5H_{10}S$	Cyclopentanethiol		-89.5	-48.1					256.9			165.2	
$C_5H_{11}Br$	1-Bromopentane		-170.2	-128.9									
$C_5H_{11}Cl$	1-Chloropentane		-213.2	-174.9									
$C_5H_{11}Cl$	2-Chloro-2-methylbutane		-235.7	-202.2									
$C_5H_{11}Cl$	1-Chloro-3-methylbutane		-216.0	-179.7									
$C_5H_{11}Cl$	2-Chloro-3-methylbutane		-226.6	-185.1									
$C_5H_{11}N$	Cyclopentylamine		-95.1	-54.9					241.0			181.2	

Molecular formula	Name	$\Delta_f H°$/kJ mol⁻¹			$\Delta_f G°$/kJ mol⁻¹			$S°$/J mol⁻¹ K⁻¹			C_p/J mol⁻¹ K⁻¹		
		Crys.	Liq.	Gas	Crys.	Liq.	Gas	Crys.	Liq.	Gas	Crys.	Liq.	Gas
C₅H₁₁N	Piperidine		-86.4	-47.1					210.0			179.9	
C₅H₁₁NO	Pentanamide	-379.5		-290.2									
C₅H₁₁NO	2,2-Dimethylpropanamide	-399.7		-313.1									
C₅H₁₁NO₂	1-Nitropentane		-215.4										
C₅H₁₁NO₂	DL-Valine	-628.9		-455.1									
C₅H₁₁NO₂	L-Valine	-617.9		-460.0									
C₅H₁₁NO₂	5-Aminopentanoic acid	-604.1											
C₅H₁₁NO₂S	L-Methionine	-577.5		-413.5									
C₅H₁₁NO₄	2-Ethyl-2-nitro-1,3-propanediol	-606.4											
C₅H₁₂	Pentane		-173.5	-146.9					260.4			167.2	
C₅H₁₂	Isopentane		-178.4	-153.6								164.8	
C₅H₁₂	Neopentane		-190.2	-168.0									
C₅H₁₂N₂O	Butylurea	-419.5											
C₅H₁₂N₂O	tert-Butylurea	-417.4											
C₅H₁₂N₂O	N,N-Diethylurea	-372.2											
C₅H₁₂N₂O	Tetramethylurea	-262.2											
C₅H₁₂N₂S	Tetramethylthiourea	-38.1		44.9									
C₅H₁₂O	1-Pentanol		-351.6	-294.6								208.1	
C₅H₁₂O	2-Pentanol		-365.2	-311.0									
C₅H₁₂O	3-Pentanol		-368.9	-314.9								239.7	
C₅H₁₂O	2-Methyl-1-butanol		-356.6	-301.4									
C₅H₁₂O	3-Methyl-1-butanol		-356.4	-300.7									
C₅H₁₂O	2-Methyl-2-butanol		-379.5	-329.3								247.1	
C₅H₁₂O	3-Methyl-2-butanol		-366.6	-313.5									
C₅H₁₂O	2,2-Dimethyl-1-propanol		-399.4										
C₅H₁₂O	Butyl methyl ether		-290.6	-258.1					295.3			192.7	
C₅H₁₂O	tert-Butyl methyl ether		-313.6	-283.7					265.3			187.5	
C₅H₁₂O	Ethyl propyl ether		-303.6	-272.0					295.0			197.2	
C₅H₁₂O₂	1,5-Pentanediol		-528.8	-450.8									
C₅H₁₂O₂	2,2-Dimethyl-1,3-propanediol	-551.2	-450.5	-414.7									
C₅H₁₂O₂	Diethoxymethane		-443.6										
C₅H₁₂O₂	1,1-Dimethoxypropane		-459.4	-429.9									
C₅H₁₂O₃	Diethylene glycol monomethyl ether											271.1	
C₅H₁₂O₃	2-(Hydroxymethyl)-2-methyl-1,3-propanediol	-744.6		-776.7									
C₅H₁₂O₄	Pentaerythritol	-920.6											
C₅H₁₂O₅	Xylitol	-1118.5											
C₅H₁₂S	1-Pentanethiol		-151.3	-110.0					290.1			198.1	
C₅H₁₂S	2-Methyl-1-butanethiol		-154.4	-114.9									
C₅H₁₂S	3-Methyl-1-butanethiol		-154.4	-114.9									
C₅H₁₂S	2-Methyl-2-butanethiol		-162.8	-127.1									
C₅H₁₂S	3-Methyl-2-butanethiol		-158.8	-121.3									

Molecular formula	Name	$\Delta_f H°$/kJ mol^{-1} Crys.	Liq.	Gas	$\Delta_f G°$/kJ mol^{-1} Crys.	Liq.	Gas	$S°$/J mol^{-1} K^{-1} Crys.	Liq.	Gas	C_p/J mol^{-1} K^{-1} Crys.	Liq.	Gas
$C_5H_{12}S$	2,2-Dimethyl-1-propanethiol		-165.4	-129.0					307.5			200.9	
$C_5H_{12}S$	Butyl methyl sulfide		-142.9	-102.4					276.1			199.9	
$C_5H_{12}S$	tert-Butyl methyl sulfide		-157.1	-121.3					309.5			198.4	
$C_5H_{12}S$	Ethyl propyl sulfide		-144.8	-104.8									
$C_5H_{12}S$	Ethyl isopropyl sulfide		-156.1	-118.3									
$C_5H_{13}N$	Pentylamine		-51.1	-18.2								218.0	
$C_5H_{14}N_2$	N,N,N',N'-Tetramethyl-methanediamine												
C_6ClF_5	Chloropentafluorobenzene	-858.4		-809.3									
C_6Cl_6	Hexachlorobenzene	-127.6		-35.5							201.2		
C_6F_6	Hexafluorobenzene		-991.3	-955.4									
C_6F_{10}	Perfluorocyclohexene		-1963.5	-1932.7									
C_6F_{12}	Perfluorocyclohexane		-2406.3	-2370.4				260.2	280.8			221.6	
C_6HF_5	Pentafluorobenzene	-852.7	-841.8	-806.5									
C_6HF_5O	Pentafluorophenol	-1024.1	-1007.7										
$C_6H_2F_4$	1,2,4,5-Tetrafluorobenzene		-683.8										
$C_6H_3Cl_3$	1,2,3-Trichlorobenzene	-70.8		3.8									
$C_6H_3Cl_3$	1,2,4-Trichlorobenzene		-63.1	-8.1									
$C_6H_3Cl_3$	1,3,5-Trichlorobenzene	-78.4		-13.4									
$C_6H_3N_3O_8$	2,4,6-Trinitro-1,3-benzenediol	-467.5											
$C_6H_4Cl_2$	o-Dichlorobenzene		-17.5	30.2									
$C_6H_4Cl_2$	m-Dichlorobenzene		-20.7	25.7									
$C_6H_4Cl_2$	p-Dichlorobenzene	-42.3		22.5				175.4			147.8	162.4	
$C_6H_4F_2$	o-Difluorobenzene		-330.0	-293.8					222.6			159.0	
$C_6H_4F_2$	m-Difluorobenzene		-343.9	-309.2					223.8			159.1	
$C_6H_4F_2$	p-Difluorobenzene		-342.3	-306.7								157.5	
$C_6H_4N_2O_5$	2,4-Dinitrophenol	-232.7		-128.1									
$C_6H_4O_2$	p-Benzoquinone	-185.7		-122.9									
C_6H_5Br	Bromobenzene		60.9						219.2		129.0	154.3	
C_6H_5Cl	Chlorobenzene		11.1	52.0								150.1	
C_6H_5ClO	m-Chlorophenol	-206.4	-189.3										
C_6H_5ClO	p-Chlorophenol	-197.7	-181.3										
C_6H_5F	Fluorobenzene		-150.6	-115.9					205.9			146.4	
C_6H_5I	Iodobenzene		117.2	164.9					205.4			158.7	
$C_6H_5NO_2$	Nitrobenzene		12.5	67.5								185.8	
$C_6H_5NO_3$	3-Pyridinecarboxylic acid	-344.9		-221.5									
$C_6H_5NO_3$	2-Nitrophenol	-202.4											
$C_6H_5N_3$	1H-Benzotriazole	236.5		335.5									
$C_6H_5N_3O_4$	2,3-Dinitroaniline	-11.7											
$C_6H_5N_3O_4$	2,4-Dinitroaniline	-67.8											
$C_6H_5N_3O_4$	2,5-Dinitroaniline	-44.3											
$C_6H_5N_3O_4$	2,6-Dinitroaniline	-50.6											
$C_6H_5N_3O_4$	3,5-Dinitroaniline	-38.9											
C_6H_6	1,5-Hexadiyne		384.2										
C_6H_6	Benzene		49.1	82.9		124.5	129.7		173.4	269.2		136.0	82.4

Molecular formula	Name	$\Delta_f H°$/kJ mol⁻¹			$\Delta_f G°$/kJ mol⁻¹			$S°$/J mol⁻¹ K⁻¹			C_p/J mol⁻¹ K⁻¹		
		Crys.	Liq.	Gas	Crys.	Liq.	Gas	Crys.	Liq.	Gas	Crys.	Liq.	Gas
C₆H₆N₂O₂	o-Nitroaniline	-26.1	-9.4	63.8							166.0		
C₆H₆N₂O₂	m-Nitroaniline	-38.3	-14.4	58.4							158.8		
C₆H₆N₂O₂	p-Nitroaniline	-42.0	-20.7	58.8							167.0		
C₆H₆O	Phenol	-165.1		-96.4							127.4		
C₆H₆O	2-Vinylfuran		-10.3	27.8									
C₆H₆O₂	p-Hydroquinone	-364.5		-265.3							136.0		
C₆H₆O₂	Pyrocatechol	-354.1		-267.5									
C₆H₆O₂	Resorcinol	-368.0		-274.7									
C₆H₆O₃	1,2,3-Benzenetriol	-551.1		-434.2									
C₆H₆O₃	1,2,4-Benzenetriol	-563.8		-444.0									
C₆H₆O₃	1,3,5-Benzenetriol	-584.6		-452.9									
C₆H₆O₃	3,4-Dimethyl-2,5-furandione	-581.4						144.0					
C₆H₆O₆	cis-1-Propene-1,2,3-tricarboxylic acid	-1224.4											
C₆H₆O₆	trans-1-Propene-1,2,3-tricarboxylic acid	-1232.7											
C₆H₆S	Benzenethiol		63.7	111.3					222.8			173.2	
C₆H₇N	Aniline		31.6	87.5			-7.0			317.9		191.9	107.9
C₆H₇N	2-Methylpyridine		56.7	99.2					216.3			158.6	
C₆H₇N	3-Methylpyridine		61.9	106.4					209.1			158.7	
C₆H₇N	4-Methylpyridine		59.2	103.8								159.0	
C₆H₇N	1-Cyclopentenecarbonitrile		111.5	156.5									
C₆H₈N₂	Adiponitrile		85.1	149.5								128.7	
C₆H₈N₂	o-Phenylenediamine	-0.3											
C₆H₈N₂	m-Phenylenediamine	-7.8						154.5			159.6		
C₆H₈N₂	p-Phenylenediamine	3.0											
C₆H₈N₂	Phenylhydrazine		141.0	202.9								217.0	
C₆H₈N₂S	Bis(2-cyanoethyl) sulfide		96.3									263.2	
C₆H₈O₄	Dimethyl maleate												
C₆H₈O₆	L-Ascorbic acid	-1164.6											
C₆H₈O₇	2-Hydroxy-1,2,3-propane-tricarboxylic acid	-1543.8											
C₆H₉Cl₃O₂	Butyl trichloroacetate		-545.8	-492.3									
C₆H₉Cl₃O₂	Isobutyl trichloroacetate		-553.4	-500.2									
C₆H₉N	Cyclopentanecarbonitrile		0.7	44.1									
C₆H₉N	2,4-Dimethylpyrrole	-422.3											
C₆H₉N	2,5-Dimethylpyrrole		-16.7	39.8									
C₆H₉NO₃	Triacetamide		-610.5	-550.1									
C₆H₉NO₆	Nitrilotriacetic acid	-1311.9											
C₆H₉N₃O₂	L-Histidine	-466.7											
C₆H₁₀	1,5-Hexadiene		54.1	84.2									
C₆H₁₀	3,3-Dimethyl-1-butyne		78.4										
C₆H₁₀	Cyclohexene		-38.5	-5.0									
C₆H₁₀	1-Methylcyclopentene		-36.4	-3.8					214.6			148.3	
C₆H₁₀	3-Methylcyclopentene		-23.7	7.4									

Molecular formula	Name	ΔfH°/kJ mol⁻¹			ΔfG°/kJ mol⁻¹			S°/J mol⁻¹ K⁻¹			C_p/J mol⁻¹ K⁻¹		
		Crys.	Liq.	Gas	Crys.	Liq.	Gas	Crys.	Liq.	Gas	Crys.	Liq.	Gas
C_6H_{10}	4-Methylcyclopentene		-17.6	14.6									
$C_6H_{10}Cl_2O_2$	Butyl dichloroacetate		-550.1	-497.8									
$C_6H_{10}O$	Cyclohexanone		-271.2	-226.1								182.2	
$C_6H_{10}O$	2-Methylcyclopentanone		-265.2									212.5	
$C_6H_{10}O$	Mesityl oxide		-420.0	-375.6									
$C_6H_{10}O_2$	Ethyl trans-2-butenoate		-395.0	-350.2									
$C_6H_{10}O_2$	Methyl cyclobutane-carboxylate												
$C_6H_{10}O_3$	Ethyl acetoacetate		-679.1	-626.5								248.0	
$C_6H_{10}O_3$	Propanoic anhydride												
$C_6H_{10}O_4$	Adipic acid	-994.3											
$C_6H_{10}O_4$	Diethyl oxalate		-805.5	-742.0									
$C_6H_{10}O_4$	Ethylene glycol diacetate											310.0	
$C_6H_{11}Cl$	Chlorocyclohexane		-207.2	-163.7									
$C_6H_{11}ClO_2$	Ethyl 4-chlorobutanoate		-566.5	-513.8									
$C_6H_{11}ClO_2$	Propyl 3-chloropropanoate		-537.6	-485.7									
$C_6H_{11}ClO_2$	Butyl chloroacetate		-538.4	-487.4									
$C_6H_{11}NO$	Caprolactam	-329.4		-239.6							156.8		
$C_6H_{11}NO$	1-Methyl-2-piperidinone		-293.0										
C_6H_{12}	1-Hexene		-74.2	-43.5					295.2			183.3	
C_6H_{12}	cis-2-Hexene		-83.9	-52.3									
C_6H_{12}	trans-2-Hexene		-85.5	-53.9									
C_6H_{12}	cis-3-Hexene		-78.9	-47.6									
C_6H_{12}	trans-3-Hexene		-86.1	-54.4									
C_6H_{12}	2-Methyl-1-pentene		-90.0	-59.4									
C_6H_{12}	3-Methyl-1-pentene		-78.2	-49.5									
C_6H_{12}	4-Methyl-1-pentene		-80.0	-51.3									
C_6H_{12}	3-Methyl-cis-2-pentene		-94.5	-62.3									
C_6H_{12}	4-Methyl-cis-2-pentene		-87.0	-57.5									
C_6H_{12}	3-Methyl-trans-2-pentene		-94.6	-63.1									
C_6H_{12}	4-Methyl-trans-2-pentene		-91.6	-61.5									
C_6H_{12}	2-Ethyl-1-butene		-87.1	-56.0									
C_6H_{12}	2,3-Dimethyl-1-butene		-93.2	-62.4									
C_6H_{12}	3,3-Dimethyl-1-butene		-87.5	-60.3									
C_6H_{12}	2,3-Dimethyl-2-butene		-101.4	-68.1									
C_6H_{12}	Cyclohexane		-156.4	-123.4					270.2			174.7	
C_6H_{12}	Methylcyclopentane		-137.9	-106.2								154.9	
C_6H_{12}	Ethylcyclobutane		-59.0	-27.5									
C_6H_{12}	1,1,2-Trimethylcyclo-propane		-96.2										
C_6H_{12}	2-Methyl-2-pentene		-98.5	-66.9									
$C_6H_{12}N_2O_4S_2$	L-Cystine	-1032.7											
$C_6H_{12}O$	Cyclohexanol		-348.2	-286.2								208.2	
$C_6H_{12}O$	cis-2-Methylcyclopentanol		-345.5										
$C_6H_{12}O$	Butyl vinyl ether		-218.8	-182.6								232.0	

Molecular formula	Name	$\Delta_f H°$/kJ mol⁻¹			$\Delta_f G°$/kJ mol⁻¹			$S°$/J mol⁻¹ K⁻¹			C_p/J mol⁻¹ K⁻¹		
		Crys.	Liq.	Gas	Crys.	Liq.	Gas	Crys.	Liq.	Gas	Crys.	Liq.	Gas
C₆H₁₂O	2-Hexanone		-322.0	-278.9								213.3	
C₆H₁₂O	3-Hexanone		-320.2	-277.6					305.3			216.9	
C₆H₁₂O	Methyl isobutyl ketone			-286.0								213.3	
C₆H₁₂O	2-Methyl-3-pentanone		-325.9	-290.6									
C₆H₁₂O	3,3-Dimethyl-2-butanone		-328.6										
C₆H₁₂O₂	Hexanoic acid		-583.8	-511.9									
C₆H₁₂O₂	Butyl acetate		-529.2	-485.3								227.8	
C₆H₁₂O₂	Isobutyl acetate											233.8	
C₆H₁₂O₂	Ethyl butanoate											228.0	
C₆H₁₂O₂	Methyl pentanoate		-514.2	-471.1								229.3	
C₆H₁₂O₂	Methyl 2,2-dimethyl-propanoate		-530.0	-491.2									
C₆H₁₂O₂	Diacetone alcohol											257.9	
C₆H₁₂O₃	Ethylene glycol ethyl ether acetate											221.3	
C₆H₁₂O₃	Paraldehyde		-673.1	-631.7								376.0	
C₆H₁₂O₆	β-D-Fructose	-1265.6											
C₆H₁₂O₆	D-Galactose	-1286.3											
C₆H₁₂O₆	α-D-Glucose	-1273.3											
C₆H₁₂O₆	D-Mannose	-1263.0											
C₆H₁₂O₆	L-Sorbose	-1271.5											
C₆H₁₂S	Cyclohexanethiol		-140.7	-96.2					255.6			192.6	
C₆H₁₂S	Cyclopentyl methyl sulfide		-109.8	-64.7					453.0			203.5	
C₆H₁₃Br	1-Bromohexane		-194.2	-148.3									
C₆H₁₃Cl	2-Chlorohexane		-246.1	-204.3									
C₆H₁₃N	Cyclohexylamine		-147.6	-104.0									
C₆H₁₃N	(±)-2-Methylpiperidine		-124.9	-84.4									
C₆H₁₃NO	Hexanamide		-397.9	-324.2									
C₆H₁₃NO	N-Butylacetamide		-380.9	-305.9									
C₆H₁₃NO₂	DL-Leucine	-640.6									200.1		
C₆H₁₃NO₂	D-Leucine	-637.3											
C₆H₁₃NO₂	L-Leucine	-637.4		-486.8									
C₆H₁₃NO₂	DL-Isoleucine	-635.3											
C₆H₁₃NO₂	L-Isoleucine	-637.8											
C₆H₁₃NO₂	Norleucine	-639.1											
C₆H₁₃NO₂	6-Aminohexanoic acid	-637.3											
C₆H₁₄	Hexane		-198.7	-166.9					290.6			195.6	
C₆H₁₄	2-Methylpentane		-204.6	-174.6					292.5			193.7	
C₆H₁₄	3-Methylpentane		-202.4	-171.9					272.5			190.7	
C₆H₁₄	2,2-Dimethylbutane		-213.8	-185.9								191.9	
C₆H₁₄	2,3-Dimethylbutane		-207.4	-178.1					287.8			189.7	
C₆H₁₄N₂	Azopropane		11.5	51.3									
C₆H₁₄N₂O₂	Lysine	-678.7						250.6			232.0		
C₆H₁₄N₄O₂	D-Arginine	-623.5											
C₆H₁₄O	1-Hexanol		-377.5	-315.9					287.4			240.4	

STANDARD THERMODYNAMIC PROPERTIES OF CHEMICAL SUBSTANCES (continued)

Molecular formula	Name	$\Delta_f H°$/kJ mol⁻¹ Crys.	Liq.	Gas	$\Delta_f G°$/kJ mol⁻¹ Crys.	Liq.	Gas	$S°$/J mol⁻¹ K⁻¹ Crys.	Liq.	Gas	C_p/J mol⁻¹ K⁻¹ Crys.	Liq.	Gas
C₆H₁₄O	2-Hexanol		-392.0	-333.5								286.2	
C₆H₁₄O	3-Hexanol		-392.4									248.0	
C₆H₁₄O	2-Methyl-1-pentanol											275.9	
C₆H₁₄O	3-Methyl-2-pentanol											273.0	
C₆H₁₄O	4-Methyl-2-pentanol		-394.7										
C₆H₁₄O	2-Methyl-3-pentanol		-396.4										
C₆H₁₄O	3-Methyl-3-pentanol											293.4	
C₆H₁₄O	Dipropyl ether		-328.8	-293.0					323.9			221.6	
C₆H₁₄O	Diisopropyl ether		-351.5	-319.2								216.8	
C₆H₁₄O	Butyl ethyl ether			-313.9									
C₆H₁₄O	Ethyl *tert*-butyl ether		-329.4	-254.9								159.0	
C₆H₁₄OS	Dipropyl sulfoxide												
C₆H₁₄O₂	1,2-Hexanediol		-577.1	-490.1									
C₆H₁₄O₂	1,6-Hexanediol	-569.9	-548.6	-461.2									
C₆H₁₄O₂	Hexylene glycol											336.0	
C₆H₁₄O₂	Ethylene glycol monobutyl ether											281.0	
C₆H₁₄O₂	1,1-Diethoxyethane		-491.4	-453.5									
C₆H₁₄O₂	1,2-Diethoxyethane		-451.4	-408.1								259.4	
C₆H₁₄O₃	Diethylene glycol monoethyl ether											301.0	
C₆H₁₄O₃	Diethylene glycol dimethyl ether											274.1	
C₆H₁₄O₃	Trimethylolpropane	-750.9		-725.0									
C₆H₁₄O₄	Triethylene glycol		-804.3	-792.0									
C₆H₁₄O₄S	Dipropyl sulfate		-859.0										
C₆H₁₄O₆	Galactitol	-1317.0											
C₆H₁₄O₆	*D*-Mannitol	-1314.5											
C₆H₁₄S	1-Hexanethiol		-175.7	-129.9									
C₆H₁₄S	2-Methyl-2-pentanethiol		-188.3	-148.3									
C₆H₁₄S	2,3-Dimethyl-2-butanethiol		-187.1	-147.9									
C₆H₁₄S	Diisopropyl sulfide		-181.6	-142.0					313.0			232.0	
C₆H₁₄S	Butyl ethyl sulfide		-172.3	-127.8									
C₆H₁₄S	Methyl pentyl sulfide		-167.1	-121.8									
C₆H₁₄S₂	Dipropyl disulfide		-171.5	-118.3									
C₆H₁₅B	Triethylborane		-194.6	-157.7		9.4	16.1		336.7	437.8		241.2	
C₆H₁₅N	Dipropylamine		-156.1	-116.0									
C₆H₁₅N	Diisopropylamine		-178.5	-143.8									
C₆H₁₅N	Triethylamine		-127.7	-92.7								219.9	
C₆H₁₅NO	2-Diethylaminoethanol		-305.9										
C₆H₁₅NO₃	Triethanolamine	-664.2		-558.3							389.0		
C₆H₁₈N₃OP	Hexamethylphosphoric triamide											321.0	
C₆H₁₈OSi₂	Hexamethyldisiloxane		-815.0	-777.7		-541.5	-534.5		433.8	535.0		311.4	238.5
C₆MoO₆	Molybdenum hexacarbonyl	-982.8		-912.1	-877.7		-856.0	325.9		490.0	242.3		205.0

Molecular formula	Name	$\Delta_f H°$/kJ mol⁻¹			$\Delta_f G°$/kJ mol⁻¹			S°/J mol⁻¹ K⁻¹			C_p/J mol⁻¹ K⁻¹		
		Crys.	Liq.	Gas	Crys.	Liq.	Gas	Crys.	Liq.	Gas	Crys.	Liq.	Gas
C₆N₄	Tetracyanoethylene	623.8		705.0									
C₇F₈	Perfluorotoluene		-1311.1						355.5			262.3	
C₇F₁₄	Perfluoromethylcyclohexane		-2931.1	-2897.2								353.1	
C₇F₁₆	Perfluoroheptane		-3420.0	-3383.6					561.8			419.0	
C₇H₃F₅	2,3,4,5,6-Pentafluorotoluene		-883.8	-842.7					306.4			225.8	
C₇H₄Cl₂O	m-Chlorobenzoyl chloride		-189.7	-103.2									
C₇H₄N₂O₆	3,5-Dinitrobenzoic acid	-409.8											
C₇H₅ClO	Benzoyl chloride		-158.0										
C₇H₅ClO₂	o-Chlorobenzoic acid	-404.5		-325.0									
C₇H₅ClO₂	o-Chlorobenzoic acid	-424.3		-342.3									
C₇H₅ClO₂	p-Chlorobenzoic acid	-428.9		-341.0							163.2		
C₇H₅F₃	(Trifluoromethyl)benzene		163.2						209.1			188.4	
C₇H₅N	Benzonitrile			215.7								165.2	
C₇H₅NO	Benzoxazole	-24.2		44.8									
C₇H₅NO₄	2-Nitrobenzoic acid	-378.8											
C₇H₅NO₄	3-Nitrobenzoic acid	-394.7											
C₇H₅NO₄	4-Nitrobenzoic acid	-392.2											
C₇H₆N₂	1H-Benzimidazole	79.5		181.7									
C₇H₆N₂	1H-Indazole	151.9		243.0				167.6			146.8		
C₇H₆O	Benzaldehyde		-87.0	-36.7					221.2			172.0	
C₇H₆O₂	Benzoic acid	-385.2		-294.0									
C₇H₆O₂	3-(2-Furanyl)-2-propenal	-182.0		-105.9									
C₇H₆O₂	Salicylaldehyde											222.0	
C₇H₆O₃	Salicylic acid	-589.9		-494.8									
C₇H₇Cl	o-Chlorotoluene		-32.5	18.9								166.8	
C₇H₇Cl	(Chloromethyl)benzene		-186.9									171.2	
C₇H₇F	p-Fluorotoluene			-147.4									
C₇H₇NO	Benzamide	-202.6		-100.9									
C₇H₇NO₂	2-Aminobenzoic acid		-380.4	-296.0									
C₇H₇NO₂	Aniline-3-carboxylic acid		-389.8	-283.6									
C₇H₇NO₂	Aniline-4-carboxylic acid		-391.9	-296.7									
C₇H₇NO₂	o-Nitrotoluene		-9.7	31.0									
C₇H₇NO₂	m-Nitrotoluene		-31.5	30.7									
C₇H₇NO₂	p-Nitrotoluene	-48.1											
C₇H₇NO₂	(Nitromethyl)benzene		-22.8										
C₇H₇NO₂	Salicylaldoxime	-183.7									172.3		
C₇H₈	Toluene		12.4	50.5								157.3	
C₇H₈N₂O	Phenylurea	-218.6						165.4					
C₇H₈O	o-Cresol	-204.6		-128.6							154.6		
C₇H₈O	m-Cresol		-194.0	-132.3					212.6			224.9	
C₇H₈O	p-Cresol	-199.3		-125.4				167.3			150.2		
C₇H₈O	Benzyl alcohol		-160.7	-100.4					216.7			217.9	
C₇H₈O	Anisole		-114.8	-67.9									
C₇H₉N	Benzylamine		34.2	94.4									
C₇H₉N	o-Methylaniline		-6.3	56.4			167.6			351.0			130.2

Molecular formula	Name	$\Delta_f H°$/kJ mol⁻¹			$\Delta_f G°$/kJ mol⁻¹			$S°$/J mol⁻¹ K⁻¹			C_p/J mol⁻¹ K⁻¹		
		Crys.	Liq.	Gas	Crys.	Liq.	Gas	Crys.	Liq.	Gas	Crys.	Liq.	Gas
C₇H₉N	*m*-Methylaniline		-8.1	54.6			165.4			352.5			125.5
C₇H₉N	*p*-Methylaniline	-23.5		55.3			167.7			347.0			126.2
C₇H₉N	*N*-Methylaniline											207.1	
C₇H₉N	1-Cyclohexenecarbonitrile		48.1	101.6									
C₇H₉N	2,3-Dimethylpyridine		19.4	67.1					243.7			189.5	
C₇H₉N	2,4-Dimethylpyridine		16.1	63.6					248.5			184.8	
C₇H₉N	2,5-Dimethylpyridine		18.7	66.5					248.8			184.7	
C₇H₉N	2,6-Dimethylpyridine		12.7	58.1					244.2			185.2	
C₇H₉N	3,4-Dimethylpyridine		18.3	68.8					240.7			191.8	
C₇H₉N	3,5-Dimethylpyridine		22.5	72.0					241.7			184.5	
C₇H₁₀O₂	Ethyl 2-pentynoate		-301.8	-250.3									
C₇H₁₀O₂	Methyl 2-hexynoate		-242.7										
C₇H₁₁Cl₃O₂	3-Methylbutyl trichloroacetate		-580.9	-523.1									
C₇H₁₁N	Cyclohexanecarbonitrile		-47.2	4.8									
C₇H₁₂	Methylenecyclohexane		-61.3	-25.2									
C₇H₁₂	Vinylcyclopentane		-34.8	-19.8									
C₇H₁₂	1-Ethylcyclopentene		-53.3										
C₇H₁₂	Bicyclo[2.2.1]heptane	-95.1		-54.8									
C₇H₁₂	1-Methylbicyclo(3,1,0)hexane		-33.2	1.7									
C₇H₁₂O	2-Methylenecyclohexanol	-277.6											
C₇H₁₂O₄	Diethyl malonate		-571.7	-517.3								285.0	
C₇H₁₃ClO₂	Butyl 2-chloropropanoate		-603.1	-549.6									
C₇H₁₃ClO₂	Isobutyl 2-chloropropanoate		-557.9	-502.3									
C₇H₁₃ClO₂	Butyl 3-chloropropanoate		-572.6	-517.3									
C₇H₁₃ClO₂	Isobutyl 3-chloropropanoate		-630.7	-578.4									
C₇H₁₃ClO₂	Propyl 2-chlorobutanoate												
C₇H₁₃N	Heptanenitrile		-82.8	-31.0									
C₇H₁₄	1-Heptene		-97.9	-62.3					327.6			211.8	
C₇H₁₄	*cis*-2-Heptene		-105.1										
C₇H₁₄	*trans*-2-Heptene		-109.5										
C₇H₁₄	*cis*-3-Heptene		-104.3										
C₇H₁₄	*trans*-3-Heptene		-109.3										
C₇H₁₄	5-Methyl-1-hexene		-100.0	-65.7									
C₇H₁₄	3-Methyl-*cis*-3-hexene		-115.9	-79.4									
C₇H₁₄	3-Methyl-*trans*-3-hexene		-112.7	-76.8									
C₇H₁₄	2,4-Dimethyl-1-pentene		-117.0	-83.8									
C₇H₁₄	4,4-Dimethyl-1-pentene		-110.6	-81.6									
C₇H₁₄	2,4-Dimethyl-2-pentene		-123.1	-88.7									
C₇H₁₄	*cis*-4,4-Dimethyl-2-pentene		-105.3	-72.6									
C₇H₁₄	*trans*-4,4-Dimethyl-2-pentene		-121.7	-88.8									
C₇H₁₄	2-Ethyl-3-methyl-1-butene		-114.1	-79.5									
C₇H₁₄	2,3,3-Trimethyl-1-butene		-117.7	-85.5									
C₇H₁₄	Cycloheptane		-156.6	-118.1									
C₇H₁₄	Methylcyclohexane		-190.1	-154.7								184.8	

Molecular formula	Name	$\Delta_f H°$/kJ mol^{-1}			$\Delta_f G°$/kJ mol^{-1}			$S°$/J mol^{-1} K^{-1}			C_p/J mol^{-1} K^{-1}		
		Crys.	Liq.	Gas	Crys.	Liq.	Gas	Crys.	Liq.	Gas	Crys.	Liq.	Gas
C$_7$H$_{14}$	Ethylcyclopentane		-163.4	-126.9					279.9				
C$_7$H$_{14}$	1,1-Dimethylcyclopentane		-172.1	-138.2									
C$_7$H$_{14}$	cis-1,2-Dimethylcyclopentane		-165.3	-129.5					269.2				
C$_7$H$_{14}$	trans-1,2-Dimethylcyclopentane		-171.2	-136.6									
C$_7$H$_{14}$	cis-1,3-Dimethylcyclopentane		-170.1	-135.8									
C$_7$H$_{14}$	trans-1,3-Dimethylcyclopentane		-168.1	-133.6									
C$_7$H$_{14}$	1,1,2,2-Tetramethylcyclopropane		-119.8										
C$_7$H$_{14}$Br$_2$	1,2-Dibromoheptane		-212.3	-157.9									
C$_7$H$_{14}$O	1-Heptanal		-311.5	-263.8					335.4			230.1	
C$_7$H$_{14}$O	2-Heptanone											232.6	
C$_7$H$_{14}$O	2,2-Dimethyl-3-pentanone		-356.1	-313.6									
C$_7$H$_{14}$O	2,4-Dimethyl-3-pentanone		-352.9	-311.3					318.0			233.7	
C$_7$H$_{14}$O	cis-2-Methylcyclohexanol		-390.2	-327.0									
C$_7$H$_{14}$O	trans-2-Methylcyclohexanol		-415.7	-352.5									
C$_7$H$_{14}$O	cis-3-Methylcyclohexanol		-416.1	-350.9									
C$_7$H$_{14}$O	trans-3-Methylcyclohexanol		-394.4	-329.1									
C$_7$H$_{14}$O	cis-4-Methylcyclohexanol		-413.2	-347.5									
C$_7$H$_{14}$O	trans-4-Methylcyclohexanol		-433.3	-367.2									
C$_7$H$_{14}$O$_2$	Heptanoic acid		-610.2	-536.2								265.4	
C$_7$H$_{14}$O$_2$	Pentyl acetate											261.0	
C$_7$H$_{14}$O$_2$	Isopentyl acetate											248.5	
C$_7$H$_{14}$O$_2$	Ethyl pentanoate		-553.0	-505.9									
C$_7$H$_{14}$O$_2$	Ethyl 3-methylbutanoate		-571.0	-527.0									
C$_7$H$_{14}$O$_2$	Ethyl 2,2-dimethylpropanoate		-577.2	-536.0									
C$_7$H$_{14}$O$_2$	Methyl hexanoate		-540.2	-492.2									
C$_7$H$_{14}$O$_6$	α-Methylglucoside	-1233.3											
C$_7$H$_{15}$Br	1-Bromoheptane		-218.4	-167.8									
C$_7$H$_{16}$	Heptane		-224.2	-187.6					323.3			224.7	
C$_7$H$_{16}$	2-Methylhexane		-229.5	-194.5								222.9	
C$_7$H$_{16}$	3-Methylhexane		-226.4	-191.3									
C$_7$H$_{16}$	3-Ethylpentane		-224.9	-189.5					314.5			219.6	
C$_7$H$_{16}$	2,2-Dimethylpentane		-238.3	-205.7					300.3			221.1	
C$_7$H$_{16}$	2,3-Dimethylpentane		-233.1	-198.7									
C$_7$H$_{16}$	2,4-Dimethylpentane		-234.6	-201.6					303.2			224.2	
C$_7$H$_{16}$	3,3-Dimethylpentane		-234.2	-201.0									
C$_7$H$_{16}$	2,2,3-Trimethylbutane		-236.5	-204.4									
C$_7$H$_{16}$O	1-Heptanol		-403.3	-336.5					292.2			213.5	
C$_7$H$_{16}$O	tert-Butyl isopropyl ether		-392.8	-358.1								272.1	
C$_7$H$_{16}$O$_2$	1,7-Heptanediol		-574.2										

Molecular formula	Name	$\Delta_f H°$/kJ mol⁻¹ Crys.	Liq.	Gas	$\Delta_f G°$/kJ mol⁻¹ Crys.	Liq.	Gas	$S°$/J mol⁻¹ K⁻¹ Crys.	Liq.	Gas	C_p/J mol⁻¹ K⁻¹ Crys.	Liq.	Gas
$C_7H_{16}O_2$	2,2-Diethoxypropane		-538.9	-506.9									
$C_7H_{16}S$	1-Heptanethiol		-200.5	-149.9									
$C_8H_4O_3$	Phthalic anhydride	-460.1		-371.4				180.0			160.0		
$C_8H_5NO_2$	1H-Indole-2,3-dione	-268.2											
$C_8H_6O_4$	Phthalic acid	-782.0						207.9			188.1		
$C_8H_6O_4$	Isophthalic acid	-803.0		-696.3									
$C_8H_6O_4$	Terephthalic acid	-816.1		-717.9									
C_8H_6S	Benzo[b]thiophene	100.6		166.3									
C_8H_7N	1H-Indole	86.6		156.5									
C_8H_8	Styrene		103.8	147.9								182.0	
C_8H_8O	Phenyl vinyl ether		-26.2	22.7									
C_8H_8O	Acetophenone		-142.5	-86.7									
$C_8H_8O_2$	o-Toluic acid	-416.5									174.9		
$C_8H_8O_2$	m-Toluic acid	-426.1									163.6		
$C_8H_8O_2$	p-Toluic acid	-429.2									169.0		
$C_8H_8O_2$	Methyl benzoate		-343.5	-287.9								221.3	
$C_8H_8O_3$	Methyl salicylate											249.0	
C_8H_{10}	1,7-Octadiyne		334.4										
C_8H_{10}	Ethylbenzene		-12.3	29.9								183.2	
C_8H_{10}	o-Xylene		-24.4	19.1								186.1	
C_8H_{10}	m-Xylene		-25.4	17.3								183.0	
C_8H_{10}	p-Xylene		-24.4	18.0								181.5	
$C_8H_{10}O$	o-Ethylphenol		-208.8	-145.2									
$C_8H_{10}O$	m-Ethylphenol		-214.3	-146.1									
$C_8H_{10}O$	p-Ethylphenol	-224.4		-144.1							206.9		
$C_8H_{10}O$	2,3-Xylenol	-241.1		-157.2									
$C_8H_{10}O$	2,4-Xylenol		-228.7	-163.8									
$C_8H_{10}O$	2,5-Xylenol	-246.6		-161.6									
$C_8H_{10}O$	2,6-Xylenol	-237.4		-162.1									
$C_8H_{10}O$	3,4-Xylenol	-242.3		-157.3									
$C_8H_{10}O$	3,5-Xylenol	-244.4		-162.4									
$C_8H_{10}O$	Phenetole		-152.6	-101.6								228.5	
$C_8H_{10}O_2$	Veratrole		-290.3	-223.3									
$C_8H_{11}N$	N-Ethylaniline		4.0	56.3									
$C_8H_{11}N$	N,N-Dimethylaniline		47.7	100.5									
$C_8H_{11}N$	2,6-Xylidine											238.9	
C_8H_{12}	1-Octen-3-yne		140.7										
C_8H_{12}	cis-1,2-Divinylcyclobutane		124.3	166.5									
C_8H_{12}	trans-1,2-Divinylcyclobutane		101.3	143.5									
$C_8H_{12}N_4$	2,2'-Azobis[isobutyronitrile]	228.9											
$C_8H_{12}O_2$	2,2,4-Tetramethyl-1,3-cyclobutanedione	-379.9		-307.6									
C_8H_{14}	Ethylidenecyclohexane		-103.5	-59.5									

Molecular formula	Name	$\Delta_f H°$/kJ mol^{-1} Crys.	Liq.	Gas	$\Delta_f G°$/kJ mol^{-1} Crys.	Liq.	Gas	$S°$/J mol^{-1} K^{-1} Crys.	Liq.	Gas	C_p/J mol^{-1} K^{-1} Crys.	Liq.	Gas
C$_8$H$_{14}$	Allylcyclopentane		-64.5	-24.1									283.7
C$_8$H$_{14}$O$_3$	Butanoic anhydride												
C$_8$H$_{15}$ClO$_2$	3-Methylbutyl 2-chloropropanoate		-627.3	-575.0									
C$_8$H$_{15}$ClO$_2$	3-Methylbutyl 3-chloropropanoate		-593.4	-539.4									
C$_8$H$_{15}$N	Octanenitrile		-107.3	-50.5								241.0	
C$_8$H$_{16}$	1-Octene		-121.8	-81.3								239.0	
C$_8$H$_{16}$	cis-2-Octene		-135.7									239.0	
C$_8$H$_{16}$	trans-2-Octene		-135.7										
C$_8$H$_{16}$	2,2-Dimethyl-cis-3-hexene		-126.4	-89.3									
C$_8$H$_{16}$	2,2-Dimethyl-trans-3-hexene		-144.9	-107.7									
C$_8$H$_{16}$	3-Ethyl-2-methyl-1-pentene		-137.9	-100.3									
C$_8$H$_{16}$	2,4,4-Trimethyl-1-pentene		-145.9	-110.5									
C$_8$H$_{16}$	2,4,4-Trimethyl-2-pentene		-142.4	-104.9									
C$_8$H$_{16}$	Cyclooctane		-167.7	-124.4									
C$_8$H$_{16}$	Ethylcyclohexane		-212.1	-171.5					280.9			211.8	
C$_8$H$_{16}$	1,1-Dimethylcyclohexane		-218.7	-180.9					267.2			209.2	
C$_8$H$_{16}$	cis-1,2-Dimethylcyclohexane		-211.8	-172.1					274.1			210.2	
C$_8$H$_{16}$	trans-1,2-Dimethylcyclohexane		-218.2	-179.9					273.2			209.4	
C$_8$H$_{16}$	cis-1,3-Dimethylcyclohexane		-222.9	-184.6					272.6			209.4	
C$_8$H$_{16}$	trans-1,3-Dimethylcyclohexane		-215.7	-176.5					276.3			212.8	
C$_8$H$_{16}$	cis-1,4-Dimethylcyclohexane		-215.6	-176.6					271.1			212.1	
C$_8$H$_{16}$	trans-1,4-Dimethylcyclohexane		-222.4	-184.5					268.0			210.2	
C$_8$H$_{16}$	Propylcyclopentane		-188.8	-147.7					310.8			216.3	
C$_8$H$_{16}$	1-Ethyl-1-methylcyclopentane		-193.8										
C$_8$H$_{16}$	cis-1-Ethyl-2-methylcyclopentane		-190.8										
C$_8$H$_{16}$	trans-1-Ethyl-2-methylcyclopentane		-195.1	-156.2									
C$_8$H$_{16}$	cis-1-Ethyl-3-methylcyclopentane		-194.4										
C$_8$H$_{16}$	trans-1-Ethyl-3-methylcyclopentane		-196.0										
C$_8$H$_{16}$O	2-Ethylhexanal		-348.5	-299.6								273.3	
C$_8$H$_{16}$O	2-Octanone		-381.6	-338.3								297.9	
C$_8$H$_{16}$O	2,2,4-Trimethyl-3-pentanone												
C$_8$H$_{16}$O$_2$	Octanoic acid		-636.0	-554.3								282.8	
C$_8$H$_{16}$O$_2$	2-Ethylhexanoic acid		-635.1	-559.5									
C$_8$H$_{16}$O$_2$	Hexyl acetate		-583.0	-533.6									
C$_8$H$_{16}$O$_2$	Propyl pentanoate		-592.2	-544.9									
C$_8$H$_{16}$O$_2$	Isopropyl pentanoate												

STANDARD THERMODYNAMIC PROPERTIES OF CHEMICAL SUBSTANCES (continued)

Molecular formula	Name	$\Delta_f H°$/kJ mol^{-1} Crys.	Liq.	Gas	$\Delta_f G°$/kJ mol^{-1} Crys.	Liq.	Gas	$S°$/J mol^{-1} K^{-1} Crys.	Liq.	Gas	C_p/J mol^{-1} K^{-1} Crys.	Liq.	Gas
C$_8$H$_{16}$O$_2$	Methyl heptanoate		-567.1	-515.5								285.1	
C$_8$H$_{17}$Br	1-Bromooctane		-245.1	-189.3									
C$_8$H$_{17}$Cl	1-Chlorooctane		-291.3	-238.9									
C$_8$H$_{17}$NO	Octanamide	-473.2		-362.7									
C$_8$H$_{18}$	Octane		-250.1	-208.5								254.6	
C$_8$H$_{18}$	2-Methylheptane		-255.0	-215.3					356.4			252.0	
C$_8$H$_{18}$	3-Methylheptane		-252.3	-212.5					362.6			250.2	
C$_8$H$_{18}$	4-Methylheptane		-251.6	-211.9								251.1	
C$_8$H$_{18}$	3-Ethylhexane		-250.4	-210.7									
C$_8$H$_{18}$	2,2-Dimethylhexane		-261.9	-224.5									
C$_8$H$_{18}$	2,3-Dimethylhexane		-252.6	-213.8									
C$_8$H$_{18}$	2,4-Dimethylhexane		-257.0	-219.2									
C$_8$H$_{18}$	2,5-Dimethylhexane		-260.4	-222.5								249.2	
C$_8$H$_{18}$	3,3-Dimethylhexane		-257.5	-219.9								246.6	
C$_8$H$_{18}$	3,4-Dimethylhexane		-251.8	-212.8									
C$_8$H$_{18}$	3-Ethyl-2-methylpentane		-249.6	-211.0									
C$_8$H$_{18}$	3-Ethyl-3-methylpentane		-252.8	-214.8									
C$_8$H$_{18}$	2,2,3-Trimethylpentane		-256.9	-220.0									
C$_8$H$_{18}$	2,2,4-Trimethylpentane		-259.2	-224.0								239.1	
C$_8$H$_{18}$	2,3,3-Trimethylpentane		-253.5	-216.3								245.6	
C$_8$H$_{18}$	2,3,4-Trimethylpentane		-255.0	-217.3								247.3	
C$_8$H$_{18}$	2,2,3,3-Tetramethylbutane	-269.0		-226.0				273.7			239.2		
C$_8$H$_{18}$N$_2$	Azobutane		-40.1	9.2									
C$_8$H$_{18}$O	1-Octanol		-426.5	-355.6									
C$_8$H$_{18}$O	2-Octanol											305.2	
C$_8$H$_{18}$O	2-Ethyl-1-hexanol		-432.8	-365.3								330.1	
C$_8$H$_{18}$O	Dibutyl ether		-377.9	-332.8								317.5	
C$_8$H$_{18}$O	Di-sec-butyl ether		-401.5	-360.6								278.2	
C$_8$H$_{18}$O	Di-tert-butyl ether		-399.6	-362.0									
C$_8$H$_{18}$O	tert-Butyl isobutyl ether		-409.1	-369.0					347.0			276.1	
C$_8$H$_{18}$O$_2$	1,8-Octanediol	-626.6											
C$_8$H$_{18}$O$_2$	2,5-Dimethyl-2,5-hexanediol	-681.7											
C$_8$H$_{18}$O$_3$	Diethylene glycol diethyl ether		-693.1	-625.3									
C$_8$H$_{18}$O$_3$S	Dibutyl sulfite		-981.7	-883.0								341.4	
C$_8$H$_{18}$O$_5$	Tetraethylene glycol								405.1			428.8	
C$_8$H$_{18}$S	Dibutyl sulfide		-220.7	-167.7								284.3	
C$_8$H$_{18}$S	Di-sec-butyl sulfide		-220.7	-167.7									
C$_8$H$_{18}$S	Di-tert-butyl sulfide		-232.6	-188.8									
C$_8$H$_{18}$S	Diisobutyl sulfide		-229.2	-180.5									
C$_8$H$_{18}$S$_2$	Dibutyl disulfide		-222.9	-160.6									
C$_8$H$_{18}$S$_2$	Di-tert-butyl disulfide		-255.2	-201.0									
C$_8$H$_{19}$N	Dibutylamine		-206.0	-156.6									
C$_8$H$_{19}$N	Diisobutylamine		-218.5	-179.2								292.9	

STANDARD THERMODYNAMIC PROPERTIES OF CHEMICAL SUBSTANCES (continued)

Molecular formula	Name	$\Delta_f H°$/kJ mol⁻¹ Crys.	Liq.	Gas	$\Delta_f G°$/kJ mol⁻¹ Crys.	Liq.	Gas	$S°$/J mol⁻¹ K⁻¹ Crys.	Liq.	Gas	C_p/J mol⁻¹ K⁻¹ Crys.	Liq.	Gas
C₈H₂₀BrN	Tetraethylammonium bromide	-342.7											
C₈H₂₀Pb	Tetraethyl lead		52.7	109.6					464.6			307.4	
C₈H₂₀Si	Tetraethylsilane											298.1	
C₉H₇N	Quinoline		141.2	200.5					216.0			196.2	
C₉H₇N	Isoquinoline		144.3	204.6					215.3			186.9	
C₉H₇NO	2-Hydroxyquinoline	-144.9		-25.5									
C₉H₈	Indene		110.6	163.4								190.2	
C₉H₈O₄	2-(Acetyloxy)benzoic acid	-815.6											
C₉H₁₀	Cyclopropylbenzene		100.3	150.5					56.0				
C₉H₁₀	Indan		11.5	60.3								148.5	
C₉H₁₀N₂	2,2'-Dipyrrolylmethane	126.2											
C₉H₁₀O₂	Benzyl acetate												
C₉H₁₀O₂	Ethyl benzoate			-312.9								246.0	
C₉H₁₁NO₂	L-Phenylalanine	-466.9						213.6			203.0		
C₉H₁₁NO₃	L-Tyrosine	-685.1						214.0			216.4		
C₉H₁₂	Propylbenzene		-38.3	7.9					287.8			214.7	
C₉H₁₂	Cumene		-41.1	4.0								210.7	
C₉H₁₂	o-Ethyltoluene		-46.4	1.3									
C₉H₁₂	m-Ethyltoluene		-48.7	-1.8									
C₉H₁₂	p-Ethyltoluene		-49.8	-3.2									
C₉H₁₂	1,2,3-Trimethylbenzene		-58.5	-9.5								216.4	
C₉H₁₂	1,2,4-Trimethylbenzene		-61.8	-13.8								215.0	
C₉H₁₂	Mesitylene		-63.4	-15.9					267.9			209.3	
C₉H₁₂O	2-Isopropylphenol		-233.7	-182.2									
C₉H₁₂O	3-Isopropylphenol		-252.5	-196.0									
C₉H₁₂O	4-Isopropylphenol		-265.9	-209.4									
C₉H₁₂O₂	Hydroperoxide, 1-methyl-1-phenylethyl		-148.3	-78.4									
C₉H₁₃NO₂	Ethyl 3,5-dimethylpyrrole-2-carboxylate	-474.5											
C₉H₁₃NO₂	Ethyl 2,4-dimethylpyrrole-3-carboxylate	-463.2											
C₉H₁₃NO₂	Ethyl 2,5-dimethylpyrrole-3-carboxylate	-478.7											
C₉H₁₃NO₂	Ethyl 4,5-dimethylpyrrole-3-carboxylate	-470.3											
C₉H₁₄O	Isophorone								458.3			253.5	
C₉H₁₄O₆	Triacetin		-1330.8	-1245.0								384.7	
C₉H₁₅N	3-Ethyl-2,4,5-trimethylpyrrole	-89.2											
C₉H₁₆O₄	Nonanedioic acid	-1054.3											
C₉H₁₇NO	2,2,6,6-Tetramethyl-4-piperidinone	-334.2		-273.4									
C₉H₁₈	Propylcyclohexane		-237.4	-192.3					311.9			242.0	

Molecular formula	Name	$\Delta_f H°$/kJ mol⁻¹ Crys.	Liq.	Gas	$\Delta_f G°$/kJ mol⁻¹ Crys.	Liq.	Gas	$S°$/J mol⁻¹ K⁻¹ Crys.	Liq.	Gas	C_p/J mol⁻¹ K⁻¹ Crys.	Liq.	Gas
C_9H_{18}	1,3,5-Trimethylcyclohexane (1',3',5')			-212.1									
$C_9H_{18}O$	2-Nonanone		-397.2	-340.7								303.6	
$C_9H_{18}O$	5-Nonanone		-398.2	-344.9								297.3	
$C_9H_{18}O$	2,6-Dimethyl-4-heptanone		-408.5	-357.6								362.4	
$C_9H_{18}O_2$	Nonanoic acid		-659.7	-577.3									
$C_9H_{18}O_2$	Butyl pentanoate	-613.3		-560.2									
$C_9H_{18}O_2$	sec-Butyl pentanoate		-624.2	-573.2									
$C_9H_{18}O_2$	Isobutyl pentanoate		-620.0	-568.6									
$C_9H_{18}O_2$	Methyl octanoate		-590.3	-533.9									
$C_9H_{19}N$	N-Butylpiperidine		-171.8										
$C_9H_{19}N$	2,2,6,6-Tetramethyl-piperidine		-206.9	-159.9									
C_9H_{20}	Nonane		-274.7	-228.2					401.4			284.4	
C_9H_{20}	2,2-Dimethylheptane		-288.1										
C_9H_{20}	2,2,3-Trimethylhexane		-282.7										
C_9H_{20}	2,2,4-Trimethylhexane		-282.8										
C_9H_{20}	2,2,5-Trimethylhexane		-293.3										
C_9H_{20}	2,3,3-Trimethylhexane		-281.1										
C_9H_{20}	2,3,5-Trimethylhexane		-284.0	-242.6									
C_9H_{20}	2,4,4-Trimethylhexane		-280.2										
C_9H_{20}	3,3,4-Trimethylhexane		-277.5										
C_9H_{20}	3,3-Diethylpentane		-275.4	-233.3								278.2	
C_9H_{20}	3-Ethyl-2,2-dimethylpentane		-272.7										
C_9H_{20}	3-Ethyl-2,4-dimethylpentane		-269.7										
C_9H_{20}	2,2,3,3-Tetramethylpentane		-278.3	-237.1								271.5	
C_9H_{20}	2,2,3,4-Tetramethylpentane		-277.7	-236.9									
C_9H_{20}	2,2,4,4-Tetramethylpentane		-280.0	-241.6								266.3	
C_9H_{20}	2,3,3,4-Tetramethylpentane		-277.9	-236.1									
$C_9H_{20}N_2O$	Tetraethylurea	-403.0		-376.5									
$C_9H_{20}O$	1-Nonanol		-453.4									212.6	
$C_9H_{20}O_2$	1,9-Nonanediol	-657.6											
$C_9H_{21}N$	Tripropylamine		-207.1	-161.0									
$C_{10}H_6N_2O_4$	1,5-Dinitronaphthalene	29.8											
$C_{10}H_6N_2O_4$	1,8-Dinitronaphthalene	39.7											
$C_{10}H_7Cl$	1-Chloronaphthalene		54.6	119.8									
$C_{10}H_7Cl$	2-Chloronaphthalene	55.4		137.4									
$C_{10}H_7I$	1-Iodonaphthalene		161.5	233.8									
$C_{10}H_7I$	2-Iodonaphthalene	144.3		235.1									
$C_{10}H_7NO_2$	1-Nitronaphthalene	42.6		111.2									
$C_{10}H_8$	Naphthalene	78.5		150.6	201.6		224.1	167.4		333.1	165.7		131.9
$C_{10}H_8$	Azulene	212.3		289.1									
$C_{10}H_8O$	1-Naphthol	-121.5		-30.4							166.9		
$C_{10}H_8O$	2-Naphthol		-124.1	-29.9									
$C_{10}H_9N$	1-Naphthalenamine	67.8		132.8									

Molecular formula	Name	$\Delta_f H°$/kJ mol⁻¹			$\Delta_f G°$/kJ mol⁻¹			S°/J mol⁻¹ K⁻¹			C_p/J mol⁻¹ K⁻¹		
		Crys.	Liq.	Gas	Crys.	Liq.	Gas	Crys.	Liq.	Gas	Crys.	Liq.	Gas
$C_{10}H_9N$	2-Naphthalenamine	60.2		134.3									
$C_{10}H_{10}$	1,2-Dihydronaphthalene		71.6										
$C_{10}H_{10}$	1,4-Dihydronaphthalene		84.2										
$C_{10}H_{10}O$	1-Tetralone	-209.6											
$C_{10}H_{10}O_4$	Dimethyl phthalate											303.1	
$C_{10}H_{10}O_4$	Dimethyl isophthalate	-730.9											
$C_{10}H_{10}O_4$	Dimethyl terephthalate	-732.6									261.1		
$C_{10}H_{12}$	1,2,3,4-Tetrahydrona-phthalene		-29.2	26.0									
$C_{10}H_{14}$	Butylbenzene		-63.2	-11.8					321.2			217.5	
$C_{10}H_{14}$	sec-Butylbenzene		-66.4	-18.4								243.4	
$C_{10}H_{14}$	tert-Butylbenzene		-70.7	-23.0									
$C_{10}H_{14}$	Isobutylbenzene		-69.8	-21.9									
$C_{10}H_{14}$	o-Cymene		-73.3										
$C_{10}H_{14}$	m-Cymene		-78.6										
$C_{10}H_{14}$	p-Cymene		-78.0									236.4	
$C_{10}H_{14}$	o-Diethylbenzene		-68.5										
$C_{10}H_{14}$	m-Diethylbenzene		-73.5										
$C_{10}H_{14}$	p-Diethylbenzene		-72.8										
$C_{10}H_{14}$	3-Ethyl-o-xylene		-80.5										
$C_{10}H_{14}$	4-Ethyl-1,2-dimethylbenzene		-86.0										
$C_{10}H_{14}$	2-Ethyl-1,3-dimethylbenzene		-80.1										
$C_{10}H_{14}$	2-Ethyl-1,4-dimethylbenzene		-84.8										
$C_{10}H_{14}$	1-Ethyl-2,4-dimethylbenzene		-84.1										
$C_{10}H_{14}$	1-Ethyl-3,5-dimethylbenzene		-87.8										
$C_{10}H_{14}$	1,2,4,5-Tetramethylbenzene	-119.9		-218.5				245.6			215.1		
$C_{10}H_{14}O$	Thymol	-309.7											
$C_{10}H_{16}$	cis,cis-2,6-Dimethyl-2,4,6-octatriene		-24.0										
$C_{10}H_{16}$	Dipentene		-50.8	-2.6									
$C_{10}H_{16}$	D-Limonene		-54.5										
$C_{10}H_{16}$	β-Myrcene		14.5										
$C_{10}H_{16}$	α-Pinene		-16.4	28.3									
$C_{10}H_{16}$	β-Pinene		-7.7	38.7									
$C_{10}H_{16}$	α-Terpinene			-20.6									
$C_{10}H_{16}N_2O_8$	Glycine, N,N-1,2-ethane-diylbis[N-(carboxymethyl)]-	-1759.5											
$C_{10}H_{16}O$	Camphor	-319.4		-267.5							271.2		
$C_{10}H_{18}$	1,1'-Bicyclopentyl		-178.9	-169.2									
$C_{10}H_{18}$	cis-Decahydronaphthalene		-219.4	-182.1					265.0			232.0	
$C_{10}H_{18}$	trans-Decahydronaphthalene		-230.6						264.9			228.5	
$C_{10}H_{18}O_4$	Sebacic acid	-1082.6		-921.9									
$C_{10}H_{19}N$	Decanenitrile		-158.4	-91.5									
$C_{10}H_{20}$	1-Decene		-173.8	-123.3					425.0			300.8	
$C_{10}H_{20}$	cis-1,2-Di-tert-butylethylene		-163.6										

Molecular formula	Name	$\Delta_f H°$/kJ mol⁻¹ Crys.	Liq.	Gas	$\Delta_f G°$/kJ mol⁻¹ Crys.	Liq.	Gas	$S°$/J mol⁻¹ K⁻¹ Crys.	Liq.	Gas	C_p/J mol⁻¹ K⁻¹ Crys.	Liq.	Gas
C₁₀H₂₀	Butylcyclohexane		-263.1	-213.7					345.0			271.0	
C₁₀H₂₀O₂	Decanoic acid	-713.7	-684.3	-594.9									
C₁₀H₂₀O₂	Methyl nonanoate		-616.2	-554.2									
C₁₀H₂₁NO₂	1-Nitrodecane		-351.5										
C₁₀H₂₂	Decane		-300.9	-249.5								314.4	
C₁₀H₂₂	2-Methylnonane		-309.8	-260.2					420.1			313.3	
C₁₀H₂₂	5-Methylnonane		-307.9	-258.6					423.8			314.4	
C₁₀H₂₂O	1-Decanol		-478.1	-396.6								370.6	
C₁₀H₂₂O	Dipentyl ether											250.0	
C₁₀H₂₂O	Diisopentyl ether											379.0	
C₁₀H₂₂O₂	1,10-Decanediol	-678.9											
C₁₀H₂₂O₂	Ethylene glycol dibutyl ether												
C₁₀H₂₂S	1-Decanethiol	-309.9	-276.5	-211.5								350.0	
C₁₀H₂₂S	Dipentyl sulfide		-266.4	-204.9					476.1			350.4	
C₁₀H₂₂S	Diisopentylsulfide		-281.8	-221.5									
C₁₁H₈O₂	1-Naphthalenecarboxylic acid	-333.5		-223.1									
C₁₁H₈O₂	2-Naphthoic acid	-346.1		-232.5									
C₁₁H₁₀	1-Methylnaphthalene		56.3	106.7					254.8			224.4	
C₁₁H₁₀	2-Methylnaphthalene	44.9						220.0			196.0		
C₁₁H₁₂N₂O₂	L-Tryptophan	-415.3						251.0			238.1		
C₁₁H₁₄	1,1-Dimethylindan		-53.6	-1.6									
C₁₁H₁₆	Pentamethylbenzene	-144.6		-67.2									
C₁₁H₂₀	Spiro[5.5]undecane		-244.5	-188.3									
C₁₁H₂₂	1-Undecenè												
C₁₁H₂₂O₂	Methyl decanoate		-640.5	-573.8								344.9	
C₁₁H₂₄	Undecane		-327.2	-270.8								344.9	
C₁₁H₂₄O	1-Undecanol		-504.8										
C₁₂F₂₇N	Trinonafluorobutylamine											418.4	
C₁₂H₈	Acenaphthylene	186.7		259.7							166.4		
C₁₂H₈N₂	Phenazine	237.0		328.8									
C₁₂H₈O	Dibenzofuran	-5.3		83.4									
C₁₂H₈S	Dibenzothiophene	120.0		205.1									
C₁₂H₈S₂	Thianthrene	182.0		286.0									
C₁₂H₉N	Carbazole	101.7		200.7									
C₁₂H₁₀	Acenaphthene	70.3		156.0				188.9			190.4		
C₁₂H₁₀	Biphenyl	99.4		181.4				209.4			198.4		
C₁₂H₁₀N₂O	trans-Azoxybenzene	243.4		342.0									
C₁₂H₁₀N₂O	N-Nitrosodiphenylamine	227.2											
C₁₂H₁₀O	Diphenyl ether	-32.1	-14.9	52.0				233.9			216.6		
C₁₂H₁₀O₂	1-Naphthaleneacetic acid	-359.2											
C₁₂H₁₀O₂	2-Naphthaleneacetic acid	-371.9											
C₁₂H₁₁N	2-Aminobiphenyl	93.8		184.4									
C₁₂H₁₁N	p-Biphenylamine	81.0											
C₁₂H₁₁N	Diphenylamine	130.2		219.3									
C₁₂H₁₂N₂	p-Benzidine	70.7											

STANDARD THERMODYNAMIC PROPERTIES OF CHEMICAL SUBSTANCES (continued)

Molecular formula	Name	$\Delta_f H°$/kJ mol⁻¹ Crys.	Liq.	Gas	$\Delta_f G°$/kJ mol⁻¹ Crys.	Liq.	Gas	$S°$/J mol⁻¹ K⁻¹ Crys.	Liq.	Gas	C_p/J mol⁻¹ K⁻¹ Crys.	Liq.	Gas
C₁₂H₁₄O₄	Diethyl phthalate		-776.6	-688.4					425.1			366.1	
C₁₂H₁₆	Cyclohexylbenzene		-76.6	-16.7									
C₁₂H₁₇NO₄	Diethyl 3,5-dimethyl-pyrrole-2,4-dicarboxylate	-916.7											
C₁₂H₁₈	3,9-Dodecadiyne		197.8										
C₁₂H₁₈	5,7-Dodecadiyne		181.5										
C₁₂H₁₈	Hexamethylbenzene	-162.4		-77.4				306.3			245.6		
C₁₂H₂₂	Cyclohexylcyclohexane		-273.7	-215.7									
C₁₂H₂₂O₄	Dodecanedioic acid	-1130.0		-976.9									
C₁₂H₂₂O₁₁	Sucrose	-2226.1											
C₁₂H₂₂O₁₁	β-Lactose	-2236.7									404.3		
C₁₂H₂₄	1-Dodecene		-226.2	-165.4					484.8			360.7	
C₁₂H₂₄O₂	Dodecanoic acid	-774.6	-737.9	-642.0									
C₁₂H₂₄O₂	Methyl undecanoate		-665.2	-593.8									
C₁₂H₂₄O₁₂	α-Lactose monohydrate	-2484.1											
C₁₂H₂₅Br	1-Bromododecane		-344.7	-269.9									
C₁₂H₂₅Cl	1-Chlorododecane		-392.3	-321.1									
C₁₂H₂₆	Dodecane		-350.9	-289.4								375.8	
C₁₂H₂₆O	1-Dodecanol		-528.5	-436.6								438.1	
C₁₂H₂₆O₃	Diethylene glycol dibutyl ether		-281.6									452.0	
C₁₂H₂₇N	Tributylamine												
C₁₃H₈O₂	Xanthone	-191.5											
C₁₃H₉N	Acridine	179.4		273.9									
C₁₃H₉N	Phenanthridine	141.9		240.5									
C₁₃H₉N	Benzo[f]quinoline	150.6		233.7									
C₁₃H₁₀N₂	9-Acridinamine	159.2											
C₁₃H₁₀O	Benzophenone	-34.5		54.9									
C₁₃H₁₁N	9-Methylcarbazole	105.5		201.0				239.3					
C₁₃H₁₂	Diphenylmethane	71.5	89.7	139.0									
C₁₃H₂₄O₄	Tridecanedioic acid	-1148.3											
C₁₃H₂₆	1-Tridecene											391.8	
C₁₃H₂₆O₂	Methyl dodecanoate		-693.0	-614.9								406.7	
C₁₃H₂₈	Tridecane												
C₁₃H₂₈O	1-Tridecanol	-599.4											
C₁₄H₈O₂	9,10-Anthracenedione	-188.5		-75.7									
C₁₄H₈O₂	9,10-Phenanthrenedione	-154.7		-46.6									
C₁₄H₈O₄	1,4-Dihydroxy-9,10-anthracenedione	-595.8		-471.7							224.8		
C₁₄H₁₀	Anthracene	129.2		230.9				207.5			210.5		
C₁₄H₁₀	Phenanthrene	116.2		207.5				215.1			220.6		
C₁₄H₁₀	Diphenylacetylene	312.4									225.9		
C₁₄H₁₀O₂	Benzil	-153.9		-55.5									
C₁₄H₁₂	cis-Stilbene		183.3	252.3									
C₁₄H₁₂	trans-Stilbene	136.9		236.1									

STANDARD THERMODYNAMIC PROPERTIES OF CHEMICAL SUBSTANCES (continued)

Molecular formula	Name	ΔfH°/kJ mol⁻¹ Crys.	Liq.	Gas	ΔfG°/kJ mol⁻¹ Crys.	Liq.	Gas	S°/J mol⁻¹ K⁻¹ Crys.	Liq.	Gas	Cp/J mol⁻¹ K⁻¹ Crys.	Liq.	Gas
$C_{14}H_{14}$	1,1-Diphenylethane		48.7										
$C_{14}H_{14}$	1,2-Diphenylethane	51.5		142.9									
$C_{14}H_{23}N_3O_{10}$	Glycine, N,N-bis[2-[bis(carboxymethyl)amino]ethyl]	-2225.2											
$C_{14}H_{27}N$	Tetradecanenitrile		-260.2	-174.9									
$C_{14}H_{28}O_2$	Tetradecanoic acid	-833.5	-788.8	-693.7							432.0		
$C_{14}H_{28}O_2$	Methyl tridecanoate		-717.9	-635.3									
$C_{14}H_{30}O$	1-Tetradecanol	-629.6	-580.6								388.0		
$C_{15}H_{16}O_2$	2,2-Bis(4-hydroxyphenyl)propane	-368.6											
$C_{15}H_{30}$	Decylcyclopentane		-367.3										
$C_{15}H_{30}O_2$	Pentadecanoic acid	-861.7	-811.7	-699.0							443.3		
$C_{15}H_{30}O_2$	Methyl tetradecanoate		-743.9	-656.9									
$C_{15}H_{32}O$	1-Pentadecanol	-658.2											
$C_{16}H_{10}$	Fluoranthene	189.9		289.0				230.6			230.2		
$C_{16}H_{10}$	Pyrene	125.5		225.7				224.9			229.7		
$C_{16}H_{22}O_4$	Dibutyl phthalate		-842.6	-750.9									
$C_{16}H_{22}O_{11}$	α-D-Glucose pentaacetate	-2249.4											
$C_{16}H_{22}O_{11}$	β-D-Glucose pentaacetate	-2232.6											
$C_{16}H_{26}$	Decylbenzene		-218.3	-138.6									
$C_{16}H_{32}$	1-Hexadecene		-328.7	-248.4					587.9				
$C_{16}H_{32}O_2$	Hexadecanoic acid	-891.5	-838.1	-737.1				452.4			460.7	488.9	
$C_{16}H_{32}O_2$	Methyl pentadecanoate		-771.0	-680.0									
$C_{16}H_{33}Br$	1-Bromohexadecane		-444.5	-350.2									
$C_{16}H_{34}$	Hexadecane		-456.1	-374.8									
$C_{16}H_{34}O$	1-Hexadecanol	-686.5		-517.0							422.0	501.6	
$C_{16}H_{36}IN$	Tetrabutyl ammonium iodide	-498.6											
$C_{17}H_{34}O_2$	Margaric acid	-924.4	-865.6								475.7		
$C_{18}H_{12}$	Benz[a]anthracene	170.8		293.0									
$C_{18}H_{12}$	Chrysene	145.3		269.8									
$C_{18}H_{15}N$	Triphenylamine	234.7		326.8									
$C_{18}H_{34}O_2$	Oleic acid												
$C_{18}H_{34}O_4$	Dibutyl sebacate												
$C_{18}H_{36}O_2$	Stearic acid	-947.7	-884.7	-781.2							501.5	577.0	
$C_{18}H_{37}Cl$	1-Chlorooctadecane		-544.1	-446.0								619.0	
$C_{18}H_{38}$	Octadecane	-567.4	-433.0	-414.6				480.2					
$C_{18}H_{39}N$	Trihexylamine												
$C_{19}H_{16}O$	Triphenylmethanol	-2.5									485.6		
$C_{19}H_{36}O_2$	Methyl oleate		-734.5	-649.9									
$C_{19}H_{36}O_2$	Methyl trans-9-octadecenoate		-737.0										
$C_{20}H_{12}$	Perylene	182.8						264.6			274.9		
$C_{20}H_{14}O_4$	1,2-Benzenedicarboxylic acid, diphenyl ester	-489.2											
$C_{20}H_{38}O_2$	Ethyl cis-9-octadecenoate		-775.8										

Molecular formula	Name	$\Delta_fH°$/kJ mol⁻¹ Crys.	Liq.	Gas	$\Delta_fG°$/kJ mol⁻¹ Crys.	Liq.	Gas	S°/J mol⁻¹ K⁻¹ Crys.	Liq.	Gas	C_p/J mol⁻¹ K⁻¹ Crys.	Liq.	Gas
$C_{20}H_{38}O_2$	Ethyl trans-9-octa-decenoate		-773.3										
$C_{20}H_{40}O_2$	Arachidic acid	-1011.9	-940.0	-812.4							545.1		
$C_{22}H_{42}O_2$	Brassidic acid	-960.7											
$C_{22}H_{42}O_2$	Butyl oleate		-816.9										
$C_{22}H_{44}O_2$	Butyl stearate												
$C_{24}H_{38}O_4$	Bis(2-ethylhexyl) phthalate											704.7	
$C_{24}H_{51}N$	Trioctylamine		-585.0										
$C_{26}H_{18}$	9,10-Diphenylanthracene	308.7		465.6									
$C_{26}H_{54}$	5-Butyldocosane		-713.5	-587.6									
$C_{26}H_{54}$	11-Butyldocosane		-716.0	-593.4									
$C_{28}H_{18}$	9,9′-Bianthracene	326.2		454.3									
$C_{31}H_{64}$	11-Decylheneicosane		-848.0	-705.8									
$C_{32}H_{66}$	Dotriacontane		-968.3	-697.2									
C_{60}	Fullerene-C_{60}	2327.0		2502.0	2302.0		2442.0	426.0		544.0	520.0		512.0
C_{70}	Fullerene-C_{70}	2555.0		2755.0	2537.0		2692.0	464.0		614.0	650.0		585.0

THERMODYNAMIC PROPERTIES AS A FUNCTION OF TEMPERATURE

L. V. Gurvich, V. S. Iorish, V. S. Yungman, and O. V. Dorofeeva

The thermodynamic properties $C_p°(T)$, $S°(T)$, $H°(T)-H°(T_r)$, $-[G°(T)-H°(T_r)]/T$ and formation properties $\Delta_f H°(T)$, $\Delta_f G°(T)$, $\log K_f°(T)$ are tabulated as functions of temperature in the range 298.15 to 1500 K for 80 substances in the standard state. The reference temperature, T_r, is equal to 298.15 K. The standard state pressure is taken as 1 bar (100,000 Pa). The tables are presented in the JANAF Thermochemical Tables format (Reference 2). The numerical data are extracted from IVTANTHERMO databases except for C_2H_4O, C_3H_6O, C_6H_6, C_6H_6O, $C_{10}H_8$, and CH_5N, which are based upon TRC Tables. See the references for information on standard states and other details.

REFERENCES

1. Gurvich, L. V., Veyts, I. V., and Alcock, C. B., Eds., *Thermodynamic Properties of Individual Substances, 4th ed.*, Hemisphere Publishing Corp., New York, 1989.
2. Chase, M. W., et al., *JANAF Thermochemical Tables, 3rd ed., J. Phys. Chem. Ref. Data*, 14, Suppl. 1, 1985.

Order of Listing of Tables

No.	Formula	Name	State	No.	Formula	Name	State
1	Ar	Argon	g	41	$CuCl_2$	Copper dichloride	cr, l
2	Br	Bromine	g	42	$CuCl_2$	Copper dichloride	g
3	Br_2	Dibromine	g	43	F	Fluorine	g
4	BrH	Hydrogen bromide	g	44	F_2	Difluorine	g
5	C	Carbon (graphite)	cr	45	FH	Hydrogen fluoride	g
6	C	Carbon (diamond)	cr	46	Ge	Germanium	cr, l
7	C_2	Dicarbon	g	47	Ge	Germanium	g
8	C_3	Tricarbon	g	48	GeO_2	Germanium dioxide	cr, l
9	CO	Carbon oxide	g	49	$GeCl_4$	Germanium tetrachloride	g
10	CO_2	Carbon dioxide	g	50	H	Hydrogen	g
11	CH_4	Methane	g	51	H_2	Dihydrogen	g
12	C_2H_2	Acetylene	g	52	HO	Hydroxyl	g
13	C_2H_4	Ethylene	g	53	H_2O	Water	l
14	C_2H_6	Ethane	g	54	H_2O	Water	g
15	C_3H_6	Cyclopropane	g	55	I	Iodine	g
16	C_3H_8	Propane	g	56	I_2	Diiodine	cr, l
17	C_6H_6	Benzene	l	57	I_2	Diiodine	g
18	C_6H_6	Benzene	g	58	IH	Hydrogen iodide	g
19	$C_{10}H_8$	Naphthalene	cr, l	59	K	Potassium	cr, l
20	$C_{10}H_8$	Naphthalene	g	60	K	Potassium	g
21	CH_2O	Formaldehyde	g	61	K_2O	Dipotassium oxide	cr, l
22	CH_4O	Methanol	g	62	KOH	Potassium hydroxide	cr, l
23	C_2H_4O	Acetaldehyde	g	63	KOH	Potassium hydroxide	g
24	C_2H_6O	Ethanol	g	64	KCl	Potassium chloride	cr, l
25	$C_2H_4O_2$	Acetic acid	g	65	KCl	Potassium chloride	g
26	C_3H_6O	Acetone	g	66	N_2	Dinitrogen	g
27	C_6H_6O	Phenol	g	67	NO	Nitric oxide	g
28	CF_4	Carbon tetrafluoride	g	68	NO_2	Nitrogen dioxide	g
29	CHF_3	Trifluoromethane	g	69	NH_3	Ammonia	g
30	$CClF_3$	Chlorotrifluoromethane	g	70	O	Oxygen	g
31	CCl_2F_2	Dichlorodifluoromethane	g	71	O_2	Dioxygen	g
32	$CHClF_2$	Chlorodifluoromethane	g	72	S	Sulfur	cr, l
33	CH_5N	Methylamine	g	73	S	Sulfur	g
34	Cl	Chlorine	g	74	S_2	Disulfur	g
35	Cl_2	Dichlorine	g	75	S_8	Octasulfur	g
36	ClH	Hydrogen chloride	g	76	SO_2	Sulfur dioxide	g
37	Cu	Copper	cr, l	77	Si	Silicon	cr
38	Cu	Copper	g	78	Si	Silicon	g
39	CuO	Copper oxide	cr	79	SiO_2	Silicon dioxide	cr
40	Cu_2O	Dicopper oxide	cr	80	$SiCl_4$	Silicon tetrachloride	g

T/K	J/K·mol			kJ/mol			
	C_p°	S°	$-(G^\circ - H^\circ(T_r))/T$	$H^\circ - H^\circ(T_r)$	$\Delta_f H^\circ$	$\Delta_f G^\circ$	Log K_f
1. ARGON	**Ar (g)**						
298.15	20.786	154.845	154.845	0.000	0.000	0.000	0.000
300	20.786	154.973	154.845	0.038	0.000	0.000	0.000
400	20.786	160.953	155.660	2.117	0.000	0.000	0.000
500	20.786	165.591	157.200	4.196	0.000	0.000	0.000
600	20.786	169.381	158.924	6.274	0.000	0.000	0.000
700	20.786	172.585	160.653	8.353	0.000	0.000	0.000
800	20.786	175.361	162.322	10.431	0.000	0.000	0.000
900	20.786	177.809	163.909	12.510	0.000	0.000	0.000
1000	20.786	179.999	165.410	14.589	0.000	0.000	0.000
1100	20.786	181.980	166.828	16.667	0.000	0.000	0.000
1200	20.786	183.789	168.167	18.746	0.000	0.000	0.000
1300	20.786	185.453	169.434	20.824	0.000	0.000	0.000
1400	20.786	186.993	170.634	22.903	0.000	0.000	0.000
1500	20.786	188.427	171.773	24.982	0.000	0.000	0.000
2. BROMINE	**Br (g)**						
298.15	20.786	175.017	175.017	0.000	111.870	82.379	−14.432
300	20.786	175.146	175.018	0.038	111.838	82.196	−14.311
400	20.787	181.126	175.833	2.117	96.677	75.460	−9.854
500	20.798	185.765	177.373	4.196	96.910	70.129	−7.326
600	20.833	189.559	179.097	6.277	97.131	64.752	−5.637
700	20.908	192.776	180.827	8.364	97.348	59.338	−4.428
800	21.027	195.575	182.499	10.461	97.568	53.893	−3.519
900	21.184	198.061	184.093	12.571	97.796	48.420	−2.810
1000	21.365	200.302	185.604	14.698	98.036	42.921	−2.242
1100	21.559	202.347	187.034	16.844	98.291	37.397	−1.776
1200	21.752	204.231	188.390	19.010	98.560	31.850	−1.386
1300	21.937	205.980	189.676	21.195	98.844	26.279	−1.056
1400	22.107	207.612	190.900	23.397	99.141	20.686	−0.772
1500	22.258	209.142	192.065	25.615	99.449	15.072	−0.525
3. DIBROMINE	**Br$_2$ (g)**						
298.15	36.057	245.467	245.467	0.000	30.910	3.105	−0.544
300	36.074	245.690	245.468	0.067	30.836	2.933	−0.511
332.25	36.340	249.387	245.671	1.235		pressure = 1 bar	
400	36.729	256.169	246.892	3.711	0.000	0.000	0.000
500	37.082	264.406	249.600	7.403	0.000	0.000	0.000
600	37.305	271.188	252.650	11.123	0.000	0.000	0.000
700	37.464	276.951	255.720	14.862	0.000	0.000	0.000
800	37.590	281.962	258.694	18.615	0.000	0.000	0.000
900	37.697	286.396	261.530	22.379	0.000	0.000	0.000
1000	37.793	290.373	264.219	26.154	0.000	0.000	0.000
1100	37.883	293.979	266.763	29.938	0.000	0.000	0.000
1200	37.970	297.279	269.170	33.730	0.000	0.000	0.000
1300	38.060	300.322	271.451	37.532	0.000	0.000	0.000
1400	38.158	303.146	273.615	41.343	0.000	0.000	0.000
1500	38.264	305.782	275.673	45.164	0.000	0.000	0.000
4. HYDROGEN BROMIDE	**HBr (g)**						
298.15	29.141	198.697	198.697	0.000	−36.290	−53.360	9.348
300	29.141	198.878	198.698	0.054	−36.333	−53.466	9.309
400	29.220	207.269	199.842	2.971	−52.109	−55.940	7.305
500	29.454	213.811	202.005	5.903	−52.484	−56.854	5.939
600	29.872	219.216	204.436	8.868	−52.844	−57.694	5.023
700	30.431	223.861	206.886	11.882	−53.168	−58.476	4.363
800	31.063	227.965	209.269	14.957	−53.446	−59.214	3.866

THERMODYNAMIC PROPERTIES AS A FUNCTION OF TEMPERATURE (continued)

	J/K·mol			kJ/mol			
T/K	C_p°	S°	$-(G^\circ-H^\circ(T_r))/T$	$H^\circ-H^\circ(T_r)$	$\Delta_f H^\circ$	$\Delta_f G^\circ$	Log K_f

4. HYDROGEN BROMIDE   HBr (g) (continued)

900	31.709	231.661	211.555	18.095	−53.677	−59.921	3.478
1000	32.335	235.035	213.737	21.298	−53.864	−60.604	3.166
1100	32.919	238.145	215.816	24.561	−54.012	−61.271	2.909
1200	33.454	241.032	217.799	27.880	−54.129	−61.925	2.696
1300	33.938	243.729	219.691	31.250	−54.220	−62.571	2.514
1400	34.374	246.261	221.499	34.666	−54.291	−63.211	2.358
1500	34.766	248.646	223.230	38.123	−54.348	−63.846	2.223

5. CARBON (GRAPHITE)   C (cr; graphite)

298.15	8.536	5.740	5.740	0.000	0.000	0.000	0.000
300	8.610	5.793	5.740	0.016	0.000	0.000	0.000
400	11.974	8.757	6.122	1.054	0.000	0.000	0.000
500	14.537	11.715	6.946	2.385	0.000	0.000	0.000
600	16.607	14.555	7.979	3.945	0.000	0.000	0.000
700	18.306	17.247	9.113	5.694	0.000	0.000	0.000
800	19.699	19.785	10.290	7.596	0.000	0.000	0.000
900	20.832	22.173	11.479	9.625	0.000	0.000	0.000
1000	21.739	24.417	12.662	11.755	0.000	0.000	0.000
1100	22.452	26.524	13.827	13.966	0.000	0.000	0.000
1200	23.000	28.502	14.968	16.240	0.000	0.000	0.000
1300	23.409	30.360	16.082	18.562	0.000	0.000	0.000
1400	23.707	32.106	17.164	20.918	0.000	0.000	0.000
1500	23.919	33.749	18.216	23.300	0.000	0.000	0.000

6. CARBON (DIAMOND)   C (cr; diamond)

298.15	6.109	2.362	2.362	0.000	1.850	2.857	−0.501
300	6.201	2.400	2.362	0.011	1.846	2.863	−0.499
400	10.321	4.783	2.659	0.850	1.645	3.235	−0.422
500	13.404	7.431	3.347	2.042	1.507	3.649	−0.381
600	15.885	10.102	4.251	3.511	1.415	4.087	−0.356
700	17.930	12.709	5.274	5.205	1.361	4.537	−0.339
800	19.619	15.217	6.361	7.085	1.338	4.993	−0.326
900	21.006	17.611	7.479	9.118	1.343	5.450	−0.316
1000	22.129	19.884	8.607	11.277	1.372	5.905	−0.308
1100	23.020	22.037	9.731	13.536	1.420	6.356	−0.302
1200	23.709	24.071	10.842	15.874	1.484	6.802	−0.296
1300	24.222	25.990	11.934	18.272	1.561	7.242	−0.291
1400	24.585	27.799	13.003	20.714	1.646	7.675	−0.286
1500	24.824	29.504	14.047	23.185	1.735	8.103	−0.282

7. DICARBON C_2 (g)

298.15	43.548	197.095	197.095	0.000	830.457	775.116	−135.795
300	43.575	197.365	197.096	0.081	830.506	774.772	−134.898
400	42.169	209.809	198.802	4.403	832.751	755.833	−98.700
500	39.529	218.924	201.959	8.483	834.170	736.423	−76.933
600	37.837	225.966	205.395	12.342	834.909	716.795	−62.402
700	36.984	231.726	208.758	16.078	835.148	697.085	−52.016
800	36.621	236.637	211.943	19.755	835.020	677.366	−44.227
900	36.524	240.943	214.931	23.411	834.618	657.681	−38.170
1000	36.569	244.793	217.728	27.065	834.012	638.052	−33.328
1100	36.696	248.284	220.349	30.728	833.252	618.492	−29.369
1200	36.874	251.484	222.812	34.406	832.383	599.006	−26.074
1300	37.089	254.444	225.133	38.104	831.437	579.596	−23.288
1400	37.329	257.201	227.326	41.824	830.445	560.261	−20.903
1500	37.589	259.785	229.405	45.570	829.427	540.997	−18.839

	J/K·mol			kJ/mol			
T/K	C_p°	S°	$-(G^\circ - H^\circ(T_r))/T$	$H^\circ - H^\circ(T_r)$	$\Delta_f H^\circ$	$\Delta_f G^\circ$	Log K_f

8. TRICARBON C_3 (g)

298.15	42.202	237.611	237.611	0.000	839.958	774.249	−135.643
300	42.218	237.872	237.611	0.078	839.989	773.841	−134.736
400	43.383	250.164	239.280	4.354	841.149	751.592	−98.147
500	44.883	260.003	242.471	8.766	841.570	729.141	−76.172
600	46.406	268.322	246.104	13.331	841.453	706.659	−61.519
700	47.796	275.582	249.807	18.042	840.919	684.230	−51.057
800	48.997	282.045	253.440	22.884	840.053	661.901	−43.217
900	50.006	287.876	256.948	27.835	838.919	639.698	−37.127
1000	50.844	293.189	260.310	32.879	837.572	617.633	−32.261
1100	51.535	298.069	263.524	37.999	836.059	595.711	−28.288
1200	52.106	302.578	266.593	43.182	834.420	573.933	−24.982
1300	52.579	306.768	269.524	48.417	832.690	552.295	−22.191
1400	52.974	310.679	272.326	53.695	830.899	530.793	−19.804
1500	53.307	314.346	275.006	59.010	829.068	509.421	−17.739

9. CARBON OXIDE   CO (g)

298.15	29.141	197.658	197.658	0.000	−110.530	−137.168	24.031
300	29.142	197.838	197.659	0.054	−110.519	−137.333	23.912
400	29.340	206.243	198.803	2.976	−110.121	−146.341	19.110
500	29.792	212.834	200.973	5.930	−110.027	−155.412	16.236
600	30.440	218.321	203.419	8.941	−110.157	−164.480	14.319
700	31.170	223.067	205.895	12.021	−110.453	−173.513	12.948
800	31.898	227.277	208.309	15.175	−110.870	−182.494	11.915
900	32.573	231.074	210.631	18.399	−111.378	−191.417	11.109
1000	33.178	234.538	212.851	21.687	−111.952	−200.281	10.461
1100	33.709	237.726	214.969	25.032	−112.573	−209.084	9.928
1200	34.169	240.679	216.990	28.426	−113.228	−217.829	9.482
1300	34.568	243.430	218.920	31.864	−113.904	−226.518	9.101
1400	34.914	246.005	220.763	35.338	−114.594	−235.155	8.774
1500	35.213	248.424	222.527	38.845	−115.291	−243.742	8.488

10. CARBON DIOXIDE CO_2 (g)

298.15	37.135	213.783	213.783	0.000	−393.510	−394.373	69.092
300	37.220	214.013	213.784	0.069	−393.511	−394.379	68.667
400	41.328	225.305	215.296	4.004	−393.586	−394.656	51.536
500	44.627	234.895	218.280	8.307	−393.672	−394.914	41.256
600	47.327	243.278	221.762	12.909	−393.791	−395.152	34.401
700	49.569	250.747	225.379	17.758	−393.946	−395.367	29.502
800	51.442	257.492	228.978	22.811	−394.133	−395.558	25.827
900	53.008	263.644	232.493	28.036	−394.343	−395.724	22.967
1000	54.320	269.299	235.895	33.404	−394.568	−395.865	20.678
1100	55.423	274.529	239.172	38.893	−394.801	−395.984	18.803
1200	56.354	279.393	242.324	44.483	−395.035	−396.081	17.241
1300	57.144	283.936	245.352	50.159	−395.265	−396.159	15.918
1400	57.818	288.196	248.261	55.908	−395.488	−396.219	14.783
1500	58.397	292.205	251.059	61.719	−395.702	−396.264	13.799

11. METHANE CH_4 (g)

298.15	35.695	186.369	186.369	0.000	−74.600	−50.530	8.853
300	35.765	186.590	186.370	0.066	−74.656	−50.381	8.772
400	40.631	197.501	187.825	3.871	−77.703	−41.827	5.462
500	46.627	207.202	190.744	8.229	−80.520	−32.525	3.398
600	52.742	216.246	194.248	13.199	−82.969	−22.690	1.975
700	58.603	224.821	198.008	18.769	−85.023	−12.476	0.931
800	64.084	233.008	201.875	24.907	−86.693	−1.993	0.130

T/K	C_p°	S°	$-(G^\circ - H^\circ(T_r))/T$	$H^\circ - H^\circ(T_r)$	$\Delta_f H^\circ$	$\Delta_f G^\circ$	Log K_f
	J/K·mol			kJ/mol			

11. METHANE   CH$_4$ (g) (continued)

T/K	C_p°	S°	$-(G^\circ - H^\circ(T_r))/T$	$H^\circ - H^\circ(T_r)$	$\Delta_f H^\circ$	$\Delta_f G^\circ$	Log K_f
900	69.137	240.852	205.773	31.571	−88.006	8.677	−0.504
1000	73.746	248.379	209.660	38.719	−88.996	19.475	−1.017
1100	77.919	255.607	213.511	46.306	−89.698	30.358	−1.442
1200	81.682	262.551	217.310	54.289	−90.145	41.294	−1.797
1300	85.067	269.225	221.048	62.630	−90.367	52.258	−2.100
1400	88.112	275.643	224.720	71.291	−90.390	63.231	−2.359
1500	90.856	281.817	228.322	80.242	−90.237	74.200	−2.584

12. ACETYLENE   C$_2$H$_2$ (g)

T/K	C_p°	S°	$-(G^\circ - H^\circ(T_r))/T$	$H^\circ - H^\circ(T_r)$	$\Delta_f H^\circ$	$\Delta_f G^\circ$	Log K_f
298.15	44.036	200.927	200.927	0.000	227.400	209.879	−36.769
300	44.174	201.199	200.927	0.082	227.397	209.770	−36.524
400	50.388	214.814	202.741	4.829	227.161	203.928	−26.630
500	54.751	226.552	206.357	10.097	226.846	198.154	−20.701
600	58.121	236.842	210.598	15.747	226.445	192.452	−16.754
700	60.970	246.021	215.014	21.704	225.968	186.823	−13.941
800	63.511	254.331	219.418	27.931	225.436	181.267	−11.835
900	65.831	261.947	223.726	34.399	224.873	175.779	−10.202
1000	67.960	268.995	227.905	41.090	224.300	170.355	−8.898
1100	69.909	275.565	231.942	47.985	223.734	164.988	−7.835
1200	71.686	281.725	235.837	55.067	223.189	159.672	−6.950
1300	73.299	287.528	239.592	62.317	222.676	154.400	−6.204
1400	74.758	293.014	243.214	69.721	222.203	149.166	−5.565
1500	76.077	298.218	246.709	77.264	221.774	143.964	−5.013

13. ETHYLENE   C$_2$H$_4$ (g)

T/K	C_p°	S°	$-(G^\circ - H^\circ(T_r))/T$	$H^\circ - H^\circ(T_r)$	$\Delta_f H^\circ$	$\Delta_f G^\circ$	Log K_f
298.15	42.883	219.316	219.316	0.000	52.400	68.358	−11.976
300	43.059	219.582	219.317	0.079	52.341	68.457	−11.919
400	53.045	233.327	221.124	4.881	49.254	74.302	−9.703
500	62.479	246.198	224.864	10.667	46.533	80.887	−8.450
600	70.673	258.332	229.441	17.335	44.221	87.982	−7.659
700	77.733	269.770	234.393	24.764	42.278	95.434	−7.121
800	83.868	280.559	239.496	32.851	40.655	103.142	−6.734
900	89.234	290.754	244.630	41.512	39.310	111.036	−6.444
1000	93.939	300.405	249.730	50.675	38.205	119.067	−6.219
1100	98.061	309.556	254.756	60.280	37.310	127.198	−6.040
1200	101.670	318.247	259.688	70.271	36.596	135.402	−5.894
1300	104.829	326.512	264.513	80.599	36.041	143.660	−5.772
1400	107.594	334.384	269.225	91.223	35.623	151.955	−5.669
1500	110.018	341.892	273.821	102.107	35.327	160.275	−5.581

14. ETHANE   C$_2$H$_6$ (g)

T/K	C_p°	S°	$-(G^\circ - H^\circ(T_r))/T$	$H^\circ - H^\circ(T_r)$	$\Delta_f H^\circ$	$\Delta_f G^\circ$	Log K_f
298.15	52.487	229.161	229.161	0.000	−84.000	−32.015	5.609
300	52.711	229.487	229.162	0.097	−84.094	−31.692	5.518
400	65.459	246.378	231.379	5.999	−88.988	−13.473	1.759
500	77.941	262.344	235.989	13.177	−93.238	5.912	−0.618
600	89.188	277.568	241.660	21.545	−96.779	26.086	−2.271
700	99.136	292.080	247.835	30.972	−99.663	46.800	−3.492
800	107.936	305.904	254.236	41.334	−101.963	67.887	−4.433
900	115.709	319.075	260.715	52.525	−103.754	89.231	−5.179
1000	122.552	331.628	267.183	64.445	−105.105	110.750	−5.785
1100	128.553	343.597	273.590	77.007	−106.082	132.385	−6.286
1200	133.804	355.012	279.904	90.131	−106.741	154.096	−6.708
1300	138.391	365.908	286.103	103.746	−107.131	175.850	−7.066
1400	142.399	376.314	292.178	117.790	−107.292	197.625	−7.373
1500	145.905	386.260	298.121	132.209	−107.260	219.404	−7.640

	J/K·mol			kJ/mol			
T/K	C_p°	S°	$-(G^\circ - H^\circ(T_r))/T$	$H^\circ - H^\circ(T_r)$	$\Delta_f H^\circ$	$\Delta_f G^\circ$	Log K_f

15. CYCLOPROPANE C_3H_6 (g)

T/K	C_p°	S°	$-(G^\circ - H^\circ(T_r))/T$	$H^\circ - H^\circ(T_r)$	$\Delta_f H^\circ$	$\Delta_f G^\circ$	Log K_f
298.15	55.571	237.488	237.488	0.000	53.300	104.514	−18.310
300	55.941	237.832	237.489	0.103	53.195	104.832	−18.253
400	76.052	256.695	239.924	6.708	47.967	122.857	−16.043
500	93.859	275.637	245.177	15.230	43.730	142.091	−14.844
600	108.542	294.092	251.801	25.374	40.405	162.089	−14.111
700	120.682	311.763	259.115	36.854	37.825	182.583	−13.624
800	130.910	328.564	266.755	49.447	35.854	203.404	−13.281
900	139.658	344.501	274.516	62.987	34.384	224.441	−13.026
1000	147.207	359.616	282.277	77.339	33.334	245.618	−12.830
1100	153.749	373.961	289.965	92.395	32.640	266.883	−12.673
1200	159.432	387.588	297.538	108.060	32.249	288.197	−12.545
1300	164.378	400.549	304.967	124.257	32.119	309.533	−12.437
1400	168.689	412.892	312.239	140.915	32.215	330.870	−12.345
1500	172.453	424.662	319.344	157.976	32.507	352.193	−12.264

16. PROPANE C_3H_8 (g)

T/K	C_p°	S°	$-(G^\circ - H^\circ(T_r))/T$	$H^\circ - H^\circ(T_r)$	$\Delta_f H^\circ$	$\Delta_f G^\circ$	Log K_f
298.15	73.597	270.313	270.313	0.000	−103.847	−23.458	4.110
300	73.931	270.769	270.314	0.136	−103.972	−22.959	3.997
400	94.014	294.739	273.447	8.517	−110.33	15.029	−0.657
500	112.591	317.768	280.025	18.872	−115.658	34.507	−3.605
600	128.700	339.753	288.162	30.955	−119.973	64.961	−5.655
700	142.674	360.668	297.039	44.540	−123.384	96.065	−7.168
800	154.766	380.528	306.245	59.427	−126.016	127.603	−8.331
900	165.352	399.381	315.555	75.444	−127.982	159.430	−9.253
1000	174.598	417.293	324.841	92.452	−129.380	191.444	−10.000
1100	182.673	434.321	334.026	110.325	−130.296	223.574	−10.617
1200	189.745	450.526	343.064	128.954	−130.802	255.770	−11.133
1300	195.853	465.961	351.929	148.241	−130.961	287.993	−11.572
1400	201.209	480.675	360.604	168.100	−130.829	320.217	−11.947
1500	205.895	494.721	369.080	188.460	−130.445	352.422	−12.272

17. BENZENE C_6H_6 (l)

T/K	C_p°	S°	$-(G^\circ - H^\circ(T_r))/T$	$H^\circ - H^\circ(T_r)$	$\Delta_f H^\circ$	$\Delta_f G^\circ$	Log K_f
298.15	135.950	173.450	173.450	0.000	49.080	124.521	−21.815
300	136.312	174.292	173.453	.252	49.077	124.989	−21.762
400	161.793	216.837	179.082	15.102	48.978	150.320	−19.630
500	207.599	257.048	190.639	33.204	50.330	175.559	−18.340

18. BENZENE C_6H_6 (g)

T/K	C_p°	S°	$-(G^\circ - H^\circ(T_r))/T$	$H^\circ - H^\circ(T_r)$	$\Delta_f H^\circ$	$\Delta_f G^\circ$	Log K_f
298.15	82.430	269.190	269.190	0.000	82.880	129.750	−22.731
300	83.020	269.700	269.190	0.153	82.780	130.040	−22.641
400	113.510	297.840	272.823	10.007	77.780	146.570	−19.140
500	139.340	326.050	280.658	22.696	73.740	164.260	−17.160
600	160.090	353.360	290.517	37.706	70.490	182.680	−15.903
700	176.790	379.330	301.360	54.579	67.910	201.590	−15.042
800	190.460	403.860	312.658	72.962	65.910	220.820	−14.418
900	201.840	426.970	324.084	92.597	64.410	240.280	−13.945
1000	211.430	448.740	335.473	113.267	63.340	259.890	−13.575
1100	219.580	469.280	346.710	134.827	62.620	277.640	−13.184
1200	226.540	488.690	357.743	157.137	62.200	299.320	−13.029
1300	232.520	507.070	368.534	180.097	62.000	319.090	−12.821
1400	237.680	524.490	379.056	203.607	61.990	338.870	−12.643
1500	242.140	541.040	389.302	227.607	62.110	358.640	−12.489

T/K	C_p°	S°	$-(G^\circ - H^\circ(T_r))/T$	$H^\circ - H^\circ(T_r)$	$\Delta_f H^\circ$	$\Delta_f G^\circ$	Log K_f
		J/K·mol			kJ/mol		

19. NAPHTHALENE $C_{10}H_8$ (cr, l)

T/K	C_p°	S°	$-(G^\circ - H^\circ(T_r))/T$	$H^\circ - H^\circ(T_r)$	$\Delta_f H^\circ$	$\Delta_f G^\circ$	Log K_f
298.15	165.720	167.390	167.390	0.000	78.530	201.585	−35.316
300	167.001	168.419	167.393	0.308	78.466	202.349	−35.232
353.43	208.722	198.948	169.833	10.290	96.099	224.543	−33.186
		PHASE TRANSITION: $\Delta_{trs} H = 18.980$ kJ/mol, $\Delta_{trs} S = 53.702$ J/K·mol, cr–l					
353.43	217.200	252.650	169.833	29.270	96.099	224.543	−33.186
400	241.577	280.916	181.124	39.917	96.067	241.475	−31.533
470	276.409	322.712	199.114	58.091	97.012	266.859	−29.658

20. NAPHTHALENE $C_{10}H_8$ (g)

T/K	C_p°	S°	$-(G^\circ - H^\circ(T_r))/T$	$H^\circ - H^\circ(T_r)$	$\Delta_f H^\circ$	$\Delta_f G^\circ$	Log K_f
298.15	131.920	333.150	333.150	0.000	150.580	224.100	−39.260
300	132.840	333.970	333.157	0.244	150.450	224.560	−39.098
400	180.070	378.800	338.950	15.940	144.190	250.270	−32.681
500	219.740	423.400	351.400	36.000	139.220	277.340	−28.973
600	251.530	466.380	367.007	59.624	135.350	305.330	−26.581
700	277.010	507.140	384.146	86.096	132.330	333.950	−24.919
800	297.730	545.520	401.935	114.868	130.050	362.920	−23.696
900	314.850	581.610	419.918	145.523	128.430	392.150	−22.759
1000	329.170	615.550	437.806	177.744	127.510	421.700	−22.027
1100	341.240	647.500	455.426	211.281	127.100	450.630	−21.398
1200	351.500	677.650	472.707	245.932	126.960	480.450	−20.913
1300	360.260	706.130	489.568	281.531	127.060	509.770	−20.482
1400	367.780	733.110	506.009	317.941	127.390	539.740	−20.137
1500	374.270	758.720	522.019	355.051	127.920	568.940	−19.812

21. FORMALDEHYDE H_2CO (g)

T/K	C_p°	S°	$-(G^\circ - H^\circ(T_r))/T$	$H^\circ - H^\circ(T_r)$	$\Delta_f H^\circ$	$\Delta_f G^\circ$	Log K_f
298.15	35.387	218.760	218.760	0.000	−108.700	−102.667	17.987
300	35.443	218.979	218.761	0.066	−108.731	−102.630	17.869
400	39.240	229.665	220.192	3.789	−110.438	−100.340	13.103
500	43.736	238.900	223.028	7.936	−112.073	−97.623	10.198
600	48.181	247.270	226.381	12.534	−113.545	−94.592	8.235
700	52.280	255.011	229.924	17.560	−114.833	−91.328	6.815
800	55.941	262.236	233.517	22.975	−115.942	−87.893	5.739
900	59.156	269.014	237.088	28.734	−116.889	−84.328	4.894
1000	61.951	275.395	240.603	34.792	−117.696	−80.666	4.213
1100	64.368	281.416	244.042	41.111	−118.382	−76.929	3.653
1200	66.453	287.108	247.396	47.655	−118.966	−73.134	3.183
1300	68.251	292.500	250.660	54.392	−119.463	−69.294	2.784
1400	69.803	297.616	253.833	61.297	−119.887	−65.418	2.441
1500	71.146	302.479	256.915	68.346	−120.249	−61.514	2.142

22. METHANOL CH_3OH (g)

T/K	C_p°	S°	$-(G^\circ - H^\circ(T_r))/T$	$H^\circ - H^\circ(T_r)$	$\Delta_f H^\circ$	$\Delta_f G^\circ$	Log K_f
298.15	44.101	239.865	239.865	0.000	−201.000	−162.298	28.434
300	44.219	240.139	239.866	0.082	−201.068	−162.057	28.216
400	51.713	253.845	241.685	4.864	−204.622	−148.509	19.393
500	59.800	266.257	245.374	10.442	−207.750	−134.109	14.010
600	67.294	277.835	249.830	16.803	−210.387	−119.125	10.371
700	73.958	288.719	254.616	23.873	−212.570	−103.737	7.741
800	79.838	298.987	259.526	31.569	−214.350	−88.063	5.750
900	85.025	308.696	264.455	39.817	−215.782	−72.188	4.190
1000	89.597	317.896	269.343	48.553	−216.916	−56.170	2.934
1100	93.624	326.629	274.158	57.718	−217.794	−40.050	1.902
1200	97.165	334.930	278.879	67.262	−218.457	−23.861	1.039
1300	100.277	342.833	283.497	77.137	−218.936	−7.624	0.306
1400	103.014	350.367	288.007	87.304	−219.261	8.644	−0.322
1500	105.422	357.558	292.405	97.729	−219.456	24.930	−0.868

	J/K·mol			kJ/mol			
T/K	C_p°	S°	$-(G^\circ - H^\circ(T_r))/T$	$H^\circ - H^\circ(T_r)$	$\Delta_f H^\circ$	$\Delta_f G^\circ$	Log K_f

23. ACETALDEHYDE C_2H_4O (g)

T/K	C_p°	S°	$-(G^\circ - H^\circ(T_r))/T$	$H^\circ - H^\circ(T_r)$	$\Delta_f H^\circ$	$\Delta_f G^\circ$	Log K_f
298.15	55.318	263.840	263.840	0.000	−166.190	−133.010	23.302
300	55.510	264.180	263.837	0.103	−166.250	−132.800	23.122
400	66.282	281.620	266.147	6.189	−169.530	−121.130	15.818
500	76.675	297.540	270.850	13.345	−172.420	−108.700	11.356
600	85.942	312.360	276.550	21.486	−174.870	−95.720	8.334
700	94.035	326.230	282.667	30.494	−176.910	−82.350	6.145
800	101.070	339.260	288.938	40.258	−178.570	−68.730	4.487
900	107.190	351.520	295.189	50.698	−179.880	−54.920	3.187
1000	112.490	363.100	301.431	61.669	−180.850	−40.930	2.138
1100	117.080	374.040	307.537	73.153	−181.560	−27.010	1.283
1200	121.060	384.400	313.512	85.065	−182.070	−12.860	0.560
1300	124.500	394.230	319.350	97.344	−182.420	1.240	−0.050
1400	127.490	403.570	325.031	109.954	−182.640	15.470	−0.577
1500	130.090	412.460	330.571	122.834	−182.750	29.580	−1.030

24. ETHANOL C_2H_5OH (g)

T/K	C_p°	S°	$-(G^\circ - H^\circ(T_r))/T$	$H^\circ - H^\circ(T_r)$	$\Delta_f H^\circ$	$\Delta_f G^\circ$	Log K_f
298.15	65.652	281.622	281.622	0.000	−234.800	−167.874	29.410
300	65.926	282.029	281.623	0.122	−234.897	−167.458	29.157
400	81.169	303.076	284.390	7.474	−239.826	−144.216	18.832
500	95.400	322.750	290.115	16.318	−243.940	−119.820	12.517
600	107.656	341.257	297.112	26.487	−247.260	−94.672	8.242
700	118.129	358.659	304.674	37.790	−249.895	−69.023	5.151
800	127.171	375.038	312.456	50.065	−251.951	−43.038	2.810
900	135.049	390.482	320.276	63.185	−253.515	−16.825	0.976
1000	141.934	405.075	328.033	77.042	−254.662	9.539	−0.498
1100	147.958	418.892	335.670	91.543	−255.454	36.000	−1.709
1200	153.232	431.997	343.156	106.609	−255.947	62.520	−2.721
1300	157.849	444.448	350.473	122.168	−256.184	89.070	−3.579
1400	161.896	456.298	357.612	138.160	−256.206	115.630	−4.314
1500	165.447	467.591	364.571	154.531	−256.044	142.185	−4.951

25. ACETIC ACID $C_2H_4O_2$ (g)

T/K	C_p°	S°	$-(G^\circ - H^\circ(T_r))/T$	$H^\circ - H^\circ(T_r)$	$\Delta_f H^\circ$	$\Delta_f G^\circ$	Log K_f
298.15	63.438	283.470	283.470	0.000	−432.249	−374.254	65.567
300	63.739	283.863	283.471	0.118	−432.324	−373.893	65.100
400	79.665	304.404	286.164	7.296	−436.006	−353.840	46.206
500	93.926	323.751	291.765	15.993	−438.875	−332.950	34.783
600	106.181	341.988	298.631	26.014	−440.993	−311.554	27.123
700	116.627	359.162	306.064	37.169	−442.466	−289.856	21.629
800	125.501	375.331	313.722	49.287	−443.395	−267.985	17.497
900	132.989	390.558	321.422	62.223	−443.873	−246.026	14.279
1000	139.257	404.904	329.060	75.844	−443.982	−224.034	11.702
1100	144.462	418.429	336.576	90.039	−443.798	−202.046	9.594
1200	148.760	431.189	343.933	104.707	−443.385	−180.086	7.839
1300	152.302	443.240	351.113	119.765	−442.795	−158.167	6.355
1400	155.220	454.637	358.105	135.146	−442.071	−136.299	5.085
1500	157.631	465.432	364.903	150.793	−441.247	−114.486	3.987

26. ACETONE C_3H_6O (g)

T/K	C_p°	S°	$-(G^\circ - H^\circ(T_r))/T$	$H^\circ - H^\circ(T_r)$	$\Delta_f H^\circ$	$\Delta_f G^\circ$	Log K_f
298.15	74.517	295.349	295.349	0.000	−217.150	−152.716	26.757
300	74.810	295.809	295.349	0.138	−217.233	−152.339	26.521
400	91.755	319.658	298.498	8.464	−222.212	−129.913	16.962
500	107.864	341.916	304.988	18.464	−226.522	−106.315	11.107
600	122.047	362.836	312.873	29.978	−230.120	−81.923	7.133
700	134.306	382.627	321.470	42.810	−233.049	−56.986	4.252
800	144.934	401.246	330.265	56.785	−235.350	−31.673	2.068

T/K	C_p°	S°	$-(G^\circ-H^\circ(T_r))/T$	$H^\circ-H^\circ(T_r)$	$\Delta_f H^\circ$	$\Delta_f G^\circ$	Log K_f
	J/K·mol			kJ/mol			

26. ACETONE C_3H_6O (g) (continued)

T/K	C_p°	S°	$-(G^\circ-H^\circ(T_r))/T$	$H^\circ-H^\circ(T_r)$	$\Delta_f H^\circ$	$\Delta_f G^\circ$	Log K_f
900	154.097	418.860	339.141	71.747	−237.149	−6.109	0.353
1000	162.046	435.513	347.950	87.563	−238.404	19.707	−1.030
1100	168.908	451.286	356.617	104.136	−239.283	45.396	−2.157
1200	174.891	466.265	365.155	121.332	−239.827	71.463	−3.110
1300	180.079	480.491	373.513	139.072	−240.120	97.362	−3.912
1400	184.556	493.963	381.596	157.314	−240.203	123.470	−4.607
1500	188.447	506.850	389.533	175.975	−240.120	149.369	−5.202

27. PHENOL C_6H_6O (g)

T/K	C_p°	S°	$-(G^\circ-H^\circ(T_r))/T$	$H^\circ-H^\circ(T_r)$	$\Delta_f H^\circ$	$\Delta_f G^\circ$	Log K_f
298.15	103.220	314.810	314.810	0.000	−96.400	−32.630	5.720
300	103.860	315.450	314.810	0.192	−96.490	−32.230	5.610
400	135.790	349.820	319.278	12.217	−100.870	−10.180	1.330
500	161.910	383.040	328.736	27.152	−104.240	12.970	−1.360
600	182.480	414.450	340.430	44.412	−106.810	36.650	−3.190
700	198.840	443.860	353.134	63.508	−108.800	60.750	−4.530
800	212.140	471.310	366.211	84.079	−110.300	85.020	−5.550
900	223.190	496.950	379.327	105.861	−111.370	109.590	−6.360
1000	232.490	520.960	392.302	128.658	−111.990	134.280	−7.010
1100	240.410	543.500	405.033	152.314	−112.280	158.620	−7.530
1200	247.200	564.720	417.468	176.703	−112.390	183.350	−7.980
1300	253.060	584.740	429.568	201.723	−112.330	208.070	−8.360
1400	258.120	603.680	441.331	227.288	−112.120	233.050	−8.700
1500	262.520	621.650	452.767	253.325	−111.780	257.540	−8.970

28. CARBON TETRAFLUORIDE CF_4 (g)

T/K	C_p°	S°	$-(G^\circ-H^\circ(T_r))/T$	$H^\circ-H^\circ(T_r)$	$\Delta_f H^\circ$	$\Delta_f G^\circ$	Log K_f
298.15	61.050	261.455	261.455	0.000	−933.200	−888.518	155.663
300	61.284	261.833	261.456	0.113	−933.219	−888.240	154.654
400	72.399	281.057	264.001	6.822	−933.986	−873.120	114.016
500	80.713	298.153	269.155	14.499	−934.372	−857.852	89.618
600	86.783	313.434	275.284	22.890	−934.490	−842.533	73.348
700	91.212	327.162	281.732	31.801	−934.431	−827.210	61.726
800	94.479	339.566	288.199	41.094	−934.261	−811.903	53.011
900	96.929	350.842	294.542	50.670	−934.024	−796.622	46.234
1000	98.798	361.156	300.695	60.460	−933.745	−781.369	40.814
1100	100.250	370.643	306.629	70.416	−933.442	−766.146	36.381
1200	101.396	379.417	312.334	80.500	−933.125	−750.952	32.688
1300	102.314	387.571	317.811	90.687	−932.800	−735.784	29.564
1400	103.059	395.181	323.069	100.957	−932.470	−720.641	26.887
1500	103.671	402.313	328.116	111.295	−932.137	−705.522	24.568

29. TRIFLUOROMETHANE CHF_3 (g)

T/K	C_p°	S°	$-(G^\circ-H^\circ(T_r))/T$	$H^\circ-H^\circ(T_r)$	$\Delta_f H^\circ$	$\Delta_f G^\circ$	Log K_f
298.15	51.069	259.675	259.675	0.000	−696.700	−662.237	116.020
300	51.258	259.991	259.676	0.095	−696.735	−662.023	115.267
400	61.148	276.113	261.807	5.722	−698.427	−650.186	84.905
500	69.631	290.700	266.149	12.275	−699.715	−637.969	66.647
600	76.453	304.022	271.368	19.593	−700.634	−625.528	54.456
700	81.868	316.230	276.917	27.519	−701.253	−612.957	45.739
800	86.201	327.455	282.542	35.930	−701.636	−600.315	39.196
900	89.719	337.818	288.116	44.732	−701.832	−587.636	34.105
1000	92.617	347.426	293.572	53.854	−701.879	−574.944	30.032
1100	95.038	356.370	298.879	63.240	−701.805	−562.253	26.699
1200	97.084	364.730	304.022	·72.849	−701.629	−549.574	23.922
1300	98.833	372.571	308.997	82.647	−701.368	−536.913	21.573
1400	100.344	379.952	313.804	92.607	−701.033	−524.274	19.561
1500	101.660	386.921	318.449	102.709	−700.635	−511.662	17.817

T/K	C_p°	S°	$-(G^\circ - H^\circ(T_r))/T$	$H^\circ - H^\circ(T_r)$	$\Delta_f H^\circ$	$\Delta_f G^\circ$	Log K_f
	J/K·mol			**kJ/mol**			

30. CHLOROTRIFLUOROMETHANE   CClF₃ (g)

T/K	C_p°	S°	$-(G^\circ - H^\circ(T_r))/T$	$H^\circ - H^\circ(T_r)$	$\Delta_f H^\circ$	$\Delta_f G^\circ$	Log K_f
298.15	66.886	285.419	285.419	0.000	−707.800	−667.238	116.896
300	67.111	285.834	285.421	0.124	−707.810	−666.986	116.131
400	77.528	306.646	288.187	7.383	−708.153	−653.316	85.313
500	85.013	324.797	293.734	15.532	−708.170	−639.599	66.818
600	90.329	340.794	300.271	24.314	−707.975	−625.901	54.489
700	94.132	355.020	307.096	33.547	−707.654	−612.246	45.686
800	96.899	367.780	313.897	43.106	−707.264	−598.642	39.087
900	98.951	379.317	320.536	52.903	−706.837	−585.090	33.957
1000	100.507	389.827	326.947	62.880	−706.396	−571.586	29.856
1100	101.708	399.465	333.108	72.993	−705.950	−558.126	26.503
1200	102.651	408.357	339.013	83.213	−705.505	−544.707	23.710
1300	103.404	416.604	344.668	93.517	−705.064	−531.326	21.349
1400	104.012	424.290	350.084	103.889	−704.628	−517.977	19.326
1500	104.512	431.484	355.273	114.316	−704.196	−504.660	17.574

31. DICHLORODIFLUOROMETHANE   CCl₂F₂ (g)

T/K	C_p°	S°	$-(G^\circ - H^\circ(T_r))/T$	$H^\circ - H^\circ(T_r)$	$\Delta_f H^\circ$	$\Delta_f G^\circ$	Log K_f
298.15	72.476	300.903	300.903	0.000	−486.000	−447.030	78.317
300	72.691	301.352	300.905	0.134	−486.002	−446.788	77.792
400	82.408	323.682	303.883	7.919	−485.945	−433.716	56.637
500	89.063	342.833	309.804	16.514	−485.618	−420.692	43.949
600	93.635	359.500	316.729	25.663	−485.136	−407.751	35.497
700	96.832	374.189	323.909	35.196	−484.576	−394.897	29.467
800	99.121	387.276	331.027	44.999	−483.984	−382.126	24.950
900	100.801	399.053	337.942	55.000	−483.388	−369.429	21.441
1000	102.062	409.742	344.596	65.146	−482.800	−356.799	18.637
1100	103.030	419.517	350.969	75.402	−482.226	−344.227	16.346
1200	103.786	428.515	357.061	85.745	−481.667	−331.706	14.439
1300	104.388	436.847	362.882	96.154	−481.121	−319.232	12.827
1400	104.874	444.602	368.445	106.618	−480.588	−306.799	11.447
1500	105.270	451.851	373.767	117.126	−480.065	−294.404	10.252

32. CHLORODIFLUOROMETHANE   CHClF₂ (g)

T/K	C_p°	S°	$-(G^\circ - H^\circ(T_r))/T$	$H^\circ - H^\circ(T_r)$	$\Delta_f H^\circ$	$\Delta_f G^\circ$	Log K_f
298.15	55.853	280.915	280.915	0.000	−475.000	−443.845	77.759
300	56.039	281.261	280.916	0.104	−475.028	−443.652	77.246
400	65.395	298.701	283.231	6.188	−476.390	−432.978	56.540
500	73.008	314.145	287.898	13.123	−477.398	−422.001	44.086
600	78.940	328.003	293.448	20.733	−478.103	−410.851	35.767
700	83.551	340.533	299.294	28.867	−478.574	−399.603	29.818
800	87.185	351.936	305.172	37.411	−478.870	−388.299	25.353
900	90.100	362.379	310.956	46.280	−479.031	−376.967	21.878
1000	92.475	371.999	316.586	55.413	−479.090	−365.622	19.098
1100	94.433	380.908	322.033	64.761	−479.068	−354.276	16.823
1200	96.066	389.196	327.289	74.289	−478.982	−342.935	14.927
1300	97.438	396.941	332.352	83.966	−478.843	−331.603	13.324
1400	98.601	404.206	337.228	93.769	−478.661	−320.283	11.950
1500	99.593	411.044	341.923	103.681	−478.443	−308.978	10.759

33. METHYLAMINE   CH₅N (g)

T/K	C_p°	S°	$-(G^\circ - H^\circ(T_r))/T$	$H^\circ - H^\circ(T_r)$	$\Delta_f H^\circ$	$\Delta_f G^\circ$	Log K_f
298.15	50.053	242.881	242.881	0.000	−22.529	32.734	−5.735
300	50.227	243.196	242.893	0.091	−22.614	33.077	−5.759
400	60.171	258.986	244.975	5.604	−26.846	52.294	−6.829
500	70.057	273.486	249.244	12.121	−30.431	72.510	−7.575
600	78.929	287.063	254.431	19.579	−33.364	93.382	−8.129
700	86.711	299.826	260.008	27.873	−35.712	114.702	−8.559
800	93.545	311.865	265.749	36.893	−37.548	136.316	−8.900

	J/K·mol			kJ/mol			
T/K	C_p°	S°	$-(G^\circ - H^\circ(T_r))/T$	$H^\circ - H^\circ(T_r)$	$\Delta_f H^\circ$	$\Delta_f G^\circ$	Log K_f

33. METHYLAMINE   CH$_5$N (g) (continued)

900	99.573	323.239	271.511	46.555	−38.949	158.138	−9.178
1000	104.886	334.006	277.220	56.786	−39.967	180.098	−9.407
1100	109.576	344.233	282.861	67.509	−40.681	201.822	−9.584
1200	113.708	353.944	288.374	78.685	−41.136	224.240	−9.761
1300	117.341	363.190	293.775	90.239	−41.376	246.364	−9.899
1400	120.542	372.012	299.061	102.131	−41.451	268.504	−10.018
1500	123.353	380.426	304.209	114.326	−41.381	290.639	−10.121

34. CHLORINE   Cl (g)

298.15	21.838	165.190	165.190	0.000	121.302	105.306	−18.449
300	21.852	165.325	165.190	0.040	121.311	105.207	−18.318
400	22.467	171.703	166.055	2.259	121.795	99.766	−13.028
500	22.744	176.752	167.708	4.522	122.272	94.203	−9.841
600	22.781	180.905	169.571	6.800	122.734	88.546	−7.709
700	22.692	184.411	171.448	9.074	123.172	82.813	−6.179
800	22.549	187.432	173.261	11.337	123.585	77.019	−5.029
900	22.389	190.079	174.986	13.584	123.971	71.175	−4.131
1000	22.233	192.430	176.615	15.815	124.334	65.289	−3.410
1100	22.089	194.542	178.150	18.031	124.675	59.368	−2.819
1200	21.959	196.458	179.597	20.233	124.996	53.416	−2.325
1300	21.843	198.211	180.963	22.423	125.299	47.439	−1.906
1400	21.742	199.826	182.253	24.602	125.587	41.439	−1.546
1500	21.652	201.323	183.475	26.772	125.861	35.418	−1.233

35. DICHLORINE   Cl$_2$ (g)

298.15	33.949	223.079	223.079	0.000	0.000	0.000	0.000
300	33.981	223.290	223.080	0.063	0.000	0.000	0.000
400	35.296	233.263	224.431	3.533	0.000	0.000	0.000
500	36.064	241.229	227.021	7.104	0.000	0.000	0.000
600	36.547	247.850	229.956	10.736	0.000	0.000	0.000
700	36.874	253.510	232.926	14.408	0.000	0.000	0.000
800	37.111	258.450	235.815	18.108	0.000	0.000	0.000
900	37.294	262.832	238.578	21.829	0.000	0.000	0.000
1000	37.442	266.769	241.203	25.566	0.000	0.000	0.000
1100	37.567	270.343	243.692	29.316	0.000	0.000	0.000
1200	37.678	273.617	246.052	33.079	0.000	0.000	0.000
1300	37.778	276.637	248.290	36.851	0.000	0.000	0.000
1400	37.872	279.440	250.416	40.634	0.000	0.000	0.000
1500	37.961	282.056	252.439	44.426	0.000	0.000	0.000

36. HYDROGEN CHLORIDE   HCl (g)

298.15	29.136	186.902	186.902	0.000	−92.310	−95.298	16.696
300	29.137	187.082	186.902	0.054	−92.314	−95.317	16.596
400	29.175	195.468	188.045	2.969	−92.587	−96.278	12.573
500	29.304	201.990	190.206	5.892	−92.911	−97.164	10.151
600	29.576	207.354	192.630	8.835	−93.249	−97.983	8.530
700	29.988	211.943	195.069	11.812	−93.577	−98.746	7.368
800	30.500	215.980	197.435	14.836	−93.879	−99.464	6.494
900	31.063	219.604	199.700	17.913	−94.149	−100.145	5.812
1000	31.639	222.907	201.858	21.049	−94.384	−100.798	5.265
1100	32.201	225.949	203.912	24.241	−94.587	−101.430	4.816
1200	32.734	228.774	205.867	27.488	−94.760	−102.044	4.442
1300	33.229	231.414	207.732	30.786	−94.908	−102.645	4.124
1400	33.684	233.893	209.513	34.132	−95.035	−103.235	3.852
1500	34.100	236.232	211.217	37.522	−95.146	−103.817	3.615

	J/K·mol			kJ/mol			
T/K	C_p°	S°	$-(G^\circ-H^\circ(T_r))/T$	$H^\circ-H^\circ(T_r)$	$\Delta_f H^\circ$	$\Delta_f G^\circ$	Log K_f

37. COPPER   Cu (cr, l)

T/K	C_p°	S°	$-(G^\circ-H^\circ(T_r))/T$	$H^\circ-H^\circ(T_r)$	$\Delta_f H^\circ$	$\Delta_f G^\circ$	Log K_f
298.15	24.440	33.150	33.150	0.000	0.000	0.000	0.000
300	24.460	33.301	33.150	0.045	0.000	0.000	0.000
400	25.339	40.467	34.122	2.538	0.000	0.000	0.000
500	25.966	46.192	35.982	5.105	0.000	0.000	0.000
600	26.479	50.973	38.093	7.728	0.000	0.000	0.000
700	26.953	55.090	40.234	10.399	0.000	0.000	0.000
800	27.448	58.721	42.322	13.119	0.000	0.000	0.000
900	28.014	61.986	44.328	15.891	0.000	0.000	0.000
1000	28.700	64.971	46.245	18.726	0.000	0.000	0.000
1100	29.553	67.745	48.075	21.637	0.000	0.000	0.000
1200	30.617	70.361	49.824	24.644	0.000	0.000	0.000
1300	31.940	72.862	51.501	27.769	0.000	0.000	0.000
1358	32.844	74.275	52.443	29.647	0.000	0.000	0.000
PHASE TRANSITION: $\Delta_{trs} H$ = 13.141 kJ/mol, $\Delta_{trs} S$ = 9.676 J/K·mol, cr–l							
1358	32.800	83.951	52.443	42.788	0.000	0.000	0.000
1400	32.800	84.950	53.403	44.166	0.000	0.000	0.000
1500	32.800	87.213	55.583	47.446	0.000	0.000	0.000

38. COPPER   Cu (g)

T/K	C_p°	S°	$-(G^\circ-H^\circ(T_r))/T$	$H^\circ-H^\circ(T_r)$	$\Delta_f H^\circ$	$\Delta_f G^\circ$	Log K_f
298.15	20.786	166.397	166.397	0.000	337.600	297.873	−52.185
300	20.786	166.525	166.397	0.038	337.594	297.626	−51.821
400	20.786	172.505	167.213	2.117	337.179	284.364	−37.134
500	20.786	177.143	168.752	4.196	336.691	271.215	−28.333
600	20.786	180.933	170.476	6.274	336.147	258.170	−22.475
700	20.786	184.137	172.205	8.353	335.554	245.221	−18.298
800	20.786	186.913	173.874	10.431	334.913	232.359	−15.171
900	20.786	189.361	175.461	12.510	334.219	219.581	−12.744
1000	20.786	191.551	176.963	14.589	333.463	206.883	−10.806
1100	20.788	193.532	178.380	16.667	332.631	194.265	−9.225
1200	20.793	195.341	179.719	18.746	331.703	181.726	−7.910
1300	20.803	197.006	180.986	20.826	330.657	169.270	−6.801
1400	20.823	198.548	182.186	22.907	316.342	157.305	−5.869
1500	20.856	199.986	183.325	24.991	315.146	145.987	−5.084

39. COPPER OXIDE   CuO (cr)

T/K	C_p°	S°	$-(G^\circ-H^\circ(T_r))/T$	$H^\circ-H^\circ(T_r)$	$\Delta_f H^\circ$	$\Delta_f G^\circ$	Log K_f
298.15	42.300	42.740	42.740	0.000	−162.000	−134.277	23.524
300	42.417	43.002	42.741	0.078	−161.994	−134.105	23.349
400	46.783	55.878	44.467	4.564	−161.487	−124.876	16.307
500	49.190	66.596	47.852	9.372	−160.775	−115.803	12.098
600	50.827	75.717	51.755	14.377	−159.973	−106.883	9.305
700	52.099	83.651	55.757	19.526	−159.124	−98.102	7.320
800	53.178	90.680	59.691	24.791	−158.247	−89.444	5.840
900	54.144	97.000	63.491	30.158	−157.356	−80.897	4.695
1000	55.040	102.751	67.134	35.617	−156.462	−72.450	3.784
1100	55.890	108.037	70.615	41.164	−155.582	−64.091	3.043
1200	56.709	112.936	73.941	46.794	−154.733	−55.812	2.429
1300	57.507	117.507	77.118	52.505	−153.940	−47.601	1.913
1400	58.288	121.797	80.158	58.295	−166.354	−39.043	1.457
1500	59.057	125.845	83.070	64.163	−165.589	−29.975	1.044

40. DICOPPER OXIDE   Cu$_2$O (cr)

T/K	C_p°	S°	$-(G^\circ-H^\circ(T_r))/T$	$H^\circ-H^\circ(T_r)$	$\Delta_f H^\circ$	$\Delta_f G^\circ$	Log K_f
298.15	62.600	92.550	92.550	0.000	−173.100	−150.344	26.339
300	62.721	92.938	92.551	0.116	−173.102	−150.203	26.152
400	67.587	111.712	95.078	6.654	−173.036	−142.572	18.618
500	70.784	127.155	99.995	13.580	−172.772	−134.984	14.101

	J/K·mol			kJ/mol			
T/K	C_p°	S°	$-(G^\circ - H^\circ(T_r))/T$	$H^\circ - H^\circ(T_r)$	$\Delta_f H^\circ$	$\Delta_f G^\circ$	Log K_f

40. DICOPPER OXIDE Cu_2O (cr) (continued)

T/K	C_p°	S°	$-(G^\circ - H^\circ(T_r))/T$	$H^\circ - H^\circ(T_r)$	$\Delta_f H^\circ$	$\Delta_f G^\circ$	Log K_f
600	73.323	140.291	105.643	20.789	−172.389	−127.460	11.096
700	75.552	151.764	111.429	28.235	−171.914	−120.009	8.955
800	77.616	161.989	117.121	35.894	−171.363	−112.631	7.354
900	79.584	171.245	122.629	43.755	−170.750	−105.325	6.113
1000	81.492	179.729	127.920	51.809	−170.097	−98.091	5.124
1100	83.360	187.584	132.992	60.052	−169.431	−90.922	4.317
1200	85.202	194.917	137.850	68.480	−168.791	−83.814	3.648
1300	87.026	201.808	142.507	77.092	−168.223	−76.756	3.084
1400	88.836	208.324	146.978	85.885	−194.030	−68.926	2.572
1500	90.636	214.515	151.276	94.858	−193.438	−60.010	2.090

41. COPPER DICHLORIDE $CuCl_2$ (cr, l)

T/K	C_p°	S°	$-(G^\circ - H^\circ(T_r))/T$	$H^\circ - H^\circ(T_r)$	$\Delta_f H^\circ$	$\Delta_f G^\circ$	Log K_f
298.15	71.880	108.070	108.070	0.000	−218.000	−173.826	30.453
300	71.998	108.515	108.071	0.133	−217.975	−173.552	30.218
400	76.338	129.899	110.957	7.577	−216.494	−158.962	20.758
500	78.654	147.204	116.532	15.336	−214.873	−144.765	15.123
600	80.175	161.687	122.884	23.282	−213.182	−130.901	11.396
675	81.056	171.183	127.732	29.329	−211.185	−120.693	9.340
PHASE TRANSITION: $\Delta_{trs} H = 0.700$ kJ/mol, $\Delta_{trs} S = 1.037$ J/K·mol, crII–crI							
675	82.400	172.220	127.732	30.029	−211.185	−120.693	9.340
700	82.400	175.216	129.375	32.089	−210.719	−117.350	8.757
800	82.400	186.219	135.808	40.329	−208.898	−104.137	6.799
871	82.400	193.226	140.207	46.179	−192.649	−94.893	5.691
PHASE TRANSITION: $\Delta_{trs} H = 15.001$ kJ/mol, $\Delta_{trs} S = 17.221$ J/K·mol, crI–l							
871	100.000	210.447	140.207	61.180	−192.649	−94.893	5.691
900	100.000	213.723	142.523	64.080	−191.640	−91.655	5.319
1000	100.000	224.259	150.179	74.080	−188.212	−80.730	4.217
1100	100.000	233.790	157.353	84.080	−184.873	−70.144	3.331
1130.75	100.000	236.547	159.470	87.155	−183.867	−66.951	3.093

42. COPPER DICHLORIDE $CuCl_2$ (g)

T/K	C_p°	S°	$-(G^\circ - H^\circ(T_r))/T$	$H^\circ - H^\circ(T_r)$	$\Delta_f H^\circ$	$\Delta_f G^\circ$	Log K_f
298.15	56.814	278.418	278.418	0.000	−43.268	−49.883	8.739
300	56.869	278.769	278.419	0.105	−43.271	−49.924	8.692
400	58.992	295.456	280.679	5.911	−43.428	−52.119	6.806
500	60.111	308.752	285.010	11.871	−43.606	−54.271	5.670
600	60.761	319.774	289.911	17.918	−43.814	−56.385	4.909
700	61.168	329.173	294.865	24.015	−44.060	−58.462	4.362
800	61.439	337.360	299.677	30.147	−44.349	−60.500	3.950
900	61.630	344.608	304.274	36.301	−44.688	−62.499	3.627
1000	61.776	351.109	308.638	42.471	−45.088	−64.457	3.367
1100	61.900	357.003	312.771	48.655	−45.566	−66.372	3.152
1200	62.022	362.394	316.685	54.851	−46.139	−68.239	2.970
1300	62.159	367.364	320.395	61.060	−46.829	−70.053	2.815
1400	62.325	371.976	323.916	67.284	−60.784	−71.404	2.664
1500	62.531	376.283	327.265	73.526	−61.613	−72.133	2.512

43. FLUORINE   F (g)

T/K	C_p°	S°	$-(G^\circ - H^\circ(T_r))/T$	$H^\circ - H^\circ(T_r)$	$\Delta_f H^\circ$	$\Delta_f G^\circ$	Log K_f
298.15	22.746	158.750	158.750	0.000	79.380	62.280	−10.911
300	22.742	158.891	158.750	0.042	79.393	62.173	−10.825
400	22.432	165.394	159.639	2.302	80.043	56.332	−7.356
500	22.100	170.363	161.307	4.528	80.587	50.340	−5.259
600	21.832	174.368	163.161	6.724	81.046	44.246	−3.852
700	21.629	177.717	165.008	8.897	81.442	38.081	−2.842
800	21.475	180.595	166.780	11.052	81.792	31.862	−2.080
900	21.357	183.117	168.458	13.193	82.106	25.601	−1.486

	J/K·mol			kJ/mol			
T/K	C_p°	S°	$-(G^\circ-H^\circ(T_r))/T$	$H^\circ-H^\circ(T_r)$	$\Delta_f H^\circ$	$\Delta_f G^\circ$	$\text{Log } K_f$

43. FLUORINE   F (g) (continued)

1000	21.266	185.362	170.039	15.324	82.391	19.308	−1.009
1100	21.194	187.386	171.525	17.447	82.654	12.986	−0.617
1200	21.137	189.227	172.925	19.563	82.897	6.642	−0.289
1300	21.091	190.917	174.245	21.675	83.123	0.278	−0.011
1400	21.054	192.479	175.492	23.782	83.335	−6.103	0.228
1500	21.022	193.930	176.673	25.886	83.533	−12.498	0.435

44. DIFLUORINE   F$_2$ (g)

298.15	31.304	202.790	202.790	0.000	0.000	0.000	0.000
300	31.337	202.984	202.790	0.058	0.000	0.000	0.000
400	32.995	212.233	204.040	3.277	0.000	0.000	0.000
500	34.258	219.739	206.453	6.643	0.000	0.000	0.000
600	35.171	226.070	209.208	10.117	0.000	0.000	0.000
700	35.839	231.545	212.017	13.669	0.000	0.000	0.000
800	36.343	236.365	214.765	17.279	0.000	0.000	0.000
900	36.740	240.669	217.409	20.934	0.000	0.000	0.000
1000	37.065	244.557	219.932	24.625	0.000	0.000	0.000
1100	37.342	248.103	222.334	28.346	0.000	0.000	0.000
1200	37.588	251.363	224.619	32.093	0.000	0.000	0.000
1300	37.811	254.381	226.794	35.863	0.000	0.000	0.000
1400	38.019	257.191	228.866	39.654	0.000	0.000	0.000
1500	38.214	259.820	230.843	43.466	0.000	0.000	0.000

45. HYDROGEN FLUORIDE   HF (g)

298.15	29.137	173.776	173.776	0.000	−273.300	−275.399	48.248
300	29.137	173.956	173.776	0.054	−273.302	−275.412	47.953
400	29.149	182.340	174.919	2.968	−273.450	−276.096	36.054
500	29.172	188.846	177.078	5.884	−273.679	−276.733	28.910
600	29.230	194.496	179.496	8.804	−273.961	−277.318	24.142
700	29.350	198.683	181.923	11.732	−274.277	−277.852	20.733
800	29.549	202.614	184.269	14.676	−274.614	−278.340	18.174
900	29.827	206.110	186.505	17.645	−274.961	−278.785	16.180
1000	30.169	209.270	188.626	20.644	−275.309	−279.191	14.583
1100	30.558	212.163	190.636	23.680	−275.652	−279.563	13.275
1200	30.974	214.840	192.543	26.756	−275.988	−279.904	12.184
1300	31.403	217.336	194.355	29.875	−276.315	−280.217	11.259
1400	31.831	219.679	196.081	33.037	−276.631	−280.505	10.466
1500	32.250	221.889	197.729	36.241	−276.937	−280.771	9.777

46. GERMANIUM   Ge (cr, l)

298.15	23.222	31.090	31.090	0.000	0.000	0.000	0.000
300	23.249	31.234	31.090	0.043	0.000	0.000	0.000
400	24.310	38.083	32.017	2.426	0.000	0.000	0.000
500	24.962	43.582	33.798	4.892	0.000	0.000	0.000
600	25.452	48.178	35.822	7.414	0.000	0.000	0.000
700	25.867	52.133	37.876	9.980	0.000	0.000	0.000
800	26.240	55.612	39.880	12.586	0.000	0.000	0.000
900	26.591	58.723	41.804	15.227	0.000	0.000	0.000
1000	26.926	61.542	43.639	17.903	0.000	0.000	0.000
1100	27.252	64.124	45.386	20.612	0.000	0.000	0.000
1200	27.571	66.509	47.048	23.353	0.000	0.000	0.000
1211.4	27.608	66.770	47.232	23.668	0.000	0.000	0.000
			PHASE TRANSITION: $\Delta_{trs} H$ = 37.030 kJ/mol, $\Delta_{trs} S$ = 30.568 J/K·mol, cr–l				
1211.4	27.600	97.338	47.232	60.698	0.000	0.000	0.000
1300	27.600	99.286	50.714	63.143	0.000	0.000	0.000

	J/K·mol			kJ/mol			
T/K	C_p°	S°	$-(G^\circ - H^\circ(T_r))/T$	$H^\circ - H^\circ(T_r)$	$\Delta_f H^\circ$	$\Delta_f G^\circ$	Log K_f

46. GERMANIUM   Ge (cr, l) (continued)

1400	27.600	101.331	54.258	65.903	0.000	0.000	0.000
1500	27.600	103.236	57.460	68.663	0.000	0.000	0.000

47. GERMANIUM   Ge (g)

298.15	30.733	167.903	167.903	0.000	367.800	327.009	−57.290
300	30.757	168.094	167.904	0.057	367.814	326.756	−56.893
400	31.071	177.025	169.119	3.162	368.536	312.959	−40.868
500	30.360	183.893	171.415	6.239	369.147	298.991	−31.235
600	29.265	189.334	173.965	9.222	369.608	284.914	−24.804
700	28.102	193.758	176.487	12.090	369.910	270.773	−20.205
800	27.029	197.439	178.882	14.845	370.060	256.598	−16.754
900	26.108	200.567	181.122	17.501	370.073	242.414	−14.069
1000	25.349	203.277	183.205	20.072	369.969	228.234	−11.922
1100	24.741	205.664	185.141	22.575	369.763	214.069	−10.165
1200	24.264	207.795	186.941	25.025	369.471	199.928	−8.703
1300	23.898	209.722	188.621	27.432	332.088	188.521	−7.575
1400	23.624	211.483	190.192	29.807	331.704	177.492	−6.622
1500	23.426	213.105	191.666	32.159	331.296	166.491	−5.798

48. GERMANIUM DIOXIDE   GeO$_2$ (cr, l)

298.15	50.166	39.710	39.710	0.000	−580.200	−521.605	91.382
300	50.475	40.021	39.711	0.093	−580.204	−521.242	90.755
400	61.281	56.248	41.850	5.759	−579.893	−501.610	65.503
500	66.273	70.519	46.191	12.164	−579.013	−482.134	50.368
600	69.089	82.872	51.299	18.943	−577.915	−462.859	40.295
700	70.974	93.671	56.597	25.952	−576.729	−443.776	33.115
800	72.449	103.247	61.841	33.125	−575.498	−424.866	27.741
900	73.764	111.857	66.928	40.436	−574.235	−406.113	23.570
1000	75.049	119.696	71.819	47.877	−572.934	−387.502	20.241
1100	76.378	126.910	76.504	55.447	−571.582	−369.024	17.523
1200	77.796	133.616	80.987	63.155	−570.166	−350.671	15.264
1300	79.332	139.903	85.279	71.010	−605.685	−329.732	13.249
1308	79.460	140.390	85.615	71.646	−584.059	−328.034	13.100

PHASE TRANSITION: $\Delta_{trs} H = 21.500$ kJ/mol, $\Delta_{trs} S = 16.437$ J/K·mol, crII–crI

1308	80.075	156.827	85.615	93.146	−584.059	−328.034	13.100
1388	81.297	161.617	89.858	99.601	−565.504	−312.415	11.757

PHASE TRANSITION: $\Delta_{trs} H = 17.200$ kJ/mol, $\Delta_{trs} S = 12.392$ J/K·mol, crI–l

1388	78.500	174.009	89.858	116.801	−565.504	−312.415	11.757
1400	78.500	174.685	90.582	117.743	−565.328	−310.228	11.575
1500	78.500	180.100	96.372	125.593	−563.882	−292.057	10.170

49. GERMANIUM TETRACHLORIDE   GeCl$_4$ (g)

298.15	95.918	348.393	348.393	0.000	−500.000	−461.582	80.866
300	96.041	348.987	348.395	0.178	−499.991	−461.343	80.326
400	100.750	377.342	352.229	10.045	−499.447	−448.540	58.573
500	103.206	400.114	359.604	20.255	−498.845	−435.882	45.536
600	104.624	419.067	367.980	30.652	−498.234	−423.347	36.855
700	105.509	435.266	376.463	41.162	−497.634	−410.914	30.662
800	106.096	449.396	384.715	51.744	−497.057	−398.565	26.023
900	106.504	461.917	392.611	62.375	−496.509	−386.287	22.419
1000	106.799	473.155	400.113	73.041	−495.993	−374.068	19.539
1100	107.020	483.344	407.224	83.733	−495.512	−361.899	17.185
1200	107.189	492.664	413.961	94.444	−495.067	−349.772	15.225
1300	107.320	501.249	420.349	105.169	−531.677	−334.973	13.459
1400	107.425	509.206	426.416	115.907	−531.265	−319.857	11.934
1500	107.509	516.621	432.185	126.654	−530.861	−304.771	10.613

THERMODYNAMIC PROPERTIES AS A FUNCTION OF TEMPERATURE (continued)

	J/K·mol			kJ/mol			
T/K	C_p°	S°	$-(G^\circ-H^\circ(T_r))/T$	$H^\circ-H^\circ(T_r)$	$\Delta_f H^\circ$	$\Delta_f G^\circ$	Log K_f

50. HYDROGEN   H (g)

298.15	20.786	114.716	114.716	0.000	217.998	203.276	−35.613
300	20.786	114.845	114.716	0.038	218.010	203.185	−35.377
400	20.786	120.824	115.532	2.117	218.635	198.149	−25.875
500	20.786	125.463	117.071	4.196	219.253	192.956	−20.158
600	20.786	129.252	118.795	6.274	219.867	187.639	−16.335
700	20.786	132.457	120.524	8.353	220.476	182.219	−13.597
800	20.786	135.232	122.193	10.431	221.079	176.712	−11.538
900	20.786	137.680	123.780	12.510	221.670	171.131	−9.932
1000	20.786	139.870	125.282	14.589	222.247	165.485	−8.644
1100	20.786	141.852	126.700	16.667	222.806	159.781	−7.587
1200	20.786	143.660	128.039	18.746	223.345	154.028	−6.705
1300	20.786	145.324	129.305	20.824	223.864	148.230	−5.956
1400	20.786	146.864	130.505	22.903	224.360	142.393	−5.313
1500	20.786	148.298	131.644	24.982	224.835	136.522	−4.754

51. DIHYDROGEN   H$_2$ (g)

298.15	28.836	130.680	130.680	0.000	0.000	0.000	0.000
300	28.849	130.858	130.680	0.053	0.000	0.000	0.000
400	29.181	139.217	131.818	2.960	0.000	0.000	0.000
500	29.260	145.738	133.974	5.882	0.000	0.000	0.000
600	29.327	151.078	136.393	8.811	0.000	0.000	0.000
700	29.440	155.607	138.822	11.749	0.000	0.000	0.000
800	29.623	159.549	141.172	14.702	0.000	0.000	0.000
900	29.880	163.052	143.412	17.676	0.000	0.000	0.000
1000	30.204	166.217	145.537	20.680	0.000	0.000	0.000
1100	30.580	169.113	147.550	23.719	0.000	0.000	0.000
1200	30.991	171.791	149.460	26.797	0.000	0.000	0.000
1300	31.422	174.288	151.275	29.918	0.000	0.000	0.000
1400	31.860	176.633	153.003	33.082	0.000	0.000	0.000
1500	32.296	178.846	154.653	36.290	0.000	0.000	0.000

52. HYDROXYL   OH (g)

298.15	29.886	183.737	183.737	0.000	39.349	34.631	−6.067
300	29.879	183.922	183.738	0.055	39.350	34.602	−6.025
400	29.604	192.476	184.906	3.028	39.384	33.012	−4.311
500	29.495	199.067	187.104	5.982	39.347	31.422	−3.283
600	29.513	204.445	189.560	8.931	39.252	29.845	−2.598
700	29.655	209.003	192.020	11.888	39.113	28.287	−2.111
800	29.914	212.979	194.396	14.866	38.945	26.752	−1.747
900	30.265	216.522	196.661	17.874	38.763	25.239	−1.465
1000	30.682	219.731	198.810	20.921	38.577	23.746	−1.240
1100	31.135	222.677	200.848	24.012	38.393	22.272	−1.058
1200	31.603	225.406	202.782	27.149	38.215	20.814	−0.906
1300	32.069	227.954	204.621	30.332	38.046	19.371	−0.778
1400	32.522	230.347	206.374	33.562	37.886	17.941	−0.669
1500	32.956	232.606	208.048	36.836	37.735	16.521	−0.575

53. WATER   H$_2$O (l)

298.15	75.300	69.950	69.950	0.000	−285.830	−237.141	41.546
300	75.281	70.416	69.951	0.139	−285.771	−236.839	41.237
373.21	76.079	86.896	71.715	5.666	−283.454	−225.160	31.513

54. WATER   H$_2$O (g)

298.15	33.598	188.832	188.832	0.000	−241.826	−228.582	40.046

THERMODYNAMIC PROPERTIES AS A FUNCTION OF TEMPERATURE (continued)

T/K	C_p°	S°	$-(G^\circ - H^\circ(T_r))/T$	$H^\circ - H^\circ(T_r)$	$\Delta_f H^\circ$	$\Delta_f G^\circ$	Log K_f
	J/K·mol			**kJ/mol**			

54. WATER H_2O (g) (continued)
300	33.606	189.040	188.833	0.062	−241.844	−228.500	39.785
400	34.283	198.791	190.158	3.453	−242.845	−223.900	29.238
500	35.259	206.542	192.685	6.929	−243.822	−219.050	22.884
600	36.371	213.067	195.552	10.509	−244.751	−214.008	18.631
700	37.557	218.762	198.469	14.205	−245.620	−208.814	15.582
800	38.800	223.858	201.329	18.023	−246.424	−203.501	13.287
900	40.084	228.501	204.094	21.966	−247.158	−198.091	11.497
1000	41.385	232.792	206.752	26.040	−247.820	−192.603	10.060
1100	42.675	236.797	209.303	30.243	−248.410	−187.052	8.882
1200	43.932	240.565	211.753	34.574	−248.933	−181.450	7.898
1300	45.138	244.129	214.108	39.028	−249.392	−175.807	7.064
1400	46.281	247.516	216.374	43.599	−249.792	−170.132	6.348
1500	47.356	250.746	218.559	48.282	−250.139	−164.429	5.726

55. IODINE   I (g)
298.15	20.786	180.787	180.787	0.000	106.760	70.172	−12.294
300	20.786	180.915	180.787	0.038	106.748	69.945	−12.178
400	20.786	186.895	181.602	2.117	97.974	58.060	−7.582
500	20.786	191.533	183.142	4.196	75.988	50.202	−5.244
600	20.786	195.323	184.866	6.274	76.190	45.025	−3.920
700	20.786	198.527	186.594	8.353	76.385	39.816	−2.971
800	20.787	201.303	188.263	10.432	76.574	34.579	−2.258
900	20.789	203.751	189.851	12.510	76.757	29.319	−1.702
1000	20.795	205.942	191.352	14.589	76.936	24.038	−1.256
1100	20.806	207.924	192.770	16.669	77.109	18.740	−0.890
1200	20.824	209.735	194.110	18.751	77.277	13.426	−0.584
1300	20.851	211.403	195.377	20.835	77.440	8.098	−0.325
1400	20.889	212.950	196.577	22.921	77.596	2.758	−0.103
1500	20.936	214.392	197.717	25.013	77.745	−2.592	0.090

56. DIIODINE I_2 (cr, l)
298.15	54.440	116.139	116.139	0.000	0.000	0.000	0.000
300	54.518	116.476	116.140	0.101	0.000	0.000	0.000
386.75	61.531	131.039	117.884	5.088	0.000	0.000	0.000

PHASE TRANSITION: $\Delta_{trs} H$ = 15.665 kJ/mol, $\Delta_{trs} S$ = 40.504 J/K·mol, cr–l

386.75	79.555	171.543	117.884	20.753	0.000	0.000	0.000
400	79.555	174.223	119.706	21.807	0.000	0.000	0.000
457.67	79.555	184.938	127.266	26.395	0.000	0.000	0.000

57. DIIODINE I_2 (g)
298.15	36.887	260.685	260.685	0.000	62.420	19.324	−3.385
300	36.897	260.913	260.685	0.068	62.387	19.056	−3.318
400	37.256	271.584	262.138	3.778	44.391	5.447	−0.711
457.67	37.385	276.610	263.652	5.931		pressure = 1 bar	
500	37.464	279.921	264.891	7.515	0.000	0.000	0.000
600	37.613	286.765	267.983	11.269	0.000	0.000	0.000
700	37.735	292.573	271.092	15.037	0.000	0.000	0.000
800	37.847	297.619	274.099	18.816	0.000	0.000	0.000
900	37.956	302.083	276.965	22.606	0.000	0.000	0.000
1000	38.070	306.088	279.681	26.407	0.000	0.000	0.000
1100	38.196	309.722	282.249	30.220	0.000	0.000	0.000
1200	38.341	313.052	284.679	34.047	0.000	0.000	0.000
1300	38.514	316.127	286.981	37.890	0.000	0.000	0.000
1400	38.719	318.989	289.166	41.751	0.000	0.000	0.000
1500	38.959	321.668	291.245	45.635	0.000	0.000	0.000

	J/K·mol			kJ/mol			
T/K	C_p°	S°	$-(G^\circ-H^\circ(T_r))/T$	$H^\circ-H^\circ(T_r)$	$\Delta_f H^\circ$	$\Delta_f G^\circ$	Log K_f

58. HYDROGEN IODIDE   HI (g)

298.15	29.157	206.589	206.589	0.000	26.500	1.700	−0.298
300	29.158	206.769	206.589	0.054	26.477	1.546	−0.269
400	29.329	215.176	207.734	2.977	17.093	−6.289	0.821
500	29.738	221.760	209.904	5.928	−5.481	−9.946	1.039
600	30.351	227.233	212.348	8.931	−5.819	−10.806	0.941
700	31.070	231.965	214.820	12.002	−6.101	−11.614	0.867
800	31.807	236.162	217.230	15.145	−6.323	−12.386	0.809
900	32.511	239.950	219.548	18.362	−6.489	−13.133	0.762
1000	33.156	243.409	221.763	21.646	−6.608	−13.865	0.724
1100	33.735	246.597	223.878	24.991	−6.689	−14.586	0.693
1200	34.249	249.555	225.896	28.391	−6.741	−15.302	0.666
1300	34.703	252.314	227.823	31.839	−6.775	−16.014	0.643
1400	35.106	254.901	229.666	35.330	−6.797	−16.723	0.624
1500	35.463	257.336	231.430	38.858	−6.814	−17.432	0.607

59. POTASSIUM   K (cr, l)

298.15	29.600	64.680	64.680	0.000	0.000	0.000	0.000
300	29.671	64.863	64.681	0.055	0.000	0.000	0.000
336.86	32.130	68.422	64.896	1.188	0.000	0.000	0.000
PHASE TRANSITION: $\Delta_{trs} H$ = 2.321 kJ/mol, $\Delta_{trs} S$ = 6.891 J/K·mol, cr–l							
336.86	32.129	75.313	64.896	3.509	0.000	0.000	0.000
400	31.552	80.784	66.986	5.519	0.000	0.000	0.000
500	30.741	87.734	70.469	8.632	0.000	0.000	0.000
600	30.158	93.283	73.824	11.675	0.000	0.000	0.000
700	29.851	97.905	76.943	14.673	0.000	0.000	0.000
800	29.838	101.887	79.818	17.655	0.000	0.000	0.000
900	30.130	105.415	82.470	20.651	0.000	0.000	0.000
1000	30.730	108.618	84.927	23.691	0.000	0.000	0.000
1039.4	31.053	109.812	85.847	24.908	0.000	0.000	0.000

60. POTASSIUM   K (g)

298.15	20.786	160.340	160.340	0.000	89.000	60.479	−10.596
300	20.786	160.468	160.340	0.038	88.984	60.302	−10.499
400	20.786	166.448	161.155	2.117	85.598	51.332	−6.703
500	20.786	171.086	162.695	4.196	84.563	42.887	−4.480
600	20.786	174.876	164.419	6.274	83.599	34.643	−3.016
700	20.786	178.080	166.148	8.353	82.680	26.557	−1.982
800	20.786	180.856	167.817	10.431	81.776	18.601	−1.215
900	20.786	183.304	169.404	12.510	80.859	10.759	−0.624
1000	20.786	185.494	170.905	14.589	79.897	3.021	−0.158
1039.4	20.786	186.297	171.474	15.408		pressure = 1 bar	
1100	20.786	187.475	172.323	16.667	0.000	0.000	0.000
1200	20.786	189.284	173.662	18.746	0.000	0.000	0.000
1300	20.789	190.948	174.929	20.825	0.000	0.000	0.000
1400	20.793	192.489	176.129	22.904	0.000	0.000	0.000
1500	20.801	193.923	177.268	24.983	0.000	0.000	0.000

61. DIPOTASSIUM OXIDE   K$_2$O (cr, l)

298.15	72.000	96.000	96.000	0.000	−361.700	−321.171	56.267
300	72.130	96.446	96.001	0.133	−361.704	−320.920	55.876
400	79.154	118.158	98.914	7.698	−366.554	−306.416	40.013
500	86.178	136.575	104.647	15.964	−366.043	−291.423	30.444
590	92.500	151.348	110.662	24.005	−364.204	−278.079	24.619
PHASE TRANSITION: $\Delta_{trs} H$ = 0.700 kJ/mol, $\Delta_{trs} S$ = 1.186 J/K·mol, crIII–crII							

	J/K·mol			kJ/mol			
T/K	C_p°	S°	$-(G^\circ-H^\circ(T_r))/T$	$H^\circ-H^\circ(T_r)$	$\Delta_f H^\circ$	$\Delta_f G^\circ$	Log K_f

61. DIPOTASSIUM OXIDE K_2O (cr, l) (continued)

590	100.000	152.534	110.662	24.705	−364.204	−278.079	24.619
600	100.000	154.215	111.374	25.705	−363.968	−276.621	24.082
645	100.000	161.447	114.618	30.205	−358.901	−270.109	21.874
PHASE TRANSITION: $\Delta_{trs} H$ = 4.000 kJ/mol, $\Delta_{trs} S$ = 6.202 J/K·mol, crII–crI							
645	100.000	167.649	114.618	34.205	−358.901	−270.109	21.874
700	100.000	175.832	119.111	39.705	−357.592	−262.592	19.595
800	100.000	189.185	127.054	49.705	−355.224	−249.183	16.270
900	100.000	200.963	134.625	59.705	−352.919	−236.067	13.701
1000	100.000	211.499	141.794	69.705	−350.732	−223.202	11.659
1013	100.000	212.791	142.697	71.005	−323.459	−221.546	11.424
PHASE TRANSITION: $\Delta_{trs} H$ = 27.000 kJ/mol, $\Delta_{trs} S$ =26.654 J/K·mol, crI–l							
1013	100.000	239.444	142.697	98.005	−323.459	−221.546	11.424
1100	100.000	247.684	150.679	106.705	−479.439	−203.633	9.670
1200	100.000	256.385	159.131	116.705	−475.371	−178.740	7.780
1300	100.000	264.389	166.924	126.705	−471.321	−154.185	6.195
1400	100.000	271.800	174.154	136.705	−467.287	−129.941	4.848
1500	100.000	278.699	180.896	146.705	−463.268	−105.986	3.691

62. POTASSIUM HYDROXIDE KOH (cr, l)

298.15	64.900	78.870	78.870	0.000	−424.580	−378.747	66.354
300	65.038	79.272	78.871	0.120	−424.569	−378.463	65.895
400	72.519	99.007	81.512	6.998	−426.094	−362.765	47.372
500	80.000	115.993	86.745	14.624	−424.572	−347.093	36.260
520	81.496	119.159	87.931	16.239	−417.725	−344.002	34.555
PHASE TRANSITION: $\Delta_{trs} H$ = 6.450 kJ/mol, $\Delta_{trs} S$ = 12.404 J/K·mol, crII–crI							
520	79.000	131.563	87.931	22.689	−417.725	−344.002	34.555
600	79.000	142.868	94.520	29.009	−416.274	−332.766	28.969
678	79.000	152.523	100.649	35.171	−405.464	−321.998	24.807
PHASE TRANSITION: $\Delta_{trs} H$ = 9.400 kJ/mol, $\Delta_{trs} S$ = 13.865 J/K·mol, crI–l							
678	83.000	166.388	100.649	44.571	−405.464	−321.998	24.807
700	83.000	169.038	102.757	46.397	−404.981	−319.297	23.826
800	83.000	180.121	111.750	54.697	−402.808	−307.206	20.058
900	83.000	189.897	119.901	62.997	−400.694	−295.383	17.143
1000	83.000	198.642	127.345	71.297	−398.668	−283.791	14.824
1100	83.000	206.553	134.192	79.597	−475.618	−267.780	12.716
1200	83.000	213.775	140.527	87.897	−472.711	−249.014	10.839
1300	83.000	220.418	146.421	96.197	−469.843	−230.490	9.261
1400	83.000	226.569	151.929	104.497	−467.011	−212.184	7.917
1500	83.000	232.296	157.098	112.797	−464.217	−194.080	6.758

63. POTASSIUM HYDROXIDE KOH (g)

298.15	49.184	238.283	238.283	0.000	−227.989	−229.685	40.239
300	49.236	238.588	238.284	0.091	−228.007	−229.696	39.993
400	51.178	253.053	240.243	5.124	−231.377	−229.667	29.991
500	52.178	264.591	243.998	10.296	−232.309	−229.129	23.937
600	52.804	274.163	248.251	15.547	−233.145	−228.413	19.885
700	53.296	282.340	252.551	20.853	−233.934	−227.562	16.981
800	53.758	289.487	256.730	26.206	−234.708	−226.599	14.795
900	54.229	295.846	260.730	31.605	−235.495	−225.538	13.090
1000	54.713	301.585	264.533	37.052	−236.322	−224.388	11.721
1100	55.203	306.823	268.143	42.548	−316.077	−218.535	10.377
1200	55.686	311.647	271.570	48.092	−315.925	−209.674	9.127
1300	56.153	316.122	274.827	53.684	−315.764	−200.826	8.069
1400	56.598	320.300	277.927	59.322	−315.595	−191.991	7.163
1500	57.016	324.220	280.884	65.003	−315.420	−183.169	6.378

T/K	J/K·mol			kJ/mol			
	C_p°	S°	$-(G^\circ - H^\circ(T_r))/T$	$H^\circ - H^\circ(T_r)$	$\Delta_f H^\circ$	$\Delta_f G^\circ$	Log K_f

64. POTASSIUM CHLORIDE   KCl (cr, l)

T/K	C_p°	S°	$-(G^\circ - H^\circ(T_r))/T$	$H^\circ - H^\circ(T_r)$	$\Delta_f H^\circ$	$\Delta_f G^\circ$	Log K_f
298.15	51.300	82.570	82.570	0.000	−436.490	−408.568	71.579
300	51.333	82.887	82.571	0.095	−436.481	−408.395	71.107
400	52.977	97.886	84.605	5.312	−438.463	−398.651	52.058
500	54.448	109.867	88.498	10.685	−437.990	−388.749	40.612
600	55.885	119.921	92.919	16.201	−437.332	−378.960	32.991
700	57.425	128.649	97.413	21.865	−436.502	−369.295	27.557
800	59.205	136.430	101.812	27.694	−435.505	−359.760	23.490
900	61.361	143.523	106.058	33.719	−434.337	−350.360	20.334
1000	64.032	150.121	110.138	39.983	−432.981	−341.100	17.817
1044	65.405	152.908	111.882	42.830	−485.450	−336.720	16.847

PHASE TRANSITION: $\Delta_{trs} H$ = 26.320 kJ/mol, $\Delta_{trs} S$ = 25.210 J/K·mol, cr–l

T/K	C_p°	S°	$-(G^\circ - H^\circ(T_r))/T$	$H^\circ - H^\circ(T_r)$	$\Delta_f H^\circ$	$\Delta_f G^\circ$	Log K_f
1044	72.000	178.118	111.882	69.150	−485.450	−336.720	16.847
1100	72.000	181.880	115.351	73.182	−483.633	−328.790	15.613
1200	72.000	188.145	121.160	80.382	−480.393	−314.856	13.705
1300	72.000	193.908	126.537	87.582	−477.158	−301.192	12.102
1400	72.000	199.244	131.542	94.782	−473.928	−287.778	10.737
1500	72.000	204.211	136.223	101.982	−470.704	−274.594	9.562

65. POTASSIUM CHLORIDE   KCl (g)

T/K	C_p°	S°	$-(G^\circ - H^\circ(T_r))/T$	$H^\circ - H^\circ(T_r)$	$\Delta_f H^\circ$	$\Delta_f G^\circ$	Log K_f
298.15	36.505	239.091	239.091	0.000	−214.575	−233.320	40.876
300	36.518	239.317	239.092	0.068	−214.594	−233.436	40.644
400	37.066	249.904	240.532	3.749	−218.112	−239.107	31.224
500	37.384	258.212	243.267	7.473	−219.287	−244.219	25.513
600	37.597	265.048	246.344	11.222	−220.396	−249.100	21.686
700	37.769	270.857	249.441	14.991	−221.461	−253.799	18.938
800	37.907	275.910	252.441	18.775	−222.509	−258.347	16.868
900	38.041	280.382	255.302	22.572	−223.568	−262.764	15.250
1000	38.162	284.397	258.014	26.383	−224.667	−267.061	13.950
1100	38.279	288.039	260.581	30.205	−304.696	−266.627	12.661
1200	38.401	291.375	263.010	34.039	−304.821	−263.161	11.455
1300	38.518	294.454	265.312	37.885	−304.941	−259.684	10.434
1400	38.639	297.313	267.496	41.743	−305.053	−256.199	9.559
1500	38.761	299.983	269.574	45.613	−305.159	−252.706	8.800

66. DINITROGEN N_2 (g)

T/K	C_p°	S°	$-(G^\circ - H^\circ(T_r))/T$	$H^\circ - H^\circ(T_r)$	$\Delta_f H^\circ$	$\Delta_f G^\circ$	Log K_f
298.15	29.124	191.608	191.608	0.000	0.000	0.000	0.000
300	29.125	191.788	191.608	0.054	0.000	0.000	0.000
400	29.249	200.180	192.752	2.971	0.000	0.000	0.000
500	29.580	206.738	194.916	5.911	0.000	0.000	0.000
600	30.109	212.175	197.352	8.894	0.000	0.000	0.000
700	30.754	216.864	199.812	11.936	0.000	0.000	0.000
800	31.433	221.015	202.208	15.046	0.000	0.000	0.000
900	32.090	224.756	204.509	18.222	0.000	0.000	0.000
1000	32.696	228.169	206.706	21.462	0.000	0.000	0.000
1100	33.241	231.311	208.802	24.759	0.000	0.000	0.000
1200	33.723	234.224	210.801	28.108	0.000	0.000	0.000
1300	34.147	236.941	212.708	31.502	0.000	0.000	0.000
1400	34.517	239.485	214.531	34.936	0.000	0.000	0.000
1500	34.842	241.878	216.275	38.404	0.000	0.000	0.000

67. NITRIC OXIDE   NO (g)

T/K	C_p°	S°	$-(G^\circ - H^\circ(T_r))/T$	$H^\circ - H^\circ(T_r)$	$\Delta_f H^\circ$	$\Delta_f G^\circ$	Log K_f
298.15	29.862	210.745	210.745	0.000	91.277	87.590	−15.345
300	29.858	210.930	210.746	0.055	91.278	87.567	−15.247
400	29.954	219.519	211.916	3.041	91.320	86.323	−11.272
500	30.493	226.255	214.133	6.061	91.340	85.071	−8.887

T/K	C_p°	S°	$-(G^\circ-H^\circ(T_r))/T$	$H^\circ-H^\circ(T_r)$	$\Delta_f H^\circ$	$\Delta_f G^\circ$	Log K_f
		J/K·mol			kJ/mol		

67. NITRIC OXIDE   NO (g) (continued)

T/K	C_p°	S°	$-(G^\circ-H^\circ(T_r))/T$	$H^\circ-H^\circ(T_r)$	$\Delta_f H^\circ$	$\Delta_f G^\circ$	Log K_f
600	31.243	231.879	216.635	9.147	91.354	83.816	−7.297
700	32.031	236.754	219.168	12.310	91.369	82.558	−6.160
800	32.770	241.081	221.642	15.551	91.386	81.298	−5.308
900	33.425	244.979	224.022	18.862	91.405	80.036	−4.645
1000	33.990	248.531	226.298	22.233	91.426	78.772	−4.115
1100	34.473	251.794	228.469	25.657	91.445	77.505	−3.680
1200	34.883	254.811	230.540	29.125	91.464	76.237	−3.318
1300	35.234	257.618	232.516	32.632	91.481	74.967	−3.012
1400	35.533	260.240	234.404	36.170	91.495	73.697	−3.012
1500	35.792	262.700	236.209	39.737	91.506	72.425	−2.750

68. NITROGEN DIOXIDE   NO$_2$ (g)

T/K	C_p°	S°	$-(G^\circ-H^\circ(T_r))/T$	$H^\circ-H^\circ(T_r)$	$\Delta_f H^\circ$	$\Delta_f G^\circ$	Log K_f
298.15	37.178	240.166	240.166	0.000	34.193	52.316	−9.165
300	37.236	240.397	240.167	0.069	34.181	52.429	−9.129
400	40.513	251.554	241.666	3.955	33.637	58.600	−7.652
500	43.664	260.939	244.605	8.167	33.319	64.882	−6.778
600	46.383	269.147	248.026	12.673	33.174	71.211	−6.199
700	48.612	276.471	251.575	17.427	33.151	77.553	−5.787
800	50.405	283.083	255.107	22.381	33.213	83.893	−5.478
900	51.844	289.106	258.555	27.496	33.334	90.221	−5.236
1000	53.007	294.631	261.891	32.741	33.495	96.534	−5.042
1100	53.956	299.729	265.102	38.090	33.686	102.828	−4.883
1200	54.741	304.459	268.187	43.526	33.898	109.105	−4.749
1300	55.399	308.867	271.148	49.034	34.124	115.363	−4.635
1400	55.960	312.994	273.992	54.603	34.360	121.603	−4.537
1500	56.446	316.871	276.722	60.224	34.604	127.827	−4.451

69. AMMONIA   NH$_3$ (g)

T/K	C_p°	S°	$-(G^\circ-H^\circ(T_r))/T$	$H^\circ-H^\circ(T_r)$	$\Delta_f H^\circ$	$\Delta_f G^\circ$	Log K_f
298.15	35.630	192.768	192.768	0.000	−45.940	−16.407	2.874
300	35.678	192.989	192.769	0.066	−45.981	−16.223	2.825
400	38.674	203.647	194.202	3.778	−48.087	−5.980	0.781
500	41.994	212.633	197.011	7.811	−49.908	4.764	−0.498
600	45.229	220.578	200.289	12.174	−51.430	15.846	−1.379
700	48.269	227.781	203.709	16.850	−52.682	27.161	−2.027
800	51.112	234.414	207.138	21.821	−53.695	38.639	−2.523
900	53.769	240.589	210.516	27.066	−54.499	50.231	−2.915
1000	56.244	246.384	213.816	32.569	−55.122	61.903	−3.233
1100	58.535	251.854	217.027	38.309	−55.589	73.629	−3.496
1200	60.644	257.039	220.147	44.270	−55.920	85.392	−3.717
1300	62.576	261.970	223.176	50.432	−56.136	97.177	−3.905
1400	64.339	266.673	226.117	56.779	−56.251	108.975	−4.066
1500	65.945	271.168	228.971	63.295	−56.282	120.779	−4.206

70. OXYGEN   O (g)

T/K	C_p°	S°	$-(G^\circ-H^\circ(T_r))/T$	$H^\circ-H^\circ(T_r)$	$\Delta_f H^\circ$	$\Delta_f G^\circ$	Log K_f
298.15	21.911	161.058	161.058	0.000	249.180	231.743	−40.600
300	21.901	161.194	161.059	0.041	249.193	231.635	−40.331
400	21.482	167.430	161.912	2.207	249.874	225.677	−29.470
500	21.257	172.197	163.511	4.343	250.481	219.556	−22.937
600	21.124	176.060	165.290	6.462	251.019	213.319	−18.571
700	21.040	179.310	167.067	8.570	251.500	206.997	−15.446
800	20.984	182.115	168.777	10.671	251.932	200.610	−13.098
900	20.944	184.584	170.399	'12.767	252.325	194.171	−11.269
1000	20.915	186.789	171.930	14.860	252.686	187.689	−9.804
1100	20.893	188.782	173.372	16.950	253.022	181.173	−8.603
1200	20.877	190.599	174.733	19.039	253.335	174.628	−7.601
1300	20.864	192.270	176.019	21.126	253.630	168.057	−6.753

T/K	C_p°	S°	$-(G^\circ-H^\circ(T_r))/T$	$H^\circ-H^\circ(T_r)$	$\Delta_f H^\circ$	$\Delta_f G^\circ$	Log K_f
		J/K·mol			kJ/mol		

70. OXYGEN O (g) (continued)

T/K	C_p°	S°	$-(G^\circ-H^\circ(T_r))/T$	$H^\circ-H^\circ(T_r)$	$\Delta_f H^\circ$	$\Delta_f G^\circ$	Log K_f
1400	20.853	193.815	177.236	23.212	253.908	161.463	−6.024
1500	20.845	195.254	178.389	25.296	254.171	154.851	−5.392

71. DIOXYGEN O_2 (g)

T/K	C_p°	S°	$-(G^\circ-H^\circ(T_r))/T$	$H^\circ-H^\circ(T_r)$	$\Delta_f H^\circ$	$\Delta_f G^\circ$	Log K_f
298.15	29.378	205.148	205.148	0.000	0.000	0.000	0.000
300	29.387	205.330	205.148	0.054	0.000	0.000	0.000
400	30.109	213.873	206.308	3.026	0.000	0.000	0.000
500	31.094	220.695	208.525	6.085	0.000	0.000	0.000
600	32.095	226.454	211.045	9.245	0.000	0.000	0.000
700	32.987	231.470	213.612	12.500	0.000	0.000	0.000
800	33.741	235.925	216.128	15.838	0.000	0.000	0.000
900	34.365	239.937	218.554	19.244	0.000	0.000	0.000
1000	34.881	243.585	220.878	22.707	0.000	0.000	0.000
1100	35.314	246.930	223.096	26.217	0.000	0.000	0.000
1200	35.683	250.019	225.213	29.768	0.000	0.000	0.000
1300	36.006	252.888	227.233	33.352	0.000	0.000	0.000
1400	36.297	255.568	229.162	36.968	0.000	0.000	0.000
1500	36.567	258.081	231.007	40.611	0.000	0.000	0.000

72. SULFUR S (cr, l)

T/K	C_p°	S°	$-(G^\circ-H^\circ(T_r))/T$	$H^\circ-H^\circ(T_r)$	$\Delta_f H^\circ$	$\Delta_f G^\circ$	Log K_f
298.15	22.690	32.070	32.070	0.000	0.000	0.000	0.000
300	22.737	32.210	32.070	0.042	0.000	0.000	0.000
368.3	24.237	37.030	32.554	1.649	0.000	0.000	0.000
		PHASE TRANSITION: $\Delta_{trs} H$ = 0.401 kJ/mol, $\Delta_{trs} S$ = 1.089 J/K·mol, crII–crI					
368.3	24.773	38.119	32.553	2.050	0.000	0.000	0.000
388.36	25.180	39.444	32.875	2.551	0.000	0.000	0.000
		PHASE TRANSITION: $\Delta_{trs} H$ = 1.722 kJ/mol, $\Delta_{trs} S$ = 4.431 J/K·mol, crI–l					
388.36	31.710	43.875	32.872	4.273	0.000	0.000	0.000
400	32.369	44.824	33.206	4.647	0.000	0.000	0.000
500	38.026	53.578	36.411	8.584	0.000	0.000	0.000
600	34.371	60.116	39.842	12.164	0.000	0.000	0.000
700	32.451	65.278	43.120	15.511	0.000	0.000	0.000
800	32.000	69.557	46.163	18.715	0.000	0.000	0.000
882.38	32.000	72.693	48.496	21.351	0.000	0.000	0.000

73. SULFUR S (g)

T/K	C_p°	S°	$-(G^\circ-H^\circ(T_r))/T$	$H^\circ-H^\circ(T_r)$	$\Delta_f H^\circ$	$\Delta_f G^\circ$	Log K_f
298.15	23.673	167.828	167.828	0.000	277.180	236.704	−41.469
300	23.669	167.974	167.828	0.044	277.182	236.453	−41.170
400	23.233	174.730	168.752	2.391	274.924	222.962	−29.115
500	22.741	179.860	170.482	4.689	273.286	210.145	−21.953
600	22.338	183.969	172.398	6.942	271.958	197.646	−17.206
700	22.031	187.388	174.302	9.160	270.829	185.352	−13.831
800	21.800	190.314	176.125	11.351	269.816	173.210	−11.309
900	21.624	192.871	177.847	13.522	215.723	162.258	−9.417
1000	21.489	195.142	179.465	15.677	216.018	156.301	−8.164
1100	21.386	197.185	180.985	17.821	216.284	150.317	−7.138
1200	21.307	199.043	182.413	19.955	216.525	144.309	−6.282
1300	21.249	200.746	183.759	22.083	216.743	138.282	−5.556
1400	21.209	202.319	185.029	24.206	216.940	132.239	−4.934
1500	21.186	203.781	186.231	26.325	217.119	126.182	−4.394

74. DISULFUR S_2 (g)

T/K	C_p°	S°	$-(G^\circ-H^\circ(T_r))/T$	$H^\circ-H^\circ(T_r)$	$\Delta_f H^\circ$	$\Delta_f G^\circ$	Log K_f
298.15	32.505	228.165	228.165	0.000	128.600	79.696	−13.962
300	32.540	228.366	228.165	0.060	128.576	79.393	−13.823
400	34.108	237.956	229.462	3.398	122.703	63.380	−8.276
500	35.133	245.686	231.959	6.863	118.296	49.031	−5.122

T/K	$C_p°$	$S°$	$-(G°-H°(T_r))/T$	$H°-H°(T_r)$	$\Delta_f H°$	$\Delta_f G°$	Log K_f
	J/K·mol			kJ/mol			

74. DISULFUR S_2 (g) (continued)

T/K	$C_p°$	$S°$	$-(G°-H°(T_r))/T$	$H°-H°(T_r)$	$\Delta_f H°$	$\Delta_f G°$	Log K_f
600	35.815	252.156	234.800	10.413	114.685	35.530	−3.093
700	36.305	257.715	237.686	14.020	111.599	22.588	−1.685
800	36.697	262.589	240.501	17.671	108.841	10.060	−0.657
882.38	36.985	266.200	242.734	20.706		pressure = 1 bar	
900	37.045	266.932	243.201	21.358	0.000	0.000	0.000
1000	37.377	270.852	245.773	25.079	0.000	0.000	0.000
1100	37.704	274.430	248.218	28.833	0.000	0.000	0.000
1200	38.030	277.725	250.541	32.620	0.000	0.000	0.000
1300	38.353	280.781	252.751	36.439	0.000	0.000	0.000
1400	38.669	283.635	254.856	40.290	0.000	0.000	0.000
1500	38.976	286.314	256.865	44.173	0.000	0.000	0.000

75. OCTASULFUR S_8 (g)

T/K	$C_p°$	$S°$	$-(G°-H°(T_r))/T$	$H°-H°(T_r)$	$\Delta_f H°$	$\Delta_f G°$	Log K_f
298.15	156.500	432.536	432.536	0.000	101.277	48.810	−8.551
300	156.768	433.505	432.539	0.290	101.231	48.484	−8.442
400	167.125	480.190	438.834	16.542	80.642	32.003	−4.179
500	173.181	518.176	451.022	33.577	66.185	21.409	−2.237
600	177.936	550.180	464.951	51.137	55.101	13.549	−1.180
700	182.441	577.948	479.152	69.157	46.349	7.343	−0.548
800	186.764	602.596	493.071	87.620	39.177	2.263	−0.148
900	190.595	624.821	506.495	106.494	−392.062	6.554	−0.380
1000	193.618	645.067	519.355	125.712	−387.728	50.614	−2.644
1100	195.684	663.625	531.639	145.185	−383.272	94.233	−4.475
1200	196.825	680.707	543.359	164.817	−378.786	137.444	−5.983
1300	197.195	696.480	554.539	184.524	−374.356	180.283	−7.244
1400	196.988	711.089	565.206	204.237	−370.048	222.785	−8.312
1500	196.396	724.662	575.389	223.909	−365.905	264.984	−9.227

76. SULFUR DIOXIDE SO_2 (g)

T/K	$C_p°$	$S°$	$-(G°-H°(T_r))/T$	$H°-H°(T_r)$	$\Delta_f H°$	$\Delta_f G°$	Log K_f
298.15	39.842	248.219	248.219	0.000	−296.810	−300.090	52.574
300	39.909	248.466	248.220	0.074	−296.833	−300.110	52.253
400	43.427	260.435	249.828	4.243	−300.240	−300.935	39.298
500	46.490	270.465	252.978	8.744	−302.735	−300.831	31.427
600	48.938	279.167	256.634	13.520	−304.699	−300.258	26.139
700	50.829	286.859	260.413	18.513	−306.308	−299.386	22.340
800	52.282	293.746	264.157	23.671	−307.691	−298.302	19.477
900	53.407	299.971	267.796	28.958	−362.075	−295.987	17.178
1000	54.290	305.646	271.301	34.345	−362.012	−288.647	15.077
1100	54.993	310.855	274.664	39.810	−361.934	−281.314	13.358
1200	55.564	315.665	277.882	45.339	−361.849	−273.989	11.926
1300	56.033	320.131	280.963	50.920	−361.763	−266.671	10.715
1400	56.426	324.299	283.911	56.543	−361.680	−259.359	9.677
1500	56.759	328.203	286.735	62.203	−361.605	−252.053	8.777

77. SILICON Si (cr)

T/K	$C_p°$	$S°$	$-(G°-H°(T_r))/T$	$H°-H°(T_r)$	$\Delta_f H°$	$\Delta_f G°$	Log K_f
298.15	19.789	18.810	18.810	0.000	0.000	0.000	0.000
300	19.855	18.933	18.810	0.037	0.000	0.000	0.000
400	22.301	25.023	19.624	2.160	0.000	0.000	0.000
500	23.610	30.152	21.231	4.461	0.000	0.000	0.000
600	24.472	34.537	23.092	6.867	0.000	0.000	0.000
700	25.124	38.361	25.006	9.348	0.000	0.000	0.000
800	25.662	41.752	26.891	11.888	0.000	0.000	0.000
900	26.135	44.802	28.715	14.478	0.000	0.000	0.000
1000	26.568	47.578	30.464	17.114	0.000	0.000	0.000
1100	26.974	50.130	32.138	19.791	0.000	0.000	0.000
1200	27.362	52.493	33.737	22.508	0.000	0.000	0.000

T/K	$C_p°$	$S°$	$-(G°-H°(T_r))/T$	$H°-H°(T_r)$	$\Delta_f H°$	$\Delta_f G°$	Log K_f
		J/K·mol			**kJ/mol**		

77. SILICON   Si (cr) (continued)

T/K	$C_p°$	$S°$	$-(G°-H°(T_r))/T$	$H°-H°(T_r)$	$\Delta_f H°$	$\Delta_f G°$	Log K_f
1300	27.737	54.698	35.265	25.263	0.000	0.000	0.000
1400	28.103	56.767	36.728	28.055	0.000	0.000	0.000
1500	28.462	58.719	38.130	30.883	0.000	0.000	0.000

78. SILICON   Si (g)

T/K	$C_p°$	$S°$	$-(G°-H°(T_r))/T$	$H°-H°(T_r)$	$\Delta_f H°$	$\Delta_f G°$	Log K_f
298.15	22.251	167.980	167.980	0.000	450.000	405.525	−71.045
300	22.234	168.117	167.980	0.041	450.004	405.249	−70.559
400	21.613	174.416	168.843	2.229	450.070	390.312	−50.969
500	21.316	179.204	170.456	4.374	449.913	375.388	−39.216
600	21.153	183.074	172.246	6.497	449.630	360.508	−31.385
700	21.057	186.327	174.032	8.607	449.259	345.682	−25.795
800	21.000	189.135	175.748	10.709	448.821	330.915	−21.606
900	20.971	191.606	177.375	12.808	448.329	316.205	−18.352
1000	20.968	193.815	178.911	14.904	447.791	301.553	−15.751
1100	20.989	195.815	180.358	17.002	447.211	286.957	−13.626
1200	21.033	197.643	181.723	19.103	446.595	272.416	−11.858
1300	21.099	199.329	183.014	21.209	445.946	257.927	−10.364
1400	21.183	200.895	184.236	23.323	445.268	243.489	−9.085
1500	21.282	202.360	185.396	25.446	444.563	229.101	−7.978

79. SILICON DIOXIDE   SiO$_2$ (cr)

T/K	$C_p°$	$S°$	$-(G°-H°(T_r))/T$	$H°-H°(T_r)$	$\Delta_f H°$	$\Delta_f G°$	Log K_f
298.15	44.602	41.460	41.460	0.000	−910.700	−856.288	150.016
300	44.712	41.736	41.461	0.083	−910.708	−855.951	149.032
400	53.477	55.744	43.311	4.973	−910.912	−837.651	109.385
500	60.533	68.505	47.094	10.705	−910.540	−819.369	85.598
600	64.452	79.919	51.633	16.971	−909.841	−801.197	69.749
700	68.234	90.114	56.414	23.590	−908.958	−783.157	58.439
800	76.224	99.674	61.226	30.758	−907.668	−765.265	49.966
848	82.967	104.298	63.533	34.569	−906.310	−756.747	46.613
	PHASE TRANSITION: $\Delta_{trs} H = 0.411$ kJ/mol, $\Delta_{trs} S = 0.484$ J/K·mol, crII–crII′						
848	67.446	104.782	63.532	34.980	−906.310	−756.747	46.613
900	67.953	108.811	66.033	38.500	−905.922	−747.587	43.388
1000	68.941	116.021	70.676	45.345	−905.176	−730.034	38.133
1100	69.940	122.639	75.104	52.289	−904.420	−712.557	33.836
1200	70.947	128.768	79.323	59.333	−901.382	−695.148	30.259
	PHASE TRANSITION: $\Delta_{trs} H = 2.261$ kJ/mol, $\Delta_{trs} S = 1.883$ J/K·mol, crII′–crI						
1200	71.199	130.651	79.323	61.594	−901.382	−695.148	30.259
1300	71.743	136.372	83.494	68.742	−900.574	−677.994	27.242
1400	72.249	141.707	87.463	75.941	−899.782	−660.903	24.658
1500	72.739	146.709	91.248	83.191	−899.004	−643.867	22.421

80. SILICON TETRACHLORIDE   SiCl$_4$ (g)

T/K	$C_p°$	$S°$	$-(G°-H°(T_r))/T$	$H°-H°(T_r)$	$\Delta_f H°$	$\Delta_f G°$	Log K_f
298.15	90.404	331.446	331.446	0.000	−662.200	−622.390	109.039
300	90.562	332.006	331.448	0.167	−662.195	−622.143	108.323
400	96.893	359.019	335.088	9.572	−661.853	−608.841	79.505
500	100.449	381.058	342.147	19.456	−661.413	−595.637	62.225
600	102.587	399.576	350.216	29.616	−660.924	−582.527	50.713
700	103.954	415.500	358.432	39.948	−660.417	−569.501	42.496
800	104.875	429.445	366.455	50.392	−659.912	−556.548	36.338
900	105.523	441.837	374.155	60.914	−659.422	−543.657	31.553
1000	105.995	452.981	381.490	71.491	−658.954	−530.819	27.727
1100	106.349	463.101	388.456	82.109	−658.515	−518.027	24.599
1200	106.620	472.366	395.068	92.758	−658.107	−505.274	21.994
1300	106.834	480.909	401.347	103.431	−657.735	−492.553	19.791
1400	107.003	488.833	407.316	114.123	−657.400	−479.860	17.904
1500	107.141	496.220	413.000	124.830	−657.104	−467.189	16.269

THERMODYNAMIC PROPERTIES OF AQUEOUS SYSTEMS

This table contains standard state thermodynamic properties of ions and neutral species in aqueous solution. It includes enthalpy and Gibbs energy of formation, entropy, and heat capacity, and thus serves as a companion to the preceding table, "Standard Thermodynamic Properties of Chemical Substances". The standard state is the hypothetical ideal solution with molality $m = 1$ mol/kg (mean ionic molality $m_{\pm}$ in the case of a species which is assumed to dissociate at infinite dilution). Further details on conventions may be found in Reference 1.

Cations are listed by formula in the first part of the table, followed by anions and finally neutral species. All values refer to standard conditions of 25°C and 100 kPa pressure.

REFERENCES

1. Wagman, D. D., Evans, W. H., Parker, V. B., Schumm, R. H., Halow, I., Bailey, S. M., Churney, K. L., and Nuttall, R. L., *The NBS Tables of Chemical Thermodynamic Properties, J. Phys. Chem. Ref. Data*, Vol. 11, Suppl. 2, 1982.
2. Zemaitis, J. F., Clark, D. M., Rafal, M., and Scrivner, N. C., *Handbook of Aqueous Electrolyte Thermodynamics*, American Institute of Chemical Engineers, New York, 1986.

Species	$\Delta_f H°/$ kJ mol^{-1}	$\Delta_f G°/$ kJ mol^{-1}	$S°/$ J mol^{-1}K^{-1}	$C_p/$ J mol^{-1}K^{-1}	Species	$\Delta_f H°/$ kJ mol^{-1}	$\Delta_f G°/$ kJ mol^{-1}	$S°/$ J mol^{-1}K^{-1}	$C_p/$ J mol^{-1}K^{-1}
Cations					InOH^{+2}	-370.3	-313.0	-88.0	
					In(OH)$_2{}^+$	-619.0	-525.0	25.0	
Ag$^+$	105.6	77.1	72.7	21.8	K$^+$	-252.4	-283.3	102.5	21.8
Al^{+3}	-531.0	-485.0	-321.7		La^{+3}	-707.1	-683.7	-217.6	-13.0
AlOH^{+2}		-694.1			Li$^+$	-278.5	-293.3	13.4	68.6
Ba^{+2}	-537.6	-560.8	9.6		Lu^{+3}	-665.0	-628.0	-264.0	25.0
BaOH$^+$		-730.5			LuF^{+2}		-931.4		
Be^{+2}	-382.8	-379.7	-129.7		Mg^{+2}	-466.9	-454.8	-138.1	
Bi^{+3}		82.8			MgOH$^+$		-626.7		
BiOH^{+2}		-146.4			Mn^{+2}	-220.8	-228.1	-73.6	50.0
Ca^{+2}	-542.8	-553.6	-53.1		MnOH$^+$	-450.6	-405.0	-17.0	
CaOH$^+$		-718.4			NH$_4{}^+$	-132.5	-79.3	113.4	79.9
Cd^{+2}	-75.9	-77.6	-73.2		N$_2$H$_5{}^+$	-7.5	82.5	151.0	70.3
CdOH$^+$		-261.1			Na$^+$	-240.1	-261.9	59.0	46.4
Ce^{+3}	-696.2	-672.0	-205.0		Nd^{+3}	-696.2	-671.6	-206.7	-21.0
Ce^{+4}	-537.2	-503.8	-301.0		Ni^{+2}	-54.0	-45.6	-128.9	
Co^{+2}	-58.2	-54.4	-113.0		NiOH$^+$	-287.9	-227.6	-71.0	
Co^{+3}	92.0	134.0	-305.0		PH$_4{}^+$		92.1		
Cr^{+2}	-143.5				Pa^{+4}	-619.0			
Cs$^+$	-258.3	-292.0	133.1	-10.5	Pb^{+2}	-1.7	-24.4	10.5	
Cu$^+$	71.7	50.0	40.6		PbOH$^+$		-226.3		
Cu^{+2}	64.8	65.5	-99.6		Pd^{+2}	149.0	176.5	-184.0	
Dy^{+3}	-699.0	-665.0	-231.0	21.0	Po^{+2}		71.0		
Er^{+3}	-705.4	-669.1	-244.3	21.0	Po^{+4}		293.0		
Eu^{+2}	-527.0	-540.2	-8.0		Pr^{+3}	-704.6	-679.1	-209.0	-29.0
Eu^{+3}	-605.0	-574.1	-222.0	8.0	Pt^{+2}		254.8		
Fe^{+2}	-89.1	-78.9	-137.7		Ra^{+2}	-527.6	-561.5	54.0	
Fe^{+3}	-48.5	-4.7	-315.9		Rb$^+$	-251.2	-284.0	121.5	
FeOH$^+$	-324.7	-277.4	-29.0		Re$^+$		-33.0		
FeOH^{+2}	-290.8	-229.4	-142.0		Sc^{+3}	-614.2	-586.6	-255.0	
Fe(OH)$_2{}^+$		-438.0			ScOH^{+2}	-861.5	-801.2	-134.0	
Ga^{+2}		-88.0			Sm^{+2}		-497.5		
Ga^{+3}	-211.7	-159.0	-331.0		Sm^{+3}	-691.6	-666.6	-211.7	-21.0
GaOH^{+2}		-380.3			Sn^{+2}	-8.8	-27.2	-17.0	
Ga(OH)$_2{}^+$		-597.4			SnOH$^+$	-286.2	-254.8	50.0	
Gd^{+3}	-686.0	-661.0	-205.9		Sr^{+2}	-545.8	-559.5	-32.6	
H$^+$	0	0	0	0	SrOH$^+$		-721.3		
Hg^{+2}	171.1	164.4	-32.2		Tb^{+3}	-682.8	-651.9	-226.0	17.0
Hg$_2{}^{+2}$	172.4	153.5	84.5		Te(OH)$_3{}^+$	-608.4	-496.1	111.7	
HgOH$^+$	-84.5	-52.3	71.0		Th^{+4}	-769.0	-705.1	-422.6	
Ho^{+3}	-705.0	-673.7	-226.8	17.0	Th(OH)$^{+3}$	-1030.1	-920.5	-343.0	
In$^+$		-12.1			Th(OH)$_2{}^{+2}$	-1282.4	-1140.9	-218.0	
In^{+2}		-50.7			Tl$^+$	5.4	-32.4	125.5	
In^{+3}	-105.0	-98.0	-151.0		Tl^{+3}	196.6	214.6	-192.0	

THERMODYNAMIC PROPERTIES OF AQUEOUS SYSTEMS (continued)

Species	$\Delta_f H°/$ kJ mol⁻¹	$\Delta_f G°/$ kJ mol⁻¹	$S°/$ J mol⁻¹K⁻¹	$C_p/$ J mol⁻¹K⁻¹	Species	$\Delta_f H°/$ kJ mol⁻¹	$\Delta_f G°/$ kJ mol⁻¹	$S°/$ J mol⁻¹K⁻¹	$C_p/$ J mol⁻¹K⁻¹
$TlOH^{+2}$		-15.9			$H_2PO_4^-$	-1296.3	-1130.2	90.4	
$Tl(OH)_2^+$		-244.7			$H_2P_2O_7^{-2}$	-2278.6	-2010.2	163.0	
Tm^{+3}	-697.9	-662.0	-243.0	25.0	I^-	-55.2	-51.6	111.3	-142.3
U^{+3}	-489.1	-476.2	-188.0		IO^-	-107.5	-38.5	-5.4	
U^{+4}	-591.2	-531.9	-410.0		IO_3^-	-221.3	-128.0	118.4	
Y^{+3}	-723.4	-693.8	-251.0		IO_4^-	-151.5	-58.5	222.0	
$Y_2(OH)_2^{+4}$		-1780.3			MnO_4^-	-541.4	-447.2	191.2	-82.0
Yb^{+2}		-527.0			MnO_4^{-2}	-653.0	-500.7	59.0	
Yb^{+3}	-674.5	-644.0	-238.0	25.0	MoO_4^{-2}	-997.9	-836.3	27.2	
$Y(OH)^{+2}$		-879.1			NO_2^-	-104.6	-32.2	123.0	-97.5
Zn^{+2}	-153.9	-147.1	-112.1	46.0	NO_3^-	-207.4	-111.3	146.4	-86.6
$ZnOH^+$		-330.1			N_3^-	275.1	348.2	107.9	
					OCN^-	-146.0	-97.4	106.7	
Anions					OH^-	-230.0	-157.2	-10.8	-148.5
					PO_4^{-3}	-1277.4	-1018.7	-220.5	
					$P_2O_7^{-4}$	-2271.1	-1919.0	-117.0	
AlO_2^-	-930.9	-830.9	-36.8		Re^-	46.0	10.1	230.0	
$Al(OH)_4^-$	-1502.5	-1305.3	102.9		S^{-2}	33.1	85.8	-14.6	
AsO_2^-	-429.0	-350.0	40.6		SCN^-	76.4	92.7	144.3	-40.2
AsO_4^{-3}	-888.1	-648.4	-162.8		SO_3^{-2}	-635.5	-486.5	-29.0	
BF_4^-	-1574.9	-1486.9	180.0		SO_4^{-2}	-909.3	-744.5	20.1	-293.0
BH_4^-	48.2	114.4	110.5		S_2^{-2}	30.1	79.5	28.5	
BO_2^-	-772.4	-678.9	-37.2		$S_2O_3^{-2}$	-652.3	-522.5	67.0	
$B_4O_7^{-2}$		-2604.8			$S_2O_4^{-2}$	-753.5	-600.3	92.0	
BeO_2^{-2}	-790.8	-640.1	-159.0		$S_2O_8^{-2}$	-1344.7	-1114.9	244.3	
Br^-	-121.6	-104.0	82.4	-141.8	Se^{-2}		129.3		
BrO^-	-94.1	-33.4	42.0		SeO_3^{-2}	-509.2	-369.8	13.0	
BrO_3^-	-67.1	18.6	161.7		SeO_4^{-2}	-599.1	-441.3	54.0	
BrO_4^-	13.0	118.1	199.6		VO_3^-	-888.3	-783.6	50.0	
$CHOO^-$	-425.6	-351.0	92.0	-87.9	VO_4^{-3}		-899.0		
CH_3COO^-	-486.0	-369.3	86.6	-6.3	WO_4^{-2}	-1075.7			
$C_2O_4^{-2}$	-825.1	-673.9	45.6						
$C_2O_4H^-$	-818.4	-698.3	149.4		**Neutral species**				
Cl^-	-167.2	-131.2	56.5	-136.4					
ClO^-	-107.1	-36.8	42.0						
ClO_2^-	-66.5	17.2	101.3		$AgBr$	-16.0	-26.9	155.2	-120.1
ClO_3^-	-104.0	-8.0	162.3		$AgCl$	-61.6	-54.1	129.3	-114.6
ClO_4^-	-129.3	-8.5	182.0		AgF	-227.1	-201.7	59.0	-84.9
CN^-	150.6	172.4	94.1		AgI	50.4	25.5	184.1	-120.5
CO_3^{-2}	-677.1	-527.8	-56.9		$AgNO_3$	-101.8	-34.2	219.2	-64.9
CrO_4^{-2}	-881.2	-727.8	50.2		Ag_2SO_4	-698.1	-590.3	165.7	-251.0
$Cr_2O_7^{-2}$	-1490.3	-1301.1	261.9		$AlBr_3$	-895.0	-799.0	-74.5	
F^-	-332.6	-278.8	-13.8	-106.7	$AlCl_3$	-1033.0	-879.0	-152.3	
$Fe(CN)_6^{-3}$	561.9	729.4	270.3		AlF_3	-1531.0	-1322.0	-363.2	
$Fe(CN)_6^{-4}$	455.6	695.1	95.0		AlI_3	-699.0	-640.0	12.1	
$HB_4O_7^-$		-2685.1			$Al_2(SO_4)_3$	-3791.0	-3205.0	-583.2	
HCO_3^-	-692.0	-586.8	91.2		$BaBr_2$	-780.7	-768.7	174.5	
HF_2^-	-649.9	-578.0	92.5		$BaCO_3$	-1214.8	-1088.6	-47.3	
HPO_3F^-		-1198.2			$BaCl_2$	-872.0	-823.2	122.6	
HPO_4^{-2}	-1292.1	-1089.2	-33.5		BaF_2	-1202.9	-1118.4	-18.0	
$HP_2O_7^{-3}$	-2274.8	-1972.2	46.0		$Ba(HCO_3)_2$	-1921.6	-1734.3	192.0	
HS^-	-17.6	12.1	62.8		BaI_2	-648.0	-663.9	232.2	
HSO_3^-	-626.2	-527.7	139.7		$Ba(NO_3)_2$	-952.4	-783.3	302.5	
HSO_4^-	-887.3	-755.9	131.8	-84.0	$BaSO_4$	-1446.9	-1305.3	29.7	
$HS_2O_4^-$		-614.5			$BeSO_4$	-1292.0	-1124.3	-109.6	
HSe^-	15.9	44.0	79.0		CCl_3COOH	-516.3			
$HSeO_3^-$	-514.6	-411.5	135.1		$CHCl_2COOH$	-512.1			
$HSeO_4^-$	-581.6	-452.2	149.4		$CHOOCs$	-683.8	-643.0	226.0	
$H_2AsO_3^-$	-714.8	-587.1	110.5		$CHOOH$	-425.6	-351.0	92.0	-87.9
$H_2AsO_4^-$	-909.6	-753.2	117.0		$CHOOK$	-677.9	-634.2	192.0	-66.1

Species	$\Delta_f H°/$ kJ mol^{-1}	$\Delta_f G°/$ kJ mol^{-1}	$S°/$ J mol^{-1}K^{-1}	$C_p/$ J mol^{-1}K^{-1}	Species	$\Delta_f H°/$ kJ mol^{-1}	$\Delta_f G°/$ kJ mol^{-1}	$S°/$ J mol^{-1}K^{-1}	$C_p/$ J mol^{-1}K^{-1}
CHOONH$_4$	-558.1	-430.4	205.0	-7.9	GdCl$_3$	-1188.0	-1059.0	-36.8	-410.0
CHOONa	-665.7	-612.9	151.0	-41.4	HBr	-121.6	-104.0	82.4	-141.8
CHOORb	-676.7	-635.1	213.0		HCN	150.6	172.4	94.1	
CH$_2$ClCOOH	-501.3				HCl	-167.2	-131.2	56.5	-136.4
CH$_3$COOCs	-744.3	-661.3	219.7		HF	-332.6	-278.8	-13.8	-106.7
CH$_3$COOH	-486.0	-369.3	86.6	-6.3	HI	-55.2	-51.6	111.3	-142.3
CH$_3$COOK	-738.4	-652.6	189.1	15.5	HNO$_3$	-207.4	-111.3	146.4	-86.6
CH$_3$COONH$_4$	-618.5	-448.6	200.0	73.6	HSCN	76.4	92.7	144.3	-40.2
CH$_3$COONa	-726.1	-631.2	145.6	40.2	H$_2$SO$_4$	-909.3	-744.5	20.1	-293.0
CH$_3$COORb	-737.2	-653.3	207.9		HoCl$_3$	-1206.7	-1067.3	-57.7	-393.0
(COOH)$_2$	-825.1	-673.9	45.6		KBr	-373.9	-387.2	184.9	-120.1
(CH$_3$)$_3$N	-76.0	93.1	133.5		KCl	-419.5	-414.5	159.0	-114.6
CaBr$_2$	-785.9	-761.5	111.7		KF	-585.0	-562.1	88.7	-84.9
CaCO$_3$	-1220.0	-1081.4	-110.0		KHCO$_3$	-944.4	-870.0	193.7	
CaCl$_2$	-877.1	-816.0	59.8		KHSO$_4$	-1139.7	-1039.2	234.3	-63.0
CaF$_2$	-1208.1	-1111.2	-80.8		KI	-307.6	-334.9	213.8	-120.5
CaI$_2$	-653.2	-656.7	169.5		KNO$_3$	-459.7	-394.5	248.9	-64.9
Ca(NO$_3$)$_2$	-957.6	-776.1	239.7		K$_2$CO$_3$	-1181.9	-1094.4	148.1	
CaSO$_4$	-1452.1	-1298.1	-33.1		K$_2$S	-471.5	-480.7	190.4	
CdBr$_2$	-319.0	-285.5	91.6		K$_2$SO$_4$	-1414.0	-1311.1	225.1	-251.0
CdCl$_2$	-410.2	-340.1	39.7		K$_2$Se		-437.2		
CdF$_2$	-741.2	-635.2	-100.8		LaCl$_3$	-1208.8	-1077.3	-50.0	-423.0
CdI$_2$	-186.3	-180.8	149.4		LiBr	-400.0	-397.3	95.8	-73.2
Cd(NO$_3$)$_2$	-490.6	-300.1	219.7		LiCl	-445.6	-424.6	69.9	-67.8
CdSO$_4$	-985.2	-822.1	-53.1		LiF	-611.1	-571.9	-0.4	-38.1
CeCl$_3$	-1197.5	-1065.6	-38.0		LiI	-333.7	-344.8	124.7	-73.6
CoBr$_2$	-301.2	-262.3	50.0		LiNO$_3$	-485.9	-404.5	160.2	-18.0
CoCl$_2$	-392.5	-316.7			Li$_2$CO$_3$	-1234.1	-1114.6	-29.7	
CoI$_2$	-168.6	-157.7	109.0		Li$_2$SO$_4$	-1466.2	-1331.2	47.3	-155.6
Co(NO$_3$)$_2$	-472.8	-276.9	180.0		LuCl$_3$	-1167.0	-1021.0	-96.0	-385.0
CoSO$_4$	-967.3	-799.1	-92.0		MgBr$_2$	-709.9	-662.7	26.8	
CsBr	-379.8	-396.0	215.5		MgCl$_2$	-801.2	-717.1	-25.1	
CsCl	-425.4	-423.2	189.5	-146.9	MgI$_2$	-577.2	-558.1	84.5	
CsF	-590.9	-570.8	119.2		Mg(NO$_3$)$_2$	-881.6	-677.3	154.8	
CsHCO$_3$	-950.3	-878.8	224.3		MgSO$_4$	-1376.1	-1199.5	-118.0	
CsHSO$_4$	-1145.6	-1047.9	264.8		MnBr$_2$	-464.0			
CsI	-313.5	-343.6	244.3	-152.7	MnCl$_2$	-555.1	-490.8	38.9	-222.0
CsNO$_3$	-465.6	-403.3	279.5	-99.0	MnI$_2$	-331.0			
Cs$_2$CO$_3$	-1193.7	-1111.9	209.2		Mn(NO$_3$)$_2$	-635.5	-450.9	218.0	-121.0
Cs$_2$S	-483.7	-498.3	251.0		MnSO$_4$	-1130.1	-972.7	-53.6	-243.0
Cs$_2$SO$_4$	-1425.8	-1328.6	286.2		NH$_4$Br	-254.1	-183.3	195.8	-61.9
Cs$_2$Se		-454.8			NH$_4$BrO$_3$	-199.6	-60.7	275.1	
Cu(NO$_3$)$_2$	-350.0	-157.0	193.3		NH$_4$CN	18.0	93.0	207.5	
CuSO$_4$	-844.5	-679.0	-79.5		NH$_4$Cl	-299.7	-210.5	169.9	-56.5
DyCl$_3$	-1197.0	-1059.0	-61.9	-389.0	NH$_4$ClO$_3$	-236.5	-87.3	275.7	
ErCl$_3$	-1207.1	-1062.7	-75.3	-389.0	NH$_4$ClO$_4$	-261.8	-87.8	295.4	
EuCl$_2$	-862.0				NH$_4$F	-465.1	-358.1	99.6	-26.8
EuCl$_3$	-1106.2	-967.7	-54.0	-402.0	NH$_4$HCO$_3$	-824.5	-666.1	204.6	
FeBr$_2$	-332.2	-286.8	27.2		NH$_4$HS	-150.2	-67.2	176.1	
FeBr$_3$	-413.4	-316.7	-68.6		NH$_4$HSO$_3$	-758.7	-607.0	253.1	
FeCl$_2$	-423.4	-341.3	-24.7		NH$_4$HSO$_4$	-1019.9	-835.2	245.2	-3.8
FeCl$_3$	-550.2	-398.3	-146.4		NH$_4$HSeO$_4$	-714.2	-531.6	262.8	
FeF$_2$	-754.4	-636.5	-165.3		NH$_4$H$_2$AsO$_3$	-847.3	-666.4	223.8	
FeF$_3$	-1046.4	-840.9	-357.3		NH$_4$H$_2$AsO$_4$	-1042.1	-832.5	230.5	
FeI$_2$	-199.6	-182.1	84.9		NH$_4$H$_2$PO$_4$	-1428.8	-1209.6	203.8	
FeI$_3$	-214.2	-159.4	18.0		NH$_4$H$_3$P$_2$O$_7$	-2409.1	-2102.6	326.0	
Fe(NO$_3$)$_3$	-670.7	-338.3	123.4		NH$_4$I	-187.7	-130.9	224.7	-62.3
FeSO$_4$	-998.3	-823.4	-117.6		NH$_4$IO$_3$	-354.0	-207.4	231.8	
Fe$_2$(SO$_4$)$_3$	-2825.0	-2242.8	-571.5		NH$_4$NO$_2$	-237.2	-111.6	236.4	-17.6

THERMODYNAMIC PROPERTIES OF AQUEOUS SYSTEMS (continued)

Species	$\Delta_f H°/$ kJ mol^{-1}	$\Delta_f G°/$ kJ mol^{-1}	$S°/$ J mol^{-1}K^{-1}	$C_p/$ J mol^{-1}K^{-1}	Species	$\Delta_f H°/$ kJ mol^{-1}	$\Delta_f G°/$ kJ mol^{-1}	$S°/$ J mol^{-1}K^{-1}	$C_p/$ J mol^{-1}K^{-1}
NH_4NO_3	-339.9	-190.6	259.8	-6.7	$RaSO_4$	-1436.8	-1306.2	75.0	
NH_4OH	-362.5	-236.5	102.5	-68.6	$RbBr$	-372.7	-387.9	203.9	
NH_4SCN	-56.1	13.4	257.7	39.7	$RbCl$	-418.3	-415.2	178.0	
$(NH_4)_2CO_3$	-942.2	-686.4	169.9		RbF	-583.8	-562.8	107.5	
$(NH_4)_2CrO_4$	-1146.2	-886.4	277.0		$RbHCO_3$	-943.2	-870.8	212.7	
$(NH_4)_2Cr_2O_7$	-1755.2	-1459.5	488.7		$RbHSO_4$	-1138.5	-1039.9	253.1	
$(NH_4)_2HAsO_4$	-1171.4	-873.2	225.1		RbI	-306.4	-335.6	232.6	
$(NH_4)_2HPO_4$	-1557.2	-1247.8	193.3		$RbNO_3$	-458.5	-395.2	267.8	
$(NH_4)_2S$	-231.8	-72.6	212.1		Rb_2CO_3	-1179.5	-1095.8	186.2	
$(NH_4)_2SO_3$	-900.4	-645.0	197.5		Rb_2S	-469.4	-482.0	228.4	
$(NH_4)_2SO_4$	-1174.3	-903.1	246.9	-133.1	Rb_2SO_4	-1411.6	-1312.5	263.2	
$(NH_4)_2SeO_4$	-864.0	-599.8	280.7		$SmCl_3$	-1193.3	-1060.2	-42.7	-431.0
$(NH_4)_3PO_4$	-1674.9	-1256.6	117.0		$SrBr_2$	-788.9	-767.4	132.2	
$NaBr$	-361.7	-365.8	141.4	-95.4	$SrCO_3$	-1222.9	-1087.3	-89.5	
$NaCl$	-407.3	-393.1	115.5	-90.0	$SrCl_2$	-880.1	-821.9	80.3	
NaF	-572.8	-540.7	45.2	-60.2	SrI_2	-656.2	-662.6	190.0	
$NaHCO_3$	-932.1	-848.7	150.2		$Sr(NO_3)_2$	-960.5	-782.0	260.2	
$NaHSO_4$	-1127.5	-1017.8	190.8	-38.0	$SrSO_4$	-1455.1	-1304.0	-12.6	
NaI	-295.3	-313.5	170.3	-95.8	$TbCl_3$	-1184.1	-1045.5	-59.0	-393.0
$NaNO_3$	-447.5	-373.2	205.4	-40.2	$TlBr$	-116.2	-136.4	207.9	
Na_2CO_3	-1157.4	-1051.6	61.1		$TlBr_3$	-168.2	-97.1	54.0	
Na_2S	-447.3	-438.1	103.3		$TlCl$	-161.8	-163.6	182.0	
Na_2SO_4	-1389.5	-1268.4	138.1	-201.0	$TlCl_3$	-305.0	-179.0	-23.0	
Na_2Se		-394.6			TlF	-327.3	-311.2	111.7	
$NdCl_3$	-1197.9	-1065.6	-37.7	-431.0	TlI	-49.8	-84.0	236.8	
$NiBr_2$	-297.1	-253.6	36.0		$TlNO_3$	-202.0	-143.7	272.0	
$NiCl_2$	-388.3	-307.9	-15.1		Tl_2SO_4	-898.6	-809.3	271.1	
NiF_2	-719.2	-603.3	-156.5		$TmCl_3$	-1199.1	-1055.6	-75.0	-385.0
NiI_2	-164.4	-149.0	93.7		UCl_4	-1259.8	-1056.8	-184.0	
$Ni(NO_3)_2$	-468.6	-268.5	164.0		UO_2CO_3	-1696.6	-1481.5	-154.4	
$NiSO_4$	-963.2	-790.3	-108.8		$UO_2(NO_3)_2$	-1434.3	-1176.0	195.4	
$PbBr_2$	-244.8	-232.3	175.3		UO_2SO_4	-1928.8	-1698.2	-77.4	
$PbCl_2$	-336.0	-286.9	123.4		$YbCl_3$	-1176.1	-1037.6	-71.0	-385.0
PbF_2	-666.9	-582.0	-17.2		$ZnBr_2$	-397.0	-355.0	52.7	-238.0
PbI_2	-112.1	-127.6	233.0		$ZnCl_2$	-488.2	-409.5	0.8	-226.0
$Pb(NO_3)_2$	-416.3	-246.9	303.3		ZnF_2	-819.1	-704.6	-139.7	-167.0
$PrCl_3$	-1206.2	-1072.7	-42.0	-439.0	ZnI_2	-264.3	-250.2	110.5	-238.0
$RaCl_2$	-861.9	-823.8	167.0		$Zn(NO_3)_2$	-568.6	-369.6	180.7	-126.0
$Ra(NO_3)_2$	-942.2	-784.0	347.0		$ZnSO_4$	-1063.2	-891.6	-92.0	-247.0

HEAT OF COMBUSTION

The heat of combustion of a substance at 25°C can be calculated from the enthalpy of formation ($\Delta_f H°$) data in the table "Standard Thermodynamic Properties of Chemical Substances" in this Section. We can write the general combustion reaction as

$$X + O_2 \rightarrow CO_2(g) + H_2O(l) + \text{other products}$$

For a compound containing only carbon, hydrogen, and oxygen, the reaction is simply

$$C_aH_bO_c + \left(a + \frac{1}{4}b - \frac{1}{2}c\right)O_2 \rightarrow a\ CO_2(g) + \frac{1}{2}b\ H_2O(l)$$

and the standard heat of combustion $\Delta_c H°$, which is defined as the negative of the enthalpy change for the reaction (i.e., the heat released in the combustion process), is given by

$$\Delta_c H° = -a\Delta_f H°\left(CO_2, g\right) - \frac{1}{2}b\ \Delta_f H°\left(H_2O, l\right) + \Delta_f H°\left(C_aH_b, O_c\right)$$

$$= 393.51\ a + 142.915\ b + \Delta_f H°\left(C_aH_b, O_c\right)$$

This equation applies if the reactants start in their standard states (25°C and one atmosphere pressure) and the products return to the same conditions. The same equation applies to a compound containing another element if that element ends in its standard reference state (e.g., nitrogen, if the product is N_2); in general, however, the exact products containing the other elements must be known in order to calculate the heat of combustion.

The following table gives the standard heat of combustion calculated in this manner for a few representative substances.

Molecular formula	Name	$\Delta_c H°/\text{kJ mol}^{-1}$
Inorganic substances		
C	Carbon (graphite)	393.5
CO	Carbon monoxide (g)	283.0
H_2	Hydrogen (g)	285.8
H_3N	Ammonia (g)	382.8
H_4N_2	Hydrazine (g)	667.1
N_2O	Nitrous oxide (g)	82.1
Hydrocarbons		
CH_4	Methane (g)	890.8
C_2H_2	Acetylene (g)	1301.1
C_2H_4	Ethylene (g)	1411.2
C_2H_6	Ethane (g)	1560.7
C_3H_6	Propylene (g)	2058.0
C_3H_6	Cyclopropane (g)	2091.3
C_3H_8	Propane (g)	2219.2
C_4H_6	1,3-Butadiene (g)	2541.5
C_4H_{10}	Butane (g)	2877.6
C_5H_{12}	Pentane (l)	3509.0
C_6H_6	Benzene (l)	3267.6
C_6H_{12}	Cyclohexane (l)	3919.6
C_6H_{14}	Hexane (l)	4163.2
C_7H_8	Toluene (l)	3910.3
C_7H_{16}	Heptane (l)	4817.0
$C_{10}H_8$	Naphthalene (s)	5156.3
Alcohols and ethers		
CH_4O	Methanol (l)	726.1
C_2H_6O	Ethanol (l)	1366.8
C_2H_6O	Dimethyl ether (g)	1460.4
$C_2H_6O_2$	Ethylene glycol (l)	1189.2

Molecular formula	Name	$\Delta_c H°/\text{kJ mol}^{-1}$
C_3H_8O	1-Propanol (l)	2021.3
$C_3H_8O_3$	Glycerol (l)	1655.4
$C_4H_{10}O$	Diethyl ether (l)	2723.9
$C_5H_{12}O$	1-Pentanol (l)	3330.9
C_6H_6O	Phenol (s)	3053.5
Carbonyl compounds		
CH_2O	Formaldehyde (g)	570.7
C_2H_2O	Ketene (g)	1025.4
C_2H_4O	Acetaldehyde (l)	1166.9
C_3H_6O	Acetone (l)	1789.9
C_3H_6O	Propanal (l)	1822.7
C_4H_8O	2-Butanone (l)	2444.1
Acids and esters		
CH_2O_2	Formic acid (l)	254.6
$C_2H_4O_2$	Acetic acid (l)	874.2
$C_2H_4O_2$	Methyl formate (l)	972.6
$C_3H_6O_2$	Methyl acetate (l)	1592.2
$C_4H_8O_2$	Ethyl acetate (l)	2238.1
$C_7H_6O_2$	Benzoic acid (s)	3226.9
Nitrogen compounds		
CHN	Hydrogen cyanide (g)	671.5
CH_3NO_2	Nitromethane (l)	709.2
CH_5N	Methylamine (g)	1085.6
C_2H_3N	Acetonitrile (l)	1247.2
C_2H_5NO	Acetamide (s)	1184.6
C_3H_9N	Trimethylamine (g)	2443.1
C_5H_5N	Pyridine (l)	2782.3
C_6H_7N	Aniline (l)	3392.8

ELECTRICAL CONDUCTIVITY OF WATER

This table gives the electrical conductivity of highly purified water over a range of temperature and pressure. The first column of conductivity data refers to water at its own vapor pressure. Equations for calculating the conductivity at any temperature and pressure may be found in the reference.

REFERENCE

Marshall, W. L., *J. Chem. Eng. Data* 32, 221, 1987.

Conductivity in µS/cm at the indicated pressure

$t/°C$	Sat. vapor	50 MPa	100 MPa	200 MPa	400 MPa	600 MPa
0	0.0115	0.0150	0.0189	0.0275	0.0458	0.0667
25	0.0550	0.0686	0.0836	0.117	0.194	0.291
100	0.765	0.942	1.13	1.53	2.45	3.51
200	2.99	4.08	5.22	7.65	13.1	19.5
300	2.41	4.87	7.80	14.1	28.9	46.5
400		1.17	4.91	14.3	39.2	71.3
600			0.134	4.65	33.8	85.7

STANDARD KCl SOLUTIONS FOR CALIBRATING CONDUCTIVITY CELLS

This table presents recommended electrolytic conductivity (κ) values for aqueous potassium chloride solutions with molalities of 0.01 mol/kg, 0.1 mol/kg and 1.0 mol/kg at temperatures from 0°C to 50°C. The values, which are based on measurements at the National Institute of Standards and Technology, provide primary standards for the calibration of conductivity cells. The measurements at 0.01 and 0.1 molal are described in Reference 1, while those at 1.0 molal are in Reference 2. Temperatures are given on the ITS-90 scale. The uncertainty in the conductivity is about 0.03% for the 0.01 molal values and about 0.04% for the 0.1 and 1.0 molal values. The conductivity of water saturated with atmospheric CO_2 is given in the last column. These values were subtracted from the original measurements to give the values in the second, third, and fourth columns. All κ values are given in units of 10^{-4} S/m (numerically equal to μS/cm).

The assistance of Kenneth W. Pratt is appreciated.

REFERENCES

1. Wu, Y.C., Koch, W.F., and Pratt, K.W., *J. Res. Natl. Inst. Stand. Technol.* 96, 191, 1991.

2. Wu, Y.C., Koch, W.F., Feng, D., Holland, L.A., Juhasz, E., Arvay, E., and Tomek, A., *J. Res. Natl. Inst. Stand. Technol.* 99, 241, 1994.

	10^4 κ/S m^{-1}			
t/°C	0.01 m KCl	0.1 m KCl	1.0 m KCl	H_2O (CO_2 sat.)
0	772.92	7 116.85	63 488	0.58
5	890.96	8 183.70	72 030	0.68
10	1 013.95	9 291.72	80 844	0.79
15	1 141.45	10 437.1	89 900	0.89
18	1 219.93	11 140.6	——	0.95
20	1 273.03	11 615.9	99 170	0.99
25	1 408.23	12 824.6	108 620	1.10
30	1 546.63	14 059.2	118 240	1.20
35	1 687.79	15 316.0	127 970	1.30
40	1 831.27	16 591.0	137 810	1.40
45	1 976.62	17 880.6	147 720	1.51
50	2 123.43	19 180.9	157 670	1.61

MOLAR CONDUCTIVITY OF AQUEOUS HF, HCl, HBr, AND HI

The molar conductivity Λ of an electrolyte solution is defined as the conductivity divided by amount-of-substance concentration. The customary unit is S cm^{2}mol^{-1} (i.e., Ω^{-1} cm^{2}mol^{-1}). The first part of this table gives the molar conductivity of the hydrohalogen acids at 25°C as a function of the concentration in mol/L. The second part gives the temperature dependence of Λ for HCl and HBr. More extensive tables and mathematical representations may be found in the reference.

REFERENCE

Hamer, W.J., and DeWane, H.J., *Electrolytic Conductance and the Conductances of the Hydrohalogen Acids in Water*, Natl. Stand. Ref. Data Sys.-Natl. Bur. Standards (U.S.), No. 33, 1970.

c/mol L^{-1}	HF	HCl	HBr	HI	c/mol L^{-1}	HF	HCl	HBr	HI
Inf. dil.	405.1	426.1	427.7	426.4	3.5		218.3	217.5	215.4
0.0001		424.5	425.9	424.6	4.0		200.0	199.4	195.1
0.0005		422.6	424.3	423.0	4.5		183.1	182.4	176.8
0.001		421.2	422.9	421.7	5.0		167.4	166.5	160.4
0.005	128.1	415.7	417.6	416.4	5.5		152.9	151.8	145.5
0.01	96.1	411.9	413.7	412.8	6.0		139.7	138.2	131.7
0.05	50.1	398.9	400.4	400.8	6.5		127.7	125.7	118.6
0.10	39.1	391.1	391.9	394.0	7.0		116.9	114.2	105.7
0.5	26.3	360.7	361.9	369.8	7.5		107.0	103.8	
1.0	24.3	332.2	334.5	343.9	8.0		98.2	94.4	
1.5		305.8	307.6	316.4	8.5		90.3	85.8	
2.0		281.4	281.7	288.9	9.0		83.1		
2.5		258.9	257.8	262.5	9.5		76.6		
3.0		237.6	236.8	237.9	10.0		70.7		

c/mol L^{-1}	–20°C	–10°C	0°C	10°C	20°C	30°C	40°C	50°C
				HCl				
0.5			228.7	283.0	336.4	386.8	436.9	482.4
1.0			211.7	261.6	312.2	359.0	402.9	445.3
1.5			196.2	241.5	287.5	331.1	371.6	410.8
2.0			182.0	222.7	262.9	303.3	342.4	378.2
2.5		131.7	168.5	205.1	239.8	277.0	315.2	347.6
3.0		120.8	154.6	188.5	219.3	253.3	289.3	319.0
3.5	85.5	111.3	139.6	172.2	201.6	232.9	263.9	292.1
4.0	79.3	102.7	129.2	158.1	185.6	214.2	242.2	268.2
4.5	73.7	94.9	119.5	145.4	170.6	196.6	222.5	246.7
5.0	68.5	87.8	110.3	133.5	156.6	180.2	204.1	226.5
5.5	63.6	81.1	101.7	122.5	143.6	165.0	187.1	207.7
6.0	58.9	74.9	93.7	112.3	131.5	151.0	171.3	190.3
6.5	54.4	69.1	86.2	103.0	120.4	138.2	156.9	174.3
7.0	50.2	63.7	79.3	94.4	110.2	126.4	143.3	159.7
7.5	46.3	58.6	73.0	86.5	100.9	115.7	131.6	146.2
8.0	42.7	54.0	67.1	79.4	92.4	106.1	120.6	134.0
8.5	39.4	49.8	61.7	72.9	84.7	97.3	110.7	123.0
9.0	36.4	45.9	56.8	67.1	77.8	89.4	101.7	112.9
9.5	33.6	42.3	52.3	61.8	71.5	82.3	93.6	103.9
10.0	31.2	39.1	48.2	57.0	65.8	75.9	86.3	95.7
10.5	28.9	36.1	44.5	52.7	60.7	70.1	79.6	88.4
11.0	26.8	33.4	41.1	48.8	56.1	64.9	73.6	81.7
11.5	24.9	31.0	38.0	45.3	51.9	60.1	68.0	75.6
12.0	23.1	28.7	35.3	42.0	48.0	55.6	62.8	70.0
12.5	21.4	26.7	32.7	39.0	44.4	51.4	57.9	64.8

c/mol L^{-1}	–20°C	–10°C	0°C	10°C	20°C	30°C	40°C	50°C
				HBr				
0.5			240.9	295.9	347.0	398.9	453.6	496.8
1.0			229.6	276.0	329.0	380.4	418.6	465.2
1.5			209.5	254.9	298.9	340.6	381.8	421.4
2.0		150.8	188.6	231.3	271.8	314.1	350.5	387.4
2.5		136.8	171.7	208.3	244.8	281.7	316.0	349.1
3.0		125.7	157.2	189.5	222.2	255.0	287.8	318.6
3.5		116.1	144.1	174.6	203.2	234.4	263.7	291.9
4.0	84.0	107.5	132.3	160.2	186.8	214.2	239.7	266.9
4.5	78.0	99.0	123.0	146.4	171.2	195.1	218.8	242.6
5.0	72.3	91.4	112.6	134.0	155.7	178.2	199.6	221.3
5.5	67.0	84.2	103.1	122.7	142.1	162.8	181.4	201.8
6.0	61.8	77.2	94.3	112.0	129.6	148.0	165.4	183.4
6.5	56.8	70.7	86.0	102.0	118.0	134.1	150.5	166.3
7.0	51.9	64.6	78.4	92.6	107.1	121.4	136.3	150.8

EQUIVALENT CONDUCTIVITY OF ELECTROLYTES IN AQUEOUS SOLUTION
Petr Vanýsek

This table gives the equivalent (molar) conductivity Λ at 25°C for some common electrolytes in aqueous solution at concentrations up to 0.1 mol/L. The units of Λ are 10^{-4} m^2 S mol^{-1}.

For very dilute solutions, the equivalent conductivity for any electrolyte of concentration c can be approximately calculated using the Debye-Hückel-Onsager equation, which can be written for a symmetrical (equal charge on cation and anion) electrolyte as

$$\Lambda = \Lambda^\circ - (A + B\Lambda^\circ)c^{1/2}$$

For a solution at 25°C and both cation and anion with charge $|1|$, the constants are $A = 60.20$ and $B = 0.229$. Λ° can be found from the next table, "Ionic Conductivity and Diffusion at Infinite Dilution". The equation is reliable for $c < 0.001$ mol/L; with higher concentration the error increases.

Compound	Infinite dilution Λ°	Concentration (mol/L) Λ						
		0.0005	0.001	0.005	0.01	0.02	0.05	0.1
$AgNO_3$	133.29	131.29	130.45	127.14	124.70	121.35	115.18	109.09
$1/2BaCl_2$	139.91	135.89	134.27	127.96	123.88	119.03	111.42	105.14
$1/2CaCl_2$	135.77	131.86	130.30	124.19	120.30	115.59	108.42	102.41
$1/2Ca(OH)_2$	258	—	—	233	226	214	—	—
$1/2CuSO_4$	133.6	121.6	115.20	94.02	83.08	72.16	59.02	50.55
HCl	425.95	422.53	421.15	415.59	411.80	407.04	398.89	391.13
KBr	151.9	149.8	148.9	146.02	143.36	140.41	135.61	131.32
KCl	149.79	147.74	146.88	143.48	141.20	138.27	133.30	128.90
$KClO_4$	139.97	138.69	137.80	134.09	131.39	127.86	121.56	115.14
$1/3K_3Fe(CN)_6$	174.5	166.4	163.1	150.7	—	—	—	—
$1/4K_4Fe(CN)_6$	184	—	167.16	146.02	134.76	122.76	107.65	97.82
$KHCO_3$	117.94	116.04	115.28	112.18	110.03	107.17	—	—
KI	150.31	148.2	143.32	144.30	142.11	139.38	134.90	131.05
KIO_4	127.86	125.74	124.88	121.18	118.45	114.08	106.67	98.2
KNO_3	144.89	142.70	141.77	138.41	132.75	132.34	126.25	120.34
$KMnO_4$	134.8	132.7	131.9	—	126.5	—	—	113
KOH	271.5	—	234	230	228	—	219	213
$KReO_4$	128.20	126.03	125.12	121.31	118.49	114.49	106.40	97.40
$1/3LaCl_3$	145.9	139.6	137.0	127.5	121.8	115.3	106.2	99.1
LiCl	114.97	113.09	112.34	109.35	107.27	104.60	100.06	95.81
$LiClO_4$	105.93	104.13	103.39	100.52	98.56	96.13	92.15	88.52
$1/2MgCl_2$	129.34	125.55	124.15	118.25	114.49	109.99	103.03	97.05
NH_4Cl	149.6	147.5	146.7	134.4	141.21	138.25	133.22	128.69
NaCl	126.39	124.44	123.68	120.59	118.45	115.70	111.01	106.69
$NaClO_4$	117.42	115.58	114.82	111.70	109.54	106.91	102.35	98.38
NaI	126.88	125.30	124.19	121.19	119.18	116.64	112.73	108.73
$NaOOCCH_3$	91.0	89.2	88.5	85.68	83.72	81.20	76.88	72.76
NaOH	247.7	245.5	244.6	240.7	237.9	—	—	—
Na picrate	80.45	78.7	78.6	75.7	73.7	—	66.3	61.8
$1/2Na_2SO_4$	129.8	125.68	124.09	117.09	112.38	106.73	97.70	89.94
$1/2SrCl_2$	135.73	131.84	130.27	124.18	120.23	115.48	108.20	102.14
$1/2ZnSO_4$	132.7	121.3	114.47	95.44	84.87	74.20	61.17	52.61

IONIC CONDUCTIVITY AND DIFFUSION AT INFINITE DILUTION
Petr Vanýsek

This table gives the molar (equivalent) conductivity λ for common ions at infinite dilution. All values refer to aqueous solutions at 25°C. It also lists the diffusion coefficient D of the ion in dilute aqueous solution, which is related to λ through the equation

$$D = \left(RT / F^2\right)\left(\lambda / |z|\right)$$

where R is the molar gas constant, T the temperature, F the Faraday constant, and z the charge on the ion. The variation with temperature is fairly sharp; for typical ions, λ and D increase by 2 to 3% per degree as the temperature increases from 25°C.

The diffusion coefficient for a salt, D_{salt}, may be calculated from the D_+ and D_- values of the constituent ions by the relation

$$D_{salt} = \frac{\left(z_+ + |z_-|\right)D_+ D_-}{z_+ D_+ + |z_-| D_-}$$

For solutions of simple, pure electrolytes (one positive and one negative ionic species), such as NaCl, equivalent ionic conductivity $\Lambda°$, which is the conductivity per unit concentration of charge, is defined as

$$\Lambda° = \lambda_+ + \lambda_-$$

where λ_+ and λ_- are equivalent ionic conductivities of the cation and anion. The more general formula is

$$\Lambda° = \nu_+ \lambda_+ + \nu_- \lambda_-$$

where ν_+ and ν_- refer to the number of moles of cations and anions to which one mole of the electrolyte gives a rise in the solution.

REFERENCES

1. Gray, D. E., Ed., *American Institute of Physics Handbook,* McGraw-Hill, New York, 1972, 2—226.
2. Robinson, R. A., and Stokes, R. H., *Electrolyte Solutions,* Butterworths, London, 1959.
3. Lobo, V. M. M., and Quaresma, J. L., *Handbook of Electrolyte Solutions,* Physical Science Data Series 41, Elsevier, Amsterdam, 1989.
4. Conway, B. E., *Electrochemical Data,* Elsevier, Amsterdam, 1952.
5. Milazzo, G., *Electrochemistry: Theoretical Principles and Practical Applications,* Elsevier, Amsterdam, 1963.

Ion	λ 10^{-4} m² S mol⁻¹	D 10^{-5} cm² s⁻¹	Ion	λ 10^{-4} m² S mol⁻¹	D 10^{-5} cm² s⁻¹
Inorganic Cations					
			$1/3Ho^{3+}$	66.3	0.589
Ag^+	61.9	1.648	K^+	73.48	1.957
$1/3Al^{3+}$	61	0.541	$1/3La^{3+}$	69.7	0.619
$1/2Ba^{2+}$	63.6	0.847	Li^+	38.66	1.029
$1/2Be^{2+}$	45	0.599	$1/2Mg^{2+}$	53.0	0.706
$1/2Ca^{2+}$	59.47	0.792	$1/2Mn^{2+}$	53.5	0.712
$1/2Cd^{2+}$	54	0.719	NH_4^+	73.5	1.957
$1/3Ce^{3+}$	69.8	0.620	$N_2H_5^+$	59	1.571
$1/2Co^{2+}$	55	0.732	Na^+	50.08	1.334
$1/3[Co(NH_3)_6]^{3+}$	101.9	0.904	$1/3Nd^{3+}$	69.4	0.616
$1/3[Co(en)_3]^{3+}$	74.7	0.663	$1/2Ni^{2+}$	49.6	0.661
$1/6[Co_2(trien)_3]^{6+}$	69	0.306	$1/4[Ni_2(trien)_3]^{4+}$	52	0.346
$1/3Cr^{3+}$	67	0.595	$1/2Pb^{2+}$	71	0.945
Cs^+	77.2	2.056	$1/3Pr^{3+}$	69.5	0.617
$1/2Cu^{2+}$	53.6	0.714	$1/2Ra^{2+}$	66.8	0.889
D^+	249.9	6.655	Rb^+	77.8	2.072
$1/3Dy^{3+}$	65.6	0.582	$1/3Sc^{3+}$	64.7	0.574
$1/3Er^{3+}$	65.9	0.585	$1/3Sm^{3+}$	68.5	0.608
$1/3Eu^{3+}$	67.8	0.602	$1/2Sr^{2+}$	59.4	0.791
$1/2Fe^{2+}$	54	0.719	Tl^+	74.7	1.989
$1/3Fe^{3+}$	68	0.604	$1/3Tm^{3+}$	65.4	0.581
$1/3Gd^{3+}$	67.3	0.597	$1/2UO_2^{2+}$	32	0.426
H^+	349.65	9.311	$1/3Y^{3+}$	62	0.550
$1/2Hg^{2+}$	68.6	0.913	$1/3Yb^{3+}$	65.6	0.582
$1/2Hg^{2+}$	63.6	0.847	$1/2Zn^{2+}$	52.8	0.703

Ion	λ 10^{-4} m^2 S mol^{-1}	D 10^{-5} cm^2 s^{-1}	Ion	λ 10^{-4} m^2 S mol^{-1}	D 10^{-5} cm^2 s^{-1}
Inorganic Anions			$1/2SeO_4^{2-}$	75.7	1.008
			$1/2WO_4^{2-}$	69	0.919
$Au(CN)_2^-$	50	1.331			
$Au(CN)_4^-$	36	0.959	**Organic Cations**		
$B(C_6H_5)_4^-$	21	0.559			
Br^-	78.1	2.080	Benzyltrimethylammonium$^+$	34.6	0.921
Br_3^-	43	1.145	Isobutylammonium$^+$	38	1.012
BrO_3^-	55.7	1.483	Butyltrimethylammonium$^+$	33.6	0.895
CN^-	78	2.077	Decylpyridinium$^+$	29.5	0.786
CNO^-	64.6	1.720	Decyltrimethylammonium$^+$	24.4	0.650
$1/2CO_3^{2-}$	69.3	0.923	Diethylammonium$^+$	42.0	1.118
Cl^-	76.31	2.032	Dimethylammonium$^+$	51.8	1.379
ClO_2^-	52	1.385	Dipropylammonium$^+$	30.1	0.802
ClO_3^-	64.6	1.720	Dodecylammonium$^+$	23.8	0.634
ClO_4^-	67.3	1.792	Dodecyltrimethylammonium$^+$	22.6	0.602
$1/3[Co(CN)_6]^{3-}$	98.9	0.878	Ethanolammonium$^+$	42.2	1.124
$1/2CrO_4^{2-}$	85	1.132	Ethylammonium$^+$	47.2	1.257
F^-	55.4	1.475	Ethyltrimethylammonium$^+$	40.5	1.078
$1/4[Fe(CN)_6]^{4-}$	110.4	0.735	Hexadecyltrimethylammonium$^+$	20.9	0.557
$1/3[Fe(CN)_6]^{3-}$	100.9	0.896	Hexyltrimethylammonium$^+$	29.6	0.788
$H_2AsO_4^-$	34	0.905	Histidyl$^+$	23.0	0.612
HCO_3^-	44.5	1.185	Hydroxyethyltrimethylarsonium$^+$	39.4	1.049
HF_2^-	75	1.997	Methylammonium$^+$	58.7	1.563
$1/2HPO_4^{2-}$	57	0.759	Octadecylpyridinium$^+$	20	0.533
$H_2PO_4^-$	36	0.959	Octadecyltributylammonium$^+$	16.6	0.442
$H_2PO_2^-$	46	1.225	Octadecyltriethylammonium$^+$	17.9	0.477
HS^-	65	1.731	Octadecyltrimethylammonium$^+$	19.9	0.530
HSO_3^-	58	1.545	Octadecyltripropylammonium$^+$	17.2	0.458
HSO_4^-	52	1.385	Octyltrimethylammonium$^+$	26.5	0.706
$H_2SbO_4^-$	31	0.825	Pentylammonium$^+$	37	0.985
I^-	76.8	2.045	Piperidinium$^+$	37.2	0.991
IO_3^-	40.5	1.078	Propylammonium$^+$	40.8	1.086
IO_4^-	54.5	1.451	Pyrilammonium$^+$	24.3	0.647
MnO_4^-	61.3	1.632	Tetrabutylammonium$^+$	19.5	0.519
$1/2MoO_4^{2-}$	74.5	1.984	Tetradecyltrimethylammonium$^+$	21.5	0.573
$N(CN)_2^-$	54.5	1.451	Tetraethylammonium$^+$	32.6	0.868
NO_2^-	71.8	1.912	Tetramethylammonium$^+$	44.9	1.196
NO_3^-	71.42	1.902	Tetraisopentylammonium$^+$	17.9	0.477
$NH_2SO_3^-$	48.3	1.286	Tetrapentylammmonium$^+$	17.5	0.466
N_3^-	69	1.837	Tetrapropylammonium$^+$	23.4	0.623
OCN^-	64.6	1.720	Triethylammonium$^+$	34.3	0.913
OD^-	119	3.169	Triethylsulfonium$^+$	36.1	0.961
OH^-	198	5.273	Trimethylammonium$^+$	47.23	1.258
PF_6^-	56.9	1.515	Trimethylhexylammonium$^+$	34.6	0.921
$1/2PO_3F^{2-}$	63.3	0.843	Trimethylsulfonium$^+$	51.4	1.369
$1/3PO_4^{3-}$	92.8	0.824	Tripropylammonium$^+$	26.1	0.695
$1/4P_2O_7^{4-}$	96	0.639			
$1/3P_3O_9^{3-}$	83.6	0.742	**Organic Anions**		
$1/5P_3O_{10}^{5-}$	109	0.581			
ReO_4^-	54.9	1.462	Acetate$^-$	40.9	1.089
SCN^-	66	1.758	p-Anisate$^-$	29.0	0.772
$1/2SO_3^{2-}$	72	0.959	$1/2$Azelate^{2-}	40.6	0.541
$1/2SO_4^{2-}$	80.0	1.065	Benzoate$^-$	32.4	0.863
$1/2S_2O_3^{2-}$	85.0	1.132	Bromoacetate$^-$	39.2	1.044
$1/2S_2O_4^{2-}$	66.5	0.885	Bromobenzoate$^-$	30	0.799
$1/2S_2O_6^{2-}$	93	1.238	Butyrate$^-$	32.6	0.868
$1/2S_2O_8^{2-}$	86	1.145	Chloroacetate$^-$	39.8	1.060
$Sb(OH)_6^-$	31.9	0.849	m-Chlorobenzoate$^-$	31	0.825
$SeCN^-$	64.7	1.723	o-Chlorobenzoate$^-$	30.2	0.804

Ion	λ 10^{-4} m^2 S mol^{-1}	D 10^{-5} cm^2 s^{-1}	Ion	λ 10^{-4} m^2 S mol^{-1}	D 10^{-5} cm^2 s^{-1}
1/3Citrate^{3-}	70.2	0.623	Iodoacetate$^-$	40.6	1.081
Crotonate$^-$	33.2	0.884	Lactate$^-$	38.8	1.033
Cyanoacetate$^-$	43.4	1.156	1/2Malate^{2-}	58.8	0.783
Cyclohexane carboxylate$^-$	28.7	0.764	1/2Maleate^{2-}	61.9	0.824
1/2 1,1-Cyclopropanedicarboxylate^{2-}	53.4	0.711	1/2Malonate^{2-}	63.5	0.845
Decylsulfate$^-$	26	0.692	Methylsulfate$^-$	48.8	1.299
Dichloroacetate$^-$	38.3	1.020	Naphthylacetate$^-$	28.4	0.756
1/2Diethylbarbiturate^{2-}	26.3	0.350	1/2Oxalate^{2-}	74.11	0.987
Dihydrogencitrate$^-$	30	0.799	Octylsulfate$^-$	29	0.772
1/2Dimethylmalonate^{2-}	49.4	0.658	Phenylacetate$^-$	30.6	0.815
3,5-Dinitrobenzoate$^-$	28.3	0.754	1/2o-Phthalate^{2-}	52.3	0.696
Dodecylsulfate$^-$	24	0.639	1/2m-Phthalate^{2-}	54.7	0.728
Ethylmalonate$^-$	49.3	1.313	Picrate$^-$	30.37	0.809
Ethylsulfate$^-$	39.6	1.055	Pivalate$^-$	31.9	0.849
Fluoroacetate$^-$	44.4	1.182	Propionate$^-$	35.8	0.953
Fluorobenzoate$^-$	33	0.879	Propylsulfate$^-$	37.1	0.988
Formate$^-$	54.6	1.454	Salicylate$^-$	36	0.959
1/2Fumarate^{2-}	61.8	0.823	1/2Suberate^{2-}	36	0.479
1/2Glutarate^{2-}	52.6	0.700	1/2Succinate^{2-}	58.8	0.783
Hydrogenoxalate$^-$	40.2	1.070	p-Sulfonate	29.3	0.780
Isovalerate$^-$	32.7	0.871	1/2Tartarate^{2-}	59.6	0.794
			Trichloroacetate$^-$	35	0.932

ACTIVITY COEFFICIENTS OF ACIDS, BASES, AND SALTS
Petr Vanýsek

This table gives mean activity coefficients at 25°C for molalities in the range 0.1 to 1.0. See the following table for definitions, references, and data over a wider concentration range.

	0.1	0.2	0.3	0.4	0.5	0.6	0.7	0.8	0.9	1.0
AgNO$_3$	0.734	0.657	0.606	0.567	0.536	0.509	0.485	0.464	0.446	0.429
AlCl$_3$	0.337	0.305	0.302	0.313	0.331	0.356	0.388	0.429	0.479	0.539
Al$_2$(SO$_4$)$_3$	0.035	0.0225	0.0176	0.0153	0.0143	0.014	0.0142	0.0149	0.0159	0.0175
BaCl$_2$	0.500	0.444	0.419	0.405	0.397	0.391	0.391	0.391	0.392	0.395
BeSO$_4$	0.150	0.109	0.0885	0.0769	0.0692	0.0639	0.0600	0.0570	0.0546	0.0530
CaCl$_2$	0.518	0.472	0.455	0.448	0.448	0.453	0.460	0.470	0.484	0.500
CdCl$_2$	0.2280	0.1638	0.1329	0.1139	0.1006	0.0905	0.0827	0.0765	0.0713	0.0669
Cd(NO$_3$)$_2$	0.513	0.464	0.442	0.430	0.425	0.423	0.423	0.425	0.428	0.433
CdSO$_4$	0.150	0.103	0.0822	0.0699	0.0615	0.0553	0.0505	0.0468	0.0438	0.0415
CoCl$_2$	0.522	0.479	0.463	0.459	0.462	0.470	0.479	0.492	0.511	0.531
CrCl$_3$	0.331	0.298	0.294	0.300	0.314	0.335	0.362	0.397	0.436	0.481
Cr(NO$_3$)$_3$	0.319	0.285	70.279	0.281	0.291	0.304	0.322	0.344	0.371	0.401
Cr$_2$(SO$_4$)$_3$	0.0458	0.0300	0.0238	0.0207	0.0190	0.0182	0.0181	0.0185	0.0194	0.0208
CsBr	0.754	0.694	0.654	0.626	0.603	0.586	0.571	0.558	0.547	0.538
CsCl	0.756	0.694	0.656	0.628	0.606	0.589	0.575	0.563	0.553	0.544
CsI	0.754	0.692	0.651	0.621	0.599	0.581	0.567	0.554	0.543	0.533
CsNO$_3$	0.733	0.655	0.602	0.561	0.528	0.501	0.478	0.458	0.439	0.422
CsOH	0.795	0.761	0.744	0.739	0.739	0.742	0.748	0.754	0.762	0.771
CsOAc	0.799	0.771	0.761	0.759	0.762	0.768	0.776	0.783	0.792	0.802
Cs$_2$SO$_4$	0.456	0.382	0.338	0.311	0.291	0.274	0.262	0.251	0.242	0.235
CuCl$_2$	0.508	0.455	0.429	0.417	0.411	0.409	0.409	0.410	0.413	0.417
Cu(NO$_3$)$_2$	0.511	0.460	0.439	0.429	0.426	0.427	0.431	0.437	0.445	0.455
CuSO$_4$	0.150	0.104	0.0829	0.0704	0.0620	0.0559	0.0512	0.0475	0.0446	0.0423
FeCl$_2$	0.5185	0.473	0.454	0.448	0.450	0.454	0.463	0.473	0.488	0.506
HBr	0.805	0.782	0.777	0.781	0.789	0.801	0.815	0.832	0.850	0.871
HCl	0.796	0.767	0.756	0.755	0.757	0.763	0.772	0.783	0.795	0.809
HClO$_4$	0.803	0.778	0.768	0.766	0.769	0.776	0.785	0.795	0.808	0.823
HI	0.818	0.807	0.811	0.823	0.839	0.860	0.883	0.908	0.935	0.963

ACTIVITY COEFFICIENTS OF ACIDS, BASES, AND SALTS (continued)

	0.1	0.2	0.3	0.4	0.5	0.6	0.7	0.8	0.9	1.0
HNO_3	0.791	0.754	0.735	0.725	0.720	0.717	0.717	0.718	0.721	0.724
H_2SO_4	0.2655	0.2090	0.1826	—	0.1557	—	0.1417	—	—	0.1316
KBr	0.772	0.722	0.693	0.673	0.657	0.646	0.636	0.629	0.622	0.617
KCl	0.770	0.718	0.688	0.666	0.649	0.637	0.626	0.618	0.610	0.604
$KClO_3$	0.749	0.681	0.635	0.599	0.568	0.541	0.518	—	—	—
K_2CrO_4	0.456	0.382	0.340	0.313	0.292	0.276	0.263	0.253	0.243	0.235
KF	0.775	0.727	0.700	0.682	0.670	0.661	0.654	0.650	0.646	0.645
$K_3Fe(CN)_6$	0.268	0.212	0.184	0.167	0.155	0.146	0.140	0.135	0.131	0.128
$K_4Fe(CN)_6$	0.139	0.0993	0.0808	0.0693	0.0614	0.0556	0.0512	0.0479	0.0454	—
KH_2PO_4	0.731	0.653	0.602	0.561	0.529	0.501	0.477	0.456	0.438	0.421
KI	0.778	0.733	0.707	0.689	0.676	0.667	0.660	0.654	0.649	0.645
KNO_3	0.739	0.663	0.614	0.576	0.545	0.519	0.496	0.476	0.459	0.443
$KOAc$	0.796	0.766	0.754	0.750	0.751	0.754	0.759	0.766	0.774	0.783
KOH	0.798	0.760	0.742	0.734	0.732	0.733	0.736	0.742	0.749	0.756
$KSCN$	0.769	0.716	0.685	0.663	0.646	0.633	0.623	0.614	0.606	0.599
K_2SO_4	0.441	0.360	0.316	0.286	0.264	0.246	0.232	—	—	—
$LiBr$	0.796	0.766	0.756	0.752	0.753	0.758	0.767	0.777	0.789	0.803
$LiCl$	0.790	0.757	0.744	0.740	0.739	0.743	0.748	0.755	0.764	0.774
$LiClO_4$	0.812	0.794	0.792	0.798	0.808	0.820	0.834	0.852	0.869	0.887
LiI	0.815	0.802	0.804	0.813	0.824	0.838	0.852	0.870	0.888	0.910
$LiNO_3$	0.788	0.752	0.736	0.728	0.726	0.727	0.729	0.733	0.737	0.743
$LiOH$	0.760	0.702	0.665	0.638	0.617	0.599	0.585	0.573	0.563	0.554
$LiOAc$	0.784	0.742	0.721	0.709	0.700	0.691	0.689	0.688	0.688	0.689
Li_2SO_4	0.468	0.398	0.361	0.337	0.319	0.307	0.297	0.289	0.282	0.277
$MgCl_2$	0.529	0.489	0.477	0.475	0.481	0.491	0.506	0.522	0.544	0.570
$MgSO_4$	0.150	0.107	0.0874	0.0756	0.0675	0.0616	0.0571	0.0536	0.0508	0.0485
$MnCl_2$	0.516	0.469	0.450	0.442	0.440	0.443	0.448	0.455	0.466	0.479
$MnSO_4$	0.150	0.105	0.0848	0.0725	0.0640	0.0578	0.0530	0.0493	0.0463	0.0439
NH_4Cl	0.770	0.718	0.687	0.665	0.649	0.636	0.625	0.617	0.609	0.603
NH_4NO_3	0.740	0.677	0.636	0.606	0.582	0.562	0.545	0.530	0.516	0.504
$(NH_4)_2SO_4$	0.439	0.356	0.311	0.280	0.257	0.240	0.226	0.214	0.205	0.196
$NaBr$	0.782	0.741	0.719	0.704	0.697	0.692	0.689	0.687	0.687	0.687
$NaCl$	0.778	0.735	0.710	0.693	0.681	0.673	0.667	0.662	0.659	0.657
$NaClO_3$	0.772	0.720	0.688	0.664	0.645	0.630	0.617	0.606	0.597	0.589
$NaClO_4$	0.775	0.729	0.701	0.683	0.668	0.656	0.648	0.641	0.635	0.629
Na_2CrO_4	0.464	0.394	0.353	0.327	0.307	0.292	0.280	0.269	0.261	0.253
NaF	0.765	0.710	0.676	0.651	0.632	0.616	0.603	0.592	0.582	0.573
NaH_2PO_4	0.744	0.675	0.629	0.593	0.563	0.539	0.517	0.499	0.483	0.468
NaI	0.787	0.751	0.735	0.727	0.723	0.723	0.724	0.727	0.731	0.736
$NaNO_3$	0.762	0.703	0.666	0.638	0.617	0.599	0.583	0.570	0.558	0.548
$NaOAc$	0.791	0.757	0.744	0.737	0.735	0.736	0.740	0.745	0.752	0.757
$NaOH$	0.766	0.727	0.708	0.697	0.690	0.685	0.681	0.679	0.678	0.678
$NaSCN$	0.787	0.750	—	0.720	0.715	0.712	0.710	0.710	0.711	0.712
Na_2SO_4	0.445	0.365	0.320	0.289	0.266	0.248	0.233	0.221	0.210	0.201
$NiCl_2$	0.522	0.479	0.463	0.460	0.464	0.471	0.482	0.496	0.515	0.563
$NiSO_4$	0.150	0.105	0.0841	0.0713	0.0627	0.0562	0.0515	0.0478	0.0448	0.0425
$Pb(NO_3)_2$	0.395	0.308	0.260	0.228	0.205	0.187	0.172	0.160	0.150	0.141
$RbBr$	0.763	0.706	0.673	0.650	0.632	0.617	0.605	0.595	0.586	0.578
$RbCl$	0.764	0.709	0.675	0.652	0.634	0.620	0.608	0.599	0.590	0.583
RbI	0.762	0.705	0.671	0.647	0.629	0.614	0.602	0.591	0.583	0.575
$RbNO_3$	0.734	0.658	0.606	0.565	0.534	0.508	0.485	0.465	0.446	0.430
$RbOAc$	0.796	0.767	0.756	0.753	0.755	0.759	0.766	0.773	0.782	0.792
Rb_2SO_4	0.451	0.374	0.331	0.301	0.279	0.263	0.249	0.238	0.228	0.219
$SrCl_2$	0.511	0.462	0.442	0.433	0.430	0.431	0.434	0.441	0.449	0.461
$TlClO_4$	0.730	0.652	0.599	0.559	0.527	—	—	—	—	—
$TlNO_3$	0.702	0.606	0.545	0.500	—	—	—	—	—	—
UO_2Cl_2	0.544	0.510	0.520	0.505	0.517	0.532	0.549	0.571	0.595	0.620
UO_2SO_4	0.150	0.102	0.0807	0.0689	0.0611	0.0566	0.0515	0.0483	0.0458	0.0439
$ZnCl_2$	0.515	0.462	0.432	0.411	0.394	0.380	0.369	0.357	0.348	0.339
$Zn(NO_3)_2$	0.531	0.489	0.474	0.469	0.473	0.480	0.489	0.501	0.518	0.535
$ZnSO_4$	0.150	0.140	0.0835	0.0714	0.0630	0.0569	0.0523	0.0487	0.0458	0.0435

MEAN ACTIVITY COEFFICIENTS OF ELECTROLYTES AS A FUNCTION OF CONCENTRATION

The mean activity coefficient γ of an electrolyte X_aY_b is defined as

$$\gamma = \left(\gamma_+^a \gamma_-^b\right)^{1/(a+b)}$$

where γ_+ and γ_- are activity coefficients of the individual ions (which cannot be directly measured). This table gives the mean activity coefficients of about 100 electrolytes in aqueous solution as a function of concentration, expressed in molality terms. All values refer to a temperature of 25°C. Substances are arranged in alphabetical order by formula.

REFERENCES

1. Hamer,W. J., and Wu, Y. C., *J. Phys. Chem. Ref. Data*, 1, 1047, 1972.
2. Staples, B. R., *J. Phys. Chem. Ref. Data*, 6, 385, 1977; 10, 767, 1981; 10, 779, 1981.
3. Goldberg, R. N. et al., *J. Phys. Chem. Ref. Data*, 7, 263, 1978; 8, 923, 1979; 8, 1005, 1979; 10, 1, 1981; 10, 671, 1981.

Mean Activity Coefficient at 25°C

m/mol kg^{-1}	AgNO$_3$	BaBr$_2$	BaCl$_2$	BaI$_2$	CaBr$_2$	CaCl$_2$	CaI$_2$
0.001	0.964	0.881	0.887	0.890	0.890	0.888	0.890
0.002	0.950	0.850	0.849	0.853	0.853	0.851	0.853
0.005	0.924	0.785	0.782	0.792	0.791	0.787	0.791
0.010	0.896	0.727	0.721	0.737	0.735	0.727	0.736
0.020	0.859	0.661	0.653	0.678	0.674	0.664	0.677
0.050	0.794	0.573	0.559	0.600	0.594	0.577	0.600
0.100	0.732	0.517	0.492	0.551	0.540	0.517	0.552
0.200	0.656	0.463	0.436	0.520	0.502	0.469	0.524
0.500	0.536	0.435	0.391	0.536	0.500	0.444	0.554
1.000	0.430	0.470	0.393	0.664	0.604	0.495	0.729
2.000	0.316	0.654		1.242	1.125	0.784	
5.000	0.181				18.7	5.907	
10.000	0.108					43.1	
15.000	0.085						

m/mol kg^{-1}	Cd(NO$_2$)$_2$	Cd(NO$_3$)$_2$	CoBr$_2$	CoCl$_2$	CoI$_2$	Co(NO$_3$)$_2$	CsBr
0.001	0.881	0.888	0.890	0.889	0.887	0.888	0.965
0.002	0.837	0.851	0.854	0.852	0.849	0.850	0.951
0.005	0.759	0.787	0.794	0.789	0.783	0.786	0.925
0.010	0.681	0.728	0.740	0.732	0.724	0.728	0.898
0.020	0.589	0.664	0.681	0.670	0.661	0.663	0.864
0.050	0.451	0.576	0.605	0.586	0.582	0.576	0.806
0.100	0.344	0.515	0.556	0.528	0.540	0.516	0.752
0.200	0.247	0.465	0.523	0.483	0.527	0.469	0.691
0.500	0.148	0.428	0.538	0.465	0.596	0.446	0.605
1.000	0.098	0.437	0.685	0.532	0.845	0.492	0.540
2.000	0.069	0.517	1.421	0.864	2.287	0.722	0.485
5.000	0.054		13.9		55.3	3.338	0.454
10.000					196		

m/mol kg^{-1}	CsCl	CsF	CsI	CsNO$_3$	CsOH	Cs$_2$SO$_4$	CuBr$_2$
0.001	0.965	0.965	0.965	0.964	0.966	0.885	0.889
0.002	0.951	0.952	0.951	0.951	0.953	0.845	0.853
0.005	0.925	0.929	0.925	0.924	0.930	0.775	0.791
0.010	0.898	0.905	0.898	0.897	0.906	0.709	0.735
0.020	0.864	0.876	0.863	0.860	0.878	0.634	0.674
0.050	0.805	0.830	0.804	0.796	0.836	0.526	0.594
0.100	0.751	0.792	0.749	0.733	0.802	0.444	0.541

m/mol kg^{-1}	CsCl	CsF	CsI	CsNO$_3$	CsOH	Cs$_2$SO$_4$	CuBr$_2$
0.200	0.691	0.755	0.688	0.655	0.772	0.369	0.504
0.500	0.607	0.721	0.601	0.529	0.755	0.285	0.503
1.000	0.546	0.726	0.534	0.421	0.782	0.233	0.591
2.000	0.496	0.803	0.470				0.859
5.000	0.474						
10.000	0.508						

m/mol kg^{-1}	CuCl$_2$	Cu(ClO$_4$)$_2$	Cu(NO$_3$)$_2$	FeCl$_2$	HBr	HCl	HClO$_4$
0.001	0.887	0.890	0.888	0.888	0.966	0.965	0.966
0.002	0.849	0.854	0.851	0.850	0.953	0.952	0.953
0.005	0.783	0.795	0.787	0.785	0.930	0.929	0.929
0.010	0.722	0.741	0.729	0.725	0.907	0.905	0.906
0.020	0.654	0.685	0.664	0.659	0.879	0.876	0.878
0.050	0.561	0.613	0.577	0.570	0.837	0.832	0.836
0.100	0.495	0.572	0.516	0.509	0.806	0.797	0.803
0.200	0.441	0.553	0.466	0.462	0.783	0.768	0.776
0.500	0.401	0.617	0.431	0.443	0.790	0.759	0.769
1.000	0.405	0.892	0.456	0.500	0.872	0.811	0.826
2.000	0.453	2.445	0.615	0.782	1.167	1.009	1.055
5.000	0.601		2.083		3.800	2.380	3.100
10.000					33.4	10.4	30.8
15.000							323

m/mol kg^{-1}	HF	HI	HNO$_3$	H$_2$SO$_4$	KBr	KCNS	KCl
0.001	0.551	0.966	0.965	0.804	0.965	0.965	0.965
0.002	0.429	0.953	0.952	0.740	0.952	0.951	0.951
0.005	0.302	0.931	0.929	0.634	0.927	0.927	0.927
0.010	0.225	0.909	0.905	0.542	0.902	0.901	0.901
0.020	0.163	0.884	0.875	0.445	0.870	0.869	0.869
0.050	0.106	0.847	0.829	0.325	0.817	0.815	0.816
0.100	0.0766	0.823	0.792	0.251	0.771	0.768	0.768
0.200	0.0550	0.811	0.756	0.195	0.772	0.716	0.717
0.500	0.0352	0.845	0.725	0.146	0.658	0.647	0.649
1.000	0.0249	0.969	0.730	0.125	0.617	0.598	0.604
2.000	0.0175	1.363	0.788	0.119	0.593	0.556	0.573
5.000	0.0110	4.760	1.063	0.197	0.626	0.525	0.593
10.000	0.0085	49.100	1.644	0.527			
15.000	0.0077		2.212	1.077			
20.000	0.0075		2.607	1.701			

m/mol kg^{-1}	KClO$_3$	K$_2$CrO$_4$	KF	KH$_2$PO$_4$*	K$_2$HPO$_4$**	KI	KNO$_3$
0.001	0.965	0.886	0.965	0.964	0.886	0.965	0.964
0.002	0.951	0.847	0.952	0.950	0.847	0.952	0.950
0.005	0.926	0.779	0.927	0.924	0.779	0.927	0.924
0.010	0.899	0.715	0.902	0.896	0.715	0.902	0.896
0.020	0.865	0.643	0.870	0.859	0.643	0.871	0.860
0.050	0.805	0.539	0.818	0.793	0.538	0.820	0.797
0.100	0.749	0.460	0.773	0.730	0.457	0.776	0.735
0.200	0.681	0.385	0.726	0.652	0.379	0.731	0.662
0.500	0.569	0.296	0.670	0.529	0.283	0.676	0.546
1.000		0.239	0.645	0.422		0.646	0.444

MEAN ACTIVITY COEFFICIENTS OF ELECTROLYTES AS A FUNCTION OF CONCENTRATION (continued)

m/mol kg^{-1}	KClO$_3$	K$_2$CrO$_4$	KF	KH$_2$PO$_4$*	K$_2$HPO$_4$**	KI	KNO$_3$
2.000		0.199	0.658			0.638	0.332
5.000			0.871				
10.000			1.715				
15.000			3.120				

m/mol kg^{-1}	KOH	K$_2$SO$_4$	LiBr	LiCl	LiClO$_4$	LiI	LiNO$_3$
0.001	0.965	0.885	0.965	0.965	0.966	0.966	0.965
0.002	0.952	0.844	0.952	0.952	0.953	0.953	0.952
0.005	0.927	0.772	0.929	0.928	0.931	0.930	0.928
0.010	0.902	0.704	0.905	0.904	0.908	0.908	0.904
0.020	0.871	0.625	0.877	0.874	0.882	0.882	0.874
0.050	0.821	0.511	0.832	0.827	0.843	0.843	0.827
0.100	0.779	0.424	0.797	0.789	0.815	0.817	0.788
0.200	0.740	0.343	0.767	0.756	0.795	0.802	0.753
0.500	0.710	0.251	0.754	0.739	0.806	0.824	0.726
1.000	0.733		0.803	0.775	0.887	0.912	0.743
2.000	0.860		1.012	0.924	1.161	1.197	0.837
5.000	1.697		2.696	2.000			1.298
10.000	6.110		20.0	9.600			2.500
15.000	19.9		147	30.9			3.960
20.000	46.4		486				4.970

m/mol kg^{-1}	LiOH	Li$_2$SO$_4$	MgBr$_2$	MgCl$_2$	MgI$_2$	MnBr$_2$	MnCl$_2$
0.001	0.964	0.887	0.889	0.889	0.889	0.889	0.888
0.002	0.950	0.847	0.852	0.852	0.853	0.853	0.850
0.005	0.923	0.780	0.790	0.790	0.791	0.791	0.786
0.010	0.895	0.716	0.733	0.734	0.736	0.735	0.727
0.020	0.858	0.645	0.672	0.672	0.677	0.674	0.662
0.050	0.794	0.544	0.593	0.590	0.602	0.595	0.574
0.100	0.735	0.469	0.543	0.535	0.556	0.543	0.513
0.200	0.668	0.400	0.512	0.493	0.535	0.508	0.464
0.500	0.579	0.325	0.540	0.485	0.594	0.519	0.437
1.000	0.522	0.284	0.715	0.577	0.858	0.650	0.477
2.000	0.484	0.270	1.590	1.065	2.326	1.224	0.661
5.000	0.493		36.1	14.40	109.8	6.697	1.539

m/mol kg^{-1}	Mn(ClO$_4$)$_2$	NH$_4$Cl	NH$_4$ClO$_4$	(NH$_4$)$_2$HPO$_4$**	NH$_4$NO$_3$	NaBr	NaBrO$_3$
0.001	0.892	0.965	0.964	0.882	0.964	0.965	0.965
0.002	0.858	0.952	0.950	0.839	0.951	0.952	0.951
0.005	0.801	0.927	0.924	0.763	0.925	0.928	0.926
0.010	0.752	0.901	0.895	0.688	0.897	0.903	0.900
0.020	0.700	0.869	0.859	0.600	0.862	0.873	0.867
0.050	0.637	0.816	0.794	0.469	0.801	0.824	0.811
0.100	0.604	0.769	0.734	0.367	0.744	0.783	0.759
0.200	0.596	0.718	0.663	0.273	0.678	0.742	0.698
0.500	0.686	0.649	0.560	0.171	0.582	0.697	0.605
1.000	1.030	0.603	0.479	0.114	0.502	0.687	0.528
2.000	3.072	0.569	0.399	0.074	0.419	0.730	0.449
5.000		0.563			0.303	1.083	
10.000					0.220		
15.000					0.179		
20.000					0.154		

MEAN ACTIVITY COEFFICIENTS OF ELECTROLYTES AS A FUNCTION
OF CONCENTRATION (continued)

m/mol kg^{-1}	Na$_2$CO$_3$	NaCl	NaClO$_3$	NaClO$_4$	Na$_2$CrO$_4$	NaF	Na$_2$HPO$_4$*
0.001	0.887	0.965	0.965	0.965	0.887	0.965	0.887
0.002	0.847	0.952	0.952	0.952	0.849	0.951	0.848
0.005	0.780	0.928	0.927	0.928	0.783	0.926	0.780
0.010	0.716	0.903	0.902	0.903	0.722	0.901	0.717
0.020	0.644	0.872	0.870	0.872	0.653	0.868	0.644
0.050	0.541	0.822	0.818	0.821	0.554	0.813	0.539
0.100	0.462	0.779	0.771	0.777	0.479	0.764	0.456
0.200	0.385	0.734	0.719	0.729	0.406	0.710	0.373
0.500	0.292	0.681	0.646	0.668	0.318	0.633	0.266
1.000	0.229	0.657	0.590	0.630	0.261	0.573	0.191
2.000	0.182	0.668	0.537	0.608	0.231		0.133
5.000		0.874		0.648			

m/mol kg^{-1}	NaI	NaNO$_3$	NaOH	Na$_2$SO$_3$	Na$_2$SO$_4$	Na$_2$WO$_4$	NiBr$_2$
0.001	0.965	0.965	0.965	0.887	0.886	0.886	0.889
0.002	0.952	0.951	0.952	0.847	0.846	0.846	0.853
0.005	0.928	0.926	0.927	0.779	0.777	0.777	0.791
0.010	0.904	0.900	0.902	0.716	0.712	0.712	0.735
0.020	0.874	0.866	0.870	0.644	0.637	0.638	0.675
0.050	0.827	0.810	0.819	0.540	0.529	0.534	0.596
0.100	0.789	0.759	0.775	0.462	0.446	0.457	0.546
0.200	0.753	0.701	0.731	0.386	0.366	0.388	0.514
0.500	0.722	0.617	0.685	0.296	0.268	0.320	0.535
1.000	0.734	0.550	0.674	0.237	0.204	0.291	0.692
2.000	0.823	0.480	0.714	0.196	0.155	0.291	1.476
5.000	1.402	0.388	1.076				
10.000	4.011	0.329	3.258				
15.000			9.796				
20.000			19.410				

m/mol kg^{-1}	NiCl$_2$	Ni(ClO$_4$)$_2$	Ni(NO$_3$)$_2$	Pb(ClO$_4$)$_2$	Pb(NO$_3$)$_2$	RbBr	RbCl
0.001	0.889	0.891	0.889	0.889	0.882	0.965	0.965
0.002	0.852	0.855	0.851	0.851	0.840	0.951	0.951
0.005	0.789	0.797	0.787	0.787	0.764	0.926	0.926
0.010	0.732	0.745	0.730	0.729	0.690	0.900	0.900
0.020	0.669	0.690	0.666	0.666	0.604	0.866	0.867
0.050	0.584	0.621	0.581	0.580	0.476	0.811	0.811
0.100	0.527	0.582	0.524	0.522	0.379	0.760	0.761
0.200	0.482	0.567	0.481	0.476	0.291	0.705	0.707
0.500	0.465	0.639	0.467	0.458	0.195	0.630	0.633
1.000	0.538	0.946	0.528	0.516	0.136	0.578	0.583
2.000	0.915	2.812	0.797	0.799		0.535	0.546
5.000	4.785			4.043		0.514	0.544
10.000				33.8			

m/mol kg^{-1}	RbF	RbI	RbNO$_3$	Rb$_2$SO$_4$	SrBr$_2$	SrCl$_2$	SrI$_2$
0.001	0.965	0.965	0.964	0.886	0.889	0.888	0.890
0.002	0.952	0.951	0.950	0.845	0.852	0.850	0.854
0.005	0.927	0.926	0.924	0.776	0.790	0.785	0.793
0.010	0.902	0.900	0.896	0.710	0.734	0.725	0.740
0.020	0.871	0.866	0.859	0.635	0.673	0.659	0.681

MEAN ACTIVITY COEFFICIENTS OF ELECTROLYTES AS A FUNCTION OF CONCENTRATION (continued)

m/mol kg^{-1}	RbF	RbI	RbNO$_3$	Rb$_2$SO$_4$	SrBr$_2$	SrCl$_2$	SrI$_2$
0.050	0.821	0.810	0.795	0.526	0.591	0.569	0.606
0.100	0.780	0.759	0.733	0.443	0.535	0.506	0.557
0.200	0.739	0.703	0.657	0.365	0.492	0.455	0.526
0.500	0.701	0.627	0.536	0.274	0.476	0.421	0.542
1.000	0.697	0.574	0.430	0.217	0.545	0.451	0.686
2.000	0.724	0.532	0.320		0.921	0.650	
5.000		0.517					

m/mol kg^{-1}	UO$_2$Cl$_2$	UO$_2$(NO$_3$)$_2$	ZnBr$_2$	ZnCl$_2$	ZnI$_2$
0.001	0.888	0.888	0.890	0.887	0.893
0.002	0.851	0.849	0.854	0.847	0.859
0.005	0.787	0.784	0.794	0.781	0.804
0.010	0.729	0.726	0.741	0.719	0.757
0.020	0.666	0.663	0.683	0.652	0.708
0.050	0.583	0.583	0.606	0.561	0.644
0.100	0.529	0.535	0.553	0.499	0.601
0.200	0.493	0.509	0.515	0.447	0.574
0.500	0.501	0.532	0.516	0.384	0.635
1.000	0.601	0.673	0.558	0.330	0.836
2.000	0.948	1.223	0.578	0.283	1.062
5.000		3.020	0.788	0.342	1.546
10.000			2.317	0.876	4.698
15.000			5.381	1.914	
20.000			7.965	2.968	

* The anion is H$_2$PO$_4^-$.
** The anion is HPO$_4^{-2}$.

ENTHALPY OF DILUTION OF ACIDS

The quantity given in this table is $-\Delta_{dil}H$, the negative of the enthalpy (heat) of dilution to infinite dilution for aqueous solutions of several common acids; i.e., the negative of the enthalpy change when a solution of molality m at a temperature of 25°C is diluted with an infinite amount of water. The tabulated numbers thus represent the heat produced (or, if the value is negative, the heat absorbed) when the acid is diluted. The initial molality m is given in the first column. The second column gives the dilution ratio, which is the number of moles of water that must be added to one mole of the acid to produce a solution of the molality in the first column.

REFERENCE

Parker, V. B., *Thermal Properties of Aqueous Uni-Univalent Electrolytes*, Natl. Stand. Ref. Data Ser. - Natl. Bur. Stand. (U.S.) 2, U.S. Government Printing Office, 1965.

$-\Delta_{dil}H$ in kJ/mol at 25°C

m	Dil. ratio	HF	HCl	$HClO_4$	HBr	HI	HNO_3	CH_2O_2	$C_2H_4O_2$
55.506	1.0		45.61		48.83		19.73	0.046	2.167
20	2.775	14.88	19.87	13.81	19.92	21.71	9.498	0.038	2.075
15	3.700	14.34	15.40	7.920	14.29	14.02	6.883	0.109	1.962
10	5.551	13.87	10.24	2.013	8.694	7.615	3.933	0.205	1.824
9	6.167	13.81	9.213	1.280	7.719	6.569	3.368	0.230	1.782
8	6.938	13.77	8.201	0.611	6.786	5.607	2.791	0.255	1.724
7	7.929	13.73	7.217	0.046	5.925	4.728	2.251	0.272	1.648
6	9.251	13.69	6.268	-0.351	5.004	3.975	1.749	0.280	1.540
5.5506	10	13.66	5.841	-0.490	4.590	3.577	1.540	0.285	1.477
5	11.10	13.62	5.318	-0.628	4.113	3.197	1.310	0.289	1.393
4.5	12.33	13.58	4.899	-0.732	3.711	2.828	1.109	0.289	1.310
4	13.88	13.53	4.402	-0.787	3.330	2.460	0.958	0.289	1.218
3.5	15.86	13.47	3.958	-0.820	2.966	2.105	0.791	0.289	1.121
3	18.50	13.45	3.506	-0.782	2.611	1.787	0.665	0.289	1.025
2.5	22.20	13.43	3.063	-0.724	2.301	1.527	0.582	0.285	0.912
2	27.75	13.40	2.623	-0.623	1.996	1.318	0.527	0.276	0.803
1.5	37.00	13.36	2.167	-0.431	1.665	1.125	0.506	0.259	0.678
1	55.51	13.30	1.695	-0.201	1.314	0.933	0.506	0.226	0.544
0.5551	100	13.22	1.234	0.050	0.983	0.736	0.502	0.184	0.423
0.5	111.0	13.20	1.172	0.075	0.941	0.711	0.498	0.176	0.406
0.2	277.5	13.09	0.761	0.247	0.649	0.536	0.439	0.146	0.331
0.1	555.1	12.80	0.556	0.272	0.498	0.439	0.372	0.134	0.289
0.0925	600	12.79	0.540	0.272	0.481	0.427	0.368	0.134	0.285
0.0793	700	12.70	0.502	0.272	0.452	0.402	0.351	0.134	0.285
0.0694	800	12.61	0.473	0.268	0.427	0.385	0.339	0.130	0.280
0.0617	900	12.50	0.448	0.264	0.406	0.368	0.326	0.126	0.276
0.05551	1000	12.42	0.427	0.259	0.385	0.351	0.318	0.121	0.272
0.05	1110	12.24	0.406	0.259	0.372	0.339	0.305	0.121	0.272
0.02775	2000	11.29	0.310	0.226	0.285	0.264	0.247	0.117	0.264
0.01850	3000	10.66	0.251	0.197	0.234	0.218	0.213	0.117	0.259
0.01388	4000	10.25	0.226	0.180	0.205	0.192	0.192	0.113	0.259
0.01110	5000	9.874	0.197	0.167	0.184	0.172	0.176	0.109	0.255
0.00555	10000	8.912	0.142	0.126	0.130	0.121	0.130	0.105	0.243
0.00278	20000	7.531	0.105	0.092	0.092	0.084	0.096	0.096	0.230
0.00111	50000	5.439	0.067	0.059	0.054	0.050	0.063	0.084	0.222
0.000555	100000	3.766	0.042	0.042	0.038	0.038	0.046	0.054	0.209
0.000111	500000	1.255	0.021	0.021	0.021	0.021	0.021	0.038	0.167
0	∞	0	0	0	0	0	0	0	0

ENTHALPY OF SOLUTION OF ELECTROLYTES

This table gives the molar enthalpy (heat) of solution at infinite dilution for some common uni-univalent electrolytes. This is the enthalpy change when 1 mol of solute in its standard state is dissolved in an infinite amount of water. Values are given in kilojoules per mole at 25°C.

REFERENCE

Parker, V. B., *Thermal Properties of Uni-Univalent Electrolytes*, Natl. Stand. Ref. Data Series — Natl. Bur. Stand.(U.S.), No.2, 1965.

Solute	State	$\Delta_{sol} H°$ kJ/mol	Solute	State	$\Delta_{sol} H°$ kJ/mol	Solute	State	$\Delta_{sol} H°$ kJ/mol
HF	g	−61.50	$LiBr \cdot 2H_2O$	c	−9.41	KCl	c	17.22
HCl	g	−74.84	$LiBrO_3$	c	1.42	$KClO_3$	c	41.38
$HClO_4$	l	−88.76	LiI	c	−63.30	$KClO_4$	c	51.04
$HClO_4 \cdot H_2O$	c	−32.95	$LiI \cdot H_2O$	c	−29.66	KBr	c	19.87
HBr	g	−85.14	$LiI \cdot 2H_2O$	c	−14.77	$KBrO_3$	c	41.13
HI	g	−81.67	$LiI \cdot 3H_2O$	c	0.59	KI	c	20.33
HIO_3	c	8.79	$LiNO_2$	c	−11.00	KIO_3	c	27.74
HNO_3	l	−33.28	$LiNO_2 \cdot H_2O$	c	7.03	KNO_2	c	13.35
HCOOH	l	−0.86	$LiNO_3$	c	−2.51	KNO_3	c	34.89
CH_3COOH	l	−1.51				$KC_2H_3O_2$	c	−15.33
			NaOH	c	−44.51	KCN	c	11.72
NH_3	g	−30.50	$NaOH \cdot H_2O$	c	−21.41	KCNO	c	20.25
NH_4Cl	c	14.78	NaF	c	0.91	KCNS	c	24.23
NH_4ClO_4	c	33.47	NaCl	c	3.88	$KMnO_4$	c	43.56
NH_4Br	c	16.78	$NaClO_2$	c	0.33			
NH_4I	c	13.72	$NaClO_2 \cdot 3H_2O$	c	28.58	RbOH	c	−62.34
NH_4IO_3	c	31.80	$NaClO_3$	c	21.72	$RbOH \cdot H_2O$	c	−17.99
NH_4NO_2	c	19.25	$NaClO_4$	c	13.88	$RbOH \cdot 2H_2O$	c	0.88
NH_4NO_3	c	25.69	$NaClO_4 \cdot H_2O$	c	22.51	RbF	c	−26.11
$NH_4C_2H_3O_2$	c	−2.38	NaBr	c	−0.60	$RbF \cdot H_2O$	c	−0.42
NH_4CN	c	17.57	$NaBr \cdot 2H_2O$	c	18.64	$RbF \cdot 1.5H_2O$	c	1.34
NH_4CNS	c	22.59	$NaBrO_3$	c	26.90	RbCl	c	17.28
CH_3NH_3Cl	c	5.77	NaI	c	−7.53	$RbClO_3$	c	47.74
$(CH_3)_3NHCl$	c	1.46	$NaI \cdot 2H_2O$	c	16.13	$RbClO_4$	c	56.74
$N(CH_3)_4Cl$	c	4.08	$NaIO_3$	c	20.29	RbBr	c	21.88
$N(CH_3)_4Br$	c	24.27	$NaNO_2$	c	13.89	$RbBrO_3$	c	48.95
$N(CH_3)_4I$	c	42.07	$NaNO_3$	c	20.50	RbI	c	25.10
			$NaC_2H_3O_2$	c	−17.32	$RbNO_3$	c	36.48
$AgClO_4$	c	7.36	$NaC_2H_3O_2 \cdot 3H_2O$	c	19.66			
$AgNO_2$	c	36.94	NaCN	c	1.21	CsOH	c	−71.55
$AgNO_3$	c	22.59	$NaCN \cdot 0.5H_2O$	c	3.31	$CsOH \cdot H_2O$	c	−20.50
			$NaCN \cdot 2H_2O$	c	18.58	CsF	c	−36.86
LiOH	c	−23.56	NaCNO	c	19.20	$CsF \cdot H_2O$	c	−10.46
$LiOH \cdot H_2O$	c	−6.69	NaCNS	c	6.83	$CsF \cdot 1.5H_2O$	c	−5.44
LiF	c	4.73				CsCl	c	17.78
LiCl	c	−37.03	KOH	c	−57.61	$CsClO_4$	c	55.44
$LiCl \cdot H_2O$	c	−19.08	$KOH \cdot H_2O$	c	−14.64	CsBr	c	25.98
$LiClO_4$	c	−26.55	$KOH \cdot 1.5H_2O$	c	−10.46	$CsBrO_3$	c	50.46
$LiClO_4 \cdot 3H_2O$	c	32.61	KF	c	−17.73	CsI	c	33.35
LiBr	c	−48.83	$KF \cdot 2H_2O$	c	6.97	$CsNO_3$	c	40.00
$LiBr \cdot H_2O$	c	−23.26						

CHEMICAL KINETIC DATA FOR STRATOSPHERIC MODELING

The present compilation of kinetic data represents the 12th evaluation prepared by the NASA Panel for Data Evaluation. The Panel was established in 1977 by the NASA Upper Atmosphere Research Program Office for the purpose of providing a critical tabulation of the latest kinetic and photochemical data for use by modelers in computer simulations of stratospheric chemistry. The recommended rate data and cross sections are based on laboratory measurements. The major use of theoretical extrapolation of data is in connection with three-body reactions, in which the required pressure or temperature dependence is sometimes unavailable from laboratory measurements, and can be estimated by use of appropriate theoretical treatment. In the case of important rate constants for which no experimental data are available, the panel may provide estimates of rate constant parameters based on analogy to similar reactions for which data are available.

Rate constants are expressed in the form $k(T) = A \exp(-E/RT)$, where A is the pre-exponential factor, E the activation energy, R the gas constant, and T the absolute temperature. Uncertainties are expressed by the factor f, e.g., a value of 4.2×10^{-10} with $f = 2$ indicates that the true value is believed to lie between 2.1×10^{-10} and 8.4×10^{-10}. The value of f at other temperatures may be calculated from $f(298)$, given in the last column, by:

$$f(T) = f(298) \exp[(\Delta E/R)(1/T - 1/298)] ,$$

where $\Delta E/R$ is the uncertainty in E/R.

Table 1 covers rate constant data on second order reactions, grouped by class, while Table 2 covers association reactions. Relevant equilibrium constant data are given in Table 3. All concentrations are measured in molecules cm^{-3}. Notes on each reaction, as well as related photochemical data, may be found in the reference.

The assistance of Robert Hampson is gratefully acknowledged.

REFERENCE

DeMore, W. B., Sander, S. P., Golden, D. M., Hampson, R. F., Kurylo, M. J., Howard, C. J., Ravishankara, A. R., Kolb, C. E., and Molina, M. J., *Chemical Kinetics and Photochemical Data for use in Atmospheric Modeling. Evaluation Number 12*, Jet Propulsion Laboratory Publication 97-4, Pasadena CA, 1997.

The report is also available at the World Wide Web site < http://remus.jpl.nasa.gov/pub/jpl97>.

Table 1. Rate Constants for Second Order Reactions

Reaction	A cm^3 molecule^{-1} s^{-1}	E/R K	k (298 K) cm^3 molecule^{-1} s^{-1}	$f(298)$
O_x Reactions				
$O + O_3 \rightarrow O_2 + O_2$	8.0×10^{-12}	2060 ± 250	8.0×10^{-15}	1.15
$O(^1D)$ Reactions				
$O(^1D) + O_2 \rightarrow O + O_2$	3.2×10^{-11}	$-(70 \pm 100)$	4.0×10^{-11}	1.2
$O(^1D) + O_3 \rightarrow O_2 + O_2$	1.2×10^{-10}	0 ± 100	1.2×10^{-10}	1.3
$\rightarrow O_2 + O + O$	1.2×10^{-10}	0 ± 100	1.2×10^{-10}	1.3
$O(^1D) + H_2 \rightarrow OH + H$	1.1×10^{-10}	0 ± 100	1.1×10^{-10}	1.1
$O(^1D) + H_2O \rightarrow OH + OH$	2.2×10^{-10}	0 ± 100	2.2×10^{-10}	1.2
$O(^1D) + N_2 \rightarrow O + N_2$	1.8×10^{-11}	$-(110 \pm 100)$	2.6×10^{-11}	1.2
$O(^1D) + N_2O \rightarrow N_2 + O_2$	4.9×10^{-11}	0 ± 100	4.9×10^{-11}	1.3
$\rightarrow NO + NO$	6.7×10^{-11}	0 ± 100	6.7×10^{-11}	1.3
$O(^1D) + NH_3 \rightarrow OH + NH_2$	2.5×10^{-10}	0 ± 100	2.5×10^{-10}	1.3
$O(^1D) + CO_2 \rightarrow O + CO_2$	7.4×10^{-11}	$-(120 \pm 100)$	1.1×10^{-10}	1.2
$O(^1D) + CH_4 \rightarrow$ products	1.5×10^{-10}	0 ± 100	1.5×10^{-10}	1.2
$O(^1D) + HCl \rightarrow$ products	1.5×10^{-10}	0 ± 100	1.5×10^{-10}	1.2
$O(^1D) + HF \rightarrow OH + F$	1.4×10^{-10}	0 ± 100	1.4×10^{-10}	2.0
$O(^1D) + HBr \rightarrow$ products	1.5×10^{-10}	0 ± 100	1.5×10^{-10}	2.0
$O(^1D) + Cl_2 \rightarrow$ products	2.8×10^{-10}	0 ± 100	2.8×10^{-10}	2.0
$O(^1D) + CCl_2O \rightarrow$ products	3.6×10^{-10}	0 ± 100	3.6×10^{-10}	2.0
$O(^1D) + CClFO \rightarrow$ products	1.9×10^{-10}	0 ± 100	1.9×10^{-10}	2.0
$O(^1D) + CF_2O \rightarrow$ products	7.4×10^{-11}	0 ± 100	7.4×10^{-11}	2.0
$O(^1D) + CCl_4 \rightarrow$ products (CFC-10)	3.3×10^{-10}	0 ± 100	3.3×10^{-10}	1.2
$O(^1D) + CH_3Br \rightarrow$ products	1.8×10^{-10}	0 ± 100	1.8×10^{-10}	1.3
$O(^1D) + CH_2Br_2 \rightarrow$ products	2.7×10^{-10}	0 ± 100	2.7×10^{-10}	1.3
$O(^1D) + CHBr_3 \rightarrow$ products	6.6×10^{-10}	0 ± 100	6.6×10^{-10}	1.5
$O(^1D) + CH_3F \rightarrow$ products (HFC-41)	1.5×10^{-10}	0 ± 100	1.5×10^{-10}	1.2

Table 1. Rate Constants for Second Order Reactions (continued)

Reaction	A cm^3 molecule^{-1} s^{-1}	E/R K	k (298 K) cm^3 molecule^{-1} s^{-1}	f(298)
O(^{1}D) + CH$_2$F$_2$ → products (HFC-32)	5.1x10^{-11}	0±100	5.1x10^{-11}	1.3
O(^{1}D) + CHF$_3$ → products (HFC-23)	9.1x10^{-12}	0±100	9.1x10^{-12}	1.2
O(^{1}D) + CHCl$_2$F → products (HCFC-21)	1.9x10^{-10}	0±100	1.9x10^{-10}	1.3
O(^{1}D) + CHClF$_2$ → products (HCFC-22)	1.0x10^{-10}	0±100	1.0x10^{-10}	1.2
O(^{1}D) + CCl$_3$F → products (CFC-11)	2.3x10^{-10}	0±100	2.3x10^{-10}	1.2
O(^{1}D) + CCl$_2$F$_2$ → products (CFC-12)	1.4x10^{-10}	0±100	1.4x10^{-10}	1.3
O(^{1}D) + CClF$_3$ → products (CFC-13)	8.7x10^{-11}	0±100	8.7x10^{-11}	1.3
O(^{1}D) + CClBrF$_2$ → products (Halon-1211)	1.5x10^{-10}	0±100	1.5x10^{-10}	1.3
O(^{1}D) + CBr$_2$F$_2$ → products (Halon-1202)	2.2x10^{-10}	0±100	2.2x10^{-10}	1.3
O(^{1}D) + CBrF$_3$ → products (Halon-1301)	1.0x10^{-10}	0±100	1.0x10^{-10}	1.3
O(^{1}D) + CF$_4$ → CF$_4$ + O (CFC-14)	-	-	2.0x10^{-14}	1.5
O(^{1}D) + CH$_3$CH$_2$F → products (HFC-161)	2.6x10^{-10}	0±100	2.6x10^{-10}	1.3
O(^{1}D) + CH$_3$CHF$_2$ → products (HFC-152a)	2.0x10^{-10}	0±100	2.0x10^{-10}	1.3
O(^{1}D) + CH$_3$CCl$_2$F → products (HCFC-141b)	2.6x10^{-10}	0±100	2.6x10^{-10}	1.3
O(^{1}D) + CH$_3$CClF$_2$ → products (HCFC-142b)	2.2x10^{-10}	0±100	2.2x10^{-10}	1.3
O(^{1}D) + CH$_3$CF$_3$ → products (HFC-143a)	1.0x10^{-10}	0±100	1.0x10^{-10}	3.0
O(^{1}D) + CH$_2$ClCClF$_2$ → products (HCFC-132b)	1.6x10^{-10}	0±100	1.6x10^{-10}	2.0
O(^{1}D) + CH$_2$ClCF$_3$ → products (HCFC-133a)	1.2x10^{-10}	0±100	1.2x10^{-10}	1.3
O(^{1}D) + CH$_2$FCF$_3$ → products (HFC-134a)	4.9x10^{-11}	0±100	4.9x10^{-11}	1.3
O(^{1}D) + CHCl$_2$CF$_3$ → products (HCFC-123)	2.0x10^{-10}	0±100	2.0x10^{-10}	1.3
O(^{1}D) + CHClFCF$_3$ → products (HCFC-124)	8.6x10^{-11}	0±100	8.6x10^{-11}	1.3
O(^{1}D) + CHF$_2$CF$_3$ → products (HFC-125)	1.2x10^{-10}	0±100	1.2x10^{-10}	2.0
O(^{1}D) + CCl$_3$CF$_3$ → products (CFC-113a)	2x10^{-10}	0±100	2x10^{-10}	2.0
O(^{1}D) + CCl$_2$FCClF$_2$ → products (CFC-113)	2x10^{-10}	0±100	2x10^{-10}	2.0
O(^{1}D) + CCl$_2$FCF$_3$ → products (CFC-114a)	1x10^{-10}	0±100	1x10^{-10}	2.0
O(^{1}D) + CClF$_2$CClF$_2$ → products (CFC-114)	1.3x10^{-10}	0±100	1.3x10^{-10}	1.3
O(^{1}D) + CClF$_2$CF$_3$ → products (CFC-115)	5x10^{-11}	0±100	5x10^{-11}	1.3
O(^{1}D) + CBrF$_2$CBrF$_2$ → products (Halon-2402)	1.6x10^{-10}	0±100	1.6x10^{-10}	1.3

Table 1. Rate Constants for Second Order Reactions (continued)

Reaction	A cm^3 molecule^{-1} s^{-1}	E/R K	k (298 K) cm^3 molecule^{-1} s^{-1}	f(298)
O(^{1}D) + CF$_3$CF$_3$ → O + CF$_3$CF$_3$ (CFC-116)	-	-	1.5×10^{-13}	1.5
O(^{1}D) + CHF$_2$CF$_2$CF$_2$CHF$_2$ → products (HFC-338pcc)	1.8×10^{-11}	0 ± 100	1.8×10^{-11}	1.5
O(^{1}D) + c-C$_4$F$_8$ → products	-	-	8×10^{-13}	1.3
O(^{1}D) + CF$_3$CHFCHFCF$_2$CF$_3$ → products (HFC-43-10mee)	2.1×10^{-10}	0 ± 100	2.1×10^{-10}	4
O(^{1}D) + C$_5$F$_{12}$ → products (CFC-41-12)	-	-	3.9×10^{-13}	2
O(^{1}D) + C$_6$F$_{14}$ → products (CFC-51-14)	-	-	1×10^{-12}	2
O(^{1}D) + 1,2-(CF$_3$)$_2$c-C$_4$F$_6$ → products	-	-	2.8×10^{-13}	2
O(^{1}D) + SF$_6$ → products	-	-	1.8×10^{-14}	1.5

Singlet O$_2$ Reactions

Reaction	A cm^3 molecule^{-1} s^{-1}	E/R K	k (298 K) cm^3 molecule^{-1} s^{-1}	f(298)
O$_2$($^1\Delta$) + O → products	-	-	$<2 \times 10^{-16}$	-
O$_2$($^1\Delta$) + O$_2$ → products	3.6×10^{-18}	220 ± 100	1.7×10^{-18}	1.2
O$_2$($^1\Delta$) + O$_3$ → O + 2O$_2$	5.2×10^{-11}	2840 ± 500	3.8×10^{-15}	1.2
O$_2$($^1\Delta$) + H$_2$O → products	-	-	4.8×10^{-18}	1.5
O$_2$($^1\Delta$) + N → NO + O	-	-	$<9 \times 10^{-17}$	-
O$_2$($^1\Delta$) + N$_2$ → products	-	-	$<10^{-20}$	-
O$_2$($^1\Delta$) + CO$_2$ → products	-	-	$<2 \times 10^{-20}$	-
O$_2$($^1\Sigma$) + O → products	-	-	8×10^{-14}	5.0
O$_2$($^1\Sigma$) + O$_2$ → products	-	-	3.9×10^{-17}	1.5
O$_2$($^1\Sigma$) + O$_3$ → products	2.2×10^{-11}	0 ± 200	2.2×10^{-11}	1.2
O$_2$($^1\Sigma$) + H$_2$O → products	-	-	5.4×10^{-12}	1.3
O$_2$($^1\Sigma$) + N → products	-	-	$<10^{-13}$	-
O$_2$($^1\Sigma$) + N$_2$ → products	2.1×10^{-15}	0 ± 200	2.1×10^{-15}	1.2
O$_2$($^1\Sigma$) + CO$_2$ → products	4.2×10^{-13}	0 ± 200	4.2×10^{-13}	1.2

HO$_x$ Reactions

Reaction	A cm^3 molecule^{-1} s^{-1}	E/R K	k (298 K) cm^3 molecule^{-1} s^{-1}	f(298)
O + OH → O$_2$ + H	2.2×10^{-11}	$-(120 \pm 100)$	3.3×10^{-11}	1.2
O + HO$_2$ → OH + O$_2$	3.0×10^{-11}	$-(200 \pm 100)$	5.9×10^{-11}	1.2
O + H$_2$O$_2$ → OH + HO$_2$	1.4×10^{-12}	2000 ± 1000	1.7×10^{-15}	2.0
H + O$_3$ → OH + O$_2$	1.4×10^{-10}	470 ± 200	2.9×10^{-11}	1.25
H + HO$_2$ → products	8.1×10^{-11}	0 ± 100	8.1×10^{-11}	1.3
OH + O$_3$ → HO$_2$ + O$_2$	1.6×10^{-12}	940 ± 300	6.8×10^{-14}	1.3
OH + H$_2$ → H$_2$O + H	5.5×10^{-12}	2000 ± 100	6.7×10^{-15}	1.1
OH + HD → products	5.0×10^{-12}	2130 ± 200	4.0×10^{-15}	1.2
OH + OH → H$_2$O + O	4.2×10^{-12}	240 ± 240	1.9×10^{-12}	1.4
OH + HO$_2$ → H$_2$O + O$_2$	4.8×10^{-11}	$-(250 \pm 200)$	1.1×10^{-10}	1.3
OH + H$_2$O$_2$ → H$_2$O + HO$_2$	2.9×10^{-12}	160 ± 100	1.7×10^{-12}	1.2
HO$_2$ + O$_3$ → OH + 2O$_2$	1.1×10^{-14}	$500 \pm$	2.0×10^{-15}	1.3
HO$_2$ + HO$_2$ → H$_2$O$_2$ + O$_2$	2.3×10^{-13}	$-(600 \pm 200)$	1.7×10^{-12}	1.3
$\quad$ H$_2$O$_2$ + O$_2$	1.7×10^{-33}[M]	$-(1000 \pm 400)$	4.9×10^{-32}[M]	1.3

NO$_x$ Reactions

Reaction	A cm^3 molecule^{-1} s^{-1}	E/R K	k (298 K) cm^3 molecule^{-1} s^{-1}	f(298)
O + NO$_2$ → NO + O$_2$	6.5×10^{-12}	$-(120 \pm 120)$	9.7×10^{-12}	1.1
O + NO$_3$ → O$_2$ + NO$_2$	1.0×10^{-11}	0 ± 150	1.0×10^{-11}	1.5
O + N$_2$O$_5$ → products			$<3.0 \times 10^{-16}$	
O + HNO$_3$ → OH + NO$_3$			$<3.0 \times 10^{-17}$	
O + HO$_2$NO$_2$ → products	7.8×10^{-11}	3400 ± 750	8.6×10^{-16}	3.0

CHEMICAL KINETIC DATA FOR STRATOSPHERIC MODELING (continued)

Table 1. Rate Constants for Second Order Reactions (continued)

Reaction	A cm³ molecule⁻¹ s⁻¹	E/R K	k (298 K) cm³ molecule⁻¹ s⁻¹	f(298)
$H + NO_2 \rightarrow OH + NO$	4.0×10^{-10}	340 ± 300	1.3×10^{-10}	1.3
$OH + NO_3 \rightarrow$ products			2.2×10^{-11}	1.5
$OH + HONO \rightarrow H_2O + NO_2$	1.8×10^{-11}	$390 \pm$	4.5×10^{-12}	1.5
$OH + HNO_3 \rightarrow H_2O + NO_3$	See reference	1.3		1.5
$OH + HO_2NO_2 \rightarrow$ products	1.3×10^{-12}	$-(380\pm)$	4.6×10^{-12}	
$OH + NH_3 \rightarrow H_2O + NH_2$	1.7×10^{-12}	710 ± 200	1.6×10^{-13}	1.5
$HO_2 + NO \rightarrow NO_2 + OH$	3.5×10^{-12}	$-(250\pm50)$	8.1×10^{-12}	1.2
$HO_2 + NO_2 \rightarrow HONO + O_2$	See reference			1.15
$HO_2 + NO_3 \rightarrow$ products			3.5×10^{-12}	
$HO_2 + NH_2 \rightarrow$ products			3.4×10^{-11}	1.5
$N + O_2 \rightarrow NO + O$	1.5×10^{-11}	3600 ± 400	8.5×10^{-17}	2.0
$N + O_3 \rightarrow NO + O_2$			$<2.0 \times 10^{-16}$	1.25
$N + NO \rightarrow N_2 + O$	2.1×10^{-11}	$-(100\pm100)$	3.0×10^{-11}	
$N + NO_2 \rightarrow N_2O + O$	5.8×10^{-12}	$-(220\pm100)$	1.2×10^{-11}	1.3
$NO + O_3 \rightarrow NO_2 + O_2$	2.0×10^{-12}	1400 ± 200	1.8×10^{-14}	1.5
$NO + NO_3 \rightarrow 2NO_2$	1.5×10^{-11}	$-(170\pm100)$	2.6×10^{-11}	1.1
$NO_2 + O_3 \rightarrow NO_3 + O_2$	1.2×10^{-13}	2450 ± 150	3.2×10^{-17}	1.3
$NO_2 + NO_3 \rightarrow NO + NO_2 + O_2$	See reference			1.15
$NO_3 + NO_3 \rightarrow 2NO_2 + O_2$	8.5×10^{-13}	2450 ± 500	2.3×10^{-16}	
$NH_2 + O_2 \rightarrow$ products			$<6.0 \times 10^{-21}$	1.5
$NH_2 + O_3 \rightarrow$ products	4.3×10^{-12}	930 ± 500	1.9×10^{-13}	
$NH_2 + NO \rightarrow$ products	4.0×10^{-12}	$-(450\pm150)$	1.8×10^{-11}	3.0
$NH_2 + NO_2 \rightarrow$ products	2.1×10^{-12}	$-(650\pm250)$	1.9×10^{-11}	1.3
$NH + NO \rightarrow$ products	4.9×10^{-11}	0 ± 300	4.9×10^{-11}	3.0
$NH + NO_2 \rightarrow$ products	3.5×10^{-13}	$-(1140\pm500)$	1.6×10^{-11}	1.5
$O_3 + HNO_2 \rightarrow O_2 + HNO_3$			$<5.0 \times 10^{-19}$	2.0
$N_2O_5 + H_2O \rightarrow 2HNO_3$			$<2.0 \times 10^{-21}$	
$N_2(A,v) + O_2 \rightarrow$ products			2.5×10^{-12}, $v=0$	1.5
$N_2(A,v) + O_3 \rightarrow$ products			4.1×10^{-11}, $v=0$	2.0

Reactions of Organic Compounds

Reaction	A cm³ molecule⁻¹ s⁻¹	E/R K	k (298 K) cm³ molecule⁻¹ s⁻¹	f(298)
$O + CH_3 \rightarrow$ products	1.1×10^{-10}	0 ± 250	1.1×10^{-10}	1.3
$O + HCN \rightarrow$ products	1.0×10^{-11}	4000 ± 1000	1.5×10^{-17}	10
$O + C_2H_2 \rightarrow$ products	3.0×10^{-11}	1600 ± 250	1.4×10^{-13}	1.3
$O + H_2CO \rightarrow$ products	3.4×10^{-11}	1600 ± 250	1.6×10^{-13}	1.25
$O + CH_3CHO \rightarrow CH_3CO + OH$	1.8×10^{-11}	1100 ± 200	4.5×10^{-13}	1.25
$O_3 + C_2H_2 \rightarrow$ products	1.0×10^{-14}	4100 ± 500	1.0×10^{-20}	3
$O_3 + C_2H_4 \rightarrow$ products	1.2×10^{-14}	2630 ± 100	1.7×10^{-18}	1.25
$O_3 + C_3H_6 \rightarrow$ products	6.5×10^{-15}	1900 ± 200	1.1×10^{-17}	1.2
$OH + CO \rightarrow$ Products	$1.5 \times 10^{-13} \times (1+0.6P_{atm})$	0 ± 300	$1.5 \times 10^{-13} \times (1+0.6P_{atm})$	1.3
$OH + CH_4 \rightarrow CH_3 + H_2O$	2.45×10^{-12}	1775 ± 100	6.3×10^{-15}	1.1
$OH + {}^{13}CH_4 \rightarrow {}^{13}CH_3 + H_2O$	See reference			
$OH + CH_3D \rightarrow$ products	3.5×10^{-12}	1950 ± 200	5.0×10^{-15}	1.15
$OH + H_2CO \rightarrow H_2O + HCO$	1.0×10^{-11}	0 ± 200	1.0×10^{-11}	1.25
$OH + CH_3OH \rightarrow$ products	6.7×10^{-12}	600 ± 300	8.9×10^{-13}	1.2
$OH + CH_3OOH \rightarrow$ Products	3.8×10^{-12}	$-(200\pm200)$	7.4×10^{-12}	1.5
$OH + HC(O)OH \rightarrow$ products	4.0×10^{-13}	0 ± 200	4.0×10^{-13}	1.3
$OH + HCN \rightarrow$ products	1.2×10^{-13}	400 ± 150	3.1×10^{-14}	3
$OH + C_2H_6 \rightarrow H_2O + C_2H_5$	8.7×10^{-12}	1070 ± 100	2.4×10^{-13}	1.1
$OH + C_3H_8 \rightarrow H_2O + C_3H_7$	1.0×10^{-11}	660 ± 100	1.1×10^{-12}	1.2
$OH + CH_3CHO \rightarrow CH_3CO + H_2O$	5.6×10^{-12}	$-(270\pm200)$	1.4×10^{-11}	1.2
$OH + C_2H_5OH \rightarrow$ products	7.0×10^{-12}	235 ± 100	3.2×10^{-12}	1.3
$OH + CH_3C(O)OH \rightarrow$ products	4.0×10^{-13}	$-(200\pm400)$	8.0×10^{-13}	1.3

Table 1. Rate Constants for Second Order Reactions (continued)

Reaction	A cm³ molecule⁻¹ s⁻¹	E/R K	k (298 K) cm³ molecule⁻¹ s⁻¹	f(298)
OH + CH₃C(O)CH₃ → CH₃C(O)CH₂ + H₂O	2.2 x 10⁻¹²	685±100	2.2x10⁻¹³	1.15
OH + CH₃CN → products	7.8x10⁻¹³	1050±200	2.3x10⁻¹⁴	1.5
OH+ CH₃ONO₂ → products	5.0x10⁻¹³	890±500	2.4x10⁻¹⁴	3
OH + CH₃C(O)O₂NO₂ (PAN)→ products			<4 x 10⁻¹⁴	
OH+ C₂H₅ONO₂ → products	8.2x10⁻¹³	450±300	1.8x10⁻¹³	3
HO₂ + CH₂O → adduct	6.7x10⁻¹⁵	-(600±600)	5.0x10⁻¹⁴	5
HO₂ + CH₃O₂ → CH₃OOH + O₂	3.8x10⁻¹³	-(800±400)	5.6x10⁻¹²	2
HO₂ + C₂H₅O₂ → C₂H₅OOH + O₂	7.5x10⁻¹³	-(700±250)	8.0x10⁻¹²	1.5
HO₂ + CH₃C(O)O₂ → products	4.5x10⁻¹³	-(1000±600)	1.3x10⁻¹¹	2
NO₃ + CO → products			<4.0x10⁻¹⁹	
NO₃ + CH₂O → products			5.8x10⁻¹⁶	1.3
NO₃ + CH₃CHO → products	1.4x10⁻¹²	1900±300	2.4x10⁻¹⁵	1.3
CH₃ + O₂ → products			<3.0x10⁻¹⁶	
CH₃ + O₃ → products	5.4x10⁻¹²	220±150	2.6x10⁻¹²	2
HCO + O₂ → CO + HO₂	3.5x10⁻¹²	-(140±140)	5.5x10⁻¹²	1.3
CH₂OH + O₂ → CH₂O + HO₂	9.1x10⁻¹²	0±200	9.1x10⁻¹²	1.3
CH₃O + O₂ → CH₂O + HO₂	3.9x10⁻¹⁴	900±300	1.9x10⁻¹⁵	1.5
CH₃O + NO → CH₂O + HNO	See reference			
CH₃O+ NO₂ → CH₂O + HONO	1.1 x 10⁻¹¹	1200±600	2.0 x 10⁻¹³	5
CH₃O₂ + O₃ → products			<3.0x10⁻¹⁷	
CH₃O₂ + CH₃O₂ → products	2.5x10⁻¹³	-(190±190)	4.7x10⁻¹³	1.5
CH₃O₂ + NO → CH₃O + NO₂	3.0x10⁻¹²	-(280±60)	7.7x10⁻¹²	1.15
CH₃O₂ + CH₃C(O)O₂ → products	1.3x10⁻¹²	-(640±200)	1.1x10⁻¹¹	1.5
C₂H₅ + O₂ → C₂H₄ + HO₂			<2.0x10⁻¹⁴	
C₂H₅O + O₂ → CH₃CHO + HO₂	6.3 x 10⁻¹⁴	550±200	1.0x10⁻¹⁴	1.5
C₂H₅O₂ + C₂H₅O₂ → products	6.8x10⁻¹⁴	0±300	6.8x10⁻¹⁴	2
C₂H₅O₂ + NO → products	2.6x10⁻¹²	-(365±150)	8.7x10⁻¹²	1.2
CH₃C(O)O₂ + CH₃C(O)O₂ → products	2.9x10⁻¹²	-(500±150)	1.5x10⁻¹¹	1.5
CH₃C(O)O₂ + NO → products	5.3x10⁻¹²	-(360±150)	1.8x10⁻¹¹	1.4

FO$_x$ Reactions

Reaction	A cm³ molecule⁻¹ s⁻¹	E/R K	k (298 K) cm³ molecule⁻¹ s⁻¹	f(298)
O + FO → F + O₂	2.7x10⁻¹¹	0±250	2.7x10⁻¹¹	3.0
O + FO₂ → FO + O₂	5.0x10⁻¹¹	0±250	5.0x10⁻¹¹	5.0
OH + CH₃F → CH₂F + H₂O (HFC-41)	3.0x10⁻¹²	1500±300	2.0x10⁻¹⁴	1.1
OH + CH₂F₂ → CHF₂ + H₂O (HFC-32)	1.9x10⁻¹²	1550±200	1.0x10⁻¹⁴	1.2
OH + CHF₃ → CF₃ + H₂O (HFC-23)	1.0x10⁻¹²	2440±200	2.8x10⁻¹⁶	1.3
OH + CF₃OH → CF₃O + H₂O			<2x10⁻¹⁷	
OH + CH₃CH₂F → products (HFC-161)	7.0x10⁻¹²	1100±300	1.7x10⁻¹³	1.4
OH + CH₃CHF₂ → products (HFC-152a)	2.4x10⁻¹²	1260±200	3.5x10⁻¹⁴	1.2
OH + CH₂FCH₂F → CHFCH₂F + H₂O (HFC-152)	1.7x10⁻¹¹	1500±500	1.1x10⁻¹³	2.0
OH + CH₃CF₃ → CH₂CF₃ + H₂O (HFC-143a)	1.8x10⁻¹²	2170±150	1.2x10⁻¹⁵	1.1
OH + CH₂FCHF₂ → products (HFC-143)	4.0x10⁻¹²	1650±300	1.6x10⁻¹⁴	1.5
OH + CH₂FCF₃ → CHFCF₃ + H₂O (HFC-134a)	1.5x10⁻¹²	1750±200	4.2x10⁻¹⁵	1.1

Table 1. Rate Constants for Second Order Reactions (continued)

Reaction	A cm^3 molecule^{-1} s^{-1}	E/R K	k (298 K) cm^3 molecule^{-1} s^{-1}	$f(298)$
OH + CHF$_2$CHF$_2$ → CF$_2$CHF$_2$ (HFC-134) + H$_2$O	1.6x10^{-12}	1680±300	5.7x10^{-15}	2.0
OH + CHF$_2$CF$_3$ → CF$_2$CF$_3$ + H$_2$O (HFC-125)	5.6x10^{-13}	1700±300	1.9x10^{-15}	1.3
OH + CH$_3$OCHF$_2$ → products (HFOC-152a)	6.0x10^{-12}	1530±150	3.5x10^{-14}	1.2
OH + CF$_3$OCH$_3$ → CF$_3$OCH$_2$ + H$_2$O (HFOC-143a)	1.5x10^{-12}	1450±150	1.2x10^{-14}	1.1
OH + CF$_2$HOCF$_2$H → CF$_2$OCF$_2$H (HFOC-134) + H$_2$O	1.9x10^{-12}	2000±150	2.3x10^{-15}	1.2
OH + CF$_3$OCHF$_2$ → CF$_3$OCF$_2$ + H$_2$O (HFOC-125)	4.7x10^{-13}	2100±300	4.1x10^{-16}	1.2
OH + CF$_3$CH$_2$CH$_3$ → products (HFC-263fb)	-	-	4.2x10^{-14}	1.5
OH + CH$_2$FCF$_2$CHF$_2$ → products (HFC-245ca)	2.4x10^{-12}	1660±150	9.1x10^{-15}	1.3
OH + CHF$_2$CHFCHF$_2$ → products (HFC-245ea)	-	-	1.6x10^{-14}	2.0
OH + CF$_3$CHFCH$_2$F → products (HFC-245eb)	-	-	1.5x10^{-14}	2.0
OH + CHF$_2$CH$_2$CF$_3$ → products (HFC-245fa)	6.1x10^{-13}	1330±150	7.0x10^{-15}	1.2
OH + CF$_3$CF$_2$CH$_2$F → CF$_3$CF$_2$CHF (HFC-236cb) +H$_2$O	1.5x10^{-12}	1750±500	4.2x10^{-15}	2.0
OH + CF$_3$CHFCHF$_2$ → products (HFC-236ea)	1.1x10^{-12}	1590±150	5.3x10^{-15}	1.1
OH + CF$_3$CH$_2$CF$_3$ → CF$_3$CHCF$_3$ (HFC-236fa) +H$_2$O	1.3x10^{-12}	2480±150	3.2x10^{-16}	1.1
OH + CF$_3$CHFCF$_3$ → CF$_3$CFCF$_3$+H$_2$O (HFC-227ea)	5.0x10^{-13}	1700±300	1.7x10^{-15}	1.1
OH + CHF$_2$OCH$_2$CF$_3$ → products (HFOC-245fa)	2.6x10^{-12}	1610±150	1.2x10^{-14}	2.0
OH + CF$_3$CH$_2$CF$_2$CH$_3$ → products (HFC-365mfc)	2.0x10^{-12}	1750±200	5.7x10^{-15}	1.3
OH + CF$_3$CH$_2$CH$_2$CF$_3$ → products (HFC-356mff)	3.0x10^{-12}	1800±300	7.1x10^{-15}	1.3
OH + CF$_3$CF$_2$CH$_2$CH$_2$F → products (HFC-356mcf)	1.7x10^{-12}	1110±200	4.2x10^{-14}	2.0
OH + CHF$_2$CF$_2$CF$_2$CF$_2$H → products (HFC-338pcc)	7.8x10^{-13}	1530±200	4.6x10^{-15}	1.5
OH + CF$_3$CH$_2$CF$_2$CH$_2$CF$_3$ → products (HFC-458mfcf)	1.2x10^{-12}	1830±200	2.6x10^{-15}	2.0
OH + CF$_3$CHFCHFCF$_2$CF$_3$ → products (HFC-43-10mee)	5.2x10^{-13}	1500±300	3.4x10^{-15}	1.3
OH + CF$_3$CF$_2$CH$_2$CH$_2$CF$_2$CF$_3$ → (HFC-55-10-mcff) products	-	-	8.3x10^{-15}	1.5
F + O$_3$ → FO + O$_2$	2.2x10^{-11}	230±200	1.0x10^{-11}	1.5
F + H$_2$ → HF + H	1.4x10^{-10}	500±200	2.6x10^{-11}	1.2
F + H$_2$O → HF + OH	1.4x10^{-11}	0±200	1.4x10^{-11}	1.3
F + HNO$_3$ → HF + NO$_3$	6.0x10^{-12}	-(400±200)	2.3x10^{-11}	1.3
F + CH$_4$ → HF + CH$_3$	1.6x10^{-10}	260±200	6.7x10^{-11}	1.4
FO + O$_3$ → products			<1 x 10^{-14}	
FO + NO → NO$_2$ + F	8.2x10^{-12}	-(300±200)	2.2x10^{-11}	1.5
FO + FO → 2 F + O$_2$	1.0x10^{-11}	0±250	1.0x10^{-11}	1.5
FO$_2$ + O$_3$ → products			<3.4x10^{-16}	
FO$_2$ + NO → FNO + O$_2$	7.5x10^{-12}	690±400	7.5x10^{-13}	2.0

Table 1. Rate Constants for Second Order Reactions (continued)

Reaction	A cm^3 molecule^{-1} s^{-1}	E/R K	k (298 K) cm^3 molecule^{-1} s^{-1}	f(298)
$FO_2 + NO_2 \rightarrow$ products	3.8×10^{-11}	2040 ± 500	4.0×10^{-14}	2.0
$FO_2 + CO \rightarrow$ products			$<5.1 \times 10^{-16}$	
$FO_2 + CH_4 \rightarrow$ products			$<2 \times 10^{-16}$	
$CF_3O + O_2 \rightarrow FO_2 + CF_2O$	$<3 \times 10^{-11}$	5000	$<1.5 \times 10^{-18}$	
$CF_3O + O_3 \rightarrow CF_3O_2 + O_2$	2×10^{-12}	1400 ± 600	1.8×10^{-14}	1.3
$CF_3O + H_2O \rightarrow OH + CF_3OH$	3×10^{-12}	>3600	$<2 \times 10^{-17}$	
$CF_3O + NO \rightarrow CF_2O + FNO$	3.7×10^{-11}	$-(110 \pm 70)$	5.4×10^{-11}	1.2
$CF_3O + NO_2 \rightarrow$ products	See reference			
$CF_3O + CO \rightarrow$ products			$<2 \times 10^{-15}$	
$CF_3O + CH_4 \rightarrow CH_3 + CF_3OH$	2.6×10^{-12}	1420 ± 200	2.2×10^{-14}	1.1
$CF_3O + C_2H_6 \rightarrow C_2H_5 + CF_3OH$	4.9×10^{-12}	400 ± 100	1.3×10^{-12}	1.2
$CF_3O_2 + O_3 \rightarrow CF_3O + 2O_2$			$<3 \times 10^{-15}$	
$CF_3O_2 + CO \rightarrow CF_3O + CO_2$			$<5 \times 10^{-16}$	
$CF_3O_2 + NO \rightarrow CF_3O + NO_2$	5.4×10^{-12}	$-(320 \pm 150)$	1.6×10^{-11}	1.1

ClO$_x$ Reactions

Reaction	A cm^3 molecule^{-1} s^{-1}	E/R K	k (298 K) cm^3 molecule^{-1} s^{-1}	f(298)
$O + ClO \rightarrow Cl + O_2$	3.0×10^{-11}	$-(70 \pm 70)$	3.8×10^{-11}	1.2
$O + OClO \rightarrow ClO + O_2$	2.4×10^{-12}	960 ± 300	1.0×10^{-13}	2.0
$O + Cl_2O \rightarrow ClO + ClO$	2.7×10^{-11}	530 ± 150	4.5×10^{-12}	1.3
$O + HCl \rightarrow OH + Cl$	1.0×10^{-11}	3300 ± 350	1.5×10^{-16}	2.0
$O + HOCl \rightarrow OH + ClO$	1.7×10^{-13}	0 ± 300	1.7×10^{-13}	3.0
$O + ClONO_2 \rightarrow$ products	2.9×10^{-12}	800 ± 200	2.0×10^{-13}	1.5
$O_3 + OClO \rightarrow$ products	2.1×10^{-12}	4700 ± 1000	3.0×10^{-19}	2.5
$O_3 + Cl_2O_2 \rightarrow$ products	-	-	$<1.0 \times 10^{-19}$	-
$OH + Cl_2 \rightarrow HOCl + Cl$	1.4×10^{-12}	900 ± 400	6.7×10^{-14}	1.2
$OH + ClO \rightarrow$ products	1.1×10^{-11}	$-(120 \pm 150)$	1.7×10^{-11}	1.5
$OH + OClO \rightarrow HOCl + O_2$	4.5×10^{-13}	$-(800 \pm 200)$	6.8×10^{-12}	2.0
$OH + HCl \rightarrow H_2O + Cl$	2.6×10^{-12}	350 ± 100	8.0×10^{-13}	1.2
$OH + HOCl \rightarrow H_2O + ClO$	3.0×10^{-12}	500 ± 500	5.0×10^{-13}	3.0
$OH + ClNO_2 \rightarrow HOCl + NO_2$	2.4×10^{-12}	1250 ± 300	3.6×10^{-14}	2.0
$OH + ClONO_2 \rightarrow$ products	1.2×10^{-12}	330 ± 200	3.9×10^{-13}	1.5
$OH + CH_3Cl \rightarrow CH_2Cl + H_2O$	4.0×10^{-12}	1400 ± 250	3.6×10^{-14}	1.2
$OH + CH_2Cl_2 \rightarrow CHCl_2 + H_2O$	3.8×10^{-12}	1050 ± 150	1.1×10^{-13}	1.4
$OH + CHCl_3 \rightarrow CCl_3 + H_2O$	2.0×10^{-12}	900 ± 150	1.0×10^{-13}	1.2
$OH + CCl_4 \rightarrow$ products	$\sim 1.0 \times 10^{-12}$	>2300	$<5.0 \times 10^{-16}$	-
$OH + CFCl_3 \rightarrow$ products (CFC-11)	$\sim 1.0 \times 10^{-12}$	>3700	$<5.0 \times 10^{-18}$	-
$OH + CF_2Cl_2 \rightarrow$ products (CFC-12)	$\sim 1.0 \times 10^{-12}$	>3600	$<6.0 \times 10^{-18}$	-
$OH + CH_2ClF \rightarrow CHClF + H_2O$ (HCFC-31)	2.8×10^{-12}	1270 ± 200	3.9×10^{-14}	1.2
$OH + CHFCl_2 \rightarrow CFCl_2 + H_2O$ (HCFC-21)	1.7×10^{-12}	1250 ± 150	2.6×10^{-14}	1.2
$OH + CHF_2Cl \rightarrow CF_2Cl + H_2O$ (HCFC-22)	1.0×10^{-12}	1600 ± 150	4.7×10^{-15}	1.1
$OH + CH_3OCl \rightarrow$ products	2.4×10^{-12}	360 ± 200	7.2×10^{-13}	3.0
$OH + CH_3CCl_3 \rightarrow CH_2CCl_3 + H_2O$ (HCC-140)	1.8×10^{-12}	1550 ± 150	1.0×10^{-14}	1.1
$OH + C_2HCl_3 \rightarrow$ products	4.9×10^{-13}	$-(450 \pm 200)$	2.2×10^{-12}	1.25
$OH + C_2Cl_4 \rightarrow$ products	9.4×10^{-12}	1200 ± 200	1.7×10^{-13}	1.25
$OH + CCl_3CHO \rightarrow H_2O + CCl_3CO$	8.2×10^{-12}	600 ± 300	1.1×10^{-12}	1.5
$OH + CH_3CFCl_2 \rightarrow CH_2CFCl_2 + H_2O$ (HCFC-141b)	1.7×10^{-12}	1700 ± 150	5.7×10^{-15}	1.2
$OH + CH_3CF_2Cl \rightarrow CH_2CF_2Cl + H_2O$ (HCFC-142b)	1.3×10^{-12}	1800 ± 150	3.1×10^{-15}	1.2

CHEMICAL KINETIC DATA FOR STRATOSPHERIC MODELING (continued)

Table 1. Rate Constants for Second Order Reactions (continued)

Reaction	A cm^3 molecule^{-1} s^{-1}	E/R K	k (298 K) cm^3 molecule^{-1} s^{-1}	$f(298)$
OH + CH$_2$ClCF$_2$Cl → CHClCF$_2$Cl (HCFC-132b) + H$_2$O	3.6x10^{-12}	1600±400	1.7x10^{-14}	2.0
OH + CHCl$_2$CF$_2$Cl → CCl$_2$CF$_2$Cl (HCFC-122) + H$_2$O	1.0x10^{-12}	900±150	4.9x10^{-14}	1.2
OH + CHFClCFCl$_2$ → CFClCFCl$_2$ (HCFC-122a) + H$_2$O	1.0x10^{-12}	1250±150	1.5x10^{-14}	1.1
OH + CH$_2$ClCF$_3$ → CHClCF$_3$ + H$_2$O (HCFC-133a)	5.2x10^{-13}	1100±300	1.3x10^{-14}	1.3
OH + CHCl$_2$CF$_3$ → CCl$_2$CF$_3$ + H$_2$O (HCFC-123)	7.0x10^{-13}	900±150	3.4x10^{-14}	1.2
OH + CHFClCF$_2$Cl → CFClCF$_2$Cl (HCFC-123a) + H2O	9.2x10^{-13}	1280±150	1.3x10^{-14}	1.2
OH + CHFClCF$_3$ → CFClCF$_3$ + H$_2$O (HCFC-124)	8.0x10^{-13}	1350±150	8.6x10^{-15}	1.2
OH + CH$_3$CF$_2$CFCl$_2$ → products (HCFC-243cc)	7.7x10^{-13}	1700±300	2.6x10^{-15}	2.0
OH + CF$_3$CF$_2$CHCl$_2$ → products (HCFC-225ca)	1.0x10^{-12}	1100±200	2.5x10^{-14}	1.3
OH + CF$_2$ClCF$_2$CHFCl → products (HCFC-225cb)	5.5x10^{-13}	1250±200	8.3x10^{-15}	1.3
HO$_2$ + Cl → HCl + O$_2$	1.8x10^{-11}	-(170±200)	3.2x10^{-11}	1.5
→ OH + ClO	4.1x10^{-11}	450±200	9.1x10^{-12}	2.0
HO$_2$ + ClO → HOCl + O$_2$	4.8x10^{-13}	-(700±)	5.0x10^{-12}	1.4
H$_2$O + ClONO$_2$ → products	-	-	<2.0x10^{-21}	-
NO + OClO → NO$_2$ + ClO	2.5x10^{-12}	600±300	3.4x10^{-13}	2.0
NO + Cl$_2$O$_2$ → products	-	-	<2.0x10^{-14}	-
NO$_3$ + HCl → HNO$_3$ + Cl	-	-	<5.0x10^{-17}	-
HO$_2$NO$_2$ + HCl → products	-	-	<1.0x10^{-21}	-
Cl + O$_3$ → ClO + O$_2$	2.9x10^{-11}	260±100	1.2x10^{-11}	1.15
Cl + H$_2$ → HCl + H	3.7x10^{-11}	2300±200	1.6x10^{-14}	1.25
Cl + H$_2$O$_2$ → HCl + HO$_2$	1.1x10^{-11}	980±500	4.1x10^{-13}	1.5
Cl + NO$_3$ → ClO + NO$_2$	2.4x10^{-11}	0±400	2.4x10^{-11}	1.5
Cl + N$_2$O → ClO + N$_2$	See reference			
Cl + HNO$_3$ → products	-	-	<2.0x10^{-16}	-
Cl + CH$_4$ → HCl + CH$_3$	1.1x10^{-11}	1400±150	1.0x10^{-13}	1.1
Cl + CH$_3$D → products	-	-	7.4x10^{-14}	2.0
Cl + H$_2$CO → HCl + HCO	8.1x10^{-11}	30±100	7.3x10^{-11}	1.15
Cl + CH$_3$O$_2$ → products	-	-	1.6x10^{-10}	1.5
Cl + CH$_3$OH → CH$_2$OH + HCl	5.4x10^{-11}	0±250	5.4x10^{-11}	1.5
Cl + C$_2$H$_6$ → HCl + C$_2$H$_5$	7.7x10^{-11}	90±90	5.7x10^{-11}	1.1
Cl + C$_2$H$_5$O$_2$ → ClO + C$_2$H$_5$O	-	-	7.4x10^{-11}	2.0
→ HCl + C$_2$H$_4$O$_2$	-	-	7.7x10^{-11}	2.0
Cl + CH$_3$CN → products	1.6x10^{-11}	2140±300	1.2x10^{-14}	2.0
Cl + CH$_3$CO$_3$NO$_2$ → products	-	-	<1x10^{-14}	
Cl + C$_3$H$_8$ → HCl + C$_3$H$_7$	1.2x10^{-10}	-(40±250)	1.4x10^{-10}	1.3
Cl + OClO → ClO + ClO	3.4x10^{-11}	-(160±200)	5.8x10^{-11}	1.25
Cl + ClOO → Cl$_2$ + O$_2$	2.3x10^{-10}	0±250	2.3x10^{-10}	3.0
→ ClO + ClO	1.2x10^{-11}	0±250	1.2x10^{-11}	3.0
Cl + Cl$_2$O → Cl$_2$ + ClO	6.2x10^{-11}	-(130±130)	9.6x10^{-11}	1.2
Cl + Cl$_2$O$_2$ → products	-	-	1.0x10^{-10}	2.0
Cl + HOCl → products	2.5x10^{-12}	130±250	1.6x10^{-12}	1.5
Cl + ClNO → NO + Cl$_2$	5.8x10^{-11}	-(100±200)	8.1x10^{-11}	1.5
Cl + ClONO$_2$ → products	6.5x10^{-12}	-(135±50)	1.0x10^{-11}	1.2
Cl + CH$_3$Cl → CH$_2$Cl + HCl	3.2x10^{-11}	1250±200	4.8x10^{-13}	1.2
Cl + CH$_2$Cl$_2$ → HCl + CHCl$_2$	3.1x10^{-11}	1350±500	3.3x10^{-13}	1.5
Cl + CHCl$_3$ → HCl + CCl$_3$	8.2x10^{-12}	1325±300	9.6x10^{-14}	1.3

Table 1. Rate Constants for Second Order Reactions (continued)

Reaction	A $cm^3\ molecule^{-1}\ s^{-1}$	E/R K	k (298 K) $cm^3\ molecule^{-1}\ s^{-1}$	$f(298)$
$Cl + CH_3F \rightarrow HCl + CH_2F$ (HFC-41)	2.0×10^{-11}	1200 ± 500	3.5×10^{-13}	1.3
$Cl + CH_2F_2 \rightarrow HCl + CHF_2$ (HFC-32)	1.2×10^{-11}	1630 ± 500	5.0×10^{-14}	1.5
$Cl + CF_3H \rightarrow HCl + CF_3$ (HFC-23)	-	-	3.0×10^{-18}	5.0
$Cl + CH_2FCl \rightarrow HCl + CHFCl$ (HCFC-31)	1.2×10^{-11}	1390 ± 500	1.1×10^{-13}	2.0
$Cl + CHFCl_2 \rightarrow HCl + CFCl_2$ (HCFC-21)	5.5×10^{-12}	1675 ± 200	2.0×10^{-14}	1.3
$Cl + CHF_2Cl \rightarrow HCl + CF_2Cl$ (HCFC-22)	5.9×10^{-12}	2430 ± 200	1.7×10^{-15}	1.3
$Cl + CH_3CCl_3 \rightarrow CH_2CCl_3 + HCl$	2.8×10^{-12}	1790 ± 400	7.0×10^{-15}	2.0
$Cl + CH_3CH_2F \rightarrow HCl + CH_3CHF$ (HFC-161)	1.8×10^{-11}	290 ± 500	6.8×10^{-12}	3.0
$\rightarrow HCl + CH_2CH_2F$	1.4×10^{-11}	880 ± 500	7.3×10^{-13}	3.0
$Cl + CH_3CHF_2 \rightarrow HCl + CH_3CF_2$ (HFC-152a)	6.4×10^{-12}	950 ± 500	2.6×10^{-13}	1.3
$\rightarrow HCl + CH_2CHF_2$	7.2×10^{-12}	2390 ± 500	2.4×10^{-15}	3.0
$Cl + CH_2FCH_2F \rightarrow HCl + CHFCH_2F$ (HFC-152)	2.6×10^{-11}	1060 ± 500	7.5×10^{-13}	3.0
$Cl + CH_3CFCl_2 \rightarrow HCl + CH_2CFCl_2$ (HCFC-141b)	1.8×10^{-12}	2000 ± 300	2.2×10^{-15}	1.2
$Cl + CH_3CF_2Cl \rightarrow HCl + CH_2CF_2Cl$ (HCFC-142b)	1.4×10^{-12}	2420 ± 500	4.2×10^{-16}	1.2
$Cl + CH_3CF_3 \rightarrow HCl + CH_2CF_3$ (HFC-143a)	1.2×10^{-11}	3880 ± 500	2.6×10^{-17}	5.0
$Cl + CH_2FCHF_2 \rightarrow HCl + CH_2FCF_2$ (HFC-143)	5.5×10^{-12}	1610 ± 500	2.5×10^{-14}	3.0
$\rightarrow HCl + CHFCHF_2$	7.7×10^{-12}	1720 ± 500	2.4×10^{-14}	3.0
$Cl + CH_2ClCF_3 \rightarrow HCl + CHClCF_3$ (HCFC-133a)	1.8×10^{-12}	1710 ± 500	5.9×10^{-15}	3.0
$Cl + CH_2FCF_3 \rightarrow HCl + CHFCF_3$ (HFC-134a)	-	-	1.5×10^{-15}	1.2
$Cl + CHF_2CHF_2 \rightarrow HCl + CF_2CHF_2$ (HCF-134)	7.5×10^{-12}	2430 ± 500	2.2×10^{-15}	1.5
$Cl + CHCl_2CF_3 \rightarrow HCl + CCl_2CF_3$ (HCFC-123)	4.4×10^{-12}	1750 ± 500	1.2×10^{-14}	1.3
$Cl + CHFClCF_3 \rightarrow HCl + CFClCF_3$ (HCFC-124)	1.1×10^{-12}	1800 ± 500	2.7×10^{-15}	1.3
$Cl + CHF_2CF_3 \rightarrow HCl + CF_2CF_3$ (HFC-125)	-	-	2.4×10^{-16}	1.3
$ClO + O_3 \rightarrow ClOO + O_2$	-	-	$<1.4 \times 10^{-17}$	-
$\rightarrow OClO + O_2$	1.0×10^{-12}	>4000	$<1.0 \times 10^{-18}$	-
$ClO + H_2 \rightarrow products$	$\sim 1.0 \times 10^{-12}$	>4800	$<1.0 \times 10^{-19}$	-
$ClO + NO \rightarrow NO_2 + Cl$	6.4×10^{-12}	$-(290 \pm 100)$	1.7×10^{-11}	1.15
$ClO + NO_3 \rightarrow ClOO + NO_2$	4.7×10^{-13}	0 ± 400	4.7×10^{-13}	1.5
$ClO + N_2O \rightarrow products$	$\sim 1.0 \times 10^{-12}$	>4300	$<6.0 \times 10^{-19}$	-
$ClO + CO \rightarrow products$	$\sim 1.0 \times 10^{-12}$	>3700	$<4.0 \times 10^{-18}$	-
$ClO + CH_4 \rightarrow products$	$\sim 1.0 \times 10^{-12}$	>3700	$<4.0 \times 10^{-18}$	-
$ClO + H_2CO \rightarrow products$	$\sim 1.0 \times 10^{-12}$	>2100	$<1.0 \times 10^{-15}$	-
$ClO + CH_3O_2 \rightarrow products$	3.3×10^{-12}	115 ± 115	2.2×10^{-12}	1.5
$ClO + ClO \rightarrow Cl_2 + O_2$	1.0×10^{-12}	1590 ± 300	4.8×10^{-15}	1.5
$\rightarrow ClOO + Cl$	3.0×10^{-11}	2450 ± 500	8.0×10^{-15}	1.5
$\rightarrow OClO + Cl$	3.5×10^{-13}	1370 ± 300	3.5×10^{-15}	1.5
$HCl + ClONO_2 \rightarrow products$	-	-	$<1.0 \times 10^{-20}$	-

Table 1. Rate Constants for Second Order Reactions (continued)

Reaction	A cm^3 molecule^{-1} s^{-1}	E/R K	k (298 K) cm^3 molecule^{-1} s^{-1}	f(298)
$CH_2ClO + O_2 \rightarrow CHClO + HO_2$	-	-	6×10^{-14}	5
$CH_2ClO_2 + HO_2 \rightarrow$				
$CH_2ClO_2H + O_2$	3.3×10^{-13}	$-(820\pm200)$	5.2×10^{-12}	1.5
$CH_2ClO_2 + NO \rightarrow CH_2ClO + NO_2$	7×10^{-12}	$-(300\pm200)$	1.9×10^{-11}	1.5
$CCl_3O_2 + NO \rightarrow CCl_2O + NO_2 + Cl$	7.3×10^{-12}	$-(270\pm200)$	1.8×10^{-11}	1.3
$CCl_2FO_2 + NO \rightarrow CClFO +$				
$NO_2 + Cl$	4.5×10^{-12}	$-(350\pm200)$	1.5×10^{-11}	1.3
$CClF_2O_2 + NO \rightarrow CF_2O +$				
$NO_2 + Cl$	3.8×10^{-12}	$-(400\pm200)$	1.5×10^{-11}	1.2

BrO_x Reactions

Reaction	A cm^3 molecule^{-1} s^{-1}	E/R K	k (298 K) cm^3 molecule^{-1} s^{-1}	f(298)
$O + BrO \rightarrow Br + O_2$	1.9×10^{-11}	$-(230\pm150)$	4.1×10^{-11}	1.5
$O + HBr \rightarrow OH + Br$	5.8×10^{-12}	1500 ± 200	3.8×10^{-14}	1.3
$O + HOBr \rightarrow OH + BrO$	1.2×10^{-10}	430 ± 300	2.8×10^{-11}	3.0
$OH + Br_2 \rightarrow HOBr + Br$	4.2×10^{-11}	0 ± 600	4.2×10^{-11}	1.3
$OH + BrO \rightarrow products$	-	-	7.5×10^{-11}	3.0
$OH + HBr \rightarrow H_2O + Br$	1.1×10^{-11}	0 ± 250	1.1×10^{-11}	1.2
$OH + CH_3Br \rightarrow CH_2Br + H_2O$	4.0×10^{-12}	1470 ± 150	2.9×10^{-14}	1.1
$OH + CH_2Br_2 \rightarrow CHBr_2 + H_2O$	2.4×10^{-12}	900 ± 300	1.2×10^{-13}	1.1
$OH + CHBr_3 \rightarrow CBr_3 + H_2O$	1.6×10^{-12}	710 ± 200	1.5×10^{-13}	2.0
$OH + CHF_2Br \rightarrow CF_2Br + H_2O$	1.1×10^{-12}	1400 ± 200	1.0×10^{-14}	1.1
$OH + CH_2ClBr \rightarrow CHClBr + H_2O$	2.3×10^{-12}	930 ± 150	1.0×10^{-13}	1.2
$OH + CF_2ClBr \rightarrow products$	-	-	$<1.5 \times 10^{-16}$	-
$OH + CF_2Br_2 \rightarrow products$	-	-	$<5.0 \times 10^{-16}$	-
$OH + CF_3Br \rightarrow products$	-	-	$<1.2 \times 10^{-16}$	-
$OH + CH_2BrCF_3 \rightarrow CHBrCF_3 + H_2O$	1.4×10^{-12}	1340 ± 200	1.6×10^{-14}	1.3
$OH + CHFBrCF_3 \rightarrow CFBrCF_3$	7.2×10^{-13}	1110 ± 150	1.8×10^{-14}	1.5
$OH + CHClBrCF_3 \rightarrow CClBrCF_3 + H_2O$	1.3×10^{-12}	995 ± 150	4.5×10^{-14}	1.5
$OH + CF_2BrCHFCl \rightarrow CF_2BrCFCl + H_2O$	9.3×10^{-13}	1250 ± 150	1.4×10^{-14}	1.5
$OH + CF_2BrCF_2Br \rightarrow products$	-	-	$<1.5 \times 10^{-16}$	-
$HO_2 + Br \rightarrow HBr + O_2$	1.5×10^{-11}	600 ± 600	2.0×10^{-12}	2.0
$HO_2 + BrO \rightarrow products$	3.4×10^{-12}	$-(540\pm200)$	2.1×10^{-11}	1.5
$NO_3 + HBr \rightarrow HNO_3 + Br$	-	-	$<1.0 \times 10^{-16}$	-
$Cl + CH_2ClBr \rightarrow HCl + CHClBr$	4.3×10^{-11}	1370 ± 500	4.3×10^{-13}	3.0
$Cl + CH_3Br \rightarrow HCl + CH_2Br$	1.5×10^{-11}	1060 ± 100	4.3×10^{-13}	1.2
$Cl + CH_2Br_2 \rightarrow HCl + CHBr_2$	6.4×10^{-12}	810 ± 100	4.2×10^{-13}	1.2
$Br + O_3 \rightarrow BrO + O_2$	1.7×10^{-11}	800 ± 200	1.2×10^{-12}	1.2
$Br + H_2O_2 \rightarrow HBr + HO_2$	1.0×10^{-11}	>3000	$<5.0 \times 10^{-16}$	-
$Br + NO_3 \rightarrow BrO + NO_2$	-	-	1.6×10^{-11}	2.0
$Br + H_2CO \rightarrow HBr + HCO$	1.7×10^{-11}	800 ± 200	1.1×10^{-12}	1.3
$Br + OClO \rightarrow BrO + ClO$	2.6×10^{-11}	1300 ± 300	3.4×10^{-13}	2.0
$Br + Cl_2O \rightarrow BrCl + ClO$	2.1×10^{-11}	470 ± 150	4.3×10^{-12}	1.3
$Br + Cl_2O_2 \rightarrow products$	-	-	3.0×10^{-12}	2.0
$BrO + O_3 \rightarrow products$	$\sim 1.0 \times 10^{-12}$	>3200	$<2.0 \times 10^{-17}$	-
$BrO + NO \rightarrow NO_2 + Br$	8.8×10^{-12}	$-(260\pm130)$	2.1×10^{-11}	1.15
$BrO + NO_3 \rightarrow products$	-	-	1.0×10^{-12}	3.0
$BrO + ClO \rightarrow Br + OClO$	1.6×10^{-12}	$-(430\pm200)$	6.8×10^{-12}	1.25
$\rightarrow Br + ClOO$	2.9×10^{-12}	$-(220\pm200)$	6.1×10^{-12}	1.25
$\rightarrow BrCl + O_2$	5.8×10^{-13}	$-(170\pm200)$	1.0×10^{-12}	1.25
$BrO + BrO \rightarrow products$	1.5×10^{-12}	$-(230\pm150)$	3.2×10^{-12}	1.15
$CH_2BrO_2 + NO \rightarrow CH_2O +$				
$NO_2 + Br$	4×10^{-12}	$-(300\pm200)$	1.1×10^{-11}	1.5

Table 1. Rate Constants for Second Order Reactions (continued)

Reaction	A cm^3 molecule^{-1} s^{-1}	E/R K	k (298 K) cm^3 molecule^{-1} s^{-1}	$f(298)$
IO$_x$ Reactions				
O + I$_2$ → IO + I	1.4x10^{-10}	0±250	1.4x10^{-10}	1.4
O + IO → O$_2$ + I			1.2x10^{-10}	2.0
OH + I$_2$ → HOI + I			1.8x10^{-10}	2.0
OH + HI → H$_2$O + I			3.0x10^{-11}	2.0
OH + CH$_3$I → H$_2$O + CH$_2$I	3.1x10^{-12}	1120±500	7.2x10^{-14}	3.0
OH + CF$_3$I → HOI + CF$_3$			3.1x10^{-14}	5.0
HO$_2$ + I → HI + O$_2$	1.5x10^{-11}	1090±500	3.8x10^{-13}	2.0
HO$_2$ + IO → HOI + O$_2$			8.4x10^{-11}	1.5
NO$_3$ + HI → HNO$_3$ + I	See reference			
I + O$_3$ → IO + O$_2$	2.3x10^{-11}	870±200	1.2x10^{-12}	1.2
I + BrO → IO + Br	-	-	1.2x10^{-11}	2.0
IO + NO → I + NO$_2$	9.1x10^{-12}	-(240±150)	2.0x10^{-11}	1.2
IO + ClO → products	5.1x10^{-12}	-(280±200)	1.3x10^{-11}	2.0
IO + BrO → products	-	-	6.9x10^{-11}	1.5
IO + IO → products	1.5x10^{-11}	-(500±500)	8.0x10^{-11}	1.5
INO + INO → I$_2$ + 2NO	8.4x10^{-11}	2620±600	1.3x10^{-14}	2.5
INO$_2$ + INO$_2$ → I$_2$ + 2NO$_2$	2.9x10^{-11}	2600±1000	4.7x10^{-15}	3.0
SO$_x$ Reactions				
O + SH → SO + H	-	-	1.6x10^{-10}	5.0
O + CS → CO + S	2.7x10^{-10}	760±250	2.1x10^{-11}	1.1
O + H$_2$S → OH + SH	9.2x10^{-12}	1800±550	2.2x10^{-14}	1.7
O + OCS → CO + SO	2.1x10^{-11}	2200±150	1.3x10^{-14}	1.2
O + CS$_2$ → CS + SO	3.2x10^{-11}	650±150	3.6x10^{-12}	1.2
O + CH$_3$SCH$_3$ → CH$_3$SO + CH$_3$	1.3x10^{-11}	-(410±100)	5.0x10^{-11}	1.1
O + CH$_3$SSCH$_3$ → CH$_3$SO + CH$_3$S	5.5x10^{-11}	-(250±100)	1.3x10^{-10}	1.3
O$_3$ + H$_2$S → products	-	-	<2.0x10^{-20}	-
O$_3$ + CH$_3$SCH$_3$ → products	-	-	<1.0x10^{-18}	-
O$_3$ + SO$_2$ → SO$_3$ + O$_2$	3.0x10^{-12}	>7000	<2.0x10^{-22}	-
OH + H$_2$S → SH + H$_2$O	6.0x10^{-12}	75±75	4.7x10^{-12}	1.2
OH + OCS → products	1.1x10^{-13}	1200±500	1.9x10^{-15}	2.0
OH + CS$_2$ → products	See reference	-	-	-
OH + CH$_3$SH → CH$_3$S + H$_2$O	9.9x10^{-12}	-(360±100)	3.3x10^{-11}	1.2
OH + CH$_3$SCH$_3$ → H$_2$O + CH$_2$SCH$_3$	1.2x10^{-11}	260±100	5.0x10^{-12}	1.15
OH + CH$_3$SSCH$_3$ → products	6.0x10^{-11}	-(400±200)	2.3x10^{-10}	1.2
OH + S → H + SO	-	-	6.6x10^{-11}	3.0
OH + SO → H + SO$_2$	-	-	8.6x10^{-11}	2.0
HO$_2$ + H$_2$S → products	-	-	<3.0x10^{-15}	-
HO$_2$ + CH$_3$SH → products	-	-	<4.0x10^{-15}	-
HO$_2$ + CH$_3$SCH$_3$ → products	-	-	<5.0x10^{-15}	-
HO$_2$ + SO$_2$ → products	-	-	<1.0x10^{-18}	-
NO$_2$ + SO$_2$ → products	-	-	<2.0x10^{-26}	-
NO$_3$+ H$_2$S → products	-	-	<8.0x10^{-16}	-
NO$_3$ + OCS → products	-	-	<1.0x10^{-16}	-
NO$_3$ + CS$_2$ → products	-	-	<4.0x10^{-16}	-
NO$_3$ + CH$_3$SH → products	4.4x10^{-13}	-(210±210)	8.9x10^{-13}	1.25
NO$_3$ + CH$_3$SCH$_3$ → CH$_3$SCH$_2$ + HNO$_3$	1.9x10^{-13}	-(500±200)	1.0x10^{-12}	1.2
NO$_3$ + CH$_3$SSCH$_3$ → products	1.3x10^{-12}	270±270	5.3x10^{-13}	1.4
NO$_3$ + SO$_2$ → products	-	-	<7.0x10^{-21}	-
N$_2$O$_5$ + CH$_3$SCH$_3$ → products	-	-	<1.0x10^{-17}	-
CH$_3$O$_2$ + SO$_2$ → products	-	-	<5.0x10^{-17}	-
F + CH$_3$SCH$_3$ → products	-	-	2.4.x10^{-10}	2.0

Table 1. Rate Constants for Second Order Reactions (continued)

Reaction	A cm^3 molecule^{-1} s^{-1}	E/R K	k (298 K) cm^3 molecule^{-1} s^{-1}	f(298)
$Cl + H_2S \rightarrow HCl + SH$	3.7×10^{-11}	$-(210 \pm 100)$	7.4×10^{-11}	1.25
$Cl + OCS \rightarrow$ products	-	-	$<1.0 \times 10^{-16}$	-
$Cl + CS_2 \rightarrow$ products	-	-	$<4.0 \times 10^{-15}$	-
$Cl + CH_3SH \rightarrow CH_3S + HCl$	1.2×10^{-10}	$-(150 \pm 50)$	2.0×10^{-10}	1.25
$Cl + CH_3SCH_3 \rightarrow$ products	See reference	-	-	-
$ClO + OCS \rightarrow$ products	-	-	$<2.0 \times 10^{-16}$	-
$ClO + CH_3SCH_3 \rightarrow$ products	-	-	9.5×10^{-15}	2.0
$ClO + SO \rightarrow Cl + SO_2$	2.8×10^{-11}	0 ± 50	2.8×10^{-11}	1.3
$ClO + SO_2 \rightarrow Cl + SO_3$	-	-	$<4.0 \times 10^{-18}$	-
$Br + H_2S \rightarrow HBr + SH$	1.4×10^{-11}	2750 ± 300	1.4×10^{-15}	2.0
$Br + CH_3SH \rightarrow CH_3S + HBr$	9.2×10^{-12}	390 ± 100	2.5×10^{-12}	2.0
$Br + CH_3SCH_3 \rightarrow$ products	See reference			
$BrO + CH_3SCH_3 \rightarrow$ products	1.5×10^{-14}	$-(850 \pm 200)$	2.6×10^{-13}	1.3
$BrO + SO \rightarrow Br + SO_2$			5.7×10^{-11}	1.4
$IO + CH_3SH \rightarrow$ products			6.6×10^{-16}	2.0
$IO + CH_3SCH_3 \rightarrow$ products			1.2×10^{-14}	1.5
$S + O_2 \rightarrow SO + O$	2.3×10^{-12}	0 ± 200	2.3×10^{-12}	1.2
$S + O_3 \rightarrow SO + O_2$			1.2×10^{-11}	2.0
$SO + O_2 \rightarrow SO_2 + O$	2.6×10^{-13}	2400 ± 500	8.4×10^{-17}	2.0
$SO + O_3 \rightarrow SO_2 + O_2$	3.6×10^{-12}	1100 ± 200	9.0×10^{-14}	1.2
$SO + NO_2 \rightarrow SO_2 + NO$	1.4×10^{-11}	0 ± 50	1.4×10^{-11}	1.2
$SO + OClO \rightarrow SO_2 + ClO$			1.9×10^{-12}	3.0
$SO_3 + H_2O \rightarrow$ products	See reference		-	-
$SO_3 + NO_2 \rightarrow$ products			1.0×10^{-19}	10.0
$SH + O_2 \rightarrow OH + SO$			$<4.0 \times 10^{-19}$	-
$SH + O_3 \rightarrow HSO + O_2$	9.0×10^{-12}	280 ± 200	3.5×10^{-12}	1.3
$SH + H_2O_2 \rightarrow$ products			$<5.0 \times 10^{-15}$	-
$SH + NO_2 \rightarrow HSO + NO$	2.9×10^{-11}	$-(240 \pm 50)$	6.5×10^{-11}	1.2
$SH + Cl_2 \rightarrow ClSH + Cl$	1.7×10^{-11}	690 ± 200	1.7×10^{-12}	2.0
$SH + BrCl \rightarrow$ products	2.3×10^{-11}	$-(350 \pm 200)$	7.4×10^{-11}	2.0
$SH + Br_2 \rightarrow BrSH + Br$	6.0×10^{-11}	$-(160 \pm 160)$	1.0×10^{-10}	2.0
$SH + F_2 \rightarrow FSH + F$	4.3×10^{-11}	1390 ± 200	4.0×10^{-13}	2.0
$HSO + O_2 \rightarrow$ products			$<2.0 \times 10^{-17}$	-
$HSO + O_3 \rightarrow$ products			1.0×10^{-13}	1.3
$HSO + NO \rightarrow$ products			$<1.0 \times 10^{-15}$	-
$HSO + NO_2 \rightarrow HSO_2 + NO$			9.6×10^{-12}	2.0
$HSO_2 + O_2 \rightarrow HO_2 + SO_2$			3.0×10^{-13}	3.0
$HOSO_2 + O_2 \rightarrow HO_2 + SO_3$	1.3×10^{-12}	330 ± 200	4.4×10^{-13}	1.2
$CS + O_2 \rightarrow OCS + O$			2.9×10^{-19}	2.0
$CS + O_3 \rightarrow OCS + O_2$			3.0×10^{-16}	3.0
$CS + NO_2 \rightarrow OCS + NO$			7.6×10^{-17}	3.0
$CH_3S + O_2 \rightarrow$ products			$<3.0 \times 10^{-18}$	-
$CH_3S + O_3 \rightarrow$ products	2.0×10^{-12}	$-(290 \pm 100)$	5.3×10^{-12}	1.15
$CH_3S + NO \rightarrow$ products			$<1.0 \times 10^{-13}$	-
$CH_3S + NO_2 \rightarrow CH_3SO + NO$	2.1×10^{-11}	$-(320 \pm 100)$	6.1×10^{-11}	1.15
$CH_2SH + O_2 \rightarrow$ products			6.5×10^{-12}	2.0
$CH_2SH + O_3 \rightarrow$ products			3.5×10^{-11}	2.0
$CH_2SH + NO \rightarrow$ products			1.9×10^{-11}	2.0
$CH_2SH + NO_2 \rightarrow$ products			5.2×10^{-11}	2.0
$CH_3SO + O_3 \rightarrow$ products			6.0×10^{-13}	1.5
$CH_3SO + NO_2 \rightarrow CH_3SO_2 + NO$			1.2×10^{-11}	1.4
$CH_3SOO + O_3 \rightarrow$ products			$<8.0 \times 10^{-13}$	-
$CH_3SOO + NO \rightarrow$ products	1.1×10^{-11}	0 ± 100	1.1×10^{-11}	2.0
$CH_3SO_2 + NO_2 \rightarrow$ products	2.2×10^{-11}	0 ± 100	2.2×10^{-11}	2.0
$CH_3SCH_2 + NO_3 \rightarrow$ products			3.0×10^{-10}	2.0
$CH_3SCH_2O_2 + NO \rightarrow CH_3SCH_2O + NO_2$			1.9×10^{-11}	2.0

Table 1. Rate Constants for Second Order Reactions (continued)

Reaction	A cm^3 molecule^{-1} s^{-1}	E/R K	k (298 K) cm^3 molecule^{-1} s^{-1}	f(298)
$CH_3SS + O_3 \rightarrow$ products			4.6×10^{-13}	2.0
$CH_3SS + NO_2 \rightarrow$ products			1.8×10^{-11}	2.0
$CH_3SSO + NO_2 \rightarrow$ products			4.5×10^{-12}	2.0

Metal Reactions

Reaction	A	E/R	k (298 K)	f(298)
$Na + O_3 \rightarrow NaO + O_2$	1.0×10^{-9}	95 ± 50	7.3×10^{-10}	1.2
$\rightarrow NaO_2 + O$	-	-	$<4.0 \times 10^{-11}$	-
$Na + N_2O \rightarrow NaO + N_2$	2.8×10^{-10}	1600 ± 400	1.3×10^{-12}	1.2
$Na + Cl_2 \rightarrow NaCl + Cl$	7.3×10^{-10}	0 ± 200	7.3×10^{-10}	1.3
$NaO + O \rightarrow Na + O_2$	3.7×10^{-10}	0 ± 400	3.7×10^{-10}	3.0
$NaO + O_3 \rightarrow NaO_2 + O_2$	1.1×10^{-9}	570 ± 300	1.6×10^{-10}	1.5
$\rightarrow Na + 2O_2$	6.0×10^{-11}	0 ± 800	6.0×10^{-11}	3.0
$NaO + H_2 \rightarrow NaOH + H$	2.6×10^{-11}	0 ± 600	2.6×10^{-11}	2.0
$NaO + H_2O \rightarrow NaOH + OH$	2.2×10^{-10}	0 ± 400	2.2×10^{-10}	2.0
$NaO + NO \rightarrow Na + NO_2$	1.5×10^{-10}	0 ± 400	1.5×10^{-10}	4.0
$NaO + HCl \rightarrow$ products	2.8×10^{-10}	0 ± 400	2.8×10^{-10}	3.0
$NaO_2 + O \rightarrow NaO + O_2$	2.2×10^{-11}	0 ± 600	2.2×10^{-11}	5.0
$NaO_2 + NO \rightarrow NaO + NO_2$	-	-	$<10^{-14}$	-
$NaO_2 + HCl \rightarrow$ products	2.3×10^{-10}	0 ± 400	2.3×10^{-10}	3.0
$NaOH + HCl \rightarrow NaCl + H_2O$	2.8×10^{-10}	0 ± 400	2.8×10^{-10}	3.0

Table 2. Rate Constants for Association Reactions

The values quoted are suitable for air as the third body, M. The integer in parentheses is the power of ten.

Reaction	Low pressure limit $k_0(T) = k_0(300) (T/300)^{-n}$ cm^6 molecule^{-2} s^{-1}		High pressure limit $k_\infty(T) = k_\infty(300) (T/300)^{-m}$ cm^3 molecule^{-1} s^{-1}	
	$k_0(300)$	n	$k_\infty(300)$	m
O_x Reactions				
$O + O_2 \rightarrow O_3$	(6.0 ± 0.5) (-34)	2.3 ± 0.5	-	-
$O(^1D)$ Reactions				
$O(^1D) + N_2 \rightarrow N_2O$	(3.5 ± 3.0) (-37)	0.6	-	-
HO_x Reactions				
$H + O_2 \rightarrow HO_2$	(5.7 ± 0.5) (-32)	1.6 ± 0.5	(7.5 ± 4.0) (-11)	0 ± 1.0
$OH + OH \rightarrow H_2O_2$	(6.2 ± 1.2) (-31)	1.0	(2.6 ± 1.0) (-11)	0 ± 0.5
NO_x Reactions				
$O + NO \rightarrow NO_2$	(9.0 ± 2.0) (-32)	1.5 ± 0.3	(3.0 ± 1.0) (-11)	0 ± 1.0
$O + NO_2 \rightarrow NO_3$	(9.0 ± 1.0) (-32)	2.0 ± 1.0	(2.2 ± 0.3) (-11)	0 ± 1.0
$OH + NO \rightarrow HONO$	(7.0 ± 1.0) (-31)	2.6 ± 0.3	(3.6 ± 1.0) (-11)	0.1 ± 0.5
$OH + NO_2 \rightarrow HNO_3$	(2.5 ± 0.1) (-30)	4.4 ± 0.3	(1.6 ± 0.2) (-11)	1.7 ± 0.2
$HO_2 + NO_2 \rightarrow HO_2NO_2$	(1.8 ± 0.3) (-31)	3.2 ± 0.4	(4.7 ± 1.0) (-12)	1.4 ± 1.4
$NO_2 + NO_3 \rightarrow N_2O_5$	(2.2 ± 0.5) (-30)	3.9 ± 1.0	(1.5 ± 0.8) (-12)	0.7 ± 0.4
$NO_3 \rightarrow NO + O_2$	See reference			
$Hydrocarbon Reactions$				
$CH_3 + O_2 \rightarrow CH_3O_2$	(4.5 ± 1.5) (-31)	3.0 ± 1.0	(1.8 ± 0.2) (-12)	1.7 ± 1.7
$C_2H_5 + O_2 \rightarrow C_2H_5O_2$	(1.5 ± 1.0) (-28)	3.0 ± 1.0	(8.0 ± 1.0) (-12)	0 ± 1.0
$OH + C_2H_2 \rightarrow HOCHCH$	(5.5 ± 2.0) (-30)	0.0 ± 0.2	(8.3 ± 1.0) (-13)	-2
$OH + C_2H_4 \rightarrow HOCH_2CH_2$	(1.0 ± 0.6) (-28)	0.8 ± 2.0	(8.8 ± 0.9) (-12)	0

Table 2. Rate Constants for Association Reactions (continued)

The values quoted are suitable for air as the third body, M. The integer in parentheses is the power of ten.

Reaction	Low pressure limit $k_0(T) = k_0(300) (T/300)^{-n}$ cm^6 molecule^{-2} s^{-1}		High pressure limit $k_\infty(T) = k_\infty(300) (T/300)^{-m}$ cm^3 molecule^{-1} s^{-1}	
	$k_0(300)$	n	$k_\infty(300)$	m
$CH_3O + NO \rightarrow CH_3ONO$	(1.4±0.5) (-29)	3.8±1.0	(3.6±1.6) (-11)	0.6±1.0
$CH_3O + NO_2 \rightarrow CH_3ONO_2$	(1.1±0.4) (-28)	4.0±2.0	(1.6±0.5) (-11)	1.0±1.0
$C_2H_5O + NO \rightarrow C_2H_5ONO$	(2.8±1.0) (-27)	4.0±2.0	(5.0±1.0) (-11)	1.0±1.0
$C_2H_5O + NO_2 \rightarrow C_2H_5ONO_2$	(2.0±1.0) (-27)	4.0±2.0	(2.8±0.4) (-11)	1.0±1.0
$CH_3O_2 + NO_2 \rightarrow CH_3O_2NO_2$	(1.5±0.8) (-30)	4.0±2.0	(6.5±3.2) (-12)	2.0±2.0
$CH_3C(O)O_2 + NO_2 \rightarrow$ $CH_3C(O)O_2NO_2$	(9.7±3.8) (-29)	5.6±2.8	(9.3±0.4)(-12)	1.5±0.3
FO_x Reactions				
$F + O_2 \rightarrow FO_2$	(4.4±0.4) (-33)	1.2±0.5	-	-
$F + NO \rightarrow FNO$	(1.8±0.3) (-31)	1.0±10	(2.8±1.4) (-10)	0.0±1.0
$F + NO_2 \rightarrow FNO_2$	(6.3±3.0) (-32)	2.0±2.0	(2.6±1.3) (-10)	0.0±1.0
$FO + NO_2 \rightarrow FONO_2$	(2.6±2.0) (-31)	1.3±1.3	(2.0±1.0) (-11)	1.5±1.5
$CF_3 + O_2 \rightarrow CF_3O_2$	(3.0±0.3) (-29)	4.0±2.0	(4.0±1.0) (-12)	1.0±1.0
$CF_3O + NO_2 \rightarrow CF_3ONO_2$	See reference			
$CF_3O_2 + NO_2 CF_3O_2NO_2$	(2.2±0.5) (-29)	5.0±1.0	(6.0±1.0) (-12)	2.5±1.0
$CF_3O + CO \rightarrow CF_3OCO$	(2.5±0.2) (-31)	-	(6.8±0.4) (-14)	-1.2
$CF_3O \rightarrow CF_2O + F$	See reference			
ClO_x Reactions				
$Cl + O_2 \rightarrow ClOO$	(2.7±1.0) (-33)	1.5±0.5	-	-
$Cl + NO \rightarrow ClNO$	(9.0±2.0) (-32)	1.6±0.5	-	-
$Cl + NO_2 ClONO \rightarrow$	(1.3±0.2) (-30)	2.0±1.0	(1.0±0.5) (-10)	1.0±1.0
$ClNO_2$	(1.8±0.3) (-31)	2.0±1.0	(1.0±0.5) (-10)	1.0±1.0
$Cl + CO \rightarrow ClCO$	(1.3±0.5) (-33)	3.8±0.5	-	-
$Cl + C_2H_2 \rightarrow ClC_2H_2$	((5.9±1.0) (-30)	2.1±1.0	(2.1±0.4) (-10)	1.0±0.5
$Cl + C_2H_4 \rightarrow ClC_2H_4$	(1.6±1) (-29)	3.3±1.0	(3.1±2) (-10)	1.0±0.5
$Cl + C_2Cl_4 \rightarrow C_2Cl_5$	(1.4±0.6) (-28)	8.5±1.0	(4.0±1.0) (-11)	1.2±0.5
$ClO + NO_2 \rightarrow ClONO_2$	(1.8±0.3) (-31)	3.4±1.0	(1.5±0.7) (-11)	1.9±1.9
$OClO + NO_3 \rightarrow O_2ClONO_2$	See reference			
$ClO + ClO \rightarrow Cl_2O_2$	(2.2±0.4) (-32)	3.1±0.5	(3.5±2) (-12)	1.0±1.0
$ClO + OClO \rightarrow Cl_2O_3$	(6.2±1.0) (-32)	4.7±0.6	(2.4±1.2) (-11)	0±1.0
$OClO + O \rightarrow ClO_3$	(1.9±0.5) (-31)	1.1±1.0	(3.1±0.8) (-11)	0±1.0
$CH_2Cl + O_2 \rightarrow CH_2ClO_2$	(1.9±0.1) (-30)	3.2±0.2	(2.9±0.2) (-12)	1.2±0.6
$CHCl_2 + O_2 \rightarrow CHCl_2O_2$	(1.3±0.1) (-30)	4.0±0.2	(2.8±0.2) (-12)	1.4±0.6
$CCl_3 + O_2 \rightarrow CCl_3O_2$	(6.9±0.2) (-31)	6.4±0.3	(2.4±0.2) (-12)	2.1±0.6
$CFCl_2 + O_2 \rightarrow CFCl_2O_2$	(5.0±0.8) (-30)	4.0±2.0	(6.0±1.0) (-12)	1.0±1.0
$CF_2Cl + O_2 \rightarrow CF_2ClO_2$	(3.0±1.5) (-30)	4.0±2.0	(3±2) (-12)	1.0±1.0
$CCl_3O_2 + NO_2 \rightarrow CCl_3O_2NO_2$	(5.0±1.0) (-29)	5.0±1.0	(6.0±1.0) (-12)	2.5±1.0
$CFCl_2O_2 + NO_2 \rightarrow CFCl_2O_2NO_2$	(3.5±0.5) (-29)	5.0±1.0	(6.0±1.0) (-12)	2.5±1.0
$CF_2ClO_2 + NO_2 \rightarrow CF_2ClO_2NO_2$	(3.3±0.7) (-29)	6.7±1.3	(4.1±1.9) (-12)	2.8±0.7
BrO_x Reactions				
$Br + NO_2 \rightarrow BrNO_2$	(4.2±0.8) (-31)	2.4±0.5	(2.7±0.5) (-11)	0±1.0
$BrO + NO_2 \rightarrow BrONO_2$	(5.2±0.6) (-31)	3.2±0.8	(6.9±1.0) (-12)	2.9±1.0
IO_x Reactions				
$I + NO \rightarrow INO$	(1.8±0.5) (-32)	1.0±0.5	(1.7±1.0) (-11)	0±1.0
$I + NO_2 \rightarrow INO_2$	(3.0±1.5) (-31)	1.0±1.0	(6.6±5.0) (-11)	0±1.0
$IO + NO_2 \rightarrow IONO_2$	(5.9±2.0) (-31)	3.5±1.0	(9.0±1.0) (-12)	1.5±1.0

Table 2. Rate Constants for Association Reactions (continued)

The values quoted are suitable for air as the third body, M. The integer in parentheses is the power of ten.

Reaction	Low pressure limit $k_0(T) = k_0(300) (T/300)^{-n}$ cm^6 molecule^{-2} s^{-1}		High pressure limit $k_\infty(T) = k_\infty(300) (T/300)^{-m}$ cm^3 molecule^{-1} s^{-1}	
	$k_0(300)$	n	$k_\infty(300)$	m
SO$_x$ Reactions				
HS + NO → HSNO	(2.4±0.4) (-31)	3.0±1.0	(2.7±0.5) (-11)	0
CH$_3$S + NO → CH$_3$SNO	(3.2±0.4) (-29)	4.0±1.0	(3.9±0.6) (-11)	2.7±1.0
O + SO$_2$ → SO$_3$	(1.3±)(-33)	-3.6±0.7		
OH + SO$_2$ → HOSO$_2$	(3.0±1.0) (-31)	3.3±1.5	(1.5±0.5) (-12)	0
CH$_3$SCH$_2$ + O$_2$ → CH$_3$SCH$_2$O$_2$	See reference			
SO$_3$ + NH$_3$ → H$_3$NSO$_3$	(3.9±0.8) (-30)	3.0±3.0	(4.7±1.3) (-11)	0±1.0
Metal Reactions				
Na + O$_2$ → NaO$_2$	(3.2±0.3) (-30)	1.4±0.3	(6.0±2.0) (-10)	0±1.0
NaO + O$_2$ → NaO$_3$	(3.5±0.7) (-30)	2.0±2.0	(5.7±3.0) (-10)	0±1.0
NaO + CO$_2$ → NaCO$_3$	(8.7±2.6) (-28)	2.0±2.0	(6.5±3.0) (-10)	0±1.0
NaOH + CO$_2$ → NaHCO$_3$	(1.3±0.3) (-28)	2.0±2.0	(6.8±4.0) (-10)	0±1.0

Table 3. Equilibrium Constants

$K(T)/\text{cm}^3 \text{ molecule}^{-1} = A \exp(B/T)$ [200 < T/K < 300]

Reaction	A/cm^3 molecule^{-1}	B/K	K (298 K)	f (298 K)
HO$_2$ + NO$_2$ → HO$_2$NO$_2$	2.1x10^{-27}	10900±1000	1.6x10^{-11}	5
NO + NO$_2$ → N$_2$O$_3$	3.3x10^{-27}	4667±100	2.1x10^{-20}	2
NO$_2$ + NO$_2$ → N$_2$O$_4$	5.2x10^{-29}	6643±250	2.5x10^{-19}	2
NO$_2$ + NO$_3$ → N$_2$O$_5$	2.7x10^{-27}	11000±500	2.9x10^{-11}	1.3
CH$_3$O$_2$ + NO$_2$ → CH$_3$O$_2$NO$_2$	1.3x10^{-28}	11200±1000	2.7x10^{-12}	2
CH$_3$C(O)O$_2$ + NO$_2$ → CH$_3$C(O)O$_2$NO$_2$	9.0x10^{-29}	14000±200	2.3x10^{-8}	2
F + O$_2$ → FOO	3.2x10^{-25}	6100±1200	2.5x10^{-16}	1.0
Cl + O$_2$ → ClOO	5.7x10^{-25}	2500±750	2.5x10^{-21}	2
Cl + CO → ClCO	1.6x10^{-25}	4000±500	1.1x10^{-19}	5
ClO + O$_2$ → ClO·O$_2$	2.9x10^{-26}	<3700	<7.2x10^{-21}	-
ClO + ClO → Cl$_2$O$_2$	1.3x10^{-27}	8744±850	7.2x10^{-15}	1.5
ClO + OClO → Cl$_2$O$_3$	1.1x10^{-24}	5455±300	9.8x10^{-17}	3
OClO + NO$_3$ → O$_2$ClONO$_2$	1x10^{-28}	9300±1000	3.6x10^{-15}	5
OH + CS$_2$ → CS$_2$OH	4.5x10^{-25}	5140±500	1.4x10^{-17}	1.4
CH$_3$S + O$_2$ → CH$_3$SO$_2$	1.8x10^{-27}	5545±300	2.2x10^{-19}	1.4

Section 6
Fluid Properties

Thermodynamic Properties of Air 6-1
Properties of Water in the Range 0-100°C 6-3
Enthalpy of Vaporization of Water 6-3
Fixed Point Properties of H_2O and D_2O 6-4
Thermal Conductivity of Saturated H_2O and D_2O 6-4
Standard Density of Water 6-5
Properties of Ice and Supercooled Water 6-6
Volumetric Properties of Aqueous Sodium Chloride Solutions 6-8
Density of D_2O 6-9
Vapor Pressure of Ice 6-9
Vapor Pressure of Water from 0 to 370°C 6-10
Boiling Point of Water at Various Pressures 6-12
Melting Point of Ice as a Function of Pressure 6-12
Properties of Water and Steam as a Function of Temperature and Pressure 6-13
Permittivity (Dielectric Constant) of Water at Various Frequencies 6-15
Thermophysical Properties of Fluids 6-16
Virial Coefficients of Selected Gases 6-25
Van der Waals Constants for Gases 6-45
Mean Free Path and Related Properties of Gases 6-46
Influence of Pressure on Freezing Points 6-47
Critical Constants 6-48
Sublimation Pressure of Solids 6-61
Vapor Pressure 6-64
Vapor Pressure of Fluids at Temperatures below 300 K 6-92
Vapor Pressure of Saturated Salt Solutions 6-103
IUPAC Recommended Data for Vapor Pressure Calibration 6-104
Enthalpy of Vaporization 6-105
Enthalpy of Fusion 6-121
Pressure and Temperature Dependence of Liquid Density 6-136
Properties of Cryogenic Fluids 6-138
Properties of Liquid Helium 6-139
Halocarbon Refrigerants 6-140
Density and Specific Volume of Mercury 6-143
Thermal Properties of Mercury 6-144
Surface Tension of Common Liquids 6-145
Permittivity (Dielectric Constant) of Liquids 6-149
Permittivity (Dielectric Constant) of Gases 6-172
Azeotropic Data for Binary Mixtures 6-174
Viscosity of Gases 6-178
Viscosity of Liquids 6-180
Thermal Conductivity of Gases 6-185
Thermal Conductivity of Liquids 6-187
Diffusion in Gases 6-189
Diffusion of Gases in Water 6-191
Diffusion Coefficients in Liquids at Infinite Dilution 6-192

THERMODYNAMIC PROPERTIES OF AIR

These tables summarize the thermodynamic properties of air in the liquid and gaseous states, as well as along the saturation line. In the table for the saturation state, P(boil) is the bubble point temperature (i.e., the pressure at which boiling begins as the temperature of the liquid is raised), and P(con) is the dew point temperature (pressure at which condensation begins as the temperature of the gas is lowered). The other properties tabulated are density (ρ), enthalpy (H), entropy (S), and isobaric heat capacity (C_p). More detailed tables may be found in the references.

REFERENCES

1. Vasserman, A.A., and Rabinovich, V.A., *Thermophysical Properties of Liquid Air and its Components*, Izdatel'stvo Komiteta, Standartov, Moscow, 1968.
2. Vasserman, A.A., et al., *Thermophysical Properties of Air and Air Components,* Izdatel'stvo Nauka, Moscow, 1966.

Properties in the saturation state:

T K	P(boil) bar	P(con) bar	ρ (liq) g/cm³	ρ (gas) g/L
65	0.1468	0.0861	0.939	0.464
70	0.3234	0.2052	0.917	1.033
75	0.6366	0.4321	0.894	2.048
80	1.146	0.8245	0.871	3.709
85	1.921	1.453	0.845	6.258
90	3.036	2.397	0.819	9.980
95	4.574	3.748	0.792	15.21
100	6.621	5.599	0.763	22.39
110	12.59	11.22	0.699	45.15
120	21.61	20.14	0.622	87.34
130	34.16	33.32	0.487	184.33
132.55	37.69	37.69	0.313	312.89

Properties of liquid air:

P bar	T K	ρ g/cm³	H J/g	S J/g K	C_p J/g K
1	75	0.8935	−131.7	2.918	1.843
5	75	0.8942	−131.4	2.916	1.840
5	80	0.8718	−122.3	3.031	1.868
5	85	0.8482	−112.9	3.143	1.901
5	90	0.8230	−103.3	3.250	1.941
5	95	0.7962	−93.5	3.356	1.991
10	75	0.8952	−131.1	2.913	1.836
10	80	0.8729	−122.0	3.028	1.863
10	90	0.8245	−103.1	3.246	1.932
10	100	0.7695	−83.2	3.452	2.041
50	75	0.9025	−128.2	2.892	1.806
50	100	0.7859	−81.8	3.415	1.939
50	125	0.6222	−28.3	3.889	2.614
50	150	0.1879	91.9	4.764	2.721
100	75	0.9111	−124.5	2.867	1.774
100	100	0.8033	−79.4	3.376	1.852
100	125	0.6746	−31.4	3.805	2.062
100	150	0.4871	32.8	4.271	2.832

Properties of air in the gaseous state:

P bar	T K	ρ g/L	H J/g	S J/g K	C_p J/g K
1	100	3.556	98.3	5.759	1.032
1	200	1.746	199.7	6.463	1.007
1	300	1.161	300.3	6.871	1.007

THERMODYNAMIC PROPERTIES OF AIR (continued)

P bar	T K	ρ g/L	H J/g	S J/g K	C_p J/g K
1	500	0.696	503.4	7.389	1.030
1	1000	0.348	1046.6	8.138	1.141
10	200	17.835	195.2	5.766	1.049
10	300	11.643	298.3	6.204	1.021
10	500	6.944	502.9	6.727	1.034
10	1000	3.471	1047.2	7.477	1.142
100	200	213.950	148.8	4.949	1.650
100	300	116.945	279.9	5.486	1.158
100	500	66.934	499.0	6.048	1.073
100	1000	33.613	1052.4	6.812	1.151

PROPERTIES OF WATER IN THE RANGE 0 — 100 °C

This table summarizes the best available values of the density, specific heat capacity at constant pressure (C_p), vapor pressure, viscosity, thermal conductivity, dielectric constant, and surface tension for liquid water in the range 0 — 100 °C. All values (except vapor pressure) refer to a pressure of 100 kPa (1 bar). The temperature scale is IPTS-68.

t °C	Density g/cm³	C_p J/g K	Vap. pres. kPa	Visc. μPa s	Ther. cond. mW/K m	Diel. const.	Surf. ten. mN/m
0	0.99984	4.2176	0.6113	1793	561.0	87.90	75.64
10	0.99970	4.1921	1.2281	1307	580.0	83.96	74.23
20	0.99821	4.1818	2.3388	1002	598.4	80.20	72.75
30	0.99565	4.1784	4.2455	797.7	615.4	76.60	71.20
40	0.99222	4.1785	7.3814	653.2	630.5	73.17	69.60
50	0.98803	4.1806	12.344	547.0	643.5	69.88	67.94
60	0.98320	4.1843	19.932	466.5	654.3	66.73	66.24
70	0.97778	4.1895	31.176	404.0	663.1	63.73	64.47
80	0.97182	4.1963	47.373	354.4	670.0	60.86	62.67
90	0.96535	4.2050	70.117	314.5	675.3	58.12	60.82
100	0.95840	4.2159	101.325	281.8	679.1	55.51	58.91
Ref.	1—3	2	1, 3	3	3	4	5

REFERENCES

1. L. Harr, J. S. Gallagher, and G. S. Kell, *NBS/NRC Steam Tables*, Hemisphere Publishing Corp., 1984.
2. K. N. Marsh, Ed., *Recommended Reference Materials for the Realization of Physicochemical Properties*, Blackwell Scientific Publications, Oxford, 1987.
3. J. V. Sengers and J. T. R. Watson, Improved international formulations for the viscosity and thermal conductivity of water substance, *J. Phys. Chem. Ref. Data*, 15, 1291, 1986.
4. D. G. Archer and P. Wang, The dielectric constant of water and Debye-Hückel limiting law slopes, *J. Phys. Chem. Ref. Data*, 19, 371, 1990.
5. N. B. Vargaftik, et al., International tables of the surface tension of water, *J. Phys. Chem. Ref. Data*, 12, 817, 1983.

ENTHALPY OF VAPORIZATION OF WATER

The enthalpy (heat) of vaporization of water is tabulated as a function of temperature on the IPTS-68 scale.

REFERENCE

Marsh, K. N., Ed., *Recommended Reference Materials for the Realization of Physicochemical Properties,* Blackwell, Oxford, 1987.

t °C	$\Delta_{vap}H$ kJ/mol	t °C	$\Delta_{vap}H$ kJ/mol
0	45.054	200	34.962
25	43.990	220	33.468
40	43.350	240	31.809
60	42.482	260	29.930
80	41.585	280	27.795
100	40.657	300	25.300
120	39.684	320	22.297
140	38.643	340	18.502
160	37.518	360	12.966
180	36.304	374	2.066

FIXED POINT PROPERTIES OF H_2O AND D_2O

Temperatures are given on the IPTS-68 scale.

REFERENCES

1. Haar, L., Gallagher, J.S., and Kell, G.S., *NBS/NRC Steam Tables*, Hemisphere Publishing Corp., 1984.
2. Levelt Sengers, J.M.H., Straub, J., Watanabe, K., and Hill, P.G., Assessment of critical parameter values for H_2O and D_2O, *J. Phys. Chem. Ref. Data,* 14, 193, 1985.
3. Kestin, J. et. al., Thermophysical properties of fluid D_2O, *J. Phys. Chem. Ref. Data,* 13, 601, 1984.
4. Kestin, J. et. al., Thermophysical properties of fluid H_2O, *J. Phys. Chem. Ref. Data,* 13, 175, 1984.
5. Hill, P.G., MacMillan, R.D.C., and Lee, V., A fundamental equation of state for heavy water, *J. Phys. Chem. Ref. Data*, 11, 1, 1982.

	Unit	H_2O	D_2O
Molar mass	g/mol	18.01528	20.02748
Melting point(101.325 kPa)	°C	0.00	3.82
Boiling point(101.325 kPa)	°C	100.00	101.42
Triple point temperature	°C	0.01	3.82
Triple point pressure	Pa	611.73	661
Triple point density(l)	g/cm^3	0.99978	1.1055
Triple point density(g)	mg/L	4.885	5.75
Critical temperature	°C	373.99	370.74
Critical pressure	MPa	22.064	21.671
Critical density	g/cm^3	0.322	0.356
Critical specific volume	cm^3/g	3.11	2.81
Maximum density(saturated liquid)	g/cm^3	0.99995	1.1053
Temperature of maximum density	°C	4.0	11.2

THERMAL CONDUCTIVITY OF SATURATED H_2O AND D_2O

This table gives the thermal conductivity λ for water (H_2O or D_2O) in equilibrium with its vapor. Values for the liquid (λ_l) and vapor (λ_v) are listed, as well as the vapor pressure.

REFERENCES

1. Sengers, J.V. and Watson, J.T.R., Improved international formulations for the viscosity and thermal conductivity of water substance, *J. Phys. Chem. Ref. Data*, 15, 1291, 1986.
2. Matsunaga, N. and Nagashima, A., Transport properties of liquid and gaseous D_2O over a wide range of temperature and pressure, *J. Phys. Chem. Ref. Data*, 12, 933, 1983.

	H_2O			D_2O		
t/°C	P/kPa	λ_l/(mW/K m)	λ_v/(mW/K m)	P/kPa	λ_l/(mW/K m)	λ_v/(mW/K m)
0	0.6	561.0	16.49			
10	1.2	580.0	17.21	1.0	575	17.0
20	2.3	598.4	17.95	2.0	589	17.8
30	4.2	615.4	18.70	3.7	600	18.5
40	7.4	630.5	19.48	6.5	610	19.3
50	12.3	643.5	20.28	11.1	618	20.2
60	19.9	654.3	21.10	18.2	625	21.0
70	31.2	663.1	21.96	28.8	629	21.9
80	47.4	670.0	22.86	44.2	633	22.8
90	70.1	675.3	23.80	66.1	635	23.8
100	101.3	679.1	24.79	96.2	636	24.8
150	476	682.1	30.77	465	625	30.8
200	1555	663.4	39.10	1546	592	39.0
250	3978	621.4	51.18	3995	541	52.0
300	8593	547.7	71.78	8688	473	75.2
350	16530	447.6	134.59	16820	391	143.0

STANDARD DENSITY OF WATER

This table gives the density ρ of standard mean ocean water (SMOW), free from dissolved salts and gases, at a pressure of 101325 Pa. SMOW is a standard water sample of high purity and known isotopic composition. Methods of correcting for different isotopic compositions are discussed in the reference. The table below is reprinted with the permission of IUPAC. Note that the temperature scale is IPTS-68.

REFERENCE

Marsh, K. N., Ed., *Recommended Reference Materials for the Realization of Physicochemical Properties,* Blackwell Scientific Publications, Oxford, 1987.

$\rho/\text{kg m}^{-3}$

$t_{68}/°C$	0.0	0.1	0.2	0.3	0.4	0.5	0.6	0.7	0.8	0.9
0	999.8426	8493	8558	8622	8683	8743	8801	8857	8912	8964
1	999.9015	9065	9112	9158	9202	9244	9284	9323	9360	9395
2	999.9429	9461	9491	9519	9546	9571	9595	9616	9636	9655
3	999.9672	9687	9700	9712	9722	9731	9738	9743	9747	9749
4	999.9750	9748	9746	9742	9736	9728	9719	9709	9696	9683
5	999.9668	9651	9632	9612	9591	9568	9544	9518	9490	9461
6	999.9430	9398	9365	9330	9293	9255	9216	9175	9132	9088
7	999.9043	8996	8948	8898	8847	8794	8740	8684	8627	8569
8	999.8509	8448	8385	8321	8256	8189	8121	8051	7980	7908
9	999.7834	7759	7682	7604	7525	7444	7362	7279	7194	7108
10	999.7021	6932	6842	6751	6658	6564	6468	6372	6274	6174
11	999.6074	5972	5869	5764	5658	5551	5443	5333	5222	5110
12	999.4996	4882	4766	4648	4530	4410	4289	4167	4043	3918
13	999.3792	3665	3536	3407	3276	3143	3010	2875	2740	2602
14	999.2464	2325	2184	2042	1899	1755	1609	1463	1315	1166
15	999.1016	0864	0712	0558	0403	0247	0090	9932*	9772*	9612*
16	998.9450	9287	9123	8957	8791	8623	8455	8285	8114	7942
17	998.7769	7595	7419	7243	7065	6886	6706	6525	6343	6160
18	998.5976	5790	5604	5416	5228	5038	4847	4655	4462	4268
19	998.4073	3877	3680	3481	3282	3081	2880	2677	2474	2269
20	998.2063	1856	1649	1440	1230	1019	0807	0594	0380	0164
21	997.9948	9731	9513	9294	9073	8852	8630	8406	8182	7957
22	997.7730	7503	7275	7045	6815	6584	6351	6118	5883	5648
23	997.5412	5174	4936	4697	4456	4215	3973	3730	3485	3240
24	997.2994	2747	2499	2250	2000	1749	1497	1244	0990	0735
25	997.0480	0223	9965*	9707*	9447*	9186*	8925*	8663*	8399*	8135*
26	996.7870	7604	7337	7069	6800	6530	6259	5987	5714	5441
27	996.5166	4891	4615	4337	4059	3780	3500	3219	2938	2655
28	996.2371	2087	1801	1515	1228	0940	0651	0361	0070	9778*
29	995.9486	9192	8898	8603	8306	8009	7712	7413	7113	6813
30	995.6511	6209	5906	5602	5297	4991	4685	4377	4069	3760
31	995.3450	3139	2827	2514	2201	1887	1572	1255	0939	0621
32	995.0302	9983*	9663*	9342*	9020*	8697*	8373*	8049*	7724*	7397*
33	994.7071	6743	6414	6085	5755	5423	5092	4759	4425	4091
34	994.3756	3420	3083	2745	2407	2068	1728	1387	1045	0703
35	994.0359	0015	9671*	9325*	8978*	8631*	8283*	7934*	7585*	7234*
36	993.6883	6531	6178	5825	5470	5115	4759	4403	4045	3687
37	993.3328	2968	2607	2246	1884	1521	1157	0793	0428	0062
38	992.9695	9328	8960	8591	8221	7850	7479	7107	6735	6361
39	992.5987	5612	5236	4860	4483	4105	3726	3347	2966	2586
40	992.2204									

* The leading figure decreases by 1.

PROPERTIES OF ICE AND SUPERCOOLED WATER

The common form of ice at ambient temperature and pressure is hexagonal ice, designated as ice I_h (see phase diagram on p. 12-174). The data given here refer to that form. Data have been taken from the references indicated; values have been interpolated and smoothed in some cases. All properties are sensitive to the method of preparation of the sample, since air or other gases are sometimes occluded. For this reason there is often disagreement among values found in the literature.

Density values (except at 0°C) and the thermal expansion coefficient were calculated from the temperature variation in the crystal lattice constants of ice (see Ref. 1). The thermal expansion coefficient appears to become negative around -200°C, but there is considerable scatter in the data.

Density of ice I_h and supercooled water in g cm^{-3}

$t/°C$	ρ (ice)	ρ (supercooled water)
0	0.9167	0.9998
-10	0.9187	0.9982
-20	0.9203	0.9935
-30	0.9216	0.9839
-40	0.9228	
-50	0.9240	
-60	0.9252	
-80	0.9274	
-100	0.9292	
-120	0.9305	
-140	0.9314	
-160	0.9331	
-180	0.9340	
Ref.	1	8

Phase transition properties:

$\Delta_{fus}H(0°C) = 333.6$ J/g (Ref. 2)

$\Delta_{sub}H(0°C) = 2838$ J/g (Ref. 2)

Other properties of ice I_h :

α_V: cubic thermal expansion coefficient, $\alpha_V = -(1/V)(\partial V/\partial t)_p$

κ : adiabatic compressibility, $\kappa = -(1/V)(\partial V/\partial p)_S$

ε : relative permittivity (dielectric constant)

k : thermal conductivity

c_p: specific heat capacity at constant pressure

$t/°C$	$\alpha_V/10^{-6}\ °C^{-1}$	$\kappa/10^{-5}\ MPa^{-1}$	ε	$k/W\ cm^{-1}\ °C^{-1}$	$c_p/J\ g^{-1}\ °C^{-1}$
0	159	13.0	91.6	0.0214	2.11
−10	155	12.8	94.4	0.023	2.03
−20	149	12.7	97.5	0.024	1.96
−30	143	12.5	99.7	0.025	1.88
−40	137	12.4	101.9	0.026	1.80
−50	130	12.2	106.9	0.028	1.72
−60	122	12.1	119.5	0.030	1.65
−80	105	11.9		0.033	1.50
−100	85	11.6		0.037	1.36
−120	77	11.4		0.042	1.23
−140	60	11.3		0.049	1.10
−160	45	11.2		0.057	0.97
−180	30	11.1		0.070	0.83
−200		11.0		0.087	0.67
−220		10.9		0.118	0.50
−240		10.9		0.20	0.29
−250		10.9		0.32	0.17
Ref.	1,2,3,5	1,5	6	7	1

REFERENCES

1. Eisenberg, D., and Kauzmann, W., *The Structure and Properties of Water*, Oxford University Press, Oxford, 1969.
2. *Landolt-Börnstein, Numerical Data and Functional Relationships in Science and Technology, New Series*, V/1b, Springer-Verlag, Heidelberg, 1982.
3. LaPlaca, S., and Post, B., *Acta Cryst.*, 13, 503, 1960. [Thermal expansion of lattice]
4. Brill, R., and Tippe, A., *Acta Cryst.*, 23, 343, 1967. [Thermal expansion of lattice]
5. Leadbetter, A. J., *Proc. Roy. Soc. A* 287, 403, 1965. [Compressibility and thermal expansion]
6. Auty, R. P., and Cole, R. H., *J. Chem. Phys.*, 20, 1309, 1952. [Dielectric constant]
7. Slack, G. A., *Phys. Rev. B*, 22, 3065, 1980. [Thermal conductivity]
8. Hare, D. E., and Sorensen, C. M., *J. Chem. Phys.*, 87, 4840, 1987. [Supercooled water]
9. Hobbs, P. V., *Ice Physics,* Clarendon Press, Oxford, 1974.

VOLUMETRIC PROPERTIES OF AQUEOUS SODIUM CHLORIDE SOLUTIONS

This table gives the following properties of aqueous solutions of NaCl as a function of temperature and concentration:

Specific volume v (reciprocal of density) in cm^3/g
Isothermal compressibility $\kappa_T = -(1/v)(\partial v/\partial P)_T$ in GPa^{-1}
Cubic expansion coefficient $\alpha_v = (1/v)(\partial v/\partial T)_P$ in kK^{-1}

All data refer to a pressure of 100 kPa (1 bar). The reference gives properties over a wider range of temperature and pressure.

REFERENCE

Rogers, P. S. Z., and Pitzer, K. S., *J. Phys. Chem. Ref. Data,* 11, 15, 1982.

Molality in mol/kg

T/°C	0.100	0.250	0.500	0.750	1.000	2.000	3.000	4.000	5.000
				Specific volume v in cm^3/g					
0	0.995732	0.989259	0.978889	0.968991	0.959525	0.925426	0.896292	0.870996	0.848646
10	0.995998	0.989781	0.979804	0.970256	0.961101	0.927905	0.899262	0.874201	0.851958
20	0.997620	0.991564	0.981833	0.972505	0.963544	0.930909	0.902565	0.877643	0.855469
25	0.998834	0.992832	0.983185	0.973932	0.965038	0.932590	0.904339	0.879457	0.857301
30	1.000279	0.994319	0.984735	0.975539	0.966694	0.934382	0.906194	0.881334	0.859185
40	1.003796	0.997883	0.988374	0.979243	0.970455	0.938287	0.910145	0.885276	0.863108
50	1.008064	1.002161	0.992668	0.983551	0.974772	0.942603	0.914411	0.889473	0.867241
60	1.0130	1.0071	0.9976	0.9885	0.9797	0.9474	0.9191	0.8940	0.8716
70	1.0186	1.0127	1.0031	0.9939	0.9851	0.9526	0.9240	0.8987	0.8762
80	1.0249	1.0188	1.0092	0.9999	0.9909	0.9581	0.9293	0.9037	0.8809
90	1.0317	1.0256	1.0157	1.0063	0.9972	0.9640	0.9348	0.9089	0.8858
100	1.0391	1.0329	1.0228	1.0133	1.0040	0.9703	0.9406	0.9144	0.8910
				Compressibility κ_T in GPa^{-1}					
0	0.503	0.492	0.475	0.459	0.443	0.389	0.346	0.315	0.294
10	0.472	0.463	0.449	0.436	0.423	0.377	0.341	0.313	0.294
20	0.453	0.446	0.433	0.422	0.411	0.371	0.338	0.313	0.294
25	0.447	0.440	0.428	0.417	0.407	0.369	0.337	0.313	0.294
30	0.443	0.436	0.425	0.414	0.404	0.367	0.337	0.313	0.294
40	0.438	0.432	0.421	0.411	0.401	0.367	0.338	0.315	0.296
50	0.438	0.431	0.421	0.411	0.402	0.369	0.340	0.317	0.299
60	0.44	0.44	0.43	0.42	0.41	0.38	0.35	0.32	0.30
70	0.45	0.44	0.43	0.42	0.42	0.38	0.36	0.33	0.31
80	0.46	0.45	0.44	0.43	0.43	0.39	0.37	0.34	0.32
90	0.47	0.47	0.46	0.45	0.44	0.41	0.38	0.35	0.33
100	0.49	0.48	0.47	0.46	0.45	0.42	0.39	0.37	0.34
				Cubic expansion coefficient α_V in kK^{-1}					
0	-0.058	-0.026	0.024	0.069	0.110	0.237	0.313	0.355	
10	0.102	0.123	0.156	0.186	0.213	0.297	0.349	0.380	
20	0.218	0.232	0.254	0.274	0.292	0.349	0.384	0.406	
25	0.267	0.278	0.296	0.312	0.327	0.373	0.401	0.420	
30	0.311	0.320	0.334	0.347	0.359	0.395	0.418	0.433	
40	0.389	0.394	0.402	0.410	0.417	0.438	0.451	0.460	
50	0.458	0.460	0.464	0.467	0.470	0.479	0.484	0.486	
60	0.52	0.52	0.52	0.52	0.52	0.52	0.52	0.52	
70	0.58	0.58	0.58	0.57	0.57	0.56	0.55	0.54	
80	0.64	0.63	0.63	0.62	0.61	0.60	0.58	0.56	
90	0.69	0.68	0.67	0.67	0.66	0.63	0.61	0.59	
100	0.74	0.73	0.72	0.71	0.70	0.66	0.64	0.61	

DENSITY OF D₂O

Density of liquid D_2O in g/cm³ at a pressure of 100 kPa (1 bar).

REFERENCE

Kirillin, V.A., Ed., *Heavy Water: Thermophysical Properties*, Gosudarstvennoe Energeticheskoe Izdatel'stvo, Moscow, 1963.

$t/°C$	3.8	5	10	15	20	25	30
Density	1.1053	1.1055	1.1057	1.1056	1.105	1.1044	1.1034

$t/°C$	35	40	45	50	55	60	65
Density	1.1019	1.1001	1.0979	1.0957	1.0931	1.0905	1.0875

$t/°C$	70	75	80	85	90	95	100
Density	1.0847	1.0815	1.0783	1.0748	1.0712	1.0673	1.0635

VAPOR PRESSURE OF ICE

The values of the vapor (sublimation) pressure of ice in this table were calculated from the equation recommended by the International Association for the Properties of Steam (IAPS) in 1993. Temperature values correspond to the ITS-90 temperature scale. The uncertainty in the pressure is estimated to be 0.1% for $t > -25°C$ and 0.5% for $t < -25°C$. The first entry in the table is the triple point of water.

REFERENCE

Wagner, W., Saul, A., and Pruss, A., *J. Phys. Chem. Ref. Data*, 23, 515, 1994.

$t/°C$	p/Pa	$t/°C$	p/Pa	$t/°C$	p/Pa
0.01	611.657	−16	150.68	−33	27.71
0	611.15	−17	137.25	−34	24.90
−1	562.67	−18	124.92	−35	22.35
−2	517.72	−19	113.62	−36	20.04
−3	476.06	−20	103.26	−37	17.96
−4	437.47	−21	93.77	−38	16.07
−5	401.76	−22	85.10	−39	14.37
−6	368.73	−23	77.16	−40	12.84
−7	338.19	−24	69.91	−45	7.202
−8	309.98	−25	63.29	−50	3.936
−9	283.94	−26	57.25	−55	2.093
−10	259.90	−27	51.74	−60	1.080
−11	237.74	−28	46.73	−65	0.540
−12	217.32	−29	42.16	−70	0.261
−13	198.52	−30	38.01	−75	0.122
−14	181.22	−31	34.24	−80	0.055
−15	165.30	−32	30.82		

VAPOR PRESSURE OF WATER FROM 0 TO 370° C

This table gives the vapor pressure of water at intervals of 1° C from the melting point to the critical point.

REFERENCE

Haar, L., Gallagher, J.S., and Kell, G.S., *NBS/NRC Steam Tables*, Hemisphere Publishing Corp., New York, 1984.

t/°C	P/kPa	t/°C	P/kPa	t/°C	P/kPa	t/°C	P/kPa
0	0.61129	53	14.303	106	125.03	159	602.11
1	0.65716	54	15.012	107	129.39	160	617.66
2	0.70605	55	15.752	108	133.88	161	633.53
3	0.75813	56	16.522	109	138.50	162	649.73
4	0.81359	57	17.324	110	143.24	163	666.25
5	0.87260	58	18.159	111	148.12	164	683.10
6	0.93537	59	19.028	112	153.13	165	700.29
7	1.0021	60	19.932	113	158.29	166	717.83
8	1.0730	61	20.873	114	163.58	167	735.70
9	1.1482	62	21.851	115	169.02	168	753.94
10	1.2281	63	22.868	116	174.61	169	772.52
11	1.3129	64	23.925	117	180.34	170	791.47
12	1.4027	65	25.022	118	186.23	171	810.78
13	1.4979	66	26.163	119	192.28	172	830.47
14	1.5988	67	27.347	120	198.48	173	850.53
15	1.7056	68	28.576	121	204.85	174	870.98
16	1.8185	69	29.852	122	211.38	175	891.80
17	1.9380	70	31.176	123	218.09	176	913.03
18	2.0644	71	32.549	124	224.96	177	934.64
19	2.1978	72	33.972	125	232.01	178	956.66
20	2.3388	73	35.448	126	239.24	179	979.09
21	2.4877	74	36.978	127	246.66	180	1001.9
22	2.6447	75	38.563	128	254.25	181	1025.2
23	2.8104	76	40.205	129	262.04	182	1048.9
24	2.9850	77	41.905	130	270.02	183	1073.0
25	3.1690	78	43.665	131	278.20	184	1097.5
26	3.3629	79	45.487	132	286.57	185	1122.5
27	3.5670	80	47.373	133	295.15	186	1147.9
28	3.7818	81	49.324	134	303.93	187	1173.8
29	4.0078	82	51.342	135	312.93	188	1200.1
30	4.2455	83	53.428	136	322.14	189	1226.9
31	4.4953	84	55.585	137	331.57	190	1254.2
32	4.7578	85	57.815	138	341.22	191	1281.9
33	5.0335	86	60.119	139	351.09	192	1310.1
34	5.3229	87	62.499	140	361.19	193	1338.8
35	5.6267	88	64.958	141	371.53	194	1368.0
36	5.9453	89	67.496	142	382.11	195	1397.6
37	6.2795	90	70.117	143	392.92	196	1427.8
38	6.6298	91	72.823	144	403.98	197	1458.5
39	6.9969	92	75.614	145	415.29	198	1489.7
40	7.3814	93	78.494	146	426.85	199	1521.4
41	7.7840	94	81.465	147	438.67	200	1553.6
42	8.2054	95	84.529	148	450.75	201	1586.4
43	8.6463	96	87.688	149	463.10	202	1619.7
44	9.1075	97	90.945	150	475.72	203	1653.6
45	9.5898	98	94.301	151	488.61	204	1688.0
46	10.094	99	97.759	152	501.78	205	1722.9
47	10.620	100	101.32	153	515.23	206	1758.4
48	11.171	101	104.99	154	528.96	207	1794.5
49	11.745	102	108.77	155	542.99	208	1831.1
50	12.344	103	112.66	156	557.32	209	1868.4
51	12.970	104	116.67	157	571.94	210	1906.2
52	13.623	105	120.79	158	586.87	211	1944.6

$t/°C$	P/kPa	$t/°C$	P/kPa	$t/°C$	P/kPa	$t/°C$	P/kPa
212	1983.6	253	4178.9	294	7881.3	335	13701
213	2023.2	254	4249.1	295	7995.2	336	13876
214	2063.4	255	4320.2	296	8110.3	337	14053
215	2104.2	256	4392.2	297	8226.8	338	14232
216	2145.7	257	4465.1	298	8344.5	339	14412
217	2187.8	258	4539.0	299	8463.5	340	14594
218	2230.5	259	4613.7	300	8583.8	341	14778
219	2273.8	260	4689.4	301	8705.4	342	14964
220	2317.8	261	4766.1	302	8828.3	343	15152
221	2362.5	262	4843.7	303	8952.6	344	15342
222	2407.8	263	4922.3	304	9078.2	345	15533
223	2453.8	264	5001.8	305	9205.1	346	15727
224	2500.5	265	5082.3	306	9333.4	347	15922
225	2547.9	266	5163.8	307	9463.1	348	16120
226	2595.9	267	5246.3	308	9594.2	349	16320
227	2644.6	268	5329.8	309	9726.7	350	16521
228	2694.1	269	5414.3	310	9860.5	351	16725
229	2744.2	270	5499.9	311	9995.8	352	16931
230	2795.1	271	5586.4	312	10133	353	17138
231	2846.7	272	5674.0	313	10271	354	17348
232	2899.0	273	5762.7	314	10410	355	17561
233	2952.1	274	5852.4	315	10551	356	17775
234	3005.9	275	5943.1	316	10694	357	17992
235	3060.4	276	6035.0	317	10838	358	18211
236	3115.7	277	6127.9	318	10984	359	18432
237	3171.8	278	6221.9	319	11131	360	18655
238	3228.6	279	6317.0	320	11279	361	18881
239	3286.3	280	6413.2	321	11429	362	19110
240	3344.7	281	6510.5	322	11581	363	19340
241	3403.9	282	6608.9	323	11734	364	19574
242	3463.9	283	6708.5	324	11889	365	19809
243	3524.7	284	6809.2	325	12046	366	20048
244	3586.3	285	6911.1	326	12204	367	20289
245	3648.8	286	7014.1	327	12364	368	20533
246	3712.1	287	7118.3	328	12525	369	20780
247	3776.2	288	7223.7	329	12688	370	21030
248	3841.2	289	7330.2	330	12852	371	21283
249	3907.0	290	7438.0	331	13019	372	21539
250	3973.6	291	7547.0	332	13187	373	21799
251	4041.2	292	7657.2	333	13357	373.98	22055
252	4109.6	293	7768.6	334	13528		

BOILING POINT OF WATER AT VARIOUS PRESSURES

Data are based on the equation of state recommended by the International Association for the Properties of Steam in 1984, as presented in Haar, Gallagher, and Kell, *NBS-NRC Steam Tables* (Hemisphere Publishing Corp., New York, 1984). The temperature scale is IPTS-68.
Note that: 1 mbar = 100 Pa = 0.000986923 atmos = 0.750062 mmHg.

P/mbar	T/°C	P/mbar	T/°C	P/mbar	T/°C	P/mbar	T/°C
50	32.88	915	97.17	1013.25	100.00	1200	104.81
100	45.82	920	97.32	1015	100.05	1250	105.99
150	53.98	925	97.47	1020	100.19	1300	107.14
200	60.07	930	97.62	1025	100.32	1350	108.25
250	64.98	935	97.76	1030	100.46	1400	109.32
300	69.11	940	97.91	1035	100.60	1450	110.36
350	72.70	945	98.06	1040	100.73	1500	111.38
400	75.88	950	98.21	1045	100.87	1550	112.37
450	78.74	955	98.35	1050	101.00	1600	113.33
500	81.34	960	98.50	1055	101.14	1650	114.26
550	83.73	965	98.64	1060	101.27	1700	115.18
600	85.95	970	98.78	1065	101.40	1750	116.07
650	88.02	975	98.93	1070	101.54	1800	116.94
700	89.96	980	99.07	1075	101.67	1850	117.79
750	91.78	985	99.21	1080	101.80	1900	118.63
800	93.51	990	99.35	1085	101.93	1950	119.44
850	95.15	995	99.49	1090	102.06	2000	120.24
900	96.71	1000	99.63	1095	102.19	2050	121.02
905	96.87	1005	99.77	1100	102.32	2100	121.79
910	97.02	1010	99.91	1150	103.59	2150	122.54

MELTING POINT OF ICE AS A FUNCTION OF PRESSURE

This table gives values of the melting temperature of ice at various pressures, as calculated from the equation for the ice I - liquid water phase boundary recommended by the International Association for the Properties of Steam (IAPS). Temperatures are on the ITS-90 scale. See the Reference for information on forms of ice that exist at higher pressures. The transition points for transformations of the various forms of ice (in each case in equilibrium with liquid water) are:

ice I - ice III	209.9 MPa	–21.985°C
ice III - ice V	350.1	–16.986
ice V - ice VI	632.4	0.16
ice VI - ice VII	2216	82

REFERENCE

Wagner, W., Saul, A., and Pruss, A., *J. Phys. Chem. Ref. Data,* 23, 515, 1994.

p/MPa	t/°C	p/MPa	t/°C	p/MPa	t/°C
0.1	0.00	40	-3.15	130	-12.07
1	-0.06	50	-4.02	140	-13.22
2	-0.14	60	-4.91	150	-14.40
3	-0.21	70	-5.83	160	-15.62
4	-0.29	80	-6.79	170	-16.85
5	-0.36	90	-7.78	180	-18.11
10	-0.74	100	-8.80	190	-19.39
20	-1.52	110	-9.86	200	-20.69
30	-2.32	120	-10.95	210	-22.00

PROPERTIES OF WATER AND STEAM AS A FUNCTION OF TEMPERATURE AND PRESSURE

This table gives properties of compressed water and superheated steam at selected pressures and temperatures. The properties included are density ρ, enthalpy H, entropy S, heat capacity at constant pressure C_p, and static dielectric constant (relative permittivity). The table was generated from the formulation approved by the International Association for the Properties of Water and Steam for general and scientific use. The reference state for this table is the liquid at the triple point, at which the internal energy and entropy are taken as zero. A duplicate entry in the temperature column indicates a phase transition (liquid-vapor) at that temperature; property values are then given for both phases. In the 100 MPa section of the table, an entry is given at the critical temperature, 647.10 K. Temperatures refer to the ITS-90 scale, on which the normal boiling point of water is 373.12 K (99.97°C).

REFERENCES

1. Release on the IAPWS Formulation 1995 for the Thermodynamic Properties of Ordinary Water Substance for General and Scientific Use, September 1996; available from Executive Secretary of IAPWS, Electric Power Research Institute, 3412 Hillview Ave., Palo Alto, CA 94304-1395.
2. NIST Chemistry WebBook, NIST Standard Reference Database Number 69, Mallard, W. G., and Linstrom, P. J., Eds., March 1998, National Institute of Standards and Technology, Gaithersburg, MD, 20899 (http://webbook.nist.gov).
3. Pruss, A. and Wagner, W., to be published.
4. Fernandez, D. P., Goodwin, A. R. H., Lemmon, E. W., Levelt Sengers, J. M. H., and Williams, R. C., *J. Phys. Chem. Ref. Data,* 26, 1125, 1997. [Dielectric constant]

p/MPa	T/K	ρ/kg m^{-3}	H/J g^{-1}	S/J g^{-1}K^{-1}	C_p/J g^{-1}K^{-1}	Diel. const.
0.1	273.16	999.84	0.10	0.0000	4.2194	87.90
0.1	300	996.56	112.65	0.3931	4.1806	77.75
0.1	325	987.19	217.15	0.7276	4.1819	69.32
0.1	350	973.73	321.84	1.0380	4.1945	61.79
0.1	372.76	958.63	417.50	1.3028	4.2152	55.61
0.1	372.76	0.59034	2674.9	7.3588	2.0784	1.006
0.1	375	0.58653	2679.6	7.3713	2.0686	1.006
0.1	400	0.54761	2730.4	7.5025	2.0078	1.005
0.1	450	0.48458	2829.7	7.7365	1.9752	1.004
0.1	500	0.43514	2928.6	7.9447	1.9813	1.003
0.1	550	0.39507	3028.1	8.1344	2.0010	1.003
0.1	600	0.36185	3128.8	8.3096	2.0268	1.002
0.1	650	0.33384	3230.8	8.4730	2.0557	1.002
0.1	700	0.30988	3334.4	8.6264	2.0867	1.002
0.1	750	0.28915	3439.5	8.7715	2.1191	1.002
0.1	800	0.27102	3546.3	8.9093	2.1525	1.001
0.1	850	0.25504	3654.8	9.0408	2.1868	1.001
0.1	900	0.24085	3765.0	9.1668	2.2216	1.001
0.1	950	0.22815	3876.9	9.2879	2.2568	1.001
0.1	1000	0.21673	3990.7	9.4045	2.2921	1.001
0.1	1050	0.20640	4106.1	9.5172	2.3273	1.001
0.1	1100	0.19701	4223.4	9.6263	2.3621	1.001
0.1	1150	0.18844	4342.3	9.7321	2.3965	1.001
0.1	1200	0.18058	4463.0	9.8348	2.4302	1.001
1	273.16	1000.3	1.02	0.0000	4.2150	87.93
1	300	996.96	113.48	0.3928	4.1781	77.78
1	325	987.58	217.93	0.7272	4.1798	69.36
1	350	974.13	322.56	1.0374	4.1925	61.82
1	375	957.43	427.64	1.3274	4.2158	55.09
1	400	937.87	533.47	1.6005	4.2535	49.06
1	450	890.39	749.20	2.1086	4.3924	38.81
1	453.03	887.13	762.51	2.1381	4.4045	38.23
1	453.03	5.1450	2777.1	6.5850	2.7114	1.042
1	500	4.5323	2891.2	6.8250	2.2795	1.034

p/MPa	T/K	ρ/kg m⁻³	H/J g⁻¹	S/J g⁻¹K⁻¹	C_p/J g⁻¹K⁻¹	Diel. const.
1	550	4.0581	3001.8	7.0359	2.1647	1.028
1	600	3.6871	3109.0	7.2224	2.1292	1.024
1	650	3.3843	3215.2	7.3925	2.1254	1.020
1	700	3.1305	3321.7	7.5504	2.1368	1.017
1	750	2.9140	3429.0	7.6984	2.1566	1.015
1	800	2.7265	3537.5	7.8384	2.1816	1.013
1	850	2.5624	3647.3	7.9715	2.2098	1.012
1	900	2.4174	3758.5	8.0986	2.2402	1.011
1	950	2.2882	3871.3	8.2206	2.2721	1.010
1	1000	2.1723	3985.7	8.3380	2.3048	1.009
1	1050	2.0678	4101.8	8.4512	2.3380	1.008
1	1100	1.9729	4219.5	8.5608	2.3713	1.007
1	1150	1.8865	4338.9	8.6669	2.4044	1.007
1	1200	1.8074	4460.0	8.7699	2.4371	1.006
10	273.16	1004.8	10.1	0.000	4.173	88.30
10	300	1001.0	121.7	0.390	4.153	78.11
10	325	991.46	225.6	0.723	4.160	69.67
10	350	978.09	329.7	1.031	4.173	62.13
10	375	961.62	434.4	1.320	4.195	55.40
10	400	942.42	539.6	1.592	4.230	49.39
10	450	896.16	753.9	2.096	4.355	39.17
10	500	838.02	977.1	2.566	4.602	30.79
10	550	761.82	1218	3.027	5.140	23.53
10	584.15	688.42	1408	3.360	6.123	18.70
10	584.15	55.463	2725	5.616	7.140	1.404
10	600	49.773	2820	5.775	5.136	1.365
10	650	40.479	3022	6.100	3.396	1.267
10	700	35.355	3177	6.330	2.874	1.214
10	750	31.810	3314	6.520	2.645	1.179
10	800	29.107	3443	6.686	2.531	1.154
10	850	26.933	3568	6.838	2.473	1.134
10	900	25.123	3691	6.978	2.445	1.118
10	950	23.580	3813	7.110	2.436	1.105
10	1000	22.241	3935	7.235	2.439	1.095
10	1050	21.063	4057	7.354	2.450	1.086
10	1100	20.017	4180	7.469	2.466	1.078
10	1150	19.078	4304	7.579	2.485	1.072
10	1200	18.230	4429	7.685	2.507	1.066
100	273.16	1045.3	95.4	-0.008	3.905	91.83
100	300	1037.2	201.4	0.362	3.979	81.22
100	325	1026.6	301.3	0.682	4.008	72.58
100	350	1013.6	401.7	0.979	4.025	64.95
100	375	998.59	502.6	1.258	4.040	58.19
100	400	981.82	603.7	1.518	4.056	52.20
100	450	943.51	807.8	1.999	4.110	42.15
100	500	899.21	1015	2.436	4.196	34.15
100	550	848.78	1228	2.842	4.323	27.67
100	600	791.49	1448	3.225	4.501	22.29
100	647.10	730.24	1665	3.573	4.733	17.97
100	650	726.21	1679	3.595	4.750	17.72
100	700	651.77	1925	3.958	5.083	13.75
100	750	568.52	2188	4.322	5.449	10.34
100	800	482.23	2466	4.681	5.610	7.562
100	850	404.66	2742	5.016	5.380	5.571
100	900	343.61	3000	5.310	4.887	4.284
100	950	298.61	3231	5.560	4.382	3.477
100	1000	265.45	3440	5.774	3.978	2.956
100	1050	240.32	3631	5.961	3.683	2.601
100	1100	220.62	3809	6.127	3.471	2.347
100	1150	204.71	3979	6.278	3.319	2.158
100	1200	191.53	4142	6.417	3.209	2.011

PERMITTIVITY (DIELECTRIC CONSTANT) OF WATER AT VARIOUS FREQUENCIES

The permittivity of liquid water in the radiofrequency and microwave regions can be represented by the Debye equation (References 1 and 2):

$$\varepsilon' = \varepsilon_\infty + \frac{\varepsilon_s - \varepsilon_\infty}{1 + \omega^2 \tau^2}$$

$$\varepsilon'' = \frac{(\varepsilon_s - \varepsilon_\infty)\omega\tau}{1 + \omega^2 \tau^2}$$

where $\varepsilon = \varepsilon' + i\,\varepsilon''$ is the (complex) relative permittivity (i.e., the absolute permittivity divided by the permittivity of free space $\varepsilon_0 = 8.854 \cdot 10^{-12}$ F m^{-1}). Here ε_s is the static permittivity (see Reference 3 and the table "Properties of Water in the Range 0—100°C" in this Section); ε_∞ is a parameter describing the permittivity in the high frequency limit; τ is the relaxation time for molecular orientation; and $\omega = 2\pi f$ is the angular frequency. The values in this table have been calculated from parameters given in Reference 2:

	0°C	25°C	50°C
ε_∞	5.7	5.2	4.0
τ/ps	17.67	8.27	4.75

Other useful quantities that can be calculated from the values in the table are the loss tangent:

$$\tan\delta = \varepsilon'' / \varepsilon'$$

and the absorption coefficient α which describes the power attentuation per unit length ($P = P_0\,e^{-\alpha l}$):

$$\alpha = \frac{\pi f\,\varepsilon''}{c\sqrt{\varepsilon'}}$$

and c is the speed of light. The last equation is valid when $\varepsilon''/\varepsilon' \ll 1$.

REFERENCES

1. Fernendez, D.P., Mulev, Y., Goodwin, A.R.H., and Levelt Sengers, J.M.H., *J. Phys. Chem. Ref. Data*, 24, 33, 1995.
2. Kaatze, U., *J. Chem. Eng. Data,* 34, 371, 1989.
3. Archer, D.G., and Wang, P., *J. Phys. Chem. Ref. Data*, 12, 817, 1983.

Frequency	0°C ε'	0°C ε''	25°C ε'	25°C ε''	50°C ε'	50°C ε''
0	87.90	0.00	78.36	0.00	69.88	0.00
1 kHz	87.90	0.00	78.36	0.00	69.88	0.00
1 MHz	87.90	0.01	78.36	0.00	69.88	0.00
10 MHz	87.90	0.09	78.36	0.04	69.88	0.02
100 MHz	87.89	0.91	78.36	0.38	69.88	0.20
200 MHz	87.86	1.82	78.35	0.76	69.88	0.39
500 MHz	87.65	4.55	78.31	1.90	69.87	0.98
1 GHz	86.90	9.01	78.16	3.79	69.82	1.96
2 GHz	84.04	17.39	77.58	7.52	69.65	3.92
3 GHz	79.69	24.64	76.62	11.13	69.36	5.85
4 GHz	74.36	30.49	75.33	14.58	68.95	7.75
5 GHz	68.54	34.88	73.73	17.81	68.45	9.62
10 GHz	42.52	40.88	62.81	29.93	64.49	18.05
20 GHz	19.56	30.78	40.37	36.55	52.57	28.99
30 GHz	12.50	22.64	26.53	33.25	40.57	32.74
40 GHz	9.67	17.62	18.95	28.58	31.17	32.43
50 GHz	8.28	14.34	14.64	24.53	24.42	30.47

THERMOPHYSICAL PROPERTIES OF FLUIDS

These tables give thermodynamic and transport properties of some important fluids, as generated from the equations of state presented in the references below. The properties tabulated are density (ρ), energy (E), enthalpy (H), entropy (S), isochoric heat capacity (C_v), isobaric heat capacity (C_p), speed of sound (v_s), viscosity (η), thermal conductivity (λ), and dielectric constant (D). All extensive properties are given on a molar basis. Not all properties are included for every substance. The references should be consulted for information on the uncertainties and the reference states for E, H, and S.

Values are given as a function of temperature for several isobars. The phase can be determined by noting the sharp decrease in density between two successive temperature entries; all lines above this point refer to the liquid phase, and all lines below refer to the gas phase. If there is no sharp discontinuity in density, all data in the table refer to the supercritical region (i.e., the isobar is above the critical pressure).

REFERENCES

1. Younglove, B.A., *Thermophysical Properties of Fluids. Part I, J. Phys. Chem. Ref. Data*, 11, Suppl. 1, 1982.
2. Younglove, B.A., and Ely, J.F., *Thermophysical Properties of Fluids. Part II, J. Phys. Chem. Ref. Data*, 16, 577, 1987.
3. McCarty, R.D., *Thermodynamic Properties of Helium, J. Phys. Chem. Ref. Data*, 2, 923, 1973.

Nitrogen (N₂)

T K	ρ mol/L	E J/mol	H J/mol	S J/mol K	C_v J/mol K	C_p J/mol K	η μPa s	λ mW/m K	D
P = 0.1 MPa (1 bar)									
70	30.017	−3828	−3824	73.8	28.5	57.2	203.9	143.5	1.45269
77.25	28.881	−3411	−3407	79.5	27.8	57.8	152.2	133.8	1.43386
77.25	0.163	1546	2161	151.6	21.6	31.4	5.3	7.6	1.00215
100	0.123	2041	2856	159.5	21.1	30.0	6.8	9.6	1.00162
200	0.060	4140	5800	179.9	20.8	29.2	12.9	18.4	1.00079
300	0.040	6223	8717	191.8	20.8	29.2	18.0	25.8	1.00053
400	0.030	8308	11635	200.2	20.9	29.2	22.2	32.3	1.00040
500	0.024	10414	14573	206.7	21.2	29.6	26.1	38.5	1.00032
600	0.020	12563	17554	212.2	21.8	30.1	29.5	44.5	1.00026
700	0.017	14770	20593	216.8	22.4	30.7	32.8	50.5	1.00023
800	0.015	17044	23698	221.0	23.1	31.4	35.8	56.3	1.00020
900	0.013	19383	26869	224.7	23.7	32.0	38.7	62.0	1.00017
1000	0.012	21786	30103	228.1	24.3	32.6	41.5	67.7	1.00016
1500	0.008	34530	47004	241.8	26.4	34.7	54.0	93.3	1.00010
P = 1 MPa									
70	30.070	−3838	−3805	73.6	28.9	56.9	205.9	144.1	1.45355
80	28.504	−3267	−3232	81.3	27.8	57.7	139.5	130.7	1.42760
90	26.721	−2685	−2648	88.2	26.7	59.4	100.1	115.3	1.39824
100	24.634	−2073	−2032	94.6	26.2	64.4	73.1	98.5	1.36417
103.75	23.727	−1828	−1786	97.1	26.2	67.8	64.8	91.8	1.34947
103.75	1.472	1788	2467	138.1	24.1	45.0	7.6	12.5	1.01954
200	0.614	4048	5675	160.3	21.0	30.4	13.2	19.3	1.00812
300	0.402	6171	8661	172.5	20.9	29.6	18.1	26.3	1.00529
400	0.300	8273	11609	180.9	20.9	29.5	22.4	32.7	1.00395
500	0.240	10389	14563	187.5	21.3	29.7	26.1	38.8	1.00315
600	0.200	12544	17554	193.0	21.8	30.2	29.6	44.8	1.00262
700	0.171	14756	20600	197.7	22.4	30.8	32.8	50.7	1.00224
800	0.150	17032	23709	201.8	23.1	31.4	35.9	56.5	1.00196
900	0.133	19374	26884	205.6	23.7	32.1	38.8	62.2	1.00174
1000	0.120	21778	30121	209.0	24.3	32.7	41.5	67.8	1.00157
1500	0.080	34527	47029	222.7	26.4	34.8	54.0	93.4	1.00104
P = 10 MPa									
65.32	31.120	−4176	−3855	68.6	31.8	53.8	275.7	153.8	1.47067
100	26.201	−2328	−1946	92.0	27.4	56.3	90.2	112.3	1.38942
200	7.117	3037	4442	136.4	22.7	45.5	17.6	30.4	1.09698
300	3.989	5667	8174	151.7	21.4	33.4	20.1	31.9	1.05347
400	2.898	7941	11392	161.0	21.3	31.3	23.7	36.7	1.03860
500	2.302	10148	14492	167.9	21.5	30.8	27.1	42.0	1.03055

T K	ρ mol/L	E J/mol	H J/mol	S J/mol K	C_v J/mol K	C_p J/mol K	η μPa s	λ mW/m K	D
600	1.918	12361	17575	173.5	21.9	30.9	30.4	47.4	1.02538
700	1.647	14613	20683	178.3	22.5	31.3	33.5	53.0	1.02175
800	1.445	16919	23837	182.5	23.2	31.8	36.4	58.6	1.01904
900	1.288	19283	27046	186.3	23.8	32.4	39.3	64.1	1.01694
1000	1.162	21705	30308	189.8	24.4	32.9	42.0	69.6	1.01526
1500	0.783	34504	47283	203.5	26.5	34.8	54.3	94.7	1.01020

Oxygen (O_2)

T K	ρ mol/L	E J/mol	H J/mol	S J/mol K	C_v J/mol K	C_p J/mol K	η μPa s	λ mW/m K	D
P = 0.1 MPa (1 bar)									
60	40.049	−5883	−5880	72.4	34.9	53.4	425.2	188.2	1.55619
80	37.204	−4814	−4812	87.7	31.0	53.6	251.7	166.1	1.51114
100	0.123	2029	2840	172.9	21.4	30.5	7.5	9.3	1.00146
120	0.102	2458	3442	178.4	21.0	29.8	9.0	11.2	1.00121
140	0.087	2881	4035	182.9	20.9	29.5	10.5	13.1	1.00103
160	0.076	3301	4624	186.9	20.9	29.4	11.9	15.0	1.00090
180	0.067	3720	5210	190.3	20.8	29.3	13.3	16.7	1.00080
200	0.060	4138	5796	193.4	20.8	29.3	14.6	18.4	1.00072
220	0.055	4556	6381	196.2	20.8	29.3	15.9	20.1	1.00065
240	0.050	4974	6966	198.8	20.9	29.3	17.2	21.7	1.00060
260	0.046	5393	7552	201.1	20.9	29.3	18.4	23.2	1.00055
280	0.043	5812	8138	203.3	21.0	29.4	19.5	24.8	1.00051
300	0.040	6234	8726	205.3	21.1	29.4	20.6	26.3	1.00048
320	0.038	6657	9316	207.2	21.2	29.5	21.7	27.8	1.00045
340	0.035	7082	9908	209.0	21.3	29.7	22.8	29.3	1.00042
360	0.033	7510	10503	210.7	21.5	29.8	23.8	30.8	1.00040
380	0.032	7941	11100	212.3	21.6	30.0	24.8	32.2	1.00038
P = 1 MPa									
60	40.084	−5887	−5863	72.3	34.9	53.3	428.5	188.4	1.55674
80	37.254	−4822	−4795	87.6	31.0	53.5	253.8	166.4	1.51192
100	34.153	−3741	−3712	99.7	28.5	55.2	155.6	137.9	1.46381
120	1.198	2163	2997	156.7	24.0	40.6	9.4	13.9	1.01429
140	0.950	2683	3735	162.4	22.2	34.4	10.8	14.9	1.01133
160	0.802	3151	4398	166.8	21.5	32.2	12.2	16.3	1.00955
180	0.698	3598	5030	170.5	21.2	31.2	13.5	17.7	1.00831
200	0.620	4035	5647	173.8	21.1	30.6	14.8	19.3	1.00738
220	0.559	4466	6255	176.7	21.0	30.3	16.1	20.8	1.00665
240	0.509	4894	6858	179.3	21.0	30.1	17.3	22.3	1.00606
260	0.468	5321	7458	181.7	21.0	29.9	18.5	23.8	1.00556
280	0.433	5748	8056	183.9	21.1	29.9	19.6	25.2	1.00515
300	0.403	6174	8654	186.0	21.1	29.9	20.7	26.7	1.00479
320	0.377	6602	9252	187.9	21.2	29.9	21.8	28.2	1.00448
340	0.355	7032	9851	189.7	21.4	30.0	22.8	29.6	1.00421
360	0.335	7463	10452	191.4	21.5	30.1	23.9	31.1	1.00397
380	0.317	7898	11056	193.1	21.7	30.2	24.9	32.6	1.00376
P = 10 MPa									
60	40.419	−5931	−5684	71.5	35.1	53.0	461.8	189.9	1.56210
80	37.727	−4893	−4628	86.7	31.6	52.7	274.4	168.6	1.51936
100	34.881	−3856	−3570	98.5	29.1	53.4	171.0	141.2	1.47500
120	31.721	−2796	−2481	108.4	27.3	55.9	113.0	115.1	1.42677
140	27.890	−1662	−1304	117.5	26.2	62.9	76.3	91.8	1.36972
160	22.379	−322	125	127.0	26.1	84.8	48.6	71.2	1.29037
180	13.232	1489	2245	139.5	26.6	105.9	26.2	46.8	1.16560
200	8.666	2681	3835	147.9	24.0	60.6	21.2	34.0	1.10650

T K	ρ mol/L	E J/mol	H J/mol	S J/mol K	C_v J/mol K	C_p J/mol K	η μPa s	λ mW/m K	D
220	6.868	3424	4880	152.9	22.6	46.4	20.5	30.8	1.08380
240	5.836	4029	5742	156.6	22.0	40.6	20.8	30.1	1.07090
260	5.134	4573	6521	159.7	21.8	37.6	21.4	30.2	1.06219
280	4.613	5086	7254	162.5	21.6	35.8	22.1	30.8	1.05575
300	4.205	5581	7959	164.9	21.6	34.7	22.9	31.6	1.05073
320	3.874	6063	8645	167.1	21.7	33.9	23.7	32.6	1.04667
340	3.598	6538	9318	169.1	21.8	33.4	24.6	33.7	1.04329
360	3.363	7009	9982	171.0	21.9	33.0	25.4	34.9	1.04043
380	3.161	7477	10641	172.8	22.0	32.8	26.3	36.1	1.03796

Hydrogen (H_2)

T K	ρ mol/L	E J/mol	H J/mol	S J/mol K	C_v J/mol K	C_p J/mol K	v_s m/s	D
P = 0.1 MPa (1 bar)								
15	37.738	−605	−603	11.2	9.7	14.4	1319	1.24827
20	35.278	−524	−521	15.8	11.3	19.1	1111	1.23093
40	0.305	491	818	75.6	12.5	21.3	521	1.00186
60	0.201	748	1244	84.3	13.1	21.6	636	1.00122
80	0.151	1030	1694	90.7	15.3	23.7	714	1.00091
100	0.120	1370	2202	96.4	18.7	27.1	773	1.00073
120	0.100	1777	2776	101.6	21.8	30.2	827	1.00061
140	0.086	2237	3401	106.4	23.8	32.2	883	1.00052
160	0.075	2723	4054	110.8	24.6	33.0	940	1.00046
180	0.067	3216	4714	114.7	24.6	32.9	998	1.00041
200	0.060	3703	5367	118.1	24.1	32.4	1054	1.00037
220	0.055	4179	6009	121.2	23.4	31.8	1110	1.00033
240	0.050	4641	6638	123.9	22.8	31.2	1163	1.00030
260	0.046	5093	7256	126.4	22.3	30.6	1214	1.00028
280	0.043	5535	7865	128.6	21.9	30.2	1263	1.00026
300	0.040	5970	8466	130.7	21.6	29.9	1310	1.00024
400	0.030	8093	11421	139.2	21.0	29.3	1518	1.00018
P = 1 MPa								
15	38.109	−609	−583	10.9	10.1	14.1	1315	1.25089
20	35.852	−532	−504	15.5	11.4	18.4	1155	1.23496
40	3.608	399	676	54.1	12.9	28.4	498	1.02209
60	2.098	697	1173	64.3	13.2	23.5	635	1.01280
80	1.523	994	1651	71.1	15.4	24.7	719	1.00928
100	1.204	1343	2174	77.0	18.8	27.7	779	1.00733
120	0.999	1756	2758	82.3	21.9	30.6	835	1.00608
140	0.854	2219	3390	87.1	23.9	32.5	891	1.00520
160	0.747	2709	4048	91.5	24.7	33.2	949	1.00454
180	0.663	3204	4712	95.4	24.6	33.1	1006	1.00404
200	0.597	3693	5368	98.9	24.1	32.5	1063	1.00363
220	0.543	4170	6012	102.0	23.5	31.9	1118	1.00330
240	0.498	4634	6643	104.7	22.9	31.2	1171	1.00303
260	0.460	5087	7263	107.2	22.3	30.7	1222	1.00279
280	0.427	5530	7873	109.5	21.9	30.3	1271	1.00259
300	0.399	5966	8475	111.5	21.6	30.0	1317	1.00242
400	0.299	8091	11433	120.1	21.0	29.4	1525	1.00182
P = 10 MPa								
20	39.669	−568	−316	13.0	10.9	15.0	1458	1.26198
40	31.344	−209	110	27.3	13.2	27.0	1171	1.20354
60	21.273	255	725	39.7	13.8	32.5	931	1.13527
80	14.830	686	1360	48.8	15.9	31.1	886	1.09303
100	11.417	1110	1986	55.8	19.3	31.9	904	1.07109

T K	ρ mol/L	E J/mol	H J/mol	S J/mol K	C_v J/mol K	C_p J/mol K	v_s m/s	D
120	9.357	1571	2640	61.8	22.4	33.5	941	1.05801
140	7.969	2068	3323	67.0	24.3	34.6	989	1.04925
160	6.963	2583	4020	71.7	25.0	34.9	1042	1.04294
180	6.195	3099	4713	75.7	24.9	34.4	1096	1.03814
200	5.588	3604	5393	79.3	24.4	33.6	1150	1.03436
220	5.094	4094	6057	82.5	23.7	32.8	1203	1.03129
240	4.683	4569	6704	85.3	23.1	32.0	1254	1.02874
260	4.336	5030	7336	87.8	22.6	31.3	1302	1.02659
280	4.038	5481	7958	90.1	22.1	30.8	1349	1.02475
300	3.780	5924	8570	92.3	21.8	30.4	1394	1.02315
400	2.869	8073	11559	100.9	21.2	29.6	1592	1.01753

Helium (He-4)

T K	ρ mol/L	E J/mol	H J/mol	S J/mol K	C_v J/mol K	C_p J/mol K	v_s m/s	η μPa s	D
P = 0.1 MPa (1 bar)									
3	35.794	−39	−36	9.8	7.6	9.4	222	3.85	1.05646
4	32.477	−27	−24	13.3	9.1	16.3	185	3.33	1.05114
5	2.935	52	86	39.1	12.7	27.1	120	1.39	1.00456
10	1.238	120	201	55.2	12.5	21.7	185	2.26	1.00192
20	0.602	247	413	69.9	12.5	21.0	264	3.58	1.00093
50	0.240	623	1039	89.0	12.5	20.8	417	6.36	1.00037
100	0.120	1247	2079	103.4	12.5	20.8	589	9.78	1.00019
200	0.060	2494	4158	117.8	12.5	20.8	833	15.14	1.00009
300	0.040	3741	6237	126.3	12.5	20.8	1020	19.93	1.00006
400	0.030	4988	8315	132.3	12.5	20.8	1177	24.29	1.00005
500	0.024	6236	10394	136.9	12.5	20.8	1316	28.36	1.00004
600	0.020	7483	12472	140.7	12.5	20.8	1441	32.22	1.00003
700	0.017	8730	14551	143.9	12.5	20.8	1557	35.89	1.00003
800	0.015	9977	16630	146.7	12.5	20.8	1664	39.43	1.00002
900	0.013	11224	18708	149.1	12.5	20.8	1765	42.85	1.00002
1000	0.012	12471	20787	151.3	12.5	20.8	1861	46.16	1.00002
1500	0.008	18707	31179	159.7	12.5	20.8	2279	61.55	1.00001
P = 1 MPa									
3	39.703	−42	−16	8.6	7.1	7.8	300	5.63	1.06274
4	38.210	−34	−7	11.2	8.3	10.9	290	5.01	1.06034
5	35.818	−22	6	14.0	9.7	15.1	269	4.38	1.05650
10	15.378	78	143	32.2	12.3	30.5	198	3.07	1.02402
20	6.067	228	393	49.8	12.6	22.9	274	3.94	1.00943
50	2.353	617	1042	69.8	12.5	21.1	428	6.53	1.00365
100	1.186	1245	2089	84.3	12.5	20.9	597	9.89	1.00184
200	0.597	2495	4170	98.7	12.5	20.8	838	15.21	1.00093
300	0.399	3742	6249	107.1	12.5	20.8	1024	19.96	1.00062
400	0.300	4990	8327	113.1	12.5	20.8	1180	24.32	1.00046
500	0.240	6237	10406	117.8	12.5	20.8	1319	28.38	1.00037
600	0.200	7485	12484	121.5	12.5	20.8	1444	32.23	1.00031
700	0.172	8732	14562	124.7	12.5	20.8	1559	35.91	1.00027
800	0.150	9979	16641	127.5	12.5	20.8	1666	39.44	1.00023
900	0.133	11227	18719	130.0	12.5	20.8	1767	42.86	1.00021
1000	0.120	12474	20798	132.2	12.5	20.8	1862	46.17	1.00019
1500	0.080	18710	31190	140.6	12.5	20.8	2280	61.55	1.00012
P = 10 MPa									
4	51.978	−24	169	6.7	6.0	7.3	586	24.27	1.08262
5	51.118	−18	177	8.5	7.9	9.3	576	18.16	1.08122
10	46.872	23	236	16.6	11.0	14.5	546	9.31	1.07432

T K	ρ mol/L	E J/mol	H J/mol	S J/mol K	C_v J/mol K	C_p J/mol K	v_s m/s	η μPa s	D
20	37.092	154	423	29.5	12.6	20.7	498	6.99	1.05854
50	19.192	572	1093	49.9	12.9	22.4	541	8.07	1.03003
100	10.525	1231	2181	65.0	12.8	21.3	674	10.93	1.01640
200	5.605	2500	4284	79.6	12.6	20.9	889	15.82	1.00871
300	3.829	3755	6367	88.0	12.6	20.8	1063	20.25	1.00595
400	2.908	5006	8445	94.0	12.6	20.8	1212	24.54	1.00452
500	2.344	6256	10522	98.6	12.5	20.8	1346	28.56	1.00364
600	1.963	7505	12599	102.4	12.5	20.8	1467	32.38	1.00305
700	1.689	8754	14676	105.6	12.5	20.8	1580	36.04	1.00262
800	1.481	10003	16753	108.4	12.5	20.8	1685	39.56	1.00230
900	1.320	11252	18830	110.9	12.5	20.8	1784	42.96	1.00205
1000	1.189	12500	20907	113.0	12.5	20.8	1877	46.26	1.00185
1500	0.797	18742	31294	121.5	12.5	20.8	2289	61.62	1.00124

Argon (Ar)

T K	ρ mol/L	E J/mol	H J/mol	S J/mol K	C_v J/mol K	C_p J/mol K	v_s m/s	η μPa s	λ mW/m K
P = 0.1 MPa (1 bar)									
85	35.243	−4811	−4808	53.6	23.1	44.7	820	278.8	132.4
90	0.138	1077	1802	129.4	13.1	22.5	174	7.5	6.0
100	0.123	1211	2024	131.8	12.9	21.9	184	8.2	6.6
120	0.102	1471	2456	135.7	12.6	21.4	203	9.8	7.8
140	0.087	1727	2881	139.0	12.6	21.1	220	11.4	9.0
160	0.076	1980	3302	141.8	12.5	21.0	235	13.0	10.2
180	0.067	2232	3722	144.3	12.5	21.0	250	14.5	11.4
200	0.060	2483	4141	146.5	12.5	20.9	263	16.0	12.5
220	0.055	2734	4559	148.5	12.5	20.9	276	17.5	13.7
240	0.050	2984	4976	150.3	12.5	20.9	289	18.9	14.8
260	0.046	3234	5394	152.0	12.5	20.9	300	20.3	15.8
280	0.043	3484	5811	153.5	12.5	20.8	312	21.6	16.9
300	0.040	3734	6227	155.0	12.5	20.8	323	22.9	17.9
320	0.038	3984	6644	156.3	12.5	20.8	333	24.2	18.9
340	0.035	4234	7060	157.6	12.5	20.8	344	25.4	19.9
360	0.033	4484	7477	158.7	12.5	20.8	354	26.6	20.8
380	0.032	4734	7893	159.9	12.5	20.8	363	27.8	21.7
P = 1 MPa									
85	35.307	−4820	−4792	53.5	23.1	44.6	823	281.3	133.0
90	34.542	−4598	−4569	56.1	21.6	44.7	808	242.7	124.2
100	32.909	−4145	−4115	60.9	19.9	46.2	753	185.0	109.2
120	1.181	1210	2057	114.3	14.7	30.1	189	10.3	9.3
140	0.945	1544	2603	118.5	13.5	25.4	212	11.8	10.1
160	0.799	1838	3089	121.8	13.0	23.6	231	13.3	11.1
180	0.697	2116	3551	124.5	12.8	22.7	247	14.8	12.1
200	0.619	2384	3999	126.9	12.7	22.2	262	16.3	13.2
220	0.559	2648	4438	128.9	12.6	21.8	275	17.7	14.2
240	0.509	2908	4873	130.8	12.6	21.6	288	19.1	15.3
260	0.468	3167	5304	132.6	12.6	21.5	301	20.4	16.3
280	0.433	3423	5732	134.2	12.6	21.4	312	21.8	17.3
300	0.403	3679	6159	135.6	12.5	21.3	324	23.1	18.3
320	0.377	3934	6583	137.0	12.5	21.2	334	24.3	19.2
340	0.355	4188	7007	138.3	12.5	21.2	345	25.5	20.2
360	0.335	4441	7429	139.5	12.5	21.1	355	26.7	21.1
380	0.317	4694	7851	140.6	12.5	21.1	365	27.9	22.0
P = 10 MPa									
90	35.208	−4694	−4410	55.0	21.9	43.2	846	265.2	129.5

T K	ρ mol/L	E J/mol	H J/mol	S J/mol K	C_v J/mol K	C_p J/mol K	v_s m/s	η μPa s	λ mW/m K
100	33.744	−4271	−3974	59.6	20.4	44.0	800	205.0	115.1
120	30.525	−3396	−3069	67.8	18.8	46.9	672	131.2	92.1
140	26.609	−2447	−2072	75.5	17.6	54.1	526	85.9	71.7
160	20.816	−1279	−799	83.9	17.4	78.6	357	51.3	52.8
180	12.296	228	1042	94.8	17.3	83.6	257	27.8	32.0
200	8.442	1118	2302	101.4	15.3	48.6	268	23.3	23.6
220	6.776	1661	3137	105.4	14.2	36.8	284	22.8	21.6
240	5.787	2087	3815	108.4	13.7	31.6	300	23.2	21.3
260	5.105	2458	4416	110.8	13.4	28.8	314	23.9	21.4
280	4.596	2798	4974	112.9	13.2	27.1	327	24.8	21.8
300	4.195	3119	5503	114.7	13.1	25.9	339	25.7	22.3
320	3.869	3427	6012	116.3	13.0	25.0	350	26.7	22.9
340	3.596	3726	6506	117.8	13.0	24.4	361	27.7	23.5
360	3.364	4017	6989	119.2	12.9	23.9	372	28.7	24.2
380	3.164	4303	7464	120.5	12.9	23.5	381	29.7	24.9

Methane (CH_4)

T K	ρ mol/L	E J/mol	H J/mol	S J/mol K	C_v J/mol K	C_p J/mol K	η μPa s	λ mW/m K	D
$P = 0.1$ MPa (1 bar)									
100	27.370	−5258	−5254	73.0	33.4	54.1	156.3	208.1	1.65504
125	0.099	3026	4039	156.5	25.4	34.6	5.0	13.4	1.00193
150	0.081	3667	4896	162.7	25.2	34.0	5.9	16.2	1.00159
175	0.069	4301	5743	168.0	25.2	33.8	6.9	19.1	1.00136
200	0.061	4935	6587	172.5	25.3	33.8	7.8	21.9	1.00119
225	0.054	5571	7434	176.5	25.5	34.0	8.7	24.8	1.00105
250	0.048	6216	8288	180.1	26.0	34.4	9.6	27.8	1.00095
275	0.044	6875	9156	183.4	26.6	35.0	10.4	30.9	1.00086
300	0.040	7552	10042	186.4	27.5	35.9	11.2	34.1	1.00079
325	0.037	8252	10951	189.4	28.5	36.9	12.0	37.6	1.00073
350	0.034	8979	11887	192.1	29.7	38.0	12.8	41.2	1.00068
375	0.032	9737	12853	194.8	30.9	39.3	13.5	45.1	1.00063
400	0.030	10528	13852	197.4	32.3	40.7	14.3	49.1	1.00059
425	0.028	11354	14886	199.9	33.7	42.1	15.0	53.3	1.00056
450	0.027	12215	15956	202.3	35.2	43.5	15.7	57.6	1.00053
500	0.024	14047	18204	207.1	38.0	46.4	17.0	66.5	1.00047
600	0.020	18111	23101	216.0	42.9	51.3	19.4	84.1	1.00039
$P = 1$ MPa									
100	27.413	−5268	−5231	72.9	33.4	54.0	158.1	208.9	1.65617
125	25.137	−3882	−3842	85.3	32.4	57.4	89.2	168.2	1.59261
150	0.969	3282	4315	140.9	27.9	45.2	6.2	18.4	1.01911
175	0.765	4041	5348	147.3	26.4	38.9	7.1	20.6	1.01507
200	0.644	4736	6289	152.3	25.9	36.8	8.0	23.1	1.01268
225	0.560	5410	7197	156.6	25.9	36.0	8.9	25.8	1.01102
250	0.497	6081	8093	160.4	26.2	35.8	9.7	28.7	1.00979
275	0.448	6758	8991	163.8	26.8	36.1	10.6	31.7	1.00882
300	0.408	7449	9901	167.0	27.6	36.7	11.4	34.9	1.00803
325	0.375	8160	10829	169.9	28.6	37.6	12.1	38.3	1.00738
350	0.347	8897	11781	172.8	29.7	38.6	12.9	41.9	1.00683
375	0.323	9662	12760	175.5	31.0	39.8	13.6	45.7	1.00636
400	0.302	10460	13770	178.1	32.4	41.1	14.4	49.6	1.00595
425	0.284	11291	14814	180.6	33.8	42.4	15.1	53.8	1.00559
450	0.268	12157	15892	183.1	35.2	43.8	15.7	58.1	1.00527
500	0.241	13997	18153	187.8	38.1	46.6	17.0	66.9	1.00474
600	0.200	18073	23070	196.8	43.0	51.4	19.5	84.5	1.00394

T K	ρ mol/L	E J/mol	H J/mol	S J/mol K	C_v J/mol K	C_p J/mol K	η μPa s	λ mW/m K	D
P = 10 MPa									
100	27.815	−5362	−5003	72.0	33.8	53.2	175.4	217	1.66668
125	25.754	−4036	−3648	84.1	32.7	55.3	100.4	178.8	1.60895
150	23.441	−2655	−2229	94.4	31.4	58.6	65.7	144.6	1.54553
175	20.613	−1175	−689	103.9	30.3	65.5	44.9	113.4	1.47021
200	16.602	542	1144	113.6	30.1	84.7	29.4	85.8	1.36789
225	10.547	2680	3628	125.3	30.8	102.2	17.6	61.0	1.22352
250	7.013	4289	5714	134.1	29.3	67.4	14.3	47.6	1.14481
275	5.530	5387	7195	139.8	28.7	53.4	13.8	44.1	1.11297
300	4.685	6320	8454	144.2	28.9	48.0	13.9	44.6	1.09513
325	4.115	7192	9622	147.9	29.6	45.8	14.3	46.6	1.08322
350	3.695	8047	10753	151.3	30.5	44.9	14.7	49.2	1.07450
375	3.366	8903	11874	154.4	31.7	44.8	15.2	52.3	1.06773
400	3.101	9774	12999	157.3	32.9	45.2	15.8	55.7	1.06227
425	2.880	10666	14138	160.0	34.3	46.0	16.3	59.4	1.05775
450	2.692	11584	15298	162.7	35.7	46.9	16.9	63.3	1.05392
500	2.389	13507	17692	167.7	38.5	48.9	18.0	71.6	1.04775
600	1.963	17700	22795	177.0	43.3	52.9	20.2	88.3	1.03911

Ethane (C_2H_6)

T K	ρ mol/L	E J/mol	H J/mol	S J/mol K	C_v J/mol K	C_p J/mol K	v_s m/s	D
P = 0.1 MPa (1 bar)								
95	21.50	−14555	−14550	80.2	47.2	68.7	1970	1.93480
100	21.32	−14210	−14205	83.8	47.1	69.3	1943	1.92500
125	20.41	−12468	−12463	99.3	45.0	69.8	1775	1.87634
150	19.47	−10717	−10712	112.1	43.4	70.4	1587	1.82726
175	18.49	−8938	-8933	123.1	42.7	72.1	1396	1.77671
200	0.062	5503	7123	210.1	34.5	43.8	258	1.00208
225	0.054	6401	8238	215.4	36.5	45.5	273	1.00183
250	0.049	7349	9401	220.3	38.9	47.7	287	1.00164
275	0.044	8360	10624	224.9	41.6	50.2	300	1.00148
300	0.040	9439	11914	229.4	44.5	53.1	312	1.00136
325	0.037	10592	13278	233.8	47.6	56.1	324	1.00125
350	0.035	11823	14719	238.1	50.7	59.2	335	1.00116
375	0.032	13133	16240	242.3	54.0	62.4	345	1.00108
400	0.030	14525	17841	246.4	57.2	65.6	355	1.00101
450	0.027	17548	21282	254.5	63.6	72.0	375	1.00090
500	0.024	20883	25035	262.4	69.7	78.1	393	1.00081
600	0.020	28429	33415	277.6	80.9	89.3	428	1.00067
P = 1 MPa								
95	21.514	−14562	−14515	80.2	47.3	68.7	1972	1.93537
100	21.334	−14217	−14170	83.7	47.2	69.3	1946	1.92560
125	20.427	−12478	−12429	99.2	45.0	69.8	1778	1.87709
150	19.494	−10731	−10679	112.0	43.4	70.3	1592	1.82823
175	18.515	−8957	−8903	123.0	42.7	72.0	1402	1.77800
200	17.464	−7127	−7070	132.7	42.9	74.9	1209	1.72513
225	16.288	−5199	−5137	141.8	43.8	80.2	1008	1.66733
250	0.564	6762	8534	198.7	41.6	57.5	260	1.01909
275	0.489	7902	9949	204.2	43.2	56.2	280	1.01650
300	0.435	9063	11363	209.0	45.5	57.2	297	1.01467
325	0.393	10273	12815	213.7	48.3	59.1	311	1.01327
350	0.360	11546	14321	218.1	51.3	61.5	325	1.01214
375	0.333	12889	15893	222.5	54.4	64.2	337	1.01121
400	0.310	14306	17534	226.7	57.5	67.1	349	1.01043

THERMOPHYSICAL PROPERTIES OF FLUIDS (continued)

T K	ρ mol/L	E J/mol	H J/mol	S J/mol K	C_v J/mol K	C_p J/mol K	v_s m/s	D
450	0.272	17367	21038	234.9	63.8	73.0	370	1.00917
500	0.244	20730	24836	242.9	69.9	78.9	390	1.00819
600	0.201	28313	33278	258.3	81.0	89.8	427	1.00677
P = 10 MPa								
95	21.624	−14626	−14163	79.5	47.4	68.5	2000	1.94104
100	21.448	−14286	−13819	83.0	47.4	69.1	1974	1.93146
125	20.570	−12572	−12086	98.5	45.5	69.3	1814	1.88436
150	19.678	−10858	−10350	111.1	43.9	69.6	1637	1.83753
175	18.758	−9130	−8596	121.9	43.3	70.8	1459	1.79010
200	17.793	−7363	−6801	131.5	43.5	73.0	1284	1.74134
225	16.760	−5535	−4938	140.3	44.3	76.4	1110	1.69017
250	15.620	−3609	−2969	148.6	45.8	81.5	935	1.63488
275	14.301	−1539	−839	156.7	47.9	89.4	758	1.57249
300	12.666	757	1547	165.0	50.8	102.7	577	1.49740
325	10.398	3443	4404	174.1	54.7	129.1	399	1.39745
350	7.292	6643	8015	184.8	58.8	150.1	290	1.26832
375	5.182	9419	11349	194.1	60.0	115.7	289	1.18570
400	4.182	11577	13968	200.8	61.4	96.9	310	1.14797
450	3.204	15379	18500	211.5	65.8	87.5	347	1.11193
500	2.677	19135	22870	220.7	71.2	88.0	378	1.09288
600	2.076	27160	31978	237.3	81.8	94.7	427	1.07142

Propane (C₃H₈)

T K	ρ mol/L	E J/mol	H J/mol	S J/mol K	C_v J/mol K	C_p J/mol K	v_s m/s	D
P = 0.1 MPa (1 bar)								
90	16.526	−21486	−21426	87.3	59.2	84.5	2126	2.07988
100	16.295	−20639	−20577	96.2	59.6	85.2	2041	2.05806
125	15.726	−18495	−18432	115.4	59.2	86.5	1856	2.00674
150	15.156	−16319	−16253	131.3	58.9	88.0	1685	1.95796
175	14.577	−14096	−14028	145.0	59.5	90.3	1521	1.91036
200	13.982	−11806	−11735	157.3	61.0	93.5	1359	1.86300
225	13.339	−9395	−9387	168.5	63.4	97.9	1197	1.81487
250	0.050	9194	11213	257.6	57.2	66.8	228	1.00238
275	0.045	10691	12930	264.1	61.6	70.7	239	1.00215
300	0.041	12297	14752	270.5	66.2	75.1	249	1.00195
325	0.037	14019	16689	276.7	71.1	79.8	259	1.00179
350	0.035	15862	18744	282.8	76.0	84.6	269	1.00166
375	0.032	17827	20921	288.8	80.9	89.5	278	1.00154
400	0.030	19912	23217	294.7	85.7	94.3	286	1.00144
450	0.027	24441	28166	306.4	95.2	103.6	303	1.00128
500	0.024	29428	33573	317.7	104.1	112.6	318	1.00115
600	0.020	40677	45658	339.7	120.4	128.8	347	1.00095
P = 1 MPa								
90	16.526	−21486	−21426	87.2	59.3	84.5	2128	2.08034
100	16.295	−20639	−20577	96.2	59.7	85.2	2043	2.05856
125	15.726	−18495	−18432	115.3	59.2	86.4	1859	2.00736
150	15.156	−16319	−16253	131.2	59.0	88.0	1690	1.95873
175	14.577	−14096	−14028	144.9	59.6	90.2	1526	1.91132
200	13.982	−11806	−11735	157.2	61.1	93.4	1365	1.86421
225	13.361	−9424	−9349	168.4	63.4	97.7	1205	1.81642
250	12.696	−6919	−6840	179.0	66.4	103.3	1045	1.76672
275	11.962	−4252	−4169	189.1	70.0	110.8	881	1.71316
300	11.102	−1360	−1270	199.2	74.1	121.9	708	1.65216
325	0.428	13278	15614	255.2	74.1	89.6	233	1.02067

THERMOPHYSICAL PROPERTIES OF FLUIDS (continued)

T K	ρ mol/L	E J/mol	H J/mol	S J/mol K	C_v J/mol K	C_p J/mol K	v_s m/s	D
350	0.383	15259	17869	261.9	78.0	91.2	248	1.01846
375	0.349	17318	20183	268.3	82.2	94.2	261	1.01678
400	0.322	19472	22582	274.4	86.7	97.8	272	1.01544
450	0.279	24092	27672	286.4	95.7	105.9	293	1.01337
500	0.248	29137	33172	298.0	104.4	114.1	312	1.01184
600	0.203	40455	45374	320.2	120.5	129.7	344	1.00968
P = 10 MPa								
90	16.590	−21553	−20951	86.5	59.9	84.4	2146	2.08489
100	16.364	−20714	−20103	95.4	60.1	85.1	2068	2.06350
125	15.810	−18595	−17962	114.5	59.6	86.1	1895	2.01342
150	15.259	−16448	−15793	130.3	59.3	87.5	1733	1.96617
175	14.705	−14261	−13581	144.0	59.9	89.5	1577	1.92048
200	14.141	−12016	−11309	156.1	61.4	92.4	1425	1.87557
225	13.562	−9692	−8955	167.2	63.7	96.1	1277	1.83076
250	12.960	−7268	−6496	177.5	66.7	100.7	1133	1.78529
275	12.322	−4721	−3909	187.4	70.2	106.4	991	1.73826
300	11.631	−2027	−1167	196.9	74.1	113.2	851	1.68849
325	10.860	843	1764	206.3	78.4	121.5	715	1.63437
350	9.973	3924	4927	215.7	82.9	132.0	582	1.57361
375	8.905	7270	8393	225.2	87.7	146.1	455	1.50271
400	7.561	10957	12279	235.3	93.0	165.7	339	1.41671
450	4.614	18845	21013	255.8	101.8	167.8	249	1.24060
500	3.241	25567	28652	272.0	107.8	142.7	276	1.16439
600	2.242	38131	42591	297.4	121.7	140.5	332	1.11122

VIRIAL COEFFICIENTS OF SELECTED GASES

Henry V. Kehiaian

This table gives second virial coefficients of about 110 inorganic and organic gases as a function of temperature. Selected data from the literature have been fitted by least squares to the equation

$$B/\text{cm}^3\,\text{mol}^{-1} = \sum_{i=1}^{n} a(i)\left[\left(T_0/T\right)-1\right]^{i-1}$$

where $T_0 = 298.15$ K. The table gives the coefficients $a(i)$ and values of B at fixed temperature increments, as calculated from this smoothing equation.

The equation may be used with the tabulated coefficients for interpolation within the indicated temperature range. It should not be used for extrapolation beyond this range.

Compounds are listed in the modified Hill order (see Introduction), with carbon-containing compounds following those compounds not containing carbon.

A useful compilation of virial coefficient data from the literature may be found in:

J. H. Dymond and E. B. Smith, *The Virial Coefficients of Pure Gases and Mixtures, A Critical Compilation*, Oxford University Press, Oxford, 1980.

Compounds Not Containing Carbon

Mol. form.	Name	T/K	$B/\text{cm}^3\,\text{mol}^{-1}$
Ar	Argon	100	−184
		120	−131
		140	−98
	$a(1) = -16$	160	−76
	$a(2) = -60$	80	−60
	$a(3) = -10$	200	−48
		300	−16
		400	−1
		500	7
		600	12
		700	15
		800	18
		900	20
		1000	22
BF$_3$	Boron trifluoride	200	−338
		240	−202
		280	−129
	$a(1) = -106$	320	−85
	$a(2) = -330$	360	−56
	$a(3) = -251$	400	−37
	$a(4) = -80$	440	−23
ClH	Hydrogen chloride	190	−451
		230	−269
		270	−181
	$a(1) = -144$	310	−132
	$a(2) = -325$	350	−102
	$a(3) = -277$	390	−81
	$a(4) = -170$	430	−66
		470	−54
Cl$_2$	Chlorine	210	−508
		220	−483

Mol. form.	Name	T/K	$B/cm^3\ mol^{-1}$
		230	−457
	a(1) = −303	240	−432
	a(2) = −555	250	−407
	a(3) = 9	260	−383
	a(4) = 329	270	−360
	a(5) = 68	280	−339
		290	−318
		300	−299
		350	−221
		400	−166
		450	−126
		500	−97
		600	−59
		700	−36
		800	−22
		900	−12
F_2	Fluorine	80	−386
		110	−171
		140	−113
	a(1) = −25	170	−73
	a(2) = 21	200	−47
	a(3) = −185	230	−32
	a(4) = 113	260	−25
F_4Si	Silicon tetrafluoride	210	−268
		240	−213
		270	−170
	a(1) = −138	300	−136
	a(2) = −312	330	−108
		360	−84
		390	−64
		420	−47
		450	−32
F_5I	Iodine pentafluoride	320	−2540
		330	−2344
		340	−2172
	a(1) = −3077	350	−2021
	a(2) = −8474	360	−1890
	a(3) = −9116	370	−1775
		380	−1674
		390	−1587
		400	−1510
		410	−1443
F_5P	Phosphorus pentafluoride	320	−162
		340	−143
		360	−127
	a(1) = −186	380	−112
	a(2) = −345	400	−98
		420	−86
		440	−75
		460	−64
F_6Mo	Molybdenum hexafluoride	300	−896
		310	−810
		320	−737

Mol. form.	Name	T/K	$B/cm^3\ mol^{-1}$
	$a(1) = -914$	330	−677
	$a(2) = -2922$	340	−627
	$a(3) = -4778$	350	−586
		360	−553
		370	−527
		380	−506
		390	−491
F_6S	Sulfur hexafluoride	200	−685
		250	−416
		300	−275
	$a(1) = -279$	350	−190
	$a(2) = -647$	400	−135
	$a(3) = -335$	450	−96
	$a(4) = -72$	500	−68
F_6U	Uranium hexafluoride	320	−1030
		340	−905
		360	−805
	$a(1) = -1204$	380	−724
	$a(2) = -2690$	400	−658
	$a(3) = -2144$	420	−604
		440	−560
F_6W	Tungsten hexafluoride	320	−641
		340	−578
		360	−523
	$a(1) = -719$	380	−473
	$a(2) = -1143$	400	−428
		420	−387
		440	−350
		460	−317
H_2	Hydrogen	15	−230
		20	−151
		25	−108
	$a(1) = 15.4$	30	−82
	$a(2) = -9.0$	35	−64
	$a(3) = -0.2$	40	−52
		45	−42
		50	−35
		60	−24
		70	−16
		80	−11
		90	−7
		100	−3
		200	11
		300	15
		400	18
H_2O	Water	300	−1126
		320	−850
		340	−660
	$a(1) = -1158$	360	−526
	$a(2) = -5157$	380	−428
	$a(3) = -10301$	400	−356
	$a(4) = -10597$	420	−301
	$a(5) = -4415$	440	−258

Mol. form.	Name	T/K	$B/cm^3\ mol^{-1}$
		460	−224
		480	−197
		500	−175
		600	−104
		700	−67
		800	−44
		900	−30
		1000	−20
		1100	−14
		1200	−11
H_3N	Ammonia	290	−302
		300	−265
		310	−236
	$a(1) = -271$	320	−213
	$a(2) = -1022$	330	−194
	$a(3) = -2715$	340	−179
	$a(4) = -4189$	350	−166
		360	−154
		370	−144
		380	−135
		400	−118
		420	−101
H_3P	Phosphine	190	−457
		200	−404
		210	−364
	$a(1) = -146$	220	−332
	$a(2) = -733$	230	−305
	$a(3) = 1022$	240	−281
	$a(4) = -1220$	250	−258
		260	−235
		270	−213
		280	−190
		290	−166
He	Helium	2	−172
		6	−48
		10	−24
	$a(1) = 12$	14	−13
	$a(2) = -1$	18	−7
		22	−3
		26	−1
		30	1
		50	6
		70	8
		90	10
		110	10
		150	11
		250	12
		650	13
		700	13
Kr	Krypton	110	−363
		120	−307
		130	−263
	$a(1) = -51$	140	−229
	$a(2) = -118$	150	−201

Mol. form.	Name	T/K	B/cm^3 mol^{-1}
	a(3) = −29	160	−178
	a(4) = −5	170	−159
		180	−143
		190	−129
		200	−117
		250	−75
		300	−51
		400	−23
		500	−8
		600	2
		700	8
NO	Nitric oxide	120	−232
		130	−176
		140	−138
	a(1) = −12	150	−113
	a(2) = −119	160	−96
	a(3) = 89	170	−83
	a(4) = −73	180	−73
		190	−65
		200	−58
		210	−52
		230	−42
		250	−32
		270	−24
N_2	Nitrogen	75	−274
		100	−161
		125	−104
	a(1) = −4	150	−71
	a(2) = −56	175	−49
	a(3) = −12	200	−34
		225	−24
		250	−15
		300	−4
		400	9
		500	16
		600	21
		700	24
N_2O	Nitrous oxide	240	−219
		260	−181
		280	−151
	a(1) = −130	300	−128
	a(2) = −307	320	−110
	a(3) = −248	340	−96
		360	−85
		380	−76
		400	−68
Ne	Neon	60	−25
		80	−13
		100	−6
	a(1) = 10.8	120	−1
	a(2) = −7.5	140	2
	a(3) = 0.4	160	4
		180	6
		200	7

Mol. form.	Name	T/K	$B/cm^3\ mol^{-1}$
		300	11
		400	13
		500	14
		600	15
O_2	Oxygen	90	−241
		110	−161
		130	−117
	a(1) = −16	150	−88
	a(2) = −62	170	−69
	a(3) = −8	190	−55
	a(4) = −3	210	−44
		230	−36
		250	−29
		270	−23
		290	−18
		310	−14
		330	−10
		350	−7
		400	−1
O_2S	Sulfur dioxide	290	−465
		320	−354
		350	−276
	a(1) = −430	380	−221
	a(2) = −1193	410	−181
	a(3) = −1029	440	−153
		470	−132
Xe	Xenon	160	−421
		170	−377
		180	−340
	a(1) = −130	190	−307
	a(2) = −262	200	−280
	a(3) = −87	210	−255
		220	−234
		230	−215
		240	−199
		250	−184
		300	−129
		350	−93
		400	−69
		500	−39
		600	−21
		650	−14

Compounds Containing Carbon

Mol. form.	Name	T/K	$B/cm^3\ mol^{-1}$
$CClF_3$	Chlorotrifluoromethane	240	−369
		290	−237
		340	−165
	a(1) = −223	390	−119
	a(2) = −504	440	−86
	a(3) = −340	490	−60
	a(4) = −291	540	−39

Mol. form.	Name		T/K	$B/cm^3\ mol^{-1}$
CCl_2F_2	Dichlorodifluoromethane		250	−769
			280	−570
			310	−441
	$a(1) = -486$		340	−353
	$a(2) = -1217$		370	−289
	$a(3) = -1188$		400	−241
	$a(4) = -698$		430	−204
			460	−174
CCl_3F	Trichlorofluoromethane		240	−1140
			280	−879
			320	−689
	$a(1) = -786$		360	−545
	$a(2) = -1428$		400	−431
	$a(3) = -142$		440	−340
			480	−265
CCl_4	Tetrachloromethane		320	−1345
			340	−1171
			360	−1040
	$a(1) = -1600$		380	−942
	$a(2) = -4059$		400	−868
	$a(3) = -4653$		420	−814
CF_4	Tetrafluoromethane		250	−137
			300	−87
			350	−55
	$a(1) = -88$		400	−32
	$a(2) = -238$		450	−16
	$a(3) = -70$		500	−4
			600	14
			700	25
			800	33
$CHClF_2$	Chlorodifluoromethane		300	−343
			325	−298
			350	−257
	$a(1) = -347$		375	−221
	$a(2) = -575$		400	−188
	$a(3) = 187$		425	−158
$CHCl_2F$	Dichlorofluoromethane		250	−728
			275	−634
			300	−557
	$a(1) = -562$		325	−491
	$a(2) = -862$		350	−434
			375	−385
			400	−343
			425	−305
			450	−271
$CHCl_3$	Trichloromethane		320	−1001
			330	−926
			340	−858
	$a(1) = -1193$		350	−797
	$a(2) = -2936$		360	−740
	$a(3) = -1751$		370	−689
			380	−642

Mol. form.	Name	T/K	$B/cm^3\ mol^{-1}$
		390	−599
		400	−559
CHF_3	Trifluoromethane	200	−433
		220	−350
		240	−288
	a(1) = −177	260	−241
	a(2) = −399	280	−204
	a(3) = −250	300	−174
		320	−151
		340	−132
		360	−116
		380	−103
		400	−91
CH_2Cl_2	Dichloromethane	320	−706
		330	−634
		340	−574
	a(1) = −913	350	−524
	a(2) = −3371	360	−482
	a(3) = −5013	370	−447
		380	−420
		400	−380
		420	−357
CH_2F_2	Difluoromethane	280	−375
		290	−343
		300	−316
	a(1) = −321	310	−294
	a(2) = −754	320	−275
	a(3) = −1300	330	−260
		340	−248
		350	−238
CH_3Br	Bromomethane	280	−645
		290	−596
		300	−551
	a(1) = −559	310	−509
	a(2) = −1324	320	−469
		340	−396
		360	−332
		380	−274
CH_3Cl	Chloromethane	280	−466
		300	−402
		320	−348
	a(1) = −407	340	−304
	a(2) = −887	360	−266
	a(3) = −385	380	−234
		400	−206
		420	−182
		440	−161
		460	−142
		480	−126
		500	−112
		600	−58
CH_3F	Fluoromethane	280	−244

Mol. form.	Name		T/K	$B/cm^3\ mol^{-1}$
			300	−205
			320	−174
	a(1) = −209		340	−150
	a(2) = −525		360	−129
	a(3) = −365		380	−112
			400	−99
			420	−87
CH_3I	Iodomethane		310	−725
			320	−646
			330	−582
	a(1) = −844		340	−531
	a(2) = −3353		350	−492
	a(3) = −6590		360	−462
			370	−441
			380	−427
CH_4	Methane		110	−328
			120	−276
			130	−237
	a(1) = −43		140	−206
	a(2) = −114		150	−181
	a(3) = −19		160	−160
	a(4) = −7		170	−143
			180	−128
			190	−116
			200	−105
			250	−66
			300	−43
			350	−27
			400	−16
			500	0
			600	10
CH_4O	Methanol		320	−1431
			330	−1299
			340	−1174
	a(1) = −1752		350	−1056
	a(2) = −4694		360	−945
			370	−840
			380	−741
			390	−646
			400	−557
CH_5N	Methylamine		300	−451
			325	−367
			350	−304
	a(1) = −459		375	−257
	a(2) = −1191		400	−220
	a(3) = −995		425	−192
			450	−170
			500	−140
			550	−122
CO	Carbon monoxide		210	−36
			240	−24
			270	−15
	a(1) = −9		300	−8

Mol. form.	Name	T/K	$B/\text{cm}^3\ \text{mol}^{-1}$
	a(2) = −58	330	−3
	a(3) = −18	360	1
		420	7
		480	11
CO_2	Carbon dioxide	220	−244
		240	−204
		260	−172
	a(1) = −127	280	−146
	a(2) = −288	300	−126
	a(3) = −118	320	−108
		340	−94
		360	−81
		380	−71
		400	−62
		500	−30
		600	−13
		700	−1
		800	7
		900	12
		1000	16
		1100	19
CS_2	Carbon disulfide	280	−932
		310	−740
		340	−603
	a(1) = −807	370	−504
	a(2) = −1829	400	−431
	a(3) = −1371	430	−375
$C_2Cl_2F_4$	1,2-Dichloro-1,1,2,2-tetrafluoroethane	300	−801
		320	−695
		340	−608
	a(1) = −812	360	−536
	a(2) = −1773	380	−475
	a(3) = −963	400	−423
		420	−379
		440	−341
		460	−307
		480	−279
		500	−253
$C_2Cl_3F_3$	1,1,2-Trichloro-1,2,2-trifluoroethane	290	−1041
		310	−943
		330	−856
	a(1) = −999	350	−780
	a(2) = −1479	370	−712
		390	−651
		410	−596
		430	−546
		450	−500
C_2H_2	Ethyne	200	−573
		210	−500
		220	−440
	a(1) = −216	230	−390
	a(2) = −375	240	−349
	a(3) = −716	250	−315

Mol. form.	Name	T/K	$B/cm^3\ mol^{-1}$
		260	−287
		270	−263
C_2H_3N	Ethanenitrile	330	−3468
		340	−2971
		350	−2563
	a(1) = −5840	360	−2233
	a(2) = −29175	370	−1970
	a(3) = −47611	380	−1765
		390	−1610
		400	−1499
		410	−1425
C_2H_4	Ethene	240	−218
		270	−172
		300	−139
	a(1) = −140	330	−113
	a(2) = −296	360	−92
	a(3) = −101	390	−76
		420	−63
		450	−52
$C_2H_4Cl_2$	1,2-Dichloroethane	370	−812
		390	−716
		410	−635
	a(1) = −1362	430	−566
	a(2) = −3240	450	−508
	a(3) = −2100	470	−458
		490	−416
		510	−379
		530	−347
		550	−319
		570	−295
C_2H_4O	Ethanal	290	−1352
		320	−927
		350	−654
	a(1) = −1217	380	−482
	a(2) = −4647	410	−375
	a(3) = −5725	440	−314
		470	−283
$C_2H_4O_2$	Methyl methanoate	320	−821
		330	−744
		340	−677
	a(1) = −1035	350	−620
	a(2) = −3425	360	−571
	a(3) = −4203	370	−528
		380	−492
		390	−461
		400	−435
C_2H_5Cl	Chloroethane	320	−634
		360	−450
		400	−330
	a(1) = −777	440	−249
	a(2) = −2205	480	−195
	a(3) = −1764	520	−157

VIRIAL COEFFICIENTS OF SELECTED GASES (continued)

Mol. form.	Name		T/K	B/cm^3 mol^{-1}
			560	−131
			600	−114
C_2H_6	Ethane		200	−409
			220	−337
			240	−284
	a(1) = −184		260	−242
	a(2) = −376		280	−209
	a(3) = −143		300	−181
	a(4) = −54		320	−159
			340	−140
			360	−123
			380	−109
			400	−96
			500	−52
			600	−24
C_2H_6O	Ethanol		320	−2710
			330	−2135
			340	−1676
	a(1) = −4475		350	−1317
	a(2) = −29719		360	−1043
	a(3) = −56716		370	−843
			380	−705
			390	−622
C_2H_6O	Dimethyl ether		275	−536
			280	−517
			285	−499
	a(1) = −455		290	−482
	a(2) = −965		295	−465
			300	−449
			305	−433
			310	−418
C_2H_7N	Dimethylamine		310	−606
			320	−563
			330	−523
	a(1) = −662		340	−487
	a(2) = −1504		350	−454
	a(3) = −667		360	−423
			370	−395
			380	−369
			390	−345
			400	−322
C_2H_7N	Ethylamine		300	−773
			310	−710
			320	−654
	a(1) = −785		330	−604
	a(2) = −2012		340	−558
	a(3) = −1397		350	−517
			360	−480
			370	−447
			380	−416
			390	−389
			400	−363
C_3H_6	Cyclopropane		300	−383

Mol. form.	Name	T/K	$B/cm^3\ mol^{-1}$
		310	−356
		320	−332
	a(1) = −388	330	−310
	a(2) = −861	340	−290
	a(3) = −538	350	−272
		360	−256
		370	−241
		380	−227
		390	−215
		400	−204
C_3H_6	Propene	280	−395
		300	−342
		320	−299
	a(1) = −347	340	−262
	a(2) = −727	360	−232
	a(3) = −325	380	−205
		400	−183
		420	−163
		440	−146
		460	−131
		480	−118
		500	−106
C_3H_6O	2-Propanone	300	−1996
		320	−1522
		340	−1198
	a(1) = −2051	360	−971
	a(2) = −8903	380	−806
	a(3) = −18056	400	−683
	a(4) = −16448	420	−586
		440	−506
		460	−437
		480	−375
C_3H_6O	Ethyl methanoate	330	−1003
		340	−916
		350	−839
	a(1) = −1371	360	−771
	a(2) = −4231	370	−712
	a(3) = −4312	380	−660
		390	−614
C_3H_6O	Methyl ethanoate	320	−1320
		330	−1186
		340	−1074
	a(1) = −1709	350	−980
	a(2) = −6348	360	−903
	a(3) = −9650	370	−840
		380	−789
		390	−749
C_3H_7Cl	1-Chloropropane	310	−1001
		340	−772
		370	−614
	a(1) = −1121	400	−501
	a(2) = −3271	430	−417
	a(3) = −3786	460	−352
	a(4) = −1974	490	−302

Mol. form.	Name	T/K	B/cm³ mol⁻¹

Mol. form.	Name	T/K	$B/\text{cm}^3\ \text{mol}^{-1}$
		520	−261
		550	−227
		580	−198
C_3H_8	Propane	240	−641
		260	−527
		280	−444
	a(1) = −386	300	−381
	a(2) = −844	320	−331
	a(3) = −720	340	−292
	a(4) = −574	360	−259
		380	−232
		400	−208
		440	−169
		480	−138
		520	−112
		560	−90
C_3H_8O	1-Propanol	380	−873
		385	−826
		390	−783
	a(1) = −2690	395	−744
	a(2) = −12040	400	−709
	a(3) = −16738	405	−679
		410	−651
		415	−627
		420	−606
C_3H_8O	2-Propanol	380	−821
		385	−766
		390	−717
	a(1) = −3165	395	−674
	a(2) = −16092	400	−636
	a(3) = −24197	405	−604
		410	−576
		415	−552
		420	−533
C_3H_9N	Trimethylamine	310	−675
		320	−628
		330	−585
	a(1) = −737	340	−547
	a(2) = −1669	350	−512
	a(3) = −986	360	−480
		370	−450
C_4H_8	1-Butene	300	−624
		320	−539
		340	−470
	a(1) = −633	360	−413
	a(2) = −1442	380	−366
	a(3) = −932	400	−327
		420	−294
C_4H_8O	2-Butanone	310	−2056
		320	−1878
		330	−1712
	a(1) = −2282	340	−1555

Mol. form.	Name	T/K	$B/cm^3\,mol^{-1}$
	a(2) = −5907	350	−1407
		360	−1267
		370	−1135
$C_4H_8O_2$	Propyl methanoate	330	−1496
		340	−1354
		350	−1231
	a(1) = −2118	360	−1126
	a(2) = −7299	370	−1035
	a(3) = −8851	380	−957
		390	−890
		400	−834
$C_4H_8O_2$	Ethyl ethanoate	330	−1543
		340	−1385
		350	−1254
	a(1) = −2272	360	−1144
	a(2) = −8818	370	−1055
	a(3) = −13130	380	−982
		390	−923
		400	−878
$C_4H_8O_2$	Methyl propanoate	330	−1588
		340	−1444
		350	−1319
	a(1) = −2216	360	−1211
	a(2) = −7339	370	−1117
	a(3) = −8658	380	−1037
		390	−968
		400	−908
C_4H_9Cl	1-Chlorobutane	330	−1224
		370	−898
		410	−691
	a(1) = −1643	450	−551
	a(2) = −4897	490	−449
	a(3) = −6178	530	−371
	a(4) = −3718	570	−309
C_4H_{10}	Butane	250	−1170
		280	−863
		310	−668
	a(1) = −735	340	−536
	a(2) = −1835	370	−442
	a(3) = −1922	400	−371
	a(4) = −1330	430	−315
		460	−270
		490	−232
		520	−199
		550	−171
C_4H_{10}	2-Methylpropane	270	−900
		300	−697
		330	−553
	a(1) = −707	360	−450
	a(2) = −1719	390	−374
	a(3) = −1282	420	−317
		450	−273

Mol. form.	Name	T/K	B/cm³ mol⁻¹

Mol. form.	Name	T/K	$B/\text{cm}^3\ \text{mol}^{-1}$
		480	−240
		510	−215
$C_4H_{10}O$	1-Butanol	350	−1693
		360	−1544
		370	−1402
	$a(1) = -2629$	380	−1268
	$a(2) = -6315$	390	−1141
		400	−1021
		420	−796
		440	−593
$C_4H_{10}O$	2-Methyl-1-propanol	390	−1076
		400	−979
		410	−887
	$a(1) = -2269$	420	−800
	$a(2) = -5065$	430	−716
		440	−636
$C_4H_{10}O$	2-Butanol	380	−1110
		390	−1005
		400	−906
	$a(1) = -2232$	410	−811
	$a(2) = -5209$	420	−721
$C_4H_{10}O$	2-Methyl-2-propanol	380	−924
		390	−827
		400	−736
	$a(1) = -1952$	410	−649
	$a(2) = -4775$	420	−567
$C_4H_{10}O$	Diethyl ether	280	−1550
		300	−1199
		320	−954
	$a(1) = -1226$	340	−776
	$a(2) = -4458$	360	−638
	$a(3) = -7746$	380	−525
	$a(4) = -10005$	400	−428
		420	−340
$C_4H_{11}N$	Diethylamine	320	−1228
		330	−1134
		340	−1056
	$a(1) = -1522$	350	−988
	$a(2) = -5204$	360	−926
	$a(3) = -15047$	370	−868
	$a(4) = -28835$	380	−812
		390	−755
		400	−697
C_5H_5N	Pyridine	350	−1257
		360	−1176
		370	−1099
	$a(1) = -1765$	380	−1026
	$a(2) = -3431$	390	−957
		400	−892
		420	−770
		440	−659

Mol. form.	Name	T/K	$B/cm^3\ mol^{-1}$
C_5H_{10}	Cyclopentane	300	−1049
		305	−1015
		310	−981
	a(1) = −1062	315	−949
	a(2) = −2116	320	−918
C_5H_{10}	1-Pentene	310	−966
		320	−898
		330	−836
	a(1) = −1055	340	−780
	a(2) = −2377	350	−729
	a(3) = −1189	360	−681
		370	−638
		380	−598
		390	−561
		400	−527
		410	−495
$C_5H_{10}O$	2-Pentanone	330	−2850
		340	−2420
		350	−2076
	a(1) = −4962	360	−1804
	a(2) = −26372	370	−1595
	a(3) = −46537	380	−1440
		390	−1332
C_5H_{12}	Pentane	300	−1234
		310	−1130
		320	−1038
	a(1) = −1254	330	−957
	a(2) = −3345	340	−884
	a(3) = −2726	350	−818
		400	−579
		450	−436
		500	−348
		550	−294
C_5H_{12}	2-Methylbutane	280	−1263
		290	−1166
		300	−1079
	a(1) = −1095	310	−1001
	a(2) = −2503	320	−931
	a(3) = −1534	330	−867
		340	−810
		350	−757
		400	−557
		450	−424
C_5H_{12}	2,2-Dimethylpropane	300	−916
		310	−843
		320	−780
	a(1) = −931	330	−724
	a(2) = −2387	340	−674
	a(3) = −2641	350	−629
	a(4) = −1810	360	−590
		370	−554
		380	−521
		390	−492

Mol. form.	Name	T/K	$B/cm^3\ mol^{-1}$
		400	−464
		450	−357
		500	−279
		550	−218
C_6H_6	Benzene	290	−1588
		300	−1454
		310	−1335
a(1) = −1477		320	−1231
a(2) = −3851		330	−1139
a(3) = −3683		340	−1056
a(4) = −1423		350	−983
		400	−712
		450	−542
		500	−429
		550	−349
		600	−291
C_6H_7N	2-Methylpyridine	360	−1656
		370	−1523
		380	−1404
a(1) = −2940		390	−1297
a(2) = −8813		400	−1202
a(3) = −7809		410	−1117
		420	−1040
		430	−972
C_6H_7N	3-Methylpyridine	380	−1819
		390	−1612
		400	−1448
a(1) = −6304		410	−1322
a(2) = −30415		420	−1230
a(3) = −44549		430	−1166
C_6H_7N	4-Methylpyridine	380	−1787
		390	−1578
		400	−1417
a(1) = −6553		410	−1297
a(2) = −32873		420	−1214
a(3) = −49874		430	−1163
C_6H_{12}	Cyclohexane	300	−1698
		320	−1391
		340	−1170
a(1) = −1733		360	−1007
a(2) = −5618		380	−883
a(3) = −9486		400	−786
a(4) = −7936		420	−707
		440	−641
		460	−584
		480	−534
		500	−488
		520	−446
		540	−406
		560	−368
C_6H_{12}	Methylcyclopentane	305	−1447
		315	−1357

Mol. form.	Name		T/K	B/cm^3 mol^{-1}
			325	−1272
	a(1) = −1512		335	−1192
	a(2) = −2910		345	−1117
C_6H_{14}	Hexane		300	−1920
			310	−1724
			320	−1561
	a(1) = −1961		330	−1424
	a(2) = −6691		340	−1309
	a(3) = −13167		350	−1209
	a(4) = −15273		360	−1123
			370	−1046
			380	−978
			390	−916
			400	−859
			410	−806
			430	−707
			450	−616
$C_6H_{15}N$	Triethylamine		330	−1562
			340	−1444
			350	−1340
	a(1) = −2061		360	−1249
	a(2) = −5735		370	−1169
	a(3) = −5899		380	−1099
			390	−1037
			400	−983
C_7H_8	Toluene		350	−1641
			360	−1511
			370	−1394
	a(1) = −2620		380	−1289
	a(2) = −7548		390	−1195
	a(3) = −6349		400	−1110
			410	−1034
			420	−965
			430	−903
C_7H_{14}	1-Heptene		340	−1781
			350	−1651
			360	−1532
	a(1) = −2491		370	−1424
	a(2) = −6230		380	−1324
	a(3) = −3780		390	−1233
			400	−1150
			410	−1073
C_7H_{16}	Heptane		300	−2782
			320	−2297
			340	−1928
	a(1) = −2834		360	−1641
	a(2) = −8523		380	−1415
	a(3) = −10068		400	−1233
	a(4) = −5051		420	−1085
			440	−963
			460	−862
			480	−775
			500	−702

Mol. form.	Name		T/K	$B/\text{cm}^3\ \text{mol}^{-1}$
			540	−583
			580	−490
			620	−416
			660	−355
			700	−304
C_8H_{10}	1,2-Dimethylbenzene		380	−2046
			390	−1848
			400	−1681
	$a(1) = -5632$		410	−1543
	$a(2) = -22873$		420	−1428
	$a(3) = -28900$		430	−1335
			440	−1261
C_8H_{10}	1,3-Dimethylbenzene		380	−2082
			390	−1865
			400	−1679
	$a(1) = -5808$		410	−1521
	$a(2) = -23244$		420	−1388
	$a(3) = -27607$		430	−1276
			440	−1184
C_8H_{10}	1,4-Dimethylbenzene		380	−2043
			390	−1851
			400	−1680
	$a(1) = -4921$		410	−1529
	$a(2) = -16843$		420	−1395
	$a(3) = -16159$		430	−1276
			440	−1171
C_8H_{16}	1-Octene		360	−2147
			370	−2000
			380	−1861
	$a(1) = -3273$		390	−1729
	$a(2) = -6557$		400	−1604
			410	−1485
C_8H_{18}	Octane		300	−4042
			350	−2511
			400	−1704
	$a(1) = -4123$		450	−1234
	$a(2) = -13120$		500	−936
	$a(3) = -16408$		550	−732
	$a(4) = -8580$		600	−583
			650	−468
			700	−375

VAN DER WAALS CONSTANTS FOR GASES

The van der Waals equation of state for a real gas is

$$(P + n^2a/V^2)(V - nb) = nRT$$

where P is the pressure, V the volume, T the temperature, n the amount of substance (in moles), and R the gas constant. The van der Waals constants a and b are characteristic of the substance and are independent of temperature. They are related to the critical temperature and pressure, T_c and P_c, by

$$a = 27R^2T_c^2/64P_c \quad b = RT_c/8P_c$$

This table gives values of a and b for some common gases. Most of the values have been calculated from the critical temperature and pressure values given in the table "Critical Constants" in this section. Van der Waals constants for other gases may easily be calculated from the data in that table.

To convert the van der Waals constants to SI units, note that 1 bar L^2/mol^2 = 0.1 Pa m^6/mol^2 and 1 L/mol = 0.001 m^3/mol.

REFERENCE

Reid, R.C, Prausnitz, J. M., and Poling, B.E., *The Properties of Gases and Liquids, Fourth Edition*, McGraw-Hill, New York, 1987.

Substance	a bar L^2/mol^2	b L/mol	Substance	a bar L^2/mol^2	b L/mol
Acetic acid	17.71	0.1065	Hydrogen sulfide	4.544	0.0434
Acetone	16.02	0.1124	Isobutane	13.32	0.1164
Acetylene	4.516	0.0522	Krypton	5.193	0.0106
Ammonia	4.225	0.0371	Methane	2.303	0.0431
Aniline	29.14	0.1486	Methanol	9.476	0.0659
Argon	1.355	0.0320	Methylamine	7.106	0.0588
Benzene	18.82	0.1193	Neon	0.208	0.0167
Bromine	9.75	0.0591	Neopentane	17.17	0.1411
Butane	13.89	0.1164	Nitric oxide	1.46	0.0289
1-Butanol	20.94	0.1326	Nitrogen	1.370	0.0387
2-Butanone	19.97	0.1326	Nitrogen dioxide	5.36	0.0443
Carbon dioxide	3.658	0.0429	Nitrogen trifluoride	3.58	0.0545
Carbon disulfide	11.25	0.0726	Nitrous oxide	3.852	0.0444
Carbon monoxide	1.472	0.0395	Octane	37.88	0.2374
Chlorine	6.343	0.0542	1-Octanol	44.71	0.2442
Chlorobenzene	25.80	0.1454	Oxygen	1.382	0.0319
Chloroethane	11.66	0.0903	Ozone	3.570	0.0487
Chloromethane	7.566	0.0648	Pentane	19.09	0.1449
Cyclohexane	21.92	0.1411	1-Pentanol	25.88	0.1568
Cyclopropane	8.34	0.0747	Phenol	22.93	0.1177
Decane	52.74	0.3043	Propane	9.39	0.0905
1-Decanol	59.51	0.3086	1-Propanol	16.26	0.1079
Diethyl ether	17.46	0.1333	2-Propanol	15.82	0.1109
Dimethyl ether	8.690	0.0774	Propene	8.442	0.0824
Dodecane	69.38	0.3758	Pyridine	19.77	0.1137
1-Dodecanol	75.70	0.3750	Pyrrole	18.82	0.1049
Ethane	5.580	0.0651	Silane	4.38	0.0579
Ethanol	12.56	0.0871	Sulfur dioxide	6.865	0.0568
Ethylene	4.612	0.0582	Sulfur hexafluoride	7.857	0.0879
Fluorine	1.171	0.0290	Tetrachloromethane	20.01	0.1281
Furan	12.74	0.0926	Tetrachlorosilane	20.96	0.1470
Helium	0.0346	0.0238	Tetrafluoroethylene	6.954	0.0809
Heptane	31.06	0.2049	Tetrafluoromethane	4.040	0.0633
1-Heptanol	38.17	0.2150	Tetrafluorosilane	5.259	0.0724
Hexane	24.84	0.1744	Tetrahydrofuran	16.39	0.1082
1-Hexanol	31.79	0.1856	Thiophene	17.21	0.1058
Hydrazine	8.46	0.0462	Toluene	24.86	0.1497
Hydrogen	0.2452	0.0265	1,1,1-Trichloroethane	20.15	0.1317
Hydrogen bromide	4.500	0.0442	Trichloromethane	15.34	0.1019
Hydrogen chloride	3.700	0.0406	Trifluoromethane	5.378	0.0640
Hydrogen cyanide	11.29	0.0881	Trimethylamine	13.37	0.1101
Hydrogen fluoride	9.565	0.0739	Water	5.537	0.0305
Hydrogen iodide	6.309	0.0530	Xenon	4.192	0.0516

MEAN FREE PATH AND RELATED PROPERTIES OF GASES

In the simplest version of the kinetic theory of gases, molecules are treated as hard spheres of diameter d which make binary collisions only. In this approximation the mean distance traveled by a molecule between successive collisions, the mean free path l, is related to the collision diameter by:

$$l = \frac{kT}{\pi\sqrt{2}\,Pd^2}$$

where P is the pressure, T the absolute temperature, and k the Boltzmann constant. At standard conditions ($P = 100\,000$ Pa and $T = 298.15$ K) this relation becomes:

$$l = \frac{9.27 \cdot 10^{-27}}{d^2}$$

where l and d are in meters.

Using the same model and the same standard pressure, the collision diameter can be calculated from the viscosity η by the kinetic theory relation:

$$\eta = \frac{2.67 \cdot 10^{-20}\,(MT)^{1/2}}{d^2}$$

where η is in units of μPa s and M is the molar mass in g/mol. Kinetic theory also gives a relation for the mean velocity $\bar{v}$ of molecules of mass m:

$$\bar{v} = \left(\frac{8kT}{\pi m}\right)^{1/2} = 145.5\,(T/M)^{1/2} \text{ m/s}$$

Finally, the mean time τ between collisions can be calculated from the relation $\tau\bar{v} = l$.

The table below gives values of l, $\bar{v}$, and τ for some common gases at 25°C and atmospheric pressure, as well as the value of d, all calculated from measured gas viscosities (see References 2 and 3 and the table "Viscosity of Gases" in this section). It is seen from the above equations that the mean free path varies directly with T and inversely with P, while the mean velocity varies as the square root of T and, in this approximation, is independent of P.

A more accurate model, in which molecular interactions are described by a Lennard-Jones potential, gives mean free path values about 5% lower than this table (see Reference 4).

REFERENCES

1. Reid, R.C., Prausnitz, J.M., and Poling, B.E., *The Properties of Gases and Liquids, Fourth Edition*, McGraw-Hill, New York, 1987.
2. Lide, D.R., and Kehiaian, H.V., *CRC Handbook of Thermophysical and Thermochemical Data*, CRC Press, Boca Raton, FL, 1994.
3. Vargaftik, N.B., *Tables of Thermophysical Properties of Liquids and Gases, Second Edition*, John Wiley, New York, 1975.
4. Kaye, G.W.C., and Laby, T.H., *Tables of Physical and Chemical Constants, 15th Edition*, Longman, London, 1986.

Gas	d	l	$\bar{v}$	τ
Air	$3.66 \cdot 10^{-10}$ m	$6.91 \cdot 10^{-8}$ m	467 m/s	148 ps
Ar	3.58	7.22	397	182
CO_2	4.53	4.51	379	119
H_2	2.71	12.6	1769	71
He	2.15	20.0	1256	159
Kr	4.08	5.58	274	203
N_2	3.70	6.76	475	142
NH_3	4.32	4.97	609	82
Ne	2.54	14.3	559	256
O_2	3.55	7.36	444	166
Xe	4.78	4.05	219	185

INFLUENCE OF PRESSURE ON FREEZING POINTS

This table illustrates the variation of the freezing point of representative types of liquids with pressure. Substances are listed in alphabetical order. Note that 1 MPa = 0.01 kbar = 9.87 atm.

REFERENCES

1. Isaacs, N.S., *Liquid Phase High Pressure Chemistry*, John Wiley, New York, 1981.
2. Merrill, L., *J. Phys. Chem. Ref. Data*, 6, 1205, 1977; 11, 1005, 1982.

Substance	Molecular formula	Freezing point in °C at:		
		0.1 MPa	100 MPa	1000 MPa
Acetic acid	$C_2H_4O_2$	16.6	37	
Acetophenone	C_8H_8O	20.0	41.2	
Aniline	C_6H_7N	–6.0	13.5	140
Benzene	C_6H_6	5.5	33.4	
Benzonitrile	C_7H_5N	–12.8	7.6	
Benzyl alcohol	C_7H_8O	–15.2	0.2	
Bromobenzene	C_6H_5Br	–30.6	–12	108
Bromoethane	C_2H_5Br	–118.6	–108	
1-Bromonaphthalene	$C_{10}H_7Br$	–1.8	6.1	
1-Bromopropane	C_3H_7Br	–110	–98	
p-Bromotoluene	C_7H_7Br	28.0	56.7	
Butanoic acid	$C_4H_8O_2$	–5.7	13.8	
1-Butanol	$C_4H_{10}O$	–89.8	–77.2	
Carbon disulfide	CS_2	–111.5	–98	
Chlorobenzene	C_6H_5Cl	–45.2	–28	84
p-Chlorotoluene	C_7H_7Cl	6.9	33.1	
o-Cresol	C_7H_8O	29.8	47.7	
m-Cresol	C_7H_8O	11.8	25.6	
p-Cresol	C_7H_8O	35.8	56.2	
Cyclohexane	C_6H_{12}	6.6	32.5	
Cyclohexanol	$C_6H_{12}O$	25.5	62.3	
1,2-Dibromoethane	$C_2H_4Br_2$	9.9	34.0	
p-Dichlorobenzene	$C_6H_4Cl_2$	52.7	79.1	
Dichloromethane	CH_2Cl_2	–95.1	–83	
N,N-Dimethylaniline	$C_8H_{11}N$	2.5	26.3	
1,4-Dioxane	$C_4H_8O_2$	11	23	
Ethanol	C_2H_6O	–114.1	–108	
Formamide	CH_3NO	–15.5	10.8	
Formic acid	CH_2O_2	8.3	20.6	
Furan	C_4H_4O	–85.6	–73	
Hexamethyldisiloxane	$C_6H_{18}OSi_2$	–66	–37	
Menthol	$C_{10}H_{20}O$	42	60	
Methyl benzoate	$C_8H_8O_2$	–15	31.8	
2-Methyl-2-butanol	$C_5H_{12}O$	–8.8	13.4	
2-Methyl-2-propanol	$C_4H_{10}O$	25.4	58.1	
Naphthalene	$C_{10}H_8$	78.2	115.7	
Nitrobenzene	$C_6H_5NO_2$	5.7	13.5	
m-Nitrotoluene	$C_7H_7NO_2$	15.5	40.6	
Pentachloroethane	C_2HCl_5	–29.0	–6.3	
Potassium	K	63.7	78	170
Potassium chloride	ClK	771		945
Propanoic acid	$C_3H_6O_2$	–20.7	–1.2	
Silver chloride	AgCl	455		545
Sodium	Na	97.8	106	167
Sodium chloride	ClNa	800.7		997
Sodium fluoride	FNa	996		1115
Tetrachloromethane	CCl_4	–23.0	14.2	
Tribromomethane	$CHBr_3$	8.1	31.5	
Trichloromethane	$CHCl_3$	–63.6	–45.2	
Water	H_2O	0.0	–9.0	
o-Xylene	C_8H_{10}	–25.2	–3.5	
m-Xylene	C_8H_{10}	–47.8	–25.2	
p-Xylene	C_8H_{10}	13.2	46.0	

CRITICAL CONSTANTS

Douglas Ambrose

This table gives the liquid - gas critical constants and normal boiling points of about 700 inorganic and organic substances. The properties tabulated are:

T_b: Normal boiling point in K at a pressure of 101.325 kPa (1 atmosphere); an "s" following the value indicates a sublimation point (temperature at which the solid is in equilibrium with the gas at a pressure of 101.325 kPa)

T_c: Critical temperature in K

P_c: Critical pressure in MPa

V_c: Critical molar volume in cm^3/mol

The number of digits given for T_b, T_c, and P_c indicates the estimated accuracy of these quantities; however, values of T_c greater than 750 K may be in error by 10 K or more. Although most V_c values are given to three figures, they cannot be assumed accurate to better than a few percent.

Many of the critical constants in this table are taken from reviews produced by the IUPAC Commission on Thermodynamics (References 2-6) and from the publications of the Design Institute for Physical Property Data (References 8-9). Compounds are listed by molecular formula in modified Hill order, with compounds not containing carbon preceding those that do contain carbon.

An asterisk following an entry indicates the value was obtained by extrapolation.

REFERENCES

1. Ambrose, D., "Vapor-Liquid Constants of Fluids", in Stevenson, R. M., and Malanowski, S., *Handbook of the Thermodynamics of Organic Compounds*, Elsevier, New York, 1987.
2. Ambrose, D., and Tsonopoulos, C., *J. Chem. Eng. Data*, 40, 531, 1995.
3. Tsonopoulos, C., and Ambrose, D., *J. Chem. Eng. Data*, 40, 547, 1995.
4. Gude, M., and Teja, A. S., *J. Chem. Eng. Data*, 40, 1025, 1995.
5. Daubert, T. E., *J. Chem. Eng. Data*, 41, 365, 1966.
6. Tsonopoulos, C., and Ambrose, D., *J. Chem. Eng. Data*, 41, 645, 1996.
7. Das, A., Frenkel, M., Gadalla, N. A. M., Kudchadker, S., Marsh, K. N., Rodgers, A. S., and Wilhoit, R. C., *J. Phys. Chem. Ref. Data*, 22, 659, 1993.
8. Wilson, L. C., Wilson, H. L., Wilding, W. V., and Wilson, G. M., *J. Chem. Eng. Data*, 41, 1252, 1997.
9. Daubert, T. E., Danner, R. P., Sibul, H. M., and Stebbins, C. C., *Physical and Thermodynamic Properties of Pure Compounds: Data Compilation*, extant 1996 (core with 6 supplements), Taylor & Francis, Bristol, PA.
10. Reid, R. C., Prausnitz, J. M., and Poling, B. E., *The Properties of Gases and Liquids, Fourth Edition,* McGraw-Hill, New York, 1987.
11. Vargaftik, N.B., *Tables of Thermophysical Properties of Liquids and Gases, Second Edition*, John Wiley, New York, 1975.

Mol. Form.	Name	T_b/K	T_c/K	P_c/MPa	V_c/cm^3 mol^{-1}
AlBr$_3$	Aluminum tribromide	528	763	2.89	310
AlCl$_3$	Aluminum trichloride	453 s	620	2.63	257
AlI$_3$	Aluminum triiodide	655	983		408
Ar	Argon	87.30	150.87	4.898	75
As	Arsenic	876 s	1673		35
AsCl$_3$	Arsenic(III) chloride	403	654		252
AsH$_3$	Arsine	210.7	373.1		
BBr$_3$	Boron tribromide	364	581		272
BCl$_3$	Boron trichloride	285.80	455	3.87	239
BF$_3$	Boron trifluoride	172	260.8	4.98	115
BI$_3$	Boron triiodide	483	773		356
B$_2$H$_6$	Diborane	180.7	289.8	4.05	
BiBr$_3$	Bismuth tribromide	726	1220		301
BiCl$_3$	Bismuth trichloride	720	1179	12.0	261
BrH	Hydrogen bromide	206.77	363.2	8.55	
BrI	Iodine bromide	389	719		139
Br$_2$	Bromine	332.0	588	10.34	127
Br$_2$Hg	Mercury(II) bromide	595	1012		
Br$_3$Ga	Gallium(III) bromide	552	806.7		303
Br$_3$HSi	Tribromosilane	382	610.0		305
Br$_3$P	Phosphorus(III) bromide	446.10	711		300
Br$_3$Sb	Antimony(III) bromide	553	904		300
Br$_4$Ge	Germanium(IV) bromide	459.50	718		392
Br$_4$Hf	Hafnium(IV) bromide	596 s	746		415

Mol. Form.	Name	T_b/K	T_c/K	P_c/MPa	V_c/cm³ mol⁻¹
Br₄Si	Tetrabromosilane	427	663		382
Br₄Sn	Tin(IV) bromide	478	744		417
Br₄Ti	Titanium(IV) bromide	503	795.7		391
Br₄Zr	Zirconium(IV) bromide	633 s	805		424
Br₅Ta	Tantalum(V) bromide	622	974		461
ClFO₃	Perchloryl fluoride	226.40	368.4	5.37	161
ClF₂N	Nitrogen chloride difluoride	207	337.5	5.15	
ClF₂P	Phosphorus(III) chloride difluoride	225.90	362.4	4.52	
ClF₂PS	Phosphorus(V) sulfide chloride difluoride	279.5	439.2	4.14	
ClF₃Si	Chlorotrifluorosilane	203.1	307.7	3.46	
ClF₅	Chlorine pentafluoride	260.0	416	5.27	233
ClF₅S	Sulfur chloride pentafluoride	254.10	390.9		
ClH	Hydrogen chloride	188	324.7	8.31	81
ClH₄N	Ammonium chloride	611 s	1155	163.5	
ClH₄P	Phosphonium chloride	246 s	322.3	7.37	
ClNO	Nitrosyl chloride	267.7	440		
ClOV	Vanadyl chloride	400	636		171
Cl₂	Chlorine	239.11	416.9	7.991	123
Cl₂FP	Phosphorus(III) dichloride fluoride	287	463.0	4.96	
Cl₂F₂Si	Dichlorodifluorosilane	241	369.0	3.5	
Cl₂Hg	Mercury(II) chloride	577	973		174
Cl₂OSe	Selenium oxychloride	450	730	7.09	235
Cl₃FSi	Trichlorofluorosilane	285.40	438.6	3.58	
Cl₃Ga	Gallium(III) chloride	474	694		263
Cl₃HSi	Trichlorosilane	306	479		268
Cl₃P	Phosphorus(III) chloride	349.10	563		264
Cl₃Sb	Antimony(III) chloride	493.5	794		272
Cl₄Ge	Germanium(IV) chloride	359.70	553.2	3.861	330
Cl₄Hf	Hafnium(IV) chloride	590 s	725.7	5.42	314
Cl₄ORe	Rhenium(VI) oxytetrachloride	496	781		362
Cl₄OW	Tungsten(VI) oxytetrachloride	500.70	782		338
Cl₄Si	Tetrachlorosilane	330.80	508.1	3.593	326
Cl₄Sn	Tin(IV) chloride	387.30	591.9	3.75	351
Cl₄Te	Tellurium tetrachloride	660	1002	8.56	310
Cl₄Ti	Titanium(IV) chloride	409.60	638	4.66	339
Cl₄Zr	Zirconium(IV) chloride	604 s	778	5.77	319
Cl₅Mo	Molybdenum(V) chloride	541	850		369
Cl₅Nb	Niobium(V) chloride	527.2	803.5	4.88	397
Cl₅P	Phosphorus(V) chloride	433 s	646		
Cl₅Ta	Tantalum(V) chloride	512.50	767		402
Cl₆W	Tungsten(VI) chloride	619.90	923		422
Cs	Cesium	944	1938	9.4	341
FH	Hydrogen fluoride	293	461	6.48	69
FNO₂	Nitryl fluoride	200.7	349.5		
F₂	Fluorine	85.03	144.13	5.172	66
F₂HN	Difluoramine	250	403		
F₂N₂	cis-Difluorodiazine	167.40	272	7.09	
F₂N₂	trans-Difluorodiazine	161.70	260	5.57	
F₂O	Fluorine monoxide	128.40	215		
F₂Xe	Xenon difluoride	387.50 s	631	9.32	148
F₃N	Nitrogen trifluoride	144.40	234.0	4.46	126
F₃NO	Trifluoramine oxide	185.6	303	6.43	147
F₃P	Phosphorus(III) fluoride	171.6	271.2	4.33	
F₃PS	Phosphorus(V) sulfide trifluoride	220.90	346.0	3.82	
F₄N₂	Tetrafluorohydrazine	199	309	3.75	
F₄S	Sulfur tetrafluoride	232.70	364		
F₄Si	Tetrafluorosilane	187	259.0	3.72	
F₄Xe	Xenon tetrafluoride	388.90 s	612	7.04	188
F₅Nb	Niobium(V) fluoride	502	737	6.28	155

Mol. Form.	Name	T_b/K	T_c/K	P_c/MPa	V_c/cm^3 mol^{-1}
F$_6$Mo	Molybdenum(VI) fluoride	307.2	473	4.75	226
F$_6$S	Sulfur hexafluoride	209.4 s	318.69	3.77	199
F$_6$Se	Selenium hexafluoride	226.6 s	345.5		
F$_6$Te	Tellurium hexafluoride	234.3 s	356		
F$_6$U	Uranium(VI) fluoride	329.7 s	505.8	4.66	250
F$_6$W	Tungsten(VI) fluoride	290.3	444	4.34	233
GaI$_3$	Gallium(III) iodide	613	951	395	
GeH$_4$	Germane	185.0	312.2	4.95	147
GeI$_4$	Germanium(IV) iodide	650	973		500
HI	Hydrogen iodide	237.60	424.0	8.31	
H$_2$	Hydrogen	20.28	32.97	1.293	65
H$_2$O	Water	373.15	647.14	22.06	56
H$_2$O$_2$	Hydrogen peroxide	423.4	728*	22*	
H$_2$S	Hydrogen sulfide	213.60	373.2	8.94	99
H$_2$Se	Hydrogen selenide	231.90	411	8.92	
H$_3$N	Ammonia	239.82	405.5	11.35	72
H$_3$P	Phosphine	185.40	324.5	6.54	
H$_4$N$_2$	Hydrazine	386.70	653	14.7	
He	Helium	4.22	5.19	0.227	57
HfI$_4$	Hafnium iodide	667 s	916		528
Hg	Mercury	629.88	1750	172.00	43
HgI$_2$	Mercury(II) iodide	627	1072		
I$_2$	Iodine	457.6	819		155
I$_3$Sb	Antimony(III) iodide	674	1102		
I$_4$Si	Tetraiodosilane	560.50	944		558
I$_4$Sn	Tin(IV) iodide	637.50	968		531
I$_4$Ti	Titanium(IV) iodide	650	1040		505
I$_4$Zr	Zirconium(IV) iodide	704 s	960		530
K	Potassium	1032	2223*	16*	209*
Kr	Krypton	119.93	209.41	5.50	91
Li	Lithium	1615	3223*	67*	66*
NO	Nitric oxide	121.41	180	6.48	58
N$_2$	Nitrogen	77.36	126.21	3.39	90
N$_2$O	Nitrous oxide	184.67	309.57	7.255	97
N$_2$O$_4$	Nitrogen tetroxide	294.30	431	10.1	167
Na	Sodium	1156	2573*	35*	116*
Ne	Neon	27.07	44.4	2.76	42
O$_2$	Oxygen	90.20	154.59	5.043	73
O$_2$S	Sulfur dioxide	263.10	430.8	7.884	122
O$_3$	Ozone	161.80	261.1	5.57	89
O$_3$S	Sulfur trioxide	318	491.0	8.2	127
O$_4$Os	Osmium(VIII) oxide	408	678		
O$_7$Re$_2$	Rhenium(VII) oxide	633	942		334
P	Phosphorus	553.7	994		
Rb	Rubidium	961	2093*	16*	247*
Rn	Radon	211.5	377	6.28	
S	Sulfur	717.75	1314	20.7	
Se	Selenium	958	1766	27.2	
Xe	Xenon	165.04	289.77	5.841	118
CBrClF$_2$	Bromochlorodifluoromethane	269.5	426.88	4.254	246
CBrF$_3$	Bromotrifluoromethane	215.3	340.2	3.97	196
CBr$_2$F$_2$	Dibromodifluoromethane	298	471.3		
CClF$_3$	Chlorotrifluoromethane	191.7	302	3.870	180
CCl$_2$F$_2$	Dichlorodifluoromethane	243.3	384.95	4.136	217
CCl$_2$O	Carbonyl chloride	281	455	5.67	190
CCl$_3$F	Trichlorofluoromethane	296.8	471.1	4.47	247
CCl$_4$	Tetrachloromethane	350.0	556.6	4.516	276
CF$_4$	Tetrafluoromethane	145.1	227.6	3.74	140
CHClF$_2$	Chlorodifluoromethane	232.5	369.5	5.035	164

Mol. Form.	Name	T_b/K	T_c/K	P_c/MPa	V_c/cm^3 mol^{-1}
CHCl$_2$F	Dichlorofluoromethane	282.0	451.58	5.18	196
CHCl$_3$	Trichloromethane	334.32	536.4	5.47	239
CHF$_3$	Trifluoromethane	191.0	299.3	4.858	133
CHN	Hydrogen cyanide	299	456.7	5.39	139
CH$_2$Cl$_2$	Dichloromethane	313	510	6.10	
CH$_2$F$_2$	Difluoromethane	221.6	351.6	5.830	121
CH$_2$O$_2$	Formic acid	374	588		
CH$_3$Cl	Chloromethane	249.06	416.25	6.679	139
CH$_3$Cl$_3$Si	Methyltrichlorosilane	338.8	517	3.28	348
CH$_3$F	Fluoromethane	194.7	317.8	5.88	113
CH$_3$I	Iodomethane	315.70	528		
CH$_3$NO$_2$	Nitromethane	374.34	588	5.87	173
CH$_4$	Methane	111.8	190.56	4.599	98.60
CH$_4$O	Methanol	337.8	512.5	8.084	117
CH$_4$S	Methanethiol	279.0	470.0	7.23	145
CH$_5$ClSi	Chloromethylsilane	280	517.8		
CH$_5$N	Methylamine	266.83	430.7	7.614	
CH$_6$N$_2$	Methylhydrazine	360.7	567	8.24	271
CH$_6$Si	Methylsilane	215.7	352.5		
CO	Carbon monoxide	81.6	132.91	3.499	93
COS	Carbon oxysulfide	223	375	5.88	137
CO$_2$	Carbon dioxide	194.8 s	304.13	7.375	94
CS$_2$	Carbon disulfide	319	552	7.90	173
C$_2$Br$_2$ClF$_3$	1,2-Dibromo-1-chloro-1,2,2-trifluoroethane	366	560.7	3.61	368
C$_2$Br$_2$F$_4$	1,2-Dibromotetrafluoroethane	320.50	487.8	3.393	341
C$_2$ClF$_3$	Chlorotrifluoroethylene	245.3	379	4.05	212
C$_2$ClF$_5$	Chloropentafluoroethane	235.20	353.2	3.229	252
C$_2$Cl$_2$F$_4$	1,1-Dichlorotetrafluoroethane	277	418.6	3.30	294
C$_2$Cl$_2$F$_4$	1,2-Dichlorotetrafluoroethane	277.0	418.78	3.252	297
C$_2$Cl$_3$F$_3$	1,1,2-Trichloro-1,2,2-trifluoroethane	320.8	487.3	3.42	325
C$_2$Cl$_4$	Tetrachloroethylene	394.5	620.2		
C$_2$Cl$_4$F$_2$	1,1,2,2-Tetrachloro-1,2-difluoroethane	366.0	551		
C$_2$F$_3$N	Trifluoroacetonitrile	204.3	311.11	3.618	202
C$_2$F$_4$	Tetrafluoroethylene	197.2	306.5	3.94	172
C$_2$F$_6$	Hexafluoroethane	195.0	293		222
C$_2$HClF$_2$	1-Chloro-2,2-difluoroethylene	254.7	400.6	4.46	197
C$_2$HClF$_4$	1-Chloro-1,1,2,2-tetrafluoroethane	263	399.9	3.72	244
C$_2$HClF$_4$	1-Chloro-1,2,2,2-tetrafluoroethane	261	395.4	3.66	
C$_2$HCl$_2$F$_3$	1,2-Dichloro-1,1,2-trifluoroethane	301	461.6		278
C$_2$HCl$_2$F$_3$	2,2-Dichloro-1,1,1-trifluoroethane	300.97	456.83	3.661	278
C$_2$HCl$_3$	Trichloroethylene	360.36	544.2	5.02	
C$_2$HF$_3$O$_2$	Trifluoroacetic acid	346	491.3	3.258	204
C$_2$HF$_5$	Pentafluoroethane	224.7	339.41	3.639	209.7
C$_2$HF$_5$O	Trifluoromethyl difluoromethyl ether	237	354.49	3.33	
C$_2$H$_2$	Acetylene	188.4 s	308.3	6.138	112.2
C$_2$H$_2$Cl$_2$	cis-1,2-Dichloroethylene	333.3	544.2		
C$_2$H$_2$Cl$_2$	trans-1,2-Dichloroethylene	321.8	516.5	5.51	
C$_2$H$_2$Cl$_4$	1,1,2,2-Tetrachloroethane	419.7	661.15		
C$_2$H$_2$F$_2$	1,1-Difluoroethylene	187.5	302.9	4.46	154
C$_2$H$_2$F$_4$	1,1,1,2-Tetrafluoroethane	247.07	374.18	4.065	198
C$_2$H$_2$F$_4$	1,1,2,2-Tetrafluoroethane	253.2	391.8		191
C$_2$H$_3$ClF$_2$	1-Chloro-1,1-difluoroethane	263.5	410.34	4.048	225
C$_2$H$_3$Cl$_2$F	1,1-Dichloro-1-fluoroethane	305.2	477.5	4.194	255
C$_2$H$_3$Cl$_3$	1,1,1-Trichloroethane	347.24	545	4.30	
C$_2$H$_3$F	Fluoroethylene	201	327.9	5.24	144
C$_2$H$_3$F$_3$	1,1,1-Trifluoroethane	225.90	346.04	3.775	194
C$_2$H$_3$F$_3$O	Methyl trifluoromethyl ether	249	378.0	3.680	228
C$_2$H$_3$N	Acetonitrile	354.80	545.5	4.85	173
C$_2$H$_4$	Ethylene	169.5	282.34	5.041	131.1

Mol. Form.	Name	T_b/K	T_c/K	P_c/MPa	V_c/cm^3 mol^{-1}
$C_2H_4Br_2$	1,2-Dibromoethane	404.8	583.0	7.2	
$C_2H_4Cl_2$	1,1-Dichloroethane	330.5	523	5.07	236
$C_2H_4Cl_2$	1,2-Dichloroethane	356.7	561	5.4	225
$C_2H_4F_2$	1,1-Difluoroethane	248.20	386.7	4.50	181
C_2H_4O	Acetaldehyde	293.3	466		154
C_2H_4O	Ethylene oxide	283.8	469	7.19	140
$C_2H_4O_2$	Acetic acid	391.0	592.71	5.786	171
$C_2H_4O_2$	Methyl formate	304.8	487.2	5.998	172
C_2H_5Br	Bromoethane	311.7	503.9	6.23	215
C_2H_5Cl	Chloroethane	285.5	460.4	5.3	
C_2H_5F	Fluoroethane	235.6	375.31	5.028	
C_2H_6	Ethane	184.5	305.32	4.872	145.5
$C_2H_6Cl_2Si$	Dichlorodimethylsilane	343.5	520.4	3.49	350
C_2H_6O	Ethanol	351.44	514.0	6.137	168
C_2H_6O	Dimethyl ether	248.3	400.0	5.37	190
$C_2H_6O_2$	Ethylene glycol	470.5	718		
C_2H_6S	Ethanethiol	308.3	499	5.49	207
C_2H_6S	Dimethyl sulfide	310.48	503.0	5.53	201
C_2H_7N	Ethylamine	289.7	456	5.62	182
C_2H_7N	Dimethylamine	280.03	437.22	5.340	
$C_2H_8N_2$	1,2-Ethanediamine	390	613.1	6.707	
C_2N_2	Cyanogen	252.1	400	5.98	
C_3ClF_5O	Chloropentafluoroacetone	281	410.6	2.878	
C_3F_6O	Perfluoroacetone	245.7	357.14	2.84	329
C_3F_6O	Perfluorooxetane	244	361.8	3.094	272
C_3F_8	Perfluoropropane	236.6	345.1	2.680	299
$C_3F_8O_2$	Perfluorodimethoxymethane	263	372.3	2.333	363
C_3HF_7O	Trifluoromethyl 1,1,2,2-tetrafluoroethyl ether	270	387.7	2.640	337
$C_3H_2F_6$	1,1,1,2,3,3-Hexafluoropropane	279.3	412.38	3.412	269
$C_3H_2F_6O$	1,2,2,2-Tetrafluoroethyl difluoromethyl ether	296.50	428.95	3.050	315
$C_3H_3F_3$	3,3,3-Trifluoropropene	256	376.2	3.80	211
$C_3H_3F_5$	1,1,1,2,2-Pentafluoropropane	255.7	380.11	3.137	273
$C_3H_3F_5O$	Perfluoroethyl methyl ether	278.74	406.80	2.887	301
C_3H_3N	Acrylonitrile	350.5	540	4.660	
C_3H_3NO	Isoxazole	368	552.0		
C_3H_4	Allene	238.7	394	5.25	
C_3H_4	Propyne	250.0	402.4	5.63	163.5
C_3H_5Cl	3-Chloropropene	318.3	514		
$C_3H_5F_3O$	2,2,2-Trifluoroethyl methyl ether	304.77	448.98	3.513	277
C_3H_5N	Propanenitrile	370.29	561.3	4.26	229
C_3H_6	Propene	225.46	364.9	4.60	184.6
C_3H_6	Cyclopropane	240.34	398.0	5.54	162
C_3H_6O	Allyl alcohol	370.2	545.1		
C_3H_6O	Propanal	321	504.4	5.27	204
C_3H_6O	Acetone	329.20	508.1	4.700	209
C_3H_6O	Methyloxirane	308	482.2	4.92	186
$C_3H_6O_2$	Propanoic acid	414.30	604	4.53	222
$C_3H_6O_2$	Ethyl formate	327.5	508.5	4.74	229
$C_3H_6O_2$	Methyl acetate	330.02	506.55	4.75	228
C_3H_7Cl	1-Chloropropane	319.7	503	4.58	
C_3H_7NO	N,N-Dimethylformamide	426	649.6		262
C_3H_8	Propane	231.1	369.83	4.248	200
C_3H_8O	1-Propanol	370.3	536.8	5.169	218
C_3H_8O	2-Propanol	355.5	508.3	4.764	222
C_3H_8O	Ethyl methyl ether	280.5	437.8	4.40	221
$C_3H_8O_2$	Ethylene glycol monomethyl ether	397.3	597.6	5.285	263
C_3H_8S	1-Propanethiol	341.0	536.6		286
C_3H_8S	Ethyl methyl sulfide	339.8	533	4.26	
$C_3H_9BO_3$	Trimethyl borate	340.7	501.7	3.59	

CRITICAL CONSTANTS (continued)

Mol. Form.	Name	T_b/K	T_c/K	P_c/MPa	V_c/cm^3 mol^{-1}
C$_3$H$_9$ClSi	Trimethylchlorosilane	333	497.8	3.20	366
C$_3$H$_9$N	Propylamine	320.37	497.0	4.72	
C$_3$H$_9$N	Isopropylamine	304.91	471.8	4.54	221
C$_3$H$_9$N	Trimethylamine	276.02	432.79	4.087	254
C$_4$Br$_2$F$_8$	1,4-Dibromooctafluorobutane	370	532.5	2.39	
C$_4$F$_8$	Perfluorocyclobutane	267.3	388.46	2.784	324
C$_4$F$_{10}$	Perfluorobutane	271.3	386.4	2.323	378
C$_4$F$_{10}$	Perfluoroisobutane	273	395.4		
C$_4$H$_2$F$_8$O	Perfluoroethyl 2,2,2-trifluoroethyl ether	301.04	421.68	2.330	409
C$_4$H$_3$F$_7$O	Perfluoropropyl methyl ether	307.38	437.70	2.481	377
C$_4$H$_3$F$_7$O	Perfluoroisopropyl methyl ether	302.56	433.30	2.553	369
C$_4$H$_4$O	Furan	304.7	490.2	5.50	218
C$_4$H$_4$S	Thiophene	357.2	579.4	5.69	219
C$_4$H$_5$F$_5$O	Perfluoroethyl ethyl ether	301.26	431.23	2.533	366
C$_4$H$_5$N	Pyrrole	402.94	639.7	6.34	200
C$_4$H$_6$	1,3-Butadiene	268.74	425	4.32	221
C$_4$H$_6$	1-Butyne	281.23	440	4.60	208
C$_4$H$_6$	2-Butyne	300.0	488.7		
C$_4$H$_6$O$_2$	γ-Butyrolactone	477	731.0	5.131	
C$_4$H$_6$O$_3$	Acetic anhydride	412.7	606	4.0	
C$_4$H$_7$N	Butanenitrile	390.8	585.4	3.88	
C$_4$H$_8$	1-Butene	266.89	419.5	4.02	240.8
C$_4$H$_8$	cis-2-Butene	276.86	435.5	4.21	233.8
C$_4$H$_8$	trans-2-Butene	274.03	428.6	4.10	237.7
C$_4$H$_8$	Isobutene	266.3	417.9	4.000	238.8
C$_4$H$_8$	Cyclobutane	285.8	460.0	4.98	210
C$_4$H$_8$O	Ethyl vinyl ether	308.7	475	4.07	
C$_4$H$_8$O	Butanal	348.0	537.2	4.32	258
C$_4$H$_8$O	2-Butanone	352.74	536.78	4.207	267
C$_4$H$_8$O	Tetrahydrofuran	338	540.1	5.19	224
C$_4$H$_8$OS	Ethyl thioacetate	389.5	590.55	4.075	319
C$_4$H$_8$O$_2$	Butanoic acid	436.90	624	4.03	290
C$_4$H$_8$O$_2$	2-Methylpropanoic acid	427.60	605	3.7	292
C$_4$H$_8$O$_2$	Propyl formate	354.0	538.0	4.06	285
C$_4$H$_8$O$_2$	Ethyl acetate	350.26	523.3	3.882	286
C$_4$H$_8$O$_2$	Methyl propanoate	353.0	530.6	4.004	282
C$_4$H$_8$O$_2$	1,4-Dioxane	374.7	587	5.21	238
C$_4$H$_8$S	Tetrahydrothiophene	394.2	632.0		
C$_4$H$_9$N	Pyrrolidine	359.71	568	6.00	238
C$_4$H$_{10}$	Butane	272.7	425.12	3.796	255
C$_4$H$_{10}$	Isobutane	261.42	407.8	3.640	259
C$_4$H$_{10}$O	1-Butanol	390.88	563.0	4.414	274
C$_4$H$_{10}$O	2-Butanol	372.66	536.2	4.202	269
C$_4$H$_{10}$O	2-Methyl-1-propanol	381.04	547.8	4.295	274
C$_4$H$_{10}$O	2-Methyl-2-propanol	355.5	506.2	3.972	275
C$_4$H$_{10}$O	Diethyl ether	307.7	466.74	3.638	280
C$_4$H$_{10}$O	Methyl propyl ether	312.3	476.25	3.801	
C$_4$H$_{10}$O	Isopropyl methyl ether	303.92	464.48	3.762	
C$_4$H$_{10}$O$_2$	1,2-Butanediol	463.7	680	5.21	303
C$_4$H$_{10}$O$_2$	1,3-Butanediol	480.7	676	4.02	305
C$_4$H$_{10}$O$_2$	1,4-Butanediol	508	727	6.22	
C$_4$H$_{10}$O$_2$	Ethylene glycol dimethyl ether	358	536	3.87	271
C$_4$H$_{10}$O$_2$	Propylene glycol monomethyl ether	392	579.8	4.113	
C$_4$H$_{10}$S	1-Butanethiol	371.7	570.1		324
C$_4$H$_{10}$S	Diethyl sulfide	365.3	557	3.96	318
C$_4$H$_{11}$N	Butylamine	350.15	531.9	4.25	277
C$_4$H$_{11}$N	sec-Butylamine	335.88	514.3	4.20	278
C$_4$H$_{11}$N	tert-Butylamine	317.19	483.9	3.84	292
C$_4$H$_{11}$N	Isobutylamine	340.90	519	4.07	278

6-53

Mol. Form.	Name	T_b/K	T_c/K	P_c/MPa	V_c/cm^3 mol^{-1}
$C_4H_{11}N$	Diethylamine	328.7	499.99	3.758	
$C_4H_{12}Si$	Tetramethylsilane	299.8	448.64	2.821	362
$C_4H_{12}Sn$	Tetramethylstannane	351	521.8	2.981	
C_5F_{12}	Perfluoropentane	302.3	420.59	2.045	473
$C_5H_2F_6O_2$	Hexafluoroacetylacetone	327.30	485.1	2.767	
$C_5H_4O_2$	Furfural	434.9	670	5.89	
C_5H_5N	Pyridine	388.38	620.0	5.67	243
$C_5H_6N_2$	2-Methylpyrazine	410	634.3	5.01	283
C_5H_6O	2-Methylfuran	338	527	4.72	247
C_5H_7N	1-Methylpyrrole	385.96	596.0	4.86	271
C_5H_7N	2-Methylpyrrole	420.7	654	5.08	266
C_5H_7N	3-Methylpyrrole	416.0	647	5.08	266
C_5H_8	1-Pentyne	313.3	493.5		
C_5H_8	Cyclopentene	317.3	506.5	4.80	245
C_5H_8O	Cyclopentanone	403.72	624.5	4.60	
C_5H_8O	3,4-Dihydro-2H-pyran	359	561.7	4.56	268
C_5H_9N	Pentanenitrile	414.5	610.3	3.58	
C_5H_9NO	N-Methyl-2-pyrrolidone	475	721.8		311
C_5H_{10}	1-Pentene	303.11	464.8	3.56	298.4
C_5H_{10}	cis-2-Pentene	310.08	475	3.69	
C_5H_{10}	trans-2-Pentene	309.49	471	3.52	
C_5H_{10}	2-Methyl-1-butene	304.3	470	3.8	
C_5H_{10}	3-Methyl-1-butene	293.3	452.7	3.53	304.9
C_5H_{10}	2-Methyl-2-butene	311.71	470	3.42	
C_5H_{10}	Cyclopentane	322.5	511.7	4.51	259
$C_5H_{10}O$	Cyclopentanol	413.57	619.5	4.9	
$C_5H_{10}O$	Allyl ethyl ether	340.8	518		
$C_5H_{10}O$	Pentanal	376	566.1	3.97	313
$C_5H_{10}O$	2-Pentanone	375.41	561.08	3.694	301
$C_5H_{10}O$	3-Pentanone	375.11	561.46	3.729	336
$C_5H_{10}O$	3-Methyl-2-butanone	367.48	553.4	3.85	310
$C_5H_{10}O$	Tetrahydropyran	361	572.2	4.77	263
$C_5H_{10}O$	2-Methyltetrahydrofuran	351	537	3.76	267
$C_5H_{10}O_2$	Pentanoic acid	459.3	643	3.58	340
$C_5H_{10}O_2$	3-Methylbutanoic acid	449.7	629	3.40	
$C_5H_{10}O_2$	Isobutyl formate	371.3	551	3.88	352
$C_5H_{10}O_2$	Propyl acetate	374.69	549.7	3.36	345
$C_5H_{10}O_2$	Isopropyl acetate	361.8	531		
$C_5H_{10}O_2$	Ethyl propanoate	372.3	546.0	3.362	345
$C_5H_{10}O_2$	Methyl butanoate	376.0	554.4	3.47	340
$C_5H_{10}O_2$	Methyl isobutanoate	365.7	540.8	3.43	339
$C_5H_{11}N$	Piperidine	379.37	594	4.94	288
C_5H_{12}	Pentane	309.21	469.7	3.370	311
C_5H_{12}	Isopentane	301.03	460.4	3.38	306
C_5H_{12}	Neopentane	282.63	433.8	3.196	307
$C_5H_{12}O$	1-Pentanol	411.13	588.1	3.897	326
$C_5H_{12}O$	2-Pentanol	392.5	560.3	3.675	329
$C_5H_{12}O$	3-Pentanol	389.40	559.6		325
$C_5H_{12}O$	2-Methyl-1-butanol	400.7	575.4	3.94	
$C_5H_{12}O$	3-Methyl-1-butanol	404.3	577.2	3.93	
$C_5H_{12}O$	2-Methyl-2-butanol	375.5	543.7	3.71	
$C_5H_{12}O$	3-Methyl-2-butanol	386.0	556.1	3.87	
$C_5H_{12}O$	Butyl methyl ether	343.31	512.78	3.371	329
$C_5H_{12}O$	Methyl tert-butyl ether	328.3	497.1	3.430	
$C_5H_{12}O$	Ethyl propyl ether	336.36	500.23	3.370	339
$C_5H_{12}O_2$	Ethylene glycol monopropyl ether	423.0	614.6		364
$C_5H_{12}O_3$	Diethylene glycol monomethyl ether	466	672	3.67	
$C_5H_{12}S$	3-Methyl-1-butanethiol	393	604		
C_6BrF_5	Bromopentafluorobenzene	410	601	3.0	

CRITICAL CONSTANTS (continued)

Mol. Form.	Name	T_b/K	T_c/K	P_c/MPa	V_c/cm^3 mol^{-1}
C$_6$ClF$_5$	Chloropentafluorobenzene	391.11	570.81	3.238	376
C$_6$Cl$_2$F$_4$	1,2-Dichlorotetrafluorobenzene	430.9	626	5.32	
C$_6$Cl$_3$F$_3$	1,3,5-Trichloro-2,4,6-trifluorobenzene	471.6	684.8	3.27	448
C$_6$F$_6$	Hexafluorobenzene	353.41	516.73	3.273	335
C$_6$F$_{10}$	Perfluorocyclohexene	325.2	461.8		
C$_6$F$_{12}$	Perfluoro-1-hexene	330.2	454.4		
C$_6$F$_{12}$	Perfluorocyclohexane	326.0	457.2	2.43	
C$_6$F$_{14}$	Perfluorohexane	329.8	448.77	1.868	606
C$_6$F$_{14}$	Perfluoro-2-methylpentane	330.8	455.3	1.923	532
C$_6$F$_{14}$	Perfluoro-3-methylpentane	331.5	450	1.69	
C$_6$F$_{14}$	Perfluoro-2,3-dimethylbutane	333.0	463	1.87	525
C$_6$HF$_5$	Pentafluorobenzene	358.89	530.97	3.531	324
C$_6$HF$_5$O	Pentafluorophenol	418.8	609	4.0	348
C$_6$HF$_{11}$	Undecafluorocyclohexane	335.2	477.7		
C$_6$H$_2$F$_4$	1,2,3,4-Tetrafluorobenzene	367.5	550.83	3.791	313
C$_6$H$_2$F$_4$	1,2,3,5-Tetrafluorobenzene	357.5	535.25	3.747	
C$_6$H$_2$F$_4$	1,2,4,5-Tetrafluorobenzene	363.3	543.35	3.801	
C$_6$H$_4$F$_2$	p-Difluorobenzene	362	556	4.40	
C$_6$H$_5$Br	Bromobenzene	429.21	670	4.52	324
C$_6$H$_5$Cl	Chlorobenzene	404.87	632.4	4.52	308
C$_6$H$_5$F	Fluorobenzene	357.88	560.09	4.551	269
C$_6$H$_5$I	Iodobenzene	461.6	721	4.52	351
C$_6$H$_6$	Benzene	353.24	562.05	4.895	256
C$_6$H$_6$O	Phenol	455.02	694.2	6.13	
C$_6$H$_7$N	Aniline	457.32	699	4.89	287
C$_6$H$_7$N	2-Methylpyridine	402.53	621.0	4.60	292
C$_6$H$_7$N	3-Methylpyridine	417.29	645.0	4.65	288
C$_6$H$_7$N	4-Methylpyridine	418.51	645.7	4.70	292
C$_6$H$_{10}$	1,5-Hexadiene	332.5	508		
C$_6$H$_{10}$	Cyclohexene	356.13	560.4		
C$_6$H$_{10}$O	Cyclohexanone	428.58	653.0	4.0	
C$_6$H$_{10}$O$_2$	Ethyl trans-2-butenoate	411	599		
C$_6$H$_{10}$S	Diallylsulfide	411.8	653		
C$_6$H$_{11}$N	Hexanenitrile	436.80	633.8	3.30	
C$_6$H$_{12}$	1-Hexene	336.63	504.0	3.21	355.1
C$_6$H$_{12}$	Cyclohexane	353.88	553.8	4.08	308
C$_6$H$_{12}$	Methylcyclopentane	345.0	532.7	3.79	318
C$_6$H$_{12}$O	Hexanal	404	591	3.46	
C$_6$H$_{12}$O	2-Hexanone	400.8	587.0	3.32	
C$_6$H$_{12}$O	3-Hexanone	396.7	582.82	3.320	
C$_6$H$_{12}$O	4-Methyl-2-pentanone	389.7	571	3.27	
C$_6$H$_{12}$O	Cyclohexanol	433.99	647.1	4.401	
C$_6$H$_{12}$O$_2$	Hexanoic acid	478.4	662	3.20	
C$_6$H$_{12}$O$_2$	Pentyl formate	403.5	576	3.46	
C$_6$H$_{12}$O$_2$	Isopentyl formate	396.7	578		
C$_6$H$_{12}$O$_2$	Butyl acetate	399.3	579		
C$_6$H$_{12}$O$_2$	Isobutyl acetate	389.7	561	3.16	
C$_6$H$_{12}$O$_2$	Propyl propanoate	395.7	578		
C$_6$H$_{12}$O$_2$	Ethyl butanoate	394.7	566	3.06	421
C$_6$H$_{12}$O$_2$	Ethyl 2-methylpropanoate	383.3	553	3.07	421
C$_6$H$_{12}$O$_2$	Methyl pentanoate	400.5	567	3.19	
C$_6$H$_{12}$O$_3$	Ethylene glycol monoethyl ether acetate	429.6	611	3.18	441
C$_6$H$_{12}$O$_3$	Paraldehyde	397.5	563		
C$_6$H$_{14}$	Hexane	341.88	507.6	3.025	368
C$_6$H$_{14}$	2-Methylpentane	333.41	497.7	3.04	368
C$_6$H$_{14}$	3-Methylpentane	336.42	504.6	3.12	368
C$_6$H$_{14}$	2,2-Dimethylbutane	322.88	489.0	3.10	358
C$_6$H$_{14}$	2,3-Dimethylbutane	331.08	500.0	3.15	361

CRITICAL CONSTANTS (continued)

Mol. Form.	Name	T_b/K	T_c/K	P_c/MPa	V_c/cm^3 mol^{-1}
C$_6$H$_{14}$O	1-Hexanol	430.8	610.3	3.417	387
C$_6$H$_{14}$O	2-Hexanol	413	585.9	3.31	384
C$_6$H$_{14}$O	3-Hexanol	408	582.4	3.36	383
C$_6$H$_{14}$O	2-Methyl-1-pentanol	422	604.4	3.45	
C$_6$H$_{14}$O	4-Methyl-1-pentanol	425.1	603.5		
C$_6$H$_{14}$O	2-Methyl-2-pentanol	394.3	559.5		
C$_6$H$_{14}$O	4-Methyl-2-pentanol	404.8	574.4		
C$_6$H$_{14}$O	2-Methyl-3-pentanol	399.7	576.0	3.46	
C$_6$H$_{14}$O	3-Methyl-3-pentanol	395.5	575.6	3.52	
C$_6$H$_{14}$O	Dipropyl ether	363.23	530.6	3.028	
C$_6$H$_{14}$O	Diisopropyl ether	341.66	500.32	2.832	386
C$_6$H$_{14}$O	Methyl pentyl ether	372	546.53	3.042	391
C$_6$H$_{14}$O$_2$	Ethylene glycol monobutyl ether	441.6	633.9		424
C$_6$H$_{14}$O$_2$	1,1-Diethoxyethane	375.40	527		
C$_6$H$_{14}$O$_3$	Diethylene glycol monoethyl ether	469	670	3.167	
C$_6$H$_{15}$N	Dipropylamine	382.5	555.8	3.63	
C$_6$H$_{15}$N	Diisopropylamine	357.0	523.1	3.02	
C$_6$H$_{15}$N	Triethylamine	362	535.6	3.032	389
C$_7$F$_8$	Perfluorotoluene	377.7	534.47	2.705	428
C$_7$F$_{14}$	Perfluoro-1-heptene	354.2	478.2		
C$_7$F$_{14}$	Perfluoromethylcyclohexane	349.5	485.91	2.019	570
C$_7$F$_{16}$	Perfluoroheptane	355.7	474.8	1.62	664
C$_7$HF$_{15}$	1H-Pentadecafluoroheptane	369.2	495.8		
C$_7$H$_3$F$_5$	2,3,4,5,6-Pentafluorotoluene	390.7	566.52	3.126	384
C$_7$H$_5$N	Benzonitrile	464.3	699.4	4.21	
C$_7$H$_6$O	Benzaldehyde	452.2	695	4.65	
C$_7$H$_8$	Toluene	383.78	591.75	4.108	316
C$_7$H$_8$O	o-Cresol	464.19	697.6	5.01	
C$_7$H$_8$O	m-Cresol	475.42	705.8	4.56	309
C$_7$H$_8$O	p-Cresol	475.13	704.6	5.15	
C$_7$H$_8$O	Benzyl alcohol	478.46	715	4.3	
C$_7$H$_8$O	Anisole	426.9	646.1	4.222	341
C$_7$H$_9$N	o-Methylaniline	473.5	707	4.37	
C$_7$H$_9$N	m-Methylaniline	476.5	707	4.28	
C$_7$H$_9$N	p-Methylaniline	473.6	706	4.58	
C$_7$H$_9$N	N-Methylaniline	469.4	701	5.20	
C$_7$H$_9$N	2,3-Dimethylpyridine	434.27	655.4	4.10	356
C$_7$H$_9$N	2,4-Dimethylpyridine	431.53	647	3.95	361
C$_7$H$_9$N	2,5-Dimethylpyridine	430.13	645	3.85	361
C$_7$H$_9$N	2,6-Dimethylpyridine	417.16	624	3.85	361
C$_7$H$_9$N	3,4-Dimethylpyridine	452.25	684	4.20	355
C$_7$H$_9$N	3,5-Dimethylpyridine	444.99	668	4.05	361
C$_7$H$_{14}$	1-Heptene	366.79	537.3	2.92	409
C$_7$H$_{14}$	Cycloheptane	391.5	604.2	3.82	353
C$_7$H$_{14}$	Methylcyclohexane	374.08	572.1	3.48	369
C$_7$H$_{14}$	Ethylcyclopentane	376.7	569.5	3.40	375
C$_7$H$_{14}$	1,1-Dimethylcyclopentane	360.7	547	3.45	
C$_7$H$_{14}$	cis-1,2-Dimethylcyclopentane	372.7	565	3.45	
C$_7$H$_{14}$	trans-1,2-Dimethylcyclopentane	365.0	553	3.45	
C$_7$H$_{14}$	cis-1,3-Dimethylcyclopentane	364.0	551	3.45	
C$_7$H$_{14}$	trans-1,3-Dimethylcyclopentane	364.8	553	3.45	
C$_7$H$_{14}$O	2-Heptanone	424.20	611.5	3.436	
C$_7$H$_{14}$O$_2$	Heptanoic acid	495.4	679	2.90	
C$_7$H$_{14}$O$_2$	Pentyl acetate	422.4	600	2.68	470
C$_7$H$_{14}$O$_2$	Isopentyl acetate	415.7	599		
C$_7$H$_{14}$O$_2$	Isobutyl propanoate	410	592		
C$_7$H$_{14}$O$_2$	Propyl butanoate	416.2	600		
C$_7$H$_{14}$O$_2$	Propyl isobutanoate	408.5	589		
C$_7$H$_{14}$O$_2$	Ethyl pentanoate	419.3	570		

Mol. Form.	Name	T_b/K	T_c/K	P_c/MPa	V_c/cm^3 mol^{-1}
C$_7$H$_{14}$O$_2$	Ethyl 3-methylbutanoate	408.2	588		
C$_7$H$_{16}$	Heptane	371.7	540.2	2.74	428
C$_7$H$_{16}$	2-Methylhexane	363.19	530.4	2.74	421
C$_7$H$_{16}$	3-Methylhexane	365	535.2	2.81	404
C$_7$H$_{16}$	3-Ethylpentane	366.7	540.6	2.89	416
C$_7$H$_{16}$	2,2-Dimethylpentane	352.3	520.5	2.77	416
C$_7$H$_{16}$	2,3-Dimethylpentane	362.93	537.3	2.91	393
C$_7$H$_{16}$	2,4-Dimethylpentane	353.64	519.8	2.74	418
C$_7$H$_{16}$	3,3-Dimethylpentane	359.21	536.4	2.95	414
C$_7$H$_{16}$	2,2,3-Trimethylbutane	354.01	531.1	2.95	398
C$_7$H$_{16}$O	1-Heptanol	449.60	632.6	3.058	435
C$_7$H$_{16}$O	2-Heptanol	432	608.3	3.021	442
C$_7$H$_{16}$O	3-Heptanol	430	605.4		434
C$_7$H$_{16}$O	4-Heptanol	429	602.6		432
C$_8$F$_{16}$O	Perfluoro-2-butyltetrahydrofuran	375.8	500.2	1.607	588
C$_8$F$_{18}$	Perfluorooctane	379.0	502	1.66	
C$_8$H$_7$N	4-Methylbenzonitrile	490.2	723		
C$_8$H$_7$N	Indole	526.7	794		
C$_8$H$_8$O	Acetophenone	475	713	4.8	356
C$_8$H$_8$O$_3$	Methyl salicylate	496.1	709	4.40	380
C$_8$H$_{10}$	Ethylbenzene	409.34	617.15	3.609	374
C$_8$H$_{10}$	o-Xylene	417.7	630.3	3.732	370
C$_8$H$_{10}$	m-Xylene	412.27	617.0	3.541	375
C$_8$H$_{10}$	p-Xylene	411.52	616.2	3.511	378
C$_8$H$_{10}$O	o-Ethylphenol	477.7	703.0		
C$_8$H$_{10}$O	m-Ethylphenol	491.6	718.8		
C$_8$H$_{10}$O	p-Ethylphenol	491.1	716.4		
C$_8$H$_{10}$O	2,3-Xylenol	490.1	722.8		
C$_8$H$_{10}$O	2,4-Xylenol	484.13	707.6		
C$_8$H$_{10}$O	2,5-Xylenol	484.3	706.9		
C$_8$H$_{10}$O	2,6-Xylenol	474.22	701.0		
C$_8$H$_{10}$O	3,4-Xylenol	500	729.8		
C$_8$H$_{10}$O	3,5-Xylenol	494.89	715.6		
C$_8$H$_{10}$O	Phenetole	442.96	647	3.42	
C$_8$H$_{11}$N	N-Ethylaniline	476.2	698		
C$_8$H$_{11}$N	N,N-Dimethylaniline	467.30	687	3.63	
C$_8$H$_{14}$O$_4$	Diethyl succinate	490.9	663		
C$_8$H$_{15}$N	Octanenitrile	478.40	674.4	2.85	
C$_8$H$_{16}$	1-Octene	394.44	567.0	2.68	468
C$_8$H$_{16}$	Cyclooctane	422	647.2	3.56	410
C$_8$H$_{16}$	Ethylcyclohexane	405.0	609	3.04	
C$_8$H$_{16}$	cis-1,2-Dimethylcyclohexane	403.0	606	2.95	
C$_8$H$_{16}$	trans-1,2-Dimethylcyclohexane	396.7	596	2.94	
C$_8$H$_{16}$	cis-1,3-Dimethylcyclohexane	393.3	591	2.94	
C$_8$H$_{16}$	trans-1,3-Dimethylcyclohexane	397.7	598	2.94	
C$_8$H$_{16}$	trans-1,4-Dimethylcyclohexane	392.5	587.7		
C$_8$H$_{16}$O$_2$	Octanoic acid	512	695	2.64	
C$_8$H$_{16}$O$_2$	2-Ethylhexanoic acid	501	674.6	2.778	528
C$_8$H$_{16}$O$_2$	Isopentyl propanoate	433.4	611		
C$_8$H$_{16}$O$_2$	Isobutyl butanoate	430.1	611		
C$_8$H$_{16}$O$_2$	Isobutyl isobutanoate	421.8	602		
C$_8$H$_{16}$O$_2$	Propyl 3-methylbutanoate	429.1	609		
C$_8$H$_{16}$O$_4$	Diethylene glycol monoethyl ether acetate	491.7	663	2.73	
C$_8$H$_{18}$	Octane	398.82	568.7	2.49	492
C$_8$H$_{18}$	2-Methylheptane	390.81	559.7	2.50	488
C$_8$H$_{18}$	3-Methylheptane	392.0	563.6	2.55	464
C$_8$H$_{18}$	4-Methylheptane	390.87	561.7	2.54	476
C$_8$H$_{18}$	3-Ethylhexane	391.8	565.5	2.61	455
C$_8$H$_{18}$	2,2-Dimethylhexane	380.01	549.8	2.53	478

Mol. Form.	Name	T_b/K	T_c/K	P_c/MPa	V_c/cm^3 mol^{-1}
C_8H_{18}	2,3-Dimethylhexane	388.77	563.5	2.63	468
C_8H_{18}	2,4-Dimethylhexane	382.7	553.5	2.56	472
C_8H_{18}	2,5-Dimethylhexane	382.27	550.0	2.49	482
C_8H_{18}	3,3-Dimethylhexane	385.12	562.0	2.65	443
C_8H_{18}	3,4-Dimethylhexane	390.88	568.8	2.69	466
C_8H_{18}	3-Ethyl-2-methylpentane	388.81	567.1	2.70	442
C_8H_{18}	3-Ethyl-3-methylpentane	391.42	576.5	2.81	455
C_8H_{18}	2,2,3-Trimethylpentane	383	563.5	2.73	436
C_8H_{18}	2,2,4-Trimethylpentane	372.37	543.8	2.57	468
C_8H_{18}	2,3,3-Trimethylpentane	388.0	573.5	2.82	455
C_8H_{18}	2,3,4-Trimethylpentane	386.7	566.4	2.73	460
C_8H_{18}	2,2,3,3-Tetramethylbutane	379.60	567.8	2.87	461
$C_8H_{18}O$	1-Octanol	468.31	652.5	2.777	497
$C_8H_{18}O$	2-Octanol	453	629.6	2.754	519
$C_8H_{18}O$	3-Octanol	444	628.5		515
$C_8H_{18}O$	4-Octanol	449.5	625.1		515
$C_8H_{18}O$	4-Methyl-3-heptanol	443	623.5		
$C_8H_{18}O$	5-Methyl-3-heptanol	445	621.2		
$C_8H_{18}O$	2-Ethyl-1-hexanol	457.8	640.6	2.8	
$C_8H_{18}O$	Dibutyl ether	413.43	584.1	3.01	
$C_8H_{18}O$	Di-tert-butyl ether	380.38	550		
$C_8H_{19}N$	Dibutylamine	432.8	607.5	3.11	
$C_8H_{19}N$	Diisobutylamine	412.8	584.4	3.20	
$C_8H_{20}Si$	Tetraethylsilane	427.9	603.7	2.602	
C_9F_{20}	Perfluorononane	398.5	524	1.56	
C_9H_7N	Quinoline	510.31	782	4.86	371
C_9H_7N	Isoquinoline	516.37	803	5.10	374
C_9H_{10}	Indan	451.12	684.9	3.95	
C_9H_{12}	Propylbenzene	432.39	638.35	3.200	440
C_9H_{12}	Isopropylbenzene	425.56	631.0	3.209	
C_9H_{12}	o-Ethyltoluene	438.4	651	3.38	
C_9H_{12}	m-Ethyltoluene	434.5	637	3.25	
C_9H_{12}	p-Ethyltoluene	435	640.2	3.23	
C_9H_{12}	1,2,3-Trimethylbenzene	449.27	664.5	3.454	
C_9H_{12}	1,2,4-Trimethylbenzene	442.53	649.1	3.232	
C_9H_{12}	1,3,5-Trimethylbenzene	437.89	637.3	3.127	
$C_9H_{13}N$	2-Methyl-N,N-dimethylaniline	467.3	668	3.12	
C_9H_{18}	1-Nonene	420.1	594.0		526
C_9H_{18}	Cyclononane	451.6	682	3.34	
C_9H_{18}	1α,3α,5α-1,3,5-Trimethylcyclohexane	413.7	602.2		
$C_9H_{18}O$	2-Nonanone	468.5	651.9	2.482	
$C_9H_{18}O$	5-Nonanone	461.60	640		
$C_9H_{18}O_2$	Nonanoic acid	527.7	711	2.40	
$C_9H_{18}O_2$	Isopentyl butanoate	452	619		
$C_9H_{18}O_2$	Isobutyl 3-methylbutanoate	441.7	621		
C_9H_{20}	Nonane	423.97	594.6	2.29	555
C_9H_{20}	2-Methyloctane	416.4	582.8	2.31	
C_9H_{20}	2,2-Dimethylheptane	405.8	576.7	2.35	
C_9H_{20}	2,2,5-Trimethylhexane	397.24	569.8		
C_9H_{20}	2,2,3,3-Tetramethylpentane	413.4	607.5	2.74	
C_9H_{20}	2,2,3,4-Tetramethylpentane	406.2	592.6	2.60	
C_9H_{20}	2,2,4,4-Tetramethylpentane	395.44	574.6	2.49	
C_9H_{20}	2,3,3,4-Tetramethylpentane	414.7	607.5	2.72	
$C_9H_{20}O$	1-Nonanol	486.52	670.7	2.528	572
$C_9H_{20}O$	2-Nonanol	466.7	649.6	2.53	575
$C_9H_{20}O$	3-Nonanol	468	648.0		577
$C_9H_{20}O$	4-Nonanol	465.7	645.1		575
$C_{10}F_8$	Perfluoronaphthalene	482	673.1		
$C_{10}F_{18}$	Perfluorodecalin	415	566	1.52	

Mol. Form.	Name	T_b/K	T_c/K	P_c/MPa	V_c/cm^3 mol^{-1}
$C_{10}F_{22}$	Perfluorodecane	417.4	542	1.45	
$C_{10}H_8$	Naphthalene	491.1	748.4	4.05	407
$C_{10}H_9N$	1-Naphthylamine	573.8	850	5.0	438
$C_{10}H_9N$	2-Naphthylamine	579.3	850	4.9	438
$C_{10}H_{12}$	1,2,3,4-Tetrahydronaphthalene	480.8	720	3.65	408
$C_{10}H_{14}$	Butylbenzene	456.46	660.5	2.89	497
$C_{10}H_{14}$	Isobutylbenzene	445.94	650	3.05	
$C_{10}H_{14}$	1-Isopropyl-4-methylbenzene	450.3	652	2.8	
$C_{10}H_{14}$	p-Diethylbenzene	456.9	657.9	2.803	
$C_{10}H_{14}$	1,2,4,5-Tetramethylbenzene	470.0	676	2.9	
$C_{10}H_{14}O$	Thymol	505.7	698		
$C_{10}H_{16}$	d-Limonene	451	653		470
$C_{10}H_{16}$	α-Pinene	429	644		454
$C_{10}H_{16}$	(+)-3-Carene	444	658		487
$C_{10}H_{18}$	1,3-Decadiene	442	615		
$C_{10}H_{18}$	cis-Decahydronaphthalene	469.0	702.3	3.20	
$C_{10}H_{18}$	trans-Decahydronaphthalene	460.5	687.1		
$C_{10}H_{20}$	1-Decene	443.7	617	2.22	584
$C_{10}H_{20}O$	Decanal	481.7	674.2		
$C_{10}H_{20}O$	Menthol	489	694		
$C_{10}H_{20}O_2$	Decanoic acid	541.8	726	2.23	
$C_{10}H_{20}O_2$	Ethyl octanoate	481.7	659		
$C_{10}H_{20}O_4$	Diethylene glycol monobutyl ether acetate	518	681	3.15	
$C_{10}H_{22}$	Decane	447.30	617.7	2.11	624
$C_{10}H_{22}$	3,3,5-Trimethylheptane	428.9	609.5	2.32	
$C_{10}H_{22}$	2,2,3,3-Tetramethylhexane	433.5	623.0	2.51	
$C_{10}H_{22}$	2,2,5,5-Tetramethylhexane	410.6	581.4	2.19	
$C_{10}H_{22}O$	1-Decanol	504.3	687.3	2.315	649
$C_{10}H_{22}O$	2-Decanol	484	668.6		646
$C_{10}H_{22}O$	3-Decanol		666.1		643
$C_{10}H_{22}O$	4-Decanol	483.7	663.7		643
$C_{10}H_{22}O$	5-Decanol	474	663.2		646
$C_{10}H_{22}S$	Diisopentylsulfide	484	664		
$C_{11}H_{10}$	1-Methylnaphthalene	517.8	772	3.60	
$C_{11}H_{10}$	2-Methylnaphthalene	514.3	761		
$C_{11}H_{22}O_2$	Ethyl nonanoate	500.2	674		
$C_{11}H_{24}$	Undecane	469.1	639	1.98	689
$C_{11}H_{24}O$	1-Undecanol	516	703.6	2.147	718
$C_{12}H_8$	Acenaphthylene	553	792	3.20	
$C_{12}H_9N$	Carbazole	627.84	901.8	3.13	454
$C_{12}H_{10}$	Biphenyl	529.3	773	3.38	497
$C_{12}H_{10}O$	Diphenyl ether	531.2	766.8		
$C_{12}H_{12}$	2,7-Dimethylnaphthalene	538	775	3.23	601
$C_{12}H_{18}$	Hexamethylbenzene	536.6	758		
$C_{12}H_{24}$	1-Dodecene	487.0	658	1.93	
$C_{12}H_{26}$	Dodecane	489.47	658	1.82	754
$C_{12}H_{26}O$	1-Dodecanol	532	719.4	1.994	
$C_{13}H_9N$	Acridine	618.01	891.1	3.21	548
$C_{13}H_9N$	Phenanthridine	622.0	895	3.6	548
$C_{13}H_{11}N$	9-Methylcarbazole	616.79	890	3.38	572
$C_{13}H_{12}$	Diphenylmethane	538.2	760	2.71	563
$C_{13}H_{26}O_2$	Methyl dodecanoate	540	712		
$C_{13}H_{28}$	Tridecane	508.62	675	1.68	823
$C_{13}H_{28}O$	1-Tridecanol	546.8	734	1.935	
$C_{14}H_{10}$	Anthracene	613.1	869.3		554
$C_{14}H_{10}$	Phenanthrene	613	869		
$C_{14}H_{30}$	Tetradecane	526.73	693	1.57	894
$C_{14}H_{30}O$	1-Tetradecanol	562	747	1.81	

CRITICAL CONSTANTS (continued)

Mol. Form.	Name	T_b/K	T_c/K	P_c/MPa	V_c/cm^3 mol^{-1}
$C_{15}H_{32}$	Pentadecane	543.8	708	1.48	966
$C_{16}H_{34}$	Hexadecane	560.01	723	1.40	1034
$C_{16}H_{34}$	2,2,4,4,6,8,8-Heptamethylnonane	519.5	692		
$C_{16}H_{34}O$	1-Hexadecanol	607	770	1.61	
$C_{17}H_{36}$	Heptadecane	575.2	736	1.34	1103
$C_{17}H_{36}O$	1-Heptadecanol	606	780	1.50	
$C_{18}H_{14}$	o-Terphenyl	605	857	2.99	731
$C_{18}H_{14}$	m-Terphenyl	636	883	2.48	724
$C_{18}H_{14}$	p-Terphenyl	649	908	2.99	729
$C_{18}H_{38}$	Octadecane	589.5	747	1.29	1189
$C_{18}H_{38}O$	1-Octadecanol	616.7	790	1.44	
$C_{19}H_{40}$	Nonadecane	603.1	755	1.16	
$C_{20}H_{42}$	Eicosane	616	768	1.07	
$C_{20}H_{42}O$	1-Eicosanol	582	809	1.30	
$C_{21}H_{44}$	Heneicosane	629.7	778	1.03	
$C_{22}H_{46}$	Docosane	641.8	786	0.98	
$C_{23}H_{48}$	Tricosane	653	790	0.92	
$C_{24}H_{50}$	Tetracosane	664.5	800	0.87	

SUBLIMATION PRESSURE OF SOLIDS

This table gives the sublimation (vapor) pressure of some representative solids as a function of temperature. Entries include simple inorganic and organic substances in their solid phase below room temperature, as well as polycyclic organic compounds which show measurable sublimation pressure only at elevated temperatures. Substances are listed by molecular formula in the Hill order. Values marked by * represent the solid-liquid-gas triple point. Note that some pressure values are in pascals (Pa) and others are in kilopascals (kPa). For conversion, 1 kPa = 7.506 mmHg = 0.0098692 atm.

REFERENCES

1. Lide, D.R. and Kehiaian, H.V., *CRC Handbook of Thermophysical and Thermochemical Data,* CRC Press, Boca Raton, FL, 1994.
2. *TRC Thermodynamic Tables,* Thermodynamic Research Center, Texas A&M University, College Station, TX.
3. Oja, V. and Suuberg, E.M., *J. Chem. Eng. Data,*43, 486, 1998.

Ar Argon								
T/K	55	60	65	70	75	80	83.81*	
p/kPa	0.2	0.8	2.8	7.7	18.7	40.7	68.8*	

BrH Hydrogen bromide								
T/K	135	140	150	160	170	180	185.1*	
p/kPa	0.1	0.3	1.1	3.3	8.7	20.1	27.4*	

Br_2 Bromine								
T/K	170	180	190	200	210	220	230	240*
p/Pa	0.069	0.416	2.04	8.45	30.3	96.0	273	710*

ClH Hydrogen chloride								
T/K	120	130	140	150	155	159.0*		
p/kPa	0.1	0.5	1.9	5.8	9.5	13.5*		

Cl_2 Chlorine								
T/K	120	130	140	150	160	170*		
p/Pa	0.144	1.52	11.2	63.1	283	1054*		

F_4Si Tetrafluorosilane								
T/K	130	140	150	160	170	175	180	186.3*
p/kPa	0.2	0.9	3.9	14.0	43.8	74.2	122.4	220.8*

F_6S Sulfur hexafluoride								
T/K	150	165	180	190	200	210	220	223.1*
p/kPa	0.4	2.6	11.3	25.9	54.5	106.1	195.1	232.7*

HI Hydrogen iodide								
T/K	160	170	180	190	200	210	220	222.4*
p/kPa	0.2	0.8	2.2	5.3	11.7	23.6	44.1	49.3*

H_2O Water								
T/K	190	210	225	240	250	260	270	273.16*
p/Pa	0.032	0.702	4.942	27.28	76.04	195.8	470.1	611.66*

H_2S Hydrogen sulfide								
T/K	140	150	160	165	170	175	180	187.6*
p/kPa	0.2	0.6	1.9	3.2	5.2	8.3	12.7	22.7*

H_3N Ammonia								
T/K	160	170	180	190	195	195.4*		
p/kPa	0.1	0.4	1.2	3.5	5.8	6.12*		

I_2 Iodine								
T/K	240	250	260	270	280	290	300	310*
p/Pa	0.081	0.297	0.971	2.89	7.92	20.1	47.9	107*

Kr Krypton								
T/K	80	90	95	100	105	110	115.8*	
p/kPa	0.4	2.7	6.0	12.1	22.8	40.4	73.1*	

NO Nitric oxide								
T/K	85	90	95	100	105	109.5*		
p/kPa	0.1	0.4	1.3	3.8	10.0	21.9*		

Xe Xenon								
T/K	110	120	130	140	150	155	160	161.4*
p/kPa	0.3	1.5	4.9	14.0	34.2	51.1	74.2	81.7*

CHN Hydrogen cyanide								
T/K	200	210	220	230	240	250	255	259.83*
p/kPa	0.2	0.4	1.0	2.2	4.8	9.7	13.6	18.62*

CH_4 Methane								
T/K	65	70	75	80	85	90.69*		
p/kPa	0.1	0.3	0.8	2.1	4.9	11.70*		

CO								
Carbon monoxide	T/K	50	55	60	65	68.13*		
	p/kPa	0.1	0.6	2.6	8.2	15.4*		
CO_2	T/K	130	140	155	170	185	194.7	205	216.58*
Carbon dioxide	p/kPa	0.032	0.187	1.674	9.987	44.02	101.3	227.1	518.0*
C_2Cl_6	T/K	275	300	325	350	375	400	425	459.9*
Hexachloroethane	p/Pa	0.004	0.056	0.383	1.62	5.30	14.8	36.4	107.4*
C_2H_2	T/K	130	140	150	160	170	180	190	192.4*
Acetylene	p/kPa	0.2	0.7	2.6	7.8	20.6	49.0	106.3	126.0*
$C_2H_4O_2$	T/K	250	260	270	280	289.7*		
Acetic acid	p/kPa	0.092	0.199	0.406	0.79	1.29*		
C_5H_{12}	T/K	200	210	220	230	240	250	255	256.58*
Neopentane	p/kPa	0.7	1.6	3.6	7.3	13.9	24.8	32.4	35.8*

$C_6H_6Cl_6$	T/K	300	320	330	340	350	360	370	380
1,2,3,4,5,6-Hexa- chlorocyclohexane (Lindane)	p/Pa	0.01	0.13	0.39	1.04	2.66	6.42	14.8	32.7

$C_6H_6O_2$	T/K	330	340	350	360	370	380
Resorcinol	p/Pa	1.03	2.78	7.09	17.2	39.6	87.6
$C_6H_6O_2$	T/K	350	360	370	380	390	400
p-Hydroquinone	p/Pa	1.20	3.18	7.96	19.0	43.4	95.1

$C_{10}H_8$	T/K	250	270	280	290	300	310	330	353.43*
Naphthalene	p/Pa	0.036	0.514	1.662	4.918	13.43	34.15	182.9	999.6*

$C_{12}H_8N_2$	T/K	290	300	310	320
Phenazine	p/Pa	0.0013	0.0046	0.0150	0.0448

$C_{12}H_8O$	T/K	300	310	320	330	340	350
Dibenzofuran	p/Pa	0.408	1.21	3.35	8.71	21.4	50.0

$C_{12}H_9N$	T/K	350	355	360
Carbazole	p/Pa	0.086	0.140	0.245

$C_{13}H_7NO_2$	T/K	330	340	350	360	370	380
Benz[g]isoquinoline- 5,10-dione	p/Pa	0.006	0.018	0.053	0.148	0.394	0.994

$C_{13}H_8O$	T/K	330	340	350
1H-Phenalen-1-one	p/Pa	0.040	0.113	0.302

$C_{13}H_8O_2$	T/K	400	410	420	430
3-Hydroxy-1H- phenalen-1-one	p/Pa	0.006	0.018	0.053	0.144

$C_{13}H_9N$	T/K	290	300	310	320
Acridine	p/Pa	0.0024	0.0085	0.0278	0.0845

$C_{13}H_9N$	T/K	310	320	330	340
Phenanthridine	p/Pa	0.020	0.066	0.206	0.603

$C_{14}H_{10}$	T/K	320	330	340	350	360	370	380	390
Anthracene	p/Pa	0.014	0.043	0.125	0.342	1.01	2.38	5.35	11.5

$C_{14}H_{10}$ Phenanthrene	T/K	300	310	320	330	340	350	360	
	p/Pa	0.025	0.085	0.270	0.796	2.02	4.89	11.2	
$C_{16}H_{10}$ Pyrene	T/K	320	330	340	350	360	370	380	390
	p/Pa	0.008	0.024	0.073	0.208	0.556	1.32	2.86	6.30
$C_{16}H_{10}O$ 1-Pyrenol	T/K	360	370	380	390	400			
	p/Pa	0.005	0.016	0.047	0.135	0.364			
$C_{16}H_{12}S$ Benzo[b]naphtho-(2,1-d)thiophene	T/K	330	340	350	360	370	380	390	
	p/Pa	0.001	0.004	0.012	0.036	0.098	0.255	0.631	
$C_{17}H_{12}$ 11H-Benzo[b]fluorene	T/K	340	350	360	370	380	390	400	
	p/Pa	0.003	0.009	0.029	0.085	0.235	0.619	1.55	
$C_{18}H_{10}O_4$ 6,11-Dihydroxy-5,12-naphthacenedione	T/K	420	430	440	450				
	p/Pa	0.008	0.022	0.055	0.131				
$C_{18}H_{12}$ Chrysene	T/K	390	400	410	420				
	p/Pa	0.087	0.221	0.539	1.26				
$C_{18}H_{12}$ Naphthacene	T/K	390	400	410	420	430	440	450	460
	p/Pa	0.005	0.014	0.035	0.084	0.194	0.432	0.928	1.929
$C_{20}H_{12}$ Perylene	T/K	390	400	410	420	430			
	p/Pa	0.006	0.015	0.040	0.102	0.246			
$C_{22}H_{14}$ Pentacene	T/K	450	460	470	480	490			
	p/Pa	0.002	0.006	0.013	0.031	0.069			
$C_{24}H_{12}$ Coronene	T/K	430	440	450	460	470	480	490	500
	p/Pa	0.004	0.010	0.021	0.046	0.097	0.197	0.389	0.747

VAPOR PRESSURE

This table gives vapor pressure data for about 1800 inorganic and organic substances. In order to accommodate elements and compounds ranging from refractory to highly volatile in a single table, the temperature at which the vapor pressure reaches specified pressure values is listed. The pressure values run in decade steps from 1 Pa (about 7.5 μm Hg) to 100 kPa (about 750 mm Hg). All temperatures are given in °C.

The data used in preparing the table came from a large number of sources; the main references used for each substance are indicated in the last column. Since the data were refit in most cases, values appearing in this table may not be identical with values in the source cited. The temperature entry in the 100 kPa column is close to, but not identical with, the normal boiling point (which is defined as the temperature at which the vapor pressure reaches 101.325 kPa). Although some temperatures are quoted to 0.1°C, uncertainties of several degrees should generally be assumed. Values followed by an "e" were obtained by extrapolating (usually with an Antoine equation) beyond the region for which experimental measurements were available and are thus subject to even greater uncertainty.

Compounds are listed by molecular formula following the Hill convention. Substances not containing carbon are listed first, followed by those that contain carbon. To locate an organic compound by name or CAS Registry Number when the molecular formula is not known, use the table *Physical Constants of Organic Compounds* in Section 3 and its indexes to determine the molecular formula. The indexes to *Physical Constants of Inorganic Compounds* in Section 4 can be used in a similar way.

More extensive and detailed vapor pressure data on selected important substances appear in other tables in this section of the *Handbook*. These substances are flagged by a symbol following the name as follows:

*   See *Vapor Pressure of Fluids below 300 K*
**   See *IUPAC Recommended Data for Vapor Pressure Calibration*
***   See *Vapor Pressure of Ice* and *Vapor Pressure of Water from 0 to 370°C*

The following notations appear after individual temperature entries:

s — Indicates the substance is a solid at this temperature.

e — Indicates an extrapolation beyond the region where experimental measurements exist.

i — Indicates the value was calculated from ideal gas thermodynamic functions, such as those in the *JANAF Thermochemical Tables* (see Reference 8).

REFERENCES

1. Lide, D.R., and Kehiaian, H.V., *CRC Handbook of Thermophysical and Thermochemical Data*, CRC Press, Boca Raton, FL, 1994.
2. Stull, D., in *American Institute of Physics Handbook, Third Edition*, Gray, D.E., Ed., McGraw Hill, New York, 1972.
3. Hultgren, R., Desai, P.D., Hawkins, D.T., Gleiser, M., Kelley, K.K., and Wagman, D.D., *Selected Values of Thermodynamic Properties of the Elements*, American Society for Metals, Metals Park, OH, 1973.
4. Stull, D., *Ind. Eng. Chem.*, 39, 517, 1947.
5. *TRCVP, Vapor Pressure Database, Version 2.2P*, Thermodynamic Research Center, Texas A&M University, College Station, TX.
6. *TRC Thermodynamic Tables,* Thermodynamic Research Center, Texas A&M University, College Station, TX.
7. Ohe, S., *Computer Aided Data Book of Vapor Pressure*, Data Book Publishing Co., Tokyo, 1976.
8. Chase, M.W., Davies, C.A., Downey, J.R., Frurip, D.J., McDonald, R.A., and Syverud, A.N., *JANAF Thermochemical Tables, Third Edition, J. Phys. Chem. Ref. Data*, Vol. 14, Suppl. 1, 1985.
9. Barin, I., *Thermochemical Data of Pure Substances*, VCH Publishers, New York, 1993.
10. Jacobsen, R.T., et. al, *International Thermodynamic Tables of the Fluid State, No. 10. Ethylene*, Blackwell Scientific Publications, Oxford, 1988.
11. Wakeham, W.A., *International Thermodynamic Tables of the Fluid State, No. 12. Methanol*, Blackwell Scientific Publications, Oxford, 1993.
12. Janz, G.J., *Molten Salts Handbook*, Academic Press, New York, 1967.
13. Ohse, R.W. *Handbook of Thermodynamic and Transport Properties of Alkali Metals*, Blackwell Scientific Publications, Oxford, 1994.
14. Gschneidner, K.A., in *CRC Handbook of Chemistry and Physics, 77th Edition*, p. 4-112, CRC Press, Boca Raton, FL, 1996.
15. Leider, H.R., Krikorian, O.H., and Young, D.A., *Carbon*, 11, 555, 1973.
16. Ruzicka, K., and Majer, V., *J. Phys. Chem. Ref. Data*, 23, 1, 1994.
17. Tillner-Roth, R., and Baehr, H.D., *J. Phys. Chem. Ref. Data*, 23, 657, 1994.
18. Younglove, B.A., and McLinden, M.O., *J. Phys. Chem. Ref. Data*, 23, 731, 1994.
19. Outcalt, S.L., and McLinden, M.O., *J. Phys. Chem. Ref. Data*, 25, 605, 1996.
20. Weber, L.A., and Defibaugh, D.R., *J. Chem. Eng. Data*, 41, 382, 1996.
21. Rodrigues, M.F., and Bernardo-Gil, M.G., *J. Chem. Eng. Data*, 41, 581, 1996.
22. Piacente, V., Gigli, G., Scardala, P., and Giustini, A., *J. Phys. Chem.*, 100, 9815, 1996.
23. Barton, J.L., and Bloom, H., *J. Phys. Chem.*, 60, 1413, 1956.
24. Sense, K.A., Alexander, C.A., Bowman, R.E., and Filbert, R.B., *J. Phys. Chem.*, 61, 337, 1957.
25. Ewing, C.T., and Stern, K.H., *J. Phys. Chem.* 78, 1998, 1974.
26. Cady, G.H., and Hargreaves, G.B., *J. Chem. Soc.*, 1961, 1563; 1961, 1568.
27. Skudlarski, K., Dudek, J., and Kapala, J., *J. Chem. Thermodynamics*, 19, 857, 1987.
28. Wagner, W., and de Reuck, K.M., *International Thermodynamic Tables of the Fluid State, No. 9. Oxygen*, Blackwell Scientific Publications, Oxford, 1987.

29. Marsh, K.N., Editor, *Recommended Reference Materials for the Realization of Physicochemical Properties*, Blackwell Scientific Publications, Oxford, 1987.
30. Alcock, C.B., Itkin, V.P., and Horrigan, M.K., *Canadian Metallurgical Quarterly*, 23, 309, 1984.
31. Stewart, R.B., and Jacobsen, R.T., *J. Phys. Chem. Ref. Data*, 18, 639, 1989.
32. Sifner, O., and Klomfar, J., *J. Phys. Chem. Ref. Data*, 23, 63, 1994.
33. Bah, A., and Dupont-Pavlovsky, N., *J. Chem. Eng. Data*, 40, 869, 1995.
34. Behrens, R.G., and Rosenblatt, G., *J. Chem. Thermodynamics*, 4, 175, 1972.
35. Behrens, R.G., and Rosenblatt, G., *J. Chem. Thermodynamics*, 5, 173, 1973.
36. Haar, L., Gallagher, J.S., and Kell, G.S., *NBS/NRC Steam Tables*, Hemisphere Publishing Corp., New York, 1984.
37. Wagner, W., Saul, A., and Pruss, A., *J. Phys. Chem. Ref. Data*, 23, 515. 1994.
38. Behrens, R.G., Lemons, R.S., and Rosenblatt, G., *J. Chem. Thermodynamics*, 6, 457, 1974.
39. Boublik, T., Fried, V., and Hala, E., *The Vapor Pressure of Pure Substances, Second Edition,* Elsevier, Amsterdam, 1984.
40. Goodwin, R.D., *J. Phys. Chem. Ref. Data*, 14, 849, 1985.
41. Younglove, B.A., and Ely, J.F., *J. Phys. Chem. Ref. Data*, 16, 577, 1987.

		Temperature in °C for the indicated pressure						
Mol. Form.	Name	1 Pa	10 Pa	100 Pa	1 kPa	10 kPa	100 kPa	Ref.
Substances not containing carbon:								
Ag	Silver	1010	1140	1302	1509	1782	2160	2
AgBr	Silver(I) bromide	569 i	656 i	765 i	905 i	1093 i	1359 i	9
AgCl	Silver(I) chloride	670	769	873	1052	1264	1561	4
AgI	Silver(I) iodide	594	686	803	959	1177	1503	4
Al	Aluminum	1209	1359	1544	1781	2091	2517	2
AlB$_3$H$_{12}$	Aluminum borohydride				-46.8	-9.4	45.5	4
AlCl$_3$	Aluminum trichloride	58.4 s	76.5 s	97.1 s	120.7 s	148.2 s	180.5 s	4
AlF$_3$	Aluminum trifluoride	744 s	819 s	906 s	1008 s	1130 s	1276 s	8
AlI$_3$	Aluminum triiodide				218	285	385	4
Al$_2$O$_3$	Aluminum oxide			2122	2351	2629	2975	4
Ar	Argon*		-226.4 s	-220.3 s	-212.4 s	-201.7 s	-186.0	1,5,31
As	Arsenic	280 s	323 s	373 s	433 s	508 s	601 s	3
AsCl$_3$	Arsenic(III) chloride			-8 e	21.3	63.1	129.4	1
AsF$_3$	Arsenic(III) fluoride					8.1	56.0	4
AsI$_3$	Arsenic(III) iodide				187	261	367 e	7
As$_2$O$_3$	Arsenic(III) oxide (arsenolite)	133.7 s	163.0 s	196.8 s	236.2 s	283.0		34
At	Astatine	88 s	119 s	156 s	202 s	258 s	334	2
Au	Gold	1373	1541	1748	2008	2347	2805	2
B	Boron	2075	2289	2549	2868	3272	3799	2
BBr$_3$	Boron tribromide			-45 e	-15 e	27.5	90.4	1
BCl$_3$	Boron trichloride*			-94.0	-70.5	-37.4	12.3	4
BF$_3$	Boron trifluoride*	-173.9 s	-166.0 s	-156.0 s	-143.0 s	-125.9	-101.1	4
B$_2$F$_4$	Tetrafluorodiborane						-34	1
B$_2$H$_6$	Diborane			-162 e	-147.0	-125.8	-92.6	1
B$_5$H$_9$	Pentaborane(9)				-34.8	3.8	57.6	4
Ba	Barium	638 s	765	912	1115	1413	1897	9
Be	Beryllium	1189 s	1335	1518	1750	2054	2469	2
BeBr$_2$	Beryllium bromide	203 s	240 s	283 s	335 s	397 s	473 s	4
BeCl$_2$	Beryllium chloride	196 s	237 s	284 s	339 s	402 s	487	4
BeF$_2$	Beryllium fluoride		686 e	767 e	869	999	1172 e	7
BeI$_2$	Beryllium iodide	188 s	229 s	276 s	333 s	402 s	487	4
Bi	Bismuth	668	768	892	1052	1265	1562	2
BiBr$_3$	Bismuth tribromide			217 s	273 i	348 i	455 i	4,9
BiCl$_3$	Bismuth trichloride				248.9	328.6	438.7	1,4
BrCs	Cesium bromide	531 s	601 s	701 i	834 i	1019 i	1293 e	9
BrH	Hydrogen bromide*		-153.3 s	-140.4 s	-123.8 s	-101.5 s	-67.0	5
BrH$_3$Si	Bromosilane				-81.0	-47.3	2.2	4
BrH$_4$N	Ammonium bromide	121 s	154 s	195 s	246 s	310.4 s	395.1 s	5
BrK	Potassium bromide	597 s	674 s	773				25
BrLi	Lithium bromide		630	733	868	1049	1308	4
BrNa	Sodium bromide			791	931	1120	1389	4
BrRb	Rubidium bromide			766	903	1087	1350	4
BrTl	Thallium(I) bromide				509	635	817	4
Br$_2$	Bromine*	-87.7 s	-71.8 s	-52.7 s	-29.3 s	2.5	58.4	1
Br$_2$Cd	Cadmium bromide	373 s	435 s	509 s				27
Br$_2$Hg	Mercury(II) bromide	71 s	98 s	132 s	174 s	227 s	318	4
Br$_2$OS	Thionyl bromide	-49 e	-29 e	-5 e	27.8	72.9	139.6	5

Mol. Form.	Name	Temperature in °C for the indicated pressure						Ref.
		1 Pa	10 Pa	100 Pa	1 kPa	10 kPa	100 kPa	
Br_2Pb	Lead(II) bromide	374	431	502	597	726	914	4
Br_2S_2	Sulfur bromide	-7 e	15 e	42 e	78.4	128.1	200.9	5
Br_3In	Indium(III) bromide			304.6 s	328.7 s	364.8 s		1
Br_3OP	Phosphorus(V) oxybromide				64 e	115.5	191.4	5
Br_3P	Phosphorus(III) bromide		-23 e	5 e	42.3	94.6	172.6	5
Br_3Sb	Antimony(III) bromide				136.5	196.9	286.5	1
Br_4Ge	Germanium(IV) bromide				51	105	188	4
Br_4Sn	Tin(IV) bromide				67	122	204	4
Br_4Zr	Zirconium(IV) bromide	136 s	167 s	203 s	245 s	295 s	356 s	4
Br_5P	Phosphorus(V) bromide		-19 s	4 s	31 s	65.5 s	110.1	5
Ca	Calcium	591 s	683 s	798 s	954	1170	1482	2
Cd	Cadmium	257 s	310 s	381	472	594	767	2
$CdCl_2$	Cadmium chloride	412 s	471 s	541 s	634	768	959	23 27
CdF_2	Cadmium fluoride				1257	1461	1742	4
CdI_2	Cadmium iodide	296 s	344 s	406	498	622	795	4,27
CdO	Cadmium oxide	770 s	866 s	983 s	1128 s	1314 s	1558 s	4
Ce	Cerium	1719	1921	2169	2481	2886	3432	14
ClCs	Cesium chloride			730	864	1043	1297	4
ClCu	Copper(I) chloride		459	543	675	914	1477	4
ClF	Chlorine fluoride*				-144.4	-122.6	-90.2	5
ClF_2P	Phosphorus(III) chloride difluoride				-119.5	-91.1	-47.6	5
ClF_3	Chlorine trifluoride				-63.7	-33.0	11.4	5
ClF_5	Chlorine pentafluoride				-88 e	-59	-14	7
ClH	Hydrogen chloride*				-138.2 s	-118.0	-85.2	1,5
$ClHO_3S$	Chlorosulfonic acid	-40 e	-20 e	5 e	38.7	85.0	153.6	5
ClH_4N	Ammonium chloride	91 s	121 s	159 s	204.7 s	263.1 s	339.5 s	5
ClK	Potassium chloride	625 s	704 s	804 s	945	1137	1411	23,25
ClLi	Lithium chloride		649 i	761 i	905 i	1101 i	1381 i	8
ClNO	Nitrosyl chloride		-116 s	-100 s	-78.7 s	-50.2	-5.7	5
$ClNO_2$	Nitryl chloride	-121 e	-113 e	-102 e	-86.1	-60.9	-15.7	5
ClNa	Sodium chloride	653 s	733 s	835	987	1182	1461	23,25
ClO_2	Chlorine dioxide*					-34.3	10.5	5
ClRb	Rubidium chloride			777	916	1105	1379	4
ClTl	Thallium(I) chloride				504	626	806	4
Cl_2	Chlorine*	-145 s	-133.7 s	-120.2 s	-103.6 s	-76.1	-34.2	1
Cl_2Co	Cobalt(II) chloride					818	1048	4
Cl_2FP	Phosphorus(III) dichloride fluoride				-71.1	-37.4	13.5	5
Cl_2F_3P	Phosphorus(V) dichloride trifluoride		-120 e	-101 e	-77.1	-44.3	3 e	7
Cl_2Fe	Iron(II) chloride				685	821	1025	4
Cl_2Hg	Mercury(II) chloride	64.4 s	94.7 s	130.8 s	174.5 s	228.5 s	304.0	4
Cl_2Mg	Magnesium chloride			762	908	1111	1414	4
Cl_2Mn	Manganese(II) chloride				760	933	1189	4
Cl_2Ni	Nickel(II) chloride	534 s	592 s	662 s	747 s	852 s	985 s	4
Cl_2OS	Thionyl chloride	-99 e	-81 e	-58 e	-27.1	14.6	75.2	5
Cl_2O_2S	Sulfuryl chloride				-27 e	11.8	69.0	5
Cl_2Pb	Lead(II) chloride			541 e	637	765	949	23
Cl_2S	Sulfur dichloride	-76 e	-61 e	-41 e	-16.7	15.3	58.7	5
Cl_2S_2	Sulfur chloride	-55 e	-36 e	-12 e	21.0	67.2	137.1	5
Cl_2Sn	Tin(II) chloride		253	308	381	479	622	4
Cl_2Zn	Zinc chloride	305 i	356 i	419 i	497 i	596 i	726 i	4,9,12
Cl_3Fe	Iron(III) chloride	118 s	153 s	190 s	229 s	268 s	319	4
Cl_3HSi	Trichlorosilane			-81 e	-56 e	-21 e	31.6	7
Cl_3N	Nitrogen trichloride				-25 e	13.2	70.6	5
Cl_3OP	Phosphorus(V) oxychloride					39.9	105.0	5
Cl_3P	Phosphorus(III) chloride	-93 e	-77 e	-55 e	-26.0	14.5	75.7	5
Cl_4Po	Polonium(IV) chloride					300.6	389.4	5
Cl_4Se	Selenium tetrachloride	23 s	45 s	71 s	102 s	141.4 s	191.1 s	5
Cl_4Si	Tetrachlorosilane*				-39 e	0 e	57.3	1
Cl_4Te	Tellurium tetrachloride				237 e	299.4	387.8	5
Cl_4Zr	Zirconium(IV) chloride	117 s	146 s	181 s	222 s	272 s	336 s	9
Cl_5P	Phosphorus(V) chloride	-2 s	19 s	44 s	74 s	111.4 s	158.9 s	5
Co	Cobalt	1517	1687	1892	2150	2482	2925	2
Cr	Chromium	1383 s	1534 s	1718 s	1950	2257	2669	2
Cs	Cesium	144.5	195.6	260.9	350.0	477.1	667.0	13,30

VAPOR PRESSURE (continued)

Mol. Form.	Name	Temperature in °C for the indicated pressure						Ref.
		1 Pa	10 Pa	100 Pa	1 kPa	10 kPa	100 kPa	
CsF	Cesium fluoride				825	999	1249	4
CsI	Cesium iodide	523 s	595 s	692	854	1029	1278	4,25
Cu	Copper	1236	1388	1577	1816	2131	2563	2
CuI	Copper(I) iodide				636	864	1331	4
Dy	Dysprosium	1105 s	1250 s	1431 i	1681 i	2031 i	2558 i	3
Er	Erbium	1231 s	1390 s	1612 i	1890 i	2279 i	2859 i	3
Eu	Europium	590 s	684 s	799 s	961	1179	1523	14
FH	Hydrogen fluoride*				-71.1	-33.7	19.2	1,5
FHO$_3$S	Fluorosulfonic acid	-14 e	4 e	28 e	59.1	101.3	162.2	5
FK	Potassium fluoride			869	1017	1216	1499	4
FLi	Lithium fluoride	801 s	896	1024	1188	1395	1672	4,12,25
FNO	Nitrosyl fluoride			-131 e	-116.1	-94.3	-60.1	5
FNO$_2$	Nitryl fluoride		-156 e	-144 e	-128.1	-106.0	-72.6	5
FNO$_3$	Fluorine nitrate	-160 e	-149 e	-135 e	-115.1	-87.4	-45.0	5
FNa	Sodium fluoride		920 s	1058	1218	1426	1702	4,12,24
FRb	Rubidium fluoride			910	1001	1145	1409	4,12
F$_2$	Fluorine*	-235 s	-229.5 s	-222.9 s	-214.8	-204.3	-188.3	1,5
F$_2$O	Fluorine monoxide*	-211.7	-204.7	-195.9	-184.2	-168.2	-144.9	5
F$_2$OS	Thionyl fluoride			-124 e	-106.5	-81.5	-44.1	5
F$_2$O$_2$Re	Rhenium(VI) dioxydifluoride				89.2	131.9	185 e	26
F$_2$Pb	Lead(II) fluoride				865	1054	1292	4
F$_2$Xe	Xenon difluoride			2.9 s	31.8 s	67.9 s	114 s	1,5
F$_2$Zn	Zinc fluoride	731 s	813 s	911 i	1048 i	1237 i	1503 i	9
F$_3$N	Nitrogen trifluoride*	-201 e	-194 e	-185 e	-172.8	-155.5	-129.2	5
F$_3$OP	Phosphorus(V) oxyfluoride	-124 s	-113 s	-100 s	-83.7 s	-64.1 s	-39.7 s	5
F$_3$P	Phosphorus(III) fluoride*				-152 e	-132.6	-101.4	5
F$_4$MoO	Molybdenum(VI) oxytetrafluoride	-21 s	3 s	33 s	69.3 s	117.3	184.1	26
F$_4$ORe	Rhenium(VI) oxytetrafluoride	5 s	26 s	50.7 s	80.1 s	117.1	171.2	26
F$_4$OW	Tungsten(VI) oxytetrafluoride	2 s	25 s	52.1 s	84.3 s	126.7	185.4	26
F$_4$S	Sulfur tetrafluoride				-110.0	-82.1	-40.3	5
F$_4$Se	Selenium tetrafluoride				13.6	51.6	104.7	5
F$_4$Si	Tetrafluorosilane*	-166 s	-157 s	-145.6 s	-132.3 s	-115.7 s	-94.9 s	4,7
F$_5$Mo	Molybdenum(V) fluoride				86.6	140.3	213 e	26
F$_5$Nb	Niobium(V) fluoride				80	140	224	4
F$_5$ORe	Rhenium(VII) oxypentafluoride	-103 s	-84 s	-59 s	-28 s	13.7 s	72.8	26
F$_5$Os	Osmium(V) fluoride			74.1	113.2	162.3	226 e	26
F$_5$P	Phosphorus(V) fluoride	-157 s	-148 s	-137 s	-124.5 s	-108.6 s	-84.8	5
F$_5$Re	Rhenium(V) fluoride			58.8	99.5	152 e	221 e	26
F$_5$Ta	Tantalum(V) fluoride					119	229	4
F$_6$Ir	Iridium(VI) fluoride	-88 s	-71 s	-51 s	-27 s	3.8 s	53.1	26
F$_6$Mo	Molybdenum(VI) fluoride	-98 s	-82 s	-64 s	-41.2 s	-13.4 s	33.5	26
F$_6$Os	Osmium(VI) fluoride	-89 s	-73 s	-54 s	-30.6 s	-1.7 s	47.4	26
F$_6$Re	Rhenium(VI) fluoride	-97 s	-82 s	-63 s	-40.2 s	-11.9 s	33.4	26
F$_6$S	Sulfur hexafluoride*	-158 s	-147 s	-133.6 s	-116.6 s	-94.4 s	-64.1 s	5
F$_6$Se	Selenium hexafluoride	-143 s	-132 s	-118 s	-100.7 s	-77.8 s	-46.5 s	5
F$_6$Te	Tellurium hexafluoride	-142 s	-130 s	-115 s	-96 s	-71.8 s	-39.1 s	5
F$_6$W	Tungsten(VI) fluoride	-107 s	-92 s	-74 s	-52.1 s	-24.8 s	16.9	26
F$_{10}$S$_2$	Sulfur decafluoride					-22.0	28.5	5
Fe	Iron	1455 s	1617	1818	2073	2406	2859	2
Fr	Francium	131 e	181 e	246 e	335 e	465 e	673 e	2
Ga	Gallium	1037	1175	1347	1565	1852	2245	2
Gd	Gadolinium	1563 i	1755 i	1994 i	2300 i	2703 i	3262 i	3
Ge	Germanium	1371	1541	1750	2014	2360	2831	2
HI	Hydrogen iodide*	-146 s	-135.2 s	-120.8 s	-101.9 s	-75.9 s	-35.9	5
HKO	Potassium hydroxide	520 e	601 e	704	842	1035	1325	4
HNO$_3$	Nitric acid			-37 e	-9 e	28.4	82.2	5
HN$_3$	Hydrazoic acid			-79 e	-54 e	-18.0	35.7	5
HNaO	Sodium hydroxide	513	605	722	874	1080	1377	4
H$_2$	Hydrogen*					-258.6	-252.8	1
H$_2$I$_2$Si	Diiodosilane				11.8	70.5	149.4	4
H$_2$O	Water***	-60.7 s	-42.2 s	-20.3 s	7.0	45.8	99.6	36,37
H$_2$O$_2$	Hydrogen peroxide			13 e	45 e	89.0	149.8	5
H$_2$O$_4$S	Sulfuric acid	72	103	140	187	248	330	4
H$_2$S	Hydrogen sulfide*		-149 s	-136 s	-118.9 s	-95.9 s	-60.5	1,5
H$_2$S$_2$	Hydrogen disulfide				-27 e	12.2	70.7	5
H$_2$Se	Hydrogen selenide	-145 s	-134 s	-120 s	-102.8 s	-78.9 s	-41.5	5

VAPOR PRESSURE (continued)

Mol. Form.	Name	Temperature in °C for the indicated pressure						Ref.
		1 Pa	10 Pa	100 Pa	1 kPa	10 kPa	100 kPa	
H_2Te	Hydrogen telluride					-46.6	-2.3	5
H_3ISi	Iodosilane				-47.7	-10.1	45.2	4
H_3N	Ammonia*	-139 s	-127 s	-112 s	-94.5 s	-71.3	-33.6	1,5,6
H_3NO	Hydroxylamine				43.7	73.3	109.8	4
H_3P	Phosphine*	-182 s	-173 s	-161 s	-145 s	-122.7	-88.0	5
H_4IN	Ammonium iodide	125 s	159 s	201 s	253 s	318.4 s	405.2 s	5
H_4N_2	Hydrazine				14.7	55.6	113 e	5
H_4Si	Silane*			-181	-165.4	-143.7	-111.8	4
He	Helium*					-270.6	-268.9	2
Hf	Hafnium	2416	2681	3004	3406	3921	4603	9
Hg	Mercury**	42.0	76.6	120.0	175.6	250.3	355.9	29,30
HgI_2	Mercury(II) iodide	85.1 s	115.6 s	152.4 s	197.8 s	255.1 s	353.6	4
Ho	Holmium	1159 s	1311 s	1502 i	1767 i	2137 i	2691 i	3
IK	Potassium iodide			731	866	1052	1322	4
ILi	Lithium iodide	545	619	710	824	972	1170	4
INa	Sodium iodide			753	883	1058	1301	4
IRb	Rubidium iodide			733	866	1045	1302	4
ITl	Thallium(I) iodide				520	644	821	4
I_2	Iodine (rhombic)	-12.8 s	9.3 s	35.9 s	68.7 s	108 s	184.0	1,2
I_2Pb	Lead(II) iodide			470	558	682	869	4
I_2Zn	Zinc iodide	301 s	351 s	409 s	488 i	598 i	750 i	9
I_3Sb	Antimony(III) iodide				214.9	292.0	401.2	4
I_4Sn	Tin(IV) iodide				167.1	242.7	347.7	4
I_4Zr	Zirconium(IV) iodide	187 s	220 s	259 s	305 s	361 s	430 s	4
In	Indium	923	1052	1212	1417	1689	2067	2
Ir	Iridium	2440 s	2684	2979	3341	3796	4386	2
K	Potassium	200.2	256.5	328	424	559	756.2	13,30
Kr	Krypton*	-214.0 s	-208.0 s	-199.4 s	-188.9 s	-174.6 s	-153.6	5
La	Lanthanum	1732 i	1935 i	2185 i	2499 i	2905 i	3453 i	3
Li	Lithium	524.3	612.3	722.1	871.2	1064.3	1337.1	13,30
Lu	Lutetium	1633 s	1829.8	2072.8	2380 i	2799 i	3390 i	3
Mg	Magnesium	428 s	500 s	588 s	698	859	1088	2
Mn	Manganese	955 s	1074 s	1220 s	1418	1682	2060	2
Mo	Molybdenum	2469 s	2721	3039	3434	3939	4606	2
MoO_3	Molybdenum(VI) oxide				801	935	1151	4
NO	Nitric oxide*	-201 s	-195 s	-188 s	-179.3 s	-168.1 s	-151.9	5
N_2	Nitrogen*	-236 s	-232 s	-226.8 s	-220.2 s	-211.1 s	-195.9	1,5
N_2O	Nitrous oxide*	-167 s	-157 s	-145.4 s	-131.1 s	-112.9 s	-88.7	5
N_2O_4	Nitrogen tetroxide	-92 s	-78 s	-61 s	-41.1 s	-16.6 s	28.7	5
N_2O_5	Nitrogen pentoxide	-71 s	-56 s	-40 s	-19.9 s	3.9 s	33.2	5
Na	Sodium	280.6	344.2	424.3	529	673	880.2	13,30
Nb	Niobium	2669	2934	3251	3637	4120	4740	2
Nd	Neodymium	1322.3	1501.2	1725.3	2023 i	2442 i	3063 i	3
Ne	Neon*	-261 s	-260 s	-258 s	-255 s	-252 s	-246.1	2
Ni	Nickel	1510	1677	1881	2137	2468	2911	2
OPb	Lead(II) oxide	724	816	928	1065	1241	1471	4
OSr	Strontium oxide	1789 s	1903 s	2047 s	2235 s	2488 s		4
O_2	Oxygen*				-211.9	-200.5	-183.1	1,28
O_2S	Sulfur dioxide*			-98 s	-80 s	-52.2	-10.3	1,5
O_2Se	Selenium dioxide	124.5 s	153.9 s	188 s	228 s	275 s	315 s	38
O_2Si	Silicon dioxide	1966 i	2149 i	2368 i				8
O_3	Ozone*	-189 e	-182 e	-172 e	-158 e	-139.7	-111.5	5
O_3P_2	Phosphorus(III) oxide				47.3	100.3	172.8	4
O_3S	Sulfur trioxide				-20 s	6.6 s	44.5	5
O_3Sb_2	Antimony(III) oxide (valentinite)	426.1 s	478 s	539 s	610 s	907	1420	4,35
O_5P_2	Phosphorus(V) oxide	285 s	328 s	377.5 s	434.4 s	500.5 s	591	4
O_7Re_2	Rhenium(VII) oxide	147 s	176 s	208 s	244 s	284 s	362	4
Os	Osmium	2887 s	3150	3478	3875	4365	4983	2
P	Phosphorus (white)	6 s	34 s	69	115	180	276	3,9
P	Phosphorus (red)	182 s	216 s	256 s	303 s	362 s	431 s	2,3
Pb	Lead	705	815	956	1139	1387	1754	2
PbS	Lead(II) sulfide	656 s	741 s	838 s	953 s	1088 s	1280	4
Pd	Palladium	1448 s	1624	1844	2122	2480	2961	2
Po	Polonium				573 e	730.2	963.3	5
Pr	Praseodymium	1497.7	1699.4	1954 i	2298 i	2781 i	3506 i	3
Pt	Platinum	2057	2277 e	2542	2870	3283	3821	2

VAPOR PRESSURE (continued)

Mol. Form.	Name	Temperature in °C for the indicated pressure						Ref.
		1 Pa	10 Pa	100 Pa	1 kPa	10 kPa	100 kPa	
Pu	Plutonium	1483	1680	1925	2238	2653	3226	2
Ra	Radium	546 s	633 s	764	936	1173	1526	2
Rb	Rubidium	160.4	212.5	278.9	368	496.1	685.3	13,30
Re	Rhenium	3030 s	3341	3736	4227	4854	5681	2
Rh	Rhodium	2015	2223	2476	2790	3132	3724	2
Rn	Radon*	-163 s	-152 s	-139 s	-121.4 s	-97.6 s	-62.3	5
Ru	Ruthenium	2315 s	2538	2814	3151	3572	4115	2
S	Sulfur	102 s	135	176	235	318	444	3
Sb	Antimony	534 s	603 s	738	946	1218	1585	2,3
Sc	Scandium	1372 s	1531 s	1733 i	1993 i	2340 i	2828 i	3
Se	Selenium	227	279	344	431	540	685	3
Si	Silicon	1635	1829	2066	2363	2748	3264	2
Sm	Samarium	728 s	833 s	967 s	1148 i	1402 i	1788 i	3
Sn	Tin	1224	1384	1582	1834	2165	2620	2
Sr	Strontium	523 s	609 s	717 s	866	1072	1373	2
Ta	Tantalum	3024	3324	3684	4122	4666	5361	2
Tb	Terbium	1516.1	1706.1	1928 i	2232 i	2640 i	3218 i	3
Tc	Technetium	2454 e	2725 e	3051 e	3453 e	3961 e	4621 e	2
Te	Tellurium			502 e	615 e	768.8	992.4	5
Th	Thorium	2360	2634	2975	3410	3986	4782	2
Ti	Titanium	1709	1898	2130 e	2419	2791	3285	2
Tl	Thallium	609	704	824	979	1188	1485	2
Tm	Thulium	844 s	962 s	1108 s	1297 s	1548 i	1944 i	3
U	Uranium	2052	2291	2586	2961	3454	4129	2
V	Vanadium	1828 s	2016	2250	2541	2914	3406	2
W	Tungsten	3204	3500	3864	4306	4854	5550	2
Xe	Xenon*	-190 s	-181 s	-170 s	-155.8 s	-136.6 s	-108.4	5,32
Y	Yttrium	1610.1	1802.3	2047 i	2354 i	2763 i	3334 i	3
Yb	Ytterbium	463 s	540 s	637 s	774 s	993 i	1192 i	3
Zn	Zinc	337 s	397 s	477	579	717	912 e	2
Zr	Zirconium	2366	2618	2924	3302	3780	4405	2

Substances containing carbon:

Mol. Form.	Name	1 Pa	10 Pa	100 Pa	1 kPa	10 kPa	100 kPa	Ref.
C	Carbon (graphite)		2566 s	2775 s	3016 s	3299 s	3635 s	15
CBrClF$_2$	Bromochloro-difluoromethane	-136 e	-123 e	-106 e	-83.4	-51.8	-4.3	1
CBrCl$_3$	Bromotrichloromethane				-6 e	38.9	104.4	5
CBrF$_3$	Bromotrifluoromethane*	-168 e	-156 e	-142 e	-122.8	-96.6	-58.1	5
CBrN	Cyanogen bromide				-13 s	17.7 s	61.0	1
CBr$_2$F$_2$	Dibromodifluoromethane		-110 e	-91 e	-66 e	-30 e	22.5	1
CBr$_4$	Tetrabromomethane			25.6 s	65.8 s	111.6	188.9	5
CClF$_3$	Chlorotrifluoromethane	-176 e	-167 e	-155 e	-139 e	-116 e	-81.7	5
CClN	Cyanogen chloride		-94.6 s	-78.1 s	-57 s	-29 s	13.0	5
CCl$_2$F$_2$	Dichlorodifluoromethane*	-150 e	-138 e	-122 e	-101.8	-73.1	-30.0	5
CCl$_2$O	Carbonyl chloride	-127 e	-113 e	-96 e	-73 e	-40.6	7.2	5
CCl$_3$F	Trichlorofluoromethane*		-107 e	-89 e	-63 e	-28.5	23.3	1,5
CCl$_3$NO$_2$	Trichloronitromethane		-59 e	-30 e	4.4	47.8	112.0	5
CCl$_4$	Tetrachloromethane*	-79.4 s	-70.8 s	-53.5 s	-24.4 s	15.8	76.2	1,5
CFN	Cyanogen fluoride		-135 s	-121.2 s	-104.1 s	-82.8 s	-46.2	1,5
CF$_4$	Tetrafluoromethane*	-199.9 s	-193 s	-183.9 s	-171.6	-153.9	-128.3	1,5
CHBrF$_2$	Bromodifluoromethane		-128 s	-111.4 s	-89.7 s	-59.7 s	-16 s	5
CHBr$_3$	Tribromomethane				30.5	78.3	148.8	1
CHClF$_2$	Chlorodifluoromethane*	-152 e	-141 e	-126 e	-107.1	-80.5	-41.1	5
CHCl$_2$F	Dichlorofluoromethane	-76 e	-70 e	-61 e	-49 e	-28.7	8.6	1
CHCl$_3$	Trichloromethane*			-61 e	-34 e	4.3	60.8	1
CHF$_3$	Trifluoromethane*			-152 e	-136 e	-114.4	-82.3	1
CHI$_3$	Triiodomethane	51.1 s	82.7 s	121 e			218.0	5
CHN	Hydrogen cyanide*			-77 s	-52.6 s	-22.7 s	25.4	1,5
CHNO	Cyanic acid			-81.1	-56.8	-23.9	23 e	5
CH$_2$BrCl	Bromochloromethane	-83 e	-69 e	-50 e	-25 e	11.4	67.7	1
CH$_2$Br$_2$	Dibromomethane			-37 e	-7 e	35.2	96.5	5
CH$_2$ClF	Chlorofluoromethane		-124 e	-108 e	-86.2	-55.7	-9.4	5
CH$_2$Cl$_2$	Dichloromethane*		-92 e	-73 e	-48 e	-12.5	39.3	1
CH$_2$F$_2$	Difluoromethane*	-156.7	-145.8	-131.9	-113.6	-88.6	-51.9	1
CH$_2$I$_2$	Diiodomethane			17 e	55 e	106.1	181.6	5
CH$_2$O	Formaldehyde*				-91 e	-61.7	-19.3	1

VAPOR PRESSURE (continued)

Mol. Form.	Name	Temperature in °C for the indicated pressure						Ref.
		1 Pa	10 Pa	100 Pa	1 kPa	10 kPa	100 kPa	
CH_2O_2	Formic acid	-56 s	-40.4 s	-22.3 s	-0.8 s	37.0	100.2	1,5
CH_3AsF_2	Methyldifluoroarsine				-15 e	22.1	76.1	5
CH_3BO	Borane carbonyl				-124	-99	-64	4
CH_3Br	Bromomethane				-77 e	-44.3	3.3	1
CH_3Cl	Chloromethane*	-140.2 s	-128.6 s	-114.7 s	-96 e	-67.1	-24.4	1,33
CH_3Cl_3Si	Methyltrichlorosilane		-83 e	-61 e	-33 e	7 e	65.7	1
CH_3F	Fluoromethane*				-130 e	-111 e	-78.6	1
CH_3I	Iodomethane				-49 e	-12.4	42.1	1
CH_3NO	Formamide		22 e	53 e	93 e	145.0	218 e	5
CH_3NO_2	Nitromethane				-2 e	40 e	100.8	1
CH_3NO_3	Methyl nitrate		-75 e	-55 e	-27 e	9.8	63 e	5
CH_4	Methane*	-220 s	-214.2 s	-206.8 s	-197 s	-183.6 s	-161.7	5,41
CH_4Cl_2Si	Dichloromethylsilane				-77 e	-51 e	40.5	1
CH_4O	Methanol*	-87 e	-69 e	-47.5	-20.4	15.2	64.2	11
CH_4S	Methanethiol		-115 e	-97 e	-74 e	-41.7	5.7	1
CH_5ClSi	Chloromethylsilane	-129 e	-115 e	-97.9	-74.4	-41.5	8.3	5
CH_5N	Methylamine				-76.7	-48.1	-6.6	1
CH_6N_2	Methylhydrazine			-31 e	-4.7	32.9	91 e	1
CH_6OSi	Methyl silyl ether				-90.2	-61.8	-18 e	1
CH_6Si	Methylsilane			-144 e	-124.6	-97.5	-57.5	5
CIN	Cyanogen iodide						153.8	5
$CNNa$	Sodium cyanide		672 e	798	961	1182	1497	4
CN_4O_8	Tetranitromethane				18.0	61.8	124 e	5
CO	Carbon monoxide*			-223 s	-216.5 s	-207.2 s	-191.7	40
COS	Carbon oxysulfide*			-136 e	-117 e	-90.0	-50.4	1
$COSe$	Carbon oxyselenide			-120	-98	-67	-22	4
CO_2	Carbon dioxide*	-159.1 s	-148.9 s	-136.7 s	-121.6 s	-103.1 s	-78.6 s	5
CS_2	Carbon disulfide		-96 e	-76	-49 e	-10.9	45.9	1
CSe_2	Carbon diselenide			-24 e	9.4	56.2	127 e	1
$C_2Br_2ClF_3$	1,2-Dibromo-1-chloro-1,2,2-trifluoroethane						92.3	5
$C_2Br_2F_4$	1,2-Dibromotetrafluoroethane		-97 e	-75 e	-46 e	-7.2	47.1	5
C_2Br_4	Tetrabromoethylene		-54.5 s	-31.7 s	-3.5 s	32.2 s	226.0	5
C_2ClF_3	Chlorotrifluoroethylene	-146 e	-134 e	-119 e	-99 e	-71 e	-28.4	1
C_2ClF_5	Chloropentafluoroethane					-80.3	-39.4	1
$C_2Cl_2F_4$	1,1-Dichlorotetrafluoroethane					-45.4	2.7	5
$C_2Cl_2F_4$	1,2-Dichlorotetrafluoroethane				-76.8	-44.9	3.2	5
$C_2Cl_3F_3$	1,1,1-Trichlorotrifluoroethane						45.6	1,5
$C_2Cl_3F_3$	1,1,2-Trichlorotrifluoroethane					-8.2	47.3	1,5
C_2Cl_3N	Trichloroacetonitrile				-16 e	25.3	85.1	1
C_2Cl_4	Tetrachloroethylene			-22 e	10 e	54.4	120.7	1
$C_2Cl_4F_2$	1,1,1,2-Tetrachloro-2,2-difluoroethane				-7 e	31.0	91.1	5
$C_2Cl_4F_2$	1,1,2,2-Tetrachloro-1,2-difluoroethane					32.3	92.5	1
C_2Cl_4O	Trichloroacetyl chloride			-25 e	7 e	51.7	117.8	1,5
C_2Cl_6	Hexachloroethane	-7.6 s	9.9 s	33.6 s	67.7 s	116.9 s	184.2 s	5
C_2F_3N	Trifluoroacetonitrile				-126.1	-102.5	-67.8	1
C_2F_4	Tetrafluoroethylene				-132.3	-109.7	-75.8	1
$C_2F_4N_2O_4$	1,1,2,2-Tetrafluoro-1,2-dinitroethane				-30 e	6.4	59.5	5
C_2F_6	Hexafluoroethane**			-155.2 s	-137.5 s	-113.4 s	-78.4 s	1,5
$C_2HBrClF_3$	2-Bromo-2-chloro-1,1,1-trifluoroethane				-41.4	-4.8	49.8	1
C_2HBr_3O	Tribromoacetaldehyde			15.0	52.7	103.0	173.5	5
C_2HClF_4	1-Chloro-1,1,2,2-tetrafluoroethane			-110 e	-87.6	-57.0	-12.1	5
$C_2HCl_2F_3$	2,2-Dichloro-1,1,1-trifluoroethane		-101.0	-82.2	-57.4	-23.3	26.7	18
C_2HCl_3	Trichloroethylene	-74 e	-59 e	-39 e	-12 e	26.7	86.8	1
C_2HCl_3O	Trichloroacetaldehyde			-41.6	-9.8	33.8	97.4	5
$C_2HCl_3O_2$	Trichloroacetic acid				83.8	130.0	197.2	1,5
C_2HCl_5	Pentachloroethane		-23 e	3 e	37.4	86.0	159.4	1
$C_2HF_3O_2$	Trifluoroacetic acid					16.8	71.4	1,5
C_2HF_5O	Trifluoromethyl difluoromethyl ether	-147 e	-136	-121 e	-102 e	-75.0	-35.4	20
C_2H_2	Acetylene*			-146.6 s	-130.7 s	-110.6 s	-84.8 s	5

Mol. Form.	Name	Temperature in °C for the indicated pressure						Ref.
		1 Pa	10 Pa	100 Pa	1 kPa	10 kPa	100 kPa	
$C_2H_2Br_2$	cis-1,2-Dibromoethylene		-45 e	-21 e	10 e	52.2	114.8	1
$C_2H_2Br_2$	trans-1,2-Dibromoethylene				-4 e	42.2	107.4	5
$C_2H_2Br_2Cl_2$	1,2-Dibromo-1,1-dichloroethane					103.6	177.8	5
$C_2H_2Br_2Cl_2$	1,2-Dibromo-1,2-dichloroethane		-11 e	22 e	64.1	119 e	193 e	5
$C_2H_2Br_4$	1,1,2,2-Tetrabromoethane	14 e	38 e	69 e	109 e	163.7	242.9	5
$C_2H_2Cl_2$	1,1-Dichloroethylene	-116 e	-101 e	-82 e	-57 e	-21.4	31.2	1
$C_2H_2Cl_2$	cis-1,2-Dichloroethylene			-62 e	-34 e	3.8	60.3	1
$C_2H_2Cl_2$	trans-1,2-Dichloroethylene				-44 e	-7.5	47.3	1
$C_2H_2Cl_2F_2$	1,2-Dichloro-1,1-difluoroethane	-101 e	-87 e	-68 e	-42.2	-6.8	46.3	5
$C_2H_2Cl_2O$	Chloroacetyl chloride			-23.7	5.6	46.1	105.6	5
$C_2H_2Cl_4$	1,1,1,2-Tetrachloroethane	-58 e	-40 e	-15 e	17 e	62.2	129.7	1
$C_2H_2Cl_4$	1,1,2,2-Tetrachloroethane		-22 e	1 e	32.4	76.9	144.7	1
$C_2H_2F_4$	1,1,1,2-Tetrafluoroethane				-94.3	-66.8	-26.4	17
$C_2H_2F_4$	1,1,2,2-Tetrafluoroethane				-96.0	-66.9	-23.3	5
C_2H_2O	Ketene		-151 e	-135 e	-115 e	-88.2	-50.0	1
C_2H_3Br	Bromoethylene	-124 e	-110 e	-92 e	-68 e	-34.5	15.4	5
C_2H_3BrO	Acetyl bromide	-78 e	-65 e	-49 e	-25 e	13.9	84 e	5
$C_2H_3Br_3$	1,1,2-Tribromoethane	-18 e	4 e	32 e	68 e	117.1	188.4	5
C_2H_3Cl	Chloroethylene	-139 e	-127 e	-110 e	-89 e	-59.0	-14.1	1
$C_2H_3ClF_2$	1-Chloro-1,1-difluoroethane		-123 e	-107 e	-85.3	-55.4	-10.5	5
C_2H_3ClO	Acetyl chloride	-100 e	-85 e	-66 e	-40 e	-3.6	50.4	1
$C_2H_3ClO_2$	Chloroacetic acid				78.4	123.9	188.9	1
$C_2H_3Cl_2F$	1,1-Dichloro-1-fluoroethane		-101 e	-83 e	-57.9	-22.7	31.4	5
$C_2H_3Cl_2F$	1,2-Dichloro-1-fluoroethane			-50 e	-23.8	14.1	73.4	5
$C_2H_3Cl_3$	1,1,1-Trichloroethane				-25.3	14.2	73.7	5
$C_2H_3Cl_3$	1,1,2-Trichloroethane			-23 e	7 e	49.9	113.4	1
C_2H_3F	Fluoroethylene			-153.3	-135.2	-109.9	-72.2	5
C_2H_3FO	Acetyl fluoride				-64.1	17.0		5
$C_2H_3F_3$	1,1,1-Trifluoroethane				-113 e	-86.6	-47.8	1
$C_2H_3F_3O$	2,2,2-Trifluoroethanol			-33 e	-8 e	26.0	74 e	5
C_2H_3I	Iodoethylene				-41 e	-3 e	55.6	5
C_2H_3IO	Acetyl iodide				-0.6	47 e	107.0	5
C_2H_3N	Acetonitrile				-20 e	21.4	81.2	1
C_2H_3NO	Methylisocyanate				-43.5	-10.2	38.8	1
C_2H_3NS	Methyl thiocyanate			-18.4	16.2	63.5	132.5	5
C_2H_4	Ethylene*				-155.6	-135.1	-104.0	1,10
C_2H_4BrCl	1-Bromo-2-chloroethane				-0.4	41.7	105.7	6
$C_2H_4Br_2$	1,1-Dibromoethane		-49 e	-26 e	5 e	46.4	107.6	5
$C_2H_4Br_2$	1,2-Dibromoethane				18 e	62.2	130.9	1
C_2H_4ClF	1-Chloro-1-fluoroethane				-69.9	-36.1	15.8	5
$C_2H_4Cl_2$	1,1-Dichloroethane		-84 e	-64 e	-36.7	1.0	56.9	1
$C_2H_4Cl_2$	1,2-Dichloroethane				-16.4	23.7	83.1	1
$C_2H_4F_2$	1,1-Difluoroethane			-115.2	-94.6	-66.1	-24.3	19
$C_2H_4N_2O_6$	Ethylene glycol dinitrate	4 e	25.6	51.0	81 e	117 e	162 e	5
C_2H_4O	Acetaldehyde		-105 e	-87 e	-62.8	-29.4	20.0	5
C_2H_4O	Ethylene oxide		-111 e	-93 e	-70 e	-37.0	10.2	1
$C_2H_4O_2$	Acetic acid	-42.8 s	-26.7 s	-8 s	14.2 s	55.9	117.5	1,5
$C_2H_4O_2$	Methyl formate		-95 e	-76 e	-51.8	-18.1	31.4	5
$C_2H_4O_3$	Peroxyacetic acid				14.4	55.3	109.7	5
$C_2H_4O_3$	Glycolic acid						99.9	5
$C_2H_5AsF_2$	Ethyldifluoroarsine			-36 e	-6.0	35.0	93.1	5
C_2H_5Br	Bromoethane	-111 e	-96 e	-77 e	-51.3	-15.5	38.0	5
C_2H_5Cl	Chloroethane	-126 e	-112 e	-94 e	-70 e	-37.0	12.0	1
C_2H_5ClO	2-Chloroethanol	-61 e	-39 e	-12 e	23 e	67.1	127.3	5
C_2H_5ClO	Chloromethyl methyl ether	-96 e	-80 e	-59 e	-32 e	6 e	61 e	5
$C_2H_5Cl_3OSi$	Trichloroethoxysilane	-78 e	-60 e	-36.0	-4.6	38.7	102.0	5
$C_2H_5Cl_3Si$	Trichloroethylsilane	-79 e	-61 e	-38 e	-8 e	34.9	98.7	5
C_2H_5F	Fluoroethane		-142 e	-127 e	-106.3	-78.7	-37.9	1
C_2H_5FO	2-Fluoroethanol			-22 e	8.3	47.5	99 e	5
C_2H_5I	Iodoethane	-94 e	-78 e	-56 e	-27.9	11.9	71.9	5
C_2H_5N	Ethyleneimine		-74 e	-55 e	-30 e	4.1	55 e	5
C_2H_5NO	Acetamide	16.7 s	39.1 s	65.2 s	102.8	150.8	218.2	5
C_2H_5NO	N-Methylformamide		13 e	41 e	78 e	127.9	199.1	1
$C_2H_5NO_2$	Nitroethane	-61 e	-44 e	-21 e	8.3	50.1	113.5	5
$C_2H_5NO_3$	Ethyl nitrate	-81 e	-63 e	-41 e	-12 e	28.2	87 e	1
C_2H_6	Ethane*	-183.3 s	-173.2	-161.3	-145.3	-122.8	-88.8	41
$C_2H_6Cl_2Si$	Dichlorodimethylsilane					11.1	70.1	5

VAPOR PRESSURE (continued)

Mol. Form.	Name	Temperature in °C for the indicated pressure						Ref.
		1 Pa	10 Pa	100 Pa	1 kPa	10 kPa	100 kPa	
C_2H_6Hg	Dimethyl mercury				-13.5	29.0	92.1	5
$C_2H_6N_2O$	N-Nitrosodimethylamine				30.7	80.5	149.8	5
C_2H_6O	Ethanol	-73 e	-56 e	-34 e	-7 e	29.2	78.0	1,5
C_2H_6O	Dimethyl ether*		-135 e	-118 e	-96.8	-67.6	-25.1	1,5
C_2H_6OS	Dimethyl sulfoxide			27.4	65.0	115.9	188.6	1
$C_2H_6O_2$	Ethylene glycol	2 e	24 e	51.1	86.1	132.5	196.9	1
$C_2H_6O_2$	Ethyl hydroperoxide	-70 e	-49 e	-25 e	6.8	47.0	101 e	5
$C_2H_6O_2S$	Dimethyl sulfone				109 e	166.8	248.9	5
C_2H_6S	Ethanethiol	-112 e	-97 e	-78 e	-53 e	-18 e	34.7	1
C_2H_6S	Dimethyl sulfide		-96 e	-77 e	-51.2	-16.0	37.0	1,5
$C_2H_6S_2$	Dimethyl disulfide	-71 e	-53 e	-29 e	1.7	45.0	109.3	5
$C_2H_7BO_2$	Dimethoxyborane	-116 e	-101.9	-83.5	-59.2	-25.4	25 e	1
C_2H_7N	Ethylamine			-71 e	-53 e	-27 e	16.4	1
C_2H_7N	Dimethylamine			-88 e	-66.9	-37.2	6.6	1
C_2H_7NO	Ethanolamine		11 e	35 e	66.2	109.0	170.6	1
$C_2H_8N_2$	1,2-Ethanediamine				17.0	57.5	116.6	1,5
$C_2H_8N_2$	1,1-Dimethylhydrazine			-52 e	-25.6	10.5	63 e	5
$C_2H_8N_2$	1,2-Dimethylhydrazine		-49 e	-33 e	-9 e	26.4	88 e	1
C_2N_2	Cyanogen	-127 s	-114.1 s	-98.5 s	-79.2 s	-54.9 s	-21.4	5
C_3ClF_5O	Chloropentafluoroacetone	-122 e	-109 e	-93 e	-71 e	-39.4	7.4	5
C_3Cl_6	Hexachloropropene	-12 e	11 e	40 e	79 e	132.8	213.6	5
C_3F_6	Perfluoropropene	-150 e	-138 e	-122 e	-101 e	-72 e	-30.6	5
C_3F_6O	Perfluoroacetone			-113 e	-94 e	-67.8	-27.6	1
C_3F_8	Perfluoropropane		-139 e	-124 e	-105 e	-77.5	-37.0	5
C_3HN	Cyanoacetylene			-58.7 s	-35.6 s	-7 s	42.0	5
$C_3H_2F_6O$	1,1,1,3,3,3-Hexafluoro-2-propanol					12.7	57.1	5
$C_3H_3F_5$	1,1,1,2,2-Pentafluoropropane					-60 e	-17.9	5
C_3H_3N	2-Propenenitrile		-72 e	-50 e	-22 e	17.7	77.0	1
C_3H_3NS	Thiazole					54.4	117.8	5
C_3H_4	Allene*		-129 e	-118 e	-101.4	-76.7	-34.7	5
C_3H_4	Propyne				-94 e	-65.3	-23.2	1
$C_3H_4ClF_3$	3-Chloro-1,1,1-trifluoropropane	-102 e	-87 e	-68 e	-43 e	-8 e	45.3	5
$C_3H_4Cl_2O$	1,1-Dichloroacetone				1 e	47.8	118.0	5
$C_3H_4Cl_2O_2$	Methyl dichloroacetate	-44 e	-25 e	0 e	33 e	77.7	142.3	5
$C_3H_4Cl_4$	1,1,1,2-Tetrachloropropane	-48 e	-28 e	-2 e	32 e	79.1	149.5	5
$C_3H_4F_4O$	2,2,3,3-Tetrafluoro-1-propanol			-10 e	17 e	53.9	107.2	5
C_3H_4O	Acrolein		-87 e	-67 e	-40 e	-3.0	52.8	1
$C_3H_4O_2$	Propenoic acid				35 e	78.0	140.7	1
$C_3H_4O_2$	Vinyl formate			-58 e	-34 e	-1.6	46.2	1
$C_3H_4O_2$	2-Oxetanone		-21 e	8 e	45.5	93.8	159.3	5
$C_3H_4O_3$	Ethylene carbonate	12.7 s	37 e				247	5
C_3H_5Br	cis-1-Bromopropene	-100 e	-84 e	-64 e	-37 e	1.0	57.4	5
C_3H_5Br	2-Bromopropene	-112 e	-95 e	-75 e	-47 e	-9 e	48.0	5
C_3H_5Br	3-Bromopropene	-98 e	-80 e	-58 e	-28 e	12 e	69.6	5
C_3H_5Cl	cis-1-Chloropropene	-114 e	-100 e	-81 e	-55 e	-20.1	32.4	5
C_3H_5Cl	trans-1-Chloropropene		-97 e	-77 e	-52 e	-16.2	37.0	5
C_3H_5Cl	2-Chloropropene	-120 e	-106 e	-87 e	-63 e	-28.7	22.3	5
C_3H_5Cl	3-Chloropropene	-107 e	-92 e	-72.4	-46.3	-9.8	44.6	5
C_3H_5ClO	Epichlorohydrin			-21 e	11 e	53.8	115.5	5
$C_3H_5ClO_2$	Methyl chloroacetate		-28 e	-5 e	25 e	66.9	129.1	5
$C_3H_5Cl_3$	1,1,3-Trichloropropane	-51 e	-31 e	-5 e	28 e	75.3	145.1	5
$C_3H_5Cl_3$	1,2,3-Trichloropropane			2 e	37 e	84.9	156.3	5
$C_3H_5Cl_3Si$	Trichloro-2-propenylsilane					53.0	116.5	5
C_3H_5I	3-Iodopropene	-80 e	-62 e	-39 e	-8 e	36 e	101.5	5
C_3H_5N	Propanenitrile	-69.4	-55.3	-36.0	-7.9	35.2	97.4	1,5
C_3H_5NO	Acrylamide				109.6	161 e		5
C_3H_5NO	3-Hydroxypropanenitrile	-11 e	18 e	53 e	96.1	150.3	220.8	5
C_3H_5NS	Ethyl thiocyanate	-39 e	-20 e	4 e	35 e	79.1	143.4	5
C_3H_5NS	Ethyl isothiocyanate				17.4	66 e	136 e	5
$C_3H_5N_3O_9$	Trinitroglycerol	48.6	75.7	118 e	191 e	353 e	1007 e	5
C_3H_6	Propene*	-160.6	-149.0	-134.3	-114.9	-88.2	-47.9	1,5
C_3H_6	Cyclopropane			-124 e	-104 e	-75.7	-33.1	1
C_3H_6BrCl	1-Bromo-3-chloropropane	-51 e	-31 e	-6 e	28 e	74.1	142.9	5
$C_3H_6Br_2$	1,2-Dibromopropane	-46 e	-26 e	-2 e	31 e	75.3	139.5	5

| Mol. Form. | Name | \multicolumn{6}{c}{Temperature in °C for the indicated pressure} | Ref. |

Mol. Form.	Name	1 Pa	10 Pa	100 Pa	1 kPa	10 kPa	100 kPa	Ref.
$C_3H_6Br_2$	1,3-Dibromopropane	-30 e	-9 e	17 e	52 e	98.7	166.8	5
$C_3H_6Cl_2$	1,1-Dichloropropane				-14 e	27.0	87.7	5
$C_3H_6Cl_2$	1,2-Dichloropropane	-78 e	-61 e	-38.1	-8.1	33.7	95.9	5
$C_3H_6Cl_2$	1,3-Dichloropropane	-65 e	-46 e	-22 e	10 e	54.0	119.9	5
$C_3H_6Cl_2$	2,2-Dichloropropane				-28 e	10.8	68.9	5
$C_3H_6Cl_2O$	1,3-Dichloro-2-propanol			21.8	59.0	107.6	173.9	5
$C_3H_6N_2O_4$	1,1-Dinitropropane	-9 e	12 e	39 e	73.2	120 e	187 e	5
C_3H_6O	Allyl alcohol	-63 e	-48 e	-21.9	6.8	44.5	96.2	5
C_3H_6O	Methyl vinyl ether			-114 e	-89 e	-52.7	4.6	1
C_3H_6O	Propanal			-69 e	-42 e	-6 e	47.7	1
C_3H_6O	Acetone	-95	-81.8	-62.8	-35.6	1.3	55.7	1,5
C_3H_6O	Methyloxirane	-109 e	-95 e	-76 e	-51.5	-17.2	33.9	5
$C_3H_6O_2$	Propanoic acid			0 e	35.1	79.9	140.8	1,5
$C_3H_6O_2$	Ethyl formate		-80 e	-61 e	-35 e	1 e	54.0	1
$C_3H_6O_2$	Methyl acetate	-95 e	-79 e	-59 e	-33 e	3.3	56.6	1
$C_3H_6O_2$	1,3-Dioxolane		-72 e	-50 e	-22 e	17.0	75.3	1
$C_3H_6O_3$	1,3,5-Trioxane				53 e	113.7		1
C_3H_6S	Thiacyclobutane		-62 e	-40 e	-9 e	32.5	94.5	5
C_3H_7Br	1-Bromopropane	-95 e	-78 e	-57 e	-28 e	11.6	70.6	1
C_3H_7Br	2-Bromopropane		-84 e	-65 e	-39.6	-1.7	59.1	1,5
C_3H_7Cl	1-Chloropropane	-106 e	-90 e	-71 e	-44.5	-8.1	46.2	1
C_3H_7Cl	2-Chloropropane		-91 e	-74 e	-51.1	-17.8	35.4	1,5
C_3H_7ClO	2-Chloro-1-propanol			23 e	63.8	125.7		5
C_3H_7F	1-Fluoropropane	-133 e	-120 e	-103 e	-80.7	-49.4	-2.8	5
C_3H_7I	1-Iodopropane	-78 e	-60 e	-37 e	-6 e	36.9	102.0	5
C_3H_7I	2-Iodopropane	-89 e	-71 e	-47 e	-16.3	26.5	89.2	5
C_3H_7N	Allylamine		-88 e	-65 e	-37 e	0.4	52 e	5
C_3H_7NO	N,N-Dimethylformamide	-39 e	-20 e	5 e	38.0	83.9	152.6	1
C_3H_7NO	N-Methylacetamide	-13.3 s	13 s	43 e	83.8	136.1	206.3	5
$C_3H_7NO_2$	1-Nitropropane	-56 e	-37 e	-13 e	20 e	64.8	130.8	1
$C_3H_7NO_2$	2-Nitropropane		-48 e	-22 e	10.7	55.6	119.8	1
$C_3H_7NO_3$	Propyl nitrate			-23.9	6.1	48.1	111 e	5
C_3H_8	Propane*	-156.9	-145.6	-130.9	-111.4	-83.8	-42.3	1,41
C_3H_8O	1-Propanol	-54 e	-38 e	-16 e	10 e	47 e	96.9	1,5
C_3H_8O	2-Propanol	-65 e	-49 e	-28 e	-1.3	33.6	82.0	1,5
C_3H_8O	Ethyl methyl ether	-98 e	-89 e	-77 e	-60 e	-34.8	7.0	5
$C_3H_8O_2$	1,2-Propylene glycol	-11 e	13 e	42 e	78 e	125.0	187.2	5
$C_3H_8O_2$	1,3-Propylene glycol	4 e	30 e	62 e	101 e	149.9	214.0	5
$C_3H_8O_2$	Ethylene glycol monomethyl ether	-57 e	-37 e	-12 e	21 e	63.8	124.3	1
$C_3H_8O_2$	Dimethoxymethane	-93 e	-81 e	-64 e	-42 e	-9.3	41.7	5
$C_3H_8O_3$	Glycerol	96 e	113 e	136 e	168 e	213.4	287 e	1
C_3H_8S	1-Propanethiol	-94 e	-78 e	-57 e	-29.1	9.6	67.4	1,5
C_3H_8S	2-Propanethiol	-102 e	-87 e	-67 e	-41 e	-3 e	52.2	1
C_3H_8S	Ethyl methyl sulfide	-94 e	-78 e	-57 e	-29.7	8.8	66.3	1
$C_3H_8S_2$	1,3-Propanedithiol	-53 e	-28 e	3 e	43 e	97 e	172.4	5
C_3H_9As	Trimethylarsine		-74 e	-45 e	-5.4	52.0		5
$C_3H_9BO_3$	Trimethyl borate			-14 e	15.6	67.9		5
C_3H_9BS	Methyl dimethylthioborane			-62 e	-30.4	11.4	70.7	5
C_3H_9ClSi	Trimethylchlorosilane			-37.8	0.4	57.3		5
C_3H_9N	Propylamine		-81 e	-63 e	-38.3	-4.1	46.9	1,5
C_3H_9N	Isopropylamine		-91 e	-74 e	-50.4	-17.6	31.5	1,5
C_3H_9N	Trimethylamine		-114 e	-97 e	-75.0	-43.8	2.6	1,5
C_3H_9NO	1-Amino-2-propanol			18 e	53.2	98.2	157.9	5
$C_3H_9O_4P$	Trimethyl phosphate	-31 e	-7 e	23.6	62.8	116.0	192.0	5
C_3H_9P	Trimethylphosphine			-81 e	-53 e	-15.0	37.1	5
C_3H_9Sb	Trimethylstibine			-56 e	-23.8	19 e	80 e	5
$C_3H_{10}N_2$	1,2-Propanediamine		-35.4	-12.0	18.8	61 e	119 e	5
C_3N_2O	Carbonyl dicyanide			-21.7	15.3	65.2		5
C_4Cl_6	Hexachloro-1,3-butadiene	-1 e	22 e	50 e	86.7	137.0	209.7	5
$C_4F_6O_3$	Trifluoroacetic acid anhydride			-63 e	-39 e	-7.1	38.8	5
C_4F_8	Perfluorocyclobutane						-6.2	1
C_4F_{10}	Perfluorobutane		-122 e	-105 e	-82 e	-49.8	-2.5	1,5
$C_4H_2Cl_2O_2$	trans-2-Butenedioyl dichloride			8.0	45.6	94.3	159.8	5
$C_4H_2Cl_2S$	2,5-Dichlorothiophene			-20 e	22 e	81.4	171 e	5

Mol. Form.	Name	\multicolumn{6}{c}{Temperature in °C for the indicated pressure}	Ref.
		1 Pa	10 Pa	100 Pa	1 kPa	10 kPa	100 kPa	
$C_4H_2O_3$	Maleic anhydride				73.7	127.9	201.7	5
C_4H_3ClS	2-Chlorothiophene		-62 e	-35 e	2 e	51.8	123 e	5
C_4H_3IS	2-Iodothiophene			-25 e	23 e	94.9	181.0	5
C_4H_4	1-Buten-3-yne			-96.1	-73.4	-41.8	4.9	5
$C_4H_4N_2$	Succinonitrile	24.8 s					266.0	5
C_4H_4O	Furan			-78 e	-54 e	-20 e	31.0	1
$C_4H_4O_2$	Diketene				19.3	63.3	126 e	5
$C_4H_4O_3$	Succinic anhydride				121 e	180.8	260.8	5
$C_4H_4O_4$	Fumaric acid	123.9 s	150 s	180 s				5
C_4H_4S	Thiophene				-17 e	23.7	83.7	5
C_4H_5Cl	2-Chloro-1,3-butadiene	-113 e	-95 e	-71 e	-41 e	0.3	59.0	5
C_4H_5ClO	2-Methyl-2-propenoyl chloride		-57 e	-35 e	-5 e	36.4	98.2	5
$C_4H_5Cl_3O_2$	Ethyl trichloroacetate			15.3	51.9	100.1	166.6	5
C_4H_5N	3-Butenenitrile	-67 e	-48 e	-23.1	9.3	53.7	118.4	5
C_4H_5N	Methylacrylonitrile				-12 e	29.0	89.8	5
C_4H_5N	Pyrrole			-8 e	24 e	66.7	129.4	1
$C_4H_5NO_2$	Methyl cyanoacetate	-3 e	19 e	48 e	84 e	134.0	204.6	5
C_4H_5NS	Allyl isothiocyanate	-45 e	-27 e	-3 e	32.1	89 e	198 e	5
C_4H_5NS	4-Methylthiazole						67.0	5
C_4H_6	1,2-Butadiene	-132 e	-117 e	-98 e	-72.8	-38.9	10.5	5
C_4H_6	1,3-Butadiene*			-106 e	-83 e	-51.9	-4.7	1
C_4H_6	1-Butyne	-125 e	-111 e	-94 e	-71.2	-39.4	7.8	1
C_4H_6	2-Butyne		-89.2 s	-73.8 s	-53.5 s	-23.9	26.6	5
$C_4H_6Cl_2O_2$	Ethyl dichloroacetate			2.6	40.1	89.1	156.3	5
C_4H_6O	Divinyl ether		-99 e	-80 e	-56 e	-22.1	28.0	5
C_4H_6O	trans-2-Butenal	-74 e	-56 e	-33 e	-3 e	39.7	102.4	5
C_4H_6O	3-Buten-2-one					21 e	81.0	5
C_4H_6O	Cyclobutanone			-34 e	-4 e	37.1	97 e	5
$C_4H_6O_2$	cis-Crotonic acid			30 e	63 e	106.7	168.9	5
$C_4H_6O_2$	trans-Crotonic acid				74 e	120.8	184.9	5
$C_4H_6O_2$	3-Butenoic acid	-19 e	2 e	27 e	61 e	105.6	168.6	5
$C_4H_6O_2$	Methacrylic acid				22 e	56 e	99.9	161.5	5
$C_4H_6O_2$	Vinyl acetate	-88 e	-71 e	-50 e	-22 e	16.2	72.2	1
$C_4H_6O_2$	Methyl acrylate		-71 e	-48 e	-18 e	22 e	79.9	5
$C_4H_6O_2$	2,3-Butanedione					30.7	84.8	5
$C_4H_6O_2$	gamma-Butyrolactone		-17 e	24 e	72 e	130.2	203 e	5
$C_4H_6O_3$	Acetic anhydride	-44 e	-25 e	-1 e	31 e	75.1	139.7	1
$C_4H_6O_3$	Propylene carbonate	-40 e	-5 e	43 e	112 e	220 e	410 e	5
$C_4H_6O_4$	Dimethyl oxalate				50.5	98.1	163.0	5
C_4H_7Br	trans-1-Bromo-1-butene	-87 e	-68 e	-43.3	-11.4	31.9	94.4	5
C_4H_7Br	2-Bromo-1-butene	-87 e	-70 e	-48 e	-20 e	20.7	80.6	5
C_4H_7Br	cis-2-Bromo-2-butene	-90 e	-72 e	-49.0	-18.5	23.5	85.2	5
C_4H_7Br	trans-2-Bromo-2-butene	-86 e	-67 e	-43.4	-12.0	31.0	93.5	5
$C_4H_7Br_3$	1,2,3-Tribromobutane	0 e	23 e	53 e	91 e	143.7	219.5	5
$C_4H_7Br_3$	1,2,4-Tribromobutane	-3 e	20 e	49 e	87 e	139.4	214.5	5
C_4H_7Cl	3-Chloro-1-butene			-64 e	-36 e	4 e	63.6	5
C_4H_7Cl	cis-2-Chloro-2-butene	-100 e	-83 e	-62 e	-34 e	6 e	66.4	5
C_4H_7Cl	trans-2-Chloro-2-butene	-102 e	-86 e	-65 e	-37 e	3 e	62.2	5
C_4H_7Cl	3-Chloro-2-methylpropene		-75 e	-54 e	-25 e	13.8	71.5	5
$C_4H_7ClO_2$	Ethyl chloroacetate			-2.6	32.6	79.1	143.8	5
C_4H_7N	Butanenitrile	-67 e	-48 e	-24 e	8 e	52.3	117.2	1
C_4H_8	1-Butene	-139.0	-125.2	-107.8	-85.3	-53.7	-6.6	1,5
C_4H_8	cis-2-Butene	-131.2	-117.4	-99.8	-76.7	-44.8	3.4	1,5
C_4H_8	trans-2-Butene			-102 e	-80 e	-47.6	0.6	1
C_4H_8	Isobutene	-139.1	-125.5	-108.2	-85.5	-54.5	-7.3	1,5
C_4H_8	Cyclobutane				-71.8	-38.1	12.1	5
C_4H_8	Methylcyclopropane	-130 e	-116 e	-99.3	-76.3	-44.2	4.2	5
$C_4H_8Br_2$	1,2-Dibromobutane	-54 e	-30 e	0.4	39.6	92.1	166.1	5
$C_4H_8Br_2$	1,4-Dibromobutane	-13 e	9 e	37 e	74 e	124.0	196.5	5
$C_4H_8Cl_2$	1,1-Dichlorobutane			-25 e	6 e	49.3	113.4	5
$C_4H_8Cl_2$	1,2-Dichlorobutane			-28.4	5.8	53.1	123.1	5
$C_4H_8Cl_2$	1,4-Dichlorobutane		-26 e	0 e	35 e	82.4	153.4	5
$C_4H_8Cl_2$	2,2-Dichlorobutane		-58 e	-35 e	-5 e	37.8	102.1	5
$C_4H_8Cl_2O$	Bis(2-chloroethyl) ether	-32 e	-9 e	19.8	56.9	106.9	177.9	5
C_4H_8O	Ethyl vinyl ether		-102 e	-81 e	-53.1	-16.5	34.7	5
C_4H_8O	1,2-Epoxybutane	-135 e	-114 e	-87 e	-53 e	-5.5	62.1	5

Mol. Form.	Name	Temperature in °C for the indicated pressure						Ref.
		1 Pa	10 Pa	100 Pa	1 kPa	10 kPa	100 kPa	
C_4H_8O	Butanal	-88 e	-72 e	-50 e	-22 e	16.6	74.5	1,5
C_4H_8O	Isobutanal		-56 e	-29 e	8 e	63.8		1
C_4H_8O	2-Butanone	-85 e	-68 e	-46 e	-18.1	21.2	79.2	1
C_4H_8O	Tetrahydrofuran	-94 e	-78 e	-57.3	-29.8	9 e	65.6	1
$C_4H_8O_2$	Butanoic acid			12.9	52.2	101.4	163.3	1,5
$C_4H_8O_2$	2-Methylpropanoic acid	-30.1	-8.2	18.1	50.5	92.9	154.0	5
$C_4H_8O_2$	Propyl formate	-78 e	-62 e	-42 e	-15.1	23.0	80.4	1,5
$C_4H_8O_2$	Isopropyl formate	-80 e	-65 e	-47 e	-22.2	13.2	67.7	5
$C_4H_8O_2$	Ethyl acetate	-83 e	-66 e	-45 e	-18 e	20.4	76.8	1
$C_4H_8O_2$	Methyl propanoate	-80 e	-64 e	-43 e	-15.8	22.2	79.0	1
$C_4H_8O_2$	cis-2-Butene-1,4-diol	17 e	44 e	77 e	117.4	168.5	234.9	5
$C_4H_8O_2$	1,3-Dioxane			-37 e	-3 e	43.4	106.0	5
$C_4H_8O_2$	1,4-Dioxane					39.6	101.0	1
$C_4H_8O_2S$	Sulfolane		49 e	87 e	135 e	198.0	283.5	5
C_4H_8S	Tetrahydrothiophene	-66 e	-47 e	-23 e	9.4	54.1	120.5	1
C_4H_9Br	1-Bromobutane	-68.4	-53.9	-34.1	-5.4	37.6	101.1	1,5
C_4H_9Br	2-Bromobutane	-86 e	-68 e	-46 e	-16 e	26.6	90.7	5
C_4H_9Br	1-Bromo-2-methylpropane	-85 e	-68 e	-46 e	-16 e	26.8	91.1	5
C_4H_9Br	2-Bromo-2-methylpropane					11.7	72.4	1,5
C_4H_9Cl	1-Chlorobutane	-87 e	-71 e	-49 e	-21 e	18.4	78.1	1
C_4H_9Cl	2-Chlorobutane	-96 e	-80 e	-59 e	-31.0	8.5	67.9	1
C_4H_9Cl	1-Chloro-2-methylpropane	-94 e	-78 e	-56.6	-28.7	10.2	68.5	5
C_4H_9Cl	2-Chloro-2-methylpropane					-4.2	50.3	5
$C_4H_9Cl_3Si$	Butyltrichlorosilane					77.2	148.4	5
C_4H_9F	1-Fluorobutane	-114 e	-99 e	-80 e	-55 e	-20.0	32.1	5
C_4H_9F	2-Fluorobutane	-117 e	-103 e	-85 e	-60.7	-26.7	24.7	5
C_4H_9I	1-Iodobutane	-62 e	-43 e	-19 e	14 e	60.5	130.0	5
C_4H_9I	2-Iodobutane	-70 e	-51 e	-27 e	5 e	50 e	119.5	5
C_4H_9I	1-Iodo-2-methylpropane		-47 e	-21.4	12.0	56.8	120.0	5
C_4H_9I	2-Iodo-2-methylpropane	-75.1 s	-58.8 s	-39.5 s	-5.2	41 e	100.0	5
C_4H_9N	Pyrrolidine		-59 e	-38 e	-10 e	28.5	86.2	1
C_4H_9NO	N-Methylpropanamide				81.1	105 e		5
C_4H_9NO	N,N-Dimethylacetamide	-8 e	8 e	28.0	56.4	98.2	165.7	1
C_4H_9NO	2-Butanone oxime		-18 e	7 e	38.9	81.9	142.9	5
C_4H_9NO	Morpholine				21 e	64.5	128.5	1
$C_4H_9NO_3$	Isobutyl nitrate			-18 e	15.1	59.2	123.0	5
C_4H_{10}	Butane*	-134.3	-121.0	-103.9	-81.1	-49.1	-0.8	1,41
C_4H_{10}	Isobutane*		-129.0	-113.0	-90.9	-59.4	-12.0	1,41
$C_4H_{10}O$	1-Butanol	-37 e	-20 e	0 e	28 e	64 e	117.4	1
$C_4H_{10}O$	2-Butanol	-50 e	-34 e	-14 e	12.6	48.2	99.2	1,5
$C_4H_{10}O$	2-Methyl-1-propanol	-39 e	-24 e	-5 e	20.9	56.0	107.6	1,5
$C_4H_{10}O$	2-Methyl-2-propanol					34.4	82.1	1,5
$C_4H_{10}O$	Diethyl ether	-111 e	-96 e	-77 e	-52.6	-17.8	34.1	1
$C_4H_{10}O$	Methyl propyl ether				-40 e	-11.3	38.7	5
$C_4H_{10}O$	Isopropyl methyl ether				-56 e	-21.2	30.4	5
$C_4H_{10}O_2$	1,3-Butanediol	-4 e	23 e	55 e	94 e	142.9	206.1	5
$C_4H_{10}O_2$	1,4-Butanediol		45 e	77 e	116 e	164.7	227.6	5
$C_4H_{10}O_2$	2,3-Butanediol		15 e	43 e	77 e	121.2	180.3	5
$C_4H_{10}O_2$	Ethylene glycol monoethyl ether	-49 e	-29 e	-3 e	30 e	73.6	135.3	1
$C_4H_{10}O_2$	Ethylene glycol dimethyl ether			-44 e	-15 e	25.2	85.2	1
$C_4H_{10}O_2$	Dimethylacetal	-89 e	-74 e	-55 e	-29 e	7.7	64.1	5
$C_4H_{10}O_2$	Diethylperoxide				-39 e	3.6	65.0	5
$C_4H_{10}O_2S$	Bis(2-hydroxyethyl) sulfide			31 e	114.2		282.0	5
$C_4H_{10}O_3$	Diethylene glycol	35 e	58 e	86 e	123 e	173.6	245.2	1
$C_4H_{10}O_4S$	Diethyl sulfate		3 e	36 e	79 e	134 e	208.3	5
$C_4H_{10}S$	1-Butanethiol	-77 e	-59 e	-37 e	-6 e	35.4	98.0	5
$C_4H_{10}S$	2-Butanethiol	-86 e	-69 e	-47 e	-17 e	23.4	84.5	5
$C_4H_{10}S$	2-Methyl-1-propanethiol		-66 e	-44 e	-15 e	26.5	88.1	5
$C_4H_{10}S$	2-Methyl-2-propanethiol					5.8	63.8	5
$C_4H_{10}S$	Diethyl sulfide	-80 e	-62 e	-40 e	-10.8	30.3	91.7	1
$C_4H_{10}S$	Methyl propyl sulfide	-78 e	-61 e	-38 e	-8 e	33.1	95.1	5
$C_4H_{10}S$	Isopropyl methyl sulfide	-85 e	-68 e	-46 e	-17 e	23.4	84.3	5
$C_4H_{10}S_2$	1,4-Butanedithiol	-17 e	5 e	32 e	69.1	119.9	195.1	5
$C_4H_{10}S_2$	Diethyl disulfide	-46 e	-26 e	0 e	35 e	82.4	153.5	5

Mol. Form.	Name	Temperature in °C for the indicated pressure						Ref.
		1 Pa	10 Pa	100 Pa	1 kPa	10 kPa	100 kPa	
$C_4H_{11}N$	Butylamine			-46 e	-18.1	20.0	75.9	5
$C_4H_{11}N$	sec-Butylamine			-55 e	-29.1	7.5	62.3	5
$C_4H_{11}N$	tert-Butylamine			-67 e	-42.4	-8.1	43.7	5
$C_4H_{11}N$	Isobutylamine	-85 e	-70 e	-50 e	-24.5	12.0	67.3	5
$C_4H_{11}N$	Diethylamine			-46 e	-26 e	5 e	55.2	1
$C_4H_{11}NO$	N,N-Dimethylethanolamine	-52 e	-31 e	-6 e	27 e	70.9	133 e	5
$C_4H_{11}NO_2$	Diethanolamine	53 e	77 e	107 e	146 e	197.3	268 e	5
$C_4H_{12}BN$	(Dimethylamino)dimethyl-borane		-81 e	-60.1	-31.9	7.0	64.2	5
$C_4H_{12}Cl_2OSi_2$	1,3-Dichloro-1,1,3,3-tetramethyldisiloxane		-33 e	-9 e	23.8	69.1	136.5	5
$C_4H_{12}O_4Si$	Tetramethyl silicate				14.4	59.3	119.7	5
$C_4H_{12}Si$	Tetramethylsilane			-83 e	-59 e	-25 e	26.7	5
$C_4H_{12}Sn$	Tetramethylstannane			-55.0	-25.6	16.6	77.7	5
$C_4H_{13}N_3$	Diethylenetriamine	-10 e	13 e	43 e	80 e	129.6	198 e	5
C_4NiO_4	Nickel carbonyl					-12	42	4
C_5F_{12}	Perfluoropentane				-54.7	-20.9	28.6	5
C_5FeO_5	Iron pentacarbonyl				0	44	105	4
C_5H_4ClN	2-Chloropyridine			7.4	45.8	97.3	169.9	5
$C_5H_4O_2$	Furfural	-26 e	-8 e	16 e	47 e	92.4	161.4	1
C_5H_5N	Pyridine			-23 e	8 e	51.0	114.9	1
C_5H_6	1,3-Cyclopentadiene			-77 e	-51 e	-14 e	39.8	5
$C_5H_6N_2$	Pentanedinitrile	24.1	52 e	85 e	126 e	178 e	245 e	5
C_5H_6O	2-Methylfuran			-66 e	-35 e	6 e	64.5	1
$C_5H_6O_2$	Furfuryl alcohol	-30 e	-5 e	25 e	62.6	109.3	169.7	5
C_5H_6S	2-Methylthiophene		-58 e	-32 e	2 e	47.9	112.2	1
C_5H_6S	3-Methylthiophene		-53 e	-28 e	6 e	50.6	115.1	1
C_5H_7N	1-Methylpyrrole				8 e	49.9	112.3	5
$C_5H_7NO_2$	Ethyl cyanoacetate	16 e	39 e	67.0	102.1	146.7	205.6	5
C_5H_8	1,2-Pentadiene	-109 e	-93 e	-73 e	-46.1	-9.7	44.5	5
C_5H_8	cis-1,3-Pentadiene	-109 e	-93 e	-73 e	-47.0	-10.5	43.7	1,5
C_5H_8	trans-1,3-Pentadiene			-75 e	-49.0	-13 e	42 e	1
C_5H_8	1,4-Pentadiene	-120 e	-105 e	-86 e	-60.9	-26.2	25.6	5
C_5H_8	2,3-Pentadiene	-106 e	-90 e	-70 e	-42.9	-6.3	47.9	5
C_5H_8	3-Methyl-1,2-butadiene	-111 e	-95 e	-75 e	-49.2	-13.1	40.4	5
C_5H_8	2-Methyl-1,3-butadiene	-115 e	-100 e	-81 e	-55.4	-19.7	33.7	1,5
C_5H_8	1-Pentyne			-75 e	-49.1	-13.5	39.9	5
C_5H_8	2-Pentyne	-100 e	-85 e	-65 e	-37.9	-0.5	55.7	5
C_5H_8	3-Methyl-1-butyne			-82 e	-57.5	-23.1	28.6	5
C_5H_8	Cyclopentene	-109 e	-94 e	-74 e	-48 e	-11.1	43.8	5
C_5H_8	Spiropentane	-110 e	-95 e	-76 e	-51 e	-15 e	38.6	5
C_5H_8O	3-Methyl-3-buten-2-one			-35 e	-5 e	36.0	97.3	5
C_5H_8O	Cyclopropyl methyl ketone		-57 e	-31 e	3 e	49 e	112 e	5
C_5H_8O	Cyclopentanone		-39 e	-14 e	19 e	64 e	130.3	1
C_5H_8O	3,4-Dihydro-2H-pyran				-22 e	22.0	84.9	5
$C_5H_8O_2$	4-Pentenoic acid	0 e	19 e	44 e	77 e	122.0	187.5	5
$C_5H_8O_2$	Vinyl propanoate					31.2	94 e	5
$C_5H_8O_2$	Ethyl acrylate		-55 e	-32.7	-2.8	38.5	99.2	5
$C_5H_8O_2$	Methyl methacrylate			-31 e	-1 e	39.7	100.0	1
$C_5H_8O_2$	2,4-Pentanedione			-5 e	24.7	67.8	137.4	1
$C_5H_8O_2$	Tetrahydro-2H-pyran-2-one		5 e	35.1	74.4	128.3	207.0	5
$C_5H_8O_3$	Methyl acetoacetate				50.1	101.1	171.3	5
$C_5H_8O_4$	Glutaric acid		121 e	153.2	191.9	240.3	302.5	5
$C_5H_8O_4$	Dimethyl malonate	-22 e	1 e	30.0	66.7	114.7	180.2	5
$C_5H_9ClO_2$	Ethyl 2-chloropropanoate			1.4	36.4	82.5	146.0	5
$C_5H_9ClO_2$	Isopropyl chloroacetate			-2 e	35.0	83.3	148.1	5
C_5H_9N	Pentanenitrile	-54 e	-34 e	-8 e	26 e	72.2	140.9	1
C_5H_9N	2,2-Dimethylpropanenitrile					41.1	104.8	5
C_5H_9NO	N-Methyl-2-pyrrolidone	1 e	24 e	53.1	92.3	147.2	229 e	5
C_5H_{10}	1-Pentene	-118.9	-103.4	-84.0	-58.8	-23.3	29.6	1,5
C_5H_{10}	cis-2-Pentene	-113.8	-98.1	-78.4	-52.7	-16.8	36.6	1,5
C_5H_{10}	trans-2-Pentene	-114.5	-98.9	-79.1	-53.3	-17.5	36.0	1,5
C_5H_{10}	2-Methyl-1-butene	-117.7	-102.2	-82.7	-57.2	-21.9	30.8	1,5
C_5H_{10}	3-Methyl-1-butene	-125.0	-110.1	-91.2	-66.7	-32.1	19.7	1,5
C_5H_{10}	2-Methyl-2-butene	-113.4	-97.6	-77.7	-51.6	-15.8	38.2	1,5

Mol. Form.	Name	1 Pa	10 Pa	100 Pa	1 kPa	10 kPa	100 kPa	Ref.
					Temperature in °C for the indicated pressure			
C_5H_{10}	Cyclopentane			-77.0	-45.4	-7.1	48.8	5
C_5H_{10}	Ethylcyclopropane	-118 e	-102 e	-83 e	-57 e	-20 e	35.5	5
C_5H_{10}	cis-1,2-Dimethylcyclo-propane	-118 e	-103 e	-83 e	-57 e	-20 e	36.6	5
C_5H_{10}	trans-1,2-Dimethylcyclo-propane	-122 e	-108 e	-89 e	-63 e	-27 e	27.8	5
$C_5H_{10}Br_2$	1,5-Dibromopentane	1 e	25 e	54 e	93 e	145.6	221.8	5
$C_5H_{10}Cl_2$	1,2-Dichloropentane				30 e	77.4	147.8	5
$C_5H_{10}Cl_2$	1,5-Dichloropentane	-31 e	-10 e	17 e	54 e	104.1	178.9	5
$C_5H_{10}N_2$	3-(Dimethylamino)-propanenitrile				51.1	101.8	171.4	5
$C_5H_{10}O$	Cyclopentanol		-13 e	11.5	42.2	82.5	140.0	5
$C_5H_{10}O$	Allyl ethyl ether			-56 e	-28.7	9.8	67.2	5
$C_5H_{10}O$	Pentanal	-71 e	-53 e	-31 e	-1 e	40.8	102.6	5
$C_5H_{10}O$	2-Pentanone				-1 e	40.3	101.9	1,5
$C_5H_{10}O$	3-Pentanone			-31 e	-1 e	40 e	101.6	1
$C_5H_{10}O$	3-Methyl-2-butanone	-69 e	-54 e	-34 e	-6.9	32.2	94.0	1,5
$C_5H_{10}O$	Tetrahydropyran				-15 e	26.0	88 e	5
$C_5H_{10}O$	2-Methyltetrahydrofuran				-20 e	19.7	79.8	5
$C_5H_{10}O_2$	Pentanoic acid	-7.4	15.3	42.7	76.3	122.1	185.7	5
$C_5H_{10}O_2$	2-Methylbutanoic acid	-10 e	10 e	36 e	69 e	112.8	175.2	5
$C_5H_{10}O_2$	3-Methylbutanoic acid	-15.8	4 e	30.0	64.7	110.6	176.1	5
$C_5H_{10}O_2$	Butyl formate			-29 e	2 e	44.4	105.7	5
$C_5H_{10}O_2$	Isobutyl formate	-69 e	-53 e	-31 e	-3 e	37.4	97.6	5
$C_5H_{10}O_2$	Propyl acetate	-69 e	-51 e	-29 e	0 e	40.9	101.2	1
$C_5H_{10}O_2$	Isopropyl acetate		-61 e	-40 e	-11 e	29.8	88.2	5
$C_5H_{10}O_2$	Ethyl propanoate	-69 e	-52 e	-30 e	-1 e	38.9	98.7	1
$C_5H_{10}O_2$	Methyl butanoate	-68 e	-50 e	-28 e	0.9	41.7	102.3	5
$C_5H_{10}O_2$	Methyl isobutanoate	-83 e	-65 e	-41 e	-11 e	31 e	92.1	5
$C_5H_{10}O_2$	Tetrahydrofurfuryl alcohol	-40 e	-16 e	15 e	55 e	106 e	176.8	5
$C_5H_{10}O_3$	Diethyl carbonate		-42 e	-17 e	17 e	61.6	125.9	5
$C_5H_{10}O_3$	Ethylene glycol monomethyl ether acetate	-47 e	-26 e	0 e	34 e	79.4	144.1	5
$C_5H_{10}S$	Thiacyclohexane				24 e	71.1	141.2	5
$C_5H_{10}S$	Cyclopentanethiol				18 e	64 e	131.7	5
$C_5H_{11}Br$	1-Bromopentane	-60 e	-41 e	-16 e	16 e	61.5	129.1	5
$C_5H_{11}Br$	2-Bromopentane	-69 e	-51 e	-27 e	5 e	49.7	116.9	5
$C_5H_{11}Br$	3-Bromopentane	-68 e	-50 e	-26 e	6 e	50.8	118.1	5
$C_5H_{11}Br$	1-Bromo-3-methylbutane	-67 e	-49 e	-25 e	8 e	52.4	119.9	5
$C_5H_{11}Cl$	1-Chloropentane	-73 e	-55 e	-32 e	-1 e	42.5	107.9	5
$C_5H_{11}Cl$	2-Chloropentane	-80 e	-62 e	-39 e	-9 e	33.2	96.1	5
$C_5H_{11}Cl$	3-Chloropentane	-77 e	-60 e	-37 e	-7 e	34.9	97.3	5
$C_5H_{11}Cl$	2-Chloro-2-methylbutane			-52 e	-21 e	21.8	85.2	5
$C_5H_{11}Cl$	1-Chloro-2,2-dimethyl-propane				-17 e	23.5	83.9	5
$C_5H_{11}F$	1-Fluoropentane	-97 e	-80 e	-60 e	-32 e	5.7	62.4	5
$C_5H_{11}I$	1-Iodopentane	-47 e	-27 e	-1 e	34 e	83.0	156.5	5
$C_5H_{11}I$	1-Iodo-3-methylbutane		-34 e	-6.6	28.8	77.3	147.8	5
$C_5H_{11}N$	Cyclopentylamine	-66 e	-48 e	-26 e	4 e	45.8	108 e	5
$C_5H_{11}N$	Piperidine				2 e	43.3	105.8	5
$C_5H_{11}N$	N-Methylpyrrolidine				-23 e	18.5	78 e	5
$C_5H_{11}NO_3$	3-Methylbutyl nitrate		-26 e	1.0	35.5	81.7	147.0	5
C_5H_{12}	Pentane**	-115.5	-99.8	-80.0	-54.0	-18.1	35.7	16
C_5H_{12}	Isopentane	-119 e	-105 e	-86 e	-61 e	-26 e	27.5	1
C_5H_{12}	Neopentane*		-107.5 s	-90.8 s	-68.8 s	-38.5 s	9.2	1,5
$C_5H_{12}N_2O$	Tetramethylurea			20.7	58.0	106.7	179.5	5
$C_5H_{12}O$	1-Pentanol	-27 e	-10 e	12 e	41 e	79.8	137.4	5
$C_5H_{12}O$	2-Pentanol	-35 e	-19 e	1 e	28.0	64.9	118.7	1
$C_5H_{12}O$	3-Pentanol	-41 e	-25 e	-4 e	24 e	61.1	114.9	5
$C_5H_{12}O$	2-Methyl-1-butanol	-27 e	-11 e	9 e	36.2	73.4	128.3	1
$C_5H_{12}O$	3-Methyl-1-butanol	-22 e	-7 e	13 e	39.1	75.7	130.1	5
$C_5H_{12}O$	2-Methyl-2-butanol			-5 e	17.7	50.6	101.7	1,5
$C_5H_{12}O$	3-Methyl-2-butanol			-3 e	22.7	58.2	111.1	5
$C_5H_{12}O$	2,2-Dimethyl-1-propanol					59.2	112.7	5
$C_5H_{12}O$	Butyl methyl ether			-54 e	-27 e	12 e	69.8	1
$C_5H_{12}O$	Methyl tert-butyl ether			-66 e	-39 e	-2 e	54.8	1
$C_5H_{12}O$	Ethyl propyl ether	-92 e	-77 e	-57 e	-30.5	6.7	63.4	1,5

VAPOR PRESSURE (continued)

Mol. Form.	Name	Temperature in °C for the indicated pressure						Ref.
		1 Pa	10 Pa	100 Pa	1 kPa	10 kPa	100 kPa	
$C_5H_{12}O_2$	1,5-Pentanediol	25 e	52 e	85 e	125 e	175.1	238.9	5
$C_5H_{12}O_2$	Ethylene glycol monopropyl ether				40 e	85.6	149.3	5
$C_5H_{12}O_2$	Diethoxymethane		-65 e	-43 e	-14 e	27.3	87.7	5
$C_5H_{12}O_3$	Diethylene glycol monomethyl ether		12 e	40 e	76 e	124.2	193.7	1
$C_5H_{12}S$	1-Pentanethiol	-60 e	-41 e	-17 e	15 e	60 e	126.2	1
$C_5H_{12}S$	2-Pentanethiol	-70 e	-52 e	-28 e	3 e	46.6	111.9	5
$C_5H_{12}S$	3-Pentanethiol	-70 e	-51 e	-28 e	4 e	47.7	113.4	5
$C_5H_{12}S$	2-Methyl-1-butanethiol				8.0	52.3	118.5	5
$C_5H_{12}S$	3-Methyl-1-butanethiol				7.8	51.9	117.9	5
$C_5H_{12}S$	2-Methyl-2-butanethiol				-8.0	34.6	98.7	5
$C_5H_{12}S$	Butyl methyl sulfide		-43 e	-19 e	13 e	57 e	123.0	1
$C_5H_{12}S$	tert-Butyl methyl sulfide				-7.8	34.7	98.4	5
$C_5H_{12}S$	Ethyl propyl sulfide	-64 e	-46 e	-23 e	9 e	52.7	118.0	5
$C_5H_{12}S$	Ethyl isopropyl sulfide	-72 e	-54 e	-31 e	0 e	42.7	106.9	5
$C_5H_{13}N$	Pentylamine		-52 e	-29 e	1 e	42.8	104.0	5
C_6BrF_5	Bromopentafluorobenzene		-10 e	23 e	68 e	136.0		5
C_6ClF_5	Chloropentafluorobenzene		-44 e	-21 e	11 e	53.8	117.6	1
$C_6Cl_3F_3$	1,3,5-Trichloro-2,4,6-trifluorobenzene	-19 e	4 e	32 e	70 e	121.7	197.9	1
C_6F_6	Hexafluorobenzene		-56.9 s	-36 s	-11.5 s	22.6	79.9	1,5
C_6F_{12}	Perfluorocyclohexane				-46.2 s	-7.6 s	48.9 s	5
C_6F_{14}	Perfluorohexane		-75 e	-57 e	-32 e	2.8	56.8	5
C_6F_{14}	Perfluoro-2-methylpentane				-33 e	2.9	57.1	5
C_6F_{14}	Perfluoro-3-methylpentane	-95 e	-80 e	-60 e	-34 e	2.8	57.9	5
C_6F_{14}	Perfluoro-2,3-dimethylbutane					4.3	59.3	5
C_6HF_5	Pentafluorobenzene			-41 e	-13 e	27 e	85.3	5
C_6HF_5O	Pentafluorophenol				39 e	82 e	145.2	5
$C_6H_2F_4$	1,2,3,4-Tetrafluorobenzene			-36 e	-7 e	33.8	94.0	1
$C_6H_2F_4$	1,2,3,5-Tetrafluorobenzene			-43 e	-14 e	25.5	84.1	1
$C_6H_2F_4$	1,2,4,5-Tetrafluorobenzene					30.7	89.9	1
$C_6H_3Cl_3O$	2,4,6-Trichlorophenol			71.8	114.0	169.5	245.7	5
$C_6H_3F_3$	1,3,5-Trifluorobenzene					18.2	75.0	5
$C_6H_4Br_2$	m-Dibromobenzene	-7 e	16 e	44 e	83 e	137.0	218.2	5
$C_6H_4ClNO_2$	1-Chloro-4-nitrobenzene	15.4 s	35.8 s		97 e	156.0	238 e	5
$C_6H_4Cl_2$	o-Dichlorobenzene		-13 e	16.3	53.9	104.6	180.0	1,5
$C_6H_4Cl_2$	m-Dichlorobenzene		-22 e	8.0	46.7	97.8	172.5	1,5
$C_6H_4Cl_2$	p-Dichlorobenzene	-45.5 s	-21.8 s	8 s	46.7 s	99.0	173.6	1,5
$C_6H_4O_2$	p-Benzoquinone	-4.1 s	17.8 s	43.5 s	74.3 s	111.6 s		5
$C_6H_5AsCl_2$	Dichlorophenylarsine	6.9	35.2	70 e	113 e	170 e	245 e	5
C_6H_5Br	Bromobenzene		-25 e	1 e	34.9	83.1	155.4	1
C_6H_5Cl	Chlorobenzene		-43 e	-17 e	16.8	62.9	131.3	1,5
C_6H_5ClO	o-Chlorophenol				45.8	97.9	173.9	5
C_6H_5ClO	m-Chlorophenol			39.7	80.2	135.1	213.4	5
C_6H_5ClO	p-Chlorophenol			45.0	86.5	142.0	219.9	5
$C_6H_5Cl_3Si$	Trichlorophenylsilane			33 e	70.2	122.6	201 e	5
C_6H_5F	Fluorobenzene				-16.9	24.2	84.4	1
C_6H_5I	Iodobenzene	-30 e	-7 e	20.9	58.5	110.6	187.8	1
$C_6H_5NO_2$	Nitrobenzene		10 e	40 e	78 e	132 e	210.3	1
$C_6H_5NO_3$	p-Nitrophenol	72.6 s	97.4 s					5
C_6H_6	1,5-Hexadien-3-yne	-82 e	-66 e	-44.3	-16.0	23.7	83.6	5
C_6H_6	Benzene**			-40 s	-15.1 s	20.0	79.7	1,5
C_6H_6ClN	o-Chloroaniline		10 e	39.0	75.2	131.4	208.3	5
C_6H_6ClN	m-Chloroaniline	-5 e	19.7	49.4	94.2	162 e	1069 e	5
$C_6H_6N_2O_2$	p-Nitroaniline	87.8 s			192.0	252.6	331.2	5
C_6H_6O	Phenol	-9.7 s	9.6 s	34.1 s	68.9	113.7	181.4	1,5
$C_6H_6O_3$	1,2,3-Benzenetriol				162.0	222.8	308.3	5
C_6H_6S	Benzenethiol		-15 e	12 e	47 e	96.0	168.6	5
C_6H_7N	Aniline		-2.5	26.7	63.5	112.5	183.5	1,5
C_6H_7N	2-Methylpyridine	-56.5	-37.8	-13.9	18.3	62.9	129.0	1,5
C_6H_7N	3-Methylpyridine			-5 e	28.8	75.2	143.7	1
C_6H_7N	4-Methylpyridine	-58.2 s	-43.1 s	-3.9 s	29.6	76.1	144.9	1,5
C_6H_8	cis-1,3,5-Hexatriene					21 e	78 e	5
C_6H_8	1,3-Cyclohexadiene	-88 e	-71 e	-50 e	-21 e	19 e	79.9	5
C_6H_8	1,4-Cyclohexadiene				-15 e	27.3	85.0	5
$C_6H_8N_2$	Adiponitrile	30 e	61 e	100 e	148.6	211.8	297 e	5

Mol. Form.	Name	Temperature in °C for the indicated pressure						Ref.
		1 Pa	10 Pa	100 Pa	1 kPa	10 kPa	100 kPa	
$C_6H_8N_2$	m-Phenylenediamine			94.5	140.2	200.8	285.0	5
$C_6H_8N_2$	Phenylhydrazine		38 e	69 e	109 e	163.9	242.5	5
$C_6H_8O_4$	Dimethyl maleate		5 e	36 e	76 e	127.3	197 e	5
C_6H_8S	2,5-Dimethylthiophene		-43 e	-16 e	20 e	67.5	134.8	5
C_6H_{10}	trans-1,3-Hexadiene	-86 e	-70 e	-51 e	-24 e	14 e	72 e	5
C_6H_{10}	trans-1,4-Hexadiene	-98 e	-81 e	-60 e	-33 e	7 e	65 e	5
C_6H_{10}	1,5-Hexadiene	-99 e	-84 e	-64 e	-37 e	0.9	59.2	5
C_6H_{10}	cis,cis-2,4-Hexadiene					18 e	79.6	5
C_6H_{10}	trans,cis-2,4-Hexadiene	-89 e	-73 e	-52 e	-23 e	18 e	79.6	5
C_6H_{10}	trans,trans-2,4-Hexadiene				-23 e	18 e	79.6	5
C_6H_{10}	trans-2-Methyl-1,3-pentadiene	-92 e	-75 e	-54 e	-26 e	14 e	75.6	5
C_6H_{10}	2,3-Dimethyl-1,3-butadiene			-59 e	-30 e	9.7	68.1	5
C_6H_{10}	1-Hexyne	-91 e	-75 e	-54 e	-26 e	12.8	71.0	5
C_6H_{10}	2-Hexyne	-84 e	-67 e	-46 e	-17 e	23.6	84.1	5
C_6H_{10}	3-Hexyne	-86 e	-69 e	-48 e	-19.1	21.0	81.0	1,5
C_6H_{10}	4-Methyl-1-pentyne	-97 e	-81 e	-61 e	-34 e	4.1	60.7	5
C_6H_{10}	4-Methyl-2-pentyne	-91 e	-74 e	-54 e	-26 e	13.8	72.7	5
C_6H_{10}	Cyclohexene	-87 e	-70 e	-49 e	-19 e	21 e	82.6	1
$C_6H_{10}Cl_2$	1,1-Dichlorocyclohexane	-39 e	-19 e	8 e	43 e	93.5	170.5	5
$C_6H_{10}Cl_2$	cis-1,2-Dichlorocyclohexane			27 e	69 e	125.7	206.2	5
$C_6H_{10}O$	4-Methyl-4-penten-2-one	-59 e	-41 e	-17 e	14 e	57.0	121.0	5
$C_6H_{10}O$	Cyclohexanone		-25 e	1 e	36 e	84 e	155.2	1
$C_6H_{10}O$	Mesityl oxide	-56 e	-37 e	-13 e	19 e	63.5	129.3	5
$C_6H_{10}O_2$	Vinyl butanoate					53 e	114.5	5
$C_6H_{10}O_2$	Ethyl methacrylate				8 e	53.2	116.8	5
$C_6H_{10}O_2$	Allyl glycidyl ether				40.1	85.7	152.8	5
$C_6H_{10}O_3$	Ethyl acetoacetate	-25 e	-3 e	25.7	62.3	111.3	180.2	5
$C_6H_{10}O_3$	Propanoic anhydride	-32 e	-15 e	6 e	36 e	77.6	142.9	5
$C_6H_{10}O_4$	Diethyl oxalate	-5 e	18 e	44.9	79.4	124.3	185.2	5
$C_6H_{10}O_4$	Dimethyl succinate			30 e	70.4	123.3	195.4	5
$C_6H_{10}O_4$	Ethylene glycol diacetate	-17 e	6 e	35.0	71.9	121.1	190.0	5
$C_6H_{10}S$	Diallylsulfide	-58 e	-38 e	-12.4	21.7	68.8	138.1	5
$C_6H_{11}Cl$	Chlorocyclohexane		-35 e	-9 e	25 e	71.6	142.1	5
$C_6H_{11}N$	Hexanenitrile	-40 e	-19 e	8 e	43 e	91.5	163.2	1,5
$C_6H_{11}N$	4-Methylpentanenitrile		-50 e	-20 e	20 e	75.2	155.2	5
$C_6H_{11}NO$	Caprolactam	36.8 s	58.9 s	86.6 s			270	5
C_6H_{12}	1-Hexene	-99.8	-82.8	-61.4	-33.7	5.2	63.1	1,5
C_6H_{12}	cis-2-Hexene	-97 e	-80 e	-58 e	-30 e	9.9	68.5	5
C_6H_{12}	trans-2-Hexene	-94 e	-78 e	-57 e	-30 e	9.3	67.5	5
C_6H_{12}	cis-3-Hexene	-96 e	-79 e	-59 e	-30.8	7.9	66.0	5
C_6H_{12}	trans-3-Hexene	-95 e	-79 e	-58 e	-30.0	8.8	66.7	5
C_6H_{12}	2-Methyl-1-pentene	-98 e	-82 e	-62 e	-34.2	4.1	61.7	5
C_6H_{12}	3-Methyl-1-pentene	-104 e	-88 e	-68 e	-41.5	-3.6	53.8	5
C_6H_{12}	4-Methyl-1-pentene	-105 e	-89 e	-69 e	-41.6	-3.6	53.5	5
C_6H_{12}	2-Methyl-2-pentene	-95 e	-78 e	-58 e	-30 e	9.0	66.9	5
C_6H_{12}	3-Methyl-cis-2-pentene	-95 e	-79 e	-58 e	-30 e	8.9	67.3	5
C_6H_{12}	3-Methyl-trans-2-pentene	-93 e	-77 e	-55 e	-27.4	11.7	70.0	5
C_6H_{12}	4-Methyl-cis-2-pentene	-102 e	-86 e	-66 e	-38.7	-0.9	56.0	5
C_6H_{12}	4-Methyl-trans-2-pentene	-100 e	-84 e	-64 e	-36.8	1.2	58.2	5
C_6H_{12}	2-Ethyl-1-butene	-98 e	-81 e	-60 e	-32 e	6.6	64.3	5
C_6H_{12}	2,3-Dimethyl-1-butene	-103 e	-87 e	-67 e	-39.9	-1.9	55.2	5
C_6H_{12}	3,3-Dimethyl-1-butene	-110 e	-95 e	-76 e	-50.8	-14.5	40.8	5
C_6H_{12}	2,3-Dimethyl-2-butene		-75 e	-54 e	-25 e	14 e	72.9	1
C_6H_{12}	Cyclohexane	-85.6 s	-68.9 s	-47.6 s	-19.8 s	19.3	80.4	1,5
C_6H_{12}	Methylcyclopentane	-97 e	-80 e	-58 e	-28.8	11.6	71.4	1,5
C_6H_{12}	Ethylcyclobutane	-99 e	-82 e	-61 e	-32 e	9 e	70.2	5
C_6H_{12}	Isopropylcyclopropane	-104 e	-88 e	-68 e	-40 e	-1 e	57.9	5
C_6H_{12}	1-Ethyl-1-methylcyclopropane	-105 e	-89 e	-69 e	-41 e	-3 e	56.3	5
C_6H_{12}	1,1,2-Trimethylcyclopropane	-109 e	-94 e	-73 e	-46 e	-7 e	52.0	5
$C_6H_{12}Cl_2$	1,2-Dichlorohexane				49 e	98.1	171.7	5
$C_6H_{12}Cl_2O$	2,2'-Dichlorodiisopropyl ether		-1 e	27.3	63.4	112.3	182.1	5
$C_6H_{12}O$	Butyl vinyl ether	-87 e	-67 e	-42 e	-9.3	33.6	93.2	5
$C_6H_{12}O$	Isobutyl vinyl ether	-87 e	-68 e	-44 e	-13 e	26.5	80.7	5
$C_6H_{12}O$	Hexanal	-56 e	-37 e	-13 e	19 e	62.6	127.8	5
$C_6H_{12}O$	2-Hexanone	-43 e	-21 e	4.2	34.5	61.9	127.2	1,5
$C_6H_{12}O$	3-Hexanone		-40 e	-16 e	15 e	58.5	123.1	1
$C_6H_{12}O$	3-Methyl-2-pentanone				8.5	52.7	117.0	5

Mol. Form.	Name	Temperature in °C for the indicated pressure						Ref.
		1 Pa	10 Pa	100 Pa	1 kPa	10 kPa	100 kPa	
$C_6H_{12}O$	4-Methyl-2-pentanone	-61 e	-43 e	-21 e	9 e	51.5	116.1	5
$C_6H_{12}O$	2-Methyl-3-pentanone					50.2	113.0	5
$C_6H_{12}O$	3,3-Dimethyl-2-butanone			-30 e	0 e	42.5	105.7	1
$C_6H_{12}O$	Cyclohexanol			34 e	61 e	99.2	160.7	1
$C_6H_{12}O_2$	Hexanoic acid		33 e	59 e	93 e	139.3	204.5	1
$C_6H_{12}O_2$	4-Methylpentanoic acid	36 e	49 e	67.1	92.9	133.6	206.8	5
$C_6H_{12}O_2$	Diethylacetic acid	-9 e	16 e	46 e	83 e	130.7	192.5	5
$C_6H_{12}O_2$	Isopentyl formate	-60 e	-41 e	-17 e	15 e	59.1	124 e	5
$C_6H_{12}O_2$	Butyl acetate	-63 e	-43 e	-19 e	14 e	61.0	125.6	1,5
$C_6H_{12}O_2$	Isobutyl acetate	-63 e	-45 e	-21 e	10 e	53.4	116 e	5
$C_6H_{12}O_2$	Propyl propanoate	-62 e	-42 e	-18 e	14 e	58.3	122.0	5
$C_6H_{12}O_2$	Ethyl butanoate	-49 e	-34 e	-14 e	14.3	55.2	121.1	5
$C_6H_{12}O_2$	Ethyl 2-methylpropanoate	-65 e	-47 e	-24.6	5.4	47.3	109.8	5
$C_6H_{12}O_2$	Methyl pentanoate				19.2	63.7	127.4	5
$C_6H_{12}O_2$	Methyl isopentanoate					53.3	116.3	5
$C_6H_{12}O_2$	Diacetone alcohol	-41 e	-17 e	13 e	50.1	98.5	164 e	5
$C_6H_{12}O_3$	Ethylene glycol monoethyl ether acetate	-25 e	-8 e	14 e	44.6	88.0	155.6	5
$C_6H_{12}O_3$	Paraldehyde				17 e	62.2	124 e	5
$C_6H_{12}S$	Cyclohexanethiol					84.8	158.3	5
$C_6H_{12}S$	cis-Tetrahydro-2,5-dimethylthiophene	-53 e	-34 e	-8 e	25 e	72.0	142.1	5
$C_6H_{12}S$	Tetrahydro-3-methyl-2H-thiopyran	-48 e	-27 e	0 e	35 e	84.1	157.5	5
$C_6H_{13}Br$	1-Bromohexane	-45 e	-25 e	2 e	36 e	83.7	154.8	5
$C_6H_{13}Cl$	1-Chlorohexane	-55 e	-36 e	-11 e	21 e	66.7	134.6	5
$C_6H_{13}F$	1-Fluorohexane	-80 e	-62 e	-40 e	-11 e	30.4	91.1	5
$C_6H_{13}I$	1-Iodohexane	-33 e	-11 e	16 e	53 e	104.0	180.8	5
$C_6H_{13}N$	Cyclohexylamine			-9 e	22 e	66.6	133.5	1
C_6H_{14}	Hexane	-96.4 s	-79.2	-57.6	-29.3	9.8	68.3	16
C_6H_{14}	2-Methylpentane	-100 e	-84 e	-64 e	-36 e	2 e	59.9	1
C_6H_{14}	3-Methylpentane	-99 e	-83 e	-62 e	-34.3	4.6	62.9	1
C_6H_{14}	2,2-Dimethylbutane		-90 e	-71.5	-45.5	-7.7	49.4	1
C_6H_{14}	2,3-Dimethylbutane	-103 e	-87 e	-66 e	-39.0	-0.4	57.6	1
$C_6H_{14}O$	1-Hexanol		5 e	28 e	56.8	97.3	157.1	1
$C_6H_{14}O$	2-Hexanol	-28 e	-10 e	12 e	41.4	81.5	139.6	1
$C_6H_{14}O$	3-Hexanol	-43 e	-23 e	1 e	33 e	75.4	135.1	1
$C_6H_{14}O$	2-Methyl-1-pentanol			14 e	45.9	88.3	147.6	5
$C_6H_{14}O$	4-Methyl-1-pentanol			24 e	53 e	92.4	151.4	5
$C_6H_{14}O$	2-Methyl-2-pentanol	-29 e	-15 e	3 e	27.1	63.0	120.9	5
$C_6H_{14}O$	3-Methyl-2-pentanol				36.5	76.1	133.8	5
$C_6H_{14}O$	4-Methyl-2-pentanol	-43 e	-24 e	0 e	30 e	71.9	131.3	5
$C_6H_{14}O$	2-Methyl-3-pentanol				29.8	68.8	126.0	5
$C_6H_{14}O$	3-Methyl-3-pentanol		-23 e	-4 e	22.9	61.1	121.1	5
$C_6H_{14}O$	2-Ethyl-1-butanol		-5 e	17 e	46 e	85.7	146.1	5
$C_6H_{14}O$	3,3-Dimethyl-1-butanol	-37 e	-16 e	9 e	42 e	84.3	142.5	5
$C_6H_{14}O$	2,3-Dimethyl-2-butanol			-5 e	23 e	61.3	118.2	5
$C_6H_{14}O$	Dipropyl ether	-80 e	-63 e	-41 e	-12 e	28.8	89.7	1
$C_6H_{14}O$	Diisopropyl ether		-76 e	-55 e	-28 e	11 e	68.1	1
$C_6H_{14}O$	Butyl ethyl ether	-78 e	-61 e	-39 e	-10 e	31.0	91.9	1
$C_6H_{14}O$	tert-Butyl ethyl ether	-90 e	-74 e	-53 e	-24.6	14.4	72.6	5
$C_6H_{14}O_2$	2-Methyl-2,4-pentanediol	-8 e	17 e	48 e	86 e	134.4	197.5	5
$C_6H_{14}O_2$	Ethylene glycol monobutyl ether	-31 e	-8 e	20 e	55 e	103.2	170.2	5
$C_6H_{14}O_2$	1,1-Diethoxyethane	-68 e	-49 e	-26 e	3.7	44.2	101.9	5
$C_6H_{14}O_2$	Ethylene glycol diethyl ether		-59 e	-35.3	-2.8	44.4	118.8	5
$C_6H_{14}O_3$	1,2,6-Hexanetriol	92 e	114.8	146.0	191 e			5
$C_6H_{14}O_3$	Dipropylene glycol				110 e	162.6	231.4	5
$C_6H_{14}O_3$	Diethylene glycol monoethyl ether			40 e	80.3	132.4	201.4	5
$C_6H_{14}O_3$	Diethylene glycol dimethyl ether	-42 e	-20 e	8.3	44.3	92.3	159.4	5
$C_6H_{14}O_3$	Trimethylolpropane	73 e	98 e	128 e	167.8	220.5	295 e	5
$C_6H_{14}O_4$	Triethylene glycol	44 e	74 e	109.0	152.6	207.2	277.9	5
$C_6H_{14}S$	1-Hexanethiol	-45 e	-25 e	1 e	35 e	81.7	152.2	5
$C_6H_{14}S$	2-Hexanethiol	-50 e	-32 e	-8 e	25 e	69.9	138.4	5
$C_6H_{14}S$	Dipropyl sulfide	-50 e	-30 e	-6 e	28 e	73.6	142.4	5

VAPOR PRESSURE (continued)

Mol. Form.	Name	1 Pa	10 Pa	100 Pa	1 kPa	10 kPa	100 kPa	Ref.
$C_6H_{14}S$	Diisopropyl sulfide	-65 e	-47 e	-23 e	9 e	53.1	119.6	5
$C_6H_{14}S$	Isopropyl propyl sulfide				18.5	63.8	131.6	5
$C_6H_{14}S$	Butyl ethyl sulfide	-49 e	-30 e	-5 e	29 e	74.8	143.8	5
$C_6H_{15}N$	Hexylamine			-10 e	22 e	66.0	130.6	5
$C_6H_{15}N$	Butylethylamine				6.1	47.7	107.0	5
$C_6H_{15}N$	Dipropylamine		-48 e	-25 e	6 e	47.5	108.8	5
$C_6H_{15}N$	Diisopropylamine			-47 e	-17.5	23.5	84.0	5
$C_6H_{15}N$	Triethylamine	-58 e	-45 e	-29 e	-5 e	29.9	88.5	1
$C_6H_{15}NO$	2-Diethylaminoethanol					97 e	160.6	5
$C_6H_{15}NO_3$	Triethanolamine	75 e	108 e	148 e	196 e	256.7	334 e	5
$C_6H_{15}O_4P$	Triethyl phosphate			34	76	132	211	4
$C_6H_{16}N_2$	Hexamethylenediamine				76.0	128.2	199.0	5
$C_6H_{16}O_2Si$	Diethoxydimethylsilane	-62 e	-44 e	-21.2	9.1	51.0	113.0	5
$C_6H_{18}Cl_2O_2Si_3$	1,5-Dichloro-1,1,3,3,5,5-hexamethyltrisiloxane	-29 e	-7 e	22.2	59.7	110.5	183.4	5
$C_6H_{18}OSi_2$	Hexamethyldisiloxane		-56 e	-34 e	-5 e	37.1	100.1	5
C_6MoO_6	Molybdenum hexacarbonyl		17.4 s	42.8 s	73.1 s	109.9 s	155.4 s	5
C_7F_{14}	Perfluoromethylcyclohexane				-21 e	18 e	75.9	1
C_7F_{16}	Perfluoroheptane		-62 e	-41 e	-14 e	24.7	82.1	1
C_7HF_{15}	1H-Pentadecafluoroheptane				-7 e	35.9	96.0	5
$C_7H_3ClF_3NO_2$	1-Chloro-2-nitro-4-(trifluoromethyl)benzene	3 e	26 e	55 e	92.8	145.2	222.0	5
$C_7H_3F_5$	2,3,4,5,6-Pentafluorotoluene			-20 e	11 e	53.6	117.0	5
$C_7H_4ClF_3$	1-Chloro-2-(trifluoromethyl)benzene			1 e	34.5	81.8	151.8	5
$C_7H_4ClF_3$	1-Chloro-3-(trifluoromethyl)benzene	-53 e	-34 e	-9 e	24.2	69.8	137.2	5
$C_7H_4ClF_3$	1-Chloro-4-(trifluoromethyl)benzene			-9 e	24.2	70.4	138.1	5
$C_7H_4Cl_2O$	o-Chlorobenzoyl chloride				93 e	149 e	237.0	5
$C_7H_4Cl_2O$	m-Chlorobenzoyl chloride				87.8	147 e	225.0	5
$C_7H_4F_3NO_2$	1-Nitro-3-(trifluoromethyl)benzene		11 e	39 e	76.2	127.3	202.2	5
$C_7H_4F_4$	1-Fluoro-4-(trifluoromethyl)benzene			-38 e·	-6 e	38.6	102.3	5
C_7H_5BrO	Benzoyl bromide	-15 e	11 e	42.6	83.9	139.5	218.0	5
C_7H_5ClO	Benzoyl chloride			27.5	67.0	120.4	196.7	5
$C_7H_5Cl_3$	(Trichloromethyl)benzene		9 e	40.6	81.5	136.2	213.0	5
$C_7H_5F_3$	(Trifluoromethyl)benzene				-3 e	39 e	101.6	5
C_7H_5N	Benzonitrile		-6 e	23.9	63.1	115.7	190.0	5
C_7H_5NS	Phenyl isothiocyanate				79.4	105 e	117 e	5
$C_7H_6Cl_2$	2,4-Dichlorotoluene		6 e	33 e	68.3	119.5	199.1	5
$C_7H_6Cl_2$	3,4-Dichlorotoluene	-13 e	9 e	38 e	76 e	129.3	208.4	5
$C_7H_6Cl_2$	(Dichloromethyl)benzene			31	72	130	213	4
C_7H_6O	Benzaldehyde		-9 e	19 e	54.6	104.6	178.3	1
$C_7H_6O_2$	Salicylaldehyde		-1 e	29 e	68 e	120.7	196.2	5
C_7H_7Br	o-Bromotoluene		-10 e	17 e	54 e	104.8	181.1	5
C_7H_7Br	m-Bromotoluene	-34 e	-11 e	19.4	58.1	109.9	183.1	5
C_7H_7Br	p-Bromotoluene				57 e	107.8	183.8	5
C_7H_7Br	(Bromomethyl)benzene			25.4	66.8	121.7	198.3	5
C_7H_7Cl	o-Chlorotoluene		-24 e	3 e	38 e	86.3	158.7	1,5
C_7H_7Cl	m-Chlorotoluene	-41 e	-21 e	6 e	41 e	89 e	161.8	5
C_7H_7Cl	p-Chlorotoluene				40 e	88.9	161.5	1,5
C_7H_7Cl	(Chloromethyl)benzene	-34 e	-11 e	17.7	55.4	106.3	178.9	5
C_7H_7ClO	1-Chloro-2-methoxy-benzene	-22 e	2 e	33 e	72 e	125.2	201 e	5
C_7H_7F	o-Fluorotoluene		-50 e	-26 e	5 e	49.0	113.9	5
C_7H_7F	m-Fluorotoluene	-67 e	-48 e	-25 e	7 e	51.0	116.1	5
C_7H_7F	p-Fluorotoluene		-48 e	-24 e	7 e	51 e	116.2	5
$C_7H_7NO_2$	o-Nitrotoluene	23 e	40 e	62 e	94 e	141.9	221.9	5
$C_7H_7NO_2$	m-Nitrotoluene			45 e	89.7	148.7	231.3	5
$C_7H_7NO_3$	2-Nitroanisole	15 e	45 e	82 e	129 e	189.4	271.8	5
C_7H_8	Toluene	-78.1	-57.1	-31.3	1.5	45.2	110.1	5
C_7H_8	Bicyclo[2.2.1]hepta-2,5-diene				-15 e	27.4	91 e	5
$C_7H_8Cl_2Si$	Dichloromethylphenylsilane			32.4	71.8	126.0	205.0	5
C_7H_8O	o-Cresol	-6.4 s	12.8 s	40.2	72.3	120.3	190.5	1,5
C_7H_8O	m-Cresol	20.8	33.6	52.4	82.6	130.6	201.8	1,5

Mol. Form.	Name	Temperature in °C for the indicated pressure						Ref.
		1 Pa	10 Pa	100 Pa	1 kPa	10 kPa	100 kPa	
C_7H_8O	p-Cresol	-0.2 s	20.7 s	52.7	83.1	130.7	201.5	1,5
C_7H_8O	Benzyl alcohol	8 e	28 e	54 e	88 e	134.7	204.9	1
C_7H_8O	Anisole		-21 e	4 e	38 e	84 e	153.2	1,5
C_7H_8S	3-Methylbenzenethiol		0 e	29 e	66 e	117.9	194.6	5
C_7H_9N	Benzylamine			25.6	62.6	112.7	183.9	5
C_7H_9N	o-Methylaniline	1.0	18.8	42.6	76.1	125.6	199.9	1,5
C_7H_9N	m-Methylaniline	3.8	22.0	46.2	80.1	128.8	202.9	1,5
C_7H_9N	p-Methylaniline			77.1		126.2	199.9	5
C_7H_9N	N-Methylaniline	-16 e	6 e	34 e	70.3	121.1	195.8	1
C_7H_9N	2-Ethylpyridine	-46 e	-26 e	-1 e	33 e	79.3	149.0	5
C_7H_9N	3-Ethylpyridine	-38 e	-17 e	9 e	44 e	92.7	166.5	5
C_7H_9N	4-Ethylpyridine	-35 e	-15 e	11 e	46 e	94.4	168.6	5
C_7H_9N	2,3-Dimethylpyridine				42 e	89.9	160.6	5
C_7H_9N	2,4-Dimethylpyridine		-25 e	3.7	40.0	87.5	157.9	1,5
C_7H_9N	2,5-Dimethylpyridine			4 e	39 e	86.2	156.6	1
C_7H_9N	2,6-Dimethylpyridine			-3 e	29.9	75.8	143.6	1
C_7H_9N	3,4-Dimethylpyridine		-9 e	19 e	55 e	104.8	178.6	5
C_7H_9N	3,5-Dimethylpyridine			11 e	48 e	98 e	171.5	1
$C_7H_{10}N_2$	Toluene-2,4-diamine			100.4	145.3	202.9	279.5	5
C_7H_{12}	1-Heptyne	-75 e	-57 e	-35 e	-5 e	37.1	99.5	5
C_7H_{12}	2-Heptyne		-51 e	-27 e	4 e	46.9	111.5	5
C_7H_{12}	3-Heptyne	-71 e	-53 e	-31 e	0 e	42.7	106.4	5
C_7H_{12}	5-Methyl-1-hexyne	-80 e	-62 e	-40 e	-11 e	30.1	91.4	5
C_7H_{12}	5-Methyl-2-hexyne	-75 e	-57 e	-34 e	-4 e	38.6	102.0	5
C_7H_{12}	2-Methyl-3-hexyne	-78 e	-61 e	-39 e	-9 e	32.6	94.8	5
C_7H_{12}	4,4-Dimethyl-1-pentyne		-73 e	-52 e	-24 e	15.9	75.6	5
C_7H_{12}	4,4-Dimethyl-2-pentyne		-70 e	-48 e	-19 e	21.4	82.6	5
C_7H_{12}	Bicyclo[4.1.0]heptane					49.9	116.3	5
C_7H_{12}	Cycloheptene			-30.0	3.4	47.5	108 e	5
C_7H_{12}	1-Methylbicyclo(3,1,0)hexane					29.8	92.6	5
C_7H_{12}	Methylenecyclohexane	-76 e	-58 e	-35 e	-5 e	38 e	103.0	5
C_7H_{12}	1-Methylcyclohexene	-72 e	-53 e	-30 e	1 e	45 e	109.8	5
C_7H_{12}	4-Methylcyclohexene	-76 e	-59 e	-36 e	-5 e	37.9	102.3	5
C_7H_{12}	1-Ethylcyclopentene	-75 e	-57 e	-34 e	-3 e	40.7	105.8	5
C_7H_{12}	1,2-Dimethylcyclopentene	-75 e	-57 e	-34 e	-3 e	40.2	105.3	5
C_7H_{12}	1,5-Dimethylcyclopentene	-77 e	-59 e	-36 e	-5.5	37.3	101.5	5
$C_7H_{12}O$	Cycloheptanone			18 e	53.7	104.0	178.7	5
$C_7H_{12}O_2$	Butyl acrylate	-52 e	-31 e	-4.5	30.4	78.0	146.9	5
$C_7H_{12}O_2$	Propyl methacrylate				26 e	73.8	139.7	5
$C_7H_{12}O_3$	Ethyl levulinate		17 e	45.3	82.6	133.2	205.7	5
$C_7H_{12}O_4$	Diethyl malonate	-23 e	4 e	36.0	76.4	128.5	198.3	5
$C_7H_{12}O_4$	Dimethyl glutarate	-11 e	15 e	47 e	87.7	139.8	209.5	5
$C_7H_{13}ClO$	Heptanoyl chloride	-17 e	4 e	29.4	59.7	96.9	144.0	5
C_7H_{14}	1-Heptene	-82.1	-63.8	-40.6	-10.7	31.1	93.2	1,5
C_7H_{14}	cis-2-Heptene	-79 e	-61 e	-38 e	-8 e	34.3	98.0	5
C_7H_{14}	trans-2-Heptene	-79 e	-61 e	-39 e	-8 e	34.0	97.5	5
C_7H_{14}	cis-3-Heptene	-80 e	-62 e	-40 e	-10 e	32.3	95.3	5
C_7H_{14}	trans-3-Heptene	-80 e	-62 e	-40 e	-10 e	32.2	95.2	5
C_7H_{14}	2-Methyl-1-hexene	-81 e	-64 e	-42 e	-12 e	29.3	91.6	5
C_7H_{14}	4-Methyl-1-hexene	-84 e	-67 e	-45 e	-16 e	25.3	86.3	5
C_7H_{14}	2-Methyl-2-hexene	-80 e	-63 e	-40 e	-10 e	32.0	95.0	5
C_7H_{14}	cis-3-Methyl-2-hexene	-79 e	-62 e	-39 e	-9 e	33.4	96.8	5
C_7H_{14}	trans-4-Methyl-2-hexene	-83 e	-66 e	-44 e	-15 e	25.9	87.1	5
C_7H_{14}	trans-5-Methyl-2-hexene	-83 e	-66 e	-44 e	-15 e	26.3	87.7	5
C_7H_{14}	trans-2-Methyl-3-hexene	-84 e	-67 e	-45 e	-16 e	24.6	85.5	5
C_7H_{14}	3-Ethyl-1-pentene	-85 e	-68 e	-46 e	-17 e	23.2	83.7	5
C_7H_{14}	2,3-Dimethyl-1-pentene	-85 e	-68 e	-46 e	-17 e	23.4	83.8	5
C_7H_{14}	2,4-Dimethyl-1-pentene	-88 e	-71 e	-50 e	-21 e	20.0	81.2	5
C_7H_{14}	3,3-Dimethyl-1-pentene	-87 e	-71 e	-50 e	-21 e	18.1	77.1	5
C_7H_{14}	4,4-Dimethyl-1-pentene	-94 e	-78 e	-57 e	-28 e	11.5	72.1	5
C_7H_{14}	2,3-Dimethyl-2-pentene	-79 e	-62 e	-39 e	-9 e	33.5	96.9	5
C_7H_{14}	2,4-Dimethyl-2-pentene	-84 e	-68 e	-46 e	-18 e	22.6	82.9	5
C_7H_{14}	cis-3,4-Dimethyl-2-pentene	-83 e	-65 e	-43 e	-14 e	27.2	88.8	5
C_7H_{14}	trans-3,4-Dimethyl-2-pentene	-82 e	-64 e	-42 e	-13 e	29.0	91.1	5
C_7H_{14}	cis-4,4-Dimethyl-2-pentene	-90 e	-73 e	-51 e	-22 e	18.6	80.0	5
C_7H_{14}	trans-4,4-Dimethyl-2-pentene	-90 e	-73 e	-52 e	-23 e	16.6	76.3	5
C_7H_{14}	2,3,3-Trimethyl-1-butene	-91 e	-75 e	-53 e	-24.2	16.3	77.5	5
C_7H_{14}	Cycloheptane				6 e	51.1	118.4	1

VAPOR PRESSURE (continued)

Mol. Form.	Name	\multicolumn{6}{c}{Temperature in °C for the indicated pressure}	Ref.
		1 Pa	10 Pa	100 Pa	1 kPa	10 kPa	100 kPa	
C_7H_{14}	Methylcyclohexane	-79 e	-62 e	-39 e	-7.9	35.5	100.5	1
C_7H_{14}	Ethylcyclopentane	-76 e	-59 e	-35 e	-5 e	38.4	103.0	5
C_7H_{14}	1,1-Dimethylcyclopentane		-69 e	-47 e	-17 e	24.8	87.4	5
C_7H_{14}	cis-1,2-Dimethylcyclopentane			-38 e	-8 e	34.9	99.0	5
C_7H_{14}	trans-1,2-Dimethylcyclopentane	-83 e	-66 e	-43 e	-13 e	28.4	91.4	5
C_7H_{14}	cis-1,3-Dimethylcyclopentane	-84 e	-66 e	-44 e	-14 e	28.2	91.1	5
C_7H_{14}	trans-1,3-Dimethylcyclopentane	-84 e	-67 e	-44 e	-14 e	27.4	90.3	5
$C_7H_{14}O$	1-Heptanal	-41 e	-21 e	4 e	37 e	83.7	152.3	5
$C_7H_{14}O$	2-Heptanone		-22 e	3 e	36 e	82.2	150.6	1
$C_7H_{14}O$	3-Heptanone		-28 e	0 e	36 e	83.2	147.0	5
$C_7H_{14}O$	4-Heptanone	-27 e	-6 e	18.8	50.2	90.3	143.4	5
$C_7H_{14}O$	5-Methyl-2-hexanone		-27 e	-2 e	31.0	76.6	144.4	5
$C_7H_{14}O$	2,4-Dimethyl-3-pentanone	-61 e	-42 e	-18 e	14 e	58.5	124.8	1
$C_7H_{14}O_2$	Heptanoic acid	24 e	46 e	72 e	107 e	154.6	222.6	5
$C_7H_{14}O_2$	Pentyl acetate	-58 e	-39 e	-14 e	20 e	70.1	149 e	5
$C_7H_{14}O_2$	Isopentyl acetate	-51 e	-30 e	-4 e	30.3	76.2	141.4	5
$C_7H_{14}O_2$	Isobutyl propanoate	-35 e	-19 e	2 e	31 e	72.0	136.1	5
$C_7H_{14}O_2$	Propyl butanoate	-35 e	-19 e	3 e	32.0	74.9	142.8	5
$C_7H_{14}O_2$	Propyl isobutanoate		-28 e	-5.7	24.5	67.5	133.3	5
$C_7H_{14}O_2$	Isopropyl isobutanoate		-44 e	-19.7	12.2	56.0	120.1	5
$C_7H_{14}O_2$	Ethyl 3-methylbutanoate	-57 e	-36 e	-10 e	23.9	69.5	134.4	5
$C_7H_{14}O_2$	Methyl hexanoate	-47 e	-26 e	2 e	36.6	83.3	149 e	5
$C_7H_{14}O_2$	4-Methoxy-4-methyl-2-pentanone				43 e	89.8	160 e	5
$C_7H_{15}Br$	1-Bromoheptane	-30 e	-9 e	18 e	54 e	104.4	178.4	5
$C_7H_{15}Cl$	1-Chloroheptane	-39 e	-19 e	7 e	41 e	88.6	159.9	5
$C_7H_{15}F$	1-Fluoroheptane	-64 e	-45 e	-22 e	10 e	53.3	117.4	5
$C_7H_{15}I$	1-Iodoheptane	-19 e	3 e	32 e	71 e	123.8	203.4	5
C_7H_{16}	Heptane	-78.6	-60.2	-37.0	-6.6	35.4	98.0	16
C_7H_{16}	2-Methylhexane	-82 e	-65 e	-43 e	-13 e	27.8	89.7	1
C_7H_{16}	3-Methylhexane	-81 e	-64 e	-42 e	-12 e	29.2	91.5	1
C_7H_{16}	3-Ethylpentane	-81 e	-63 e	-41 e	-11 e	30.5	93.1	1
C_7H_{16}	2,2-Dimethylpentane	-90 e	-73 e	-52 e	-22.9	17.6	78.8	1
C_7H_{16}	2,3-Dimethylpentane	-87 e	-68.4	-45.3	-14.9	26.8	89.3	5
C_7H_{16}	2,4-Dimethylpentane	-89 e	-72 e	-50 e	-21.3	19.2	80.1	1
C_7H_{16}	3,3-Dimethylpentane	-88 e	-71 e	-49 e	-18.8	22.9	85.6	1
C_7H_{16}	2,2,3-Trimethylbutane				-23.2	18.1	80.4	5
$C_7H_{16}O$	1-Heptanol		17 e	40 e	70.1	112.5	176 e	1
$C_7H_{16}O$	2-Heptanol	-9 e	7 e	27 e	55.0	95.2	158.7	5
$C_7H_{16}O$	3-Heptanol	-8 e	7 e	27 e	54.5	93.9	156.3	5
$C_7H_{16}O$	4-Heptanol	-16 e	1 e	22 e	51 e	91.9	154.6	5
$C_7H_{16}O$	2,2-Dimethyl-3-pentanol				9 e	35 e	73.1	5
$C_7H_{16}S$	1-Heptanethiol	-30 e	-9 e	18 e	53 e	102.7	176.4	5
$C_7H_{17}N$	Heptylamine			5 e	39 e	86.7	156.4	5
$C_7H_{18}N_2$	N,N-Diethyl-1,3-propanediamine				50.1	99.9	167.7	5
C_8F_{18}	Perfluorooctane				5 e	45.0	105.6	5
$C_8H_4O_3$	Phthalic anhydride	48.2 s	72.4 s			192.7	284.2	5
C_8H_6O	Benzofuran		-16 e	12 e	47.9	97.7	170.7	5
C_8H_7Cl	o-Chlorostyrene	-33 e	-10 e	20 e	58 e	110.8	188 e	5
C_8H_7N	2-Methylbenzonitrile		1 e	32.1	72.2	126.6	204.7	5
C_8H_7N	4-Methylbenzonitrile			40.1	78.7	134.3	221.3	5
C_8H_7N	Benzeneacetonitrile	-3 e	23 e	55.3	97.4	153.7	233.1	5
C_8H_7N	Indole	20.6 s	44.5 s				254.0	5
$C_8H_7NO_4$	Methyl 2-nitrobenzoate	17 e	49 e	89 e	140 e	208 e	302 e	5
C_8H_8	Styrene		-31 e	-5 e	28.6	75.4	144.7	1
C_8H_8	1,3,5,7-Cyclooctatetraene				24.3	71.0	140.1	5
C_8H_8O	Acetophenone			36 e	73 e	125.3	201.5	5
$C_8H_8O_2$	Phenyl acetate		3 e	33.1	72.2	123.9	195.5	5
$C_8H_8O_2$	Methyl benzoate		-1 e	29 e	68 e	121.2	198.9	5
$C_8H_8O_2$	4-Methoxybenzaldehyde	9 e	35 e	68.1	110.8	167.9	248.5	5
$C_8H_8O_3$	Methyl salicylate	-1 e	22 e	51 e	88.8	141.8	219.9	5
C_8H_9Cl	1-Chloro-2-ethylbenzene	-30 e	-9 e	18 e	54 e	103.7	177.9	5
C_8H_9Cl	1-Chloro-4-ethylbenzene	-27 e	-6 e	22 e	58 e	108.7	183.9	5
$C_8H_9NO_2$	1-Ethyl-4-nitrobenzene	10 e	36 e	69 e	111.6	168 e	245 e	5
C_8H_{10}	Ethylbenzene	-56.2	-36.8	-12.0	21.1	67.1	135.7	1
C_8H_{10}	o-Xylene			-7 e	27 e	74.2	143.9	1
C_8H_{10}	m-Xylene		-35 e	-10 e	23.4	69.8	138.7	1
C_8H_{10}	p-Xylene				22.4	68.9	137.9	1

Mol. Form.	Name	Temperature in °C for the indicated pressure						Ref.
		1 Pa	10 Pa	100 Pa	1 kPa	10 kPa	100 kPa	
$C_8H_{10}O$	o-Ethylphenol		16.9	44.5	81.1	130.9	204.0	5
$C_8H_{10}O$	m-Ethylphenol	5.6	29.2	57.5	91.9	144.8	217.9	5
$C_8H_{10}O$	p-Ethylphenol			60 e	95.5	144.6	217.5	5
$C_8H_{10}O$	2,3-Xylenol	14.3 s	34.3 s	57.2 s	91.4	141.7	216.4	1,5
$C_8H_{10}O$	2,4-Xylenol			50.2	85.5	137.2	210.5	1,5
$C_8H_{10}O$	2,5-Xylenol	13.4 s	33.2 s	55.9 s	87.4	137.0	210.6	5
$C_8H_{10}O$	2,6-Xylenol	-3.1 s	16.7 s	39.6 s	75.3	125.9	200.6	1,5
$C_8H_{10}O$	3,4-Xylenol	19.7 s	40.2 s	63.7 s	102.1	152.3	226.4	1,5
$C_8H_{10}O$	3,5-Xylenol	16.5 s	37.2 s	61.1 s	98.0	147.9	221.3	1,5
$C_8H_{10}O$	Benzeneethanol	2 e	25 e	54 e	92 e	143.6	217.7	5
$C_8H_{10}O$	Phenetole		-9 e	17 e	51 e	99 e	169.3	5
$C_8H_{10}O_2$	2-Phenoxyethanol	21 e	46 e	75.9	115.4	168.7	244.8	5
$C_8H_{10}O_2$	1,3-Dimethoxybenzene	18 e	34 e	56 e	86.7	135.5	223 e	5
$C_8H_{11}N$	p-Ethylaniline	-2 e	21 e	49 e	87 e	139.4	216.7	5
$C_8H_{11}N$	N-Ethylaniline	-15 e	8 e	38 e	76.4	128.8	204.2	5
$C_8H_{11}N$	N,N-Dimethylaniline			28 e	66 e	118.1	193.6	1
$C_8H_{11}N$	2,4-Xylidine	-2 e	21 e	51 e	88 e	139.1	210.9	5
$C_8H_{11}N$	2,6-Xylidine			37 e	80 e	137.7	217.7	5
$C_8H_{11}N$	5-Ethyl-2-picoline	-33 e	-9.3	20 e			178.0	5
$C_8H_{11}NO$	o-Phenetidine	0 e	27 e	60 e	102.2	156.0	228.1	5
C_8H_{12}	1,5-Cyclooctadiene		-37 e	-8 e	30 e	80.2	150 e	5
C_8H_{12}	4-Vinylcyclohexene	-62 e	-43 e	-19 e	14.1	59.9	129 e	5
$C_8H_{12}O_4$	Diethyl maleate	-6 e	20 e	52.2	93.5	148.4	224.8	5
C_8H_{14}	2,5-Dimethyl-1,5-hexadiene	-38 e	-26 e	-10 e	14 e	50.8	115.1	5
C_8H_{14}	1-Octyne	-59 e	-40 e	-16 e	16 e	60.3	125.8	1
C_8H_{14}	2-Octyne	-52 e	-33 e	-8 e	25 e	70.6	137.8	1
C_8H_{14}	3-Octyne	-55 e	-35 e	-11 e	22 e	66.8	132.8	1
C_8H_{14}	4-Octyne	-56 e	-36 e	-12 e	21 e	65.6	131.4	1
C_8H_{14}	1-Ethylcyclohexene	-55 e	-35 e	-11 e	22 e	68 e	136.5	5
$C_8H_{14}O_2$	Cyclohexyl acetate					103.1	172.9	5
$C_8H_{14}O_2$	Butyl methacrylate				47 e	93.3	159.0	5
$C_8H_{14}O_3$	Butanoic anhydride	-28 e	-2 e	30 e	71 e	123.8	196.5	5
$C_8H_{14}O_4$	Ethyl succinate	-6 e	20 e	51.0	91.1	143.7	216.1	5
$C_8H_{14}O_4$	Dipropyl oxalate	-4 e	20 e	49.9	88.6	140.4	213.0	5
$C_8H_{14}O_4$	Dimethyl adipate		28 e	61 e	103 e	156.1	227.3	5
$C_8H_{15}Br$	(2-Bromoethyl)cyclohexane	-14 e	8 e	36.9	75.3	129.7	212.5	5
$C_8H_{15}ClO$	Octanoyl chloride	1 e	22 e	46 e	74.7	109 e	150 e	5
$C_8H_{15}N$	Octanenitrile	-15 e	8 e	37 e	75 e	127.7	204.4	5
C_8H_{16}	1-Octene	-65.7	-46.1	-21.4	10.5	54.9	120.9	1,5
C_8H_{16}	cis-2-Octene	-59 e	-41 e	-17 e	15 e	59 e	125.2	5
C_8H_{16}	trans-2-Octene	-59 e	-41 e	-17 e	14 e	59 e	124.5	5
C_8H_{16}	cis-3-Octene	-65 e	-46 e	-22 e	10 e	55.1	122.4	5
C_8H_{16}	trans-3-Octene	-61 e	-43 e	-19 e	13 e	57 e	122.8	5
C_8H_{16}	cis-4-Octene	-63 e	-44 e	-20 e	11 e	56 e	122.1	5
C_8H_{16}	trans-4-Octene	-65 e	-46 e	-22 e	10 e	54.6	121.8	5
C_8H_{16}	2-Methyl-1-heptene	-66 e	-48 e	-24 e	8 e	52.3	118.7	5
C_8H_{16}	2,2-Dimethyl-cis-3-hexene	-74 e	-56 e	-33 e	-3 e	40.1	105.0	5
C_8H_{16}	2,3-Dimethyl-2-hexene	-65 e	-47 e	-23 e	10 e	54.3	121.3	5
C_8H_{16}	2,3,3-Trimethyl-1-pentene		-53 e	-30 e	1 e	43.8	107.9	5
C_8H_{16}	2,4,4-Trimethyl-1-pentene	-79 e	-61 e	-38 e	-7 e	36.2	101.0	5
C_8H_{16}	2,3,4-Trimethyl-2-pentene	-68 e	-49 e	-26 e	6 e	50.0	115.8	5
C_8H_{16}	2,4,4-Trimethyl-2-pentene	-73 e	-56 e	-33 e	-2 e	40.4	104.5	5
C_8H_{16}	Cyclooctane				30 e	78 e	150.7	1
C_8H_{16}	Ethylcyclohexane	-61 e	-42 e	-17 e	15.8	61.9	131.3	5
C_8H_{16}	1,1-Dimethylcyclohexane			-27 e	5 e	50.6	119.1	5
C_8H_{16}	cis-1,2-Dimethylcyclohexane		-44 e	-20 e	14 e	59.7	129.2	5
C_8H_{16}	trans-1,2-Dimethylcyclohexane	-68 e	-49 e	-25 e	8 e	53.9	122.9	5
C_8H_{16}	cis-1,3-Dimethylcyclohexane	-68 e	-48 e	-23 e	10 e	55.6	123.1	5
C_8H_{16}	trans-1,3-Dimethylcyclohexane	-62 e	-45 e	-23 e	8 e	51.5	120.9	5
C_8H_{16}	cis-1,4-Dimethylcyclohexane	-66 e	-47 e	-23 e	10 e	55.3	123.8	5
C_8H_{16}	trans-1,4-Dimethylcyclohexane			-27 e	5 e	50.6	118.9	5
C_8H_{16}	Propylcyclopentane	-60 e	-41 e	-16 e	16.5	62.1	130.5	5
C_8H_{16}	Isopropylcyclopentane	-65 e	-46 e	-21 e	12 e	57.3	125.9	5
C_8H_{16}	1-Ethyl-1-methylcyclopentane	-67 e	-49 e	-24 e	8 e	53.2	121.0	5
C_8H_{16}	cis-1-Ethyl-2-methylcyclopentane	-63 e	-44 e	-19 e	13.3	59.1	127.6	5
C_8H_{16}	1,1,2-Trimethylcyclopentane				2 e	46.2	113.2	5
C_8H_{16}	1,1,3-Trimethylcyclopentane	-77 e	-59 e	-36 e	-5 e	38.7	104.4	5

VAPOR PRESSURE (continued)

Mol. Form.	Name	Temperature in °C for the indicated pressure						Ref.
		1 Pa	10 Pa	100 Pa	1 kPa	10 kPa	100 kPa	
C_8H_{16}	1′,2′,4a-1,2,4-Trimethylcyclopentane	-70 e	-52 e	-28 e	4 e	48.9	116.2	5
C_8H_{16}	1′,2a,4′-1,2,4-Trimethylcyclopentane	-74 e	-56 e	-33 e	-1 e	42.8	108.8	5
$C_8H_{16}O$	1-Propylcyclopentanol	9 e	24 e	43 e	69.0	108.4	173.5	5
$C_8H_{16}O$	Octanal			6 e	45.7	97.8	170.2	5
$C_8H_{16}O$	2-Octanone		-3 e	23 e	57 e	103.8	172.1	5
$C_8H_{16}O$	3-Octanone			8 e	47.7	97 e	161 e	5
$C_8H_{16}O$	2,2,4-Trimethyl-3-pentanone			11.3	42.1	81.7	134.6	5
$C_8H_{16}O_2$	Octanoic acid	37 e	58 e	85 e	120 e	165.5	238.4	1,5
$C_8H_{16}O_2$	2-Ethylhexanoic acid				108 e	159.6	226.6	5
$C_8H_{16}O_2$	Hexyl acetate	-37 e	-13 e	16 e	52.8	100.4	164 e	5
$C_8H_{16}O_2$	Isopentyl propanoate				3.1	40.7	90.6	159.8	5
$C_8H_{16}O_2$	Isobutyl isobutanoate	-47 e	-26 e	0.4	34.8	81.1	147.0	5
$C_8H_{16}O_2$	Propyl 3-methylbutanoate			1.8	38.9	87.9	155.6	5
$C_8H_{16}O_2$	Ethyl hexanoate	-31 e	-9 e	18.7	53.9	100.7	166.2	5
$C_8H_{16}O_2$	Methyl heptanoate	-30 e	-9 e	19 e	54.2	102.4	172 e	5
$C_8H_{16}O_4$	Diethylene glycol monoethyl ether acetate	-16 e	10.6	43.9	86.2	141.3	216.6	5
$C_8H_{17}Br$	1-Bromooctane	-17 e	6 e	34 e	72 e	123.8	200.3	5
$C_8H_{17}Cl$	1-Chlorooctane	-25 e	-4 e	23 e	59 e	108.8	182.9	5
$C_8H_{17}Cl$	3-(Chloromethyl)heptane					100.3	172.4	5
$C_8H_{17}F$	1-Fluorooctane				29 e	74.6	141.8	5
$C_8H_{17}I$	1-Iodooctane	-6 e	18 e	48 e	87 e	142.5	224.5	5
C_8H_{18}	Octane		-42.6	-17.9	14.4	58.9	125.3	16
C_8H_{18}	2-Methylheptane	-69 e	-49.1	-24.5	7.6	51.6	117.2	1,5
C_8H_{18}	3-Methylheptane	-67 e	-48.1	-23.6	8.5	52.7	118.5	1,5
C_8H_{18}	4-Methylheptane	-65 e	-47 e	-24 e	7.8	51.6	117.2	5
C_8H_{18}	3-Ethylhexane				8 e	52.1	118.1	5
C_8H_{18}	2,2-Dimethylhexane	-73 e	-55 e	-32 e	-1.5	41.6	106.4	5
C_8H_{18}	2,3-Dimethylhexane				5 e	49.2	115.1	5
C_8H_{18}	2,4-Dimethylhexane				0.6	43.9	109.0	5
C_8H_{18}	2,5-Dimethylhexane	-71 e	-53 e	-30 e	0.7	43.8	108.6	5
C_8H_{18}	3,3-Dimethylhexane	-72 e	-54 e	-30 e	1.4	45.4	111.5	5
C_8H_{18}	3,4-Dimethylhexane				7 e	50.9	117.3	5
C_8H_{18}	3-Ethyl-2-methylpentane	-69 e	-50 e	-27 e	5 e	48.9	115.2	5
C_8H_{18}	3-Ethyl-3-methylpentane	-70 e	-51 e	-27 e	5 e	50.2	117.8	5
C_8H_{18}	2,2,3-Trimethylpentane	-74 e	-56 e	-32 e	-0.8	43.1	109.4	5
C_8H_{18}	2,2,4-Trimethylpentane	-81.9	-63.4	-39.8	-8.9	34.0	98.8	5
C_8H_{18}	2,3,3-Trimethylpentane	-72 e	-54 e	-30 e	2.1	46.9	114.3	5
C_8H_{18}	2,3,4-Trimethylpentane	-74 e	-54.5	-30.0	2.2	46.7	113.1	1,5
C_8H_{18}	2,2,3,3-Tetramethylbutane	-62.5 s	-44 s	-20.9 s	8.9 s	48.8 s	105.8	5
$C_8H_{18}O$	1-Octanol	12 e	30 e	53 e	84 e	128.2	194.8	1,39
$C_8H_{18}O$	2-Octanol			40 e	69.9	112.5	179.4	1,39
$C_8H_{18}O$	3-Octanol	12 e	24 e	40 e	64 e	102.8	174.1	1
$C_8H_{18}O$	4-Octanol			40 e	66.9	107.3	176.0	1,39
$C_8H_{18}O$	4-Methyl-3-heptanol	-52 e	-28 e	1 e	39 e	87.6	155.0	5
$C_8H_{18}O$	5-Methyl-3-heptanol	-35 e	-16 e	8 e	40 e	84.8	153.0	5
$C_8H_{18}O$	4-Methyl-4-heptanol	-17 e	1 e	24 e	55 e	97.2	160.7	5
$C_8H_{18}O$	2-Ethyl-1-hexanol			45 e	75 e	118.3	184.2	1
$C_8H_{18}O$	2-Ethyl-2-hexanol	-13 e	4 e	26 e	55 e	96.3	160.3	5
$C_8H_{18}O$	2,4,4-Trimethyl-2-pentanol		-7 e	13 e	40 e	79.8	146.1	5
$C_8H_{18}O$	2,2,4-Trimethyl-3-pentanol	-2 e	9 e	24 e	47 e	82.6	150.4	5
$C_8H_{18}O$	Dibutyl ether	-55 e	-35 e	-8 e	26 e	73.0	141.2	5
$C_8H_{18}O$	Di-sec-butyl ether			-19 e	12.1	55.4	120.6	5
$C_8H_{18}O$	Di-tert-butyl ether			-33 e	-2 e	41.7	106.8	1
$C_8H_{18}O_2$	Ethylene glycol monohexyl ether	-13 e	14 e	46 e	86 e	137.7	206.9	5
$C_8H_{18}O_2$	1,2-Dipropoxyethane			-44.2	-2.0	63.6	179.2	5
$C_8H_{18}O_2$	Di-tert-butyl peroxide			-26 e	4.3	46.6	110.5	5
$C_8H_{18}O_3$	Diethylene glycol monobutyl ether	14 e	37 e	66.8	104.9	153 e	230.4	5
$C_8H_{18}O_3$	Diethylene glycol diethyl ether	-32 e	-7 e	25 e	64.9	117.1	189 e	5
$C_8H_{18}O_5$	Tetraethylene glycol	89 e	117 e	151.1	192.2	242.9	307.3	5
$C_8H_{18}S$	1-Octanethiol	-15 e	6 e	34 e	71 e	122.1	198.5	5
$C_8H_{18}S$	Dibutyl sulfide	-22 e	0 e	27 e	63 e	113.5	188.4	5

Mol. Form.	Name	Temperature in °C for the indicated pressure						Ref.
		1 Pa	10 Pa	100 Pa	1 kPa	10 kPa	100 kPa	
$C_8H_{19}N$	Dibutylamine	-37 e	-16 e	10 e	44 e	90.8	159.1	5
$C_8H_{19}N$	Diisobutylamine	-57 e	-36 e	-9.0	25.5	72.2	139.0	5
$C_8H_{20}O_4Si$	Ethyl silicate	-77 e	-52 e	-21 e	21.6	80.5	164.1	5
$C_8H_{20}Si$	Tetraethylsilane		-6.5	30.5	80.6	152.6		5
C_9F_{20}	Perfluorononane					40 e	114.7	5
$C_9H_6N_2O_2$	Toluene-2,4-diisocyanate		39 e	72 e	113.9	169.7	247 e	5
C_9H_7N	Quinoline	-1.3	23.7	55.4	96.8	153.4	236.5	1,5
C_9H_7N	Isoquinoline		30.2	60.7	101.3	157.9	242.7	1,5
C_9H_8	Indene			12 e	53.0	106.8	181.0	5
C_9H_{10}	cis-1-Propenylbenzene	-38 e	-15.4	13.3	51.4	103.7	178.4	5
C_9H_{10}	trans-1-Propenylbenzene		-16 e	13.3	51.6	103.7	178.4	5
C_9H_{10}	Isopropenylbenzene			3.2	41.5	92.8	164.9	5
C_9H_{10}	Indan	-33 e	-12 e	16 e	52 e	102.3	177.5	1
$C_9H_{10}O$	2,4-Dimethylbenzaldehyde	-3 e	23 e	54 e	93.2	144.6	214.5	5
$C_9H_{10}O_2$	Ethyl benzoate	-18 e	8 e	39 e	80.1	135.1	212.8	5
$C_9H_{10}O_2$	Benzyl acetate	-11 e	15 e	46.6	86.9	139.5	211 e	5
$C_9H_{11}Br$	1-Bromo-4-isopropylbenzene	-8 e	15 e	45 e	84 e	138.1	218.5	5
$C_9H_{11}Cl$	1-Chloro-2-isopropylbenzene	-23 e	-1 e	27 e	64 e	114.6	190.5	5
$C_9H_{11}Cl$	1-Chloro-4-isopropylbenzene		3 e	31 e	69 e	120.5	197.8	5
C_9H_{12}	Propylbenzene	-43 e	-23 e	4 e	38 e	86.7	158.8	1
C_9H_{12}	Isopropylbenzene	-46 e	-26 e	-1 e	33 e	80.9	152.0	1
C_9H_{12}	o-Ethyltoluene	-40 e	-19 e	8 e	43 e	92.1	164.7	5
C_9H_{12}	m-Ethyltoluene	-42 e	-21 e	5 e	40.4	88.9	160.8	5
C_9H_{12}	p-Ethyltoluene	-41 e	-21 e	6 e	41 e	89.2	161.5	5
C_9H_{12}	1,2,3-Trimethylbenzene		-12 e	15 e	52 e	101.5	175.6	1
C_9H_{12}	1,2,4-Trimethylbenzene	-37 e	-16 e	11 e	47 e	95.9	168.9	1
C_9H_{12}	1,3,5-Trimethylbenzene	-39 e	-18 e	9 e	43.7	92.4	164.3	1
$C_9H_{12}O$	Benzyl ethyl ether		-10 e	20.4	59.3	111.3	184.5	5
$C_9H_{12}O$	Phenyl propyl ether		-10 e	21 e	61 e	113.9	189.3	5
$C_9H_{12}O$	Phenyl isopropyl ether	-20 e	-1 e	23 e	56 e	103.7	176.9	5
$C_9H_{13}N$	2,4,6-Trimethylaniline	12 e	36 e	66 e	104.1	154.9	226 e	5
$C_9H_{13}N$	N,N-Dimethyl-o-toluidine	-25 e	-3 e	24.4	60.6	110.7	184.5	5
$C_9H_{13}N$	Amphetamine			33 e	70.1	118 e	202.0	5
$C_9H_{14}O$	Isophorone		1 e	33.1	75.1	132.4	215.1	5
$C_9H_{14}O_6$	Triacetin	37.6	62 e	90 e	124 e	165 e	214 e	5
$C_9H_{16}O_4$	Diethyl glutarate	-1 e	26 e	60.2	103.3	159.6	236.5	5
$C_9H_{17}N$	Nonanenitrile	-3 e	21 e	50.9	90.7	145.4	225.1	5
C_9H_{18}	1-Nonene	-50.1	-29.4	-3.3	30.4	77.1	146.4	1,5
C_9H_{18}	2-Methyl-1-octene	-53 e	-34 e	-9 e	25 e	72 e	144.1	5
C_9H_{18}	Butylcyclopentane	-45 e	-24 e	1 e	36 e	84 e	156.1	5
C_9H_{18}	Propylcyclohexane	-46 e	-26 e	0 e	35.1	83.6	156.2	5
C_9H_{18}	Isopropylcyclohexane	-48 e	-28 e	-2 e	33 e	81.3	154.0	5
C_9H_{18}	trans-1-Ethyl-4-methylcyclo-hexane	-53 e	-33 e	-8 e	25 e	71.8	141.5	5
C_9H_{18}	1,1,2-Trimethylcyclohexane			-12 e	23 e	71.5	145.5	5
C_9H_{18}	1,1,3-Trimethylcyclohexane	-60 e	-41 e	-16 e	18 e	65.2	136.1	5
C_9H_{18}	1',2a,4a-1,2,4-Trimethylcyclo-hexane	-71 e	-50 e	-22 e	15 e	65.7	140.7	5
C_9H_{18}	1',3',5'-1,3,5-Trimethylcyclo-hexane	-72 e	-50 e	-22 e	14 e	65.1	140.0	5
C_9H_{18}	Isobutylcyclopentane	-105 e	-88 e	-64 e	-28 e	31 e	147.0	5
C_9H_{18}	cis-1-Methyl-2-propylcyclo-pentane	-52 e	-33 e	-7 e	28 e	77 e	152.0	5
C_9H_{18}	trans-1-Methyl-2-propylcyclo-pentane	-56 e	-36 e	-11 e	23 e	72 e	145.8	5
C_9H_{18}	1,1,3,3-Tetramethylcyclo-pentane	-72 e	-54 e	-30 e	2 e	47 e	117.4	5
$C_9H_{18}O$	Nonanal		-3 e	27.4	65.5	115.6	184.6	5
$C_9H_{18}O$	2-Nonanone		8 e	35 e	71 e	121.0	194.0	5
$C_9H_{18}O$	5-Nonanone			-1 e	39.1	94 e	188 e	5
$C_9H_{18}O$	2,6-Dimethyl-4-heptanone	-32 e	-12 e	14 e	48 e	96.2	167.7	5
$C_9H_{18}O_2$	Nonanoic acid	48 e	69 e	97 e	133 e	182.7	255.1	5
$C_9H_{18}O_2$	Heptyl acetate	-16 e	6 e	34 e	70 e	119.9	191.9	5
$C_9H_{18}O_2$	Isopentyl butanoate				55 e	105.6	178.4	5
$C_9H_{18}O_2$	Isobutyl 3-methylbutanoate			11.3	48.3	97.9	168.3	5
$C_9H_{18}O_2$	Propyl hexanoate	-26 e	-2 e	28 e	65.1	113.4	178 e	5
$C_9H_{18}O_2$	Methyl octanoate	-26 e	-9 e	13 e	40 e	76 e	127.9	5

Mol. Form.	Name	Temperature in °C for the indicated pressure						Ref.
		1 Pa	10 Pa	100 Pa	1 kPa	10 kPa	100 kPa	
$C_9H_{19}Cl$	1-Chlorononane	-11 e	11 e	39 e	76 e	127.8	204.7	5
C_9H_{20}	Nonane	-46.8	-26.0	0.0	34.0	80.8	150.3	16
C_9H_{20}	2-Methyloctane	-49 e	-30 e	-5 e	28 e	73.9	142.8	5
C_9H_{20}	3-Methyloctane	-49 e	-29 e	-5 e	29 e	74.7	143.7	5
C_9H_{20}	4-Methyloctane	-50 e	-30 e	-6 e	27 e	73.2	141.9	5
C_9H_{20}	2,2-Dimethylheptane	-58 e	-39 e	-15 e	18 e	63.6	132.3	5
C_9H_{20}	2,3-Dimethylheptane	-53 e	-33 e	-9 e	25 e	70.8	140.0	5
C_9H_{20}	2,6-Dimethylheptane	-55 e	-36 e	-12 e	21 e	66.4	134.7	5
C_9H_{20}	3-Ethyl-4-methylhexane			-9 e	24 e	70.6	139.9	5
C_9H_{20}	2,2,4-Trimethylhexane	-66.1	-46.4	-21.3	11.8	57.7	126.0	5
C_9H_{20}	2,2,5-Trimethylhexane	-65.1	-45.8	-21.2	11.2	56.2	123.7	1,5
C_9H_{20}	2,3,3-Trimethylhexane	-58 e	-38 e	-13 e	20 e	66.7	137.2	5
C_9H_{20}	2,3,5-Trimethylhexane	-60 e	-41 e	-16 e	17 e	62.3	130.9	5
C_9H_{20}	2,4,4-Trimethylhexane	-62 e	-43 e	-18 e	15 e	61.0	130.2	5
C_9H_{20}	3,3,4-Trimethylhexane	-53 e	-33 e	-7 e	28 e	76.3	148.9	5
C_9H_{20}	3,3-Diethylpentane			-9 e	26 e	73.7	145.7	1
C_9H_{20}	3-Ethyl-2,4-dimethylpentane	-58 e	-38 e	-13 e	20 e	66.7	136.2	5
C_9H_{20}	2,2,3,3-Tetramethylpentane				21 e	68.5	139.8	1
C_9H_{20}	2,2,3,4-Tetramethylpentane	-61 e	-42 e	-17 e	16 e	62.5	132.6	1
C_9H_{20}	2,2,4,4-Tetramethylpentane		-49 e	-25 e	8 e	53.2	121.8	1
C_9H_{20}	2,3,3,4-Tetramethylpentane	-57 e	-37 e	-12 e	22 e	69.7	141.1	1
$C_9H_{20}O$	1-Nonanol		40 e	64 e	96.9	141.0	213.0	5,39
$C_9H_{20}O$	3-Nonanol		24 e	47 e	78 e	123.0	194.2	5
$C_9H_{20}O$	4-Nonanol			45 e	76.4	121.3	192.0	5
$C_9H_{20}O$	5-Nonanol	13 e	31 e	54 e	84.5	128.1	194.7	5
$C_9H_{20}O$	2,2,4,4-Tetramethyl-3-pentanol				58	100	167	5
$C_9H_{20}S$	1-Nonanethiol	-2 e	21 e	49 e	87 e	140.4	219.2	5
$C_9H_{21}BO_3$	Triisopropyl borate					73.1	139.0	5
$C_9H_{21}N$	Nonylamine		9 e	37 e	75 e	126.2	202.1	5
$C_9H_{21}N$	Tripropylamine	-39 e	-18 e	8 e	42 e	88.2	156.0	5
$C_{10}F_8$	Perfluoronaphthalene	5.2 s	25.1 s	48.1 s				5
$C_{10}F_{22}$	Perfluorodecane					52 e	132.9	5
$C_{10}H_7Br$	1-Bromonaphthalene	17 e	45 e	80.3	126.7	189.8	280.5	5
$C_{10}H_7Cl$	1-Chloronaphthalene	14 e	39 e	70.5	112.8	171.6	258.6	5
$C_{10}H_8$	Naphthalene**	3.2 s	24.1 s	49.3 s	80.7	135.6	217.5	1,5
$C_{10}H_8$	Azulene	24.1 s	46 s	71.5 s	103.3	162.6	244.0	5
$C_{10}H_8O$	1-Naphthol				137.2	196.7	281.8	5
$C_{10}H_8O$	2-Naphthol				140.7	200.5	286.8	5
$C_{10}H_9N$	1-Naphthalenamine		62 e	99.0	146.9	210.7	300.1	5
$C_{10}H_9N$	2-Naphthalenamine	36.3 s	65.9 s	103 s	150.9	215.1	305.5	5
$C_{10}H_9N$	2-Methylquinoline	5.3	31.9	63.8	102.9	165.8	247.2	5
$C_{10}H_9N$	4-Methylquinoline	29 e	54 e	85 e	127 e	183.0	265.1	5
$C_{10}H_9N$	6-Methylquinoline	27 e	51 e	81 e	122 e	179.2	264.5	5
$C_{10}H_9N$	8-Methylquinoline	15 e	40 e	70 e	111 e	166.1	247.3	5
$C_{10}H_{10}$	m-Divinylbenzene	-29 e	-4 e	27.1	67.6	122.1	199 e	5
$C_{10}H_{10}O_4$	Dimethyl phthalate	27 e	56 e	92.7	137.8	195.8	272.7	5
$C_{10}H_{10}O_4$	Dimethyl isophthalate			85 e	129.5	189.2	273 e	5
$C_{10}H_{10}O_4$	Dimethyl terephthalate	56.6 s	79.4 s	106.1 s	137.9 s	197.9	282 e	5
$C_{10}H_{12}$	1,2,3,4-Tetrahydronaphthalene	-21 e	3 e	33.2	74.1	127.4	207.8	5
$C_{10}H_{12}$	2-Ethylstyrene	-31 e	-8 e	21 e	60 e	111.7	187 e	5
$C_{10}H_{12}$	3-Ethylstyrene	-28 e	-5.3	24.1	62.6	116 e	193 e	5
$C_{10}H_{12}$	4-Ethylstyrene	-31 e	-8.2	21.3	60.5	115 e	196 e	5
$C_{10}H_{12}O$	Estragole			48.5	88.0	140.7	214.6	5
$C_{10}H_{12}O$	4-Isopropylbenzaldehyde			54.1	96.0	152.2	231.5	5
$C_{10}H_{12}O_2$	4-Allyl-2-methoxyphenol	9 e	37 e	72 e	115.9	173.8	252.9	5
$C_{10}H_{12}O_2$	2-Phenylethyl acetate	-4 e	22 e	54 e	96 e	152.3	232.0	5
$C_{10}H_{12}O_2$	Propyl benzoate	-8 e	18 e	50.2	92.3	149.2	230.5	5
$C_{10}H_{12}O_2$	Ethyl phenylacetate	-9 e	19 e	52 e	95 e	150.2	225 e	5
$C_{10}H_{12}O_2$	Isoeugenol				125 e	185.3	267.1	5
$C_{10}H_{14}$	Butylbenzene	-28 e	-7 e	21 e	56.9	107.6	182.8	1,5
$C_{10}H_{14}$	sec-Butylbenzene	-35 e	-14 e	13 e	48 e	98.3	172.8	5
$C_{10}H_{14}$	tert-Butylbenzene	-37 e	-16 e	10 e	46 e	94.9	168.6	5
$C_{10}H_{14}$	Isobutylbenzene	-36 e	-15 e	12 e	47.9	97.8	172.3	5
$C_{10}H_{14}$	o-Cymene	-39 e	-16 e	13 e	51 e	103.1	177.8	5
$C_{10}H_{14}$	m-Cymene	-34 e	-13 e	14 e	50 e	99.9	174.6	5
$C_{10}H_{14}$	p-Cymene	-33 e	-12 e	16 e	52 e	102.2	176.6	5
$C_{10}H_{14}$	o-Diethylbenzene	-28 e	-6 e	21 e	58 e	107.9	182.9	5

Mol. Form.	Name	Temperature in °C for the indicated pressure						Ref.
		1 Pa	10 Pa	100 Pa	1 kPa	10 kPa	100 kPa	
$C_{10}H_{14}$	m-Diethylbenzene	-28 e	-7 e	20 e	56 e	106.2	180.6	5
$C_{10}H_{14}$	p-Diethylbenzene	-28 e	-6 e	21 e	57 e	108.1	183.3	5
$C_{10}H_{14}$	3-Ethyl-1,2-dimethylbenzene	-22 e	0 e	28 e	66 e	117.2	193.4	5
$C_{10}H_{14}$	4-Ethyl-1,2-dimethylbenzene	-24 e	-2 e	26 e	63 e	113.6	189.2	5
$C_{10}H_{14}$	2-Ethyl-1,3-dimethylbenzene		-2 e	26 e	63 e	113.7	189.5	5
$C_{10}H_{14}$	2-Ethyl-1,4-dimethylbenzene	-27 e	-5 e	23 e	60 e	110.6	186.4	5
$C_{10}H_{14}$	1-Ethyl-2,4-dimethylbenzene	-25 e	-4 e	24 e	61 e	112.2	187.9	5
$C_{10}H_{14}$	1-Ethyl-3,5-dimethylbenzene	-28 e	-6 e	21 e	58 e	108.3	183.2	5
$C_{10}H_{14}$	1-Methyl-2-propylbenzene	-27 e	-6 e	22 e	58.2	108.9	184.3	5
$C_{10}H_{14}$	1-Methyl-3-propylbenzene	-29 e	-8 e	20 e	56.1	106.5	181.3	5
$C_{10}H_{14}$	1-Methyl-4-propylbenzene	-29 e	-7 e	20 e	56.6	107.4	182.8	5
$C_{10}H_{14}$	1,2,3,4-Tetramethylbenzene		7 e	36 e	74 e	126.6	204.5	5
$C_{10}H_{14}$	1,2,3,5-Tetramethylbenzene	-19 e	3 e	32 e	69 e	120.9	197.5	5
$C_{10}H_{14}$	1,2,4,5-Tetramethylbenzene					119.9	196.3	5
$C_{10}H_{14}O$	2-Butylphenol	7 e	31 e	61 e	101 e	155.2	234.4	5
$C_{10}H_{14}O$	Butyl phenyl ether	-16 e	8 e	38 e	77 e	131.3	209.7	5
$C_{10}H_{14}O$	Thymol	18.9 s	37.9 s	59.5	101.2	155.0	230.4	5
$C_{10}H_{15}N$	2-Methyl-5-isopropylaniline	19 e	43 e	72 e	107.4	150 e	204 e	5
$C_{10}H_{15}N$	N-Butylaniline	11 e	35 e	66 e	106 e	160.9	241.0	5
$C_{10}H_{15}N$	N,N-Diethylaniline	-11 e	14 e	44.3	84.2	138.4	216.3	5
$C_{10}H_{16}$	Dipentene	-42 e	-19 e	10.6	48.7	100.2	173.9	5
$C_{10}H_{16}$	d-Limonene	-45 e	-21 e	9.1	48.0	100.4	174.5	5
$C_{10}H_{16}$	l-Limonene	-33 e	-12 e	16 e	52.0	102.3	177.0	21
$C_{10}H_{16}$	β-Myrcene			9.4	47.3	98.3	171.0	5
$C_{10}H_{16}$	α-Pinene	-48 e	-27 e	-1 e	33.6	82.2	155.1	21
$C_{10}H_{16}$	β-Pinene	-43 e	-22 e	5.0	40.6	90.5	165.5	21
$C_{10}H_{16}$	Camphene					90.7	160.1	4
$C_{10}H_{16}$	Terpinolene			26.5	64.9	115.4	184.6	5
$C_{10}H_{16}$	β-Phellandrene			16 e	53.2	104 e	171.0	5
$C_{10}H_{16}O$	(+)-Camphor	-15.8 s	10 s	41.5 s	80.8 s	131.4 s	207.6	5
$C_{10}H_{16}O$	Pulegone	37 e	49.1	66.4	92.2	135.1	220.2	5
$C_{10}H_{18}$	1-Decyne	-34 e	-13 e	14 e	51 e	100.3	173.5	5
$C_{10}H_{18}$	cis-Decahydronaphthalene	-26 e	-4 e	24 e	62.4	115.5	195.3	1
$C_{10}H_{18}$	trans-Decahydronaphthalene		-10 e	18 e	55.3	107.9	186.8	1
$C_{10}H_{18}O$	α-Terpineol			48	89	142	217	4
$C_{10}H_{18}O$	Eucalyptol			10.6	48.5	100.3	175.4	5
$C_{10}H_{18}O$	trans-Geraniol	4 e	31 e	63.2	104.3	157.7	229.6	5
$C_{10}H_{18}O_4$	Sebacic acid	125.9 s						5
$C_{10}H_{18}O_4$	Dipropyl succinate	11 e	38 e	72.1	115.4	172.3	250.4	5
$C_{10}H_{18}O_4$	Diethyl adipate	4 e	35 e	72 e	116.6	171.2	239.5	5
$C_{10}H_{19}N$	Decanenitrile	13 e	36 e	66 e	105.8	160.6	241.6	5
$C_{10}H_{20}$	1-Decene	-35.5	-13.7	13.7	49.0	97.9	170.1	1,5
$C_{10}H_{20}$	Cyclodecane			29 e	68 e	121.3	201.8	1
$C_{10}H_{20}$	Butylcyclohexane	-31 e	-9 e	18 e	54 e	104.7	180.4	5
$C_{10}H_{20}$	Isobutylcyclohexane	-37 e	-16 e	10 e	46 e	95.9	170.8	5
$C_{10}H_{20}$	tert-Butylcyclohexane	-39 e	-18 e	9 e	45 e	95.3	171.1	5
$C_{10}H_{20}O$	Decanal		16 e	47.2	86.3	137.7	208.0	5
$C_{10}H_{20}O_2$	Decanoic acid	58 e	80 e	108 e	145 e	195.2	269.5	5
$C_{10}H_{20}O_2$	Octyl acetate	-26 e	-3 e	27 e	66.3	120.0	198.2	5
$C_{10}H_{20}O_2$	2-Ethylhexyl acetate	-11 e	5 e	26 e	57.6	107.1	197.2	5
$C_{10}H_{20}O_2$	Isopentyl isopentanoate			22 e	62.8	116.9	193.6	5
$C_{10}H_{20}O_2$	Ethyl octanoate	-17 e	9 e	41 e	81.4	133.2	203 e	5
$C_{10}H_{20}O_4$	Diethylene glycol monobutyl ether acetate	6 e	34 e	69 e	112.6	169.2	245.4	5
$C_{10}H_{21}Br$	1-Bromodecane	9 e	33 e	63 e	104 e	159.2	240.0	5
$C_{10}H_{21}Cl$	1-Chlorodecane	2 e	25 e	54 e	92 e	145.7	225.3	5
$C_{10}H_{21}F$	1-Fluorodecane	-22 e	0 e	27 e	64 e	113.3	185.7	5
$C_{10}H_{22}$	Decane		-10.6	16.7	52.3	101.1	173.7	16
$C_{10}H_{22}$	2-Methylnonane	-34 e	-14 e	12 e	47 e	94.8	166.5	5
$C_{10}H_{22}$	3-Methylnonane	-34 e	-14 e	12 e	47 e	95.1	167.3	5
$C_{10}H_{22}$	4-Methylnonane	-36 e	-16 e	10 e	45 e	93.1	165.2	5
$C_{10}H_{22}$	5-Methylnonane	-36 e	-16 e	10 e	45 e	92.6	164.6	5
$C_{10}H_{22}$	2,4-Dimethyloctane				38 e	84.9	155.4	5
$C_{10}H_{22}$	2,7-Dimethyloctane	-39 e	-19 e	7 e	41 e	88.4	159.4	5
$C_{10}H_{22}$	2,2,6-Trimethylheptane	-46 e	-27 e	-2 e	32 e	78.5	148.4	5
$C_{10}H_{22}$	3,3,5-Trimethylheptane			0 e	35 e	82.7	155.2	5
$C_{10}H_{22}$	2,2,3,3-Tetramethylhexane	-46 e	-25 e	1 e	36 e	85.6	159.8	5

| Mol. Form. | Name | \multicolumn{7}{c}{Temperature in °C for the indicated pressure} | | | | | | |
		1 Pa	10 Pa	100 Pa	1 kPa	10 kPa	100 kPa	Ref.
$C_{10}H_{22}$	2,2,5,5-Tetramethylhexane			-10 e	22 e	68.3	137.0	5
$C_{10}H_{22}$	2,4-Dimethyl-3-isopropylpentane	-46 e	-26 e	0 e	35 e	83.2	156.5	5
$C_{10}H_{22}$	2,2,3,3,4-Pentamethylpentane		-24 e	3 e	39 e	89.1	165.5	5
$C_{10}H_{22}$	2,2,3,4,4-Pentamethylpentane		-29 e	-3 e	33 e	82.8	158.7	5
$C_{10}H_{22}O$	1-Decanol	30 e	50 e	75 e	109 e	157.3	230.6	1,39
$C_{10}H_{22}O$	4-Decanol	18 e	37 e	61 e	93 e	139 e	210 e	5
$C_{10}H_{22}O$	Dipentyl ether	-31 e	-8 e	22 e	60 e	111.6	186.2	5
$C_{10}H_{22}O$	Diisopentyl ether			14.0	51.5	101.8	172.8	5
$C_{10}H_{22}O_2$	Ethylene glycol dibutyl ether	0 e	20 e	44 e	78.4	127.1	202.9	5
$C_{10}H_{22}O_5$	Tetraethylene glycol dimethyl ether				138 e	200.9	275.3	5
$C_{10}H_{22}S$	1-Decanethiol	11 e	34 e	64 e	103 e	157.5	238.6	5
$C_{10}H_{22}S$	Diisopentylsulfide			7 e	82 e	118 e	139 e	5
$C_{10}H_{23}N$	Dipentylamine				77 e	127.7	202.0	5
$C_{10}H_{30}O_3Si_4$	Decamethyltetrasiloxane	-31 e	-6 e	26 e	66.8	118.8	193.9	5
$C_{10}H_{30}O_5Si_5$	Decamethylcyclopentasiloxane	-2 e	19 e	46 e	82 e	132.9	210.4	5
$C_{11}H_8O_2$	1-Naphthalenecarboxylic acid				191.9	239.3	299.6	5
$C_{11}H_8O_2$	2-Naphthalenecarboxylic acid				197.9	246.0	308.1	5
$C_{11}H_{10}$	1-Methylnaphthalene	5 e	29 e	60 e	102 e	159.1	244.1	1
$C_{11}H_{10}$	2-Methylnaphthalene			57 e	99 e	156.0	240.5	1
$C_{11}H_{12}O_2$	Ethyl *trans*-cinnamate			79	125	187	271	4
$C_{11}H_{12}O_3$	Myristicin	23 e	53 e	88.9	135.2	196.0	279.4	5
$C_{11}H_{14}$	4-Isopropylstyrene	-25 e	-1 e	30.2	70.3	124.5	202.1	5
$C_{11}H_{14}$	1,2,3,4-Tetrahydro-5-methylnaphthalene	9 e	31 e	60 e	99 e	153.1	233.8	5
$C_{11}H_{14}$	1,2,3,4-Tetrahydro-6-methylnaphthalene	17 e	36 e	62 e	97 e	147.8	228.5	5
$C_{11}H_{14}O_2$	Butyl benzoate	6 e	34 e	67.9	110.3	165 e	237 e	5
$C_{11}H_{16}$	Pentylbenzene	-14 e	8 e	37 e	74 e	126.7	204.9	5
$C_{11}H_{16}$	*p-tert*-Butyltoluene	-24 e	-2 e	27 e	64.1	115.5	190.8	5
$C_{11}H_{16}$	1,3-Diethyl-5-methylbenzene	-26 e	-1 e	29.5	69.5	123.5	200.2	5
$C_{11}H_{16}$	2-Ethyl-1,3,5-trimethylbenzene		6 e	36 e	75.7	129.6	207.6	5
$C_{11}H_{16}$	1-Ethyl-2,4,5-trimethylbenzene	-13 e	11 e	40 e	79.4	132.1	207.7	5
$C_{11}H_{20}$	1-Undecyne	-22 e	0 e	29 e	67 e	118.5	194.5	5
$C_{11}H_{20}$	2-Undecyne	-17 e	6 e	35 e	74 e	127.4	205.4	5
$C_{11}H_{20}O_2$	10-Undecenoic acid	35 e	67 e	105 e	150.0	205.4	274.5	5
$C_{11}H_{20}O_4$	Ethyl diethylmalonate			74 e	105 e	149.4	219 e	5
$C_{11}H_{21}N$	Undecanenitrile			78.6	120.3	177.3	259.9	5
$C_{11}H_{22}$	1-Undecene	-21.6	1.2	29.7	66.4	117.1	192.2	5
$C_{11}H_{22}$	*cis*-2-Undecene	-14 e	7 e	34 e	70.2	120.6	196 e	5
$C_{11}H_{22}$	*trans*-2-Undecene	-14 e	7 e	33 e	69.3	119.6	195 e	5
$C_{11}H_{22}$	*cis*-4-Undecene	-19 e	3 e	30 e	66.6	117.1	192 e	5
$C_{11}H_{22}$	*trans*-4-Undecene	-17 e	4 e	31 e	67.1	117.4	193 e	5
$C_{11}H_{22}$	*cis*-5-Undecene	-19 e	2 e	30 e	66.2	116.7	191 e	5
$C_{11}H_{22}$	*trans*-5-Undecene	-18 e	3 e	31 e	67.0	117.4	192 e	5
$C_{11}H_{22}$	Pentylcyclohexane	-17 e	6 e	34 e	72 e	124.2	202.7	5
$C_{11}H_{22}$	Hexylcyclopentane	-15 e	7 e	36 e	73 e	125.0	202.5	5
$C_{11}H_{22}O$	2-Undecanone	17 e	37 e	64.3	103.0	153.6	232.6	1,5
$C_{11}H_{22}O$	6-Undecanone		28 e	57 e	95 e	148.4	226.9	1
$C_{11}H_{22}O_2$	Undecanoic acid	68 e	90 e	118 e	156 e	207.2	283.6	5
$C_{11}H_{22}O_2$	Heptyl butanoate	2 e	29 e	62 e	102.6	155.1	224.7	5
$C_{11}H_{22}O_2$	Propyl octanoate	-2 e	23 e	55 e	94.0	145.2	215 e	5
$C_{11}H_{22}O_2$	Methyl decanoate	10 e	33 e	62 e	100.9	154.0	232 e	5
$C_{11}H_{24}$	Undecane	-18.4	4.3	32.6	69.5	120.2	195.4	16
$C_{11}H_{24}$	2-Methyldecane	-20 e	1 e	28 e	64 e	114.0	188.7	5
$C_{11}H_{24}$	3-Methyldecane	-35 e	-10 e	22 e	61.9	115.6	190.4	5
$C_{11}H_{24}$	4-Methyldecane	-38 e	-12 e	20 e	60.8	113.9	186.4	5
$C_{11}H_{24}$	2,4,7-Trimethyloctane				43 e	94 e	170.4	5
$C_{11}H_{24}O$	1-Undecanol	52.2	80.0	82 e	118 e	167.6	244.1	5
$C_{11}H_{24}S$	1-Undecanethiol	23 e	47 e	77 e	118 e	173.6	256.8	5
$C_{12}F_{27}N$	Trinonafluorobutylamine		3 e	29.0	63.3	109.9	176.8	5
$C_{12}H_8$	Acenaphthylene	24 s	49.8 s	80.6 s				5
$C_{12}H_9N$	Carbazole					254.7	354.0	5
$C_{12}H_{10}$	Acenaphthene				126.2	187 e	276 e	1
$C_{12}H_{10}$	Biphenyl			69.0	111.1	169.5	254.7	1
$C_{12}H_{10}N_2$	Azobenzene			98.1	144.8	206.7	292.7	4
$C_{12}H_{10}O$	Diphenyl ether		44 e	75 e	116 e	173 e	257.4	5

Mol. Form.	Name	Temperature in °C for the indicated pressure						Ref.
		1 Pa	10 Pa	100 Pa	1 kPa	10 kPa	100 kPa	
$C_{12}H_{10}O$	1-Acetonaphthone	37 e	69 e	107.0	154.6	215.2	294.9	5
$C_{12}H_{10}O$	2-Acetonaphthone	48.3 s		118.7	163.0	221.1	300.3	5
$C_{12}H_{10}S$	Diphenyl sulfide	20 e	51 e	88.7	137.5	202.2	291.8	5
$C_{12}H_{11}N$	Diphenylamine	48 s		102.8	150.5	213.7	301.4	5
$C_{12}H_{12}$	1-Ethylnaphthalene	16 e	41 e	72 e	114 e	171.8	257.7	5
$C_{12}H_{12}$	2-Ethylnaphthalene	14 e	39 e	71 e	113 e	171.2	257.3	5
$C_{12}H_{12}$	1,2-Dimethylnaphthalene	26 e	51 e	82 e	123 e	180.5	265.7	5
$C_{12}H_{12}$	2,7-Dimethylnaphthalene	31.5 s	53.1 s	78.8 s	115.9	175 e	260 e	5
$C_{12}H_{14}O_4$	Diethyl phthalate	12 e	51 e	96 e	150.5	215.9	296.2	5
$C_{12}H_{16}$	p-Isopropenylisopropylbenzene	-11 e	15 e	46 e	87 e	142.4	221 e	5
$C_{12}H_{16}$	Cyclohexylbenzene		28 e	58 e	98 e	154.7	239.5	5
$C_{12}H_{16}O_2$	3-Methylbutyl benzoate			66 e	115.0	177.7	261.4	5
$C_{12}H_{18}$	Hexylbenzene	-2 e	22 e	51 e	90 e	144.5	225.5	5
$C_{12}H_{18}$	1,2-Diisopropylbenzene	-14 e	9 e	37 e	74 e	125.9	203.2	5
$C_{12}H_{18}$	1,3-Diisopropylbenzene	-14 e	8 e	36 e	74 e	125.5	202.6	5
$C_{12}H_{18}$	1,4-Diisopropylbenzene	-6 e	18 e	49 e	90 e	148.8	238 e	5
$C_{12}H_{18}$	Hexamethylbenzene	46.3 s	72.5 s	81.7 s	121.8 s	178.3	263.7	5
$C_{12}H_{18}$	1,5,9-Cyclododecatriene	-14 e	11 e	44 e	87 e	145.0	229.8	5
$C_{12}H_{20}O_2$	Geranyl acetate			67.7	110.8	166.9	242.9	5
$C_{12}H_{20}O_4$	Dibutyl maleate	12.3	50.4	94.0	144.2	203 e	272 e	5
$C_{12}H_{22}$	1-Dodecyne	-11 e	13 e	43 e	82 e	135.8	214.4	5
$C_{12}H_{22}$	Cyclohexylcyclohexane		20 e	53.1	96.0	154.1	237.2	5
$C_{12}H_{22}O_2$	Methyl 10-undecenoate	10 e	38 e	73 e	116 e	172.2	247.1	5
$C_{12}H_{22}O_4$	Dimethyl sebacate		53 e	97	150	214	293	4
$C_{12}H_{23}N$	Dodecanenitrile	36 e	60 e	92 e	133 e	190.5	275.5	5
$C_{12}H_{24}$	1-Dodecene	-8.3	15.2	44.8	82.9	135.4	212.8	5
$C_{12}H_{24}$	Hexylcyclohexane	-3 e	20 e	50 e	89 e	143.1	224.2	5
$C_{12}H_{24}$	Heptylcyclopentane	-1 e	22 e	51 e	90 e	143.5	223.5	5
$C_{12}H_{24}O$	Dodecanal			70 e	116.2	175.9	256.6	5
$C_{12}H_{24}O_2$	Dodecanoic acid	78 e	100 e	128 e	166 e	219.1	298.1	5
$C_{12}H_{24}O_2$	Decyl acetate	12 e	40 e	74 e	115.1	168.1	238 e	5
$C_{12}H_{24}O_2$	Ethyl decanoate	8 e	35 e	69 e	111.8	166.1	238 e	5
$C_{12}H_{25}Br$	1-Bromododecane	31 e	57 e	90 e	132 e	190.8	275.3	5
$C_{12}H_{25}Cl$	1-Chlorododecane	27 e	51 e	81 e	122 e	178.7	262.6	5
$C_{12}H_{26}$	Dodecane	-5.4	18.2	47.6	85.8	138.2	215.8	16
$C_{12}H_{26}O$	1-Dodecanol			133 e	185.0	264.1		1
$C_{12}H_{26}O_3$	Diethylene glycol dibutyl ether	5 e	34.4	70.2	115.3	174.1	253.8	5
$C_{12}H_{27}N$	Tributylamine	-26 e	1 e	35 e	77.7	134.5	213.4	5
$C_{12}H_{27}N$	Triisobutylamine		1 e	28.9	64.9	112.5	178.5	5
$C_{12}H_{27}O_4P$	Tributyl phosphate					205 e	288.3	5
$C_{12}H_{36}O_6Si_6$	Dodecamethylcyclohexasiloxane	18 e	41 e	69 e	108 e	162.2	244.7	5
$C_{13}H_9N$	Acridine			124.4	176.2	246.0	345.4	5
$C_{13}H_9N$	Phenanthridine	79 s						5
$C_{13}H_{10}$	Fluorene	48.4 s			137.4	205.4	295 e	5
$C_{13}H_{10}O_2$	Phenyl benzoate			102.3	151.4	217.9	313.3	5
$C_{13}H_{10}O_3$	Phenyl salicylate				166.0	224.8	312.4	5
$C_{13}H_{12}$	Diphenylmethane		45 e	77 e	119.3	177.7	263.6	1,5
$C_{13}H_{13}N$	Methyldiphenylamine	35 e	63 e	98.4	143.1	201.6	281.6	5
$C_{13}H_{14}$	1-Isopropylnaphthalene	27 e	51 e	82 e	123.2	180.8	267.3	5
$C_{13}H_{20}$	Heptylbenzene	12 e	36 e	66 e	107 e	162.7	246.2	5
$C_{13}H_{24}O_2$	Ethyl 10-undecenoate	32 e	55 e	86 e	125.2	179.5	258.4	5
$C_{13}H_{26}$	1-Tridecene	4.1	28.5	59.0	98.3	152.5	232.3	5
$C_{13}H_{26}$	Heptylcyclohexane	11 e	34 e	65 e	105 e	160.9	244.3	5
$C_{13}H_{26}$	Octylcyclopentane	13 e	36 e	66 e	106 e	160.9	243.1	5
$C_{13}H_{26}O_2$	Tridecanoic acid	87 e	109 e	138 e	176 e	230.3	311.5	5
$C_{13}H_{26}O_2$	Methyl dodecanoate	38 e	61 e	90 e	130 e	184.9	269 e	5
$C_{13}H_{28}$	Tridecane	7.2	31.5	61.8	101.1	155.1	234.9	16
$C_{13}H_{28}O$	1-Tridecanol	71.6	101.0	103 e	140 e	192.3	273.1	5
$C_{14}H_{10}$	Anthracene	89.2 s	125.9 s	151.5 s	165 s	238.8	340.2	1,5
$C_{14}H_{10}$	Phenanthrene	53 s	83 s	120.8	170.4	238.4	337.7	5
$C_{14}H_{10}O_2$	Benzil			123	175	246	346	4
$C_{14}H_{12}$	cis-Stilbene	26 e	54 e	88 e	130.4	183 e	253 e	5
$C_{14}H_{12}$	trans-Stilbene				155.6	218.1	305.8	5
$C_{14}H_{12}O_2$	Benzoin				181	248	342	4
$C_{14}H_{14}$	1,1-Diphenylethane	19 e	47 e	82.0	125.3	181 e	254 e	5
$C_{14}H_{15}N$	Dibenzylamine	48 e	77 e	113.1	158.9	218.5	299.4	5
$C_{14}H_{16}$	1-Butylnaphthalene	67 e	82 e	103 e	135 e	186.7	288.6	5

Mol. Form.	Name	Temperature in °C for the indicated pressure						Ref.
		1 Pa	10 Pa	100 Pa	1 kPa	10 kPa	100 kPa	
$C_{14}H_{16}$	2-Butylnaphthalene	44 e	67 e	98 e	139 e	197.5	287.4	5
$C_{14}H_{22}$	Octylbenzene	20.1	46.2	79.1	121.9	178.1	263.8	5
$C_{14}H_{26}O_4$	Diethyl sebacate		83 e	120	166	225	305	4
$C_{14}H_{27}N$	Tetradecanenitrile	52 e	79 e	114.0	159.0	219.7	306.3	5
$C_{14}H_{28}$	1-Tetradecene	16.1	41.3	72.7	113.2	168.7	250.6	5
$C_{14}H_{28}$	Octylcyclohexane	16.9	44.3	77.8	120.0	177.6	263.2	5
$C_{14}H_{28}$	Nonylcyclopentane	25 e	49 e	80 e	120 e	177.2	261.5	5
$C_{14}H_{28}O_2$	Tetradecanoic acid	96 e	118 e	147 e	186 e	241.3	325.6	5
$C_{14}H_{30}$	Tetradecane	19.1	44.1	75.3	115.7	171.1	253.0	16
$C_{14}H_{30}O$	1-Tetradecanol	80.0	110.5	149.6	152 e	205.3	286.7	5
$C_{14}H_{31}N$	Tetradecylamine			104 e	147 e	206.1	290.9	5
$C_{14}H_{42}O_5Si_6$	Tetradecamethylhexasiloxane	6 e	36 e	72 e	117 e	176.0	259.1	5
$C_{15}H_{18}$	1-Pentylnaphthalene	34 e	62 e	96 e	141.3	202.2	289 e	5
$C_{15}H_{24}$	Nonylbenzene	33.0	58.9	92.0	135.4	193.7	281.4	5
$C_{15}H_{30}$	Nonylcyclohexane	35 e	60 e	92 e	134 e	193.4	280.9	5
$C_{15}H_{30}$	Decylcyclopentane	37 e	61 e	93 e	134 e	192.5	278.8	5
$C_{15}H_{30}O_2$	Methyl tetradecanoate		75 e	110	155	214	295	4
$C_{15}H_{32}$	Pentadecane	30.5	56.1	88.1	129.6	186.3	270.1	16
$C_{16}H_{22}O_4$	Dibutyl phthalate		104.0	142.7	191.5	254.5	339.4	4
$C_{16}H_{32}$	1-Hexadecene	38.4	65.0	98.1	140.5	198.8	284.3	5
$C_{16}H_{32}O_2$	Hexadecanoic acid		136 e	165 e	205 e	261.9	350.2	5
$C_{16}H_{34}$	Hexadecane	41.1	67.4	100.3	142.7	200.7	286.3	16
$C_{16}H_{34}O$	1-Hexadecanol	99.5	130.6	171.9	175 e	229.0	311.7	5
$C_{16}H_{35}N$	Hexadecylamine	63 e	91 e	126 e	171 e	232.6	320.5	5
$C_{17}H_{10}O$	Benzanthrone		184 e	229.3	290.3	377.2	511 e	5
$C_{17}H_{34}O_2$	Methyl hexadecanoate	65 e	93	129	177			4
$C_{17}H_{36}$	Heptadecane	51.5	78.5	112.0	155.3	214.5	302 e	16
$C_{17}H_{36}O$	1-Heptadecanol	94 e	117 e	146 e	185 e	240.1	323.3	5
$C_{18}H_{14}$	o-Terphenyl	66 e	94 e	129 e	176 e	241.3	336.3	5
$C_{18}H_{14}$	m-Terphenyl	87 e	118 e	156 e	206.6	275.3	374.6	5
$C_{18}H_{14}$	p-Terphenyl	127.1 s	154.7 s		217.2	284.0	383.0	5
$C_{18}H_{30}$	Hexaethylbenzene				144.1	206.8	297.5	5
$C_{18}H_{34}O_2$	Oleic acid	94 e	126 e	165.5	214.5	277.0	359.7	5
$C_{18}H_{34}O_2$	Elaidic acid		124 e	166	216	280	361	4
$C_{18}H_{36}O$	Stearaldehyde			142 e	186 e	246.9	336.7	5
$C_{18}H_{36}O_2$	Stearic acid		153 e	183 e	223 e	281.6	374.5	5
$C_{18}H_{38}$	Octadecane	61.5	89.0	123.1	167.3	227.6	316 e	16
$C_{18}H_{38}O$	1-Octadecanol	106 e	130 e	160 e	200.5	257.3	343.0	5
$C_{19}H_{16}$	Triphenylmethane	81 s		112 e	175 e	254.6	360.0	5
$C_{19}H_{36}O_2$	Methyl oleate	85 e	114 e	149.7	195.6	256 e	340 e	5
$C_{19}H_{40}$	Nonadecane	71.1	99.1	133.8	178.8	240.1	330 e	16
$C_{20}H_{42}$	Eicosane	80.4	108.9	144.2	189.8	252.1	344 e	16
$C_{20}H_{42}O$	1-Eicosanol	119 e	143 e	173 e	213 e	270.0	355.1	5
$C_{20}H_{60}O_8Si_9$	Eicosamethylnonasiloxane			141 e	183.1	236.7	307.1	5
$C_{21}H_{21}O_4P$	Tri-o-cresyl phosphate	119.0	156.1	201.0	256.3	326.3	418 e	5
$C_{21}H_{21}O_4P$	Tri-m-cresyl phosphate	147.8	177.3	211.4	251.3	298 e	355 e	5
$C_{21}H_{21}O_4P$	Tri-p-cresyl phosphate	140.6	174 e	214 e	262 e	320 e	392 e	5
$C_{21}H_{44}$	Heneicosane	82.3	113.5	152.2	201.6	263.8	355.9	5
$C_{22}H_{42}O_2$	Brassidic acid	134 e	166 e	203.6	249.8	307.6	382.0	5
$C_{22}H_{42}O_2$	Erucic acid	126 e	160 e	199.4	247.4	306.5	381.1	5
$C_{22}H_{42}O_2$	Butyl oleate	95.5	124.2	158 e	198 e	245 e	304 e	5
$C_{22}H_{44}O_2$	Behenic acid	145.4	176.5	213.7	259.3	316.2	390 e	5
$C_{22}H_{44}O_2$	Butyl stearate	99.6	128 e	162 e	201 e	249 e	307 e	5
$C_{22}H_{46}$	Docosane	83.5	115.0	154.0	203.6	274.8	368.0	5
$C_{23}H_{48}$	Tricosane	102.9	135.1	174.8	221 e	285.3	379.5	5
$C_{24}H_{38}O_4$	Dioctyl phthalate	130 e	163.7	203.8	252 e	311 e	385 e	5
$C_{24}H_{38}O_4$	Bis(2-ethylhexyl) phthalate	122.0	153.2	189.2	231.3	281.1	341.1	5
$C_{24}H_{50}$	Tetracosane	115.0	148.1	188.5	239.1	295.4	390.6	5
$C_{25}H_{52}$	Pentacosane	119.7	152.7	193.2	244.4	305.0	401.1	5
$C_{26}H_{54}$	Hexacosane	125.1	158.8	200.1	252.1	314.3	411.3	5
$C_{27}H_{56}$	Heptacosane	136.7	168.8	206.5	255.8	323.3	421.2	5
$C_{28}H_{58}$	Octacosane	136.5	169.8	210.9	263.1	332.0	430.6	5
$C_{29}H_{60}$	Nonacosane	148.2	182.8	221.2	271.5	340.2	439.7	5
$C_{30}H_{62}$	Squalane	66 e	84 e	105.8	131.9	163.7	203.2	5
C_{70}	Carbon (fullerene-C_{70})	598 s	662 s					22

VAPOR PRESSURE OF FLUIDS AT TEMPERATURES BELOW 300 K

This table gives vapor pressures of 67 important fluids in the temperature range 2 to 300 K. Helium (^{4}He), hydrogen (H_2), and neon (Ne) are covered on this page. The remaining fluids are listed on subsequent pages by molecular formula in the Hill order (see Introduction). The data have been taken from evaluated sources; references are listed at the end of the table.

Pressures are given in kilopascals (kPa). Note that:

1 kPa = 7.50062 Torr
100 kPa = 1 bar
101.325 kPa = 1 atmos

s following an entry indicates that the compound is solid at that temperature.

Helium		Hydrogen		Neon	
T/K	P/kPa	T/K	P/kPa	T/K	P/kPa
2.2	5.3	14.0	7.90	25.0	51.3
2.3	6.7	14.5	10.38	26.0	71.8
2.4	8.3	15.0	13.43	27.0	98.5
2.5	10.2	15.5	17.12	28.0	132.1
2.6	12.4	16.0	21.53	29.0	173.5
2.7	14.8	16.5	26.74	30.0	223.8
2.8	17.5	17.0	32.84	31.0	284.0
2.9	20.6	17.5	39.92	32.0	355.2
3.0	24.0	18.0	48.08	33.0	438.6
3.1	27.8	18.5	57.39	34.0	535.2
3.2	32.0	19.0	67.96	35.0	646.2
3.3	36.5	19.5	79.89	36.0	772.8
3.4	41.5	20.0	93.26	37.0	916.4
3.5	47.0	20.5	108.2	38.0	1078
3.6	52.9	21.0	124.7	39.0	1260
3.7	59.3	21.5	143.1	40.0	1462
3.8	66.1	22.0	163.2	41.0	1688
3.9	73.5	22.5	185.3	42.0	1939
4.0	81.5	23.0	209.4	43.0	2216
4.1	90.0	23.5	235.7	44.0	2522
4.2	99.0	24.0	264.2		
4.3	108.7	24.5	295.1		
4.4	119.0	25.0	328.5		
4.5	129.9	25.5	364.3		
4.6	141.6	26.0	402.9		
4.7	153.9	26.5	444.3		
4.8	167.0	27.0	488.5		
4.9	180.8	27.5	535.7		
5.0	195.4	28.0	586.1		
5.1	210.9	28.5	639.7		
		29.0	696.7		
		29.5	757.3		
		30.0	821.4		
		30.5	889.5		
		31.0	961.5		
		31.5	1038.0		
		32.0	1119.0		
		32.5	1204.0		
Ref.	17,18		1		13

T/K	Ar Argon		BCl₃ Boron trichloride	BF₃ Boron trifluoride	BrH Hydrogen bromide		Br₂ Bromine		ClF Chlorine fluoride	ClH Hydrogen chloride	
50	0.1	s									
55	0.2	s									
60	0.8	s									
65	2.8	s									
70	7.7	s									
75	18.7	s									
80	40.7	s									
85	79.0										
90	134										
95	213										
100	324										
105	473										
110	666										
115	910								0.1		
120	1214								0.3	0.1	s
125	1584								0.6	0.3	s
130	2027								1.2	0.5	s
135	2553				0.1	s			2.1	1.0	s
140	3170				0.3	s			3.6	1.9	s
145	3892			7.7	0.6	s			6.0	3.4	s
150	4736			13.4	1.1	s			9.5	5.8	s
155				22.3	1.9	s			14.6	9.5	s
160				35.2	3.3	s			21.8	14.7	
165				53.7	5.4	s			31.7	22.0	
170				79.1	8.7	s			44.8	31.9	
175				113	13.4	s			62.0	45.1	
180			0.1	157	20.1	s			84.2	62.5	
185			0.2	214	29.5	s			112	84.7	
190			0.3	285	37.9				147	113	
195			0.5	372	51.8				190	148	
200			0.8	479	69.5				242	190	
205			1.2	608	91.8				304	242	
210			1.8	762	119				378	304	
215			2.6	944	153				464	377	
220			3.8	1160	194		0.1	s	564	463	
225			5.2	1413	242		0.2	s	680	563	
230			7.2	1709	299		0.3	s	812	678	
235			9.7	2056	366		0.4	s	961	811	
240			12.9	2460	443		0.7	s	1130	961	
245			17.0	2913	532		1.1	s	1319	1132	
250			22.0	3481	633		1.7	s	1529	1325	
255			28.1	4123	748		2.6	s	1762	1542	
260			35.6	4874	878		3.8	s	2019	1784	
265			44.5		1023		5.5	s	2301	2054	
270			55.1		1185		7.3		2608	2354	
275			67.6		1364		9.5		2941	2686	
280			82.2		1562		12.3		3303	3053	
285			99.1		1780		15.6		3693	3457	
290			119		2018		19.7		4111	3901	
295			141		2278		24.6		4560	4388	
300			166		2561		30.5		5039	4921	
Ref.	8,15		12	12	12		12		12	12	

T/K	ClO_2 Chlorine dioxide	Cl_2 Chlorine	Cl_4Si Silicon tetrachloride	FH Hydrogen fluoride	F_2 Fluorine	F_2O Difluorine oxide	F_3N Nitrogen trifluoride
50							
55					0.4		
60					1.5		
65					4.8		
70					12.3		
75					27.6	0.1	
80					55.3	0.2	
85					101	0.5	0.1
90					172	1.2	0.2
95					276	2.6	0.4
100					420	5.3	0.9
105					615	10.1	2.0
110					870	18.0	4.0
115					1196	30.5	7.3
120					1605	49.3	12.8
125					2108	76.7	21.1
130					2721	115	33.5
135					3458	168	51.1
140					4339	237	75.4
145						328	108
150						444	150
155						588	205
160						766	273
165						981	357
170						1238	459
175		1.8				1541	581
180		2.8				1895	726
185		4.2				2303	896
190		6.1		0.3		2771	1092
195	0.1	8.7		0.5		3302	1319
200	0.3	12.3		0.8		3899	1578
205	0.5	16.9		1.2		4567	1871
210	0.9	22.9	0.1	1.7		5308	2203
215	1.4	30.5	0.2	2.3			2577
220	2.3	40.1	0.3	3.2			2995
225	3.5	51.9	0.5	4.4			3464
230	5.3	66.4	0.7	5.9			3991
235	7.6	84.0	1.0	7.9			
240	10.8	105	1.5	10.3			
245	14.9	130	2.0	13.4			
250	20.1	160	2.8	17.2			
255	26.6	194	3.8	21.8			
260	34.6	234	5.0	27.4			
265	44.4	280	6.6	34.2			
270	56.1	332	8.6	42.2			
275	69.9	392	11.1	51.8			
280	86.2	459	14.2	63.1			
285	105	535	17.9	76.3			
290	127	619	22.3	91.7			
295	151	714	27.7	110			
300	179	818	34.0	130			
Ref.	12	5	12	12	12	12	1

T/K	F₃P Phosphorous trifluoride	F₄Si Silicon tetrafluoride	F₆S Sulfur hexafluoride	HI Hydrogen iodide	H₂S Hydrogen sulfide	H₃N Ammonia	H₃P Phosphine
50							
55							
60							
65							
70							
75							
80							
85							
90							
95							
100							
105	0.1						
110	0.2						0.1
115	0.5						0.2
120	1.0						0.4
125	1.9	0.1 s					0.7
130	3.5	0.2 s					1.3
135	5.9	0.4 s			0.1 s		2.3
140	9.5	0.9 s	0.1 s		0.2 s		3.9
145	14.9	1.9 s	0.2 s		0.3 s		6.2
150	22.5	3.8 s	0.4 s		0.6 s		9.6
155	33.1	7.5 s	0.8 s	0.1 s	1.1 s		14.5
160	47.3	14.0 s	1.5 s	0.2 s	1.9 s	0.1 s	21.1
165	66.0	25.2 s	2.6 s	0.4 s	3.2 s	0.2 s	30.0
170	90.1	43.8 s	4.4 s	0.8 s	5.2 s	0.3 s	41.6
175	121	74.2 s	7.1 s	1.3 s	8.3 s	0.6 s	56.6
180	159	122 s	11.3 s	2.2 s	12.7 s	1.2 s	75.6
185	206	197 s	17.3 s	3.4 s	18.9 s	2.1 s	99.2
190	262	280	25.9 s	5.3 s	26.6	3.5 s	128
195	330	376	38.0 s	8.0 s	36.7	5.8 s	163
200	410	488	54.4 s	11.7 s	49.8	8.7	205
205	503	618	76.6 s	16.8 s	66.4	12.6	254
210	611	766	106 s	23.6 s	87.1	17.9	312
215	736	932	145 s	32.5 s	113	24.9	379
220	877	1117	195 s	44.0 s	144	34.1	456
225	1037	1324	249	56.2	182	45.9	544
230	1217	1555	305	71.4	227	60.8	644
235	1418	1816	371	89.7	281	79.6	756
240	1640	2111	448	112	344	103	881
245	1885	2449	536	137	416	131	1019
250	2154	2841	636	168	500	165	1172
255	2448	3301	750	203	597	207	1341
260	2767		878	244	706	256	1525
265	3112		1021	290	830	313	1725
270			1181	343	969	381	1942
275			1358	404	1124	460	2176
280			1554	472	1297	552	2428
285			1768	548	1488	655	2699
290			2003	633	1698	774	2987
295			2258	727	1929	909	3295
300			2534	831	2181	1062	3621
Ref.	12	12	12,15	12	12,15	11	12

T/K	H_4Si Silane	Kr Krypton		NO Nitric oxide		N_2 Nitrogen		N_2O Nitrous oxide	O_2 Oxygen	O_2S Sulfur dioxide
50						0.4	s		0.2	
55						1.8	s		0.7	
60						6.3	s		2.3	
65						17.4			6.3	
70						38.6			14.5	
75		0.1	s			76.1			30.1	
80		0.4	s			137			56.8	
85		1.1	s	0.1	s	229			99.3	
90		2.7	s	0.4	s	361			163	
95	0.1	6.0	s	1.3	s	541			254	
100	0.2	12.1	s	3.8	s	779			379	
105	0.4	22.8	s	10.0	s	1084			543	
110	1.0	40.4	s	23.5		1467			756	
115	1.9	68.0	s	46.8		1939		0.1	1022	
120	3.5	103		86.5		2513		0.1	1351	
125	6.1	150		151		3209		0.3	1749	
130	10.0	211		248				0.7	2225	
135	15.8	290		391				1.3	2788	
140	24.1	388		592				2.5	3448	
145	35.3	509		867				4.3	4219	
150	50.3	655		1231				7.1		
155	69.8	830		1703				11.4		
160	94.6	1037		2302				17.6		
165	126	1278		3050				26.4		0.1
170	164	1557		3971				38.5		0.2
175	210	1877		5089				54.7		0.3
180	265	2241		6433				75.9		0.5
185	331	2655						103		0.8
190	408	3120						138		1.3
195	498	3641						181		2.0
200	602	4223						234		3.0
205	722	4870						298		4.4
210	859							374		6.3
215	1017							465		9.0
220	1196							571		12.6
225	1398							694		17.3
230	1628							835		23.3
235	1888							996		31.1
240	2180							1179		40.9
245	2509							1385		53.2
250	2880							1615		68.3
255	3296							1870		86.7
260	3763							2152		109
265	4288							2462		136
270								2802		168
275								3172		205
280								3573		249
285								4006		300
290								4473		359
295								4973		426
300								5508		
Ref.	12	13, 15		12, 15		1		12	3	12

T/K	O_3 Ozone	Rn Radon	Xe Xenon		$CBrF_3$ Bromotri-fluoromethane	$CClF_3$ Chlorotri-fluoromethane	CCl_2F_2 Dichlorodi-fluoromethane	CCl_3F Trichloro-fluoromethane
50								
55								
60								
65								
70								
75								
80								
85								
90								
95								
100	0.1		0.1	s				
105	0.2		0.1	s				
110	0.4		0.3	s				
115	1.0		0.7	s		0.1		
120	2.0		1.5	s		0.2		
125	3.8		2.7	s		0.3		
130	6.8	0.1	4.9	s		0.6		
135	11.5	0.3	8.5	s	0.1	1.1		
140	18.7	0.5	14.0	s	0.3	2.0		
145	29.1	0.9	22.2	s	0.5	3.3		
150	43.7	1.5	34.2	s	0.9	5.3		
155	63.6	2.4	51.1	s	1.5	8.3	0.1	
160	89.9	3.8	74.2	s	2.5	12.6	0.3	
165	124	5.8	101		3.9	18.6	0.5	
170	168	8.6	134		5.9	26.8	0.8	
175	222	12.5	173		8.8	37.6	1.3	
180	289	17.7	222		12.8	51.7	2.1	
185	367	24.5	280		18.1	69.7	3.2	
190	468	33.2	348		25.1	92.3	4.8	
195	584	44.4	428		34.1	120	6.9	0.2
200	721	58.2	521		45.6	155	9.9	0.3
205	881	75.3	628		60.0	196	13.7	0.4
210	1068	96	750		77.8	246	18.8	0.6
215	1285	121	889		99.5	304	25.2	1.0
220	1536	151	1045		126	372	33.3	1.4
225	1824	185	1220		157	451	43.3	2.0
230	2155		1416		194	542	55.5	2.9
235	2534		1633		237	646	70.4	4.1
240	2968		1872		287	763	88.1	5.6
245	3464		2136		344	896	109	7.6
250	4031		2425		410	1044	134	10.1
255	4678		2742		485	1210	163	13.3
260	5417		3087		570	1394	196	17.2
265			3462		665	1598	234	22.1
270			3869		771	1823	278	28.0
275			4310		889	2071	327	35.1
280			4786		1021	2343	383	43.7
285			5299		1166	2641	445	53.8
290					1325	2968	515	65.7
295					1501	3325	593	79.6
300					1692	3716	679	95.6
								114.1
Ref.	12	15	12,13		12	12	12	12

T/K	CCl_4 Tetrachloro- methane	CF_4 Tetrafluoro- methane	CO Carbon monoxide		COS Carbon oxysulfide	CO_2 Carbon dioxide		CHClF_2 Chlorodifluo- methane	$CHCl_3$ Trichloro- methane
50			0.1	s					
55			0.6	s					
60			2.6	s					
65			8.2	s					
70			21.0						
75			44.4						
80			83.7						
85			147						
90		0.1	239						
95		0.3	371						
100		0.8	545						
105		1.7	771						
110		3.4	1067						
115		6.5	1428						
120		11.5	1877						
125		19.3	2400						
130		30.8	3064						
135		47.4				0.1	s		
140		70.2			0.1	0.2	s		
145		101			0.2	0.4	s		
150		141			0.4	0.8	s	0.1	
155		191			0.8	1.7	s	0.3	
160		254			1.3	3.1	s	0.5	
165		332			2.2	5.7	s	0.8	
170		425			3.4	9.9	s	1.4	
175		537			5.2	16.8	s	2.3	
180		669			7.8	27.6	s	3.6	
185		824			11.3	44.0	s	5.5	
190		1005			15.9	68.4	s	8.1	
195		1216			22.1	104	s	11.8	
200		1460			30.0	155	s	16.7	
205		1743			40.1	227	s	23.1	
210		2073			52.7	327	s	31.5	
215		2457			68.2	465	s	42.1	0.1
220		2907			87.2	600		55.3	0.2
225		3438			110	735		71.7	0.3
230					137	894		91.6	0.4
235					169	1075		116	0.7
240					207	1283		144	1.0
245					250	1519		178	1.4
250					301	1786		218	2.0
255	1.5				358	2085		264	2.7
260	2.1				423	2419		317	3.7
265	2.8				497	2790		377	5.0
270	3.7				580	3203		446	6.6
275	4.9				673	3658		525	8.7
280	6.4				777	4160		613	11.3
285	8.2				892	4712		711	14.4
290	10.5				1019	5315		821	18.3
295	13.2				1159	5984		944	22.9
300	16.5				1313	6710		1080	28.5
Ref.	12	12	9		12	6		12	12

T/K	CHF₃ Trifluoro-methane	CHN Hydrogen cyanide		CH₂Cl₂ Dichloro-methane	CH₂F₂ Difluoro methane	CH₂O Formaldehyde	CH₃Cl Chloromethane	CH₃F Fluoromethane
50								
55								
60								
65								
70								
75								
80								
85								
90								
95								
100								
105								
110								
115								
120	0.1							
125	0.2							
130	0.4							
135	0.7							0.6
140	1.4				0.1			1.2
145	2.5				0.2			2.1
150	4.3				0.3			3.6
155	7.1				0.6			5.9
160	11.1				1.0			9.3
165	17.0				1.7			14.1
170	25.3				2.8			20.9
175	36.5				4.4			29.9
180	51.4				6.8			42.0
185	70.9				10.2	1.3	2.1	57.6
190	95.8				14.8	2.0	3.1	77.4
195	127				21.2	3.0	4.6	102
200	166	0.1	s	0.1	29.5	4.4	6.7	133
205	214	0.2	s	0.2	40.5	6.4	9.5	171
210	271	0.4	s	0.3	54.5	9.1	13.1	216
215	340	0.6	s	0.4	72.1	12.7	17.9	270
220	421	1	s	0.6	94.1	17.4	24.0	333
225	516	1.5	s	0.9	121	23.4	31.8	408
230	626	2.2	s	1.4	154	31.0	41.4	495
235	754	3.3	s	2.0	193	40.6	53.3	595
240	900	4.7	s	2.8	240	52.5	67.7	711
245	1067	6.8	s	3.8	295	67.0	85.1	843
250	1257	9.7	s	5.3	360	84.6	106	993
255	1472	13.6	s	7.1	434	106	131	1163
260	1713	18.8		9.5	521	131	159	1355
265	1984	24.1		12.4	620	161	193	1571
270	2287	30.5		16.1	732	196	232	1813
275	2624	38.3		20.7	860	236	277	2084
280	3000	47.7		26.3	1004	283	327	2387
285	3418	58.8		33.0	1165	337	385	2724
290	3881	72.1		41.1	1346	399	450	3099
295	4393	87.6		50.8	1547	470	524	3516
300		105.9		62.1	1770	549	606	3978
Ref.	12	12,16		12	12	12	12	12

T/K	CH_4 Methane	CH_4O Methanol	C_2H_2 Acetylene	C_2H_4 Ethylene	C_2H_6 Ethane	C_2H_6O Dimethyl ether	C_3H_4 Propadiene
50							
55							
60							
65	0.1						
70	0.3						
75	0.8						
80	2.1						
85	4.9						
90	10.6						
95	20.0						
100	34.5						
105	57.0						
110	88.4			0.3			
115	133			0.8	0.1		
120	192			1.4	0.4		
125	269			2.7	0.7		
130	368		0.1 s	4.5	1.3		
135	491		0.3 s	7.7	2.2		
140	642		0.7 s	11.9	3.8		
145	824		1.3 s	18.3	6.0		
150	1041		2.6 s	27.5	9.7		0.1
155	1297		4.6 s	39.9	15.0	0.1	0.2
160	1594		7.8 s	56.4	21.5	0.2	0.3
165	1937		12.8 s	77.9	31.0	0.3	0.6
170	2331		20.6 s	105	42.9	0.5	1.0
175	2779		32.2 s	140	59.0	0.9	1.7
180	3288		49.0 s	182	78.7	1.4	2.7
185	3865		72.9 s	234	104	2.1	4.1
190	4520		106 s	296	135	3.2	6.1
195			146	369	172	4.7	8.9
200			190	456	217	6.8	12.5
205			244	557	271	9.6	17.4
210			309	673	334	13.3	23.7
215			385	806	407	18.1	31.6
220			475	958	492	24.3	41.4
225			579	1128	590	32.1	53.5
230		0.1	699	1321	700	41.9	68.2
235		0.2	837	1535	826	53.9	85.8
240		0.4	993	1774	967	68.6	107
245		0.5	1170	2039	1125	86.3	131
250		0.8	1370	2331	1301	108	160
255		1.2	1593	2652	1496	133	193
260		1.7	1843	3005	1712	162	230
265		2.4	2121	3391	1949	197	273
270		3.3	2429	3813	2210	237	322
275		4.5	2771	4275	2495	283	376
280		6.2	3150		2806	335	438
285		8.3	3567		3146	395	506
290		11	4028		3515	463	582
295		14.4	4535		3917	538	666
300		18.7	5093		4355	623	759
Ref.	2,16	12	12,16	4	2	12	12

VAPOR PRESSURE OF FLUIDS AT TEMPERATURES BELOW 300 K (continued)

T/K	C_3H_6 Propylene	C_3H_8 Propane	C_4H_6 Buta-1,3-diene	C_4H_{10} n-Butane	C_4H_{10} Isobutane	C_5H_{12} n-Pentane	C_5H_{12} Neopentane	
50								
55								
60								
65								
70								
75								
80								
85								
90								
95								
100								
105								
110								
115								
120								
125								
130								
135								
140	0.1							
145	0.2							
150	0.4							
155	0.7							
160	1.2	0.8			0.1			
165	2.0	1.4			0.1			
170	3.1	2.2	0.1	0.1	0.3			
175	4.7	3.3	0.2	0.2	0.4			
180	7.0	5.0	0.4	0.3	0.7			
185	10.1	7.3	0.6	0.5	1.1			
190	14.2	10.5	1.0	0.8	1.7		0.1	s
195	19.7	15.0	1.5	1.3	2.5		0.2	s
200	26.9	20.1	2.3	1.9	3.7		0.4	s
205	35.9	27.0	3.4	2.8	5.3		0.7	s
210	47.3	36.0	4.8	4.0	7.4		1.1	s
215	61.3	47.0	6.7	5.7	10.2		1.6	s
220	78.5	60.0	9.2	7.8	13.8	1.0	2.4	s
225	99.2	77.0	12.5	10.6	18.3	1.5	3.6	s
230	124	97.0	16.7	14.1	24.0	2.1	5.2	s
235	153	120	21.9	18.5	31.1	3.0	7.3	s
240	188	148	28.4	24.1	39.8	4.2	10.2	s
245	228	180	36.3	30.9	50.3	5.7	13.9	s
250	274	218	46.0	39.1	62.9	7.6	18.7	s
255	327	261	57.6	49.1	77.8	10.0	24.8	s
260	387	311	71.3	61.0	95.4	13.0	32.4	s
265	456	367	87.6	75.0	116	16.6	41.6	
270	533	431	107	91.5	140	21.1	51.4	
275	619	502	129	111	167	26.6	63.0	
280	715	582	154	133	198	33.1	76.6	
285	822	671	184	159	234	40.8	92.3	
290	940	769	217	188	274	50.0	111	
295	1069	878	255	221	319	60.7	131	
300	1212	998	297	258	370	73.2	155	
							182	
Ref.	7	2	12	2	2	14	12,16	

VAPOR PRESSURE OF FLUIDS AT TEMPERATURES BELOW 300 K (continued)

REFERENCES

1. B. A. Younglove, Thermophysical properties of fluids. I. Ethylene, parahydrogen, nitrogen trifluoride, and oxygen, *J. Phys. Chem. Ref. Data*, 11, Supp. 1, 1982.
2. B. A. Younglove and J. F. Ely, Thermophysical properties of fluids. II. Methane, ethane, propane, isobutane, and normal butane, *J. Phys. Chem. Ref. Data*, 16, 577, 1987.
3. W. Wagner, et al., *International Tables for the Fluid State: Oxygen*, Blackwell Scientific Publications, Oxford, 1987.
4. R. T. Jacobsen, et al., *International Tables for the Fluid State: Ethylene*, Blackwell Scientific Publications, Oxford, 1988.
5. S. Angus, et al., *International Tables for the Fluid State: Chlorine*, Pergamon Press, Oxford, 1985.
6. S. Angus, et al., *International Tables for the Fluid State: Carbon Dioxide*, Pergamon Press, Oxford, 1976.
7. S. Angus, et al., *International Tables for the Fluid State: Propylene*, Pergamon Press, Oxford, 1980.
8. R. B. Stewart and R. T. Jacobsen, Thermophysical properties of argon, *J. Phys. Chem. Ref. Data*, 18, 639, 1989.
9. R. D. Goodwin, Carbon monoxide thermophysical properties, *J. Phys. Chem. Ref. Data*, 14, 849, 1985.
10. R. D. Goodwin, Methanol thermophysical properties, *J. Phys. Chem. Ref. Data*, 16, 799, 1987.
11. L. Haar, Thermodynamic properties of ammonia, *J. Phys. Chem. Ref. Data*, 7, 635, 1978.
12. DIPPR Data Compilation of Pure Compound Properties, Design Institute for Physical Properties Data, American Institute of Chemical Engineers, 1987.
13. V. A. Rabinovich, et al., *Thermophysical Properties of Neon, Argon, Krypton, and Xenon*, Hemisphere Publishing Corp., New York, 1987.
14. K. N. Marsh, *Recommended Reference Methods for the Realization of Physicochemical Properties*, Blackwell Scientific Publications, Oxford, 1987.
15. TRC Thermodynamic Tables: Non-Hydrocarbons, Thermodynamic Research Center, Texas A & M University, College Station, Texas, 1985.
16. R. M. Stevenson and S. Malanowski, *Handbook of the Thermodynamics of Organic Compounds*, Elsevier, New York, 1987.
17. S. Angus and K. M. de Reuck, *International Tables of the Fluid State: Helium-4*, Pergamon Press, Oxford, 1977.
18. R. D. McCarty, *J. Phys. Chem. Ref. Data*, 2, 923, 1973.

VAPOR PRESSURE OF SATURATED SALT SOLUTIONS

This table gives the vapor pressure of water above saturated solutions of some common salts at ambient temperatures. Data on pure water are given on the last line for comparison.

The references provide additional information on water activity, osmotic coefficient, and enthalpy of vaporization.

REFERENCES

1. Apelblat, A., *J. Chem. Thermodynamics,* 24, 619, 1992.
2. Apelblat, A., *J. Chem. Thermodynamics,* 25, 63, 1993.
3. Apelblat, A., *J. Chem. Thermodynamics,* 25, 1513, 1993.
4. Apelblat, A. and Korin, E., *J. Chem. Thermodynamics,* 30, 59, 1998.

Vapor Pressure in kPa

Salt	10°C	15°C	20°C	25°C	30°C	35°C	40°C	Ref.
$BaCl_2$	0.971	1.443	2.073	2.887	3.903	5.133	6.576	1
$Ca(NO_3)_2$	0.701	1.015	1.381	1.772	2.154	2.487		1
$CuSO_4$	1.113	1.574	2.189	2.996	4.037	5.363		3
$FeSO_4$	0.978	1.516	2.208	3.035	3.950	4.884		3
KBr	0.953	1.338	1.853	2.533	3.419	4.563		3
KIO_3	1.100	1.564	2.177	2.970	3.979	5.236	6.778	4
K_2CO_3	0.541	0.802	1.134	1.536	1.997	2.499	3.016	1
LiCl	0.128	0.193	0.279	0.384				2
$Mg(NO_3)_2$	0.726	0.999	1.339	1.749	2.231	2.782	3.397	1
$MnCl_2$	0.697	1.064	1.515	2.020	2.535	3.002		3
NH_4Cl	0.971	1.328	1.836	2.481				2
NH_4NO_3	0.853	1.152	1.524	1.972				2
$(NH_4)_2SO_4$	0.901	1.319	1.871	2.573	3.439	4.474		3
NaBr	0.722	1.004	1.376	1.858	2.475	3.255	4.229	4
NaCl	0.921	1.285	1.768	2.401	3.218	4.262	5.581	4
$NaNO_2$	0.703	0.994	1.381	1.888	2.540	3.368	4.403	4
$NaNO_3$	0.884	1.244	1.719	2.335	3.121	4.109	5.333	4
RbCl	0.862	1.215	1.684	2.298	3.088	4.089	5.343	4
$ZnSO_4$	0.945	1.401	1.986	2.698	3.523	4.431	5.382	1
Water	1.228	1.706	2.339	3.169	4.246	5.627	7.381	

IUPAC RECOMMENDED DATA FOR VAPOR PRESSURE CALIBRATION

Vapor pressures are given in kPa (1 kPa = 0.0098692 atmos = 7.5006 Torr). Reprinted with permission if IUPAC.

REFERENCE

Marsh, K.N., Ed., *Recommended Reference Materials for the Realization of Physicochemical Properties*, Blackwell Scientific Publications, Oxford (1987).

T/K	$CO_2(s)$	$H_2O(s)$	$C_{10}H_8(s)$	$n\text{-}C_5H_{12}$	C_6H_6	C_6F_6	H_2O	Hg
180	27.62							
190	68.44							
200	155.11	0.0002						
210	327.17	0.0007						
220		0.0026						
230		0.0089						
240		0.0273						
250		0.0760		7.60				
260		0.1958	0.0001	12.98			0.485	
270		0.4701	0.0005	21.15			0.991	
280			0.0017	33.11	5.148	4.322	1.919	
290			0.0049	50.01	8.606	7.463	3.535	
300			0.0134	73.17	13.816	12.328	6.228	
310			0.0341	104.07	21.389	19.576	10.540	
320			0.0814	144.3	32.054	30.009	17.202	
330			0.1829	195.7	46.656	44.578	27.167	
340			0.3899	260.1	66.152	64.380	41.647	
350			0.7920	339.4	91.609	90.664	62.139	
360				435.9	124.192	124.816	90.453	
370				551.5	165.2	168.4	128.74	
380				688.8	215.9	223.0	179.48	
390				850.2	277.7	290.4	245.54	0.138
400				1038	353.2	372.6	330.15	0.215
410				1256	441.0	471.5	436.89	0.329
420				1507	545.5	589.3	569.73	0.493
430				1793	667.6	728.3	732.99	0.724
440				2120	808.8	890.9	931.36	1.045
450				2490	971.1	1080	1169.9	1.485
460				2910	1156	1297	1453.9	2.078
470					1366	1547	1789.0	2.866
480					1602	1833	2181.4	3.899
490					1868	2159	2637.3	5.239
500					2164	2530	3163.3	6.955
510					2494	2954	3766.4	9.131
520					2861		4453.9	11.861
530					3267		5233.5	15.256
540					3717		6113.4	19.438
550					4216		7102.0	25.547
560					4770		8208.6	30.74
570							9443.0	38.19
580							10816	47.09
590							12339	57.64
600								

ENTHALPY OF VAPORIZATION

The molar enthalpy (heat) of vaporization $\Delta_{vap}H$, which is defined as the enthalpy change in the conversion of one mole of liquid to gas at constant temperature, is tabulated here for approximately 850 inorganic and organic compounds. Values are given, when available, both at the normal boiling point t_b, referred to a pressure of 101.325 kPa (760 mmHg), and at 25°C. Substances are listed by molecular formula in the modified Hill order (see Preface).

The values in this table were measured either by calorimetric techniques or by application of the Claperyon equation to the variation of vapor pressure with temperature. See Reference 1 for a discussion of the accuracy of different experimental techniques and for methods of estimating enthalpy of vaporization at other temperatures.

REFERENCES

1. Majer, V. and Svoboda, V., *Enthalpies of Vaporization of Organic Compounds,* Blackwell Scientific Publications, Oxford, 1985.
2. Chase, M. W., Davies, C. A., Downey, J. R., Frurip, D. J., McDonald, R. A., and Syverud, A. N., *JANAF Thermochemical Tables, Third Edition, J. Phys. Chem. Ref. Data,* Vol. 14, Suppl. 1, 1985.
3. *Landolt-Börnstein, Numerical Data and Functional Relationships in Science and Technology, Sixth Edition,* II/4, *Caloric Quantities of State,* Springer-Verlag, Heidelberg, 1961.
4. Daubert, T. E., Danner, R. P., Sibul, H. M., and Stebbins, C. C., *Physical and Thermodynamic Properties of Pure Compounds: Data Compilation,* extant 1994 (core with 4 supplements), Taylor & Francis, Bristol, PA.
5. Ruzicka, K. and Majer, V., "Simultaneous Treatment of Vapor Pressures and Related Thermal Data Between the Triple and Normal Boiling Temperatures for *n*-Alkanes C_5 - C_{20}", *J. Phys. Chem. Ref. Data,* 23, 1, 1994.
6. Verevkin, S. P., "Thermochemistry of Amines: Experimental Standard Molar Enthalpies of Formation of Some Aliphatic and Aromatic Amines", *J. Chem. Thermodynamics,* 29, 891, 1997.
7. Cady, G. H. and Hargreaves, G. B., "The Vapor Pressure of Some Heavy Transition Metal Hexafluorides", *J. Chem. Soc.,* 1961, 1563; 1961, 1578.
8. Steele, W. V., Chirico, R. D., Knipmeyer, S. E., and Nguyen, A., *J. Chem. Eng. Data,* 41, 1255, 1996.

Mol. Form.	Name	t_b/°C	$\Delta_{vap}H(t_b)$ kJ/mol	$\Delta_{vap}H(25°C)$ kJ/mol
AgBr	Silver(I) bromide	1502	198	
AgCl	Silver(I) chloride	1547	199	
AgI	Silver(I) iodide	1506	143.9	
Al	Aluminum	2519	294	
AlB_3H_{12}	Aluminum borohydride	44.5	30	
$AlBr_3$	Aluminum tribromide	255	23.5	
AlI_3	Aluminum triiodide	382	32.2	
Ar	Argon	-185.85	6.43	
$AsBr_3$	Arsenic(III) bromide	221	41.8	
$AsCl_3$	Arsenic(III) chloride	130	35.01	
AsF_3	Arsenic(III) fluoride	57.8	29.7	
AsF_5	Arsenic(V) fluoride	-52.8	20.8	
AsH_3	Arsine	-62.5	16.69	
AsI_3	Arsenic(III) iodide	424	59.3	
Au	Gold	2856	324	
B	Boron	4000	480	
BBr_3	Boron tribromide	91	30.5	
BCl_3	Boron trichloride	12.65	23.77	23.1
BF_3	Boron trifluoride	-101	19.33	
BI_3	Boron triiodide	210	40.5	
B_2F_4	Tetrafluorodiborane	-34	28	
B_2H_6	Diborane	-92.4	14.28	
B_4H_{10}	Tetraborane	18	27.1	
B_5H_{11}	Pentaborane(11)	63	31.8	
Ba	Barium	1897	140	
$BeCl_2$	Beryllium chloride	482	105	
BeI_2	Beryllium iodide	487	70.5	
Bi	Bismuth	1564	151	
$BiBr_3$	Bismuth tribromide	453	75.4	
$BiCl_3$	Bismuth trichloride	447	72.61	
BrF	Bromine fluoride	20	25.1	
BrF_3	Bromine trifluoride	125.8	47.57	

Mol. Form.	Name	t_b/°C	$\Delta_{vap}H(t_b)$ kJ/mol	$\Delta_{vap}H(25°C)$ kJ/mol
BrF_5	Bromine pentafluoride	40.76	30.6	
BrH	Hydrogen bromide	-66.38		12.69
BrH_3Si	Bromosilane	1.9	24.4	
BrIn	Indium(I) bromide	656	92	
BrTl	Thallium(I) bromide	819	99.56	
Br_2	Bromine	58.8	29.96	30.91
Br_2Cd	Cadmium bromide	844	115	
Br_2H_2Si	Dibromosilane	66	31	
Br_2Hg	Mercury(II) bromide	322	58.89	
Br_2Pb	Lead(II) bromide	892	133	
Br_2Sn	Tin(II) bromide	639	102	
Br_2Zn	Zinc bromide	697	118	
Br_3Ga	Gallium(III) bromide	279	38.9	
Br_3HSi	Tribromosilane	109	34.8	
Br_3OP	Phosphorus(V) oxybromide	191.7	38	
Br_3P	Phosphorus(III) bromide	172.95	38.8	
Br_3Sb	Antimony(III) bromide	280	59	
Br_4Ge	Germanium(IV) bromide	186.35	41.4	
Br_4Si	Tetrabromosilane	154	37.9	
Br_4Sn	Tin(IV) bromide	205	43.5	
Br_4Ti	Titanium(IV) bromide	230	44.37	
Br_5Ta	Tantalum(V) bromide	349	62.3	
Cd	Cadmium	767	99.87	
$CdCl_2$	Cadmium chloride	960	124.3	
CdF_2	Cadmium fluoride	1748	214	
CdI_2	Cadmium iodide	742	115	
ClF	Chlorine fluoride	-101.1	24	
$ClFO_3$	Perchloryl fluoride	-46.75	19.33	
ClF_2P	Phosphorus(III) chloride difluoride	-47.25	17.6	
ClF_3	Chlorine trifluoride	11.75	27.53	
ClF_3Si	Chlorotrifluorosilane	-70.0	18.7	
ClH	Hydrogen chloride	-85	16.15	9.08
ClH_3Si	Chlorosilane	-30.4	21	
ClNO	Nitrosyl chloride	-5.5	25.78	
$ClNO_2$	Nitryl chloride	-15	25.7	
ClO_2	Chlorine dioxide	11	30	
ClTl	Thallium(I) chloride	720	102.2	
Cl_2	Chlorine	-34.04	20.41	17.65
Cl_2Cr	Chromium(II) chloride	1300	197	
Cl_2CrO_2	Chromyl chloride	117	35.1	
Cl_2FP	Phosphorus(III) dichloride fluoride	14	24.9	
Cl_2F_2Si	Dichlorodifluorosilane	-32	21.2	
Cl_2H_2Si	Dichlorosilane	8.3	25	24.2
Cl_2Hg	Mercury(II) chloride	304	58.9	
Cl_2O	Chlorine monoxide	2.2	25.9	
Cl_2OS	Thionyl chloride	75.6	31.7	31
Cl_2O_2S	Sulfuryl chloride	69.4	31.4	30.1
Cl_2Pb	Lead(II) chloride	951	127	
Cl_2Sn	Tin(II) chloride	623	86.8	
Cl_2Ti	Titanium(II) chloride	1500	232	
Cl_2Zn	Zinc chloride	732	126	
Cl_3Ga	Gallium(III) chloride	201	23.9	
Cl_3HSi	Trichlorosilane	33		25.7
Cl_3OP	Phosphorus(V) oxychloride	105.5	34.35	38.6
Cl_3OV	Vanadyl trichloride	127	36.78	
Cl_3P	Phosphorus(III) chloride	75.95	30.5	32.1
Cl_3Sb	Antimony(III) chloride	220.3	45.19	
Cl_3Ti	Titanium(III) chloride	960	124	

Mol. Form.	Name	$t_b/°C$	$\Delta_{vap}H(t_b)$ kJ/mol	$\Delta_{vap}H(25°C)$ kJ/mol
Cl_4Ge	Germanium(IV) chloride	86.55	27.9	
Cl_4OW	Tungsten(VI) oxytetrachloride	227.55	67.8	
Cl_4Si	Tetrachlorosilane	57.65	28.7	29.7
Cl_4Sn	Tin(IV) chloride	114.15	34.9	
Cl_4Te	Tellurium tetrachloride	387	77	
Cl_4Th	Thorium(IV) chloride	921	146.4	
Cl_4Ti	Titanium(IV) chloride	136.45	36.2	
Cl_4V	Vanadium(IV) chloride	148	41.4	42.5
Cl_5Mo	Molybdenum(V) chloride	268	62.8	
Cl_5Nb	Niobium(V) chloride	254.0	52.7	
Cl_5Ta	Tantalum(V) chloride	239.35	54.8	
Cl_6W	Tungsten(VI) chloride	346.75	52.7	
FH_3Si	Fluorosilane	-98.6	18.8	
FLi	Lithium fluoride	1673	147	
FNO	Nitrosyl fluoride	-59.9	19.28	
FNO_2	Nitryl fluoride	-72.4	18.05	
FNS	Thionitrosyl fluoride (NSF)	4.8	22.2	
F_2	Fluorine	-188.12	6.62	
F_2H_2Si	Difluorosilane	-77.8	16.3	
F_2O	Fluorine monoxide	-144.75	11.09	
F_2OS	Thionyl fluoride	-43.8	21.8	
F_2O_2	Fluorine dioxide	-57	19.1	
F_2Pb	Lead(II) fluoride	1293	160.4	
F_2Zn	Zinc fluoride	1500	190.1	
F_3HSi	Trifluorosilane	-95	16.2	
F_3N	Nitrogen trifluoride	-128.75	11.56	
F_3O_2Re	Rhenium(VII) dioxytrifluoride	185.4	65.7	
F_3P	Phosphorus(III) fluoride	-101.5	16.5	
F_3PS	Phosphorus(V) sulfide trifluoride	-52.25	19.6	
F_4MoO	Molybdenum(VI) oxytetrafluoride	186.0	50.6	
F_4N_2	Tetrafluorohydrazine	-74	13.27	
F_4ORe	Rhenium(VI) oxytetrafluoride	171.7	61.0	
F_4OW	Tungsten(VI) oxytetrafluoride	185.9	59.5	
F_4S	Sulfur tetrafluoride	-40.45	26.44	
F_4Se	Selenium tetrafluoride	106	47.2	
F_4Th	Thorium(IV) fluoride	1680	258	
F_5I	Iodine pentafluoride	100.5	41.3	
F_5Mo	Molybdenum(V) fluoride	213.6	51.8	
F_5Nb	Niobium(V) fluoride	229	52.3	
F_5Os	Osmium(V) fluoride	225.9	65.6	
F_5P	Phosphorus(V) fluoride	-84.6	17.2	
F_5Re	Rhenium(V) fluoride	221.3	58.1	
F_5Ta	Tantalum(V) fluoride	229.2	56.9	
F_5V	Vanadium(V) fluoride	48.3	44.52	
F_6Ir	Iridium(VI) fluoride	53.6	30.9	
F_6Mo	Molybdenum(VI) fluoride	34.0	29.0	
F_6Os	Osmium(VI) fluoride	47.5	28.1	
F_6Re	Rhenium(VI) fluoride	33.8	28.7	
F_6S	Sulfur hexafluoride			8.99
F_6W	Tungsten(VI) fluoride	17.1	26.5	
Ga	Gallium	2204	254	
GaI_3	Gallium(III) iodide	340	56.5	
Ge	Germanium	2833	334	
GeH_4	Germane	-88.1	14.06	
Ge_2H_6	Digermane	30.8	25.1	
Ge_3H_8	Trigermane	110.5	32.2	
HI	Hydrogen iodide	-35.55	19.76	17.36
$HLiO$	Lithium hydroxide	1626	188	

Mol. Form.	Name	$t_b/°C$	$\Delta_{vap}H(t_b)$ kJ/mol	$\Delta_{vap}H(25°C)$ kJ/mol
HNO_3	Nitric acid	83		39.1
HN_3	Hydrazoic acid	35.7	30.5	
$HNaO$	Sodium hydroxide	1388	175	
H_2	Hydrogen	-252.87	0.90	
H_2O	Water	100.0	40.65	43.98
H_2O_2	Hydrogen peroxide	150.2		51.6
H_2S	Hydrogen sulfide	-59.55	18.67	14.08
H_2S_2	Hydrogen disulfide	70.7		33.78
H_2Se	Hydrogen selenide	-41.25	19.7	
H_2Te	Hydrogen telluride	-2	19.2	
H_3N	Ammonia	-33.33	23.33	19.86
H_3P	Phosphine	-87.75	14.6	
H_3Sb	Stibine	-17	21.3	
H_4N_2	Hydrazine	113.55	41.8	44.7
H_4P_2	Diphosphine	63.5	28.8	
H_4Si	Silane	-111.9	12.1	
H_4Sn	Stannane	-51.8	19.05	
H_6Si_2	Disilane	-14.3	21.2	
H_8Si_3	Trisilane	52.9	28.5	
He	Helium	-268.93	0.08	
Hg	Mercury	356.73	59.11	
HgI_2	Mercury(II) iodide	354	59.2	
IIn	Indium(I) iodide	712	90.8	
ITl	Thallium(I) iodide	824	104.7	
I_2	Iodine	184.4	41.57	
I_2Pb	Lead(II) iodide	872	104	
I_2Sn	Tin(II) iodide	714	105	
I_3P	Phosphorus(III) iodide	227	43.9	
I_3Sb	Antimony(III) iodide	401	68.6	
I_4Si	Tetraiodosilane	287.35	50.2	
I_4Sn	Tin(IV) iodide	364.35	56.9	
I_4Ti	Titanium(IV) iodide	377	58.4	
Kr	Krypton	-153.22	9.08	
MoO_3	Molybdenum(VI) oxide	1155	138	
NO	Nitric oxide	-151.74	13.83	
N_2	Nitrogen	-195.79	5.57	
N_2O	Nitrous oxide	-88.48	16.53	
N_2O_4	Nitrogen tetroxide	21.15	38.12	
Ne	Neon	-246.08	1.71	
O_2	Oxygen	-182.95	6.82	
O_2S	Sulfur dioxide	-10.05	24.94	22.92
O_3S	Sulfur trioxide	45	40.69	43.14
P	Phosphorus	280.5	12.4	14.2
Pb	Lead	1749	179.5	
S	Sulfur	444.60	45	
STl_2	Thallium(I) sulfide	1367	154	
Se	Selenium	685	95.48	
Te	Tellurium	988	114.1	
Xe	Xenon	-108.11	12.57	
$CClF_3$	Chlorotrifluoromethane	-81.4	15.8	
CCl_2F_2	Dichlorodifluoromethane	-29.8	20.1	
CCl_3F	Trichlorofluoromethane	23.7	25.1	
CCl_4	Tetrachloromethane	76.8	29.82	32.43
$CHBr_3$	Tribromomethane	149.1	39.66	46.05
$CHClF_2$	Chlorodifluoromethane	-40.7	20.2	
$CHCl_2F$	Dichlorofluoromethane	8.9	25.2	
$CHCl_3$	Trichloromethane	61.17	29.24	31.28
CH_2BrCl	Bromochloromethane	68.0	30.0	

Mol. Form.	Name	$t_b/°C$	$\Delta_{vap}H(t_b)$ kJ/mol	$\Delta_{vap}H(25°C)$ kJ/mol
CH_2Br_2	Dibromomethane	97	32.92	36.97
CH_2Cl_2	Dichloromethane	40	28.06	28.82
CH_2I_2	Diiodomethane	182	42.5	
CH_2O_2	Formic acid	101	22.69	20.10
CH_3Br	Bromomethane	3.5	23.91	22.81
CH_3Cl	Chloromethane	-24.09	21.40	18.92
CH_3I	Iodomethane	42.55	27.34	27.97
CH_3NO	Formamide	220		60.15
CH_3NO_2	Nitromethane	101.19	33.99	38.27
CH_4	Methane	-161.48	8.19	
CH_4O	Methanol	64.6	35.21	37.43
CH_5N	Methylamine	-6.32	25.60	23.37
CH_6N_2	Methylhydrazine	87.5	36.12	40.37
CN_4O_8	Tetranitromethane	126.1	40.74	49.93
CO	Carbon monoxide	-191.5	6.04	
CS_2	Carbon disulfide	46	26.74	27.51
$C_2Br_2ClF_3$	1,2-Dibromo-1-chloro-1,2,2-trifluoroethane	93	31.17	35.04
$C_2Br_2F_4$	1,2-Dibromotetrafluoroethane	47.35	27.03	28.39
C_2ClF_5	Chloropentafluoroethane	-37.95	19.41	
$C_2Cl_2F_4$	1,2-Dichlorotetrafluoroethane	3.8	23.3	
$C_2Cl_3F_3$	1,1,1-Trichlorotrifluoroethane	46.1	26.85	28.08
$C_2Cl_3F_3$	1,1,2-Trichloro-1,2,2-trifluoroethane	47.7	27.04	28.40
C_2Cl_4	Tetrachloroethylene	121.3	34.68	39.68
C_2F_6	Hexafluoroethane	-78.1	16.15	
$C_2HBrClF_3$	2-Bromo-2-chloro-1,1,1-trifluoroethane	50.2	28.08	29.61
C_2HCl_3	Trichloroethylene	87.21	31.40	34.54
C_2HCl_5	Pentachloroethane	159.8	36.9	
$C_2HF_3O_2$	Trifluoroacetic acid	73	33.3	
$C_2H_2Br_4$	1,1,2,2-Tetrabromoethane	243.5	48.7	
$C_2H_2Cl_2$	1,1-Dichloroethylene	31.6	26.14	26.48
$C_2H_2Cl_2$	cis-1,2-Dichloroethylene	60.1	30.2	
$C_2H_2Cl_2$	trans-1,2-Dichloroethylene	48.7	28.9	
$C_2H_2Cl_4$	1,1,2,2-Tetrachloroethane	146.5	37.64	45.71
C_2H_3Br	Bromoethylene	15.8	23.4	
C_2H_3Cl	Chloroethylene	-13.3	20.8	
$C_2H_3Cl_2F$	1,1-Dichloro-1-fluoroethane	32.0	26.06	26.48
$C_2H_3Cl_3$	1,1,1-Trichloroethane	74.09	29.86	32.50
$C_2H_3Cl_3$	1,1,2-Trichloroethane	113.8	34.82	40.24
$C_2H_3F_3$	1,1,1-Trifluoroethane	-47.25	18.99	
C_2H_3N	Acetonitrile	81.65	29.75	32.94
C_2H_4	Ethylene	-103.77	13.53	
$C_2H_4Br_2$	1,2-Dibromoethane	131.6	34.77	41.73
$C_2H_4Cl_2$	1,1-Dichloroethane	57.4	28.85	30.62
$C_2H_4Cl_2$	1,2-Dichloroethane	83.5	31.98	35.16
$C_2H_4F_2$	1,1-Difluoroethane	-24.95	21.56	19.08
C_2H_4O	Acetaldehyde	20.1	25.76	25.47
C_2H_4O	Ethylene oxide	10.6	25.54	24.75
$C_2H_4O_2$	Acetic acid	117.9	23.70	23.36
$C_2H_4O_2$	Methyl formate	31.7	27.92	28.35
C_2H_5Br	Bromoethane	38.5	27.04	28.03
C_2H_5Cl	Chloroethane	12.3	24.65	
C_2H_5ClO	2-Chloroethanol	128.6	41.4	
C_2H_5I	Iodoethane	72.5	29.44	31.93
C_2H_5NO	N-Methylformamide	199.51		56.19
$C_2H_5NO_2$	Nitroethane	114.0	38.0	
C_2H_6	Ethane	-88.6	14.69	5.16
C_2H_6O	Ethanol	78.29	38.56	42.32
C_2H_6O	Dimethyl ether	-24.8	21.51	18.51

Mol. Form.	Name	$t_b/°C$	$\Delta_{vap}H(t_b)$ kJ/mol	$\Delta_{vap}H(25°C)$ kJ/mol
C_2H_6OS	Dimethyl sulfoxide	189	43.1	
$C_2H_6O_2$	Ethylene glycol	197.3	50.5	
C_2H_6S	Ethanethiol	35.1	26.79	27.30
C_2H_6S	Dimethyl sulfide	37.33	27.0	27.65
$C_2H_6S_2$	1,2-Ethanedithiol	146.1	37.93	44.68
$C_2H_6S_2$	Dimethyl disulfide	109.8	33.78	37.86
C_2H_7N	Dimethylamine	6.88	26.40	25.05
C_2H_7NO	Ethanolamine	171	49.83	
$C_2H_8N_2$	1,2-Ethanediamine	117	37.98	44.98
$C_2H_8N_2$	1,1-Dimethylhydrazine	63.9	32.55	35.0
C_2N_2	Cyanogen	-21.1	23.33	19.75
$C_3Cl_2F_6$	1,2-Dichlorohexafluoropropane	34.1	26.28	26.93
$C_3H_3Cl_3O_2$	Methyl trichloroacetate	153.8		48.33
C_3H_3N	Acrylonitrile	77.3	32.6	
$C_3H_4Cl_2O_2$	Methyl dichloroacetate	142.9	39.28	47.72
C_3H_4O	Acrolein	52.6	28.3	
$C_3H_4O_2$	2-Oxetanone	162		47.03
C_3H_5Br	3-Bromopropene	70.1	30.24	32.73
C_3H_5Cl	3-Chloropropene	45.1	29.0	
$C_3H_5ClO_2$	Methyl chloroacetate	129.5	39.23	46.73
$C_3H_5Cl_3$	1,2,3-Trichloropropane	157	37.1	
C_3H_5N	Propanenitrile	97.14	31.81	36.03
C_3H_6	Propene	-47.69	18.42	14.24
C_3H_6	Cyclopropane	-32.81	20.05	16.93
$C_3H_6Br_2$	1,2-Dibromopropane	141.9	35.61	41.67
$C_3H_6Br_2$	1,3-Dibromopropane	167.3		47.45
$C_3H_6Cl_2$	1,3-Dichloropropane	120.9	35.18	40.75
C_3H_6O	Allyl alcohol	97.0	40.0	
C_3H_6O	Propanal	48	28.31	29.62
C_3H_6O	Acetone	56.05	29.10	30.99
C_3H_6O	Methyloxirane	35	27.35	27.89
C_3H_6O	Oxetane	47.6	28.67	29.85
$C_3H_6O_2$	Propanoic acid	141.15		32.14
$C_3H_6O_2$	Ethyl formate	54.4	29.91	31.96
$C_3H_6O_2$	Methyl acetate	56.87	30.32	32.29
C_3H_6S	Thiacyclobutane	95	32.32	35.97
C_3H_7Br	1-Bromopropane	71.1	29.84	32.01
C_3H_7Br	2-Bromopropane	59.5	28.33	30.17
C_3H_7Cl	1-Chloropropane	46.5	27.18	28.35
C_3H_7Cl	2-Chloropropane	35.7	26.30	26.90
C_3H_7I	1-Iodopropane	102:6	32.08	36.25
C_3H_7I	2-Iodopropane	89.5	30.68	34.06
C_3H_7NO	N-Ethylformamide	198		58.44
C_3H_7NO	N,N-Dimethylformamide	153		46.89
$C_3H_7NO_2$	1-Nitropropane	131.1	38.5	
$C_3H_7NO_2$	2-Nitropropane	120.2	36.8	
C_3H_8	Propane	-42.1	19.04	14.79
C_3H_8O	1-Propanol	97.2	41.44	47.45
C_3H_8O	2-Propanol	82.3	39.85	45.39
$C_3H_8O_2$	1,2-Propylene glycol	187.6	52.4	
$C_3H_8O_2$	1,3-Propylene glycol	214.4	57.9	
$C_3H_8O_2$	Ethylene glycol monomethyl ether	124.1	37.54	45.17
$C_3H_8O_3$	Glycerol	290	61.0	
C_3H_8S	1-Propanethiol	67.8	29.54	31.89
C_3H_8S	2-Propanethiol	52.6	27.91	29.45
C_3H_8S	Ethyl methyl sulfide	66.7	29.53	31.85
$C_3H_8S_2$	1,3-Propanedithiol	172.9		49.66
C_3H_9N	Propylamine	47.22	29.55	31.27

Mol. Form.	Name	$t_b/°C$	$\Delta_{vap}H(t_b)$ kJ/mol	$\Delta_{vap}H(25°C)$ kJ/mol
C_3H_9N	Isopropylamine	31.76	27.83	28.36
C_3H_9N	Trimethylamine	2.87	22.94	21.66
$C_3H_{10}N_2$	1,3-Propanediamine	139.8	40.85	50.16
C_4F_8	Perfluorocyclobutane	-5.9	23.2	
C_4F_{10}	Perfluorobutane	-1.9	22.9	
$C_4H_4N_2$	Succinonitrile	266	48.5	
$C_4H_4N_2$	Pyrimidine	123.8	43.09	49.79
$C_4H_4N_2$	Pyridazine	208		53.47
C_4H_4O	Furan	31.5	27.10	27.45
$C_4H_4O_2$	Diketene	126.1	36.80	42.89
C_4H_4S	Thiophene	84.0	31.48	34.70
$C_4H_5Cl_3O_2$	Ethyl trichloroacetate	167.5		50.97
C_4H_5N	2-Methylacrylonitrile	90.3	31.8	
C_4H_5N	Pyrrole	129.79	38.75	45.09
C_4H_5N	Cyclopropanecarbonitrile	135.1	35.55	41.94
$C_4H_5NO_2$	Methyl cyanoacetate	200.5	48.2	
C_4H_5NS	4-Methylthiazole	133.3	37.58	43.85
C_4H_6	1,2-Butadiene	10.9	24.02	23.21
C_4H_6	1,3-Butadiene	-4.41	22.47	20.86
C_4H_6	1-Butyne	8.08	24.52	23.35
$C_4H_6Cl_2O_2$	Ethyl dichloroacetate	155		50.60
$C_4H_6O_2$	Vinyl acetate	72.5	34.6	
$C_4H_6O_2$	Methyl acrylate	80.7	33.1	
$C_4H_6O_2$	γ-Butyrolactone	204	52.2	
$C_4H_6O_3$	Acetic anhydride	139.5	38.2	
C_4H_6S	2,3-Dihydrothiophene	112.1	33.24	37.74
C_4H_6S	2,5-Dihydrothiophene	122.4	34.83	39.95
$C_4H_7ClO_2$	Ethyl chloroacetate	144.3	40.43	49.47
C_4H_7N	Butanenitrile	117.6	33.68	39.33
C_4H_7N	2-Methylpropanenitrile	103.9	32.39	37.13
C_4H_8	1-Butene	-6.26	22.07	20.22
C_4H_8	cis-2-Butene	3.71	23.34	22.16
C_4H_8	trans-2-Butene	0.88	22.72	21.40
C_4H_8	Cyclobutane	12.6	24.19	23.51
$C_4H_8Br_2$	1,4-Dibromobutane	197		53.09
$C_4H_8Cl_2$	1,2-Dichlorobutane	124.1	33.90	39.58
$C_4H_8Cl_2$	1,4-Dichlorobutane	161		46.36
$C_4H_8Cl_2O$	Bis(2-chloroethyl) ether	178.5	45.2	
C_4H_8O	Ethyl vinyl ether	35.5	26.2	
C_4H_8O	1,2-Epoxybutane	63.4	30.3	
C_4H_8O	Butanal	74.8	31.5	
C_4H_8O	2-Butanone	79.59	31.30	34.79
C_4H_8O	Tetrahydrofuran	65	29.81	31.99
$C_4H_8O_2$	Butanoic acid	163.75		40.45
$C_4H_8O_2$	2-Methylpropanoic acid	154.45		35.30
$C_4H_8O_2$	Propyl formate	80.9	33.61	37.53
$C_4H_8O_2$	Ethyl acetate	77.11	31.94	35.60
$C_4H_8O_2$	Methyl propanoate	79.8	32.24	35.85
$C_4H_8O_2$	1,3-Dioxane	106.1	34.37	39.09
$C_4H_8O_2$	1,4-Dioxane	101.5	34.16	38.60
C_4H_8S	Tetrahydrothiophene	121.0	34.66	39.43
C_4H_9Br	1-Bromobutane	101.6	32.51	36.64
C_4H_9Br	2-Bromobutane	91.3	30.77	34.41
C_4H_9Br	1-Bromo-2-methylpropane	91.1	31.33	34.82
C_4H_9Br	2-Bromo-2-methylpropane	73.3	29.23	31.81
C_4H_9Cl	1-Chlorobutane	78.6	30.39	33.51
C_4H_9Cl	2-Chlorobutane	68.2	29.17	31.53
C_4H_9Cl	1-Chloro-2-methylpropane	68.5	29.22	31.67

Mol. Form.	Name	$t_b/°C$	$\Delta_{vap}H(t_b)$ kJ/mol	$\Delta_{vap}H(25°C)$ kJ/mol
C_4H_9Cl	2-Chloro-2-methylpropane	50.9	27.55	28.98
C_4H_9I	1-Iodobutane	130.6	34.66	40.63
C_4H_9I	2-Iodobutane	120.1	33.27	38.46
C_4H_9I	1-Iodo-2-methylpropane	121.1	33.54	38.83
C_4H_9I	2-Iodo-2-methylpropane	100.1	31.43	35.41
C_4H_9N	Pyrrolidine	86.56	33.01	37.52
C_4H_9NO	N-Ethylacetamide	205		64.89
C_4H_9NO	N,N-Dimethylacetamide	165		50.24
C_4H_9NO	Morpholine	128	37.1	
C_4H_{10}	Butane	-0.5	22.44	21.02
C_4H_{10}	Isobutane	-11.73	21.30	19.23
$C_4H_{10}O$	1-Butanol	117.73	43.29	52.35
$C_4H_{10}O$	2-Butanol	99.51	40.75	49.72
$C_4H_{10}O$	2-Methyl-1-propanol	107.89	41.82	50.82
$C_4H_{10}O$	2-Methyl-2-propanol	82.4	39.07	46.69
$C_4H_{10}O$	Diethyl ether	34.5	26.52	27.10
$C_4H_{10}O$	Methyl propyl ether	39.1	26.75	27.60
$C_4H_{10}O$	Isopropyl methyl ether	30.77	26.05	26.41
$C_4H_{10}O_2$	1,2-Butanediol	190.5	52.84	71.55
$C_4H_{10}O_2$	1,3-Butanediol	207.5	54.31	74.46
$C_4H_{10}O_2$	Ethylene glycol monoethyl ether	135	39.22	48.21
$C_4H_{10}O_2$	Ethylene glycol dimethyl ether	85	32.42	36.39
$C_4H_{10}O_3$	Diethylene glycol	245.8	52.3	
$C_4H_{10}S$	1-Butanethiol	98.5	32.23	36.63
$C_4H_{10}S$	2-Butanethiol	85	30.59	33.99
$C_4H_{10}S$	2-Methyl-1-propanethiol	88.5	31.01	34.63
$C_4H_{10}S$	2-Methyl-2-propanethiol	64.3	28.45	30.78
$C_4H_{10}S$	Diethyl sulfide	92.1	31.77	35.80
$C_4H_{10}S$	Methyl propyl sulfide	95.6	32.08	36.24
$C_4H_{10}S$	Isopropyl methyl sulfide	84.8	30.71	34.15
$C_4H_{10}S_2$	1,4-Butanedithiol	195.5		55.10
$C_4H_{10}S_2$	Diethyl disulfide	154.1	37.58	45.18
$C_4H_{11}N$	Butylamine	77.00	31.81	35.72
$C_4H_{11}N$	sec-Butylamine	62.73	29.92	32.85
$C_4H_{11}N$	tert-Butylamine	44.04	28.27	29.64
$C_4H_{11}N$	Isobutylamine	67.75	30.61	33.85
$C_4H_{11}N$	Diethylamine	55.5	29.06	31.31
$C_4H_{11}N$	Isopropylmethylamine	50.4	28.71	30.69
$C_4H_{11}NO$	2-Amino-2-methyl-1-propanol	165.5	50.6	
$C_4H_{11}NO_2$	Diethanolamine	268.8	65.2	
$C_5H_2F_6O_2$	Hexafluoroacetylacetone	54.15	27.05	30.58
$C_5H_4O_2$	Furfural	161.7	43.2	
C_5H_5N	Pyridine	115.23	35.09	40.21
$C_5H_6O_2$	Furfuryl alcohol	171	53.6	
C_5H_6S	2-Methylthiophene	112.6	33.90	38.87
C_5H_6S	3-Methylthiophene	115.5	34.24	39.43
C_5H_7N	trans-3-Pentenenitrile	142.6	37.09	44.77
C_5H_7N	Cyclobutanecarbonitrile	149.6	36.88	44.34
C_5H_8	Spiropentane	39	26.76	27.49
C_5H_8O	Cyclopropyl methyl ketone	111.3	34.07	39.41
C_5H_8O	Cyclopentanone	130.57	36.35	42.72
$C_5H_8O_2$	Methyl cyclopropanecarboxylate	114.9	35.25	41.27
$C_5H_8O_2$	Allyl acetate	103.5	36.3	
$C_5H_8O_2$	Ethyl acrylate	99.4	34.7	
$C_5H_8O_2$	Methyl methacrylate	100.5	36.0	
$C_5H_8O_2$	2,4-Pentanedione	138	34.30	41.77
C_5H_9N	Pentanenitrile	141.3	36.09	43.60
C_5H_9N	3-Methylbutanenitrile	127.5	35.10	41.64

Mol. Form.	Name	$t_b/°C$	$\Delta_{vap}H(t_b)$ kJ/mol	$\Delta_{vap}H(25°C)$ kJ/mol
C_5H_9N	2,2-Dimethylpropanenitrile	106.1	32.40	37.35
C_5H_{10}	1-Pentene	29.96	25.20	25.47
C_5H_{10}	cis-2-Pentene	36.93		26.86
C_5H_{10}	trans-2-Pentene	36.34		26.76
C_5H_{10}	2-Methyl-1-butene	31.2	25.50	25.92
C_5H_{10}	3-Methyl-1-butene	20.1		23.77
C_5H_{10}	2-Methyl-2-butene	38.56	26.31	27.06
C_5H_{10}	Cyclopentane	49.3	27.30	28.52
$C_5H_{10}Cl_2$	1,2-Dichloropentane	148.3	36.45	43.89
$C_5H_{10}Cl_2$	1,5-Dichloropentane	179		50.71
$C_5H_{10}O$	Cyclopentanol	140.42		57.05
$C_5H_{10}O$	2-Pentanone	102.26	33.44	38.40
$C_5H_{10}O$	3-Pentanone	101.96	33.45	38.52
$C_5H_{10}O$	3-Methyl-2-butanone	94.33	32.35	36.78
$C_5H_{10}O$	3,3-Dimethyloxetane	80.6	30.85	33.94
$C_5H_{10}O$	Tetrahydropyran	88	31.17	34.58
$C_5H_{10}O_2$	Pentanoic acid	186.1	44.1	
$C_5H_{10}O_2$	2-Methylbutanoic acid	177		46.91
$C_5H_{10}O_2$	Butyl formate	106.1	36.58	41.11
$C_5H_{10}O_2$	Isobutyl formate	98.2	33.6	
$C_5H_{10}O_2$	Propyl acetate	101.54	33.92	39.72
$C_5H_{10}O_2$	Isopropyl acetate	88.6	32.93	37.20
$C_5H_{10}O_2$	Ethyl propanoate	99.1	33.88	39.21
$C_5H_{10}O_2$	Methyl butanoate	102.8	33.79	39.28
$C_5H_{10}O_2$	Methyl isobutanoate	92.5	32.61	37.32
$C_5H_{10}O_2$	Tetrahydrofurfuryl alcohol	178	45.2	
$C_5H_{10}O_3$	Diethyl carbonate	126		43.60
$C_5H_{10}O_3$	Ethylene glycol monomethyl ether acetate	143	43.9	
$C_5H_{10}S$	Thiacyclohexane	141.8	35.96	42.58
$C_5H_{10}S$	Cyclopentanethiol	132.1	35.32	41.42
$C_5H_{11}Br$	1-Bromopentane	129.8	35.01	41.28
$C_5H_{11}Cl$	1-Chloropentane	107.8	33.15	38.24
$C_5H_{11}Cl$	2-Chloropentane	97.0	31.79	36.03
$C_5H_{11}Cl$	1-Chloro-3-methylbutane	98.9	32.02	36.24
$C_5H_{11}I$	1-Iodopentane	155		45.27
$C_5H_{11}N$	Piperidine	106.22		39.29
C_5H_{12}	Pentane	36.06	25.79	26.43
C_5H_{12}	Isopentane	27.88	24.69	24.85
C_5H_{12}	Neopentane	9.48	22.74	21.84
$C_5H_{12}O$	1-Pentanol	137.98	44.36	57.02
$C_5H_{12}O$	2-Pentanol	119.3	41.40	54.21
$C_5H_{12}O$	3-Pentanol	116.25		54.0
$C_5H_{12}O$	2-Methyl-1-butanol	128		55.16
$C_5H_{12}O$	3-Methyl-1-butanol	131.1	44.07	55.61
$C_5H_{12}O$	2-Methyl-2-butanol	102.4	39.04	50.10
$C_5H_{12}O$	3-Methyl-2-butanol	112.9		53.0
$C_5H_{12}O$	Butyl methyl ether	70.16	29.55	32.37
$C_5H_{12}O$	sec-Butyl methyl ether	59.1	28.09	30.23
$C_5H_{12}O$	Methyl tert-butyl ether	55.2	27.94	29.82
$C_5H_{12}O$	Isobutyl methyl ether	58.6	28.02	30.13
$C_5H_{12}O$	Ethyl propyl ether	63.21	28.94	31.43
$C_5H_{12}O$	Ethyl isopropyl ether	54.1	28.21	30.08
$C_5H_{12}O_2$	1-Ethoxy-2-methoxyethane	102.1	34.33	39.83
$C_5H_{12}O_2$	1,5-Pentanediol	239	60.7	
$C_5H_{12}O_2$	Ethylene glycol monopropyl ether	149.8	41.40	52.12
$C_5H_{12}O_2$	Diethoxymethane	88	31.33	35.65
$C_5H_{12}O_3$	Diethylene glycol monomethyl ether	193	46.6	
$C_5H_{12}S$	1-Pentanethiol	126.6	34.88	41.24

Mol. Form.	Name	$t_b/°C$	$\Delta_{vap}H(t_b)$ kJ/mol	$\Delta_{vap}H(25°C)$ kJ/mol
$C_5H_{12}S$	2-Methyl-1-butanethiol	119.1	33.79	39.45
$C_5H_{12}S$	2-Methyl-2-butanethiol	99.1	31.37	35.67
$C_5H_{12}S$	Butyl methyl sulfide	123.5	34.47	40.46
$C_5H_{12}S$	tert-Butyl methyl sulfide	99	31.47	35.84
$C_5H_{12}S$	Ethyl propyl sulfide	118.6	34.24	39.97
$C_5H_{12}S$	Ethyl isopropyl sulfide	107.5	32.74	37.78
$C_5H_{13}N$	Pentylamine	104.3	34.01	40.08
$C_5H_{13}N$	Ethylisopropylamine	69.6	29.94	33.13
C_6ClF_5	Chloropentafluorobenzene	117.96	34.76	41.07
C_6F_6	Hexafluorobenzene	80.26	31.66	35.71
C_6HF_5	Pentafluorobenzene	85.74	32.15	36.27
$C_6H_4Cl_2$	o-Dichlorobenzene	180	39.66	50.21
$C_6H_4Cl_2$	m-Dichlorobenzene	173	38.62	48.58
$C_6H_4Cl_2$	p-Dichlorobenzene	174	38.79	49.0
$C_6H_4F_2$	o-Difluorobenzene	94	32.21	36.18
$C_6H_4F_2$	m-Difluorobenzene	82.6	31.10	34.59
$C_6H_4F_2$	p-Difluorobenzene	89	31.77	35.54
C_6H_5Br	Bromobenzene	156.06		44.54
C_6H_5Cl	Chlorobenzene	131.72	35.19	40.97
C_6H_5F	Fluorobenzene	84.73	31.19	34.58
C_6H_5I	Iodobenzene	188.4	39.5	
$C_6H_5NO_2$	Nitrobenzene	210.8		55.01
C_6H_6	Benzene	80.09	30.72	33.83
C_6H_6ClN	o-Chloroaniline	208.8	44.4	
C_6H_6O	Phenol	181.87	45.69	57.82
C_6H_6S	Benzenethiol	169.1	39.93	47.56
C_6H_7N	Aniline	184.17	42.44	55.83
C_6H_7N	2-Methylpyridine	129.38	36.17	42.48
C_6H_7N	3-Methylpyridine	144.14	37.35	44.44
C_6H_7N	4-Methylpyridine	145.36	37.51	44.56
C_6H_7N	1-Cyclopentenecarbonitrile			44.98
C_6H_9N	Cyclopentanecarbonitrile			43.43
$C_6H_9NO_3$	Triacetamide			60.41
C_6H_{10}	Cyclohexene	82.98	30.46	33.47
$C_6H_{10}O$	Cyclohexanone	155.43		45.06
$C_6H_{10}O$	Mesityl oxide	130	36.1	
$C_6H_{10}O_2$	Methyl cyclobutanecarboxylate	135.5	37.13	44.72
$C_6H_{10}O_3$	Propanoic anhydride	170	41.7	
$C_6H_{10}O_4$	Diethyl oxalate	185.7	42.0	
$C_6H_{10}O_4$	Ethylene glycol diacetate	190		61.44
$C_6H_{11}N$	Hexanenitrile	163.65		47.91
C_6H_{12}	1-Hexene	63.48		30.61
C_6H_{12}	cis-2-Hexene	68.8		32.19
C_6H_{12}	trans-2-Hexene	67.9		31.60
C_6H_{12}	cis-3-Hexene	66.4		31.23
C_6H_{12}	trans-3-Hexene	67.1		31.55
C_6H_{12}	2-Methyl-1-pentene	62.1		30.48
C_6H_{12}	3-Methyl-1-pentene	54.2		28.62
C_6H_{12}	4-Methyl-1-pentene	53.9		28.71
C_6H_{12}	2-Methyl-2-pentene	67.3		31.60
C_6H_{12}	3-Methyl-cis-2-pentene	67.7		32.09
C_6H_{12}	3-Methyl-trans-2-pentene	70.4		31.35
C_6H_{12}	4-Methyl-cis-2-pentene	56.3		29.48
C_6H_{12}	4-Methyl-trans-2-pentene	58.6		29.97
C_6H_{12}	2-Ethyl-1-butene	64.7		31.13
C_6H_{12}	2,3-Dimethyl-1-butene	55.6		29.18
C_6H_{12}	3,3-Dimethyl-1-butene	41.2		26.61
C_6H_{12}	2,3-Dimethyl-2-butene	73.3	29.64	32.51

Mol. Form.	Name	t_b/°C	$\Delta_{vap}H(t_b)$ kJ/mol	$\Delta_{vap}H(25°C)$ kJ/mol
C_6H_{12}	Cyclohexane	80.73	29.97	33.01
C_6H_{12}	Methylcyclopentane	71.8	29.08	31.64
C_6H_{12}	Ethylcyclobutane	70.8	28.67	31.24
$C_6H_{12}Cl_2$	1,2-Dichlorohexane	173		48.16
$C_6H_{12}O$	Butyl vinyl ether	94	31.58	36.17
$C_6H_{12}O$	2-Hexanone	127.6	36.35	43.14
$C_6H_{12}O$	3-Hexanone	123.5	35.36	42.47
$C_6H_{12}O$	3-Methyl-2-pentanone	117.5	34.16	40.53
$C_6H_{12}O$	4-Methyl-2-pentanone	116.5	34.49	40.61
$C_6H_{12}O$	2-Methyl-3-pentanone	113.5	33.84	39.79
$C_6H_{12}O$	3,3-Dimethyl-2-butanone	106.1	33.39	37.91
$C_6H_{12}O$	Cyclohexanol	160.84		62.01
$C_6H_{12}O_2$	Butyl acetate	126.1	36.28	43.86
$C_6H_{12}O_2$	tert-Butyl acetate	95.1	33.07	38.03
$C_6H_{12}O_2$	Isobutyl acetate	116.5	35.9	
$C_6H_{12}O_2$	Propyl propanoate	122.5	35.54	43.45
$C_6H_{12}O_2$	Ethyl butanoate	121.5	35.47	42.68
$C_6H_{12}O_2$	Ethyl 2-methylpropanoate	110.1	33.67	39.83
$C_6H_{12}O_2$	Methyl pentanoate	127.4	35.36	43.10
$C_6H_{12}O_2$	Methyl 2,2-dimethylpropanoate	101.1	33.42	38.76
$C_6H_{12}O_3$	Ethylene glycol monoethyl ether acetate	156.4	40.76	52.61
$C_6H_{12}S$	Cyclohexanethiol	158.9	37.06	44.57
$C_6H_{13}Br$	1-Bromohexane	155.3		45.89
$C_6H_{13}Cl$	1-Chlorohexane	135	35.67	42.83
$C_6H_{13}I$	1-Iodohexane	181		49.75
$C_6H_{13}N$	Cyclohexylamine	134	36.14	43.67
C_6H_{14}	Hexane	68.73	28.85	31.56
C_6H_{14}	2-Methylpentane	60.26	27.79	29.89
C_6H_{14}	3-Methylpentane	63.27	28.06	30.28
C_6H_{14}	2,2-Dimethylbutane	49.73	26.31	27.68
C_6H_{14}	2,3-Dimethylbutane	57.93	27.38	29.12
$C_6H_{14}N_2$	Azopropane	114		39.88
$C_6H_{14}O$	1-Hexanol	157.6	44.50	61.61
$C_6H_{14}O$	2-Hexanol	140	41.01	58.46
$C_6H_{14}O$	2-Methyl-1-pentanol	149	50.2	
$C_6H_{14}O$	4-Methyl-1-pentanol	151.9	44.46	60.47
$C_6H_{14}O$	2-Methyl-2-pentanol	121.1	39.59	54.77
$C_6H_{14}O$	4-Methyl-2-pentanol	131.6	44.2	
$C_6H_{14}O$	2-Ethyl-1-butanol	147	43.2	
$C_6H_{14}O$	Dipropyl ether	90.08	31.31	35.69
$C_6H_{14}O$	Diisopropyl ether	68.51	29.10	32.12
$C_6H_{14}O$	Butyl ethyl ether	92.3	31.63	36.32
$C_6H_{14}O$	Methyl pentyl ether	99	32.02	36.85
$C_6H_{14}O_2$	2-Methyl-2,4-pentanediol	197.1	57.3	
$C_6H_{14}O_2$	Ethylene glycol monobutyl ether	168.4		56.59
$C_6H_{14}O_2$	1,1-Diethoxyethane	102.25	36.28	43.20
$C_6H_{14}O_2$	Ethylene glycol diethyl ether	119.4	36.28	43.20
$C_6H_{14}O_3$	Bis(ethoxymethyl) ether	140.6	36.17	44.69
$C_6H_{14}O_3$	Diethylene glycol monoethyl ether	196	47.5	
$C_6H_{14}O_3$	Diethylene glycol dimethyl ether	162	36.17	44.69
$C_6H_{14}O_4$	Triethylene glycol	285	71.4	
$C_6H_{14}S$	Dipropyl sulfide	142.9	36.60	44.21
$C_6H_{14}S$	Diisopropyl sulfide	120.1	33.80	39.60
$C_6H_{14}S$	Isopropyl propyl sulfide	132.1	35.11	41.78
$C_6H_{14}S$	Butyl ethyl sulfide	144.3	37.01	44.51
$C_6H_{14}S$	Methyl pentyl sulfide	145.1	37.41	45.24
$C_6H_{15}N$	Hexylamine	132.8	36.54	45.10
$C_6H_{15}N$	Butylethylamine	107.5	33.97	40.15

Mol. Form.	Name	t_b/°C	$\Delta_{vap}H(t_b)$ kJ/mol	$\Delta_{vap}H(25°C)$ kJ/mol
$C_6H_{15}N$	Dipropylamine	109.3	33.47	40.04
$C_6H_{15}N$	Diisopropylamine	83.9	30.40	34.61
$C_6H_{15}N$	Isopropylpropylamine	96.9	32.14	37.23
$C_6H_{15}N$	Triethylamine	89	31.01	34.84
C_6MoO_6	Molybdenum hexacarbonyl	701	72.51	
$C_7H_3F_5$	2,3,4,5,6-Pentafluorotoluene	117.5	34.75	41.12
$C_7H_5F_3$	(Trifluoromethyl)benzene	102.1	32.63	37.60
C_7H_5N	Benzonitrile	191.1	45.9	
C_7H_6O	Benzaldehyde	179.0	42.5	
$C_7H_6O_2$	Salicylaldehyde	197	38.2	
C_7H_7Cl	o-Chlorotoluene	159.0	37.5	
C_7H_7Cl	p-Chlorotoluene	162.4	38.7	
C_7H_7F	o-Fluorotoluene	115	35.4	
C_7H_7F	p-Fluorotoluene	116.6	34.08	39.42
C_7H_8	Toluene	110.63	33.18	38.01
C_7H_8O	o-Cresol	191.04	45.19	
C_7H_8O	m-Cresol	202.27	47.40	61.71
C_7H_8O	p-Cresol	201.98	47.45	
C_7H_8O	Benzyl alcohol	205.31	50.48	
C_7H_8O	Anisole	153.7	38.97	46.90
C_7H_9N	Benzylamine	185		60.16
C_7H_9N	o-Methylaniline	200.3	44.6	
C_7H_9N	m-Methylaniline	203.3	44.9	
C_7H_9N	p-Methylaniline	200.4	44.3	
C_7H_9N	1-Cyclohexenecarbonitrile			53.55
C_7H_9N	2,3-Dimethylpyridine	161.12	39.08	47.82
C_7H_9N	2,4-Dimethylpyridine	158.38	38.53	47.49
C_7H_9N	2,5-Dimethylpyridine	156.98	38.68	47.04
C_7H_9N	2,6-Dimethylpyridine	144.01	37.46	45.34
C_7H_9N	3,4-Dimethylpyridine	179.10	39.99	50.50
C_7H_9N	3,5-Dimethylpyridine	171.84	39.46	49.33
$C_7H_{10}O$	Dicyclopropyl ketone	161		53.70
$C_7H_{11}N$	Cyclohexanecarbonitrile			51.92
C_7H_{12}	1-Methylbicyclo(3,1,0)hexane	93.1	31.07	34.77
$C_7H_{12}O_4$	Diethyl malonate	200	54.8	
C_7H_{14}	1-Heptene	93.64		35.49
C_7H_{14}	cis-2-Heptene	98.4		36.26
C_7H_{14}	trans-2-Heptene	98		36.27
C_7H_{14}	cis-3-Heptene	95.8		35.81
C_7H_{14}	trans-3-Heptene	95.7		35.84
C_7H_{14}	cis-3-Methyl-3-hexene	95.4		36.31
C_7H_{14}	trans-3-Methyl-3-hexene	93.5		35.70
C_7H_{14}	2,4-Dimethyl-1-pentene	81.6		33.03
C_7H_{14}	4,4-Dimethyl-1-pentene	72.5		31.13
C_7H_{14}	2,4-Dimethyl-2-pentene	83.4		34.19
C_7H_{14}	cis-4,4-Dimethyl-2-pentene	80.4		32.56
C_7H_{14}	trans-4,4-Dimethyl-2-pentene	76.7		32.81
C_7H_{14}	2-Ethyl-3-methyl-1-butene	89		34.35
C_7H_{14}	2,3,3-Trimethyl-1-butene	77.9		32.09
C_7H_{14}	Methylcyclohexane	100.93	31.27	35.36
C_7H_{14}	Ethylcyclopentane	103.5	31.96	36.40
C_7H_{14}	cis-1,3-Dimethylcyclopentane	90.8	30.40	34.20
$C_7H_{14}O$	2-Heptanone	151.05		47.24
$C_7H_{14}O$	2,2-Dimethyl-3-pentanone	125.6	36.09	42.34
$C_7H_{14}O$	2,4-Dimethyl-3-pentanone	125.4	34.64	41.51
$C_7H_{14}O$	1-Methylcyclohexanol	155	79.0	
$C_7H_{14}O$	cis-2-Methylcyclohexanol	165	48.5	
$C_7H_{14}O$	trans-2-Methylcyclohexanol	167.5	53.0	

Mol. Form.	Name	$t_b/°C$	$\Delta_{vap}H(t_b)$ kJ/mol	$\Delta_{vap}H(25°C)$ kJ/mol
$C_7H_{14}O_2$	Pentyl acetate	149.2	38.42	48.56
$C_7H_{14}O_2$	Isopentyl acetate	142.5	37.5	
$C_7H_{14}O_2$	Ethyl pentanoate	146.1	36.96	47.01
$C_7H_{14}O_2$	Ethyl 3-methylbutanoate	135.0	37.0	
$C_7H_{14}O_2$	Ethyl 2,2-dimethylpropanoate	118.4	34.51	41.25
$C_7H_{14}O_2$	Methyl hexanoate	149.5	38.55	48.04
$C_7H_{15}Br$	1-Bromoheptane	179		50.60
$C_7H_{15}Cl$	1-Chloroheptane	159		47.66
C_7H_{16}	Heptane	98.5	31.77	36.57
C_7H_{16}	2-Methylhexane	90.04	30.62	34.87
C_7H_{16}	3-Methylhexane	92	30.9	
C_7H_{16}	3-Ethylpentane	93.5	31.12	35.22
C_7H_{16}	2,2-Dimethylpentane	79.2	29.23	32.42
C_7H_{16}	2,3-Dimethylpentane	89.78	30.46	34.26
C_7H_{16}	2,4-Dimethylpentane	80.49	29.55	32.88
C_7H_{16}	3,3-Dimethylpentane	86.06	29.62	33.03
C_7H_{16}	2,2,3-Trimethylbutane	80.86	28.90	32.05
$C_7H_{16}O$	Hexyl methyl ether	126.1	34.93	42.07
$C_7H_{16}O$	1-Heptanol	176.45		66.81
$C_7H_{16}O$	3-Heptanol	157	42.5	
$C_7H_{16}O$	Butyl propyl ether	118.1	33.72	40.22
$C_7H_{16}O$	Ethyl pentyl ether	117.6	34.41	41.01
$C_7H_{17}N$	Heptylamine	156		49.96
C_8F_{18}	Perfluorooctane	105.9	33.38	41.13
C_8H_8	Styrene	145	38.7	
C_8H_8O	Acetophenone	202	43.98	55.40
$C_8H_8O_2$	Methyl benzoate	199		55.57
$C_8H_8O_3$	Methyl salicylate	222.9	46.7	
C_8H_{10}	Ethylbenzene	136.19	35.57	42.24
C_8H_{10}	o-Xylene	144.5	36.24	43.43
C_8H_{10}	m-Xylene	139.12	35.66	42.65
C_8H_{10}	p-Xylene	138.37	35.67	42.40
$C_8H_{10}O$	2,4-Xylenol	210.98		64.96
$C_8H_{10}O$	2,5-Xylenol	211.1	46.9	
$C_8H_{10}O$	2,6-Xylenol	201.07		75.31
$C_8H_{10}O$	3,4-Xylenol	227		85.03
$C_8H_{10}O$	3,5-Xylenol	221.74		82.01
$C_8H_{10}O$	Phenetole	169.81		51.04
$C_8H_{11}N$	N-Ethylaniline	203.0		58.3
$C_8H_{11}N$	N,N-Dimethylaniline	194.15		52.83
$C_8H_{11}N$	2,4-Dimethylaniline	214		61.3
$C_8H_{11}N$	2,5-Dimethylaniline	214		61.7
$C_8H_{11}N$	2,3,6-Trimethylpyridine	171.6	39.95	50.61
$C_8H_{11}N$	2,4,6-Trimethylpyridine	170.6	39.87	50.33
C_8H_{14}	1-Octyne	126.3	35.83	42.30
C_8H_{14}	2-Octyne	137.6	37.26	44.49
C_8H_{14}	3-Octyne	133.1	36.94	43.92
C_8H_{14}	4-Octyne	131.6	36.0	42.73
$C_8H_{14}O_3$	Butanoic anhydride	200	50.0	
$C_8H_{15}N$	Octanenitrile	205.25		56.80
C_8H_{16}	1-Octene	121.29	34.07	40.34
C_8H_{16}	cis-2,2-Dimethyl-3-hexene	105.5		36.86
C_8H_{16}	trans-2,2-Dimethyl-3-hexene	100.8		37.03
C_8H_{16}	3-Ethyl-2-methyl-1-pentene	109.5		37.27
C_8H_{16}	2,4,4-Trimethyl-1-pentene	101.4		35.59
C_8H_{16}	2,4,4-Trimethyl-2-pentene	104.9		37.23
C_8H_{16}	Ethylcyclohexane	131.9	34.04	40.56
C_8H_{16}	1,1-Dimethylcyclohexane	119.6	32.51	37.92

Mol. Form.	Name	$t_b/°C$	$\Delta_{vap}H(t_b)$ kJ/mol	$\Delta_{vap}H(25°C)$ kJ/mol
C_8H_{16}	cis-1,2-Dimethylcyclohexane	129.8	33.47	39.70
C_8H_{16}	trans-1,2-Dimethylcyclohexane	123.5	32.96	38.36
C_8H_{16}	cis-1,3-Dimethylcyclohexane	120.1	32.91	38.26
C_8H_{16}	trans-1,3-Dimethylcyclohexane	124.5	33.39	39.16
C_8H_{16}	cis-1,4-Dimethylcyclohexane	124.4	33.28	39.02
C_8H_{16}	trans-1,4-Dimethylcyclohexane	119.4	32.56	37.90
C_8H_{16}	Propylcyclopentane	131	34.70	41.08
C_8H_{16}	Isopropylcyclopentane	126.5	33.56	39.44
C_8H_{16}	1-Ethyl-1-methylcyclopentane	121.6	33.20	38.85
$C_8H_{16}O$	2,2,4-Trimethyl-3-pentanone	135.1	35.64	43.30
$C_8H_{16}O_2$	Octanoic acid	239	58.5	
$C_8H_{16}O_2$	2-Ethylhexanoic acid	228		75.60
$C_8H_{16}O_2$	Isobutyl isobutanoate	148.6	38.2	
$C_8H_{16}O_2$	Ethyl hexanoate	167		51.72
$C_8H_{16}O_2$	Methyl heptanoate	174		51.62
$C_8H_{17}Br$	1-Bromooctane	200		55.77
$C_8H_{17}Cl$	1-Chlorooctane	181.5		52.42
$C_8H_{17}F$	1-Fluorooctane	142.4	40.43	49.65
C_8H_{18}	Octane	125.67	34.41	41.49
C_8H_{18}	2-Methylheptane	117.66	33.26	39.67
C_8H_{18}	3-Methylheptane	118.9	33.66	39.83
C_8H_{18}	4-Methylheptane	117.72	33.35	39.69
C_8H_{18}	3-Ethylhexane	118.6	33.59	39.64
C_8H_{18}	2,2-Dimethylhexane	106.86	32.07	37.28
C_8H_{18}	2,3-Dimethylhexane	115.62	33.17	38.78
C_8H_{18}	2,4-Dimethylhexane	109.5	32.51	37.76
C_8H_{18}	2,5-Dimethylhexane	109.12	32.54	37.85
C_8H_{18}	3,3-Dimethylhexane	111.97	32.31	37.53
C_8H_{18}	3,4-Dimethylhexane	117.73	33.24	38.97
C_8H_{18}	3-Ethyl-2-methylpentane	115.66	32.93	38.52
C_8H_{18}	3-Ethyl-3-methylpentane	118.27	32.78	37.99
C_8H_{18}	2,2,3-Trimethylpentane	110	31.94	36.91
C_8H_{18}	2,2,4-Trimethylpentane	99.22	30.79	35.14
C_8H_{18}	2,3,3-Trimethylpentane	114.8	32.12	37.27
C_8H_{18}	2,3,4-Trimethylpentane	113.5	32.36	37.75
C_8H_{18}	2,2,3,3-Tetramethylbutane	106.45		42.90
$C_8H_{18}N_2$	Azobutane			49.31
$C_8H_{18}O$	1-Octanol	195.16		70.98
$C_8H_{18}O$	2-Octanol	180	44.4	
$C_8H_{18}O$	2-Ethyl-1-hexanol	184.6	54.2	
$C_8H_{18}O$	Dibutyl ether	140.28	36.49	44.97
$C_8H_{18}O$	Di-sec-butyl ether	121.1	34.06	40.84
$C_8H_{18}O$	Di-tert-butyl ether	107.23	32.15	37.61
$C_8H_{18}O_2$	1,2-Dipropoxyethane			50.62
$C_8H_{18}O_3$	Diethylene glycol diethyl ether	188		58.40
$C_8H_{18}S$	Dibutyl sulfide	185		52.96
$C_8H_{18}S$	Di-tert-butyl sulfide	149.1	33.26	43.76
$C_8H_{18}S$	Diisobutyl sulfide	171		48.71
$C_8H_{19}N$	Dibutylamine	159.6	38.44	49.45
$C_8H_{19}N$	2-Ethylhexylamine	169.2	40.0	
C_9H_7N	Quinoline	237.16	49.7	59.30
C_9H_7N	Isoquinoline	243.22	49.0	60.26
C_9H_{10}	Cyclopropylbenzene	173.6		50.22
C_9H_{10}	Indan	177.97	39.63	48.79
$C_9H_{10}O_2$	Benzyl acetate	213	49.4	
C_9H_{12}	Propylbenzene	159.24		46.22
C_9H_{12}	Isopropylbenzene	152.41		45.13
C_9H_{12}	1,2,3-Trimethylbenzene	176.12		49.05

Mol. Form.	Name	t_b/°C	$\Delta_{vap}H(t_b)$ kJ/mol	$\Delta_{vap}H(25°C)$ kJ/mol
C_9H_{12}	1,2,4-Trimethylbenzene	169.38		47.93
C_9H_{12}	1,3,5-Trimethylbenzene	164.74		47.50
$C_9H_{14}O_6$	Triacetin	259		85.74
C_9H_{18}	Butylcyclopentane	156.6	36.16	45.89
C_9H_{18}	Propylcyclohexane	156.7		45.08
C_9H_{18}	Isopropylcyclohexane	154.8		44.02
$C_9H_{18}O$	2-Nonanone	195.3		56.44
$C_9H_{18}O$	5-Nonanone	188.45		53.30
$C_9H_{18}O$	2,6-Dimethyl-4-heptanone	169.4		50.92
$C_9H_{18}O_2$	Methyl octanoate	192.9		56.41
C_9H_{20}	Nonane	150.82	37.18	46.55
C_9H_{20}	2,2,5-Trimethylhexane	124.09	33.65	40.16
C_9H_{20}	2,3,5-Trimethylhexane	131.4	34.43	41.41
C_9H_{20}	3,3-Diethylpentane	146.3	34.61	42.0
C_9H_{20}	2,2,4,4-Tetramethylpentane	122.29	32.51	38.49
$C_9H_{20}O$	1-Nonanol	213.37		76.86
$C_{10}H_7Br$	1-Bromonaphthalene	281	39.3	
$C_{10}H_7Cl$	1-Chloronaphthalene	259	52.1	
$C_{10}H_8$	Naphthalene	217.9	43.2	
$C_{10}H_9N$	2-Methylquinoline	246.5		66.1
$C_{10}H_9N$	4-Methylquinoline	262		67.6
$C_{10}H_9N$	6-Methylquinoline	258.6		67.7
$C_{10}H_9N$	8-Methylquinoline	247.5		65.7
$C_{10}H_{12}$	1,2,3,4-Tetrahydronaphthalene	207.6	43.9	
$C_{10}H_{14}$	Butylbenzene	183.31	38.87	51.36
$C_{10}H_{14}$	sec-Butylbenzene	173.3		47.98
$C_{10}H_{14}$	tert-Butylbenzene	169.1		47.71
$C_{10}H_{14}$	Isobutylbenzene	172.79		47.86
$C_{10}H_{14}$	1-Isopropyl-4-methylbenzene	177.1	38.2	
$C_{10}H_{16}O$	(+)-Camphor	207.4	59.5	
$C_{10}H_{18}$	cis-Decahydronaphthalene	195.8	41.0	
$C_{10}H_{18}$	trans-Decahydronaphthalene	187.3	40.2	
$C_{10}H_{19}N$	Decanenitrile	243		66.84
$C_{10}H_{20}$	1-Decene	170.5		50.43
$C_{10}H_{20}$	Butylcyclohexane	180.9		49.36
$C_{10}H_{20}O_2$	2-Ethylhexyl acetate	199	43.5	
$C_{10}H_{20}O_2$	Isopentyl isopentanoate	190.4	45.9	
$C_{10}H_{22}$	Decane	174.15	39.58	51.42
$C_{10}H_{22}$	2-Methylnonane	167.1	38.23	49.63
$C_{10}H_{22}$	3-Methylnonane	167.9	38.26	49.71
$C_{10}H_{22}$	5-Methylnonane	165.1	38.14	49.34
$C_{10}H_{22}$	2,4-Dimethyloctane	156	36.47	47.13
$C_{10}H_{22}O$	1-Decanol	231.1		81.50
$C_{10}H_{22}O$	Diisopentyl ether	172.5	35.1	
$C_{10}H_{22}S$	1-Decanethiol	240.6		65.48
$C_{11}H_{10}$	1-Methylnaphthalene	244.7	45.5	
$C_{11}H_{21}N$	Undecanenitrile	253		71.14
$C_{11}H_{22}$	Pentylcyclohexane	203.7		53.88
$C_{11}H_{24}$	Undecane	195.9	41.91	56.58
$C_{11}H_{24}$	2-Methyldecane	189.3	40.25	54.28
$C_{11}H_{24}$	4-Methyldecane	187	40.70	53.76
$C_{11}H_{24}$	2,4,7-Trimethyloctane	168.1	38.22	49.91
$C_{12}F_{27}N$	Tris(perfluorobutyl)amine	178	46.4	
$C_{12}H_{10}O$	Diphenyl ether	258.0	48.2	
$C_{12}H_{16}$	Cyclohexylbenzene	240.1		59.94
$C_{12}H_{22}$	Cyclohexylcyclohexane	238		57.98
$C_{12}H_{23}N$	Dodecanenitrile	277		76.12
$C_{12}H_{24}$	1-Dodecene	213.8		60.78

Mol. Form.	Name	$t_b/°C$	$\Delta_{vap}H(t_b)$ kJ/mol	$\Delta_{vap}H(25°C)$ kJ/mol
$C_{12}H_{26}$	2,2,4,6,6-Pentamethylheptane	177.8		48.97
$C_{12}H_{26}$	Dodecane	216.32	44.09	61.52
$C_{12}H_{26}O$	1-Dodecanol	259		91.96
$C_{12}H_{27}BO_3$	Tributyl borate	234	56.1	
$C_{12}H_{27}N$	Tributylamine	216.5	46.9	
$C_{13}H_{13}N$	N-Benzylaniline	306.5		79.6
$C_{13}H_{26}O_2$	Methyl dodecanoate	267		77.17
$C_{13}H_{28}$	Tridecane	235.47	46.20	66.68
$C_{14}H_{10}$	Phenanthrene	340		75.50
$C_{14}H_{12}O_2$	Benzyl benzoate	323.5	53.6	
$C_{14}H_{27}N$	Tetradecanenitrile			85.29
$C_{14}H_{30}$	Tetradecane	253.58	48.16	71.73
$C_{14}H_{30}O$	1-Tetradecanol	289		102.20
$C_{15}H_{32}$	Pentadecane	270.6	50.08	76.77
$C_{16}H_{22}O_4$	Dibutyl phthalate	340	79.2	
$C_{16}H_{32}$	1-Hexadecene	284.9		80.25
$C_{16}H_{34}$	Hexadecane	286.86	51.84	81.35
$C_{17}H_{36}$	Heptadecane	302.0	53.58	86.47
$C_{18}H_{34}O_2$	Oleic acid	360	67.4	
$C_{18}H_{38}$	Octadecane	316.3	55.23	91.44
$C_{19}H_{40}$	Nonadecane	329.9	56.93	96.4
$C_{20}H_{42}$	Eicosane	343	58.49	101.81

ENTHALPY OF FUSION

This table lists the molar enthalpy (heat) of fusion, $\Delta_{fus}H$, of over 800 inorganic and organic compounds. All values refer to the enthalpy change at equilibrium between the liquid phase and the most stable solid phase at the transition temperature. Most values of $\Delta_{fus}H$ are given at the normal melting point t_m. However, a "t" following the entry in the melting point column indicate a triple-point temperature, where the solid, liquid, and gas phases are in equilibrium. Substances are listed by molecular formula in the Hill order, with substances containing carbon (except graphite) following those that do not contain carbon.

All temperatures are given on the ITS-90 scale.

A * following an entry indicates that the value includes the enthalpy of transition between crystalline phases whose transformation occurs within 1°C of the melting point.

REFERENCES

1. Chase, M. W., Davies, C. A., Downey, J. R., Frurip, D. J., McDonald, R. A., and Syverud, A. N., *JANAF Thermochemical Tables, Third Edition, J. Phys. Chem. Ref. Data*, Vol. 14, Suppl. 1, 1985.
2. Gurvich, L. V., Veyts, I. V., and Alcock, C. B., *Thermodynamic Properties of Individual Substances, Fourth Edition*; Vol. 2, Hemisphere Publishing Corp., New York, 1991; Vol. 3, CRC Press, Boca Raton, FL, 1994.
3. Dinsdale, A. T., *CALPHAD*, 15, 317, 1991
4. *Landolt-Börnstein, Numerical Data and Functional Relationships in Science and Technology, New Series*, IV/8A, *Enthalpies of Fusion and Transition of Organic Compounds*, Springer-Verlag, Heidelberg, 1995.
5. *Landolt-Börnstein, Numerical Values and Functions for Physics, Chemistry, Astronomy, Geophysics, and Technology, Sixth Edition*, Vol. 2, Part 4, Springer-Verlag, Heidelberg, 1961.
6. Janz, G. J., et al., *Physical Properties Data Compilations Relevant to Energy Storage. II. Molten Salts*, Nat. Stand. Ref. Data Sys.- Nat. Bur. Standards (U.S.), No. 61, Part 2, 1979.
7. *TRC Thermodynamic Tables*, Thermodynamic Research Center, Texas A&M University, College Station, TX.

Molecular formula	Name	$t_m/°C$	$\Delta_{fus}H/kJ\ mol^{-1}$
Ag	Silver	961.78	11.30
AgBr	Silver(I) bromide	432	9.12
AgCl	Silver(I) chloride	455	13.2
AgI	Silver(I) iodide	558	9.41
AgNO$_3$	Silver(I) nitrate	212	11.5
Ag$_2$S	Silver(I) sulfide	825	14.1
Al	Aluminum	660.32	10.71
AlBr$_3$	Aluminum bromide	97.5	11.25
AlCl$_3$	Aluminum chloride	192.6	35.4
AlF$_3$	Aluminum fluoride	2250 t	98
AlI$_3$	Aluminum iodide	188.28	15.9
Al$_2$O$_3$	Aluminum oxide (corundum)	2053	111.4
Al$_2$S$_3$	Aluminum sulfide	1100	55
Am	Americium	1176	14.39
Ar	Argon	-189.36 t	1.18
As	Arsenic	817 t	24.44
AsBr$_3$	Arsenic(III) bromide	31.1	11.7
AsCl$_3$	Arsenic(III) chloride	-16	10.1
AsF$_3$	Arsenic(III) fluoride	-5.9	10.4
Au	Gold	1064.18	12.55
B	Boron	2075	50.2
BCl$_3$	Boron trichloride	-107	2.10
BF$_3$	Boron trifluoride	-126.8	4.20
BHO$_2$	Metaboric acid (<gamma> form)	236	14.3
BH$_3$O$_3$	Boric acid (orthoboric acid)	170.9	22.3
BN	Boron nitride	2966	81

Molecular formula	Name	$t_m/°C$	$\Delta_{fus}H/kJ\ mol^{-1}$
$BNaO_2$	Sodium metaborate	966	36.2
B_2O_3	Boron oxide	450	24.56
Ba	Barium	727	7.12
$BaBr_2$	Barium bromide	857	32.2
$BaCl_2$	Barium chloride	962	15.85
BaF_2	Barium fluoride	1368	17.8
BaH_2	Barium hydride	1200	25
BaH_2O_2	Barium hydroxide	408	16
BaI_2	Barium iodide	711	26.5
BaO	Barium oxide	1972	46
BaO_4S	Barium sulfate	1580	40
BaS	Barium sulfide	2229	63
Be	Beryllium	1287	7.895
$BeBr_2$	Beryllium bromide	508	18
$BeCl_2$	Beryllium chloride	415	8.66
BeF_2	Beryllium fluoride	552	4.77
BeI_2	Beryllium iodide	470	18
BeO	Beryllium oxide	2577	86
BeO_4S	Beryllium sulfate	1127	6
Bi	Bismuth	271.40	11.30
$BiCl_3$	Bismuth trichloride	230	10.9
BrF_5	Bromine pentafluoride	-60.5	5.67
BrH	Hydrogen bromide	-86.80	2.41
BrIn	Indium(I) bromide	290	15
BrK	Potassium bromide	734	25.5
BrLi	Lithium bromide	552	17.6
BrNa	Sodium bromide	747	26.11
$BrNaO_3$	Sodium bromate	381	28.11
BrRb	Rubidium bromide	682	15.5
BrTl	Thallium(I) bromide	460	16.4
Br_2	Bromine	-7.2	10.57
Br_2Ca	Calcium bromide	742	29.1
Br_2Cd	Cadmium bromide	568	20.9
Br_2Fe	Iron(II) bromide	691	50.2
Br_2Hg	Mercury(II) bromide	236	17.9
Br_2Mg	Magnesium bromide	711	39.3
Br_2Pb	Lead(II) bromide	371	16.44
Br_2Sr	Strontium bromide	657	10.1
Br_2Zn	Zinc bromide	394	16.7
Br_3Ga	Gallium(III) bromide	121.5	12.1
Br_3In	Indium(III) bromide	420	26
Br_3Pu	Plutonium(III) bromide	681	55.2
Br_3U	Uranium(III) bromide	727	43.9
Br_4Sn	Tin(IV) bromide	29.1	12.2
Br_4Th	Thorium(IV) bromide	679	66.9
Br_4Ti	Titanium(IV) bromide	39	12.9
Br_4U	Uranium(IV) bromide	519	55.2
Br_5Ta	Tantalum(V) bromide	265	45.6
C	Carbon (graphite)	4489 t	117
Ca	Calcium	842	8.54
$CaCl_2$	Calcium chloride	775	28.05
CaF_2	Calcium fluoride	1418	30
CaH_2	Calcium hydride	1000	6.7
CaI_2	Calcium iodide	783	41.8
CaO	Calcium oxide	2898	80
CaO_4S	Calcium sulfate	1460	28
CaS	Calcium sulfide	2524	70
Cd	Cadmium	321.07	6.19

ENTHALPY OF FUSION (continued)

Molecular formula	Name	$t_m/°C$	$\Delta_{fus}H/\text{kJ mol}^{-1}$
$CdCl_2$	Cadmium chloride	564	48.58
CdF_2	Cadmium fluoride	1110	22.6
CdI_2	Cadmium iodide	387	15.3
Ce	Cerium	798	5.46
$CeCl_3$	Cerium(III) chloride	817	54.4
ClCs	Cesium chloride	645	15.9
ClCu	Copper(I) chloride	430	10.2
ClH	Hydrogen chloride	-114.17	2.00
ClI	Iodine chloride	27.39	11.6
ClIn	Indium(I) chloride	211	21.3
ClK	Potassium chloride	771	26.53
ClLi	Lithium chloride	610	19.9
$ClLiO_4$	Lithium perchlorate	236	29
ClNa	Sodium chloride	800.7	28.16
$ClNaO_3$	Sodium chlorate	248	22.1
ClRb	Rubidium chloride	715	18.4
ClTl	Thallium(I) chloride	430	15.56
Cl_2	Chlorine	-101.5	6.40
Cl_2Co	Cobalt(II) chloride	740	45
Cl_2Cr	Chromium(II) chloride	814	32.2
Cl_2Cu	Copper(II) chloride	630	20.4
Cl_2Fe	Iron(II) chloride	677	43.01
Cl_2Hg	Mercury(II) chloride	276	19.41
Cl_2Mg	Magnesium chloride	714	43.1
Cl_2Mn	Manganese(II) chloride	650	30.7
Cl_2Ni	Nickel(II) chloride	1009	71.2
Cl_2Pb	Lead(II) chloride	501	21.75
Cl_2Sn	Tin(II) chloride	247.1	14.52
Cl_2Sr	Strontium chloride	874	17.5
Cl_3Fe	Iron(III) chloride	304	43.1
Cl_3Ga	Gallium(III) chloride	77.9	11.13
Cl_3In	Indium(III) chloride	583	27
Cl_3La	Lanthanum chloride	859	43.1
Cl_3OP	Phosphorus(V) oxychloride	1.18	13.1
Cl_3P	Phosphorus(III) chloride	-112	7.10
Cl_3Sb	Antimony(III) chloride	73.4	12.7
Cl_4OW	Tungsten(VI) oxytetrachloride	211	45
Cl_4Si	Tetrachlorosilane	-68.74	7.60
Cl_4Sn	Tin(IV) chloride	-34.07	9.20
Cl_4Th	Thorium(IV) chloride	770	40.2
Cl_4Ti	Titanium(IV) chloride	-24.12	9.97
Cl_4U	Uranium(IV) chloride	590	45
Cl_4V	Vanadium(IV) chloride	-25.7	2.30
Cl_4Zr	Zirconium(IV) chloride	437 t	50
Cl_5Mo	Molybdenum(V) chloride	194	19
Cl_5Nb	Niobium(V) chloride	204.7	38.3
Cl_5Ta	Tantalum(V) chloride	216	41.6
Cl_6W	Tungsten(VI) chloride	275	6.60
Co	Cobalt	1495	16.2
CoF_2	Cobalt(II) fluoride	1127	59
Cr	Chromium	1907	21.0
Cr_2O_3	Chromium(III) oxide	2329	130
Cs	Cesium	28.5	2.09
CsF	Cesium fluoride	703	21.7
CsHO	Cesium hydroxide	342.3	7.78
Cs_2O_4S	Cesium sulfate	1005	35.7
Cu	Copper	1084.62	13.26
CuF_2	Copper(II) fluoride	836	55

ENTHALPY OF FUSION (continued)

Molecular formula	Name	$t_m/°C$	$\Delta_{fus}H/kJ\ mol^{-1}$
CuO	Copper(II) oxide	1446	11.8
Dy	Dysprosium	1412	11.06
Er	Erbium	1529	19.9
Eu	Europium	822	9.21
FH	Hydrogen fluoride	-83.35	4.58
FK	Potassium fluoride	858	27.2
FLi	Lithium fluoride	848.2	27.09
FNa	Sodium fluoride	996	33.35
FRb	Rubidium fluoride	833	17.3
FTl	Thallium(I) fluoride	326	13.87
F_2	Fluorine	-219.66	0.51
F_2Fe	Iron(II) fluoride	1100	52
F_2HK	Potassium hydrogen fluoride	238.9	6.62
F_2Mg	Magnesium fluoride	1263	58.5
F_2Pb	Lead(II) fluoride	830	14.7
F_2Sr	Strontium fluoride	1477	28.5
F_3In	Indium(III) fluoride	1170	64
F_3Pu	Plutonium(III) fluoride	1396	59.8
F_4Pu	Plutonium(IV) fluoride	1027	65.3
F_4Th	Thorium(IV) fluoride	1110	44.0
F_4U	Uranium(IV) fluoride	1036	42.7
F_4Zr	Zirconium(IV) fluoride	932 t	64.2
F_5Nb	Niobium(V) fluoride	80	12.2
F_5V	Vanadium(V) fluoride	19.5	49.96
F_6Ir	Iridium(VI) fluoride	44	8.40
F_6Mo	Molybdenum(VI) fluoride	17.5	4.33
F_6Pu	Plutonium(VI) fluoride	52	17.6
F_6S	Sulfur hexafluoride	-50.7 t	5.02
F_6U	Uranium(VI) fluoride	64.0 t	19.1
F_6W	Tungsten(VI) fluoride	2.3	4.10
Fe	Iron	1538	13.81
FeI_2	Iron(II) iodide	587	45
FeO	Iron(II) oxide	1377	24
FeS	Iron(II) sulfide	1188	31.5
Fe_3O_4	Iron(II,III) oxide	1597	138
Ga	Gallium	29.76	5.59
GaI_3	Gallium(III) iodide	212	12.9
GaSb	Gallium antimonide	712	25.1
Ga_2O_3	Gallium(III) oxide	1806	100
Gd	Gadolinium	1313	10.0
Ge	Germanium	938.25	36.94
HI	Hydrogen iodide	-50.76	2.87
HKO	Potassium hydroxide	406	7.9
HLi	Lithium hydride	688.7	22.59
HLiO	Lithium hydroxide	471.1	20.88
HNO_3	Nitric acid	-41.6	10.5
HNaO	Sodium hydroxide	323	6.60
HORb	Rubidium hydroxide	382	8.0
H_2	Hydrogen	-259.34	0.12
H_2Mg	Magnesium hydride	327	14
H_2O	Water	0.00	6.01
H_2O_2	Hydrogen peroxide	-0.43	12.50
H_2O_2Sr	Strontium hydroxide	535	23
H_2O_4S	Sulfuric acid	10.31	10.71
H_2S	Hydrogen sulfide	-85.5	2.38
H_2Sr	Strontium hydride	1050	23
H_3N	Ammonia	-77.73	5.66
H_3O_2P	Hypophosphorous acid	26.5	9.7

Molecular formula	Name	$t_m/°C$	$\Delta_{fus}H/kJ\ mol^{-1}$
H_3O_3P	Phosphorous acid	74.4	12.8
H_3O_4P	Phosphoric acid	42.4	13.4
H_4IN	Ammonium iodide	551	21
H_4N_2	Hydrazine	1.4	12.6
$H_4N_2O_3$	Ammonium nitrate	210	6.40
Hf	Hafnium	2233	27.2
Hg	Mercury	-38.83	2.29
HgI_2	Mercury(II) iodide	259	18.9
Hg_2I_2	Mercury(I) iodide	290	27
Ho	Holmium	1474	17.0
IIn	Indium(I) iodide	364.4	17.26
IK	Potassium iodide	681	24
ILi	Lithium iodide	469	14.6
INa	Sodium iodide	660	23.6
IRb	Rubidium iodide	642	12.5
ITl	Thallium(I) iodide	441.7	14.73
I_2	Iodine (rhombic)	113.7	15.52
I_2Mg	Magnesium iodide	634	26
I_2Pb	Lead(II) iodide	410	23.4
I_2Sr	Strontium iodide	538	19.7
I_3In	Indium(III) iodide	207	18.48
I_4Si	Tetraiodosilane	120.5	19.7
I_4Th	Thorium(IV) iodide	570	61.4
I_4Ti	Titanium(IV) iodide	150	19.8
I_4U	Uranium(IV) iodide	506	70.7
In	Indium	156.60	3.28
InSb	Indium antimonide	525	25.5
In_2O_3	Indium(III) oxide	1912	105
Ir	Iridium	2446	41.12
K	Potassium	63.5	2.33
KNO_3	Potassium nitrate	337	10.1
K_2O_4S	Potassium sulfate	1069	36.4
K_2S	Potassium sulfide	948	16.15
Kr	Krypton	-157.38 t	1.64
La	Lanthanum	918	6.20
Li	Lithium	180.50	3.00
$LiNO_3$	Lithium nitrate	253	24.9
Li_2O_3Si	Lithium metasilicate	1201	28
Li_2O_4S	Lithium sulfate	859	7.50
Lu	Lutetium	1663	22
Mg	Magnesium	650	8.48
MgO	Magnesium oxide	2825	77
MgO_4S	Magnesium sulfate	1127	14.6
MgS	Magnesium sulfide	2226	63
Mg_2O_4Si	Magnesium orthosilicate	1897	71
Mn	Manganese	1246	12.91
MnO	Manganese(II) oxide	1839	54.4
Mo	Molybdenum	2623	37.48
MoO_3	Molybdenum(VI) oxide	801	48
$NNaO_3$	Sodium nitrate	307	15
NO	Nitric oxide	-163.6	2.30
NO_3Rb	Rubidium nitrate	305	5.60
NO_3Tl	Thallium(I) nitrate	206	9.6
N_2	Nitrogen	-210.0	0.71
N_2O	Nitrous oxide	-90.8	6.54
N_2O_4	Nitrogen tetroxide	-9.3	14.65
Na	Sodium	97.80	2.60
Na_2O	Sodium oxide	1132	48

Molecular formula	Name	$t_m/°C$	$\Delta_{fus}H/\text{kJ mol}^{-1}$
Na_2O_3Si	Sodium metasilicate	1089	52
Na_2O_4S	Sodium sulfate	884	23.6
Na_2S	Sodium sulfide	1172	19
Nb	Niobium	2477	30
NbO	Niobium(II) oxide	1936	85
NbO_2	Niobium(IV) oxide	1901	92
Nb_2O_5	Niobium(V) oxide	1512	104.3
Nd	Neodymium	1021	7.14
Ne	Neon	-248.61 t	0.328
Ni	Nickel	1455	17.48
NiS	Nickel(II) sulfide	976	30.1
Np	Neptunium	644	3.20
OSr	Strontium oxide	2531	81
OTl_2	Thallium(I) oxide	579	30.3
OV	Vanadium(II) oxide	1789	63
OZn	Zinc oxide	1974	52.3
O_2	Oxygen	-218.79	0.44
O_2Si	Silicon dioxide (cristobalite)	1722	9.6
O_2Zr	Zirconium(IV) oxide	2709	87
O_3S	Sulfur trioxide	16.8	8.60
O_3Tl_2	Thallium(III) oxide	834	53
O_3W	Tungsten(VI) oxide	1472	73
O_3Y_2	Yttrium oxide	2438	105
O_4Os	Osmium(VIII) oxide	41	9.8
O_4SSr	Strontium sulfate	1606	36
O_4STl_2	Thallium(I) sulfate	632	23
O_5P_2	Phosphorus(V) oxide	562	27.2
O_5Ta_2	Tantalum(V) oxide	1784	120
O_5V_2	Vanadium(V) oxide	670	64.5
O_7Re_2	Rhenium(VII) oxide	297	64.2
Os	Osmium	3033	57.85
P	Phosphorus (white)	44.15	0.66
Pa	Protactinium	1572	12.34
Pb	Lead	327.46	4.77
PbS	Lead(II) sulfide	1113	49.4
Pd	Palladium	1554.9	16.74
Pr	Praseodymium	931	6.89
Pt	Platinum	1768.4	22.17
Pu	Plutonium	640	2.82
Rb	Rubidium	39.3	2.19
Re	Rhenium	3186	60.43
Rh	Rhodium	1964	26.59
Ru	Ruthenium	2334	38.59
S	Sulfur (monoclinic)	115.21	1.72
SSr	Strontium sulfide	2226	63
STl_2	Thallium(I) sulfide	448	12
Sb	Antimony	630.63	19.87
Sc	Scandium	1541	14.1
Se	Selenium	220.5	6.69
Si	Silicon	1414	50.21
Sm	Samarium	1074	8.62
Sn	Tin (white)	231.93	7.03
Sr	Strontium	777	7.43
Ta	Tantalum	3017	36.57
Tb	Terbium	1356	10.15
Tc	Technetium	2157	33.29
Te	Tellurium	449.51	17.49
Th	Thorium	1750	13.81

Molecular formula	Name	$t_m/°C$	$\Delta_{fus}H/\text{kJ mol}^{-1}$
Ti	Titanium	1668	14.15
Tl	Thallium	304	4.14
Tm	Thulium	1545	16.84
U	Uranium	1135	9.14
V	Vanadium	1910	21.5
W	Tungsten	3422	52.31
Xe	Xenon	-111.79 t	2.27
Y	Yttrium	1522	11.4
Yb	Ytterbium	819	7.66
Zn	Zinc	419.53	7.32
Zr	Zirconium	1855	21.00
$CBaO_3$	Barium carbonate	1555	40
$CBrCl_3$	Bromotrichloromethane	-5.65	2.53
CBr_4	Tetrabromomethane	92.3	3.76
$CCaO_3$	Calcium carbonate	1330	36
CCl_2O	Carbonyl chloride	-127.78	5.74
CCl_3F	Trichlorofluoromethane	-110.44	6.89
CCl_4	Tetrachloromethane	-22.62	2.56
CF_4	Tetrafluoromethane	-183.60	0.704
$CHBr_3$	Tribromomethane	8.69	11.05
$CHClF_2$	Chlorodifluoromethane	-157.42	4.12
$CHCl_3$	Trichloromethane	-63.41	9.5
CHF_3	Trifluoromethane	-155.2	4.06
CHI_3	Triiodomethane	121.2	16.44
CHN	Hydrogen cyanide	-13.29	8.41
$CHNaO_2$	Sodium formate	257.3	17.7
CHO_2Tl	Thallium(I) formate	101	10.9
CH_2Cl_2	Dichloromethane	-97.2	4.60
CH_2N_2	Cyanamide	45.56	7.27
CH_2N_4	Tetrazole	157.3	18.2
CH_2O_2	Formic acid	8.3	12.68
CH_3Br	Bromomethane	-93.68	5.98
CH_3Cl	Chloromethane	-97.7	6.43
CH_3NO	Formamide	2.49	8.44
CH_3NO_2	Nitromethane	-28.38	9.70
CH_3NO_3	Methyl nitrate	-83.0	8.24
CH_4	Methane	-182.47	0.94
CH_4N_2O	Urea	133.3	13.9
CH_4N_2S	Thiourea	178	14.0
CH_4O	Methanol	-97.53	3.215
CH_4S	Methanethiol	-123	5.91
CH_5N	Methylamine	-93.5	6.13
CH_6N_2	Methylhydrazine	-52.36	10.42
CK_2O_3	Potassium carbonate	898	27.6
CLi_2O_3	Lithium carbonate	723	41
$CMgO_3$	Magnesium carbonate	990	59
CNa_2O_3	Sodium carbonate	858.1	29.7
CO	Carbon monoxide	-205.02	0.833
COS	Carbon oxysulfide	-138.8	4.73
CO_2	Carbon dioxide	-56.56 t	9.02
CO_3Sr	Strontium carbonate	1494	40
CO_3Tl_2	Thallium(I) carbonate	272	18.4
CS_2	Carbon disulfide	-112.1	4.39
CSe_2	Carbon diselenide	-43.7	6.36
$C_2Br_2F_4$	1,2-Dibromotetrafluoroethane	-110.32	7.04
C_2ClF_3	Chlorotrifluoroethylene	-158.2	5.55
C_2ClF_5	Chloropentafluoroethane	-99.4	1.86
$C_2Cl_2F_4$	1,2-Dichloro-1,1,2,2-tetrafluoroethane	-92.53	1.51

Molecular formula	Name	$t_m/°C$	$\Delta_{fus}H/kJ\ mol^{-1}$
$C_2Cl_3F_3$	1,1,2-Trichloro-1,2,2-trifluoroethane	-36.22	2.47
C_2Cl_4	Tetrachloroethylene	-22.3	10.88
$C_2Cl_4F_2$	1,1,2,2-Tetrachloro-1,2-difluoroethane	24.8	3.67
C_2Cl_6	Hexachloroethane	186.8 t	9.75
C_2F_4	Tetrafluoroethylene	-131.15	7.72
C_2F_6	Hexafluoroethane	-100.05	2.69
C_2HCl_3	Trichloroethylene	-84.7	8.45
$C_2HCl_3O_2$	Trichloroacetic acid	59.2	5.90
C_2HCl_5	Pentachloroethane	-28.78	11.3
$C_2H_2Cl_2$	1,1-Dichloroethylene	-122.56	6.51
$C_2H_2Cl_2$	cis-1,2-Dichloroethylene	-80.0	7.2
$C_2H_2Cl_4$	1,1,2,2-Tetrachloroethane	-42.4	9.17
C_2H_3Br	Bromoethylene	-139.54	5.12
C_2H_3Cl	Chloroethylene	-153.84	4.92
$C_2H_3ClO_2$	Chloroacetic acid	63	12.28
$C_2H_3Cl_3$	1,1,1-Trichloroethane	-30.01	2.35
$C_2H_3Cl_3$	1,1,2-Trichloroethane	-36.3	11.46
$C_2H_3F_3$	1,1,1-Trifluoroethane	-111.3	6.19
$C_2H_3KO_2$	Potassium acetate	309	7.65
C_2H_3N	Acetonitrile	-43.82	8.16
$C_2H_3NaO_2$	Sodium acetate	328.2	17.9
C_2H_4	Ethylene	-169.15	3.35
$C_2H_4Br_2$	1,2-Dibromoethane	9.84	10.89
$C_2H_4Cl_2$	1,1-Dichloroethane	-96.9	7.87
$C_2H_4Cl_2$	1,2-Dichloroethane	-35.7	8.84
C_2H_4O	Acetaldehyde	-123.37	2.31
C_2H_4O	Ethylene oxide	-112.5	5.17
$C_2H_4O_2$	Acetic acid	16.64	11.73
C_2H_5Br	Bromoethane	-118.6	7.47
C_2H_5Cl	Chloroethane	-138.4	4.45
C_2H_5NO	Acetamide	80.16	15.59
$C_2H_5NO_2$	Nitroethane	-89.5	9.85
C_2H_6	Ethane	-182.79	2.72*
$C_2H_6N_2O$	N-Methylurea	104.9	14.0
C_2H_6O	Ethanol	-114.14	4.931
C_2H_6O	Dimethyl ether	-141.5	4.94
C_2H_6OS	Dimethyl sulfoxide	17.89	14.37
$C_2H_6O_2$	Ethylene glycol	-12.69	9.96
$C_2H_6O_2S$	Dimethyl sulfone	108.9	18.30
C_2H_6S	Ethanethiol	-147.88	4.98
C_2H_6S	Dimethyl sulfide	-98.24	7.99
$C_2H_6S_2$	Dimethyl disulfide	-84.67	9.19
C_2H_6Zn	Dimethyl zinc	-43.0	6.83
C_2H_7N	Dimethylamine	-92.18	5.94
$C_2H_8N_2$	1,2-Ethanediamine	11.14	22.58
$C_2H_8N_2$	1,1-Dimethylhydrazine	-57.20	10.07
$C_2H_8N_2$	1,2-Dimethylhydrazine	-8.9	13.64
C_2N_2	Cyanogen	-27.83	8.11
C_3F_6O	Perfluoroacetone	-125.45	8.38
C_3F_8	Perfluoropropane	-147.70	0.477
C_3H_3N	Acrylonitrile	-83.48	6.23
C_3H_3NS	Thiazole	-33.62	9.57
$C_3H_3N_3$	1,3,5-Triazine	80.3	14.56
C_3H_4	Allene	-136.6	4.40
$C_3H_4N_2$	1H-Pyrazole	70.7	14.0
$C_3H_4N_2$	Imidazole	89.5	12.82
$C_3H_4O_2$	Acrylic acid	12.5	9.51
C_3H_5N	Propanenitrile	-92.78	5.03

Molecular formula	Name	$t_m/°C$	$\Delta_{fus}H/kJ\ mol^{-1}$
$C_3H_5N_3O_9$	Trinitroglycerol	13.5	21.87
C_3H_6	Propene	-185.24	3.003
C_3H_6	Cyclopropane	-127.58	5.44
$C_3H_6Br_2$	1,2-Dibromopropane	-55.49	8.94
$C_3H_6Br_2$	1,3-Dibromopropane	-34.5	14.6
$C_3H_6Cl_2$	1,2-Dichloropropane, (Ò)	-100.53	6.40
$C_3H_6Cl_2$	2,2-Dichloropropane	-33.9	2.30
C_3H_6O	Acetone	-94.7	5.77
C_3H_6O	Methyloxirane	-111.9	6.53
C_3H_6O	Oxetane	-97	6.5
$C_3H_6O_2$	Propanoic acid	-20.5	10.66
$C_3H_6O_2$	Methyl acetate	-98.25	7.49
$C_3H_6O_2$	1,3-Dioxolane	-97.22	6.57
$C_3H_6O_3$	1,3,5-Trioxane	60.29	15.11
C_3H_6S	Thiacyclobutane	-73.24	8.25
C_3H_7Br	1-Bromopropane	-110.3	6.44
C_3H_7Br	2-Bromopropane	-89.0	6.53
C_3H_7Cl	1-Chloropropane	-122.9	5.54
C_3H_7Cl	2-Chloropropane	-117.18	7.39
C_3H_7N	Cyclopropylamine	-35.39	13.18
C_3H_7NO	N,N-Dimethylformamide	-60.48	7.90
C_3H_8	Propane	-187.63	3.50
$C_3H_8N_2O$	N,N-Dimethylurea	182.1	23.0
$C_3H_8N_2O$	N,N'-Dimethylurea	106.6	13.0
C_3H_8O	1-Propanol	-124.39	5.37
C_3H_8O	2-Propanol	-87.9	5.41
$C_3H_8O_2$	1,3-Propylene glycol	-27.7	7.1
$C_3H_8O_2$	Dimethoxymethane	-105.1	8.33
$C_3H_8O_3$	Glycerol	18.1	18.3
C_3H_8S	1-Propanethiol	-113.13	5.48
C_3H_8S	2-Propanethiol	-130.5	5.74
C_3H_8S	Ethyl methyl sulfide	-105.93	9.76
C_3H_9N	Propylamine	-84.75	10.97
C_3H_9N	Isopropylamine	-95.13	7.33
C_3H_9N	Trimethylamine	-117.1	7
C_3H_9NO	3-Amino-1-propanol	12.4	19.7
C_4F_8	Perfluorocyclobutane	-40.19	2.77
C_4F_{10}	Perfluorobutane	-129.1	7.66
$C_4H_2O_3$	Maleic anhydride	52.56	13.60
$C_4H_4N_2$	Succinonitrile	57.98	3.70
$C_4H_4N_2$	Pyrazine	51.0	12.9
C_4H_4O	Furan	-85.61	3.80
$C_4H_4O_3$	Succinic anhydride	119	20.4
C_4H_4S	Thiophene	-38.21	5.07
C_4H_5N	Pyrrole	-23.39	7.91
C_4H_6	1,2-Butadiene	-136.2	6.96
C_4H_6	1,3-Butadiene	-108.91	7.98
C_4H_6	1-Butyne	-125.7	6.03
C_4H_6	2-Butyne	-32.2	9.23
C_4H_6O	Divinyl ether	-100.6	7.9
$C_4H_6O_2$	cis-Crotonic acid	15	12.6
$C_4H_6O_2$	trans-Crotonic acid	71.5	13.0
$C_4H_6O_2$	<gamma>-Butyrolactone	-43.61	9.57
$C_4H_6O_3$	Acetic anhydride	-74.1	10.5
$C_4H_6O_4$	Succinic acid	187.9	32.4
$C_4H_6O_4$	Dimethyl oxalate	54.8	21.1
C_4H_8	1-Butene	-185.34	3.96
C_4H_8	cis-2-Butene	-138.88	7.31

Molecular formula	Name	$t_m/°C$	$\Delta_{fus}H/\text{kJ mol}^{-1}$
C_4H_8	*trans*-2-Butene	-105.52	9.76
C_4H_8	Isobutene	-140.7	5.92
C_4H_8	Cyclobutane	-90.7	1.09
C_4H_8	Methylcyclopropane	-177.6	2.8
C_4H_8O	Butanal	-96.86	10.77
C_4H_8O	2-Butanone	-86.64	8.39
C_4H_8O	Tetrahydrofuran	-108.44	8.54
$C_4H_8O_2$	Butanoic acid	-5.1	11.59
$C_4H_8O_2$	Ethyl acetate	-83.8	10.48
$C_4H_8O_2$	1,4-Dioxane	11.85	12.84
C_4H_8S	Tetrahydrothiophene	-96.2	7.35
C_4H_9Br	1-Bromobutane	-112.6	9.23
C_4H_9Br	2-Bromobutane, (Ò)	-112.65	6.89
C_4H_9Cl	2-Chloro-2-methylpropane	-25.60	2.07
C_4H_9N	Pyrrolidine	-57.79	8.58
C_4H_9NO	Morpholine	-4.8	14.5
C_4H_{10}	Butane	-138.3	4.66
C_4H_{10}	Isobutane	-159.4	4.54
$C_4H_{10}O$	1-Butanol	-88.6	9.37
$C_4H_{10}O$	2-Butanol	-88.5	5.97
$C_4H_{10}O$	2-Methyl-1-propanol	-101.9	6.32
$C_4H_{10}O$	2-Methyl-2-propanol	25.69	6.70
$C_4H_{10}O$	Diethyl ether	-116.2	7.19
$C_4H_{10}O_2$	1,4-Butanediol	20.4	18.70
$C_4H_{10}O_2$	Ethylene glycol dimethyl ether	-69.20	12.6
$C_4H_{10}S$	1-Butanethiol	-115.7	10.46
$C_4H_{10}S$	Diethyl sulfide	-103.91	10.90
$C_4H_{11}N$	*tert*-Butylamine	-66.94	0.882
$C_4H_{12}Pb$	Tetramethyl lead	-30.2	10.80
$C_4H_{12}Si$	Tetramethylsilane	-99.06	6.87
$C_4H_{12}Sn$	Tetramethylstannane	-55.1	9.30
$C_5H_4O_2$	Furfural	-38.1	14.37
C_5H_5N	Pyridine	-41.70	8.28
C_5H_6O	2-Methylfuran	-91.3	8.55
$C_5H_6O_2$	Furfuryl alcohol	-14.6	13.13
C_5H_8	*cis*-1,3-Pentadiene	-140.8	5.64
C_5H_8	*trans*-1,3-Pentadiene	-87.4	7.14
C_5H_8	1,4-Pentadiene	-148.2	6.12
C_5H_8	2-Methyl-1,3-butadiene	-145.9	4.93
C_5H_8	Cyclopentene	-135.0	3.36
C_5H_8	Spiropentane	-107.0	6.43
$C_5H_8O_2$	Methyl methacrylate	-47.55	14.4
$C_5H_8O_3$	4-Oxopentanoic acid	33	9.22
$C_5H_8O_4$	Glutaric acid	97.8	20.3
C_5H_9N	Pentanenitrile	-96.2	9
C_5H_{10}	1-Pentene	-165.12	5.94
C_5H_{10}	*cis*-2-Pentene	-151.36	7.11
C_5H_{10}	*trans*-2-Pentene	-140.21	8.35
C_5H_{10}	2-Methyl-1-butene	-137.53	7.91
C_5H_{10}	3-Methyl-1-butene	-168.43	5.36
C_5H_{10}	2-Methyl-2-butene	-133.72	7.60
C_5H_{10}	Cyclopentane	-93.4	0.61
$C_5H_{10}O$	Cyclopentanol	-17.5	1.535
$C_5H_{10}O$	2-Pentanone	-76.8	10.63
$C_5H_{10}O$	3-Pentanone	-39	11.59
$C_5H_{10}O$	3-Methyl-2-butanone	-93.1	9.34
$C_5H_{10}O$	Tetrahydropyran	-49.1	1.8
$C_5H_{10}O_2$	Pentanoic acid	-33.6	14.16

Molecular formula	Name	$t_m/°C$	$\Delta_{fus}H/kJ\ mol^{-1}$
$C_5H_{11}Br$	1-Bromopentane	-88.0	14.37
$C_5H_{11}N$	Cyclopentylamine	-82.7	8.31
$C_5H_{11}N$	Piperidine	-11.02	14.85
C_5H_{12}	Pentane	-129.67	8.40
C_5H_{12}	Isopentane	-159.77	5.15
C_5H_{12}	Neopentane	-16.4	3.10
$C_5H_{12}O$	1-Pentanol	-77.6	10.50
$C_5H_{12}O$	2-Methyl-2-butanol	-9.1	4.46
$C_5H_{12}O$	Butyl methyl ether	-115.7	10.85
$C_5H_{12}O$	Methyl *tert*-butyl ether	-108.6	7.60
$C_5H_{12}O_4$	Pentaerythritol	258	4.8
$C_5H_{12}S$	1-Pentanethiol	-75.65	17.53
C_6Cl_6	Hexachlorobenzene	228.83	25.2
C_6F_6	Hexafluorobenzene	5.03	11.59
C_6F_{14}	Perfluorohexane	-88.2	6.84
C_6HF_5	Pentafluorobenzene	-47.4	10.87
C_6HF_5O	Pentafluorophenol	37.5	16.41
$C_6H_2F_4$	1,2,3,5-Tetrafluorobenzene	-46.25	6.36
$C_6H_2F_4$	1,2,4,5-Tetrafluorobenzene	3.88	15.05
$C_6H_3Cl_3$	1,2,3-Trichlorobenzene	51.3	17.9
$C_6H_3Cl_3$	1,2,4-Trichlorobenzene	16.92	16.4
$C_6H_3Cl_3$	1,3,5-Trichlorobenzene	62.8	18.1
$C_6H_3N_3O_6$	1,3,5-Trinitrobenzene	122.9	15.4
$C_6H_4ClNO_2$	1-Chloro-2-nitrobenzene	32.1	17.9
$C_6H_4ClNO_2$	1-Chloro-3-nitrobenzene	44.4	19.4
$C_6H_4ClNO_2$	1-Chloro-4-nitrobenzene	82	14.1
$C_6H_4Cl_2$	*o*-Dichlorobenzene	-17.0	12.4
$C_6H_4Cl_2$	*m*-Dichlorobenzene	-24.8	12.6
$C_6H_4Cl_2$	*p*-Dichlorobenzene	53.09	18.19
$C_6H_4F_2$	*o*-Difluorobenzene	-47.1	11.05
$C_6H_4F_2$	*m*-Difluorobenzene	-69.12	8.58
$C_6H_4O_2$	*p*-Benzoquinone	115	18.5
C_6H_5Br	Bromobenzene	-30.72	10.70
C_6H_5Cl	Chlorobenzene	-45.31	9.6
C_6H_5ClO	*o*-Chlorophenol	9.4	13.0
C_6H_5ClO	*m*-Chlorophenol	32.6	14.9
C_6H_5ClO	*p*-Chlorophenol	42.8	14.1
C_6H_5F	Fluorobenzene	-42.18	11.31
C_6H_5I	Iodobenzene	-31.3	9.75
C_6H_5NO	Benzene, nitroso-	67	31.0
$C_6H_5NO_2$	Nitrobenzene	5.7	12.12
$C_6H_5NO_3$	*o*-Nitrophenol	44.8	17.7
$C_6H_5NO_3$	*m*-Nitrophenol	96.8	20.6
$C_6H_5NO_3$	*p*-Nitrophenol	113.6	18.8
C_6H_6	Benzene	5.49	9.87
C_6H_6ClN	*o*-Chloroaniline	-1.9	11.9
C_6H_6ClN	*m*-Chloroaniline	-10.28	10.15
C_6H_6ClN	*p*-Chloroaniline	70.5	20.0
$C_6H_6N_2O_2$	*o*-Nitroaniline	71.0	16.1
$C_6H_6N_2O_2$	*m*-Nitroaniline	113.4	23.6
$C_6H_6N_2O_2$	*p*-Nitroaniline	147.5	21.2
C_6H_6O	Phenol	40.89	11.51
$C_6H_6O_2$	*p*-Hydroquinone	172.4	26.8
$C_6H_6O_2$	Pyrocatechol	104.6	22.8
$C_6H_6O_2$	Resorcinol	109.4	20.4
C_6H_6S	Benzenethiol	-14.93	11.48
C_6H_7N	Aniline	-6.02	10.54
C_6H_7N	2-Methylpyridine	-66.68	9.72

Molecular formula	Name	$t_m/°C$	$\Delta_{fus}H/kJ\ mol^{-1}$
C_6H_7N	3-Methylpyridine	-18.14	14.18
C_6H_7N	4-Methylpyridine	3.67	12.58
$C_6H_8N_2$	o-Phenylenediamine	102.1	23.1
$C_6H_8N_2$	m-Phenylenediamine	66.0	15.57
$C_6H_8N_2$	p-Phenylenediamine	141.1	23.8
$C_6H_8N_2$	Phenylhydrazine	20.6	14.05
C_6H_{10}	Cyclohexene	-103.5	3.29
$C_6H_{10}O$	Cyclohexanone	-27.9	1.328
$C_6H_{10}O_2$	2-Oxepanone	-1.0	13.83
$C_6H_{10}O_4$	Adipic acid	152.5	36.3
$C_6H_{11}Cl$	Chlorocyclohexane	-43.81	2.043
C_6H_{12}	1-Hexene	-139.76	9.35
C_6H_{12}	cis-2-Hexene	-141.11	8.88
C_6H_{12}	2,3-Dimethyl-2-butene	-74.19	6.45
C_6H_{12}	Cyclohexane	6.59	2.68
C_6H_{12}	Methylcyclopentane	-142.42	6.93
$C_6H_{12}O$	Hexanal	-56	13.3
$C_6H_{12}O$	2-Hexanone	-55.5	14.9
$C_6H_{12}O$	3-Hexanone	-55.4	13.49
$C_6H_{12}O$	Cyclohexanol	25.93	1.78
$C_6H_{12}O_3$	Paraldehyde	12.6	13.5
$C_6H_{13}Br$	1-Bromohexane	-83.7	18.1
$C_6H_{13}N$	Cyclohexylamine	-17.8	17.5
C_6H_{14}	Hexane	-95.35	13.08
C_6H_{14}	2-Methylpentane	-153.6	6.27
C_6H_{14}	3-Methylpentane	-162.90	5.30
C_6H_{14}	2,2-Dimethylbutane	-98.8	0.58
C_6H_{14}	2,3-Dimethylbutane	-128.10	0.79
$C_6H_{14}O$	1-Hexanol	-47.4	15.38
$C_6H_{14}O$	Dipropyl ether	-114.8	10.8
$C_6H_{14}O$	Diisopropyl ether	-85.4	12.04
$C_6H_{14}O_2$	1,6-Hexanediol	41.5	22.2
C_7F_8	Perfluorotoluene	-65.49	11.54
C_7F_{16}	Perfluoroheptane	-51.2	6.95
$C_7H_3F_5$	2,3,4,5,6-Pentafluorotoluene	-29.78	13.1
C_7H_5ClO	Benzoyl chloride	-0.4	19.2
$C_7H_5ClO_2$	o-Chlorobenzoic acid	140.2	25.6
C_7H_5N	Benzonitrile	-13.99	9.1
$C_7H_5N_3O_6$	2,4,6-Trinitrotoluene	80.5	22.9
$C_7H_6O_2$	Benzoic acid	122.35	18.02
$C_7H_6O_3$	o-Hydroxybenzoic acid	159.0	14.2
C_7H_7Cl	o-Chlorotoluene	-35.8	9.6
C_7H_7NO	Benzamide	127.3	19.5
$C_7H_7NO_2$	p-Nitrotoluene	51.63	16.81
C_7H_8	Toluene	-94.95	6.64
C_7H_8O	o-Cresol	31.03	15.82
C_7H_8O	m-Cresol	12.24	10.71
C_7H_8O	p-Cresol	34.77	12.71
C_7H_8O	Benzyl alcohol	-15.4	8.97
C_7H_8O	Anisole	-37.13	12.9
C_7H_9N	o-Methylaniline	-14.41	11.66
C_7H_9N	m-Methylaniline	-31.3	7.9
C_7H_9N	p-Methylaniline	43.6	18.9
C_7H_{14}	1-Heptene	-118.9	12.41
C_7H_{14}	Cycloheptane	-8.46	1.88
C_7H_{14}	Methylcyclohexane	-126.6	6.75
$C_7H_{14}O$	1-Heptanal	-43.4	23.2
$C_7H_{14}O$	Cycloheptanol	7.2	1.60

Molecular formula	Name	$t_m/°C$	$\Delta_{fus}H$/kJ mol^{-1}
$C_7H_{14}O_2$	Heptanoic acid	-7.17	15.13
$C_7H_{15}Br$	1-Bromoheptane	-56.1	21.8
C_7H_{16}	Heptane	-90.55	14.03
C_7H_{16}	2-Methylhexane	-118.2	9.19
C_7H_{16}	3-Ethylpentane	-118.55	9.55
C_7H_{16}	2,2-Dimethylpentane	-123.7	5.82
C_7H_{16}	2,4-Dimethylpentane	-119.2	6.85
C_7H_{16}	3,3-Dimethylpentane	-134.4	6.85
C_7H_{16}	2,2,3-Trimethylbutane	-24.6	2.26
$C_7H_{16}O$	1-Heptanol	-33.2	18.17
C_8H_8	Styrene	-30.65	10.9
$C_8H_8O_2$	o-Toluic acid	103.5	19.5
$C_8H_8O_2$	m-Toluic acid	109.9	15.7
$C_8H_8O_2$	p-Toluic acid	179.6	22.7
$C_8H_8O_2$	Benzeneacetic acid	76.5	16.3
$C_8H_8O_2$	Methyl benzoate	-12.4	9.74
C_8H_{10}	Ethylbenzene	-94.96	9.18
C_8H_{10}	o-Xylene	-25.2	13.6
C_8H_{10}	m-Xylene	-47.8	11.6
C_8H_{10}	p-Xylene	13.25	17.12
$C_8H_{10}O$	2,3-Xylenol	72.5	21.0
$C_8H_{10}O$	2,5-Xylenol	74.8	23.4
$C_8H_{10}O$	2,6-Xylenol	45.8	18.9
$C_8H_{10}O$	3,4-Xylenol	65.1	18.1
$C_8H_{10}O$	3,5-Xylenol	63.4	17.4
C_8H_{16}	1-Octene	-101.7	15.31
C_8H_{16}	Cyclooctane	14.59	2.41
C_8H_{16}	Ethylcyclohexane	-111.3	8.33
C_8H_{16}	1,1-Dimethylcyclohexane	-33.3	2.07
C_8H_{16}	cis-1,2-Dimethylcyclohexane	-49.8	1.64
C_8H_{16}	trans-1,2-Dimethylcyclohexane	-88.15	10.49
C_8H_{16}	cis-1,3-Dimethylcyclohexane	-75.53	10.82
C_8H_{16}	trans-1,3-Dimethylcyclohexane	-90.07	9.87
C_8H_{16}	cis-1,4-Dimethylcyclohexane	-87.39	9.31
C_8H_{16}	trans-1,4-Dimethylcyclohexane	-36.93	12.33
$C_8H_{16}O_2$	Octanoic acid	16.5	21.35
$C_8H_{17}Br$	1-Bromooctane	-55.0	24.7
C_8H_{18}	Octane	-56.82	20.73
C_8H_{18}	2-Methylheptane	-109.02	11.92
C_8H_{18}	3-Methylheptane	-120.48	11.69
C_8H_{18}	4-Methylheptane	-121.0	10.8
C_8H_{18}	2,2,4-Trimethylpentane	-107.3	9.20
$C_8H_{18}O$	1-Octanol	-14.8	23.7
C_9H_7N	Quinoline	-14.78	10.66
C_9H_7N	Isoquinoline	26.47	13.54
C_9H_8	Indene	-1.5	10.20
C_9H_{10}	Indan	-51.38	8.60
C_9H_{12}	Propylbenzene	-99.6	9.27
C_9H_{12}	Isopropylbenzene	-96.02	7.33
C_9H_{12}	o-Ethyltoluene	-79.83	9.96
C_9H_{12}	m-Ethyltoluene	-95.6	7.6
C_9H_{12}	p-Ethyltoluene	-62.35	12.7
C_9H_{12}	1,2,3-Trimethylbenzene	-25.4	8.18
C_9H_{12}	1,2,4-Trimethylbenzene	-43.77	13.19
C_9H_{12}	1,3,5-Trimethylbenzene	-44.72	9.51
C_9H_{18}	Propylcyclohexane	-94.9	10.37
$C_9H_{18}O$	Nonanal	-19.3	30.5
$C_9H_{18}O$	5-Nonanone	-3.8	24.93

Molecular formula	Name	$t_m/°C$	$\Delta_{fus}H/kJ\ mol^{-1}$
$C_9H_{18}O_2$	Nonanoic acid	12.4	19.82
C_9H_{20}	Nonane	-53.46	15.47
C_9H_{20}	3,3-Diethylpentane	-33.1	10.09
C_9H_{20}	2,2,3,3-Tetramethylpentane	-9.75	2.33
C_9H_{20}	2,2,4,4-Tetramethylpentane	-66.54	9.74
$C_{10}H_7Br$	1-Bromonaphthalene	6.1	15.2
$C_{10}H_7Br$	2-Bromonaphthalene	55.9	14.4
$C_{10}H_7Cl$	1-Chloronaphthalene	-2.5	12.9
$C_{10}H_7Cl$	2-Chloronaphthalene	58.0	14.0
$C_{10}H_8$	Naphthalene	80.26	19.01
$C_{10}H_8O$	1-Naphthol	95.0	23.1
$C_{10}H_8O$	2-Naphthol	121.5	18.1
$C_{10}H_{14}$	Butylbenzene	-87.85	11.22
$C_{10}H_{14}$	1-Isopropyl-4-methylbenzene	-67.94	9.66
$C_{10}H_{14}$	1,2,4,5-Tetramethylbenzene	79.3	21
$C_{10}H_{14}O$	Thymol	49.5	21.3
$C_{10}H_{18}$	cis-Decahydronaphthalene	-42.9	9.49
$C_{10}H_{18}$	trans-Decahydronaphthalene	-30.4	14.41
$C_{10}H_{18}O_4$	Sebacic acid	130.9	40.8
$C_{10}H_{20}$	1-Decene	-66.3	13.81
$C_{10}H_{20}$	Butylcyclohexane	-74.73	14.16
$C_{10}H_{20}O$	Decanal	-4.0	34.5
$C_{10}H_{20}O_2$	Decanoic acid	31.4	27.8
$C_{10}H_{22}$	Decane	-29.6	28.72
$C_{10}H_{22}O$	1-Decanol	6.9	43
$C_{11}H_{10}$	1-Methylnaphthalene	-30.43	6.95
$C_{11}H_{10}$	2-Methylnaphthalene	34.6	12.13
$C_{11}H_{24}$	Undecane	-25.5	22.2
$C_{12}H_8$	Acenaphthylene	91.8	6.9
$C_{12}H_9N$	Carbazole	246.3	24.1
$C_{12}H_{10}$	Acenaphthene	93.4	21.49
$C_{12}H_{10}$	Biphenyl	68.93	18.57
$C_{12}H_{10}N_2$	Azobenzene	67.88	22.52
$C_{12}H_{10}N_2O$	trans-Azoxybenzene	34.6	17.9
$C_{12}H_{10}O$	Diphenyl ether	26.87	17.22
$C_{12}H_{11}N$	Diphenylamine	53.2	18.5
$C_{12}H_{16}$	Cyclohexylbenzene	7.07	15.6
$C_{12}H_{18}$	Hexamethylbenzene	165.5	20.6
$C_{12}H_{24}$	1-Dodecene	-35.2	19.9
$C_{12}H_{24}O_2$	Dodecanoic acid	43.8	36.3
$C_{12}H_{26}$	Dodecane	-9.57	36.8
$C_{12}H_{26}O$	1-Dodecanol	23.9	40.2
$C_{13}H_{10}$	9H-Fluorene	114.77	19.58
$C_{13}H_{10}O$	Benzophenone	47.9	18.19
$C_{13}H_{12}$	Diphenylmethane	25.4	18.6
$C_{13}H_{28}$	Tridecane	-5.4	28.50
$C_{13}H_{28}O$	1-Tridecanol	31.7	41.4
$C_{14}H_{10}$	Anthracene	215.76	29.4
$C_{14}H_{10}$	Phenanthrene	99.24	16.46
$C_{14}H_{10}O_2$	Benzil	94.87	23.5
$C_{14}H_{12}$	trans-Stilbene	124.2	27.7
$C_{14}H_{12}O_2$	‡-Phenylbenzeneacetic acid	147.29	31.3
$C_{14}H_{28}O_2$	Tetradecanoic acid	54.2	45.1
$C_{14}H_{30}$	Tetradecane	5.82	45.07
$C_{14}H_{30}O$	1-Tetradecanol	38.2	25.1*
$C_{15}H_{32}$	Pentadecane	9.95	34.6
$C_{16}H_{10}$	Fluoranthene	110.19	18.69
$C_{16}H_{10}$	Pyrene	150.62	17.36

Molecular formula	Name	$t_m/°C$	$\Delta_{fus}H/kJ\ mol^{-1}$
$C_{16}H_{32}O_2$	Hexadecanoic acid	62.5	53.7
$C_{16}H_{34}$	Hexadecane	18.12	53.36
$C_{16}H_{34}O$	1-Hexadecanol	49.2	33.6
$C_{17}H_{36}$	Heptadecane	22.0	40.16
$C_{18}H_{12}$	Benz[a]anthracene	160.5	21.4
$C_{18}H_{12}$	Benzo[c]phenanthrene	68	16.3
$C_{18}H_{12}$	Chrysene	255.5	26.2
$C_{18}H_{12}$	Triphenylene	197.8	24.74
$C_{18}H_{14}$	o-Terphenyl	56.20	17.19
$C_{18}H_{14}$	p-Terphenyl	213.9	35.3
$C_{18}H_{15}N$	Triphenylamine	126.5	24.9
$C_{18}H_{36}O_2$	Stearic acid	69.3	61.2
$C_{18}H_{38}$	Octadecane	28.2	61.7
$C_{18}H_{38}O$	1-Octadecanol	57.9	45
$C_{19}H_{40}$	Nonadecane	32.0	45.8
$C_{20}H_{12}$	Perylene	277.76	31.9
$C_{20}H_{12}$	Benzo[a]pyrene	181.1	17.3
$C_{20}H_{12}$	Benzo[e]pyrene	181.4	16.6
$C_{20}H_{14}$	2,2'-Binaphthalene	187.9	38.9
$C_{20}H_{42}$	Eicosane	36.6	69.9
$C_{20}H_{42}O$	1-Eicosanol	65.4	42
$C_{24}H_{12}$	Coronene	437.4	19.2

PRESSURE AND TEMPERATURE DEPENDENCE OF LIQUID DENSITY

This table gives data on the variation of the density of some common liquids with pressure and temperature. The pressure dependence is described to first order by the isothermal compressibility coefficient κ defined as

$$\kappa = -(1/V)\,(\partial V/\partial P)_T$$

where V is the volume, and the temperature dependence by the cubic expansion coefficient α,

$$\alpha = (1/V)\,(\partial V/\partial T)_P$$

Substances are listed by molecular formula in the Hill order. More precise data on the variation of density with temperature over a wide temperature range can be found in Reference 1.

REFERENCES

1. Lide, D. R., and Kehiaian, H. V., *CRC Handbook of Thermophysical and Thermochemical Data*, CRC Press, Boca Raton, FL, 1994.
2. Le Neindre, B., *Effets des Hautes et Très Hautes Pressions*, in *Techniques de l'Ingénieur*, Paris, 1991.
3. *Landolt-Börnstein, Numerical Data and Functional Relationships in Science and Technology, New Series*, IV/4, *High-pressure Properties of Matter*, Springer-Verlag, Heidelberg, 1980.
4. Riddick, J.A., Bunger, W.B., and Sakano, T.K., *Organic Solvents, Fourth Edition*, John Wiley & Sons, New York, 1986.
5. Isaacs, N. S., *Liquid Phase High Pressure Chemistry*, John Wiley, New York, 1981.

Molecular formula	Name	Isothermal Compressibility		Cubic Thermal Expansion	
		$t/°C$	$\kappa \times 10^4/\mathrm{MPa}^{-1}$	$t/°C$	$\alpha \times 10^3/°C^{-1}$
Cl_3P	Phosphorus trichloride	20	9.45	20	1.9
H_2O	Water	20	4.591	20	0.206
		25	4.524	25	0.256
		30	4.475	30	0.302
Hg	Mercury	20	0.401	20	1.811
CCl_4	Tetrachloromethane	20	10.50	20	1.14
		40	12.20	40	1.21
		70	15.6	70	1.33
$CHBr_3$	Tribromomethane	50	8.76	25	0.91
$CHCl_3$	Trichloromethane	20	9.96	20	1.21
		50	12.9	50	1.33
CH_2Br_2	Dibromomethane	27	6.85		
CH_2Cl_2	Dichloromethane	25	10.3	25	1.39
CH_3I	Iodomethane	27	10.3	25	1.26
CH_4O	Methanol	20	12.14	20	1.49
		40	13.83	40	1.59
CS_2	Carbon disulfide	20	9.38	20	1.12
		40	10.6	35	1.16
C_2Cl_4	Tetrachloroethylene	25	7.56	25	1.02
C_2HCl_3	Trichloroethylene	25	8.57	25	1.17
$C_2H_2Cl_2$	trans-1,2-Dichloroethylene	25	11.2	25	1.36
$C_2H_4Cl_2$	1,1-Dichloroethane	20	7.97	25	0.93
$C_2H_4Cl_2$	1,2-Dichloroethane	30	8.46	20	1.14
$C_2H_4O_2$	Acetic acid	20	9.08	20	1.08
		80	13.7	80	1.38
C_2H_5Br	Bromoethane	20	11.53	20	1.31
C_2H_5I	Iodoethane	20	9.82	25	1.17
C_2H_6O	Ethanol	20	11.19	20	1.40
		70	15.93	70	1.67
$C_2H_6O_2$	Ethylene glycol	20	3.64	20	0.626
C_3H_6O	Acetone	20	12.62	20	1.46
		40	15.6	40	1.57
C_3H_7Br	1-Bromopropane	0	10.22	25	1.2
C_3H_7Cl	1-Chloropropane	0	12.09	20	1.4
C_3H_7I	1-Iodopropane	0	10.22	25	1.09
C_3H_8O	1-Propanol	0	8.43	0	1.22
C_3H_8O	2-Propanol	40	13.32	40	1.55
$C_3H_8O_2$	1,2-Propanediol	0	4.45	20	0.695

Molecular formula	Name	Isothermal Compressibility		Cubic Thermal Expansion	
		$t/°C$	$\kappa \times 10^4/MPa^{-1}$	$t/°C$	$\alpha \times 10^3/°C^{-1}$
$C_3H_8O_2$	1,3-Propanediol	0	4.09	20	0.61
$C_3H_8O_3$	Glycerol	0	2.54	20	0.520
$C_4H_8O_2$	Ethyl acetate	20	11.32	20	1.35
		60	16.2	60	1.54
C_4H_9Br	1-Bromobutane	25	10.26	20	1.13
C_4H_9I	1-Iodobutane	0	7.73	25	1.02
$C_4H_{10}O$	1-Butanol	0	8.10	0	1.12
$C_4H_{10}O$	Diethyl ether	20	18.65	20	1.65
		30	20.85	30	1.72
$C_4H_{10}O_3$	Diethylene glycol	0	3.34	20	0.635
C_5H_{10}	Cyclopentane	20	13.31	20	1.35
$C_5H_{11}Br$	1-Bromopentane	0	8.42	25	1.04
$C_5H_{11}I$	1-Iodopentane	0	7.56		
C_5H_{12}	Pentane	25	21.80	25	1.64
$C_5H_{12}O$	1-Pentanol	0	7.71	0	1.02
C_6H_5Br	Bromobenzene	20	6.46	20	0.86
C_6H_5Cl	Chlorobenzene	20	7.45	20	0.94
$C_6H_5NO_2$	Nitrobenzene	20	4.93	25	0.833
C_6H_6	Benzene	25	9.66	25	1.14
		45	11.28	45	1.21
C_6H_6O	Phenol	60	6.05	60	0.82
C_6H_7N	Aniline	20	4.53	20	0.81
		80	6.32	80	0.91
C_6H_{12}	Cyclohexane	20	11.30	20	1.15
		60	15.2	60	1.29
C_6H_{14}	Hexane	25	16.69	25	1.41
		45	20.27	45	1.52
C_6H_{14}	2-Methylpentane	0	13.97	25	1.43
C_6H_{14}	3-Methylpentane	0	14.57	25	1.40
C_6H_{14}	2,3-Dimethylbutane	20	17.97	25	1.39
$C_6H_{14}O$	1-Hexanol	25	8.24	25	1.03
$C_6H_{15}NO_3$	Triethanolamine	0	3.61	55	0.53
C_7H_8	Toluene	20	8.96	20	1.05
		50	11.0	50	1.13
C_7H_8O	Anisole	20	6.60	20	0.951
C_7H_{14}	Cycloheptane	20	9.22		
C_7H_{16}	Heptane	25	14.38	25	1.26
C_8H_{10}	o-Xylene	25	8.10	25	0.96
C_8H_{10}	m-Xylene	20	8.46	20	0.99
C_8H_{10}	p-Xylene	25	8.59	25	1.00
C_8H_{16}	Cyclooctane	20	8.03		
C_8H_{18}	Octane	25	12.82	25	1.16
		45	15.06	45	1.23
$C_8H_{18}O$	1-Octanol	25	7.64	25	0.827
C_9H_{12}	Mesitylene	25	8.14	25	0.94
$C_9H_{14}O_6$	Triacetin	0	4.49	25	0.94
C_9H_{20}	Nonane	25	11.75	25	1.08
$C_{10}H_{22}$	Decane	25	10.94	25	1.02
$C_{11}H_{24}$	Undecane	25	10.31	25	0.97
$C_{12}H_{26}$	Dodecane	25	9.88	25	0.93
$C_{13}H_{28}$	Tridecane	25	9.48	25	0.90
$C_{14}H_{30}$	Tetradecane	25	9.10	25	0.87
$C_{15}H_{32}$	Pentadecane	25	8.82		
$C_{16}H_{22}O_4$	Butyl phthalate	0	5.0	25	0.86
$C_{16}H_{34}$	Hexadecane	25	8.57		
		45	9.78		
$C_{19}H_{36}O_2$	Methyl oleate	0	6.18	60	0.85

PROPERTIES OF CRYOGENIC FLUIDS

This table gives physical and thermodynamic properties of eight cryogenic fluids. The properties are:

M	Molar mass in grams per mole
T_t	Triple point temperature in kelvins
P_t	Triple point pressure in kilopascals
$\rho_t(l)$	Liquid density at the triple point in grams per milliliter
$\Delta_{fus}H$ @ T_t	Enthalpy of fusion at the triple point in joules per gram
T_b	Normal boiling point in kelvins at a pressure of 101325 pascals (760 mmHg)
$\Delta_{vap}H$ @ T_b	Enthalpy of vaporization at the normal boiling point in joules per gram
$\rho(l)$ @ T_b	Liquid density at the normal boiling point in grams per milliliter
$\rho(g)$ @ T_b	Vapor density at the normal boiling point in grams per liter
$C_p(l)$ @ T_b	Liquid heat capacity at constant pressure at the normal boiling point in joules per gram kelvin
$C_p(g)$ @ T_b	Vapor heat capacity at constant pressure at the normal boiling point in joules per gram kelvin
T_c	Critical temperature in kelvins
P_c	Critical pressure in megapascals
ρ_c	Critical density in grams per milliliter

In the case of air, the value given for the triple point temperature is the incipient solidification temperature, and the normal boiling point value is the incipient boiling (bubble) point. See Reference 3 for more details.

Property	Units	Air	N₂	O₂	H₂	He	Ne	Ar	Kr	Xe	CH₄
M	g/mol	28.96	28.014	31.999	2.0159	4.0026	20.180	39.948	83.800	131.290	16.043
T_t	K	59.75	63.15	54.3584	13.8		24.5561	83.8058	115.8	161.4	90.694
P_t	kPa		12.463	0.14633	7.042		50	68.95	72.92	81.59	11.696
$\rho_t(l)$	g/mL	0.959	0.870	1.306	0.0770		1.251	1.417	2.449	2.978	0.4515
$\Delta_{fus}H$ @ T_t	J/g		25.3	13.7	59.5		16.8	28.0	16.3	13.8	58.41
T_b	K	78.67	77.35	90.188	20.28	4.2221	27.07	87.293	119.92	165.10	111.668
$\Delta_{vap}H$ @ T_b	J/g	198.7	198.8	213.1	445	20.7	84.8	161.0	108.4	96.1	510.83
$\rho(l)$ @ T_b	g/mL	0.8754	0.807	1.141	0.0708	0.124901	1.204	1.396	2.418	2.953	0.4224
$\rho(g)$ @ T_b	g/L	3.199	4.622	4.467	1.3390	16.89	9.51	5.79	8.94		1.816
$C_p(l)$ @ T_b	J/g K	1.865	2.042	1.699	9.668	4.545	1.877	1.078	0.533	0.340	3.481
$C_p(g)$ @ T_b	J/g K		1.341	0.980	12.24	9.78		0.570	0.248	0.158	2.218
T_c	K	132.5	126.20	154.581	32.98	5.1953	44.40	150.663	209.40	289.73	190.56
P_c	MPa	3.766	3.390	5.043	1.293	0.227460	2.760	4.860	5.500	5.840	4.592
ρ_c	g/mL	0.316	0.313	0.436	0.031	0.06964	0.484	0.531	0.919	1.110	0.1627

REFERENCES

1. Younglove, B. A., *J. Phys. Chem. Ref. Data*, 11, Suppl. 1, 1982.
2. Daubert, T. E., Danner, R. P., Sibul, H. M., and Stebbins, C. C., *Physical and Thermodynamic Properties of Pure Compounds: Data Compilation*, extant 1994 (core with 4 supplements), Taylor & Francis, Bristol, PA (also available as database).
3. Sytchev, V. V., et al., *Thermodynamic Properties of Air*, Hemisphere Publishing, New York, 1987.
4. Jacobsen, R. T., Stewart, R. B., and Jahangiri, M., *J. Phys. Chem. Ref. Data*, 15, 735, 1986. [Nitrogen]
5. Stewart, R. B., Jacobsen, R. T., and Wagner, W., *J. Phys. Chem. Ref. Data*, 20, 917, 1991. [Oxygen]
6. McCarty, R. D., *J. Phys. Chem. Ref. Data*, 2, 923, 1973. [Helium] Also, Donnelly, R. J., private communication.
7. Stewart, R. B. and Jacobsen, R. T., *J. Phys. Chem. Ref. Data*, 18, 639, 1989. [Argon]
8. Setzmann, U. and Wagner, W., *J. Phys. Chem. Ref. Data*, 20, 1061, 1991. [Methane]
9. Vargaftik, N. B., *Thermophysical Properties of Liquids and Gases*, 2nd ed., John Wiley, New York, 1975.

PROPERTIES OF LIQUID HELIUM

The following data were obtained by a critical evaluation of all existing experimental measurements on liquid helium, using a fitting procedure described in the reference. All values refer to liquid helium at saturated vapor pressure; temperatures are on the ITS-90 scale. Several properties show a singularity at the lambda point (2.1768 K).

p : vapor pressure
ρ : density
C_s : molar heat capacity
$\Delta_{vap}H$: molar enthalpy of vaporization
ε : relative permittivity (dielectric constant)

σ : surface tension
α : coefficient of linear expansion
η : viscosity
λ : thermal conductivity

REFERENCE

Donnelly, R. J., and Barenghi, C. F., *J. Phys. Chem. Reference Data* 27, 1217, 1998.

T/K	p/kPa	ρ/g cm^{-3}	C_s/J mol^{-1}K^{-1}	$\Delta_{vap}H$/J mol^{-1}	ε	σ/mN m^{-1}	$10^3\alpha$/K^{-1}	η/μPa s	λ/W cm^{-1}K^{-1}
0.0		0.1451397	0	59.83	1.057255		0.000		
0.5		0.1451377	0.010	70.24	1.057254	0.3530	0.107		
1.0	0.01558	0.1451183	0.415	80.33	1.057246	0.3471	0.309	3.873	
1.5	0.4715	0.1451646	4.468	89.35	1.057265	0.3322	-2.36	1.346	
2.0	3.130	0.1456217	21.28	93.07	1.057449	0.3021	-12.2	1.468	
2.5	10.23	0.1448402	9.083	92.50	1.057135	0.2623	39.4	3.259	0.1497
3.0	24.05	0.1412269	9.944	94.11	1.055683	0.2161	61.5	3.517	0.1717
3.5	47.05	0.1360736	12.37	92.84	1.053615	0.1626	88.7	3.509	0.1868
4.0	81.62	0.1289745	15.96	87.00	1.050770	0.1095	129	3.319	0.1965
4.5	130.3	0.1188552	21.8	75.86	1.046725	0.0609	211		
5.0	196.0		44.7	47.67		0.0157			

HALOCARBON REFRIGERANTS

Halogen derivatives of the lower aliphatic hydrocarbons, especially fluorocarbons and chlorocarbons, are widely used as refrigerants, solvents, and cleaning agents. This table lists the most important halocarbons in current use or being considered as possible substitutes for those compounds that are potential environmental hazards. The code number for each compound appears in the first column; this number is frequently used in expressions like R 11 or CFC 11 or HFC 134.

In addition to name, molecular formula, CAS Registry Number, and molecular weight, the table lists the following properties:

t_m: normal melting point
t_b: normal boiling point (at 101.324 kPa)
t_c: critical temperature
TLV: Threshold Limit Value, which is the maximum safe concentration in air in the workplace, expressed as the time-weighted average (TWA) in parts per million by volume over an 8-hr workday and 40-hr workweek (see Reference 3). An * following the TLV indicates that the substance in a confirmed or suspected human carcinogen.

REFERENCES

1. *1989 ASHRAE Handbook: Fundamentals*, American Society of Heating, Refrigerating, and Air Conditioning Engineers, Atlanta, GA, 1989.
2. Platzer, B., Polt, A.. and Mauer,G.,*Thermophysical Properties of Refrigerants*, Springer, Berlin, 1990.
3. 1991-1992 Threshold Limit Values and Biological Exposure Indexes, American Conference of Government Industrial Hygienists, Cincinnati, OH, 1991.

Code no.	Name	Molecular formula	CAS reg. no.	Molecular weight	$t_m/°C$	$t_b/°C$	$t_c/°C$	TLV ppm
10	Tetrachloromethane	CCl_4	56-23-5	153.822	-23	76.8	283.5	5*
11	Trichlorofluoromethane	CCl_3F	75-69-4	137.368	-111.11	23.75	198.1	1000
12	Dichlorodifluoromethane	CCl_2F_2	75-71-8	120.913	-158	-29.8	111.80	1000
12B1	Bromochlorodifluoromethane	$CBrClF_2$	353-59-3	165.365	-159.5	-3.72	153.73	
13	Chlorotrifluoromethane	$CClF_3$	75-72-9	104.459	-181	-81.4	29	
13B1	Bromotrifluoromethane	$CBrF_3$	75-63-8	148.910	-172	-57.89	67.1	
14	Tetrafluoromethane	CF_4	75-73-0	88.005	-183.59	-128.02	-45.5	
20	Trichloromethane	$CHCl_3$	67-66-3	119.377	-63.6	61.17	263.3	10*
21	Dichlorofluoromethane	$CHCl_2F$	75-43-4	102.923	-135	8.95	178.43	10
22	Chlorodifluoromethane	$CHClF_2$	75-45-6	86.468	-157.42	-40.75	96.2	1000
23	Trifluoromethane	CHF_3	75-46-7	70.014	-155.18	-82.1	26.2	
30	Dichloromethane	CH_2Cl_2	75-09-2	84.932	-95.14	40	237	50*
31	Chlorofluoromethane	CH_2ClF	593-70-4	68.478	-133	-9.1	78.5	
32	Difluoromethane	CH_2F_2	75-10-5	52.024	-136	-51.69	143.10	50*
40	Chloromethane	CH_3Cl	74-87-3	50.488	-97.7	-24.09	44.7	
41	Fluoromethane	CH_3F	593-53-3	34.033	-141.8	-78.41		1*
110	Hexachloroethane	C_2Cl_6	67-72-1	236.738	187	187		
111	Pentachlorofluoroethane	C_2Cl_5F	354-56-3	220.284			278	500
112	Tetrachloro-1,2-difluoroethane	$C_2Cl_4F_2$	76-12-0	203.830	26	93		
112a	1,1,1,2-Tetrachloro-2,2-difluoroethane	$C_2Cl_4F_2$	76-11-9	203.830	40.6	91.5	214.2	1000
113	1,1,2-Trichlorotrifluoroethane	$C_2Cl_3F_3$	76-13-1	187.375	-35	47.7		
113a	1,1,1-Trichlorotrifluoroethane	$C_2Cl_3F_3$	354-58-5	187.375	14.2	46.1	145.63	1000
114	1,2-Dichlorotetrafluoroethane	$C_2Cl_2F_4$	76-14-2	170.921	-94	3.8	145.5	1000
114a	1,1-Dichlorotetrafluoroethane	$C_2Cl_2F_4$	374-07-2	170.921	-56.6	4		

HALOCARBON REFRIGERANTS (continued)

Code no.	Name	Molecular formula	CAS reg. no.	Molecular weight	t_m/°C	t_b/°C	t_c/°C	TLV ppm
114B2	1,2-Dibromotetrafluoroethane	$C_2Br_2F_4$	124-73-2	259.824	-110.4	47.35	214.7	1000
115	Chloropentafluoroethane	C_2ClF_5	76-15-3	154.467	-99.44	-37.95	80.1	
116	Hexafluoroethane	C_2F_6	76-16-4	138.012	-100.7	-78.1	20	
120	Pentachloroethane	C_2HCl_5	76-01-7	202.293	-29	159.88		
121	1,1,2,2-Tetrachloro-1-fluoroethane	C_2HCl_4F	354-14-3	185.839	-82.6	116.6		
121a	1,1,1,2-Tetrachloro-2-fluoroethane	C_2HCl_4F	354-11-0	185.840	-95.35	116.5		
122	1,2,2-Trichloro-1,1-difluoroethane	$C_2HCl_3F_2$	354-21-2	169.385	-140	71.9		
122a	1,1,2-Trichloro-1,2-difluoroethane	$C_2HCl_3F_2$	354-15-4	169.385		72.5		
122b	1,1,1-Trichloro-2,2-difluoroethane	$C_2HCl_3F_2$	354-12-1	169.385		73		
123	2,2-Dichloro-1,1,1-trifluoroethane	$C_2HCl_2F_3$	306-83-2	152.931	-107	27.1		
123a	1,2-Dichloro-1,2,2-trifluoroethane	$C_2HCl_2F_3$	354-23-4	152.931		28.2		
124	2-Chloro-1,1,1,2-tetrafluoroethane	C_2HClF_4	2837-89-0	136.476		-12	126.8	
124a	1-Chloro-1,1,2,2-tetrafluoroethane	C_2HClF_4	354-25-6	136.476		-10.2		
125	Pentafluoroethane	C_2HF_5	354-33-6	120.022	-103	-48.5		1*
130	1,1,2,2-Tetrachloroethane	$C_2H_2Cl_4$	79-34-5	167.849	-43.8	146.5	388.00	
131	1,1,2-Trichloro-2-fluoroethane	$C_2H_2Cl_3F$	359-28-4	151.394		102.5		
132	1,2-Dichloro-1,2-difluoroethane	$C_2H_2Cl_2F_2$	431-06-1	134.940	-155	59		
132b	1,2-Dichloro-1,1-difluoroethane	$C_2H_2Cl_2F_2$	1649-08-7	134.940				
133	1-Chloro-1,2,2-trifluoroethane	$C_2H_2ClF_3$	431-07-2	118.486		17		
133a	2-Chloro-1,1,1-trifluoroethane	$C_2H_2ClF_3$	75-88-7	118.486	-105.5	6.1		
133b	1-Chloro-1,1,2-trifluoroethane	$C_2H_2ClF_3$	421-04-5	118.486		12		
134	1,1,2,2-Tetrafluoroethane	$C_2H_2F_4$	359-35-3	102.032	-89	-19.9	119	
134a	1,1,1,2-Tetrafluoroethane	$C_2H_2F_4$	811-97-2	102.032	-101	-26.5	101.2	
140	1,1,2-Trichloroethane	$C_2H_3Cl_3$	79-00-5	133.404	-36.6	113.8	272	10
140a	1,1,1-Trichloroethane	$C_2H_3Cl_3$	71-55-6	133.404	-30.4	74.09		350
141	1,2-Dichloro-1-fluoroethane	$C_2H_3Cl_2F$	430-57-9	116.950		75.7		
141b	1,1-Dichloro-1-fluoroethane	$C_2H_3Cl_2F$	1717-00-6	116.950	-103.5	32.11		
142	2-Chloro-1,1-difluoroethane	$C_2H_3ClF_2$	338-65-8	100.495		35.1		
142b	1-Chloro-1,1-difluoroethane	$C_2H_3ClF_2$	75-68-3	100.495	-130.8	-9.75	137.14	
143	1,1,2-Trifluoroethane	$C_2H_3F_3$	430-66-0	84.041	-84	5		
143a	1,1,1-Trifluoroethane	$C_2H_3F_3$	420-46-2	84.041	-111.3	-47.55	73.2	
150	1,2-Dichloroethane	$C_2H_4Cl_2$	107-06-2	98.959	-35.5	83.5	288	10*
150a	1,1-Dichloroethane	$C_2H_4Cl_2$	75-34-3	98.959	-96.96	57.4	250	100
151	1-Chloro-2-fluoroethane	C_2H_4ClF	762-50-5	82.505		53.2		
151a	1-Chloro-1-fluoroethane	C_2H_4ClF	1615-75-4	82.505		16.1		
152	1,2-Difluoroethane	$C_2H_4F_2$	624-72-6	66.051		30.7		
152a	1,1-Difluoroethane	$C_2H_4F_2$	75-37-6	66.051	-117	-24.95	113.6	1000
160	Chloroethane	C_2H_5Cl	75-00-3	64.514	-138.7	13.1	187.3	
161	Fluoroethane	C_2H_5F	353-36-6	48.060	-143.2	-37.65	102.16	
218	Perfluoropropane	C_3F_8	76-19-7	188.020	-147.69	-36.65	72.0	
1112a	1,1-Dichloro-2,2-difluoroethylene	$C_2Cl_2F_2$	79-35-6	132.924				
1113	Chlorotrifluoroethylene	C_2ClF_3	79-38-9	116.470	-158	-27.85	106	

HALOCARBON REFRIGERANTS (continued)

Code no.	Name	Molecular formula	CAS reg. no.	Molecular weight	t_m/°C	t_b/°C	t_c/°C	TLV ppm
1114	Tetrafluoroethylene	C_2F_4	116-14-3	100.016	-142.5	-75.95	33.4	
1120	Trichloroethylene	C_2HCl_3	79-01-6	131.388	-84.75	87.21	271.1	50*
1130	cis-1,2-Dichloroethylene	$C_2H_2Cl_2$	156-59-2	96.943	-80	60.19	271.1	200
1130	trans-1,2-Dichloroethylene	$C_2H_2Cl_2$	156-60-5	96.943	-49.8	48.73	243.4	200
1132a	1,1-Difluoroethylene	$C_2H_2F_2$	75-38-7	64.035	-144	-85.7	29.8	
1140	Chloroethylene	C_2H_3Cl	75-01-4	62.499	-153.79	-13.37		5*
1141	Fluoroethylene	C_2H_3F	75-02-5	46.044	-160.5	-72	54.8	
C316	1,2-Dichlorohexafluorocyclobutane	$C_4Cl_2F_6$	356-18-3	232.940				
C317	Chloroheptafluorocyclobutane	C_4ClF_7	377-41-3	216.486				
C318	Perfluorocyclobutane	C_4F_8	115-25-3	200.031	-40.19	-5.99	115.31	

* Confirmed or suspected human carcinogen.

DENSITY AND SPECIFIC VOLUME OF MERCURY

The data in this table have been adjusted to the ITS-90 temperature scale. The uncertainty in density values is 0.0003 g/mL between −20 and −10°C; 0.0001 or less between −10 and 200°C; and 0.0002 between 200 and 300°C.

REFERENCE

Ambrose, D., *Metrologia*, 27, 245, 1990.

$t/°C$	$\rho/(g/mL)$	$v/(mL/kg)$	$t/°C$	$\rho/(g/mL)$	$v/(mL/kg)$	$t/°C$	$\rho/(g/mL)$	$v/(mL/kg)$
−20	13.64461	73.2890	27	13.52869	73.9170	74	13.41423	74.5477
−19	13.64212	73.3024	28	13.52624	73.9304	75	13.41181	74.5612
−18	13.63964	73.3157	29	13.52379	73.9438	76	13.40939	74.5746
−17	13.63716	73.3291	30	13.52134	73.9572	77	13.40697	74.5881
−16	13.63468	73.3424	31	13.51889	73.9705	78	13.40455	74.6016
−15	13.63220	73.3558	32	13.51645	73.9839	79	13.40213	74.6150
−14	13.62972	73.3691	33	13.51400	73.9973	80	13.39971	74.6285
−13	13.62724	73.3824	34	13.51156	74.0107	81	13.39729	74.6420
−12	13.62476	73.3958	35	13.50911	74.0241	82	13.39487	74.6554
−11	13.62228	73.4091	36	13.50667	74.0375	83	13.39245	74.6689
−10	13.61981	73.4225	37	13.50422	74.0509	84	13.39003	74.6824
−9	13.61733	73.4358	38	13.50178	74.0643	85	13.38762	74.6959
−8	13.61485	73.4492	39	13.49934	74.0777	86	13.38520	74.7094
−7	13.61238	73.4625	40	13.49690	74.0911	87	13.38278	74.7229
−6	13.60991	73.4759	41	13.49446	74.1045	88	13.38037	74.7364
−5	13.60743	73.4892	42	13.49202	74.1179	89	13.37795	74.7498
−4	13.60496	73.5026	43	13.48958	74.1313	90	13.37554	74.7633
−3	13.60249	73.5160	44	13.48714	74.1447	91	13.37313	74.7768
−2	13.60002	73.5293	45	13.48470	74.1581	92	13.37071	74.7903
−1	13.59755	73.5427	46	13.48226	74.1715	93	13.36830	74.8038
0	13.59508	73.5560	47	13.47982	74.1850	94	13.36589	74.8173
1	13.59261	73.5694	48	13.47739	74.1984	95	13.36347	74.8308
2	13.59014	73.5827	49	13.47495	74.2118	96	13.36106	74.8443
3	13.58768	73.5961	50	13.47251	74.2252	97	13.35865	74.8579
4	13.58521	73.6095	51	13.47008	74.2386	98	13.35624	74.8714
5	13.58275	73.6228	52	13.46765	74.2520	99	13.35383	74.8849
6	13.58028	73.6362	53	13.46521	74.2655	100	13.35142	74.8984
7	13.57782	73.6495	54	13.46278	74.2789	110	13.3273	75.0337
8	13.57535	73.6629	55	13.46035	74.2923	120	13.3033	75.1693
9	13.57289	73.6763	56	13.45791	74.3057	130	13.2793	75.3052
10	13.57043	73.6896	57	13.45548	74.3192	140	13.2553	75.4413
11	13.56797	73.7030	58	13.45305	74.3326	150	13.2314	75.5778
12	13.56551	73.7164	59	13.45062	74.3460	160	13.2075	75.7147
13	13.56305	73.7297	60	13.44819	74.3594	170	13.1836	75.8519
14	13.56059	73.7431	61	13.44576	74.3729	180	13.1597	75.9895
15	13.55813	73.7565	62	13.44333	74.3863	190	13.1359	76.1274
16	13.55567	73.7698	63	13.44090	74.3998	200	13.1120	76.2659
17	13.55322	73.7832	64	13.43848	74.4132	210	13.0882	76.4047
18	13.55076	73.7966	65	13.43605	74.4266	220	13.0644	76.5440
19	13.54831	73.8100	66	13.43362	74.4401	230	13.0406	76.6838
20	13.54585	73.8233	67	13.43120	74.4535	240	13.0167	76.8241
21	13.54340	73.8367	68	13.42877	74.4670	250	12.9929	76.9650
22	13.54094	73.8501	69	13.42635	74.4804	260	12.9691	77.1064
23	13.53849	73.8635	70	13.42392	74.4939	270	12.9453	77.2484
24	13.53604	73.8769	71	13.42150	74.5073	280	12.9214	77.3909
25	13.53359	73.8902	72	13.41908	74.5208	290	12.8975	77.5341
26	13.53114	73.9036	73	13.41665	74.5342	300	12.8736	77.6779

THERMAL PROPERTIES OF MERCURY
Lev R. Fokin

The first of these tables gives the molar heat capacity at constant pressure of liquid and gaseous mercury as a function of temperature. To convert to specific heat in units of J/g K, divide these values by 200.59, the atomic weight of mercury.

REFERENCE

Douglas, T. B., Ball, A. T., and Ginnings, D. C., *J. Res. Natl. Bur. Stands.,* 46, 334, 1951.

t/°C	C_p/(J/mol K) Liquid	Gas	t/°C	C_p/(J/mol K) Liquid	Gas	t/°C	C_p/(J/mol K) Liquid	Gas
−38.84	28.2746	20.786	140	27.3675	20.786	340	27.1500	20.836
−20	28.1466	20.786	160	27.3090	20.786	356.73	27.1677	20.849
0	28.0190	20.786	180	27.2588	20.790	360	27.1709	20.853
20	27.9002	20.786	200	27.2169	20.790	380	27.1981	20.870
25	27.8717	20.786	220	27.1834	20.794	400	27.2324	20.891
40	27.7897	20.786	240	27.1583	20.794	420	27.2738	20.916
60	27.6880	20.786	260	27.1412	20.799	440	27.3207	20.941
80	27.5952	20.786	280	27.1320	20.807	460	27.3742	20.974
100	27.5106	20.786	300	27.1303	20.815	480	27.4332	21.008
120	27.4349	20.786	320	27.1366	20.824	500	27.4985	21.046

The second table gives the molar heat capacity of solid mercury in its rhombohedral (α-mercury) form.

REFERENCES

1. Busey and Giaque, *J. Am. Chem. Soc.,* 75, 806, 1953.
2. Amitin, Lebedeva, and Paukov, *Rus. J. Phys. Chem.,* 2666, 1979.

t/°C	C_p/J mol^{-1}	t/°C	C_p/J mol^{-1}	t/°C	C_p/J mol^{-1}	t/°C	C_p/J mol^{-1}
−268.99	0.99*	−248.15	12.74	−193.15	23.16	−113.15	26.15
−268.99	0.97**	−243.15	14.78	−183.15	23.76	−93.15	26.69
−268.15	1.6	−233.15	17.90	−173.15	24.24	−73.15	27.28
−263.15	4.6	−223.15	19.94	−153.15	25.00	−53.15	27.96
−258.15	7.6	−213.15	21.40	−133.15	25.61	−38.87	28.5
−253.15	10.33	−203.15	22.42				

*   Superconducting state
**   Normal state

The final table gives the cubic thermal expansion coefficient α, the isothermal compressibility coefficient κ_T, and the speed of sound U for liquid mercury as a function of temperature. These properties are defined as follows:

$$\alpha = \frac{1}{v}\left(\frac{\partial v}{\partial T}\right)_p \qquad \kappa_T = -\frac{1}{v}\left(\frac{\partial v}{\partial P}\right)_T \qquad U^2 = \left(\frac{\partial P}{\partial \rho}\right)_S \qquad \rho = v^{-1}$$

where v is the specific volume (given in the table on the preceding page).

REFERENCE

Vukalovich, M. P., et al., *Thermophysical Properties of Mercury,* Moscow Standard Press, 1971.

t/°C	α × 10^4/K^{-1}	κ_T × 10^6/bar^{-1} At 1 bar	At 1000 bar	U/m s^{-1}	t/°C·	α × 10^4/K^{-1}	κ_T × 10^6/bar^{-1} At 1 bar	At 1000 bar	U/m s^{-1}
−20	1.818	3.83		1470	120	1.8058	4.513	4.33	1404.7
0	1.8144	3.918	3.78	1460.8	140	1.8074	4.622		1395.4
20	1.8110	4.013	3.87	1451.4	160	1.8100	4.731	4.53	1386.1
40	1.8083	4.109	3.96	1442.0	180	1.8136	4.844		1376.7
60	1.8064	4.207		1432.7	200	1.818	4.96		1367
80	1.8053	4.308	4.14	1423.4	250	1.834	5.26		1344
100	1.8051	4.410		1414.1	300	1.856	5.59		1321

SURFACE TENSION OF COMMON LIQUIDS

The surface tension γ of about 200 liquids is tabulated here as a function of temperature. Values of γ are given in units of millinewtons per meter (mN/m), which is equivalent to dyn/cm in cgs units. The values refer to a nominal pressure of one atmosphere (about 100 kPa) except in cases where the indicated temperature is above the normal boiling point of the substance; in those cases, the applicable pressure is the saturation vapor pressure at the temperature in question.

The uncertainty of the values is 0.1 to 0.2 mN/m or less in most cases. Values at temperatures between the points tabulated can be obtained by linear interpolation to a good approximation.

Substances are listed by molecular formula in the modified Hill order, with substances not containing carbon appearing before those that do contain carbon. A more extensive compilation of surface tension may be found in the Reference.

REFERENCE

Jasper, J. J., *J. Phys. Chem. Ref. Data*, 1, 841, 1972.

Mol. form.	Name	γ in mN/m				
		10°C	25°C	50°C	75°C	100°C
Br_2	Bromine	43.68	40.95	36.40		
Cl_2O_2S	Sulfuryl chloride		28.78			
Cl_3OP	Phosphoryl chloride		32.03	28.85	25.66	
Cl_3P	Phosphorus trichloride		27.98	24.81		
Cl_4Si	Silicon tetrachloride	19.78	18.29	15.80		
H_2O	Water	74.23	71.99	67.94	63.57	58.91
H_4N_2	Hydrazine		66.39			
Hg	Mercury	488.55	485.48	480.36	475.23	470.11
CCl_4	Tetrachloromethane		26.43	23.37	20.31	17.25
CS_2	Carbon disulfide	33.81	31.58	27.87		
$CHBr_3$	Tribromomethane		44.87	41.60	38.33	
$CHCl_3$	Trichloromethane		26.67	23.44	20.20	
CH_2Br_2	Dibromomethane		39.05	35.33	31.61	
CH_2Cl_2	Dichloromethane		27.20			
CH_2O_2	Formic acid		37.13	34.38	31.64	
CH_3I	Iodomethane	32.19	30.34			
CH_3NO	Formamide		57.03	54.92	52.82	50.71
CH_3NO_2	Nitromethane	39.04	36.53	32.33		
CH_4O	Methanol	23.23	22.07	20.14		
CH_5N	Methylamine		19.15			
C_2HCl_5	Pentachloroethane		34.15	31.20	28.26	
$C_2HF_3O_2$	Trifluoroacetic acid		13.53	11.42		
$C_2H_2Cl_4$	1,1,2,2-Tetrachloroethane		35.58	32.41	29.24	26.07
$C_2H_3Cl_3$	1,1,1-Trichloroethane		25.18	22.07		
$C_2H_3Cl_3$	1,1,2-Trichloroethane		34.02	30.65	27.27	23.89
C_2H_3N	Acetonitrile		28.66	25.51		
$C_2H_4Br_2$	1,2-Dibromoethane		39.55	36.25	32.95	
$C_2H_4Cl_2$	1,1-Dichloroethane		24.07			
$C_2H_4Cl_2$	1,2-Dichloroethane		31.86	28.29	24.72	
C_2H_4O	Acetaldehyde	22.54	20.50	17.10		
$C_2H_4O_2$	Acetic acid		27.10	24.61	22.13	
$C_2H_4O_2$	Methyl formate	26.72	24.36	20.43	16.50	12.57
C_2H_5Br	Bromoethane	25.36	23.62			
C_2H_5I	Iodoethane	30.38	28.46	25.24		
$C_2H_5NO_2$	Nitroethane	34.02	32.13	29.00		
C_2H_6O	Ethanol	23.22	21.97	19.89		
C_2H_6OS	Dimethyl sulfoxide		42.92	40.06		
$C_2H_6O_2$	Ethylene glycol		47.99	45.76	43.54	41.31
C_2H_6S	Dimethyl sulfide	25.27	24.06			
C_2H_6S	Ethanethiol		23.08			
$C_2H_6S_2$	Dimethyl disulfide		33.39	30.04		
C_2H_7N	Dimethylamine		26.34			
C_2H_7N	Ethylamine		19.20			

Mol. form.	Name	γ in mN/m				
		10°C	25°C	50°C	75°C	100°C
C_2H_7NO	Ethanolamine		48.32	45.53	42.73	
C_3H_5Br	3-Bromopropene		26.31	23.17		
C_3H_5Cl	3-Chloropropene		23.14			
C_3H_5ClO	Epichlorohydrin	38.40	36.36	32.96	29.56	26.16
C_3H_5N	Propanenitrile		26.75	23.87		
$C_3H_6Cl_2$	1,2-Dichloropropane		28.32	25.22	22.12	
C_3H_6O	Acetone		23.46	20.66		
C_3H_6O	Allyl alcohol	26.63	25.28	23.02	20.77	
$C_3H_6O_2$	Ethyl formate	25.16	23.18			
$C_3H_6O_2$	Methyl acetate	26.66	24.73	21.51		
$C_3H_6O_2$	Propanoic acid		26.20	23.72	21.23	
C_3H_7Br	1-Bromopropane	27.08	25.26	22.21		
C_3H_7Br	2-Bromopropane	25.03	23.25	20.30		
C_3H_7Cl	1-Chloropropane	23.16	21.30			
C_3H_7Cl	2-Chloropropane	20.49	19.16			
$C_3H_7NO_2$	2-Nitropropane	31.02	29.29	26.39		
C_3H_8O	1-Propanol	24.48	23.32	21.38	19.43	
C_3H_8O	2-Propanol	22.11	20.93	18.96	16.98	
$C_3H_8O_2$	2-Methoxyethanol	32.32	30.84	28.38	25.92	23.46
C_3H_8S	1-Propanethiol		24.20	21.02		
C_3H_8S	2-Propanethiol		21.33	18.39		
C_3H_9N	Propylamine		21.75			
C_3H_9N	Trimethylamine		13.41			
$C_4H_4N_2$	Pyridazine	49.51	47.96	45.37	42.78	40.19
$C_4H_4N_2$	Pyrimidine		30.33	27.80	25.28	22.75
C_4H_4S	Thiophene		30.68	27.36		
C_4H_5N	Pyrrole	38.71	37.06	34.31		
$C_4H_6O_3$	Acetic anhydride	34.08	31.93	28.34	24.75	21.16
C_4H_7N	Butanenitrile		26.92	24.33	21.73	
C_4H_8O	2-Butanone		23.97	21.16		
$C_4H_8O_2$	1,4-Dioxane		32.75	29.28	25.80	22.32
$C_4H_8O_2$	Ethyl acetate	25.13	23.39	20.49	17.58	14.68
$C_4H_8O_2$	Methyl propanoate	26.32	24.44	21.29		
$C_4H_8O_2$	Butanoic acid		26.05	23.75	21.45	
C_4H_9Br	1-Bromobutane	27.58	25.90	23.08	20.27	17.45
C_4H_9Cl	1-Chlorobutane	24.85	23.18	20.39		
C_4H_9I	1-Iodobutane	29.79	28.24	25.67	23.09	20.51
C_4H_9N	Pyrrolidine	30.58	29.23	26.98		
$C_4H_{10}O$	1-Butanol	26.28	24.93	22.69	20.44	18.20
$C_4H_{10}O$	2-Butanol	23.74	22.54	20.56	18.57	16.58
$C_4H_{10}O$	2-Methyl-2-propanol		19.96	17.71		
$C_4H_{10}O$	Diethyl ether		16.65			
$C_4H_{10}O_2$	2-Ethoxyethanol		28.35	26.11	23.86	21.62
$C_4H_{10}O_3$	Diethylene glycol		44.77	42.57	40.37	38.17
$C_4H_{10}S$	Diethyl sulfide	26.22	24.57	21.80		
$C_4H_{11}N$	Butylamine		23.44	20.63		
$C_4H_{11}N$	Isobutylamine		21.75	19.02		
$C_4H_{11}N$	*tert*-Butylamine		16.87			
$C_4H_{11}N$	Diethylamine		19.85			
$C_5H_4O_2$	Furfural	45.08	43.09	39.78	36.46	33.14
C_5H_5N	Pyridine		36.56	33.29	30.03	
C_5H_8	Cyclopentene	24.45	22.20			
C_5H_8O	Cyclopentanone	34.45	32.80	30.05	27.30	24.55
C_5H_{10}	1-Pentene	17.10	15.45			
C_5H_{10}	2-Methyl-2-butene	18.61	17.15			
C_5H_{10}	Cyclopentane	24.07	21.88	18.22		
$C_5H_{10}O$	2-Pentanone		23.25	21.62		

Mol. form.	Name	γ in mN/m				
		10°C	25°C	50°C	75°C	100°C
$C_5H_{10}O$	3-Pentanone		24.74	22.13		
$C_5H_{10}O$	Pentanal	26.95	25.44	22.91		
$C_5H_{10}O_2$	Butyl formate	26.05	24.52	21.95	19.39	16.82
$C_5H_{10}O_2$	Propyl acetate	25.48	23.80	21.00	18.20	15.40
$C_5H_{10}O_2$	Isopropyl acetate	23.37	21.76	19.08	16.40	
$C_5H_{10}O_2$	Ethyl propanoate	25.55	23.80	20.88	17.96	
$C_5H_{10}O_2$	Methyl butanoate	26.34	24.62	21.76	18.89	16.03
$C_5H_{11}Cl$	1-Chloropentane	26.01	24.40	21.71	19.02	16.33
$C_5H_{11}N$	Piperidine	30.64	28.91	26.03	23.14	20.26
C_5H_{12}	Pentane	17.15	15.49			
$C_5H_{12}O$	1-Pentanol	26.67	25.36	23.17	20.99	18.80
$C_5H_{12}O$	2-Pentanol	24.96	23.45	20.94	18.43	15.92
$C_5H_{12}O$	3-Methyl-1-butanol	24.94	23.71	21.66	19.61	17.56
$C_5H_{13}N$	Pentylamine		24.69	22.14	19.58	
$C_6H_4Cl_2$	m-Dichlorobenzene	37.15	35.43	32.57	29.70	26.83
C_6H_5Br	Bromobenzene	36.98	35.24	32.34	29.44	26.54
C_6H_5Cl	Chlorobenzene	34.78	32.99	30.02	27.04	24.06
C_6H_5ClO	o-Chlorophenol		39.70	36.89	34.09	31.28
C_6H_5ClO	m-Chlorophenol		41.18	38.66	36.13	33.61
C_6H_5F	Fluorobenzene	28.47	26.66	23.65	20.64	
C_6H_5I	Iodobenzene	40.40	38.71	35.91	33.10	30.29
$C_6H_5NO_2$	Nitrobenzene			40.56	37.66	34.77
C_6H_6	Benzene		28.22	25.00	21.77	
C_6H_6O	Phenol			38.20	35.53	32.86
C_6H_7N	Aniline		42.12	39.41	36.69	
C_6H_7N	2-Methylpyridine		33.00	29.90	26.79	
$C_6H_8N_2$	Adiponitrile		45.45	43.02	40.58	
C_6H_{10}	Cyclohexene	28.01	26.17	23.12		
$C_6H_{10}O$	Cyclohexanone	36.43	34.57	31.46	28.36	25.25
$C_6H_{11}N$	Hexanenitrile		27.37	25.11	22.84	
C_6H_{12}	Cyclohexane	26.43	24.65	21.68		
C_6H_{12}	Methylcyclopentane	23.47	21.72	18.82		
C_6H_{12}	1-Hexene	19.44	17.90	15.33		
$C_6H_{12}O$	Cyclohexanol		32.92	30.50	28.09	25.67
$C_6H_{12}O$	2-Hexanone		25.45	22.72		
$C_6H_{12}O_2$	Butyl acetate	26.48	24.88	22.21	19.54	16.87
$C_6H_{12}O_2$	Isobutyl acetate	24.58	23.06	20.53	17.99	15.46
$C_6H_{12}O_2$	Ethyl butanoate	25.51	23.94	21.33	18.71	16.10
$C_6H_{12}O_3$	Paraldehyde	27.22	25.63	22.97	20.32	17.66
$C_6H_{13}Cl$	1-Chlorohexane	27.28	25.73	23.13	20.54	17.94
$C_6H_{13}N$	Cyclohexylamine		31.22	28.25	25.28	
C_6H_{14}	Hexane	19.42	17.89	15.33		
C_6H_{14}	2-Methylpentane	18.37	16.88	14.39		
C_6H_{14}	3-Methylpentane	19.20	17.61	14.96		
$C_6H_{14}O$	Diisopropyl ether		17.27	14.65		
$C_6H_{14}O$	1-Hexanol		25.81	23.81	21.80	19.80
$C_6H_{14}O_2$	1,1-Diethoxyethane		20.89	18.31	15.74	
$C_6H_{14}O_2$	2-Butoxyethanol	27.36	26.14	24.10	22.06	20.02
$C_6H_{15}N$	Triethylamine		20.22	17.74		
$C_6H_{15}N$	Dipropylamine		22.31	19.75	17.20	
$C_6H_{15}N$	Diisopropylamine		19.14	16.45		
C_7H_5N	Benzonitrile		38.79	35.90	33.00	
C_7H_6O	Benzaldehyde	39.63	38.00	35.27	32.55	29.82
C_7H_8	Toluene	29.71	27.93	24.96	21.98	19.01
C_7H_8O	o-Cresol		36.90	34.38	31.85	29.32
C_7H_8O	m-Cresol		35.69	33.38	31.07	28.76
C_7H_8O	Benzyl alcohol				27.89	24.44

Mol. form.	Name	γ in mN/m				
		10°C	25°C	50°C	75°C	100°C
C_7H_8O	Anisole		35.10	32.09	29.08	
C_7H_9N	N-Methylaniline		36.90	34.47	32.05	
C_7H_9N	2,3-Dimethylpyridine		32.71	30.04	27.36	
C_7H_9N	Benzylamine		39.30	36.27	33.23	
C_7H_{14}	Methylcyclohexane	24.98	23.29	20.46		
C_7H_{14}	1-Heptene	21.29	19.80	17.33	14.85	
$C_7H_{14}O$	2-Heptanone		26.12	23.48		
$C_7H_{14}O_2$	Pentyl acetate	26.67	25.17	22.69	20.20	17.72
$C_7H_{14}O_2$	Heptanoic acid		27.76	25.64		
C_7H_{16}	Heptane	21.12	19.65	17.20	14.75	
C_7H_{16}	3-Methylhexane	20.76	19.31	16.88	14.46	
C_8H_8O	Acetophenone		39.04	36.15	33.27	
$C_8H_8O_2$	Methyl benzoate		37.17	34.25	31.32	
$C_8H_8O_3$	Methyl salicylate	40.98	39.22	36.28	33.35	30.41
C_8H_{10}	Ethylbenzene	30.39	28.75	26.01	23.28	20.54
C_8H_{10}	o-Xylene	31.41	29.76	27.01	24.25	21.50
C_8H_{10}	m-Xylene	30.13	28.47	25.71	22.95	20.19
C_8H_{10}	p-Xylene		28.01	25.32	22.64	19.95
$C_8H_{10}O$	Phenetole		32.41	29.65	26.89	
$C_8H_{11}N$	N,N-Dimethylaniline		35.52	32.90	30.27	
$C_8H_{11}N$	N-Ethylaniline		36.33	33.65	30.98	
C_8H_{16}	Ethylcyclohexane	26.73	25.15	22.51		
C_8H_{18}	Octane	22.57	21.14	18.77	16.39	14.01
C_8H_{18}	2,5-Dimethylhexane	20.77	19.40	17.12	14.84	12.56
$C_8H_{18}O$	1-Octanol	28.30	27.10	25.12		
$C_8H_{19}N$	Dibutylamine		24.12	21.74	19.36	
$C_8H_{19}N$	Diisobutylamine		21.72	19.44	17.16	
C_9H_7N	Quinoline	44.19	42.59	39.94	37.28	34.62
C_9H_{12}	Cumene	29.27	27.69	25.05	22.42	19.78
C_9H_{12}	1,2,4-Trimethylbenzene	30.74	29.20	26.64	24.07	21.51
C_9H_{12}	Mesitylene	28.89	27.55	25.31	23.07	20.82
$C_9H_{18}O$	5-Nonanone		26.28	23.85		
C_9H_{20}	Nonane	23.79	22.38	20.05	17.71	15.37
$C_9H_{20}O$	1-Nonanol	29.03	27.89	26.00	24.10	22.20
$C_{10}H_{12}$	1,2,3,4-Tetrahydronaphthalene		33.17	30.78	28.40	
$C_{10}H_{22}$	Decane	24.75	23.37	21.07	18.77	16.47
$C_{10}H_{22}O$	1-Decanol	29.61	28.51	26.68	24.85	23.02
$C_{11}H_{24}$	Undecane	25.56	24.21	21.96	19.70	17.45
$C_{12}H_{10}O$	Diphenyl ether		26.75	24.80		
$C_{12}H_{27}N$	Tributylamine		24.39	22.32	20.24	
$C_{13}H_{28}$	Tridecane	26.86	25.55	23.37	21.19	19.01
$C_{14}H_{12}O_2$	Benzyl benzoate	44.47	42.82	40.06	37.31	34.55
$C_{14}H_{30}$	Tetradecane	27.43	26.13	23.96	21.78	19.61
$C_{16}H_{34}$	Hexadecane		27.05	24.91	22.78	20.64
$C_{18}H_{38}$	Octadecane		27.87	25.77	23.66	21.55

PERMITTIVITY (DIELECTRIC CONSTANT) OF LIQUIDS

Christian Wohlfarth

The permittivity of a substance (often called the dielectric constant) is the ratio of the electric displacement D to the electric field strength E when an external field is applied to the substance. The quantity tabulated here is the relative permittivity, which is the ratio of the actual permittivity to the permittivity of a vacuum; it is a dimensionless number.

The table gives the static relative permittivity ε_r, i.e., the relative permittivity measured in static fields or at low frequencies where no relaxation effects occur. The fourth column of the table lists the value of ε_r at the temperature specified in the third column, usually 293.15 or 298.15 K. Otherwise, the temperature closest to 293.15 K was chosen, or (as it is the case for many of the substances included here) ε_r is given at the only temperature for which data are available.

The static permittivity refers to nominal atmospheric pressure as long as the corresponding temperature is below the normal boiling point. Otherwise, at temperatures above the normal boiling point, the pressure is understood to be the saturated vapor pressure of the substance considered.

For substances where information on the temperature dependence of the permittivity is available, the table gives the coefficients of a simple polynomial fitting of permittivity to temperature with an equation of the form

$$\varepsilon_r(T) = a + bT + cT^2 + dT^3$$

where T is the absolute temperature in K. Since the parameter d was used in only a few cases where the quadratic fit was not satisfactory, only a, b, and c are listed as columns in the table, while the d values are given at the end of this introduction. For all other substances, $d = 0$. The temperature range of the fit is given in the last column. The coefficients of the fitting equation can be used to calculate dielectric constants within the fitted temperature range but should not be used for extrapolation outside this range. The user who needs dielectric constant data with more accuracy than can be provided by this equation is referred to Reference 1, which gives the original data together with their literature source.

Substances are listed by molecular formula in modified Hill order, with substances not containing carbon preceding those that do contain carbon.

* Indicates that the isomer was not specified in the original reference.
** Indicates a compound for which the cubic term is needed:

Ethanol	$d = -0.15512E-05$
N-Methylacetamide	$d = -0.12998E-04$
1,2-Propylene glycol	$d = -0.32544E-05$
1-Butanol	$d = -0.48841E-06$
2-Butanol	$d = -0.89512E-06$
2-Methyl-1-propanol	$d = -0.45229E-06$
2-Methyl-2-propanol	$d = -0.25968E-05$
N-Butylacetamide	$d = -0.48716E-05$

REFERENCES

1. Wohlfarth, Ch., "Static Dielectric Constants of Pure Liquids and Binary Liquid Mixtures", *Landolt-Börnstein, Numerical Data and Functional Relationships in Science and Technology*, New Series, Editor in Chief, O. Madelung, Group IV, Macroscopic and Technical Properties of Matter, Volume 6, Springer-Verlag, Berlin, Heidelberg, New York, 1991.
2. Marsh, K. N., Ed., *Recommended Reference Materials for the Realization of Physicochemical Properties*, Blackwell Scientific Publications, Oxford, 1987.

Mol. Form.	Name	T/K	ε	a	b	c	Range/K
$AlBr_3$	Aluminum tribromide	373.2	3.38				
Ar	Argon	140.00	1.3247	0.12408E+01	0.68755E-02	-0.45344E-04	87-149
AsH_3	Arsine	200.9	2.40	0.37674E+01	-0.97454E-02	0.14537E-04	157-201
BBr_3	Boron tribromide	273.2	2.58				
B_2H_6	Diborane	180.66	1.8725	0.23848E+01	-0.29501E-02	0.64189E-06	108-181
B_5H_9	Pentaborane(9)	298.2	21.1	0.40952E+03	-0.24414E+01	0.38225E-02	226-298
BrF_3	Bromine trifluoride	298.2	106.8				
BrF_5	Bromine pentafluoride	297.7	7.91	0.11428E+02	-0.11822E-01		262-298
BrH	Hydrogen bromide	186.8	8.23				
BrNO	Nitrosyl bromide	288.4	13.4				
Br_2	Bromine	297.9	3.1484	0.32701E+01	-0.12535E-03		273-327
Br_2OS	Thionyl bromide	293.2	9.06				
Br_3OV	Vanadyl tribromide	298.2	3.6	0.61112E+01	-0.84211E-02		203-298
Br_4Ge	Germanium(IV) bromide	299.9	2.955	0.34450E+01	-0.16083E-02		300-316
Br_4Sn	Tin(IV) bromide	303.45	3.169	0.50001E+01	-0.60383E-02		304-316
$ClFO_3$	Perchloryl fluoride	150.2	2.194	0.23808E+01	-0.38629E-03	-0.57143E-05	125-150
ClF_3	Chlorine trifluoride	293.2	4.394	0.96716E+01	-0.18000E-01		273-313

Mol. Form.	Name	T/K	ε	a	b	c	Range/K
ClF_5	Chlorine pentafluoride	193.2	4.28	0.78192E+01	-0.20860E-01	0.13132E-04	193-256
ClH	Hydrogen chloride	158.9	14.3	0.47316E+02	-0.28455E+00	0.48650E-03	159-258
ClNO	Nitrosyl chloride	285.2	18.2				
Cl_2	Chlorine	208.0	2.147	0.29440E+01	-0.44649E-02	0.30388E-05	208-240
Cl_2F_3P	Phosphorus(V) dichloride trifluoride	228.63	2.8129	0.46501E+01	-0.80358E-02		172-229
Cl_2OS	Thionyl chloride	298.2	8.675				
Cl_2OSe	Selenium oxychloride	293.2	46.2				
Cl_2O_2S	Sulfuryl chloride	293.2	9.1				
Cl_2S	Sulfur dichloride	298.2	2.915				
Cl_2S_2	Sulfur chloride	288.2	4.79				
Cl_3F_2P	Phosphorus(V) trichloride difluoride	268.0	2.3752	0.28905E+01	-0.19228E-02		215-268
Cl_3OP	Phosphorus(V) oxychloride	293.2	14.1				
Cl_3OV	Vanadyl trichloride	298.2	3.4				
Cl_3P	Phosphorus(III) chloride	290.2	3.498	0.59098E+01	-0.83322E-02		290-333
Cl_3PS	Phosphorus(V) sulfide trichloride	298.2	4.94				
Cl_4FP	Phosphorus(V) tetrachloride fluoride	272.64	2.6499	0.33503E+01	-0.29651E-02		244-273
Cl_4Ge	Germanium(IV) chloride	273.2	2.463	-0.55078E+01	0.64881E-01	-0.13091E-03	246-273
Cl_4Pb	Lead(IV) chloride	293.2	2.78				
Cl_4Si	Tetrachlorosilane	273.2	2.248	0.58041E+01	-0.27129E-01	0.51678E-04	207-273
Cl_4Sn	Tin(IV) chloride	273.2	3.014	0.43951E+01	-0.48805E-02		234-273
Cl_4Ti	Titanium(IV) chloride	257.4	2.843	0.33668E+01	-0.19675E-02		237-257
Cl_4V	Vanadium(IV) chloride	298.2	3.05				
Cl_5P	Phosphorus(V) chloride	433.2	2.85				
Cl_5Sb	Antimony(V) chloride	293.0	3.222	0.45413E+01	-0.45078E-02		276-320
FH	Hydrogen fluoride	273.2	83.6	0.50352E+03	-0.19297E+01	0.14372E-02	200-273
F_2	Fluorine	53.48	1.4913	0.14144E+01	0.26387E-02	-0.28356E-04	54-144
F_5I	Iodine pentafluoride	293.2	37.13	0.95184E+02	-0.19800E+00		273-313
F_6S	Sulfur hexafluoride	223.2	1.81				
F_6Xe	Xenon hexafluoride	398.2	4.10				
F_7I	Iodine heptafluoride	298.2	1.75				
$F_{10}S_2$	Sulfur decafluoride	293.2	2.0202				
HI	Hydrogen iodide	220.2	3.87	0.51557E+03	-0.44552E+01	0.96795E-02	220-236
H_2	Hydrogen	13.52	1.2792	0.13327E+01	-0.51946E-02		14-19
H_2O	Water	293.2	80.100	0.24921E+03	-0.79069E+00	0.72997E-03	273-372
H_2O_2	Hydrogen peroxide	290.2	74.6	0.48511E+03	-0.23145E+01	0.31020E-02	233-303
H_2S	Hydrogen sulfide	283.2	5.93	0.14736E+02	-0.33675E-01	0.96740E-05	212-363
H_3N	Ammonia	293.2	16.61	0.66756E+02	-0.24696E+00	0.25913E-03	238-323
H_4N_2	Hydrazine	298.2	51.7	0.22061E+03	-0.89633E+00	0.11066E-02	278-323
He	Helium	2.055	1.0555	0.10640E+01	-0.35584E-02		2-4
I_2	Iodine	391.25	11.08	0.64730E+02	-0.29266E+00	0.39759E-03	391-441
Kr	Krypton	119.80	1.664				
Mn_2O_7	Manganese(VII) oxide	293.2	3.28	0.37655E+01	-0.16463E-02		283-312
NO	Nitric oxide		1.997				
N_2	Nitrogen	63.15	1.4680	0.12550E+01	0.67949E-02	-0.56704E-04	63-126
N_2O_3	Nitrogen trioxide	203.2	31.13	0.92287E+02	-0.43306E+00	0.65000E-03	203-243
N_2O_4	Nitrogen tetroxide	293.2	2.44	0.28212E+01	-0.13000E-02		253-293
Ne	Neon	26.11	1.1907	0.12667E+01	-0.29064E-02		26-29
O_2	Oxygen	54.478	1.5684	0.15434E+01	0.14615E-02	-0.21964E-04	55-154
O_2S	Sulfur dioxide	298.2	16.3	0.52045E+02	-0.16125E+00	0.11042E-03	213-449
O_3	Ozone	90.2	4.75	0.86344E+01	-0.54807E-01	0.12596E-03	90-185
O_3S	Sulfur trioxide	291.2	3.11				
P	Phosphorus	307.2	4.096	0.79018E+00	0.23911E-01	-0.42826E-04	307-358
S	Sulfur	407.2	3.4991	0.51651E+01	-0.77381E-02	0.89120E-05	407-479
Se	Selenium	510.65	5.44	0.67569E+01	-0.25829E-02		511-575
Xe	Xenon	161.35	1.880				
$CBrClF_2$	Bromochlorodifluoromethane	123.2	3.920	0.52442E+01	-0.11000E-01		123-223
$CBrCl_3$	Bromotrichloromethane	293.2	2.405	0.29249E+01	-0.17650E-02		273-333
$CBrF_3$	Bromotrifluoromethane	123.2	3.730	0.54154E+01	-0.13680E-01		123-173
CBr_2Cl_2	Dibromodichloromethane	298.2	2.542	0.32330E+01	-0.23162E-02		298-333

Mol. Form.	Name	T/K	ε	a	b	c	Range/K
CBr_2F_2	Dibromodifluoromethane	273.2	2.939	0.67296E+01	-0.22133E-01	0.30213E-04	139-273
CBr_3Cl	Tribromochloromethane	333.2	2.601				
CBr_3F	Tribromofluoromethane	293.2	3.00	0.53203E+01	-0.11061E-01	0.10688E-04	206-323
CBr_3NO_2	Tribromonitromethane	298.2	9.034	0.16079E+02	-0.23630E-01		298-328
$CClF_3$	Chlorotrifluoromethane	123.2	3.010	0.43677E+01	-0.11020E-01		123-173
CCl_2F_2	Dichlorodifluoromethane	123.2	3.500	0.46984E+01	-0.97600E-02		123-223
CCl_2O	Carbonyl chloride	295.2	4.30				
CCl_3D	Trichloromethane-d	298.2	4.67				
CCl_3F	Trichlorofluoromethane	293.2	3.00	0.53203E+01	-0.11061E-01	0.10688E-04	206-323
CCl_3NO_2	Trichloronitromethane	293.2	7.319	0.14403E+02	-0.24178E-01		276-333
CCl_4	Tetrachloromethane	293.2	2.2379	0.28280E+01	-0.20339E-02	0.71795E-07	283-333
CF_4	Tetrafluoromethane	126.3	1.685	0.20350E+01	-0.27616E-02		126-142
$CHBr_3$	Tribromomethane	283.2	4.404	0.71707E+01	-0.98000E-02		283-343
$CHCl_3$	Trichloromethane	293.2	4.8069	0.15115E+02	-0.51830E-01	0.56803E-04	218-323
CHF_3	Trifluoromethane	294.0	5.2	0.11442E+03	-0.75600E+00	0.13562E-02	130-263
CHN	Hydrogen cyanide	293.2	114.9	0.37331E+04	-0.23180E+02	0.36963E-01	258-299
CH_2Br_2	Dibromomethane	283.2	7.77	0.18060E+02	-0.36333E-01		283-313
CH_2Cl_2	Dichloromethane	298.0	8.93	0.40452E+02	-0.17748E+00	0.23942E-03	184-306
CH_2F_2	Difluoromethane	152.2	53.74	0.19428E+03	-0.12939E+01	0.24280E-02	152-224
CH_2I_2	Diiodomethane	298.2	5.32				
CH_2O_2	Formic acid	298.2	51.1	0.14040E+03	-0.24673E+00	-0.17151E-03	287-358
CH_3Br	Bromomethane	275.7	9.71	0.40580E+02	-0.18418E+00	0.26219E-03	195-276
CH_3Cl	Chloromethane	295.2	10.0	0.42775E+02	-0.16175E+00	0.17108E-03	190-392
CH_3ClO_2S	Methanesulfonyl chloride	293.2	34.0	0.10384E+03	-0.33838E+00	0.34156E-03	293-373
CH_3DO	Methan-d_1-ol	297.5	31.68	0.20839E+03	-0.10318E+01	0.14740E-02	176-298
CH_3F	Fluoromethane	131.0	51.0	0.11338E+03	-0.63979E+00	0.96983E-03	150-299
CH_3I	Iodomethane	293.2	6.97	0.24264E+02	-0.93914E-01	0.11926E-03	223-303
CH_3NO	Formamide	293.2	111.0	0.26076E+03	-0.61145E+00	0.34296E-03	278-333
CH_3NO_2	Nitromethane	293.2	37.27	0.11227E+03	-0.35591E+00	0.34206E-03	288-343
CH_3NO_2	Methyl nitrite	200.0	20.77	0.11071E+03	-0.73428E+00	0.14054E-02	110-260
CH_3NO_3	Methyl nitrate	293.2	23.9				
CH_4	Methane	91.0	1.6761	0.15996E+01	0.27434E-02	-0.22086E-04	91-184
CH_4O	Methanol	293.2	33.0	0.19341E+03	-0.92211E+00	0.12839E-02	177-293
CH_5N	Methylamine	215.2	16.7	0.34398E+02	-0.73630E-01	-0.41279E-04	198-258
CN_4O_8	Tetranitromethane	293.2	2.317				
COS	Carbon oxysulfide	185.0	4.47	0.84702E+01	-0.21488E-01		143-185
$COSe$	Carbon oxyselenide	283.2	3.47	0.48740E+01	-0.49425E-02		219-283
CO_2	Carbon dioxide	295.0	1.4492	0.79062E+00	0.10639E-01	-0.28510E-04	220-300
CS_2	Carbon disulfide	293.2	2.6320	0.45024E+01	-0.12054E-01	0.19147E-04	154-319
$C_2Br_2F_4$	1,2-Dibromotetrafluoroethane	298.2	2.34				
$C_2Cl_2F_4$	1,2-Dichlorotetrafluoroethane	273.2	2.4842	0.36663E+01	-0.42271E-01	-0.36255E-06	193-273
$C_2Cl_2O_2$	Oxalyl chloride	294.35	3.470				
C_2Cl_3N	Trichloroacetonitrile	292.2	7.85				
C_2Cl_4	Tetrachloroethylene	303.2	2.268				
$C_2Cl_4F_2$	1,1,2,2-Tetrachloro-1,2-difluoroethane	308.2	2.52				
C_2HBr_3O	Tribromoacetaldehyde	293.2	7.6				
C_2HCl_3	Trichloroethylene	301.5	3.390	0.58319E+01	-0.80828E-02		302-338
$C_2HCl_3F_2$	1,2,2-Trichloro-1,1-difluoroethane	303.2	4.01	0.75423E+01	-0.11667E-01		303-333
C_2HCl_3O	Trichloroacetaldehyde	298.2	6.8				
$C_2HCl_3O_2$	Trichloroacetic acid	333.2	4.34	0.13412E+01	0.90000E-02	-0.24130E-14	333-393
C_2HCl_5	Pentachloroethane	298.2	3.716	0.65972E+01	-0.96800E-02		298-338
$C_2HF_3O_2$	Trifluoroacetic acid	293.2	8.42	0.21652E+02	-0.68146E-01	0.78571E-04	263-323
C_2H_2	Acetylene	195.0	2.4841				
$C_2H_2Br_2$	cis-1,2-Dibromoethylene	298.2	7.08				
$C_2H_2Br_2$	trans-1,2-Dibromoethylene	298.2	2.88				
$C_2H_2Br_4$	1,1,2,2-Tetrabromoethane	303.2	6.72	0.16246E+02	-0.31500E-01		303-333
$C_2H_2Cl_2$	1,1-Dichloroethylene	293.2	4.60				
$C_2H_2Cl_2$	cis-1,2-Dichloroethylene	298.2	9.20				
$C_2H_2Cl_2$	trans-1,2-Dichloroethylene	293.2	2.14				

Mol. Form.	Name	T/K	ε	a	b	c	Range/K
C₂H₂Cl₂O₂	Dichloroacetic acid	293.2	8.33	0.11014E+02	-0.10859E-01	0.49242E-05	284-363
C₂H₂Cl₄	1,1,1,2-Tetrachloroethane	207.2	9.22	0.19606E+02	-0.49847E-01		207-233
C₂H₂Cl₄	1,1,2,2-Tetrachloroethane	293.2	8.50				
C₂H₂I₂	cis-1,2-Diiodoethylene	345.65	4.46				
C₂H₃ClO	Acetyl chloride	295.2	15.8				
C₂H₃ClO₂	Chloroacetic acid	338.2	12.35	0.17310E+02	-0.14674E-01		338-393
C₂H₃Cl₂NO₂	1,1-Dichloro-1-nitroethane	303.2	16.3	0.37576E+02	-0.70400E-01		303-333
C₂H₃Cl₃	1,1,1-Trichloroethane	293.2	7.243	0.27705E+02	-0.10621E+00	0.12424E-03	258-318
C₂H₃Cl₃	1,1,2-Trichloroethane	298.2	7.1937	0.17147E+02	-0.33371E-01		288-318
C₂H₃F₃O	2,2,2-Trifluoroethanol	293.2	27.68	0.90593E+02	-0.21421E+00		293-318
C₂H₃N	Acetonitrile	293.2	36.64	0.29724E+03	-0.15508E+01	0.22591E-02	288-333
C₂H₃NO	Methyl isocyanate	288.7	21.75				
C₂H₄	Ethylene	270.0	1.4833	0.13546E+01	0.62614E-02	-0.21374E-04	200-270
C₂H₄BrCl	1-Bromo-2-chloroethane	283.2	7.41	0.19493E+02	-0.59054E-01	0.58036E-04	263-363
C₂H₄Br₂	1,2-Dibromoethane	293.2	4.9612	0.67142E+01	-0.59800E-02		293-313
C₂H₄Cl₂	1,1-Dichloroethane	298.2	10.10	0.24429E+02	-0.48000E-01		288-318
C₂H₄Cl₂	1,2-Dichloroethane	293.2	10.42	0.24404E+02	-0.47892E-01		293-343
C₂H₄Cl₂O	Bis(chloromethyl) ether	293.2	3.51				
C₂H₄N₂O₆	Ethylene glycol dinitrate	293.2	28.26				
C₂H₄O	Acetaldehyde	291.2	21.0				
C₂H₄O	Ethylene oxide	293.2	12.42	0.52661E+02	-0.21337E+00	0.25947E-03	293-243
C₂H₄OS	Thioacetic acid	298.2	14.30				
C₂H₄O₂	Acetic acid	293.2	6.20	-0.15731E+02	0.12662E+00	-0.17738E-03	293-363
C₂H₄O₂	Methyl formate	288.2	9.20	0.19699E+02	-0.36429E-01		288-302
C₂H₄O₃S	Ethylene glycol sulfite	298.2	39.6	0.85483E+02	-0.15400E+00		298-328
C₂H₅Br	Bromoethane	298.2	9.01	0.28473E+02	-0.85495E-01	0.67971E-04	243-308
C₂H₅Cl	Chloroethane	293.2	9.45	0.60693E+02	-0.31290E+00	0.47154E-03	237-293
C₂H₅ClO	2-Chloroethanol	293.2	25.80	0.11155E+03	-0.30149E+00		140-175
C₂H₅I	Iodoethane	293.2	7.82	0.25598E+02	-0.94367E-01	0.11424E-03	183-343
C₂H₅N	Ethyleneimine	298.2	18.3	0.61405E+02	-0.14474E+00		273-298
C₂H₅NO	Acetamide	363.7	67.6	-0.20055E+03	0.15515E+01	-0.22392E-02	364-448
C₂H₅NO	N-Methylformamide	293.2	189.0	0.10383E+04	-0.43165E+01	0.48398E-02	276-353
C₂H₅NO	Acetaldoxime	298.2	4.70				
C₂H₅NO₂	Nitroethane	288.2	29.11	0.57406E+02	-0.97657E-01		276-333
C₂H₅NO₂	Methyl carbamate	328.2	18.48	0.36773E+02	-0.55700E-01		328-368
C₂H₅NO₃	Ethyl nitrate	293.2	19.7				
C₂H₆	Ethane	95.0	1.9356	0.20185E+01	-0.51493E-03	-0.48148E-05	95-295
C₂H₆O	Ethanol	293.2	25.3	0.15145E+03	-0.87020E+00	0.19570E-02	163-523
C₂H₆O	Dimethyl ether	258.0	6.18	0.22389E+02	-0.86524E-01	0.91291E-04	155-258
C₂H₆OS	Dimethyl sulfoxide	293.2	47.24	0.38478E+02	0.16939E+00	-0.47423E-03	288-343
C₂H₆O₂	Ethylene glycol	293.2	41.4	0.14355E+03	-0.48573E+00	0.46703E-03	293-423
C₂H₆O₂S	Dimethyl sulfone	383.2	47.39	0.10830E+03	-0.15900E+00		383-398
C₂H₆O₄S	Dimethyl sulfate	298.2	55.0				
C₂H₆S	Ethanethiol	298.2	6.667				
C₂H₆S	Dimethyl sulfide	294.2	6.70				
C₂H₆S₂	1,2-Ethanedithiol	293.2	7.26	0.11228E+02	-0.13500E-01		293-333
C₂H₆S₂	Dimethyl disulfide	298.2	9.6	0.19109E+02	-0.32000E-01		298-323
C₂H₇N	Ethylamine	273.2	8.7	0.30163E+02	-0.79000E-01		233-273
C₂H₇NO	Ethanolamine	293.2	31.94	0.14890E+03	-0.62491E+00	0.77143E-03	253-293
C₂H₈N₂	1,2-Ethanediamine	293.2	13.82	0.48922E+02	-0.17021E+00	0.17262E-03	273-333
C₃Cl₆O	Hexachloroacetone	291.9	3.925	0.76423E+01	-0.15838E-01	0.10618E-04	269-303
C₃F₆O	Perfluoroacetone	202.2	2.104	0.34809E+01	-0.92883E-02	0.12282E-04	151-238
C₃HN	Cyanoacetylene	291.9	72.3	0.91803E+03	-0.49149E+01	0.69104E-02	281-314
C₃H₂F₆O	1,1,1,3,3,3-Hexafluoro-2-propanol	293.2	16.70				
C₃H₃ClO₃	4-Chloro-1,3-dioxolan-2-one	313.2	62.0				
C₃H₃N	Acrylonitrile	293.2	33.0	0.11109E+03	-0.36806E+00	0.34879E-03	233-413
C₃H₃NO₂	Cyanoacetic acid	277.2	33.4				
C₃H₄	Allene	269.0	2.025	0.26049E+01	-0.44147E-03	-0.63420E-05	156-269
C₃H₄	Propyne	246.0	3.218	0.60871E+01	-0.11730E-01		185-246

Mol. Form.	Name	T/K	ε	a	b	c	Range/K
$C_3H_4ClF_3$	3-Chloro-1,1,1-trifluoropropane	295.2	7.32	0.22361E+02	-0.68840E-01	0.60594E-04	275-313
C_3H_4ClNO	2-Chloroethyl isocyanate	288.2	29.1	0.64311E+02	-0.12217E+00		288-403
$C_3H_4Cl_2O$	1,1-Dichloroacetone	293.2	14.6				
$C_3H_4F_4O$	2,2,3,3-Tetrafluoro-1-propanol	298.2	21.03				
C_3H_4O	Propargyl alcohol	293.2	20.8	0.99895E+02	-0.38911E+00	0.40776E-03	213-293
$C_3H_4O_3$	Ethylene carbonate	313.2	89.78	0.20746E+03	-0.37610E+00		313-343
C_3H_5Br	3-Bromopropene	293.2	7.0				
$C_3H_5BrO_2$	2-Bromopropanoic acid	294.2	11.0				
$C_3H_5Br_3$	1,2,3-Tribromopropane	303.2	6.00	0.11024E+02	-0.16596E-01		303-358
C_3H_5Cl	2-Chloropropene	299.25	8.92				
C_3H_5Cl	3-Chloropropene	293.2	8.2				
$C_3H_5ClN_2O_6$	3-Chloro-1,2-propanediol dinitrate	293.2	17.50				
C_3H_5ClO	Epichlorohydrin	293.2	22.6				
$C_3H_5ClO_2$	Ethyl chloroformate	308.7	9.736	0.15356E+02	-0.18250E-01		309-349
$C_3H_5ClO_2$	Methyl chloroacetate	293.2	12.0				
$C_3H_5Cl_3$	1,2,3-Trichloropropane	293.2	7.5				
C_3H_5I	3-Iodopropene	292.2	6.1				
C_3H_5N	Propanenitrile	293.2	29.7	0.82222E+02	-0.22937E+00	0.17424E-03	213-473
C_3H_5NO	Ethyl isocyanate	293.2	19.7				
C_3H_5NS	Ethyl isothiocyanate	293.2	19.6				
$C_3H_5N_3O_9$	Trinitroglycerol	293.2	19.25				
C_3H_6	Propene	220.0	2.1365	0.29623E+01	-0.37564E-02		220-250
$C_3H_6Br_2$	1,2-Dibromopropane	283.2	4.60	0.54973E+01	-0.31695E-02		283-333
$C_3H_6Br_2$	1,3-Dibromopropane	293.2	9.482	0.29193E+02	-0.94450E-01	0.92800E-04	293-368
$C_3H_6ClNO_2$	2-Chloro-2-nitropropane	250.4	31.90				
$C_3H_6Cl_2$	1,2-Dichloropropane	293.2	8.37	0.18915E+02	-0.35907E-01		281-323
$C_3H_6Cl_2$	1,3-Dichloropropane	303.2	10.27	0.21609E+02	-0.37333E-01		303-333
$C_3H_6Cl_2$	2,2-Dichloropropane	293.2	11.37	0.32421E+02	-0.72188E-01		245-293
$C_3H_6N_2O_4$	2,2-Dinitropropane	325.1	42.4				
C_3H_6O	Allyl alcohol	293.2	19.7	0.62714E+02	-0.14771E+00	0.37879E-05	213-303
C_3H_6O	Propanal	290.2	18.5				
C_3H_6O	Acetone	293.2	21.01	0.88157E+02	-0.34300E+00	0.38925E-03	273-323
$C_3H_6O_2$	Propanoic acid	298.2	3.44	0.18793E+01	0.46841E-02	0.19983E-05	289-408
$C_3H_6O_2$	Ethyl formate	288.2	8.57	0.15884E+02	-0.25333E-01		288-318
$C_3H_6O_2$	Methyl acetate	288.2	7.07	0.13190E+02	-0.21226E-01		276-318
$C_3H_6O_3$	3-Hydroxypropanoic acid	296.2	30.0				
$C_3H_6O_3$	Dimethyl carbonate	298.2	3.087				
$C_3H_6O_3$	1,3,5-Trioxane	338.2	15.55				
C_3H_7Br	1-Bromopropane	293.2	8.09	0.17769E+02	-0.32599E-01		274-328
C_3H_7Br	2-Bromopropane	293.2	9.46	0.26195E+02	-0.72995E-01	0.55454E-04	186-328
C_3H_7Cl	1-Chloropropane	293.2	8.588	0.21214E+02	-0.43130E-01		273-313
C_3H_7ClO	3-Chloro-1-propanol	215.2	36.0	0.12436E+03	-0.60841E+00	0.92060E-03	145-215
C_3H_7ClO	1-Chloro-2-propanol	153.2	59.0	-0.19169E+02	0.13605E+01	-0.55567E-02	153-177
$C_3H_7ClO_2$	3-Chloro-1,2-propanediol	293.2	31.0				
C_3H_7I	1-Iodopropane	293.2	7.07	0.13744E+02	-0.22745E-01		293-323
C_3H_7I	2-Iodopropane	298.2	8.19				
C_3H_7NO	N-Ethylformamide	298.2	102.7	0.64764E+03	-0.28499E+01	0.34286E-02	298-338
C_3H_7NO	N,N-Dimethylformamide	293.2	38.25	0.15364E+03	-0.60367E+00	0.71505E-03	213-353
C_3H_7NO	N-Methylacetamide	303.2	179.0	0.15975E+04	-0.90451E+01	0.18345E-01	303-473
$C_3H_7NO_2$	1-Nitropropane	288.2	24.70	0.94999E+02	-0.38358E+00	0.48480E-03	276-333
$C_3H_7NO_2$	2-Nitropropane	288.2	26.74	0.60138E+02	-0.11566E+00		276-303
$C_3H_7NO_2$	Propyl nitrite	250.0	12.35	0.70552E+02	-0.40362E+00	0.66687E-03	110-310
$C_3H_7NO_2$	Isopropyl nitrite	260.0	13.92	0.74578E+02	-0.38283E+00	0.57071E-03	150-300
$C_3H_7NO_2$	Ethyl carbamate	328.2	14.14	0.32431E+02	-0.65097E-01	0.28571E-04	328-368
C_3H_8	Propane	293.19	1.6678	0.22883E+01	-0.23276E-02	0.84710E-06	90-300
C_3H_8O	1-Propanol	293.2	20.8	0.98045E+02	-0.36860E+00	0.36422E-03	193-493
C_3H_8O	2-Propanol	293.2	20.18	0.10416E+03	-0.41011E+00	0.42049E-03	193-493
$C_3H_8O_2$	1,2-Propylene glycol	303.2	27.5	0.24546E+03	-0.15738E+01	0.38068E-02	193-403
$C_3H_8O_2$	1,3-Propylene glycol	293.2	35.1	0.11365E+03	-0.36680E+00	0.33766E-03	288-328

Mol. Form.	Name	T/K	ε	a	b	c	Range/K
$C_3H_8O_2$	Ethylene glycol monomethyl ether	298.2	17.2	0.11803E+03	-0.58000E+00	0.81001E-03	254-318
$C_3H_8O_2$	Dimethoxymethane	293.2	2.644	0.25877E+01	-0.93019E-03	0.38472E-05	171-293
$C_3H_8O_3$	Glycerol	293.2	46.53	0.77503E+02	-0.37984E-01	-0.23107E-03	288-343
C_3H_8S	1-Propanethiol	288.2	5.937	0.11602E+02	-0.19580E-01		273-318
C_3H_8S	2-Propanethiol	298.2	5.952				
$C_3H_8S_2$	1,2-Propanedithiol	293.2	7.24	0.14667E+02	-0.32660E-01	0.25000E-04	293-333
$C_3H_8S_2$	1,3-Propanedithiol	303.2	8.11	0.66607E+01	0.31310E-01	-0.87500E-04	303-343
$C_3H_9BO_3$	Trimethyl borate	293.2	2.2762				
C_3H_9ClSi	Trimethylchlorosilane	273.2	10.21	-0.19492E+02	0.29806E+00	-0.69284E-03	223-273
C_3H_9N	Propylamine	296.2	5.08	0.17719E+02	-0.59022E-01	0.54780E-04	204-296
C_3H_9N	Isopropylamine	293.2	5.6268	0.40429E+02	-0.21441E+00	0.32634E-03	213-298
C_3H_9N	Trimethylamine	298.2	2.440	0.39745E+01	-0.51331E-02		273-298
$C_3H_9O_4P$	Trimethyl phosphate	293.2	20.6				
C_4Cl_6	Hexachloro-1,3-butadiene	293.2	2.55				
$C_4Cl_6O_3$	Trichloroacetic anhydride	298.2	5.0				
$C_4F_6O_3$	Trifluoroacetic acid anhydride	298.2	2.7				
$C_4H_2Cl_4O_3$	Dichloroacetic anhydride	298.2	15.8				
$C_4H_2O_3$	Maleic anhydride	326.2	52.75				
$C_4H_3F_7O$	2,2,3,3,4,4,4-Heptafluoro-1-butanol	298.2	14.4				
$C_4H_4N_2$	Succinonitrile	298.2	62.6	0.17724E+03	-0.54654E+00	0.54046E-03	236-351
$C_4H_4N_2$	Pyrazine	323.2	2.80				
C_4H_4O	Furan	277.1	2.88	0.13636E+01	0.12864E-01	-0.22701E-04	188-277
C_4H_4S	Thiophene	293.2	2.739	0.32941E+01	-0.19019E-02		253-293
C_4H_5Cl	2-Chloro-1,3-butadiene	293.2	4.914				
$C_4H_5Cl_3O_2$	Ethyl trichloroacetate	293.2	8.428				
C_4H_5N	Pyrrole	293.0	8.00	0.12672E+02	-0.14075E-01	-0.62671E-05	293-357
C_4H_5NO	Allyl isocynate	288.2	15.15	0.34299E+02	-0.66444E-01		288-333
C_4H_6	1,3-Butadiene	265.0	2.050	0.27674E+01	-0.26738E-02		185-265
C_4H_6O	Divinyl ether	288.2	3.94				
C_4H_6O	Ethoxyacetylene	298.2	8.05				
C_4H_6O	Cyclobutanone	298.2	14.27	0.43974E+02	-0.15712E+00	0.19264E-03	220-317
$C_4H_6O_2$	Methyl acrylate	303.2	7.03	0.11968E+02	-0.16500E-01		303-333
$C_4H_6O_2$	2,3-Butanedione	298.2	4.04	0.46907E+01	-0.22302E-02		278-348
$C_4H_6O_2$	γ-Butyrolactone	293.2	39.0				
$C_4H_6O_3$	Acetic anhydride	293.2	22.45				
$C_4H_6O_3$	Propylene carbonate	293.0	66.14	0.15940E+03	-0.39530E+00	0.26284E-03	273-333
C_4H_7Br	cis-2-Bromo-2-butene	293.2	5.38				
C_4H_7Br	trans-2-Bromo-2-butene	293.2	6.76				
$C_4H_7BrO_2$	2-Bromobutanoic acid	293.2	7.2				
$C_4H_7BrO_2$	Ethyl bromoacetate	303.2	9.75	0.15627E+02	-0.19600E-01		303-333
$C_4H_7BrO_2$	Methyl 3-bromopropanoate	303.2	5.81	0.36001E+01	0.72500E-02		303-343
$C_4H_7ClO_2$	Propyl chlorocarbonate	293.2	11.2				
$C_4H_7ClO_2$	Methyl 2-chloropropanoate	303.2	11.45	0.22449E+02	-0.36250E-01		303-343
C_4H_7N	Butanenitrile	293.2	24.83	0.53884E+02	-0.99257E-01		293-333
C_4H_7N	2-Methylpropanenitrile	293.2	24.42	0.52554E+02	-0.96000E-01		293-313
C_4H_7NO	2-Pyrrolidone	298.2	28.18	0.11054E+03	-0.47945E+00	0.68182E-03	298-338
C_4H_8	1-Butene	220.0	2.2195	0.29354E+01	-0.32580E-02		220-250
C_4H_8	cis-2-Butene	296.0	1.960	0.28802E+01	-0.31064E-02		197-296
C_4H_8	Isobutene	288.7	2.1225	0.33701E+01	-0.43295E-02		220-289
$C_4H_8Br_2$	1,2-Dibromobutane	293.2	4.74	0.11199E+03	-0.63334E+00	0.91250E-03	293-333
$C_4H_8Br_2$	1,3-Dibromobutane	293.2	9.14	0.34031E+02	-0.13254E+00	0.16250E-03	293-333
$C_4H_8Br_2$	1,4-Dibromobutane	303.2	8.68	0.20944E+02	-0.55620E-01	0.50000E-04	303-333
$C_4H_8Br_2$	2,3-Dibromobutane	298.2	6.245	0.23849E+02	-0.96300E-01	0.12500E-03	293-333
$C_4H_8Br_2$	1,2-Dibromo-2-methylpropane	293.2	4.1				
$C_4H_8Cl_2$	1,2-Dichlorobutane	293.2	7.74	0.31925E+02	-0.13232E+00	0.17007E-03	293-356
$C_4H_8Cl_2$	1,4-Dichlorobutane	308.2	9.30	0.59766E+01	0.49300E-01	-0.12500E-03	308-338
$C_4H_8Cl_2$	1,2-Dichloro-2-methylpropane	296.0	7.15	0.39429E+02	-0.20028E+00	0.30917E-03	165-296
$C_4H_8Cl_2O$	Bis(2-chloroethyl) ether	293.2	21.20				
C_4H_8O	Butanal	298.2	13.45				

Mol. Form.	Name	T/K	ε	a	b	c	Range/K
C_4H_8O	2-Butanone	293.2	18.56	0.15457E+02	0.90152E-01	-0.27100E-03	293-333
C_4H_8O	Tetrahydrofuran	295.2	7.52	0.30739E+02	-0.12946E+00	0.17195E-03	224-295
$C_4H_8O_2$	Butanoic acid	287.2	2.98	0.15010E+01	0.50046E-02		287-403
$C_4H_8O_2$	2-Methylpropanoic acid	293.2	2.58				
$C_4H_8O_2$	Propyl formate	303.2	6.92				
$C_4H_8O_2$	Ethyl acetate	293.2	6.0814	0.15646E+02	-0.44066E-01	0.39137E-04	293-433
$C_4H_8O_2$	Methyl propanoate	293.2	6.200	0.12798E+02	-0.22540E-01		293-333
$C_4H_8O_2$	1,4-Dioxane	293.2	2.2189	0.27299E+01	-0.17440E-02		293-313
$C_4H_8O_3$	2-Hydroxybutanoic acid	296.2	37.7				
$C_4H_8O_3$	3-Hydroxybutanoic acid	296.2	31.5				
$C_4H_8O_3$	Ethyl methyl carbonate	293.2	2.985				
$C_4H_8O_3$	Ethylene glycol monoacetate	303.2	12.95				
C_4H_9Br	1-Bromobutane	283.2	7.315	0.22542E+02	-0.79306E-01	0.89867E-04	183-363
C_4H_9Br	2-Bromobutane	298.2	8.64	0.18461E+02	-0.32933E-01		274-328
C_4H_9Br	1-Bromo-2-methylpropane	273.2	7.70	0.37558E+02	-0.20571E+00	0.35496E-03	112-273
C_4H_9Br	2-Bromo-2-methylpropane	293.0	10.98	0.35085E+02	-0.14075E+00	0.19960E-03	258-293
C_4H_9Cl	1-Chlorobutane	293.2	7.276	0.13565E+02	-0.10161E-01	-0.38750E-04	273-323
C_4H_9Cl	2-Chlorobutane	293.2	8.564	0.30376E+02	-0.11377E+00	0.13429E-03	273-323
C_4H_9Cl	1-Chloro-2-methylpropane	293.2	7.027	0.14945E+02	-0.33747E-01	0.23036E-04	273-323
C_4H_9Cl	2-Chloro-2-methylpropane	293.2	9.663	0.35077E+02	-0.12867E+00	0.14304E-03	273-323
C_4H_9I	1-Iodobutane	293.2	6.27	0.16493E+02	-0.50262E-01	0.52485E-04	293-323
C_4H_9I	2-Iodobutane	293.2	7.873	0.10883E+02	-0.14680E-02	-0.30000E-04	293-323
C_4H_9I	2-Iodo-2-methylpropane	283.2	6.65	0.76780E+01	0.69900E-02	-0.37500E-04	283-323
C_4H_9N	Pyrrolidine	293.0	8.30	0.38191E+02	-0.15462E+00	0.17941E-03	274-333
C_4H_9NO	N-Methylpropanamide	293.2	170.0				
C_4H_9NO	N-Ethylacetamide	293.2	135.0	0.74494E+03	-0.31400E+01	0.36131E-02	213-353
C_4H_9NO	N,N-Dimethylacetamide	294.2	38.85	0.15420E+03	-0.57506E+00	0.61911E-03	294-433
C_4H_9NO	2-Butanone oxime	293.2	3.4				
C_4H_9NO	Morpholine	298.2	7.42				
$C_4H_9NO_2$	tert-Butyl nitrite	298.2	11.47				
$C_4H_9NO_2$	Propyl carbamate	338.2	12.06	0.24356E+02	-0.36400E-01		338-378
$C_4H_9NO_2$	Ethyl-N-methyl carbamate	298.2	21.10	0.11477E+03	-0.47568E+00	0.54127E-03	298-373
$C_4H_9NO_2$	N-Acetylethanolamine	298.2	96.6	0.37016E+03	-0.13113E+01	0.13214E-02	298-348
$C_4H_9NO_3$	Butyl nitrate	293.2	13.10				
C_4H_{10}	Butane	295.0	1.7697	0.22379E+01	-0.13884E-02	-0.66711E-06	135-303
C_4H_{10}	Isobutane	295.0	1.7518	0.23295E+01	-0.19953E-02	0.14197E-06	115-303
$C_4H_{10}O$	1-Butanol	293.2	17.84	0.10578E+03	-0.50587E+00	0.84733E-03	193-553
$C_4H_{10}O$	2-Butanol	293.2	17.26	0.13850E+03	-0.75146E+00	0.14086E-02	172-533
$C_4H_{10}O$	2-Methyl-1-propanol	293.2	17.93	0.10762E+03	-0.51398E+00	0.83702E-03	173-533
$C_4H_{10}O$	2-Methyl-2-propanol	298.2	12.47	0.22541E+03	-0.14990E+01	0.34050E-02	298-503
$C_4H_{10}O$	Diethyl ether	293.2	4.2666	0.79725E+01	-0.12519E-01		283-301
$C_4H_{10}O_2$	1,2-Butanediol	298.2	22.4	0.63702E+02	-0.13807E+00		278-323
$C_4H_{10}O_2$	1,3-Butanediol	298.2	28.8	0.72883E+02	-0.14770E+00		278-323
$C_4H_{10}O_2$	1,4-Butanediol	298.2	31.9	0.13079E+03	-0.46985E+00	0.46320E-03	288-328
$C_4H_{10}O_2$	Ethylene glycol monoethyl ether	298.2	13.38				
$C_4H_{10}O_2$	Ethylene glycol dimethyl ether	296.7	7.30	0.48832E+02	-0.24218E+00	0.34413E-03	256-318
$C_4H_{10}O_2S$	Bis(2-hydroxyethyl) sulfide	293.2	28.61	0.13128E+03	-0.52719E+00	0.60465E-03	253-333
$C_4H_{10}O_3$	Diethylene glycol	293.2	31.82	0.13973E+03	-0.54725E+00	0.61149E-03	288-343
$C_4H_{10}O_3S$	Diethyl sulfite	293.2	15.6				
$C_4H_{10}O_4$	1,2,3,4-Butanetetrol	393.2	28.2				
$C_4H_{10}O_4S$	Diethyl sulfate	293.2	29.2				
$C_4H_{10}S$	1-Butanethiol	288.2	5.204	0.11201E+02	-0.20767E-01		273-318
$C_4H_{10}S$	2-Butanethiol	288.2	5.645	0.10866E+02	-0.17993E-01		273-318
$C_4H_{10}S$	2-Methyl-1-propanethiol	298.2	4.961				
$C_4H_{10}S$	2-Methyl-2-propanethiol	293.2	5.475	0.10597E+02	-0.17500E-01		283-313
$C_4H_{10}S$	Diethyl sulfide	298.2	5.723				
$C_4H_{11}N$	Butylamine	293.2	4.71	0.13322E+02	-0.44176E-01	0.50250E-04	223-333
$C_4H_{11}N$	Diethylamine	293.2	3.680	0.26462E+02	-0.13750E+00	0.20373E-03	243-323
$C_4H_{11}NO_2$	Diethanolamine	293.2	25.75	0.73435E+02	-0.21377E+00	0.17500E-03	273-323

PERMITTIVITY (DIELECTRIC CONSTANT) OF LIQUIDS (continued)

Mol. Form.	Name	T/K	ε	a	b	c	Range/K
$C_4H_{12}O_2Si$	Dimethoxydimethylsilane	298.2	3.663				
$C_4H_{12}O_3Si$	Trimethoxymethylsilane	298.2	4.9				
$C_4H_{12}O_4Si$	Tetramethyl silicate	293.2	6.0				
$C_4H_{12}Si$	Diethylsilane	293.2	2.544				
$C_4H_{12}Si$	Tetramethylsilane	293.2	1.921				
$C_4H_{13}N_3$	Diethylenetriamine	293.2	12.62	0.57840E+02	-0.23873E+00	0.28841E-03	213-333
C_5FeO_5	Iron pentacarbonyl	293.2	2.602				
C_5H_4BrN	2-Bromopyridine	298.2	23.18	0.73391E+02	-0.23678E+00	0.22930E-03	298-398
C_5H_4ClN	2-Chloropyridine	298.2	27.32	0.98702E+02	-0.34237E+00	0.34502E-03	298-398
$C_5H_4F_8O$	2,2,3,3,4,4,5,5-Octafluoro-1-pentanol	298.2	15.30				
$C_5H_4O_2$	Furfural	293.2	42.1				
C_5H_5N	Pyridine	293.2	13.260	0.43991E+02	-0.15150E+00	0.15925E-03	293-323
C_5H_5NO	Pyridine-1-oxide	343.0	35.94	0.20878E+02	0.16450E+00	-0.35269E-03	343-398
C_5H_6O	2-Methylfuran	293.2	2.76				
$C_5H_6O_2$	Furfuryl alcohol	298.2	16.85				
$C_5H_7Cl_3O_2$	Propyl trichloroacetate	298.2	8.32				
$C_5H_7NO_2$	Ethyl cyanoacetate	263.2	31.62				
C_5H_8	1,3-Pentadiene*	298.2	2.319				
C_5H_8	1,4-Pentadiene	294.0	2.054	0.29994E+01	-0.34578E-02	0.85300E-06	178-294
C_5H_8	2-Methyl-1,3-butadiene	293.2	2.098	0.28170E+01	-0.23147E-02	-0.43975E-06	198-293
C_5H_8	Cyclopentene	295.0	2.083	0.28177E+01	-0.27597E-02	0.89346E-06	171-319
C_5H_8O	Cyclopentanone	298.2	13.58	0.24083E+02	-0.30286E-01	-0.16802E-04	219-298
$C_5H_8O_2$	Ethyl acrylate	303.2	6.05	0.47827E+02	-0.24394E+00	0.35000E-03	303-343
$C_5H_8O_2$	Methyl trans-2-butenoate	293.2	6.6645				
$C_5H_8O_2$	Methyl methacrylate	303.2	6.32	0.32098E+02	-0.14568E+00	0.20000E-03	303-343
$C_5H_8O_2$	2,4-Pentanedione	303.2	26.524				
$C_5H_8O_4$	Dimethyl malonate	293.2	9.82	0.26470E+02	-0.76656E-01	0.67888E-04	293-433
$C_5H_9BrO_2$	Ethyl 2-bromopropanoate	293.2	9.4				
$C_5H_9ClO_2$	Isobutyl chlorocarbonate	293.2	9.1				
$C_5H_9ClO_2$	Ethyl 2-chloropropanoate	303.2	11.95	0.25965E+02	-0.46250E-01		303-343
$C_5H_9ClO_2$	Ethyl 3-chloropropanoate	303.2	10.19	0.21951E+02	-0.38750E-01		303-343
$C_5H_9ClO_2$	Methyl 4-chlorobutanoate	303.2	9.51	0.17127E+02	-0.25000E-01		303-343
C_5H_9N	Pentanenitrile	293.2	20.04	0.55793E+02	-0.15750E+00	0.12432E-03	183-333
C_5H_9N	2,2-Dimethylpropanenitrile	293.2	21.1	0.58418E+02	-0.16884E+00	0.14131E-03	293-453
C_5H_9NO	Isobutyl isocyanate	293.2	11.638	0.38026E+02	-0.12714E+00	0.12679E-03	293-353
C_5H_9NO	N-Methyl-2-pyrrolidone	293.2	32.55				
C_5H_{10}	1-Pentene	293.2	2.011	-0.11438E+01	0.25420E-01	-0.50000E-04	273-293
C_5H_{10}	2-Methyl-1-butene	293.2	2.180				
C_5H_{10}	2-Methyl-2-butene	296.0	1.979	0.26064E+01	-0.19578E-02	-0.53908E-06	225-296
C_5H_{10}	Cyclopentane	293.2	1.9687	0.24287E+01	-0.15304E-02	-0.13095E-06	278-313
C_5H_{10}	Ethylcyclopropane	293.2	1.933				
$C_5H_{10}Br_2$	1,2-Dibromopentane	298.2	4.39				
$C_5H_{10}Br_2$	1,4-Dibromopentane	293.2	9.05	0.26443E+02	-0.88640E-01	0.10000E-03	293-333
$C_5H_{10}Br_2$	1,5-Dibromopentane	303.2	9.14	0.38192E+02	-0.15648E+00	0.20000E-03	303-333
$C_5H_{10}Cl_2$	1,2-Dichloropentane	293.2	6.89	0.19016E+02	-0.57954E-01	0.56801E-04	293-356
$C_5H_{10}Cl_2$	1,5-Dichloropentane	298.2	9.92				
$C_5H_{10}O$	Cyclopentanol	288.2	18.5	0.10565E+03	-0.44244E+00	0.48657E-03	258-323
$C_5H_{10}O$	Pentanal	293.2	10.00				
$C_5H_{10}O$	2,2-Dimethylpropanal	293.2	9.051	0.18645E+02	-0.32395E-01	-0.16157E-05	280-333
$C_5H_{10}O$	2-Pentanone	293.2	15.45	0.40893E+02	-0.10423E+00	0.60557E-04	204-353
$C_5H_{10}O$	3-Pentanone	293.2	17.00	0.12690E+02	0.95177E-01	-0.27321E-03	233-353
$C_5H_{10}O$	3-Methyl-2-butanone	293.2	10.37	0.30695E+02	-0.10962E+00	0.13810E-03	293-328
$C_5H_{10}O$	Tetrahydropyran	293.2	5.66	0.19793E+02	-0.76071E-01	0.94852E-04	234-333
$C_5H_{10}O$	2-Methyltetrahydrofuran	298.2	6.97				
$C_5H_{10}O_2$	Pentanoic acid	294.4	2.661	0.33491E+01	-0.75156E-02	0.17820E-04	250-344
$C_5H_{10}O_2$	Butyl formate	303.2	6.10	0.21532E+02	-0.84106E-01	0.10952E-03	288-323
$C_5H_{10}O_2$	Isobutyl formate	293.2	6.41				
$C_5H_{10}O_2$	Propyl acetate	293.2	5.62	0.17677E+02	-0.61404E-01	0.69196E-04	253-353
$C_5H_{10}O_2$	Ethyl propanoate	293.2	5.76				

Mol. Form.	Name	T/K	ε	a	b	c	Range/K
$C_5H_{10}O_2$	Methyl butanoate	301.2	5.48	0.38604E+02	-0.19171E+00	0.27128E-03	301-343
$C_5H_{10}O_2$	Tetrahydrofurfuryl alcohol	303.2	13.48				
$C_5H_{10}O_2S$	3-Methyl sulfolane	298.2	29.4	0.53158E+02	-0.93730E-01	0.47275E-04	298-398
$C_5H_{10}O_3$	Diethyl carbonate	297.2	2.820				
$C_5H_{10}O_3$	Ethyl lactate	303.2	15.4	0.31225E+02	-0.43531E-01	-0.28571E-04	273-373
$C_5H_{10}O_4$	1,2,3-Propanetriol-1-acetate	242.2	38.57	0.10653E+03	-0.26439E+00	-0.62371E-04	215-242
$C_5H_{11}Br$	2-Bromo-2-methylbutane	298.2	9.21				
$C_5H_{11}Br$	1-Bromopentane	299.2	6.31	0.20954E+02	-0.78743E-01	0.98908E-04	183-328
$C_5H_{11}Br$	3-Bromopentane	298.2	8.37				
$C_5H_{11}Br$	1-Bromo-3-methylbutane	291.5	6.33	0.27743E+02	-0.13927E+00	0.22627E-03	123-292
$C_5H_{11}Cl$	1-Chloropentane	293.2	6.654	0.18626E+02	-0.54719E-01	0.47143E-04	273-323
$C_5H_{11}Cl$	1-Chloro-3-methylbutane	292.0	6.10	0.22228E+02	-0.93189E-01	0.12991E-03	171-297
$C_5H_{11}Cl$	2-Chloro-2-methylbutane	222.75	12.31	0.55104E+02	-0.29866E+00	0.47840E-03	201-223
$C_5H_{11}F$	1-Fluoropentane	293.2	3.931				
$C_5H_{11}I$	1-Iodopentane	293.2	5.78	0.15753E+02	-0.50543E-01	0.56401E-04	293-323
$C_5H_{11}I$	3-Iodopentane	293.2	7.432				
$C_5H_{11}I$	1-Iodo-3-methylbutane	292.2	5.6				
$C_5H_{11}I$	2-Iodo-2-methylbutane	293.2	8.192				
$C_5H_{11}N$	Piperidine	293.0	4.33	0.82317E+01	-0.11229E-01	-0.71429E-05	293-333
$C_5H_{11}N$	N-Methylpyrrolidine	298.2	32.2				
$C_5H_{11}NO$	2,2-Dimethylpropanamide	298.2	20.13	0.10400E+03	-0.46017E+00	0.60000E-03	298-328
$C_5H_{11}NO$	N,N-Diethylformamide	293.2	29.6				
$C_5H_{11}NO$	2-Pentanone oxime	293.2	3.3				
$C_5H_{11}NO_2$	Pentyl nitrite	298.2	7.21				
C_5H_{12}	Pentane	293.2	1.8371				
C_5H_{12}	Isopentane	293.2	1.845	0.22384E+01	-0.12985E-02	-0.16182E-06	143-293
C_5H_{12}	Neopentane	296.0	1.769	0.10949E+02	-0.63057E-01	0.10835E-03	251-296
$C_5H_{12}N_2O$	Tetramethylurea	293.2	23.10				
$C_5H_{12}O$	1-Pentanol	298.2	15.13	0.73397E+02	-0.28165E+00	0.28427E-03	213-513
$C_5H_{12}O$	2-Pentanol	298.2	13.71	0.16437E+03	-0.86506E+00	0.11955E-02	273-323
$C_5H_{12}O$	3-Pentanol	298.2	13.35	0.12838E+03	-0.60980E+00	0.75000E-03	288-318
$C_5H_{12}O$	2-Methyl-1-butanol	298.2	15.63	0.14020E+02	0.13948E+00	-0.45000E-03	288-318
$C_5H_{12}O$	3-Methyl-1-butanol	293.2	15.63	0.79733E+02	-0.31272E+00	0.32014E-03	173-513
$C_5H_{12}O$	2-Methyl-2-butanol	298.2	5.78	0.11662E+03	-0.69756E+00	0.10920E-02	268-318
$C_5H_{12}O$	3-Methyl-2-butanol	298.2	12.1				
$C_5H_{12}O$	2,2-Dimethyl-1-propanol	333.2	8.35	0.92350E+02	-0.41870E+00	0.50000E-03	333-373
$C_5H_{12}O_2$	1,2-Pentanediol	296.8	17.31	0.18436E+03	-0.10682E+01	0.17037E-02	197-297
$C_5H_{12}O_2$	1,4-Pentanediol	295.7	26.74	0.13568E+03	-0.59198E+00	0.75398E-03	193-318
$C_5H_{12}O_2$	1,5-Pentanediol	293.2	26.2	0.11858E+03	-0.45920E+00	0.49341E-03	243-343
$C_5H_{12}O_2$	2,3-Pentanediol	296.9	17.37	0.95876E+02	-0.46463E+00	0.67434E-03	238-297
$C_5H_{12}O_2$	2,4-Pentanediol	294.2	24.69	0.11914E+03	-0.52569E+00	0.69607E-03	224-294
$C_5H_{12}O_2$	Diethoxymethane	293.2	2.527	0.25294E+01	0.73988E-04	-0.28331E-06	227-293
$C_5H_{12}O_4$	Tetramethoxymethane	293.2	2.40				
$C_5H_{12}O_5$	Xylitol	293.2	40.0				
$C_5H_{12}S$	1-Pentanethiol	293.2	4.847	0.71131E+01	-0.30228E-02	-0.16414E-04	273-333
$C_5H_{12}S$	2-Methyl-2-butanethiol	293.2	5.087	0.15116E+02	-0.50700E-01	0.56250E-04	273-333
$C_5H_{12}S_4$	Tetrakis(methylthio)methane	343.2	2.818				
$C_5H_{13}N$	Pentylamine	293.2	4.27	0.11274E+02	-0.34965E-01	0.37706E-04	223-353
$C_5H_{13}N_3$	1,1,3,3-Tetramethylguanidine	298.2	11.5				
$C_5H_{14}OSi$	Ethoxytrimethylsilane	298.2	3.013				
C_6F_6	Hexafluorobenzene	298.2	2.029	0.24041E+01	-0.83086E-03	-0.14286E-05	298-338
C_6F_{14}	Perfluorohexane	298.2	1.76				
$C_6H_3N_3O_7$	2,4,6-Trinitrophenol	294.2	4.0				
C_6H_4BrF	1-Bromo-2-fluorobenzene	298.2	4.72				
C_6H_4BrF	1-Bromo-3-fluorobenzene	298.2	4.85				
C_6H_4BrF	1-Bromo-4-fluorobenzene	298.2	2.60				
$C_6H_4BrNO_2$	1-Bromo-3-nitrobenzene	328.2	20.2	0.81413E+02	-0.27645E+00	0.27367E-03	328-413
$C_6H_4Br_2$	o-Dibromobenzene	293.2	7.86	-0.81849E-02	0.62671E-01	-0.12222E-03	293-353
$C_6H_4Br_2$	m-Dibromobenzene	293.2	4.81	0.93214E+01	-0.20273E-01	0.16667E-04	293-353

Mol. Form.	Name	T/K	ε	a	b	c	Range/K
$C_6H_4Br_2$	p-Dibromobenzene	368.2	2.57				
C_6H_4ClF	1-Chloro-2-fluorobenzene	298.2	6.10				
C_6H_4ClF	1-Chloro-3-fluorobenzene	298.2	4.96				
C_6H_4ClF	1-Chloro-4-fluorobenzene	298.2	3.34				
$C_6H_4ClNO_2$	1-Chloro-2-nitrobenzene	323.2	37.7	0.16800E+03	-0.59708E+00	0.59957E-03	323-436
$C_6H_4ClNO_2$	1-Chloro-3-nitrobenzene	323.2	20.9	0.77193E+02	-0.25118E+00	0.23798E-03	323-433
$C_6H_4ClNO_2$	1-Chloro-4-nitrobenzene	393.2	8.09				
$C_6H_4Cl_2$	o-Dichlorobenzene	293.2	10.12	0.13629E+02	0.10622E-02	-0.44444E-04	293-353
$C_6H_4Cl_2$	m-Dichlorobenzene	293.2	5.02	0.77565E+01	-0.93333E-02	-0.26880E-14	293-353
$C_6H_4Cl_2$	p-Dichlorobenzene	328.2	2.3943	0.26999E+01	-0.35325E-03	-0.17619E-05	328-363
C_6H_4FI	1-Fluoro-2-iodobenzene	298.2	8.22				
C_6H_4FI	1-Fluoro-4-iodobenzene	298.2	3.12				
$C_6H_4F_2$	o-Difluorobenzene	301.2	13.38	0.59107E+02	-0.23611E+00	0.27987E-03	273-323
$C_6H_4F_2$	m-Difluorobenzene	301.2	5.01	0.14448E+02	-0.46982E-01	0.51948E-04	273-323
$C_6H_4I_2$	o-Diiodobenzene	323.2	5.41	0.31150E+02	-0.14428E+00	0.20000E-03	323-353
$C_6H_4I_2$	m-Diiodobenzene	323.2	4.11				
$C_6H_4I_2$	p-Diiodobenzene	393.2	2.88				
$C_6H_4N_2$	2-Pyridinecarbonitrile	303.2	93.77	0.45596E+03	-0.17746E+01	0.19105E-02	303-398
$C_6H_4N_2$	3-Pyridinecarbonitrile	323.2	20.54	0.60484E+02	-0.17280E+00	0.15218E-03	323-398
$C_6H_4N_2$	4-Pyridinecarbonitrile	353.2	5.23	0.12533E+02	-0.30115E-01	0.26674E-04	353-398
$C_6H_4N_2O_4$	1,3-Dinitrobenzene	365.2	22.9	0.10406E+03	-0.34133E+00	0.32609E-03	365-413
C_6H_5Br	Bromobenzene	293.2	5.45	0.94100E+01	-0.12537E-01	-0.31127E-05	234-333
C_6H_5Cl	Chlorobenzene	293.2	5.6895	0.19471E+02	-0.70786E-01	0.82466E-04	293-430
C_6H_5ClO	o-Chlorophenol	296.2	7.40	0.29755E+02	-0.11256E+00	0.12390E-03	296-448
C_6H_5ClO	m-Chlorophenol	293.2	6.255				
C_6H_5ClO	p-Chlorophenol	314.2	11.18	0.31997E+02	-0.94241E-01	0.88392E-04	314-453
$C_6H_5ClO_2S$	Benzenesulfonyl chloride	323.2	28.90	0.83886E+02	-0.23405E+00	0.19713E-03	323-473
C_6H_5ClS	4-Chlorobenzenethiol	338.2	3.59				
C_6H_5F	Fluorobenzene	293.2	5.465				
C_6H_5I	Iodobenzene	293.2	4.59	0.89442E+01	-0.20008E-01	0.17641E-04	243-323
C_6H_5NOS	N-Sulfinylaniline	298.2	6.97				
$C_6H_5NO_2$	Nitrobenzene	293.0	35.6	0.11212E+03	-0.35211E+00	0.31128E-03	279-533
$C_6H_5NO_3$	o-Nitrophenol	323.2	16.50	0.33827E+02	-0.62123E-01	0.26774E-04	323-453
$C_6H_5NO_3$	m-Nitrophenol	373.2	35.45	0.18967E+03	-0.66144E+00	0.66532E-03	373-458
$C_6H_5NO_3$	p-Nitrophenol	393.2	42.20	0.22901E+03	-0.74264E+00	0.68006E-03	393-463
C_6H_6	Benzene	293.2	2.2825	0.26706E+01	-0.91648E-03	-0.14257E-05	293-513
C_6H_6BrN	m-Bromoaniline	293.2	13.0				
C_6H_6ClN	o-Chloroaniline	293.2	13.40				
C_6H_6ClN	m-Chloroaniline	293.2	13.3				
$C_6H_6N_2O_2$	o-Nitroaniline	353.0	47.3	0.18900E+03	-0.56977E+00	0.47484E-03	353-468
$C_6H_6N_2O_2$	m-Nitroaniline	398.0	35.6	0.20352E+03	-0.66582E+00	0.61310E-03	398-468
$C_6H_6N_2O_2$	p-Nitroaniline	428.0	78.5	0.48673E+03	-0.15040E+01	0.12857E-02	428-468
C_6H_6O	Phenol	303.2	12.40	0.63391E+02	-0.24988E+00	0.26930E-03	303-433
$C_6H_6O_2$	Pyrocatechol	388.2	17.57	0.74930E+02	-0.22142E+00	0.18919E-03	388-463
$C_6H_6O_2$	Resorcinol	393.2	13.55	0.30252E+02	-0.56443E-01	0.35578E-04	393-463
C_6H_6S	Benzenethiol	303.2	4.26	0.57155E+01	-0.70336E-02	0.73617E-05	303-358
C_6H_7N	Aniline	293.2	7.06	0.89534E+01	0.38990E-02	-0.36310E-04	293-413
C_6H_7N	2-Methylpyridine	293.2	10.18	0.34560E+02	-0.11980E+00	0.12500E-03	293-333
C_6H_7N	3-Methylpyridine	303.0	11.10	0.19643E+03	-0.11167E+01	0.16667E-02	303-333
C_6H_7N	4-Methylpyridine	293.0	12.2	0.33765E+02	-0.10113E+00	0.93860E-04	274-333
C_6H_7NO	2-Methylpyridine-1-oxide	323.2	36.4	0.11705E+03	-0.35301E+00	0.32000E-03	323-398
C_6H_7NO	3-Methylpyridine-1-oxide	318.2	28.26	0.59851E+02	-0.12682E+00	0.86622E-04	318-398
C_6H_8	1,3-Cyclohexadiene	184.2	2.68				
C_6H_8	1,4-Cyclohexadiene	296.0	2.211	0.27459E+01	-0.16975E-02	-0.36461E-06	232-356
$C_6H_8N_2$	Phenylhydrazine	293.2	7.15				
$C_6H_8N_2$	2,5-Dimethylpyrazine	293.2	2.436				
$C_6H_8N_2$	2,6-Dimethylpyrazine	308.2	2.653				
$C_6H_8O_2$	1,4-Cyclohexanedione	351.2	4.40				
$C_6H_9Cl_3O_2$	Butyl trichloroacetate	293.2	7.480				

PERMITTIVITY (DIELECTRIC CONSTANT) OF LIQUIDS (continued)

Mol. Form.	Name	T/K	ε	a	b	c	Range/K
$C_6H_9Cl_3O_2$	Isobutyl trichloroacetate	293.2	7.667				
C_6H_9N	Cyclopentanecarbonitrile	293.2	22.68	0.69830E+02	-0.25303E+00	0.31491E-03	201-293
C_6H_{10}	1,5-Hexadiene	294.0	2.125	0.30014E+01	-0.28668E-02	-0.31026E-06	151-294
C_6H_{10}	cis,cis-2,4-Hexadiene	297.0	2.163	0.27284E+01	-0.17178E-02	-0.62926E-06	234-351
C_6H_{10}	trans,trans-2,4-Hexadiene	297.0	2.123	0.26774E+01	-0.16977E-02	-0.55637E-06	232-353
C_6H_{10}	2-Methyl-1,3-pentadiene*	298.2	2.422				
C_6H_{10}	3-Methyl-1,3-pentadiene	298.2	2.426				
C_6H_{10}	4-Methyl-1,3-pentadiene	293.2	2.599	0.51328E+01	-0.12774E-01	0.14215E-04	198-323
C_6H_{10}	2,3-Dimethyl-1,3-butadiene	293.2	2.102	0.26258E+01	-0.17990E-02	0.12035E-06	223-323
C_6H_{10}	1-Hexyne	296.0	2.621	0.58591E+01	-0.17099E-01	0.20856E-04	184-296
C_6H_{10}	Cyclohexene	293.2	2.2176	0.30598E+01	-0.39841E-02	0.37554E-05	141-313
$C_6H_{10}O$	Butoxyacetylene	298.2	6.62				
$C_6H_{10}O$	Cyclohexanone	293.0	16.1	0.41577E+02	-0.11463E+00	0.92454E-04	253-423
$C_6H_{10}O$	Mesityl oxide	273.2	15.6				
$C_6H_{10}O_2$	Ethyl 2-butenoate	293.2	5.4				
$C_6H_{10}O_2$	Ethyl methacrylate	303.2	5.68	0.40962E+02	-0.20520E+00	0.29286E-03	303-343
$C_6H_{10}O_3$	Ethyl acetoacetate	293.2	14.0				
$C_6H_{10}O_3$	Propanoic anhydride	293.2	18.30				
$C_6H_{10}O_4$	Monomethyl glutarate	293.2	8.37	0.16779E+02	-0.39839E-01	0.38095E-04	293-363
$C_6H_{10}O_4$	Diethyl oxalate	293.2	8.266	0.21938E+02	-0.66226E-01	0.66800E-04	293-368
$C_6H_{10}O_4$	Dimethyl succinate	293.2	7.19	0.13551E+02	-0.23109E-01	0.55440E-05	293-433
$C_6H_{10}O_4$	Ethylene glycol diacetate	290.2	7.7	0.25093E+02	-0.95171E-01	0.12224E-03	223-290
$C_6H_{11}Br$	Bromocyclohexane	303.2	8.0026				
$C_6H_{11}BrO_2$	Ethyl 2-bromobutanoate	303.2	8.57	0.49005E+02	-0.23193E+00	0.32500E-03	303-333
$C_6H_{11}BrO_2$	Ethyl 2-bromo-2-methylpropanoate	303.2	8.55	0.77044E+02	-0.40784E+00	0.60000E-03	303-333
$C_6H_{11}Cl$	Chlorocyclohexane	303.2	7.9505				
$C_6H_{11}N$	Hexanenitrile	298.2	17.26				
$C_6H_{11}N$	4-Methylpentanenitrile	295.2	17.5				
$C_6H_{11}NO$	Cyclohexanone oxime	362.2	3.04				
C_6H_{12}	1-Hexene	294.0	2.077	0.31476E+01	-0.50003E-02	0.46673E-05	149-294
C_6H_{12}	trans-2-Hexene	295.0	1.978	0.24338E+01	-0.11323E-02	-0.13720E-05	157-295
C_6H_{12}	cis-3-Hexene	296.0	2.069	0.30691E+01	-0.45458E-02	0.39898E-05	155-296
C_6H_{12}	trans-3-Hexene	293.2	1.954				
C_6H_{12}	Cyclohexane	293.2	2.0243	0.24293E+01	-0.12095E-02	-0.58741E-06	283-333
C_6H_{12}	Methylcyclopentane	293.2	1.9853	0.21587E+01	-0.22450E-03	-0.12500E-05	293-323
C_6H_{12}	Ethylcyclobutane	293.2	1.965				
$C_6H_{12}Br_2$	1,6-Dibromohexane	298.2	8.52	-0.55185E+01	0.11746E+00	-0.23658E-03	274-328
$C_6H_{12}Br_2$	3,4-Dibromohexane	298.2	6.732				
$C_6H_{12}Cl_2$	1,6-Dichlorohexane	308.2	8.60	0.11277E+02	0.67200E-02	-0.50000E-04	308-338
$C_6H_{12}O$	1-Methylcyclopentanol	310.1	7.11	0.75444E+02	-0.36617E+00	0.47021E-03	310-333
$C_6H_{12}O$	Isobutyl vinyl ether	293.2	3.34	0.48060E+01	-0.50000E-02	-0.41495E-14	293-323
$C_6H_{12}O$	2-Hexanone	293.2	14.56	0.70378E+02	-0.29385E+00	0.35289E-03	243-293
$C_6H_{12}O$	4-Methyl-2-pentanone	293.2	13.11	0.36341E+02	-0.97119E-01	0.61896E-04	204-373
$C_6H_{12}O$	3,3-Dimethyl-2-butanone	293.2	12.73	0.66857E+02	-0.28552E+00	0.34422E-03	243-293
$C_6H_{12}O$	Cyclohexanol	293.2	16.40	0.10173E+03	-0.43072E+00	0.47926E-03	293-423
$C_6H_{12}O_2$	Hexanoic acid	298.2	2.600	0.21730E+01	0.14840E-02	-0.16526E-06	298-433
$C_6H_{12}O_2$	2-Ethylbutanoic acid	296.2	2.72				
$C_6H_{12}O_2$	tert-Butylacetic acid	296.2	2.85				
$C_6H_{12}O_2$	Pentyl formate	292.2	5.7				
$C_6H_{12}O_2$	Isopentyl formate	288.2	5.44	0.29257E+02	-0.14028E+00	0.20000E-03	288-323
$C_6H_{12}O_2$	Butyl acetate	293.2	5.07	0.13825E+02	-0.43994E-01	0.48214E-04	253-353
$C_6H_{12}O_2$	sec-Butyl acetate	293.2	5.135	0.12427E+02	-0.32035E-01	0.24286E-04	273-323
$C_6H_{12}O_2$	tert-Butyl acetate	293.2	5.672	0.55435E+02	-0.30494E+00	0.46107E-03	273-323
$C_6H_{12}O_2$	Isobutyl acetate	293.2	5.068	0.14323E+02	-0.46048E-01	0.49286E-04	273-323
$C_6H_{12}O_2$	Propyl propanoate	293.2	5.249				
$C_6H_{12}O_2$	Ethyl butanoate	301.2	5.18	0.48698E+02	-0.25660E+00	0.37237E-03	301-343
$C_6H_{12}O_2$	Methyl pentanoate	293.2	4.992				
$C_6H_{12}O_2$	Diacetone alcohol	298.2	18.2				
$C_6H_{12}O_3$	Ethylene glycol monoethyl ether acetate	303.2	7.567	0.23290E+02	-0.71566E-01	0.65000E-04	303-323

Mol. Form.	Name	T/K	ε	a	b	c	Range/K
$C_6H_{12}S$	Cyclohexanethiol	298.2	5.420				
$C_6H_{13}Br$	1-Bromohexane	298.2	5.82	0.15233E+02	-0.44385E-01	0.43039E-04	274-328
$C_6H_{13}Cl$	1-Chlorohexane	293.2	6.104	0.15994E+02	-0.43647E-01	0.33393E-04	273-323
$C_6H_{13}ClO$	6-Chloro-1-hexanol	242.2	21.6	-0.73364E+01	0.46377E+00	-0.14202E-02	195-242
$C_6H_{13}I$	1-Iodohexane	293.3	5.35	0.16685E+02	-0.61309E-01	0.77262E-04	293-323
$C_6H_{13}N$	Cyclohexylamine	293.2	4.547				
$C_6H_{13}NO$	N-Propylpropanamide	298.2	118.1	0.58846E+03	-0.22012E+01	0.20870E-02	298-328
$C_6H_{13}NO$	N-Butylacetamide	293.2	104.0	0.70739E+03	-0.37369E+01	0.71585E-02	253-493
$C_6H_{13}NO$	N,N-Diethylacetamide	293.2	32.1				
C_6H_{14}	Hexane	293.2	1.8865	0.19768E+01	0.70933E-03	-0.34470E-05	293-473
C_6H_{14}	2-Methylpentane	293.2	1.886	0.20745E+01	0.50871E-03	-0.39286E-05	273-323
C_6H_{14}	3-Methylpentane	293.2	1.886	0.24739E+01	-0.23190E-02	0.10714E-05	273-323
C_6H_{14}	2,2-Dimethylbutane	293.2	1.869	0.22740E+01	-0.96229E-03	-0.14286E-05	273-313
C_6H_{14}	2,3-Dimethylbutane	293.2	1.889	0.24305E+01	-0.20081E-02	0.53571E-06	273-323
$C_6H_{14}O$	1-Hexanol	293.2	13.03	0.62744E+02	-0.24214E+00	0.24704E-03	233-513
$C_6H_{14}O$	2-Hexanol	298.2	11.06				
$C_6H_{14}O$	3-Hexanol	298.2	9.66				
$C_6H_{14}O$	3-Methyl-1-pentanol	298.2	15.2				
$C_6H_{14}O$	3-Methyl-3-pentanol	293.2	4.322				
$C_6H_{14}O$	2-Ethyl-1-butanol	362.2	6.19				
$C_6H_{14}O$	2,2-Dimethyl-1-butanol	293.2	10.5	0.14054E+03	-0.72925E+00	0.97821E-03	243-393
$C_6H_{14}O$	Dipropyl ether	297.0	3.38	0.14600E+02	-0.72670E-01	0.11742E-03	161-297
$C_6H_{14}O$	Diisopropyl ether	303.2	3.805				
$C_6H_{14}OS$	Dipropyl sulfoxide	303.2	30.37	0.84868E+02	-0.23486E+00	0.18198E-03	303-373
$C_6H_{14}O_2$	2-Methyl-2,4-pentanediol	293.2	25.86	0.14531E+03	-0.65285E+00	0.83503E-03	203-333
$C_6H_{14}O_2$	Ethylene glycol diethyl ether	293.2	3.90	0.99099E+01	-0.33403E-01	0.44048E-04	223-303
$C_6H_{14}O_2S$	Dipropyl sulfone	303.2	32.62	0.70195E+02	-0.15008E+00	0.86506E-04	303-398
$C_6H_{14}O_3$	1,2,6-Hexanetriol	285.3	31.5	0.26127E+03	-0.14552E+01	0.22765E-02	261-285
$C_6H_{14}O_3$	Diethylene glycol dimethyl ether	298.2	7.23	0.28291E+02	-0.11236E+00	0.14000E-03	298-333
$C_6H_{14}O_4$	Triethylene glycol	293.2	23.69	0.91845E+02	-0.33827E+00	0.36062E-03	253-333
$C_6H_{14}O_6$	D-Glucitol	353.2	35.5				
$C_6H_{14}O_6$	D-Mannitol	443.2	24.6				
$C_6H_{14}S$	1-Hexanethiol	293.2	4.436	0.11774E+02	-0.37298E-01	0.41875E-04	273-333
$C_6H_{15}B$	Triethylborane	293.2	1.974				
$C_6H_{15}N$	Hexylamine	293.2	4.08	0.80244E+01	-0.16627E-01	0.10874E-04	253-373
$C_6H_{15}N$	Dipropylamine	293.2	2.923	0.11376E+02	-0.49796E-01	0.71792E-04	243-323
$C_6H_{15}N$	Triethylamine	293.2	2.418	0.29205E+01	-0.14007E-02	-0.13469E-05	233-323
$C_6H_{15}OP$	Triethylphosphine oxide	323.2	35.5				
$C_6H_{15}O_4P$	Triethyl phosphate	298.2	13.20	0.61230E+02	-0.26047E+00	0.33333E-03	298-333
$C_6H_{15}PS$	Triethylphosphine sulfide	371.2	39.0				
$C_6H_{16}O_2Si$	Diethoxydimethylsilane	298.2	3.216				
$C_6H_{16}Si$	Triethylsilane	293.2	2.323				
$C_6H_{18}N_3OP$	Hexamethylphosphoric triamide	293.2	31.3	0.95666E+02	-0.29769E+00	0.26407E-03	283-363
$C_6H_{18}N_4$	N,N'-Bis(2-aminoethyl)-1,2-ethanediamine	293.2	10.76	0.50699E+02	-0.21730E+00	0.27582E-03	213-333
$C_6H_{18}OSi_2$	Hexamethyldisiloxane	293.2	2.179	0.34537E+01	-0.61530E-02	0.61544E-05	213-313
$C_6H_{18}O_3Si_3$	Hexamethylcyclotrisiloxane	343.2	2.139				
$C_6H_{19}NSi_2$	Hexamethyldisilazane	294.2	2.273	0.23358E+01	0.16127E-02	-0.62078E-05	294-333
C_7F_{14}	Perfluoromethylcyclohexane	298.2	1.82				
C_7F_{16}	Perfluoroheptane	289.2	1.847				
$C_7H_3Cl_5$	2,3,4,5,6-Pentachlorotoluene	293.2	4.8				288-348
C_7H_4ClNO	4-Chlorophenyl isocyanate	288.2	3.177	0.40896E+01	-0.31667E-02		
C_7H_5BrO	Benzoyl bromide	293.2	21.33	0.84231E+02	-0.31089E+00	0.32857E-03	283-313
C_7H_5ClO	Benzoyl chloride	293.2	23.0				
C_7H_5FO	Benzoyl fluoride	293.2	22.7				
$C_7H_5F_3$	(Trifluoromethyl)benzene	298.2	9.22				
C_7H_5N	Benzonitrile	293.2	25.9	0.57605E+02	-0.13354E+00	0.87767E-04	273-453
C_7H_5NO	Phenyl isocyanate	293.2	8.940	0.17541E+02	-0.29790E-01	0.15476E-05	293-353
$C_7H_6ClNO_2$	4-Chloro-3-nitrotoluene	301.2	28.07				

Mol. Form.	Name	T/K	ε	a	b	c	Range/K
C$_7$H$_6$Cl$_2$	2,4-Dichlorotoluene	301.2	5.68				
C$_7$H$_6$Cl$_2$	2,6-Dichlorotoluene	301.2	3.36				
C$_7$H$_6$Cl$_2$	3,4-Dichlorotoluene	301.2	9.39				
C$_7$H$_6$Cl$_2$	(Dichloromethyl)benzene	293.2	6.9				
C$_7$H$_6$O	Benzaldehyde	293.2	17.85	0.35046E+02	-0.61271E-01	0.16222E-04	301-346
C$_7$H$_6$O$_2$	Salicylaldehyde	293.2	18.35	0.51315E+02	-0.15379E+00	0.14111E-03	289-453
C$_7$H$_7$Br	o-Bromotoluene	293.2	4.641	0.10229E+02	-0.25050E-01	0.20357E-04	273-323
C$_7$H$_7$Br	m-Bromotoluene	293.2	5.566	0.11522E+02	-0.24946E-01	0.15714E-04	273-323
C$_7$H$_7$Br	p-Bromotoluene	293.2	5.503	0.10014E+02	-0.13918E-01	-0.50000E-05	273-293
C$_7$H$_7$Br	(Bromomethyl)benzene	293.2	6.658	0.18482E+02	-0.57207E-01	0.57321E-04	273-323
C$_7$H$_7$BrO	o-Bromoanisole	303.2	8.96	0.12023E+02	-0.59116E-02	-0.13787E-04	303-358
C$_7$H$_7$BrO	p-Bromoanisole	303.2	7.40	0.74367E+01	0.12648E-01	-0.42128E-04	303-358
C$_7$H$_7$Cl	o-Chlorotoluene	293.2	4.721	0.11507E+02	-0.31148E-01	0.27143E-04	273-323
C$_7$H$_7$Cl	m-Chlorotoluene	293.2	5.763	0.13921E+02	-0.37186E-01	0.31786E-04	273-323
C$_7$H$_7$Cl	p-Chlorotoluene	293.2	6.25	0.20265E+01	0.40060E-01	-0.87500E-04	293-333
C$_7$H$_7$Cl	(Chloromethyl)benzene	293.2	6.854	0.17108E+02	-0.45285E-01	0.35000E-04	273-323
C$_7$H$_7$ClO	p-Chloroanisole	293.2	7.84	0.64019E+01	0.30560E-01	-0.87500E-04	293-333
C$_7$H$_7$ClO$_2$S	p-Toluenesulfonyl chloride	343.2	22.6				
C$_7$H$_7$ClO$_3$S	4-Methoxybenzenesulfonyl chloride	314.2	27.2				
C$_7$H$_7$F	o-Fluorotoluene	298.2	4.23				
C$_7$H$_7$F	m-Fluorotoluene	298.2	5.41				
C$_7$H$_7$F	p-Fluorotoluene	298.2	5.88				
C$_7$H$_7$I	p-Iodotoluene	308.2	4.4				
C$_7$H$_7$N	2-Vinylpyridine	293.2	9.126				
C$_7$H$_7$N	4-Vinylpyridine	293.2	10.50				
C$_7$H$_7$NO$_2$	Benzyl nitrite	298.2	7.78				
C$_7$H$_7$NO$_2$	o-Nitrotoluene	293.0	26.26	0.10420E+03	-0.41726E+00	0.51607E-03	273-323
C$_7$H$_7$NO$_2$	m-Nitrotoluene	303.2	24.95	0.62492E+02	-0.16235E+00	0.12844E-03	303-403
C$_7$H$_7$NO$_2$	p-Nitrotoluene	331.2	22.2				
C$_7$H$_7$NO$_2$S	4-Nitrothioanisole	346.0	21.7				
C$_7$H$_7$NO$_3$	2-Nitroanisole	293.2	45.75	0.16684E+03	-0.58196E+00	0.57382E-03	293-423
C$_7$H$_7$NO$_3$	3-Nitroanisole	318.2	25.7	0.65402E+02	-0.16460E+00	0.12560E-03	318-443
C$_7$H$_7$NO$_3$	4-Nitroanisole	338.2	26.95	0.59811E+02	-0.10955E+00	0.36042E-04	338-443
C$_7$H$_8$	Toluene	296.35	2.379	0.32584E+01	-0.34410E-02	0.15937E-05	207-316
C$_7$H$_8$O	o-Cresol	298.2	6.76	0.21633E+02	-0.71069E-01	0.70590E-04	298-453
C$_7$H$_8$O	m-Cresol	298.2	12.44	0.81716E+02	-0.35039E+00	0.39878E-03	274-463
C$_7$H$_8$O	p-Cresol	298.2	13.05	0.70253E+02	-0.28870E+00	0.31979E-03	298-453
C$_7$H$_8$O	Benzyl alcohol	303.2	11.916	0.13661E+03	-0.72127E+00	0.10225E-02	303-333
C$_7$H$_8$O	Anisole	294.2	4.30	0.10887E+02	-0.32372E-01	0.33629E-04	294-413
C$_7$H$_8$O$_2$	2-Methoxyphenol	298.2	11.95	0.31751E+02	-0.88173E-01	0.72953E-04	291-448
C$_7$H$_8$O$_2$	3-Methoxyphenol	298.2	11.59	0.37279E+02	-0.12113E+00	0.11698E-03	298-433
C$_7$H$_8$O$_2$	4-Methoxyphenol	333.7	11.05	0.39483E+02	-0.12142E+00	0.10841E-03	334-453
C$_7$H$_8$O$_2$S	Ethyl thiophene-2-carboxylate	293.2	6.18				
C$_7$H$_8$O$_2$S	Methyl phenyl sulfone	373.2	37.9				
C$_7$H$_8$S	Benzenemethanethiol	298.2	4.705	0.16628E+02	-0.68276E-01	0.94636E-04	298-358
C$_7$H$_8$S	4-Methylbenzenethiol	323.2	4.74	0.87052E+01	-0.15347E-01	0.95238E-05	323-358
C$_7$H$_8$S	(Methylthio)benzene	303.2	4.88	0.21841E+02	-0.97630E-01	0.13750E-03	303-343
C$_7$H$_9$N	Benzylamine	293.2	5.18				
C$_7$H$_9$N	o-Methylaniline	298.2	6.138	0.10988E+02	-0.18976E-01	0.91958E-05	298-398
C$_7$H$_9$N	m-Methylaniline	298.2	5.816	0.13477E+02	-0.35551E-01	0.33135E-04	298-398
C$_7$H$_9$N	p-Methylaniline	333.2	5.058	0.78897E+01	-0.10196E-01	0.51190E-05	333-403
C$_7$H$_9$N	N-Methylaniline	293.2	5.96				
C$_7$H$_9$N	2-Ethylpyridine	293.2	8.33	0.36397E+02	-0.15070E+00	0.18750E-03	293-333
C$_7$H$_9$N	4-Ethylpyridine	293.2	10.98	-0.73831E+01	0.14326E+00	-0.27500E-03	293-333
C$_7$H$_9$N	2,4-Dimethylpyridine	293.2	9.60	0.25895E+02	-0.73900E-01	0.62500E-04	293-333
C$_7$H$_9$N	2,6-Dimethylpyridine	293.2	7.33	0.17714E+02	-0.39080E-01	0.12500E-04	293-333
C$_7$H$_9$NO	2,6-Dimethylpyridine-1-oxide	298.2	46.11	0.22765E+03	-0.90760E+00	0.10011E-02	298-398
C$_7$H$_9$NO	o-Methoxyaniline	303.2	5.230	0.79911E+01	-0.92183E-02	0.37879E-06	303-393
C$_7$H$_9$NO	m-Methoxyaniline	298.2	8.76	0.28179E+02	-0.97840E-01	0.11027E-03	289-393

Mol. Form.	Name	T/K	ε	a	b	c	Range/K
C_7H_9NO	p-Methoxyaniline	333.2	7.85	0.30149E+02	-0.10523E+00	0.11467E-03	333-453
$C_7H_{10}N_2$	1-Methyl-1-phenylhydrazine	292.2	7.3				
$C_7H_{11}Cl_3O_2$	Isopentyl trichloroacetate	293.2	7.287				
C_7H_{12}	1,6-Heptadiene	293.0	2.161	0.30815E+01	-0.36095E-02	0.16354E-05	184-293
C_7H_{12}	Cycloheptene	295.0	2.265	0.32309E+01	-0.42373E-02	0.32572E-05	227-363
$C_7H_{12}O$	Cycloheptanone	298.2	13.16	0.17511E+03	-0.11221E+01	0.19417E-02	258-298
$C_7H_{12}O$	2-Methylcyclohexanone	293.2	14.0				
$C_7H_{12}O$	3-Methylcyclohexanone	293.2	12.4				
$C_7H_{12}O$	4-Methylcyclohexanone	293.2	12.35				
$C_7H_{12}O_2$	Cyclohexanecarboxylic acid	304.2	2.67				
$C_7H_{12}O_2$	Cyclohexyl formate	293.2	6.47				
$C_7H_{12}O_2$	Butyl acrylate	301.2	5.25	0.38296E+02	-0.19109E+00	0.27006E-03	301-343
$C_7H_{12}O_4$	Monomethyl adipate	293.2	6.69	0.11962E+02	-0.23973E-01	0.20608E-04	293-433
$C_7H_{12}O_4$	Diethyl malonate	304.2	7.550	0.14809E+02	-0.31207E-01	0.24066E-04	304-393
$C_7H_{12}O_4$	Dimethyl glutarate	293.2	7.87	0.20697E+02	-0.57794E-01	0.48405E-04	293-433
$C_7H_{12}O_5$	1,2,3-Propanetriol-1,3-diacetate	288.2	9.80	0.28321E+02	-0.89073E-01	0.86891E-04	258-374
C_7H_{14}	1-Heptene	293.2	2.092	0.21755E+01	0.13896E-02	-0.57049E-05	273-323
C_7H_{14}	2-Methyl-2-hexene	293.2	2.962				
C_7H_{14}	3-Ethyl-2-pentene	293.2	2.051				
C_7H_{14}	Cycloheptane	293.2	2.0784	0.25136E+01	-0.15089E-02	0.84915E-07	278-333
C_7H_{14}	Methylcyclohexane	293.2	2.024				
$C_7H_{14}Br_2$	1,2-Dibromoheptane	298.2	3.77				
$C_7H_{14}Br_2$	2,3-Dibromoheptane	298.2	5.08				
$C_7H_{14}Br_2$	3,4-Dibromoheptane	298.2	4.70				
$C_7H_{14}Cl_2$	1,7-Dichloroheptane	298.2	8.34				
$C_7H_{14}O$	1-Heptanal	295.2	9.07				
$C_7H_{14}O$	2-Heptanone	293.2	11.95	0.38348E+02	-0.12531E+00	0.12005E-03	253-413
$C_7H_{14}O$	3-Heptanone	293.2	12.7				
$C_7H_{14}O$	4-Heptanone	293.2	12.60	0.41520E+02	-0.13839E+00	0.13497E-03	253-393
$C_7H_{14}O$	5-Methyl-2-hexanone	293.2	13.53	0.52353E+02	-0.17695E+00	0.15195E-03	293-333
$C_7H_{14}O$	Cyclohexanemethanol	333.2	9.70	0.10164E+03	-0.45839E+00	0.54762E-03	333-368
$C_7H_{14}O$	2-Methylcyclohexanol*	293.2	9.375	0.17315E+03	-0.98794E+00	0.14634E-02	273-323
$C_7H_{14}O$	3-Methylcyclohexanol*	293.2	13.79	0.65896E+02	-0.21954E+00	0.14107E-03	273-323
$C_7H_{14}O$	4-Methylcyclohexanol*	293.2	13.45	0.65021E+02	-0.22896E+00	0.17946E-03	273-323
$C_7H_{14}O_2$	Heptanoic acid	288.2	3.04	0.36423E+01	-0.31996E-02	0.39362E-05	288-423
$C_7H_{14}O_2$	Pentyl acetate	293.2	4.79	0.12091E+02	-0.36536E-01	0.39732E-04	253-353
$C_7H_{14}O_2$	Isopentyl acetate	293.2	4.72				
$C_7H_{14}O_2$	Butyl propanoate	293.2	4.838				
$C_7H_{14}O_2$	Propyl butanoate	293.2	4.3				
$C_7H_{14}O_2$	Ethyl pentanoate	291.2	4.71				
$C_7H_{14}O_2$	Ethyl 3-methylbutanoate	293.2	4.71				
$C_7H_{14}O_2$	Methyl hexanoate	293.2	4.615				
$C_7H_{15}Br$	1-Bromoheptane	303.2	5.255	0.15289E+02	-0.50621E-01	0.57753E-04	203-343
$C_7H_{15}Br$	2-Bromoheptane	295.2	6.46				
$C_7H_{15}Br$	4-Bromoheptane	295.2	6.81				
$C_7H_{15}Cl$	1-Chloroheptane	293.2	5.521	0.14279E+02	-0.39431E-01	0.32321E-04	273-323
$C_7H_{15}Cl$	2-Chloroheptane	295.2	6.52				
$C_7H_{15}Cl$	3-Chloroheptane	295.2	6.70				
$C_7H_{15}Cl$	4-Chloroheptane	295.2	6.54				
$C_7H_{15}I$	1-Iodoheptane	298.2	4.92	0.11856E+02	-0.33493E-01	0.34368E-04	294-323
$C_7H_{15}I$	3-Iodoheptane	295.2	6.39				
C_7H_{16}	Heptane	293.2	1.9209	0.24740E+01	-0.22577E-02	0.12428E-05	273-373
C_7H_{16}	2-Methylhexane	293.2	1.9221	0.24759E+01	-0.22535E-02	0.12500E-05	293-323
C_7H_{16}	3-Methylhexane	293.2	1.920	0.27089E+01	-0.37908E-02	0.37500E-05	273-323
C_7H_{16}	3-Ethylpentane	293.2	1.942	0.23771E+01	-0.15140E-02	0.10093E-06	163-363
C_7H_{16}	2,2-Dimethylpentane	293.2	1.915	0.23414E+01	-0.14362E-02	-0.51322E-07	153-353
C_7H_{16}	2,3-Dimethylpentane	293.2	1.929	0.25637E+01	-0.26328E-02	0.16071E-05	273-323
C_7H_{16}	2,4-Dimethylpentane	293.2	1.902	0.23979E+01	-0.17436E-02	0.17857E-06	273-323
C_7H_{16}	3,3-Dimethylpentane	291.3	1.9419	0.24007E+01	-0.16802E-02	0.36069E-06	291-322

Mol. Form.	Name	T/K	ε	a	b	c	Range/K
C_7H_{16}	2,2,3-Trimethylbutane	293.2	1.930				
$C_7H_{16}O$	1-Heptanol	293.2	11.75	0.60662E+02	-0.24049E+00	0.25155E-03	239-513
$C_7H_{16}O$	2-Heptanol	293.7	9.72	0.10050E+03	-0.49793E+00	0.64504E-03	207-365
$C_7H_{16}O$	3-Heptanol	296.1	7.07	0.19586E+03	-0.11465E+01	0.17175E-02	248-349
$C_7H_{16}O$	4-Heptanol	296.2	6.18	0.28995E+03	-0.18499E+01	0.30109E-02	270-301
$C_7H_{16}O$	2-Methyl-2-hexanol	297.0	3.257				
$C_7H_{16}O$	3-Methyl-2-hexanol	297.2	4.990	0.59724E+02	-0.32417E+00	0.47058E-03	244-372
$C_7H_{16}O$	3-Methyl-3-hexanol	298.2	3.248				
$C_7H_{16}O$	3-Ethyl-3-pentanol	293.2	3.158				
$C_7H_{16}O$	2,2-Dimethyl-1-pentanol	293.2	6.020	0.37318E+02	-0.17095E+00	0.22022E-03	283-393
$C_7H_{16}O$	Ethyl pentyl ether	296.2	3.6				
$C_7H_{16}O$	Ethyl isopentyl ether	293.2	3.955	0.66541E+01	-0.55450E-02	-0.12500E-04	293-323
$C_7H_{16}O_3$	Triethoxymethane	293.2	4.779				
$C_7H_{16}S$	1-Heptanethiol	293.2	4.194	0.71333E+01	-0.97320E-02	-0.12500E-05	273-333
$C_7H_{17}N$	Heptylamine	293.2	3.81	0.87794E+01	-0.24363E-01	0.25325E-04	253-373
$C_7H_{18}O_3Si$	Triethoxymethylsilane	298.2	3.845				
$C_8H_4F_6$	1,3-Bis(trifluoromethyl)benzene	303.2	5.98				
C_8H_6	Phenylacetylene	298.2	2.98				
$C_8H_6Cl_2$	2,5-Dichlorostyrene	298.2	2.58				
$C_8H_6Cl_4$	1,2,3,4-Tetrachloro-5,6-dimethylbenzene	293.2	8.0				
$C_8H_6Cl_4$	1,2,3,5-Tetrachloro-4,6-dimethylbenzene	293.2	5.4				
C_8H_6O	Phenoxyacetylene	298.2	4.76				
C_8H_7N	Benzeneacetonitrile	299.2	17.87	0.82175E+02	-0.37416E+00	0.53220E-03	299-343
$C_8H_7NO_2$	4-Methoxyphenyl isocyanate	333.2	10.26	0.20780E+02	-0.31571E-01		333-403
$C_8H_7NO_4$	Methyl 2-nitrobenzoate	300.1	27.76				
C_8H_8	Styrene	293.2	2.4737	0.44473E+01	-0.11422E-01	0.16000E-04	293-313
C_8H_8O	Acetophenone	298.2	17.44	0.26099E+02	0.64048E-02	-0.11905E-03	298-333
$C_8H_8O_2$	Benzeneacetic acid	353.2	3.47	0.24104E+01	0.30000E-02		353-393
$C_8H_8O_2$	Benzyl formate	303.2	6.34	0.26162E+02	-0.11026E+00	0.14787E-03	303-358
$C_8H_8O_2$	Phenyl acetate	298.2	5.403	0.11327E+02	-0.26707E-01	0.22938E-04	298-404
$C_8H_8O_2$	Methyl benzoate	302.7	6.642	0.17486E+02	-0.51027E-01	0.50222E-04	303-393
$C_8H_8O_2$	(Hydroxyacetyl)benzene	298.2	21.33	0.42286E+02	-0.69215E-01	-0.35714E-05	298-368
$C_8H_8O_2$	4-Methoxybenzaldehyde	303.2	22.0				
$C_8H_8O_3$	Methyl salicylate	314.4	8.80	0.20501E+02	-0.39045E-01	0.68298E-05	223-398
C_8H_9Br	1-Bromo-2-ethylbenzene	298.2	5.55				
C_8H_9Br	1-Bromo-3-ethylbenzene	298.2	5.56				
C_8H_9Br	1-Bromo-4-ethylbenzene	298.2	5.42				
C_8H_9BrO	1-Bromo-2-ethoxybenzene	313.2	7.04	0.23146E+02	-0.75753E-01	0.77778E-04	313-358
C_8H_9Cl	1-Chloro-2-ethylbenzene	298.2	4.36				
C_8H_9Cl	1-Chloro-3-ethylbenzene	298.2	5.18				
C_8H_9Cl	1-Chloro-4-ethylbenzene	298.2	5.16				
$C_8H_9NO_2$	1-Ethyl-2-nitrobenzene	273.4	21.9				
$C_8H_9NO_2$	Methyl 2-aminobenzoate	298.2	21.9				
$C_8H_9NO_2$	Ethyl 4-pyridinecarboxylate	293.2	8.95				
C_8H_{10}	Ethylbenzene	293.2	2.4463	0.35969E+01	-0.53169E-02	0.47500E-05	293-323
C_8H_{10}	o-Xylene	293.2	2.562	0.36163E+01	-0.40177E-02	0.14286E-05	273-323
C_8H_{10}	m-Xylene	293.2	2.359	0.28421E+01	-0.10191E-02	-0.21429E-05	273-323
C_8H_{10}	p-Xylene	293.2	2.2735	0.23140E+01	0.97221E-03	-0.37500E-05	293-363
$C_8H_{10}O$	2,3-Xylenol	343.2	4.81	0.14399E+02	-0.41438E-01	0.39244E-04	343-433
$C_8H_{10}O$	2,4-Xylenol	303.2	5.060	0.22125E+02	-0.85543E-01	0.96548E-04	303-363
$C_8H_{10}O$	2,5-Xylenol	338.2	5.36	0.18049E+02	-0.54991E-01	0.51656E-04	338-455
$C_8H_{10}O$	2,6-Xylenol	313.2	4.90	0.12284E+02	-0.32996E-01	0.29867E-04	313-453
$C_8H_{10}O$	3,4-Xylenol	333.2	9.02	0.54423E+02	-0.21153E+00	0.22508E-03	333-453
$C_8H_{10}O$	3,5-Xylenol	323.2	9.06	0.54251E+02	-0.21647E+00	0.23542E-03	323-453
$C_8H_{10}O$	Benzeneethanol	293.2	12.31	0.12170E+03	-0.63124E+00	0.87776E-03	278-333
$C_8H_{10}O$	1-Phenylethanol	293.2	8.77	0.32971E+02	-0.12042E+00	0.12809E-03	293-423
$C_8H_{10}O$	Phenetole	293.2	4.216	-0.15043E+02	0.13752E+00	-0.24500E-03	293-313

Mol. Form.	Name	T/K	ε	a	b	c	Range/K
$C_8H_{10}O$	2-Methylanisole	293.2	3.502	0.50825E+01	-0.62297E-02	0.28571E-05	293-333
$C_8H_{10}O$	3-Methylanisole	293.2	3.967	0.12830E+02	-0.49701E-01	0.66429E-04	293-333
$C_8H_{10}O$	4-Methylanisole	293.2	3.914	0.86608E+01	-0.23510E-01	0.25000E-04	293-333
$C_8H_{10}O_2$	1,2-Dimethoxybenzene	293.2	4.45	0.74604E+01	-0.13445E-01	0.10737E-04	293-443
$C_8H_{10}O_2$	1,3-Dimethoxybenzene	298.2	5.363	0.11911E+02	-0.30804E-01	0.29643E-04	298-358
$C_8H_{10}O_2$	1,4-Dimethoxybenzene	333.7	5.60	0.11289E+02	-0.20765E-01	0.11987E-04	334-463
$C_8H_{10}O_2S$	Ethyl phenyl sulfone	348.2	39.0				
$C_8H_{10}S$	(Ethylthio)benzene	298.2	4.95				
$C_8H_{11}N$	p-Ethylaniline	298.2	4.84				
$C_8H_{11}N$	N-Ethylaniline	293.2	5.87				
$C_8H_{11}N$	N,N-Dimethylaniline	298.2	4.90	0.84052E+01	-0.13549E-01	0.62835E-05	289-453
$C_8H_{11}N$	2,4,6-Trimethylpyridine	298.2	7.807	0.20990E+02	-0.57419E-01	0.44286E-04	298-358
$C_8H_{11}NO$	4-Ethoxyaniline	298.2	7.43				
$C_8H_{12}N_2O_2$	Hexamethylene diisocyanate	288.2	14.41	0.26715E+02	-0.42696E-01		288-403
$C_8H_{12}O_4$	Diethyl maleate	298.2	7.560	0.13953E+02	-0.21969E-01	0.17817E-05	298-343
$C_8H_{12}O_4$	Diethyl fumarate	296.2	6.56				
C_8H_{14}	1,7-Octadiene	293.0	2.186	0.28376E+01	-0.17442E-02	-0.16141E-05	214-293
C_8H_{14}	cis-Cyclooctene	296.0	2.306	0.31115E+01	-0.32058E-02	0.16713E-05	269-406
C_8H_{14}	1,2-Dimethylcyclohexene	296.0	2.144	0.26443E+01	-0.17973E-02	0.35815E-06	211-374
C_8H_{14}	1,3-Dimethylcyclohexene	296.0	2.182	0.29951E+01	-0.34615E-02	0.24026E-05	213-373
$C_8H_{14}O_2$	Methyl cyclohexanecarboxylate	293.2	4.87				
$C_8H_{14}O_2$	Cyclohexyl acetate	293.2	5.08				
$C_8H_{14}O_3$	Butanoic anhydride	293.2	12.8				
$C_8H_{14}O_3$	2-Methylpropanoic anhydride	292.2	13.6				
$C_8H_{14}O_4$	Diisopropyl oxalate	293.2	6.403	0.10709E+02	-0.16328E-01	0.56000E-05	293-368
$C_8H_{14}O_4$	Diethyl succinate	293.2	6.098	0.80213E+01	0.11810E-02	-0.26400E-04	293-343
$C_8H_{14}O_4$	Dimethyl adipate	293.2	6.84	0.11739E+02	-0.17281E-01	0.11447E-05	293-433
$C_8H_{15}N$	Octanenitrile	293.2	13.90				
C_8H_{16}	1-Octene	293.2	2.113	0.24348E+01	0.34200E-03	-0.50000E-05	273-323
C_8H_{16}	cis-3-Octene	298.2	2.062				
C_8H_{16}	trans-3-Octene	298.2	2.002				
C_8H_{16}	cis-4-Octene	298.2	2.053				
C_8H_{16}	trans-4-Octene	298.2	2.004				
C_8H_{16}	3-Methyl-2-heptene*	293.2	2.436				
C_8H_{16}	2,5-Dimethyl-2-hexene	293.2	2.431				
C_8H_{16}	2,4,4-Trimethyl-1-pentene	298.2	2.0908				
C_8H_{16}	Cyclooctane	295.0	2.116	0.25036E+01	-0.12460E-02	-0.23175E-06	295-411
$C_8H_{16}Br_2$	1,8-Dibromooctane	298.2	7.43	0.94117E+00	0.61520E-01	-0.13333E-03	298-328
$C_8H_{16}Cl_2$	1,8-Dichlorooctane	298.2	7.64				
$C_8H_{16}O$	2-Octanone	293.2	9.51	-0.16219E+02	0.18799E+00	-0.34156E-03	293-333
$C_8H_{16}O$	3-Octanone	303.2	10.50				
$C_8H_{16}O_2$	Octanoic acid	288.2	2.85	0.29391E+01	-0.38721E-03		288-423
$C_8H_{16}O_2$	2-Ethylhexanoic acid	296.2	2.64				
$C_8H_{16}O_2$	Hexyl acetate	293.2	4.42				
$C_8H_{16}O_2$	Pentyl propanoate	293.2	4.552				
$C_8H_{16}O_2$	Isopentyl propanoate	273.2	5.21	0.17665E+02	-0.71718E-01	0.95635E-04	273-373
$C_8H_{16}O_2$	Butyl butanoate	298.2	4.39	0.79684E+01	-0.12000E-01	0.15266E-13	298-318
$C_8H_{16}O_2$	Propyl pentanoate	292.2	4.0				
$C_8H_{16}O_2$	Ethyl hexanoate	293.2	4.45	0.11007E+02	-0.32800E-01	0.35714E-04	253-353
$C_8H_{16}O_2$	Methyl heptanoate	293.2	4.355				
$C_8H_{16}O_3$	Isopentyl lactate	273.2	11.2	0.48649E+02	-0.21253E+00	0.27619E-03	273-373
$C_8H_{17}Br$	1-Bromooctane	293.2	5.0957	0.12404E+02	-0.35050E-01	0.34542E-04	283-353
$C_8H_{17}Br$	2-Bromooctane	293.2	5.44				
$C_8H_{17}Cl$	1-Chlorooctane	298.2	5.05	0.11346E+02	-0.25120E-01	0.13450E-04	274-328
$C_8H_{17}Cl$	2-Chlorooctane	293.2	5.42				
$C_8H_{17}F$	1-Fluorooctane	293.2	3.89				
$C_8H_{17}I$	1-Iodooctane	293.2	4.67	0.12452E+02	-0.41229E-01	0.50108E-04	233-313
$C_8H_{17}NO_2$	1-Nitrooctane	293.2	11.46				
C_8H_{18}	Octane	293.2	1.948	0.22590E+01	-0.84212E-03	-0.75758E-06	233-393

Mol. Form.	Name	T/K	ε	a	b	c	Range/K
C_8H_{18}	2-Methylheptane	293.2	1.9519				
C_8H_{18}	3-Ethylhexane	293.2	1.9617				
C_8H_{18}	2,2-Dimethylhexane	293.2	1.9498				
C_8H_{18}	2,5-Dimethylhexane	293.95	1.9619	0.25821E+01	-0.26804E-02	0.19404E-05	294-324
C_8H_{18}	3,3-Dimethylhexane	293.2	1.9645				
C_8H_{18}	3,4-Dimethylhexane	292.1	1.9814	0.26849E+01	-0.33712E-02	0.32949E-05	292-324
C_8H_{18}	3-Ethyl-3-methylpentane	291.49	1.9869	0.25983E+01	-0.28027E-02	0.24195E-05	292-324
C_8H_{18}	2,2,3-Trimethylpentane	293.2	1.960				
C_8H_{18}	2,2,4-Trimethylpentane	293.2	1.943	0.23677E+01	-0.14768E-02	0.94261E-07	173-373
C_8H_{18}	2,3,3-Trimethylpentane	293.2	1.9780				
C_8H_{18}	2,3,4-Trimethylpentane	293.2	1.9738				
$C_8H_{18}O$	1-Octanol	293.2	10.30	0.51647E+02	-0.20371E+00	0.21320E-03	258-513
$C_8H_{18}O$	2-Octanol	293.2	8.13	0.63760E+02	-0.27643E+00	0.31075E-03	213-513
$C_8H_{18}O$	3-Octanol	293.2	5.55	0.12505E+03	-0.70646E+00	0.10245E-02	223-383
$C_8H_{18}O$	4-Octanol	293.2	4.48	0.51049E+02	-0.26664E+00	0.37280E-03	243-403
$C_8H_{18}O$	2-Methyl-1-heptanol	293.1	5.16	0.61698E+02	-0.33647E+00	0.49066E-03	236-328
$C_8H_{18}O$	3-Methyl-1-heptanol	290.3	2.884	0.84687E+01	-0.33712E-01	0.49793E-04	241-316
$C_8H_{18}O$	4-Methyl-1-heptanol	290.6	4.63	0.48612E+02	-0.26773E+00	0.39972E-03	237-332
$C_8H_{18}O$	5-Methyl-1-heptanol	290.4	7.68	0.54581E+02	-0.24772E+00	0.29734E-03	235-328
$C_8H_{18}O$	6-Methyl-1-heptanol	290.3	10.54	0.57997E+02	-0.23517E+00	0.24663E-03	265-328
$C_8H_{18}O$	2-Methyl-2-heptanol	292.2	3.43				
$C_8H_{18}O$	3-Methyl-2-heptanol	289.6	7.47	0.39178E+02	-0.17976E+00	0.24218E-03	229-329
$C_8H_{18}O$	4-Methyl-2-heptanol	290.0	3.59	0.39715E+02	-0.23115E+00	0.36771E-03	240-333
$C_8H_{18}O$	5-Methyl-2-heptanol	278.5	7.5	0.68568E+02	-0.40706E+00	0.67433E-03	230-279
$C_8H_{18}O$	6-Methyl-2-heptanol	290.1	6.41	0.77520E+02	-0.41724E+00	0.59448E-03	239-329
$C_8H_{18}O$	2-Methyl-3-heptanol	293.2	3.260	-0.59739E+01	0.56700E-01	-0.83125E-04	343-403
$C_8H_{18}O$	3-Methyl-3-heptanol	293.2	3.013	-0.38440E+01	0.42327E-01	-0.61250E-04	343-403
$C_8H_{18}O$	4-Methyl-3-heptanol	293.2	3.312	-0.48003E+01	0.50740E-01	-0.75000E-04	343-403
$C_8H_{18}O$	5-Methyl-3-heptanol	293.2	3.832	0.61967E+01	-0.63750E-02		343-383
$C_8H_{18}O$	6-Methyl-3-heptanol	293.2	4.992	0.23037E+02	-0.98029E-01	0.12479E-03	283-383
$C_8H_{18}O$	2-Methyl-4-heptanol	296.3	3.338	0.42102E+00	0.10427E-01	-0.20438E-05	230-333
$C_8H_{18}O$	3-Methyl-4-heptanol	290.0	7.46	0.33354E+02	-0.14077E+00	0.17750E-03	230-330
$C_8H_{18}O$	4-Methyl-4-heptanol	296.2	2.902				
$C_8H_{18}O$	2-Ethyl-1-hexanol	298.2	7.58	0.86074E+02	-0.42636E+00	0.55078E-03	208-318
$C_8H_{18}O$	2,2-Dimethyl-1-hexanol	293.2	4.50	0.91244E+01	-0.21785E-01	0.21018E-04	283-393
$C_8H_{18}O$	Dibutyl ether	293.2	3.0830	0.65383E+01	-0.16172E-01	0.14969E-04	293-314
$C_8H_{18}OS$	Dibutyl sulfoxide	313.2	24.73	0.67156E+02	-0.16448E+00	0.92275E-04	313-393
$C_8H_{18}O_2$	2-Ethyl-1,3-hexanediol	293.2	18.73	0.57919E+02	-0.17128E+00	0.12949E-03	233-333
$C_8H_{18}O_2S$	Dibutyl sulfone	323.2	25.72	0.66248E+02	-0.16417E+00	0.12001E-03	323-398
$C_8H_{18}O_4$	Triethylene glycol dimethyl ether	298.2	7.62				
$C_8H_{18}O_5$	Tetraethylene glycol	293.2	20.44	0.83547E+02	-0.31691E+00	0.34689E-03	253-333
$C_8H_{18}S$	1-Octanethiol	293.2	3.949	0.63667E+01	-0.87920E-02	0.18750E-05	273-333
$C_8H_{18}S$	Dibutyl sulfide	298.2	4.29				
$C_8H_{19}N$	Octylamine	293.2	3.58	0.77931E+01	-0.20015E-01	0.19347E-04	273-373
$C_8H_{19}N$	Dibutylamine	293.2	2.765	0.52504E+01	-0.10538E-01	0.71485E-05	243-323
$C_8H_{20}O_4Si$	Ethyl silicate	293.2	2.50				
$C_8H_{20}Si$	Tetraethylsilane	293.2	2.090				
$C_8H_{20}Sn$	Tetraethylstannane	293.2	2.241				
$C_8H_{23}N_5$	Tetraethylenepentamine	293.2	9.40	0.40553E+02	-0.16681E+00	0.20659E-03	213-333
$C_8H_{24}O_4Si_4$	Octamethylcyclotetrasiloxane	296.2	2.390	0.36286E+01	-0.56885E-02	0.50874E-05	296-333
$C_9H_6N_2O_2$	Toluene-2,4-diisocyanate	293.2	8.433	0.22174E+02	-0.66982E-01	0.68571E-04	293-353
$C_9H_6O_2$	2H-1-Benzopyran-2-one	343.2	34.04	0.11311E+03	-0.33804E+00	0.31324E-03	343-423
C_9H_7N	Quinoline	293.2	9.16	0.33432E+02	-0.13497E+00	0.17788E-03	258-323
C_9H_7N	Isoquinoline	298.2	11.0	0.14412E+03	-0.79935E+00	0.11839E-02	298-323
C_9H_8O	Cinnamaldehyde	305.8	17.72	0.41837E+02	-0.11060E+00	0.10401E-03	306-354
$C_9H_8O_4$	2-(Acetyloxy)benzoic acid	333.2	6.55	0.69994E+01	-0.14553E-02		333-416
C_9H_{10}	1-Propenylbenzene	293.2	2.73				
C_9H_{10}	Allylbenzene	293.2	2.63				
C_9H_{10}	Isopropenylbenzene	293.2	2.28				

Mol. Form.	Name	T/K	ε	a	b	c	Range/K
$C_9H_{10}OS$	4-Acetylthioanisole	355.2	11.34				
$C_9H_{10}O_2$	Ethyl benzoate	293.2	6.20	0.18216E+02	-0.62361E-01	0.72884E-04	288-343
$C_9H_{10}O_2$	Methyl 4-methylbenzoate	306.2	4.3				
$C_9H_{10}O_2$	Benzyl acetate	303.2	5.34	0.11727E+02	-0.30869E-01	0.32340E-04	303-358
$C_9H_{10}O_2$	Phenyl propanoate	293.2	4.77				
$C_9H_{10}O_2$	4-Acetylanisole	313.2	17.3				
$C_9H_{10}O_3$	Ethyl salicylate	308.2	8.48	0.18910E+02	-0.35623E-01	0.46529E-05	225-321
$C_9H_{10}O_3$	Methyl 2-methoxybenzoate	294.2	7.7				
$C_9H_{11}Br$	(3-Bromopropyl)benzene	302.2	5.41	0.11360E+02	-0.27471E-01	0.25775E-04	302-358
$C_9H_{11}NO$	N-Ethylbenzamide	352.7	42.6	-0.20109E+03	0.17866E+01	-0.31065E-02	353-389
$C_9H_{11}NO$	N,N-Dimethylbenzamide	318.2	20.77	0.76725E+02	-0.26908E+00	0.29409E-03	318-443
$C_9H_{11}NO_2$	Ethyl 2-aminobenzoate	298.2	4.14				
C_9H_{12}	Propylbenzene	293.2	2.370	0.26933E+01	0.21679E-03	-0.44643E-05	273-323
C_9H_{12}	Isopropylbenzene	293.2	2.381	0.31149E+01	-0.30801E-02	0.19643E-05	273-323
C_9H_{12}	o-Ethyltoluene	293.2	2.595				
C_9H_{12}	m-Ethyltoluene	293.2	2.365				
C_9H_{12}	p-Ethyltoluene	293.2	2.265				
C_9H_{12}	1,2,3-Trimethylbenzene	293.2	2.656	0.76006E+01	-0.29118E-01	0.41786E-04	273-323
C_9H_{12}	1,2,4-Trimethylbenzene	293.2	2.377	0.31517E+01	-0.30634E-02	0.14286E-05	273-323
C_9H_{12}	1,3,5-Trimethylbenzene	293.2	2.279	0.38998E+01	-0.88072E-02	0.11149E-04	288-358
$C_9H_{12}O$	Benzenepropanol	293.2	11.97	0.94482E+02	-0.45540E+00	0.59307E-03	213-303
$C_9H_{12}O$	α-Ethylbenzenemethanol	293.2	6.68	0.44520E+02	-0.21505E+00	0.29443E-03	233-373
$C_9H_{12}O$	α,α-Dimethylbenzenemethanol	303.2	5.61	0.57072E+01	0.86568E-02	-0.29580E-04	303-373
$C_9H_{12}O$	1-Phenyl-2-propanol	293.2	9.35	0.10762E+03	-0.56026E+00	0.76915E-03	233-373
$C_9H_{12}O$	Benzyl ethyl ether	298.2	3.90				
$C_9H_{12}O$	2,6-Dimethylanisole	293.2	3.780	0.76700E+01	-0.18298E-01	0.17143E-04	293-333
$C_9H_{12}O$	3,5-Dimethylanisole	293.2	3.711	0.54981E+01	-0.56651E-02	-0.14286E-05	293-333
$C_9H_{12}O_2S$	Butyl thiophene-2-carboxylate	293.2	6.40				
$C_9H_{12}S$	Benzenepropanethiol	303.2	4.36	0.82411E+01	-0.15034E-01	0.73617E-05	303-358
$C_9H_{13}N$	Benzylethylamine	293.2	4.3				
$C_9H_{13}N$	N-Propylaniline	293.2	5.48				
$C_9H_{13}N$	2-Methyl-N,N-dimethylaniline	293.2	3.4				
$C_9H_{13}N$	4-Methyl-N,N-dimethylaniline	293.2	3.9				
$C_9H_{14}OSi$	Trimethylphenoxysilane	298.2	3.3953				
$C_9H_{14}O_6$	Triacetin	293.6	7.11	0.17819E+02	-0.53656E-01	0.57759E-04	219-304
$C_9H_{14}Si$	Trimethylphenylsilane	298.2	2.3533	0.21463E+01	0.32711E-02	-0.86264E-05	288-323
$C_9H_{16}O_2$	2-Nonenoic acid	296.2	2.5				
$C_9H_{16}O_2$	Cyclohexyl propanoate	293.2	4.82				
$C_9H_{16}O_2$	Ethyl cyclohexanecarboxylate	293.2	4.64				
$C_9H_{16}O_4$	Diethyl glutarate	303.2	6.659				
$C_9H_{17}N$	Nonanenitrile	293.2	12.08				
C_9H_{18}	1-Nonene	293.2	2.180	0.22710E+01	0.15797E-02	-0.64286E-05	273-323
$C_9H_{18}Br_2$	1,9-Dibromononane	293.2	7.153	0.18931E+02	-0.57764E-01	0.60000E-04	293-343
$C_9H_{18}O$	2-Nonanone	295.2	9.14				
$C_9H_{18}O$	5-Nonanone	293.2	10.6				
$C_9H_{18}O$	Di-tert-butyl ketone	287.65	10.0				
$C_9H_{18}O$	2,6-Dimethyl-4-heptanone	293.2	9.91	0.33178E+02	-0.11290E+00	0.11454E-03	273-393
$C_9H_{18}O_2$	Nonanoic acid	294.9	2.475	0.25039E+01	0.67274E-03	-0.24180E-05	295-365
$C_9H_{18}O_2$	2-Methyloctanoic acid	293.2	2.39				
$C_9H_{18}O_2$	2-Ethylheptanoic acid	293.2	1.98				
$C_9H_{18}O_2$	Heptyl acetate	293.2	4.2				
$C_9H_{18}O_2$	Pentyl butanoate	301.2	4.08	0.59029E+01	-0.49905E-02	-0.34292E-05	301-343
$C_9H_{18}O_2$	Isopentyl butanoate	293.2	4.0				
$C_9H_{18}O_2$	Isobutyl pentanoate	292.2	3.8				
$C_9H_{18}O_2$	Methyl octanoate	293.2	4.101				
$C_9H_{19}Br$	1-Bromononane	298.2	4.74	0.79870E+01	-0.10488E-01	-0.13450E-05	274-328
$C_9H_{19}Cl$	1-Chlorononane	293.2	4.803	0.95528E+01	-0.16200E-01	-0.16365E-13	293-323
$C_9H_{19}NO$	N,N-Dibutylformamide	293.2	18.4				
C_9H_{20}	Nonane	293.2	1.9722	0.23894E+01	-0.14830E-02	0.14881E-06	253-393

Mol. Form.	Name	T/K	ε	a	b	c	Range/K
C_9H_{20}	2-Methyloctane	293.2	1.967				
C_9H_{20}	4-Methyloctane	293.2	1.967				
C_9H_{20}	2,4-Dimethylheptane	293.2	1.89				
C_9H_{20}	2,5-Dimethylheptane	293.2	1.89				
C_9H_{20}	2,6-Dimethylheptane	293.2	1.987				
$C_9H_{20}N_2O$	Tetraethylurea	296.8	14.29	0.52820E+02	-0.18790E+00	0.19580E-03	205-411
$C_9H_{20}O$	1-Nonanol	293.2	8.83	0.97467E+02	-0.51103E+00	0.71429E-03	288-343
$C_9H_{20}O$	2-Nonanol	298.2	6.66	0.10136E+03	-0.55612E+00	0.80000E-03	288-308
$C_9H_{20}O$	3-Nonanol	298.2	4.49	0.55214E+02	-0.31920E+00	0.50000E-03	288-308
$C_9H_{20}O$	4-Nonanol	298.2	3.69	0.27954E+01	0.30000E-02	-0.52375E-13	288-308
$C_9H_{20}O$	5-Nonanol	298.2	3.54	-0.25463E+01	0.35320E-01	-0.50000E-04	288-308
$C_9H_{21}B$	Tripropylborane	293.2	2.026				
$C_9H_{21}N$	Nonylamine	293.2	3.42	0.53575E+01	-0.71982E-02	0.19481E-05	293-373
$C_9H_{21}N$	Tripropylamine	293.2	2.380	0.33380E+01	-0.86332E-02	0.18322E-04	243-293
$C_9H_{21}O_4P$	Tripropyl phosphate	293.2	10.93	0.33166E+02	-0.10514E+00	0.10000E-03	293-373
$C_{10}H_7Br$	1-Bromonaphthalene	298.2	4.768	0.10561E+02	-0.27671E-01	0.27655E-04	293-323
$C_{10}H_7Cl$	1-Chloronaphthalene	298.2	5.04	0.84861E+01	-0.12357E-01	0.26899E-05	274-328
$C_{10}H_7NO_2$	1-Nitronaphthalene	333.2	19.68	0.36267E+02	-0.41283E-01	-0.25595E-04	333-403
$C_{10}H_8$	Naphthalene	363.2	2.54				
$C_{10}H_8O$	1-Naphthol	373.0	5.03	0.16489E+02	-0.46700E-01	0.42857E-04	373-453
$C_{10}H_8O$	2-Naphthol	413.0	4.95	0.92865E+01	-0.10500E-01	0.42501E-15	413-453
$C_{10}H_9N$	1-Naphthylamine	333.2	5.20	0.10577E+02	-0.22114E-01	0.17857E-04	333-453
$C_{10}H_9N$	2-Naphthylamine	393.0	5.26	0.19722E+02	-0.60679E-01	0.60714E-04	393-473
$C_{10}H_9N$	2-Methylquinoline	293.2	7.24	0.11688E+02	-0.78400E-02	-0.25000E-04	293-333
$C_{10}H_9N$	4-Methylquinoline	293.2	9.31	0.17788E+02	-0.32580E-01	0.12500E-04	293-333
$C_{10}H_9N$	6-Methylquinoline	293.2	8.48	0.21696E+02	-0.63400E-01	0.62500E-04	293-333
$C_{10}H_9N$	8-Methylquinoline	293.2	6.58	0.19356E+02	-0.61900E-01	0.62500E-04	293-333
$C_{10}H_{10}O_4$	Methyl 2-(acetyloxy)benzoate	328.9	5.31	0.19579E+02	-0.69970E-01	0.80889E-04	329-371
$C_{10}H_{10}O_4$	Dimethyl phthalate	293.2	8.66				
$C_{10}H_{12}$	1,2,3,4-Tetrahydronaphthalene	298.2	2.771	0.29172E+01	0.12832E-02	-0.59453E-05	298-343
$C_{10}H_{12}$	4-Ethylstyrene	298.2	3.350				
$C_{10}H_{12}$	Dicyclopentadiene	313.2	2.43	0.30564E+01	-0.20000E-02	0.82443E-15	313-373
$C_{10}H_{12}O$	Tetrahydro-2-naphthol*	293.2	11.70	0.98978E+02	-0.48267E+00	0.63008E-03	293-363
$C_{10}H_{12}O$	4-Isopropylbenzaldehyde	288.2	10.68				
$C_{10}H_{12}O_2$	4-Allyl-2-methoxyphenol	293.2	9.55	0.52377E+02	-0.24380E+00	0.33333E-03	273-323
$C_{10}H_{12}O_2$	2-Phenylethyl acetate	297.2	4.93				
$C_{10}H_{12}O_2$	Benzyl propanoate	303.0	5.11	0.42301E+01	0.13962E-01	-0.36426E-04	303-358
$C_{10}H_{12}O_2$	Phenyl butanoate	293.2	4.48				
$C_{10}H_{12}O_2$	Propyl benzoate	303.2	5.78	0.10927E+02	-0.20535E-01	0.11745E-04	303-358
$C_{10}H_{12}O_2$	Ethyl phenylacetate	293.2	5.320				
$C_{10}H_{14}$	Butylbenzene	293.2	2.359				
$C_{10}H_{14}$	sec-Butylbenzene	293.2	2.357	0.28348E+01	-0.68586E-03	-0.32143E-05	273-323
$C_{10}H_{14}$	tert-Butylbenzene	293.2	2.359	0.27924E+01	-0.38350E-03	-0.37500E-05	273-323
$C_{10}H_{14}$	Isobutylbenzene	293.2	2.318	0.28055E+01	-0.92614E-03	-0.25000E-05	273-323
$C_{10}H_{14}$	1-Isopropyl-4-methylbenzene	298.2	2.2322	0.25266E+01	-0.25121E-03	-0.24867E-05	277-333
$C_{10}H_{14}$	o-Diethylbenzene	293.2	2.594				
$C_{10}H_{14}$	m-Diethylbenzene	293.2	2.369				
$C_{10}H_{14}$	p-Diethylbenzene	293.2	2.259				
$C_{10}H_{14}$	1-Ethyl-3,5-dimethylbenzene	293.2	2.275				
$C_{10}H_{14}$	1,2,3,4-Tetramethylbenzene	296.0	2.538	0.33822E+01	-0.33630E-02	0.17475E-05	273-412
$C_{10}H_{14}$	1,2,4,5-Tetramethylbenzene	356.0	2.223	0.26834E+01	-0.10327E-02	-0.73533E-06	356-430
$C_{10}H_{14}N_2$	L-Nicotine	293.2	8.937	0.21347E+02	-0.57177E-01	0.50655E-04	293-363
$C_{10}H_{14}O$	1-Phenyl-2-methyl-2-propanol	298.2	5.71	0.21922E+02	-0.84231E-01	0.99475E-04	298-423
$C_{10}H_{14}O$	Butyl phenyl ether	293.2	3.734				
$C_{10}H_{14}O$	Thymol	333.2	4.259				
$C_{10}H_{15}N$	N,N-Diethylaniline	303.2	5.15	0.50773E+01	0.15399E-01	-0.50000E-04	303-328
$C_{10}H_{16}$	γ-Terpinene	298.2	2.2738				
$C_{10}H_{16}$	d-Limonene	298.2	2.3746				
$C_{10}H_{16}$	l-Limonene	298.2	2.3738				

Mol. Form.	Name	T/K	ε	a	b	c	Range/K
$C_{10}H_{16}$	Terpinolene	298.2	2.2918				
$C_{10}H_{16}$	α-Pinene	298.2	2.1787				
$C_{10}H_{16}$	β-Pinene	298.2	2.4970				
$C_{10}H_{16}$	α-Terpinene	298.2	2.4526				
$C_{10}H_{16}$	β-Myrcene	298.2	2.3				
$C_{10}H_{16}O$	Carvenone	293.2	18.8				
$C_{10}H_{16}O$	d-Fenchone	294.2	12.8				
$C_{10}H_{17}Cl$	2-Chlorobornane	368.2	5.21				
$C_{10}H_{18}$	Pinane	298.2	2.1456				
$C_{10}H_{18}$	cis-Decahydronaphthalene	293.2	2.219	0.25410E+01	-0.11420E-02	0.15092E-06	293-373
$C_{10}H_{18}$	trans-Decahydronaphthalene	293.2	2.184	0.26615E+01	-0.21241E-02	0.16864E-05	293-373
$C_{10}H_{18}O$	Eucalyptol	298.2	4.57				
$C_{10}H_{18}O_2$	Cyclohexyl butanoate	293.2	4.58				
$C_{10}H_{18}O_4$	Diethyl adipate	293.2	6.109	0.14824E+02	-0.40749E-01	0.37600E-04	293-343
$C_{10}H_{20}$	1-Decene	293.2	2.136	0.19091E+01	0.33442E-02	-0.87500E-05	273-323
$C_{10}H_{20}$	cis-5-Decene	298.2	2.071				
$C_{10}H_{20}$	trans-5-Decene	298.2	2.030				
$C_{10}H_{20}$	5-Methyl-4-nonene	293.2	2.175				
$C_{10}H_{20}$	2,4,6-Trimethyl-3-heptene	293.2	2.293				
$C_{10}H_{20}Br_2$	1,10-Dibromodecane	303.2	6.56	0.17350E+02	-0.50328E-01	0.48633E-04	303-368
$C_{10}H_{20}Cl_2$	1,10-Dichlorodecane	308.2	6.68	-0.57423E+01	0.94220E-01	-0.17500E-03	308-338
$C_{10}H_{20}O$	2-Decanone	287.2	8.3				
$C_{10}H_{20}O$	Menthol	309.3	3.90	0.68202E+01	-0.15894E-01	0.20837E-04	309-358
$C_{10}H_{20}O_2$	2,2-Dimethyloctanoic acid	296.2	2.8				
$C_{10}H_{20}O_2$	Octyl acetate	288.2	4.18	-0.34691E+01	0.58106E-01	-0.10952E-03	288-323
$C_{10}H_{20}O_2$	2-Methylheptyl acetate	288.2	4.27	0.23285E+02	-0.11538E+00	0.17143E-03	288-323
$C_{10}H_{20}O_2$	Pentyl pentanoate	305.6	4.076	0.77641E+01	-0.14335E-01	0.73740E-05	306-393
$C_{10}H_{20}O_2$	Isopentyl pentanoate	292.2	3.6				
$C_{10}H_{20}O_2$	Isopentyl isopentanoate	288.2	4.39	0.14698E+02	-0.57726E-01	0.76190E-04	288-323
$C_{10}H_{20}O_2$	Methyl nonanoate	293.2	3.943				
$C_{10}H_{21}Br$	1-Bromodecane	298.2	4.44	0.11202E+02	-0.33491E-01	0.36314E-04	274-328
$C_{10}H_{21}Cl$	1-Chlorodecane	293.2	4.581	0.68741E+01	-0.12210E-02	-0.22500E-04	293-323
$C_{10}H_{21}NO$	N,N-Dibutylacetamide	293.2	19.1				
$C_{10}H_{22}$	Decane	293.2	1.9853	0.24054E+01	-0.15445E-02	0.44643E-06	253-393
$C_{10}H_{22}$	2,7-Dimethyloctane	293.2	1.98				
$C_{10}H_{22}$	4-Propylheptane	293.2	1.9955				
$C_{10}H_{22}O$	1-Decanol	293.2	7.93	0.47195E+02	-0.20740E+00	0.24942E-03	293-343
$C_{10}H_{22}O$	2-Decanol	298.2	5.82	0.13621E+03	-0.81000E+00	0.12500E-02	288-308
$C_{10}H_{22}O$	3-Decanol	298.2	4.05	0.52090E+02	-0.31020E+00	0.50000E-03	288-308
$C_{10}H_{22}O$	4-Decanol	298.2	3.42	-0.11260E+02	0.93960E-01	-0.15000E-03	288-308
$C_{10}H_{22}O$	5-Decanol	298.2	3.24	-0.25832E+01	0.31456E-01	-0.40000E-04	288-308
$C_{10}H_{22}O$	2,2-Dimethyl-1-octanol	293.2	7.86	0.69536E+02	-0.34596E+00	0.46250E-03	293-333
$C_{10}H_{22}O$	Dipentyl ether	298.2	2.798				
$C_{10}H_{22}O$	Diisopentyl ether	293.2	2.817	0.44690E+01	-0.63710E-02	0.25000E-05	293-323
$C_{10}H_{22}OS$	Dipentyl sulfoxide	348.2	18.8				
$C_{10}H_{22}O_5$	Tetraethylene glycol dimethyl ether	298.2	7.68				
$C_{10}H_{22}S$	Dipentyl sulfide	298.2	3.826				
$C_{10}H_{23}N$	Decylamine	293.2	3.31	0.61497E+01	-0.12801E-01	0.10606E-04	293-373
$C_{10}H_{30}O_3Si_4$	Decamethyltetrasiloxane	293.2	2.370				
$C_{10}H_{30}O_5Si_5$	Decamethylcyclopentasiloxane	293.2	2.50				
$C_{11}H_{10}$	1-Methylnaphthalene	293.2	2.915	0.45126E+01	-0.76480E-02	0.75000E-05	293-333
$C_{11}H_{10}$	2-Methylnaphthalene	313.2	2.747				
$C_{11}H_{10}O$	1-Methoxynaphthalene	293.2	4.020	0.71885E+01	-0.14838E-01	0.13750E-04	293-333
$C_{11}H_{10}O$	2-Methoxynaphthalene	353.2	3.563	0.56702E+01	-0.69754E-02	0.28571E-05	353-373
$C_{11}H_{12}O_2$	Ethyl trans-cinnamate	293.2	5.63				
$C_{11}H_{12}O_3$	Ethyl benzoylacetate	303.2	13.50	0.93644E+01	0.74280E-01	-0.20000E-03	303-323
$C_{11}H_{14}O_2$	Benzyl butanoate	301.2	4.55				
$C_{11}H_{14}O_2$	Phenyl pentanoate	293.2	4.30				
$C_{11}H_{14}O_2$	Butyl benzoate	303.2	5.52	0.77854E+01	-0.34972E-02	-0.13149E-04	303-358

Mol. Form.	Name	T/K	ε	a	b	c	Range/K
$C_{11}H_{14}O_2$	Isobutyl benzoate	291.2	5.39				
$C_{11}H_{16}$	1,3-Diethyl-5-methylbenzene	293.2	2.264				
$C_{11}H_{16}$	Pentamethylbenzene	334.0	2.358	0.30196E+01	-0.22619E-02	0.83831E-06	334-413
$C_{11}H_{22}$	1-Undecene	293.2	2.137	0.22132E+01	0.13121E-02	-0.53571E-05	273-323
$C_{11}H_{22}O$	2-Undecanone	285.3	8.3				
$C_{11}H_{22}O_2$	Nonyl acetate	293.2	3.87				
$C_{11}H_{22}O_2$	Pentyl hexanoate	288.2	4.22	0.83503E+01	-0.18449E-01	0.14286E-04	288-323
$C_{11}H_{23}Br$	1-Bromoundecane	272.6	4.61				
$C_{11}H_{24}$	Undecane	293.2	1.9972	0.23637E+01	-0.12500E-02	-0.85869E-16	283-363
$C_{11}H_{24}O$	1-Undecanol	313.2	5.98				
$C_{11}H_{25}N$	Undecylamine	293.2	3.25	0.54945E+01	-0.96161E-02	0.66017E-05	293-373
$C_{12}F_{27}N$	Tris(perfluorobutyl)amine	293.2	2.15				
$C_{12}H_8O$	Dibenzofuran	373.2	3.00				
$C_{12}H_{10}$	Biphenyl	348.2	2.53	0.26869E+01	0.63072E-03	-0.30995E-05	348-428
$C_{12}H_{10}N_2O$	trans-Azoxybenzene	311.2	5.2				
$C_{12}H_{10}O$	Diphenyl ether	283.2	3.726				
$C_{12}H_{10}O$	2-Acetonaphthone	333.2	13.03	0.14538E+03	-0.73040E+00	0.10000E-02	333-363
$C_{12}H_{10}OS$	Diphenyl sulfoxide	344.7	16.6				
$C_{12}H_{10}O_2S$	Diphenyl sulfone	406.2	21.1				
$C_{12}H_{10}S$	Diphenyl sulfide	298.2	5.43				
$C_{12}H_{11}N$	Diphenylamine	323.2	3.73				
$C_{12}H_{11}NO$	N-1-Naphthylenylacetamide	433.2	24.3	0.84739E+02	-0.12391E+00	-0.35714E-04	433-533
$C_{12}H_{12}$	1,6-Dimethylnaphthalene	293.2	2.7250				
$C_{12}H_{12}O$	1-Ethoxynaphthalene	292.2	3.3				
$C_{12}H_{14}O_2$	Propyl cinnamate	293.2	5.45				
$C_{12}H_{14}O_4$	Diethyl phthalate	293.2	7.86				
$C_{12}H_{16}O$	2-Cyclohexylphenol	328.2	3.97				
$C_{12}H_{16}O$	4-Cyclohexylphenol	404.2	4.42				
$C_{12}H_{16}O_2$	Pentyl benzoate	293.2	5.07				
$C_{12}H_{16}O_3$	Pentyl salicylate	301.2	6.25				
$C_{12}H_{16}O_3$	Isopentyl salicylate	293.12	7.26	0.13129E+02	-0.19190E-01	-0.36060E-05	225-397
$C_{12}H_{17}NO$	N-Butyl-N-phenylacetamide	298.2	11.66				
$C_{12}H_{18}$	Hexylbenzene	293.2	2.3				
$C_{12}H_{18}$	1,3,5-Triethylbenzene	293.2	2.256				
$C_{12}H_{18}$	Hexamethylbenzene	449.0	2.172	0.35710E+01	-0.46912E-02	0.35088E-05	449-489
$C_{12}H_{20}O_2$	l-Bornyl acetate	303.2	4.46	0.60791E+01	0.98200E-02	-0.50000E-04	303-323
$C_{12}H_{22}O$	Dicyclohexyl ether	293.2	3.45	0.95324E+01	-0.31740E-01	0.37500E-04	293-333
$C_{12}H_{22}O$	Cyclododecanone	303.2	11.4	0.39327E+02	-0.13248E+00	0.13298E-03	303-423
$C_{12}H_{22}O_6$	Dibutyl tartrate	314.2	9.4				
$C_{12}H_{24}$	1-Dodecene	293.2	2.152	0.22581E+01	0.11106E-02	-0.50000E-05	273-323
$C_{12}H_{24}O_2$	Decyl acetate	293.2	3.75				
$C_{12}H_{24}O_2$	Ethyl decanoate	293.2	3.75	0.70969E+01	-0.15080E-01	0.12500E-04	293-353
$C_{12}H_{24}O_2$	Methyl undecanoate	293.2	3.671				
$C_{12}H_{25}Br$	1-Bromododecane	298.2	4.07	0.86103E+01	-0.20891E-01	0.18994E-04	274-328
$C_{12}H_{25}Cl$	1-Chlorododecane	298.2	4.17	0.10002E+02	-0.27798E-01	0.27559E-04	274-328
$C_{12}H_{25}I$	1-Iodododecane	298.2	3.91	0.34641E+01	0.97404E-02	-0.27602E-04	293-323
$C_{12}H_{26}$	Dodecane	293.2	2.0120	0.23697E+01	-0.12200E-02	-0.36375E-16	283-363
$C_{12}H_{26}O$	1-Dodecanol	303.2	5.82	0.18518E+02	-0.44859E-01	0.99900E-05	303-358
$C_{12}H_{26}O$	2-Butyl-1-octanol	363.2	3.28				
$C_{12}H_{27}BO_3$	Tributyl borate	293.2	2.23				
$C_{12}H_{27}N$	Dodecylamine	303.2	3.07	0.27999E+01	0.44810E-02	-0.11905E-04	303-373
$C_{12}H_{27}N$	Tributylamine	293.2	2.340	0.19846E+01	0.28108E-02	-0.54545E-05	233-293
$C_{12}H_{27}O_4P$	Tributyl phosphate	293.2	8.34	0.26304E+02	-0.88480E-01	0.92857E-04	293-373
$C_{12}H_{28}O_4Si$	Tetrapropoxysilane	298.2	3.21				
$C_{12}H_{28}Sn$	Tetrapropylstannane	293.2	2.267				
$C_{12}H_{30}OSi_2$	Hexaethyldisiloxane	298.2	2.259	0.36559E+01	-0.72406E-02	0.85714E-05	298-333
$C_{13}H_{10}O$	Benzophenone	300.2	12.62	0.34130E+02	-0.10249E+00	0.10268E-03	300-420
$C_{13}H_{10}O_3$	Phenyl salicylate	290.2	6.92	0.26545E+02	-0.11180E+00	0.15220E-03	290-358
$C_{13}H_{12}$	Diphenylmethane	303.2	2.540	0.30638E+01	-0.17286E-02		303-333

Mol. Form.	Name	T/K	ε	a	b	c	Range/K
$C_{13}H_{12}O$	Benzyl phenyl ether	313.2	3.748				
$C_{13}H_{18}O_2$	Hexyl benzoate	293.2	4.80				
$C_{13}H_{20}$	Heptylbenzene	293.2	2.26				
$C_{13}H_{20}O$	α-Ionone*	292.4	10.78				
$C_{13}H_{20}O$	β-Ionone*	297.65	11.66				
$C_{13}H_{24}O_4$	Diethyl nonanedioate	303.2	5.133				
$C_{13}H_{26}$	1-Tridecene	293.2	2.139	0.14154E+01	0.66514E-02	-0.14286E-04	273-323
$C_{13}H_{26}O$	7-Tridecanone	303.2	7.6				
$C_{13}H_{26}O_2$	Ethyl undecanoate	293.2	3.55				
$C_{13}H_{26}O_2$	Methyl dodecanoate	293.2	3.539				
$C_{13}H_{27}Br$	1-Bromotridecane	281.15	4.19				
$C_{13}H_{28}$	Tridecane	293.2	2.0213	0.23731E+01	-0.12000E-02	-0.21841E-15	283-363
$C_{13}H_{28}$	5-Butylnonane	293.2	2.0319				
$C_{13}H_{28}O$	1-Tridecanol	333.2	4.02				
$C_{14}H_{10}$	Anthracene	502.0	2.649	0.20571E+02	-0.69169E-01	0.66667E-04	502-516
$C_{14}H_{10}$	Phenanthrene	383.2	2.72				
$C_{14}H_{10}O_2$	Benzil	368.2	13.04	-0.23599E+02	0.22715E+00	-0.34667E-03	368-393
$C_{14}H_{12}O_2$	Benzyl benzoate	303.2	5.26	0.76856E+01	-0.80000E-02	-0.80361E-15	303-358
$C_{14}H_{12}O_3$	Benzyl salicylate	301.2	4.12				
$C_{14}H_{14}$	1,2-Diphenylethane	331.2	2.47	0.31178E+01	-0.21572E-02	0.59800E-06	331-451
$C_{14}H_{14}O$	Dibenzyl ether	293.2	3.821	0.80154E+01	-0.20536E-01	0.21250E-04	293-333
$C_{14}H_{15}N$	Dibenzylamine	293.2	3.446				
$C_{14}H_{16}O_2Si$	Dimethyldiphenoxysilane	298.2	3.500	0.51669E+01	-0.77001E-02	0.70156E-05	283-353
$C_{14}H_{18}O_2$	Pentyl cinnamate	293.2	4.89				
$C_{14}H_{22}$	Octylbenzene	293.2	2.26				
$C_{14}H_{26}O_4$	Diisobutyl adipate	293.2	5.19				
$C_{14}H_{26}O_4$	Diethyl sebacate	303.2	4.995	0.39143E+02	-0.20965E+00	0.32000E-03	303-313
$C_{14}H_{28}O_2$	Dodecyl acetate	293.2	3.6				
$C_{14}H_{28}O_2$	Ethyl laurate	273.2	3.94				
$C_{14}H_{28}O_2$	Methyl tridecanoate	293.2	3.442				
$C_{14}H_{29}Br$	1-Bromotetradecane	293.2	3.84	0.10058E+02	-0.33905E-01	0.43528E-04	274-328
$C_{14}H_{30}$	Tetradecane	293.2	2.0343	0.23832E+01	-0.11900E-02	-0.51229E-16	283-363
$C_{14}H_{30}O$	1-Tetradecanol	318.2	4.42	0.12272E+02	-0.24667E-01	-0.13168E-13	318-358
$C_{14}H_{31}N$	Tetradecylamine	312.55	2.90				
$C_{15}H_{12}O_4$	Phenyl 2-(acetyloxy)benzoate	384.2	4.33				
$C_{15}H_{26}O_6$	Tributyrin	282.8	5.72	0.13152E+02	-0.36684E-01	0.36795E-04	199-283
$C_{15}H_{30}O_2$	Methyl tetradecanoate	293.2	3.352				
$C_{15}H_{31}Br$	1-Bromopentadecane	293.35	3.88				
$C_{15}H_{32}$	Pentadecane	293.2	2.0391	0.23792E+01	-0.11600E-02	-0.71069E-16	283-363
$C_{15}H_{32}O$	1-Pentadecanol	333.2	3.70				
$C_{15}H_{33}N$	Pentadecylamine	313.25	2.85				
$C_{15}H_{33}N$	Triisopentylamine	294.2	2.29				
$C_{16}H_{22}O_4$	Dibutyl phthalate	293.2	6.58	0.12444E+02	-0.20000E-01		293-333
$C_{16}H_{32}O_2$	Hexadecanoic acid	338.2	2.417				
$C_{16}H_{32}O_2$	Ethyl myristate	293.2	3.50	0.52642E+01	-0.60000E-02	-0.47358E-15	293-353
$C_{16}H_{32}O_2$	Methyl pentadecanoate	293.2	3.296	0.58668E+01	-0.73333E-02	-0.52666E-14	298-328
$C_{16}H_{33}Br$	1-Bromohexadecane	298.2	3.68	0.79531E+01	-0.22859E-01	0.26955E-04	293-323
$C_{16}H_{33}I$	1-Iodohexadecane	293.2	3.57				
$C_{16}H_{34}$	Hexadecane	293.2	2.0460	0.23861E+01	-0.11600E-02	0.25555E-15	293-363
$C_{16}H_{34}O$	1-Hexadecanol	333.2	3.69	0.85935E+01	-0.14714E-01	-0.45533E-13	333-363
$C_{16}H_{35}N$	Hexadecylamine	328.35	2.71				
$C_{16}H_{36}Sn$	Tetrabutylstannane	293.2	9.74	0.56115E+02	-0.24812E+00	0.30682E-03	293-313
$C_{17}H_{12}O_3$	2-Naphthyl salicylate	293.0	6.30	0.11229E+02	-0.18857E-01	0.70332E-05	293-353
$C_{17}H_{34}O$	9-Heptadecanone	328.2	5.43	0.44176E+02	-0.21183E+00	0.28571E-03	328-363
$C_{17}H_{34}O_2$	Methyl palmitate	313.2	3.124				
$C_{17}H_{36}$	Heptadecane	293.2	2.0578	0.23627E+01	-0.10400E-02	-0.10397E-12	293-308
$C_{17}H_{36}O$	1-Heptadecanol	333.2	3.41				
$C_{18}H_{26}O_4$	Dipentyl phthalate	293.2	6.00				
$C_{18}H_{28}O_2$	Phenyl laurate	293.2	3.28				

Mol. Form.	Name	T/K	ε	a	b	c	Range/K
$C_{18}H_{30}O_2$	Linolenic acid	293.2	2.825	0.33867E+01	-0.19181E-02		274-368
$C_{18}H_{30}O_4$	Dicyclohexyl adipate	308.2	4.84				
$C_{18}H_{32}O_2$	Linoleic acid	293.2	2.754	0.32073E+01	-0.15477E-02		275-368
$C_{18}H_{34}O_2$	Oleic acid	293.2	2.336	0.25385E+01	-0.69448E-03		275-368
$C_{18}H_{34}O_4$	Dibutyl sebacate	293.2	4.54				
$C_{18}H_{36}O_2$	Stearic acid	293.2	2.314	0.27159E+01	-0.13300E-02		293-373
$C_{18}H_{36}O_2$	Hexadecyl acetate	308.2	3.19	0.47310E+01	-0.50000E-02	0.41338E-14	308-348
$C_{18}H_{36}O_2$	Ethyl palmitate	303.2	3.07	0.57938E+01	-0.12294E-01	0.10919E-04	303-455
$C_{18}H_{36}O_2$	Methyl heptadecanoate	313.2	3.07				
$C_{18}H_{37}Br$	1-Bromooctadecane	303.35	3.53	0.46790E+01	-0.30355E-02	-0.24798E-05	303-332
$C_{18}H_{38}O$	1-Octadecanol	333.2	3.38	0.73784E+01	-0.12000E-01	-0.22871E-13	333-363
$C_{18}H_{39}BO_3$	Trihexyl borate	293.2	2.22				
$C_{18}H_{39}N$	Octadecylamine	326.35	2.67				
$C_{19}H_{16}$	Triphenylmethane	367.2	2.46	0.40201E+01	-0.66507E-02	0.65329E-05	367-448
$C_{19}H_{18}O_3Si$	Methyltriphenoxysilane	298.2	3.628				
$C_{19}H_{32}O_2$	Methyl linolenate	293.2	3.355				
$C_{19}H_{34}O_2$	Methyl linoleate	293.2	3.466				
$C_{19}H_{36}O_2$	Methyl oleate	293.2	3.211				
$C_{19}H_{38}O$	10-Nonadecanone	353.2	5.37				
$C_{19}H_{38}O_2$	Methyl stearate	313.2	3.021				
$C_{19}H_{40}$	Nonadecane	293.2	2.0706				
$C_{20}H_{30}O_4$	Dihexyl phthalate	293.2	5.62				
$C_{20}H_{38}O_2$	Ethyl oleate	301.2	3.17	0.57033E+01	-0.11223E-01	0.93447E-05	301-423
$C_{20}H_{40}O_2$	Octadecyl acetate	308.2	3.07	0.44569E+01	-0.45000E-02	0.33923E-14	308-348
$C_{20}H_{40}O_2$	Ethyl stearate	313.2	2.958	0.70930E+01	-0.19081E-01	0.19555E-04	331-440
$C_{20}H_{40}O_2$	Methyl nonadecanoate	313.2	2.982				
$C_{20}H_{42}O$	1-Eicosanol	338.2	3.13	0.21700E+01	0.12497E-01	-0.28571E-04	338-363
$C_{20}H_{42}O$	Didecyl ether	293.2	2.644	0.41465E+01	-0.62240E-02	0.37500E-05	293-333
$C_{20}H_{60}O_8Si_9$	Eicosamethylnonasiloxane	293.2	2.645	0.57840E+01	-0.16568E-01	0.20000E-04	293-323
$C_{21}H_{21}O_4P$	Tricresyl phosphate*	298.2	6.7				
$C_{21}H_{38}O_6$	1,2,3-Propanetriyl hexanoate	293.2	4.476				
$C_{22}H_{42}O_2$	Butyl oleate	298.2	4.00				
$C_{22}H_{44}O_2$	Butyl stearate	298.2	3.120	0.73894E+02	-0.46261E+00	0.75500E-03	298-343
$C_{22}H_{46}$	Docosane	293.2	2.0840				
$C_{22}H_{46}O$	1-Docosanol	348.2	2.94	0.82062E+01	-0.25069E-01	0.28571E-04	348-373
$C_{24}H_{20}O_4Si$	Tetraphenoxysilane	333.2	3.4915				
$C_{24}H_{38}O_4$	Dioctyl phthalate	293.2	5.22				
$C_{26}H_{50}O_4$	Dioctyl sebacate	299.2	4.01				
$C_{27}H_{50}O_6$	1,2,3-Propanetriyl octanoate	293.2	3.931				
$C_{30}H_{58}O_4$	Ethylene glycol ditetradecanoate	343.2	2.98				
$C_{30}H_{62}$	Triacontane	373.2	1.9112				
$C_{30}H_{62}$	2,6,10,15,19,23-Hexamethyltetracosane	373.2	1.9106				
$C_{34}H_{66}O_4$	Ethylene glycol dipalmitate	348.2	2.89				
$C_{34}H_{68}O_2$	Hexadecyl stearate	333.2	2.61				
$C_{38}H_{74}O_4$	Ethylene glycol distearate	353.2	2.79				
$C_{39}H_{74}O_6$	Glycerol trilaurate	313.2	3.287				
$C_{51}H_{98}O_6$	Glycerol tripalmitate	328.2	2.901	-0.29131E+01	0.32206E-01	-0.44154E-04	328-393
$C_{57}H_{104}O_6$	Glycerol trioleate	293.2	3.109				
$C_{57}H_{104}O_6$	Glycerol trielaidate	313.2	2.980				
$C_{57}H_{110}O_6$	Glycerol tristearate	353.2	2.740				

PERMITTIVITY (DIELECTRIC CONSTANT) OF GASES

This table gives the permittivity ε (often called the dielectric constant) of some common gases at a temperature of 20°C and pressure of one atmosphere (101.325 kPa). Values of the permanent dipole moment μ in Debye Units (1 D = 3.33564×10^{-30} C m) are also included.

The density dependence of the permittivity is given by the equation

$$\frac{\varepsilon-1}{\varepsilon+2} = \rho_m \left(\frac{4\pi N \alpha}{3} + \frac{4\pi N \mu^2}{9kT} \right)$$

where ρ_m is the molar density, N is Avogadro's number, k is the Boltzmann constant, T is the temperature, and α is the molecular polarizability. Therefore, in regions where the gas can be considered ideal, $\varepsilon - 1$ is approximately proportional to the pressure at constant temperature. For nonpolar gases ($\mu = 0$), $\varepsilon - 1$ is inversely proportional to temperature at constant pressure.

The number of significant figures indicates the accuracy of the values given. The values of ε for air, Ar, H_2, He, N_2, O_2, and CO_2 are recommended as reference values; these are accurate to 1 ppm or better.

The second part of the table gives the permittivity of water vapor in equilibrium with liquid water as a function of temperature (derived from Reference 4).

REFERENCE

1. A. A. Maryott and F. Buckley, *Table of Dielectric Constants and Electric Dipole Moments of Substances in the Gaseous State*, National Bureau of Standards Circular 537, 1953.
2. B. A. Younglove, *J. Phys. Chem. Ref. Data*, 11, Suppl. 1, 1982; 16, 577, 1987 (for data on N_2, H_2, O_2, and hydrocarbons over a range of pressure and temperature).
3. Landolt-Börnstein, *Numerical Data and Functional Relationships in Science and Technology*, New Series, Group IV, Vol. 4, Springer-Verlag, Heidelberg, 1980 (for data at high pressures).
4. G. Birnbaum and S. K. Chatterjee, *J. Appl. Phys.*, 23, 220, 1952 (for data on water vapor).

Mol. form.	Name	ε	μ/D
	Compounds not containing carbon		
	Air (dry, CO_2 free)	1.0005364	
Ar	Argon	1.0005172	0
BF_3	Boron trifluoride	1.0011	0
BrH	Hydrogen bromide	1.00279	0.827
ClH	Hydrogen chloride	1.00390	1.109
F_3N	Nitrogen trifluoride	1.0013	0.235
F_6S	Sulfur hexafluoride	1.00200	0
HI	Hydrogen iodide	1.00214	0.448
H_2	Hydrogen	1.0002538	0
H_2S	Hydrogen sulfide	1.00344	0.97
H_3N	Ammonia	1.00622	1.471
He	Helium	1.0000650	0
Kr	Krypton	1.00078	0
NO	Nitric oxide	1.00060	0.159
N_2	Nitrogen	1.0005480	0
N_2O	Nitrous oxide	1.00104	0.161
Ne	Neon	1.00013	0
O_2	Oxygen	1.0004947	0
O_2S	Sulfur dioxide	1.00825	1.633
O_3	Ozone	1.0017	0.534
Xe	Xenon	1.00126	0
	Compounds containing carbon		
CF_4	Tetrafluoromethane	1.00121	0
CO	Carbon monoxide	1.00262	0.110
CO_2	Carbon dioxide	1.000922	0

PERMITTIVITY (DIELECTRIC CONSTANT) OF GASES (continued)

Mol. form.	Name	ε	μ/D
CH_3Br	Bromomethane	1.01028	1.822
CH_3Cl	Chloromethane	1.01080	1.892
CH_3F	Fluoromethane	1.00973	1.858
CH_3I	Iodomethane	1.00914	1.62
CH_4	Methane	1.00081	0
C_2H_2	Acetylene	1.00124	0
C_2H_3Cl	Chloroethylene	1.0075	1.45
C_2H_4	Ethylene	1.00134	0
C_2H_5Cl	Chloroethane	1.01325	2.05
C_2H_6	Ethane	1.00140	0
C_2H_6O	Dimethyl ether	1.0062	1.30
C_3H_6	Propene	1.00228	0.366
C_3H_6	Cyclopropane	1.00178	0
C_3H_8	Propane	1.00200	0.084
C_4H_{10}	Butane	1.00258	0
C_4H_{10}	Isobutane	1.00260	0.132

PERMITTIVITY OF SATURATED WATER VAPOR

$t/°C$	ε	$t/°C$	ε
0	1.00007	60	1.00144
10	1.00012	70	1.00213
20	1.00022	80	1.00305
30	1.00037	90	1.00428
40	1.00060	100	1.00587
50	1.00095		

AZEOTROPIC DATA FOR BINARY MIXTURES

Liquid mixtures having an extremum (maximum or minimum) vapor pressure at constant temperature, as a function of composition, are called azeotropic mixtures, or simply azeotropes. Mixtures that do not show a maximum or minimum are called zeotropic. Azeotropes in which the pressure is a maximum are often called positive azeotropes, while pressure-minimum azeotropes are called negative azeotropes. The coordinates of an azeotropic point are the azeotropic temperature t_{az}, pressure P_{az}, and liquid-phase composition, usually expressed as mole fractions. At the azeotropic point, the vapor-phase composition is the same as the liquid-phase composition.

This table gives azeotropic data for a number of binary mixtures at normal atmospheric pressure ($P_{az} = 101.3$ kPa). Component 1 of each mixture is given in bold face. The temperature t_{az} and mole fraction x_1 of component 1 are listed for each choice of component 2.

The components are arranged in a modified Hill order, with substances that do not contain carbon preceding those that do contain carbon.

REFERENCES

1. Lide, D.R., and Kehiaian, H.V., *CRC Handbook of Thermophysical and Thermochemical Data*, CRC Press, Boca Raton, FL, 1994.
2. Horsley, L.H., *Azeotropic Data, III*, American Chemical Society, Washington, D.C., 1973.

Molecular formula	Name	$t_{az}/°C$	x_1
Water H$_2$O			
CHCl$_3$	Trichloromethane	56.1	0.160
CH$_2$O$_2$	Formic acid	107.2	0.427
CH$_3$NO$_2$	Nitromethane	83.6	0.511
CS$_2$	Carbon disulfide	42.6	0.109
C$_2$H$_3$N	Acetonitrile	76.5	0.307
C$_2$H$_5$NO$_2$	Nitroethane	87.2	0.624
C$_2$H$_6$O	Ethanol	78.2	0.096
C$_4$H$_8$O$_2$	Ethyl acetate	70.4	0.312
C$_4$H$_{10}$O	1-Butanol	92.7	0.753
C$_4$H$_{10}$O	2-Butanol	87	0.601
C$_5$H$_5$N	Pyridine	93.6	0.755
C$_5$H$_{11}$N	Piperidine	92.8	0.718
C$_5$H$_{12}$	Pentane	34.6	0.054
C$_6$H$_5$Cl	Chlorobenzene	90.2	0.712
C$_6$H$_6$	Benzene	69.3	0.295
C$_6$H$_6$O	Phenol	99.5	0.981
C$_6$H$_{10}$	Cyclohexene	70.8	0.308
C$_6$H$_{12}$	Cyclohexane	69.5	0.300
C$_6$H$_{14}$	Hexane	61.6	0.221
C$_7$H$_8$	Toluene	84.1	0.444
C$_7$H$_{16}$	Heptane	79.2	0.452
C$_8$H$_{10}$	1,3-Dimethylbenzene	92	0.767
C$_8$H$_{10}$	Ethylbenzene	92	0.744
C$_8$H$_{18}$	Octane	89.6	0.673
C$_8$H$_{18}$O	Dibutyl ether	92.9	0.781
C$_9$H$_{20}$	Nonane	94.8	0.970
C$_{12}$H$_{27}$N	Tributylamine	99.7	0.976
Tetrachloromethane CCl$_4$			
CH$_2$O$_2$	Formic acid	66.7	0.569
CH$_3$NO$_2$	Nitromethane	71.3	0.660
CH$_4$O	Methanol	55.7	0.445
C$_2$H$_3$N	Acetonitrile	65.1	0.566
C$_2$H$_6$O	Ethanol	65.0	0.615
C$_3$H$_6$O	Acetone	56.1	0.047
C$_3$H$_8$O	1-Propanol	73.4	0.820
C$_4$H$_{10}$O	1-Butanol	76.6	0.951
Formic acid CH$_2$O$_2$			
CS$_2$	Carbon disulfide	42.6	0.253
Nitromethane CH$_3$NO$_2$			
CS$_2$	Carbon disulfide	41.2	0.845
Methanol CH$_4$O			
C$_3$H$_6$O	Acetone	55.5	0.198

AZEOTROPIC DATA FOR BINARY MIXTURES (continued)

Molecular formula	Name	$t_{az}/°C$	x_1
$C_3H_6O_2$	Methyl acetate	53.5	0.352
C_5H_{10}	Cyclopentane	38.8	0.263
C_5H_{12}	Pentane	30.9	0.145
$C_5H_{12}O$	*tert*-Butyl methyl ether	51.3	0.315
C_6H_6	Benzene	57.5	0.610
C_6H_{12}	Cyclohexane	53.9	0.601
C_7H_8	Toluene	63.5	0.883
C_7H_{16}	Heptane	59.1	0.769
C_8H_{18}	Octane	62.8	0.881
C_9H_{20}	Nonane	64.1	0.953
Carbon disulfide CS$_2$			
C_2H_6O	Ethanol	42.6	0.860
C_3H_6O	Acetone	39.3	0.608
C_3H_8O	1-Propanol	45.7	0.931
$C_4H_8O_2$	Ethyl acetate	46.1	0.974
Acetonitrile C$_2$H$_3$N			
C_2H_6O	Ethanol	72.5	0.469
C_7H_8	Toluene	81.4	0.900
Acetic acid C$_2$H$_4$O$_2$			
$C_4H_8O_2$	1,4-Dioxane	119.5	0.831
C_5H_5N	Pyridine	138.1	0.579
C_6H_6	Benzene	80.1	0.026
C_6H_{12}	Cyclohexane	78.8	0.130
C_6H_{14}	Hexane	68.3	0.084
$C_6H_{15}N$	Triethylamine	163	0.774
C_7H_8	Toluene	100.7	0.375
C_7H_{16}	Heptane	91.7	0.451
C_8H_{10}	Ethylbenzene	114.7	0.774
C_8H_{18}	Octane	105.7	0.688
C_9H_{20}	Nonane	112.9	0.826
Iodoethane C$_2$H$_5$I			
C_6H_{14}	Hexane	64.7	0.420
Ethanol C$_2$H$_6$O			
C_5H_{10}	Cyclopentane	44.7	0.110
C_5H_{12}	Pentane	34.3	0.076
C_6H_6	Benzene	67.9	0.440
C_6H_{12}	Cyclohexane	64.8	0.430
C_6H_{14}	Hexane	58.7	0.332
C_7H_8	Toluene	76.7	0.810
C_8H_{18}	Octane	77	0.898
Ethylene glycol C$_2$H$_6$O$_2$			
C_7H_8	Toluene	110.1	0.034
C_7H_{16}	Heptane	97.9	0.048
$C_8H_{18}O$	Dibutyl ether	139.5	0.125
$C_{10}H_{22}$	Decane	161	0.406
Dimethyl sulfide C$_2$H$_6$S			
C_5H_{12}	Pentane	31.8	0.503
1,2-Ethanediamine C$_2$H$_8$N$_2$			
C_7H_8	Toluene	104	0.406
Propanenitrile C$_3$H$_5$N			
C_6H_{14}	Hexane	63.5	0.134
Acetone C$_3$H$_6$O			
$C_3H_6O_2$	Methyl acetate	55.8	0.544
C_5H_{10}	Cyclopentane	41	0.404
C_6H_{12}	Cyclohexane	53	0.751
Ethyl formate C$_3$H$_6$O$_2$			
C_5H_{12}	Pentane	32.5	0.294
Methyl acetate C$_3$H$_6$O$_2$			
C_6H_{12}	Cyclohexane	55.5	0.801

AZEOTROPIC DATA FOR BINARY MIXTURES (continued)

Molecular formula	Name	$t_{az}/°C$	x_1
C_6H_{14}	Hexane	51.8	0.642
Propanoic acid $C_3H_6O_2$			
C_5H_5N	Pyridine	148.6	0.686
C_7H_{16}	Heptane	97.8	0.027
C_9H_{12}	Propylbenzene	139.5	0.830
1-Nitropropane $C_3H_7NO_2$			
C_3H_8O	1-Propanol	97.0	0.061
C_7H_{16}	Heptane	96.6	0.149
1-Propanol C_3H_8O			
$C_4H_8O_2$	1,4-Dioxane	95.3	0.642
C_6H_6	Benzene	77.1	0.209
C_6H_{12}	Cyclohexane	74.7	0.241
C_7H_{16}	Heptane	84.6	0.470
2-Propanol C_3H_8O			
$C_4H_{11}N$	Butylamine	74.7	0.646
C_5H_{12}	Pentane	35.5	0.071
C_6H_{12}	Cyclohexane	69.4	0.397
C_7H_8	Toluene	80.6	0.773
Ethyl methyl sulfide C_3H_8S			
C_6H_{12}	Methylcyclopentane	65.6	0.664
C_7H_{16}	2,2-Dimethylpentane	66.4	0.908
1-Propanethiol C_3H_8S			
C_6H_{12}	Cyclohexane	67.8	0.978
C_6H_{14}	Hexane	64.4	0.557
$C_6H_{14}O$	Diisopropyl ether	65.9	0.714
Thiophene C_4H_4S			
C_6H_{12}	Cyclohexane	77.9	0.412
C_6H_{14}	Hexane	68.5	0.114
Butanal C_4H_8O			
C_6H_{14}	Hexane	60	0.296
2-Butanone C_4H_8O			
C_4H_9Cl	1-Chlorobutane	77	0.440
$C_4H_{11}N$	Butylamine	74	0.353
C_6H_6	Benzene	78.3	0.460
C_6H_{12}	Cyclohexane	71.8	0.438
C_7H_{16}	Heptane	77	0.764
Butanoic acid $C_4H_8O_2$			
C_5H_5N	Pyridine	163.2	0.912
C_6H_5Cl	Chlorobenzene	131.8	0.035
C_8H_{10}	1,2-Dimethylbenzene	143	0.118
1,4-Dioxane $C_4H_8O_2$			
C_4H_9Br	1-Bromobutane	98	0.580
Ethyl acetate $C_4H_8O_2$			
C_6H_{14}	Hexane	65.2	0.394
Methyl propanoate $C_4H_8O_2$			
C_4H_9Cl	1-Chlorobutane	76.8	0.392
Propyl formate $C_4H_8O_2$			
C_4H_9Cl	1-Chlorobutane	76.1	0.392
C_6H_6	Benzene	78.5	0.440
C_6H_{12}	Cyclohexane	75	0.469
1-Butanol $C_4H_{10}O$			
C_5H_5N	Pyridine	118.6	0.704
C_6H_5Cl	Chlorobenzene	115.3	0.659
C_6H_{10}	Cyclohexene	82	0.055
C_7H_8	Toluene	105.5	0.324
C_7H_{16}	Heptane	93.9	0.229
C_8H_{10}	1,2-Dimethylbenzene	116.8	0.811
$C_8H_{18}O$	Dibutyl ether	117.7	0.892

AZEOTROPIC DATA FOR BINARY MIXTURES (continued)

Molecular formula	Name	$t_{az}/°C$	x_1
	2-Butanol $C_4H_{10}O$		
C_6H_6	Benzene	78.5	0.161
C_7H_{16}	Heptane	88.1	0.439
	Diethyl ether $C_4H_{10}O$		
C_5H_{12}	Pentane	33.7	0.553
	***tert*-Butyl alcohol $C_4H_{10}O$**		
C_6H_6	Benzene	74.0	0.378
C_7H_{16}	Heptane	78	0.688
	Methyl propyl ether $C_4H_{10}O$		
C_5H_{12}	Pentane	35.6	0.215
	2-Ethoxyethanol $C_4H_{10}O_2$		
C_7H_{16}	Heptane	96.5	0.153
C_9H_{12}	Propylbenzene	134.6	0.842
	2-Furaldehyde $C_5H_4O_2$		
C_7H_{16}	Heptane	98.3	0.055
C_9H_{12}	Propylbenzene	151.4	0.475
	Pyridine C_5H_5N		
C_7H_8	Toluene	110.1	0.249
	Benzene C_6H_6		
C_6H_{10}	Cyclohexene	78.9	0.635
C_6H_{12}	Cyclohexane	77.6	0.538
	Phenol C_6H_6O		
C_6H_7N	2-Methylpyridine	185.5	0.752
C_7H_9N	2,4-Dimethylpyridine	193.4	0.601
C_9H_{12}	1,3,5-Trimethylbenzene	163.5	0.253
$C_{10}H_{22}$	Decane	168	0.449
	Aniline C_6H_7N		
C_9H_{12}	1,3,5-Trimethylbenzene	164.4	0.150
$C_{10}H_{22}O$	Dipentyl ether	177.5	0.675
$C_{12}H_{26}$	Dodecane	180.4	0.821
	2-Methylpyridine C_6H_7N		
C_8H_{18}	Octane	121.1	0.470
	Cyclohexanol $C_6H_{12}O$		
C_8H_{10}	1,2-Dimethylbenzene	143	0.147

VISCOSITY OF GASES

The following table gives the viscosity of some common gases as a function of temperature. Unless otherwise noted, the viscosity values refer to a pressure of 100 kPa (1 bar). The notation $P=0$ indicates the low pressure limiting value is given. The difference between the viscosity at 100 kPa and the limiting value is generally less than 1%. Viscosity is given in units of μPa s; note that 1 μPa s = 10^{-5} poise. Substances are listed in the modified Hill order (see Introduction).

		Viscosity in micropascal seconds (μPa s)						
		100 K	200 K	300 K	400 K	500 K	600 K	Ref.
	Air	7.1	13.3	18.6	23.1	27.1	30.8	1
Ar	Argon	8.0	15.9	22.9	28.8	34.2	39.0	2,8
BF$_3$	Boron trifluoride		12.3	17.1	21.7	26.1	30.2	13
ClH	Hydrogen chloride			14.6	19.7	24.3		13
F$_6$S	Sulfur hexafluoride							
	($P=0$)			15.3	19.8	23.9	27.7	10
H$_2$	Hydrogen ($P=0$)	4.2	6.8	9.0	10.9	12.7	14.4	4
D$_2$	Deuterium ($P=0$)	5.9	9.6	12.6	15.4	17.9	20.3	11
H$_2$O	Water			10.0	13.3	17.3	21.4	6
D$_2$O	Deuterium oxide			11.1	13.7	17.7	22.0	7
He	Helium ($P=0$)	9.7	15.3	20.0	24.4	28.4	32.3	8
Kr	Krypton ($P=0$)	8.8	17.1	25.6	33.1	39.8	45.9	8
NO	Nitric oxide		13.8	19.2	23.8	28.0	31.9	13
N$_2$	Nitrogen ($P=0$)		12.9	17.9	22.2	26.1	29.6	12
N$_2$O	Nitrous oxide		10.0	15.0	19.4	23.6	27.4	13
Ne	Neon ($P=0$)	14.4	24.3	32.1	38.9	45.0	50.8	8
O$_2$	Oxygen ($P=0$)	7.5	14.6	20.8	26.1	30.8	35.1	12
O$_2$S	Sulfur dioxide		8.6	12.9	17.5	21.7		13
Xe	Xenon ($P=0$)	8.3	15.4	23.2	30.7	37.6	44.0	8
CO	Carbon monoxide	6.7	12.9	17.8	22.1	25.8	29.1	13
CO$_2$	Carbon dioxide		10.0	15.0	19.7	24.0	28.0	9,10
CHCl$_3$	Chloroform			10.2	13.7	16.9	20.1	13
CH$_4$	Methane		7.7	11.2	14.3	17.0	19.4	10
CH$_4$O	Methanol				13.2	16.5	19.6	13
C$_2$H$_2$	Acetylene			10.4	13.5	16.5		13
C$_2$H$_4$	Ethylene		7.0	10.4	13.6	16.5	19.1	3
C$_2$H$_6$	Ethane		6.4	9.5	12.3	14.9	17.3	5
C$_2$H$_6$O	Ethanol				11.6	14.5	17.0	13
C$_3$H$_8$	Propane			8.3	10.9	13.4	15.8	5
C$_4$H$_{10}$	Butane			7.5	10.0	12.3	14.6	5
C$_4$H$_{10}$	Isobutane			7.6	10.0	12.3	14.6	5
C$_4$H$_{10}$O	Diethyl ether			7.6	10.1	12.4		13
C$_5$H$_{12}$	Pentane			6.7	9.2	11.4	13.4	13
C$_6$H$_{14}$	Hexane				8.6	10.8	12.8	13

REFERENCES

1. K. Kadoya, N. Matsunaga, and A. Nagashima, Viscosity and thermal conductivity of dry air in the gaseous phase, *J. Phys. Chem. Ref. Data*, 14, 947, 1985.
2. B. A. Younglove and H. J. M. Hanley, The viscosity and thermal conductivity coefficients of gaseous and liquid argon, *J. Phys. Chem. Ref. Data*, 15, 1323, 1986.
3. P. M. Holland, B. E. Eaton, and H. J. M. Hanley, A Correlation of the viscosity and thermal conductivity data of gaseous and liquid ethylene, *J. Phys. Chem. Ref. Data*, 12, 917, 1983.
4. M. J. Assael, S. Mixafendi, and W. A. Wakeham, The viscosity and thermal conductivity of normal hydrogen in the limit zero density, *J. Phys. Chem. Ref. Data*, 15, 1315, 1986.
5. B. A. Younglove and J. F. Ely, Thermophysical properties of fluids. II. Methane, ethane, propane, isobutane, and normal butane, *J. Phys. Chem. Ref. Data*, 16, 577, 1987.

6. J. V. Sengers and J. T. R. Watson, Improved international formulations for the viscosity and thermal conductivity of water substance, *J. Phys. Chem. Ref. Data*, 15, 1291, 1986.

7. N. Matsunaga and A. Nagashima, Transport properties of liquid and gaseous D_2O over a wide range of temperature and pressure, *J. Phys. Chem. Ref. Data*, 12, 933, 1983.

8. J. Kestin, et al., Equilibrium and transport properties of the noble gases and their mixtures at low density, *J. Phys. Chem. Ref. Data*, 13, 299, 1984.

9. V. Vescovic, et al., The transport properties of carbon dioxide, *J. Phys. Chem. Ref. Data*, 19, 1990.

10. R. D. Trengove and W. A. Wakeham, The viscosity of carbon dioxide, methane, and sulfur hexafluoride in the limit of zero density, *J. Phys. Chem. Ref. Data*, 16, 175, 1987.

11. M. J. Assael, S. Mixafendi, and W. A. Wakeham, The viscosity of normal deuterium in the limit of zero density, *J. Phys. Chem. Ref. Data*, 16, 189, 1987.

12. W. A. Cole and W. A. Wakeham, The viscosity of nitrogen, oxygen, and their binary mixtures in the limit of zero density, *J. Phys. Chem. Ref. Data*, 14, 209, 1985.

13. C. Y. Ho, Ed., *Properties of Inorganic and Organic Fluids, CINDAS Data Series on Materials Properties*, Vol. V-1, Hemisphere Publishing Corp., New York, 1988.

VISCOSITY OF LIQUIDS

The absolute viscosity of some common liquids at temperatures between –25 and 100°C is given in this table. Values were derived by fitting experimental data to suitable expressions for the temperature dependence. The substances are arranged by molecular formula in the modified Hill order (see Preface). All values are given in units of millipascal seconds (mPa s); this unit is identical to centipoise (cp).

Viscosity values correspond to a nominal pressure of 1 atmosphere. If a value is given at a temperature above the normal boiling point, the applicable pressure is understood to be the vapor pressure of the liquid at that temperature. A few values are given at a temperature slightly below the normal freezing point; these refer to the supercooled liquid.

The accuracy ranges from 1% in the best cases to 5 to 10% in the worst cases. Additional significant figures are included in the table to facilitate interpolation.

REFERENCES

1. Viswanath, D. S. and Natarajan, G., *Data Book on the Viscosity of Liquids*, Hemisphere Publishing Corp., New York, 1989.
2. Daubert, T. E., Danner, R. P., Sibul, H. M., and Stebbins, C. C., *Physical and Thermodynamic Properties of Pure Compounds: Data Compilation*, extant 1994 (core with 4 supplements), Taylor & Francis, Bristol, PA (also available as database).
3. Ho, C. Y., Ed., *CINDAS Data Series on Material Properties*, Vol. V-1, *Properties of Inorganic and Organic Fluids*, Hemisphere Publishing Corp., New York, 1988.
4. Stephan, K. and Lucas, K., *Viscosity of Dense Fluids*, Plenum Press, New York, 1979.
5. Vargaftik, N. B., *Tables of Thermophysical Properties of Liquids and Gases*, 2nd ed., John Wiley, New York, 1975.

Molecular formula	Name	Viscosity in mPa s					
		–25°C	0°C	25°C	50°C	75°C	100°C
		Compounds not containing carbon					
Br_2	Bromine		1.252	0.944	0.746		
Cl_3HSi	Trichlorosilane		0.415	0.326			
Cl_3P	Phosphorous trichloride	0.870	0.662	0.529	0.439		
Cl_4Si	Tetrachlorosilane			99.4	96.2		
H_2O	Water		1.793	0.890	0.547	0.378	0.282
H_4N_2	Hydrazine			0.876	0.628	0.480	0.384
Hg	Mercury			1.526	1.402	1.312	1.245
NO_2	Nitrogen dioxide		0.532	0.402			
		Compounds containing carbon					
CCl_3F	Trichlorofluoromethane	0.740	0.539	0.421			
CCl_4	Tetrachloromethane		1.321	0.908	0.656	0.494	
CS_2	Carbon disulfide		0.429	0.352			
$CHBr_3$	Tribromomethane			1.857	1.367	1.029	
$CHCl_3$	Trichloromethane	0.988	0.706	0.537	0.427		
CHN	Hydrogen cyanide		0.235	0.183			
CH_2Br_2	Dibromomethane	1.948	1.320	0.980	0.779	0.652	
CH_2Cl_2	Dichloromethane	0.727	0.533	0.413			
CH_2O_2	Formic acid			1.607	1.030	0.724	0.545
CH_3I	Iodomethane		0.594	0.469			
CH_3NO	Formamide		7.114	3.343	1.833		
CH_3NO_2	Nitromethane	1.311	0.875	0.630	0.481	0.383	0.317
CH_4O	Methanol	1.258	0.793	0.544			
CH_5N	Methylamine	0.319	0.231				
$C_2Cl_3F_3$	1,1,2-Trichlorotrifluoro-ethane	1.465	0.945	0.656	0.481		
C_2Cl_4	Tetrachloroethylene		1.114	0.844	0.663	0.535	0.442
C_2HCl_3	Trichloroethylene		0.703	0.545	0.444	0.376	
C_2HCl_5	Pentachloroethane		3.761	2.254	1.491	1.061	
$C_2HF_3O_2$	Trifluoroacetic acid			0.808	0.571		
$C_2H_2Cl_2$	cis-1,2-Dichloroethylene	0.786	0.575	0.445			
$C_2H_2Cl_2$	trans-1,2-Dichloroethylene	0.522	0.398	0.317	0.261		
$C_2H_2Cl_4$	1,1,1,2-Tetrachloroethane	3.660	2.200	1.437	1.006	0.741	0.570
$C_2H_3ClF_2$	1-Chloro-1,1-difluoro-ethane	0.477	0.376				

Molecular formula	Name	Viscosity in mPa s					
		−25°C	0°C	25°C	50°C	75°C	100°C
C$_2$H$_3$ClO	Acetyl chloride			0.368	0.294		
C$_2$H$_3$Cl$_3$	1,1,1-Trichloroethane	1.847	1.161	0.793	0.578	0.428	
C$_2$H$_3$N	Acetonitrile		0.400	0.369	0.284	0.234	
C$_2$H$_4$Br$_2$	1,2-Dibromoethane			1.595	1.116	0.837	0.661
C$_2$H$_4$Cl$_2$	1,1-Dichloroethane			0.464	0.362		
C$_2$H$_4$Cl$_2$	1,2-Dichloroethane		1.125	0.779	0.576	0.447	
C$_2$H$_4$O$_2$	Acetic acid			1.056	0.786	0.599	0.464
C$_2$H$_4$O$_2$	Methyl formate		0.424	0.325			
C$_2$H$_5$Br	Bromoethane	0.635	0.477	0.374			
C$_2$H$_5$Cl	Chloroethane	0.416	0.319				
C$_2$H$_5$I	Iodoethane		0.723	0.556	0.444	0.365	
C$_2$H$_5$NO	N-Methylformamide		2.549	1.678	1.155	0.824	0.606
C$_2$H$_5$NO$_2$	Nitroethane	1.354	0.940	0.688	0.526	0.415	0.337
C$_2$H$_6$O	Ethanol	3.262	1.786	1.074	0.694	0.476	
C$_2$H$_6$OS	Dimethyl sulfoxide			1.987	1.290		
C$_2$H$_6$O$_2$	Ethylene glycol			16.1	6.554	3.340	1.975
C$_2$H$_6$S	Dimethyl sulfide		0.356	0.284			
C$_2$H$_6$S	Ethanethiol		0.364	0.287			
C$_2$H$_7$N	Dimethylamine	0.300	0.232				
C$_2$H$_7$NO	Ethanolamine			21.1	8.560	3.935	1.998
C$_3$H$_5$Br	3-Bromopropene		0.620	0.471	0.373		
C$_3$H$_5$Cl	3-Chloropropene		0.408	0.314			
C$_3$H$_5$ClO	Epichlorohydrin	2.492	1.570	1.073	0.781	0.597	0.474
C$_3$H$_5$N	Propanenitrile			0.294	0.240	0.202	
C$_3$H$_6$O	Acetone	0.540	0.395	0.306	0.247		
C$_3$H$_6$O	Allyl alcohol			1.218	0.759	0.505	
C$_3$H$_6$O	Propanal			0.321	0.249		
C$_3$H$_6$O$_2$	Ethyl formate		0.506	0.380	0.300		
C$_3$H$_6$O$_2$	Methyl acetate		0.477	0.364	0.284		
C$_3$H$_6$O$_2$	Propanoic acid		1.499	1.030	0.749	0.569	0.449
C$_3$H$_7$Br	1-Bromopropane		0.645	0.489	0.387		
C$_3$H$_7$Br	2-Bromopropane		0.612	0.458	0.359		
C$_3$H$_7$Cl	1-Chloropropane		0.436	0.334			
C$_3$H$_7$Cl	2-Chloropropane		0.401	0.303			
C$_3$H$_7$I	1-Iodopropane		0.970	0.703	0.541	0.436	0.363
C$_3$H$_7$I	2-Iodopropane		0.883	0.653	0.506	0.407	
C$_3$H$_7$NO	N,N-Dimethylformamide		1.176	0.794	0.624		
C$_3$H$_7$NO$_2$	1-Nitropropane	1.851	1.160	0.798	0.589	0.460	0.374
C$_3$H$_8$O	1-Propanol	8.645	3.815	1.945	1.107	0.685	
C$_3$H$_8$O	2-Propanol		4.619	2.038	1.028	0.576	
C$_3$H$_8$O$_2$	1,2-Propylene glycol		248	40.4	11.3	4.770	2.750
C$_3$H$_8$O$_3$	Glycerol			934	152	39.8	14.8
C$_3$H$_8$S	1-Propanethiol		0.503	0.385			
C$_3$H$_8$S	2-Propanethiol		0.477	0.357	0.280		
C$_3$H$_9$N	Propylamine			0.376			
C$_3$H$_9$N	Isopropylamine		0.454	0.325			
C$_4$H$_4$O	Furan	0.661	0.475	0.361			
C$_4$H$_5$N	Pyrrole		2.085	1.225	0.828	0.612	
C$_4$H$_6$O$_3$	Acetic anhydride		1.241	0.843	0.614	0.472	0.377
C$_4$H$_7$N	Butanenitrile			0.553	0.418	0.330	0.268
C$_4$H$_8$O	2-Butanone	0.720	0.533	0.405	0.315	0.249	
C$_4$H$_8$O	Tetrahydrofuran	0.849	0.605	0.456	0.359		
C$_4$H$_8$O$_2$	1,4-Dioxane			1.177	0.787	0.569	
C$_4$H$_8$O$_2$	Ethyl acetate		0.578	0.423	0.325	0.259	
C$_4$H$_8$O$_2$	Methyl propionate		0.581	0.431	0.333	0.266	
C$_4$H$_8$O$_2$	Propyl formate		0.669	0.485	0.370	0.293	
C$_4$H$_8$O$_2$	Butanoic acid		2.215	1.426	0.982	0.714	0.542

Molecular formula	Name	Viscosity in mPa s					
		−25°C	0°C	25°C	50°C	75°C	100°C
$C_4H_8O_2$	2-Methylpropanoic acid		1.857	1.226	0.863	0.639	0.492
$C_4H_8O_2S$	Sulfolane				6.280	3.818	2.559
C_4H_8S	Tetrahydrothiophene			0.973	0.912		
C_4H_9Br	1-Bromobutane		0.815	0.606	0.471	0.379	
C_4H_9Cl	1-Chlorobutane		0.556	0.422	0.329	0.261	
C_4H_9N	Pyrrolidine	1.914	1.071	0.704	0.512		
C_4H_9NO	N,N-Dimethylacetamide			1.956	1.279	0.896	0.661
C_4H_9NO	Morpholine			2.021	1.247	0.850	0.627
$C_4H_{10}O$	1-Butanol	12.19	5.185	2.544	1.394	0.833	0.533
$C_4H_{10}O$	2-Butanol			3.096	1.332	0.698	0.419
$C_4H_{10}O$	2-Methyl-2-propanol			4.312	1.421	0.678	
$C_4H_{10}O$	Diethyl ether		0.283	0.224			
$C_4H_{10}O_3$	Diethylene glycol			30.200	11.130	4.917	2.505
$C_4H_{10}S$	Diethyl sulfide		0.558	0.422	0.331	0.267	
$C_4H_{11}N$	Butylamine		0.830	0.574	0.409	0.298	
$C_4H_{11}N$	Isobutylamine		0.770	0.571	0.367		
$C_4H_{11}N$	Diethylamine			0.319	0.239		
$C_4H_{11}NO_2$	Diethanolamine				109.5	28.7	9.100
$C_5H_4O_2$	Furfural		2.501	1.587	1.143	0.906	0.772
C_5H_5N	Pyridine		1.361	0.879	0.637	0.497	0.409
C_5H_{10}	1-Pentene	0.313	0.241	0.195			
C_5H_{10}	2-Methyl-2-butene		0.255	0.203			
C_5H_{10}	Cyclopentane		0.555	0.413	0.321		
$C_5H_{10}O$	Mesityl oxide	1.291	0.838	0.602	0.465	0.381	0.326
$C_5H_{10}O$	2-Pentanone		0.641	0.470	0.362	0.289	0.238
$C_5H_{10}O$	3-Pentanone		0.592	0.444	0.345	0.276	0.227
$C_5H_{10}O_2$	Butyl formate		0.937	0.644	0.472	0.362	0.289
$C_5H_{10}O_2$	Propyl acetate		0.768	0.544	0.406	0.316	0.255
$C_5H_{10}O_2$	Ethyl propanoate		0.691	0.501	0.380	0.299	0.242
$C_5H_{10}O_2$	Methyl butanoate		0.759	0.541	0.406	0.318	0.257
$C_5H_{10}O_2$	Methyl isobutanoate		0.672	0.488	0.373	0.296	
$C_5H_{11}N$	Piperidine			1.573	0.958	0.649	0.474
C_5H_{12}	Pentane	0.351	0.274	0.224			
C_5H_{12}	Isopentane	0.376	0.277	0.214			
$C_5H_{12}O$	1-Pentanol	25.4	8.512	3.619	1.820	1.035	0.646
$C_5H_{12}O$	2-Pentanol			3.470	1.447	0.761	0.465
$C_5H_{12}O$	3-Pentanol			4.149	1.473	0.727	0.436
$C_5H_{12}O$	2-Methyl-1-butanol			4.453	1.963	1.031	0.612
$C_5H_{12}O$	3-Methyl-1-butanol		8.627	3.692	1.842	1.031	0.631
$C_5H_{13}N$	Pentylamine		1.030	0.702	0.493	0.356	
C_6F_6	Hexafluorobenzene			2.789	1.730	1.151	
$C_6H_4Cl_2$	o-Dichlorobenzene		1.958	1.324	0.962	0.739	0.593
$C_6H_4Cl_2$	m-Dichlorobenzene		1.492	1.044	0.787	0.628	0.525
C_6H_5Br	Bromobenzene		1.560	1.074	0.798	0.627	0.512
C_6H_5Cl	Chlorobenzene	1.703	1.058	0.753	0.575	0.456	0.369
C_6H_5ClO	o-Chlorophenol			3.589	1.835	1.131	0.786
C_6H_5ClO	m-Chlorophenol			4.041			
C_6H_5F	Fluorobenzene		0.749	0.550	0.423	0.338	
C_6H_5I	Iodobenzene		2.354	1.554	1.117	0.854	0.683
$C_6H_5NO_2$	Nitrobenzene		3.036	1.863	1.262	0.918	0.704
C_6H_6	Benzene			0.604	0.436	0.335	
C_6H_6ClN	o-Chloroaniline			3.316	1.913	1.248	0.887
C_6H_6O	Phenol				3.437	1.784	1.099
C_6H_7N	Aniline			3.847	2.029	1.247	0.850
$C_6H_8N_2$	Phenylhydrazine			13.0	4.553	1.850	0.848
C_6H_{10}	Cyclohexene		0.882	0.625	0.467	0.364	
$C_6H_{10}O$	Cyclohexanone			2.017	1.321	0.919	0.671

Molecular formula	Name	Viscosity in mPa s					
		−25°C	0°C	25°C	50°C	75°C	100°C
$C_6H_{11}N$	Hexanenitrile			0.912	0.650	0.488	0.382
C_6H_{12}	Cyclohexane			0.894	0.615	0.447	
C_6H_{12}	Methylcyclopentane	0.927	0.653	0.479	0.364		
C_6H_{12}	1-Hexene	0.441	0.326	0.252	0.202		
$C_6H_{12}O$	Cyclohexanol			57.5	12.3	4.274	1.982
$C_6H_{12}O$	2-Hexanone	1.300	0.840	0.583	0.429	0.329	0.262
$C_6H_{12}O$	4-Methyl-2-pentanone			0.545	0.406		
$C_6H_{12}O_2$	Butyl acetate		1.002	0.685	0.500	0.383	0.305
$C_6H_{12}O_2$	Isobutyl acetate			0.676	0.493	0.370	0.286
$C_6H_{12}O_2$	Ethyl butanoate			0.639	0.453		
$C_6H_{12}O_2$	Diacetone alcohol	28.7	6.621	2.798	1.829	1.648	
$C_6H_{12}O_3$	Paraldehyde			1.079	0.692	0.485	0.362
$C_6H_{13}N$	Cyclohexylamine			1.944	1.169	0.782	0.565
C_6H_{14}	Hexane		0.405	0.300	0.240		
C_6H_{14}	2-Methylpentane		0.372	0.286	0.226		
C_6H_{14}	3-Methylpentane		0.395	0.306			
$C_6H_{14}O$	Dipropyl ether		0.542	0.396	0.304	0.242	
$C_6H_{14}O$	1-Hexanol			4.578	2.271	1.270	0.781
$C_6H_{15}N$	Triethylamine		0.455	0.347	0.273	0.221	
$C_6H_{15}N$	Dipropylamine		0.751	0.517	0.377	0.288	0.228
$C_6H_{15}N$	Diisopropylamine			0.393	0.300	0.237	
$C_6H_{15}NO_3$	Triethanolamine			609	114	31.5	11.7
C_7H_5N	Benzonitrile			1.267	0.883	0.662	0.524
C_7H_7Cl	o-Chlorotoluene		1.390	0.964	0.710	0.547	0.437
C_7H_7Cl	m-Chlorotoluene		1.165	0.823	0.616	0.482	0.391
C_7H_7Cl	p-Chlorotoluene			0.837	0.621	0.483	0.390
C_7H_8	Toluene	1.165	0.778	0.560	0.424	0.333	0.270
C_7H_8O	o-Cresol				3.035	1.562	0.961
C_7H_8O	m-Cresol			12.9	4.417	2.093	1.207
C_7H_8O	Benzyl alcohol			5.474	2.760	1.618	1.055
C_7H_8O	Anisole			1.056	0.747	0.554	0.427
C_7H_9N	N-Methylaniline		4.120	2.042	1.222	0.825	0.606
C_7H_9N	o-Methyl aniline		10.3	3.823	1.936	1.198	0.839
C_7H_9N	m-Methyl aniline		8.180	3.306	1.679	1.014	0.699
C_7H_9N	Benzylamine			1.624	1.080	0.769	0.577
C_7H_{14}	Methylcyclohexane		0.991	0.679	0.501	0.390	0.316
C_7H_{14}	1-Heptene		0.441	0.340	0.273	0.226	
$C_7H_{14}O$	2-Heptanone			0.714	0.407	0.297	
$C_7H_{14}O_2$	Heptanoic acid			3.840	2.282	1.488	1.041
C_7H_{16}	Heptane	0.757	0.523	0.387	0.301	0.243	
C_7H_{16}	3-Methylhexane			0.350			
$C_7H_{16}O$	1-Heptanol			5.810	2.603	1.389	0.849
$C_7H_{16}O$	2-Heptanol			3.955	1.799	0.987	0.615
$C_7H_{16}O$	3-Heptanol				1.957	0.976	0.584
$C_7H_{16}O$	4-Heptanol			4.207	1.695	0.882	0.539
$C_7H_{17}N$	Heptylamine			1.314	0.865	0.600	0.434
C_8H_8	Styrene		1.050	0.695	0.507	0.390	0.310
C_8H_8O	Acetophenone			1.681			0.634
$C_8H_8O_2$	Methyl benzoate			1.857			
$C_8H_8O_3$	Methyl salicylate					1.102	0.815
C_8H_{10}	Ethylbenzene		0.872	0.631	0.482	0.380	0.304
C_8H_{10}	o-Xylene		1.084	0.760	0.561	0.432	0.345
C_8H_{10}	m-Xylene		0.795	0.581	0.445	0.353	0.289
C_8H_{10}	p-Xylene			0.603	0.457	0.359	0.290
$C_8H_{10}O$	Phenetole			1.197	0.817	0.594	0.453
$C_8H_{11}N$	N,N-Dimethylaniline		1.996	1.300	0.911	0.675	0.523
$C_8H_{11}N$	N-Ethylaniline		3.981	2.047	1.231	0.825	0.596

VISCOSITY OF LIQUIDS (continued)

Molecular formula	Name	Viscosity in mPa s					
		−25°C	0°C	25°C	50°C	75°C	100°C
C_8H_{16}	Ethylcyclohexane		1.139	0.784	0.579		
$C_8H_{16}O_2$	Octanoic acid			5.020	2.656	1.654	1.147
C_8H_{18}	Octane		0.700	0.508	0.385	0.302	0.243
$C_8H_{18}O$	1-Octanol			7.288	3.232	1.681	0.991
$C_8H_{18}O$	4-Methyl-3-heptanol		1.904	1.085	0.702	0.497	0.375
$C_8H_{18}O$	5-Methyl-3-heptanol		2.052	1.178	0.762	0.536	0.401
$C_8H_{18}O$	2-Ethyl-1-hexanol		20.7	6.271	2.631	1.360	0.810
$C_8H_{18}O$	Dibutyl ether	1.417	0.918	0.637	0.466	0.356	0.281
$C_8H_{19}N$	Dibutylamine		1.509	0.918	0.619	0.449	0.345
$C_8H_{19}N$	Diisobutylamine		1.115	0.723	0.511	0.384	0.303
C_9H_7N	Quinoline			3.337	1.892	1.201	0.833
C_9H_{10}	Indane		2.230	1.357	0.931	0.692	0.545
C_9H_{12}	Cumene		1.075	0.737	0.547		
$C_9H_{14}O$	Isophorone		4.201	2.329	1.415	0.923	0.638
$C_9H_{18}O$	5-Nonanone			1.199	0.834	0.619	0.484
$C_9H_{18}O_2$	Nonanoic acid			7.011	3.712	2.234	1.475
C_9H_{20}	Nonane		0.964	0.665	0.488	0.375	0.300
$C_9H_{20}O$	1-Nonanol			9.123	4.032		
$C_{10}H_{10}O_4$	Dimethyl phthalate		63.2	14.4	5.309	2.824	1.980
$C_{10}H_{14}$	Butylbenzene			0.950	0.683	0.515	
$C_{10}H_{18}$	cis-Decahydronaphthalene	12.8	5.645	3.042	1.875	1.271	0.924
$C_{10}H_{18}$	trans-Decahydronaphthalene	6.192	3.243	1.948	1.289	0.917	0.689
$C_{10}H_{20}O_2$	Decanoic acid			4.327	2.651		
$C_{10}H_{22}$	Decane	2.188	1.277	0.838	0.598	0.453	0.359
$C_{10}H_{22}O$	1-Decanol			10.9	4.590		
$C_{11}H_{24}$	Undecane		1.707	1.098	0.763	0.562	0.433
$C_{12}H_{10}O$	Diphenyl ether			2.130	1.407	1.023
$C_{12}H_{26}$	Dodecane		2.277	1.383	0.930	0.673	0.514
$C_{13}H_{12}$	Diphenylmethane					1.265	0.929
$C_{13}H_{28}$	Tridecane		2.909	1.724	1.129	0.796	0.594
$C_{14}H_{30}$	Tetradecane			2.128	1.376	0.953	0.697
$C_{16}H_{22}O_4$	Dibutyl phthalate	483	66.4	16.6	6.470	3.495	2.425
$C_{16}H_{34}$	Hexadecane			3.032	1.879	1.260	0.899
$C_{18}H_{38}$	Octadecane				2.487	1.609	1.132

THERMAL CONDUCTIVITY OF GASES

This table gives the thermal conductivity of several gases as a function of temperature. Unless otherwise noted, the values refer to a pressure of 100 kPa (1 bar) or to the saturation vapor pressure if that is less than 100 kPa. The notation $P = 0$ indicates the low pressure limiting value is given. In general, the $P = 0$ and $P = 100$ kPa values differ by less than 1%. Units are milliwatts per meter kelvin. Substances are listed in the modified Hill order.

| MF | Name | Thermal conductivity in mW/m K | | | | | | Ref. |
		100 K	200 K	300 K	400 K	500 K	600 K	
	Air	9.4	18.4	26.2	33.3	39.7	45.7	1
Ar	Argon	6.2	12.4	17.9	22.6	26.8	30.6	2,8
BF_3	Boron trifluoride			19.0	24.6			11
ClH	Hydrogen chloride		9.2	14.5	19.5	24.0	28.1	11
F_6S	Sulfur hexafluoride ($P = 0$)			13.0	20.6	27.5	33.8	16
H_2	Hydrogen ($P = 0$)	68.6	131.7	186.9	230.4			4
H_2O	Water			18.7	27.1	35.7	47.1	6
	Deuterium oxide				27.0	36.5	47.6	7
H_2S	Hydrogen sulfide			14.6	20.5	26.4	32.4	11
H_3N	Ammonia			24.4	37.4	51.6	66.8	11
He	Helium ($P = 0$)	75.5	119.3	156.7	190.6	222.3	252.4	8
Kr	Krypton ($P = 0$)	3.3	6.4	9.5	12.3	14.8	17.1	8
NO	Nitric oxide		17.8	25.9	33.1	39.6	46.2	11
N_2	Nitrogen	9.8	18.7	26.0	32.3	38.3	44.0	12
N_2O	Nitrous oxide		9.8	17.4	26.0	34.1	41.8	11
Ne	Neon ($P = 0$)	22.3	37.6	49.8	60.3	69.9	78.7	8
O_2	Oxygen	9.3	18.4	26.3	33.7	41.0	48.1	10
O_2S	Sulfur dioxide			9.6	14.3	20.0	25.6	11
Xe	Xenon ($P = 0$)	2.0	3.6	5.5	7.3	8.9	10.4	8
CCl_2F_2	Dichlorodifluoromethane			9.9	15.0	20.1	25.2	13
CF_4	Tetrafluoromethane ($P = 0$)			16.0	24.1	32.2	39.9	16
CO	Carbon monoxide ($P = 0$)			25.0	32.3	39.2	45.7	14
CO_2	Carbon dioxide		9.6	16.8	25.1	33.5	41.6	9
$CHCl_3$	Trichloromethane			7.5	11.1	15.1		11
CH_4	Methane		22.5	34.1	49.1	66.5	84.1	5,15
CH_4O	Methanol				26.2	38.6	53.0	11
$C_2Cl_2F_4$	1,2-Dichlorotetrafluoro-ethane			10.25	15.7	21.1		13
$C_2Cl_3F_3$	1,1,2-Trichlorotrifluoro-ethane			9.0	13.6	18.3		13
C_2H_2	Acetylene			21.4	33.3	45.4	56.8	11
C_2H_4	Ethylene		11.1	20.5	34.6	49.9	68.6	3
C_2H_6	Ethane		11.0	21.3	35.4	52.2	70.5	5
C_2H_6O	Ethanol			14.4	25.8	38.4	53.2	11
C_3H_6O	Acetone			11.5	20.2	30.6	42.7	11
C_3H_8	Propane			18.0	30.6	45.5	61.9	5
C_4F_8	Perfluorocyclobutane			12.5	19.5			13
C_4H_{10}	Butane			16.4	28.4	43.0	59.1	5
C_4H_{10}	Isobutane			16.1	27.9	42.1	57.6	5
$C_4H_{10}O$	Diethyl ether			15.1	25.0	37.1		11
C_5H_{12}	Pentane			14.4	24.9	37.8	52.7	11
C_6H_{14}	Hexane				23.4	35.4	48.7	11

REFERENCES

1. Kadoya, K. Matsunaga, N., and Nagashima, A., Viscosity and thermal conductivity of dry air in the gaseous phase, *J. Phys. Chem. Ref. Data*, 14, 947, 1985.
2. Younglove, B. A. and Hanley, H. J. M., The viscosity and thermal conductivity coefficients of gaseous and liquid argon, *J. Phys. Chem. Ref. Data*, 15, 1323, 1986.
3. Holland, P. M., Eaton, B. E., and Hanley, H. J. M., A correlation of the viscosity and thermal conductivity data of gaseous and liquid ethylene, *J. Phys. Chem. Ref. Data*, 12, 917, 1983.

4. Assael, M. J., Mixafendi, S., and Wakeham, W. A., The viscosity and thermal conductivity of normal hydrogen in the limit of zero density, *J. Phys. Chem. Ref. Data*, 15, 1315, 1986.
5. Younglove, B. A. and Ely, J. F., Thermophysical properties of fluids. II. Methane, ethane, propane, isobutane, and normal butane, *J. Phys. Chem. Ref. Data*, 16, 577, 1987.
6. Sengers, J. V. and Watson, J. T. R., Improved international formulations for the viscosity and thermal conductivity of water substance, *J. Phys. Chem. Ref. Data*, 15, 1291, 1986.
7. Matsunaga, N. and Nagashima, A., Transport properties of liquid and gaseous D_2O over a wide range of temperature and pressure, *J. Phys. Chem. Ref. Data*, 12, 933, 1983.
8. Kestin, J. et al., Equilibrium and transport properties of the noble gases and their mixtures at low density, *J. Phys. Chem. Ref. Data*, 13, 229, 1984.
9. Vescovic, V. et al., The transport properties of carbon dioxide, *J. Phys. Chem. Ref. Data*, 19, 1990.
10. Younglove, B. A., Thermophysical properties of fluids. I. Argon, ethylene, parahydrogen, nitrogen, nitrogen trifluoride, and oxygen, *J. Phys. Chem. Ref. Data*, 11, Suppl. 1, 1982.
11. Ho, C. Y., Ed., *Properties of Inorganic and Organic Fluids, CINDAS Data Series on Materials Properties*, Volume V-1, Hemisphere Publishing Corp., New York, 1988.
12. Stephen, K., Krauss, R., and Laesecke, A., Viscosity and thermal conductivity of nitrogen for a wide range of fluid states, *J. Phys. Chem. Ref. Data*, 16, 993, 1987.
13. Krauss, R. and Stephan, K., Thermal conductivity of refrigerants in a wide range of temperature and pressure, *J. Phys. Chem. Ref. Data*, 18, 43, 1989.
14. Millat, J. and Wakeham, W. A., The thermal conductivity of nitrogen and carbon monoxide in the limit of zero density, *J. Phys. Chem. Ref. Data*, 18, 565, 1989.
15. Friend, D. G., Ely, J. F., and Ingham, H., Thermophysical properties of methane, *J. Phys. Chem. Ref. Data*, 18, 583, 1989.
16. Uribe, F. J., Mason, E. A., and Kestin, J., Thermal conductivity of nine polyatomic gases at low density, *J. Phys. Chem. Ref. Data*, 19, 1123, 1990.

THERMAL CONDUCTIVITY OF LIQUIDS

This table gives the thermal conductivity of some common liquids at temperatures between -25 and 100°C. All values are given in units of watts per meter kelvin (W/m K). Values refer to nominal atmospheric pressure (about 100 kPa); when an entry is given at a temperature above the normal boiling point of the substance, the pressure is understood to be the saturation vapor pressure at that temperature.

Substances are arranged by molecular formula in the modified Hill order, with compounds not containing carbon preceding those that do contain carbon.

The values for water, benzene, toluene, heptane, and dimethyl phthalate are particularly well determined and can be used for calibration purposes.

REFERENCES

1. Daubert, T. E., Danner, R. P., Sibul, H. M., and Stebbins, C. C., *Physical and Thermodynamic Properties of Pure Compounds: Data Compilation*, extant 1994 (core with 4 supplements), Taylor & Francis, Bristol, PA (also available as database).
2. Marsh, K. N., Ed., *Recommended Reference Materials for the Realization of Physicochemical Properties*, Blackwell Scientific Publications, Oxford, 1987.

Molecular formula	Name	Thermal conductivity in W/m K					
		–25°C	0°C	25°C	50°C	75°C	100°C
Cl_4Si	Silicon tetrachloride			0.099	0.096		
H_2O	Water		0.5610	0.6071	0.6435	0.6668	0.6791
Hg	Mercury	7.25	7.77	8.25	8.68	9.07	9.43
CCl_4	Tetrachloromethane		0.104	0.099	0.093	0.088	
CS_2	Carbon disulfide		0.154	0.149			
$CHCl_3$	Trichloromethane	0.127	0.122	0.117	0.112	0.107	0.102
CH_2Br_2	Dibromomethane	0.120	0.114	0.108	0.103	0.097	
CH_4O	Methanol	0.214	0.207	0.200	0.193		
C_2Cl_4	Tetrachloroethylene		0.117	0.110	0.104	0.097	0.091
C_2HCl_3	Trichloroethylene	0.133	0.124	0.116	0.108	0.100	
$C_2H_3Cl_3$	1,1,1-Trichloroethane		0.106	0.101	0.096		
C_2H_3N	Acetonitrile	0.208	0.198	0.188	0.178	0.168	
$C_2H_4O_2$	Acetic acid			0.158	0.153	0.149	0.144
C_2H_5Cl	Chloroethane	0.145	0.132	0.119	0.106	0.093	
C_2H_5NO	N-Methylformamide			0.203	0.201	0.199	0.196
C_2H_6O	Ethanol		0.176	0.169	0.162		
$C_2H_6O_2$	Ethylene glycol		0.256	0.256	0.256	0.256	0.256
C_2H_7NO	Ethanolamine			0.299	0.286	0.274	0.261
C_3H_5ClO	Epichlorohydrin	0.142	0.137	0.131	0.125	0.119	0.114
C_3H_6O	Acetone		0.169	0.161			
$C_3H_6O_2$	Methyl acetate	0.174	0.164	0.153	0.143	0.133	0.122
C_3H_7NO	N,N-Dimethylformamide			0.184	0.178	0.171	0.165
C_3H_8O	1-Propanol	0.162	0.158	0.154	0.149	0.145	0.141
C_3H_8O	2-Propanol	0.146	0.141	0.135	0.129	0.124	0.118
$C_3H_8O_2$	1,2-Propanediol		0.202	0.200	0.199	0.198	0.197
$C_3H_8O_3$	Glycerol			0.292	0.295	0.297	0.300
C_3H_9N	Trimethylamine	0.143	0.133				
C_4H_4O	Furan	0.142	0.134	0.126			
C_4H_4S	Thiophene			0.199	0.195	0.191	0.186
C_4H_6	2-Butyne	0.137	0.129	0.121			
C_4H_8O	2-Butanone	0.158	0.151	0.145	0.139	0.133	
C_4H_8O	Tetrahydrofuran	0.132	0.126	0.120	0.114		
$C_4H_8O_2$	1,4-Dioxane			0.159	0.147	0.135	0.123
$C_4H_8O_2$	Ethyl acetate	0.162	0.153	0.144	0.135	0.126	
$C_4H_{10}O$	1-Butanol		0.158	0.154	0.149		
$C_4H_{10}O$	Diethyl ether	0.150	0.140	0.130	0.120	0.110	0.100
C_5H_5N	Pyridine		0.169	0.165	0.161	0.158	
C_5H_8	Cyclopentene	0.143	0.136	0.129			
C_5H_{10}	1-Pentene	0.131	0.124	0.116			
C_5H_{10}	Cyclopentane	0.140	0.133	0.126			
C_5H_{12}	Pentane	0.132	0.122	0.113	0.103	0.095	0.087
$C_5H_{12}O$	1-Pentanol		0.157	0.153	0.149	0.145	
C_6H_5Cl	Chlorobenzene	0.136	0.131	0.127	0.122	0.117	0.112

THERMAL CONDUCTIVITY OF LIQUIDS (continued)

Molecular formula	Name	Thermal conductivity in W/m K					
		−25°C	0°C	25°C	50°C	75°C	100°C
C_6H_6	Benzene			0.1411	0.1329	0.1247	
C_6H_6O	Phenol				0.156	0.153	0.151
C_6H_{10}	Cyclohexene	0.142	0.136	0.130	0.124	0.118	
$C_6H_{10}O$	Mesityl oxide	0.170	0.163	0.156	0.149	0.142	0.134
C_6H_{12}	Cyclohexane			0.123	0.117	0.111	
C_6H_{12}	1-Hexene	0.137	0.129	0.121	0.113		
$C_6H_{12}O$	Cyclohexanol			0.134	0.131		
$C_6H_{12}O$	2-Hexanone	0.151	0.145	0.139	0.133	0.127	0.121
C_6H_{14}	Hexane	0.137	0.128	0.120	0.111	0.102	0.093
$C_6H_{14}O$	1-Hexanol	0.159	0.154	0.150	0.145	0.141	0.137
C_7H_6O	Benzaldehyde			0.151	0.141	0.131	0.121
C_7H_8	Toluene	0.1461	0.1386	0.1311	0.1236	0.1161	
C_7H_8O	Anisole	0.170	0.163	0.156	0.150	0.143	0.136
C_7H_{16}	Heptane	0.1378	0.1303	0.1228	0.1152	0.1077	
$C_7H_{16}O$	1-Heptanol		0.166	0.159	0.153	0.147	0.141
C_8H_8	Styrene	0.148	0.142	0.137	0.131	0.126	0.120
C_8H_{10}	Ethylbenzene			0.130	0.124	0.118	0.112
C_8H_{10}	o-Xylene			0.131	0.126	0.120	0.114
C_8H_{10}	m-Xylene			0.130	0.124	0.118	0.113
C_8H_{10}	p-Xylene			0.130	0.124	0.118	0.112
C_8H_{18}	Octane	0.143	0.135	0.128	0.120	0.113	0.106
$C_8H_{18}O$	1-Octanol		0.168	0.161	0.154	0.147	0.141
C_9H_{12}	Cumene			0.128	0.120	0.112	0.107
C_9H_{12}	Mesitylene	0.147	0.141	0.136	0.130	0.124	0.118
C_9H_{20}	Nonane	0.144	0.138	0.131	0.124	0.118	0.111
$C_9H_{20}O$	1-Nonanol		0.166	0.161	0.155	0.149	0.143
$C_{10}H_{10}O_4$	Dimethyl phthalate		0.1501	0.1473	0.1443	0.1409	0.1373
$C_{10}H_{14}$	p-Cymene	0.132	0.127	0.122	0.117	0.112	0.107
$C_{10}H_{22}$	Decane	0.144	0.138	0.132	0.126	0.119	0.113
$C_{10}H_{22}O$	1-Decanol			0.162	0.156	0.150	0.145
$C_{11}H_{24}$	Undecane			0.140	0.135	0.129	0.123
$C_{12}H_{10}O$	Diphenyl ether				0.139	0.135	0.131
$C_{12}H_{26}$	Dodecane		0.157	0.152	0.146	0.140	0.135
$C_{12}H_{26}O$	1-Dodecanol			0.146	0.142	0.139	0.135
$C_{13}H_{28}$	Tridecane			0.137	0.132	0.127	0.122
$C_{14}H_{30}$	Tetradecane			0.136	0.131	0.126	0.121
$C_{14}H_{30}O$	1-Tetradecanol				0.167	0.162	0.157
$C_{16}H_{22}O_4$	Dibutyl phthalate	0.144	0.140	0.136	0.133	0.129	0.125
$C_{16}H_{34}$	Hexadecane			0.140	0.135	0.130	0.125
$C_{18}H_{38}$	Octadecane				0.146	0.142	0.137

DIFFUSION IN GASES

This table gives binary diffusion coefficients D_{12} for a number of common gases as a function of temperature. Values refer to atmospheric pressure. The diffusion coefficient is inversely proportional to pressure as long as the gas is in a regime where binary collisions dominate. See Reference 1 for a discussion of the dependence of D_{12} on temperature and composition.

The first part of the table gives data for several gases in the presence of a large excess of air. The remainder applies to equimolar mixtures of gases. Each gas pair is ordered alphabetically according to the most common way of writing the formula. The listing of pairs then follows alphabetical order by the first constituent.

REFERENCES

1. Marrero, T. R., and Mason, E. A., *J. Phys. Chem. Ref. Data*, 1, 1, 1972.
2. Kestin, J., et al., *J. Phys. Chem. Ref. Data*, 13, 229, 1984.

$D_{12}/cm^2\,s^{-1}$ for $p = 101.325$ kPa and the Specified T/K

System	200	273.15	293.15	373.15	473.15	573.15	673.15
Large Excess of Air							
Ar-air		0.167	0.148	0.289	0.437	0.612	0.810
CH_4-air			0.106	0.321	0.485	0.678	0.899
CO-air			0.208	0.315	0.475	0.662	0.875
CO_2-air			0.160	0.252	0.390	0.549	0.728
H_2-air		0.668	0.627	1.153	1.747	2.444	3.238
H_2O-air			0.242	0.399	0.638	0.873	1.135
He-air		0.617	0.580	1.057	1.594	2.221	2.933
SF_6-air				0.150	0.233	0.329	0.438
Equimolar Mixture							
Ar-CH_4				0.306	0.467	0.657	0.876
Ar-CO		0.168	0.187	0.290	0.439	0.615	0.815
Ar-CO_2		0.129	0.078	0.235	0.365	0.517	0.689
Ar-H_2		0.698	0.794	1.228	1.876	2.634	3.496
Ar-He	0.381	0.645	0.726	1.088	1.617	2.226	2.911
Ar-Kr	0.064	0.117	0.134	0.210	0.323	0.456	0.605
Ar-N_2		0.168	0.190	0.290	0.439	0.615	0.815
Ar-Ne	0.160	0.277	0.313	0.475	0.710	0.979	1.283
Ar-O_2		0.166	0.189	0.285	0.430	0.600	0.793
Ar-SF_6				0.128	0.202	0.290	0.389
Ar-Xe	0.052	0.095	0.108	0.171	0.264	0.374	0.498
CH_4-H_2			0.782	1.084	1.648	2.311	3.070
CH_4-He			0.723	0.992	1.502	2.101	2.784
CH_4-N_2			0.220	0.317	0.480	0.671	0.890
CH_4-O_2			0.210	0.341	0.523	0.736	0.978
CH_4-SF_6				0.167	0.257	0.363	0.482
CO-CO_2			0.162	0.250	0.384		
CO-H_2	0.408	0.686	0.772	1.162	1.743	2.423	3.196
CO-He	0.365	0.619	0.698	1.052	1.577	2.188	2.882
CO-Kr		0.131	0.581	0.227	0.346	0.485	0.645
CO-N_2	0.133	0.208	0.231	0.336	0.491	0.673	0.878
CO-O_2			0.202	0.307	0.462	0.643	0.849
CO-SF_6				0.144	0.226	0.323	0.432
CO_2-C_3H_8			0.084	0.133	0.209		
CO_2-H_2	0.315	0.552	0.412	0.964	1.470	2.066	2.745
CO_2-H_2O			0.162	0.292	0.496	0.741	1.021
CO_2-He	0.300	0.513	0.400	0.878	1.321		
CO_2-N_2			0.160	0.253	0.392	0.553	0.733
CO_2-N_2O	0.055	0.099	0.113	0.177	0.276		
CO_2-Ne	0.131	0.227	0.199	0.395	0.603	0.847	

DIFFUSION IN GASES (continued)

System	200	273.15	293.15	373.15	473.15	573.15	673.15
CO_2-O_2			0.159	0.248	0.380	0.535	0.710
CO_2-SF_6			0.099	0.155			
D_2-H_2	0.631	1.079	1.219	1.846	2.778	3.866	5.103
H_2-He	0.775	1.320	1.490	2.255	3.394	4.726	6.242
H_2-Kr	0.340	0.601	0.682	1.053	1.607	2.258	2.999
H_2-N_2	0.408	0.686	0.772	1.162	1.743	2.423	3.196
H_2-Ne	0.572	0.982	0.317	1.684	2.541	3.541	4.677
H_2-O_2		0.692	0.756	1.188	1.792	2.497	3.299
H_2-SF_6			0.208	0.649	0.998	1.400	1.851
H_2-Xe		0.513	0.122	0.890	1.349	1.885	2.493
H_2O-N_2			0.242	0.399			
H_2O-O_2			0.244	0.403	0.645	0.882	1.147
He-Kr	0.330	0.559	0.629	0.942	1.404	1.942	2.550
He-N_2	0.365	0.619	0.698	1.052	1.577	2.188	2.882
He-Ne	0.563	0.948	1.066	1.592	2.362	3.254	4.262
He-O_2		0.641	0.697	1.092	1.640	2.276	2.996
He-SF_6			1.109	0.592	0.871	1.190	1.545
He-Xe	0.282	0.478	0.538	0.807	1.201	1.655	2.168
Kr-N_2		0.131	0.149	0.227	0.346	0.485	0.645
Kr-Ne	0.131	0.228	0.258	0.392	0.587	0.812	1.063
Kr-Xe	0.035	0.064	0.073	0.116	0.181	0.257	0.344
N_2-Ne			0.258	0.483	0.731	1.021	1.351
N_2-O_2			0.202	0.307	0.462	0.643	0.849
N_2-SF_6			0.148	0.231	0.328	0.436	
N_2-Xe		0.107	0.123	0.188	0.287	0.404	0.539
Ne-Xe	0.111	0.193	0.219	0.332	0.498	0.688	0.901
O_2-SF_6			0.097	0.154	0.238	0.334	0.441

DIFFUSION OF GASES IN WATER

This table gives values of the diffusion coefficient, D, for diffusion of several common gases in water at various temperatures. For simple one-dimensional transport, the diffusion coefficient describes the time-rate of change of concentration, dc/dt, through the equation

$$dc/dt = D\, d^2c/dx^2$$

where x is, for example, the perpendicular distance from a gas-liquid interface. The values below have been selected from the references indicated; in some cases data have been refitted to permit interpolation in temperature.

Gas-liquid diffusion coefficients are difficult to measure, and large differences are found between values obtained by different authors and through different experimental methods. See References 1 and 2 for a discussion of measurement techniques.

REFERENCES

1. Jähne, B., Heinz, G., and Dietrich, W., *J. Geophys. Res.*, 92, 10767, 1987.
2. Himmelblau, D. M., *Chem. Rev.* 64, 527, 1964.
3. Boerboom, A. J. H., and Kleyn, G., *J. Chem. Phys.*, 50, 1086, 1969.
4. O'Brien, R. N., and Hyslop, W. F., *Can. J. Chem.*, 55, 1415, 1977.
5. Maharajh, D. M., and Walkley, J., *Can. J. Chem.*, 51, 944, 1973.
6. *Landolt-Börnstein, Numerical Data and Functional Relationships in Science and Technology, Sixth Edition,* II/5a, *Transport Phenomena I (Viscosity and Diffusion)*, Springer-Verlag, Heidelberg, 1969.

$D/10^{-5}$ cm^2 s^{-1}

	10°C	15°C	20°C	25°C	30°C	35°C	Ref.
Ar				2.5			3,4
CHCl$_2$F				1.80			5
CH$_3$Br				1.35			5
CH$_3$Cl				1.40			5
CH$_4$	1.24	1.43	1.62	1.84	2.08	2.35	1
CO$_2$	1.26	1.45	1.67	1.91	2.17	2.47	1
C$_2$H$_2$	1.43	1.59	1.78	1.99	2.23		2
Cl$_2$		1.13	1.5	1.89			2,6
HBr				3.15			6
HCl				3.07			6
H$_2$	3.62	4.08	4.58	5.11	5.69	6.31	1
H$_2$S				1.36			2,6
He	5.67	6.18	6.71	7.28	7.87	8.48	1,3
Kr	1.20	1.39	1.60	1.84	2.11	2.40	1,3
NH$_3$		1.3	1.5				2
NO$_2$			1.23	1.4	1.59		2,6
N$_2$				2.0			2
N$_2$O		1.62	2.11	2.57			2,6
Ne	2.93	3.27	3.64	4.03	4.45	4.89	1,3
O$_2$		1.67	2.01	2.42			2,6
Rn	0.81	0.96	1.13	1.33	1.55	1.80	1
SO$_2$			1.62	1.83	2.07	2.32	2
Xe	0.93	1.08	1.27	1.47	1.70	1.95	1,3

DIFFUSION COEFFICIENTS IN LIQUIDS AT INFINITE DILUTION

This table lists diffusion coefficients D_{AB} at infinite dilution for some binary liquid mixtures. Although values are given to two decimal places, measurements in the literature are often in poor agreement. Therefore most values in the table cannot be relied upon to better than 10%.

Solvents are listed in alphabetical order, as are the solutes within each solvent group.

REFERENCE

Landolt-Börnstein, *Numerical Data and Functional Relationships in Science and Technology*, Sixth Edition, Vol. II/5a, 1969.

Solute	Solvent	$t/°C$	D_{AB} 10^{-5} cm^2 s^{-1}	Solute	Solvent	$t/°C$	D_{AB} 10^{-5} cm^2 s^{-1}
Acetic acid	Acetone	25	3.31	Acetone	Tetrachloromethane	25	1.75
Benzoic acid	Acetone	25	2.62	Benzene	Tetrachloromethane	25	1.42
Formic acid	Acetone	25	3.77	Cyclohexane	Tetrachloromethane	25	1.30
Nitrobenzene	Acetone	20	2.94	Ethanol	Tetrachloromethane	25	1.90
Tetrachloromethane	Acetone	25	3.29	Iodine	Tetrachloromethane	30	1.63
Trichloromethane	Acetone	25	3.64	Trichloromethane	Tetrachloromethane	25	1.66
Water	Acetone	25	4.56	Acetic acid	Toluene	25	2.26
Acetic acid	Benzene	25	2.09	Benzene	Toluene	25	2.54
Aniline	Benzene	25	1.96	Benzoic acid	Toluene	25	1.49
Benzoic acid	Benzene	25	1.38	Cyclohexane	Toluene	25	2.42
Bromobenzene	Benzene	8	1.45	Formic acid	Toluene	25	2.65
2-Butanone	Benzene	30	2.09	Water	Toluene	25	6.19
Chloroethylene	Benzene	8	1.77	Acetone	Trichloromethane	25	2.55
Cyclohexane	Benzene	25	2.25	Benzene	Trichloromethane	25	2.89
Ethanol	Benzene	25	3.02	2-Butanone	Trichloromethane	25	2.13
Formic acid	Benzene	25	2.28	Butyl acetate	Trichloromethane	25	1.71
Heptane	Benzene	25	1.78	Diethyl ether	Trichloromethane	25	2.15
Methanol	Benzene	25	3.80	Ethanol	Trichloromethane	15	2.20
Toluene	Benzene	25	1.85	Ethyl acetate	Trichloromethane	25	2.02
1,2,4-Trichlorobenzene	Benzene	8	1.34	Acetic acid	Water	25	1.29
Trichloromethane	Benzene	25	2.26	Acetone	Water	25	1.28
Adipic acid	1-Butanol	30	0.40	Acetonitrile	Water	15	1.26
Benzene	1-Butanol	25	1.00	Alanine	Water	25	0.91
Biphenyl	1-Butanol	25	0.63	Allyl alcohol	Water	15	0.90
Butyric acid	1-Butanol	30	0.51	Aniline	Water	20	0.92
p-Dichlorobenzene	1-Butanol	25	0.82	Arabinose	Water	20	0.69
Methanol	1-Butanol	30	0.59	Benzene	Water	20	1.02
Oleic acid	1-Butanol	30	0.25	1-Butanol	Water	25	0.56
Propane	1-Butanol	25	1.57	Caprolactam	Water	25	0.87
Water	1-Butanol	25	0.56	Chloroethylene	Water	25	1.34
Benzene	Cyclohexane	25	1.41	Cyclohexane	Water	20	0.84
Tetrachloromethane	Cyclohexane	25	1.49	Diethylamine	Water	20	0.97
Toluene	Cyclohexane	25	1.57	Ethanol	Water	25	1.24
Allyl alcohol	Ethanol	20	0.98	Ethanolamine	Water	25	1.08
Benzene	Ethanol	25	1.81	Ethyl acetate	Water	20	1.00
Iodine	Ethanol	25	1.32	Ethylbenzene	Water	20	0.81
Iodobenzene	Ethanol	20	1.00	Ethylene glycol	Water	25	1.16
3-Methyl-1-butanol	Ethanol	20	0.81	Glucose	Water	25	0.67
Pyridine	Ethanol	20	1.10	Glycerol	Water	25	1.06
Tetrachloromethane	Ethanol	25	1.50	Glycine	Water	25	1.05
Water	Ethanol	25	1.24	Lactose	Water	15	0.38
Acetic acid	Ethyl acetate	20	2.18	Maltose	Water	15	0.38
Acetone	Ethyl acetate	20	3.18	Mannitol	Water	15	0.50
2-Butanone	Ethyl acetate	30	2.93	Methane	Water	25	1.49
Ethyl benzoate	Ethyl acetate	20	1.85	Methanol	Water	15	1.28
Nitrobenzene	Ethyl acetate	20	2.25	3-Methyl-1-butanol	Water	10	0.69
Water	Ethyl acetate	25	3.20	Methylcyclopentane	Water	20	0.85
Benzene	Heptane	25	3.91	Phenol	Water	20	0.89
Toluene	Heptane	25	3.72	1-Propanol	Water	15	0.87
Bromobenzene	Hexane	8	2.60	Propene	Water	25	1.44
2-Butanone	Hexane	30	3.74	Pyridine	Water	25	0.58
Dodecane	Hexane	25	2.73	Raffinose	Water	15	0.33
Iodine	Hexane	25	4.45	Sucrose	Water	25	0.52
Methane	Hexane	25	0.09	Toluene	Water	20	0.85
Propane	Hexane	25	4.87	Urea	Water	25	1.38
Tetrachloromethane	Hexane	25	3.70	Urethane	Water	15	0.80
Toluene	Hexane	25	4.21				

Section 7
Biochemistry

Properties of Amino Acids 7-1
Structures of Common Amino Acids 7-3
Properties of Purine and Pyrimidine Bases 7-4
The Genetic Code 7-5
Properties of Fatty Acids 7-6
Carbohydrate Names and Symbols 7-7
Biological Buffers 7-9
Typical pH Values of Biological Materials and Foods 7-10
Chemical Composition of the Human Body 7-11

PROPERTIES OF AMINO ACIDS

This table gives selected properties of some important amino acids and closely related compounds. The first part of the table lists the 20 "standard" amino acids that are the basic constituents of proteins (structures of these amino acids may be found in the following table). The second part includes other amino acids and related compounds of biochemical importance. Within each part of the table the compounds are listed by name in alphabetical order.

Symbol — Three-letter symbol for the standard amino acids
M_r — Molecular weight
t_m — Melting point
pK_a, pK_b, pK_c, pK_d — Negative of the logarithm of the acid dissociation constants for the COOH and NH_2 groups (and, in some cases, other groups) in the molecule (at 25°C)
pI — pH at the isoelectric point
S — Solubility in water at 25°C in units of grams of compound per kilogram of water; when quantitative data are not available, the notations sl.s. (for slightly soluble) and v.s. (for very soluble) are used.

Data on the enthalpy of formation of many of these compounds are included in the table "Standard Thermodynamic Properties of Chemical Substances" in Section 5 of this Handbook. Absorption spectra and optical rotation data can be found in Reference 3. Partial molar volume and other thermodynamic properties, including solubility as a function of temperature, are given in References 3 and 5. Most of the pK values come from Reference 7.

REFERENCES

1. Dawson, R. M. C., Elliott, D. C., Elliott, W. H., and Jones, K. M., *Data for Biochemical Research*, 3rd ed., Clarendon Press, Oxford, 1986.
2. Budavari, S., Ed., *The Merck Index, Twelfth Edition*, Merck & Co., Rahway, NJ, 1996.
3. Sober, H. A., Ed., *CRC Handbook of Biochemistry. Selected Data for Molecular Biology*, CRC Press, Boca Raton, FL, 1968.
4. Voet, D. and Voet, J. G., *Biochemistry, Second Edition*, John Wiley & Sons, New York, 1995.
5. Hinz, H. J., Ed., *Thermodynamic Data for Biochemistry and Biotechnology,* Springer-Verlag, Heidelberg, 1986.
6. Fasman, G. D., Ed., *Practical Handbook of Biochemistry and Molecular Biology,* CRC Press, Boca Raton, FL, 1989.
7. Smith, R. M., and Martell, A. E., *NIST Standard Reference Database 46: Critically Selected Stability Constants of Metal Complexes Database, Version 3.0*, National Institute of Standards and Technology, Gaithersburg, MD, 1997.
8. Jin, Z. and Chao, K. C., *J. Chem. Eng. Data*, 37, 199, 1992.

The standard amino acids:

Symbol	Name	Mol. form.	M_r	t_m/°C	pK_a	pK_b	pK_c	pI	S/g kg^{-1}
Ala	Alanine	$C_3H_7NO_2$	89.09	297	2.33	9.71		6.00	165.0
Arg	Arginine	$C_6H_{14}N_4O_2$	174.20	244	2.03	9.00	12.10	10.76	182.6
Asn	Asparagine	$C_4H_8N_2O_3$	132.12	235	2.16	8.73		5.41	25.1
Asp	Aspartic acid	$C_4H_7NO_4$	133.10	270	1.95	9.66	3.71	2.77	4.95
Cys	Cysteine	$C_3H_7NO_2S$	121.16	240	1.91	10.28	8.14	5.07	v.s.
Glu	Glutamic acid	$C_5H_9NO_4$	147.13	160	2.16	9.58	4.15	3.22	8.61
Gln	Glutamine	$C_5H_{10}N_2O_3$	146.15	185	2.18	9.00		5.65	42
Gly	Glycine	$C_2H_5NO_2$	75.07	290	2.34	9.58		5.97	250.9
His	Histidine	$C_6H_9N_3O_2$	155.16	287	1.70	9.09	6.04	7.59	43.5
Ile	Isoleucine	$C_6H_{13}NO_2$	131.17	284	2.26	9.60		6.02	34.2
Leu	Leucine	$C_6H_{13}NO_2$	131.17	293	2.32	9.58		5.98	22.0
Lys	Lysine	$C_6H_{14}N_2O_2$	146.19	224	2.15	9.16	10.67	9.74	5.8
Met	Methionine	$C_5H_{11}NO_2S$	149.21	281	2.16	9.08		5.74	56
Phe	Phenylalanine	$C_9H_{11}NO_2$	165.19	283	2.18	9.09		5.48	27.9
Pro	Proline	$C_5H_9NO_2$	115.13	221	1.95	10.47		6.30	1623
Ser	Serine	$C_3H_7NO_3$	105.09	228	2.13	9.05		5.68	421.7
Thr	Threonine	$C_4H_9NO_3$	119.12	256	2.20	8.96		5.60	98.1
Trp	Tryptophan	$C_{11}H_{12}N_2O_2$	204.23	289	2.38	9.34		5.89	13.2
Tyr	Tyrosine	$C_9H_{11}NO_3$	181.19	343	2.24	9.04	10.10	5.66	0.46
Val	Valine	$C_5H_{11}NO_2$	117.15	315	2.27	9.52		5.96	88.5

Other amino acids and related compounds:

Name	Mol. form.	M_r	$t_m/°C$	pK_a	pK_b	pK_c	pK_d	$S/\text{g kg}^{-1}$
N-Acetylglutamic acid	$C_7H_{11}NO_5$	189.17	199					
*N*6-Acetyl-*L*-lysine	$C_8H_{16}N_2O_3$	188.23	265	2.12	9.51			
β-Alanine	$C_3H_7NO_2$	89.09	200	3.51	10.08			891
DL-2-Aminobutanoic acid	$C_4H_9NO_2$	103.12	304	2.30	9.63			210
DL-3-Aminobutanoic acid	$C_4H_9NO_2$	103.12	194	3.43	10.05			1250
4-Aminobutanoic acid	$C_4H_9NO_2$	103.12	203	4.02	10.35			v.s.
5-Aminolevulinic acid	$C_5H_9NO_3$	131.13	118	4.05	8.90			
L-3-Amino-2-methylpropanoic acid	$C_4H_9NO_2$	103.12	183					
Azaserine	$C_5H_7N_3O_4$	173.13	150		8.55			v.s.
L-γ-Carboxyglutamic acid	$C_6H_9NO_6$	191.14	167	1.70	9.90	4.75	3.20	
Carnosine	$C_9H_{14}N_4O_3$	226.24	260	2.51	9.35	6.76		322
Citrulline	$C_6H_{13}N_3O_3$	175.19	222	2.32	9.30			
Creatine	$C_4H_9N_3O_2$	131.13	303	2.63	14.30			16
L-Cysteic acid	$C_3H_7NO_5S$	169.16	260	1.89	8.70	1.30		v.s.
L-Cystine	$C_6H_{12}N_2O_4S_2$	240.30	260	1.50	8.80	2.05	8.03	0.11
L-3,5-Diiodotyrosine	$C_9H_9I_2NO_3$	432.98	213	2.12	9.10	6.16		0.62
Dopamine	$C_8H_{11}NO_2$	153.18			10.36	8.88		
L-Ethionine	$C_6H_{13}NO_2S$	163.24	273	2.18	9.05	13.10		
Glycocyamine	$C_3H_7N_3O_2$	117.11	282	2.82				5
N-Glycylglycine	$C_4H_8N_2O_3$	132.12	263	3.13	8.10			
Histamine	$C_5H_9N_3$	111.15	83		9.83	6.11		v.s.
Homocysteine	$C_4H_9NO_2S$	135.19	232	2.15	8.57	10.38		
Homocystine	$C_8H_{16}N_2O_4S_2$	268.36	264	1.59	9.44	2.54	8.52	0.2
L-Homoserine	$C_4H_9NO_3$	119.12	203	2.27	9.28			1100
5-Hydroxylysine	$C_6H_{14}N_2O_3$	162.19		2.13	8.85	9.83		
trans-4-Hydroxyproline	$C_5H_9NO_3$	131.13	274	1.82	9.47			361
L-3-Iodotyrosine	$C_9H_{10}INO_3$	307.09	205	2.20	9.10	8.70		sl.s.
L-Kynurenine	$C_{10}H_{12}N_2O_3$	208.22	185					sl.s.
L-Lanthionine	$C_6H_{12}N_2O_4S$	208.24	295					1.5
Levodopa	$C_9H_{11}NO_4$	197.19	277	2.20	8.75	9.81	13.40	1650
2-Methylalanine	$C_4H_9NO_2$	103.12	335	2.36	10.21			137
L-1-Methylhistidine	$C_7H_{11}N_3O_2$	169.18	250	1.69	8.85	6.48		
L-Norleucine	$C_6H_{13}NO_2$	131.17	301	2.31	9.68			15
L-Norvaline	$C_5H_{11}NO_2$	117.15	305	2.31	9.65			107
L-Ornithine	$C_5H_{12}N_2O_2$	132.16	140	1.94	8.78	10.52		v.s.
O-Phosphoserine	$C_3H_8NO_6P$	185.07	166	2.14	9.80	5.70		
L-Pyroglutamic acid	$C_5H_7NO_3$	129.12	162	3.32				
Sarcosine	$C_3H_7NO_2$	89.09	212	2.18	9.97			428
L-Thyroxine	$C_{15}H_{11}I_4NO_4$	776.87	235	2.20	10.01	6.45		sl.s.

STRUCTURES OF COMMON AMINO ACIDS

Alanine
(Ala, A)

$$H_3C-\underset{\underset{H}{|}}{\overset{\overset{NH_2}{|}}{C}}-COOH$$

Leucine
(Leu, L)

$$\underset{CH_3}{\overset{CH_3}{}}CH-CH_2-\underset{\underset{H}{|}}{\overset{\overset{NH_2}{|}}{C}}-COOH$$

Arginine
(Arg, R)

$$H_2N-\overset{\overset{NH}{\|}}{C}-\underset{\overset{|}{H}}{N}-CH_2-CH_2-CH_2-\underset{\underset{H}{|}}{\overset{\overset{NH_2}{|}}{C}}-COOH$$

Lysine
(Lys, K)

$$H_2N-CH_2-CH_2-CH_2-CH_2-\underset{\underset{H}{|}}{\overset{\overset{NH_2}{|}}{C}}-COOH$$

Asparagine
(Asn, N)

$$H_2N-\overset{\overset{O}{\|}}{C}-CH_2-\underset{\underset{H}{|}}{\overset{\overset{NH_2}{|}}{C}}-COOH$$

Methionine
(Met, M)

$$CH_3-S-CH_2-CH_2-\underset{\underset{H}{|}}{\overset{\overset{NH_2}{|}}{C}}-COOH$$

Aspartic acid
(Asp, D)

$$HOOC-CH_2-\underset{\underset{H}{|}}{\overset{\overset{NH_2}{|}}{C}}-COOH$$

Phenylalanine
(Phe, F)

Cysteine
(Cys, C)

$$HS-CH_2-\underset{\underset{H}{|}}{\overset{\overset{NH_2}{|}}{C}}-COOH$$

Proline
(Pro, P)

Glutamine
(Gln, Q)

$$H_2N-\overset{\overset{O}{\|}}{C}-CH_2-CH_2-\underset{\underset{H}{|}}{\overset{\overset{NH_2}{|}}{C}}-COOH$$

Serine
(Ser, S)

$$HO-CH_2-\underset{\underset{H}{|}}{\overset{\overset{NH_2}{|}}{C}}-COOH$$

Glutamic acid
(Glu, E)

$$HOOC-CH_2-CH_2-\underset{\underset{H}{|}}{\overset{\overset{NH_2}{|}}{C}}-COOH$$

Threonine
(Thr, T)

$$CH_3-\underset{\underset{OH}{|}}{CH}-\underset{\underset{H}{|}}{\overset{\overset{NH_2}{|}}{C}}-COOH$$

Glycine
(Gly, G)

$$H-\underset{\underset{H}{|}}{\overset{\overset{NH_2}{|}}{C}}-COOH$$

Tryptophan
(Trp, W)

Histidine
(His, H)

Tyrosine
(Tyr, Y)

Isoleucine
(Ile, I)

$$\underset{CH_3}{\overset{CH_3-CH_2}{}}CH-\underset{\underset{H}{|}}{\overset{\overset{NH_2}{|}}{C}}-COOH$$

Valine
(Val, V)

$$\underset{CH_3}{\overset{CH_3}{}}CH-\underset{\underset{H}{|}}{\overset{\overset{NH_2}{|}}{C}}-COOH$$

PROPERTIES OF PURINE AND PYRIMIDINE BASES

This table lists some of the important purine and pyrimidine bases that occur in nucleic acids. The pK_a values (negative logarithm of the acid dissociation constant) are given for each ionization stage. The last column gives the aqueous solubility S at the indicated temperature in units of grams per 100 grams of solution.

The numbering system in the rings is:

Purine   Pyrimidine

REFERENCES

1. R. M. C. Dawson, et al., *Data for Biochemical Research*, 3rd Ed., Clarendon Press, Oxford, 1986.
2. S. Budavari, Ed., *The Merck Index*, 11th Ed., Merk and Co., Rahway, NJ., 1989.

Common name	Systematic name	Mol form.	Mol. wt.	pK_a values			S/mass % (temp.)
Pyrimidines							
Cytosine	4-Amino-2-hydroxypyrimidine	$C_4H_5N_3O$	111.10	4.5	12.2		0.76 (25°C)
5-Methylcytosine	4-Amino-2-hydroxy-5-methylpyrimidine	$C_5H_7N_3O$	125.13	4.6	12.4		0.45 (25°C)
5-Hydroxymethyl-cytosine	4-Amino-2-hydroxy-5-hydroxy-methylpyrimidine	$C_5H_7N_3O_2$	141.13	4.3	13		
Uracil	2,4-Dihydroxypyrimidine	$C_4H_4N_2O_2$	112.09	0.5	9.5	>13	0.36 (25°C)
Thymine	5-Methyluracil	$C_5H_6N_2O_2$	126.11	9.9	>13		0.4 (25°C)
Orotic acid	Uracil-6-carboxylic acid	$C_5H_4N_2O_4$	156.10	2.4	9.5	>13	0.18 (18°C)
Purines							
Adenine	6-Aminopurine	$C_5H_5N_5$	135.14	<1	4.1	9.8	0.09 (25°C)
Guanine	2-Amino-6-hydroxypurine	$C_5H_5N_5O$	151.13	3.3	9.2	12.3	0.004 (40°C)
7-Methylguanine	7-Methyl-2-amino-6-hydroxypurine	$C_6H_7N_5O$	165.16	3.5	9.9		
Isoguanine	6-Amino-2-hydroxypurine	$C_5H_5N_5O$	151.13	4.5	9.0		0.006 (25°C)
Xanthine	2,6-Dioxopurine	$C_5H_4N_4O_2$	152.11	0.8	7.4	11.1	0.05 (20°C)
Hypoxanthine	6-Hydroxypurine	$C_5H_4N_4O$	136.11	2.0	8.9	12.1	0.07 (19°C)
Uric acid	2,6,8-Trihydroxypurine	$C_5H_4N_4O_3$	168.11	5.4	11.3		0.002 (20°C)

THE GENETIC CODE

This table gives the correspondence between a messenger RNA codon and the amino acid which it specifies. The symbols for bases in the codon are:

U: uracil
C: cytosine
A: adenine
G: guanine

The amino acid symbols are given in the table entitled "Structures of Common Amino Acids". A chain-initiating codon is indicated by **init** and a chain-terminating codon by **term**.

Example: UCA codes for **Ser**, UAC codes for **Tyr**, etc.

First position	Second position				Third position
	U	**C**	**A**	**G**	
U	Phe	Ser	Tyr	Cys	U
	Phe	Ser	Tyr	Cys	C
	Leu	Ser	**term**	**term**	A
	Leu	Ser	**term**	Trp	G
C	Leu	Pro	His	Arg	U
	Leu	Pro	His	Arg	C
	Leu	Pro	Gln	Arg	A
	Leu	Pro	Gln	Arg	G
A	Ile	Thr	Asn	Ser	U
	Ile	Thr	Asn	Ser	C
	Ile	Thr	Lys	Arg	A
	Met (**init**)	Thr	Lys	Arg	G
G	Val	Ala	Asp	Gly	U
	Val	Ala	Asp	Gly	C
	Val	Ala	Glu	Gly	A
	Val (**init**)	Ala	Glu	Gly	G

PROPERTIES OF FATTY ACIDS

This table gives the systematic names and selected properties of some of the more important fatty acids of five or more carbon atoms. Compounds are listed first by degree of saturation and, secondly, by number of carbon atoms. The following data are included:

M_r: Molecular weight
t_m: Melting point in °C

S: Aqueous solubility at 20°C in units of grams of solute per 100 grams of water

REFERENCES

1. Dawson, R. M. C., Elliott, D. C., Elliott, W. H., and Jones, K. M., *Data for Biochemical Research*, Third Edition, Clarendon Press, Oxford, 1986.
2. Fasman, G. D., Ed., *Practical Handbook of Biochemistry and Molecular Biology*, CRC Press, Boca Raton, FL, 1989.

Common name	Systematic name	Mol. form.	M_r	$t_m/°C$	S
Saturated					
Valeric acid	Pentanoic acid	$C_5H_{10}O_2$	102.13	-34	2.5
Isovaleric acid	3-Methylbutanoic acid	$C_5H_{10}O_2$	102.13	-29.3	4.3
Caproic acid	Hexanoic acid	$C_6H_{12}O_2$	116.16	-3	0.967
Enanthic acid	Heptanoic acid	$C_7H_{14}O_2$	130.19	-7.5	0.24
Caprylic acid	Octanoic acid	$C_8H_{16}O_2$	144.21	16.3	0.080
Pelargonic acid	Nonanoic acid	$C_9H_{18}O_2$	158.24	12.3	0.0284
Capric acid	Decanoic acid	$C_{10}H_{20}O_2$	172.27	31.9	0.015
Lauric acid	Dodecanoic acid	$C_{12}H_{24}O_2$	200.32	43.2	0.0055
Tridecylic acid	Tridecanoic acid	$C_{13}H_{26}O_2$	214.35	41.5	0.0033
Myristic acid	Tetradecanoic acid	$C_{14}H_{28}O_2$	228.38	53.9	0.0020
Pentadecylic acid	Pentadecanoic acid	$C_{15}H_{30}O_2$	242.40	52.3	0.0012
Palmitic acid	Hexadecanoic acid	$C_{16}H_{32}O_2$	256.43	63.1	0.00072
Margaric acid	Heptadecanoic acid	$C_{17}H_{34}O_2$	270.46	61.3	0.00042
Stearic acid	Octadecanoic acid	$C_{18}H_{36}O_2$	284.48	69.6	0.00029
Arachidic acid	Eicosanoic acid	$C_{20}H_{40}O_2$	312.54	76.5	
Phytanic acid	3,7,11,15-Tetramethylhexadecanoic acid	$C_{20}H_{40}O_2$	312.54	-65	
Behenic acid	Docosanoic acid	$C_{22}H_{44}O_2$	340.59	81.5	
Lignoceric acid	Tetracosanoic acid	$C_{24}H_{48}O_2$	368.64	87.5	
Cerotic acid	Hexacosanoic acid	$C_{26}H_{52}O_2$	396.70	88.5	
Montanic acid	Octacosanoic acid	$C_{28}H_{56}O_2$	424.75	90.9	
Monounsaturated					
Caproleic acid	9-Decenoic acid	$C_{10}H_{18}O_2$	170.25	26.5	
Palmitoleic acid	*cis*-9-Hexadecenoic acid	$C_{16}H_{30}O_2$	254.41	-0.1	
Oleic acid	*cis*-9-Octadecenoic acid	$C_{18}H_{34}O_2$	282.47	13.4	
Elaidic acid	*trans*-9-Octadecenoic acid	$C_{18}H_{34}O_2$	282.47	45	
Vaccenic acid	*trans*-11-Octadecenoic acid	$C_{18}H_{34}O_2$	282.47	44	
Erucic acid	*cis*-13-Docosenoic acid	$C_{22}H_{42}O_2$	338.57	34.7	
Brassidic acid	*trans*-13-Docosenoic acid	$C_{22}H_{42}O_2$	338.57	61.9	
Nervonic acid	*cis*-15-Tetracosenoic acid	$C_{24}H_{46}O_2$	366.63	43	
Diunsaturated					
Linoleic acid	*cis,cis*-9,12-Octadecadienoic acid	$C_{18}H_{32}O_2$	280.45	-12	
Triunsaturated					
cis-Eleostearic acid	*trans,cis,trans*-9,11,13-Octadecatrienoic acid	$C_{18}H_{30}O_2$	278.44	49	
trans-Eleostearic acid	*trans,trans,trans*-9,11,13-Octadecatrienoic acid	$C_{18}H_{30}O_2$	278.44	71.5	
Linolenic acid	*cis,cis,cis*-9,12,15-Octadecatrienoic acid	$C_{18}H_{30}O_2$	278.44	-11	
Tetraunsaturated					
Arachidonic acid	5,8,11,14-Eicosatetraenoic acid, (all-*trans*)	$C_{20}H_{32}O_2$	304.47	-49.5	

CARBOHYDRATE NAMES AND SYMBOLS

The following table lists the systematic names and symbols for selected carbohydrates and some of their derivatives. The symbols for monosaccharide residues and derivatives are recommended by IUPAC for use in describing the structures of oligosaccharide chains. A more complete list can be found in the reference.

REFERENCE

McNaught, A. D., *Pure Appl. Chem.*, 68, 1919-2008, 1996.

Common Name	Symbol	Systematic Name
Abequose	Abe	3,6-Dideoxy-D-*xylo*-hexose
N-Acetyl-2-deoxyneur-2-enaminic acid	Neu2en5Ac	
N-Acetylgalactosamine	GalNAc	
N-Acetylglucosamine	GlcNAc	
N-Acetylneuraminic acid	Neu5Ac	
Allose	All	*allo*-Hexose
Altrose	Alt	*altro*-Hexose
Apiose	Api	3-*C*-(Hydroxymethyl)-*glycero*-tetrose
Arabinitol	Ara-ol	Arabinitol
Arabinose	Ara	*arabino*-Pentose
Arcanose		2,6-Dideoxy-3-*C*-methyl-3-*O*-methyl-*xylo*-hexose
Ascarylose		3,6-Dideoxy-L-*arabino*-hexose
Boivinose		2,6-Dideoxy-D-gulose
Chalcose		4,6-Dideoxy-3-*O*-methyl-D-*xylo*-hexose
Cladinose		2,6-Dideoxy-3-*C*-methyl-3-*O*-methyl-L-*ribo*-hexose
Colitose		3,6-Dideoxy-L-*xylo*-hexose
Cymarose		6-Deoxy-3-*O*-methyl-*ribo*-hexose
3-Deoxy-D-*manno*-oct-2-ulosonic acid	Kdo	
2-Deoxyribose	dRib	2-Deoxy-*erythro*-pentose
2,3-Diamino-2,3-dideoxy-D-glucose	GlcN3N	
Diginose		2,6-Dideoxy-3-*O*-methyl-*lyxo*-hexose
Digitalose		6-Deoxy-3-*O*-methyl-D-galactose
Digitoxose		2,6-Dideoxy-D-*ribo*-hexose
3,4-Di-*O*-methylrhamnose	Rha3,4Me$_2$	
Ethyl glucopyranuronate	Glc*p*A6Et	
Evalose		6-Deoxy-3-*C*-methyl-D-mannose
Fructose	Fru	*arabino*-Hex-2-ulose
Fucitol	Fuc-ol	6-Deoxy-D-galactitol
Fucose	Fuc	6-Deoxygalactose
β-D-Galactopyranose 4-sulfate	β-D-Gal*p*4*S*	
Galactosamine	GalN	2-Amino-2-deoxygalactose
Galactose	Gal	*galacto*-Hexose
Glucitol	Glc-ol	
Glucosamine	GlcN	2-Amino-2-deoxyglucose
Glucose	Glc	*gluco*-Hexose
Glucuronic acid	GlcA	
N-Glycoloylneuraminic acid	Neu5Gc	
Gulose	Gul	*gulo*-Hexose
Hamamelose		2-*C*-(Hydroxymethyl)-D-ribose
Idose	Ido	*ido*-Hexose
Iduronic acid	IdoA	
Lactose	Lac	β-D-Galactopyranosyl-(1→4)-D-glucose
Lyxose	Lyx	*lyxo*-Pentose
Maltose		α-D-Glucopyranosyl-(1→4)-D-glucose
Mannose	Man	*manno*-Hexose
2-*C*-Methylxylose	Xyl2*C*Me	
Muramic acid	Mur	2-Amino-3-*O*-[(R)-1-carboxyethyl]-2-deoxy-D-glucose
Mycarose		2,6-Dideoxy-3-*C*-methyl-L-*ribo*-hexose
Mycinose		6-Deoxy-2,3-di-*O*-methyl-D-allose
Neuraminic acid	Neu	5-Amino-3,5-dideoxy-D-*glycero*-D-*galacto*-non-2-ulosonic acid

CARBOHYDRATE NAMES AND SYMBOLS (continued)

Common Name	Symbol	Systematic Name
Panose		α-D-Glucopyranosyl-(1→6)-α-D-glucopyranosyl-(1→4)-D-glucose
Paratose		3,6-Dideoxy-D-*ribo*-hexose
Primeverose		β-D-Xylopyranosyl-(1→6)-D-glucose
Psicose	Psi	*ribo*-Hex-2-ulose
Quinovose	Qui	6-Deoxyglucose
Raffinose		β-D-Fructofuranosyl-α-D-galactopyranosyl-(1→6)-α-D-glucopyranoside
Rhamnose	Rha	6-Deoxymannose
Rhodinose		2,3,6-Trideoxy-L-*threo*-hexose
Ribose	Rib	*ribo*-Pentose
Ribose 5-phosphate	Rib5*P*	
Ribulose	Ribulo (Rul)	*erythro*-Pent-2-ulose
Rutinose		α-L-Rhamnopyranosyl-(1→6)-D-glucose
Sarmentose		2,6-Dideoxy-3-*O*-methyl-D-*xylo*-hexose
Sedoheptulose		D-*altro*-Hept-2-ulose
Sorbose	Sor	*xylo*-Hex-2-ulose
Streptose		5-Deoxy-3-*C*-formyl-L-lyxose
Sucrose		β-D-Fructofuranosyl-α-D-glucopyranoside
Tagatose	Tag	*lyxo*-Hex-2-ulose
Talose	Tal	*talo*-Hexose
Turanose		α-D-Glucopyranosyl-(1→3)-D-fructose
Tyvelose	Tyv	3,6-Dideoxy-D-*arabino*-hexose
Xylose	Xyl	*xylo*-Pentose
Xylulose	Xylulo (Xul)	*threo*-Pent-2-ulose

BIOLOGICAL BUFFERS

This table of frequently used buffers gives the pK_a value at 25°C and the useful pH range of each buffer. The buffers are listed in order of increasing pH.

The table is reprinted with permission of Sigma Chemical Company, St. Louis, Mo.

Acronym	Name	Mol. wt.	pK_a	Useful pH range
MES	2-(N-Morpholino)ethanesulfonic acid	195.2	6.1	5.5—6.7
BIS TRIS	Bis(2-hydroxyethyl)iminotris(hydroxymethyl)methane	209.2	6.5	5.8—7.2
ADA	N-(2-Acetamido)-2-iminodiacetic acid	190.2	6.6	6.0—7.2
ACES	2-[(2-Amino-2-oxoethyl)amino]ethanesulfonic acid	182.2	6.8	6.1—7.5
PIPES	Piperazine-N,N'-bis(2-ethanesulfonic acid)	302.4	6.8	6.1—7.5
MOPSO	3-(N-Morpholino)-2-hydroxypropanesulfonic acid	225.3	6.9	6.2—7.6
BIS TRIS PROPANE	1,3-Bis[tris(hydroxymethyl)methylamino]propane	282.3	6.8[a]	6.3—9.5
BES	N,N-Bis(2-hydroxyethyl)-2-aminoethanesulfonic acid	213.2	7.1	6.4—7.8
MOPS	3-(N-Morpholino)propanesulfonic acid	209.3	7.2	6.5—7.9
HEPES	N-(2-Hydroxyethyl)piperazine-N'-(2-ethanesulfonic acid)	238.3	7.5	6.8—8.2
TES	N-Tris(hydroxymethyl)methyl-2-aminoethanesulfonic acid	229.2	7.5	6.8—8.2
DIPSO	3-[N,N-Bis(2-hydroxyethyl)amino]-2-hydroxypropanesulfonic acid	243.3	7.6	7.0—8.2
TAPSO	3-[N-Tris(hydroxymethyl)methylamino]-2-hydroxypropanesulfonic acid	259.3	7.6	7.0—8.2
TRIZMA	Tris(hydroxymethyl)aminomethane	121.1	8.1	7.0—9.1
HEPPSO	N-(2-hydroxyethyl)piperazine-N'-(2-hydroxypropanesulfonic acid)	268.3	7.8	7.1—8.5
POPSO	Piperazine-N,N'-bis(2-hydroxypropanesulfonic acid)	362.4	7.8	7.2—8.5
EPPS	N-(2-Hydroxyethyl)piperazine-N'-(3-propanesulfonic acid)	252.3	8.0	7.3—8.7
TEA	Triethanolamine	149.2	7.8	7.3—8.3
TRICINE	N-Tris(hydroxymethyl)methylglycine	179.2	8.1	7.4—8.8
BICINE	N,N-Bis(2-hydroxyethyl)glycine	163.2	8.3	7.6—9.0
TAPS	N-Tris(hydroxymethyl)methyl-3-aminopropanesulfonic acid	243.3	8.4	7.7—9.1
AMPSO	3-[(1,1-Dimethyl-2-hydroxyethyl)amino]-2-hydroxypropanesulfonic acid	227.3	9.0	8.3—9.7
CHES	2-(N-Cyclohexylamino)ethanesulfonic acid	207.3	9.3	8.6—10.0
CAPSO	3-(Cyclohexylamino)-2-hydroxy-1-propanesulfonic acid	237.3	9.6	8.9—10.3
AMP	2-Amino-2-methyl-1-propanol	89.1	9.7	9.0—10.5
CAPS	3-(Cyclohexylamino)-1-propanesulfonic acid	221.3	10.4	9.7—11.1

[a] $pK_a = 9.0$ for the second dissociation stage.

TYPICAL pH VALUES OF BIOLOGICAL MATERIALS AND FOODS

This table gives typical pH ranges for various biological fluids and common foods. All values refer to 25°C.

Biological Materials

Blood, human	7.35-7.45
Blood, dog	6.9-7.2
Spinal fluid, human	7.3-7.5
Saliva, human	6.5-7.5
Gastric contents, human	1.0-3.0
Duodenal contents, human	4.8-8.2
Feces, human	4.6-8.4
Urine, human	4.8-8.4
Milk, human	6.6-7.6
Bile, human	6.8-7.0

Foods

Apples	2.9-3.3
Apricots	3.6-4.0
Asparagus	5.4-5.8
Bananas	4.5-4.7
Beans	5.0-6.0
Beers	4.0-5.0
Beets	4.9-5.5
Blackberries	3.2-3.6
Bread, white	5.0-6.0
Butter	6.1-6.4
Cabbage	5.2-5.4
Carrots	4.9-5.3
Cheese	4.8-6.4
Cherries	3.2-4.0
Cider	2.9-3.3
Corn	6.0-6.5
Crackers	6.5-8.5
Dates	6.2-6.4
Eggs, fresh white	7.6-8.0
Flour, wheat	5.5-6.5
Gooseberries	2.8-3.0
Grapefruit	3.0-3.3

Grapes	3.5-4.5
Hominy (lye)	6.8-8.0
Jams, fruit	3.5-4.0
Jellies, fruit	2.8-3.4
Lemons	2.2-2.4
Limes	1.8-2.0
Maple syrup	6.5-7.0
Milk, cows	6.3-6.6
Olives	3.6-3.8
Oranges	3.0-4.0
Oysters	6.1-6.6
Peaches	3.4-3.6
Pears	3.6-4.0
Peas	5.8-6.4
Pickles, dill	3.2-3.6
Pickles, sour	3.0-3.4
Pimento	4.6-5.2
Plums	2.8-3.0
Potatoes	5.6-6.0
Pumpkin	4.8-5.2
Raspberries	3.2-3.6
Rhubarb	3.1-3.2
Salmon	6.1-6.3
Sauerkraut	3.4-3.6
Shrimp	6.8-7.0
Soft drinks	2.0-4.0
Spinach	5.1-5.7
Squash	5.0-5.4
Strawberries	3.0-3.5
Sweet potatoes	5.3-5.6
Tomatoes	4.0-4.4
Tuna	5.9-6.1
Turnips	5.2-5.6
Vinegar	2.4-3.4
Water, drinking	6.5-8.0
Wines	2.8-3.8

CHEMICAL COMPOSITION OF THE HUMAN BODY

The elemental composition of the "standard man" of mass 70 kg is given below.

REFERENCES

1. Padikal, T.N., and Fivozinsky, S.P., *Medical Physics Data Book, National Bureau of Standards Handbook 138*, U. S. Government Printing Office, Washington, DC, 1981.
2. Snyde, W.S., et al., *Reference Man: Anatomical, Physiological, and Metabolic Characteristics*, Pergamon, New York, 1975.

Element	Amount (g)	Percent of total body mass
Oxygen	43,000	61
Carbon	16,000	23
Hydrogen	7000	10
Nitrogen	1800	2.6
Calcium	1000	1.4
Phosphorus	780	1.1
Sulfur	140	0.20
Potassium	140	0.20
Sodium	100	0.14
Chlorine	95	0.12
Magnesium	19	0.027
Silicon	18	0.026
Iron	4.2	0.006
Fluorine	2.6	0.0037
Zinc	2.3	0.0033
Rubidium	0.32	0.00046
Strontium	0.32	0.00046
Bromine	0.20	0.00029
Lead	0.12	0.00017
Copper	0.072	0.00010
Aluminum	0.061	0.00009
Cadmium	0.050	0.00007
Boron	<0.048	0.00007
Barium	0.022	0.00003
Tin	<0.017	0.00002
Manganese	0.012	0.00002
Iodine	0.013	0.00002
Nickel	0.010	0.00001
Gold	<0.010	0.00001
Molybdenum	<0.0093	0.00001
Chromium	<0.0018	0.000003
Cesium	0.0015	0.000002
Cobalt	0.0015	0.000002
Uranium	0.00009	0.0000001
Beryllium	0.000036	
Radium	$3.1 \cdot 10^{-11}$	

Section 8
Analytical Chemistry

Preparation of Special Analytical Reagents 8-1
Standard Solutions of Acids, Bases, and Salts 8-5
Standard Solutions of Oxidation and Reduction Reagents 8-7
Organic Analytical Reagents for the Determination of Inorganic Substances 8-8
Acid-Base Indicators 8-16
Fluorescent Indicators 8-19
Conversion Formulas for Concentration of Solutions 8-20
Electrochemical Series 8-21
Reduction and Oxidation Potentials for Certain Ion Radicals 8-32
pH Scale for Aqueous Solutions 8-35
Practical pH Measurements on Natural Waters 8-41
Buffer Solutions Giving Round Values of pH at 25°C 8-43
Dissociation Constants of Inorganic Acids and Bases 8-44
Dissociation Constants of Organic Acids and Bases 8-46
Concentrative Properties of Aqueous Solutions: Density, Refractive Index, Freezing Point
   Depression, and Viscosity 8-57
Ion Product of Water Substance 8-84
Ionization Constant of Normal and Heavy Water 8-85
Solubility of Selected Gases in Water 8-86
Solubility of Carbon Dioxide in Water at Various Temperatures and Pressures 8-90
Aqueous Solubility and Henry's Law Constants of Organic Compounds 8-91
Aqueous Solubility of Inorganic Compounds at Various Temperatures 8-102
Solubility Product Constants 8-111
Solubility of Common Salts at Ambient Temperatures 8-115
Solubility Chart 8-116
Reduction of Weighings in Air to Vacuo 8-119
Volume of One Gram of Water 8-119
Properties of Carrier Gases for Gas Chromatography 8-120
Solvents for Ultraviolet Spectrophotometry 8-121
^{13}C Chemical Shifts of Useful NMR Solvents 8-122
Mass Spectral Peaks of Common Organic Solvents 8-123

PREPARATION OF SPECIAL ANALYTICAL REAGENTS

Aluminon (qualitative test for aluminum). Aluminon is a trade name for the ammonium salt of aurintricarboxylic acid. Dissolve 1 g of the salt in 1 L of distilled water. Shake the solution well to insure thorough mixing.

Bang's reagent (for glucose estimation). Dissolve 100 g of K_2CO_3, 66 g of KCl and 160 g of $KHCO_3$ in the order given in about 700 mL of water at 30°C. Add 4.4 g of $CuSO_4$ and dilute to 1 L after the CO_2 is evolved. This solution should be shaken only in such a manner as not to allow entry of air. After 24 hours 300 mL are diluted to 1 L with saturated KCl solution, shaken gently and used after 24 hours; 50 mL is equivalent to 10 mg glucose.

Barfoed's reagent (test for glucose). See Cupric acetate.

Baudisch's reagent. See Cupferron.

Benedict's solution (qualitative reagent for glucose). With the aid of heat, dissolve 173 g of sodium citrate and 100 g of Na_2CO_3 in 800 mL of water. Filter, if necessary, and dilute to 850 mL. Dissolve 17.3 g of $CuSO_4·5H_2O$ in 100 mL of water. Pour the latter solution, with constant stirring, into the carbonate-citrate solution, and dilute to 1 L.

Benzidine hydrochloride solution (for sulfite determination). Make a paste of 8 g of benzidine hydrochloride ($C_{12}H_8(NH_3)_2·2HCl$) and 20 mL of water, add 20 mL of HCl (sp. gr. 1.12) and dilute to 1 L with water. Each mL of this solution is equivalent to 0.00357 g of H_2SO_4.

Bertrand's reagent (glucose estimation). Consists of the following solutions:
1. Dissolve 200 g of Rochelle salt and 150 g of NaOH in sufficient water to make 1 L of solution.
2. Dissolve 40 g of $CuSO_4$ in enough water to make 1 L of solution.
3. Dissolve 50 g of $Fe_2(SO_4)_3$ and 200 g of H_2SO_4 (sp. gr. 1.84) in sufficient water to make 1 L of solution.
4. Dissolve 5 g of $KMnO_4$ in sufficient water to make 1 L of solution.

Bial's reagent (for pentose). Dissolve 1 g of orcinol (5-methyl-1,3-benzenediol) in 500 mL of 30% HCl to which 30 drops of a 10% solution of $FeCl_3$ has been added.

Boutron — Boudet soap solution:
1. Dissolve 100 g of pure castile soap in about 2.5 L of 56% ethanol.
2. Dissolve 0.59 g of $Ba(NO_3)_2$ in 1 L of water.

Adjust the castile soap solution so that 2.4 mL of it will give a permanent lather with 40 mL of solution (b). When adjusted, 2.4 mL of soap solution is equivalent to 220 parts per million of hardness (as $CaCO_3$) for a 40 mL sample. See also Soap solution.

Brucke's reagent (protein precipitation). See Potassium iodide-mercuric iodide.

Clarke's soap solution (estimation of hardness in water).
1. Dissolve 100 g of pure powdered castile soap in 1 L of 80% ethanol and allow to stand over night.
2. Prepare a solution of $CaCl_2$ by dissolving 0.5 g of $CaCO_3$ in HCl (sp. gr. 1.19), neutralize with NH_4OH and make slightly alkaline to litmus, and dilute to 500 mL. One mL is equivalent to 1 mg of $CaCO_3$.

Titrate (1) against (2) and dilute (1) with 80% ethanol until 1 mL of the resulting solution is equivalent to 1 mL of (2) after making allowance for the lather factor (the amount of standard soap solution required to produce a permanent lather in 50 mL of distilled water). One mL of the adjusted solution after subtracting the lather factor is equivalent to 1 mg of $CaCO_3$. See also Soap solution.

Cobalticyanide paper (Rinnmann's test for Zn). Dissolve 4 g of $K_3Co(CN)_6$ and 1 g of $KClO_3$ in 100 mL of water. Soak filter paper in solution and dry at 100°C. Apply drop of zinc solution and burn in an evaporating dish. A green disk is obtained if zinc is present.

Cochineal. Extract 1 g of cochineal for 4 days with 20 mL of alcohol and 60 mL of distilled water. Filter.

Congo red. Dissolve 0.5 g of congo red in 90 mL of distilled water and 10 mL of alcohol.

Cupferron (Baudisch's reagent for iron analysis). Dissolve 6 g of the ammonium salt of N-hydroxy-N-nitrosoaniline (cupferron) in 100 mL of H_2O. Reagent good for 1 week only and must be kept in the dark.

Cupric acetate (Barfoed's reagent for reducing monosaccharides). Dissolve 66 g of cupric acetate and 10 mL of glacial acetic acid in water and dilute to 1 L.

Cupric oxide, ammoniacal; Schweitzer's reagent (dissolves cotton, linen, and silk, but not wool).
1. Dissolve 5 g of cupric sulfate in 100 mL of boiling water, and add sodium hydroxide until precipitation is complete. Wash the precipitate well, and dissolve it in a minimum quantity of ammonium hydroxide.
2. Bubble a slow stream of air through 300 mL of strong ammonium hydroxide containing 50 g of fine copper turnings. Continue for 1 hour.

Cupric sulfate in glycerin-potassium hydroxide (reagent for silk). Dissolve 10 g of cupric sulfate, $CuSO_4·5H_2O$, in 100 mL of water and add 5 g of glycerol. Add KOH solution slowly until a deep blue solution is obtained.

Cupron (precipitates copper). Dissolve 5 g of benzoinoxime in 100 mL of 95% ethanol.

Cuprous chloride, acidic (reagent for CO in gas analysis).
1. Cover the bottom of a 2-L flask with a layer of cupric oxide about 0.5 inch deep, suspend a coil of copper wire so as to reach from the bottom to the top of the solution, and fill the flask with hydrochloric acid (sp. gr. 1.10). Shake occasionally. When the solution becomes nearly colorless, transfer to reagent bottles, which should also contain copper wire. The stock bottle may be refilled with dilute hydrochloric acid until either the cupric oxide or the copper wire is used up. Copper sulfate may be substituted for copper oxide in the above procedure.
2. Dissolve 340 g of $CuCl_2·2H_2O$ in 600 mL of conc. HCl and reduce the cupric chloride by adding 190 mL of a saturated solution of stannous chloride or until the solution is colorless. The stannous chloride is prepared by treating 300 g of metallic tin in a 500 mL flask with conc. HCl until no more tin goes into solution.
3. (Winkler method). Add a mixture of 86 g of CuO and 17 g of finely divided metallic Cu, made by the reduction of CuO with hydrogen, to a solution of HCl, made by diluting 650 mL of conc. HCl with 325 mL of water. After the mixture has been added slowly and with frequent stirring, a spiral of copper wire is suspended in the bottle, reaching all the way to the bottom. Shake occasionally, and when the solution becomes colorless, it is ready for use.

Cuprous chloride, ammoniacal (reagent for CO in gas analysis).

    1. The acid solution of cuprous chloride as prepared above is neutralized with ammonium hydroxide until an ammonia odor persists. An excess of metallic copper must be kept in the solution.

    2. Pour 800 mL of acidic cuprous chloride, prepared by the Winkler method, into about 4 L of water. Transfer the precipitate to a 250 mL graduate. After several hours, siphon off the liquid above the 50 mL mark and refill with 7.5% NH_4OH solution which may be prepared by diluting 50 mL of conc. NH_4OH with 150 mL of water. The solution is well shaken and allowed to stand for several hours. It should have a faint odor of ammonia.

Dichlorofluorescein indicator. Dissolve 1 g in 1 L of 70% alcohol or 1 g of the sodium salt in 1 L of water.

Dimethyglyoxime, 0.01 N. Dissolve 0.6 g of dimethylglyoxime (2,3-butanedione oxime) in 500 mL of 95% ethanol. This is an especially sensitive test for nickel, a very definite crimson color being produced.

Diphenylamine (reagent for rayon). Dissolve 0.2 g in 100 mL of concentrated sulfuric acid.

Diphenylamine sulfonate (for titration of iron with $K_2Cr_2O_7$). Dissolve 0.32 g of the barium salt of diphenylamine sulfonic acid in 100 mL of water, add 0.5 g of sodium sulfate and filter off the precipitate of $BaSO_4$.

Diphenylcarbazide. Dissolve 0.2 g of diphenylcarbazide in 10 mL of glacial acetic acid and dilute to 100 mL with 95% ethanol.

Esbach's reagent (estimation of protein). To a water solution of 10 g of picric acid and 20 g of citric acid, add sufficient water to make 1 L of solution.

Eschka's compound. Two parts of calcined ("light") magnesia are thoroughly mixed with 1 part of anhydrous sodium carbonate.

Fehling's solution (reagent for reducing sugars.)

    1. Copper sulfate solution. Dissolve 34.66 g of $CuSO_4 \cdot 5H_2O$ in water and dilute to 500 mL.

    2. Alkaline tartrate solution. Dissolve 173 g of potassium sodium tartrate (Rochelle salt, $KNaC_4H_4O_6 \cdot 4H_2O$) and 50 g of NaOH in water and dilute when cold to 500 mL.

Mix equal volumes of the two solutions at the time of using.

Ferric-alum indicator. Dissolve 140 g of ferric ammonium sulfate crystals in 400 mL of hot water. When cool, filter, and make up to a volume of 500 mL with dilute nitric acid.

Folin's mixture (for uric acid). To 650 mL of water add 500 g of $(NH_4)_2SO_4$, 5 g of uranium acetate, and 6 g of glacial acetic acid. Dilute to 1 L.

Formaldehyde — sulfuric acid (Marquis' reagent for alkaloids). Add 10 mL of formaldehyde solution to 50 mL of sulfuric acid.

Froehde's reagent. See Sulfomolybdic acid.

Fuchsin (reagent for linen). Dissolve 1 g of fuchsin in 100 mL of alcohol.

Fuchsin — sulfurous acid (Schiff's reagent for aldehydes). Dissolve 0.5 g of fuchsin and 9 g of sodium bisulfite in 500 mL of water, and add 10 mL of HCl. Keep in well-stoppered bottles and protect from light.

Gunzberg's reagent (detection of HCl in gastric juice). Prepare as needed a solution containing 4 g of phloroglucinol (1,3,5-benzenetriol) and 2 g of vanillin in 100 mL of absolute ethanol.

Hager's reagent. See Picric acid.

Hanus solution (for iodine number). Dissolve 13.2 g of resublimed iodine in 1 L of glacial acetic acid which will pass the dichromate test for reducible matter. Add sufficient bromine to double the halogen content, determined by titration (3 mL is about the proper amount). The iodine may be dissolved by the aid of heat, but the solution should be cold when the bromine is added.

Iodine, tincture of. To 50 mL of water add 70 g of I_2 and 50 g of KI. Dilute to 1 L with alcohol.

Iodo-potassium iodide (Wagner's reagent for alkaloids). Dissolve 2 g of iodine and 6 g of KI in 100 mL of water.

Litmus (indicator). Extract litmus powder three times with boiling alcohol, each treatment consuming an hour. Reject the alcoholic extract. Treat residue with an equal weight of cold water and filter; then exhaust with five times its weight of boiling water, cool and filter. Combine the aqueous extracts.

Magnesia mixture (reagent for phosphates and arsenates). Dissolve 55 g of magnesium chloride and 105 g of ammonium chloride in water, barely acidify with hydrochloric acid, and dilute to 1 L. The ammonium hydroxide may be omitted until just previous to use. The reagent, if completely mixed and stored for any period of time, becomes turbid

Magnesium uranyl acetate. Dissolve 100 g of $UO_2(C_2H_3O_2)_2 \cdot 2H_2O$ in 60 mL of glacial acetic acid and dilute to 500 mL. Dissolve 330 g of $Mg(C_2H_3O_2)_2 \cdot 4H_2O$ in 60 mL of glacial acetic acid and dilute to 200 mL. Heat solutions to the boiling point until clear, pour the magnesium solution into the uranyl solution, cool and dilute to 1 L. Let stand over night and filter if necessary.

Marme's reagent. See Potassium-cadmium iodide.

Marquis' reagent. See Formaldehyde-sulfuric acid.

Mayer's reagent (white precipitate with most alkaloids in slightly acid solutions). Dissolve 1.358 g of $HgCl_2$ in 60 mL of water and pour into a solution of 5 g of KI in 10 mL of H_2O. Add sufficient water to make 100 mL.

Methyl orange indicator. Dissolve 1 g of methyl orange in 1 L of water. Filter, if necessary.

Methyl orange, modified. Dissolve 2 g of methyl orange and 2.8 g of xylene cyanole FF in 1 L of 50% alcohol.

Methyl red indicator. Dissolve 1 g of methyl red in 600 mL of alcohol and dilute with 400 mL of water.

Methyl red, modified. Dissolve 0.50 g of methyl red and 1.25 g of xylene cyanole FF in 1 L of 90% alcohol. Or, dissolve 1.25 g of methyl red and 0.825 g of methylene blue in 1 L of 90% alcohol.

Millon's reagent (for albumins and phenols). Dissolve 1 part of mercury in 1 part of cold fuming nitric acid. Dilute with twice the volume of water and decant the clear solution after several hours.

Molisch's reagent. See 1-Naphthol.

1-Naphthol (Molisch's reagent for wool). Dissolve 15 g of 1-naphthol in 100 mL of alcohol or chloroform.

Nessler's reagent (for ammonia). Dissolve 50 g of KI in the smallest possible quantity of cold water (50 mL). Add a saturated solution of mercuric chloride (about 22 g in 350 mL of water will be needed) until an excess is indicated by the formation of a precipitate. Then add 200 mL of 5 N NaOH and dilute to 1 L. Let settle, and draw off the clear liquid.

PREPARATION OF SPECIAL ANALYTICAL REAGENTS (continued)

Nickel oxide, ammoniacal (reagent for silk). Dissolve 5 g of nickel sulfate in 100 mL of water, and add sodium hydroxide solution until nickel hydroxide is completely precipitated. Wash the precipitate well and dissolve in 25 mL of concentrated ammonium hydroxide and 25 mL of water.

Nitron (detection of nitrate radical). Dissolve 10 g of nitron (1,4-diphenyl-3-(phenylamino)-1,2,4-triazolium hydroxide) in 5 mL of glacial acetic acid and 95 mL of water. The solution may be filtered with slight suction through an alumdum crucible and kept in a dark bottle.

1-Nitroso-2-naphthol. Make a saturated solution in 50% acetic acid (1 part of glacial acetic acid with 1 part of water). Does not keep well.

Nylander's solution (carbohydrates). Dissolve 20 g of bismuth subnitrate and 40 g of Rochelle salt in 1 L of 8% NaOH solution. Cool and filter.

Obermayer's reagent (for indoxyl in urine). Dissolve 4 g of $FeCl_3$ in 1 L of HCl (sp. gr. 1.19).

Oxine. Dissolve 14 g of 8-hydroxyquinoline in 30 mL of glacial acetic acid. Warm slightly, if necessary. Dilute to 1 L.

Oxygen absorbent. Dissolve 300 g of ammonium chloride in 1 L of water and add 1 L of concentrated ammonium hydroxide solution. Shake the solution thoroughly. For use as an oxygen absorbent, a bottle half full of copper turnings is filled nearly full with the NH_4Cl-NH_4OH solution and the gas passed through.

Pasteur's salt solution. To 1 L of distilled water add 2.5 g of potassium phosphate, 0.25 g of calcium phosphate, 0.25 g of magnesium sulfate, and 12.00 g of ammonium tartrate.

Pavy's solution (glucose reagent). To 120 mL of Fehling's solution, add 300 mL of NH_4OH (sp. gr. 0.88) and dilute to 1 L with water.

Phenanthroline ferrous ion indicator. Dissolve 1.485 g of 1,10-phenanthroline monohydrate in 100 mL of 0.025 M ferrous sulfate solution.

Phenolphthalein. Dissolve 1 g of phenolphthalein in 50 mL of alcohol and add 50 mL of water.

Phenolsulfonic acid (determination of nitrogen as nitrate). Dissolve 25 g of phenol in 150 mL of conc. H_2SO_4, add 75 mL of fuming H_2SO_4 (15% SO_3), stir well and heat for 2 hours at 100°C.

Phloroglucinol solution (pentosans). Make a 3% phloroglucinol (1,3,5-benzenetriol) solution in alcohol. Keep in a dark bottle.

Phosphomolybdic acid (Sonnenschein's reagent for alkaloids).

1. Prepare ammonium phosphomolybdate and after washing with water, boil with nitric acid and expel NH_3; evaporate to dryness and dissolve in 2 M nitric acid.
2. Dissolve ammonium molybdate in HNO_3 and treat with phosphoric acid. Filter, wash the precipitate, and boil with aqua regia until the ammonium salt is decomposed. Evaporate to dryness. The residue dissolved in 10 % HNO_3 constitutes Sonnenschein's reagent.

Phosphoric acid — sulfuric acid mixture. Dilute 150 mL of conc. H_2SO_4 and 100 mL of conc. H_3PO_4 (85%) with water to a volume of 1 L.

Phosphotungstic acid (Schcibicr's reagent for alkaloids).

1. Dissolve 20 g of sodium tungstate and 15 g of sodium phosphate in 100 mL of water containing a little nitric acid.
2. The reagent is a 10% solution of phosphotungstic acid in water. Thc.phosphotungstic acid is prepared by evaporating a mixture of 10 g of sodium tungstate dissolved in 5 g of phosphoric acid (sp. gr. 1.13) and enough boiling water to effect solution. Crystals of phosphotungstic acid separate.

Picric acid (Hager's reagent for alkaloids, wool and silk). Dissolve 1 g of picric acid in 100 mL of water.

Potassium antimonate (reagent for sodium). Boil 22 g of potassium antimonate with 1 L of water until nearly all of the salt has dissolved, cool quickly, and add 35 mL of 10% potassium hydroxide. Filter after standing overnight.

Potassium-cadmium iodide (Marme's reagent for alkaloids). Add 2 g of CdI_2 to a boiling solution of 4 g of KI in 12 mL of water, and then mix with 12 mL of saturated KI solution.

Potassium hydroxide (for CO_2 absorption). Dissolve 360 g of KOH in water and dilute to 1 L.

Potassium iodide — mercuric iodide (Brucke's reagent for proteins). Dissolve 50 g of KI in 500 mL of water, and saturate with mercuric iodide (about 120 g). Dilute to 1 L.

Potassium pyrogallate (for oxygen absorption). For mixtures of gases containing less than 28% oxygen, add 100 mL of KOH solution (50 g of KOH to 100 mL of water) to 5 g of pyrogallol. For mixtures containing more than 28% oxygen the KOH solution should contain 120 g of KOH to 100 mL of water.

Pyrogallol, alkaline.

1. Dissolve 75 g of pyrogallic acid in 75 mL of water.
2. Dissolve 500 g of KOH in 250 mL of water. When cool, adjust until sp. gr. is 1.55.

For use, add 270 mL of solution (2) to 30 mL of solution (1).

Rosolic acid (indicator). Dissolve 1 g of rosolic acid in 10 mL of alcohol and add 100 mL of water.

Scheibler's reagent. See Phosphotungstic acid.

Schiff's reagent. See Fuchsin-sulfurous acid.

Schweitzer's reagent. See Cupric oxide, ammoniacal.

Soap solution (reagent for hardness in water). Dissolve 100 g of dry castile soap in 1 L of 80% alcohol (5 parts alcohol to 1 part water). Allow to stand several days and dilute with 70% to 80% alcohol until 6.4 mL produces a permanent lather with 20 mL of standard calcium solution. The latter solution is made by dissolving 0.2 g of $CaCO_3$ in a small amount of dilute HCl, evaporating to dryness and making up to 1 L.

Sodium bismuthate (oxidation of manganese). Heat 20 parts of NaOH nearly to redness in an iron or nickel crucible and add slowly 10 parts of basic bismuth nitrate which has been previously dried. Add 2 parts of sodium peroxide, and pour the brownish-yellow fused mass onto an iron plate to cool. When cold, break up in a mortar, extract with water, and collect on an asbestos filter.

Sodium hydroxide (for CO_2 absorption). Dissolve 330 g of NaOH in water and dilute to 1 L.

Sodium nitroprusside (reagent for hydrogen sulfide and wool). Use a freshly prepared solution of 1 g of sodium nitroferricyanide in 10 mL of water.

Sodium oxalate (primary standard). Dissolve 30 g of the commercial salt in 1 L of water, make slightly alkaline with sodium hydroxide, and let stand until perfectly clear. Filter and evaporate the filtrate to 100 mL. Cool and filter. Pulverize the residue and wash it several times with small volumes of water. The procedure is repeated until the mother liquor is free from sulfate and is neutral to phenolphthalein.

Sodium plumbite (reagent for wool). Dissolve 5 g of sodium hydroxide in 100 mL of water. Add 5 g of litharge (PbO) and boil until dissolved.

Sodium polysulfide. Dissolve 480 g of $Na_2S \cdot 9H_2O$ in 500 mL of water, add 40 g of NaOH and 18 g of sulfur. Stir thoroughly and dilute to 1 L with water.

Sonnenschein's reagent. See Phosphomolybdic acid.

Starch solution.

    1. Make a paste with 2 g of soluble starch and 0.01 g of HgI_2 with a small amount of water. Add the mixture slowly to 1 L of boiling water and boil for a few minutes. Keep in a glass stoppered bottle. If other than soluble starch is used, the solution will not clear on boiling; it should be allowed to stand and the clear liquid decanted.

    2. A solution of starch which keeps indefinitely is made as follows: Mix 500 mL of saturated NaCl solution (filtered), 80 mL of glacial acetic acid, 20 mL of water and 3 g of starch. Bring slowly to a boil and boil for 2 minutes.

    3. Make a paste with 1 g of soluble starch and 5 mg of HgI_2, using as little cold water as possible. Then pour about 200 mL of boiling water on the paste and stir immediately. This will give a clear solution if the paste is prepared correctly and the water actually boiling. Cool and add 4 g of KI. Starch solution decomposes on standing due to bacterial action, but this solution will keep well if stored under a layer of toluene.

Stoke's reagent. Dissolve 30 g of $FeSO_4$ and 20 g of tartaric acid in water and dilute to 1 L. Just before using, add concentrated NH_4OH until the precipitate first formed is redissolved.

Sulfanilic acid (reagent for nitrites). Dissolve 0.5 g of sulfanilic acid in a mixture of 15 mL of glacial acetic acid and 135 mL of recently boiled water.

Sulfomolybdic acid (Froehde's reagent for alkaloids and glucosides). Dissolve 10 g of molybdic acid or sodium molybdate in 100 mL of conc. H_2SO_4.

Tannic acid (reagent for albumin, alkaloids, and gelatin). Dissolve 10 g of tannic acid in 10 mL of alcohol and dilute with water to 100 mL.

Titration mixture. (residual chlorine in water analyasis). Prepare 1 L of dilute HCl (100 mL of HCl (sp. gr. 1.19) in sufficient water to make 1 L). Dissolve 1 g of *o*-tolidine in 100 mL of the dilute HCl and dilute to 1 L with dilute HCl solution.

Trinitrophenol solution. See Picric acid.

Turmeric tincture (reagent for borates). Digest ground turmeric root with several quantities of water which are discarded. Dry the residue and digest it several days with six times its weight of alcohol. Filter.

Uffelmann's reagent (turns yellow in presence of lactic acid). To a 2% solution of pure phenol in water, add a water solution of $FeCl_3$ until the phenol solution becomes violet in color.

Wagner's reagent. See Iodo-potassium iodide.

Wagner's solution (used in phosphate rock analysis to prevent precipitation of iron and aluminum). Dissolve 25 g of citric acid and 1 g of salicylic acid in water and dilute to 1 L. Use 50 mL of the reagent.

Wij's iodine monochloride solution (for iodine number). Dissolve 13 g of resublimed iodine in 1 L of glacial acetic acid which will pass the dichromate test for reducible matter. Set aside 25 mL of this solution. Pass into the remainder of the solution dry chlorine gas (dried and washed by passing through H_2SO_4 (sp. gr. 1.84)) until the characteristic color of free iodine has been discharged. Now add the iodine solution which was reserved, until all free chlorine has been destroyed. A slight excess of iodine does little or no harm, but an excess of chlorine must be avoided. Preserve in well stoppered, amber colored bottles. Avoid use of solutions which have been prepared for more than 30 days.

Wij's special solution (for iodine number). To 200 mL of glacial acetic acid that will pass the dichromate test for reducible matter, add 12 g of dichloramine T (*N,N*-dichloro-4-methyl-benzenesulfonamide), and 16.6 g of dry KI (in small quantities with continual shaking until all the KI has dissolved). Make up to 1 L with the same quality of acetic acid used above and preserve in a dark colored bottle.

Zimmermann-Reinhardt reagent (determination of iron). Dissolve 70 g of $MnSO_4 \cdot 4H_2O$ in 500 mL of water, add 125 mL of conc. H_2SO_4 and 125 mL of 85% H_3PO_4, and dilute to 1 L.

Zinc chloride solution, basic (reagent for silk). Dissolve 1000 g of zinc chloride in 850 mL of water, and add 40 g of zinc oxide. Heat until solution is complete.

Zinc uranyl acetate (reagent for sodium). Dissolve 10 g of $UO_2(C_2H_3O_2)_2 \cdot 2H_2O$ in 6 g of 30% acetic acid with heat, if necessary, and dilute to 50 mL. Dissolve 30 g of $Zn(C_2H_3O_2)_2 \cdot H_2O$ in 3 g of 30% acetic acid and dilute to 50 mL. Mix the two solutions, add 50 mg of NaCl, allow to stand overnight and filter.

STANDARD SOLUTIONS OF ACIDS, BASES, AND SALTS

For each compound listed, the last column of this table gives the mass in grams which is contained in 1 liter of a solution whose amount-of-substance concentration divided by the equivalence factor of the compound equals 0.1 mol/L. In the older literature such a solution is often referred to as a "decinormal solution" (0.1 N).

REFERENCE

Compendium of Analytical Nomenclature (IUPAC), Pergamon Press, Oxford, 1978.

Name	Formula	Atomic or molecular weight	Equivalence factor	Mass in grams
Acetic acid	$HC_2H_3O_2$	60.0530	1	6.0053
Ammonia	NH_3	17.0306	1	1.7031
Ammonium ion	NH_4^+	18.0386	1	1.8039
Ammonium chloride	NH_4Cl	53.4916	1	5.3492
Ammonium sulfate	$(NH_4)_2SO_4$	132.1388	1/2	6.6069
Ammonium thiocyanate	NH_4CNS	76.1204	1	7.6120
Barium	Ba	137.34	1/2	6.867
Barium carbonate	$BaCO_3$	197.3494	1/2	9.8675
Barium chloride hydrate	$BaCl_2 \cdot 2H_2O$	244.2767	1/2	12.2138
Barium hydroxide	$Ba(OH)_2$	171.3547	1/2	8.5677
Barium oxide	BaO	153.3394	1/2	7.6670
Bromine	Br	79.909	1	7.9909
Calcium	Ca	40.08	1/2	2.004
Calcium carbonate	$CaCO_3$	100.0894	1/2	5.0045
Calcium chloride	$CaCl_2$	110.9860	1/2	5.5493
Calcium chloride hydrate	$CaCl_2 \cdot 6H_2O$	219.0150	1/2	10.9508
Calcium hydroxide	$Ca(OH)_2$	74.0947	1/2	3.7047
Calcium oxide	CaO	56.0794	1/2	2.8040
Chlorine	Cl	35.453	1	3.5453
Citric acid	$C_6H_8O_7 \cdot H_2O$	210.1418	1/3	7.0047
Cobalt	Co	58.9332	1/2	2.9466
Copper	Cu	63.54	1/2	3.177
Copper oxide (cupric)	CuO	79.5394	1/2	3.9770
Copper sulfate hydrate	$CuSO_4 \cdot 5H_2O$	249.6783	1/2	12.4839
Hydrochloric acid	HCl	36.4610	1	3.6461
Hydrocyanic acid	HCN	27.0258	1	2.7026
Iodine	I	126.9044	1	12.6904
Lactic acid	$C_3H_6O_3$	90.0795	1	9.0080
Malic acid	$C_4H_6O_5$	134.0894	1/2	6.7045
Magnesium	Mg	24.312	1/2	1.2156
Magnesium carbonate	$MgCO_3$	84.3214	1/2	4.2161
Magnesium chloride	$MgCl_2$	95.2180	1/2	4.7609
Magnesium chloride hydrate	$MgCl_2 \cdot 6H_2O$	203.2370	1/2	10.1623
Magnesium oxide	MgO	40.3114	1/2	2.0156
Manganese	Mn	54.938	1/2	2.7469
Manganese sulfate	$MnSO_4$	150.9996	1/2	7.5500
Mercuric chloride	$HgCl_2$	271.4960	1/2	13.5748
Nickel	Ni	58.71	1/2	2.9356
Nitric acid	HNO_3	63.0129	1	6.3013
Oxalic acid	$H_2C_2O_4$	90.0358	1/2	4.5018
Oxalic acid hydrate	$H_2C_2O_4 \cdot 2H_2O$	126.0665	1/2	6.3033
Oxalic acid anhydride	C_2O_3	72.0205	1/2	3.6010
Phosphoric acid	H_3PO_4	97.9953	1/3	3.2665
Potassium	K	39.102	1	3.9102
Potassium bicarbonate	$KHCO_3$	100.1193	1	10.0119
Potassium carbonate	K_2CO_3	138.2134	1/2	6.9106
Potassium chloride	KCl	74.5550	1	7.4555

Name	Formula	Atomic or molecular weight	Equivalence factor	Mass in grams
Potassium cyanide	KCN	65.1199	1	6.5120
Potassium hydroxide	KOH	56.1094	1	5.6109
Potassium oxide	K_2O	94.2034	1/2	4.7102
Potassium tartrate	$K_2H_4C_4O_6$	226.2769	1/2	11.3139
Silver	Ag	107.87	1	10.787
Silver nitrate	$AgNO_3$	169.8749	1	16.9875
Sodium	Na	22.9898	1	2.2990
Sodium bicarbonate	$NaHCO_3$	84.0071	1	8.4007
Sodium carbonate	Na_2CO_3	105.9890	1/2	5.2995
Sodium chloride	NaCl	58.4428	1	5.8443
Sodium hydroxide	NaOH	39.9972	1	3.9997
Sodium oxide	Na_2O	61.9790	1/2	3.0990
Sodium sulfide	Na_2S	78.0436	1/2	3.9022
Succinic acid	$H_2C_4H_4O_4$	118.0900	1/2	5.9045
Sulfuric acid	H_2SO_4	98.0775	1/2	4.9039
Tartaric acid	$C_4H_6O_6$	150.0888	1/2	7.5044
Zinc	Zn	65.37	1/2	3.269
Zinc sulfate hydrate	$ZnSO_4 \cdot 7H_2O$	287.5390	1/2	14.3769

STANDARD SOLUTIONS OF OXIDATION AND REDUCTION REAGENTS

For each reagent listed, the last column of this table gives the mass in grams which is contained in a solution whose amount-of-substance concentration divided by the equivalence factor of the compound equals 0.1 mol/L. The equivalence factor given refers to the most common reactions of the reagent. In the older literature such a solution is often called a "decinormal solution" (0.1 N).

REFERENCE

Compendium of Analytical Nomenclature (IUPAC), Pergamon Press, Oxford, 1978.

Name	Formula	Atomic or molecular weight	Equivalence factor	Mass in grams
Antimony	Sb	121.75	1/2	6.0875
Arsenic	As	74.9216	1/2	3.7461
Arsenic trisulfide	As_2S_3	246.0352	1/4	6.1509
Arsenous oxide	As_2O_3	197.8414	1/4	4.9460
Barium peroxide	BaO_2	169.3388	1/2	8.4669
Barium peroxide hydrate	$BaO_2 \cdot 8H_2O$	313.4615	1/2	15.6730
Calcium	Ca	40.08	1/2	2.004
Calcium carbonate	$CaCO_3$	100.0894	1/2	5.0045
Calcium hypochlorite	$Ca(OCl)_2$	142.9848	1/4	3.5746
Calcium oxide	CaO	56.0794	1/2	2.8040
Chlorine	Cl	35.453	1	3.5453
Chromium trioxide	CrO_3	99.9942	1/3	3.3331
Ferrous ammonium sulfate	$FeSO_4(NH_4)SO_4 \cdot 6H_2O$	392.0764	1	39.2076
Hydroferrocyanic acid	$H_4Fe(CN)_6$	215.9860	1	21.5986
Hydrogen peroxide	H_2O_2	34.0147	1/2	1.7007
Hydrogen sulfide	H_2S	34.0799	1/2	1.7040
Iodine	I	126.9044	1	12.6904
Iron	Fe	55.847	1	5.5847
Iron oxide (ferrous)	FeO	71.8464	1	7.1846
Iron oxide (ferric)	Fe_2O_3	159.6922	1/2	7.9846
Lead peroxide	PbO_2	239.1888	1/2	11.9594
Manganese dioxide	MnO_2	86.9368	1/2	4.3468
Nitric acid	HNO_3	63.0129	1/3	2.1004
Nitrogen trioxide	N_2O_3	76.0116	1/4	1.9002
Nitrogen pentoxide	N_2O_5	108.0104	1/6	1.8001
Oxalic acid	$C_2H_2O_4$	90.0358	1/2	4.5018
Oxalic acid hydrate	$C_2H_2O_4 \cdot 2H_2O$	126.0665	1/2	6.3033
Oxygen	O	15.9994	1/2	0.8000
Potassium dichromate	$K_2Cr_2O_7$	294.1918	1/6	4.9032
Potassium chlorate	$KClO_3$	122.5532	1/6	2.0425
Potassium chromate	K_2CrO_4	194.1076	1/3	6.4733
Potassium ferrocyanide	$K_4Fe(CN)_6$	368.3621	1	36.8362
Potassium ferrocyanide hydrate	$K_4Fe(CN)_6 \cdot 3H_2O$	422.4081	1	42.2408
Potassium iodide	KI	166.0064	1	16.6006
Potassium nitrate	KNO_3	101.1069	1/3	3.3702
Potassium perchlorate	$KClO_4$	138.5526	1/8	1.7319
Potassium permanganate	$KMnO_4$	158.0376	1/5	3.1608
Sodium chlorate	$NaClO_3$	106.4410	1/6	1.7740
Sodium nitrate	$NaNO_3$	84.9947	1/3	2.8332
Sodium thiosulfate hydrate	$Na_2S_2O_3 \cdot 5H_2O$	248.1825	1	24.8183
Stannous chloride	$SnCl_2$	189.5960	1/2	9.4798
Stannous oxide	SnO	134.6894	1/2	6.7345
Sulfur dioxide	SO_2	64.0628	1/2	3.2031
Tin	Sn	118.69	1/2	5.935

ORGANIC ANALYTICAL REAGENTS FOR THE DETERMINATION OF INORGANIC SUBSTANCES

G. Ackermann, L. Sommer, and D. Thorburn Burns

Determination	Reagents	Ref.
Aluminium	Alizarin Red S	Onishi, Part II a, p 28. (5), Snell, *Metals I,* p 587. (7)
	Aluminon	Fries/Getrost, p 16. (2), Onishi, IIa, p 21. (5), Snell, *Metals I,* p 590. (7)
	Aluminon + Cetyltrimethylammonium bromide	Huaxue Shiji, *8,* 85, (1986)
	Chrome Azurol S	Onishi, Part IIa, p 26. (5), Snell, *Metals I,* p 605. (7)
	Chrome Azurol S + Cetyltrimethylammonium bromide	Marczenko, p 133 (3), Snell, *Metals I,* p 606. (7)
	Chromazol KS + Cetylpyridinium bromide	*Analyst, 107,* 428, (1982).
	Eriochrome Cyanine R	Fries/Getrost, p 19 (2), Onishi, Part IIa p 25. (5), Snell, *Metals I,* p 611. (7)
	Eriochrome Cyanine R + Cetyltrimethylammonium bromide	Snell, *Metals I,* p 613. (7), *Analyst, 107,* 1431, (1982).
	8-Hydroxyquinoline	Fries/Getrost, p 22 (2), Marczenko, p 131 (3), Onishi, Part IIa, p 31. (5), Snell, *Metals I, p* 622 (7)
Ammonia	Phenol + Sodium hypochlorite	Boltz, p 210 (1), Marczenko, p 413 (3), Snell, *Nonmetals,* p 604 (9)
Antimony	Brilliant Green	Onishi, Part IIa, p 102. (5), Snell, *Metals I,* p 384. (7)
	Bromopyrogallol Red	*Talanta, 13,* 507, (1966).
	Rhodamine B	Fries/Getrost, p 32, (2), Marczenko, p 141. (3), Onishi, Part IIa, p 93. (5), Snell, *Metals I,* p 404. (7)
	Silver diethyldithiocarbamate	Fries/Getrost, p 36. (2)
Arsenic	Silver diethyldithiocarbamate	Fries/Getrost, p 41. (2), Marczenko, p 153. (3), Onishi, Part IIa, p 153. (5), Snell, *Metals I,* p 370. (7)
Barium	Sulfonazo III	Fries/Getrost, p 46. (2), Snell, *Metals II,* p 1782. (8), Onishi, Part IIa, p 202. (5)
Beryllium	Beryllon II	Snell, *Metals I,* p 667. (7)
	Chrome Azurol S	Marczenko, p 163. (3), Snell, *Metals I,* p 672. (7)
	Chrome Azurol S + Cetyltrimethylammonium bromide	Marczenko, p 164. (3), Snell, *Metals I,* p 673. (7)
	Eriochrome Cyanine R	Snell, *Metals I,* p 675. (7), *Talanta, 31,* 249, (1984).
	Eriochrome Cyanine R + Cetyltrimethylammonium bromide	Zh, *Anal. Khim., 33,* 1298, (1978).
Bismuth	Dithizone	Onishi, Part IIa, p 262. (5), Snell, *Metals I,* p 303. (7)
	Pyrocatechol Violet	Fres. Z. *Anal. Chem., 186,* 418, (1962).
	Pyrocatechol Violet + Cetyltrimethylammonium bromide	Zh. *Anal. Khim., 38,* 216, (1983).
	Thiourea	Onishi, Part IIa, p 260. (5), Snell, *Metals I,* p 317. (7)
	Xylenol Orange	Friez/Getrost, p 57. (2), Marczenko, p 172. (3), Snell, *Metals I,* p 320. (7)
Boron	Azomethine H	Snell, Nonmetals, p 165. (9)
	Carminic acid	Boltz, p 14. (1), Fries/Getrost, p 65. (2), Snell, Nonmetals, p 170. (9), Williams, p 35. (11)
	Curcumin	Boltz, p 8. (1), Fries/Getrost, p 68. (2), Marczenko, p 180. (3), Snell, Nonmetals, p 180. (9), Fres. Z. *Anal. Chem., 323,* 266, (1986).

ORGANIC ANALYTICAL REAGENTS FOR THE DETERMINATION OF
INORGANIC SUBSTANCES (continued)

Determination	Reagents	Ref.
	Methylene Blue	Boltz, p 21. (1), Marczenko, p 183. (3), Snell, Nonmetals, p 205. (9), T*alanta, 31,* 547, (1984).
Bromide	Fluorescein	Boltz, p 48. (1), Snell, Nonmetals, p 276., Fres. Z. *Anal. Chem., 301,* 28 (1980).
	Phenol Red	Boltz, p 44. (1), Marczenko, p 190. (3), Snell, Nonmetals, p 28. (9)
Cadmium	2-(5-Bromo-2-pyridylazo)-5-diethylaminophenol	Marczenko, p 197. (3)
	Cadion	Onishi, Part IIa, p 323. (5)
	Dithizone	Fries/Getrost, p 78. (2), Onishi, Part IIa, p 315. (5), Snell, *Metals I,* p 279. (7), West, p 25. (10).
	4–(2-Pyridylazo)resorcinol	Fres. Z. *Anal. Chem., 310,* 51, (1982).
Calcium	Chlorophosphonazo III	Marczenko, p 207. (3), Snell, *Metals II,* p 1744. (8)
	Glyoxal-bis(2-hydroxyanil)	Fries/Getrost, p 86. (2), Onishi, Part IIa, p 352. (5), Snell, *Metals I,* p 1762. (8)
	Murexide	Onishi, Part IIa, p 357. (5), Snell, *Metals II,* p 1769. (8)
	Phthalein Purple	*Anal. Chim. Acta, 34,* 71 (1966).
Cerium	N-benzoyl-N-phenylhydroxylamine	*Anal. Chim. Acta, 48,* 155, (1969).
	8-Hydroxyquinoline	Fries/Getrost, p 93. (2), Marczenko, p 220. (3), Onishi, Part IIa, p 383. (7)
Chlorine	N,N-Diethyl-1,4-phenylenediamine	Boltz, p 92. (1), Fries/Getrost, p 101. (2), Snell, Nonmetals, p 225. (9), *Analyst, 90,* 187, (1965).
Chromium	1,5-Diphenylcarbazide	Fries/Getrost, p 105. (2), Onishi, Part IIa, p 412. (5), Snell, *Metals I,* p 714. (7), West, p 12. (10)
	4-(2-Pyridylazo)resorcinol	Snell, *Metals I,* p 736. (7), West, p 17. (10)
	4-(2-Pyridylazo)resorcinol + Tetradecyldimethyl-benzylammonium chloride	West, p 17. (10), *Anal. Chim. Acta, 67,* 297, (1973).
	4-(2-Pyridylazo)resorcinol + Hydrogen peroxide	Fres. Z. *Anal. Chem., 304,* 382, (1980).
Cobalt	Nitroso-R salt	Fries/Getrost, p 118. (2), Onishi, Part IIa, p 454. (5), Snell, *Metals I,* p 953. (7)
	1-Nitroso-2-naphthol	Fries/Getrost, p 111. (2), Marczenko, p 246. (3), Snell, *Metals I,* p 947. (5)
	2-Nitroso-1-naphthol	Fries/Getrost, p 113. (2), Onishi, Part IIa, p 459. (5), Snell, *Metals I,* p 949. (7), West, p 45. (10)
	4-(2-Pyridylazo)resorcinol	Snell, *Metals I,* p 969. (7), West, p 44. (10)
	4-(2-Pyridylazo)resorcinol + Diphenylguanidine	Zh. *Anal. Khim., 35,* 1306, (1980).
Copper	Bathocuproine	Fries/Getrost, p 135. (2), Snell, *Metals I,* p 148. (7)
	Bathocuproine disulfonic acid	Fries/Getrost, p 137. (2), West, p 52. (10)
	Dithizone	Marczenko, p 258. (3), Onishi, Part IIa, p 529. (5), Snell, *Metals I,* p 199. (7)
	Neocuproine	Snell, *Metals I,* p 217. (5), West, p 51. (10)
	Cuprizone	Onishi, Part IIa, p 534. (5), Snell, *Metals I,* p 157. (7), West, p 53. (10)
	4-(2-pyridylazo)resorcinol + Tetradecyldimethyl-benzylammonium chloride	*Anal. Chim.* Acta, *138,* 321, (1982).
Cyanide	Barbituric Acid + Pyridine	Fries/Getrost, p 153. (2), Snell, Nonmetals, p 653. (9)
	Barbituric Acid + Pyridine-4-carboxylic acid	*Anal. Chim. Acta, 99,* 197, (1978).

Determination	Reagents	Ref.
Fluoride	Alizarin Fluorine blue + Lanthanum(III) ion	Boltz, p 129. (1), Fries/Getrost, p 158. (2), Snell, *Nonmetals,* p 333. (9), Williams, p 354. (11)
	Eriochrome Cyanine R + Zirconium(IV) ion	Boltz, p 119. (1), Snell, *Nonmetals*, p 359. (2), Williams, p 357. (10)
Gallium	Pyrocatechol violet + Diphenylguanidine	Snell, *Metals I,* p 500. (7)
	8-Hydroxyquinoline	Onishi Pt IIa, p 582. (5), Snell, *Metals I,* p 505. (7)
	1-(2-Pyridylazo)-2-naphthol	Snell, *Metals I,* p 512. (7)
	4-(2-Pyridylazo)resorcinol	Snell, *Metals I,* p 513. (7)
	Rhodamine B	Marczenko, p 284. (3), Onishi, Part IIa, p 578. (5), Snell, *Metals I,* p 515. (7)
	Xylenol Orange	Fries/Getrost, p 166. (2), Snell, *Metals I,* p 523. (7)
	Xylenol Orange + 8-Hydroxyquinoline	*Zh. Anal. Khim.*, *26*, 75, (1971).
Germanium	Brilliant Green + Molybdate	Snell, *Metals I,* p 562. (7)
	Phenylfluorone	Fries/Getrost, p 168. (2), Marczenko, p 292. (3), Onishi, Part IIa, p 607. (5), Snell, *Metals I,* p 570. (7)
Gold	5-(4-Diethylaminobenzylidene) rhodanine	Fries/Getrost, p 173. (2), Onishi, Part IIa, p 631. (5), Snell, *Metals II,* p 1516. (8)
	Rhodamine B	Fries/Getrost, p 175. (2), Marczenko, p 301. (3), Onishi, Part IIa, p 637. (5), Snell, *Metals II,* p 513. (8)
Hafnium	Arsenazo III	Snell, *Metals II,* p 1184. (8), *Talanta, 19,* 807, (1972).
Indium	Bromopyrogallol Red	Snell, *Metals I,* p 469. (7)
	Chrome Azurol S	Snell, *Metals I,* p 474. (7)
	Chrome Azurol S + Cetyltrimethylammonium bromide	*Anal. Chim. Acta, 67,* 107, (1973).
	Dithizone	Fries/Getrost, p 179. (2), Onishi, Part IIa, p 672. (5), Snell, *Metals I,* p 474. (7)
	8-Hydroxyquinoline	Onishi, Part IIa, p 670. (5), Snell, *Metals I,* p 475. (7)
	1-(2-Pyridylazo)-2-naphthol	Snell, *Metals I,* p 480. (7)
	4-(2-Pyridylazo)resorcinol	Marczenko, p 309. (3), Snell, Metals I, p 480. (7)
Iodide	Neocuproine + Copper(II)	*Anal. Chim. Acta, 69,* 321, (1974).
Iodine	Starch	Boltz, p 162. (1), Marczenko, p 316. (3), Snell, *Nonmetals,* p 307. (9)
Iridium	Rhodamine 6G + Tin(II)	Marczenko, p 323. (3)
	N,N-Dimethyl-4-nitrosoaniline	*Anal. Chem.*, *27*, 1776, (1955).
Iron	Bathophenanthroline	Fries/Getrost, p 189. (2), Onishi, Part IIa, p 729. (5), Snell, *Metals I,* p 763. (7)
	Bathophenanthroline disulfonic acid	Fries/Getrost, p 191. (2), Snell, *Metals I,* p 772. (7)
	2,2′-Bipyridyl	Snell, *Metals I,* p 750. (7)
	Chrome Azurol S + Cetyltrimethylammonium bromide	Snell, *Metals I,* p 757. (7), *Coll. Czech. Chem. Comm., 45,* 2656, (1980).
	1,10-Phenanthroline	Fries/Getrost, p 199. (2), Marczenko, p 331. (3), Onishi, Part IIa, p 725. (5), Snell, *Metals I,* p 795. (7)
	1,10-Phenanthroline + Bromothymol Blue	*Zh. Anal. Khim.*, *25*, 1348, (1970).
	Ferrozine	Onishi, Part IIa, p 730. (5), Snell, *Metals I,* p 783. (7)
Lanthanum	Arsenazo III	Marczenko, p 468. (3), Snell, *Metals II,* p 1910. (8)

Determination	Reagents	Ref.
Lead	Dithizone	Fries/Getrost, p 207. (2), Onishi, Part IIa, p 824. (5), Snell, *Metals I*, p 2. (7), West, p 34. (10)
	Sodium diethyldithiocarbamate	Fries/Getrost, p 214. (2), Snell, *Metals I*, p 27. (7)
	4-(2-Pyridylazo)resorcinol	Fries/Getrost, p 220. (2), Marczenko, p 347. (3), Snell, *Metals I*, p 34. (7)
Lithium	Thoron	Onishi, Part IIa, p 863. (5), Snell, *Metals II*, p 1726. (8), T*alanta, 30*, 587, (1983).
Magnesium	Eriochrome Black T	Fries/Getrost, p 226. (2), Marczenko, p 355. (3), Onishi, Part IIb, p 13. (6), Snell, *Metals II*, p 1932. (8)
	8-Hydroxyquinoline	Onishi, Part IIb, p 11. (6), Snell, *Metals II*, p 1938. (8)
	8-Hydroxyquinoline + Butylamine	Fries/Getrost, p 228. (2), Snell, *Metals II*, p 1938. (8)
	Titan Yellow	Fries/Getrost, p 234. (2), Marczenko, p 352. (3), Snell, *Metals II*, p 1945. (8)
	Xylidyl Blue	Fries/Getrost, p 231. (2), Onishi, Part IIb, p 14. (6), Snell, *Metals II*, p 1950. (8)
Manganese	Formaldoxime	Fries/Getrost, p 236. (2), Marczenko, p 364. (3), Onishi Part IIb, p 38. (6), Snell, *Metals II*, 1010. (8)
Mercury	Dithizone	Fries/Getrost, p 243. (2), Marczenko, p 373. (3), Onishi, Part IIb, p 66. (6), Snell, *Metals I*, p 107. (7), West, p 29. (10)
	Michler's thioketone	Marczenko, p 375. (3), Snell, *Metals I*, p 126. (7)
Molybdenum	Xylenol Orange	*Talanta, 16*, 1023, (1969)
	Bromopyrogallol Red + Cetylpyridium chloride	West, p 58. (10)
	Phenylfluorone	Snell, *Metals II*, p 1311., *Microchem. J.*, *31*, 56, (1985).
	Toluene-3,4-dithiol	Fries/Getrost, p 251. (2), Marczenko, p 384. (3), Onishi, Part IIb, p 96. (6), Snell, *Metals II*, p 1301. (8)
Nickel	2-(5-Bromo-2-pyridylazo)-5-diethylaminophenol	Marczenko, p 397. (3), *Talanta 28*, 189, (1981).
	Dimethylglyoxime	Fries/Getrost, p 263. (2), Marczenko, p 393. (3), Onishi, Part IIb, p 125. (6), Snell, *Metals I*, p 887. (7)
	Dimethylglyoxime + Oxidant	Fries/Getrost, p 263. (2), Onishi, Part IIb, p 125. (6), Snell, *Metals I*, p 887. (7)
	2,2′-Furildioxime	Marczenko, p 396. (3), Snell, *Metals I*, p 904. (7)
	2-(2-Pyridylazo)-2-naphthol	Snell, *Metals I*, p 910. (7)
	4-(2-Pyridylazo)resorcinol	Snell, *Metals I*, p 911. (7), West, p 39. (10), *Anal. Chim. Acta, 82*, 431, (1976).
Niobium	N-Benzoyl-N-phenylhydroxylamine	Snell, *Metals II*, p 1425. (8)
	Pyrocatechol + EDTA or 2,2′Bipyridyl or 1-(2-thenoyl)-3,3,3,-trifluoroacetone	Snell, *Metals II*, p 1427. (8)
	Bromopyrogallol red	Marczenko, p 407. (3), Snell, *Metals II*, p 1426. (8)
	Bromopyrogallol red + Cetylpyridinium chloride	*Talanta, 32*, 189, (1985).
	4-(2-Pyridylazo)resorcinol	Fries/Getrost, p 274. (2), Marczenko, p 406. (3), Onishi, Part IIb, p 160. (7), Snell, *Metals II*, p 1447. (8)

Determination	Reagents	Ref.
	Sulfochlorophenol S	Onishi, Part IIb, p 161. (7), Snell, *Metals II*, p 1430. (8)
		Onishi, Part IIb, p 164. (7)
Nitrate	Xylenol Orange	Boltz, p 227. (1), Fries/Getrost, p 280. (2),
	Brucine	Snell, *Nonmetals*, p 546. (9)
	Chromotropic acid	Boltz, p 229. (1), Fries/Getrost, p 281. (2), Snell, *Nonmetals*, p 548. (9), Williams, p 132. (11), *Fres. Z. Anal. Chem.*, *320*, 490, (1985).
	Sulfanilamide + *N*-(1-Naphthyl)ethylenediamine dihydrochloride	Fries/Getrost, p 279. (2), Snell, *Nonmetals*, p 559. (9)
Nitrite	Sulfanilamide + *N*-(1-Naphthyl)ethylenediamine dihydrochlorine	Boltz, p 241. (1), Snell, *Nonmetals*, *p* 585. (8), *Analyst*, *109*, 1281, (1984).
	Sulfanilic acid + 1-Naphthylamine	Boltz, p 237. (1), Fries/Getrost, p 285. (2), Marczenko, p 419. (3), Snell, *Nonmetals*, p 586. (9)
Osmium	1,5-Diphenylcarbazide	Marczenko, p 428. (3)
Palladium	2-(5 Bromo-2-pyridylazo)-5-diethylaminophenol	*Talanta*, *33*, 939, (1986).
	Dithizone	Marczenko, p 440. (3), Onishi, Part IIb, p 227. (6), Snell, *Metals II*, p 1577. (8)
	2-Nitroso-1-naphthol	Fries/Getrost, p 294. (2), Onishi, Part IIb, p 226. (6), Snell, *Metals II*, p 1581. (8)
	4-(2-Pyridylazo)resorcinol	Snell, *Metals II*, p 1583. (8) *Analyst*, *107*, 708, (1982).
Phosphate	Rhodamine B + Molybdate	Snell, *Nonmetals*, p 103. (9)
	Malachite Green + Molybdate	Snell, *Nonmetals*, p 12. (9), Analyst, *108*, 361, (1983).
Platinum	Sulfochlorophenolazorhodamine	Onishi, Part IIb, p 253. (6), *Talanta*, *34*, 87, (1987).
	Dithizone	Fries/Getrost, p 300. (2), Onishi, Part IIb, p 253. (6), Snell, *Metals II*, p 1534. (8)
	2-Mercaptobenzothiazole	Fries/Getrost, p 302. (2), *Zh. Anal. Khim.*, *24*, 1172, (1969).
Rare Earths	Arsenazo I	Marczenko, p 470. (3), Onishi, Part IIa, p 785. (5), Snell, *Metals II*, p 1857. (8)
	Arsenazo III	Fries/Getrost, p 309. (2), Marczenko, p 468. (3), Onishi, Part IIa, p 786. (5), Snell, *Metals II*, p 1862. (8)
	Xylenol Orange	Onishi, Part IIa, p 787. (5), Snell, *Metals II*, p 1874. (8)
Rhenium	2,2′-Furildioxime	Fries/Getrost, p 310. (2), Marczenko, p 481. (3), Onishi, Part IIb, p 288. (6), Snell, *Metals II*, p 1659. (8)
Rhodium	1-(2-Pyridylazo)-2-naphthol	Fries/Getrost, p 311. (2), Snell, *Metals II*, p 1553. (8)
Ruthenium	1,10-Phenanthroline	Onishi, Part IIb, p 331. (6), Snell, *Metals II*, p 1623. (8)
	Thiourea	Fries/Getrost, p 318. (2), Onishi, Part IIb, 329. (6), Snell, *Metals II*, p 1626. (8)
	1,4-Diphenylthiosemicarbazide	Marczenko, p 493. (3), Onishi, Part IIb, p 330. (8)
Scandium	Alizarin red S	Fries/Getrost, p 319. (2), Onishi, Part IIb, p 360. (6), Snell, *Metals I*, p 536. (7)
	Arsenazo III	Onishi, Part IIb, p 359. (6), Snell, *Metals I*, p 539. (7)

Determination	Reagents	Ref.
	Chrome Azurol S	Snell, Metals I, p 551. (7), *Anal. Chim. Acta*, 159, 309, (1984).
	Xylenol Orange	Marczenko, p 501. (3), Onishi, Part IIb, p 357. (6), Snell, *Metals I*, p 547. (7)
Selenium	3,3′-Diaminobenzidine	Boltz, p 391. (1), Fries/Getrost, p 323. (2), Marczenko, p 508. (3), Snell, *Nonmetals*, p 490. (9), West, p 4. (10).
	2,3-Diaminonaphthaline	Snell, *Nonmetals*, p 501. (9)
Silver	Dithizone	Fries/Getrost, p 328. (2), Marczenko, p 524. (3), Onishi, Part IIb, p 379. (6), Snell, *Metals I*, p 82. (7)
	Eosin + 1,10-Phenanthroline	Snell, *Metals I*, p 93. (7)
Sulfate	Methylthymol blue + Barium (II)	Snell, *Nonmetals*, p 457. (9)
Sulfide	N,N,-Dimethyl-1,4-phenylenediamine	Boltz, p 483. (1), Fries/Getrost, p 344. (2), Snell, *Nonmetals*, p 400. (9), Williams, p 578. (11)
Sulfite	Pararosaniline + Formaldehyde	Boltz, p 478. (1), Marczenko, p 540. (3), Snell, *Nonmetals*, p 430. (9), Williams, p 591. (11)
Tantalum	Methyl Violet	Marczenko, p 551. (3), Snell, *Metals II*, p 1485. (8)
	4-(2-Pyridylazo)resorcinol	Snell, *Metals II*, p 1488. (8)
	Phenylfluorone	Onishi, Part IIb, p 166. (6), Snell, *Metals II*, p 1486. (8)
Tellurium	Diethyldithiocarbamate	Boltz, p 402. (1), Fries/Getrost, p 348. (2), Snell, *Nonmetals*, p 533. (9), Williams, p 220. (10)
	Bismuthiol II	Boltz, p 401. (1), Marczenko, p 557. (3), Snell, *Nonmetals*, p 524. (9)
Thallium	Brilliant green	Fries/Getrost, p 352. (2), Marczenko, p 567. (3), Onishi, Part IIb, p 426. (6), Snell, *Metals I*, p 45. (7)
	Dithizone	Fries/Getrost, p 355. (2), Onishi, Part IIb, p 426. (6), Snell, *Metals I*, p 54. (7)
	Rhodamine B	Fries/Getrost, p 354. (2), Marczenko, p 566. (3), Onishi, Part IIb, p 424. (6), Snell, *Metals I*, p 63. (7)
Thorium	Arsenazo III	Fries/Getrost, p 360. (2), Marczenko, p 575. (3), Onishi, Part IIb, p 460. (6), Snell, *Metals II*, p 1820. (8)
	Thoron	Marczenko, p 574. (3), Onishi, Part IIb, p 463. (6), Snell, *Metals I*, p 1835. (7)
	Xylenol Orange	Snell, *Metals I*, p 1852. (7)
	Xylenol Orange + Cetyltrimethylammonium bromide	*Talanta*, 26, 499, (1979).
Tin	Pyrocatechol violet (and + Cetyltrimethylammonium bromide)	Marczenko, p 585. (3), Onishi, Part IIb, p 501. (6), Snell, *Metals I, p* 422. (7)
	Gallein	Onishi, Part IIb, p 507, 510. (6), Snell, *Metals I,* p 432. (7)
	Phenylfluorone	Fries/Getrost, p 368. (2), Marczenko, p 582. (3), Onishi, Part IIb, p 497. (6), Snell, *Metals I,* p 444. (7)
	Toluene-3,4-dithiol + Dispersant	Fries/Getrost, p 366. (2), Onishi, Part IIb, p 502. (6), Snell, *Metals I,* p 427. (7)
Titanium	Chromotropic acid	Marczenko, p 593. (3), Onishi, Part IIb, p 551. (6), Snell, *Metals II,* p 1080. (8)

Determination	Reagents	Ref.
	Diantipyrinylmethane	Onishi, Part IIb, p 545. (6), Snell, *Metals II,* 1085. (8)
	Tiron	Fries/Getrost, p 376. (2), Onishi, Part IIb, p 549. (6), Snell, *Metals II,* p 1114. (8)
Tungsten	Pyrocatechol Violet	Snell, Metals II, p 1265. (8)
	Tetraphenylarsonium chloride + Thiocyanate	Onishi, Part IIb, p 596. (6), Snell, *Metals II,* p 1278. (8)
	Toluene-3,5-dithiol	Marczenko, p 605. (3), Onishi, Part IIb, p 590. (6), Snell, *Metals II,* 1267. (8)
Uranium	Arsenazo III	Marczenko, p 611. (3), Onishi, Part IIb, p 627. (6), Snell, *Metals II,* p 1356. (8)
	2-(5-Bromo-2-pyridylazo)diethylaminophenol	Fries/Getrost, p 388. (2), Onishi, Part IIb, p 625. (6)
	Chlorophosphonazo III	Snell, Metals II, p 1367. (8), *Fres. Z. Anal. Chem., 306,* 110, (1981).
	1-(2-Pyridylazo)-2-naphthol	Fries/Getrost, p 386. (2), Onishi, Part IIb, p 625. (6), Snell, *Metals II,* p 1387. (8)
Vanadium	*N*-Benzoyl-*N*-phenylhydroxylamine	Fries/Getrost, p 395. (2), Marczenko, p 625. (3), Snell, *Metals II,* p 1196. (8)
	8-Hydroxyquinoline	Marczenko, p 623. (3), Snell, *Metals II,* p 1209. (8)
	4-(2-pyridylazo)resorcinol	Fries/Getrost, p 404. (23), Marczenko, p 628. (3), Onishi, Part IIb, p 625. (6), Snell, *Metals II,* p 1226. (8)
Yttrium	Alizarin Red S	Fries/Getrost, p 406. (2), Onishi, Part IIa, p 784. (5), Snell, *Metals II,* p 1919. (8)
	Arsenazo III	Marczenko, p 468. (3), Onishi, Part IIa, p 786. (5), Snell, *Metals II,* p 1921. (8)
	Xylenol Orange	Fries/Getrost, p 406. (2), Onishi, Part IIa, p 787. (5), Snell, *Metals II,* p 1923. (8)
Zinc	Dithizone	Fries/Getrost, p 408. (2), Marczenko, p 637. (3), Onishi, Part IIb, p 708. (6), Snell, *Metals II,* p 1042. (8)
	1-(2-Pyridylazo)-2-naphthol	Marczenko, p 639. (3), Onishi, Part IIb, p 719. (6), Snell, *Metals II,* p 1056. (8)
	Xylenol Orange	Fries/Getrost, p 417. (2), Snell, *Metals II,* 1062. (8), Talanta, *26,* 693, (1979).
	Zircon	Fries/Getrost, p 412. (2), Onishi, Part IIb, p 719. (6), Snell, *Metals II,* 1063. (8), West, p 23. (10)
Zirconium	Alizarin Red S	Fries/Getrost, p 421. (2), Marczenko, p 647. (3), Onishi, Part IIb, p 763. (6), Snell, *Metals II,* p 1136. (8)
	Arsenazo III	Fries/Getrost, p 421. (2), Onishi, Part IIb, p 770. (6), Snell, *Metals II,* p 1143. (8)
	Pyrocatechol Violet	Onishi, Part IIb, p 771. (6), Snell, *Metals II,* p 1149. (8)
	Morin	Fries/Getrost, p 424. (2), Onishi, Part IIb, p 765. (6), Snell, *Metals II,* p 1158. (8)
	Xylenol Orange	Fries/Getrost, p 419. (2), Marczenko, p 648. (3), Onishi, Part IIb, p 767. (6), Snell, *Metals II,* p 1167. (8)

ORGANIC ANALYTICAL REAGENTS FOR THE DETERMINATION OF INORGANIC SUBSTANCES (continued)

REVIEWS

Sommer, L, Ackermann, G., Thorburn Burns, D., and Savvin, S. B., *Pure and Applied Chem.,* 62, 2147, 1990.

Sommer, L., Ackermann, G., and Thorburn Burns, D., *Pure and Applied Chem.,* 62, 2323, 1990)

Sommer, L., Komarek, J., and Thorburn Burns, D., *Pure and Applied Chem., 64,* 213, 1992.

Savvin, S. B., *Crit. Rev. Anal. Chem.,* 8, 55, 1979.

MONOGRAPHS

1. Boltz, D. F., and Howell, J. A., *Colorimetric Determination of Nonmetals,* 2nd ed, Wiley, New York, 1978.

2. Fries, J. and Getrost, H., *Organic Reagents for Trace Analysis,* E Merck, Darmstadt, 1977.

3. Marczenko, Z., *Separation and Spectrophotometric Determination of Elements,* Ellis Horwood, Chichester, 1986.

4. Sandell, E. B. and Onishi, H., *Photometric Determination of Traces of Metals. General Aspects, Part I,* 4th ed, J. Wiley, New York, 1978.

5. Onishi, H., *Photometric Determination of Traces of Metals. Part IIa: Individual Metals, Aluminium to Lithium,* 4th ed, J. Wiley, New York, 1986.

6. Onishi, H., *Photometric Determination of Traces of Metals. Part IIb: Individual Metals, Magnesium to Zinc,* 4th ed, J. Wiley, New York, 1989.

7. Snell, F. D., *Photometric and Fluorimetric Methods of Analysis, Metals Part 1,* J. Wiley, New York, 1978.

8. Snell, F. D., *Photometric and Fluorimetric Methods of Analysis, Metals Part 2,* J. Wiley, New York, 1978.

9. Snell, F. D., *Photometric and Fluorimetric Methods of Analysis, Nonmetals,* J. Wiley, New York, 1981.

10. West, T. S. and Nürnberg, H. W., Eds., *The Determination of Trace Metals in Natural Waters,* Blackwell, Oxford, 1988.

11. Williams, W. J.,, *Handbook of Anion Determination,* Butterworth, London, 1979.

12. Townshend, A., Burns, D. T., Guilbault, G. G., Lobinski, R., Mavenzenko, Z., Newman, E., and Onishi, H., *Dictionary of Analytical Reagents,* Chapman & Hall, London, 1993.

ACID-BASE INDICATORS

A. K. Covington

The first part of this table lists some common acid-base indicators in alphabetical order along with the approximate pH range(s) at which a color change occurs. Following this is a table of the same indicators ordered by pH range, which includes the nature of the color change, instructions on preparation of the indicator solution, and the acid dissociation constant pK, when available. The color code is:

C = colorless   A = amber   B/G = blue-green   Pk = pink   Y = yellow   V = violet   R = red   B = blue

P = purple   O = orange

REFERENCE

Bishop, E., Ed., *Indicators*, Pergamon, Oxford, 1972.

Indicator	pH Range	Indicator	pH Range
Alizarin	5.6-7.2; 11.0-12.4	Erythrosin, disodium salt	2.2-3.6
Alizarin Red S	4.6-6.0	4-(p-Ethoxyphenylazo)-m-	
Alizarin Yellow R	10.1-12.0	phenylene-diamine	
Benzopurpurine 4B	2.2-4.2	monohydrochloride	4.4-5.8
4,4'-Bis(2-amino-1-		Ethyl bis(2,4-dimethylphenyl)	
naphthylazo)-2,2'-		ethanoate	8.4-9.6
stilbenedisulfonic acid	3.0-4.0	Ethyl Orange	3.4-4.8
4,4'-Bis(4-amino-1-		Ethyl Red	4.0-5.8
naphthylazo)-2,2'-		Ethyl Violet	0.0-2.4
stilbenedisulfonic acid	8.0-9.0	5,5'-Indigodisulfonic acid,	
Brilliant Yellow	6.6-7.8	disodium salt	11.4-13.0
Bromocresol Green	3.8-5.4	Malachite Green	0.2-1.8
Bromocresol Purple	5.2-6.8	Metacresol Purple	1.2-2.8; 7.4-9.0
Bromophenol Blue	3.0-4.6	Metanil Yellow	1.2-2.4
Bromothymol Blue	6.0-7.6	Methyl Green	0.2-1.8
Chlorophenol Red	5.2-6.8	Methyl Orange	3.2-4.4
Clayton Yellow	12.2-13.2	Methyl Red	4.8-6.0
Congo Red	3.0-5.0	Methyl Violet	0.0-1.6
o-Cresolphthalein	8.2-9.8	p-Naphtholbenzein	8.2-10.0
Cresol Red	0.0-1.0; 7.0-8.8	Neutral Red	6.8-8.0
Crystal Violet	0.0-1.8	p-Nitrophenol	5.4-6.6
Curcumin (Turmaric)	7.4-8.6	m-Nitrophenol	6.8-8.6
p-(2,4-Dihydroxyphenylazo)		Orange IV	1.4-2.8
benzenesulfonic acid,		Paramethyl Red	1.0-3.0
sodium salt	11.4-12.6	Phenolphthalein	8.2-10.0
p-Dimethylaminoazobenzene	2.8-4.4	Phenol Red	6.6-8.0
4-(4-Dimethylamino-1-		4-Phenylazodiphenylamine	1.2-2.6
naphylazo)-3-		4-Phenylazo-1-naphthylamine	4.0-5.6
methoxybenzenesulfonic		Propyl Red	4.8-6.6
acid	3.5-4.8	Quinaldine Red	1.4-3.2
2-(p-Dimethylamino-		Resazurin	3.8-6.4
phenylazo)pyridine	0.2-1.8; 4.4-5.6	Resorcin Blue	4.4-6.2
N,N-Dimethyl-p-		Tetrabromophenolphthalein	
(m-tolylazo)aniline	2.6-4.8	ethyl ester, potassium salt	3.0-4.2
2,4-Dinitrophenol	2.0-4.7	Thymol Blue	1.2-2.8; 8.0-9.6
2-(2,4 Dinitrophenylazo)-1-		Thymolphthalein	9.4-10.6
naphthol-3,6-disulfonic		4-o-Tolylazo-o-toluidine	1.4-2.8
acid, disodium salt	6.0-7.0	1,3,5-Trinitrobenzene	12.0-14.0
6,8-Dinitro-2,4-		2,4,6-Trinitrotoluene	11.5-13.0
(1H)quinazolinedione	6.4-8.0	Turmaric	7.4-8.6

pH range	Color change	Indicator	pK	Preparation
0.0-1.0	R-Y	Cresol Red		0.1 g in 26.2 mL 0.01 M NaOH + 223.8 mL water
0.0-1.6	Y-B	Methyl Violet		0.01-0.05% in water
0.0-1.8	Y-B	Crystal Violet		0.02% in water
0.0-2.4	Y-B	Ethyl Violet		0.1 g in 50 mL 50% v/v methanol-water
0.2-1.8	Y-B/G	Malachite Green	1.3	water
0.2-1.8	Y-B	Methyl Green		0.1% in water
0.2-1.8	Y-R	2-(p-Dimethylaminophenylazo)pyridine		0.1% in ethanol
1.0-3.0	R-Y	Paramethyl Red		ethanol
1.2-2.4	R-Y	Metanil Yellow		0.01% in water
1.2-2.6	R-Y	4-Phenylazodiphenylamine		0.01 g in 1 mL 1 M HCl + 50 mL ethanol + 49 mL water
1.2-2.8	R-Y	Thymol Blue	1.65	0.1 g in 21.5 mL 0.01 M NaOH + 228.5 mL water
1.2-2.8	R-Y	Metacresol Purple	1.51	0.1 g in 26.2 mL 0.01 M NaOH + 223.8 mL water
1.4-2.8	R-Y	Orange IV		0.01% in water
1.4-2.8	O-Y	4-o-Tolylazo-o-toluidine		water
1.4-3.2	C-R	Quinaldine Red	2.63	1% in ethanol
2.0-4.7	C-Y	2,4-Dinitrophenol	3.96	sat. solution in water
2.2-3.6	O-R	Erythrosin, disodium salt		0.1% in water
2.2-4.2	V-R	Benzopurpurine 4B		0.1% in water
2.6-4.8	R-Y	N,N-Dimethyl-p-(m-tolylazo)aniline		0.1% in water
2.8-4.4	R-Y	p-Dimethylaminoazobenzene		0.1 g in 100 mL 90% v/v ethanol-water
3.0-4.0	P-R	4,4'-Bis(2-amino-1-naphthylazo)-2,2'-stilbenedisulfonic acid		0.1 g in 5.9 mL 0.05 M NaOH + 94.1 mL water
3.0-4.2	Y-B	Tetrabromophenolphthalein ethyl ester, potassium salt		0.1% in ethanol
3.0-4.6	Y-B	Bromophenol Blue	4.10	0.1 g in 14.9 mL 0.01 M NaOH + 235.1 mL water
3.0-5.0	B-R	Congo Red		0.1% in water
3.2-4.4	R-Y	Methyl Orange	3.46	0.1% in water
3.4-4.8	R-Y	Ethyl Orange	4.34	0.05-0.2% in water or aqueous ethanol
3.5-4.8	V-Y	4-(4-Dimethylamino-1-naphylazo)-3-methoxybenzenesulfonic acid		0.1% in 60% ethanol-water
3.8-5.4	Y-B	Bromocresol Green	4.90	0.1 g in 14.3 mL 0.01 M NaOH + 235.7 mL water
3.8-6.4	O-V	Resazurin		water
4.0-5.6	R-Y	4-Phenylazo-1-naphthylamine		0.1% in ethanol
4.0-5.8	C-R	Ethyl Red	5.42	0.1 g in 100 mL 50% v/v methanol-water
4.4-5.6	R-Y	2-(p-Dimethylaminophenylazo)pyridine		0.1% in ethanol
4.4-5.8	O-Y	4-(p-Ethoxyphenylazo)-m-phenylene-diamine monohydrochloride		0.1% in water
4.4-6.2	R-B	Resorcin Blue		0.2% in ethanol
4.6-6.0	Y-R	Alizarin Red S		water
4.8-6.0	R-Y	Methyl Red	5.00	0.02 g in 100 mL 60% v/v ethanol-water
4.8-6.6	R-Y	Propyl Red	5.48	ethanol
5.2-6.8	Y-P	Bromocresol Purple	6.40	0.1 g in 18.5 mL 0.01 M NaOH + 231.5 mL water
5.2-6.8	Y-R	Chlorophenol Red	6.25	0.1 g in 23.6 mL 0.01 M NaOH + 226.4 mL water
5.4-6.6	C-Y	p-Nitrophenol	7.15	0.1% in water
5.6-7.2	Y-R	Alizarin		0.1% in methanol
6.0-7.0	Y-B	2-(2,4-Dinitrophenylazo)-1-naphthol-3,6-disulfonic acid, disodium salt		0.1% in water
6.0-7.6	Y-B	Bromothymol Blue	7.30	0.1 g in 16 mL 0.01 M NaOH + 234 mL water
6.4-8.0	C-Y	6,8-Dinitro-2,4-(1H)quinazolinedione		25 g in 115 mL 1 M NaOH + 50 mL water at 100°C
6.6-7.8	Y-R	Brilliant Yellow		1% in water
6.6-8.0	Y-R	Phenol Red	8.00	0.1 g in 28.2 mL 0.01 M NaOH + 221.8 mL water
6.8-8.0	R-A	Neutral Red		0.01 g in 100 mL 50% v/v ethanol-water
6.8-8.6	C-Y	m-Nitrophenol	8.28	0.3% in water
7.0-8.8	Y-R	Cresol Red	8.46	0.1 g in 26.2 mL 0.01 M NaOH + 223.8 mL water
7.4-8.6	Y-R	Turmaric (Curcumin)		ethanol
7.4-9.0	Y-P	Metacresol Purple	8.3	0.1 g in 26.2 mL 0.01 M NaOH + 223.8 mL water

pH range	Color change	Indicator	pK	Preparation
8.0-9.0	B-R	4,4'-Bis(4-amino-1-naphthylazo)-2,2'-stilbenedisulfonic acid		0.1 g in 5.9 mL 0.05 M NaOH + 94.1 mL water
8.0-9.6	Y-B	Thymol Blue	9.20	0.1 g in 21.5 mL 0.01 M NaOH + 228.5 mL water
8.2-10.0	O-B	p-Naphtholbenzein		1% in dil. alkali
8.2-10.0	C-Pk	Phenolphthalein	9.5	0.5 g in 100 mL 50% v/v ethanol-water
8.2-9.8	C-R	o-Cresolphthalein		0.04% in ethanol
8.4-9.6	C-B	Ethyl bis(2,4-dimethylphenyl)ethanoate		sat. solution in 50% acetone-ethanol
9.4-10.6	C-B	Thymolphthalein		0.04 g in 100 mL 50% v/v ethanol-water
10.1-12.0	Y-R	Alizarin Yellow R		0.01% in water
11.0-12.4	R-P	Alizarin		0.1% in methanol
11.4-12.6	Y-O	p-(2,4-Dihydroxyphenylazo) benzenesulfonic acid, sodium salt		0.1% in water
11.4-13.0	B-Y	5,5'-Indigodisulfonic acid, disodium salt		water
11.5-13.0	C-O	2,4,6-Trinitrotoluene		0.1-0.5% in ethanol
12.0-14.0	C-O	1,3,5-Trinitrobenzene		0.1-0.5% in ethanol
12.2-13.2	Y-A	Clayton Yellow		0.1% in water

FLUORESCENT INDICATORS

Jack DeMent

Fluorescent indicators are substances which show definite changes in fluorescence with change in pH. Some fluorescent materials are not suitable for indicators since their change in fluorescence is too gradual. Fluorescent indicators find greatest utility in the titration of opaque, highly turbid or deeply colored solutions. A long wavelength ultraviolet ("black light") lamp in a dimly lighted room provides the best environment for titrations involving fluorescent indicators, although bright daylight is sometimes sufficient to evoke a response in the bright green, yellow and orange fluorescent indicators. Titrations are carried out in non-fluorescent glassware. One should check the glassware prior to use to make certain that it does not fluoresce due to the wavelengths of light involved in the titration. The meniscus of the liquid in the burette can be followed when a few particles of an insoluble fluorescent solid are dropped onto its surface.

In this table the indicators are arranged by approximate pH range covered. In the case of some of the dyestuffs the end point may vary slightly with the source or manufacturer.

pH 0 to 2

Indicator	C.I.	From pH	To pH
Benzoflavine	—	0.3, yellow fl.	1.7, green fl.
3,6-Dioxyphthalimide	—	0, blue fl.	2.4, green fl.
Eosine YS	768	0, yellow colored	3.0, yellow fl.
Erythrosine	772	0, yellow colored	3.6, yellow fl.
Esculin	—	1.5, colorless	2, blue fl.
4-Ethoxyacridone	—	1.2, green fl.	3.2, blue fl.
3,6-Tetramethyldiaminooxanthone	—	1.2, green fl.	3.4, blue fl.

pH 2 to 4

Indicator	C.I.	From pH	To pH
Chromotropic acid	—	3.5, colorless	4.5, blue fl.
Fluorescein	766	4, colorless	4.5, green fl.
Magdala Red	—	3.0, purple colored	4.0, fl.
α-Naphthylamine	—	3.4, colorless	4.8, blue fl.
β-Naphthylamine	—	2.8, colorless	4.4, violet fl.
Phloxine	774	3.4, colorless	5.0, bright yellow fl.
Salicylic acid	—	2.5, colorless	3.5, blue fl.

pH 4 to 6

Indicator	C.I.	From pH	To pH
Acridine	788	4.9, green fl.	5.1, violet colored
Dichlorofluorescein	—	4.0, colorless	5.0, green fl.
3,6-Dioxyxanthone	—	5.4, colorless	7.6, blue-violet fl.
Erythrosine	772	4.0, colorless	4.5, yellow-green fl.
β-Methylesculetin	—	4.0, colorless	6.2, blue fl.
Neville-Winther acid	—	6.0, colorless	6.5, blue fl.
Resorufin	—	4.4, yellow fl.	6.4, weak orange fl.
Quininic acid	—	4.0, yellow colored	5.0, blue fl.
Quinine [first end point]	—	5.0, blue fl.	6.1, violet fl.

pH 6 to 8

Indicator	C.I.	From pH	To pH
Acid R Phosphine	—	(claimed for range pH 6.0–7.0)	
Brilliant Diazol Yellow	—	6.5, colorless	7.5, violet fl.
Cleves acid	—	6.5, colorless	7.5, green fl.
Coumaric acid	—	7.2, colorless	9.0, green fl.
3,6-Dioxyphthalic dinitrile	—	5.8, blue fl.	8.2, green fl.
Magnesium 8-hydroxyquinolinate	—	6.5, colorless	7.5, golden fl.
β-Methylumbelliferone	—	7.0, colorless	7.5, blue fl.
1-Naphthol-4-sulfonic acid	—	6.0, colorless	6.5, blue fl.
Orcinaurine	—	6.5, colorless	8.0, green fl.
Patent Phosphine	789	(for the range pH 6.0–7.0, green-yellow fl.)	
Thioflavine	816	(for the region pH 6.5–7.0, yellow fl.)	
Umbelliferone	—	6.5, colorless	7.6, blue fl.

pH 8 to 10

Indicator	C.I.	From pH	To pH
Acridine Orange	788	8.4, orange colored	10.4, green fl.
Ethoxyphenylnaphthostilbazonium chloride	—	9, green fl.	11, non-fl.
G Salt	—	9.0, dull blue fl.	9.5, bright blue fl.
Naphthazol derivatives	—	8.2, colorless	10.0, yellow or green fl.
α-Naphthionic acid	—	9, blue fl.	11, green fl.
2-Naphthol-3,6-disulfonic acid	—	9.5, dark blue fl.	Light blue fl. at higher pH
β-Naphthol	—	8.6, colorless	Blue fl. at higher pH
α-Naphtholsulfonic acid	—	8.0, dark blue fl.	9.0, bright violet fl.
1,4-Naphtholsulfonic acid	—	8.2, dark blue fl.	Light blue fl. at higher pH
Orcinsulfonphthalein	—	8.6, yellow colored	10.0 fl.
Quinine [second end point]	—	9.5, violet fl.	10.0, colorless
R-Salt	—	9.0, dull blue fl.	9.5, bright blue fl.
Sodium 1-naphthol-2-sulfonate	—	9.0, dark blue fl.	10.0, bright violet fl.

pH 10 to 12

Indicator	C.I.	From pH	To pH
Coumarin	—	9.8, deep green fl.	12, light green fl.
Eosine BN	771	10.5, colorless	14.0, yellow fl.
Papaverine (permanganate oxidized)	—	9.5, yellow fl.	11.0, blue fl.
Schaffers Salt	—	5.0, violet fl.	11.0, green-blue fl.
SS-Acid (sodium salt)	—	10.0, violet fl.	12.0, yellow colored

pH 12 to 14

Indicator	C.I.	From pH	To pH
Cotarnine	—	12.0, yellow fl.	13.0, white fl.
α-Naphthionic acid	—	12, blue fl.	13, green fl.
β-Naphthionic acid	—	12, blue fl.	13, violet fl.

CONVERSION FORMULAS FOR CONCENTRATION OF SOLUTIONS

A = Weight per cent of solute G = Molality

B = Molecular weight of solvent M = Molarity

E = Molecular weight of solute N = Mole fraction

F = Grams of solute per liter of solution R = Density of solution in grams per milliliter

Concentration of solute—SOUGHT	Concentration of solute—GIVEN				
	A	N	G	M	F
A	—	$\dfrac{100N \times E}{N \times E + (1 - N)B}$	$\dfrac{100G \times E}{1000 + G \times E}$	$\dfrac{M \times E}{10R}$	$\dfrac{F}{10R}$
N	$\dfrac{\dfrac{A}{E}}{\dfrac{A}{E} + \dfrac{100 - A}{B}}$	—	$\dfrac{B \times G}{B \times G + 1000}$	$\dfrac{B \times M}{M(B - E) + 1000R}$	$\dfrac{B \times F}{F(B - E) + 1000R \times E}$
G	$\dfrac{1000A}{E(100 - A)}$	$\dfrac{1000N}{B - N \times B}$	—	$\dfrac{1000M}{1000R - (M \times E)}$	$\dfrac{1000F}{E(1000R - F)}$
M	$\dfrac{10R \times A}{E}$	$\dfrac{1000R \times N}{N \times E + (1 - N)B}$	$\dfrac{1000R \times G}{1000 + E \times G}$	—	$\dfrac{F}{E}$
F	$10AR$	$\dfrac{1000R \times N \times E}{N \times E + (1 - N)B}$	$\dfrac{1000R \times G \times E}{1000 + G \times E}$	$M \times E$	—

ELECTROCHEMICAL SERIES

Petr Vanýsek

There are three tables for this electrochemical series. Each table lists standard reduction potentials, $E°$ values, at 298.15 K (25°C), and at a pressure of 101.325 kPa (1 atm). Table 1 is an alphabetical listing of the elements, according to the symbol of the elements. Thus, data for silver (Ag) precedes those for aluminum (Al). Table 2 lists only those reduction reactions which have $E°$ values positive in respect to the standard hydrogen electrode. In Table 2, the reactions are listed in the order of increasing positive potential, and they range from 0.0000 V to + 3.4 V. Table 3 lists only those reduction potentials which have $E°$ negative with respect to the standard hydrogen electrode. In Table 3, the reactions are listed in the order of decreasing potential and range from 0.0000 V to –4.10 V. The reliability of the potentials is not the same for all the data. Typically, the values with fewer significant figures have lower reliability. The values of reduction potentials, in particular those of less common reactions, are not definite; they are subject to occasional revisions.

Abbreviations: ac = acetate; bipy = 2,2′-dipyridine, or bipyridine; en = ethylenediamine; phen = 1,10-phenanthroline.

REFERENCES

1. G. Milazzo, S. Caroli, and V. K. Sharma, *Tables of Standard Electrode Potentials*, Wiley, Chichester, 1978.
2. A. J. Bard, R. Parsons, and J. Jordan, *Standard Potentials in Aqueous Solutions*, Marcel Dekker, New York, 1985.
3. S. G. Bratsch, *J. Phys. Chem. Ref. Data*, 18, 1—21, 1989.

TABLE 1
Alphabetical Listing

Reaction	$E°/V$	Reaction	$E°/V$
$Ac^{3+} + 3\,e \rightleftharpoons Ac$	−2.20	$Al(OH)_4^- + 3\,e \rightleftharpoons Al + 4\,OH^-$	−2.328
$Ag^+ + e \rightleftharpoons Ag$	0.7996	$H_2AlO_3^- + H_2O + 3\,e \rightleftharpoons Al + 4\,OH^-$	−2.33
$Ag^{2+} + e \rightleftharpoons Ag^+$	1.980	$AlF_6^{3-} + 3\,e \rightleftharpoons Al + 6\,F^-$	−2.069
$Ag(ac) + e \rightleftharpoons Ag + (ac)^-$	0.643	$Am^{4+} + e \rightleftharpoons Am^{3+}$	2.60
$AgBr + e \rightleftharpoons Ag + Br^-$	0.07133	$Am^{2+} + 2\,e \rightleftharpoons Am$	−1.9
$AgBrO_3 + e \rightleftharpoons Ag + BrO_3^-$	0.546	$Am^{3+} + 3\,e \rightleftharpoons Am$	−2.048
$Ag_2C_2O_4 + 2\,e \rightleftharpoons 2\,Ag + C_2O_4^{2-}$	0.4647	$Am^{3+} + e \rightleftharpoons Am^{2+}$	−2.3
$AgCl + e \rightleftharpoons Ag + Cl^-$	0.22233	$As + 3\,H^+ + 3\,e \rightleftharpoons AsH_3$	−0.608
$AgCN + e \rightleftharpoons Ag + CN^-$	−0.017	$As_2O_3 + 6\,H^+ + 6\,e \rightleftharpoons 2\,As + 3\,H_2O$	0.234
$Ag_2CO_3 + 2\,e \rightleftharpoons 2\,Ag + CO_3^{2-}$	0.47	$HAsO_2 + 3\,H^+ + 3\,e \rightleftharpoons As + 2\,H_2O$	0.248
$Ag_2CrO_4 + 2\,e \rightleftharpoons 2\,Ag + CrO_4^{2-}$	0.4470	$AsO_2^- + 2\,H_2O + 3\,e \rightleftharpoons As + 4\,OH^-$	−0.68
$AgF + e \rightleftharpoons Ag + F^-$	0.779	$H_3AsO_4 + 2\,H^+ + 2\,e^- \rightleftharpoons HAsO_2 + 2\,H_2O$	0.560
$Ag_4[Fe(CN)_6] + 4\,e \rightleftharpoons 4\,Ag + [Fe(CN)_6]^{4-}$	0.1478	$AsO_4^{3-} + 2\,H_2O + 2\,e \rightleftharpoons AsO_2^- + 4\,OH^-$	−0.71
$AgI + e \rightleftharpoons Ag + I^-$	−0.15224	$At_2 + 2\,e \rightleftharpoons 2\,At^-$	0.3
$AgIO_3 + e \rightleftharpoons Ag + IO_3^-$	0.354	$Au^+ + e \rightleftharpoons Au$	1.692
$Ag_2MoO_4 + 2\,e \rightleftharpoons 2\,Ag + MoO_4^{2-}$	0.4573	$Au^{3+} + 2\,e \rightleftharpoons Au^+$	1.401
$AgNO_2 + e \rightleftharpoons Ag + 2\,NO_2^-$	0.564	$Au^{3+} + 3\,e \rightleftharpoons Au$	1.498
$Ag_2O + H_2O + 2\,e \rightleftharpoons 2\,Ag + 2\,OH^-$	0.342	$Au^{2+} + e^- \rightleftharpoons Au^+$	1.8
$Ag_2O_3 + H_2O + 2\,e \rightleftharpoons 2\,AgO + 2\,OH^-$	0.739	$AuOH^{2+} + H^+ + 2\,e \rightleftharpoons Au^+ + H_2O$	1.32
$Ag^{3+} + 2\,e \rightleftharpoons Ag^+$	1.9	$AuBr_2^- + e \rightleftharpoons Au + 2\,Br^-$	0.959
$Ag^{3+} + e \rightleftharpoons Ag^{2+}$	1.8	$AuBr_4^- + 3\,e \rightleftharpoons Au + 4\,Br^-$	0.854
$Ag_2O_2 + 4\,H^+ + e \rightleftharpoons 2\,Ag + 2\,H_2O$	1.802	$AuCl_4^- + 3\,e \rightleftharpoons Au + 4\,Cl^-$	1.002
$2\,AgO + H_2O + 2\,e \rightleftharpoons Ag_2O + 2\,OH^-$	0.607	$Au(OH)_3 + 3\,H^+ + 3\,e \rightleftharpoons Au + 3\,H_2O$	1.45
$AgOCN + e \rightleftharpoons Ag + OCN^-$	0.41	$H_2BO_3^- + 5\,H_2O + 8\,e \rightleftharpoons BH_4^- + 8\,OH^-$	−1.24
$Ag_2S + 2\,e \rightleftharpoons 2\,Ag + S^{2-}$	−0.691	$H_2BO_3^- + H_2O + 3\,e \rightleftharpoons B + 4\,OH^-$	−1.79
$Ag_2S + 2\,H^+ + 2\,e \rightleftharpoons 2\,Ag + H_2S$	−0.0366	$H_3BO_3 + 3\,H^+ + 3\,e \rightleftharpoons B + 3\,H_2O$	−0.8698
$AgSCN + e \rightleftharpoons Ag + SCN^-$	0.08951	$B(OH)_3 + 7\,H^+ + 8\,e \rightleftharpoons BH_4^- + 3\,H_2O$	−0.481
$Ag_2SeO_3 + 2\,e \rightleftharpoons 2\,Ag + SeO_4^{2-}$	0.3629	$Ba^{2+} + 2\,e \rightleftharpoons Ba$	−2.912
$Ag_2SO_4 + 2\,e \rightleftharpoons 2\,Ag + SO_4^{2-}$	0.654	$Ba^{2+} + 2\,e \rightleftharpoons Ba(Hg)$	−1.570
$Ag_2WO_4 + 2\,e \rightleftharpoons 2\,Ag + WO_4^{2-}$	0.4660	$Ba(OH)_2 + 2\,e \rightleftharpoons Ba + 2\,OH^-$	−2.99
$Al^{3+} + 3\,e \rightleftharpoons Al$	−1.662	$Be^{2+} + 2\,e \rightleftharpoons Be$	−1.847
$Al(OH)_3 + 3\,e \rightleftharpoons Al + 3\,OH^-$	−2.31	$Be_2O_3^{2-} + 3\,H_2O + 4\,e \rightleftharpoons 2\,Be + 6\,OH^-$	−2.63

TABLE 1
Alphabetical Listing (continued)

Reaction	$E°/V$	Reaction	$E°/V$
p–benzoquinone + 2 H$^+$ + 2 e $\rightleftharpoons$ hydroquinone	0.6992	$HClO_2 + 3 H^+ + 4 e \rightleftharpoons Cl^- + 2 H_2O$	1.570
$Bi^+ + e \rightleftharpoons Bi$	0.5	$ClO_2^- + H_2O + 2 e \rightleftharpoons ClO^- + 2 OH^-$	0.66
$Bi^{3+} + 3 e \rightleftharpoons Bi$	0.308	$ClO_2^- + 2 H_2O + 4 e \rightleftharpoons Cl^- + 4 OH^-$	0.76
$Bi^{3+} + 2 e \rightleftharpoons Bi^+$	0.2	$ClO_2(aq) + e \rightleftharpoons ClO_2^-$	0.954
$Bi + 3 H^+ + 3 e \rightleftharpoons BiH_3$	−0.8	$ClO_3^- + 2 H^+ + e \rightleftharpoons ClO_2 + H_2O$	1.152
$BiCl_4^- + 3 e \rightleftharpoons Bi + 4 Cl^-$	0.16	$ClO_3^- + 3 H^+ + 2 e \rightleftharpoons HClO_2 + H_2O$	1.214
$Bi_2O_3 + 3 H_2O + 6 e \rightleftharpoons 2 Bi + 6 OH^-$	−0.46	$ClO_3^- + 6 H^+ + 5 e \rightleftharpoons 1/2 Cl_2 + 3 H_2O$	1.47
$Bi_2O_4 + 4 H^+ + 2 e \rightleftharpoons 2 BiO^+ + 2 H_2O$	1.593	$ClO_3^- + 6 H^+ + 6 e \rightleftharpoons Cl^- + 3 H_2O$	1.451
$BiO^+ + 2 H^+ + 3 e \rightleftharpoons Bi + H_2O$	0.320	$ClO_3^- + H_2O + 2 e \rightleftharpoons ClO_2^- + 2 OH^-$	0.33
$BiOCl + 2 H^+ + 3 e \rightleftharpoons Bi + Cl^- + H_2O$	0.1583	$ClO_3^- + 3 H_2O + 6 e \rightleftharpoons Cl^- + 6 OH^-$	0.62
$Bk^{4+} + e \rightleftharpoons Bk^{3+}$	1.67	$ClO_4^- + 2 H^+ + 2 e \rightleftharpoons ClO_3^- H_2O$	1.189
$Bk^{2+} + 2 e \rightleftharpoons Bk$	−1.6	$ClO_4^- + 8 H^+ + 7 e \rightleftharpoons 1/2 Cl_2 + 4 H_2O$	1.39
$Bk^{3+} + e \rightleftharpoons Bk^{2+}$	−2.8	$ClO_4^- + 8 H^+ + 8 e \rightleftharpoons Cl^- + 4 H_2O$	1.389
$Br_2(aq) + 2 e \rightleftharpoons 2 Br^-$	1.0873	$ClO_4^- + H_2O + 2 e \rightleftharpoons ClO_3^- + 2 OH^-$	0.36
$Br_2(1) + 2 e \rightleftharpoons 2 Br^-$	1.066	$Cm^{4+} + e \rightleftharpoons Cm^{3+}$	3.0
$HBrO + H^+ + 2 e \rightleftharpoons Br^- + H_2O$	1.331	$Cm^{3+} + 3 e \rightleftharpoons Cm$	−2.04
$HBrO + H^+ + e \rightleftharpoons 1/2 Br_2(aq) + H_2O$	1.574	$Co^{2+} + 2 e \rightleftharpoons Co$	−0.28
$HBrO + H^+ + e \rightleftharpoons 1/2 Br_2(1) + H_2O$	1.596	$Co^{3+} + e \rightleftharpoons Co^{2+}$	1.92
$BrO^- + H_2O + 2 e \rightleftharpoons Br^- + 2 OH^-$	0.761	$[Co(NH_3)_6]^{3+} + e \rightleftharpoons [Co(NH_3)_6]^{2+}$	0.108
$BrO_3^- + 6 H^+ + 5 e \rightleftharpoons 1/2 Br_2 + 3 H_2O$	1.482	$Co(OH)_2 + 2 e \rightleftharpoons Co + 2 OH^-$	−0.73
$BrO_3^- + 6 H^+ + 6 e \rightleftharpoons Br^- + 3 H_2O$	1.423	$Co(OH)_3 + e \rightleftharpoons Co(OH)_2 + OH^-$	0.17
$BrO_3^- + 3 H_2O + 6 e \rightleftharpoons Br^- + 6 OH^-$	0.61	$Cr^{2+} + 2 e \rightleftharpoons Cr$	−0.913
$(CN)_2 + 2 H^+ + 2 e \rightleftharpoons 2 HCN$	0.373	$Cr^{3+} + e \rightleftharpoons Cr^{2+}$	−0.407
$2 HCNO + 2 H^+ + 2 e \rightleftharpoons (CN)_2 + 2 H_2O$	0.330	$Cr^{3+} + 3 e \rightleftharpoons Cr$	−0.744
$(CNS)_2 + 2 e \rightleftharpoons 2 CNS^-$	0.77	$Cr_2O_7^{2-} + 14 H^+ + 6 e \rightleftharpoons 2 Cr^{3+} + 7 H_2O$	1.232
$CO_2 + 2 H^+ + 2 e \rightleftharpoons HCOOH$	−0.199	$CrO_2^- + 2 H_2O + 3 e \rightleftharpoons Cr + 4 OH^-$	−1.2
$Ca^+ + e \rightleftharpoons Ca$	−3.80	$HCrO_4^- + 7 H^+ + 3 e \rightleftharpoons Cr^{3+} + 4 H_2O$	1.350
$Ca^{2+} + 2 e \rightleftharpoons Ca$	−2.868	$CrO_2 + 4 H^+ + e \rightleftharpoons Cr^{3+} + 2H_2O$	1.48
$Ca(OH)_2 + 2 e \rightleftharpoons Ca + 2 OH^-$	−3.02	$Cr(V) + e \rightleftharpoons Cr(IV)$	1.34
Calomel electrode, 1 molal KCl	0.2800	$CrO_4^{2-} + 4 H_2O + 3 e \rightleftharpoons Cr(OH)_3 + 5 OH^-$	−0.13
Calomel electrode, 1 molar KCl (NCE)	0.2801	$Cr(OH)_3 + 3 e \rightleftharpoons Cr + 3 OH^-$	−1.48
Calomel electrode, 0.1 molar KCl	0.3337	$Cs^+ + e \rightleftharpoons Cs$	−3.026
Calomel electrode, saturated KCl (SCE)	0.2412	$Cu^+ + e \rightleftharpoons Cu$	0.521
Calomel electrode, saturated NaCl (SSCE)	0.2360	$Cu^{2+} + e \rightleftharpoons Cu^+$	0.153
$Cd^{2+} + 2 e \rightleftharpoons Cd$	−0.4030	$Cu^{2+} + 2 e \rightleftharpoons Cu$	0.3419
$Cd^{2+} + 2 e \rightleftharpoons Cd(Hg)$	−0.3521	$Cu^{2+} + 2 e \rightleftharpoons Cu(Hg)$	0.345
$Cd(OH)_2 + 2 e \rightleftharpoons Cd(Hg) + 2 OH^-$	−0.809	$Cu^{3+} + e \rightleftharpoons Cu^{2+}$	2.4
$CdSO_4 + 2 e \rightleftharpoons Cd + SO_4^{2-}$	−0.246	$Cu_2O_3 + 6 H^+ + 2e \rightleftharpoons 2Cu^{2+} + 3 H_2O$	2.0
$Cd(OH)_4^{2-} + 2 e \rightleftharpoons Cd + 4 OH^-$	−0.658	$Cu^{2+} + 2 CN^- + e \rightleftharpoons [Cu(CN)_2]^-$	1.103
$CdO + H_2O + 2 e \rightleftharpoons Cd + 2 OH^-$	−0.783	$CuI_2^- + e \rightleftharpoons Cu + 2 I^-$	0.00
$Ce^{3+} + 3 e \rightleftharpoons Ce$	−2.336	$Cu_2O + H_2O + 2 e \rightleftharpoons 2 Cu + 2 OH^-$	−0.360
$Ce^{3+} + 3 e \rightleftharpoons Ce(Hg)$	−1.4373	$Cu(OH)_2 + 2 e \rightleftharpoons Cu + 2 OH^-$	−0.222
$Ce^{4+} + e \rightleftharpoons Ce^{3+}$	1.72	$2 Cu(OH)_2 + 2 e \rightleftharpoons Cu_2O + 2 OH^- + H_2O$	−0.080
$CeOH^{3+} + H^+ + e \rightleftharpoons Ce^{3+} + H_2O$	1.715	$2 D^+ + 2 e \rightleftharpoons D_2$	−0.013
$Cf^{4+} + e \rightleftharpoons Cf^{3+}$	3.3	$Dy^{2+} + 2 e \rightleftharpoons Dy$	−2.2
$Cf^{3+} + e \rightleftharpoons Cf^{2+}$	−1.6	$Dy^{3+} + 3 e \rightleftharpoons Dy$	−2.295
$Cf^{3+} + 3 e \rightleftharpoons Cf$	−1.94	$Dy^{3+} + e \rightleftharpoons Dy^{2+}$	−2.6
$Cf^{2+} + 2 e \rightleftharpoons Cf$	−2.12	$Er^{2+} + 2 e \rightleftharpoons Er$	−2.0
$Cl_2(g) + 2 e \rightleftharpoons 2 Cl^-$	1.35827	$Er^{3+} + 3 e \rightleftharpoons Er$	−2.331
$HClO + H^+ + e \rightleftharpoons 1/2 Cl_2 + H_2O$	1.611	$Er^{3+} + e \rightleftharpoons Er^{2+}$	−3.0
$HClO + H^+ + 2 e \rightleftharpoons Cl^- + H_2O$	1.482	$Es^{3+} + e \rightleftharpoons Es^{2+}$	−1.3
$ClO^- + H_2O + 2 e \rightleftharpoons Cl^- + 2 OH^-$	0.81	$Es^{3+} + 3 e \rightleftharpoons Es$	−1.91
$ClO_2 + H^+ + e \rightleftharpoons HClO_2$	1.277	$Es^{2+} + 2 e \rightleftharpoons Es$	−2.23
$HClO_2 + 2 H^+ + 2 e \rightleftharpoons HClO + H_2O$	1.645	$Eu^{2+} + 2 e \rightleftharpoons Eu$	−2.812
$HClO_2 + 3 H^+ + 3 e \rightleftharpoons 1/2 Cl_2 + 2 H_2O$	1.628	$Eu^{3+} + 3 e \rightleftharpoons Eu$	−1.991

TABLE 1
Alphabetical Listing (continued)

Reaction	$E°/V$	Reaction	$E°/V$
$Eu^{3+} + e \rightleftharpoons Eu^{2+}$	-0.36	$Ho^{3+} + 3e \rightleftharpoons Ho$	-2.33
$F_2 + 2H^+ + 2e \rightleftharpoons 2HF$	3.053	$Ho^{3+} + e \rightleftharpoons Ho^{2+}$	-2.8
$F_2 + 2e \rightleftharpoons 2F^-$	2.866	$I_2 + 2e \rightleftharpoons 2I^-$	0.5355
$F_2O + 2H^+ + 4e \rightleftharpoons H_2O + 2F^-$	2.153	$I_3^- + 2e \rightleftharpoons 3I^-$	0.536
$Fe^{2+} + 2e \rightleftharpoons Fe$	-0.447	$H_3IO_6^{2-} + 2e \rightleftharpoons IO_3^- + 3OH^-$	0.7
$Fe^{3+} + 3e \rightleftharpoons Fe$	-0.037	$H_5IO_6 + H^+ + 2e \rightleftharpoons IO_3^- + 3H_2O$	1.601
$Fe^{3+} + e \rightleftharpoons Fe^{2+}$	0.771	$2HIO + 2H^+ + 2e \rightleftharpoons I_2 + 2H_2O$	1.439
$2HFeO_4^- + 8H^+ + 6e \rightleftharpoons Fe_2O_3 + 5H_2O$	2.09	$HIO + H^+ + 2e \rightleftharpoons I^- + H_2O$	0.987
$HFeO_4^- + 4H^+ + 3e \rightleftharpoons FeOOH + 2H_2O$	2.08	$IO^- + H_2O + 2e \rightleftharpoons I^- + 2OH^-$	0.485
$HFeO_4^- + 7H^+ + 3e \rightleftharpoons Fe^{3+} + 4H_2O$	2.07	$2IO_3^- + 12H^+ + 10e \rightleftharpoons I_2 + 6H_2O$	1.195
$Fe_2O_3 + 4H^+ + 2e \rightleftharpoons 2FeOH^+ + H_2O$	0.16	$IO_3^- + 6H^+ + 6e \rightleftharpoons I^- + 3H_2O$	1.085
$[Fe(CN)_6]^{3-} + e \rightleftharpoons [Fe(CN)_6]^{4-}$	0.358	$IO_3^- + 2H_2O + 4e \rightleftharpoons IO^- + 4OH^-$	0.15
$FeO_4^{2-} + 8H^+ + 3e \rightleftharpoons Fe^{3+} + 4H_2O$	2.20	$IO_3^- + 3H_2O + 6e \rightleftharpoons IO^- + 6OH^-$	0.26
$[Fe(bipy)_2]^{3+} + e \rightleftharpoons Fe(bipy)_2]^{2+}$	0.78	$In^+ + e \rightleftharpoons In$	-0.14
$[Fe(bipy)_3]^{3+} + e \rightleftharpoons Fe(bipy)_3]^{2+}$	1.03	$In^{2+} + e \rightleftharpoons In^+$	-0.40
$Fe(OH)_3 + e \rightleftharpoons Fe(OH)_2 + OH^-$	-0.56	$In^{3+} + e \rightleftharpoons In^{2+}$	-0.49
$[Fe(phen)_3]^{3+} + e \rightleftharpoons [Fe(phen)_3]^{2+}$	1.147	$In^{3+} + 2e \rightleftharpoons In^+$	-0.443
$[Fe(phen)_3]^{3+} + e \rightleftharpoons [Fe(phen)_3]^{2+} (1 \text{ molar } H_2SO_4)$	1.06	$In^{3+} + 3e \rightleftharpoons In$	-0.3382
$[Ferricinium]^+ + e \rightleftharpoons ferrocene$	0.400	$In(OH)_3 + 3e \rightleftharpoons In + 3OH^-$	-0.99
$Fm^{3+} + e \rightleftharpoons Fm^{2+}$	-1.1	$In(OH)_4^- + 3e \rightleftharpoons In + 4OH^-$	-1.007
$Fm^{3+} + 3e \rightleftharpoons Fm$	-1.89	$In_2O_3 + 3H_2O + 6e \rightleftharpoons 2In + 6OH^-$	-1.034
$Fm^{2+} + 2e \rightleftharpoons Fm$	-2.30	$Ir^{3+} + 3e \rightleftharpoons Ir$	1.156
$Fr^+ + e \rightleftharpoons Fr$	-2.9	$[IrCl_6]^{2-} + e \rightleftharpoons [IrCl_6]^{3-}$	0.8665
$Ga^{3+} + 3e \rightleftharpoons Ga$	-0.549	$[IrCl_6]^{3-} + 3e \rightleftharpoons Ir + 6Cl^-$	0.77
$Ga^+ + e \rightleftharpoons Ga$	-0.2	$Ir_2O_3 + 3H_2O + 6e \rightleftharpoons 2Ir + 6OH^-$	0.098
$GaOH^{2+} + H^+ + 3e \rightleftharpoons Ga + H_2O$	-0.498	$K^+ + e \rightleftharpoons K$	-2.931
$H_2GaO_3^- + H_2O + 3e \rightleftharpoons Ga + 4OH^-$	-1.219	$La^{3+} + 3e \rightleftharpoons La$	-2.379
$Gd^{3+} + 3e \rightleftharpoons Gd$	-2.279	$La(OH)_3 + 3e \rightleftharpoons La + 3OH^-$	-2.90
$Ge^{2+} + 2e \rightleftharpoons Ge$	0.24	$Li^+ + e \rightleftharpoons Li$	-3.0401
$Ge^{4+} + 4e \rightleftharpoons Ge$	0.124	$Lr^{3+} + 3e \rightleftharpoons Lr$	-1.96
$Ge^{4+} + 2e \rightleftharpoons Ge^{2+}$	0.00	$Lu^{3+} + 3e \rightleftharpoons Lu$	-2.28
$GeO_2 + 2H^+ + 2e \rightleftharpoons GeO + H_2O$	-0.118	$Md^{3+} + e \rightleftharpoons Md^{2+}$	-0.1
$H_2GeO_3 + 4H^+ + 4e \rightleftharpoons Ge + 3H_2O$	-0.182	$Md^{3+} + 3e \rightleftharpoons Md$	-1.65
$2H^+ + 2e \rightleftharpoons H_2$	0.00000	$Md^{2+} + 2e \rightleftharpoons Md$	-2.40
$H_2 + 2e \rightleftharpoons 2H^-$	-2.23	$Mg^+ + e \rightleftharpoons Mg$	-2.70
$HO_2 + H^+ + e \rightleftharpoons H_2O_2$	1.495	$Mg^{2+} + 2e \rightleftharpoons Mg$	-2.372
$2H_2O + 2e \rightleftharpoons H_2 + 2OH^-$	-0.8277	$Mg(OH)_2 + 2e \rightleftharpoons Mg + 2OH^-$	-2.690
$H_2O_2 + 2H^+ + 2e \rightleftharpoons 2H_2O$	1.776	$Mn^{2+} + 2e \rightleftharpoons Mn$	-1.185
$Hf^{4+} + 4e \rightleftharpoons Hf$	-1.55	$Mn^{3+} + 3e \rightleftharpoons Mn^{2+}$	1.5415
$HfO^{2+} + 2H^+ + 4e \rightleftharpoons Hf + H_2O$	-1.724	$MnO_2 + 4H^+ + 2e \rightleftharpoons Mn^{2+} + 2H_2O$	1.224
$HfO_2 + 4H^+ + 4e \rightleftharpoons Hf + 2H_2O$	-1.505	$MnO_4^- + e \rightleftharpoons MnO_4^{2-}$	0.558
$HfO(OH)_2 + H_2O + 4e \rightleftharpoons Hf + 4OH^-$	-2.50	$MnO_4^- + 4H^+ + 3e \rightleftharpoons MnO_2 + 2H_2O$	1.679
$Hg^{2+} + 2e \rightleftharpoons Hg$	0.851	$MnO_4^- + 8H^+ + 5e \rightleftharpoons Mn^{2+} + 4H_2O$	1.507
$2Hg^{2+} + 2e \rightleftharpoons Hg_2^{2+}$	0.920	$MnO_4^- + 2H_2O + 3e \rightleftharpoons MnO_2 + 4OH^-$	0.595
$Hg_2^{2+} + 2e \rightleftharpoons 2Hg$	0.7973	$MnO_4^{2-} + 2H_2O + 2e \rightleftharpoons MnO_2 + 4OH^-$	0.60
$Hg_2(ac)_2 + 2e \rightleftharpoons 2Hg + 2(ac)^-$	0.51163	$Mn(OH)_2 + 2e \rightleftharpoons Mn + 2OH^-$	-1.56
$Hg_2Br_2 + 2e \rightleftharpoons 2Hg + 2Br^-$	0.13923	$Mn(OH)_3 + e \rightleftharpoons Mn(OH)_2 + OH^-$	0.15
$Hg_2Cl_2 + 2e \rightleftharpoons 2Hg + 2Cl^-$	0.26808	$Mn_2O_3 + 6H^+ + e \rightleftharpoons 2Mn^{2+} + 3H_2O$	1.485
$Hg_2HPO_4 + 2e \rightleftharpoons 2Hg + HPO_4^{2-}$	0.6359	$Mo^{3+} + 3e \rightleftharpoons Mo$	-0.200
$Hg_2I_2 + 2e \rightleftharpoons 2Hg + 2I^-$	-0.0405	$MoO_2 + 4H^+ + 4e \rightleftharpoons Mo + 4H_2O$	-0.152
$Hg_2O + H_2O + 2e \rightleftharpoons 2Hg + 2OH^-$	0.123	$H_3Mo_7O_{24}^{3-} + 45H^+ + 42e \rightleftharpoons 7Mo + 24H_2O$	0.082
$HgO + H_2O + 2e \rightleftharpoons Hg + 2OH^-$	0.0977	$MoO_3 + 6H^+ + 6e \rightleftharpoons Mo + 3H_2O$	0.075
$Hg(OH)_2 + 2H^+ + 2e \rightleftharpoons Hg + 2H_2O$	1.034	$N_2 + 2H_2O + 6H^+ + 6e \rightleftharpoons 2NH_4OH$	0.092
$Hg_2SO_4 + 2e \rightleftharpoons 2Hg + SO_4^{2-}$	0.6125	$3N_2 + 2H^+ + 2e \rightleftharpoons 2HN_3$	-3.09
$Ho^{2+} + 2e \rightleftharpoons Ho$	-2.1	$N_5^+ + 3H^+ + 2e \rightleftharpoons 2NH_4^+$	1.275

TABLE 1
Alphabetical Listing (continued)

Reaction	$E°/V$	Reaction	$E°/V$
$N_2O + 2\,H^+ + 2\,e \rightleftharpoons N_2 + H_2O$	1.766	$H_2P_2^- + e \rightleftharpoons P + 2\,OH^-$	-1.82
$H_2N_2O_2 + 2\,H^+ + 2\,e \rightleftharpoons N_2 + 2\,H_2O$	2.65	$H_3PO_2 + H^+ + e \rightleftharpoons P + 2\,H_2O$	-0.508
$N_2O_4 + 2\,e \rightleftharpoons 2\,NO_2^-$	0.867	$H_3PO_3 + 2\,H^+ + 2\,e \rightleftharpoons H_3PO_2 + H_2O$	-0.499
$N_2O_4 + 2\,H^+ + 2\,e \rightleftharpoons 2\,HNO_2$	1.065	$H_3PO_3 + 3\,H^+ + 3\,e \rightleftharpoons P + 3\,H_2O$	-0.454
$N_2O_4 + 4\,H^+ + 4\,e \rightleftharpoons 2\,NO + 2\,H_2O$	1.035	$HPO_3^{2-} + 2\,H_2O + 2\,e \rightleftharpoons H_2PO_2^- + 3\,OH^-$	-1.65
$2\,NH_3OH^+ + H^+ + 2\,e \rightleftharpoons N_2H_5^+ + 2\,H_2O$	1.42	$HPO_3^{2-} + 2\,H_2O + 3\,e \rightleftharpoons P + 5\,OH^-$	-1.71
$2\,NO + 2\,H^+ + 2\,e \rightleftharpoons N_2O + H_2O$	1.591	$H_3PO_4 + 2\,H^+ + 2\,e \rightleftharpoons H_3PO_3 + H_2O$	-0.276
$2\,NO + H_2O + 2\,e \rightleftharpoons N_2O + 2\,OH^-$	0.76	$PO_4^{3-} + 2\,H_2O + 2\,e \rightleftharpoons HPO_3^{2-} + 3\,OH^-$	-1.05
$HNO_2 + H^+ + e \rightleftharpoons NO + H_2O$	0.983	$Pa^{3+} + 3\,e \rightleftharpoons Pa$	-1.34
$2\,HNO_2 + 4\,H^+ + 4\,e \rightleftharpoons H_2N_2O_2 + 2\,H_2O$	0.86	$Pa^{4+} + 4\,e \rightleftharpoons Pa$	-1.49
$2\,HNO_2 + 4\,H^+ + 4\,e \rightleftharpoons N_2O + 3\,H_2O$	1.297	$Pa^{4+} + e \rightleftharpoons Pa^{3+}$	-1.9
$NO_2^- + H_2O + e \rightleftharpoons NO + 2\,OH^-$	-0.46	$Pb^{2+} + 2\,e \rightleftharpoons Pb$	-0.1262
$2\,NO_2^- + 2\,H_2O + 4\,e \rightleftharpoons N_2O_2^{2-} + 4\,OH^-$	-0.18	$Pb^{2+} + 2\,e \rightleftharpoons Pb(Hg)$	-0.1205
$2\,NO_2^- + 3\,H_2O + 4\,e \rightleftharpoons N_2O + 6\,OH^-$	0.15	$PbBr_2 + 2\,e \rightleftharpoons Pb + 2\,Br^-$	-0.284
$NO_3^- + 3\,H^+ + 2\,e \rightleftharpoons HNO_2 + H_2O$	0.934	$PbCl_2 + 2\,e \rightleftharpoons Pb + 2\,Cl^-$	-0.2675
$NO_3^- + 4\,H^+ + 3\,e \rightleftharpoons NO + 2\,H_2O$	0.957	$PbF_2 + 2\,e \rightleftharpoons Pb + 2\,F^-$	-0.3444
$2\,NO_3^- + 4\,H^+ + 2\,e \rightleftharpoons N_2O_4 + 2\,H_2O$	0.803	$PbHPO_4 + 2\,e \rightleftharpoons Pb + HPO_4^{2-}$	-0.465
$NO_3^- + H_2O + 2\,e \rightleftharpoons NO_2^- + 2\,OH^-$	0.01	$PbI_2 + 2\,e \rightleftharpoons Pb + 2\,I^-$	-0.365
$2\,NO_3^- + 2\,H_2O + 2\,e \rightleftharpoons N_2O_4 + 4\,OH^-$	-0.85	$PbO + H_2O + 2\,e \rightleftharpoons Pb + 2\,OH^-$	-0.580
$Na^+ + e \rightleftharpoons Na$	-2.71	$PbO_2 + 4\,H^+ + 2\,e \rightleftharpoons Pb^{2+} + 2\,H_2O$	1.455
$Nb^{3+} + 3\,e \rightleftharpoons Nb$	-1.099	$HPbO_2^- + H_2O + 2\,e \rightleftharpoons Pb + 3\,OH^-$	-0.537
$NbO_2 + 2\,H^+ + 2\,e \rightleftharpoons NbO + H_2O$	-0.646	$PbO_2 + H_2O + 2\,e \rightleftharpoons PbO + 2\,OH^-$	0.247
$NbO_2 + 4\,H^+ + 4\,e \rightleftharpoons Nb + 2\,H_2O$	-0.690	$PbO_2 + SO_4^{2-} + 4\,H^+ + 2\,e \rightleftharpoons PbSO_4 + 2\,H_2O$	1.6913
$NbO + 2\,H^+ + 2\,e \rightleftharpoons Nb + H_2O$	-0.733	$PbSO_4 + 2\,e \rightleftharpoons Pb + SO_4^{2-}$	-0.3588
$Nb_2O_5 + 10\,H^+ + 10\,e \rightleftharpoons 2\,Nb + 5\,H_2O$	-0.644	$PbSO_4 + 2\,e \rightleftharpoons Pb(Hg) + SO_4^{2-}$	-0.3505
$Nd^{3+} + 3\,e \rightleftharpoons Nd$	-2.323	$Pd^{2+} + 2\,e \rightleftharpoons Pd$	0.951
$Nd^{2+} + 2\,e \rightleftharpoons Nd$	-2.1	$[PdCl_4]^{2-} + 2\,e \rightleftharpoons Pd + 4\,Cl^-$	0.591
$Nd^{3+} + e \rightleftharpoons Nd^{2+}$	-2.7	$[PdCl_6]^{2-} + 2\,e \rightleftharpoons [PdCl_4]^{2-} + 2\,Cl^-$	1.288
$Ni^{2+} + 2\,e \rightleftharpoons Ni$	-0.257	$Pd(OH)_2 + 2\,e \rightleftharpoons Pd + 2\,OH^-$	0.07
$Ni(OH)_2 + 2\,e \rightleftharpoons Ni + 2\,OH^-$	-0.72	$Pm^{2+} + 2\,e \rightleftharpoons Pm$	-2.2
$NiO_2 + 4\,H^+ + 2\,e \rightleftharpoons Ni^{2+} + 2\,H_2O$	1.678	$Pm^{3+} + 3\,e \rightleftharpoons Pm$	-2.30
$NiO_2 + 2\,H_2O + 2\,e \rightleftharpoons Ni(OH)_2 + 2\,OH^-$	-0.490	$Pm^{3+} + e \rightleftharpoons Pm^{2+}$	-2.6
$No^{3+} + e \rightleftharpoons No^{2+}$	1.4	$Po^{4+} + 2\,e \rightleftharpoons Po^{2+}$	0.9
$No^{3+} + 3\,e \rightleftharpoons No$	-1.20	$Po^{4+} + 4\,e \rightleftharpoons Po$	0.76
$No^{2+} + 2\,e \rightleftharpoons No$	-2.50	$Pr^{4+} + e \rightleftharpoons Pr^{3+}$	3.2
$Np^{3+} + 3\,e \rightleftharpoons Np$	-1.856	$Pr^{2+} + 2\,e \rightleftharpoons Pr$	-2.0
$Np^{4+} + e \rightleftharpoons Np^{3+}$	0.147	$Pr^{3+} + 3\,e \rightleftharpoons Pr$	-2.353
$NpO_2 + H_2O + H^+ + e \rightleftharpoons Np(OH)_3$	-0.962	$Pr^{3+} + e \rightleftharpoons Pr^{2+}$	-3.1
$O_2 + 2\,H^+ + 2\,e \rightleftharpoons H_2O_2$	0.695	$Pt^{2+} + 2\,e \rightleftharpoons Pt$	1.18
$O_2 + 4\,H^+ + 4\,e \rightleftharpoons 2\,H_2O$	1.229	$[PtCl_4]^{2-} + 2\,e \rightleftharpoons Pt + 4\,Cl^-$	0.755
$O_2 + H_2O + 2\,e \rightleftharpoons HO_2^- + OH^-$	-0.076	$[PtCl_6]^{2-} + 2\,e \rightleftharpoons [PtCl_4]^{2-} + 2\,Cl^-$	0.68
$O_2 + 2\,H_2O + 2\,e \rightleftharpoons H_2O_2 + 2\,OH^-$	-0.146	$Pt(OH)_2 + 2\,e \rightleftharpoons Pt + 2\,OH^-$	0.14
$O_2 + 2\,H_2O + 4\,e \rightleftharpoons 4\,OH^-$	0.401	$PtO_3 + 2\,H^+ + 2\,e \rightleftharpoons PtO_2 + H_2O$	1.7
$O_3 + 2\,H^+ + 2\,e \rightleftharpoons O_2 + H_2O$	2.076	$PtO_3 + 4\,H^+ + 2\,e \rightleftharpoons Pt(OH)_2^{2+} + H_2O$	1.5
$O_3 + H_2O + 2\,e \rightleftharpoons O_2 + 2\,OH^-$	1.24	$PtOH^+ + H^+ + 2\,e \rightleftharpoons Pt + H_2O$	1.2
$O(g) + 2\,H^+ + 2\,e \rightleftharpoons H_2O$	2.421	$PtO_2 + 2\,H^+ + 2\,e \rightleftharpoons PtO + H_2O$	1.01
$OH + e \rightleftharpoons OH^-$	2.02	$PtO_2 + 4\,H^+ + 4\,e \rightleftharpoons Pt + 2\,H_2O$	1.00
$HO_2^- + H_2O + 2\,e \rightleftharpoons 3\,OH^-$	0.878	$Pu^{3+} + 3\,e \rightleftharpoons Pu$	-2.031
$OsO_4 + 8\,H^+ + 8\,e \rightleftharpoons Os + 4\,H_2O$	0.838	$Pu^{4+} + e \rightleftharpoons Pu^{3+}$	1.006
$OsO_4 + 4\,H^+ + 4\,e \rightleftharpoons OsO_2 + 2\,H_2O$	1.02	$Pu^{5+} + e \rightleftharpoons Pu^{4+}$	1.099
$[Os(bipy)_2]^{3+} + e \rightleftharpoons [Os(bipy)_2]^{2+}$	0.81	$PuO_2(OH)_2 + 2\,H^+ + 2\,e \rightleftharpoons Pu(OH)_4$	1.325
$[Os(bipy)_3]^{3+} + e \rightleftharpoons [Os(bipy)_3]^{2+}$	0.80	$PuO_2(OH)_2 + H^+ + e \rightleftharpoons PuO_2OH + H_2O$	1.062
$P(red) + 3\,H^+ + 3\,e \rightleftharpoons PH_3(g)$	-0.111	$Ra^{2+} + 2\,e \rightleftharpoons Ra$	-2.8
$P(white) + 3\,H^+ + 3\,e \rightleftharpoons PH_3(g)$	-0.063	$Rb^+ + e \rightleftharpoons Rb$	-2.98
$P + 3\,H_2O + 3\,e \rightleftharpoons PH_3(g) + 3\,OH^-$	-0.87	$Re^{3+} + 3\,e \rightleftharpoons Re$	0.300

TABLE 1
Alphabetical Listing (continued)

Reaction	$E°/V$	Reaction	$E°/V$
$ReO_4^- + 4 H^+ + 3 e \rightleftharpoons ReO_2 + 2 H_2O$	0.510	$SiO_2 \text{ (quartz)} + 4 H^+ + 4 e \rightleftharpoons Si + 2 H_2O$	0.857
$ReO_2 + 4 H^+ + 4 e \rightleftharpoons Re + 2 H_2O$	0.2513	$SiO_3^{2-} + 3 H_2O + 4 e \rightleftharpoons Si + 6 OH^-$	−1.697
$ReO_4^- + 2 H^+ + e \rightleftharpoons ReO_3 + H_2O$	0.768	$Sm^{3+} + e \rightleftharpoons Sm^{2+}$	−1.55
$ReO_4^- + 4 H_2O + 7 e \rightleftharpoons Re + 8 OH^-$	−0.584	$Sm^{3+} + 3 e \rightleftharpoons Sm$	−2.304
$ReO_4^- + 8 H^+ + 7 e \rightleftharpoons Re + 4 H_2O$	0.368	$Sm^{2+} + 2 e \rightleftharpoons Sm$	−2.68
$Rh^+ + e \rightleftharpoons Rh$	0.600	$Sn^{2+} + 2 e \rightleftharpoons Sn$	−0.1375
$Rh^+ + 2e \rightleftharpoons Rh$	0.600	$Sn^{4+} + 2 e \rightleftharpoons Sn^{2+}$	0.151
$Rh^{3+} + 3 e \rightleftharpoons Rh$	0.758	$Sn(OH)_3^+ + 3 H^+ + 2 e \rightleftharpoons Sn^{2+} + 3 H_2O$	0.142
$[RhCl_6]^{3-} + 3 e \rightleftharpoons Rh + 6 Cl^-$	0.431	$SnO_2 + 4 H^+ + 2 e^- \rightleftharpoons Sn^{2+} + 2 H_2O$	−0.094
$RhOH^{2+} + H^+ + 3 e \rightleftharpoons Rh + H_2O$	0.83	$SnO_2 + 4 H^+ + 4 e \rightleftharpoons Sn + 2 H_2O$	−0.117
$Ru^{2+} + 2 e \rightleftharpoons Ru$	0.455	$SnO_2 + 3 H^+ + 2 e \rightleftharpoons SnOH^+ + H_2O$	−0.194
$Ru^{3+} + e \rightleftharpoons Ru^{2+}$	0.2487	$SnO_2 + 2 H_2O + 4 e \rightleftharpoons Sn + 4 OH^-$	−0.945
$RuO_2 + 4 H^+ + 2 e \rightleftharpoons Ru^{2+} + 2 H_2O$	1.120	$HSnO_2^- + H_2O + 2 e \rightleftharpoons Sn + 3 OH^-$	−0.909
$RuO_4^- + e \rightleftharpoons RuO_4^{2-}$	0.59	$Sn(OH)_6^{2-} + 2 e \rightleftharpoons HSnO_2^- + 3 OH^- + H_2O$	−0.93
$RuO_4 + e \rightleftharpoons RuO_4^-$	1.00	$Sr^+ + e \rightleftharpoons Sr$	−4.10
$RuO_4 + 6 H^+ + 4 e \rightleftharpoons Ru(OH)_2^{2+} + 2 H_2O$	1.40	$Sr^{2+} + 2 e \rightleftharpoons Sr$	−2.899
$RuO_4 + 8 H^+ + 8 e \rightleftharpoons Ru + 4 H_2O$	1.038	$Sr^{2+} + 2 e \rightleftharpoons Sr(Hg)$	−1.793
$[Ru(bipy)_3]^{3+} + e^- \rightleftharpoons [Ru(bipy)_3]^{2+}$	1.24	$Sr(OH)_2 + 2 e \rightleftharpoons Sr + 2 OH^-$	−2.88
$[Ru(H_2O)_6]^{3+} + e^- \rightleftharpoons [Ru(H_2O)_6]^{2+}$	0.23	$Ta_2O_5 + 10 H^+ + 10 e \rightleftharpoons 2 Ta + 5 H_2O$	−0.750
$[Ru(NH_3)_6]^{3+} + e^- \rightleftharpoons [Ru(NH_3)_6]^{2+}$	0.10	$Ta^{3+} + 3 e \rightleftharpoons Ta$	−0.6
$[Ru(en)_3]^{3+} + e^- \rightleftharpoons [Ru(en)_3]^{2+}$	0.210	$Tc^{2+} + 2 e \rightleftharpoons Tc$	0.400
$[Ru(CN)_6]^{3-} + e^- \rightleftharpoons [Ru(CN)_6]^{4-}$	0.86	$TcO_4^- + 4 H^+ + 3 e \rightleftharpoons TcO_2 + 2 H_2O$	0.782
$S + 2 e \rightleftharpoons S^{2-}$	−0.47627	$Tc^{3+} + e \rightleftharpoons Tc^{2+}$	0.3
$S + 2 H^+ + 2 e \rightleftharpoons H_2S(aq)$	0.142	$TcO_4^- + 8 H^+ + 7 e \rightleftharpoons Tc + 4 H_2O$	0.472
$S + H_2O + 2 e \rightleftharpoons SH^- + OH^-$	−0.478	$Tb^{4+} + e \rightleftharpoons Tb^{3+}$	3.1
$2 S + 2 e \rightleftharpoons S_2^{2-}$	−0.42836	$Tb^{3+} + 3 e \rightleftharpoons Tb$	−2.28
$S_2O_6^{2-} + 4 H^+ + 2 e \rightleftharpoons 2 H_2SO_3$	0.564	$Te + 2 e \rightleftharpoons Te^{2-}$	−1.143
$S_2O_8^{2-} + 2 e \rightleftharpoons 2 SO_4^{2-}$	2.010	$Te + 2 H^+ + 2 e \rightleftharpoons H_2Te$	−0.793
$S_2O_8^{2-} + 2 H^+ + 2 e \rightleftharpoons 2 HSO_4^-$	2.123	$Te^{4+} + 4 e \rightleftharpoons Te$	0.568
$S_4O_6^{2-} + 2 e \rightleftharpoons 2 S_2O_3^{2-}$	0.08	$TeO_2 + 4 H^+ + 4 e \rightleftharpoons Te + 2 H_2O$	0.593
$2 H_2SO_3 + H^+ + 2 e \rightleftharpoons HS_2O_4^- + 2 H_2O$	−0.056	$TeO_3^{2-} + 3 H_2O + 4 e \rightleftharpoons Te + 6 OH^-$	−0.57
$H_2SO_3 + 4 H^+ + 4 e \rightleftharpoons S + 3 H_2O$	0.449	$TeO_4^- + 8 H^+ + 7 e \rightleftharpoons Te + 4 H_2O$	0.472
$2 SO_3^{2-} + 2 H_2O + 2 e \rightleftharpoons S_2O_4^{2-} + 4 OH^-$	−1.12	$H_6TeO_6 + 2 H^+ + 2 e \rightleftharpoons TeO_2 + 4 H_2O$	1.02
$2 SO_3^{2-} + 3 H_2O + 4 e \rightleftharpoons S_2O_3^{2-} + 6 OH^-$	−0.571	$Th^{4+} + 4 e \rightleftharpoons Th$	−1.899
$SO_4^{2-} + 4 H^+ + 2 e \rightleftharpoons H_2SO_3 + H_2O$	0.172	$ThO_2 + 4 H^+ + 4 e \rightleftharpoons Th + 2 H_2O$	−1.789
$2 SO_4^{2-} + 4 H^+ + 2 e \rightleftharpoons S_2O_6^{2-} + H_2O$	−0.22	$Th(OH)_4 + 4 e \rightleftharpoons Th + 4 OH^-$	−2.48
$SO_4^{2-} + H_2O + 2 e \rightleftharpoons SO_3^{2-} + 2 OH^-$	−0.93	$Ti^{2+} + 2 e \rightleftharpoons Ti$	−1.630
$Sb + 3 H^+ + 3 e \rightleftharpoons SbH_3$	−0.510	$Ti^{3+} + e \rightleftharpoons Ti^{2+}$	−0.9
$Sb_2O_3 + 6 H^+ + 6 e \rightleftharpoons 2 Sb + 3 H_2O$	0.152	$TiO_2 + 4 H^+ + 2 e \rightleftharpoons Ti^{2+} + 2 H_2O$	−0.502
$Sb_2O_5 \text{ (senarmontite)} + 4 H^+ + 4 e \rightleftharpoons Sb_2O_3 + 2 H_2O$	0.671	$Ti^{3+} + 3 e \rightleftharpoons Ti$	−1.37
$Sb_2O_5 \text{ (valentinite)} + 4 H^+ + 4 e \rightleftharpoons Sb_2O_3 + 2 H_2O$	0.649	$TiOH^{3+} + H^+ + e \rightleftharpoons Ti^{3+} + H_2O$	−0.055
$Sb_2O_5 + 6 H^+ + 4 e \rightleftharpoons 2 SbO^+ + 3 H_2O$	0.581	$Tl^+ + e \rightleftharpoons Tl$	−0.336
$SbO^+ + 2 H^+ + 3 e \rightleftharpoons Sb + 2 H_2O$	0.212	$Tl^+ + e \rightleftharpoons Tl(Hg)$	−0.3338
$SbO_2^- + 2 H_2O + 3 e \rightleftharpoons Sb + 4 OH^-$	−0.66	$Tl^{3+} + 2 e \rightleftharpoons Tl^+$	1.252
$SbO_3^- + H_2O + 2 e \rightleftharpoons SbO_2^- + 2 OH^-$	−0.59	$Tl^{3+} + 3 e \rightleftharpoons Tl$	0.741
$Sc^{3+} + 3 e \rightleftharpoons Sc$	−2.077	$TlBr + e \rightleftharpoons Tl + Br^-$	−0.658
$Se + 2 e \rightleftharpoons Se^{2-}$	−0.924	$TlCl + e \rightleftharpoons Tl + Cl^-$	−0.5568
$Se + 2 H^+ + 2 e \rightleftharpoons H_2Se(aq)$	−0.399	$TlI + e \rightleftharpoons Tl + I^-$	−0.752
$H_2SeO_3 + 4 H^+ + 4 e \rightleftharpoons Se + 3 H_2O$	0.74	$Tl_2O_3 + 3 H_2O + 4 e \rightleftharpoons 2 Tl^+ + 6 OH^-$	0.02
$Se + 2 H^+ + 2 e \rightleftharpoons H_2Se$	−0.082	$TlOH + e \rightleftharpoons Tl + OH^-$	−0.34
$SeO_3^{2-} + 3 H_2O + 4 e \rightleftharpoons Se + 6 OH^-$	−0.366	$Tl(OH)_3 + 2 e \rightleftharpoons TlOH + 2 OH^-$	−0.05
$SeO_4^{2-} + 4 H^+ + 2 e \rightleftharpoons H_2SeO_3 + H_2O$	1.151	$Tl_2SO_4 + 2 e \rightleftharpoons Tl + SO_4^{2-}$	−0.4360
$SeO_4^{2-} + H_2O + 2 e \rightleftharpoons SeO_3^{2-} + 2 OH^-$	0.05	$Tm^{3+} + e \rightleftharpoons Tm^{2+}$	−2.2
$SiF_6^{2-} + 4 e \rightleftharpoons Si + 6 F^-$	−1.24	$Tm^{3+} + 3 e \rightleftharpoons Tm$	−2.319
$SiO + 2 H^+ + 2 e \rightleftharpoons Si + H_2O$	−0.8	$Tm^{2+} + 2 e \rightleftharpoons Tm$	−2.4

TABLE 1
Alphabetical Listing (continued)

Reaction	$E°$/V	Reaction	$E°$/V
$U^{3+} + 3\,e \rightleftharpoons U$	−1.798	$2\,WO_3 + 2\,H^+ + 2\,e \rightleftharpoons W_2O_5 + H_2O$	−0.029
$U^{4+} + e \rightleftharpoons U^{3+}$	−0.607	$H_4XeO_6 + 2\,H^+ + 2\,e \rightleftharpoons XeO_3 + 3\,H_2O$	2.42
$UO_2^+ + 4\,H^+ + e \rightleftharpoons U^{4+} + 2\,H_2O$	0.612	$XeO_3 + 6\,H^+ + 6\,e \rightleftharpoons Xe + 3\,H_2O$	2.10
$UO_2^{2+} + e \rightleftharpoons UO_2^+$	0.062	$XeF + e \rightleftharpoons Xe + F^-$	3.4
$UO_2^{2+} + 4\,H^+ + 2\,e \rightleftharpoons U^{4+} + 2\,H_2O$	0.327	$Y^{3+} + 3\,e \rightleftharpoons Y$	−2.372
$UO_2^{2+} + 4\,H^+ + 6\,e \rightleftharpoons U + 2\,H_2O$	−1.444	$Yb^{3+} + e \rightleftharpoons Yb^{2+}$	−1.05
$V^{2+} + 2\,e \rightleftharpoons V$	−1.175	$Yb^{3+} + 3\,e \rightleftharpoons Yb$	−2.19
$V^{3+} + e \rightleftharpoons V^{2+}$	−0.255	$Yb^{2+} + 2\,e \rightleftharpoons Yb$	−2.76
$VO^{2+} + 2\,H^+ + e \rightleftharpoons V^{3+} + H_2O$	0.337	$Zn^{2+} + 2\,e \rightleftharpoons Zn$	−0.7618
$VO_2^+ + 2\,H^+ + e \rightleftharpoons VO^{2+} + H_2O$	0.991	$Zn^{2+} + 2\,e \rightleftharpoons Zn(Hg)$	−0.7628
$V_2O_5 + 6\,H^+ + 2\,e \rightleftharpoons 2\,VO^{2+} + 3\,H_2O$	0.957	$ZnO_2^{2-} + 2\,H_2O + 2\,e \rightleftharpoons Zn + 4\,OH^-$	−1.215
$V_2O_5 + 10\,H^+ + 10\,e \rightleftharpoons 2\,V + 5\,H_2O$	−0.242	$ZnSO_4 \cdot 7\,H_2O + 2\,e = Zn(Hg) + SO_4^{2-} + 7\,H_2O$	−0.7993
$V(OH)_4^+ + 2\,H^+ + e \rightleftharpoons VO^{2+} + 3\,H_2O$	1.00	(Saturated $ZnSO_4$)	
$V(OH)_4^+ + 4\,H^+ + 5\,e \rightleftharpoons V + 4\,H_2O$	−0.254	$ZnOH^+ + H^+ + 2\,e \rightleftharpoons Zn + H_2O$	−0.497
$[V(phen)_3]^{3+} + e \rightleftharpoons [V(phen)_3]^{2+}$	0.14	$Zn(OH)_4^{2-} + 2\,e \rightleftharpoons Zn + 4\,OH^-$	−1.199
$W^{3+} + 3\,e \rightleftharpoons W$	0.1	$Zn(OH)_2 + 2\,e \rightleftharpoons Zn + 2\,OH^-$	−1.249
$W_2O_5 + 2\,H^+ + 2\,e \rightleftharpoons 2\,WO_2 + H_2O$	−0.031	$ZnO + H_2O + 2\,e \rightleftharpoons Zn + 2\,OH^-$	−1.260
$WO_2 + 4\,H^+ + 4\,e \rightleftharpoons W + 2\,H_2O$	−0.119	$ZrO_2 + 4\,H^+ + 4\,e \rightleftharpoons Zr + 2\,H_2O$	−1.553
$WO_3 + 6\,H^+ + 6\,e \rightleftharpoons W + 3\,H_2O$	−0.090	$ZrO(OH)_2 + H_2O + 4\,e \rightleftharpoons Zr + 4\,OH^-$	−2.36
$WO_3 + 2\,H^+ + 2\,e \rightleftharpoons WO_2 + H_2O$	0.036	$Zr^{4+} + 4\,e \rightleftharpoons Zr$	−1.45

TABLE 2
Reduction Reactions Having $E°$ Values More Positive than that of the Standard Hydrogen Electrode

Reaction	$E°$/V	Reaction	$E°$/V
$2\,H^+ + 2\,e \rightleftharpoons H_2$	0.00000	$Sn(OH)_3^+ + 3\,H^+ + 2\,e \rightleftharpoons Sn^{2+} + 3\,H_2O$	0.142
$CuI_2^- + e \rightleftharpoons Cu + 2\,I^-$	0.00	$Np^{4+} + e \rightleftharpoons Np^{3+}$	0.147
$Ge^{4+} + 2\,e \rightleftharpoons Ge^{2+}$	0.00	$Ag_4[Fe(CN)_6] + 4\,e \rightleftharpoons 4\,Ag + [Fe(CN)_6]^{4-}$	0.1478
$NO_3^- + H_2O + 2\,e \rightleftharpoons NO_2^- + 2\,OH^-$	0.01	$IO_3^- + 2\,H_2O + 4\,e \rightleftharpoons IO^- + 4\,OH^-$	0.15
$Tl_2O_3 + 3\,H_2O + 4\,e \rightleftharpoons 2\,Tl^+ + 6\,OH^-$	0.02	$Mn(OH)_3 + e \rightleftharpoons Mn(OH)_2 + OH^-$	0.15
$SeO_4^{2-} + H_2O + 2\,e \rightleftharpoons SeO_3^{2-} + 2\,OH^-$	0.05	$2\,NO_2^- + 3\,H_2O + 4\,e \rightleftharpoons N_2O + 6\,OH^-$	0.15
$WO_3 + 2\,H^+ + 2\,e \rightleftharpoons WO_2 + H_2O$	0.036	$Sn^{4+} + 2\,e \rightleftharpoons Sn^{2+}$	0.151
$UO_2^{2+} + e = UO_2^+$	0.062	$Sb_2O_3 + 6\,H^+ + 6\,e \rightleftharpoons 2\,Sb + 3\,H_2O$	0.152
$Pd(OH)_2 + 2\,e \rightleftharpoons Pd + 2\,OH^-$	0.07	$Cu^{2+} + e \rightleftharpoons Cu^+$	0.153
$AgBr + e \rightleftharpoons Ag + Br^-$	0.07133	$BiOCl + 2\,H^+ + 3\,e \rightleftharpoons Bi + Cl^- + H_2O$	0.1583
$MoO_3 + 6\,H^+ + 6\,e \rightleftharpoons Mo + 3\,H_2O$	0.075	$BiCl_4^- + 3\,e \rightleftharpoons Bi + 4\,Cl^-$	0.16
$S_4O_6^{2-} + 2\,e \rightleftharpoons 2\,S_2O_3^{2-}$	0.08	$Fe_2O_3 + 4\,H^+ + 2\,e \rightleftharpoons 2\,FeOH^+ + H_2O$	0.16
$H_3Mo_7O_{24}^{3-} + 45\,H^+ + 42\,e \rightleftharpoons 7\,Mo + 24\,H_2O$	0.082	$Co(OH)_3 + e \rightleftharpoons Co(OH)_2 + OH^-$	0.17
$AgSCN + e \rightleftharpoons Ag + SCN^-$	0.8951	$SO_4^{2-} + 4\,H^+ + 2\,e \rightleftharpoons H_2SO_3 + H_2O$	0.172
$N_2 + 2\,H_2O + 6\,H^+ + 6\,e \rightleftharpoons 2\,NH_4OH$	0.092	$Bi^{3+} + 2\,e \rightleftharpoons Bi^+$	0.2
$HgO + H_2O + 2\,e \rightleftharpoons Hg + 2\,OH^-$	0.0977	$[Ru(en)_3]^{3+} + e \rightleftharpoons [Ru(en)_3]^{2+}$	0.210
$Ir_2O_3 + 3\,H_2O + 6\,e \rightleftharpoons 2\,Ir + 6\,OH^-$	0.098	$SbO^+ + 2\,H^+ + 3\,e \rightleftharpoons Sb + 2\,H_2O$	0.212
$2\,NO + 2\,e \rightleftharpoons N_2O_2^{2-}$	0.10	$AgCl + e \rightleftharpoons Ag + Cl^-$	0.22233
$[Ru(NH_3)_6]^{3+} + e \rightleftharpoons [Ru(NH_3)_6]^{2+}$	0.10	$[Ru(H_2O)_6]^{3+} + e \rightleftharpoons [Ru(H_2O)_6]^{2+}$	0.23
$W^{3+} + 3\,e \rightleftharpoons W$	0.1	$As_2O_3 + 6\,H^+ + 6\,e \rightleftharpoons 2\,As + 3\,H_2O$	0.234
$[Co(NH_3)_6]^{3+} + e \rightleftharpoons [Co(NH_3)_6]^{2+}$	0.108	Calomel electrode, saturated NaCl (SSCE)	0.2360
$Hg_2O + H_2O + 2\,e \rightleftharpoons 2\,Hg + 2\,OH^-$	0.123	$Ge^{2+} + 2\,e \rightleftharpoons Ge$	0.24
$Ge^{4+} + 4\,e \rightleftharpoons Ge$	0.124	$Ru^{3+} + e \rightleftharpoons Ru^{2+}$	0.24
$Hg_2Br_2 + 2\,e \rightleftharpoons 2\,Hg + 2\,Br^-$	0.13923	Calomel electrode, saturated KCl	0.2412
$Pt(OH)_2 + 2\,e \rightleftharpoons Pt + 2\,OH^-$	0.14	$PbO_2 + H_2O + 2\,e \rightleftharpoons PbO + 2\,OH^-$	0.247
$[V(phen)_3]^{3+} + e \rightleftharpoons [V(phen)_3]^{2+}$	0.14	$HAsO_2 + 3\,H^+ + 3\,e \rightleftharpoons As + 2\,H_2O$	0.248
$S + 2\,H^+ + 2\,e \rightleftharpoons H_2S(aq)$	0.142	$Ru^{3+} + e \rightleftharpoons Ru^{2+}$	0.2487

TABLE 2
Reduction Reactions Having $E°$ Values More Positive than that of the Standard Hydrogen Electrode (continued)

Reaction	$E°$/V	Reaction	$E°$/V
$ReO_2 + 4\,H^+ + 4\,e \rightleftharpoons Re + 2\,H_2O$	0.2513	$[PdCl_4]^{2-} + 2\,e \rightleftharpoons Pd + 4\,Cl^-$	0.591
$IO_3^- + 3\,H_2O + 6\,e \rightleftharpoons I^- + OH^-$	0.26	$TeO_2 + 4\,H^+ + 4\,e \rightleftharpoons Te + 2\,H_2O$	0.593
$Hg_2Cl_2 + 2\,e \rightleftharpoons 2\,Hg + 2\,Cl^-$	0.26808	$MnO_4^- + 2\,H_2O + 3\,e \rightleftharpoons MnO_2 + 4\,OH^-$	0.595
Calomel electrode, 1 molal KCl	0.2800	$Rh^{2+} + 2\,e \rightleftharpoons Rh$	0.600
Calomel electrode, 1 molar KCl (NCE)	0.2801	$Rh^+ + e \rightleftharpoons Rh$	0.600
$At_2 + 2\,e \rightleftharpoons 2\,At^-$	0.3	$MnO_4^{2-} + 2\,H_2O + 2\,e \rightleftharpoons MnO_2 + 4\,OH^-$	0.60
$Re^{3+} + 3\,e \rightleftharpoons Re$	0.300	$2\,AgO + H_2O + 2\,e \rightleftharpoons Ag_2O + 2\,OH^-$	0.607
$Tc^{3+} + e \rightleftharpoons Tc^{2+}$	0.3	$BrO_3^- + 3\,H_2O + 6\,e \rightleftharpoons Br^- + 6\,OH^-$	0.61
$Bi^{3+} + 3\,e \rightleftharpoons Bi$	0.308	$UO_2^+ + 4\,H^+ + e \rightleftharpoons U^{4+} + 2\,H_2O$	0.612
$BiO^+ + 2\,H^+ + 3\,e \rightleftharpoons Bi + H_2O$	0.320	$Hg_2SO_4 + 2\,e \rightleftharpoons 2\,Hg + SO_4^{2-}$	0.6125
$UO_2^{2+} + 4\,H^+ + 2\,e \rightleftharpoons U^{4+} + 2\,H_2O$	0.327	$ClO_3^- + 3\,H_2O + 6\,e \rightleftharpoons Cl^- + 6\,OH^-$	0.62
$ClO_3^- + H_2O + 2\,e \rightleftharpoons ClO_2^- + 2\,OH^-$	0.33	$Hg_2HPO_4 + 2\,e \rightleftharpoons 2\,Hg + HPO_4^{2-}$	0.6359
$2\,HCNO + 2\,H^+ + 2\,e \rightleftharpoons (CN)_2 + 2\,H_2O$	0.330	$Ag(ac) + e \rightleftharpoons Ag + (ac)^-$	0.643
Calomel electrode, 0.1 molar KCl	0.3337	$Sb_2O_5(valentinite) + 4\,H^+ + 4\,e \rightleftharpoons Sb_2O_3 + 2\,H_2O$	0.649
$VO^{2+} + 2\,H^+ + e \rightleftharpoons V^{3+} + H_2O$	0.337	$Ag_2SO_4 + 2\,e \rightleftharpoons 2\,Ag + SO_4^{2-}$	0.654
$Cu^{2+} + 2\,e \rightleftharpoons Cu$	0.3419	$ClO_2^- + H_2O + 2\,e \rightleftharpoons ClO^- + 2\,OH^-$	0.66
$Ag_2O + H_2O + 2\,e \rightleftharpoons 2\,Ag + 2\,OH^-$	0.342	$Sb_2O_5(senarmontite) + 4\,H^+ + 4\,e \rightleftharpoons Sb_2O_5 + 2\,H_2O$	0.671
$Cu^{2+} + 2\,e \rightleftharpoons Cu(Hg)$	0.345	$[PtCl_6]^{2-} + 2\,e \rightleftharpoons [PtCl_4]^{2-} + 2\,Cl^-$	0.68
$AgIO_3 + e \rightleftharpoons Ag + IO_3^-$	0.354	$O_2 + 2\,H^+ + 2\,e \rightleftharpoons H_2O_2$	0.695
$[Fe(CN)_6]^{3-} + e \rightleftharpoons [Fe(CN)_6]^{4-}$	0.358	$p\text{-benzoquinone} + 2\,H^+ + 2\,e \rightleftharpoons \text{hydroquinone}$	0.6992
$ClO_4^- + H_2O + 2\,e \rightleftharpoons ClO_3^- + 2\,OH^-$	0.36	$H_3IO_6^{2-} + 2\,e \rightleftharpoons IO_3^- + 3\,OH^-$	0.7
$Ag_2SeO_3 + 2\,e \rightleftharpoons 2\,Ag + SeO_3^{2-}$	0.3629	$Ag_2O_3 + H_2O + 2\,e \rightleftharpoons 2\,AgO + 2\,OH^-$	0.739
$ReO_4^- + 8\,H^+ + 7\,e \rightleftharpoons Re + 4\,H_2O$	0.368	$Tl^{3+} + 3\,e \rightleftharpoons Tl$	0.741
$(CN)_2 + 2\,H^+ + 2\,e \rightleftharpoons 2\,HCN$	0.373	$[PtCl_4]^{2-} + 2\,e \rightleftharpoons Pt + 4\,Cl^-$	0.755
$[Ferricinium]^+ + e \rightleftharpoons ferrocene$	0.400	$Rh^{3+} + 3\,e \rightleftharpoons Rh$	0.758
$Tc^{2+} + 2\,e \rightleftharpoons Tc$	0.400	$ClO_2^- + 2\,H_2O + 4\,e \rightleftharpoons Cl^- + 4\,OH^-$	0.76
$O_2 + 2\,H_2O + 4\,e \rightleftharpoons 4\,OH^-$	0.401	$2\,NO + H_2O + 2\,e \rightleftharpoons N_2O + 2\,OH^-$	0.76
$AgOCN + e \rightleftharpoons Ag + OCN^-$	0.41	$Po^{4+} + 4\,e \rightleftharpoons Po$	0.76
$[RhCl_6]^{3-} + 3\,e \rightleftharpoons Rh + 6\,Cl^-$	0.431	$BrO^- + H_2O + 2\,e \rightleftharpoons Br^- + 2\,OH^-$	0.761
$Ag_2CrO_4 + 2\,e \rightleftharpoons 2\,Ag + CrO_4^{2-}$	0.4470	$ReO_4^- + 2\,H^+ + e \rightleftharpoons ReO_3 + H_2O$	0.768
$H_2SO_3 + 4\,H^+ + 4\,e \rightleftharpoons S + 3\,H_2O$	0.449	$(CNS)_2 + 2\,e \rightleftharpoons 2\,CNS^-$	0.77
$Ru^{2+} + 2\,e \rightleftharpoons Ru$	0.455	$[IrCl_6]^{3-} + 3\,e \rightleftharpoons Ir + 6\,Cl^-$	0.77
$Ag_2MoO_4 + 2\,e \rightleftharpoons 2\,Ag + MoO_4^{2-}$	0.4573	$Fe^{3+} + e \rightleftharpoons Fe^{2+}$	0.771
$Ag_2C_2O_4 + 2\,e \rightleftharpoons 2\,Ag + C_2O_4^{2-}$	0.4647	$AgF + e \rightleftharpoons Ag + F^-$	0.779
$Ag_2WO_4 + 2\,e \rightleftharpoons 2\,Ag + WO_4^{2-}$	0.4660	$[Fe(bipy)_2]^{3+} + e \rightleftharpoons [Fe(bipy)_2]^{2+}$	0.78
$Ag_2CO_3 + 2\,e \rightleftharpoons 2\,Ag + CO_3^{2-}$	0.47	$TcO_4^- + 4\,H^+ + 3\,e \rightleftharpoons TcO_2 + 2\,H_2O$	0.782
$TcO_4^- + 8\,H^+ + 7\,e \rightleftharpoons Tc + 4\,H_2O$	0.472	$Hg_2^{2+} + 2\,e \rightleftharpoons 2\,Hg$	0.7973
$TeO_4^- + 8\,H^+ + 7\,e \rightleftharpoons Te + 4\,H_2O$	0.472	$Ag^+ + e \rightleftharpoons Ag$	0.7996
$IO^- + H_2O + 2\,e \rightleftharpoons I^- + 2\,OH^-$	0.485	$[Os(bipy)_3]^{3+} + e \rightleftharpoons [Os(bipy)_3]^{2+}$	0.80
$NiO_2 + 2\,H_2O + 2\,e \rightleftharpoons Ni(OH)_2 + 2\,OH^-$	0.490	$2\,NO_3^- + 4\,H^+ + 2\,e \rightleftharpoons N_2O_4 + 2\,H_2O$	0.803
$Bi^+ + e \rightleftharpoons Bi$	0.5	$[Os(bipy)_2]^{3+} + e \rightleftharpoons [Os(bipy)_2]^{2+}$	0.81
$ReO_4^- + 4\,H^+ + 3\,e \rightleftharpoons ReO_2 + 2\,H_2O$	0.510	$RhOH^{2+} + H + 3\,e \rightleftharpoons Rh + H_2O$	0.83
$Hg_2(ac)_2 + 2\,e \rightleftharpoons 2\,Hg + 2(ac)^-$	0.51163	$OsO_4 + 8\,H^+ + 8\,e \rightleftharpoons Os + 4\,H_2O$	0.838
$Cu^+ + e \rightleftharpoons Cu$	0.521	$ClO^- + H_2O + 2\,e \rightleftharpoons Cl^- + 2\,OH^-$	0.841
$I_2 + 2\,e \rightleftharpoons 2\,I^-$	0.5355	$Hg^{2+} + 2\,e \rightleftharpoons Hg$	0.851
$I_3^- + 2\,e \rightleftharpoons 3\,I^-$	0.536	$AuBr_4^- + 3\,e \rightleftharpoons Au + 4\,Br^-$	0.854
$AgBrO_3 + e \rightleftharpoons Ag + BrO_3^-$	0.546	$SiO_2(quartz) + 4\,H^+ + 4\,e \rightleftharpoons Si + 2\,H_2O$	0.857
$MnO_4^- + e \rightleftharpoons MnO_4^{2-}$	0.558	$2\,HNO_2 + 4\,H^+ + 4\,e \rightleftharpoons H_2N_2O_2 + H_2O$	0.86
$H_3AsO_4 + 2\,H^+ + 2\,e \rightleftharpoons HAsO_2 + 2\,H_2O$	0.560	$[Ru(CN)_6]^{3-} + e^- \rightleftharpoons [Ru(CN)_6]^{4-}$	0.86
$S_2O_6^{2-} + 4\,H^+ + 2\,e \rightleftharpoons 2\,H_2SO_3$	0.564	$[IrCl_6]^{2-} + e \rightleftharpoons [IrCl_6]^{3-}$	0.8665
$AgNO_2 + e \rightleftharpoons Ag + NO_2^-$	0.564	$N_2O_4 + 2\,e \rightleftharpoons 2\,NO_2^-$	0.867
$Te^{4+} + 4\,e \rightleftharpoons Te$	0.568	$HO_2^- + H_2O + 2\,e \rightleftharpoons 3\,OH^-$	0.878
$Sb_2O_5 + 6\,H^+ + 4\,e \rightleftharpoons 2\,SbO^+ + 3\,H_2O$	0.581	$Po^{4+} + 2\,e \rightleftharpoons Po^{2+}$	0.9
$RuO_4^- + e \rightleftharpoons RuO_4^{2-}$	0.59	$2\,Hg^+ + 2\,e \rightleftharpoons Hg_2^{2+}$	0.920

TABLE 2
Reduction Reactions Having $E°$ Values More Positive than that of the Standard Hydrogen Electrode
(continued)

Reaction	$E°$/V	Reaction	$E°$/V
$NO_3^- + 3 H^+ + 2 e \rightleftharpoons HNO_2 + H_2O$	0.934	$Cl_2(g) + 2 e \rightleftharpoons 2 Cl^-$	1.35827
$Pd^{2+} + 2 e \rightleftharpoons Pd$	0.951	$ClO_4^- + 8 H^+ + 8 e \rightleftharpoons Cl^- + 4 H_2O$	1.389
$ClO_2(aq) + e \rightleftharpoons ClO_2^-$	0.954	$ClO_4^- + 8 H^+ + 7 e \rightleftharpoons 1/2 Cl_2 + 4 H_2O$	1.39
$NO_3^- + 4 H^+ + 3 e \rightleftharpoons NO + 2 H_2O$	0.957	$No^{3+} + e \rightleftharpoons No^{2+}$	1.4
$V_2O_5 + 6 H^+ + 2 e \rightleftharpoons 2 VO^{2+} + 3 H_2O$	0.957	$RuO_4 + 6 H^+ + 4 e \rightleftharpoons Ru(OH)_2^{2+} + 2 H_2O$	1.40
$AuBr_2^- + e \rightleftharpoons Au + 2 Br^-$	0.959	$Au^{3+} + 2 e \rightleftharpoons Au^+$	1.401
$HNO_2 + H^+ + e \rightleftharpoons NO + H_2O$	0.983	$2 NH_3OH^+ + H^+ + 2 e \rightleftharpoons N_2H_5^+ + 2 H_2O$	1.42
$HIO + H^+ + 2 e \rightleftharpoons I^- + H_2O$	0.987	$BrO_3^- + 6 H^+ + 6 e \rightleftharpoons Br^- + 3 H_2O$	1.423
$VO_2^+ + 2 H^+ + e \rightleftharpoons VO^{2+} + H_2O$	0.991	$2 HIO + 2 H^+ + 2 e \rightleftharpoons I_2 + 2 H_2O$	1.439
$PtO_2 + 4 H^+ + 4 e \rightleftharpoons Pt + 2 H_2O$	1.00	$Au(OH)_3 + 3 H^+ + 3 e \rightleftharpoons Au^- + 3 H_2O$	1.45
$RuO_4 + e \rightleftharpoons RuO_4^-$	1.00	$3 IO_3^- + 6 H^+ + 6 e \rightleftharpoons Cl^- + 3 H_2O$	1.451
$V(OH)_4^+ + 2 H^+ + e \rightleftharpoons VO^{2+} + 3 H_2O$	1.00	$PbO_2 + 4 H^+ + 2 e \rightleftharpoons Pb^{2+} + 2 H_2O$	1.455
$AuCl_4^- + 3 e \rightleftharpoons Au + 4 Cl^-$	1.002	$ClO_3^- + 6 H^+ + 5 e \rightleftharpoons 1/2 Cl_2 + 3 H_2O$	1.47
$Pu^{4+} + e \rightleftharpoons Pu^{3+}$	1.006	$CrO_2 + 4 H^+ + e \rightleftharpoons Cr^{3+} + 2 H_2O$	1.48
$PtO_2 + 2 H^+ + 2 e \rightleftharpoons PtO + H_2O$	1.01	$BrO_3^- + 6 H^+ + 5 e \rightleftharpoons 1/2 Br_2 + 3 H_2O$	1.482
$OsO_4 + 4 H + 4 e \rightleftharpoons OsO_2 + 2 H_2O$	1.02	$HClO + H^+ + 2 e \rightleftharpoons Cl^- + H_2O$	1.482
$H_6TeO_6 + 2 H^+ + 2 e \rightleftharpoons TeO_2 + 4 H_2O$	1.02	$Mn_2O_3 + 6 H^+ + e \rightleftharpoons 2 Mn^{2+} + 3 H_2O$	1.485
$[Fe(bipy)_3]^{3+} + e \rightleftharpoons [Fe(bipy)_3]^{2+}$	1.03	$HO_2 + H^+ + e \rightleftharpoons H_2O_2$	1.495
$Hg(OH)_2 + 2 H^+ + 2 e \rightleftharpoons Hg + 2 H_2O$	1.034	$Au^{3+} + 3 e \rightleftharpoons Au$	1.498
$N_2O_4 + 4 H^+ + 4 e \rightleftharpoons 2 NO + 2 H_2O$	1.035	$PtO_3 + 4 H^+ + 2 e \rightleftharpoons Pt(OH)_2^{2+} + H_2O$	1.5
$RuO_4 + 8 H^+ + 8 e \rightleftharpoons Ru + 4 H_2O$	1.038	$MnO_4^- + 8 H^+ + 5 e \rightleftharpoons Mn^{2+} + 4 H_2O$	1.507
$[Fe(phen)_3]^{3+} + e \rightleftharpoons [Fe(phen)_3]^{2+}$ (1 molar H_2SO_4)	1.06	$Mn^{3+} + e \rightleftharpoons Mn^{2-}$	1.5415
$PuO_2(OH)_2 + H^+ + e \rightleftharpoons PuO_2OH + H_2O$	1.062	$HClO_2 + 3 H^+ + 4 e \rightleftharpoons Cl^- + 2 H_2O$	1.570
$N_2O_4 + 2 H^+ + 2 e \rightleftharpoons 2 HNO_2$	1.065	$HBrO + H^+ + e \rightleftharpoons 1/2 Br_2(aq) + H_2O$	1.574
$Br_2(l) + 2 e \rightleftharpoons 2 Br^-$	1.066	$2 NO + 2 H^+ + 2 e \rightleftharpoons N_2O + H_2O$	1.591
$IO_3^- + 6 H^+ + 6 e \rightleftharpoons I^- + 3 H_2O$	1.085	$Bi_2O_4 + 4 H^+ + 2 e \rightleftharpoons 2 BiO^+ + 2 H_2O$	1.593
$Br_2(aq) + 2 e \rightleftharpoons 2 Br^-$	1.0873	$HBrO + H^+ + e \rightleftharpoons 1/2 Br_2(l) + H_2O$	1.596
$Pu^{5+} + e \rightleftharpoons Pu^{4+}$	1.099	$H_5IO_6 + H^+ + 2 e \rightleftharpoons IO_3^- + 3 H_2O$	1.601
$Cu^{2+} + 2 CN^- + e \rightleftharpoons [Cu(CN)_2]^-$	1.103	$HClO + H^+ + e \rightleftharpoons 1/2 Cl_2 + H_2O$	1.611
$RuO_2 + 4 H^+ + 2 e \rightleftharpoons Ru^{2+} + 2 H_2O$	1.120	$HClO_2 + 3 H^+ + 3 e \rightleftharpoons 1/2 Cl_2 + 2 H_2O$	1.628
$[Fe(phen)_3]^{3+} + e \rightleftharpoons [Fe(phen)_3]^{2+}$	1.147	$HClO_2 + 2 H^+ + 2 e \rightleftharpoons HClO + H_2O$	1.645
$SeO_4^{2-} + 4 H^+ + 2 e \rightleftharpoons H_2SeO_3 + H_2O$	1.151	$Bk^{4+} + e \rightleftharpoons Bk^{3+}$	1.67
$ClO_3^- + 2 H^+ + e \rightleftharpoons ClO_2 + H_2O$	1.152	$NiO_2 + 4 H^+ + 2 e \rightleftharpoons Ni^{2+} + 2 H_2O$	1.678
$Ir^{3+} + 3 e \rightleftharpoons Ir$	1.156	$MnO_4^- + 4 H^+ + 3 e \rightleftharpoons MnO_2 + 2 H_2O$	1.679
$Pt^{2+} + 2 e \rightleftharpoons Pt$	1.18	$PbO_2 + SO_4^{2-} + 4 H^+ + 2 e \rightleftharpoons PbSO_4 + 2 H_2O$	1.6913
$ClO_4^- + 2 H^+ + 2 e \rightleftharpoons ClO_3^- + H_2O$	1.189	$Au^+ + e \rightleftharpoons Au$	1.692
$2 IO_3^- + 12 H^+ + 10 e \rightleftharpoons I_2 + 6 H_2O$	1.195	$PtO_3 + 2 H^+ + 2 e \rightleftharpoons PtO_2 + H_2O$	1.7
$PtOH^+ + H^+ + 2 e \rightleftharpoons Pt + H_2O$	1.2	$CeOH^{3+} + H^+ + e \rightleftharpoons Ce^{3+} + H_2O$	1.715
$ClO_3^- + 3 H^+ + 2 e \rightleftharpoons HClO_2 + H_2O$	1.214	$Ce^{4+} + e \rightleftharpoons Ce^{3+}$	1.72
$MnO_2 + 4 H^+ + 2 e \rightleftharpoons Mn^{2+} + 2 H_2O$	1.224	$N_2O + 2 H^+ + 2 e \rightleftharpoons N_2 + H_2O$	1.766
$O_2 + 4 H^+ + 4 e \rightleftharpoons 2 H_2O$	1.229	$H_2O_2 + 2 H^+ + 2 e \rightleftharpoons 2 H_2O$	1.776
$Cr_2O_7^{2-} + 14 H^+ + 6 e \rightleftharpoons 2 Cr^{3+} + 7 H_2O$	1.232	$Ag^{3+} + e \rightleftharpoons Ag^{2+}$	1.8
$O_3 + H_2O + 2 e \rightleftharpoons O_2 + 2 OH^-$	1.24	$Au^{2+} + e \rightleftharpoons Au^+$	1.8
$[Ru(bipy)_3]^{3+} + e \rightleftharpoons [Ru(bipy)_3]^{2+}$	1.24	$Ag_2O_2 + 4 H^+ + e \rightleftharpoons 2 Ag + 2 H_2O$	1.802
$Tl^{3+} + 2 e \rightleftharpoons Tl^+$	1.252	$Co^{3+} + e \rightleftharpoons Co^{2-}$ (2 molar H_2SO_4)	1.83
$N_2H_5^+ + 3 H^+ + 2 e \rightleftharpoons 2 NH_4^+$	1.275	$Ag^{3+} + 2 e \rightleftharpoons Ag^+$	1.9
$ClO_2 + H^+ + e \rightleftharpoons HClO_2$	1.277	$Co^{3+} + e \rightleftharpoons Co^{2+}$	1.92
$[PdCl_6]^{2-} + 2 e \rightleftharpoons [PdCl_4]^{2-} + 2 Cl^-$	1.288	$Ag^{2+} + e \rightleftharpoons Ag^+$	1.980
$2 HNO_2 + 4 H^+ + 4 e \rightleftharpoons N_2O + 3 H_2O$	1.297	$Cu_2O_3 + 6 H^+ + 2 e \rightleftharpoons 2 Cu^{2+} + 3 H_2O$	2.0
$AuOH^{2+} + H^+ + 2 e \rightleftharpoons Au^+ + H_2O$	1.32	$S_2O_8^{2-} + 2 e \rightleftharpoons 2 SO_4^{2-}$	2.010
$PuO_2(OH)_2 + 2 H^- + 2 e \rightleftharpoons Pu(OH)_4$	1.325	$OH + e \rightleftharpoons OH^-$	2.02
$HBrO + H^+ + 2 e \rightleftharpoons Br^- + H_2O$	1.331	$HFeO_4^- + 7 H^+ + 3 e \rightleftharpoons Fe^{3+} + 4 H_2O$	2.07
$Cr(V) + e \rightleftharpoons Cr(IV)$	1.34	$O_3 + 2 H^+ + 2 e \rightleftharpoons O_2 + H_2O$	2.076
$HCrO_4^- + 7 H^+ + 3 e \rightleftharpoons Cr^{3+} + 4 H_2O$	1.350	$HFeO_4^- + 4 H^+ + 3 e \rightleftharpoons FeOOH + 2 H_2O$	2.08

TABLE 2
Reduction Reactions Having $E°$ Values More Positive than that of the Standard Hydrogen Electrode (continued)

Reaction	$E°/V$	Reaction	$E°/V$
$2\,HFeO_4^- + 8\,H^+ + 6\,e \rightleftharpoons Fe_2O_3 + 5\,H_2O$	2.09	$H_2N_2O_2 + 2\,H^+ + 2\,e \rightleftharpoons N_2 + 2\,H_2O$	2.65
$XeO_3 + 6\,H^+ + 6\,e \rightleftharpoons Xe + 3\,H_2O$	2.10	$F_2 + 2\,e \rightleftharpoons 2\,F^-$	2.866
$S_2O_8^{2-} + 2\,H^+ + 2\,e \rightleftharpoons 2\,HSO_4^-$	2.123	$Cm^{4+} + e \rightleftharpoons Cm^{3+}$	3.0
$F_2O + 2\,H^+ + 4\,e \rightleftharpoons H_2O + 2\,F^-$	2.153	$F_2 + 2\,H^+ + 2\,e \rightleftharpoons 2\,HF$	3.053
$FeO_4^{2-} + 8\,H^+ + 3\,e \rightleftharpoons Fe^{3+} + 4\,H_2O$	2.20	$Tb^{4+} + e \rightleftharpoons Tb^{3+}$	3.1
$Cu^{3+} + e \rightleftharpoons Cu^{2+}$	2.4	$Pr^{4+} + e \rightleftharpoons Pr^{3+}$	3.2
$H_4XeO_6 + 2\,H^+ + 2\,e \rightleftharpoons XeO_3 + 3\,H_2O$	2.42	$Cf^{4+} + e \rightleftharpoons Cf^{3+}$	3.3
$O(g) + 2\,H^+ + 2\,e \rightleftharpoons H_2O$	2.421	$XeF + e \rightleftharpoons Xe + F^-$	3.4
$Am^{4+} + e \rightleftharpoons Am^{3+}$	2.60		

TABLE 3
Reduction Reactions Having $E°$ Values More Negative than that of the Standard Hydrogen Electrode

Reaction	$E°/V$	Reaction	$E°/V$
$2\,H^+ + 2\,e \rightleftharpoons H_2$	0.00000	$Cu(OH)_2 + 2\,e \rightleftharpoons Cu + 2\,OH^-$	−0.222
$2\,D^+ + 2\,e \rightleftharpoons D_2$	−0.013	$V_2O_5 + 10\,H^+ + 10\,e \rightleftharpoons 2\,V + 5\,H_2O$	−0.242
$AgCN + e \rightleftharpoons Ag + CN^-$	−0.017	$CdSO_4 + 2\,e \rightleftharpoons Cd + SO_4^{2-}$	−0.246
$2\,WO_3 + 2\,H^+ + 2\,e \rightleftharpoons W_2O_5 + H_2O$	−0.029	$V(OH)_4^+ + 4\,H^+ + 5\,e \rightleftharpoons V + 4\,H_2O$	−0.254
$W_2O_5 + 2\,H^+ + 2\,e \rightleftharpoons 2\,WO_2 + H_2O$	−0.031	$V^{3+} + e \rightleftharpoons V^{2+}$	−0.255
$Ag_2S + 2\,H^+ + 2\,e \rightleftharpoons 2\,Ag + H_2S$	−0.0366	$Ni^{2+} + 2\,e \rightleftharpoons Ni$	−0.257
$Fe^{3+} + 3\,e \rightleftharpoons Fe$	−0.037	$PbCl_2 + 2\,e \rightleftharpoons Pb + 2\,Cl^-$	−0.2675
$Hg_2I_2 + 2\,e \rightleftharpoons 2\,Hg + 2\,I^-$	−0.0405	$H_3PO_4 + 2\,H^+ + 2\,e \rightleftharpoons H_3PO_3 + H_2O$	−0.276
$Tl(OH)_3 + 2\,e \rightleftharpoons TlOH + 2\,OH^-$	−0.05	$Co^{2+} + 2\,e \rightleftharpoons Co$	−0.28
$TiOH^{3+} + H^+ + e \rightleftharpoons Ti^{3+} + H_2O$	−0.055	$PbBr_2 + 2\,e \rightleftharpoons Pb + 2\,Br^-$	−0.284
$2\,H_2SO_3 + H^+ + 2\,e \rightleftharpoons HS_2O_4^- + 2\,H_2O$	−0.056	$Tl^+ + e \rightleftharpoons Tl(Hg)$	−0.3338
$P(white) + 3\,H^+ + 3\,e \rightleftharpoons PH_3(g)$	−0.063	$Tl^+ + e \rightleftharpoons Tl$	−0.336
$O_2 + H_2O + 2\,e \rightleftharpoons HO_2^- + OH^-$	−0.076	$In^{3+} + 3\,e \rightleftharpoons In$	−0.3382
$2\,Cu(OH)_2 + 2\,e \rightleftharpoons Cu_2O + 2\,OH^- + H_2O$	−0.080	$TlOH + e \rightleftharpoons Tl + OH^-$	−0.34
$Se + 2\,H^+ + 2\,e \rightleftharpoons H_2Se$	−0.082	$PbF_2 + 2\,e \rightleftharpoons Pb + 2\,F^-$	−0.3444
$WO_3 + 6\,H^+ + 6\,e \rightleftharpoons W + 3\,H_2O$	−0.090	$PbSO_4 + 2\,e \rightleftharpoons Pb(Hg) + SO_4^{2-}$	−0.3505
$SnO_2 + 4\,H^+ + 2\,e \rightleftharpoons Sn^{2+} + 2\,H_2O$	−0.094	$Cd^{2+} + 2\,e \rightleftharpoons Cd(Hg)$	−0.3521
$Md^{3+} + e \rightleftharpoons Md^{2+}$	−0.1	$PbSO_4 + 2\,e \rightleftharpoons Pb + SO_4^{2-}$	−0.3588
$P(red) + 3\,H^+ + 3\,e \rightleftharpoons PH_3(g)$	−0.111	$Cu_2O + H_2O + 2\,e \rightleftharpoons 2\,Cu + 2\,OH^-$	−0.360
$SnO_2 + 4\,H^+ + 4\,e \rightleftharpoons Sn + 2\,H_2O$	−0.117	$Eu^{3+} + e \rightleftharpoons Eu^{2+}$	−0.36
$GeO_2 + 2\,H^+ + 2\,e \rightleftharpoons GeO + H_2O$	−0.118	$PbI_2 + 2\,e \rightleftharpoons Pb + 2\,I^-$	−0.365
$WO_2 + 4\,H^+ + 4\,e \rightleftharpoons W + 2\,H_2O$	−0.119	$SeO_3^{2-} + 3\,H_2O + 4\,e \rightleftharpoons Se + 6\,OH^-$	−0.366
$Pb^{2+} + 2\,e \rightleftharpoons Pb(Hg)$	−0.1205	$Se + 2\,H^+ + 2\,e \rightleftharpoons H_2Se(aq)$	−0.399
$Pb^{2+} + 2\,e \rightleftharpoons Pb$	−0.1262	$In^{2+} + e \rightleftharpoons In^+$	−0.40
$CrO_4^{2-} + 4\,H_2O + 3\,e \rightleftharpoons Cr(OH)_3 + 5\,OH^-$	−0.13	$Cd^{2+} + 2\,e \rightleftharpoons Cd$	−0.4030
$Sn^{2-} + 2\,e \rightleftharpoons Sn$	−0.1375	$Cr^{3+} + e \rightleftharpoons Cr^{2+}$	−0.407
$In^+ + e \rightleftharpoons In$	−0.14	$2\,S + 2\,e \rightleftharpoons S_2^{2-}$	−0.42836
$O_2 + 2\,H_2O + 2\,e \rightleftharpoons H_2O_2 + 2\,OH^-$	−0.146	$Tl_2SO_4 + 2\,e \rightleftharpoons Tl + SO_4^{2-}$	−0.4360
$MoO_2 + 4\,H^+ + 4\,e \rightleftharpoons Mo + 4\,H_2O$	−0.152	$In^{3+} + 2\,e \rightleftharpoons In^+$	−0.443
$AgI + e \rightleftharpoons Ag + I^-$	−0.15224	$Fe^{2+} + 2\,e \rightleftharpoons Fe$	−0.447
$2\,NO_2^- + 2\,H_2O + 4\,e \rightleftharpoons N_2O_2^{2-} + 4\,OH^-$	−0.18	$H_3PO_3 + 3\,H^+ + 3\,e \rightleftharpoons P + 3\,H_2O$	−0.454
$H_2GeO_3 + 4\,H^+ + 4\,e \rightleftharpoons Ge + 3\,H_2O$	−0.182	$Bi_2O_3 + 3\,H_2O + 6\,e \rightleftharpoons 2\,Bi + 6\,OH^-$	−0.46
$SnO_2 + 3\,H^+ + 2\,e \rightleftharpoons SnOH^+ + H_2O$	−0.194	$NO_2^- + H_2O + e \rightleftharpoons NO + 2\,OH$	−0.46
$CO_2 + 2\,H^+ + 2\,e \rightleftharpoons HCOOH$	−0.199	$PbHPO_4 + 2\,e \rightleftharpoons Pb + HPO_4^{2-}$	−0.465
$Mo^{3+} + 3\,e \rightleftharpoons Mo$	−0.200	$S + 2\,e \rightleftharpoons S^{2-}$	−0.47627
$Ga^+ + e \rightleftharpoons Ga$	−0.2	$S + H_2O + 2\,e \rightleftharpoons HS^- + OH^-$	−0.478
$2\,SO_2^{2-} + 4\,H^+ + 2\,e \rightleftharpoons S_2O_6^{2-} + H_2O$	−0.22	$B(OH)_3 + 7\,H^+ + 8\,e \rightleftharpoons BH_4^- + 3\,H_2O$	−0.481

TABLE 3
Reduction Reactions Having $E°$ Values More Negative than that of the Standard Hydrogen Electrode
(continued)

Reaction	$E°/V$	Reaction	$E°/V$
$In^{3+} + e \rightleftharpoons In^{2+}$	−0.49	$SnO_2 + 2 H_2O + 4 e \rightleftharpoons Sn + 4 OH^-$	−0.945
$ZnOH^+ + H^+ + 2 e \rightleftharpoons Zn + H_2O$	−0.497	$In(OH)_3 + 3 e \rightleftharpoons In + 3 OH^-$	−0.99
$GaOH^{2+} + H^+ + 3 e \rightleftharpoons Ga + H_2O$	−0.498	$NpO_2 + H_2O + H^+ + e \rightleftharpoons Np(OH)_3$	−0.962
$H_3PO_3 + 2 H^+ + 2 e \rightleftharpoons H_3PO_2 + H_2O$	−0.499	$In(OH)_4^- + 3 e \rightleftharpoons In + 4 OH^-$	−1.007
$TiO_2 + 4 H^+ + 2 e \rightleftharpoons Ti^{2+} + 2 H_2O$	−0.502	$In_2O_3 + 3 H_2O + 6 e \rightleftharpoons 2 In + 6 OH^-$	−1.034
$H_3PO_2 + H^+ + e \rightleftharpoons P + 2 H_2O$	−0.508	$PO_4^{3-} + 2 H_2O + 2 e \rightleftharpoons HPO_3^{2-} + 3 OH^-$	−1.05
$Sb + 3 H^+ + 3 e \rightleftharpoons SbH_3$	−0.510	$Yb^{3+} + e \rightleftharpoons Yb^{2+}$	−1.05
$HPbO_2^- + H_2O + 2 e \rightleftharpoons Pb + 3 OH^-$	−0.537	$Nb^{3+} + 3 e \rightleftharpoons Nb$	−1.099
$Ga^{3+} + 3 e \rightleftharpoons Ga$	−0.549	$Fm^{3+} + e \rightleftharpoons Fm^{2+}$	−1.1
$TlCl + e \rightleftharpoons Tl + Cl^-$	−0.5568	$2 SO_3^{2-} + 2 H_2O + 2 e \rightleftharpoons S_2O_4^{2-} + 4 OH^-$	−1.12
$Fe(OH)_3 + e \rightleftharpoons Fe(OH)_2 + OH^-$	−0.56	$Te + 2 e \rightleftharpoons Te^{2-}$	−1.143
$TeO_3^{2-} + 3 H_2O + 4 e \rightleftharpoons Te + 6 OH^-$	−0.57	$V^{2+} + 2 e \rightleftharpoons V$	−1.175
$2 SO_3^{2-} + 3 H_2O + 4 e \rightleftharpoons S_2O_3^{2-} + 6 OH^-$	−0.571	$Mn^{2+} + 2 e \rightleftharpoons Mn$	−1.185
$PbO + H_2O + 2 e \rightleftharpoons Pb + 2 OH^-$	−0.580	$Zn(OH)_4^{2-} + 2 e \rightleftharpoons Zn + 4 OH^-$	−1.199
$ReO_2^- + 4 H_2O + 7 e \rightleftharpoons Re + 8 OH^-$	−0.584	$CrO_2 + 2 H_2O + 3 e \rightleftharpoons Cr + 4 OH^-$	−1.2
$SbO_3^- + H_2O + 2 e \rightleftharpoons SbO_2^- + 2 OH^-$	−0.59	$No^{3+} + 3 e \rightleftharpoons No$	−1.20
$Ta^{3+} + 3 e \rightleftharpoons Ta$	−0.6	$ZnO_2^- + 2 H_2O + 2 e \rightleftharpoons Zn + 4 OH^-$	−1.215
$U^{4+} + e \rightleftharpoons U^{3+}$	−0.607	$H_2GaO_3^- + H_2O + 3 e \rightleftharpoons Ga + 4 OH^-$	−1.219
$As + 3 H^+ + 3 e \rightleftharpoons AsH_3$	−0.608	$H_2BO_3^- + 5 H_2O + 8 e \rightleftharpoons BH_4^- + 8 OH^-$	−1.24
$Nb_2O_5 + 10 H^+ + 10 e \rightleftharpoons 2 Nb + 5 H_2O$	−0.644	$SiF_6^{2-} + 4 e \rightleftharpoons Si + 6 F^-$	−1.24
$NbO_2 + 2 H^+ + 2 e \rightleftharpoons NbO + H_2O$	−0.646	$Zn(OH)_2 + 2 e \rightleftharpoons Zn + 2 OH^-$	−1.249
$Cd(OH)_4^{2-} + 2 e \rightleftharpoons Cd + 4 OH^-$	−0.658	$ZnO + H_2O + 2 e \rightleftharpoons Zn + 2 OH^-$	−1.260
$TlBr + e \rightleftharpoons Tl + Br^-$	−0.658	$Es^{3+} + e \rightleftharpoons Es^{2+}$	−1.3
$SbO_2^- + 2 H_2O + 3 e \rightleftharpoons Sb + 4 OH^-$	−0.66	$Pa^{3+} + 3 e \rightleftharpoons Pa$	−1.34
$AsO_2^- + 2 H_2O + 3 e \rightleftharpoons As + 4 OH^-$	−0.68	$Ti^{3+} + 3 e \rightleftharpoons Ti$	−1.37
$NbO_2 + 4 H^+ + 4 e \rightleftharpoons Nb + 2 H_2O$	−0.690	$Ce^{3+} + 3 e \rightleftharpoons Ce(Hg)$	−1.4373
$Ag_2S + 2 e \rightleftharpoons 2 Ag + S^{2-}$	−0.691	$UO_2^{2+} + 4 H^+ + 6 e \rightleftharpoons U + 2 H_2O$	−1.444
$AsO_4^{3-} + 2 H_2O + 2 e \rightleftharpoons AsO_2^- + 4 OH^-$	−0.71	$Zr^{4+} + 4 e \rightleftharpoons Zr$	−1.45
$Ni(OH)_2 + 2 e \rightleftharpoons Ni + 2 OH^-$	−0.72	$Cr(OH)_3 + 3 e \rightleftharpoons Cr + 3 OH^-$	−1.48
$Co(OH)_2 + 2 e \rightleftharpoons Co + 2 OH^-$	−0.73	$Pa^{4+} + 4 e \rightleftharpoons Pa$	−1.49
$NbO + 2 H^+ + 2 e \rightleftharpoons Nb + H_2O$	−0.733	$HfO_2 + 4 H^+ + 4 e \rightleftharpoons Hf + 2 H_2O$	−1.505
$H_2SeO_3 + 4 H^+ + 4 e \rightleftharpoons Se + 3 H_2O$	−0.74	$Hf^{4+} + 4 e \rightleftharpoons Hf$	−1.55
$Cr^{3+} + 3 e \rightleftharpoons Cr$	−0.744	$Sm^{3+} + e \rightleftharpoons Sm^{2+}$	−1.55
$Ta_2O_5 + 10 H^+ + 10 e \rightleftharpoons 2 Ta + 5 H_2O$	−0.750	$ZrO_2 + 4 H^+ + 4 e \rightleftharpoons Zr + 2 H_2O$	−1.553
$TlI + e \rightleftharpoons Tl + I^-$	−0.752	$Mn(OH)_2 + 2 e \rightleftharpoons Mn + 2 OH^-$	−1.56
$Zn^{2+} + 2 e \rightleftharpoons Zn$	−0.7618	$Ba^{2+} + 2 e \rightleftharpoons Ba(Hg)$	−1.570
$Zn^{2+} + 2 e \rightleftharpoons Zn(Hg)$	−0.7628	$Bk^{2+} + 2 e \rightleftharpoons Bk$	−1.6
$CdO + H_2O + 2 e \rightleftharpoons Cd + 2 OH^-$	−0.783	$Cf^{3+} + e \rightleftharpoons Cf^{2+}$	−1.6
$Te + 2 H^+ + 2 e \rightleftharpoons H_2Te$	−0.793	$Ti^{2+} + 2 e \rightleftharpoons Ti$	−1.630
$ZnSO_4 \cdot 7H_2O + 2 e \rightleftharpoons Zn(Hg) + SO_4^{2-} + 7 H_2O$	−0.7993	$Md^{3+} + 3 e \rightleftharpoons Md$	−1.65
(Saturated $ZnSO_4$)		$HPO_3^{2-} + 2 H_2O + 2 e \rightleftharpoons H_2PO_2^- + 3 OH^-$	−1.65
$Bi + 3 H^+ + 3 e \rightleftharpoons BiH_3$	−0.8	$Al^{3+} + 3 e \rightleftharpoons Al$	−1.662
$SiO + 2 H^+ + 2 e \rightleftharpoons Si + H_2O$	−0.8	$SiO_3^{2-} + H_2O + 4 e \rightleftharpoons Si + 6 OH^-$	−1.697
$Cd(OH)_2 + 2 e \rightleftharpoons Cd(Hg) + 2 OH^-$	−0.809	$HPO_3^{2-} + 2 H_2O + 3 e \rightleftharpoons P + 5 OH^-$	−1.71
$2 H_2O + 2 e \rightleftharpoons H_2 + 2 OH^-$	−0.8277	$HfO^{2+} + 2 H^+ + 4 e \rightleftharpoons Hf + H_2O$	−1.724
$2 NO_3^- + 2 H_2O + 2 e \rightleftharpoons N_2O_4 + 4 OH^-$	−0.85	$ThO_2 + 4 H^+ + 4 e \rightleftharpoons Th + 2 H_2O$	−1.789
$H_3BO_3 + 3 H^+ + 3 e \rightleftharpoons B + 3 H_2O$	−0.8698	$H_2BO_3^- + H_2O + 3 e \rightleftharpoons B + 4 OH^-$	−1.79
$P + 3 H_2O + 3 e \rightleftharpoons PH_3(g) + 3 OH^-$	−0.87	$Sr^{2+} + 2 e \rightleftharpoons Sr(Hg)$	−1.793
$Ti^{3+} + e \rightleftharpoons Ti^{2+}$	−0.9	$U^{3+} + 3 e \rightleftharpoons U$	−1.798
$HSnO_2^- + H_2O + 2 e \rightleftharpoons Sn + 3 OH^-$	−0.909	$H_2PO_2^- + e \rightleftharpoons P + 2 OH^-$	−1.82
$Cr^{2+} + 2 e \rightleftharpoons Cr$	−0.913	$Be^{2+} + 2 e \rightleftharpoons Be$	−1.847
$Se + 2 e \rightleftharpoons Se^{2-}$	−0.924	$Np^{3+} + 3 e \rightleftharpoons Np$	−1.856
$SO_4^{2-} + H_2O + 2 e \rightleftharpoons SO_3^{2-} + 2 OH^-$	−0.93	$Fm^{3+} + 3 e \rightleftharpoons Fm$	−1.89
$Sn(OH)_6^{2-} + 2 e \rightleftharpoons HSnO_2^- + 3 OH^- + H_2O$	−0.93	$Th^{4+} + 4 e \rightleftharpoons Th$	−1.899

TABLE 3
Reduction Reactions Having $E°$ Values More Negative than that of the Standard Hydrogen Electrode (continued)

Reaction	$E°/V$	Reaction	$E°/V$
$Am^{2+} + 2\,e \rightleftharpoons Am$	−1.9	$ZrO(OH)_2 + H_2O + 4\,e \rightleftharpoons Zr + 4\,OH^-$	−2.36
$Pa^{4+} + e \rightleftharpoons Pa^{3+}$	−1.9	$Mg^{2+} + 2\,e \rightleftharpoons Mg$	−2.372
$Es^{3+} + 3\,e \rightleftharpoons Es$	−1.91	$Y^{3+} + 3\,e \rightleftharpoons Y$	−2.372
$Cf^{3+} + 3\,e \rightleftharpoons Cf$	−1.94	$La^{3+} + 3\,e \rightleftharpoons La$	−2.379
$Lr^{3+} + 3\,e \rightleftharpoons Lr$	−1.96	$Tm^{2+} + 2\,e \rightleftharpoons Tm$	−2.4
$Eu^{3+} + 3\,e \rightleftharpoons Eu$	−1.991	$Md^{2+} + 2\,e \rightleftharpoons Md$	−2.40
$Er^{2+} + 2\,e \rightleftharpoons Er$	−2.0	$Th(OH)_4 + 4\,e \rightleftharpoons Th + 4\,OH^-$	−2.48
$Pr^{2+} + 2\,e \rightleftharpoons Pr$	−2.0	$HfO(OH)_2 + H_2O + 4\,e \rightleftharpoons Hf + 4\,OH^-$	−2.50
$Pu^{3+} + 3\,e \rightleftharpoons Pu$	−2.031	$No^{2+} + 2\,e \rightleftharpoons No$	−2.50
$Cm^{3+} + 3\,e \rightleftharpoons Cm$	−2.04	$Dy^{3+} + e \rightleftharpoons Dy^{2+}$	−2.6
$Am^{3+} + 3\,e \rightleftharpoons Am$	−2.048	$Pm^{3+} + e \rightleftharpoons Pm^{2+}$	−2.6
$AlF_6^{3-} + 3\,e \rightleftharpoons Al + 6\,F^-$	−2.069	$Be_2O_3^{2-} + 3\,H_2O + 4\,e \rightleftharpoons 2\,Be + 6\,OH^-$	−2.63
$Sc^{3+} + 3\,e \rightleftharpoons Sc$	−2.077	$Sm^{2+} + 2\,e \rightleftharpoons Sm$	−2.68
$Ho^{2+} + 2\,e \rightleftharpoons Ho$	−2.1	$Mg(OH)_2 + 2\,e \rightleftharpoons Mg + 2\,OH^-$	−2.690
$Nd^{2+} + 2\,e \rightleftharpoons Nd$	−2.1	$Nd^{3+} + e \rightleftharpoons Nd^{2+}$	−2.7
$Cf^{2+} + 2\,e \rightleftharpoons Cf$	−2.12	$Mg^+ + e \rightleftharpoons Mg$	−2.70
$Yb^{3+} + 3\,e \rightleftharpoons Yb$	−2.19	$Na^+ + e \rightleftharpoons Na$	−2.71
$Ac^{3+} + 3\,e \rightleftharpoons Ac$	−2.20	$Yb^{2+} + 2\,e \rightleftharpoons Yb$	−2.76
$Dy^{2+} + 2\,e \rightleftharpoons Dy$	−2.2	$Bk^{3+} + e \rightleftharpoons Bk^{2+}$	−2.8
$Tm^{3+} + e \rightleftharpoons Tm^{2+}$	−2.2	$Ho^{3+} + e \rightleftharpoons Ho^{2+}$	−2.8
$Pm^{2+} + 2\,e \rightleftharpoons Pm$	−2.2	$Ra^{2+} + 2\,e \rightleftharpoons Ra$	−2.8
$Es^{2+} + 2\,e \rightleftharpoons Es$	−2.23	$Eu^{2+} + 2\,e \rightleftharpoons Eu$	−2.812
$H_2 + 2\,e \rightleftharpoons 2\,H^-$	−2.23	$Ca^{2+} + 2\,e \rightleftharpoons Ca$	−2.868
$Gd^{3+} + 3\,e \rightleftharpoons Gd$	−2.279	$Sr(OH)_2 + 2\,e \rightleftharpoons Sr + 2\,OH^-$	−2.88
$Tb^{3+} + 3\,e \rightleftharpoons Tb$	−2.28	$Sr^{2+} + 2\,e \rightleftharpoons Sr$	−2.89
$Lu^{3+} + 3\,e \rightleftharpoons Lu$	−2.28	$Fr^+ + e \rightleftharpoons Fr$	−2.9
$Dy^{3+} + 3\,e \rightleftharpoons Dy$	−2.295	$La(OH)_3 + 3\,e \rightleftharpoons La + 3\,OH^-$	−2.90
$Am^{3+} + e \rightleftharpoons Am^{2+}$	−2.3	$Ba^{2+} + 2\,e \rightleftharpoons Ba$	−2.912
$Fm^{2+} + 2\,e \rightleftharpoons Fm$	−2.30	$K^+ + e \rightleftharpoons K$	−2.931
$Pm^{3+} + 3\,e \rightleftharpoons Pm$	−2.30	$Rb^+ + e \rightleftharpoons Rb$	−2.98
$Sm^{3+} + 3\,e \rightleftharpoons Sm$	−2.304	$Ba(OH)_2 + 2\,e \rightleftharpoons Ba + 2\,OH^-$	−2.99
$Al(OH)_3 + 3\,e \rightleftharpoons Al + 3\,OH^-$	−2.31	$Er^{3+} + e \rightleftharpoons Er^{2+}$	−3.0
$Tm^{3+} + 3\,e \rightleftharpoons Tm$	−2.319	$Ca(OH)_2 + 2\,e \rightleftharpoons Ca + 2\,OH^-$	−3.02
$Nd^{3+} + 3\,e \rightleftharpoons Nd$	−2.323	$Cs^+ + e \rightleftharpoons Cs$	−3.026
$Al(OH)_4^- + 3\,e \rightleftharpoons Al + 4\,OH^-$	−2.328	$Li^+ + e \rightleftharpoons Li$	−3.0401
$H_2AlO_3^- + H_2O + 3\,e \rightleftharpoons Al + 4\,OH^-$	−2.33	$3\,N_2 + 2\,H^+ + 2\,e \rightleftharpoons 2\,HN_3$	−3.09
$Ho^{3+} + 3\,e \rightleftharpoons Ho$	−2.33	$Pr^{3+} + e \rightleftharpoons Pr^{2+}$	−3.1
$Er^{3+} + 3\,e \rightleftharpoons Er$	−2.331	$Ca^+ + e \rightleftharpoons Ca$	−3.80
$Ce^{3+} + 3\,e \rightleftharpoons Ce$	−2.336	$Sr^+ + e \rightleftharpoons Sr$	−4.10
$Pr^{3+} + 3\,e \rightleftharpoons Pr$	−2.353		

REDUCTION AND OXIDATION POTENTIALS FOR CERTAIN ION RADICALS

Petr Vanýsek

There are two tables for ion radicals. The first table lists reduction potentials for organic compounds which produce anion radicals during reduction, a process described as $A + e^- \rightleftharpoons A^-$. The second table lists oxidation potentials for organic compounds which produce cation radicals during oxidation, a process described as $A \rightleftharpoons A^+ + e^-$. To obtain reduction potential for a reverse reaction, the sign for the potential is changed.

Unlike the table of the Electrochemical Series, which lists *standard* potentials, values for radicals are experimental values with experimental conditions given in the second column. Since the measurements leading to potentials for ion radicals are very dependent on conditions, an attempt to report standard potentials for radicals would serve no useful purpose. For the same reason, the potentials are also reported as experimental values, usually a half-wave potential ($E_{1/2}$ in polarography) or a peak potential (E_p in cyclic voltammetry). Unless otherwise stated, the values are reported vs. SCE (saturated calomel electrode). To obtain a value vs. normal hydrogen electrode, 0.241 V has to be added to the SCE values. All the ion radicals chosen for inclusion in the tables result from electrochemically reversible reactions. More detailed data on ion radicals can be found in the *Encyclopedia of Electrochemistry of Elements*, (A. J. Bard, Ed.), Vol. XI and XII in particular, Marcel Dekker, New York, 1978.

Abbreviations are: CV — cyclic voltammetry; DMF — *N,N*-Dimethylformamide; E swp — potential sweep; E° — standard potential; E_p — peak potential; $E_{p/2}$ — half-peak potential; $E_{1/2}$ — half wave potential; M — mol/L; MeCN — acetonitrile; pol — polarography; rot Pt dsk — rotated Pt disk; SCE — saturated calomel electrode; TBABF$_4$ — tetrabutylammonium tetrafluoroborate; TBAI — tetrabutylammonium iodide; TBAP — tetrabutylammonium perchlorate; TEABr — tetraethylammonium bromide; TEAP — tetraethylammonium perchlorate; THF — tetrahydrofuran; TPACF$_3$SO$_3$ — tetrapropylammonium trifluoromethanesulfite; TPAP — tetrapropylammonium perchlorate; and wr — wire.

Reduction Potentials (Products are Anion Radicals)

Substance	Conditions/electrode/technique	Potential V (vs. SCE)
Acetone	DMF, 0.1 M TEABr/Hg/pol	$E_{1/2} = -2.84$
1-Naphthyphenylacetylene	DMF, 0.03 M TBAI/Hg/pol	$E_{1/2} = -1.91$
1-Naphthalenecarboxyaldehyde	-/Hg/pol	$E_{1/2} = -0.91$
2-Naphthalenecarboxyaldehyde	-/Hg/pol	$E_{1/2} = -0.96$
2-Phenanthrenecarboxaldehyde	-/Hg/pol	$E_{1/2} = -1.00$
3-Phenanthrenecarboxaldehyde	-/Hg/pol	$E_{1/2} = -0.94$
9-Phenanthrenecarboxaldehyde	-/Hg/pol	$E_{1/2} = -0.83$
1-Anthracenecarboxaldehyde	-/Hg/pol	$E_{1/2} = -0.75$
1-Pyrenecarboxaldehyde	-/Hg/pol	$E_{1/2} = -0.76$
2-Pyrenecarboxaldehyde	-/Hg/pol	$E_{1/2} = -1.00$
Anthracene	DMF, 0.1 M TBAP/Pt dsk/CV	$E_p = -2.00$
	DMF, 0.5 M TBABF$_4$/Hg/CV	$E_{1/2} = -1.93$
	MeCN, 0.1 M TEAP/Hg/CV	$E_{1/2} = -2.07$
	DMF, 0.1 M TBAI/Hg/pol	$E_{1/2} = -1.92$
9,10-Dimethylanthracene	DMF, 0.1 M TBAP/Pt/CV	$E_p = -2.08$
	MeCN, 0.1 M TBAP/Pt/CV	$E_p = -2.10$
1-Phenylanthracene	DMF, 0.5 M TBABF$_2$/Hg/CV	$E_{1/2} = -1.91$
	DMF, 0.1 M TBAI/Hg/pol	$E_{1/2} = -1.878$
2-Phenylanthracene	DMF, 0.1 M TBAI/Hg/pol	$E_{1/2} = -1.875$
8-Phenylanthracene	DMF, 0.5 M TBABF$_4$/Hg/CV	$E_{1/2} = -1.91$
9-Phenylanthracene	DMF, 0.5 M TBABF$_4$/Hg/CV	$E_{1/2} = -1.93$
	DMF, 0.1 M TBAI/Hg/pol	$E_{1/2} = -1.863$
1,8-Diphenylanthracene	DMF, 0.5 M TBABF$_4$/Hg/CV	$E_{1/2} = -1.88$
1,9-Diphenylanthracene	DMF, 0.1 M TBAI/Hg/pol	$E_{1/2} = -1.846$
1,10-Diphenylanthracene	DMF, 0.1 M TBAI/Hg/pol	$E_{1/2} = -1.786$
8,9-Diphenylanthracene	DMF, 0.5 M TBABF$_4$/Hg/CV	$E_{1/2} = -1.90$
9,10-Diphenylanthracene	MeCN, 0.1 M TBAP/rot Pt/E swp	$E_{1/2} = -1.83$
	DMF, 0.1 M TBAI/Hg/pol	$E_{1/2} = -1.835$
1,8,9-Triphenylanthracene	DMF, 0.5 M TBABF$_4$/Hg/CV	$E_{1/2} = -1.85$
1,8,10-Triphenylanthracene	DMF, 0.5 M TBABF$_4$/Hg/CV	$E_{1/2} = -1.81$
9,10-Dibiphenylanthracene	MeCN, 0.1 M TBAP/rot Pt/E swp	$E_{1/2} = -1.94$
Benz(a)anthracene	MeCN, 0.1 M TEAP/Hg/CV	$E_{1/2} = -2.11$
	MeCN, 0.1 M TEAP/Hg/pol	$E_{1/2} = -2.40$[a]
Azulene	DMF, 0.1 M TBAI/Hg/pol	$E_{1/2} = -1.10$[c]
Annulene	DMF, 0.5 M TBAP 0°C/Hg/pol	$E_{1/2} = -1.23$
Benzaldehyde	DMF, 0.1 M TBAP/Hg/pol	$E_{1/2} = -1.67$
Benzil	DMSO, 0.1 M TBAP/Hg/pol	$E_{1/2} = -1.04$

Reduction Potentials (Products are Anion Radicals) (continued)

Substance	Conditions/electrode/technique	Potential V (vs. SCE)
Benzophenone	-/Hg/pol	$E_{1/2} = -1.80$
	DMF/Pt dsk/CV	$E° = -1.72$
Chrysene	MeCN, 0.1 M TEAP/Hg/pol	$E_{1/2} = -2.73$[a]
Fluoranthrene	DMF, 0.1 M TBAP/Pt dsk/CV	$E_p = -1.76$
Cyclohexanone	DMF, 0.1 M TEABr/Hg/pol	$E_{1/2} = -2.79$
5,5-Dimethyl-3-phenyl-2-cyclohexen-1-one	DMF, 0.5 M/Hg/pol	$E_{1/2} = -1.71$
1,2,3-Indanetrione hydrate (ninhydrin)	DMF, 0.2 M NaNO$_3$/Hg/pol	$E_{1/2} = -0.039$
Naphthacene	DMF, 0.1 M TBAI/Hg/pol	$E_{1/2} = -1.53$
Naphthalene	DMF, 0.1 M TBAP/Pt dsk/CV	$E_p = -2.55$
	DMF, 0.5 M TBABF$_4$/Hg/CV	$E_{1/2} = -2.56$
	DMF, MeCN, 0.1 M TEAP/Hg/CV	$E_{1/2} = -2.63$
	DMF, 0.1 M TBAI/Hg/pol	$E_{1/2} = -2.50$
1-Phenylnaphthalene	DMF, 0.5 M TBABF$_4$/Hg/CV	$E_{1/2} = -2.36$
1,2-Diphenylnaphthalene	DMF, 0.5 M TBABF$_4$/Hg/CV	$E_{1/2} = -2.25$
Cyclopentanone	DMF, 0.1 M TEABr/Hg/pol	$E_{1/2} = -2.82$
Phenanthrene	MeCN, 0.1 M TBAP/Pt wr/CV	$E_{1/2} = -2.47$
	MeCN, 0.1 M TEAP/Hg/pol	$E_{1/2} = -2.88$[a]
Pentacene	THF, 0.1 M TBAP/rot Pt dsk/E swp	$E_{1/2} = -1.40$
Perylene	MeCN, 0.1 M TEAP/Hg/CV	$E_{1/2} = -1.73$
1,3-Diphenyl-1,3-propanedione	DMSO, 0.2 M TBAP/Hg/CV	$E_{1/2} = -1.42$
2,2-Dimethyl-1,3-diphenyl-1,3 propanedione	DMSO, TBAP/Hg/CV	$E_{1/2} = -1.80$
Pyrene	DMF, 0.1 M TBAP/Pt/CV	$E_p = -2.14$
	MeCN, 0.1 M TEAP/Hg/pol	$E_{1/2} = -2.49$[a]
Diphenylsulfone	DMF, TEABr	$E_{1/2} = -2.16$
Triphenylene	MeCN, 0.1 M TEAP/Hg/pol	$E_{1/2} = -2.87$[a]
9,10-Anthraquinone	DMF, 0.5 M TBAP, 20°/Pt dsk/CV	$E_{1/2} = -1.01$
1,4-Benzoquinone	MeCN, 0.1 M TEAP/Pt/CV	$E_p = -0.54$
1,4-Naphthohydroquinone, dipotassium salt	DMF, 0.5 M TBAP, 20°/Pt dsk/CV	$E_{1/2} = -1.55$
Rubrene	DMF, 0.1 M TBAP/Pt dsk/CV	$E_p = -1.48$
	DMF, 0.1 M TBAI/Hg/pol	$E_{1/2} = -1.410$
Benzocyclooctatetraene	THF, 0.1 M TBAP/Hg/pol	$E_{1/2} = -2.13$
sym-Dibenzocyclooctatetraene	THF, 0.1 M TBAP/Hg/pol	$E_{1/2} = -2.29$
Ubiquinone-6	MeCN, 0.1 M TEAP/Pt/CV	$E_p = -1.05$[e]
(9-Phenyl-fluorenyl)$^+$	10.2 M H$_2$SO$_4$/Hg/CV	$E_p = -0.01$[b]
(Triphenylcyclopropenyl)$^+$	MeCN, 0.1 M TEAP/Hg/CV	$E_p = -1.87$
(Triphenylmethyl)$^+$	MeCN, 0.1 M TBAP/Hg/pol	$E_{1/2} = 0.27$
	H$_2$SO$_4$, 10.2 M/Hg/CV	$E_p = -0.58$[b]
(Tribiphenylmethyl)$^+$	MeCN, 0.1 M TBAP/Hg/pol	$E_{1/2} = 0.19$
(Tri-4-t-butyl-5-phenylmethyl)$^+$	MeCN, 0.1 M TBAP/Hg/pol	$E_{1/2} = 0.13$
(Tri-4-isopropylphenylmethyl)$^+$	MeCN, 0.1 M TBAP/Hg/pol	$E_{1/2} = 0.07$
(Tri-4-methylphenylmethyl)$^+$	MeCN, 0.1 M TBAP/Hg/pol	$E_{1/2} = 0.05$
(Tri-4-cyclopropylphenylmethyl)$^+$	MeCN, 0.1 M TBAP/Hg/pol	$E_{1/2} = 0.01$
(Tropylium)$^+$	MeCN, 0.1 M TBAP/Hg/pol	$E_{1/2} = -0.17$
	DMF, 0.15 M TBAI/Hg/pol	$E_{1/2} = -1.55$
	DMF, 0.15 M TBAI/Hg/pol	$E_{1/2} = -1.55$
	DMF, 0.15 M TBAI/Hg/pol	$E_{1/2} = -1.57$
	DMF, 0.15 M TBAI/Hg/pol	$E_{1/2} = -1.60$
	DMF, 0.15 M TBAI/Hg/pol	$E_{1/2} = -1.87$
	DMF, 0.15 M TBAI/Hg/pol	$E_{1/2} = -1.96$
	DMF, 0.15 M TBAI/Hg/pol	$E_{1/2} = -2.05$

Oxidation Potentials (Products are Cation Radicals)

Substance	Conditions/electrode/technique	Potential V (vs. SCE)
Anthracene	CH_2Cl_2, 0.2 M TBABF$_4$, −70°C/Pt dsk/CV	E_p = +0.73[d]
9,10-Dimethylanthracene	MeCN, 0.1 M LiClO$_4$/Pt wr/CV	E_p = +1.0
9,10-Dipropylanthracene	MeCN, 0.1 M TEAP/Pt/CV	E_p = +1.08
1,8-Diphenylanthracene	CH_2Cl_2, 0.2 M TPrACF$_3$SO$_3$/rot Pt wr/E swp	$E_{1/2}$ = +1.34
8,9-Diphenylanthracene	CH_2Cl_2, 0.2 M TPrACF$_3$SO$_3$/rot Pt wr/E swp	$E_{1/2}$ = +1.30
9,10-Diphenylanthracene	MeCN/Pt/CV	E_p = +1.22
Perylene	MeCN, 0.1 M TBAP/Pt/CV	E_p = +1.34
Pyrene	DMF, 0.1 M TBAP/Pt dsk/CV	E_p = +1.25
Rubrene	DMF, 0.1 M TBAP/Pt dsk/CV	E_p = +1.10
Tetracene	CH_2Cl_2, 0.2 M TBABF$_4$, −70°C/Pt wr/CV	E_p = +0.35[d]
1,4-Dithiabenzene	MeCN, 0.1 M TEAP/Pt dsk/rot	$E_{1/2}$ = +0.69
1,4-Dithianaphthalene	MeCN, 0.1 M TEAP/Pt dsk/rot	$E_{1/2}$ = +0.80
Thianthrene	0.1 M TPAP/Pt/CV	$E_{1/2}$ = +1.28

[a] vs 0.01 M Ag/AgClO$_4$

[b] vs. Hg/Hg$_2$SO$_4$, 17 M H$_2$SO$_4$

[c] vs Hg pool

[d] vs Ag/saturated AgNO$_3$

[e] vs Ag/0.01 M Ag$^+$

pH SCALE FOR AQUEOUS SOLUTIONS

A. K. Covington

The pH value is the negative decadic logarithm of the (relative) ion activity of the hydrogen ion in the solution.

$$pH = -\log a_H \tag{1}$$

This is only a notional definition since Equation 1 involves a single ion activity, which is immeasurable, and has to be attained through a nonthermodynamic assumption such as that described in Equation 5 below. In terms of substance concentration, molarity, Equation 1 may be rewritten

$$pH = -\log (c_H y_H/c^o) \tag{2}$$

where c^o is an arbitrary constant representing the standard state condition and equal to 1 mol dm^{-3}, c_H is the concentration of hydrogen ion and y_H is the single ion activity of the hydrogen ion. In terms of molality, Equation 1 may be rewritten

$$pH = -\log (m_H \gamma_H/m^o) \tag{3}$$

where m^o is an arbitrary constant representing the standard state condition and equal to 1 mol kg^{-1}, m_H is the concentration of hydrogen ion and γ_H is the single ion activity of the hydrogen ion. For most purposes the difference between these two scales can be ignored for dilute aqueous solutions; the difference is 0.001 at 25° C and 0.02 at 100° C. Arising from the nonexperimental determinability of single ion activities, the definition and determination of pH have an operational basis, and depend on the assignment of pH values to a standard solution (or solutions) together with the determination of pH difference by a cell with liquid junction called the operational cell.

The Operational Definition of pH Difference[1,2]

The electromotive force, EMF, $E(X)$ of the cell with liquid junction:

$$\text{Reference electrode} \mid \text{KCl (aq., concentrated)} \parallel \text{Solution X} \mid \text{H}_2 \mid \text{Pt} \tag{I}$$

is measured, and likewise that, $E(S)$, of the cell:

$$\text{Reference electrode} \mid \text{KCl (aq., concentrated)} \parallel \text{Solution S} \mid \text{H}_2 \mid \text{Pt} \tag{II}$$

The temperature of both cells (I and II) must be equal and uniform throughout, and the hydrogen gas pressures identical. The two bridge solutions may be any molality of KCl not less than 3.5 mol kg^{-1} provided they are the same.

The pH of the solution X, pH(X), is then related to the assigned pH of the solution S, pH(S) by the definition:

$$pH(X) = pH(S) + [E(S) - E(X)]/[(RT/F) \ln 10] \tag{4}$$

where R is the gas constant, T the thermodynamic temperature, F the Faraday constant. The quantity $k = (RT/F) \ln 10$ is called the slope factor whose values are given as a function of temperature in Table 1. As a consequence of this definition any difference in liquid junction potential between cells I and II is subsumed into the value of pH(X).

The pH Scale

The pH scale at a particular temperature is defined by Equation 4 as a straight line, on the plot of pH against $E(X)$, having a slope of k drawn through the pH value assigned to the Reference Value Standard (RVS) solution (as given in Table 2) and the value of $E(S)$ for cell II when it contains the Reference Value Standard solution. The solution chosen for the RVS is 0.05 mol kg^{-1} aqueous potassium hydrogen phthalate. The procedure by which pH(RVS) values have been assigned to the Reference Value Standard (RVS) is the cell III without transference:[1,3]

$$\text{Pt(Pd)} \mid \text{H}_2 \text{ (g, } p=1 \text{ atm} = 101\,325 \text{ Pa)} \mid \text{RVS, Cl}^- \mid \text{AgCl} \mid \text{Ag} \tag{III}$$

The palladised-platinum hydrogen electrode is used to reduce the catalytised chemical reduction of the phthalate by hydrogen gas. The calculation involves a non-thermodynamic assumption, the Bates-Guggenheim Convention, for the single ion activity of the chloride ion[1,2] as

$$\log (\gamma_{Cl})^o = -A(I/m^o)^{1/2}/[1 + 1.5 \, (I/m^o)^{1/2}] \tag{5}$$

where I is the ionic strength = $(1/2)\Sigma m_i z_i^2 = 0.0534$ mol kg^{-1} for the RVS solution and A is a known function of temperature (Table 1).

To prepare the RVS solution, dry the sample at 110° C for 2 h before use. The water should have a conductivity of less than 0.1 mS m^{-1}. The required solution contains 10.211 g kg^{-1} water. It can be prepared on a volume basis by dissolving 10.138 g potassium hydrogen phthalate in water and making up to 1 L at 20° C. This solution is 0.04964 mol/L with a density of 1.00300 g/L at 20° C.

pH SCALE FOR AQUEOUS SOLUTIONS (continued)

Primary Standards[2]

pH values may be assigned by the cell without transference method (cell III) to six other buffer solutions which meet certain criteria of reproducibility of preparation and properties. These solutions are called primary pH standards (PS), and details and the pH(PS) values assigned to them are given in Table 3. When these PS solutions are used in the operational cell I, the experimental value of the slope will not be in accord with the slope factor values of Table 1, and, moreover, the experimental value could change if additional primary solutions were to be defined. Hence the pH value determined for an unknown solution can be slightly dependent (±0.02) on the choice of primary standard.[2,4,5] Some useful data for standard buffers are given in Table 5.

Operational Standards[2,6]

Operational standards (OS) are also defined which are traceable to the Reference Value Standard (RVS). Values are assigned by means of the operational cells I and II where the liquid junctions are the free diffusion type reproducibly formed in 1 mm vertical capillary tubes. These operational standards are not restricted in number provided certain preparation criteria are met, and pH(OS) values for 16 solutions are given in Table 4.[2] These OS represent an alternative procedure and are in no way to be regarded as inferior to the primary standards. As a consequence of their definition, all pH(OS) values fall on the line with slope given by the slope factor value for the appropriate temperature in Table 1. Any difference in liquid junction potential between the solutions of cells I and II and KCl is subsumed into the assigned value of pH(OS).

Measurement of pH. Choice of Standard Reference Solution

1a. If pH is not required to better than ±0.05 any standard reference solution may be selected.

1b. If pH is required to ±0.002 and interpretation in terms of hydrogen ion concentration or activity is desired, choose a standard reference solution, pH(PS) or pH(OS), to match X as closely as possible in terms of pH, composition and ionic strength.

2. Alternatively, a bracketting procedure may be adopted whereby two standard reference solutions are chosen whose pH values, pH(S1), pH(S2) are on either side of pH(X). Then if the corresponding potential difference measurements are $E(S1)$, $E(S2)$, $E(X)$, then pH(X) is obtained from

$$pH(X) = pH(S1) + [E(X) - E(S1)]/ \%k$$

where $\%k = 100[E(S2) - E(S1)]/[pH(S2) - pH(S1)]$ is the apparent percentage slope. This procedure is very easily done on some pH meters simply by adjusting downwards the slope factor control with the electrodes in S2. The purpose of the bracketting procedure is to compensate for deficiencies in the electrodes and measuring system.

Information to be Given about the Measurement of pH(X)

The standard solutions selected for calibration of the pH meter system should be reported with the measurement as follows,

1. System calibrated with pH(RVS) = ... at ...K.
2. System calibrated with two primary standards pH(PS1) = ... and pH(PS2) = ... at ... K.
3. System calibrated with two operational standards pH(OS1) =..... and pH(OS2) = ... at .... K.

Interpretation of pH(X) in Terms of Hydrogen Ion Concentration

The operationally defined pH has no simple interpretation in terms of hydrogen ion concentration but the mean ionic activity coefficient of a typical 1:1 electrolyte can be substituted into equation 2 or 3 to obtain hydrogen ion concentration subject to an uncertainty of 3.9% in concentration corresponding to 0.02 in pH.

REFERENCES

1. R. G. Bates, *Measurement of pH. Theory and Practice*, 2nd ed., John Wiley & Sons,, New York, 1973.
2. A. K. Covington, R. G. Bates, and R. A. Durst, *Pure Appl. Chem.*, 57, 531, 1985.
3. H. P. Butikofer and A. K. Covington, *Anal. Chim. Acta*, 108, 179, 1979.
4. R. G. Bates, *Crit. Rev. Anal. Chem.*, 10, 247, 1981.
5. A. K. Covington, *Anal. Chim. Acta*, 127, 1, 1981.
6. A. K. Covington and M. J. Rebelo, *Anal. Chim. Acta.*, 200, 245, 1987.
7. V. E. Bower and R. G. Bates, *J. Res. Natl. Bur. Stand.*, 39, 263, 1954.

TABLE 1
Standard EMF, Slope Factor and Debye-Huckel Constant A
(Unit Weight of Solvent) as Functions of Temperature

Temperature/°C	$E°$/mV[7]	Slope Factor k/mV	A^1
0	236.55	54.199	0.4918
5	234.13	55.191	0.4952
10	231.42	56.183	0.4988
15	228.57	57.175	0.5026
20	225.57	58.167	0.5066
25	222.34	59.159	0.5108
30	219.04	60.152	0.5150
35	215.65	61.144	0.5196
40	212.08	62.136	0.5242
45	208.35	63.128	0.5291
50	204.49	64.120	0.5341
55	200.56	65.112	0.5393
60	196.49	66.104	0.5448
70	187.82	68.088	0.5562
80	178.73	70.073	0.5685
90	169.52	72.057	0.5817
95	165.11	73.049	0.5886

TABLE 2
Values of pH(RVS) for the Reference Value Standard of 0.05 mol kg^{-1} Potassium
Hydrogen Phthalate at Various Temperatures

t/°C	pH(RVS)	t/°C	pH(RVS)	t/°C	pH(RVS)
0	4.000	35	4.018	65	4.097
5	3.998	37	4.022	70	4.116
10	3.997	40	4.027	75	4.137
15	3.998	45	4.038	80	4.159
20	4.001	50	4.050	85	4.183
25	4.005	55	4.064	90	4.21
30	4.011	60	4.080	95	4.24

TABLE 3
Values of pH(PS) for Primary Standard Reference Solutions

Primary ref. standard	t/°C															
	0	5	10	15	20	25	30	35	37	40	50	60	70	80	90	95
Saturated (at 25° C) Potassium hydrogen tartrate	—	—	—	—	—	3.557	3.552	3.549	3.548	3.547	3.549	3.560	3.580	3.610	3.650	3.674
0.1 mol/kg Potassium dihydrogen citrate	3.863	3.840	3.820	3.802	3.788	3.776	3.766	3.759	3.756	3.754	3.749	—	—	—	—	—
0.025 mol/kg Disodium hydrogen phosphate +0.025 mol/kg Potassium dihydrogen phosphate	6.984	6.951	6.923	6.900	6.881	6.865	6.853	6.844	6.841	6.838	6.833	6.836	6.845	6.859	6.876	6.886
0.03043 mol/kg Disodium hydrogen phosphate +0.008695 mol/kg Potassium dihydrogen phosphate	7.534	7.500	7.472	7.448	7.429	7.413	7.400	7.389	7.386	7.380	7.367					
0.01 mol/kg Disodium tetraborate	9.464	9.395	9.332	9.276	9.225	9.180	9.139	9.102	9.088	9.068	9.011	8.962	8.921	8.884	8.850	8.833
0.025 mol/kg Sodium hydrogen carbonate +0.025 mol/kg sodium carbonate	10.317	10.245	10.179	10.118	10.062	10.012	9.966	9.926	9.910	9.889	9.828	—	—	—	—	—

Note: Based on an uncertainty of ±0.2 mV in determined (E-E^0), the uncertainty is ±0.003 in pH in the range 0—50° C.

pH SCALE FOR AQUEOUS SOLUTIONS (continued)

TABLE 4

pH (OS) Values for Operational Reference Solutions

Operational standard ref. solution	\(t/°C\)														
	0	5	10	15	20	25	30	37	40	50	60	70	80	90	95
0.1 mol/kg Potassium tetroxalate[a]	—	—	—	—	1.475	1.479	1.483	1.490	1.493	1.503	1.513	1.52	1.53	1.53	1.53
0.05 mol/kg potassium tetroxalate[a]	—	—	1.638	1.642	1.644	1.646	1.648	1.649	1.650	1.653	1.660	1.671	1.689	1.72	1.73
0.05 mol/kg sodium hydrogen diglycolate[b]	—	3.466	3.470	3.476	3.484	3.492	3.502	3.519	3.527	3.558	3.595	—	—	—	—
Saturated (at 25° C) potassium hydrogen tartrate	—	—	—	—	—	3.556	3.549	3.544	3.542	3.544	3.553	3.570	3.596	3.627	3.649
0.05 mol/kg Potassium hydrogen phthalate (RVS)	4.000	3.998	3.997	3.998	4.000	4.005	4.011	4.022	4.027	4.050	4.080	4.115	4.159	4.21	4.24
0.1 mol/dm³ Acetic acid + 0.1 mol/dm³ sodium acetate	4.664	4.657	4.652	4.647	4.645	4.644	4.643	4.647	4.650	4.663	4.684	4.713	4.75	4.80	4.83
0.01 mol/dm³ Acetic acid + 0.1 mol/dm³ sodium acetate	4.729	4.722	4.717	4.714	4.712	4.713	4.715	4.722	4.726	4.743	4.768	4.800	4.839	4.88	4.91
0.02 mol/kg Piperazine phosphate[c]	—	6.477	6.419	6.364	6.310	6.259	6.209	6.143	6.116	6.030	5.952	—	—	—	—
0.025 mol/kg Disodium hydrogen phosphate + 0.025 mol/kg potassium dihydrogen phosphate	6.961	6.935	6.912	6.891	6.873	6.857	6.843	6.828	6.823	6.814	6.817	6.830	6.85	6.90	6.92
0.03043 mol/kg Disodium hydrogen phosphate + 0.008695 mol/kg potassium disodium phosphate	7.506	7.482	7.460	7.441	7.423	7.406	7.390	7.369	—	—	—	—	—	—	—
0.04 mol/kg Disodium hydrogen phosphate + 0.01 mol/kg potassium dihydrogen phosphate	—	7.512	7.488	7.466	7.445	7.428	7.414	7.404	—	—	—	—	—	—	—
0.05 mol/kg Tris hydrochloride + 0.01667 mol/kg Tris[d]	8.399	8.238	8.083	7.933	7.788	7.648	7.513	7.332	7.257	7.018	6.794	—	—	—	—
0.05 mol/kg Disodium tetraborate (Na₂B₄O₇)	9.475	9.409	9.347	9.288	9.233	9.182	9.134	9.074	9.051	8.983	8.932	8.898	8.88	8.84	8.89
0.01 mol/kg Disodium tetraborate (Na₂B₄O₇)	9.451	9.388	9.329	9.275	9.225	9.179	9.138	9.086	9.066	9.009	8.965	8.932	8.91	8.90	8.89

TABLE 4
pH(OS) Values for Operational Standard Reference Solutions (continued)

Operational standard ref. solution	t/°C														
	0	5	10	15	20	25	30	37	40	50	60	70	80	90	95
0.025 mol/kg Sodium hydrogen carbonate + 0.025 mol/kg sodium carbonate	10.273	10.212	10.154	10.098	10.045	9.995	9.948	9.889	9.866	9.800	9.753	9.728	9.725	9.75	9.77
Saturated (at 20°C) calcium hydroxide	13.360	13.159	12.965	12.780	12.602	12.431	12.267	12.049	11.959	11.678	11.423	11.192	10.984	10.80	10.71

Note: Uncertainty is ±0.003 in pH between 0 and 60° C rising to ±0.01 above 70° C.

a Potassium trihydrogen dioxalate ($KH_3C_4O_8$).
b Sodium hydrogen 2,2'-oxydiethanoate.
c $C_4H_{10}N_2 \cdot H_3PO_4$.
d 2-Amino-2-(hydroxymethyl)-1,3 propanediol or tris(hydroxymethyl)aminomethane.

TABLE 5
Useful Data on Some Standard Buffer Solutions

	Molecular formula	Molality (mol/kg)	Relative molar mass	Density at 20° C (g/cm³)	Molarity at at 20° C (mol/L)	Mass of 1 L at 20° C (g)	Mass tolerance for ±0.001 pH[a] (g)	Mass tolerance expressed as a percentage (%)
Potassium tetraoxalate	$KH_3C_4O_8 \cdot 2H_2O$	0.1	254.1913	1.0091	0.09875	25.1017	0.07	0.27
Potassium tetraoxalate	$KH_3C_4O_8 \cdot 2H_2O$	0.05	254.1913	1.0038	0.04965	12.6202	0.034	0.26
Disodium hydrogen orthophosphate	Na_2HPO_4	0.025	141.9588	1.0038	0.02492	3.5379	0.02	0.56
Potassium dihydrogen orthophosphate	KH_2PO_4	0.025	136.0852			3.3912	0.02	0.58
Disodium tetraborate	$Na_2B_4O_7 \cdot 10H_2O$	0.05	381.367	1.0075	0.04985	19.0117	0.9	4.73
Disodium tetraborate	$Na_2B_4O_7 \cdot 10H_2O$	0.01	381.367	1.0001	0.009981	3.8064	0.19	0.49
Sodium carbonate	Na_2CO_3	0.025	105.9887	1.0021	0.02494	2.6428	0.017	0.064
Sodium hydrogen carbonate	$NaHCO_3$	0.025	84.0069			2.0947	0.013	0.62

a Calculated from known dilution value of solution.

PRACTICAL pH MEASUREMENTS ON NATURAL WATERS

A. K. Covington and W. Davison

(1) Dilute solutions and freshwater including 'acid-rain' samples ($I < 0.02$ mol kg^{-1})

Major problems could be encountered due to errors associated with the liquid junction. It is recommended that either a free diffusion junction is used or it is verified that the junction is working correctly using dilute solutions as follows. For commercial electrodes calibrated with IUPAC aqueous RVS or PS standards, the pH(X) of dilute solutions should be within ±0.02 of those given in Table 1. The difference in determined pH(X) between a stirred and unstirred dilute solution should be < 0.02. The characteristics of glass electrodes are such that below pH 5 the readings should be stable within 2 min, but for pH 5 to 8, 8 or so minutes may be necessary to attain stability. Interpretation of pH(X) measured in this way in terms of activity of hydrogen ion, a_{H+} is subject[1] to an uncertainty of ±0.02 in pH.

(2) Seawater

Measurements made by calibration of electrodes with IUPAC aqueous RVS or PS standards to obtain pH(X) are perfectly valid. However, the interpretation of pH(X) in terms of the activity of hydrogen ion is complicated by the non zero residual liquid junction potential as well as by systematic differences between electrode pairs, principally attributable to the reference electrode. For 35‰ salinity seawater ($S = 0.035$) a_{H+} calculated from pH(X) is typically 12% too low. Special seawater pH scales have been devised to overcome this problem:
(i) The total hydrogen ion scale, pH_T, is defined in terms of the sum of free and complexed (total) hydrogen ion concentrations, where

$$^{T}C_H = [H^+] + [HSO_4^-] + [HF].$$

$$\text{So, } pH_T = -\log {}^{T}C_H$$

Calibration of the electrodes with a buffer having a composition similar to that of seawater, to which pH_T has been assigned, results in values of pHT(X) (Tables 2, 3) which are accurately interpretable in terms of $^{T}C_H$.
(ii) The free hydrogen ion scale, pH_F, is defined, and fully interpretable, in terms of the concentration of free hydrogen ions.

$$pH_F = -\log [H^+]$$

Values of pH_F as a function of temperature have been assigned to the same set of pH_T seawater buffers, and so alternatively can be used for calibration (Tables 2, 3) [2,3]

(3) Estuarine water

Prescriptions for seawater scale buffers are available for a range of salinities. Reliable estuarine pH measurements can be made by calibrating with a buffer of the same salinity as the sample. However, these buffers are difficult to prepare and their use presumes prior knowledge of salinity of the sample. Interpretable measurements of estuarine pH can be made by calibration with IUPAC aqueous RVS or PS standards if the electrode pair is additionally calibrated using a 20‰ salinity seawater buffer.[4] The difference between the assigned pH_{SWS} of the seawater buffer and its measured pH(X) value using RVS or PS standards is

$$\Delta pH = pH_{SWS} - pH(X)$$

Values of ΔpH should be in the range of 0.08 to 0.18. It empirically corrects for differences between the two pH scales and for measurement errors associated with the electrode pair. The pH(X) of samples measured using IUPAC aqueous buffers, can be converted to pH_T or pH_F using the appropriate measured ΔpH:

$$pH_T = pH(X) - \Delta pH$$
$$\text{or } pH_F = pH(X) - \Delta pH$$

This simple procedure is appropriate to pH measurement at salinities from 2‰ to 35‰. For salinities lower than 2‰ the procedures for freshwaters should be adopted.

REFERENCES

1. Davison, W. and Harbinson, T. R., *Analyst*, 113, 709, 1988.
2. Culberson, C. H., in *Marine Electrochemistry*, Whitfield, M. and Jagner, D., Eds., Wiley, 1981.
3. Millero, F. J., *Limnol. Oceanogr.*, 31, 839, 1986.
4. Covington, A. K., Whalley, P. D., Davison, W., and Whitfield, M., in *The Determination of Trace Metals in Natural Waters*, West, T. S. and Nurnberg, H. W., Eds., Blackwell, Oxford, 1988.
5. Koch, W. F., Marinenko, G., and Paule, R. C., *J. Res. NBS*, 91, 33, 1986.

Table 1
pH of Dilute Solutions at 25°C, Degassed and Equilibrated with Air, Suitable as Quality Control Standards

	Ionic strength mmol kg^{-1}	Concentration(x) mmol kg^{-1}	pH $p_{CO_2} = 0$	pH $p_{CO_2} =$ air
Potassium hydrogen phthalate	10.7	10	4.12	4.12
	1.1	1	4.33	4.33
xKH$_2$PO$_4$ + xNa$_2$HPO$_4$	9.9	2.5	7.07	7.05
xKH$_2$PO$_4$ + 3.5xNa$_2$HPO$_4$	10	0.87	7.61	7.58
Na$_2$B$_4$O$_7 \cdot$ 10H$_2$O	10	5	9.20	—
HCl	0.1	0.1	4.03	4.03
SRM2694-I[a]	—	—	4.30	—
SRM2694-II[a]	—	—	3.59	—

Note: The pH of solutions near to pH 4 is virtually independent of temperature over the range of 5 to 30°C.

[a] Simulated rainwater samples are available (Reference 5) from NIST containing sulfate, nitrate, chloride, fluoride, sodium, potassium, calcium and magnesium

Table 2
Composition of Seawater Buffer of Salinity $S = 35‰$ at 25°C
(Reference 3)

Solute	mol dm^{-3}	mol kg^{-1}	g kg^{-1}	g dm^{-3}
NaCl	0.3666	0.3493	20.416	20.946
Na$_2$SO$_4$	0.02926	0.02788	3.96	4.063
KCl	0.01058	0.01008	0.752	0.772
CaCl$_2$	0.01077	0.01026	1.139	1.169
MgCl$_2$	0.05518	0.05258	5.006	5.139
Tris	0.06	0.05717	6.926	7.106
Tris $\cdot$ HCl	0.06	0.05717	9.010	9.244

Tris = tris(hydroxymethyl)aminomethane (HOCH$_2$)$_3$CNH$_2$.
A 20‰ buffer is made by diluting the 35‰ in the ratio 20:35.

Table 3
Assigned Values of 20‰ and 35‰ Buffers on Free and Total Hydrogen Ion Scales. Calculated from Equations Provided by Millero (Reference 3)

Temp (°C)	pH$_T$ $S = 20‰$	pH$_T$ $S = 35‰$	pH$_F$ $S = 20‰$	pH$_F$ $S = 35‰$
5	8.683	8.718	8.759	8.81
10	8.513	8.542	8.597	8.647
15	8.351	8.374	8.442	8.491
20	8.195	8.212	8.292	8.341
25	8.045	8.057	8.149	8.197
30	7.901	7.908	8.011	8.059
35	7.762	7.764	7.879	7.926

BUFFER SOLUTIONS GIVING ROUND VALUES OF pH AT 25°C

A		B		C		D		E	
pH	x	pH	x	pH	x	pH	x	pH	x
1.00	67.0	2.20	49.5	4.10	1.3	5.80	3.6	7.00	46.6
1.10	52.8	2.30	45.8	4.20	3.0	5.90	4.6	7.10	45.7
1.20	42.5	2.40	42.2	4.30	4.7	6.00	5.6	7.20	44.7
1.30	33.6	2.50	38.8	4.40	6.6	6.10	6.8	7.30	43.4
1.40	26.6	2.60	35.4	4.50	8.7	6.20	8.1	7.40	42.0
1.50	20.7	2.70	32.1	4.60	11.1	6.30	9.7	7.50	40.3
1.60	16.2	2.80	28.9	4.70	13.6	6.40	11.6	7.60	38.5
1.70	13.0	2.90	25.7	4.80	16.5	6.50	13.9	7.70	36.6
1.80	10.2	3.00	22.3	4.90	19.4	6.60	16.4	7.80	34.5
1.90	8.1	3.10	18.8	5.00	22.6	6.70	19.3	7.90	32.0
2.00	6.5	3.20	15.7	5.10	25.5	6.80	22.4	8.00	29.2
2.10	5.10	3.30	12.9	5.20	28.8	6.90	25.9	8.10	26.2
2.20	3.9	3.40	10.4	5.30	31.6	7.00	29.1	8.20	22.9
		3.50	8.2	5.40	34.1	7.10	32.1	8.30	19.9
		3.60	6.3	5.50	36.6	7.20	34.7	8.40	17.2
		3.70	4.5	5.60	38.8	7.30	37.0	8.50	14.7
		3.80	2.9	5.70	40.6	7.40	39.1	8.60	12.2
		3.90	1.4	5.80	42.3	7.50	40.9	8.70	10.3
		4.00	0.1	5.90	43.7	7.60	42.4	8.80	8.5
						7.70	43.5	8.90	7.0
						7.80	44.5	9.00	5.7
						7.90	45.3		
						8.00	46.1		

F		G		H		I		J	
pH	x	pH	x	pH	x	pH	x	pH	x
8.00	20.5	9.20	0.9	9.60	5.0	10.90	3.3	12.00	6.0
8.10	19.7	9.30	3.6	9.70	6.2	11.00	4.1	12.10	8.0
8.20	18.8	9.40	6.2	9.80	7.6	11.10	5.1	12.20	10.2
8.30	17.7	9.50	8.8	9.90	9.1	11.20	6.3	12.30	12.8
8.40	16.6	9.60	11.1	10.00	10.7	11.30	7.6	12.40	16.2
8.50	15.2	9.70	13.1	10.10	12.2	11.40	9.1	12.50	20.4
8.60	13.5	9.80	15.0	10.20	13.8	11.50	11.1	12.60	25.6
8.70	11.6	9.90	16.7	10.30	15.2	11.60	13.5	12.70	32.2
8.80	9.6	10.00	18.3	10.40	16.5	11.70	16.2	12.80	41.2
8.90	7.1	10.10	19.5	10.50	17.8	11.80	19.4	12.90	53.0
9.00	4.6	10.20	20.5	10.60	19.1	11.90	23.0	13.00	66.0
9.10	2.0	10.30	21.3	10.70	20.2	12.00	26.9		
		10.40	22.1	10.80	21.2				
		10.50	22.7	10.90	22.0				
		10.60	23.3	11.00	22.7				
		10.70	23.8						
		10.80	24.25						

A. 25 ml of 0.2 molar KCl + x ml of 0.2 molar HCl.
B. 50 ml of 0.1 molar potassium hydrogen phthalate + x ml of 0.1 molar HCl.
C. 50 ml of 0.1 molar potassium hydrogen phthalate + x ml of 0.1 molar NaOH.
D. 50 ml of 0.1 molar potassium dihydrogen phosphate + x ml of 0.1 molar NaOH.
E. 50 ml of 0.1 molar tris(hydroxymethyl)aminomethane + x ml of 0.1 M HCl.
F. 50 ml of 0.025 molar borax + x ml of 0.1 molar HCl.
G. 50 ml of 0.025 molar borax + x ml of 0.1 molar NaOH.
H. 50 ml of 0.05 molar sodium bicarbonate + x ml of 0.1 molar NaOH.
I. 50 ml of 0.05 molar disodium hydrogen phosphate + x ml of 0.1 molar NaOH.
J. 25 ml of 0.2 molar KCl + x ml of 0.2 molar NaOH.

Final volume of mixtures = 100 ml.

REFERENCES

1. Bower, V.E., and Bates, R.G., *J. Res. Natl. Bur. Stand.*, 55, 197, 1955 (A–D).
2. Bates, R.G., and Bower, V.E., *Anal. Chem.*, 28, 1322, 1956 (E–J).

DISSOCIATION CONSTANTS OF INORGANIC ACIDS AND BASES

The data in this table are presented as values of pK_a, defined as the negative logarithm of the acid dissociation constant K_a for the reaction

$$BH \rightleftharpoons B^- + H^+$$

Thus $pK_a = -\log K_a$, and the hydrogen ion concentration $[H^+]$ can be calculated from

$$K_a = \frac{[H^+][B^-]}{[BH]}$$

In the case of bases, the entry in the table is for the conjugate acid; e.g., ammonium ion for ammonia. The OH⁻ concentration in the system

$$NH_3 + H_2O \rightleftharpoons NH_4^+ + OH^-$$

can be calculated from the equation

$$K_b = K_{water} / K_a = \frac{[OH^-][NH_4^+]}{[NH_3]}$$

where $K_{water} = 1.01 \times 10^{-14}$ at 25 °C. Note that $pK_a + pK_b = pK_{water}$.

All values refer to dilute aqueous solutions at zero ionic strength at the temperature indicated. The table is arranged alphabetically by compound name.

REFERENCE

1. Perrin, D. D., *Ionization Constants of Inorganic Acids and Bases in Aqueous Solution, Second Edition,* Pergamon, Oxford, 1982.

Name	Formula	Step	t/°C	pKₐ
Aluminum(III) ion	Al^{+3}		25	5.0
Ammonia	NH_3		25	9.25
Arsenic acid	H_3AsO_4	1	25	2.26
		2	25	6.76
		3	25	11.29
Arsenious acid	H_2AsO_3		25	9.29
Barium(II) ion	Ba^{+2}		25	13.4
Boric acid	H_3BO_3	1	20	9.27
		2	20	>14
Calcium(II) ion	Ca^{+2}		25	12.6
Carbonic acid	H_2CO_3	1	25	6.35
		2	25	10.33
Chlorous acid	$HClO_2$		25	1.94
Chromic acid	H_2CrO_4	1	25	0.74
		2	25	6.49
Cyanic acid	$HCNO$		25	3.46
Germanic acid	H_2GeO_3	1	25	9.01
		2	25	12.3
Hydrazine	N_2H_4		25	8.1
Hydrazoic acid	HN_3		25	4.6
Hydrocyanic acid	HCN		25	9.21
Hydrofluoric acid	HF		25	3.20
Hydrogen peroxide	H_2O_2		25	11.62
Hydrogen selenide	H_2Se	1	25	3.89
		2	25	11.0
Hydrogen sulfide	H_2S	1	25	7.05
		2	25	19
Hydrogen telluride	H_2Te	1	18	2.6
		2	25	11
Hydroxylamine	NH_2OH		25	5.94
Hypobromous acid	$HBrO$		25	8.55

Name	Formula	Step	$t/°C$	pK_a
Hypochlorous acid	HClO		25	7.40
Hypoiodous acid	HIO		25	10.5
Iodic acid	HIO_3		25	0.78
Lithium ion	Li^+		25	13.8
Magnesium(II) ion	Mg^{+2}		25	11.4
Nitrous acid	HNO_2		25	3.25
Perchloric acid	$HClO_4$		20	-1.6
Periodic acid	HIO_4		25	1.64
Phosphoric acid	H_3PO_4	1	25	2.16
		2	25	7.21
		3	25	12.32
Phosphorous acid	H_3PO_3	1	20	1.3
		2	20	6.70
Pyrophosphoric acid	$H_4P_2O_7$	1	25	0.91
		2	25	2.10
		3	25	6.70
		4	25	9.32
Selenic acid	H_2SeO_4	2	25	1.7
Selenious acid	H_2SeO_3	1	25	2.62
		2	25	8.32
Silicic acid	H_4SiO_4	1	30	9.9
		2	30	11.8
		3	30	12
		4	30	12
Sodium ion	Na^+		25	14.8
Strontium(II) ion	Sr^{+2}		25	13.2
Sulfamic acid	NH_2SO_3H		25	1.05
Sulfuric acid	H_2SO_4	2	25	1.99
Sulfurous acid	H_2SO_3	1	25	1.85
		2	25	7.2
Telluric acid	H_2TeO_4	1	18	7.68
		2	18	11.0
Tellurous acid	H_2TeO_3	1	25	6.27
		2	25	8.43
Tetrafluoroboric acid	HBF_4		25	0.5
Thiocyanic acid	HSCN		25	-1.8
Water	H_2O		25	13.995

DISSOCIATION CONSTANTS OF ORGANIC ACIDS AND BASES

This table lists the dissociation (ionization) constants of over 1070 organic acids, bases, and amphoteric compounds. All data apply to dilute aqueous solutions and are presented as values of pK_a, which is defined as the negative of the logarithm of the equilibrium constant K_a for the reaction

$$HA \rightleftharpoons H^+ + A^-$$

i.e.,

$$K_a = [H^+][A^-]/[HA]$$

where $[H^+]$, etc. represent the concentrations of the respective species in mol/L. It follows that $pK_a = pH + \log[HA] - \log[A^-]$, so that a solution with 50% dissociation has pH equal to the pK_a of the acid.

Data for bases are presented as pK_a values for the conjugate acid, i.e., for the reaction

$$BH^+ \rightleftharpoons H^+ + B$$

In older literature, an ionization constant K_b was used for the reaction $B + H_2O \rightleftharpoons BH^+ + OH^-$. This is related to K_a by

$$pK_a + pK_b = pK_{water} = 14.00 \quad (at\ 25°C)$$

Compounds are listed by molecular formula in Hill order.

REFERENCES

1. Perrin, D.D., *Dissociation Constants of Organic Bases in Aqueous Solution*, Butterworths, London, 1965; Supplement, 1972.
2. Serjeant, E.P., and Dempsey, B., *Ionization Constants of Organic Acids in Aqueous Solution*, Pergamon, Oxford, 1979.
3. Albert, A., "Ionization Constants of Heterocyclic Substances", in Katritzky, A.R., Ed., *Physical Methods in Heterocyclic Chemistry*, Academic Press, New York, 1963.
4. Sober, H.A., Ed., *CRC Handbook of Biochemistry*, CRC Press, Boca Raton, FL, 1968.
5. Perrin, D.D., Dempsey, B., and Serjeant, E.P., pK_a *Prediction for Organic Acids and Bases*, Chapman and Hall, London, 1981.
6. Albert, A., and Serjeant, E. P., *The Determination of Ionization Constants, Third Edition*, Chapman and Hall, London, 1984.
7. Budavari, S., Editor, *The Merck Index, Twelfth Edition*, Merck & Co., Whitehouse Station, NJ, 1996.

Mol. Form.	Name	Step	$t/°C$	pK_a	Mol. Form.	Name	Step	$t/°C$	pK_a
CHNO	Cyanic acid		25	3.7	$C_2H_4N_2$	Aminoacetonitrile		25	5.34
CH_2N_2	Cyanamide		29	1.1	C_2H_4O	Acetaldehyde		25	13.57
CH_2O	Formaldehyde		25	13.27	C_2H_4OS	Thioacetic acid		25	3.33
CH_2O_2	Formic acid		25	3.75	$C_2H_4O_2$	Acetic acid		25	4.756
CH_3NO_2	Nitromethane		25	10.21	$C_2H_4O_2S$	Thioglycolic acid		25	3.68
CH_3NS_2	Carbamodithioic acid		25	2.95	$C_2H_4O_3$	Glycolic acid		25	3.83
CH_4N_2O	Urea		25	0.10	C_2H_5N	Ethyleneimine		25	8.04
CH_4N_2S	Thiourea		25	-1	C_2H_5NO	Acetamide		25	15.1
CH_4O	Methanol		25	15.5	$C_2H_5NO_2$	Acetohydroxamic acid			8.70
CH_4S	Methanethiol		25	10.33	$C_2H_5NO_2$	Nitroethane		25	8.46
CH_5N	Methylamine		25	10.66	$C_2H_5NO_2$	Glycine	1	25	2.35
CH_5NO	O-Methylhydroxylamine			12.5			2	25	9.78
CH_5N_3	Guanidine		25	13.6	$C_2H_6N_2$	Ethanimidamide		25	12.1
C_2HCl_3O	Trichloroacetaldehyde		25	10.04	C_2H_6O	Ethanol		25	15.5
$C_2HCl_3O_2$	Trichloroacetic acid		20	0.66	C_2H_6OS	2-Mercaptoethanol		25	9.72
$C_2HF_3O_2$	Trifluoroacetic acid		25	0.52	$C_2H_6O_2$	Ethyleneglycol		25	15.1
$C_2H_2Cl_2O_2$	Dichloroacetic acid		25	1.35	$C_2H_7AsO_2$	Dimethylarsinic acid	1	25	1.57
$C_2H_2O_3$	Glyoxylic acid		25	3.18			2	25	6.27
$C_2H_2O_4$	Oxalic acid	1	25	1.25	C_2H_7N	Ethylamine		25	10.65
		2	25	3.81	C_2H_7N	Dimethylamine		25	10.73
$C_2H_3BrO_2$	Bromoacetic acid		25	2.90	C_2H_7NO	Ethanolamine		25	9.50
$C_2H_3ClO_2$	Chloroacetic acid		25	2.87	$C_2H_7NO_3S$	2-Aminoethanesulfonic	1	25	1.5
$C_2H_3Cl_3O$	2,2,2-Trichloroethanol		25	12.24		acid	2	25	9.06
$C_2H_3FO_2$	Fluoroacetic acid		25	2.59	C_2H_7NS	Cysteamine	1	25	8.27
$C_2H_3F_3O$	2,2,2-Trifluoroethanol		25	12.37			2	25	10.53
$C_2H_3IO_2$	Iodoacetic acid		25	3.18	$C_2H_7N_5$	Biguanide	1		11.52
$C_2H_3NO_4$	Nitroacetic acid		24	1.48			2		2.93
$C_2H_3N_3$	1H-1,2,3-Triazole		20	1.17	$C_2H_8N_2$	1,2-Ethanediamine	1	25	9.92
$C_2H_3N_3$	1H-1,2,4-Triazole		20	2.27			2	25	6.86

Mol. Form.	Name	Step	$t/°C$	pK_a	Mol. Form.	Name	Step	$t/°C$	pK_a
$C_2H_8O_7P_2$	1-Hydroxy-1,1-	1		1.35	C_3H_9NO	2-Methoxyethylamine		25	9.40
	diphosphonoethane	2		2.87	C_3H_9NO	Trimethylamine oxide		20	4.65
		3		7.03	$C_3H_{10}N_2$	1,2-Propanediamine, (±)	1	25	9.82
		4		11.3			2	25	6.61
$C_3H_2O_2$	2-Propynoic acid		25	1.84	$C_3H_{10}N_2$	1,3-Propanediamine	1	25	10.55
C_3H_3NO	Oxazole		33	0.8			2	25	8.88
C_3H_3NO	Isoxazole		25	-2.0	$C_3H_{10}N_2O$	1,3-Diamino-2-propanol	1	20	9.69
$C_3H_3NO_2$	Cyanoacetic acid		25	2.47			2	20	7.93
C_3H_3NS	Thiazole		25	2.52	$C_3H_{11}N_3$	1,2,3-Triaminopropane	1	20	9.59
$C_3H_3N_3O_3$	Cyanuric acid	1		6.88			2	20	7.95
		2		11.40	$C_4H_4FN_3O$	Flucytosine			3.26
		3		13.5	$C_4H_4N_2$	Pyrazine		20	0.65
$C_3H_4N_2$	1H-Pyrazole		25	2.49	$C_4H_4N_2$	Pyrimidine		20	1.23
$C_3H_4N_2$	Imidazole		25	6.99	$C_4H_4N_2$	Pyridazine		20	2.24
$C_3H_4N_2S$	2-Thiazolamine		20	5.36	$C_4H_4N_2O_2$	Uracil		25	9.45
C_3H_4O	Propargyl alcohol		25	13.6	$C_4H_4N_2O_3$	Barbituric acid		25	4.01
$C_3H_4O_2$	Acrylic acid		25	4.25	$C_4H_4N_2O_5$	Alloxanic acid		25	6.64
$C_3H_4O_3$	Pyruvic acid		25	2.39	$C_4H_4N_4O_2$	5-Nitropyrimidinamine		20	0.35
$C_3H_4O_4$	Malonic acid	1	25	2.85	$C_4H_4O_2$	2-Butynoic acid		25	2.62
		2	25	5.70	$C_4H_4O_4$	Maleic acid	1	25	1.92
$C_3H_4O_5$	Hydroxypropanedioic	1		2.42			2	25	6.23
	acid	2		4.54	$C_4H_4O_4$	Fumaric acid	1	25	3.02
$C_3H_5BrO_2$	3-Bromopropanoic acid		25	4.00			2	25	4.38
$C_3H_5ClO_2$	2-Chloropropanoic acid		25	2.83	$C_4H_4O_5$	Oxaloacetic acid	1	25	2.55
$C_3H_5ClO_2$	3-Chloropropanoic acid		25	3.98			2	25	4.37
$C_3H_6N_2$	3-Aminopropanenitrile		20	7.80			3	25	13.03
$C_3H_6N_6$	1,3,5-Triazine-2,4,6-		25	5.00	C_4H_5N	Pyrrole		25	-3.8
	triamine				$C_4H_5NO_2$	Succinimide		25	9.62
C_3H_6O	Allyl alcohol		25	15.5	$C_4H_5N_3$	2-Pyrimidinamine		20	3.45
$C_3H_6O_2$	Propanoic acid		25	4.87	$C_4H_5N_3$	4-Pyrimidinamine		20	5.71
$C_3H_6O_2S$	(Methylthio)acetic acid		25	3.66	$C_4H_5N_3O$	Cytosine	1		4.60
$C_3H_6O_3$	Lactic acid		25	3.86			2		12.16
$C_3H_6O_3$	3-Hydroxypropanoic acid		25	4.51	$C_4H_5N_3O_2$	6-Methyl-1,2,4-triazine-			7.6
$C_3H_6O_4$	Glyceric acid		25	3.52		3,5(2H,4H)-dione			
C_3H_7N	Allylamine		25	9.49	$C_4H_6N_2$	1-Methylimidazol		25	6.95
C_3H_7N	Azetidine		25	11.29	$C_4H_6N_4O_3$	Allantoin		25	8.96
C_3H_7NO	2-Propanone oxime		25	12.42	$C_4H_6N_4O_3S_2$	Acetazolamide			7.2
$C_3H_7NO_2$	L-Alanine	1	25	2.34	$C_4H_6O_2$	trans-Crotonic acid		25	4.69
		2	25	9.87	$C_4H_6O_2$	3-Butenoic acid		25	4.34
$C_3H_7NO_2$	β-Alanine	1	25	3.55	$C_4H_6O_2$	Cyclopropanecarboxylic acid		25	4.83
		2	25	10.24	$C_4H_6O_3$	2-Oxobutanoic acid		25	2.50
$C_3H_7NO_2$	Sarcosine	1	25	2.21	$C_4H_6O_3$	Acetoacetic acid		25	3.6
		2	25	10.1	$C_4H_6O_4$	Succinic acid	1	25	4.21
$C_3H_7NO_2S$	L-Cysteine	1	25	1.5			2	25	5.64
		2	25	8.7	$C_4H_6O_4$	Methylmalonic acid	1	25	3.07
		3	25	10.2			2	25	5.76
$C_3H_7NO_3$	L-Serine	1	25	2.19	$C_4H_6O_5$	Malic acid	1	25	3.40
		2	25	9.21			2	25	5.11
$C_3H_7NO_5S$	DL-Cysteic acid	1	25	1.3	$C_4H_6O_6$	DL-Tartaric acid	1	25	3.03
		2	25	1.9			2	25	4.37
		3	25	8.70	$C_4H_6O_6$	meso-Tartaric acid	1	25	3.17
$C_3H_7N_3O_2$	Glycocyamine		25	2.82			2	25	4.91
$C_3H_8O_2$	Ethylene glycol		25	14.8	$C_4H_6O_6$	L-Tartaric acid	1	25	2.98
	monomethyl ether						2	25	4.34
$C_3H_8O_3$	Glycerol		25	14.15	$C_4H_6O_8$	Dihydroxytartaric acid		25	1.92
C_3H_9N	Propylamine		25	10.54	$C_4H_7ClO_2$	2-Chlorobutanoic acid			2.86
C_3H_9N	Isopropylamine		25	10.63	$C_4H_7ClO_2$	3-Chlorobutanoic acid			4.05
C_3H_9N	Trimethylamine		25	9.80	$C_4H_7ClO_2$	4-Chlorobutanoic acid			4.52

Mol. Form.	Name	Step	t/°C	pKa	Mol. Form.	Name	Step	t/°C	pKa
$C_4H_7NO_2$	4-Cyanobutanoic acid		25	2.42	$C_5H_4N_4O$	Hypoxanthine		25	8.7
$C_4H_7NO_3$	N-Acetylglycine		25	3.67	$C_5H_4N_4O$	Allopurinol			10.2
$C_4H_7NO_4$	Iminodiacetic acid	1		2.98	$C_5H_4N_4O_3$	Uric acid		12	3.89
		2		9.89	$C_5H_4N_4S$	1,7-Dihydro-6H-	1		7.77
$C_4H_7NO_4$	L-Aspartic acid	1	25	1.99		purine-6-thione	2		11.17
		2	25	3.90	$C_5H_4O_2S$	2-Thiophenecarboxylic acid		25	3.49
		3	25	9.90	$C_5H_4O_2S$	3-Thiophenecarboxylic acid		25	4.1
$C_4H_7N_3O$	Creatinine	1	25	4.8	$C_5H_4O_3$	2-Furancarboxylic acid		25	3.16
		2		9.2	$C_5H_4O_3$	3-Furancarboxylic acid		25	3.9
$C_4H_7N_5$	2,4,6-Pyrimidinetriamine		20	6.84	C_5H_5N	Pyridine		25	5.23
$C_4H_8N_2O_3$	L-Asparagine	1	20	2.1	C_5H_5NO	2-Pyridinol	1	20	0.75
		2	20	8.80			2	20	11.65
$C_4H_8N_2O_3$	N-Glycylglycine	1	25	3.14	C_5H_5NO	3-Pyridinol	1	20	4.79
		2		8.17			2	20	8.75
$C_4H_8O_2$	Butanoic acid		25	4.83	C_5H_5NO	4-Pyridinol	1	20	3.20
$C_4H_8O_2$	2-Methylpropanoic acid		20	4.84			2	20	11.12
$C_4H_8O_3$	3-Hydroxybutanoic acid, (±)		25	4.70	C_5H_5NO	2(1H)-Pyridinone	1	20	0.75
$C_4H_8O_3$	4-Hydroxybutanoic acid		25	4.72			2	20	11.65
$C_4H_8O_3$	Ethoxyacetic acid		18	3.65	C_5H_5NO	Pyridine-1-oxide		24	0.79
C_4H_9N	Pyrrolidine		25	11.31	$C_5H_5NO_2$	1H-Pyrrole-2-carboxylic		20	4.45
C_4H_9NO	Morpholine		25	8.50		acid			
$C_4H_9NO_2$	2-Methylalanine	1	25	2.36	$C_5H_5NO_2$	1H-Pyrrole-3-carboxylic		20	5.00
		2	25	10.21		acid			
$C_4H_9NO_2$	N,N-Dimethylglycine		25	9.89	$C_5H_5N_3O$	Pyrazinecarboxamide			0.5
$C_4H_9NO_2$	DL-2-Aminobutanoic acid	1	25	2.29	$C_5H_5N_5$	Adenine	1		4.3
		2	25	9.83			2		9.83
$C_4H_9NO_2$	4-Aminobutanoic acid	1	25	4.031	$C_5H_5N_5O$	Guanine		40	9.92
		2	25	10.556	$C_5H_6N_2$	2-Pyridinamine		20	6.82
$C_4H_9NO_2S$	DL-Homocysteine	1	25	2.22	$C_5H_6N_2$	3-Pyridinamine		25	6.04
		2	25	8.87	$C_5H_6N_2$	4-Pyridinamine		25	9.11
		3	25	10.86	$C_5H_6N_2$	2-Methylpyrazine		27	1.45
$C_4H_9NO_3$	L-Threonine	1	25	2.09	$C_5H_6N_2O_2$	Thymine		25	9.94
		2	25	9.10	$C_5H_6O_4$	1,1-Cyclopropanedi-	1	25	1.82
$C_4H_9NO_3$	L-Homoserine	1	25	2.71		carboxylic acid	2	25	7.43
		2	25	9.62	$C_5H_6O_4$	trans-1-Propene-1,2-	1	25	3.09
$C_4H_9N_3O_2$	Creatine	1	25	2.63		dicarboxylic acid	2	25	4.75
		2	25	14.3	$C_5H_6O_4$	1-Propene-2,3-	1	25	3.85
$C_4H_{10}N_2$	Piperazine	1	25	9.73		dicarboxylic acid	2	25	5.45
		2	25	5.33	$C_5H_6O_5$	2-Oxoglutaric acid	1	25	2.47
$C_4H_{10}N_2O_2$	2,4-Diaminobutanoic acid	1	25	1.85			2	25	4.68
		2	25	8.24	$C_5H_7NO_3$	5,5-Dimethyl-2,4-		37	6.13
		3	25	10.44		oxazolidinedione			
$C_4H_{10}O_4$	1,2,3,4-Butanetetrol			13.9	$C_5H_7NO_3$	L-Pyroglutamic acid		25	3.32
$C_4H_{11}N$	Butylamine		25	10.60	$C_5H_7N_3$	2,5-Pyridinediamine		20	6.48
$C_4H_{11}N$	sec-Butylamine		25	10.56	$C_5H_7N_3$	Methylaminopyrazine		25	3.39
$C_4H_{11}N$	tert-Butylamine		25	10.68	$C_5H_7N_3O_4$	Azaserine			8.55
$C_4H_{11}N$	Diethylamine		25	10.84	$C_5H_8N_2$	2,4-Dimethylimidazole		25	8.36
$C_4H_{11}NO_3$	Tris(hydroxymethyl)		20	8.3	$C_5H_8N_4O_3S_2$	Methazolamide			7.30
	methylamine				$C_5H_8O_2$	trans-3-Pentenoic acid		25	4.51
$C_4H_{12}N_2$	1,4-Butanediamine	1	25	10.80	$C_5H_8O_4$	Dimethylmalonic acid		25	3.15
		2	25	9.63	$C_5H_8O_4$	Glutaric acid	1	18	4.32
C_5H_4BrN	3-Bromopyridine		25	2.84			2	25	5.42
C_5H_4ClN	2-Chloropyridine		25	0.49	$C_5H_8O_4$	Methylsuccinic acid	1	25	4.13
C_3H_4ClN	3-Chloropyridine		25	2.81			2	25	5.64
C_5H_4ClN	4-Chloropyridine		25	3.83	$C_5H_9NO_2$	L-Proline	1	25	1.95
C_5H_4FN	2-Fluoropyridine		25	-0.44			2	25	10.64
$C_5H_4N_2O_2$	4-Nitropyridine		25	1.61	$C_5H_9NO_3$	5-Amino-4-oxopentanoic	1	25	4.05
$C_5H_4N_4$	1H-Purine	1	20	2.30		acid	2	25	8.90
		2	20	8.96	$C_5H_9NO_3$	trans-4-Hydroxyproline	1	25	1.82

DISSOCIATION CONSTANTS OF ORGANIC ACIDS AND BASES (continued)

Mol. Form.	Name	Step	$t/°C$	pK_a	Mol. Form.	Name	Step	$t/°C$	pK_a
		2	25	9.66	C_6H_5ClO	2-Chlorophenol		25	8.56
$C_5H_9NO_4$	L-Glutamic acid	1	25	2.13	C_6H_5ClO	3-Chlorophenol		25	9.12
		2	25	4.31	C_6H_5ClO	4-Chlorophenol		25	9.41
		3		9.67	$C_6H_5Cl_2N$	2,4-Dichloroaniline		22	2.05
$C_5H_9N_3$	Histamine	1	25	6.04	C_6H_5FO	2-Fluorophenol		25	8.73
		2	25	9.75	C_6H_5FO	3-Fluorophenol		25	9.29
$C_5H_{10}N_2O_3$	Glycylalanine		25	3.15	C_6H_5FO	4-Fluorophenol		25	9.89
$C_5H_{10}N_2O_3$	L-Glutamine	1	25	2.17	C_6H_5IO	2-Iodophenol		25	8.51
		2	25	9.13	C_6H_5IO	3-Iodophenol		25	9.03
$C_5H_{10}N_2O_4$	Glycylserine	1	25	2.98	C_6H_5IO	4-Iodophenol		25	9.33
		2	25	8.38	C_6H_5NO	2-Pyridinecarboxaldehyde		25	12.68
$C_5H_{10}O_2$	Pentanoic acid		20	4.83	C_6H_5NO	4-Pyridinecarboxaldehyde		30	12.05
$C_5H_{10}O_2$	2-Methylbutanoic acid		25	4.80	$C_6H_5NO_2$	Nitrobenzene		0	3.98
$C_5H_{10}O_2$	3-Methylbutanoic acid		25	4.77	$C_6H_5NO_2$	2-Pyridinecarboxylic acid	1	20	0.99
$C_5H_{10}O_2$	2,2-Dimethylpropanoic acid		20	5.03			2	20	5.39
$C_5H_{10}O_4$	D-2-Deoxyribose		25	12.61	$C_6H_5NO_2$	3-Pyridinecarboxylic acid	1	25	2.00
$C_5H_{10}O_5$	L-Ribose		25	12.22			2	25	4.82
$C_5H_{10}O_5$	D-Xylose		18	12.14	$C_6H_5NO_2$	4-Pyridinecarboxylic acid	1	25	1.77
$C_5H_{11}N$	Piperidine		25	11.123			2	25	4.84
$C_5H_{11}N$	N-Methylpyrrolidine		25	10.46	$C_6H_5NO_3$	2-Nitrophenol		25	7.23
$C_5H_{11}NO$	4-Methylmorpholine		25	7.38	$C_6H_5NO_3$	3-Nitrophenol		25	8.36
$C_5H_{11}NO_2$	L-Valine	1	25	2.29	$C_6H_5NO_3$	4-Nitrophenol		25	7.15
		2	25	9.74	$C_6H_5N_3$	1H-Benzotriazole		20	1.6
$C_5H_{11}NO_2$	DL-Norvaline	1		2.36	$C_6H_5N_5O$	2-Amino-4-	1	20	2.27
		2		9.72		hydroxypteridine	2	20	7.96
$C_5H_{11}NO_2$	L-Norvaline	1	25	2.32	$C_6H_5N_5O_2$	Xanthopterin	2	20	6.59
		2	25	9.81			3	20	9.31
$C_5H_{11}NO_2$	N-Propylglycine	1	25	2.35	C_6H_6BrN	2-Bromoaniline		25	2.53
		2	25	10.19	C_6H_6BrN	3-Bromoaniline		25	3.53
$C_5H_{11}NO_2$	5-Aminopentanoic acid	1	25	4.27	C_6H_6BrN	4-Bromoaniline		25	3.89
		2	25	10.77	C_6H_6ClN	2-Chloroaniline		25	2.66
$C_5H_{11}NO_2$	Betaine		0	1.83	C_6H_6ClN	3-Chloroaniline		25	3.52
$C_5H_{11}NO_2S$	L-Methionine	1	25	2.13	C_6H_6ClN	4-Chloroaniline		25	3.98
		2	25	9.27	C_6H_6FN	2-Fluoroaniline		25	3.20
$C_5H_{12}N_2O$	Tetramethylurea			2	C_6H_6FN	3-Fluoroaniline		25	3.59
$C_5H_{12}N_2O_2$	L-Ornithine	1	25	1.71	C_6H_6FN	4-Fluoroaniline		25	4.65
		2	25	8.69	C_6H_6IN	2-Iodoaniline		25	2.54
		3	25	10.76	C_6H_6IN	3-Iodoaniline		25	3.58
$C_5H_{13}N$	Pentylamine		25	10.63	C_6H_6IN	4-Iodoaniline		25	3.81
$C_5H_{13}N$	3-Pentanamine		17	10.59	$C_6H_6N_2O$	3-Pyridinecarboxamide		20	3.3
$C_5H_{13}N$	3-Methyl-1-butanamine		25	10.60	$C_6H_6N_2O$	2-Pyridinecarbox-	1	20	3.59
$C_5H_{13}N$	2-Methyl-2-butanamine		19	10.85		aldehyde oxime	2	20	10.18
$C_5H_{13}N$	2,2-Dimethylpropylamine		25	10.15	$C_6H_6N_2O_2$	2-Nitroaniline		25	-0.25
$C_5H_{13}N$	Diethylmethylamine		25	10.35	$C_6H_6N_2O_2$	3-Nitroaniline		25	2.46
$C_5H_{14}NO$	Choline		25	13.9	$C_6H_6N_2O_2$	4-Nitroaniline		25	1.02
$C_5H_{14}N_2$	1,5-Pentanediamine	1	25	10.05	C_6H_6O	Phenol		25	9.99
		2	25	10.93	$C_6H_6O_2$	p-Hydroquinone	1	25	9.85
$C_6H_3Cl_3N_2O_2$	4-Amino-3,5,6-trichloro-			3.6			2	25	11.4
	2-pyridinecarboxlic acid				$C_6H_6O_2$	Pyrocatechol	1	25	9.34
$C_6H_3N_3O_7$	2,4,6-Trinitrophenol		24	0.42			2	25	12.6
$C_6H_4Cl_2O$	2,3-Dichlorophenol		25	7.44	$C_6H_6O_2$	Resorcinol	1	25	9.32
$C_6H_4N_2O_5$	2,4-Dinitrophenol		25	4.07			2	25	11.1
$C_6H_4N_2O_5$	2,5-Dinitrophenol		15	5.15	$C_6H_6O_2S$	Benzenesulfinic acid		20	1.3
$C_6H_4N_4$	Pteridine		20	4.05	$C_6H_6O_3S$	Benzenesulfonic acid		25	0.70
C_6H_5BrO	2-Bromophenol		25	8.45	$C_6H_6O_4$	5-Hydroxy-2-(hydroxy-			7.9
C_6H_5BrO	3-Bromophenol		25	9.03		methyl)-4H-pyran-4-one			
C_6H_5BrO	4-Bromophenol		25	9.37	$C_6H_6O_4S$	3-Hydroxybenzene-		25	9.07
$C_6H_5Br_2N$	3,5-Dibromoaniline		25	2.34		sulfonic acid			

Mol. Form.	Name	Step	$t/°C$	pK_a	Mol. Form.	Name	Step	$t/°C$	pK_a
$C_6H_6O_4S$	4-Hydroxybenzene-sulfonic acid		25	9.11	$C_6H_{10}O_2$	Cyclopentanecarboxylic acid		25	4.99
$C_6H_6O_6$	cis-1-Propene-1,2,3-tricarboxylic acid		25	1.95	$C_6H_{10}O_3$	Ethyl acetoacetate		25	10.68
					$C_6H_{10}O_4$	3-Methylglutaric acid		25	4.24
$C_6H_6O_6$	trans-1-Propene-1,2,3-tricarboxylic acid	1	25	2.80	$C_6H_{10}O_4$	Adipic acid	1	18	4.41
		2	25	4.46			2	18	5.41
C_6H_6S	Benzenethiol		25	6.62	$C_6H_{11}NO_2$	2-Piperidinecarboxylic acid	1	25	2.28
$C_6H_7BO_2$	Benzeneboronic acid			8.83			2	25	10.72
C_6H_7N	Aniline		25	4.87	$C_6H_{11}NO_3$	Adipamic acid		25	4.63
C_6H_7N	2-Methylpyridine		25	6.00	$C_6H_{11}NO_4$	2-Aminoadipic acid	1	25	2.14
C_6H_7N	3-Methylpyridine		25	5.70			2	25	4.21
C_6H_7N	4-Methylpyridine		25	5.99			3	25	9.77
C_6H_7NO	2-Aminophenol	1	20	4.78	$C_6H_{11}N_3O_4$	N-(N-Glycylglycyl)glycine	1	25	3.225
		2	20	9.97			2	25	8.09
C_6H_7NO	3-Aminophenol	1	20	4.37	$C_6H_{11}N_3O_4$	Glycylasparagine	1	25	2.942
		2	20	9.82			2	18	8.44
C_6H_7NO	4-Aminophenol	1	25	5.48	$C_6H_{12}N_2$	Triethylenediamine	1		3.0
		2	25	10.30			2		8.7
C_6H_7NO	2-Methoxypyridine		20	3.28	$C_6H_{12}N_2O_4S_2$	L-Cystine	1		1
C_6H_7NO	3-Methoxypyridine		25	4.78			2		2.1
C_6H_7NO	4-Methoxypyridine		25	6.58			3		8.02
$C_6H_7NO_3S$	2-Aminobenzenesulfonic acid		25	2.46			4		8.71
					$C_6H_{12}O_2$	Hexanoic acid		25	4.85
$C_6H_7NO_3S$	3-Aminobenzenesulfonic acid		25	3.74	$C_6H_{12}O_2$	4-Methylpentanoic acid		18	4.84
					$C_6H_{12}O_6$	β-D-Fructose		25	12.27
$C_6H_7NO_3S$	4-Aminobenzenesulfonic acid		25	3.23	$C_6H_{12}O_6$	α-D-Glucose		25	12.46
					$C_6H_{12}O_6$	D-Mannose		25	12.08
$C_6H_8N_2$	N-Methylpyridinamine		20	9.65	$C_6H_{13}N$	Cyclohexylamine		25	10.64
$C_6H_8N_2$	o-Phenylenediamine	1	20	4.57	$C_6H_{13}N$	1-Methylpiperidine		25	10.38
		2	20	0.80	$C_6H_{13}N$	1,2-Dimethylpyrrolidine		26	10.20
$C_6H_8N_2$	m-Phenylenediamine	1	20	5.11	$C_6H_{13}NO$	N-Ethylmorpholine		25	7.67
		2	20	2.50	$C_6H_{13}NO_2$	L-Leucine	1	25	2.33
$C_6H_8N_2$	p-Phenylenediamine	1	20	6.31			2	25	9.74
		2	20	2.97	$C_6H_{13}NO_2$	L-Isoleucine	1	25	2.32
$C_6H_8N_2$	Phenylhydrazine		15	8.79			2	25	9.76
$C_6H_8O_2$	2,4-Hexadienoic acid		25	4.76	$C_6H_{13}NO_2$	L-Norleucine	1	25	2.34
$C_6H_8O_2$	1,3-Cyclohexanedione		25	5.26			2	25	9.83
$C_6H_8O_4$	2,2-Dimethyl-1,3-dioxane-4,6-dione			5.1	$C_6H_{13}NO_2$	6-Aminohexanoic acid	1	25	4.37
							2	25	10.80
$C_6H_8O_6$	L-Ascorbic acid	1	25	4.04	$C_6H_{13}NO_4$	N,N-Bis(2-hydroxy-ethyl)glycine	2	20	8.35
		2	16	11.7					
$C_6H_8O_7$	Citric acid	1	25	3.13	$C_6H_{13}N_3O_3$	Citrulline	1	25	2.43
		2	25	4.76			2	25	9.69
		3	25	6.40	$C_6H_{14}N_2$	cis-1,2-Cyclohexane-diamine	1	20	9.93
$C_6H_8O_7$	Isocitric acid	1	25	3.29			2	20	6.13
		2	25	4.71	$C_6H_{14}N_2$	trans-1,2-Cyclohexane-diamine	1	20	9.94
		3	25	6.40			2	20	6.47
$C_6H_9NO_6$	Nitrilotriacetic acid	1	20	3.03	$C_6H_{14}N_2$	cis-2,5-Dimethyl-piperazine	1	25	9.66
		2	20	3.07			2	25	5.20
		3	20	10.70	$C_6H_{14}N_2O_2$	L-Lysine	1	25	2.16
$C_6H_9NO_6$	L-γ-Carboxyglutamic acid	1	25	1.7			2	25	9.06
		2	25	3.2			3	25	10.54
		3	25	4.75	$C_6H_{14}N_4O_2$	L-Arginine	1	25	1.82
		4	25	9.9			2	25	8.99
$C_6H_9N_3$	4,6-Dimethylpyrimi-dinamine		20	4.82			3	25	12.5
					$C_6H_{14}O_6$	D-Mannitol		18	13.5
$C_6H_9N_3O_2$	L-Histidine	1	25	1.80	$C_6H_{15}N$	Hexylamine		25	10.56
		2	25	6.04	$C_6H_{15}N$	Diisopropylamine		25	11.05
		3	25	9.33	$C_6H_{15}N$	Triethylamine		25	10.75

DISSOCIATION CONSTANTS OF ORGANIC ACIDS AND BASES (continued)

Mol. Form.	Name	Step	$t/°C$	pK_a	Mol. Form.	Name	Step	$t/°C$	pK_a
$C_6H_{15}NO_3$	Triethanolamine		25	7.76			2	25	9.46
$C_6H_{16}N_2$	1,6-Hexanediamine	1	0	11.86	$C_7H_6O_4$	2,4-Dihydroxybenzoic acid	1	25	3.11
		2	0	10.76			2	25	8.55
$C_6H_{16}N_2$	*N,N,N',N'*-Tetramethyl-	1	25	10.40			3	25	14.0
	1,2-ethanediamine	2	25	8.26	$C_7H_6O_4$	2,5-Dihydroxybenzoic acid	1	25	2.97
$C_6H_{19}NSi_2$	Hexamethyldisilazane			7.55	$C_7H_6O_4$	3,4-Dihydroxybenzoic acid	1	25	4.48
$C_7HF_5O_2$	Pentafluorobenzoic acid		25	1.75			2	25	8.83
$C_7H_3Br_2NO$	3,5-Dibromo-4-			4.06			3	25	12.6
	hydroxybenzonitrile				$C_7H_6O_4$	3,5-Dihydroxybenzoic acid	1	25	4.04
$C_7H_3N_3O_8$	2,4,6-Trinitrobenzoic acid		25	0.65	$C_7H_6O_5$	2,4,6-Trihydroxybenzoic		25	1.68
$C_7H_4Cl_3NO_3$	Triclopyr			2.68		acid			
$C_7H_4N_2O_6$	2,4-Dinitrobenzoic acid		25	1.43	$C_7H_6O_5$	3,4,5-Trihydroxybenzoic		25	4.41
$C_7H_5BrO_2$	2-Bromobenzoic acid		25	2.85		acid			
$C_7H_5BrO_2$	3-Bromobenzoic acid		25	3.81	C_7H_7NO	Benzamide		25	¯13
$C_7H_5BrO_2$	4-Bromobenzoic acid		25	3.96	$C_7H_7NO_2$	Aniline-2-carboxylic acid	1	25	2.17
$C_7H_5ClO_2$	2-Chlorobenzoic acid		25	2.90			2	25	4.85
$C_7H_5ClO_2$	3-Chlorobenzoic acid		25	3.84	$C_7H_7NO_2$	Aniline-3-carboxylic acid	1	25	3.07
$C_7H_5ClO_2$	4-Chlorobenzoic acid		25	4.00			2	25	4.79
$C_7H_5FO_2$	2-Fluorobenzoic acid		25	3.27	$C_7H_7NO_2$	Aniline-4-carboxylic acid	1	25	2.50
$C_7H_5FO_2$	3-Fluorobenzoic acid		25	3.86			2	25	4.87
$C_7H_5FO_2$	4-Fluorobenzoic acid		25	4.15	$C_7H_7NO_3$	4-Amino-2-hydroxy-			3.25
$C_7H_5F_3O$	2-(Trifluoromethyl)phenol		25	8.95		benzoic acid			
$C_7H_5F_3O$	3-(Trifluoromethyl)phenol		25	8.68	$C_7H_8ClN_3O_4S_2$	Hydrochlorothiazide	1		7.9
$C_7H_5IO_2$	2-Iodobenzoic acid		25	2.86			2		9.2
$C_7H_5IO_2$	3-Iodobenzoic acid		25	3.87	$C_7H_8N_4O_2$	Theobromine		18	7.89
$C_7H_5IO_2$	4-Iodobenzoic acid		25	4.00	$C_7H_8N_4O_2$	Theophylline	1	25	8.77
C_7H_5NO	2-Hydroxybenzonitrile		25	6.86	C_7H_8O	*o*-Cresol		25	10.29
C_7H_5NO	3-Hydroxybenzonitrile		25	8.61	C_7H_8O	*m*-Cresol		25	10.09
C_7H_5NO	4-Hydroxybenzonitrile		25	7.97	C_7H_8O	*p*-Cresol		25	10.26
$C_7H_5NO_3S$	Saccharin		18	11.68	C_7H_8OS	4-(Methylthio)phenol		25	9.53
$C_7H_5NO_4$	2-Nitrobenzoic acid		25	2.17	$C_7H_8O_2$	2-Methoxyphenol		25	9.98
$C_7H_5NO_4$	3-Nitrobenzoic acid		25	3.46	$C_7H_8O_2$	3-Methoxyphenol		25	9.65
$C_7H_5NO_4$	4-Nitrobenzoic acid		25	3.43	$C_7H_8O_2$	4-Methoxyphenol		25	10.21
$C_7H_5NO_4$	2,3-Pyridinedicarboxylic	1	25	2.43	C_7H_8S	Benzenemethanethiol		25	9.43
	acid	2	25	4.78	C_7H_9N	Benzylamine		25	9.34
$C_7H_5NO_4$	2,4-Pyridinedicarboxylic	1	25	2.15	C_7H_9N	2-Methylaniline		25	4.45
	acid				C_7H_9N	3-Methylaniline		25	4.71
$C_7H_5NO_4$	2,6-Pyridinedicarboxylic	1	25	2.16	C_7H_9N	4-Methylaniline		25	5.08
	acid	2	25	4.76	C_7H_9N	*N*-Methylaniline		25	4.85
$C_7H_5NO_4$	3,5-Pyridinedicarboxylic	1	25	2.80	C_7H_9N	2-Ethylpyridine		25	5.89
	acid				C_7H_9N	2,3-Dimethylpyridine		25	6.57
$C_7H_6ClN_3O_4S_2$	Chlorothiazide	1		6.85	C_7H_9N	2,4-Dimethylpyridine		25	6.99
		2		9.45	C_7H_9N	2,5-Dimethylpyridine		25	6.40
$C_7H_6F_3N$	3-(Trifluoromethyl)aniline		25	3.49	C_7H_9N	2,6-Dimethylpyridine		25	6.65
$C_7H_6F_3N$	4-(Trifluoromethyl)aniline		25	2.45	C_7H_9N	3,4-Dimethylpyridine		25	6.46
$C_7H_6N_2$	1*H*-Benzimidazole		25	5.53	C_7H_9N	3,5-Dimethylpyridine		25	6.15
$C_7H_6N_2$	2-Aminobenzonitrile		25	0.77	C_7H_9NO	2-Methoxyaniline		25	4.53
$C_7H_6N_2$	3-Aminobenzonitrile		25	2.75	C_7H_9NO	3-Methoxyaniline		25	4.20
$C_7H_6N_2$	4-Aminobenzonitrile		25	1.74	C_7H_9NO	4-Methoxyaniline		25	5.36
C_7H_6O	Benzaldehyde		25	14.90	C_7H_9NS	2-(Methylthio)aniline		25	3.45
$C_7H_6O_2$	Benzoic acid		25	4.204	C_7H_9NS	4-(Methylthio)aniline		25	4.35
$C_7H_6O_2$	Salicylaldehyde		25	8.37	$C_7H_9N_5$	2-Dimethylaminopurine	1	20	4.00
$C_7H_6O_2$	3-Hydroxybenzaldehyde		25	8.98			2	20	10.24
$C_7H_6O_2$	4-Hydroxybenzaldehyde		25	7.61	$C_7H_{11}N_3O_2$	*L*-1-Methylhistidine	1	25	1.69
$C_7H_6O_3$	2-Hydroxybenzoic acid	1	20	2.98			2	25	6.48
		2	20	13.6			3	25	8.85
$C_7H_6O_3$	3-Hydroxybenzoic acid	1	25	4.08	$C_7H_{11}N_3O_2$	*L*-3-Methylhistidine	1	25	1.92
		2	19	9.92			2	25	6.56
$C_7H_6O_3$	4-Hydroxybenzoic acid	1	25	4.57			3	25	8.73

Mol. Form.	Name	Step	$t/°C$	pK_a
$C_7H_{12}O_2$	Cyclohexanecarboxylic acid		25	4.91
$C_7H_{12}O_4$	Heptanedioic acid	1	25	4.71
		2	25	5.58
$C_7H_{12}O_4$	Butylpropanedioic acid	1	5	2.96
$C_7H_{13}NO_4$	α-Ethylglutamic acid	1	25	3.846
		2	25	7.838
$C_7H_{14}O_2$	Heptanoic acid		25	4.89
$C_7H_{14}O_6$	α-Methylglucoside		25	13.71
$C_7H_{15}N$	1-Ethylpiperidine		23	10.45
$C_7H_{15}N$	1,2-Dimethylpiperidine,(±)		25	10.22
$C_7H_{15}NO_3$	Carnitine		25	3.80
$C_7H_{17}N$	Heptylamine		25	10.67
$C_7H_{17}N$	2-Heptanamine		19	10.7
$C_8H_5NO_2$	3-Cyanobenzoic acid		25	3.60
$C_8H_5NO_2$	4-Cyanobenzoic acid		25	3.55
$C_8H_6N_2$	Cinnoline		20	2.37
$C_8H_6N_2$	Quinazoline		29	3.43
$C_8H_6N_2$	Quinoxaline		20	0.56
$C_8H_6N_2$	Phthalazine		20	3.47
$C_8H_6N_4O_5$	Nitrofurantoin			7.2
$C_8H_6O_3$	3-Formylbenzoic acid		25	3.84
$C_8H_6O_3$	4-Formylbenzoic acid		25	3.77
$C_8H_6O_4$	Phthalic acid	1	25	2.943
		2	25	5.432
$C_8H_6O_4$	Isophthalic acid	1	25	3.70
		2	25	4.60
$C_8H_6O_4$	Terephthalic acid	1	25	3.54
		2	25	4.34
$C_8H_7ClO_2$	2-Chlorobenzeneacetic acid		25	4.07
$C_8H_7ClO_2$	3-Chlorobenzeneacetic acid		25	4.14
$C_8H_7ClO_2$	4-Chlorobenzeneacetic acid		25	4.19
$C_8H_7ClO_3$	2-Chlorophenoxyacetic acid		25	3.05
$C_8H_7ClO_3$	3-Chlorophenoxyacetic acid		25	3.10
$C_8H_7NO_4$	2-Nitrobenzeneacetic acid		25	4.00
$C_8H_7NO_4$	3-Nitrobenzeneacetic acid		25	3.97
$C_8H_7NO_4$	4-Nitrobenzeneacetic acid		25	3.85
$C_8H_8F_3N_3O_4S_2$	Hydroflumethiazide	1		8.9
		2		9.7
$C_8H_8N_2$	2-Methyl-1H-benzimidazole		25	6.19
$C_8H_8O_2$	o-Toluic acid		25	3.91
$C_8H_8O_2$	m-Toluic acid		25	4.25
$C_8H_8O_2$	p-Toluic acid		25	4.37
$C_8H_8O_2$	Benzeneacetic acid		25	4.31
$C_8H_8O_2$	1-(2-Hydroxyphenyl)ethanone		25	10.06
$C_8H_8O_2$	1-(3-Hydroxyphenyl)ethanone		25	9.19
$C_8H_8O_2$	1-(4-Hydroxyphenyl)ethanone		25	8.05
$C_8H_8O_3$	2-Methoxybenzoic acid		25	4.08
$C_8H_8O_3$	3-Methoxybenzoic acid		25	4.10
$C_8H_8O_3$	4-Methoxybenzoic acid		25	4.50
$C_8H_8O_3$	Phenoxyacetic acid		25	3.17
$C_8H_8O_3$	Mandelic acid		25	3.37
$C_8H_8O_4$	2,5-Hydroxybenzeneacetic acid		25	4.40
C_8H_9NO	Acetanilide		25	0.5
$C_8H_9NO_2$	2-(Methylamino)benzoic acid		25	5.34
$C_8H_9NO_2$	3-(Methylamino)benzoic acid		25	5.10
$C_8H_9NO_2$	4-(Methylamino)benzoic acid		25	5.04
$C_8H_9NO_2$	N-Phenylglycine	1	25	1.83
		2		4.39
$C_8H_{10}BrN$	4-Bromo-N,N-dimethylaniline		25	4.23
$C_8H_{10}ClN$	3-Chloro-N,N-dimethylaniline		20	3.83
$C_8H_{10}ClN$	4-Chloro-N,N-dimethylaniline		20	4.39
$C_8H_{10}N_2O_2$	N,N-Dimethyl-3-nitroaniline		25	2.62
$C_8H_{11}N$	N-Ethylaniline		25	5.12
$C_8H_{11}N$	N,N-Dimethylaniline		25	5.07
$C_8H_{11}N$	2,6-Dimethylaniline		25	3.89
$C_8H_{11}N$	Benzeneethanamine		25	9.83
$C_8H_{11}N$	2,4,6-Trimethylpyridine		25	7.43
$C_8H_{11}NO$	2-Ethoxyaniline		28	4.43
$C_8H_{11}NO$	3-Ethoxyaniline		25	4.18
$C_8H_{11}NO$	4-Ethoxyaniline		28	5.20
$C_8H_{11}NO$	4-(2-Aminoethyl)phenol	1	25	9.74
		2	25	10.52
$C_8H_{11}NO$	2-(2-Methoxyethyl)pyridine			5.5
$C_8H_{11}NO_2$	Dopamine	1	25	8.9
		2	25	10.6
$C_8H_{11}NO_3$	Norepinephrine	1	25	8.64
		2	25	9.70
$C_8H_{11}N_3O_6$	6-Azauridine			6.70
$C_8H_{11}N_5$	Phenylbiguanide	1		10.76
		2		2.13
$C_8H_{12}N_2O_3$	Barbital		25	7.43
$C_8H_{12}O_2$	5,5-Dimethyl-1,3-cyclohexanedione		25	5.15
$C_8H_{13}NO_2$	Arecoline			6.84
$C_8H_{14}O_2S_2$	Thioctic acid			5.4
$C_8H_{14}O_4$	Octanedioic acid	1	25	4.52
$C_8H_{15}NO$	Tropine		15	3.80
$C_8H_{15}NO$	Pseudotropine		15	3.80
$C_8H_{16}N_2O_3$	N-Glycylleucine		25	3.18
$C_8H_{16}N_2O_3$	N-Leucylglycine	1	25	3.25
		2	25	8.2
$C_8H_{16}N_2O_4S_2$	Homocystine	1	25	1.59
		2	25	2.54
		3	25	8.52
		4	25	9.44
$C_8H_{16}O_2$	Octanoic acid		25	4.89
$C_8H_{16}O_2$	2-Propylpentanoic acid			4.6
$C_8H_{17}N$	2-Propylpiperidine,(S)			10.9
$C_8H_{17}N$	2,2,4-Trimethylpiperidine		30	11.04
$C_8H_{17}NO$	trans-6-Propyl-3-piperidinol,(3S)			10.3
$C_8H_{19}N$	Octylamine		25	10.65
$C_8H_{19}N$	N-Methyl-2-heptanamine		17	10.99
$C_8H_{19}N$	Dibutylamine		21	11.25
$C_8H_{20}N_2$	1,8-Octanediamine	1	20	11.00
		2	20	10.1
C_9H_6BrN	3-Bromoquinoline		25	2.69
$C_9H_7ClO_2$	trans-o-Chlorocinnamic acid		25	4.23

Mol. Form.	Name	Step	$t/°C$	pK_a
$C_9H_7ClO_2$	trans-m-Chlorocinnamic acid		25	4.29
$C_9H_7ClO_2$	trans-p-Chlorocinnamic acid		25	4.41
C_9H_7N	Quinoline		20	4.90
C_9H_7N	Isoquinoline		20	5.40
C_9H_7NO	2-Quinolinol	1	20	-0.31
		2	20	11.76
C_9H_7NO	3-Quinolinol	1	20	4.28
		2	20	8.08
C_9H_7NO	4-Quinolinol	1	20	2.23
		2	20	11.28
C_9H_7NO	6-Quinolinol	1	20	5.15
		2	20	8.90
C_9H_7NO	8-Quinolinol	1	25	4.91
		2	25	9.81
C_9H_7NO	7-Isoquinolinol	1	20	5.68
		2	20	8.90
$C_9H_7NO_3$	2-Cyanophenoxyacetic acid		25	2.98
$C_9H_7NO_3$	3-Cyanophenoxyacetic acid		25	3.03
$C_9H_7NO_3$	4-Cyanophenoxyacetic acid		25	2.93
$C_9H_7N_7O_2S$	Azathioprine			8.2
$C_9H_8N_2$	2-Quinolinamine		20	7.34
$C_9H_8N_2$	3-Quinolinamine		20	4.91
$C_9H_8N_2$	4-Quinolinamine		20	9.17
$C_9H_8N_2$	1-Isoquinolinamine		20	7.62
$C_9H_8N_2$	3-Isoquinolinamine		20	5.05
$C_9H_8O_2$	cis-Cinnamic acid		25	3.88
$C_9H_8O_2$	trans-Cinnamic acid		25	4.44
$C_9H_8O_2$	α-Methylenebenzene-acetic acid			4.35
$C_9H_8O_4$	2-(Acetyloxy)benzoic acid		25	3.48
$C_9H_9Br_2NO_3$	3,5-Dibromo-L-tyrosine	1		2.17
		2		6.45
		3		7.60
$C_9H_9ClO_2$	3-(2-Chlorophenyl)-propanoic acid		25	4.58
$C_9H_9ClO_2$	3-(3-Chlorophenyl)-propanoic acid		25	4.59
$C_9H_9ClO_2$	3-(4-Chlorophenyl)-propanoic acid		25	4.61
$C_9H_9I_2NO_3$	L-3,5-Diiodotyrosine	1	25	2.12
		2	25	5.32
		3	25	9.48
$C_9H_9NO_3$	N-Benzoylglycine		25	3.62
$C_9H_9NO_4$	3-(2-Nitrophenyl)-propanoic acid		25	4.50
$C_9H_9NO_4$	3-(4-Nitrophenyl)-propanoic acid		25	4.47
$C_9H_9N_3O_2$	Carbendazim			4.48
$C_9H_9N_3O_2S_2$	Sulfathiazole			7.2
$C_9H_{10}INO_3$	L-3-Iodotyrosine	1	25	2.2
		2	25	8.7
		3	25	9.1
$C_9H_{10}N_2$	2-Ethylbenzimidazole		25	6.18
$C_9H_{10}O_2$	3,5-Dimethylbenzoic acid		25	4.32
$C_9H_{10}O_2$	Benzenepropanoic acid		25	4.66
$C_9H_{10}O_2$	α-Methylbenzeneacetic acid		25	4.64
$C_9H_{10}O_3$	α-Hydroxy-α-methyl-benzeneacetic acid		25	3.47

Mol. Form.	Name	Step	$t/°C$	pK_a
$C_9H_{11}Cl_2N_3O_4S_2$	Methylclothiazide			9.4
$C_9H_{11}N$	N-Allylaniline		25	4.17
$C_9H_{11}N$	1-Indanamine		22	9.21
$C_9H_{11}NO_2$	4-(Dimethylamino)-benzoic acid	1		6.03
		2		11.49
$C_9H_{11}NO_2$	Ethyl 4-aminobenzoate			2.5
$C_9H_{11}NO_2$	L-Phenylalanine	1	25	2.20
		2	25	9.31
$C_9H_{11}NO_3$	L-Tyrosine	1	25	2.20
		2	25	9.11
		3	25	10.1
$C_9H_{11}NO_4$	Levodopa	1	25	2.32
		2	25	8.72
		3	25	9.96
		4	25	11.79
$C_9H_{12}N_2O_2$	Tyrosineamide		25	7.33
$C_9H_{13}N$	N-Isopropylaniline		25	5.77
$C_9H_{13}NO_3$	Epinephrine	1	25	8.66
		2	25	9.95
$C_9H_{13}N_2O_9P$	5'-Uridylic acid	1		6.4
		2		9.5
$C_9H_{13}N_3O_5$	Cytidine	1		4.22
		2		12.5
$C_9H_{14}ClNO$	Phenylpropanolamine hydrochloride			9.44
$C_9H_{14}N_2O_3$	Metharbital			8.45
$C_9H_{14}N_3O_8P$	3'-Cytidylic acid	1		0.8
		2		4.28
		3		6.0
$C_9H_{14}N_4O_3$	Carnosine	1	20	2.73
		2	20	6.87
		3	20	9.73
$C_9H_{15}NO_3S$	Captopril	1		3.7
		2		9.8
$C_9H_{15}N_5O$	Minoxidil			4.61
$C_9H_{16}O_4$	Nonanedioic acid	1	25	4.53
		2	25	5.33
$C_9H_{18}O_2$	Nonanoic acid		25	4.96
$C_9H_{19}N$	N-Butylpiperidine		23	10.47
$C_9H_{19}N$	2,2,6,6-Tetramethyl-piperidine		25	11.07
$C_9H_{21}N$	Nonylamine		25	10.64
$C_{10}H_7NO_2$	8-Quinolinecarboxylic acid		25	1.82
$C_{10}H_8O$	1-Naphthol		25	9.39
$C_{10}H_8O$	2-Naphthol		25	9.63
$C_{10}H_9N$	1-Naphthylamine		25	3.92
$C_{10}H_9N$	2-Naphthylamine		25	4.16
$C_{10}H_9N$	2-Methylquinoline		20	5.83
$C_{10}H_9N$	4-Methylquinoline		20	5.67
$C_{10}H_9N$	5-Methylquinoline		20	5.20
$C_{10}H_9NO$	5-Amino-1-naphthol		25	3.97
$C_{10}H_9NO$	6-Methoxyquinoline		20	5.03
$C_{10}H_9NO_2$	1H-Indole-3-acetic acid			4.75
$C_{10}H_{10}O_2$	o-Methylcinnamic acid		25	4.50
$C_{10}H_{10}O_2$	m-Methylcinnamic acid		25	4.44
$C_{10}H_{10}O_2$	p-Methylcinnamic acid		25	4.56
$C_{10}H_{12}N_2$	Tryptamine		25	10.2
$C_{10}H_{12}N_2O$	5-Hydroxytryptamine	1	25	9.8
		2	25	11.1

Mol. Form.	Name	Step	$t/°C$	pK_a	Mol. Form.	Name	Step	$t/°C$	pK_a
$C_{10}H_{12}N_2O_5$	Dinoseb			4.62	$C_{11}H_{17}NO_3$	Isoproterenol			8.64
$C_{10}H_{12}N_4O_3$	Dideoxyinosine			9.12	$C_{11}H_{17}N_3O_8$	Tetrodotoxin			8.76
$C_{10}H_{12}O$	5,6,7,8-Tetrahydro-2-naphthalenol		25	10.48	$C_{11}H_{18}ClNO_3$	Methoxamine hydrochloride		25	9.2
					$C_{11}H_{18}N_2O_3$	Amobarbital		25	8.0
$C_{10}H_{12}O_2$	Benzenebutanoic acid		25	4.76	$C_{11}H_{25}N$	Undecylamine		25	10.63
$C_{10}H_{12}O_5$	Propyl 3,4,5-trihydroxy-benzoate			8.11	$C_{11}H_{26}NO_2PS$	Methylphosphonothioic acid S[2-[bis(1-isopropyl)amino]-ethyl],O-ethylester			7.9
$C_{10}H_{13}N_5O_4$	Adenosine	1	25	3.6					
		2	25	12.4	$C_{12}H_6Cl_4O_2S$	Bithionol	1		4.82
$C_{10}H_{14}N_2$	L-Nicotine	1		8.02			2		10.50
		2		3.12	$C_{12}H_8N_2$	1,10-Phenanthroline		25	4.84
$C_{10}H_{14}N_5O_7P$	5'-Adenylic acid	1		3.8	$C_{12}H_8N_2$	Phenazine		20	1.20
		2		6.2	$C_{12}H_{10}O$	2-Hydroxybiphenyl		25	10.01
$C_{10}H_{14}O$	2-$tert$-Butylphenol		25	10.62	$C_{12}H_{10}O$	3-Hydroxybiphenyl		25	9.64
$C_{10}H_{14}O$	3-$tert$-Butylphenol		25	10.12	$C_{12}H_{10}O$	4-Hydroxybiphenyl		25	9.55
$C_{10}H_{14}O$	4-$tert$-Butylphenol		25	10.23	$C_{12}H_{11}N$	Diphenylamine		25	0.79
$C_{10}H_{15}N$	N-$tert$-Butylaniline		25	7.00	$C_{12}H_{11}N$	2-Aminobiphenyl		25	3.83
$C_{10}H_{15}N$	N,N-Diethylaniline		25	6.57	$C_{12}H_{11}N$	3-Aminobiphenyl		18	4.25
$C_{10}H_{15}NO$	d-Ephedrine		10	10.139	$C_{12}H_{11}N$	4-Aminobiphenyl		18	4.35
$C_{10}H_{15}NO$	l-Ephedrine		10	9.958	$C_{12}H_{11}N$	2-Benzylpyridine		25	5.13
$C_{10}H_{17}N_3O_6S$	l-Glutathione	1	25	2.12	$C_{12}H_{11}N_3$	4-Aminoazobenzene		25	2.82
		2	25	3.59	$C_{12}H_{12}N_2$	p-Benzidine	1	20	4.65
		3	25	8.75			2	20	3.43
		4	25	9.65	$C_{12}H_{12}N_2O_3$	Phenobarbital	1		7.3
$C_{10}H_{18}N_4O_5$	L-Argininosuccinic acid	1	25	1.62			2		11.8
		2	25	2.70	$C_{12}H_{13}I_3N_2O_3$	Iocetamic acid			4
		3	25	4.26	$C_{12}H_{13}N$	N,N-Dimethyl-1-naphthylamine		25	4.83
		4	25	9.58					
$C_{10}H_{18}O_4$	Sebacic acid	1		4.59	$C_{12}H_{13}N$	N,N-Dimethyl-2-naphthylamine		25	4.566
		2		5.59					
$C_{10}H_{19}N$	Bornylamine		25	10.17	$C_{12}H_{14}N_4O_2S$	Sulfamethazine	1		7.4
$C_{10}H_{19}N$	Neobornylamine		25	10.01			2		2.65
$C_{10}H_{21}N$	Butylcyclohexylamine		25	11.23	$C_{12}H_{14}N_4O_3S$	Sulfacytine			6.9
$C_{10}H_{21}N$	1,2,2,6,6-Pentamethyl-piperidine		30	11.25	$C_{12}H_{17}N_3O_4$	Agaritine	1		3.4
							2		8.86
$C_{10}H_{23}N$	Decylamine		25	10.64	$C_{12}H_{20}N_2O_2$	Aspergillic acid			5.5
$C_{11}H_8N_2$	1H-Perimidine		20	6.35	$C_{12}H_{21}N_5O_2S_2$	Nizatidine	1		2.1
$C_{11}H_8O_2$	1-Naphthalenecarboxylic acid		25	3.69			2		6.8
					$C_{12}H_{22}O_{11}$	Sucrose		25	12.7
$C_{11}H_8O_2$	2-Naphthalenecarboxylic acid		25	4.16	$C_{12}H_{22}O_{11}$	α-Maltose		21	12.05
					$C_{12}H_{23}N$	Dicyclohexylamine			10.4
$C_{11}H_{11}N$	Methyl-1-naphthylamine		27	3.67	$C_{12}H_{27}N$	Dodecylamine		25	10.63
$C_{11}H_{12}I_3NO_2$	Iopanoic acid			4.8	$C_{13}H_9N$	Acridine		20	5.58
$C_{11}H_{12}N_2O_2$	L-Tryptophan	1	25	2.46	$C_{13}H_9N$	Phenanthridine		20	5.58
		2	25	9.41	$C_{13}H_{10}N_2$	9-Acridinamine		20	9.99
$C_{11}H_{12}N_4O_3S$	Sulfamethoxypyridazine			6.7	$C_{13}H_{10}N_2$	2-Phenylbenzimidazole	1	25	5.23
$C_{11}H_{13}F_3N_2O_3S$	Mefluidide			4.6			2	25	11.91
$C_{11}H_{13}NO_3$	Hydrastinine			11.38	$C_{13}H_{10}O_2$	2-Phenylbenzoic acid		25	3.46
$C_{11}H_{13}N_3O_3S$	Sulfisoxazole			5	$C_{13}H_{10}O_3$	2-Phenoxybenzoic acid		25	3.53
$C_{11}H_{14}N_2O$	Cytisine	1		6.11	$C_{13}H_{10}O_3$	3-Phenoxybenzoic acid		25	3.95
		2		13.08	$C_{13}H_{10}O_3$	4-Phenoxybenzoic acid		25	4.57
$C_{11}H_{14}O_2$	2-$tert$-Butylbenzoic acid		25	3.54	$C_{13}H_{11}N_3$	3,6-Acridinediamine		20	9.65
$C_{11}H_{14}O_2$	3-$tert$-Butylbenzoic acid		25	4.20	$C_{13}H_{12}Cl_2O_4$	Ethacrynic acid			3.50
$C_{11}H_{14}O_2$	4-$tert$-Butylbenzoic acid		25	4.38	$C_{13}H_{12}N_2O$	Harmine			7.70
$C_{11}H_{16}N_2O_2$	Pilocarpine	1	25	1.6	$C_{13}H_{12}N_2O_3S$	Sulfabenzamide		25	4.57
		2	25	6.9	$C_{13}H_{13}N$	4-Benzylaniline		25	2.17
$C_{11}H_{16}N_4O_4$	Pentostatin			5.2	$C_{13}H_{14}N_2O_{13}$	Harmaline			4.2
$C_{11}H_{17}N$	N,N-Diethyl-2-methyl-aniline		25	7.24	$C_{13}H_{15}N_3O_3$	Imazapyr	1		1.9
							2		3.6

Mol. Form.	Name	Step	$t/^\circ C$	pK_a	Mol. Form.	Name	Step	$t/^\circ C$	pK_a
$C_{13}H_{16}ClNO$	Ketamine			7.5	$C_{18}H_{33}ClN_2O_5S$	Clindamycin			7.6
$C_{13}H_{19}NO_4S$	4-[(Dipropylamino)-sulfonyl]benzoic acid			5.8	$C_{18}H_{39}N$	Octadecylamine		25	10.60
$C_{13}H_{21}N$	2,6-Di-*tert*-butylpyridine			3.58	$C_{19}H_{10}Br_4O_5S$	Bromophenol Blue			4.0
$C_{13}H_{29}N$	(Tridecyl)amine		25	10.63	$C_{19}H_{14}O_5S$	Phenol Red			7.9
$C_{14}H_{12}F_3NO_4S_2$	Perfluidone			2.5	$C_{19}H_{16}ClNO_4$	Indomethacin			4.5
$C_{14}H_{12}O_2$	α-Phenylbenzeneacetic acid		25	3.94	$C_{19}H_{17}N_3O_4S_2$	Cephaloridine			3.2
$C_{14}H_{12}O_3$	α-Hydroxy-α-phenyl-benzeneacetic acid		25	3.04	$C_{19}H_{20}N_2O_2$	Phenylbutazone			4.5
					$C_{19}H_{21}N$	Protriptyline			8.2
$C_{14}H_{18}N_4O_3$	Trimethoprim			6.6	$C_{19}H_{21}NO_3$	Thebaine		15	6.05
$C_{14}H_{19}NO_2$	Methylphenidate			8.9	$C_{19}H_{22}N_2O$	Cinchonine	1		5.85
$C_{14}H_{21}N_3O_3S$	Tolazamide		25	3.6			2		9.92
$C_{14}H_{22}N_2O_3$	Atenolol			9.6	$C_{19}H_{22}N_2O$	Cinchonidine	1		5.80
$C_{14}H_{31}N$	Tetradecylamine		25	10.62			2		10.03
$C_{15}H_{10}ClN_3O_3$	Clonazepam	1		1.5	$C_{19}H_{22}N_2O_2$	Cupreine			6.57
		2		10.5	$C_{19}H_{22}O_6$	Gibberellic acid			4.0
$C_{15}H_{11}I_4NO_4$	*L*-Thyroxine	1	25	2.2	$C_{19}H_{23}N_3O_2$	Ergometrinine			7.3
		2	25	6.45	$C_{19}H_{23}N_3O_2$	Ergonovine			6.8
		3	25	10.1	$C_{20}H_{14}O_4$	Phenolphthalein		25	9.7
$C_{15}H_{14}O_3$	Fenoprofen			7.3	$C_{20}H_{21}NO_4$	Papaverine			6.4
$C_{15}H_{15}NO_2$	Mefenamic acid			4.2	$C_{20}H_{23}N$	Amitriptyline			9.4
$C_{15}H_{15}N_3O_2$	Methyl Red	1		2.5	$C_{20}H_{23}N_7O_7$	Folinic acid	1		3.1
		2		9.5			2		4.8
$C_{15}H_{17}ClN_4$	NeutralRed			6.7			3		10.4
$C_{15}H_{19}NO_2$	Tropacocaine		15	4.32	$C_{20}H_{24}N_2O_2$	Quinine	1	25	8.52
$C_{15}H_{19}N_3O_3$	Imazethapyr	1		2.1			2	25	4.13
		2		3.9	$C_{20}H_{24}N_2O_2$	Quinidine	1	20	5.4
$C_{15}H_{21}N_3O_2$	Physostigmine	1		6.12			2	20	10.0
		2		12.24	$C_{20}H_{26}N_2O_2$	Hydroquinine			5.33
$C_{15}H_{26}N_2$	Sparteine	1	20	2.24	$C_{21}H_{14}Br_4O_5S$	Bromocresol Green			4.7
		2	20	9.46	$C_{21}H_{16}Br_2O_5S$	Bromocresol Purple			6.3
$C_{15}H_{33}N$	Pentadecylamine		25	10.61	$C_{21}H_{18}O_5S$	CresolRed			8.3
$C_{16}H_{13}ClN_2O$	Valium			3.4	$C_{21}H_{21}NO_6$	Hydrastine			7.8
$C_{16}H_{14}ClN_3O$	Chlorodiazepoxide			4.8	$C_{21}H_{22}N_2O_2$	Strychnine		25	8.26
$C_{16}H_{16}N_2O_2$	Lysergic acid	1		3.44	$C_{21}H_{23}ClFNO_2$	Haloperidol			8.3
		2		7.68	$C_{21}H_{31}NO_4$	Furethidine			7.48
$C_{16}H_{17}N_3O_4S$	Cephalexin	1		5.2	$C_{21}H_{35}N_3O_7$	Lisinopril	1		2.5
		2		7.3			2		4.0
$C_{16}H_{19}N_3O_4S$	Cephradine	1		2.63			3		6.7
		2		7.27			4		10.1
$C_{16}H_{22}N_2$	Lycodine	1		3.97	$C_{22}H_{18}O_4$	*o*-Cresolphthalein			9.4
		2		8.08	$C_{22}H_{22}FN_3O_2$	Droperidol			7.64
$C_{16}H_{35}N$	Hexadecylamine		25	10.61	$C_{22}H_{23}NO_7$	Noscapine			7.8
$C_{17}H_{17}NO_2$	Apomorphine	1		7.0	$C_{22}H_{25}NO_6$	Colchicine		20	12.36
		2		8.92	$C_{22}H_{25}N_3O$	Benzpiperylon	1		6.73
$C_{17}H_{19}NO_3$	Piperine		18	12.22			2		9.13
$C_{17}H_{19}NO_3$	Morphine	1	25	8.21	$C_{22}H_{33}NO_2$	Atisine			12.2
		2	20	9.85	$C_{23}H_{26}N_2O_4$	Brucine	1		6.04
$C_{17}H_{20}N_4O_6$	Riboflavin	1		1.7			2		11.07
		2	25	9.69	$C_{24}H_{40}O_4$	Deoxycholic acid			6.58
$C_{17}H_{20}O_6$	Mycophenolic acid			4.5	$C_{24}H_{40}O_5$	Cholic acid			6.4
$C_{17}H_{23}NO_3$	Hyoscyamine		21	9.7	$C_{25}H_{29}I_2NO_3$	Amiodarone		25	6.56
$C_{17}H_{27}NO_4$	Nadolol			9.67	$C_{25}H_{41}NO_9$	Aconine			9.52
$C_{18}H_{19}ClN_4$	Clozapine	1		3.70	$C_{26}H_{43}NO_6$	Glycocholic acid			4.4
		2		7.60	$C_{26}H_{45}NO_7S$	Taurocholic acid			1.4
$C_{18}H_{21}NO_3$	Codeine			8.21	$C_{27}H_{28}Br_2O_5S$	Bromothymol Blue			7.0
$C_{18}H_{21}N_3O$	Dibenzepin			8.25	$C_{27}H_{38}N_2O_4$	Verapamil			8.6
$C_{18}H_{32}O_2$	Linoleic acid			7.6	$C_{29}H_{32}O_{13}$	Etoposide			9.8
					$C_{29}H_{40}N_2O_4$	Emetine	1		5.77
							2		6.64

DISSOCIATION CONSTANTS OF ORGANIC ACIDS AND BASES (continued)

Mol. Form.	Name	Step	$t/°C$	pK_a	Mol. Form.	Name	Step	$t/°C$	pK_a
$C_{30}H_{23}BrO_4$	Bromadiolone		21	4.04	$C_{36}H_{51}NO_{11}$	Veratridine			9.54
$C_{30}H_{48}O_3$	Oleanolic acid			2.52	$C_{37}H_{67}NO_{13}$	Erythromycin			8.8
$C_{31}H_{36}N_2O_{11}$	Novobiocin	1		4.3	$C_{43}H_{58}N_4O_{12}$	Rifampin	1		1.7
		2		9.1			2		7.9
$C_{32}H_{32}O_{13}S$	Teniposide			10.13	$C_{45}H_{73}NO_{15}$	Solanine		15	6.66
$C_{33}H_{40}N_2O_9$	Reserpine			6.6	$C_{46}H_{56}N_4O_{10}$	Vincristine			5.4
$C_{34}H_{47}NO_{11}$	Aconitine			5.88	$C_{46}H_{58}N_4O_9$	Vinblastine	1		5.4
							2		7.4

CONCENTRATIVE PROPERTIES OF AQUEOUS SOLUTIONS: DENSITY, REFRACTIVE INDEX, FREEZING POINT DEPRESSION, AND VISCOSITY

This table gives properties of aqueous solutions of 66 substances as a function of concentration. All data refer to a temperature of 20°C. The properties are:

Mass %: Mass of solute divided by total mass of solution, expressed as percent.
m Molality (moles of solute per kg of water).
c Molarity (moles of solute per liter of solution).
ρ Density of solution in g/cm^3.
n Index of refraction, relative to air, at a wavelength of 589 nm (sodium D line); the index of pure water at 20°C is 1.3330.
Δ Freezing point depression in °C relative to pure water.
η Absolute (dynamic) viscosity in mPa s (equal to centipoise, cP); the viscosity of pure water at 20°C is 1.002 mPa s.

Density data for aqueous solutions over a wider range of temperatures and pressures (and for other compounds) may be found in Reference 2. Solutes are listed in the following order:

Acetic acid	Lithium chloride	2-Propanol
Acetone	Magnesium chloride	Silver nitrate
Ammonia	Magnesium sulfate	Sodium acetate
Ammonium chloride	Maltose	Sodium bicarbonate
Ammonium sulfate	Manganese(II) sulfate	Sodium bromide
Barium chloride	D-Mannitol	Sodium carbonate
Calcium chloride	Methanol	Sodium chloride
Cesium chloride	Nitric acid	Sodium citrate
Citric acid	Oxalic acid	Sodium hydroxide
Copper sulfate	Phosphoric acid	Sodium nitrate
Disodium ethylenediamine	Potassium bicarbonate	Sodium phosphate
tetraacetate (EDTA sodium)	Potassium bromide	Sodium hydrogen phosphate
Ethanol	Potassium carbonate	Sodium dihydrogen phosphate
Ethylene glycol	Potassium chloride	Sodium sulfate
Ferric chloride	Potassium hydroxide	Sodium thiosulfate
Formic acid	Potassium iodide	Strontium chloride
D-Fructose	Potassium nitrate	Sucrose
D-Glucose	Potassium permanganate	Sulfuric acid
Glycerol	Potassium hydrogen phosphate	Trichloroacetic acid
Hydrochloric acid	Potassium dihydrogen phosphate	Tris(hydroxymethyl)methylamine
Lactic acid	Potassium sulfate	Urea
Lactose	1-Propanol	Zinc sulfate

REFERENCES

1. Wolf, A. V., Aqueous Solutions and Body Fluids, Hoeber, 1966.
2. Söhnel, O., and Novotny, P., *Densities of Aqueous Solutions of Inorganic Substances*, Elsevier, Amsterdam, 1985.

Solute	Mass %	m/mol kg^{-1}	c/mol L^{-1}	ρ/g cm^{-3}	n	Δ/°C	η/mPa s
Acetic acid	0.5	0.084	0.083	0.9989	1.3334	0.16	1.012
CH$_3$COOH	1.0	0.168	0.166	0.9996	1.3337	0.32	1.022
	2.0	0.340	0.333	1.0011	1.3345	0.63	1.042
	3.0	0.515	0.501	1.0025	1.3352	0.94	1.063
	4.0	0.694	0.669	1.0038	1.3359	1.26	1.084
	5.0	0.876	0.837	1.0052	1.3366	1.58	1.105
	6.0	1.063	1.006	1.0066	1.3373	1.90	1.125
	7.0	1.253	1.175	1.0080	1.3381	2.23	1.143
	8.0	1.448	1.345	1.0093	1.3388	2.56	1.162
	9.0	1.647	1.515	1.0107	1.3395	2.89	1.186
	10.0	1.850	1.685	1.0121	1.3402	3.23	1.210
	12.0	2.271	2.028	1.0147	1.3416	3.91	1.253
	14.0	2.711	2.372	1.0174	1.3430	4.61	1.298
	16.0	3.172	2.718	1.0200	1.3444	5.33	1.341
	18.0	3.655	3.065	1.0225	1.3458	6.06	1.380
	20.0	4.163	3.414	1.0250	1.3472	6.81	1.431

Solute	Mass %	m/mol kg^{-1}	c/mol L^{-1}	ρ/g cm^{-3}	n	Δ/°C	η/mPa s
	22.0	4.697	3.764	1.0275	1.3485	7.57	1.478
	24.0	5.259	4.116	1.0299	1.3498	8.36	1.525
	26.0	5.851	4.470	1.0323	1.3512	9.17	1.572
	28.0	6.476	4.824	1.0346	1.3525	10.00	1.613
	30.0	7.137	5.180	1.0369	1.3537	10.84	1.669
	32.0	7.837	5.537	1.0391	1.3550	11.70	1.715
	34.0	8.579	5.896	1.0413	1.3562	12.55	1.762
	36.0	9.367	6.255	1.0434	1.3574	13.38	1.812
	38.0	10.207	6.615	1.0454	1.3586		1.852
	40.0	11.102	6.977	1.0474	1.3598		1.912
	50.0	16.653	8.794	1.0562	1.3653		2.158
	60.0	24.979	10.620	1.0629	1.3700		2.409
	70.0	38.857	12.441	1.0673	1.3738		2.629
	80.0	66.611	14.228	1.0680	1.3767		2.720
	90.0	149.875	15.953	1.0644	1.3771		2.386
	92.0	191.507	16.284	1.0629	1.3766		2.240
	94.0	260.894	16.602	1.0606	1.3759		2.036
	96.0	399.667	16.911	1.0578	1.3748		1.813
	98.0	815.987	17.198	1.0538	1.3734		1.535
	100.0		17.447	1.0477	1.3716		1.223
Acetone	0.5	0.087	0.086	0.9975	1.3334	0.16	1.013
$(CH_3)_2CO$	1.0	0.174	0.172	0.9968	1.3337	0.32	1.024
	2.0	0.351	0.343	0.9954	1.3344	0.65	1.047
	3.0	0.533	0.513	0.9940	1.3352	0.97	1.072
	4.0	0.717	0.684	0.9926	1.3359	1.30	1.099
	5.0	0.906	0.853	0.9912	1.3366	1.63	1.125
	6.0	1.099	1.023	0.9899	1.3373	1.96	1.150
	7.0	1.296	1.191	0.9886	1.3381	2.29	1.174
	8.0	1.497	1.360	0.9874	1.3388	2.62	1.198
	9.0	1.703	1.528	0.9861	1.3395	2.95	1.221
	10.0	1.913	1.696	0.9849	1.3402	3.29	1.244
Ammonia	0.5	0.295	0.292	0.9960	1.3332	0.55	1.009
NH_3	1.0	0.593	0.584	0.9938	1.3335	1.14	1.015
	2.0	1.198	1.162	0.9895	1.3339	2.32	1.029
	3.0	1.816	1.736	0.9853	1.3344	3.53	1.043
	4.0	2.447	2.304	0.9811	1.3349	4.78	1.057
	5.0	3.090	2.868	0.9770	1.3354	6.08	1.071
	6.0	3.748	3.428	0.9730	1.3359	7.43	1.085
	7.0	4.420	3.983	0.9690	1.3365	8.95	1.099
	8.0	5.106	4.533	0.9651	1.3370	10.34	1.113
	9.0	5.807	5.080	0.9613	1.3376	11.90	1.127
	10.0	6.524	5.622	0.9575	1.3381	13.55	1.141
	12.0	8.007	6.695	0.9502	1.3393	17.13	1.169
	14.0	9.558	7.753	0.9431	1.3404	21.13	1.195
	16.0	11.184	8.794	0.9361	1.3416	25.63	1.218
	18.0	12.889	9.823	0.9294	1.3428	30.70	1.237
	20.0	14.679	10.837	0.9228	1.3440	36.42	1.254
	22.0	16.561	11.838	0.9164	1.3453	43.36	1.268
	24.0	18.542	12.826	0.9102	1.3465	51.38	1.280
	26.0	20.630	13.801	0.9040	1.3477	60.77	1.288
	28.0	22.834	14.764	0.8980	1.3490	71.66	
	30.0	25.164	15.713	0.8920	1.3502	84.06	
Ammonium	0.5	0.094	0.093	0.9998	1.3340	0.32	0.999
chloride	1.0	0.189	0.187	1.0014	1.3349	0.64	0.996
NH_4Cl	2.0	0.382	0.376	1.0045	1.3369	1.27	0.992

Solute	Mass %	m/mol kg^{-1}	c/mol L^{-1}	ρ/g cm^{-3}	n	$\Delta/°C$	η/mPa s
	3.0	0.578	0.565	1.0076	1.3388	1.91	0.988
	4.0	0.779	0.756	1.0107	1.3407	2.57	0.985
	5.0	0.984	0.948	1.0138	1.3426	3.25	0.982
	6.0	1.193	1.141	1.0168	1.3445	3.94	0.979
	7.0	1.407	1.335	1.0198	1.3464	4.66	0.976
	8.0	1.626	1.529	1.0227	1.3483	5.40	0.974
	9.0	1.849	1.726	1.0257	1.3502	6.16	0.972
	10.0	2.077	1.923	1.0286	1.3521	6.95	0.970
	12.0	2.549	2.320	1.0344	1.3559	8.60	0.969
	14.0	3.043	2.722	1.0401	1.3596		0.969
	16.0	3.561	3.128	1.0457	1.3634		0.971
	18.0	4.104	3.537	1.0512	1.3671		0.973
	20.0	4.674	3.951	1.0567	1.3708		0.978
	22.0	5.273	4.368	1.0621	1.3745		0.986
	24.0	5.903	4.789	1.0674	1.3782		0.996
Ammonium	0.5	0.038	0.038	1.0012	1.3338	0.17	1.008
sulfate	1.0	0.076	0.076	1.0042	1.3346	0.33	1.014
$(NH_4)_2SO_4$	2.0	0.154	0.153	1.0101	1.3363	0.63	1.027
	3.0	0.234	0.231	1.0160	1.3379	0.92	1.041
	4.0	0.315	0.309	1.0220	1.3395	1.21	1.057
	5.0	0.398	0.389	1.0279	1.3411	1.49	1.073
	6.0	0.483	0.469	1.0338	1.3428	1.77	1.090
	7.0	0.570	0.551	1.0397	1.3444	2.05	1.108
	8.0	0.658	0.633	1.0456	1.3460	2.33	1.127
	9.0	0.748	0.716	1.0515	1.3476	2.61	1.147
	10.0	0.841	0.800	1.0574	1.3492	2.89	1.168
	12.0	1.032	0.971	1.0691	1.3523	3.47	1.210
	14.0	1.232	1.145	1.0808	1.3555	4.07	1.256
	16.0	1.441	1.323	1.0924	1.3586	4.69	1.305
	18.0	1.661	1.504	1.1039	1.3616		1.359
	20.0	1.892	1.688	1.1154	1.3647		1.421
	22.0	2.134	1.876	1.1269	1.3677		1.490
	24.0	2.390	2.067	1.1383	1.3707		1.566
	26.0	2.659	2.262	1.1496	1.3737		1.650
	28.0	2.943	2.460	1.1609	1.3766		1.743
	30.0	3.243	2.661	1.1721	1.3795		1.847
	32.0	3.561	2.866	1.1833	1.3824		1.961
	34.0	3.898	3.073	1.1945	1.3853		2.086
	36.0	4.257	3.284	1.2056	1.3881		2.222
	38.0	4.638	3.499	1.2166	1.3909		2.371
	40.0	5.045	3.716	1.2277	1.3938		2.530
Barium	0.5	0.024	0.024	1.0026	1.3337	0.12	1.009
chloride	1.0	0.049	0.048	1.0070	1.3345	0.23	1.016
$BaCl_2$	2.0	0.098	0.098	1.0159	1.3360	0.46	1.026
	3.0	0.149	0.148	1.0249	1.3375	0.69	1.037
	4.0	0.200	0.199	1.0341	1.3391	0.93	1.049
	5.0	0.253	0.251	1.0434	1.3406	1.18	1.062
	6.0	0.307	0.303	1.0528	1.3422	1.44	1.075
	7.0	0.361	0.357	1.0624	1.3438	1.70	1.087
	8.0	0.418	0.412	1.0721	1.3454	1.98	1.101
	9.0	0.475	0.468	1.0820	1.3470	2.27	1.114
	10.0	0.534	0.524	1.0921	1.3487	2.58	1.129
	12.0	0.655	0.641	1.1128	1.3520	3.22	1.161
	14.0	0.782	0.763	1.1342	1.3555	3.92	1.195
	16.0	0.915	0.889	1.1564	1.3591	4.69	1.234

Solute	Mass %	m/mol kg^{-1}	c/mol L^{-1}	ρ/g cm^{-3}	n	Δ/$^\circ$C	η/mPa s
	18.0	1.054	1.019	1.1793	1.3627		1.277
	20.0	1.201	1.156	1.2031	1.3664		1.325
	22.0	1.355	1.297	1.2277	1.3703		1.378
	24.0	1.517	1.444	1.2531	1.3741		1.437
	26.0	1.687	1.597	1.2793	1.3781		1.503
Calcium	0.5	0.045	0.045	1.0024	1.3342	0.22	1.015
chloride	1.0	0.091	0.091	1.0065	1.3354	0.44	1.028
CaCl$_2$	2.0	0.184	0.183	1.0148	1.3378	0.88	1.050
	3.0	0.279	0.277	1.0232	1.3402	1.33	1.078
	4.0	0.375	0.372	1.0316	1.3426	1.82	1.110
	5.0	0.474	0.469	1.0401	1.3451	2.35	1.143
	6.0	0.575	0.567	1.0486	1.3475	2.93	1.175
	7.0	0.678	0.667	1.0572	1.3500	3.57	1.208
	8.0	0.784	0.768	1.0659	1.3525	4.28	1.242
	9.0	0.891	0.872	1.0747	1.3549	5.04	1.279
	10.0	1.001	0.976	1.0835	1.3575	5.86	1.319
	12.0	1.229	1.191	1.1014	1.3625	7.70	1.408
	14.0	1.467	1.413	1.1198	1.3677	9.83	1.508
	16.0	1.716	1.641	1.1386	1.3730	12.28	1.625
	18.0	1.978	1.878	1.1579	1.3784	15.11	1.764
	20.0	2.253	2.122	1.1775	1.3839	18.30	1.930
	22.0	2.541	2.374	1.1976	1.3895	21.70	2.127
	24.0	2.845	2.634	1.2180	1.3951	25.30	2.356
	26.0	3.166	2.902	1.2388	1.4008	29.70	2.645
	28.0	3.504	3.179	1.2600	1.4066	34.70	3.000
	30.0	3.862	3.464	1.2816	1.4124	41.00	3.467
	32.0	4.240	3.759	1.3036	1.4183	49.70	4.035
	34.0	4.642	4.062	1.3260	1.4242		4.820
	36.0	5.068	4.375	1.3488	1.4301		5.807
	38.0	5.522	4.698	1.3720	1.4361		7.321
	40.0	6.007	5.030	1.3957	1.4420		8.997
Cesium	0.5	0.030	0.030	1.0020	1.3334	0.10	1.000
chloride	1.0	0.060	0.060	1.0058	1.3337	0.20	0.997
CsCl	2.0	0.121	0.120	1.0135	1.3345	0.40	0.992
	3.0	0.184	0.182	1.0214	1.3353	0.61	0.988
	4.0	0.247	0.245	1.0293	1.3361	0.81	0.984
	5.0	0.313	0.308	1.0374	1.3369	1.02	0.980
	6.0	0.379	0.373	1.0456	1.3377	1.22	0.977
	7.0	0.447	0.438	1.0540	1.3386	1.43	0.974
	8.0	0.516	0.505	1.0625	1.3394	1.64	0.971
	9.0	0.587	0.573	1.0711	1.3403	1.85	0.969
	10.0	0.660	0.641	1.0798	1.3412	2.06	0.966
	12.0	0.810	0.782	1.0978	1.3430	2.51	0.961
	14.0	0.967	0.928	1.1163	1.3448	2.97	0.955
	16.0	1.131	1.079	1.1355	1.3468	3.46	0.950
	18.0	1.304	1.235	1.1552	1.3487	3.96	0.945
	20.0	1.485	1.397	1.1756	1.3507	4.49	0.939
	22.0	1.675	1.564	1.1967	1.3528		0.934
	24.0	1.876	1.737	1.2185	1.3550		0.930
	26.0	2.087	1.917	1.2411	1.3572		0.926
	28.0	2.310	2.103	1.2644	1.3594		0.924
	30.0	2.546	2.296	1.2885	1.3617		0.922
	32.0	2.795	2.497	1.3135	1.3641		0.922
	34.0	3.060	2.705	1.3393	1.3666		0.924
	36.0	3.341	2.921	1.3661	1.3691		0.926

Solute	Mass %	m/mol kg^{-1}	c/mol L^{-1}	ρ/g cm^{-3}	n	Δ/°C	η/mPa s
	38.0	3.640	3.146	1.3938	1.3717		0.930
	40.0	3.960	3.380	1.4226	1.3744		0.934
	42.0	4.301	3.624	1.4525	1.3771		0.940
	44.0	4.667	3.877	1.4835	1.3800		0.947
	46.0	5.060	4.142	1.5158	1.3829		0.956
	48.0	5.483	4.418	1.5495	1.3860		0.967
	50.0	5.940	4.706	1.5846	1.3892		0.981
	60.0	8.910	6.368	1.7868	1.4076		1.120
	64.0	10.560	7.163	1.8842	1.4167		1.238
Citric acid	0.5	0.026	0.026	1.0002	1.3336	0.05	1.013
(HO)C(COOH)$_3$	1.0	0.053	0.052	1.0022	1.3343	0.11	1.024
	2.0	0.106	0.105	1.0063	1.3356	0.21	1.048
	3.0	0.161	0.158	1.0105	1.3368	0.32	1.073
	4.0	0.217	0.211	1.0147	1.3381	0.43	1.098
	5.0	0.274	0.265	1.0189	1.3394	0.54	1.125
	6.0	0.332	0.320	1.0232	1.3407	0.65	1.153
	7.0	0.392	0.374	1.0274	1.3420	0.76	1.183
	8.0	0.453	0.430	1.0316	1.3433	0.88	1.214
	9.0	0.515	0.485	1.0359	1.3446	1.00	1.247
	10.0	0.578	0.541	1.0402	1.3459	1.12	1.283
	12.0	0.710	0.655	1.0490	1.3486	1.38	1.357
	14.0	0.847	0.771	1.0580	1.3514	1.66	1.436
	16.0	0.991	0.889	1.0672	1.3541	1.95	1.525
	18.0	1.143	1.008	1.0764	1.3569	2.26	1.625
	20.0	1.301	1.130	1.0858	1.3598	2.57	1.740
	22.0	1.468	1.254	1.0953	1.3626	2.88	1.872
	24.0	1.644	1.380	1.1049	1.3655	3.21	2.017
	26.0	1.829	1.508	1.1147	1.3684	3.55	2.178
	28.0	2.024	1.639	1.1246	1.3714	3.89	2.356
	30.0	2.231	1.772	1.1346	1.3744	4.25	2.549
Copper	0.5	0.031	0.031	1.0033	1.3339	0.08	1.017
sulfate	1.0	0.063	0.063	1.0085	1.3348	0.14	1.036
CuSO$_4$	2.0	0.128	0.128	1.0190	1.3367	0.26	1.084
	3.0	0.194	0.194	1.0296	1.3386	0.37	1.129
	4.0	0.261	0.261	1.0403	1.3405	0.48	1.173
	5.0	0.330	0.329	1.0511	1.3424	0.59	1.221
	6.0	0.400	0.399	1.0620	1.3443	0.70	1.276
	7.0	0.472	0.471	1.0730	1.3462	0.82	1.336
	8.0	0.545	0.543	1.0842	1.3481	0.93	1.400
	9.0	0.620	0.618	1.0955	1.3501	1.05	1.469
	10.0	0.696	0.694	1.1070	1.3520	1.18	1.543
	12.0	0.854	0.850	1.1304	1.3560	1.45	1.701
	14.0	1.020	1.013	1.1545	1.3601	1.75	1.889
	16.0	1.193	1.182	1.1796	1.3644		2.136
	18.0	1.375	1.360	1.2059	1.3689		2.449
Disodium	0.5	0.015	0.015	1.0009	1.3339	0.07	1.017
ethylenediamine	1.0	0.030	0.030	1.0036	1.3348	0.14	1.032
tetraacetate	1.5	0.045	0.045	1.0062	1.3356	0.21	1.046
(EDTA sodium)	2.0	0.061	0.060	1.0089	1.3365	0.27	1.062
Na$_2$C$_{10}$H$_{14}$N$_2$O$_8$	2.5	0.076	0.075	1.0115	1.3374	0.33	1.077
	3.0	0.092	0.090	1.0142	1.3383	0.40	1.093
	3.5	0.108	0.106	1.0169	1.3392	0.46	1.109
	4.0	0.124	0.121	1.0196	1.3400	0.52	1.125
	4.5	0.140	0.137	1.0223	1.3409	0.58	1.142

CONCENTRATIVE PROPERTIES OF AQUEOUS SOLUTIONS:
DENSITY, REFRACTIVE INDEX, FREEZING POINT DEPRESSION, AND VISCOSITY (continued)

Solute	Mass %	m/mol kg^{-1}	c/mol L^{-1}	ρ/g cm^{-3}	n	Δ/°C	η/mPa s
	5.0	0.157	0.152	1.0250	1.3418	0.65	1.160
	5.5	0.173	0.168	1.0277	1.3427	0.71	1.178
	6.0	0.190	0.184	1.0305	1.3436	0.77	1.197
Ethanol	0.5	0.109	0.108	0.9973	1.3333	0.20	1.023
CH$_3$CH$_2$OH	1.0	0.219	0.216	0.9963	1.3336	0.40	1.046
	2.0	0.443	0.432	0.9945	1.3342	0.81	1.095
	3.0	0.671	0.646	0.9927	1.3348	1.23	1.140
	4.0	0.904	0.860	0.9910	1.3354	1.65	1.183
	5.0	1.142	1.074	0.9893	1.3360	2.09	1.228
	6.0	1.385	1.286	0.9878	1.3367	2.54	1.279
	7.0	1.634	1.498	0.9862	1.3374	2.99	1.331
	8.0	1.887	1.710	0.9847	1.3381	3.47	1.385
	9.0	2.147	1.921	0.9833	1.3388	3.96	1.442
	10.0	2.412	2.131	0.9819	1.3395	4.47	1.501
	12.0	2.960	2.551	0.9792	1.3410	5.56	1.627
	14.0	3.534	2.967	0.9765	1.3425	6.73	1.761
	16.0	4.134	3.382	0.9739	1.3440	8.01	1.890
	18.0	4.765	3.795	0.9713	1.3455	9.40	2.019
	20.0	5.427	4.205	0.9687	1.3469	10.92	2.142
	22.0	6.122	4.613	0.9660	1.3484	12.60	2.259
	24.0	6.855	5.018	0.9632	1.3498	14.47	2.370
	26.0	7.626	5.419	0.9602	1.3511	16.41	2.476
	28.0	8.441	5.817	0.9571	1.3524	18.43	2.581
	30.0	9.303	6.212	0.9539	1.3535	20.47	2.667
	32.0	10.215	6.601	0.9504	1.3546	22.44	2.726
	34.0	11.182	6.987	0.9468	1.3557	24.27	2.768
	36.0	12.210	7.370	0.9431	1.3566	25.98	2.803
	38.0	13.304	7.747	0.9392	1.3575	27.62	2.829
	40.0	14.471	8.120	0.9352	1.3583	29.26	2.846
	42.0	15.718	8.488	0.9311	1.3590	30.98	2.852
	44.0	17.055	8.853	0.9269	1.3598	32.68	2.850
	46.0	18.490	9.213	0.9227	1.3604	34.36	2.843
	48.0	20.036	9.568	0.9183	1.3610	36.04	2.832
	50.0	21.706	9.919	0.9139	1.3616	37.67	2.813
	60.0	32.559	11.605	0.8911	1.3638	44.93	2.547
	70.0	50.648	13.183	0.8676	1.3652		2.214
	80.0	86.824	14.649	0.8436	1.3658		1.881
	90.0	195.355	15.980	0.8180	1.3650		1.542
	92.0	249.620	16.225	0.8125	1.3646		1.475
	94.0	340.062	16.466	0.8070	1.3642		1.407
	96.0	520.946	16.697	0.8013	1.3636		1.342
	98.0		16.920	0.7954	1.3630		1.273
	100.0		17.133	0.7893	1.3614		1.203
Ethylene	0.5	0.081	0.080	0.9988	1.3335	0.15	1.010
glycol	1.0	0.163	0.161	0.9995	1.3339	0.30	1.020
(CH$_2$OH)$_2$	2.0	0.329	0.322	1.0007	1.3348	0.61	1.048
	3.0	0.498	0.484	1.0019	1.3358	0.92	1.074
	4.0	0.671	0.646	1.0032	1.3367	1.24	1.099
	5.0	0.848	0.809	1.0044	1.3377	1.58	1.125
	6.0	1.028	0.972	1.0057	1.3386	1.91	1.153
	7.0	1.213	1.136	1.0070	1.3396	2.26	1.182
	8.0	1.401	1.299	1.0082	1.3405	2.62	1.212
	9.0	1.593	1.464	1.0095	1.3415	2.99	1.243
	10.0	1.790	1.628	1.0108	1.3425	3.37	1.277
	12.0	2.197	1.959	1.0134	1.3444	4.16	1.348

Solute	Mass %	m/mol kg^{-1}	c/mol L^{-1}	ρ/g cm^{-3}	n	Δ/°C	η/mPa s
	14.0	2.623	2.292	1.0161	1.3464	5.01	1.424
	16.0	3.069	2.626	1.0188	1.3484	5.91	1.500
	18.0	3.537	2.962	1.0214	1.3503	6.89	1.578
	20.0	4.028	3.300	1.0241	1.3523	7.93	1.661
	24.0	5.088	3.981	1.0296	1.3564	10.28	1.843
	28.0	6.265	4.669	1.0350	1.3605	13.03	2.047
	32.0	7.582	5.364	1.0405	1.3646	16.23	2.280
	36.0	9.062	6.067	1.0460	1.3687	19.82	2.537
	40.0	10.741	6.776	1.0514	1.3728	23.84	2.832
	44.0	12.659	7.491	1.0567	1.3769	28.32	3.166
	48.0	14.872	8.212	1.0619	1.3811	33.30	3.544
	52.0	17.453	8.939	1.0670	1.3851	38.81	3.981
	56.0	20.505	9.671	1.0719	1.3892	44.83	4.475
	60.0	24.166	10.406	1.0765	1.3931	51.23	5.026
Ferric	0.5	0.031	0.031	1.0025	1.3344	0.21	1.024
chloride	1.0	0.062	0.062	1.0068	1.3358	0.39	1.047
FeCl$_3$	2.0	0.126	0.125	1.0153	1.3386	0.75	1.093
	3.0	0.191	0.189	1.0238	1.3413	1.15	1.139
	4.0	0.257	0.255	1.0323	1.3441	1.56	1.187
	5.0	0.324	0.321	1.0408	1.3468	2.00	1.238
	6.0	0.394	0.388	1.0493	1.3496	2.48	1.292
	7.0	0.464	0.457	1.0580	1.3524	2.99	1.350
	8.0	0.536	0.526	1.0668	1.3552	3.57	1.412
	9.0	0.610	0.597	1.0760	1.3581	4.19	1.480
	10.0	0.685	0.669	1.0853	1.3611	4.85	1.553
	12.0	0.841	0.817	1.1040	1.3670	6.38	1.707
	14.0	1.004	0.969	1.1228	1.3730	8.22	1.879
	16.0	1.174	1.126	1.1420		10.45	2.080
	18.0	1.353	1.289	1.1615		13.08	2.311
	20.0	1.541	1.457	1.1816		16.14	2.570
	24.0	1.947	1.810	1.2234		23.79	3.178
	28.0	2.398	2.189	1.2679		33.61	4.038
	32.0	2.901	2.595	1.3153		49.16	5.274
	36.0	3.468	3.030	1.3654			7.130
	40.0	4.110	3.496	1.4176			9.674
Formic acid	0.5	0.109	0.109	0.9994	1.3333	0.21	1.006
HCOOH	1.0	0.219	0.217	1.0006	1.3336	0.42	1.011
	2.0	0.443	0.436	1.0029	1.3342	0.82	1.017
	3.0	0.672	0.655	1.0053	1.3348	1.24	1.195
	4.0	0.905	0.876	1.0077	1.3354	1.67	1.032
	5.0	1.143	1.097	1.0102	1.3359	2.10	1.039
	6.0	1.387	1.320	1.0126	1.3365	2.53	1.046
	7.0	1.635	1.544	1.0150	1.3371	2.97	1.052
	8.0	1.889	1.768	1.0175	1.3376	3.40	1.058
	9.0	2.149	1.994	1.0199	1.3382	3.84	1.064
	10.0	2.414	2.221	1.0224	1.3387	4.27	1.070
	12.0	2.962	2.678	1.0273	1.3397	5.19	1.082
	14.0	3.537	3.139	1.0322	1.3408	6.11	1.094
	16.0	4.138	3.605	1.0371	1.3418	7.06	1.106
	18.0	4.769	4.074	1.0419	1.3428	8.08	1.119
	20.0	5.431	4.548	1.0467	1.3437	9.11	1.132
	28.0	8.449	6.481	1.0654	1.3475	13.10	1.179
	36.0	12.220	8.477	1.0839	1.3511	17.65	1.227
	44.0	17.070	10.529	1.1015	1.3547	22.93	1.281
	52.0	23.535	12.633	1.1183	1.3581	29.69	1.340

Solute	Mass %	m/mol kg^{-1}	c/mol L^{-1}	ρ/g cm^{-3}	n	$\Delta/°C$	η/mPa s
	60.0	32.587	14.813	1.1364	1.3612	38.26	1.410
	68.0	46.166	17.054	1.1544	1.3641		1.490
D-Fructose	0.5	0.028	0.028	1.0002	1.3337	0.05	1.015
$C_6H_{12}O_6$	1.0	0.056	0.056	1.0021	1.3344	0.10	1.028
	2.0	0.113	0.112	1.0061	1.3358	0.21	1.054
	3.0	0.172	0.168	1.0101	1.3373	0.32	1.080
	4.0	0.231	0.225	1.0140	1.3387	0.43	1.106
	5.0	0.292	0.283	1.0181	1.3402	0.54	1.134
	6.0	0.354	0.340	1.0221	1.3417	0.66	1.165
	7.0	0.418	0.399	1.0262	1.3431	0.78	1.198
	8.0	0.483	0.458	1.0303	1.3446	0.90	1.232
	9.0	0.549	0.517	1.0344	1.3461	1.03	1.270
	10.0	0.617	0.576	1.0385	1.3476	1.16	1.309
	12.0	0.757	0.697	1.0469	1.3507	1.43	1.391
	14.0	0.904	0.820	1.0554	1.3538	1.71	1.483
	16.0	1.057	0.945	1.0640	1.3569	2.01	1.587
	18.0	1.218	1.072	1.0728	1.3601	2.32	1.703
	20.0	1.388	1.201	1.0816	1.3634	2.64	1.837
	22.0	1.566	1.332	1.0906	1.3667	3.05	1.986
	24.0	1.753	1.465	1.0996	1.3700	3.43	2.154
	26.0	1.950	1.600	1.1089	1.3734	3.82	2.348
	28.0	2.159	1.738	1.1182	1.3768	4.20	2.562
	30.0	2.379	1.878	1.1276	1.3803		2.817
	32.0	2.612	2.020	1.1372	1.3839		3.112
	34.0	2.859	2.164	1.1469	1.3874		3.462
	36.0	3.122	2.312	1.1568	1.3911		3.899
	38.0	3.402	2.461	1.1668	1.3948		4.418
	40.0	3.700	2.613	1.1769	1.3985		5.046
	42.0	4.019	2.767	1.1871	1.4023		5.773
	44.0	4.361	2.925	1.1975	1.4062		6.644
	46.0	4.728	3.084	1.2080	1.4101		7.753
	48.0	5.124	3.247	1.2187	1.4141		9.060
D-Glucose	0.5	0.028	0.028	1.0001	1.3337	0.05	1.010
$C_6H_{12}O_6$	1.0	0.056	0.056	1.0020	1.3344	0.11	1.021
	2.0	0.113	0.112	1.0058	1.3358	0.21	1.052
	3.0	0.172	0.168	1.0097	1.3373	0.32	1.083
	4.0	0.231	0.225	1.0136	1.3387	0.43	1.113
	5.0	0.292	0.282	1.0175	1.3402	0.55	1.145
	6.0	0.354	0.340	1.0214	1.3417	0.67	1.179
	7.0	0.418	0.398	1.0254	1.3432	0.79	1.214
	8.0	0.483	0.457	1.0294	1.3447	0.91	1.250
	9.0	0.549	0.516	1.0334	1.3462	1.04	1.289
	10.0	0.617	0.576	1.0375	1.3477	1.17	1.330
	12.0	0.757	0.697	1.0457	1.3508	1.44	1.416
	14.0	0.904	0.819	1.0540	1.3539	1.73	1.512
	16.0	1.057	0.944	1.0624	1.3571	2.03	1.625
	18.0	1.218	1.070	1.0710	1.3603	2.35	1.757
	20.0	1.388	1.199	1.0797	1.3635	2.70	1.904
	22.0	1.566	1.329	1.0884	1.3668	3.07	2.063
	24.0	1.753	1.462	1.0973	1.3702	3.48	2.242
	26.0	1.950	1.597	1.1063	1.3736	3.90	2.458
	28.0	2.159	1.734	1.1154	1.3770	4.34	2.707
	30.0	2.379	1.873	1.1246	1.3805	4.79	2.998
	32.0	2.612	2.014	1.1340	1.3840		3.324
	34.0	2.859	2.158	1.1434	1.3876		3.704

Solute	Mass %	m/mol kg^{-1}	c/mol L^{-1}	ρ/g cm^{-3}	n	Δ/°C	η/mPa s
	36.0	3.122	2.304	1.1529	1.3912		4.193
	38.0	3.402	2.452	1.1626	1.3949		4.786
	40.0	3.700	2.603	1.1724	1.3986		5.493
	42.0	4.019	2.756	1.1823	1.4024		6.288
	44.0	4.361	2.912	1.1924	1.4062		7.235
	46.0	4.728	3.071	1.2026	1.4101		8.454
	48.0	5.124	3.232	1.2130	1.4141		9.883
	50.0	5.551	3.396	1.2235	1.4181		11.884
	52.0	6.013	3.562	1.2342	1.4222		14.489
	54.0	6.516	3.732	1.2451	1.4263		17.916
	56.0	7.064	3.905	1.2562	1.4306		22.886
	58.0	7.665	4.081	1.2676	1.4349		29.389
	60.0	8.326	4.261	1.2793	1.4394		37.445
Glycerol	0.5	0.055	0.054	0.9994	1.3336	0.07	1.011
CH$_2$OHCHOHCH$_2$OH	1.0	0.110	0.109	1.0005	1.3342	0.18	1.022
	2.0	0.222	0.218	1.0028	1.3353	0.41	1.048
	3.0	0.336	0.327	1.0051	1.3365	0.63	1.074
	4.0	0.452	0.438	1.0074	1.3376	0.85	1.100
	5.0	0.572	0.548	1.0097	1.3388	1.08	1.127
	6.0	0.693	0.659	1.0120	1.3400	1.32	1.157
	7.0	0.817	0.771	1.0144	1.3412	1.56	1.188
	8.0	0.944	0.883	1.0167	1.3424	1.81	1.220
	9.0	1.074	0.996	1.0191	1.3436	2.06	1.256
	10.0	1.207	1.109	1.0215	1.3448	2.32	1.291
	12.0	1.481	1.337	1.0262	1.3472	2.88	1.365
	14.0	1.768	1.568	1.0311	1.3496	3.47	1.445
	16.0	2.068	1.800	1.0360	1.3521	4.09	1.533
	18.0	2.384	2.035	1.0409	1.3547	4.76	1.630
	20.0	2.715	2.271	1.0459	1.3572	5.46	1.737
	24.0	3.429	2.752	1.0561	1.3624	7.01	1.988
	28.0	4.223	3.242	1.0664	1.3676	8.77	2.279
	32.0	5.110	3.742	1.0770	1.3730	10.74	2.637
	36.0	6.108	4.252	1.0876	1.3785	12.96	3.088
	40.0	7.239	4.771	1.0984	1.3841	15.50	3.653
	44.0	8.532	5.300	1.1092	1.3897		4.443
	48.0	10.024	5.838	1.1200	1.3954		5.413
	52.0	11.764	6.385	1.1308	1.4011		6.666
	56.0	13.820	6.944	1.1419	1.4069		8.349
	60.0	16.288	7.512	1.1530	1.4129		10.681
	64.0	19.305	8.092	1.1643	1.4189		13.657
	68.0	23.075	8.680	1.1755	1.4249		18.457
	72.0	27.923	9.277	1.1866	1.4310		27.625
	76.0	34.387	9.884	1.1976	1.4370		40.571
	80.0	43.436	10.498	1.2085	1.4431		59.900
	84.0	57.009	11.121	1.2192	1.4492		84.338
	88.0	79.632	11.753	1.2299	1.4553		147.494
	92.0	124.878	12.392	1.2404	1.4613		384.467
	96.0	260.615	13.039	1.2508	1.4674		780.458
	100.0		13.694	1.2611	1.4735		
Hydrochloric	0.5	0.138	0.137	1.0007	1.3341	0.49	1.008
acid	1.0	0.277	0.275	1.0031	1.3353	0.99	1.015
HCl	2.0	0.560	0.553	1.0081	1.3376	2.08	1.015
	3.0	0.848	0.833	1.0130	1.3399	3.28	1.029
	4.0	1.143	1.117	1.0179	1.3422	4.58	1.044
	5.0	1.444	1.403	1.0228	1.3445	5.98	1.075

CONCENTRATIVE PROPERTIES OF AQUEOUS SOLUTIONS:
DENSITY, REFRACTIVE INDEX, FREEZING POINT DEPRESSION, AND VISCOSITY (continued)

Solute	Mass %	m/mol kg^{-1}	c/mol L^{-1}	ρ/g cm^{-3}	n	Δ/°C	η/mPa s
	6.0	1.751	1.691	1.0278	1.3468	7.52	1.091
	7.0	2.064	1.983	1.0327	1.3491	9.22	1.108
	8.0	2.385	2.277	1.0377	1.3515	11.10	1.125
	9.0	2.713	2.574	1.0426	1.3538	13.15	1.143
	10.0	3.047	2.873	1.0476	1.3561	15.40	1.161
	12.0	3.740	3.481	1.0576	1.3607	20.51	1.199
	14.0	4.465	4.099	1.0676	1.3653		1.239
	16.0	5.224	4.729	1.0777	1.3700		1.282
	18.0	6.020	5.370	1.0878	1.3746		1.326
	20.0	6.857	6.023	1.0980	1.3792		1.374
	22.0	7.736	6.687	1.1083	1.3838		1.426
	24.0	8.661	7.362	1.1185	1.3884		1.483
	26.0	9.636	8.049	1.1288	1.3930		1.547
	28.0	10.666	8.748	1.1391	1.3976		1.620
	30.0	11.754	9.456	1.1492	1.4020		1.705
	32.0	12.907	10.175	1.1594	1.4066		1.799
	34.0	14.129	10.904	1.1693	1.4112		1.900
	36.0	15.427	11.642	1.1791	1.4158		2.002
	38.0	16.810	12.388	1.1886	1.4204		2.105
	40.0	18.284	13.140	1.1977	1.4250		
Lactic acid	0.5	0.056	0.055	0.9992	1.3335	0.10	1.014
$CH_3CHOHCOOH$	1.0	0.112	0.111	1.0002	1.3340	0.19	1.027
	2.0	0.227	0.223	1.0023	1.3350	0.38	1.056
	3.0	0.343	0.334	1.0043	1.3360	0.57	1.084
	4.0	0.463	0.447	1.0065	1.3370	0.76	1.110
	5.0	0.584	0.560	1.0086	1.3380	0.95	1.138
	6.0	0.709	0.673	1.0108	1.3390	1.16	1.167
	7.0	0.836	0.787	1.0131	1.3400	1.36	1.198
	8.0	0.965	0.902	1.0153	1.3410	1.57	1.229
	9.0	1.098	1.017	1.0176	1.3420	1.79	1.262
	10.0	1.233	1.132	1.0199	1.3430	2.02	1.296
	12.0	1.514	1.365	1.0246	1.3450	2.49	1.366
	14.0	1.807	1.600	1.0294	1.3470	2.99	1.441
	16.0	2.115	1.837	1.0342	1.3491	3.48	1.522
	18.0	2.437	2.076	1.0390	1.3511	3.96	1.607
	20.0	2.775	2.318	1.0439	1.3532	4.44	1.699
	24.0	3.506	2.807	1.0536	1.3573		1.902
	28.0	4.317	3.305	1.0632	1.3615		2.136
	32.0	5.224	3.811	1.0728	1.3657		2.414
	36.0	6.244	4.325	1.0822	1.3700		2.730
	40.0	7.401	4.847	1.0915	1.3743		3.114
	44.0	8.722	5.377	1.1008	1.3786		3.566
	48.0	10.247	5.917	1.1105	1.3828		4.106
	52.0	12.026	6.466	1.1201	1.3871		4.789
	56.0	14.129	7.023	1.1297	1.3914		5.579
	60.0	16.652	7.588	1.1392	1.3958		6.679
	64.0	19.736	8.161	1.1486	1.4001		8.024
	68.0	23.590	8.741	1.1579	1.4045		9.863
	72.0	28.546	9.328	1.1670	1.4088		12.866
	76.0	35.154	9.922	1.1760	1.4131		16.974
	80.0	44.405	10.522	1.1848	1.4173		22.164
Lactose	0.5	0.015	0.015	1.0002	1.3337	0.03	1.013
$C_{12}H_{22}O_{11}$	1.0	0.030	0.029	1.0021	1.3345	0.06	1.026
	2.0	0.060	0.059	1.0061	1.3359	0.11	1.058
	3.0	0.090	0.089	1.0102	1.3375	0.17	1.089
	4.0	0.122	0.119	1.0143	1.3390	0.23	1.120

Solute	Mass %	m/mol kg^{-1}	c/mol L^{-1}	ρ/g cm^{-3}	n	Δ/°C	η/mPa s
	5.0	0.154	0.149	1.0184	1.3406	0.29	1.154
	6.0	0.186	0.179	1.0225	1.3421	0.35	1.191
	7.0	0.220	0.210	1.0267	1.3437	0.42	1.232
	8.0	0.254	0.241	1.0308	1.3453	0.50	1.276
	9.0	0.289	0.272	1.0349	1.3468		1.321
	10.0	0.325	0.304	1.0390	1.3484		1.370
	12.0	0.398	0.367	1.0473	1.3515		1.476
	14.0	0.476	0.432	1.0558	1.3548		1.593
	16.0	0.556	0.498	1.0648	1.3582		1.724
	18.0	0.641	0.565	1.0746	1.3619		1.869
Lithium	0.5	0.119	0.118	1.0012	1.3341	0.42	1.019
chloride	1.0	0.238	0.237	1.0041	1.3351	0.84	1.037
LiCl	2.0	0.481	0.476	1.0099	1.3373	1.72	1.072
	3.0	0.730	0.719	1.0157	1.3394	2.68	1.108
	4.0	0.983	0.964	1.0215	1.3415	3.73	1.146
	5.0	1.241	1.211	1.0272	1.3436	4.86	1.185
	6.0	1.506	1.462	1.0330	1.3457	6.14	1.226
	7.0	1.775	1.715	1.0387	1.3478	7.56	1.269
	8.0	2.051	1.971	1.0444	1.3499	9.11	1.313
	9.0	2.333	2.230	1.0502	1.3520	10.79	1.360
	10.0	2.621	2.491	1.0560	1.3541	12.61	1.411
	12.0	3.217	3.022	1.0675	1.3583	16.59	1.522
	14.0	3.840	3.564	1.0792	1.3625	21.04	1.647
	16.0	4.493	4.118	1.0910	1.3668		1.787
	18.0	5.178	4.683	1.1029	1.3711		1.942
	20.0	5.897	5.260	1.1150	1.3755		2.128
	22.0	6.653	5.851	1.1274	1.3799		2.341
	24.0	7.449	6.453	1.1399	1.3844		2.600
	26.0	8.288	7.069	1.1527	1.3890		2.925
	28.0	9.173	7.700	1.1658	1.3936		3.318
	30.0	10.109	8.344	1.1791	1.3983		3.785
Magnesium	0.5	0.053	0.053	1.0022	1.3343	0.26	1.024
chloride	1.0	0.106	0.106	1.0062	1.3356	0.52	1.046
MgCl$_2$	2.0	0.214	0.213	1.0144	1.3381	1.06	1.091
	3.0	0.325	0.322	1.0226	1.3406	1.65	1.139
	4.0	0.438	0.433	1.0309	1.3432	2.30	1.188
	5.0	0.553	0.546	1.0394	1.3457	3.01	1.241
	6.0	0.670	0.660	1.0479	1.3483		1.298
	7.0	0.791	0.777	1.0564	1.3508		1.358
	8.0	0.913	0.895	1.0651	1.3534		1.423
	9.0	1.039	1.015	1.0738	1.3560		1.493
	10.0	1.167	1.137	1.0826	1.3587		1.570
	12.0	1.432	1.387	1.1005	1.3641		1.745
	14.0	1.710	1.645	1.1189	1.3695		1.956
	16.0	2.001	1.911	1.1372	1.3749		2.207
	18.0	2.306	2.184	1.1553	1.3804		2.507
	20.0	2.626	2.467	1.1742	1.3859		2.867
	22.0	2.962	2.758	1.1938	1.3915		3.323
	24.0	3.317	3.060	1.2140	1.3972		3.917
	26.0	3.690	3.371	1.2346	1.4030		4.694
	28.0	4.085	3.692	1.2555	1.4089		5.709
	30.0	4.501	4.022	1.2763	1.4148		7.017
Magnesium	0.5	0.042	0.042	1.0033	1.3340	0.10	1.027
sulfate	1.0	0.084	0.084	1.0084	1.3350	0.19	1.054
MgSO$_4$	2.0	0.170	0.169	1.0186	1.3371	0.36	1.112

Solute	Mass %	m/mol kg^{-1}	c/mol L^{-1}	ρ/g cm^{-3}	n	Δ/°C	η/mPa s
	3.0	0.257	0.256	1.0289	1.3391	0.52	1.177
	4.0	0.346	0.345	1.0392	1.3411	0.69	1.249
	5.0	0.437	0.436	1.0497	1.3431	0.87	1.328
	6.0	0.530	0.528	1.0602	1.3451	1.05	1.411
	7.0	0.625	0.623	1.0708	1.3471	1.24	1.498
	8.0	0.722	0.719	1.0816	1.3492	1.43	1.593
	9.0	0.822	0.817	1.0924	1.3512	1.64	1.702
	10.0	0.923	0.917	1.1034	1.3532	1.85	1.829
	12.0	1.133	1.122	1.1257	1.3572	2.31	2.104
	14.0	1.352	1.336	1.1484	1.3613	2.86	2.412
	16.0	1.582	1.557	1.1717	1.3654	3.67	2.809
	18.0	1.824	1.788	1.1955	1.3694		3.360
	20.0	2.077	2.027	1.2198	1.3735		4.147
	22.0	2.343	2.275	1.2447	1.3776		5.199
	24.0	2.624	2.532	1.2701	1.3817		6.498
	26.0	2.919	2.800	1.2961	1.3858		8.066
Maltose	0.5	0.015	0.015	1.0003	1.3337	0.03	1.016
$C_{12}H_{22}O_{11}$	1.0	0.030	0.029	1.0023	1.3345	0.06	1.030
	2.0	0.060	0.059	1.0063	1.3359	0.11	1.060
	3.0	0.090	0.089	1.0104	1.3374	0.17	1.092
	4.0	0.122	0.119	1.0144	1.3389	0.23	1.126
	5.0	0.154	0.149	1.0184	1.3404	0.29	1.162
	6.0	0.186	0.179	1.0224	1.3420	0.35	1.200
	7.0	0.220	0.210	1.0265	1.3435	0.42	1.239
	8.0	0.254	0.241	1.0305	1.3450	0.48	1.281
	9.0	0.289	0.272	1.0345	1.3466	0.55	1.325
	10.0	0.325	0.303	1.0385	1.3482	0.62	1.372
	12.0	0.398	0.367	1.0465	1.3513	0.77	1.474
	14.0	0.476	0.431	1.0545	1.3546	0.92	1.588
	16.0	0.556	0.497	1.0629	1.3578	1.08	1.715
	18.0	0.641	0.564	1.0716	1.3612	1.25	1.859
	20.0	0.730	0.631	1.0801	1.3644	1.43	2.021
	22.0	0.824	0.700	1.0894	1.3678	1.64	2.216
	24.0	0.923	0.770	1.0984	1.3714	1.85	2.463
	26.0	1.026	0.842	1.1080	1.3749	2.08	2.753
	28.0	1.136	0.914	1.1171	1.3785	2.34	3.066
	30.0	1.252	0.988	1.1269	1.3821	2.62	3.427
	40.0	1.948	1.375	1.1769	1.4013	4.41	6.926
	50.0	2.921	1.797	1.2304	1.4217		17.786
	52.0	3.165	1.886	1.2416	1.4260		22.034
	54.0	3.429	1.976	1.2528	1.4308		28.757
	56.0	3.718	2.068	1.2638	1.4350		38.226
	58.0	4.034	2.159	1.2740	1.4394		49.298
	60.0	4.382	2.253	1.2855	1.4440		
Manganese(II)	1.0	0.067	0.067	1.0080	1.3348	0.16	1.046
sulfate	2.0	0.135	0.135	1.0178	1.3366	0.31	1.090
$MnSO_4$	3.0	0.205	0.204	1.0277	1.3384	0.44	1.137
	4.0	0.276	0.275	1.0378	1.3402	0.57	1.187
	5.0	0.349	0.347	1.0480	1.3420	0.70	1.242
	6.0	0.423	0.421	1.0583	1.3438	0.84	1.301
	7.0	0.498	0.495	1.0688	1.3457	0.98	1.363
	8.0	0.576	0.572	1.0794	1.3475	1.12	1.431
	9.0	0.655	0.650	1.0902	1.3494	1.28	1.505
	10.0	0.736	0.729	1.1012	1.3513	1.44	1.587
	12.0	0.903	0.893	1.1236	1.3551	1.80	1.779

CONCENTRATIVE PROPERTIES OF AQUEOUS SOLUTIONS:
DENSITY, REFRACTIVE INDEX, FREEZING POINT DEPRESSION, AND VISCOSITY (continued)

Solute	Mass %	m/mol kg^{-1}	c/mol L^{-1}	ρ/g cm^{-3}	n	Δ/°C	η/mPa s
	14.0	1.078	1.063	1.1467	1.3589	2.21	2.005
	16.0	1.261	1.240	1.1705	1.3629	2.67	2.272
	18.0	1.454	1.424	1.1950	1.3668	3.19	2.580
	20.0	1.656	1.616	1.2203	1.3708	3.80	2.938
D-Mannitol	0.5	0.028	0.027	1.0000	1.3337	0.05	1.019
$CH_2(CHOH)_4CH_2OH$	1.0	0.055	0.055	1.0017	1.3345	0.10	1.032
	2.0	0.112	0.110	1.0053	1.3359	0.21	1.057
	3.0	0.170	0.166	1.0088	1.3374	0.32	1.081
	4.0	0.229	0.222	1.0124	1.3389	0.43	1.107
	5.0	0.289	0.279	1.0159	1.3403	0.54	1.135
	6.0	0.350	0.336	1.0195	1.3418	0.66	1.166
	7.0	0.413	0.393	1.0230	1.3433	0.77	1.200
	8.0	0.477	0.451	1.0266	1.3447	0.90	1.236
	9.0	0.543	0.509	1.0302	1.3462	1.02	1.275
	10.0	0.610	0.567	1.0338	1.3477	1.15	1.314
	11.0	0.678	0.626	1.0375	1.3491	1.28	1.355
	12.0	0.749	0.686	1.0412	1.3506	1.41	1.398
	13.0	0.820	0.746	1.0450	1.3521	1.55	1.443
	14.0	0.894	0.806	1.0489	1.3536	1.69	1.489
	15.0	0.969	0.867	1.0529	1.3552	1.84	1.537
Methanol	0.5	0.157	0.156	0.9973	1.3331	0.28	1.022
CH_3OH	1.0	0.315	0.311	0.9964	1.3332	0.56	1.040
	2.0	0.637	0.621	0.9947	1.3334	1.14	1.070
	3.0	0.965	0.930	0.9930	1.3336	1.75	1.100
	4.0	1.300	1.238	0.9913	1.3339	2.37	1.131
	5.0	1.643	1.544	0.9896	1.3341	3.02	1.163
	6.0	1.992	1.850	0.9880	1.3343	3.71	1.196
	7.0	2.349	2.155	0.9864	1.3346	4.41	1.229
	8.0	2.714	2.459	0.9848	1.3348	5.13	1.264
	9.0	3.087	2.762	0.9832	1.3351	5.85	1.297
	10.0	3.468	3.064	0.9816	1.3354	6.60	1.329
	12.0	4.256	3.665	0.9785	1.3359	8.14	1.389
	14.0	5.081	4.262	0.9755	1.3365	9.72	1.446
	16.0	5.945	4.856	0.9725	1.3370	11.36	1.501
	18.0	6.851	5.447	0.9695	1.3376	13.13	1.554
	20.0	7.803	6.034	0.9666	1.3381	15.02	1.604
	22.0	8.803	6.616	0.9636	1.3387	16.98	1.652
	24.0	9.856	7.196	0.9606	1.3392	19.04	1.697
	26.0	10.966	7.771	0.9576	1.3397	21.23	1.735
	28.0	12.138	8.341	0.9545	1.3402	23.59	1.769
	30.0	13.376	8.908	0.9514	1.3407	25.91	1.795
	32.0	14.688	9.470	0.9482	1.3411	28.15	1.814
	34.0	16.078	10.028	0.9450	1.3415	30.48	1.827
	36.0	17.556	10.580	0.9416	1.3419	32.97	1.835
	38.0	19.129	11.127	0.9382	1.3422	35.60	1.839
	40.0	20.807	11.669	0.9347	1.3425	38.60	1.837
	50.0	31.211	14.288	0.9156	1.3431	54.50	1.761
	60.0	46.816	16.749	0.8944	1.3426	74.50	1.600
	70.0	72.826	19.040	0.8715	1.3411		1.368
	80.0	124.844	21.144	0.8468	1.3385		1.128
	90.0	280.899	23.045	0.8204	1.3348		0.861
	100.0		24.710	0.7917	1.3290		0.586
Nitric acid	0.5	0.080	0.079	1.0009	1.3336	0.28	1.004
HNO_3	1.0	0.160	0.159	1.0037	1.3343	0.56	1.005
	2.0	0.324	0.320	1.0091	1.3356	1.12	1.007

Solute	Mass %	m/mol kg^{-1}	c/mol L^{-1}	ρ/g cm^{-3}	n	Δ/°C	η/mPa s
	3.0	0.491	0.483	1.0146	1.3368	1.70	1.010
	4.0	0.661	0.648	1.0202	1.3381	2.32	1.014
	5.0	0.835	0.814	1.0257	1.3394	2.96	1.018
	6.0	1.013	0.982	1.0314	1.3407	3.63	1.022
	7.0	1.194	1.152	1.0370	1.3421	4.33	1.027
	8.0	1.380	1.324	1.0427	1.3434	5.05	1.032
	9.0	1.570	1.498	1.0485	1.3447	5.81	1.038
	10.0	1.763	1.673	1.0543	1.3460	6.60	1.044
	12.0	2.164	2.030	1.0660	1.3487	8.27	1.058
	14.0	2.583	2.395	1.0780	1.3514	10.08	1.075
	16.0	3.023	2.768	1.0901	1.3541	12.04	1.094
	18.0	3.484	3.149	1.1025	1.3569	14.16	1.116
	20.0	3.967	3.539	1.1150	1.3596		1.141
	22.0	4.476	3.937	1.1277	1.3624		1.169
	24.0	5.011	4.344	1.1406	1.3652		1.199
	26.0	5.576	4.760	1.1536	1.3680		1.233
	28.0	6.172	5.185	1.1668	1.3708		1.271
	30.0	6.801	5.618	1.1801	1.3736		1.311
	32.0	7.468	6.060	1.1934	1.3763		1.354
	34.0	8.175	6.512	1.2068	1.3790		1.400
	36.0	8.927	6.971	1.2202	1.3817		1.450
	38.0	9.727	7.439	1.2335	1.3842		1.504
	40.0	10.580	7.913	1.2466	1.3867		1.561
Oxalic acid	0.5	0.056	0.056	1.0006	1.3336	0.16	1.013
$(COOH)_2$	1.0	0.112	0.111	1.0030	1.3342	0.30	1.023
	1.5	0.169	0.167	1.0054	1.3347	0.44	1.033
	2.0	0.227	0.224	1.0079	1.3353	0.57	1.044
	2.5	0.285	0.281	1.0103	1.3359	0.71	1.055
	3.0	0.343	0.337	1.0126	1.3364	0.84	1.065
	3.5	0.403	0.395	1.0150	1.3370	0.97	1.076
	4.0	0.463	0.452	1.0174	1.3375	1.09	1.086
	4.5	0.523	0.510	1.0197	1.3381		1.097
	5.0	0.585	0.568	1.0220	1.3386		1.108
	6.0	0.709	0.684	1.0265	1.3397		1.129
	7.0	0.836	0.802	1.0310	1.3407		1.150
	8.0	0.966	0.920	1.0355	1.3418		1.172
Phosphoric	0.5	0.051	0.051	1.0010	1.3335	0.12	1.010
acid	1.0	0.103	0.102	1.0038	1.3340	0.24	1.020
H_3PO_4	2.0	0.208	0.206	1.0092	1.3349	0.46	1.050
	3.0	0.316	0.311	1.0146	1.3358	0.69	1.079
	4.0	0.425	0.416	1.0200	1.3367	0.93	1.108
	5.0	0.537	0.523	1.0254	1.3376	1.16	1.138
	6.0	0.651	0.631	1.0309	1.3385	1.38	1.169
	7.0	0.768	0.740	1.0363	1.3394	1.62	1.200
	8.0	0.887	0.850	1.0418	1.3403	1.88	1.232
	9.0	1.009	0.962	1.0474	1.3413	2.16	1.267
	10.0	1.134	1.075	1.0531	1.3422	2.45	1.303
	12.0	1.392	1.304	1.0647	1.3441	3.01	1.382
	14.0	1.661	1.538	1.0765	1.3460	3.76	1.469
	16.0	1.944	1.777	1.0885	1.3480	4.45	1.565
	18.0	2.240	2.022	1.1009	1.3500	5.25	1.671
	20.0	2.551	2.273	1.1135	1.3520	6.23	1.788
	22.0	2.878	2.529	1.1263	1.3540	7.38	1.914
	24.0	3.223	2.791	1.1395	1.3561	8.69	2.049
	26.0	3.585	3.059	1.1528	1.3582	10.12	2.198

Solute	Mass %	m/mol kg^{-1}	c/mol L^{-1}	ρ/g cm^{-3}	n	Δ/°C	η/mPa s
	28.0	3.968	3.333	1.1665	1.3604	11.64	2.365
	30.0	4.373	3.614	1.1804	1.3625	13.23	2.553
	32.0	4.802	3.901	1.1945	1.3647	14.94	2.766
	34.0	5.257	4.194	1.2089	1.3669	16.81	3.001
	36.0	5.740	4.495	1.2236	1.3691	18.85	3.260
	38.0	6.254	4.803	1.2385	1.3713	21.09	3.544
	40.0	6.803	5.117	1.2536	1.3735	23.58	3.856
Potassium bicarbonate $KHCO_3$	0.5	0.050	0.050	1.0014	1.3335	0.18	1.009
	1.0	0.101	0.100	1.0046	1.3341	0.34	1.015
	2.0	0.204	0.202	1.0114	1.3353	0.67	1.027
	3.0	0.309	0.305	1.0181	1.3365	0.98	1.040
	4.0	0.416	0.409	1.0247	1.3376	1.29	1.053
	5.0	0.526	0.515	1.0310	1.3386	1.60	1.067
	6.0	0.638	0.622	1.0379	1.3397	1.91	1.081
	7.0	0.752	0.730	1.0446	1.3409	2.22	1.096
	8.0	0.869	0.840	1.0514	1.3419	2.53	1.112
	9.0	0.988	0.951	1.0581	1.3430	2.84	1.128
	10.0	1.110	1.064	1.0650	1.3441	3.16	1.145
	12.0	1.362	1.293	1.0788	1.3462	3.79	1.183
	14.0	1.626	1.528	1.0929	1.3484	4.41	1.224
	16.0	1.903	1.770	1.1073	1.3506		1.270
	18.0	2.193	2.017	1.1221	1.3528		1.319
	20.0	2.497	2.272	1.1372	1.3550		1.373
	22.0	2.817	2.533	1.1527	1.3572		1.432
	24.0	3.154	2.801	1.1685	1.3595		1.497
Potassium bromide KBr	0.5	0.042	0.042	1.0018	1.3336	0.15	1.000
	1.0	0.085	0.084	1.0054	1.3342	0.29	0.998
	2.0	0.171	0.170	1.0127	1.3354	0.59	0.994
	3.0	0.260	0.257	1.0200	1.3366	0.88	0.990
	4.0	0.350	0.345	1.0275	1.3379	1.18	0.985
	5.0	0.442	0.435	1.0350	1.3391	1.48	0.981
	6.0	0.536	0.526	1.0426	1.3403	1.78	0.977
	7.0	0.633	0.618	1.0503	1.3416	2.10	0.974
	8.0	0.731	0.711	1.0581	1.3429	2.42	0.970
	9.0	0.831	0.806	1.0660	1.3441	2.74	0.967
	10.0	0.934	0.903	1.0740	1.3454	3.07	0.964
	12.0	1.146	1.099	1.0903	1.3481	3.76	0.958
	14.0	1.368	1.302	1.1070	1.3507	4.49	0.953
	16.0	1.601	1.512	1.1242	1.3535	5.25	0.949
	18.0	1.845	1.727	1.1419	1.3562	6.04	0.946
	20.0	2.101	1.950	1.1601	1.3591	6.88	0.944
	22.0	2.370	2.179	1.1788	1.3620	7.76	0.943
	24.0	2.654	2.416	1.1980	1.3650	8.70	0.943
	26.0	2.952	2.661	1.2179	1.3680	9.68	0.944
	28.0	3.268	2.914	1.2383	1.3711	10.72	0.947
	30.0	3.601	3.175	1.2593	1.3743	11.82	0.952
	32.0	3.954	3.445	1.2810	1.3776	12.98	0.959
	34.0	4.329	3.724	1.3033	1.3809		0.968
	36.0	4.727	4.012	1.3263	1.3843		0.979
	38.0	5.150	4.311	1.3501	1.3878		0.993
	40.0	5.602	4.620	1.3746	1.3914		1.010
Potassium carbonate K_2CO_3	0.5	0.036	0.036	1.0027	1.3339	0.18	1.013
	1.0	0.073	0.073	1.0072	1.3347	0.34	1.025
	2.0	0.148	0.147	1.0163	1.3365	0.66	1.048

Solute	Mass %	m/mol kg^{-1}	c/mol L^{-1}	ρ/g cm^{-3}	n	Δ/°C	η/mPa s
	3.0	0.224	0.223	1.0254	1.3382	0.99	1.071
	4.0	0.301	0.299	1.0345	1.3399	1.32	1.094
	5.0	0.381	0.378	1.0437	1.3416	1.67	1.119
	6.0	0.462	0.457	1.0529	1.3433	2.03	1.146
	7.0	0.545	0.538	1.0622	1.3450	2.40	1.174
	8.0	0.629	0.620	1.0715	1.3467	2.77	1.204
	9.0	0.716	0.704	1.0809	1.3484	3.17	1.235
	10.0	0.804	0.789	1.0904	1.3501	3.57	1.269
	12.0	0.987	0.963	1.1095	1.3535	4.45	1.339
	14.0	1.178	1.144	1.1291	1.3569	5.39	1.414
	16.0	1.378	1.330	1.1490	1.3603	6.42	1.497
	18.0	1.588	1.523	1.1692	1.3637	7.55	1.594
	20.0	1.809	1.722	1.1898	1.3671	8.82	1.707
	24.0	2.285	2.139	1.2320	1.3739	11.96	1.978
	28.0	2.814	2.584	1.2755	1.3807	16.01	2.331
	32.0	3.405	3.057	1.3204	1.3874	21.46	2.834
	36.0	4.070	3.559	1.3665	1.3940	28.58	3.503
	40.0	4.824	4.093	1.4142	1.4006	37.55	4.360
	50.0	7.236	5.573	1.5404	1.4168		9.369
Potassium chloride KCl	0.5	0.067	0.067	1.0014	1.3337	0.23	1.000
	1.0	0.135	0.135	1.0046	1.3343	0.46	0.999
	2.0	0.274	0.271	1.0110	1.3357	0.92	0.999
	3.0	0.415	0.409	1.0174	1.3371	1.38	0.998
	4.0	0.559	0.549	1.0239	1.3384	1.85	0.997
	5.0	0.706	0.691	1.0304	1.3398	2.32	0.996
	6.0	0.856	0.835	1.0369	1.3411	2.80	0.994
	7.0	1.010	0.980	1.0434	1.3425	3.29	0.992
	8.0	1.166	1.127	1.0500	1.3438	3.80	0.990
	9.0	1.327	1.276	1.0566	1.3452	4.30	0.989
	10.0	1.490	1.426	1.0633	1.3466	4.81	0.988
	12.0	1.829	1.733	1.0768	1.3493	5.88	0.990
	14.0	2.184	2.048	1.0905	1.3521		0.994
	16.0	2.555	2.370	1.1043	1.3549		0.999
	18.0	2.944	2.701	1.1185	1.3577		1.004
	20.0	3.353	3.039	1.1328	1.3606		1.012
	22.0	3.783	3.386	1.1474	1.3635		1.024
	24.0	4.236	3.742	1.1623	1.3665		1.040
Potassium hydroxide KOH	0.5	0.090	0.089	1.0025	1.3340	0.30	1.010
	1.0	0.180	0.179	1.0068	1.3350	0.61	1.019
	2.0	0.364	0.362	1.0155	1.3369	1.24	1.038
	3.0	0.551	0.548	1.0242	1.3388	1.89	1.058
	4.0	0.743	0.736	1.0330	1.3408	2.57	1.079
	5.0	0.938	0.929	1.0419	1.3427	3.36	1.102
	6.0	1.138	1.124	1.0509	1.3445	4.14	1.126
	7.0	1.342	1.322	1.0599	1.3464	4.92	1.151
	8.0	1.550	1.524	1.0690	1.3483		1.177
	9.0	1.763	1.729	1.0781	1.3502		1.205
	10.0	1.980	1.938	1.0873	1.3520		1.233
	12.0	2.431	2.365	1.1059	1.3558		1.294
	14.0	2.902	2.806	1.1246	1.3595		1.361
	16.0	3.395	3.261	1.1435	1.3632		1.436
	18.0	3.913	3.730	1.1626	1.3670		1.521
	20.0	4.456	4.213	1.1818	1.3707		1.619
	22.0	5.027	4.711	1.2014	1.3744		1.732
	24.0	5.629	5.223	1.2210	1.3781		1.861

Solute	Mass %	m/mol kg^{-1}	c/mol L^{-1}	ρ/g cm^{-3}	n	$\Delta/°C$	η/mPa s
	26.0	6.262	5.750	1.2408	1.3818		2.006
	28.0	6.931	6.293	1.2609	1.3854		2.170
	30.0	7.639	6.851	1.2813	1.3889		2.357
	40.0	11.882	9.896	1.3881	1.4068		3.879
	50.0	17.824	13.389	1.5024	1.4247		7.892
Potassium	0.5	0.030	0.030	1.0019	1.3337	0.11	0.999
iodide	1.0	0.061	0.061	1.0056	1.3343	0.22	0.997
KI	2.0	0.123	0.122	1.0131	1.3357	0.43	0.991
	3.0	0.186	0.184	1.0206	1.3370	0.64	0.986
	4.0	0.251	0.248	1.0282	1.3384	0.86	0.981
	5.0	0.317	0.312	1.0360	1.3397	1.08	0.976
	6.0	0.385	0.377	1.0438	1.3411	1.30	0.969
	7.0	0.453	0.443	1.0517	1.3425	1.53	0.963
	8.0	0.524	0.511	1.0598	1.3440	1.77	0.957
	9.0	0.596	0.579	1.0679	1.3454	2.01	0.951
	10.0	0.669	0.648	1.0762	1.3469	2.26	0.946
	12.0	0.821	0.790	1.0931	1.3498	2.77	0.937
	14.0	0.981	0.937	1.1105	1.3529	3.30	0.929
	16.0	1.147	1.088	1.1284	1.3560	3.87	0.921
	18.0	1.322	1.244	1.1469	1.3593	4.46	0.915
	20.0	1.506	1.405	1.1659	1.3626	5.09	0.910
	22.0	1.699	1.571	1.1856	1.3661	5.76	0.905
	24.0	1.902	1.744	1.2060	1.3696	6.46	0.901
	26.0	2.117	1.922	1.2270	1.3733	7.21	0.898
	28.0	2.343	2.106	1.2487	1.3771	8.01	0.895
	30.0	2.582	2.297	1.2712	1.3810	8.86	0.892
	32.0	2.835	2.495	1.2944	1.3851	9.76	0.891
	34.0	3.103	2.700	1.3185	1.3893	10.72	0.890
	36.0	3.388	2.913	1.3434	1.3936	11.73	0.890
	38.0	3.692	3.134	1.3692	1.3981	12.81	0.893
	40.0	4.016	3.364	1.3959	1.4027	13.97	0.897
Potassium	0.5	0.050	0.050	1.0014	1.3335	0.17	0.999
nitrate	1.0	0.100	0.099	1.0045	1.3339	0.33	0.996
KNO$_3$	2.0	0.202	0.200	1.0108	1.3349	0.64	0.990
	3.0	0.306	0.302	1.0171	1.3358	0.94	0.986
	4.0	0.412	0.405	1.0234	1.3368	1.22	0.983
	5.0	0.521	0.509	1.0298	1.3377	1.50	0.980
	6.0	0.631	0.615	1.0363	1.3386	1.76	0.977
	7.0	0.744	0.722	1.0428	1.3396	2.02	0.975
	8.0	0.860	0.830	1.0494	1.3405	2.27	0.973
	9.0	0.978	0.940	1.0560	1.3415	2.52	0.971
	10.0	1.099	1.051	1.0627	1.3425	2.75	0.970
	12.0	1.349	1.277	1.0762	1.3444		0.970
	14.0	1.610	1.509	1.0899	1.3463		0.972
	16.0	1.884	1.747	1.1039	1.3482		0.976
	18.0	2.171	1.991	1.1181	1.3502		0.982
	20.0	2.473	2.240	1.1326	1.3521		0.990
	22.0	2.790	2.497	1.1473	1.3541		0.999
	24.0	3.123	2.759	1.1623	1.3561		1.010
Potassium	0.5	0.032	0.032	1.0017		0.11	1.001
permanganate	1.0	0.064	0.064	1.0051		0.22	1.000
KMnO$_4$	1.5	0.096	0.096	1.0085		0.32	0.999
	2.0	0.129	0.128	1.0118		0.43	0.998
	3.0	0.196	0.193	1.0186			0.995

Solute	Mass %	m/mol kg^{-1}	c/mol L^{-1}	ρ/g cm^{-3}	n	Δ/°C	η/mPa s
	4.0	0.264	0.260	1.0254			0.992
	5.0	0.333	0.327	1.0322			0.989
	6.0	0.404	0.394	1.0390			0.985
Potassium hydrogen phosphate K_2HPO_4	0.5	0.029	0.029	1.0025	1.3338	0.13	1.013
	1.0	0.058	0.058	1.0068	1.3345	0.25	1.023
	1.5	0.087	0.087	1.0110	1.3353	0.37	1.034
	2.0	0.117	0.117	1.0153	1.3361	0.49	1.046
	2.5	0.147	0.146	1.0195	1.3368	0.61	1.057
	3.0	0.178	0.176	1.0238	1.3376	0.73	1.069
	3.5	0.208	0.207	1.0281	1.3384	0.86	1.081
	4.0	0.239	0.237	1.0324	1.3392	0.97	1.094
	4.5	0.271	0.268	1.0368	1.3399	1.10	1.107
	5.0	0.302	0.299	1.0412	1.3407	1.22	1.120
	6.0	0.366	0.362	1.0500	1.3422	1.46	1.147
	7.0	0.432	0.426	1.0590	1.3438	1.70	1.177
	8.0	0.499	0.491	1.0680	1.3453	1.95	1.209
Potassium dihydrogen phosphate KH_2PO_4	0.5	0.037	0.037	1.0018	1.3336	0.13	1.010
	1.0	0.074	0.074	1.0053	1.3342	0.25	1.019
	1.5	0.112	0.111	1.0089	1.3348	0.37	1.028
	2.0	0.150	0.149	1.0125	1.3354	0.49	1.038
	3.0	0.227	0.225	1.0197	1.3365	0.72	1.060
	4.0	0.306	0.302	1.0269	1.3377	0.96	1.083
	5.0	0.387	0.380	1.0342	1.3388	1.19	1.108
	6.0	0.469	0.459	1.0414	1.3400	1.41	1.133
	7.0	0.553	0.539	1.0486	1.3411	1.63	1.160
	8.0	0.639	0.621	1.0558	1.3422	1.84	1.187
	9.0	0.727	0.703	1.0630	1.3434	2.04	1.215
	10.0	0.816	0.786	1.0703	1.3445	2.23	1.245
Potassium sulfate K_2SO_4	0.5	0.029	0.029	1.0022	1.3336	0.14	1.006
	1.0	0.058	0.058	1.0062	1.3343	0.26	1.011
	2.0	0.117	0.116	1.0143	1.3355	0.50	1.021
	3.0	0.177	0.176	1.0224	1.3368	0.73	1.033
	4.0	0.239	0.237	1.0306	1.3380	0.95	1.045
	5.0	0.302	0.298	1.0388	1.3393	1.17	1.058
	6.0	0.366	0.360	1.0470	1.3405		1.072
	7.0	0.432	0.424	1.0553	1.3417		1.087
	8.0	0.499	0.488	1.0637	1.3428		1.102
	9.0	0.568	0.554	1.0721	1.3440		1.117
	10.0	0.638	0.620	1.0806	1.3452		1.132
1-Propanol $CH_3CH_2CH_2OH$	1.0	0.168	0.166	0.9963	1.3339	0.31	1.051
	2.0	0.340	0.331	0.9946	1.3348	0.61	1.100
	3.0	0.515	0.496	0.9928	1.3357	0.93	1.152
	4.0	0.693	0.660	0.9911	1.3366	1.24	1.208
	5.0	0.876	0.823	0.9896	1.3376	1.57	1.267
	6.0	1.062	0.987	0.9882	1.3385	1.91	1.325
	7.0	1.252	1.149	0.9868	1.3394	2.26	1.387
	8.0	1.447	1.312	0.9855	1.3404	2.61	1.449
	9.0	1.646	1.474	0.9842	1.3414	2.99	1.514
	10.0	1.849	1.635	0.9829	1.3423	3.36	1.577
	12.0	2.269	1.958	0.9804	1.3442	4.09	1.710
	14.0	2.709	2.278	0.9779	1.3460	4.91	1.849
	16.0	3.169	2.595	0.9749	1.3477	5.78	1.986
	18.0	3.652	2.911	0.9719	1.3494	6.67	2.106

CONCENTRATIVE PROPERTIES OF AQUEOUS SOLUTIONS:
DENSITY, REFRACTIVE INDEX, FREEZING POINT DEPRESSION, AND VISCOSITY (continued)

Solute	Mass %	m/mol kg^{-1}	c/mol L^{-1}	ρ/g cm^{-3}	n	Δ/°C	η/mPa s
	20.0	4.160	3.223	0.9686	1.3510	7.76	2.218
	24.0	5.254	3.838	0.9612	1.3539	9.12	2.432
	28.0	6.471	4.441	0.9533	1.3566	10.17	2.612
	32.0	7.830	5.033	0.9452	1.3592	10.66	2.765
	36.0	9.359	5.613	0.9370	1.3614		2.900
	40.0	11.093	6.182	0.9288	1.3635		3.010
	60.0	24.958	8.860	0.8875	1.3734		3.186
	80.0	66.556	11.275	0.8470	1.3812		2.822
	100.0		13.368	0.8034	1.3852		2.227
2-Propanol	1.0	0.168	0.166	0.9960	1.3338	0.30	1.056
CH$_3$CHOHCH$_3$	2.0	0.340	0.331	0.9939	1.3346	0.60	1.112
	3.0	0.515	0.495	0.9920	1.3355	0.93	1.166
	4.0	0.693	0.659	0.9902	1.3364	1.26	1.225
	5.0	0.876	0.822	0.9884	1.3373	1.61	1.287
	6.0	1.062	0.985	0.9871	1.3382	1.96	1.352
	7.0	1.252	1.148	0.9855	1.3392	2.32	1.417
	8.0	1.447	1.310	0.9843	1.3400	2.68	1.485
	9.0	1.646	1.472	0.9831	1.3410	3.06	1.553
	10.0	1.849	1.633	0.9816	1.3420	3.48	1.629
	12.0	2.269	1.955	0.9793	1.3439	4.43	1.794
	14.0	2.709	2.276	0.9772	1.3459	5.29	1.970
	16.0	3.169	2.596	0.9751	1.3478	6.36	2.160
	18.0	3.652	2.913	0.9725	1.3496	7.40	2.352
	20.0	4.160	3.227	0.9696	1.3514	8.52	2.550
	40.0	11.093	6.191	0.9302	1.3642		
	60.0	24.958	8.809	0.8824	1.3717		
	80.0	66.556	11.103	0.8341	1.3765		
	100.0		13.058	0.7848	1.3742		
Silver	0.5	0.030	0.030	1.0027	1.3336	0.10	1.003
nitrate	1.0	0.059	0.059	1.0070	1.3342	0.20	1.005
AgNO$_3$	2.0	0.120	0.120	1.0154	1.3352	0.40	1.009
	3.0	0.182	0.181	1.0239	1.3363	0.59	1.013
	4.0	0.245	0.243	1.0327	1.3374	0.78	1.016
	5.0	0.310	0.307	1.0417	1.3385	0.96	1.020
	6.0	0.376	0.371	1.0506	1.3396	1.15	1.024
	7.0	0.443	0.437	1.0597	1.3407	1.33	1.027
	8.0	0.512	0.503	1.0690	1.3419	1.51	1.031
	9.0	0.582	0.571	1.0785	1.3431	1.69	1.035
	10.0	0.654	0.641	1.0882	1.3443	1.87	1.039
	12.0	0.803	0.783	1.1079	1.3467	2.21	1.049
	14.0	0.958	0.930	1.1284	1.3493	2.55	1.060
	16.0	1.121	1.083	1.1496	1.3519	2.86	1.072
	18.0	1.292	1.241	1.1715	1.3546		1.086
	20.0	1.472	1.406	1.1942	1.3574		1.101
	22.0	1.660	1.577	1.2177	1.3602		1.117
	24.0	1.859	1.755	1.2420	1.3632		1.135
	26.0	2.068	1.940	1.2672	1.3662		1.154
	28.0	2.289	2.132	1.2933	1.3694		1.176
	30.0	2.523	2.332	1.3204	1.3726		1.200
	32.0	2.770	2.541	1.3487	1.3760		1.227
	34.0	3.033	2.758	1.3780	1.3795		1.257
	36.0	3.311	2.985	1.4087	1.3832		1.290
	38.0	3.608	3.223	1.4407	1.3871		1.326
	40.0	3.925	3.472	1.4743	1.3911		1.366

Solute	Mass %	m/mol kg^{-1}	c/mol L^{-1}	ρ/g cm^{-3}	n	Δ/°C	η/mPa s
Sodium	0.5	0.061	0.061	1.0008	1.3337	0.22	1.021
acetate	1.0	0.123	0.122	1.0034	1.3344	0.43	1.040
CH$_3$COONa	2.0	0.249	0.246	1.0085	1.3358	0.88	1.080
	3.0	0.377	0.371	1.0135	1.3372	1.34	1.124
	4.0	0.508	0.497	1.0184	1.3386	1.82	1.171
	5.0	0.642	0.624	1.0234	1.3400	2.32	1.222
	6.0	0.778	0.752	1.0283	1.3414	2.85	1.278
	7.0	0.918	0.882	1.0334	1.3428	3.40	1.337
	8.0	1.060	1.013	1.0386	1.3442	3.98	1.401
	9.0	1.206	1.145	1.0440	1.3456	4.57	1.468
	10.0	1.354	1.279	1.0495	1.3470		1.539
	12.0	1.662	1.552	1.0607	1.3498		1.688
	14.0	1.984	1.829	1.0718	1.3526		1.855
	16.0	2.322	2.112	1.0830	1.3554		2.054
	18.0	2.676	2.400	1.0940	1.3583		2.284
	20.0	3.047	2.694	1.1050	1.3611		2.567
	22.0	3.438	2.993	1.1159	1.3639		2.948
	24.0	3.849	3.297	1.1268	1.3666		3.400
	26.0	4.283	3.606	1.1377	1.3693		3.877
	28.0	4.741	3.921	1.1488	1.3720		4.388
	30.0	5.224	4.243	1.1602	1.3748		4.940
Sodium	0.5	0.060	0.060	1.0018	1.3337	0.20	1.015
bicarbonate	1.0	0.120	0.120	1.0054	1.3344	0.40	1.028
NaHCO$_3$	1.5	0.181	0.180	1.0089	1.3351	0.59	1.042
	2.0	0.243	0.241	1.0125	1.3357	0.78	1.057
	2.5	0.305	0.302	1.0160	1.3364	0.98	1.071
	3.0	0.368	0.364	1.0196	1.3370	1.16	1.086
	3.5	0.432	0.426	1.0231	1.3377	1.35	1.102
	4.0	0.496	0.489	1.0266	1.3383	1.54	1.118
	4.5	0.561	0.552	1.0301	1.3390	1.72	1.134
	5.0	0.627	0.615	1.0337	1.3396	1.90	1.151
	5.5	0.693	0.679	1.0372	1.3403	2.08	1.168
	6.0	0.760	0.743	1.0408	1.3409	2.26	1.185
Sodium	0.5	0.049	0.049	1.0021	1.3337	0.17	1.004
bromide	1.0	0.098	0.098	1.0060	1.3344	0.34	1.007
NaBr	2.0	0.198	0.197	1.0139	1.3358	0.69	1.012
	3.0	0.301	0.298	1.0218	1.3372	1.04	1.017
	4.0	0.405	0.400	1.0298	1.3386	1.39	1.022
	5.0	0.512	0.504	1.0380	1.3401	1.76	1.028
	6.0	0.620	0.610	1.0462	1.3415	2.14	1.034
	7.0	0.732	0.717	1.0546	1.3430	2.53	1.040
	8.0	0.845	0.826	1.0630	1.3445	2.93	1.046
	9.0	0.961	0.937	1.0716	1.3460	3.34	1.053
	10.0	1.080	1.050	1.0803	1.3475	3.77	1.060
	12.0	1.325	1.281	1.0981	1.3506	4.67	1.077
	14.0	1.582	1.519	1.1164	1.3538	5.65	1.096
	16.0	1.851	1.765	1.1352	1.3570	6.74	1.119
	18.0	2.133	2.020	1.1546	1.3604		1.144
	20.0	2.430	2.283	1.1745	1.3638		1.174
	22.0	2.741	2.555	1.1951	1.3673		1.207
	24.0	3.069	2.837	1.2163	1.3708		1.244
	26.0	3.415	3.129	1.2382	1.3745		1.287
	28.0	3.780	3.431	1.2608	1.3783		1.336
	30.0	4.165	3.744	1.2842	1.3822		1.395
	32.0	4.574	4.069	1.3083	1.3862		1.465

Solute	Mass %	$m/\text{mol kg}^{-1}$	$c/\text{mol L}^{-1}$	$\rho/\text{g cm}^{-3}$	n	$\Delta/^\circ\text{C}$	$\eta/\text{mPa s}$
	34.0	5.007	4.406	1.3333	1.3903		1.546
	36.0	5.467	4.755	1.3592	1.3946		1.639
	38.0	5.957	5.119	1.3860	1.3990		1.745
	40.0	6.479	5.496	1.4138	1.4035		1.866
Sodium	0.5	0.047	0.047	1.0034	1.3341	0.22	1.025
carbonate	1.0	0.095	0.095	1.0086	1.3352	0.43	1.049
Na_2CO_3	2.0	0.193	0.192	1.0190	1.3375	0.75	1.102
	3.0	0.292	0.291	1.0294	1.3397	1.08	1.159
	4.0	0.393	0.392	1.0398	1.3419	1.42	1.222
	5.0	0.497	0.495	1.0502	1.3440	1.77	1.292
	6.0	0.602	0.600	1.0606	1.3462	2.13	1.367
	7.0	0.710	0.707	1.0711	1.3483		1.448
	8.0	0.820	0.816	1.0816	1.3504		1.538
	9.0	0.933	0.927	1.0922	1.3525		1.638
	10.0	1.048	1.041	1.1029	1.3547		1.754
	11.0	1.166	1.156	1.1136	1.3568		1.884
	12.0	1.287	1.273	1.1244	1.3589		2.028
	13.0	1.410	1.392	1.1353	1.3610		2.186
	14.0	1.536	1.514	1.1463	1.3631		2.361
	15.0	1.665	1.638	1.1574	1.3652		2.551
Sodium	0.5	0.086	0.086	1.0018	1.3339	0.30	1.011
chloride	1.0	0.173	0.172	1.0053	1.3347	0.59	1.020
NaCl	2.0	0.349	0.346	1.0125	1.3365	1.19	1.036
	3.0	0.529	0.523	1.0196	1.3383	1.79	1.052
	4.0	0.713	0.703	1.0268	1.3400	2.41	1.068
	5.0	0.901	0.885	1.0340	1.3418	3.05	1.085
	6.0	1.092	1.069	1.0413	1.3435	3.70	1.104
	7.0	1.288	1.256	1.0486	1.3453	4.38	1.124
	8.0	1.488	1.445	1.0559	1.3470	5.08	1.145
	9.0	1.692	1.637	1.0633	1.3488	5.81	1.168
	10.0	1.901	1.832	1.0707	1.3505	6.56	1.193
	12.0	2.333	2.229	1.0857	1.3541	8.18	1.250
	14.0	2.785	2.637	1.1008	1.3576	9.94	1.317
	16.0	3.259	3.056	1.1162	1.3612	11.89	1.388
	18.0	3.756	3.486	1.1319	1.3648	14.04	1.463
	20.0	4.278	3.928	1.1478	1.3684	16.46	1.557
	22.0	4.826	4.382	1.1640	1.3721	19.18	1.676
	24.0	5.403	4.847	1.1804	1.3757		1.821
	26.0	6.012	5.326	1.1972	1.3795		1.990
Sodium	1.0	0.039	0.039	1.0049	1.3348	0.20	1.043
citrate	2.0	0.079	0.078	1.0120	1.3366	0.39	1.081
$(HO)C(COONa)_3$	3.0	0.120	0.118	1.0186	1.3383	0.59	1.122
	4.0	0.161	0.159	1.0260	1.3401	0.79	1.166
	5.0	0.204	0.200	1.0331	1.3419	0.97	1.210
	6.0	0.247	0.242	1.0405	1.3437	1.17	1.263
	7.0	0.292	0.284	1.0482	1.3455	1.36	1.314
	8.0	0.337	0.327	1.0557	1.3473	1.57	1.371
	9.0	0.383	0.371	1.0632	1.3491	1.77	1.427
	10.0	0.431	0.415	1.0708	1.3509	1.96	1.499
	12.0	0.528	0.505	1.0861	1.3546	2.38	1.649
	14.0	0.631	0.598	1.1019	1.3583	2.82	1.832
	16.0	0.738	0.693	1.1173	1.3618	3.27	2.045
	18.0	0.851	0.790	1.1327	1.3656	3.82	2.290
	20.0	0.969	0.891	1.1492	1.3693	4.39	2.596

Solute	Mass %	m/mol kg^{-1}	c/mol L^{-1}	ρ/g cm^{-3}	n	$\Delta/°C$	η/mPa s
	24.0	1.224	1.099	1.1813	1.3767		3.409
	28.0	1.507	1.318	1.2151	1.3845		4.586
	32.0	1.823	1.548	1.2487	1.3923		6.541
	36.0	2.180	1.792	1.2843	1.4001		9.788
Sodium	0.5	0.126	0.125	1.0039	1.3344	0.43	1.027
hydroxide	1.0	0.253	0.252	1.0095	1.3358	0.86	1.054
NaOH	2.0	0.510	0.510	1.0207	1.3386	1.74	1.112
	3.0	0.773	0.774	1.0318	1.3414	2.64	1.176
	4.0	1.042	1.043	1.0428	1.3441	3.59	1.248
	5.0	1.316	1.317	1.0538	1.3467	4.57	1.329
	6.0	1.596	1.597	1.0648	1.3494	5.60	1.416
	7.0	1.882	1.883	1.0758	1.3520	6.69	1.510
	8.0	2.174	2.174	1.0869	1.3546	7.87	1.616
	9.0	2.473	2.470	1.0979	1.3572	9.12	1.737
	10.0	2.778	2.772	1.1089	1.3597	10.47	1.882
	12.0	3.409	3.393	1.1309	1.3648	13.42	2.201
	14.0	4.070	4.036	1.1530	1.3697	16.76	2.568
	15.0	4.412	4.365	1.1640	1.3722		2.789
	16.0	4.762	4.701	1.1751	1.3746		3.043
	18.0	5.488	5.387	1.1971	1.3793		3.698
	20.0	6.250	6.096	1.2192	1.3840		4.619
	22.0	7.052	6.827	1.2412	1.3885		5.765
	24.0	7.895	7.579	1.2631	1.3929		7.100
	26.0	8.784	8.352	1.2848	1.3971		8.744
	28.0	9.723	9.145	1.3064	1.4012		10.832
	30.0	10.715	9.958	1.3277	1.4051		13.517
	32.0	11.766	10.791	1.3488	1.4088		16.844
	34.0	12.880	11.643	1.3697	1.4123		20.751
	36.0	14.064	12.512	1.3901	1.4156		25.290
	38.0	15.324	13.398	1.4102	1.4186		30.461
	40.0	16.668	14.300	1.4299	1.4215		36.312
Sodium	0.5	0.059	0.059	1.0016	1.3336	0.20	1.004
nitrate	1.0	0.119	0.118	1.0050	1.3341	0.40	1.007
NaNO$_3$	2.0	0.240	0.238	1.0117	1.3353	0.79	1.012
	3.0	0.364	0.359	1.0185	1.3364	1.18	1.018
	4.0	0.490	0.483	1.0254	1.3375	1.56	1.025
	5.0	0.619	0.607	1.0322	1.3387	1.94	1.032
	6.0	0.751	0.734	1.0392	1.3398	2.32	1.040
	7.0	0.886	0.862	1.0462	1.3409	2.70	1.049
	8.0	1.023	0.991	1.0532	1.3421	3.08	1.059
	9.0	1.164	1.123	1.0603	1.3432	3.46	1.069
	10.0	1.307	1.256	1.0674	1.3443	3.84	1.081
	12.0	1.604	1.527	1.0819	1.3466	4.60	1.107
	14.0	1.915	1.806	1.0967	1.3489	5.37	1.138
	18.0	2.583	2.387	1.1272	1.3536	6.98	1.215
	20.0	2.941	2.689	1.1429	1.3559	7.81	1.263
	30.0	5.042	4.326	1.2256	1.3678		1.609
	40.0	7.844	6.200	1.3175	1.3802		2.226
Sodium	0.5	0.031	0.031	1.0042	1.3343	0.19	1.033
phosphate	1.0	0.062	0.062	1.0100	1.3356	0.37	1.064
Na$_3$PO$_4$	1.5	0.093	0.093	1.0158	1.3369	0.53	1.094
	2.0	0.124	0.125	1.0216	1.3381	0.67	1.126
	2.5	0.156	0.157	1.0275	1.3394	0.79	1.161
	3.0	0.189	0.189	1.0335	1.3406		1.198

Solute	Mass %	m/mol kg^{-1}	c/mol L^{-1}	ρ/g cm^{-3}	n	$\Delta/°C$	η/mPa s
	3.5	0.221	0.222	1.0395	1.3419		1.238
	4.0	0.254	0.255	1.0456	1.3432		1.281
	4.5	0.287	0.289	1.0517	1.3444		1.327
	5.0	0.321	0.323	1.0579	1.3457		1.375
	5.5	0.355	0.357	1.0642	1.3470		1.426
	6.0	0.389	0.392	1.0705	1.3482		1.480
	6.5	0.424	0.427	1.0768	1.3495		1.538
	7.0	0.459	0.462	1.0832	1.3507		1.598
	7.5	0.495	0.498	1.0896	1.3519		1.662
	8.0	0.530	0.535	1.0961	1.3532		1.729
Sodium hydrogen phosphate Na$_2$HPO$_4$	0.5	0.035	0.035	1.0032	1.3340	0.17	1.021
	1.0	0.071	0.071	1.0082	1.3349	0.32	1.042
	1.5	0.107	0.107	1.0131	1.3358	0.46	1.064
	2.0	0.144	0.143	1.0180	1.3368		1.088
	2.5	0.181	0.180	1.0229	1.3377		1.113
	3.0	0.218	0.217	1.0279	1.3386		1.138
	3.5	0.255	0.255	1.0328	1.3396		1.165
	4.0	0.293	0.292	1.0378	1.3405		1.193
	4.5	0.332	0.331	1.0428	1.3414		1.223
	5.0	0.371	0.369	1.0478	1.3424		1.254
	5.5	0.410	0.408	1.0528	1.3433		1.286
Sodium dihydrogen phosphate NaH$_2$PO$_4$	0.5	0.042	0.042	1.0019	1.3336	0.14	1.018
	1.0	0.084	0.084	1.0056	1.3343	0.28	1.035
	1.5	0.127	0.126	1.0094	1.3349	0.42	1.051
	2.0	0.170	0.169	1.0131	1.3356	0.56	1.068
	2.5	0.214	0.212	1.0168	1.3362	0.70	1.085
	3.0	0.258	0.255	1.0206	1.3369	0.84	1.103
	3.5	0.302	0.299	1.0244	1.3375	0.98	1.121
	4.0	0.347	0.343	1.0281	1.3382	1.12	1.140
	4.5	0.393	0.387	1.0319	1.3388	1.25	1.160
	5.0	0.439	0.432	1.0358	1.3395	1.39	1.180
	6.0	0.532	0.522	1.0434	1.3408	1.65	1.223
	7.0	0.627	0.613	1.0511	1.3421	1.89	1.270
	8.0	0.725	0.706	1.0589	1.3434	2.12	1.319
	9.0	0.824	0.800	1.0668	1.3447	2.35	1.371
	10.0	0.926	0.896	1.0747	1.3460	2.58	1.428
	12.0	1.137	1.091	1.0907	1.3486	3.06	1.552
	14.0	1.357	1.292	1.1070	1.3512	3.53	1.694
	16.0	1.588	1.499	1.1236	1.3538	4.03	1.861
	18.0	1.830	1.711	1.1404	1.3565	4.55	2.050
	20.0	2.084	1.930	1.1576	1.3592	5.10	2.283
	22.0	2.351	2.155	1.1752	1.3618		2.550
	24.0	2.632	2.387	1.1931	1.3646		2.850
	26.0	2.929	2.625	1.2113	1.3673		3.214
	28.0	3.242	2.870	1.2299	1.3700		3.682
	30.0	3.572	3.123	1.2488	1.3728		4.300
	32.0	3.923	3.383	1.2682	1.3756		5.079
	34.0	4.294	3.650	1.2879	1.3784		6.008
	36.0	4.689	3.925	1.3080	1.3812		7.098
	38.0	5.109	4.208	1.3285	1.3840		8.363
	40.0	5.557	4.499	1.3493	1.3869		9.814
Sodium sulfate Na$_2$SO$_4$	0.5	0.035	0.035	1.0027	1.3338	0.17	1.013
	1.0	0.071	0.071	1.0071	1.3345	0.32	1.026
	2.0	0.144	0.143	1.0161	1.3360	0.61	1.058

Solute	Mass %	m/mol kg^{-1}	c/mol L^{-1}	ρ/g cm^{-3}	n	Δ/°C	η/mPa s
	3.0	0.218	0.217	1.0252	1.3376	0.87	1.091
	4.0	0.293	0.291	1.0343	1.3391	1.13	1.126
	5.0	0.371	0.367	1.0436	1.3406	1.36	1.163
	6.0	0.449	0.445	1.0526	1.3420	1.56	1.202
	7.0	0.530	0.523	1.0619	1.3435		1.244
	8.0	0.612	0.603	1.0713	1.3449		1.289
	9.0	0.696	0.685	1.0808	1.3464		1.337
	10.0	0.782	0.768	1.0905	1.3479		1.390
	12.0	0.960	0.938	1.1101	1.3509		1.508
	14.0	1.146	1.114	1.1301	1.3539		1.646
	16.0	1.341	1.296	1.1503	1.3567		1.812
	18.0	1.545	1.483	1.1705	1.3595		2.005
	20.0	1.760	1.677	1.1907	1.3620		2.227
	22.0	1.986	1.875	1.2106	1.3643		2.481
Sodium thiosulfate $Na_2S_2O_3$	0.5	0.032	0.032	1.0024	1.3340	0.14	1.012
	1.0	0.064	0.064	1.0065	1.3351	0.28	1.023
	2.0	0.129	0.128	1.0148	1.3371	0.57	1.044
	3.0	0.196	0.194	1.0231	1.3392	0.84	1.066
	4.0	0.264	0.261	1.0315	1.3413	1.09	1.090
	5.0	0.333	0.329	1.0399	1.3434	1.34	1.115
	6.0	0.404	0.398	1.0483	1.3454	1.59	1.141
	7.0	0.476	0.468	1.0568	1.3475	1.83	1.169
	8.0	0.550	0.539	1.0654	1.3496	2.06	1.199
	9.0	0.626	0.611	1.0740	1.3517	2.30	1.231
	10.0	0.703	0.685	1.0827	1.3538	2.55	1.267
	12.0	0.862	0.835	1.1003	1.3581	3.06	1.345
	14.0	1.030	0.990	1.1182	1.3624	3.60	1.435
	16.0	1.205	1.150	1.1365	1.3667	4.17	1.537
	18.0	1.388	1.315	1.1551	1.3711	4.76	1.657
	20.0	1.581	1.485	1.1740	1.3756	5.37	1.798
	30.0	2.711	2.417	1.2739	1.3987		2.903
	40.0	4.216	3.498	1.3827	1.4229		5.758
Strontium chloride $SrCl_2$	0.5	0.032	0.032	1.0027	1.3339	0.16	1.012
	1.0	0.064	0.064	1.0071	1.3348	0.31	1.021
	2.0	0.129	0.128	1.0161	1.3366	0.62	1.039
	3.0	0.195	0.194	1.0252	1.3384	0.93	1.057
	4.0	0.263	0.261	1.0344	1.3402	1.26	1.076
	5.0	0.332	0.329	1.0437	1.3421	1.61	1.096
	6.0	0.403	0.399	1.0532	1.3440	1.98	1.116
	7.0	0.475	0.469	1.0628	1.3459	2.38	1.136
	8.0	0.549	0.541	1.0726	1.3478	2.80	1.157
	9.0	0.624	0.615	1.0825	1.3498	3.25	1.180
	10.0	0.701	0.689	1.0925	1.3518	3.74	1.204
	12.0	0.860	0.843	1.1131	1.3558	4.81	1.258
	14.0	1.027	1.002	1.1342	1.3599	6.03	1.317
	16.0	1.202	1.167	1.1558	1.3641	7.41	1.383
	18.0	1.385	1.338	1.1780	1.3684	8.98	1.460
	20.0	1.577	1.515	1.2008	1.3728	10.74	1.549
	22.0	1.779	1.699	1.2241	1.3772	12.74	1.650
	24.0	1.992	1.890	1.2481	1.3817	14.99	1.765
	26.0	2.216	2.087	1.2728	1.3864		1.897
	28.0	2.453	2.293	1.2983	1.3911		2.056
	30.0	2.703	2.507	1.3248	1.3961		2.245
	32.0	2.968	2.730	1.3523	1.4013		2.527

Solute	Mass %	m/mol kg^{-1}	c/mol L^{-1}	ρ/g cm^{-3}	n	Δ/°C	η/mPa s
	34.0	3.250	2.962	1.3811	1.4067		2.846
	36.0	3.548	3.205	1.4114	1.4124		3.206
Sucrose	0.5	0.015	0.015	1.0002	1.3337	0.03	1.015
$C_{12}H_{22}O_{11}$	1.0	0.030	0.029	1.0021	1.3344	0.06	1.028
	2.0	0.060	0.059	1.0060	1.3359	0.11	1.055
	3.0	0.090	0.089	1.0099	1.3373	0.17	1.084
	4.0	0.122	0.118	1.0139	1.3388	0.23	1.114
	5.0	0.154	0.149	1.0178	1.3403	0.29	1.146
	6.0	0.186	0.179	1.0218	1.3418	0.35	1.179
	7.0	0.220	0.210	1.0259	1.3433	0.42	1.215
	8.0	0.254	0.241	1.0299	1.3448	0.49	1.254
	9.0	0.289	0.272	1.0340	1.3463	0.55	1.294
	10.0	0.325	0.303	1.0381	1.3478	0.63	1.336
	12.0	0.398	0.367	1.0465	1.3509	0.77	1.429
	14.0	0.476	0.431	1.0549	1.3541	0.93	1.534
	16.0	0.556	0.497	1.0635	1.3573	1.10	1.653
	18.0	0.641	0.564	1.0722	1.3606	1.27	1.790
	20.0	0.730	0.632	1.0810	1.3639	1.47	1.945
	22.0	0.824	0.700	1.0899	1.3672	1.67	2.124
	24.0	0.923	0.771	1.0990	1.3706	1.89	2.331
	26.0	1.026	0.842	1.1082	1.3741	2.12	2.573
	28.0	1.136	0.914	1.1175	1.3776	2.37	2.855
	30.0	1.252	0.988	1.1270	1.3812	2.64	3.187
	32.0	1.375	1.063	1.1366	1.3848	2.94	3.762
	34.0	1.505	1.139	1.1464	1.3885	3.27	4.052
	36.0	1.643	1.216	1.1562	1.3922	3.63	4.621
	38.0	1.791	1.295	1.1663	1.3960	4.02	5.315
	40.0	1.948	1.375	1.1765	1.3999	4.45	6.162
	42.0	2.116	1.456	1.1868	1.4038	4.93	7.234
	44.0	2.295	1.539	1.1972	1.4078		8.596
	46.0	2.489	1.623	1.2079	1.4118		10.301
	48.0	2.697	1.709	1.2186	1.4159		12.515
	50.0	2.921	1.796	1.2295	1.4201		15.431
	60.0	4.382	2.255	1.2864	1.4419		58.487
	70.0	6.817	2.755	1.3472	1.4654		481.561
	80.0	11.686	3.299	1.4117	1.4906		
Sulfuric acid	0.5	0.051	0.051	1.0016	1.3336	0.21	1.010
H_2SO_4	1.0	0.103	0.102	1.0049	1.3342	0.42	1.019
	2.0	0.208	0.206	1.0116	1.3355	0.80	1.036
	3.0	0.315	0.311	1.0183	1.3367	1.17	1.059
	4.0	0.425	0.418	1.0250	1.3379	1.60	1.085
	5.0	0.537	0.526	1.0318	1.3391	2.05	1.112
	6.0	0.651	0.635	1.0385	1.3403	2.50	1.136
	7.0	0.767	0.746	1.0453	1.3415	2.95	1.159
	8.0	0.887	0.858	1.0522	1.3427	3.49	1.182
	9.0	1.008	0.972	1.0591	1.3439	4.08	1.206
	10.0	1.133	1.087	1.0661	1.3451	4.64	1.230
	12.0	1.390	1.322	1.0802	1.3475	5.93	1.282
	14.0	1.660	1.563	1.0947	1.3500	7.49	1.337
	16.0	1.942	1.810	1.1094	1.3525	9.26	1.399
	18.0	2.238	2.064	1.1245	1.3551	11.29	1.470
	20.0	2.549	2.324	1.1398	1.3576	13.64	1.546
	22.0	2.876	2.592	1.1554	1.3602	16.48	1.624
	24.0	3.220	2.866	1.1714	1.3628	19.85	1.706
	26.0	3.582	3.147	1.1872	1.3653	24.29	1.797
	28.0	3.965	3.435	1.2031	1.3677	29.65	1.894

CONCENTRATIVE PROPERTIES OF AQUEOUS SOLUTIONS:
DENSITY, REFRACTIVE INDEX, FREEZING POINT DEPRESSION, AND VISCOSITY (continued)

Solute	Mass %	m/mol kg^{-1}	c/mol L^{-1}	ρ/g cm^{-3}	n	Δ/°C	η/mPa s
	30.0	4.370	3.729	1.2191	1.3701	36.21	2.001
	32.0	4.798	4.030	1.2353	1.3725	44.76	2.122
	34.0	5.252	4.339	1.2518	1.3749	55.28	2.255
	36.0	5.735	4.656	1.2685	1.3773		2.392
	38.0	6.249	4.981	1.2855	1.3797		2.533
	40.0	6.797	5.313	1.3028	1.3821		2.690
	42.0	7.383	5.655	1.3205	1.3846		2.872
	44.0	8.011	6.005	1.3386	1.3870		3.073
	46.0	8.685	6.364	1.3570	1.3895		3.299
	48.0	9.411	6.734	1.3759	1.3920		3.546
	50.0	10.196	7.113	1.3952	1.3945		3.826
	52.0	11.045	7.502	1.4149	1.3971		4.142
	54.0	11.969	7.901	1.4351	1.3997		4.499
	56.0	12.976	8.312	1.4558	1.4024		4.906
	58.0	14.080	8.734	1.4770	1.4050		5.354
	60.0	15.294	9.168	1.4987	1.4077		5.917
	70.0	23.790	11.494	1.6105			
	80.0	40.783	14.088	1.7272			
	90.0	91.762	16.649	1.8144			
	92.0	117.251	17.109	1.8240			
	94.0	159.734	17.550	1.8312			
	96.0	244.698	17.966	1.8355			
	98.0	499.592	18.346	1.8361			
	100.0		18.663	1.8305			
Trichloroacetic	0.5	0.031	0.031	1.0008	1.3337	0.11	1.011
acid	1.0	0.062	0.061	1.0034	1.3343	0.21	1.021
CCl$_3$COOH	2.0	0.125	0.123	1.0083	1.3356	0.42	1.044
	3.0	0.189	0.186	1.0133	1.3369	0.64	1.069
	4.0	0.255	0.249	1.0182	1.3381	0.86	1.096
	5.0	0.322	0.313	1.0230	1.3394	1.08	1.123
	6.0	0.391	0.377	1.0279	1.3406	1.30	1.150
	7.0	0.461	0.442	1.0328	1.3418	1.53	1.177
	8.0	0.532	0.508	1.0378	1.3431	1.76	1.204
	9.0	0.605	0.574	1.0428	1.3444	1.99	1.233
	10.0	0.680	0.641	1.0479	1.3456	2.23	1.263
	12.0	0.835	0.777	1.0583	1.3483	2.73	1.326
	14.0	0.996	0.916	1.0692	1.3510	3.26	1.393
	16.0	1.166	1.058	1.0806	1.3539	3.82	1.462
	18.0	1.343	1.203	1.0921	1.3568		1.533
	20.0	1.530	1.351	1.1035	1.3597		1.608
	24.0	1.933	1.654	1.1260	1.3652		1.768
	28.0	2.380	1.968	1.1485	1.3705		1.935
	32.0	2.880	2.294	1.1713	1.3759		2.118
	36.0	3.443	2.632	1.1947	1.3813		2.320
	40.0	4.080	2.984	1.2188	1.3868		1.543
	44.0	4.809	3.349	1.2435	1.3923		2.797
	48.0	5.650	3.726	1.2682	1.3977		3.076
Tris	0.5	0.041	0.041	0.9994	1.3337	0.08	1.014
(hydroxymethyl)-	1.0	0.083	0.083	1.0006	1.3344	0.16	1.027
methylamine	2.0	0.168	0.166	1.0030	1.3359	0.31	1.054
H$_2$NC(CH$_2$OH)$_3$	3.0	0.255	0.249	1.0054	1.3374	0.47	1.083
	4.0	0.344	0.333	1.0078	1.3388	0.64	1.115
	5.0	0.434	0.417	1.0103	1.3403	0.80	1.148
	6.0	0.527	0.502	1.0128	1.3418	0.97	1.182

Solute	Mass %	m/mol kg^{-1}	c/mol L^{-1}	ρ/g cm^{-3}	n	Δ/°C	η/mPa s
	7.0	0.621	0.587	1.0153	1.3433	1.15	1.218
	8.0	0.718	0.672	1.0179	1.3448	1.33	1.256
	9.0	0.816	0.758	1.0204	1.3463	1.51	1.295
	10.0	0.917	0.844	1.0230	1.3478	1.70	1.337
	12.0	1.126	1.019	1.0282	1.3508	2.08	1.427
	14.0	1.344	1.194	1.0335	1.3539	2.47	1.527
	16.0	1.572	1.372	1.0389	1.3570	2.90	1.642
	18.0	1.812	1.552	1.0443	1.3601	3.36	1.772
	20.0	2.064	1.733	1.0498	1.3633	3.85	1.920
	30.0	3.538	2.670	1.0781	1.3797		2.998
	40.0	5.503	3.657	1.1076	1.3970		5.208
Urea	0.5	0.084	0.083	0.9995	1.3337	0.16	1.007
$(NH_2)_2CO$	1.0	0.168	0.167	1.0007	1.3344	0.31	1.010
	2.0	0.340	0.334	1.0033	1.3358	0.62	1.012
	3.0	0.515	0.502	1.0058	1.3372	0.93	1.017
	4.0	0.694	0.672	1.0085	1.3387	1.24	1.025
	5.0	0.876	0.842	1.0111	1.3401	1.55	1.033
	6.0	1.063	1.013	1.0138	1.3416	1.88	1.041
	7.0	1.253	1.185	1.0165	1.3431	2.22	1.049
	8.0	1.448	1.358	1.0192	1.3446	2.56	1.057
	9.0	1.647	1.531	1.0220	1.3461	2.91	1.065
	10.0	1.850	1.706	1.0248	1.3476	3.26	1.074
	12.0	2.270	2.059	1.0304	1.3506	3.95	1.091
	14.0	2.710	2.415	1.0360	1.3537	4.66	1.109
	16.0	3.171	2.775	1.0417	1.3568	5.40	1.130
	18.0	3.655	3.139	1.0473	1.3599	6.19	1.153
	20.0	4.163	3.506	1.0530	1.3629	7.00	1.178
	22.0	4.696	3.878	1.0586	1.3661	7.81	1.205
	24.0	5.258	4.253	1.0643	1.3692	8.64	1.235
	26.0	5.850	4.632	1.0699	1.3723	9.52	1.266
	28.0	6.475	5.014	1.0756	1.3754	10.45	1.298
	30.0	7.136	5.401	1.0812	1.3785	11.40	1.332
	32.0	7.835	5.791	1.0869	1.3817	12.34	1.371
	34.0	8.577	6.185	1.0926	1.3848	13.27	1.413
	36.0	9.366	6.584	1.0984	1.3881	14.20	1.459
	38.0	10.205	6.988	1.1044	1.3913	15.11	1.509
	40.0	11.100	7.397	1.1106	1.3947	15.99	1.565
	42.0	12.057	7.812	1.1171	1.3982	16.83	1.629
	44.0	13.082	8.234	1.1239	1.4018	17.62	1.700
	46.0	14.183	8.665	1.1313	1.4056		1.780
Zinc sulfate	0.5	0.031	0.031	1.0034	1.3339	0.08	1.021
$ZnSO_4$	1.0	0.063	0.062	1.0085	1.3348	0.15	1.040
	2.0	0.126	0.126	1.0190	1.3366	0.28	1.081
	3.0	0.192	0.191	1.0296	1.3384	0.41	1.126
	4.0	0.258	0.258	1.0403	1.3403	0.53	1.175
	5.0	0.326	0.326	1.0511	1.3421	0.65	1.227
	6.0	0.395	0.395	1.0620	1.3439	0.77	1.283
	7.0	0.466	0.465	1.0730	1.3457	0.89	1.341
	8.0	0.539	0.537	1.0842	1.3475	1.01	1.403
	9.0	0.613	0.611	1.0956	1.3494	1.14	1.470
	10.0	0.688	0.686	1.1071	1.3513	1.27	1.545
	12.0	0.845	0.840	1.1308	1.3551	1.55	1.716
	14.0	1.008	1.002	1.1553	1.3590	1.89	1.918
	16.0	1.180	1.170	1.1806	1.3630	2.31	2.152

ION PRODUCT OF WATER SUBSTANCE
William L. Marshall and E. U. Franck

Pressure (bars)	Temperature (°C)								
	0	25	50	75	100	150	200	250	300
Saturated vapor	14.938	13.995	13.275	12.712	12.265	11.638	11.289	11.191	11.406
250	14.83	13.90	13.19	12.63	12.18	11.54	11.16	11.01	11.14
500	14.72	13.82	13.11	12.55	12.10	11.45	11.05	10.85	10.86
750	14.62	13.73	13.04	12.48	12.03	11.36	10.95	10.72	10.66
1,000	14.53	13.66	12.96	12.41	11.96	11.29	10.86	10.60	10.50
1,500	14.34	13.53	12.85	12.29	11.84	11.16	10.71	10.43	10.26
2,000	14.21	13.40	12.73	12.18	11.72	11.04	10.57	10.27	10.08
2,500	14.08	13.28	12.62	12.07	11.61	10.92	10.45	10.12	9.91
3,000	13.97	13.18	12.53	11.98	11.53	10.83	10.34	9.99	9.76
3,500	13.87	13.09	12.44	11.90	11.44	10.74	10.24	9.88	9.63
4,000	13.77	13.00	12.35	11.82	11.37	10.66	10.16	9.79	9.52
5,000	13.60	12.83	12.19	11.66	11.22	10.52	10.00	9.62	9.34
6,000	13.44	12.68	12.05	11.53	11.09	10.39	9.87	9.48	9.18
7,000	13.31	12.55	11.93	11.41	10.97	10.27	9.75	9.35	9.04
8,000	13.18	12.43	11.82	11.30	10.86	10.17	9.64	9.24	8.93
9,000	13.04	12.31	11.71	11.20	10.77	10.07	9.54	9.13	8.82
10,000	12.91	12.21	11.62	11.11	10.68	9.98	9.45	9.04	8.71

Pressure (bars)	Temperature (°C)								
	350	400	450	500	600	700	800	900	1000
Saturated vapor	12.30	—	—	—	—	—	—	—	—
250	11.77	19.43	21.59	22.40	23.27	23.81	24.23	24.59	24.93
500	11.14	11.88	13.74	16.13	18.30	19.29	19.92	20.39	20.80
750	10.79	11.17	11.89	13.01	15.25	16.55	17.35	17.93	18.39
1,000	10.54	10.77	11.19	11.81	13.40	14.70	15.58	16.22	16.72
1,500	10.22	10.29	10.48	10.77	11.59	12.50	13.30	13.97	14.50
2,000	9.98	9.98	10.07	10.23	10.73	11.36	11.98	12.54	12.97
2,500	9.79	9.74	9.77	9.86	10.18	10.63	11.11	11.59	12.02
3,000	9.61	9.54	9.53	9.57	9.78	10.11	10.49	10.89	11.24
3,500	9.47	9.37	9.33	9.34	9.48	9.71	10.02	10.35	10.62
4,000	9.34	9.22	9.16	9.15	9.23	9.41	9.65	9.93	10.13
5,000	9.13	8.99	8.90	8.85	8.85	8.95	9.11	9.30	9.42
6,000	8.96	8.80	8.69	8.62	8.57	8.61	8.72	8.86	8.97
7,000	8.81	8.64	8.51	8.42	8.34	8.34	8.40	8.51	8.64
8,000	8.68	8.50	8.36	8.25	8.13	8.10	8.13	8.21	8.38
9,000	8.57	8.37	8.22	8.10	7.95	7.89	7.89	7.95	8.12
10,000	8.46	8.25	8.09	7.96	7.78	7.70	7.68	7.70	7.85

Data in this table were calculated from the equation, $\log_{10} K_w^* = A + B/T + C/T^2 + D/T^3 + (E + F/T + G/T^2) \log_{10} \rho_w^*$, where $K_w^* = K_w/(\text{mol kg}^{-1})$, and $\rho_w^* = \rho_w/(\text{g cm}^{-3})$. The parameters are:

$A = -4.098$ $E = +13.957$
$B = -3245.2$ K $F = 1262.3$ K
$C = +2.2362 \times 10^5$ K^2 $G = +8.5641 \times 10^5$ K^2
$D = -3.984 \times 10^7$ K^3

Reprinted with permission from W. L. Marshall and E. U. Franck, *J. Phys. Chem. Ref. Data*, 10, 295, 1981.

IONIZATION CONSTANT OF NORMAL AND HEAVY WATER

This table gives the ionization constant in molality terms for H_2O and D_2O at temperatures from 0 to 100°C at the saturated vapor pressure. The quantity tabulated is $-\log K_W$, where K_W is defined by

$$K_W = m_+ \times m_-$$

and m_+ and m_- are the molalities, in mol/kg of water, for H^+ and OH^-, respectively.

REFERENCES

1. W.L. Marshall and E.U. Franck, *J. Phys. Chem. Ref. Data*, 10, 295, 1981.
2. R.E. Mesmer and D.L. Herting, *J. Solution Chem.*, 7, 901, 1978.

	$-\log K_W$	
$t/°C$	H_2O	D_2O
0	14.938	15.972
5	14.727	15.743
10	14.528	15.527
15	14.340	15.324
20	14.163	15.132
25	13.995	14.951
30	13.836	14.779
35	13.685	14.616
40	13.542	14.462
45	13.405	14.316
50	13.275	14.176
55	13.152	14.044
60	13.034	13.918
65	12.921	13.798
70	12.814	13.683
75	12.712	13.574
80	12.613	13.470
85	12.520	13.371
90	12.428	13.276
95	12.345	13.186
100	12.265	13.099

SOLUBILITY OF SELECTED GASES IN WATER

L. H. Gevantman

The values in this table are taken almost exclusively from the International Union of Pure and Applied Chemistry "Solubility Data Series". Unless noted, they comprise evaluated data fitted to a smoothing equation. The data at each temperature are then derived from the smoothing equation which expresses the mole fraction solubility X_1 of the gas in solution as:

$$\ln X_1 = A + B/T^* + C \ln T^*$$

where

$$T^* = T/100 \text{ K}$$

All values refer to a partial pressure of the gas of 101.325 kPa (one atmosphere).

The equation constants, the standard deviation for $\ln X_1$ (except where noted), and the temperature range over which the equation applies are given in the column headed Equation constants. There are two exceptions. The equation for methane has an added term, DT^*. The equation for H_2Se and H_2S takes the form,

$$\ln X_1 = A + B/T + C \ln T + DT$$

where T is the temperature in kelvin.

Solubilities given for those gases which react with water, namely ozone, nitrogen oxides, chlorine and its oxides, carbon dioxide, hydrogen sulfide, hydrogen selenide and sulfur dioxide, are recorded as bulk solubilities; i.e., all chemical species of the gas and its reaction products with water are included.

Gas	T/K	Solubility (X_1)	Equation constants	Ref.
Hydrogen (H_2)	288.15	1.510×10^{-5}	$A = -48.1611$	1
$M_r = 2.01588$	293.15	1.455×10^{-5}	$B = 55.2845$	
	298.15	1.411×10^{-5}	$C = 16.8893$	
	303.15	1.377×10^{-5}	Std. dev. $= \pm 0.54\%$	
	308.15	1.350×10^{-5}	Temp.range $= 273.15-353.15$	
Deuterium (D_2)	283.15	$1.675 \times 10^{-5} \pm 0.57\%$	Averaged experimental	1
$M_r = 4.0282$	288.15	$1.595 \times 10^{-5} \pm 0.57\%$	values	
	293.15	$1.512 \times 10^{-5} \pm 0.78\%$	Temp. range $= 278.15-303.15$	
	298.15	$1.460 \times 10^{-5} \pm 0.52\%$		
	303.15	$1.395 \times 10^{-5} \pm 0.37\%$		
Helium (He)	288.15	7.123×10^{-6}	$A = -41.4611$	2
$A_r = 4.0026$	293.15	7.044×10^{-6}	$B = 42.5962$	
	298.15	6.997×10^{-6}	$C = 14.0094$	
	303.15	6.978×10^{-6}	Std. dev. $= \pm 0.54\%$	
	308.15	6.987×10^{-6}	Temp.range $= 273.15-348.15$	
Neon (Ne)	288.15	8.702×10^{-6}	$A = -52.8573$	2
$A_r = 20.1797$	293.15	8.395×10^{-6}	$B = 61.0494$	
	298.15	8.152×10^{-6}	$C = 18.9157$	
	303.15	7.966×10^{-6}	Std. dev. $= \pm 0.47\%$	
	308.15	7.829×10^{-6}	Temp.range $= 273.15-348.15$	
Argon (Ar)	288.15	3.025×10^{-5}	$A = -57.6661$	3
$A_r = 39.948$	293.15	2.748×10^{-5}	$B = 74.7627$	
	298.15	2.519×10^{-5}	$C = 20.1398$	
	303.15	2.328×10^{-5}	Std. dev. $= \pm 0.26\%$	
	308.15	2.169×10^{-5}	Temp.range $= 273.15-348.15$	
Krypton (Kr)	288.15	5.696×10^{-5}	$A = -66.9928$	4
$A_r = 83.80$	293.15	5.041×10^{-5}	$B = 91.0166$	
	298.15	4.512×10^{-5}	$C = 24.2207$	

Gas	T/K	Solubility (X_1)	Equation constants	Ref.
	303.15	4.079×10^{-5}	Std. dev. = ±0.32%	
	308.15	3.725×10^{-5}	Temp.range = 273.15—353.15	
Xenon (Xe)	288.15	10.519×10^{-5}	$A = -74.7398$	4
$A_r = 131.29$	293.15	9.051×10^{-5}	$B = 105.210$	
	298.15	7.890×10^{-5}	$C = 27.4664$	
	303.15	6.961×10^{-5}	Std. dev. = ±0.35%	
	308.15	6.212×10^{-5}	Temp.range = 273.15—348.15	
Radon-222(^{222}Rn)	288.15	2.299×10^{-4}	$A = -90.5481$	
$A_r = 222$	293.15	1.945×10^{-4}	$B = 130.026$	
	298.15	1.671×10^{-4}	$C = 35.0047$	
	303.15	1.457×10^{-4}	Std. dev. = ±1.02%	
	308.15	1.288×10^{-4}	Temp.range = 273.15—373.15	
Oxygen (O_2)	288.15	2.756×10^{-5}	$A = -66.7354$	5
$M_r = 31.9988$	293.15	2.501×10^{-5}	$B = 87.4755$	
	298.15	2.293×10^{-5}	$C = 24.4526$	
	303.15	2.122×10^{-5}	Std. dev. = ±0.36%	
	308.15	1.982×10^{-5}	Temp.range = 273.15—348.15	
Ozone (O_3)	293.15	$1.885 \times 10^{-6} \pm 10\%$	Experimental value derived	5
$M_r = 47.9982$		pH = 7.0	from Henry's Law Constant	
Nitrogen (N_2)	288.15	1.386×10^{-5}	$A = -67.3877$	6
$M_r = 28.0134$	293.15	1.274×10^{-5}	$B = 86.3213$	
	298.15	1.183×10^{-5}	$C = 24.7981$	
	303.15	1.108×10^{-5}	Std. dev. = ±0.72%	
	308.15	1.047×10^{-5}	Temp.range = 273.15—348.15	
Nitrous oxide (N_2O)	288.15	5.948×10^{-4}	$A = -60.7467$	7
$M_r = 44.0129$	293.15	5.068×10^{-4}	$B = 88.8280$	
	298.15	4.367×10^{-4}	$C = 21.2531$	
	303.15	3.805×10^{-4}	Std. dev. = ±1.2%	
	308.15	3.348×10^{-4}	Temp.range = 273.15—313.15	
Nitric oxide (NO)	288.15	4.163×10^{-5}	$A = -62.8086$	7
$M_r = 30.0061$	293.15	3.786×10^{-5}	$B = 82.3420$	
	298.15	3.477×10^{-5}	$C = 22.8155$	
	303.15	3.222×10^{-5}	Std. dev. = ±0.76%	
	308.15	3.012×10^{-5}	Temp.range = 273.15—358.15	
Carbon monoxide (CO)	288.15	2.095×10^{-5}	Derived from Henry's	8
$M_r = 28.0104$	293.15	1.918×10^{-5}	Law Constant Equation	
	298.15	1.774×10^{-5}	Std. dev. = ±0.043%	
	303.15	1.657×10^{-5}	Temp.range = 273.15—328.15	
	308.15	1.562×10^{-5}		
Carbon dioxide (CO_2)	288.15	8.21×10^{-4}	Derived from Henry's	9
$M_r = 44.0098$	293.15	7.07×10^{-4}	Law Constant Equation	
	298.15	6.15×10^{-4}	Std. dev. = ±1.1%	
	303.15	5.41×10^{-4}	Temp.range = 273.15—353.15	
	308.15	4.80×10^{-4}		
Hydrogen selenide (H_2Se)	288.15	1.80×10^{-3}	$A = 9.15$	10
$M_r = 80.976$	298.15	1.49×10^{-3}	$B = 974$	
	308.15	1.24×10^{-3}	$C = -3.542$	
			$D = 0.0042$	

SOLUBILITY OF SELECTED GASES IN WATER (continued)

Gas	T/K	Solubility (X_1)	Equation constants	Ref.
			Std. dev. = $\pm 2.3 \times 10^{-5}$	
			Temp. range = 288.15—343.15	
Hydrogen sulfide (H$_2$S)	288.15	2.335×10^{-3}	$A = -24.912$	10
$M_r = 34.082$	293.15	2.075×10^{-3}	$B = 3477$	
	298.15	1.85×10^{-3}	$C = 0.3993$	
	303.15	1.66×10^{-3}	$D = 0.0157$	
	308.15	1.51×10^{-3}	Std. dev. = $\pm 6.5 \times 10^{-5}$	
			Temp. range = 283.15—603.15	
Sulfur dioxide (SO$_2$)	288.15	3.45×10^{-2}	$A = -25.2629$	11
$M_r = 64.0648$	293.15	2.90×10^{-2}	$B = 45.7552$	
	298.15	2.46×10^{-2}	$C = 5.6855$	
	303.15	2.10×10^{-2}	Std. dev. = $\pm 1.8\%$	
	308.15	1.80×10^{-2}	Temp.range = 278.15—328.15	
Chlorine (Cl$_2$)	283.15	$2.48 \times 10^{-3} \pm 2\%$	Experimental data	11
$M_r = 70.9054$	293.15	$1.88 \times 10^{-3} \pm 2\%$	Temp.range = 283.15—333.15	
	303.15	$1.50 \times 10^{-3} \pm 2\%$		
	313.15	$1.23 \times 10^{-3} \pm 2\%$		
Chlorine monoxide (Cl$_2$O)	273.15	$5.25 \times 10^{-1} \pm 1\%$	Experimental data	11
$M_r = 86.9048$	276.61	$4.54 \times 10^{-1} \pm 1\%$	Temp. range = 273.15—293.15	
	283.15	$4.273 \times 10^{-1} \pm 1\%$		
	293.15	$3.353 \times 10^{-1} \pm 1\%$		
Chlorine dioxide (ClO$_2$)	288.15	2.67×10^{-2}	$A = 7.9163$	11
$M_r = 67.4515$	293.15	2.20×10^{-2}	$B = 0.4791$	
	298.15	1.823×10^{-2}	$C = 11.0593$	
	303.15	1.513×10^{-2}	Std. dev. = $\pm 4.6\%$	
	308.15	1.259×10^{-2}	Temp.range = 283.15—333.15	
Methane (CH$_4$)	288.15	3.122×10^{-5}	$A = -115.6477$	12
$M_r = 16.0428$	293.15	2.806×10^{-5}	$B = 155.5756$	
	298.15	2.552×10^{-5}	$C = 65.2553$	
	303.15	2.346×10^{-5}	$D = -6.6170$	
	308.15	2.180×10^{-5}	Std. dev. = $\pm 0.056\%$	
			Temp.range = 273.15—328.15	
Ethane (C$_2$H$_6$)	288.15	4.556×10^{-5}	$A = -90.8225$	13
$M_r = 30.0696$	293.15	3.907×10^{-5}	$B = 126.9559$	
	298.15	3.401×10^{-5}	$C = 34.7413$	
	303.15	3.002×10^{-5}	Std. dev. = $\pm 0.13\%$	
	308.15	2.686×10^{-5}	Temp.range = 273.15—323.15	
Propane (C$_3$H$_8$)	288.15	3.813×10^{-5}	$A = -102.044$	14
$M_r = 44.097$	293.15	3.200×10^{-5}	$B = 144.345$	
	298.15	2.732×10^{-5}	$C = 39.4740$	
	303.15	2.370×10^{-5}	Std. dev. = $\pm 0.012\%$	
	308.15	2.088×10^{-5}	Temp.range = 273.15—347.15	
Butane (C$_4$H$_{10}$)	288.15	3.274×10^{-5}	$A = -102.029$	14
$M_r = 58.123$	293.15	2.687×10^{-5}	$B = 146.040$	
	298.15	2.244×10^{-5}	$C = 38.7599$	
	303.15	1.906×10^{-5}	Std. dev. = $\pm 0.026\%$	
	308.15	1.645×10^{-5}	Temp.range = 273.15—349.15	
2-Methyl propane (Isobutane)	288.15	2.333×10^{-5}	$A = -129.714$	14

SOLUBILITY OF SELECTED GASES IN WATER (continued)

Gas	T/K	Solubility (X_1)	Equation constants	Ref.
(C_4H_{10})	293.15	1.947×10^{-5}	$B = 183.044$	
$M_r = 58.123$	298.15	1.659×10^{-5}	$C = 53.4651$	
	303.15	1.443×10^{-5}	Std. dev. = $\pm 0.034\%$	
	308.15	1.278×10^{-5}	Temp.range = 278.15—318.15	

REFERENCES

1. C. L. Young, Ed., *IUPAC Solubility Data Series*, Vol. 5/6, Hydrogen and Deuterium, Pergamon Press, Oxford, England, 1981.
2. H. L. Clever, Ed., *IUPAC Solubility Data Series*, Vol. 1, Helium and Neon, Pergamon Press, Oxford, England, 1979.
3. H. L. Clever, Ed., *IUPAC Solubility Data Series*, Vol. 4, Argon, Pergamon Press, Oxford, England, 1980.
4. H. L. Clever, Ed., *IUPAC Solubility Data Series*, Vol. 2, Krypton, Xenon and Radon, Pergamon Press, Oxford, England, 1979.
5. R. Battino, Ed., *IUPAC Solubility Data Series*, Vol. 7, Oxygen and Ozone, Pergamon Press, Oxford, England, 1981.
6. R. Battino, Ed., *IUPAC Solubility Data Series*, Vol. 10, Nitrogen and Air, Pergamon Press, Oxford, England, 1982.
7. C. L. Young, Ed., *IUPAC Solubility Data Series*, Vol. 8, Oxides of Nitrogen, Pergamon Press, Oxford, England, 1981.
8. R. W. Cargill, Ed., *IUPAC Solubility Data Series*, Vol. 43, Carbon Monoxide, Pergamon Press, Oxford, England, 1990.
9. R. Crovetto, Evaluation of Solubility Data for the System CO_2-H_2O, *J. Phys. Chem. Ref. Data*, 20, 575, 1991.
10. P. G. T. Fogg and C. L. Young, Eds., *IUPAC Solubility Data Series*, Vol. 32, Hydrogen Sulfide, Deuterium Sulfide, and Hydrogen Selenide, Pergamon Press, Oxford, England, 1988.
11. C. L. Young, Ed., *IUPAC Solubility Data Series*, Vol. 12, Sulfur Dioxide, Chlorine, Fluorine and Chlorine Oxides, Pergamon Press, Oxford, England, 1983.
12. H. L. Clever and C. L. Young, Eds., *IUPAC Solubility Data Series*, Vol. 27/28, Methane, Pergamon Press, Oxford, England, 1987.
13. W. Hayduk, Ed., *IUPAC Solubility Data Series*, Vol. 9, Ethane, Pergamon Press, Oxford, England, 1982.
14. W. Hayduk, Ed., *IUPAC Solubility Data Series*, Vol. 24, Propane, Butane and 2-Methylpropane, Pergamon Press, Oxford, England, 1986.

SOLUBILITY OF CARBON DIOXIDE IN WATER AT VARIOUS TEMPERATURES AND PRESSURES

The solubility of CO_2 in water, expressed as mole fraction of CO_2 in the liquid phase, is given for pressures up to atmospheric and temperatures of 0 to 100°C. Note that 1 standard atmosphere equals 101.325 kPa. The references give data over a wider range of temperature and pressure. The estimated accuracy is about 2%.

REFERENCES

1. Carroll, J. J., Slupsky, J. D., and Mather, A. E., *J. Phys. Chem. Ref. Data*, 20, 1201, 1991.
2. Fernandez-Prini, R. and Crovetto, R., *J. Phys. Chem. Ref. Data*, 18, 1231, 1989.
3. Crovetto, R., *J. Phys. Chem. Ref. Data*, 20, 575, 1991

$1000 \times$ mole fraction of CO_2 in liquid phase

$t/°C$	Partial pressure of CO_2 in kPa						
	5	10	20	30	40	50	100
0	0.067	0.135	0.269	0.404	0.538	0.671	1.337
5	0.056	0.113	0.226	0.338	0.451	0.564	1.123
10	0.048	0.096	0.191	0.287	0.382	0.477	0.950
15	0.041	0.082	0.164	0.245	0.327	0.409	0.814
20	0.035	0.071	0.141	0.212	0.283	0.353	0.704
25	0.031	0.062	0.123	0.185	0.247	0.308	0.614
30	0.027	0.054	0.109	0.163	0.218	0.271	0.541
35	0.024	0.048	0.097	0.145	0.193	0.242	0.481
40	0.022	0.043	0.087	0.130	0.173	0.216	0.431
45	0.020	0.039	0.078	0.117	0.156	0.196	0.389
50	0.018	0.036	0.071	0.107	0.142	0.178	0.354
55	0.016	0.033	0.065	0.098	0.131	0.163	0.325
60	0.015	0.030	0.060	0.090	0.121	0.150	0.300
65	0.014	0.028	0.056	0.084	0.112	0.140	0.279
70	0.013	0.026	0.052	0.079	0.105	0.131	0.261
75	0.012	0.025	0.049	0.074	0.099	0.123	0.245
80	0.012	0.023	0.047	0.070	0.093	0.116	0.232
85	0.011	0.022	0.044	0.067	0.089	0.111	0.221
90	0.011	0.021	0.042	0.064	0.085	0.106	0.211
95	0.010	0.020	0.041	0.061	0.082	0.102	0.203
100	0.010	0.020	0.039	0.059	0.079	0.098	0.196

AQUEOUS SOLUBILITY AND HENRY'S LAW CONSTANTS OF ORGANIC COMPOUNDS

The solubility in water of about 580 organic compounds, including many compounds of environmental interest, is tabulated here. Values are given at 25°C or at the nearest temperature to this where data are available. In some cases solubility values are given at other temperatures as well.

Solubility of solids is defined as the concentration of the compound in a solution that is in equilibrium with the solid phase at the specified temperature and one atmosphere pressure. For liquids whose water mixtures separate into two phases, the solubility given here is the concentration of the compound in the water-rich phase at equilibrium. In the case of gases (i.e., compounds whose vapor pressure at the specified temperature exceeds one atmosphere) the solubility is defined here as the concentration in the water phase when the partial pressure of the compound above the solution is 101.325 kPa (1 atm). Values for gases are marked with an asterisk.

All solubility values are expressed as mass percent of solute, $S = 100\,w_2$, where the mass fraction w_2 is given by $w_2 = m_2/(m_1 + m_2)$. In these equations m_2 is the mass of solute and m_1 the mass of water. This quantity is related to other common measures of solubility as follows:

Molality: $m_2 = 1000 w_2/M_2(1-w_2)$

Mole fraction: $x_2 = (w_2/M_2)/\{(w_2/M_2) + (1-w_2)/M_1\}$

Mass of solute per 100 g of H_2O: $r_2 = 100w_2/(1-w_2)$

Here M_2 is the molar mass of the solute and $M_1 = 18.015$ g/mol is the molar mass of water. For small values of S the amount of substance concentration c_2 in moles per liter is approximately $10S/M_2$.

Data have been selected from evaluated sources wherever possible, in particular the *IUPAC Solubility Data Series* (References 1-4). The primary source for each value is listed in the column following the solubility values. The user is cautioned that wide variations of data are found in the literature for the lower solubility compounds.

The table also contains values of the Henry's Law constant k_H, which provides a measure of the partition of a substance between the atmosphere and the aqueous phase. Here k_H is defined as the limit of p_2/c_2 as the concentration approaches zero, where p_2 is the partial pressure of the solute above the solution and c_2 is the concentration in the solution at equilibrium (other formulations of Henry's Law are often used; see Reference 5). The values of k_H listed here are based on direct experimental measurement whenever available, but many of them are simply calculated as the ratio of the pure compound vapor pressure to the solubility. This approximation is reliable only for compounds of very low solubility. In fact, values of k_H found in the literature frequently differ by a factor of two or three, and variations over an order of magnitude are not unusual (Reference 5). Therefore the data given here should be taken only as a rough indication of the true Henry's Law constant, which is difficult to measure precisely.

All values of k_H refer to 25°C. If the vapor pressure of the compound at 25°C is greater than one atmosphere, it can be assumed that the k_H value has been calculated as $101.325/c_2$ kPa m^3/mol. The source of the Henry's Law data is given in the last column. The air-water partition coefficient (i.e., ratio of air concentration to water concentration when both are expressed in the same units) is equal to k_H/RT or $k_H/2.48$ in the units used here.

Compounds are listed by molecular formula following the Hill convention. To locate a compound by name or CAS Registry Number when the molecular formula is not known, use the table "Physical Constants of Organic Compounds" in Section 3 and its indexes to determine the molecular formula.

* Indicates a value of S for a gas at a partial pressure of 101.325 kPa (1 atm) in equilibrium with the solution.

REFERENCES

1. *Solubility Data Series, International Union of Pure and Applied Chemistry, Vol. 15,* Pergamon Press, Oxford, 1982.
2. *Solubility Data Series, International Union of Pure and Applied Chemistry, Vol. 20,* Pergamon Press, Oxford, 1985.
3. *Solubility Data Series, International Union of Pure and Applied Chemistry, Vol. 37,* Pergamon Press, Oxford, 1988.
4. *Solubility Data Series, International Union of Pure and Applied Chemistry, Vol. 38,* Pergamon Press, Oxford, 1988.
5. Mackay, D., and Shiu, W.Y., *J. Phys. Chem. Ref. Data,* 10, 1175, 1981.
6. Pearlman, R.S., and Yalkowsky, S.H., *J. Phys. Chem. Ref. Data,* 13, 975, 1984.
7. Shiu, W.Y., and Mackay, D., *J. Phys. Chem. Ref. Data,* 15, 911, 1986.
8. Varhanickova, D., Lee, S.C., Shiu, W. Y., and Mackay, D., *J. Chem. Eng. Data,* 40, 620, 1995.
9. Miller, M.M., Ghodbane, S., Wasik, S. P., Tewari, Y.B., and Martire, D.E., *J. Chem. Eng. Data,* 29, 184, 1984.
10. Riddick, J.A., Bunger, W.B., and Sakano, T.K., *Organic Solvents, Fourth Edition*, John Wiley & Sons, New York, 1986.
11. Mackay, D., Shiu, W.Y., and Ma, K.C., *Illustrated Handbook of Physical-Chemical Properties and Environmental Fate for Organic Chemicals*, Vol. I, Lewis Publishers/CRC Press, Boca Raton, FL, 1992.
12. Mackay, D., Shiu, W.Y., and Ma, K.C., *Illustrated Handbook of Physical-Chemical Properties and Environmental Fate for Organic Chemicals*, Vol. II, Lewis Publishers/CRC Press, Boca Raton, FL, 1992.
13. Mackay, D., Shiu, W.Y., and Ma, K.C., *Illustrated Handbook of Physical-Chemical Properties and Environmental Fate for Organic Chemicals*, Vol. III, Lewis Publishers/CRC Press, Boca Raton, FL, 1993.
14. Horvath, A.L., *Halogenated Hydrocarbons*, Marcel Dekker, New York, 1982.
15. Howard, P.H., *Handbook of Environmental Fate and Exposure Data for Organic Chemicals, Vol. I,* Lewis Publishers/CRC Press, Boca Raton, FL, 1989.
16. Howard, P.H., *Handbook of Environmental Fate and Exposure Data for Organic Chemicals, Vol. II,* Lewis Publishers/CRC Press, Boca Raton, FL, 1990.
17. Banergee, S., Yalkowsky, S.H., and Valvani, S.C., *Environ. Sci. Technol.,* 14, 1227, 1980.
18. Gevantman, L., in *CRC Handbook of Chemistry and Physics, 77th Edition*, p. 6-3, CRC Press, Boca Raton, FL, 1996.
19. Wilhelm, E., Battino, R., and Wilcock, R.J., *Chem. Rev.,* 77, 219, 1977.
20. Stephenson, R.M., *J. Chem. Eng. Data,* 37, 80, 1992.
21. Stephenson, R.M., Stuart, J., and Tabak, M., *J. Chem. Eng. Data,* 29, 287, 1984.

Mol. Form.	Name	$t/°C$	S/mass %	Ref.	k_H/kPa m^3 mol^{-1}	Ref.
CBrF$_3$	Bromotrifluoromethane	25	0.032*	14		
CBr$_3$F	Tribromofluoromethane	25	0.040	14		
CBr$_4$	Tetrabromomethane	30	0.024	14		
CClF$_3$	Chlorotrifluoromethane	25	0.009*	10	6.9	13
CCl$_2$F$_2$	Dichlorodifluoromethane	20	0.028*	5	41	13
CCl$_3$F	Trichlorofluoromethane	20	0.11	5	10.2	13
CCl$_4$	Tetrachloromethane	25	0.065	20	2.99	13
CF$_4$	Tetrafluoromethane	25	0.00187*	19		
CHBr$_3$	Tribromomethane	25	0.30	5	0.047	13
CHClF$_2$	Chlorodifluoromethane	25	0.30*	10	3.0	13
CHCl$_2$F	Dichlorofluoromethane	25	0.95*	10		
CHCl$_3$	Trichloromethane	25	0.80	20	0.43	13
CHF$_3$	Trifluoromethane	25	0.09*	14		
CHI$_3$	Triiodomethane	25	0.012	14		
CH$_2$BrCl	Bromochloromethane	25	1.7	10	0.18	13
CH$_2$Br$_2$	Dibromomethane	25	1.14	14	0.086	13
CH$_2$ClF	Chlorofluoromethane	25	1.05*	14		
CH$_2$Cl$_2$	Dichloromethane	25	1.73	20	0.30	13
CH$_2$I$_2$	Diiodomethane	30	0.124	10	0.032	13
CH$_3$Br	Bromomethane	20	1.80*	5	0.63	13
CH$_3$Cl	Chloromethane	25	0.535*	5	0.98	13
CH$_3$F	Fluoromethane	30	0.177*	5		
CH$_3$I	Iodomethane	20	1.4	10	0.54	13
CH$_3$NO$_2$	Nitromethane	25	11.1	10		
CH$_4$	Methane	25	0.00227*	18	67.4	5
CO	Carbon monoxide	25	0.00276*	18		
CO$_2$	Carbon dioxide	25	0.1501	18		
CS$_2$	Carbon disulfide	20	0.210	10		
C$_2$ClF$_5$	Chloropentafluoroethane	25	0.006*	10	260	13
C$_2$Cl$_2$F$_4$	1,2-Dichlorotetrafluoroethane	25	0.013*	10	127	13
C$_2$Cl$_3$F$_3$	1,1,2-Trichlorotrifluoroethane	25	0.017	10	32	13
C$_2$Cl$_4$	Tetrachloroethylene	25	0.026	20	1.73	13
C$_2$Cl$_4$F$_2$	1,1,2,2-Tetrachloro-1,2-difluoroethane	25	0.012	10		
C$_2$Cl$_6$	Hexachloroethane	25	0.0050	5	0.85	13
C$_2$F$_4$	Tetrafluoroethylene	25	0.0158*	19		
C$_2$HCl$_3$	Trichloroethylene	25	0.11	5	1.03	13
C$_2$HCl$_5$	Pentachloroethane	25	0.048	5	0.25	13
C$_2$H$_2$	Acetylene	25	0.1081*	19		
C$_2$H$_2$Br$_4$	1,1,2,2-Tetrabromoethane	30	0.0651	10		
C$_2$H$_2$Cl$_2$	1,1-Dichloroethylene	25	0.040	5	2.62	13
C$_2$H$_2$Cl$_2$	cis-1,2-Dichloroethylene	25	0.35	5	0.46	13
C$_2$H$_2$Cl$_2$	trans-1,2-Dichloroethylene	25	0.63	5	0.96	13
C$_2$H$_2$Cl$_4$	1,1,1,2-Tetrachloroethane	25	0.11	5	0.24	13
C$_2$H$_2$Cl$_4$	1,1,2,2-Tetrachloroethane	25	0.30	5	0.026	13
C$_2$H$_3$Cl	Chloroethylene	25	0.27*	5	2.68	13
C$_2$H$_3$Cl$_3$	1,1,1-Trichloroethane	25	0.072	5	1.76	13
C$_2$H$_3$Cl$_3$	1,1,2-Trichloroethane	25	0.442	5	0.092	13
C$_2$H$_4$	Ethylene	25	0.01336*	19	21.7	5
C$_2$H$_4$BrCl	1-Bromo-2-chloroethane	30	0.683	10		
C$_2$H$_4$Br$_2$	1,2-Dibromoethane	25	0.17	5	0.066	13
C$_2$H$_4$Cl$_2$	1,1-Dichloroethane	25	0.51	5	0.63	13
C$_2$H$_4$Cl$_2$	1,2-Dichloroethane	25	0.75	20	0.14	13
C$_2$H$_4$O$_2$	Methyl formate	25	23	10		
C$_2$H$_5$Br	Bromoethane	20	0.91	10	1.23	13
C$_2$H$_5$Cl	Chloroethane	20	0.57*	5	1.02	13
C$_2$H$_5$F	Fluoroethane	25	0.216*	14		
C$_2$H$_5$I	Iodoethane	30	3.88	10	0.52	13
C$_2$H$_5$NO	Acetamide	20	40.8	10		
C$_2$H$_5$NO$_2$	Nitroethane	25	4.68	10		
C$_2$H$_6$	Ethane	25	0.00568*	18	50.6	5

Mol. Form.	Name	t/°C	S/mass %	Ref.	k_H/kPa m^3 mol^{-1}	Ref.
C_2H_6O	Dimethyl ether	24	35.3*	10	0.077	13
C_2H_6OS	Dimethyl sulfoxide	25	25.3	10		
C_2H_6S	Dimethyl sulfide	25	2	10		
C_3Cl_6	Hexachloropropene	25	0.00170	14		
C_3F_6	Perfluoropropene	25	0.0194*	14		
C_3F_8	Perfluoropropane	15	0.0015*	14		
C_3H_3N	2-Propenenitrile	20	7.35	10		
C_3H_4	Propyne	25	0.364*	5	1.11	5
$C_3H_4Cl_2$	cis-1,3-Dichloropropene	25	0.27	5	0.24	5
$C_3H_4Cl_2$	trans-1,3-Dichloropropene	20	0.28	5	0.18	5
$C_3H_4Cl_2$	2,3-Dichloropropene	25	0.215	5	0.36	5
C_3H_4O	Acrolein	20	20.8	10		
C_3H_5Cl	3-Chloropropene	25	0.33	5	1.10	5
C_3H_5ClO	Epichlorohydrin	20	6.58	10	0.003	13
$C_3H_5Cl_3$	1,2,3-Trichloropropane	20	0.19	10	0.038	13
C_3H_5N	Propanenitrile	25	10.3	10		
C_3H_6	Propene	25	0.0200*	5	21.3	5
C_3H_6	Cyclopropane	25	0.0484*	19		
$C_3H_6Br_2$	1,2-Dibromopropane	25	0.143	10		
$C_3H_6Cl_2$	1,2-Dichloropropane	25	0.247	10	0.29	13
$C_3H_6N_6O_6$	Hexahydro-1,3,5-trinitro-1,3,5-triazine	25	0.0060	17		
C_3H_6O	Propanal	25	30.6	10		
C_3H_6O	Methyloxirane	20	40.5	10	0.0087	13
$C_3H_6O_2$	Ethyl formate	25	11.8	10		
$C_3H_6O_2$	Methyl acetate	20	24.5	10		
C_3H_7Br	1-Bromopropane	30	0.230	10	3.8	13
C_3H_7Br	2-Bromopropane	18	0.286	10	1.27	13
C_3H_7Cl	1-Chloropropane	20	0.271	10	1.41	13
C_3H_7Cl	2-Chloropropane	12	0.342	10		
C_3H_7F	1-Fluoropropane	14	0.386*	14		
C_3H_7F	2-Fluoropropane	15	0.366	14		
C_3H_7I	1-Iodopropane	30	0.104	10	0.93	13
C_3H_7I	2-Iodopropane	20	0.140	10		
$C_3H_7NO_2$	1-Nitropropane	25	1.50	10		
$C_3H_7NO_2$	2-Nitropropane	25	1.71	10		
C_3H_8	Propane	25	0.00669*	18	71.6	5
$C_3H_8O_2$	Dimethoxymethane	16	24.4	10		
C_4F_8	Perfluorocyclobutane	21	0.014*	14		
$C_4H_4N_2$	Succinonitrile	25	11.5	10		
C_4H_4O	Furan	25	1	10	0.54	13
C_4H_5N	Methylacrylonitrile	20	2.57	10		
C_4H_5N	Pyrrole	25	4.5	10		
C_4H_6	1,3-Butadiene	25	0.0735*	5	20.7	13
C_4H_6	1-Butyne	25	0.287*	5	1.91	5
C_4H_6O	trans-2-Butenal	20	15.6	10		
$C_4H_6O_2$	Methacrylic acid	20	8.9	10		
$C_4H_6O_2$	trans-Crotonic acid	25	9	10		
$C_4H_6O_2$	Vinyl acetate	20	2.0	10		
$C_4H_6O_2$	Methyl acrylate	25	4.94	10		
C_4H_7Cl	1-Chloro-2-methylpropene	25	0.916	5	0.12	5
C_4H_7N	Butanenitrile	20	3.3	10		
C_4H_8	1-Butene	25	0.0222*	5	25.6	13
C_4H_8	Isobutene	25	0.0263*	5	21.6	13
$C_4H_8Cl_2O$	Bis(2-chloroethyl) ether	25	1.03	20	0.003	13
C_4H_8O	cis-Crotonyl alcohol	20	16.6	10		
C_4H_8O	Ethyl vinyl ether	20	0.9	10		
C_4H_8O	Butanal	25	7.1	10		
C_4H_8O	Isobutanal	20	9.1	10		
C_4H_8O	2-Butanone	25	25.9	20		
$C_4H_8O_2$	2-Methylpropanoic acid	20	22.8	10		

Mol. Form.	Name	$t/°C$	S/mass %	Ref.	k_H/kPa m^3 mol^{-1}	Ref.
$C_4H_8O_2$	Propyl formate	22	2.05	10		
$C_4H_8O_2$	Ethyl acetate	25	8.08	10		
C_4H_9Br	1-Bromobutane	30	0.0608	10	1.2	13
C_4H_9Cl	1-Chlorobutane	20	0.11	10	1.54	13
C_4H_9Cl	1-Chloro-2-methylpropane	12	0.092	10		
C_4H_9I	1-Iodobutane	17	0.012	10	1.87	13
C_4H_{10}	Butane	25	0.00724*	18	95.9	5
C_4H_{10}	Isobutane	25	0.00535*	18	120	5
$C_4H_{10}O$	Diethyl ether	25	6.04	10	0.088	13
$C_4H_{10}O$	1-Butanol	0	10.4	1		
		25	7.4	1		
		50	6.4	1		
$C_4H_{10}O$	2-Butanol	10	23.9	1		
		25	18.1	1		
		50	14.0	1		
$C_4H_{10}O$	2-Methyl-1-propanol	0	11.5	1		
		25	8.1	1		
		50	6.5	1		
$C_4H_{10}S$	1-Butanethiol	20	0.0597	10		
$C_4H_{11}NO_2$	Diethanolamine	20	95.4	10		
$C_4H_{12}Si$	Tetramethylsilane	25	0.00196	10		
$C_5H_4O_2$	Furfural	20	8.2	10		
C_5H_6	1,3-Cyclopentadiene	25	0.068	3		
$C_5H_7NO_2$	Ethyl cyanoacetate	20	25.9	10		
C_5H_8	1,4-Pentadiene	25	0.056	3	12	5
C_5H_8	2-Methyl-1,3-butadiene	25	0.061	3	7.78	5
		50	0.076*	3		
C_5H_8	1-Pentyne	25	0.157	3	2.5	5
C_5H_8	Cyclopentene	25	0.054	3	6.56	13
$C_5H_8O_2$	Ethyl acrylate	25	1.50	10		
$C_5H_8O_2$	Methyl methacrylate	20	1.56	10		
$C_5H_8O_2$	2,4-Pentanedione	20	16.6	10		
C_5H_{10}	1-Pentene	25	0.0148	3	40.3	5
C_5H_{10}	cis-2-Pentene	25	0.0203	3	22.8	5
C_5H_{10}	3-Methyl-1-butene	25	0.013*	3	54.7	5
C_5H_{10}	2-Methyl-2-butene	25	0.041	3		
C_5H_{10}	Cyclopentane	25	0.0157	3	19.1	13
$C_5H_{10}O$	2-Pentanone	25	5.5	20		
$C_5H_{10}O$	3-Pentanone	25	4.72	20		
$C_5H_{10}O$	2-Methyltetrahydrofuran	25	13.9	10	0.67	13
$C_5H_{10}O_2$	Pentanoic acid	20	2.4	10		
$C_5H_{10}O_2$	3-Methylbutanoic acid	21	4.1	10		
$C_5H_{10}O_2$	Isobutyl formate	22	1.0	10		
$C_5H_{10}O_2$	Propyl acetate	20	2.3	10		
$C_5H_{10}O_2$	Isopropyl acetate	20	2.9	10		
$C_5H_{10}O_2$	Ethyl propanoate	20	1.92	10		
$C_5H_{11}Br$	1-Bromopentane	25	0.0127	10		
$C_5H_{11}Cl$	1-Chloropentane	25	0.02	10	2.37	13
C_5H_{12}	Pentane	25	0.0041	3	128	13
C_5H_{12}	Isopentane	25	0.00485	3	479	13
C_5H_{12}	Neopentane	25	0.00332*	3	220	13
$C_5H_{12}O$	1-Pentanol	0	3.1	1		
		25	2.20	1		
		50	1.8	1		
$C_5H_{12}O$	2-Pentanol	25	4.3	21		
$C_5H_{12}O$	3-Pentanol	25	5.6	21		
$C_5H_{12}O$	2-Methyl-1-butanol	25	3.0	3		
$C_5H_{12}O$	3-Methyl-1-butanol	25	2.7	1		
$C_5H_{12}O$	2-Methyl-2-butanol	25	11.0	1		
$C_5H_{12}O$	3-Methyl-2-butanol	25	5.6	1		

Mol. Form.	Name	$t/°C$	S/mass %	Ref.	k_H/kPa m^3 mol^{-1}	Ref.
$C_5H_{12}O$	2,2-Dimethyl-1-propanol	25	3.5	1		
$C_5H_{12}O$	Methyl *tert*-butyl ether	25	3.62	20	0.070	13
C_6Cl_6	Hexachlorobenzene	25	0.0000005	2	0.131	11
C_6HCl_5	Pentachlorobenzene	25	0.000055	2	0.085	11
C_6HCl_5O	Pentachlorophenol	25	0.0010	2		
$C_6H_2Br_4$	1,2,4,5-Tetrabromobenzene	25	0.00000434	2		
$C_6H_2Cl_4$	1,2,3,4-Tetrachlorobenzene	25	0.0000433	2	0.144	11
$C_6H_2Cl_4$	1,2,3,5-Tetrachlorobenzene	25	0.000346	2	0.59	11
$C_6H_2Cl_4$	1,2,4,5-Tetrachlorobenzene	25	0.0000606	2	0.122	11
$C_6H_2Cl_4O$	2,3,4,6-Tetrachlorophenol	25	1.8	2		
$C_6H_2Cl_4O_2$	3,4,5,6-Tetrachloro-1,2-benzenediol	25	0.071	8		
$C_6H_3Br_3$	1,2,4-Tribromobenzene	25	0.0010	2		
$C_6H_3Br_3$	1,3,5-Tribromobenzene	25	0.0000789	2		
$C_6H_3Br_3O$	2,4,6-Tribromophenol	15	0.0007	2		
$C_6H_3Cl_3$	1,2,3-Trichlorobenzene	25	0.00309	2	0.242	11
$C_6H_3Cl_3$	1,2,4-Trichlorobenzene	25	0.00379	2	0.277	11
$C_6H_3Cl_3$	1,3,5-Trichlorobenzene	25	0.000655	2	1.1	11
$C_6H_3Cl_3O$	2,4,5-Trichlorophenol	25	0.1	2		
$C_6H_3Cl_3O$	2,4,6-Trichlorophenol	25	0.04	2		
$C_6H_3Cl_3O_2$	3,4,5-Trichloro-1,2-benzenediol	25	0.051	8		
C_6H_4BrCl	1-Bromo-2-chlorobenzene	25	0.0124	2		
C_6H_4BrCl	1-Bromo-3-chlorobenzene	25	0.0118	2		
C_6H_4BrCl	1-Bromo-4-chlorobenzene	25	0.00442	2		
C_6H_4BrI	1-Bromo-4-iodobenzene	25	0.000794	2		
$C_6H_4Br_2$	*o*-Dibromobenzene	25	0.00748	2		
$C_6H_4Br_2$	*m*-Dibromobenzene	25	0.0064	2		
$C_6H_4Br_2$	*p*-Dibromobenzene	25	0.0020	2		
$C_6H_4Br_2O$	2,4-Dibromophenol	25	0.2	2		
C_6H_4ClI	1-Chloro-2-iodobenzene	25	0.00689	2		
C_6H_4ClI	1-Chloro-3-iodobenzene	25	0.00674	2		
C_6H_4ClI	1-Chloro-4-iodobenzene	25	0.00311	2		
$C_6H_4Cl_2$	*o*-Dichlorobenzene	0	0.0142	2		
		25	0.0147	2	0.244	11
		50	0.0212	2		
$C_6H_4Cl_2$	*m*-Dichlorobenzene	10	0.0103	2		
		25	0.0106	2	0.376	11
		50	0.0165	2		
$C_6H_4Cl_2$	*p*-Dichlorobenzene	10	0.00512	2		
		25	0.00829	2	0.160	11
		50	0.0167	2		
$C_6H_4Cl_2O$	2,4-Dichlorophenol	20	0.45	2		
$C_6H_4Cl_2O_2$	3,5-Dichloro-1,2-benzenediol	25	0.78	8		
$C_6H_4Cl_2O_2$	4,5-Dichloro-1,2-benzenediol	25	1.19	8		
$C_6H_4F_2$	*o*-Difluorobenzene	25	0.114	2		
$C_6H_4F_2$	*m*-Difluorobenzene	25	0.114	2		
$C_6H_4F_2$	*p*-Difluorobenzene	25	0.122	2		
$C_6H_4I_2$	*o*-Diiodobenzene	25	0.00192	2		
$C_6H_4I_2$	*m*-Diiodobenzene	25	0.000185	2		
$C_6H_4I_2$	*p*-Diiodobenzene	25	0.000893	2		
C_6H_5Br	Bromobenzene	10	0.0387	2		
		25	0.0445	2	0.21	5
		40	0.0516	2		
C_6H_5BrO	*p*-Bromophenol	25	1.86	2		
C_6H_5Cl	Chlorobenzene	10	0.0387	2		
		25	0.0495	2	0.38	11
		50	0.0882	2		
C_6H_5ClO	*o*-Chlorophenol	25	2.0	2		
C_6H_5ClO	*m*-Chlorophenol	25	2.2	2		
C_6H_5ClO	*p*-Chlorophenol	25	2.7	2		
C_6H_5F	Fluorobenzene	27	0.154	2	0.70	11

Mol. Form.	Name	$t/°C$	S/mass %	Ref.	k_H/kPa m^3 mol^{-1}	Ref.
C_6H_5I	Iodobenzene	10	0.0193	2		
		25	0.0226	2	0.078	11
		45	0.0279	2		
$C_6H_5NO_2$	Nitrobenzene	25	0.21	17		
C_6H_6	Benzene	10	0.178	3		
		25	0.177	3	0.557	11
		50	0.208	3		
C_6H_6ClN	o-Chloroaniline	25	0.876	10		
C_6H_6O	Phenol	25	8.66	10		
C_6H_7N	Aniline	25	3.38	10	14	15
C_6H_8	1,4-Cyclohexadiene	25	0.08	3	1.03	13
$C_6H_8N_2$	Adiponitrile	20	0.80	16		
$C_6H_8O_4$	Dimethyl maleate	25	8.0	10		
C_6H_{10}	1,5-Hexadiene	25	0.017	3		
C_6H_{10}	1-Hexyne	25	0.036	3	4.14	13
C_6H_{10}	Cyclohexene	25	0.016	3	4.57	13
$C_6H_{10}O$	Cyclohexanone	25	8.8	20		
$C_6H_{10}O$	Mesityl oxide	20	2.89	10		
$C_6H_{10}O_3$	Ethyl acetoacetate	25	12	10		
$C_6H_{11}NO$	Caprolactam	25	84.0	10		
C_6H_{12}	1-Hexene	25	0.0053	3	41.8	5
C_6H_{12}	trans-2-Hexene	25	0.0067	3		
C_6H_{12}	2-Methyl-1-pentene	25	0.0078	3	28.1	5
C_6H_{12}	4-Methyl-1-pentene	25	0.0048	3	63.2	5
C_6H_{12}	2,3-Dimethyl-1-butene	30	0.046	3		
C_6H_{12}	Cyclohexane	25	0.0058	3	19.4	13
C_6H_{12}	Methylcyclopentane	25	0.0043	3	36.7	5
$C_6H_{12}O$	1-Hexen-3-ol	25	2.52	1		
$C_6H_{12}O$	4-Hexen-2-ol	25	3.81	1		
$C_6H_{12}O$	Cyclohexanol	10	4.62	1		
		25	3.8	1		
		40	3.30	1		
$C_6H_{12}O$	Butyl vinyl ether	20	0.3	10		
$C_6H_{12}O$	2-Hexanone	20	1.75	10		
$C_6H_{12}O$	4-Methyl-2-pentanone	25	1.7	10		
$C_6H_{12}O_2$	Hexanoic acid	20	0.958	10		
$C_6H_{12}O_2$	Butyl acetate	20	0.68	10		
$C_6H_{12}O_2$	Isobutyl acetate	20	0.63	10		
$C_6H_{12}O_2$	sec-Butyl acetate	20	0.62	10		
$C_6H_{12}O_2$	Ethyl butanoate	20	0.49	10		
C_6H_{14}	Hexane	25	0.0011	3	183	13
		60	0.00136	3		
C_6H_{14}	2-Methylpentane	25	0.00137	3	176	13
C_6H_{14}	3-Methylpentane	25	0.00129	3	170	13
C_6H_{14}	2,2-Dimethylbutane	25	0.0021	3	199	13
C_6H_{14}	2,3-Dimethylbutane	25	0.0021	3	144	13
$C_6H_{14}O$	1-Hexanol	0	0.79	1		
		25	0.60	1		
		50	0.51	1		
$C_6H_{14}O$	2-Hexanol	25	1.4	1		
$C_6H_{14}O$	3-Hexanol	25	1.6	1		
$C_6H_{14}O$	Dipropyl ether	25	0.49	10	0.26	13
$C_6H_{14}O$	2-Methyl-1-pentanol	25	0.81	1		
$C_6H_{14}O$	4-Methyl-1-pentanol	25	0.76	1		
$C_6H_{14}O$	2-Methyl-2-pentanol	25	3.2	1		
$C_6H_{14}O$	3-Methyl-2-pentanol	25	1.9	1		
$C_6H_{14}O$	4-Methyl-2-pentanol	27	1.5	1		
$C_6H_{14}O$	2-Methyl-3-pentanol	25	2.0	1		
$C_6H_{14}O$	3-Methyl-3-pentanol	25	4.3	1		
$C_6H_{14}O$	2-Ethyl-1-butanol	25	1.0	1		

Mol. Form.	Name	$t/°C$	S/mass %	Ref.	k_H/kPa m^3 mol^{-1}	Ref.
$C_6H_{14}O$	2,2-Dimethyl-1-butanol	25	0.8	1		
$C_6H_{14}O$	Diisopropyl ether	20	1.2	10	0.26	13
$C_6H_{14}O$	2,3-Dimethyl-2-butanol	25	4.2	1		
$C_6H_{14}O$	3,3-Dimethyl-2-butanol	25	2.4	1		
$C_6H_{14}O_2$	1,1-Diethoxyethane	25	5	10		
$C_6H_{15}N$	Dipropylamine	20	2.5	10		
$C_6H_{15}N$	Triethylamine	20	5.5	10		
$C_7H_4Cl_4O$	2,3,4,6-Tetrachloro-5-methylphenol	25	0.00061	2		
$C_7H_5Cl_3$	(Trichloromethyl)benzene	5	0.0053	10		
$C_7H_5Cl_3O$	2,4,6-Trichloro-3-methylphenol	25	0.0112	2		
C_7H_5N	Benzonitrile	25	0.2	10		
C_7H_5NO	Benzoxazole	20	0.834	6		
$C_7H_6Cl_2$	(Dichloromethyl)benzene	30	0.025	10		
$C_7H_6Cl_2O$	2,6-Dichloro-4-methylphenol	25	0.0673	2		
$C_7H_6Cl_2O$	2,4-Dichloro-6-methylphenol	25	0.0283	2		
$C_7H_6N_2$	1*H*-Benzimidazole	20	0.201	6		
$C_7H_6N_2$	1*H*-Indazole	20	0.0827	6		
C_7H_6O	Benzaldehyde	20	0.3	10		
$C_7H_6O_2$	Salicylaldehyde	86	1.68	10		
C_7H_7Br	*p*-Bromotoluene	25	0.011	2		
C_7H_7Cl	(Chloromethyl)benzene	20	0.0493	10		
C_7H_7ClO	4-Chloro-2-methylphenol	25	0.68	2		
C_7H_7ClO	4-Chloro-3-methylphenol	25	0.40	2		
C_7H_7ClO	2-Chloro-6-methylphenol	25	0.36	2		
$C_7H_7NO_3$	2-Nitroanisole	30	0.619	10		
C_7H_8	Toluene	5	0.063	3		
		25	0.053	3	0.680	13
C_7H_8	1,3,5-Cycloheptatriene	25	0.064	3	0.47	13
C_7H_8	1,6-Heptadiyne	25	0.125	3		
C_7H_8O	Benzyl alcohol	20	0.08	10		
C_7H_8O	*o*-Cresol	40	3.08	10		
C_7H_8O	*m*-Cresol	40	2.51	10		
C_7H_8O	*p*-Cresol	40	2.26	10		
C_7H_8O	Anisole	25	0.19	20	0.025	13
C_7H_9N	*o*-Methylaniline	20	1.66	10		
C_7H_9N	*p*-Methylaniline	21	7.35	10		
C_7H_{12}	1-Heptyne	25	0.0094	3	4.47	13
C_7H_{12}	Cycloheptene	25	0.0066	3	4.9	13
C_7H_{12}	1-Methylcyclohexene	25	0.0052	3		
C_7H_{14}	1-Heptene	25	0.032	3	40.3	13
C_7H_{14}	*trans*-2-Heptene	25	0.015	3	42.2	13
C_7H_{14}	Cycloheptane	25	0.0030	3	9.59	13
C_7H_{14}	Methylcyclohexane	25	0.00151	3	43.3	13
		50	0.0019	3		
C_7H_{14}	Ethylcyclopentane	20	0.012	3		
$C_7H_{14}O$	2-Heptanone	25	0.43	10		
$C_7H_{14}O$	3-Heptanone	20	1.43	10		
$C_7H_{14}O$	2,4-Dimethyl-3-pentanone	20	0.59	10		
$C_7H_{14}O_2$	Pentyl acetate	20	0.17	10		
$C_7H_{14}O_2$	Isopentyl acetate	20	0.2	10		
$C_7H_{14}O_2$	Ethyl 3-methylbutanoate	20	0.2	10		
C_7H_{16}	Heptane	0	0.0003	3		
		25	0.00024	3	209	13
		40	0.00025	3		
C_7H_{16}	2-Methylhexane	25	0.00025	3	346	5
C_7H_{16}	3-Methylhexane	25	0.00026	3	249	13
C_7H_{16}	2,2-Dimethylpentane	25	0.00044	3	318	5
C_7H_{16}	2,3-Dimethylpentane	25	0.00052	3	175	5
C_7H_{16}	2,4-Dimethylpentane	25	0.00042	3	323	13
C_7H_{16}	3,3-Dimethylpentane	25	0.00059	3	186	5

Mol. Form.	Name	$t/°C$	S/mass %	Ref.	k_H/kPa m^3 mol^{-1}	Ref.
$C_7H_{16}O$	1-Heptanol	10	0.25	1		
		25	0.174	1		
		50	0.12	1		
$C_7H_{16}O$	2-Heptanol	30	0.33	1		
$C_7H_{16}O$	3-Heptanol	25	0.43	1		
$C_7H_{16}O$	4-Heptanol	25	0.47	1		
$C_7H_{16}O$	2-Methyl-2-hexanol	25	1.0	1		
$C_7H_{16}O$	5-Methyl-2-hexanol	25	0.49	1		
$C_7H_{16}O$	3-Methyl-3-hexanol	25	1.2	1		
$C_7H_{16}O$	2,3-Dimethyl-2-pentanol	25	1.5	1		
$C_7H_{16}O$	2,4-Dimethyl-2-pentanol	25	1.3	1		
$C_7H_{16}O$	3-Ethyl-3-pentanol	25	1.7	1		
$C_7H_{16}O$	2,2-Dimethyl-3-pentanol	25	0.82	1		
$C_7H_{16}O$	2,3-Dimethyl-3-pentanol	25	1.6	1		
$C_7H_{16}O$	2,4-Dimethyl-3-pentanol	25	0.70	1		
$C_7H_{16}O$	2,3,3-Trimethyl-2-butanol	40	2.2	1		
$C_8H_4F_6$	1,3-Bis(trifluoromethyl)benzene	25	0.0041	2		
$C_8H_6N_2$	Quinoxaline	50	54	6		
C_8H_6S	Benzo[b]thiophene	20	0.0130	6		
$C_8H_7ClO_3$	3-Chloro-4-hydroxy-5-methoxybenzaldehyde	25	0.093	8		
$C_8H_7ClO_3$	2-Chloro-4-hydroxy-5-methoxybenzaldehyde	25	0.013	8		
$C_8H_7Cl_3O$	2,4,6-Trichloro-3,5-dimethylphenol	25	0.00050	2		
C_8H_7N	Indole	20	0.187	6		
C_8H_8	Styrene	25	0.025	4	0.3	15
		50	0.046	4	0.30	13
$C_8H_8N_2$	2-Methyl-1H-benzimidazole	20	0.145	6		
C_8H_8O	Acetophenone	25	0.55			
$C_8H_8O_2$	Methyl benzoate	20	0.21	10		
$C_8H_8O_3$	Methyl salicylate	30	0.74	10		
$C_8H_8O_3$	4-Hydroxy-3-methoxybenzaldehyde	25	0.247	8		
C_8H_9ClO	4-Chloro-2,5-dimethylphenol	25	0.89	2		
C_8H_9ClO	4-Chloro-2,6-dimethylphenol	25	0.52	2		
C_8H_9ClO	4-Chloro-3,5-dimethylphenol	25	0.34	2		
C_8H_{10}	Ethylbenzene	0	0.020	4		
		25	0.0169	4	0.887	11
		40	0.0200	4		
C_8H_{10}	o-Xylene	25	0.0173	4	0.565	13
		45	0.021	4		
C_8H_{10}	m-Xylene	0	0.0203	4		
		25	0.016	4	0.730	13
		40	0.022	4		
C_8H_{10}	p-Xylene	0	0.0160	4		
		25	0.018	4	0.578	13
		40	0.022	4		
$C_8H_{10}O$	Phenetole	25	0.12	10		
$C_8H_{10}O$	2,4-Xylenol	25	0.787	10		
$C_8H_{10}O$	3,5-Xylenol	29	0.62	10		
C_8H_{12}	4-Vinylcyclohexene	25	0.005	4		
C_8H_{14}	1-Octyne	25	0.0024	4	7.87	13
C_8H_{16}	1-Octene	25	0.00027	4	96.3	13
C_8H_{16}	Cyclooctane	25	0.00079	4	10.7	13
C_8H_{16}	Ethylcyclohexane	40	0.00066	4		
C_8H_{16}	cis-1,2-Dimethylcyclohexane	25	0.00060	4	36	5
C_8H_{16}	trans-1,4-Dimethylcyclohexane	25	0.000384	4	88.2	5
C_8H_{16}	Propylcyclopentane	25	0.00020	4	90.2	5
C_8H_{16}	1,1,3-Trimethylcyclopentane	25	0.00037	4	159	5
$C_8H_{16}O$	2-Octanone	25	0.113	10		
$C_8H_{16}O_2$	Octanoic acid	25	0.0798	10		

Mol. Form.	Name	$t/°C$	S/mass %	Ref.	k_H/kPa m^3 mol^{-1}	Ref.
$C_8H_{16}O_2$	Hexyl acetate	20	0.02	10		
$C_8H_{16}O_2$	sec-Hexyl acetate	20	0.13	10		
$C_8H_{16}O_2$	Isobutyl isobutanoate	20	0.5	10		
$C_8H_{17}Cl$	3-(Chloromethyl)heptane	20	0.01	10		
C_8H_{18}	Octane	25	0.000071	4	311	13
		50	0.00010	4		
C_8H_{18}	3-Methylheptane	25	0.000079	4	376	5
C_8H_{18}	2,2,4-Trimethylpentane	25	0.00022	4	307	13
C_8H_{18}	2,3,4-Trimethylpentane	25	0.00018	4	206	13
$C_8H_{18}O$	Dibutyl ether	20	0.03	10	0.48	13
$C_8H_{18}O$	1-Octanol	25	0.054	1		
$C_8H_{18}O$	2-Octanol	25	0.4	1		
$C_8H_{18}O$	2-Methyl-2-heptanol	30	0.25	1		
$C_8H_{18}O$	2-Ethyl-1-hexanol	25	0.01	1		
$C_8H_{19}N$	Dibutylamine	20	0.47	10		
$C_8H_{19}N$	2-Ethylhexylamine	20	0.25	10		
$C_8H_{20}Si$	Tetraethylsilane	25	0.0000325	10		
C_9H_7N	Quinoline	20	0.633	6		
C_9H_7N	Isoquinoline	20	0.452	6		
C_9H_9N	3-Methyl-1H-indole	20	0.050	6		
C_9H_{10}	Indan	25	0.010	4		
$C_9H_{10}O_2$	Ethyl benzoate	25	0.083	20		
C_9H_{12}	1,8-Nonadiyne	25	0.0125	4		
C_9H_{12}	Propylbenzene	25	0.0055	4	1.04	11
C_9H_{12}	Isopropylbenzene	25	0.0056	4	1.47	11
C_9H_{12}	o-Ethyltoluene	25	0.0093	5	0.529	13
C_9H_{12}	p-Ethyltoluene	25	0.0094	5	0.500	13
C_9H_{12}	1,2,3-Trimethylbenzene	25	0.0069	4	0.343	11
C_9H_{12}	1,2,4-Trimethylbenzene	25	0.0056	4	0.569	11
C_9H_{12}	1,3,5-Trimethylbenzene	25	0.00489	4	0.781	11
$C_9H_{14}O_6$	Triacetin	25	5.8	10		
C_9H_{16}	1-Nonyne	25	0.00072	4		
C_9H_{18}	1,1,3-Trimethylcyclohexane	25	0.000177	4	105	13
$C_9H_{18}O$	Diisobutyl ketone	25	0.043	10		
$C_9H_{18}O_2$	Nonanoic acid	20	0.0284	10		
C_9H_{20}	Nonane	25	0.000017	4	333	13
		50	0.000022	4		
C_9H_{20}	4-Methyloctane	25	0.0000115	4	1000	5
C_9H_{20}	2,2,5-Trimethylhexane	25	0.00008	4	246	13
$C_9H_{20}O$	3,5-Dimethyl-4-heptanol	15	0.072	1		
$C_9H_{20}O$	1-Nonanol	25	0.014	1		
$C_9H_{20}O$	2-Nonanol	15	0.026	1		
$C_9H_{20}O$	3-Nonanol	15	0.032	1		
$C_9H_{20}O$	4-Nonanol	15	0.0026	1		
$C_9H_{20}O$	5-Nonanol	15	0.0032	1		
$C_{10}H_7Cl$	1-Chloronaphthalene	25	0.00224	5		
$C_{10}H_7Cl$	2-Chloronaphthalene	25	0.00117	5		
$C_{10}H_8$	Naphthalene	10	0.0019	4		
		25	0.0031	4	0.0430	5
		50	0.0082	4		
$C_{10}H_9N$	3-Methylisoquinoline	20	0.092	6		
$C_{10}H_{10}O_4$	Dimethyl phthalate	25	0.40	15		
$C_{10}H_{14}$	Butylbenzene	25	0.0015	4	1.33	11
$C_{10}H_{14}$	sec-Butylbenzene	25	0.0014	4	1.89	11
$C_{10}H_{14}$	tert-Butylbenzene	25	0.0032	4	1.28	11
$C_{10}H_{14}$	Isobutylbenzene	25	0.0010	4	3.32	11
$C_{10}H_{14}$	p-Cymene	25	0.00234	4	0.80	5
$C_{10}H_{14}$	1,2,4,5-Tetramethylbenzene	25	0.000348	4	2.55	11
$C_{10}H_{16}$	d-Limonene	0	0.00097	4		
		25	0.00138	4		

Mol. Form.	Name	$t/°C$	S/mass %	Ref.	k_H/kPa m^3 mol^{-1}	Ref.
$C_{10}H_{16}O$	Camphor	20	0.01	10		
$C_{10}H_{18}$	trans-Decahydronaphthalene	25	0.000089	4	3	13
$C_{10}H_{20}$	Pentylcyclopentane	25	0.0000115	4	185	5
$C_{10}H_{20}$	1-Decene	25	0.00057	4		
$C_{10}H_{22}$	Decane	0	0.0000015	4	479	13
$C_{10}H_{22}O$	Diisopentyl ether	20	0.02	10		
$C_{10}H_{22}O$	1-Decanol	25	0.0037	1		
$C_{11}H_{10}$	1-Methylnaphthalene	25	0.0028	4	0.0450	12
$C_{11}H_{10}$	2-Methylnaphthalene	25	0.0025	4	0.051	12
$C_{11}H_{16}$	Pentylbenzene	25	0.00105	5	1.69	11
$C_{12}Cl_{10}$	Decachlorobiphenyl	25	0.00000000012	7	0.0208	7
$C_{12}HCl_9$	2,2',3,3',4,5,5',6,6'-Nonachlorobiphenyl	25	0.0000000018	7		
$C_{12}H_2Cl_8$	2,2',3,3'5,5',6,6'-Octachlorobiphenyl	25	0.0000003	7	0.0381	7
$C_{12}H_3Cl_7$	2,2',3,3',4,4',6-Heptachlorobiphenyl	25	0.0000002	7	0.0054	7
$C_{12}H_4Cl_6$	2,2',3,3',4,4'-Hexachlorobiphenyl	25	0.00000006	7	0.0119	7
$C_{12}H_4Cl_6$	2,2',4,4',6,6'-Hexachlorobiphenyl	25	0.00000007	7	0.818	7
$C_{12}H_4Cl_6$	2,2',3,3',6,6'-Hexachlorobiphenyl	25	0.00000008	7		
$C_{12}H_5Cl_5$	2,3,4,5,6-Pentachlorobiphenyl	25	0.0000008	7		
$C_{12}H_5Cl_5$	2,2',4,5,5'-Pentachlorobiphenyl	25	0.000001	7	0.0355	7
$C_{12}H_6Cl_4$	2,3,4,5-Tetrachlorobiphenyl	25	0.000002	7		
$C_{12}H_6Cl_4$	2,2',4',5-Tetrachlorobiphenyl	25	0.0000016	9		
$C_{12}H_7Cl_3$	2,4,5-Trichlorobiphenyl	25	0.000014	7	0.0243	7
$C_{12}H_7Cl_3$	2,4,6-Trichlorobiphenyl	25	0.00002	7	0.0495	7
$C_{12}H_8Cl_2$	2,5-Dichlorobiphenyl	25	0.0002	7	0.0201	7
$C_{12}H_8Cl_2$	2,6-Dichlorobiphenyl	25	0.00014	7		
$C_{12}H_8O$	Dibenzofuran	25	0.000656	6	0.011	12
$C_{12}H_8S$	Dibenzothiophene	25	0.000103	6		
$C_{12}H_9Cl$	2-Chlorobiphenyl	25	0.00055	7	0.0701	7
$C_{12}H_9N$	Carbazole	22	0.000120	6		
$C_{12}H_{10}$	Acenaphthene	0	0.00015	4		
		25	0.00038	4	0.012	12
		50	0.00092	4		
$C_{12}H_{10}$	Biphenyl	0	0.000272	4		
		25	0.00072	4	0.028	5
		50	0.0022	4		
$C_{12}H_{10}N_2O$	N-Nitrosodiphenylamine	25	0.0035	17		
$C_{12}H_{10}O$	Diphenyl ether	25	0.00180	6	0.027	13
$C_{12}H_{12}$	1-Ethylnaphthalene	25	0.00101	4	0.039	12
$C_{12}H_{12}$	2-Ethylnaphthalene	25	0.00080	4	0.078	12
$C_{12}H_{12}$	1,3-Dimethylnaphthalene	25	0.0008	4		
$C_{12}H_{12}$	1,4-Dimethylnaphthalene	25	0.00114	4		
$C_{12}H_{12}$	1,5-Dimethylnaphthalene	25	0.00031	4		
$C_{12}H_{12}$	2,3-Dimethylnaphthalene	25	0.00025	4		
$C_{12}H_{12}$	2,6-Dimethylnaphthalene	25	0.00017	4		
$C_{12}H_{18}$	Hexylbenzene	25	0.00021	4		
$C_{12}H_{26}$	Dodecane	25	0.00000037	4	750	5
$C_{12}H_{26}O$	1-Dodecanol	25	0.0004	1		
$C_{12}H_{27}O_4P$	Tributyl phosphate	25	0.039	10		
$C_{13}H_9N$	Acridine	25	0.00466	6		
$C_{13}H_9N$	Benzo[f]quinoline	25	0.0079	6		
$C_{13}H_{10}$	9H-Fluorene	0	0.00007	4		
		25	0.00019	4	0.0079	12
		50	0.00063	4		
$C_{13}H_{12}$	Diphenylmethane	25	0.000141	4	0.001	12
$C_{13}H_{14}$	1,4,5-Trimethylnaphthalene	25	0.00021	4		
$C_{14}H_{10}$	Anthracene	0	0.0000022	4		
		25	0.0000062	4	0.0040	12
$C_{14}H_{10}$	Phenanthrene	10	0.000050	4		
		25	0.00011	4	0.0032	12
		50	0.00041	4		

Mol. Form.	Name	$t/°C$	S/mass %	Ref.	k_H/kPa m^3 mol^{-1}	Ref.
$C_{14}H_{12}$	*trans*-Stilbene	25	0.000029	4	0.040	12
$C_{14}H_{14}$	1,2-Diphenylethane	25	0.00044	6	0.017	12
$C_{14}H_{14}O$	Dibenzyl ether	35	0.0040	10		
$C_{14}H_{30}$	Tetradecane	25	0.000012	5		
$C_{14}H_{30}O$	1-Tetradecanol	25	0.000031	1		
$C_{15}H_{12}$	2-Methylanthracene	25	0.0000030	4		
$C_{15}H_{12}$	9-Methylanthracene	25	0.000026	4		
$C_{15}H_{12}$	1-Methylphenanthrene	25	0.0000269	4		
$C_{15}H_{32}O$	1-Pentadecanol	25	0.000010	1		
$C_{16}H_{10}$	Fluoranthene	25	0.000024	4	0.0010	12
$C_{16}H_{10}$	Pyrene	25	0.0000132	4	0.00092	12
		50	0.00009	4		
$C_{16}H_{14}$	9,10-Dimethylanthracene	25	0.0000056	4		
$C_{16}H_{22}O_4$	Dibutyl phthalate	25	0.00112	15		
$C_{16}H_{34}O$	1-Hexadecanol	25	0.000003	1		
$C_{17}H_{12}$	11H-Benzo[a]fluorene	25	0.0000045	4		
$C_{17}H_{12}$	11H-Benzo[b]fluorene	25	0.0000002	4		
$C_{18}H_{12}$	Benz[a]anthracene	25	0.0000011	4	0.00058	12
$C_{18}H_{12}$	Chrysene	25	0.00000019	4	0.006	12
$C_{18}H_{12}$	Naphthacene	25	0.00000006	4	0.000004	12
$C_{18}H_{12}$	Triphenylene	25	0.0000041	4	0.00001	12
$C_{18}H_{12}N_2$	2,2′-Biquinoline	24	0.000102	6		
$C_{18}H_{34}O_4$	Dibutyl sebacate	20	0.004	10		
$C_{19}H_{14}$	5-Methylchrysene	27	0.0000062	4		
$C_{19}H_{14}$	9-Methylbenz[a]anthracene	27	0.0000066	4		
$C_{19}H_{14}$	10-Methylbenz[a]anthracene	25	0.0000055	4		
$C_{20}H_{12}$	Benzo[a]pyrene	25	0.00000038	4	0.000046	12
$C_{20}H_{12}$	Perylene	25	0.00000004	4	0.000003	12
$C_{20}H_{12}$	Benzo[e]pyrene	20	0.00000046	4	0.00002	12
$C_{20}H_{13}N$	13H-Dibenzo[a,i]carbazole	24	0.00000104	6		
$C_{20}H_{14}$	1,2-Dihydrobenz[j]aceanthrylene	25	0.00000036	6		
$C_{20}H_{42}$	Eicosane	25	0.00000019	4		
$C_{21}H_{13}N$	Dibenz[a,j]acridine	25	0.000016	6		
$C_{21}H_{16}$	1,2-Dihydro-3-methylbenz[j]aceanthrylene	25	0.00000022	6		
$C_{22}H_{12}$	Benzo[ghi]perylene	25	0.000000026	4	0.000075	12
$C_{22}H_{14}$	Picene	25	0.00000025	4		
$C_{22}H_{14}$	Benzo[b]triphenylene	25	0.0000027	4		
$C_{22}H_{14}$	Dibenz[a,h]anthracene	25	0.00000006	4		
$C_{22}H_{14}$	Dibenz[a,j]anthracene	25	0.0000012	4		
$C_{22}H_{44}O_2$	Butyl stearate	25	0.2	10		
$C_{24}H_{12}$	Coronene	25	0.000000014	4		

AQUEOUS SOLUBILITY OF INORGANIC COMPOUNDS AT VARIOUS TEMPERATURES

The solubility of over 300 common inorganic compounds in water is tabulated here as a function of temperature. Solubility is defined as the concentration of the compound in a solution that is in equilibrium with a solid phase at the specified temperature. In this table the solid phase is generally the most stable crystalline phase at the temperature in question. An asterisk * on solubility values in adjacent columns indicates that the solid phase changes between those two temperatures (usually from one hydrated phase to another or from a hydrate to the anhydrous solid). In such cases the slope of the solubility vs. temperature curve may show a discontinuity.

All solubility values are expressed as mass percent of solute, $100 \cdot w_2$, where

$$w_2 = m_2/(m_1 + m_2)$$

and m_2 is the mass of solute and m_1 the mass of water. This quantity is related to other common measures of solubility as follows:

Molality: $m_2 = 1000w_2/M_2(1-w_2)$
Mole fraction: $x_2 = (w_2/M_2)/\{(w_2/M_2) + (1-w_2)/M_1\}$
Mass of solute per 100 g of H_2O: $r_2 = 100w_2/(1-w_2)$

Here M_2 is the molar mass of the solute and $M_1 = 18.015$ g/mol is the molar mass of water.

The data in the table have been derived from the references indicated; in many cases the data have been refitted or interpolated in order to present solubility at rounded values of temperature. Where available, values were taken from the IUPAC *Solubility Data Series* (Reference 1) or the related papers in the *Journal of Physical and Chemical Reference Data* (References 2 to 5), which present carefully evaluated data.

The solubility of sparingly soluble compounds that do not appear in this table may be calculated from the data in the table "Solubility Product Constants". Solubility of inorganic gases may be found in the table "Solubility of Selected Gases in Water".

Compounds are listed alphabetically by chemical formula in the most commonly used form (e.g., NaCl, NH_4NO_3, etc.).

REFERENCES

1. *Solubility Data Series*, International Union of Pure and Applied Chemistry. Volumes 1 to 53 were published by Pergamon Press, Oxford, from 1979 to 1994; subsequent volumes were published by Oxford University Press, Oxford. The number following the colon is the volume number in the series.
2. Clever, H.L., and Johnston, F.J., *J. Phys. Chem. Ref. Data*, 9, 751, 1980.
3. Marcus, Y., *J. Phys. Chem. Ref. Data*, 9, 1307, 1980.
4. Clever, H.L., Johnson, S.A., and Derrick, M.E., *J. Phys. Chem. Ref. Data*, 14, 631, 1985.
5. Clever, H.L., Johnson, S.A., and Derrick, M.E., *J. Phys. Chem. Ref. Data*, 21, 941, 1992.
6. Söhnel, O., and Novotny, P., *Densities of Aqueous Solutions of Inorganic Substances*, Elsevier, Amsterdam, 1985.
7. Krumgalz, B.S., *Mineral Solubility in Water at Various Temperatures*, Israel Oceanographic and Limnological Research Ltd., Haifa, 1994.
8. Potter, R.W., and Clynne, M.A., *J. Research U.S. Geological Survey*, 6, 701, 1978; Clynne, M.A., and Potter, R.W., *J. Chem. Eng. Data*, 24, 338, 1979.
9. Marshal, W.L., and Slusher, R., *J. Phys. Chem.*, 70, 4015, 1966; Knacke, O., and Gans, W., *Zeit. Phys. Chem.*, NF, 104, 41, 1977.
10. Stephen, H., and Stephen, T., *Solubilities of Inorganic and Organic Compounds, Vol. 1*, Macmillan, New York, 1963.

AQUEOUS SOLUBILITY OF INORGANIC COMPOUNDS AT VARIOUS TEMPERATURES (continued)

Compound	0°C	10°C	20°C	25°C	30°C	40°C	50°C	60°C	70°C	80°C	90°C	100°C	Ref.
AgBrO$_3$	0.17	0.31	0.47	0.193	0.64	0.82	1.02	1.22	1.44	1.66	1.32		7
AgClO$_2$				0.55							1.88	2.11	7
AgClO$_3$				15									7
AgClO$_4$	81.6	83.0	84.2	84.8	85.3	86.3	86.9	87.5	87.9	88.3	88.6	88.8	6
AgNO$_2$	0.155			0.413									7
AgNO$_3$	55.9	62.3	67.8	70.1	72.3	76.1	79.2	81.7	83.8	85.4	86.7	87.8	6
Ag$_2$SO$_4$	0.56	0.67	0.78	0.83	0.88	0.97	1.05	1.13	1.20	1.26	1.32	1.39	7
AlCl$_3$	30.84	30.91	31.03	31.10	31.18	31.37	31.60	31.87	32.17	32.51	32.90	33.32	7
Al(ClO$_4$)$_3$	54.9			44.2							64.4		7
AlF$_3$	0.25	0.34	0.44	0.50	0.56	0.68	0.81	0.96	1.11	1.28	1.45	1.64	7
Al(NO$_3$)$_3$	37.0	38.2	39.9	40.8	42.0	44.5	47.3	50.4	53.8*			61.5*	6
Al$_2$(SO$_4$)$_3$	27.5			27.8	28.2	29.2	30.7	32.6	34.9	37.6	40.7	44.2	7
As$_2$O$_3$	1.19	1.48	1.80	2.01	2.27	2.86	3.43	4.11	4.89	5.77	6.72	7.71	10
BaBr$_2$	47.6	48.5	49.5	50.0	50.4	51.4	52.5	53.5	54.5	55.5	56.6	57.6	6
Ba(BrO$_3$)$_2$	0.285	0.442	0.656	0.788	0.935	1.30	1.74	2.27	2.90	3.61	4.40	5.25	1:14
Ba(C$_2$H$_3$O$_2$)$_2$	37.0			44.2									7
BaCl$_2$	23.30	24.88	26.33	27.03	27.70	29.00	30.27	31.53	32.81	34.14	35.54	37.05	8
Ba(ClO$_2$)$_2$	30.5			31.3								44.7	7
Ba(ClO$_3$)$_2$	16.90	21.23	23.66	27.50	29.43	33.16	36.69	40.05	43.04	45.90	48.70	51.17	1:14
Ba(ClO$_4$)$_2$	67.30	70.96	74.30	75.75	77.05	79.23	80.92	82.21	83.16	83.88	84.43	84.90	7
BaF$_2$	0.158	0.158		0.161									7
BaI$_2$	62.5	64.7	67.3	68.8	69.1	69.5	70.1	70.7	71.3	72.0	72.7	73.4	6
Ba(IO$_3$)$_2$	0.0182	0.0262	0.0342	0.0396	0.045*	0.058*	0.073	0.090	0.109	0.131	0.156	0.182	1:14
Ba(NO$_2$)$_2$	31.1	36.6	41.8	44.3	46.8	51.6	56.2	60.5	64.6	68.5	72.1	75.6	10
Ba(NO$_3$)$_2$	4.7	6.3	8.2	9.3	10.2	12.4	14.7	17.0	19.3	21.5	23.5	25.5	6
Ba(OH)$_2$	1.67			4.68	8.4	19	33	52	74	101			7
BaS	2.79	4.78	6.97	8.21	9.58	12.67	16.18	20.05	24.19	28.55	33.04	37.61	7
Ba(SCN)$_2$				62.6									1:26
BaSO$_3$				0.0011									7
BeCl$_2$	40.5			41.7									7
Be(ClO$_4$)$_2$				59.5									7
BeSO$_4$	26.69	27.58	28.61	29.22	29.90	31.51	33.39	35.50	37.78	40.21	42.72	45.28	7
CaBr$_2$	55	56	59	61	63	68	71	73					10
CaCl$_2$	36.70	39.19	42.13	44.83*	49.12*	52.85*	56.05*	56.73	57.44	58.21	59.04	59.94	8
Ca(ClO$_3$)$_2$	63.2	64.2	65.5	66.3	67.2	69.0	71.0	73.2	75.5*	77.4*	77.7	78.0	1:14
Ca(ClO$_4$)$_2$				65.3									7
CaF$_2$	0.0013			0.0016									10
CaI$_2$	64.6	66.0	67.6	68.3	69.0	70.8	72.4	74.0	76.0	78.0	79.6	81.0	7
Ca(IO$_3$)$_2$	0.082	0.155	0.243	0.305	0.384*	0.517*	0.590	0.652	0.811*	0.665*	0.668		1:14
Ca(NO$_2$)$_2$	38.6	39.5	44.5	48.6									7
Ca(NO$_3$)$_2$	50.1	53.1	56.7	59.0	60.9	65.4	77.8	78.1	78.2	78.3	78.4	78.5	6
CaSO$_3$			0.0059	0.0054	0.0049	0.0041	0.0035	0.0030	0.0026	0.0023	0.0020	0.0019	1:26
CaSO$_4$	0.174	0.191	0.202	0.205	0.208	0.210	0.207	0.201	0.193	0.184	0.173	0.163	9

AQUEOUS SOLUBILITY OF INORGANIC COMPOUNDS AT VARIOUS TEMPERATURES (continued)

Compound	0°C	10°C	20°C	25°C	30°C	40°C	50°C	60°C	70°C	80°C	90°C	100°C	Ref.
CdBr$_2$	36.0	43.0	49.9	53.4	56.4	60.3*	60.3*	60.5	60.7	60.9	61.3	61.6	6
CdC$_2$O$_4$				0.0060									5
CdCl$_2$	47.2	50.1	53.2	54.6	56.3*	57.3*	57.5	57.8	58.1	58.51	58.98	59.5	6
Cd(ClO$_4$)$_2$				58.7								66.9	7
CdF$_2$		5.82	4.65	4.18	3.76								5
CdI$_2$	44.1	44.9	45.8	46.3	46.8	47.9	49.0	50.2	51.5	52.7	54.1	55.4	6
Cd(IO$_3$)$_2$				0.091									5
Cd(NO$_3$)$_2$	55.4	57.1	59.6	61.0	62.8	66.5	70.6	86.1	86.5	86.8	87.1	87.4	6
CdSO$_4$	43.1	43.1	43.2	43.4	43.6	44.1	43.5	42.5	41.4	40.2	38.5	36.7	6
CdSeO$_4$	42.04	40.59	39.02	38.18	37.29	35.35	33.15	30.65	27.84	24.69	21.24	17.49	5
Ce(NO$_3$)$_3$	57.99	59.80	61.89	63.05	64.31*	67.0*	68.6	71.1*	74.9*	79.2	80.9	83.1	1:13
CoCl$_2$	30.30	32.60	34.87	35.99	37.10	39.27	41.38	43.46	45.50	47.51	49.51	51.50	7
Co(ClO$_4$)$_2$	50.0			53.0									7
CoF$_2$				1.4									7
CoI$_2$	58.00	61.78	65.35	66.99	68.51	71.17	73.41	75.29	76.89	78.28	79.52	80.70	7
Co(NO$_2$)$_2$	0.076			0.49									7
Co(NO$_3$)$_2$	45.5	47.0	49.4	50.8	52.4	56.0	60.1	62.6	64.9	67.7			6
CoSO$_4$	19.9	23.0	26.1	27.7	29.2	32.3	34.4	35.9	35.5	33.2	30.6	27.8	6
Co(SCN)$_2$				50.7									6
CrO$_3$	62.2	62.3	62.6	62.8	63.0	63.5	64.1	64.7	65.5	66.2	67.1	67.9	6
CsBr				55.2									7
CsBrO$_3$	1.16	1.93	3.01	3.69	4.46	6.32	8.60	11.32	14.45	17.96	21.83	25.98	1:30
CsCl	61.83	63.48	64.96	65.64	66.29	67.50	68.60	69.61	70.54	71.40	72.21	72.96	1:47
CsClO$_3$	2.40	3.87	5.94	7.22	8.69	12.15	16.33	21.14	26.45	32.10	37.89	43.42	1:30
CsClO$_4$	0.79	1.01	1.51	1.96	2.57	4.28	6.55	9.29	12.41	15.80	19.39	23.07	7
CsI	30.9	37.2	43.2	45.9	48.6	53.3	57.3	60.7	63.6	65.9	67.7	69.2	6
CsIO$_3$	1.08	1.58	2.21	2.59	3.02	3.96	5.06	6.29	7.70	9.20	10.79	12.45	1:30
CsNO$_3$	8.46	13.0	18.6	21.8	25.1	32.0	39.0	45.7	51.9	57.3	62.1	66.2	6
CsOH					75								7
Cs$_2$SO$_4$	62.6	63.4	64.1	64.5	64.8	65.5	66.1	66.7	67.3	67.8	68.3	68.8	6
CuBr$_2$				55.8									7
CuCl$_2$	40.8	41.7	42.6	43.1	43.7	44.8	46.0	47.2	48.5	49.9	51.3	52.7	6
Cu(ClO$_4$)$_2$	54.3				59.3								7
CuF$_2$				0.075									7
Cu(NO$_3$)$_2$	45.2	49.8	56.3	59.2	61.1	62.0	63.1	64.5	65.9	67.5	69.2	71.0	6
CuSO$_4$	12.4	14.4	16.7	18.0	19.3	22.2	25.4	28.8	32.4	36.3	40.3	43.5	6
CuSeO$_4$	10.6			16.0									7
Dy(NO$_3$)$_3$	58.79	59.99	61.49	62.35	63.29	65.43	68.04	71.58	77.75				1:13
Er(NO$_3$)$_3$	61.58	63.15	64.84	65.75	66.69	68.70	70.96	73.64					1:13
Eu(NO$_3$)$_3$	55.2	56.7	58.5	59.4	60.4	62.5	64.6						1:13
FeBr$_2$	33.2*			54.6								64.8*	7
FeCl$_2$				39.4*								48.7*	7
FeCl$_3$	42.7	44.9	47.9	47.7	51.6	74.8	76.7	84.6	84.3	84.3	84.4	84.7	6

AQUEOUS SOLUBILITY OF INORGANIC COMPOUNDS AT VARIOUS TEMPERATURES (continued)

Compound	0°C	10°C	20°C	25°C	30°C	40°C	50°C	60°C	70°C	80°C	90°C	100°C	Ref.
Fe(ClO$_4$)$_2$	63.39			67.76									7
FeF$_3$				5.59									7
Fe(NO$_3$)$_3$	40.15			46.57									7
Fe(NO$_3$)$_2$	41.44			46.67									7
FeSO$_4$	13.5	17.0	20.8	22.8	24.8	28.8	32.8	35.5	33.6	30.4	27.1	24.0	6
Gd(NO$_3$)$_3$	56.3	57.7	59.2	60.1	61.0	62.9	65.2	67.9	71.5				1:13
HIO$_3$	73.45	74.10	74.98	75.48	76.03	77.20	78.46	79.78	81.13	82.48	83.82	85.14	1:30
H$_3$BO$_3$	2.61	3.57	4.77	5.48	6.27	8.10	10.3	12.9	15.9	19.3	23.1	27.3	6
HgBr$_2$	0.26	0.37	0.52	0.61	0.72	0.96	1.26	1.63	2.08	2.61	3.23	3.95	4
Hg(CN)$_2$	6.57	7.83	9.33	10.2	11.1	13.1	15.5	18.2	21.2	24.6	28.3	32.3	6
HgCl$_2$	4.24	5.05	6.17	6.81	7.62	9.53	12.02	15.18	19.16	24.06	29.90	36.62	4
HgI$_2$			0.0041	0.0055	0.0072	0.0122	0.0199						4
Hg(SCN)$_2$				0.070									4
Hg$_2$Cl$_2$				0.0004									4
Hg$_2$(ClO$_4$)$_2$	73.8			79.8*								85.3*	3
Hg$_2$SO$_4$	0.038	0.043	0.048	0.051	0.054	0.059	0.065	0.070	0.076	0.082	0.088	0.093	7
Ho(NO$_3$)$_3$				63.8									4
KBF$_4$	0.28	0.34	0.45	0.55	0.75	1.38	2.09	2.82	3.58	4.34	5.12	5.90	1:13
KBr	35.0	37.3	39.4	40.4	41.4	43.2	44.8	46.2	47.6	48.8	49.8	50.8	10
KBrO$_3$	2.97	4.48	6.42	7.55	8.79	11.57	14.71	18.14	21.79	25.57	29.42	33.28	6
KC$_2$H$_3$O$_2$	68.40	70.29	72.09	72.92	73.70	75.08	76.27	77.31	78.22	79.04	79.80	80.55	1:30
KCl	21.74	23.61	25.39	26.22	27.04	28.59	30.04	31.40	32.66	33.86	34.99	36.05	7
KClO$_3$	3.03	4.67	6.74	7.93	9.21	12.06	15.26	18.78	22.65	26.88	31.53	36.65	1:47
KClO$_4$	0.70	1.10	1.67	2.04	2.47	3.54	4.94	6.74	8.99	11.71	14.94	18.67	1:30
KF	30.90	39.8	47.3	50.41	53.2					60.0			6
KHCO$_3$	18.62	21.73	24.92	26.6	28.13	31.32	34.46	37.51	40.45				7
KHSO$_4$	27.1	29.7	32.3	33.6	35.0	37.8	40.5	43.4	46.2	49.02	51.82	54.6	6
KH$_2$PO$_4$	11.74	14.91	18.25	19.97	21.77	25.28	28.95	32.76	36.75	40.96	45.41	50.12	6
KI	56.0	57.6	59.0	59.7	60.4	61.6	62.8	63.8	64.8	65.7	66.6	67.4	1:31
KIO$_3$	4.53	5.96	7.57	8.44	9.34	11.09	13.22	15.29	17.41	19.58	21.78	24.03	6
KIO$_4$	0.16	0.22	0.37	0.51	0.70	1.24	1.96	2.83	3.82	4.89	6.02	7.17	1:30
KMnO$_4$	2.74	4.12	5.96	7.06	8.28	11.11	14.42	18.16					7
KNO$_2$	73.7	74.6	75.3	75.7	76.0	76.7	77.4	78.0	78.5	79.1	79.6	80.1	6
KNO$_3$	12.0	17.6	24.2	27.7	31.3	38.6	45.7	52.2	58.0	63.0	67.3	70.8	6
KOH	48.7	50.8	53.2	54.7	56.1	57.9	58.6	59.5	60.6	61.8	63.1	64.6	6
KSCN	63.8	66.4	69.1	70.4	71.6	74.1	76.5	78.9	81.1	83.3	85.3	87.3	6
K$_2$CO$_3$	51.3	51.7	52.3	52.7	53.1	54.0	54.9	56.0	57.2	58.4	59.6	61.0	6
K$_2$CrO$_4$	37.1	38.1	38.9	39.4	39.8	40.5	41.3	41.9	42.6	43.2	43.8	44.3	6
K$_2$Cr$_2$O$_7$	4.30	7.12	10.9	13.1	15.5	20.8	26.3	31.7	36.9	41.5	45.5	48.9	6
K$_2$HAsO$_4$	48.5*			63.6*		67.7*		72.7*				79.8*	7
K$_2$HPO$_4$	57.0	59.1	61.5	62.7	64.1								1:31
K$_2$MoO$_4$				64.7							66.5		7
K$_2$SO$_3$	51.30	51.39	51.49	51.55	51.62	51.76	51.93	52.11	52.32	52.54	52.79	53.06	1:26

AQUEOUS SOLUBILITY OF INORGANIC COMPOUNDS AT VARIOUS TEMPERATURES (continued)

Compound	0°C	10°C	20°C	25°C	30°C	40°C	50°C	60°C	70°C	80°C	90°C	100°C	Ref.
K_2SO_4	7.11	8.46	9.95	10.7	11.4	12.9	14.2	15.5	16.7	17.7	18.6	19.3	6
$K_2S_2O_3$	49.0*			62.3*							75.7*		7
$K_2S_2O_5$	22.1	26.7	31.1	33.1	35.2	39.0	42.6	46.0	49.1	52.0	54.6		1:26
K_2SeO_3	68.4*			68.5*								68.5*	7
K_2SeO_4	52.70	52.93	53.17	53.30	53.43	53.70	53.99	54.30	54.61	54.94	55.26	55.60	7
K_3AsO_4	51.5*			55.6*								73*	7
$K_3Fe(CN)_6$	23.9	27.6	31.1	32.8	34.3	37.2	39.6	41.7	43.5	45.0	46.1	47.0	6
K_3PO_4	44.3			51.4									7
$K_4Fe(CN)_6$	12.5	17.3	22.0	23.9	25.6	29.2	32.5	35.5	38.2	40.6	41.4	43.1	6
$LaCl_3$	49.0	48.5	48.6	48.9	49.3	50.5	52.1	54.0	56.3	58.9	61.7		6
$La(NO_3)_3$	55.0	56.9	58.9	60.0	61.1	63.6	66.3	69.9*	74.1*				1:13
$LiBr$	58.4	60.1	62.7	64.4	65.9	67.8	68.3	69.0	69.8	70.7	71.7	72.8	6
$LiBrO_3$	61.03	62.62	64.44	65.44	66.51	68.90	71.68*	73.24*	74.43	75.66	76.93	78.32	1:30
$LiC_2H_3O_2$	23.76	26.49	29.42	31.02	32.72	36.48	40.65	45.15	49.93	54.91	60.04	65.26	7
$LiCl$	40.45	42.46*	45.29*	45.81	46.25	47.30	48.47	49.78	51.27	52.98	54.98*	56.34*	1:47
$LiClO_3$	73.2	75.6*	80.8*	82.1	83.4	85.9*	87.1*	88.2	89.6	91.3	93.4	95.7	1:30
$LiClO_4$	30.1	32.6	35.5	37.0	38.6	41.9	45.5	49.2	53.2	57.2	61.3	71.4	6
LiF	0.120	0.126	0.131	0.134									7
LiH_2PO_4	55.8												6
LiI	59.4	60.5	61.7	62.3	63.0	64.3	65.8	67.3	68.8	81.3	81.7	82.6	7
$LiIO_3$				43.8									6
$LiNO_2$	41	45	49	51	53	56	60	63	66	68			1:30
$LiNO_3$	34.8	37.6	42.7	50.5	57.9	60.1	62.2	64.0	65.7	67.2	68.5	69.7	10
$LiOH$	10.8	10.8	11.0	11.1	11.3	11.7	12.2	12.7	13.4	14.2	15.1	16.1	6
$LiSCN$				54.5									6
Li_2CO_3	1.54	1.43	1.33	1.28	1.24	1.15	1.07	0.99	0.92	0.85	0.78	0.72	7
$Li_2C_2O_4$				5.87									7
Li_2HPO_3	9.07	8.40	7.77	7.47	7.18	6.64	6.16	5.71	5.30	4.91	4.53	4.16	7
Li_2SO_4	26.3	25.9	25.6	25.5	25.3	25.0	24.8	24.5	24.3	24.0	23.8	23.6	6
Li_3PO_4				0.027									1:31
$Lu(NO_3)_3$				71.1									1:13
$MgBr_2$	49.3	49.8	50.3	50.6	50.9	51.5	52.1	52.8	53.5	54.2	55.0	55.7	6
$Mg(BrO_3)_2$	43.0	45.2	48.0	49.4	51.0	54.3	57.9	61.6	65.3	69.0*	70.9*	71.7	1:14
$Mg(C_2H_3O_2)_2$	36.18	37.55	38.92	39.61									7
MgC_2O_4				0.038									
$MgCl_2$	33.96	34.85	35.58	35.90	36.20	36.77	37.34	37.97	38.71	39.62	40.75	42.15	8
$Mg(ClO_3)_2$	53.35	54.40	56.81	58.66	60.91*	65.46*	67.33	69.27	71.01	72.44	73.48		1:14
$Mg(ClO_4)_2$	47.8	48.7	49.6	50.1	50.5	51.3	52.1						6
$MgCrO_4$	32.06*			35.39*									7
$MgCr_2O_7$				58.9						67.0			7
MgF_2				0.013									7
MgI_2	54.7	56.1	58.2	59.4	60.8	63.9	65.0	65.0	65.0	65.0	65.1	65.2	6
$Mg(IO_3)_2$	3.19*	6.70*	7.92	8.52	9.11	10.45	11.99	13.7	15.6	17.6	19.6		1:14

AQUEOUS SOLUBILITY OF INORGANIC COMPOUNDS AT VARIOUS TEMPERATURES (continued)

Compound	0°C	10°C	20°C	25°C	30°C	40°C	50°C	60°C	70°C	80°C	90°C	100°C	Ref.
$Mg(NO_2)_2$				47									7
$Mg(NO_3)_2$	38.4	39.5	40.8	41.6	42.4	44.1	45.9	47.9	50.0	52.2	70.6	72.0	6
$MgSO_3$	0.32	0.37	0.46	0.52	0.61	0.87*	0.85*	0.76	0.69	0.64	0.62	0.60	1:26
$MgSO_4$	18.2	21.7	25.1	26.3	28.2	30.9	33.4	35.6	36.9	35.9	34.7	33.3	6
MgS_2O_3	30.7			34.1									7
$MgSeO_4$	31.4*			35.7*								47*	7
$MnBr_2$	56.00	57.72	59.39	60.19	60.96	62.41	63.75	65.01	66.19	67.32	68.42	69.50	7
$MnCl_2$	38.7	40.6	42.5	43.6	44.7	47.0	49.4	54.1	54.7	55.2	55.7	56.1	7
MnF_2	0.80*			1.01*									6
$Mn(IO_3)_2$				0.27							0.34	0.48	7
$Mn(NO_3)_2$	50.5			61.7									7
$MnSO_4$	34.6	37.3	38.6	38.9	38.9	37.7	36.3	34.6	32.8	30.8	28.8	26.7	7
NH_4Br	37.5	40.2	42.7	43.9	45.1	47.3	49.4	51.3	53.0	54.6	56.1	57.4	6
NH_4Cl	22.92	25.12	27.27	28.34	29.39	31.46	33.50	35.49	37.46	39.40	41.33	43.24	7
NH_4ClO_4	10.8	14.1	17.8	19.7	21.7	25.8	29.8	33.6	37.3	40.7	43.8	46.6	1:47
NH_4F	41.7	43.2	44.7	45.5	46.3	47.8	49.3	50.9	52.5	54.1			6
NH_4HCO_3	10.6	13.7	17.6	19.9	22.4	27.9	34.2	41.4	49.3	58.1	67.6	78.0	7
$NH_4H_2AsO_4$	25.2	29.0	32.7	34.5	36.3	39.7	43.1	46.2	49.3	52.2	55.0		7
$NH_4H_2PO_4$	17.8	22.0	26.4	28.8	31.2	36.2	41.6	47.2	53.0	59.2	65.7	72.4	7
NH_4I	60.7	62.1	63.4	64.0	64.6	65.8	66.8	67.8	68.7	69.6	70.4	71.1	6
NH_4IO_3				3.70	4.20	5.64	7.63						1:30
NH_4NO_2	55.7	59.0	64.9	68.8	70.3	74.3	77.7	80.8	83.4	85.8	88.2	90.3	7
NH_4NO_3	54.0	60.1	65.5	68.0					81.1				6
NH_4SCN				64.4									7
$(NH_4)_2C_2O_4$	2.31	3.11	4.25	4.94	5.73	7.56	9.73	12.2	15.1	18.3	21.8	25.7	7
$(NH_4)_2HPO_4$	36.4	38.2	40.0	41.0	42.0	44.1	46.2	48.5	50.9	53.3	55.9	58.6	7
$(NH_4)_2S_2O_5$	65.5	67.9	69.8	70.5	71.3	72.3	72.9	73.1					1:26
$(NH_4)_2S_2O_8$	37.00	40.45	43.84	45.49	47.11	50.25	53.28	56.23	59.13	62.00			7
$(NH_4)_2SO_3$	32.2	34.9	37.7	39.1	40.6	43.7	47.0	50.6	54.5	58.9			1:26
$(NH_4)_2SO_4$	41.3	42.1	42.9	43.3	43.8	44.7	45.6	46.6	47.5	48.5			6
$(NH_4)_2SeO_3$	49.0	51.1	53.4	54.7	56.0	58.9	62.0	65.4	69.1				7
$(NH_4)_2SeO_4$				54.02							49.5	50.5	1:26
$(NH_4)_3PO_4$				15.5									7
$NaBr$	44.4	45.9	47.7	48.6	49.6	51.6	53.7	54.1	54.3	54.5	54.7	54.9	7
$NaBrO_3$	20.0	23.22	26.65	28.28	29.86	32.83	35.55	38.05	40.37	42.52			6
$NaCHO_2$	30.8	37.9	45.7	48.7	50.6	52.0	53.5	55.0					1:30
$NaC_2H_3O_2$	26.5	28.8	31.8	33.5	35.5	39.9	45.1	58.3	59.3	60.5	61.7	62.9	6
$NaCl$	26.28	26.32	26.41	26.45	26.52	26.67	26.84	27.03	27.25	27.50	27.78	28.05	6
$NaClO$	22.7			44.4									1:47
$NaClO_2$				97.0*				95.3*					7
$NaClO_3$	44.27	46.67	49.3	50.1	51.2	53.6	55.5	57.0	58.5	60.5	63.3	67.1	1:30
$NaClO_4$	61.9	64.1	66.2	67.2	68.3	70.4	72.5	74.1	74.7	75.4	76.1	76.7	6
NaF	3.52	3.72	3.89	3.97	4.05	4.20	4.34	4.46	4.57	4.66	4.75	4.82	6

AQUEOUS SOLUBILITY OF INORGANIC COMPOUNDS AT VARIOUS TEMPERATURES (continued)

Compound	0°C	10°C	20°C	25°C	30°C	40°C	50°C	60°C	70°C	80°C	90°C	100°C	Ref.
NaHCO$_3$	6.48	7.59	8.73	9.32	9.91	11.13	12.40	13.70	15.02	16.37	17.73	19.10	7
NaHSO$_4$				22.2								33.3	10
NaH$_2$PO$_4$	36.54	41.07	46.00	48.68	51.54	57.89*	61.7*	62.3*	65.9	68.7			1:31
NaI	61.2	62.4	63.9	64.8	65.7	67.7	69.8	72.0	74.7	74.8	74.9	75.1	6
NaIO$_3$	2.43	4.40	7.78*	8.65*	9.60	11.67	13.99	16.52	19.25*	21.1*	22.9	24.7	1:30
NaIO$_4$				12.62									7
NaNO$_2$	41.9	43.4	45.1	45.9	46.8	48.7	50.7	52.8	55.0	57.2	59.5	61.8	6
NaNO$_3$	42.2	44.4	46.6	47.7	48.8	51.0	53.2	55.3	57.5	59.6	61.7	63.8	6
NaOH	30	39	46	50	53	58	63	67	71	74	76	79	10
NaSCN		52.9	57.1	60.2	62.7	63.5	64.2	65.0	65.9	66.9	67.9	69.0	6
Na$_2$B$_4$O$_7$	1.23	1.71	2.50	3.07	3.82	6.02	9.7	14.9	17.1	19.9	23.5	28.0	6
Na$_2$CO$_3$	6.44	10.8	17.9	23.5	28.7	32.8	32.2	31.7	31.3	31.1	30.9	30.9	6
Na$_2$C$_2$O$_4$	2.62	2.95	3.30	3.48	3.65	4.00	4.36	4.71	5.06	5.41	5.75	6.08	6
Na$_2$CrO$_4$	22.6	32.3	44.6	46.7	46.9	48.9	51.0	53.4	55.3	55.5	55.8	56.1	6
Na$_2$Cr$_2$O$_7$	62.1	63.1	64.4	65.2	66.1	68.0	70.1	72.3	74.6	77.0	79.6	80.7	6
Na$_2$HAsO$_4$	5.6*			29.3*								67*	7
Na$_2$HPO$_4$	1.66	4.19	7.51	10.55	16.34*	35.17*	44.64*	45.20	46.81	48.78	50.52	51.53	1:31
Na$_2$MoO$_4$	30.6	38.8	39.4	39.4	39.8	40.3	41.0	41.7	42.6	43.5	44.5	45.5	6
Na$_2$S	11.1	13.2	15.7	17.1	18.6	22.1	26.7	28.1	30.2	33.0	36.4	41.0	6
Na$_2$SO$_3$	12.0	16.1	20.9	23.5	26.3*	27.3*	25.9	24.8	23.7	22.8	22.1	21.5	1:26
Na$_2$SO$_4$			16.13	21.94	29.22*	32.35*	31.55	30.90	30.39	30.02	29.79	29.67	8
Na$_2$S$_2$O$_3$	33.1	36.3	40.6	43.3	45.9	52.0	62.3	65.7	68.8	69.4	70.1	71.0	6
Na$_2$S$_2$O$_5$		38.4	39.5	40.0	40.6	41.8	43.0	44.2	45.5	46.8	48.1	49.5	1:26
Na$_2$SeO$_3$				47.3*								45*	7
Na$_2$SeO$_4$	11.7			36.9*								42.1*	7
Na$_2$WO$_4$	41.6	41.9	42.3	42.6	42.9	43.6	44.4	45.3	46.2	47.3	48.4	49.5	6
Na$_3$PO$_4$	4.28	7.30	10.8	12.6	14.1	16.6	22.9	28.4	32.4	37.6	40.4	43.5	6
Na$_4$P$_2$O$_7$	2.23	3.28	4.81	6.62	7.00	10.10	14.38	20.07	27.31	36.03	32.37	30.67	6
NdCl$_3$	49.0	49.3	49.7	50.0	50.4	51.2	52.2	53.3	54.5	55.8	57.1	58.5	6
Nd(NO$_3$)$_3$	55.76	57.49	59.37	60.38	61.43	63.69	66.27	69.47					1:13
NiCl$_2$	34.7	36.1	38.5	40.3	41.7	42.1	43.2	45.0	46.1	46.2	46.4	46.6	6
Ni(ClO$_4$)$_2$	51.1			52.8									7
NiF$_2$				2.50							2.52		7
NiI$_2$	55.40	57.68	59.78	60.69	61.50	62.80	63.73	64.38	64.80	65.09	65.30		7
Ni(NO$_3$)$_2$	44.1	46.0	48.4	49.8	51.3	54.6	58.3	61.0	63.1	65.6	67.9	69.0	6
NiSO$_4$	21.4	24.4	27.4	28.8	30.3*	32.0*	34.1	35.8	37.7	39.9	42.3	44.8	6
Ni(SCN)$_2$				35.48									7
NiSeO$_4$	21.6		26.2*									45.6*	7
PbBr$_2$	0.449	0.620	0.841	0.966	1.118	1.46	1.89						2
PbCl$_2$	0.66	0.81	0.98	1.07	1.17	1.39	1.64	1.93	2.24	2.60	2.99	3.42	2
Pb(ClO$_4$)$_2$				81.5									7
PbF$_2$	0.041	0.0603	0.0649	0.0670	0.0693								2
PbI$_2$		0.052	0.067	0.076	0.086	0.112	0.144	0.187	0.243	0.315			2

AQUEOUS SOLUBILITY OF INORGANIC COMPOUNDS AT VARIOUS TEMPERATURES (continued)

Compound	0°C	10°C	20°C	25°C	30°C	40°C	50°C	60°C	70°C	80°C	90°C	100°C	Ref.
$Pb(IO_3)_2$				0.0025									7
$Pb(NO_3)_2$	28.46	32.13	35.67	37.38	39.05	42.22	45.17	47.90	50.42	52.72	54.82	56.75	2
$PbSO_4$	0.0033	0.0038	0.0042	0.0044	0.0047	0.0052	0.0058						2
$PrCl_3$	48.0	48.1	48.6	49.0	49.5	50.8	52.3	54.1	56.1	58.3			6
$Pr(NO_3)_3$	57.50	59.20	61.16	62.24	63.40*	65.7*	67.8	70.2	73.4				1:13
$RbBr$	47.4	50.1	52.6	53.8	54.9	57.0	58.8	60.6	62.1	63.5	64.8	65.9	6
$RbBrO_3$	0.97	1.55	2.36	2.87	3.45	4.87	6.64	8.78	11.29	14.15	17.32	20.76	1:30
$RbCl$	43.58	45.65	47.53	48.42	49.27	50.86	52.34	53.67	54.92	56.08	57.16	58.15	1:47
$RbClO_3$	2.10	3.38	5.14	6.22	7.45	10.35	13.85	17.93	22.53	27.57	32.96	38.60	1:30
$RbClO_4$	1			1.5								17	1:30
RbF			75										7
$RbHCO_3$			53.7										7
RbI	55.8	58.6	61.1	62.3	63.4	65.4	67.2	68.8	70.3	71.6	72.7	73.8	7
$RbIO_3$	1.09	1.53	2.07	2.38	2.74	3.52	4.41	5.42	6.52	7.74	9.00	10.36	7
$RbNO_3$	16.4	25.0	34.6	39.4	44.2	53.1	60.8	67.2	72.2	76.1	79.0	81.2	6
$RbOH$				63.4									1:30
Rb_2CrO_4	38.27			43.26									6
Rb_2SO_4	27.3	30.0	32.5	33.7	34.8	36.9	38.7	40.3	41.8	43.0	44.1	44.9	7
$SbCl_3$	85.7			90.8									7
SbF_3	79.4			83.1									7
$Sc(NO_3)_3$	57.0	59.3	61.6	62.8	63.9	66.2	68.5	68.1*					7
$Sm(NO_3)_3$	54.83	56.33	58.08	59.05	60.08	62.38	65.05*						1:13
$SmCl_3$		48.0	48.2	48.4	48.6	49.2	50.0						1:13
$SnCl_2$	46	64											6
SnI_2			0.97										7
$SrBr_2$	46.0	48.3	50.6	51.7	52.9	55.2	57.6	59.9	62.3	64.6	66.8	69.0	7
$Sr(BrO_3)_2$	18.53	22.00	25.39	27.02	28.59	31.55	34.21	36.57	38.64*	40.2*	40.8	41.0	6
$SrCl_2$	31.94	32.93	34.43	35.37	36.43	38.93	41.94	45.44*	46.81*	47.69	48.70	49.87	1:14
$Sr(ClO_2)_2$	13.0	13.6	14.1	14.3	14.5	14.9	15.3	15.6	15.9				8
$Sr(ClO_3)_2$	63.29	63.42	63.64	63.77	63.93	64.29	64.70	65.16	65.65	66.18	66.74	67.31	1:14
$Sr(ClO_4)_2$	70.04*			75.35*		78.44*							7
SrF_2	0.011			0.021									7
SrI_2	62.5	62.8	63.5	63.9	64.5	65.8	67.3	69.0	70.8	72.7	74.7	79.2	7
$Sr(IO_3)_2$	0.102	0.126	0.152	0.165	0.179	0.206	0.233	0.259	0.284	0.307	0.328	0.346	6
$Sr(MnO_4)_2$	2.5												1:14
$Sr(NO_2)_2$					41.9	44.3						58.6	7
$Sr(NO_3)_2$	28.2	34.6	41.0	44.5	47.0	47.4	47.9	48.4	48.9	49.5	50.1	50.7	6
$Sr(OH)_2$	0.9			2.2									7
$SrSO_3$				0.0015									7
$SrSO_4$				0.0135									7
SrS_2O_3	8.8	13.2	17.7	20.0	22.2	26.8							1:26
$Tb(NO_3)_3$			60.6	61.02									1:13
Tl_2SO_4	2.65	3.56	4.61	5.19	5.80	7.09	8.46	9.89	11.33	12.77	14.18	15.53	6

AQUEOUS SOLUBILITY OF INORGANIC COMPOUNDS AT VARIOUS TEMPERATURES (continued)

Compound	0°C	10°C	20°C	25°C	30°C	40°C	50°C	60°C	70°C	80°C	90°C	100°C	Ref.
$Tm(NO_3)_3$				67.9									1:13
$UO_2(NO_3)_2$	49.52	51.82	54.42	55.85	57.55	61.59	67.07						1:55
$Y(NO_3)_3$	55.57	56.93	58.75	59.86	61.11*	63.3*	64.9	67.9	72.5				1:13
$Yb(NO_3)_3$				70.5									1:13
$ZnBr_2$	79.3	80.1	81.8	83.0	84.1	85.6	85.8	86.1	86.3	86.6	86.8	87.1	6
ZnC_2O_4		0.0010	0.0019	0.0026									5
$ZnCl_2$		76.6	79.0	80.3	81.4	81.8	82.4	83.0	83.7	84.4	85.2	86.0	6
$Zn(ClO_4)_2$	44.29*			46.27*			48.70						7
ZnF_2				1.53									5
ZnI_2	81.1	81.2	81.3	81.4	81.5	81.7	82.0	82.3	82.6	83.0	83.3	83.7	6
$Zn(IO_3)_2$			0.58	0.64	0.69	0.77	0.82						5
$Zn(NO_3)_2$	47.8	50.8	54.4	54.6	58.5	79.1	80.1	87.5	89.9				6
$ZnSO_3$			0.1786	0.1790	0.1794	0.1803	0.1812						5
$ZnSO_4$	29.1	32.0	35.0	36.6	38.2	41.3	43.0	42.1	41.0	39.9	38.8	37.6	6
$ZnSeO_4$	33.06	34.98	37.38	38.79	40.34								5

SOLUBILITY PRODUCT CONSTANTS

The solubility product constant K_{sp} is a useful parameter for calculating the aqueous solubility of sparingly soluble compounds under various conditions. It may be determined by direct measurement or calculated from the standard Gibbs energies of formation $\Delta_f G°$ of the species involved at their standard states. Thus if $K_{sp} = [M^+]^m [A^-]^n$ is the equilibrium constant for the reaction

$$M_m A_n(s) \rightleftharpoons mM^+(aq) + nA^-(aq),$$

where $M_m A_n$ is the slightly soluble substance and M^+ and A^- are the ions produced in solution by the dissociation of $M_m A_n$, then the Gibbs energy change is

$$\Delta G° = m\, \Delta_f G° (M^+,aq) + n\, \Delta_f G° (A^-,aq) - \Delta_f G° (M_m A_n, s)$$

The solubility product constant is calculated from the equation

$$\ln K_{sp} = -\Delta G°/RT$$

The first table below gives selected values of K_{sp} at 25°C. Many of these have been calculated from standard state thermodynamic data in References 1 and 2; other values are taken from publications of the IUPAC Solubility Data Project (References 3 to 7).

The above formulation is not convenient for treating sulfides because the S^{-2} ion is usually not present in significant concentrations (see Reference 8). This is due to the hydrolysis reaction

$$S^{-2} + H_2O \rightleftharpoons HS^- + OH^-$$

which is strongly shifted to the right except in very basic solutions. Furthermore, the equilibrium constant for this reaction, which depends on the second ionization constant of H_2S, is poorly known. Therefore it is more useful in the case of sulfides to define a different solubility product K_{spa} based on the reaction

$$M_m S_n(s) + 2H^+ \rightleftharpoons mM^+ + nH_2S \text{ (aq)}$$

Values of K_{spa}, taken from Reference 8, are given for several sulfides in the auxiliary table following the main table. Additional discussion of sulfide equilibria may be found in References 7 and 9.

REFERENCES

1. Wagman, D.D., Evans, W.H., Parker, V.B., Schumm, R.H., Halow, I., Bailey, S.M., Churney, K.L., and Nuttall, R L., *The NBS Tables of Chemical Thermodynamic Properties, J. Phys. Chem. Ref. Data*, Vol. 11, Suppl. 2, 1982.
2. Garvin, D., Parker, V.B., and White, H.J., *CODATA Thermodynamic Tables*, Hemisphere, New York, 1987.
3. *Solubility Data Series* (53 Volumes), International Union of Pure and Applied Chemistry, Pergamon Press, Oxford, 1979—1992.
4. Clever, H.L., and Johnston, F.J., *J. Phys. Chem. Ref. Data*, 9, 751, 1980.
5. Marcus, Y., *J. Phys. Chem. Ref. Data*, 9, 1307, 1980.
6. Clever, H.L., Johnson, S.A., and Derrick, M.E., *J. Phys. Chem. Ref. Data*, 14, 631, 1985.
7. Clever, H.L., Johnson, S.A., and Derrick, M.E., *J. Phys. Chem. Ref. Data*, 21, 941, 1992.
8. Myers, R.J., *J. Chem. Educ.*, 63, 687, 1986.
9. Licht, S., *J. Electrochem. Soc.*, 135, 2971, 1988.

Compound	Formula	K_{sp}
Aluminum phosphate	$AlPO_4$	$9.84 \cdot 10^{-21}$
Barium bromate	$Ba(BrO_3)_2$	$2.43 \cdot 10^{-4}$
Barium carbonate	$BaCO_3$	$2.58 \cdot 10^{-9}$
Barium chromate	$BaCrO_4$	$1.17 \cdot 10^{-10}$
Barium fluoride	BaF_2	$1.84 \cdot 10^{-7}$
Barium hydroxide octahydrate	$Ba(OH)_2 \cdot 8H_2O$	$2.55 \cdot 10^{-4}$
Barium iodate	$Ba(IO_3)_2$	$4.01 \cdot 10^{-9}$
Barium iodate monohydrate	$Ba(IO_3)_2 \cdot H_2O$	$1.67 \cdot 10^{-9}$
Barium molybdate	$BaMoO_4$	$3.54 \cdot 10^{-8}$
Barium nitrate	$Ba(NO_3)_2$	$4.64 \cdot 10^{-3}$
Barium selenate	$BaSeO_4$	$3.40 \cdot 10^{-8}$
Barium sulfate	$BaSO_4$	$1.08 \cdot 10^{-10}$
Barium sulfite	$BaSO_3$	$5.0 \cdot 10^{-10}$
Beryllium hydroxide	$Be(OH)_2$	$6.92 \cdot 10^{-22}$
Bismuth arsenate	$BiAsO_4$	$4.43 \cdot 10^{-10}$

SOLUBILITY PRODUCT CONSTANTS (continued)

Compound	Formula	K_{sp}
Bismuth iodide	BiI_3	$7.71 \cdot 10^{-19}$
Cadmium arsenate	$Cd_3(AsO_4)_2$	$2.2 \cdot 10^{-33}$
Cadmium carbonate	$CdCO_3$	$1.0 \cdot 10^{-12}$
Cadmium fluoride	CdF_2	$6.44 \cdot 10^{-3}$
Cadmium hydroxide	$Cd(OH)_2$	$7.2 \cdot 10^{-15}$
Cadmium iodate	$Cd(IO_3)_2$	$2.5 \cdot 10^{-8}$
Cadmium oxalate trihydrate	$CdC_2O_4 \cdot 3H_2O$	$1.42 \cdot 10^{-8}$
Cadmium phosphate	$Cd_3(PO_4)_2$	$2.53 \cdot 10^{-33}$
Calcium carbonate (calcite)	$CaCO_3$	$3.36 \cdot 10^{-9}$
Calcium fluoride	CaF_2	$3.45 \cdot 10^{-11}$
Calcium hydroxide	$Ca(OH)_2$	$5.02 \cdot 10^{-6}$
Calcium iodate	$Ca(IO_3)_2$	$6.47 \cdot 10^{-6}$
Calcium iodate hexahydrate	$Ca(IO_3)_2 \cdot 6H_2O$	$7.10 \cdot 10^{-7}$
Calcium molybdate	$CaMoO_4$	$1.46 \cdot 10^{-8}$
Calcium oxalate monohydrate	$CaC_2O_4 \cdot H_2O$	$2.32 \cdot 10^{-9}$
Calcium phosphate	$Ca_3(PO_4)_2$	$2.07 \cdot 10^{-33}$
Calcium sulfate	$CaSO_4$	$4.93 \cdot 10^{-5}$
Calcium sulfate dihydrate	$CaSO_4 \cdot 2H_2O$	$3.14 \cdot 10^{-5}$
Calcium sulfite hemihydrate	$CaSO_3 \cdot 0.5H_2O$	$3.1 \cdot 10^{-7}$
Cesium perchlorate	$CsClO_4$	$3.95 \cdot 10^{-3}$
Cesium periodate	$CsIO_4$	$5.16 \cdot 10^{-6}$
Cobalt(II) arsenate	$Co_3(AsO_4)_2$	$6.80 \cdot 10^{-29}$
Cobalt(II) hydroxide (blue)	$Co(OH)_2$	$5.92 \cdot 10^{-15}$
Cobalt(II) iodate dihydrate	$Co(IO_3)_2 \cdot 2H_2O$	$1.21 \cdot 10^{-2}$
Cobalt(II) phosphate	$Co_3(PO_4)_2$	$2.05 \cdot 10^{-35}$
Copper(I) bromide	$CuBr$	$6.27 \cdot 10^{-9}$
Copper(I) chloride	$CuCl$	$1.72 \cdot 10^{-7}$
Copper(I) cyanide	$CuCN$	$3.47 \cdot 10^{-20}$
Copper(I) iodide	CuI	$1.27 \cdot 10^{-12}$
Copper(I) thiocyanate	$CuSCN$	$1.77 \cdot 10^{-13}$
Copper(II) arsenate	$Cu_3(AsO_4)_2$	$7.95 \cdot 10^{-36}$
Copper(II) iodate monohydrate	$Cu(IO_3)_2 \cdot H_2O$	$6.94 \cdot 10^{-8}$
Copper(II) oxalate	CuC_2O_4	$4.43 \cdot 10^{-10}$
Copper(II) phosphate	$Cu_3(PO_4)_2$	$1.40 \cdot 10^{-37}$
Europium(III) hydroxide	$Eu(OH)_3$	$9.38 \cdot 10^{-27}$
Gallium(III) hydroxide	$Ga(OH)_3$	$7.28 \cdot 10^{-36}$
Iron(II) carbonate	$FeCO_3$	$3.13 \cdot 10^{-11}$
Iron(II) fluoride	FeF_2	$2.36 \cdot 10^{-6}$
Iron(II) hydroxide	$Fe(OH)_2$	$4.87 \cdot 10^{-17}$
Iron(III) hydroxide	$Fe(OH)_3$	$2.79 \cdot 10^{-39}$
Iron(III) phosphate dihydrate	$FePO_4 \cdot 2H_2O$	$9.91 \cdot 10^{-16}$
Lanthanum iodate	$La(IO_3)_3$	$7.50 \cdot 10^{-12}$
Lead(II) bromide	$PbBr_2$	$6.60 \cdot 10^{-6}$
Lead(II) carbonate	$PbCO_3$	$7.40 \cdot 10^{-14}$
Lead(II) chloride	$PbCl_2$	$1.70 \cdot 10^{-5}$
Lead(II) fluoride	PbF_2	$3.3 \cdot 10^{-8}$
Lead(II) hydroxide	$Pb(OH)_2$	$1.43 \cdot 10^{-20}$
Lead(II) iodate	$Pb(IO_3)_2$	$3.69 \cdot 10^{-13}$
Lead(II) iodide	PbI_2	$9.8 \cdot 10^{-9}$
Lead(II) selenate	$PbSeO_4$	$1.37 \cdot 10^{-7}$
Lead(II) sulfate	$PbSO_4$	$2.53 \cdot 10^{-8}$
Lithium carbonate	Li_2CO_3	$8.15 \cdot 10^{-4}$
Lithium fluoride	LiF	$1.84 \cdot 10^{-3}$
Lithium phosphate	Li_3PO_4	$2.37 \cdot 10^{-11}$
Magnesium carbonate	$MgCO_3$	$6.82 \cdot 10^{-6}$
Magnesium carbonate trihydrate	$MgCO_3 \cdot 3H_2O$	$2.38 \cdot 10^{-6}$
Magnesium carbonate pentahydrate	$MgCO_3 \cdot 5H_2O$	$3.79 \cdot 10^{-6}$
Magnesium fluoride	MgF_2	$5.16 \cdot 10^{-11}$
Magnesium hydroxide	$Mg(OH)_2$	$5.61 \cdot 10^{-12}$
Magnesium oxalate dihydrate	$MgC_2O_4 \cdot 2H_2O$	$4.83 \cdot 10^{-6}$

Compound	Formula	K_{sp}
Magnesium phosphate	$Mg_3(PO_4)_2$	$1.04 \cdot 10^{-24}$
Manganese(II) carbonate	$MnCO_3$	$2.24 \cdot 10^{-11}$
Manganese(II) iodate	$Mn(IO_3)_2$	$4.37 \cdot 10^{-7}$
Manganese(II) oxalate dihydrate	$MnC_2O_4 \cdot 2H_2O$	$1.70 \cdot 10^{-7}$
Mercury(I) bromide	Hg_2Br_2	$6.40 \cdot 10^{-23}$
Mercury(I) carbonate	Hg_2CO_3	$3.6 \cdot 10^{-17}$
Mercury(I) chloride	Hg_2Cl_2	$1.43 \cdot 10^{-18}$
Mercury(I) fluoride	Hg_2F_2	$3.10 \cdot 10^{-6}$
Mercury(I) iodide	Hg_2I_2	$5.2 \cdot 10^{-29}$
Mercury(I) oxalate	$Hg_2C_2O_4$	$1.75 \cdot 10^{-13}$
Mercury(I) sulfate	Hg_2SO_4	$6.5 \cdot 10^{-7}$
Mercury(I) thiocyanate	$Hg_2(SCN)_2$	$3.2 \cdot 10^{-20}$
Mercury(II) bromide	$HgBr_2$	$6.2 \cdot 10^{-20}$
Mercury(II) iodide	HgI_2	$2.9 \cdot 10^{-29}$
Neodymium carbonate	$Nd_2(CO_3)_3$	$1.08 \cdot 10^{-33}$
Nickel(II) carbonate	$NiCO_3$	$1.42 \cdot 10^{-7}$
Nickel(II) hydroxide	$Ni(OH)_2$	$5.48 \cdot 10^{-16}$
Nickel(II) iodate	$Ni(IO_3)_2$	$4.71 \cdot 10^{-5}$
Nickel(II) phosphate	$Ni_3(PO_4)_2$	$4.74 \cdot 10^{-32}$
Palladium(II) thiocyanate	$Pd(SCN)_2$	$4.39 \cdot 10^{-23}$
Potassium hexachloroplatinate	K_2PtCl_6	$7.48 \cdot 10^{-6}$
Potassium perchlorate	$KClO_4$	$1.05 \cdot 10^{-2}$
Potassium periodate	KIO_4	$3.71 \cdot 10^{-4}$
Praseodymium hydroxide	$Pr(OH)_3$	$3.39 \cdot 10^{-24}$
Radium iodate	$Ra(IO_3)_2$	$1.16 \cdot 10^{-9}$
Radium sulfate	$RaSO_4$	$3.66 \cdot 10^{-11}$
Rubidium perchlorate	$RbClO_4$	$3.00 \cdot 10^{-3}$
Scandium fluoride	ScF_3	$5.81 \cdot 10^{-24}$
Scandium hydroxide	$Sc(OH)_3$	$2.22 \cdot 10^{-31}$
Silver(I) acetate	$AgCH_3COO$	$1.94 \cdot 10^{-3}$
Silver(I) arsenate	Ag_3AsO_4	$1.03 \cdot 10^{-22}$
Silver(I) bromate	$AgBrO_3$	$5.38 \cdot 10^{-5}$
Silver(I) bromide	$AgBr$	$5.35 \cdot 10^{-13}$
Silver(I) carbonate	Ag_2CO_3	$8.46 \cdot 10^{-12}$
Silver(I) chloride	$AgCl$	$1.77 \cdot 10^{-10}$
Silver(I) chromate	Ag_2CrO_4	$1.12 \cdot 10^{-12}$
Silver(I) cyanide	$AgCN$	$5.97 \cdot 10^{-17}$
Silver(I) iodate	$AgIO_3$	$3.17 \cdot 10^{-8}$
Silver(I) iodide	AgI	$8.52 \cdot 10^{-17}$
Silver(I) oxalate	$Ag_2C_2O_4$	$5.40 \cdot 10^{-12}$
Silver(I) phosphate	Ag_3PO_4	$8.89 \cdot 10^{-17}$
Silver(I) sulfate	Ag_2SO_4	$1.20 \cdot 10^{-5}$
Silver(I) sulfite	Ag_2SO_3	$1.50 \cdot 10^{-14}$
Silver(I) thiocyanate	$AgSCN$	$1.03 \cdot 10^{-12}$
Strontium arsenate	$Sr_3(AsO_4)_2$	$4.29 \cdot 10^{-19}$
Strontium carbonate	$SrCO_3$	$5.60 \cdot 10^{-10}$
Strontium fluoride	SrF_2	$4.33 \cdot 10^{-9}$
Strontium iodate	$Sr(IO_3)_2$	$1.14 \cdot 10^{-7}$
Strontium iodate monohydrate	$Sr(IO_3)_2 \cdot H_2O$	$3.77 \cdot 10^{-7}$
Strontium iodate hexahydrate	$Sr(IO_3)_2 \cdot 6H_2O$	$4.55 \cdot 10^{-7}$
Strontium sulfate	$SrSO_4$	$3.44 \cdot 10^{-7}$
Thallium(I) bromate	$TlBrO_3$	$1.10 \cdot 10^{-4}$
Thallium(I) bromide	$TlBr$	$3.71 \cdot 10^{-6}$
Thallium(I) chloride	$TlCl$	$1.86 \cdot 10^{-4}$
Thallium(I) chromate	Tl_2CrO_4	$8.67 \cdot 10^{-13}$
Thallium(I) iodate	$TlIO_3$	$3.12 \cdot 10^{-6}$
Thallium(I) iodide	TlI	$5.54 \cdot 10^{-8}$
Thallium(I) thiocyanate	$TlSCN$	$1.57 \cdot 10^{-4}$
Thallium(III) hydroxide	$Tl(OH)_3$	$1.68 \cdot 10^{-44}$
Tin(II) hydroxide	$Sn(OH)_2$	$5.45 \cdot 10^{-27}$

Compound	Formula	K_{sp}
Yttrium carbonate	$Y_2(CO_3)_3$	$1.03 \cdot 10^{-31}$
Yttrium fluoride	YF_3	$8.62 \cdot 10^{-21}$
Yttrium hydroxide	$Y(OH)_3$	$1.00 \cdot 10^{-22}$
Yttrium iodate	$Y(IO_3)_3$	$1.12 \cdot 10^{-10}$
Zinc arsenate	$Zn_3(AsO_4)_2$	$2.8 \cdot 10^{-28}$
Zinc carbonate	$ZnCO_3$	$1.46 \cdot 10^{-10}$
Zinc carbonate monohydrate	$ZnCO_3 \cdot H_2O$	$5.42 \cdot 10^{-11}$
Zinc fluoride	ZnF_2	$3.04 \cdot 10^{-2}$
Zinc hydroxide	$Zn(OH)_2$	$3 \cdot 10^{-17}$
Zinc iodate dihydrate	$Zn(IO_3)_2 \cdot 2H_2O$	$4.1 \cdot 10^{-6}$
Zinc oxalate dihydrate	$ZnC_2O_4 \cdot 2H_2O$	$1.38 \cdot 10^{-9}$
Zinc selenide	$ZnSe$	$3.6 \cdot 10^{-26}$
Zinc selenite monohydrate	$ZnSeO_3 \cdot H_2O$	$1.59 \cdot 10^{-7}$

Sulfides

Compound	Formula	K_{spa}
Cadmium sulfide	CdS	$8 \cdot 10^{-7}$
Copper(II) sulfide	CuS	$6 \cdot 10^{-16}$
Iron(II) sulfide	FeS	$6 \cdot 10^{2}$
Lead(II) sulfide	PbS	$3 \cdot 10^{-7}$
Manganese(II) sulfide (green)	MnS	$3 \cdot 10^{7}$
Mercury(II) sulfide (red)	HgS	$4 \cdot 10^{-33}$
Mercury(II) sulfide (black)	HgS	$2 \cdot 10^{-32}$
Silver(I) sulfide	Ag_2S	$6 \cdot 10^{-30}$
Tin(II) sulfide	SnS	$1 \cdot 10^{-5}$
Zinc sulfide (sphalerite)	ZnS	$2 \cdot 10^{-4}$
Zinc sulfide (wurtzite)	ZnS	$3 \cdot 10^{-2}$

SOLUBILITY OF COMMON SALTS AT AMBIENT TEMPERATURES

This table gives the aqueous solubility of selected salts at temperatures from 10°C to 40°C. Values are given in molality terms.

REFERENCES

1. Apelblat, A., *J. Chem. Thermodynamics*, 24, 619, 1992.
2. Apelblat, A., *J. Chem. Thermodynamics*, 25, 63, 1993.
3. Apelblat, A., *J. Chem. Thermodynamics*, 25, 1513, 1993.
4. Apelblat, A. and Korin, E., *J. Chem. Thermodynamics*, 30, 59, 1998.

Salt	10°C	15°C	20°C	25°C	30°C	35°C	40°C	Ref.
$BaCl_2$	1.603	1.659	1.716	1.774	1.834	1.895	1.958	1
$Ca(NO_3)_2$	6.896	7.398	7.986	8.675	9.480	10.421		1
$CuSO_4$	1.055	1.153	1.260	1.376	1.502	1.639		3
$FeSO_4$	1.352	1.533	1.729	1.940	2.165	2.405		3
KBr	5.002	5.237	5.471	5.703	5.932	6.157		3
KIO_3	0.291	0.333	0.378	0.426	0.478	0.534	0.593	4
K_2CO_3	7.756	7.846	7.948	8.063	8.191	8.331	8.483	1
$LiCl$	19.296	19.456	19.670	19.935				2
$Mg(NO_3)_2$	4.403	4.523	4.656	4.800	4.958	5.130	5.314	1
$MnCl_2$	5.421	5.644	5.884	6.143	6.422	6.721		3
NH_4Cl	6.199	6.566	6.943	7.331				2
NH_4NO_3	18.809	21.163	23.721	26.496				2
$(NH_4)_2SO_4$	5.494	5.589	5.688	5.790	5.896	6.005		3
$NaBr$	8.258	8.546	8.856	9.191	9.550	9.937	10.351	4
$NaCl$	6.110	6.121	6.136	6.153	6.174	6.197	6.222	4
$NaNO_2$	11.111	11.484	11.883	12.310	12.766	13.253	13.772	4
$NaNO_3$	9.395	9.819	10.261	10.723	11.204	11.706	12.230	4
$RbCl$	6.911	7.180	7.449	7.717	7.986	8.253	8.520	4
$ZnSO_4$	2.911	3.116	3.336	3.573	3.827	4.099	4.194	1

SOLUBILITY CHART

Abbreviations: **W**, soluble in water; **A**, insoluble in water but soluble in acids; **w**, sparingly soluble in water and only sparingly soluble in acids; **I**, insoluble in water and acids; **a**, insoluble in water and acids; **d**, decomposes in water. * Indicates two modifications of the salt

No.		Al	NH₄	Sb	Ba	Bi	Cd	Ca	Cr	Co	Cu	Au (I)	Au (II)	H	Fe (II)	Fe (III)
1	Acetate —(C₂H₃O₂)	W $Al(—)_3$	W $NH_4(—)$		W $Ba(—)_2$	W $Bi(—)_3$	W $Cd(—)_2$	W $Ca(—)_2$	W $Cr(—)_3$	W $Co(—)_2$	W $Cu(—)_2$	W	W	W $C_2H_4O_2$	W $Fe(—)_2$	W $Fe_2(—)_6$
2	Arsenate —(AsO₄)	a $Al(—)$	W $(NH_4)_3(—)$	A $Sb(—)$	w $Ba_3(—)_2$	A $Bi(—)$	A $Cd_3(—)_2$	w $Ca_3(—)_2$		A $Co_3(—)_2$	A $Cu_3(—)_2$			H_3AsO_4	A $Fe_3(—)_2$	A $Fe(—)$
3	Arsenite —(AsO₃)	$Al(—)$	W NH_4AsO_2	A $Sb(—)$				w		A $Co_3(—)_2$	A $CuH(—)$					
4	Benzoate —(C₇H₅O₂)	W	W $NH_4(—)$		W $Ba(—)_2$	A $Bi(—)_3$	W $Cd(—)_2$	W $Ca(—)_2$	A	W $Co(—)_2$	w $Cu(—)_2$			W $C_7H_6O_2$	W $Fe(—)_2$	A $Fe_2(—)_6$
5	Bromide	W $AlBr_3$	W NH_4Br	d $SbBr_3$	W $BaBr_2$	d $BiBr_3$	W $CdBr_2$	W $CaBr_2$	W(I)* $CrBr_3$	W $CoBr_2$	W $CuBr_2$	w $AuBr$	W $AuBr_3$	W HBr	W $FeBr_2$	W $FeBr_3$
6	Carbonate		W $(NH_4)_2CO_3$		w $BaCO_3$		A $CdCO_3$	A $CaCO_3$	W $CrCO_3$	A $CoCO_3$	W $CuCO_3$				w $FeCO_3$	
7	Chlorate —(ClO₃)	W $Al(—)_3$	W $NH_4(—)$		W $Ba(—)_2$	W $Bi(—)_3$	W $Cd(—)_2$	W $Ca(—)_2$	W	W $Co(—)_2$	W $Cu(—)_2$			W $HClO_3$	W $Fe(—)_2$	W $Fe(—)_3$
8	Chloride	W $AlCl_3$	W NH_4Cl	W $SbCl_3$	W $BaCl_2$	A $BiCl_3$	W $CdCl_2$	W $CaCl_2$	I $CrCl_3$	A $CoCl_2$	W $CuCl_2$	w $AuCl$	W $AuCl_3$	W HCl	W $FeCl_2$	A $FeCl_3$
9	Chromate —(CrO₄)		W $(NH_4)_2(—)$		A $Ba(—)$	A $Bi(—)$	A $Cd(—)$	W $Ca(—)$		A $Co(—)$						A $Fe_2(—)_3$
10	Citrate —(C₆H₅O₇)	W $Al(—)$	W $(NH_4)_3(—)$		w $Ba_3(—)_2$	A $Bi(—)$	W $Cd_3(—)_2$	w $Ca_3(—)_2$		w $Co_3(—)_2$				W $C_6H_8O_7$	w $Fe_3(—)_2$	W $Fe(—)$
11	Cyanide —(CN)	$Al(—)$	W NH_4CN		W $Ba(CN)_2$	A $Bi(CN)_3$	W $Cd(CN)_2$	W $Ca(CN)_2$	A $Cr(CN)_3$	W $Co(CN)_2$	A $Cu(CN)_2$	w $AuCN$	W $Au(CN)_3$	W HCN	a $Fe(CN)_2$	
12	Ferricy'de —[Fe(CN)₆]		W $(NH_4)_3(—)$		w $Ba_3(—)$		A $Cd_3(—)_2$	w $Ca_3(—)_2$		I $Co_3(—)_2$	I $Cu_3(—)$			W $H_3(—)$	I	a $Fe_4(—)_3$
13	Ferrocy'de —[Fe(CN)₆]	w $Al_4(—)_3$	W $(NH_4)_4(—)$		W $Ba_2(—)$		W $Cd_2(—)$	w $Ca_2(—)$		I $Co_2(—)$	I $Cu_2(—)$			W $H_4(—)$	I $Fe_2(—)$	W $Fe_4(—)_3$
14	Fluoride —F	W AlF_3	W NH_4F	W SbF_3	W BaF_2	W BiF_3	W CdF_2	A CaF_2	W(a)* CrF_3	W CoF_2	w CuF_2			W HF	W FeF_2	W FeF_3
15	Formate —(CHO₂)	W $Al(—)_3$	W $NH_4(—)$		W $Ba(—)_2$	W $Bi(—)_3$	W $Cd(—)_2$	W $Ca(—)_2$	W	W $Co(—)_2$	W $Cu(—)_2$			W CH_2O_2	W $Fe(—)_2$	W $Fe(—)_3$
16	Hydroxide —(OH)	A $Al(OH)_3$	W NH_4OH	A $Bi(OH)_3$	W $Ba(OH)_2$	A $Bi(OH)_3$	A $Cd(OH)_2$	W $Ca(OH)_2$	A $Cr(OH)_3$	W $Co(OH)_2$	A $Cu(OH)_2$	W $AuOH$	A $Au(OH)_3$	W H_2O	W $Fe(OH)_2$	A $Fe(OH)_3$
17	Iodide —I	W AlI_3	W NH_4I	d SbI_3	W BaI_2	A BiI_3	W CdI_2	W CaI_2	W CrI_3	W CoI_2	a CuI	a AuI	a AuI_3	W HI	W FeI_2	W FeI_3
18	Nitrate —NO₃	W $Al(NO_3)_3$	W NH_4NO_3	A SbI_3	W $Ba(NO_3)_2$	d $Bi(NO_3)_3$	W $Cd(NO_3)_2$	W $Ca(NO_3)_2$	W $Cr(NO_3)_2$	W $Co(NO_3)_2$	W $Cu(NO_3)_2$			W HNO_3	W $Fe(NO_3)_2$	W $Fe(NO_3)_3$
19	Oxalate —(C₂O₄)	A $Al_2(—)_3$	W $(NH_4)_2(—)$		A $Ba(—)$	A $Bi_2(—)_3$	A $Cd(—)_2$	A $Ca(—)$	a $Cr(—)$	A $Co(—)$	A $Cu(—)$		A Au_2O_3	W $C_2H_2O_4$	A $Fe(—)$	A $Fe_2(—)_3$
20	Oxide —O	a Al_2O_3	W	w Sb_2O_3	W BaO	A Bi_2O_3	A CdO	W CaO	A Cr_2O_3	A CoO	A CuO	Au_2O	A Au_2O_3	W H_2O_2	A FeO	A Fe_2O_3
21	Phosphate —(PO₄)	A $AlPO_4$	W $NH_4H_2PO_4$		A $Ba_3(PO_4)_2$	A $BiPO_4$	A $Cd_3(PO_4)_2$	A $Ca_3(PO_4)_2$	a $Cr_2(PO_4)*$	A $Co_3(PO_4)_2$	A $Cu_3(PO_4)_2$		H_3PO_4	I $Fe_3(PO_4)$	A $FePO_4$	w $FePO_4$
22	Silicate, —(SiO₃)	I $Al_2(—)_3$	W		W $Ba(—)$		W	w $Ca(—)$	W(I)* Cr_2SiO_5	W Co_2SiO_4	A $Cu(—)$			W H_2SiO_3	W Fe_2SiO_4	W
23	Sulfate —(SO₄)	W $Al_2(SO_4)_3$	W $(NH_4)_2SO_4$	A $Sb_2(SO_4)_3$	a $BaSO_4$	d $Bi_2(SO_4)_3$	W $CdSO_4$	w $CaSO_4$	d $Cr_2(SO_4)_3$	W $CoSO_4$	W $CuSO_4$	I	I	W H_2SO_4	W $FeSO_4$	A $Fe_2(SO_4)_3$
24	Sulfide —S	d Al_2S_3	W $(NH_4)_2S$	W Sb_2S_3	w BaS	A Bi_2S_3	A CdS	w CaS	d Cr_2S_3	A CoS	A CuS	I Au_2S_3	I Au_2S_3	W H_2S	A FeS	d Fe_2S_3
25	Tartrate —(C₄H₄O₆)	w $Al_2(—)_3$	W $(NH_4)_2(—)$	W $Sb_2(—)_3$	W $Ba(—)$	A $Bi_2(—)_3$	W $Cd(—)_2$	w $Ca(—)$	d	W $Co(—)$	A $Cu(—)$	I	I	W $C_4H_6O_6$	W $Fe(—)$	w $Fe_2(—)_3$

No.		Al	NH₄	Sb	Ba	Bi	Cd	Cr	Co	Cu	Au (I)	Au (II)	H	Fe (II)	Fe (III)	Pb	Mg	Mn	Hg (I)	Hg (II)	Ni	Ca	K	Pt	Ag	Na	Sn (IV)	Sn (II)	Sr	Zn
26	Thiocy'te		W NH₄CNS		W Ba(CNS)₂				W Co(CNS)₂	d CuCNS			W CNSH	W Fe(CNS)₂	W Fe(CNS)₃							W Ca(CNS)				W		d	W	W
1	Acetate —(C₂H₃O₂)															W Pb(—)₂	W Mg(—)₂	W Mn(—)₂	w Hg(—)₂	W Hg(—)₂	W Ni(—)₂		W K(—)		w Ag(—)	W Na(—)	W Sn(—)₄	W Sn(—)₂	W Sr(—)₂	W Zn(—)₂
2	Arsenate —(AsO₄)															A PbH(—)	A Mg₃(—)₂	w MnH(—)	A Hg₃(—)	A Hg₃(—)₂	A Ni₃(—)₂		W K₃(—)		A Ag₃(—)	W Na₃(—)		A Sn₃(—)₂	A SrH(—)	A Zn₃(—)₂
3	Arsenite —(AsO₃)												SrH(—)			W	W Mg₃(—)₂	A	w Hg₃(—)	A Hg₃(—)₂	A		W K₃(—)		A Ag₃(—)	W Na₃(—)		A Sn₃(—)₂	w Sr₃(—)₂	
4	Benzoate —(C₇H₅O₂)															w Pb(—)₂	W Mg(—)₂	W Mn₃H₆(—)₄	w Hg₂(—)₂	W Hg₃(—)₂	W Ni₃H₆(—)₄		W K₃AsO₃		A Ag₃(—)	W Na₂H(—)			W	W Zn(—)₂
5	Bromide															W PbBr₂	W MgBr₂	W MnBr₂	W HgBr	W HgBr₂	w NiBr₂		W KBr	W PtBr₄	a AgBr	W NaBr	W SnBr₄	W SnBr₂	W SrBr₂	W ZnBr₂
6	Carbonate															A PbCO₃	W MgCO₃	W MnCO₃	W Hg₂CO₃	A	w NiCO₃		W K₂CO₃		A Ag₂CO₃	W Na₂CO₃			W SrCO₃	W ZnCO₃
7	Chlorate —(ClO₃)															W	W Mg(—)₂	W Mn(—)₂	W Hg(—)₂	W Hg(—)₂	W Ni(—)₂		W K(—)		W Ag(—)	W Na(—)	W Sn(—)₂	W Sn(—)₂	W Sr(—)₂	W Zn(—)₂
8	Chloride															W PbCl₂	W MgCl₂	W MnCl₂	W HgCl	W HgCl₂	W NiCl₂		W KCl	W PtCl₄	a AgCl	W NaCl	W SnCl₄	W SnCl₂	W SrCl₂	W ZnCl₂
9	Chromate —(CrO₄)															A Pb(—)	W Mg(—)		w Hg₂(—)	w Hg(—)	A Ni(—)		W K₂(—)		W Ag₂(—)	W Na₂(—)		A Sn(—)	A Sr(—)	W Zn(—)
10	Citrate —(C₆H₅O₇)															W	W Mg(—)₂	W Mn(—)₂	W Hg₃(—)₂	W Hg(—)₂	W Ni₃(—)₂		W K₃(—)		A Ag₃(—)	W Na₃(—)		W Sn(—)	W Sr(—)₂	W Zn(—)₂
11	Cyanide												SrH(—)			w Pb(CN)₂	W Mg(CN)₂	w MnH(—)	W Hg₃(—)₂	W Hg(CN)₂	a Ni₃(—)₂		W K₃(—)	I Pt(CN)₂	a AgCN	W Na₃(—)		A Sn₃(CN)₂	W Sr(CN)₂	A Zn(CN)₂
12	Ferricy'de —Fe(CN)₆															w Pb₃(—)₂	W Mg₃(—)₂		W Hg₃(—)	W Hg₃(—)₂	W Ni(CN)₂		W K₃(—)		I Ag₃(—)	W Na₃(—)		A Sn₃(—)₂	W Sr(CN)₂	A Zn₃(CN)₂
13	Ferrocy'de —Fe(CN)₆															a Pb₂(—)	W Mg₃(—)₂	A Mn₂(—)	d	I Hg₃(—)₂	I Ni₃(—)₂		W K₄(—)		I Ag₄(—)	W Na₄(—)		a Sn₂(—)	w Sr₃(—)₂	I Zn₃(—)₂
14	Fluoride															W PbF₂	W MgF₂	W MnF₂	W HgF	d HgF₂	W NiF₂		W KF	W PtF₄	W AgF	W NaF	W SnF₄	W SnF₂	A SrF₂	w ZnF₂
15	Formate —(CHO₂)															W Pb(—)₂	W Mg(—)₂	W Mn(—)₂	w Hg(—)	W Hg(—)₂	W Ni(—)₂		W K(—)		W Ag(—)	W Na(—)		W	W Sr(—)₂	W Zn(—)₂
16	Hydroxide												SrH(—)			w Pb(OH)₂	A Mg(OH)₂	A Mn(OH)₂	A Hg(—)	A Hg(OH)₂	A Ni(OH)₂		W KOH	A Pt(OH)₄	A Ag(—)	W NaOH	w Sn(OH)₄	A Sn(OH)₂	W Sr(OH)₂	A Zn(OH)₂
17	Iodide															w PbI₂	W MgI₂	W MnI₂	A HgI	w HgI₂	W NiI₂		W KI	I PtI₂	I AgI	W NaI	d SnI₄	W SnI₂	W SrI₂	W ZnI₂
18	Nitrate															W Pb(NO₃)₂	W Mg(NO₃)₂	W Mn(NO₃)₂	W HgNO₃	W Hg(NO₃)₂	W Ni(NO₃)₂		W KNO₃	W Pt(NO₃)₄	W AgNO₃	W NaNO₃		d Sn(NO₃)₂	W Sr(NO₃)₂	W Zn(NO₃)₂
19	Oxalate —(C₂O₄)															A Pb(—)	W Mg(—)	A Mn(—)	a	A Hg(—)	A Ni(—)		W K₂(—)		a Ag₂(—)	W Na₂(—)	A SnO₂	A Sn(—)	w Sr(—)	A Zn(—)₂
20	Oxide															w PbO	A MgO	A MnO	w Hg₂O	w HgO	A NiO		W K₂O	A PtO	A Ag₂O	d Na₂O		w SnO	W SrO	W ZnO
21	Phosphate															A Pb₃(PO₄)₂	W Mg₃(PO₄)₂	W Mn₃(PO₄)₂	w Hg₃PO₄	W Hg₃(PO₄)₂	W Ni₃(PO₄)₂		W K₃PO₄		A Ag₃PO₄	W Na₃PO₄		A Sn₃(PO₄)₂	A Sr₃(PO₄)₂	A Zn₃(PO₄)₂
22	Silicate —(SiO₃)															A Pb(—)	A Mg(—)	W Mn(—)	w	A Hg(—)₂			W K₂(—)			W Na₂(—)		A Sn(—)	W Sr(—)	W Zn(—)
23	Sulfate															w PbSO₄	W MgSO₄	W MnSO₄	w Hg₂SO₄	d HgSO₄	W NiSO₄		W K₂SO₄	W Pt(SO₄)₂	w Ag₂SO₄	W Na₂SO₄	W Sn(SO₄)₂	W SnSO₄	W SrSO₄	W ZnSO₄

SOLUBILITY CHART (continued)

No.		Pb	Mg	Mn	Hg (I)	Hg (II)	Ni	K	Pt	Ag	Na	Sn (IV)	Sn (II)	Sr	Zn
24	Sulfide	A PbS	d MgS	A MnS	I Hg$_2$S	I HgS	A NiS	W K$_2$S	I PtS	A Ag$_2$S	W Na$_2$S	A SnS$_2$	A SnS	W SrS	A ZnS
25	Tartrate —(C$_4$H$_4$O$_6$)	A Pb(—)	w Mg(—)	w Mn(—)	I Hg$_2$(—)	I	A Ni(—)	W K$_2$(—)	I	A Ag$_2$(—)	W Na$_2$(—)	A	A Sn(—)	W Sr(—)	A Zn(—)
26	Thiocy'te	w Pb(CNS)$_2$	W Mg(CNS)$_2$	W Mn(CNS)$_2$	A HgCNS	w Hg(CNS)$_2$	W	W KCNS		I AgCNS	W NaCNS			W Sr(CNS)$_2$	W Zn(CNS)$_2$

REDUCTION OF WEIGHINGS IN AIR TO VACUO

When the mass M of a body is determined in air, a correction is necessary for the buoyancy of the air. The corrected mass is given by $M + kM/1000$, where k is a function of the material used for the weights, given by

$$k = 1000\rho_{air}(1/\rho_{body} - 1/\rho_{weight})$$

and ρ is density. The table below is computed for an air density of 0.0012 g/cm^3 and for densities of three common weights: platinum-iridium (21.6 g/cm^3), brass (8.5 g/cm^3), and aluminum or quartz (2.65 g/cm^3).

REFERENCES

1. Kaye, G. W. C., and Laby, T. H., *Tables of Physical and Chemical Constants, 16th Edition,* pp. 25-28, Longman, London, 1995.
2. Giacomo, P., *Metrologia* 18, 33, 1982.
3. Davis, R. S., *Metrologia* 29, 67, 1992.

Density of body (g/cm^3)	Value of k for weights of: Pt-Ir	Brass	Quartz or Al	Density of body (g/cm^3)	Value of k for weights of: Pt-Ir	Brass	Quartz or Al
0.5	2.34	2.26	1.95	1.8	0.61	0.53	0.21
0.6	1.94	1.86	1.55	1.9	0.58	0.49	0.18
0.7	1.66	1.57	1.26	2.0	0.54	0.46	0.15
0.8	1.44	1.36	1.05	2.5	0.42	0.34	0.03
0.9	1.28	1.19	0.88	3.0	0.34	0.26	-0.05
1.0	1.14	1.06	0.75	4.0	0.24	0.16	-0.15
1.1	1.04	0.95	0.64	6.0	0.14	0.06	-0.25
1.2	0.94	0.86	0.55	8.0	0.09	0.01	-0.30
1.3	0.87	0.78	0.47	10.0	0.06	-0.02	-0.33
1.4	0.80	0.72	0.40	15.0	0.02	-0.06	-0.37
1.5	0.74	0.66	0.35	20.0	0.00	-0.08	-0.39
1.6	0.69	0.61	0.30	22.0	0.00	-0.09	-0.40
1.7	0.65	0.56	0.25				

For a more accurate calculation, use the following values of the density of air (assuming 50% relative humidity and 0.04% CO_2):

P/kPa	Air temperature 10°C	20°C	30°C
85	0.001043	0.001005	0.000968
90	0.001105	0.001065	0.001025
95	0.001166	0.001124	0.001083
100	0.001228	0.001184	0.001140
105	0.001290	0.001243	0.001198

Formulas for calculating the density of air over more extended ranges of temperature, pressure, and humidity may be found in the references.

VOLUME OF ONE GRAM OF WATER

The following table, which is designed for gravimetric calibration of volumetric apparatus, gives the specific volume of water at standard atmospheric pressure as a function of temperature.

REFERENCE

Marsh, K. N., Editor, *Recommended Reference Materials for the Realization of Physicochemical Properties*, pp. 25-27, Blackwell Scientific Publications, Oxford, 1987.

t/°C	Volume of 1 g H_2O in cm^3	t/°C	Volume of 1 g H_2O in cm^3	t/°C	Volume of 1 g H_2O in cm^3
10	1.0002980	17	1.0012246	24	1.0027079
11	1.0003928	18	1.0014044	25	1.0029607
12	1.0005007	19	1.0015952	26	1.0032234
13	1.0006212	20	1.0017969	27	1.0034956
14	1.0007542	21	1.0020092	28	1.0037771
15	1.0008992	22	1.0022320	29	1.0040679
16	1.0010561	23	1.0024649	30	1.0043679

PROPERTIES OF CARRIER GASES FOR GAS CHROMATOGRAPHY

The following is a list of carrier gases sometimes used in gas chromatography, with properties relevant to the design of chromatographic systems. All data refer to normal atmospheric pressure (101.325 kPa).

M_r : Molecular weight (relative molar mass)
ρ_{25} : Density at 25°C in g/L
λ : Thermal conductivity in mW/m °C
η : Viscosity in μPa s (equal to 10^{-3} cp)
c_p : Specific heat at 25°C in J/g °C

REFERENCES

1. Lide, D. R., and Kehiaian, H. V., *CRC Handbook of Thermophysical and Thermochemical Data*, CRC Press, Boca Raton, FL, 1994.
2. Bruno, T. J., and Svoronos, P. D. N., *CRC Handbook of Basic Tables for Chemical Analysis*, CRC Press, Boca Raton, FL, 1989

Gas	M_r	ρ_{25} g L^{-1}	At 25°C λ mW/m °C	At 25°C η μPa s	At 250°C λ mW/m °C	At 250°C η μPa s	$c_p(25\ °C)$ J/g °C
Hydrogen	2.016	0.0824	185.9	8.9	280	13.1	14.3
Helium	4.003	0.1636	154.6	19.9	230	29.5	5.20
Argon	39.95	1.6329	17.8	22.7	27.7	35.3	0.521
Nitrogen	28.01	1.1449	25.9	17.9	39.6	26.8	1.039
Oxygen	32.00	1.3080	26.2	20.7	42.6	31.8	0.919
Carbon monoxide	28.01	1.1449	24.8	17.8	40.7	26.5	1.039
Carbon dioxide	44.01	1.7989	16.7	14.9	35.5	24.9	0.843
Sulfur hexafluoride	146.05	5.9696	13.1	28.1	15.3	24.8	0.664
Methane	16.04	0.6556	34.5	11.1	75.0	17.6	2.23
Ethane	30.07	1.2291	20.9	9.4	57.7	15.5	1.75
Ethylene	28.05	1.1465	20.5	10.3	53.8	17.2	1.53
Propane	44.10	1.8025	17.9	8.3	49.2	14.0	1.67

SOLVENTS FOR ULTRAVIOLET SPECTROPHOTOMETRY

This table lists some solvents commonly used for sample preparation for ultraviolet spectrophotometry. The properties given are:

λ_c: cutoff wavelength, below which the solvent absorption becomes excessive.
ε: dielectric constant (relative permittivity); the temperature in °C is given as a superscript.
t_b: normal boiling point.

REFERENCES

1. Bruno, T. J., and Svoronos, P. D. N., *CRC Handbook of Basic Tables for Chemical Analysis*, CRC Press, Boca Raton, FL, 1989.
2. *Landolt-Börnstein, Numerical Data and Functional Relationships in Science and Technology, New Series, IV/6, Static Dielectric Constants of Pure Liquids and Binary Liquid Mixtures,* Springer-Verlag, Heidelberg, 1991.

Name	λ_c/nm	ε	t_b/°C
Acetic acid	260	6.20[20]	117.9
Acetone	330	21.01[20]	56.0
Acetonitrile	190	36.64[20]	81.6
Benzene	280	2.28[20]	80.0
2-Butanol	260	17.26[20]	99.5
Butyl acetate	254	5.07[20]	126.1
Carbon disulfide	380	2.63[20]	46
Carbon tetrachloride	265	2.24[20]	76.8
1-Chlorobutane	220	7.28[20]	78.6
Chloroform	245	4.81[20]	61.1
Cyclohexane	210	2.02[20]	80.7
1,2-Dichloroethane	226	10.42[20]	83.5
Dichloromethane	235	8.93[25]	40
Diethyl ether	218	4.27[20]	34.5
N,N-Dimethylacetamide	268	38.85[21]	165
N,N-Dimethylformamide	270	38.25[20]	153
Dimethyl sulfoxide	265	47.24[20]	189
1,4-Dioxane	215	2.22[20]	101.5
Ethanol	210	25.3[20]	78.2
Ethyl acetate	255	6.08[20]	77.1
Ethylene glycol dimethyl ether	240	7.30[24]	85
Ethylene glycol monoethyl ether	210	13.38[25]	135
Ethylene glycol monomethyl ether	210	17.2[25]	124.1
Glycerol	207	46.53[20]	290
Heptane	197	1.92[20]	98.5
Hexadecane	200	2.05[20]	286.8
Hexane	210	1.89[20]	68.7
Methanol	210	33.0[20]	64.6
Methylcyclohexane	210	2.02[20]	100.9
Methyl ethyl ketone	330	18.56[20]	79.5
Methyl isobutyl ketone	335	13.11[20]	116.5
2-Methyl-1-propanol	230	17.93[20]	107.8
N-Methyl-2-pyrrolidone	285	32.55[20]	202
Nitromethane	380	37.27[20]	101.1
Pentane	210	1.84[20]	36.0
Pentyl acetate	212	4.79[20]	149.2
1-Propanol	210	20.8[20]	97.2
2-Propanol	210	20.18[20]	82.3
Pyridine	330	13.26[20]	115.2
Tetrachloroethylene	290	2.27[30]	121.3
Tetrahydrofuran	220	7.52[22]	65
Toluene	286	2.38[23]	110.6
1,1,2-Trichloro-1,2,2-trifluoroethane	231	2.41[25]	47.7
2,2,4-Trimethylpentane	215	1.94[20]	99.2
Water	191	80.10[20]	100.0
o-Xylene	290	2.56[20]	144.5
m-Xylene	290	2.36[20]	139.1
p-Xylene	290	2.27[20]	138.3

^{13}C CHEMICAL SHIFTS OF USEFUL NMR SOLVENTS

The following table gives the expected carbon-13 chemical shifts, relative to tetramethylsilane, for various useful NMR solvents. In some solvents, slight changes can occur with change of concentration.[2,3]

REFERENCES

1. Bruno, T. J. and Svoronos, P. D. N., *CRC Handbook of Basic Tables for Chemical Analysis*, CRC Press, Boca Raton, FL, 1989.
2. Silverstein, R. M., Bassler, G. C., and Morrill, T. C., *Spectrometric Identification of Organic Compounds*, John Wiley & Sons, Now York, 1981.
3. Rahman, A. U., *Nuclear Magnetic Resonance. Basic Principles*, Springer-Verlag, New York, 1986.
4. Pretsch, E., Clerc, T., Seibl, J., and Simon, W., *Spectral Data for Structure Determination of Organic Compounds, Second Edition*, Springer-Verlag, Heidelberg, 1989.

Solvent	Formula	Chemical shift (ppm)
Acetic acid-d_4	CD_3COOD	20.0 (CD_3) 205.8 (C=O)
Acetone	$(CH_3)_2C=O$	30.7 (CH_3) 206.7 (C=O)
Acetone-d_6	$(CD_3)_2C=O$	29.2 (CD_3) 204.1 (C=O)
Acetonitrile-d_3	$CD_3C\equiv N$	1.3 (CD_3) 117.1 (C$\equiv$N)
Benzene	C_6H_6	128.5
Benzene-d_6	C_6D_6	128.4
Carbon disulfide	CS_2	192.3
Carbon tetrachloride	CCl_4	96.0
Chloroform	$CHCl_3$	77.2
Chloroform-d_3	$CDCl_3$	77.05
Cyclohexane-d_{12}	C_6D_{12}	27.5
Dichloromethane-d_2	CD_2Cl_2	53.6
Dimethylformamide-d_7	$(CD_3)_2NCDO$	31 (CD_3) 36 (CD_3) 162.4 (C=O)
Dimethylsulfoxide-d_6	$(CD_3)_2S=O$	39.6
Dioxane-d_8	$C_4D_3O_2$	67.4
Formic acid-d_2	DCOOD	165.5
Methanol-d_4	CD_3OD	49.3
Nitromethane-d_3	CD_3NO_2	57.3
Pyridine	C_5H_5N	123.6 (C_3) 135.7 (C_4) 149.8 (C_2)
Pyridine-d_5	C_5D_5N	123.9 (C_3) 135.9 (C_4) 150.2 (C_2)
1,1,2,2-Tetrachloroethane-d_2	$CDCl_2CDCl_2$	75.5
Tetrahydrofuran-d_8	C_4D_8O	25.8 (C_2) 67.9 (C_1)
Trichlorofluoromethane	$CFCl_3$	117.6

MASS SPECTRAL PEAKS OF COMMON ORGANIC SOLVENTS

The strongest peaks in the mass spectra of 200 important solvents are listed in this table. The e/m value for each peak is followed by the relative intensity in parentheses, with the strongest peak assigned an intensity of 100. The peaks for each compound are listed in order of decreasing intensity. Solvents are listed in alphabetical order by common name.

Data on the physical properties of the same compounds may be found in Section 15 in the table "Properties of Common Laboratory Solvents".

REFERENCES

1. NIST/EPA/NIH Mass Spectral Database, National Institute of Standards and Technology, Gaithersburg, MD, 20899.
2. Lide, D.R., and Milne, G.W.A., Editors, *Handbook of Data on Organic Compounds, Third Edition*, CRC Press, Boca Raton, FL, 1994. (Also available as a CD ROM database.)

MASS SPECTRAL PEAKS OF COMMON ORGANIC SOLVENTS (continued)

Compound	e/m (intensity)									
Acetal (1,1-Diethoxyethane)	44(100)	43(92)	29(77)	31(76)	45(74)	27(52)	72(48)	73(23)	28(17)	46(15)
Acetic acid	43(100)	45(87)	60(57)	15(42)	42(14)	29(13)	14(13)	28(7)	18(6)	16(6)
Acetone	43(100)	15(34)	58(23)	27(9)	14(9)	42(8)	26(7)	29(5)	28(5)	39(4)
Acetonitrile	41(100)	40(46)	39(13)	14(9)	38(6)	28(4)	26(4)	25(3)	42(2)	27(2)
Acetylacetone	43(100)	85(31)	100(20)	27(12)	42(10)	29(10)	41(7)	39(7)	31(5)	26(5)
Acrylonitrile	53(100)	26(85)	52(79)	51(34)	27(13)	50(8)	25(7)	38(5)	54(3)	37(3)
Adiponitrile	41(100)	68(50)	54(42)	40(21)	55(20)	27(17)	39(16)	28(13)	52(7)	42(6)
Allyl alcohol.	57(100)	31(34)	29(32)	28(31)	58(25)	39(22)	27(20)	30(16)	32(14)	26(11)
Allylamine	30(100)	56(80)	28(76)	57(33)	39(21)	29(20)	27(18)	26(13)	41(8)	18(8)
2-Aminoisobutanol	58(100)	41(18)	18(17)	42(13)	28(11)	56(10)	30(10)	29(8)	43(6)	59(5)
Benzal chloride	125(100)	127(32)	160(14)	89(13)	162(9)	63(9)	126(8)	62(7)	105(5)	39(5)
Benzaldehyde	51(100)	77(81)	50(55)	106(44)	105(43)	52(26)	78(16)	39(13)	27(10)	74(8)
Benzene	78(100)	77(20)	52(19)	51(17)	50(15)	39(12)	79(6)	76(5)	74(4)	38(4)
Benzonitrile	103(100)	76(34)	50(13)	104(9)	75(7)	51(7)	77(5)	52(4)	39(4)	74(3)
Benzyl chloride	91(100)	126(20)	65(14)	92(9)	39(9)	63(8)	128(6)	45(6)	89(5)	125(3)
Bromochloromethane	49(100)	130(67)	128(52)	51(31)	93(23)	81(20)	79(20)	95(17)	132(16)	47(8)
Bromoform (Tribromomethane)	173(100)	171(50)	175(49)	93(22)	91(22)	79(18)	81(17)	94(13)	92(13)	254(11)
Butyl acetate	43(100)	56(34)	41(17)	27(16)	29(15)	73(11)	61(10)	28(7)	55(6)	39(6)
Butyl alcohol	31(100)	56(81)	41(62)	43(60)	27(50)	42(31)	29(31)	28(17)	39(16)	55(12)
sec-Butyl alcohol	45(100)	31(22)	27(22)	59(20)	29(18)	43(13)	41(12)	44(8)	18(8)	28(5)
tert-Butyl alcohol	59(100)	31(33)	41(22)	43(18)	29(13)	27(11)	57(10)	42(4)	60(3)	28(3)
Butylamine	30(100)	73(10)	28(5)	41(3)	27(3)	18(3)	44(2)	42(2)	31(2)	29(2)
tert-Butylamine	58(100)	41(21)	42(15)	18(9)	30(8)	15(8)	39(7)	57(6)	28(6)	59(4)
Butyl methyl ketone	43(100)	58(60)	57(17)	100(16)	29(15)	41(13)	85(8)	27(8)	71(7)	59(5)
p-tert-Butyltoluene	133(100)	105(38)	41(23)	148(18)	93(16)	91(14)	115(13)	134(11)	39(11)	116(10)
γ-Butyrolactone	28(100)	42(74)	29(48)	27(33)	41(27)	56(25)	86(24)	26(18)	85(10)	39(10)
Caprolactam	55(100)	113(87)	30(81)	56(66)	84(60)	85(57)	42(51)	41(33)	28(26)	43(17)
Carbon disulfide	76(100)	32(22)	44(17)	78(9)	38(6)	28(5)	77(3)	64(1)	46(1)	39(1)
Carbon tetrachloride	117(100)	119(98)	121(31)	82(24)	47(23)	84(16)	35(14)	49(8)	28(8)	36(6)
1-Chloro-1,1-difluoroethane	65(100)	45(31)	85(14)	31(10)	64(8)	44(7)	35(6)	26(6)	87(5)	81(4)
Chlorobenzene	112(100)	77(63)	114(33)	51(29)	50(14)	75(8)	113(7)	78(5)	76(5)	28(4)
Chloroform	83(100)	85(64)	47(35)	35(19)	48(16)	49(12)	87(10)	37(6)	50(5)	84(4)
Chloropentafluoroethane	85(100)	69(61)	31(38)	87(32)	50(17)	35(8)	119(6)	66(4)	100(3)	47(3)
Cumene (Isopropylbenzene)	105(100)	120(25)	77(13)	51(12)	79(10)	106(9)	39(9)	27(8)	103(6)	91(5)
Cyclohexane	56(100)	84(71)	41(70)	27(37)	55(36)	39(35)	42(30)	69(23)	28(18)	43(14)
Cyclohexanol	57(100)	44(68)	41(68)	39(51)	32(40)	43(38)	31(32)	42(22)	67(18)	82(16)
Cyclohexanone	55(100)	42(85)	41(34)	27(33)	98(31)	39(27)	69(26)	70(20)	43(14)	28(14)
Cyclohexylamine	56(100)	43(23)	28(17)	99(10)	70(8)	57(6)	30(6)	93(5)	54(4)	41(4)
Cyclopentane	42(100)	70(30)	55(29)	41(29)	39(22)	27(15)	40(7)	29(5)	28(4)	43(3)
Cyclopentanone	55(100)	28(50)	84(42)	41(38)	56(29)	27(24)	39(19)	42(15)	26(9)	29(7)
p-Cymene (1-Methyl-4-isopropyl-benzene)	119(100)	91(42)	134(33)	39(27)	41(20)	117(18)	65(18)	77(17)	27(16)	120(15)
cis-Decalin	67(100)	81(87)	41(81)	138(67)	96(62)	82(62)	39(50)	55(45)	27(44)	95(42)
trans-Decalin	41(100)	68(91)	67(88)	82(67)	27(65)	96(61)	95(55)	138(51)	81(51)	29(51)

MASS SPECTRAL PEAKS OF COMMON ORGANIC SOLVENTS (continued)

e/m (intensity)

Compound	Peaks — e/m (intensity)
Diacetone alcohol	43(100) 59(41) 58(17) 101(10) 41(9) 31(9) 83(6) 56(6) 55(6) 29(6)
1,2-Dibromoethane	27(100) 107(77) 109(72) 58(17) 26(24) 28(10) 81(5) 79(5) 25(5) 93(4)
Dibromofluoromethane	111(100) 113(98) 192(29) 43(16) 41(16) 190(15) 194(14) 81(9) 79(9) 122(7)
Dibromomethane	174(100) 93(96) 95(84) 172(53) 176(50) 91(11) 81(9) 79(9) 65(5)
1,2-Dibromotetrafluoroethane	179(100) 181(97) 129(34) 131(33) 100(17) 31(13) 260(12) 50(8) 262(6)
Dibutylamine	86(100) 72(52) 30(48) 44(40) 57(24) 41(21) 100(17) 73(15) 69(7) 43(13)
o-Dichlorobenzene	146(100) 148(64) 111(38) 75(23) 113(12) 74(12) 50(11) 73(9) 147(7)
1,1-Dichloroethane (Ethylidene dichloride)	63(100) 27(71) 65(31) 26(19) 83(11) 85(7) 61(7) 35(6) 62(5)
1,2-Dichloroethane (Ethylene dichloride)	62(100) 27(91) 49(40) 64(32) 26(31) 63(19) 51(13) 85(7) 100(9)
1,1-Dichloroethylene	61(100) 96(61) 98(38) 64(32) 63(32) 26(16) 60(15) 51(13) 35(6)
cis-1,2-Dichloroethylene	61(100) 96(73) 98(47) 63(32) 26(16) 60(15) 25(7) 62(7) 100(6)
trans-1,2-Dichloroethylene	61(100) 96(67) 98(43) 63(32) 26(30) 60(21) 25(13) 62(9) 100(8)
Dichloroethyl ether	93(100) 63(74) 95(32) 27(38) 65(24) 62(10) 100(7) 31(9) 47(7)
Dichloromethane (Methylene chloride)	49(100) 84(64) 86(39) 51(31) 47(14) 48(8) 88(6) 50(3) 28(4) 62(3)
1,2-Dichloropropane	63(100) 62(71) 27(57) 41(49) 51(31) 76(27) 39(32) 88(6) 85(2) 83(2)
1,2-Dichlorotetrafluoroethane	85(100) 135(52) 87(33) 137(17) 101(9) 31(9) 65(31) 50(5) 49(13) 77(12)
Diethanolamine	30(100) 74(82) 28(77) 56(69) 18(50) 42(46) 31(9) 100(6) 103(6) 69(4)
Diethylamine	30(100) 58(81) 44(28) 73(18) 29(18) 28(17) 42(46) 27(34) 45(30) 43(19)
Diethyl carbonate	29(100) 45(70) 31(53) 73(18) 91(24) 27(39) 72(12) 63(11) 42(11) 59(4)
Diethylene glycol	45(100) 75(23) 31(20) 44(16) 27(14) 76(12) 63(11) 26(10) 30(6) 43(5)
Diethylene glycol dimethyl ether (Diglyme)	59(100) 58(43) 29(32) 45(28) 89(15) 28(19) 43(9) 29(12) 42(9) 41(4)
Diethylene glycol monoethyl ether (Carbitol)	45(100) 59(56) 72(37) 73(22) 60(14) 28(19) 89(15) 43(9) 27(5) 60(4)
Diethylene glycol monoethyl ether acetate	43(100) 31(42) 87(33) 137(17) 101(9) 75(11) 44(9) 104(8) 103(7)
Diethylene glycol monomethyl ether	43(100) 31(42) 59(24) 45(40) 72(18) 44(10) 73(9) 42(9) 30(6)
Diethylenetriamine	44(100) 31(42) 59(41) 29(38) 28(32) 58(21) 43(14) 27(13) 44(11) 32(10)
Diethyl ether	31(100) 73(59) 29(63) 30(35) 19(18) 56(16) 27(16) 42(11) 99(8) 43(8)
Diisobutyl ketone (Isovalerone)	57(100) 29(63) 59(40) 45(33) 43(39) 74(23) 15(17) 43(9) 28(9) 26(9)
Diisopropyl ether	45(100) 85(82) 41(46) 43(39) 58(33) 28(30) 26(30) 39(22) 42(12) 142(11)
N,N-Dimethylacetamide	44(100) 43(39) 87(69) 41(12) 59(10) 27(8) 39(4) 69(3) 42(3) 31(3)
Dimethylamine	44(100) 87(69) 43(46) 28(30) 42(19) 72(15) 15(11) 30(8) 28(5) 88(4)
Dimethyl disulfide	94(100) 45(81) 18(32) 28(30) 43(19) 42(15) 15(9) 46(5) 41(5) 27(5)
N,N-Dimethylformamide	73(100) 45(63) 79(59) 46(38) 28(30) 47(26) 15(18) 61(12) 64(11) 96(9)
Dimethyl sulfoxide	63(100) 44(86) 42(36) 30(22) 45(35) 28(20) 29(8) 48(14) 72(6) 74(4)
1,4-Dioxane	28(100) 78(70) 15(40) 45(35) 29(16) 61(13) 43(7) 31(11) 58(5) 47(10)
1,3-Dioxolane	73(100) 29(37) 88(31) 58(24) 31(17) 15(17) 46(12) 30(13) 48(10) 26(9)
Dipentene	68(100) 44(53) 45(28) 28(21) 43(20) 27(15) 31(7) 27(13) 43(11) 42(3)
Epichlorohydrin	57(100) 93(50) 94(22) 39(22) 28(21) 107(18) 92(18) 53(18) 31(7) 79(16)
	27(39) 49(25) 31(22) 62(18) 28(16) 92(1) 74(5) 136(16)

8-125

MASS SPECTRAL PEAKS OF COMMON ORGANIC SOLVENTS (continued)

Compound	e/m (intensity)									
Ethanolamine (Glycinol)	30(100)	18(30)	28(15)	42(7)	31(6)	17(6)	61(5)	15(5)	43(3)	29(3)
Ethyl acetate	43(100)	29(46)	27(33)	45(32)	61(28)	28(25)	42(18)	73(11)	88(10)	70(10)
Ethyl acetoacetate	43(100)	29(24)	88(18)	28(16)	85(14)	27(12)	42(11)	60(9)	130(6)	45(6)
Ethyl alcohol	31(100)	45(44)	46(18)	27(18)	29(15)	43(14)	30(6)	42(3)	19(3)	14(3)
Ethylamine	30(100)	28(32)	44(20)	45(19)	27(13)	15(10)	42(9)	29(8)	41(5)	40(5)
Ethylbenzene	91(100)	106(31)	51(14)	39(10)	77(8)	65(8)	105(7)	92(7)	78(7)	27(6)
Ethyl bromide (Bromoethane)	108(100)	110(97)	29(62)	27(51)	28(35)	26(14)	93(6)	32(6)	95(5)	81(5)
Ethyl chloride (Cloroethane)	64(100)	28(91)	29(84)	27(75)	66(32)	26(28)	49(25)	51(8)	63(6)	65(4)
Ethylene carbonate	29(100)	44(62)	43(54)	88(40)	30(16)	28(11)	45(7)	58(6)	42(6)	73(4)
Ethylenediamine (1,2-Ethanediamine)	30(100)	18(13)	42(6)	43(5)	27(5)	44(4)	29(4)	17(4)	15(4)	41(3)
Ethylene glycol	31(100)	33(35)	29(13)	32(11)	43(6)	27(5)	28(4)	62(3)	30(3)	44(2)
Ethylene glycol diethyl ether	31(100)	59(71)	29(58)	45(43)	27(33)	74(27)	43(15)	15(14)	28(12)	44(10)
Ethylene glycol dimethyl ether	45(100)	60(13)	29(13)	90(7)	58(6)	31(5)	28(5)	43(4)	59(3)	46(2)
Ethylene glycol monobutyl ether	57(100)	45(38)	29(35)	41(31)	87(16)	27(12)	56(11)	31(9)	75(7)	28(7)
Ethylene glycol monoethyl ether (Cellosolve)	31(100)	29(52)	59(50)	45(26)	72(14)	43(14)	15(14)	28(8)	26(6)	
Ethylene glycol monoethyl ether acetate	43(100)	31(34)	59(31)	72(28)	44(25)	29(24)	45(12)	27(11)	15(11)	87(7)
Ethylene glycol monomethyl ether	45(100)	31(15)	29(14)	28(11)	47(9)	76(6)	43(6)	58(4)	46(4)	27(4)
Ethylene glycol monomethyl ether acetate	43(100)	45(48)	58(42)	29(10)	42(4)	31(4)	73(3)	27(3)	59(2)	26(2)
Ethyl formate	31(100)	28(73)	27(51)	29(38)	45(34)	26(17)	74(11)	43(9)	47(8)	56(4)
Furan	68(100)	39(64)	40(9)	38(9)	42(6)	29(6)	37(5)	69(4)	34(2)	67(1)
Furfural	39(100)	96(55)	95(52)	38(38)	29(35)	37(29)	40(11)	97(9)	50(7)	42(7)
Furfuryl alcohol	98(100)	41(65)	39(59)	81(55)	53(53)	97(51)	42(49)	69(39)	70(36)	29(28)
Glycerol	61(100)	43(90)	31(57)	44(54)	29(38)	18(32)	27(12)	42(11)	60(10)	45(10)
Heptane	43(100)	41(56)	29(49)	57(47)	27(46)	71(45)	56(27)	42(26)	39(23)	70(18)
1-Heptanol	41(100)	70(87)	56(86)	31(78)	43(72)	29(70)	55(67)	27(65)	42(54)	69(41)
Hexane	57(100)	43(78)	41(77)	29(61)	27(57)	56(45)	42(39)	39(27)	28(16)	86(14)
1-Hexanol (Caproyl alcohol)	56(100)	43(78)	31(74)	41(71)	27(64)	29(59)	55(58)	42(53)	39(37)	69(27)
Hexylene glycol	59(100)	43(61)	56(25)	45(17)	41(16)	57(13)	42(13)	85(11)	61(10)	31(10)
Hexyl methyl ketone	43(100)	58(79)	41(56)	59(52)	71(49)	27(46)	29(36)	39(27)	57(18)	55(17)
Isobutyl acetate	43(100)	56(26)	73(15)	41(10)	29(5)	71(3)	57(3)	39(3)	27(3)	86(2)
Isobutyl alcohol	43(100)	33(73)	31(72)	41(66)	42(60)	27(43)	29(18)	39(17)	28(8)	74(6)
Isobutylamine	30(100)	28(9)	41(6)	73(5)	27(5)	39(4)	29(3)	15(3)	58(2)	56(2)
Isopentyl acetate	43(100)	70(49)	55(38)	61(15)	42(15)	41(14)	27(12)	87(11)	29(10)	73(9)
Isophorone	82(100)	39(20)	138(17)	54(13)	27(12)	41(10)	53(8)	83(7)	29(7)	55(6)
Isopropyl acetate	43(100)	61(17)	41(14)	87(9)	59(8)	27(8)	42(7)	39(4)	45(3)	44(2)
Isopropyl alcohol	45(100)	43(19)	27(17)	29(12)	41(7)	31(6)	19(6)	42(5)	44(4)	59(3)
Isoquinoline	129(100)	102(26)	51(20)	128(18)	50(11)	130(10)	75(10)	76(9)	103(8)	74(7)
d-Limonene (Citrene)	68(100)	93(50)	67(49)	41(22)	94(21)	79(21)	39(21)	136(20)	53(19)	121(16)
2,6-Lutidine (2,6-Dimethylpyridine)	107(100)	39(39)	106(29)	66(22)	92(18)	65(18)	38(12)	27(11)	79(9)	63(9)

MASS SPECTRAL PEAKS OF COMMON ORGANIC SOLVENTS (continued)

Compound	e/m (intensity)									
Mesitylene (1,3,5-Trimethyl-benzene)	105(100)	120(64)	119(15)	77(13)	39(11)	106(9)	91(9)	51(8)	27(7)	121(6)
Mesityl oxide	55(100)	83(89)	43(73)	29(42)	98(36)	39(32)	27(28)	53(11)	41(10)	56(5)
Methyl acetate	43(100)	74(52)	28(38)	42(19)	59(17)	44(8)	32(8)	29(6)	31(4)	75(2)
Methylal (Dimethoxymethane)	45(100)	75(61)	29(59)	31(13)	30(6)	15(6)	47(5)	76(2)	46(2)	44(2)
Methyl alcohol	31(100)	29(72)	32(67)	15(42)	28(12)	14(10)	30(9)	13(6)	12(3)	16(2)
Methylamine	30(100)	31(87)	28(56)	29(19)	32(15)	15(12)	27(9)	—	—	—
Methyl benzoate	105(100)	77(81)	51(45)	136(24)	50(18)	106(8)	78(6)	28(6)	39(5)	27(5)
Methylcyclohexane	83(100)	55(82)	41(60)	98(44)	42(35)	56(30)	27(29)	39(27)	69(23)	70(22)
Methyl ethyl ketone	43(100)	72(24)	29(19)	27(12)	57(7)	42(5)	26(4)	28(3)	44(2)	39(2)
N-Methylformamide	59(100)	30(54)	28(34)	29(13)	58(8)	15(7)	60(3)	41(3)	27(3)	31(2)
Methyl formate	31(100)	29(63)	32(34)	60(28)	30(7)	28(7)	44(2)	18(2)	61(1)	59(1)
Methyl iodide (Iodomethane)	142(100)	127(38)	141(14)	15(13)	139(5)	140(4)	128(3)	14(1)	13(1)	71(0)
Methyl isobutyl ketone	43(100)	58(84)	29(65)	41(56)	57(44)	27(42)	39(31)	85(19)	100(14)	42(14)
Methyl isopentyl ketone	43(100)	58(34)	27(14)	41(13)	15(13)	57(11)	39(9)	71(8)	59(8)	29(8)
2-Methylpentane	43(100)	42(53)	41(35)	27(31)	71(29)	39(20)	29(18)	57(11)	15(10)	70(7)
4-Methyl-2-pentanol	45(100)	43(47)	69(30)	41(27)	27(19)	39(13)	29(12)	87(11)	84(10)	57(10)
Methyl pentyl ketone	43(100)	58(60)	71(14)	41(11)	27(11)	59(9)	39(8)	29(8)	42(5)	114(4)
Methyl propyl ketone	43(100)	41(17)	86(12)	42(12)	27(11)	39(8)	71(7)	58(7)	45(7)	44(3)
N-Methyl-2-pyrrolidone	99(100)	44(89)	98(80)	42(60)	41(38)	43(17)	28(17)	71(13)	39(11)	70(10)
Morpholine	57(100)	29(100)	87(69)	28(69)	30(38)	56(33)	86(28)	31(28)	27(12)	15(7)
Nitrobenzene	77(100)	51(59)	123(42)	50(25)	30(15)	65(14)	39(10)	93(9)	74(7)	78(6)
Nitroethane	29(100)	30(12)	28(11)	26(9)	27(8)	43(5)	41(5)	14(5)	15(3)	46(2)
Nitromethane	30(100)	61(64)	46(39)	28(30)	45(8)	27(8)	44(7)	29(7)	60(5)	43(4)
1-Nitropropane	43(100)	27(93)	41(90)	39(34)	30(25)	44(20)	42(20)	26(20)	28(13)	54(12)
2-Nitropropane	43(100)	41(73)	27(71)	39(30)	30(18)	15(11)	42(9)	28(8)	26(8)	38(6)
Octane	43(100)	57(30)	85(25)	41(25)	71(19)	29(17)	56(14)	70(10)	42(10)	27(10)
1-Octanol	41(100)	56(85)	43(82)	55(81)	31(69)	27(69)	29(68)	42(62)	70(53)	69(48)
Pentachloroethane	167(100)	165(91)	117(90)	119(89)	83(58)	169(54)	130(43)	132(42)	60(40)	85(37)
Pentamethylene glycol (1,5-Pentanediol)	31(100)	56(85)	41(67)	57(59)	55(51)	44(45)	29(37)	43(31)	68(29)	27(26)
Pentane	43(100)	42(55)	41(45)	27(42)	29(26)	39(19)	57(13)	28(9)	15(9)	72(8)
1-Pentanol (Amyl alcohol)	42(100)	70(72)	55(65)	41(56)	31(47)	29(41)	27(26)	57(22)	28(22)	43(21)
Pentyl acetate (Amyl acetate)	43(100)	70(90)	42(52)	28(51)	61(50)	55(41)	73(21)	41(20)	29(14)	69(11)
2-Picoline (2-Methylpyridine)	93(100)	66(41)	39(31)	92(20)	78(19)	51(19)	65(16)	38(13)	50(12)	52(11)
α-Pinene	93(100)	92(30)	39(24)	41(23)	77(22)	91(21)	27(21)	79(18)	121(13)	53(10)
β-Pinene	93(100)	41(64)	69(47)	39(33)	27(31)	79(20)	77(18)	53(14)	94(13)	91(13)
Piperidine (Hexahydropyridine)	84(100)	85(53)	56(46)	57(43)	28(41)	29(37)	44(34)	42(30)	30(30)	43(25)
Propanenitrile	28(100)	54(63)	26(20)	27(17)	52(11)	55(10)	51(9)	15(9)	53(7)	25(7)
Propyl acetate	43(100)	61(19)	31(18)	27(15)	42(11)	73(9)	41(9)	29(9)	59(5)	39(5)
Propyl alcohol	31(100)	27(19)	29(18)	59(11)	42(9)	60(7)	41(7)	41(7)	43(3)	32(3)
Propylamine	30(100)	28(13)	59(8)	27(7)	41(5)	42(3)	39(3)	29(3)	26(3)	18(3)
Propylbenzene	91(100)	120(21)	92(10)	38(10)	65(9)	78(6)	51(6)	27(5)	63(4)	105(3)
1,2-Propylene glycol	45(100)	18(46)	29(21)	43(19)	31(18)	27(17)	28(11)	19(8)	44(6)	61(5)

MASS SPECTRAL PEAKS OF COMMON ORGANIC SOLVENTS (continued)

e/m (intensity)

Compound										
Pseudocumene (1,2,4-Trimethyl-benzene)	105(100)	120(56)	119(17)	77(15)	39(15)	51(11)	91(10)	27(10)	106(9)	79(7)
Pyridine	79(100)	52(62)	51(31)	50(19)	78(11)	53(7)	39(7)	80(6)	27(3)	77(2)
Pyrrole	67(100)	41(58)	39(58)	40(51)	28(42)	38(20)	37(12)	66(7)	68(5)	27(3)
Pyrrolidine	43(100)	28(52)	70(33)	71(26)	42(22)	41(20)	27(16)	39(15)	29(10)	30(9)
2-Pyrrolidone	85(100)	42(43)	41(36)	28(33)	30(29)	56(16)	84(14)	40(12)	27(12)	29(9)
Quinoline	129(100)	51(28)	76(25)	128(24)	44(24)	50(20)	32(19)	75(18)	74(12)	103(11)
Styrene	104(100)	103(41)	78(32)	51(28)	77(23)	105(12)	50(12)	52(11)	39(11)	102(10)
Sulfolane	41(100)	28(94)	56(82)	55(72)	120(37)	27(32)	39(19)	29(17)	26(11)	48(5)
α-Terpinene	121(100)	93(85)	136(43)	91(40)	77(34)	39(33)	27(33)	79(27)	41(26)	43(18)
1,1,1,2-Tetrachloro-2,2-difluoroethane	167(100)	169(96)	119(82)	171(31)	85(29)	121(26)	82(14)	47(14)	101(13)	
Tetrachloro-1,2-difluoroethane	101(100)	103(64)	167(54)	169(52)	117(19)	119(18)	171(17)	105(11)	31(11)	132(9)
1,1,1,2-Tetrachloroethane	131(100)	133(96)	117(76)	119(73)	95(34)	135(31)	121(23)	97(23)	61(19)	60(18)
1,1,2,2-Tetrachloroethane	83(100)	85(63)	95(11)	87(10)	168(8)	133(8)	131(8)	96(8)	61(8)	60(8)
Tetrachloroethylene	166(100)	164(82)	131(71)	129(71)	168(45)	94(38)	47(31)	96(24)	133(20)	59(17)
Tetraethylene glycol	45(100)	89(10)	44(8)	43(6)	31(6)	29(6)	27(6)	101(5)	75(5)	28(5)
Tetrahydrofuran	42(100)	41(52)	27(33)	72(29)	71(27)	39(24)	43(22)	29(22)	40(13)	15(10)
1,2,3,4-Tetrahydronaphthalene	104(100)	132(53)	91(43)	51(17)	39(17)	131(15)	117(15)	115(14)	78(13)	77(13)
Tetrahydropyran	41(100)	28(64)	56(57)	45(57)	29(51)	27(49)	86(42)	85(42)	39(28)	55(23)
Tetramethylsilane	73(100)	43(14)	45(12)	74(8)	29(7)	15(5)	75(4)	44(4)	42(4)	31(4)
Toluene	91(100)	92(73)	39(20)	65(14)	63(11)	51(11)	50(7)	27(6)	93(5)	90(5)
o-Toluidine	106(100)	107(83)	77(17)	79(13)	39(12)	53(10)	52(10)	54(9)	51(9)	28(9)
Triacetin	43(100)	103(44)	145(34)	116(17)	115(13)	44(10)	86(9)	28(8)	73(7)	42(7)
Tributylamine	142(100)	100(19)	143(11)	29(8)	185(7)	57(6)	44(6)	41(6)	30(5)	86(4)
1,1,1-Trichloroethane	97(100)	99(64)	61(58)	26(31)	27(24)	117(19)	63(19)	119(18)	35(17)	62(11)
1,1,2-Trichloroethane	97(100)	83(95)	99(62)	85(60)	61(58)	26(23)	96(21)	63(19)	27(17)	98(15)
Trichloroethylene	95(100)	130(90)	132(85)	60(65)	97(64)	35(40)	134(27)	47(26)	62(21)	59(13)
Trichlorofluoromethane	101(100)	103(66)	66(13)	105(11)	35(11)	47(9)	31(8)	82(4)	68(4)	37(4)
1,1,2-Trichlorotrifluoroethane	101(100)	151(68)	103(64)	85(45)	31(45)	153(44)	35(20)	66(19)	47(18)	87(14)
Triethanolamine	118(100)	56(69)	45(60)	42(56)	44(27)	43(25)	41(14)	116(8)	57(8)	86(7)
Triethylamine	86(100)	30(68)	58(37)	28(24)	29(23)	27(19)	44(18)	101(17)	42(16)	56(8)
Triethylene glycol	45(100)	58(11)	89(9)	31(8)	29(8)	75(7)	44(7)	43(7)	27(7)	28(5)
Triethyl phosphate	99(100)	81(71)	155(56)	82(45)	45(45)	109(44)	127(41)	43(24)	125(16)	111(14)
Trimethylamine	58(100)	59(47)	30(29)	42(26)	44(17)	15(14)	28(10)	18(10)	43(8)	57(7)
Trimethylene glycol (1,3-Propanediol)	28(100)	58(93)	31(76)	57(70)	29(40)	27(26)	45(24)	43(23)	19(18)	30(17)
Trimethyl phosphate	110(100)	109(35)	79(34)	95(25)	80(23)	15(20)	140(18)	47(10)	31(7)	139(5)
Veratrole	138(100)	95(65)	77(48)	123(44)	52(42)	41(33)	65(30)	51(29)	39(19)	63(17)
o-Xylene	91(100)	106(40)	39(21)	105(17)	51(17)	77(15)	27(12)	65(10)	92(8)	79(8)
m-Xylene	91(100)	106(65)	105(29)	39(18)	51(15)	77(14)	27(10)	92(8)	79(8)	78(8)
p-Xylene	91(100)	106(62)	105(30)	51(16)	39(16)	77(13)	27(11)	92(7)	78(7)	65(7)

Section 9
Molecular Structure and Spectroscopy

Bond Lengths in Crystalline Organic Compounds 9-1
Bond Lengths in Organometallic Compounds 9-15
Bond Lengths and Angles in Gas-Phase Molecules 9-17
Dipole Moments 9-44
Strengths of Chemical Bonds 9-51
Electronegativity 9-75
Force Constants for Bond Stretching 9-76
Fundamental Vibrational Frequencies of Small Molecules 9-77
Spectroscopic Constants of Diatomic Molecules 9-81
Infrared Correlation Charts 9-87
Nuclear Spins, Moments, and Other Data Related to NMR Spectroscopy 9-92
Proton NMR Chemical Shifts for Characteristic Organic Structures 9-95
^{13}C NMR Absorptions of Major Functional Groups 9-96

BOND LENGTHS IN CRYSTALLINE ORGANIC COMPOUNDS

The following table gives average interatomic distances for bonds between the elements H, B, C, N, O, F, Si, P, S, Cl, As, Se, Br, Te, and I as determined from X-ray and neutron diffraction measurements on organic crystals. The table has been derived from an analysis of high-precision structure data on about 10,000 crystals contained in the 1985 version of the Cambridge Structural Database, which is maintained by the Cambridge Crystallographic Data Center. The explanation of the columns is:

Column 1: Specification of elements in the bond, with coordination number given in parentheses, and bond type (single, double, etc.). For carbon, the hybridization state is given.

Column 2: Substructure in which the bond is found. The target bond is set in boldface. Where X is not specified, it denotes any element type. C# indicates any sp^3 carbon atom, and C* denotes an sp^3 carbon whose bonds, in addition to those specified in the linear formulation, are to C and H atoms only.

Column 3: d is the unweighted mean in Å units of all the values for that bond length found in the sample.

Column 4: m is the median in Å units of all values.

Column 5: σ is the standard deviation in the sample.

Column 6: q_1 is the lower quartile for the sample (i.e., 25% of values are less than q_1 and 75% exceed it).

Column 7: q_2 is the upper quartile for the sample.

Column 8: n is number of observations in the sample.

Column 9: Notes refer to the footnotes in Appendix 1.

References to special cases are given in a shorthand form and listed in Appendix 2. Further information on the method of analysis of the data may be found in the reference cited below.

The table is reprinted with permission of the authors, the Royal Society of Chemistry, and the International Union of Crystallography.

REFERENCE

Frank H. Allen, Olga Kennard, David G. Watson, Lee Brammer, A. Guy Orpen, and Robin Taylor, *J. Chem. Soc. Perkin Trans. II*, S1—S19, 1987.

Bond	Substructure	d	m	σ	q_1	q_u	n	Note
As(3)–As(3)	X_2–As–As–X_2	2.459	2.457	0.011	2.456	2.466	8	
As–B	see CUDLOC (2.065), CUDLUI (2.041)							
As–Br	see CODDEE, CODDII (2.346—3.203)							
As(4)–C	X_3–As–CH_3	1.903	1.907	0.016	1.893	1.916	12	
	$(X)_2$(C,O,S=)As–Csp^3	1.927	1.929	0.017	1.921	1.937	16	
	As–C*ar* in Ph_4As^+	1.905	1.909	0.012	1.897	1.912	108	
	$(X)_2$(C,O,S=)As–C*ar*	1.922	1.927	0.016	1.908	1.934	36	
As(3)–C	X_2–As–Csp^3	1.963	1.965	0.017	1.948	1.978	6	
	X_2–As–C*ar*	1.956	1.956	0.015	1.944	1.964	41	
As(3)–Cl	X_2–As–Cl	2.268	2.256	0.039	2.247	2.281	10	
As(6)–F	in AsF_6^-	1.678	1.676	0.020	1.659	1.695	36	
As(3)–I	see OPIMAS (2.579, 2.590)							
As(3)–N(3)	X_2–As–N–X_2	1.858	1.858	0.029	1.839	1.873	19	
As(4)=N(2)	see TPASSN (1.837)							
As(4)–O	$(X)_2$(O=)As–OH	1.710	1.712	0.017	1.695	1.726	6	
As(3)–O	see ASAZOC, PHASOC01 (1.787—1.845)							
As(4)=O	X_3–As=O	1.661	1.661	0.016	1.652	1.667	9	
As(3)=P(3)	see BELNIP (2.350, 2.362)							†
As(3)–P(3)	see BUTHAZ10 (2.124)							†
As(3)–S	X_2–As–S	2.275	2.266	0.032	2.247	2.298	14	
As(4)=S	X_3–As=S	2.083	2.082	0.004	2.080	2.086	9	
As(3)–Se(2)	see COSDIX, ESEARS (2.355—2.401)							†
As(3)–Si(4)	see BICGEZ, MESIAD (2.351—2.365)							†
As(3)–Te(2)	see ETEARS (2.571, 2.576)							†
B(n)–B(n)	$n = 5$—7 in boron cages	1.775	1.773	0.031	1.763	1.786	688	
B(4)–B(4)	see CETTAW (2.041)							
B(4)–B(3)	see COFVOI (1.698)							
B(3)–B(3)	X_2–B–B–X_2	1.701	1.700	0.014	1.691	1.712	8	
B(6)–Br		1.967	1.971	0.014	1.954	1.979	7	†
B(4)–Br		2.017	2.008	0.031	1.990	2.044	15	†
B(n)–C	$n = 5$—7: B–C in cages	1.716	1.717	0.020	1.707	1.728	96	
	$n = 3$—4: B–Csp^3 not cages	1.597	1.599	0.022	1.585	1.611	29	1
	$n = 4$: B–C*ar*	1.606	1.607	0.012	1.596	1.615	41	
	$n = 4$: B–C*ar* in Ph_4B^-	1.643	1.643	0.006	1.641	1.645	16	
B(n)–C	$n = 3$: B–C*ar*	1.556	1.552	0.015	1.546	1.566	24	
B(n)–Cl	B(5)–Cl and B(3)–Cl	1.751	1.751	0.011	1.743	1.761	14	
	B(4)–Cl	1.833	1.833	0.013	1.821	1.843	22	
B(4)–F	B–F (B neutral)	1.366	1.368	0.017	1.356	1.375	25	
	B^-–F in BF_4^-	1.365	1.372	0.029	1.352	1.390	84	
B(4)–I	see TMPBTI (2.220, 2.253)							
B(4)–N(3)	X_3–B–N(=C)(X)	1.611	1.617	0.013	1.601	1.625	8	
	in pyrazaboles	1.549	1.552	0.015	1.536	1.560	10	

Bond	Substructure	d	m	σ	q_1	q_u	n	Note
B(3)–N(3)	X_2–B–N–C_2: all coplanar	1.404	1.404	0.014	1.389	1.408	40	2
	for τ(BN) > 30° see BOGSUL, BUSHAY, CILRUK (1.434—1.530)							
	S_2–B–N–X_2	1.447	1.443	0.013	1.435	1.470	14	
B(4)–O	B^-–O in BO_4^-	1.468	1.468	0.022	1.453	1.479	24	3
	for neutral B–O see Note 3							
B(3)–O(2)	X_2–B–O–X	1.367	1.367	0.024	1.349	1.382	35	
B(n)–P	$n = 4$: B–P	1.922	1.927	0.027	1.900	1.954	10	
	$n = 3$: see BUPSIB10 (1.892, 1.893)							
B(4)–S	B(4)–S(3)	1.930	1.927	0.009	1.925	1.934	10	
	B(4)–S(2)	1.896	1.896	0.004	1.893	1.899	6	
B(3)–S	N–B–S_2	1.806	1.806	0.010	1.799	1.816	28	
	(=X–)(N–)B–S	1.851	1.854	0.013	1.842	1.859	10	
Br–Br	see BEPZEB, TPASTB	2.542	2.548	0.015	2.526	2.551	4	
Br–C	Br–C*	1.966	1.967	0.029	1.951	1.983	100	4
	Br–Csp^3 (cyclopropane)	1.910	1.910	0.010	1.900	1.914	8	
	Br–Csp^2	1.883	1.881	0.015	1.874	1.894	31	4
	Br–Car (mono-Br + m,p-Br_2)	1.899	1.899	0.012	1.892	1.906	119	4
	Br–Car (o-Br_2)	1.875	1.872	0.011	1.864	1.884	8	4
$^-$Br(2)–Cl	see TEACBR (2.362—2.402)							†
Br–I	see DTHIBR10 (2.646), TPHOSI (2.695)							
Br–N	see NBBZAM (1.843)							
Br–O	see CIYFOF	1.581	1.581	0.007	1.574	1.587	4	
Br–P	see CISTED (2.366)							†
Br–S(2)	see BEMLIO (2.206)							†
Br–S(3)	see CIWYIQ (2.435, 2.453)							†
Br–S(3)$^+$	see THINBR (2.321)							†
Br–Se	see CIFZUM (2.508, 2.619)							
Br–Si	see BIZJAV (2.284)							
Br–Te	In Br_6Te^{2-} see CUGBAH (2.692—2.716)							
	Br–Te(4) see BETUTE10 (3.079, 3.015)							
	Br–Te(3) see BTUPTE (2.835)							
Csp^3–Csp^3	C#–CH_2–CH_3	1.513	1.514	0.014	1.507	1.523	192	
	(C#)$_2$–CH–CH_3	1.524	1.526	0.015	1.518	1.534	226	
	(C#)$_3$–C–CH_3	1.534	1.534	0.011	1.527	1.541	825	
	C#–CH_2–CH_2–C#	1.524	1.524	0.014	1.516	1.532	2 459	
	(C#)$_2$–CH–CH_2–C#	1.531	1.531	0.012	1.524	1.538	1 217	
	(C#)$_3$–C–CH_2–C#	1.538	1.539	0.010	1.533	1.544	330	
	(C#)$_2$–CH–CH–(C#)$_2$	1.542	1.542	0.011	1.536	1.549	321	
	(C#)$_3$–C–CH–(C#)$_2$	1.556	1.556	0.011	1.549	1.562	215	
	(C#)$_3$–C–C–(C#)$_3$	1.588	1.580	0.025	1.566	1.610	21	
	C*–C* (overall)	1.530	1.530	0.015	1.521	1.539	5 777	5,6
	in cyclopropane (any subst.)	1.510	1.509	0.026	1.497	1.523	888	7
	in cyclobutane (any subst.)	1.554	1.553	0.021	1.540	1.567	679	8
	in cyclopentane (C,H-subst.)	1.543	1.543	0.018	1.532	1.554	1 641	
	in cyclohexane (C,H-subst.)	1.535	1.535	0.016	1.525	1.545	2 814	
	cyclopropyl-C* (exocyclic)	1.518	1.518	0.019	1.505	1.531	366	7
	cyclobutyl-C* (exocyclic)	1.529	1.529	0.016	1.519	1.539	376	8
	cyclopentyl-C* (exocyclic)	1.540	1.541	0.017	1.527	1.549	956	
	cyclohexyl-C* (exocyclic)	1.539	1.538	0.016	1.529	1.549	2 682	
	in cyclobutene (any subst.)	1.573	1.574	0.017	1.566	1.586	25	8
	in cyclopentene (C,H-subst.)	1.541	1.539	0.015	1.532	1.549	208	
	in cyclohexene (C,H-subst.)	1.541	1.541	0.020	1.528	1.554	586	
	in oxirane (epoxide)	1.466	1.466	0.015	1.458	1.474	249	9
	in aziridine	1.480	1.481	0.021	1.465	1.496	67	9
	in oxetane	1.541	1.541	0.019	1.527	1.557	16	
	in azetidine	1.548	1.543	0.018	1.536	1.558	22	
	oxiranyl-C* (exocyclic)	1.509	1.507	0.018	1.497	1.519	333	9
	aziridinyl-C* (exocyclic)	1.512	1.512	0.018	1.496	1.526	13	9
Csp^3–Csp^2	CH_3–C=C	1.503	1.504	0.011	1.497	1.509	215	
	C#–CH_2–C=C	1.502	1.502	0.013	1.494	1.510	483	
	(C#)$_2$–CH–C=C	1.510	1.510	0.014	1.501	1.518	564	
	(C#)$_3$–C–C=C	1.522	1.522	0.016	1.511	1.533	193	
Csp^3–Csp^2	C*–C=C (overall)	1.507	1.507	0.015	1.499	1.517	1 456	5
	C*–C=C (endocyclic)							
	in cyclopropene	1.509	1.508	0.016	1.500	1.516	20	10
	in cyclobutene	1.513	1.512	0.018	1.500	1.525	50	8
	in cyclopentene	1.512	1.512	0.014	1.502	1.521	208	
	in cyclohexene	1.506	1.505	0.016	1.495	1.516	391	
	in cyclopentadiene	1.502	1.503	0.019	1.490	1.515	18	
	in cyclohexa-1,3-diene	1.504	1.504	0.017	1.491	1.517	56	
	C*–C=C (exocyclic):							
	cyclopropenyl-C*	1.478	1.475	0.012	1.470	1.485	7	10
	cyclobutenyl-C*	1.489	1.483	0.015	1.479	1.496	11	8

Bond	Substructure	d	m	σ	q_1	q_u	n	Note
	cyclopentenyl-C*	1.504	1.506	0.012	1.495	1.512	115	
	cyclohexenyl-C*	1.511	1.511	0.013	1.502	1.519	292	
	C*–CH=O in aldehydes	1.510	1.510	0.008	1.501	1.518	7	
	(C*)₂–C=O							
	in ketones	1.511	1.511	0.015	1.501	1.521	952	11
	in cyclobutanone	1.529	1.530	0.016	1.514	1.545	18	
	in cyclopentanone	1.514	1.514	0.016	1.505	1.523	312	
	acyclic and 6+ rings	1.509	1.509	0.016	1.499	1.519	626	
	C*–COOH in carboxylic acids	1.502	1.502	0.014	1.495	1.510	176	
	C*–COO⁻ in carboxylate anions	1.520	1.521	0.011	1.516	1.528	57	
	C*–C(=O)(–OC*)							
	in acyclic esters	1.497	1.496	0.018	1.484	1.509	553	12
	in β-lactones	1.519	1.519	0.020	1.500	1.538	4	13
	in γ-lactones	1.512	1.512	0.015	1.501	1.521	110	12
	in δ-lactones	1.504	1.502	0.013	1.495	1.517	27	12
	cyclopropyl (C)–C=O in ketones, acids							
	and esters	1.486	1.485	0.018	1.474	1.497	105	7
	C*–C(=O)(–NH₂) in acyclic amides	1.514	1.512	0.016	1.506	1.526	32	14
	C*–C(=O)(–NHC*) in acyclic amides	1.506	1.505	0.012	1.498	1.515	78	14
	C*–C(=O)[–N(C*)₂] in acyclic amides	1.505	1.505	0.011	1.496	1.517	15	14
Csp³–Car	CH₃–Car	1.506	1.507	0.011	1.501	1.513	454	
	C#–CH₂–Car	1.510	1.510	0.009	1.505	1.516	674	
	(C#)₂–CH–Car	1.515	1.515	0.011	1.508	1.522	363	
	(C#)₃–C–Car	1.527	1.530	0.016	1.517	1.539	308	
	C*–Car (overall)	1.513	1.513	0.014	1.505	1.521	1 813	
	cyclopropyl (C)–Car	1.490	1.490	0.015	1.479	1.503	90	7
Csp³–Csp¹	C*–C≡C	1.466	1.465	0.010	1.460	1.469	21	15
	C#–C≡C	1.472	1.472	0.012	1.464	1.481	88	15
	C*–C≡N	1.470	1.469	0.013	1.463	1.479	106	7b
	cyclopropyl (C)–C≡N	1.444	1.447	0.010	1.436	1.451	38	7
Csp²–Csp²	C=C–C=C							
	(conjugated)	1.455	1.455	0.011	1.447	1.463	30	16,18
	(unconjugated)	1.478	1.476	0.012	1.470	1.479	8	17,18
	(overall)	1.460	1.460	0.015	1.450	1.470	38	
	C=C–C=C–C=C	1.443	1.445	0.013	1.431	1.454	29	18
	C=C–C=C (endocyclic in TCNQ)	1.432	1.433	0.012	1.424	1.441	280	19
	C=C–C(=O)(–C*)							
	(conjugated)	1.464	1.462	0.018	1.453	1.476	211	16,18
	(unconjugated)	1.484	1.486	0.017	1.475	1.497	14	17,18
	(overall)	1.465	1.462	0.018	1.453	1.478	226	
	C=C–C(=O)–C=C							
	in benzoquinone (C,H-subst. only)	1.478	1.476	0.011	1.469	1.488	28	
	in benzoquinone (any subst.)	1.478	1.478	0.031	1.464	1.498	172	
	non-quinonoid	1.456	1.455	0.012	1.447	1.464	28	
	C=C–COOH	1.475	1.476	0.015	1.461	1.488	22	
	C=C–COOC*	1.488	1.489	0.014	1.478	1.497	113	
	C=C–COO⁻	1.502	1.499	0.017	1.488	1.510	11	
	HOOC–COOH	1.538	1.537	0.007	1.535	1.541	9	
	HOOC–COO⁻	1.549	1.552	0.009	1.546	1.553	13	
	⁻OOC–COO⁻	1.564	1.559	0.022	1.554	1.568	9	
	formal Csp²–Csp² single bond in selected							
	non-fused heterocycles:							
	in 1H-pyrrole (C3–C4)	1.412	1.410	0.016	1.401	1.427	29	
	in furan (C3–C4)	1.423	1.423	0.016	1.412	1.433	62	
	in thiophene (C3–C4)	1.424	1.425	0.015	1.415	1.433	40	
	in pyrazole (C3–C4)	1.410	1.412	0.016	1.400	1.418	20	
	in isoxazole (C3–C4)	1.425	1.425	0.016	1.413	1.438	9	
	in furazan (C3–C4)	1.428	1.427	0.007	1.422	1.435	6	
	in furoxan (C3–C4)	1.417	1.417	0.006	1.412	1.422	14	
Csp²–Car	C=C–Car							
	(conjugated)	1.470	1.470	0.015	1.463	1.480	37	16,18
Csp²–Car		1.488	1.490	0.012	1.480	1.496	87	17,18
	(overall)	1.483	1.483	0.015	1.472	1.494	124	
	cyclopropenyl (C=C)–Car	1.447	1.448	0.006	1.441	1.452	8	10
	Car–C(=O)–C*	1.488	1.489	0.016	1.478	1.500	84	
	Car–C(=O)–Car	1.480	1.481	0.017	1.468	1.494	58	
	Car–COOH	1.484	1.485	0.014	1.474	1.491	75	
	Car–C(=O)(–OC*)	1.487	1.487	0.012	1.480	1.494	218	
	Car–COO⁻	1.504	1.509	0.014	1.495	1.512	26	
	Car–C(=O)–NH₂	1.500	1.503	0.020	1.498	1.510	19	
	Car–C=N–C#							
	(conjugated)	1.476	1.478	0.014	1.466	1.486	27	16
	(unconjugated)	1.491	1.490	0.008	1.485	1.496	48	17
	(overall)	1.485	1.487	0.013	1.481	1.493	75	

Bond	Substructure	d	m	σ	q_l	q_u	n	Note
	in indole (C3–C3a)	1.434	1.434	0.011	1.428	1.439	40	
Csp^2–Csp^1	C=C–C≡C	1.431	1.427	0.014	1.425	1.441	11	7b
	C=C–C≡N in TCNQ	1.427	1.427	0.010	1.420	1.433	280	19
Car–Car	in biphenyls (*ortho* subst. all H)	1.487	1.488	0.007	1.484	1.493	30	
	(≥1 non-H *ortho*-subst.)	1.490	1.491	0.010	1.486	1.495	212	
Car–Csp^1	Car–C≡C	1.434	1.436	0.006	1.430	1.437	37	
	Car–C≡N	1.443	1.444	0.008	1.436	1.448	31	
Csp^1–Csp^1	C≡C–C≡C	1.377	1.378	0.012	1.374	1.384	21	
Csp^2=Csp^2	C*–CH=CH₂	1.299	1.300	0.027	1.280	1.311	42	
	(C*)₂–C=CH₂	1.321	1.321	0.013	1.313	1.328	77	
	C*–CH=CH–C*							
	(*cis*)	1.317	1.318	0.013	1.310	1.323	106	
	(*trans*)	1.312	1.311	0.011	1.304	1.320	19	
	(overall)	1.316	1.317	0.015	1.309	1.323	127	
	(C*)₂–C=CH–C*	1.326	1.328	0.011	1.319	1.334	168	
	(C*)₂–C=C–(C*)₂	1.331	1.330	0.009	1.326	1.334	89	
	(C*,H)₂–C=C–(C*,H)₂ (overall)	1.322	1.323	0.014	1.315	1.331	493	5
	in cyclopropene (any subst.)	1.294	1.288	0.017	1.284	1.302	10	10
	in cyclobutene (any subst.)	1.335	1.335	0.019	1.324	1.347	25	8
	in cyclopentene (C,H-subst.)	1.323	1.324	0.013	1.314	1.331	104	
	in cyclohexene (C,H-subst.)	1.326	1.325	0.012	1.318	1.334	196	
	C=C=C (allenes, any subst.)	1.307	1.307	0.005	1.303	1.310	18	
	C=C–C=C (C,H subst., conjugated)	1.330	1.330	0.014	1.322	1.338	76	16
	C=C–C=C–C=C (C,H subst., conjugated)	1.345	1.345	0.012	1.337	1.350	58	16
	C=C–Car (C,H subst., conjugated)	1.339	1.340	0.011	1.334	1.346	124	16
	C=C in cyclopenta-1,3-diene (any subst.)	1.341	1.341	0.017	1.328	1.356	18	
	C=C in cyclohexa-1,3-diene (any subst.)	1.332	1.332	0.013	1.323	1.341	56	
	in C=C–C=O							
	(C,H subst., conjugated)	1.340	1.340	0.013	1.332	1.348	211	16,18
	(C,H subst., unconjugated)	1.331	1.330	0.008	1.326	1.339	14	17,18
	(C,H subst., overall)	1.340	1.339	0.013	1.332	1.348	226	
	in cyclohexa-2,5-dien-1-ones	1.329	1.327	0.011	1.321	1.335	28	
	in *p*-benzoquinones							
	(C*,H subst.)	1.333	1.337	0.011	1.325	1.338	14	
	(any subst.)	1.349	1.339	0.030	1.330	1.364	86	
	in TCNQ							
	(endocyclic)	1.352	1.353	0.010	1.345	1.358	142	19
	(exocyclic)	1.392	1.391	0.017	1.379	1.405	139	19
	C=C–OH in enol tautomers	1.362	1.360	0.020	1.349	1.370	54	
	in heterocycles (any subst.):							
	1*H*-pyrrole (C2–C3, C4–C5)	1.375	1.377	0.018	1.361	1.388	58	
	furan (C2–C3, C4–C5)	1.341	1.342	0.021	1.329	1.351	125	
	thiophene (C2–C3, C4–C5)	1.362	1.359	0.025	1.346	1.377	60	
	pyrazole (C4–C5)	1.369	1.372	0.019	1.362	1.383	20	
	imidazole (C4–C5)	1.360	1.361	0.014	1.352	1.367	44	
	isoxazole (C4–C5)	1.341	1.336	0.012	1.331	1.355	9	
	indole (C2–C3)	1.364	1.363	0.012	1.355	1.371	40	
$Car \simeq Car$	in phenyl rings with C*,H subst. only							
	H–C≃C–H	1.380	1.381	0.013	1.372	1.388	2 191	
	C*–C≃C–H	1.387	1.388	0.010	1.382	1.393	891	
	C*–C≃C–C*	1.397	1.397	0.009	1.392	1.403	182	
	C≃C (overall)	1.384	1.384	0.013	1.375	1.391	3 264	
	F–C≃C–F	1.372	1.374	0.011	1.366	1.380	84	4
	Cl–C≃C–Cl	1.388	1.389	0.014	1.380	1.398	152	4
	in naphthalene (D_{2h}, any subst.)							
	C1–C2	1.364	1.364	0.014	1.356	1.373	440	
	C2–C3	1.406	1.406	0.014	1.397	1.415	218	
	C1–C8a	1.420	1.419	0.012	1.412	1.426	440	
	C4a–C8a	1.422	1.424	0.011	1.417	1.429	109	
$Car \simeq Car$	in anthracene (D_{2h}, any subst.)							
	C1–C2	1.356	1.356	0.009	1.350	1.360	56	
	C2–C3	1.410	1.410	0.010	1.401	1.416	34	
	C1–C9a	1.430	1.430	0.006	1.426	1.434	56	
	C4a–C9a	1.435	1.436	0.007	1.429	1.440	34	
	C9–C9a	1.400	1.402	0.009	1.395	1.406	68	
	in pyridine (C,H subst.)	1.379	1.381	0.012	1.371	1.387	276	20
	(any subst.)	1.380	1.380	0.015	1.371	1.389	537	20
	in pyridinium cation							
	(N⁺–H; C,H subst. on C)							
	C2–C3	1.373	1.375	0.012	1.368	1.380	30	
	C3–C4	1.379	1.380	0.011	1.371	1.388	30	
	(N⁺–X; C,H subst. on C)							
	C2–C3	1.373	1.372	0.019	1.362	1.382	151	
	C3–C4	1.383	1.385	0.019	1.372	1.394	151	

BOND LENGTHS IN CRYSTALLINE ORGANIC COMPOUNDS (continued)

Bond	Substructure	d	m	σ	q_1	q_u	n	Note
	in pyrazine (H subst. on C)	1.379	1.377	0.010	1.370	1.388	10	
	(any subst. on C)	1.405	1.405	0.024	1.388	1.420	60	
	in pyrimidine (C,H subst. on C)	1.387	1.389	0.018	1.379	1.400	28	
$Csp^1\equiv Csp^1$	X–C≡C–X	1.183	1.183	0.014	1.174	1.193	119	15
	C,H–C≡C–C,H	1.181	1.181	0.014	1.173	1.192	104	15
	in C≡C–C(sp^2,ar)	1.189	1.193	0.010	1.181	1.195	38	15
	in C≡C–C≡C	1.192	1.192	0.010	1.187	1.197	42	15
	in CH≡C–C#	1.174	1.174	0.011	1.167	1.180	42	15
Csp^3–Cl	Omitting 1,2-dichlorides:							
	C–CH$_2$–Cl	1.790	1.790	0.007	1.783	1.795	13	4
	C$_2$–CH–Cl	1.803	1.802	0.003	1.800	1.807	8	4
	C$_3$–C–Cl	1.849	1.856	0.011	1.837	1.858	5	4
	X–CH$_2$–Cl (X = C,H,N,O)	1.790	1.791	0.011	1.783	1.797	37	4
	X$_2$–CH–Cl (X = C,H,N,O)	1.805	1.803	0.014	1.800	1.812	26	4
	X$_3$–C–Cl (X = C,H,N,O)	1.843	1.838	0.014	1.835	1.858	7	4
	X$_2$–C–Cl$_2$ (X = C,H,N,O)	1.779	1.776	0.015	1.769	1.790	18	4
	X–C–Cl$_3$ (X = C,H,N,O)	1.768	1.765	0.011	1.761	1.776	33	4
	Cl–CH(–C)–CH(–C)–Cl	1.793	1.793	0.013	1.786	1.800	66	4
	Cl–C(–C$_2$)–C(–C$_2$)–Cl	1.762	1.760	0.010	1.757	1.765	54	4
	cyclopropyl–Cl	1.755	1.756	0.011	1.749	1.763	64	
Csp^2–Cl	C=C–Cl (C,H,N,O subst. on C)	1.734	1.729	0.019	1.719	1.748	63	4
	C=C–Cl$_2$ (C,H,N,O subst. on C)	1.720	1.716	0.013	1.708	1.729	20	4
	Cl–C=C–Cl	1.713	1.711	0.011	1.705	1.720	80	4
Car–Cl	Car–Cl (mono-Cl + m,p-Cl$_2$)	1.739	1.741	0.010	1.734	1.745	340	4
	Car–Cl (o-Cl$_2$)	1.720	1.720	0.010	1.713	1.717	364	4
Csp^1–Cl	see HCLENE10 (1.634, 1.646)							
Csp^3–F	Omitting 1,2-difluorides							
	C-CH$_2$–F and C$_2$-CH–F	1.399	1.399	0.017	1.389	1.408	25	4
	C$_3$–C–F	1.428	1.431	0.009	1.421	1.435	11	4
	(C*,H)$_2$–C–F$_2$	1.349	1.347	0.012	1.342	1.356	58	4
	C*–C–F$_3$	1.336	1.334	0.007	1.330	1.344	12	4
	F–C*–C*–F	1.371	1.374	0.007	1.362	1.375	26	4
	X$_3$–C–F (X = C,H,N,O)	1.386	1.389	0.033	1.373	1.408	70	4
	X$_2$–C–F$_2$ (X = C,H,N,O)	1.351	1.349	0.013	1.342	1.356	58	4
	X–C–F$_3$ (X = C,H,N,O)	1.322	1.323	0.015	1.314	1.332	309	4
	F–C(–X)$_2$–C(–X)$_2$–F (X = C,H,N,O)	1.373	1.374	0.009	1.362	1.377	30	4
	F–C(–X)$_2$–NO$_2$ (X = any subst.)	1.320	1.319	0.009	1.312	1.327	18	
Csp^2–F	C=C–F (C,H,N,O subst. on C)	1.340	1.340	0.013	1.334	1.346	34	4
Car–F	Car–F (mono-F + m,p-F$_2$)	1.363	1.362	0.008	1.357	1.368	38	4
	Car–F (o-F$_2$)	1.340	1.340	0.009	1.336	1.344	167	4
Csp^3–H	C–C–H$_3$ (methyl)	1.059	1.061	0.030	1.039	1.083	83	21
	C$_2$–C–H$_2$ (primary)	1.092	1.095	0.013	1.088	1.099	100	21
	C$_3$–C–H (secondary)	1.099	1.097	0.004	1.095	1.103	14	21
	C$_{2,3}$–C–H (primary and secondary)	1.093	1.095	0.012	1.089	1.100	118	21
	X–C–H$_3$ (methyl)	1.066	1.074	0.028	1.049	1.087	160	21
	X$_2$–C–H$_2$ (primary)	1.092	1.095	0.012	1.088	1.099	230	21
	X$_3$–C–H (secondary)	1.099	1.099	0.007	1.095	1.103	117	21
	X$_{2,3}$–C–H (primary and secondary)	1.094	1.096	0.011	1.091	1.100	348	21
Csp^2–H	C–C=C–H	1.077	1.079	0.012	1.074	1.085	14	21
Car–H	Car–H	1.083	1.083	0.011	1.080	1.087	218	21
Csp^3–I	C*–I	2.162	2.159	0.015	2.149	2.179	15	4
Car–I	Car–I	2.095	2.095	0.015	2.089	2.104	51	4
Csp^3–N(4)	C*–NH_3^+	1.488	1.488	0.013	1.482	1.495	298	
	(C*)$_2$–NH$_2^+$	1.494	1.493	0.016	1.484	1.503	249	
	(C*)$_3$–NH$^+$	1.502	1.502	0.015	1.491	1.512	509	
	(C*)$_4$–N$^+$	1.510	1.509	0.020	1.496	1.523	319	
	C*–N$^+$ (overall)	1.499	1.498	0.018	1.488	1.510	1 370	
Csp^3–N(3)	C*–N$^+$ in N-subst. pyridinium	1.485	1.484	0.009	1.477	1.490	32	
	C*–NH$_2$ (Nsp^3: pyramidal)	1.469	1.470	0.010	1.462	1.474	19	22
	(C*)$_2$–NH (Nsp^3: pyramidal)	1.469	1.467	0.012	1.461	1.477	152	5,22
	(C*)$_3$–N (Nsp^3: pyramidal)	1.469	1.468	0.014	1.460	1.476	1 042	5,22
	C*–Nsp^3 (overall)	1.469	1.468	0.014	1.460	1.476	1 201	
	Csp^3–Nsp^3							
	in aziridine	1.472	1.471	0.016	1.464	1.482	134	
	in azetidine	1.484	1.481	0.018	1.472	1.495	21	
	in tetrahydropyrrole	1.475	1.473	0.016	1.464	1.483	66	
	in piperidine	1.473	1.473	0.013	1.460	1.479	240	
	Csp^3–Nsp^2 (N planar) in:							23
	acyclic amides C*–NH–C=O	1.454	1.451	0.011	1.446	1.461	78	14
	β-lactams C*–N(–X)–C=O (endo)	1.464	1.465	0.012	1.458	1.475	23	13
	γ-lactams							
	C*–NH–C=O (endo)	1.457	1.458	0.011	1.449	1.465	20	13
	C*–N(–C*)–C=O (endo)	1.462	1.461	0.010	1.453	1.466	15	13
	C*–N(–C*)–C=O (exo)	1.458	1.456	0.014	1.448	1.465	15	13

Bond	Substructure	d	m	σ	q_1	q_u	n	Note
	δ-lactams							
	C*–NH–C=O (endo)	1.478	1.472	0.016	1.467	1.491	6	14
	C*–N(–C*)–C=O (endo)	1.479	1.476	0.007	1.475	1.482	15	14
	C*–N(–C*)–C=O (exo)	1.468	1.471	0.009	1.462	1.477	15	14
	nitro compounds (1,2-dinitro omitted):							
	C–CH₂–NO₂	1.485	1.483	0.020	1.478	1.502	8	
	C₂–CH–NO₂	1.509	1.509	0.011	1.502	1.511	12	
	C₃–C–NO₂	1.533	1.533	0.013	1.530	1.539	17	
	C₂–C–(NO₂)₂	1.537	1.536	0.016	1.525	1.550	19	
	1,2-dinitro: NO₂–C*–C*–NO₂	1.552	1.550	0.023	1.536	1.572	32	
Csp^3–N(2)	C#–N=N	1.493	1.493	0.020	1.477	1.506	54	
	C*–N=C–Car	1.465	1.468	0.011	1.461	1.472	75	
Csp^2–N(3)	C=C–NH₂ Nsp^2 planar	1.336	1.344	0.017	1.317	1.348	10	23
	C=C–NH–C# Nsp^2 planar	1.339	1.340	0.016	1.327	1.351	17	23
	C=C–N–(C#)₂							
	Nsp^2 planar	1.355	1.358	0.014	1.341	1.363	22	23
	Nsp^3 pyramidal	1.416	1.418	0.018	1.397	1.432	18	22
	Csp^2–Nsp^2 (N planar) in:							23
	acyclic amides							
	NH₂–C=O	1.325	1.323	0.009	1.318	1.331	32	14
	C*–NH–C=O	1.334	1.333	0.011	1.326	1.343	78	14
	(C*)₂–N–C=O	1.346	1.342	0.011	1.339	1.356	5	14
	β-lactams C*–NH–C=O	1.385	1.388	0.019	1.374	1.396	23	13
	γ-lactams							
	C*–NH–C=O	1.331	1.331	0.011	1.326	1.337	20	13
	C*–N(–C*)–C=O	1.347	1.344	0.014	1.335	1.359	15	13
	δ-lactams							
	C*–NH–C=O	1.334	1.334	0.006	1.330	1.339	6	14
	C*–N(–C*)–C=O	1.352	1.353	0.010	1.344	1.356	15	14
	peptides C#–N(–X)–C(–C#)(=O)	1.333	1.334	0.013	1.326	1.340	380	24
	ureas							
	(NH₂)₂–C=O	1.334	1.334	0.008	1.329	1.339	48	25,26
	(C#–NH)₂–C=O	1.347	1.345	0.010	1.341	1.354	26	25
	[(C#)ₙ–N]₂–C=O	1.363	1.359	0.014	1.354	1.370	40	25,27
	thioureas							
	(X₂N)₂–C=S	1.346	1.343	0.023	1.328	1.361	192	
	imides							
	[C#–C(=O)]₂–NH	1.376	1.377	0.012	1.369	1.383	64	
	[C#–C(=O)]₂–N–C#	1.389	1.383	0.017	1.376	1.404	38	
	[Csp^2–C(=O)]₂–N–C#	1.396	1.396	0.010	1.389	1.403	46	
	[Csp^2–C(=O)]₂–N–Csp^2	1.409	1.406	0.020	1.391	1.419	28	
	guanidinium [C–(NH₂)₃]⁺ (unsubst.)	1.321	1.320	0.008	1.314	1.327	39	
	(any subst.)	1.328	1.325	0.015	1.317	1.333	140	
	in heterocyclic systems (any subst.)							
	1H-pyrrole (N1–C2, N1–C5)	1.372	1.374	0.016	1.363	1.384	58	
	indole (N1–C2)	1.370	1.370	0.012	1.364	1.377	40	
	pyrazole (N1–C5)	1.357	1.359	0.012	1.347	1.365	20	
	imidazole (N1–C2)	1.349	1.349	0.018	1.338	1.358	44	
	imidazole (N1–C5)	1.370	1.370	0.010	1.365	1.377	44	
Csp^2–N(2)	in imidazole (N3–C4)	1.376	1.377	0.011	1.369	1.384	44	
Car–N(4)	Car–N⁺–(C,H)₃	1.465	1.466	0.007	1.461	1.470	23	
Car–N(3)	Car–NH₂							
	(Nsp^2: planar)	1.355	1.360	0.020	1.340	1.372	33	23
	(Nsp^3: pyramidal)	1.394	1.396	0.011	1.385	1.403	25	22
	(overall)	1.375	1.377	0.025	1.363	1.394	98	28
Car–N(3)	Car–NH–C#							
	(Nsp^2: planar)	1.353	1.353	0.007	1.347	1.359	16	23
	(Nsp^3: pyramidal)	1.419	1.423	0.017	1.412	1.432	8	22
	(overall)	1.380	1.364	0.032	1.353	1.412	31	28
	Car–N–(C#)₂							
	(Nsp^2: planar)	1.371	1.370	0.016	1.363	1.382	41	23
	(Nsp^3: pyramidal)	1.426	1.425	0.011	1.421	1.431	22	22
	(overall)	1.390	1.385	0.030	1.366	1.420	69	28
	in indole (N1–C7a)	1.372	1.372	0.007	1.367	1.376	40	
	Car–NO₂	1.468	1.469	0.014	1.460	1.476	556	
Car–N(2)	Car–N=N	1.431	1.435	0.020	1.422	1.442	26	
Csp^2=N(3)	in furoxan (⁺N2=C3)	1.316	1.316	0.009	1.311	1.324	14	
Csp^2=N(2)	Car–C=N–C#	1.279	1.279	0.008	1.275	1.285	75	
	(C,H)₂–C=N–OH in oximes	1.281	1.280	0.013	1.273	1.288	67	
	S–C=N–X	1.302	1.302	0.021	1.285	1.319	36	
	in pyrazole (N2=C3)	1.329	1.331	0.014	1.315	1.339	20	
	in imidazole (C2=N3)	1.313	1.314	0.011	1.307	1.319	44	
	in isoxazole (N2=C3)	1.314	1.315	0.009	1.305	1.320	9	

BOND LENGTHS IN CRYSTALLINE ORGANIC COMPOUNDS (continued)

Bond	Substructure	d	m	σ	q_1	q_u	n	Note
	in furazan (N2=C3, C4=N5)	1.298	1.299	0.006	1.294	1.303	12	
	in furoxan (C4=N5)	1.304	1.306	0.008	1.300	1.308	14	
$Car \simeq N(3)$	$C \simeq N^+ - H$ (pyrimidinium)	1.335	1.334	0.015	1.325	1.342	30	
	$C \simeq N^+ - C^*$ (pyrimidinium)	1.346	1.346	0.010	1.340	1.352	64	
	$C \simeq N^+ - O^-$ (pyrimidinium)	1.362	1.359	0.013	1.353	1.369	56	
$Car \simeq N(2)$	$C \simeq N$ (pyridine)	1.337	1.338	0.012	1.330	1.344	269	
	$C \simeq N$ (pyrazine)	1.336	1.335	0.022	1.319	1.347	120	
	$C \simeq N \simeq C$ (pyrimidine)	1.339	1.338	0.015	1.333	1.342	28	
	$N \simeq C \simeq N$ (pyrimidine)	1.333	1.335	0.013	1.326	1.337	28	
	$C \simeq N$ (pyrimidine) (overall)	1.336	1.337	0.014	1.331	1.339	56	
	in any 6-membered N-containing aromatic ring:							
	$H-C \simeq N \simeq C-H$	1.334	1.334	0.014	1.327	1.341	146	
	$H-C \simeq N \simeq C-C^*$	1.339	1.341	0.013	1.336	1.345	38	
	$C^*-C \simeq N \simeq C-C^*$	1.345	1.345	0.008	1.342	1.348	24	
	$C \simeq N \simeq C$ (overall)	1.336	1.337	0.014	1.329	1.344	204	
$Csp^1 \equiv N(2)$	$X-S-N \equiv C^-$ (isothiocyanide)	1.144	1.147	0.006	1.140	1.148	6	
$Csp^1 \equiv N(1)$	$C^*-C \equiv N$	1.136	1.137	0.010	1.131	1.142	140	
	$C=C-C \equiv N$ in TCNQ	1.144	1.144	0.008	1.139	1.149	284	19
	$Car-C \equiv N$	1.138	1.138	0.007	1.133	1.143	31	
	$X-C \equiv N$	1.144	1.141	0.012	1.138	1.151	10	
	$(S-C \equiv N)^-$	1.155	1.156	0.012	1.147	1.165	14	
$Csp^3 - O(2)$	in alcohols							
	$CH_3 - OH$	1.413	1.414	0.018	1.395	1.425	17	
	$C-CH_2 - OH$	1.426	1.426	0.011	1.420	1.431	75	
	$C_2 - CH - OH$	1.432	1.431	0.011	1.425	1.439	266	
	$C_3 - C - OH$	1.440	1.440	0.012	1.432	1.449	106	
	$C^* - OH$ (overall)	1.432	1.431	0.013	1.424	1.441	464	
	in dialkyl ethers							29
	$CH_3 - O - C^*$	1.416	1.418	0.016	1.405	1.426	110	
	$C-CH_2 - O - C^*$	1.426	1.424	0.011	1.418	1.435	34	
	$C_2 - CH - O - C^*$	1.429	1.430	0.010	1.420	1.437	53	
	$C_3 - C - O - C^*$	1.452	1.450	0.011	1.445	1.458	39	
	$C^* - O - C^*$ (overall)	1.426	1.425	0.019	1.414	1.437	236	5
	in aryl alkyl ethers							29
	$CH_3 - O - Car$	1.424	1.424	0.012	1.417	1.431	616	
	$C-CH_2 - O - Car$	1.431	1.430	0.013	1.422	1.438	188	
	$C_2 - CH - O - Car$	1.447	1.446	0.020	1.435	1.466	58	
	$C_3 - C - O - Car$	1.470	1.469	0.018	1.456	1.483	55	
	$C^* - O - Car$ (overall)	1.429	1.427	0.018	1.419	1.436	917	
	in alkyl esters of carboxylic acids							12,29
	$CH_3 - O - C(=O) - C^*$	1.448	1.449	0.010	1.442	1.455	200	
	$C-CH_2 - O - C(=O) - C^*$	1.452	1.453	0.009	1.445	1.458	32	
	$C_2 - CH - O - C(=O) - C^*$	1.460	1.460	0.010	1.454	1.465	78	
	$C_3 - C - O - C(=O) - C^*$	1.477	1.475	0.008	1.472	1.484	6	
	$C^* - O - C(=O) - C^*$ (overall)	1.450	1.451	0.014	1.442	1.459	314	
	in alkyl esters of α,β-unsaturated acids:							
	$C^* - O - C(=O) - C=C$ (overall)	1.453	1.452	0.013	1.444	1.459	112	
	in alkyl esters of benzoic acid							
	$C^* - O - C(=O) - C$(phenyl) (overall)	1.454	1.454	0.012	1.446	1.463	219	
	in ring systems							
	oxirane (epoxides) (any subst.)	1.446	1.446	0.014	1.438	1.456	498	9
	oxetane (any subst.)	1.463	1.460	0.015	1.451	1.474	16	
	tetrahydrofuran (C,H subst.)	1.442	1.441	0.017	1.430	1.451	154	
$Csp^3 - O(2)$	tetrahydropyran (C,H subst.)	1.441	1.442	0.015	1.431	1.451	22	
	β-lactones: $C^* - O - C(=O)$	1.492	1.494	0.010	1.481	1.501	4	16
	γ-lactones: $C^* - O - C(=O)$	1.464	1.464	0.012	1.455	1.473	110	12
	δ-lactones: $C^* - O - C(=O)$	1.461	1.464	0.017	1.452	1.473	27	12
	O–C–O system in gem-diols, and pyranose and furanose sugars:							30,31
	$HO-C^*-OH$	1.397	1.401	0.012	1.388	1.405	18	
	$C_5 - O_5 - C_1 - O_1 H$ in pyranoses							
	O_1 axial (α):							
	$C_5 - O_5$	1.439	1.440	0.008	1.432	1.445	29	
	$O_5 - C_1$	1.427	1.426	0.012	1.421	1.432	29	
	$C_1 - O_1$	1.403	1.400	0.012	1.391	1.412	29	
	O_1 equatorial (β):							
	$C_5 - O_5$	1.435	1.436	0.008	1.429	1.440	17	
	$O_5 - C_1$	1.430	1.431	0.010	1.424	1.436	17	
	$C_1 - O_1$	1.393	1.393	0.007	1.386	1.399	17	
	$\alpha + \beta$ (overall):							
	$C_5 - O_5$	1.439	1.440	0.008	1.432	1.446	60	
	$O_5 - C_1$	1.430	1.429	0.012	1.421	1.436	60	
	$C_1 - O_1$	1.401	1.399	0.011	1.392	1.407	60	

Bond	Substructure	d	m	σ	q_1	q_u	n	Note
	$C_4-O_4-C_1-O_1H$ in furanoses							
	(overall values)							
	C_4-O_4	1.442	1.446	0.012	1.436	1.449	18	
	O_4-C_1	1.432	1.432	0.012	1.421	1.443	18	
	C_1-O_1	1.404	1.405	0.013	1.397	1.409	18	
	$C_5-O_5-C_1-O_1-C^*$ in pyranoses							
	O_1 axial (α):							
	C_5-O_5	1.439	1.438	0.010	1.433	1.446	67	
	O_5-C_1	1.417	1.417	0.009	1.410	1.424	67	
	C_1-O_1	1.409	1.409	0.014	1.401	1.417	67	
	O_1-C^*	1.435	1.435	0.013	1.427	1.443	67	
	O_1 equatorial (β):							
	C_5-O_5	1.434	1.435	0.006	1.429	1.439	39	
	O_5-C_1	1.424	1.424	0.008	1.418	1.431	39	
	C_1-O_1	1.390	1.390	0.011	1.381	1.400	39	
	O_1-C^*	1.437	1.438	0.013	1.428	1.445	39	
	$\alpha + \beta$ (overall):							
	C_5-O_5	1.436	1.436	0.009	1.431	1.442	126	
	O_5-C_1	1.419	1.419	0.011	1.412	1.426	126	
	C_1-O_1	1.402	1.403	0.016	1.391	1.413	126	
	O_1-C^*	1.436	1.436	0.013	1.428	1.445	126	
	$C_4-O_4-C_1-O_1-C^*$ in furanoses							
	(overall values)							
	C_4-O_4	1.443	1.445	0.013	1.429	1.453	23	
	O_4-C_1	1.421	1.418	0.012	1.413	1.431	23	
	C_1-O_1	1.410	1.409	0.014	1.401	1.420	23	
	O_1-C^*	1.439	1.437	0.014	1.429	1.449	23	
	Miscellaneous:							
	$C\#-O-SiX_3$	1.416	1.416	0.017	1.405	1.428	29	
	C^*-O-SO_2-C	1.465	1.461	0.014	1.454	1.475	33	
$Csp^2-O(2)$	in enols: $C=C-OH$	1.333	1.331	0.017	1.324	1.342	53	
	in enol esters: $C=C-O-C^*$	1.354	1.353	0.016	1.341	1.363	40	
	in acids:							
	$C^*-C(=O)-OH$	1.308	1.311	0.019	1.298	1.320	174	
	$C=C-C(=O)-OH$	1.293	1.295	0.019	1.279	1.307	22	
	$Car-C(=O)-OH$	1.305	1.311	0.020	1.291	1.317	75	
	in esters:							
	$C^*-C(=O)-O-C^*$	1.336	1.337	0.014	1.328	1.346	551	12,29
	$C=C-C(=O)-O-C^*$	1.332	1.331	0.011	1.324	1.339	112	
	$Car-C(=O)-O-C^*$	1.337	1.335	0.013	1.329	1.344	219	12
	$C^*-C(=O)-O-C=C$	1.362	1.359	0.018	1.351	1.374	26	
	$C^*-C(=O)-O-C=C$	1.407	1.405	0.017	1.394	1.420	26	
	$C^*-C(=O)-O-Car$	1.360	1.359	0.011	1.355	1.367	40	12
	in anhydrides: $O=C-O-C=O$	1.386	1.386	0.011	1.379	1.393	70	
	in ring systems:							
	furan (O1–C2, O1–C5)	1.368	1.369	0.015	1.359	1.377	125	
	isoxazole (O1–C5)	1.354	1.354	0.010	1.345	1.360	9	
	β-lactones: $C^*-C(=O)-O-C^*$	1.359	1.359	0.013	1.348	1.371	4	13
	γ-lactones: $C^*-C(=O)-O-C^*$	1.350	1.349	0.012	1.342	1.359	110	12
	δ-lactones: $C^*-C(=O)-O-C^*$	1.339	1.339	0.016	1.332	1.347	27	12
$Car-O(2)$	in phenols: $Car-OH$	1.362	1.364	0.015	1.353	1.373	551	
	in aryl alkyl ethers: $Car-O-C^*$	1.370	1.370	0.011	1.363	1.377	920	29,32
$Car-O(2)$	in diaryl ethers: $Car-O-Car$	1.384	1.381	0.014	1.375	1.391	132	
	in esters: $Car-O-C(=O)-C^*$	1.401	1.401	0.010	1.394	1.408	40	12
$Csp^2=O(1)$	in aldehydes and ketones:							
	$C^*-CH=O$	1.192	1.192	0.005	1.188	1.197	7	
	$(C^*)_2-C=O$	1.210	1.210	0.008	1.206	1.215	474	5
	$(C\#)_2-C=O$							
	in cyclobutanones	1.198	1.198	0.007	1.194	1.204	12	
	in cyclopentanones	1.208	1.208	0.007	1.203	1.212	155	
	in cyclohexanones	1.211	1.211	0.009	1.207	1.216	312	
	$C=C-C=O$	1.222	1.222	0.010	1.216	1.229	225	
	$(C=C)_2-C=O$	1.233	1.229	0.010	1.226	1.242	28	
	$Car-C=O$	1.221	1.218	0.014	1.212	1.229	85	
	$(Car)_2-C=O$	1.230	1.226	0.015	1.220	1.238	66	
	$C=O$ in benzoquinones	1.222	1.220	0.013	1.211	1.231	86	
	delocalized double bonds in carboxylate anions:							
	$H-C\simeq O_2^-$ (formate)	1.242	1.243	0.012	1.234	1.252	24	
	$C^*-C\simeq O_2^-$	1.254	1.253	0.010	1.247	1.261	114	
	$C=C-C\simeq O_2^-$	1.250	1.248	0.017	1.238	1.261	52	
	$Car-C\simeq O_2^-$	1.255	1.253	0.010	1.249	1.262	22	
	$HOOC-C\simeq O_2^-$ (hydrogen oxalate)	1.243	1.247	0.015	1.232	1.256	26	
	$^-O_2\simeq C-C\simeq O_2-$ (oxalate)	1.251	1.251	0.007	1.248	1.254	18	

Bond	Substructure	d	m	σ	q_1	q_u	n	Note
	in carboxylic acids (X–COOH)							
	C*–C(=O)–OH	1.214	1.214	0.019	1.203	1.224	175	
	C=C–C(=O)–OH	1.229	1.226	0.017	1.218	1.237	22	
	Car–C(=O)–OH	1.226	1.223	0.020	1.211	1.241	75	
	in esters:							
	C*–C(=O)–O–C*	1.196	1.196	0.010	1.190	1.202	551	12
	C=C–C(=O)–O–C*	1.199	1.198	0.009	1.193	1.203	113	
	Car–C(=O)–O–C*	1.202	1.201	0.009	1.196	1.207	218	12
	C*–C(=O)–O–C=C	1.190	1.190	0.014	1.184	1.198	26	
	C*–C(=O)–O–Car	1.187	1.188	0.011	1.181	1.195	40	12
	in anhydrides: O=C–O–C=O	1.187	1.187	0.010	1.184	1.193	70	
	in β-lactones: C*–C(=O)–O–C*	1.193	1.193	0.006	1.187	1.198	4	13
	γ-lactones: C*–C(=O)–O–C*	1.201	1.202	0.009	1.196	1.206	109	12
	δ-lactones: C*–C(=O)–O–C*	1.205	1.207	0.008	1.201	1.209	27	12
	in amides:							
	NH$_2$–C(–C*)=O	1.234	1.233	0.012	1.225	1.243	32	14
	(C*–)(C*,H–)N–C(–C*)=O	1.231	1.231	0.012	1.224	1.238	378	14
	β-lactams: C*–NH–C=O	1.198	1.200	0.012	1.193	1.204	23	13
	γ-lactams:							
	C*–NH–C=O	1.235	1.235	0.008	1.232	1.240	20	13
	C*–N(–C*)–C=O	1.225	1.226	0.011	1.217	1.233	15	13
	δ-lactams:							
	C*–NH–C=O	1.240	1.241	0.003	1.237	1.243	6	14
	C*–N(–C*)–C=O	1.233	1.233	0.007	1.229	1.239	15	14
	in ureas:							
	(NH)$_2$)$_2$–C=O	1.256	1.256	0.007	1.249	1.261	24	25,26
	(C#–NH)$_2$–C=O	1.241	1.237	0.011	1.235	1.245	13	25
	[(C#)$_n$–N]$_2$–C=O	1.230	1.230	0.007	1.224	1.234	20	25,27
Csp^3–P(4)	C$_3$–P$^+$–C*	1.800	1.802	0.015	1.790	1.812	35	33
	C$_2$–P(=O)–CH$_3$	1.791	1.790	0.006	1.786	1.795	10	
	C$_2$–P(=O)–CH$_2$–C	1.806	1.806	0.009	1.801	1.813	45	
	C$_2$–P(=O)–CH–C$_2$	1.821	1.821	0.009	1.815	1.828	15	
	C$_2$–P(=O)–C–C$_3$	1.841	1.842	0.008	1.835	1.847	14	
	C$_2$–P(=O)–C* (overall)	1.813	1.811	0.017	1.800	1.822	84	
Csp^3–P(3)	C$_2$–P–C*	1.855	1.857	0.019	1.840	1.870	23	
Car–P(4)	C$_3$–P$^+$–Car	1.793	1.792	0.011	1.786	1.800	276	
	C$_2$–P(=O)–Car	1.801	1.802	0.011	1.796	1.807	98	
	Ph$_3$–P=N$^+$=P–Ph$_3$	1.795	1.795	0.008	1.789	1.800	197	
Car–P(3)	C$_2$–P–Car	1.836	1.837	0.010	1.830	1.844	102	
	(N$\simeq$)$_2$P-Car (P$\simeq$N aromatic)	1.795	1.793	0.011	1.788	1.803	43	
Csp^3–S(4)	C*–SO$_2$–C (C* = CH$_3$ excluded)	1.786	1.782	0.018	1.774	1.797	75	
	C*–SO$_2$–C (overall)	1.779	1.778	0.020	1.764	1.790	94	
	C*–SO$_2$–O–X	1.745	1.744	0.009	1.738	1.754	7	34
	C*–SO$_2$–N–X$_2$	1.758	1.756	0.018	1.746	1.773	17	34
Csp^3–S(3)	C*–S(=O)–C (C* = CH$_3$ excluded)	1.818	1.814	0.024	1.802	1.829	69	
	C*–S(=O)–C (overall)	1.809	1.806	0.025	1.793	1.820	88	
	CH$_3$–S$^+$–X$_2$	1.786	1.787	0.007	1.779	1.792	21	
	C*–S$^+$–X$_2$ (C* = CH$_3$ excluded)	1.823	1.820	0.016	1.812	1.834	18	
	C*–S$^+$–X$_2$ (overall)	1.804	1.794	0.025	1.788	1.820	41	
Csp^3–S(2)	C*–SH	1.808	1.805	0.010	1.800	1.819	6	
	CH$_3$–S–C*	1.789	1.787	0.008	1.784	1.794	9	
Csp^3–S(2)	C–CH$_2$–S–C*	1.817	1.816	0.013	1.808	1.824	92	
	C$_2$–CH–S–C*	1.819	1.819	0.011	1.811	1.825	32	
	C$_3$–C–S–C*	1.856	1.860	0.011	1.854	1.863	26	
	C*–S–C* (overall)	1.819	1.817	0.019	1.809	1.827	242	
	in thiirane	1.834	1.835	0.025	1.810	1.858	4	9
	in thietane: see ZCMXSP (1.817, 1.844)							
	in tetrahydrothiophene	1.827	1.826	0.018	1.811	1.837	20	
	in tetrahydrothiopyran	1.823	1.821	0.014	1.812	1.832	24	
	C–CH$_2$–S–S–X	1.823	1.820	0.014	1.813	1.832	41	
	C$_3$–C–S–S–X	1.863	1.865	0.015	1.848	1.878	11	
	C*–S–S–X (overall)	1.833	1.828	0.022	1.818	1.848	59	
Csp^2–S(2)	C=C–S–C*	1.751	1.755	0.017	1.740	1.764	61	
	C=C–S–C=C (in tetrathiafulvalene)	1.741	1.741	0.011	1.733	1.750	88	
	C=C–S–C=C (in thiophene)	1.712	1.712	0.013	1.703	1.722	60	
	O=C–S–C#	1.762	1.759	0.018	1.747	1.778	20	
Car–S(4)	Car–SO$_2$–C	1.763	1.764	0.009	1.756	1.769	96	
	Car–SO$_2$–O–X	1.752	1.750	0.008	1.749	1.756	27	
	Car–SO$_2$–N–X$_2$	1.758	1.759	0.013	1.749	1.765	106	35
Car–S(3)	Car–S(=O)–C	1.790	1.790	0.010	1.783	1.798	41	
	Car–S$^+$–X$_2$	1.778	1.779	0.010	1.771	1.787	10	
Car–S(2)	Car–S–C*	1.773	1.774	0.009	1.765	1.779	44	
	Car–S–Car	1.768	1.767	0.010	1.762	1.774	158	
	Car–S–Car (in phenothiazine)	1.764	1.764	0.008	1.760	1.769	48	

Bond	Substructure	d	m	σ	q_1	q_u	n	Note
	Car–S–S–X	1.777	1.777	0.012	1.767	1.785	47	
Csp^1–S(2)	N≡C–S–X	1.679	1.683	0.026	1.645	1.698	10	
Csp^1–S(1)	(N≡C–S)⁻	1.630	1.630	0.014	1.619	1.641	14	
Csp^2=S(1)	(C*)₂–C=S: see IPMUDS (1.599)							
	(Car)₂–C=S: see CELDOM (1.611)							
	(X)₂–C=S (X = C,N,O,S)	1.671	1.675	0.024	1.656	1.689	245	
	X₂N–C(=S)–S–X	1.660	1.660	0.016	1.648	1.674	38	
	(X₂N)₂–C=S (thioureas)	1.681	1.684	0.020	1.669	1.693	96	
	N–C(≃S)₂	1.720	1.721	0.012	1.709	1.731	20	
Csp^3–Se	C#–Se	1.970	1.967	0.032	1.948	1.998	21	
Csp^2–Se(2)	C=C–Se–C=C (in tetraselenafulvalene)	1.893	1.895	0.013	1.882	1.902	32	
Car–Se(3)	Ph₃–Se⁺	1.930	1.929	0.006	1.924	1.936	13	
Csp^3–Si(5)	C#–Si⁻–X₄	1.874	1.876	0.015	1.859	1.884	9	
Csp^3–Si(4)	CH₃–Si–X₃	1.857	1.857	0.018	1.848	1.869	552	
	C*–Si–X₃ (C* = CH₃ excluded)	1.888	1.887	0.023	1.872	1.905	124	
	C*–Si–X₃ (overall)	1.863	1.861	0.024	1.850	1.875	681	
Car–Si(4)	Car–Si–X₃	1.868	1.868	0.014	1.857	1.878	178	
Csp^1–Si(4)	C≡C–Si–X₃	1.837	1.840	0.012	1.824	1.849	8	
Csp^3–Te	C#–Te	2.158	2.159	0.030	2.128	2.177	13	
Car–Te	Car–Te	2.116	2.115	0.020	2.104	2.130	72	
Csp^2=Te	see CEDCUJ (2.044)							
Cl–Cl	see PHASCL (2.306, 2.227)							
Cl–I	see CMBIDZ (2.563), HXPASC (2.541, 2.513), METAMM (2.552), BQUINI (2.416, 2.718)							
Cl–N	see BECTAE (1.743—1.757), BOGPOC (1.705)							
Cl–O(1)	in ClO₄⁻	1.414	1.419	0.026	1.403	1.431	252	
Cl–P	(N≃)₂P–Cl (N ≃ P aromatic)	1.997	1.994	0.015	1.989	2.004	46	
	Cl–P (overall)	2.008	2.001	0.035	1.986	2.028	111	
Cl–S	Cl–S (overall)	2.072	2.079	0.023	2.047	2.091	6	
	see also longer bonds in CILSAR (2.283), BIHXIZ (2.357), CANLUY (2.749)							
Cl–Se	see BIRGUE10, BIRHAL10, CTCNSE (2.234—2.851)							
Cl–Si(4)	Cl–Si–X₃ (monochloro)	2.072	2.075	0.009	2.066	2.078	5	
	Cl₂–Si–X₂ and Cl₃–Si–X	2.020	2.012	0.015	2.007	2.036	5	
Cl–Te	Cl–Te in range 2.34—2.60	2.520	2.515	0.034	2.493	2.537	22	36
	see also longer bonds in BARRIV, BOJPUL, CETUTE, EPHTEA, OPNTEC10 (2.73—2.94)							
F–N(3)	F–N–C₂ and F₂–N–C	1.406	1.404	0.016	1.395	1.416	9	
F–P(6)	in hexafluorophosphate, PF₆⁻	1.579	1.587	0.025	1.563	1.598	72	
F–P(3)	(N≃)₂P–F (N ≃ P aromatic)	1.495	1.497	0.016	1.481	1.510	10	
F–S	43 observations in range 1.409—1.770 in a wide variety of environments; F–S(6) in F₂–SO₂–C₂ (see FPSULF10, BETJOZ)	1.640	1.646	0.011	1.626	1.649	6	
	F–S(4) in F₂–S(=O)–N (see BUDTEZ)	1.527	1.528	0.004	1.524	1.530	24	37
F–Si(6)	in SiF₆²⁻	1.694	1.701	0.013	1.677	1.703	6	
F–Si(5)	F–Si⁻–X₄	1.636	1.639	0.035	1.602	1.657	10	
F–Si(4)	F–Si–X₃	1.588	1.587	0.014	1.581	1.599	24	
F–Te	see CUCPIZ (F–Te(6) = 1.942, 1.937), FPHTEL (F–Te(4) = 2.006)							
H–N(4)	X₃–N⁺–H	1.033	1.036	0.022	1.026	1.045	87	21
H–N(3)	X₂–N–H	1.009	1.010	0.019	0.997	1.023	95	21
H–O(2)	in alcohols C*–O–H	0.967	0.969	0.010	0.959	0.974	63	21
	C#–O–H	0.967	0.970	0.010	0.959	0.974	73	21
	in acids O=C–O–H	1.015	1.017	0.017	1.001	1.031	16	21,38
I–I	in I₃⁻	2.917	2.918	0.011	2.907	2.927	6	
I–N	see BZPRIB, CMBIDZ, HMTITI, HMTNTI, IFORAM, IODMAM (2.042—2.475)							
I–O	X–I–O (see BZPRIB, CAJMAB, IBZDAC11)	2.144	2.144	0.028	2.127	2.164	6	
	for IO₆⁻ see BOVMEE (1.829—1.912)							
I–P(3)	see CEHKAB (2.490—2.493)							†
I–S	see DTHIBR10 (2.687), ISUREA10 (2.629), BZTPPI (3.251)							
I–Te(4)	I–Te–X₃	2.926	2.928	0.026	2.902	2.944	8	
N(4)–N(3)	X₃–N⁺–N⁰–X₂ (N⁰ planar)	1.414	1.414	0.005	1.412	1.418	13	
N(3)–N(3)	(C)(C,H)–Nₐ–N_b–(C)(C,H)							5,39
	Nₐ, N_b pyramidal	1.454	1.452	0.021	1.444	1.457	44	40
	Nₐ pyramidal, N_b planar	1.420	1.420	0.015	1.407	1.433	68	40
	Nₐ, N_b planar	1.401	1.401	0.018	1.384	1.418	40	40
	overall	1.425	1.425	0.027	1.407	1.443	139	
N(3)–N(2)	in pyrazole (N1–N2)	1.366	1.366	0.019	1.350	1.375	20	
	in pyridazinium (N1⁺ ≃ N2)	1.350	1.349	0.010	1.345	1.361	7	

Bond	Substructure	d	m	σ	q_1	q_u	n	Note
$N(2) \simeq N(2)$	$N \simeq N$ (aromatic) in pyridazine							
	with C,H as *ortho* substituents	1.304	1.300	0.019	1.287	1.326	6	
	with N,Cl as *ortho* substituents	1.368	1.373	0.011	1.362	1.375	9	
$N(2)=N(2)$	$C\#-N=N-C\#$							
	cis	1.245	1.244	0.009	1.239	1.252	21	
	trans	1.222	1.222	0.006	1.218	1.227	6	
	(overall)	1.240	1.241	0.012	1.230	1.251	27	
	$Car-N=N-Car$	1.255	1.253	0.016	1.247	1.262	13	
	$X-N=N=N$ (azides)	1.216	1.226	0.028	1.202	1.237	19	
$N(2)=N(1)$	$X-N=N=N$ (azides)	1.124	1.128	0.015	1.114	1.137	19	
$N(3)-O(2)$	$(C,H)_2-N-OH$ (Nsp^2: planar)	1.396	1.394	0.012	1.390	1.401	28	
	$C_2-N-O-C$							
	(Nsp^3: pyramidal)	1.463	1.465	0.012	1.457	1.468	22	
	(Nsp^2: planar)	1.397	1.394	0.011	1.388	1.409	12	
	in furoxan (N2–O1)	1.438	1.436	0.009	1.430	1.447	14	
$N(3)-O(1)$	$(C\simeq)_2N^+-O^-$ in pyridine *N*-oxides	1.304	1.299	0.015	1.291	1.316	11	
	in furoxan ($^+$N2–O6$^-$)	1.234	1.234	0.008	1.228	1.240	14	
$N(2)-O(2)$	in oximes							
	$(C\#)_2-C=N-OH$	1.416	1.418	0.006	1.416	1.420	7	
	$(H)(Csp^2)-C=N-OH$	1.390	1.390	0.011	1.380	1.401	20	
	$(C\#)(Csp^2)-C=N-OH$	1.402	1.403	0.010	1.393	1.410	18	
	$(Csp^2)_2-C=N-OH$	1.378	1.377	0.017	1.365	1.393	16	
	$(C,H)_2-C=N-OH$ (overall)	1.394	1.395	0.018	1.379	1.408	67	
	in furazan (O1–N2, O1–N5)	1.385	1.383	0.013	1.378	1.392	12	
	in furoxan (O1–N5)	1.380	1.380	0.011	1.370	1.388	14	
	in isoxazole (O1–N2)	1.425	1.425	0.010	1.417	1.434	9	
$N(3)=O(1)$	in nitrate ions NO_3^-	1.239	1.240	0.020	1.227	1.251	105	
	in nitro groups							
	C^*-NO_2	1.212	1.214	0.012	1.206	1.221	84	
	$C\#-NO_2$	1.210	1.210	0.011	1.203	1.218	251	
	$Car-NO_2$	1.217	1.218	0.011	1.211	1.215	1 116	
	$C-NO_2$ (overall)	1.218	1.219	0.013	1.210	1.226	1 733	
$N(3)-P(4)$	$X_2-P(=X)-NX_2$							
	Nsp^2: planar	1.652	1.651	0.024	1.634	1.670	205	
	Nsp^3: pyramidal	1.683	1.683	0.005	1.680	1.686	6	
	(overall)	1.662	1.662	0.029	1.639	1.682	358	
	subsets of this group are:							
	$O_2-P(=S)-NX_2$	1.628	1.624	0.015	1.615	1.634	9	
	$C-P(=S)-(NX_2)_2$	1.691	1.694	0.018	1.678	1.703	28	
	$O-P(=S)-(NX_2)_2$	1.652	1.654	0.014	1.642	1.664	28	
	$P(=O)-(NX_2)_3$	1.663	1.668	0.026	1.640	1.679	78	
$N(3)-P(3)$	$-NX-P(-X)-NX-P(-X)-$ (P_2N_2 ring)	1.730	1.721	0.017	1.716	1.748	20	
	$-NX-P(=S)-NX-P(=S)-$ (P_2N_2 ring)	1.697	1.697	0.015	1 690	1.703	44	
	in *P*-substituted phosphazenes:							
	$(N\simeq)_2P-N$ (amino)	1.637	1.638	0.014	1.625	1.651	16	
	(aziridinyl)	1.672	1.674	0.010	1.665	1.676	15	
$N(2)=P(4)$	$Ph_3-P=N^+=P-Ph_3$	1.571	1.573	0.013	1.563	1.580	66	
$N(2)=P(3)$	$Ph_3-P=N-C,S$	1.599	1.597	0.018	1.580	1.615	7	
$N(2) \simeq P(3)$	$N \simeq P$ aromatic							
	in phosphazenes	1.582	1.582	0.019	1.571	1.594	126	
	in $P \simeq N \simeq S$	1.604	1.606	0.009	1.594	1.612	36	
$N(3)-S(4)$	$C-SO_2-NH_2$	1.600	1.601	0.012	1.591	1.610	14	35
	$C-SO_2-NH-C\#$	1.633	1.633	0.019	1.615	1.652	47	35
	$C-SO_2-N-C(\#)_2$	1.642	1.641	0.024	1.623	1.659	38	35
$N(3)-S(2)$	$C-S-NX_2$ Nsp^2: planar	1.710	1.707	0.019	1.698	1.722	22	23
	(for Nsp^3 pyramidal see MODIAZ: 1.765)							
	$X-S-NX_2$ Nsp^2: planar	1.707	1.705	0.012	1.699	1.715	30	23
$N(2)-S(2)$	$C=N-S-X$	1.656	1.663	0.027	1.632	1.677	36	
$N(2) \simeq S(2)$	$N \simeq S$ aromatic in $P \simeq N \simeq S$	1.560	1.558	0.011	1.554	1.563	37	
$N(2)=S(2)$	$N=S$ in $N=S=N$ and $N=S=S$	1.541	1.546	0.022	1.521	1.558	37	
$N(3)-Se$	see COJCUZ (1.830), DSEMOR10 (1.846, 1.852), MORTRS10 (1.841)							
$N(2)-Se$	see SEBZQI (1.805), NAPSEZ10 (1.809, 1.820)							
$N(2)=Se$	see CISMUM (1.790, 1.791)							
$N(3)-Si(5)$	see DMESIP01, BOJLER, CASSAQ, CASYOK, CECXEN, CINTEY, CIPBUY, FMESIB, MNPSIL, PNPOSI (1.973—2.344)							
$N(3)-Si(4)$	$X_3-Si-NX_2$ (overall)	1.748	1.746	0.022	1.735	1.757	170	
	subsets of this group are:							
	$X_3-Si-NHX$	1.714	1.719	0.014	1.702	1.727	16	
	$X_3-Si-NX-Si-X_3$ acyclic	1.743	1.744	0.016	1.731	1.755	45	
	N–Si–N in 4-membered rings	1.742	1.742	0.009	1.735	1.748	53	
	N–Si–N in 5-membered rings	1.741	1.742	0.019	1.726	1.749	33	
$N(2)-Si(4)$	$X_3-Si-N^--Si-X_3$	1.711	1.712	0.019	1.693	1.729	15	

BOND LENGTHS IN CRYSTALLINE ORGANIC COMPOUNDS (continued)

Bond	Substructure	d	m	σ	q_l	q_u	n	Note
N–Te	see ACLTEP (2.402), BIBLAZ (1.980), CESSAU (2.023)							
O(2)–O(2)	C*–O–O–C*,H							
	τ(OO) = 70—85°	1.464	1.464	0.009	1.458	1.472	12	
	τ(OO) *ca.* 180°	1.482	1.480	0.005	1.478	1.486	5	
	overall	1.469	1.471	0.012	1.461	1.478	17	
	O=C–O–O–C=O see ACBZPO01 (1.446), CEYLUN (1.452), CIMHIP (1.454)							
	Si–O–O–Si	1.496	1.499	0.005	1.490	1.499	10	
O(2)–P(5)	X–P–(OX)$_4$							41
	trigonal bipyramidal:							
	axial	1.689	1.685	0.024	1.675	1.712	20	
	equatorial	1.619	1.622	0.024	1.604	1.628	20	
	square pyramidal	1.662	1.661	0.020	1.649	1.673	28	
O(2)–P(4)	C–O–P($\simeq$O)$_3$$^{2-}$	1.621	1.622	0.007	1.615	1.628	12	
	(H–O)$_2$–P($\simeq$O)$_2$$^-$	1.560	1.561	0.009	1.555	1.566	16	
	(C–O)$_2$–P($\simeq$O)$_2$$^-$	1.608	1.607	0.013	1.599	1.615	16	
	(C#–O)$_3$–P=O	1.558	1.554	0.011	1.550	1.564	30	
	(Car–O)$_3$–P=O	1.587	1.588	0.014	1.572	1.599	19	
	X–O–P(=O)–(C,N)$_2$	1.590	1.585	0.016	1.577	1.601	33	
	(X–O)$_2$–P(=O)–(C,N)	1.571	1.572	0.013	1.563	1.579	70	
O(2)–P(3)	(N$\simeq$)$_2$P–O–C (N$\simeq$P aromatic)	1.573	1.573	0.011	1.563	1.584	16	
O(1)=P(4)	C–O–P($\simeq$O)$_3$$^{2-}$ (delocalized)	1.513	1.512	0.008	1.508	1.518	42	
	(H–O)$_2$–P($\simeq$O)$_2$$^-$ (delocalized)	1.503	1.503	0.005	1.499	1.508	16	
	(C–O)$_2$–P($\simeq$O)$_2$$^-$ (delocalized)	1.483	1.485	0.008	1.474	1.490	16	
	(C–O)$_3$–P=O	1.449	1.448	0.007	1.446	1.452	18	
	C$_3$–P=O	1.489	1.486	0.010	1.481	1.496	72	
	N$_3$–P=O	1.461	1.462	0.014	1.449	1.470	26	
	(C)$_2$(N)–P=O	1.487	1.489	0.007	1.479	1.493	5	
	(C,N)$_2$(O)–P=O	1.467	1.465	0.007	1.462	1.472	33	
	(C,N)(O)$_2$–P=O	1.457	1.458	0.009	1.454	1.462	35	
O(2)–S(4)	C–O–SO$_2$–C	1.577	1.576	0.015	1.566	1.584	41	
	C–O–SO$_2$–CH$_3$	1.569	1.569	0.013	1.556	1.582	7	
	C–O–SO$_2$–Car	1.580	1.578	0.015	1.571	1.588	27	
O(1)=S(4)	C–SO$_2$–C	1.436	1.437	0.010	1.431	1.442	316	42
	X–SO$_2$–NX$_2$	1.428	1.428	0.010	1.422	1.434	326	
	C–SO$_2$–N–(C,H)$_2$	1.430	1.430	0.009	1.425	1.435	206	
	C–SO$_2$–O–C	1.423	1.423	0.008	1.418	1.428	82	
	in SO$_4$$^{2-}$	1.472	1.473	0.013	1.463	1.481	104	
O(1)=S(3)	C–S(=O)–C	1.497	1.498	0.013	1.489	1.505	90	5
O–Se	see BAPPAJ, BIRGUE10, BIRHAL10, CXMSEO, DGLYSE, SPSEBU (1.597 for O=Se to 1.974 for O–Se)							
O(2)–Si(5)	(X–O)$_3$–Si–(N)(C)	1.663	1.658	0.023	1.650	1.665	21	
O(2)–Si(4)	X$_3$–Si–O–X (overall)	1.631	1.630	0.022	1.617	1.646	191	
O(2)–Si(4)	subsets of this group are:							
	X$_3$–Si–O–C#	1.645	1.647	0.012	1.634	1.652	29	
	X$_3$–Si–O–Si–X$_3$	1.622	1.625	0.014	1.614	1.631	70	
	X$_3$–Si–O–O–Si–X$_3$	1.680	1.676	0.008	1.673	1.688	10	
O(2)–Te(6)	(X–O)$_6$–Te	1.927	1.927	0.020	1.908	1.942	16	
O(2)–Te(4)	(X–O)$_2$–Te–X$_2$	2.133	2.136	0.054	2.078	2.177	12	
P(4)–P(4)	X$_3$–P–P–X$_3$	2.256	2.259	0.025	2.243	2.277	6	
P(4)–P(3)	see CECHEX (2.197), COZPIQ (2.249)							
P(3)–P(3)	X$_2$–P–P–X$_2$	2.214	2.210	0.022	2.200	2.224	41	
P(4)=P(4)	see BUTSUE (2.054)							
P(3)=P(3)	see BALXOB (2.034)							
P(4)=S(1)	C$_3$–P=S	1.954	1.952	0.005	1.950	1.957	13	
	(N,O)$_2$(C)–P=S	1.922	1.924	0.014	1.913	1.927	26	
	(N,O)$_3$–P=S	1.913	1.914	0.014	1.906	1.921	50	
P(4)=Se(1)	X$_3$–P=Se	2.093	2.099	0.019	2.075	2.108	12	
P(3)–Si(4)	X$_2$–P–Si–X$_3$: 3- and 4-rings	2.264	2.260	0.019	2.249	2.283	22	
	excluded (see BOPFER, BOPFIV, CASTOF10, COZVIW: 2.201—2.317)							
P(4)=Te(1)	see MOPHTE (2.356), TTEBPZ (2.327)							
S(2)–S(2)	C–S–S–C							
	τ(SS) = 75—105°	2.031	2.029	0.015	2.021	2.038	46	
	τ(SS) = 0—20°	2.070	2.068	0.022	2.057	2.077	28	
	(overall)	2.048	2.045	0.026	2.028	2.068	99	
	in polysulphide chain–S–S–S–	2.051	2.050	0.022	2.037	2.065	126	
S(2)–S(1)	X–N=S–S	1.897	1.896	0.012	1.887	1.908	5	
S–Se(4)	see BUWZUO (2.264, 2.269)							
S–Se(2)	X–Se–S (any)	2.193	2.195	0.015	2.174	2.207	9	
S(2)–Si(4)	X$_3$–Si–S–X	2.145	2.138	0.020	2.130	2.158	19	

BOND LENGTHS IN CRYSTALLINE ORGANIC COMPOUNDS (continued)

Bond	Substructure	d	m	σ	q_1	q_u	n	Note
S(2)–Te	X–S–Te (any)	2.405	2.406	0.022	2.383	2.424	10	
	X=S–Te (any)	2.682	2.686	0.035	2.673	2.694	28	
Se(2)–Se(2)	X–Se–Se–X	2.340	2.340	0.024	2.315	2.361	15	
Se(2)–Te(2)	see BAWFUA, BAWGAH (2.524—2.561)							†
Si(4)–Si(4)	X₃–Si–Si–X₃ 3-membered rings excluded: see CIHRAM (2.511)	2.359	2.359	0.012	2.349	2.366	42	
Te–Te	see CAHJOK (2.751, 2.704)							

Appendix 1. (Footnotes to Table)

1. Sample dominated by B–CH₃. For longer bonds in B⁻–CH₃ see LITMEB10 [B(4)–CH₃ = 1.621—1.644Å].
2. $p(\pi)$–$p(\pi)$ Bonding with Bsp^2 and Nsp^2 coplanar (τBN = 0 ± 15°) predominates. See G. Schmidt, R. Boese, and D. Bläser, *Z. Naturforsch.*, 1982, **37b**, 1230.
3. 84 observations range from 1.38 to 1.61 Å and individual values depend on substituents on B and O. For a discussion of borinic acid adducts see S. J. Rettig and J. Trotter, *Can. J. Chem.*, 1982, **60**, 2957.
4. See M. Kaftory in 'The Chemistry of Functional Groups. Supplement D: The Chemistry of Halides, Pseudohalides, and Azides' eds. S. Patai and Z. Rappoport, Wiley: New York, 1983, Part 2, ch. 24.
5. Bonds which are endocyclic or exocyclic to any 3- or 4-membered rings have been omitted from all averages in this section.
6. The overall average given here is for Csp^3–Csp^3 bonds which carry only C or H substituents. The value cited reflects the relative abundance of each 'substitution' group. The 'mean of means' for the 9 subgroups is 1.538 (σ = 0.022) Å.
7. See F. H. Allen, (a) *Acta Crystallogr.*, 1980, **B36**, 81; (b) 1981, **B37**, 890.
8. See F. H. Allen, *Acta Crystallogr.*, 1984, **B40**, 64.
9. See F. H. Allen, *Tetrahedron*, 1982, **38**, 2843.
10. See F. H. Allen, *Tetrahedron*, 1982, **38**, 645.
11. Cyclopropanones and cyclobutanones excluded.
12. See W. B. Schweizer and J. D. Dunitz, *Helv. Chim. Acta*, 1982, **65**, 1547.
13. See L. Norskov-Lauritsen, H.-B. Bürgi, P. Hoffmann, and H. R. Schmidt, *Helv. Chim. Acta*, 1985, **68**, 76.
14. See P. Chakrabarti and J. D. Dunitz, *Helv. Chim. Acta*, 1982, **65**, 1555.
15. See J. L. Hencher in 'The Chemistry of the C≡C Triple Bond,' ed. S. Patai, Wiley, New York, 1978, ch. 2.
16. Conjugated: torsion angle about central C–C single bond is 0 ± 20° (*cis*) or 180 ± 20° (*trans*).
17. Unconjugated: torsion angle about central C–C single bond is 20—160°.
18. Other conjugative substituents excluded.
19. TCNQ is tetracyanoquinodimethane.
20. No difference detected between C2≃C3 and C3≃C4 bonds.
21. Derived from neutron diffraction results only.
22. Nsp^3: pyramidal; mean valence angle at N is in range 108—114°.
23. Nsp^2: planar; mean valence angle at N is ≥117.5°.
24. Cyclic and acyclic peptides.
25. See R. H. Blessing, *J. Am. Chem. Soc.*, 1983, **105**, 2776.
26. See L. Lebioda, *Acta Crystallogr.*, 1980, **B36**, 271.
27. n = 3 or 4, *i.e.* tri- or tetra-substituted ureas.
28. Overall value also includes structures with mean valence angle at N in the range 115—118°.
29. See F. H. Allen and A. J. Kirby, *J. Am. Chem. Soc.*, 1984, **106**, 6197.
30. See A. J. Kirby, 'The Anomeric Effect and Related Stereoelectronic Effects at Oxygen,' Springer, Berlin, 1983.
31. See B. Fuchs, L. Schleifer, and E. Tartakovsky, *Nouv. J. Chim.*, 1984, **8**, 275.
32. See S. C. Nyburg and C. H. Faerman, *J. Mol. Struct.*, 1986, **140**, 347.
33. Sample dominated by P–CH₃ and P–CH₂–C.
34. Sample dominated by C* = methyl.
35. See A. Kalman, M. Czugler, and G. Argay, *Acta Crystallogr.*, 1981, **B37**, 868.
36. Bimodal distribution resolved into 22 'short' bonds and 5 longer outliers.
37. All 24 observations come from BUDTEZ.
38. 'Long' O–H bonds in centrosymmetric O – – – H – – – O H-bonded dimers are excluded.
39. N–N bond length also dependent on torsion angle about N–N bond and on nature of substituent C atoms; these effects are ignored here.
40. N pyramidal has average angle at N in range 100—113.5°; N planar has average angle of ≥117.5°.
41. See R. R. Holmes and J. A. Deiters, *J. Amer. Chem. Soc.*, 1977, **99**, 3318.
42. No detectable variation in S=O bond length with type of C-substituent.

Appendix 2.

Short-form references to individual CSD entries cited by reference code in the Table. A full list of CSD bibliographic entries is given in SUP 56701.

ACBZPO01	*J. Am. Chem. Soc.*, 1975, **97**, 6729.	BIBLAZ	*Zh. Strukt. Khim.*, 1981, **22**, 118.
ACLTEP	*J. Organomet. Chem.*, 1980, **184**, 417.	BICGEZ	*Z. Anorg. Allg. Chem.*, 1982, **486**, 90.
ASAZOC	*Dokl. Akad. Nauk SSSR*, 1979, **249**, 120.	BIHXIZ	*J. Chem. Soc., Chem. Commun.*, 1982, 982.
BALXOB	*J. Am. Chem. Soc.*, 1981, **103**, 4587.	BIRGUE10	*Z. Naturforsch., Teil B*, 1983, **38**, 20.
BAPPAJ	*Inorg. Chem.*, 1981, **20**, 3071.	BIRHAL10	*Z. Naturforsch., Teil B*, 1982, **37**, 1410.
BARRIV	*Acta Chem. Scand., Ser. A*, 1981, **35**, 443.	BIZJAV	*J. Organomet. Chem.*, 1982, **238**, C1.
BAWFUA	*Cryst. Struct. Commun.*, 1981, **10**, 1345.	BOGPOC	*Z. Naturforsch., Teil B*, 1982, **37**, 1402.
BAWGAH	*Cryst. Struct. Commun.*, 1981, **10**, 1353.	BOGSUL	*Z. Naturforsch., Teil B*, 1982, **37**, 1230.
BECTAE	*J. Org. Chem.*, 1981, **46**, 5048, 1981.	BOJLER	*Z. Anorg. Allg. Chem.*, 1982, **493**, 53.
BELNIP	*Z. Naturforsch., Teil B*, 1982, **37**, 299.	BOJPUL	*Acta Chem. Scand., Ser. A*, 1982, **36**, 829.
BEMLIO	*Chem. Ber.*, 1982, **115**, 1126.	BOPFER	*Chem. Ber.*, 1983, **116**, 146.
BEPZEB	*Cryst. Struct. Commun.*, 1982, **11**, 175.	BOPFIV	*Chem. Ber.*, 1983, **116**, 146.
BETJOZ	*J. Am. Chem. Soc.*, 1982, **104**, 1683.	BOVMEE	*Acta Crystallogr., Sect. B*, 1982, **38**, 1048.
BETUTE10	*Acta Chem. Scand., Ser. A*, 1976, **30**, 719.	BQUINI	*Acta Crystallogr., Sect. B*, 1979, **35**, 1930.

BTUPTE	*Acta Chem. Scand., Ser. A*, 1975, **29**, 738.
BUDTEZ	*Z. Naturforsch., Teil B*, 1983, **38**, 454.
BUPSIB10	*Z. Anorg. Allg. Chem.*, 1981, **474**, 31.
BUSHAY	*Z. Naturforsch., Teil. B*, 1983, **38**, 692.
BUTHAZ10	*Inorg. Chem.*, 1984, **23**, 2582.
BUTSUE	*J. Chem. Soc., Chem. Commun.*, 1983, 862.
BUWZUO	*Acta Chem. Scand., Ser A*, 1983, **37**, 219.
BZPRIB	*Z. Naturforsch., Teil B*, 1981, **36**, 922.
BZTPPI	*Inorg. Chem.*, 1978, **17**, 894.
CAHJOK	*Inorg. Chem.*, 1983, **22**, 1809.
CAJMAB	*Chem. Z*, 1983, **107**, 169.
CANLUY	*Tetrahedron Lett.*, 1983, **24**, 4337.
CASSAQ	*J. Struct. Chem.*, 1983, **2**, 101.
CASTOF10	*Acta Crystallogr., Sect. C*, 1984, **40**, 1879.
CASYOK	*J. Struct. Chem.*, 1983, **2**, 107.
CECHEX	*Z. Anorg. Allg. Chem.*, 1984, **508**, 61.
CECXEN	*J. Struct. Chem.*, 1983, **2**, 207.
CEDCUJ	*J. Org. Chem.*, 1983, **48**, 5149.
CEHKAB	*Z. Naturforsch., Teil B*, 1984, **39**, 139.
CELDOM	*Acta Crystallogr., Sect. C*, 1984, **40**, 556.
CESSAU	*Acta Crystallogr., Sect. C*, 1984, **40**, 653.
CETTAW	*Chem. Ber.*, 1984, **117**, 1089.
CETUTE	*Acta Chem. Scand., Ser A*, 1975, **29**, 763.
CEYLUN	*Izv. Akad. Nauk SSSR, Ser. Khim.*, 1983, 2744.
CIFZUM	*Acta Chem. Scand., Ser A*, 1984, **38**, 289.
CIHRAM	*Angew. Chem., Int. Ed. Engl.*, 1984, **23**, 302.
CILRUK	*J. Chem. Soc., Chem. Commun.*, 1984, 1023.
CILSAR	*J. Chem. Soc., Chem. Commun.*, 1984, 1021.
CIMHIP	*Acta Crystallogr., C*, 1984, **40**, 1458.
CINTEY	*Dokl. Akad. Nauk SSSR*, 1984, **274**, 615.
CIPBUY	*J. Struct. Chem.*, 1983, **2**, 281.
CISMUM	*Z. Naturforsch., Teil B*, 1984, **39**, 485.
CISTED	*Z. Anorg. Allg. Chem.*, 1984, **511**, 95.
CIWYIQ	*Inorg. Chem.*, 1984, **23**, 1946.
CIYFOF	*Inorg. Chem.*, 1984, **23**, 1790.
CMBIDZ	*J. Org. Chem.*, 1979, **44**, 1447.
CODDEE	*Z. Naturforsch., Teil B*, 1984, **39**, 1257.
CODDII	*Z. Naturforsch., Teil B*, 1984, **39**, 1257.
COFVOI	*Z. Naturforsch., Teil B*, 1984, **39**, 1027.
COJCUZ	*Chem. Ber.*, 1984, **117**, 2686.
COSDIX	*Z. Naturforsch., Teil B*, 1984, **39**, 1344.
COZPIQ	*Chem. Ber.*, 1984, **117**, 2063.
COZVIW	*Z. Anorg. Allg. Chem.*, 1984, **515**, 7.
CTCNSE	*J. Am. Chem. Soc.*, 1980, **102**, 5430.
CUCPIZ	*J. Am. Chem. Soc.*, 1984, **106**, 7529.
CUDLOC	*J. Cryst. Spectrosc.*, 1985, **15**, 53.
CUDLUI	*J. Cryst. Spectrosc.*, 1985, **15**, 53.
CUGBAH	*Acta Crystallogr., Sect. C*, 1985, **41**, 476.
CXMSEO	*Acta Crystallogr., Sect. B*, 1973, **29**, 595.
DGLYSE	*Acta Crystallogr., Sect. B*, 1975, **31**, 1785.
DMESIP01	*Acta Crystallogr., Sect. C*, 1984, **40**, 895.
DSEMOR10	*J. Chem. Soc., Dalton Trans.*, 1980, 628.
DTHIBR10	*Inorg. Chem.*, 1971, **10**, 697.
EPHTEA	*Inorg. Chem.*, 1980, **19**, 2487.
ESEARS	*J. Chem. Soc. C*, 1971, 1511.
ETEARS	*J. Chem. Soc. C*, 1971, 1511.
FMESIB	*J. Organomet. Chem.*, 1980, **197**, 275.
FPHTEL	*J. Chem. Soc., Dalton Trans.*, 1980, 2306.
FPSULF10	*J. Am. Chem. Soc.*, 1982, **104**, 1683.
HCLENE10	*Acta Crystallogr., Sect. B*, 1982, **38**, 3139.
HMTITI	*Acta Crystallogr., Sect. B*, 1975, **31**, 1505.
HMTNTI	*Z. Anorg. Allg. Chem.*, 1974, **409**, 237.
HXPASC	*J. Chem. Soc., Dalton Trans.*, 1975, 1381.
IBZDAC11	*J. Chem. Soc., Dalton Trans.*, 1979, 854.
IFORAM	*Monatsh. Chem.*, 1974, **105**, 621.
IODMAM	*Acta Crystallogr., Sect. B*, 1977, **33**, 3209.
IPMUDS	*Acta Crystallogr., Sect. B*, 1973, **29**, 2128.
ISUREA10	*Acta Crystallogr., Sect. B*, 1972, **28**, 643.
LITMEB10	*J. Am. Chem. Soc.*, 1975, **97**, 6401.
MESIAD	*Z. Naturforsch., Teil B*, 1980, **35**, 789.
METAMM	*Acta Crystallogr.*, 1964, **17**, 1336.
MNPSIL	*J. Am. Chem. Soc.*, 1969, **91**, 4134.
MODIAZ	*J. Heterocycl. Chem.*, 1980, **17**, 1217.
MOPHTE	*Acta Chem. Scand., Ser. A*, 1980, **34**, 333.
MORTRS10	*J. Chem. Soc., Dalton Trans.*, 1980, 628.
NAPSEZ10	*J. Am. Chem. Soc.*, 1980, **102**, 5070.
NBBZAM	*Z. Naturforsch., Teil B*, 1977, **32**, 1416.
OPIMAS	*Aust. J. Chem.*, 1977, **30**, 2417.
OPNTEC10	*J. Chem. Soc., Dalton Trans.*, 1982, 251.
PHASCL	*Acta Crystallogr., Sect. B*, 1981, **37**, 1357.
PHASOC01	*Aust. J. Chem.*, 1975, **28**, 15.
PNPOSI	*J. Am. Chem. Soc.*, 1968, **90**, 5102.
SEBZQI	*J. Chem. Soc., Chem. Commun.*, 1977, 325.
SPSEBU	*Acta Chem. Scand., Ser. A*, 1979, **33**, 403.
TEACBR	*Cryst. Struct. Commun.*, 1974, **3**, 753.
THINBR	*J. Am. Chem. Soc.*, 1970, **92**, 4002.
TMPBTI	*Acta Crystallogr., Sect. B*, 1975, **31**, 1116.
TPASSN	*J. Chem. Soc., Dalton Trans.*, 1977, 514.
TPASTB	*Cryst. Struct. Commun.*, 1976, **5**, 39.
TPHOSI	*Z. Naturforsch., Teil B*, 1979, **34**, 1064.
TTEBPZ	*Z. Naturforsch., Teil B*, 1979, **34**, 256.
ZCMXSP	*Cryst. Struct. Commun.*, 1977, **6**, 93.

BOND LENGTHS IN ORGANOMETALLIC COMPOUNDS

This table summarizes the average values of interatomic distances of representative metal-ligand bonds. Sigma bonds between d- and f-block metals and the elements C, N, O, P, S, and As are included. The values are extracted from a much larger list in Reference 1. The tabulated values are the unweighted means of reported measurements on compounds in each category. If four or more measurements are available, the standard deviation is given in parentheses. All values are in Ångstrom units (10^{-10} m).

The first part of the table covers metal-carbon bonds in different ligand categories, while the second part covers metal bonds to other elements. R stands for any alkyl group; Me for a CH_3 group; C_6R_5 indicates an aryl group; and $C(=O)R$ an acyl group. Metals are listed in atomic number order.

REFERENCE

1. Orpen, A. G., Brammer, L., Allen, F.H., Kennard, O., Watson, D. G., and Taylor, R., *J. Chem. Soc. Dalton Trans.*, 1989, S1-S83.

M	M-CH$_3$	M-CH$_2$R	M-CR=CR$_2$	M-C$_6$R$_5$	M-C(=O)R
Ti		2.167	2.215(0.042)	2.148	
V				2.114(0.012)	
Cr	2.168		2.035(0.009)	2.075(0.019)	
Mn	2.095(0.030)	2.176(0.024)	2.007	2.064(0.021)	2.044
Fe	2.074	2.091(0.030)	1.991(0.039)	2.031(0.062)	1.997(0.033)
Co	2.014(0.023)	2.039(0.032)	1.934(0.019)	1.974	1.990
Ni	2.029	1.964	1.892(0.017)	1.917(0.038)	1.850(0.059)
Cu				2.020	
Zn		1.964			
Zr	2.292(0.049)		2.257		
Nb	2.336	1.319			
Mo	2.254(0.065)	2.250(0.061)	2.204(0.049)	2.193(0.054)	2.109
Ru	2.179(0.045)	2.036(0.010)	2.063	2.092(0.057)	2.091
Rh	2.092(0.027)	2.100	2.040(0.054)	2.011(0.026)	1.995(0.031)
Pd		2.028	2.000(0.024)	1.981(0.032)	1.982(0.029)
Hf	2.275(0.049)		2.205		
Ta	2.217(0.035)	2.225(0.056)		2.199(0.073)	
W	2.189(0.039)	2.175	2.224		
Re	2.173(0.051)	2.290		2.027	2.190(0.027)
Os		2.221	2.052	2.090(0.032)	2.161
Ir	2.175		2.071(0.044)	2.070(0.038)	2.019
Pt	2.083(0.045)	2.062(0.031)	2.024(0.037)	2.049(0.046)	1.991(0.025)
Au	2.066(0.045)		2.042	2.059(0.024)	
Hg	2.072(0.026)	2.125		2.086(0.040)	
Th	2.567				

M	M-NH$_3$	M-OH$_2$	M-PMe$_3$	M-SR	M-AsR$_3$
Ti		2.066(0.052)		2.369	2.686
V		2.129(0.131)	2.510(0.010)	2.378(0.007)	
Cr	2.069(0.008)	1.997(0.070)	2.389(0.069)	2.362	2.460(0.040)
Mn		2.189(0.040)	2.455(0.164)	2.366(0.054)	2.400(0.013)
Fe		2.085(0.066)	2.246(0.042)	2.271(0.028)	2.352(0.043)
Co	1.965(0.021)	2.085(0.064)	2.217(0.043)	2.254(0.025)	2.323(0.021)
Ni	2.074(0.093)	2.079(0.038)	2.204(0.031)	2.187(0.007)	2.333(0.035)
Cu	1.987(0.017)	2.186(0.215)			2.367(0.016)
Zn	2.044	2.090(0.061)		2.295	
Y		2.398(0.068)			
Zr			2.692		
Nb		2.248(0.137)			2.741(0.008)
Mo	2.217	2.201(0.094)	2.462(0.046)	2.401(0.050)	2.582(0.036)
Ru	2.126(0.024)	2.074(0.051)	2.307(0.050)		2.446(0.031)
Rh	2.114(0.018)	2.190(0.096)	2.266(0.036)		2.416(0.039)
Pd	2.032	2.200	2.287(0.018)		2.386(0.052)
Ag		2.350			
Cd		2.318(0.065)		2.444	

M	M-NH$_3$	M-OH$_2$	M-PMe$_3$	M-SR	M-AsR$_3$
La		2.556(0.062)			
Ce		2.565(0.063)			
Pr		2.518(0.038)			
Nd		2.533(0.058)			
Sm		2.459(0.050)			
Eu		2.441(0.055)			
Gd		2.443(0.074)			
Tb		2.455			
Dy		2.409(0.074)			
Ho		2.407(0.069)			
Er		2.404(0.083)			
Yb		2.353(0.066)			
Lu		2.404(0.116)			
Ta			2.589(0.044)		
W		2.115(0.065)	2.485(0.039)		
Re	2.253	2.199(0.091)	2.369(0.065)		2.575(0.006)
Os	2.136	2.166	2.328(0.029)		
Ir	2.050(0.021)		2.323(0.028)	2.461	
Pt			2.295(0.036)	2.320(0.015)	2.366(0.058)
Au		2.157		2.293	
Hg		2.690(0.083)		2.402(0.065)	
Th		2.483(0.032)			
U		2.455(0.047)			

BOND LENGTHS AND ANGLES IN GAS-PHASE MOLECULES

This table is reprinted from *Kagaku Benran, 3rd Edition*, Vol. II, pp. 649—661 (1984), with permission of the publisher, Maruzen Company, LTD. (Copyright 1984 by the Chemical Society of Japan). Translation was carried out by Kozo Kuchitsu.

Internuclear distances and bond angles are represented in units of Å (1 Å = 10^{-10} m) and degrees, respectively. The same but inequivalent atoms are discriminated by subscripts a, b, etc. In some molecules ax for axial and eq for equatorial are also used. All measurements were made in the gas phase. The methods used are abbreviated as follows. UV: ultraviolet (including visible) spectroscopy; IR: infrared spectroscopy; R: Raman spectroscopy; MW: microwave spectroscopy; ED: electron diffraction; NMR: nuclear magnetic resonance; LMR: laser magnetic resonance; EPR: electron paramagnetic resonance; MBE: molecular beam electric resonance. If two methods were used jointly for structure determination, they are listed together, as (ED, MW). If the numerical values listed refer to the equilibrium values, they are specified by r_e and θ_e. In other cases the listed values represent various average values in vibrational states; it is frequently the case that they represent the r_s structure derived from several isotopic species for MW or the r_g structure (i.e., the average internuclear distances at thermal equilibrium) for ED. These internuclear distances for the same atom pair with different definitions may sometimes differ as much as 0.01 Å. Appropriate comments are made on the symmetry and conformation in the equilibrium structure.

In general, the numerical values listed in the following tables contain uncertainties in the last digits. However, for certain molecules such as diatomic molecules, with experimental uncertainties of the order of 10^{-5} Å or smaller, numerical values are listed to four decimal places.

REFERENCES

1. L. E. Sutton, ed., *Tables of Interatomic Distances and Configuration in Molecules and Ions*, The Chemical Society Special Publication, No. 11, 18, The Chemical Society (London) (1958, 1965).
2. K.-H. Hellwege, ed., *Landolt-Börnstein Numerical Data and Functional Relations in Science and Technology*, New Series, II/7, J. H. Callomon, E. Hirota, K. Kuchitsu, W. J. Lafferty, A. G. Maki, C. S. Pote, with assistance of I. Buck and B. Starck, *Structure Data of Free Polyatomic Molecules*, Springer-Verlag (1976).
3. K. P. Huber and G. Herzberg, *Molecular Spectra and Molecular Structure IV. Constants of Diatomic Molecules*, Van Nostrand Reinhold Co., London (1979).
4. B. Starck, *Microwave Catalogue and Supplements*.
5. B. Starck, *Electron Diffraction Catalogue and Supplements*.

STRUCTURES OF ELEMENTS AND INORGANIC COMPOUNDS
Compounds are Arranged in Alphabetical Order by their Chemical Formulas
(Lengths in Å and Angles in Degrees)

Compound	Structure			Method
AgBr	Ag—Br (r_e)	2.3931		MW
AgCl	Ag—Cl (r_e)	2.2808		MW
AgF	Ag—F (r_e)	1.9832		MW
AgH	Ag—H (r_e)	1.617		UV
AgI	Ag—I (r_e)	2.5446		MW
AgO	Ag—O (r_e)	2.0030		UV
AlBr	Al—Br (r_e)	2.295		UV
AlCl	Al—Cl (r_e)	2.1301		MW
AlF	Al—F (r_e)	1.6544		MW
AlH	Al—H (r_e)	1.6482		UV
AlI	Al—I (r_e)	2.5371		MW
AlO	Al—O (r_e)	1.6176		UV
Al_2Br_6		Al—Br$_a$	2.22	ED
		Al—Br$_b$	2.38	
		∠Br$_b$AlBr$_b$	82	
		∠Br$_a$AlBr$_a$	118	
		(D$_{2h}$)		
Al_2Cl_6		Al—Cl$_a$	2.04	ED
		Al—Cl$_b$	2.24	
		∠Cl$_b$AlCl$_b$	87	
		∠Cl$_a$AlCl$_a$	122	
		(D$_{2h}$)		
AsBr$_3$	As—Br	2.324	∠BrAsBr	99.6	ED
AsCl$_3$	As—Cl	2.165	∠ClAsCl	98.6	ED, MW
AsF$_3$	As—F	1.710	∠FAsF	95.9	ED, MW

Compound	Structure						Method
AsF$_5$	(structure diagram)	As—F$_a$ 1.711		As—F$_b$ 1.656 (D_{3h})			
AsH$_3$	As—H (r_e)	1.511		∠HAsH (θ_e)	92.1		MW, IR
AsI$_3$	As—I	2.557		∠IAsI	100.2		ED
AuH	Au—H (r_e)	1.5237					UV
BBr$_3$	B—Br	1.893		(D_{3h})			ED
BCl$_3$	B—Cl	1.742		(D_{3h})			ED
BF	B—F (r_e)	1.2626					UV
BF$_2$H	B—H	1.189	B—F 1.311	∠FBF	118.3		MW
BF$_2$OH	B—F	1.32	B—O 1.34	O—H	0.941		MW
	∠FBF	118	∠FBO 123	∠BOH	114.1		
BF$_3$	B—F	1.313		(D_{3h})			ED, IR
BH	B—H (r_e)	1.2325					UV
BH$_3$PH$_3$	B—P	1.937	B—H 1.212	P—H	1.399		MW
	∠PBH	103.6	∠BPH 116.9	∠HBH	114.6		
	∠HPH	101.3	staggered form				
BI$_3$	B—I	2.118		(D_{3h})			ED
BN	B—N (r_e)	1.281					UV
BO	B—O (r_e)	1.2045					EPR
BO$_2$	B—O	1.265		linear			UV
BS	B—S	1.6091					UV
B$_2$H$_6$	(structure diagram)			B—H$_a$	1.19		IR, ED
				B—H$_b$	1.33		
				B···B	1.77		
				∠H$_a$BH$_a$	122		
				∠H$_b$BH$_b$	97		
B$_3$H$_3$O$_3$	B—O	1.376		∠BOB≅∠OBO	120		ED
B$_3$H$_6$N$_3$	B—N	1.435	B—H 1.26	N—H	1.05		ED
	∠NBN	118	∠BNB 121	(C_2)			
BaH	Ba—H (r_e)	2.2318					UV
BaO	Ba—O (r_e)	1.9397					MW
BaS	Ba—S (r_e)	2.5074					MBE
BeF	Be—F (r_e)	1.3609					UV
BeH	Be—H (r_e)	1.3431					UV
BeO	Be—O (r_e)	1.3308					UV
BiBr	Bi—Br (r_e)	2.6095					MW
BiBr$_3$	Bi—Br	2.63		∠BrBiBr	90	(C_{3v})	ED
BiCl	Bi—Cl (r_e)	2.4716					MW
BiCl$_3$	Bi—Cl	2.423		∠ClBiCl	100	(C_{3v})	ED
BiF	Bi—F (r_e)	2.0516					MW
BiH	Bi—H (r_e)	1.805					UV
BiI	Bi—I (r_e)	2.8005					MW
BiO	Bi—O (r_e)	1.934					UV
BrCN	C—N (r_e)	1.157		C—Br (r_e)	1.790		IR
BrCl	Br—Cl (r_e)	2.1361					MW
BrF	Br—F (r_e)	1.7590					MW
BrF$_3$	(structure diagram) F$_a$—Br—F$_a$, F$_b$			Br—F$_a$ 1.810	Br—F$_b$ 1.721		MW
				∠F$_a$BrF$_b$ 86.2	(C_{2v})		
BrF$_5$	Br—F (average)	1.753					ED, MW
	(Br—F$_{eq}$) – (Br—F$_{ax}$) = 0.069						
	∠F$_{ax}$BrF$_{eq}$	85.1		(C_{4v})			
BrO	Br—O (r_e)	1.7172					MW
Br$_2$	Br—Br (r_e)	2.2811					R
CBr$_4$	C—Br	1.935		(T_d)			ED
CCl	C—Cl	1.6512					UV

Compound	Structure						Method
$CClF_3$	C—Cl	1.752	C—F	1.325	∠FCF	108.6	ED, MW
CCl_3F	C—Cl	1.754	C—F	1.362	∠ClCCl	111	MW
					(C_{3v})		
CCl_4	C—Cl		1.767		(T_d)		ED
CF	C—F (r_e)		1.2718				EPR
CF_3I	C—I	2.138	C—F	1.330	∠FCF	108.1	ED, MW
CF_4	C—F		1.323		(T_d)		ED
CH	C—H (r_e)		1.1199				UV
CI_4	C—I		2.15		(T_d)		ED
CN	C—N (r_e)		1.1718				MW
CO	C—O (r_e)		1.1283				MW
$COBr_2$	C—O		1.178		C—Br	1.923	ED, MW
	∠BrCBr		112.3				
COClF	C—F	1.334	C—O	1.173	C—Cl	1.725	ED, MW
	∠FCCl	108.8	∠ClCO	127.5			
$COCl_2$	C—O		1.179		C—Cl	1.742	ED, MW
	∠ClCCl		111.8				
COF_2	C—F		1.3157		C—O	1.172	ED, MW
	∠FCF		107.71				
CO_2	C—O (r_e)		1.1600				IR
CP	C—P (r_e)		1.562				UV
CS	C—S (r_e)		1.5349				MW
CS_2	C—S (r_e)		1.5526				IR
C_2	C—C (r_e)		1.2425				UV
C_3O_2	C—O		1.163		C—C	1.289	ED
	linear (large-amplitude bending vibration)						
CaH	Ca—H (r_e)		2.002				UV
CaO	Ca—O (r_e)		1.8221				UV
CaS	Ca—S (r_e)		2.3178				UV
CdH	Cd—H (r_e)		1.781				EPR
$CdBr_2$	Cd—Br		2.35		linear		ED
$CdCl_2$	Cd—Cl		2.24		linear		ED
CdI_2	Cd—I		2.56		linear		ED
ClCN	C—Cl (r_e)		1.629		C—N (r_e)	1.160	MW
ClF	Cl—F (r_e)		1.6283				MW
ClF_3	F_a—Cl—F_a			Cl—F_a	1.698	Cl—F_b 1.598	MW
	F_b			∠F_aClF_b	87.5	(C_{2v})	
ClO	Cl—O (r_e)		1.5696				MW, UV
ClOH	O—Cl		1.690	O—H	0.975	∠HOCl 102.5	MW, IR
ClO_2	Cl—O		1.470		∠OClO	117.38	MW
$ClO_3(OH)$	O_a—Cl		1.407	O_b—Cl	1.639		ED
				∠O_aClO_a	114.3	∠O_aClO_b 104.1	

Compound	Structure						Method
Cl_2	Cl—Cl (r_e)		1.9878				UV
Cl_2O	Cl—O		1.6959		∠ClOCl	110.89	MW
CoH	Co—H (r_e)		1.542				UV
$Cr(CO)_6$	C—O		1.16		Cr—C	1.92	ED
	∠CrCO		180				
CrO	Cr—O (r_e)		1.615				UV
CsBr	Cs—Br (r_e)		3.0723				MW
CsCl	Cs—Cl (r_e)		2.9063				MW
CsF	Cs—F (r_e)		2.3454				MW
CsH	Cs—H (r_e)		2.4938				UV
CsI	Cs—I (r_e)		3.3152				MW

Compound	Structure						Method
CsOH	Cs—O (r_e)	2.395	O—H (r_e)	0.97			MW
CuBr	Cu—Br (r_e)	2.1734					MW
CuCl	Cu—Cl (r_e)	2.0512					MW
CuF	Cu—F (r_e)	1.7449					MW
CuH	Cu—H (r_e)	1.4626					UV
CuI	Cu—I (r_e)	2.3383					MW
FCN	C—F	1.262	C—N	1.159			MW
FOH	O—H	0.96	O—F	1.442	∠HOF	97.2	MW
F_2	F—F (r_e)	1.4119					R
$Fe(CO)_5$	Fe—C (average)	1.821					ED
	$(Fe—C)_{eq} - (Fe—C)_{ax}$	0.020					
	C—O (average)	1.153	(D_{3h})				
GaBr	Ga—Br (r_e)	2.3525					MW
GaCl	Ga—Cl (r_e)	2.2017					MW
GaF	Ga—F (r_e)	1.7744					MW
GaF_3	Ga—F	1.88	(D_{3h})				ED
GaI	Ga—I (r_e)	2.5747					MW
GaI_3	Ga—I	2.458	(D_{3h})				ED
GdI_3	Gd—I	2.841	∠IGdI	108	(C_{3v})		ED
$GeBrH_3$	Ge—H	1.526	Ge—Br	2.299			MW, IR
	∠HGeH	106.2					
$GeBr_4$	Ge—Br	2.272	(T_d)				ED
$GeClH_3$	Ge—H	1.537	Ge—Cl	2.150			IR, MW
	∠HGeH	111.0					
$GeCl_2$	Ge—Cl	2.183	∠ClGeCl	100.3			ED
$GeCl_4$	Ge—Cl	2.113	(T_d)				ED
$GeFH_3$	Ge—H	1.522	Ge—F	1.732			MW, IR
	∠HGeH	113.0					
GeF_2	Ge—F (r_e)	1.7321	∠FGeF (θ_e)	97.17			MW
GeH	Ge—H (r_e)	1.5880					UV
GeH_4	Ge—H	1.5251	(T_d)				IR, R
GeO	Ge—O (r_e)	1.6246					MW
GeS	Ge—S (r_e)	2.0121					MW
GeSe	Ge—Se (r_e)	2.1346					MW
GeTe	Ge—Te (r_e)	2.3402					MW
Ge_2H_6	Ge—H	1.541	Ge—Ge	2.403			ED
	∠HGeH	106.4	∠GeGeH	112.5			
HBr	H—Br (r_e)	1.4145					MW
HCN	C—H (r_e)	1.0655	C—N (r_e)	1.1532			MW, IR
			linear				
HCNO	H—C	1.027	C—N	1.161	N—O	1.207	MW
					linear		
HCl	H-Cl (r_e)	1.2746					MW
HF	H—F (r_e)	0.9169					MW
HI	H—I (r_e)	1.6090					MW
HNCO	N—H	0.986	N—C	1.209	C—O	1.166	MW
	∠HNC	128.0					
HNCS	N—H	0.989	N—C	1.216	C—S	1.561	MW
	∠HNC	135.0	∠NCS	180			
HNO	N—H	1.063	N—O	1.212	∠HNO	108.6	UV
HNO_2			s-trans conformer		s-cis conformer		MW
	$O_a=N—O_bH$		O_b—H	0.958	0.98		
			N—O_b	1.432	1.39		
			N—O_a	1.170	1.19		
			∠$O_a NO_b$	110.7	114		
			∠$NO_b H$	102.1	104		
HNO_3			O_c—H	0.96	N—O_c	1.41	MW
			N—O_a	1.20	N—O_b	1.21	
			∠$HO_c N$	102.2	∠$O_c NO_a$	113.9	
			∠$O_c NO_b$	115.9	planar		

Compound	Structure							Method

Compound			Structure					Method
HNSO	N—H	1.029	N—S	1.512	S—O	1.451		MW
	∠HNS	115.8	∠NSO	120.4				
			planar					
H_2	H—H (r_e)	0.7414						UV
H_2O	O—H (r_e)	0.9575		∠HOH (θ_e)	104.51			MW, IR
H_2O_2	O—O	1.475		∠OOH	94.8			IR
	dihedral angle of internal rotation			119.8	(C_2)			
H_2S	H—S (r_e)	1.3356		∠HSH (θ_e)	92.12			MW, IR
H_2SO_4			O—H	0.97	S—O_a	1.574		MW
			S—O_c	1.422	∠H_aO_aS	108.5		
			∠O_aSO_b	101.3	∠O_cSO_d	123.3		
			∠O_aSO_c	108.6	∠O_aSO_d	106.4		
		dihedral angle between the H_aO_aS and O_aSO_c planes			20.8			
		dihedral angle between the H_aO_aS and O_aSO_b planes			90.9			
		dihedral angle between the H_aSO_b and O_cSO_d planes			88.4	(C_2)		
H_2S_2	S—S	2.055	S—H	1.327	∠SSH	91.3		ED, MW
	dihedral angle of internal rotation		90.6	(C_2)				
$HfCl_4$	Hf—Cl	2.33	(T_d)					ED
$HgCl_2$	Hg—Cl	2.252	linear					ED
HgH	Hg—H (r_e)	1.7404						UV
HgI_2	Hg—I	2.553	linear					ED
IBr	I—Br (r_e)	2.4691						MW
ICN	C—I	1.995	C—N	1.159				MW
ICl	I—Cl (r_e)	2.3210						MW
IF_5	I—F (average)	1.860	(I—F)$_{eq}$ – (I—F)$_{ax}$	0.03				ED, MW
	∠$F_{ax}IF_{eq}$	82.1	(C_{4v})					
IO	I—O (r_e)	1.8676						MW
I_2	I—I (r_e)	2.6663						R
InBr	In—Br (r_e)	2.5432						MW
InCl	In—Cl (r_e)	2.4012						MW
InF	In—F (r_e)	1.9854						MW
InH	In—H (r_e)	1.8376						UV
InI	In—I (r_e)	2.7537						MW
IrF_6	Ir—F	1.830	(O_h)					ED
KBr	K—Br (r_e)	2.8208						MW
KCl	K—Cl (r_e)	2.6667						MW
KF	K—F (r_e)	2.1716						MW
KH	K—H (r_e)	2.244						UV
KI	K—I (r_e)	3.0478						MW
KOH	O—H	0.91	K—O	2.212 linear				MW
K_2	K—K (r_e)	3.9051						UV
KrF_2	Kr—F	1.89	linear					ED
LiBr	Li—Br (r_e)	2.1704						MW
LiCl	Li—Cl (r_e)	2.0207						MW
LiF	Li—F (r_e)	1.5639						MW
LiH	Li—H (r_e)	1.5949						MW
LiI	Li—I (r_e)	2.3919						MW
Li_2	Li—Li (r_e)	2.6729						UV
Li_2Cl_2			Li—Cl	2.23				ED
			Cl—Cl	3.61				
			∠ClLiCl	108				
$LuCl_3$	Lu—Cl	2.417	∠ClLuCl	112	(C_{3v})			ED
MgF	Mg—F (r_e)	1.7500						UV
MgH	Mg—H (r_e)	1.7297						UV
MgO	Mg—O (r_e)	1.749						UV
MnH	Mn—H (r_e)	1.7308						UV
$Mo(CO)_6$	Mo—C	2.063	C—O	1.145	(O_h)			ED
$MoCl_4O$	Mo—Cl	2.279	Mo—O	1.658				ED
	∠ClMoCl	87.2	(C_{4v})					

Compound	Structure						Method
MoF_6	Mo—F	1.820		(O_h)			ED
$NClH_2$	N—H	1.017		N—Cl	1.748		MW, IR
	∠HNCl	103.7		∠HNH	107		
NCl_3	N—Cl	1.759		∠ClNCl	107.1		ED
NF_2	N—F	1.3528		∠FNF	103.18		MW
NH_2	N—H	1.024		∠HNH	103.3		UV
NH_2CN	N—H	1.00		N_a—C	1.35		MW

NH_2CN structure: C—N_b 1.160; ∠HNH 114; angle between the NH_2 plane and the N—C bond 142

H—N_a—C≡N_b—H (with H atoms on N_a)

Compound	Structure						Method
NH_2NO_2	N—N	1.427		N—H	1.005		MW
	∠HNH	115.2		∠ONO	130.1		
	dihedral angle between the NH_2 and NNO_2 planes			128.2			
NH_3	N—H (r_e)	1.012		∠HNH (θ_e)	106.7		IR
NH_4Cl	N—H	1.22	N—Cl	2.54	(C_{3v})		ED
NF_2CN	F_2N_b—C≡N_a		C—N_a	1.158	C—N_b	1.386	MW
	N_b—F	1.399	∠N_aCN_b	174			
	∠CN_bF	105.4	∠FN_bF	102.8			
NH	N—H (r_e)	1.0362					LMR
NH_2OH	N—H	1.02	N—O	1.453	O—H	0.962	MW
	∠HNH	107	∠HNO	103.3	∠NOH	101.4	
	The bisector of H—N—H angle is *trans* to the O—H bond						
NO	N—O (r_e)	1.1506					IR
NOCl	N—Cl	1.975	N—O	1.14	∠ONCl	113	MW
NOF	O—N	1.136	N—F	1.512	∠FNO	110.1	MW
NO_2	N—O	1.193		∠ONO	134.1		MW
NO_2Cl	N—Cl	1.840		N—O	1.202		MW
	∠ONO	130.6		(C_{2v})			
NO_2F	N—O	1.1798		N—F	1.467		MW
	∠ONO	136		(C_{2v})			
NS	N—S (r_e)	1.4940					IR
N_2	N—N (r_e)	1.0977					UV
N_2H_4	N—H	1.021		N—N	1.449		ED, MW
	∠HNH	106.6 (assumed)		∠NNH_a	112		
	∠NNH_b 106	dihedral angle of internal rotation		91			
	H_a: the H atom closer to the C_2 axis, H_b: the H atom farther from the C_2 axis						
N_2O	N—N (r_e)	1.1284		N—O (r_e)	1.1841		MW, IR
N_2O_3			N_a—N_b	1.864	N_a—O_a	1.142	MW
			N_b—O_b	1.202	N_b—O_c	1.217	
			∠$O_aN_aN_b$	105.05			
			∠$N_aN_bO_b$	112.72			
			∠$N_aN_bO_c$	117.47			
N_2O_4			N—N	1.782	N—O	1.190	ED
			∠ONO	135.4	(D_{2h})		
NaBr	Na—Br (r_e)	2.5020					MW
NaCl	Na—Cl (r_e)	2.3609					MW
NaF	Na—F (r_e)	1.9260					MW
NaH	Na—H (r_e)	1.8873					UV
NaI	Na—I (r_e)	2.7115					MW
Na_2	Na—Na (r_e)	3.0789					UV
$NbCl_5$	Nb—Cl_{eq} 2.241		Nb—Cl_{ax} 2.338 (D_{3h})		ED		
NbO	Nb—O (r_e)	1.691					UV
$Ni(CO)_4$	Ni—C	1.838	C—O	1.141	(T_d)		ED
NiH	Ni—H (r_e)	1.476					UV
NpF_6	Np—F	1.981		(O_h)			ED
OCS	C—O (r_e)	1.1578		C—S (r_e)	1.5601		MW

BOND LENGTHS AND ANGLES IN GAS-PHASE MOLECULES (continued)

Compound	Structure						Method
OCSe	C—O	1.159		C—Se	1.709		MW
OF	O—F (r_e)	1.3579					LMR
OF$_2$	O—F (r_e)	1.4053		∠FOF (θ_e)	103.07	(C$_{2v}$)	MW
O(SiH$_3$)$_2$	Si—H	1.486		Si—O	1.634		ED
	∠SiOSi	144.1					
O$_2$	O—O (r_e)	1.2074					MW
O$_2$F$_2$	O—O	1.217		F—O	1.575		MW
	∠OOF	109.5	dihedral angle of internal rotation	87.5	(C$_2$)		
O$_3$	O—O (r_e)	1.2716		∠OOO (θ_e)	117.47	(C$_{2v}$)	MW
OsF$_6$	Os—F	1.831		(O$_h$)			ED
OsO$_4$	Os—O	1.712		(T$_d$)			ED
PBr$_3$	P—Br	2.220		∠BrPBr	101.0		ED
PCl$_3$	P—Cl	2.039		∠ClPCl	100.27		ED
PCl$_5$	(structure)		P—Cl$_a$	2.124	P—Cl$_h$	2.020	ED
					(D$_{3h}$)		
PF	P—F (r_e)	1.5896					UV
PF$_3$	P—F	1.570		∠FPF	97.8		ED, MW
PF$_5$	P—F$_{ax}$	1.577	P—F$_{eq}$	1.534	(D$_{3h}$)		ED
PH	P—H (r_e)	1.4223					LMR
PH$_2$	P—H	1.418		∠HPH	91.70		UV
PH$_3$	P—H	1.4200		∠HPH	93.345		MW
PN	N—P (r_e)	1.4909					MW
PO	O—P (r_e)	1.4759					UV
POCl$_3$	P—O	1.449		P—Cl	1.993		ED
	∠ClPCl	103.3					
POF$_3$	P—O	1.436	P—F	1.524	∠FPF	101.3	ED, MW
P$_2$	P—P (r_e)	1.8931					UV
P$_2$F$_4$	P—F	1.587		P—P	2.281		ED
	∠PPF	95.4		∠FPF	99.1		
	The two PF$_2$ planes are *trans* to each other (the *gauche* conformer is less than 10%)						
P$_4$	P—P	2.21		(T$_d$)			ED
P$_4$O$_6$	P—O	1.638	∠POP	126.4	(T$_d$)		ED
PbH	Pb—H (r_e)	1.839					UV
PbO	Pb—O (r_e)	1.9218					MW
PbS	Pb—S (r_e)	2.2869					MW
PbSe	Pb—Se (r_e)	2.4022					MW
PbTe	Pb—Te (r_e)	2.5950					MW
PrI$_3$	Pr—I	2.904	∠IPrI	113	(C$_{3v}$)		ED
PtO	Pt—O (r_e)	1.7273					UV
PuF$_6$	Pu—F	1.971		(O$_h$)			ED
RbBr	Rb—Br (r_e)	2.9447					MW
RbCl	Rb—Cl (r_e)	2.7869					MW
RbF	Rb—F (r_e)	2.2703					MW
RbH	Rb—H (r_e)	2.367					UV
RbI	Rb—I (r_e)	3.1768					MW
RbOH	Rb—O	2.301	O—H	0.957	linear		MW
ReClO$_3$	Re—O	1.702		Re—Cl	2.229		MW
	∠ClReO	109.4		(C$_{3v}$)			
ReF$_6$	Re—F	1.832		(O$_h$)			ED
RuO$_4$	Ru—O	1.706		(T$_d$)			ED
SCSe	C—Se	1.693		C—S	1.553		MW
SCTe	C—S	1.557		C—Te	1.904		MW
SCl$_2$	S—Cl	2.006	∠ClSCl	103.0	(C$_{2v}$)		ED
SF	S—F (r_e)	1.6006					MW
SF$_2$	S—F	1.5921		∠FSF	98.20		MW
SF$_6$	S—F	1.561		(O$_h$)			ED

9-23

Compound	Structure					Method
SO	S—O (r_e)	1.4811				MW
SOCl$_2$	S—O	1.44	S—Cl	2.072		MW
	∠ClSCl	97.2	∠OSCl	108.0		
SOF$_2$	S—O	1.420	S—F	1.583		ED
	∠OSF	106.2	∠FSF	92.2		
SOF$_4$	F$_b$ F$_b$ / F$_a$—S—F$_a$ / O		S—O	1.403	S—F$_a$ 1.575	ED
			S—F$_b$	1.552	∠OSF$_a$ 90.7	
			∠OSF$_b$	124.9	∠F$_a$SF$_b$ 89.6	
			∠F$_b$SF$_b$	110.2	(C$_{2v}$)	
SO$_2$	S—O (r_e)	1.4308	∠OSO (θ_e)	119.329		MW
SO$_2$Cl$_2$	S—O	1.404	S—Cl	2.011	∠OSO 123.5	ED
	∠ClSCl	100.0	(C$_{2v}$)			
SO$_2$F$_2$	S—O	1.397	S—F	1.530	∠OSO 123	ED
	∠FSF	97	(C$_{2v}$)			
SO$_3$	S—O	1.4198	(D$_{3h}$)			IR
S(SiH$_3$)$_2$	Si—H	1.494	Si—S	2.136	∠SiSSi 97.4	ED
S$_2$	S—S (r_e)	1.8892				R
S$_2$Br$_2$	S—Br	2.24	S—S	1.98		ED
	∠SSBr	105	dihedral angle of internal rotation	83.5		
S$_2$Cl$_2$	S—Cl	2.057	S—S	1.931		ED
	∠SSCl	108.2	dihedral angle of internal rotation	84.1 (C$_2$)		
S$_2$O$_2$	S—O	1.458	S—S 2.025		∠OSS 112.8	MW
			planar *cis* form			
S$_8$	(ring structure)		S—S	2.07		ED
			∠SSS	105		
			(D$_{4d}$)			
SbCl$_3$	Sb—Cl	2.333	∠ClSbCl	97.2		ED
SbH$_3$	Sb—H	1.704	∠HSbH	91.6		MW
SeF	Se—F	1.742				MW
SeF$_6$	Se—F	1.69	(O$_h$)			ED
SeO	Se—O (r_e)	1.6393				MW
SeOF$_2$	Se—O	1.576	Se—F	1.730		MW
	∠OSeF	104.82	∠FSeF	92.22		
SeO$_2$	Se—O (r_e)	1.6076	∠OSeO (θ_e)	113.83		MW
SeO$_3$	Se—O	1.69	(D$_{3h}$)			ED
Se$_2$	Se—Se (r_e)	2.1660				UV
Se$_6$	Se—Se	2.34	∠SeSeSe	102		ED
			six-membered ring with chair conformation			
SiBrF$_3$	Si—F	1.560	Si—Br	2.153		MW
	∠FSiBr	108.5	(C$_{3v}$)			
SiBrH$_3$	Si—H	1.485	Si—Br	2.210		MW
	∠HSiBr	107.8	(C$_{3v}$)			
SiClH$_3$	Si—H	1.482	Si—Cl	2.048		MW
	∠HSiCl	107.9	(C$_{3v}$)			
SiCl$_4$	Si—Cl	2.019	(T$_4$)			ED
SiF	Si—F	1.6008				UV
SiFH$_3$	Si—H	1.484	Si—F	1.593		MW, IR
	∠HSiH	110.63	(C$_{3v}$)			
SiF$_2$	Si—F (r_e)	1.590	∠FSiF (θ_e)	100.8		MW
SiF$_3$H	Si—H (r_e)	1.4468	Si—F (r_e)	1.5624		MW
	∠HSiF (θ_e)	110.64				
SiF$_4$	Si—F	1.553	(T$_d$)			ED
SiH	Si—H (r_e)	1.5201				UV
SiH$_3$I	Si—H	1.485	Si—I	2.437		MW
	∠HSH	107.8				
SiH$_4$	Si—H	1.4798	(T$_d$)			IR
SiN	N—Si (r_e)	1.572				UV
SiO	Si—O (r_e)	1.5097				MW

Compound	Structure						Method
SiS	Si—S (r_e)	1.9293					MW
SiSe	Se—Si (r_e)	2.0583					MW
Si$_2$	Si—Si (r_e)	2.246					UV
Si$_2$Cl$_6$	Si—Si	2.32		Si—Cl	2.009		ED
	∠ClSiCl	109.7					
Si$_2$F$_6$	Si—Si	2.317		Si—F	1.564		ED
	∠FSiF	108.6					
Si$_2$H$_6$	Si—H	1.492		Si—Si	2.331		ED
	∠SiSiH	110.3		∠HSiH	108.6		
				staggered form (assumed)			
SnCl$_4$	Sn—Cl	2.280		(T$_d$)			ED
SnH	Sn—H (r_e)	1.7815					UV
SnH$_4$	Sn—H	1.711		(T$_d$)			R, IR
SnO	Sn—O	1.8325					MW
SnS	S—Sn (r_e)	2.2090					MW
SnSe	Se—Sn (r_e)	2.3256					MW
SnTe	Sn—Te (r_e)	2.5228					MW
SrH	Sr—H (r_e)	2.1455					UV
SrO	Sr—O (r_e)	1.9198					MW
SrS	S—Sr (r_e)	2.4405					UV
TaCl$_5$	Ta—Cl$_{eq}$	2.227		Ta—Cl$_{ax}$	2.369	(D$_{3h}$)	ED
TaO	Ta—O (r_e)	1.6875					UV
TeF$_6$	Te—F	1.815		(O$_h$)			ED
Te$_2$	Te—Te (r_e)	2.5574					UV
ThCl$_4$	Th—Cl	2.58		(T$_d$)			ED
ThF$_4$	Th—F	2.14		(T$_d$)			ED
TlBr	Tl—Br (r_e)	2.6182					MW
TlCl	Tl—Cl (r_e)	2.4848					MW
TlF	Tl—F (r_e)	2.0844					MW
TlH	Tl—H (r_e)	1.870					UV
TlI	Tl—I (r_e)	2.8137					MW
TiBr$_4$	Ti—Br	2.339		(T$_d$)			ED
TiCl$_4$	Ti—Cl	2.170		(T$_d$)			ED
TiO	Ti—O (r_e)	1.620					UV
TiS	Ti—S (r_e)	2.0825					UV
UF$_6$	U—F	1.996		(O$_h$)			ED
V(CO)$_6$	V—C	2.015		C—O	1.138		ED
	(O$_h$, involving dynamic Jahn-Teller effect)						
VCl$_3$O	V—O	1.570		V—Cl	2.142		ED, MW
	∠ClVCl	111.3					
VCl$_4$	V—Cl	2.138		(T$_d$, involving dynamic Jahn-Teller effect)			ED
VF$_5$	V—F (average)	1.71					ED
VO	V—O (r_e)	1.5893					UV
W(CO)$_6$	W—C	2.059	C—O	1.149	(O$_h$)		ED
WClF$_5$				W—Cl	2.251		MW
				W—F (average)	1.836		
				∠F$_a$WF$_b$	88.7		
WF$_4$O	W—O	1.666		W—F	1.847		ED
	∠FWF	86.2		(C$_{4v}$)			
WF$_6$	W—F	1.832		(O$_h$)			ED
XeF$_2$	Xe—F	1.977		linear			IR
XeF$_4$	Xe—F	1.94		(D$_{4h}$)			ED
XeF$_6$	Xe—F	1.890		(large-amplitude bending vibration around the O$_h$ structure)			ED
XeO$_4$	Xe—O	1.736		(T$_d$)			ED
ZnH	Zn—H (r_e)	1.5949					UV
ZrCl$_4$	Zr—Cl	2.32		(T$_d$)			ED
ZrF$_4$	Zr—F	1.902		(T$_d$)			ED
ZrO	Zr—O (r_e)	1.7116					UV

For WClF$_5$ the structure diagram shows:

```
      Cl
      |   F_b
      |  /
F_b —W— F_b
   /  |
 F_b  F_a
```

STRUCTURES OF ORGANIC MOLECULES

Compounds are Arranged in Alphabetical Order by Chemical Name; Cross References are Given for Common Synonyms (Lengths in Å and Angles in Degrees)

Compound	Structure				Method
Acetaldehyde			C_a—O	1.210	ED, MW
			C_b—H	1.107	
			C_a—H	1.128	
	C_a—C_b	1.515	$\angle C_bC_aO$	124.1	
	$\angle HC_bH$	109.8	$\angle C_bC_aH$	115.3	
Acetamide	C—O	1.220	C—N	1.380	ED
CH_3CONH_2	C—C	1.519	N—H	1.022	
	C—H	1.124	$\angle NCO$	122.0	
	$\angle CCN$	115.1			
Acetic acid			C—C	1.520	ED
			C—O_a	1.214	
			C—O_b	1.364	
	C—H	1.10	$\angle CCO_a$	126.6	
	$\angle CCO_b$	110.6			
Acetone	C—C	1.520	C—O	1.213	ED, MW
$(CH_3)_2CO$	C—H	1.103	$\angle CCC$	116.0	
	$\angle HCH$	108.5	symmetry axis of each methyl group is tilted 2° from the C—C bond		
Acetonitrile	C—H	1.107	C—C	1.468	ED, MW
CH_3CN	C—N	1.159	$\angle CCH$	109.7	
Acetonitrile oxide	C—C	1.442	C—N	1.169	MW
CH_3CNO	N—O	1.217	(C_{3v})		
Acetyl chloride	C—H	1.105	C—O	1.187	ED, MW
CH_3COCl	C—C	1.506	C—Cl	1.798	
	$\angle HCH$	108.6	$\angle OCCl$	121.2	
	$\angle CCCl$	111.6			
Acetyl cyanide → Pyruvonitrile					
Acetylene	C—H (r_e)	1.060	C—C (r_e)	1.203	IR
HC≡CH					
Acrolein → Acrylaldehyde					
Acrylaldehyde			C_b—C_c	1.484	ED, MW
			C_a—C_b	1.345	
			C_c—O	1.217	
			C_a—H	1.10	
			C_c—H	1.13	
	$\angle C_aC_bC_c$	120.3	$\angle C_bC_cO$	123.3	
	$\angle HC_cC_b$	114	other CCH angles (average)	122	
	planar s-trans form				
Acrylonitrile			C_a—C_b	1.343	ED, MW
			C_b—C_c	1.438	
			C_c—N	1.167	
			C_a—H	1.114	
			$\angle C_aC_bC_c$	121.7	
	$\angle C_bC_cN$	178	$\angle HCC$	120	
Acryloyl chloride	C—H	1.086 (assumed)	C_b—C_c	1.48	MW
	C_c—Cl	1.82	C_a—C_b	1.35	
			C_c—O	1.19	
			$\angle C_aC_bH$	120 (assumed)	
			$\angle C_bC_aH$	121.5 (assumed)	
			$\angle C_aC_bC_c$	123	
			$\angle C_bC_cCl$	116	
			$\angle C_bC_cO$	127	

Compound	Structure				Method

Allene
$CH_2=C=CH_2$

C—C	1.3084	C—H	1.087	IR
∠HCH	118.2			

Allyl chloride

		cis conformer	C—Cl 1.811	MW
$H_2C_a=C_bH—C_cH_2$			∠CCCl 115.2	
		skew conformer	C—Cl 1.809	
∠CCCl	109.6	CCCCl	dihedral angle of internal rotation 122.4	

Aniline
$C_6H_5NH_2$

C—C	1.392	C—N	1.431	MW
N—H	0.998	∠HNH	113.9	
dihedral angle between the NH_2 plane and the N—C bond			140.6	

Azetidine

$CH_2—CH_2$
$|\qquad|$
$CH_2—NH$

		C—N	1.482	ED
		C—C	1.553	
		C—H	1.107	
N—H	1.03	∠CNC	92.2	
∠CCC	86.9	∠CCN	85.8	
dihedral angle between the CCC and CNC planes			147	

Aziridine

		N—H	1.016	MW
		N—C	1.475	
		C—C	1.481	
		C—H	1.084	
		∠CNC	60.3	
		∠H_aNC	109.3	
∠H_bCH_c	115.7	∠H_bCC	117.8	
∠H_bCN	118.3	∠H_cCC	119.3	
∠H_cCN	114.3			

Azomethane
$CH_3N=NCH_3$

C—N	1.482	N—N	1.247	ED
∠CNN	112.3	*trans* conformer		

Benzene
C_6H_6

C—C	1.399	C—H	1.101	ED, IR

***p*-Benzoquinone**

C_a—O	1.225		ED
C_b—C_b	1.344		
C_a—C_b	1.481		
∠$C_b$$C_a$$C_b$	118.1		

Biacetyl
$CH_3COCOCH_3$

C—O	1.215	C—C (average)	1.524	ED
C—H	1.108	∠CCO	119.5	
∠CCC	116.2	*trans* conformer		

Bicyclo[1.1.0]butane

		C_a—C_a	1.497	MW
		C_a—C_b	1.498	
		C_a—H_a	1.071	
		C_b—H_b, C_b—H_c	1.093	
		∠$H_b$$C_b$$H_c$	115.6	
∠$C_b$$C_a$$H_a$	130.4	∠$C_a$$C_b$$C_a$	60.0	
∠$C_a$$C_a$$H_a$	128.4	dihedral angle between the two $C_a$$C_a$$C_b$ planes	121.7	

Bicyclo[2.2.1]hepta-2,5-diene

		C_a—C_b	1.535	ED
		C_b—C_b	1.343	
		C_a—C_c	1.573	
		C—H	1.12	
		∠$C_a$$C_c$$C_a$	94	
dihedral angle between the two $C_a$$C_b$$C_b$$C_a$ planes			115.6 (C_{2v})	

Bicyclo[2.2.1]heptane
C_7H_{12}

See the preceding molecule for the labels of the C atoms				ED
C_a—C_b	1.54	C_b—C_b	1.56	
C_a—C_c	1.56	C—C (average)	1.549	
∠$C_a$$C_c$$C_a$	93.1	dihedral angle between the two $C_a$$C_b$$C_b$$C_a$ planes	113.1	

Compound	Structure			Method

Bicyclo[2.2.0]hexa-2,5-diene

C_b—C_b	1.345
C_a—C_a	1.574
C_a—C_b	1.524
dihedral angle between the two $C_aC_bC_bC_a$ planes	117.3

Method: ED

Bicyclo[2.2.2]octane

$HC_a(C_bH_2C_bH_2)_3C_aH$

		C_a—C_b	1.54
C_b—C_b	1.55	$\angle C_aC_bC_b$	109.7
C—C (average)	1.542		

large-amplitude torsional motion about the D_{3h} symmetry axis

Method: ED

Bicyclo[1.1.1]pentane C_5H_8

C—C	1.557	$\angle CCC$	74.2

Method: ED

Bicyclo[2.1.0]pentane

C_a—C_a	1.536
C_b—C_b	1.565
C_a—C_b	1.528
C_a—C_c	1.507

Dihedral angle between the $C_aC_aC_bC_b$ and $C_aC_aC_c$ planes 112.7

Method: MW

Biphenyl

C—C (intra-ring)	1.396
(inter-ring)	1.49
torsional dihedral angle between the two rings	~40

Method: ED

4,4'-Bipyridyl

C—C, C—N (intra-ring)

	1.375
C—C (inter-ring)	1.465
torsional dihedral angle between the two rings	~37

Method: ED

Bis (cyclopentadienyl) beryllium $(C_5H_5)_2Be$

Be—(cyclopentadienyl plane) 1.470, 1.92
C—C 1.423 (C_{5v}) (The Be atom has two equilibrium positions)

Method: ED

Bis (cyclopentadienyl) iron → Ferrocene

Bis (cyclopentadienyl) lead $(C_5H_5)_2Pb$

C—C	1.430	Pb—C	2.79

dihedral angle between the two C_5H_5 planes 40~50 (The two rings are not parallel.)

Method: ED

Bis (cyclopentadienyl) manganese $(C_5H_5)_2Mn$

Mn—C	2.383	C—C	1.429	(D_{5h})

Method: ED

Bis (cyclopentadienyl) nickel $(C_5H_5)_2Ni$

Ni—C	2.196	C—C	1.430
			(D_{5h})

Method: ED

Bis (cyclopentadienyl) ruthenium $(C_5H_5)_2Ru$

C—C	1.439	Ru—C	2.196

Method: ED

Bis (cyclopentadienyl) tin $(C_5H_5)_2Sn$

C—C	1.431	Sn—C	2.71
C—H	1.14	(D_{5h})	

Method: ED

Bis (trifluoromethyl) peroxide CF_3OOCF_3

O—O	1.42	C—O	1.399
C—F	1.320	$\angle COO$	107
$\angle FCF$	109.0	COOC dihedral angle of internal rotation	123

Method: ED

Borine carbonyl BH_3CO

B—H	1.194	B—C	1.540
C—O	1.131	$\angle HBH$	113.9
$\angle BCO$	180	(C_{3v})	

Method: MW

Bromobenzene

C—H	1.072
C_c—C_d	1.401
C_b—C_c	1.375
C—Br	1.85
C_a—C_b	1.42
$\angle C_bC_aC_b$	117.4

Method: MW

Bromoform $CHBr_3$

C—Br	1.924	C—H	1.11
$\angle BrCBr$	111.7	(C_{3v})	

Method: ED, MW

Bromoiodoacetylene $IC\equiv CBr$

C—I	1.972	C—C	1.206
C—Br	1.795		

Method: ED

Compound	Structure				Method

1,3-Butadiene

C_aH_2
$\quad C_bH—C_bH$
$\qquad\qquad C_aH_2$

$\angle C_bC_aH$	120.9	
	$C_b—C_b$	1.467
	$C_a—C_b$	1.349
	C—H (average)	1.108
	$\angle CCC$	124.4
	anti conformer	(C_{2h})

Method: ED

1,3-Butadiyne

$HC_a{\equiv}C_b—C_b{\equiv}C_aH$

$C_a—C_b$	1.218	
C—H	1.09	
$C_b—C_b$	1.384	
linear		

Method: ED

Butane
$CH_3CH_2CH_2CH_3$

C—C	1.531		C—H	1.117
$\angle CCC$	113.8		$\angle CCH$	111.0
trans conformer	54%	dihedral angle for the *gauche* conformer		65

Method: ED

2-Butanone → Ethyl methyl ketone

Butatriene

$H_2C_a{=}C_b{=}C_b{=}C_aH_2$

$C_a—C_b$	1.32		C—H	1.08
			$C_b—C_b$	1.28 (D_{2h})

Method: ED

2-Butene

$C_aH_3—C_bH{=}C_bH—C_aH_3$

$C_a—C_b$	*cis* conformer	ED	
		1.506	*trans* conformer 1.508
$C_b—C_b$	1.346	1.347	
$\angle C_aC_bC_b$	125.4	123.8	

3-Buten-1-yne → Vinylacetylene

tert-Butyl chloride
$(CH_3)_3CCl$

C—H	1.102		C—C	1.528
C—Cl	1.828		$\angle CCCl$	107.3
$\angle CCH$	110.8		$\angle CCC$	111.6

Method: ED, MW

tert-Butyl cyanide → Pivalonitrile

2-Butyne

$C_aH_3—C_b{\equiv}C_b—C_aH_3$

C—H	1.116	ED	
$C_b—C_b$	1.214	$C_a—C_b$	1.468
$\angle C_bC_aH$	110.7		

Carbon C₂

C—C (r_e)	1.3119

Method: UV

Carbon C₃

C—C	1.277	linear

Method: UV

Carbon suboxide → Tricarbon dioxide

Carbon tetrabromide
CBr_4

C—Br	1.935	(T_d)

Method: ED

Carbon tetrachloride
CCl_4

C—Cl	1.767	(T_d)

Method: ED

Carbon tetrafluoride
CF_4

C—F	1.323	(T_d)

Method: ED

Carbon tetraiodide
CI_4

C—I	2.15	(T_d)

Method: ED

Carbonyl cyanide
$CO(CN)_2$

C—O	1.209		C—C	1.466
C—N	1.153		$\angle CCC$	115
$\angle CCN$	180			

Method: ED, MW

Chloroacetylene
$HC{\equiv}CCl$

C—H	1.0550		C—C	1.2033
C—Cl	1.6368			

Method: MW

Chlorobenzene
C_6H_5Cl

C—C	1.400		C—Cl	1.737
C—H	1.083		$\angle CC(Cl)C$	121.7
$\angle CC(H)C$	120			

Method: ED

Chlorobromoacetylene
$ClC{\equiv}CBr$

Cl—C	1.636	C—C	1.206
C—Br	1.784		

Method: ED

Chlorocyanoacetylene
$ClC{\equiv}CCN$

C—Cl	1.624	C—C	1.205
C—CN	1.362	C—N	1.160

Method: ED

Chloroethane → Ethyl chloride

2-Chloroethanol
$ClCH_2CH_2OH$

C—O	1.413		C—C	1.519
C—Cl	1.801		C—H	1.093
O—H	1.033	$\angle CCCl$ 110.7	$\angle CCO$	113.8

fraction of the *gauche* conformer at 37°C is 92 ~ 94%,
dihedral angle of internal rotation 62.4

Method: ED

Chloroethylene → Vinyl chloride

Chloroform
$CHCl_3$

C—H	1.100		C—Cl	1.758
$\angle ClCCl$	111.3		(C_{3v})	

Method: MW

Compound	Structure				Method

Chloroiodoacetylene
$ClC \equiv CI$

C—Cl	1.63	C—I	1.99		MW
C—C	1.209 (assumed)				

Chloromethane → Methyl chloride

3-Chloropropene → Allyl chloride

Cyanamide
$H_2N_aCN_b$

N_a—C	1.346	C—N_b	1.160	MW
N—H	1.00	∠HNH	114	
dihedral angle between the NH_2 plane and the N—C bond			142	

Cyanoacetylene
$H—C_b \equiv C_a—C_c \equiv N$

C_b—H	1.058	C_a—C_b	1.205	MW
C_a—C_c	1.378	C_c—N	1.159	

Cyanocyclopropane
$C_3H_5C_aN$

C—C (ring)	1.513	C—C_a	1.472	MW
C—H	1.107	C_a—N	1.157	
∠HCH	114.6	∠C_aCH	119.6	

Cyanogen
$(CN)_2$

C—N	1.163	C—C	1.393	ED
			linear	

Cyclobutane
$(CH_2)_4$

C—H	1.113	C—C	1.555	ED
dihedral angle between the two CCC planes 145				

Cyclobutanone

$$C_cH_2 \quad \overset{C_bH_2}{\underset{C_bH_2}{\diamondsuit}} \quad C_a{=}O$$

C_a—C_b	1.527	MW
C_b—C_c	1.556	
∠$C_bC_aC_b$	93.1	
∠$C_aC_bC_c$	88.0	

Cyclobutene

$$\begin{array}{c} H_2C_a{-}C_aH_2 \\ HC_b{=}C_bH \end{array}$$

C_b—C_b	1.342	C_a—C_a	1.566	MW
C_a—C_b	1.517	C_a—H	1.094	
C_b—H	1.083			
∠$C_aC_bC_b$	94.2	∠C_bC_bH	133.5	
∠C_aC_aH	114.5	∠$C_aC_aC_b$	85.8	
∠HC_aH	109.2	dihedral angle between the CH_2 plane and the C_a—C_a bond 135.8		

Cyclohexane
C_6H_{12}

C—C	1.536	C—H	1.119	ED
∠CCC	111.3	chair form		

Cyclohexene

$$\begin{array}{c} HC_a{=}C_aH \\ H_2C_b \qquad C_bH_2 \\ C_cH_2{-}C_cH_2 \end{array}$$

		C_a—C_a	1.334	ED
		C_a—C_b	1.50	
		C_b—C_c	1.52	
		C_c—C_c	1.54	
∠$C_aC_aC_b$	123.4	∠$C_aC_bC_c$	112.0	
∠$C_bC_cC_c$	110.9	(C_2) half-chair form		

Cyclooctatetraene

C_a—C_b	1.476	ED	
C—H	1.100		
C_a—C_a, C_b—C_b	1.340		
∠$C_bC_aC_a$, ∠$C_aC_bC_b$	126.1		
dihedral angle between the $C_aC_aC_aC_a$ and $C_aC_bC_bC_a$ planes	136.9		
tub form (D_{2d})			

1,3-Cyclopentadiene

$$\begin{array}{c} C_aH_2 \\ HC_b \qquad C_bH \\ HC_c{-}C_cH \end{array}$$

		C_a—C_b	1.509	MW
		C_b—C_c	1.342	
		C_c—C_c	1.469	
		∠$C_aC_bC_c$	109.3	
∠$C_bC_cC_c$	109.4	∠$C_bC_aC_b$	102.8	

Cyclopentadienylindium

$$\begin{array}{c} In \\ HC{+}CH \\ HC \qquad CH \\ CH \end{array}$$

In—C	2.621	ED	
C—C	1.426		
(C_{5v})			

Cyclopentane
$(CH_2)_5$

C—H	1.114	C—C	1.546	ED
∠CCH	111.7			
(The out-of-plane vibration of the C atoms is essentially free pseudorotation; average value of the displacements of the C atoms from the molecular plane 0.43)				

Compound	Structure		Method

Cyclopentene

	C_a—C_b	1.546	ED
	C_b—C_c	1.519	
	C_c—C_c	1.342	
	$\angle C_b C_a C_b$	104.0	
$\angle C_b C_c C_c$ 110.0	$\angle C_a C_b C_c$	103.0	

dihedral angle between the $C_b C_a C_b$ and $C_b C_c C_c C_b$ planes   151.2

Cyclopropane
(CH$_2$)$_3$

C—C	1.512	C—H	1.083
$\angle$HCH	114.0		

Method: R

Cyclopropanone

C—H	1.086	C_a—C_b	1.475
C_b—C_b	1.575	C_a—O	1.191
$\angle C_a C_b C_b$	57.7		

Method: MW

$\angle HC_b H$ 114   dihedral angle between the CH$_2$ plane and the C_b—C_b bond   151

Cyclopropene

	C_b—C_b	1.304	ED
	C_a—C_b	1.519	
	C_b—H	1.077	
C_a—H 1.112	$\angle C_b C_b H$	133	
$\angle HC_a H$ 118			

Decalin
C$_{10}$H$_{18}$

C—C (average)	1.530	C—H (average)	1.113
$\angle$CCC (average)	111.4		

Method: ED

Dewer benzene → Bicyclo[2.2.0]
  hexa-2,5-diene

Diacetylene → 1,3-Butadiyne

1,4-Diazabicyclo[2.2.2]octane

C—N	1.472	ED
C—C	1.562	
$\angle$NCC	110.2	
$\angle$CNC	108.7	

large-amplitude torsional motion about the D_{3h} symmetry axis

2,3-Diaza-1,3-butadiene → Formalde-
  hyde azine

Diazirine

C—H	1.09	MW
C—N	1.482	
N—N	1.228	
$\angle$HCH	117	

Diazoacetonitrile

C_b—N_b	1.280	MW
N_b—N_c	1.132	
C_a—N_a	1.165	
C—H	1.082	
C_a—C_b	1.424	

$\angle C_a C_b H$	117	$\angle C_a C_b N_b$	119.5

Diazomethane
CH$_2$N$_2$

C—H	1.075	C—N	1.32
N—N	1.12	$\angle$HCH	126.0 linear

Method: MW, IR

1,2-Dibromoethane
CH$_2$BrCH$_2$Br

C—C	1.506	C—Br	1.950
C—H	1.108	$\angle$CCBr	109.5
$\angle$CCH 110	fraction of the *trans* conformer at 25°C	95%	

Method: ED

Dibromomethane
CH$_2$Br$_2$

C—H	1.08	C—Br	1.924
$\angle$HCBr	109	$\angle$BrCBr	113.2

Method: ED

2,2'-Dichlorobiphenyl
C$_6$H$_4$Cl—C$_6$H$_4$Cl

C—C	1.398	C—C inter-ring	1.495
C—Cl	1.732	C—H	1.10
$\angle$CCCl	121.4	$\angle$CCH	126

Method: ED

dihedral angle between the two aromatic rings 74 (defined to be 0 for that of the *cis* conformer)

trans-1,4-Dichlorocyclohexane
C$_6$H$_{10}$Cl$_2$

C—H	1.102	C—Cl	1.810
C—C	1.530	$\angle$CCC	111.5
$\angle$CCCl (*ee*)	108.6	$\angle$CCCl (*aa*)	110.6
$\angle$HCCl (*ee*)	111.5	$\angle$HCCl (*aa*)	107.6
ee 49%	*aa* 51%	e: equatorial, a: axial	

Method: ED

Compound	Structure				Method
1,1-Dichloroethane	C—Cl	1.766	C—C	1.540	MW
CHCl$_2$CH$_3$	∠ClCCl	112.0	∠CCCl	111.0	
1,2-Dichloroethane	C—C	1.531	C—Cl	1.790	ED
CH$_2$ClCH$_2$Cl	C—H	1.11	∠CCCl	109.0	
	∠CCH	113			
	fraction of the *trans* conformer at room temperature 73%, that of the *gauche* conformer 27%				
1,1-Dichloroethylene	C—C	1.32 (assumed)	C—Cl	1.73	MW
CH$_2$=CCl$_2$	∠ClCC	123	(C$_{2v}$)		
cis-1,2-Dichloroethylene	C—Cl	1.718	C—C	1.354	ED
CHCl=CHCl	∠ClCC	123.8			
Dichloromethane	C—H (r_e)	1.087	C—Cl (r_e)	1.765	MW, IR
CH$_2$Cl$_2$	∠HCH (θ_e)	111.5	∠ClCCl (θ_e)	112.0	
1,1-Difluoroethane	C—C	1.498	C—H (average)	1.081	ED
CH$_3$CHF$_2$	C—F	1.364	∠CCH (average)	111.0	
	∠CCF	110.7	dihedral angle between the two CCF planes	118.9	
1,2-Difluoroethane	C—F	1.389	C—C	1.503	ED
CH$_2$FCH$_2$F	C—H	1.103	∠CCF	110.3	
	∠CCH	111	dihedral angle of internal rotation	109	
	fraction of the *gauche* conformer at 22°C		94%		
1,1-Difluoroethane	C—C	1.340	C—F	1.315	ED, MW
CH$_2$=CF$_2$	C—H	1.091	∠CCF	124.7	
	∠CCH	119.0			
cis-1,2-Difluoroethylene	C—C	1.33	C—F	1.342	ED, MW
CHF=CHF	C—H	1.099	∠CCF	122.0	
	∠CCH	124.1			
Difluoromethane	C—H	1.093	C—F	1.357	MW
CH$_2$F$_2$	∠HCH	113.7	∠FCF	108.3	
Dimethoxymethane			C$_a$—O	1.432	ED
			C$_b$—O	1.382	
			C—H (average)	1.108	
	∠COC	114.6	∠OCO	114.3	
	∠OCH	110.3			
Dimethylacetylene → 2-Butyne					
Dimethylamine	C—H	1.106	N—H	1.00	ED
(CH)$_2$NH	C—N	1.455	∠CNC	111.8	
	∠CNH	107	∠NCH	112	
	∠HCH	107			
Dimethylberyllium	Be—C	1.698	C—H	1.127	ED
(CH$_3$)$_2$Be	∠BeCH	113.9	CBeC linear		
Dimethylcadmium	C—Cd	2.112	∠HCH	108.4	R
(CH$_3$)$_2$Cd					
Dimethyl carbonate			C$_b$—O$_b$	1.209	ED
			C$_b$—O$_a$	1.34	
			C$_a$—O$_a$	1.42	
	∠O$_a$C$_b$O$_a$	107	∠C$_b$O$_a$C$_a$	114.5	
Dimethylcyanamide	C$_b$—N$_b$	1.161	C$_b$—N$_a$	1.338	ED
(C$_a$H$_3$)$_2$N$_a$—C$_b$≡N$_b$	C$_a$—N$_a$	1.463	∠C$_a$NC$_a$	115.5	
	∠C$_a$NC$_b$	116.0			
1,2-Dimethyldiborane			B—B	1.799	ED
			B—C	1.580	
			B—H$_b$	1.358 (*cis*), 1.365 (*trans*)	
	B—H$_t$	1.24			
	∠BBC	122.6 (*cis*), 121.8 (*trans*)			
Dimethyl diselenide	C—H	1.13	C—Se	1.95	ED
(CH$_3$)$_2$Se$_2$	Se—Se	2.326	∠CSeSe	98.9	
	∠HCSe	108	dihedral angle between the CSeSe and SeSeC planes	88	

9-32

Compound	Structure				Method
Dimethyl disulfide	C—S	1.816	S—S	2.029	ED
$(CH_3)_2S_2$	C—H	1.105	∠SSC	103.2	
	∠SCH	111.3	CSSC dihedral angle of internal rotation	85	
S,S'-Dimethyl dithiocarbonate	$C_aH_3SC_bSC_aH_3$		C_b—O	1.206	ED
	‖		C_b—S	1.777	
	O		C_a—S	1.802	
	∠OCS	124.9	∠CSC	99.3	
			syn-syn conformer		
Dimethyl ether	C—O	1.416	C—H	1.121	ED
$(CH_3)_2O$	∠COC	112	∠HCH	108	
Dimethylglyoxal → Biacetyl					
N,N'-Dimethylhydrazine	N—N	1.42	C—N	1.46	ED
CH_3NH—$NHCH_3$	N—H	1.03	C—H	1.12	
	∠NNC	112	CNNC dihedral angle of internal rotation	90	
Dimethylmercury	C—Hg	2.083	C—H	1.160 (assumed)	ED
$(CH_3)_2Hg$	Hg···H	2.71			
Dimethylphosphine	C—P	1.848	P—H	1.419	MW
$(CH_3)_2PH$	∠CPC	99.7	∠CPH	97.0	
Dimethyl selenide	C—H	1.093	Se—C	1.943	MW
$(CH_3)_2Se$	∠CSeC	96.2	∠SeCH	108.7	
	∠HCH	110.3			
Dimethyl sulfide	C—S	1.807	C—H	1.116	ED, MW
$(CH_3)_2S$	∠CSC	99.05	∠HCH	109.3	
Dimethyl sulfone	C—H	1.114	S—O	1.435	ED
$(CH_3)_2SO_2$	S—C	1.771	∠CSC	102	
	∠OSO	121			
Dimethyl sulfoxide	C—H	1.081	C—S	1.799	MW
$(CH_3)_2SO$	S—O	1.485	∠CSC	96.6	
	∠CSO	106.7	∠HCH	110.3	
	dihedral angle between the SCC plane and the S—O bond			115.5	
Dimethylzinc	Zn—C	1.929	∠HCH	107.7	R
$(CH_3)_2Zn$					
1,4-Dioxane	C—C	1.523	C—O	1.423	ED
	C—H	1.112	∠COC	112.45	
	∠CCO	109.2	chair form		
Ethanal → Acetaldehyde					
Ethane	C—C	1.5351	C—H	1.0940	MW
C_2H_6	∠CCH	111.17	staggered conformation		
Ethanethiol	C_bH_3—C_aH_2—SH		C_a—H	1.090	MW
	C_b—H	1.093	C_a—C_b	1.530	
	C_a—S	1.829	S—H	1.350	
	∠C_bC_aH	109.6	∠C_aC_bH	109.7	
	∠C_bC_aS	108.3	∠C_aSH	96.4	
Ethanol	$C_bH_3C_aH_2OH$		C—C	1.512	MW
	C—O	1.431	O—H	0.971	
	C_a—H	1.10	C_b—H	1.09	
	∠CCO	107.8	∠COH	105	
	∠C_bC_aH	111	∠C_aC_bH	110	
	staggered conformation				
Ethyl chloride			C—C	1.528	ED, MW
			C—Cl	1.802	
			C—H	1.103	
			C_a—H_a=C_b—H_b (assumed)		
			∠CCCl	110.7	
	∠$H_bC_bH_b$	109.8	∠$H_aC_aH_a$	109.2	
	∠$C_bC_aH_a$	110.6			
Ethylene	C—H	1.087	C—C	1.339	MW
$CH_2=CH_2$	∠CCH	121.3			

Compound	Structure				Method
Ethylenediamine	C—N	1.469	C—C	1.545	ED
$H_2NCH_2CH_2NH_2$	C—H	1.11	∠CCN	110.2	
	gauche conformer	dihedral angle between the NCC and CCN planes		64	
Ethylene dibromide → 1,2-Dibromoethane					
Ethylene dichloride → 1,2-Dichloroethane					
Ethyleneimine → Aziridine					
Ethylene oxide	C—C	1.466	C—H	1.085	MW
	C—O	1.431	∠HCH	116.6	
	dihedral angle between the NH_2 plane and the N—C bond			158.0	
Ethylene sulfide → Thiirane					
Ethyl methyl ether	C—O (average)	1.418	C—C	1.520	ED
$C_2H_5OCH_3$	C—H (average)	1.118	∠COC	111.9	
	∠OCC	109.4	∠HCH	109.0	
	fraction of the *trans* conformer at 20°C		80%		
Ethyl methyl ketone			C—C (average)	1.518	ED
			C_c—O	1.219	
			C—H (average)	1.102	
			∠$C_aC_bC_c$	113.5	
	∠C_bC_cO, ∠C_dC_cO	121.9	*trans* conformer	95%	
Ethyl methyl sulfide	C—S (average)	1.813	C—C	1.536	ED
$C_2H_5SCH_3$	C—H	1.111	∠CSC	97	
	∠SCC	114.0	∠HCH	110	
	fraction of the *gauche* conformer at 20°C		75%		
Ferrocene	C—C	1.440	C—H	1.104	ED
$(C_5H_5)_2Fe$	Fe—C	2.064	(D_{5h})		
Fluoroform	C—H	1.098	C—F	1.332	MW
CHF_3	∠FCF	108.8	(C_{3v})		
Formaldehyde	C—H	1.116	C—O	1.208	MW
H_2CO	∠HCH	116.5			
Formaldehyde azine	$H_2C=N—N=CH_2$	N—N	1.418		ED
	C—N	1.277	C—H	1.094	
	∠CNN	111.4	∠HCN	120.7	
	fraction of the *trans* conformer at −30°C		91%		
Formaldehyde dimethylacetal → Dimethoxy-methane					
Formaldoxime			C—H_a	1.085	MW
			C—H_b	1.086	
			C—N	1.276	
	N—O	1.408	O—H_c	0.956	
	∠H_bCN	115.6	∠CNO	110.2	
	∠H_aCN	121.8	∠NOH_c	102.7	
Formamide			C—H_a	1.125	ED, MW
			N—H	1.027	
			C—N	1.368	
	C—O	1.212	∠NCO	125.0	
	∠CNH (average)	119.2			
Formic acid			C—O_a	1.202	MW
			C—O_b	1.343	
	O_b—H	0.972	C—H	1.097	
	∠HCO_a	124.1	∠O_aCO_b	124.9	
	∠CO_bH	106.3	planar		

Compound	Structure				Method

Formic acid dimer

$O_a \cdots O_b$	2.703	ED
$C\!-\!O_a$	1.220	
$C\!-\!O_b$	1.323	
$\angle O_aCO_b$	126.2	
$\angle CO_aO_b$	108.5	

Formyl radical

$C\!-\!H$	1.110	$C\!-\!O$	1.1712	MW
$\angle HCO$	127.43			

Fulvene

$C_a\!-\!C_d$	1.349	MW
$C_a\!-\!C_b$	1.470	
$C_b\!-\!C_c$	1.355	
$C_c\!-\!C_c$	1.476	
$C_b\!-\!H$	1.078	
$C_c\!-\!H$	1.080	

$C_d\!-\!H$	1.13	$\angle C_bC_aC_b$	106.6
$\angle C_aC_bC_c$	107.7	$\angle C_bC_cC_c$	109
$\angle C_aC_bH$	124.7	$\angle C_bC_cH$	126.4
$\angle HC_dH$	117		

2-Furaldehyde

$C_a\!-\!C_e$	1.458	MW
$C_e\!-\!O_b$	1.250	
$C_e\!-\!H$	1.088	
$\angle C_eC_aC_b$	133.9	

$\angle C_aC_eH$	116.9	$\angle C_aC_eO$	121.6

trans conformer (with respect to the O_a and O_b atoms)

Furan

$C_b\!-\!C_b$	1.431	MW
$C_a\!-\!C_b$	1.361	
$C_a\!-\!O$	1.362	
$C_a\!-\!H_a$	1.075	
$C_b\!-\!H_b$	1.077	

$\angle C_aC_bC_b$	106.1	$\angle C_aOC_a$	106.6
$\angle C_bC_aO$	110.7	$\angle OC_aH_a$	115.9
$\angle C_bC_bH_b$	128.0		

Furfural → 2-Furaldehyde

Glycolaldehyde

$C_b\!-\!O_b$	1.209	MW
$C_a\!-\!O_a$	1.437	
$C_a\!-\!C_b$	1.499	
$O_a\!-\!H_a$	1.051	
$C_b\!-\!H_c$	1.102	
$C_a\!-\!H_b$	1.093	

$\angle C_aC_bO_b$	122.7	$\angle C_aC_bH_c$	115.3
$\angle C_bC_aO_a$	111.5	$\angle C_aO_aH_a$	101.6
$\angle C_bC_aH_b$	109.2	$\angle H_bC_aH_b$	107.6
$\angle H_bC_aO_a$	109.7		

Glyoxal
CHOCHO

$C\!-\!C$	1.526	$C\!-\!O$	1.212	ED, UV
$C\!-\!H$	1.132	$\angle CCO$	121.2	
$\angle HCO$	112	*trans* conformer	(C_{2h} (assumed))	

Hexachloroethane
Cl_3CCCl_3

$C\!-\!C$	1.56	$C\!-\!Cl$	1.769	ED
$\angle CCCl$	110.0			

2,4-Hexadiyne

$C_aH_3\!-\!C_b\!\equiv\!C_c\!-\!C_c\!\equiv\!C_b\!-\!C_aH_3$

				ED
$C_a\!-\!C_b$	1.450	$C_b\!-\!C_c$	1.208	
$C_c\!-\!C_c$	1.377	$C_a\!-\!H$	1.09	

Hexafluoroethane
F_3CCF_3

$C\!-\!C$	1.545	$C\!-\!F$	1.326	ED
$\angle CCF$	109.8	staggered conformation		

Hexafluoropropene
$CF_2\!=\!CFCF_3$

average value of the C=C and C—F distances			1.329	ED
$C\!-\!C$	1.513	$\angle CCC$	127.8	
$\angle FCC\ (CF_2)$	124	$\angle FCC\ (CF)$	120	
$\angle FCC\ (CF_3)$	110			

Compound	Structure				Method
1,3,5-Hexatriene	$H_2C_a=C_bH-C_cH=C_cH-C_bH=C_aH_2$				ED
	C_a-C_b	1.337	C_b-C_c	1.458	
	C_c-C_c	1.368	$\angle C_aC_bC_c$	121.7	
	$\angle C_bC_cC_c$	124.4			
Iminocyanide radical	N—H	1.034	N···N	2.470	UV
$H\dot{N}CN$	$\angle HNC$	116.5	$\angle NCN$	~180	
Iodocyanoacetylene	$I-C_a$	1.985	C_a-C_b	1.207	MW
$I-C_a\equiv C_b-C_c\equiv N$	C_b-C_c	1.370	C_c-N	1.160	
Isobutane	C_a-H	1.122	C_b-H	1.113	ED, MW
$(C_bH_3)_3C_aH$	C_a-C_b	1.535	$\angle C_bC_aC_b$	110.8	
	$\angle C_aC_bH$	111.4			
Isobutylene → 2-Methylpropene					
Ketene	C—C	1.317	C—O	1.161	MW
$CH_2=C=O$	C—H	1.080	$\angle HCH$	123.0	
Malononitrile	C—H	1.091	C—C	1.480	MW
$C_aH_2(C_bN)_2$	C—N	1.147	$\angle CCC$	110.4	
	$\angle HCH$	108.4	$\angle CCN$	176.6	
	(The two N atoms are bent away from each other in the plane of $C_b-C_a-C_b$)				
Methane	C—H (r_e)	1.0870	(T_d)		MW
CH_4					
Methanethiol	C—H	1.09	C—S	1.819	MW
CH_3SH	S—H	1.34	$\angle HSC$	96.5	
	$\angle HCH$	109.8			
	angle between the CH_3 symmetry axis and the C—S bond 2.2.				
	(The axis of the CH_3 group is tilted away from the H atom with respect to the C—S bond.)				
Methanol	C—H	1.0936	C—O	1.4246	MW
CH_3OH	O—H	0.9451	$\angle HCH$	108.63	
	$\angle COH$	108.53			
	angle between the CH_3 symmetry axis and the C—O bond				
	(The axis of the CH_3 group is tilted away from the H atom with respect to the C—O bond.)			3.27	
Methyl radical	C—H	1.08	planar		UV
$\cdot CH_3$					
N-Methylacetamide			C_a-C_b	1.520	ED
			$N-C_c$	1.469	
			C—H	1.107	
	C_b-N	1.386	C_b-O	1.225	
	$\angle C_bNC_c$	119.7			
	$\angle NC_bO$	121.8			
	$\angle C_aC_bN$	114.1			
Methylacetylene → Propyne					
Methylal → Dimethoxymethane					
Methylamine	N—H	1.010	C—N	1.471	MW
CH_3NH_2	C—H	1.099	$\angle NHN$	107.1	
	$\angle HNC$	110.3	$\angle HCH$	108.0	
	dihedral angle between the CH_3 symmetry axis and the C—N bond (The axis of the CH_3 group is tilted away from the NH_2 group with respect to the C—N bond.)			2.9	
Methyl azide			C—H	1.09	ED
			$C-N_a$	1.468	
			N_a-N_b	1.216	
	N_b-N_c	1.113	$\angle CN_aN_b$	116.8	
	NNN linear				
Methyl bromide	C—H (r_e)	1.086	C—Br (r_e)	1.933	MW, IR
CH_3Br	$\angle HCH$ (θ_e)	111.2	(C_{3v})		
Methyl chloride	C—H	1.090	C—Cl	1.785	MW, IR
CH_3Cl	$\angle HCH$	110.8			

Compound	Structure					Method

Methyldiazirine

$CH_3CH\diagdown\substack{N\\ \parallel\\ N}$

C—N	1.481	C—C	1.501	MW
N—N	1.235	∠NCN	49.3	

dihedral angle between the CNN plane and the C—C bond   122.3

Methylene
:CH_2

C—H	1.078	∠HCH	130	LMR

Methylenecyclopropane

$\substack{C_cH_2\\ |\\ C_cH_2}\diagup C_b{=}C_aH_2$

		C_a—C_b	1.332	MW
		C_b—C_c	1.457	
		C_c—C_c	1.542	
C_c—H	1.09	∠$C_cC_bC_c$	63.9	
∠HC_aH	114.3	∠HC_cH	113.5	

dihedral angle between the C_cH_2 plane and the C_c—C_c bond   150.8

3-Methyleneoxetane

$O\diagup\substack{C_cH_2\\ \\ C_cH_2}C_b{=}C_aH_2$

		C_b—C_c	1.52	MW
		C_c—O	1.45	
		C_a—C_b	1.33	
C—H	1.09 (assumed)	∠$C_cC_bC_c$	87	
∠HC_cH	114 (assumed)	∠HC_aH	120 (assumed)	

Methyl fluoride
CH_3F

C—H (r_e)	1.095			MW, IR
C—F (r_e)	1.382	∠HCH (θ_e)	110.45 (C_{3v})	

Methyl formate

$\substack{C_aH_3\\ \diagdown}O_a{-}C_b\diagup\substack{O_b\\ \\ H_b}$

		C_a—H	1.08	ED
		C_b—O_b	1.206	
		C—O (average)	1.393	
		C_b—H	1.101 (assumed)	
∠$O_aC_bO_b$	127	∠COC	114	
∠O_aC_aH	110			

Methylgermane
CH_3GeH_3

C—H	1.083	Ge—H	1.529	MW
C—Ge	1.945	∠HCH	108.4	
∠HGeH	109.3			

Methyl hypochlorite
CH_3OCl

C—H	1.103	O—Cl	1.674	MW
O—C	1.389	∠HCH	109.6	
∠COCl	112.8			

Methylidyne radical
:ĊH

C—H (r_e)	1.1198			UV

Methylidyne phosphide
HCP

H—C (r_e)	1.0692	C—P (r_e)	1.5398	MW

Methyl iodide
CH_3I

C—H (r_e)	1.084	C—I (r_e)	2.132 (C_{3v})	MW, IR
∠HCH (θ_e)	111.2			

Methyl isocyanide

C_aH_3—N≡C_b	C_a—H	1.102	C_a—N	1.424	MW
N—C_b	1.166	∠NC_aH	109.12	

Methylketene

$\substack{C_cH_3\\ \diagdown\\ H}C_b{=}C_a{=}O$

		O—C_a	1.171	MW
		C_b—C_c	1.518	
		C_c—H	1.10	
C_a—C_b	1.306	C_b—H	1.083	
∠OC_aC_b	180.5	∠$C_aC_bC_c$	122.6	
∠C_aC_bH	113.7	∠C_cC_bH	123.7	
∠HCH	109.2			

Methylmercury chloride
CH_3HgCl

Hg—Cl	2.282	C—H	1.15 (C_{3v})	MW, NMR
Hg—C	1.99			

Methyl nitrate

$\substack{H_a\\ \diagup}C\substack{\diagdown\\ H_b}O\diagdown N\substack{\diagup O_a\\ \diagdown O_b}$

		C—H_a	1.10	MW
		C—H_b	1.09	
		C—O	1.437	
		O—N	1.402	
N—O_a	1.205	N—O_b	1.208	
∠OCH_a	110	∠OCH_b	103	
∠CON	112.7	∠ONO$_a$	118.1	
∠ONO$_b$	112.4			

Compound	Structure				Method

Methylphosphine CH_3PH_2

C—P 1.858   C—H 1.094   ED

2-Methylpropane → Isobutane

2-Methylpropene   ED, MW

C_aH_3, H_c, C_b=C_c, C_aH_3, H_c

C_a—H	1.119	
C_c—H_c	1.10	
C_a—C_b	1.508	
C_b—C_c	1.342	

∠HC$_a$C$_b$ (average) 111.4   ∠H$_c$C$_c$H$_c$ 118.5
∠C$_a$C$_b$C$_a$ 115.6   ∠C$_a$C$_b$C$_c$ 122.2
∠HC$_a$H 107.9   ∠C$_b$C$_c$H 121

Methylsilane CH_3SiH_3

C—H 1.093   C—Si 1.867   MW
Si—H 1.485   ∠HCH 107.7
∠HSiH 108.3   (C_{3v})

Methylstannane CH_3SnH_3

C—Sn 2.143   Sn—H 1.700   MW
(C_{3v})

Methyl thiocyanate

S—C_b 1.684   S—C_a 1.824   MW
C_b—N 1.170   C—H 1.081

C_aH_3
S—C_b—N

∠C_aSC_b 99.0   ∠HCH 110.6   ∠HCS 108.3

Naphthalene   ED

C_a—C_b	1.37
C_b—C_b	1.41
C_a—C_c	1.42
C_c—C_c	1.42
C—C (average)	1.40
∠$C_a$$C_c$$C_c$	119.4

Neopentane $C(CH_3)_4$

C—C 1.537   C—H 1.114   ED
∠CCH 112

Nickelocene → Bis (cyclopentadienyl) nickel

Nitromethane CH_3NO_2

C—H 1.088 (assumed)   C—N 1.489   MW
N—O 1.224   ∠NCH 107
∠ONO 125.3

N-Nitrosodimethylamine $(CH_3)_2NNO$

N—O 1.235   N—N 1.344   ED
C—N 1.461   ∠ONN 113.6
∠CNC 123.2   ∠CNN 116.4

Nitrosomethane CH_3NO

C—N 1.49   N—O 1.22   MW
C—H 1.084   ∠CNO 112.6
∠NCH 109.0

Norbornane → Bicyclo[2.2.1]heptane

Norbornadiene → Bicyclo[2.2.1]hepta-2,5-diene

1,2,5-Oxadiazole

O—N 1.380   ∠NON 110.4   MW
C—N 1.300   ∠ONC 105.8
C—C 1.421   ∠CCN 109.0
C—H 1.076   ∠CCH 130.2
∠NCH 120.9   planar

1,3,4-Oxadiazole

O—C 1.348   ∠COC 102.0   MW
C—N 1.297   ∠OCN 113.4
N—N 1.399   ∠CNN 105.6
C—H 1.075   ∠OCH 118.1
∠NCH 128.5   planar

Oxalic acid   ED

C—C	1.544
C—O_a	1.205
C—O_b	1.336
O_b—H	1.05
∠CCO_a	123.1
∠O_aCO_b	125.0
∠CO_bH	104

BOND LENGTHS AND ANGLES IN GAS-PHASE MOLECULES (continued)

Compound	Structure						Method

Oxalyl dichloride

C—O	1.182	ED
C—C	1.534	
C—Cl	1.744	
∠CCO	124.2	∠CCCl 111.7

fraction of the *trans* conformer at 0°C 68%, that of the *gauche* conformer 32%

Oxetane

C—O	1.448	MW
C—C	1.546	
C—H (average)	1.090	
∠COC	92	
∠CCC 85	∠OCC 92	
∠HCH (average) 109.9		

Oxirane → Ethylene oxide

Phenol

C—C (average)	1.397	MW
C_b—H	1.084	
C_c—H	1.076	
C_d—H	1.082	
C_a—O	1.364	
O—H	0.956	
∠COH	109.0	

Phosphirane

C—P	1.867	P—H	1.43	MW
C—C	1.502	C—H	1.09	
∠CPC	47.4	∠HCH	114.4	

∠HPC 95.2   ∠CCH 118

dihedral angle between the PCC plane and the PH bond   95.7

Piperazine

C—C	1.540	ED
C—N	1.467	
C—H	1.110	
∠CNC 109.0	∠CCN 110.4 (C_{2h})	

Pivalonitrile

$(C_cH_3)_3C_b—C_a≡N$

$C_a—C_b$	1.495	$C_a—N$	1.159	MW
$C_b—C_c$	1.536	∠$C_cC_bC_c$	110.5	

Propadiene → Allene

Propane
C_3H_8

C—C	1.532	C—H	1.107	ED
∠CCC	112	∠HCH	107	

Propenal → Acrylaldehyde

Propene

$C_a—H_a$	1.104	ED, MW
$C_a—C_b$	1.341	
$C_c—H_d$	1.117	
$C_b—C_c$	1.506	
∠$C_bC_aH_{a,b,c}$	121.3	
∠$C_bC_cH_d$ 110.7	∠$C_aC_bC_c$ 124.3	

1-Propenyl chloride
$CH_3—C_bH=C_aH—Cl$ $C_a—Cl$ 1.728   MW
∠C_bC_aCl 121.9   *trans* conformer

Propiolaldehyde
$H_aC_a≡C_b—C_cH_cO$

		$C_a—H_a$	1.085	ED, MW
$C_a—C_b$	1.211	$C_b—C_c$	1.453	
$C_c—H_c$	1.130	$C_c—O$	1.214	
∠C_bC_cO	124.2	∠$C_bC_cH_c$	113.7	
∠$C_aC_bC_c$	178.6	planar		

Propylene → Propene

Propylene oxide
$C_aH_3C_bH—C_cH_2$ (over O)

$C_a—C_b$	1.51	MW
∠$C_aC_bC_c$	121.0	

dihedral angle between the C_bC_cO plane
and the C_aC_b bond   123.8

Compound	Structure					Method

Propynal → Propiolaldehyde

Propyne

$H_3C_c—C_b≡C_aH$

$C_c—C_b$	1.459	$C_c—H$	1.105
$C_a—H$	1.056	$C_b—C_a$	1.206
		$∠HC_cC_b$	110.2

Method: MW

Pyrazine

C—C	1.339	C—N	1.403
C—H	1.115	∠CCN	115.6
∠CCH	123.9		

Method: ED

Pyridazine

$N—C_a$	1.341
$C_a—C_b$	1.393
N—N	1.330
$C_b—C_b$	1.375

Method: ED, MW

∠NCC	123.7	∠NNC	119.3

Pyridine

$N—C_a$	1.340
$C_b—C_c$	1.394
$C_b—H_b$	1.081
$C_a—C_b$	1.395
$C_a—H_a$	1.084
$C_c—H_c$	1.077

Method: MW

$∠C_aNC_a$	116.8	$∠NC_aC_b$	123.9
$∠C_aC_bC_c$	118.5	$∠C_bC_cC_b$	118.3
$∠NC_aH_a$	115.9	$∠C_cC_bH_b$	121.3

Pyrimidine

N—C	1.340	C—C	1.393
∠NCN	127.6	∠CNC	115.5
(C_{2v} assumed)			

Method: ED

Pyrrole

$N—C_a$	1.370
$C_b—C_b$	1.417
$C_a—C_b$	1.382
N—H	0.996
$C_a—H_a$	1.076

Method: MW

$C_b—H_b$	1.077	$∠C_aNC_a$	109.8
$∠NC_aC_b$	107.7	$∠C_aC_bC_b$	107.4
$∠NC_aH_a$	121.5	$∠C_bC_bH$	127.1

Pyruvonitrile

$C_aH_3—C_b$

C—H	1.12
C—N	1.17
C—O	1.208
$C_b—C_c$	1.477

Method: ED, MW

$C_a—C_b$	1.518	∠HCH	109.2
$∠C_aC_bO$	124.5	$∠C_aC_bC_c$	114.2
∠CCN	179		

**Ruthenocene → Bis (cyclopentadienyl)
 ruthenium**

Silacyclobutane

$CH_2—CH_2$
$|\quad\quad|$
$CH_2—SiH_2$

Si—C	1.892
C—C	1.600
Si—H	1.47

Method: ED

C—H	1.14	∠CSiC	80.7
∠SiCC	84.8	∠CCC	99.8
dihedral angle between the CCC and CSiC planes			146

Spiropentane

$C_b—C_b$	1.52
$C_a—C_b$	1.47
C—H	1.09
$∠C_bC_aC_b$	62

Method: ED

∠HCH	118	(D_{2d})

Compound	Structure						Method

Succinonitrile

CH₂CN
|
CH₂CN

C—C	1.561	C—C(N)	1.465	ED
C—N	1.161	C—H	1.09	
∠CCC	110.4			

fraction of the *anti* conformer at 170°C 74%,
dihedral angle of CCCC for the *gauche* conformer   75

Tetrachloroethylene
CCl₂=CCl₂

C—Cl	1.718	C—C	1.354	ED
∠ClCCl	115.7			

Tetracyanoethylene
(CN)₂C=C(CN)₂

C—N	1.162	C—C	1.435	ED
C=C	1.357	∠CC=C	121.1	

Tetrafluoro-1,3-dithietane

F₂C⟨S⟩CF₂ (with S at top and bottom)

C—S	1.785	ED
C—F	1.314	
∠CSC	83.2	
∠FCS	113.7	(D₂ₕ assumed)

Tetrafluoroethylene
CF₂=CF₂

C—C	1.31	C—F	1.319	ED
∠CCF	123.8	(D₂ₕ assumed)		

Tetrahydrofuran

CH₂CH₂
| ⟩O
CH₂CH₂

C—H	1.115	C—O	1.428	ED
C—C	1.536			

The skeletal bending vibration of the molecular plane is essentially free pseudorotation

Tetrahydropyran

C—O	1.420	ED
C—C	1.531	
C—H	1.116	
∠COC	111.5	
∠OCC	111.8	

∠CCC (C)	108	∠CCC (O)	111

chair form

Tetrahydrothiophene

CH₂CH₂
| ⟩S
CH₂CH₂

C—S	1.839	C—H	1.120
C—C	1.536	∠CSC	93.4
∠SCC	106.1	∠CCC	105.0

Tetramethylgermane
(CH₃)₄Ge

Ge—C	1.945	C—H	1.12	ED
∠GeCH	108	(T_d excluding the H atoms)		

Tetramethyllead
(CH₃)₄Pb

Pb—C	2.238	(T_d excluding the H atoms)	ED

Tetramethylsilane
(CH₃)₄Si

C—H	1.115	C—Si	1.875	ED
∠HCH	109.8	(T_d excluding the H atoms)		

Tetramethylstannane
(CH₃)₄Sn

C—Sn	2.144		ED
C—H	1.12	(T_d excluding the H atoms)	

1,2,5-Thiadiazole

S—N	1.631	∠NSN	99.6	MW
C—N	1.328	∠CCN	113.8	
C—C	1.420	∠CCH	126.2	
C—H	1.079	planar		

1,3,4-Thiadiazole

S—C	1.721	∠CSC	86.4	MW
N—N	1.371	∠SCN	114.6	
C—N	1.302	∠CCN	112.2	
C—H	1.08	∠SCH	121.9	
∠NCH	123.5	planar		

Thietane

CH₂—CH₂
| |
CH₂ — S

C—S	1.847	ED, MW
C—C	1.549	
C—H (average)	1.100	
∠CSC	76.8	∠HCH (average)	112

dihedral angle between the CCC and CSC planes   154

Thiirane

H₂C⟨S⟩ (three-membered ring with S)
H₂C

C—C	1.484	∠HCH	116	MW
C—H	1.083	∠CSC	48.3	
C—S	1.815	∠CCS	65.9	

dihedral angle between the CH₂ plane and the
C—C bond   152

Compound	Structure					Method
Thioformaldehyde	C—S	1.611		C—H	1.093	MW
CH_2S	∠HCH	116.9				
Thioformamide				N—H_a	1.002	MW
				N—H_b	1.007	
				C—N	1.358	
	C—S	1.626		C—H_c	1.10	
	∠$H_a NH_b$	121.7		∠$H_a NC$	117.9	
	∠$H_b NC$	120.4		∠NCS	125.3	
	∠NCH_c	108		∠SCH_c	127	
Thiolane → Tetrahydrothiophene						
Thiophene				C_a—H_a	1.078	MW
				C_b—H_b	1.081	
				C_a—S	1.714	
				C_a—C_b	1.370	
				C_b—C_b	1.423	
				∠$C_a SC_a$	92.2	
	∠$SC_a C_b$	115.5		∠$C_a C_b C_b$	112.5	
	∠$SC_a H_a$	119.9		∠$C_b C_b H_b$	124.3	
Toluene		C—C (ring)	1.399	C—CH_3	1.524	ED
		C—H (average)	1.11			
		the difference between the C—$H(CH_3)$ and C—H(ring): about 0.01				
1,1,1-Tribromoethane	C—Br	1.93		C—H	1.095 (assumed)	MW
$CH_3 CBr_3$	C—C	1.51 (assumed)		∠CCBr	108	
	∠BrCBr	111		∠CCH	109.0 (assumed)	
Tribromomethane → Bromoform						
Tri-*tert*-butyl methane	C_a—C_b	1.611		C—H	1.111	ED
$HC_a[C_b(C_c H_3)_3]_3$	C_b—C_c	1.548		∠$C_a C_b C_c$	113.0	
Tricarbon dioxide	C—O	1.163		C—C	1.289	ED
OCCCO	linear (with a large-amplitude bending vibration)					
Trichloroacetonitrile	C—N	1.165		C—C	1.460	ED
$CCl_3 CN$	C—Cl	1.763		∠ClCCl	110.0	
1,1,1-Trichloroethane	C—H	1.090		C—C	1.541	MW
$CH_3 CCl_3$	C—Cl	1.771		∠HCH	110.0	
	∠CCH	108.9		∠ClCCl	109.4	
	∠CCCl	109.6				
Trichloro(methyl)germane	Ge—Cl	2.132		Ge—C	1.89	ED, MW
$CH_3 GeCl_3$	C—H	1.103 (assumed)		∠ClGeCl	106.4	
	∠GeCH	110.5 (assumed)				
Trichloro(methyl)silane	C—Si	1.876		Si—Cl	2.021	MW
$CH_3 SiCl_3$				(C_{3v})		
Trichloro(methyl)stannane	Sn—Cl	2.304		Sn—C	2.10	ED
$CH_3 SnCl_3$	C—H	1.100		∠CSnCl	113.9	
	∠ClSnCl	104.7		∠SnCH	108	
Triethylenediamine → 1,4-Diazabicyclo [2.2.2]octane						
Trifluoroacetic acid				C—F	1.325	ED
				C—C	1.546	
				C—O_a	1.192	
	C—O_b	1.35		O—H	0.96 (assumed)	
	∠CCO_a	126.8		∠CCO_b	111.1	
	∠CCF	109.5				
1,1,1-Trifluoroethane	C—C	1.494		C—F	1.340	ED
$CH_3 CF_3$	C—H	1.081		∠CCF	119.2	
	∠CCH	112				
Trifluoromethane → Fluoroform						

Compound	Structure				Method
1,1,1-Trifluoro-2,2,2-trichloroethane CF_3CCl_3	C—C C—Cl ∠CCCl	1.54 1.77 109.6	C—F ∠CCF staggered conformation	1.33 110	MW
Trimethylaluminium $(CH_3)_3Al$	C—H ∠AlCH	1.113 111.7	Al—C ∠CAlC	1.957 120	ED
Trimethylamine $(CH_3)_3N$	C—N ∠CNC	1.458 110.9	C—H ∠HCH	1.100 110	ED
Trimethylarsine $(CH_3)_3As$	C—As ∠AsCH	1.979 111.4	∠CAsC	98.8	ED
Trimethylbismuth $(CH_3)_3Bi$	Bi—C ∠CBiC	2.263 97.1	C—H	1.07	ED
Trimethylborane $(CH_3)_3B$	C—B ∠CBC	1.578 120.0	C—H ∠BCH	1.114 112.5	ED
Trimethyleneimine → Azetidine					
Trimethylphosphine $(CH_3)_3P$	C—P ∠CPC	1.847 98.6	C—H ∠PCH	1.091 110.7	ED
1,3,5-Trioxane	(structure diagram)		C—O ∠OCO ∠COC	1.422 112.2 110.3	MW
Triphenylamine $(C_6H_5)_3N$	C—C ∠CNC	1.392 116	C—N (C₃)	1.42	ED

torsional dihedral angle of the two phenyl rings 47° (defined to be 0 when the symmetry axis is contained in the phenyl planes)

Tropone	(structure diagram)		C_a—O C_a—C_b C_b—C_c C_c—C_d C_d—C_d ∠$C_bC_aC_b$ ∠$C_aC_bC_c$	1.23 1.45 1.36 1.46 1.34 122 133	ED
	∠$C_bC_cC_d$	126	∠$C_cC_dC_d$ (C₂ᵥ)	130	
Vinylacetylene	(structure diagram)		C_b—C_c C_a—C_b C_c—C_d C_a—H_a C_d—H_d ∠$C_aC_bC_c$	1.434 1.344 1.215 1.11 1.09 123.1	ED, MW
	∠$C_bC_cC_d$ ∠$H_bC_aC_b$ ∠$C_cC_dH_d$	178 122 182	∠$H_aC_aC_b$ ∠$H_cC_bC_a$	119 122	
Vinyl chloride	(structure diagram)		C—C C—Cl C—H	1.342 1.730 1.09	ED, MW
	∠CCCl ∠CCH_b	122.5 120	∠CCH_a ∠CCH_c	124 121.1	

DIPOLE MOMENTS

This table gives values of the electric dipole moment for about 800 molecules. When available, values determined by microwave spectroscopy, molecular beam electric resonance, and other high-resolution spectroscopic techniques were selected. Otherwise, the values come from measurements of the dielectric constant in the gas phase or, if these do not exist, in the liquid phase. Compounds are listed by molecular formula in Hill order; compounds not containing carbon are listed first, followed by compounds containing carbon.

The dipole moment μ is given in debye units (D). The conversion factor to SI units is $1\ D = 3.33564 \times 10^{-30}\ C\ m$.

Dipole moments of individual conformers (rotational isomers) are given when they have been measured. The conformers are designated as *gauche*, *trans, axial*, etc. The meaning of these terms can be found in the references. In some cases an average value, obtained from measurements on the bulk gas, is also given. Other information on molecules that have been studied by spectroscopy, such as the components of the dipole moment in the molecular framework and the variation with vibrational state and isotopic species, is given in References 1 and 2.

When the accuracy of a value is explicitly stated (i.e., 1.234 ± 0.005), the stated uncertainty generally indicates two or three standard deviations. When no uncertainty is given, the value may be assumed to be precise to a few units in the last decimal place. However, if more than three decimal places are given, the exact interpretation of the final digits may require analysis of the vibrational averaging.

Values measured in the gas phase that are questionable because of undetermined error sources are indicated as approximate ($\approx$). Values obtained by liquid phase measurements, which sometimes have large errors because of association effects, are enclosed in brackets, e.g., [1.8].

REFERENCES

1. Nelson, R. D., Lide, D. R., and Maryott, A. A., *Selected Values of Electric Dipole Moments for Molecules in the Gas Phase*, Natl. Stand. Ref. Data Ser. - Nat. Bur. Stnds. 10, 1967.
2. *Landolt-Börnstein, Numerical Data and Functional Relationships in Science and Technology*, New Series, II/6 (1974), II/14a (1982), II/14b (1983), II/19c (1992), Springer-Verlag, Heidelberg.
3. Riddick, J. A., Bunger, W. B., and Sakano, T. K., *Organic Solvents, Fourth Edition*, John Wiley & Sons, New York, 1986.

Mol. Form.	Name	μ/D	Mol. Form.	Name	μ/D
Compounds not containing carbon					
AgBr	Silver(I) bromide	5.62 ± 0.03	ClF$_3$Si	Chlorotrifluorosilane	0.636 ± 0.004
AgCl	Silver(I) chloride	6.08 ± 0.06	ClGeH$_3$	Chlorogermane	2.13 ± 0.02
AgF	Silver(I) fluoride	6.22 ± 0.30	ClH	Hydrogen chloride	1.1086 ± 0.0003
AgI	Silver(I) iodide	4.55 ± 0.05	ClHO	Hypochlorous acid	≈ 1.3
AlF	Aluminum monofluoride	1.53 ± 0.15	ClH$_3$Si	Chlorosilane	1.31 ± 0.01
AsCl$_3$	Arsenic(III) chloride	1.59 ± 0.08	ClI	Iodine chloride	1.24 ± 0.02
AsF$_3$	Arsenic(III) fluoride	2.59 ± 0.05	ClIn	Indium(I) chloride	3.79 ± 0.19
AsH$_3$	Arsine	0.217 ± 0.003	ClK	Potassium chloride	10.269 ± 0.001
BClH2	Chloroborane	0.75 ± 0.05	ClLi	Lithium chloride	7.12887
BF	Fluoroborane(1)	≈ 0.5	ClNO$_2$	Nitryl chloride	0.53
BF$_2$H	Difluoroborane	0.971 ± 0.010	ClNS	Thionitrosyl chloride	1.87 ± 0.02
B$_4$H$_{10}$	Tetraborane	0.486 ± 0.002	ClNa	Sodium chloride	9.00117
B$_5$H$_9$	Pentaborane(9)	2.13 ± 0.04	ClO	Chlorine oxide	1.297 ± 0.001
B$_6$H$_{10}$	Hexaborane	2.50 ± 0.05	ClRb	Rubidium chloride	10.510 ± 0.005
BaO	Barium oxide	7.954 ± 0.003	ClTl	Thallium(I) chloride	4.54299
BaS	Barium sulfide	10.86 ± 0.02	Cl$_2$H$_2$Si	Dichlorosilane	1.17 ± 0.02
BrCl	Bromine chloride	0.519 ± 0.004	Cl$_2$OS	Thionyl chloride	1.45 ± 0.03
BrF	Bromine fluoride	1.422 ± 0.016	Cl$_2$O$_2$S	Sulfuryl chloride	1.81 ± 0.04
BrF$_3$Si	Bromotrifluorosilane	0.83 ± 0.01	Cl$_2$S	Sulfur dichloride	0.36 ± 0.01
BrF$_5$	Bromine pentafluoride	1.51 ± 0.15	Cl$_3$FSi	Trichlorofluorosilane	0.49 ± 0.01
BrH	Hydrogen bromide	0.8272 ± 0.0003	Cl$_3$HSi	Trichlorosilane	0.86 ± 0.01
BrH$_3$Si	Bromosilane	1.319	Cl$_3$N	Nitrogen trichloride	0.39 ± 0.01
BrI	Iodine bromide	0.726 ± 0.003	Cl$_3$OP	Phosphorus(V) oxychloride	2.54 ± 0.05
BrK	Potassium bromide	10.628 ± 0.001	Cl$_3$P	Phosphorus(III) chloride	0.56 ± 0.02
BrLi	Lithium bromide	7.268 ± 0.001	CrO	Chromium monoxide	3.88 ± 0.13
BrNO	Nitrosyl bromide	≈ 1.8	CsF	Cesium fluoride	7.884 ± 0.001
BrNa	Sodium bromide	9.1183 ± 0.0006	CsNa	Cesium sodium	4.75 ± 0.20
BrO	Bromine monoxide	1.76 ± 0.04	CuF	Copper(I) fluoride	5.77 ± 0.29
BrO$_2$	Bromine dioxide	2.8 ± 0.3	CuO	Copper(II) oxide	4.5 ± 0.5
BrRb	Rubidium bromide	≈ 10.9	FGa	Gallium monofluoride	2.45 ± 0.05
BrTl	Thallium(I) bromide	4.49 ± 0.05	FGeH$_3$	Fluorogermane	2.33 ± 0.12
CaCl	Calcium monochloride	≈ 3.6	FH	Hydrogen fluoride	1.826178
ClCs	Cesium chloride	10.387 ± 0.004	FHO	Hypofluorous acid	2.23 ± 0.11
ClF	Chlorine fluoride	0.888061	FH$_2$N	Fluoramide	2.27 ± 0.18
ClFO$_3$	Perchloryl fluoride	0.023 ± 0.001	FH$_3$Si	Fluorosilane	1.2969 ± 0.0006
ClF$_3$	Chlorine trifluoride	0.6 ± 0.10	FI	Iodine fluoride	1.948 ± 0.020

Mol. Form.	Name	μ/D
FIn	Indium(I) fluoride	3.40 ± 0.07
FK	Potassium fluoride	8.585 ± 0.003
FLi	Lithium fluoride	6.3274 ± 0.0002
FNO	Nitrosyl fluoride	1.730 ± 0.003
FNO_2	Nitryl fluoride	0.466 ± 0.005
FNS	Thionitrosyl fluoride (NSF)	1.902 ± 0.012
FN_3	Fluorine azide	≈1.3
FNa	Sodium fluoride	8.156 ± 0.001
FO	Fluorine oxide	0.0043 ± 0.0004
FRb	Rubidium fluoride	8.5465 ± 0.0005
FS	Sulfur monofluoride	0.794 ± 0.02
FTl	Thallium(I) fluoride	4.2282 ± 0.0008
F_2Ge	Germanium(II) fluoride	2.61 ± 0.02
F_2HN	Difluoramine	1.92 ± 0.02
F_2H_2Si	Difluorosilane	1.55 ± 0.02
F_2N_2	cis-Difluorodiazine	0.16 ± 0.01
F_2O	Fluorine monoxide	0.308180
F_2OS	Thionyl fluoride	1.63 ± 0.01
F_2O_2	Fluorine dioxide	1.44 ± 0.07
F_2O_2S	Sulfuryl fluoride	1.12 ± 0.02
F_2S	Sulfur difluoride	1.05 ± 0.05
F_2Si	Difluorosilylene	1.23 ± 0.02
F_3HSi	Trifluorosilane	1.27 ± 0.03
$F_3H_3Si_2$	1,1,1-Trifluorodisilane	2.03 ± 0.10
F_3ISi	Trifluoroiodosilane	1.11 ± 0.03
F_3N	Nitrogen trifluoride	0.235 ± 0.004
F_3NO	Trifluoramine oxide	0.0390 ± 0.0004
F_3OP	Phosphorus(V) oxyfluoride	1.8685 ± 0.0001
F_3P	Phosphorus(III) fluoride	1.03 ± 0.01
F_3PS	Phosphorus(V) sulfide trifluoride	0.64 ± 0.02
F_4N_2	Tetrafluorohydrazine (gauche)	0.257 ± 0.002
F_4S	Sulfur tetrafluoride	0.632 ± 0.003
F_4Se	Selenium tetrafluoride	1.78 ± 0.09
F_5I	Iodine pentafluoride	2.18 ± 0.11
GeH_3N_3	Germylazide	2.579 ± 0.003
GeO	Germanium(II) oxide	3.2823 ± 0.0001
GeS	Germanium(II) sulfide	2.00 ± 0.06
GeSe	Germanium(II) selenide	1.65 ± 0.05
GeTe	Germanium(II) telluride	1.06 ± 0.07
HI	Hydrogen iodide	0.448 ± 0.001
HKO	Potassium hydroxide	7.415 ± 0.002
HLi	Lithium hydride	5.884 ± 0.001
HLiO	Lithium hydroxide	4.754 ± 0.002
HN	Imidogen	1.39 ± 0.07
HNO	Nitrosyl hydride	1.62 ± 0.03
HNO_2	Nitrous acid (cis)	1.423 ± 0.005
HNO_2	Nitrous acid (trans)	1.855 ± 0.016
HNO_3	Nitric acid	2.17 ± 0.02
HN_3	Hydrazoic acid	1.70 ± 0.09
HO	Hydroxyl	1.655 ± 0.001
HS	Mercapto	0.7580 ± 0.0001
H_2O	Water	1.8546 ± 0.0040
H_2O_2	Hydrogen peroxide	1.573 ± 0.001
H_2S	Hydrogen sulfide	0.97833
H_3N	Ammonia	1.4718 ± 0.0002
H_3NO	Hydroxylamine	0.59 ± 0.05
H_3P	Phosphine	0.5740 ± 0.0003
H_3Sb	Stibine	0.12 ± 0.05
H_4N_2	Hydrazine	1.75 ± 0.09
H_6OSi_2	Disiloxane	0.24 ± 0.02
IK	Potassium iodide	≈10.8

Mol. Form.	Name	μ/D
ILi	Lithium iodide	7.428 ± 0.001
INa	Sodium iodide	9.236 ± 0.003
IO	Iodine monoxide	2.45 ± 0.05
IRb	Rubidium iodide	≈11.5
ITl	Thallium(I) iodide	4.61 ± 0.07
KLi	Lithium potassium	3.45 ± 0.20
KNa	Potassium sodium	2.693 ± 0.014
LaO	Lanthanum monoxide	3.207 ± 0.011
LiNa	Lithium sodium	0.463 ± 0.002
LiO	Lithium monoxide	6.84 ± 0.03
LiRb	Lithium rubidium	4.0 ± 0.1
MgO	Magnesium oxide	6.2 ± 0.6
NO	Nitric oxide	0.15872
NO_2	Nitrogen dioxide	0.316 ± 0.010
NP	Phosphorus nitride	2.7470 ± 0.0001
NS	Nitrogen sulfide	1.81 ± 0.02
N_2O	Nitrous oxide	0.16083
N_2O_3	Nitrogen trioxide	2.122 ± 0.010
NaRb	Rubidium sodium	3.1 ± 0.3
OP	Phosphorus monoxide	1.88 ± 0.07
OPb	Lead(II) oxide	4.64 ± 0.50
OS	Sulfur monoxide	1.55 ± 0.02
OS_2	Sulfur oxide (SSO)	1.47 ± 0.03
OSi	Silicon monoxide	3.0982
OSn	Tin(II) oxide	4.32 ± 0.22
OSr	Strontium oxide	8.900 ± 0.003
OTi	Titanium(II) oxide	2.96 ± 0.05
OY	Yttrium monoxide	4.524 ± 0.007
OZr	Zirconium(II) oxide	2.55 ± 0.01
O_2S	Sulfur dioxide	1.63305
O_2Se	Selenium dioxide	2.62 ± 0.05
O_2Zr	Zirconium(IV) oxide	7.80 ± 0.02
O_3	Ozone	0.53373
PbS	Lead(II) sulfide	3.59 ± 0.18
SSi	Silicon monosulfide	1.73 ± 0.09
SSn	Tin(II) sulfide	3.18 ± 0.16

Compounds containing carbon

Mol. Form.	Name	μ/D
$CBrF_3$	Bromotrifluoromethane	0.65 ± 0.05
CBr_2F_2	Dibromodifluoromethane	0.66 ± 0.05
$CClF_3$	Chlorotrifluoromethane	0.50 ± 0.01
CClN	Cyanogen chloride	2.8331 ± 0.0002
CCl_2F_2	Dichlorodifluoromethane	0.51 ± 0.05
CCl_2O	Carbonyl chloride	1.17 ± 0.01
CCl_3F	Trichlorofluoromethane	0.46 ± 0.02
CF	Fluoromethylidyne	0.645 ± 0.005
CFN	Cyanogen fluoride	2.120 ± 0.001
CF_2	Difluoromethylene	0.47 ± 0.02
CF_2O	Carbonyl fluoride	0.95 ± 0.01
CF_3I	Trifluoroiodomethane	1.048 ± 0.003
CH	Methylidyne	≈1.46
CHBrClF	Bromochlorofluoromethane	1.5 ± 0.3
$CHBr_3$	Tribromomethane	0.99 ± 0.02
$CHClF_2$	Chlorodifluoromethane	1.42 ± 0.03
$CHCl_2F$	Dichlorofluoromethane	1.29 ± 0.03
$CHCl_3$	Trichloromethane	1.04 ± 0.02
CHFO	Formyl fluoride	2.081 ± 0.001
CHF_2N	Carboimidic difluoride	1.393 ± 0.001
CHF_3	Trifluoromethane	1.65150
CHN	Hydrogen cyanide	2.985188

Mol. Form.	Name	μ/D
CHN	Hydrogen isocyanide	3.05 ± 0.15
CHNO	Isocyanic acid (HNCO)	≈1.6
CHNO	Fulminic acid	3.09934
CH₂BrCl	Bromochloromethane	[1.66]
CH₂Br₂	Dibromomethane	1.43 ± 0.03
CH₂ClF	Chlorofluoromethane	1.82 ± 0.04
CH₂Cl₂	Dichloromethane	1.60 ± 0.03
CH₂F₂	Difluoromethane	1.9785 ± 0.02
CH₂I₂	Diiodomethane	[1.08]
CH₂N₂	Diazomethane	1.50 ± 0.01
CH₂N₂	Cyanamide	4.28 ± 0.10
CH₂N₄	1H-Tetrazole	2.19 ± 0.05
CH₂O	Formaldehyde	2.332 ± 0.002
CH₂O₂	Formic acid	1.425 ± 0.002
CH₂S	Thioformaldehyde	1.6491 ± 0.0004
CH₂Se	Selenoformaldehyde	1.41 ± 0.01
CH₃BCl₂	Dichloromethylborane	1.419 ± 0.013
CH₃BF₂	Difluoromethylborane	1.668 ± 0.003
CH₃BO	Borane carbonyl	1.698 ± 0.020
CH₃Br	Bromomethane	1.8203 ± 0.0004
CH₃Cl	Chloromethane	1.8963 ± 0.0002
CH₃Cl₃Si	Methyltrichlorosilane	1.91 ± 0.01
CH₃F	Fluoromethane	1.858 ± 0.002
CH₃F₂OP	Methylphosphonic difluoride	3.69 ± 0.26
CH₃F₂P	Methyldifluorophosphine	2.056 ± 0.006
CH₃F₃Si	Trifluoromethylsilane	2.3394 ± 0.0002
CH₃F₃Si	(Trifluoromethyl)silane	2.32 ± 0.02
CH₃I	Iodomethane	1.6406 ± 0.0004
CH₃NO	Formamide	3.73 ± 0.07
CH₃NO₂	Nitromethane	3.46 ± 0.02
CH₃N₃	Methyl azide	2.17 ± 0.04
CH₄O	Methanol	1.70 ± 0.02
CH₄O₂	Methylhydroperoxide	≈0.65
CH₄S	Methanethiol	1.52 ± 0.08
CH₅FSi	Fluoromethylsilane	1.700 ± 0.008
CH₅ISi	Iodomethylsilane	1.862 ± 0.005
CH₅N	Methylamine	1.31 ± 0.03
CH₆OSi	Methyl silyl ether	1.15 ± 0.02
CH₆Si	Methylsilane	0.73456
CH₈B₂	Methyldiborane(6)	0.566 ± 0.006
CIN	Cyanogen iodide	3.67 ± 0.02
CO	Carbon monoxide	0.10980
COS	Carbon oxysulfide	0.715189
COSe	Carbon oxyselenide	0.73 ± 0.02
CS	Carbon monosulfide	1.958 ± 0.005
CSe	Carbon monoselenide	1.99 ± 0.04
C₂BrF	Bromofluoroacetylene	0.448 ± 0.002
C₂ClF₃	Chlorotrifluoroethene	0.40 ± 0.10
C₂ClF₅	Chloropentafluoroethane	0.52 ± 0.05
C₂Cl₂F₂	1,1-Dichloro-2,2-difluoroethene	0.50
C₂Cl₂F₄	1,2-Dichloro-1,1,2,2-tetrafluoroethane	≈0.5
C₂F₃N	Trifluoroacetonitrile	1.262 ± 0.010
C₂F₃N	Trifluoroisocyanomethane	1.153 ± 0.010
C₂HBr	Bromoacetylene	0.22962
C₂HCl	Chloroacetylene	0.44408
C₂HCl₃	Trichloroethene	[0.8]
C₂HCl₅	Pentachloroethane	0.92 ± 0.05
C₂HF	Fluoroacetylene	0.7207 ± 0.0003
C₂HF₃	Trifluoroethene	1.32 ± 0.03
C₂HF₃O₂	Trifluoroacetic acid	2.28 ± 0.25

Mol. Form.	Name	μ/D
C₂HI	Iodoacetylene	0.02525
C₂H₂Br₄	1,1,2,2-Tetrabromoethane	[1.38]
C₂H₂Cl₂	1,1-Dichloroethene	1.34 ± 0.01
C₂H₂Cl₂	cis-1,2-Dichloroethene	1.90 ± 0.04
C₂H₂Cl₂O	Chloroacetyl chloride	2.23 ± 0.11
C₂H₂Cl₄	1,1,2,2-Tetrachloroethane	1.32 ± 0.07
C₂H₂F₂	1,1-Difluoroethene	1.3893 ± 0.0002
C₂H₂F₂	cis-1,2-Difluoroethene	2.42 ± 0.02
C₂H₂F₄	1,1,1,2-Tetrafluoroethane	1.80 ± 0.22
C₂H₂N₂S	1,2,5-Thiadiazole	1.579 ± 0.007
C₂H₂O	Ketene	1.42215
C₂H₂O₂	Glyoxal (cis)	4.8 ± 0.2
C₂H₃Br	Bromoethene	1.42 ± 0.03
C₂H₃Cl	Chloroethene	1.45 ± 0.03
C₂H₃ClF₂	1-Chloro-1,1-difluoroethane	2.14 ± 0.04
C₂H₃ClO	Acetyl chloride	2.72 ± 0.14
C₂H₃Cl₃	1,1,1-Trichloroethane	1.755 ± 0.015
C₂H₃Cl₃	1,1,2-Trichloroethane	[1.4]
C₂H₃F	Fluoroethene	1.468 ± 0.003
C₂H₃FO	Acetyl fluoride	2.96 ± 0.03
C₂H₃F₃	1,1,1-Trifluoroethane	2.3470 ± 0.005
C₂H₃HgN	Cyanomethylmercury	4.7 ± 0.1
C₂H₃I	Iodoethene	1.311 ± 0.005
C₂H₃N	Acetonitrile	3.92519
C₂H₃NO	Methyl cyanate	4.26 ± 0.18
C₂H₃NO	Methyl isocyanate	≈2.8
C₂H₃NS	Methyl isothiocyanate	3.453 ± 0.003
C₂H₃N₃	1H-1,2,4-Triazole	2.7 ± 0.1
C₂H₄BrCl	1-Bromo-2-chloroethane	[1.2]
C₂H₄Br₂	1,2-Dibromoethane	[1.19]
C₂H₄ClF	1-Chloro-1-fluoroethane	2.068 ± 0.014
C₂H₄Cl₂	1,1-Dichloroethane	2.06 ± 0.04
C₂H₄Cl₂	1,2-Dichloroethane	[1.83]
C₂H₄F₂	1,1-Difluoroethane	2.27 ± 0.05
C₂H₄F₂	1,2-Difluoroethane (gauche)	2.67 ± 0.13
C₂H₄O	Acetaldehyde	2.750 ± 0.006
C₂H₄O	Ethylene oxide	1.89 ± 0.01
C₂H₄O₂	Acetic acid	1.70 ± 0.03
C₂H₄O₂	Methyl formate	1.77 ± 0.04
C₂H₄O₂	Glycolaldehyde	2.73 ± 0.05
C₂H₅Br	Bromoethane	2.04 ± 0.02
C₂H₅Cl	Chloroethane	2.05 ± 0.02
C₂H₅ClO	2-Chloroethanol	1.78 ± 0.09
C₂H₅Cl₃Si	Trichloroethylsilane	[2.04]
C₂H₅F	Fluoroethane	1.937 ± 0.007
C₂H₅I	Iodoethane	1.976 ± 0.002
C₂H₅N	Ethyleneimine	1.90 ± 0.01
C₂H₅NO	Acetamide	3.68 ± 0.03
C₂H₅NO	N-Methylformamide	3.83 ± 0.08
C₂H₅NO₂	Nitroethane	3.23 ± 0.03
C₂H₆O	Ethanol (gauche)	1.68 ± 0.03
C₂H₆O	Ethanol (trans)	1.44 ± 0.03
C₂H₆O	Ethanol (average)	1.69 ± 0.03
C₂H₆O	Dimethyl ether	1.30 ± 0.01
C₂H₆OS	Dimethyl sulfoxide	3.96 ± 0.04
C₂H₆O₂	Ethylene glycol (average)	2.36 ± 0.10
C₂H₆S	Ethanethiol (gauche)	1.61 ± 0.08
C₂H₆S	Ethanethiol (trans)	1.58 ± 0.08
C₂H₆S	Dimethyl sulfide	1.554 ± 0.004
C₂H₆S₂	1,2-Ethanedithiol	2.03 ± 0.08
C₂H₆S₂	Dimethyl disulfide	[1.85]

Mol. Form.	Name	μ/D	Mol. Form.	Name	μ/D
C_2H_6Si	Vinylsilane	0.657 ± 0.002	$C_3H_6O_2$	Ethyl formate (gauche)	1.81 ± 0.02
C_2H_7N	Ethylamine (gauche)	1.210 ± 0.015	$C_3H_6O_2$	Ethyl formate (trans)	1.98 ± 0.02
C_2H_7N	Ethylamine (trans)	1.304 ± 0.011	$C_3H_6O_2$	Ethyl formate (average)	1.93
C_2H_7N	Ethylamine (average)	1.22 ± 0.10	$C_3H_6O_2$	Methyl acetate	1.72 ± 0.09
C_2H_7N	Dimethylamine	1.01 ± 0.02	$C_3H_6O_2$	1,3-Dioxolane	1.19 ± 0.06
C_2H_7NO	Ethanolamine	[2.27]	$C_3H_6O_2S$	Thietane 1,1-dioxide	4.8 ± 0.1
$C_2H_8N_2$	1,2-Ethanediamine	1.99 ± 0.10	$C_3H_6O_3$	1,3,5-Trioxane	2.08 ± 0.02
C_3HF_3	3,3,3-Trifluoro-1-propyne	2.317 ± 0.013	C_3H_6S	Thiacyclobutane	1.85 ± 0.09
C_3HN	Cyanoacetylene	3.73172	C_3H_7Br	1-Bromopropane	2.18 ± 0.11
$C_3H_2F_2$	3,3-Difluorocyclopropene	2.98 ± 0.02	C_3H_7Br	2-Bromopropane	2.21 ± 0.11
C_3H_2O	2-Propynal	2.78 ± 0.02	C_3H_7Cl	1-Chloropropane (gauche)	2.02 ± 0.03
$C_3H_3Cl_2F$	1,1-Dichloro-2-fluoropropene	2.43 ± 0.02	C_3H_7Cl	1-Chloropropane (trans)	1.95 ± 0.02
C_3H_3F	3-Fluoropropyne	1.73 ± 0.02	C_3H_7Cl	1-Chloropropane (average)	2.05 ± 0.04
$C_3H_3F_3$	3,3,3-Trifluoropropene	2.45 ± 0.05	C_3H_7Cl	2-Chloropropane	2.17 ± 0.11
C_3H_3N	Acrylonitrile	3.92 ± 0.07	C_3H_7F	1-Fluoropropane (gauche)	1.90 ± 0.10
C_3H_3NO	Oxazole	1.503 ± 0.030	C_3H_7F	1-Fluoropropane (trans)	2.05 ± 0.04
C_3H_3NO	Isoxazole	2.95 ± 0.04	C_3H_7F	2-Fluoropropane	1.958 ± 0.001
C_3H_4	Propyne	0.784 ± 0.001	C_3H_7I	1-Iodopropane	2.04 ± 0.10
C_3H_4	Cyclopropene	0.454 ± 0.010	C_3H_7I	2-Iodopropane	[1.95]
$C_3H_4F_2$	1,1-Difluoro-1-propene	0.889 ± 0.007	C_3H_7N	Allylamine	≈ 1.2
$C_3H_4N_2$	1H-Pyrazole	2.20 ± 0.01	C_3H_7N	Cyclopropylamine	1.19 ± 0.01
$C_3H_4N_2$	Imidazole	3.8 ± 0.4	C_3H_7N	Propyleneimine (cis)	1.77 ± 0.09
C_3H_4O	Propargyl alcohol	1.13 ± 0.06	C_3H_7N	Propyleneimine (trans)	1.57 ± 0.03
C_3H_4O	Acrolein (trans)	3.117 ± 0.004	C_3H_7NO	N,N-Dimethylformamide	3.82 ± 0.08
C_3H_4O	Acrolein (cis)	2.552 ± 0.003	C_3H_7NO	N-Methylacetamide	[4.3]
C_3H_4O	Cyclopropanone	2.67 ± 0.13	$C_3H_7NO_2$	1-Nitropropane	3.66 ± 0.07
$C_3H_4O_2$	Vinyl formate	1.49 ± 0.01	$C_3H_7NO_2$	2-Nitropropane	3.73 ± 0.07
$C_3H_4O_2$	2-Oxetanone	4.18 ± 0.03	C_3H_8	Propane	0.084 ± 0.001
$C_3H_4O_2$	3-Oxetanone	0.887 ± 0.005	C_3H_8O	1-Propanol (gauche)	1.58 ± 0.03
$C_3H_4O_3$	Ethylene carbonate	[4.9]	C_3H_8O	1-Propanol (trans)	1.55 ± 0.03
C_3H_5Br	2-Bromopropene	[1.51]	C_3H_8O	2-Propanol (trans)	1.58 ± 0.03
C_3H_5Br	3-Bromopropene	≈ 1.9	C_3H_8O	Ethyl methyl ether (trans)	1.17 ± 0.02
C_3H_5Cl	cis-1-Chloropropene	1.67 ± 0.08	$C_3H_8O_2$	1,2-Propylene glycol	[2.25]
C_3H_5Cl	trans-1-Chloropropene	1.97 ± 0.10	$C_3H_8O_2$	1,3-Propylene glycol	[2.55]
C_3H_5Cl	2-Chloropropene	1.647 ± 0.010	$C_3H_8O_2$	Ethylene glycol monomethyl ether (gauche)	2.36 ± 0.05
C_3H_5Cl	3-Chloropropene	1.94 ± 0.10	$C_3H_8O_2$	Dimethoxymethane	[0.74]
C_3H_5ClO	Epichlorohydrin	[1.8]	$C_3H_8O_3$	Glycerol	[2.56]
C_3H_5F	cis-1-Fluoropropene	1.46 ± 0.03	C_3H_8S	1-Propanethiol (gauche)	1.683 ± 0.010
C_3H_5F	trans-1-Fluoropropene	≈ 1.9	C_3H_8S	1-Propanethiol (trans)	1.60 ± 0.08
C_3H_5F	2-Fluoropropene	1.61 ± 0.03	C_3H_8S	2-Propanethiol (gauche)	1.53 ± 0.03
C_3H_5F	3-Fluoropropene (gauche)	1.939 ± 0.015	C_3H_8S	2-Propanethiol (trans)	1.61 ± 0.03
C_3H_5F	3-Fluoropropene (cis)	1.765 ± 0.014	C_3H_8S	Ethyl methyl sulfide (gauche)	1.593 ± 0.004
C_3H_5N	Propanenitrile	4.05 ± 0.03	C_3H_8S	Ethyl methyl sulfide (trans)	1.56 ± 0.03
C_3H_5NO	Ethyl cyanate	4.72 ± 0.09	C_3H_9N	Propylamine	1.17 ± 0.06
C_3H_5NO	3-Hydroxypropanenitrile (gauche)	3.17 ± 0.02	C_3H_9N	Isopropylamine	1.19 ± 0.06
C_3H_6	Propene	0.366 ± 0.001	C_3H_9N	Trimethylamine	0.612 ± 0.003
$C_3H_6Br_2$	1,2-Dibromopropane	[1.2]	$C_3H_9O_4P$	Trimethyl phosphate	[3.18]
$C_3H_6Cl_2$	1,2-Dichloropropane	[1.85]	C_4H_4	1-Buten-3-yne	0.22 ± 0.02
$C_3H_6Cl_2$	1,3-Dichloropropane	2.08 ± 0.04	C_4H_4	Methylenecyclopropene	1.90 ± 0.01
C_3H_6O	Acetone	2.88 ± 0.03	$C_4H_4N_2$	Succinonitrile	[3.7]
C_3H_6O	Propanal (gauche)	2.86 ± 0.01	$C_4H_4N_2$	Pyrimidine	2.334 ± 0.010
C_3H_6O	Propanal (cis)	2.52 ± 0.05	$C_4H_4N_2$	Pyridazine	4.22 ± 0.02
C_3H_6O	Propanal (average)	2.72	C_4H_4O	Furan	0.66 ± 0.01
C_3H_6O	Allyl alcohol (gauche)	1.55 ± 0.08	$C_4H_4O_2$	Diketene	3.53 ± 0.07
C_3H_6O	Allyl alcohol (average)	1.60 ± 0.08	C_4H_4S	Thiophene	0.55 ± 0.01
C_3H_6O	Methyl vinyl ether	0.965 ± 0.002	C_4H_5N	2-Methylacrylonitrile	3.69 ± 0.18
C_3H_6O	Methyloxirane	2.01 ± 0.02	C_4H_5N	Pyrrole	1.767 ± 0.001
C_3H_6O	Oxetane	1.94 ± 0.01	C_4H_5N	Isocyanocyclopropane	4.03 ± 0.10
$C_3H_6O_2$	Propanoic acid (cis)	1.46 ± 0.07	C_4H_5NO	2-Methyloxazole	1.37 ± 0.07
$C_3H_6O_2$	Propanoic acid (average)	1.75 ± 0.09	C_4H_5NO	4-Methyloxazole	1.08 ± 0.05

Mol. Form.	Name	μ/D	Mol. Form.	Name	μ/D
C_4H_5NO	5-Methyloxazole	2.16 ± 0.04	C_4H_9Cl	2-Chlorobutane	2.04 ± 0.10
C_4H_5NO	4-Methylisoxazole	3.583 ± 0.005	C_4H_9Cl	1-Chloro-2-methylpropane	2.00 ± 0.10
C_4H_6	1,2-Butadiene	0.403 ± 0.002	C_4H_9Cl	2-Chloro-2-methylpropane	2.13 ± 0.04
C_4H_6	1-Butyne	0.782 ± 0.004	C_4H_9I	1-Iodobutane	[1.93]
C_4H_6	Cyclobutene	0.132 ± 0.001	C_4H_9I	2-Iodobutane	2.12 ± 0.11
C_4H_6O	Divinyl ether	0.78 ± 0.05	C_4H_9I	1-Iodo-2-methylpropane	[1.87]
C_4H_6O	3-Methoxy-1,2-propadiene	0.963 ± 0.020	C_4H_9N	Pyrrolidine	[1.57]
C_4H_6O	trans-2-Butenal	3.67 ± 0.07	C_4H_9NO	N-Methylpropanamide	3.61
C_4H_6O	2-Methylpropenal	2.68 ± 0.13	C_4H_9NO	N,N-Dimethylacetamide	[3.7]
C_4H_6O	Cyclobutanone	2.89 ± 0.03	C_4H_9NO	Morpholine	1.55 ± 0.03
C_4H_6O	2,3-Dihydrofuran	1.32 ± 0.03	C_4H_{10}	Isobutane	0.132 ± 0.002
C_4H_6O	2,5-Dihydrofuran	1.63 ± 0.01	$C_4H_{10}O$	1-Butanol	1.66 ± 0.03
$C_4H_6O_2$	trans-Crotonic acid	[2.13]	$C_4H_{10}O$	2-Butanol	[1.8]
$C_4H_6O_2$	Methacrylic acid	[1.65]	$C_4H_{10}O$	2-Methyl-1-propanol	1.64 ± 0.08
$C_4H_6O_2$	Vinyl acetate	[1.79]	$C_4H_{10}O$	2-Methyl-2-propanol	[1.66]
$C_4H_6O_2$	Methyl acrylate	[1.77]	$C_4H_{10}O$	Diethyl ether	1.15 ± 0.02
$C_4H_6O_2$	γ-Butyrolactone	4.27 ± 0.03	$C_4H_{10}O$	Methyl propyl ether (trans-trans)	1.107 ± 0.013
$C_4H_6O_2$	2,3-Dihydro-1,4-dioxin	0.939 ± 0.008	$C_4H_{10}O$	Isopropyl methyl ether	1.247 ± 0.003
$C_4H_6O_2$	3,6-Dihydro-1,2-dioxin	2.329 ± 0.001	$C_4H_{10}O_2$	1,4-Butanediol	[2.58]
$C_4H_6O_3$	Acetic anhydride	≈2.8	$C_4H_{10}O_2$	Ethylene glycol monoethyl ether	[2.08]
$C_4H_6O_3$	Propylene carbonate	[4.9]	$C_4H_{10}O_3$	Diethylene glycol	[2.31]
C_4H_6S	2,3-Dihydrothiophene	1.61 ± 0.20	$C_4H_{10}S$	1-Butanethiol	[1.53]
C_4H_6S	2,5-Dihydrothiophene	1.75 ± 0.01	$C_4H_{10}S$	2-Methyl-2-propanethiol	1.66 ± 0.03
C_4H_7N	Butanenitrile (gauche)	3.91 ± 0.04	$C_4H_{10}S$	Diethyl sulfide	1.54 ± 0.08
C_4H_7N	Butanenitrile (anti)	3.73 ± 0.06	$C_4H_{11}N$	Butylamine	≈1.0
C_4H_7N	2-Methylpropanenitrile	4.29 ± 0.09	$C_4H_{11}N$	sec-Butylamine	[1.28]
C_4H_7N	2-Isocyanopropane	4.055 ± 0.001	$C_4H_{11}N$	tert-Butylamine	[1.29]
C_4H_7NO	2-Pyrrolidone	[3.5]	$C_4H_{11}N$	Isobutylamine	[1.27]
C_4H_8	1-Butene (cis)	0.438 ± 0.007	$C_4H_{11}N$	Diethylamine	0.92 ± 0.05
C_4H_8	1-Butene (skew)	0.359 ± 0.011	$C_4H_{11}NO_2$	Diethanolamine	[2.8]
C_4H_8	cis-2-Butene	0.253 ± 0.005	$C_4H_{13}N_3$	Diethylenetriamine	[1.89]
C_4H_8	Isobutene	0.503 ± 0.010	C_5F_5N	Perfluoropyridine	0.98 ± 0.08
C_4H_8	Methylcyclopropane	0.139 ± 0.004	C_5H_3NS	2-Thiophenecarbonitrile	4.59 ± 0.02
$C_4H_8Cl_2$	1,4-Dichlorobutane	2.22 ± 0.11	C_5H_3NS	3-Thiophenecarbonitrile	4.13 ± 0.02
$C_4H_8Cl_2O$	Bis(2-chloroethyl) ether	[2.58]	C_5H_4	1,3-Pentadiyne	1.207 ± 0.001
C_4H_8O	cis-2-Buten-1-ol	1.96 ± 0.03	C_5H_4ClN	4-Chloropyridine	0.756 ± 0.005
C_4H_8O	trans-2-Buten-1-ol	1.90 ± 0.02	C_5H_4FN	3-Fluoropyridine	2.09 ± 0.26
C_4H_8O	2-Methyl-2-propenol (skew)	1.295 ± 0.022	C_5H_4O	2,4-Cyclopentadien-1-one	3.132 ± 0.007
C_4H_8O	Ethyl vinyl ether	[1.26]	C_5H_4OS	4H-Pyran-4-thione	3.95 ± 0.05
C_4H_8O	1,2-Epoxybutane	1.891 ± 0.011	$C_5H_4O_2$	Furfural	[3.54]
C_4H_8O	Butanal	2.72 ± 0.05	$C_5H_4O_2$	4H-Pyran-4-one	3.79 ± 0.02
C_4H_8O	Isobutanal (gauche)	2.69 ± 0.01	$C_5H_4S_2$	4H-Thiopyran-4-thione	3.9 ± 0.2
C_4H_8O	Isobutanal (trans)	2.86 ± 0.01	C_5H_5N	Pyridine	2.215 ± 0.010
C_4H_8O	2-Butanone	2.779 ± 0.015	C_5H_6	1,2,3-Pentatriene	0.51 ± 0.05
C_4H_8O	Tetrahydrofuran	1.75 ± 0.04	C_5H_6	1-Penten-3-yne	0.66 ± 0.02
C_4H_8OS	1,4-Oxathiane	0.295 ± 0.003	C_5H_6	cis-3-Penten-1-yne	0.78 ± 0.02
$C_4H_8O_2$	Butanoic acid	[1.65]	C_5H_6	trans-3-Penten-1-yne	1.06 ± 0.05
$C_4H_8O_2$	2-Methylpropanoic acid	[1.08]	C_5H_6	2-Methyl-1-buten-3-yne	0.513 ± 0.02
$C_4H_8O_2$	Propyl formate	[1.89]	C_5H_6	1,3-Cyclopentadiene	0.419 ± 0.004
$C_4H_8O_2$	Ethyl acetate	1.78 ± 0.09	$C_5H_6N_2$	2-Methylpyrimidine	1.676 ± 0.010
$C_4H_8O_2$	cis-2-Butene-1,4-diol	[2.48]	$C_5H_6N_2$	5-Methylpyrimidine	2.881 ± 0.006
$C_4H_8O_2$	trans-2-Butene-1,4-diol	[2.45]	C_5H_6O	2-Methylfuran	0.65 ± 0.05
$C_4H_8O_2$	1,3-Dioxane	2.06 ± 0.04	C_5H_6O	3-Methylfuran	1.03 ± 0.02
$C_4H_8O_2S$	Sulfolane	[4.8]	C_5H_6O	3-Cyclopenten-1-one	2.79 ± 0.03
C_4H_8S	3-Methylthietane	2.046 ± 0.009	$C_5H_6O_2$	5-Methyl-2(3H)-furanone	4.08 ± 0.02
C_4H_8S	Tetrahydrothiophene	[1.90]	$C_5H_6O_2$	Furfuryl alcohol	[1.92]
$C_4H_8S_2$	1,3-Dithiane	2.14 ± 0.04	C_5H_6S	2-Methylthiophene	0.674 ± 0.005
C_4H_9Br	1-Bromobutane	2.08 ± 0.10	C_5H_6S	3-Methylthiophene	0.914 ± 0.015
C_4H_9Br	2-Bromobutane	2.23 ± 0.11	C_5H_7N	3-Methyl-2-butenenitrile	4.61 ± 0.13
C_4H_9Br	2-Bromo-2-methylpropane	[2.17]	C_5H_7N	Cyclobutanecarbonitrile	4.04 ± 0.04
C_4H_9Cl	1-Chlorobutane	2.05 ± 0.04	$C_5H_7NO_2$	Ethyl cyanoacetate	[2.17]

Mol. Form.	Name	μ/D	Mol. Form.	Name	μ/D
C_5H_8	cis-1,3-Pentadiene	0.500 ± 0.015	$C_6H_4ClNO_2$	1-Chloro-2-nitrobenzene	4.64 ± 0.09
C_5H_8	trans-1,3-Pentadiene	0.585 ± 0.010	$C_6H_4ClNO_2$	1-Chloro-3-nitrobenzene	3.73 ± 0.07
C_5H_8	2-Methyl-1,3-butadiene	0.25 ± 0.01	$C_6H_4ClNO_2$	1-Chloro-4-nitrobenzene	2.83 ± 0.06
C_5H_8	1-Pentyne (gauche)	0.769 ± 0.028	$C_6H_4Cl_2$	o-Dichlorobenzene	2.50 ± 0.05
C_5H_8	1-Pentyne (trans)	0.842 ± 0.010	$C_6H_4Cl_2$	m-Dichlorobenzene	1.72 ± 0.09
C_5H_8	Cyclopentene	0.20 ± 0.02	$C_6H_4FNO_2$	1-Fluoro-4-nitrobenzene	2.87 ± 0.06
C_5H_8	3,3-Dimethylcyclopropene	0.287 ± 0.003	$C_6H_4F_2$	o-Difluorobenzene	2.46 ± 0.05
C_5H_8O	Cyclopropyl methyl ketone	2.62 ± 0.25	$C_6H_4F_2$	m-Difluorobenzene	1.51 ± 0.02
C_5H_8O	Cyclopentanone	≈3.3	$C_6H_4N_2$	2-Pyridinecarbonitrile	5.78 ± 0.11
C_5H_8O	3,4-Dihydro-2H-pyran	1.400 ± 0.008	$C_6H_4N_2$	3-Pyridinecarbonitrile	3.66 ± 0.11
C_5H_8O	3,6-Dihydro-2H-pyran	1.283 ± 0.005	$C_6H_4N_2$	4-Pyridinecarbonitrile	1.96 ± 0.03
$C_5H_8O_2$	Ethyl acrylate	[1.96]	$C_6H_4O_2$	3,5-Cyclohexadiene-1,2-dione	4.23 ± 0.02
$C_5H_8O_2$	Methyl methacrylate	[1.67]	C_6H_5Br	Bromobenzene	1.70 ± 0.03
$C_5H_8O_2$	2,4-Pentanedione	[2.78]	C_6H_5Cl	Chlorobenzene	1.69 ± 0.03
$C_5H_8O_2$	Dihydro-3-methyl-2(3H)-furanone	4.56 ± 0.02	C_6H_5ClO	p-Chlorophenol	2.11 ± 0.11
$C_5H_8O_2$	Dihydro-5-methyl-2(3H)-furanone	4.71 ± 0.05	C_6H_5F	Fluorobenzene	1.60 ± 0.08
$C_5H_8O_2$	Tetrahydro-4H-pyran-4-one	1.720 ± 0.003	C_6H_5I	Iodobenzene	1.70 ± 0.09
C_5H_9N	Pentanenitrile	4.12 ± 0.08	C_6H_5NO	2-Pyridinecarboxaldehyde	3.56 ± 0.07
C_5H_9N	2,2-Dimethylpropanenitrile	3.95 ± 0.04	C_6H_5NO	3-Pyridinecarboxaldehyde	1.44
C_5H_9N	1,2,5,6-Tetrahydropyridine	1.007 ± 0.003	C_6H_5NO	4-Pyridinecarboxaldehyde	1.66
C_5H_9NO	N-Methyl-2-pyrrolidone	[4.1]	$C_6H_5NO_2$	Nitrobenzene	4.22 ± 0.08
C_5H_{10}	1-Pentene	≈0.5	C_6H_6	Fulvene	0.4236 ± 0.013
C_5H_{10}	3-Methyl-1-butene (gauche)	0.398 ± 0.004	C_6H_6ClN	o-Chloroaniline	[1.77]
C_5H_{10}	3-Methyl-1-butene (trans)	0.320 ± 0.010	C_6H_6O	Phenol	1.224 ± 0.008
C_5H_{10}	1,1-Dimethylcyclopropane	0.142 ± 0.001	C_6H_6O	2-Vinylfuran	0.69 ± 0.07
$C_5H_{10}O$	2,2-Dimethylpropanal	2.66 ± 0.05	$C_6H_6O_2$	p-Hydroquinone	2.38 ± 0.05
$C_5H_{10}O$	2-Pentanone	[2.70]	C_6H_6S	Benzenethiol	[1.23]
$C_5H_{10}O$	3-Pentanone	[2.82]	C_6H_7N	Aniline	1.13 ± 0.02
$C_5H_{10}O$	Tetrahydropyran (chair)	1.58 ± 0.03	C_6H_7N	2-Methylpyridine	1.85 ± 0.04
$C_5H_{10}O_2$	Pentanoic acid	[1.61]	C_6H_7N	3-Methylpyridine	[2.40]
$C_5H_{10}O_2$	3-Methylbutanoic acid	[0.63]	C_6H_7N	4-Methylpyridine	2.70 ± 0.02
$C_5H_{10}O_2$	Butyl formate	[2.03]	C_6H_8O	3-Methyl-2-cyclopenten-1-one	4.33 ± 0.002
$C_5H_{10}O_2$	Isobutyl formate	[1.88]	$C_6H_8O_4$	Dimethyl maleate	[2.48]
$C_5H_{10}O_2$	Propyl acetate	[1.78]	C_6H_8Si	Phenylsilane	0.845 ± 0.012
$C_5H_{10}O_2$	Ethyl propanoate	[1.74]	C_6H_9F	1-Fluorocyclohexene	1.942 ± 0.010
$C_5H_{10}O_2$	Tetrahydrofurfuryl alcohol	[2.1]	C_6H_{10}	1-Hexyne	0.83 ± 0.05
$C_5H_{10}O_3$	Diethyl carbonate	1.10 ± 0.06	C_6H_{10}	3,3-Dimethyl-1-butyne	0.661 ± 0.004
$C_5H_{10}O_3$	Ethylene glycol monomethyl ether acetate	[2.13]	C_6H_{10}	Cyclohexene (half-chair)	0.332 ± 0.012
			$C_6H_{10}F_2$	1,1-Difluorocyclohexane	2.556 ± 0.010
$C_5H_{10}O_3$	Ethyl lactate	[2.4]	$C_6H_{10}O$	3-Methylcyclopentanone	3.14 ± 0.03
$C_5H_{10}S$	Thiacyclohexane	1.781 ± 0.010	$C_6H_{10}O$	Cyclohexanone	3.246 ± 0.006
$C_5H_{11}Br$	1-Bromopentane	2.20 ± 0.11	$C_6H_{10}O$	Mesityl oxide	[2.79]
$C_5H_{11}Cl$	1-Chloropentane	2.16 ± 0.11	$C_6H_{10}O_4$	Diethyl oxalate	[2.49]
$C_5H_{11}Cl$	1-Chloro-3-methylbutane	[1.92]	$C_6H_{10}O_4$	Ethylene glycol diacetate	[2.34]
$C_5H_{11}N$	Piperidine (equitorial)	0.82 ± 0.02	$C_6H_{11}Cl$	Chlorocyclohexane (equitorial)	2.44 ± 0.07
$C_5H_{11}N$	Piperidine (axial)	1.19 ± 0.02	$C_6H_{11}Cl$	Chlorocyclohexane (axial)	1.91 ± 0.02
$C_5H_{11}N$	Piperidine (average)	[1.19]	$C_6H_{11}F$	Fluorocyclohexane (equitorial)	2.11 ± 0.04
$C_5H_{11}N$	N-Methylpyrrolidine	0.572 ± 0.003	$C_6H_{11}F$	Fluorocyclohexane (axial)	1.81 ± 0.04
C_5H_{12}	Isopentane	0.13 ± 0.05	$C_6H_{11}N$	4-Methylpentanenitrile	[3.5]
$C_5H_{12}N_2O$	Tetramethylurea	[3.5]	$C_6H_{11}NO$	Caprolactam	[3.9]
$C_5H_{12}O$	1-Pentanol	[1.7]	$C_6H_{12}O$	Butyl vinyl ether	[1.25]
$C_5H_{12}O$	2-Pentanol	[1.66]	$C_6H_{12}O$	2-Hexanone	[2.66]
$C_5H_{12}O$	3-Pentanol	[1.64]	$C_6H_{12}O_2$	Hexanoic acid	[1.13]
$C_5H_{12}O$	2-Methyl-1-butanol	[1.88]	$C_6H_{12}O_2$	Pentyl formate	1.90 ± 0.10
$C_5H_{12}O$	2-Methyl-2-butanol	[1.82]	$C_6H_{12}O_2$	Butyl acetate	[1.87]
$C_5H_{12}O_2$	1,5-Pentanediol	[2.5]	$C_6H_{12}O_2$	sec-Butyl acetate	[1.87]
$C_5H_{12}O_3$	Diethylene glycol monomethyl ether	[1.6]	$C_6H_{12}O_2$	Isobutyl acetate	[1.86]
			$C_6H_{12}O_2$	Ethyl butanoate	[1.74]
$C_6H_2F_4$	1,2,3,4-Tetrafluorobenzene	2.42 ± 0.05	$C_6H_{12}O_2$	Diacetone alcohol	[3.24]
$C_6H_2F_4$	1,2,3,5-Tetrafluorobenzene	1.46 ± 0.06	$C_6H_{12}O_3$	Ethylene glycol monoethyl ether acetate	[2.25]
$C_6H_3F_3$	1,2,4-Trifluorobenzene	1.402 ± 0.009			

Mol. Form.	Name	μ/D	Mol. Form.	Name	μ/D
$C_6H_{12}O_3$	Paraldehyde	1.43 ± 0.07	C_8H_8	Styrene	0.123 ± 0.003
$C_6H_{13}N$	Cyclohexylamine	[1.26]	C_8H_8O	Acetophenone	3.02 ± 0.06
$C_6H_{14}O$	Dipropyl ether	1.21 ± 0.06	$C_8H_8O_2$	Methyl benzoate	[1.94]
$C_6H_{14}O$	Diisopropyl ether	1.13 ± 0.10	$C_8H_8O_3$	Methyl salicylate	[2.47]
$C_6H_{14}O$	Butyl ethyl ether	[1.24]	C_8H_{10}	Ethylbenzene	0.59 ± 0.05
$C_6H_{14}O_2$	2-Methyl-2,4-pentanediol	[2.9]	C_8H_{10}	o-Xylene	0.640 ± 0.005
$C_6H_{14}O_2$	Ethylene glycol monobutyl ether	[2.08]	$C_8H_{10}O$	2,4-Xylenol	[1.4]
$C_6H_{14}O_2$	1,1-Diethoxyethane	[1.38]	$C_8H_{10}O$	2,5-Xylenol	[1.45]
$C_6H_{14}O_3$	Diethylene glycol monoethyl ether	[1.6]	$C_8H_{10}O$	2,6-Xylenol	[1.40]
$C_6H_{14}O_3$	Diethylene glycol dimethyl ether	[1.97]	$C_8H_{10}O$	3,4-Xylenol	[1.56]
$C_6H_{15}N$	Dipropylamine	[1.03]	$C_8H_{10}O$	3,5-Xylenol	[1.55]
$C_6H_{15}N$	Diisopropylamine	[1.15]	$C_8H_{10}O$	Phenetole	1.45 ± 0.15
$C_6H_{15}N$	Triethylamine	0.66 ± 0.05	$C_8H_{10}O_2$	1,2-Dimethoxybenzene	[1.29]
$C_6H_{15}NO_3$	Triethanolamine	[3.57]	$C_8H_{11}N$	N,N-Dimethylaniline	1.68 ± 0.17
$C_6H_{15}O_4P$	Triethyl phosphate	[3.12]	$C_8H_{11}N$	2,4-Dimethylaniline	[1.40]
$C_6H_{18}N_3OP$	Hexamethylphosphoric triamide	[5.5]	$C_8H_{11}N$	2,6-Dimethylaniline	[1.63]
$C_7H_5Cl_3$	(Trichloromethyl)benzene	[2.03]	$C_8H_{11}N$	2,4,6-Trimethylpyridine	[2.05]
$C_7H_5F_3$	(Trifluoromethyl)benzene	2.86 ± 0.06	$C_8H_{16}O$	2-Octanone	[2.70]
C_7H_5N	Benzonitrile	4.18 ± 0.08	$C_8H_{16}O_2$	Octanoic acid	[1.15]
C_7H_5N	Isocyanobenzene	4.018 ± 0.003	$C_8H_{16}O_2$	sec-Hexyl acetate	[1.9]
$C_7H_6Cl_2$	2,4-Dichlorotoluene	[1.70]	$C_8H_{16}O_2$	Isobutyl isobutanoate	[1.9]
$C_7H_6Cl_2$	3,4-Dichlorotoluene	[2.95]	$C_8H_{16}O_4$	Diethylene glycol monoethyl ether acetate	[1.8]
$C_7H_6Cl_2$	(Dichloromethyl)benzene	[2.07]			
C_7H_6O	2,4,6-Cycloheptatrien-1-one	4.1 ± 0.3	$C_8H_{17}Cl$	1-Chlorooctane	[2.00]
C_7H_6O	Benzaldehyde	[3.0]	$C_8H_{18}O$	1-Octanol	[1.76]
$C_7H_6O_2$	Salicylaldehyde	[2.86]	$C_8H_{18}O$	2-Octanol	[1.71]
C_7H_7Cl	o-Chlorotoluene	1.56 ± 0.08	$C_8H_{18}O$	2-Ethyl-1-hexanol	[1.74]
C_7H_7Cl	m-Chlorotoluene	[1.82]	$C_8H_{18}O$	Dibutyl ether	1.17 ± 0.06
C_7H_7Cl	p-Chlorotoluene	2.21 ± 0.04	$C_8H_{18}S$	Dibutyl sulfide	[1.61]
C_7H_7Cl	(Chloromethyl)benzene	[1.82]	$C_8H_{19}N$	Dibutylamine	[0.98]
C_7H_7F	o-Fluorotoluene	1.37 ± 0.07	C_9H_7N	Quinoline	2.29 ± 0.11
C_7H_7F	m-Fluorotoluene	1.82 ± 0.04	C_9H_7N	Isoquinoline	2.73 ± 0.14
C_7H_7F	p-Fluorotoluene	2.00 ± 0.10	$C_9H_{10}O_2$	Ethyl benzoate	2.00 ± 0.10
$C_7H_7NO_3$	2-Nitroanisole	[5.0]	$C_9H_{10}O_2$	Benzyl acetate	[1.22]
C_7H_8	Toluene	0.375 ± 0.010	C_9H_{12}	Isopropylbenzene	≈0.79
C_7H_8	2,5-Norbornadiene	0.0587 ± 0.0001	$C_9H_{18}O$	2,6-Dimethyl-4-heptanone	[2.66]
C_7H_8O	o-Cresol	[1.45]	$C_9H_{18}O_2$	Nonanoic acid	[0.79]
C_7H_8O	m-Cresol	[1.48]	$C_{10}H_7Br$	1-Bromonaphthalene	[1.55]
C_7H_8O	p-Cresol	[1.48]	$C_{10}H_7Cl$	1-Chloronaphthalene	[1.57]
C_7H_8O	Benzyl alcohol	1.71 ± 0.09	$C_{10}H_8$	Azulene	0.80 ± 0.02
C_7H_8O	Anisole	1.38 ± 0.07	$C_{10}H_{14}$	tert-Butylbenzene	≈0.83
C_7H_9N	o-Methylaniline	[1.60]	$C_{10}H_{16}O$	Camphor, (+)	[3.1]
C_7H_9N	m-Methylaniline	[1.45]	$C_{10}H_{20}O_2$	2-Ethylhexyl acetate	[1.8]
C_7H_9N	p-Methylaniline	[1.52]	$C_{10}H_{21}Br$	1-Bromodecane	[1.93]
C_7H_9N	2,4-Dimethylpyridine	[2.30]	$C_{10}H_{22}O$	Dipentyl ether	[1.20]
C_7H_9N	2,6-Dimethylpyridine	[1.66]	$C_{10}H_{22}O$	Diisopentyl ether	[1.23]
C_7H_{10}	1,3-Cycloheptadiene	0.740	$C_{11}H_{12}O_2$	Ethyl trans-cinnamate	[1.84]
C_7H_{12}	Methylenecyclohexane	0.62 ± 0.01	$C_{12}H_{10}$	Acenaphthene	≈0.85
$C_7H_{12}O_4$	Diethyl malonate	[2.54]	$C_{12}H_{10}O$	Diphenyl ether	≈1.3
$C_7H_{14}O$	2-Heptanone	[2.59]	$C_{12}H_{27}BO_3$	Tributyl borate	[0.77]
$C_7H_{14}O$	3-Heptanone	[2.78]	$C_{12}H_{27}N$	Tributylamine	[0.78]
$C_7H_{14}O$	2,4-Dimethyl-3-pentanone	[2.74]	$C_{12}H_{27}O_4P$	Tributyl phosphate	[3.07]
$C_7H_{14}O$	cis-3-Methylcyclohexanol	[1.91]	$C_{14}H_{12}O_2$	Benzyl benzoate	[2.06]
$C_7H_{14}O$	trans-3-Methylcyclohexanol	[1.75]	$C_{16}H_{22}O_4$	Dibutyl phthalate	[2.82]
$C_7H_{14}O_2$	Pentyl acetate	1.75 ± 0.10	$C_{18}H_{34}O_2$	Oleic acid	[1.18]
$C_7H_{14}O_2$	Isopentyl acetate	[1.86]	$C_{18}H_{34}O_4$	Dibutyl sebacate	[2.48]
$C_7H_{15}Br$	1-Bromoheptane	2.16 ± 0.11	$C_{21}H_{21}O_4P$	Tri-o-cresyl phosphate	[2.87]
$C_7H_{16}O$	2-Heptanol	[1.71]	$C_{21}H_{21}O_4P$	Tri-m-cresyl phosphate	[3.05]
$C_7H_{16}O$	3-Heptanol	[1.71]	$C_{21}H_{21}O_4P$	Tri-p-cresyl phosphate	[3.18]
C_8H_6	Phenylacetylene	0.656 ± 0.005	$C_{22}H_{44}O_2$	Butyl stearate	[1.88]
C_8H_7N	Benzeneacetonitrile	[3.5]	$C_{24}H_{38}O_4$	Bis(2-ethylhexyl) phthalate	[2.84]

STRENGTHS OF CHEMICAL BONDS*
J. Alistair Kerr and David W. Stocker

The strength of a chemical bond, $D°$(R-X), often known as the bond dissociation energy, is defined as the standard enthalpy change of the reaction in which the bond is broken: RX → R + X. It is given by the thermochemical equation, $D°$(R-X) = $\Delta_f H°$(R) + $\Delta_f H°$(X) - $\Delta_f H°$(RX). Some authors list bond strengths at a temperature of absolute zero but here the values at 298 K are given because more thermodynamic data are available for this temperature. Bond strengths or bond dissociation energies are not equal to, and may differ considerably from, mean bond energies determined solely from thermochemical data on atoms and molecules.

Table 1
BOND STRENGTHS IN DIATOMIC MOLECULES

These have usually been measured spectroscopically or by mass spectrometric analysis of hot gases effusing a Knudsen cell. Excellent accounts of these and other methods are given in (i) *Dissociation Energies and Spectra of Diatomic Molecules*, by A. G. Gaydon, 3rd. ed., Chapman & Hall, London, 1968 and (ii) "Mass Spectrometric Determination of Bond Energies of High-Temperature Molecules", K. A. Gingerich, *Chimia*, 26, 619, 1972. The errors quoted in the table are those given in the original paper or review article. The references have been chosen primarily as a key to the literature. It should not be assumed that the author referred to was responsible for the value quoted, as the reference may be to a review article.

Bond strengths reported at a temperature of absolute zero, $D°_0$, have been converted to $D°_{298}$ by the use of enthalpy functions taken mainly from the JANAF Thermochemical Tables, Third Edition, *J. Phys. Chem. Ref. Data*, 14, Suppl. 1, 1985, wherever possible. For most bonds, however, this data is not available and the conversion has been made by the approximate relation:

$$D°_{298} = D°_0 + (3/2)RT$$

The list below does not include the increasing number of bond strengths of diatomic molecules now being calculated by *ab initio* methods. The Table has been arranged in an alphabetical order of the atoms.

Molecule	$D°_{298}$/kJ mol^{-1}	Ref.	Molecule	$D°_{298}$/kJ mol^{-1}	Ref.	Molecule	$D°_{298}$/kJ mol^{-1}	Ref.
Ag-Ag	160.3 ± 3.4	314	Al-Cl	511.3 ± 0.8	312	As-Ga	209.6 ± 1.2	83
Ag-Al	183.7 ± 9.2	79	Al-Co	181.6 ± 0.2	22	As-H	274.0 ± 2.9	29
Ag-Au	202.9 ± 9.2	4	Al-Cr	223.6 ± 0.6	19	As-I	296.6 ± 28.0	325
Ag-Bi	193 ± 42	246	Al-Cu	227.1 ± 1.2	21	As-In	201	297
Ag-Br	293 ± 29	120	Al-D	290.8	246	As-N	489 ± 2	310
Ag-Cl	314.2	184	Al-F	663.6 ± 6.3	80	As-O	481 ± 8	262
Ag-Cu	174.1 ± 9.2	136	Al-H	284.9 ± 6.3	80	As-P	433.5 ± 12.6	137
Ag-D	226.8	205	Al-I	369.9 ± 2.1	267	As-S	379.5 ± 6.3	262
Ag-Dy	130 ± 19	203	Al-Kr	6.047 ± 0.001	180	As-Sb	330.5 ± 5.4	94
Ag-Eu	129.7 ± 12.6	66	Al-Li	76.5	39	As-Se	96	288
Ag-F	354.4 ± 16.3	120	Al-N	297 ± 96	120	As-Tl	198.3 ± 14.6	300
Ag-Ga	180 ± 15	44	Al-Ni	225 ± 5	20	At-At	~80	89
Ag-Ge	174.5 ± 20.9	135	Al-O	511 ± 3	56,74	Au-Au	226.2 ± 0.5	210
Ag-H	215.1 ± 8	217	Al-P	216.7 ± 12.6	80	Au-B	367.8 ± 10.5	145
Ag-Ho	123.4 ± 16.7	62	Al-Pd	254.4 ± 12.1	64	Au-Ba	254.8 ± 10.0	135
Ag-I	234 ± 29	120	Al-S	373.6 ± 7.9	376	Au-Be	285 ± 8	120
Ag-In	166.5 ± 4.9	13	Al-Sb	216.3 ± 5.9	293	Au-Bi	297 ± 8.4	135
Ag-Li	173.6 ± 6.3	276,303	Al-Se	337.6 ± 10.0	376	Au-Ca	243	135
Ag-Mn	100 ± 21	246	Al-Si	229.3 ± 30.1	51	Au-Ce	339 ± 21	135
Ag-Na	138.1 ± 8.4	291,298	Al-Te	267.8 ± 10.0	376	Au-Cl	343 ± 9.6	120
Ag-Nd	<209	221	Al-U	326 ± 29	123	Au-Co	222 ± 17	135
Ag-O	220.1 ± 20.9	287	Al-V	147.4 ± 1.0	22	Au-Cr	213 ± 17	135
Ag-S	217.1	349	Al-Xe	7.43 ± 0.69	43	Au-Cs	255 ± 3.3	41
Ag-Se	202.5	349	Ar-Ar	4.73 ± 0.04	181	Au-Cu	228.0 ± 5.0	34,135
Ag-Si	177.8 ± 10.0	320	Ar-He	3.89	246	Au-D	318.4	205
Ag-Sn	136.0 ± 20.9	3	Ar-Hg	6.15	246	Au-Dy	259 ± 21	203
Ag-Te	195.8	349	Ar-I	10.0	40	Au-Eu	241.0 ± 10.5	66
Al-Al	133 ± 6	118	Ar-K	4.2	205	Au-Fe	187.0 ± 16.7	220
Al-Ar	5.182 ± 0.005	180	As-As	382.0 ± 10.5	247	Au-Ga	234 ± 38	135
Al-As	202.9 ± 7.1	294,301	As-Cl	448	80	Au-Ge	274.1 ± 5.0	135
Al-Au	325.9 ± 6.3	124	As-D	270.3	205	Au-H	292.0 ± 8	217
Al-Br	429 ± 6	186	As-F	410	205	Au-Ho	267.4 ± 16.7	62,250

* Revised to November 1998.

Table 1
BOND STRENGTHS IN DIATOMIC MOLECULES (continued)

Molecule	$D°_{298}$/kJ mol^{-1}	Ref.	Molecule	$D°_{298}$/kJ mol^{-1}	Ref.	Molecule	$D°_{298}$/kJ mol^{-1}	Ref.
Au-In	286.0 ± 5.7	13	Ba-H	176 ± 14.6	120	Br-Ni	360 ± 13	120
Au-La	336.4 ± 20.9	127	Ba-I	320.8 ± 6.3	188	Br-O	235.5 ± 2.4	46
Au-Li	284.5 ± 6.7	276	Ba-O	561.9 ± 13.4	287	Br-Pb	247 ± 38	120
Au-Lu	332.2 ± 16.7	132	Ba-Pd	221.8 ± 5.0	125	Br-Rb	380.7 ± 4	362
Au-Mg	243 ± 42	246	Ba-Rh	259.4 ± 25.1	125	Br-Sb	314 ± 59	120
Au-Mn	185.4 ± 12.6	342	Ba-S	400.0 ± 18.8	70	Br-Sc	444 ± 63	246
Au-Na	215.1 ± 12.6	298	Be-Be	59	35,87	Br-Se	297 ± 84	246
Au-Nd	299.2 ± 20.9	127	Be-Br	381 ± 84	246	Br-Si	367.8 ± 10.0	106
Au-Ni	247 ± 16	352	Be-Cl	388.3 ± 9.2	108,191,382	Br-Sn	≥552	286
Au-O	221.8 ± 20.9	287	Be-D	203.05	205	Br-Sr	333.0 ± 9.2	199
Au-Pb	130 ± 42	246	Be-F	577 ± 42	80,108	Br-Th	364	187
Au-Pd	155 ± 21	135	Be-H	200.0 ± 1.3	67	Br-Ti	439	246
Au-Pr	310 ± 21	135	Be-O	434.7 ± 13.4	287	Br-Tl	333.9 ± 1.7	28
Au-Rb	243 ± 2.9	41	Be-S	372 ± 59	120	Br-U	377.4 ± 6.3	260
Au-Rh	231.0 ± 29	66	Bi-Bi	200.4 ± 7.5	307,324	Br-V	439 ± 42	246
Au-S	418 ± 25	131	Bi-Br	267.4 ± 4.2	76	Br-W	329.3	223
Au-Sc	280.3 ± 16.7	128	Bi-Cl	301 ± 4	78	Br-Xe	5.94 ± 0.02	58
Au-Se	243.1	349	Bi-D	283.7	266	Br-Y	485 ± 84	246
Au-Si	305.4 ± 5.9	139	Bi-F	367 ± 13	400	Br-Zn	142 ± 29	246
Au-Sn	254.8 ± 7.1	233	Bi-Ga	159 ± 17	296	C-C	610 ± 2.0	373
Au-Sr	264 ± 42	246	Bi-H	≤283.3	266	C-Ce	444 ± 13	236
Au-Tb	289.5 ± 33.5	152,250	Bi-I	218.0 ± 4.6	77	C-Cl	397 ± 29	283
Au-Te	317.6	349	Bi-In	153.6 ± 1.7	321	C-D	341.4	205
Au-U	318 ± 29	123	Bi-Li	154.0 ± 5.0	277,305	C-F	552	193
Au-V	240.6 ± 12.1	172	Bi-O	337.2 ± 12.6	287	C-Ge	460 ± 21	120
Au-Y	307.1 ± 8.4	177	Bi-P	280 ± 13	137	C-H	338.4 ± 1.2	205
B-B	297 ± 21	80	Bi-Pb	141.8 ± 14.6	324	C-Hf	540 ± 25	357
B-Br	396	32	Bi-S	315.5 ± 4.6	375	C-I	209 ± 21	120
B-C	448 ± 29	246	Bi-Sb	251 ± 4	244	C-Ir	632 ± 4	171
B-Ce	305 ± 21	246	Bi-Se	280.3 ± 5.9	375	C-La	462 ± 20	290
B-Cl	511.3 ± 4	195	Bi-Sn	210.0 ± 8.4	135	C-Mo	481 ± 15.9	167
B-D	341.0 ± 6.3	246	Bi-Te	232.2 ± 11.3	375	C-N	748.0 ± 10	42
B-F	757	257	Bi-Tl	121 ± 13	84	C-Nb	569 ± 13.0	167
B-H	340	302	Br-Br	192.807	1	C-O	1076.5 ± 0.4	80
B-I	220.5 ± 0.8	315	Br-C	280 ± 21	120	C-Os	≥594	126
B-Ir	514.2 ± 17.2	381	Br-Ca	310.9 ± 9.2	319	C-P	513.4 ± 8	350
B-La	339 ± 63	246	Br-Cd	159 ± 96	120	C-Pt	598 ± 5.9	171,379
B-N	389 ± 21	80	Br-Cl	217.53 ± 0.29	59	C-Rh	580.0 ± 3.8	332
B-O	808.8 ± 20.9	287	Br-Co	331 ± 42	246	C-Ru	616.2 ± 10.5	333
B-P	346.9 ± 16.7	147	Br-Cr	328.0 ± 24.3	120	C-S	714.1 ± 1.2	71,354
B-Pd	329.3 ± 20.9	381	Br-Cs	389.1 ± 4	285,362	C-Sc	≤444	148
B-Pt	477.8 ± 16.7	268	Br-Cu	331 ± 25	120	C-Se	590.4 ± 5.9	343
B-Rh	475.7 ± 20.9	381	Br-D	370.74	205	C-Si	451.5	91,387
B-Ru	446.9 ± 20.9	381	Br-F	280 ± 12	211	C-Tc	565 ± 29	322
B-S	580.7 ± 9.2	374	Br-Fe	247 ± 96	120	C-Th	453 ± 17	166,357
B-Sc	276 ± 63	246	Br-Ga	444 ± 17	80	C-Ti	423 ± 29	162,357
B-Se	461.9 ± 14.6	374	Br-Ge	255 ± 29	120	C-U	454.8 ± 15.1	165,169
B-Si	317 ± 7	390	Br-H	366.35	205	C-V	427 ± 23.8	167
B-Te	354.4 ± 20.1	374	Br-Hg	72.8 ± 4	80	C-Y	418 ± 14	338
B-Th	297	140	Br-I	179.1 ± 0.4	120,309	C-Zr	561 ± 25	357
B-Ti	276 ± 63	246	Br-In	414 ± 21	80	Ca-Ca	~17	153
B-U	322 ± 33	246	Br-K	379.9 ± 0.8	362,378	Ca-Cl	409 ± 9	269
B-Y	293 ± 63	246	Br-Li	418.8 ± 4	362	Ca-D	≤169.9	205
Ba-Br	362.8 ± 8.4	104,199,230	Br-Mg	≤327.2	205	Ca-F	527 ± 21	101,190
Ba-Cl	436.0 ± 8.4	197,199	Br-Mn	314.2 ± 9.6	120	Ca-H	167.8	120
Ba-D	≤193.7	205	Br-N	276 ± 21	120	Ca-I	284.7 ± 8.4	188
Ba-F	587.0 ± 6.7	101,196	Br-Na	367.4 ± 0.8	362,378	Ca-Li	84.9 ± 8.4	397

Table 1
BOND STRENGTHS IN DIATOMIC MOLECULES (continued)

Molecule	D°_{298}/kJ mol^{-1}	Ref.	Molecule	D°_{298}/kJ mol^{-1}	Ref.	Molecule	D°_{298}/kJ mol^{-1}	Ref.
Ca-O	402.1 ± 16.7	287,326	Cl-Si	406	308,385	Cu-I	197 ± 21	120
Ca-S	337.6 ± 18.8	70,205	Cl-Sm	≥423	399	Cu-In	187.4 ± 7.9	13
Cd-Cd	7.36	251	Cl-Sn	414 ± 17	246	Cu-Li	192.9 ± 8.8	276
Cd-Cl	208.4	205	Cl-Sr	406 ± 13	197,199	Cu-Mn	158.6 ± 17	222
Cd-F	305 ± 21	31	Cl-Ta	544	23	Cu-Na	176.1 ± 16.7	299
Cd-H	69.0 ± 0.4	120	Cl-Th	489	261	Cu-Ni	202 ± 10	117
Cd-I	97.23	215	Cl-Ti	405.4 ± 10.5	192	Cu-O	269.0 ± 20.9	287
Cd-In	138	246	Cl-Tl	372.8 ± 2.1	28	Cu-S	276	349
Cd-O	235.6 ± 83.7	158,287	Cl-U	452 ± 8	259	Cu-Se	251	349
Cd-S	208.4 ± 20.9	158	Cl-V	477 ± 63	246	Cu-Si	221.3 ± 6.3	320
Cd-Se	127.6 ± 25.1	158	Cl-W	423 ± 42	246	Cu-Sn	169.5 ± 6.7	3,237
Cd-Te	100.0 ± 15.1	158	Cl-Xe	6.7	205	Cu-Tb	193 ± 19	203
Ce-Ce	245.2	127	Cl-Y	527 ± 84	246	Cu-Te	278.7	1
Ce-F	582 ± 42	246	Cl-Yb	~322	113	D-D	443.533	205
Ce-Ir	586	149	Cl-Zn	228.9 ± 19.7	73	D-F	576.6	205
Ce-N	519 ± 21	146	Cm-O	736	341	D-Ga	<272.8	253
Ce-O	795 ± 8	95	Co-Co	167 ± 25	218	D-Ge	≤322	205
Ce-Os	506 ± 33	126	Co-Cu	167 ± 17	135	D-H	439.433	205
Ce-Pd	322.2	63	Co-F	435 ± 63	246	D-Hg	42.05	205
Ce-Pt	556	149	Co-Ge	234 ± 21	135	D-In	246.0	205
Ce-Rh	548	149	Co-H	226 ± 42	369	D-Li	240.1892 ± 0.0046	207,360
Ce-Ru	531 ± 25	126	Co-I	285 ± 21	246	D-Lu	302	308
Ce-S	569	24	Co-Nb	267.0 ± 0.1	8	D-Mg	135.1	205
Ce-Se	494.5 ± 14.6	271	Co-O	384.5 ± 13.4	287	D-Ni	≤302.9	205
Ce-Te	189.4 ± 12.8	252	Co-S	331	349	D-Pt	≤350.2	205
Cl-Cl	242.580 ± 0.004	205	Co-Si	276 ± 17	380	D-S	351	205
Cl-Co	337.6 ± 6.7	194	Co-Ti	235.4 ± 0.1	353	D-Si	302.5	205
Cl-Cr	377.8 ± 6.7	194	Co-Y	253.7 ± 0.1	8	D-Sr	≥275.7	205
Cl-Cs	448 ± 8	285,363	Co-Zr	306.4 ± 0.1	8	D-Zn	88.7	205
Cl-Cu	377.8 ± 7.5	184	Cr-Cr	142.9 ± 5.4	201	Dy-F	531	406
Cl-D	436.47	205	Cr-Cu	155 ± 21	222	Dy-O	607 ± 17	95
Cl-Eu	~326	113	Cr-F	444.8 ± 19.7	227	Dy-S	414 ± 42	246
Cl-F	256.23	205,279	Cr-Ge	154 ± 7	202	Dy-Se	322 ± 42	246
Cl-Fe	329.7 ± 6.7	194	Cr-H	190.3 ± 7.0	53	Dy-Te	234 ± 42	246
Cl-Ga	481 ± 13	80	Cr-I	287.0 ± 24.3	120	Er-F	565 ± 17	406
Cl-Ge	~431	205	Cr-N	377.8 ± 18.8	152,355	Er-O	615 ± 13	95
Cl-H	431.62	205	Cr-O	461 ± 9	179	Er-S	418 ± 42	246
Cl-Hg	100 ± 8	120	Cr-Pb	105 ± 2	202	Er-Se	326 ± 42	246
Cl-I	211.3 ± 0.4	120	Cr-S	331	93	Er-Te	238 ± 42	246
Cl-In	439 ± 8	80	Cr-Sn	141 ± 3	202	Eu-Eu	33.5 ± 17	66
Cl-K	433.0 ± 8	363	Cs-Cs	43.919 ± 0.010	394	Eu-F	544	242
Cl-Li	469 ± 13	80	Cs-F	519 ± 8	285	Eu-Li	66.9 ± 2.9	275
Cl-Mg	327.6 ± 2.1	105,197,382	Cs-H	175.364	401	Eu-O	479 ± 10	95
Cl-Mn	338.5 ± 6.7	194	Cs-Hg	8	205	Eu-Rh	233.9 ± 33	66
Cl-N	333.9 ± 9.6	54	Cs-I	337.2 ± 2.1	285,361	Eu-S	362.3 ± 13.0	271,347
Cl-Na	412.1 ± 8	363	Cs-Na	63.2 ± 1.3	86	Eu-Se	301 ± 14.6	25,178,271
Cl-Ni	377.0 ± 6.7	194	Cs-O	295.8 ± 62.8	287	Eu-Te	243 ± 14.6	25,271
Cl-O	268.85 ± 0.10	2	Cs-Rb	49.57 ± 0.01	174	F-F	158.78	205
Cl-P	289 ± 42	246	Cu-Cu	176.52 ± 2.38	135,323	F-Ga	577 ± 14.6	270
Cl-Pb	301 ± 29	120	Cu-D	270.3	205	F-Gd	590.4 ± 27.2	405
Cl-Ra	343 ± 75	120	Cu-Dy	142 ± 21	203	F-Ge	485 ± 21	98
Cl-Rb	427.6 ± 8	363	Cu-F	413.4 ± 13	99	F-H	569.87 ± 0.06	402
Cl-S	277.0	224	Cu-Ga	215.9 ± 15.1	44	F-Hf	650 ± 15	16
Cl-Sb	360 ± 50	120	Cu-Ge	208.8 ± 21	273	F-Hg	~180	205
Cl-Sc	331	386	Cu-H	277.8	217,318	F-Ho	540	406
Cl-Se	322	246	Cu-Ho	142 ± 21	203	F-I	≤271.5	7,33,60,75

Table 1
BOND STRENGTHS IN DIATOMIC MOLECULES (continued)

Molecule	$D°_{298}$/kJ mol⁻¹	Ref.	Molecule	$D°_{298}$/kJ mol⁻¹	Ref.	Molecule	$D°_{298}$/kJ mol⁻¹	Ref.
F-In	506 ± 14.6	270	Gd-S	526.8 ± 10.5	116,345	Hg-S	217.1 ± 22.2	158
F-K	497.5 ± 2.5	17	Gd-Se	431 ± 14.6	25	Hg-Se	144.3 ± 30.1	158
F-La	598 ± 42	246	Gd-Te	343 ± 14.6	25	Hg-Te	≤142	246
F-Li	577 ± 21	80	Ge-Ge	263.6 ± 7.1	238	Hg-Tl	4	183
F-Lu	333.5	208	Ge-H	≤321.7	243	Ho-Ho	84 ± 17	62
F-Mg	461.9 ± 5.0	101,196	Ge-Ni	290.3 ± 10.9	335	Ho-O	611 ± 17	95
F-Mn	423.4 ± 14.6	228	Ge-O	659.4 ± 12.6	226,287	Ho-S	428.4 ± 14.6	345
F-Mo	464.8	198	Ge-Pd	254.7 ± 10.5	334	Ho-Se	335 ± 17	25
F-N	343	205	Ge-S	534 ± 3	280	Ho-Te	259 ± 17	25
F-Na	519	205	Ge-Sc	271.0 ± 11	234	I-I	151.088	205,371
F-Nd	545.2 ± 12.6	403	Ge-Se	484.7 ± 1.7	282	I-In	331	384
F-Ni	430 ± 20	85	Ge-Si	296.4 ± 8.6	391	I-K	325.1 ± 0.8	361,378
F-O	219.54 ± 10	48	Ge-Te	397 ± 3	282	I-Li	345.2 ± 4.2	361
F-P	439 ± 96	120	Ge-Y	279.8 ± 11.4	235	I-Mg	~285	26
F-Pb	356 ± 8	408	H-H	435.990	205	I-Mn	282.8 ± 9.6	120
F-Pm	540 ± 42	246	H-Hg	39.844	205	I-N	159 ± 17	246
F-Pr	582 ± 46	246	H-I	298.407	205	I-Na	304.2 ± 2.1	361,378
F-Pu	538.5 ± 29	229	H-In	243.1	205	I-Ni	293 ± 21	120
F-Rb	494 ± 21	80	H-K	174.576	206,401	I-O	230	47
F-Ru	402	185	H-Li	238.049 ± 0.004	393	I-Pb	193 ± 4	340
F-S	342.7 ± 5.0	30,200,231	H-Mg	126.4 ± 2.9	14,15,100	I-Rb	318.8 ± 2.1	361
F-Sb	439 ± 96	120	H-Mn	234 ± 29	120	I-Si	293	205
F-Sc	589.1 ± 13	404	H-N	≤339	205	I-Sn	234 ± 42	246
F-Se	339 ± 42	246	H-Na	185.69 ± 0.25	274,316	I-Sr	269.9 ± 5.9	239
F-Si	552.7 ± 2.1	109	H-Ni	252.3 ± 8	217	I-Te	192 ± 42	246
F-Sm	565	242	H-O	427.6	205	I-Ti	310 ± 42	246
F-Sn	466.5 ± 13	408	H-P	297	205	I-Tl	272 ± 8	26
F-Sr	541.8 ± 6.7	101,196	H-Pb	≤157	205	I-Zn	108.29	215
F-Ta	573 ± 13	256	H-Pd	234 ± 25	369	I-Zr	305	241
F-Tb	561 ± 42	246	H-Pt	≤335	205	In-In	100 ± 8	246
F-Th	652	263	H-Rb	167 ± 21	120	In-Li	92.5 ± 14.6	160
F-Ti	569 ± 33	407	H-Rh	247 ± 21	369	In-O	<320.1	287
F-Tl	445.2 ± 19.2	28	H-Ru	234 ± 21	369	In-P	197.9 ± 8.4	291
F-Tm	510	242	H-S	344.3 ± 12.1	212	In-S	289 ± 17	69
F-U	659.0 ± 10.5	157,258	H-Sc	~180	329	In-Sb	151.9 ± 10.5	81
F-V	590 ± 63	246	H-Se	314.47 ± 0.96	122	In-Se	247 ± 17	69
F-W	548 ± 63	246	H-Si	≤299.2	205	In-Te	218 ± 17	69
F-Xe	15.77	317,368	H-Sn	264 ± 17	120	Ir-La	577 ± 13	176
F-Y	605.0 ± 20.9	404	H-Sr	163 ± 8	120	Ir-O	414.6 ± 42.3	287
F-Yb	≥521.3	18,113,399	H-Te	268 ± 2.1	119	Ir-Si	462.8 ± 20.9	381
F-Zn	368 ± 63	246	H-Ti	204.6 ± 8.8	52	Ir-Th	573	133
F-Zr	616 ± 15	16	H-Tl	188 ± 8	120	Ir-Ti	422 ± 13	289
Fe-Fe	75 ± 17	330	H-V	208.7 ± 7.0	53	Ir-Y	456.1 ± 16.7	177
Fe-Ge	210.9 ± 29	219	H-Yb	159 ± 38	120	K-K	54.63 ± 0.02	6,265
Fe-H	180 ± 25	369	H-Zn	85.8 ± 2.1	120	K-Kr	4.6	205
Fe-O	390.4 ± 17.2	287	He-He	3.8	205	K-Li	82.0 ± 4.2	103,410
Fe-S	322	93	He-Hg	6.61	246	K-Na	65.994 ± 0.008	36,410
Fe-Si	297 ± 25	380	Hf-C	548 ± 63	246	K-O	277.8 ± 20.9	287
Ga-Ga	112.1 ± 7	337	Hf-N	536 ± 29	152,245	K-Xe	5.0	205
Ga-H	<274.1	253	Hf-O	801.7 ± 13.4	287	Kr-Kr	5.23	50,205
Ga-I	339 ± 9.6	120	Hg-Hg	8 ± 2	204	Kr-O	<8	246
Ga-Li	133.1 ± 14.6	160	Hg-I	34.69 ± 0.96	388	Kr-Xe	5.505 ± 0.002	9
Ga-O	353.5 ± 41.8	287	Hg-K	8.24 ± 0.21	246	La-La	247 ± 21	386
Ga-P	229.7 ± 12.6	130	Hg-Li	13.8	205	La-N	519 ± 42	246
Ga-Sb	192.0 ± 12.6	295	Hg-Na	9.2	205,411	La-O	799 ± 4	95
Ga-Te	251 ± 25	377	Hg-O	220.9 ± 33.1	158	La-Pt	502 ± 21	272
Gd-O	719 ± 10	95	Hg-Rb	8.4	205	La-Rh	527 ± 17	65

Table 1
BOND STRENGTHS IN DIATOMIC MOLECULES (continued)

Molecule	$D°_{298}$/kJ mol^{-1}	Ref.	Molecule	$D°_{298}$/kJ mol^{-1}	Ref.	Molecule	$D°_{298}$/kJ mol^{-1}	Ref.
La-S	573.2 ± 1.7	214,359	Nd-Se	385 ± 17	25,156,271	P-Tl	209 ± 13	293
La-Se	477 ± 17	25,271	Nd-Te	305 ± 17	25	P-U	297 ± 21	246
La-Te	381 ± 17	25,154	Ne-Ne	3.93	365	P-W	305 ± 4	150
La-Y	202.1	386	Ni-Ni	200.7 ± 0.2	304	Pb-Pb	86.6 ± 0.8	138,305
Li-Li	110.21 ± 4	383,398	Ni-O	382.0 ± 16.7	287	Pb-S	346.0 ± 1.7	375
Li-Mg	67.4 ± 6.3	396	Ni-Pt	273.7 ± 0.3	367	Pb-Sb	161.5 ± 10.5	409
Li-Na	87.181 ± 0.001	102,111	Ni-S	344.3	93	Pb-Se	302.9 ± 4	375
Li-O	333.5 ± 8.4	287	Ni-Si	318 ± 17	380	Pb-Te	251 ± 13	375
Li-Pb	78.7 ± 7.9	277	Ni-V	206.3 ± 0.1	353	Pd-Pd	100 ± 15	331
Li-S	312.5 ± 7.5	232	Ni-Y	283.9 ± 0.1	8	Pd-Si	261 ± 12	336
Li-Sb	172.8 ± 10.0	278	Ni-Zr	279.7 ± 0.1	8	Pd-Y	238 ± 17	313
Li-Sm	49.0 ± 4.2	275	Np-O	718.4 ± 41.8	287	Pm-S	423 ± 63	246
Li-Tm	69.0 ± 3.3	275	O-O	498.36 ± 0.17	38,205	Pm-Se	339 ± 63	246
Li-Yb	37.2 ± 2.9	275	O-Os	575	189	Pm-Te	255 ± 63	246
Lu-Lu	142 ± 33	246	O-P	599.1 ± 12.6	287	Po-Po	187.0	205
Lu-O	678 ± 8	95	O-Pa	788.3 ± 17.2	240	Pr-S	492.5 ± 4.6	112
Lu-Pt	402 ± 33	141	O-Pb	382.0 ± 12.6	287	Pr-Se	446.4 ± 23.0	155,271
Lu-S	507.1 ± 14.6	114,345	O-Pd	380.7 ± 83.7	287	Pr-Te	326 ± 42	246
Lu-Se	418 ± 17	25	O-Pm	674 ± 63	246	Pt-Pt	307 ± 2	366
Lu-Te	326 ± 17	25	O-Pr	753 ± 13	95	Pt-Si	501.2 ± 18.0	381
Mg-Mg	8.552 ± 0.004	264,397	O-Pt	391.6 ± 41.8	287	Pt-Th	552	133
Mg-O	363.2 ± 12.6	287	O-Pu	715.9 ± 33.9	287	Pt-Ti	397 ± 13	173
Mg-S	234	70	O-Rb	255 ± 84	37	Pt-Y	474.0 ± 12.1	170
Mn-Mn	25.9	221	O-Re	626.8 ± 83.7	287	Rb-Rb	48.898 ± 0.005	5
Mn-O	402.9 ± 41.8	225,287	O-Rh	405.0 ± 41.8	287	Rh-Rh	285.3 ± 0.05	255
Mn-S	301 ± 17	395	O-Ru	528.4 ± 41.8	287	Rh-Sc	443.9 ± 10.5	175
Mn-Se	239.3 ± 9.2	351	O-S	517.90 ± 0.05	57	Rh-Si	395.0 ± 18.0	381
Mo-Mo	406 ± 21	168	O-Sb	434.3 ± 41.8	287	Rh-Th	515 ± 21	129
Mo-Nb	456 ± 25	163	O-Sc	681.6 ± 11.3	287	Rh-Ti	390.8 ± 14.6	61
Mo-O	560.2 ± 20.9	287	O-Se	464.8 ± 21.3	287,344	Rh-U	519 ± 17	129
N-N	945.33 ± 0.59	205	O-Si	799.6 ± 13.4	287	Rh-V	364 ± 29	135
N-O	630.57 ± 0.13	205	O-Sm	565 ± 13	95	Rh-Y	445.2 ± 10.5	175
N-P	617.1 ± 20.9	72,151	O-Sn	531.8 ± 12.6	287	Ru-Si	397.1 ± 20.9	381
N-Pu	473 ± 63	246	O-Sr	426.3 ± 6.3	327	Ru-Th	591.6 ± 42	134
N-S	464 ± 21	246	O-Ta	799.1 ± 12.6	287	Ru-V	414 ± 29	135
N-Sb	301 ± 50	120	O-Tb	711 ± 13	95	S-S	425.30	205
N-Sc	469 ± 84	246	O-Te	376.1 ± 20.9	287	S-Sb	378.7	110
N-Se	370 ± 11	254	O-Th	878.6 ± 12.1	287	S-Sc	477 ± 13	359,372
N-Si	470 ± 15	311	O-Ti	672.4 ± 9.2	287	S-Se	371.1 ± 6.7	90
N-Ta	611 ± 84	246	O-Tm	502 ± 13	95	S-Si	623	205
N-Th	577.4 ± 33.1	144,152	O-U	759.4 ± 13.4	287	S-Sm	389	112
N-Ti	476.1 ± 33.1	152,356	O-V	626.8 ± 18.8	12,287	S-Sn	464 ± 3.3	88
N-U	531.4 ± 2.1	142	O-W	672.0 ± 41.8	287	S-Sr	339	45
N-V	477.4 ± 17.2	107,152	O-Xe	36.4	246	S-Tb	515 ± 42	246
N-Xe	23.0	182	O-Y	719.6 ± 11.3	209,287	S-Te	339 ± 21	88
N-Y	481 ± 63	246	O-Yb	397 ± 17	95	S-Ti	418 ± 3	96,292
N-Zr	564.8 ± 25.1	143,152	O-Zn	159 ± 4	55	S-Tm	368 ± 42	246
Na-Na	73.0813 ± 0.0001	213	O-Zr	776.1 ± 13.4	287	S-U	522.6 ± 9.6	359
Na-O	256.1 ± 16.7	287	P-P	489.5 ± 10.5	151	S-V	450	97,205
Na-Rb	63.25	392	P-Pt	≤416.7	348	S-Y	528.4 ± 10.5	358
Nb-Nb	510 ± 10.0	164	P-Rh	353.1 ± 17	348	S-Yb	167	246
Nb-Ni	271.9 ± 0.1	8	P-S	444 ± 8	92	S-Zn	205 ± 13	82,158
Nb-O	771.5 ± 25.1	287	P-Sb	356.9	249	S-Zr	575.3 ± 16.7	359
Nb-Ti	302.0 ± 0.1	254	P-Se	363.6 ± 10.0	92	Sb-Sb	299.2 ± 6.3	81,248
Nd-Nd	<163	246	P-Si	363.6	346	Sb-Te	277.4 ± 3.8	306,364
Nd-O	703 ± 13	95	P-Te	297.9 ± 10.0	92	Sb-Tl	126.8 ± 10.5	11,293
Nd-S	471.5	25	P-Th	550.2 ± 42	135	Sc-Sc	162.8 ± 21	136

Molecule	$D°_{298}$/kJ mol^{-1}	Ref.	Molecule	$D°_{298}$/kJ mol^{-1}	Ref.	Molecule	$D°_{298}$/kJ mol^{-1}	Ref.
Sc-Se	385 ± 17	246	Se-Zn	170.7 ± 25.9	82,158	Te-Zn	117.6 ± 18.0	158
Sc-Si	228.7 ± 14	234	Si-Si	325 ± 7	328	Th-Th	≤289	140
Sc-Te	289 ± 17	246	Si-Te	452 ± 8	205,281	Ti-Ti	141.4 ± 21	216
Se-Se	332.6 ± 0.4	90,375	Si-Y	258.8 ± 17.3	235	Ti-V	203.2 ± 0.1	353
Se-Si	538 ± 13	370	Sm-Te	272.4 ± 14.6	271	Ti-Zr	214.3 ± 0.1	254
Se-Sm	331.0 ± 14.6	271	Sn-Sn	187.1 ± 0.3	284	Tl-Tl	64.4 ± 17	10
Se-Sn	401.2 ± 5.9	68	Sn-Te	359.8	205	U-U	222 ± 21	246
Se-Sr	~285	27	Sr-Sr	15.5 ± 0.4	121	V-V	269.3 ± 0.1	353
Se-Tb	423 ± 42	246	Tb-Tb	131.4 ± 25.1	250	V-Zr	260.6 ± 0.3	254
Se-Te	291.6 ± 4	88,90,159	Tb-Te	339 ± 42	246	Xe-Xe	6.138 ± 0.001	49,115
Se-Ti	381 ± 42	246	Te-Te	257.6 ± 4.1	389	Y-Y	159 ± 21	246
Se-Tm	276 ± 42	246	Te-Ti	289 ± 17	246	Yb-Yb	20.5 ± 17	161
Se-V	347 ± 21	246	Te-Tm	276 ± 42	246	Zn-Zn	29	339
Se-Y	435 ± 13	246	Te-Y	339 ± 13	246	Zr-Zr	298.2 ± 0.1	8

REFERENCES

1. Abbasov, A. S., Azizov, T. Kh., Alleva, N. A., Aliev, I. Ya, Mustafaev, F. M., and Mamedov, A. N., *Zh. Fiz. Khim.*, 50, 2172, 1976.
2. Abramowitz, S. and Chase, M. L., Pure Appl. Chem., 63, 1449, 1991.
3. Ackerman, M., Drowart, J., Stafford, F. E., and Verhaegen, G., *J. Chem. Phys.*, 36, 1557, 1962.
4. Ackerman, M., Stafford, F. E., and Drowart, J., *J. Chem. Phys.*, 33, 1784, 1960.
5. Amiot, C., *J. Chem. Phys.*, 93, 8591, 1990.
6. Amiot, C., *J. Mol. Spectrosc.*, 147, 370, 1991.
7. Appelman, E. H. and Clyne, M. A. A., *J. Chem. Soc. Faraday Trans. 1*, 71, 2072, 1975.
8. Arrington, C. A., Blume, T., Morse, M. D., Doverstal, M., and Sassenberg, U., J. Phys. Chem., 98, 1398, 1994.
9. Balakrishnan, A., Jones, W. J., Mahajan, C. G., and Stoicheff, B. P., *Chem. Phys. Lett.*, 155, 43, 1989.
10. Balducci, G. and Piacente, V., *J. Chem. Soc. Chem. Commun.*, 1287, 1980.
11. Balducci, G., Ferro, D., and Piacente, V., *High Temp. Sci.*, 14, 207, 1981.
12. Balducci, G., Gigli, G., and Guido, M., *J. Chem. Phys.*, 79, 5616, 1983.
13. Balducci, G., Nunzio, P. E., Gigli, G., and Guido, M., *J. Chem. Phys.*, 90, 406, 1989.
14. Balfour, W. J. and Cartwright, H. M., *Astron. Astrophys. Suppl. Ser.*, 26, 389, 1976.
15. Balfour, W. J. and Lingren, B., Can. J. Chem., 56, 767, 1978.
16. Barkovskii, N. V., Tsirel'nikov, V. I., Emel'yanov, A. M., and Khodeev, Yu. S., *Teplofiz. Vys. Temp.*, 29, 474, 1991.
17. Barrow, R. F. and Caunt, A. D., *Proc. R. Soc. London Ser. A*, 219, 120, 1953.
18. Barrow, R. F. and Chojnicki, A. H., *J. Chem. Soc. Faraday Trans. 2*, 71, 728, 1975.
19. Behm, J. M. and Morse, M. D., J. Chem. Phys., 101, 6500, 1994.
20. Behm, J. M., Arrington, C. A., and Morse, M. D., *J. Chem. Phys.*, 99, 6409, 1993.
21. Behm, J. M., Arrington, C. A., Langenberg, J. D., and Morse, M. D., *J. Chem. Phys.*, 99, 6394, 1993.
22. Behm, J. M., Brugh, D. J., and Morse, M. D., J. Chem. Phys., 101, 6487, 1994.
23. Behrens, R. G. and Feber, R. C., *J. Less-Common Met.*, 75, 281, 1980.
24. Bergman, C. and Gingerich, K. A., *J. Phys. Chem.*, 76, 2332, 1972.
25. Bergman, C., Coppens P., Drowart, J., and Smoes, S., *Trans. Faraday Soc.*, 66, 800, 1970.
26. Berkowitz, J. and Chupka, W. A., *J. Chem. Phys.*, 45, 1287, 1966.
27. Berkowitz, J. and Chupka, W. A., *J. Chem. Phys.*, 45, 4289, 1966.
28. Berkowitz, J. and Walter, T., *J. Chem. Phys.*, 49, 1184, 1968.
29. Berkowitz, J., *J. Chem. Phys.*, 89, 7065, 1988.
30. Berneike ,W., Kreuttle, U., and Neuert, H., *Chem. Phys. Lett.*, 76, 525, 1980.
31. Besenbruch, G., Kana'an, A. S., and Margrave, J. L., *J. Phys. Chem.*, 69, 3174, 1965.
32. Bharate, N. S., Bhartiya, J. B., and Behere, S. H., *Proc. Indian Natl. Sci. Acad. Part A*, 57, 419, 1991.
33. Birks, J. W., Gabelnick, S. D., and Johnston, H. S., *J. Mol. Spectrosc.*, 57, 23, 1975.
34. Bishea, G. A. and Morse, M. D., *Chem. Phys. Lett.*, 171, 430, 1990.
35. Bondybey, V. E., *Chem. Phys. Lett.*, 109, 436, 1984.
36. Breford, E. J. and Engelke, F., *J. Chem. Phys.*, 71, 1994, 1979.
37. Brewer, L. and Rosenblatt, G. M., *Adv. High Temp. Sci.*, 2, 1, 1969.
38. Brix, P. and Herzberg, G., *J. Chem. Phys.*, 21, 2240, 1953.
39. Brock, L. R., Pilgrim, J. S., and Duncan, M. A., Chem. Phys. Lett., 230, 93, 1994.
40. Burns, G., LeRoy, L. J., Morris, D. J., and Blake, J. A., *Proc. R. Soc. London Ser. A*, 316, 81, 1970.
41. Busse, V. B. and Weil, K. G., *Ber. Bunsenges. Phys. Chem.*, 85, 309, 1981.

42. Calculated from $\Delta_f H°$(CN) = 441.4 ± 4.6 kJ mol^{-1} (Table 4), but see also Costes, M., Naulin, C., and Dorthe, G., *Astron. Astronphys.*, 232, 270, 1990.
43. Callender, C. L., Mitchell, S. A., and Hackett, P. A., *J. Chem. Phys.*, 90, 5252, 1989.
44. Carbonel, M., Bergman, C., and Laffite, M., *Colloq. Int. Cent. Nat. Rech. Sci.*, 201, 311, 1972.
45. Cater, E. D. and Johnson, E. W., *J. Chem. Phys.*, 47, 5353, 1967.
46. Chase, M. L., *J. Phys. Chem. Ref. Data*, 25, 1069, 1996.
47. Chase, M. L., *J. Phys. Chem. Ref. Data*, 25, 1297, 1996.
48. Chase, M. L., *J. Phys. Chem. Ref. Data*, 25, 551, 1996.
49. Chashchina, G. I. and Shreider, E. Ya., *Zh. Prikl. Spektrosk.*, 21, 696, 1974.
50. Chashchina, G. I. and Shreider, E. Ya., *Zh. Prikl. Spektrosk.*, 25, 163, 1976.
51. Chatillon, C., Allibert, M., and Pattoret, A., *C. R. Acad. Sci. Ser. C*, 280, 1505, 1975.
52. Chen, Y. M., Clemmer, D. E., and Armentrout, P. B., *J. Chem. Phys.*, 95, 1228, 1991.
53. Chen, Y.-M., Clemmer, D. E., and Armentrout, P. B., *J. Chem. Phys.*, 98, 4929, 1993.
54. Clarke, T. C. and Clyne, M. A. A., *Trans. Faraday Soc.*, 66, 877, 1970.
55. Clemmer, D. E., Daltaska, N. F., and Armentrout, P. B., *J. Chem. Phys.*, 95, 7263, 1991.
56. Clemmer, D. E., Weber, M. E., and Armentrout, P. B., *J. Phys. Chem.*, 96, 10888, 1992.
57. Clerbaux, C. and Colin, R., *J. Mol. Spectrosc.* 165, 334, 1994.
58. Clevenger, J. O. and Tellinghuisen, J., *J. Chem. Phys.,* 103, 9611, 1995.
59. Clyne, M. A. A. and McDermid, I. S., *Faraday Discuss. Chem. Soc.,* 67, 316, 1979.
60. Clyne, M. A. A. and McDermid, I. S., *J. Chem. Soc. Faraday Trans. 2*, 72, 2252, 1976.
61. Cocke, D. L. and Gingerich, K. A., *J. Chem. Phys.*, 60, 1958, 1974.
62. Cocke, D. L. and Gingerich, K. A., *J. Phys. Chem.*, 75, 3264, 1971.
63. Cocke, D. L. and Gingerich, K. A., *J. Phys. Chem.,* 76, 2332, 1972.
64. Cocke, D. L., Gingerich, K. A., and Chang, C. A., *J. Chem. Soc. Faraday Trans. 1,* 72, 268, 1976.
65. Cocke, D. L., Gingerich, K. A., and Kordis, J., *High Temp. Sci.*, 5, 474, 1973.
66. Cocke, D. L., Gingerich, K. A., and Kordis, J., *High Temp. Sci.*, 7, 61, 1975.
67. Colin, R. and De Greef, D., *Can. J. Phys.*, 53, 2142, 1975.
68. Colin, R. and Drowart, J., *Trans. Faraday Soc.*, 60, 673, 1964.
69. Colin, R. and Drowart, J., *Trans. Faraday Soc.*, 64, 2611, 1968.
70. Colin, R., Goldfinger, P., and Jeunehomme, M., *Trans. Faraday Soc.*, 60, 306, 1964.
71. Coppens, P., Reynaert, J. C., and Drowart, J., *J. Chem. Soc. Faraday Trans. 2*, 75, 292, 1979.
72. Coquart, B. and Prudhomme, J. C., *J. Mol. Spectrosc.*, 87, 75, 1981.
73. Corbett, J. D. and Lynde, R. A., *Inorg. Chem.,* 6, 2199, 1967.
74. Costes, M., Naulin, C., Dorthe, G., Vaucamps, C., Nouchi, G., *Faraday Discuss. Chem. Soc.*, 84, 75, 1987.
75. Coxon, J. A., *Chem. Phys. Lett.,* 33, 136, 1975.
76. Cubicciotti, D., *Inorg. Chem.,* 7, 208, 1968.
77. Cubicciotti, D., *Inorg. Chem.*, 7, 211, 1968.
79. Cubicciotti, D., *J. Phys. Chem.*, 71, 3066, 1967.
80. Cuthill, A. M., Fabian, D. J., and Shu-Shou-Shen, S., *J. Phys. Chem.*, 77, 2008, 1973.
81. De Maria, G., Drowart, J., and Inghram, M. G., *J. Chem. Phys.*, 31, 1076, 1959.
82. De Maria, G., Goldfinger, P., Malaspina, L., and Piacente, V., *Trans. Faraday Soc.*, 61, 2146, 1965.
83. De Maria, G., Malaspina, L., and Piacente, V., *J. Chem. Phys.*, 52, 1019, 1970.
84. De Maria, G., Malaspina, L., and Piacente, V., *J. Chem. Phys.*, 56, 1978, 1972.
85. Devore, T. C., McQuaid, M., and Gole, J. L., *High Temp. Sci.,* 30, 83, 1990.
86. Diemer, U., Weickenmeier, H., Wahl, M., and Demtroeder, W., *Chem. Phys. Lett.*, 104, 489, 1984.
87. Drowart, J. and Goldfinger, P., *Angew. Chem.*, 6, 581, 1967.
88. Drowart, J. and Goldfinger, P., *Q. Rev. (London),* 20, 545, 1966.
89. Drowart, J. and Honig, R. E., *J. Phys. Chem.*, 61, 980, 1957.
90. Drowart, J. and Smoes, S., *J. Chem. Soc. Faraday Trans. 2*, 73, 1755, 1977.
91. Drowart, J., De Maria, G., and Inghram, M. G., *J. Chem. Phys.*, 29, 1015, 1958.
92. Drowart, J., Myers, C. E., Szwarc, R.,Vander Auwera-Mahieu, A., and Uy, O. M., *High Temp. Sci.*, 5, 482, 1973.
93. Drowart, J., Pattoret, A., and Smoes, S., *Proc. Br. Ceramic Soc.*, No. 8, 67, 1967.
94. Drowart, J., Smoes, S., and Vander Auwera-Mahieu, A., *J. Chem. Thermodyn.*, 10, 453, 1978.
95. Dulick, M., Murad, E., and Barrow, R. F., *J. Chem. Phys.*, 85, 385, 1986.
96. Edwards, J. G., Franklin, H. F., and Gilles, P. W., *J. Chem. Phys.*, 54, 545, 1971.
97. Edwards, J. G., *J. Chem. Phys.*, 96, 866, 1992
98. Ehlert, T. C. and Margrave, J. L., *J. Chem. Phys.*, 41, 1066, 1964.
99. Ehlert, T. C. and Wang, J. S., *J. Phys. Chem.*, 81, 2069, 1977.
100. Ehlert, T. C., Hilmer, R. M., and Beauchamp, E. A., *J. Inorg. Nucl. Chem.*, 30, 3112, 1968.
101. Engelke, F., *Chem. Phys.*, 39, 279, 1979.
102. Engelke, F., Ennen, G., and Meiwes, K. H., *Chem. Phys.,* 66, 391, 1982.

103. Engelke, F., Hage, H., and Sprick, U., *Chem. Phys.*, 88, 443, 1984.
104. Estler, C. and Zare, R. N., *Chem. Phys.*, 28, 253, 1978.
105. Farber, M. and Srivastava, R. D., *Chem. Phys. Lett.*, 42, 567, 1976.
106. Farber, M. and Srivastava, R. D., *High Temp. Sci.*, 12, 21, 1980.
107. Farber, M. and Srivastava, R. D., *J. Chem. Soc. Faraday Trans. 1,* 69, 390, 1973.
108. Farber, M. and Srivastava, R. D., *J. Chem. Soc. Faraday Trans. 1,* 70, 1581, 1974.
109. Farber, M. and Srivastava, R. D., *J. Chem. Soc. Faraday Trans. 1,* 74, 1089, 1978.
110. Faure, F. M., Mitchell, M. J., and Bartlett, R. W., *High Temp. Sci.*, 4, 181, 1972.
111. Fellows, C. E., *J. Chem. Phys.,* 94, 5855, 1991.
112. Fenochka, B. V. and Gorkienko, S. P., *Zh. Fiz. Khim*, 47, 2445, 1973.
113. Filippenko, N. V., Morozov, E. V., Giricheva, N. L., and Krasnev, K. S., *Izv. Vyssh. Ucheb. Zaved Khim.Technol.*, 15, 1416, 1972.
114. Franzen, H. and Hariharan, A. V., *J. Chem. Phys.*, 70, 4907, 1979.
115. Freeman, D. E., Yoshino, K., and Tanaka, Y., *J. Chem. Phys.,* 61, 4880, 1974.
116. Fries, J. A. and Cater, E. D., *J. Chem. Phys.*, 68, 3978, 1978.
117. Fu, Z. and Morse, M. D., *J. Chem. Phys.*, 90, 3417, 1989.
118. Fu, Z., Lemire, G. W., Bishea, G. A., and Morse, M. D., *J. Chem. Phys.*, 93, 8420, 1990.
119. Gal, J. F., Maria, P. C., and Decouzon, M., *Int. J. Mass Spectrom. Ion Processes*, 93, 87, 1989.
120. Gaydon, A. G., *Dissociation Energies and Spectra of Diatomic Molecules, 3rd ed.*, Chapman & Hall, London, 1968.
121. Gerber, G. and Moeller, R., *Contrib. Symp. At. Surf. Phys.*, 168, 1982.
122. Gibson, S. T., Greene, J. P., and Berkowitz, J., *J. Chem. Phys.,* 85, 4815, 1986.
123. Gingerich, K. A. and Blue, G. D., *J. Chem. Phys.,* 47, 5447, 1967.
124. Gingerich, K. A. and Blue, G. D., *J. Chem. Phys.,* 59, 186, 1973.
125. Gingerich, K. A. and Choudary, U. V., *J. Chem. Phys.,* 68, 3265, 1978.
126. Gingerich, K. A. and Cocke, D. L., *Inorg. Chim. Acta*, 28, L171, 1978.
127. Gingerich, K. A. and Finkbeiner, H. C., *J. Chem. Phys.*, 54, 2621, 1971.
128. Gingerich, K. A. and Finkbeiner, H. C., *Proc. 9th Rare Earth Res. Conf.*, 2, 795, 1971.
129. Gingerich, K. A. and Gupta, S. K., *J. Chem. Phys.*, 69, 505, 1978.
130. Gingerich, K. A. and Piacente, V., *J. Chem. Phys.*, 54, 2498, 1971.
131. Gingerich, K. A., *Chem. Commun.*, 580, 1970.
132. Gingerich, K. A., *Chem. Phys. Lett.*, 13, 262, 1972.
133. Gingerich, K. A., *Chem. Phys. Lett.*, 23, 270, 1973.
134. Gingerich, K. A., *Chem. Phys. Lett.,* 25, 523, 1974.
135. Gingerich, K. A., *Chem. Soc. Faraday, Symp.*, No.14, 109, 1980.
136. Gingerich, K. A., *Chimia*, 26, 619, 1972.
137. Gingerich, K. A., Cocke, D. L., and Kordis, J., *J. Phys. Chem.*, 78, 603, 1974.
138. Gingerich, K. A., Cocke, D. L., and Miller, F., *J. Chem. Phys.*, 64, 4027, 1976.
139. Gingerich, K. A., Haque, R., and Kingcade, J. E., *Thermochim. Acta*, 30, 61, 1979.
140. Gingerich, K. A., *High Temp. Sci.*, 1, 258, 1969.
141. Gingerich, K. A., *High Temp. Sci.*, 3, 415, 1971.
142. Gingerich, K. A., *J. Chem. Phys.,* 47, 2192, 1967.
143. Gingerich, K. A., *J. Chem. Phys.,* 49, 14, 1968.
144. Gingerich, K. A., *J. Chem. Phys.,* 49, 19, 1968.
145. Gingerich, K. A., *J. Chem. Phys.*, 54, 2646, 1971.
146. Gingerich, K. A., *J. Chem. Phys.*, 54, 3720, 1971.
147. Gingerich, K. A., *J. Chem. Phys.*, 56, 4239, 1972.
148. Gingerich, K. A., *J. Chem. Phys.*, 74, 6407, 1981.
149. Gingerich, K. A., *J. Chem. Soc. Faraday Trans. 2,* 70, 471, 1974.
150. Gingerich, K. A., *J. Phys. Chem.*, 68, 768, 1964.
151. Gingerich, K. A., *J. Phys. Chem.*, 73, 2734, 1969.
152. Gingerich, K. A., *NBS Spec. Publ. (U. S.)*, 561, 289, 1979.
153. Gondal, M. A., Khan, M. A., and Rais, M. H., Chem. Phys. Lett., 243, 94, 1995.
154. Gordienko, S. P. and Fenochka, B. V., *Izv. Akad. Nauk. SSSR Neorg. Mater.*, 18, 1811, 1982.
155. Gordienko, S. P., Fenochka, B. V., Viksman, G. Sh., Klockkova, L. A., and Mikhlina, T. M., *Izv. Akad. Nauk. SSSR Neorg. Mater.*, 18, 18, 1982.
156. Gordienko, S. P., *Izv. Akad. Nauk. SSSR Neorg. Mater.*, 20, 1472, 1984.
157. Gorokhov, L. N., Smirnov, V. K., and Khodeev, Yu. S., *Zh. Fiz. Khim.,* 58, 1603, 1984.
158. Grade, M. and Hirschwald, W., *Ber. Bunsenges. Phys. Chem.*, 86, 899, 1982.
159. Grade, M., Wienecke, J., Rosinger, W., and Hirschwald, W., *Ber. Bunsenges. Phys. Chem.*, 87, 355, 1983.
160. Guggi, D. J., Neubert, A., and Zmbov, K. F., *Conf. Int. Thermodyn. Chim. [C. R.] 4th*, 3, 124, 1975.
161. Guido, M. and Balducci, G., *J. Chem. Phys.,* 57, 5611, 1972.
162. Gupta, S. K. and Gingerich, K. A., *High Temp. - High Pressures*, 12, 273, 1980.

STRENGTHS OF CHEMICAL BONDS (continued)

Table 1
BOND STRENGTHS IN DIATOMIC MOLECULES (continued)

163. Gupta, S. K. and Gingerich, K. A., *J. Chem. Phys.*, 69, 4318, 1978.
164. Gupta, S. K. and Gingerich, K. A., *J. Chem. Phys.*, 70, 5350, 1979.
165. Gupta, S. K. and Gingerich, K. A., *J. Chem. Phys.*, 71, 3072, 1979.
166. Gupta, S. K. and Gingerich, K. A., *J. Chem. Phys.*, 72, 2795, 1980.
167. Gupta, S. K. and Gingerich, K. A., *J. Chem. Phys.*, 74, 3584, 1981.
168. Gupta, S. K., Atkins, R. M., and Gingerich, K. A., *Inorg. Chem.*, 17, 3211, 1978.
169. Gupta, S. K., Kingcade, J. E., and Gingerich, K. A., *Adv. Mass Spectrom.*, 8A, 445, 1980.
170. Gupta, S. K., Nappi, B. M., and Gingerich, K. A., *Inorg. Chem.*, 20, 966, 1981.
171. Gupta, S. K., Nappi, B. M., and Gingerich, K. A., *J. Phys. Chem.*, 85, 971, 1981.
172. Gupta, S. K., Pelino, M., and Gingerich, K. A., *J. Chem. Phys.*, 70, 2044, 1979.
173. Gupta, S. K., Pelino, M., and Gingerich, K. A., *J. Phys. Chem.*, 83, 2335, 1979.
174. Gustavsson, T., Amiot, C., Verges, J., *Mol. Phys.*, 64, 279, 1988.
175. Haque, R. and Gingerich, K. A., *J. Chem. Thermodyn.*, 12, 439, 1980.
176. Haque, R., Pelino, M., and Gingerich, K. A., *J. Chem. Phys.*, 71, 2929, 1979.
177. Haque, R., Pelino, M., and Gingerich, K. A., *J. Chem. Phys.*, 73, 4045, 1980.
178. Hariharan, A. V. and Eick, H. A., *J. Chem. Thermodyn.*, 6, 373, 1974.
179. Hedgecock, I. M., Naulin, C., and Costes, M., Chem. Phys., 207, 379, 1996.
180. Heidecke, S. A., Fu, Z., Colt, J. R., and Morse, M. D., J. Chem. Phys., 97, 1692, 1992.
181. Herman, P. R., La Rocque, P. E., and Stoicheff, B., *J. Chem. Phys.*, 89, 4535, 1988.
182. Herman, R. and Herman, L., *J. Phys. Radium.*, 24, 73, 1963.
183. Herzberg, G., *Molecular Spectra and Molecular Structure. I. Spectra of Diatomic Molecules*, 2nd ed.,Van Nostrand, New York, 1950.
184. Hildenbrand, D. L. and Lau, K. H., *High Temp. Mater. Sci.*, 35, 11, 1996.
185. Hildenbrand, D. L. and Lau, K. H., *J. Chem. Phys.*, 89, 5825, 1988.
186. Hildenbrand, D. L. and Lau, K. H., *J. Chem. Phys.*, 91, 4909, 1989.
187. Hildenbrand, D. L. and Lau, K. H., *J. Chem. Phys.*, 93, 5983, 1990.
188. Hildenbrand, D. L. and Lau, K. H., *J. Chem. Phys.*, 96, 3830, 1992.
189. Hildenbrand, D. L. and Lau, K. H., *J. Phys. Chem.*, 96, 2325, 1992.
190. Hildenbrand, D. L. and Murad, E., *J. Chem. Phys.*, 44, 1524, 1966.
191. Hildenbrand, D. L. and Theard, L. P., *J. Chem. Phys.*, 50, 5350, 1969.
192. Hildenbrand, D. L., *High Temp. Mater. Sci.*, 35, 151, 1996.
193. Hildenbrand, D. L., *Chem. Phys. Lett.*, 32, 523, 1975.
194. Hildenbrand, D. L., *J. Chem. Phys.*, 103, 2634, 1995.
195. Hildenbrand, D. L., *J. Chem. Phys.*, 105, 10507, 1996.
196. Hildenbrand, D. L., *J. Chem. Phys.*, 48, 3657, 1968.
197. Hildenbrand, D. L., *J. Chem. Phys.*, 52, 5751, 1970.
198. Hildenbrand, D. L., *J. Chem. Phys.*, 65, 614, 1976.
199. Hildenbrand, D. L., *J. Chem. Phys.*, 66, 3526, 1977.
200. Hildenbrand, D. L., *J. Phys. Chem.*, 77, 897, 1973.
201. Hilpert, K. and Ruthardt, K., *Ber. Bunsenges. Phys. Chem.*, 91, 724, 1987.
202. Hilpert, K. and Ruthardt, K., *Ber. Bunsenges. Phys. Chem.*, 93, 1070, 1989.
203. Hilpert, K., *Ber. Kernforschungsanlage Juelich, JUEL-1744, 272, 1981.
204. Hilpert, K., *J. Phys. Chem.*, 77, 1425, 1982.
205. Huber, K. P. and Herzberg, G., *Molecular Spectra and Molecular Structure Constants of Diatomic Molecules*, Van Nostrand, New York, 1979.
206. Hussein, K., Effantin, C., D'Incan, J., Verges, J., and Barrow, R. F., *Chem. Phys. Lett.*, 124, 105, 1986.
207. Ihle, H. R. and Wu, C. H., *J. Chem. Phys.*, 63, 1605, 1975.
208. Ishwar, N. B., Varma, M. P., and Jha, B. L., *Acta Phys. Pol. A*, A61, 503, 1982.
209. Ishwar, N. B., Varma, M. P., and Jha, B. L., *Indian J. Pure Appl. Phys.*, 20, 992, 1982.
210. James, A. M., Kowalczyk, P., Simard, B., Pinegar, J. C., Morse, M. D., J. Mol. Spectrosc., 168, 248, 1994.
211. Jeyagopal, T., Rajavel, S. R. K., Ramakrishnan, M., and Rajamanickam, N., *Acta Phys. Hung.*, 68, 145, 1990.
212. Johns, J. W. C. and Ramsey, D. A., *Can. J. Phys.*, 39, 210, 1961.
213. Jones, K. M., Maleki, S., Bize, S., Lett, P. D., Williams, C. J., Richling, H., Knöckel, H., Tiemann, E., Wang, H., Gould, P. L., and Stwalley, W. C., Phys. Rev., A, 54, R1006, 1996.
214. Jones, R. W. and Gole, J. L., *Chem. Phys.*, 20, 311, 1977.
215. Jordan, K. J., Lipson, R. H., McDonald, N. A., and Le Roy, R. J., *J. Phys. Chem.*, 96, 4778, 1992.
216. Kant, A. and Lin, S.-S., *J. Chem. Phys.*, 51, 1644, 1969.
217. Kant, A. and Moon, K. A., *High Temp. Sci.*, 11, 55, 1979.
218. Kant, A. and Strauss, B. H., *J. Chem. Phys.*, 41, 3806, 1964.
219. Kant, A. and Strauss, B., *J. Chem. Phys.*, 49, 3579, 1968.
220. Kant, A., *J. Chem. Phys.*, 49, 5144, 1968.
221. Kant, A., Lin, S.-S, and Strauss, B., *J. Chem. Phys.*, 49, 1983, 1968.

222. Kant, A., Strauss, B., and Lin, S.-S., *J. Chem. Phys.*, 52, 2384, 1970.
223. Kaposi, O., *Magy. Kem. Foly.*, 83, 356, 1977.
224. Kaufel, R., Vahl, G., Nunkwitz, R., and Baumgaertel, H., *Z. Anorg. Allg. Chem.*, 481, 207, 1981.
225. Kazenas ,E., Tagirov, V. K., and Zviadadze, G. N., *Izv. Akad. Nauk. SSSR Met.*, 58, 1984.
226. Kazenas, E. K., Bol'shikh, M. A., and Petrov, A. A., *Izvestiya Rossiiskoi Akademii Nauk. Metally,* (3), 29, 1996.
227. Kent, R. A. and Margrave, J. L., *J. Am. Chem. Soc.*, 87, 3582, 1965.
228. Kent, R. A., Ehlert, T. C., and Margrave, J. L., *J. Am. Chem. Soc.*, 86, 5090, 1964.
229. Kent, R. A., *J. Am. Chem. Soc.*, 90, 5657, 1968.
230. Khitrov, A. N., Ryabova, V. G., and Gurvich, L. V., *Teplofiz. Vys. Tempo.*, 11, 1126, 1973.
231. Kiang, T. and Zare, R. N., *J. Am. Chem. Soc.*, 102, 4024, 1980.
232. Kimura, H., Asano, M., and Kubo, K., *J. Nucl. Mater.*, 97, 259, 1981.
233. Kingcade, J. E. Jr. and Gingerich, K. A., *J. Chem. Phys.*, 84, 3432, 1986.
234. Kingcade, J. E. Jr. and Gingerich, K. A., *J. Chem. Soc. Faraday Trans. 2*, 85, 195, 1989.
235. Kingcade, J. E. Jr., and Gingerich, K. A., *J. Chem. Phys.*, 84, 4574, 1986.
236. Kingcade, J. E., Cocke, D. L., and Gingerich, K. A., *High Temp. Sci.*, 16, 89, 1983,
237. Kingcade, J. E., Dufner, D. C., Gupta, S. K., and Gingerich, K. A., *High Temp. Sci.*, 10, 213, 1978.
238. Kingcade, J. E., Nagarathna, H. M., Shim, I., and Gingerich, K. A., *J. Phys. Chem.*, 90, 2830, 1986.
239. Kleinschmidt, P. D. and Hildenbrand, D. L., *J. Chem. Phys.*, 68, 2819, 1978.
240. Kleinschmidt, P. D. and Ward, J. W., *J. Less-Common. Met.*, 121, 61, 1986.
241. Kleinschmidt, P. D., Cubicciotti, D., and Hildenbrand, D. L., *J. Electrochem. Soc.*, 125, 1543, 1978; *Proc. Electrochem. Soc.*, 78, 217, 1978.
242. Kleinschmidt, P. D., Lau, K. H., and Hildenbrand, D. L., *J. Chem. Phys.*, 74, 653, 1981.
243. Klynning, L. and Lindgren, B., *Arkiv. Fysik.*, 32, 575, 1966.
244. Kohl, F. J. and Carlson, K. D., *J. Am. Chem. Soc.*, 90, 4814, 1968.
245. Kohl, F. J. and Stearns, C. A., *J. Phys. Chem.*, 78, 273, 1974.
246. Kondratiev, V. N., *Bond Dissociation Energies, Ionization Potentials and Electron Affinities*, Mauka Publishing House, Moscow, 1974.
247. Kordis, J. and Gingerich, K. A., *J. Chem. Eng. Data,* 18, 135, 1973.
248. Kordis, J. and Gingerich, K. A., *J. Chem. Phys.*, 58, 5141, 1973.
249. Kordis, J. and Gingerich, K. A., *J. Phys. Chem.,* 76, 2336, 1972.
250. Kordis, J., Gingerich, K. A., and Seyse, R. J., *J. Chem. Phys.*, 61, 5114, 1974.
251. Kowalski, A., Czaikowski, M., and Breckenridge, W. H., *Chem. Phys. Lett.*, 119, 368, 1985.
252. Koyama, T. and Yamawaki, M., *J. Nucl. Mater.*, 152, 30, 1988.
253. Kronekvist, M., Lagerqvist, A., and Neuhaus, H., *J. Mol. Spectrosc.*, 39, 516, 1971.
254. Langenberg, J. D. and Morse, M. D., *Chem. Phys. Lett.,* 239, 25, 1995.
255. Langenberg, J. D. and Morse, M. D., *J. Chem. Phys.,* 100, 2331, 1998.
256. Lau, K. H. and Hildenbrand, D. L., *J. Chem. Phys.*, 71, 1572, 1979.
257. Lau, K. H. and Hildenbrand, D. L., *J. Chem. Phys.*, 72, 4928, 1980.
258. Lau, K. H. and Hildenbrand, D. L., *J. Chem. Phys.*, 76, 2646, 1982.
259. Lau, K. H. and Hildenbrand, D. L., *J. Chem. Phys.*, 80, 1312, 1984.
260. Lau, K. H. and Hildenbrand, D. L., *J. Chem. Phys.*, 86, 2949, 1987.
261. Lau, K. H. and Hildenbrand, D. L., *J. Chem. Phys.*, 92, 6124, 1990.
262. Lau, K. H., Brittain, R. D., and Hildenbrand, D. L., *Chem. Phys. Lett.*, 81, 227, 1981; *J. Phys. Chem.*, 86, 4429, 1982.
263. Lau, K. H., Brittain, R. D., and Hildenbrand, D. L., *J. Chem. Phys.*, 90, 1158, 1989.
264. Li, K. C. and Stwalley, W. C., *J. Chem. Phys.*, 59, 4423, 1973.
265. Li, L., Lyyra, A. M., Luh, W. T., and Stwalley, W. C., *J. Chem. Phys.*, 93, 8452, 1990.
266. Lindgren, B. and Nilsson, Ch., *J. Mol. Spectrosc.*, 55, 407, 1975.
267. Martin, E. and Barrow, R. F., *Phys. Scr.,* 17, 501, 1978.
268. McIntyre, N. S., Vander Auwera-Mahieu, A., and Drowart, J., *Trans. Faraday Soc.,* 64, 3006, 1968.
269. Menendez, M., Garay, M., Verdasco, E., and Gonzalez, U. A., *J. Chem. Phys.*, 99, 2760, 1993.
270. Murad, E., Hildenbrand, D. L., and Main, R. P., *J. Chem. Phys.*, 45, 263, 1966.
271. Nagai, S., Shinmei, M., and Yokokawa, T., *J. Inorg. Nucl. Chem.*, 36, 1904, 1974.
272. Nappi, B. M. and Gingerich, K. A., *Inorg. Chem.*, 20, 522, 1981.
273. Neckel, A. and Sodeck, G., *Monatsch. Chem.*, 103, 367, 1972.
274. Nedelec, O. and Giroud, M., *J. Chem. Phys.*, 79, 2121, 1983.
275. Neubert, A. and Zmbov, K. F., *Chem. Phys.*, 76, 469, 1983.
276. Neubert, A. and Zmbov, K. F., *J. Chem. Soc. Faraday Trans. 1*, 70, 2219, 1974.
277. Neubert, A., Ihle, H. R., and Gingerich, K. A., *J. Chem. Phys.*, 73, 1406, 1980.
278. Neubert, A., Zmbov, K. F., Gingerich, K. A., and Ihle, H. R., *J. Chem. Phys.*, 77, 5218, 1982.
279. Nordine, P. C., *J. Chem. Phys.*, 61, 224, 1974.
280. O'Hare, P. A. G. and Curtiss, L. A., *J. Chem. Thermodyn.,* 27, 643, 1995.
281. O'Hare, P. A. G., *J. Phys. Chem. Ref. Data*, 22, 1455, 1993.

282. O'Hare, P. A. G., Zywocinski, A., and Curtiss, L. A., *J. Chem. Thermodyn.*, 28, 459, 1996.
283. Ovcharenko, I. E., Ya, Kuzyankow, Y., and Tatevaskii, V. M., *Opt. Spectrosk.*, 19, 528, 1965.
284. Pak, K., Cai, M. F., Dzugan, T. P., and Bondybey, V. E., *Faraday Discuss. Chem. Soc.*, 86, 153, 1988.
285. Parks, E. K. and Wexler, S., *J. Phys. Chem.*, 88, 4492, 1984.
286. Parr, T. P., Behrens, R., Freedman, A., and Heron, R. R., *Chem. Phys. Lett.*, 56, 71, 1978.
287. Pedley, J. B. and Marshall, E. M., *J. Phys. Chem. Ref. Data*, 12, 967, 1984.
288. Pelevin, O. V., Mil'vidskii, M. G., Belyaev, A. I., and Khotin, B. A., *Izv. Akad. Nauk. SSSR Neorg. Mater.*, 2, 924, 1966.
289. Pelino, M. and Gingerich, K. A., *J. Chem. Phys.*, 90, 1286, 1989.
290. Pelino, M., and Gingerich, K. A., *J. Chem. Phys.*, 93, 1581, 1989.
291. Pelino, M., Piacente, V., and Ascenzo, G., *Thermochim. Acta*, 31, 383, 1979.
292. Pelino, M., Viswanadham, P., and Edwards, J. G., *J. Phys. Chem.*, 83, 2964, 1979.
293. Piacente, V. and Balducci, G., *Adv. Mass Spectrom.*, 7A, 626, 1978.
294. Piacente, V. and Balducci, G., *Dyn. Mass Spectrom.*, 4, 295, 1976.
295. Piacente, V. and Balducci, G., *High Temp. Sci.*, 6, 254, 1974.
296. Piacente, V. and Desideri, A., *J. Chem. Phys.*, 57, 2213, 1972.
297. Piacente, V. and Gigli, R., *J. Chem. Phys.*, 77, 4790, 1982.
298. Piacente, V. and Gingerich, K. A., *High Temp. Sci.*, 9, 189, 1977.
299. Piacente, V. and Gingerich, K. A., *Z. Naturforsch. Teil A*, 28, 316, 1973.
300. Piacente, V. and Malaspina, L., *J. Chem. Phys.*, 56, 1780, 1972.
301. Piacente, V., *J. Chem. Phys.*, 70, 5911, 1979.
302. Pianalto, F. S., O'Brien, L. C., Keller, P. C., and Bernath, P. F., *J. Mol. Spectrosc.*, 129, 348, 1988.
303. Pilgrim, J. C. and Duncan, M. A., *Chem. Phys. Lett.*, 232, 335, 1995.
304. Pinegar, J. C., Langenberg, J. D., Arrington, C.A., Spain, E. M., and Morse, M. D., *J. Chem. Phys.*, 102, 666, 1995.
305. Pitzer, K. S., *J. Chem. Phys.*, 74, 3078, 1981.
306. Porter, R. F. and Spencer, C. W. J., *J. Chem. Phys.*, 32, 943, 1960.
307. Prasad, R., Venugopal, V., and Sood, D. D., *J. Chem. Thermodyn.*, 9, 593, 1977.
308. Rajamanickam, N., Dhuvaragaikannan, N., and Raja Mohamed, K., *Acta Physica Hungarica*, 74, 385, 1994.
309. Rajamanickam, N., Palaniselvam, K., Rajavel, S. R. K., Rajesh, M., and Sureshkumar, G., *Acta Phys. Hung.*, 70, 141, 1991.
310. Rajamanickam, N., Senthilkumar, R. N., Ganesan, S., Gopalakrishnan, N., Rajkumer, J., Jegadesan, V., and Dandapani, C., *Acta Phys. Hung.*, 70, 71, 1991.
311. Rajamanickan, N., *Acta Ciencia Indica Phys.*, 14, 18, 1988.
312. Ram, R. S., Rai, S. B., Ram, R. S., Upadhya, K. N., *J. Chim. Phys. Phys-Chim. Biol.*, 76, 560, 1979.
313. Ramakrishnan, E. S., Shim, I., and Gingerich, K. A., *J. Chem. Soc. Faraday Trans. 2*, 80, 395, 1984.
314. Ran, Q., Schmude, R. W., Gingerich, K. A., Wilhite, D. W., and Kincade, J. E., *J. Phys. Chem.*, 97, 8535, 1993.
315. Rao, P. S. and Rao, T. V. R., *J. Quant. Spectrosc. Radiat. Transfer*, 27, 207, 1982.
316. Rao, S. P. and Rao, T. V. R., *Acta Ciencia Indica Phys.*, 7, 58, 1981.
617. Rao, T. V. R., Reddy, R. R., and Rao, P. S., *Indian J. Pure Appl. Phys.*, 19, 1219, 1981.
318. Rao, V. M., Rao, M. L. P., and Rao, P. T., *J. Quant. Spectrosc. Radiat. Transfer*, 25, 547, 1981.
319. Reddy, R. R., Reddy, A. S. R., and Rao, T. V. R., *Acta Phys. Slovaca*, 36, 273, 1986.
320. Riekert, G., Lamparter, P., and Steeb, S., *Z. Metallkd.*, 72, 765, 1981.
321. Riekert, G., Rainer-Harbach, G., Lamparter, P., and Steeb, S., *Z. Metallkd.*, 76, 406, 1981.
322. Rinehart, G. H. and Behrens, R. G., *J. Phys. Chem.*, 83, 2052, 1979.
323. Rohlfing, E. A. and Valentine, J. J., *J. Chem. Phys.*, 84, 6560, 1986.
324. Rovner, L., Drowart, A., and Drowart, J., *Trans. Faraday Soc.*, 63, 2910, 1967.
325. Rusin, A. D., Zhukov, E., Agamirova, L. M., and Kalinnikov, V. T., *Zh. Neorg. Khim.*, 24, 1457, 1979.
326. Samoilova, I. O. and Kazenas, E. K., *Izvestiya Rossiiskoi Akademii Nauk. Metally*, (1), 33, 1995.
327. Samoilova, I. O. and Kazenas, E. K., *Izvestiya Rossiiskoi Akademii Nauk. Metally*, (3), 36, 1994.
328. Schumde, R. W. Jr., Ran, Q., Gingerich, K. A., and Kincade, J. E. Jr., *J. Chem. Phys.*, 102, 2574, 1995.
329. Scott, P. R. and Richards, W. G., *J. Phys. B*, 7, 1679, 1974.
330. Shim, I. and Gingerich, K. A., *J. Chem. Phys.*, 77, 2490, 1982.
331. Shim, I. and Gingerich, K. A., *J. Chem. Phys.*, 80, 5107, 1984.
332. Shim, I., and Gingerich, K. A., *J. Chem. Phys.*, 81, 5937, 1984.
333. Shim, I., Finkbeiner, H. C., and Gingerich, K. A., *J. Phys. Chem.*, 91, 3171, 1987.
334. Shim, I., Kingcade, J. E. Jr., and Gingerich, K. A., *J. Chem. Phys.*, 85, 6629, 1986.
335. Shim, I., Kingcade, J. E. Jr., and Gingerich, K. A., *J. Chem. Phys.*, 89, 3104, 1988.
336. Shim, I., Kingcade, J. E. Jr., and Gingerich, K. A., *Z. Phys. D - Atoms Molecules and Clusters*, 7, 261, 1987.
337. Shim, I., Mandix, K., and Gingerich, K. A., *J. Phys. Chem.*, 95, 5435, 1991.
338. Shim, I., Pelino, M., and Gingerich, K. A., *J. Chem. Phys.*, 97, 9240, 1992.
339. Siegel, B., *Q. Rev. (London)*, 19, 77, 1965.
340. Simons, J. W., Oldenberg, R. C., and Baughaim, S. L., *J. Phys. Chem.*, 91, 3840, 1987.

341. Smith, P. K. and Peterson, D. E., *J .Chem. Phys.*, 52, 4963, 1970.
342. Smoes, S. and Drowart, J., *Chem. Commun.*, p.534, 1968.
343. Smoes, S. and Drowart, J., *J. Chem. Soc. Faraday Trans. 2*, 73, 1746, 1977.
344. Smoes, S. and Drowart, J., *J. Chem. Soc. Faraday Trans. 2*, 80, 1171, 1984.
345. Smoes, S., Coppens, P., Bergman, C., and Drowart, J., *Trans. Faraday Soc.*, 65, 682, 1969.
346. Smoes, S., Depiere, D., and Drowart, J., *Rev. Int. Hautes Temp. Refractaires Paris*, 9, 171, 1972.
347. Smoes, S., Drowart, J., Welter, J. M., *J. Chem. Thermodyn.*, 9, 275, 1977; *Adv. Mass Spectrom.*, 7A, 622, 1978.
348. Smoes, S., Huguet, R., and Drowart, J., *Z. Naturforsch. Teil A*, 26, 1934, 1971.
349. Smoes, S., Mandy, F., Vander Auwera-Mahieu, A., and Drowart, *J., Bull. Soc. Chim. Belg.*, 81, 45, 1972.
350. Smoes, S., Myers, C. E., and Drowart, J., *Chem. Phys. Lett.*, 8, 10, 1971.
351. Smoes, S., Pattje, W. R., and Drowart, J., *High Temp. Sci.*, 10, 109, 1978.
352. Spain, E. M. and Morse, M. D., *J. Chem. Phys.*, 97, 4605, 1992.
353. Spain, E. M. and Morse, M. D., *J. Phys. Chem.*, 96, 2479, 1992.
354. Sreedhara Murthy, N., *Indian J. Phys.*, 62B, 92, 1988.
355. Srivastara, R. D. and Farber, M., *High Temp. Sci.*, 5, 489, 1973.
356. Stearns, C. A. and Kohl, F. J., *High Temp. Sci.*, 2, 146, 1970.
357. Stearns, C. A. and Kohl, F. J., *High Temp. Sci.*, 6, 284, 1974.
358. Steiger, R. A. and Cater, E. D., *High Temp. Sci.*, 7, 204, 1975.
359. Steiger, R. P. and Cater, E. D., *High Temp. Sci.*, 7, 288, 1975.
360. Stwalley, W. C., Way, K. R., and Velasco, R., *J. Chem. Phys.*, 60, 3611, 1974.
361. Su, T.-M. R. and Riley, S. J., *J. Chem. Phys.*, 71, 3194, 1979.
362. Su, T.-M. R. and Riley, S. J., *J. Chem. Phys.*, 72, 1614, 1980.
363. Su, T.-M. R. and Riley, S. J., *J. Chem. Phys.*, 72, 6632, 1980.
364. Sullivan, C. L., Zehe, M. J., and Carlson, K. D., *High Temp. Sci.*, 6, 80, 1974.
365. Tanaka, Y., Yushina, K., and Freeman, D. E., *J. Chem. Phys.*, 59, 564, 1973.
366. Taylor, S., Lemire, G. W., Hamrick, Y. M., Fu, Z., and Morse, M. D., *J. Chem. Phys.*, 89, 5517, 1988.
367. Taylor, S., Spain, E. M., and Morse, M. D., *J. Chem. Phys.*, 92, 2698, 1990.
368. Tellinghuisen, J., Tisone, G. C., Hoffmann, J. M., and Hays, A. K., *J. Chem. Phys.*, 64, 4796, 1976.
369. Tolbert, M. A. and Beauchamp, J. L., *J. Phys. Chem.*, 90, 5015, 1986.
370. Tomaszkiewicz, I., Susman, S., Volin, K. J., and O'Hare, P. A. G., *J. Chem. Thermodyn.*, 26, 1081, 1994.
371. Tromp, J. W., LeRoy, R. J., Gerstenkorn, S., and Luc, P., *J. Mol. Spectrosc.*, 100, 82, 1983.
372. Tuenge, R. T., Laabs, F., and Franzen, H. F., *J. Chem. Phys.*, 65, 2400, 1976.
373. Urdahl, R. S., Bao, Y., and Jackson, W. M., *Chem. Phys. Lett.*, 178, 425, 1991.
374. Uy, O. M. and Drowart, J., *High Temp. Sci.*, 2, 293, 1970.
375. Uy, O. M. and Drowart, J., *Trans. Faraday Soc.*, 65, 3221, 1969.
376. Uy, O. M. and Drowart, J., *Trans. Faraday Soc.*, 67, 1293, 1971.
377. Uy, O. M., Muenow, D. W., Ficalora, P, J., and Margrave, J. L., *Trans. Faraday Soc.*, 64, 2998, 1968.
378. Van Veen, N. J. A., DeVries, M., and DeVries, A. E., *Chem. Phys. Lett.*, 64, 213, 1979.
379. Vander Auwera-Mahieu, A. and Drowart, J., *Chem. Phys. Lett.*, 1, 311, 1967.
380. Vander Auwera-Mahieu, A., McIntyre, N. S., and Drowart, J., *Chem. Phys. Lett.*, 4, 198, 1969.
381. Vander Auwera-Mahieu, A., Peeters, R., McIntyre, N. S., and Drowart, J., *Trans. Faraday Soc.*, 66, 809, 1970.
382. Varma, M. P., Ishwar, N. B., and Jha, B. L., *Indian J. Pure Appl. Phys.*, 20, 828, 1982.
383. Velasco, R., Ottinger, C., and Zare, R. N., *J. Chem. Phys.*, 51, 5522, 1969.
384. Vempati, S. N. and Jones, W. E., *J. Mol. Spectrosc.*, 127, 232, 1988.
385. Venkataramanaiah, M. and Lakshman, S. V. J., *J. Quant. Spectrosc. Radiat. Transfer*, 26, 11, 1981.
386. Verhaegen, G., Smoes, S., and Drowart, J., *J. Chem. Phys.*, 40, 239, 1964.
387. Verhaegen, G., Stafford, F. E., and Drowart, J., *J. Chem. Phys.*, 40, 1622, 1964.
388. Viswanathan, K. S. and Tellinghuisen, J., *J. Mol. Spectrosc.*, 98, 185, 1983.
389. Viswanathan, R., Baba, M. S., Raj, D. D. A., Balasubramanian, R., Narasimhan, T. S. L., and Mathews, C. K., *Spectrochim. Acta*, Part B, 49B, 243, 1994.
390. Viswanathan, R., Schmude, R. W., and Gingerich, K. A., *J. Phys. Chem.*, 100, 10784, 1996.
391. Viswanathan, R., Schumde, R. W., and Gingerich, K. A., *J. Chem. Thermodyn.*, 27, 763, 1995.
392. Wang, Y. C., Kajitani, M., Kasahara, S., Bata, M., Ishikawa, K., and Kato, H., *J. Chem. Phys.*, 95, 6229, 1991.
393. Way, K. R. and Stwalley, W. C., *J. Chem. Phys.*, 59, 5298, 1973.
394. Weickenmeier, W., Diemer, U., Wahl, M., Raab, M., Demtroeder, W., and Mueller, W., *J. Phys.Chem.*, 82, 5354, 1985.
395. Wiedemeier, H. and Gilles, P. W., *J. Chem. Phys.*, 42, 2765, 1965.
396. Wu, C. H. and Ihle, H. R., *Adv. Mass Spectrom.*, 8A, 374, 1980.
397. Wu, C. H., Ihle, H. R., and Gingerich, K. A., *Int. J. Mass Spectrom. Ion Phys.*, 47, 235, 1983.
398. Wu, C. H., *J. Chem. Phys.*, 65, 3181, 1976; 65, 2040, 1976.

Table 1
BOND STRENGTHS IN DIATOMIC MOLECULES (continued)

399. Yokozeki, A. and Menzinger, M., *Chem. Phys.*, 14, 427, 1976.
400. Yoo, R. K., Ruscic, B., and Berkowitz, J., *Chem. Phys.*, 166, 215, 1992.
401. Zemke, W. T. and Stwalley, W. C., *Chem. Phys. Lett.*, 143, 84, 1988.
402. Zemke, W. T., Stwalley, W. C., Langhoff, S. R., Valderrama, G. L., and Berry, M. J., *J. Chem. Phys.*, 95, 7846, 1991.
403. Zmbov, K. F. and Margrave, J. L., *J. Chem. Phys.*, 45, 3167, 1966.
404. Zmbov, K. F. and Margrave, J. L., *J. Chem. Phys.*, 47, 3122, 1967.
405. Zmbov, K. F. and Margrave, J. L., *J. Inorg. Nucl. Chem.*, 29, 59, 1967.
406. Zmbov, K. F. and Margrave, J. L., *J. Phys. Chem.*, 70, 3379, 1966.
407. Zmbov, K. F. and Margrave, J. L., *J. Phys. Chem.*, 71, 2893, 1967.
408. Zmbov, K. F., Hastie, J. W., and Margrave, J. L., *Trans. Faraday Soc.*, 64, 861, 1968.
409. Zmbov, K. F., Neubert, A., and Ihle, H. R., *Z. Naturforsch., A*, 36A, 914, 1981.
410. Zmbov, K. F., Wu, C. H., and Ihle, H. R., *J. Chem. Phys.*, 67, 4603, 1977.
411. Zollweg, R. J., *Contrib. Pap. Int. Conf. Phenom. Ioniz. Gases.* 11th, 402, 1973.

Table 2
ENTHALPY OF FORMATION OF GASEOUS ATOMS FROM ELEMENTS IN THEIR STANDARD STATES

For elements that are diatomic gases in their standard states these are readily obtained from the bond strength. For elements that are crystalline in their standard states they are derived from vapor pressure data.

Atom	$\Delta_f H^\circ_{298}$/kJ mol^{-1}	Ref.	Atom	$\Delta_f H^\circ_{298}$/kJ mol^{-1}	Ref.	Atom	$\Delta_f H^\circ_{298}$/kJ mol^{-1}	Ref.
			Hf	619 ± 4	1	Re	774 ± 6.3	1
Ag	284.9 ± 0.8	2	Hg	61.38 ± 0.04	2	Rh	556 ± 4	1
Al	330.0 ± 4.0	2	I	106.76 ± 0.04	2	Ru	650.6 ± 6.3	1
As	302.5 ± 13	1	In	243 ± 4	1	S	277.17 ± 0.15	2
Au	368.2 ± 2.1	1	Ir	669 ± 4	1	Sb	264.4 ± 2.5	1
B	565 ± 5	2	K	89.0 ± 0.8	2	Sc	377.8 ± 4	1
Ba	177.8 ± 4	1	Li	159.3 ± 1.0	2	Se	227.2 ± 4	1
Be	324 ± 5	2	Mg	147.1 ± 0.8	2	Si	450 ± 8	2
Bi	209.6 ± 2.1	1	Mn	283.3 ± 4	1	Sn	301.2 ± 1.5	2
Br	111.87 ± 0.12	2	Mo	658.1 ± 2.1	1	Sr	163.6 ± 2.1	1
C	716.68 ± 0.45	2	N	472.68 ± 0.40	2	Ta	782.0 ± 2.5	1
Ca	177.8 ± 0.8	2	Na	107.5 ± 0.7	2	Te	196.6 ± 2.1	1
Cd	111.80 ± 0.20	2	Nb	721.3 ± 4	1	Th	602 ± 6	2
Ce	423 ± 13	1	Ni	430.1 ± 2.1	1	Ti	473 ± 3	2
Cl	121.301 ± 0.008	2	O	249.18 ± 0.10	2	Tl	182.21 ± 0.4	1
Co	428.4 ± 4	1	Os	787 ± 6.3	1	U	533 ± 8	2
Cr	397 ± 4	1	P	316.5 ± 1.0	2	V	514.2 ± 1.3	1
Cs	76.5 ± 1.0	2	Pb	195.2 ± 0.8	2	W	849.8 ± 4	1
Cu	337.4 ± 1.2	2	Pd	376.6 ± 2.1	1	Y	424.7 ± 2.1	1
Er	317.1 ± 4	1	Pt	565.7 ± 1.3	1	Yb	152.09 ± 0.8	1
F	79.38 ± 0.30	2	Pu	364.4 ± 17	1	Zn	130.40 ± 0.40	2
Ge	372 ± 3	2	Rb	80.9 ± 0.8	2	Zr	608.8 ± 4	1
H	217.998 ± 0.006	2						

REFERENCES

1. Brewer, L. and Rosenblatt, G. M., *Adv. High Temp. Chem.*, 2, 1, 1969.
2. Cox, J. D., Wagman, D. D., and Medvedev, V. A., Eds., *CODATA Key Values for Thermodynamics*, Hemisphere Publishing Corporation, New York, 1989.

Table 3
BOND STRENGTHS IN POLYATOMIC MOLECULES

The values below refer to a temperature of 298 K and have mostly been determined by kinetic methods (see (i) S. W. Benson, *J. Chem. Educ.*, 42, 502, 1965, (ii) J. A. Kerr, *Chem. Rev.*, 66, 465, 1966 and (iii) D. F. McMillen and D. M. Golden, *Ann. Rev. Phys. Chem.*, 33, 493, 1982, for a full description of the methods). An increasing number of bond strengths are being determined from gas-phase acidity cycles and from photoionization mass spectrometry (see J. Berkowitz, G. B. Ellison and D. Gutman, *J. Phys. Chem.*, 98, 2744, 1994).

Bond strengths in polyatomic molecules are notoriously difficult to measure accurately since the mechanisms of the kinetic systems involved in many of the measurements are seldom straightforward. Thus much controversy has taken place in the literature over the past 15 years concerning C-H bond strengths in simple alkanes, for which we recommend data based largely on kinetic studies involving time-resolved flow tube experiments with mass spectrometric determination of reactant radical concentrations (see Berkowitz, J., Ellison, G. B., and Gutman, D., *J. Phys. Chem.*, 98, 2744, 1994.). These alkane bond strengths and the enthalpies of formation of the corresponding radicals are significantly larger than values derived from experiments in very low pressure reactors (see Dobis, O. and Benson, S. W., *J. Phys. Chem.*, 101, 6030, 1997; and Benson, S. W. and Dobis, O., *J. Phys. Chem.*, 102, 5175, 1998). Other examples illustrating the difficulties involved are concerned with the C-H bond strengths in ethene and methanol or the corresponding enthalpies of formation of the vinyl and hydroxymethyl radicals and changes to the recommendations could well arise.

Some of the bond strengths have been calculated from the enthalpies of formation of the species involved according to the equations:

$$D°(R–X) = \Delta_f H°(R) + \Delta_f H°(X) − \Delta_f H°(RX)$$
$$D°(R–R) = 2 \Delta_f H°(R) − \Delta_f H°(RR)$$

The enthalpies of formation of the atoms and radicals are taken from Tables 2 and 4 and for the molecules from the appropriate References following Table 3.

An attempt has been made to list all the important values obtained by methods that are considered to be valid. The references are intended to serve as a guide to the literature.

Bond	$D°_{298}$/kJ mol^{-1}	Ref.	Bond	$D°_{298}$/kJ mol^{-1}	Ref.
H-CH	424.0 ± 4.2	1	H-C$_6$H$_5$	473.1 ± 3.0	46
H-CH$_2$	462.0 ± 4.0	1	H-Cyclohexa-1,3-dien-5-yl	305 ± 21	63
H-CH$_3$	438.9 ± 0.4	16	H-Cyclohexa-1,4-dien-3-yl	305.4 ± 8.4	45
H-CCH	556.1 ± 2.9	4,37	H-Cyclohexyl	399.6 ± 4	63
H-CHCH$_2$	465.3 ± 3.4	37,81	H-C(CH$_3$)$_2$CCCH$_3$	344.3 ± 11.3	63
H-C$_2$H$_5$	423.0 ± 1.6	16	H-CH$_2$C(CH$_3$)C(CH$_3$)$_2$	326.4 ± 4.6	63
H-Cycloprop-2-en-1-yl	379.1 ± 17	63	H-C(CH$_3$)$_2$C(CH$_3$)CH$_2$	319.2 ± 4.6	63
H-CH$_2$CCH	374.0 ± 8	63	H-CH$_2$C$_6$H$_5$	375.7 ± 1.7	35
H-CH$_2$CHCH$_2$	361.9 ± 8.8	16,35	H-Cyclohepta-1,3,5-trien-7-yl	305.4 ± 8	63
H-Cyclopropyl	444.8 ± 1.3	63	H-Norbornyl	404.6 ± 10.5	63
H-n-C$_3$H$_7$	423.3 ± 2.1	82	H-Cycloheptyl	387.0 ± 4	63
H-i-C$_3$H$_7$	409.1 ± 2.0	82	H-CH(CH$_3$)C$_6$H$_5$	357.3 ± 6.3	63
H-CH$_2$CCCH$_3$	364.8 ± 8	63	H-Inden-1-yl	351 ± 13	63
H-CH(CH$_3$)CCH	347.7 ± 9.2	63	H-C(CH$_3$)$_2$C$_6$H$_5$	353.1 ± 6.3	63
H-Cyclobutyl	403.8 ± 4	63	H-1-Naphthylmethyl	356.1 ± 6.3	63
H-Cyclopropylmethyl	407.5 ± 6.7	63	H-CH(C$_6$H$_5$)$_2$	340.6	80
H-CH(CH$_3$)CHCH$_2$	345.2 ± 5.4	63	H-9,10-Dihydroanthracen-9-yl	315.1 ± 6.3	63
H-CH$_2$CHCHCH$_3$	358.2 ± 6.3	63	H-C(CH$_3$)(C$_6$H$_5$)$_2$	339 ± 8	63
H-CH$_2$C(CH$_3$)CH$_2$	358.2 ± 4	91,95	H-9-Anthracenylmethyl	342.3 ± 6.3	63
H-n-C$_4$H$_9$	425.4 ± 2.1	82	H-9-Phenanthrenylmethyl	356.1 ± 6.3	63
H-i-C$_4$H$_9$	425.2 ± 2.1	82	H-CN	527.6 ± 1.7	16
H-s-C$_4$H$_9$	411.2 ± 2.0	82	H-CH$_2$CN	392.9 ± 8.4	16
H-t-C$_4$H$_9$	404.3 ± 1.3	82	H-CH$_2$NC	380.7 ± 8.8	16
H-Cyclopenta-1,3-dien-5-yl	346.7	1,12	H-CH(CH$_3$)CN	376.1 ± 9.6	63
H-Spiropentyl	413.4 ± 4	63	H-C(CH$_3$)$_2$CN	361.9 ± 8.4	63
H-Cyclopent-1-en-3-yl	344.3 ± 4	63	H-CH$_2$NH$_2$	390.4 ± 8.4	63
H-CH$_2$CHCHCHCH$_2$	347 ± 13	63	H-CH$_2$NHCH$_3$	364 ± 8	63
H-CH(C$_2$H$_3$)$_2$	319.7	63,92	H-CH$_2$N(CH$_3$)$_2$	351 ± 8	63
H-CH(CH$_3$)CCCH$_3$	365.3 ± 11.3	63	H-CHO	368.5 ± 1.0	23
H-C(CH$_3$)$_2$CCH	338.9 ± 9.6	63	H-CHCO	440.6 ± 8.8	16
H-C(CH$_3$)$_2$CHCH$_2$	323.0 ± 6.3	63	H-COCH$_3$	373.8 ± 1.5	67
H-Cyclopentyl	403.5 ± 2.5	22,74	H-COCHCH$_2$	364.4 ± 4.2	63
H-CH$_2$C(CH$_3$)$_3$	418 ± 8	63	H-COC$_2$H$_5$	371.3	1,12
H-C(CH$_3$)$_2$C$_2$H$_5$	404.0 ± 6.3	1,74,89	H-COC$_6$H$_5$	363.6 ± 4	63

Table 3
BOND STRENGTHS IN POLYATOMIC MOLECULES (continued)

Bond	D°_{298}/kJ mol^{-1}	Ref.	Bond	D°_{298}/kJ mol^{-1}	Ref.
H-COCF$_3$	380.7 ± 8	63	H-N(CH$_3$)C$_6$H$_5$	366.1 ± 8	63
H-CH$_2$CHO	394.6 ± 9.2	16	H-NO	195.35 ± 0.25	32
H-CH$_2$COCH$_3$	411.3 ± 7.5	63	H-NO$_2$	327.6 ± 2.1	63
H-CH(CH$_3$)COCH$_3$	386.2 ± 5.9	63	H-NF$_2$	316.7 ± 10.5	63
H-CH$_2$OCH$_3$	402.2	1,12	H-NHNH$_2$	366.1	44
H-CH(CH$_3$)OC$_2$H$_5$	383.7 ± 1.7	55	H-NH$_3$	385 ± 21	63
H-Tetrahydrofuran-2-yl	385 ± 4	63	H-OH	498 ± 4	63
H-2-Furylmethyl	361.9 ± 8	63	H-OCH$_3$	436.0 ± 3.8	16
H-CH$_2$OH	401.8 ± 1.5	50	H-OC$_2$H$_5$	437.7 ± 3.4	16
H-CH(CH$_3$)OH	401.4	1,12	H-OC(CH$_3$)$_3$	439.7 ± 4	63
H-CH(OH)CHCH$_2$	341.4 ± 7.5	63	H-OCH$_2$C(CH$_3$)$_3$	428.0 ± 6.3	63
H-C(CH$_3$)$_2$OH	381 ± 4	63	H-OC$_6$H$_5$	361.9 ± 8	63
H-CH$_2$OCOC$_6$H$_5$	419.2 ± 5.4	63	H-O$_2$H	369.0 ± 4.2	88
H-COOCH$_3$	387.9 ± 4	63	H-O$_2$CH$_3$	370.3 ± 2.1	56
H-CH$_2$F	423.8 ± 4	77	H-O$_2$C(CH$_3$)$_3$	374.0 ± 0.8	47
H-CHF$_2$	431.8 ± 4	77	H-OCOCH$_3$	442.7 ± 8	63
H-CF$_3$	449.5	3,61	H-OCOC$_2$H$_5$	445.2 ± 8	63
H-CHFCl	421.7 ± 5.4	97	H-OCO-n-C$_3$H$_7$	443.1 ± 8	63
H-CF$_2$Cl	421.3 ± 8.3	64	H-ONO	327.6 ± 2.1	15
H-CHFCl$_2$	413.8 ± 5.0	97	H-ONO$_2$	423.4 ± 2.1	15
H-CH$_2$Cl	419.0 ± 2.3	84	H-SiH	351	63
H-CHCl$_2$	402.5 ± 2.7	84	H-SiH$_2$	268	63
H-CH$_2$CH$_2$Cl	423.1 ± 2.4	85	H-SiH$_3$	384.1 ± 2.0	83
H-CH(CH$_3$)Cl	406.6 ± 1.5	84	H-SiH$_2$CH$_3$	374.9	99
H-C(CH$_3$)Cl$_2$	390.6 ± 1.5	84	H-SiH(CH$_3$)$_2$	374.0	99
H-CCl$_3$	392.5 ± 2.5	48	H-Si(CH$_3$)$_3$	377.8	63,99
H-CH$_2$Br	425.1 ± 4.2	97	D-Si(CH$_3$)$_3$	389 ± 7.1	36
H-CHBr$_2$	417.2 ± 7.5	97	H-SiH$_2$C$_6$H$_5$	369.0	63,99
H-CBr$_3$	401.7 ± 6.7	63	H-SiF$_3$	418.8	63,99
H-CH$_2$I	431 ± 8	63	H-SiCl$_3$	382.0	63,99
H-CHI$_2$	431 ± 8	63	H-Si$_2$H$_5$	361.1	63
H-CHCF$_2$	448 ± 8	90	H-Si(CH$_3$)$_2$Si(CH$_3$)$_3$	356.9 ± 8.4	45
H-CFCHF	448 ± 8	90	H-Si(Si(CH$_3$)$_3$)$_3$	330.5 ± 8.4	45
H-CFCF$_2$	452 ± 8	90	H-PH$_2$	351.0 ± 2.1	16
H-CH$_2$CF$_3$	446.4 ± 4.6	63	H-SH	381.6 ± 2.9	65
H-CF$_2$CH$_3$	416.3 ± 10.5	63	H-SCH$_3$	365.3 ± 2.5	65
H-C$_2$F$_5$	429.7 ± 2.1	63	H-SC$_6$H$_5$	348.5 ± 8	63
H-CFCFCl	444 ± 8	90	H-SO	172.8	100
H-CHClCF$_3$	425.9 ± 6.3	63	H-GeH$_3$	349.0 ± 8	16
H-CClCFCl	439 ± 8	90	H-GeH$_2$I	331 ± 8	68
H-CClCH$_2$	>433.5	81	H-Ge(CH$_3$)$_3$	339 ± 8	34
H-CClCHCl	435 ± 8	90	H-AsH$_2$	319.2 ± 0.8	16
H-CCl$_2$CHCl$_2$	393 ± 8	63	H-SeH	334.93 ± 0.75	16
H-C$_2$Cl$_5$	393.5 ± 6.0	66	H-Sn(n-C$_4$H$_9$)$_3$	308.4 ± 8.4	19
H-CClBrCF$_3$	404.2 ± 6.3	63	H-SbH$_2$	288.3 ± 2.1	16
H-n-C$_3$F$_7$	435 ± 8	63	H-TeH	277.0 ± 5.0	16
H-i-C$_3$F$_7$	433.5 ± 2.5	38	HC⌡CH	965 ± 8	1,24,74
H-CHClCHCH$_2$	370.7 ± 5.9	63	H$_2$C=CH$_2$	728.3 ± 6	1,74
H-C$_6$F$_5$	476.6	63	CH$_3$-CH$_3$	376.0 ± 2.1	1,74,86
H-CH$_2$Si(CH$_3$)$_3$	415.1 ± 4	99	CH$_3$-CH$_2$CCH	318.0 ± 8	63
H-CSH	399.6 ± 5.0	16	CH$_3$-CH$_2$CCCH$_3$	308.4 ± 6.3	63
H-CH$_2$SH	392.9 ± 8.4	16	CH$_3$-CH(CH$_3$)CCH	305.4	63
H-CH$_2$SCH$_3$	384.9 ± 5.9	49	CH$_3$-C(CH$_3$)CCH$_2$	320.1 ± 9.2	63
H-NH$_2$	452.7 ± 1.3	16	CH$_3$-CH$_2$CHCHCH$_3$	305.0 ± 3.3	63
H-NHCH$_3$	418.4 ± 10.5	63	CH$_3$-CH$_2$C(CH$_3$)CH$_2$	301.2 ± 3.3	91
H-N(CH$_3$)$_2$	382.8 ± 8	63	CH$_3$-CH(CH$_3$)CCCH$_3$	320.9 ± 6.3	63
H-NHC$_6$H$_5$	368.2 ± 8	63	CH$_3$-C(CH$_3$)$_2$CCH	295.8 ± 6.3	63

Table 3
BOND STRENGTHS IN POLYATOMIC MOLECULES (continued)

Bond	D°_{298}/kJ mol^{-1}	Ref.	Bond	D°_{298}/kJ mol^{-1}	Ref.
n-C$_3$H$_7$-CH$_2$CCH	306.3 ± 6.3	63	s-C$_4$H$_9$-N$_2$s-C$_4$H$_9$	195.4	13
CH$_3$-C(CH$_3$)$_2$CHCH$_2$	284.9 ± 6.3	63	t-C$_4$H$_9$-N$_2$t-C$_4$H$_9$	182.0	13
n-C$_3$H$_7$-CH$_2$CHCH$_2$	295.8	96	C$_6$H$_5$CH$_2$-N$_2$CH$_2$C$_6$H$_5$	157.3	13
CH$_3$-C(CH$_3$)$_2$CCCH$_3$	303.3 ± 6.3	63	CF$_3$-N$_2$CF$_3$	231.0	13
CHCCH$_2$-s-C$_4$H$_9$	300.0 ± 6.3	63	CH$_3$-NO	167.4 ± 3.3	63
CH$_3$-CH$_2$C$_6$H$_5$	332.2 ± 4	63	i-C$_3$H$_7$-NO	152.7 ± 13	63
CH$_3$-CH(CH$_3$)C$_6$H$_5$	312.1 ± 6.3	63	t-C$_4$H$_9$-NO	165.3 ± 6.3	63
C$_2$H$_5$-CH$_2$C$_6$H$_5$	294.1 ± 4	63	C$_6$H$_5$-NO	212.5 ± 4	63
CH$_3$-1-Naphthylmethyl	305.0 ± 6.3	63	NC-NO	120.5 ± 10.5	43
CH$_3$-C(CH$_3$)$_2$C$_6$H$_5$	308.4 ± 6.3	63	CF$_3$-NO	179.1 ± 8	63
CHCCH$_2$-CH$_2$C$_6$H$_5$	256.9 ± 8	63	C$_6$F$_5$-NO	208.4 ± 4	63
n-C$_3$H$_7$-CH$_2$C$_6$H$_5$	292.9 ± 4	63	CCl$_3$-NO	134 ± 13	63
CH$_3$-9-Anthracenylmethyl	282.8 ± 6.3	63	t-C$_4$H$_9$-NOt-C$_4$H$_9$	121	21
CH$_3$-9-Phenanthrenylmethyl	305.0 ± 6.3	63	CH$_3$-NO$_2$	254.4	63
CH$_3$-CH(C$_6$H$_5$)$_2$	301 ± 8	63	CH$_2$C(CH$_3$)-NO$_2$	245.2	63
CH$_3$-C(CH$_3$)(C$_6$H$_5$)$_2$	289 ± 8	63	i-C$_3$H$_7$-NO$_2$	246.9	63
CH$_3$-CN	509.6 ± 8	63	t-C$_4$H$_9$-NO$_2$	244.8	63
C$_2$H-CN	602 ± 4	70	C$_6$H$_5$-NO$_2$	298.3 ± 4	63
C$_2$H$_5$-CH$_2$NH$_2$	332.2 ± 8	63	C(NO$_2$)$_3$-NO$_2$	169.5 ± 4	63
CH$_3$-CH$_2$CN	336.4 ± 4	93	CH$_3$-OC(CH$_3$)$_2$	277.4	101
C$_2$H$_5$-CH$_2$CN	321.7 ± 7.1	63	CH$_3$-OC$_6$H$_5$	238 ± 8	73
CH$_3$-CH(CH$_3$)CN	329.7 ± 8	63	CH$_3$-OCH$_2$C$_6$H$_5$	280.3	26
C$_2$H$_5$-CH$_2$CN	321.7 ± 7.1	63	C$_2$H$_5$-OC$_6$H$_5$	264 ± 6.3	63
CH$_3$-C(CH$_3$)$_2$CN	312.5 ± 6.7	63	CH$_2$CHCH$_2$-OC$_6$H$_5$	208.4 ± 8	63
CH$_3$-C(CH$_3$)(CN)C$_6$H$_5$	250.6	63	O=CO	532.2 ± 0.4	29
C$_6$H$_5$CH$_2$-CH$_2$NH$_2$	284.5 ± 8	63	CH$_3$-O$_2$	137.0 ± 3.8	53
C$_6$H$_5$CH$_2$-C$_5$H$_4$N	362.8	80	C$_2$H$_5$-O$_2$	148.4 ± 8.4	53
CN-CN	536 ± 4	30	CH$_2$CHCH$_2$-O$_2$	76.2 ± 2.1	62
CH$_3$-2-Furylmethyl	314 ± 8	63	i-C$_3$H$_7$-O$_2$	155.4 ± 9.6	53
CH$_3$-COC$_6$H$_5$	355.6 ± 9.2	102	t-C$_4$H$_9$-O$_2$	152.8 ± 7.4	53
C$_6$H$_5$CH$_2$-COCH$_2$C$_6$H$_5$	273.6 ± 8	63	C$_6$H$_5$CH$_2$-O$_2$CCH$_3$	280 ± 8	63
CH$_3$CO-COCH$_3$	282.0 ± 9.6	63	C$_6$H$_5$CH$_2$-O$_2$CC$_6$H$_5$	289	13
C$_6$H$_5$CH$_2$-COOH	280	63	CH$_3$-O$_2$SCH$_3$	279.5	63
C$_6$H$_5$CO-COC$_6$H$_5$	277.8	63	CH$_2$CHCH$_2$-O$_2$SCH$_3$	207.5	63
(C$_6$H$_5$)$_2$CH-COOH	248.5 ± 13	63	C$_6$H$_5$CH$_2$-O$_2$SCH$_3$	221.3	63
CF$_3$-COC$_6$H$_5$	308.8 ± 8	63	CF$_3$-O$_2$CF$_3$	361.5	10
CF$_2$=CF$_2$	319.2 ± 13	103	CH$_2$Cl-O$_2$	122.4 ± 10.5	53
CH$_2$F-CH$_2$F	368 ± 8	51	CHCl$_2$-O$_2$	108.2 ± 8.2	53
CH$_3$-CF$_3$	423.4 ± 4.6	79	CCl$_3$-O$_2$	92.0 ± 6.4	53
CF$_3$-CF$_3$	413.0 ± 10.5	63	CH$_3$CHCl-O$_2$	131.2 ± 1.8	53
C$_6$F$_5$-C$_6$F$_5$	487.9 ± 24.7	78	CH$_3$CCl$_2$-O$_2$	112.2 ± 2.2	53
CH$_3$-BF$_2$	~473	63	(CH$_3$)$_2$CCl-O$_2$	136.0 ± 3.8	53
C$_6$H$_5$-BCl$_2$	~510	63	CH$_3$-SH	312.5 ± 4.2	65
CH$_2$CHCH$_2$-Si(CH$_3$)$_3$	293	63	t-C$_4$H$_9$-SH	286.2 ± 6.3	63
s-C$_4$H$_9$-Si(CH$_3$)$_3$	414	63	C$_6$H$_5$-SH	361.9 ± 8	63
CH$_3$-NHC$_6$H$_5$	298.7 ± 8	63	CH$_3$-SCH$_3$	307.9 ± 3.3	65
C$_6$H$_5$CH$_2$-NH$_2$	297.5 ± 4	63	CH$_3$-SC$_6$H$_5$	290.4 ± 8	63
CH$_3$-N(CH$_3$)C$_6$H$_5$	296.2 ± 8	63	C$_6$H$_5$CH$_2$-SCH$_3$	256.9 ± 8	63
C$_6$H$_5$CH$_2$-NHCH$_3$	287.4 ± 8	63	S-CS	430.5 ± 13	63
C$_6$H$_5$CH$_2$-N(CH$_3$)$_2$	259.8 ± 8	63	F-ĊH$_3$	472	1,61
CH$_2$=N$_2$	<175	58	F-CN	469.9 ± 5.0	63
CH$_3$-N$_2$CH$_3$	219.7	13	F-COF	535 ± 12	18
C$_2$H$_5$-N$_2$C$_2$H$_5$	209.2	13	F-CHFCl	465.3 ± 9.6	97
i-C$_3$H$_7$-N$_2$i-C$_3$H$_7$	198.7	13	F-CF$_2$Cl	490 ± 25	41
n-C$_4$H$_9$-N$_2$n-C$_4$H$_9$	209.2	13	F-CFCl$_2$	462.3 ± 10.0	97
i-C$_4$H$_9$-N$_2$i-C$_4$H$_9$	205.0	13	F-CF$_2$CH$_3$	522.2 ± 8	63

Table 3
BOND STRENGTHS IN POLYATOMIC MOLECULES (continued)

Bond	$D°_{298}$/kJ mol^{-1}	Ref.	Bond	$D°_{298}$/kJ mol^{-1}	Ref.
F-C$_2$F$_5$	530.5 ± 7.5	63	CH$_3$-In(CH$_3$)$_2$	205 ± 17	63
Cl-CN	421.7 ± 5.0	63	CH$_3$-Sn(CH$_3$)$_3$	297 ± 17	63
Cl-COC$_6$H$_5$	310 ± 13	63	C$_2$H$_5$-Sn(C$_2$H$_5$)$_3$	264 ± 17	63
Cl-CSCl	265.3 ± 2.1	71	CH$_3$-Sb(CH$_3$)$_2$	255 ± 17	63
Cl-CF$_3$	360.2 ± 3.3	27	C$_2$H$_5$-Sb(C$_2$H$_5$)$_2$	243 ± 17	63
Cl-CHFCl	354.4 ± 11.7	97	CH$_3$-HgCH$_3$	255 ± 17	63
Cl-CF$_2$Cl	346.0 ± 13.4	97	C$_2$H$_5$-HgC$_2$H$_5$	205 ± 17	63
Cl-CFCl$_2$	305 ± 8	40	CH$_3$-Tl(CH$_3$)$_2$	167 ± 17	63
Cl-CH$_2$Cl	350.2 ± 0.8	97	CH$_3$-Pb(CH$_3$)$_3$	238 ± 17	63
Cl-CHCl$_2$	338.5 ± 4.2	97	C$_2$H$_5$-Pb(C$_2$H$_5$)$_3$	230 ± 17	63
Cl-CCl$_3$	305.9 ± 7.5	63	CH$_3$-Bi(CH$_3$)$_2$	218 ± 17	63
Cl-C$_2$F$_5$	346.0 ± 7.1	27	CO-Cr(CO)$_5$	155 ± 8	60
Cl-CF$_2$CF$_2$Cl	326 ± 8	63	CO-Fe(CO)$_4$	172 ± 8	60
Cl-SiCl$_3$	464	99	CO-Mo(CO)$_5$	167 ± 8	60
Br-CH$_3$	292.9 ± 5.0	39	CO-W(CO)$_5$	192 ± 8	60
Br-C$_6$H$_5$	336.8 ± 8	63	BH$_3$-BH$_3$	146	13
Br-CN	367.4 ± 5.0	63	NH$_2$-NH$_2$	275.3	63
Br-CH$_2$COCH$_3$	261.5	101	NH$_2$-NHCH$_3$	268.2 ± 8	63
Br-COC$_6$H$_5$	268.6	13	NH$_2$-N(CH$_3$)$_2$	246.9 ± 8	63
Br-CHF$_2$	289 ± 8	63	NH$_2$-NHC$_6$H$_5$	218.8 ± 8	63
Br-CF$_3$	296.2 ± 1.3	3	ON-NO$_2$	40.6 ± 2.1	63
Br-CF$_2$CH$_3$	287.0 ± 5.4	76	O$_2$N-NO$_2$	56.9	63
Br-C$_2$F$_5$	287.4 ± 6.3	63	NF$_2$-NF$_2$	88 ± 4	63
Br-n-C$_3$F$_7$	278.2 ± 10.5	63	O-N$_2$	167	1,14
Br-i-C$_3$F$_7$	274.1 ± 4.6	63	O-NO	305	1,14
Br-CH$_2$C$_6$F$_5$	225 ± 6	54	O-NO$_2$	208.7 ± 1.1	31
Br-CHClCF$_3$	274.9 ± 6.3	63	HO-NO	206.3	63
Br-CCl$_3$	231.4 ± 4	63	HO-NO$_2$	206.7	63
Br-CClBrCF$_3$	251.0 ± 6.3	63	HO$_2$-NO$_2$	96 ± 8	63
Br-CH$_2$Br	296.7 ± 1.3	97	CH$_3$O-NO	174.9 ± 3.8	9,11
Br-CHBr$_2$	292.0 ± 8	97	C$_2$H$_5$O-NO	175.7 ± 5.4	8,11
Br-CBr$_3$	235.1 ± 7.5	63	CH$_3$COO$_2$-NO$_2$	118.8 ± 3.0	17
Br-NO$_2$	82.0 ± 7.1	57	n-C$_3$H$_7$O-NO	167.8 ± 7.5	11
Br-NF$_2$	≤222	25	i-C$_3$H$_7$O-NO	171.5 ± 5.4	7,11
I-CHCH$_2$	259.0 ± 4.2	20	n-C$_4$H$_9$O-NO	177.8 ± 6.3	11
I-n-C$_4$H$_9$	205.0 ± 4	63	i-C$_4$H$_9$O-NO	175.7 ± 6.3	11
I-Norbornyl	261.5 ± 10.5	69	s-C$_4$H$_9$O-NO	173.6 ± 3.3	5,11
I-CN	305 ± 4	30	t-C$_4$H$_9$O-NO	171.1 ± 3.3	6,11
I-CF$_3$	227.2 ± 1.3	3	HO-NCHCH$_3$	207.9	13
I-CF$_2$CH$_3$	218.0 ± 4.2	63	Cl-NF$_2$	~134	1,75
I-CH$_2$CF$_3$	235.6 ± 4	63	I-NO	77.8 ± 0.4	42
I-C$_2$F$_5$	218.8 ± 2.9	2	I-NO$_2$	76.6 ± 4	98
I-n-C$_3$F$_7$	208.4 ± 4.2	63	HO-OH	213 ± 4	63
I-i-C$_3$F$_7$	215.1 ± 2.9	2	HO-OCH$_2$C(CH$_3$)$_3$	193.7 ± 7.9	63
I-n-C$_4$F$_9$	205.0 ± 4.2	72	CH$_3$O-OCH$_3$	157.3 ± 8	63
I-C(CF$_3$)$_3$	206	33	C$_2$H$_5$O-OC$_2$H$_5$	158.6 ± 4	63
I-C$_6$H$_5$	273.6 ± 8	63	n-C$_3$H$_7$O-On-C$_3$H$_7$	155.2 ± 4	63
I-C$_6$F$_5$	277.0	63	i-C$_3$H$_7$O-Oi-C$_3$H$_7$	157.7 ± 4	63
C$_5$H$_5$-FeC$_5$H$_5$	381 ± 13	59	s-C$_4$H$_9$O-Os-C$_4$H$_9$	152.3 ± 4	63
CH$_3$-ZnCH$_3$	285 ± 17	63	t-C$_4$H$_9$O-Ot-C$_4$H$_9$	159.0 ± 4	63
C$_2$H$_5$-ZnC$_2$H$_5$	238 ± 17	63	C$_2$H$_5$C(CH$_3$)$_2$O-OC(CH$_3$)$_2$C$_2$H$_5$	164.4 ± 4	63
CH$_3$-Ga(CH$_3$)$_2$	264 ± 17	63	(CH$_3$)$_3$CCH$_2$O-OCH$_2$C(CH$_3$)$_3$	152.3 ± 4	63
C$_2$H$_5$-Ga(C$_2$H$_5$)$_2$	209 ± 17	63	CF$_3$O-OCF$_3$	193.3	63
CH$_3$-Ge(CH$_3$)$_3$	347 ± 17	63	(CF$_3$)$_3$CO-OC(CF$_3$)$_3$	148.5 ± 4.6	63
CH$_3$-As(CH$_3$)$_2$	280 ± 17	63	t-C$_4$H$_9$O-OSi(CH$_3$)$_3$	197	63
CH$_3$-CdCH$_3$	251 ± 17	63	SF$_5$O-OSF$_5$	155.6	63

Table 3
BOND STRENGTHS IN POLYATOMIC MOLECULES (continued)

Bond	$D°_{298}$/kJ mol^{-1}	Ref.	Bond	$D°_{298}$/kJ mol^{-1}	Ref.
t-C_4H_9O-$OGe(C_2H_5)_3$	192	63	O=PF_3	544 ± 21	52
t-C_4H_9O-$OSn(C_2H_5)_3$	192	63	O=PCl_3	510 ± 21	52
$FClO_2$-O	244.3	13	O=PBr_3	498 ± 21	52
CF_3O-O_2CF_3	126.8 ± 8	63	HO-$Si(CH_3)_3$	536	63
SF_5O-O_2SF_5	126.8	63	HS-SH	276 ± 8	63
CH_3CO_2-O_2CCH_3	127.2 ± 8	63	CH_3S-SCH_3	272.8 ± 3.8	65
$C_2H_5CO_2$-$O_2CC_2H_5$	127.2 ± 8	63	F-SF_5	420 ± 10	94
n-$C_3H_7CO_2$-O_2Cn-C_3H_7	127.2 ± 8	63	I-SH	206.7 ± 8	63
O-SO	552 ± 8	29	I-SO	180	63
F-OCF_3	182.0 ± 2.1	28	I-SCH_3	206.3 ± 7.1	87
HO-Cl	251 ± 13	52	I-$Si(CH_3)_3$	322	99
O-ClO	247 ± 13	29	H_3Si-SiH_3	310	63,99
HO-Br	234 ± 13	52	$(CH_3)_3Si$-$Si(CH_3)_3$	336.8	63,99
HO-I	234 ± 13	52	$(C_6H_5)_3Si$-$Si(C_6H_5)_3$	368 ± 29	63,99

REFERENCES

1. A value calculated from one of the thermochemical equations above, taking enthalpy data fro atoms from Table 2 and for radicals from Table 4.
2. Ahonkhai, S. I. and Whittle, E., *Int. J. Chem. Kinet.,* 16, 543, 1984.
3. Asher, R. L. and Ruscic, B., *J. Chem. Phys.,* 106, 210, 1997.
4. Baldwin, D. P., Buntine, M. A., and Chandler, D.W., *J. Chem. Phys.*, 93, 6578, 1990.
5. Batt, L. and McCulloch, R. D., *Int. J. Chem. Kinet.,* 8, 911, 1976.
6. Batt, L. and Milne, R. T. , *Int. J. Chem. Kinet.,* 8, 59, 1976.
7. Batt, L. and Milne, R. T. , *Int. J. Chem. Kinet.,* 9, 141, 1977.
8. Batt, L. and Milne, R. T. , *Int. J. Chem. Kinet.,* 9, 549, 1977.
9. Batt, L. and Milne, R. T., and McCulloch, R. D., *Int. J. Chem. Kinet.,* 9, 567, 1977.
10. Batt, L. and Walsh, R., *Int. J. Chem. Kinet.,* 15, 605, 1983.
11. Batt, L., Christie, K., Milne, R. T., and Summers, A. J., *Int. J. Chem. Kinet.,* 6, 877, 1974.
12. Baulch, D. L., Bowman, C. T., Cobos, C. J., Cox, R. A., Just, Th., Kerr, J. A., Pilling, M. J., Stocker, D., Troe, J., Tsang, W., Walker, R. W., and Warnatz, J., *J. Phys. Chem. Ref. Data,* in press.
13. Benson, S. W. and O'Neal, H. E., *Kinetic data on Gas Phase Unimolecular Reactions,* National Bureau of Standards, NSRDS-NBS, Washington, D. C., 21, 1970.
14. Benson, S. W., *J. Chem. Educ.,* 42, 502, 1965.
15. Benson, S. W., *Thermochemical Kinetics, 2nd ed.*, John Wiley & Sons, New York, 1976.
16. Berkowitz, J., Ellison, G. B., and Gutman, D., *J. Phys. Chem.,* 98, 2744, 1994.
17. Bridier, I., Caralp, F., Loirat, H., Lesclaux, R., Veyret, B., Becker, K. H., Reimer, A., and Zabel, F., *J. Phys. Chem.*, 95, 3594, 1991.
18. Buckley, T. J., Johnson, R. D., Huie, R. E., Zhang, Z., Kuo, S. C., and Klemm, B., *J. Phys. Chem.*, 99, 4879, 1995
19. Burkey, T. J., Majewski, M., and Griller, D., *J. Am. Chem. Soc.,* 108, 2218, 1986.
20. Cao, J. R., Zhang, J. M., Zhong, X., Huang, Y. H., Fang, W. Q., Wu, X. J., and Zhu, Q. H., *Chem. Phys.,* 138, 377, 1989.
21. Carmichael, P. J., Gowenlock, B. G., and Johnson, C. A. F., *J. Chem. Soc. Perkin Trans. 2,* 1853, 1973.
22. Castelhano, A. L. and Griller, D., *J. Am. Chem. Soc.,* 104, 3655, 1982.
23. Chuang, M.-C., Foltz, M. F., and Moore, C. B., *J. Chem. Phys.,* 87, 3855, 1987.
24. Chupka, W. A. and Liftshitz, C., *J. Chem. Phys.,* 48, 1109, 1968.
25. Clyne, M. A. A. and Connor, J., *J. Chem. Soc. Faraday Trans. 1,* 68, 1220, 1972.
26. Colussi, A. J., Zabel, F., and Benson, S. W., *Int. J. Chem. Kinet.,* 9, 161, 1977.
27. Coomber, J. W. and Whittle, E., *Trans. Faraday Soc.,* 63, 2656, 1967.
28. Czarnarski, J., Castellano, E., and Schumacher, H. J., *Chem. Comm.,* p. 1255, 1968.
29. Darwent, D. deB., *Bond Dissociation Energies in Simple Molecules,* National Bureau of Standards, NSRDS-NBS, 31 Washington, D. C., 1970.
30. Davis, D. D. and Okabe, H., *J. Chem. Phys.,* 49, 5526, 1968.
31. Davis, H. F., Kim, B., Johnston, H. S., and Lee, Y. T., *J. Chem. Phys.,* 97, 2172, 1993.
32. Dixon, R. N., *J. Chem. Phys.,* 104, 6905, 1996.
33. Dobychin, S. L., Mashendzhinov, V. I., Mishin, V. I., Semenov, V. N., and Shpak, V. S., *Doklady Akademii Nauk SSSR,* 312, 1166 1991.
34. Doncaster, A. M. and Walsh, R., *J. Phys. Chem.,* 83, 578, 1979.
35. Ellison, G. B., Davico, G. E., Bierbaum, V. M., and DePuy, C. H., *Int. J. Mass Spectrom. Ion Proc.,* 156, 109, 1997.

Table 3
BOND STRENGTHS IN POLYATOMIC MOLECULES (continued)

36. Ellul, E., Potzinger, P., Reimann, B., and Camilleri, P., *Ber. Bunsenges. Phys. Chem.*, 85, 407, 1981.
37. Ervin, K. M., Gronert, S., Barlow, S. E., Gilles, M. K., Harrison, A. G., Bierbaum, V. M., DePuy, C. H., Lineberger, W. C., and Ellison, G. B., *J. Am. Chem. Soc.*, 112, 5750, 1990.
38. Evans, B. S., Weeks, I., and Whittle, E., *J. Chem. Soc. Faraday Trans. 1*, 79, 1471, 1983.
39. Ferguson, K. C., Okafo, E. N., and Whittle, E., *J. Chem. Soc. Faraday Trans. 1*, 69, 295, 1973.
40. Foon, R. and Tait, K. B., *J. Chem. Soc. Faraday Trans. 1*, 68, 104, 1972.
41. Foon, R. and Tait, K. B., *J. Chem. Soc. Faraday Trans. 1*, 68, 1121, 1972.
42. Forte, E., Hippler, H., and van den Bergh, H., *Int. J. Chem. Kinet.*, 13, 1227, 1981.
43. Gowenlock, B. G., Jonhson, C. A. F., Keary, C. M., and Pfaf, J., *J. Chem. Soc. Perkin Trans. 2*, 71, 351, 1975.
44. Grela, M. A. and Colussi, A. J., *Int. J. Chem. Kinet.*, 20, 713, 1988.
45. Griller, D. and Wayner, D. D. M., *Pure Appl. Chem.*, 61, 717, 1989.
46. Heckmann, E., Hippler, H., and Troe, J., *26th Symp. (Int.) Combust.*, Combustion Institute, Pittsburgh, Pennsylvania, pp 543, 1996.
47. Heneghan, S. P. and Benson, S. W., *Int. J. Chem. Kinet.*, 15, 815, 1983.
48. Hudgens, J. W., Johnson, R. D., Timonen, R. S., Seetula, J. A., and Gutman, D., *J. Phys. Chem.*, 95, 4400, 1991.
49. Jefferson, A., Nicovich, J. M., and Wine, P. H., *J. Phys. Chem.*, 98, 7128, 1994.
50. Johnson, R. D. and Hudgens, J. W., *J. Phys Chem.*, 100, 19874, 1996.
51. Kerr, J. A. and Timlin, D. M., *Int. J. Chem. Kinet.*, 3, 427, 1971.
52. Kerr, J. A., *Chem. Rev.*, 66, 465, 1966.
53. Knyazev, V. D. and Slagle, I. R., *J. Phys. Chem.*, 102, 1770, 1998.
54. Kominar, R. J., Krech, M. J., and Price, S. J. W., *Can. J. Chem.*, 58, 1906, 1980.
55. Kondo, O. and Benson, S. W., *Int. J. Chem. Kinet.*, 16, 949, 1984.
56. Kondo, O. and Benson, S. W., *J. Phys. Chem.*, 88, 6675, 1984.
57. Kreutter, K. D., Nicovich, J. M., and Wine, P. H., *J. Phys. Chem.*, 95, 4020, 1991.
58. Laufer, A. H. and Okabe, H., *J. Am. Chem. Soc.*, 93, 4137, 1971.
59. Lewis, K. E. and Smith, G. P., *J. Am. Chem. Soc.*, 106, 4650, 1984.
60. Lewis, K. E., Golden, D. M., and Smith, G. P., *J. Am. Chem. Soc.*, 106, 3905, 1984.
61. Lias, S. G., Bartmess, J. E., Liebman, J. F., Holmes, J. L., Levin, R. D., and Mallard, W. G., *J. Phys. Chem. Ref. Data 17*, Suppl. 1, 1988.
62. Lightfoot, P. D., Cox, R. A., Crowley, J. N., Destriau, M., Hayman G. D., Jenkin, M. E., Mootrgat, G. K., and Zabel, F., *Atmos. Environ.*, 26A, 1805, 1992.
63. McMillen, D. F. and Golden, D. M., *Ann. Rev. Phys. Chem.*, 33, 493, 1982.
64. Miyokawa, K. and Tschuikow-Roux, E., *J. Phys. Chem.*, 96, 7328, 1992.
65. Nicovich, J. M., Kreutter, K. D., van Dijk, C. A., and Wine, P. H., *J. Phys. Chem.*, 96, 2518, 1992.
66. Nicovich, J. M., Wang, S., McKee, M. L., and Wine, P. H., *J. Phys. Chem.*, 100, 680, 1996.
67. Niiranen, J. T., Gutman, D., and Krasnoperov, L. N., *J. Phys. Chem.*, 96, 5881, 1992.
68. Noble, P. N. and Walsh, R., *Int. J. Chem. Kinet.*, 15, 561, 1983.
69. O'Neal, H. E., Bagg, J. W., and Richardson, W. H., *Int. J. Chem. Kinet.*, 2, 493, 1970.
70. Okabe, H. and Dibeler, V. H., *J. Chem. Phys.*, 59, 2430, 1973.
71. Okabe, H., *J. Chem. Phys.*, 66, 2058, 1977.
72. Okafo, E. N. and Whittle, E., *Int. J. Chem. Kinet.*, 7, 287, 1975.
73. Paul, S. and Back, M. H., *Can J. Chem.*, 53, 3330, 1975.
74. Pedley, J. B. and Rylance, J., "Sussex - N.P.L. Computer Analysed Thermochemical Data; Organic and Organometallic Compounds", University of Sussex, 1977.
75. Petry, R. C., *J. Am. Chem. Soc.*, 89, 4600, 1967.
76. Pickard, J. M. and Rodgers, A. S., *Int. J. Chem. Kinet.*, 9, 759, 1977.
77. Pickard, J. M. and Rodgers, A. S., *Int. J. Chem. Kinet.*, 15, 569, 1983.
78. Price, S. J. W. and Sapiano, H. J., *Can. J. Chem.*, 57, 1468, 1979.
79. Rogers, A. S. and Ford, W. G. F., *Int. J. Chem. Kinet.*, 5, 965, 1973.
80. Rossi, M., McMillen, D. F., and Golden, D. M., *J. Phys. Chem.*, 88, 5031, 1984.
81. Russell, J. J., Senkan, S. M., Seetula, J. A., and Gutman, D., *J. Phys. Chem.*, 93, 5184, 1989.
82. Seetula, J. A., and Slagle, I. R., *J. Chem. Soc. Faraday Trans.*, 93, 1709, 1997.
83. Seetula, J. A., Feng, Y., Gutman, D., Seakins, P. W., and Pilling, M. J., *J. Phys. Chem.*, 95, 1658, 1991.
84. Seetula, J. A., *J. Chem. Soc. Faraday Trans.*, 92, 3069, 1996.
85. Seetula, J. A., *J. Chem. Soc. Faraday Trans.*, 94, 891, 1998.
86. Seetula, J. A., Russell, J. J., and Gutman, D., *J. Am. Chem. Soc.*, 112, 1347, 1990.
87. Shum, L. G. S. and Benson, S. W., *Int. J. Chem.*, 15, 433, 1983.
88. Shum, L. G. S. and Benson, S. W., *J. Phys. Chem.*, 87, 3479, 1983.
89. Stein, S. E., SRD Thermochemical Database, 25. N.I.S.T. Structures and Properties Database and Estimation Program, U.S. Department of Commerce, 1992.

Table 3
BOND STRENGTHS IN POLYATOMIC MOLECULES (continued)

90. Steinkruger, F. J. and Rowland, F. S., *J. Phys. Chem.*, 85, 136, 1981.
91. Trenwith, A. B. and Wrigley, S. P., *J. Chem. Soc. Faraday Trans. 1*, 73, 817, 1977.
92. Trenwith, A. B., *J. Chem. Soc. Faraday Trans. 1*, 78, 3131, 1982.
93. Trenwith, A. B., *J. Chem. Soc. Faraday Trans. 1*, 79, 2755, 1983.
94. Tsang, W. and Herron, J. T., *J. Chem. Phys.*, 96, 4272, 1992.
95. Tsang, W., *Int. J. Chem. Kinet.*, 5, 929, 1973.
96. Tsang, W., *Int. J. Chem. Kinet.*, 10, 1119, 1978.
97. Tschuikow-Roux, E. and Paddison, S., *Int. J. Chem. Kinet.*, 19, 15, 1987.
98. van den Bergh, H. and Troe, J., *J. Chem. Phys.*, 64, 736, 1976.
99. Walsh, R., *Acc. Chem. Res.*, 14, 246, 1981.
100. White, J. N. and Gardiner, W. C., *Chem. Phys. Lett.*, 58, 470, 1978.
101. Zabel, F., Benson, S. W., and Golden, D. M., *Int. J. Chem. Kinet.*, 10, 295, 1978.
102. Zhao, H.-Q., Cheung, Y.-S., Liao, C.-L., Liao, C.-X., Ng, C. Y., and Li, W.-K., *J. Chem. Phys.*, 107, 7230, 1997.
103. Zmbov, K. F., Uy, O. M., and Margrave, J. L., *J. Am. Chem. Soc.*, 90, 5090, 1968.

Table 4
ENTHALPIES OF FORMATION OF FREE RADICALS

The enthalpies of formation of the free radicals are related to the corresponding bond strengths by the equations

$$D°(R\text{-}X) = \Delta_f H°(R) + \Delta_f H°(X) - \Delta_f H°(RX)$$

or

$$D°(R\text{-}R) = 2\Delta_f H°(R) - \Delta_f H°(RR)$$

For an excellent review of the methods of determining the enthalpies of formation of free radicals the reader is referred to "Thermochemistry of Free Radicals" by H. E. O'Neal and S. W. Benson in *Free Radicals*, Kochi, J. K., Ed., John Wiley & Sons, New York, 1973, 275 and the article by J. Berkowitz, G. B. Ellison, and D. Gutman, *J. Phys. Chem.*, 98, 2744, 1994.

Radical	$\Delta_f H°_{298}$/kJ mol^{-1}	Ref.	Radical	$\Delta_f H°_{298}$/kJ mol^{-1}	Ref.
CH	596.4 ± 1.2	6	Spiropentyl	380.7 ± 4	48
CH_2(triplet)	390.4 ± 4	15,34	Cyclopent-1-en-3-yl	160.7 ± 4	48
CH_2(singlet)	428.3 ± 4	15	CH_2=CHCH=CHCH$_2$	205 ± 13	48
CH_3	146.4 ± 0.4	10	$(C_2H_3)_2$CH	205 ± 13	48
CH≡C	566.1 ± 2.9	5,27	CH_3C≡CCHCH$_3$	272.8 ± 9.6	48
CH_2=CH	300.0 ± 3.4	27,64	CH≡CC(CH$_3$)$_2$	257.3 ± 8.4	48
C_2H_5	120.9 ± 1.6	10	CH_2=CHC(CH$_3$)$_2$	77.4 ± 6.3	48
Cycloprop-2-en-1-yl	439.7 ± 17.2	48	Cyclopentyl	107.1 ± 2.5	17
CH≡CCH$_2$	340.6 ± 8.4	48	$(CH_3)_3$CCH$_2$	36.4 ± 8	48
CH_2=CHCH$_2$	170.7 ± 8.8	10,26	C_2H_5C(CH$_3$)$_2$	32.2 ± 6.3	71
CH_3CH=CH	262.7	77	C_6H_5	338 ± 3	35
Cyclopropyl	279.9 ± 1.1	48	Cyclohexa-1,3-dien-5-yl	197 ± 21	48
n-C$_3$H$_7$	100.8 ± 2.1	65	Cyclohexyl	58.2 ± 4	48
i-C$_3$H$_7$	86.6 ± 2.0	65	CH_3C≡CC(CH$_3$)$_2$	221.8 ± 9.6	48
CH_3C≡CCH$_2$	293.7	48	$(CH_3)_2$C=C(CH$_3$)CH$_2$	39.8 ± 6.3	48
CH_2=CHCHCH$_3$	125.5 ± 6.3	48	CH_2=C(CH$_3$)C(CH$_3$)$_2$	37.7 ± 6.3	48
CH≡CCHCH$_3$	295.0 ± 9.2	48	C_6H_5CH$_2$	208.0 ± 2.5	26
Cyclobutyl	214.2 ± 4.2	48	Cyclohepta-1,3,5-trien-7-yl	271.1 ± 8	48
Cyclopropylmethyl	213.8 ± 6.7	48	CH_3CH$_2$CH$_2$C(CH$_3$)$_2$	3.4 ± 8.4	69
CH_2=C(CH$_3$)CH$_2$	127.2 ± 5.4	48	Norbornyl	136.4 ± 10.5	48
CH_3CH=CHCH$_2$	125.5 ± 6.3	48	Cycloheptyl	51.1 ± 4	48
n-C$_4$H$_9$	80.9 ± 2.2	65	C_6H_5CHCH$_3$	169.0	48
i-C$_4$H$_9$	72.7 ± 2.2	65	C_6H_5C(CH$_3$)$_2$	134.7	48
s-C$_4$H$_9$	66.7 ± 2.1	65	1-Naphthylmethyl	252.7	48
t-C$_4$H$_9$	51.8 ± 1.3	65	$(C_6H_5)_2$CH	289	63
Cyclopenta-1,3-dien-5-yl	263.0	8	9,10-Dihydroanthracen-9-yl	256.9 ± 6.3	48

Table 4
ENTHALPIES OF FORMATION OF FREE RADICALS (continued)

Radical	$\Delta_f H°_{298}$/kJ mol^{-1}	Ref.	Radical	$\Delta_f H°_{298}$/kJ mol^{-1}	Ref.
9-Anthracenylmethyl	337.6	48	CF_3CH_2	-517.1 ± 5.0	48
9-Phenanthrenylmethyl	311.3	48	C_2F_5	-892.9 ± 4	48
CH_2CN	243.1 ± 11.3	10	$CCl=CH_2$	>251	64
CH_2NC	326.4 ± 11.3	10	CH_3CHCl	76.5 ± 1.6	67
CH_3CHCN	209.2 ± 9.6	48	CH_3CCl_2	48.4 ± 7.6	40,67
$(CH_3)_2CCN$	166.5 ± 8.4	48	CH_2CH_2Cl	93.0 ± 3.4	68
$C_6H_5C(CH_3)CN$	248.5	48	$CHCl_2CCl_2$	23.4 ± 8	48
CH_2NH_2	149.4 ± 8	48	CF_2ClCF_2	-686 ± 17	48
CH_3NHCH_2	126 ± 8	48	C_2Cl_5	33.5 ± 5.4	54
$(CH_3)_2NCH_2$	109 ± 8	48	C_6F_5	-547.7 ± 8	48
CN	441.4 ± 4.6	10	$(CH_3)_3SiCH_2$	-34.7	48
CHN_2	494.42	30	CS	278.5 ± 3.8	62
CHO	43.1	9,22	HCS	300.4 ± 8.4	10
CHCO	175.3 ± 8.4	10	CH_2SH	151.9 ± 8.4	10
CH_3CO	-10.0 ± 1.2	56	CH_3SCH_2	136.8 ± 5.9	38
$CH_2=CHCO$	72.4	48	NH	352.3 ± 9.6	60
C_2H_5CO	-32.3	8	NH_2	188.7 ± 1.3	10
C_6H_5CO	123.0 ± 9.6	78	HNO	112.9	25
CH_2CHO	10.5 ± 9.2	10	NF_2	34 ± 4	48
CH_3COCH_2	-23.9 ± 10.9	48	N_2H_3	243.5	33
$CH_3COCHCH_3$	-70.3 ± 7.1	48	N_3	469 ± 21	48
CH_3OCH_2	-0.1	8	CH_2NH	104.6 ± 13	33
$C_2H_5OCHCH_3$	-84.5	42	CH_3NH	177.4 ± 8	48
Tetrahydrofuran-2-yl	-18.0 ± 6.3	48	$(CH_3)_2N$	145.2 ± 8	48
CH_2OH	-17.8 ± 1.3	39	C_6H_5NH	237.2 ± 8	48
CH_2CH_2OH	-36.0	31	$C_6H_5NCH_3$	233.5 ± 8	48
$CH=CHOH$	113.0	31	NCO	127.0	13,79
CH_3CHOH	-51.6	8	CNO	407.01	41
$CH_2=CHCHOH$	0.0	48	CH_3N_2	215.5 ± 7.5	2
$(CH_3)_2COH$	-111.3 ± 4.6	48	$C_2H_5N_2$	187.4 ± 10.5	2
COOH	-217 ± 10	32	$i\text{-}C_3H_7N_2$	158.6 ± 9.2	2
$COOCH_3$	-169.0 ± 4	48	OH	39.3	48
$C_6H_5COOCH_2$	-69.9 ± 8.4	48	CH_3O	17.2 ± 3.8	10
CF	261.5 ± 4.6	3	C_2H_5O	-15.5 ± 3.4	10
CHF	143.1 ± 12.6	61	$n\text{-}C_3H_7O$	-41.4	48
CH_2F	-31.8 ± 8.4	59	$i\text{-}C_3H_7O$	-52.3	48
FCO	-152.1 ± 12	14	$n\text{-}C_4H_9O$	-62.8	48
CHF_2	-238.9 ± 4	59	$s\text{-}C_4H_9O$	-69.5 ± 3.3	48
CF_2	-184.1 ± 8.4	61	$t\text{-}C_4H_9O$	-90.8	48
CF_3	-466.1 ± 3.8	3	C_6H_5O	47.7	48
CHCl	336.4 ± 11.7	61	CF_3O	-655.6	7
CH_2Cl	117.3 ± 3.1	67	FO	109 ± 10	21
CFCl	31.0 ± 13.4	61	ClO	101.63 ± 0.1	1
CHFCl	-60.7 ± 10.0	73	BrO	125.8 ± 2.4	19
CF_2Cl	-279.1 ± 8.3	49	IO	126 ± 18	20
ClCO	-21.8	55	HO_2	14.6	70
$CHCl_2$	89.0 ± 3.0	67	CH_3O_2	9.0 ± 5.1	40
$CFCl_2$	-89.1 ± 10.0	73	$C_2H_5O_2$	-27.4 ± 9.9	40
CCl_2	230.1 ± 8.4	61	$CH_2=CHCH_2O_2$	87.9 ± 5.5	43
CCl_3	71.1 ± 2.5	37	$i\text{-}C_3H_7O_2$	-68.8 ± 11.3	16,40
CH_2Br	169.0 ± 4.2	73	$t\text{-}C_4H_9O_2$	-101.0 ± 9.2	40
$CHBr_2$	188.3 ± 9.2	73	$HOCH_2O_2$	-162.1 ± 2.1	43
CBr_3	207.1 ± 8	73	CF_3O_2	-614.0	43
CH_2I	230.1 ± 6.7	48	CF_2ClO_2	-406.5	43
CHI_2	333.9 ± 9.2	48	$CFCl_2O_2$	-213.7	43
CH_3CF_2	-302.5 ± 8	48	CH_2ClO_2	-5.1 ± 13.6	40

Table 4
ENTHALPIES OF FORMATION OF FREE RADICALS (continued)

Radical	$\Delta_f H°_{298}$/kJ mol^{-1}	Ref.	Radical	$\Delta_f H°_{298}$/kJ mol^{-1}	Ref.
CHCl$_2$O$_2$	-19.2 ± 11.2	40	SiF$_3$	-1025	48,76
CCl$_3$O$_2$	-20.9 ± 8.9	40	SiCl	195.8	48,76
CH$_3$CHClO$_2$	-54.7 ± 3.4	40	SiCl$_2$	-163.6	48,76
CH$_3$CCl$_2$O$_2$	-63.8 ± 9.8	40	SiCl$_3$	-318	48,76
CH$_3$CO$_2$	-207.5 ± 4	48	SiH$_3$SiH	269.9 ± 14.6	74
CH$_3$COO$_2$	-172 ± 20	12	Si$_2$H$_5$	223.0	48,76
C$_2$H$_5$CO$_2$	-228.5 ± 4	48	PH$_2$	138.5 ± 2.5	10
n-C$_3$H$_7$CO$_2$	-249.4 ± 4	48	HS	143.0 ± 2.8	53
FO$_2$	26.1	45,58	CH$_3$S	124.6 ± 1.8	53
ClO$_2$	97.5	4,52	C$_6$H$_5$S	229.7 ± 8	48
OClO	95.6	28,51	SO	5.0	18
NO$_3$	73.7 ± 1.4	24	HSO	-4	44
sym-ClO$_3$	232.6	23	HSO$_2$	-222	11
SiH	377	48,76	CH$_3$SO$_2$	-239.3	57
SiH$_2$	269.0 ± 1.3	29,47	HOSO$_2$	-385	46
SiH$_3$	200.5 ± 2.5	66	SO$_3$	-395.7	75
CH$_3$Si	310	76	SF$_4$	-746 ± 12	72
CH$_3$SiH	213.0 ± 14.6	74	SF$_5$	-879.9 ± 20	72
CH$_3$SiH$_2$	152.7	48,76	CH$_3$S$_2$	68.6 ± 8	36
(CH$_3$)$_2$Si	109	76	C$_2$H$_5$S$_2$	43.5 ± 8	36
(CH$_3$)$_2$SiH	59.8	48,76	i-C$_3$H$_7$S$_2$	13.8 ± 8	36
(CH$_3$)$_3$Si	-3.3	48,76	t-C$_4$H$_9$S$_2$	-19.3 ± 8	36
C$_6$H$_5$Si(CH$_3$)$_2$	163	57	HOCS$_2$	110.5	50
(C$_6$H$_5$)$_2$SiCH$_3$	326	57	GeH$_3$	222 ± 8	10
(C$_6$H$_5$)$_3$Si	486.2	57	AsH$_2$	167.8 ± 1.3	10
SiF	-19.3	48,76	HSe	144.8 ± 2.1	10
SiF$_2$	-587.9	48,76	SbH$_2$	215.5 ± 2.5	10
			HTe	158.6 ± 5.0	10

REFERENCES

1. Abramowitz, S. and Chase, M. W., *Pure Appl. Chem.*, 63, 1448, 1991.
2. Acs, G. and Peter, A., *Int. J. Chem. Kinet.*, 19, 929, 1987.
3. Asher, R. L. and Ruscic, B., *J. Chem. Phys.*, 106, 210, 1997.
4. Baer, S., Hippler, H., Rabu, R., Siefke, M., Seitzinger, N., and Troe, J., *J. Chem. Phys.*, 95, 6463, 1991.
5. Baldwin, D. P., Buntine, M. A., and Chandler, D. W., *J. Chem. Phys.*, 93, 6578, 1990.
6. Based on $D°$(C-H), see Table 1 and $\Delta_f H°$(C) and $D_f H°$(H), see Table 2.
7. Batt, L. and Walsh, R., Int. J. Chem. Kinet., 14, 933, 1982.
8. Baulch, D. L., Bowman, C. T., Cobos, C. J., Cox, R. A., Just, Th., Kerr, J. A., Pilling, M. J., Stocker, D., Troe, J., Tsang, W., Walker, R. W., and Warnatz, J., *J. Phys. Chem. Ref. Data,* in press.
9. Becerra,R., Carpenter, I. W., and Walsh, R., *J. Phys. Chem.*, A 101, 4185, 1997.
10. Berkowitz, J., Ellison, G. B., and Gutman, D., *J. Phys. Chem.*, 98, 2744, 1994.
11. Boyd, R. J., Gupta, A., Langler, R. F., Lownie, S. P., and Pincock, J. A., *Can. J. Chem.*, 58, 331, 1980.
12. Bridier, I., Caralp, F., Loirat, H., Lesclaux, R., Veyret, B., Becker, K. H., Reimer, A., and Zabel, F., *J. Phys. Chem.*, 95, 3594, 1991.
13. Brown, S. S., Berghout, H. L., and Crim, F. F., *J. Chem. Phys.*, 105, 8103, 1996.
14. Buckley, T. J., Johnson, R. D., Huie, R. E., Zhang, Z., Kuo, S. C., and Klemm, B., *J. Phys. Chem.*, 99, 4879, 1995.
15. Bunker, P. R. and Sears, T. J., *J. Chem. Phys.,* 83, 4866, 1985.
16. Calculated taking $\Delta_f H°$(i-C$_3$H$_7$) = 86.6 kJ mol^{-1}.
17. Castelhano, A. L. and Griller, D., *J. Am. Chem. Soc.*, 104, 3655, 1982; value adjusted to $\Delta_f H°$(CH$_3$) = 146 kJ mol^{-1} and error limits assigned here.
18. Chase, M. W. Jr., Davies, C. A., Downey, J.R. Jr., Frurip, D. J., McDonald, R. A., and Syverud, A. N., *J. Phys. Chem. Ref. Data,* 14, Suppl. 1, 1985.
19. Chase, M. W., *J. Phys. Chem. Ref. Data,* 25, 1069, 1996.
20. Chase, M. W., *J. Phys. Chem. Ref. Data,* 25, 1297, 1996.
21. Chase, M. W., *J. Phys. Chem. Ref. Data,* 25, 551, 1996.
22. Chuang, M.-C., Foltz, M. F., and Moore, C. B., *J. Chem. Phys.*, 87, 3855, 1987.
23. Colussi, A. J., *J. Phys. Chem.*, 94, 8922, 1990.
24. Davis, H. F., Kim, B., Johnston, H. S., and Lee, Y. T., *J. Phys. Chem.*, 97, 2172, 1993.

Table 4
ENTHALPIES OF FORMATION OF FREE RADICALS (continued)

25. Dixon, R. N., *J. Chem. Phys.*, 104, 6905, 1996.
26. Ellison, G. B., Davico, G. E., Bierbaum, V. M., and DePuy, C. H., *Int. J. Mass Spectrom. Ion Proc.*, 156, 109, 1996.
27. Ervin, K. M., Gronert, S., Barlow, S. E., Gilles, M. K., Harrison, A. G., Bierbaum, V. M., DePuy, C. H., Lineberger, W. C., and Ellison, G. B., *J. Am. Chem. Soc.*, 112, 5750, 1990.
28. Flesch, R., E. Rühl, K. Hottmann, and H. Baumgartel, *J. Phys. Chem.*, 97, 837 (1993).
29. Frey, H. M., Walsh, R., and Watts, I. M., *J. Chem. Soc. Chem. Comm.*, 1189, 1986.
30. Fulle, D. and Hippler, H., *J. Chem. Phys.,* 105, 5423, 1996.
31. Fulle, D., Hamann, H. F., Hippler, H., and Jansch, C. P., *Ber. Bunsenges. Phys. Chem.,* 101, 1433, 1997.
32. Fulle, D., Hamann, H. F., Hippler, H., and Troe, *J. Chem. Phys.,* 105, 983, 1996.
33. Grela, M. A. and Colussi, A. J., *Int. J. Chem. Kinet.*, 20, 713, 1988.
34. Gurvich, l.V., Veyts, I.V., Alcock, C.B., *Thermodynamic Properties of Individual Substances, 4th ed.*, Hemisphere, New York, 1991, Vol. 2.
35. Heckmann, E., Hippler, H., and Troe, J., *26th Symp. (Int.) Combust.,* The Combustion Institute, Pittsburgh, Pennsylvania, pp 543, 1996.
36. Howari, J. A. Griller, D., and Lossing, F. P., *J. Am. Chem. Soc.*, 108, 3273, 1986.
37. Hudgens, J. W., Johnson, R. D., Timonen, R. S., Seetula, J. A., and Gutman, D., *J. Phys. Chem.*, 95, 4400, 1991.
38. Jefferson, A., Nicovich, J. M., and Wine, P. H., *J. Phys. Chem.*, 98, 7128, 1994.
39. Johnson, R. D. and Hudgens, J. W., *J. Phys. Chem.,* 100, 19874, 1996.
40. Knyazev, V. D., Bencsura, A., Dubinsky, I. A., Gutman, D., Melius, C. F., and Senkan, S. M., *J. Phys. Chem.*, 99, 230, 1995.
41. Koch, W. and Frenking, G., *J. Phys. Chem.,* 91, 49, 1987.
42. Kondo, O. and Benson, S. W., *Int. J. Chem. Kinet.*, 16, 949, 1984.
43. Lightfoot, P. D., Cox, R. A., Crowley, J. N., Destriau, M., Hayman, G.D., Jenkin, M. E., Moortgat, G. K., and Zabel, F., *Atmos Environ.*, 26A, 1805, 1992.
44. Lovejoy, E. R., Wang, N. S., and Howard, C. J., *J. Phys. Chem.,* 91, 5749, 1987.
45. Lyman, J. L. and Holland, R., *J. Phys. Chem.*, 92, 7232, 1988.
46. Margitan, J. J., *J. Phys. Chem.*, 88, 3314, 1984.
47. Martin, J. G., Ring, M. A., and O'Neal, H. E., *Int. J. Chem. Kinet.*, 19, 715, 1987.
48. McMillen, D. F. and Golden, D. M., *Ann. Rev. Phys. Chem.*, 33, 493, 1982.
49. Miyokawa, K. and Tschuikow-Roux, E., *J. Phys. Chem.,* 96, 7328, 1992.
50. Murrells, T. P., Lovejoy, E.R., and Ravishankara, A. R., *J. Phys. Chem.*, 80, 4065, 1984.
51. Nickolaisen, S. L., R. R. Friedl, and S. P. Sander, *J. Phys. Chem.*, 98, 155 (1994).
52. Nicovich, J. M., Kreutter, K. D., Shockelford, C. J., and Wine, P. H., *Chem. Phys. Lett.*, 179, 367, 1991.
53. Nicovich, J. M., Kreutter, K. D., van Dijk, C. A., and Wine, P. H., *J. Phys. Chem.*, 96, 2518, 1992.
54. Nicovich, J. M., Wang, S., McKee, M. L., and Wine, P. H., *J. Phys. Chem.*, 100, 680, 1996.
55. Nicovich, J.M., Kreutter, K. D., and Wine, P.H., *J. Chem. Phys.*, 92, 3539 (1990).
56. Niiranen, J. T., Gutman, D., and Krasnoperov, L. N., *J. Phys. Chem.*, 96, 5881, 1992.
57. O'Neal, H. E. and Benson, S. W., in *Free Radicals*, Kochi, J. K., Ed., John Wiley & Sons, New York, 1973, 275.
58. Pagsberg, P., Ratajczak, E., Sillesen, A., and Jodkowski, J. T., *Chem. Phys. Lett.*, 141, 88, 1987.
59. Pickard, J. M. and Rodgers, A. S., *Int. J. Chem. Kinet.*, 15, 569, 1983.
60. Piper, L. G., *J. Chem. Phys.*, 70, 3417, 1979.
61. Poutsma, J. C., Paulino, J. A., and Squires, R. R., *J. Phys. Chem. A*, 101, 5327, 1997.
62. Prinslow, D. A. and Armentrout, P. B., *J. Chem. Phys.*, 94, 3563, 1991.
63. Rossi, M., McMillen, D. F., and Golden, D. M., *J. Phys. Chem.*, 88, 5031, 1984.
64. Russell, J. R., Senkan, S. M., Seetula, J. A., and Gutman, D., *J. Phys. Chem.*, 93, 5184, 1989.
65. Seetula, J. A. and Slagle I. R., *J. Chem. Soc. Faraday Trans.*, 93, 1709, 1997.
66. Seetula, J. A., Feng, Y., Gutman, D., Seakins, P. W., and Pilling, M. J., *J. Phys. Chem.,* 95, 1658, 1991.
67. Seetula, J. A., *J. Chem. Soc. Faraday Trans.,* 92, 3069, 1996.
68. Seetula, J. A., *J. Chem. Soc. Faraday Trans.,* 94, 891, 1998.
69. Seres, L., Gorgenyi, M., and Farkas, J., *Int. J. Chem. Kinet.*, 15, 1133, 1983.
70. Shum, L. G. S. and Benson, S. W., *J. Phys. Chem.*, 87, 3479, 1983.
71. Stein, S. E., SRD Thermochemical Database, 25. N.I.S.T. Structures and Properties Database and Estimation Program, U.S. Department of Commerce, 1992.
72. Tsang, W. and Herron, J. T., *J. Chem. Phys.*, 96, 4272, 1992.
73. Tschuikow-Roux, E. and Paddison, S., *Int. J. Chem. Kinet.*, 19, 15, 1987.
74. Vanderwielen, A. J., Ring, M. A., and O'Neal, H. E., *J. Am. Chem. Soc.,* 97, 993, 1975.
75. Wagman, D. D., Evans, W. H., Parker, V. B., Schumm, R. H., Halow, I., Bailey, S. M., Churney, K. L., and Nuttall, R.L., *J. Phys. Chem. Ref. Data*, 11, Suppl. 2, 1978.
76. Walsh, R., *Acc. Chem. Res.*, 14, 246, 1981.
77. Wu, C. H. and Kern, R. D., *J. Phys. Chem.,* 91, 6291, 1987.
78. Zhao, H.-Q., Cheung, Y.-S., Liao, C.-L., Liao, C.-X., Ng, C. Y., and Li, W.-K., *J. Chem. Phys.,* 107, 7230, 1997.
79. Zyrianov, M., Droz-Georget, T., Sanov, A., and Reisler, H., *J. Chem. Phys.*, 105, 8111, 1996.

Table 5
BOND STRENGTHS OF SOME ORGANIC MOLECULES

Bond strengths at 298 K expressed in kJ/mol^{-1} for some organic molecules of the general formula R-X are presented below. Some are experimental values taken from the preceding tables; the remainder are calculated from the enthalpies of formation of atoms (Table 2) and of radicals (Table 4), and the enthalpies of formation of the parent compounds from sources indicated by the references below. The table also includes bond strengths for the inorganic molecules, hydrogen, the hydrogen halides, water and ammonia.

	H	F	Cl	Br	I	OH	NH$_2$	CH$_3$O	CH$_3$	CH$_3$CO	NO	CF$_3$	CCl$_3$
H	435.990	569.87	431.62	366.35	298.407	498	453	436	439	374	195	450	393
CH$_3$	439	472	350[e]	293	239[e]	387[e]	358[e]	348[e]	377[e]	354[e]	167	423	362[e]
C$_2$H$_5$	423	463[d]	354[e]	295[e]	236[e]	395[e]	357[e]	355[e]	371[e]	349[e]	—	—	—
i-C$_3$H$_7$	409	460[e]	353[e]	298[e]	234[e]	399[e]	359[e]	356[e]	367[e]	339[e]	153	—	—
t-C$_4$H$_9$	404	—	355[e]	296[e]	231[e]	404[e]	362[e]	353[e]	366[e]	332[e]	165	—	—
C$_6$H$_5$	473	533[e]	407[e]	346[f]	280[e]	474[e]	439[e]	423[e]	434[e]	415[e]	213	—	—
C$_6$H$_5$CH$_2$	376	—	310[e]	256[f]	215[f]	348[e]	302[e]	—	332	297[e]	—	—	—
CCl$_3$	393	419[e]	288[e]	224[e]	167[e]	—	—	—	362[e]	—	134	335[b]	286[e]
CF$_3$	450	547[e]	362[e]	294[e]	227	—	—	—	423	—	179	413	335[b]
C$_2$F$_5$	430	531[e]	347[e]	283[e]	219	—	—	—	—	—	—	—	—
CH$_3$CO	374	512[e]	354[f]	292[e]	223[e]	462[e]	417[e]	421[e]	354[e]	282	—	—	—
CN	528	470	422	367	305	—	—	—	514[e]	—	121	—	—
C$_6$F$_5$	473	487[e]	383[e]	—	277[a]	499[e]	—	—	441[e]	—	208[a]	—	—

REFERENCES

[a] Choo, K. Y., Mendenhall, G. D., Golden, D.M., and Benson, S. W., *Int. J. Chem. Kinet.*, 6, 813, 1974.

[b] Kolesov, V. P. and Papina, T. S., *Russ. Chem. Rev.*, 52, 425, 1983.

[c] Kudchadker, S. A. and Kudchadker, A. P., *J. Phys. Chem. Ref. Data*, 1, 1285, 1978.

[d] Lias, S. G., Bartmess, J. E., Liebman, J. F., Holmes, J. L., Levin, R. D., and Mallard, W. G., *J. Phys. Chem. Ref. Data*, 17, Suppl. No. 1, 1988.

[e] Lide, D. R., Ed., *Handbook of Chemistry and Physics, 80th Edition,* CRC Press, Boca Raton, FL, 1999.

[f] Pedley, J. B. and Rylance, J., *Sussex.N.P.L. Computer Analysed Thermochemical Data: Organic and Organometallic Compounds*, University of Sussex, 1977.

ELECTRONEGATIVITY

Electronegativity is a parameter originally introduced by Pauling which describes, on a relative basis, the tendency of an atom in a molecule to attract bonding electrons. While electronegativity is not a precisely defined molecular property, the electronegativity difference between two atoms provides a useful measure of the polarity and ionic character of the bond between them. This table gives the electronegativity X, on the Pauling scale, for the most common oxidation state. Other scales are described in the references.

REFERENCES

1. Pauling, L., *The Nature of the Chemical Bond, Third Edition,* Cornell University Press, Ithaca, New York, 1960.
2. Allen, L.C., *J. Am. Chem. Soc.,* 111, 9003, 1989.
3. Allred, A.L., *J. Inorg. Nucl. Chem.,* 17, 215, 1961.

Z	Symbol	X	Z	Symbol	X	Z	Symbol	X
1	H	2.20	33	As	2.18	65	Tb	—
2	He	—	34	Se	2.55	66	Dy	1.22
3	Li	0.98	35	Br	2.96	67	Ho	1.23
4	Be	1.57	36	Kr	—	68	Er	1.24
5	B	2.04	37	Rb	0.82	69	Tm	1.25
6	C	2.55	38	Sr	0.95	70	Yb	—
7	N	3.04	39	Y	1.22	71	Lu	1.0
8	O	3.44	40	Zr	1.33	72	Hf	1.3
9	F	3.98	41	Nb	1.6	73	Ta	1.5
10	Ne	—	42	Mo	2.16	74	W	1.7
11	Na	0.93	43	Tc	2.10	75	Re	1.9
12	Mg	1.31	44	Ru	2.2	76	Os	2.2
13	Al	1.61	45	Rh	2.28	77	Ir	2.2
14	Si	1.90	46	Pd	2.20	78	Pt	2.2
15	P	2.19	47	Ag	1.93	79	Au	2.4
16	S	2.58	48	Cd	1.69	80	Hg	1.9
17	Cl	3.16	49	In	1.78	81	Tl	1.8
18	Ar	—	50	Sn	1.96	82	Pb	1.8
19	K	0.82	51	Sb	2.05	83	Bi	1.9
20	Ca	1.00	52	Te	2.1	84	Po	2.0
21	Sc	1.36	53	I	2.66	85	At	2.2
22	Ti	1.54	54	Xe	2.60	86	Rn	—
23	V	1.63	55	Cs	0.79	87	Fr	0.7
24	Cr	1.66	56	Ba	0.89	88	Ra	0.9
25	Mn	1.55	57	La	1.10	89	Ac	1.1
26	Fe	1.83	58	Ce	1.12	90	Th	1.3
27	Co	1.88	59	Pr	1.13	91	Pa	1.5
28	Ni	1.91	60	Nd	1.14	92	U	1.7
29	Cu	1.90	61	Pm	—	93	Np	1.3
30	Zn	1.65	62	Sm	1.17	94	Pu	1.3
31	Ga	1.81	63	Eu	—			
32	Ge	2.01	64	Gd	1.20			

FORCE CONSTANTS FOR BOND STRETCHING

Representative force constants (*f*) for stretching of chemical bonds are listed in this table. Except where noted, all force constants are derived from values of the harmonic vibrational frequencies ω_e. Values derived from the observed vibrational fundamentals ν, which are noted by a, are lower than the harmonic force constants, typically by 2 to 3% in the case of heavy atoms (often by 5 to 10% if one of the atoms is hydrogen). Values are given in the SI unit newton per centimeter (N/cm), which is identical to the commonly used cgs unit mdyn/Å.

REFERENCES

1. Huber, K. P., and Herzberg, G., *Molecular Spectra and Molecular Structure. IV. Constants of Diatomic Molecules*, Van Nostrand Reinhold, New York, 1979.
2. Shimanouchi, T., *The Molecular Force Field*, in Eyring, H., Henderson, D., and Yost, W., Eds., *Physical Chemistry: An Advanced Treatise*, Vol. IV, Academic Press, New York, 1970.
3. Tasumi, M., and Nakata, M., *Pure and Appl. Chem.*, 57, 121—147, 1985.

Bond	Molecule	f/(N/cm)	Note	Bond	Molecule	f/(N/cm)	Note
H-H	H_2	5.75			OCS	7.44	
Be-H	BeH	2.27		C-N	CN	16.29	
B-H	BH	3.05			HCN	18.78	
C-H	CH	4.48			CH_3CN	18.33	
	CH_4	5.44	b		CH_3NH_2	5.12	a,c
	C_2H_6	4.83	a,b,c	C-P	CP	7.83	
	CH_3CN	5.33	b	Si-Si	Si_2	2.15	
	CH_3Cl	5.02	a,b,c	Si-O	SiO	9.24	
	$CCl_2=CH_2$	5.57	b	Si-F	SiF	4.90	
	HCN	6.22		Si-Cl	SiCl	2.63	
N-H	NH	5.97		N-N	N_2	22.95	
O-H	OH	7.80			N_2O	18.72	
	H_2O	8.45		N-O	NO	15.95	
P-H	PH	3.22			N_2O	11.70	
S-H	SH	4.23		P-P	P_2	5.56	
	H_2S	4.28		P-O	PO	9.45	
F-H	HF	9.66		O-O	O_2	11.77	
Cl-H	HCl	5.16			O_3	5.74	a
Br-H	HBr	4.12		S-O	SO	8.30	
I-H	HI	3.14			SO_2	10.33	a
Li-H	LiH	1.03		S-S	S_2	4.96	
Na-H	NaH	0.78		F-F	F_2	4.70	
K-H	KH	0.56		Cl-F	ClF	4.48	
Rb-H	RbH	0.52		Br-F	BrF	4.06	
Cs-H	CsH	0.47		Cl-Cl	Cl_2	3.23	
C-C	C_2	12.16		Br-Cl	BrCl	2.82	
	$CCl_2=CH_2$	8.43		Br-Br	Br_2	2.46	
	C_2H_6	4.50	a,c	I-I	I_2	1.72	
	CH_3CN	5.16		Li-Li	Li_2	0.26	
C-F	CF	7.42		Li-Na	LiNa	0.21	
	CH_3F	5.71	a,c	Na-Na	Na_2	0.17	
C-Cl	CCl	3.95		Li-F	LiF	2.50	
	CH_3Cl	3.44	a,c	Li-Cl	LiCl	1.43	
	$CCl_2=CH_2$	4.02	b	Li-Br	LiBr	1.20	
C-Br	CH_3Br	2.89	a,c	Li-I	LiI	0.97	
C-I	CH_3I	2.34	a,c	Na-F	NaF	1.76	
C-O	CO	19.02		Na-Cl	NaCl	1.09	
	CO_2	16.00		Na-Br	NaBr	0.94	
	OCS	16.14		Na-I	NaI	0.76	
	CH_3OH	5.42	a,c	Be-O	BeO	7.51	
C-S	CS	8.49		Mg-O	MgO	3.48	
	CS_2	7.88		Ca-O	CaO	3.61	

a Derived from fundamental frequency, without anharmonicity correction.

b Average of symmetric and antisymmetric (or degenerate) modes.

c Calculated from Local Symmetry Force Field (see Reference 2).

This table lists the fundamental vibrational frequencies of selected three-, four-, and five-atom molecules. Both stable molecules and transient free radicals are included. The data have been taken from evaluated sources. In general, the selected values are based on gas-phase infrared, Raman, or ultraviolet spectra; when these were not available, liquid-phase or matrix-isolation spectra were used.

Molecules are grouped by structural type. Within each group, related molecules appear together for convenient comparison.

The vibrational modes are described by their approximate character in terms of stretching, bending, deformation, etc. However, it should be emphasized that most such descriptions are only approximate, and that the true normal mode usually involves a mixture of motions. Abbreviations are:

sym.	symmetric
antisym.	antisymmetric
str.	stretch
deform.	deformation
scis.	scissors
rock.	rocking
deg.	degenerate

In the case of free radicals, strong interactions may exist between the electronic and bending vibrational motions. Details can be found in References 3 and 4. The references should be consulted for information on the accuracy of the data and for data on other molecules not listed here.

All fundamental frequencies (more precisely, wavenumbers) are given in units of cm^{-1}.

XY_2 Molecules
Point groups $D_{\infty h}$(linear) and C_{2v}(bent)

Molecule	Structure	Sym. str.	Bend	Antisym. str.
CO_2	Linear	1333	667	2349
CS_2	Linear	658	397	1535
C_3	Linear	1224	63	2040
CNC	Linear		321	1453
NCN	Linear	1197	423	1476
BO_2	Linear	1056	447	1278
BS_2	Linear	510	120	1015
KrF_2	Linear	449	233	590
XeF_2	Linear	515	213	555
$XeCl_2$	Linear	316		481
H_2O	Bent	3657	1595	3756
D_2O	Bent	2671	1178	2788
F_2O	Bent	928	461	831
Cl_2O	Bent	639	296	686
O_3	Bent	1103	701	1042
H_2S	Bent	2615	1183	2626
D_2S	Bent	1896	855	1999
SF_2	Bent	838	357	813
SCl_2	Bent	525	208	535
SO_2	Bent	1151	518	1362
H_2Se	Bent	2345	1034	2358
D_2Se	Bent	1630	745	1696
NH_2	Bent	3219	1497	3301
NO_2	Bent	1318	750	1618
NF_2	Bent	1075	573	942
ClO_2	Bent	945	445	1111
CH_2	Bent		963	
CD_2	Bent		752	
CF_2	Bent	1225	667	1114
CCl_2	Bent	721	333	748
CBr_2	Bent	595	196	641
SiH_2	Bent	2032	990	2022
SiD_2	Bent	1472	729	1468
SiF_2	Bent	855	345	870

Molecule	Structure	Sym. str.	Bend	Antisym. str.
SiCl$_2$	Bent	515		505
SiBr$_2$	Bent	403		400
GeH$_2$	Bent	1887	920	1864
GeCl$_2$	Bent	399	159	374
SnF$_2$	Bent	593	197	571
SnCl$_2$	Bent	352	120	334
SnBr$_2$	Bent	244	80	231
PbF$_2$	Bent	531	165	507
PbCl$_2$	Bent	314	99	299
ClF$_2$	Bent	500		576

XYZ Molecules
Point Groups $C_{\infty v}$ (linear) and C_s(bent)

Molecule	Structure	XY str.	Bend	YZ str.
HCN	Linear	3311	712	2097
DCN	Linear	2630	569	1925
FCN	Linear	1077	451	2323
ClCN	Linear	744	378	2216
BrCN	Linear	575	342	2198
ICN	Linear	486	305	2188
CCN	Linear	1060	230	1917
CCO	Linear	1063	379	1967
HCO	Bent	2485	1081	1868
HCC	Linear	3612		1848
OCS	Linear	2062	520	859
NCO	Linear	1270	535	1921
NNO	Linear	2224	589	1285
HNB	Linear	3675		2035
HNC	Linear	3653		2032
HNSi	Linear	3583	523	1198
HBO	Linear		754	1817
FBO	Linear		500	2075
ClBO	Linear	676	404	1958
BrBO	Linear	535	374	1937
FNO	Bent	766	520	1844
ClNO	Bent	596	332	1800
BrNO	Bent	542	266	1799
HNF	Bent		1419	1000
HNO	Bent	2684	1501	1565
HPO	Bent	2095	983	1179
HOF	Bent	3537	886	1393
HOCl	Bent	3609	1242	725
HOO	Bent	3436	1392	1098
FOO	Bent	579	376	1490
ClOO	Bent	407	373	1443
BrOO	Bent			1487
HSO	Bent		1063	1009
NSF	Bent	1372	366	640
NSCl	Bent	1325	273	414
HCF	Bent		1407	1181
HCCl	Bent		1201	815
HSiF	Bent	1913	860	834
HSiCl	Bent		808	522
HSiBr	Bent	1548	774	408

Symmetric XY$_3$ Molecules
Point Groups D$_{3h}$ (planar) and C$_{3v}$ (pyramidal)

Molecule	Structure	Sym. str.	Sym. deform.	Deg. str.	Deg. deform.
NH$_3$	Pyram.	3337	950	3444	1627
ND$_3$	Pyram.	2420	748	2564	1191
PH$_3$	Pyram.	2323	992	2328	1118
AsH$_3$	Pyram.	2116	906	2123	1003
SbH$_3$	Pyram.	1891	782	1894	831
NF$_3$	Pyram.	1032	647	907	492
PF$_3$	Pyram.	892	487	860	344
AsF$_3$	Pyram.	741	337	702	262
PCl$_3$	Pyram.	504	252	482	198
PI$_3$	Pyram.	303	111	325	79
AsI$_3$	Pyram.	219	94	224	71
AlCl$_3$	Pyram.	375	183	595	150
SO$_3$	Planar	1065	498	1391	530
BF$_3$	Planar	888	691	1449	480
BH$_3$	Planar		1125	2808	1640
CH$_3$	Planar		606	3161	1396
CD$_3$	Planar		453	2369	1029
CF$_3$	Pyram.	1090	701	1260	510
SiF$_3$	Pyram.	830	427	937	290

Linear XYYX Molecules
Point Group D$_{\infty h}$

Molecule	Sym. XY str.	Antisym. XY str.	YY str.	Bend	Bend
C$_2$H$_2$	3374	3289	1974	612	730
C$_2$D$_2$	2701	2439	1762	505	537
C$_2$N$_2$	2330	2158	851	507	233

Planar X$_2$YZ Molecules
Point Group C$_{2v}$

Molecule	Sym.XY str.	YZ str.	YX$_2$ scis.	Antisym. XY str.	YX$_2$ rock	YX$_2$ wag
H$_2$CO	2783	1746	1500	2843	1249	1167
D$_2$CO	2056	1700	1106	2160	990	938
F$_2$CO	965	1928	584	1249	626	774
Cl$_2$CO	567	1827	285	849	440	580
O$_2$NF	1310	822	568	1792	560	742
O$_2$NCl	1286	793	370	1685	408	652

Tetrahedral XY$_4$ Molecules
Point Group T$_d$

Molecule	Sym. str.	Deg. deform.(e)	Deg. str.(f)	Deg. deform.(f)
CH$_4$	2917	1534	3019	1306
CD$_4$	2109	1092	2259	996
CF$_4$	909	435	1281	632
CCl$_4$	459	217	776	314

Molecule	Sym. str.	Deg. deform.(e)	Deg. str.(f)	Deg. deform.(f)
CBr_4	267	122	672	182
CI_4	178	90	555	125
SiH_4	2187	975	2191	914
SiD_4	1558	700	1597	681
SiF_4	800	268	1032	389
$SiCl_4$	424	150	621	221
GeH_4	2106	931	2114	819
GeD_4	1504	665	1522	596
$GeCl_4$	396	134	453	172
$SnCl_4$	366	104	403	134
$TiCl_4$	389	114	498	136
$ZrCl_4$	377	98	418	113
$HfCl_4$	382	102	390	112
RuO_4	885	322	921	336
OsO_4	965	333	960	329

REFERENCES

1. T. Shimanouchi, Tables of Molecular Vibrational Frequencies, Consolidated Volume I, Natl. Stand. Ref. Data Ser. Natl. Bur. Stand. (U.S.), 39, 1972.
2. T. Shimanouchi, Tables of Molecular Vibrational Frequencies, Consolidated Volume II, *J. Phys. Chem. Ref. Data*, 6, 993, 1977.
3. G. Herzberg, *Electronic Spectra and Electronic Structure of Polyatomic Molecules*, D. Van Nostrand Co., Princeton, 1966.
4. M. E. Jacox, Ground state vibrational energy levels of polyatomic transient molecules, *J. Phys. Chem. Ref. Data*, 13, 945, 1984.

SPECTROSCOPIC CONSTANTS OF DIATOMIC MOLECULES

This table lists the leading spectroscopic constants and equilibrium internuclear distance r_e in the ground electronic state for selected diatomic molecules. The constants are those describing the vibrational and rotational energy through the expressions:

$$E_{vib}/hc = \omega_e(v+1/2) - \omega_e x_e(v+1/2)^2 + \cdots$$

$$E_{rot}/hc = B_v J(J+1) - D_v[J(J+1)]^2 + \cdots$$

where

$$B_v = B_e - \alpha_e(v+1/2) + \cdots$$

$$D_v = D_e + \cdots$$

Here v and J are the vibrational and rotational quantum numbers, respectively, h is Planck's constant, and c is the speed of light. In this customary formulation the constants ω_e, B_e, etc. have dimensions of inverse length; in this table they are given in units of cm^{-1}.

Users should note that higher order terms in the above energy expressions are required for very precise calculations; constants for many of these terms can be found in the references. Also, if the ground electronic state is not $^1\Sigma$, additional terms are needed to account for the interaction between electronic and pure rotational angular momentum. For some molecules in the table the data have been analyzed in terms of the Dunham series expansion:

$$E/hc = \Sigma_{lm} Y_{lm}(v+1/2)^l J^m(J+1)^m$$

In such cases it has been assumed that $Y_{10} = \omega_e$, $Y_{01} = B_e$, etc., although in the highest approximations these identities are not precisely correct. Some of the values of r_e in the table have been corrected for breakdown of the Born-Oppenheimer approximation, which can affect the last decimal place. Because of differences in the method of data analysis and limitations in the model, care should be taken in comparing r_e values for different molecules to a precision beyond 0.001 Å.

Molecules are listed in alphabetical order by formula as written in the most common form. In most cases this form places the more electropositive element first, but there are exceptions such as OH, NH, CH, etc.

* Indicates a value for the interval between $v = 0$ and $v = 1$ states instead of a value of ω_e.

REFERENCES

1. Huber, K. P., and Herzberg, G., *Molecular Spectra and Molecular Structure IV. Constants of Diatomic Molecules*, Van Nostrand Reinhold, New York, 1979.
2. Lovas, F. J., and Tiemann, E., *J. Phys. Chem. Ref. Data*, 3, 609, 1974.
3. *Landolt-Börnstein, Numerical Data and Functional Relationships in Science and Technology, New Series*, II/6 (1974), II/14a (1982), II/14b (1983), II/19a (1992), II/19d-1 (1995), *Molecular Constants*, Springer-Verlag, Heidelberg.

Molecule	State	ω_e cm^{-1}	$\omega_e x_e$ cm^{-1}	B_e cm^{-1}	α_e cm^{-1}	D_e 10^{-6}cm^{-1}	r_e Å
^{107}Ag^{79}Br	$^1\Sigma^+$	249.57	0.63	0.064833	0.0002361	0.0175	2.39311
^{107}Ag^{35}Cl	$^1\Sigma^+$	343.49	1.17	0.12298388	0.00059541	0.06305	2.28079
^{107}Ag^{19}F	$^1\Sigma^+$	513.45	2.59	0.2657020	0.0019206	0.284	1.98318
^{107}Ag^{1}H	$^1\Sigma^+$	1759.9	34.06	6.449	0.201	344	1.618
^{107}Ag^{2}H	$^1\Sigma^+$	1250.70	17.17	3.2572	0.0722	85.9	1.6180
^{107}Ag^{127}I	$^1\Sigma^+$	206.50	0.46	0.04486821	0.0001414	0.00847	2.54463
^{107}Ag^{16}O	$^2\Pi_{1/2}$	490.2	3.1	0.3020	0.0025	0.45	2.003
^{27}Al$_2$	$^3\Sigma_g^-$	350.01	2.02	0.2054	0.0012	0.31	2.466
^{27}Al^{79}Br	$^1\Sigma^+$	378.0	1.28	0.15919713	0.00086045	0.11285	2.29481
^{27}Al^{35}Cl	$^1\Sigma^+$	481.30	1.95	0.24393012	0.00161113	0.2503	2.13011
^{27}Al^{19}F	$^1\Sigma^+$	802.3	4.77	0.5524798	0.0049841	1.0464	1.65437
^{27}Al^{1}H	$^1\Sigma^+$	1682.56	29.09	6.3907	0.1858	356.5	1.6478
^{27}Al^{2}H	$^1\Sigma^+$	1211.95	15.14	3.3186	0.0697	97	1.6463
^{27}Al^{127}I	$^1\Sigma^+$	316.1	1.0	0.11769985	0.00055859		2.53710
^{27}Al^{16}O	$^2\Sigma^+$	979.23	6.97	0.6414	0.0058	1.08	1.6179
^{27}Al^{32}S	$^2\Sigma^+$	617.1	3.33	0.2799	0.0018	0.22	2.029
^{75}As$_2$	$^1\Sigma_g^+$	429.55	1.12	0.10179	0.000333		2.1026
^{75}As^{1}H	$^3\Sigma^-$	2130*		7.3067	0.2117	327	1.52315
^{75}As^{2}H	$^3\Sigma^-$	1484*		3.6688		90	1.5306
^{75}As^{14}N	$^1\Sigma^+$	1068.54	5.41	0.54551	0.003366	0.53	1.6184
^{75}As^{16}O	$^2\Pi_{1/2}$	967.08	4.85	0.48482	0.003299	0.49	1.6236

Molecule	State	ω_e cm^{-1}	$\omega_e x_e$ cm^{-1}	B_e cm^{-1}	α_e cm^{-1}	D_e 10^{-6}cm^{-1}	r_e Å
$^{197}Au_2$	$^1\Sigma_g^+$	190.9	0.42	0.028013	0.0000723	0.00250	2.4719
$^{197}Au^1H$	$^1\Sigma^+$	2305.01	43.12	7.2401	0.2136	279	1.5239
$^{197}Au^2H$	$^1\Sigma^+$	1634.98	21.65	3.6415	0.07614	70.9	1.5238
$^{11}B_2$	$^3\Sigma_g^-$	1051.3	9.35	1.212	0.014		1.590
$^{11}B^{79}Br$	$^1\Sigma^+$	684.31	3.52	0.4894	0.0035	1.00	1.888
$^{11}B^{35}Cl$	$^1\Sigma^+$	840.29	5.49	0.684282	0.006812	1.84	1.71528
$^{11}B^{19}F$	$^1\Sigma^+$	1402.1	11.8	1.516950	0.019056	7.105	1.26267
$^{11}B^1H$	$^1\Sigma^+$	2366.9	49.40	12.021	0.412	1242	1.2324
$^{11}B^2H$	$^1\Sigma^+$	1703.3	28	6.54	0.17	400	1.2324
$^{11}B^{14}N$	$^3\Pi$	1514.6	12.3	1.666	0.025	8.1	1.281
$^{11}B^{16}O$	$^2\Sigma^+$	1885.69	11.81	1.7820	0.0166	6.32	1.2045
$^{11}B^{32}S$	$^2\Sigma^+$	1180.17	6.31	0.7949	0.0061	1.40	1.6092
$^{138}Ba^{79}Br$	$^2\Sigma^+$	193.77	0.41	0.0415082	0.0001219	0.00762	2.84449
$^{138}Ba^{35}Cl$	$^2\Sigma^+$	279.92	0.82	0.08396717	0.00033429	0.03022	2.68276
$^{138}Ba^{19}F$	$^2\Sigma^+$	468.9	1.79	0.2159	0.0012	0.175	2.163
$^{138}Ba^1H$	$^2\Sigma^+$	1168.31	14.50	3.38285	0.06599	112.67	2.23175
$^{138}Ba^2H$	$^2\Sigma^+$	829.77	7.32	1.7071	0.02363	28.77	2.2304
$^{138}Ba^{127}I$	$^2\Sigma^+$	152.14	0.27	0.02680587	0.00006634	0.00333	3.08476
$^{138}Ba^{16}O$	$^1\Sigma^+$	669.76	2.03	0.3126140	0.0013921	0.2724	1.93969
$^{138}Ba^{32}S$	$^1\Sigma^+$	379.42	0.88	0.10331	0.0003188	0.0306	2.5074
$^9Be^{19}F$	$^2\Sigma^+$	1247.36	9.12	1.4889	0.0176	8.28	1.3610
$^9Be^1H$	$^2\Sigma^+$	2060.78	36.31	10.3164	0.3030	1022.1	1.3426
$^9Be^2H$	$^2\Sigma^+$	1530.32	20.71	5.6872	0.1225	313.8	1.3419
$^9Be^{16}O$	$^1\Sigma^+$	1487.32	11.83	1.6510	0.0190	8.20	1.3309
$^9Be^{32}S$	$^1\Sigma^+$	997.94	6.14	0.79059	0.00664	2.00	1.7415
$^{209}Bi_2$	$^1\Sigma_g^+$	172.71	0.34	0.022781	0.000055	0.00150	2.6596
$^{209}Bi^1H$	$^3\Sigma^-$	1635.73	31.6	5.137	0.148	183	1.805
$^{209}Bi^2H$	$^3\Sigma^-$	1173.32	16.1	2.592	0.054	50.6	1.804
$^{79}Br_2$	$^1\Sigma_g^+$	325.32	1.08	0.082107	0.0003187	0.02092	2.2811
$^{79}Br^{35}Cl$	$^1\Sigma^+$	444.28	1.84	0.152470	0.000770	0.07183	2.13607
$^{79}Br^{19}F$	$^1\Sigma^+$	670.75	4.05	0.35584	0.00261	0.401	1.75894
$^{79}Br^{16}O$	$^2\Pi_{3/2}$	779	6.8	0.429598	0.003639	0.523	1.717
$^{12}C_2$	$^1\Sigma_g^+$	1854.71	13.34	1.8198	0.0177	6.92	1.2425
$^{12}C^{35}Cl$	$^2\Pi_{1/2}$	866.72*	6.2	0.6936	0.00672	1.9	1.6450
$^{12}C^{19}F$	$^2\Pi_{1/2}$	1308.1	11.10	1.4172	0.0184	6.5	1.2718
$^{12}C^1H$	$^2\Pi_{1/2}$	2858.5	63.0	14.457	0.534	1450	1.1199
$^{12}C^2H$	$^2\Pi_{1/2}$	2099.8	34.02	7.806	0.208	420	1.1190
$^{12}C^{14}N$	$^2\Sigma^+$	2068.59	13.09	1.8997830	0.0173717	6.4034	1.17181
$^{12}C^{16}O$	$^1\Sigma^+$	2169.81	13.29	1.93128075	0.01750390	6.1216	1.12823
$^{12}C^{31}P$	$^2\Sigma^+$	1239.67	6.86	0.7986	0.00597	1.33	1.562
$^{12}C^{32}S$	$^1\Sigma^+$	1285.15	6.50	0.8200434	0.0059182	1.336	1.53482
$^{12}C^{80}Se$	$^1\Sigma^+$	1035.36	4.86	0.5750	0.00379	0.71	1.67609
$^{40}Ca^{35}Cl$	$^2\Sigma^+$	367.53	1.31	0.1522302	0.0007990	0.1029	2.43676
$^{40}Ca^{19}F$	$^2\Sigma^+$	581.1	2.74	0.339	0.0026	0.45	1.967
$^{40}Ca^1H$	$^2\Sigma^+$	1298.34	19.10	4.2766	0.0970	183.7	2.0025
$^{40}Ca^2H$	$^2\Sigma^+$	910*		2.1769	0.035	47.9	2.002
$^{40}Ca^{127}I$	$^2\Sigma^+$	238.70	0.63	0.0693263	0.0002634	0.0234	2.82859
$^{40}Ca^{16}O$	$^1\Sigma^+$	732.03	4.83	0.444441	0.003282	0.6541	1.8221
$^{40}Ca^{32}S$	$^1\Sigma^+$	462.23	1.78	0.1766757	0.0008270	0.1032	2.31775
$^{114}Cd^1H$	$^2\Sigma^+$	1337.1*		5.323		314	1.781
$^{114}Cd^2H$	$^2\Sigma^+$			2.704		76	1.775
$^{35}Cl_2$	$^1\Sigma_g^+$	559.7	2.68	0.2440	0.0015	0.186	1.988
$^{35}Cl^{19}F$	$^1\Sigma^+$	786.15	6.16	0.516479	0.004358	0.88	1.62831
$^{35}Cl^{16}O$	$^2\Pi_{3/2}$	853.8	5.5	0.62345	0.0058	1.33	1.56963
$^{52}Cr^1H$	$^6\Sigma^+$	1581*	32	6.220	0.179	347	1.656
$^{52}Cr^2H$	$^6\Sigma^+$	1182*		3.14		88.8	1.664
$^{52}Cr^{16}O$	$^5\Pi$	898.4	6.8	0.5231	0.0070		1.615
$^{133}Cs_2$	$^1\Sigma_g^+$	42.02	0.08	0.0127	0.0000264	0.00464	4.47

SPECTROSCOPIC CONSTANTS OF DIATOMIC MOLECULES (continued)

Molecule	State	ω_e cm^{-1}	$\omega_e x_e$ cm^{-1}	B_e cm^{-1}	α_e cm^{-1}	D_e 10^{-6}cm^{-1}	r_e Å
^{133}Cs^{79}Br	$^1\Sigma^+$	149.66	0.37	0.03606925	0.00012401	0.00838	3.07225
^{133}Cs^{35}Cl	$^1\Sigma^+$	214.17	0.73	0.07209149	0.00033756	0.03268	2.90627
^{133}Cs^{19}F	$^1\Sigma^+$	352.56	1.62	0.18436969	0.0011756	0.20168	1.34535
^{133}Cs^{1}H	$^1\Sigma^+$	891.0	12.9	2.7099	0.0579	113	2.4938
^{133}Cs^{2}H	$^1\Sigma^+$	619.1*		1.354		20	2.505
^{133}Cs^{127}I	$^1\Sigma^+$	119.18	0.25	0.02362736	0.00006826	0.00371	3.31519
^{133}Cs^{16}O	$^2\Sigma^+$	357.5*		0.223073	0.001303	0.348	2.3007
^{63}Cu$_2$	$^1\Sigma_g^+$	264.55	1.02	0.10874	0.000614	0.0716	2.2197
^{63}Cu^{79}Br	$^1\Sigma^+$	314.8	0.96	0.10192625	0.00045214	0.04274	2.17344
^{65}Cu^{35}Cl	$^1\Sigma^+$	415.29	1.58	0.17628802	0.00099647	0.12706	2.05118
^{63}Cu^{19}F	$^1\Sigma^+$	622.7	3.95	0.3794029	0.0032298	0.563	1.74493
^{63}Cu^{1}H	$^1\Sigma^+$	1941.26	37.51	7.9441	0.2563	520	1.46263
^{63}Cu^{2}H	$^1\Sigma^+$	1384.14	18.97	4.0381	0.0917	136.2	1.4626
^{63}Cu^{127}I	$^1\Sigma^+$	264.5	0.60	0.07328742	0.00028390	0.02244	2.33832
^{63}Cu^{16}O	$^2\Pi_{3/2}$	640.17	4.43	0.44454	0.00456	0.85	1.7244
^{63}Cu^{32}S	$^2\Pi_{3/2}$	415.0	1.75	0.1891		0.18	2.051
^{19}F$_2$	$^1\Sigma_g^+$	916.64	11.24	0.89019	0.013847	3.3	1.41193
^{56}Fe^{16}O	$^5\Delta$	965*		0.650		0.72	1.444
^{69}Ga^{81}Br	$^1\Sigma^+$	263.0	0.81	0.081839	0.0003207	0.032	2.35248
^{69}Ga^{35}Cl	$^1\Sigma^+$	365.67	1.25	0.1499046	0.0007936	0.1008	2.20169
^{69}Ga^{19}F	$^1\Sigma^+$	622.2	3.2	0.3595161	0.0028642	0.50	1.77437
^{69}Ga^{1}H	$^1\Sigma^+$	1604.52	28.77	6.137	0.181	342	1.663
^{69}Ga^{2}H	$^1\Sigma^+$			3.083	0.06	84	1.663
^{69}Ga^{127}I	$^1\Sigma^+$	216.38	0.47	0.0569359	0.0001897	0.015770	2.57464
^{69}Ga^{16}O	$^2\Sigma$	767.5	6.24	0.4271		0.37	1.744
^{74}Ge^{79}Br	$^2\Pi_{1/2}$	295	0.7				
^{74}Ge^{35}Cl	$^2\Pi_{1/2}$	407.6	1.36				
^{72}Ge^{1}H	$^2\Pi_{1/2}$	1833.77	37	6.726	0.192	326	1.5880
^{72}Ge^{2}H	$^2\Pi_{1/2}$	1320.09	19	3.415	0.070	83.2	1.5874
^{74}Ge^{16}O	$^1\Sigma^+$	986.49	4.47	0.4856981	0.0030787	0.4709	1.62464
^{74}Ge^{32}S	$^1\Sigma^+$	575.8	1.80	0.18656576	0.00074910	0.07883	2.01209
^{74}Ge^{80}Se	$^1\Sigma^+$	408.7	1.36	0.09634051	0.00028904	0.02207	2.13463
^{74}Ge^{130}Te	$^1\Sigma^+$	323.9	0.75	0.06533821	0.00017246	0.012	2.34017
^{1}H$_2$	$^1\Sigma_g^+$	4401.21	121.34	60.853	3.062	47100	0.74144
^{2}H$_2$	$^1\Sigma_g^+$	3115.50	61.82	30.444	1.0786	11410	0.74152
^{3}H$_2$	$^1\Sigma_g^+$	2546.5	41.23	20.335	0.5887		0.74142
^{1}H^{81}Br	$^1\Sigma^+$	2648.97	45.22	8.46488	0.23328	345.8	1.41444
^{2}H^{81}Br	$^1\Sigma^+$	1884.75	22.72	4.245596	0.084	88.32	1.4145
^{1}H^{35}Cl	$^1\Sigma^+$	2990.95	52.82	10.59342	0.30718	531.94	1.27455
^{2}H^{35}Cl	$^1\Sigma^+$	2145.16	27.18	5.448796	0.113292	140	1.27458
^{1}H^{19}F	$^1\Sigma^+$	4138.32	89.88	20.9557	0.798	2151	0.91681
^{2}H^{19}F	$^1\Sigma^+$	2998.19	45.76	11.0102	0.3017	594	0.91694
^{1}H^{127}I	$^1\Sigma^+$	2309.01	39.64	6.4263650	0.1689	206.9	1.60916
^{202}Hg^{1}H	$^2\Sigma^+$	1203.24*		5.3888		395.3	1.7662
^{202}Hg^{2}H	$^2\Sigma^+$	896.12*		2.739		91	1.757
^{127}I$_2$	$^1\Sigma_g^+$	214.50	0.61	0.03737	0.000114	0.0043	2.666
^{127}I^{79}Br	$^1\Sigma^+$	268.64	0.81	0.0568325	0.0001969	0.0102	2.46899
^{127}I^{35}Cl	$^1\Sigma^+$	384.29	1.50	0.1141587	0.0005354	0.0403	2.32088
^{127}I^{19}F	$^1\Sigma^+$	610.24	3.12	0.2797111	0.0018738	0.2356	1.90976
^{127}I^{16}O	$^2\Pi_{3/2}$	681.5	4.3	0.34026	0.00270	0.36	1.8676
^{115}In^{81}Br	$^1\Sigma^+$	221.0	0.65	0.05489468	0.00018672	0.01350	2.54315
^{115}In^{35}Cl	$^1\Sigma^+$	317.39	1.03	0.1090583	0.0005177	0.0515	2.40117
^{115}In^{19}F	$^1\Sigma^+$	535.4	2.6	0.2623241	0.0018798	0.252	1.98540
^{115}In^{1}H	$^1\Sigma^+$	1476.0	25.61	4.995	0.143	223	1.8380
^{115}In^{2}H	$^1\Sigma^+$	1048.2	12.4	2.523	0.051	58	1.837
^{115}In^{127}I	$^1\Sigma^+$	177.08	0.34	0.03686702	0.00010411	0.00639	2.75364
^{39}K$_2$	$^1\Sigma_g^+$	92.02	0.28	0.056743	0.000165	0.0863	3.9051
^{39}K^{79}Br	$^1\Sigma^+$	213	0.80	0.08122109	0.00040481	0.04462	2.82078

Molecule	State	ω_e cm^{-1}	$\omega_e x_e$ cm^{-1}	B_e cm^{-1}	α_e cm^{-1}	D_e 10^{-6}cm^{-1}	r_e Å
^{39}K^{35}Cl	$^1\Sigma^+$	281	1.30	0.1286348	0.0007899	0.1087	2.66665
^{39}K^{19}F	$^1\Sigma^+$	426.26	2.45	0.27993741	0.00233492	0.4829	2.17146
^{39}K^{1}H	$^1\Sigma^+$	983.6	14.3	3.416400	0.085313	163.55	2.243
^{39}K^{2}H	$^1\Sigma^+$	707	7.7	1.754	0.0318	50	2.240
^{39}K^{127}I	$^1\Sigma^+$	186.53	0.57	0.06087473	0.00026776	0.02593	3.04784
^{139}La^{16}O	$^2\Sigma^+$	812.8	2.22	0.35252001	0.00142365	0.2626	1.82591
^{7}Li$_2$	$^1\Sigma_g^+$	351.43	2.61	0.67264	0.00704	9.87	2.6729
^{7}Li^{79}Br	$^1\Sigma^+$	563.2	3.5	0.555399	0.005644	2.159	2.17043
^{7}Li^{35}Cl	$^1\Sigma^+$	642.95	4.47	0.7065225	0.0080102	3.409	2.02067
^{7}Li^{19}F	$^1\Sigma^+$	910.57	8.21	1.3452583	0.0202887	11.745	1.56386
^{7}Li^{1}H	$^1\Sigma^+$	1405.65	23.20	7.51373	0.21665	862	1.59490
^{7}Li^{2}H	$^1\Sigma^+$	1054.80	12.94	4.23310	0.09155	276	1.5941
^{7}Li^{127}I	$^1\Sigma^+$	496.85	2.85	0.4431766	0.0040862	1.4104	2.39192
^{7}Li^{16}O	$^2\Pi$	814.62	7.78	1.212830	0.017899	0.1079	1.68822
^{24}Mg$_2$	$^1\Sigma_g^+$	51.12	1.64	0.09287	0.00378	1.22	3.891
^{24}Mg^{35}Cl	$^2\Sigma^+$	462.12*	2.1	0.2456154	0.0016204	0.2723	2.19639
^{24}Mg^{19}F	$^2\Sigma^+$	711.69*	4.9	0.51922	0.00470	1.080	1.7500
^{24}Mg^{1}H	$^2\Sigma^+$	1495.20	31.89	5.8257	0.1859	344	1.7297
^{24}Mg^{2}H	$^2\Sigma^+$	1077.9	16.1	3.0306	0.06289	92	1.7302
^{24}Mg^{16}O	$^1\Sigma^+$	784.78	5.26	0.57470436	0.00532377	1.2328	1.74838
^{55}Mn^{1}H	$^7\Sigma$	1548.0	28.8	5.6841	0.1570	303.9	1.7311
^{55}Mn^{2}H	$^7\Sigma$	1103	13.9	2.8957	0.051	79.5	1.7310
^{14}N$_2$	$^1\Sigma_g^+$	2358.57	14.32	1.99824	0.017318	5.76	1.09769
^{14}N^{79}Br	$^3\Sigma^-$	691.75	4.72	0.444	0.0040		1.79
^{14}N^{35}Cl	$^3\Sigma^-$	827.96	5.30	0.649770	0.006414	1.598	1.61071
^{14}N^{19}F	$^3\Sigma^-$	1141.37	8.99	1.2057	0.01492	5.39	1.3170
^{14}N^{1}H	$^3\Sigma^-$	3282.3	78.4	16.6993	0.6490	1709.7	1.0362
^{14}N^{2}H	$^3\Sigma^-$	2398	42	8.7913	0.2531	490.4	1.0361
^{14}N^{16}O	$^2\Pi_{1/2}$	1904.20	14.07	1.67195	0.0171	0.5	1.15077
^{14}N^{32}S	$^2\Pi_{1/2}$	1218.7	7.28	0.769602	0.0064	1.2	1.4940
^{23}Na$_2$	$^1\Sigma_g^+$	159.13	0.72	0.154707	0.008736	0.581	3.0789
^{23}Na^{79}Br	$^1\Sigma^+$	302	1.5	0.1512533	0.0009410	0.1554	2.50204
^{23}Na^{35}Cl	$^1\Sigma^+$	366	2.05	0.2180631	0.0016248	0.3120	2.36080
^{23}Na^{19}F	$^1\Sigma^+$	535.66	3.57	0.4369011	0.0045580	1.163	1.92595
^{23}Na^{1}H	$^1\Sigma^+$	1172.2	19.72	4.9033634	0.1370919	343.40	1.88654
^{23}Na^{2}H	$^1\Sigma^+$	826.1*		2.557089	0.051600	93.46	1.88654
^{23}Na^{127}I	$^1\Sigma^+$	258	1.1	0.1178056	0.0006478	0.0973	2.71145
^{23}Na^{16}O	$^2\Pi$	492.3		0.424630	0.004506	1.2638	2.05155
^{93}Nb^{16}O	$^4\Sigma^-$	989.0	3.8	0.4321	0.0021	0.22	1.691
^{58}Ni^{1}H	$^2\Delta_{5/2}$	1926.6	38	7.700	0.23	481	1.476
^{58}Ni^{2}H	$^2\Delta_{5/2}$	1390.1	19	3.992	0.092	130	1.465
^{16}O$_2$	$^3\Sigma_g^-$	1580.19	11.98	1.44563	0.0159	4.839	1.20752
^{16}O^{1}H	$^2\Pi_{3/2}$	3737.76	84.88	18.911	0.7242	1938	0.96966
^{16}O^{2}H	$^2\Pi_{3/2}$	2720.24	44.05	10.021	0.276	537.4	0.9698
^{31}P$_2$	$^1\Sigma_g^+$	780.77	2.84	0.30362	0.00149	0.188	1.8934
^{31}P^{35}Cl	$^3\Sigma^-$	551.38	2.23	0.2528748	0.0015119	0.2124	2.01461
^{31}P^{19}F	$^3\Sigma^-$	846.75	4.49	0.5665	0.00456		1.58938
^{31}P^{1}H	$^3\Sigma^-$	2365.2	44.5	8.5371	0.2514	436	1.42140
^{31}P^{2}H	$^3\Sigma^-$	1699.2	23.0	4.4081	0.0928	116	1.4220
^{31}P^{14}N	$^1\Sigma^+$	1337.24	6.98	0.7864854	0.0055364	1.091	1.49087
^{31}P^{16}O	$^2\Pi_{1/2}$	1233.34	6.56	0.7337	0.0055	1.3	1.4759
^{208}Pb$_2$		110.5	0.35				
^{208}Pb^{79}Br	$^2\Pi_{1/2}$	207.5	0.50				
^{208}Pb^{35}Cl	$^2\Pi_{1/2}$	303.9	0.88				
^{208}Pb^{19}F	$^2\Pi_{1/2}$	502.73	2.28	0.22875	0.001473	0.183	2.0575
^{208}Pb^{1}H	$^2\Pi_{1/2}$	1564.1	29.75	4.971	0.144	201	1.839
^{208}Pb^{16}O	$^1\Sigma^+$	720.96	3.52	0.30730373	0.00190977	0.2138	1.92181
^{208}Pb^{32}S	$^1\Sigma^+$	429.17	1.26	0.11632307	0.00043510	0.03418	2.28678

Molecule	State	ω_e cm^{-1}	$\omega_e x_e$ cm^{-1}	B_e cm^{-1}	α_e cm^{-1}	D_e 10^{-6}cm^{-1}	r_e Å
^{208}Pb^{80}Se	$^1\Sigma^+$	277.6	0.51	0.05059953	0.00012993	0.0070	2.40218
^{208}Pb^{130}Te	$^1\Sigma^+$	212.0	0.43	0.03130774	0.00006743	0.0027	2.59492
^{195}Pt^{12}C	$^1\Sigma^+$	1051.13	4.86	0.53044	0.003273	0.546	1.6767
^{195}Pt^{1}H	$^2\Delta_{5/2}$	2294.68*	46	7.1963	0.1996	261	1.52852
^{195}Pt^{2}H	$^2\Delta_{5/2}$	1644.3*	23	3.640	0.071	66	1.524
^{85}Rb^{79}Br	$^1\Sigma^+$	169.46	0.46	0.04752798	0.00018596	0.01496	2.94474
^{85}Rb^{35}Cl	$^1\Sigma^+$	228	0.92	0.0876404	0.0004537	0.04947	2.78673
^{85}Rb^{19}F	$^1\Sigma^+$	376	1.9	0.2106640	0.0015228	0.2684	2.27033
^{85}Rb^{1}H	$^1\Sigma^+$	936.9	14.21	3.020	0.072	123	2.367
^{85}Rb^{127}I	$^1\Sigma^+$	138.51	0.33	0.03283293	0.00010946	0.00738	3.17688
^{85}Rb^{16}O	$^2\Sigma^+$	388.4*		0.246481	0.002174	0.397	2.25420
^{32}S$_2$	$^3\Sigma_g^-$	725.65	2.84	0.2955	0.001570	0.19	1.8892
^{32}S^{19}F	$^2\Pi_{3/2}$			0.552174			1.60058
^{32}S^{1}H	$^2\Pi_{3/2}$	2711.6	59.9	9.5995	0.2785	480.6	1.34066
^{32}S^{2}H	$^2\Pi_{3/2}$	1885	31	4.95130	0.10308	130	1.34049
^{32}S^{16}O	$^3\Sigma^-$	1149.2	5.6	0.7208171	0.005737	1.134	1.48109
^{121}Sb^{35}Cl	$^3\Sigma^-$	374.7	0.6				
^{121}Sb^{19}F	$^3\Sigma^-$	605.0	2.6	0.2792	0.0020	0.23	1.918
^{121}Sb^{1}H	$^3\Sigma^-$			5.684		240	1.723
^{121}Sb^{2}H	$^3\Sigma^-$			2.8782		45	1.7194
^{121}Sb^{14}N	$^1\Sigma^+$	942.0	5.6				
^{121}Sb^{16}O	$^2\Pi_{1/2}$	816	4.2	0.3580	0.0022	0.270	1.826
^{45}Sc^{19}F	$^1\Sigma^+$	735.6	3.8	0.3950	0.00266		1.788
^{80}Se$_2$	$^3\Sigma_g^-$	385.30	0.96	0.08992	0.000288	0.024	2.166
^{80}Se^{1}H	$^2\Pi_{3/2}$	2400*		8.02	0.23	330	1.48
^{80}Se^{2}H	$^2\Pi_{3/2}$	1708*		3.94			1.48
^{80}Se^{16}O	$^3\Sigma^-$	914.69	4.52	0.4655	0.00323	0.5	1.648
^{28}Si$_2$	$^3\Sigma_g^-$	510.98	2.02	0.2390	0.0014	0.21	2.246
^{28}Si^{35}Cl	$^2\Pi_{1/2}$	535.60	2.17	0.2561	0.0016	0.25	2.058
^{28}Si^{19}F	$^2\Pi_{1/2}$	857.19	4.73	0.5812	0.00494	1.07	1.6011
^{28}Si^{1}H	$^2\Pi_{1/2}$	2041.80	35.51	7.4996	0.2190	397	1.5201
^{28}Si^{2}H	$^2\Pi_{1/2}$	1469.32	18.23	3.8840	0.0781	105.4	1.5199
^{28}Si^{14}N	$^2\Sigma^+$	1151.4	6.47	0.7311	0.00565	1.2	1.572
^{28}Si^{16}O	$^1\Sigma^+$	1241.54	5.97	0.7267521	0.0050379	0.9923	1.50975
^{28}Si^{32}S	$^1\Sigma^+$	749.64	2.58	0.30352788	0.00147308	0.201	1.92926
^{28}Si^{80}Se	$^1\Sigma^+$	580.0	1.78	0.1920117	0.0007767	0.0842	2.05832
^{120}Sn^{79}Br	$^2\Pi_{1/2}$	247.2	0.6				
^{120}Sn^{35}Cl	$^2\Pi_{1/2}$	351.1	1.06	0.1117	0.0004		2.361
^{118}Sn^{19}F	$^2\Pi_{1/2}$	577.6	2.69	0.2727	0.0014	0.26	1.944
^{120}Sn^{1}H	$^2\Pi_{1/2}$			5.31488		207.5	1.78146
^{120}Sn^{2}H	$^2\Pi_{1/2}$	1188.0*		2.6950	0.049	53.4	1.7770
^{120}Sn^{127}I	$^2\Pi_{1/2}$	199.0	0.6				
^{120}Sn^{16}O	$^1\Sigma^+$	822.13	3.72	0.35571998	0.00214432	0.26638	1.83251
^{120}Sn^{32}S	$^1\Sigma^+$	487.26	1.36	0.13686139	0.00050563	0.0424	2.20898
^{120}Sn^{80}Se	$^1\Sigma^+$	331.2	0.74	0.0649978	0.0001705	0.011	2.32557
^{120}Sn^{130}Te	$^1\Sigma^+$	259.5	0.50	0.04247917	0.00009543	0.0055	2.52280
^{88}Sr^{79}Br	$^2\Sigma^+$	216.60	0.52	0.0541847	0.0001827	0.01356	2.73522
^{88}Sr^{35}Cl	$^2\Sigma^+$	302.3	0.95				
^{88}Sr^{19}F	$^2\Sigma^+$	502.4	2.3	0.2505346	0.0015513	0.2498	2.07537
^{88}Sr^{1}H	$^2\Sigma^+$	1206.2	17.0	3.6751	0.0814	135	2.1456
^{88}Sr^{2}H	$^2\Sigma^+$	841	8.6	1.8609	0.0292	34.7	2.1449
^{88}Sr^{127}I	$^2\Sigma^+$	173.77	0.35	0.0367097	0.0001060	0.00655	2.94364
^{88}Sr^{16}O	$^1\Sigma^+$	653.5	3.96	0.33798	0.00219	0.36	1.91983
^{181}Ta^{16}O	$^2\Delta_{3/2}$	1028.69	3.51	0.40284	0.00182	0.2450	1.68746
^{130}Te$_2$	$^3\Sigma_g^-$	247.07	0.51	0.039681	0.000106	0.0044	2.5574
^{130}Te^{1}H	$^2\Pi_{3/2}$			5.56			1.74
^{130}Te^{16}O	0^+	797.11	4.00	0.3554	0.00237	0.27	1.825
^{232}Th^{16}O	$^1\Sigma^+$	895.77	2.39	0.332644	0.001302	0.1833	1.84032

SPECTROSCOPIC CONSTANTS OF DIATOMIC MOLECULES (continued)

Molecule	State	ω_e cm^{-1}	$\omega_e x_e$ cm^{-1}	B_e cm^{-1}	α_e cm^{-1}	D_e 10^{-6}cm^{-1}	r_e Å
^{48}Ti^{16}O	$^3\Delta_1$	1009.02	4.50	0.53541	0.00301	0.603	1.6202
^{205}Tl^{81}Br	$^1\Sigma^+$	192.10	0.39	0.0423899	0.0001276	0.0083	1.61817
^{205}Tl^{35}Cl	$^1\Sigma^+$	284.71	0.86	0.09139702	0.00039784	0.0377	2.48483
^{205}Tl^{19}F	$^1\Sigma^+$	476.86	2.24	0.22315014	0.00150380	0.1955	2.08439
^{205}Tl^{1}H	$^1\Sigma^+$	1390.7	22.7	4.806	0.154	254	1.870
^{205}Tl^{2}H	$^1\Sigma^+$	987.7	12.04	2.419	0.057	60	1.869
^{205}Tl^{127}I	$^1\Sigma^+$	150*		0.0271676	0.0000664	0.0036	2.81361
^{51}V^{16}O	$^4\Sigma^-$	1011.3	4.86	0.54825	0.00352	0.6	1.5893
^{89}Y^{35}Cl	$^1\Sigma$	380.7	1.3	0.1160	0.0003	0.09	2.41
^{89}Y^{19}F	$^1\Sigma^+$	631.29	2.50	0.29042	0.00163	0.237	1.9257
^{89}Y^{16}O	$^2\Sigma^+$	861.0	2.9	0.3881	0.0018	0.32	1.790
^{174}Yb^{1}H	$^2\Sigma^+$	1249.54	21.06	3.9931	0.0957	161.8	2.0526
^{174}Yb^{2}H	$^2\Sigma^+$	886.6	10.57	2.01162	0.03425	41.60	2.0516
^{64}Zn^{35}Cl	$^2\Sigma$	390.5	1.6				
^{64}Zn^{19}F	$^2\Sigma$	628	3.5				
^{64}Zn^{1}H	$^2\Sigma^+$	1607.6	55.14	6.6794	0.2500	466	1.5949
^{64}Zn^{2}H	$^2\Sigma^+$	1072	28	3.350		124	1.6054
^{64}Zn^{127}I	$^2\Sigma$	223.4	0.6				
^{90}Zr^{16}O	$^1\Sigma^+$	969.8	4.9	0.42263	0.0023	0.319	1.7116

INFRARED CORRELATION CHARTS

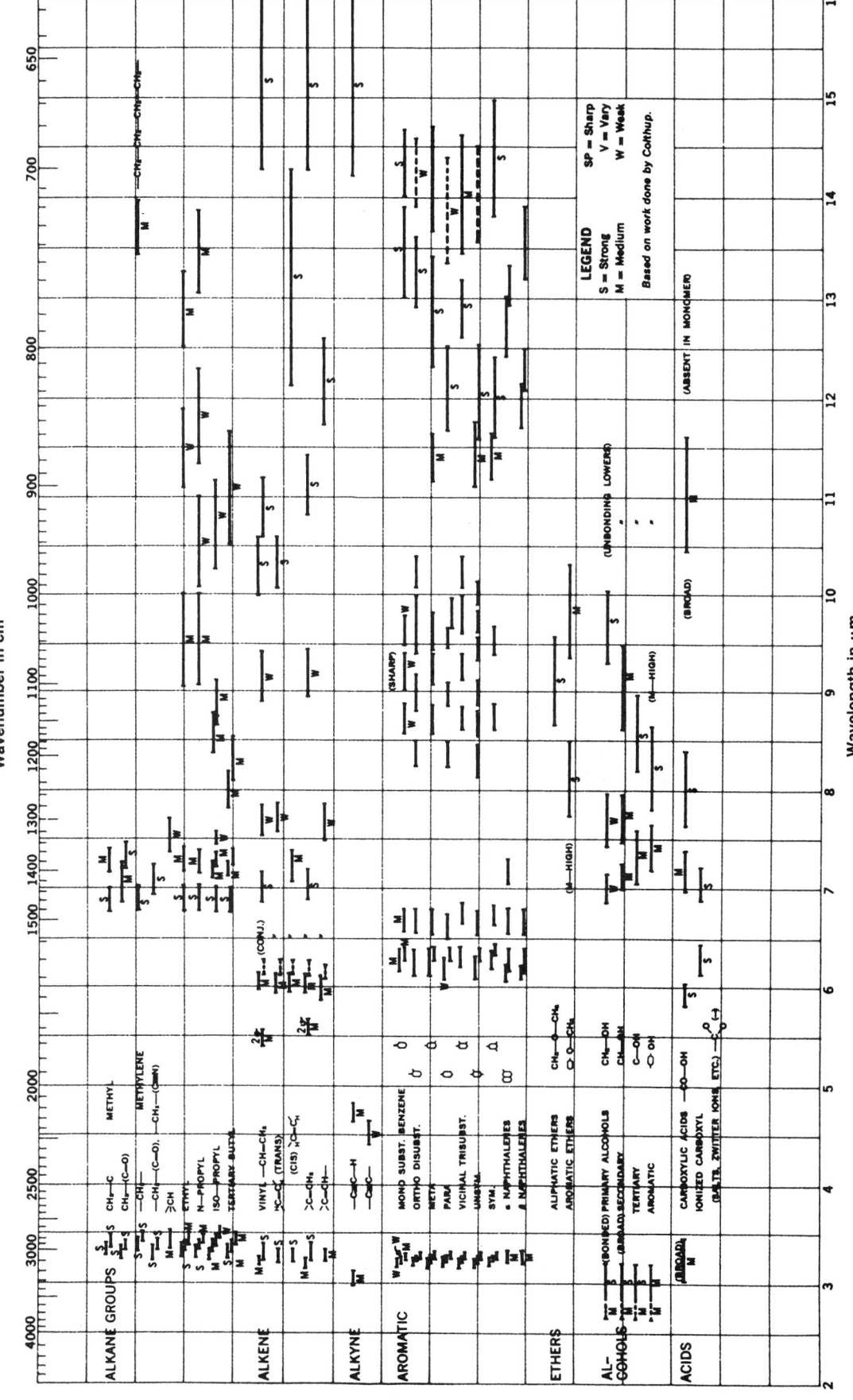

INFRARED CORRELATION CHARTS (continued)

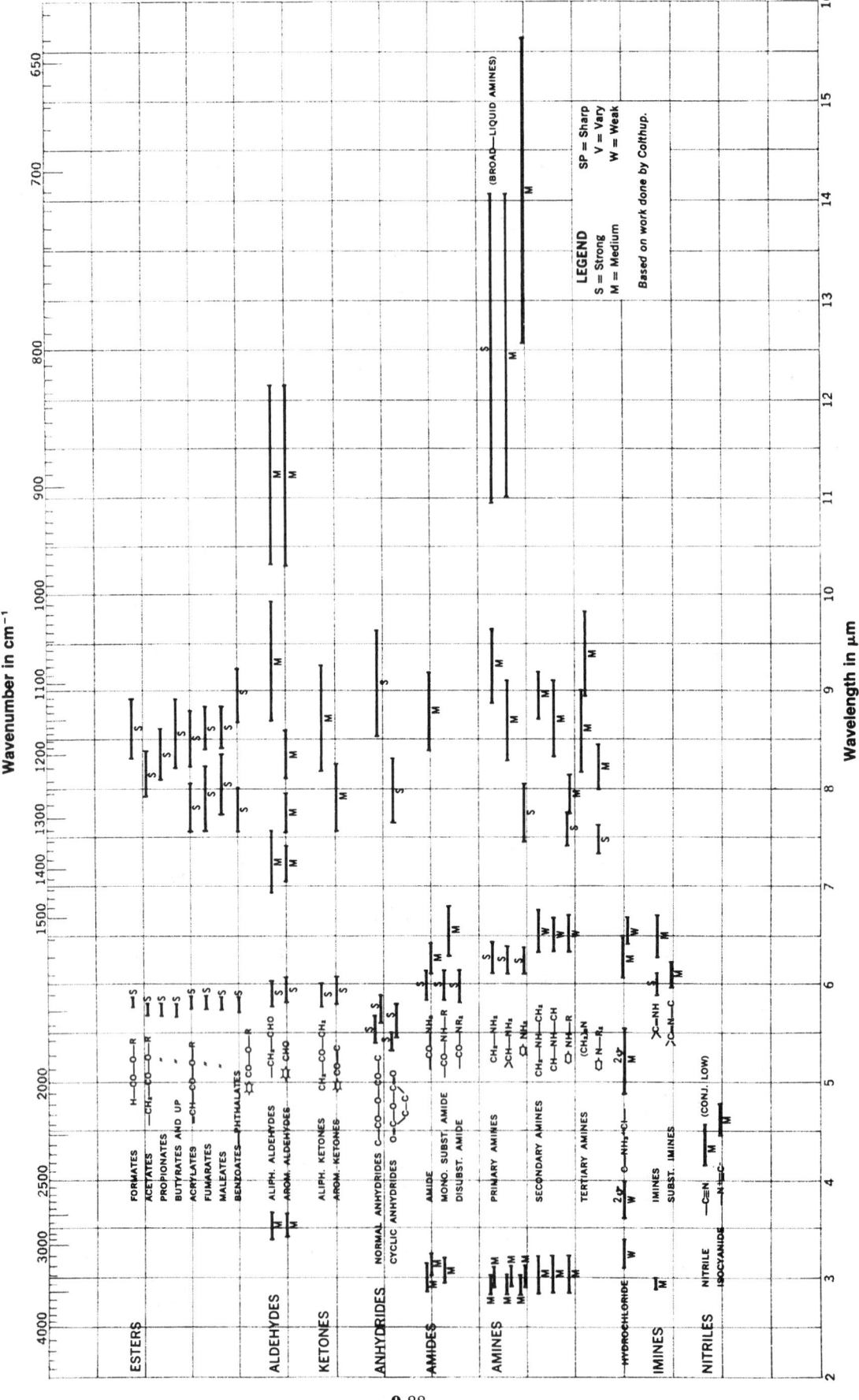

INFRARED CORRELATION CHARTS (continued)

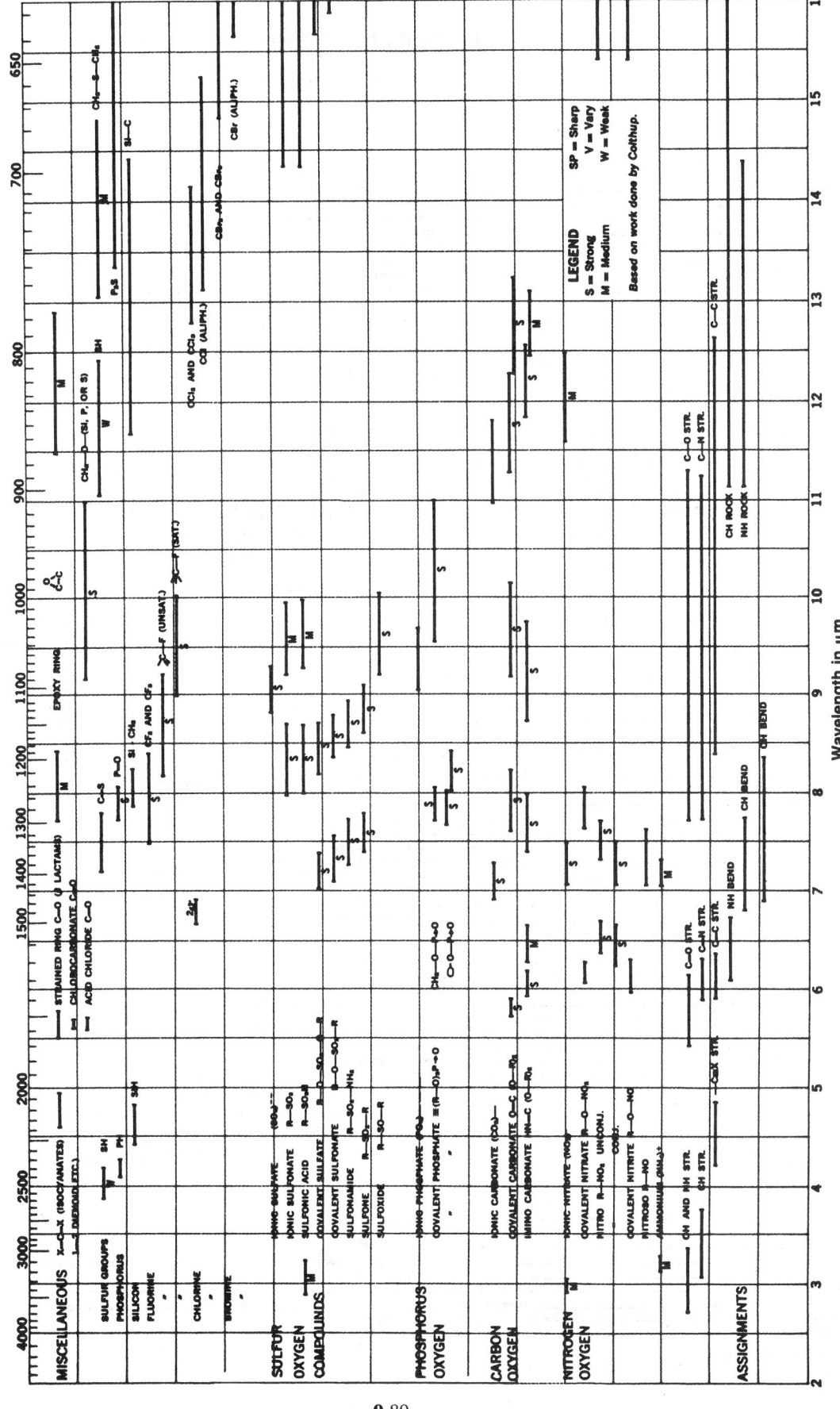

INFRARED CORRELATION CHARTS (continued)

Far Infrared Region

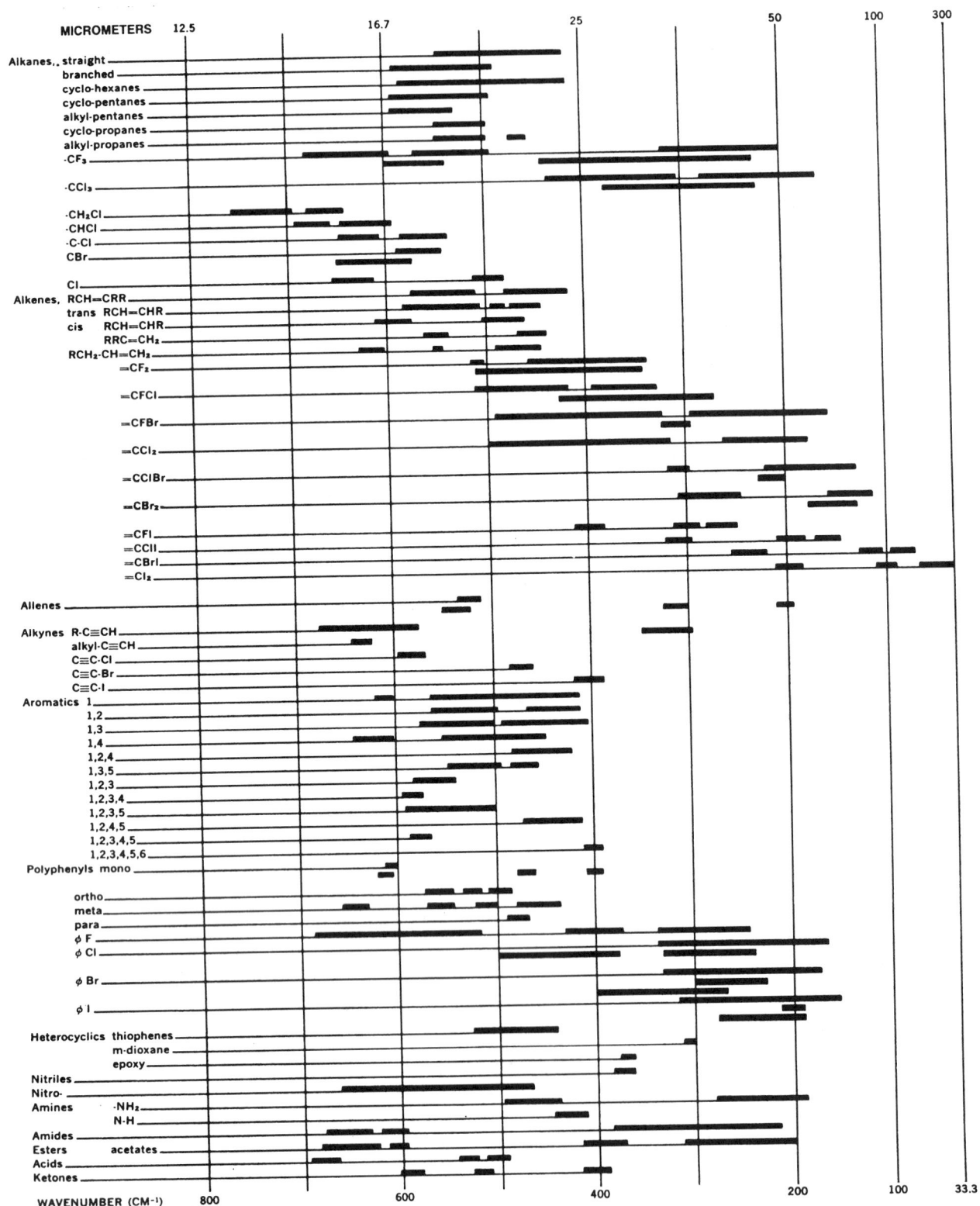

INFRARED CORRELATION CHARTS (continued)

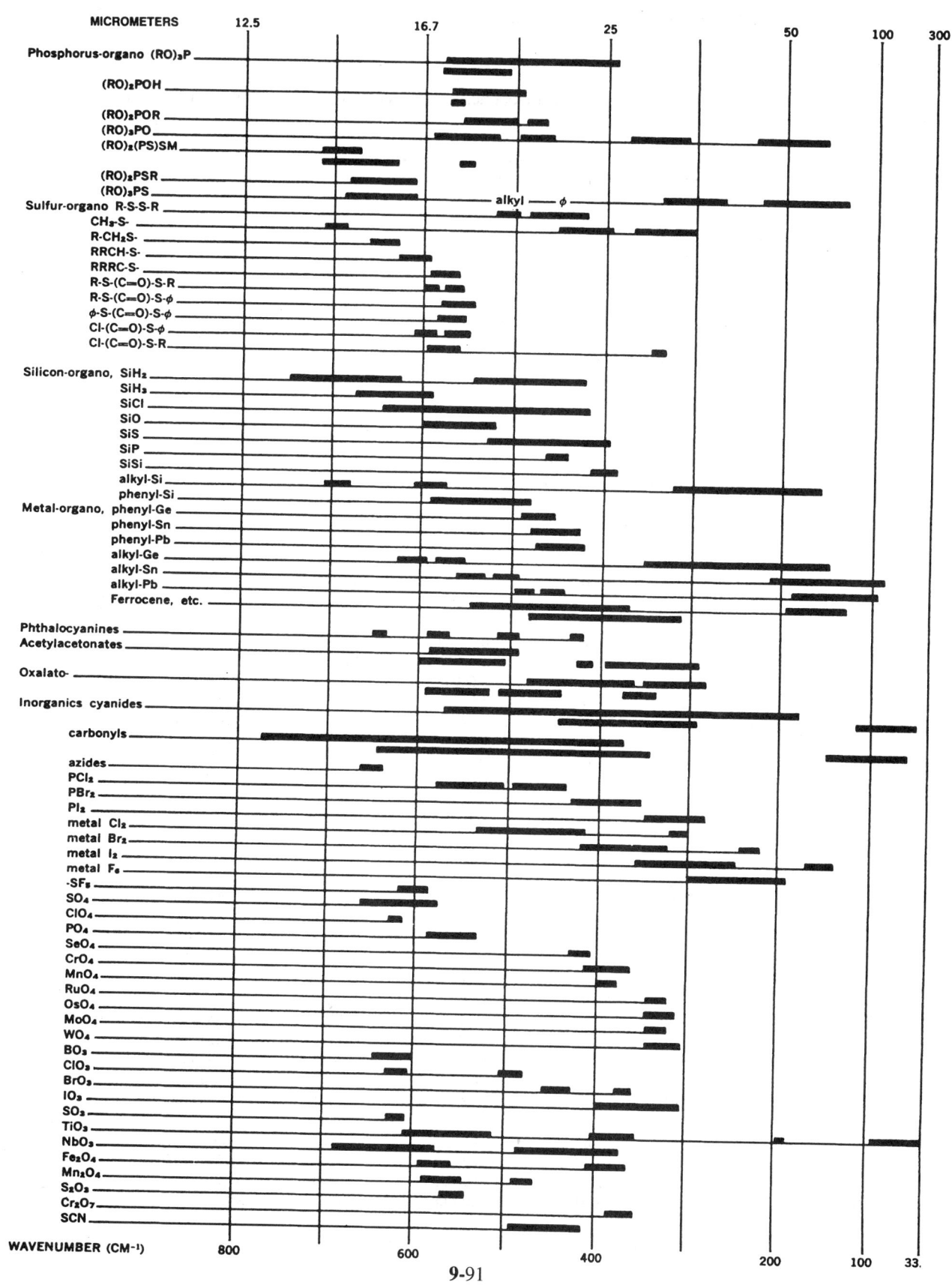

NUCLEAR SPINS, MOMENTS, AND OTHER DATA RELATED TO NMR SPECTROSCOPY

This table presents the following data relevant to nuclear magnetic resonance spectroscopy:

Z: Atomic number

Isotope: Element symbol and mass number

Abundance: Natural abundance of the isotope in percent. An * indicates a radioactive nuclide; if no value is given, the nuclide is not present in nature or its abundance is highly variable.

I: Nuclear spin

ν: Resonant frequency in megahertz for an applied field H_o of 1 tesla (in cgs units, 10 kilogauss)

Relative sensitivity: Sensitivity relative to ^1H (=1) assuming an equal number of nuclei and constant temperature. Values were calculated from the expressions:

      For constant H_o: $0.0076508(\mu/\mu_N)^3(I+1)/I^2$

      For constant ν: $0.23871(\mu/\mu_N)(I+1)$

μ: Nuclear magnetic moment in units of the nuclear magneton μ_N

Q: Nuclear quadrupole moment in units of femtometers squared (1 $fm^2 = 10^{-2}$ barn)

The table includes all stable nuclides of non-zero spin for which spin and magnetic moment values have been measured, as well as selected radioactive nuclides of current or potential interest. At least one isotope is included for each element through $Z = 95$ for which data are available. See Reference 2 for a complete listing of spins and moments.

The assistance of P. Pyykkö in providing recent data on nuclear quadrupole moments is gratefully acknowledged.

REFERENCES

1. Raghavan, P., "Table of Nuclear Moments", *At. Data Nuc. Data Tables,* 42, 189, 1989.
2. Holden, N. E., "Table of the Isotopes", in Lide, D. R., Ed., *CRC Handbook of Chemistry and Physics,* 74th Ed., CRC Press, Boca Raton, FL, 1993.
3. Pyykkö, P., *Z. Naturforsch.*, 47a, 189, 1992.

Z	Isotope	Abundance %	I	ν/MHz for $H_o = 1$ T	Relative sensitivity Const. H_o	Const. ν	μ/μ_N	Q/fm²
1	^{1}n	*	1/2	29.1639	0.32139	0.6850	-1.91304275	
1	^{1}H	99.985	1/2	42.5764	1.00000	1.0000	+2.7928474	
1	^{2}H	0.015	1	6.53573	0.00965	0.4093	+0.8574382	+0.2860
1	^{3}H	*	1/2	45.4137	1.21354	1.0666	+2.9789625	
2	^{3}He	0.0001	1/2	32.4352	0.44212	0.7618	-2.1276248	
3	^{6}Li	7.5	1	6.2660	0.00850	0.3925	+0.8220467	-0.082
3	^{7}Li	92.5	3/2	16.5478	0.29355	1.9433	+3.256427	-4.01
4	^{9}Be	100	3/2	5.986	0.01389	0.7029	-1.1779	+5.288
5	^{10}B	19.9	3	4.5751	0.01985	1.7193	+1.800645	+8.459
5	^{11}B	80.1	3/2	13.6626	0.16522	1.6045	+2.688649	+4.059
6	^{13}C	1.10	1/2	10.7081	0.01591	0.2515	+0.7024118	
7	^{14}N	99.634	1	3.0776	0.00101	0.1928	+0.4037610	+2.02
7	^{15}N	0.366	1/2	4.3172	0.00104	0.1014	-0.2831888	
8	^{17}O	0.038	5/2	5.7741	0.02910	1.5822	-1.89379	-2.558
9	^{19}F	100	1/2	40.0765	0.83400	0.9413	+2.628868	
10	^{21}Ne	0.27	3/2	3.3630	0.00246	0.3949	-0.661797	+10.155
11	^{23}Na	100	3/2	11.2686	0.09270	1.3233	+2.217522	+10.89
12	^{25}Mg	10.00	5/2	2.6082	0.00268	0.7147	-0.85545	+19.94
13	^{27}Al	100	5/2	11.1028	0.20689	3.0424	+3.641507	+14.03
14	^{29}Si	4.67	1/2	8.4653	0.00786	0.1988	-0.55529	
15	^{31}P	100	1/2	17.2510	0.06652	0.4052	+1.13160	
16	^{33}S	0.75	3/2	3.2716	0.00227	0.3842	+0.6438212	-6.78
17	^{35}Cl	75.77	3/2	4.1764	0.00472	0.4905	+0.8218743	-8.165
17	^{37}Cl	24.23	3/2	3.4764	0.00272	0.4083	+0.6841236	-6.435
18	^{37}Ar	*	3/2	5.818	0.01276	0.6833	+1.145	
18	^{39}Ar	*	7/2	2.8	0.00617	1.3964	-1.3	
19	^{39}K	93.2581	3/2	1.9893	0.00051	0.2336	+0.3914662	+6.01
19	^{40}K	0.0117	4	2.4737	0.00523	1.5493	-1.298100	-7.49
19	^{41}K	6.7302	3/2	1.0919	0.00008	0.1282	+0.2148701	+7.33

Z	Isotope	Abundance %	I	ν/MHz for $H_0 = 1$ T	Relative sensitivity Const. H_0	Const. ν	μ/μ_N	Q/fm^2
20	^{43}Ca	0.135	7/2	2.8688	0.00642	1.4150	-1.31726	-4.08
21	^{45}Sc	100	7/2	10.3588	0.30244	5.1093	+4.756487	-22
22	^{47}Ti	7.3	5/2	2.4040	0.00210	0.6587	-0.78848	+29
22	^{49}Ti	5.5	7/2	2.4047	0.00378	1.1861	-1.10417	+24
23	^{50}V	0.250	6	4.2504	0.05571	5.5904	+3.345689	+21
23	^{51}V	99.750	7/2	11.2130	0.38360	5.5306	+5.1487057	-5.2
24	^{53}Cr	9.501	3/2	2.4114	0.00091	0.2832	-0.47454	-15
25	^{55}Mn	100	5/2	10.5760	0.17881	2.8980	+3.46872	+33
26	^{57}Fe	2.1	1/2	1.3815	0.00003	0.0324	+0.0906230	
27	^{59}Co	100	7/2	10.077	0.27841	4.9702	+4.627	+42
28	^{61}Ni	1.140	3/2	3.8113	0.00359	0.4476	-0.75002	+16.2
29	^{63}Cu	69.17	3/2	11.2979	0.09342	1.3268	+2.22329	-22.0
29	^{65}Cu	30.83	3/2	12.1027	0.11484	1.4213	+2.38167	-20.4
30	^{67}Zn	4.1	5/2	2.6693	0.00287	0.7314	+0.875479	+15.0
31	^{69}Ga	60.108	3/2	10.2475	0.06971	1.2034	+2.01659	+17.0
31	^{71}Ga	39.892	3/2	13.0204	0.14300	1.5291	+2.56227	+10.0
32	^{73}Ge	7.73	9/2	1.4897	0.00141	1.1546	-0.8794677	-17.3
33	^{75}As	100	3/2	7.3148	0.02536	0.8590	+1.439475	+31.4
34	^{77}Se	7.63	1/2	8.1566	0.00703	0.1916	+0.535042	
35	^{79}Br	50.69	3/2	10.7039	0.07945	1.2570	+2.106400	+33.1
35	^{81}Br	49.31	3/2	11.5381	0.09951	1.3550	+2.270562	+27.6
36	^{83}Kr	11.5	9/2	1.6442	0.00190	1.2744	-0.970669	+25.3
37	^{85}Rb	72.165	5/2	4.1253	0.01061	1.1304	+1.35303	+27.4
37	^{87}Rb	27.835	3/2	13.9807	0.17703	1.6418	+2.75124	+13.2
38	^{87}Sr	7.00	9/2	1.8524	0.00272	1.4358	-1.093603	+33.5
39	^{89}Y	100	1/2	2.0949	0.00012	0.0492	-0.1374154	
40	^{91}Zr	11.22	5/2	3.9747	0.00949	1.0891	-1.30362	-20.6
41	^{93}Nb	100	9/2	10.4520	0.48821	8.1011	+6.1705	-32
42	^{95}Mo	15.92	5/2	2.7874	0.00327	0.7638	-0.9142	-2.2
42	^{97}Mo	9.55	5/2	2.8462	0.00349	0.7799	-0.9335	+25.5
43	^{99}Tc	*	9/2	9.63	0.38174	7.4633	+5.6847	-12.9
44	^{99}Ru	12.7	5/2	1.9553	0.00113	0.5358	-0.6413	+7.9
44	^{101}Ru	17.0	5/2	2.192	0.00159	0.6005	-0.7188	+45.7
45	^{103}Rh	100	1/2	1.3476	0.00003	0.0317	-0.08840	
46	^{105}Pd	22.33	5/2	1.957	0.00113	0.5364	-0.642	+66.0
47	^{107}Ag	51.839	1/2	1.7330	0.00007	0.0407	-0.1136796	
47	^{109}Ag	48.161	1/2	1.9924	0.00010	0.0468	-0.1306906	
48	^{111}Cd	12.80	1/2	9.0689	0.00966	0.2130	-0.5948861	
48	^{113}Cd	12.22	1/2	9.4868	0.01106	0.2228	-0.6223009	
49	^{113}In	4.3	9/2	9.3652	0.35121	7.2588	+5.5289	+79.9
49	^{115}In	95.7	9/2	9.3854	0.35348	7.2744	+5.5408	+81
50	^{115}Sn	0.34	1/2	14.0074	0.03561	0.3290	-0.91883	
50	^{117}Sn	7.68	1/2	15.2606	0.04605	0.3584	-1.00104	
50	^{119}Sn	8.59	1/2	15.9656	0.05273	0.3750	-1.04728	
51	^{121}Sb	57.36	5/2	10.2549	0.16302	2.8100	+3.3634	-36
51	^{123}Sb	42.64	7/2	5.5530	0.04659	2.7389	+2.5498	-49
52	^{123}Te	0.908	1/2	11.2346	0.01837	0.2639	-0.7369478	
52	^{125}Te	7.139	1/2	13.5451	0.03220	0.3181	-0.8885051	
53	^{127}I	100	5/2	8.5776	0.09540	2.3504	+2.813273	-78.9
54	^{129}Xe	26.4	1/2	11.8601	0.02162	0.2786	-0.7779763	
54	^{131}Xe	21.2	3/2	3.5158	0.00282	0.4129	+0.6918619	-12.0
55	^{133}Cs	100	7/2	5.6232	0.04838	2.7735	+2.582025	-0.37
56	^{135}Ba	6.592	3/2	4.2581	0.00500	0.5001	+0.837943	+16.0
56	^{137}Ba	11.23	3/2	4.7633	0.00700	0.5594	+0.937365	+24.5
57	^{138}La	* 0.0902	5	5.6614	0.09404	5.3188	+3.713646	+45
57	^{139}La	99.9098	7/2	6.0610	0.06058	2.9895	+2.7830455	+20
58	^{137}Ce	*	3/2	4.62	0.00641	0.5431	0.91	

Z	Isotope	Abundance %	I	ν/MHz for $H_0 = 1$ T	Relative sensitivity Const. H_0	Const. ν	μ/μ_N	Q/fm²
58	^{139}Ce	*	3/2	4.62	0.00641	0.5431	0.91	
58	^{141}Ce	*	7/2	2.37	0.00364	1.1708	1.09	
59	^{141}Pr	100	5/2	13.0355	0.33483	3.5720	+4.2754	-5.9
60	^{143}Nd	12.18	7/2	2.319	0.00339	1.1440	-1.065	-63
60	^{145}Nd	8.30	7/2	1.429	0.00079	0.7047	-0.656	-33
61	^{143}Pm	*	5/2	11.6	0.23510	3.1748	3.8	
61	^{147}Pm	*	7/2	5.7	0.04940	2.7928	+2.6	+70
62	^{147}Sm	15.0	7/2	1.7747	0.00152	0.8753	-0.8149	-26
62	^{149}Sm	13.8	7/2	1.4631	0.00085	0.7216	-0.6718	+9.4
63	^{151}Eu	47.8	5/2	10.5854	0.17929	2.9006	+3.4718	+90.3
63	^{153}Eu	52.2	5/2	4.6744	0.01544	1.2809	+1.5331	+241
64	^{155}Gd	14.80	3/2	1.317	0.00015	0.1546	-0.2591	+127
64	^{157}Gd	15.65	3/2	1.727	0.00033	0.2028	-0.3399	+135
65	^{159}Tb	100	3/2	10.23	0.06945	1.2019	+2.014	+143.2
66	^{161}Dy	18.9	5/2	1.4653	0.00048	0.4015	-0.4806	+247
66	^{163}Dy	24.9	5/2	2.0507	0.00130	0.5619	+0.6726	+265
67	^{165}Ho	100	7/2	9.0881	0.20423	4.4825	+4.173	+358
68	^{167}Er	22.95	7/2	1.2281	0.00050	0.6057	-0.5639	+357
69	^{169}Tm	100	1/2	3.531	0.00057	0.0829	-0.2316	
70	^{171}Yb	14.3	1/2	7.5259	0.00552	0.1768	+0.49367	
70	^{173}Yb	16.12	5/2	2.0730	0.00135	0.5680	-0.67989	+280
71	^{175}Lu	97.41	7/2	4.8624	0.03128	2.3983	+2.2327	+349
71	^{176}Lu	* 2.59	7	3.451	0.03975	6.0516	+3.169	+497
72	^{177}Hf	18.606	7/2	1.7281	0.00140	0.8524	+0.7935	+336
72	^{179}Hf	13.629	9/2	1.0856	0.00055	0.8414	-0.6409	+379
73	^{180}Ta	0.012	9	4.04	0.10251	11.3862	4.77	
73	^{181}Ta	99.988	7/2	5.1625	0.03744	2.5463	+2.3705	+317
74	^{183}W	14.3	1/2	1.7956	0.00008	0.0422	+0.1177847	
75	^{185}Re	37.40	5/2	9.717	0.13870	2.6627	+3.1871	+218
75	^{187}Re	*62.60	5/2	9.817	0.14300	2.6900	+3.2197	+207
76	^{187}Os	1.6	1/2	0.9856	0.00001	0.0231	+0.06465189	
76	^{189}Os	16.1	3/2	3.3535	0.00244	0.3938	+0.659933	+85.6
77	^{191}Ir	37.3	3/2	0.766	0.00003	0.0899	+0.1507	+81.6
77	^{193}Ir	62.7	3/2	0.832	0.00004	0.0977	+0.1637	+75.1
78	^{195}Pt	33.8	1/2	9.2920	0.01039	0.2182	+0.60952	
79	^{197}Au	100	3/2	0.7406	0.00003	0.0870	+0.145746	+54.7
80	^{199}Hg	16.87	1/2	7.7121	0.00594	0.1811	+0.5058855	
80	^{201}Hg	13.18	3/2	2.8468	0.00149	0.3343	-0.5602257	+38.6
81	^{203}Tl	29.524	1/2	24.7310	0.19598	0.5809	+1.6222579	
81	^{205}Tl	70.476	1/2	24.9742	0.20182	0.5866	+1.6382146	
82	^{207}Pb	22.1	1/2	9.0338	0.00955	0.2122	+0.59258	
83	^{209}Bi	100	9/2	6.9628	0.14433	5.3967	+4.1106	-50
84	^{209}Po	*	1/2	11.7	0.02096	0.2757	+0.77	
86	^{211}Rn	*	1/2	9.16	0.00997	0.2152	+0.601	
87	^{223}Fr	*	3/2	5.94	0.01362	0.6982	+1.17	+117
88	^{223}Ra	*	3/2	1.3746	0.00017	0.1614	+0.2705	+119
88	^{225}Ra	*	1/2	11.187	0.01814	0.2627	-0.7338	
89	^{227}Ac	*	3/2	5.6	0.01131	0.6564	+1.1	+170
90	^{229}Th	*	5/2	1.40	0.00042	0.3843	+0.46	+430
91	^{231}Pa	*	3/2	10.2	0.06903	1.1995	2.01	-172
92	^{235}U	* 0.7200	7/2	0.83	0.00015	0.4082	-0.38	+493.6
93	^{237}Np	*	5/2	9.57	0.13264	2.6234	+3.14	+388.6
94	^{239}Pu	*	1/2	3.09	0.00038	0.0727	+0.203	
95	^{243}Am	*	5/2	4.91	0.01788	1.3451	+1.61	+421

PROTON NMR CHEMICAL SHIFTS FOR CHARACTERISTIC ORGANIC STRUCTURES

The chart below summarizes the range of chemical shifts for protons in several classes or organic compounds and substituent groups. The chemical shifts δ are given in parts per million relative to tetramethylsilane.

REFERENCE

Mohacsi, E., *J. Chem. Edu.*, 41, 38, 1964 (with permission).

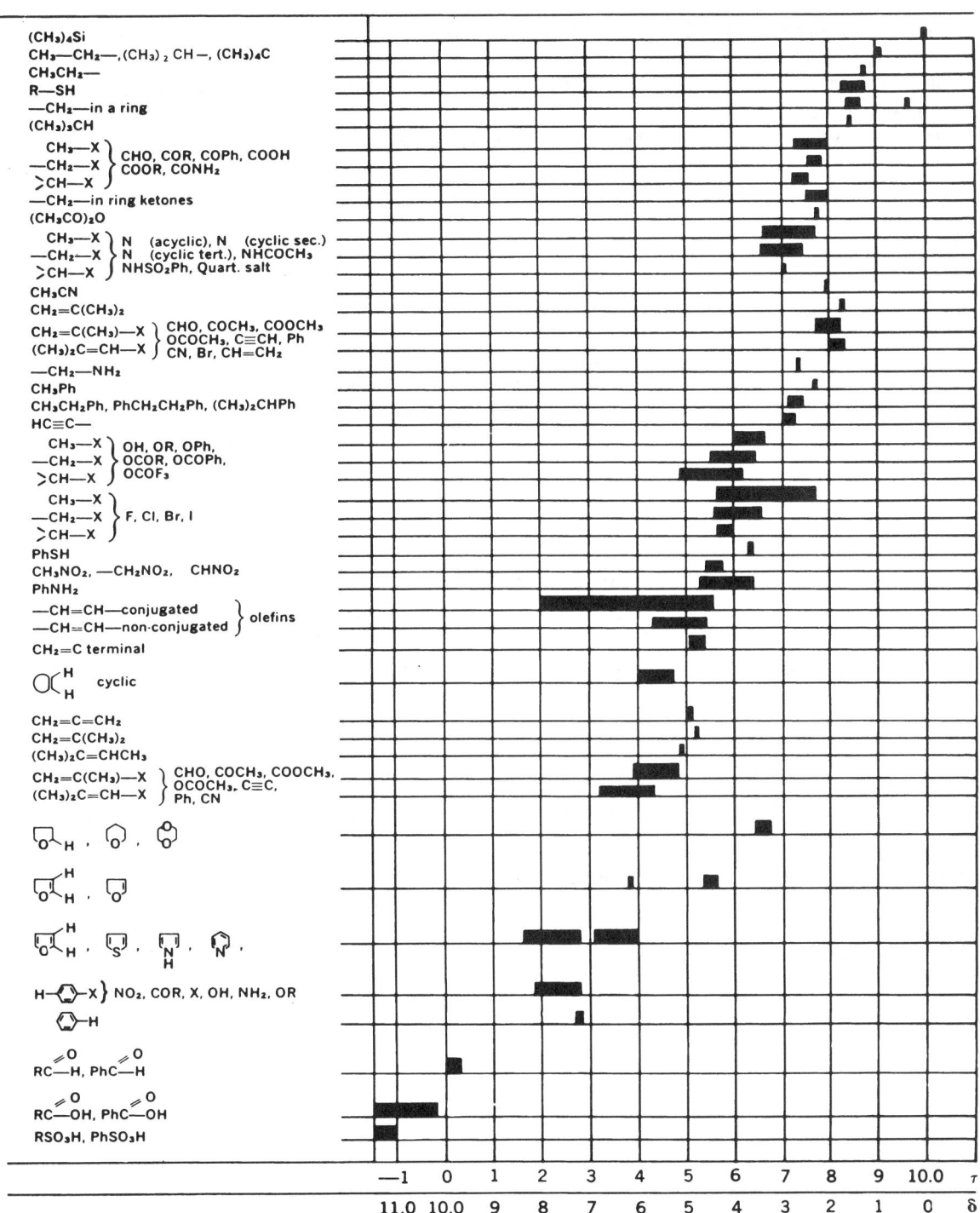

^{13}C-NMR ABSORPTIONS OF MAJOR FUNCTIONAL GROUPS

The table below lists the range of ^{13}C chemical shifts δ in parts per million relative to tetramethylsilane, in descending order, for various functional groups. Examples of simple compounds for each family are given to illustrate the correlations. The shifts for the carbons of interest, which are italicized, are given in parentheses; when two or more values appear, they refer to the sequence of italicized carbon atoms from left to right in the formula.

REFERENCES

1. Yoder, C. H. and Schaeffer, C. D., Jr., *Introduction to Multinuclear NMR: Theory and Application*, Benjamin/ Cummings, Menlo Part, CA, 1987.
2. Silverstein, R. M., Bassler, G. C., and Morrill, T. C., *Spectrometric Identification of Organic Compounds*, John Wiley & Sons, New York, 1981.
3. Brown, D. W., A Short Set of ^{13}C NMR Correlation Tables, *J. Chem. Educ.*, 62, 209, 1985.

δ (ppm)	Group	Family	Example (δ of italicized carbon)	
220-165	>C=O	Ketones	$(CH_3)_2CO$	(206.0)
			$(CH_3)_2CHCOCH_3$	(212.1)
		Aldehydes	CH_3CHO	(199.7)
		α,β-Unsaturated carbonyls	$CH_3CH=CHCHO$	(192.4)
			$CH_2=CHCOCH_3$	(169.9)
		Carboxylic acids	HCO_2H	(166.0)
			CH_3CO_2H	(178.1)
		Amides	$HCONH_2$	(165.0)
			CH_3CONH_2	(172.7)
		Esters	$CH_3CO_2CH_2CH_3$	(170.3)
			$CH_2=CHCO_2CH_3$	(165.5)
140-120	>C=C<	Aromatic	C_6H_6	(128.5)
		Alkenes	$CH_2=CH_2$	(123.2)
			$CH_2=CHCH_3$	(115.9, 136.2)
			$CH_2=CHCH_2Cl$	(117.5, 133.7)
			$CH_3CH=CHCH_2CH_3$	(132.7)
125-115	-CN	Nitriles	CH_3-CN	(117.7)
80-70	-CC-	Alkynes	$HCCH$	(71.9)
			CH_3CCH_3	(73.9)
70-45	-C-O	Esters	$CH_3OOCH_2CH_3$	(57.6, 67.9)
		Alcohols	$HOCH_3$	(49.0)
			$HOCH_2CH_3$	(57.0)
40-20	-C-NH$_2$	Amines	CH_3NH_2	(26.9)
			$CH_3CH_2NH_2$	(35.9)
30-15	-S-CH$_3$	Sulfides (thioethers)	$C_6H_5-S-CH_3$	15.6
30-(-2.3)	-C-H	Alkanes, cycloalkanes	CH_4	(-2.3)
			CH_3CH_3	(5.7)
			$CH_3CH_2CH_3$	(15.8, 16.3)
			$CH_3CH_2CH_2CH_3$	(13.4, 25.2)
			$CH_3CH_2CH_2CH_2CH_3$	(13.9, 22.8, 34.7)
		Cyclohexane		(26.9)

Section 10
Atomic, Molecular, and Optical Physics

Line Spectra of the Elements **10**-1
NIST Atomic Transition Probability Tables **10**-88
Electron Affinities **10**-147
Atomic and Molecular Polarizabilities **10**-160
Ionization Potentials of Atoms and Atomic Ions **10**-175
Ionization Energies of Gas-Phase Molecules **10**-178
X-Ray Atomic Energy Levels **10**-196
Electron Binding Energies of the Elements **10**-200
Natural Width of X-Ray Lines **10**-206
Photon Attenuation Coefficients **10**-207
Classification of Electromagnetic Radiation **10**-212
Black Body Radiation **10**-214
Characteristics of Infrared Detectors **10**-216
Refractive Index and Transmittance of Representative Glasses **10**-217
Index of Refraction of Water **10**-218
Index of Refraction of Liquids for Calibration Purposes **10**-219
Index of Refraction of Air **10**-220
Characteristics of Laser Sources **10**-221
Infrared Laser Frequencies **10**-228
Infrared and Far-Infrared Absorption Frequency Standards **10**-236

LINE SPECTRA OF THE ELEMENTS

Joseph Reader and Charles H. Corliss

The original tables from which this table was derived were prepared under the auspices of the Committee on Line Spectra of the Elements of the National Academy of Sciences-National Research Council. The table contains the outstanding spectral lines of neutral (I) and singly ionized (II) atoms of the elements from hydrogen through plutonium (Z = 1-94); selected strong lines from doubly ionized (III), triply ionized (IV), and quadruply ionized (V) atoms are also included. Listed are lines that appear in emission from the vacuum ultraviolet to the far infrared. These lines were selected from much larger lists in such a way as to include the stronger observed lines in each spectral region. A more extensive list may be found in Reference 1.

The data were compiled by the following contributors.

J. G. Conway — Lawrence Berkeley Laboratory
C. H. Corliss — National Bureau of Standards
R. D. Cowan — Los Alamos Scientific Laboratory
C. R. Cowley — University of Michigan
Henry M. and Hannah Crosswhite — Argonne National Laboratory
S. P. Davis — University of California, Berkeley
V. Kaufman — National Bureau of Standards
R. L. Kelly — Naval Postgraduate School
J. F. Kielkopf — University of Louisville
W. C. Martin — National Bureau of Standards
T. K. McCubbin — Pennsylvania State University

L. J. Radziemski — Los Alamos Scientific Laboratory
J. Reader — National Bureau of Standards
C. J. Sansonetti — National Bureau of Standards
G. V. Shalimoff — Lawrence Berkeley Laboratory
R. W. Stanley — Purdue University
J. O. Stoner, Jr. — University of Arizona
H. H. Stroke — New York University
D. R. Wood — Wright State University
E. F. Worden — Lawrence Livermore Laboratory
J. J. Wynne — International Business Machines Corporation
R. Zalubas — National Bureau of Standards

All wavelengths are given in Ångstrom units (10^{-10} m). Below 2000 Å the wavelengths are in vacuum; above 2000 Å the wavelengths are in air. Wavelengths given to three decimal places have an uncertainty of less than 0.001 Å and are therefore suitable for calibration purposes. In the air region, the elements used most commonly for calibration are Ne, Ar, Kr, Fe, Th, and Hg; in the vacuum region, the most common are C, N, O, Si, Cu.

All data refer to natural isotopic abundance of the elements except that Kr I and Kr II lines below 11,000 Å given to three decimal places are for ^{86}Kr. A separate table for ^{198}Hg contains accurately known wavelengths that are frequently used for calibration.

A large number of the lines for neutral and singly ionized atoms were extracted from the National Bureau of Standards (NBS) *Tables of Spectral-Line Intensities* (Reference 2). The intensities of these lines represent quantitative estimates of relative line strengths that take account of varying detection sensitivity at different wavelengths. They are on a linear scale. For nearly all of the other lines the intensities represent qualitative estimates of the relative strengths of lines not greatly separated in wavelength. Because different observers frequently use different scales for their intensity estimates, these intensities are useful only as a rough indication of the appearance of a spectrum. In some cases the intensity scale is not intended to be linear. In the first and second spectra the intensities of the lines of the singly ionized atom (II) relative to those of the neutral atom (I) should be used with caution, inasmuch as the concentration of ions in a light source depends greatly on the excitation conditions.

Descriptive symbols that follow the wavelength have the following meanings:

c — complex
d — line consists of two unresolved lines
h — hazy
l — shaded to longer wavelengths
s — shaded to shorter wavelengths
p — perturbed by a close line
r — easily reversed
w — wide

The table is arranged alphabetically by element name (not symbol); for each element the lines are listed by wavelength. References to the sources of data for each element are given at the end of the table, starting on page **10**-82.

GENERAL REFERENCES

1. Reader, J., Corliss, C. H., Wiese, W. L., and Martin, G. A., *Tables of Line Spectra of the Elements, Part 1. Wavelengths and Intensities*, Nat. Stand. Ref. Data Sys.- Nat. Bur. Standards (U.S.), No. 68, 1980.
2. Meggers, W. F., Corliss, C. H., and Scribner, B. F., *Tables of Spectral Line Intensities, Part 1. Arranged by Elements*, Nat. Bur. Stand. (U.S.), Monograph 145, 1975.
3. Fuhr, J. R., Martin, W. C., Musgrove, A., Sugar, J., and Wiese, W. L., "NIST Atomic Spectroscopic Database" ver. 1.1, January 1996. *NIST Physical Reference Data*, National Institute of Standards and Technology, Gaithersburg, MD. Available at the WWW address: http://physics.nist.gov/PhysRefData/contents.html

Intensity	Wavelength/Å	Intensity	Wavelength/Å	Intensity	Wavelength/Å	Intensity	Wavelength/Å
	Actinium	100	1611.814 III	110	2637.70 II	360	6696.02 I
	Ac Z = 89	800	1611.874 III	150	2652.48 I	230	6698.67 I
2000 h	2952.55 III	150	1625.627 II	200	2660.39 I	110	7361.57 I
2000 h	3392.78 III	800	1639.06 IV	160	2669.17 II	140	7362.30 I
3000	3487.59 III	100	1644.235 II	650	2816.19 II	230	7835.31 I
2000 s	3863.12 II	100	1644.809 II	150	3041.28 II	290	7836.13 I
3000 s	4088.44 II	1000	1670.787 II	360	3050.07 I	110	8075.35 I
3000 s	4168.40 II	100	1686.250 II	450	3057.14 I	290	8640.70 II
100	4179.98 I	800	1719.440 II	150	3074.64 II	360	8772.87 I
20	4183.12 I	500	1721.244 II	4500 r	3082.153 I	450	8773.90 I
20	4194.40 I	900	1721.271 II	7200 r	3092.710 I	110	8828.91 I
20 l	4384.53 I	500	1724.952 II	1800 r	3092.839 I	180	8841.28 I
20	4396.71 I	900	1724.984 II	150	3428.92 II	140	8923.56 I
2000 h	4413.09 III	350	1760.104 II	150	3443.64 I	150	9290.65 II
20	4462.73 I	300	1761.975 II	900	3492.23 IV	110	9290.75 II
3000 h	4569.87 III	290	1763.00 I	800	3508.46 IV	150	10076.29 II
1000	5910.85 II	500	1763.869 II	450	3586.56 II	110	10768.36 I
20	6359.86 I	700	1763.952 II	360	3587.07 II	140	10782.04 I
20 l	6691.27 I	450	1765.64 I	290	3587.45 II	110	10872.98 I
	Aluminum	300	1765.815 II	870	3601.63 III	230	10891.73 I
	Al Z = 13	450	1766.38 I	220	3651.06 II	450	11253.19 I
900	125.53 V	400	1767.731 II	110	3651.10 II	570	11254.88 I
800	126.07 V	450	1769.14 I	150	3654.98 II	570	13123.41 I
800	130.41 V	1000	1818.56 IV	290	3655.00 II	450	13150.76 I
1000	130.85 V	600	1828.588 II	450	3900.68 II	230	16718.96 I
900	131.00 V	400	1832.837 II	4500 r	3944.006 I	300	16750.56 I
900	131.44 V	250	1834.808 II	9000 r	3961.520 I	140	16763.36 I
800	160.07 IV	1000	1854.716 III	110	3995.86 II	300	21093.04 I
1000	278.69 V	300	1855.929 II	290	4226.81 II	360	21163.75 I
900	281.39 V	700	1858.026 II	870	4529.19 III		**Antimony**
70	486.884 III	120	1859.980 II	150	4585.82 II		**Sb Z = 51**
30	486.912 III	1000	1862.311 II	110	4588.19 II	15	722.86 III
250	511.138 III	600	1862.790 III	550	4666.80 II	15	732.33 III
150	511.191 III	200	1929.978 II	110	4898.76 II		861.5 IV
500	560.317 III	150	1931.048 II	110	4902.77 II	4	876.84 II
200	560.433 III	200	1932.377 II	150	5280.21 II	4	921.07 II
100	670.068 III	400	1934.503 II	290	5283.77 II	6	983.57 II
200	671.118 III	150	1934.713 II	150	5285.85 II	15	999.62 III
500	695.829 III	300	1935.840 III	110	5312.32 II	6	1001.13 II
400	696.217 III	200	1935.949 III	220	5316.07 II	6	1009.43 II
200	725.683 III	150	1936.907 II	150	5371.84 II	40	1011.94 III
300	726.915 III	220	1939.261 II	180	5557.06 I	6	1052.21 II
400	855.034 III	700	1990.531 II	110	5557.95 I	8	1056.27 II
500	856.746 III	150	2016.052 II	450	5593.23 II	8	1057.32 II
400	892.024 III	450	2016.234 II	1200	5696.60 III	40	1065.90 III
50	893.887 III	100	2016.368 II	1000	5722.73 III	6	1073.81 II
450	893.897 III	200	2074.008 II	110	5853.62 II	30	1075.82 III
800	1042.17 IV	700	2094.264 II	220	5971.94 II		1087.6 IV
50	1191.812 II	150	2094.744 II	290	6001.76 II	8	1104.32 V
900	1237.19 IV	300	2094.791 II	220	6001.88 II	30	1151.49 III
900	1257.62 IV	100	2095.104 II	450	6006.42 II	40	1157.74 III
800	1264.18 IV	200	2095.141 II	150	6061.11 II		1199.1 IV
1000	1272.76 IV	400	2269.10 I	290	6068.43 II	50	1205.20 III
150	1350.18 II	120	2269.22 I	110	6068.53 II	50	1210.64 III
800	1384.13 III	140	2321.56 I	450	6073.23 II	12	1226.00 V
800	1447.51 IV	460	2367.05 I	110	6181.57 II	6	1230.30 II
800	1494.79 IV	110	2367.61 I	150	6181.68 II	8	1274.98 II
1000	1526.14 V	110	2368.11 I	290	6182.28 II	20	1306.69 III
800	1537.54 IV	180	2369.30 I	220	6182.45 II	8	1327.40 II
800	1539.830 II	140	2370.22 I	450 h	6183.42 II	6	1358.04 II
1000	1557.25 IV	160	2372.07 I	450	6201.52 II	8	1384.70 II
100	1569.385 II	850	2373.12 I	360	6201.70 II	20	1404.18 III
900	1582.04 IV	170	2373.35 I	290	6226.18 II	6	1407.83 II
800	1584.46 IV	110	2373.57 I	360	6231.78 II	8	1436.49 II
125	1596.059 II	240	2567.98 I	450	6243.36 II	20 r	1486.57 I
700	1605.766 III	480	2575.10 I	450	6335.74 II	40 h	1491.36 I

Line Spectra of the Elements (continued): Antimony—Argon

Intensity	Wavelength/Å		Intensity	Wavelength/Å		Intensity	Wavelength/Å		Intensity	Wavelength/Å	
	1499.2	IV	400	2478.32	I	30 h	7648.28	I	200	666.011	II
12	1505.70	V	150	2480.44	I	80	7844.44	I	1000	670.946	II
50 r	1512.57	I	100	2510.54	I	200	7924.65	I	3000	671.851	II
12	1524.47	V	2000 r	2528.52	I	60	8411.69	I	70	676.242	II
120 r	1532.74	I	15	2528.54	II	150	8572.64	I	30	677.952	II
80 r	1535.06	I	10	2567.75	II	100	8619.55	I	30	679.218	II
6	1565.51	II	150	2574.06	I	400	9518.68	I	200	679.401	II
8	1576.11	II	15	2590.13	III	400	9949.14	I	10	683.28	IV
7	1581.36	II	1500 r	2598.05	I	200	10078.49	I	7	688.39	IV
80 r	1599.96	I	500 r	2598.09	I	300	10261.01	I	12 p	689.01	IV
10	1606.98	II	300 r	2612.31	I	200	10585.60	I	6	699.41	IV
200 w	1612.8	I	12	2617.17	III	1000	10677.41	I	8	700.28	IV
100 w	1623.3	I	200 r	2652.60	I	800	10741.94	I	3	705.35	V
20	1657.04	II	20	2669.39	III	80	10794.11	I	5	709.20	V
100 w	1662.6	I	300 r	2670.64	I	600	10839.73	I	4	715.60	V
15	1673.89	III	200 r	2682.76	I	200	10868.58	I	3	715.65	V
15	1711.84	III	120	2692.25	I	400	10879.55	I	200	718.090	II
80 r	1716.93	I	150 r	2718.90	I	300	11012.79	I	3000	723.361	II
150 r	1717.45	I	400 r	2769.95	I	150	11266.23	I	2	725.11	V
150 r	1723.43	I	1000 r	2877.92	I	5	12116.06	I	500	725.548	II
15	1725.33	III	15	2980.96	II		**Argon**		70	730.930	II
100 r	1736.19	I	500 r	3029.83	I		**Ar Z = 18**		200	740.269	II
100 h	1765.76	I	600 r	3232.52	I	3	336.56	V	200	744.925	II
100 r	1780.87	I	20	3241.28	II	3	337.56	V	70	745.322	II
100 r	1788.24	I	700 r	3267.51	I	6	338.00	V	4	754.20	IV
150	1800.18	I	15	3498.46	II	2	338.43	V	5	761.47	IV
50 r	1810.50	I	25	3637.80	II	2	339.01	V	12	769.15	III
80 r	1814.20	I	250	3637.83	I	3	339.89	V	5	800.57	IV
100	1829.50	I	20	3722.78	II	3	350.88	V	10	801.09	IV
50 r	1868.17	I	200 r	3722.79	I	4	396.87	IV	10	801.41	IV
300 r	1871.15	I	20	3850.22	II	4	398.55	IV	5	801.91	IV
150 r	1882.56	I	200	4033.55	I	2	436.67	V	20	802.859	I
100	1927.08	I	20	4033.56	II	5	446.00	V	100	806.471	I
200 r	1950.39	I	20	4133.63	II	8	446.95	V	60	806.869	I
60 r	2029.49	I	15	4140.54	II	4	447.53	V	30	807.218	I
70 r	2039.77	I	15	4195.17	II	18	449.06	V	40	807.653	I
150 r	2049.57	I	20	4219.07	II	4	449.49	V	50	809.927	I
1000 r	2068.33	I	20	4314.32	II	3	458.12	V	120	816.232	I
100	2079.56	I	15	4514.50	II	2	458.98	V	70	816.464	I
50 r	2098.41	I	30	4596.90	II	6 p	461.23	V	80	820.124	I
80 r	2118.48	I	20	4599.09	II	3	462.42	V	4	822.16	V
100 r	2127.39	I	15	4604.77	II	7	463.94	V	120	825.346	I
50 r	2137.05	I	30	4647.32	II	30	487.227	II	120	826.365	I
100 r	2139.69	I	20	4675.74	II	50	490.650	II	5	827.05	V
10	2141.80	II	40	4711.26	II	30	490.701	II	3	827.35	V
50 r	2141.83	I	20	4757.81	II	30	519.327	II	150	834.392	I
100 r	2144.86	I	20	4765.36	II	3	522.09	V	4 p	834.88	V
1500 r	2175.81	I	30	4784.03	II	5	524.19	V	100	835.002	I
250 r	2179.19	I	20	4802.01	II	6	527.69	V	2	836.13	V
200 r	2201.32	I	20	4832.82	II	30	542.912	II	15	840.03	IV
300 r	2208.45	I	20	4877.24	II	200	543.203	II	100	842.805	I
150 r	2220.73	I	15	4947.40	II	70	547.461	II	20	843.77	IV
100	2221.98	I	15	5044.56	II	2	554.50	V	25	850.60	IV
120 r	2224.93	I	20	5238.94	II	70	556.817	II	180	866.800	I
300 r	2262.51	I	20	5354.24	II	5	558.48	V	150	869.754	I
120	2288.98	I	40 h	5556.10	I	70	573.362	II	10	871.10	III
150 r	2293.44	I	100 l	5632.02	I	30	576.736	II	9	875.53	III
300 r	2306.46	I	30	5639.75	II	70	580.263	II	180 r	876.058	I
2500 r	2311.47	I	60 h	5830.34	I	30	583.437	II	12	878.73	III
150	2315.89	I	100	6005.21	I	70	597.700	II	8	879.62	III
400 h	2373.67	I	20	6053.41	II	30	602.858	II	180 r	879.947	I
300 h	2383.64	I	30	6079.80	II	30	612.372	II	9	883.18	III
100	2395.22	I	50	6130.04	II	6	623.77	IV	10	887.40	III
150	2422.13	I	20	6154.94	II	3	635.12	V	150	894.310	I
250	2426.35	I	20	6611.49	I	500	661.867	II	5	900.36	IV
400 r	2445.51	I	30	6647.44	II	30	664.562	II	9	901.17	IV

Intensity	Wavelength/Å		Intensity	Wavelength/Å		Intensity	Wavelength/Å		Intensity	Wavelength/Å	
1000	919.781	II	15	2640.34	IV	8	3511.12	III	100	4228.158	II
1000	932.054	II	10	2654.63	III	70	3514.388	II	100	4237.220	II
1000 r	1048.220	I	8	2674.02	III	70	3545.596	II	25	4251.185	I
500 r	1066.660	I	9	2678.38	III	70	3545.845	II	200	4259.362	I
7	1669.67	III	9	2682.63	IV	7	3554.306	I	100	4266.286	I
7	1673.42	III	10	2724.84	III	100	3559.508	II	70	4266.527	II
7	1675.48	III	14	2757.92	IV	100	3561.030	II	150	4272.169	I
9	1914.40	III	7	2762.23	III	70	3576.616	II	550	4277.528	II
7	1915.56	III	10	2776.26	IV	25	3581.608	II	20	4282.898	II
10	2125.16	III	12	2784.47	IV	50	3582.355	II	100	4300.101	I
15	2133.87	III	14	2788.96	IV	70	3588.441	II	25	4300.650	II
10	2138.59	III	7	2797.11	IV	7	3606.522	I	70	4309.239	II
10	2148.73	III	16	2809.44	IV	25	3622.138	II	200	4331.200	II
15	2166.19	III	10	2830.25	IV	20	3639.833	II	50	4332.030	II
10	2168.26	III	7	2842.88	III	35	3718.206	II	100	4333.561	I
20	2170.23	III	8	2855.29	III	70	3729.309	II	50	4335.338	I
25	2177.22	III	6	2874.40	IV	50	3737.889	II	25	4345.168	I
8	2184.06	III	9	2884.12	III	150	3765.270	II	800	4348.064	II
10	2188.22	III	25	2891.612	II	50	3766.119	II	50	4352.205	II
15	2192.06	III	12	2913.00	IV	20	3770.369	I	25	4362.066	II
7	2248.73	III	11	2926.33	IV	20	3770.520	II	50	4367.832	II
10	2279.10	III	200	2942.893	II	25	3780.840	II	200	4370.753	II
7	2281.22	III	100	2979.050	II	20	3795.37	III	70	4371.329	II
7	2282.21	III	10	3010.02	III	25	3803.172	II	50	4375.954	II
12	2293.03	III	12	3024.05	III	50	3809.456	II	150	4379.667	II
4	2299.72	IV	50	3033.508	II	7	3834.679	I	50	4385.057	II
10	2300.85	III	6	3037.98	IV	70	3850.581	II	70	4400.097	II
15	2302.17	III	12	3054.82	III	10	3858.32	III	200	4400.986	II
9	2317.00	III	10	3064.77	III	35	3868.528	II	400	4426.001	II
15	2317.47	III	8	3077.40	IV	7	3907.84	III	150	4430.189	II
12	2318.04	III	10	3078.15	III	35	3925.719	II	50	4430.996	II
10	2319.13	III	50	3093.402	II	50	3928.623	II	50	4433.838	II
10	2319.37	III	7	3110.41	III	25	3932.547	II	20	4439.461	II
9	2345.17	III	7	3127.90	III	70	3946.097	II	35	4448.879	II
7	2351.67	III	8	3200.37	I	7	3947.505	I	100	4474.759	II
9	2360.26	III	20	3243.689	II	35	3948.979	I	200	4481.811	II
10	2395.63	III	25	3285.85	III	8	3960.53	III	100	4510.733	I
12	2399.15	III	25	3293.640	II	20	3979.356	II	20	4522.323	I
10	2413.20	III	20	3301.88	II	35	3994.792	II	20	4530.552	II
7	2415.61	III	20	3307.228	II	50	4013.857	II	400	4545.052	II
10	2418.82	III	15	3311.25	III	6	4023.60	III	20	4564.405	II
5	2420.456	II	7	3319.34	I	50	4033.809	II	400	4579.350	II
12	2423.52	III	7	3323.59	III	20	4035.460	II	400	4589.898	II
12	2423.93	III	25	3336.13	III	150	4042.894	II	15	4596.097	I
7	2443.69	III	20	3344.72	III	50	4044.418	I	550	4609.567	II
8	2447.71	IV	25	3350.924	II	100	4052.921	II	7	4628.441	I
8	2472.95	III	15	3358.49	III	200	4072.005	II	35	4637.233	II
7	2476.10	III	7	3361.28	III	70	4072.385	II	400	4657.901	II
12	2488.86	III	7	3373.47	I	25	4076.628	II	15	4702.316	I
12	2513.28	IV	25	3376.436	II	35	4079.574	II	20	4721.591	II
10	2516.789	II	25	3388.531	II	25	4082.387	II	550	4726.868	II
6	2518.40	IV	15	3391.85	II	150	4103.912	II	50	4732.053	II
9	2525.69	IV	7	3393.73	I	300	4131.724	II	300	4735.906	II
10	2534.709	II	7	3417.49	III	5	4146.70	III	800	4764.865	II
15	2562.087	II	9	3424.25	III	35	4156.086	II	550	4806.020	II
12	2562.17	IV	8	3438.04	III	400	4158.590	I	150	4847.810	II
10	2568.07	IV	7	3461.07	I	50	4164.180	I	50	4865.910	II
7	2569.53	IV	9	3471.32	III	35	4179.297	II	800	4879.864	II
12	2599.47	IV	70	3476.747	II	50	4181.884	I	70	4889.042	II
10	2608.06	IV	20	3478.232	II	100	4190.713	I	20	4904.752	II
7	2608.44	IV	20	3480.55	III	50	4191.029	I	35	4933.209	II
12	2615.68	IV	50	3491.244	II	200	4198.317	I	200	4965.080	II
6	2619.98	IV	100	3491.536	II	400	4200.674	I	50	5009.334	II
12	2621.36	IV	12	3499.67	III	25	4218.665	II	70	5017.163	II
12	2624.92	IV	15	3503.58	III	25	4222.637	II	70	5062.037	II
7	2631.90	III	70	3509.778	II	25	4226.988	II	20	5090.495	II

Intensity	Wavelength/Å		Intensity	Wavelength/Å		Intensity	Wavelength/Å		Intensity	Wavelength/Å	
100	5141.783	II	7	6960.250	I	11	11078.869	I	800	1211.17	II
70	5145.308	II	10000	6965.431	I	30	11106.46	I	800	1218.10	II
5	5151.391	I	150	7030.251	I	12	11441.832	I	340	1223.15	II
15	5162.285	I	10000	7067.218	I	400	11488.109	I	760	1241.31	II
25	5165.773	II	100	7068.736	I	200	11668.710	I	965	1243.08	II
20	5187.746	I	25	7107.478	I	12	11719.488	I	870	1245.67	II
20	5216.814	II	25	7125.820	I	200	12112.326	I	800	1258.58	II
7	5221.271	I	1000	7147.042	I	50	12139.738	I	965	1263.77	II
5	5421.352	I	15	7158.839	I	50	12343.393	I	800	1266.34	II
10	5451.652	I	70	7206.980	I	200	12402.827	I	800	1267.59	II
25	5495.874	I	15	7265.172	I	200	12439.321	I	715	1280.99	II
5	5506.113	I	7	7270.664	I	100	12456.12	I	715	1287.54	II
25	5558.702	I	2000	7272.936	I	200	12487.663	I	715	1305.70	II
10	5572.541	I	35	7311.716	I	150	12702.281	I	340	1307.74	II
35	5606.733	I	25	7316.005	I	30	12733.418	I	760	1333.15	II
20	5650.704	I	5	7350.814	I	12	12746.232	I	965	1341.55	II
10	5739.520	I	70	7353.293	I	200	12802.739	I	760	1355.93	II
5	5834.263	I	200	7372.118	I	50	12933.195	I	965	1369.77	II
10	5860.310	I	20	7380.426	II	500	12956.659	I	800	1373.65	II
15	5882.624	I	10000	7383.980	I	200	13008.264	I	1000	1375.07	II
25	5888.584	I	20	7392.980	I	200	13213.99	I	760	1375.78	II
50	5912.085	I	15	7412.337	I	200	13228.107	I	800	1394.64	II
15	5928.813	I	10	7425.294	I	100	13230.90	I	800	1400.31	II
5	5942.669	I	25	7435.368	I	500	13272.64	I	500	1448.59	II
7	5987.302	I	10	7436.297	I	1000	13313.210	I	500	1558.88	II
5	5998.999	I	20000	7503.869	I	1000	13367.111	I	500	1570.99	II
5	6025.150	I	15000	7514.652	I	30	13499.41	I	100 r	1593.60	I
70	6032.127	I	25000	7635.106	I	1000	13504.191	I	500	1660.55	II
35	6043.223	I	15000	7723.761	I	11	13573.617	I	340	1860.34	II
10	6052.723	I	10000	7724.207	I	30	13599.333	I	1000 r	1890.42	I
20	6059.372	I	10	7891.075	I	400	13622.659	I	500	1912.94	II
7	6098.803	I	20000	7948.176	I	200	13678.550	I	800 r	1937.59	I
10	6105.635	I	20000	8006.157	I	1000	13718.577	I	585 r	1972.62	I
100	6114.923	II	25000	8014.786	I	10	13825.715	I	170 r	1990.35	I
10	6145.441	I	7	8053.308	I	10	13907.478	I	100 r	1991.13	I
7	6170.174	I	20000	8103.693	I	200	14093.640	I	100 r	1995.43	I
150	6172.278	II	35000	8115.311	I	100	15046.50	I	230 r	2003.34	I
10	6173.096	I	10000	8264.522	I	25	15172.69	I	100 r	2009.19	I
10	6212.503	I	20	8392.27	I	10	15329.34	I	200	2263.2	IV
5	6215.938	I	15000	8408.210	I	30	15989.49	I	350 r	2288.12	I
25	6243.120	II	20000	8424.648	I	30	16519.86	I	200	2301.0	IV
7	6296.872	I	15000	8521.442	I	500	16940.58	I	350 r	2349.84	I
15	6307.657	I	7	8605.776	I	12	18427.76	I	100 r	2370.77	I
7	6369.575	I	4500	8667.944	I	50	20616.23	I	135 r	2381.18	I
20	6384.717	I	20	8771.860	II	30	20986.11	I	250	2417.5	IV
70	6416.307	I	180	8849.91	I	20	23133.20	I	250	2454.0	IV
25	6483.082	II	20	9075.394	I	20	23966.52	I	170 r	2456.53	I
15	6538.112	I	35000	9122.967	I		**Arsenic**		200	2461.4	IV
15	6604.853	I	550	9194.638	I		**As Z = 33**		340	2602.00	II
25	6638.221	II	15000	9224.499	I	510	871.7	III	170 r	2780.22	I
20	6639.740	II	400	9291.531	I	325	889.0	III	300	2830.359	II
50	6643.698	II	1600	9354.220	I	325	927.5	III	300	2831.164	II
5	6660.676	I	25000	9657.786	I	325	937.2	III	100 r	2860.44	I
5	6664.051	I	4500	9784.503	I	325	953.6	III	300	2884.406	II
25	6666.359	II	180	10052.06	I	325	963.8	III	80	2926.3	III
100	6677.282	I	30	10332.72	I	250	987.7	V	615	2959.572	II
35	6684.293	II	100	10467.177	II	340	1021.96	II	300	3003.819	II
150	6752.834	I	1600	10470.054	I	250	1029.5	V	300	3116.516	II
5	6756.163	I	13	10478.034	I	340	1082.35	II	340	3842.60	II
15	6766.612	I	180	10506.50	I	500	1139.40	II	325	3922.6	III
20	6861.269	II	200	10673.565	I	615	1149.31	II	715	4190.082	II
150	6871.289	I	11	10681.773	I	555	1181.51	II	615	4197.40	II
5	6879.582	I	7	10683.034	II	555	1189.87	II	615	4242.982	II
10	6888.174	I	30	10733.87	I	615	1196.38	II	500	4315.657	II
50	6937.664	I	30	10759.16	I	615	1196.56	II	500	4323.867	II
7	6951.478	I	7	10812.896	II	340	1207.44	II	500	4336.64	II

Intensity	Wavelength/Å		Intensity	Wavelength/Å		Intensity	Wavelength/Å		Intensity	Wavelength/Å	
500	4352.145	II		1771.03	II	100	3576.28	II	400	4899.97	II
425	4352.864	II		1786.93	II	30	3577.62	I	15	4902.90	I
375	4371.17	II	100	1904.15	II	80 h	3579.67	I	20000	4934.09	II
615	4427.106	II	500	1924.70	II	200	3596.57	II	8	4947.35	I
615	4431.562	II		1985.60	II	40	3630.64	I	1000	4957.15	II
715	4458.469	II	300	1999.54	II	40 h	3636.83	I	300	4997.81	II
340	4461.075	II	10	2001.30	III	20 h	3688.47	I	1000	5013.00	II
715	4466.348	II		2009.20	II	400	3735.75	II	20 h	5159.94	I
500	4474.46	II	400	2023.95	II	200	3816.69	II	20	5267.03	I
800	4494.230	II		2052.68	II	200	3842.80	II	800	5361.35	II
850	4507.659	II		2054.57	II	100	3854.76	II	1000	5391.60	II
615	4539.74	II	500	2214.7	II	20	3889.33	I	200	5421.05	II
715	4543.483	II	800	2245.61	II	1400 l	3891.78	II	100	5424.55	I
615	4602.427	II	1000	2254.73	II	20	3892.65	I	200	5428.79	II
340	4629.787	II	1400	2304.24	II	40	3909.91	I	300	5480.30	II
340	4707.586	II	60	2331.10	III	500	3914.73	II	200	5519.05	I
340	4730.67	II	2000	2335.27	II	25	3926.85	III	1000 r	5535.48	I
340	4888.557	II	190	2347.58	II	50	3935.72	I	20 h	5620.40	I
340	5105.58	II	40	2512.28	III	20	3937.87	I	10	5680.18	I
500	5107.55	II	40	2523.83	III	200	3939.67	II	400	5777.62	I
425	5231.38	II	60	2528.51	II	500	3949.51	II	800	5784.18	II
500	5331.23	II	50	2559.54	III	25	3993.06	III	100	5800.23	I
340	5497.727	II	8 h	2596.64	I	80	3993.40	I	20	5805.69	I
425	5558.09	II	100	2634.78	II	30	3995.66	I	150	5826.28	I
425	5651.32	II	40	2681.89	III	300	4036.26	II	2800	5853.68	II
425	6110.07	II	8	2702.63	I	200	4083.77	II	15	5907.64	I
500	6170.27	II	18	2771.36	II	30 h	4084.86	I	100	5971.70	I
300	6511.74	II	15	2785.28	I	1500 h	4130.66	II	800	5981.25	II
300	7092.27	II	100 r	3071.58	I	20	4132.43	I	100	5997.09	I
300	7102.72	II	40	3079.14	III	200	4166.00	II	300	5999.85	II
340	7990.53	II	10 h	3108.21	I	500	4216.04	II	100	6019.47	I
300	8174.51	II	8	3132.60	I	800	4267.95	II	200	6063.12	I
200	9300.61	I	8 h	3135.72	I	100	4283.10	I	300	6110.78	I
230	9597.95	I	10	3137.70	I	300	4287.80	II	400	6135.83	II
290	9626.70	I	10	3155.34	I	200	4297.60	II	20000	6141.72	II
230	9833.76	I	10	3155.67	I	800	4309.32	II	150	6341.68	I
170	9915.71	I	12	3158.05	I	20 h	4323.00	I	500	6378.91	II
290	9923.05	I	12 h	3158.54	I	600	4325.73	II	10	6383.76	III
290	10024.04	I	25	3165.60	I	200	4326.74	II	90	6450.85	I
170	10614.07	I	15 h	3173.69	I	300	4329.62	II	150	6482.91	I
Astatine			30	3183.16	I	80	4350.33	I	12000	6496.90	II
At Z = 85			15	3183.96	I	60	4402.54	I	300	6498.76	I
8	2162.25	I	10	3193.91	I	400	4405.23	II	150	6527.31	I
10	2244.01	I	25 h	3203.70	I	40	4431.89	I	3000	6595.33	I
Barium			30	3221.63	I	60 h	4488.98	I	150	6654.10	I
Ba Z = 56			40	3222.19	I	50 h	4493.64	I	1500	6675.27	I
14	555.48	III	50	3261.96	I	40	4505.92	I	1800	6693.84	I
14	587.57	III	60 r	3262.34	I	200	4509.63	II	1000	6769.62	II
18	647.27	III	40	3281.50	I	60 h	4523.17	I	600	6865.69	I
300	719.86	V	15	3281.77	I	130	4524.93	II	300 h	6867.85	I
150	721.85	V	50	3322.80	I	65000	4554.03	II	1000	6874.09	II
1000	766.87	V	80 h	3356.80	I	40	4573.85	I	6000	7059.94	I
40000	794.89	IV	50	3368.18	III	80	4579.64	I	2400 hs	7120.33	I
300	877.41	V	60 r	3377.08	I	30	4599.75	I	600	7195.24	I
50000	923.74	IV	20	3377.39	I	20 h	4619.92	I	600 hl	7228.84	I
200	946.26	V	70 r	3420.32	I	25 h	4628.33	I	3000	7280.30	I
200	1486.72	II	25	3421.01	I	300	4644.10	II	1200	7392.41	I
400	1504.01	II	30 h	3421.48	I	30	4673.62	I	300	7417.53	I
300	1554.38	II	40	3463.74	I	35	4691.62	I	900 hl	7459.78	I
200	1572.73	II	200 r	3501.11	I	20	4700.43	I	600	7488.08	I
	1573.92	II	80 h	3524.97	I	800	4708.94	II	450 hl	7636.90	I
	1630.40	II	30 h	3531.35	I	40	4726.44	I	600 hl	7642.91	I
100	1674.51	II	80 h	3544.66	I	800	4843.46	II	1800	7672.09	I
400	1694.37	II	20 h	3547.68	I	300	4847.14	II	1200	7780.48	I
	1697.16	II	100	3552.45	II	200	4850.84	II	180 h	7839.57	I
	1761.75	II	200	3567.73	II	30 h	4877.65	I	1500	7905.75	I

Intensity	Wavelength/Å		Intensity	Wavelength/Å		Intensity	Wavelength/Å		Intensity	Wavelength/Å	
600	7911.34	I		93.93	II		1909.0	II	10	3046.52	II
900 h	8210.24	I		94.78	II		1912.	I	30	3046.69	II
8	8308.69	III		95.76	II	3	1917.03	III		3090.3	I
1800 h	8559.97	I		96.29	I		1919.	I	10	3110.81	I
100	8710.74	II		97.24	I	5	1929.67	I	10	3110.92	I
100	8737.71	II		97.44	I	10	1943.68	I	20	3110.99	I
300 h	8799.76	I		97.86	I	60 h	1954.97	III		3120.	I
300	8860.98	I		97.97	I		1956.	I	480	3130.42	II
450	8914.99	I		98.12	I	50	1964.59	I	320	3131.07	II
300	9219.69	I		98.37	I	5	1985.13	I		3136.	I
300	9308.08	I		98.66	I		1997.95	I		3150.	I
300 h	9324.58	I		98.94	I		1997.98	I		3160.6	I
1500	9370.06	I		99.19	I	60	1998.01	I		3163.	I
300	9455.92	I	100	100.25	III		2033.25	I		3168.	I
8	9521.76	III		100.86	I		2033.28	I		3180.7	II
450	9589.37	I		101.20	I		2033.38	I		3187.	I
900	9608.88	I		102.13	I	50	2055.90	I	20	3193.81	I
300 h	9645.72	I		102.49	II	100	2056.01	I	20	3197.10	II
1500 hl	9830.37	I		104.40	I	75 h	2076.94	I	30	3197.15	II
900	10001.08	I		104.67	I	60 h	2080.38	III	20	3208.60	I
600	10032.10	I		105.80	I	25	2118.56	III		3220.	I
1200 h	10233.23	I		107.26	I	15 h	2122.27	III	60	3229.63	I
300	10471.26	I		107.38	I	10	2125.57	I	2	3233.52	II
120 hl	10791.25	I	3	509.99	III	20	2125.68	I	10	3241.62	II
180 hl	11012.69	I	2	549.31	III	15 h	2127.20	III	30	3241.83	II
150 h	11114.42	I	6	582.08	III	5	2137.25	III	15	3269.02	I
240	11303.04	I	4	661.32	III	25	2145.	I	100	3274.58	II
120 h	11697.45	I	8	675.59	III	55	2174.99	I	30	3274.67	II
120	13207.30	I		714.0	II	55	2175.10	I	30	3282.91	I
120	13810.50	I	4	725.59	III	5	2191.57	III	30	3321.01	I
120	14077.90	I	5	725.71	II		2273.5	II	30	3321.09	I
120	15000.40	I	5	743.58	III		2324.6	II	220	3321.34	I
120	20712.00	I	7	746.23	III		2337.0	I	20	3345.43	I
150	25515.70	I	2	767.75	III	950	2348.61	I	60	3367.63	I
150	29223.90	I	8	775.37	II	20	2350.66	I		3405.6	II
Beryllium			20	842.06	II	60	2350.71	I	5	3451.37	I
Be Z = 4				865.3	II	200	2350.83	I	300	3455.18	I
	58.13	IV	2	925.25	II	2	2413.34	II	20	3476.56	I
	58.57	IV	10	943.56	II	16	2413.46	II	300	3515.54	I
	59.32	IV	10	973.27	II	20	2453.84	I	10	3555.	I
	60.74	IV		981.4	II		2480.6	I	100	3720.36	III
	64.06	IV		1020.1	II	35	2494.54	I		3720.92	III
	75.93	IV	8	1026.93	II	35	2494.58	I		3722.98	III
1 h	76.10	III	5	1036.32	II	100	2494.73	I	100	3736.30	I
2	76.48	III	15	1048.23	II	16	2507.43	II	700	3813.45	I
3	78.53	III	1	1114.69	III	5	2617.99	II	40	3865.13	I
4	78.66	III	20	1143.03	II	20	2618.13	II	80	3865.42	I
1 h	78.92	III		1155.9	II	100	2650.45	I	1	3865.51	I
5	81.89	III	60	1197.19	II	60	2650.55	I	6	3865.72	I
10	82.38	III	2	1213.12	III	200	2650.62	I	100	3866.03	I
	82.58	II	1	1214.32	III	60	2650.69	I	90 h	4249.14	III
20	83.20	III	2	1362.25	III	100	2650.76	I	100	4253.05	I
	83.66	II	1	1401.52	III	5	2697.46	I	60	4253.76	I
30	84.76	III	10	1421.26	III	20	2697.58	II	300	4360.66	II
50	88.31	III	5	1422.86	III	20	2728.88	II	500	4360.99	II
	89.16	I		1426.12	I	30	2738.05	I	400	4407.94	I
	89.80	II	1	1435.17	III		2764.2	II	2	4485.52	III
	90.04	II	2	1440.77	III	20	2898.13	I	100 h	4487.30	III
	90.21	I		1491.76	I	10	2898.19	I	1	4495.09	III
	90.67	I	20	1512.30	II	20	2898.25	I	140 h	4497.8	III
	91.06	II	60	1512.43	II	30	2986.06	I		4526.6	I
	91.36	II	100	1661.49	I	10	2986.42	I		4548.	I
	91.74	II	2 h	1754.69	III	60	3019.33	I	12	4572.66	I
	92.19	I	15	1776.12	II	30	3019.49	I	700	4673.33	II
	92.61	II	20	1776.34	II	30	3019.53	I	1000	4673.42	II
	93.14	II		1907.	I	20	3019.60	I	6	4709.37	I
	93.42	II									

Intensity	Wavelength/Å		Intensity	Wavelength/Å		Intensity	Wavelength/Å		Intensity	Wavelength/Å	
200	4828.16	II	30	12098.18	II	35	1538.06	II	2800	2989.03	I
40	4849.16	I	100	14643.92	I	20	1563.67	II	700	2993.34	I
2 h	4858.22	II	60	14644.75	I	40	1573.70	II	100	3012.	IV
80	5087.75	I	200	16157.72	I	60	1591.79	II	2400	3024.64	I
8	5218.12	II	80	17855.38	I	25	1601.58	II	60	3034.87	I
20	5218.33	II	120	17856.63	I	60 h	1606.40	III	9000 c	3067.72	I
3	5255.86	II	100	18143.54	I	40	1609.70	II	140	3076.66	I
64	5270.28	II	160	31775.05	I	40	1611.38	II	35	3115.0	III
500	5270.81	II	200	31778.70	I	20	1652.81	II	100	3239.	IV
20	5403.04	II		**Bismuth**		20	1749.29	II	550 c	3397.21	I
20	5410.21	II		**Bi Z = 83**		80	1777.11	II	10	3430.83	II
	5558.	I	6	420.7	IV	60	1787.47	II	12	3431.23	II
140 h	6142.01	III	6	431.2	IV	70	1791.93	II	40 h	3451.0	III
10	6229.11	I	2	488.39	V	70	1823.80	II	40	3473.8	III
16	6279.43	II	3	563.62	V	100	1902.41	II	35	3485.5	III
30	6279.73	II	5	670.76	III	9000	1954.53	I	500 c	3510.85	I
30	6473.54	I	6	686.88	V	7000	1960.13	I	380 c	3596.11	I
60	6547.89	II	5	730.71	V	25	1989.35	II	45	3613.4	III
60	6558.36	II	10	738.17	V	7000	2021.21	I	100	3643.	IV
30	6564.52	I	4	775.16	III	9000	2061.70	I	12	3654.2	II
2 h	6636.44	II	6	790.5	IV	45 h	2068.9	II	20	3682.	IV
1	6756.72	II	6	790.6	IV	4600	2110.26	I	10	3695.32	III
2	6757.13	II	8	792.5	IV	2500	2133.63	I	50	3695.68	III
30	6786.56	I	10	820.3	IV	15	2143.40	II	100	3734.	IV
1 h	6884.22	I	9	822.9	IV	15	2143.46	II	70 h	3792.5	II
6 h	6884.44	I	12	824.9	IV	60	2186.9	II	12	3811.1	II
100	6982.75	I	15 d	864.45	V	40 h	2214.0	II	20	3815.8	II
6 h	7154.40	I	15	872.6	IV	360	2228.25	I	10	3845.8	II
40 h	7154.65	I	12	923.9	IV	1700	2230.61	I	30	3863.9	II
100	7209.13	I	15	943.3	IV	340	2276.58	I	100	3868.	IV
3	7401.20	II	25	1039.99	III	100	2311.	IV	40 h	4079.1	II
2	7401.43	II	50 h	1045.76	III	100	2326.	IV	10	4097.2	II
10	7551.90	I	30	1051.81	III	16	2368.12	II	140	4121.53	I
10 h	7618.68	I	15	1058.88	II	12	2368.25	II	140	4121.86	I
20 h	7618.88	I	20	1085.47	II	100	2376.	IV	75 h	4259.4	II
60	8090.06	I	10	1099.20	II	190	2400.88	I	25	4272.0	II
5 h	8158.99	I	24	1103.4	IV	75 h	2414.6	III	70 h	4301.7	II
10 h	8159.24	I	20	1139.01	III	10	2501.0	II	12 h	4339.8	II
4	8254.07	I	50	1224.64	III	25	2515.69	I	25 h	4340.5	II
10 h	8287.07	I	10	1225.43	II	70	2524.49	I	12 h	4379.4	II
30	8547.36	I	15	1232.78	II	20 h	2544.5	II	25 h	4476.8	II
60	8547.67	I	10	1241.05	II	700	2627.91	I	60 h	4705.3	II
300	8801.37	I	10	1265.35	II	100	2629.	IV	600 c	4722.52	I
6	8882.18	I	15	1283.73	II	100	2677.	IV	30	4730.3	II
40	9190.45	I	10	1306.18	II	12	2693.0	II	20	4749.7	II
20 h	9243.92	I	60	1317.0	IV	280 c	2696.76	I	40 h	4797.4	III
1 h	9343.89	II	20	1325.46	II	20	2713.3	II	12	4908.2	II
40	9392.74	I	40	1326.84	III	140 d	2730.50	I	10	4916.6	II
2	9476.43	II	20	1329.47	II	100	2767.	IV	12	4969.7	II
16	9477.03	II	60	1346.12	III	100	2772.	IV	20	4993.6	II
20	9847.32	I	20	1350.07	II	360	2780.52	I	45 h	5079.3	III
10 h	9895.63	I	25	1372.61	II	100	2786.	IV	10	5091.6	II
20 h	9895.96	I	15	1376.02	II	15	2803.42	II	50 h	5124.3	II
80	9939.78	I	20	1393.92	II	11	2803.70	II	60 h	5144.3	II
16	10095.52	II	35	1423.33	III	12	2805.3	II	20	5201.5	II
20	10095.73	II	35	1423.52	III	140 c	2809.62	I	75 h	5209.2	II
60	10119.92	II	45	1436.83	II	100	2842.	IV	40 h	5270.3	II
80	10331.03	I	25	1447.94	II	80 h	2855.6	III	10	5397.8	II
30	11066.46	I	50	1455.11	II	4000	2897.98	I	10 c	5552.35	I
	11173.	II	60 h	1461.00	III	100	2924.	IV	3	5599.41	II
1	11173.73	II	25	1462.14	II	100	2933.	IV	20	5655.2	II
120	11496.39	I	35	1486.93	II	100	2936.	IV	40 h	5719.2	II
2 h	11625.16	II	20	1502.50	II	15	2936.7	II	6	5742.55	I
	11659.	II	40	1520.57	II	3200	2938.30	I	12	5818.3	II
2	11660.25	II	40	1533.17	II	20	2950.4	II	20	5860.2	II
100	12095.36	II	30	1536.77	II	12	2963.4	II			

Intensity	Wavelength/Å		Intensity	Wavelength/Å		Intensity	Wavelength/Å		Intensity	Wavelength/Å	
20	5973.0	II	160	677.14	III	40	4242.98	III	7500	1232.43	I
15	6059.1	II	40	693.95	II	70	4243.61	III	1200	1243.90	I
15	6128.0	II	40	731.36	II	110	4472.10	II	1500	1251.66	I
6	6134.82	I	40	731.44	II	110	4472.85	II	1000	1255.80	I
3	6475.73	I		749.74	V	220	4487.05	III	1500	1259.20	I
3	6476.24	I	40	758.48	III	360	4497.73	III	1200	1261.66	I
15	6497.7	II	70	758.67	III	70	4784.21	II	1200	1266.20	I
10	6577.2	II	110	882.54	II	110	4940.38	II	1000	1279.48	I
40 h	6600.2	II	110	882.68	II	110	6080.44	II	1000	1286.26	I
50 h	6808.6	II	40	984.67	II	70	6285.47	II	3000	1309.91	I
4 h	6991.12	I	110	1081.88	II	70	7030.20	II	3000	1316.74	I
12	7033.	II	110	1082.07	II	40	7031.90	II	1000	1317.37	I
2	7036.15	I	70	1112.2	IV	110	7835.25	III	2000	1317.70	I
10 h	7381.	II	450	1168.9	IV	70	7841.41	III	12000	1384.60	I
2	7502.33	I	70	1170.9	IV	20	8667.22	I	3000	1449.90	I
10 h	7637.	II	110	1230.16	II	70	8668.57	I	50000	1488.45	I
10	7750.	II	220	1362.46	II	800	11660.04	I	30000	1531.74	I
3	7838.70	I	70	1600.46	I	570	11662.47	I	25000	1540.65	I
2	7840.33	I	120	1600.73	I	125	15629.08	I	30000	1574.84	I
20	7965.	II	160	1623.58	II	200	16240.38	I	20000	1576.39	I
40	8008.	III	110	1623.77	II	250	16244.67	I	25000	1582.31	I
12 h	8050.	II	220	1624.02	II	235	18994.33	I	75000	1633.40	I
50	8070.	III	70	1624.16	II		**Bromine**		1000	2133.79	IV
15	8328.	II	160	1624.34	II		**Br Z = 35**		1000	2145.02	IV
15	8388.	II	100	1663.04	I	700	379.73	IV	1000	2257.21	IV
30	8532.	II	150	1666.87	I	700	400.37	IV	1000	2272.73	IV
2	8544.54	I	200	1667.29	I	800	482.11	V	1000	2307.40	IV
1	8579.74	I	150	1817.86	I	900	531.97	V	1000	2408.16	IV
25	8653.	II	200	1818.37	I	1000	545.43	IV	1000	2411.58	IV
2	8754.88	I	300	1825.91	I	1000	547.90	V	700	2491.14	IV
3	8761.54	I	300	1826.41	I	1000	559.76	IV	1000	2581.19	IV
25	8863.	II	110	1842.81	II	1000	569.19	IV	600	2661.40	IV
2	8907.81	I	20	1953.83	III	1000	576.59	IV	1000	2842.88	IV
2000 d	9657.04	I	550	2065.78	III	1000	585.10	IV	1100 h	2907.71	IV
40	9827.78	I	250	2066.38	I	1000	586.71	IV	500 h	2972.26	II
20	10104.5	I	250	2066.65	I	1000	597.51	IV	500	3041.18	IV
15	10138.8	I	100	2066.93	I	1000	600.09	IV	500	3074.42	III
20	10300.6	I	300	2067.19	I	1000	601.27	IV	500	3349.64	III
20	10536.19	I	450	2067.23	III	1000	607.03	IV	500	3380.56	IV
50	11072.44	I	160	2077.09	III	1000	617.85	IV	500	3540.16	III
1500 d	11710.37	I	500	2088.91	I	1000	619.87	IV	500	3562.43	III
40	11999.49	I	500	2089.57	I	1000	630.14	IV	1200	3815.65	I
200	12165.08	I	70	2220.30	II	1000	642.23	IV	1500	3992.36	I
200	12690.04	I	40	2234.09	III	1000	661.53	IV	1000	4223.89	II
100	12817.8	I	70	2234.59	III	1000	683.51	IV	2000	4365.14	I
200	14330.5	I	40	2323.03	II	1000	697.72	IV	1000	4365.60	II
50	16001.5	I	40	2328.67	II	1000	715.39	IV	1500	4425.14	I
60	22551.6	I	40	2393.20	II	1000	731.00	IV	10000	4441.74	I
	Boron		220	2395.05	II	1000	800.12	IV	10000	4472.61	I
	B Z = 5		40	2459.69	II	700	812.95	V	20000	4477.72	I
	41.00	V	40	2459.90	II	1000	813.66	IV	1000	4490.42	I
30	48.59	V	1000	2496.77	I	1000	850.81	V	3000	4513.44	I
10	52.68	IV	1000	2497.73	I	1000	889.23	II	15000	4525.59	I
30	60.31	IV	70	2524.7	IV	1000	948.97	II	3000	4575.74	I
	194.37	V	160	2530.3	IV	1000	1015.54	II	2500	4614.58	I
	262.37	V	450	2821.68	IV	1000	1049.00	II	2500	4752.28	I
160	344.0	IV	70	2824.57	IV	1000	1069.15	V	4000	4780.31	I
450	385.0	IV	285	2825.85	IV	900	1112.13	V	1600	4785.19	I
40	411.80	III	160	2918.08	II	1000	1143.56	V	4000	4979.76	I
285	418.7	IV	110	3032.26	II	1000	1189.28	I	1200	5395.48	I
20	510.77	III	70	3179.33	II	1000	1189.50	I	1200	5466.22	I
40	510.85	III	110	3323.18	II	1000	1210.73	I	1800	5852.08	I
	512.53	V	110	3323.60	II	1000	1221.13	I	1600	5940.48	I
150	518.24	III	450	3451.29	II	1000	1223.24	I	2400	6122.14	I
75	518.27	III	285	4121.93	II	1200	1224.41	I	40000	6148.60	I
110	677.00	III	110	4194.79	II	1200	1226.90	I	2000	6177.39	I

Intensity	Wavelength/Å		Intensity	Wavelength/Å		Intensity	Wavelength/Å		Intensity	Wavelength/Å	
1500	6335.48	I	9000	8964.00	I	60	567.01	IV	50	2628.979	I
60000	6350.73	I	30000	9166.06	I	150	1118.16	IV	40	2632.190	I
2500	6410.32	I	15000	9173.63	I	100	1164.65	IV	75	2639.420	I
1800	6483.56	I	20000	9178.16	I	100	1183.40	IV	40	2659.23	II
1000	6514.62	I	40000	9265.42	I	100	1256.00	II	50 h	2660.325	I
20000	6544.57	I	15000	9320.86	I	150	1296.43	II	25	2668.20	II
1500	6548.09	I	6000	9793.48	I	100	1326.50	II	50	2672.62	II
50000 c	6559.80	I	10000	9896.40	I	60	1370.48	IV	100	2677.540	I
1000	6571.31	I	3000	10140.08	I	150	1370.91	II	25	2677.748	I
1800	6579.14	I	6000	10237.74	I	60	1418.89	IV	50	2707.00	II
20000	6582.17	I	1000	10299.62	I	200	1514.26	II	75	2712.505	I
1500	6620.47	I	1500	10377.65	I	50	1545.17	III	50	2733.820	I
50000 c	6631.62	I	30000	10457.96	I	200	1571.58	II	1000	2748.54	II
20000	6682.28	I	1000	10742.14	I	100	1668.60	II	100 h	2763.894	I
10000	6692.13	I	3000	10755.92	I	50	1702.47	II	50 h	2764.230	I
8000	6728.28	I	1700	13217.17	I	40	1707.16	III	50	2774.958	I
2000	6760.06	I	1800	14354.57	I	40	1722.95	III	30	2823.19	II
2000	6779.48	I	1250	14888.70	I	50	1724.41	II	200	2836.900	I
2200	6786.74	I	1800	16731.19	I	40	1747.67	III	25	2856.46	II
6500	6790.04	I	1200	18568.31	I	40	1773.06	III	100	2868.180	I
1600 c	6791.48	I	3500	19733.62	I	100	1785.84	II	200 r	2880.767	I
1800	6861.15	I	1000	20281.73	I	75	1793.40	III	50 r	2881.224	I
10000	7005.19	I	1000	20624.67	I	40	1823.41	III	200	2914.67	II
2000	7260.45	I	1200	21787.24	I	100	1827.70	II	50	2927.87	II
10000	7348.51	I	4000	22865.65	I	50	1844.66	III	200	2929.27	II
40000	7512.96	I	1000	23513.15	I	40	1851.13	III	1000 r	2980.620	I
1600	7591.61	I	500	28346.50	I	40	1855.85	III	200 r	2981.362	I
1800	7595.07	I	500	30380.85	I	200	1856.67	III	50	2981.845	I
2000	7616.41	I	600	31630.13	I	150	1874.08	III	50	3030.60	II
30000	7803.02	I	150	38345.75	I	300	1922.23	II	150	3080.822	I
1200	7827.23	I	120	39964.36	I	100	1943.54	II	25	3081.48	II
2500 s	7881.45	I		**Cadmium**		40	1965.54	II	30	3082.593	I
2500	7881.57	I		**Cd Z = 48**		30	1986.89	II	100	3092.34	II
2500	7925.81	I	50	427.01	IV	200	1995.43	II	200	3133.167	I
30000 c	7938.68	I	50	447.85	IV	100	2007.49	II	50	3146.79	II
3000	7947.94	I	60	480.90	IV	50	2032.45	II	150	3250.33	II
3000	7950.18	I	70	493.00	IV	75	2036.23	II	300	3252.524	I
8000	7978.44	I	70	495.13	IV	40	2039.83	III	300	3261.055	I
10000	7978.57	I	70	498.14	IV	50	2045.61	II	50	3343.21	II
30000	7989.94	I	70	498.53	IV	75	2087.91	III	50	3385.49	II
2000	8026.35	I	80	504.09	IV	150	2096.00	II	30	3388.88	II
2500	8026.54	I	70	504.20	IV	50	2111.60	III	800	3403.652	I
30000	8131.52	I	70	504.50	IV	1000 r	2144.41	II	50	3417.49	II
1000 c	8152.65	I	80	506.31	IV	50	2155.06	II	50	3442.42	II
10000	8153.75	I	60	508.01	IV	100	2187.79	II	100	3464.43	II
25000	8154.00	I	50	508.95	IV	1000	2194.56	II	1000	3466.200	I
5000	8246.86	I	70	509.55	IV	1000	2265.02	II	800	3467.655	I
15000	8264.96	I	70	511.40	IV	1500 r	2288.022	I	25	3483.08	II
75000 c	8272.44	I	80	513.00	IV	1000	2312.77	II	150	3495.44	II
20000	8334.70	I	70	514.50	IV	200	2321.07	II	25	3499.952	I
10000	8343.70	I	60	519.42	IV	40	2376.82	II	100	3524.11	II
1200	8384.04	I	80	524.41	IV	50	2418.69	II	100	3535.69	II
40000	8446.55	I	70	524.47	IV	50	2469.73	II	1000	3610.508	I
4000	8477.45	I	70	525.10	IV	40	2487.93	II	800	3612.873	I
1500	8513.38	I	60	525.19	IV	40	2495.58	II	60	3614.453	I
1000	8557.73	I	70	527.07	IV	50	2509.11	II	20	3649.558	I
1000	8566.28	I	80	531.09	IV	30	2516.22	II	10	3981.926	I
20000	8638.66	I	80	531.51	IV	25 h	2525.196	I	100	4029.12	II
4000	8698.53	I	70	534.29	IV	50	2544.613	I	200	4134.77	II
10000 c	8793.47	I	70	536.77	IV	50	2551.98	II	50	4141.49	II
15000	8819.96	I	60	540.90	IV	25	2553.465	I	100	4285.08	II
25000	8825.22	I	70	541.74	IV	3	2565.789	I	8	4306.672	I
4000	8888.98	I	80	542.60	IV	500	2572.93	II	100	4412.41	II
30000	8897.62	I	80	546.55	IV	50	2580.106	I	3	4412.989	I
6000	8932.40	I	60	553.06	IV	30	2592.026	I	1000	4415.63	II
1800	8949.39	I	80	554.05	IV	25 h	2602.048	I	30	4440.45	II

Intensity	Wavelength/Å	Intensity	Wavelength/Å	Intensity	Wavelength/Å	Intensity	Wavelength/Å
8	4662.352 I	400	637.93 V	20	4456.61 I	20	8020.50 II
200	4678.149 I	300	643.12 V	20	4472.04 II	70	8133.05 II
30	4744.69 II	400	646.57 V	20	4489.18 II	100	8201.72 II
300	4799.912 I	750	656.00 IV	19	4499.88 III	110	8248.80 II
50	4881.72 II	300	656.76 V	23	4526.94 I	70	8254.73 II
50	5025.50 II	500	669.70 IV	22	4578.55 I	130	8498.02 II
1000 h	5085.822 I	24	1341.89 II	23	4581.40 I	170	8542.09 II
6	5154.660 I	12	1342.54 II	23	4581.47 I	160	8662.14 II
100	5268.01 II	20	1433.75 II	24	4585.87 I	100	8912.07 II
100	5271.60 II	20	1545.29 III	24	4585.96 I	110	8927.36 II
1000	5337.48 II	60	1649.86 II	20	4685.27 I	110	9213.90 II
1000	5378.13 II	20	1807.34 II	30	4716.74 II	90	9312.00 II
200	5381.89 II	40	1814.50 II	40	4721.03 II	100	9319.56 II
40	5843.30 II	40	1838.01 II	40	4799.97 II	110	9320.65 II
50	5880.22 II	60	1840.06 II	25	4878.13 I	25	9416.97 I
300	6099.142 I	20	1843.09 II	70	5001.48 II	100	9567.97 II
100	6111.49 I	40	1850.69 II	80	5019.97 II	110	9599.24 II
100	6325.166 I	17	2123.03 III	40	5021.14 II	80	9601.82 II
30	6330.013 I	16	2152.43 III	23	5041.62 I	80	9854.74 II
400	6354.72 II	16	2687.76 III	25	5188.85 I	110	9890.63 II
500	6359.98 II	19	2881.78 III	22	5261.71 I	90	9931.39 II
2000	6438.470 I	21	2899.79 III	23	5262.24 I	100	10223.04 II
400	6464.94 II	19	2924.33 III	22	5264.24 I	20	10343.81 I
25	6567.65 II	20	2988.63 III	24	5265.56 I	20	11838.99 II
500	6725.78 II	10	3006.86 I	25	5270.27 I	25	12816.04 I
100	6759.19 II	15	3028.59 III	60	5285.27 II	24	12823.86 I
30	6778.116 I	3	3055.32 I	70	5307.22 II	25	12909.10 I
50	7237.01 II	19	3119.67 III	50	5339.19 II	30	13033.57 I
100	7284.38 II	170	3158.87 II	27	5349.47 I	21	13086.44 I
1000	7345.670 I	180	3179.33 II	23	5512.98 I	24	13134.95 I
50	8066.99 II	150	3181.28 II	25	5581.97 I	20	16150.77 I
5	8200.309 I	20	3316.51 II	27	5588.76 I	22	16157.36 I
20	9289. I	12	3361.92 I	24	5590.12 I	21	16197.04 I
15	11652. I	19	3372.67 III	26	5594.47 I	20	18925.47 I
35	14487. I	20	3461.87 II	25	5598.49 I	24	18970.14 I
80	15708. I	13	3487.60 I	24	5601.29 I	30	19046.14 I
55 d	19120. I	18	3537.77 III	24	5602.85 I	48	19309.20 I
25	24371. I	20	3644.41 I	30	5857.45 I	49	19452.99 I
35	25448. I	30	3683.70 II	27	6102.72 I	47	19505.72 I
Calcium		40	3694.11 II	29	6122.22 I	50	19776.79 I
Ca Z = 20		170	3706.03 II	22	6161.29 I	35	19853.10 I
250	190.46 V	180	3736.90 II	30	6162.17 I	34	19862.22 I
250	196.97 V	20	3755.67 II	22	6163.76 I	23	19917.19 I
300	199.55 V	30	3758.39 II	24	6166.44 I	24	19933.70 I
250	200.51 V	230	3933.66 II	26	6169.06 I	25	22624.93 I
265	257.98 V	220	3968.47 II	28	6169.56 I	30	22651.23 I
400	267.77 V	50	4097.10 II	35	6439.07 I	**Carbon**	
300	270.31 V	60	4109.82 II	30	6449.81 I	**C Z = 6**	
400	280.99 V	30	4110.28 II	22	6455.60 I	110	34.973 V
300	284.98 V	40	4206.18 II	80	6456.87 II	450	40.268 V
450 c	286.96 V	50	4220.07 II	34	6462.57 I	110	227.19 V
500	322.17 V	50	4226.73 I	29	6471.66 I	250	244.91 IV
300	323.22 V	24	4283.01 I	32	6493.78 I	160	248.66 V
300	330.94 V	22	4289.36 I	28	6499.65 I	160	248.74 V
300	334.55 V	22	4298.99 I	23	6572.78 I	200	289.14 IV
250 c	342.45 IV	25	4302.53 I	30	6717.69 I	250	289.23 IV
250	343.93 IV	20	4302.81 III	33	7148.15 I	570	312.42 IV
450	352.92 V	23	4307.74 I	31	7202.19 I	500	312.46 IV
250	377.18 V	22	4318.65 I	33	7326.15 I	250	371.69 III
200	387.08 V	20	4355.08 I	30	7575.81 II	250	371.75 III
750	425.00 V	19	4399.59 III	60	7581.11 II	150	371.78 III
600	434.57 IV	25	4425.44 I	80	7601.30 II	650	384.03 IV
250	437.77 IV	26	4434.96 I	20	7602.32 II	700	384.18 IV
750	443.82 IV	25	4435.69 I	40	7820.78 II	500	386.203 III
500	450.57 IV	30	4454.78 I	60	7843.38 II	400	419.52 IV
500	558.60 V	28	4455.89 I	20	8017.50 II	500	419.71 IV

Intensity	Wavelength/Å		Intensity	Wavelength/Å		Intensity	Wavelength/Å		Intensity	Wavelength/Å	
200	450.734	III	900	1550.774	IV	300	5380.34	I	12	16559.66	I
400	459.46	III	150	1560.310	I	250	5648.07	II	50	16890.38	I
500	459.52	III	400	1560.683	I	350	5662.47	II	10	17338.56	I
570	459.63	III	400	1560.708	I	450	5695.92	III	11	17448.60	I
250	511.522	III	100	1561.341	I	250	5801.33	IV	13	18139.80	I
250	535.288	III	400	1561.438	I	200	5811.98	IV	23	19721.99	I
300	538.080	III	150	1656.266	I	150	5826.42	III		**Cerium**	
350	538.149	III	120	1656.928	I	570	5889.77	II		**Ce Z = 58**	
400	538.312	III	300	1657.008	I	350	5891.59	II	300	399.36	V
350	574.281	III	120	1657.380	I	200	6001.13	I	200	482.96	V
9	595.022	II	120	1657.907	I	250	6006.03	I	40	741.79	IV
30	687.053	II	150	1658.122	I	110	6007.18	I	30	754.60	IV
50	687.345	II	500	1751.823	I	150	6010.68	I	75	1332.16	IV
10	858.092	II	1000	1930.905	I	300	6013.22	I	75	1372.72	IV
20	858.559	II	250	2162.94	III	250	6014.84	I	100	2000.42	IV
30	903.624	II	40	2270.91	V	800	6578.05	II	100	2009.94	IV
60	903.962	II	5	2277.25	V	570	6582.88	II	10000	2318.64	III
150	904.142	II	20	2277.92	V	200	6587.61	I	10000	2372.34	III
30	904.480	II	800	2296.87	III	150	6744.38	III	10000	2380.12	III
800	977.03	III	800	2478.56	I	250	6783.90	II	10000	2431.45	III
9	1009.86	II	250	2509.12	II	150 h	7037.25	III	15000	2439.80	III
10	1010.08	II	350	2512.06	II	250	7113.18	I	10000	2454.32	III
10	1010.37	II	200 l	2524.41	IV	250	7115.19	I	10000	2469.95	III
80	1036.337	II	300 s	2529.98	IV	250	7115.63	II	10000	2483.82	III
150	1037.018	II	250 h	2574.83	II	200	7116.99	I	10000	2497.50	III
150	1157.910	I	150	2697.75	III	350	7119.90	I	20000	2531.99	III
150	1158.019	I	110 l	2724.85	III	800	7231.32	II	10000	2603.59	III
150	1158.035	I	150 l	2725.30	III	1000	7236.42	II	340	2651.01	II
370	1174.93	III	150 l	2725.90	III	150	7612.65	III	270	2830.90	II
350	1175.26	III	350 l	2741.28	II	90 w	7726.2	IV	250	2874.14	II
330	1175.59	III	250	2746.49	II	200	7860.89	I	10000	2923.81	III
500	1175.71	III	1000	2836.71	II	200	8058.62	I	10000	2931.54	III
350	1175.99	III	800	2837.60	II	300 h	8196.48	III	400	2976.91	II
370	1176.37	III	200	2982.11	III	150	8332.99	III	10000	3022.75	III
150	1188.992	I	800 h	2992.62	II	520	8335.15	I	50000	3031.58	III
150	1189.447	I	350	3876.19	II	300	8500.32	III	95000	3055.59	III
200	1189.631	I	350	3876.41	II	250	9061.43	I	20000	3056.56	III
300	1193.009	I	350	3876.66	II	200	9062.47	I	40000	3057.23	III
300	1193.031	I	570	3918.98	II	200	9078.28	I	20000	3057.58	III
300	1193.240	I	800	3920.69	II	250	9088.51	I	680	3063.01	II
300	1193.264	I	150	4056.06	III	450	9094.83	I	40000	3085.10	III
100	1193.393	I	200	4067.94	III	300	9111.80	I	20000	3106.98	III
150	1193.649	I	250	4068.91	III	800	9405.73	I	30000	3110.53	III
150	1193.679	I	250	4070.26	III	150	9603.03	I	30000	3121.56	III
100	1194.064	I	250	4074.52	I	250	9620.80	I	20000	3141.29	III
100	1194.488	I	350 l	4075.85	II	300	9658.44	I	20000	3143.96	III
100	1261.552	I	150	4162.86	III	200	10683.08	I	20000	3147.06	III
250	1277.245	I	250 h	4186.90	III	300	10691.25	I	710	3194.83	II
250	1277.282	I	800	4267.00	II	12	11619.29	I	990	3201.71	II
300	1277.513	I	1000	4267.26	II	23	11628.83	I	710	3218.94	II
300	1277.550	I	200	4325.56	III	13	11658.85	I	880	3221.17	II
200	1280.333	I	600	4647.42	III	47	11659.68	I	710	3227.11	II
100	1311.363	I	520	4650.25	III	24	11669.63	I	20000	3228.57	III
9	1323.951	II	375	4651.47	III	85	11748.22	I	710	3234.16	II
120	1329.578	I	200 w	4658.30	IV	142	11753.32	I	990	3272.25	II
120	1329.600	I	200	4665.86	III	114	11754.76	I	20000	3353.29	III
150	1334.532	II	200	4771.75	I	11	11777.54	I	10000	3395.77	III
300	1335.708	II	200	4932.05	I	17	11892.91	I	30000	3427.36	III
100	1354.288	I	5	4943.88	V	30	11895.75	I	40000	3443.63	III
150	1355.84	I	5	4944.56	V	26	12614.10	I	30000	3454.39	III
120	1364.164	I	200	5052.17	I	20	13502.27	I	40000	3459.39	III
100	1459.032	I	350	5132.94	II	38	14399.65	I	60000	3470.92	III
200	1463.336	I	350	5133.28	II	16	14403.25	I	710	3485.05	II
120	1467.402	I	350	5143.49	II	61	14420.12	I	50000	3497.81	III
150	1481.764	I	570	5145.16	II	12	14429.03	I	60000	3504.64	III
1000	1548.202	IV	400	5151.09	II	13	14442.24	I	770	3539.08	II

Intensity	Wavelength/Å		Intensity	Wavelength/Å		Intensity	Wavelength/Å		Intensity	Wavelength/Å	
50000	3544.07	III	770	4227.75	II	35	6343.95	II	330	2076.43	III
1200	3560.80	II	980	4239.92	II	35	6371.11	II	540	2077.30	III
1000	3577.45	II	1100	4248.68	II	28	6386.84	I	410	2088.68	III
1800	3655.85	II	2000	4289.94	II	23	6393.02	II	210	2101.63	III
880	3660.64	II	1500	4296.67	II	35	6430.07	I	200	2141.47	III
880	3667.98	II	770	4300.33	II	23	6436.40	I	1000	2316.88	III
1000	3709.29	II	770	4306.72	II	35	6458.03	I	230	2325.95	III
1000	3709.93	II	980	4337.77	II	28	6467.39	I	390	2340.49	III
1400	3716.37	II	700	4349.79	II	35	6473.72	I	1600	2455.81	III
800	3728.42	II	910	4364.66	II	23	6513.59	II	1600	2477.57	III
860	3786.63	II	910	4382.17	II	45	6555.65	I	890	2485.45	III
2500	3801.52	II	700	4386.84	II	23	6579.10	I	410	2495.07	III
800	3803.09	II	1700	4391.66	II	22	6612.06	I	1400	2525.67	III
1000	3808.11	II	980	4418.78	II	30	6628.93	I	430	2573.05	III
1100	3838.54	II	770	4449.34	II	22	6652.72	II	16000	2596.86	III
860	3848.59	II	2400	4460.21	II	26	6700.66	I	390	2610.12	III
860	3853.15	II	1400	4471.24	II	35	6704.27	I	6200	2630.51	III
1200	3854.18	II	700	4479.36	II	30	6774.28	II	370	2700.32	III
1200	3854.31	II	700	4483.90	II	35	6775.59	I	710	2701.20	III
1100	3878.36	II	840	4486.91	II	30	6924.81	I	390	2776.44	III
1500	3882.45	II	770	4523.08	II	30	6986.02	I	270	2810.87	III
1000	3889.98	II	840	4527.35	II	35	7061.75	II	630	2845.70	III
770	3907.29	II	840	4528.47	II	35	7086.35	II	3100	2859.32	III
980	3912.44	II	840	4539.75	II	22	7238.36	II	200	2893.85	III
770	3918.28	II	2100	4562.36	II	25	7252.75	I	180	2921.13	III
770	3931.09	II	1100	4572.28	II	25	7329.91	I	3200	2976.86	III
770	3940.34	II	840	4593.93	II	25	7397.77	I	210	3001.28	III
2000	3942.15	II	1700	4628.16	II	25	7616.11	II	1700	3066.59	III
2700	3942.75	II	310	4737.28	II	25	7689.17	I	1100 c	3149.36	III
770	3943.89	II	470	5079.68	II	22	7844.94	II	1400	3152.36	III
3100	3952.54	II	280	5159.69	I	22	7857.54	II	8400	3268.32	III
980	3956.28	II	280	5161.48	I	30	8025.56	II	1300	3315.51	III
770	3960.91	II	370	5187.46	II	25	8772.14	II	550	3340.60	III
770	3967.05	II	260	5223.46	I	30	8891.20	II	430	3344.02	III
770	3978.65	II	260	5245.92	I		**Cesium**		1200	3349.46	III
770	3984.68	II	340	5274.23	II		**Cs Z = 55**		400	3463.45	III
700	3992.39	II	450	5353.53	II	10000	614.01	III	580	3476.83	III
910	3993.82	II	300	5393.40	II	2000	638.17	III	480	3559.82	III
2800	3999.24	II	280	5409.23	II	2500	666.25	III	7200	3597.45	III
910	4003.77	II	260	5512.08	II	5000	691.60	III	1300	3608.31	III
2700	4012.39	II	300	5696.99	I	3500	703.89	III	2300	3618.19	III
910	4014.90	II	370	5699.23	I	15000	718.14	II	300 c	3641.34	III
840	4024.49	II	240	5719.03	I	20000	721.79	III	520	3651.08	III
840	4028.41	II	230	5940.86	I	20000	722.20	III	4800	3661.40	III
840	4031.34	II	55	6001.90	I	5000	731.56	III	640	3699.50	III
2100	4040.76	II	55	6005.86	I	12000	740.29	III	430	3837.46	III
910	4042.58	II	55	6006.82	I	15000	808.76	II	2100 c	3876.15	I
700	4053.51	II	75	6013.42	I	15000	813.84	II	2900	3888.37	III
1100	4071.81	II	110	6024.20	I	7500	830.39	III	600 c	3888.61	I
1800	4073.48	II	10000	6032.54	III	35000	901.27	II	2700	3925.60	III
1500	4075.71	II	110	6043.39	II	15000	920.35	III	680 c	4001.70	III
1500	4075.85	II	55	6047.40	I	40000	926.66	II	3100	4006.55	III
910	4083.23	II	10000	6060.91	III	25000 c	1054.79	III	420	4006.78	III
770	4118.14	II	45	6098.34	II	17 c	1673.99	III	520	4043.42	III
980	4123.87	II	45	6123.67	I	12	1705.25	III	14000	4264.70	II
980	4127.37	II	35	6143.36	II	10	1801.83	III	18000 w	4277.13	II
2700	4133.80	II	35	6186.17	I	20 c	1822.40	III	370	4403.86	III
2000	4137.65	II	35	6208.98	I	11	1823.93	III	1200	4410.22	III
770	4142.40	II	35	6228.94	I	12	1824.70	III	940	4425.68	III
980	4149.94	II	23	6232.45	II	12	1841.80	III	530	4471.48	III
1400	4151.97	II	28	6237.45	I	25	1915.50	III	12000	4501.55	II
1300	4165.61	II	45	6272.05	II	25 c	1923.29	III	1200	4506.72	III
3500	4186.60	II	35	6295.58	I	12	1961.33	III	590	4522.86	III
840	4198.72	II	28	6299.51	II	17	1996.56	III	20000	4526.74	II
910	4202.94	II	23	6300.21	I	710	2035.11	III	1000 c	4555.28	I
1500	4222.60	II	35	6310.01	I	120	2056.43	III	460 c	4593.17	I

Line Spectra of the Elements (continued): Cesium—Chlorine

Intensity	Wavelength/Å		Intensity	Wavelength/Å		Intensity	Wavelength/Å		Intensity	Wavelength/Å	
99900	4603.79	II	18000	9172.32	I	1000	659.811	II	25000	1389.693	I
420 h	4620.61	III	5200	9208.53	I	1300	661.841	II	20000	1389.957	I
210	4665.52	III	19000	10024.36	I	2000	663.074	II	12000	1396.527	I
25000	4830.19	II	4800	10123.41	I	1500	682.053	II	500	1441.470	II
140	4851.59	III	26000	10123.60	I	1500	687.656	II	500	1528.569	II
19000	4870.04	II	2900	13424.31	I	1500	693.594	II	500	1542.942	II
37000	4952.85	II	38000 c	13588.29	I	2000	725.271	II	500	1558.144	II
370	5035.72	III	8400	13602.56	I	2500	728.951	II	500	1565.050	II
27000	5043.80	II	5700	13758.81	I	2000	777.562	II	600	1822.50	III
75000	5227.04	II	55000 c	14694.91	I	5000	787.580	II	500	1828.40	III
29000	5249.38	II	820	16535.63	I	5000	788.740	II	500	1857.488	II
11000	5274.05	II	1500	17012.32	I	5000	793.342	II	500	1901.61	III
10000 c	5349.13	II	760	20138.47	I	500	834.84	IV	500	1983.61	III
22000	5370.99	II	880	22811.86	I	500	834.97	IV	450 h	1997.370	II
230	5380.79	III	1100	23037.98	I	6000	839.297	II	450	2032.116	II
60 c	5465.94	I	3900	23344.47	I	8000	839.599	II	350 h	2088.583	II
37	5502.88	I	4400	24251.21	I	600	840.93	IV	350 h	2091.458	II
39000	5563.02	II	850	24374.96	I	5000	851.691	II	700	2253.07	III
100	5635.21	I	890 d	25763.51	I	2000	888.026	II	500	2268.95	III
210 c	5664.02	I	500	25764.73	I	2000	893.549	II	500	2278.34	III
27	5745.72	I	680 c	29310.06	I	2000	961.499	II	700	2283.93	III
24000	5831.14	II	2800	30103.27	I	500	973.21	IV	600	2323.50	III
59 c	5838.83	I	610 c	30953.06	I	600	977.56	IV	500	2336.45	III
300	5845.14	I	1100	34900.13	I	40	978.284	I	600	2340.64	III
51000	5925.63	I	190	36131.00	I	700	984.95	IV	600	2359.67	III
140	5950.14	III	2 c	39177.28	I	25	998.372	I	600	2370.37	III
110	5979.97	III	2 d	39421.25	I	25	998.432	I	700	2416.42	III
640 c	6010.49	I	1	39424.11	I	75	1002.346	I	600	2447.14	III
86	6034.09	I		**Chlorine**		500	1005.28	III	600	2448.58	III
150	6043.99	III		**Cl Z = 17**		600	1008.78	III	500	2486.91	III
870	6079.86	III	500	392.43	V	150	1013.664	I	500	2532.48	III
9800	6128.61	II	800	486.17	IV	700	1015.02	III	600	2580.67	III
330	6150.42	III	800	534.73	IV	90	1025.553	I	500	2603.59	III
1000	6213.10	I	700	535.67	IV	6000	1063.831	II	500	2632.67	III
170	6217.60	I	600	536.15	IV	3000	1067.945	II	500	2633.18	III
450	6242.96	III	900	537.61	IV	9000	1071.036	II	600	2665.54	III
320 c	6354.55	I	500	538.03	V	6000	1071.767	II	700	2710.37	III
510	6456.33	III	600	538.12	IV	5000	1075.230	II	500	2724.03	IV
8300	6495.53	II	800	542.23	V	5000	1079.080	II	500	2751.23	IV
10000 w	6536.44	II	600	542.30	V	200	1084.667	I	700	2782.47	IV
490	6586.51	I	1000	545.11	V	200	1085.171	I	600	2965.56	III
97	6628.66	I	600	546.33	V	250	1085.304	I	500	3063.13	IV
8800	6646.57	II	1000	547.63	V	400	1088.06	I	600	3076.68	IV
3300 c	6723.28	I	500	549.22	IV	350	1090.271	I	600	3104.46	III
9600	6724.47	II	700	552.02	IV	250	1090.982	I	800	3139.34	III
400	6753.12	III	600	553.30	IV	250	1092.437	I	900	3191.45	III
200	6824.65	I	700	554.62	IV	400	1094.769	I	700	3289.80	III
300	6870.45	I	600	556.23	III	350	1095.148	I	700	3320.57	III
37000	6955.50	II	700	556.61	III	350	1095.662	I	700	3329.06	III
4800	6973.30	I	700	557.12	III	400	1095.797	I	800	3340.42	III
16000	6979.67	II	350	559.305	II	250	1096.810	I	900	3392.89	III
980	6983.49	I	700	561.53	III	300	1097.369	I	800	3393.45	III
13000 w	7149.54	II	700	561.68	III	200	1098.068	I	900	3530.03	III
1900 c	7219.60	III	700	561.74	III	200	1099.523	I	800	3560.68	III
790	7228.53	I	400	571.904	II	500	1107.528	I	900	3602.10	III
130	7279.90	I	800	574.406	II	800	1139.214	II	800	3612.85	III
1100	7279.96	I	500	601.50	IV	800	1167.148	I	800	3622.69	III
2600 c	7608.90	I	500	604.59	IV	3000	1179.293	I	700	3656.95	III
3300	7943.88	I	500	606.35	III	1200	1188.774	I	700	3670.28	III
22000	7997.44	II	700	618.057	II	900	1201.353	I	700	3682.05	III
3500	8015.73	I	600	619.982	II	3000	1335.726	I	600	3705.45	III
510	8078.94	I	800	620.298	II	10000	1347.240	I	600	3707.34	III
4500	8079.04	I	700	626.735	II	5000	1351.657	I	800	3720.45	III
59000 c	8521.13	I	800	635.881	II	12000	1363.447	I	800	3748.81	III
15000 c	8761.41	I	1000	636.626	II	2500	1373.116	I	500	3779.35	III
61000 c	8943.47	I	1000	650.894	II	20000	1379.528	I	10000	3850.99	II

Intensity	Wavelength/Å		Intensity	Wavelength/Å		Intensity	Wavelength/Å		Intensity	Wavelength/Å	
25000	3860.83	II	600	7980.60	I	259	16198.5	I	280	2668.71	II
500	3925.87	III	2900	7997.85	I	717	19755.3	I	350	2671.81	II
700	3991.50	III	2200	8015.61	I	100	24470.0	I	280	2672.83	II
600	4018.50	III	1100	8023.33	I		39716.0	I	1800	2677.16	II
600	4059.07	III	400	8051.07	I		40085.5	I	320	2678.79	II
500	4104.23	III	1700	8084.51	I		40089.5	I	230	2687.09	II
500	4106.83	III	2200	8085.56	I		40532.2	I	280	2691.04	II
10000 h	4132.50	II	3000	8086.67	I	**Chromium**			180	2698.41	II
500	4608.21	III	1300	8087.73	I	**Cr Z = 24**			180	2698.69	II
40	4623.938	I	2500	8194.42	I	100	438.62	V	110	2701.99	I
50	4654.040	I	2200	8199.13	I	100	464.02	V	140	2712.31	II
80	4661.208	I	2200	8200.21	I	100	620.66	IV	170	2722.75	II
45	4691.523	I	800	8203.78	I	100	629.26	IV	420 h	2726.51	I
40	4721.255	I	18000	8212.04	I	80	630.30	IV	280 h	2731.91	I
45	4740.729	I	3000	8220.45	I	100	666.55	IV	170 h	2736.47	I
13000	4781.32	II	20000	8221.74	I	100	693.92	IV	250	2743.64	II
99000	4794.55	II	18000	8333.31	I	60	1030.47	III	110 h	2748.29	I
29000	4810.06	II	99900	8375.94	I	100	1033.69	III	330	2748.98	II
16000	4819.47	II	400	8406.199	I	100	1036.03	III	390	2750.73	II
81000	4896.77	II	15000	8428.25	I	80	1055.89	IV	280	2751.87	II
47000	4904.78	II	2200	8467.34	I	80	1068.41	III	110 h	2752.88	I
26000	4917.73	II	2200	8550.44	I	100	1116.48	V	150	2757.10	I
10000	4995.48	II	20000	8575.24	I	150	1121.07	V	350	2757.72	II
26000	5078.26	II	750	8578.02	I	150	1127.63	V	750	2762.59	II
30	5099.789	I	75000	8585.97	I	100	1263.50	V	750	2766.54	II
56000	5217.94	II	450	8628.54	I	100	1417.42	IV	250 h	2769.92	I
23000	5221.36	II	300	8641.71	I	150	1465.86	V	610	2780.70	I
15000	5392.12	II	3500	8686.26	I	150	1497.97	V	180	2822.37	II
99000	5423.23	II	2200	8912.92	I	170	1519.03	V	180	2830.47	II
10000	5423.51	II	3000	8948.06	I	220	1579.70	V	2500	2835.63	II
19000	5443.37	II	2000	9038.982	I	170	1591.72	V	110	2840.02	II
10000	5444.21	II	2500	9045.43	I	150	1603.19	V	1700	2843.25	II
40	5532.162	I	1000	9069.656	I	120	1672.66	IV	1200	2849.84	II
50 d	5796.305	I	2000	9073.17	I	120	1758.51	IV	120	2851.36	II
45	5799.914	I	7500	9121.15	I	140	1802.72	IV	880	2855.68	II
30	5856.742	I	3000	9191.731	I	130	1812.41	IV	610	2858.91	II
50	6019.812	I	500	9197.596	I	200	1837.44	V	440	2860.93	II
200	6140.245	I	4000	9288.86	I	140	1873.89	IV	790	2862.57	II
160	6194.757	I	1500	9393.862	I	140	1967.18	IV	750	2865.11	II
150	6434.833	I	3500	9452.10	I	120	1972.07	IV	610	2866.74	II
300	6932.903	I	500	9486.964	I	19000	2055.52	II	480	2867.65	II
300	6981.886	I	1000	9584.801	I	14000	2061.49	II	210	2870.44	II
600	7086.814	I	3500	9592.22	I	8900	2065.42	II	110	2871.63	I
7500	7256.62	I	250	9632.509	I	200	2226.72	III	160	2873.48	II
5000	7414.11	I	1000	9702.439	I	200	2235.91	III	320	2875.99	II
550	7462.370	I	250	9744.426	I	150	2237.59	III	230	2876.24	II
550	7489.47	I	200	9807.057	I	150	2244.10	III	180	2877.98	II
700	7492.118	I	400	9875.970	I	150	2284.44	III	120	2879.27	I
11000	7547.072	I	331	10392.549	I	150	2324.88	III	170	2887.00	I
2300	7672.42	I	300	11123.05	I	130	2383.33	I	700	2889.29	I
450	7702.828	I	269	11409.69	I	140	2408.62	I	370	2893.25	I
7000	7717.581	I	1000	11436.33	I	170	2496.31	I	190	2894.17	I
10000	7744.97	I	350	13243.8	I	110	2502.53	I	210	2896.75	I
2200	7769.16	I	310	13296.0	I	190	2504.31	I	180	2905.49	I
650	7771.09	I	550	13346.8	I	110	2516.92	I	260	2909.05	I
2200	7821.36	I	525	13821.7	I	390	2519.52	I	260	2910.90	I
1700	7830.75	I	294	14931.7	I	190	2527.12	I	250	2911.14	I
3000	7878.22	I	269	15108.0	I	160	2549.54	I	480	2967.64	I
2300	7899.31	I	381	15465.1	I	130	2560.69	I	480	2971.11	I
1800	7915.08	I	1094	15520.3	I	150	2571.74	I	210	2971.91	II
3000	7924.645	I	1487	15730.1	I	100	2577.65	I	480	2975.48	I
2100	7933.89	I	2780	15869.7	I	380	2591.85	I	190	2979.74	II
1700	7935.012	I	277	15883.3	I	250	2653.59	II	350	2980.79	I
650	7952.52	I	342	15928.9	I	250	2658.59	II	110	2985.32	II
1500	7974.72	I	735	15960.0	I	320	2663.42	II	480	2985.85	I
1300	7976.97	I	283	15970.5	I	440	2666.02	II	1500	2986.00	I

Intensity	Wavelength/Å		Intensity	Wavelength/Å		Intensity	Wavelength/Å		Intensity	Wavelength/Å	
2100	2986.47	I	130	3573.64	I	190	3984.34	I	60	5013.32	I
660	2988.65	I	330 h	3574.80	I	160	3989.99	I	70	5166.23	I
160	2989.19	II	19000	3578.69	I	960	3991.12	I	70	5184.59	I
480	2991.89	I	160 h	3584.33	I	160	3991.67	I	70	5192.00	I
230	2994.07	I	130	3585.30	II	190	3992.84	I	85	5196.44	I
300	2995.10	I	17000	3593.49	I	160	4001.44	I	5300	5204.52	I
700	2996.58	I	350	3601.67	I	120	4012.47	II	8400	5206.04	I
210	2998.79	I	13000	3605.33	I	120	4026.17	I	11000	5208.44	I
1100	3000.89	I	130	3632.84	I	190	4039.10	I	85	5224.94	I
750	3005.06	I	350	3636.59	I	160	4048.78	I	290	5247.56	I
140	3013.03	I	630	3639.80	I	120	4058.77	I	530	5264.15	I
710	3013.71	I	220	3641.83	I	140	4126.52	I	180	5265.72	I
710	3014.76	I	220	3649.00	I	120	4153.82	I	95 h	5275.17	I
1400	3014.92	I	170	3653.91	I	140	4163.62	I	70 h	5276.03	I
710	3015.19	I	220	3656.26	I	170	4174.80	I	340	5296.69	I
2800	3017.57	I	130	3663.21	I	170	4179.26	I	70 h	5297.36	I
430	3018.50	I	120	3685.55	I	110	4209.37	I	660	5298.27	I
240	3018.82	I	130	3686.80	I	20000	4254.35	I	85	5300.75	I
430	3020.67	I	130	3687.25	I	110	4263.14	I	340 h	5328.34	I
2800	3021.56	I	130	3730.81	I	16000	4274.80	I	70 h	5329.17	I
1100	3024.35	I	150	3732.03	I	10000	4289.72	I	780	5345.81	I
170	3029.16	I	480	3743.58	I	780	4337.57	I	380	5348.32	I
710	3030.24	I	570	3743.88	I	1100	4339.45	I	40	5400.61	I
140	3031.35	I	340	3749.00	I	380	4339.72	I	1400	5409.79	I
390	3034.19	I	230	3757.66	I	1900	4344.51	I	24	5628.64	I
550	3037.04	I	260	3768.24	I	380	4351.05	I	7	5642.36	I
550	3040.85	I	130	3791.38	I	2300	4351.77	I	24	5664.04	I
110	3050.14	II	130	3792.14	I	570	4359.63	I	24	5694.73	I
710	3053.88	I	120	3793.29	I	530	4371.28	I	40	5698.33	I
240	3118.65	II	130	3793.88	I	110	4374.16	I	24	5702.31	I
430	3120.37	II	140	3797.13	I	530	4384.98	I	24	5712.78	I
470	3124.94	II	200	3797.72	I	110	4458.54	I	24 h	5783.11	I
120	3128.70	II	530	3804.80	I	660	4496.86	I	30 h	5783.93	I
590	3132.06	II	110	3806.83	I	380	4526.47	I	24 h	5785.00	I
140	3136.68	II	110	3807.93	I	380	4530.74	I	19 h	5785.82	I
140	3147.23	II	180	3815.43	I	240	4535.72	I	60 h	5787.99	I
100	3155.15	I	180	3819.56	I	240	4540.50	I	180 h	5791.00	I
100	3163.76	I	130	3826.42	I	240	4540.72	I	35	6330.10	I
240	3180.70	II	130	3830.03	I	140	4544.62	I	22	6362.87	I
220	3197.08	II	380	3841.28	I	600	4545.96	I	19	6661.08	I
170	3209.18	II	190	3848.98	I	120	4565.51	I	21 h	6883.03	I
140	3217.40	II	140	3849.36	I	120	4571.68	I	27 h	6924.13	I
120	3245.54	I	290	3850.04	I	360	4580.06	I	30 h	6978.48	I
130	3251.84	I	140	3852.22	I	360	4591.39	I	85	7355.90	I
130	3257.82	I	190	3854.22	I	480	4600.75	I	130	7400.21	I
130	3339.80	II	110	3855.29	I	240	4613.37	I	150	7462.31	I
110	3342.59	II	140	3855.57	I	600	4616.14	I	40	8947.15	I
170	3358.50	II	260	3857.63	I	550	4626.19	I	19	8976.83	I
160	3360.30	II	660	3883.29	I	1600	4646.17	I		**Cobalt**	
430	3368.05	II	570	3885.22	I	570	4651.28	I		**Co Z = 27**	
140	3382.68	II	380	3886.79	I	840	4652.16	I	20	355.52	V
170	3403.32	II	260	3894.04	I	240 d	4698.46	I	18	355.88	V
360	3408.76	II	360	3902.92	I	190	4708.04	I	12	356.06	V
210	3421.21	II	960	3908.76	I	240	4718.43	I	66	609.16	IV
270	3422.74	II	120 hd	3911.82	I	120	4730.71	I	70	609.21	IV
140	3433.31	II	120	3915.84	I	140	4737.35	I	64	609.28	IV
270	3433.60	I	190	3916.24	I	340	4756.11	I	10	1018.36	V
160	3436.19	I	1900	3919.16	I	190	4789.32	I	10	1021.14	V
140	3441.44	I	600	3921.02	I	120	4801.03	I	15	1231.73	V
170	3445.62	I	600	3928.64	I	110	4829.38	I	50	1277.01	V
170	3447.43	I	410	3941.49	I	140	4870.80	I	80	1299.58	II
190	3453.33	I	1900	3963.69	I	130	4887.01	I	80	1306.95	II
130	3455.60	I	120	3969.06	I	260	4922.27	I	50	1345.67	V
100	3460.43	I	1600	3969.75	I	110	4936.33	I	1000	1696.01	III
120	3550.64	I	1600	3976.66	I	70	4942.50	I	800	1697.99	III
130	3566.16	I	960	3983.91	I	110	4954.81	I	1000	1707.35	III

Intensity	Wavelength/Å		Intensity	Wavelength/Å		Intensity	Wavelength/Å		Intensity	Wavelength/Å	
5000	1760.35	III	200	2291.98	II	200	2546.74	II	8800	3443.64	I
5000	1773.57	III	300 d	2293.38	II	340	2548.34	I	50	3446.39	II
2000	1780.05	III	300	2301.40	II	310	2553.37	I	4100	3449.17	I
3000	1782.97	III	800 d	2307.85	II	310	2555.07	I	2100	3449.44	I
1000	1787.08	III	2600	2309.02	I	300	2559.41	II	21000	3453.50	I
1000	1789.07	III	500	2311.60	II	200	2560.03	II	1000	3455.23	I
1000	1823.08	III	500	2314.05	II	960	2562.15	I	5100	3462.80	I
2000	1830.09	III	300	2314.96	II	500	2564.04	II	5100	3465.80	I
2000	1831.44	III	200 p	2317.06	II	1100	2567.35	I	8000	3474.02	I
5000	1835.00	III	2400	2323.14	I	960	2574.35	I	1900	3483.41	I
1500	1842.34	I	300 p	2324.31	II	800	2580.32	II	4800	3489.40	I
1800	1847.89	I	200 d	2326.11	II	300 d	2582.22	II	2400	3495.69	I
1800	1852.71	I	500	2326.47	II	500	2587.22	II	50	3501.72	II
2400	1855.05	I	1400	2335.99	I	500	2587.52	II	9600	3502.28	I
2000	1863.83	III	1600	2338.67	I	200	2588.91	II	7000	3506.32	I
1500	1878.28	I	200	2347.39	II	100 p	2605.71	II	50	3507.77	II
1800	1936.58	I	1600	2352.85	I	100	2612.50	II	2900	3509.84	I
1500	1946.79	I	200 d	2353.41	II	100	2614.36	II	1400	3510.43	I
1500	1951.90	I	2000	2353.42	I	100 p	2628.77	II	4800	3512.64	I
1800	1954.22	I	500	2363.80	II	100	2632.26	II	3800	3513.48	I
1800	1955.17	I	400	2378.62	II	100	2636.07	II	4800	3518.35	I
1500	1958.55	I	1400	2380.48	I	310	2646.42	I	1300	3520.08	I
1500	1961.59	I	200	2381.76	II	770	2648.64	I	2700	3521.57	I
1500 h	1968.69	I	300 p	2383.45	II	100	2653.72	II	3800	3523.43	I
1500 h	1968.93	I	1400	2384.86	I	100	2663.53	II	60	3523.51	II
3000	1970.71	I	200	2386.36	II	200	2666.73	II	6400	3526.85	I
1800 h	1971.16	I	500	2388.92	II	100	2675.85	II	2700	3529.03	I
1800 h	1972.52	I	200	2397.38	II	100	2684.42	II	7300	3529.81	I
1500	1973.85	I	1100 d	2402.06	I	100	2702.02	II	1900	3533.36	I
1800	1976.97	I	200 p	2404.16	II	200	2706.62	II	50	3545.03	II
2400 h	1980.89	I	5300	2407.25	I	200	2707.35	II	1100	3560.89	I
1500	1989.80	I	5300	2411.62	I	190	2715.99	I	80	3561.07	II
1800	1990.34	I	1600	2412.76	I	100	2727.78	II	8800	3569.38	I
1500 l	1998.49	I	4800	2414.46	I	80	2734.54	II	50	3574.95	II
1500	2002.32	I	4800	2415.30	I	190	2745.10	I	1600	3574.96	I
900	2008.04	I	300	2417.65	II	100	2753.22	II	60	3575.32	II
50	2011.51	II	4100	2424.93	I	190	2764.19	I	2500	3575.36	I
1200 h	2014.58	I	3300	2432.21	I	100	2766.70	II	60	3577.96	II
900	2016.17	I	2900	2436.66	I	100	2774.97	II	1000	3585.16	I
50	2022.35	II	2400	2439.05	I	100	2791.00	II	6700	3587.19	I
50	2027.04	II	200	2442.63	II	100	2793.73	II	1900	3594.87	I
900	2031.96	I	200 d	2446.03	II	150	2815.56	I	1600	3602.08	I
1500	2039.95	I	200 p	2447.69	II	80	2835.63	II	100	3621.21	II
1200	2041.11	I	200	2450.00	II	80	2847.35	II	1000	3627.81	I
50	2065.54	II	200	2464.20	II	80	2871.22	II	80	3643.61	II
1500 h	2077.76	I	200	2486.44	II	190	2886.44	I	60	3681.35	II
900	2085.67	I	200	2498.82	II	100	2918.38	II	1100	3745.50	I
900	2087.55	I	570	2504.52	I	100	2930.24	II	1400	3842.05	I
900	2089.35	I	500	2506.46	II	100	2954.73	II	6900	3845.47	I
900	2093.40	I	360	2506.88	I	690	2987.16	I	5500	3873.12	I
900	2094.86	I	200	2511.16	II	690	2989.59	I	2800	3873.96	I
900	2095.77	I	860	2517.87	I	60	3022.59	II	7900	3894.08	I
1200	2097.51	I	500	2519.82	II	3100	3044.00	I	1500	3935.97	I
1500	2104.73	I	4300	2521.36	I	1700	3061.82	I	80 h	3963.10	II
1500	2106.80	I	200 h	2524.65	II	80	3387.70	II	6000	3995.31	I
900	2108.98	I	300	2524.97	II	1100	3388.17	I	970	3997.91	I
900 s	2117.68	I	500	2528.62	II	2200	3395.38	I	350	4020.90	I
900	2137.78	I	2900	2528.97	I	11000	3405.12	I	370	4045.39	I
900	2138.97	I	200 p	2530.09	II	4500	3409.18	I	350	4066.37	I
900	2163.03	I	720	2530.13	I	6700	3412.34	I	830	4092.39	I
1100	2174.60	I	860	2532.18	I	2200	3412.63	I	550	4110.54	I
200	2193.60	II	200 d	2533.82	II	2700	3417.16	I	2800	4118.77	I
200	2256.73	II	2900	2535.96	I	50	3423.84	II	4400	4121.32	I
150	2260.00	II	860	2536.49	I	2500	3431.58	I	90	4190.71	I
200	2283.52	II	300	2541.94	II	4500	3433.04	I	90	4469.56	I
1000	2286.15	II	1700	2544.25	I	1600	3442.93	I	690	4530.96	I

Intensity	Wavelength/Å	Intensity	Wavelength/Å	Intensity	Wavelength/Å	Intensity	Wavelength/Å
90	4549.66 I	300	1004.055 II	250	1450.304 II	400	1606.834 II
140	4565.59 I	300	1008.569 II	200	1452.294 II	250	1608.639 II
190	4581.60 I	300	1008.728 II	300	1458.002 II	150	1610.296 II
120	4629.38 I	300	1010.269 II	250	1459.412 II	200	1617.915 II
85	4663.41 I	250	1012.597 II	200	1463.752 II	600	1621.426 II
110	4792.86 I	500	1018.707 II	400	1463.838 II	400	1622.428 II
100	4840.27 I	500	1027.831 II	200	1466.070 II	250	1630.268 II
150	4867.88 I	250	1028.328 II	400	1470.697 II	100	1636.605 II
80 h	4964.18 II	200	1030.263 II	200	1472.395 II	1000 r	1642.21 III
50	5212.71 I	600	1036.470 II	250	1473.978 II	250	1649.458 II
50	5230.22 I	600	1039.348 II	200	1474.935 II	30 r	1655.32 I
50	5247.93 I	600	1039.582 II	150	1476.059 II	200	1656.322 II
50	5342.71 I	800	1044.519 II	300 r	1481.23 III	200	1660.001 II
50	5352.05 I	800	1044.744 II	200	1481.544 II	300	1663.002 II
	Copper	500	1049.755 II	200	1485.328 II	100	1672.776 II
	Cu Z = 29	600	1054.690 II	750	1488.831 II	30	1688.09 I
80	685.141 II	400	1055.797 II	300	1492.834 II	30	1691.08 I
100	709.313 II	600	1056.955 II	250	1493.366 II	30 r	1703.84 I
100	718.179 II	400	1058.799 II	250	1495.430 II	50 r	1713.36 I
150	724.489 II	600	1059.096 II	350	1496.687 II	150	1717.721 II
200	735.520 II	600	1060.634 II	150	1503.368 II	50 r	1725.66 I
250	736.032 II	600	1063.005 II	250	1504.757 II	100	1736.551 II
80	779.295 II	200	1065.782 II	200	1505.388 II	50 r	1741.57 I
100	797.455 II	200	1066.134 II	300	1508.632 II	150	1753.281 II
150	810.998 II	500	1069.195 II	350	1510.506 II	200 r	1774.82 I
200	813.883 II	300	1073.745 II	200	1512.465 II	100 r	1825.35 I
300	826.996 II	200	1088.395 II	200	1513.366 II	250	1929.751 II
150	848.808 II	300	1094.402 II	500	1514.492 II	250	1944.597 II
250	851.303 II	250	1097.053 II	200	1517.631 II	100	1946.493 II
250	858.487 II	150	1119.947 II	500	1519.492 II	200	1957.518 II
400	861.994 II	200	1142.640 II	600	1519.837 II	150	1970.495 II
400	865.390 II	300	1144.856 II	200	1520.540 II	150	1977.027 II
250	869.336 II	100	1250.048 II	200	1524.860 II	500	1979.956 II
150	873.263 II	150	1265.506 II	150	1525.764 II	300	1989.855 II
200	876.723 II	300	1275.572 II	500	1531.856 II	250	1999.698 II
250	877.012 II	150	1282.455 II	300	1532.131 II	270	2035.854 II
200	877.555 II	150	1287.468 II	250	1533.986 II	250	2037.127 II
500	878.699 II	150	1298.395 II	250	1535.002 II	350	2043.802 II
100	884.133 II	300	1308.297 II	500	1537.559 II	300	2054.980 II
250	885.847 II	300	1314.337 II	200	1540.239 II	100	2078.663 II
600	886.943 II	100	1320.686 II	300	1540.389 II	110	2098.398 II
600	890.567 II	100	1326.395 II	300	1540.588 II	320	2104.797 II
500	892.414 II	150	1350.594 II	750	1541.703 II	300	2112.100 II
800	893.678 II	250	1351.837 II	400	1544.677 II	320	2117.310 II
400	894.227 II	150	1355.305 II	100	1547.958 II	350	2122.980 II
600	896.759 II	300	1358.773 II	300	1550.653 II	350	2126.044 II
400	896.976 II	200	1359.009 II	300	1551.389 II	420	2134.341 II
600	901.073 II	200	1362.600 II	500	1552.646 II	900	2135.981 II
400	906.113 II	250	1367.951 II	250	1553.896 II	400	2148.984 II
800	914.213 II	200	1371.840 II	400	1555.134 II	150	2161.320 II
600	922.019 II	300 r	1376.79 III	500	1555.703 II	1300 r	2165.09 I
500	924.239 II	200 r	1377.49 III	300	1558.345 II	250	2174.982 II
400	935.232 II	100	1393.128 II	400	1565.924 II	1600 r	2178.94 I
600	935.898 II	100	1398.642 II	400	1566.415 II	700	2179.410 II
600	943.335 II	150	1402.777 II	100	1569.416 II	1700 r	2181.72 I
600	945.525 II	150	1407.169 II	300	1579.492 II	700	2189.630 II
500	945.965 II	100	1414.898 II	300	1580.626 II	900	2192.268 II
200	954.383 II	250	1418.426 II	400	1581.995 II	400	2195.683 II
250	956.290 II	250	1421.759 II	500	1583.682 II	1700 r	2199.58 I
400	958.154 II	200	1427.829 II	400	1590.165 II	1300 r	2199.75 I
200	960.414 II	400	1430.243 II	600	1593.556 II	100	2200.509 II
250	968.042 II	250	1434.904 II	500 r	1593.75 III	200	2209.806 II
200	974.759 II	150	1436.236 II	400	1598.402 II	750	2210.268 II
250	977.567 II	150	1442.139 II	400	1602.388 II	1600 r	2214.58 I
100	987.657 II	200	1445.984 II	200	1604.848 II	250	2215.106 II
250	992.953 II	200	1449.058 II	300	1605.281 II	1000 r	2215.65 I

Intensity	Wavelength/Å		Intensity	Wavelength/Å		Intensity	Wavelength/Å		Intensity	Wavelength/Å	
750	2218.108	II	10000 r	3273.96	I	100	4758.433	II	550	6305.972	II
2100 r	2225.70	I	1400 h	3282.72	I	400	4812.948	II	400	6312.492	II
150	2226.780	II	400	3290.418	II	120	4851.262	II	120	6326.466	II
1600 r	2227.78	I	1500 h	3290.54	I	300	4854.988	II	400	6373.268	II
350	2228.868	II	110	3300.881	II	100	4873.304	II	750	6377.840	II
2500 r	2230.08	I	250	3301.229	II	150	4901.427	II	400	6403.384	II
1100 r	2238.45	I	2500 h	3307.95	I	1000	4909.734	II	850	6423.884	II
900	2242.618	II	200	3316.276	II	500	4918.376	II	200	6442.965	II
2300 r	2244.26	I	1500	3337.84	I	200	4926.424	II	750	6448.559	II
1000	2247.002	II	150	3338.648	II	900	4931.698	II	170	6466.246	II
1300 r	2260.53	I	200	3365.648	II	120	4943.026	II	950	6470.168	II
2200 r	2263.08	I	450	3370.454	II	700	4953.724	II	750	6481.437	II
150	2263.786	II	300	3374.952	II	500	4985.506	II	400	6484.421	II
200	2276.258	II	200	3380.712	II	400	5006.801	II	220	6517.317	II
100	2286.645	II	100	3384.945	II	350	5009.851	II	400	6530.083	II
2500 r	2293.84	I	1250 h	3483.76	I	400	5012.620	II	120	6551.286	II
170	2294.368	II	1250	3524.23	I	350	5021.279	II	200	6577.080	II
1000	2303.12	I	2000	3530.38	I	200	5039.016	II	750	6624.292	II
150	2369.890	II	1400	3599.13	I	300	5047.348	II	800	6641.396	II
2500 r	2392.63	I	1400	3602.03	I	900	5051.793	II	450	6660.962	II
120	2403.337	II	1000	3686.555	II	400	5058.910	II	100	6770.362	II
1500	2406.66	I	150	3786.270	II	500	5065.459	II	300	6806.216	II
1000 r	2441.64	I	170	3797.849	II	450	5067.094	II	400	6809.647	II
100	2485.792	II	100	3818.879	II	350	5072.302	II	320	6823.202	II
2000 r	2492.15	I	140	3826.921	II	450	5088.277	II	250	6844.157	II
150	2506.273	II	160	3864.137	II	420	5093.816	II	320	6868.791	II
120	2526.593	II	280	3884.131	II	350	5100.067	II	270	6872.231	II
300	2544.805	II	150	3892.924	II	1500	5105.54	I	270	6879.404	II
100	2571.756	II	170	3903.177	II	250	5124.476	II	220	6937.553	II
150	2590.529	II	140	3920.654	II	2000	5153.24	I	150	6952.871	II
200	2600.270	II	120	3933.268	II	100	5158.093	II	150	6977.572	II
2500 r	2618.37	I	120	3987.024	II	100	5183.367	II	200	7022.860	II
200	2666.291	II	150	3993.302	II	2500	5218.20	I	300	7194.896	II
750	2689.300	II	140	4003.476	II	100	5269.991	II	400	7326.008	II
700	2700.962	II	1250	4022.63	I	100	5276.525	II	300	7331.694	II
650	2703.184	II	100	4032.647	II	1650	5292.52	I	250	7382.277	II
700	2713.508	II	600	4043.484	II	100	5368.383	II	1000	7404.354	II
650	2718.778	II	500	4043.751	II	1500	5700.24	I	270	7434.156	II
300	2721.677	II	2000	4062.64	I	1500	5782.13	I	500	7562.015	II
120	2737.342	II	120	4068.106	II	150	5805.989	II	700	7652.333	II
270	2745.271	II	500	4131.363	II	100	5833.515	II	1000	7664.648	II
2500 r	2766.37	I	200	4143.017	II	200	5897.971	II	150	7681.788	II
800	2769.669	II	300	4153.623	II	120	5937.577	II	450	7744.097	II
200	2791.795	II	500	4161.140	II	400	5941.196	II	800	7778.738	II
170	2799.528	II	370	4164.284	II	100	5993.260	II	750	7805.184	II
100	2810.804	II	400	4171.851	II	650	6000.120	II	1500	7807.659	II
1250 r	2824.37	I	500	4179.512	II	100	6023.264	II	1000	7825.654	II
350	2837.368	II	500	4211.866	II	250	6072.218	II	350	7860.577	II
100	2857.748	II	320	4230.449	II	150	6080.343	II	300	7890.567	II
600	2877.100	II	200	4255.635	II	150	6099.990	II	700	7902.553	II
270	2884.196	II	950	4275.11	I	160	6107.412	II	1500	7933.13	I
2500 r	2961.16	I	300	4279.962	II	300	6114.493	II	400	7944.438	II
100	2986.335	II	500	4292.470	II	600	6150.384	II	400	7972.033	II
2000	2997.36	I	400	4365.370	II	750	6154.222	II	1200	7988.163	II
2000	3010.84	I	100	4444.831	II	500	6172.037	II	2000	8092.63	I
2500	3036.10	I	400	4506.002	II	550	6186.884	II	500	8277.560	II
2500	3063.41	I	150	4516.049	II	400	6188.676	II	800	8283.160	II
1400	3073.80	I	150	4541.032	II	300	6198.092	II	250	8503.396	II
1500	3093.99	I	500	4555.920	II	470	6204.261	II	750	8511.061	II
1250	3099.93	I	100	4596.906	II	450	6208.457	II	200	8609.134	II
2000	3108.60	I	120	4649.271	II	750	6216.939	II	500	9813.213	II
1400 h	3126.11	I	2000	4651.12	I	700	6219.844	II	250	9827.978	II
1500	3194.10	I	120	4661.363	II	500	6261.848	II	200	9830.798	II
1400	3208.23	I	320	4671.702	II	1000	6273.349	II	600	9861.280	II
1500 h	3243.16	I	300	4673.577	II	350	6288.696	II	600	9864.137	II
10000 r	3247.54	I	450	4681.994	II	900	6301.009	II	200	9883.969	II

Intensity	Wavelength/Å		Intensity	Wavelength/Å		Intensity	Wavelength/Å		Intensity	Wavelength/Å	
550	9916.419	II	560	3517.26	II	580	3806.27	II	4400	4218.09	I
500	9917.954	II	4400	3523.98	II	470	3812.27	I	4400	4221.11	I
550	9925.594	II	22000	3531.70	II	470	3813.67	II	2700	4225.16	I
450	9938.998	II	4400	3534.96	II	1400	3816.76	II	1000	4308.63	II
500	9960.354	II	5500	3536.02	II	700	3825.68	II	540	4409.38	II
450	10006.588	II	4400	3538.52	II	2300	3836.50	II	740	4449.70	II
550	10022.969	II	1700	3542.33	II	1400	3841.31	II	420	4577.78	I
550	10038.093	II	1400	3546.83	II	420	3846.34	II	2100	4589.36	I
650	10054.938	II	4400	3550.22	II	420	3847.02	I	990	4612.26	I
450	10080.354	II	2200	3551.62	II	1200	3853.03	II	170	4731.84	II
Dysprosium			440 h	3558.23	II	420	3858.40	I	120 h	4775.79	I
Dy Z = 66			440	3559.30	II	560	3868.45	II	480	4957.34	II
260	2356.91	II	2200	3563.15	II	1600	3868.81	I	70	5022.12	I
240	2410.01	II	560	3563.69	II	820	3869.86	II	160	5042.63	I
260	2439.84	II	780	3573.83	II	7000	3872.11	II	95	5070.68	I
220	2585.30	I	1400	3574.15	II	1200	3873.99	II	120	5077.67	I
440	2634.80	II	4400	3576.24	II	470	3879.11	II	80	5090.38	II
220	2755.75	II	1700	3576.87	II	5800	3898.53	II	80	5110.32	I
300	2816.39	II	830	3577.98	II	540	3914.87	II	130 h	5120.04	I
390	2913.95	II	440	3580.04	II	540	3915.59	II	190	5139.60	II
610	3038.28	II	3300	3585.06	II	540 d	3917.29	I	110	5169.69	II
830	3135.38	II	1400	3585.78	II	420	3927.86	I	80	5185.30	I
500	3141.14	II	560	3586.11	II	540	3930.14	I	290	5192.86	II
1200	3156.52	II	1100	3591.41	II	2100	3931.52	II	95	5197.66	II
670	3162.83	II	560	3591.81	II	10000	3944.68	II	70	5259.88	I
1000	3169.99	II	560	3592.11	II	800	3957.79	II	130	5260.56	I
470	3215.19	II	1800	3595.04	II	14000	3968.39	II	65	5267.11	I
830	3216.63	II	560	3600.38	II	2700	3978.57	II	55	5282.07	I
490	3235.89	II	1800	3606.12	II	1400	3981.92	II	160	5301.58	I
490	3245.12	II	440	3618.51	II	1600	3983.65	II	65	5340.30	I
1200	3251.27	II	560	3620.16	II	800	3984.21	II	85	5389.58	II
890	3280.09	II	470	3624.27	II	540	3991.32	II	80	5419.13	I
490	3282.77	II	1100	3629.42	II	1600	3996.69	II	70	5423.32	I
1100	3308.88	II	4000	3630.24	II	8000	4000.45	II	95	5451.11	I
780	3316.32	II	440	3632.78	II	420	4005.84	I	65	5547.27	I
1000	3319.88	II	1100	3640.25	II	540	4011.29	II	100	5639.50	I
780	3341.00	II	11000	3645.40	II	540	4013.82	I	55 h	5645.99	I
510	3353.58	II	1000	3648.78	II	540	4014.70	II	80	5652.01	I
510	3368.11	II	700	3664.62	II	420	4027.78	II	70 h	5718.46	I
5300	3385.02	II	990	3672.30	II	520 d	4028.32	II	55	5745.53	I
610	3388.85	II	420	3672.70	II	520	4032.47	II	55 h	5868.11	II
3800	3393.57	II	1400	3674.08	II	420	4033.65	II	70	5945.80	I
1300	3396.16	II	2200	3676.59	II	420	4036.32	II	120	5974.49	I
5300	3407.80	II	640	3678.51	I	12000	4045.97	I	140	5988.56	I
1300	3413.78	II	820	3684.85	I	1600	4050.56	II	140	6088.26	I
530	3414.82	II	1300	3685.78	I	520	4055.14	II	100	6168.43	I
780	3419.63	II	4700	3694.81	II	2500	4073.12	II	270	6259.09	I
530	3425.06	II	990	3698.21	II	7400	4077.96	II	160	6579.37	I
1900	3434.37	II	540	3701.63	II	3900	4103.30	II	75	6667.86	I
560	3440.93	II	440	3707.57	II	860	4103.87	I	180	6835.42	I
1300	3441.45	II	440	3708.22	II	1500	4111.34	II	80	6852.96	I
3800	3445.57	II	420	3710.07	II	490	4124.63	II	65	6899.32	II
830	3446.99	II	1600	3724.45	II	990	4129.42	II	55	7426.86	II
2700	3454.32	II	930	3739.34	I	1200	4143.10	II	55	7543.73	I
1300	3456.56	II	1200	3747.82	II	990	4146.06	I	80	7662.36	I
4400	3460.97	II	1400	3753.51	II	5700	4167.97	I	100	8201.57	II
720	3468.43	II	1400	3753.75	II	930	4183.72	I	45	8791.39	II
560	3471.14	II	1200	3757.05	I	12000	4186.82	I	**Erbium**		
560 d	3471.53	II	4700	3757.37	II	2200	4191.64	I	**Er Z = 68**		
1300	3477.07	II	640	3767.63	I	6800	4194.84	I	600	2277.65	III
4400	3494.49	II	640	3773.05	I	800	4198.02	I	290	2586.73	II
560	3496.34	II	420	3781.47	I	680	4201.30	I	490	2670.26	II
830	3498.71	II	3300	3786.18	II	680	4202.24	I	500	2739.27	III
830	3504.53	II	1600	3788.44	II	16000	4211.72	I	610	2755.63	II
830	3505.45	II	700	3791.87	II	1800	4213.18	I	1000	2904.47	II
1300	3506.81	II	510	3804.14	II	3700	4215.16	I	1500	2910.36	II

Intensity	Wavelength/Å		Intensity	Wavelength/Å		Intensity	Wavelength/Å		Intensity	Wavelength/Å	
1500	2964.52	II	3200	3973.58	I	55	6326.13	I	120	3058.98	I
1200	3002.41	II	1400	3974.72	II	55	6492.35	I	220	3077.36	II
1000	3055.10	III	810	3977.02	I	60	6583.48	I	120	3097.45	II
1000	3070.40	III	1100	3982.33	I	70	6601.11	I	320	3106.18	I
610	3073.34	II	810	3987.66	I	70	6759.87	I	950	3111.43	I
720	3082.08	II	14000	4007.96	I	35	6790.92	I	120	3130.73	II
610	3084.02	II	1100	4012.58	I	70	6848.10	I	50 c	3171.00	III
770	3122.72	II	3000	4020.51	I	55	6865.13	I	50 c	3183.78	III
1500	3166.25	III	1000	4046.96	I	55	7459.55	I	420	3210.57	I
870	3181.92	II	940	4055.47	II	120	7469.51	I	1000	3212.81	I
870	3220.73	II	690	4059.78	II	35	7680.01	I	420	3213.75	I
610	3223.31	II	3500	4087.63	I	35	7797.47	I	150	3272.77	II
2300	3230.58	II	1100	4098.10	I	35	7921.85	I	210	3277.78	II
2700	3264.78	II	6900	4151.11	I	30	7937.84	I	150	3301.95	II
720	3279.33	II	1000	4190.70	I	35	8312.82	I	140	3308.02	II
720	3280.22	II	1400	4218.43	I	55	8409.90	I	140	3313.33	II
2000	3301.23	III	690	4286.56	I	9	8866.84	II	950	3334.33	I
2300	3312.42	II	40000	4290.06	III		**Europium**		110	3350.40	I
770	3323.19	II	20000	4386.86	III		**Eu Z = 63**		140	3369.06	II
770	3332.70	II	810	4409.34	I	30	2124.69	III	190	3391.99	II
1300	3346.04	II	1000	4606.61	I	200	2350.51	III	280	3396.58	II
1400	3364.08	II	570	4675.62	II	4000	2375.46	III	150	3425.02	II
1400 d	3368.02	II	15000	4735.56	III	100 d	2435.14	III	150	3441.00	II
7700	3372.71	II	2000	4783.12	III	1000	2444.38	III	130	3461.38	II
970	3374.17	II	250	5007.25	I	4000	2445.99	III	470 cw	3521.09	II
1700	3385.08	II	200	5035.94	I	2000	2513.76	III	150	3542.15	II
2300	3392.00	II	210	5042.05	II	200	2522.14	III	180	3552.52	II
770	3441.13	II	120	5124.56	I	160	2564.17	II	150	3603.20	II
970	3471.71	II	130	5127.41	II	110	2568.17	II	6400	3688.42	II
610	3479.41	II	120	5131.53	I	230	2577.14	II	20000 cw	3724.94	II
970	3485.85	II	130	5133.83	I	1000	2638.77	II	350	3741.31	II
6700	3499.10	II	170	5164.77	II	380	2641.27	II	260	3761.12	II
610	3502.78	I	130	5172.78	I	640	2668.34	II	39000 cw	3819.67	II
610	3524.91	II	160	5188.90	II	110	2673.42	II	140	3844.23	II
820	3549.84	II	150	5206.52	I	250	2678.29	II	190	3865.57	I
1500	3558.02	I	140	5255.93	II	250	2685.66	II	150	3884.75	I
1000	3559.90	II	80	5272.91	I	550	2692.03	II	28000 cw	3907.10	II
920	3570.75	II	90	5348.06	I	700	2701.14	II	32000 cw	3930.48	II
1000	3580.52	II	60	5414.63	II	800	2701.90	II	30000 cw	3971.96	II
610	3590.76	I	180	5456.62	I	240	2705.28	II	180	4011.69	II
610	3599.50	II	90	5468.32	I	180	2709.99	I	150	4017.58	II
1000	3599.83	II	80	5485.97	II	700	2716.98	II	120	4039.19	I
3100	3616.56	II	80	5593.46	I	4200	2727.78	II	120	4085.38	II
720	3628.04	I	60	5611.82	I	160	2740.62	II	33000 cw	4129.70	II
1000	3633.54	II	70	5622.01	I	120	2744.26	II	60000 cw	4205.05	II
1600	3638.68	I	80	5626.53	II	480	2781.89	II	150	4298.73	I
900	3645.94	II	90	5640.36	I	1900	2802.84	II	240	4355.09	II
7900	3692.65	II	70	5664.95	I	220	2811.75	II	14000 cw	4435.56	II
1300	3729.52	II	70	5719.55	I	3400	2813.94	II	3000	4522.57	II
900	3742.64	II	100	5739.19	I	550	2816.18	II	11000	4594.03	I
900	3747.43	I	290	5762.80	I	2000	2820.78	II	9800	4627.22	I
1800	3786.84	II	70	5784.66	I	400 cw	2828.72	II	8300	4661.88	I
1600	3810.33	I	70	5800.79	I	260	2859.67	II	110	4867.62	I
4000	3816.78	III	430	5826.79	I	280	2862.57	II	150	4907.18	I
3600	3830.48	II	100	5850.07	I	200	2892.54	I	180	4911.40	I
680	3855.90	I	120	5855.31	I	140	2893.03	I	180	5013.17	I
7500	3862.85	I	140	5872.35	I	360	2893.83	I	170	5022.91	I
1500	3880.61	II	120	5881.14	I	3200	2906.68	II	110	5029.54	I
1200	3882.89	II	8000	5903.30	III	160	2908.99	I	170	5114.37	I
4200	3892.68	I	70	6022.56	I	850	2925.04	II	170	5129.10	I
5200	3896.23	II	70	6061.25	I	200 cw	2952.68	II	210	5133.52	I
11000	3906.31	II	60	6076.45	II	260	2960.21	II	270	5160.07	I
3200	3937.01	I	360	6221.02	I	300	2991.33	II	210	5166.70	I
2100	3938.63	II	55	6262.56	I	100 c	3023.93	III	200	5199.85	I
3200	3944.42	I	60	6268.87	I	200 c	3026.79	III	110	5200.96	I
2700	3973.04	I	130	6308.77	I	320 cw	3054.94	II	120	5206.44	I

Intensity	Wavelength/Å		Intensity	Wavelength/Å		Intensity	Wavelength/Å		Intensity	Wavelength/Å	
750	5215.10	I	35	7887.99	I	750	955.55	I	150	2835.63	III
300	5223.49	I	24 cw	8209.80	I	500	958.52	I	150	2860.33	III
120	5239.24	I	21 cw	8642.67	I	20	972.40	I	120	2862.86	III
200	5266.40	I	18	8870.30	I	350	973.90	I	140	2887.58	III
390	5271.96	I		**Fluorine**		100	976.22	I	150	2889.45	III
110	5272.48	I		**F Z = 9**		40	976.51	I	120	2905.30	III
150	5282.82	I	50	148.00	V	100	977.75	I	140	2913.29	III
120	5291.26	I	50	163.56	V	60	1082.31	V	160	2916.34	III
120	5294.64	I	90	165.98	V	70	1088.39	V	140	2932.49	III
540	5357.61	I	100	166.18	V	80	1219.03	III	140	2994.28	III
120	5361.61	I	50	186.84	V	80	1266.87	III	120 h	2997.21	III
110	5376.94	I	60	190.57	V	90	1267.71	III	130	2997.53	III
120	5392.94	I	70	190.84	V	70	1297.54	III	120	2999.47	III
450	5402.77	I	50	196.39	IV	70	1359.92	III	130	3039.25	III
380	5451.51	I	60	196.45	IV	110	1498.93	III	120	3039.75	III
260	5452.94	I	70	200.09	IV	120	1502.01	III	160	3042.80	III
120	5488.65	I	80	201.16	IV	110	1504.18	III	150	3049.14	III
120	5510.52	I	90	208.25	IV	140	1504.79	III	140	3113.62	III
200	5547.44	I	90	240.08	IV	130	1506.30	III	160	3115.70	III
150	5570.33	I	100	251.03	IV	110	1506.77	III	180	3121.54	III
200	5577.14	I	140	419.65	IV	100	1553.02	III	140	3124.79	III
120	5580.03	I	150	420.05	IV	110	1557.59	III	140	3134.23	III
210	5645.80	I	160	420.73	IV	100	1563.73	III	140	3146.99	III
330	5765.20	I	100	429.51	III	100	1565.54	III	180	3174.17	III
180	5783.69	I	110	430.15	IV	100	1623.40	III	170	3174.76	III
170	5818.74	II	150	430.76	IV	100	1650.76	III	120	3214.00	III
600 cw	5830.98	I	90	464.29	III	130	1670.39	III	200	3501.45	II
330	5966.07	II	120	465.98	V	140	1677.40	III	200	3501.57	II
480 cw	5967.10	I	130	490.57	IV	100	1716.99	III	200	3502.96	II
170	5972.75	I	160	491.00	IV	120	1770.09	III	6	3594.10	I
240	5992.83	I	50	497.38	IV	150	1770.67	III	12	3668.17	I
110	6012.56	I	60	497.83	IV	110	1772.93	III	270	3847.09	II
420	6018.15	I	70	498.80	IV	140	1773.36	III	260	3849.99	II
170	6029.00	I	90	506.16	V	160	1791.65	III	250	3851.67	II
420	6049.51	II	100	508.08	V	110	1803.03	III	5	3898.48	I
140	6057.36	I	120	508.39	III	170	1805.90	III	8	3930.69	I
240	6083.84	I	60	514.08	V	110	1839.30	III	5	3934.26	I
240	6099.35	I	90	525.29	V	120	1839.97	III	5	3948.56	I
120	6118.78	I	100	526.30	V	110	1840.14	III	240	4024.73	II
330	6173.05	II	120	567.69	III	100	2027.44	III	220	4025.01	II
110	6178.76	I	110	567.75	III	120	2030.32	III	230	4025.49	II
260 cw	6188.13	I	140	570.64	IV	120	2217.17	III	200	4103.51	II
140	6195.07	I	140	571.30	IV	50	2298.29	IV	200	4246.23	II
240	6262.25	I	150	571.39	IV	40	2451.58	IV	200	4299.17	II
170	6299.77	I	160	572.66	IV	120	2452.07	III	140 h	4420.30	III
230	6303.41	II	90	605.67	II	50	2456.92	IV	120 h	4427.35	III
120 cw	6350.04	I	100	606.80	II	130	2464.85	III	120 h	4432.32	III
120 cw	6400.93	I	90	630.20	III	130	2470.29	III	140 h	4479.99	III
180	6410.04	I	100	647.77	V	120	2478.73	III	6	4960.65	I
140	6411.32	I	110	647.87	V	150	2484.37	III	150	5012.54	III
830	6437.64	II	130	654.03	V	120	2542.77	III	160	5110.99	III
120	6457.96	I	120	656.12	III	120	2580.04	III	15	5230.41	I
1400	6645.11	II	130	656.87	III	130	2583.81	III	12	5279.01	I
50	6666.35	III	110	657.23	V	120	2593.23	III	18	5540.52	I
140	6802.72	I	140	657.33	V	130	2595.53	III	12	5552.43	I
360	6864.54	I	140	658.33	III	140	2599.28	III	10	5577.33	I
120	7040.20	I	140	676.12	IV	130	2625.01	III	20	5624.06	I
330	7077.10	II	130	677.15	IV	140	2629.70	III	12	5626.93	I
570	7194.81	II	150	677.22	IV	120	2656.44	III	15	5659.15	I
570	7217.55	II	130	678.99	IV	130	2755.55	III	40	5667.53	I
540	7301.17	II	160	679.21	IV	160	2759.63	III	90	5671.67	I
720	7370.22	II	60	757.04	V	120	2788.15	III	18	5689.14	I
300	7426.57	II	150	806.96	I	160	2811.45	III	25	5700.82	I
160	7583.91	I	125	809.60	I	40	2820.74	IV	25	5707.31	I
60 cw	7742.57	I	500	951.87	I	50	2826.13	IV	140	5753.17	III
70	7746.19	I	1000	954.83	I	140	2833.99	III	120	5761.20	III

Intensity	Wavelength/Å		Intensity	Wavelength/Å		Intensity	Wavelength/Å		Intensity	Wavelength/Å	
12	5950.15	I	1500	8274.62	I	2200	3481.28	II	1100	4306.34	I
25	5959.19	I	2000	8298.58	I	1700	3481.80	II	1800	4313.84	I
70	5965.28	I	600	8302.40	I	1700	3494.40	II	2600 d	4325.57	II
50	5994.43	I	900	8807.58	I	1400	3505.51	II	1900	4327.12	I
150	6015.83	I	1000	8900.92	I	1100	3512.50	II	1000	4344.30	II
80	6038.04	I	300	8912.78	I	4300	3545.80	II	2200	4346.46	I
900	6047.54	I	350	9025.49	I	3900	3549.36	II	1400	4401.86	I
100	6080.11	I	400	9042.10	I	1400	3557.05	II	1400	4422.41	I
150	6091.82	III	350	9178.68	I	5400	3584.96	II	1100	4430.63	I
140	6125.50	III	200	9433.67	I	1100	3592.71	II	1100	4519.66	I
800	6149.76	I	25	9505.30	I	1100	3604.87	I	910	4537.81	I
400	6210.87	I	12	9662.04	I	6100	3646.19	II	520	4614.50	I
130	6233.57	III	25	9734.34	I	3900	3654.62	II	700	4694.33	I
13000	6239.65	I	15	9822.11	I	3100	3656.15	II	410	4743.65	I
10000	6348.51	I	12	9902.65	I	1400	3662.26	II	470	4767.24	I
140	6363.05	III	80 h	10047.98	II	2700	3664.60	II	300	4784.62	I
8000	6413.65	I	15	10285.45	I	2000	3671.20	II	320	4821.69	I
450	6569.69	I	20	10862.31	I	2000	3684.13	I	280	4934.12	I
300	6580.39	I		Francium		3100	3687.74	II	750	5015.04	I
400	6650.41	I		Fr Z = 87		2000	3697.73	II	75	5039.09	I
1800	6690.48	I		7177.	I	1300	3699.73	II	5000	5091.70	III
400	6708.28	I		Gadolinium		2700	3712.70	II	130	5098.38	II
7000	6773.98	I		Gd Z = 64		2000	3713.57	I	910	5103.45	I
1500	6795.53	I	1200	1007.24	IV	1400	3716.36	II	180	5108.91	II
9000	6834.26	I	1200	1063.84	IV	2000	3717.48	I	120	5125.56	II
50000	6856.03	I	1600	1476.98	IV	1800 d	3719.45	II	860	5155.84	I
8000	6870.22	I	1500	1705.03	IV	1500	3730.84	II	190	5176.28	II
15000	6902.48	I	1600	1706.01	IV	4500	3743.47	II	410	5197.77	I
6000	6909.82	I	2000	1736.24	IV	1400	3758.31	II	280	5219.40	I
4000	6966.35	I	1500	1815.32	IV	8700	3768.39	II	130	5233.93	I
45000	7037.47	I	2200	1975.24	III	1400	3770.69	II	320	5251.18	I
30000	7127.89	I	3400	2018.07	III	2900	3783.05	I	120	5252.14	II
15000	7202.36	I	2800	2359.31	III	5100	3796.37	II	140	5255.80	I
1000	7309.03	I	1400	2397.87	IV	3700	3813.97	II	280	5283.08	I
15000	7311.02	I	2800	2697.39	III	3300	3850.69	II	280	5301.67	I
700	7314.30	I	2800	2703.28	III	5100	3850.97	II	220	5302.76	I
5000	7331.96	I	2700	2727.89	III	4300	3852.45	II	280	5307.30	I
120	7336.77	III	9000	2904.73	III	1600	3866.99	I	130	5321.50	I
130	7354.94	III	9500	2955.53	III	1500	3894.70	II	280	5321.78	I
10000	7398.69	I	1200	2999.04	II	2200	3916.51	II	110	5327.32	I
4000	7425.65	I	2100	3010.13	II	1200	3934.79	I	170	5333.30	I
2200	7482.72	I	1900	3027.60	II	1400	3945.54	I	300	5343.00	I
2500	7489.16	I	2100	3032.84	II	1200	3957.67	II	200	5348.67	I
900	7514.92	I	1600	3034.05	II	1100	4023.14	I	300	5350.38	I
5000	7552.24	I	2100	3081.99	II	1100	4028.15	I	240	5353.26	I
5000	7573.38	I	3500	3100.50	II	1400	4037.33	II	3000	5365.96	III
7000	7607.17	I	930	3145.00	II	1600	4045.01	I	150	5370.63	I
18000	7754.70	I	980	3156.53	II	1300	4049.43	II	4000	5553.30	III
15000	7800.21	I	980	3161.37	II	2200	4049.86	II	3000	5587.88	III
300	7879.18	I	4000	3176.66	III	2600	4053.64	I	190	5617.91	I
500	7898.59	I	1400	3331.38	II	2600	4058.22	I	110	5632.25	I
350	7936.31	I	1100	3336.18	II	1900	4063.39	II	260	5643.24	I
300	7956.32	I	5400	3350.47	II	1300	4078.44	II	3000	5658.98	III
80	8016.01	II	4300	3358.62	II	2800	4078.70	I	390	5696.22	I
1000	8040.93	I	5400	3362.23	II	1500	4085.56	II	120	5733.86	II
900	8075.52	I	1100	3392.53	II	1100	4092.71	I	240	5791.38	I
350	8077.52	I	1100 d	3407.56	II	2600	4098.61	II	220	5851.63	I
350	8126.56	I	6900	3422.47	II	2200	4130.37	II	280	5856.22	I
600	8129.26	I	1700	3439.21	II	1100	4132.28	II	110	5904.56	I
300	8159.51	I	2700	3439.99	II	2400	4175.54	I	170	5911.45	II
600	8179.34	I	1400	3450.38	II	2400	4184.25	II	85	5930.29	I
300	8191.24	I	1100	3451.23	II	2200	4190.78	I	85	5936.84	I
350	8208.63	I	2700	3463.98	II	1300	4212.00	II	65	5937.71	I
2500	8214.73	I	1700	3467.27	II	4800	4225.85	I	110 h	5988.02	I
3000	8230.77	I	1700	3468.99	II	1700	4251.73	II	430	6114.07	I
500	8232.19	I	1400	3473.22	II	1600	4262.09	I	75	6305.15	II

Line Spectra of the Elements (continued): Gadolinium—Germanium

Intensity	Wavelength/Å		Intensity	Wavelength/Å		Intensity	Wavelength/Å		Intensity	Wavelength/Å	
40	6331.35	I	80	1054.56	V	76	1351.06	IV	60	746.88	V
40	6380.95	II	90	1058.12	V	70	1353.92	III	60	760.05	V
40 h	6538.15	I	80	1066.69	V	74	1364.63	IV	10	862.234	II
55	6564.78	I	60	1069.60	V	60	1395.54	IV	15	875.493	II
50	6634.36	II	80	1073.77	V	77	1402.55	IV	15	905.977	II
35	6681.23	II	90	1078.83	V	70	1405.32	IV	20	920.554	II
85	6730.73	I	110	1079.60	V	73	1465.87	IV	300	971.35	V
50	6752.67	II	60	1080.99	V	90	1495.07	III	300	990.66	V
26	6786.33	II	80	1085.00	III	50	1534.46	III	50	999.101	II
100	6828.25	I	250	1085.01	V	10	1813.98	II	300	1004.38	V
100	6916.57	I	80	1087.37	V	15	1845.30	II	100	1016.638	II
50	6985.89	II	90	1091.71	V	20	2091.34	II	900	1045.71	V
75	6991.92	I	100	1094.36	V	90	2417.70	III	700	1072.66	V
60	6996.76	II	80	1095.10	V	90	2423.98	III	100	1075.072	II
45	7006.16	II	160	1102.83	V	15	2424.36	III	300	1085.51	II
35	7122.57	I	140	1103.03	V	10	2632.66	I	40	1088.45	III
170	7168.37	I	60	1105.61	III	10	2665.05	I	800	1089.49	V
28	7189.57	II	75	1105.62	V	20	2700.47	II	200	1098.71	II
28	7262.66	I	70	1106.17	V	15	2780.15	II	500	1106.74	II
35	7441.85	I	80	1118.34	V	50	3521.77	III	1000	1116.94	V
40	7464.36	I	120	1126.40	V	80	3581.19	III	500	1120.46	II
55	7562.97	I	130	1128.10	V	100	3589.34	III	700	1163.39	V
80	7733.50	I	120	1128.53	V	10	3731.10	III	200	1164.27	II
35	7846.35	II	100	1129.94	V	10	3806.60	III	500	1181.19	II
35	7856.93	I	130	1136.07	V	10	4032.99	I	500	1181.65	II
25	7930.25	II	67	1137.06	IV	10	4172.04	I	200	1188.73	II
18	8146.15	I	130	1150.23	V	15	4251.16	II	20	1188.99	IV
21	8668.63	I	90	1150.27	III	10 h	4254.04	II	300	1191.26	II
21 h	8832.06	II	70	1156.10	IV	10	4255.77	II	700	1222.30	V
14 h	8849.14	I	120	1156.51	V	40	4262.00	II	20	1229.81	IV
18 h	8867.31	I	35	1157.74	V	100	4380.69	III	500	1237.059	II
5000	14332.88	III	70	1163.60	IV	150	4381.76	III	500	1261.905	II
Gallium			25	1169.40	V	100	4863.00	III	100	1264.710	II
Ga Z = 31			75	1170.58	IV	150	4993.78	III	200	1401.24	II
14	294.53	IV	48	1171.71	IV	10	5808.28	III	200	1538.091	II
61	295.67	IV	40	1178.95	V	20	5848.25	III	500	1576.855	II
30	298.44	V	68	1185.23	IV	15	5993.51	III	75	1581.070	II
30	300.01	V	40	1186.06	IV	10	6334.2	II	100	1602.486	II
30	302.86	V	73	1190.89	IV	2000	6396.56	I	3 r	1615.57	I
41	304.99	IV	73	1193.02	IV	1000	6413.44	I	2 r	1624.130	I
30	307.03	V	75	1195.02	IV	10 h	7251.4	I	2 r	1630.173	I
30	308.26	V	69	1201.54	IV	20 h	7403.0	I	3 r	1636.31	I
30	311.79	V	72	1206.89	IV	30 h	7464.0	I	4 r	1639.730	I
30	313.68	V	80	1213.17	V	10 h	7620.5	I	2	1647.531	I
40	319.41	V	63	1216.15	IV	50 h	7734.77	I	200	1649.194	II
40	322.31	V	50	1228.03	IV	100 h	7800.01	I	2	1651.528	I
50	322.99	V	60	1236.38	IV	15 h	8002.55	I	4 r	1651.955	I
30	323.10	V	60	1238.59	IV	20 h	8074.25	I	3	1661.345	I
40	324.25	V	75	1245.53	IV	100 h	8311.86	I	4 r	1663.539	I
40	324.95	V	83	1258.77	IV	200 h	8386.49	I	10 h	1665.275	I
40	326.14	V	81	1264.66	IV	200 h	9492.92	I	4	1667.802	I
30	326.77	V	82	1267.15	IV	200 h	9493.12	I	3 r	1670.608	I
30	328.65	V	90	1267.16	III	300 h	9589.36	I	100 r	1691.090	I
25	423.18	IV	81	1279.24	IV	100 h	10905.95	I	200 r	1716.784	I
16	439.92	IV	15	1283.64	V	400	11949.12	I	100 h	1739.102	I
50	620.00	III	80	1285.33	IV	200	12109.78	I	100	1742.195	I
40	622.01	III	80	1293.46	III	60	14996.64	I	50	1746.065	I
90	806.51	III	60	1295.36	III	60	22016.81	I	200	1750.043	I
90	817.30	III	82	1295.86	IV	70	22568.71	I	100	1758.279	I
50	828.70	III	83	1299.46	IV	**Germanium**			100 h	1764.185	I
20	878.17	V	82	1303.53	IV	**Ge Z = 32**			100 h	1765.284	I
40	973.21	V	80	1309.68	IV	700	294.51	V	50 h	1766.433	I
40	989.75	V	80	1314.82	IV	1000	295.64	V	200	1774.176	I
90	1014.47	V	60	1323.15	III	200	304.98	V	200	1785.046	I
90	1019.71	V	85	1338.09	IV	20	621.52	V	100 h	1793.071	I
120	1050.48	V	77	1347.03	IV	50	724.21	V	75 h	1801.432	I

Intensity	Wavelength/Å		Intensity	Wavelength/Å		Intensity	Wavelength/Å		Intensity	Wavelength/Å	
200 h	1841.328	I	1200	2651.172	I	4	9492.559	I	300	1428.93	III
200 h	1842.410	I	500	2651.568	I	7	9625.664	I	80	1429.19	I
100 h	1844.410	I	500	2691.341	I	5	10039.436	I	250	1430.06	III
100 h	1845.872	I	850	2709.624	I	4	10200.952	I	275	1433.37	III
100 h	1846.958	I	400	2729.78	II	10	10382.427	I	50	1435.79	I
200	1853.134	I	40	2740.426	I	10	10404.913	I	250	1435.81	III
500 r	1860.086	I	650	2754.588	I	8	10734.068	I	300	1439.12	III
100	1865.052	I	70	2793.925	I	8	10947.416	I	200	1441.21	III
300 r	1874.256	I	80	2829.008	I	10	11125.130	I	150	1446.37	III
100	1895.197	I	1000	2831.843	II	230	11252.83	I	250	1448.42	III
500 r	1904.702	I	1000	2845.527	II	600	11714.76	I	250	1454.95	III
50 h	1908.434	I	750	3039.067	I	1300	12069.20	I	100	1464.72	III
30	1912.409	I	600	3067.021	I	1050	12391.58	I	150	1471.28	III
300 r	1917.592	I	20	3124.816	I	235	13107.61	I	100	1474.73	III
100 h	1923.467	I	35	3211.86	III	470	14822.38	I	150	1481.10	III
500 r	1929.826	I	40	3255.05	III	150	16759.79	I	100	1481.76	I
10 h	1934.048	I	110	3269.489	I	135	17214.34	I	300	1487.15	III
100 r	1937.483	I	40	3434.03	III	70	18811.86	I	250	1487.91	III
500	1938.008	II	300	3499.21	II	62	19279.24	I	200	1489.47	III
100 r	1938.300	I	60	3554.19	IV	28	20673.64	I	250	1500.37	III
500	1938.891	II	50	3676.65	IV		**Gold**		200	1502.47	III
30 s	1944.116	I	200	4178.96	III		**Au Z = 79**		200	1503.74	III
200	1944.731	I	70	4226.562	I	100	843.44	III	100	1542.00	III
200	1955.115	I	200	4260.85	III	100	845.14	III	100	1548.50	III
500	1962.013	I	150	4291.71	III	200	945.10	III	200	1567.54	III
30 h	1963.373	I	10	4685.829	I	100	1040.63	III	200	1574.85	III
30	1965.383	I	1000	4741.806	II	80	1044.49	III	200	1579.44	III
200	1970.880	I	1000	4814.608	II	80	1046.81	III	150	1584.10	III
200	1979.274	II	50	4824.097	II	100 h	1239.96	III	200	1587.16	I
300 h	1987.849	I	100	5131.752	II	100	1278.51	III	200	1589.56	III
300	1988.267	I	200	5178.648	II	100	1314.84	III	150	1593.41	III
500 r	1998.887	I	3	5194.583	I	100 h	1328.37	I	70	1598.24	I
200	2011.29	I	6	5265.892	I	200	1336.72	III	200	1600.51	III
1700	2019.068	I	6	5513.263	I	180	1341.68	III	250	1617.16	III
2400 r	2041.712	I	8	5564.741	I	100	1348.89	III	100	1617.78	III
1600 r	2043.770	I	8	5607.010	I	150	1350.32	III	500	1621.93	III
420	2054.461	I	6	5616.135	I	150	1355.61	III	100	1624.34	I
220 h	2057.238	I	7	5621.426	I	150	1356.13	III	300 d	1629.13	III
750 r	2065.215	I	6	5664.226	I	50	1363.98	I	250	1638.88	III
2600 r	2068.656	I	5	5664.842	I	500	1365.40	III	50	1639.90	I
420	2086.021	I	9	5691.954	I	200	1367.17	III	100	1644.17	III
2000 r	2094.258	I	6	5701.776	I	60	1368.62	I	150	1646.67	I
25	2104.45	III	5	5717.877	I	70	1374.82	I	250	1652.74	III
240	2105.824	I	6	5801.029	I	50	1375.76	I	250	1664.77	III
95 h	2124.744	I	9	5802.093	I	180	1377.73	III	100	1665.76	I
50 h	2186.451	I	1000	5893.389	II	150	1378.69	III	100	1668.11	III
340 r	2198.714	I	500	6021.041	II	150	1379.98	III	125	1673.93	III
15	2220.375	I	75	6283.452	II	125	1380.53	III	1000	1693.94	III
18	2256.001	I	100	6336.377	II	200	1381.36	III	150	1697.09	III
18	2314.201	I	100	6484.181	II	80	1382.75	I	200	1698.98	III
24	2327.918	I	6	6557.488	I	50	1385.33	I	200	1699.34	I
15	2359.233	I	50	7049.369	II	300	1385.79	III	200	1700.00	III
20	2379.144	I	30	7145.390	II	100	1389.41	III	200	1702.25	III
10	2389.472	I	5	7353.334	I	180	1391.46	III	100	1707.53	III
15	2397.885	I	7	7384.208	I	60	1392.27	I	250	1710.16	III
130	2417.367	I	10	7833.575	I	180	1396.00	III	200	1715.69	III
30	2436.412	I	10	8031.039	I	50	1402.12	I	100	1716.71	III
30	2488.25	IV	6	8044.165	I	100	1402.91	III	300	1717.83	III
90	2497.962	I	10	8256.013	I	70	1407.38	I	500	1727.31	III
500	2500.54	II	10	8482.21	I	100	1408.45	I	100 d	1733.17	III
70	2533.230	I	9	8700.60	I	225	1409.50	III	300	1738.48	III
20	2542.44	IV	5	9068.785	I	250	1413.80	III	150	1744.39	III
3	2556.298	I	5	9095.957	I	100	1414.27	III	500	1746.10	III
28	2589.188	I	6	9398.868	I	100	1417.09	III	500	1756.92	III
500	2592.534	I	20	9474.993	II	125	1417.39	III	500	1761.95	III
8	2644.184	I	20	9475.645	II	150	1427.42	III	300	1767.42	III

Intensity	Wavelength/Å		Intensity	Wavelength/Å		Intensity	Wavelength/Å		Intensity	Wavelength/Å	
100	1774.42	III	100	3122.50	II	270	921.67	V	450	2393.36	II
800	1775.17	III	1600	3122.78	I	245	951.62	V	670	2393.83	II
200	1776.40	III	100	3194.72	I	180	960.12	V	540	2405.42	II
100	1780.57	III	100	3227.99	III	180	964.74	V	370	2410.14	II
60	1783.22	II	300	3230.63	I	160	971.51	V	320	2417.69	II
300	1786.11	III	300	3308.30	I	160	1092.76	V	390	2447.25	II
150	1792.65	III	300	3309.64	I	160	1201.76	V	450	2460.49	II
500	1793.76	III	100	3309.86	III	270	1232.03	V	400	2461.74	III
200	1801.98	III	100	3320.12	I	200	1233.59	V	430	2464.19	II
400	1805.24	III	100	3355.15	I	160	1237.42	V	210	2469.18	II
100	1809.81	III	100 h	3391.31	I	160	1239.53	V	2000	2495.16	III
400	1821.17	III	100	3395.40	I	160	1244.46	V	290	2496.99	II
400	1844.89	III	100	3467.21	I	440	1396.66	V	580	2512.69	II
150	1848.83	III	300	3557.36	I	270	1400.09	V	580	2513.03	II
500	1861.80	III	300	3586.73	I	160	1401.70	V	1000	2515.16	III
150	1871.92	III	100	3611.57	I	370	1407.17	V	890	2516.88	II
100	1879.83	I	100 h	3631.31	I	370	1408.38	V	340	2531.19	II
150	1918.28	III	300	3637.90	I	270	1412.28	V	200	2537.33	II
100	1932.04	III	100 h	3645.02	I	270	1413.51	V	320	2551.40	II
100	1935.42	III	100	3650.74	I	160	1421.96	V	400 h	2560.74	III
200	1948.79	III	100	3709.62	I	220	1422.53	V	250	2563.61	II
100	1958.47	III	100	3796.01	I	370	1433.43	V	300 h	2567.46	III
400	1989.63	III	100	3874.73	I	370	1437.27	V	890	2571.67	II
150	1996.85	III	100 h	3892.26	I	500	1437.73	V	320	2573.90	II
11000	2012.00	I	400	3897.86	I	370	1445.40	V	320	2576.82	II
2600	2021.38	I	300	3909.38	I	270	1457.91	V	300	2578.14	II
150	2082.09	II	100	3927.69	I	100	1717.21	IV	320	2582.54	II
300	2083.09	III	400	4040.93	I	270	1719.32	V	390	2606.37	II
60	2110.68	II	700	4065.07	I	550	1729.08	V	450	2607.03	II
100	2159.08	III	100	4084.10	I	750	1731.83	V	230	2613.60	II
80	2167.33	III	100	4241.80	I	750	1733.96	V	450	2622.74	II
200	2172.20	III	200	4315.11	I	440	1741.74	V	160	2637.00	I
100	2184.11	III	120 h	4437.27	I	1000	1749.11	V	1100	2638.71	II
500	2188.97	III	250	4488.25	I	1000	1750.19	V	1100	2641.41	II
70	2248.56	II	900 h	4607.51	I	500	1760.89	V	160	2642.75	I
80	2263.62	II	100 h	4620.56	I	370	1765.62	V	670	2647.29	II
300	2322.27	III	500	4792.58	I	270	1774.02	V	160	2657.84	II
180	2352.65	I	100	4811.60	I	6200	2012.78	II	210	2661.88	II
100	2382.40	III	100	5147.44	I	8500	2028.18	II	290	2683.35	II
120	2387.75	I	300	5230.26	I	300	2070.94	III	200	2687.22	III
150	2402.71	III	100 h	5261.76	I	1200	2096.18	II	670	2705.61	I
150	2405.12	III	100	5655.77	I	200 h	2099.30	III	210	2712.42	II
2600	2427.95	I	100 h	5721.36	I	200 h	2110.31	III	250	2718.59	I
60	2533.52	II	300	5837.37	I	200	2155.66	III	710	2738.76	II
250	2641.48	I	100 h	5862.93	I	200	2183.50	III	200	2743.64	I
3400	2675.95	I	300 h	5956.96	I	200	2195.44	III	360	2751.81	II
1100	2748.25	I	600	6278.17	I	540	2210.82	II	500	2753.60	III
100	2780.82	I	100	6562.68	I	200	2234.59	III	450	2761.63	I
1000	2802.04	II	600	7510.73	I	320	2254.01	II	160	2766.96	I
300	2819.79	II	10	8145.06	I	160	2255.15	II	170	2773.02	I
100	2822.55	II	10	9254.28	I	250	2266.83	II	980	2773.36	II
100 h	2825.44	II		**Hafnium**		620	2277.16	II	180	2774.02	II
300	2837.85	II		**Hf Z = 72**		200 h	2313.44	III	390	2779.37	I
100	2846.92	II	220	545.41	V	230	2321.14	II	230	2808.00	II
100	2856.74	II	180	600.00	V	580	2322.47	II	230	2813.86	II
300	2883.45	I	200	618.27	IV	300	2323.25	II	170	2814.48	II
300	2891.96	I	200	644.54	IV	300	2324.89	II	230	2817.68	I
100	2893.25	II	400	647.39	IV	200	2332.97	II	200	2819.74	I
100	2907.04	II	600	665.65	IV	300	2336.47	III	1200	2820.22	I
300	2913.52	II	200	673.49	IV	200	2337.33	II	490	2822.68	II
300	2918.24	II	270	867.25	V	230	2343.32	II	180	2833.28	I
100	2954.22	II	180	875.88	V	320	2347.44	II	410	2845.83	I
100 h	2990.27	II	180	885.58	V	540	2351.22	II	270	2849.21	II
300	2994.80	II	180	896.14	V	250	2380.30	II	270	2850.96	I
320	3029.20	I	180	901.54	V	250	2383.540	III	180	2851.21	II
300	3065.42	I	180	919.10	V	170	2393.18	II	180	2860.56	I

Intensity	Wavelength/Å		Intensity	Wavelength/Å		Intensity	Wavelength/Å		Intensity	Wavelength/Å	
760	2861.01	II	270	3255.28	II	1400	3777.64	I	85	6789.27	I
760	2861.70	II	180	3273.66	II	1400	3785.46	I	160	6818.94	I
2100	2866.37	I	200 h	3279.67	III	650	3793.37	II	160	7063.83	I
210	2887.14	I	270	3279.98	II	850 d	3800.38	I	570	7131.81	I
800	2889.62	I	160	3291.05	I	320	3811.78	I	650	7237.10	I
1800	2898.26	I	210	3306.12	I	1300	3820.73	I	410	7240.87	I
1200	2904.41	I	340	3310.27	I	280	3830.02	I	360	7624.40	I
890	2904.75	I	670	3312.86	I	800	3849.18	I	110	7740.17	I
2000	2916.48	I	180	3317.99	II	600	3858.31	I	310	7845.35	I
580	2918.58	I	890	3332.73	I	230	3860.91	I	130	7920.71	I
320	2919.59	II	370	3352.06	II	200	3872.55	II	250	7994.73	I
180	2924.62	I	230	3358.91	I	160	3877.10	II	130	8204.58	I
490	2929.63	II	180	3360.06	I	380	3880.82	II	150	8546.48	I
450	2929.90	I	180	3378.93	I	200	3882.52	I	160	8640.06	I
710	2937.80	II	230	3384.70	II	200	3889.23	I	40	8711.24	I
2000	2940.77	I	170	3386.21	I	200	3889.33	I	65	9004.73	I
160	2944.71	I	800	3389.83	II	620	3899.94	I		**Helium**	
1200	2950.68	I	230	3392.81	I	620	3918.09	II		He Z = 2	
1100	2954.20	I	230	3394.59	II	200	3923.90	II	15	231.454	II
540	2958.02	I	230	3397.26	I	320	3931.38	I	20	232.584	II
1400	2964.88	I	230	3397.60	I	410	3951.83	I	30	234.347	II
620	2966.93	I	2300	3399.80	II	160	3968.01	I	50	237.331	II
710	2968.81	I	170	3400.21	I	200	3973.48	I	100	243.027	II
890	2975.88	II	180	3402.51	I	180	4032.27	I	300	256.317	II
1100	2980.81	I	230	3410.17	II	230	4062.84	I	1000	303.780	II
210	2982.72	I	230	3417.34	I	180	4083.35	I	500	303.786	II
170	3000.10	II	410	3419.18	I	540	4093.16	II	10	320.293	I
800	3005.56	I	200	3428.37	II	1100	4174.34	I	2	505.500	I
1100	3012.90	II	250	3438.24	II	160	4206.58	II	3	505.684	I
540	3016.78	I	710	3472.40	I	190	4209.70	I	4	505.912	I
1100	3016.94	II	200	3478.99	II	170	4228.08	I	5	506.200	I
980	3018.31	I	480	3479.28	II	170	4232.44	II	7	506.570	I
1200	3020.53	I	250	3495.75	II	170	4260.98	I	10	507.058	I
410	3031.16	II	250	3497.16	I	200	4263.39	I	15	507.718	I
710	3050.76	I	980	3497.49	I	170	4272.85	II	20	508.643	I
1100	3057.02	I	1200	3505.23	II	320	4294.79	I	25	509.998	I
850	3067.41	I	980	3523.02	I	160	4330.27	I	35	512.098	I
2100	3072.88	I	980	3535.54	II	180	4336.66	II	50	515.616	I
170	3074.10	I	760	3536.62	I	250	4356.33	I	100	522.213	I
250	3074.79	I	180	3548.81	I	180	4370.97	II	400	537.030	I
430	3080.84	I	540	3552.70	II	160	4417.91	I	1000	584.334	I
200	3096.76	I	1300	3561.66	II	200	4438.04	I	50	591.412	I
340	3101.40	II	270	3567.36	I	250	4565.94	I	5	958.70	I
710	3109.12	II	1100	3569.04	II	500 d	4598.80	I	6	972.11	II
710	3131.81	I	210	3597.42	II	230	4620.86	I	8	992.36	II
850	3134.72	II	540	3599.87	I	210	4655.19	I	15	1025.27	II
170	3139.65	II	800	3616.89	I	120	4699.01	I	30	1084.94	II
220	3145.32	II	320	3630.87	II	160	4782.74	I	35	1215.09	II
220	3148.41	I	800	3644.36	II	310	4800.50	I	50	1215.17	II
450	3156.63	I	320	3649.10	I	130	4859.24	I	120	1640.34	II
270	3159.82	I	200	3651.84	I	120	4975.25	I	180	1640.47	II
710	3162.61	II	220	3665.35	II	95	5018.20	I	7	2385.40	II
450	3168.39	I	200	3672.27	I	95	5047.45	I	9	2511.20	II
890	3172.94	I	480	3675.74	I	75	5170.18	I	50	2577.6	I
450	3176.86	II	2200	3682.24	I	230	5181.86	I	1	2723.19	I
220	3181.01	I	280	3696.51	I	110	5243.99	I	12	2733.30	II
360	3193.53	II	240	3699.72	I	120	5294.87	I	2	2763.80	I
670	3194.19	II	340	3701.15	II	110	5354.73	I	10	2818.2	I
200	3196.93	I	1000	3717.80	I	110	5373.86	I	4	2829.08	I
310	3206.11	I	650	3719.28	II	75	5452.92	I	10	2945.11	I
180	3210.98	I	160	3729.10	I	230	5550.60	I	40	3013.7	I
180	3217.30	II	460	3733.79	I	230	5552.12	I	20	3187.74	I
180	3220.61	II	160	3737.88	II	95	5613.27	I	3	3202.96	II
360	3247.66	I	400	3746.80	I	160	5719.18	I	15	3203.10	II
220	3249.53	I	170	3766.92	II	95	6098.67	I	1	3354.55	I
890	3253.70	II	200	3768.25	I	95	6185.13	I	2	3447.59	I

Line Spectra of the Elements (continued): Helium—Hydrogen

Intensity	Wavelength/Å		Intensity	Wavelength/Å		Intensity	Wavelength/Å		Intensity	Wavelength/Å	
1	3587.27	I	200	17002.47	I	410	3580.75	II	220	4979.97	I
3	3613.64	I	1	18555.55	I	410	3581.83	II	90	4995.05	I
2	3634.23	I	6	18636.8	II	630 c	3592.23	II	130	5042.37	I
3	3705.00	I	500	18685.34	I	1100 cw	3598.77	II	80	5093.07	I
1	3732.86	I	200	18697.23	I	540 c	3600.95	II	140	5127.81	I
10	3819.607	I	100	19089.38	I	410	3618.43	I	130	5142.59	II
1	3819.76	I	20	19543.08	I	430 c	3626.69	II	110	5143.22	II
500	3888.65	I	1000	20581.30	I	490	3627.25	II	160	5149.59	II
20	3964.729	I	80	21120.07	I	430 c	3631.76	II	90 c	5167.88	I
1	4009.27	I	10	21121.43	I	430 c	3638.30	II	130 c	5182.11	I
50	4026.191	I	20	21132.03	I	1600 c	3662.29	I	90	5190.11	II
5	4026.36	I	3	30908.5	II	1400	3667.97	I	65	5251.82	I
12	4120.82	I	4	40478.90	I	720	3679.19	I	90	5301.25	I
2	4120.99	I		**Holmium**		670	3679.70	I	80	5330.11	I
3	4143.76	I		**Ho Z = 67**		720	3682.65	I	90	5359.99	I
10	4387.929	I	170	2502.91	II	580	3690.65	I	100	5407.08	I
3	4437.55	I	170	2533.80	I	410	3700.04	I	70	5566.52	I
200	4471.479	I	190	2605.86	II	490 c	3702.35	II	65	5627.60	I
25	4471.68	I	270	2750.35	II	430	3712.88	I	140	5659.58	I
6	4685.4	II	270	2769.89	II	450	3720.72	I	140 c	5691.47	I
30	4685.7	II	300	2824.20	II	1100	3731.40	I	140 c	5696.57	I
30	4713.146	I	270 c	2831.69	II	810	3736.35	I	140 c	5860.28	I
4	4713.38	I	270	2849.10	II	3200 cw	3748.17	II	70 c	5882.99	I
20	4921.931	I	360	2880.26	II	8900 c	3796.75	II	70	5921.76	I
100	5015.678	I	460	2880.98	II	8900 c	3810.73	II	70 cw	5948.03	I
10	5047.74	I	570 c	2909.41	II	410 c	3835.35	II	70	5972.76	I
5	5411.52	II	410 c	2979.63	II	1300 cw	3837.51	II	90	5973.52	I
500	5875.62	I	410	2987.64	II	410 c	3842.05	II	230 c	5982.90	I
100	5875.97	I	480 c	3049.38	II	1100	3843.86	II	120	6081.79	I
8	6560.10	II	410 c	3054.00	II	490 c	3846.73	II	70	6208.65	I
100	6678.15	I	500 c	3057.45	II	1800 c	3854.07	II	70 c	6305.36	I
3	6867.48	I	500 c	3082.34	II	2700 c	3861.68	II	70	6550.97	I
200	7065.19	I	910	3084.36	II	3000 c	3888.96	II	260	6604.94	I
30	7065.71	I	430 c	3086.54	II	13000 c	3891.02	II	120	6628.99	I
50	7281.35	I	760	3118.50	II	1300 cw	3905.68	II	55 cw	6694.32	I
1	7816.15	I	580 c	3166.62	II	580	3955.73	I	55 c	6785.43	I
2	8361.69	I	810	3173.78	II	490	3959.68	I	40 cw	6939.49	I
2	9063.27	I	810 c	3181.50	II	2700	4040.81	I	45 cw	6950.39	I
2	9210.34	I	980 c	3281.97	II	5400 c	4045.44	II	140	7555.09	I
10	9463.61	I	630 c	3337.23	II	8100	4053.93	I	40 c	7628.42	I
4	9516.60	I	980 c	3343.58	II	1700	4065.09	II	50 c	7693.15	I
3	9526.17	I	8100 c	3398.98	II	720	4068.05	I	60 cw	7815.48	I
1	9529.27	I	810 c	3410.26	II	8900	4103.84	I	60	7894.64	I
1	9603.42	I	1400 c	3414.90	II	2900	4108.62	I	50	8512.94	I
3	9702.60	I	5400	3416.46	II	1500	4120.20	I	40	8670.19	I
6	10027.73	I	1200	3421.63	II	1300	4125.65	I	90	8915.98	II
2	10031.16	I	2000 c	3425.34	II	4300	4127.16	I		**Hydrogen**	
15	10123.6	II	2000 c	3428.13	II	1500	4136.22	I		**H Z = 1**	
1	10138.50	I	3200	3453.14	II	980 cw	4152.61	II	15	926.226	I
10	10311.23	I	16000 c	3456.00	II	8100	4163.03	I	20	930.748	I
2	10311.54	I	1600	3461.97	II	2500	4173.23	I	30	937.803	I
3	10667.65	I	810 c	3473.91	II	540	4194.35	I	50	949.743	I
300	10829.09	I	5400 c	3474.26	II	2000	4227.04	I	100	972.537	I
1000	10830.25	I	6300	3484.84	II	1300 cw	4254.43	I	300	1025.722	I
2000	10830.34	I	2500 c	3494.76	II	490	4264.05	I	1000	1215.668	I
9	10913.05	I	810 c	3498.88	II	1300	4350.73	I	500	1215.674	I
3	10917.10	I	810	3510.73	I	300	4477.64	II	5	3835.384	I
4	11626.4	II	4100 c	3515.59	II	290	4629.10	II	6	3889.049	I
30	11969.12	I	410 c	3519.94	II	80	4701.69	II	8	3970.072	I
20	12527.52	I	630	3540.76	II	130	4709.84	II	15	4101.74	I
50	12784.99	I	1600	3546.05	II	130 c	4717.52	I	30	4340.47	I
20	12790.57	I	1100 c	3556.78	II	290	4742.04	II	80	4861.33	I
7	12845.96	I	410	3560.15	II	100 c	4757.01	I	120	6562.72	I
10	12968.45	I	410 c	3573.24	II	65	4782.92	I	180	6562.852	I
2	12984.89	I	630 c	3574.80	II	290	4939.01	I	5	9545.97	I
12	15083.64	I	810	3579.12	I	250 c	4967.21	II	7	10049.4	I

Intensity	Wavelength/Å		Intensity	Wavelength/Å		Intensity	Wavelength/Å		Intensity	Wavelength/Å	
12	10938.1	I	40	2281.64	II	250 w	3962.35	II	180 w	6115.9	II
20	12818.1	I	100 c	2306.05	II	120 c	4004.66	II	230 w	6128.7	II
40	18751.0	I	90 d	2313.21	II	140 d	4013.92	II	240 w	6129.4	II
5	21655.3	I	70 d	2327.95	II	100	4023.77	III	320 w	6132.1	II
8	26251.5	I	80 h	2334.57	II	150	4032.32	III	150 c	6140.0	II
15	40511.6	I	110 d	2382.63	II	410 w	4056.94	II	90	6143.23	II
4	46525.1	I	70 h	2427.20	II	100	4071.57	III	140 c	6148.10	II
6	74578.	I	100	2447.90	II	100	4072.93	III	190 w	6149.5	II
Indium			110 d	2488.62	II	17000	4101.76	I	80	6161.15	II
In Z = 49			90	2488.95	II	140 c	4205.14	II	180 w	6162.45	II
17	378.61	V	80	2498.59	II	100 d	4213.04	II	200	6197.72	III
17	386.21	V	100	2499.60	II	110 d	4219.66	II	100 c	6224.28	II
14	388.91	V	90 d	2500.99	II	100	4252.68	III	280 w	6228.3	II
25	393.89	V	110 d	2512.31	II	150 d	4372.87	II	140 w	6231.1	II
25	400.57	V	100	2521.37	I	150 c	4500.78	II	270 w	6304.8	II
25	402.39	V	100	2527.41	III	18000	4511.31	I	290 w	6362.3	II
622	472.71	IV	160 d	2554.44	II	110 c	4549.01	II	300 w	6469.0	II
689	479.39	IV	1100	2560.15	I	140 c	4570.85	II	210 c	6541.20	II
709	498.62	IV	200	2601.76	I	180 w	4578.02	II	190 c	6751.88	II
10	882.24	III	90 d	2654.70	II	180 w	4578.40	II	180 c	6765.9	II
10	890.84	III	100 d	2662.63	II	140 c	4616.08	II	100 c	6783.72	II
10	915.87	III	140 d	2668.65	II	170 c	4617.17	II	320 w	6891.5	II
85	954.67	IV	140 d	2674.56	II	250 c	4620.14	II	380 w	7182.9	II
87	973.50	IV	80	2683.12	II	150 w	4620.70	II	180 c	7255.0	II
86	991.60	IV	1600	2710.26	I	170 c	4627.30	II	210 c	7276.5	II
89	1024.68	IV	300	2713.94	I	140 c	4637.04	II	180 c	7303.4	II
85	1024.79	IV	80	2726.15	III	380 c	4638.16	II	320 c	7350.6	II
88	1031.45	IV	130 d	2749.75	II	220 c	4644.58	II	100 c	7632.7	II
82	1031.98	IV	700	2753.88	I	360 c	4655.62	II	100 c	7682.9	II
80	1054.43	IV	90 d	2818.97	II	320 w	4656.74	II	210 c	7740.7	II
84	1063.03	IV	180 c	2836.92	I	190 c	4681.11	II	100 c	7776.96	II
83	1068.25	IV	80	2865.68	II	450 w	4684.8	II	180 c	7789.0	II
82	1069.82	IV	120 d	2890.18	II	90 d	4907.06	II	90 c	7840.9	II
86	1077.64	IV	1100	2932.63	I	150 c	4973.77	II	240 c	8227.0	II
90	1082.10	IV	100	2941.05	II	80 h	5109.36	II	100 w	8813.5	II
83	1086.33	IV	100	2982.80	III	100 w	5115.14	II	80 c	8832.6	II
82	1096.81	IV	110 c	2999.40	II	140 c	5117.40	II	120 c	9197.7	II
84	1097.18	IV	100	3008.08	III	270 c	5120.80	II	120 c	9202.0	II
85	1116.10	IV	8000	3039.36	I	200 w	5121.75	II	220 w	9213.0	II
80	1124.06	IV	110 d	3099.80	II	80 d	5129.85	II	160 d	9241.1	II
90	1131.46	IV	180 c	3101.8	II	240 c	5175.42	II	100	9977.86	I
85	1144.43	IV	130 c	3138.60	II	140 c	5184.44	II	200	10257.03	I
80	1145.41	IV	80 c	3142.75	II	200	5248.77	III	100 h	10744.31	I
89	1146.62	IV	130 d	3146.70	II	150 c	5309.45	II	20	11334.72	I
83	1154.11	IV	150	3155.77	II	80	5411.41	II	20	11731.48	I
84	1154.60	IV	100 c	3158.40	II	140 c	5418.45	II	10	12912.59	I
90	1157.71	IV	90 c	3176.30	II	220 w	5436.70	II	9	13429.96	I
90	1157.82	IV	90 d	3198.11	II	130 c	5497.50	II	7	14719.08	I
85	1159.78	IV	13000	3256.09	I	140 c	5507.08	II	6	22291.06	I
88	1176.50	IV	3000	3258.56	I	320 c	5513.00	II	7	23879.13	I
85	1191.58	IV	90 c	3338.50	II	250 w	5523.28	II	**Iodine**		
83	1204.87	IV	100 c	3404.28	II	130 c	5536.50	II	**I Z = 53**		
90	1206.55	IV	110 d	3438.40	II	190 w	5555.45	II	30	363.78	V
88	1221.50	IV	180 c	3693.91	II	240 c	5576.90	II	36	380.74	V
85	1221.90	IV	95 c	3708.13	II	200 w	5636.70	II	45	565.53	V
85	1233.58	IV	380 w	3716.14	II	100	5645.15	III	50	607.57	V
87	1235.84	IV	120 c	3718.30	II	160 c	5708.50	II	6	612.46	IV
90	1373.20	IV	160 c	3718.72	II	100 c	5721.80	II	6	666.81	III
88	1398.77	IV	160 c	3723.40	II	100	5819.50	III	8	705.11	III
81	1412.09	IV	170 w	3795.21	II	210 c	5853.15	II	7	784.64	III
100	1625.42	III	230 c	3799.21	II	490 w	5903.4	II	7	784.80	III
100	1748.83	III	250 c	3834.65	II	260 w	5915.4	II	8	795.52	III
30	1842.41	III	200 c	3842.18	II	120 c	5918.78	II	7	919.28	IV
40	1850.30	III	100	3852.82	III	130 c	6062.9	II	10000	1034.66	II
30	2154.08	III	100	3889.78	II	250 c	6095.95	II	8	1094.20	III
50	2205.28	II	100 c	3902.07	II	210 c	6108.66	II	10000	1139.80	II

Intensity	Wavelength/Å		Intensity	Wavelength/Å		Intensity	Wavelength/Å		Intensity	Wavelength/Å	
10000	1160.56	II	8	2519.74	IV	2000 c	7700.20	I	10	30361.93	I
20000	1166.48	II	8	2545.67	IV	600	7897.98	I	8	30383.88	I
10000	1178.65	II	7	2545.71	III	500	7969.48	I	10	34295.73	I
15000	1187.34	II	8	2652.23	IV	1000	8003.63	I	9	34513.11	I
10000	1190.85	II	5000	3078.75	II	99000	8043.74	I	3	40228.54	I
200	1218.41	I	200	4102.23	I	300 d	8065.70	I	2	41633.80	I
20000	1220.89	II	200	4129.21	I	1000	8090.76	I	**Iridium**		
600	1224.05	I	100 d	4134.15	I	800 c	8169.38	I	**Ir Z = 77**		
600	1224.08	I	500 d	4321.84	I	500 d	8222.57	I	9900	2010.65	I
500	1228.89	I	250	4763.31	I	4000	8240.05	I	8700	2022.35	I
20000	1234.06	II	1000	4862.32	I	10000 c	8393.30	I	15000	2033.57	I
600	1251.34	I	200	4916.94	I	1000	8486.11	I	6200	2052.22	I
8	1252.35	III	10000	5119.29	I	1500 c	8664.95	I	5000	2060.64	I
2500	1259.15	I	3000 c	5161.20	II	500 c	8700.80	I	3700	2083.22	I
3000	1259.51	I	1000	5234.57	I	250 d	8748.22	I	3100	2085.74	I
800	1261.27	I	3000 c	5245.71	II	1000	8853.24	I	17000	2088.82	I
600	1267.57	I	10000	5338.22	II	2000	8853.80	I	14000	2092.63	I
600	1267.60	I	5000 c	5345.15	II	3000	8857.50	I	2700	2112.68	I
1500	1275.26	I	600 c	5427.06	I	1000 d	8898.50	I	1800	2119.54	I
3000	1289.40	I	3000	5435.83	II	400	8964.69	I	2000	2125.44	I
10000	1300.34	I	2000 c	5464.62	II	400	8993.13	I	4500	2126.81	II
3000	1302.98	I	10000	5625.69	II	5000	9022.40	I	2000	2127.52	I
3000	1313.95	I	2000 c	5690.91	II	15000	9058.33	I	4500	2127.94	I
3000	1317.54	I	4000 c	5710.53	II	1000	9098.86	I	3700	2148.22	I
2000	1330.19	I	1000 d	5764.33	I	12000	9113.91	I	2500	2150.54	I
20000	1336.52	II	2000	5894.03	I	600	9128.03	I	3500	2152.68	II
5000	1355.10	I	5000	5950.25	II	600	9227.74	I	2900	2155.81	I
3000	1357.97	I	300	5984.86	I	1000	9335.05	I	7900	2158.05	I
5000	1360.97	I	2000 d	6024.08	I	4000	9426.71	I	2100	2162.88	I
3000	1361.11	I	2000 c	6074.98	II	3000	9427.15	I	5800	2169.42	II
2500	1367.71	I	1000	6082.43	I	2000	9598.22	I	4500	2175.24	I
2500	1368.22	I	2000 c	6127.49	II	2000	9649.61	I	2700	2178.17	I
4000	1383.23	I	800	6191.88	I	3000 d	9653.06	I	1600	2187.43	II
3000	1390.75	I	500	6213.10	I	5000	9731.73	I	1100	2190.38	II
2000	1392.90	I	800	6244.48	I	500	10003.05	I	740	2191.64	I
2000	1400.01	I	1000	6293.98	I	750	10131.16	I	910	2208.09	II
8000	1425.49	I	500	6313.13	I	1000	10238.82	I	1300	2220.37	I
5000	1446.26	I	800	6330.37	I	400	10375.20	I	790	2221.07	II
5000	1453.18	I	400	6333.50	I	400	10391.74	I	2500	2242.68	II
5000	1457.39	I	2000	6337.85	I	5000	10466.54	I	620	2245.76	II
5000	1457.47	I	1000	6339.44	I	400	11236.56	I	2100	2253.38	I
10000	1457.98	I	500	6359.16	I	350	11558.46	I	2100	2255.10	I
2500	1458.79	I	1000	6566.49	I	320	11778.34	I	1400	2255.81	I
4000	1459.15	I	2000	6583.75	I	450	11996.86	I	350	2258.51	I
2500	1465.83	I	1000	6585.27	I	300	12033.69	I	1400	2258.86	I
1000	1485.92	I	5000	6619.66	I	150	12304.58	I	830	2264.61	I
5000	1492.89	I	500	6661.11	I	60	13149.16	I	1100	2266.33	I
5000	1507.04	I	500 c	6697.29	I	140	13958.27	I	1000	2268.90	I
5000	1514.68	I	400	6732.03	I	200	14287.02	I	660	2280.00	I
15000	1518.05	I	4000	6812.57	II	100	14460.00	I	950	2281.02	II
2500	1526.45	I	500	6989.78	I	225	15032.57	I	660	2281.91	I
5000	1593.58	I	500	7120.05	I	105	15528.65	I	330	2284.60	I
5000	1617.60	I	1200	7122.05	I	150	16037.33	I	330	2295.08	I
2500	1640.78	I	2000	7142.06	I	15	18275.71	I	790	2298.05	I
15000	1702.07	I	1000	7164.79	I	20	18348.52	I	460	2299.53	I
12000	1782.76	I	400 d	7191.66	I	15	18982.41	I	910	2300.50	I
5000	1799.09	I	700	7227.30	I	35	19070.17	I	2700	2304.22	I
75000	1830.38	I	1000	7236.78	I	110	19105.12	I	410	2305.47	I
15000	1844.45	I	500	7237.84	I	50	19370.02	I	210	2307.27	I
2000	2061.63	I	5000	7402.06	I	10	20648.69	I	910	2308.93	I
7	2361.13	IV	1000	7410.50	I	220	22183.03	I	460	2315.38	I
6	2372.45	IV	500	7416.48	I	150	22226.53	I	410	2321.45	I
7	2376.46	IV	5000	7468.99	I	30	22309.21	I	410	2321.58	I
8	2387.11	IV	500 c	7490.52	I	32	24420.82	I	210	2327.98	I
9	2426.10	IV	2000	7554.18	I	12	27365.42	I	540	2333.30	I
8	2475.35	IV	500 d	7556.65	I	9	27573.05	I	740	2333.84	I

Intensity	Wavelength/Å		Intensity	Wavelength/Å		Intensity	Wavelength/Å		Intensity	Wavelength/Å	
580	2334.50	I	330	2617.78	I	3400	3133.32	I	45	5625.55	I
1600	2343.18	I	210	2619.88	I	490	3168.88	I	35	5894.06	I
740	2343.61	I	250	2625.32	I	370	3177.58	I	20	6110.67	I
580	2355.00	I	700	2634.17	I	370	3198.92	I	12	6288.28	I
230	2357.53	II	250	2639.42	I	610	3212.12	I	10	6686.08	I
410	2358.16	I	3500	2639.71	I	370	3219.51	I	6	7834.32	I
500	2360.73	I	210	2644.19	I	5100	3220.78	I		**Iron**	
2500	2363.04	I	1800	2661.98	I	300	3229.28	I		**Fe Z = 26**	
370	2368.04	II	350	2662.63	I	470	3241.52	I	350	386.16	V
3500	2372.77	I	2700	2664.79	I	200	3262.01	I	350	386.88	V
290	2375.09	II	520	2669.91	I	390	3266.44	I	400	387.20	V
250	2377.28	I	520	2671.84	I	200	3322.60	I	400	387.50	V
250	2377.98	I	330	2673.61	I	560	3368.48	I	400	387.76	V
500	2379.38	I	270	2692.34	I	660	3437.02	I	400	387.78	V
540	2381.62	I	3000	2694.23	I	410	3448.97	I	400	395.90	V
210	2383.17	I	330	2772.46	I	3200	3513.64	I	400	404.62	V
1300	2386.89	I	250	2775.55	I	220	3515.95	I	400	405.50	V
2500	2390.62	I	520	2781.29	I	410	3522.03	I	800	407.42	V
2700	2391.18	I	330	2785.22	I	320	3558.99	I	600	407.44	V
230	2407.59	I	540	2797.35	I	1200	3573.72	I	400	407.49	V
290	2409.37	I	1600	2797.70	I	320	3594.39	I	500	407.75	V
290	2410.17	I	380	2798.18	I	220	3609.77	I	400	409.71	V
290	2410.73	I	410	2800.82	I	660	3628.67	I	400	410.20	V
540	2413.31	I	680	2823.18	I	220	3636.20	I	600	411.55	V
370	2415.86	I	1200	2824.45	I	300	3661.71	I	700	417.39	V
620	2418.11	I	820	2836.40	I	300	3664.62	I	700	418.04	V
210	2424.89	I	1100	2839.16	I	320	3674.98	I	500	418.47	V
370	2424.99	I	820	2840.22	I	200	3687.08	I	700	421.06	V
290	2425.66	I	3800	2849.72	II	200	3731.36	II	500	421.78	V
540	2427.61	I	380	2875.60	I	530	3747.20	I	500	422.31	V
540	2431.24	I	380	2875.98	I	3100	3800.12	I	500	426.06	V
1300	2431.94	I	270	2877.68	I	230	3817.24	I	500	426.11	V
270	2435.14	I	820	2882.64	I	480	3902.51	I	350	426.97	V
250	2445.34	I	650	2897.15	I	480	3915.38	I	17	525.69	IV
250	2447.76	I	260	2901.95	I	400	3934.84	I	15	526.29	IV
910	2452.81	I	260	2904.80	I	590	3976.31	I	13	526.63	IV
1300	2455.61	I	200	2907.24	I	460	3992.12	I	14	536.61	IV
230	2455.87	I	440	2916.36	I	350	4033.76	I	15	537.10	IV
210	2457.03	I	230	2918.57	I	370	4069.92	I	13	537.26	IV
210	2457.23	I	4400	2924.79	I	150	4070.68	I	14	537.79	IV
870	2467.30	I	1200	2934.64	I	100	4092.61	I	13	537.94	IV
3300	2475.12	I	880	2936.68	I	140	4115.78	I	13	552.14	IV
210	2478.11	I	250	2938.47	I	90	4172.56	I	14	607.53	IV
2100	2481.18	I	2700	2943.15	I	260	4268.10	I	13	608.80	IV
620	2493.08	I	230	2946.97	I	220	4311.50	I	10	813.38	III
210	2496.27	I	200	2949.76	I	160	4399.47	I	10	844.28	III
250	2502.63	I	1200	2951.22	I	65	4403.78	I	10 p	861.83	III
4100	2502.98	I	200	2974.95	I	110	4426.27	I	10	891.17	III
210	2513.71	I	440	2980.65	I	55	4478.48	I	10	950.33	III
990	2533.13	I	300	2996.08	I	55	4545.68	I	10	981.37	III
1100	2534.46	I	220	3002.25	I	30	4548.48	I	10 w	983.88	III
580	2537.22	I	600	3003.63	I	35	4568.09	I	12	1055.27	II
580	2542.02	I	270	3017.31	I	75	4616.39	I	15	1068.36	II
7900	2543.97	I	380	3029.36	I	26	4656.18	I	15	1071.60	II
790	2546.03	I	330	3039.26	I	50	4728.86	I	15	1096.89	II
210	2551.40	I	300	3047.16	I	26	4756.46	I	12	1099.12	II
210	2555.35	I	300	3049.44	I	65	4778.16	I	18	1112.09	II
910	2564.18	I	300	3057.28	I	30	4795.67	I	12	1121.99	II
210	2569.88	I	1600	3068.89	I	50	4938.09	I	12	1122.86	II
230	2572.70	I	320	3083.22	I	26	4970.48	I	12	1128.07	II
740	2577.26	I	240	3086.44	I	25	4999.74	I	12	1130.43	II
740	2592.06	I	390	3088.04	I	25	5002.74	I	15	1133.41	II
740	2599.04	I	510	3100.29	I	30	5014.98	I	12	1133.68	II
700	2608.25	I	510	3100.45	I	30	5123.66	I	12	1138.64	II
1800	2611.30	I	340	3120.76	I	35	5364.32	I	12	1142.33	II
210	2614.98	I	200	3121.78	I	75	5449.50	I	12	1143.23	II

Intensity	Wavelength/Å		Intensity	Wavelength/Å		Intensity	Wavelength/Å		Intensity	Wavelength/Å	
18	1144.95	II	13	1606.98	IV	18	1788.07	II	300	2178.118	I
12	1147.41	II	17	1609.10	IV	13	1792.10	IV	12	2180.41	III
15	1148.29	II	14	1609.83	IV	13	1796.93	IV	250	2186.486	I
12	1151.16	II	13	1610.47	IV	13	1827.98	IV	60	2186.892	I
12	1267.44	II	13	1611.20	IV	10	1869.83	III	120	2187.195	I
12	1272.00	II	13	1613.64	IV	12	1877.99	III	250	2191.839	I
400	1317.86	V	15	1614.02	IV	10	1882.05	III	150	2196.043	I
400	1323.27	V	13	1614.64	IV	12	1886.76	III	80	2200.390	I
400	1330.40	V	13	1615.00	IV	13	1890.67	III	80	2200.724	I
400	1359.01	V	16	1616.68	IV	11	1893.98	III	15	2208.41	II
600	1361.82	V	14	1617.68	IV	20	1895.46	III	10 p	2208.85	III
700	1373.59	V	12	1618.47	II	10 s	1907.58	III	20	2213.65	II
600	1373.67	V	14	1619.02	IV	19	1914.06	III	12	2218.26	II
500	1376.34	V	13	1621.16	IV	15	1915.08	III	20	2220.38	II
500	1378.56	V	14	1621.57	IV	15	1922.79	III	10	2221.83	III
800	1387.94	V	13	1623.38	IV	10 p	1926.01	III	10	2229.27	III
400	1397.97	V	13	1623.53	IV	18	1926.30	III	10	2232.43	III
600	1400.24	V	15	1626.47	IV	15	1930.39	III	10	2232.69	III
800	1402.39	V	14	1626.90	IV	14	1931.51	III	10	2235.91	III
400	1406.67	V	13	1628.54	IV	30	1934.538	I	10	2238.16	III
500	1406.82	V	13	1630.18	IV	25	1937.269	I	12 p	2241.54	III
400	1407.25	V	17	1631.08	IV	14	1937.34	III	50	2250.790	I
600	1409.45	V	15	1631.12	II	10 l	1938.90	III	60	2251.874	I
400	1415.20	V	14	1632.40	IV	14 s	1943.48	III	300	2259.511	I
600	1420.46	V	13	1634.01	IV	12	1945.34	III	12	2261.59	III
800	1430.57	V	18	1635.40	II	50	1946.988	I	60	2264.389	I
13	1431.43	IV	15	1636.32	II	10	1950.33	III	80	2267.085	I
800	1440.53	V	15	1639.40	II	12	1951.01	III	10	2267.42	III
400	1442.22	V	14	1639.40	IV	25	1951.571	I	80	2267.469	I
800	1446.62	V	16	1640.04	IV	30	1952.59	I	50	2270.862	I
700	1448.85	V	14	1640.16	IV	11	1952.65	III	150	2272.070	I
400	1449.93	V	12	1641.76	II	30	1953.005	I	150	2276.026	I
700	1456.16	V	15	1641.87	IV	13	1953.32	III	80	2279.937	I
500	1459.83	V	15	1647.09	IV	10	1953.49	III	150	2284.086	I
400	1460.73	V	12	1647.16	II	10 w	1954.22	III	150	2287.250	I
500	1462.63	V	15	1651.58	IV	60	1957.823	I	300	2292.524	I
700	1464.68	V	15	1652.90	IV	11	1958.58	III	10	2293.06	III
500	1465.38	V	13	1653.41	IV	60	1960.144	I	15	2295.86	III
400	1466.65	V	13	1656.11	IV	13	1960.32	III	200	2297.787	I
500	1469.00	V	15	1656.65	IV	30	1961.25	I	600	2298.169	I
500	1479.47	V	14	1660.10	IV	50	1962.111	I	80	2299.220	I
10 h	1505.17	III	13	1662.32	IV	12	1963.11	II	300	2300.142	I
13	1526.60	IV	13	1662.52	IV	15	1987.50	III	50	2301.684	I
13	1530.26	IV	13	1663.54	IV	14	1991.61	III	100	2303.424	I
14	1532.63	IV	13	1668.09	IV	13	1994.07	III	150	2303.581	I
13	1532.91	IV	12	1670.74	II	12	1995.56	III	120	2308.999	I
15	1533.86	IV	14	1671.04	IV	12	1996.42	III	150	2313.104	I
13	1533.95	IV	13	1673.68	IV	10	2061.55	III	10 p	2317.70	III
14	1536.58	IV	14	1675.66	IV	12	2068.24	III	10	2319.22	III
10 h	1538.63	III	13	1681.36	IV	14	2078.99	III	200	2320.358	I
13	1542.16	IV	15	1687.69	IV	100	2084.122	I	10 p	2321.71	III
14	1542.70	IV	15	1698.88	IV	10	2084.35	III	10	2326.95	III
12 h	1550.20	III	12	1702.04	II	12	2090.14	III	100	2327.40	II
13	1566.26	IV	13	1709.81	IV	15	2097.48	III	100	2331.31	II
14	1568.27	IV	15	1711.41	IV	12	2097.69	III	300	2332.80	II
13	1591.51	IV	14	1712.76	IV	10	2103.80	III	10 p	2336.77	III
13	1592.05	IV	14	1717.90	IV	12	2107.32	III	200	2338.01	III
13	1598.01	IV	14	1718.16	IV	15	2151.78	III	10	2338.96	III
13	1600.58	IV	14	1719.46	IV	12	2157.71	III	600	2343.49	II
10 h	1601.21	III	14	1722.71	IV	50	2157.794	I	80	2343.96	II
13	1601.67	IV	14	1724.06	IV	12	2158.47	III	150	2344.28	II
13	1603.18	IV	16	1725.63	IV	10	2161.27	III	200	2348.11	II
13	1603.73	IV	13	1761.08	IV	40	2166.773	I	250	2348.30	II
13	1604.88	IV	12	1761.38	II	12	2166.95	III	200	2359.12	II
13	1605.68	IV	20	1785.26	II	12	2171.04	III	150	2360.00	II
15	1605.97	IV	20	1786.74	II	15	2174.66	III	120	2360.29	II

Line Spectra of the Elements (continued): Iron

Intensity	Wavelength/Å		Intensity	Wavelength/Å		Intensity	Wavelength/Å		Intensity	Wavelength/Å	
200	2364.83	II	50	2485.990	I	80	2549.39	II	250	2742.254	I
80	2365.76	II	800	2486.373	I	600	2549.613	I	800	2742.405	I
80	2368.59	II	100	2486.691	I	400	2562.53	II	200	2743.20	II
80	2369.456	I	100	2487.066	I	200	2563.48	II	150	2743.565	I
80	2369.95	II	120	2487.370	I	150	2574.36	II	200	2744.068	I
120	2371.430	I	4000	2488.143	I	300	2576.691	I	80	2744.527	I
300	2373.624	I	100	2488.945	I	100	2582.58	II	300	2746.48	II
150	2373.74	II	80	2489.48	II	1500	2584.54	I	100	2749.32	II
120	2374.518	I	1000	2489.750	I	650	2585.88	II	500	2749.48	II
120	2376.43	II	50	2489.913	I	90	2591.54	II	1200	2750.140	I
80	2379.27	II	3000	2490.644	I	90	2593.73	II	80	2753.29	II
120	2380.76	II	100	2490.71	II	650	2598.37	II	150	2754.032	I
150	2381.835	I	2000	2491.155	I	2000	2599.40	II	100	2754.426	I
1000	2382.04	II	100	2491.40	II	300	2599.57	I	800	2755.73	II
300	2388.63	II	100	2493.18	II	60	2605.657	I	250	2756.328	I
200	2389.973	I	500	2493.26	II	300	2606.51	II	100	2757.316	I
1000	2395.62	II	60	2494.000	I	800	2606.827	I	120	2761.780	I
300	2399.24	II	50	2494.251	I	650	2607.09	II	150	2761.81	II
800	2404.88	II	100	2495.87	I	600	2611.87	II	150	2762.026	I
250	2406.66	II	600	2496.533	I	320	2613.82	II	120	2762.772	I
80	2406.97	II	150	2498.90	I	320	2617.62	II	120	2763.109	I
300	2410.52	II	1000	2501.132	I	250	2618.018	I	80	2766.910	I
200	2411.07	II	50	2501.693	I	90	2620.41	II	250	2767.522	I
150	2413.31	II	80	2506.09	II	400	2623.53	I	300	2772.07	I
80	2417.87	II	500	2507.900	I	200	2625.67	II	600	2778.220	I
60	2420.396	I	50	2508.753	I	150	2628.29	II	3000	2788.10	I
60	2423.089	I	1000	2510.835	I	250	2631.05	II	200	2797.78	I
150	2424.14	II	120	2511.76	II	250	2631.32	II	400	2804.521	I
120	2428.36	II	80	2512.275	I	100	2632.237	I	1500	2806.98	I
120	2430.08	II	400	2512.365	I	300	2635.809	I	10 p	2813.24	III
80	2432.26	II	80	2516.570	I	50	2641.646	I	2500	2813.287	I
60	2438.182	I	300	2517.661	I	200	2643.998	I	300	2823.276	I
150	2439.30	II	800	2518.102	I	300	2666.812	I	600	2825.56	I
150	2439.74	I	150	2519.629	I	60	2666.965	I	50	2825.687	I
100	2442.57	I	50	2522.480	I	600	2679.062	I	120	2828.808	I
250	2443.872	I	4000	2522.849	I	500	2684.75	II	1500	2832.436	I
100	2444.51	II	200	2523.66	I	400	2689.212	I	120	2835.950	I
50	2445.212	I	500	2524.293	I	10 h	2695.13	III	200	2838.119	I
100	2445.57	II	100	2525.02	I	200	2699.106	I	200	2843.631	I
60	2447.709	I	200	2525.39	II	80	2706.012	I	1000	2843.977	I
100	2453.476	I	300	2526.29	II	400	2706.582	I	100	2845.594	I
1500	2457.598	I	2000	2527.435	I	60	2708.571	I	800	2851.797	I
150	2458.78	II	800	2529.135	I	200	2711.655	I	50	2869.307	I
80	2461.28	II	250	2529.55	II	80	2714.41	II	50	2872.334	I
100	2461.86	II	150	2529.836	I	50	2716.257	I	80	2874.172	I
100	2462.181	I	200	2530.687	I	50	2717.786	I	50	2894.504	I
1500	2462.647	I	120	2533.63	II	250	2718.436	I	12	2904.43	III
50	2463.730	I	100	2534.42	II	4000	2719.027	I	10	2907.50	III
800	2465.149	I	120	2535.49	II	100	2719.420	I	12	2907.70	III
60	2467.732	I	400	2535.607	I	50	2720.197	I	120	2912.157	I
600	2468.879	I	200	2536.792	I	1500	2720.903	I	120	2929.007	I
80	2470.67	II	200	2536.80	II	400	2723.578	I	1200	2936.903	I
80	2470.965	I	100	2538.80	II	150	2724.953	I	60	2941.343	I
800	2472.336	I	100	2538.91	II	50	2726.235	I	1000	2947.876	I
1000	2472.895	I	150	2538.99	II	80	2727.54	II	600	2953.940	I
200	2473.16	I	50	2539.357	I	200	2728.020	I	250	2957.364	I
600	2474.814	I	200	2540.66	II	50	2728.820	I	150	2965.254	I
60	2476.657	I	600	2540.972	I	80	2728.90	II	1500	2966.898	I
120	2479.480	I	80	2541.10	II	1000	2733.581	I	120	2969.36	I
1200	2479.776	I	300	2542.10	I	60	2734.005	I	800	2970.099	I
100	2480.16	II	250	2543.92	I	50	2734.268	I	1200	2973.132	I
80	2482.12	II	150	2544.70	I	500	2735.475	I	500	2973.235	I
100	2482.66	II	800	2545.978	I	50	2735.612	I	600	2981.445	I
10000	2483.271	I	80	2546.67	II	500	2737.310	I	1000	2983.570	I
300	2483.533	I	100	2548.74	II	120	2737.83	I	1000	2994.427	I
1000	2484.185	I	80	2549.08	II	400	2739.55	II	250	2994.502	I

Intensity	Wavelength/Å		Intensity	Wavelength/Å		Intensity	Wavelength/Å		Intensity	Wavelength/Å	
500	2999.512	I	11	3276.08	III	200	3605.454	I	100	3786.68	I
120	3000.451	I	150	3286.75	I	500	3606.680	I	250	3787.880	I
800	3000.948	I	10	3288.81	III	1500	3608.859	I	250	3790.092	I
12	3001.62	III	120	3305.97	I	250	3610.16	I	150	3794.34	I
60	3001.655	I	200	3306.343	I	60	3612.068	I	400	3795.002	I
12 h	3007.28	III	400	3355.227	I	150	3617.788	I	120	3797.518	I
200	3007.282	I	80	3355.517	I	1500	3618.768	I	250	3798.511	I
500	3008.14	I	60	3369.546	I	200	3621.462	I	400	3799.547	I
120	3009.569	I	120	3370.783	I	150	3622.004	I	200	3805.345	I
15	3013.17	III	50	3378.678	I	150	3623.19	I	80	3806.696	I
60	3017.627	I	50	3380.110	I	100	3631.096	I	600	3812.964	I
60	3018.983	I	60	3383.978	I	1200	3631.463	I	60	3813.059	I
500	3020.491	I	50	3392.304	I	60	3632.041	I	1500	3815.840	I
1500	3020.639	I	150	3392.651	I	100	3638.298	I	2500	3820.425	I
600	3021.073	I	150	3399.333	I	200	3640.389	I	150	3821.179	I
500	3024.032	I	80	3404.353	I	80	3643.717	I	80	3824.306	I
150	3025.638	I	500	3407.458	I	1500	3647.842	I	2500	3824.444	I
500	3025.842	I	250	3413.131	I	250	3649.506	I	1500	3825.880	I
80	3030.148	I	60	3424.284	I	80	3650.279	I	1200	3827.823	I
60	3031.214	I	500	3427.119	I	200	3651.467	I	1000	3834.222	I
60	3034.484	I	60	3428.748	I	120	3670.024	I	120	3839.257	I
800	3037.389	I	6000	3440.606	I	150	3670.089	I	500	3840.437	I
80	3041.637	I	2500	3440.989	I	100	3676.311	I	800	3841.047	I
800	3047.604	I	1000	3443.876	I	150	3677.629	I	120	3843.256	I
600	3057.446	I	200	3445.149	I	1500	3679.913	I	80	3846.800	I
1000	3059.086	I	1200	3465.860	I	200	3682.242	I	200	3849.96	I
250	3067.244	I	2000	3475.450	I	120	3683.054	I	120	3850.817	I
120	3075.719	I	500	3476.702	I	150	3684.107	I	2500	3856.372	I
120	3091.577	I	2500	3490.574	I	120	3685.998	I	150	3859.212	I
80	3098.189	I	500	3497.840	I	500	3687.456	I	10000	3859.911	I
100	3099.895	I	10	3501.76	III	120	3689.477	I	150	3865.523	I
100	3099.968	I	250	3513.817	I	150	3694.008	I	60	3867.215	I
60	3100.303	I	300	3521.261	I	120	3695.051	I	250	3872.501	I
100	3100.665	I	400	3526.040	I	150	3701.086	I	150	3873.761	I
10 p	3136.43	III	100	3526.166	I	80	3704.462	I	250	3878.018	I
10	3174.09	III	60	3526.237	I	1200	3705.566	I	2000	3878.573	I
80	3175.445	I	60	3526.381	I	60	3707.041	I	4000	3886.282	I
10	3175.99	III	60	3526.467	I	150	3707.821	I	200	3887.048	I
10	3178.01	III	100	3533.199	I	300	3707.919	I	300	3888.513	I
150	3184.895	I	200	3536.556	I	600	3709.246	I	800	3895.656	I
250	3191.659	I	300	3541.083	I	120	3716.442	I	1200	3899.707	I
500	3193.226	I	250	3542.075	I	8000	3719.935	I	400	3902.945	I
800	3193.299	I	80	3553.739	I	1500	3722.563	I	250	3906.479	I
200	3196.928	I	400	3554.925	I	120	3724.377	I	80	3916.731	I
80	3199.500	I	200	3556.878	I	60	3725.491	I	600	3920.258	I
50	3205.398	I	400	3558.515	I	60	3727.093	I	1200	3922.911	I
100	3211.88	I	1000	3565.379	I	500	3727.619	I	1200	3927.920	I
200	3214.011	I	1200	3570.097	I	150	3732.396	I	2000	3930.296	I
200	3214.396	I	800	3570.25	I	1200	3733.317	I	60	3948.774	I
60	3215.938	I	120	3571.996	I	5000	3734.864	I	60	3949.953	I
50	3217.377	I	100	3573.393	I	120	3735.324	I	50	3951.164	I
80	3219.583	I	60	3573.829	I	6000	3737.131	I	50	3952.601	I
60	3219.766	I	60	3573.888	I	100	3738.306	I	16	3954.33	III
300	3222.045	I	4000	3581.19	I	400	3743.362	I	60	3956.454	I
600	3225.78	I	150	3582.199	I	6000	3745.561	I	250	3956.68	I
80	3227.796	I	150	3584.660	I	1200	3745.899	I	60	3966.614	I
50	3233.967	I	120	3584.929	I	3000	3748.262	I	11	3968.72	III
120	3234.613	I	300	3585.319	I	80	3748.964	I	100	3969.257	I
300	3236.222	I	150	3585.705	I	3000	3749.485	I	80	3977.741	I
100	3239.433	I	10	3586.04	III	1500	3758.232	I	10 w	3979.42	III
80	3244.187	I	200	3586.103	I	400	3760.05	I	40	3981.771	I
80	3246.005	I	400	3586.984	I	1500	3763.788	I	50	3983.956	I
80	3265.046	I	100	3594.633	I	400	3765.54	I	60	3994.114	I
50	3265.617	I	11	3600.94	III	600	3767.191	I	200	3997.392	I
13	3266.88	III	150	3603.204	I	60	3776.452	I	40	3998.053	I
50	3271.000	I	11	3603.88	III	250	3785.95	I	400	4005.241	I

Intensity	Wavelength/Å		Intensity	Wavelength/Å		Intensity	Wavelength/Å		Intensity	Wavelength/Å	
60	4009.713	I	12 h	4273.40	III	150	5216.274	I	16	6032.59	III
100	4021.867	I	12	4279.72	III	60	5226.862	I	13	6036.56	III
10	4035.42	III	1200	4282.402	I	1000	5227.150	I	11	6048.72	III
50	4040.638	I	14 h	4286.16	III	250	5232.939	I	11	6054.18	III
4000	4045.813	I	80	4291.462	I	10	5235.66	III	40	6065.482	I
11	4053.11	III	16 h	4296.85	III	18	5243.31	III	30	6102.159	I
1500	4063.594	I	250	4299.234	I	13 1	5260.34	III	40	6136.614	I
50	4066.975	I	18 h	4304.78	III	100	5266.555	I	40	6137.694	I
50	4067.977	I	1200	4307.901	I	1200	5269.537	I	40	6191.558	I
1200	4071.737	I	20 h	4310.36	III	800	5270.357	I	30	6213.429	I
40	4076.629	I	150	4315.084	I	14	5272.98	III	30	6219.279	I
12	4081.00	III	1500	4325.761	I	15	5276.48	III	40	6230.726	I
40	4100.737	I	80	4352.734	I	30	5281.789	I	20	6246.317	I
40	4107.489	I	80	4369.771	I	16	5282.30	III	80	6247.56	II
150	4118.544	I	11 h	4372.31	III	60	5283.621	I	30	6252.554	I
10	4120.90	III	14 h	4372.53	III	12	5284.83	III	20	6393.602	I
11	4122.02	III	18 h	4372.81	III	11	5298.12	III	30	6399.999	I
11	4122.78	III	800	4375.929	I	12	5299.93	III	20	6411.647	I
40	4127.608	I	3000	4383.544	I	25	5302.299	I	20	6421.349	I
400	4132.058	I	1200	4404.750	I	14 w	5302.60	III	30	6430.844	I
80	4134.676	I	300	4415.122	I	10	5306.76	III	200	6456.38	II
40	4136.997	I	12	4419.60	III	10	5322.74	III	60	6494.981	I
15	4137.76	III	600	4427.299	I	150	5324.178	I	20	6546.239	I
13	4139.35	III	400	4461.652	I	800	5328.038	I	20	6592.913	I
200	4143.415	I	120	4466.551	I	300	5328.531	I	40	6677.989	I
800	4143.869	I	80	4476.017	I	100	5332.899	I	25	7164.443	I
40	4153.898	I	80	4482.169	I	80	5339.928	I	80	7187.313	I
50	4154.500	I	200	4482.252	I	500	5341.023	I	30	7207.381	I
60	4156.799	I	50	4489.739	I	11	5346.88	III	30	7445.746	I
18	4164.73	III	50	4528.613	I	12	5353.77	III	40	7495.059	I
13	4166.84	III	30	4647.433	I	12	5363.76	III	60	7511.045	I
50	4172.744	I	30	4736.771	I	10	5368.06	III	80	7937.131	I
13	4174.26	III	50	4859.741	I	400	5371.489	I	60	7945.984	I
60	4174.912	I	120	4871.317	I	11 1	5375.47	III	80	7998.939	I
50	4175.635	I	60	4872.136	I	40	5393.167	I	60	8046.047	I
50	4177.593	I	30	4878.208	I	300	5397.127	I	50	8085.176	I
120	4181.754	I	100	4890.754	I	250	5405.774	I	150	8220.41	I
50	4184.891	I	250	4891.492	I	250	5429.695	I	120	8327.053	I
120	4187.038	I	30	4903.309	I	100	5434.523	I	20	8331.908	I
120	4187.795	I	150	4918.992	I	200	5446.871	I	120	8387.770	I
80	4191.430	I	500	4920.502	I	120	5455.609	I	30	8468.404	I
40	4195.329	I	1500	4957.597	I	25	5497.516	I	15	8514.069	I
150	4198.304	I	80	5001.862	I	20	5501.464	I	60	8661.898	I
40	4199.095	I	30	5005.711	I	30	5506.778	I	150	8688.621	I
300	4202.029	I	100	5006.117	I	30	5569.618	I	52	11422.32	I
40	4203.984	I	60	5012.067	I	60	5572.841	I	87	11439.12	I
80	4206.696	I	30	5014.941	I	120	5586.755	I	91	11593.59	I
80	4210.343	I	150	5041.755	I	200	5615.644	I	255	11607.57	I
400	4216.183	I	30	5049.819	I	20	5624.541	I	160	11638.26	I
100	4219.360	I	30	5051.634	I	50	5662.515	I	230	11689.98	I
50	4222.212	I	25	5074.748	I	11	5719.88	III	160	11783.26	I
11	4222.27	III	150	5110.357	I	10	5756.38	III	580	11882.84	I
50	4225.956	I	40	5139.251	I	20	5762.990	I	225	11884.08	I
200	4227.423	I	100	5139.462	I	18	5833.93	III	1030	11973.05	I
100	4233.602	I	25	5151.910	I	10	5854.62	III	96	14400.56	I
13	4235.56	III	12	5156.12	III	30	5862.353	I	72	14512.23	I
250	4235.936	I	80	5166.281	I	15	5891.91	III	50	14555.06	I
50	4238.809	I	2500	5167.487	I	30	5914.114	I	40	14826.43	I
12	4243.75	III	80	5168.897	I	10 p	5920.13	III	94	15294.58	I
50	4247.425	I	500	5171.595	I	18 p	5929.69	III	41	15769.42	I
200	4250.118	I	50	5191.454	I	10	5952.31	III	105	18856.65	I
300	4250.787	I	80	5192.343	I	14	5953.62	III			
40	4258.315	I	200	5194.941	I	12	5979.32	III		**Krypton**	
800	4260.473	I	10	5199.08	III	30	5986.956	I		**Kr Z = 36**	
250	4271.153	I	30	5204.582	I	12 h	5989.08	III	30	467.35	III
1200	4271.759	I	25	5215.179	I	18	5999.54	III	150	472.16	V
									100	484.39	V

Line Spectra of the Elements (continued): Krypton

Intensity	Wavelength/Å		Intensity	Wavelength/Å		Intensity	Wavelength/Å		Intensity	Wavelength/Å	
250	496.25	V	50	837.66	III	6	2774.70	IV	40 h	3868.70	III
120	500.77	V	22	842.04	IV	3	2829.60	IV	150 h	3875.44	II
200	507.20	V	100	844.06	II	100	2833.00	II	150	3906.177	II
30	540.86	III	50	854.73	III	3	2836.08	IV	200	3920.081	II
60	548.04	V	60	862.58	III	30	2841.00	III	5	3934.29	IV
30	565.64	III	60	864.82	II	30	2851.16	III	100	3994.840	II
30	569.16	III	60	868.87	II	5	2853.0	IV	100 h	3997.793	II
30	571.98	III	40	870.84	III	3	2859.3	IV	300	4057.037	II
30	579.83	III	50	876.08	III	50	2870.61	III	300	4065.128	II
30	585.14	III	200	884.14	II	100	2892.18	III	50	4067.37	III
30	585.96	III	1000	886.30	II	30	2909.17	III	500	4088.337	II
30	593.70	III	400	891.01	II	50	2952.56	III	250	4098.729	II
30	594.10	III	75	897.81	III	60	2992.22	III	100	4109.248	II
30	596.41	III	200	911.39	II	50	3022.30	III	40	4131.33	III
40	600.17	III	2000	917.43	II	80	3024.45	III	250	4145.122	II
30	603.67	III	50	945.44	I	50	3046.93	III	40	4154.46	III
50	605.86	III	50	946.54	I	30	3056.72	III	150	4250.580	II
35	606.47	III	20	951.06	I	60	3063.13	III	1000	4273.969	I
50	611.12	III	50	953.40	I	40	3097.16	III	100	4282.967	I
35	616.72	III	50	963.37	I	60	3112.25	III	600	4292.923	II
40	621.45	III	2000	964.97	II	30	3120.61	III	200	4300.49	III
45	622.80	III	50	987.29	III	100	3124.39	III	500 h	4317.81	II
50	625.02	III	100	1001.06	I	60	3141.35	III	400	4318.551	I
30	625.76	III	100	1003.55	I	3	3142.01	IV	1000	4319.579	I
45	628.59	III	100	1030.02	I	100	3189.11	III	150 h	4322.98	II
50	630.04	III	30	1158.74	III	80	3191.21	III	100	4351.359	I
35	633.09	III	200	1164.87	I	6	3224.99	IV	3000	4355.477	II
120	637.87	V	650	1235.84	I	40	3239.52	III	500	4362.641	I
50	639.98	III	6	1638.82	III	40	3240.44	III	200	4369.69	II
60	646.41	III	6	1914.09	III	300	3245.69	III	800	4376.121	I
50	651.20	III	3	2237.34	IV	3	3261.70	IV	300 h	4386.54	II
50	659.72	III	6	2291.26	IV	150	3264.81	III	200	4399.965	I
30	664.86	III	3	2329.3	IV	100	3268.48	III	100	4425.189	I
40	672.34	III	4	2336.75	IV	30	3271.65	III	500	4431.685	II
35	672.85	III	4	2348.27	IV	30	3285.89	III	600	4436.812	II
35	676.57	III	3	2358.5	IV	30	3304.75	III	600	4453.917	I
35	680.13	III	3	2388.05	IV	50	3311.47	III	800	4463.689	I
35	683.68	III	40	2393.94	III	200	3325.75	III	800	4475.014	II
45	686.25	III	4	2416.9	IV	60	3330.76	III	400 h	4489.88	II
45	687.98	III	3	2428.04	IV	50	3342.48	III	600	4502.353	I
	690.86	V	5	2442.68	IV	100	3351.93	III	400 h	4523.14	II
	691.75	V	4	2451.7	IV	40	3374.96	III	200 h	4556.61	II
45	691.93	III	6	2459.74	IV	100	3439.46	III	800	4577.209	II
50	695.61	III	100 h	2464.77	II	70	3474.65	III	300	4582.978	II
30	698.05	III	5	2474.06	IV	100	3488.59	III	150 h	4592.80	II
50	708.36	III	60	2492.48	II	200	3507.42	III	500	4615.292	II
600	708.85	V	40	2494.01	III	100	3564.23	III	1000	4619.166	II
50	714.00	III	4	2517.0	IV	100 h	3607.88	II	800	4633.885	II
100 p	722.04	III	5	2518.02	IV	200	3631.889	II	2000	4658.876	II
60	729.40	II	6	2519.38	IV	30	3641.34	III	500	4680.406	II
30	746.70	III	5	2524.5	IV	250	3653.928	II	100	4691.301	II
200	761.18	II	5	2546.0	IV	80	3665.324	I	200	4694.360	II
100	763.98	II	6	2547.0	IV	150	3669.01	II	3000	4739.002	II
60	766.20	II	4	2558.08	IV	100	3679.559	I	300	4762.435	II
200	771.03	II	30	2563.25	III	80	3686.182	II	1000	4765.744	II
60 p	773.69	II	3	2586.9	IV	30	3690.65	III	300	4811.76	II
200	782.10	II	5	2606.17	IV	300 h	3718.02	II	300	4825.18	II
100	783.72	II	10	2609.5	IV	200	3718.595	II	800	4832.077	II
60	785.97	III	8	2615.3	IV	150	3721.350	II	700	4846.612	II
	793.44	IV	7	2621.11	IV	200	3741.638	II	150	4857.20	II
	794.11	IV	60	2639.76	III	150	3744.80	II	300	4945.59	II
7	805.76	IV	30	2680.32	III	80	3754.245	II	20 h	5016.45	III
	810.70	V	40	2681.19	III	500	3778.089	II	200	5022.40	II
18	816.82	IV	80 h	2712.40	II	500	3783.095	II	250	5086.52	II
60	818.15	II	3	2730.55	IV	3	3809.30	IV	400 h	5125.73	II
60	830.38	II	8	2748.18	IV	5	3860.58	IV	500	5208.32	II

Intensity	Wavelength/Å		Intensity	Wavelength/Å		Intensity	Wavelength/Å		Intensity	Wavelength/Å	
200	5308.66	II	100	9362.082	I	140	20446.971	I	30000 c	2962.58	IV
500	5333.41	II	200 h	9402.82	II	600	21165.471	I	70000 w	3009.51	IV
200	5468.17	II	200 h	9470.93	II	1800	21902.513	I	90000 c	3056.68	IV
10	5501.43	III	500	9577.52	II	120	22485.775	I	1000	3171.63	III
500	5562.224	I	500 h	9605.80	II	180	23340.416	I	1500	3171.74	III
2000	5570.288	I	400 h	9619.61	II	120	24260.506	I	510	3245.13	II
80	5580.386	I	200	9663.34	II	180	24292.221	I	550	3265.67	II
100	5649.561	I	200 h	9711.60	II	600	25233.820	I	800	3303.11	II
400	5681.89	II	2000	9751.758	I	180	28610.55	I	1500	3337.49	II
200 h	5690.35	II	500	9803.14	II	1000	28655.72	I	870	3344.56	II
100	5832.855	I	500	9856.314	I	150	28769.71	I	1500	3380.91	II
3000	5870.914	I	1000	10221.46	II	140	28822.49	I	320	3628.83	II
200	5992.22	II	100	11187.108	I	300	29236.69	I	1000	3645.42	II
60	5993.849	I	200	11257.711	I	300	30663.54	I	550	3713.54	II
10 h	6037.17	III	150	11259.126	I	300	30979.16	I	2400	3759.08	II
60	6056.125	I	500	11457.481	I	500	39300.6	I	3700	3790.83	II
10 h	6078.38	III	150	11792.425	I	1100	39486.52	I	3900	3794.78	II
10	6310.22	III	1500	11819.377	I	220	39557.25	I	600	3840.72	II
300	6420.18	II	600	11997.105	I	100	39572.60	I	1600	3849.02	II
100	6421.026	I	160	12077.224	I	1400	39588.4	I	3400	3871.64	II
200	6456.288	I	100	12861.892	I	1100	39589.6	I	1700	3886.37	II
150	6570.07	II	1100	13177.412	I	500	39954.8	I	1300	3916.05	II
60	6699.228	I	1000	13622.415	I	300	39966.6	I	1100	3921.54	II
100	6904.678	I	2400	13634.220	I	1300	40306.1	I	2200	3929.22	II
250	7213.13	II	800	13658.394	I	250	40685.16	I	9000	3949.10	II
100	7224.104	I	200	13711.036	I		**Lanthanum**		4400	3988.52	II
80	7287.258	I	600	13738.851	I		**La Z = 57**		3600	3995.75	II
400	7289.78	II	150	13974.027	I	100	344.12	IV	2800	4031.69	II
400	7407.02	II	550	14045.657	I	400	390.72	V	3000	4042.91	II
60	7425.541	I	140	14104.298	I	1000	432.11	V	850	4067.39	II
200	7435.78	II	180	14402.22	I	2500	435.28	V	2800	4077.35	II
100	7486.862	I	2000	14426.793	I	10000	463.14	IV	5500	4086.72	II
300	7524.46	II	100	14517.84	I	5000	482.16	V	4400	4123.23	II
1000	7587.411	I	1600	14734.436	I	7000	498.08	V	550	4141.74	II
2000	7601.544	I	550	14762.672	I	15000	499.54	IV	1100	4151.97	II
150	7641.16	II	450	14765.472	I	10000	503.58	V	1500	4196.55	II
1000	7685.244	I	400	14961.894	I	12000	526.76	V	1600	4238.38	II
1200	7694.538	I	120	15005.307	I	10000	531.07	V	480	4269.50	II
250	7735.69	II	140	15209.526	I	15000	533.23	V	600	4286.97	II
150	7746.827	I	1700	15239.615	I	8000	547.44	V	600	4296.05	II
800	7854.821	I	130	15326.480	I	40000	552.02	IV	440	4322.51	II
200	7913.423	I	1500	15334.958	I	5000	600.24	V	4600	4333.74	II
180	7928.597	I	700	15372.037	I	30000	631.26	IV	550	4354.40	II
200	7933.22	II	200	15474.026	I	400	796.99	III	2000	4429.90	II
120	7973.62	II	180	15681.02	I	2000	870.40	III	850	4522.37	II
100	7982.401	I	120	15820.09	I	1000	882.34	III	420	4526.12	II
1500	8059.503	I	200	16726.513	I	400	942.86	III	400	4558.46	II
4000	8104.364	I	2000	16785.128	I	50000	1081.61	III	400	4574.88	II
6000	8112.899	I	1000	16853.488	I	95000	1099.73	III	410	4613.39	II
60	8132.967	I	2400	16890.441	I	2000	1255.63	III	410	4619.88	II
3000	8190.054	I	1600	16896.753	I	10000	1349.18	III	540	4655.50	II
200	8202.72	II	1800	16935.806	I	25000	1368.04	IV	360	4662.51	II
80	8218.365	I	600	17098.771	I	20000	1463.47	IV	230	4692.50	II
3000	8263.240	I	700	17367.606	I	15000	1507.87	IV	230	4728.42	II
100	8272.353	I	120	17404.443	I	10000	1523.79	III	500	4740.28	II
1500	8281.050	I	150	17616.854	I	4000	1808.66	IV	390	4743.09	II
5000	8298.107	I	650	17842.737	I	5000	1902.97	IV	320	4748.73	II
100	8412.430	I	700	18002.229	I	4000 c	2197.45	IV	320	4860.91	II
3000	8508.870	I	2600	18167.315	I	770	2256.76	II	850	4899.92	II
150	8764.110	I	100	18399.786	I	25000 w	2417.58	IV	1000	4920.98	II
6000	8776.748	I	150	18580.896	I	50000	2532.75	IV	1000	4921.79	II
2000	8928.692	I	300	18696.294	I	45000	2582.05	IV	370	4949.77	I
500	9238.48	II	170	18785.460	I	95000 w	2597.50	IV	340	4970.39	II
500 hl	9293.82	II	200	18797.703	I	70000 w	2662.75	IV	370	4986.83	II
200 h	9320.99	II	140	20209.878	I	420	2808.39	II	720	4999.47	II
300	9361.95	II	300	20423.964	I	50000 w	2848.30	IV	210	5050.57	I

Line Spectra of the Elements (continued): Lanthanum—Lead

Intensity	Wavelength/Å		Intensity	Wavelength/Å		Intensity	Wavelength/Å		Intensity	Wavelength/Å	
470	5114.56	II	10	827.41	IV	10	1348.37	II	4	2986.876	II
470	5122.99	II	12	832.60	IV	16	1388.94	IV	10	3043.85	III
450	5145.42	I	12	845.94	IV	18	1400.26	IV	150	3118.894	I
290	5158.69	I	18	857.64	IV	10	1404.34	IV	10	3137.81	III
580	5177.31	I	16	862.33	IV	10	1433.96	II	10	3176.50	III
850	5183.42	II	20	863.97	V	10	1512.42	II	600	3220.528	I
260	5188.22	II	14	870.44	IV	14	1535.71	IV	100	3229.613	I
720	5211.86	I	6	873.71	II	20	1553.1	III	400	3240.186	I
520	5234.27	I	12	879.96	IV	10	1671.53	II	200	3262.355	I
340	5253.46	I	18	883.90	V	10	1682.15	II	35000	3572.729	I
370	5271.19	I	14	884.96	IV	20	1726.75	II	50000 r	3639.568	I
370	5301.98	II	14	884.99	IV	10	1796.670	II	20000	3671.491	I
180	5303.55	II	14	888.37	V	10	1822.050	II	70000 r	3683.462	I
500	5455.15	I	8	889.68	II	10	1904.77	I	10	3713.982	II
470	5501.34	I	16	890.72	IV	7	1921.471	II	25000	3739.935	I
240	5648.25	I	14	894.40	V	12	1959.34	IV	12	3854.08	III
180	5740.66	I	12	896.08	V	16	1973.16	IV	15000	4019.632	I
370	5769.34	I	12	908.51	IV	10	1998.83	V	95000	4057.807	I
320	5789.24	I	14	915.71	V	5 r	2022.02	I	14000	4062.136	I
450	5791.34	I	12	917.90	IV	10	2042.58	IV	10	4157.814	I
140	5821.99	I	12	918.09	V	12	2049.34	IV	10000	4168.033	I
320	5930.62	I	12	920.28	V	8 r	2053.28	I	8	4272.66	III
720	6249.93	I	12	920.66	V	12	2079.22	IV	200	4340.413	I
260 d	6262.30	II	10	922.12	IV	6	2111.758	I	10	4496.15	IV
450	6394.23	I	12	922.49	IV	10	2115.066	I	6	4499.34	III
250	6455.99	I	10	927.64	IV	15	2154.01	IV	16	4534.60	IV
180	6709.50	I	14	932.20	IV	500 r	2170.00	I	7	4571.21	III
110	7045.96	I	12	954.35	V	7	2175.580	I	10	4579.051	II
160	7066.23	II	10	967.23	II	12	2177.46	IV	6	4761.12	III
50	7161.25	I	10	986.71	II	7	2187.888	I	1000	5005.416	I
110 w	7282.34	II	10	995.89	II	8	2189.603	I	100	5006.572	I
110 w	7334.18	I	10	1016.61	II	10	2203.534	II	50	5089.484	I
75 cw	7483.50	II	14	1028.61	IV	20	2237.425	I	10	5107.242	I
50	7498.83	I	20	1032.05	IV	20	2246.86	I	2000	5201.437	I
85	7539.23	I	16	1041.24	IV	25	2246.89	I	10	5372.099	II
40	7964.83	I	18	1044.14	IV	20	2259.01	V	40	5692.346	I
75	8086.05	I	12	1048.9	III	150	2332.418	I	200	5895.624	I
85	8324.69	I	10	1049.82	II	16	2359.53	IV	2000	6001.862	I
95	8346.53	I	10	1050.77	II	180	2388.797	I	500	6011.667	I
65	8545.44	I	10	1051.26	V	550 r	2393.792	I	500	6059.356	I
300	8583.45	III	15	1056.53	IV	140	2399.597	I	40	6081.409	II
40	8674.43	I	10	1060.66	II	320 r	2401.940	I	50	6110.520	I
35	8825.82	I	12	1072.09	IV	320 r	2411.734	I	100	6235.266	I
120	9184.38	III	18	1080.81	IV	16	2417.61	IV	50 c	6660.20	II
100	9212.63	III	20	1084.17	IV	15	2424.81	V	20000	7228.965	I
140	10284.79	III	10	1088.86	V	150 r	2443.829	I	10	7346.676	I
Lead			10	1103.94	II	160 r	2446.181	I	20	7809.259	I
Pb Z = 82			10	1108.43	II	130 r	2476.378	I	5	7896.737	I
10	496.38	IV	10	1109.84	II	80 r	2577.260	I	10	8168.001	I
12	499.94	IV	20	1116.08	IV	500 r	2613.655	I	6	8191.886	I
14	529.78	IV	10	1119.57	II	900 r	2614.175	I	5	8217.711	I
20	570.16	IV	10	1121.36	II	160	2628.262	I	40	8272.690	I
10	648.50	IV	10	1133.14	II	4	2634.256	II	20	8409.384	I
20	703.73	V	18	1137.84	IV	10	2657.094	I	10	8478.492	I
12	749.46	V	14	1144.93	IV	700	2663.154	I	5	8722.810	I
10	752.52	V	12	1157.88	V	10	2697.541	I	10	8857.457	I
10	761.09	IV	14	1185.43	V	25000 r	2801.995	I	8	9293.476	I
18	767.45	V	20	1189.95	IV	100	2822.58	I	15	9438.05	I
18	769.49	V	10	1203.63	II	14000 r	2823.189	I	15	9604.297	I
14	771.42	V	10	1231.20	II	35000 r	2833.053	I	15	9674.351	I
14	782.79	V	11	1233.50	V	6	2840.557	II	200	10290.458	I
15	797.02	V	10	1291.10	IV	14000 r	2873.311	I	100	10498.965	I
18	802.07	IV	20	1313.05	IV	3	2914.442	II	50	10649.249	I
12	802.82	IV	10	1331.65	II	15	2966.460	I	15	10886.688	I
18	809.63	V	10	1335.20	II	15	2972.991	I	40	10969.53	I
10	812.59	IV	12	1343.06	IV	15	2980.157	I		13512.6	I

Intensity	Wavelength/Å	Intensity	Wavelength/Å	Intensity	Wavelength/Å	Intensity	Wavelength/Å
	14743.0 I		2164. II		2846. I		5037.92 II
	15349.6 I		2173.4 I		2868. I		5271. I
	39039.4 I		2183. II		2895. I		5315. I
	Lithium		2214. II	2	2934.02 II		5395. I
	Li Z = 3		2222. II	2	2934.07 II		5440. I
	102.9 III		2237. II	5	2934.12 II	600 c	5483.55 II
	103.4 III	h	2249.21 III	1	2934.25 II	600 c	5485.65 II
	104.1 III		2286.82 II		2968. I	320	6103.54 I
	105.5 III		2302.57 II	3	3029.12 II	320	6103.65 I
	108.0 III		2303.33 II	3	3029.14 II	3600	6707.76 I
	113.9 III		2304.59 I		3144. I	3600	6707.91 I
	125.5 II		2304.92 I	3	3155.31 II	48	8126.23 I
	135.0 III		2305.36 I	4	3155.33 II	48	8126.45 I
	136.5 II		2305.83 I	1	3196.26 II		8517.37 I
	140.5 II		2306.29 I	9	3196.33 II		9581.42 II
	167.21 II		2306.82 I	4	3196.36 II		10120. II
	168.74 II		2307.44 I	5	3199.33 II		12232. I
	171.58 II		2308.97 I	2	3199.43 II		12782. I
	178.02 II		2309.88 I	17	3232.66 I		13566. I
	199.28 II		2310.94 I		3249.87 II		17552. I
	207.5 II		2312.11 I		3306.28 II		18697. I
	456. II		2313.49 I		3488. I		19290. I
	483. II		2315.08 I		3579.8 I		24467. I
	540. II		2316.95 I		3618. I		40475. I
	540.0 III		2319.18 I		3662. I		**Lutetium**
	729. II		2321.88 I		3684.32 II		**Lu Z = 71**
	729.1 III		2325.11 I	1	3714.00 II	100	563.72 V
	800. II		2329.02 I	5	3714.16 II	500	810.73 III
	820. II		2329.84 II	6 d	3714.27 II	2000	832.28 III
	861. II		2333.94 I	8	3714.29 II	100	861.92 V
	905.5 II	3	2336.88 II	7 d	3714.40 II	400	876.80 IV
	917.5 II	5	2336.91 II	10	3714.41 II	100	880.32 V
	936. II	2	2337.00 II	1	3714.51 II	100	891.81 V
	945. II		2340.15 I		3714.58 II	100	914.72 V
	965. II		2348.22 I	3	3718.7 I	400	1001.18 III
	972. II		2358.93 I	6	3794.72 I	100	1272.42 IV
	988. II		2373.54 I	20	3915.30 I	800	1333.79 IV
	1018. II		2381.54 II	20	3915.35 I	400	1429.38 IV
	1032. II		2383.20 II	10	3985.48 I	200	1441.76 V
	1036. II	1	2394.39 I	10	3985.54 I	200	1453.35 V
	1093. II		2402.33 I	40	4132.56 I	200	1468.99 V
	1103. II		2410.84 II	40	4132.62 I	400	1472.12 V
	1109. II	3	2425.43 I		4196. I	200	1473.71 V
	1116. II		2429.81 II	20	4273.07 I	200	1485.58 V
	1132.1 II		2460.2 I	20	4273.13 I	400	1511.26 IV
	1141. II	10	2475.06 I	5	4325.42 II	600	1772.57 IV
	1166.4 II		2506.94 II	5	4325.47 II	100 c	1786.25 V
	1198.09 II		2508.78 II	1	4325.54 II	1000	1854.57 III
	1215. II		2518. I		4516.45 II	1500	2065.35 III
	1238. II		2539.49 II	13	4602.83 I	1500 c	2070.56 III
	1253.8 II	24	2551.7 II	13	4602.89 I	600 c	2086.47 IV
	1420.89 II		2559. II		4607.34 II	1000 c	2104.41 IV
	1424. II	15	2562.31 I		4671.51 II	1000 c	2108.31 IV
3	1492.93 II		2605.08 II	6	4671.65 II	1700 h	2195.54 II
5	1492.97 II	2	2657.29 II	2	4671.70 II	1000	2236.14 III
1	1493.04 II	3	2657.30 II	3	4678.06 II	2000	2236.22 III
	1555. II		2674.46 II	1	4678.29 II	95	2276.94 II
3	1653.08 II		2728.24 II		4760. I	190	2297.41 II
5	1653.13 II	5	2728.29 II		4763. II	1300	2392.19 II
1	1653.21 II	2	2728.32 II		4788.36 II	120	2399.14 II
	1681.66 II	3	2730.47 II		4843.0 II	80	2419.21 II
	1755.33 II	1	2730.55 II	4	4881.32 II	130	2459.64 II
	2009. II	5	2741.20 I	4	4881.39 II	370	2536.95 II
	2039. I		2766.99 II	1	4881.49 II	930	2571.23 II
	2068. II		2790.31 II	8	4971.66 I	1700	2578.79 II
	2131. II		2801. I	8	4971.75 I	4500 c	2603.35 III

Intensity	Wavelength/Å		Intensity	Wavelength/Å		Intensity	Wavelength/Å		Intensity	Wavelength/Å	
1800	2613.40	II	1600	4184.25	II	150	857.29	IV	3	2646.21	I
18000	2615.42	II	150	4277.50	I	50	919.03	IV	4	2649.06	I
1800	2619.26	II	250	4281.03	I	250	1037.41	IV	8	2660.76	II
2700	2657.80	II	330 d	4295.97	I	300	1210.99	IV	8	2660.82	II
570 h	2685.08	I	150	4309.57	I	300	1342.19	IV	6	2668.12	I
4200	2701.71	II	190 c	4430.48	I	800	1346.57	IV	8	2669.55	I
180 d	2719.09	I	190	4450.81	I	300	1346.68	IV	10	2672.46	I
480 h	2728.95	I	3300	4518.57	I	600	1352.05	IV	3	2693.72	I
3600	2754.17	II	100 h	4648.21	I	900	1384.46	IV	5	2695.18	I
750 h	2765.74	I	1000	4658.02	I	500	1385.77	IV	6	2698.14	I
2000	2772.55	III	85 h	4659.03	I	800	1387.53	IV	8	2731.99	I
2700	2796.63	II	150	4785.42	II	300	1404.68	IV	10	2733.49	I
270 c	2834.35	II	85	4815.05	I	1000	1409.36	IV	12	2736.53	I
330 h	2845.13	I	460	4904.88	I	500	1437.53	IV	5	2765.22	I
3000	2847.51	II	180	4942.34	I	1000	1437.64	IV	7	2768.34	I
570 h	2885.14	I	800	4994.13	II	300	1447.42	IV	38	2776.69	I
6300	2894.84	II	800	5001.14	I	300	1459.54	IV	32	2778.27	I
4500	2900.30	II	140	5134.05	I	400	1459.62	IV	90	2779.83	I
300	2903.05	I	2700	5135.09	I	400	1481.51	IV	8	2781.29	I
9000	2911.39	II	170	5196.61	I	350	1490.45	IV	32	2781.42	I
270 h	2949.73	I	500	5402.57	I	300	1495.50	IV	36	2782.97	I
1200	2951.69	II	140 c	5421.90	I	300	1607.11	IV	1000	2795.53	II
4200	2963.32	II	100	5437.88	I	5	1668.43	I	600	2802.70	II
2400	2969.82	II	2100	5476.69	II	500	1683.02	IV	3	2809.76	I
1800	2989.27	I	550	5736.55	I	10	1683.41	I	2	2811.11	I
3000	3020.54	II	80	5800.59	I	400	1698.81	IV	1	2811.78	I
2100	3056.72	II	690 cw	5983.9	II	15	1707.06	I	12	2846.72	I
1000	3057.86	III	140	5997.13	I	40	1734.84	II	12	2846.75	I
7500	3077.60	II	1400	6004.52	I	50	1737.62	II	14	2848.34	I
390	3080.11	I	440	6055.03	I	20	1747.80	I	14	2848.42	I
5100 h	3081.47	I	150	6159.94	II	40	1750.65	II	16	2851.65	I
3000	3118.43	I	600	6198.13	III	50	1753.46	II	16	2851.66	I
2400	3171.36	I	160	6199.66	II	30	1827.93	I	6000	2852.13	I
260	3191.80	II	2100	6221.87	II	300	1844.17	IV	2	2902.92	I
1400	3198.12	I	80	6235.36	II	9	2025.82	I	4	2906.36	I
4800	3254.31	II	160	6242.34	II	25	2064.90	III	3	2915.45	I
3800	3278.97	I	70 h	6345.35	I	20	2091.96	III	10	2936.74	I
7600	3281.74	I	1100	6463.12	II	20	2177.70	III	12	2938.47	I
6200	3312.11	I	29	6477.67	I	3	2329.58	II	2	2942.00	I
7600	3359.56	I	55 c	6523.18	I	20	2395.15	III	13	2942.00	I
6200	3376.50	I	35 cw	6611.28	II	6	2449.57	II	20	3091.08	I
950	3385.50	I	23 c	6677.14	I	1	2557.23	I	22	3092.99	I
160 h	3391.55	I	30 c	6793.77	I	1	2560.94	I	14	3096.90	I
1400	3396.82	I	45	6917.31	I	1	2562.26	I	9	3104.71	II
4100	3397.07	.II	23	7031.24	I	1	2564.94	I	8	3104.81	II
4800	3472.48	II	45	7125.84	II	1	2570.91	I	6	3168.98	II
8300 c	3507.39	II	14 ch	7237.98	I	1	2572.25	I	6	3172.71	II
1600	3508.42	I	11 c	7441.52	I	2	2574.94	I	7	3175.78	II
4800	3554.43	II	9 c	8178.16	I	1	2577.89	I	2	3197.62	I
4800	3567.84	I	17	8382.08	I	1	2580.59	I	17	3329.93	I
340	3596.34	I	35	8459.19	II	1	2584.22	I	6	3332.15	I
800	3623.99	II	10 d	8478.50	I	2	2585.56	I	9	3336.68	I
680	3636.25	I	29 c	8508.08	I	3	2588.28	I	7	3535.04	II
2600	3647.77	I	35 c	8610.98	I	1	2591.89	I	8	3538.86	II
110	3756.70	I		**Magnesium**		1	2593.23	I	7	3549.52	II
110	3756.79	I		**Mg Z = 12**		2	2595.97	I	8	3553.37	II
150	3800.67	I	400	146.95	IV	2	2602.50	I	140	3829.30	I
2700	3841.18	I	20	186.51	III	4	2603.85	I	300	3832.30	I
530	3876.65	II	20	187.20	III	5	2606.62	I	500	3838.29	I
50	3918.86	I	10	188.53	III	1	2613.36	I	8	3848.24	II
480	3968.46	I	100	231.73	III	2	2614.73	I	7	3850.40	II
670	4054.45	I	80	234.26	III	3	2617.51	I	3	3878.31	I
310	4122.49	I	35	276.58	V	3	2628.66	I	3	3895.57	I
3100	4124.73	I	4000	320.99	IV	6	2630.05	I	4	3903.86	I
150 c	4131.79	I	3000	323.31	IV	8	2632.87	I	6	3938.40	I
460	4154.08	I	30	353.09	V	2	2644.80	I	8	3986.75	I

Intensity	Wavelength/Å		Intensity	Wavelength/Å		Intensity	Wavelength/Å		Intensity	Wavelength/Å	
10	4057.50	I	17	9429.81	I	80	1795.65	IV	50	2427.72	II
15	4167.27	I	19	9432.76	I	80	1795.79	IV	30	2427.94	II
20	4351.91	I	20	9438.78	I	30	1853.27	II	30	2437.37	II
9	4384.64	II	12	9631.89	II	20	1857.92	II	20	2437.84	II
10	4390.59	II	11	9632.43	II	50	1902.95	II	30	2452.49	II
8	4428.00	II	15	9953.20	I	20	1907.84	II	50	2499.00	II
9	4433.99	II	15	9983.20	I	75	1910.25	IV	30	2507.60	II
14	4481.16	II	17	9986.47	I	30	1911.41	II	20	2516.60	II
13	4481.33	II	18	9993.21	I	20 d	1914.68	II	30	2516.74	II
28	4571.10	I	14	10092.16	II	100	1915.10	II	20	2521.66	II
10	4730.03	I	35	10811.08	I	20	1918.64	II	20	2530.72	II
7	4851.10	II	11	10914.23	II	30	1919.64	II	20	2531.80	II
75	5167.33	I	10	10951.78	II	80	1921.25	II	50	2532.78	II
220	5172.68	I	25	10953.32	I	20	1923.07	II	50	2533.33	II
400	5183.61	I	27	10957.30	I	20	1923.34	II	30	2534.10	II
8	5264.21	II	28	10965.45	I	30	1925.52	II	80	2534.22	II
7	5264.37	II	15	11032.10	I	50	1926.59	II	100	2535.66	II
9	5401.54	II	14	11033.66	I	30	1931.40	II	30	2535.98	II
6	5528.41	I	45	11828.18	I	500	1941.28	III	100	2537.92	II
30	5711.09	I	30	12083.66	I	800	1943.21	III	50	2541.11	II
10	6318.72	I	28	14877.62	I	20	1945.15	II	80	2542.92	II
9	6319.24	I	35	15024.99	I	20	1947.93	II	50	2543.45	II
7	6319.49	I	30	15040.24	I	20	1950.14	II	100	2548.75	II
10	6346.74	II	25	15047.70	I	500	1952.36	III	50	2551.85	II
9	6346.96	II	10	15765.84	I	1000	1952.52	III	30	2553.27	II
11	6545.97	II	30	17108.66	I	30	1953.23	II	75	2556.57	II
7	6781.45	II	5	26392.90	I	20 d	1954.81	II	30	2556.89	II
8	6787.85	II	**Manganese**			30	1959.25	II	95	2558.59	II
7	6812.86	II	**Mn Z = 25**			20	1969.24	II	30	2559.41	II
8	6819.27	II	600	410.30	V	500	1978.95	III	150	2563.65	II
10	7193.17	I	600	410.60	V	30	1994.23	II	30	2565.22	II
10	7291.06	I	600	415.62	V	9700	1996.06	I	580	2572.76	I
12	7387.69	I	650	415.98	V	14000	1999.51	I	480	2575.51	I
20	7657.60	I	600	428.59	V	18000	2003.85	I	12000	2576.10	II
19	7659.15	I	600	435.67	V	1000 w	2027.83	III	550	2584.31	I
17	7659.90	I	1000	441.72	V	500 w	2028.14	III	30	2588.97	II
15	7691.55	I	850	442.49	V	50	2037.31	II	45	2589.71	II
12	7877.05	II	60	579.79	IV	40	2037.64	II	250	2592.94	I
13	7896.37	II	60	581.44	IV	40	2039.97	II	6200	2593.73	II
10	8098.72	I	60	581.65	IV	500	2049.68	III	250	2595.76	I
9	8115.22	II	60	585.21	IV	500	2066.38	III	95	2598.90	II
8	8120.43	II	90	1242.25	IV	1000	2069.02	III	30	2602.72	II
10	8209.84	I	90	1244.50	IV	30	2076.21	II	45	2603.72	II
20	8213.03	I	95	1251.93	IV	900	2077.38	III	4300	2605.69	II
10	8213.99	II	95	1257.28	IV	800	2084.23	III	190	2610.20	II
11	8234.64	II	90	1264.41	IV	600	2090.05	III	500	2618.14	II
10	8310.26	I	500	1283.58	III	1500	2092.16	I	140	2622.90	I
15	8346.12	I	400	1287.59	III	500	2094.78	III	150	2624.04	I
10	8710.18	I	300	1291.62	III	20	2097.46	II	40	2624.80	II
12	8712.69	I	1000	1360.72	III	500	2097.93	III	200	2625.58	II
13	8717.83	I	800	1365.20	III	500	2099.97	III	190	2632.35	II
10	8734.99	II	500 h	1609.17	III	20	2102.50	II	130	2638.17	II
17	8736.02	I	1000	1614.14	III	1700	2109.58	I	80	2639.84	II
11	8745.66	II	2000	1620.60	III	30	2113.96	II	27	2650.99	II
14	8806.76	I	500	1633.80	III	1000	2169.78	III	60	2655.91	II
10	8824.32	II	80	1667.00	IV	700	2174.15	III	30	2666.77	II
11	8835.08	II	80	1698.30	IV	900	2176.87	III	30	2667.03	II
20	8923.57	I	20	1726.47	II	800	2181.86	III	110	2672.59	II
10	8997.16	I	30	1732.70	II	800	2184.87	III	55	2673.37	II
14	9218.25	II	50	1733.55	II	290	2208.81	I	55	2674.43	II
13	9244.27	II	40	1734.49	II	540	2213.85	I	45	2680.34	II
12	9246.50	I	30	1737.93	II	900	2220.55	III	30	2680.68	II
30	9255.78	I	20	1740.16	II	770	2221.84	I	30	2681.25	II
10	9327.54	II	20	1742.00	II	1000	2227.42	III	55	2684.55	II
10	9340.54	II	85	1742.10	IV	20	2373.36	II	55	2685.94	II
25	9414.96	I	85	1766.27	IV	20	2427.38	II	110	2688.25	II

Intensity	Wavelength/Å		Intensity	Wavelength/Å		Intensity	Wavelength/Å		Intensity	Wavelength/Å	
27	2693.19	II	200	3330.78	II	730	4063.53	I	40	5481.40	I
55	2695.36	II	720	3441.99	II	290	4070.28	I	30	5505.87	I
27	2698.97	II	50	3460.03	II	730	4079.24	I	50	5516.77	I
85	2701.00	II	360	3460.33	II	730	4079.42	I	40	5537.76	I
50	2701.17	II	360 h	3474.04	II	1100	4082.94	I	21	5551.98	I
160	2701.70	II		3474.13	II	1100	4083.63	I	200	5946.65	III
100	2703.98	II	290	3482.91	II	200	4110.90	I	140	6013.50	I
130	2705.74	II	180	3488.68	II	150	4131.12	I	200	6016.64	I
80	2707.53	II	140	3495.84	II	120	4135.04	I	290	6021.80	I
110	2708.45	II	50	3496.81	II	150	4176.60	I	200	6231.21	III
45	2709.96	II	100	3497.54	II	120	4189.99	I	17	6440.97	I
80	2710.33	II	360	3531.85	I	370	4235.14	I	24	6491.71	I
110	2711.58	II	1100	3532.12	I	510	4235.29	I	14 h	6942.52	I
30	2716.80	II	1300	3547.80	I	190	4239.72	I	12	6989.96	I
30	2717.53	II	1100	3548.03	I	290	4257.66	I	14	7069.84	I
30	2719.01	II	390	3548.20	I	290	4265.92	I	12	7184.25	I
50	2719.74	II	2200	3569.49	I	270	4281.10	I	24 h	7283.82	I
30	2722.10	II	720	3569.80	I	50	4323.63	II	35 h	7302.89	I
30	2724.46	II	1400	3577.88	I	350	4414.88	I	50	7326.51	I
55	2728.61	II	720	3586.54	I	210	4436.35	I	12	7680.20	I
6200	2794.82	I	290	3595.12	I	800	4451.59	I	12 h	8672.06	I
5100	2798.27	I	150	3601.72	III	160	4453.00	I	12 h	8701.05	I
220	2799.84	I	420	3607.54	I	130	4455.01	I	17 h	8703.76	I
3700	2801.06	I	420	3608.49	I	160	4455.32	I	30 h	8740.93	I
110	2809.11	I	360	3610.30	I	110	4455.82	I	**Mercury 198**		
60	2815.02	II	290	3619.28	I	210	4457.55	I	**Hg Z = 80**		
30	2816.33	II	220	3623.79	I	270	4458.26	I	80	1250.564	I
60	2870.08	II	140	3629.74	I	150	4461.08	I	8	1259.242	I
30	2872.94	II	100	3660.40	I	510	4462.02	I	100	1268.825	I
80	2879.49	II	280	3693.67	I	290	4464.68	I	5	1307.751	I
70	2886.68	II	180	3696.57	I	200	4470.14	I	20	1402.619	I
160	2889.58	II	210	3706.08	I	130	4472.79	I	10	1435.503	I
55	2892.39	II	130	3718.93	I	170	4490.08	I	1000	1849.492	I
50	2898.70	II	130	3731.93	I	240	4498.90	I	60	2262.210	II
80	2900.16	II	260	3790.22	I	240	4502.22	I	20	2302.065	I
140 h	2914.60	I	110	3800.55	I	160	4709.72	I	20	2345.440	I
190 h	2925.57	I	3200	3806.72	I	180	4727.48	I	100	2378.325	I
1100	2933.06	II	700	3809.59	I	130	4739.11	I	20	2380.004	I
1500	2939.30	II	2100	3823.51	I	1000	4754.04	I	40	2399.349	I
250 h	2940.39	I	390	3823.89	I	180	4761.53	I	20	2399.729	I
1900	2949.20	II	200	3829.68	I	750	4762.38	I	20	2446.900	I
30	3019.92	II	480	3833.86	I	300	4765.86	I	15	2464.064	I
55	3031.06	II	1300	3834.36	I	500	4766.43	I	40	2481.999	I
30	3035.35	II	350	3839.78	I	940	4783.42	I	30	2482.713	I
330	3044.57	I	670	3841.08	I	1000	4823.52	I	40	2483.821	I
120	3045.59	I	350	3843.98	I	19	5004.91	I	90	2534.769	I
200	3047.04	I	120	3926.47	I	30	5074.79	I	15000	2536.506	I
30	3050.65	II	130	3982.58	I	200	5079.20	III	25	2563.861	I
250	3054.36	I	150	3985.24	I	150	5100.03	III	25	2576.290	I
140	3062.12	I	190	3986.83	I	60	5117.94	I	250	2652.043	I
170	3066.02	I	150	3987.10	I	50	5150.89	I	400	2653.683	I
170	3070.27	I	1500	4018.10	I	50	5196.59	I	100	2655.130	I
160	3073.13	I	150	4026.44	I	85	5255.32	I	50	2698.831	I
140 h	3178.50	I	27000	4030.76	I	160	5341.06	I	80	2752.783	I
220	3212.88	I	19000	4033.07	I	19	5349.88	I	20	2759.710	I
1000	3228.09	I	11000	4034.49	I	95	5377.63	I	40	2803.471	I
300	3230.72	I	1500	4035.73	I	95	5394.67	I	30	2804.438	I
850	3236.78	I	5600	4041.36	I	50	5399.49	I	750	2847.675	II
330	3243.78	I	210 d	4045.13	I	95	5407.42	I	50	2856.939	I
650	3248.52	I	1100	4048.76	I	35	5413.69	I	150	2893.598	I
100	3251.14	I	150	4055.21	I	85	5420.36	I	150	2916.227	II
310	3252.95	I	1900	4055.54	I	35	5432.55	I	60	2925.413	I
310	3256.14	I	210	4057.95	I	150	5454.07	III	1200	2967.283	I
220	3258.41	I	1100	4058.93	I	12	5457.47	I	300	3021.500	I
180	3260.23	I	150	4059.39	I	60	5470.64	I	120	3023.476	I
180	3264.71	I	730	4061.74	I	200	5474.68	III	30	3025.608	I

Intensity	Wavelength/Å	Intensity	Wavelength/Å	Intensity	Wavelength/Å	Intensity	Wavelength/Å
50	3027.490 I	9	1681.40 III	3	2670.49 III	100	4398.62 II
400	3125.670 I	100	1702.73 II	5	2674.91 I	15	4470.58 III
320	3131.551 I	100	1707.40 II	50	2698.83 I	12	4552.84 III
320	3131.842 I	120	1727.18 II	50	2699.38 I	90	4660.28 II
80	3341.481 I	250	1732.14 II	80	2705.36 II	50	4797.01 III
2800	3650.157 I	15	1759.75 III	70	2724.43 III	80	4855.72 II
300	3654.839 I	20	1775.68 I	80	2752.78 I	10	4869.85 III
80	3662.883 I	40	1783.70 II	20	2759.71 I	5	4883.00 I
240	3663.281 I	30	1796.22 II	6	2769.22 III	5	4889.91 I
30	3701.432 I	200	1796.90 II	40	2803.46 I	80	4916.07 I
35	3704.170 I	60	1798.74 II	30	2804.43 I	5	4970.37 I
30	3801.660 I	30	1803.89 II	2	2805.34 I	80	4973.57 III
20	3901.867 I	40	1808.29 II	2	2806.77 I	5	4980.64 I
60	3906.372 I	400	1820.34 II	150	2814.93 II	20	5102.70 I
200	3983.839 II	5	1832.74 I	3	2844.76 I	40	5120.64 I
1800	4046.572 I	1000	1849.50 I	750	2847.68 II	100	5128.45 II
150	4077.838 I	160	1869.23 II	50	2856.94 I	20	5137.94 I
40	4108.057 I	300	1870.55 II	150	2893.60 I	30	5210.82 III
250	4339.224 I	200	1875.54 II	150	2916.27 II	20	5290.74 I
400	4347.496 I	1	1894.77 III	60	2925.41 I	5	5316.78 I
4000	4358.337 I	20	1900.28 II	150	2935.94 II	60	5354.05 I
80	4916.068 I	30	1927.60 II	400	2947.08 II	30	5384.63 I
1100	5460.753 I	300	1942.27 II	1200	2967.28 I	1100	5460.74 I
160	5675.922 I	100	1972.94 II	300	3021.50 I	30	5549.63 I
240	5769.598 I	200	1973.89 II	120	3023.47 I	160	5675.86 I
280	5790.663 I	150	1987.98 II	30	3025.61 I	6	5695.71 III
20	6072.713 I	90	2026.97 II	50	3027.49 I	240	5769.60 I
30	6234.402 I	90	2052.93 II	15	3090.05 III	100	5789.66 I
160	6716.429 I	70	2148.00 II	400	3125.67 I	280	5790.66 I
250	6907.461 I	5	2247.55 I	320	3131.55 I	140	5803.78 I
240	11287.407 I	60	2262.23 II	320	3131.84 I	60	5859.25 I
Mercury		20	2302.06 I	400	3208.20 II	60	5871.73 II
Hg Z = 80		7	2314.15 III	400	3264.06 II	20	5871.98 I
3	621.44 III	15	2323.20 I	5	3283.02 III	20	6072.72 I
2	679.68 III	5	2340.57 I	12	3312.28 III	1000	6149.50 I
2	878.59 III	20	2345.43 I	80	3341.48 I	25	6220.35 III
1	886.48 III	20	2352.48 I	100	3385.25 II	30	6234.40 I
400	893.08 II	100	2378.32 I	8	3389.01 III	35	6418.98 III
300	915.83 II	20	2380.00 I	5	3450.77 III	40	6501.38 III
150	923.39 II	4	2380.55 III	400	3451.69 II	80	6521.13 II
200	940.80 II	40	2399.38 I	3	3500.35 III	10	6584.26 III
100	962.74 II	20	2399.73 I	4	3538.88 III	6	6610.12 III
50	969.13 II	10	2400.49 I	200	3549.42 II	30	6709.29 III
1	988.89 III	60	2407.35 II	5	3557.24 III	160	6716.43 I
2	1009.29 III	50	2414.13 II	2800	3650.15 I	250	6907.52 I
5	1068.03 III	8	2431.65 III	300	3654.84 I	250	7081.90 I
800	1099.26 II	5	2441.06 I	80	3662.88 I	200	7091.86 I
2	1161.95 III	20	2446.90 I	240	3663.28 I	40	7346.37 II
80	1250.58 I	15	2464.06 I	30	3701.44 I	100	7485.87 I
8	1259.24 I	5	2480.56 III	35	3704.17 I	12	7517.46 III
100	1268.82 I	40	2482.00 I	30	3801.66 I	20	7728.82 I
5	1307.75 I	30	2482.72 I	15	3803.51 III	7	7808.10 III
300	1307.93 II	40	2483.82 I	100	3806.38 II	100	7944.66 II
400	1321.71 II	7	2484.50 III	20	3901.87 I	25	7946.75 III
400	1331.74 II	90	2534.77 I	60	3906.37 I	50	7984.51 III
80	1350.07 II	15000	2536.52 I	100	3918.92 II	5	8151.64 III
200	1361.27 II	25	2563.86 I	200	3983.96 II	2000	10139.75 I
20	1402.62 I	25	2576.29 I	1800	4046.56 I	240	11287.40 I
200	1414.43 II	5	2578.91 I	150	4077.83 I	120	13209.95 I
10	1435.51 I	2	2612.92 III	40	4108.05 I	140	13426.57 I
15	1619.46 II	4	2617.97 III	70	4122.07 III	60	13468.38 I
120	1623.95 II	15	2625.19 I	10	4140.34 III	80	13505.58 I
20	1628.25 II	5	2639.78 I	100	4216.74 III	500	13570.21 I
150	1649.94 II	250	2652.04 I	250	4339.22 I	450	13673.51 I
50	1653.64 II	400	2653.69 I	400	4347.49 I	200	13950.55 I
200	1672.41 II	100	2655.13 I	4000	4358.33 I	500	15295.82 I

Intensity	Wavelength/Å		Intensity	Wavelength/Å		Intensity	Wavelength/Å		Intensity	Wavelength/Å	
100	16881.48	I	200	2506.19	III	290	2903.07	II	1300	3344.75	I
400	16920.16	I	440	2538.46	II	80	2907.12	II	95	3346.40	II
300	16942.00	I	330	2542.67	II	600	2909.12	II	1600	3358.12	I
500	17072.79	I	80	2558.88	II	1100	2911.92	II	950	3363.78	I
400	17109.93	I	85	2564.34	II	120	2918.83	II	950	3379.97	I
20	17116.75	I	250	2593.70	II	1300	2923.39	II	1900	3384.62	I
20	17198.67	I	250	2602.80	II	140	2924.32	II	130	3395.36	II
20	17213.20	I	400	2616.78	II	1100	2930.50	II	640	3404.34	I
70	17329.41	I	440	2629.85	I	800	2934.30	II	1300	3405.94	I
30	17436.18	I	330	2636.67	II	95	2940.10	II	640	3437.22	I
50	18130.38	I	720	2638.76	II	110	2941.22	II	130	3446.08	II
40	19700.17	I	410	2640.99	I	150	2944.82	II	3200	3447.12	I
	22493.28	I	600	2644.35	II	140	2946.69	II	640	3449.07	I
250	23253.07	I	370	2646.49	II	95	2947.28	II	950	3456.39	I
	32148.06	I	640	2649.46	II	125	2947.32	III	640	3460.78	I
	36303.03	I	480	2653.35	II	95	2955.84	II	800	3504.41	I
	Molybdenum		560 h	2655.03	I	240	2956.06	II	560	3508.12	I
	Mo Z = 42		640	2660.58	II	70	2956.90	II	480	3521.41	I
50	867.92	IV	720	2672.84	II	95	2960.24	II	640	3537.28	I
100	884.19	IV	250	2673.27	II	250	2963.79	II	520	3558.10	I
60	886.05	IV	1000	2679.85	I	210	2965.27	II	400	3563.14	I
50	891.74	IV	95	2681.36	II	70	2971.91	II	1400	3581.89	I
100	1169.33	III	640	2683.23	II	250	2972.61	II	1400	3624.46	I
100	1254.93	III	880	2684.14	II	80	2975.40	II	1000	3635.43	I
100	1258.52	III	560	2687.99	II	95	2992.84	II	400	3657.35	I
100	1262.21	III	480	2701.42	II	95	3027.77	II	540	3664.81	I
100	1263.74	III	190	2713.51	II	100	3060.78	II	590	3672.82	I
100	1274.37	III	290	2717.35	II	800	3064.28	I	1300	3680.60	I
100	1276.40	III	85	2726.97	II	250	3065.04	II	65	3688.31	II
200	1277.40	III	140	2729.68	II	800	3074.37	I	180	3692.64	II
200	1277.58	III	80	2730.20	II	85	3077.66	II	1400	3694.94	I
200	1278.40	III	330	2732.88	II	800	3085.62	I	500	3727.69	I
150	1281.90	III	160	2736.96	II	270	3087.62	II	80	3744.37	II
150	1283.60	III	80 h	2737.88	II	190	3092.07	II	29000	3798.25	I
100	1854.73	III	290	2746.30	II	560	3094.66	I	520	3826.70	I
80	1926.26	IV	110	2756.07	II	560	3101.34	I	940	3828.87	I
100	1929.24	IV	220	2763.62	II	1400	3112.12	I	1700	3833.75	I
80	1971.06	IV	240	2769.76	II	290	3122.00	II	29000	3864.11	I
70	2010.92	IV	160	2773.78	II	14000	3132.59	I	580	3869.08	I
19000	2015.11	II	190	2774.39	II	110	3138.72	II	580	3886.82	I
40000	2020.30	II	1700	2775.40	II	220	3152.82	II	19000	3902.96	I
21000	2038.44	II	65	2777.86	II	55	3155.64	II	65	3941.48	II
17000	2045.98	II	880	2780.04	II	6000	3158.16	I	1400	4062.08	I
50	2060.38	IV	400	2784.99	II	8700	3170.35	I	2300	4069.88	I
4800	2081.68	II	100	2807.74	III	95	3172.03	II	1300	4081.44	I
2400	2089.52	II	400	2807.76	II	160	3172.74	II	940	4084.38	I
2200	2092.50	II	1700	2816.15	II	120 d	3187.59	II	730	4107.47	I
4000	2093.11	II	220	2817.44	II	7600	3193.97	I	630	4120.10	I
2700	2100.84	II	80	2827.74	II	880	3205.88	I	2900	4143.55	I
1500	2104.29	II	80	2834.39	II	3000	3208.83	I	480	4185.82	I
1400	2108.02	II	80	2835.33	II	560	3215.07	I	2500	4188.32	I
100	2184.37	III	160	2842.15	II	880	3228.22	I	1500	4232.59	I
100	2211.02	III	1700	2848.23	II	600	3229.79	I	890	4276.91	I
400	2269.69	II	370	2853.23	II	1100	3233.14	I	1200	4277.24	I
150	2269.71	III	370	2863.81	II	950	3237.08	I	1400	4288.64	I
200	2294.97	III	220	2866.69	II	65	3240.71	II	680	4292.13	I
160	2304.25	II	1700	2871.51	II	950	3256.21	I	890	4293.21	I
160	2306.97	II	85	2872.88	II	480	3264.40	I	840	4326.14	I
150	2330.93	III	220	2879.05	II	800	3270.90	I	1900	4381.64	I
110	2332.12	II	65	2888.15	II	'200	3271.69	III	2500	4411.57	I
190	2341.59	II	1300	2890.99	II	1100	3289.02	I	990	4434.95	I
100	2359.76	III	95	2891.28	II	950	3290.82	I	480	4457.36	I
110	2389.20	II	190	2892.81	II	190	3292.31	II	630	4474.56	I
140	2403.61	II	950	2894.45	II	100	3313.62	I	400	4536.80	I
120	2413.01	II	140	2897.63	II	190	3320.90	II	460	4626.47	I
85	2498.28	II	70	2900.80	II	640	3323.95	I	640	4707.26	I

Intensity	Wavelength/Å		Intensity	Wavelength/Å		Intensity	Wavelength/Å		Intensity	Wavelength/Å	
700	4731.44	I	50	5869.33	I	780	3723.50	II	2000	3951.16	II
770	4760.19	I	820	5888.33	I	410	3724.87	II	810	3952.20	II
410	4819.25	I	50 h	5893.38	I	710	3728.13	II	590	3958.00	II
410	4830.51	I	160 h	5928.88	I	470	3730.58	II	510	3962.21	II
180	5014.60	I	35	6025.49	I	1000 d	3735.54	II	1400	3963.12	II
80	5029.00	I	1300	6030.66	I	440	3737.10	II	1100	3973.30	II
65	5030.78	I	40	6101.87	I	1000	3738.06	II	740	3973.69	II
100	5047.71	I	40	6357.22	I	580	3752.49	II	740	3976.85	II
50	5055.00	I	35	6401.07	I	510	3757.82	II	740	3979.49	II
200	5059.88	I	100	6424.37	I	930	3758.95	II	470	3986.25	II
100	5080.02	I	230	6619.13	I	930	3763.47	II	1400	3990.10	II
100	5096.65	I	50	6650.38	I	510	3769.65	II	1000	3991.74	II
130	5097.52	I	110	6733.98	I	1400	3775.50	II	1100	3994.68	II
130	5109.71	I	50	6746.27	I	710	3779.47	II	410	4000.50	II
80	5114.97	I	35	6753.97	I	580	3780.40	II	540	4004.02	II
150	5145.38	I	40	6838.88	I	510	3781.32	II	410	4007.43	II
110	5147.39	I	35	6914.01	I	2400	3784.25	II	3700	4012.25	II
80	5163.19	I	110	7109.87	I	370	3801.12	II	540	4012.70	II
100	5167.76	I	150	7242.50	I	1200	3803.47	II	1000	4020.87	II
160 d	5171.08	I	40	7245.85	I	2500	3805.36	II	1000	4021.34	II
230 h	5172.94	I	40	7391.36	I	470	3807.23	II	1000	4021.78	II
160 h	5174.18	I	140	7485.74	I	540	3808.77	II	1200	4023.00	II
110	5200.17	I	27	7720.77	I	440	3809.06	II	410	4030.47	II
50	5200.74	I	40 h	8328.44	I	580	3810.49	II	1200	4031.82	II
50	5211.86	I	45 h	8389.32	I	710	3814.73	II	3000	4040.80	II
80	5219.40	I	45 h	8483.39	I	410	3822.47	II	410	4043.59	II
65	5231.06	I		**Neodymium**		1200	3826.42	II	410	4048.81	II
100	5234.26	I		**Nd Z = 60**		540	3828.85	II	850	4051.15	II
460 h	5238.20	I	75	2764.98	I	440	3829.16	II	850	4059.96	II
230 h	5240.88	I	80	2993.20	II	510	3830.47	II	4700	4061.09	II
110 h	5242.81	I	95	3007.97	II	740	3836.54	II	1100	4069.28	II
100	5245.51	I	95	3014.19	II	1700	3838.98	II	710	4075.12	II
150	5259.04	I	95	3018.35	II	410 d	3841.82	II	470	4075.28	II
65	5261.14	I	140	3056.71	II	1700 d	3848.24	II	470	4080.23	II
65	5279.65	I	130	3069.73	II	1500	3848.52	II	1400	4109.08	II
210	5280.86	I	160	3075.38	II	470	3850.22	II	2500	4109.46	II
55	5292.08	I	240	3092.92	II	2400 d	3851.66	II	510	4110.48	II
55	5295.47	I	260	3115.18	II	3700 d	3863.33	II	410	4123.88	II
55	5313.89	I	290	3133.60	II	850	3869.07	II	470	4133.36	II
80	5354.88	I	220	3134.90	II	470	3875.87	II	510	4135.33	II
65	5356.48	I	170	3141.46	II	1100	3878.58	II	3000	4156.08	II
560 hl	5360.56	I	170	3142.44	II	1000	3879.55	II	510	4156.26	II
110 hl	5364.28	I	150	3203.47	II	780	3880.38	II	410	4168.00	II
65	5394.52	I	220	3259.24	II	1200	3880.78	II	810	4175.61	II
50	5400.47	I	220	3265.12	II	540	3887.87	II	2400	4177.32	II
55	5435.68	I	320	3275.22	II	1300	3889.93	II	640	4179.59	II
65	5437.75	I	290	3285.10	II	1300	3890.58	II	470	4205.60	II
50	5501.54	I	410	3328.28	II	1300	3890.94	II	470	4211.29	II
7800	5506.49	I	320	3353.59	II	580	3891.51	II	440	4227.73	II
5200	5533.05	I	410	3560.75	II	470	3892.06	II	1300	4232.38	II
50	5543.12	I	470	3587.51	II	810	3894.63	II	2000	4247.38	II
55	5556.28	I	370	3615.82	II	440	3897.63	II	850	4252.44	II
2500	5570.45	I	410	3653.15	II	2000	3900.21	II	410	4261.84	II
100	5610.93	I	470	3662.26	II	1300	3901.84	II	470	4282.44	II
330	5632.47	I	540	3665.18	II	1700	3905.89	II	710	4284.52	II
50	5634.86	I	540	3672.36	II	510	3907.84	II	5400	4303.58	II
230	5650.13	I	580	3673.54	II	2000	3911.16	II	470	4314.52	II
55	5674.47	I	1200	3685.80	II	850	3912.23	II	1100	4325.76	II
460	5689.14	I	440	3687.30	II	440	3915.13	II	510	4327.93	II
80	5705.72	I	410	3689.69	II	610	3915.95	II	540	4338.70	II
210	5722.74	I	470	3697.56	II	1100	3920.96	II	680	4351.29	II
620	5751.40	I	470	3713.70	II	510	3927.10	II	850	4358.17	II
520	5791.85	I	640 d	3714.73	II	610	3934.82	II	470 d	4374.93	II
55 h	5849.73	I	470	3715.68	II	410	3936.11	II	710	4385.66	II
50 h	5851.52	I	410	3718.54	II	510	3938.86	II	540	4400.83	II
520	5858.27	I	410	3721.35	II	2000	3941.51	II	510	4411.06	II

Intensity	Wavelength/Å		Intensity	Wavelength/Å		Intensity	Wavelength/Å		Intensity	Wavelength/Å	
580	4446.39	II	55	6310.49	I	100	208.48	IV	80	541.13	IV
1400	4451.57	II	65	6385.20	II	100	208.73	IV	100	542.07	IV
740	4462.99	II	45	6630.14	I	80	208.90	IV	150	543.89	IV
410	4501.82	II	45	6650.57	II	150	212.56	IV	400	568.42	V
250	4516.36	II	40	6740.11	II	140	223.24	IV	250	569.76	V
340	4541.27	II	40	6900.43	II	120	223.60	IV	500	569.83	V
340	4542.61	II	35	7037.30	II	140	234.32	IV	250	572.11	V
340	4563.22	II	40	7066.89	II	120	234.70	IV	800	572.34	V
300	4621.94	I	29	7129.35	II	20	251.14	III	35	587.213	I
510	4634.24	I	24	7189.42	II	20	251.56	III	35	589.179	I
340	4641.10	I	20	7192.01	II	20	251.73	III	35	589.911	I
250	4645.77	II	15	7236.54	II	40	267.06	III	70	591.830	I
300	4649.67	I	12	7316.81	II	40	267.52	III	100	595.920	I
310	4683.45	I	10	7406.62	II	20	267.71	III	75	598.706	I
470	4706.54	II	10	7418.18	II	40	283.18	III	35	598.891	I
240	4719.02	I	12	7511.16	II	160	283.21	III	70	600.036	I
240	4811.34	II	17	7513.73	II	110	283.69	III	170	602.726	I
350	4825.48	II	12	7528.99	II	40	283.89	III	170	615.628	I
280	4859.02	II	10	7538.26	II	220	301.12	III	170	618.672	I
350	4883.81	I	12	7696.56	II	220	313.05	III	120	619.102	I
220	4890.70	II	10	7750.95	II	220	313.68	III	200	626.823	I
240	4891.07	I	10	7808.47	II	40	313.95	III	200	629.739	I
280	4896.93	I	12	7863.04	II	90	352.956	I	1000	735.896	I
210	4901.84	I	12	7917.01	II	60	354.962	I	400	743.720	I
330	4920.68	II	12	7958.95	I	50	357.83	IV	60	993.88	I
470	4924.53	I	12	7965.73	II	400	357.96	V	70	1068.65	I
260	4944.83	I	15	7982.09	II	500	358.47	V	90	1131.72	I
290	4954.78	I	12	7982.68	II	200	358.72	IV	100	1131.85	II
290	4959.13	II	12	8000.76	II	500	359.38	V	90	1229.83	I
250	4989.94	II	10	8120.93	II	90	361.433	II	20	1255.03	III
360	5076.59	II	12	8122.07	II	60	362.455	II	110	1255.68	III
360	5092.80	II	12	8141.75	II	1000	365.59	V	160	1257.19	III
360	5107.59	II	12	8143.27	II	220	379.31	III	90	1418.38	I
340	5123.79	II	10	8231.52	II	125	387.14	IV	90	1428.58	I
680	5130.60	II	10	8307.72	II	100	388.22	IV	90	1436.09	I
500	5191.45	II	12	8346.36	II	150	405.854	II	120	1681.68	II
630	5192.62	II	17	8839.10	II	120	407.138	II	180	1688.36	II
330	5200.12	II		**Neon**		800	416.20	V	100	1888.11	II
310	5212.37	II		**Ne Z = 10**		150	421.61	IV	100	1889.71	II
450	5234.20	II	66	119.01	V	200	445.040	II	200	1907.49	II
250	5239.79	II	200	122.52	V	300	446.256	II	500	1916.08	II
720	5249.59	II	66	125.12	V	250	446.590	II	300	1930.03	II
360	5255.51	II	45	131.99	V	180	447.815	II	200	1938.83	II
590	5273.43	II	50	132.04	V	150	454.654	II	100 c	1945.46	II
680	5293.17	II	150	140.76	V	200	455.274	II	80	2007.01	II
220	5311.46	II	150	140.79	V	10	456.275	II	65	2018.44	IV
500	5319.82	II	100	142.44	V	120	456.348	II	110	2022.19	IV
290	5361.47	II	100	142.50	V	90	456.896	II	80	2025.56	II
160	5431.53	II	150	142.72	V	1000	460.728	II	150	2085.47	II
240	5594.43	II	100	143.27	V	500	462.391	II	200	2086.96	III
220	5620.54	I	150	143.34	V	140	469.77	IV	300	2089.43	III
140 d	5675.97	I	150	147.13	V	200	469.82	IV	240	2092.44	III
220	5688.53	II	66	151.23	V	180	469.87	IV	400	2095.54	III
130	5702.24	II	120	151.42	V	140	469.92	IV	180	2096.11	II
160	5708.28	II	15	151.82	IV	250	480.41	V	120	2096.25	II
100	5729.29	I	15	152.23	IV	150	481.28	V	200	2161.22	III
160	5804.02	II	45	154.50	V	250	481.36	V	300	2163.77	III
80	5811.57	II	15	158.65	IV	500	482.99	V	200	2180.89	III
70	5825.87	II	15	158.82	IV	285	488.10	III	30	2203.88	IV
80	5842.39	II	100	164.02	V	220	488.87	III	200	2209.35	III
55	5858.91	I	100	164.14	V	450	489.50	III	200	2211.85	III
45	6007.67	I	80	172.62	IV	70	489.64	III	240	2213.76	III
45	6034.24	II	500	173.93	V	220	490.31	III	300	2216.07	III
55	6066.03	I	80	177.16	IV	360	491.05	III	10	2220.81	IV
45	6178.59	I	150	186.58	IV	120	521.74	IV	75	2227.42	V
45	6223.39	I	100	194.28	IV	140	521.82	IV	110	2232.41	V

Intensity	Wavelength/Å		Intensity	Wavelength/Å		Intensity	Wavelength/Å		Intensity	Wavelength/Å	
65	2245.48	V	300	3028.86	II	40	3369.908	I	150 p	4430.90	II
250	2258.02	IV	100	3030.79	II	100	3371.80	II	150 p	4430.94	II
65	2259.57	V	120	3034.46	II	500	3378.22	II	120	4457.05	II
175	2262.08	IV	100	3035.92	II	150	3388.42	II	100	4522.72	II
240	2263.21	III	100	3037.72	II	120	3388.94	II	10	4537.754	I
65	2263.39	V	100	3039.59	II	300	3392.80	II	10	4540.380	I
110	2264.54	IV	100	3044.09	II	100	3404.82	II	100	4569.06	II
200	2264.91	III	100	3045.56	II	120	3406.95	II	15	4704.395	I
250	2265.71	V	120	3047.56	II	100	3413.15	II	12	4708.862	I
550	2285.79	IV	100	3054.34	II	120	3416.91	II	10	4710.067	I
30	2293.14	IV	100	3054.68	II	120	3417.69	II	10	4712.066	I
250	2293.49	IV	100	3059.11	II	50	3417.904	I	15	4715.347	I
250	2350.84	IV	100	3062.49	II	15	3418.006	I	10	4752.732	I
450	2352.52	IV	100	3063.30	II	120	3428.69	II	12	4788.927	I
700	2357.96	IV	100	3070.89	II	60	3447.703	I	10	4790.22	I
250	2362.68	IV	100	3071.53	II	50	3454.195	I	10	4827.344	I
250	2363.28	IV	100	3075.73	II	100	3456.61	II	10	4884.917	I
110	2365.49	IV	120	3088.17	II	100	3459.32	II	4	5005.159	I
350	2372.16	IV	100	3092.09	II	25	3460.524	I	10	5037.751	I
65	2384.20	IV	120	3092.90	II	30	3464.339	I	10	5144.938	I
350	2384.95	IV	100	3094.01	II	30	3466.579	I	25	5330.778	I
300	2412.73	III	100	3095.10	II	60	3472.571	I	20	5341.094	I
240	2412.94	III	100	3097.13	II	150	3479.52	II	8	5343.283	I
200	2413.78	III	100	3117.98	II	200	3480.72	II	60	5400.562	I
200	2473.40	III	120	3118.16	II	200	3481.93	II	5	5562.766	I
80 p	2562.12	II	10	3126.199	I	25	3498.064	I	10	5656.659	I
90 w	2567.12	II	300	3141.33	II	30	3501.216	I	5	5719.225	I
800	2590.04	III	100	3143.72	II	25	3515.191	I	12	5748.298	I
600	2593.60	III	100 p	3148.68	II	150	3520.472	I	80	5764.419	I
400	2595.68	III	100	3164.43	II	120	3542.85	II	12	5804.450	I
300	2610.03	III	100	3165.65	II	120	3557.80	II	40	5820.156	I
240	2613.41	III	100	3188.74	II	100	3561.20	II	500	5852.488	I
200	2615.87	III	120	3194.58	II	250	3568.50	II	100	5872.828	I
80	2623.11	II	500	3198.59	II	100	3574.18	II	100	5881.895	I
80	2629.89	II	60	3208.96	II	200	3574.61	II	60	5902.462	I
90 w	2636.07	II	120	3209.36	II	50	3593.526	I	60	5906.429	I
80	2638.29	II	120	3213.74	II	30	3593.640	I	100	5944.834	I
200	2638.70	III	150	3214.33	II	15	3600.169	I	100	5965.471	I
200	2641.07	III	150	3218.19	II	20	3633.665	I	100	5974.627	I
80	2644.10	II	120	3224.82	II	150	3643.93	II	120	5975.534	I
600	2677.90	III	120	3229.57	II	200	3664.07	II	80	5987.907	I
500	2678.64	III	200	3230.07	II	20	3682.243	I	100	6029.997	I
80	2762.92	II	120	3230.42	II	12	3685.736	I	100	6074.338	I
90	2792.02	II	120	3232.02	II	200	3694.21	II	80	6096.163	I
80	2794.22	II	150	3232.37	II	10	3701.225	I	60	6128.450	I
100	2809.48	II	100	3243.40	II	150	3709.62	II	100	6143.063	I
80	2906.59	II	100	3244.10	II	250	3713.08	II	120	6163.594	I
80	2906.82	II	100	3248.34	II	250	3727.11	II	250	6182.146	I
90	2910.06	II	100	3250.36	II	800	3766.26	II	150	6217.281	I
90	2910.41	II	150	3297.73	II	1000	3777.13	II	150	6266.495	I
80	2911.14	II	150	3309.74	II	100	3818.43	II	60	6304.789	I
80	2915.12	II	300	3319.72	II	120	3829.75	II	100	6334.428	I
80	2925.62	II	1000	3323.74	II	150	4219.74	II	120	6382.992	I
80 w	2932.10	II	150	3327.15	II	100	4233.85	II	200	6402.246	I
80	2940.65	II	100	3329.16	II	120	4250.65	II	150	6506.528	I
90	2946.04	II	200	3334.84	II	120	4369.86	II	60	6532.882	I
150	2955.72	II	150	3344.40	II	70	4379.40	II	150	6598.953	I
150	2963.24	II	300	3345.45	II	150	4379.55	II	70	6652.093	I
150	2967.18	II	150	3345.83	II	100	4385.06	II	90	6678.276	I
100	2973.10	II	200	3355.02	II	200	4391.99	II	20	6717.043	I
15	2974.72	I	120	3357.82	II	150	4397.99	II	100	6929.467	I
100	2979.46	II	200	3360.60	II	150	4409.30	II	90	7024.050	I
12	2982.67	I	120	3362.16	II	100	4413.22	II	100	7032.413	I
150	3001.67	II	100	3362.71	II	100	4421.39	II	50	7051.292	I
120 p	3017.31	II	120	3367.22	II	100 p	4428.52	II	80	7059.107	I
300	3027.02	II	12	3369.808	I	100 p	4428.63	II	100	7173.938	I

Intensity	Wavelength/Å		Intensity	Wavelength/Å		Intensity	Wavelength/Å		Intensity	Wavelength/Å	
150	7213.20	II	3000	9313.97	I	1100	23709.2	I	300 s	7370.60	I
150	7235.19	II	6000	9326.51	I	1800	23951.42	I	300 l	7381.03	I
100	7245.167	I	2000	9373.31	I	600	23956.46	I	300 l	7381.65	I
150	7343.94	II	5000	9425.38	I	1000	23978.12	I	300 l	7402.70	I
40	7472.439	I	3000	9459.21	I	200	24098.54	I	300 s	7512.22	I
90	7488.871	I	5000	9486.68	I	500	24161.42	I	300 l	7515.15	I
100	7492.10	II	5000	9534.16	I	600	24249.64	I	300 l	7546.05	I
150	7522.82	II	3000	9547.40	I	1500	24365.05	I	300 l	7624.83	I
80	7535.774	I	120	9577.01	II	800	24371.60	I	300	7626.85	I
60	7544.044	I	1000	9665.42	I	400	24447.85	I	300 s	7681.01	I
100	7724.628	I	100	9808.86	II	700	24459.4	I	300 s	7685.25	I
120	7740.74	II	800	10295.42	I	300	24776.46	I	1000 l	7735.14	I
300	7839.055	I	2000	10562.41	I	550	24928.88	I	300 l	7761.61	I
120	7926.20	II	1500	10798.07	I	250	25161.69	I	1000 l	7765.75	I
400	7927.118	I	2000	10844.48	I	650	25524.37	I	300 s	7776.07	I
700	7936.996	I	3000	11143.020	I	125	28386.21	I	300	7787.46	I
2000	7943.181	I	3500	11177.528	I	150	30200.	I	1000 l	7791.38	I
2000	8082.458	I	1600	11390.434	I	250	33173.09	I	300 l	7851.44	I
100	8084.34	II	1100	11409.134	I	450	33352.35	I	300 l	7887.88	I
1000	8118.549	I	3000	11522.746	I	1300	33901.	I	300 l	7901.71	I
600	8128.911	I	1500	11525.020	I	2200	33912.10	I	300 l	7975.98	I
3000	8136.406	I	950	11536.344	I	600	34131.31	I	300 h	8080.32	I
2500	8259.379	I	500	11601.537	I	100	34471.44	I	300 s	8124.59	I
100	8264.81	II	1200	11614.081	I	120	35834.78	I	300	8155.11	I
2500	8266.077	I	300	11688.002	I		**Neptunium**		300 l	8167.42	I
800	8267.117	I	2000	11766.792	I		**Np Z = 93**		300 l	8183.06	I
6000	8300.326	I	1500	11789.044	I	300	3481.93	I	300 l	8188.61	I
100	8315.00	II	500	11789.889	I	300 h	3501.50	I	300 l	8247.82	I
1500	8365.749	I	1000	11984.912	I	300 l	3986.89	I	300 l	8287.11	I
100	8372.11	II	3000	12066.334	I	300 s	5044.66	I	300 s	8287.75	I
8000	8377.606	I	800	12459.389	I	300 l	5601.70	I	300 l	8306.22	I
1000	8417.159	I	1000	12689.201	I	300 l	5652.75	I	300 s	8313.66	I
4000	8418.427	I	1100	12912.014	I	300 l	5784.39	I	1000 l	8339.12	I
1500	8463.358	I	700	13219.241	I	300 l	5878.04	I	300	8356.79	I
800	8484.444	I	800	15230.714	I	300 s	6011.22	I	300 l	8367.11	I
5000	8495.360	I	400	17161.930	I	300	6056.09	I	3000	8372.88	I
600	8544.696	I	400	18035.80	I	300 s	6073.90	I	3000	8529.96	I
1000	8571.352	I	1000	18083.21	I	300 s	6080.05	I	1000 s	8696.23	I
4000	8591.259	I	350	18221.11	I	300 l	6120.49	I	1000 s	8906.02	I
6000	8634.647	I	250	18227.02	I	300	6188.59	I	1000	8942.70	I
3000	8647.041	I	2500	18276.68	I	300 l	6200.00	I	1000 s	9004.75	I
15000	8654.383	I	2000	18282.62	I	300 s	6215.90	I	1000 l	9006.31	I
4000	8655.522	I	1200	18303.97	I	300 s	6317.84	I	10000 l	9016.18	I
100	8668.26	II	250	18359.12	I	300 l	6341.38	I	3000 l	9141.30	I
5000	8679.492	I	1200	18384.85	I	300 l	6566.11	I	3000 s	9379.33	I
5000	8681.921	I	2000	18389.95	I	300 l	6720.68	I	3000 l	9468.66	I
2000	8704.112	I	1000	18402.84	I	300 s	6751.32	I	3000 s	9679.13	I
4000	8771.656	I	1200	18422.39	I	300 s	6795.21	I	3000 l	9930.55	I
12000	8780.621	I	300	18458.65	I	300 l	6802.62	I	10000 l	10091.99	I
10000	8783.753	I	400	18475.79	I	300 l	6805.81	I	10000 s	10817.45	I
500	8830.907	I	900	18591.55	I	300 s	6816.44	I	10000 l	11695.15	I
7000	8853.867	I	1600	18597.70	I	300 l	6865.45	I	10000 l	11776.64	I
1000	8865.306	I	350	18618.96	I	300 s	6907.13	I	10000 s	12148.18	I
1000	8865.755	I	550	18625.16	I	300 h	6912.91	I	10000 s	12377.42	I
3000	8919.501	I	1200	21041.295	I	1000 s	6930.31	I	10000 l	12407.99	I
2000	8988.57	I	750	21708.145	I	300 l	6963.63	I	10000 l	13834.33	I
100	9079.46	II	300	22247.35	I	3000 s	6972.09	I		**Nickel**	
6000	9148.67	I	350	22428.13	I	300	7014.02	I		**Ni Z = 28**	
6000	9201.76	I	2250	22530.40	I	300 l	7018.91	I	55	315.24	V
4000	9220.06	I	400	22661.81	I	300 s	7039.14	I	56	315.71	V
2000	9221.58	I	600	23100.51	I	300 s	7080.01	I	72	354.18	V
2000	9226.69	I	1000	23260.30	I	300 l	7174.83	I	76	354.42	V
1000	9275.52	I	1050	23373.00	I	300 l	7184.93	I	68	354.49	V
200	9287.56	II	850	23565.36	I	300 l	7284.28	I	500	630.71	III
6000	9300.85	I	3500	23636.52	I	300 l	7292.29	I	500	676.94	III
1500	9310.58	I	300	23701.64	I	300 l	7332.52	I	300	713.33	III

Line Spectra of the Elements (continued): Nickel—Niobium

Intensity	Wavelength/Å		Intensity	Wavelength/Å		Intensity	Wavelength/Å		Intensity	Wavelength/Å	
300	713.38	III	2000	2174.67	II	660	3500.85	I	13	7797.59	I
500	718.48	III	1500	2175.15	II	2600	3510.34	I	1000	8096.75	II
300	722.09	III	2500	2185.50	II	6600	3515.05	I	700	8121.48	II
500	729.82	III	3000	2192.09	II	660	3519.77	I	9	8862.55	I
400	731.70	III	5000	2205.55	II	8200	3524.54	I	500 w	9900.92	II
300	732.16	III	4000	2206.72	II	5000	3566.37	I	**Niobium**		
300	747.99	III	6000	2216.48	II	990	3571.87	I	**Nb Z = 41**		
300	750.05	III	1000	2264.46	II	1300	3597.70	I	80	464.55	V
300	757.80	III	2000	2270.21	II	1300	3610.46	I	80	468.32	V
400	770.22	III	1600	2289.98	I	530	3612.74	I	80	763.77	V
500	778.81	III	630	2300.78	I	6600	3619.39	I	80	774.02	V
300	788.04	III	1000	2303.00	II	200	3664.10	I	60	993.54	IV
500	811.57	III	2000	2310.96	I	130	3669.24	I	400	1005.72	IV
500	826.14	III	1700	2312.34	I	180	3670.43	I	500	1007.05	IV
500	842.14	III	1400	2313.66	I	260	3674.15	I	500	1010.19	IV
400	845.24	III	1400	2313.98	I	160	3688.42	I	100	1116.08	IV
300	847.43	III	1000	2316.04	II	80	3693.93	I	150	1120.02	IV
300	860.64	III	1400	2317.16	I	120	3722.48	I	100	1258.87	V
300	862.88	III	2600	2320.03	I	150	3736.81	I	60	1314.56	III
300	863.22	III	1900	2321.38	I	60	3739.23	I	80	1445.43	III
300	867.51	III	1400	2325.79	I	600	3775.57	I	80	1445.98	III
300	973.79	III	940	2329.96	I	700	3783.53	I	80	1447.09	III
400	979.59	III	1200	2345.54	I	700	3807.14	I	100	1456.68	III
500	1317.22	II	400	2347.52	I	110	3831.69	I	80	1484.73	III
76	1398.19	IV	1000	2375.42	II	1200	3858.30	I	100	1495.94	III
74	1411.45	IV	240	2386.58	I	110	3973.56	I	80	1498.02	III
70	1438.82	IV	1000	2394.52	II	110	4401.55	I	80	1499.45	III
73	1449.01	IV	2000	2416.13	II	85	4459.04	I	100	1501.99	III
76	1452.22	IV	240	2419.31	I	55	4470.48	I	60	1502.30	IV
73	1482.25	IV	160	2472.06	I	65	4605.00	I	80	1513.81	III
72	1489.83	IV	150	2798.65	I	75	4648.66	I	60	1524.36	IV
75	1525.31	IV	250	2821.29	I	110	4714.42	I	100	1524.91	III
74	1527.68	IV	500	2943.91	I	45	4786.54	I	100	1590.21	III
74	1527.80	IV	570	2981.65	I	45	4855.41	I	80	1598.86	III
76	1534.71	IV	500	2992.60	I	40	4904.41	I	80	1604.72	III
73	1537.25	IV	1000	2994.46	I	45	4980.16	I	80	1639.51	III
75	1543.41	IV	4000	3002.49	I	45	4984.13	I	100	1682.77	III
74	1546.23	IV	2200	3003.63	I	50	5017.59	I	100	1705.44	III
300	1604.54	III	3700	3012.00	I	100	5035.37	I	100	1707.14	III
300	1652.87	III	1700	3037.94	I	100	5080.52	I	100	1758.33	V
400	1687.90	III	3500	3050.82	I	65	5081.11	I	100	1877.34	V
1000	1692.51	III	1500	3054.32	I	40 h	5146.48	I	100	1892.92	III
800	1709.90	III	1900	3057.64	I	40 h	5155.76	I	60	1922.41	IV
650	1715.30	III	500	3064.62	I	180	5476.91	I	100	1938.84	III
500	1719.46	III	2600	3101.55	I	23	5709.56	I	60	1978.22	IV
400	1722.28	III	1300	3101.88	I	16	5754.68	I	3300	2029.32	II
500	1738.25	III	2900	3134.11	I	10	5857.76	I	65	2032.53	IV
300	1739.78	III	1100	3232.96	I	10	5892.88	I	3000	2032.99	II
1000	1741.55	II	600	3243.06	I	10	6108.12	I	2000	2109.42	II
300	1741.96	III	660	3315.66	I	10	6176.81	I	1700	2125.21	II
550	1747.01	III	2000	3331.88	II	10	6191.18	I	1100	2126.54	II
300	1752.43	III	2900	3369.57	I	13	6256.36	I	80 h	2130.24	III
400	1753.01	III	3300	3380.57	I	16	6643.64	I	1500	2131.18	II
800	1764.69	III	1300	3391.05	I	22	6767.77	I	80	2273.92	III
500	1767.94	III	3300	3392.99	I	10	6914.56	I	100	2275.23	III
2000	1769.64	III	8200	3414.76	I	26	7122.20	I	80	2279.36	III
400	1776.07	III	1600	3423.71	I	16	7393.60	I	100	2281.51	III
300	1807.24	III	2600	3433.56	I	16	7409.35	I	80	2284.40	III
300	1819.28	III	990	3437.28	I	23	7422.28	I	100	2290.36	III
800	1823.06	III	4800	3446.26	I	13	7522.76	I	370	2295.68	II
400	1830.01	III	1300	3452.89	I	19	7555.60	I	280	2302.08	II
650	1847.28	III	5000	3458.47	I	23	7617.00	I	100	2313.30	III
800	1854.15	III	5000	3461.65	I	16	7714.32	I	100	2338.09	III
300	1858.75	III	1600	3472.54	I	19	7727.61	I	80	2344.12	III
1000	2165.55	II	550	3483.77	I	19	7748.89	I	90	2349.21	III
2000	2169.10	II	5500	3492.96	I	10	7788.94	I	80	2355.54	III

Intensity	Wavelength/Å		Intensity	Wavelength/Å		Intensity	Wavelength/Å		Intensity	Wavelength/Å	
100	2362.06	III	470	2716.62	II	390	3215.60	II	2700	3726.24	I
80	2362.50	III	470	2721.98	II	800	3225.48	II	2700	3739.80	I
80	2365.70	III	310	2733.26	II	140	3229.56	II	670	3740.73	II
100	2372.73	III	110	2737.09	II	400	3236.40	II	1700	3742.39	I
170	2376.40	II	240	2768.13	II	200	3247.47	II	530	3763.49	I
110	2387.09	II	310	2773.20	I	120	3248.94	II	350	3765.08	I
100	2387.41	III	270	2780.24	II	320	3254.07	II	530	3771.85	I
140	2387.52	II	110	2793.05	II	230	3260.56	II	870	3781.01	I
80	2388.23	III	190	2827.08	II	160	3263.37	II	1700	3787.06	I
45	2388.27	II	250	2841.15	II	200	3283.46	II	1300	3790.15	I
160	2398.48	II	280	2842.65	II	160	3292.02	II	3500	3791.21	I
80	2404.89	III	160	2846.28	II	320	3296.01	I	2700	3798.12	I
55	2405.34	II	240	2861.09	II	400	3312.60	I	2700	3802.92	I
55	2405.85	II	100	2865.61	II	120	3319.58	II	670	3803.88	I
140	2412.46	II	500	2868.52	II	130	3341.60	II	530	3804.74	I
100	2413.94	III	800	2875.39	II	1300	3341.97	I	670	3810.49	I
160	2416.99	II	270	2876.95	II	1300	3343.71	I	530	3811.03	I
140	2418.69	II	530	2877.03	II	1700	3349.06	I	530	3815.51	I
100	2421.91	III	100	2880.72	II	420	3349.52	I	210	3818.86	II
75	2433.80	II	570	2883.18	II	340	3354.74	I	670	3824.88	I
40	2435.95	II	280	2888.83	II	1700	3358.42	I	350	3835.18	I
45	2437.42	II	470	2897.81	II	130	3365.58	II	350	3863.38	I
40	2442.14	II	400	2899.24	II	340	3366.96	I	530	3877.56	I
28	2442.68	II	470	2908.24	II	130	3369.16	II	870	3878.82	I
65	2451.87	II	670	2910.59	II	350	3374.92	I	670	3883.14	I
65	2453.95	II	470	2911.74	II	170	3386.24	II	1100	3885.44	I
100	2456.99	III	1100	2927.81	II	350	3392.34	I	670	3885.68	I
55	2458.09	II	110	2931.47	II	230	3408.68	II	580	3891.30	I
65	2462.89	I	870	2941.54	II	180	3409.19	II	670	3914.70	I
80	2468.72	II	110 h	2945.88	II	230	3412.94	II	530	3920.20	I
80	2475.87	III	110	2946.12	II	230	3425.42	II	670	3937.44	I
110	2477.38	II	110	2946.90	II	230	3426.57	II	520	3943.67	I
65	2478.29	II	1100	2950.88	II	180	3432.70	II	910 d	3966.09	I
65	2479.94	II	400	2972.57	II	180	3440.59	II	1100	4032.52	I
35	2483.88	II	320	2974.10	II	200	3479.56	II	16000 c	4058.94	I
100	2499.73	III	210	2977.68	II	100	3484.05	II	350	4060.79	I
110	2511.00	II	200	2982.11	II	500	3498.63	I	12000	4079.73	I
110	2521.40	II	330	2990.26	II	460	3507.96	I	440	4100.40	I
390	2544.80	II	470	2994.73	II	200	3510.26	II	6700	4100.92	I
100	2545.64	III	80	3001.84	III	200	3515.42	II	310	4116.90	I
110	2551.38	II	140	3024.74	II	200	3517.67	II	5300	4123.81	I
130	2556.94	II	350	3028.44	II	2000	3535.30	I	670	4129.43	I
80	2557.94	III	300	3032.77	II	1300	3537.48	I	770	4129.93	I
130	2562.41	II	100	3044.76	II	250	3540.96	II	2300	4137.10	I
110	2571.33	II	100	3055.52	II	500	3544.02	I	440	4139.44	I
390	2583.99	II	220	3064.53	II	300	3550.45	I	2700	4139.71	I
390	2590.94	II	110	3069.68	II	1000	3554.66	I	350	4143.21	I
80	2598.86	III	100	3070.90	II	630	3563.50	I	870	4150.12	I
80	2633.17	III	110	3071.56	II	630	3563.62	I	4400	4152.58	I
200	2642.24	II	100	3073.24	II	1500	3575.85	I	870	4163.47	I
320	2646.26	II	400	3076.87	II	5000	3580.27	I	4400	4163.66	I
330	2647.50	I	110	3080.35	II	500	3584.97	I	4000	4164.66	I
330	2654.45	I	1800	3094.18	II	750	3589.11	I	3500	4168.13	I
310	2656.08	II	140	3099.19	II	500	3589.36	I	310	4184.44	I
80	2657.99	III	270	3127.53	II	500	3593.97	I	1200	4190.88	I
110	2665.25	II	1500	3130.79	II	500	3602.56	I	870	4192.07	I
110	2666.59	II	80	3142.26	III	300	3619.51	II	870	4195.09	I
110	2667.30	II	390	3145.40	II	420	3649.85	I	1300	4195.66	I
400	2671.93	II	1200	3163.40	II	400	3651.19	II	310	4198.51	I
200	2673.57	II	150	3175.78	II	200	3659.61	II	350	4201.52	I
200	2675.94	II	390	3180.29	II	630	3660.37	I	870	4205.31	I
160	2691.77	II	300	3191.10	II	900	3664.70	I	350	4214.73	I
1000	2697.06	II	150	3191.43	II	1500	3697.85	I	420	4217.94	I
320	2698.86	II	1000	3194.98	II	330	3711.34	I	420	4229.15	I
320	2702.20	II	120	3203.35	II	3300	3713.01	I	770	4262.05	I
150	2702.52	II	300	3206.34	II	480	3716.99	I	420	4266.02	I

Intensity	Wavelength/Å		Intensity	Wavelength/Å		Intensity	Wavelength/Å		Intensity	Wavelength/Å	
400	4286.99	I	210 cw	6660.84	I	600 w	297.7	IV	550	916.012	II
580	4299.60	I	150 cw	6677.33	I	700	297.82	IV	650	916.701	II
580	4300.99	I	130 c	6723.62	I	650	300.32	IV	520	921.992	IV
390	4311.27	I	85	6828.11	I	90	303.123	IV	500	922.519	IV
350	4326.33	I	85	6990.32	I	500	303.28	IV	480	923.057	IV
390	4331.37	I	190 c	7046.81	I	150	314.715	III	520	924.283	IV
330	4410.21	I	130	7159.43	I	200	314.850	III	90	953.415	I
150	4503.04	I	190 cw	7372.50	I	90	314.877	III	100	953.655	I
530	4523.41	I	65	7515.93	I	150	315.053	III	130	953.970	I
480	4546.82	I	170 c	7574.58	I	120	322.503	IV	1000	955.335	IV
370	4564.53	I	75 c	7726.68	I	150	322.570	IV	130	963.990	I
720	4573.08	I	35	7885.31	I	200	322.724	IV	115	964.626	I
480	4581.62	I	40	8135.20	I	120	323.175	IV	70	965.041	I
1200	4606.77	I	29 cw	8320.93	I	600	323.26	III	650	979.842	III
170	4616.17	I	29	8346.08	I	300	335.050	IV	700	979.919	III
450	4630.11	I	35	8905.78	I	500	338.35	III	900	989.790	III
450	4648.95	I		**Nitrogen**		500	340.20	III	700	991.514	III
450	4663.83	I		**N Z = 7**		500 w	351.93	IV	1000	991.579	III
340	4666.24	I	400	181.75	IV	500	351.98	III	150 w	1036.16	IV
240	4667.22	I	52	186.069	V	700	353.06	IV	90	1067.614	I
580	4672.09	I	62	186.153	V	120	362.833	III	60	1068.612	I
530	4675.37	I	400	191.7	IV	150	362.881	III	90	1078.71	IV
320	4685.14	I	400	192.9	IV	150	362.946	III	450	1083.990	II
130 c	4706.14	I	500	196.87	IV	90	362.985	III	600	1084.580	II
260	4708.29	I	500	197.23	IV	300	374.204	III	430	1085.546	II
150	4713.50	I	500	202.60	IV	350	374.441	III	650	1085.701	II
220 c	4749.70	I	500	205.94	IV	500	387.48	III	175	1097.237	I
130 c	4967.78	I	500	205.97	IV	500	420.77	IV	115	1098.095	I
190	4988.97	I	500	206.03	IV	250	451.869	III	115	1098.260	I
230	5017.75	I	90	209.303	V	300	452.226	III	105	1100.360	I
150	5026.36	I	500	217.20	IV	650	463.74	IV	40	1100.465	I
210	5039.04	I	500 d	217.90	IV	285	644.634	II	90	1101.291	I
170	5058.01	I	500 d	223.4	IV	360	644.837	II	360	1134.165	I
130	5065.25	I	800 w	225.12	IV	450	645.178	II	385	1134.415	I
750	5078.96	I	800	225.21	IV	140	647.50	I	410	1134.980	I
420	5095.30	I	600 w	234.12	IV	360	660.286	II	105	1143.65	I
170	5100.16	I	600 w	234.20	IV	170	671.016	II	130	1163.884	I
170	5120.30	I	600 w	234.25	IV	285	671.386	II	60	1164.206	I
210	5134.75	I	550	236.07	IV	150	671.630	II	105	1164.325	I
250	5160.33	I	500	237.99	IV	160	671.773	II	270	1167.448	I
250	5164.38	I	500 w	238.7	IV	170	672.001	II	105	1168.334	I
230	5180.31	I	600	238.80	IV	500	684.996	III	60	1168.417	I
190	5189.20	I	500 w	239.62	IV	570	685.513	III	195	1168.536	I
170	5193.08	I	900	247.20	IV	650	685.816	III	230	1176.510	I
150	5195.84	I	90	247.561	V	500	686.335	III	105	1176.630	I
150	5232.81	I	120	247.706	V	350	692.70	I	195	1177.695	I
150 d	5251.62	I	500 w	248.43	IV	90	713.518	V	500	1183.031	III
270	5271.53	I	500 w	248.46	IV	150	713.860	V	570	1184.550	III
130 c	5276.20	I	500 w	248.48	IV	285	746.984	II	90	1188.01	IV
250	5318.60	I	500	257.95	III	150	748.195	V	410	1199.550	I
460	5344.17	I	650	258.50	III	200	748.291	V	385	1200.223	I
340	5350.74	I	700	259.19	III	500	763.336	III	360	1200.710	I
110	5437.27	I	800	260.09	III	570	764.359	III	175	1225.026	I
85	5551.35	I	600	260.45	IV	570	765.148	IV	160	1225.37	I
170	5642.11	I	800	261.28	III	250	771.544	III	130	1228.41	I
130	5664.71	I	500	262.91	III	300	771.901	III	160	1228.79	I
170	5665.63	I	500	265.23	III	350	772.385	III	1000	1238.821	V
130	5729.19	I	500	265.27	III	200	772.891	III	900	1242.804	V
110	5760.34	I	150	266.196	V	150	772.975	III	360	1243.179	I
110	5819.43	I	200	266.379	V	650	775.965	II	315	1243.306	I
130 d	5838.64	I	500	268.70	III	90	885.67	I	290	1310.540	I
190 cw	5900.62	I	650	270.99	IV	90	909.697	I	250	1310.95	I
150	5983.22	I	250	283.42	IV	80	910.278	I	230	1319.00	I
75	6221.96	I	300	283.48	IV	40	910.645	I	315	1319.68	I
85 c	6430.46	I	350	283.58	IV	450	915.612	II	115	1326.57	I
65	6544.61	I	600	285.56	IV	450	915.962	II	115	1327.92	I

Line Spectra of the Elements (continued): Nitrogen

Intensity	Wavelength/Å		Intensity	Wavelength/Å		Intensity	Wavelength/Å		Intensity	Wavelength/Å	
150	1387.371	III	160	2885.27	II	160	4950.23	I	160	7406.12	I
360	1411.94	I	90 1	2974.52	V	350	4963.98	I	265	7406.24	I
700	1492.625	I	150 w	2980.78	V	285	4987.37	II	685	7423.64	I
490	1492.820	I	250 w	2981.31	V	450	4994.36	II	785	7442.29	I
640	1494.675	I	60 w	2998.43	V	650	5001.48	II	900	7468.31	I
90	1549.336	V	220	3006.83	II	360	5002.70	II	185	7608.80	I
200 1	1616.33	V	90	3078.25	IV	870	5005.15	II	60 w	7618.46	V
350 1	1619.69	V	120	3367.34	III	550	5007.32	II	450	7762.24	II
1000	1718.55	IV	360	3437.15	II	450	5010.62	II	400	8184.87	I
250	1729.945	III	90	3463.37	IV	360	5016.39	II	400	8188.02	I
775	1742.729	I	570	3478.71	IV	360	5025.66	II	250	8200.36	I
700	1745.252	I	500	3482.99	IV	550	5045.10	II	300	8210.72	I
570	1747.848	III	400	3484.96	IV	185	5281.20	I	570	8216.34	I
350	1751.218	III	90	3747.54	IV	140	5292.68	I	400	8223.14	I
650	1751.657	III	90	3754.67	III	90	5314.35	III	400	8242.39	I
150	1804.486	III	120	3771.05	III	200	5320.82	III	550	8438.74	II
200	1805.669	III	285	3838.37	II	150	5327.18	III	500	8567.74	I
150	1846.42	III	360	3919.00	II	450	5495.67	II	570	8594.00	I
90 w	1860.37	V	90	3938.52	III	285	5535.36	II	650	8629.24	I
350	1885.06	III	450	3955.85	II	650	5666.63	II	500	8655.89	I
400	1885.22	III	1000	3995.00	II	550	5676.02	II	220	8676.08	II
200	1907.99	III	150	3998.63	III	870	5679.56	II	700	8680.28	I
150	1919.55	III	200	4003.58	III	450	5686.21	II	650	8683.40	I
150	1919.77	III	360	4035.08	II	450	5710.77	II	500	8686.15	I
300	1920.65	III	550	4041.31	II	285	5747.30	II	110	8687.43	II
150	1920.84	III	360	4043.53	II	700	5752.50	I	110 h	8699.00	II
200	1921.30	III	150	4057.76	IV	240	5764.75	I	500	8703.25	I
200	2064.01	III	250	4097.33	III	265	5829.54	I	160 h	8710.54	II
250	2064.42	III	140	4099.94	I	235	5854.04	I	570	8711.70	I
120	2068.68	III	200	4103.43	III	360	5927.81	II	500	8718.83	I
90	2071.09	III	185	4109.95	I	550	5931.78	II	250	8728.89	I
90	2080.34	IV	285	4176.16	II	285	5940.24	II	200	8747.36	I
160	2095.53	II	120	4195.76	III	650	5941.65	II	500	9386.80	I
70	2096.20	II	150	4200.10	III	285	5952.39	II	570	9392.79	I
110	2096.86	II	285	4227.74	II	160	5999.43	I	250	9460.68	I
90	2117.59	III	285	4236.91	II	210	6008.47	I	200	9863.33	I
90	2121.50	III	220	4237.05	II	285	6167.76	II	160 h	9865.41	II
110	2130.18	II	450	4241.78	II	360	6379.62	II	110 h	9868.21	II
160	2142.78	II	90	4332.91	III	150	6380.77	IV	160 h	9887.39	II
90	2147.31	III	120	4345.68	III	185	6411.65	I	220 h	9891.09	II
200	2188.20	III	300	4379.11	III	210	6420.64	I	160 h	9961.86	II
150	2188.38	III	285	4432.74	II	210	6423.02	I	220 h	9969.34	II
160	2206.09	II	650	4447.03	II	210	6428.32	I	285 h	10023.27	II
160	2286.69	II	90	4510.91	III	185	6437.68	I	220 h	10035.45	II
110	2288.44	II	120	4514.86	III	235	6440.94	I	220 h	10065.15	II
220	2316.49	II	360	4530.41	II	90	6454.11	III	160 h	10070.12	II
160	2316.69	II	550	4601.48	II	185	6457.90	I	250	10105.13	I
285	2317.05	II	350	4603.73	V	120	6467.02	III	300	10108.89	I
90 w	2318.09	IV	90	4606.33	IV	300	6468.44	I	350	10112.48	I
160	2461.27	II	450	4607.16	II	265	6481.71	I	400	10114.64	I
150	2477.69	IV	360	4613.87	II	750	6482.05	II	110 h	10126.27	II
110	2496.83	II	250	4619.98	V	360	6482.70	I	250	10539.57	I
70	2496.97	II	450	4621.39	II	300	6483.75	I	200	12074.51	I
110	2520.22	II	870	4630.54	II	325	6484.80	I	380	12186.82	I
160	2520.79	II	90	4634.14	III	160	6491.22	I	225	12288.97	I
220	2522.23	II	120	4640.64	III	210	6499.54	I	290	12328.76	I
110	2590.94	II	550	4643.08	II	185	6506.31	I	310	12381.65	I
250	2645.65	IV	285	4788.13	II	750	6610.56	II	180	12438.40	I
300	2646.18	IV	450	4803.29	II	185	6622.54	I	510	12461.25	I
350	2646.96	IV	180	4847.38	I	185	6636.94	I	920	12469.62	I
250 w	2682.18	III	90	4858.82	III	235	6644.96	I	500	13429.61	I
90	2689.20	III	150	4867.15	III	185	6646.50	I	840	13581.33	I
160	2709.84	II	285	4895.11	II	235	6653.46	I	180	13587.73	I
110	2799.22	II	160	4914.94	I	210	6656.51	I	180	13602.27	I
110	2823.64	II	210	4935.12	I	185	6722.62	I	290	13624.18	I
60 1	2859.16	V	200 w	4944.56	V	210	7398.64	I	250	14757.07	I

Intensity	Wavelength/Å		Intensity	Wavelength/Å		Intensity	Wavelength/Å		Intensity	Wavelength/Å	
100	14868.87	I	5100	2838.63	I	28	5417.51	I	250	215.245	V
160	14966.60	I	2300	2844.40	I	55	5443.31	I	250	216.018	V
180	15582.27	I	1500	2850.76	I	22	5446.93	I	520	220.352	V
120 s	17516.58	I	1500	2860.96	I	22	5457.30	I	80	227.372	V
100 l	17584.86	I	9600	2909.06	I	28	5470.00	I	80	227.469	V
100	17878.26	I	2100	2912.33	I	22	5509.33	I	150	227.511	V
Osmium			2100	2919.79	I	270	5523.53	I	80	227.549	V
Os Z = 76			1100 h	2948.23	I	22	5546.82	I	80	227.634	V
9600	2001.45	I	1400	2949.53	I	80	5584.44	I	80	227.689	V
13000	2003.73	I	4400	3018.04	I	35	5620.08	I	150	231.823	V
17000	2010.15	I	1100	3030.70	I	22	5642.56	I	140	233.46	IV
29000	2018.14	I	2900	3040.90	I	28	5645.25	I	150	233.50	IV
14000	2022.76	I	120	3042.74	II	28	5680.88	I	110	233.52	IV
14000	2028.23	I	8600	3058.66	I	170	5721.93	I	200	233.56	IV
18000	2034.44	I	1100	3077.72	I	22	5765.05	I	110	233.60	IV
26000	2045.36	I	3100	3156.25	I	170	5780.82	I	90	238.36	IV
8600	2058.69	I	180	3173.93	II	40	5800.60	I	180	238.57	IV
13000	2061.69	I	150	3213.31	II	110	5857.76	I	110	248.459	V
7800	2067.21	II	1900	3232.06	I	28	5860.64	I	110	252.56	IV
4200	2070.67	II	3100	3262.29	I	65	5996.00	I	110	252.95	IV
7200	2076.95	I	3100	3267.94	I	35	6227.70	I	150	253.08	IV
14000	2079.97	I	1200	3290.26	I	22	6269.41	I	300	260.39	IV
2900	2082.54	I	7600	3301.56	I	22	6403.15	I	250	260.56	IV
2900	2089.03	I	960	3336.15	I	27	6729.56	I	80 d	264.34	III
2900	2089.21	I	960	3370.59	I	22	7145.54	I	110	264.48	III
6000	2097.60	I	620	3387.84	I	26	7602.95	I	110	266.97	III
5300	2100.63	I	620	3401.86	I	7	8041.29	I	150	266.98	III
2100	2117.66	I	620	3504.66	I	**Oxygen**			150	267.03	III
4800	2117.96	I	1200	3528.60	I	**O Z = 8**			150	277.38	III
5300	2137.11	I	1200	3560.86	I	80	124.616	V	300	279.63	IV
2600	2154.59	I	620	3598.11	I	110	135.523	V	375	279.94	IV
1300	2157.84	I	95	3604.48	II	80	138.109	V	110	285.71	IV
1200	2158.53	I	480	3670.89	I	110	139.029	V	150	285.84	IV
3100	2166.90	I	3700	3752.52	I	80	151.447	V	110	286.448	V
1100	2167.75	I	2100	3782.20	I	110	151.477	V	80	295.62	III
2100	2171.65	I	730	3876.77	I	150	151.546	V	110	295.66	III
1100	2234.61	I	1000	3963.63	I	80	164.574	V	120	295.72	III
1300	2252.15	I	730	3977.23	I	110	164.657	V	150	303.41	III
2000	2255.85	II	960	4066.69	I	80	164.709	V	150	303.46	III
1400	2264.60	I	1200	4112.02	I	80	166.235	V	140	303.52	III
1400	2282.26	II	2500	4135.78	I	150	167.99	V	160	303.62	III
500	2367.35	II	1200	4173.23	I	110	170.219	V	160	303.69	III
2600	2377.03	I	1200	4211.86	I	450	172.169	V	250	303.80	III
1700	2387.29	I	4900	4260.85	I	250	185.745	V	200	305.60	III
1100	2395.88	I	560	4293.95	I	375	192.751	V	250	305.66	III
200	2423.07	II	560	4311.40	I	450	192.799	V	190	305.70	III
1400	2424.97	I	4900	4420.47	I	520	192.906	V	300	305.77	III
110	2454.91	II	540	4550.41	I	80	193.003	V	190	305.84	III
1800	2461.42	I	670	4793.99	I	200	194.593	V	200	306.62	IV
110	2468.90	II	55	5031.83	I	150	195.86	IV	150	306.88	IV
530	2486.24	II	45	5039.12	I	200	196.01	IV	450	320.979	III
4500	2488.55	I	35	5072.88	I	80	202.161	V	300	328.45	III
2600	2498.41	I	35	5074.77	I	80	202.224	V	250	328.74	III
2400	2513.25	I	35	5079.09	I	80	202.283	V	300	345.31	III
780	2538.00	II	90	5103.50	I	80	202.334	V	110	355.14	III
1000	2542.51	I	55	5110.81	I	150	202.393	V	90	355.33	III
1000	2590.76	I	140	5149.74	I	110	203.78	V	80	355.47	III
1800	2613.06	I	40	5193.52	I	150	203.82	V	200	359.02	III
3800	2637.13	I	270	5202.63	I	100	203.85	V	190	359.22	III
1900	2644.11	I	35	5203.23	I	200	203.89	V	150	359.38	III
1900	2658.60	I	45	5255.82	I	100	203.94	V	210	373.80	III
2100	2689.82	I	55	5265.15	I	110	207.18	IV	200	374.00	III
3000	2714.64	I	40	5298.78	I	150	207.24	IV	300	374.08	III
1300	2720.04	I	110	5376.79	I	300	207.794	V	190	374.16	III
960	2770.71	I	120	5416.34	I	150	215.040	V	200	374.33	III
2800	2806.91	I	45	5416.69	I	200	215.103	V	210	374.44	III

Line Spectra of the Elements (continued): Oxygen

Intensity	Wavelength/Å		Intensity	Wavelength/Å		Intensity	Wavelength/Å		Intensity	Wavelength/Å	
450	395.558	III	80	805.810	I	30 d	2283.42	II	160	3305.15	II
300	434.98	III	240	832.762	II	30 d	2284.89	II	160	3306.60	II
800	507.391	III	600	832.927	III	110	2293.32	II	80	3312.30	III
900	507.683	III	450	833.332	II	200	2300.35	II	110	3340.74	III
1000	508.182	III	780	833.742	III	30 d	2313.05	II	230	3348.08	IV
1000	525.795	III	600	834.467	II	30 d	2316.12	II	270	3349.11	IV
250	537.83	II	600	835.096	III	30 d	2316.79	II	160	3354.27	IV
300	538.26	II	800	835.292	III	50 d	2319.68	II	200	3375.40	IV
220	539.09	II	40	877.879	I	30 d	2322.15	II	220	3377.20	II
200	539.55	II	130	921.296	IV	30 d	2339.31	II	130	3378.06	IV
150	539.85	II	160	921.366	IV	200 d	2390.44	III	360	3381.20	IV
700	553.330	IV	80	922.008	I	80	2394.33	III	360	3385.21	IV
775	554.075	IV	200	923.367	IV	110	2411.60	II	285	3390.25	II
850	554.514	IV	130	923.433	IV	80	2422.84	III	270	3396.79	IV
700	555.261	IV	90	935.193	I	80	2425.55	II	360	3403.52	IV
700	597.818	III	40	948.686	I	250	2433.56	II	220	3407.38	II
1000	599.598	III	90	971.738	I	80 d	2436.06	II	230	3409.66	IV
580	608.398	IV	40	976.448	I	80 d	2438.83	III	160	3409.84	II
110	609.70	III	160	988.773	I	80	2444.26	II	410	3411.69	IV
640	609.829	IV	40	990.204	I	300	2445.55	II	230	3413.64	IV
160	610.04	III	250	1025.762	I	200	2449.372	IV	80	3444.10	III
200	610.75	III	90	1027.431	I	200	2450.040	IV	80	3455.12	III
100	610.85	III	160	1039.230	I	200	2454.99	III	285	3470.81	II
270	616.952	IV	60	1040.942	I	200	2493.44	IV	200	3489.83	IV
150	617.005	IV	40	1152.152	I	200	2493.77	IV	160	3492.24	IV
200	617.036	IV	900	1302.168	I	200	2507.73	IV	230	3560.39	IV
520	624.617	IV	600	1304.858	I	230	2509.19	IV	270	3563.33	IV
580	625.130	IV	300	1306.029	I	200	2517.2	IV	80	3698.70	III
640	625.852	IV	200	1338.612	IV	200	2558.06	III	80	3702.75	III
1000	629.730	V	130	1342.992	IV	80	2687.53	III	80	3703.37	III
150	644.148	II	230	1343.512	IV	110	2695.49	III	110	3707.24	III
200	672.95	II	640	1371.292	V	300	2733.34	II	220	3712.75	II
150	673.77	II	160	1476.89	III	110	2747.46	II	110	3715.08	III
230	681.272	V	160 w	1506.72	V	1000	2781.01	V	315 w	3725.93	IV
70	685.544	I	285	1590.01	III	920	2786.99	V	285	3727.33	II
800	702.332	III	160	1591.33	III	775	2789.85	V	360	3729.03	IV
800	702.822	III	315 w	1643.68	V	160	2836.26	IV	410	3736.85	IV
900	702.899	III	160	1707.996	V	160	2921.45	IV	160	3739.92	II
1000	703.850	III	220	1760.12	III	200	2941.33	V	110	3744.00	III
900	718.484	II	110	1760.42	III	210	2941.65	V	230	3744.89	IV
600	718.562	II	220	1763.22	III	80	2959.68	III	360	3749.49	II
70	744.794	I	220	1764.48	III	265	2972.29	I	150	3754.67	III
700	758.678	V	750	1767.78	III	250	2983.78	III	80	3757.21	III
640	759.441	V	550	1768.24	III	80	3017.63	III	250	3759.87	III
580	760.228	V	360	1771.67	III	80	3023.45	III	110	3791.26	III
775	760.445	V	110	1773.00	III	80	3043.02	III	160	3803.14	II
640	761.128	V	110	1773.85	III	80	3047.13	III	120	3823.41	I
700	762.003	V	220	1779.16	III	110	3059.30	III	450	3911.96	II
70	770.793	I	160	1781.03	III	460	3063.42	IV	160	3919.29	II
90	771.056	I	160	1784.85	III	410	3071.61	IV	185	3947.29	I
520	774.518	V	220	1789.66	III	80	3121.71	III	160	3947.48	I
70	775.321	I	110	1848.26	III	160	3122.62	II	140	3947.59	I
200	779.734	IV	110	1856.62	III	220	3129.44	II	220	3954.37	II
315	779.821	IV	285	1872.78	III	110	3132.86	III	100	3954.61	I
360	779.912	IV	285	1872.87	III	450	3134.82	II	200	3961.59	III
200	779.997	IV	285	1874.94	III	285	3138.44	II	450	3973.26	II
640	787.711	IV	160	1920.04	III	160	3144.66	V	220	3982.20	II
520	790.109	IV	110	1920.75	III	160	3209.66	IV	160	4069.90	II
700	790.199	IV	110	1921.52	III	80	3238.57	III	285	4072.16	II
70	791.973	I	220	1923.49	III	200	3260.98	III	450	4075.87	II
300	796.66	II	110	1923.82	III	300	3265.46	III	80 d	4083.91	II
200	802.200	IV	110	1926.94	III	80	3267.31	III	50 d	4087.14	II
160	802.255	IV	360	2013.27	III	220	3270.98	II	150 d	4089.27	II
90	804.267	I	160	2026.96	III	220	3273.52	II	110	4097.24	II
70	804.848	I	220	2045.67	III	220	3277.69	II	220	4105.00	II
70	805.295	I	160	2052.74	III	360	3287.59	II	285	4119.22	II

Intensity	Wavelength/Å		Intensity	Wavelength/Å		Intensity	Wavelength/Å		Intensity	Wavelength/Å	
100	4123.99	V	130	6500.24	V	100	9622.13	I	4000	1914.62	III
160	4132.81	II	80	6604.91	I	120	9625.29	I	1000	1930.33	III
50	4146.06	II	100	6653.83	I	160	9677.38	I	2000	1941.64	III
220	4153.30	II	360	7001.92	I	80	9694.66	I	800	2002.16	III
285	4185.46	II	450	7002.23	I	65	9694.91	I	1000	2004.47	III
450	4189.79	II	210	7156.70	I	235	9741.50	I	500	2055.11	III
80	4233.27	I	400	7254.15	I	235	9760.65	I	500	2149.82	III
50 d	4253.74	II	450	7254.45	I	120	9909.05	I	500	2177.55	III
50 d	4253.98	II	320	7254.53	I	140	9936.98	I	500	2177.63	III
50 d	4275.47	II	210	7476.44	I	120	9940.41	I	100 r	2231.59	II
50 d	4303.78	II	100	7477.24	I	160	9995.31	I	200 r	2296.53	II
285	4317.14	II	120	7479.08	I	120 d	10421.18	I	100	2426.87	II
160	4336.86	II	120	7480.67	I	590	11286.34	I	100	2430.94	II
220	4345.56	II	100	7706.75	I	640	11286.91	I	100	2433.11	II
285	4349.43	II	870	7771.94	I	490	11287.02	I	100	2435.32	II
220	4366.90	II	810	7774.17	I	490	11287.32	I	150	2446.17	II
100	4368.25	I	750	7775.39	I	490	11295.10	I	1100	2447.91	I
220	4395.95	II	80	7886.27	I	540	11297.68	I	100	2457.29	II
450	4414.91	II	100	7943.15	I	590	11302.38	I	150	2469.29	II
285	4416.98	II	100	7947.17	I	265	11358.69	I	100	2471.18	II
160	4448.21	II	235	7947.55	I	490	12464.02	I	1700	2476.42	I
160	4452.38	II	210	7950.80	I	450	12570.04	I	250	2486.52	II
50	4465.45	II	185	7952.16	I	120	12990.77	I	300	2488.92	II
50 d	4466.28	II	110	7981.94	I	160	13076.91	I	200	2498.81	II
50	4467.83	II	135	7982.40	I	700	13163.89	I	150	2505.73	II
50	4469.41	II	190	7986.98	I	750	13164.85	I	150	2551.84	II
360	4590.97	II	135	7987.33	I	640	13165.11	I	150	2565.51	II
285	4596.17	II	250	7995.07	I	160	16212.06	I	100	2569.56	II
80 d	4609.39	II	400	8221.82	I	120	17966.70	I	150	2658.75	II
160	4638.85	II	265	8227.65	I	590	18021.21	I	1900	2763.09	I
360	4641.81	II	265	8230.02	I	120	18041.48	I	150 h	2776.85	II
450	4649.14	II	325	8233.00	I	120	18042.19	I	100 h	2787.92	II
160	4650.84	II	120	8235.35	I	120	18046.23	I	200	2854.59	II
360	4661.64	II	120	8426.16	I	140	18229.23	I	100 h	2871.37	II
285	4676.23	II	810	8446.25	I	540	18243.63	I	100 h	2878.01	II
220	4699.21	II	1000	8446.36	I	140	26173.56	I	520	2922.49	I
285	4705.36	II	935	8446.76	I	**Palladium**			650	3002.65	I
160	4924.60	II	325	8820.43	I	**Pd Z = 46**			1500	3027.91	I
230 w	4930.27	V	160 d	9057.01	I	200	705.49	III	1100	3065.31	I
220	4943.06	II	120	9118.29	I	200	727.72	III	2600	3114.04	I
135	5329.10	I	80	9134.71	I	500	763.06	III	11000	3242.70	I
160	5329.68	I	80	9150.14	I	500	766.42	III	2700	3251.64	I
190	5330.74	I	80	9151.48	I	2000	781.02	III	3500	3258.78	I
90	5435.18	I	235	9156.01	I	500	794.08	III	3600	3302.13	I
110	5435.78	I	450	9260.81	I	500	797.52	III	5000	3373.00	I
135	5436.86	I	490	9260.84	I	500	800.03	III	24000	3404.58	I
120	5577.34	I	450	9260.94	I	500	800.10	III	13000	3421.24	I
110	5592.37	III	400	9262.58	I	500	803.67	III	5000	3433.45	I
130	5597.91	V	540	9262.67	I	500	825.35	III	6400	3441.40	I
160	5958.39	I	590	9262.77	I	500	840.58	III	7700	3460.77	I
190	5958.58	I	490	9265.94	I	500	856.47	III	10000	3481.15	I
80	5995.28	I	640	9266.01	I	500	864.04	III	2000	3489.77	I
160	6046.23	I	185	9399.19	I	500	880.59	III	12000	3516.94	I
190	6046.44	I	120	9481.16	I	500	888.84	III	12000	3553.08	I
110	6046.49	I	120 d	9482.88	I	1000	889.29	III	4500	3571.16	I
100	6106.27	I	235	9487.43	I	300	1596.89	III	20000	3609.55	I
400	6155.98	I	140	9492.71	I	500	1741.62	III	20000	3634.70	I
450	6156.77	I	265	9497.97	I	4000	1782.55	III	5500	3690.34	I
490	6158.18	I	160	9499.30	I	400	1843.49	III	1400	3718.91	I
80	6256.83	I	235	9505.59	I	1500	1851.59	III	1500	3799.19	I
100	6261.55	I	210	9521.96	I	2000	1852.27	III	1500	3832.29	I
100	6366.34	I	120	9523.36	I	1000	1859.21	III	2200	3894.20	I
100	6374.32	I	120	9523.96	I	1500	1874.63	III	1500	3958.64	I
320	6453.60	I	100	9528.28	I	2000	1885.83	III	290	4087.34	I
360	6454.44	I				1000	1887.40	III	2500	4212.95	I
400	6455.98	I				1500	1891.34	III	180	4473.59	I

Intensity	Wavelength/Å		Intensity	Wavelength/Å		Intensity	Wavelength/Å		Intensity	Wavelength/Å	
160	5163.84	I	15	1373.500	I	520	3204.04	V	150	6992.690	III
120	5295.63	I	10	1374.732	I	300	3219.307	III	100	7102.200	I
55	5542.80	I	15	1377.080	I	400	3233.602	III	100	7158.367	I
75	5670.07	I	15	1377.937	I	650	3347.736	IV	180	7165.465	I
55 h	5695.09	I	25	1379.429	I	570	3364.467	IV	180	7175.102	I
65	6784.52	I	25	1381.469	I	400	3371.122	IV	180	7176.660	I
75	7368.12	I	15	1381.637	I	300	3957.641	III	200	7443.657	IV
120	7764.03	I	500	1484.507	IV	350	3978.307	III	250	7845.63	II
45	7915.80	I	400	1487.788	IV	400	4059.312	III	100	8046.801	I
55	8132.82	I	350	1502.228	III	300	4080.084	III	150	8113.528	III
45	8300.83	I	80	1532.51	I	500	4222.195	III	140	8278.058	I
65	8761.35	I	120	1535.90	II	350	4246.720	III	100	8367.856	I
Phosphorus			450	1610.50	V	400	4420.71	II	140	8531.475	I
P Z = 15			150	1618.632	III	250	4479.776	III	140	8613.835	I
250	328.78	V	200	1618.907	III	250	4540.288	IV	180	8637.578	I
150	359.899	IV	140	1671.070	I	250	4541.112	IV	400	8741.529	I
500	388.318	IV	100	1671.510	I	500	4588.04	II	100	8872.174	I
250	389.50	V	180	1671.680	I	500	4589.86	II	180	9175.819	I
300	390.70	V	140	1672.035	I	600	4602.08	II	950	9193.85	I
300	445.158	IV	140	1672.474	I	300	4626.70	II	600	9278.88	I
375	475.60	V	600	1674.591	I	300	4658.31	II	1250	9304.94	I
120	498.180	III	600	1679.695	I	500	4943.53	II	500	9323.50	I
520	542.57	V	140	1685.976	I	300	4954.39	II	950	9435.069	I
600	544.92	V	100	1694.028	I	300	4969.71	II	950	9441.86	I
200	569.853	III	100	1694.486	I	100	5079.381	I	600	9452.83	I
200	581.831	III	100	1706.376	I	100	5098.221	I	1250	9493.56	I
350	629.008	IV	100	1707.553	I	100	5100.974	I	1700	9525.73	I
400	629.914	IV	600	1774.951	I	140	5109.628	I	1500	9545.18	I
500	631.779	IV	500	1782.838	I	140	5154.844	I	280	9556.81	I
450	673.90	V	400	1787.656	I	180	5162.290	I	1700	9563.439	I
10	810.24	II	140	1834.801	I	300	5253.52	II	280	9593.50	I
650	823.179	IV	140	1847.165	I	140	5293.539	I	750	9609.04	I
700	824.730	IV	100	1849.820	I	400	5296.13	II	400	9638.939	I
800	827.932	IV	140	1851.194	I	250	5316.07	II	500	9676.24	I
300	847.669	III	100	1852.069	I	300	5344.75	II	180	9706.533	I
350	855.624	III	500	1858.886	I	180	5345.851	I	1500	9734.750	I
500	859.652	III	400	1859.393	I	100	5364.631	I	280	9736.680	I
10	865.44	II	140	1864.348	I	250	5378.20	II	1500	9750.77	I
450	865.45	V	650	1888.523	IV	300	5386.88	II	600	9790.21	I
600	871.39	V	180	1905.481	I	400	5425.91	II	1700	9796.85	I
700	877.476	IV	140	1906.403	I	100	5428.094	I	280	9834.80	I
300	913.971	III	280	1907.665	I	400	5450.74	II	400	9903.68	I
300	917.120	III	280	2023.489	I	140	5458.305	I	280	9976.67	I
350	918.665	III	180	2024.516	I	180	5477.672	I	229	10084.27	I
1000	950.655	IV	400	2032.432	I	140	5477.860	I	458	10511.58	I
250	1003.598	III	400	2033.477	I	140	5478.267	I	962	10529.52	I
570	1025.563	IV	400	2135.465	I	100	5514.774	I	1235	10581.57	I
500	1028.096	IV	400	2136.182	I	100	5516.997	I	415	10596.90	I
570	1030.517	IV	400	2149.145	I	250	5588.34	II	435	10681.40	I
500	1033.111	IV	280	2152.940	I	500	6024.18	II	265	10813.13	I
500	1035.517	IV	500	2154.080	I	400	6034.04	II	764	11183.23	I
900	1117.98	V	180	2235.732	I	500	6043.12	II	402	11186.75	I
570	1118.551	IV	450	2440.93	V	250	6055.50	II	479	14241.64	I
700	1128.01	V	250	2478.256	IV	150	6083.409	III	256	14307.83	I
20	1249.82	II	750	2533.976	I	350	6087.82	II	714	15711.52	I
20	1301.87	II	950	2535.603	I	180	6097.690	I	228	15962.53	I
20	1304.47	II	750	2553.262	I	350	6165.59	II	296	16254.77	I
15	1304.68	II	500	2554.915	I	500	6199.024	I	203	16292.97	I
35	1305.48	II	250	2605.506	IV	180	6210.499	I	1627	16482.92	I
60	1310.70	II	300	2632.713	III	140	6375.681	I	588	16590.07	I
500	1334.808	III	400	2644.295	IV	100	6388.579	I	225	16613.05	I
650	1344.327	III	400	2728.770	IV	250	6435.32	II	221	16738.68	I
300	1344.845	III	500	2739.309	IV	600	6459.99	II	419	16803.39	I
500	1366.695	IV	250	2739.872	IV	600	6503.46	II	471	17112.48	I
15	1372.033	I	450	2978.55	V	600	6507.97	II	289	17286.91	I
400	1372.674	IV	700	3175.09	V	100	6717.411	I	299	17423.67	I

Intensity	Wavelength/Å		Intensity	Wavelength/Å		Intensity	Wavelength/Å		Intensity	Wavelength/Å	
287	23844.97	I	100	2418.06	I	40 h	2865.05	II	14	4445.55	I
311	29097.16	I	50	2424.87	II	40 h	2875.85	II	25	4498.76	I
Platinum			80	2428.04	I	100 h	2877.52	II	12	4520.90	I
Pt Z = 78			50	2428.20	I	25	2888.20	I	35	4552.42	I
30	1621.66	II	25	2429.10	I	25	2893.22	I	12	4879.53	I
30	1723.13	II	180	2436.69	I	600	2893.86	I	14	5044.04	I
30	1751.70	II	650	2440.06	I	300	2897.87	I	30	5059.48	I
50 r	1777.09	II	60	2450.97	I	60	2905.90	I	35	5227.66	I
30	1781.86	II	440	2467.44	I	120	2912.26	I	40	5301.02	I
30	1879.09	II	35	2471.01	I	120	2913.54	I	12	5368.99	I
40	1883.05	II	1000	2487.17	I	70	2919.34	I	12	5390.79	I
50	1889.52	II	25	2488.74	II	30	2921.38	I	14	5475.77	I
50	1911.70	II	200	2490.12	I	1700	2929.79	I	14	5478.50	I
30	1929.25	II	160	2495.82	I	30	2942.76	I	6	5763.57	I
30	1929.68	II	240	2498.50	I	30	2944.75	I	20	5840.12	I
30	1939.80	II	50	2505.93	I	25	2959.10	I	8	5844.84	I
30	1949.90	II	120	2508.50	I	60	2960.75	I	6	6026.04	I
30	1983.74	II	50	2514.07	I	1800	2997.97	I	7	6318.37	I
40	2014.93	II	60	2515.03	I	35	3001.17	II	8	6326.58	I
3200	2030.63	I	240	2515.58	I	220	3002.27	I	9	6523.45	I
4400	2032.41	I	140	2524.30	I	30	3017.88	I	10	6710.42	I
100	2036.46	II	40	2529.41	I	30 h	3031.22	II	20	6760.02	I
40	2041.57	II	50	2536.49	I	130	3036.45	I	60	6842.60	I
5500	2049.37	I	160	2539.20	I	800	3042.64	I	20	7113.73	I
1500	2067.50	I	18	2549.46	I	3200	3064.71	I	10	8224.74	I
3000	2084.59	I	50	2552.25	I	30	3071.94	I	**Plutonium**		
1000	2103.33	I	50	2596.00	I	130	3100.04	I	**Pu Z = 94**		
30	2115.57	II	70	2603.14	I	320	3139.39	I	10000	2806.11	II
950	2128.61	I	30	2616.76	II	140	3156.56	I	10000	2950.06	II
30	2130.69	II	50	2619.57	I	120	3200.71	I	10000	3000.31	II
1900	2144.23	I	30	2625.34	II	320	3204.04	I	10000	3200.23	II
100	2144.24	II	1100	2628.03	I	30	3230.29	I	10000	3418.88	II
600	2165.17	I	130	2639.35	I	20	3233.42	I	10000	3805.93	I
1500	2174.67	I	1000	2646.89	I	20	3250.36	I	10000	4097.12	I
30	2190.32	II	500	2650.86	I	40	3251.98	I	10000	4170.95	I
400	2202.22	I	20	2658.17	I	160	3255.92	I	10000	4367.41	I
50 h	2202.58	II	2800	2659.45	I	25	3268.42	I	10000	5590.54	I
320	2222.61	I	40	2674.57	I	25	3281.97	I	10000	7068.90	I
50 h	2233.11	II	440	2677.15	I	120	3290.22	I	10000	8691.94	I
30 h	2240.99	II	200	2698.43	I	500	3301.86	I	3000	9533.07	I
100	2245.52	II	2000	2702.40	I	60	3315.05	I	3000	12144.46	I
150	2249.30	I	1600	2705.89	I	35	3323.80	I	3000	16897.38	I
30	2251.52	II	60	2713.13	I	340	3408.13	I	**Polonium**		
30 h	2251.92	II	1300	2719.04	I	35	3427.93	I	**Po Z = 84**		
190	2268.84	I	130	2729.92	I	60	3483.43	I	1500 w	2450.08	I
30 h	2271.72	II	1800	2733.96	I	160	3485.27	I	1500 w	2558.01	I
280	2274.38	I	70	2738.48	I	120	3628.11	I	2500 w	3003.21	I
50 h	2287.50	II	70	2747.61	I	70	3638.79	I	1200	4170.52	I
30	2288.20	II	80	2753.86	I	70	3643.17	I	800	4493.21	I
150	2289.27	I	200	2754.92	I	50	3663.10	I	500	8618.26	I
150	2292.40	I	30	2769.84	I	80	3671.99	I	**Potassium**		
240	2308.04	I	500	2771.67	I	80	3674.04	I	**K Z = 19**		
50	2310.96	II	40	2773.24	I	35	3699.91	I	100	214.35	V
90	2315.50	I	20	2774.00	I	18	3706.53	I	150	271.82	IV
220	2318.29	I	50	2774.77	II	80	3818.69	I	100	273.06	IV
100	2326.10	I	50	2793.27	I	40	3900.73	I	150	282.35	V
170	2340.18	I	100	2794.21	II	110	3922.96	I	150	293.33	V
280	2357.10	I	40 h	2799.98	II	35	3948.40	I	300	294.84	V
180	2368.28	I	140	2803.24	I	100	3966.36	I	200	296.17	V
50	2377.28	II	10	2808.51	I	20	3996.57	I	200	297.06	V
130	2383.64	I	50	2818.25	I	110	4118.69	I	200	300.25	V
40	2386.81	I	30 h	2822.27	II	80	4164.56	I	200	300.50	V
120	2389.53	I	1400	2830.30	I	40	4192.43	I	200	311.24	V
35	2396.17	I	70	2834.71	I	18	4327.06	I	250	312.77	V
70	2401.87	I	16	2853.11	I	18	4391.83	I	200	315.18	V
200	2403.09	I	80 h	2860.68	II	80	4442.55	I	250	327.38	V

Line Spectra of the Elements (continued): Potassium—Praseodymium

The last (fifth) Intensity/Wavelength pair lists **Praseodymium, Pr Z = 59**.

Intensity	Wavelength/Å	Intensity	Wavelength/Å	Intensity	Wavelength/Å	Intensity	Wavelength/Å	Intensity	Wavelength/Å (Pr, Z = 59)
25	330.68 III	75	434.72 III	15	874.04 III	11	8503.45 I	7000	865.90 V
300	340.46 IV	50	435.68 III	6	2550.02 III	10	8505.11 I	5000	869.17 V
150	340.74 IV	250	438.02 V	5	2635.11 III	4	8763.96 I	2000	1228.59 IV
30	341.92 III	25	441.81 II	5	2689.90 III	3	8767.05 I	5000	1293.22 IV
15	348.00 III	200	442.30 IV	5	2938.45 III	13	8902.19 I	5000	1295.28 IV
200	349.50 V	300	443.57 IV	5	2986.20 III	12	8904.02 I	5000	1321.36 IV
300	354.93 IV	75	444.34 III	6	2992.42 III	5	8923.31 I	5000	1333.57 IV
150	356.26 IV	200	445.61 IV	6	3052.07 III	4	8925.44 I	5000	1354.66 IV
300	359.73 IV	250	446.83 IV	5	3056.84 III	7	9347.24 I	2000	1360.64 IV
200	359.91 IV	75	448.60 III	5	3062.18 II	3	9349.25 I	2000	1365.77 IV
250	362.08 IV	750	448.60 IV	4	3101.79 I	6	9351.59 I	5000	1374.41 IV
150	362.15 IV	200	449.71 V	3	3102.04 I	15	9595.70 I	5000	1435.56 IV
150	363.02 IV	200	452.90 V	7	3217.16 I	14	9597.83 I	2000	1520.98 IV
500	372.15 V	250	455.67 V	6	3217.62 I	6	9949.67 I	5000	1574.55 IV
200	372.46 V	400	456.33 IV	11	3446.37 I	5	9954.14 I	5000	1575.10 IV
200	372.77 V	400	456.33 V	10	3447.38 I	9	10479.63 I	3000	1578.38 IV
300	375.96 IV	75	466.79 III	3	3648.84 I	5	10482.15 I	2000	1622.30 IV
300	375.96 V	100	470.09 III	4	3648.98 I	8	10487.11 I	10000	1884.87 IV
250	377.76 V	75	471.57 III	18	4044.14 I	17	11019.87 I	2000	2083.23 IV
30	379.12 III	45	474.92 III	17	4047.21 I	16	11022.67 I	3300	2246.20 V
300	379.12 V	10	476.03 II	10	4641.88 I	17	11690.21 I	2000 c	2378.98 IV
300	379.88 IV	40	479.18 III	11	4642.37 I	16	11769.62 I	40 h	2598.04 II
25	380.48 III	10	482.11 III	4	4740.91 I	17	11772.83 I	100 h	2707.37 II
250	380.48 IV	10	482.41 III	6	4744.35 I		12522.11 I	60	2760.35 II
200	381.70 IV	200	482.71 V	5	4753.93 I		13377.86 I	270	3168.24 II
30	382.23 III	200	483.75 V	7	4757.39 I		13397.09 I	200 d	3195.99 II
300	382.23 IV	30	495.14 II	5	4786.49 I		15163.08 I	190	3219.48 II
150	382.49 IV	75	497.10 III	7	4791.05 I		15168.40 I	200	3584.21 II
200	382.65 IV	10	514.94 III	6	4799.75 I		40158.37 I	250	3645.66 II
300	382.91 IV	50	520.61 III	8	4804.35 I			250	3646.30 II
250	384.10 IV	250	523.00 IV	7	4849.86 I			370	3668.83 II
200	386.61 IV	25	523.79 III	8	4856.09 I			290	3714.05 II
300	387.80 V	200	526.45 IV	8	4863.48 I			410	3739.18 II
250	388.92 IV	150	527.62 IV	9	4869.76 I			680	3761.87 II
250	389.07 IV	40	529.80 III	8	4942.02 I			680	3800.30 II
250	389.07 V	15	539.71 III	9	4950.82 I				
250	390.11 V	15	546.12 III	9	4956.15 I				
250	390.42 IV	750	580.32 V	10	4965.03 I				
300	390.57 IV	250	585.51 V	10	5084.23 I				
200	391.46 IV	500	586.32 V	11	5097.17 I				
200	392.47 IV	30	600.77 II	11	5099.20 I				
500	393.14 IV	250	602.27 V	12	5112.25 I				
250	395.40 V	400	603.43 V	12	5323.28 I				
200	398.36 V	25	607.93 II	13	5339.69 I				
15	398.63 III	30	612.62 II	12	5342.97 I				
200	398.88 V	250	638.67 V	14	5359.57 I				
200	399.75 V	750	646.19 IV	16	5782.38 I				
400	400.21 IV	300	687.50 V	17	5801.75 I				
20	402.10 III	20	708.84 III	15	5812.15 I				
300	402.91 IV	300	720.43 V	17	5831.89 I				
250	403.97 IV	400	724.42 V	8	6120.27 II				
150	404.41 IV	600	731.86 V	7	6307.29 II				
30	406.48 III	500	737.14 IV	19	6911.08 I				
250	408.08 IV	500	741.95 IV	12	6936.28 I				
40	408.96 III	500	745.26 IV	20	6938.77 I				
50	413.79 III	400	746.35 IV	7	6964.18 I				
30	414.87 III	300	749.99 IV	12	6964.67 I				
250	415.05 V	150	754.19 IV	25	7664.90 I				
200	415.79 V	400	754.67 IV	24	7698.96 I				
30	416.00 III	20	765.31 III	5	7955.37 I				
150	417.28 IV	30	765.64 III	4	7956.83 I				
30	417.54 III	150	770.29 V	7	8078.11 I				
30	418.62 III	150	771.46 V	6	8079.62 I				
400	422.18 V	35	778.53 III	9	8250.18 I				
300	425.16 V	20	872.31 III	8	8251.74 I				
500	425.59 V	10	873.86 III	3	8390.22 I				

Intensity	Wavelength/Å		Intensity	Wavelength/Å		Intensity	Wavelength/Å		Intensity	Wavelength/Å	
390	3811.84	II	560	4096.82	II	680	5259.73	II	1000 r	3998.96	II
1300 h	3816.02	II	380	4098.40	II	340 c	5292.02	II	1000	4417.96	II
680	3818.28	II	2900 c	4100.72	II	340	5292.62	II	900 r	4728.36	I
310	3821.80	II	1700 c	4118.46	II	430	5322.76	II	900	6100.21	I
960	3830.72	II	340	4130.77	II	65	5509.15	II	1000 d	6520.45	I
480	3840.99	II	1500 c	4141.22	II	150	5535.17	II	**Protactinium**		
580	3846.59	II	2700	4143.11	II	110	5623.05	II	**Pa Z = 91**		
1200	3850.79	II	1700 c	4164.16	II	90	5624.45	II	3000	2599.16	II
720 c	3851.55	II	620	4171.82	II	90	5756.17	II	3000	2699.22	II
960	3852.80	II	730	4172.25	II	90	5779.28	I	3000	2822.79	II
480 c	3865.45	II	5200	4179.39	II	160 d	5815.17	II	3000 h	2871.42	II
480	3876.19	II	2500	4189.48	II	90	5823.72	II	3000 h	2891.14	II
1700 c	3877.18	II	560 c	4191.60	II	90	5859.68	II	3000 l	3011.10	II
680	3880.47	II	2500 c	4206.72	II	160	5939.90	II	3000 s	3033.59	II
440 c	3885.19	II	500	4208.32	II	7000 w	5956.05	III	3000 l	3071.24	II
440 c	3889.34	II	320	4211.86	II	90	5956.60	II	3000 l	3093.23	II
770 c	3908.05	II	320	4217.81	II	110	5967.82	II	3000 l	3126.23	II
630	3912.90	II	3800	4222.93	II	90	6006.33	II	3000 l	3146.28	II
310	3913.55	II	3800	4225.35	II	150	6017.80	II	3000 l	3170.89	II
1300 c	3918.85	II	320	4233.11	II	150	6025.72	II	3000 l	3171.54	II
420	3919.63	II	320 c	4236.15	II	140	6055.13	I	3000 l	3240.58	II
960	3925.47	II	960	4241.01	II	65	6087.52	II	3000	3274.46	II
480	3927.46	II	340	4243.51	II	9000 w	6090.02	III	3000 l	3332.69	II
370	3929.29	II	840 c	4247.63	II	65	6114.38	II	3000 s	3346.66	II
370	3935.82	II	500	4254.40	II	65	6148.23	I	3000 l	3452.82	II
730 c	3947.63	II	320	4269.09	II	5000	6160.24	III	3000	3504.97	I
900 c	3949.43	II	790 c	4272.27	II	190	6161.18	II	3000 s	3530.65	II
900 c	3953.51	II	470 c	4280.07	II	270	6165.94	II	3000	3570.56	I
380	3956.75	II	790 c	4282.42	II	45	6244.35	II	3000	3571.82	I
470	3962.45	II	450 c	4298.98	II	110	6281.28	II	3000	3618.07	I
560	3964.26	II	1500	4305.76	II	55 c	6359.03	I	10000	3636.52	I
1600 c	3964.81	II	1300	4333.97	II	55	6411.23	I	3000	3702.74	I
560 c	3966.57	II	360	4338.70	II	45	6429.63	II	3000	3752.67	I
500	3971.16	II	620 cw	4344.30	II	45	6431.84	II	3000	3873.35	I
320	3971.67	II	470 c	4347.49	II	45	6486.55	I	3000	3931.83	I
620 c	3972.14	II	340	4350.40	II	45	6566.77	I	3000 s	3952.62	II
320	3974.85	II	450	4354.91	II	55	6616.67	I	10000 l	3957.85	II
1300 c	3989.68	II	410 c	4359.79	II	75	6656.83	II	3000 s	3970.07	II
340	3992.16	II	1200	4368.33	II	55	6673.41	II	3000	3981.82	I
1600	3994.79	II	320	4371.62	II	75	6673.78	II	10000	3982.23	I
560 c	3997.04	II	430	4405.83	II	35 c	6747.09	I	3000 l	4012.96	II
320	3999.12	II	1700	4408.82	II	55 cw	6798.60	I	3000 s	4018.21	II
620 c	4000.17	II	410	4413.77	II	35 cw	6827.60	II	3000	4030.16	II
730	4004.70	II	1200 c	4429.13	II	7000	6910.14	III	3000 s	4046.93	II
1900	4008.69	II	730	4449.83	II	40	7021.51	II	10000 s	4056.20	II
620	4010.60	II	960	4468.66	II	5000	7030.39	III	10000 s	4070.40	II
730	4015.39	II	1100	4496.46	II	4500	7076.62	III	3000 l	4176.18	II
620	4020.96	II	790	4510.15	II	20	7114.55	I	10000 l	4217.23	II
470	4022.71	II	340 c	4534.15	II	24	7227.70	II	10000 s	4248.08	II
360	4025.54	II	340	4535.92	II	16	7407.56	II	3000 s	4291.34	II
360 c	4029.72	II	270 c	4628.74	II	20 c	7451.74	II	3000 s	4601.43	II
730 c	4031.75	II	270 c	4672.09	II	14	7541.02	II	3000 l	6035.78	I
960	4033.83	II	290	4695.77	I	20	7645.66	II	3000	6162.56	I
730	4038.45	II	250	4736.69	I	16	7721.84	I	3000 l	6358.61	I
470	4039.34	II	200	4924.60	I	14	7871.67	I	3000	6379.25	I
1300	4044.81	II	320	4939.74	I	14	8067.44	I	3000 l	6438.97	I
340	4047.08	II	380	4951.37	I	10 cw	8122.78	II	3000 h	6792.75	I
450	4051.13	II	270	5034.41	II	11	8141.10	I	10000	6945.72	I
2200	4054.88	II	320	5045.52	I	5000 w	8602.74	III	3000	6960.09	I
2200	4056.54	II	360	5110.38	II	10	8714.59	II	3000 h	6961.78	I
450	4058.80	II	560	5110.76	II	**Promethium**			3000 s	6992.73	I
3400	4062.81	II	410	5129.52	II	**Pm Z = 61**			3000	7076.27	I
500 c	4079.77	II	620	5173.90	II	1000	3892.15	II	3000 h	7100.94	I
500 c	4080.98	II	360	5206.55	II	1000	3910.26	II	10000 s	7114.89	I
790	4081.85	II	360	5219.05	II	1000	3919.10	II	3000 h	7171.55	I
500	4083.34	II	560	5220.11	II	1000	3957.74	II	3000	7227.13	I

Intensity	Wavelength/Å		Intensity	Wavelength/Å		Intensity	Wavelength/Å		Intensity	Wavelength/Å	
3000	7318.79	I	680	2306.54	I	270	2649.05	I	440	3177.71	I
10000 l	7368.25	I	800	2322.49	I	660	2651.90	I	260	3178.61	I
3000 h	7471.89	I	300	2328.66	I	400	2654.12	I	600	3182.87	I
10000 h	7493.15	I	860	2344.78	I	220	2663.63	I	1100	3184.76	I
3000 h	7558.26	I	230	2349.39	I	940	2674.34	I	1100	3185.57	I
10000 h	7608.20	I	680	2352.07	I	220	2688.53	I	260	3190.78	I
10000	7626.79	I	250	2356.50	I	1300	2715.47	I	260	3192.36	I
10000 s	7635.18	I	1200	2365.90	I	220	2732.21	I	220	3198.58	I
10000	7669.34	I	570	2367.68	I	610	2733.04	II	1100 c	3204.25	I
3000	7679.20	I	520	2369.27	I	220	2758.00	I	380	3235.94	I
10000 h	7749.19	I	220	2370.76	II	210	2763.79	I	600	3258.85	I
3000	7872.95	I	320	2375.07	I	310	2767.74	I	600	3259.55	I
3000 l	7945.56	I	370	2379.77	I	220	2768.85	I	300	3268.89	I
10000	8039.34	I	340	2388.57	I	220	2769.32	I	280	3296.70	I
10000 h	8099.84	I	230	2393.65	I	350	2770.42	I	280	3296.99	I
10000	8199.04	I	320	2394.37	I	550	2783.57	I	280	3301.60	I
10000	8271.87	I	320	2396.79	I	220	2791.29	I	240	3302.23	I
3000 s	8358.98	I	210 d	2400.72	I	220	2814.68	I	320	3303.21	II
3000 s	8369.60	I	210	2401.68	I	880	2819.95	I	280	3303.75	I
3000 h	8441.04	I	1500	2405.06	I	310	2834.08	I	240	3313.95	I
10000 h	8532.66	I	740	2405.60	I	220	2843.00	I	600	3322.48	I
10000 s	8572.96	I	320	2406.70	I	270	2850.98	I	2000	3338.18	I
3000 h	8639.91	I	270	2410.37	I	240	2867.19	I	1600	3342.24	I
3000 h	8653.51	I	1200	2419.81	I	2900	2887.68	I	810	3344.32	I
10000	8735.27	I	300	2421.73	I	490	2896.01	I	320	3346.20	I
3000	10923.32	I	300	2421.88	I	830 c	2902.48	I	240 d	3356.33	I
10000	11791.73	I	2500	2428.58	I	210	2905.58	I	240	3377.74	I
10000	14344.76	I	490	2431.54	I	550	2909.82	I	320	3379.06	II
3000	18478.61	I	420	2432.18	I	830 c	2927.42	I	320	3379.70	I
Radium			340 c	2441.47	I	270	2930.61	I	240	3389.43	I
Ra Z = 88			230	2442.51	I	440	2943.14	I	4000	3399.30	I
100	3649.55	II	250	2444.94	I	270	2962.27	I	650	3404.72	I
200	3814.42	II	610	2446.98	I	720	2965.11	I	650	3405.89	I
100	4340.64	II	610	2449.71	I	1500	2965.76	I	240	3408.67	I
100	4682.28	II	390	2461.20	I	310	2976.29	I	320	3409.83	I
100	4825.91	I	800 c	2461.84	II	210	2978.15	I	320	3417.77	I
50	5660.81	I	1200	2483.92	I	220	2980.82	I	810	3419.41	I
50	7141.21	I	390	2485.81	I	220	2982.19	I	8000	3424.62	I
50	8019.70	II	980	2487.33	I	220	2988.47	I	400	3426.19	I
Radon			370	2496.04	I	1800	2992.36	I	300	3427.61	I
Rn Z = 86			370	2501.72	I	5500	2999.60	I	320	3437.71	I
100	4349.60	I	570	2502.35	II	350	3001.14	I	400	3449.37	I
200	7055.42	I	230	2504.60	II	220	3004.14	I	240	3453.50	I
100	7268.11	I	270	2505.94	I	500	3016.02	I	16000 c	3451.88	I
300	7450.00	I	1800 c	2508.99	I	300	3016.49	I	55000 c	3460.46	I
100	7809.82	I	570	2520.01	I	380	3030.45	I	40000 c	3464.73	I
100	8099.51	I	540	2521.50	I	240	3047.25	I	400	3467.96	I
100	8270.96	I	370	2534.80	I	1600	3067.40	I	240	3476.44	I
100	8600.07	I	570	2540.51	I	320	3069.94	I	400	3480.38	I
Rhenium			740 d	2544.74	I	260	3071.16	I	320	3480.85	I
Re = 75			370	2545.48	I	550	3082.43	I	240	3482.23	I
25000	2003.53	I	300	2552.02	I	340	3088.76	I	560	3503.06	I
16000	2017.87	I	370	2554.63	II	700	3100.67	I	320	3516.65	I
27000	2049.08	I	1000	2556.51	I	700	3108.81	I	320	3517.33	I
10000	2085.59	I	250	2559.08	I	340	3110.86	I	320	3537.46	I
9800	2097.12	I	340	2564.19	I	340 c	3118.19	I	240	3549.89	I
3400	2139.04	II	540	2568.64	II	340	3121.36	I	240	3570.26	I
3700	2156.67	I	370	2571.81	II	420	3128.94	I	360	3579.12	I
4900	2167.94	I	380	2586.79	I	260	3134.02	I	810 c	3580.15	II
3400	2176.21	I	290	2599.86	I	250	3141.38	I	650	3580.97	I
4200 c	2214.26	II	290	2603.89	I	440	3151.64	I	810	3583.02	I
5200 c	2275.25	II	660	2608.50	II	330	3153.79	I	320	3617.08	I
2900	2287.51	I	610 d	2611.54	I	360 c	3158.31	I	810	3637.84	I
2700	2294.49	I	310	2635.83	II	220	3164.52	I	440	3651.97	I
390	2298.09	II	550	2636.64	I	700	3168.37	I	320	3670.53	I
610	2302.99	I	270	2642.75	I	220	3174.61	I	860 c	3689.50	I

Intensity	Wavelength/Å		Intensity	Wavelength/Å		Intensity	Wavelength/Å		Intensity	Wavelength/Å	
1500 c	3691.48	I	800	1880.66	III	160	2728.94	I	880 d	3538.14	I
520	3703.24	I	500	1884.91	III	100	2771.51	I	280	3541.91	I
240	3709.93	I	500	1887.36	III	130	2783.03	I	1200	3543.95	I
360 c	3717.28	I	700	1888.62	III	150	2826.43	I	1800	3549.54	I
4000	3725.76	I	800	1901.32	III	180	2826.68	I	1200	3570.18	I
240 c	3735.01	I	500	1910.16	III	280	2862.94	I	4700	3583.10	I
810	3735.31	I	600	1919.37	III	110	2878.66	I	4700	3596.19	I
910	3740.10	I	500	1927.07	III	140	2882.37	I	5900	3597.15	I
300 cw	3745.44	I	700	1931.79	III	160	2907.21	I	3100	3612.47	I
700	3787.52	I	500	1954.25	III	65	2910.17	II	1800	3626.59	I
240	3869.94	I	500	1994.26	III	180	2924.02	I	8200	3657.99	I
240	3875.26	I	800	2013.71	III	130	2929.11	I	1300	3666.22	I
240	3876.86	I	500	2017.47	III	130	2931.94	I	560	3681.04	I
380 c	3917.27	I	500	2028.53	III	230	2968.66	I	1900	3690.70	I
550	3929.85	I	800	2036.72	III	160	2977.68	I	9400	3692.36	I
280	3961.04	I	600	2037.61	III	450	2986.20	I	940	3695.52	I
350 c	3962.48	I	1000	2040.18	III	110	3004.46	I	280	3698.26	I
220	4033.31	I	3000	2048.67	III	50	3006.43	III	380	3698.60	I
240	4081.43	I	2000	2064.11	III	130	3023.91	I	7600	3700.91	I
240 c	4110.89	I	800	2076.84	III	50	3052.44	III	940	3713.02	I
240 cw	4133.42	I	1000	2118.53	III	180	3083.96	I	650	3735.28	I
1800	4136.45	I	1000	2118.63	III	140	3121.76	I	420	3737.27	I
700	4144.36	I	1000	2139.44	III	240	3123.70	I	420	3744.17	I
220	4182.90	I	1000	2152.23	III	130	3155.78	I	1200	3748.22	I
220	4183.06	I	3000	2158.17	III	140	3189.05	I	240	3754.12	I
650	4221.08	I	3000	2163.19	III	470	3191.19	I	380	3754.27	I
3600 c	4227.46	I	3000	2167.33	III	190	3197.13	I	490	3755.58	I
260 c	4257.60	I	150	2276.21	II	520	3263.14	I	1000	3760.40	I
380	4358.69	I	140	2288.57	I	520	3271.61	I	2300	3765.08	I
360 cw	4394.38	I	110	2309.82	I	2300	3280.55	I	490	3769.97	I
2600	4513.31	I	350	2322.58	I	2300	3283.57	I	380	3778.13	I
260	4516.64	I	140	2326.47	I	280	3289.14	I	1000	3788.47	I
500	4522.73	I	190	2334.77	II	210	3294.28	I	1300	3792.18	I
2200 cw	4889.14	I	300	2361.92	I	260	3300.46	I	3800	3793.22	I
220	4923.90	I	110	2368.34	I	50	3310.69	III	4900	3799.31	I
1300	5270.95	I	270	2382.89	I	4200	3323.09	I	760	3805.92	I
1600 cw	5275.56	I	230	2383.40	I	330	3338.54	I	1300	3806.76	I
100	5667.88	I	270	2386.14	II	280	3360.80	I	470	3815.01	I
110 c	5752.93	I	80	2415.84	II	420	3368.38	I	760	3816.47	I
110 cw	5776.83	I	130	2427.68	I	1100	3372.25	I	1300	3818.19	I
550	5834.31	I	230	2429.52	I	110	3377.14	I	3800	3822.26	I
200	6307.70	I	110	2437.90	I	110	3385.78	I	2300	3828.48	I
200	6321.90	I	330	2440.34	I	5600	3396.82	I	2000	3833.89	I
100 cw	6605.19	I	90	2461.04	II	820	3399.70	I	5900	3856.52	I
180 c	6813.41	I	130	2473.09	I	160	3406.55	I	490	3870.01	I
260	6829.90	I	150	2487.47	I	820	3412.27	I	380	3877.34	I
50 cw	7640.94	I	100	2490.77	II	330	3421.22	I	120	3913.51	I
65 cw	7912.94	I	130	2502.46	I	120 d	3424.38	I	240	3922.19	I
Rhodium			300	2504.29	II	8200	3434.89	I	2000	3934.23	I
Rh Z = 45			150	2505.67	I	1400	3440.53	I	590	3942.72	I
50	813.44	III	350	2509.70	I	120	3447.74	I	3800	3958.86	I
80	882.51	III	300	2511.03	II	120	3450.29	I	45	3964.54	II
100	925.75	III	200	2515.75	I	400	3455.22	I	380	3975.31	I
150	937.28	III	130	2520.53	II	180	3457.07	I	240	3984.40	I
500	991.62	III	110	2537.04	II	220	3457.93	I	240	3995.61	I
400	992.48	III	350	2545.70	I	5900	3462.04	I	380	3996.15	I
500 d	1009.60	III	550	2555.36	I	180	3469.62	I	120	4023.14	I
200	1012.22	III	150	2622.58	I	4700	3470.66	I	560	4082.78	I
200	1015.17	III	230	2625.88	I	120	3472.25	I	140	4097.52	I
200	1073.87	III	100	2630.42	I	4700	3474.78	I	120	4119.68	I
150	1784.24	III	110	2647.28	I	2100	3478.91	I	1100	4121.68	I
200	1784.94	III	400	2652.66	I	110	3494.44	I	1500	4128.87	I
150	1796.50	III	100	2680.63	I	1200	3498.73	I	2100	4135.27	I
200	1816.03	III	400	2703.73	I	5900	3502.52	I	240	4154.37	I
1000	1832.05	III	100	2715.31	II	2800	3507.32	I	330	4196.50	I
500	1859.85	III	180	2718.54	I	8800	3528.02	I	3300	4211.14	I

Intensity	Wavelength/Å		Intensity	Wavelength/Å		Intensity	Wavelength/Å		Intensity	Wavelength/Å	
820	4288.71	I	1200	598.49	III	5	6299.224	I	370	2456.57	II
4200	4374.80	I	1500	643.878	II	10000	6458.33	II	280	2478.93	II
130	4569.00	I	25	663.76	IV	5000	6560.81	II	140	2498.42	II
150	4675.03	I	3000	697.049	II	100 l	7279.997	I	140	2498.57	II
70	4745.11	I	6000	711.187	II	150	7408.173	I	260	2507.01	II
70	5090.63	I	25	716.24	IV	200 l	7618.933	I	110	2513.32	II
60	5155.54	I	50	740.85	IV	300	7757.651	I	110	2517.32	II
60	5175.97	I	10000	741.456	II	60	7759.436	I	150	2535.59	II
95	5193.14	I	5000	769.04	III	90000 c	7800.27	I	550	2549.58	I
130	5354.40	I	25	776.89	IV	45000 c	7947.60	I	370	2609.06	I
95	5390.44	I	2500	815.28	III	40 l	8271.41	I	830	2612.07	I
160	5599.42	I	15	850.18	IV	30	8271.71	I	460	2642.96	I
40	5686.38	I	1000	1604.12	II	40 l	8868.512	I	330	2651.84	I
29	5792.66	I	5000	1760.50	II	30	8868.852	I	400	2659.62	I
40	5806.91	I	2000	2068.92	II	30 l	9522.65	I	330	2661.61	II
35	5831.58	I	10000	2075.95	II	20 l	9540.18	I	690	2678.76	II
130	5983.60	I	30000	2143.83	II	2000 c	9689.05	II	330	2692.06	II
35	6102.72	I	10000	2217.08	II	35 l	10075.282	I	200	2712.41	II
14	6199.99	I	5000	2291.71	II	30 l	10075.708	I	690	2719.52	I
16	6253.72	I	50000	2472.20	II	100	13235.17	I	140	2725.47	II
29	6319.53	I	1000	2631.75	III	20	13442.81	I	310	2734.35	II
40	6752.35	I	2000	2956.07	III	30	13443.57	I	1800	2735.72	I
13	6827.33	I	500	3086.84	III	75	13665.01	I	100	2778.38	II
11	6857.68	I	500	3111.36	III	1000	14752.41	I	110	2787.83	II
20	6879.94	I	5000 c	3148.90	II	800	15288.43	I	350	2810.03	I
65	6965.67	I	25	3157.54	I	150	15289.48	I	1700	2810.55	I
16	6979.15	I	50	3227.98	I	20	22529.65	I	350	2818.36	I
16	7001.58	I	500	3286.41	III	4	27314.31	I	400	2829.16	I
18	7101.64	I	60	3348.72	I	**Ruthenium**			640	2854.07	I
15	7104.45	I	75	3350.82	I	**Ru Z = 44**			420	2861.41	I
18	7268.18	I	100	3587.05	I	250	850.09	III	550	2866.64	I
35	7270.82	I	40	3591.57	I	200	850.30	III	1800	2874.98	I
18 h	7442.39	I	5000	3600.60	II	250	919.74	III	740	2886.54	I
12	7475.74	I	10000	3600.64	II	500	940.09	III	370	2908.88	I
12	7495.24	I	25000	3940.51	II	500	966.54	III	1100	2916.26	I
11	7557.67	I	1000	4201.80	I	750	974.14	III	180	2945.67	II
29	7791.61	I	500	4215.53	I	900	979.43	III	370	2949.50	I
55	7824.91	I	90000	4244.40	II	500	981.35	III	550	2965.16	I
21	8029.91	I	15000	4273.14	II	900	986.84	III	170	2965.55	II
29	8045.36	I	20000	4571.77	II	900	994.56	III	140	2976.59	II
15	8136.20	I	10000	4648.57	II	300	1001.65	III	550	2976.92	I
8	8425.59	I	30000	4775.95	II	500	1009.13	III	1400	2988.95	I
Rubidium			2	5087.987	I	900	1009.87	III	460	2994.96	I
Rb Z = 37			2	5132.471	I	500	1014.68	III	440	3006.59	I
30	465.85	III	10	5150.134	I	800	1190.51	III	330	3017.24	I
40	481.118	II	10000	5152.08	II	500	1200.07	III	310	3020.88	I
500	482.83	III	1	5165.023	I	500	1207.17	III	390	3064.84	I
500	489.66	III	2	5165.142	I	500	1209.77	III	330	3096.57	I
600	493.48	III	15	5195.278	I	300	1211.31	III	830	3099.28	I
90	497.430	II	2	5233.968	I	500	1941.35	III	740	3100.84	I
20	508.434	II	20	5260.034	I	500	2009.28	III	490	3294.11	I
150	513.266	II	1	5260.228	I	2400	2076.43	I	370	3301.59	I
300	530.173	II	3	5322.380	I	2600	2083.77	I	930	3339.55	I
75	533.801	II	40	5362.601	I	2400	2090.89	I	3100	3417.35	I
1200	535.86	III	4	5390.568	I	690	2255.52	I	4900	3428.31	I
40	542.887	II	75	5431.532	I	780	2272.09	I	6400	3436.74	I
200	555.036	II	3	5431.830	I	780	2279.57	I	8300	3498.94	I
1200	556.19	III	6	5578.788	I	480	2317.80	I	640	3514.49	I
1500	566.71	III	40	5647.774	I	120	2334.96	II	790	3539.37	I
1000	572.82	III	20	5653.750	I	190 h	2342.85	,II	690	3570.59	I
1500	576.65	III	60	5724.121	I	310	2351.33	I	6400	3589.22	I
2500	579.63	III	3	5724.614	I	170	2357.91	II	6900	3593.02	I
1500	581.26	III	75	6070.755	I	780	2402.72	II	6400	3596.18	I
2500	589.419	II	30 c	6159.626	I	150	2407.92	II	1300	3599.76	I
1000	594.94	III	75 c	6206.309	I	180	2455.53	II	3100	3634.93	I
1300	595.88	III	120 c	6298.325	I	150	2456.44	II	6200	3661.35	I

Intensity	Wavelength/Å		Intensity	Wavelength/Å		Intensity	Wavelength/Å		Intensity	Wavelength/Å	
830	3663.37	I	930	4307.60	I	18	7722.87	I	1500	3990.00	II
650	3669.49	I	550	4319.87	I	22	7791.86	I	1400	4064.58	II
550	3726.10	I	550	4342.07	I	30	7847.80	I	1000	4092.27	II
8700	3726.93	I	710	4354.13	I	80	7881.49	I	1900	4118.55	II
11000	3728.03	I	870	4361.21	I	18	8264.96	I	1200	4152.21	II
7100	3730.43	I	2400	4372.21	I	22	8710.84	I	1000	4188.13	II
3500	3742.28	I	870	4385.39	I		**Samarium**		1100	4203.05	II
870	3742.78	I	1300	4385.65	I		**Sm Z = 62**		1000	4225.33	II
2800	3745.59	I	1700	4390.44	I	150	2789.38	II	1200	4236.74	II
760	3753.54	I	1600	4410.03	I	410	3152.52	II	2100	4256.39	II
870	3755.93	I	1100	4460.04	I	720	3183.92	II	1300	4262.68	II
1200	3759.84	I	5400	4554.51	I	600	3211.73	II	1200	4279.68	II
600	3761.51	I	1700	4584.44	I	530	3216.85	II	2200	4280.79	II
600	3767.35	I	720	4647.61	I	600	3218.61	II	710	4282.21	I
1500	3777.59	I	1400	4709.48	I	720	3230.56	II	470	4282.83	I
600	3782.74	I	500	4757.84	I	720	3236.64	II	1600	4296.74	I
3900	3786.06	I	550	4869.15	I	720	3239.66	II	1900	4318.94	II
6000	3790.51	I	160	5011.23	I	720	3250.37	II	470	4319.53	I
760	3798.05	I	450	5057.33	I	850	3254.38	II	1800	4329.02	II
7600	3798.90	I	120	5076.32	I	1700	3306.39	II	440	4330.02	I
7600	3799.35	I	200	5093.83	I	1200	3321.18	II	1300	4334.15	II
600	3812.72	I	530	5136.55	I	1200	3365.86	II	880	4336.14	I
760	3817.27	I	170	5142.76	I	1200	3382.40	II	1100	4347.80	II
760	3819.03	I	250	5147.24	I	4200	3568.27	II	440	4362.91	I
650	3822.09	I	110	5151.07	I	4200	3592.60	II	530	4380.42	I
550	3824.93	I	500	5155.14	I	1700	3604.28	II	1600	4390.86	II
760	3831.80	I	920	5171.03	I	3400	3609.49	II	410	4401.17	I
930	3839.70	I	180	5195.02	I	1700	3621.23	II	470	4419.33	I
760	3850.43	I	130	5284.08	I	3400	3634.29	II	1500	4420.53	II
1300	3857.55	I	260	5309.27	I	2200	3661.36	II	2900	4424.34	II
650	3862.69	I	110	5335.93	I	2200	3670.84	II	470	4429.66	I
1300	3867.84	I	130	5361.77	I	1100	3693.99	II	1600	4433.88	II
650	3892.21	I	110 h	5401.04	I	1600	3728.47	II	1800	4434.32	II
760	3909.08	I	80	5484.32	I	2100	3731.26	II	530	4441.81	I
1500	3923.47	I	130	5510.71	I	1600	3735.98	II	440	4442.28	I
3300	3925.92	I	90	5559.75	I	2900	3739.12	II	710	4445.15	I
600	3931.76	I	290	5636.24	I	1200	3743.87	II	1300	4452.73	II
760	3945.57	I	180	5699.05	I	930	3745.46	I	1200	4454.63	II
600	3978.44	I	65	5814.98	I	800	3756.41	I	1000	4458.52	I
600	3979.42	I	55	5919.34	I	1200	3757.53	II	2200	4467.34	II
870	3984.86	I	80	5921.45	I	1900	3760.69	II	810	4470.89	I
1500	4022.16	I	21 h	5973.38	I	1100	3764.37	II	370	4499.11	I
600	4023.83	I	16	5988.67	I	370 d	3773.33	I	440	4581.73	I
1400	4051.40	I	35	5993.65	I	1100	3778.14	II	380	4649.49	I
710	4054.05	I	18	6116.77	I	1500	3788.12	II	470 d	4670.75	I
760	4068.37	I	26	6199.42	I	1600	3793.97	II	1100	4674.60	II
980	4076.73	I	26	6225.20	I	1600	3797.73	II	370	4688.73	I
6000	4080.60	I	18	6295.22	I	1600	3826.20	II	530	4704.40	II
930	4097.79	I	16	6390.23	I	1100	3831.50	II	730	4716.10	I
1900	4112.74	I	26 h	6444.84	I	560	3834.48	I	770	4728.42	I
2000	4144.16	I	21	6663.14	I	1600	3843.50	II	470	4745.68	II
650	4145.74	I	55	6690.00	I	530	3853.30	I	730	4760.27	I
870	4167.51	I	21	6766.95	I	2700	3854.21	II	580	4783.10	I
550	4197.58	I	30	6775.02	I	480	3854.56	I	350	4785.86	I
550	4198.88	I	21	6824.17	I	400	3858.74	I	970	4841.70	I
7600	4199.90	I	26	6911.48	I	3700	3885.29	II	730	4883.97	I
1500	4206.02	I	110	6923.23	I	1600	3896.98	II	630	4910.40	I
5400	4212.06	I	26	6982.01	I	1300	3903.42	II	350	4913.25	II
760	4214.44	I	26	7027.98	I	2500	3922.40	II	430	4918.99	I
930	4217.27	I	35	7238.92	I	1900	3928.28	II	400	5044.28	I
550	4230.31	I	16	7393.93	I	1300	3941.87	II	540	5071.20	I
760	4241.05	I	18	7468.91	I	470	3951.89	I	510	5117.16	I
760	4243.06	I	26	7485.79	I	1500	3963.00	II	350	5122.14	I
760	4284.33	I	70	7499.75	I	1500	3971.40	II	360	5155.03	II
550	4295.93	I	26	7559.61	I	620	3974.66	I	470	5175.42	I
3700	4297.71	I	18	7621.50	I	1000	3976.43	II	250	5200.59	I

Intensity	Wavelength/Å		Intensity	Wavelength/Å		Intensity	Wavelength/Å		Intensity	Wavelength/Å	
260	5251.92	I	500	252.85	V	130 d	3469.65	I	160 h	4573.99	I
400	5271.40	I	500	253.73	V	110	3471.13	I	350	4670.40	II
250	5282.91	I	900	283.91	V	200	3498.91	I	120	4706.97	I
220	5453.00	I	800	284.45	V	2700	3535.73	II	120	4709.34	I
230	5493.72	I	600	288.29	V	6600	3558.55	II	200	4728.77	I
230	5516.09	I	900	289.59	V	6100	3567.70	II	490	4729.23	I
140	5550.40	I	15	289.85	IV	13000	3572.53	II	590	4734.10	I
140	5659.86	I	1000 d	291.93	V	9900	3576.35	II	690	4737.65	I
120	5696.73	I	800	293.25	V	7700	3580.94	II	790	4741.02	I
85	5706.20	I	15	296.31	IV	4000	3589.64	II	1200	4743.81	I
70	5773.77	I	15	299.04	IV	4000	3590.48	II	200	4753.16	I
60	5778.33	I	700	300.00	V	28000	3613.84	II	220	4779.35	I
70 d	5786.98	II	1000	573.36	V	110	3617.43	I	170	4839.44	I
60	5788.38	I	600	587.94	V	20000	3630.75	II	90	4909.76	I
60	5800.52	I	10	785.12	IV	13000	3642.79	II	90	4922.84	I
65	5802.84	I	25	1168.61	III	6600	3645.31	II	90	4934.25	I
65	5867.79	I	15	1550.80	IV	110	3646.90	I	170	4954.06	I
50	5874.21	I	180	1603.06	III	5300	3651.80	II	120	4973.66	I
50	5898.96	I	150	1610.19	III	110	3664.25	II	150	4980.37	I
65	5965.71	II	160	2010.42	III	290	3666.54	II	140	4991.92	I
50	6045.00	I	12	2118.97	IV	75 h	3717.10	I	530	5031.02	II
50	6070.06	I	11	2185.43	IV	270	3833.07	II	250	5064.32	I
45	6084.12	I	11	2205.46	IV	610	3843.03	II	530	5070.23	I
45 h	6159.56	I	14	2222.22	IV	20000	3907.49	I	250	5075.81	I
45	6256.54	I	11	2271.33	IV	23000	3911.81	I	2100	5081.56	I
100	6267.28	II	110	2438.62	I	4400	3933.38	I	1200	5083.72	I
140	6569.31	II	560	2545.22	II	5500	3996.61	I	1100	5085.55	I
110	6589.72	II	2900	2552.37	II	530	4014.49	II	750	5086.95	I
50	6671.51	I	560	2555.82	II	20000	4020.40	I	390	5087.14	I
120 d	6731.84	II	2300	2560.25	II	20000	4023.69	I	270	5089.89	I
95	6794.20	II	1100	2563.21	II	220	4030.67	I	390	5096.73	I
120	6860.93	I	11	2586.93	IV	140	4031.39	I	620	5099.23	I
120	6955.29	II	120	2692.78	I	220	4043.80	I	370	5101.12	I
90	7020.44	II	350	2699.07	III	200	4046.48	I	180	5109.06	I
90	7039.22	II	360	2706.77	I	2700	4047.79	I	150	5112.86	I
90	7042.24	II	210	2707.95	I	120	4049.95	I	320	5116.69	I
90	7051.52	II	580	2711.35	I	5500	4054.55	I	390	5210.52	I
90	7082.37	II	230	2734.05	III	220	4056.59	I	280	5219.67	I
26	7088.30	I	340	2965.86	I	100	4068.66	III	350	5239.82	II
30	7095.50	I	1200	2974.01	I	160 h	4074.97	I	280	5258.33	I
30	7104.54	I	1400	2980.75	I	160	4078.57	I	210	5285.76	I
26	7115.96	I	340	2988.95	I	6100	4082.40	I	120	5341.05	I
85 d	7149.60	II	2200	3015.36	I	200	4086.67	I	350	5349.30	I
23	7213.82	I	2700	3019.34	I	400	4087.16	I	120	5349.71	I
60	7240.90	II	360	3030.76	I	440 h	4133.00	I	210	5355.75	I
26	7347.30	I	120 h	3056.31	I	530 h	4140.30	I	530	5356.10	I
30	7444.56	I	130	3065.11	II	720	4152.36	I	270	5375.35	I
26	7445.41	I	990	3251.32	II	1100 h	4165.19	I	370	5392.08	I
13	7470.76	I	1500	3255.69	I	110 h	4218.26	I	270	5446.20	I
45	7645.09	II	4400	3269.91	I	110 h	4219.73	I	120	5451.34	I
12	7645.61	I	5500	3273.63	I	180	4231.93	I	750	5481.99	I
40 w	7835.08	II	110 d	3343.28	II	200	4233.61	I	530	5484.62	I
16	7895.96	I	270	3352.05	II	400	4238.05	I	570	5514.22	I
90	7928.14	II	9900	3353.73	II	15000	4246.83	II	660	5520.50	I
40	8048.70	II	2000	3359.68	II	290	4294.77	II	660	5526.82	II
16	8065.16	I	1700	3361.27	II	350	4305.71	II	70	5564.86	I
45	8068.46	II	1700	3361.94	II	4200	4314.09	II	110	5591.33	I
40 w	8305.79	II	4000	3368.95	II	3300	4320.74	II	80	5640.98	II
19	8383.71	I	6600	3372.15	II	2400	4325.01	II	250	5657.88	II
45 w	8485.99	II	130	3418.51	I	180	4354.61	II	1500	5671.81	I
45 w	8708.43	II	200	3429.21	I	110	4358.64	I	1200	5686.84	I
95	8913.66	II	200	3429.48	I	2000	4374.46	II	1100	5700.21	I
Scandium			270	3431.36	I	130	4384.81	II	10	5706.82	IV
Sc Z = 21			530	3435.56	I	1100	4400.37	II	190	5708.61	I
350	180.14	V	270	3457.45	I	880	4415.56	II	880	5711.75	I
500	243.87	V	180	3462.19	I	120 h	4557.24	I	230	5717.28	I

Intensity	Wavelength/Å		Intensity	Wavelength/Å		Intensity	Wavelength/Å		Intensity	Wavelength/Å	
180	5724.08	I	200	1580.0	I	150	7013.875	I	10	993.52	III
14	5771.63	IV	150	1587.5	I	300	7062.065	I	13	994.79	III
620	6210.68	I	150	1593.2	I	200	7575.1	I	16	997.39	III
320	6239.78	I	250	1606.5	I	250	7583.4	I	50	1023.69	II
120	6245.63	II	200	1617.4	I	150	7592.2	I	8	1066.63	IV
110	6249.96	I	150	1621.2	I	300	8001.0	I	14	1108.37	III
80	6256.01	III	150	1643.4	I	200	8036.4	I	16	1109.97	III
250	6258.96	I	250	1671.2	I	150	8093.2	I	18	1113.23	III
750	6305.67	I	250	1675.3	I	150	8094.7	I	8	1122.49	IV
60	6378.82	I	250	1690.7	I	180	8149.3	I	10	1128.34	IV
90	6413.35	I	250	1793.3	I	150	8152.0	I	100	1190.42	II
60	6604.60	II	300	1795.3	I	200	8157.7	I	200	1193.28	II
65	6737.87	I	250	1855.2	I	180	8163.1	I	250	1194.50	II
50	6819.52	I	400	1898.6	I	150	8182.9	I	100	1197.39	II
50	6835.03	I	350	1913.8	I	150	8440.47	I	30	1206.51	III
90	7449.16	III	300	1919.2	I	150	8450.38	I	30	1206.53	III
55 h	7741.17	I	500	1960.9	I	150	8742.33	I	9	1207.52	III
30	7800.44	I	500	2039.8	I	300	8918.86	I	10	1210.46	III
19 h	8761.40	I	285	2057.5	III	200	9001.97	I	50	1226.81	II
50	8829.78	III	500	2074.8	I	200	9038.61	I	100	1227.60	II
30 h	8834.35	I	285	2136.6	IV	100	9432.50	I	150	1228.75	II
400	22051.86	I	500	2164.2	I	200	10217.25	I	200	1229.39	II
150	22065.05	I	600	2413.5	I	377	10307.45	I	100	1246.74	II
	Selenium		300	2548.0	I	900	10327.26	I	150	1248.43	II
	Se Z = 34		360	2665.5	IV	640	10386.36	I	100	1250.09	II
360	613.0	V	285	2724.3	IV	275	11946.87	I	150	1250.43	II
360	652.7	IV	285	2767.2	III	170	11952.64	I	200	1251.16	II
450	670.1	IV	220	2773.8	III	205	11972.93	I	40	1256.49	I
360	724.3	III	160	2951.6	IV	315	14817.93	I	50	1258.80	I
450	746.4	IV	450	3387.2	III	410	14917.47	I	1000	1260.42	II
450	759.1	V	450	3413.9	III	500	15151.44	I	2000	1264.73	II
360	808.7	V	450	3457.8	III	320	15471.00	I	200	1265.02	II
360	830.3	V	450	3637.6	III	265	15520.97	I	17	1294.54	III
360	832.7	II	450	3738.7	III	395	15618.40	I	14	1296.73	III
450	839.5	V	450	3800.9	III	360	16659.44	I	15	1298.89	III
360	843.0	III	450	4169.1	III	505	16813.78	I	18	1298.96	III
360	845.8	V	360	4175.3	II	205	16866.54	I	14	1301.15	III
360	912.9	II	450	4180.9	II	235	21374.24	I	16	1303.32	III
360	959.6	IV	285	4382.9	II	680	21442.56	I	100	1304.37	II
360	974.8	III	285	4446.0	II	415	21473.48	I	50 h	1305.59	II
450	996.7	IV	220	4449.2	II	270	21716.36	I	200	1309.27	II
360	1013.4	II	285	4467.6	II	240	21730.60	I	13	1312.59	III
360	1014.0	II	500	4730.8	I	150	23388.85	I	100	1346.87	II
450	1033.6	II	400	4739.0	I	265	24148.18	I	100	1348.54	II
450	1049.6	II	300	4742.2	I	375	24385.99	I	150	1350.06	II
360	1057.4	II	285	4840.6	II	255	25017.51	I	100	1352.64	II
360	1094.7	V	360	4845.0	II	510	25127.43	I	100	1353.72	II
360	1099.1	III	450	5227.5	II		**Silicon**		15	1393.76	IV
450	1119.2	III	360	5305.4	II		**Si Z = 14**		12	1402.77	IV
360	1141.9	II	100	5365.5	I	10	85.18	V	13	1417.24	III
450	1192.3	II	120	5369.9	I	15	96.44	V	90 h	1485.02	II
450	1227.6	V	110	5374.1	I	10	97.14	V	100 h	1485.51	II
285	1291.0	II	285	5522.4	II	20	117.86	V	12	1500.24	III
285	1308.9	II	285	5566.9	II	20	118.97	V	10	1501.19	III
285	1314.4	IV	285	5866.3	II	4	457.82	IV	9	1501.87	III
120	1435.3	I	450	6056.0	II	8	566.61	III	100 h	1509.10	II
120	1435.8	I	285	6303.8	III	8	653.33	III	50 h	1512.07	II
150	1449.2	I	200	6325.6	I	7	815.05	IV	60 p	1516.91	II
150	1500.9	I	360	6444.2	II	8	818.13	IV	500	1526.72	II
250	1530.4	I	285	6490.5	II	9	823.41	III	1000	1533.45	II
150	1531.3	I	285	6535.0	II	40 h	845.77	II	150	1594.55	I
200	1531.8	I	150	6831.3	I	100	889.72	II	100	1622.87	I
150	1575.3	I	120	6990.690	I	200	892.00	II	300	1629.43	I
150	1577.6	I	100	6991.792	I	9	967.95	III	200	1629.92	I
150	1577.9	I	200	7010.809	I	100	989.87	II	100	1667.62	I
150	1579.5	I				200	992.68	II	100	1668.52	I

Line Spectra of the Elements (continued): Silicon

Intensity	Wavelength/Å		Intensity	Wavelength/Å		Intensity	Wavelength/Å		Intensity	Wavelength/Å	
100	1672.59	I	300	2904.28	II	16	4819.72	III	1000	6347.10	II
200	1675.20	I	500	2905.69	II	18	4828.97	III	1000	6371.36	II
200	1696.20	I	55	2970.355	I	30	4947.607	I	45	6526.609	I
200	1697.94	I	150	2987.645	I	40	5006.061	I	45	6527.199	I
100 h	1770.92	I	50	3006.739	I	1000	5041.03	II	45	6555.462	I
100 h	1776.83	I	75	3020.004	I	1000	5055.98	II	50 h	6660.52	II
100 h	1799.12	I	100 h	3030.00	II	10 h	5091.42	III	100	6671.88	II
150	1808.00	II	9	3040.93	III	100	5181.90	II	7	6701.21	IV
500 h	1814.07	I	100 h	3043.69	II	100 h	5185.25	II	50 h	6717.04	II
200	1816.92	II	50 h	3048.30	II	200 h	5192.86	II	100	6721.853	I
200	1836.51	I	150 h	3053.18	II	500 h	5202.41	II	50	6829.82	I
200	1841.44	I	25	3086.24	III	100 h	5405.34	II	30	6848.568	I
9	1842.55	III	20	3093.42	III	100 h	5438.62	II	80	6976.523	I
200	1843.77	I	16	3096.83	III	100 h	5456.45	II	180	7003.567	I
300	1845.51	I	9	3165.71	IV	500 h	5466.43	II	180	7005.883	I
400	1847.47	I	16	3185.13	III	500 h	5466.87	II	90	7017.646	I
200	1848.14	I	13	3186.02	III	100 h	5469.21	II	250	7034.903	I
500	1850.67	I	150	3188.97	II	200 h	5496.45	II	6 h	7047.94	IV
200	1852.46	I	150	3193.09	II	35	5517.535	I	200	7165.545	I
500 h	1874.84	I	100	3195.41	II	100 h	5540.74	II	100	7226.206	I
200	1881.85	I	14	3196.50	III	150 h	5576.66	II	100	7235.326	I
200	1887.70	I	200	3199.51	II	30	5622.221	I	180	7250.625	I
200 h	1893.25	I	100 h	3203.87	II	100 h	5632.97	II	160	7275.294	I
1000 h	1901.33	I	200 h	3210.03	II	200 h	5639.48	II	400	7289.173	I
100 h	1902.46	II	15	3210.55	III	90	5645.611	I	375	7405.774	I
50 h	1910.62	II	75	3214.66	II	150 h	5660.66	II	200	7409.082	I
50	1941.67	II	12	3230.50	III	80	5665.554	I	275	7415.946	I
100	1949.56	II	14	3233.95	III	1000 h	5669.56	II	425	7423.497	I
100	1954.97	I	15	3241.62	III	120	5684.484	I	9 h	7466.32	III
50	2058.65	II	12	3258.66	III	300 h	5688.81	II	12 h	7612.36	III
50	2059.01	II	10	3276.26	III	100	5690.425	I	100	7680.267	I
200	2072.02	II	300	3333.14	II	90	5701.105	I	6 h	7723.82	IV
200	2072.70	II	500	3339.82	II	200 h	5701.37	II	30	7800.008	I
100	2124.12	I	15	3486.91	III	100 h	5706.37	II	400	7848.80	II
50 h	2136.56	II	9	3525.94	III	160	5708.397	I	500	7849.72	II
110	2207.98	I	20	3590.47	III	20	5739.73	III	30	7849.967	I
115	2210.89	I	8	3762.44	IV	45	5747.667	I	90	7918.386	I
110	2211.74	I	20 c	3791.41	III	45	5753.625	I	120	7932.349	I
120	2216.67	I	25	3796.11	III	45	5754.220	I	140	7944.001	I
120	2218.06	I	30	3806.54	III	45	5762.977	I	35	7970.306	I
10	2296.87	III	100 h	3853.66	II	70	5772.145	I	35	8035.619	I
10	2308.19	III	500 h	3856.02	II	70	5780.384	I	70	8093.241	I
100 h	2356.30	II	200 h	3862.60	II	90	5793.071	I	9 h	8102.86	III
30 h	2357.18	II	300	3905.523	I	100	5797.859	I	11 h	8103.45	III
50 h	2357.97	II	20	3924.47	III	150 h	5800.47	II	35	8230.642	I
300	2435.15	I	10	4088.85	IV	200	5806.74	II	9 h	8262.57	III
11	2449.48	III	70	4102.936	I	50	5846.13	II	40	8443.982	I
425	2506.90	I	9	4116.10	IV	300 h	5868.40	II	40	8501.547	I
375	2514.32	I	300 h	4128.07	II	40	5873.764	I	60	8502.221	I
500	2516.113	I	500 h	4130.89	II	10 h	5898.79	III	40	8536.165	I
7	2517.51	IV	100 h	4190.72	II	150	5915.22	II	120	8556.780	I
350	2519.202	I	50	4198.13	II	200	5948.545	I	50	8648.462	I
425	2524.108	I	9	4338.50	III	500	5957.56	II	40	8728.011	I
450	2528.509	I	30	4552.62	III	500	5978.93	II	75	8742.451	I
110	2532.381	I	25	4567.82	III	90	6125.021	I	100	8752.009	I
25	2541.82	III	20	4574.76	III	85	6131.574	I	35	8790.389	I
10	2546.09	III	100	4621.42	II	90	6131.850	I	100	9412.72	II
14	2559.21	III	150	4621.72	II	100	6142.487	I	100	9413.506	I
30	2563.679	I	9 h	4631.24	IV	100	6145.015	I	30	10371.269	I
85	2568.641	I	10 h	4654.32	IV	160	6155.134	I	120	10585.141	I
45	2577.151	I	9	4683.02	III	160	6237.320	I	120	10603.431	I
190	2631.282	I	16	4716.65	III	40	6238.287	I	120	10660.975	I
11	2640.79	III	50	4782.991	I	125	6243.813	I	30	10694.251	I
14	2655.51	III	35	4792.212	I	125	6244.468	I	30	10727.408	I
9	2817.11	III	80	4792.324	I	180	6254.188	I	60	10749.384	I
1000	2881.579	I	15	4813.33	III	45	6331.954	I	30	10784.550	I

Line Spectra of the Elements (continued): Silicon—Sodium

Intensity	Wavelength/Å		Intensity	Wavelength/Å		Intensity	Wavelength/Å		Intensity	Wavelength/Å	
80	10786.856	I	600	1917.08	III	100	4212.82	I	15	410.372	IV
140	10827.091	I	700	1957.62	III	50	4311.07	I	10	411.334	IV
60	10843.854	I	100	1967.38	II	50 h	4476.04	I	13	412.242	IV
130	10869.541	I	600	1975.92	III	30 h	4615.69	I	11	1582.18	IV
30	10882.802	I	500	1977.03	III	80	4620.04	II	11 d	1583.98	IV
30	10885.336	I	600	2000.24	III	50	4620.46	II	12	1584.14	IV
80	10979.308	I	150	2015.96	II	60 h	4668.48	I	12 d	1587.05	IV
30	10982.061	I	150	2033.98	II	30 h	4677.60	I	11	1615.92	IV
80	11017.965	I	200	2061.17	I	100	4788.40	II	12	1618.57	IV
370	11984.19	I	100	2069.85	I	30 h	4847.82	I	11	1655.47	IV
220	11991.57	I	80 r	2113.82	II	100	4874.10	I	15 c	1701.97	IV
440	12031.51	I	60	2145.60	II	80	5027.35	II	20 d	1887.47	III
190	15888.39	I	600	2161.89	III	1000	5209.08	I	12	1960.76	IV
95	16060.03	I	50	2186.76	II	1000	5465.50	I	11	1965.08	IV
110	19722.50	I	60	2229.53	II	100	5471.55	I	12 d	2106.33	IV
Silver			100 r	2246.43	II	100	5667.34	I	30	2230.33	III
Ag Z = 47			500	2246.51	III	10 h	6268.50	I	16	2232.19	III
25	730.83	II	75 r	2248.74	II	320	7687.78	I	20 h	2246.70	III
30	752.80	II	75	2280.03	II	25	8005.4	II	300	2315.65	II
400	799.41	III	30 h	2309.56	I	500	8273.52	I	18	2386.99	III
15	1005.32	II	700	2310.04	III	25	8403.8	II	17	2394.03	III
10	1065.49	II	70 r	2317.05	II	30 h	8645.70	I	300	2420.99	II
12	1072.23	II	80 r	2320.29	II	10 h	8704.85	I	300	2424.73	II
250	1074.22	II	70 r	2324.68	II	12	8747.6	II	25	2459.31	III
150	1107.03	II	80 r	2331.40	II	15	9000.9	II	18	2468.85	II
150	1112.46	II	70	2357.92	II	10	12551.0	I	20	2474.73	III
60	1195.83	II	50 h	2375.02	I	60	16819.5	I	1000	2493.15	II
50	1223.33	II	75	2411.41	II	20	17416.7	I	25	2497.03	III
50	1240.80	II	90 r	2413.23	II	15	18307.9	I	17	2510.26	III
50	1246.87	II	100 r	2437.81	II	15	18382.3	I	20	2543.84	I
55	1256.81	II	80	2447.93	II	**Sodium**			10	2543.87	I
55	1257.55	II	80	2473.84	II	**Na Z = 11**			70	2593.87	I
50	1266.63	II	60	2506.63	II	7	142.232	IV	35	2593.92	I
70	1273.67	II	50 h	2575.63	I	8	146.064	IV	850	2611.81	II
65	1297.51	II	60	2660.49	II	9	150.298	IV	850	2661.00	II
85	1311.20	II	60	2721.77	I	8	150.687	IV	1000	2671.83	II
55	1313.81	II	75	2767.54	II	8	155.510	IV	200	2680.34	I
50	1314.61	II	100 h	2824.39	I	8	156.537	IV	100	2680.43	I
60	1323.84	II	30 h	3130.02	I	12	162.448	IV	1000	2841.72	II
60	1342.09	II	90	3180.70	II	10	163.190	IV	400	2852.81	I
50	1342.57	II	100	3267.35	II	12	168.411	IV	200	2853.01	I
70	1346.62	II	55000 r	3280.68	I	10	168.546	IV	2	2893.62	I
50	1353.54	II	28000 r	3382.89	I	5	183.95	III	1100	2904.92	II
150	1364.50	II	30	3469.16	I	10	190.445	IV	1100	2917.52	II
100	1396.00	II	70	3475.82	I	10	199.772	IV	1100	2919.05	II
100	1410.93	II	80	3495.28	II	8	202.49	III	1200	2919.85	II
90	1419.72	II	50	3542.61	I	8	202.76	III	1300	2920.95	II
95	1432.60	II	50 h	3624.68	I	8 p	203.06	III	1000	2923.49	II
100	1464.72	II	75	3682.46	II	8	203.28	III	1200	2951.24	II
50	1466.23	II	30	3682.50	I	8	203.33	III	1100	2952.40	II
50 r	1507.37	I	80	3683.34	II	15	229.87	III	1000	2977.13	II
100 r	1515.63	I	50 h	3709.20	I	50 c	250.52	III	1100	2979.66	II
50 r	1548.58	I	200	3810.94	I	30	251.37	III	1100	2980.63	II
100	1555.16	II	50	3811.78	I	25	266.90	III	1300	2984.19	II
100	1644.50	II	100 h	3840.74	I	70	267.65	III	1700	3124.42	II
60	1651.52	I	50 h	3907.41	I	50	267.87	III	2500	3135.48	II
50	1652.10	I	50	3909.31	II	50	268.63	III	1700	3137.86	II
700	1656.18	III	50 h	3914.40	I	20 p	272.08	III	2000	3149.28	II
120	1682.82	II	70	3920.10	II	20	272.45	III	2000	3163.74	II
500	1693.51	III	60	3949.43	II	10	319.644	IV	1000	3179.06	II
10	1708.11	I	100 h	3981.58	I	300	372.08	II	1700	3189.79	II
50	1709.27	I	70	3985.19	II	350	376.38	II	1600	3212.19	II
125	1736.44	II	100 h	4055.48	I	100	378.14	III	1500	3257.96	II
750	1751.03	III	80	4085.91	II	70	380.10	III	1700	3285.60	II
10 h	1766.14	I	100	4185.48	II	12	408.684	IV	1700	3301.35	II
75	1790.37	II	90 h	4210.96	I	10	409.614	IV	1200	3302.37	I

Intensity	Wavelength/Å		Intensity	Wavelength/Å		Intensity	Wavelength/Å		Intensity	Wavelength/Å	
600	3302.98	I	400	12679.17	I	340	4305.45	II	**Sulfur**		
1500	3304.96	II	60	14767.48	I	65000	4607.33	I	**S Z = 16**		
1000	3318.04	II	100	14779.73	I	9	4685.08	IV	5	437.4	V
50	3426.86	I	60	16373.85	I	3200	4722.28	I	5	438.2	V
1500	3533.05	II	100	16388.85	I	2200	4741.92	I	5	439.6	V
1200	3631.27	II	400	18465.25	I	1400	4784.32	I	20	519.3	IV
6	4238.99	I	50	22056.44	I	4800	4811.88	I	20	520.1	IV
10	4242.08	I	25	22083.67	I	3600	4832.08	I	40	520.8	IV
1	4249.41	I	60	23348.41	I	3000	4872.49	I	20	522.0	IV
2	4252.52	I	100	23379.13	I	2000	4876.32	I	20	522.5	IV
15	4273.64	I	**Strontium**			1000	4891.98	I	20	551.2	IV
20	4276.79	I	**Sr Z = 38**			8000	4962.26	I	40	652.5	IV
2	4287.84	I	15	298.12	IV	1300	4967.94	I	40	653.0	IV
3	4291.01	I	15	300.12	IV	800 h	5156.07	I	70	653.6	IV
30	4321.40	I	125	330.67	III	1400	5222.20	I	40	654.0	IV
40	4324.62	I	500	351.62	III	2000	5225.11	I	70	655.6	IV
3	4341.49	I	75	358.80	III	2000	5229.27	I	20	655.9	IV
5	4344.74	I	250	363.49	III	2800	5238.55	I	110	657.3	IV
40	4390.03	I	150	371.21	III	4800	5256.90	I	40	658.3	V
60	4393.34	I	20	378.53	IV	40	5257.71	III	70	659.8	V
5	4419.88	I	75	392.44	IV	40	5443.48	III	40	660.9	IV
8	4423.25	I	50	393.00	IV	1500	5450.84	I	160	661.4	IV
60	4494.18	I	50	396.22	IV	7000	5480.84	I	110	663.2	V
100	4497.66	I	1000	437.24	III	3500	5504.17	I	40	663.7	IV
10	4541.63	I	1875	491.79	III	2600	5521.83	I	40	664.8	IV
15	4545.19	I	1250	507.04	III	2000	5534.81	I	70	666.1	IV
120	4664.811	I	3750	514.38	III	2000	5540.05	I	20	678.1	V
200	4668.560	I	10	517.28	V	1000	6380.75	I	40	680.3	V
20	4747.941	I	2500	562.75	III	900 h	6386.50	I	110	680.9	V
30	4751.822	I	25	578.01	V	600 h	6388.24	I	40	681.6	V
200	4978.541	I	30	624.93	V	9000	6408.47	I	20	693.5	V
400	4982.813	I	25	642.23	V	5500	6504.00	I	70	729.5	III
40	5148.838	I	50	649.21	V	1000	6546.79	I	110	732.42	III
80	5153.402	I	25	660.94	V	1700	6550.26	I	70	735.2	III
280	5682.633	I	200	664.43	IV	3000	6617.26	I	70	738.5	III
70	5688.193	I	35	686.23	V	1800	6791.05	I	110	744.9	IV
560	5688.205	I	100	710.35	IV	4800	6878.38	I	110	748.4	IV
80000	5889.950	I	12	747.82	V	1200	6892.59	I	110	750.2	IV
40000	5895.924	I	50	1025.23	III	5500	7070.10	I	110	753.8	IV
120	6154.225	I	35	1125.49	III	2500	7309.41	I	285	786.5	V
240	6160.747	I	50	1236.23	III	500	7621.50	I	70	789.0	III
130	6530.70	II	1400	2152.84	II	400 h	7673.06	I	70	796.7	III
130	6544.04	II	1400	2165.96	II	200 h	8422.80	I	70	800.5	IV
130	6545.75	II	100	2273.71	III	120	8505.69	II	70	804.0	IV
20	7373.23	I	100	2340.13	III	200	8688.91	II	70	809.7	IV
10	7373.49	I	50	2346.97	IV	100	9294.10	I	110	816.0	IV
50	7809.78	I	160	2428.10	I	400 h	9448.95	I	70	824.9	III
25	7810.24	I	100	2486.52	III	600	9596.00	I	70	836.3	III
4400	8183.256	I	40	2555.60	IV	300	9624.70	I	160	849.2	V
800	8194.790	I	40	2571.04	IV	100	9638.10	I	110	852.2	V
8800	8194.824	I	100	3002.61	III	100 h	9647.70	II	220	854.8	V
100	8649.92	I	200	3012.32	III	300	10036.66	II	110	857.9	V
60	8650.89	I	10	3019.29	IV	1000	10327.31	II	110	860.5	V
25	8942.96	I	100	3021.73	III	200	10914.88	II	40	906.9	II
40	9153.88	I	50	3061.43	III	700	11241.25	I	40	910.5	II
60	9465.94	I	50	3182.61	III	100	12014.76	II	40	912.7	II
80	9961.28	I	100	3235.39	III	60	12445.90	II	40	937.4	II
20	10566.00	I	400	3351.25	I	40	12495.00	I	40	937.7	II
60	10572.28	I	650	3380.71	II	75	12974.70	II	160	1062.7	IV
200	10746.44	I	50	3430.76	III	100	13123.80	II	160	1073.0	IV
80	10749.29	I	950	3464.46	II	50	17447.40	I	70	1073.5	IV
120	10834.87	I	600	3969.26	I	230	20261.40	I	285	1077.1	III
35	11190.19	I	1300	4030.38	I	120	20700.70	I	40	1102.3	II
50	11197.21	I	46000	4077.71	II	30	26023.60	I	70	1194.0	III
400	11381.45	I	32000	4215.52	II				70	1201.0	III
1000	11403.78	I	9	4298.57	IV				40	1234.1	II

Intensity	Wavelength/Å		Intensity	Wavelength/Å		Intensity	Wavelength/Å		Intensity	Wavelength/Å	
40	1250.5	II	110	3097.5	IV	110	8882.5	I	1500	2199.67	II
110	1253.8	II	110	3497.3	III	220	8884.2	I	90	2207.64	IV
110	1259.5	II	160	3632.0	III	160	9035.9	I	1400 d	2210.03	II
275	1270.782	I	110	3709.4	III	450	9212.9	I	1400	2239.48	II
250	1277.216	I	160	3717.8	III	450	9228.1	I	1200	2250.76	II
280	1295.653	I	160	3838.3	III	450	9237.5	I	840	2261.42	II
275	1302.337	I	285	3867.6	I	285	9413.5	I	990	2262.30	II
235	1302.863	I	285	3902.0	I	285	9421.9	I	990	2272.59	II
235	1303.110	I	160	3928.6	III	285	9437.1	I	790	2285.25	II
245	1303.430	I	360	3933.3	II	650	9649.9	I	600	2286.59	II
260	1305.883	I	450	4120.8	I	450	9672.3	I	990	2289.16	II
265	1310.194	I	280	4142.3	I	450	9680.8	I	440	2302.24	II
355	1316.542	I	360	4145.1	II	450	9693.7	I	440	2302.93	II
290	1316.618	I	450	4153.1	II	285	9697.3	I	440	2312.60	II
375	1323.515	I	450	4162.7	II	285	9739.7	I	420	2315.46	II
355	1326.643	I	360	4253.6	III	285	9932.3	I	690	2331.98	II
775	1381.552	I	450	4694.1	I	285	9949.8	I	550	2332.19	II
710	1385.510	I	285	4695.4	I	285	9958.9	I	250	2357.30	I
960	1388.435	I	160	4696.2	I	285	10455.5	I	260	2361.09	I
640	1389.154	I	280	4716.2	II	285	10459.5	I	600	2364.24	II
775	1392.588	I	450	4815.5	II		**Tantalum**		320	2371.58	I
1000	1396.112	I	360	4924.1	II		**Ta Z = 73**		1400	2387.06	II
300	1409.337	I	450	4925.3	II	60	493.07	V	2400	2400.63	II
510	1425.030	I	285	4993.5	I	1000	890.87	V	320	2416.89	II
425	1433.280	I	360	5428.6	II	500	947.30	V	360	2427.64	I
300	1436.968	I	650	5432.8	II	67	999.34	IV	360	2429.71	II
300	1448.229	I	1000	5453.8	II	79	1116.10	IV	480	2432.70	II
425	1472.972	I	1000	5473.6	II	78	1136.17	IV	380	2470.90	II
550	1473.995	I	1000	5509.7	II	85	1175.51	IV	600	2474.62	I
300	1474.380	I	280	5564.9	II	80	1189.28	IV	500	2484.95	I
355	1481.665	I	1000	5606.1	II	80	1192.67	IV	600	2488.70	II
485	1483.039	I	450	5640.0	II	85	1213.09	IV	500	2490.46	I
300	1483.233	I	450	5640.3	II	500	1213.42	V	600	2504.45	I
330	1485.622	I	280	5647.0	II	85	1215.53	IV	600	2507.45	I
390	1487.150	I	650	5659.9	II	90	1223.73	IV	1200 d	2526.35	I
20	1624.0	IV	450	5664.7	II	88	1238.12	IV	600	2532.12	II
20	1629.2	IV	160	5706.1	I	95	1240.06	IV	1200	2559.43	I
680	1666.688	I	450	5819.2	II	87	1258.34	IV	460	2562.10	I
640	1687.530	I	450	6052.7	I	94	1264.91	IV	600	2577.37	II
710	1807.311	I	280	6286.4	II	98	1272.42	IV	600	2603.49	II
680	1820.343	I	450	6287.1	II	94	1275.48	IV	1400	2608.63	I
640	1826.245	I	450	6305.5	II	86	1275.94	IV	1200	2635.58	II
710	1900.286	I	450	6312.7	II	92	1308.51	IV	860	2636.90	I
550	1914.698	I	280	6384.9	II	87	1315.58	IV	2400	2647.47	I
20	2387.0	IV	280	6397.3	II	92	1332.38	IV	2600	2653.27	I
40	2398.9	IV	280	6398.0	II	86	1343.30	IV	1900	2656.61	I
110	2460.5	III	360	6413.7	II	92	1365.88	IV	1500	2661.34	I
110	2489.6	III	160	6743.6	I	5000	1392.56	V	770	2675.90	II
160	2496.2	III	285	6748.8	I	91	1398.78	IV	1500	2685.17	II
160	2499.1	III	450	6757.2	I	93	1413.40	IV	470	2694.52	II
220	2508.2	III	450	7579.0	I	91	1454.32	IV	1000	2698.30	I
70	2636.9	III	450	7629.8	I	92	1464.41	IV	1200	2710.13	I
220	2665.4	III	285	7686.1	I	93	1469.82	IV	2600	2714.67	I
110	2691.8	III	450	7696.7	I	90	1495.25	IV	470	2727.44	II
110	2702.8	III	1000	7924.0	I	95	1514.19	IV	1200	2748.78	I
220	2718.9	III	160	7928.8	I	85	1607.70	IV	860	2749.83	I
110	2721.4	III	285	7930.3	I	7000	1709.10	V	410	2752.49	II
220	2726.8	III	450	7931.7	I	85	1712.16	IV	1000	2758.31	I
220	2731.1	III	450	7967.4	I	85	1716.13	IV	430	2761.68	II
110	2741.0	III	450	7967.4	II	85	2055.75	IV	770	2775.88	I
285	2756.9	III	450	8314.7	I	1100	2140.13	II	680	2796.34	I
110	2775.2	III	450	8314.7	II	1500	2146.87	II	680	2797.76	II
160	2785.5	III	450	8585.6	I	1200	2182.71	II	510	2806.58	I
110	2863.5	III	285	8680.5	I	1100	2193.88	II	640	2844.25	I
160	2904.3	III	450	8694.7	I	1500	2196.03	II	560	2848.52	I
160	2986.0	III	360	8874.5	I	90	2199.58	IV	1500	2850.49	I

Intensity	Wavelength/Å		Intensity	Wavelength/Å		Intensity	Wavelength/Å		Intensity	Wavelength/Å	
1900	2850.98	I	150	5645.91	I	20000	4262.27	I	200	4686.91	II
360	2861.98	I	130	5664.90	I	30000	4297.06	I	100	4696.38	II
470	2871.42	I	130	5776.77	I	20000	4853.59	I	100	4706.53	II
380	2880.02	I	90	5780.71	I	**Tellurium**			100	4766.05	II
770	2891.84	I	130	5811.10	I	**Te Z = 52**			100	4784.87	II
560	2902.05	I	240	5877.36	I	8	802.28	II	100	4827.14	II
310	2915.49	I	130	5882.30	I	8	1059.51	II	150	4831.28	II
410	2925.19	I	90	5901.91	I	8	1077.66	II	150	4842.90	II
310	2932.70	I	90	5918.95	I	10	1161.42	II	130	4865.12	II
1700	2933.55	I	130	5939.76	I	10	1174.34	II	200	4866.24	II
470	2940.06	I	240	5944.02	I	12	1175.79	II	8	5083.0	I
1200	2940.22	I	190 c	5997.23	I	9	1208.54	II	50	5449.84	II
510	2951.92	I	100	6020.72	I	9	1220.98	II	50	5487.95	II
340	2953.56	I	250	6045.39	I	9	1253.62	II	150	5576.35	II
1500	2963.32	I	100	6047.25	I	9	1270.52	II	150	5649.26	II
770	2965.13	II	100	6101.58	I	10	1324.92	II	100	5666.20	II
770	2965.54	I	65	6144.56	I	9	1363.24	II	200	5708.12	II
340	2969.47	I	130	6154.50	I	8	1366.73	II	150	5755.85	II
430	2975.56	I	150	6256.68	I	10	1374.80	II	100	5974.68	II
1800	3012.54	II	150	6268.70	I	10	1608.41	II	50	6367.13	II
290 d	3027.48	I	150	6309.58	I	10	1613.15	II	10 h	6790.0	I
530	3049.56	I	75	6325.08	I	5	1655.4	I	20 h	6837.6	I
530	3069.24	I	65	6341.17	I	5	1688.5	I	20 h	6854.7	I
360	3077.24	I	75	6356.16	I	6	1700.0	I	15 h	7191.1	I
560	3103.25	I	65	6360.84	I	5	1708.0	I	20 h	7263.5	I
380	3124.97	I	90	6389.45	I	10	1822.4	I	12	7460.98	II
380	3130.58	I	65	6428.60	I	26000	2002.02	I	15	7468.75	II
270	3132.64	I	250	6430.79	I	6500	2081.16	I	15	7921.69	II
320	3170.29	I	200	6450.36	I	18000	2142.81	I	15	7943.14	II
270	3173.59	I	380	6485.37	I	3200	2147.25	I	10	7950.34	II
600	3180.95	I	65	6505.52	I	500	2259.02	I	30 h	8061.4	I
300	3223.83	I	100	6514.39	I	1200	2383.26	I	10	8122.44	II
1100	3311.16	I	100	6516.10	I	1500	2385.78	I	20	8186.44	II
680	3318.84	I	100	6574.84	I	50	2438.69	II	15	8273.53	II
330 d	3330.99	II	110	6611.95	I	120	2530.72	I	15	8672.95	II
640	3371.54	I	75	6621.30	I	100	2649.66	II	10	8733.81	II
360	3385.05	I	100	6673.73	I	80	2661.10	II	205	8758.18	I
450	3406.94	I	180	6675.53	I	110	2677.13	I	81	9004.37	I
490	3480.52	I	75 c	6740.73	I	100	2858.29	II	5660	9722.74	I
380	3497.85	I	75	6771.74	I	150	2895.41	II	532	9868.92	I
490	3511.04	I	160 c	6813.25	I	70	2967.29	II	689	9956.30	I
750	3607.41	I	210	6866.23	I	70	3047.00	II	325	9977.13	I
980	3626.62	I	180	6875.27	I	100	3175.14	I	5950	10051.41	I
500	3642.06	I	150	6902.10	I	60	3256.80	II	4097	10091.01	I
210	3918.51	I	140	6927.38	I	60	3329.22	II	381	10118.08	I
210	3970.10	I	140	6928.54	I	150	3406.79	II	397	10300.56	I
210	3996.17	I	65	6951.26	I	50	3442.25	II	745	10493.57	I
410	4061.40	I	180	6966.13	I	50	3521.11	II	1880	10918.34	I
310	4067.91	I	110 d	6995.22	I	50	3552.19	II	10200	11089.56	I
300	4205.88	I	75	7006.96	I	100	3611.78	II	508	11163.74	I
360 c	4510.98	I	150	7148.63	I	50	3617.57	II	6620	11487.23	I
340	4574.31	I	110	7172.90	I	50	4006.52	II	1580	13247.75	I
260	4619.51	I	140	7301.74	I	70	4127.32	II	1050	14513.51	I
450	4681.88	I	160	7346.41	I	100	4169.77	II	1480	15452.45	I
200	5037.37	I	140 c	7352.86	I	80	4225.73	II	2430	15546.23	I
100	5067.87	I	100	7356.96	I	100	4261.11	II	3760	16403.90	I
110	5115.84	I	90 cw	7369.09	I	60	4273.43	II	1960	17303.54	I
100	5141.62	I	160	7407.89	I	80	4285.85	II	2780	18291.59	I
100	5143.69	I	100	7882.37	I	150	4364.00	II	1020	21043.73	I
330	5156.56	I	75	8026.50	I	75	4385.10	II	464	21602.50	I
110	5212.74	I	75	8281.62	I	170	4478.63	II	74	22555.29	I
110 d	5218.45	I	**Technetium**			80	4537.07	II	38	26539.17	I
140	5341.05	I	**Tc Z = 43**			100	4557.78	II	**Terbium**		
200	5402.51	I	10000 c	3636.07	I	70	4630.62	II	**Tb Z = 65**		
130	5419.19	I	20000 c	4031.63	I	100	4641.12	II	1000	1259.40	IV
90	5518.91	I	15000	4095.67	I	180	4654.37	II	1000	1327.67	IV

Intensity	Wavelength/Å		Intensity	Wavelength/Å		Intensity	Wavelength/Å		Intensity	Wavelength/Å	
1000	1373.86	IV	1700	3765.14	I	65	4632.07	I	30	7737.63	I
5000	1595.39	IV	2100	3776.49	II	65 h	4636.59	I	30	7855.79	II
2000	1633.19	IV	600	3783.53	I	85	4641.00	II	27	7927.90	II
2000	2027.79	IV	410	3789.92	I	210	4641.98	II	30	8025.42	II
1000	2089.98	IV	760 d	3806.85	II	260 cw	4645.31	II	30	8085.06	II
1000	2332.54	IV	1500	3830.26	I	80	4647.23	I	65	8194.82	II
110	2584.61	II	540	3833.42	I	80	4662.79	I	95	8212.57	I
110	2608.57	II	920 d	3842.50	II	80	4676.90	I	40	8450.06	II
130	2628.69	II	3700	3848.73	II	70 c	4681.87	I	30 h	8511.80	I
140	2669.29	II	3500 w	3874.17	II	80	4688.63	II	45	8583.45	II
190	2704.07	II	480	3888.22	I	80	4693.11	II	30	8603.40	I
270	2769.53	II	490	3894.64	I	200	4702.41	II	65	8765.74	II
320	2897.44	II	2400	3899.20	II	110	4707.94	II		**Thallium**	
250	2956.21	II	1600	3901.33	I	80	4739.93	I		**Tl Z = 81**	
230	3010.59	II	480	3908.06	I	70	4747.80	I	10	570.49	IV
230	3016.18	II	650	3915.43	I	410 cw	4752.53	II	5 r	670.87	II
460	3053.55	II	760	3925.45	II	180	4786.78	I	15 r	696.30	II
460	3070.05	II	810 d	3939.52	II	100	4813.77	I	5 r	709.23	II
670	3078.86	II	2200 d	3976.84	II	80	4875.57	II	10 r	817.18	II
480	3082.36	II	1800	3981.87	II	80	4881.15	II	5 r	836.34	II
480	3089.58	II	970	4002.59	II	95	4915.90	I	8 r	1018.85	II
480	3102.96	II	1900	4005.47	II	65	4931.79	I	30	1028.69	IV
440	3139.64	II	760	4012.75	II	85	4993.82	II	20	1034.73	IV
480	3187.26	II	870	4032.28	I	110	5078.25	I	20	1036.61	IV
480	3199.56	II	2100	4033.03	II	75	5089.12	II	10 r	1049.73	II
1100	3218.93	II	430	4054.12	I	85	5186.13	I	8 r	1050.30	II
1200	3219.98	II	410	4060.37	I	120	5228.12	I	5 r	1074.97	II
480	3252.32	II	1300	4061.58	I	75	5248.71	I	30	1079.68	IV
760	3280.31	II	650	4105.37	I	75 w	5262.11	II	10 r	1130.17	II
760	3281.40	II	1100	4144.41	II	75	5281.05	I	15 r	1162.55	II
1000	3285.04	II	350	4158.53	I	65	5304.72	I	10 r	1167.43	II
1500	3293.07	II	390	4196.74	I	110	5319.23	I	10 r	1183.41	II
3800	3324.40	II	650	4203.74	I	65 w	5337.90	I	12 r	1194.84	II
760	3349.42	II	600	4206.49	I	160	5354.88	I	5 r	1246.00	II
760	3364.93	II	480	4215.09	I	75	5369.72	I	10	1266.33	III
810	3454.06	II	480	4232.82	I	75	5375.98	I	15 r	1307.50	II
810 d	3472.79	II	650	4266.34	I	50	5424.10	II	8 r	1310.20	II
810	3500.84	II	760 cw	4278.52	II	55	5459.81	I	25 r	1321.71	II
5700	3509.17	II	450	4310.42	I	55	5509.61	I	8 r	1330.40	II
1300	3523.66	II	2200	4318.83	I	50	5514.54	I	10 r	1373.52	II
1100	3540.24	II	600	4322.23	I	65	5524.12	I	10	1477.14	III
810	3543.89	II	600	4325.83	II	85 c	5747.58	I	8 r	1489.65	I
3200	3561.74	II	3000	4326.43	I	75	5795.64	I	10 r	1499.30	II
810	3567.35	II	600	4332.12	I	75	5803.13	II	10 r	1507.82	II
4200	3568.52	II	870	4336.43	I	65	5815.36	I	15 r	1561.58	II
1600	3568.98	II	600	4337.64	I	65	5851.07	I	10 r	1568.57	II
1100	3579.20	II	1700	4338.41	I	65	5870.62	I	7 r	1593.26	II
710	3585.03	II	700	4340.62	I	65 c	5920.78	I	5 h	1616.	I
810	3596.38	II	870	4356.81	I	75	5967.34	II	5	1685.40	I
1600	3600.44	II	330	4382.45	I	35	6331.68	II	10 r	1792.76	II
810	3625.54	II	300	4388.23	I	35 cw	6518.68	I	12 r	1814.85	II
2300	3650.40	II	260	4390.91	I	35	6581.82	I	25 r	1908.64	II
810	3654.88	II	350	4423.10	I	90	6677.94	II	100 r	2007.56	I
2000	3658.88	II	240	4436.12	I	40 cw	6702.61	I	100 r	2210.71	I
3800	3676.35	II	240	4448.04	I	130	6794.58	II	30	2298.04	II
810	3682.26	II	430	4493.07	I	55	6896.37	II	140	2315.98	I
450	3693.58	I	75	4514.31	II	45 h	6899.95	I	900 h	2379.69	I
450	3700.12	I	110	4549.07	I	40	6901.98	I	20	2530.86	II
4700	3702.86	II	110	4550.45	I	65	7204.28	I	700	2580.14	I
2400	3703.92	II	110	4556.46	I	40	7257.73	I	420	2709.23	I
1000 d	3711.76	II	110	4563.69	II	45	7348.88	II	4400 d	2767.87	I
650	3745.04	I	210	4578.69	II	45	7496.12	I	10	2849.80	II
870	3747.17	II	65	4584.84	II	27 h	7582.03	II	2800	2918.32	I
870	3747.34	II	65	4591.56	II	45	7590.24	I	20	3091.56	II
1100	3755.24	II	75 d	4626.32	II	65	7596.44	I	15	3185.51	II
650	3759.35	I	95	4626.94	II	30	7627.81	I	15	3186.56	II

Intensity	Wavelength/Å		Intensity	Wavelength/Å		Intensity	Wavelength/Å		Intensity	Wavelength/Å	
15	3187.74	II	510	3119.526	II	310	3747.539	I	30	7045.795	II
1200	3229.75	I	510	3122.963	II	650	3752.569	II	30	7084.171	I
15	3291.01	II	480	3125.507	II	180	3770.056	I	30	7168.896	I
15	3369.15	II	100	3136.216	I	590	3803.075	I	40	7191.132	II
9	3456.34	III	420	3139.306	II	450	3828.384	I	35	7208.006	I
20000	3519.24	I	420	3142.835	II	840	3839.746	II	50	7525.508	II
5000	3529.43	I	420	3175.726	II	450	3863.405	II	30	7647.380	I
8	3540.08	II	1100	3180.193	II	210	3875.374	I	30	8330.451	I
9	3560.68	II	770	3188.233	II	340	3895.419	I	40	8967.641	I
12000 w	3775.72	I	560	3221.292	II	590	3929.669	II	20	9833.42	I
10	3832.30	II	560	3229.009	II	200	3932.911	I	20	10726.93	I
10	3887.15	II	480	3235.84	II	390	3967.392	I	20	10942.24	II
7	4109.85	III	590	3238.116	II	200	3972.155	I	30	11230.259	I
6	4269.81	III	910	3256.274	II	150	3980.089	I	20	11984.67	II
20	4274.98	II	180	3257.366	I	530	3994.549	II	20	17208.22	II
40	4306.80	II	910	3262.668	II	220	4008.210	I	15	18811.88	I
20	4737.05	II	620	3287.789	II	220	4009.056	I	10	22264.35	II
15	4981.35	II	910	3291.739	II	280	4012.495	I		**Thulium**	
25	5078.54	II	620	3292.520	II	4200	4019.129	II		**Tm Z = 69**	
25	5152.14	II	240	3301.650	I	250	4030.842	I	5000	2185.94	III
18000	5350.46	I	480	3304.238	I	250	4036.047	I	360	2284.79	II
15 d	5384.85	II	510	3321.450	II	250	4063.407	I	20000	2296.21	III
10	5410.97	II	840	3325.120	II	700	4086.520	II	5000	2305.03	III
25	5949.48	II	250	3330.476	I	700	4094.747	II	20000	2311.16	III
10	6179.98	II	620	3334.604	II	150	4100.341	I	5000	2312.72	III
10	6378.32	II	620	3337.870	II	840	4108.421	II	5000	2326.19	III
16 h	6549.84	I	310	3348.768	I	240	4112.754	I	6000	2328.50	III
10	6966.5	II	980	3351.228	II	280	4115.758	I	6000	2329.29	III
10	7815.80	I	620	3358.602	II	1100	4116.713	II	3000	2331.80	III
20	8373.6	I	250	3374.974	I	200	4127.411	I	3000	2357.05	III
10	8474.27	I	1300	3392.035	II	200	4134.067	I	4000	2406.63	III
10	8664.1	II	200	3396.727	I	450	4149.986	II	450	2409.02	II
20	9130.	II	250	3398.544	I	620	4178.060	II	450	2426.17	II
20	9130.5	I	200	3405.558	I	620	4208.890	II	770	2480.13	II
40	9509.4	I	250	3413.012	I	110	4253.538	I	30000	2489.44	III
20	9930.4	I	390	3421.210	I	110	4260.333	I	2000	2504.71	III
30	10011.9	I	270	3423.989	I	480	4277.313	II	1300	2509.08	II
40	10488.80	I	980	3433.998	II	700	4282.042	II	3000	2519.78	III
1000	11512.82	I	770	3435.976	II	130	4337.277	I	130	2527.02	I
150	12736.4	I	1300	3469.920	II	1300	4381.860	II	10000	2552.46	III
700	13013.2	I	170	3471.218	I	1100	4391.110	II	360	2552.76	I
	Thorium		200	3486.552	I	110	4498.940	I	540	2561.65	II
	Th Z = 90		670	3539.587	II	280	4510.527	II	430	2588.27	II
150	1707.37	IV	180	3544.018	I	90	4723.438	I	170 h	2596.49	I
200	1959.02	IV	170	3549.595	I	50	4840.843	I	810	2607.06	II
200	2002.34	IV	200	3555.013	I	280	4863.163	II	730	2624.33	II
200	2413.50	III	530	3559.451	II	260	5017.255	II	5000	2682.32	III
200	2427.94	III	200	3576.557	I	110	5067.974	I	2000	2707.03	III
200	2431.68	III	270	3592.780	I	120	5148.211	II	3000	2719.47	III
200	2441.24	III	270	3598.120	I	95	5216.596	II	540	2721.19	II
500	2565.593	II	980	3609.445	II	110	5231.160	I	3000	2724.44	II
480	2692.415	II	200	3612.427	I	95	5247.654	II	4000	2727.56	III
520	2747.156	II	480	3615.133	II	60	5343.581	I	680	2794.60	II
410	2752.166	II	270	3635.943	I	60	5587.026	I	730	2797.27	II
800	2832.315	II	210	3642.248	I	95	5707.103	II	2000	2806.77	III
1200	2837.295	II	170	3649.735	I	70	5760.551	I	580	2827.92	II
100	2848.084	I	220	3663.202	I	85	5989.044	II	200	2854.17	I
550	2870.406	II	280	3669.968	I	60	6169.822	I	1600	2869.23	II
100	2936.086	I	700	3675.567	II	50	6182.622	I	1000	2947.72	III
100	2943.729	I	150	3682.486	I	50	6274.116	II	490	2973.22	I
420	3049.092	II	170	3692.566	I	50	6274.117	II	1000	2998.28	III
450	3067.729	II	180	3698.105	I	50	6355.911	II	1500	3015.30	II
670	3078.828	II	340	3706.767	I	60	6457.283	I	360	3081.12	I
480	3080.217	II	590	3719.435	I	50	6462.614	I	7400	3131.26	II
510	3108.296	II	770	3721.825	II	50 h	6531.342	I	2300	3133.89	II
100	3116.263	I	1300	3741.183	II	55	6989.656	I	1900	3151.04	II

Intensity	Wavelength/Å		Intensity	Wavelength/Å		Intensity	Wavelength/Å		Intensity	Wavelength/Å	
1500	3157.34	II	1500	3958.10	II	40	7856.08	I	70	2058.31	I
450	3172.65	I	1800	3996.52	II	55	7927.51	I	80	2068.58	I
2300	3172.83	II	220	4024.23	I	110	7930.84	I	100	2072.89	I
1200	3236.81	II	380	4044.47	I	95	8017.90	I	100	2073.08	I
1600	3240.23	II	10000	4094.19	I	27	8472.01	II	200	2096.39	I
2300	3241.54	II	9500	4105.84	I		**Tin**		100	2100.93	I
320	3246.96	I	1100	4138.33	I		**Sn Z = 50**		100 r	2113.93	I
1900	3258.05	II	8800	4187.62	I	7	169.47	II	50	2121.26	I
1600	3266.64	II	6000	4203.73	I	150	361.01	V	40 r	2148.73	I
1200	3267.40	II	380	4222.67	I	100	753.01	III	20 r	2151.43	I
1100	3276.81	II	3000	4242.15	II	200	910.92	III	30	2151.54	II
1200	3283.40	II	270	4271.71	I	500	956.25	IV	80	2171.32	I
1200	3285.61	II	150	4298.36	I	7	985.13	II	150 r	2194.49	I
2300	3291.00	II	2700	4359.93	I	500	1019.72	IV	300 r	2199.34	I
2000	3302.46	II	1400	4386.43	I	1000	1044.49	IV	400 r	2209.65	I
1200	3309.80	II	200	4394.42	I	1000	1073.41	IV	80 r	2231.72	I
230	3349.99	I	140	4396.50	I	200	1089.35	V	400 r	2246.05	I
4000	3362.61	II	120	4454.03	I	8	1108.19	II	60	2251.17	I
1700	3397.50	II	540	4481.26	II	1000	1119.34	IV	400 r	2268.91	I
850	3410.05	I	150	4519.60	I	1000	1139.29	III	200 r	2286.68	I
340	3412.59	I	260	4522.57	II	1000	1158.33	III	600 r	2317.23	I
340	3416.59	I	110	4548.60	I	200	1160.74	V	300 r	2334.80	I
6400	3425.08	II	270	4599.02	I	10	1161.43	II	1000 r	2354.84	I
340	3429.33	I	300	4615.94	II	1000	1184.25	III	22	2368.33	II
4900	3441.50	II	80	4626.33	II	2000	1210.52	III	100	2408.15	I
4900	3453.66	II	95	4626.56	II	9	1219.07	II	800 r	2421.70	I
8500	3462.20	II	110	4634.26	II	13	1223.70	II	1000 r	2429.49	I
210	3467.51	I	120	4655.09	I	11	1243.00	II	15	2448.98	II
340	3476.69	I	160	4681.92	I	2000	1251.38	V	300	2483.39	I
340	3480.98	I	120	4691.11	I	1000	1259.92	III	13	2483.48	II
420	3487.38	I	110	4724.26	I	20	1290.86	II	10	2486.99	II
340	3499.95	I	680	4733.34	I	200	1294.36	V	200	2495.70	I
250	3517.60	I	70	4759.90	I	1000	1305.97	III	400	2546.55	I
1700	3535.52	II	80	4831.20	II	1000	1314.55	IV	500 r	2571.58	I
420	3537.91	I	140	4957.18	I	20	1316.59	II	200	2594.42	I
210	3555.82	I	160	5009.77	II	1000	1327.34	III	200 r	2661.24	I
340	3560.92•	I	160	5034.22	II	1000	1347.65	III	700 r	2706.51	I
420	3563.88	I	150	5060.90	I	1000	1386.74	III	150	2779.81	I
1300	3566.47	II	95	5113.97	I	25	1400.52	II	1400 r	2839.99	I
420	3567.36	I	80	5213.38	I	1000	1437.52	IV	1000 r	2863.32	I
280	3586.07	I	650	5307.12	I	20	1475.15	II	700 r	3009.14	I
2100	3608.77	II	80	5346.49	II	9	1489.22	II	850 r	3034.12	I
1000	3629.09	III	270	5631.41	I	1000	1570.36	III	12	3047.50	II
380	3638.41	I	520	5675.84	I	10 r	1737.21	I	550 r	3175.05	I
1100	3668.09	II	40	5684.76	II	15 r	1751.46	I	550 r	3262.34	I
4800	3700.26	II	35	5709.97	II	20 r	1764.98	I	50	3283.21	II
3800	3701.36	II	190	5764.29	I	30 r	1790.75	I	110	3330.62	I
7700	3717.91	I	35	5838.76	II	80 r	1804.60	I	60	3351.97	II
2400	3734.12	II	240	5895.63	I	15	1811.34	II	10	3472.46	II
5000	3744.06	I	140	5971.26	I	500	1811.71	III	11	3575.45	II
1700	3751.81	I	200	6460.26	I	40 r	1815.74	I	280 r	3801.02	I
6000	3761.33	II	95	6604.96	I	120 r	1823.00	I	10	5332.36	II
4800	3761.91	II	110	6779.77	I	9	1831.89	II	20	5561.95	II
7100	3795.75	II	120	6844.26	I	50 r	1848.75	I	25	5588.92	II
770	3798.54	I	80	6845.76	I	200 r	1860.32	I	500	5631.71	I
600	3807.72	I	10	6937.37	I	80	1886.05	I	15	5799.18	II
290	3826.39	I	10	7017.90	I	100	1891.40	I	50	5925.44	I
1300	3838.20	II	12	7034.34	I	12	1899.91	II	100	5970.30	I
290	3840.87	I	10	7106.14	I	50	1909.30	I	150	6037.70	I
8900	3848.02	II	17	7272.62	I	80	1925.31	I	250	6069.00	I
6800	3883.13	I	14	7310.51	I	500	1941.86	III	100	6073.46	I
1800	3883.44	II	14	7432.18	I	150	1952.15	I	400	6149.71	I
5400	3887.35	I	75	7481.08	I	50 h	1977.6	I	200	6154.60	I
440	3896.62	I	75	7490.20	I	80	1984.20	I	150	6171.50	I
3500	3916.48	I	140	7558.33	I	50	2040.66	I	100	6310.78	I
1500	3949.27	I	80	7731.53	I	50	2054.03	I	70	6453.50	II

Intensity	Wavelength/Å		Intensity	Wavelength/Å		Intensity	Wavelength/Å		Intensity	Wavelength/Å	
25	6844.05	II	24	2563.44	III	600	3461.50	II	6000	4533.24	I
20	7191.40	II	23	2565.42	III	600	3477.18	II	240	4533.97	II
10	7387.79	II	22	2567.56	III	480	3491.05	II	3600	4534.78	I
13	7741.80	II	270	2599.92	I	890	3504.89	II	2400	4535.58	I
100	7754.97	I	340	2605.15	I	600	3510.84	II	1200	4535.92	I
100 h	8030.5	I	510	2611.28	I	17	3576.44	IV	1200	4536.05	I
200	8114.09	I	300	2619.94	I	600	3610.16	I	720	4544.69	I
80	8357.04	I	640	2641.10	I	4800	3635.46	I	950	4548.77	I
300	8422.72	I	800	2644.26	I	6600	3642.68	I	240	4549.63	II
400	8552.60	I	950	2646.64	I	7200	3653.50	I	15	4549.84	III
50 h	8681.7	I	250	2742.32	I	600	3671.67	I	950	4552.46	I
50 h	9410.86	I	250	2802.50	I	3100	3685.20	II	720	4555.49	I
80 h	9415.37	I	190	2841.94	II	600	3689.91	I	240	4571.98	II
150	9616.40	I	180	2877.44	II	2900	3729.82	I	15 d	4572.20	III
50	9741.1	I	280	2884.11	II	3300	3741.06	I	950	4617.27	I
100 h	9742.8	I	450	2912.08	I	330	3741.64	II	480	4623.09	I
300 h	9805.38	I	340	2928.34	I	5200	3752.86	I	720	4656.47	I
500	9850.52	I	1100	2942.00	I	3300	3759.30	II	840	4667.59	I
54	10894.00	I	1300	2948.26	I	2900	3761.32	II	950	4681.92	I
70	11191.85	I	1600	2956.13	I	840	3786.04	I	470	4840.87	I
56	11277.66	I	22	2984.75	III	500	3882.89	I	400	4885.08	I
200	11454.59	I	1300 d	3066.22	II	530	3900.54	II	380	4899.91	I
200	11616.26	I	1100	3072.97	II	2600	3904.78	I	5800	4981.73	I
258	11739.78	I	1600	3075.22	II	500	3913.46	II	4600	4991.07	I
96	11825.18	I	2300	3078.64	II	500	3914.34	I	4000	4999.51	I
106	11835.82	I	3600	3088.02	II	15	3915.47	III	3600	5007.21	I
254	11932.99	I	720	3119.72	I	1100	3924.53	I	3200 d	5014.19	I
48	12009.50	I	500	3161.20	II	890	3929.88	I	840	5020.03	I
111	12313.24	I	780	3161.77	II	1100	3947.78	I	840	5022.87	I
42	12530.87	I	1000	3162.57	II	4500	3948.67	I	1200	5035.91	I
42	12536.5	I	1600	3168.52	II	4500	3956.34	I	840	5036.47	I
89	12888.5	I	2400	3186.45	I	5200	3958.21	I	740	5038.40	I
187	12981.7	I	1000	3190.87	II	950	3962.85	I	1200	5039.95	I
187	13018.5	I	3100	3191.99	I	950	3964.27	I	1400	5064.66	I
68	13081.5	I	3800	3199.92	I	4800	3981.76	I	1100	5173.75	I
378	13460.2	I	780	3202.54	II	570	3982.48	I	1300	5192.98	I
144	13608.2	I	1100	3217.06	II	5700	3989.76	I	1400	5210.39	I
40	20861.7	I	1300	3222.84	II	7800	3998.64	I	17	5278.12	III
4	24738.2	I	6600	3234.52	II	950	4008.93	I	20	5398.93	IV
	Titanium		5200	3236.57	II	1200	4024.57	I	340	5512.53	I
	Ti Z = 22		4100	3239.04	II	840	4078.47	I	270	5514.35	I
17	252.96	V	2600	3241.99	II	890	4286.01	I	320	5514.54	I
15	498.26	V	1200	3248.60	II	840	4287.40	I	250	5644.14	I
14	502.08	V	1200	3252.91	II	950	4289.07	I	130	5675.44	I
13	526.57	V	1200	3254.25	II	840	4290.94	I	95	5689.47	I
18	779.07	IV	1200	3261.60	II	840	4295.76	I	95	5715.13	I
20	1298.66	III	840	3314.42	I	2000	4298.66	I	85	5739.51	I
20	1298.97	III	2900	3322.94	II	200	4300.05	II	400	5866.46	I
23	1455.19	III	2100	3329.46	II	2900	4300.56	I	230	5899.32	I
20	1467.34	IV	1800	3335.20	II	4100	4301.09	I	120	5918.55	I
11	1717.40	V	1100	3340.34	II	6000	4305.92	I	150	5922.12	I
10	1841.49	V	5700	3341.88	II	1200	4314.80	I	120	5941.76	I
20	2067.56	IV	4300	3349.04	II	330	4395.04	II	300	5953.17	I
18	2103.16	IV	12000	3349.41	II	890	4427.10	I	200	5965.84	I
180	2273.28	I	4100	3354.64	I	230	4443.80	II	270	5978.56	I
190	2279.96	I	7200	3361.21	II	840	4449.15	I	340	5999.04	I
190	2305.67	I	1100	3370.44	I	550	4450.90	I	110	6064.63	I
22	2413.99	III	4300	3371.45	I	840	4453.32	I	120	6085.23	I
25	2516.05	III	5700	3372.80	II	950	4455.33	I	120	6091.17	I
360	2525.60	II	2900 d	3377.48	I	1100	4457.43	I	120	6126.22	I
24	2527.84	III	1400	3380.28	II	240	4468.50	II	17	6246.65	IV
210	2529.85	I	5700	3383.76	II	530	4481.26	I	380	6258.10	I
190	2531.25	II	1400	3385.95	I	780	4512.74	I	380	6258.70	I
190	2534.62	II	1400	3387.84	II	1000	4518.03	I	300	6261.10	I
130	2535.87	II	1100	3394.58	II	1000	4522.80	I	55	6546.28	I
23	2540.06	III	890	3444.31	II	780	4527.31	I	65	6554.23	I

Intensity	Wavelength/Å		Intensity	Wavelength/Å		Intensity	Wavelength/Å		Intensity	Wavelength/Å	
75	6556.07	I	1400	2466.85	I	810	2848.02	I	1000	4102.70	I
18	6621.58	III	480	2472.51	I	1500	2896.44	I	540	4137.46	I
18	6667.99	III	1200	2474.15	I	690	2935.00	I	450	4171.17	I
80	6743.12	I	870	2480.13	I	2400	2944.40	I	220	4207.05	I
20	7072.64	III	1500	2481.44	I	2400	2946.99	I	250	4219.37	I
18	7084.57	III	480 d	2482.10	I	730 d	2979.71	I	540	4244.36	I
260	7209.44	I	580	2484.74	I	360	3013.79	I	1400	4269.38	I
130	7244.86	I	390	2487.50	I	520	3016.47	I	4100	4294.61	I
130	7251.72	I	390	2489.23	II	770	3017.44	I	2200	4302.11	I
120	7344.72	I	630	2495.26	I	210	3024.93	I	200	4378.48	I
90	7357.74	I	680	2504.70	I	310 d	3026.67	I	180	4384.85	I
60	7364.11	I	75	2510.47	II	440 d	3041.73	I	200	4408.28	I
60	7978.88	I	310	2520.46	I	270	3043.80	I	640	4484.19	I
55	8024.84	I	780	2521.32	I	440	3046.44	I	170	4588.73	I
75	8364.24	I	270	2522.04	II	810	3049.69	I	640	4659.87	I
100	8377.85	I	780	2523.41	I	180	3073.28	I	640	4680.51	I
100	8382.54	I	430	2527.76	I	180 d	3084.83	I	790	4843.81	I
75	8396.87	I	780	2533.64	I	370	3093.50	I	380	4886.90	I
120	8412.36	I	1200	2547.14	I	240	3107.23	I	220	4982.59	I
170	8426.52	I	780	2550.38	I	240	3108.02	I	820	5053.28	I
490	8434.94	I	2700	2551.35	I	230	3117.57	I	770	5224.66	I
240	8435.70	I	730	2561.97	I	260	3120.18	I	220	5514.68	I
20	8466.87	III	870	2580.49	I	290	3163.42	I	65	5648.37	I
90	8675.39	I	390	2584.39	I	320	3176.60	I	55	5735.09	I
	Tungsten		390	2589.17	II	190	3181.82	I	45	5804.85	I
	W Z = 74		370	2601.96	I	390	3191.57	I	40	5902.64	I
5800	2001.71	II	680	2606.39	I	390	3198.84	I	55	5947.57	I
13000	2008.07	II	370	2608.32	I	520	3207.25	I	55	5965.86	I
5100	2009.98	II	970	2613.08	I	1000	3215.56	I	55	6012.78	I
4100	2010.23	II	480	2613.82	I	190	3232.49	I	40	6021.52	I
4100	2014.23	II	400	2620.25	I	210	3254.36	I	45	6292.02	I
7300	2026.08	II	400	2622.21	I	210	3259.66	I	35	6404.21	I
15000	2029.98	II	400	2625.22	I	210 d	3266.62	I	40	6445.12	I
5300	2049.63	II	400	2632.48	I	730	3300.82	I	17	6611.62	I
9700	2079.11	II	400	2632.70	I	440	3311.38	I	13	6678.42	I
6100	2094.75	II	810	2633.13	I	440	3326.20	I	15	6693.08	I
2100	2118.87	II	400 d	2638.62	I	440	3331.69	I	13	6984.27	I
2400	2121.59	II	650	2646.18	I	390	3373.75	I	15	7140.52	I
1500	2166.32	II	400	2646.73	I	230	3429.59	I	9	7162.64	I
1300	2204.48	II	1600	2656.54	I	240	3443.00	I	11	7200.16	I
460	2249.80	I	810	2662.84	I	400	3495.24	I	10	7278.24	I
510	2277.58	I	810	2671.47	I	650	3545.22	I	15	7285.81	I
530 d	2294.49	I	650	2677.28	I	240	3570.65	I	15	7296.55	I
340	2309.02	I	2100	2681.42	I	1900	3617.52	I	10	7509.00	I
440	2313.17	I	650	2695.67	I	650	3682.08	I	17	7569.92	I
460	2321.63	I	650	2699.59	I	400	3683.30	I	17	7614.15	I
390 d	2326.56	I	400	2700.01	I	570	3688.06	I	13	7688.97	I
320	2354.61	I	400	2706.58	I	810	3707.92	I	11	7784.15	I
580	2360.44	I	400	2708.59	I	510	3757.92	I	22	8017.19	I
850	2363.07	I	400 d	2708.80	I	680	3760.13	I	22	8055.64	I
510	2374.47	I	400	2715.50	I	1000	3768.45	I	13	8123.82	I
670	2384.82	I	2100	2718.91	I	340	3773.71	I	10	8338.08	I
730	2397.09	II	2600	2724.35	I	1000	3780.77	I	27	8585.11	I
560	2397.73	I	400	2725.03	I	290	3809.22	I	10	8594.42	I
560	2397.98	I	650	2748.84	I	190	3810.38	I	13	8865.53	I
1700 d	2405.58	I	400	2762.34	I	260	3810.79	I		**Uranium**	
610	2415.68	I	400	2764.27	II	1400	3817.48	I		**U Z = 92**	
870	2424.21	I	400	2769.74	I	1100	3835.06	I	440	2565.41	II
1800	2435.96	I	810	2770.88	I	730	3846.22	I	610	2635.53	II
580	2444.06	I	810	2774.00	I	1800	3867.99	I	830	2793.94	II
780	2451.48	II	810	2774.48	I	730	3881.41	I	870	2802.56	II
870	2452.00	I	810	2792.70	I	8600	4008.75	I	630	2807.05	II
630	2454.98	I	400	2799.93	I	540	4015.22	I	630	2817.96	II
780	2455.51	I	810	2818.06	I	910	4045.59	I	870	2821.12	II
780	2456.53	I	1600	2831.38	I	730	4069.95	I	680	2828.90	II
1100	2459.30	I	810	2833.63	I	5000	4074.36	I	920	2832.06	II

Line Spectra of the Elements (continued): Uranium—Vanadium

Intensity	Wavelength/Å		Intensity	Wavelength/Å		Intensity	Wavelength/Å		Intensity	Wavelength/Å	
970	2865.68	II	1000	3881.45	II	100	1680.20	V	490	3592.02	II
1200	2889.62	II	2200	3890.36	II	1000	1694.78	III	560	3592.53	I
780	2906.80	II	2000	3932.02	II	1000	1760.07	III	100	3679.86	III
780	2908.28	II	1200	3943.82	I	1000	1788.26	III	1300	3688.07	I
580	2931.41	II	1200	3985.79	II	1000	1794.60	III	1000	3690.28	I
530 p	2940.37	II	1000	4042.75	I	1000	1812.19	III	1500	3692.22	I
1300	2941.92	II	1600	4050.04	II	300	1861.56	IV	1000	3695.86	I
830	2943.90	II	880	4062.54	II	500	1939.06	IV	3800	3703.58	I
580	2956.06	II	2200	4090.13	II	400	1951.43	IV	1800	3704.70	I
580	2967.94	II	810	4116.10	II	500	1997.72	IV	320	3715.47	II
580	2971.06	II	880	4153.97	I	2100	2092.44	I	250	3727.34	II
530	2984.61	II	1400	4171.59	II	500	2268.30	IV	280	3732.76	II
630	3022.21	II	1000	4241.67	II	1000	2292.86	III	520	3790.32	I
630	3031.99	II	600	4472.33	II	2500	2330.42	III	1100	3794.96	I
580	3050.20	II	620	4543.63	II	2500	2371.06	III	570	3799.91	I
630	3057.91	II	170	4689.07	II	1000	2382.46	III	570	3803.47	I
630	3062.54	II	150	4756.81	I	240	2507.78	I	1000	3813.49	I
580	3072.78	II	110	5008.21	II	410	2526.22	I	1300	3818.24	I
580	3093.01	II	170	5027.38	I	210	2527.90	II	1700	3828.56	I
580	3102.39	II	80	5160.32	II	80 h	2570.72	IV	2600	3840.75	I
970	3111.62	II	70	5280.38	I	230	2574.02	I	1200	3855.37	I
530	3119.35	II	80	5475.70	II	250	2593.05	III	3000	3855.84	I
680	3124.95	II	70	5480.26	II	250	2595.10	III	1300	3864.86	I
530	3139.61	II	70	5481.20	II	80 h	2645.54	IV	1500	3875.08	I
680	3149.24	II	160	5492.95	II	180	2661.42	I	700	3890.18	I
530	3153.11	II	70	5780.59	I	1100	2687.96	II	2400	3902.25	I
730	3229.50	II	70	5798.53	II	680	2700.94	II	700	3909.89	I
680	3232.16	II	230	5915.39	I	530	2706.17	II	540	3990.57	I
730	3291.33	II	100	5976.32	I	640	2715.69	II	430	3998.73	I
1100	3305.89	II	90	6077.29	I	180	2731.35	I	170	4005.71	II
730	3390.38	I	55	6372.46	I	240	2864.36	I	1100	4090.58	I
580	3424.56	II	90	6395.42	I	900	2891.64	II	1800	4092.69	I
580	3435.49	I	110	6449.16	I	900	2892.66	II	890	4095.49	I
630	3466.30	I	90	6826.92	I	1400	2893.32	II	2800	4099.80	I
680	3482.49	II	45	7533.93	I	900	2906.46	II	590	4102.16	I
1600	3489.37	I	50	7881.94	I	2400	2908.82	II	2800	4105.17	I
530	3496.41	II	35	8445.39	I	710	2923.62	I	2300	4109.79	I
630	3500.08	I	75	8607.95	I	2400	2924.02	II	8900	4111.78	I
780	3507.34	I	30	8757.76	I	1700	2924.64	II	4300	4115.18	I
1600	3514.61	I	100	10554.93	I	900	2941.37	II	1800	4116.47	I
630	3533.57	II	75	11167.84	I	1100	2944.57	II	2000	4123.57	I
530	3540.47	II	100	11384.13	I	410	2962.77	I	3100	4128.07	I
1200	3550.82	II	100	11859.42	I	600	2968.38	II	3100	4132.02	I
680	3555.32	I	100	11908.83	I	1200	3056.33	I	2300	4134.49	I
1200	3561.80	I	100	12250.46	I	1400	3060.46	I	20	4200.32	V
2300	3566.59	I	100	13185.16	I	2400	3066.38	I	360	4232.46	I
530	3569.08	I	75	13306.23	I	3800	3093.11	II	560	4268.64	I
630	3578.72	II	100	13961.58	I	3000	3102.30	II	460	4271.55	I
3200	3584.88	I	75	18634.43	I	2600	3110.71	II	460	4276.96	I
840	3638.20	I	75	21910.22	I	2000	3118.38	II	430	4284.06	I
2800	3670.07	II		**Vanadium**		1500	3125.28	II	460	4330.02	I
1100	3701.52	II		**V Z = 23**		3200	3183.41	I	510	4332.82	I
600	3738.04	II	20	225.46	V	5300	3183.98	I	760	4341.01	I
680	3746.42	II	20	251.66	V	3800	3185.40	I	1000	4352.87	I
950	3748.68	II	20	286.84	V	410	3187.71	II	12000	4379.24	I
600	3751.17	I	35	483.01	V	530	3188.51	II	7000	4384.72	I
1900	3782.84	II	50	633.94	III	750	3190.68	II	4800	4389.97	I
570	3793.10	II	200	677.34	IV	1100	3267.70	II	3600	4395.23	I
1900	3811.99	I	500	684.37	IV	900	3271.12	II	1400	4400.58	I
750	3826.51	II	400	737.85	IV	750	3276.12	II	2300	4406.64	I
2000	3831.46	II	100	864.27	III	80 h	3514.25	IV	2800	4407.64	I
1200	3839.63	I	500	1006.46	III	560	3517.30	II	3600	4408.20	I
2400	3854.64	II	500	1149.94	III	560	3533.68	I	4600	4408.51	I
4900	3859.57	II	100	1426.65	IV	560	3545.20	II	640	4416.47	I
1900	3865.92	II	1000	1643.03	III	560	3556.80	II	640	4421.57	I
1500	3871.03	I	1000	1650.14	III	560	3589.76	II	640	4437.84	I

Intensity	Wavelength/Å		Intensity	Wavelength/Å		Intensity	Wavelength/Å		Intensity	Wavelength/Å	
830	4441.68	I	29 c	8027.39	I	30	2827.45	III	10	3669.91	I
640	4444.21	I	120 w	8116.80	I	40	2847.65	III	50	3676.67	III
610	4452.01	I	70 c	8161.07	I	30	2862.40	III	40	3685.90	I
1000	4459.76	I	60 c	8919.85	I	200	2864.73	II	40	3693.49	I
2000	4460.29	I		**Xenon**		80 w	2871.10	III	40	3776.3	III
610	4462.36	I		**Xe Z = 54**		60 w	2871.24	III	300	3781.02	III
510	4577.17	I	8	657.8	III	30	2871.7	III	100	3841.5	III
640	4580.40	I	8	660.1	III	150 h	2895.22	II	200	3877.8	III
830	4586.36	I	9	673.8	III	30	2896.62	III	60	3880.5	III
1300	4594.11	I	9	674.0	III	50	2906.6	III	100 1	3907.91	II
230	4619.77	I	9	676.6	III	40	2911.89	III	500	3922.55	III
100	4635.18	I	10	694.0	III	80 w	2912.36	III	300	3950.59	III
130	4646.40	I	20	698.5	III	40	2940.2	III	100	4037.59	II
160	4670.49	I	12	705.1	III	60	2945.2	III	200	4050.07	III
130	4776.36	I	10	721.2	III	40	2947.5	III	200 1	4057.46	II
110	4786.51	I	15	731.0	III	40	2948.1	III	60	4060.4	III
130	4796.92	I	10	733.3	III	80 w	2970.47	III	100 h	4098.89	II
130	4807.53	I	350	740.41	II	400	2979.32	II	100	4109.1	III
130	4827.45	I	15	742.6	III	40	2992.87	III	100	4145.7	III
150	4831.64	I	10	756.0	III	30	3004.25	III	200 1	4158.04	II
120	4832.43	I	10	761.5	III	100 h	3017.43	II	1000 h	4180.10	II
320	4851.48	I	10	769.1	III	100	3023.81	III	500 h	4193.15	II
480	4864.74	I	25	779.1	III	40	3083.5	III	300 h	4208.48	II
620	4875.48	I	15	792.9	III	50	3091.1	III	100 h	4209.47	II
740	4881.56	I	12	796.1	III	30	3106.46	III	300 h	4213.72	II
110	5128.53	I	15	802.0	III	300	3128.87	II	100	4215.60	II
110	5138.42	I	350	803.07	II	100 w	3138.3	III	300 h	4223.00	II
110	5192.99	I	25	823.2	III	80 c	3150.82	III	400 h	4238.25	II
110	5194.83	I	30	824.9	III	40	3185.2	III	500 h	4245.38	II
110	5234.07	I	25	853.0	III	100	3242.86	III	100 1	4251.57	II
110	5240.87	I	600	880.80	II	80	3268.98	III	30	4285.9	III
100	5401.93	I	350	885.54	II	30	3287.82	III	500 h	4296.40	II
140	5415.26	I	15	889.3	III	80 w.	3301.55	III	500 h	4310.51	II
140	5584.50	I	20	894.0	III	40	3331.6	III	1000 1	4330.52	II
100	5592.42	I	20	896.0	III	30	3358.0	III	200 h	4369.20	II
200	5624.60	I	600	925.87	II	200 h	3366.72	II	100 1	4373.78	II
400	5627.64	I	250	935.40	II	80	3384.12	III	500 h	4393.20	II
110	5657.44	I	10	965.5	III	2	3400.07	I	500 1	4395.77	II
110	5668.36	I	800	972.77	II	2	3418.37	I	200 1	4406.88	II
310	5670.85	I	700	976.68	II	2	3420.00	I	150 1	4416.07	II
1200	5698.52	I	35	1003.4	III	3	3442.66	I	50	4434.2	III
920	5703.56	I	35	1017.7	III	60	3444.2	III	500 h	4448.13	II
570	5706.98	I	500	1032.44	II	70	3454.2	III	100 w	4462.1	III
850	5727.03	I	700	1037.68	II	100 w	3458.7	III	1000 h	4462.19	II
230	5731.25	I	1100	1041.31	II	100 h	3461.26	II	500 1	4480.86	II
230	5737.06	I	10	1047.8	III	40	3468.22	III	100 1	4521.86	II
450	6039.73	I	1000	1048.27	II	4	3469.81	I	100 w	4569.1	III
480	6081.44	I	1200	1051.92	II	4	3472.36	I	100 w	4570.1	III
1300	6090.22	I	12	1066.4	III	5	3506.74	I	100 w	4641.4	III
600	6119.52	I	2000	1074.48	II	80	3522.83	III	30	4673.7	III
450	6199.19	I	600	1083.86	II	50	3542.3	III	60	4683.57	III
450	6216.37	I	1200	1100.43	II	10	3549.86	I	30	4723.60	III
430	6230.74	I	30	1130.3	III	50	3552.1	III	600	4734.152	I
710	6243.10	I	600	1158.47	II	10	3554.04	I	100 w	4757.3	III
280	6251.82	I	250	1169.63	II	40	3561.4	III	150	4792.619	I
130	6268.82	I	800 p	1183.05	II	100	3579.7	III	500	4807.02	I
170	6274.65	I	250	1192.04	I	80	3583.6	III	400	4829.71	I
200	6285.16	I	25	1232.1	III	100 w	3595.4	III	300	4843.29	I
200	6292.83	I	600	1244.76	II	100	3606.06	III	40	4869.5	III
170	6296.49	I	250	1250.20	I	40	3607.0	III	500	4916.51	I
110	6531.43	I	1000	1295.59	I	15	3610.32	I	500	4923.152	I
65 c	6753.00	I	600	1469.61	I	8	3613.06	I	200 1	4971.71	II
50 c	6766.49	I	80	2668.98	III	100 w	3615.9	III	400	4972.71	II
40	6784.98	I	100	2717.33	III	40	3623.1	III	300	4988.77	II
40	7338.92	I	30	2814.45	III	600	3624.08	III	100 1	4991.17	II
35	7356.54	I	40	2815.91	III	6	3633.06	I	200	5028.280	I

Line Spectra of the Elements (continued): Xenon—Ytterbium

Intensity	Wavelength/Å		Intensity	Wavelength/Å		Intensity	Wavelength/Å		Intensity	Wavelength/Å	
200	5044.92	II	100	6198.26	I	100	8101.98	I	175	25145.84	I
1000	5080.62	II	60	6205.97	III	150 h	8151.80	II	2000	26269.08	I
300	5122.42	II	100	6220.02	II	100	8171.02	I	2500	26510.86	I
100	5125.70	II	25	6221.7	III	700	8206.34	I	250	28381.54	I
100	5178.82	II	60	6238.2	III	10000	8231.635	I	750	28582.25	I
300	5188.04	II	60	6259.05	III	500	8266.52	I	300	29384.41	I
400	5191.37	II	500	6270.82	II	7000	8280.116	I	150	29448.06	I
100	5192.10	II	400	6277.54	II	2000	8346.82	I	100	29649.58	I
60	5239.0	III	100	6284.41	II	100	8347.24	II	100	29813.62	I
500	5260.44	II	100	6286.01	I	2000	8409.19	I	600	30253.14	I
500	5261.95	II	250	6300.86	II	50 h	8515.19	II	1500	30475.46	I
2000	5292.22	II	500	6318.06	I	200	8576.01	I	100	30504.12	I
300	5309.27	II	400	6343.96	II	50 h	8604.23	II	500	30794.18	I
1000	5313.87	II	600	6356.35	II	250	8648.54	I	6000	31069.23	I
2000	5339.33	II	200	6375.28	II	100	8692.20	I	125	31336.01	I
200	5363.20	II	100	6397.99	II	200	8696.86	I	550	31607.91	I
30	5367.1	III	300	6469.70	I	50 h	8716.19	II	100	32293.08	I
200	5368.07	II	150	6472.84	I	300	8739.39	I	1800	32739.26	I
500	5372.39	II	120	6487.76	I	100	8758.20	I	3500	33666.69	I
100	5392.80	I	100	6498.72	I	5000	8819.41	I	150	34014.67	I
50	5401.0	III	200 h	6504.18	I	300	8862.32	I	450	34335.27	I
3000	5419.15	II	300	6512.83	II	200	8908.73	I	170	34744.00	I
800	5438.96	II	200	6528.65	II	200	8930.83	I	5000	35070.25	I
300	5445.45	II	100	6533.16	I	1000	8952.25	I	110	35246.92	I
200	5450.45	II	1000	6595.01	II	100	8981.05	I	250	36209.21	I
400	5460.39	II	100	6595.56	I	200	8987.57	I	150	36231.74	I
1000	5472.61	II	400	6597.25	II	400	9045.45	I	450	36508.36	I
100 l	5494.86	II	100	6598.84	II	500	9162.65	I	850	36788.83	I
40	5524.4	III	150	6668.92	I	100	9167.52	I	140	38685.98	I
200	5525.53	II	300	6694.32	II	100	9374.76	I	175	38737.82	I
600	5531.07	II	200	6728.01	I	200	9513.38	I	270	38939.60	I
100	5566.62	I	150	6788.71	II	50 h	9591.35	II	120	39955.14	I
300	5616.67	II	100	6790.37	II	150	9685.32	I		**Ytterbium**	
300	5659.38	II	1000	6805.74	II	50 l	9698.68	II		**Yb Z = 70**	
600	5667.56	II	200	6827.32	I	100	9718.16	I	1000	1050.24	IV
150	5670.91	II	100	6872.11	I	2000	9799.70	I	1000	1054.46	IV
100	5695.75	I	300	6882.16	I	3000	9923.19	I	5000	1134.43	IV
200	5699.61	II	80	6910.22	II	100	10838.37	I	900	1316.04	IV
200	5716.10	II	100	6925.53	I	90	11742.01	I	800	1326.36	IV
500	5726.91	II	800 h	6942.11	II	375	12235.24	I	900	1350.26	IV
500	5751.03	II	100	6976.18	I	100	12257.76	I	80	1561.42	III
300	5758.65	II	2000	6990.88	II	300	12590.20	I	80 h	1765.21	III
300	5776.39	II	150	7082.15	II	2500	12623.391	I	800	1791.06	IV
100	5815.96	II	500	7119.60	II	250	13544.15	I	100	1863.32	III
300	5823.89	I	50 s	7147.50	II	2000	13657.055	I	800	1873.91	III
150	5824.80	I	200	7149.03	II	1250	14142.444	I	500	1898.25	III
100	5875.02	I	500	7164.83	II	800	14240.96	I	500	1998.82	III
300	5893.29	II	100	7284.34	II	375	14364.99	I	900	2116.65	IV
100	5894.99	I	200	7301.80	II	140	14660.81	I	2500	2116.67	II
200	5905.13	II	200	7339.30	II	3000	14732.806	I	800	2123.32	IV
100	5934.17	I	100	7386.00	I	100	15099.72	I	3000	2126.74	II
500	5945.53	II	150	7393.79	I	2500	15418.394	I	800	2139.99	IV
300	5971.13	II	300	7548.45	II	150	15557.13	I	20000	2144.77	IV
2000	5976.46	II	200	7584.68	I	250	15979.54	I	15000	2154.18	IV
200	6008.92	II	80	7618.57	II	100	16039.90	I	370	2161.60	II
1000	6036.20	II	500	7642.02	I	1000	16053.28	I	850	2185.71	II
2000	6051.15	II	100	7643.91	I	125	16554.49	I	640	2224.46	III
600	6093.50	II	200	7670.66	II	1500	16728.15	I	300	2240.11	III
1500	6097.59	II	60	7787.04	II	1500	17325.77	I	300	2305.32	III
400	6101.43	II	100	7802.65	I	350	18788.13	I	140	2320.81	II
100	6115.08	II	100	7881.32	I	150	20187.19	I	170	2390.74	II
100	6146.45	II	300	7887.40	I	3000	20262.242	I	460	2464.50	I
150	6178.30	I	500	7967.34	I	250	21470.09	I	140	2512.06	II
120	6179.66	I	100	8029.67	I	1250	23193.33	I	270	2538.67	II
300	6182.42	I	200	8057.26	I	110	23279.54	I	2000	2567.61	III
500	6194.07	II	150	8061.34	I	1800	24824.71	I	1000	2579.57	III

Intensity	Wavelength/Å		Intensity	Wavelength/Å		Intensity	Wavelength/Å		Intensity	Wavelength/Å	
800	2599.14	III	2000	3384.01	III	40	4837.46	I	300	403.45	V
600	2621.11	III	140	3387.50	I	40 h	4894.60	I	300	420.74	V
1000	2642.56	III	50	3412.45	I	27	4912.36	I	600	425.03	IV
1000	2651.74	III	140	3418.39	I	710	4935.50	I	300	473.10	IV
700	2652.25	III	360	3426.04	I	140	4966.90	I	4000	584.98	V
990	2653.75	II	240	3431.11	I	30	5067.80	I	2000	630.97	V
200	2665.04	II	85	3452.40	I	70	5069.14	I	5000	805.20	III
2000	2666.13	III	500	3454.08	II	220	5074.34	I	7000	809.92	III
2000	2666.99	III	190 d	3458.29	II	50	5076.74	I	15000	989.21	III
390	2671.96	I	360	3460.27	I	60	5196.08	I	25000	996.37	III
390	2672.66	II	2400	3464.37	I	85	5211.60	I	5000	1314.51	III
170	2718.35	II	500	3476.30	II	100	5244.11	I	4000	1334.04	III
230	2748.66	II	500	3478.84	II	150 h	5277.04	I	4000	2068.98	III
1300	2750.48	II	50	3517.00	I	170	5335.15	II	10000	2127.98	III
170	2776.28	II	230	3520.29	II	30 h	5351.29	I	16000	2191.16	III
600	2795.60	III	35	3559.03	I	150	5352.95	II	350	2243.06	II
1000	2803.43	III	200	3560.33	I	30	5363.66	I	10000	2284.34	III
600	2816.92	III	170	3560.70	I	40	5449.27	II	10000	2327.31	III
1000	2818.72	III	360	3585.47	II	60	5481.92	I	50	2354.20	I
140	2821.15	II	200	3619.80	II	40	5505.49	I	50000	2367.23	III
190	2830.99	II	240	3637.76	II	17	5524.54	I	40000	2414.64	III
230 h	2847.18	II	70	3648.15	I	85 h	5539.05	I	560	2422.20	II
360	2851.13	II	90	3655.73	I	2400	5556.47	I	60	2694.21	I
430	2859.80	II	240	3669.69	II	60	5651.98	I	95	2723.00	I
140	2861.21	II	140	3675.08	II	220	5719.99	I	70	2742.53	I
200	2867.06	II	32000	3694.19	II	27	5771.66	II	140	2760.10	I
45	2873.49	I	70	3700.58	I	35	5833.99	II	90000	2817.04	III
200	2888.04	II	400	3711.91	III	35	5837.14	II	45	2822.56	I
3600	2891.38	II	180	3734.69	I	27	5854.51	I	70	2854.43	II
600	2898.30	III	550	3770.10	I	17	5989.33	I	95	2886.48	I
1000	2906.31	II	80	3774.32	I	40	5991.51	I	160	2919.05	I
170	2914.21	II	60 h	3791.74	I	60	6152.57	II	99000	2946.01	III
140	2915.28	II	170	3839.91	I	60	6274.78	II	390	2948.40	I
280	2919.35	II	340	3872.85	I	200	6328.52	III	350	2964.96	I
35	2934.36	I	340	3900.85	I	35 h	6400.35	I	480	2974.59	I
140	2945.91	II	140	3911.27	I	35 h	6417.91	I	750	2984.26	I
2000	2970.56	II	500	3931.23	III	340	6489.06	I	140	2996.94	I
200	2983.99	II	32000	3987.99	I	180	6667.82	I	130	3021.73	I
170	2994.80	II	930	3990.88	I	25	6727.61	II	190	3045.37	I
800	2998.00	III	50	4007.36	I	690	6799.60	I	95	3095.88	II
310	3005.77	II	2000	4028.14	III	9 h	7244.41	I	110	3173.06	II
160	3017.56	II	70	4052.28	I	8 h	7305.22	I	220	3179.41	II
160	3026.67	II	440	4089.68	I	10 h	7313.05	I	70	3191.31	I
2000	3029.49	III	470	4149.07	I	16 h	7350.04	I	2300	3195.62	II
920	3031.11	II	120	4174.56	I	25	7448.28	I	2200	3200.27	II
3000	3092.50	III	340	4180.81	II	30 h	7527.46	I	2200	3203.32	II
28	3100.74	I	300	4213.64	III	750	7699.48	I	3900	3216.69	II
170	3107.90	II	150 d	4218.56	II	100	7971.46	III	6200	3242.28	II
190	3117.81	II	120	4231.97	I	70 h	8922.56	II	4700	3327.89	II
4000	3126.01	III	70	4277.74	I	200	10110.60	III	85	3388.59	I
1000	3138.58	II	120	4305.97	I	100	10830.36	III	85	3412.47	I
230	3140.94	II	60 h	4393.69	I	**Yttrium**			170	3485.73	I
28	3162.29	I	60 h	4430.21	I	**Y Z = 39**			1700	3496.09	II
800	3191.35	III	440	4439.19	I	150	264.64	IV	3900	3549.01	II
390	3192.88	II	85 h	4482.42	I	150	273.03	IV	130	3551.80	I
240	3201.16	II	100	4517.58	III	900	333.09	V	540	3552.69	I
2000	3228.58	III	85 h	4563.95	I	500	333.80	V	170	3558.76	I
35	3239.58	I	640	4576.21	I	400	335.14	V	190	3571.43	I
18000	3289.37	I	200	4582.36	I	500	336.62	V	260	3576.05	I
130	3305.25	I	70	4589.21	I	500	339.02	V	3300	3584.52	II
140	3305.73	II	140	4590.83	I	500	344.59	V	300	3587.75	I
80	3319.41	I	40	4684.27	I	900	355.86	IV	100	3589.69	I
2000	3325.51	III	190	4726.08	II	300	370.42	IV	2800	3592.92	I
240	3337.17	II	170 h	4781.87	I	300	372.05	V	10000	3600.73	I
280 d	3342.93	II	170	4786.61	II	400	379.96	V	6200	3601.92	II
240	3375.48	II	35	4816.43	I	500	386.82	IV	7800	3611.05	II

Intensity	Wavelength/Å		Intensity	Wavelength/Å		Intensity	Wavelength/Å		Intensity	Wavelength/Å	
4300	3620.94	I	170	4786.89	I	24 h	6950.31	I	200	2670.53	I
1900	3628.71	II	180	4799.30	I	24	6979.88	I	300	2684.16	I
7800	3633.12	II	140	4819.64	I	29	7052.94	I	300	2712.49	I
3000	3664.61	II	120	4822.13	I	35	7191.66	I	200	2756.45	I
170	3692.53	I	770	4839.87	I	35	7264.17	II	300	2770.86	I
13000	3710.30	II	550	4845.68	I	50	7346.46	I	300	2770.98	I
1200	3747.55	II	410	4852.69	I	29	7450.30	II	400	2800.87	I
10000	3774.33	II	120	4854.25	I	9000	7558.71	III	100	2801.06	I
1400	3776.56	II	890	4854.87	II	35	7563.13	I	200	3035.78	I
7400	3788.70	II	330	4859.84	I	29	7855.52	I	200	3072.06	I
1300	3818.35	II	1900	4883.69	II	110	7881.90	II	300	3196.31	II
4000	3832.88	II	95	4893.44	I	10000	7991.43	III	500 r	3282.33	I
80	3876.82	I	1100	4900.12	II	24	8344.43	I	800	3302.58	I
480	3878.28	II	100	4906.11	I	10000	8796.21	III	700 r	3302.94	I
4400	3950.36	II	150	4921.87	I	95	8800.62	I	800	3345.02	I
3600	3982.60	II	120	4974.30	I	19 h	8835.85	II	500	3345.57	I
940	4039.83	I	100	5006.97	I		**Zinc**		50	3883.34	I
2400	4047.64	I	75	5070.21	I		**Zn Z = 30**		300	4680.14	I
9400	4077.38	I	75	5072.19	I	200	425.90	IV	400	4722.15	I
2000	4083.71	I	1100	5087.42	II	200	428.54	IV	400	4810.53	I
9900	4102.38	I	180	5135.20	I	200	430.59	IV	800	4911.62	II
8900	4128.31	I	960	5200.41	II	1000	677.63	III	500	4924.03	II
7500	4142.85	I	1500	5205.72	II	750	677.96	III	200	5181.98	I
100 h	4157.63	I	10000	5238.10	III	200	713.90	IV	500	5894.33	II
2400	4167.52	I	180	5240.81	I	60	1193.23	II	500	6021.18	II
2000	4174.14	I	75	5380.62	I	50	1239.12	IV	500	6102.49	II
8000	4177.54	II	220	5402.78	II	50	1249.69	IV	500	6214.61	II
160	4217.80	I	90	5424.37	I	500	1265.74	IV	1000 h	6362.34	I
280 h	4220.63	I	190	5438.24	I	500	1306.66	IV	300	7588.5	II
600	4235.73	II	710	5466.46	I	200	1456.72	III	300	7732.5	II
2200	4235.94	I	100	5468.47	I	200	1459.98	IV	100	11054.25	I
300	4251.20	I	240	5497.41	II	300	1499.42	III	100	13053.63	I
360 h	4302.30	I	300	5503.45	I	300	1500.42	III	100	13150.59	I
2800	4309.63	II	250	5509.90	II	300	1505.92	III	100	14038.70	I
110	4330.78	I	120	5521.63	I	300	1515.85	III	20	16483.45	I
440 h	4348.79	I	740	5527.54	I	300	1552.30	III	20	16491.98	I
120	4357.73	I	120	5544.50	I	90	1572.99	II	20	16505.23	I
800	4358.73	II	180	5577.42	I	200	1629.19	III	10	24375.02	I
120	4366.03	I	620	5581.87	I	200	1639.33	III		**Zirconium**	
12000	4374.94	II	120	5606.33	I	200	1673.05	III		**Zr Z = 40**	
150 h	4375.61	I	560	5630.13	I	80 d	1735.61	II	500	304.01	V
100	4387.74	I	120	5644.69	I	100	1767.69	III	60	480.66	IV
1800	4398.02	II	120	5648.47	I	100	1797.64	II	60	497.23	IV
890	4422.59	II	740	5662.94	II	100 d	1811.05	II	60	500.22	IV
100	4443.66	I	90	5675.27	I	100 d	1833.57	II	600	628.66	IV
130	4446.63	I	160	5706.73	I	100	1864.12	II	500	633.56	IV
170	4475.72	I	90	5743.85	I	100	1866.08	II	50	690.39	III
180	4476.96	I	75	5765.64	I	100	1872.13	II	2000	740.61	V
160	4477.45	I	100	5781.69	II	100 d	1918.96	II	10000	800.00	V
110	4487.28	I	120	6009.19	I	100 d	1929.67	II	10000	806.89	V
300	4487.47	I	120	6023.41	I	100	1969.40	II	10000	812.05	V
500	4505.95	I	120	6135.04	I	100	1982.11	II	3000	841.40	V
890	4527.25	I	150	6138.43	I	100	1986.99	II	300	863.65	IV
440	4527.80	I	1200	6191.73	I	500	2025.48	II	500	864.59	IV
100	4544.32	I	300	6222.59	I	500	2062.00	II	9000	1183.97	IV
100	4559.37	I	1000	6435.00	I	200	2064.23	II	9000	1201.77	IV
130	4596.55	I	90	6538.60	I	120	2079.08	I	10000	1219.86	IV
95	4604.80	I	70	6557.39	I	300	2099.94	II	500	1303.93	V
2000	4643.70	I	95	6613.75	II	,200	2102.18	II	500 p	1323.81	V
200 h	4658.32	I	40	6650.61	I	800 r	2138.56	I	1000	1469.47	IV
2000	4674.84	I	150	6687.58	I	1000	2501.99	II	10000	1546.17	IV
180	4696.81	I	70	6700.71	I	150	2515.81	I	10000	1598.95	IV
170	4728.53	I	190	6793.71	I	1000	2557.95	II	5000	1607.95	IV
160	4752.79	I	21	6815.16	I	300	2582.49	I	100	1612.38	III
410	4760.98	I	45	6845.24	I	200	2608.56	I	700	1725.02	V
120	4781.04	I	29	6887.22	I	300	2608.64	I	200	1790.19	III

Intensity	Wavelength/Å		Intensity	Wavelength/Å		Intensity	Wavelength/Å		Intensity	Wavelength/Å	
150	1793.56	III	350	3120.74	I	3500	3601.19	I	660	4187.56	I
125	1798.13	III	500	3129.18	II	690	3611.89	II	400	4194.76	I
600	1860.86	V	500	3129.76	II	1100	3613.10	II	610	4199.09	I
200	1940.25	III	350	3132.07	I	1100	3614.77	II	610	4201.46	I
600	2028.54	V	690	3138.68	II	1100	3623.86	I	610	4208.98	II
125	2070.43	III	540	3164.31	II	1100	3663.65	I	400	4213.86	I
200	2086.78	III	880	3165.97	II	390	3671.27	II	2000	4227.76	I
10000	2091.49	IV	880	3182.86	II	800	3674.72	II	2000	4239.31	I
10000	2092.36	IV	540	3191.21	I	390	3697.46	II	770	4240.34	I
600	2132.42	V	540	3212.01	I	960	3698.17	I	770	4241.20	I
10000	2163.68	IV	760	3214.19	II	720	3709.26	II	1200	4241.69	I
100	2175.80	III	630	3231.69	II	560	3745.98	II	550	4282.20	I
100	2191.15	III	630	3234.12	I	880	3751.60	II	550	4294.79	I
10000	2286.67	IV	760	3241.05	II	480	3764.39	I	550	4341.13	I
100	2301.60	III	1000	3273.05	II	480	3766.72	I	1000	4347.89	I
90	2539.65	I	1300	3279.26	II	340	3766.82	II	290	4359.74	II
570	2567.64	II	880	3284.71	II	720	3780.54	I	310	4360.81	I
1600	2568.87	II	540	3305.15	II	560	3791.40	I	350	4366.45	I
2100	2571.39	II	880	3306.28	II	560	3822.41	I	550	4507.12	I
250	2620.56	III	380	3322.99	II	2200	3835.96	I	610	4535.75	I
200	2643.79	III	380	3326.80	II	1300	3836.76	II	490	4542.22	I
150	2664.26	III	380	3334.25	II	550	3843.02	II	490	4575.52	I
1800	2678.63	II	760	3340.56	II	550	3847.01	I	350	4602.57	I
90	2687.75	I	380	3344.79	II	550	3849.25	I	700	4633.98	I
750	2700.13	II	760	3356.09	II	2900	3863.87	I	2300	4687.80	I
1300	2722.61	II	540	3357.26	II	770	3864.34	I	510	4688.45	I
800	2726.49	II	380	3374.73	II	990	3877.60	I	1900	4710.08	I
1400	2734.86	II	570	3387.87	II	1500	3885.42	I	1400	4739.48	I
1100	2742.56	II	760	3388.30	II	2900	3890.32	I	870	4772.31	I
660	2745.86	II	5700	3391.98	II	2000	3891.38	I	700	4815.63	I
660	2752.21	II	570	3393.12	II	610	3921.79	I	250	5046.58	I
530	2758.81	II	570	3404.83	II	1200	3929.53	I	360	5064.91	I
620	2814.90	I	760	3410.25	II	940	3958.22	II	470	5078.25	I
390	2818.74	II	380	3414.66	I	490	3966.66	I	300	5155.45	I
530	2825.56	II	1000	3430.53	II	990	3968.26	I	200	5158.00	I
710	2837.23	I	4700	3438.23	II	660	3973.50	I	100	5191.60	II
660	2844.58	II	600	3447.36	II	770	3991.13	II	270	5385.14	I
350	2848.52	I	410	3457.56	II	770	3998.97	II	160	5664.51	I
350	2851.97	II	820	3463.02	II	400	4023.98	I	160	5797.74	I
340	2869.81	II	600	3471.19	I	770	4024.92	I	340	5879.80	I
490	2875.98	I	1200	3479.39	II	990	4027.20	I	170	6045.85	I
300	2915.99	II	1300	3481.15	II	400	4029.68	II	170	6121.91	I
270	2918.24	II	4100	3496.21	II	490	4030.04	I	680	6127.44	I
320	2926.99	II	820	3505.67	II	400	4035.89	I	340	6134.55	I
320	2948.94	II	1000	3509.32	I	610	4043.58	I	440	6143.20	I
320	2955.78	II	2000	3519.60	I	490	4044.56	I	300	6313.02	I
320	2960.87	I	440	3525.81	II	400	4045.61	II	150	6953.84	I
320	2962.68	II	440	3533.22	I	610	4048.67	II	150	6990.84	I
320	2968.96	II	630	3542.62	II	770	4055.03	I	540	7097.70	I
320	2978.05	II	1800	3547.68	I	600	4055.71	I	280	7102.91	I
820	2985.39	I	630	3550.46	I	1500	4064.16	I	170	7103.72	I
320	3003.74	II	1800	3551.95	II	2000	4072.70	I	590	7169.09	I
820	3011.75	I	2100	3556.60	II	240	4078.31	I	160	7944.61	I
350	3020.47	II	1100	3566.10	I	2000	4081.22	I	160	8005.27	I
500	3028.04	II	2100	3572.47	II	400	4121.46	I	150	8063.09	I
880	3029.52	I	1100	3575.79	I	1200	4149.20	II	790	8070.08	I
350 d	3036.39	II	1300	3576.85	II	400	4161.21	II	390	8132.99	I
690	3054.84	II	880	3586.29	I	400	4166.36	I	280	8212.53	I
690	3106.58	II									

SOURCES OF DATA FOR EACH ELEMENT

Numbers following the element name refer to the references on the following pages.

Actinium: 193
Aluminum: 6,8,81,89,127,144,146,227,228,282
Americium: 92
Antimony: 164,167,194,386,406
Argon: 190,203,204,219,367,368,372,373,374,375,414,421
Arsenic: 163,168,197,244,280
Astatine: 188
Barium: 1,78,111,252,259,277,279
Berkelium: 53,339
Beryllium: 15,44,73,102,115,134,135,171,175,198,335
Bismuth: 1,357,358,359,360,361
Boron: 66,69,74,94,104,171,221,222
Bromine: 42,122,124,139,142,240,243,246,248,249,250,316
Cadmium: 44,285,296,353,399
Calcium: 16,25,70,150,270
Californium: 52,331
Carbon: 22,66,211
Cerium: 1,136,166,261,305
Cesium: 78,82,154,155,200,201,259,263,325
Chlorine: 11,28,30,31,85,233,238,239
Chromium: 1,379,380,412
Cobalt: 1,100,125,159,236,276,291
Copper: 199,273,290,295,324
Curium: 51,332
Dysprosium: 1
Einsteinium: 333
Erbium: 1,301
Europium: 1,312
Fluorine: 68,169,224,225,226
Francium: 408
Gadolinium: 1,46,137,151,152
Gallium: 2,19,62,132,140,141,143,195,281
Germanium: 5,119,293,340,341,342
Gold: 38,72,234,393,395
Hafnium: 1,369,404,410,425
Helium: 16,94,173,183,317
Holmium: 1
Hydrogen: 214
Indium: 1,132,348,349,350,351,352,353,435,436
Iodine: 20,21,58,84,124,153,161,176,184
Iridium: 1
Iron: 56,63,71,101,105,138,174,278,381,382
Krypton: 61,121,123,147,208,232,366,390,409,417,421
Lanthanum: 1,78,79,220,309
Lead: 54,64,106,256,274,297,283,329,330
Lithium: 3,15,17,18,37,44,112,284,321,335
Lutetium: 1,148,310,401
Magnesium: 4,7,49,83,103,128,129,177,217,269,315,335
Manganese: 1,126,385,405,433
Mercury (198): 43,50,69,145,229,242

Mercury (Natural): 34,45,90,117,133,189,235,304,327,328,343
Molybdenum: 1,383,420
Neodymium: 1
Neon: 56,58,69,118,150,230,364,365,371,388,389,400,402,413,430
Neptunium: 93
Nickel: 1,294,415,416,422
Niobium: 1,392,407,431
Nitrogen: 66,107,108,212,213,318
Osmium: 1
Oxygen: 23,24,36,66,69,209,210,215
Palladium: 1,287,424
Phosphorus: 179,180,182,336
Platinum: 1,288
Plutonium: 91
Polonium: 47,48
Potassium: 32,59,60,75,76,86,150,160,172,268,314,322
Praseodymium: 1,149,306,308,337,338
Promethium: 196,260
Protactinium: 96
Radium: 253,254
Radon: 251
Rhenium: 1
Rhodium: 1,396
Rubidium: 12,109,130,241,257,258,262,264
Ruthenium: 1,423
Samarium: 1
Scandium: 1,88,150,298,323
Selenium: 9,80,181,216,245,247,275
Silicon: 87,170,237,292,319,320
Silver: 13,99,255,286,289,363,387,398
Sodium: 178,205,206,207,268,299,334
Strontium: 1,109,110,218,231,265,279,313
Sulfur: 29,144,202,209,210,266
Tantalum: 1,411,426
Technetium: 35
Tellurium: 1,344,345,346,347
Terbium: 1,302
Thallium: 1,195,348,354,355,356
Thorium: 1,97,98,156,157,165,434
Thulium: 1,307
Tin: 187,191,399,423
Titanium: 1,378,427,428
Tungsten: 1
Uranium: 1,303
Vanadium: 1,394,397,432
Xenon: 33,116,118,120,232,384,391,429
Ytterbium: 1,40,192,311
Yttrium: 1,77,265,419
Zinc: 39,55,113,131,185,186,370,376,377
Zirconium: 1,362,403,418

REFERENCES

1. Meggers, W. F., Corliss, C. H., and Scribner, B. F., *Natl. Bur. Stand. (U.S.) Monogr.*, 145, Washington, D.C., 1975.
2. Aksenov, V. P. and Ryabtsev, A. N., *Opt. Spectrosc.*, 37, 860, 1970.
3. Andersen, N., Bickel, W. S., Carriveau, G. W., Jensen, K., and Veje, E., *Phys. Scr.*, 4, 113, 1971.
4. Andersson, E. and Johannesson, G. A., *Phys. Scr.*, 3, 203, 1971.
5. Andrew, K. L. and Meissner, K. W., *J. Opt. Soc. Am.*, 49, 146, 1959.
6. Artru, M. C. and Brillet, W. U. L., *J. Opt. Soc. Am.*, 64, 1063, 1974.
7. Artru, M. C. and Kaufman, V., *J. Opt. Soc. Am.*, 62, 949, 1972.
8. Artru, M. C. and Kaufman, V., *J. Opt. Soc. Am.*, 65, 594, 1975.
9. Badami, J. S. and Rao, K. R., *Proc. R. Soc. London*, 140(A), 387, 1933.
10. Baird, K. M. and Smith, D. S., *J. Opt. Soc. Am.*, 48, 300, 1958.
11. Bashkin, S. and Martinson, I., *J. Opt. Soc. Am.*, 61, 1686, 1971.
12. Beacham, J. R., Ph.D. thesis, Purdue University, 1970.
13. Benschop, H., Joshi, Y. N., and van Kleef, T. A. M., *Can. J. Phys.*, 53, 700, 1975.
14. Berry, H. G., Bromander, J., and Buchta, R., *Phys. Scr.*, 1, 181, 1970.
15. Berry, H. G., Bromander, J., Martinson, I., and Buchta, R., *Phys. Scr.*, 3, 63, 1971.
16. Berry, H. G., Desesquelles, J., and Dufay, M., *Phys. Rev. Sect. A.*, 6, 600, 1972.
17. Berry, H. G., Desesquelles, J., and Dufay, M., *Nucl. Instrum. Methods*, 110, 43, 1973.
18. Berry, H. G., Pinnington, E. H., and Subtil, J. L., *J. Opt. Soc. Am.*, 62, 767, 1972.
19. Bidelman, W. P. and Corliss, C. H., *Astrophys. J.*, 135, 968, 1962.
20. Bloch, L. and Bloch, E., *Ann. Phys.* (Paris), 10(11), 141, 1929.
21. Bloch, L., Bloch, E., and Felici, N., *J. Phys. Radium*, 8, 355, 1937.
22. Bockasten, K., *Ark. Fys.*, 9, 457, 1955.
23. Bockasten, K., Hallin, R., Johansson, K. B., and Tsui, P., *Phys. Lett.* (Netherlands), 8, 181, 1964.
24. Bockasten, K. and Johansson, K. B., *Ark. Fys.*, 38, 563, 1969.
25. Borgstrom, A., *Ark. Fys.*, 38, 243, 1968.
26. Borgstrom, A., *Phys. Scr.*, 3, 157, 1971.
27. Bowen, I. S., *Phys. Rev.*, 29, 231, 1927.
28. Bowen, I. S., *Phys. Rev.*, 31, 34, 1928.
29. Bowen, I. S., *Phys. Rev.*, 39, 8, 1932.
30. Bowen, I. S., *Phys. Rev.*, 45, 401, 1934.
31. Bowen, I. S., *Phys. Rev.*, 46, 377, 1934.
32. Bowen, I. S., *Phys. Rev.*, 46, 791, 1934.
33. Boyce, J. C., *Phys. Rev.*, 49, 730, 1936.
34. Boyce, J. C. and Robinson, H. A., *J. Opt. Soc. Am.*, 26, 133, 1936.
35. Bozman, W. R., Meggers, W. F., and Corliss, C. H., *J. Res. Natl. Bur. Stand. Sect. A*, 71, 547, 1967.
36. Bromander, J., *Ark. Fys.*, 40, 257, 1969.
37. Bromander, J. and Buchta, R., *Phys. Scr.*, 1, 184, 1970.
38. Brown, C. M. and Ginter, M. L., *J. Opt. Soc. Am.*, 68, 243, 1978.
39. Brown, C. M., Tilford, S. G., and Ginter, M. L., *J. Opt. Soc. Am.*, 65, 1404, 1975.
40. Bryant, B. W., *Johns Hopkins Spectroscopic Report* No. 21, 1961.
41. Buchet, J. P., Buchet-Poulizac, M. C., Berry, H. G., and Drake, G. W. F., *Phys. Rev. Sect. A*, 7, 922, 1973.
42. Budhiraja, C. J. and Joshi, Y. N., *Can. J. Phys.*, 49, 391, 1971.
43. Burns, K. and Adams, K. B., *J. Opt. Soc. Am.*, 42, 56, 1952.
44. Burns, K. and Adams, K. B., *J. Opt. Soc. Am.*, 46, 94, 1956.
45. Burns, K., Adams, K. B., and Longwell, J., *J. Opt. Soc. Am.*, 40, 339, 1950.
46. Callahan, W. R., Ph.D. thesis, Johns Hopkins University, 1962.
47. Charles, G. W., *J. Opt. Soc. Am.*, 56, 1292, 1966.
48. Charles, G. W., Hunt, D. J., Pish, G., and Timma, D. L., *J. Opt. Soc. Am.*, 45, 869, 1955.
49. Codling, K., *Proc. Phys. Soc.*, 77, 797, 1961.
50. Comite Consulatif Pour La Definition du Metre, *J. Phys. Chem. Ref. Data*, 3, 852, 1974.
51. Conway, J. G., Blaise, J., and Verges, J., *Spectrochim. Acta Part B*, 31, 31, 1976.
52. Conway, J. G., Worden, E. F., Blaise, J., and Verges, J., *Spectrochim. Acta Part B*, 32, 97, 1977.
53. Conway, J. G., Worden, E. F., Blaise, J., Camus, P., and Verges, J., *Spectrochim. Acta Part B*, 32, 101, 1977.
54. Crooker, A. M., *Can. J. Res. Sect. A*, 14, 115, 1936.
55. Crooker, A. M. and Dick, K. A., *Can. J. Phys.*, 46, 1241, 1968.
56. Crosswhite, H. M., *J. Res. Natl. Bur. Stand. Sect. A*, 79, 17, 1975.
58. Crosswhite, H. M. and Dieke, G. H., *American Institute of Physics Handbook*, Section 7, 1972.
59. de Bruin, T. L., *Z. Phys.*, 38, 94, 1926.
60. de Bruin, T. L., *Z. Phys.*, 53, 658, 1929.
61. de Bruin, T. L., Humphreys, C. J., and Meggers, W. F., *J. Res. Natl. Bur. Stand.*, 11, 409, 1933.
62. Dick, K. A., *J. Opt. Soc. Am.*, 64, 702, 1973.
63. Dobbie, J. C., *Ann. Solar Phys. Observ.* (Cambridge), 5, 1, 1938.
64. Earls, L. T. and Sawyer, R. A., *Phys. Rev.*, 47, 115, 1935.
65. Edlen, B., *Z. Phys.*, 85, 85, 1933.
66. Edlen, B., *Nova Acta Reglae Soc. Sci. Ups.*, (IV) 9, No. 6, 1934.
67. Edlen, B., *Z. Phys.*, 93, 726, 1935.
68. Edlen, B., *Z. Phys.*, 94, 47, 1935.
69. Edlen, B., *Rep. Prog. Phys.*, 26, 181, 1963.
70. Edlen, B. and Risberg, P., *Ark. Fys.*, 10, 553, 1956.
71. Edlen, B. and Swings, P., *Astrophys. J.*, 95, 532, 1942.
72. Ehrhardt, J. C. and Davis, S. P., *J. Opt. Soc. Am.*, 61, 1342, 1971.
73. Eidelsberg, M., *J. Phys. B*, 5, 1031, 1972.
74. Eidelsberg, M., *J. Phys. B*, 7, 1476, 1974.
75. Ekberg, J. O. and Svensson, L. A., *Phys. Scr.*, 2, 283, 1970.
76. Ekefors, E., *Z. Phys.*, 71, 53, 1931.
77. Epstein, G. L. and Reader, J., *J. Opt. Soc. Am.*, 65, 310, 1975.
78. Epstein, G. L. and Reader, J., *J. Opt. Soc. Am.*, 66, 590, 1976.
79. Epstein, G. L. and Reader, J., unpublished.
80. Eriksson, K. B. S., *Phys. Lett. A.*, 41, 97, 1972.
81. Eriksson, K. B. S. and Isberg, H. B. S., *Ark. Fys.*, 23, 527, 1963.
82. Eriksson, K. B. S. and Wenaker, I., *Phys. Scr.*, 1, 21, 1970.
83. Esteva, J. M. and Mehlman, G., *Astrophys. J.*, 193, 747, 1974.
84. Even-Zohar, M. and Fraenkel, B. S., *J. Phys. B*, 5, 1596, 1972.
85. Fawcett, B. C., *J. Phys. B*, 3, 1732, 1970.
86. Fawcett, B. C., Culham Laboratory Report ARU-R4, 1971.
87. Ferner, E., *Ark. Mat. Astron. Fys.*, 28(A), 4, 1941.
88. Fischer, R. A., Knopf, W. C., and Kinney, F. E., *Astrophys. J.*, 130, 683, 1959.

89. Fowler, A., *Report on Series in Line Spectra*, Fleetway Press, London, 1922.
90. Fowles, G. R., *J. Opt. Soc. Am.*, 44, 760, 1954.
91. Fred, M., *Argonne Natl. Lab.*, unpublished, 1977.
92. Fred, M. and Tomkins, F. S., *J. Opt. Soc. Am.*, 47, 1076, 1957.
93. Fred, M., Tomkins, F. S., Blaise, J. E., Camus, P., and Verges, J., Argonne National Laboratory Report No. 76-68, 1976.
94. Garcia, J. D. and Mack, J. E., *J. Opt. Soc. Am.*, 55, 654, 1965.
96. Giacchetti, A., *Argonne Natl. Lab.*, unpublished, 1975.
97. Giacchetti, A., Blaise, J., Corliss, C. H., and Zalubas, R., *J. Res. Natl. Bur. Stand. Sect. A*, 78, 247, 1974.
98. Giacchetti, A., Stanley, R. W., and Zalubas, R., *J. Opt. Soc. Am.*, 69, 474, 1970.
99. Gilbert, W. P., *Phys. Rev.*, 47, 847, 1935.
100. Gilroy, H. T., *Phys. Rev.*, 38, 2217, 1931.
101. Glad, S., *Ark. Fys.*, 10, 291, 1956.
102. Goldsmith, S., *J. Phys. B*, 2, 1075, 1969.
103. Goorvitch, D., Mehlmam-Balloffet, G., and Valero, F. P. J., *J. Opt. Soc. Am.*, 60, 1458, 1970.
104. Goorvitch, D. and Valero, F. P. J., *Astrophys. J.*, 171, 643, 1972.
105. Green, L. C., *Phys. Rev.*, 55, 1209, 1939.
106. Gutman, F., *Diss. Abstr. Int. B*, 31, 363, 1970.
107. Hallin, R., *Ark. Fys.*, 31, 511, 1966.
108. Hallin, R., *Ark. Fys.*, 32, 201, 1966.
109. Hansen, J. E. and Persson, W., *J. Opt. Soc. Am.*, 64, 696, 1974.
110. Hansen, J. E. and Persson, W., *Phys. Scr.*, 13, 166, 1976.
111. Hellintin, P., *Phys. Scr.*, 13, 155, 1976.
112. Herzberg, G. and Moore, H. R., *Can. J. Phys.*, 37, 1293, 1959.
113. Hetzler, C. W., Boreman, R. W., and Burns, K., *Phys. Rev.*, 48, 656, 1935.
114. Holmstrom, J. E. and Johansson, L., *Ark. Fys.*, 40, 133, 1969.
115. Hontzeas, S., Martinson, I., Erman, P., and Buchta, R., *Nucl. Instrum. Methods*, 110, 51, 1973.
116. Humphreys, C. J., *J. Res. Natl. Bur. Stand.*, 22, 19, 1939.
117. Humphreys, C. J., *J. Opt. Soc. Am.*, 43, 1027, 1953.
118. Humphreys, C. J., *J. Phys. Chem. Ref. Data*, 2, 519, 1973.
119. Humphreys, C. J. and Andrew, K. L., *J. Opt. Soc. Am.*, 54, 1134, 1964.
120. Humphreys, C. J. and Meggers, W. F., *J. Res. Natl. Bur. Stand.*, 10, 139, 1933.
121. Humphreys, C. J. and Paul, E., Jr., *J. Opt. Soc. Am.*, 60, 200, 1970.
122. Humphreys, C. J. and Paul, E., Jr., *J. Opt. Soc. Am.*, 62, 432, 1972.
123. Humphreys, C. J., Paul, E., Jr., Cowan, R. D., and Andrew, K. L., *J. Opt. Soc. Am.*, 57, 855, 1967.
124. Humphreys, C. J., Paul, E., Jr., and Minnhagen, L., *J. Opt. Soc. Am.*, 61, 110, 1971.
125. Iglesias, L., Inst. of Optics, Madrid, unpublished, 1977.
126. Iglesias, L. and Velasco, R., *Publ. Inst. Opt. Madrid*, No. 23, 1964.
127. Isberg, B., *Ark. Fys.*, 35, 551, 1967.
128. Johannesson, G. A., Lundstrom, T., and Minnhagen, L., *Phys. Scr.*, 6, 129, 1972.
129. Johannesson, G. A. and Lundstrom, T., *Phys. Scr.*, 8, 53, 1973.
130. Johansson, I. *Ark. Fys.*, 20, 135, 1961.
131. Johansson, I. and Contreras, R., *Ark. Fys.*, 37, 513, 1968.
132. Johansson, I. and Litzen, U., *Ark. Fys.*, 34, 573, 1967.
133. Johansson, I. and Svensson, K. F., *Ark. Fys.*, 16, 353, 1960.
134. Johansson, L., *Ark. Fys.*, 20, 489, 1961.
135. Johansson, L., *Ark. Fys.*, 23, 119, 1963.
136. Johansson, S. and Litzen, U., *Phys. Scr.*, 6, 139, 1972.
137. Johansson, S. and Litzen, U., *Phys. Scr.*, 8, 43, 1973.
138. Johansson, S. and Litzen, U., *Phys. Scr.*, 10, 121, 1974.
139. Joshi, Y. N., St. Francis Xavier Univ., Nova Scotia, unpublished.
140. Joshi, Y. N., Bhatia, K. S., and Jones, W. E., *Sci. Light Tokyo*, 21, 113, 1972.
141. Joshi, Y. N., Bhatia, K. S., and Jones, W. E., *Spectrochim. Acta Part B*, 28, 149, 1973.
142. Joshi, Y. N. and Budhiraja, C. J., *Can. J. Phys.*, 49, 670, 1971.
143. Joshi, Y. N. and van Kleef, T. A. M., *Can. J. Phys.*, 52, 1891, 1974.
144. Kaufman, V., *Natl. Bur. Stand.*, unpublished.
145. Kaufman, V., *J, Opt. Soc. Am.*, 52, 866, 1962.
146. Kaufman, V., Artru, M. C., and Brillet, W. U. L., *J. Opt. Soc. Am.*, 64, 197, 1974.
147. Kaufman, V. and Humphreys, C. J., *J. Opt. Soc. Am.*, 59, 1614, 1969.
148. Kaufman, V. and Sugar, J., *J. Opt. Soc. Am.*, 61, 1693, 1971.
149. Kaufman, V. and Sugar, J., *J. Res. Natl. Bur. Stand. Sect. A*, 71, 583, 1967.
150. Kelly, R. L. and Palumbo, L. J., *Naval Research Laboratory Report 7599*, Washington, DC., 1973.
151. Kielkopf, J. F., *Univ. of Louisville*, unpublished. 1975.
152. Kielkopf, J. F., *Univ. of Louisville*, unpublished, 1976.
153. Kiess, C. C. and Corliss, C. H., *J. Res. Natl. Bur. Stand. Sect. A*, 63, 1, 1959.
154. Kleiman, H., *J. Opt. Soc. Am.*, 52, 441, 1962.
155. Eriksson, K. B., Johansson, I., and Norlen, G., *Ark. Fys.*, 28, 233, 1964.
156. Klinkenberg, P. F. A., *Physica*, 15, 774, 1949.
157. Klinkenberg, P. F. A., *Physica*, 16, 618, 1950.
158. Krishnamurty, S. G., *Proc. Phys. Soc. London*, 48, 277, 1936.
159. Kruger, P. G. and Gilroy, H. T., *Phys. Rev.*, 48, 720, 1935.
160. Kruger, P. G. and Pattin, H. S., *Phys. Rev.*, 52, 621, 1937.
161. Lacroute, P., *Ann. Phys.* (Paris), 3, 5, 1935.
162. Lang, R. J., *Phys. Rev.*, 30, 762, 1927.
163. Lang, R. J., *Phys. Rev.*, 32, 737, 1928.
164. Lang, R. J., *Phys. Rev.*, 35, 445, 1930.
165. Lang, R. J., *Can. J. Res. Sect. A*, 14, 43, 1936.
166. Lang, R. J., *Can. J. Res. Sect. A*, 14, 127, 1936.
167. Lang, R. J. and Vestine, E. H., *Phys. Rev.*, 42, 233, 1932.
168. Li, H. and Andrew, K. L., *J. Opt. Soc. Am.*, 61, 96, 1971.
169. Liden, K., *Ark. Fys.*, 1, 229, 1949.
170. Litzen, U., *Ark. Fys.*, 28, 239, 1965.
171. Litzen, U., *Phys. Scr.*, 1, 251, 1970.
172. Litzen, U., *Phys. Scr.*, 1, 253, 1970.
173. Litzen, U., *Phys. Scr.*, 2, 103, 1970.
174. Litzen, U. and Verges, I., *Phys. Scr.*, 13, 240, 1976.
175. Lofstrand, B., *Phys. Scr.*, 8, 57, 1973.
176. Luc-Koenig, E., Morillon, C., and Verges, J., *Phys. Scr.*, 12, 199, 1975.
177. Lundstrom, T., *Phys. Scr.*, 7, 62, 1973.
178. Lundstrom, T. and Minnhagen, L., *Phys. Scr.*, 5, 243, 1972.
179. Magnusson, C. E. and Zetterberg, P. O., *Phys. Scr.*, 10, 177, 1974.
180. Magnusson, C. E., and Zetterberg, P. O., *Phys. Scr.*, 15, 237, 1977.
181. Martin, D. C., *Phys. Rev.*, 48. 938, 1935.
182. Svendenius, N., *Phys. Scr.*, 22, 240, 1980.
183. Martin, W. C., *J. Res. Natl. Bur. Stand. Sect. A*, 64, 19, 1960.
184. Martin, W. C. and Corliss, C. H., *J. Res. Natl. Bur. Stand. Sect. A*, 64, 443, 1960.
185. Martin, W. C. and Kaufman, V., *J. Res. Natl. Bur. Stand. Sect. A*, 74, 11, 1970.
186. Martin, W. C. and Kaufman, V., *J. Opt. Soc. Am.*, 60, 1096, 1970.

187. McCormick, W. W. and Sawyer, R. A., *Phys. Rev.*, 54, 71, 1938.
188. McLaughlin, R., *J. Opt. Soc. Am.*, 54, 965, 1964.
189. McLennan, J. C., McLay, A. B., and Crawford, M. F., *Proc. R. Soc. London Ser. A*, 134, 41, 1931.
190. Meissner, K. W., *Z. Phys.*, 39, 172, 1926.
191. Meggers, W. F., *J. Res. Natl. Bur. Stand.*, 24, 153, 1940.
192. Meggers, W. F. and Corliss, C. H., *J. Res. Natl. Bur. Stand. Sect. A*, 70, 63, 1966.
193. Meggers, W. F., Fred, M., and Tomkins, F. S., *J. Res. Natl. Bur. Stand.*, 58, 297, 1957.
194. Meggers, W. F. and Humphreys, C. J., *J. Res. Natl. Bur. Stand.*, 28, 463, 1942.
195. Meggers, W. F. and Murphy, R. J., *J. Res. Natl. Bur. Stand.*, 48, 334, 1952.
196. Meggers, W. F., Scribner, B. F., and Bozman, W. R., *J. Res. Natl. Bur. Stand.*, 46, 85, 1951.
197. Meggers, W. F., Shenstone, A. G., and Moore, C. E., *J. Res. Natl. Bur. Stand.*, 45, 346, 1950.
198. Mehlman, G. and Esteva, J. M., *Astrophys. J.*, 188, 191, 1974.
199. Meinders, E., *Physica*, 84(C), 117, 1976.
200. Sansonetti, C. J., Dissertation, Purdue University, 1981.
201. Sansonetti, C. J., *Natl. Bur. Stand. (U.S.)*, unpublished.
202. Millikan, R. A. and Bowen, I. S., *Phys. Rev.*, 25, 600, 1925.
203. Minnhagen, L., *J. Opt. Soc. Am.*, 61, 1257, 1925.
204. Minnhagen, L., *J. Opt. Soc. Am.*, 63, 1185, 1973.
205. Minnhagen, L., *Phys. Scr.*, 11, 38, 1975.
206. Minnhagen, L., *J. Opt. Soc. Am.*, 66, 659, 1976.
207. Minnhagen, L. and Nietsche, H., *Phys. Scr.*, 5, 237, 1972.
208. Minnhagen. L., Strihed, H., and Petersson, B., *Ark. Fys.*, 39, 471, 1969.
209. Moore, C. E., *Natl. Bur. Stand. (U.S.) Circ.*, 488, 1950.
210. Moore, C. E., *Revised Multiplet Table*, Princeton University Observatory No. 20, 1945.
211. Moore, C. E., National Standard Reference Data Series - National Bureau of Standards 3, Sect. 3, 1970.
212. Moore, C. E., National Standard Reference Data Series - National Bureau of Standards 3, Sect. 4, 1971.
213. Moore, C. E., National Standard Reference Data Series - National Bureau of Standards 3, Sect. 5, 1975.
214. Moore, C. E., National Standard Reference Data Series - National Bureau of Standards 3, Sect. 6, 1972.
215. Moore, C. E., National Standard Reference Data Series - National Bureau of Standards 3, Sect. 7, 1975.
216. Morillon, C. and Verges, J., *Phys. Scr.*, 10, 227, 1974.
217. Newsom, G. H., *Astrophys. J.*, 166, 243, 1971.
218. Newsom, G. H., O'Connor, S., and Learner, R. C. M., *J. Phys. B*, 6, 2162, 1973.
219. Norlen, G., *Phys. Scr.*, 8, 249, 1973.
220. Odabasi, H., *J. Opt. Soc. Am.*, 57, 1459, 1967.
221. Olme, A., *Ark. Fys.*, 40, 35, 1969.
222. Olme, A., *Phys. Scr.*, 1, 256, 1970.
223. Johansson, S., and Litzen, U., *J. Opt. Soc. Am.*, 61, 1427, 1971.
224. Palenius, H. P., *Ark. Fys.*, 39, 15, 1969.
225. Palenius, H. P., *Phys. Scr.*, 1, 113, 1970.
226. Palenius, H. P., *Univ. of Lund, Sweden*, unpublished.
227. Paschen, F., *Ann. Phys.*, Series 5, 12, 509, 1932.
228. Paschen, F. and Ritschl, R., *Ann. Phys.*, Series 5, 18, 867, 1933.
229. Peck, E. R., Khanna, B. N., and Anderholm, N. C., *J. Opt. Soc. Am.*, 52, 53, 1962.
230. Persson, W., *Phys. Scr.*, 3, 133, 1971.
231. Persson, W. and Valind, S., *Phys. Scr.*, 5, 187, 1972.
232. Petersson, B., *Ark. Fys.*, 27, 317, 1964.
233. Phillips, L. W. and Parker, W. L., *Phys. Rev.*, 60, 301, 1941.
234. Platt, J. R. and Sawyer, R. A., Phys. Rev., 60, 866, 1941.
235. Plyer, E. K., Blaine, L. R., and Tidwell, E., *J. Res. Natl. Bur. Stand.*, 55, 279, 1955.
236. Poppe, R., van Kleef, T. A. M., and Raassen, A. J. J., *Physica*, 77, 165, 1974.
237. Radziemski, L. J., Jr. and Andrew, K. L., *J. Opt. Soc. Am.*, 55, 474, 1965.
238. Radziemski, L. J., Jr. and Kaufman, V., *J. Opt. Soc. Am.*, 59, 424, 1969.
239. Radziemski, L. J., Jr. and Kaufman, V., *J. Opt. Soc. Am.*, 64, 366, 1974.
240. Ramanadham, R. and Rao, K. R., *Indian J. Phys.*, 18, 317, 1944.
241. Ramb, R., *Ann. Phys.*, 10, 311, 1931.
242. Rank, D. H., Bennett, J. M., and Bennett, H. E., *J. Opt. Soc. Am.*, 40, 477, 1950.
243. Rao, A. S. and Krishnamurty, S. G., *Proc. Phys. Soc. London*, 46, 531, 1943.
244. Rao, K. R., *Proc. R. Soc. London, Ser. A*, 134, 604, 1932.
245. Rao, K. R. and Badami, J. S., *Proc. R. Soc. London Ser. A*, 131, 154, 1931.
246. Rao, K. R. and Krishnamurty, S. G., *Proc. R. Soc. London Ser. A*, 161, 38, 1937.
247. Rao, K. R. and Murti, S. G. K., *Proc. R. Soc. London Ser. A*, 145, 681, 1934.
248. Rao, Y. B., *Indian J. Phys.*, 32, 497, 1958.
249. Rao, Y. B., *Indian J. Phys.*, 33, 546, 1959.
250. Rao, Y. B., *Indian J. Phys.*, 35, 386, 1961.
251. Rasmussen, E., *Z. Phys.*, 80, 726, 1933.
252. Rasmussen, E., *Z. Phys.*, 83, 404, 1933.
253. Rasmussen, E., *Z. Phys.*, 86, 24, 1934.
254. Rasmussen, E., *Z. Phys.*, 87, 607, 1934.
255. Rasmussen, E., *Phys. Rev.*, 57, 840, 1940.
256. Rau, A. S. and Narayan, A. L., *Z. Phys.*, 59, 687, 1930.
257. Reader, J., *J. Opt. Soc. Am.*, 65, 286, 1975.
258. Reader, J., *J. Opt. Soc. Am.*, 65, 988, 1975.
259. Reader, J., *J. Opt Soc. Am.*, 73, 349, 1983.
260. Reader, J. and Davis, S., *J. Res. Natl. Bur. Stand. Sect. A*, 71, 587, 1967, and unpublished.
261. Reader, J. and Ekberg, J. O., *J. Opt. Soc. Am.*, 62, 464, 1972.
262. Reader, J. and Epstein, G. L., *J. Opt. Soc. Am.*, 62, 1467, 1972.
263. Reader, J. and Epstein, G. L., *J. Opt. Soc. Am.*, 65, 638, 1975.
264. Reader, J. and Epstein, G. L., *Natl. Bur. Stand.*, unpublished.
265. Reader, J., Epstein, G. L., and Ekberg, J. O., *J. Opt. Soc. Am.*, 62, 273, 1972.
266. Kaufman, V., *Phys. Scr.*, 26, 439, 1982.
267. Ricard, R., Givord, M., and George, F., *C. R. Acad. Sci. Paris*, 205, 1229, 1937.
268. Risberg, P., *Ark. Fys.*, 10, 583, 1956.
269. Risberg, G., *Ark. Fys.*, 28, 381, 1965.
270. Risberg, G., *Ark. Fys.*, 37, 231, 1968.
271. Robinson, H. A., *Phys. Rev.*, 49, 297, 1936.
272. Robinson, H. A., *Phys. Rev.*, 50, 99, 1936.
273. Ross, C. B., Jr., Doctoral dissertation, Purdue University, 1969.
274. Ross, C. B., Wood, D. R., and Scholl, P. S., *J. Opt. Soc. Am.*, 66, 36, 1976.
275. Ruedy, J. E. and Gibbs, R. C., *Phys. Rev.*, 46, 880, 1934.
276. Russell, H. N., King, R. B., and Moore, C. E., *Phys. Rev.*, 58, 407, 1940.
277. Russell, H. N. and Moore, C. E., *J. Res. Natl. Bur. Stand.*, 55, 299, 1955.
278. Russell, H. N., Moore, C. E., and Weeks, D. W., *Trans. Am. Philos. Soc.*, 34(2), 111, 1944.

279. Saunders, F., Schneider, E., and Buckingham, E., *Proc. Natl. Acad. Sci.*, 20, 291, 1934.
280. Sawyer, R. A. and Humphreys, C. J., *Phys. Rev.*, 32, 583, 1928.
281. Sawyer, R. A. and Lang, R. J., *Phys. Rev.*, 34, 712, 1929.
282. Sawyer, R. A. and Paschen, F., *Ann. Phys.*, 84(4),1, 1927.
283. Scholl, P. S., M.S. thesis, Wright State Univ., 1975.
284. Schurmann, D., *Z. Phys.*, 17, 4, 1975.
285. Seguier, J., *C. R. Acad. Sci. Paris*, 256, 1703, 1963.
286. Shenstone, A. G., *Phys. Rev.*, 31, 317, 1928.
287. Shenstone, A. G., *Phys. Rev.*, 32, 30, 1928.
288. Shenstone, A. G., *Trans. R. Soc. London*, 237(A), 57, 1938.
289. Shenstone, A. G., *Phys. Rev.*, 57, 894, 1940.
290. Shenstone, A. G. *Philos. Trans. R. Soc. London Ser. A*, 241, 297, 1948.
291. Shenstone, A. G., *Can. J. Phys.*, 38, 677, 1960.
292. Shenstone, A. G., *Proc. R. Soc. London*, 261(A), 153, 1961.
293. Shenstone, A. G., *Proc. R. Soc. London*, 276(A), 293, 1963.
294. Shenstone, A. G., *J. Res. Natl. Bur. Stand. Sect. A*, 74, 801, 1970.
295. Shenstone, A. G., *J. Res. Natl. Bur. Stand. Sect. A*, 79, 497, 1975.
296. Shenstone, A. G. and Pittenger, J. T., *J. Opt. Soc. Am.*, 39, 219, 1949.
297. Smith, S., *Phys. Rev.*, 36, 1, 1930.
298. Smitt, R., *Phys. Scr.*, 8, 292, 1973.
299. Soderqvist, J., *Ark. Mat. Astronom. Fys.*, 32(A), 1, 1946.
300. Sommer, L. A., *Ann. Phys.*, 75, 163, 1924.
301. Spector, N., *J. Opt. Soc. Am.*, 63, 358, 1973.
302. Spector, N. and Sugar, J., *J. Opt. Soc. Am.*, 66, 436, 1976.
303. Steinhaus, D. W., Radziemski, L. J., Jr., and Blaise, J., *Los Alamos Sci. Lab.*, unpublished, 1975.
304. Subbaraya, T. S., *Z. Phys.*, 78, 541, 1932.
305. Sugar, J., *J. Opt. Soc. Am.*, 55, 33, 1965.
306. Sugar, J., *J. Res. Natl. Bur. Stand. Sect. A*, 73, 333, 1969.
307. Sugar, J., *J. Opt. Soc. Am.*, 60, 454, 1970.
308. Sugar, J., *J. Res. Natl. Bur. Stand. Sect. A*, 78, 555, 1974.
309. Sugar, J. and Kaufman, V., *J. Opt. Soc. Am.*, 55, 1283, 1965.
310. Sugar, J. and Kaufman, V., *J. Opt. Soc. Am.*, 62, 562, 1972.
311. Sugar, J., Kaufman, V., and Spector, N., *J. Res. Natl. Bur. Stand., Sect. A*, 83, 233, 1978.
312. Sugar, J. and Spector, N., *J. Opt. Soc. Am.*, 64, 1484, 1974.
313. Sullivan, F. J. *Univ. Pittsburgh Bull.*, 35, 1, 1938.
314. Svensson, L. A. and Ekberg, J. O., *Ark. Fys.*, 37, 65, 1968.
315. Swensson, J. W. and Risberg, G., *Ark. Fys.*, 31, 237, 1966.
316. Tech, J. L., *J. Res. Natl. Bur. Stand. Sect. A*, 67, 505, 1963.
317. Tech, J. L. and Ward, J. F., *Phys. Rev. Lett.*, 27, 367, 1971.
318. Tilford, S. G., *J. Opt. Soc. Am.*, 53, 1051, 1963.
319. Toresson, Y. G., *Ark. Fys.*, 17, 179, 1960.
320. Toresson, Y. G., *Ark. Fys.*, 18, 389, 1960.
321. Toresson, Y. G. and Edlen, B., *Ark. Fys.*, 23, 117, 1963.
322. Tsien, W. Z., *Chin. J. Phys.*, Peiping, 3, 117, 1939.
323. van Deurzen, C. H. H., Conway, J., and Davis, S. P., *J. Opt. Soc. Am.*, 63, 158, 1973.
324. van Kleef, T. A. M., Raassen, A. J. J., and Joshi, Y. N., *Physica*, 84(C), 401, 1976.
325. Sansonetti, C. J., Andrew, K. L., and Verges, J., *J. Opt. Soc. Am.*, 71, 423, 1981.
326. Wheatley, M. A. and Sawyer, R. A., *Phys. Rev.*, 61, 591, 1942.
327. Wilkinson, P. G., *J. Opt. Soc. Am.*, 45, 862, 1955.
328. Wilkinson, P. G. and Andrew, K. L., *J. Opt. Soc. Am.*, 53, 710, 1963.
329. Wood, D. and Andrew, K. L., *J. Opt. Soc. Am.*, 58, 818, 1968.
330. Wood, D. R., Ron, C. B., Scholl, P. S., and Hoke, M., *J. Opt. Soc. Am.*, 64, 1159, 1974.
331. Worden, E. F. and Conway, J. G., *Lawrence Livermore Lab.*, unpublished, 1977.
332. Worden, E. F., Hulet, E. K., Gutmacher, R. G., Conway, J. G., *At. Data Nucl. Data Tables*, 18, 459, 1976.
333. Worden, E. F., Lougheed, R. W., Gutmacher, R. G., and Conway, J. G., *J. Opt. Soc. Am.*, 64, 77, 1974.
334. Wu, C. M., Ph.D. thesis, University of British Columbia, 1971.
335. Zaidel, A. N., Prokofev, V. K., Raiskii, S. M., Slavnyi, V. A., and Schreider, E. Y., *Tables of Spectral Lines*, 3rd ed., Plenum, New York, 1970.
336. Zetterberg, P. O. and Magnusson, C. E., *Phys. Scr.*, 15, 189, 1977.
337. Sugar, J., *J. Opt. Soc. Am.*, 55, 1058, 1965.
338. Sugar, J., *J. Opt. Soc. Am.*, 61, 727, 1971.
339. Worden, E. F., and Conway, J. G., *At. Data Nucl. Data Tables*, 22, 329, 1978.
340. Kaufman, V. and Edlen, B., *J. Phys. Chem. Ref. Data*, 3, 825, 1974.
341. Lang, R. J., *Phys. Rev.*, 34, 697, 1929.
342. Ryabtsev, A. N., *Opt. Spectros.*, 39, 455, 1975.
343. Foster, E. W., *Proc. R. Soc. London*, 200(A), 429, 1950.
344. Morillon, C. and Verges, J., *Phys. Scr.*, 12, 129, 1975.
345. Ruedy, J. E., *Phys. Rev.*, 41, 588, 1932.
346. McLennan, J. C., McLay, A. B., and McLeod, J. H., *Philos. Mag.*, 4 ,486, 1927.
347. Handrup, M. B. and Mack, J. E., *Physica*, 30, 1245, 1964.
348. Clearman, H. E., *J. Opt. Soc. Am.*, 42, 373, 1952.
349. Paschen, F., *Ann. Physik*, 424, 148, 1938.
350. Paschen, F. and Campbell, J. S., *Ann. Phys.*, 31(5), 29, 1938.
351. Nodwell, R., *Univ. of British Columbia, Vancouver*, unpublished, 1955.
352. Gibbs, R. C. and White, H. E., *Phys. Rev.*, 31, 776, 1928.
353. Green, M., *Phys. Rev.*, 60, 117, 1941.
354. Ellis, C. B. and Sawyer, R. A., *Phys. Rev.*, 49, 145, 1936.
355. McLennan, J. C., McLay, A. B., and Crawford, M. F., *Proc. R. Soc. London Ser. A*, 125, 50, 1929.
356. Mack, J. E. and Fromer, M., *Phys. Rev.*, 48, 346, 1935.
357. Humphreys, C. J. and Paul, E., U.S. Nav. Ord. Lab., Navord Rep. 4589, 25, 1956.
358. Walters, F. M., *Sci. Pap. Bur. Stand.*, 17, 161, 1921.
359. Crawford, M. F. and McLay, A. B., *Proc. R. Soc. London Ser. A*, 143, 540, 1934.
360. McLay, A. D. and Crawford, M. F., Phys. Rev., 44, 986, 1933.
361. Schoepfle, G. K., *Phys. Rev.*, 47, 232, 1935.
362. Acquista, N., and Reader, J., *J. Opt. Soc. Am.*, 70, 789, 1980.
363. Benschop, H., Joshi, Y. N., and van Kleef, T. A. M., *Can. J. Phys.*, 53, 498, 1975.
364. Bockasten, K., Hallin, R., and Hughes, T. P., *Proc. Phys. Soc.*, 81, 522, 1963.
365. Boyce, J. C., *Phys. Rev.*, 46, 378, 1934.
366. Boyce, J. C., *Phys. Rev.*, 47, 718, 1935.
367. Boyce, J. C., *Phys. Rev.*, 48, 396, 1935.
368. Boyce, J. C., *Phys. Rev.*, 49, 351, 1936.
369. Corliss, C. H. and Meggers, W. F., *J. Res. Natl. Bur. Stand.*, 61, 269, 1958.
370. Crooker, A. M. and Dick, K. A., *Can. J. Phys.*, 42, 766, 1964.
371. De Bruin, T. L., *Z. Physik*, 77, 505, 1932.
372. De Bruin, T. L., *Proc. Roy. Acad. Amsterdam*, 36, 727, 1933.
373. De Bruin, T. L., *Zeeman Verhandelingen*, (The Hague), 1935, p. 415.
374. De Bruin, T. L., *Physica*, 3, 809, 1936.
375. De Bruin, T. L., *Proc. Roy. Acad. Amsterdam*, 40, 339, 1937.
376. Dick, K. A., *Can. J. Phys.*, 46, 1291, 1968.
377. Dick, K. A., unpublished, 1978.

378. Edlen, B. and Swensson, J. W., *Phys. Scr.*, 12, 21, 1975.
379. Ekberg, J. O., *Phys. Scr.*, 7, 55, 1973.
380. Ekberg, J. O., *Phys. Scr.*, 7, 59, 1973.
381. Ekberg, J. O., *Phys. Scr.*, 12, 42, 1975.
382. Ekberg, J. O. and Edlen, B., *Phys. Scr.*, 18, 107, 1978.
383. Eliason, A. Y., *Phys. Rev.*, 43, 745, 1933.
384. Gallardo, M., Massone, C. A., Tagliaferri, A. A., Garavaglia, M., and Persson, W., *Phys. Scr.*, 19, 538, 1979.
385. Garcia-Riquelme, O., *Optica Pura Y Aplicada*, 1, 53, 1968.
386. Gibbs, R. C., Vieweg, A. M., and Gartlein, C. W., *Phys. Rev.*, 34, 406, 1929.
387. Gilbert, W. P., *Phys. Rev.*, 48, 338, 1935.
388. Goldsmith, S. and Kaufman, A. S., *Proc. Phys. Soc.*, 81, 544, 1963.
389. Hermansdorfer, H., *J. Opt. Soc. Am.*, 62, 1149, 1972.
390. Humphreys, C. J., *Phys. Rev.*, 47, 712, 1935.
391. Humphreys, C. J., *J. Res. Natl. Bur. Stand.*, 16, 639, 1936.
392. Iglesias, L., *J. Opt. Soc. Am.*, 45, 856, 1955.
393. Iglesias, L., *J. Res. Natl. Bur. Stand.*, 64A, 481, 1960.
394. Iglesias, L., *Anales Fisica Y Quimica*, 58A, 191, 1962.
395. Iglesias, L., *J. Res. Natl. Bur. Stand.*, 70A, 465, 1966.
396. Iglesias, L., *Can. J. Phys.*, 44, 895, 1966.
397. Iglesias, L., *J. Res. Natl. Bur. Stand.*, 72A, 295, 1968.
398. Joshi, Y. N., *Can. Spectrosc.*, 15, 96, 1970.
399. Joshi, Y. N. and van Kleef, T. A. M., *Can. J. Phys.*, 55, 714, 1977.
400. Kaufman, A. S., Hughes, T. P., and Williams, R. V., *Proc. Phys. Soc.*, 76, 17, 1960.
401. Kaufman, V. and Sugar, J., *J. Opt. Soc. Am.*, 68, 1529, 1978.
402. Keussler, V., *Z. Physik*, 85, 1, 1933.
403. Kiess, C. C., *J. Res. Natl. Bur. Stand.*, 56, 167, 1956.
404. Klinkenberg, P. F. A., van Kleef, T. A. M., and Noorman, P. E., *Physica*, 27, 1177, 1961.
405. Kovalev, V. I., Romanos, A. A., and Ryabtsev, A. N., *Opt. Spectrosc.*, 43, 10, 1977.
406. Lang, R. J., *Proc. Natl. Acad. Sci.*, 13, 341, 1927.
407. Lang, R. J., *Zeeman Verhandelingen*, (The Hague), 44, 1935.
408. Liberman, S., et al., *C. R. Acad. Sci.* (Paris), 286, 253, 1978.
409. Livingston, A. E., *J. Phys.*, B9, L215, 1976.
410. Meijer, F. G., *Physica*, 72, 431, 1974.
411. Meijer, F. G. and Metsch, B. C., *Physica*, 94C, 259, 1978.
412. Moore, F. L., thesis, Princeton, 1949.
413. Paul, F. W. and Polster, H. D., *Phys. Rev.*, 59, 424, 1941.
414. Phillips, L. W. and Parker, W. L., *Phys. Rev.*, 60, 301, 1941.
415. Poppe, R., *Physica*, 81C, 351, 1976.
416. Raassen, A. J. J., van Kleef, T. A. M., and Metsch, B. C., *Physica*, 84C, 133, 1976.
417. Rao, A. B. and Krishnamurty, S. G., *Proc. Phys. Soc. (London)*, 51, 772, 1939.
418. Reader, J. and Acquista, N., *J. Opt. Soc. Am.*, 69, 239, 1979.
419. Reader, J. and Epstein, G. L., *J. Opt. Soc. Am.*, 62, 619, 1972.
420. Rico, F. R., *Anales, Real Soc. Esp. Fis. Quim.*, 61, 103, 1965.
421. Schonheit, E., *Optik*, 23, 409, 1966.
422. Shenstone, A. G., *J. Opt. Soc. Am.*, 44, 749, 1954.
423. Shenstone, A. G., unpublished, 1958.
424. Shenstone, A. G., *J. Res. Natl. Bur. Stand.*, 67A, 87, 1963.
425. Sugar, J. and Kaufman, V., *J. Opt. Soc. Am.*, 64, 1656, 1974.
426. Sugar, J. and Kaufman, V., *Phys. Rev.*, C12, 1336, 1975.
427. Svensson, L. A., *Phys. Scr.*, 13, 235, 1976.
428. Swensson, J. W. and Edlen, B., *Phys. Scr.*, 9, 335, 1974.
429. Tagliaferri, A. A., Gallego Lluesma, E., Garavaglia, M., Gallardo, M., and Massone, C. A., *Optica Pura Y Aplica*, 7, 89.
430. Tilford, S. G. and Giddings, L. E., *Astrophys. J.*, 141, 1222, 1965.
431. Trawick, M. W., *Phys. Rev.*, 46, 63, 1934.
432. Van Deurzen, C. H. H., *J. Opt. Soc. Am.*, 67, 476, 1977.
433. Yarosewick, S. L. and Moore, F. L., *J. Opt. Soc. Am.*, 57, 1381, 1967.
434. Zalubas, R., unpublished, 1979.
435. Bhatia, K. S., Jones, W. E., and Crooker, A. M., *Can. J. Phys.*, 50, 2421, 1972.
436. van Kleef, T. A. M. and Joshi, Y. N., *Phys. Scr.*, 24, 557, 1981.

NIST ATOMIC TRANSITION PROBABILITY TABLES

J.R. Fuhr and W.L. Wiese

These tables substantially update and enlarge our earlier tables in this *Handbook*. The new tables contain critically evaluated atomic transition probabilities for about 9000 selected lines of all elements for which reliable data are available on an absolute scale. The material is largely for neutral and singly ionized spectra, but also includes a number of prominent lines of more highly charged ions of important elements.

Many of the data are obtained from comprehensive compilations of the Data Center on Atomic Transition Probabilities at the National Institute of Standards and Technology (formerly the National Bureau of Standards). Specifically, data have been taken from three recent comprehensive critical compilations on C, N and O,[1] on Sc through Mn,[2] on Fe through Ni.[3] Material from earlier compilations for the elements H through Ne[4] and Na through Ca[5] was supplemented by more recent material taken directly from the original literature. For the highly charged ions, some of the data were derived from studies of the systematic behavior of transition probabilities.[6-8] Most of the original literature is cited in the above tables and in recent bibliographies[9,10]; for lack of space, individual literature references are not cited here.

The wavelength range for the neutral species is normally the visible spectrum or shorter wavelengths; only the very prominent near infrared lines are included. For the higher ions, most of the strong lines are located in the far UV. The tabulation is limited to electric dipole — including intercombination — lines and comprises essentially the fairly strong transitions with estimated uncertainties of 50% or less. With the exception of hydrogen, helium, and the alkalis, most transitions are between states with low principal quantum numbers.

The transition probability, A, is given in units of 10^8 s^{-1} and is listed to as many digits as is consistent with the indicated accuracy. The power of 10 is indicated by the E notation (i.e., E-02 means 10^{-2}). Generally, the estimated uncertainties of the A-values are ± 25 to 50% for two-digit numbers, ± 10 to 25% for three-digit numbers and $\pm 1\%$ or better for four- and five-digit numbers.

Each transition is identified by the wavelength, λ, in angstroms; and the statistical weights, g_i and g_k, of the lower (i) and upper (k) states [the product $g_k A$ (or $g_i f$) is needed for many applications]. Whenever the wavelengths of individual lines within a multiplet are extremely close, only an average wavelength for the multiplet as well as the multiplet A-value are given, and this is indicated by an asterisk (*) to the left of the wavelength. This also has been done when the transition probability for an entire multiplet has been taken from the literature and values for individual lines cannot be determined because of insufficient knowledge of the coupling of electrons. The wavelength data have been taken either from recent compilations or from the original literature cited in bibliographies published by the Atomic Energy Levels Data Center[11,12] at the National Institute of Standards and Technology. Wavelength values are consistent with those given in the table "Line Spectra of the Elements", which appears elsewhere in this *Handbook*.

The transition probabilities for hydrogen and hydrogen-like ions are known precisely. Because of the hydrogen degeneracy, a "transition" is actually the sum of all fine-structure transitions between the principal quantum numbers listed in the transition column; therefore, the special hydrogen table which appears below gives weighted average A-values.

In addition to the transition probability A, the atomic oscillator strength f and the line strength S are often used in the literature. The conversion factors between these quantities are (for electric-dipole transitions):

$$g_i f = 1.499 \times 10^{-8} \lambda^2 g_k A = 303.8 \lambda^{-1} S$$

where λ is in angstroms, A is in 10^8 s^{-1}, and S is in atomic units, which are $a_0^2 e^2 = 7.188 \times 10^{-59}$ m^2 C^2.

After the special table for hydrogen, the tables for other elements appear in alphabetical sequence by element name (not symbol). Within each element, the tables are ordered by increasing ionization stage (e.g., Al I, Al II, etc.).

REFERENCES

1. Wiese, W.L., Fuhr, J.R., and Deters, T.M., *Atomic Transition Probabilities of Carbon, Nitrogen and Oxygen, J. Phys. Chem. Ref. Data, Monograph 7*, 1996.
2. Martin, G.A., Fuhr, J.R., and Wiese, W.L., *Atomic Transition Probabilities—Scandium through Manganese, J. Phys. Chem. Ref. Data, 17, Suppl.* 3, 1988.
3. Fuhr, J.R., Martin, G.A., and Wiese, W.L., *Atomic Transition Probabilities—Iron through Nickel, J. Phys. Chem. Ref. Data, 17, Suppl.* 4, 1988.
4. Wiese, W.L., Smith, M.W., and Glennon, B.M., *Atomic Transition Probabilities (H through Ne—A Critical Data Compilation)*, National Standard Reference Data Series, National Bureau of Standards 4, Vol. I, U.S. Government Printing Office, Washington, D.C., 1966.
5. Wiese, W.L., Smith, M.W., and Miles, B.M., *Atomic Transition Probabilities (Na through Ca—A Critical Data Compilation)*, National Standard Reference Data Series, National Bureau of Standards 22, Vol. II, U. S. Government Printing Office, Washington, D.C., 1969.
6. Wiese, W.L. and Weiss, A.W., *Phys. Rev.*, 175, 50, 1968.
7. Smith, M.W. and Wiese,M.L., *Astrophys. J., Suppl. Ser.*, 23, No. 196, 103, 1971.
8. Martin, G.A., and Wiese,W.L., *J. Phys. Chem. Ref. Data*, 5, 537, 1976.
9. Fuhr, J.R., Miller, B.J., and Martin, G.A., *Bibliography on Atomic Transition Probabilities (1914 through October 1997)*, National Bureau of Standards Special Publication 505, 1978; Miller, B.J., Fuhr, J.R., and Martin, G.A., *Bibliography on Atomic Transition Probabilities (November 1977 through February 1980)*, National Bureau of Standards Special Publication 505, Supplement 1, 1980.
10. Wiese, W.L., Reports on Astronomy, *Trans. Int. Astron. Union*, 18A, 116,—123, 1982; 19A, 122—138, 1985.; 20A, 117—123, 1988, Reidel, D., Ed., Kluwer, Dordrecht, Holland.
11. Moore, C.E., *Bibliography on the Analyses of Optical Atomic Spectra*, National Bureau of Standards Special Publication 306—Section 1, 1968; Sections 2—4, 1969.

12. Hagan, L. and Martin, W.C., *Bibliography on Atomic Energy Levels and Spectra (July 1968 through June 1971)*, National Bureau of Standards Special Publication 363, 1972; Hagan, L., *Bibliography on Atomic Energy Levels and Spectra (July 1971 through June 1975)*, National Bureau of Standards Special Publication 363, Supplement 1, 1977; Zalubas, R. and Albright, A., *Bibliography on Atomic Energy Levels and Spectra (July 1975 through June 1979)*, National Bureau of Standards Special Publication 363, Supplement 2, 1980; Musgrove, A. and Zalubas, R., *Bibliography on Atomic Energy Levels and Spectra (July 1979 through December 1983)*, National Bureau of Standards Special Publication 363, Supplement 3, 1985.
13. Younger, S.M. and Weiss, A., *J. Res. Natl. Bur. Stand.*, 79A, 629, 1975.

Transition Probabilities for Allowed Lines of Hydrogen

λ Å	g_i	g_k	A 10^8 s^{-1}	λ Å	g_i	g_k	A 10^8 s^{-1}	λ Å	g_i	g_k	A 10^8 s^{-1}
Hydrogen				3664.68	8	1568	4.022E-06	8598.40	18	392	9.211E-05
H I				3666.10	8	1458	4.826E-06	8665.02	18	338	1.343E-04
912.768	2	1800	5.167E-06	3667.68	8	1352	5.830E-06	8750.48	18	288	2.021E-04
912.839	2	1682	6.122E-06	3669.46	8	1250	7.096E-06	8862.79	18	242	3.156E-04
912.918	2	1568	7.297E-06	3671.48	8	1152	8.707E-06	9014.91	18	200	5.156E-04
913.006	2	1458	8.753E-06	3673.76	8	1058	1.078E-05	9229.02	18	162	8.905E-04
913.104	2	1352	1.057E-05	3676.36	8	968	1.347E-05	9545.97	18	128	1.651E-03
913.215	2	1250	1.286E-05	3679.35	8	882	1.700E-05	10049.4	18	98	3.358E-03
913.339	2	1152	1.578E-05	3682.81	8	800	2.172E-05	10938.1	18	72	7.783E-03
913.480	2	1058	1.952E-05	3686.83	8	722	2.809E-05	12818.1	18	50	2.201E-02
913.641	2	968	2.438E-05	3691.55	8	648	3.685E-05	16407.2	32	288	1.620E-04
913.826	2	882	3.077E-05	3697.15	8	578	4.910E-05	16806.5	32	242	2.556E-04
914.039	2	800	3.928E-05	3703.85	8	512	6.658E-05	17362.1	32	200	4.235E-04
914.286	2	722	5.077E-05	3711.97	8	450	9.210E-05	18174.1	32	162	7.459E-04
914.576	2	648	6.654E-05	3721.94	8	392	1.303E-04	18751.0	18	32	8.986E-02
914.919	2	578	8.858E-05	3734.37	8	338	1.893E-04	19445.6	32	128	1.424E-03
915.329	2	512	1.200E-04	3750.15	8	288	2.834E-04	21655.3	32	98	3.041E-03
915.824	2	450	1.657E-04	3770.63	8	242	4.397E-04	26251.5	32	72	7.711E-03
916.429	2	392	2.341E-04	3797.90	8	200	7.122E-04	27575	50	288	1.402E-04
917.181	2	338	3.393E-04	3835.38	8	162	1.216E-03	28722	50	242	2.246E-04
918.129	2	288	5.066E-04	3889.05	8	128	2.215E-03	30384	50	200	3.800E-04
919.351	2	242	7.834E-04	3970.07	8	98	4.389E-03	32961	50	162	6.908E-04
920.963	2	200	1.263E-03	4101.73	8	72	9.732E-03	37395	50	128	1.388E-03
923.150	2	162	2.143E-03	4340.46	8	50	2.530E-02	40511.5	32	50	2.699E-02
926.226	2	128	3.869E-03	4861.32	8	32	8.419E-02	43753	72	288	1.288E-04
930.748	2	98	7.568E-03	6562.80	8	18	4.410E-01	46525	50	98	3.253E-03
937.803	2	72	1.644E-02	8392.40	18	800	1.517E-05	46712	72	242	2.110E-04
949.743	2	50	4.125E-02	8413.32	18	722	1.964E-05	51273	72	200	3.688E-04
972.537	2	32	1.278E-01	8437.96	18	648	2.580E-05	59066	72	162	7.065E-04
1025.72	2	18	5.575E-01	8467.26	18	578	3.444E-05	74578	50	72	1.025E-02
1215.67	2	8	4.699E+00	8502.49	18	512	4.680E-05	75004	72	128	1.561E-03
3662.26	8	1800	2.847E-06	8545.39	18	450	6.490E-05	123680	72	98	4.561E-03
3663.40	8	1682	3.374E-06								

For hydrogen-like ions of nuclear charge Z, the following scaling laws hold:

$$A_Z = Z^4 A_{Hydrogen}; f_Z = f_H; S_Z = Z^{-2} S_H$$
$$\text{(For wavelengths, } \lambda_Z = Z^{-2} \lambda_H)$$

For very highly charged hydrogen-like ions, starting at about Z>25, relativistic corrections[13] must be applied.

Transition Probabilities for Other Elements

λ Å	Weights g_i	g_k	A 10^8 s^{-1}	λ Å	Weights g_i	g_k	A 10^8 s^{-1}	λ Å	Weights g_i	g_k	A 10^8 s^{-1}
Aluminum				1384.1	4	2	9.1E+00	*4761	2	6	2.55E-01
Al I				1605.8	2	4	1.22E+01	5172	2	4	3.95E-02
2263.5	2	4	6.6E-01	1611.8	4	4	2.42E+00	5551	4	6	3.85E-02
2269.1	4	6	7.9E-01	1611.9	4	6	1.45E+01	5687	4	4	6.0E-03
2269.2	4	4	1.3E-01	1854.7	2	4	5.40E+00				
2367.1	2	4	7.2E-01	1862.8	2	2	5.33E+00	**Argon**			
2373.1	4	6	8.6E-01	*1935.9	10	14	1.22E+01	**Ar I**			
2373.4	4	4	1.4E-01	3601.6	6	4	1.34E+00	1048.22	1	3	5.36E+00
2568.0	2	4	2.3E-01	3601.9	4	4	1.49E-01	1066.66	1	3	1.29E+00
2575.1	4	6	2.8E-01	3612.4	4	2	1.5E+00	3406.18	3	1	3.9E-03
2575.4	4	4	4.4E-02					3461.08	3	5	6.7E-04
2652.5	2	2	1.33E-01	**Al X**				3554.30	5	5	2.7E-03
2660.4	4	2	2.64E-01	39.925	1	3	2.22E+03	3563.29	1	3	1.2E-03
3082.2	2	4	6.3E-01	51.979	1	3	4.8E+03	3567.66	5	7	1.1E-03
3092.7	4	6	7.4E-01	55.227	1	3	5.2E+03	3572.30	3	1	5.1E-03
3092.8	4	4	1.2E-01	55.272	3	5	7.2E+03	3606.52	3	1	7.6E-03
3944.0	2	2	4.93E-01	55.376	5	7	9.5E+03	3632.68	3	5	6.6E-04
3961.5	4	2	9.8E-01	59.107	3	5	4.6E+03	3634.46	3	3	1.3E-03
6696.0	2	4	1.69E-02	332.78	1	3	5.6E+01	3643.12	3	5	2.4E-04
6698.7	2	2	1.69E-02	394.83	3	1	8.3E+01	3649.83	3	1	8.0E-03
7835.3	4	6	5.7E-02	395.36	3	5	1.2E+01	3659.53	3	3	4.4E-04
7836.1	6	8	6.2E-02	397.76	1	3	1.7E+01	3670.67	3	5	3.1E-04
				400.43	3	3	1.3E+01	3675.23	3	3	4.9E-04
Al II				401.12	5	5	3.6E+01	3770.37	1	3	7.0E-04
1047.9	1	3	3.6E-01	403.55	3	1	4.9E+01	3834.68	3	1	7.5E-03
1048.6	3	5	4.8E-01	406.31	5	3	1.9E+01	3894.66	3	3	5.7E-04
1539.8	3	5	8.8E+00	670.06	3	5	9.8E+00	3947.50	5	5	5.6E-04
1670.8	1	3	1.46E+01	2535	1	3	3.8E-01	3948.98	5	3	4.55E-03
1719.4	1	3	6.79E+00					4044.42	3	5	3.33E-03
1764.0	5	5	9.8E+00	**Al XI**				4045.96	3	3	4.1E-04
1772.8	1	3	9.5E+00	*36.675	2	6	1.5E+03	4054.53	3	3	2.7E-04
1777.0	5	7	1.7E+01	39.091	2	4	2.6E+03	4158.59	5	5	1.40E-02
*1819.0	15	15	5.6E+00	39.180	4	6	3.1E+03	4164.18	5	3	2.88E-03
1855.9	1	3	8.32E-01	39.530	2	2	1.8E+02	4181.88	1	3	5.61E-03
1858.0	3	3	2.48E+00	39.623	4	2	3.7E+02	4190.71	5	5	2.80E-03
1862.3	5	3	4.12E+00	48.298	2	4	3.09E+03	4191.03	1	3	5.39E-03
1931.0	3	1	1.08E+01	48.338	2	2	3.08E+03	4198.32	3	1	2.57E-02
1990.5	3	5	1.47E+01	52.299	2	4	8.1E+03	4200.67	5	7	9.67E-03
2816.2	3	1	3.83E+00	52.446	4	6	9.6E+03	4251.18	5	3	1.11E-03
4663.1	5	3	5.3E-01	52.458	4	4	1.6E+03	4259.36	3	1	3.98E-02
6226.2	1	3	6.2E-01	54.217	2	2	4.8E+02	4266.29	3	5	3.12E-03
6231.8	3	5	8.4E-01	54.388	4	2	9.6E+02	4272.17	3	3	7.97E-03
6243.4	5	7	1.1E+00	*99.083	2	6	2.2E+02	4300.10	3	5	3.77E-03
6335.7	5	3	1.4E-01	103.6	2	4	4.2E+02	4333.56	3	5	5.68E-03
6823.4	3	3	3.4E-01	103.8	4	6	5.0E+02	4335.34	3	3	3.87E-03
6837.1	5	3	5.7E-01	*141.6	2	6	4.07E+02	4345.17	3	3	2.97E-03
6920.3	3	1	9.6E-01	150.31	2	4	8.5E+01	4363.79	3	3	1.2E-04
7042.1	3	5	5.9E-01	150.61	4	6	9.9E+02	4424.00	1	3	7.3E-05
7056.7	3	3	5.8E-01	157.0	2	2	1.3E+02	4510.73	3	1	1.18E-02
7471.4	5	7	9.4E-01	157.4	4	2	2.6E+02	4522.32	1	3	8.98E-04
				*205.0	2	6	6.3E+01	4544.75	3	3	8.3E-04
Al III				*308.6	2	6	9.9E+01	4554.32	3	5	3.8E-04
*560.36	2	6	4.0E-01	*341.3	6	2	1.3E+02	4584.96	3	5	1.6E-03
695.83	2	4	7.4E-01	550.05	2	4	8.55E+00	4586.61	3	3	2.3E-03
696.22	2	2	7.2E-01	568.12	2	2	7.73E+00	4587.21	3	1	4.9E-03
*1352.8	10	14	4.40E+00	1997	2	4	1.07E+00	4589.29	3	5	6.2E-05
1379.7	2	2	4.59E+00	2069	2	2	9.7E-01	4596.10	3	3	9.47E-04

λ Å	Weights g_i	g_k	A 10^8 s^{-1}	λ Å	Weights g_i	g_k	A 10^8 s^{-1}	λ Å	Weights g_i	g_k	A 10^8 s^{-1}
4628.44	3	5	3.83E-04	5473.46	5	3	2.0E-03	6025.15	5	3	9.0E-03
4642.15	3	5	9.6E-04	5490.12	5	5	8.5E-04	6043.22	5	7	1.47E-02
4647.49	3	3	1.2E-03	5492.09	3	1	5.6E-03	6052.73	3	5	1.9E-03
4702.32	3	3	1.09E-03	5495.87	7	9	1.69E-02	6064.76	5	7	5.8E-04
4746.82	3	1	3.6E-03	5506.11	5	7	3.6E-03	6081.25	3	3	7.5E-04
4752.94	3	3	4.5E-03	5524.96	7	7	1.7E-03	6085.86	3	3	9.0E-05
4768.68	3	5	8.6E-04	5528.97	1	3	1.2E-03	6090.79	1	3	3.0E-03
4798.74	7	9	8.8E-04	5534.49	5	3	2.7E-03	6098.81	3	3	5.2E-03
4835.97	7	9	9.3E-04	5540.87	7	5	4.1E-04	6101.16	3	3	3.3E-03
4836.70	3	5	1.02E-03	5552.77	3	3	7.9E-04	6104.58	3	1	3.4E-03
4876.26	3	5	7.8E-03	5558.70	3	5	1.42E-02	6105.64	3	5	1.21E-02
4886.29	7	9	1.2E-03	5559.66	3	5	2.2E-03	6113.46	3	5	4.7E-04
4887.95	3	3	1.3E-02	5572.54	5	7	6.6E-03	6119.66	3	3	5.1E-04
4894.69	3	1	1.8E-02	5574.22	3	5	4.6E-04	6121.86	3	5	1.3E-04
4921.04	5	7	5.9E-04	5581.87	7	5	5.6E-04	6127.42	5	3	1.1E-03
4937.72	7	5	3.6E-04	5588.72	5	5	1.5E-03	6128.73	3	5	8.6E-04
4956.75	7	9	1.8E-03	5597.48	5	7	4.2E-03	6145.44	5	7	7.6E-03
4989.95	5	7	1.1E-03	5606.73	3	3	2.20E-02	6155.24	5	3	5.1E-03
5032.03	7	5	8.2E-04	5618.01	3	3	2.1E-03	6165.12	5	5	9.89E-04
5048.81	3	5	4.6E-03	5620.92	3	1	3.6E-03	6170.17	5	5	5.0E-03
5054.18	3	3	4.5E-03	5623.78	5	5	1.4E-03	6173.10	3	5	6.7E-03
5056.53	3	1	5.7E-03	5635.58	3	5	9.6E-04	6179.41	5	3	6.6E-04
5060.08	7	9	3.7E-03	5637.33	1	3	9.1E-04	6212.50	5	7	3.9E-03
5070.99	5	3	2.6E-03	5639.12	1	3	2.1E-03	6215.94	5	5	5.7E-03
5073.08	3	5	5.9E-04	5641.39	3	5	8.7E-04	6230.93	5	5	1.2E-04
5078.03	7	7	4.7E-04	5648.69	5	3	1.2E-03	6243.40	3	1	1.3E-03
5087.09	5	7	1.6E-03	5650.70	3	1	3.20E-02	6244.73	3	5	2.0E-04
5104.74	3	5	8.7E-04	5659.13	5	5	2.6E-03	6248.41	3	5	6.8E-04
5118.21	5	7	2.7E-03	5681.90	5	7	2.0E-03	6278.65	5	7	2.0E-04
5127.80	5	5	3.3E-04	5683.73	5	5	2.0E-03	6296.87	3	5	9.0E-03
5151.39	3	1	2.39E-02	5700.87	5	7	5.9E-03	6307.66	5	5	6.0E-03
5152.30	3	5	1.1E-03	5712.51	1	3	8.7E-04	6309.14	3	3	7.6E-04
5162.29	3	3	1.90E-02	5739.52	3	5	8.7E-03	6364.89	3	1	5.6E-03
5177.54	7	5	2.4E-03	5772.11	5	7	2.0E-03	6369.58	5	3	4.2E-03
5192.72	7	7	1.2E-04	5773.99	5	5	1.1E-03	6384.72	3	3	4.21E-03
5194.02	3	1	7.8E-03	5783.54	3	5	8.1E-04	6416.31	3	5	1.16E-02
5210.49	7	7	1.1E-03	5789.48	5	5	4.6E-04	6431.56	5	3	5.1E-04
5214.77	5	3	2.1E-03	5790.40	5	3	3.4E-04	6466.55	1	3	1.5E-03
5216.28	5	3	1.3E-03	5802.08	5	3	4.2E-03	6481.14	1	3	9.4E-04
5221.27	7	9	8.8E-03	5843.77	3	5	3.3E-04	6513.85	3	3	5.4E-04
5241.09	5	5	1.3E-03	5882.62	3	1	1.23E-02	6538.11	7	7	1.1E-03
5246.24	5	7	1.2E-03	5888.58	7	5	1.29E-02	6596.12	7	5	2.3E-04
5249.20	5	5	7.9E-04	5916.58	5	3	5.9E-04	6598.68	5	5	3.6E-04
5252.79	5	7	5.4E-03	5927.11	7	7	3.7E-04	6604.02	7	5	2.8E-03
5254.47	3	5	3.6E-03	5928.81	5	3	1.1E-02	6604.85	5	7	1.3E-04
5286.07	5	7	9.6E-04	5940.86	1	3	1.2E-03	6632.09	3	3	5.3E-04
5290.00	5	3	9.0E-04	5942.67	5	5	1.8E-03	6656.88	3	3	3.1E-04
5309.52	5	5	1.2E-03	5943.89	7	5	3.6E-04	6660.68	3	1	7.8E-03
5317.73	5	7	2.6E-03	5949.26	3	3	1.5E-03	6664.05	5	5	1.5E-03
5373.50	3	5	2.7E-03	5964.48	1	3	7.7E-04	6677.28	3	1	2.36E-03
5393.27	5	5	9.6E-04	5968.32	3	3	1.8E-03	6684.73	3	5	3.9E-04
5410.48	5	7	2.0E-03	5971.60	3	1	1.1E-02	6698.47	3	3	2.5E-04
5421.35	7	5	6.0E-03	5981.90	5	7	1.2E-04	6698.88	5	3	1.6E-03
5439.99	3	3	1.9E-03	5987.30	7	7	1.2E-03	6719.22	1	3	2.4E-03
5442.24	7	7	9.3E-04	5988.13	3	5	6.1E-04	6722.88	5	7	3.2E-04
5451.65	3	5	4.7E-03	5994.66	3	5	2.6E-04	6752.84	3	5	1.93E-02
5457.42	5	3	3.6E-03	5999.00	5	5	1.4E-03	6754.37	3	3	2.1E-03
5459.65	7	7	3.8E-04	6005.73	5	3	1.4E-03	6756.10	5	5	3.6E-03
5467.16	5	5	7.6E-04	6013.68	7	5	1.4E-03	6766.61	5	3	4.0E-03

λ Å	Weights g_i	g_k	A 10^8 s^{-1}	λ Å	Weights g_i	g_k	A 10^8 s^{-1}	λ Å	Weights g_i	g_k	A 10^8 s^{-1}
6779.93	1	3	1.21E-03	8014.79	5	5	9.28E-02	13622.4	3	5	7.3E-02
6818.29	3	1	2.0E-03	8037.23	1	3	3.59E-03	13678.5	3	5	6.2E-02
6827.25	5	3	2.4E-03	8046.13	3	1	1.12E-02	14093.6	1	3	4.3E-02
6851.88	3	5	6.7E-04	8053.31	5	3	8.6E-03	14739.1	5	7	8.8E-04
6871.29	3	3	2.78E-02	8066.60	5	5	1.4E-03	15046.4	1	3	5.2E-02
6879.59	3	5	1.8E-03	8103.69	3	3	2.5E-01	15172.3	1	3	1.3E-02
6887.10	5	7	1.3E-03	8115.31	5	7	3.31E-01	15329.6	5	5	1.2E-03
6888.17	3	5	2.5E-03	8264.52	3	3	1.53E-01	15555.5	5	7	9.8E-05
6925.01	3	3	1.2E-03	8384.73	5	7	2.4E-03	15734.9	5	3	2.9E-04
6937.67	3	1	3.08E-02	8408.21	3	5	2.23E-01	15816.8	5	3	8.7E-04
6951.46	5	5	2.2E-03	8424.65	3	5	2.15E-01	15989.3	1	3	1.9E-02
6960.23	5	5	2.4E-03	8490.30	3	5	9.6E-04	16122.7	5	3	3.9E-04
6965.43	5	3	6.39E-02	8521.44	3	3	1.39E-01	16180.0	5	5	1.2E-03
6992.17	3	1	7.5E-03	8605.78	5	5	1.04E-02	16264.1	3	3	3.0E-04
7030.25	7	5	2.67E-02	8620.46	1	3	9.2E-03	16520.1	3	5	2.6E-03
7067.22	5	5	3.80E-02	8667.94	1	3	2.43E-02	16739.8	3	5	3.1E-03
7068.73	5	3	2.0E-02	8761.69	3	5	9.5E-03	16940.4	5	5	2.5E-02
7086.70	1	3	1.5E-03	8784.61	3	1	2.4E-03	20317.0	1	3	1.6E-03
7107.48	5	5	4.5E-03	8799.08	5	3	4.6E-03	20616.5	5	5	3.9E-03
7125.83	3	3	6.0E-03	8962.19	3	3	1.6E-03	20812.0	5	7	7.6E-04
7147.04	5	3	6.25E-03	9075.42	3	1	1.2E-02	21332.2	3	3	3.2E-04
7158.83	3	1	2.1E-02	9122.97	5	3	1.89E-01	21534.9	3	5	1.1E-03
7162.57	1	3	5.8E-04	9194.64	3	3	1.76E-02	22039.2	3	1	1.2E-03
7206.98	5	3	2.48E-02	9224.50	3	5	5.03E-02	22077.4	5	3	1.4E-03
7229.93	5	5	6.6E-04	9291.53	3	1	3.26E-02	23133.4	3	3	1.7E-03
7265.17	3	3	1.7E-03	9354.22	3	3	1.06E-02	23844.8	9	7	1.1E-02
7270.66	7	7	1.1E-03	9657.78	3	3	5.43E-02	23967.5	3	1	3.6E-03
7272.93	3	3	1.83E-02	9784.50	3	5	1.47E-02				
7285.44	5	3	1.2E-03	10470.05	1	3	9.8E-03	**Ar II**			
7311.72	3	3	1.7E-02	10478.0	3	3	2.44E-02	2317.7	6	4	1.4E-01
7316.01	3	3	9.6E-03	10950.7	5	3	3.96E-03	2891.6	4	2	1.82E-01
7350.78	3	1	1.2E-02	11078.9	5	5	8.3E-03	2942.9	4	4	5.3E-01
7353.32	5	7	9.6E-03	11393.7	3	1	2.22E-02	2979.1	2	2	4.16E-01
7372.12	7	9	1.9E-02	11441.8	5	3	1.39E-02	3033.5	2	4	9.9E-02
7383.98	3	5	8.47E-02	11467.5	3	5	3.69E-03	3139.0	6	6	5.2E-01
7392.97	5	3	7.2E-03	11488.11	3	3	1.9E-03	3169.7	4	6	4.9E-01
7412.33	3	5	3.9E-03	11668.7	5	5	3.76E-02	3181.0	6	4	3.7E-01
7422.26	3	5	6.6E-04	11719.5	5	3	9.52E-03	3212.5	4	4	5.2E-02
7425.29	5	7	3.1E-03	12026.6	1	3	4.2E-03	3221.6	6	6	1.8E-02
7435.33	5	5	9.0E-03	12112.2	7	7	3.1E-02	3226.0	4	4	2.1E-02
7436.25	7	5	2.7E-03	12139.8	3	3	4.5E-02	3243.7	4	2	1.06E+00
7471.17	3	3	2.2E-04	12343.7	5	7	2.0E-02	3249.8	2	4	6.3E-01
7484.24	3	5	3.4E-03	12402.9	3	3	1.1E-01	3263.6	2	4	1.55E-01
7503.84	3	1	4.45E-01	12439.2	3	5	4.9E-02	3281.7	2	2	4.2E-01
7510.42	5	5	4.5E-03	12456.1	5	3	8.9E-02	3430.4	6	8	6.2E-02
7514.65	3	1	4.02E-01	12487.6	7	5	1.1E-01	3454.1	6	4	3.14E-01
7618.33	3	5	2.9E-03	12554.4	7	5	1.2E-03	3466.3	8	6	3.0E-02
7628.86	3	5	2.9E-03	12702.4	3	3	7.1E-02	3476.7	6	6	1.25E+00
7635.11	5	5	2.45E-01	12733.6	5	5	1.1E-01	3491.2	4	4	1.79E+00
7670.04	5	3	2.8E-03	12746.3	3	3	2.0E-02	3491.5	6	8	2.31E+00
7704.81	5	7	6.3E-04	12802.7	5	5	5.7E-02	3509.8	2	2	2.55E+00
7723.76	5	3	5.18E-02	12933.3	3	1	1.0E-01	3514.4	4	6	1.36E+00
7724.21	1	3	1.17E-01	12956.6	3	3	7.4E-02	3520.0	6	6	5.2E-01
7798.55	3	5	8.7E-04	13008.5	5	3	8.9E-02	3521.3	8	8	2.27E-01
7868.20	1	3	3.50E-03	13214.7	3	1	8.1E-02	3535.3	2	4	5.7E-01
7891.08	5	5	9.5E-03	13273.1	5	7	1.5E-01	3548.5	4	4	8.7E-01
7916.45	3	3	1.2E-03	13313.4	3	5	1.3E-01	3550.0	6	6	2.6E-02
7948.18	1	3	1.86E-01	13504.0	5	7	1.1E-01	3556.9	2	2	5.0E-02
8006.16	3	5	4.90E-02	13599.2	5	5	2.2E-02	3559.5	6	8	2.88E+00

λ Å	Weights g_i	g_k	A 10^8 s^{-1}	λ Å	Weights g_i	g_k	A 10^8 s^{-1}	λ Å	Weights g_i	g_k	A 10^8 s^{-1}
3565.0	2	4	5.5E-01	4228.2	4	6	1.31E-01	6483.1	4	2	1.06E-01
3576.6	6	8	2.75E+00	4237.2	4	4	1.12E-01	6638.2	6	4	1.37E-01
3581.6	2	4	1.76E+00	4266.5	6	6	1.64E-01	6639.7	4	2	1.69E-01
3582.4	4	6	2.53E+00	4277.5	6	4	8.0E-01	6643.7	10	8	1.47E-01
3588.4	8	10	3.03E+00	4282.9	4	2	1.32E-01	6666.4	2	2	8.8E-02
3605.9	4	6	4.4E-02	4300.6	6	6	5.7E-02	6684.3	8	6	1.07E-01
3656.0	6	6	7.6E-02	4331.2	4	4	5.74E-01	6756.6	4	4	2.0E-02
3682.5	4	2	1.7E-02	4332.0	4	2	1.92E-01	6863.5	6	6	2.5E-02
3709.9	4	4	4.7E-02	4348.1	6	8	1.17E+00	7233.5	2	4	3.7E-02
3717.2	6	8	5.2E-02	4352.2	2	2	2.12E-01	7380.4	4	4	5.6E-02
3729.3	6	4	4.80E-01	4362.1	4	6	5.5E-02	7589.3	6	4	1.07E-01
3746.9	4	6	2.1E-02	4370.8	4	4	6.6E-01				
3763.5	8	6	1.78E-01	4371.3	6	4	2.21E-01	**Ar III**			
3766.1	4	4	7.4E-02	4376.0	4	2	2.05E-01	769.15	5	3	6.0E+00
3777.5	2	2	1.1E-02	4379.7	2	2	1.00E+00	871.10	5	3	1.59E+00
3780.8	8	8	7.7E-01	4383.8	4	4	1.1E-02	875.53	3	1	3.74E+00
3786.4	8	6	1.5E-02	4400.1	4	4	1.60E-01	878.73	5	5	2.79E+00
3799.4	6	4	1.7E-01	4401.0	8	6	3.04E-01	879.62	3	3	9.2E-01
3808.6	6	6	1.0E-02	4412.9	6	8	6.1E-02	883.18	1	3	1.22E+00
3826.8	6	6	2.81E-01	4420.9	2	4	3.1E-02	887.40	3	5	9.0E-01
3841.5	4	2	2.69E-01	4426.0	4	6	8.17E-01	3024.1	5	7	2.6E+00
3844.7	6	8	4.8E-02	4430.2	2	4	5.69E-01	3027.2	5	5	6.4E-01
3845.4	6	4	1.6E-02	4431.0	6	6	1.09E-01	3054.8	3	5	1.9E+00
3850.6	4	4	3.87E-01	4460.6	4	6	1.5E-02	3064.8	3	3	1.0E+00
3868.5	4	6	1.4E+00	4474.8	4	2	2.90E-01	3078.2	1	3	1.4E+00
3872.1	4	4	1.5E-01	4481.8	6	6	4.55E-01	3285.9	5	7	2.0E+00
3875.3	4	2	8.2E-02	4491.0	6	4	4.6E-02	3301.9	5	5	2.0E+00
3880.3	2	2	2.32E-01	4530.5	6	4	2.1E-02	3311.3	5	3	2.0E+00
3891.4	2	2	4.3E-02	4545.1	4	4	4.71E-01	3336.1	7	9	2.0E+00
3892.0	6	4	6.3E-02	4579.4	2	2	8.0E-01	3344.7	5	7	1.8E+00
3900.6	4	6	7.2E-02	4589.9	4	6	6.64E-01	3352.1	7	7	2.2E-01
3911.6	2	4	7.7E-02	4598.8	4	4	6.7E-02	3358.5	3	5	1.6E+00
3914.8	4	4	3.7E-02	4609.6	6	8	7.89E-01	3361.3	5	5	3.0E-01
3928.6	2	4	2.44E-01	4637.2	6	6	7.1E-02	3472.6	5	7	2.0E+00
3931.2	2	4	2.0E-02	4657.9	4	2	8.92E-01	3480.6	7	7	1.6E+00
3932.5	4	4	9.3E-01	4726.9	4	4	5.88E-01	3499.7	3	3	1.3E+00
3944.3	8	6	4.1E-02	4732.1	6	4	6.7E-02	3500.6	3	5	2.6E-01
3952.7	4	4	2.08E-01	4735.9	6	4	5.80E-01	3502.7	5	3	4.3E-01
3958.4	6	4	3.8E-02	4764.9	2	4	6.4E-01	3503.6	5	5	1.2E+00
3968.4	6	6	4.8E-02	4806.0	6	6	7.80E-01	3511.7	7	5	2.6E-01
3979.4	4	2	9.8E-01	4847.8	4	2	8.49E-01				
3988.2	6	6	4.1E-02	4879.9	4	6	8.23E-01	**Ar IV**			
3992.1	4	6	1.6E-02	4889.0	2	2	1.9E-01	840.03	4	2	2.73E+00
4013.9	8	8	1.05E-01	4904.8	6	8	3.7E-02	843.77	4	4	2.70E+00
4031.4	4	2	7.5E-02	4933.2	4	4	1.44E-01	850.60	4	6	2.63E+00
4035.5	4	6	4.4E-02	4965.1	2	4	3.94E-01				
4038.8	6	8	1.2E-02	4972.2	2	2	9.7E-02	**Ar VI**			
4042.9	4	4	4.06E-01	5009.3	4	6	1.51E-01	292.15	2	2	6.9E+01
4045.7	4	4	1.6E-02	5017.2	4	6	2.07E-01	294.05	4	2	1.36E+02
4052.9	2	4	6.7E-01	5017.6	4	4	1.1E-02				
4065.1	4	4	1.1E-02	5062.0	2	4	2.23E-01	**Ar VII**			
4072.0	6	6	5.8E-01	5141.8	6	8	8.1E-02	*250.41	9	3	2.78E+02
4079.6	6	4	1.19E-01	5145.3	4	6	1.06E-01	*477.54	9	15	9.92E+01
4082.4	6	6	2.9E-02	5176.2	6	6	1.7E-02	585.75	1	3	7.83E+01
4112.8	4	4	1.1E-02	6103.5	2	2	1.7E-02	*637.30	9	9	6.7E+01
4128.6	8	6	1.4E-02	6114.9	10	8	2.00E-01				
4131.7	4	2	8.5E-01	6138.7	6	4	1.2E-02	**Ar VIII**			
4178.4	6	4	1.2E-02	6172.3	8	6	2.00E-01	158.92	2	4	1.1E+02
4202.0	2	4	2.1E-02	6243.1	8	6	3.0E-02	159.18	2	2	1.11E+02

λ Å	Weights g_i	Weights g_k	A 10^8 s^{-1}	λ Å	Weights g_i	Weights g_k	A 10^8 s^{-1}	λ Å	Weights g_i	Weights g_k	A 10^8 s^{-1}
229.44	2	2	1.12E+02	2860.4	2	2	5.5E-01	5777.6	5	7	6.5E-01
230.88	4	2	2.21E+02	2898.7	4	2	9.9E-02	5784.0	3	5	2.1E-01
337.09	4	4	1.2E+01					5800.2	5	5	9.9E-02
337.26	6	4	1.0E+02	**Barium**				5805.7	7	7	1.1E-02
338.22	4	2	1.1E+02	**Ba I**				5826.3	5	3	5.6E-01
519.43	2	4	6.3E+01	2409.2	1	3	8.6E-04	5907.6	3	5	1.5E-02
526.46	4	6	7.2E+01	2414.1	1	3	1.5E-03	5971.7	5	5	1.8E-01
526.87	4	4	1.2E+01	2420.1	1	3	2.3E-03	5997.1	3	3	2.7E-01
700.24	2	4	2.55E+01	2427.4	1	3	5.6E-03	6019.5	3	1	1.4E+00
713.81	2	2	2.4E+01	2432.5	1	3	7.2E-03	6063.1	5	3	5.7E-01
				2438.8	1	3	1.4E-03	6083.4	3	1	1.1E-01
Ar IX				2444.6	1	3	4.5E-03	6110.8	7	5	5.5E-01
48.739	1	3	1.69E+03	2452.4	1	3	8.1E-04	6129.2	3	1	6.0E-02
				2473.2	1	3	4.6E-03	6341.7	5	7	1.9E-01
Ar XIII				2500.2	1	3	1.5E-02	6450.9	3	5	1.1E-01
162.96	5	3	3.4E+02	2543.2	1	3	4.1E-02	6482.9	5	7	4.4E-01
*163.08	9	3	5.3E+02	2596.6	1	3	1.2E-01	6498.8	7	7	8.6E-01
184.90	5	5	1.66E+02	2646.5	1	3	1.1E-02	6527.3	5	5	5.9E-01
186.38	1	3	8.8E+01	2702.6	1	3	2.5E-02	6595.3	3	3	3.9E-01
*207.89	9	9	9.5E+01	2739.2	1	3	9.1E-03	6675.3	5	3	1.9E-01
*245.10	9	15	3.7E+01	2785.3	1	3	2.8E-02	6693.8	7	5	2.8E-01
				3071.6	1	3	4.1E-01	6865.7	5	5	2.3E-02
Ar XIV				3501.1	1	3	1.9E-01	7059.9	7	9	7.1E-01
180.29	2	4	4.5E+01	3889.3	1	3	8.8E-03	7120.3	3	5	2.1E-01
183.41	2	2	1.69E+02	3909.9	3	5	4.9E-01	7195.2	1	3	2.4E-01
187.95	4	4	1.97E+02	3935.7	5	7	4.7E-01	7280.3	5	7	5.3E-01
191.35	4	2	7.5E+01	3937.9	5	5	1.1E-01	7392.4	3	3	5.0E-01
194.39	2	2	4.6E+01	3993.4	7	9	5.5E-01	7417.5	7	5	2.5E-02
203.35	4	2	7.8E+01	3995.7	7	7	8.8E-02	7488.1	7	7	1.0E-01
				4132.4	1	3	7.1E-03	7528.2	5	5	2.7E-02
Ar XV				4239.6	5	3	2.4E-01	7672.1	3	5	3.1E-01
25.05	1	3	1.7E+04	4242.6	3	5	5.6E-02	7780.5	5	5	1.3E-01
221.10	1	3	9.55E+01	4264.4	1	3	1.5E-01	7905.8	5	3	6.3E-01
*265.3	9	9	8.1E+01	4283.1	5	7	6.4E-01	7911.3	1	3	2.98E-03
				4323.0	3	5	1.5E-01	8147.7	5	5	6.3E-02
Ar XVI				4325.2	5	7	7.1E-01	9645.6	7	5	1.1E-01
*23.52	2	6	1.43E+04	4332.9	3	3	1.5E-01	9704.3	3	1	1.6E-01
*24.96	6	10	4.4E+04	4350.3	3	5	6.0E-01	9821.5	3	1	5.5E-02
353.88	2	4	1.5E+01	4402.5	3	5	2.7E-01	10370.3	3	5	1.3E-02
389.11	2	2	1.1E+01	4406.8	5	5	1.0E-01	10649.1	5	5	2.7E-02
1268	2	4	1.9E+00	4431.9	1	3	1.2E+00	11075.7	3	3	3.6E-05
1401	2	2	1.4E+00	4467.1	5	7	6.6E-02	11303.1	5	3	1.2E-03
2975	2	4	9.0E-02	4489.0	5	7	4.2E-01	11373.8	3	1	1.3E-01
3514	4	6	6.5E-02	4493.6	5	5	3.6E-01	14158.4	9	7	2.0E-03
				4505.9	3	3	1.1E+00	14723.2	3	5	8.6E-03
Arsenic				4523.2	5	5	9.6E-01	14999.9	5	3	2.8E-03
As I				4573.9	3	1	1.21E+00	17123.7	7	7	3.3E-03
1890.4	4	6	2.0E+00	4579.6	5	5	7.0E-01	17187.1	3	1	2.7E-02
1937.6	4	4	2.0E+00	4591.8	5	5	1.6E-02	20563.9	5	7	2.6E-03
1972.6	4	2	2.0E+00	4599.7	3	1	4.07E-01				
2288.1	6	4	2.8E+00	4605.0	3	1	7.7E-02	**Ba II**			
2344.0	2	4	3.5E-01	4619.9	1	3	9.3E-02	1413.4	6	8	1.7E-02
2349.8	4	2	3.1E+00	4628.3	5	3	6.0E-02	1417.1	4	6	3.8E-02
2369.7	4	4	6.0E-01	4673.6	7	5	6.5E-02	1444.9	4	6	8.1E-02
2370.8	4	6	4.2E-01	4691.6	5	3	1.6E+00	1461.5	6	8	8.7E-02
2456.5	6	4	7.2E-02	4700.4	3	3	2.4E-01	1487.0	4	6	1.4E-01
2492.9	4	2	1.2E-01	4726.4	5	3	4.6E-01	1503.9	6	8	1.5E-01
2745.0	2	4	2.6E-01	5519.1	3	5	5.0E-01	1554.4	4	6	2.6E-01
2780.2	4	4	7.8E-01	5535.5	1	3	1.19E+00	1572.7	6	8	2.4E-01

λ Å	Weights g_i	g_k	A 10^8 s^{-1}	λ Å	Weights g_i	g_k	A 10^8 s^{-1}	λ Å	Weights g_i	g_k	A 10^8 s^{-1}
1573.9	6	6	1.6E-02	5428.8	6	4	2.3E-02	2061.7	4	6	9.9E-01
1630.4	2	2	1.7E-02	5480.3	8	6	1.8E-02	2110.3	4	2	9.1E-01
1674.5	4	6	2.2E-01	5784.2	2	4	2.0E-01	2177.3	4	2	2.6E-02
1694.4	6	8	2.1E-01	5853.7	4	4	4.8E-02	2228.3	4	4	8.9E-01
1697.2	6	6	1.7E-02	5981.3	4	6	1.6E-01	2230.6	4	6	2.6E+00
1761.8	4	4	3.9E-03	5999.9	4	4	2.6E-02	2276.6	4	4	2.5E-01
1771.0	4	2	3.4E-02	6135.8	2	2	8.5E-02	2515.7	4	6	4.3E-02
1786.9	6	4	4.4E-02	6141.7	6	4	3.7E-01	2627.9	4	4	4.7E-01
1892.7	2	4	9.0E-02	6363.2	6	4	2.9E-03	2696.8	4	6	6.4E-02
1904.2	4	6	1.1E-02	6372.9	4	4	6.7E-04	2780.5	4	2	3.09E-01
1906.8	2	2	5.1E-02	6378.9	4	2	9.9E-02	2798.7	6	6	3.6E-02
1924.7	6	8	3.1E-02	6457.7	6	4	3.0E-03	2898.0	4	2	1.53E+00
1954.2	4	6	1.3E-01	6496.9	4	2	3.32E-01	2938.3	6	4	1.23E+00
1955.1	4	4	1.8E-02	7556.8	6	4	1.6E-03	2989.0	4	4	5.5E-01
1970.2	4	2	6.7E-02	7678.2	8	6	6.6E-04	2993.3	4	6	1.6E-01
1985.6	2	4	2.5E-01	8710.7	6	8	8.0E-01	3024.6	6	6	8.8E-01
1999.5	2	4	1.0E-01	8737.7	4	6	9.3E-01	3067.7	4	2	2.07E+00
2009.2	2	2	8.6E-02					3076.7	4	4	3.5E-02
2052.7	4	6	2.0E-01	**Beryllium**				3397.2	6	4	1.81E-01
2054.6	4	4	2.9E-02	**Be I**				3402.9	6	6	1.6E-02
2080.0	4	2	1.0E-01	1491.8	1	3	1.3E-02	3510.9	6	4	6.8E-02
2153.9	2	4	5.3E-01	1661.5	1	3	2.0E-01	3596.1	2	4	1.98E-01
2200.9	2	2	2.0E-01	2348.6	1	3	5.55E+00	3888.2	2	2	6.9E-02
2232.8	4	6	2.9E-01	*2494.7	9	15	1.6E+00	4121.5	2	2	1.64E-01
2235.4	4	4	4.4E-02	*2650.6	9	9	4.24E+00	4308.5	2	4	1.6E-02
2286.0	4	2	1.3E-01	4572.7	3	5	7.9E-01	4493.0	2	4	1.5E-02
2528.5	2	4	7.1E-01					4722.5	4	2	1.17E-01
2634.8	4	6	7.6E-01	**Be II**				6134.8	4	4	1.8E-02
2641.4	4	4	1.2E-01	1197.1	2	2	4.7E-01				
2647.3	2	2	2.0E-01	1197.2	4	2	9.4E-01	**Boron**			
2771.4	4	2	4.0E-01	1512.3	2	4	9.2E+00	**B I**			
3816.7	4	6	2.3E-03	1512.4	4	6	1.1E+01	1378.6	2	4	3.50E+00
3842.8	6	8	2.2E-03	1776.1	2	2	1.4E+00	1378.9	2	2	1.40E+01
3891.8	2	4	1.67E+00	1776.3	4	2	2.9E+00	1378.9	4	4	1.75E+01
4024.1	6	4	5.3E-03	*2453.8	2	6	1.42E-01	1379.2	4	2	7.0E+00
4057.5	8	6	1.2E-02	3046.5	2	4	4.8E-01	1465.5	2	4	3.34E+00
4130.7	4	6	1.80E+00	3046.7	4	6	5.9E-01	1465.7	4	4	6.7E+00
4166.0	4	4	3.7E-01	3130.4	2	4	1.14E+00	1465.8	6	4	1.00E+01
4216.0	2	4	5.8E-02	3131.1	2	2	1.15E+00	1825.9	2	4	1.76E+00
4287.8	2	2	2.4E-02	3241.6	2	2	1.41E-01	1826.4	4	6	2.11E+00
4325.7	4	6	5.9E-02	3241.8	4	2	2.8E-01	2088.9	2	4	2.8E-01
4329.6	4	4	8.8E-03	3274.6	2	4	1.9E-01	2089.6	4	6	3.3E-01
4405.2	4	2	3.9E-02	3274.7	2	2	1.9E-01	2496.8	2	2	8.64E-01
4470.7	6	4	1.4E-02	4360.7	2	4	9.2E-01	2497.7	4	2	1.73E+00
4509.6	8	6	1.2E-02	4361.0	4	6	1.1E+00				
4524.9	2	2	7.2E-01	*5255.9	2	6	2.56E-02	**Bromine**			
4554.0	2	4	1.17E+00	5270.3	2	2	3.30E-01	**Br I**			
4708.9	2	4	9.7E-02	5270.8	4	2	6.6E-01	1488.5	4	4	1.2E+00
4843.5	4	6	9.3E-02	6279.4	2	4	1.2E-01	1540.7	4	4	1.4E+00
4847.1	2	2	4.1E-02	6279.7	4	6	1.43E-01	1574.8	2	4	2.0E-01
4850.8	4	4	1.4E-02	6756.7	2	2	5.1E-02	1576.4	4	6	2.1E-02
4900.0	4	2	7.75E-01	6757.1	4	2	1.02E-01	1633.4	2	4	8.1E-02
4934.1	2	2	9.55E-01	7401.2	2	4	3.0E-02	4365.1	2	4	7.5E-03
4997.8	4	2	6.1E-02	7401.4	2	2	3.0E-02	4425.1	4	2	4.2E-03
5185.0	2	4	1.8E-02					4441.7	6	4	7.5E-03
5361.4	4	6	4.8E-02	**Bismuth**				4472.6	4	4	9.3E-03
5391.6	6	8	5.2E-02	**Bi I**				4477.7	6	8	1.3E-02
5413.6	6	6	8.4E-04	1954.5	4	6	1.2E+00	4513.4	6	4	2.8E-03
5421.1	6	6	1.9E-03	2021.2	4	4	6.0E-02				

λ Å	Weights g_i	g_k	A 10^8 s^{-1}	λ Å	Weights g_i	g_k	A 10^8 s^{-1}	λ Å	Weights g_i	g_k	A 10^8 s^{-1}
4525.6	6	6	7.2E-03	3361.9	5	7	2.23E-01	6493.8	3	5	4.4E-01
4575.7	4	4	1.6E-02	3624.1	1	3	2.12E-01	6499.7	5	5	8.1E-02
4614.6	4	6	5.4E-03	3630.8	3	5	2.97E-01				
4979.8	4	4	2.6E-03	3631.0	3	3	1.53E-01	**Ca II**			
5245.1	2	4	3.1E-03	3644.4	5	7	3.55E-01	1341.9	2	4	1.5E-02
5345.4	2	4	7.6E-03	3644.8	5	5	9.4E-02	1342.5	2	2	1.5E-02
7348.5	4	6	1.2E-01	3870.5	3	5	7.2E-02	1649.9	2	4	3.2E-03
7513.0	6	4	1.2E-01	3957.1	3	3	9.8E-02	1652.0	2	2	3.1E-03
7803.0	2	4	5.3E-02	3973.7	5	3	1.75E-01	1673.9	2	4	2.24E-01
7938.7	6	6	1.9E-01	4092.6	3	5	1.1E-01	1680.1	4	6	2.65E-01
8131.5	2	4	3.8E-02	4094.9	5	7	1.2E-01	1680.1	4	4	4.41E-02
8343.7	2	2	2.2E-02	4098.5	7	9	1.3E-01	1807.3	2	4	3.54E-01
8446.6	4	4	1.2E-01	4108.5	5	7	9.0E-01	1814.5	4	6	4.2E-01
8638.7	6	4	9.7E-02	4226.7	1	3	2.18E+00	1814.7	4	4	7.0E-02
				4283.0	3	5	4.34E-01	1843.1	2	2	1.6E-01
Br II				4289.4	1	3	6.0E-01	1850.7	4	2	3.08E-01
4704.9	5	7	1.1E+00	4299.0	3	3	4.66E-01	2103.2	2	4	8.2E-01
4785.5	5	5	9.4E-01	4302.5	5	5	1.36E+00	2112.8	4	6	9.7E-01
4816.7	5	3	1.1E+00	4307.7	3	1	1.99E+00	2113.2	4	4	1.6E-01
				4318.7	5	3	7.4E-01	2197.8	2	2	3.1E-01
Cadmium				4355.1	5	7	1.9E-01	2208.6	4	2	6.2E-01
Cd I				4425.4	1	3	4.98E-01	3158.9	2	4	3.1E+00
2288.0	1	3	5.3E+00	4435.0	3	5	6.7E-01	3179.3	4	6	3.6E+00
2836.9	1	3	2.8E-01	4435.7	3	3	3.42E-01	3181.3	4	4	5.8E-01
2880.8	3	5	4.2E-01	4454.8	5	7	8.7E-01	3706.0	2	2	8.8E-01
2881.2	3	3	2.4E-01	4455.9	5	5	2.0E-01	3736.9	4	2	1.7E+00
2980.6	5	7	5.9E-01	4526.9	5	3	4.1E-01	3933.7	2	4	1.47E+00
2981.4	5	5	1.5E-01	4578.6	3	5	1.76E-01	3968.5	2	2	1.4E+00
3261.1	1	3	4.06E-03	4581.4	5	7	2.09E-01				
3403.7	1	3	7.7E-01	4585.9	7	9	2.29E-01	**Ca III**			
3466.2	3	5	1.2E+00	4685.3	3	5	8.0E-02	357.97	1	3	8.8E+02
3467.7	3	3	6.7E-01	4878.1	5	7	1.88E-01	439.69	1	3	1.9E-01
3610.5	5	7	1.3E+00	5041.6	5	3	3.3E-01	490.55	1	3	1.6E-02
3612.9	5	5	3.5E-01	5188.9	3	5	4.0E-01				
4140.5	3	5	4.7E-02	5261.7	3	3	1.5E-01	**Ca V**			
4662.4	3	5	5.5E-02	5262.2	3	1	6.0E-01	558.60	5	3	2.2E+01
4678.1	1	3	1.3E-01	5264.2	5	5	9.1E-02	637.93	5	3	3.9E+00
4799.9	3	3	4.1E-01	5265.6	5	3	4.4E-01	643.12	3	1	9.1E+00
5085.8	5	3	5.6E-01	5270.3	7	5	5.0E-01	646.57	5	5	6.9E+00
6438.5	3	5	5.9E-01	5582.0	5	7	6.0E-02	647.88	3	3	2.3E+00
				5588.8	7	7	4.9E-01	651.55	1	3	2.9E+00
Cd II				5590.1	3	5	8.3E-02	656.76	3	5	2.1E+00
2144.4	2	4	2.8E+00	5594.5	5	5	3.8E-01				
2265.0	2	2	3.0E+00	5598.5	3	3	4.3E-01	**Ca VII**			
2572.9	2	2	1.7E+00	5601.3	7	5	8.6E-02	550.20	5	5	1.8E+01
2748.5	4	2	2.8E+00	5602.9	5	3	1.4E-01	624.39	1	3	3.3E+00
4415.6	4	6	1.4E-02	5857.5	3	5	6.6E-01	630.54	3	5	4.5E+00
				6102.7	1	3	9.6E-02	630.79	3	3	2.2E+00
Calcium				6122.2	3	3	2.87E-01	639.15	5	7	5.7E+00
Ca I				6161.3	5	5	3.3E-02	640.41	5	5	1.3E+00
2275.5	1	3	3.01E-01	6162.2	5	3	3.54E-01				
2995.0	1	3	3.67E-01	6163.8	3	3	5.6E-02	**Ca VIII**			
2997.3	3	5	2.41E-01	6166.4	3	1ʳ	2.2E-01	182.71	2	2	1.6E+02
2999.6	3	3	2.79E-01	6169.1	5	3	1.7E-01	184.16	4	2	3.2E+02
3000.9	3	1	1.58E+00	6169.6	7	5	1.9E-01				
3006.9	5	5	7.5E-01	6439.1	7	9	5.3E-01	**Ca IX**			
3009.2	5	3	4.30E-01	6449.8	3	5	9.0E-02	163.23	5	3	3.76E+02
3344.5	1	3	1.51E-01	6462.6	5	7	4.7E-01	371.89	1	3	8.8E+01
3350.2	3	5	1.78E-01	6471.7	7	7	5.9E-02	373.81	3	5	1.16E+02

λ Å	Weights g_i	g_k	A 10^8 s^{-1}	λ Å	Weights g_i	g_k	A 10^8 s^{-1}	λ Å	Weights g_i	g_k	A 10^8 s^{-1}
378.08	5	7	1.5E+02	1261.00	3	3	4.42E-01	4817.37	3	3	8.76E-04
395.03	3	5	2.2E+02	1261.12	3	5	3.71E-01	4826.80	5	3	6.28E-04
466.24	1	3	1.12E+02	1261.43	5	3	7.06E-01	4932.05	3	1	6.02E-02
498.01	3	5	2.49E+01	1261.55	5	5	1.27E+00	5023.84	7	9	1.81E-03
506.18	5	5	7.2E+01	1274.11	5	7	1.03E-02	5039.06	7	9	4.73E-03
515.57	5	3	3.75E+01	1277.25	1	3	1.27E+00	5041.48	3	5	5.25E-03
				1277.28	3	5	1.73E+00	5041.79	5	7	3.28E-03
Ca X				1277.51	3	3	9.12E-01	5052.17	3	5	2.60E-02
110.96	2	4	2.9E+02	1277.55	5	7	2.31E+00	5380.34	3	3	1.86E-02
111.20	2	2	2.92E+02	1277.72	5	5	6.35E-01	5545.05	3	3	3.04E-03
151.84	2	2	2.3E+02	1277.95	5	3	5.56E-02	5668.94	3	3	2.35E-02
153.02	4	2	4.5E+02	1279.23	5	7	1.10E-01	5793.12	7	5	3.44E-03
206.57	4	4	2.9E+01	1279.89	3	5	3.08E-01	5794.47	5	5	6.44E-04
206.75	6	4	2.6E+02	1280.14	1	3	3.11E-01	5800.23	3	3	1.04E-03
207.39	4	2	2.8E+02	1280.33	5	5	5.77E-01	5800.60	5	3	3.04E-03
411.70	2	4	8.3E+01	1280.40	3	3	1.73E-01	5805.20	3	1	4.12E-03
419.75	4	6	9.5E+01	1280.60	3	1	8.22E-01	6001.12	5	5	3.22E-03
420.47	4	4	1.6E+01	1280.85	5	3	3.33E-01	6006.02	7	5	1.79E-02
557.76	2	4	3.50E+01	1328.83	1	3	7.95E-01	6007.18	3	3	5.34E-03
574.01	2	2	3.2E+01	1329.09	3	1	2.41E+00	6010.68	3	1	2.13E-02
				1329.58	5	5	1.79E+00	6013.17	7	5	1.79E-02
Ca XI				1329.60	5	3	1.00E+00	6013.21	7	9	4.35E-03
30.448	1	3	6.2E+03	1355.84	5	7	1.04E+00	6014.83	5	3	1.60E-02
30.867	1	3	4.9E+04	1364.16	5	5	1.57E-01	6016.45	5	7	3.86E-03
35.212	1	3	2.0E+03	1431.60	5	7	2.11E+00	6587.61	3	3	5.09E-02
				1432.10	5	5	2.01E+00	6655.52	3	3	5.03E-03
Ca XII				1432.53	5	3	2.11E+00	6828.12	3	5	9.89E-03
140.05	4	2	3.7E+02	1459.03	5	3	4.76E-01	7111.47	3	5	2.17E-02
147.27	2	2	1.6E+02	1463.34	5	7	1.88E+00	7113.18	7	9	2.47E-02
				1467.40	5	3	5.49E-01	7115.17	5	7	2.19E-02
Ca XV				1468.41	5	3	3.90E-02	7115.18	3	1	4.43E-02
141.69	5	3	4.08E+02	1470.09	5	7	1.37E-02	7116.99	7	5	3.26E-02
*142.23	9	3	6.3E+02	1472.23	5	3	8.01E-03	7119.66	5	3	3.12E-02
161.00	5	5	1.9E+02	1481.76	5	5	3.92E-01	7860.88	5	5	1.53E-02
				1560.31	1	3	6.57E-01	8058.62	5	5	1.09E-02
Ca XVII				1561.34	5	5	2.94E-01	8335.15	3	1	3.51E-01
19.558	1	3	3.8E+04	1561.44	5	7	1.18E+00	9061.44	3	5	7.31E-02
21.198	3	5	4.9E+04	1656.27	3	5	8.58E-01	9062.49	1	3	9.48E-02
192.82	1	3	1.21E+02	1656.93	1	3	1.13E+00	9078.29	3	3	7.07E-02
218.82	3	5	2.76E+01	1657.01	5	5	2.52E+00	9088.51	3	1	3.00E-01
223.02	1	3	3.44E+01	1657.38	3	3	8.64E-01	9094.83	5	5	2.28E-01
228.72	3	3	2.37E+01	1657.91	3	1	3.43E+00	9111.81	5	3	1.35E-01
232.83	5	5	6.5E+01	1658.12	5	3	1.44E+00	9405.73	3	5	2.91E-01
244.06	5	3	3.28E+01	1751.83	1	3	9.07E-01	9603.03	1	3	3.06E-02
				1763.91	1	3	3.59E-02	9620.78	3	3	8.62E-02
Ca XVIII				1765.37	1	3	1.04E-02	9658.43	5	3	1.25E-01
*18.71	2	6	2.31E+04	1930.90	5	3	3.51E+00				
*19.74	6	10	7.0E+04	2478.56	1	3	3.40E-01	**C II**			
302.19	2	4	2.0E+01	2902.23	1	3	4.32E-03	687.345	4	6	2.84E+01
344.76	2	2	1.3E+01	2903.27	3	3	1.29E-02	858.092	2	2	1.18E+00
				2905.00	5	3	2.15E-02	858.559	4	2	2.35E+00
Carbon				4371.37	3	3	1.27E-02	903.623	2	4	6.85E+00
C I				4762.31	1	3	3.37E-03	903.962	2	2	2.74E+01
945.191	1	3	3.79E+00	4762.53	3	5	2.72E-03	904.142	4	4	3.42E+01
945.338	3	3	1.14E+01	4766.67	3	3	2.36E-03	904.480	4	2	1.37E+01
945.579	5	3	1.89E+01	4770.03	3	1	1.07E-02	1009.86	2	4	5.71E+00
1193.24	5	7	1.22E+00	4771.74	5	5	7.97E-03	1010.08	4	4	1.14E+01
1260.74	1	3	5.32E-01	4775.90	5	3	4.84E-03	1010.37	6	4	1.71E+01
1260.93	3	1	1.70E+00	4812.92	1	3	4.03E-04	1036.34	2	2	7.61E+00

λ Å	Weights g_i	g_k	A 10^8 s^{-1}	λ Å	Weights g_i	g_k	A 10^8 s^{-1}	λ Å	Weights g_i	g_k	A 10^8 s^{-1}
1037.02	4	2	1.52E+01	7046.25	4	2	3.20E-01	5695.92	3	5	4.27E-01
1323.91	4	4	4.33E+00	7053.09	4	4	3.19E-01	5858.34	3	1	1.34E-01
1323.95	6	6	4.49E+00	7063.68	4	6	3.17E-01	5863.25	3	3	3.35E-02
1334.53	2	4	2.37E+00	7112.48	2	4	2.94E-01	5871.68	5	3	1.00E-01
1335.71	4	6	2.84E+00	7113.04	4	6	3.15E-01	5880.56	5	5	1.99E-02
2091.14	2	4	1.00E-01	7115.63	6	8	3.60E-01	5894.07	7	5	1.11E-01
2091.19	4	6	1.69E-01	7119.76	4	4	1.17E-01	6727.48	1	3	1.12E-01
2091.65	6	8	2.41E-01	7119.91	8	10	4.19E-01	6731.04	3	5	1.50E-01
2093.16	6	6	7.20E-02	7125.72	6	6	1.02E-01	6742.15	3	3	8.32E-02
2173.85	2	4	2.31E-01	7132.47	6	4	8.33E-03	6744.39	5	7	1.99E-01
2174.17	2	2	2.31E-01	7134.10	8	8	5.93E-02	6762.17	5	5	4.95E-02
2509.13	2	4	4.53E-01	7231.33	2	4	3.52E-01	6773.39	5	3	5.47E-03
2511.74	4	4	9.04E-02	7236.42	4	6	4.22E-01	6851.18	3	5	7.60E-03
2512.06	4	6	5.42E-01	7237.17	4	4	7.03E-02	6853.68	5	7	5.64E-03
2727.31	2	4	6.63E-02	8028.85	2	2	1.71E-02	6857.24	3	3	3.79E-02
2728.72	4	4	3.31E-01	8037.73	2	4	4.26E-02	6862.69	5	5	3.51E-02
2729.21	2	2	2.65E-01	8039.40	4	2	8.51E-02	6868.78	5	3	1.26E-02
2730.63	4	2	1.32E-01	8048.31	4	4	1.36E-02	6872.04	7	7	4.46E-02
5132.95	2	4	3.89E-01	8062.10	4	6	3.04E-02	6881.10	7	5	7.80E-03
5133.28	4	6	2.80E-01	8062.80	6	4	4.56E-02	7353.88	5	3	3.09E-02
5137.26	2	2	1.55E-01	8076.64	6	6	7.05E-02	7707.43	3	5	1.30E-01
5139.17	4	4	1.24E-01	9238.30	4	6	3.34E-02	7771.76	3	1	1.77E-01
5143.49	4	2	7.73E-01	9251.01	2	4	2.77E-02	7780.41	3	3	1.76E-01
5145.16	6	6	6.49E-01	9863.06	2	4	5.56E-02	7796.00	3	5	1.75E-01
5151.08	6	4	4.16E-01	9870.78	4	6	9.31E-02	8500.32	1	3	1.01E-01
5640.55	2	4	9.89E-02	9882.68	6	8	1.33E-01	9593.32	3	3	5.32E-03
5648.07	4	4	1.97E-01					9651.47	5	5	1.57E-02
5662.46	6	4	2.93E-01	**C III**				9696.48	5	7	7.53E-03
5818.31	2	2	3.38E-02	310.170	1	3	6.56E+00	9696.54	3	5	7.12E-03
5822.98	2	4	3.38E-03	386.203	1	3	3.46E+01	9699.57	7	9	8.47E-03
5823.18	4	2	3.38E-02	459.466	1	3	5.91E+01	9701.10	1	3	4.40E-02
5827.85	4	4	2.16E-02	459.514	3	5	7.97E+01	9705.41	3	5	5.93E-02
5836.37	6	4	4.22E-02	459.627	5	7	1.06E+02	9706.44	3	3	3.29E-02
5843.62	6	6	1.20E-02	574.281	3	5	6.24E+01	9715.09	5	7	7.88E-02
5856.06	8	6	5.31E-02	977.020	1	3	1.767E+01	9717.75	5	5	1.97E-02
6095.29	2	4	4.20E-01	1174.93	3	5	3.293E+00	9718.79	5	3	2.19E-03
6098.51	4	6	5.03E-01	1175.26	1	3	4.385E+00				
6102.56	4	4	8.37E-02	1175.59	3	3	3.287E+00	**C IV**			
6578.05	2	4	3.63E-01	1175.71	5	5	9.856E+00	*312.43	2	6	4.63E+01
6582.88	2	2	3.62E-01	1175.99	3	1	1.313E+01	*384.13	6	10	1.76E+02
6724.56	2	4	3.17E-02	1176.37	5	3	5.468E+00	1548.19	2	4	2.65E+00
6727.07	2	2	6.34E-02	1247.38	3	1	2.082E+01	1550.77	2	2	2.64E+00
6727.26	4	6	2.96E-02	2296.87	3	5	1.376E+00	5801.31	2	4	3.17E-01
6731.07	4	4	5.06E-02	2849.05	3	1	1.95E-01	5811.97	2	2	3.16E-01
6733.58	4	2	6.32E-02	3703.70	3	3	5.90E-01				
6734.00	6	8	1.80E-02	4325.56	3	5	1.24E-01	**C V**			
6738.61	6	6	7.23E-02	4647.42	3	5	7.26E-01	34.9728	1	3	2.554E+03
6742.43	6	4	4.41E-02	4650.25	3	3	7.25E-01	40.2678	1	3	8.873E+03
6750.54	8	8	1.08E-01	4651.02	3	5	2.28E-01	*227.19	3	9	1.363E+02
6755.16	8	6	2.38E-02	4651.47	3	1	7.24E-01	247.315	1	3	1.278E+02
6779.94	4	6	2.56E-01	4652.05	1	3	3.04E-01	*248.71	9	15	4.247E+02
6780.59	2	4	1.52E-01	4659.06	3	3	2.27E-01	*260.19	9	3	6.680E+01
6783.91	6	8	3.65E-01	4663.64	3	1	9.05E-01	267.267	3	5	3.947E+02
6787.21	2	2	3.04E-01	4665.86	5	5	6.78E-01	*2273.9	3	9	5.646E-01
6791.47	4	4	1.94E-01	4673.95	5	3	3.75E-01	3526.66	1	3	1.663E-01
6798.10	4	2	6.04E-02	5244.66	1	3	5.30E-02	8420.72	3	5	6.898E-02
6800.69	6	6	1.09E-01	5253.58	3	3	1.58E-01	*8433.2	3	9	6.868E-02
6812.28	6	4	1.80E-02	5272.52	5	3	2.61E-01	8448.12	3	1	6.832E-02

λ Å	Weights g_i	Weights g_k	A 10^8 s^{-1}	λ Å	Weights g_i	Weights g_k	A 10^8 s^{-1}	λ Å	Weights g_i	Weights g_k	A 10^8 s^{-1}
8449.19	3	3	6.829E-02	7256.6	6	4	1.5E-01	3530.0	6	8	1.8E+00
				7414.1	6	4	4.7E-02	3560.7	4	6	1.7E+00
Cesium				7547.1	4	4	1.2E-01	3602.1	6	8	1.7E+00
Cs I				7717.6	4	4	3.0E-02	3612.9	4	6	1.2E+00
3203.5	2	4	7.6E-06	7745.0	2	4	6.3E-02	3720.5	4	6	1.7E+00
3205.3	2	4	7.9E-06	7769.2	6	6	6.0E-02				
3207.5	2	4	8.5E-06	7821.4	6	8	9.8E-02	**Chromium**			
3210.0	2	4	9.4E-06	7830.8	4	4	9.7E-02	**Cr I**			
3212.8	2	4	1.19E-05	7878.2	6	6	1.8E-02	1999.95	9	9	1.4E+00
3216.2	2	4	1.49E-05	7899.3	4	6	5.1E-02	2383.30	9	11	4.1E-01
3220.1	2	4	1.7E-05	7924.6	2	4	2.1E-02	2389.21	3	5	2.3E-01
3220.2	2	2	1.07E-07	7935.0	6	8	3.9E-02	2408.60	9	7	6.7E-01
3224.8	2	4	2.0E-05	7997.9	4	4	2.1E-02	2408.72	7	5	2.9E-01
3225.0	2	2	1.43E-07					2492.57	3	5	4.5E-01
3230.5	2	4	2.5E-05	**Cl II**				2495.08	3	3	2.7E-01
3230.7	2	2	1.97E-07	3329.1	5	7	1.5E+00	2496.30	5	7	5.6E-01
3237.4	2	4	2.8E-05	3522.1	7	7	1.4E+00	2502.55	7	9	2.2E-01
3237.6	2	2	2.63E-07	3798.8	5	7	1.6E+00	2504.31	7	9	4.5E-01
3245.9	2	4	3.45E-05	3805.2	7	9	1.8E+00	2508.11	5	5	2.1E-01
3246.2	2	2	3.7E-07	3809.5	3	5	1.5E+00	2508.97	5	3	3.8E-01
3256.7	2	4	4.25E-05	3851.0	5	7	1.8E+00	2527.11	9	9	5.3E-01
3257.1	2	2	7.0E-07	3851.4	5	5	1.6E+00	2549.55	3	3	4.8E-01
3270.5	2	4	5.6E-05	3854.7	3	5	2.2E+00	2560.70	5	5	4.3E-01
3271.0	2	2	9.8E-07	3861.9	5	7	2.4E+00	2571.74	7	5	6.4E-01
3288.6	2	4	1.0E-04	3868.6	7	9	2.7E+00	2577.66	7	7	2.6E-01
3289.3	2	2	2.7E-06	3913.9	9	9	8.2E-01	2591.84	9	7	6.5E-01
3313.1	2	4	1.6E-04	3990.2	5	7	8.4E-01	2620.48	5	3	1.9E-01
3314.0	2	2	5.2E-06	4132.5	5	5	1.6E+00	2673.64	3	3	1.8E-01
3347.5	2	4	2.2E-04	4276.5	9	7	7.6E-01	2701.99	9	11	2.1E-01
3348.8	2	2	1.1E-05	4768.7	3	5	7.7E-01	2726.50	5	7	7.5E-01
3397.9	2	4	4.0E-04	4781.3	5	7	1.0E+00	2731.90	5	5	7.8E-01
3400.0	2	2	2.4E-05	4794.6	5	7	1.04E+00	2736.46	5	3	7.5E-01
3476.8	2	4	6.6E-04	4810.1	5	5	9.9E-01	2752.85	3	3	8.7E-01
3480.0	2	2	6.6E-05	4819.5	5	3	1.00E+00	2757.09	5	5	6.8E-01
3611.4	2	4	1.5E-03	4904.8	5	7	8.1E-01	2761.74	5	3	6.8E-01
3617.3	2	2	2.5E-04	4917.7	3	5	7.5E-01	2764.36	7	7	3.7E-01
3876.1	2	4	3.8E-03	5078.3	7	7	7.7E-01	2769.90	7	5	1.1E+00
3888.6	2	2	9.7E-04	5219.1	3	9	8.6E-01	2780.70	9	7	1.4E+00
4555.3	2	4	1.88E-02	5392.1	5	7	1.0E+00	2879.27	5	7	2.1E-01
4593.2	2	2	8.0E-03					2887.00	3	5	2.7E-01
8521.1	2	4	3.276E-01	**Cl III**				2889.22	9	9	6.6E-01
8943.5	2	2	2.87E-01	2298.5	4	4	4.2E+00	2893.25	7	7	5.2E-01
				2340.6	6	6	4.2E+00	2894.17	1	3	3.3E-01
Chlorine				2370.4	8	6	2.8E+00	2896.76	5	5	3.0E-01
Cl I				2531.8	2	4	4.4E+00	2905.48	3	1	1.3E+00
1188.8	4	6	2.33E+00	2532.5	4	6	5.3E+00	2909.05	5	3	6.8E-01
1188.8	4	4	2.71E-01	2577.1	4	6	4.3E+00	2910.89	7	5	3.4E-01
1201.4	2	4	2.39E+00	2580.7	6	8	4.7E+00	2911.15	9	7	2.6E-01
1335.7	4	2	1.74E+00	2601.2	2	4	4.6E+00	2967.64	7	9	3.9E-01
1347.2	4	4	4.19E+00	2603.6	4	6	5.0E+00	2971.10	5	7	7.1E-01
1351.7	2	2	3.23E+00	2609.5	6	8	5.7E+00	2975.48	3	5	8.9E-01
1363.4	2	4	7.5E-01	2617.0	8	10	6.6E+00	2980.78	1	3	5.10E-01
4323.3	4	4	1.1E-02	2661.6	4	6	3.4E+00	2988.64	5	7	5.2E-01
4363.3	4	6	6.8E-03	2665.5	6	8	4.8E+00	2991.88	3	1	3.0E+00
4379.9	4	4	1.4E-02	2691.5	4	4	3.5E+00	2994.06	5	5	2.5E-01
4389.8	6	8	1.4E-02	2710.4	4	6	3.5E+00	2995.09	5	5	4.3E-01
4526.2	4	4	5.1E-02	3340.4	6	6	1.5E+00	2996.57	5	3	2.0E+00
4601.0	2	2	4.2E-02	3392.9	4	4	1.9E+00	2998.78	5	3	4.07E-01
4661.2	2	4	1.2E-02	3393.5	6	6	1.9E+00	3000.88	7	5	1.6E+00

λ Å	Weights g_i	g_k	A 10^8 s^{-1}	λ Å	Weights g_i	g_k	A 10^8 s^{-1}	λ Å	Weights g_i	g_k	A 10^8 s^{-1}
3005.06	9	7	9.2E-01	4432.77	15	15	4.9E-01	2857.40	6	8	2.8E-01
3013.72	3	5	8.3E-01	4443.72	3	1	4.5E-01	2860.92	2	4	6.9E-01
3015.20	1	3	1.63E+00	4482.88	3	3	3.0E-01	2862.57	8	8	6.3E-01
3020.67	3	3	1.5E+00	4490.55	9	7	3.9E-01	2866.72	4	4	1.2E+00
3021.58	9	11	2.91E+00	4492.31	5	3	4.47E-01	2867.09	4	4	1.1E+00
3024.36	5	5	1.27E+00	4495.28	9	7	2.0E-01	2867.65	2	2	1.1E+00
3029.17	5	3	3.8E-01	4500.29	7	7	2.1E-01	2870.43	6	6	1.3E+00
3030.25	7	7	1.1E+00	4506.84	13	11	2.7E-01	2873.81	4	2	8.8E-01
3031.35	5	3	3.1E-01	4540.72	11	11	3.14E-01	2880.86	6	4	7.9E-01
3034.19	7	7	3.5E-01	4564.17	11	13	5.1E-01	2898.53	10	12	1.2E+00
3037.05	9	9	5.4E-01	4595.60	13	13	4.7E-01	2921.81	8	10	9.0E-01
3040.84	7	5	7.4E-01	4622.47	7	7	4.1E-01	2930.83	2	4	1.1E+00
3053.87	9	7	7.97E-01	4663.33	3	3	2.0E-01	2935.12	6	8	1.8E+00
3148.44	9	11	5.6E-01	4665.90	3	3	3.0E-01	2953.34	2	2	1.8E+00
3155.16	11	13	5.7E-01	4689.38	7	5	2.3E-01	2966.03	10	8	5.4E-01
3163.76	13	15	6.0E-01	4698.46	9	7	2.2E-01	2971.90	14	14	2.0E+00
3237.73	9	9	1.3E+00	4708.02	11	9	4.31E-01	2979.73	12	12	1.8E+00
3238.09	11	11	2.0E-01	4718.43	13	11	3.4E-01	2985.32	10	10	2.2E+00
3578.68	7	9	1.48E+00	4730.69	7	5	3.83E-01	2989.18	8	8	2.2E+00
3593.48	7	7	1.50E+00	4737.33	9	7	3.38E-01	3118.64	2	4	1.7E+00
3605.32	7	5	1.62E+00	4741.09	3	5	2.2E-01	3120.36	4	6	1.5E+00
3639.80	13	11	1.8E+00	4752.07	13	13	6.2E-01	3122.59	12	12	4.4E-01
3743.89	13	13	7.61E-01	4756.09	11	9	4.0E-01	3128.69	4	4	8.1E-01
3757.66	7	7	4.13E-01	4792.49	7	5	2.6E-01	3136.68	6	6	6.4E-01
3768.24	5	5	5.10E-01	4801.02	9	7	3.06E-01	4588.22	8	6	1.2E-01
3804.80	9	9	6.9E-01	4816.13	9	9	1.8E-01				
3963.69	13	15	1.3E+00	4870.79	7	9	3.5E-01	**Cr V**			
3969.75	11	13	1.2E+00	4887.01	9	11	3.2E-01	434.306	9	9	1.5E+01
3983.90	7	9	1.05E+00	4922.28	11	13	4.0E-01	436.351	9	7	2.4E+01
3991.12	5	7	1.07E+00	4966.80	3	1	3.0E-01	436.601	7	5	2.1E+01
4001.44	9	11	6.8E-01	5204.51	5	3	5.09E-01	437.420	7	7	1.4E+01
4039.10	15	15	6.7E-01	5206.02	5	5	5.14E-01	437.655	5	5	1.3E+01
4048.78	13	13	6.4E-01	5208.42	5	7	5.06E-01	441.056	5	3	2.3E+01
4058.78	11	11	6.7E-01	5243.38	5	3	2.19E-01	456.357	1	3	9.5E+00
4065.71	9	11	3.5E-01	5297.37	7	9	3.88E-01	456.637	3	1	3.3E+01
4165.52	11	13	7.5E-01	5297.99	7	7	3.0E-01	456.743	3	3	9.1E+00
4204.48	13	11	3.1E-01	5328.36	9	11	6.2E-01	457.028	5	5	2.7E+01
4254.33	7	9	3.15E-01	5329.17	9	9	2.25E-01	457.504	5	3	1.2E+01
4263.15	15	17	6.4E-01	5783.11	3	3	2.1E-01	464.015	9	7	3.6E+01
4274.81	7	7	3.07E-01	5783.89	5	5	2.02E-01	469.634	5	5	2.3E+01
4275.98	11	11	2.2E-01	5787.97	5	7	2.35E-01	1106.25	7	9	1.2E+01
4280.42	13	15	4.7E-01					1121.07	7	9	2.1E+01
4289.73	7	5	3.16E-01	**Cr II**				1127.63	9	11	3.5E+01
4291.97	7	5	2.4E-01	2653.57	4	6	3.5E-01	1465.86	5	3	1.1E+01
4297.75	11	13	4.9E-01	2658.59	2	4	5.8E-01	1481.65	3	1	1.0E+01
4298.05	9	9	2.6E-01	2666.02	6	8	5.9E-01	1519.03	5	7	9.5E+00
4300.52	9	7	1.9E-01	2668.71	4	2	1.4E+00	1579.70	7	9	8.6E+00
4301.19	11	9	2.6E-01	2671.80	6	4	1.0E+00				
4302.78	11	11	2.5E-01	2672.83	8	6	5.5E-01	**Cr VI**			
4319.66	5	3	1.8E-01	2744.97	4	6	8.5E-01	161.687	6	6	1.7E+02
4337.25	5	7	2.0E-01	2787.61	6	6	1.5E+00	168.088	4	6	2.0E+02
4373.65	9	9	2.8E-01	2822.38	14	16	2.3E+00	201.007	4	4	2.5E+03
4376.80	13	13	3.2E-01	2835.63	10	12	2.0E+00	201.224	4	6	1.8E+02
4413.86	7	5	2.7E-01	2840.01	10	12	2.7E+00	201.388	6	4	2.7E+02
4422.70	5	5	2.7E-01	2843.24	8	10	6.4E-01	201.606	6	6	2.6E+03
4424.29	9	7	2.1E-01	2849.83	6	8	9.2E-01	202.442	6	4	1.0E+03
4429.93	3	3	2.4E-01	2851.35	8	10	2.2E+00	202.739	4	2	1.2E+03
4432.16	1	3	1.8E-01	2856.77	4	6	4.3E-01	226.241	6	8	7.2E+02

λ Å	g_i	g_k	A 10^8 s^{-1}
227.202	4	6	6.6E+02
Cr X			
216.72	6	8	9.0E+02
223.86	4	2	7.7E+02
224.74	4	4	7.6E+02
226.24	4	6	7.3E+02
227.42	4	4	5.2E+02
227.50	4	6	1.8E+01
228.63	6	4	8.1E+01
228.71	6	6	4.5E+02
231.21	2	4	1.2E+02
232.96	4	4	4.4E+02
242.20	2	4	5.0E+01
244.19	4	6	5.8E+01
395.984	4	4	2.4E+01
398.150	6	6	2.1E+01
Cr XI			
214.31	5	7	1.4E+01
226.45	5	7	6.0E+02
232	3	1	4.1E+02
235.53	5	7	5.5E+02
240.76	1	3	4.8E+02
250.28	5	7	1.0E+01
366.491	3	3	1.2E+01
366.942	3	1	3.0E+01
374.927	5	5	2.3E+01
422.083	3	5	1.0E+01
Cr XII			
216	4	6	2.4E+02
218	6	8	2.4E+02
239	2	2	1.6E+02
244.70	2	4	3.0E+02
247	4	2	2.4E+02
247	2	2	3.3E+02
248	6	8	1.4E+02
250	6	8	3.5E+02
250	6	6	2.2E+02
251.52	4	6	3.4E+02
252	4	6	2.0E+02
256	2	2	1.5E+02
259	2	4	3.2E+02
269	2	2	2.1E+02
300.32	2	2	1.4E+02
305.81	4	4	2.76E+02
309	4	2	2.7E+02
309	6	6	1.6E+02
311.55	4	2	1.6E+02
324	4	6	2.2E+02
327	6	8	2.2E+02
332.06	6	4	1.4E+02
Cr XIII			
49.59	1	3	9.9E+02
67.01	1	3	1.67E+03
228	5	7	1.8E+02
267.73	5	7	1.9E+02
270	3	1	1.7E+02
276.4	5	7	2.2E+02
277	1	3	2.1E+02
279.32	3	5	3.5E+02
286	3	1	4.6E+02
328.29	1	3	1.86E+02
345	7	9	1.74E+02
Cr XIV			
*38.036	2	6	2.47E+02
39.796	2	4	3.05E+02
40.018	4	6	3.6E+02
40.782	2	4	3.9E+02
40.800	2	2	3.9E+02
41.556	2	4	4.5E+02
41.788	4	6	5.3E+02
44.597	2	4	7.1E+02
44.869	4	6	8.3E+02
46.125	4	2	3.1E+02
46.468	2	4	6.6E+02
46.527	2	2	6.7E+02
48.300	4	6	5.9E+02
48.338	6	8	6.3E+02
50.821	2	4	1.2E+03
51.172	4	6	1.4E+03
51.180	4	4	2.3E+02
52.321	4	6	1.0E+03
52.363	6	8	1.1E+03
53.760	2	2	3.0E+02
54.164	4	2	5.9E+02
60.699	4	6	2.05E+03
60.756	6	8	2.19E+03
63.324	2	4	1.07E+03
63.539	2	2	1.13E+03
68.594	2	4	1.98E+03
69.213	4	6	2.31E+03
69.247	4	4	3.8E+02
86.060	4	6	5.3E+03
86.169	6	8	5.9E+03
86.185	6	6	3.9E+02
101.05	6	4	4.4E+02
101.42	4	2	4.83E+02
104.4	4	6	3.0E+02
104.5	6	8	3.1E+02
109.8	2	4	2.3E+02
110.4	4	6	2.8E+02
118.3	4	2	2.1E+02
125.2	4	6	5.0E+02
125.3	6	8	5.4E+02
148.5	2	4	2.18E+02
149.1	2	2	2.1E+02
157.1	2	4	3.3E+02
158.4	4	6	3.7E+02
187.02	4	6	9.3E+02
187.30	6	8	9.6E+02
189.1	2	2	2.13E+02
191.0	4	2	4.11E+02
222.9	4	2	2.2E+02
346.3	4	6	2.4E+02
346.5	6	8	2.5E+02
Cr XV			
18.497	1	3	1.62E+05
18.782	1	3	2.8E+04
19.015	1	3	6.3E+02
20.863	1	3	6.0E+03
21.153	1	3	5.6E+03
102	3	3	1.6E+02
102.18	5	3	7.0E+02
103	3	1	3.8E+02
105	7	5	5.3E+02
111.27	3	3	1.7E+02
Cr XVI			
17.073	4	6	1.2E+04
17.242	2	4	8.6E+04
17.299	4	4	2.5E+04
17.372	4	4	1.4E+05
17.438	4	2	1.1E+05
17.514	2	4	1.1E+05
17.587	2	4	2.0E+04
17.656	2	2	2.0E+04
19.442	4	2	9.9E+03
19.714	2	2	1.1E+04
Cr XVII			
16.31	5	3	9.6E+03
16.32	5	7	3.2E+04
16.37	3	1	9.7E+04
16.44	5	7	1.3E+05
16.59	3	1	5.7E+04
16.65	5	5	1.1E+04
16.66	1	3	1.8E+05
16.68	5	7	6.8E+04
16.80	5	7	4.4E+04
16.97	1	3	2.63E+04
16.97	3	3	1.5E+04
17.968	5	3	8.6E+03
18.336	5	3	1.7E+04
18.336	5	5	1.6E+04
18.389	1	3	9.2E+03
Cr XVIII			
95.77	4	2	3.08E+02
102.32	4	4	1.54E+02
104.98	6	4	8.7E+02
106.84	4	2	3.4E+02
110.41	4	2	7.9E+02
112.27	4	2	4.24E+02
119.62	2	2	3.2E+02
123.87	6	4	3.9E+02
125.51	4	4	3.4E+02
128.10	6	6	2.8E+02
136.52	4	2	1.66E+02
139.87	4	4	1.49E+02
140.82	4	2	2.66E+02
155.46	2	2	2.84E+02
157.40	4	4	2.83E+02

λ Å	Weights g_i	Weights g_k	A 10^8 s^{-1}	λ Å	Weights g_i	Weights g_k	A 10^8 s^{-1}	λ Å	Weights g_i	Weights g_k	A 10^8 s^{-1}
Cr XIX				14.04	3	5	1.2E+05	2467.69	6	8	7.0E-02
14.73	3	3	7.1E+04	14.24	1	3	1.41E+05	2470.27	10	12	1.5E-01
14.80	1	3	1.3E+05					2476.64	10	8	2.2E-01
14.81	5	3	3.4E+04	**Cr XXII**				2504.52	10	8	1.8E-01
14.84	5	7	1.3E+05	2.190	4	2	1.7E+06	2511.02	10	10	9.2E-01
109.64	3	3	2.46E+02	2.191	2	2	2.5E+06	2521.36	10	8	3.0E+00
110.37	5	3	6.0E+02	2.198	4	4	4.5E+06	2528.97	8	6	2.8E+00
113.97	5	3	5.5E+02	2.199	2	4	2.3E+06	2530.13	6	6	7.1E-02
118.31	3	1	3.29E+02	2.202	4	6	1.6E+06	2535.96	6	4	1.9E+00
118.67	5	3	2.1E+02	2.203	4	2	1.3E+06	2536.50	8	8	3.0E-01
118.83	3	3	1.35E+02	13.149	2	4	1.29E+05	2544.25	4	2	3.0E+00
126.30	1	3	1.56E+02	13.292	4	6	1.54E+05	2562.12	4	4	3.9E-01
126.33	5	5	4.35E+02					2567.34	6	6	3.0E-01
130.99	7	5	2.9E+02	**Cr XXIII**				2574.35	8	8	1.7E-01
134.89	3	1	1.98E+02	1.7632	1	3	3.68E+05	2685.34	6	8	5.5E-02
138.15	3	1	1.75E+02	1.8557	1	3	8.97E+05	3017.55	8	6	6.9E-02
138.45	5	5	1.71E+02	2.095	3	1	3.5E+06	3044.00	10	10	1.9E-01
140.92	5	3	1.38E+02	2.101	1	3	2.0E+06	3048.89	6	4	7.5E-02
143.57	3	1	7.2E+02	2.101	5	5	7.9E+06	3061.82	8	8	1.6E-01
163.94	5	5	3.1E+02	2.102	3	5	2.1E+06	3072.34	6	6	1.5E-01
179.18	3	1	1.45E+02	2.103	3	5	1.2E+06	3086.78	4	4	1.9E-01
				2.104	1	3	1.4E+06	3354.37	8	6	1.1E-01
Cr XX				2.105	3	3	9.6E+05	3367.11	10	8	6.0E-02
14.13	2	4	1.1E+05	2.106	3	3	2.0E+06	3385.22	8	6	1.1E-01
14.26	4	6	1.3E+05	2.107	5	5	2.3E+06	3388.16	6	4	2.4E-01
128.42	4	4	3.8E+02	2.107	3	5	3.3E+06	3395.37	6	8	2.9E-01
131.31	6	4	1.27E+02	2.109	5	3	1.7E+06	3405.12	10	10	1.0E+00
133.82	2	4	8.3E+01	2.113	3	5	5.9E+05	3409.17	8	8	4.2E-01
135.26	4	2	2.41E+02	2.119	3	1	2.7E+05	3412.34	8	10	6.1E-01
140.75	4	4	1.35E+02	2.129	3	1	5.1E+05	3412.63	10	8	1.2E-01
148.99	6	4	1.75E+02	2.1818	1	3	3.37E+06	3414.74	4	4	8.8E-02
156.00	2	4	8.4E+01	2.1923	1	3	2.34E+05	3417.15	6	6	3.2E-01
167.97	6	6	1.12E+02					3431.58	8	6	1.1E-01
180.85	4	4	1.6E+02	**Cobalt**				3433.05	4	4	1.0E+00
				Co I				3442.92	6	4	1.2E-01
Cr XXI				2287.80	8	8	8.6E-01	3443.64	8	8	6.9E-01
12.97	3	1	4.8E+04	2295.22	10	8	2.2E-01	3449.17	6	6	7.6E-01
12.98	5	5	3.9E+04	2309.03	10	10	5.6E-01	3449.44	10	10	1.8E-01
13.02	3	5	3.8E+04	2323.13	8	8	5.0E-01	3453.51	10	12	1.1E+00
13.02	5	7	3.9E+04	2325.53	6	8	1.1E-01	3455.24	4	2	1.9E-01
13.08	1	3	5.2E+04	2335.98	6	6	5.1E-01	3462.80	4	6	7.9E-01
13.22	3	1	4.6E+04	2338.66	4	4	7.7E-01	3465.79	10	12	9.2E-02
13.34	3	5	5.2E+04	2353.36	8	10	1.5E-01	3474.02	6	8	5.6E-01
13.49	1	3	9.0E+04	2355.48	6	8	1.3E-01	3483.41	8	10	5.5E-02
13.53	3	3	6.6E+04	2358.18	4	6	1.4E-01	3489.40	8	6	1.3E+00
13.55	3	5	1.2E+05	2365.06	10	10	1.3E-01	3491.32	4	4	5.0E-02
13.65	5	7	1.5E+05	2371.85	6	8	7.3E-02	3495.68	4	6	4.9E-01
13.66	3	1	1.2E+05	2384.86	10	8	2.4E-01	3502.28	10	8	8.0E-01
13.67	5	5	3.9E+04	2392.03	6	6	4.0E-01	3502.63	6	6	5.2E-02
13.68	3	3	8.2E+04	2402.06	8	6	5.1E-01	3506.32	8	6	8.2E-01
13.75	5	3	4.5E+04	2407.25	10	12	3.6E+00	3509.84	6	8	3.2E-01
13.75	5	5	9.5E+04	2412.76	4	6	6.5E-01	3512.64	6	4	1.0E+00
13.76	1	3	1.51E+05	2414.46	6	8	3.4E+00	3513.48	8	10	7.8E-02
13.78	5	7	1.7E+05	2415.29	4	6	3.6E+00	3518.34	6	4	1.6E+00
13.84	5	7	2.59E+05	2424.93	10	10	3.2E+00	3521.58	10	8	1.8E-01
13.87	3	5	8.5E+04	2432.21	8	8	2.6E+00	3523.42	4	2	9.8E-01
13.92	3	5	8.5E+04	2436.66	6	6	2.6E+00	3526.85	10	10	1.3E-01
13.93	5	7	4.2E+04	2439.04	4	4	2.7E+00	3529.03	6	8	8.8E-02
13.95	5	5	3.8E+04	2460.80	4	6	1.2E-01	3529.82	8	10	4.6E-01

λ Å	Weights g_i	Weights g_k	A 10^8 s^{-1}	λ Å	Weights g_i	Weights g_k	A 10^8 s^{-1}	λ Å	Weights g_i	Weights g_k	A 10^8 s^{-1}
3533.36	4	6	9.1E-02	**Copper**				3868.8	17	17	3.1E+00
3560.89	4	4	2.3E-01	**Cu I**				3967.5	17	19	8.7E-01
3564.95	6	8	7.0E-02	*2024.3	2	6	9.8E-02	4046.0	17	15	1.5E+00
3569.37	8	8	1.6E+00	2165.1	2	4	5.1E-01	4103.9	13	11	1.7E+00
3574.97	6	6	1.5E-01	2178.9	2	4	9.13E-01	4186.8	17	17	1.32E+00
3575.36	8	8	9.6E-02	2181.7	2	2	1.0E+00	4194.8	17	17	7.2E-01
3585.15	8	8	7.1E-02	2225.7	2	2	4.6E-01	4211.7	17	19	2.08E+00
3587.19	6	6	1.4E+00	2244.3	2	4	1.19E-02	4218.1	15	15	1.85E+00
3594.87	6	6	9.2E-02	2441.6	2	2	2.0E-02	4221.1	15	17	1.52E+00
3602.08	4	4	1.0E-01	2492.2	2	4	3.11E-02	4225.2	13	15	4.5E+00
3704.06	6	8	1.2E-01	2618.4	6	4	3.07E-01	4268.3	15	15	3.6E-02
3745.49	8	8	7.5E-02	2766.4	4	4	9.6E-02	4276.7	13	13	7.3E-01
3842.05	8	6	1.3E-01	2824.4	6	6	7.8E-02	4292.0	15	15	5.8E-02
3845.47	8	10	4.6E-01	2961.2	6	8	3.76E-02	4577.8	17	19	2.2E-02
3861.16	6	4	1.4E-01	3063.4	4	4	1.55E-02	4589.4	17	15	1.3E-01
3873.12	10	8	1.2E-01	3194.1	4	4	1.55E-02	4612.3	17	15	8.2E-02
3873.95	8	6	1.0E-01	3247.5	2	4	1.39E+00	5077.7	17	17	5.7E-03
3881.87	6	4	8.2E-02	3274.0	2	2	1.37E+00	5301.6	17	15	1.1E-02
3894.07	6	8	6.9E-01	3337.8	6	8	3.8E-03	5547.3	17	17	2.7E-03
3894.98	4	2	8.8E-02	4022.6	2	4	1.90E-01	5639.5	17	19	4.7E-03
3935.96	8	10	6.2E-02	4062.6	4	6	2.10E-01	5974.5	17	17	4.0E-03
3995.31	8	10	2.5E-01	4249.0	2	2	1.95E-01	5988.6	17	15	5.3E-03
3997.90	6	8	7.0E-02	4275.1	6	8	3.45E-01	6010.8	15	15	2.6E-02
4092.39	8	8	5.7E-02	4480.4	2	2	3.0E-02	6088.3	15	13	3.5E-02
4110.53	6	6	5.5E-02	4509.4	4	2	2.75E-01	6168.4	15	17	2.5E-02
4118.77	6	8	1.6E-01	4530.8	4	2	8.4E-02	6259.1	17	19	8.5E-03
4121.32	8	10	1.9E-01	4539.7	6	4	2.12E-01	6579.4	17	15	7.5E-03
5146.75	8	8	1.5E-01	4587.0	8	6	3.20E-01				
5212.70	10	10	1.9E-01	4651.1	10	8	3.80E-01	**Erbium**			
5265.79	6	8	5.0E-02	4704.6	8	8	5.5E-02	**Er I**			
5280.63	10	8	2.8E-01	5105.5	6	4	2.0E-02	3862.9	13	13	2.5E+00
5352.05	12	10	2.7E-01	5153.2	2	4	6.0E-01	4008.0	13	15	2.6E+00
5477.09	6	8	6.8E-02	5218.2	4	6	7.5E-01	4151.1	13	11	1.8E+00
5483.96	8	10	7.3E-02	5220.1	4	4	1.50E-01				
6082.43	10	10	5.4E-02	5292.5	8	8	1.09E-01	**Europium**			
6455.00	8	10	9.0E-02	5700.2	4	4	2.4E-03	**Eu I**			
7838.12	8	10	5.4E-02	5782.1	4	2	1.65E-02	2372.9	8	6	1.9E-01
8093.93	12	10	2.0E-01					2375.3	8	8	2.0E-01
8372.79	10	10	8.7E-02	**Cu II**				2379.7	8	10	2.0E-01
				2489.7	5	5	1.5E-02	2619.3	8	10	7.0E-03
Co II				2544.8	9	7	1.1E+00	2643.8	8	8	6.6E-03
2286.15	11	13	3.3E+00	2689.3	7	7	4.1E-01	2659.4	8	10	1.2E-02
2307.85	9	11	2.6E+00	2701.0	5	5	6.7E-01	2682.6	8	6	1.2E-02
2311.61	7	9	2.8E+00	2703.2	3	3	1.2E+00	2710.0	8	10	1.4E-01
2314.05	5	7	2.8E+00	2713.5	5	5	6.8E-01	2724.0	8	8	1.2E-01
2314.97	3	5	2.7E+00	2769.7	7	7	6.1E-01	2731.4	8	8	3.1E-02
2330.36	5	3	1.32E+00					2732.6	8	6	3.7E-02
2344.28	3	3	1.5E+00	**Dysprosium**				2735.3	8	10	4.7E-02
2353.41	7	7	1.9E+00	**Dy I**				2738.6	8	10	1.3E-02
2363.80	9	9	2.1E+00	2862.7	17	15	6.5E-02	2743.3	8	6	1.1E-01
2378.62	11	9	1.9E+00	2964.6	17	17	6.5E-02	2745.6	8	6	5.0E-02
2383.45	9	7	1.8E+00	3147.7	15	17	1.1E-01	2747.8	8	8	5.2E-02
2388.92	11	11	2.8E+00	3263.2	15	13	1.4E-01	2772.9	8	6	1.0E-02
2389.54	5	3	1.5E+00	3511.0	15	13	3.1E-01	2878.9	8	10	2.8E-02
2404.17	3	3	1.5E+00	3571.4	15	13	2.0E-01	2892.5	8	8	1.0E-01
2417.66	9	9	8.5E-01	3757.1	17	19	3.0E+00	2893.0	8	6	1.0E-01

λ (Å)	g_i	g_k	A (10^8 s⁻¹)
2909.0	8	10	6.9E-02
2958.9	8	6	1.6E-02
3059.0	8	8	3.8E-02
3067.0	8	10	9.1E-03
3106.2	8	10	5.5E-02
3111.4	8	10	3.0E-01
3168.3	8	10	6.9E-02
3185.5	8	10	5.8E-03
3210.6	8	8	1.1E-01
3212.8	8	8	2.9E-01
3213.8	8	6	1.8E-01
3235.1	8	10	1.0E-02
3241.4	8	8	2.3E-02
3246.0	8	6	1.4E-02
3247.6	8	8	2.3E-02
3322.3	8	6	3.5E-02
3334.3	8	6	3.4E-01
3350.4	8	10	1.5E-02
3353.7	8	8	5.8E-03
3457.1	8	8	8.4E-03
3467.9	8	8	1.0E-02
3589.3	8	6	6.9E-03
4594.0	8	10	1.4E+00
4627.2	8	8	1.3E+00
4661.9	8	6	1.3E+00
5645.8	8	6	5.4E-03
5765.2	8	8	1.1E-02
6018.2	8	10	8.5E-03
6291.3	8	6	1.8E-03
6864.5	8	10	5.8E-03
7106.5	8	8	2.6E-03

Fluorine
F I

λ (Å)	g_i	g_k	A (10^8 s⁻¹)
806.96	4	6	3.3E+00
809.60	2	4	2.8E+00
951.87	4	2	2.6E+00
954.83	4	4	5.77E+00
955.55	2	2	5.1E+00
958.52	2	4	1.3E+00
6239.7	6	4	2.5E-01
6348.5	4	4	1.8E-01
6413.7	2	4	1.1E-01
6708.3	6	4	1.4E-02
6774.0	6	6	1.0E-01
6795.5	4	2	5.2E-02
6834.3	4	4	2.1E-01
6856.0	6	8	4.94E-01
6870.2	2	2	3.8E-01
6902.5	4	6	3.2E-01
6909.8	2	4	2.2E-01
6966.4	4	2	1.1E-01
7037.5	4	4	3.0E-01
7127.9	2	2	3.8E-01
7309.0	6	8	4.7E-01
7311.0	4	2	3.9E-01
7314.3	4	6	4.8E-01
7332.0	6	4	3.1E-01
7398.7	6	6	2.85E-01
7425.7	4	2	3.4E-01
7482.7	4	4	5.6E-02
7489.2	2	2	1.1E-01
7514.9	2	2	5.2E-02
7552.2	4	6	7.8E-02
7573.4	2	4	1.0E-01
7607.2	4	4	7.0E-02
7754.7	4	6	3.82E-01
7800.2	2	4	2.1E-01

Gallium
Ga I

λ (Å)	g_i	g_k	A (10^8 s⁻¹)
2195.4	2	2	1.9E-02
2199.7	4	2	3.3E-02
2214.4	4	6	1.2E-02
2235.9	4	2	4.3E-02
2255.0	2	2	3.1E-02
2259.2	4	6	3.1E-02
2294.2	2	4	7.0E-02
2297.9	4	2	5.8E-02
2338.2	4	6	9.8E-02
2371.3	2	2	5.7E-02
2418.7	4	2	1.0E-01
2450.1	2	4	2.8E-01
2500.2	4	6	3.4E-01
2659.9	2	2	1.2E-01
2719.7	4	2	2.3E-01
2874.2	2	4	1.2E+00
2943.6	4	6	1.4E+00
2944.2	4	4	2.7E-01
4033.0	2	2	4.9E-01
4172.0	4	2	9.2E-01

Ga II

λ (Å)	g_i	g_k	A (10^8 s⁻¹)
829.60	1	3	2.2E-01
1414.4	1	3	1.88E+01

Germanium
Ge I

λ (Å)	g_i	g_k	A (10^8 s⁻¹)
1944.7	3	1	7.0E-01
1955.1	3	3	2.8E-01
1988.3	5	3	2.5E-01
1998.9	5	5	5.5E-01
2041.7	1	3	1.1E+00
2065.2	3	3	8.5E-01
2068.7	3	5	1.2E+00
2086.0	3	5	4.0E-01
2094.3	5	7	9.7E-01
2105.8	5	5	1.7E-01
2256.0	5	5	3.2E-02
2417.4	5	5	9.6E-01
2498.0	1	3	1.3E-01
2533.2	3	3	1.0E-01
2589.2	5	3	5.1E-02
2592.5	3	5	7.1E-01
2651.2	5	5	2.0E+00
2651.6	1	3	8.5E-01
2691.3	3	3	6.1E-01
2709.6	3	1	2.8E+00
2754.6	5	3	1.1E+00
3039.1	5	3	2.8E+00
3124.8	5	5	3.1E-02
3269.5	5	3	2.9E-01
4226.6	1	3	2.1E-01
4685.8	1	3	9.5E-02

Ge II

λ (Å)	g_i	g_k	A (10^8 s⁻¹)
999.10	2	4	1.9E+00
1016.6	4	6	2.1E+00
1017.1	4	4	3.5E-01
1055.0	2	2	6.9E-01
1237.1	2	4	1.9E+01
1261.9	4	6	2.2E+01
1264.7	4	4	3.5E+00
1602.5	2	2	3.4E+00
1649.2	4	2	6.5E+00
4741.8	2	4	4.6E-01
4814.6	4	6	5.1E-01
4824.1	4	4	8.6E-02
5131.8	4	6	1.9E+00
5178.5	6	6	1.3E-01
5178.6	6	8	2.0E+00
5893.4	2	4	9.2E-01
6021.0	2	2	8.4E-01
6336.4	2	2	4.4E-01
6484.2	4	2	8.5E-01

Gold
Au I

λ (Å)	g_i	g_k	A (10^8 s⁻¹)
2427.95	2	4	1.99E+00
2675.95	2	2	1.64E+00
3122.78	6	4	1.90E-01
6278.30	4	2	3.4E-02

Helium
He I

λ (Å)	g_i	g_k	A (10^8 s⁻¹)
510.00	1	3	4.6224E-01
512.10	1	3	7.3174E-01
515.62	1	3	1.2582E+00
522.21	1	3	2.4356E+00
537.03	1	3	5.6634E+00
584.33	1	3	1.7989E+01
*2677.1	3	9	4.4174E-03
*2696.1	3	9	6.0234E-03
*2723.2	3	9	8.4996E-03
*2763.8	3	9	1.2508E-02
*2829.1	3	9	1.9389E-02
*2945.1	3	9	3.2006E-02
*3187.7	3	9	5.6361E-02
3231.3	1	3	5.1015E-03
3258.3	1	3	6.9627E-03
3296.8	1	3	9.8432E-03
3354.6	1	3	1.4537E-02
3447.6	1	3	2.2691E-02
*3554.4	9	15	7.5971E-03
*3563.0	9	3	4.8362E-03
*3587.3	9	15	1.8107E-02

λ Å	g_i	g_k	A 10^8 s^-1	λ Å	g_i	g_k	A 10^8 s^-1	λ Å	g_i	g_k	A 10^8 s^-1
3613.6	1	3	3.8022E-02	**In II**				2276.03	9	7	1.7E-01
*3634.2	9	15	2.6062E-02	2941.1	3	1	1.4E+00	2277.11	7	5	3.7E+01
*3652.0	9	3	9.7444E-03					2287.25	5	3	3.4E-01
*3705.0	9	15	3.9528E-02	**Iodine**				2292.52	7	9	4.3E-02
*3819.6	9	15	6.4351E-02	**I I**				2294.41	3	1	6.1E-01
3833.6	3	5	9.6470E-03	1782.8	4	4	2.71E+00	2300.14	5	7	8.0E-02
*3867.5	9	3	2.4465E-02	1830.4	4	6	1.6E-01	2301.68	1	3	1.3E-01
3871.8	3	5	1.3386E-02					2303.42	1	3	9.4E-02
*3888.7	3	9	9.4746E-02	**Iridium**				2303.58	3	5	7.6E-02
3926.5	3	5	1.9371E-02	**Ir I**				2309.00	3	5	1.5E-01
3935.9	3	1	7.4475E-03	2475.12	10	10	2.1E-01	2313.10	5	7	1.4E-01
3964.7	1	3	6.9507E-02	2502.98	10	12	3.2E-01	2320.36	7	9	1.2E-01
4009.3	3	5	2.9612E-02	2639.71	10	10	4.7E-01	2371.43	5	5	5.2E-02
4024.0	3	1	1.1281E-02	2661.98	10	10	2.5E-01	2373.62	7	7	6.7E-02
*4026.2	9	15	1.1600E-01	2664.79	10	8	4.0E-01	2374.52	1	3	2.9E-01
*4120.8	9	3	4.4529E-02	2694.23	10	12	4.8E-01	2381.83	3	5	5.4E-02
4143.8	3	5	4.8812E-02	2849.72	10	10	2.2E-01	2389.97	5	7	5.0E-02
4169.0	3	1	1.8298E-02	2853.31	10	10	2.0E-03	2462.18	7	5	1.5E-01
4387.9	3	5	8.9889E-02	2882.64	10	8	7.2E-02	2462.65	9	9	5.8E-01
4437.6	3	1	3.2689E-02	2924.79	10	12	1.42E-01	2479.78	5	5	1.8E+00
*4471.5	9	15	2.4578E-01	2934.64	8	10	2.0E-01	2483.27	9	11	4.9E+00
*4713.2	9	3	9.5209E-02	2951.22	10	8	2.8E-02	2488.14	7	9	4.7E+00
4921.9	3	5	1.9863E-01	3003.63	8	10	5.9E-02	2490.64	5	7	3.8E+00
5015.7	1	3	1.3372E-01	3168.88	8	10	5.47E-02	2491.15	3	5	3.0E+00
5047.7	3	1	6.7712E-02	3220.78	10	8	2.4E-01	2501.13	9	7	6.8E-01
*5875.7	9	15	7.0703E-01	3558.99	6	8	1.5E-02	2510.83	7	5	1.3E+00
6678.2	3	5	6.3705E-01	3573.72	8	10	5.4E-02	2518.10	5	3	1.9E+00
*7065.2	9	3	2.7853E-01	3617.21	6	8	2.0E-02	2522.85	9	9	2.9E+00
7281.4	3	1	1.8299E-01	3628.67	8	8	2.8E-02	2524.29	3	1	3.4E+00
*8361.7	3	9	3.8126E-03	3661.71	8	10	4.0E-02	2527.43	7	7	1.9E+00
*9463.6	3	9	5.6868E-03	3734.77	8	8	2.7E-02	2529.13	5	5	9.8E-01
9603.4	1	3	5.8286E-03	4033.76	8	10	2.7E-02	2535.61	1	3	9.7E-01
*9702.6	9	3	8.6511E-03	4069.92	6	8	3.6E-02	2540.97	3	5	9.2E-01
*10311	9	15	1.9945E-02	4913.35	12	12	3.3E-02	2545.98	5	7	6.7E-01
*10668	9	3	1.4471E-02	4939.24	10	12	2.5E-03	2549.61	7	9	3.6E-01
*10830	3	9	1.0216E-01					2584.54	11	13	4.6E-01
*10913	15	21	1.9801E-02	**Iron**				2606.83	9	11	4.2E-01
10917	5	7	1.6083E-02	**Fe I**				2618.02	7	7	4.0E-01
*10997	15	9	1.4253E-03	1934.54	9	7	2.5E-01	2623.53	7	9	3.3E-01
11013	1	3	9.2496E-03	1937.27	9	7	2.2E-01	2656.15	13	15	2.8E-01
11045	3	5	1.8457E-02	1940.66	7	5	2.6E-01	2669.49	11	13	1.7E-01
11226	3	1	1.1168E-02	2084.12	9	7	3.7E-01	2679.06	11	11	1.9E-01
*11969	9	15	3.4781E-02	2102.35	7	7	8.8E-02	2719.03	9	7	1.4E+00
*12528	3	9	7.0932E-03	2112.97	1	3	1.9E-01	2720.90	7	5	1.1E+00
12756	5	3	1.2754E-03	2132.02	9	9	7.6E-02	2723.58	5	3	6.4E-01
*12785	15	21	4.1339E-02	2145.19	7	7	5.7E-02	2733.58	11	9	8.6E-01
12791	5	7	3.2475E-02	2153.01	5	5	6.9E-02	2735.48	9	7	6.2E-01
*12846	9	3	2.7317E-02	2161.58	3	5	5.0E-02	2737.31	3	3	8.5E-01
12968	3	5	3.3615E-02	2166.77	9	7	2.7E+00	2742.41	5	5	6.3E-01
*12985	15	9	2.7292E-03	2171.30	5	7	5.1E-02	2744.07	1	3	3.5E-01
				2173.21	3	5	8.3E-02	2750.14	7	7	3.9E-01
Indium				2176.84	1	3	1.0E-01	2756.33	3	5	2.0E-01
In I				2191.20	1	3	7.3E-02	2788.10	11	13	6.3E-01
2560.2	2	4	4.0E-01	2191.84	5	5	1.2E+00	2894.50	5	5	6.2E-01
2710.3	4	6	4.0E-01	2196.04	3	3	1.2E+00	2899.42	5	3	5.9E-01
3039.4	2	4	1.3E+00	2200.72	3	5	2.8E-01	2920.69	5	5	5.2E-02
3256.1	4	6	1.3E+00	2259.51	9	11	7.0E-02	2923.29	11	11	1.6E+00
4101.8	2	2	5.6E-01	2267.08	7	5	7.1E-02	2925.36	7	9	1.8E-01
4511.3	4	2	1.02E+00	2272.07	7	9	3.8E-02	2929.01	7	5	7.3E-02

λ Å	Weights g_i	g_k	A 10^8 s^{-1}	λ Å	Weights g_i	g_k	A 10^8 s^{-1}	λ Å	Weights g_i	g_k	A 10^8 s^{-1}
2936.90	9	9	1.3E-01	3215.94	5	5	8.0E-01	3442.36	5	5	4.55E-02
2941.34	5	3	5.6E-02	3217.38	11	9	2.2E-01	3443.88	5	3	6.2E-02
2947.88	7	7	2.0E-01	3219.58	7	9	6.2E-01	3445.15	5	7	2.8E-01
2953.94	5	5	1.89E-01	3222.07	11	11	3.3E-01	3447.28	5	5	9.1E-02
2954.65	5	7	1.0E-01	3225.79	11	13	8.8E-01	3450.33	3	3	2.0E-01
2957.36	3	3	1.77E-01	3227.80	9	7	1.4E+00	3476.70	1	3	5.4E-02
2965.25	1	3	1.16E-01	3228.25	5	3	4.5E-01	3477.85	3	1	4.2E-02
2966.90	9	11	2.72E-01	3229.99	9	11	4.5E-01	3485.34	5	3	1.4E-01
2969.36	3	1	3.66E-02	3230.21	5	5	1.9E-01	3495.29	9	7	9.46E-02
2973.13	5	7	1.35E-01	3230.96	7	5	3.9E-01	3497.10	7	7	1.4E-01
2973.24	7	9	1.83E-01	3233.05	13	15	5.4E-01	3505.07	5	3	9.9E-02
2980.53	7	7	2.2E-01	3233.97	9	9	2.0E-01	3506.50	5	5	7.1E-02
2981.45	7	5	6.54E-02	3246.96	5	3	9.9E-02	3508.49	9	11	5.7E-02
2983.57	9	7	2.80E-01	3248.20	7	7	2.2E-01	3510.44	1	3	4.4E-02
2987.29	9	7	6.6E-02	3253.60	7	9	1.8E-01	3516.56	7	5	3.7E-02
2990.39	9	11	3.9E-01	3254.36	11	13	5.1E-01	3521.84	3	5	9.6E-02
2994.43	7	5	4.4E-02	3257.59	7	5	1.4E-01	3523.31	5	3	7.6E-02
2996.39	3	5	1.6E-01	3265.62	7	5	3.8E-01	3524.08	7	5	7.5E-02
2999.51	11	11	2.3E-01	3268.23	3	3	5.9E-02	3524.24	5	7	4.2E-02
3000.95	5	3	6.42E-01	3271.00	5	3	6.6E-02	3527.79	9	9	2.0E-01
3008.14	3	1	1.07E+00	3280.26	9	11	5.4E-01	3529.82	3	3	7.6E-02
3009.09	13	11	6.7E-02	3282.89	3	5	3.0E-01	3536.56	5	7	7.8E-01
3009.57	9	9	1.7E-01	3284.59	5	5	5.4E-02	3537.73	5	3	1.1E-01
3011.48	7	9	4.7E-01	3290.99	3	5	6.0E-02	3537.90	11	11	8.4E-02
3015.92	11	9	5.9E-02	3292.02	7	9	6.1E-01	3540.12	7	9	1.2E-01
3016.18	5	3	8.5E-02	3292.59	3	3	2.6E-01	3541.08	9	11	6.2E-01
3017.63	3	3	6.82E-02	3298.13	3	5	8.1E-02	3542.08	7	9	7.4E-01
3018.98	7	7	1.3E-01	3305.97	5	7	4.7E-01	3543.67	3	5	1.8E-01
3021.07	7	7	4.56E-01	3306.36	3	5	6.1E-01	3548.02	5	3	9.7E-02
3024.03	3	5	4.88E-02	3307.23	13	13	2.0E-01	3552.11	3	5	4.5E-02
3025.84	1	3	3.48E-01	3314.74	5	7	6.9E-01	3552.83	5	5	1.5E-01
3026.46	5	5	1.1E-01	3322.47	9	11	6.2E-02	3553.74	11	9	8.1E-01
3031.63	3	3	1.5E-01	3323.74	5	5	3.0E-01	3556.88	9	11	4.4E-01
3037.39	3	5	3.2E-01	3328.87	11	11	2.7E-01	3559.50	3	3	1.9E-01
3042.02	3	5	4.9E-01	3337.66	11	9	5.7E-02	3560.70	7	9	6.5E-02
3042.66	5	7	5.7E-02	3347.93	5	5	4.0E-02	3565.38	7	9	3.8E-01
3047.60	5	7	2.84E-01	3354.06	1	3	7.7E-02	3567.03	5	7	6.5E-02
3053.07	3	5	1.5E-01	3355.23	9	9	3.2E-01	3568.42	5	3	5.3E-02
3057.45	11	9	4.4E-01	3369.55	9	9	2.4E-01	3568.82	7	9	5.6E-02
3059.09	7	9	1.7E-01	3370.78	11	11	3.3E-01	3570.10	9	11	6.77E-01
3067.24	9	7	3.4E-01	3380.11	7	7	2.4E-01	3572.00	11	11	2.4E-01
3068.17	5	3	9.8E-02	3383.98	7	7	9.3E-02	3573.39	5	7	7.5E-02
3075.72	7	5	2.9E-01	3392.65	7	7	2.6E-01	3576.76	11	9	9.6E-02
3083.74	5	3	3.0E-01	3394.58	5	3	9.9E-02	3578.38	1	3	6.3E-02
3091.58	3	1	5.4E-01	3399.33	5	5	3.8E-01	3581.19	11	13	1.02E+00
3098.19	11	11	1.1E-01	3402.26	13	13	2.8E-01	3582.20	13	11	2.5E-01
3100.67	7	7	1.4E-01	3406.44	3	5	3.0E-01	3583.33	1	3	2.3E-01
3119.49	11	9	8.2E-02	3407.46	7	9	5.8E-01	3585.32	7	7	1.3E-01
3120.43	9	7	8.9E-02	3410.17	3	5	4.7E-01	3585.71	9	9	3.75E-02
3156.27	7	7	5.4E-01	3411.35	9	9	5.5E-02	3586.98	5	5	1.6E-01
3160.66	9	9	1.9E-01	3413.13	5	7	3.6E-01	3591.48	1	3	6.0E-02
3161.95	11	13	1.2E-01	3417.84	3	3	5.1E-01	3592.67	7	5	4.0E-02
3166.44	9	7	1.14E-01	3418.51	3	1	1.3E+00	3594.63	9	9	2.7E-01
3168.85	5	7	5.7E-02	3424.28	7	7	2.0E-01	3595.30	5	5	5.4E-02
3175.45	11	11	1.3E-01	3425.01	9	7	2.8E-01	3597.02	5	3	1.7E-01
3176.36	5	3	9.2E-02	3427.12	7	9	5.5E-01	3599.62	11	9	1.8E-01
3196.93	9	11	9.0E-01	3428.19	5	5	2.1E-01	3603.20	11	11	2.6E-01
3199.53	9	9	2.6E-01	3428.75	7	5	2.7E-01	3603.82	3	3	1.7E-01
3205.40	3	3	1.2E+00	3440.99	7	5	8.4E-02	3605.45	9	9	6.4E-01

λ (Å)	g_i	g_k	A (10^8 s^{-1})	λ (Å)	g_i	g_k	A (10^8 s^{-1})	λ (Å)	g_i	g_k	A (10^8 s^{-1})
3606.68	11	13	8.2E-01	3724.38	5	7	1.3E-01	3820.43	11	9	6.68E-01
3608.86	3	5	8.14E-01	3726.93	5	5	4.6E-01	3821.18	11	13	7.0E-01
3610.16	13	13	4.8E-01	3727.09	9	7	2.0E-01	3821.83	5	5	7.8E-02
3610.70	5	3	7.1E-02	3727.62	7	5	2.25E-01	3825.88	9	7	5.98E-01
3612.07	11	13	7.5E-02	3730.39	9	11	1.3E-01	3827.82	7	5	1.05E+00
3613.45	7	7	6.7E-02	3730.95	5	7	3.8E-02	3833.31	9	9	4.69E-02
3615.19	3	3	5.8E-02	3732.40	5	5	2.8E-01	3834.22	7	5	4.53E-01
3617.79	5	7	6.5E-01	3733.32	3	3	6.2E-02	3836.33	5	5	3.7E-01
3618.77	5	7	7.3E-01	3734.86	11	11	9.02E-01	3839.26	9	9	2.8E-01
3621.46	9	11	5.1E-01	3735.32	9	9	2.4E-01	3839.61	3	5	3.9E-01
3622.00	7	7	5.1E-01	3737.13	7	9	1.42E-01	3840.44	5	3	4.70E-01
3623.19	13	13	7.4E-02	3738.31	11	13	3.8E-02	3841.05	5	3	1.3E+00
3624.06	5	3	5.4E-02	3740.24	7	9	1.4E-01	3843.26	9	7	4.7E-01
3630.35	9	7	7.6E-02	3742.62	9	9	1.0E-01	3845.17	3	3	6.8E-02
3631.46	7	9	5.17E-01	3743.36	5	3	2.60E-01	3845.69	5	7	4.9E-02
3632.04	3	5	4.8E-01	3744.10	5	3	3.6E-01	3846.00	9	7	4.3E-02
3632.55	11	9	5.2E-02	3745.56	5	7	1.15E-01	3846.41	11	9	1.9E-01
3635.19	5	3	1.4E-01	3745.90	1	3	7.33E-02	3846.80	7	7	6.6E-01
3637.86	9	9	5.5E-02	3746.93	7	7	2.2E-01	3849.96	3	1	6.06E-01
3638.30	7	9	2.6E-01	3748.26	3	5	9.15E-02	3856.37	7	5	4.64E-02
3640.39	9	11	3.8E-01	3749.48	9	9	7.64E-01	3859.21	13	11	8.5E-02
3644.80	7	5	7.8E-02	3753.61	7	5	9.3E-02	3859.91	9	9	9.70E-02
3645.82	1	3	5.7E-01	3756.94	11	11	2.4E-01	3865.52	3	3	1.55E-01
3647.84	9	11	2.92E-01	3757.45	5	3	1.2E-01	3867.22	5	5	3.4E-01
3649.51	11	9	4.2E-01	3758.23	7	7	6.34E-01	3871.75	11	11	6.7E-02
3650.03	7	7	9.9E-02	3760.05	13	15	4.47E-02	3872.50	5	5	1.05E-01
3651.47	7	9	6.2E-01	3760.53	3	5	4.8E-02	3873.76	11	9	8.0E-02
3655.46	5	5	1.0E-01	3763.79	5	5	5.44E-01	3878.02	7	7	7.72E-02
3659.52	9	9	5.8E-02	3765.54	13	15	9.8E-01	3878.57	5	3	6.6E-02
3667.25	9	7	1.4E-01	3766.67	5	3	9.7E-02	3883.28	7	7	1.6E-01
3669.15	9	7	7.4E-02	3767.19	3	3	6.40E-01	3884.36	11	9	3.5E-02
3669.52	9	7	3.0E-01	3768.03	3	1	8.4E-02	3885.51	3	5	5.8E-02
3670.09	11	13	7.6E-02	3774.82	3	3	4.7E-02	3886.28	7	7	5.30E-02
3674.77	5	3	6.7E-02	3778.51	7	5	1.2E-01	3887.05	9	9	3.52E-02
3676.31	9	11	4.63E-02	3781.94	5	7	3.7E-02	3888.51	5	5	2.6E-01
3677.31	5	7	3.1E-01	3785.95	11	13	4.2E-02	3888.82	5	3	2.7E-01
3677.63	7	5	8.0E-01	3786.19	5	5	1.2E-01	3891.93	3	3	4.0E-01
3678.86	3	5	4.1E-02	3787.16	5	5	1.0E-01	3893.39	11	11	1.3E-01
3682.24	5	5	1.7E+00	3787.88	3	5	1.29E-01	3895.66	3	1	9.40E-02
3684.11	9	7	3.4E-01	3789.82	9	7	3.9E-02	3900.52	7	7	7.5E-02
3686.00	9	11	2.6E-01	3791.73	5	3	6.3E-02	3902.95	7	7	2.14E-01
3686.26	3	1	1.2E-01	3793.87	3	3	7.4E-02	3903.90	9	9	9.6E-02
3687.46	11	9	8.01E-02	3794.34	9	11	3.8E-02	3906.75	5	7	6.7E-02
3688.48	7	9	6.9E-02	3795.00	5	7	1.15E-01	3907.93	7	5	6.7E-02
3690.73	11	11	2.7E-01	3799.55	7	9	7.32E-02	3909.66	3	5	5.3E-02
3694.01	5	7	6.8E-01	3801.68	5	7	6.6E-02	3909.83	3	3	6.5E-02
3697.43	7	7	2.1E-01	3802.00	11	13	3.5E-02	3914.27	3	3	5.4E-02
3698.60	5	7	3.8E-02	3802.28	5	5	5.0E-02	3916.73	13	11	1.2E-01
3699.15	5	7	4.5E-02	3804.01	11	9	4.7E-02	3919.07	9	9	3.9E-02
3701.09	7	9	4.8E-01	3805.35	9	11	9.8E-01	3925.20	1	3	5.7E-02
3702.03	3	1	3.5E-01	3806.22	3	3	2.3E-01	3931.12	5	7	4.5E-02
3703.69	9	11	5.3E-02	3806.70	11	11	5.4E-01	3941.28	5	5	8.4E-02
3703.82	1	3	1.2E-01	3807.54	3	5	8.0E-02	3942.44	3	5	9.0E-02
3704.46	11	9	1.3E-01	3808.73	9	9	3.54E-02	3946.99	9	11	4.4E-02
3709.25	9	7	1.56E-01	3810.76	5	3	2.0E-01	3948.77	11	9	2.2E-01
3711.41	3	5	7.3E-02	3813.88	13	11	8.7E-02	3949.14	3	3	3.9E-02
3718.41	7	7	5.3E-02	3815.84	9	7	1.3E+00	3949.95	7	5	5.9E-02
3719.93	9	11	1.62E-01	3817.64	11	11	8.3E-02	3951.16	3	5	3.6E-01
3722.56	5	5	4.97E-02	3819.50	7	5	4.6E-02	3952.60	11	11	4.1E-02

NIST ATOMIC TRANSITION PROBABILITY TABLES (continued)

λ Å	g_i	g_k	A 10^8 s^{-1}	λ Å	g_i	g_k	A 10^8 s^{-1}	λ Å	g_i	g_k	A 10^8 s^{-1}
3953.15	7	9	3.7E-02	4109.07	1	3	4.5E-02	4250.79	7	7	1.0E-01
3955.34	3	3	1.4E-01	4109.80	3	3	1.6E-01	4260.47	11	11	3.2E-01
3955.96	3	3	5.7E-02	4112.96	11	13	1.4E-01	4267.83	1	3	9.4E-02
3956.45	13	11	2.1E-01	4114.45	5	5	4.7E-02	4268.75	5	3	4.2E-02
3957.02	5	7	1.6E-01	4118.54	11	13	5.8E-01	4271.15	7	9	1.82E-01
3960.28	5	7	4.2E-02	4126.18	11	11	3.9E-02	4271.76	9	11	2.28E-01
3963.10	3	5	1.7E-01	4127.61	1	3	1.3E-01	4282.40	7	5	1.1E-01
3967.42	9	7	2.3E-01	4132.06	5	7	1.2E-01	4300.83	5	5	4.7E-02
3967.96	7	9	6.3E-02	4132.90	3	5	9.4E-02	4305.45	5	3	6.0E-02
3969.26	9	7	2.3E-01	4134.68	5	7	1.8E-01	4307.90	7	9	3.4E-01
3970.39	3	1	3.5E-01	4137.00	3	5	2.2E-01	4315.08	5	5	7.7E-02
3971.32	11	9	5.7E-02	4137.42	5	7	6.1E-02	4325.76	5	7	5.0E-01
3973.65	5	7	6.6E-02	4142.63	3	5	7.4E-02	4327.09	5	5	7.8E-02
3976.61	3	5	1.8E-01	4143.87	7	9	1.5E-01	4352.73	3	5	3.9E-02
3977.74	5	5	7.0E-02	4149.37	11	13	3.6E-02	4369.77	9	9	7.2E-02
3981.77	9	9	3.9E-02	4150.25	3	3	7.1E-02	4383.54	9	11	5.00E-01
3983.96	9	7	7.6E-02	4153.90	7	9	2.3E-01	4387.89	3	3	3.9E-02
3985.39	5	5	6.7E-02	4154.54	9	11	1.5E-01	4388.41	7	7	1.3E-01
3989.86	5	7	5.0E-02	4156.80	5	5	1.9E-01	4401.29	7	7	5.9E-02
3996.97	9	9	6.7E-02	4158.79	3	5	1.6E-01	4404.75	7	9	2.75E-01
3997.39	9	11	1.5E-01	4170.90	5	5	6.1E-02	4415.12	5	7	1.19E-01
3998.05	11	9	6.6E-02	4172.12	7	5	9.7E-02	4422.57	3	3	8.8E-02
4003.76	3	3	7.1E-02	4175.64	3	5	1.6E-01	4430.61	3	1	7.45E-02
4005.24	7	5	2.04E-01	4181.75	5	7	3.6E-01	4433.22	5	3	2.3E-01
4006.31	11	9	4.7E-02	4182.38	5	5	4.9E-02	4438.34	3	1	7.9E-02
4007.27	7	5	4.2E-02	4184.89	5	5	1.1E-01	4442.34	5	5	3.76E-02
4009.71	3	5	5.2E-02	4187.04	7	5	2.15E-01	4443.19	1	3	1.1E-01
4014.53	11	11	2.4E-01	4187.79	9	7	1.52E-01	4446.83	3	3	5.3E-02
4017.15	9	11	4.5E-02	4191.68	1	3	4.8E-02	4447.72	3	3	5.11E-02
4021.87	7	9	1.0E-01	4196.21	7	7	9.8E-02	4454.38	5	5	3.8E-02
4024.72	7	9	8.9E-02	4198.30	11	9	8.03E-02	4455.03	9	7	3.9E-02
4031.96	3	5	7.1E-02	4198.64	5	5	1.3E-01	4466.55	5	7	1.2E-01
4040.64	5	7	4.4E-02	4199.09	9	11	6.1E-01	4469.37	5	7	2.6E-01
4044.61	5	3	1.1E-01	4200.09	7	7	4.0E-02	4481.61	3	3	4.2E-02
4045.81	9	9	8.63E-01	4200.92	7	9	4.2E-02	4484.22	7	9	7.0E-02
4054.87	5	3	1.6E-01	4202.03	9	9	8.22E-02	4485.67	3	3	1.1E-01
4058.22	9	7	4.9E-02	4203.67	7	9	8.6E-02	4528.61	7	9	5.44E-02
4059.73	5	3	8.1E-02	4203.94	13	13	1.3E-01	4533.13	3	1	3.7E-02
4062.44	3	3	2.2E-01	4205.54	5	5	3.6E-02	4547.85	5	7	7.6E-02
4063.59	7	7	6.8E-01	4207.13	5	3	4.3E-02	4619.29	7	5	4.7E-02
4065.40	3	1	1.9E-01	4210.34	3	3	1.7E-01	4669.17	5	3	4.0E-02
4067.98	9	9	1.7E-01	4213.65	3	1	1.9E-01	4673.16	5	7	4.6E-02
4070.77	7	5	1.3E-01	4217.55	3	5	2.3E-01	4678.85	7	9	7.4E-02
4071.74	5	5	7.65E-01	4219.36	11	13	3.8E-01	4704.95	3	1	8.1E-02
4073.76	5	3	1.6E-01	4220.34	3	1	1.9E-01	4736.77	9	11	4.9E-02
4074.79	9	9	4.8E-02	4222.21	7	7	5.77E-02	4789.65	5	5	7.2E-02
4076.63	9	9	1.9E-01	4224.17	9	11	1.3E-01	4859.74	5	3	1.3E-01
4078.35	5	3	4.2E-02	4224.51	3	5	7.1E-02	4871.32	7	5	2.2E-01
4079.18	5	5	5.1E-02	4225.45	5	7	1.7E-01	4872.14	3	3	2.4E-01
4079.84	1	3	6.3E-02	4226.42	3	3	3.7E-02	4878.21	1	3	9.1E-02
4080.21	3	1	2.4E-01	4233.60	3	5	1.85E-01	4890.75	5	5	2.1E-01
4082.44	3	3	3.8E-02	4235.94	9	9	1.88E-01	4891.49	9	7	2.9E-01
4084.49	11	9	1.1E-01	4238.81	7	9	2.2E-01	4892.87	3	3	4.8E-02
4085.00	3	5	4.2E-02	4240.37	5	3	5.7E-02	4903.31	3	5	4.7E-02
4085.30	7	7	1.1E-01	4245.26	1	3	8.3E-02	4917.23	5	3	6.1E-02
4085.98	7	5	5.0E-02	4246.08	7	5	5.7E-02	4918.01	1	3	4.0E-02
4088.57	5	3	3.9E-02	4247.43	9	11	2.0E-01	4918.99	7	7	1.7E-01
4098.18	7	7	6.8E-02	4248.22	3	5	3.5E-02	4920.50	11	9	3.5E-01
4107.49	5	3	2.5E-01	4250.12	5	7	2.08E-01	4930.31	3	3	4.1E-02

λ Å	Weights g_i	g_k	A 10^8 s^{-1}	λ Å	Weights g_i	g_k	A 10^8 s^{-1}	λ Å	Weights g_i	g_k	A 10^8 s^{-1}
4969.92	3	3	1.8E-01	5679.02	5	7	3.6E-02	2370.50	4	4	1.4E-01
4973.10	3	3	1.0E-01	5686.53	9	11	4.4E-02	2373.74	10	10	3.3E-01
4978.60	5	3	1.1E-01	5691.51	3	1	6.2E-02	2375.19	4	2	9.8E-01
4988.95	7	7	4.9E-02	5705.99	7	9	6.7E-02	2379.27	8	8	1.5E-01
4991.27	5	7	8.2E-02	5717.85	1	3	5.0E-02	2380.76	6	8	3.1E-01
5001.86	9	7	3.9E-01	5753.12	3	5	7.0E-02	2382.04	10	12	3.8E+00
5004.04	5	3	3.5E-02	5762.99	5	7	1.0E-01	2382.90	12	14	2.2E-01
5014.94	7	5	3.0E-01	5816.36	9	11	3.7E-02	2383.25	6	6	3.4E-01
5022.24	5	3	2.6E-01	5905.67	5	3	1.2E-01	2384.39	4	4	2.3E-01
5074.75	9	11	1.5E-01	5927.80	5	3	5.1E-02	2388.37	10	12	2.2E-01
5090.78	7	5	2.0E-01	5930.17	5	7	1.6E-01	2388.63	8	8	1.0E+00
5109.65	3	5	5.4E-02	6020.17	7	9	1.1E-01	2390.10	14	16	5.5E+00
5121.64	5	5	7.9E-02	6024.07	9	11	1.3E-01	2390.77	6	6	9.3E-01
5125.11	9	7	2.6E-01	6055.99	7	9	7.0E-02	2395.42	6	4	3.3E-01
5133.69	11	13	2.7E-01	6170.49	5	5	1.3E-01	2395.62	8	10	2.5E+00
5137.38	11	9	1.1E-01	6336.84	3	3	4.9E-02	2399.24	6	6	1.4E+00
5159.06	5	3	1.3E-01	6338.90	5	3	4.8E-02	2400.06	12	14	5.2E+00
5162.27	11	11	2.4E-01	6400.00	7	9	5.5E-02	2401.29	6	8	2.5E+00
5184.26	5	7	3.5E-02	6411.65	5	7	3.5E-02	2404.43	4	2	7.1E-01
5208.59	7	5	5.2E-02	6419.98	7	7	1.3E-01	2404.89	6	8	1.7E+00
5232.94	9	11	1.4E-01	6469.21	3	3	9.0E-02	2406.66	4	4	1.6E+00
5263.30	5	5	5.2E-02	6495.78	3	3	6.0E-02	2410.52	4	6	1.5E+00
5266.55	7	9	8.6E-02	6496.46	5	5	8.5E-02	2411.07	2	2	2.4E+00
5283.62	7	7	8.0E-02	6569.23	7	9	6.5E-02	2413.31	2	4	1.1E+00
5302.30	3	5	6.3E-02	6633.76	7	7	3.6E-02	2416.45	8	10	1.6E+00
5324.18	9	9	1.5E-01	6733.16	3	1	3.9E-02	2418.44	6	8	1.6E+00
5339.93	5	7	7.0E-02	6841.35	5	7	3.6E-02	2423.21	4	6	1.4E+00
5353.39	9	7	4.8E-02	7130.94	3	5	4.3E-02	2428.36	8	10	2.7E+00
5364.87	5	7	5.5E-01					2432.87	14	14	3.2E+00
5367.47	7	9	5.8E-01	**Fe II**				2434.06	8	6	7.0E-01
5369.96	9	11	4.7E-01	1144.94	10	12	4.8E+00	2434.24	8	10	2.0E+00
5373.71	7	9	3.5E-02	1635.40	8	6	2.4E+00	2434.73	12	12	3.2E+00
5383.37	11	13	5.6E-01	1641.76	6	4	1.8E+00	2439.30	12	14	2.8E+00
5389.48	7	7	1.3E-01	1647.16	6	6	5.2E-01	2445.11	12	12	1.9E+00
5398.29	5	5	9.8E-02	2208.41	10	10	1.8E+00	2445.80	4	6	1.5E+00
5400.50	9	9	1.8E-01	2213.66	14	14	4.4E-01	2446.47	12	14	2.9E-01
5410.91	7	9	4.8E-01	2218.27	8	10	1.9E+00	2447.20	6	6	1.2E+00
5415.20	11	13	5.6E-01	2327.40	6	4	5.9E-01	2453.98	8	10	7.3E-01
5424.07	13	15	5.0E-01	2331.31	10	8	2.9E-01	2455.71	8	8	1.0E+00
5432.95	5	5	4.1E-02	2332.80	8	6	1.5E+00	2458.78	10	12	2.7E+00
5445.04	11	11	2.0E-01	2338.01	4	4	1.1E+00	2458.97	6	4	2.0E+00
5463.27	9	9	3.2E-01	2343.49	10	8	1.7E+00	2460.44	10	12	5.3E+00
5466.39	9	7	7.5E-02	2343.96	8	6	2.9E-01	2461.28	6	8	2.6E+00
5473.90	7	7	5.5E-02	2344.28	2	4	8.2E-01	2461.86	8	10	2.6E+00
5480.87	3	1	1.2E-01	2348.11	10	8	5.1E-01	2466.52	2	4	2.1E+00
5487.74	7	5	8.6E-02	2348.30	6	6	1.2E+00	2469.51	8	6	2.8E+00
5554.89	9	9	8.7E-02	2351.67	6	6	1.7E+00	2472.61	8	10	3.7E+00
5569.62	5	3	2.1E-01	2352.31	2	4	4.2E+00	2475.12	4	6	3.9E+00
5572.84	7	5	2.1E-01	2353.68	8	8	1.3E+00	2475.54	6	8	3.5E+00
5576.09	3	1	2.1E-01	2354.89	6	4	2.4E-01	2481.05	12	12	1.9E-01
5586.76	9	7	1.9E-01	2360.00	10	10	2.4E-01	2484.44	8	8	2.3E+00
5598.30	5	5	1.8E-01	2360.29	8	6	5.9E-01	2492.34	10	12	1.6E-01
5615.64	11	9	1.7E-01	2362.02	8	8	1.3E-01	2493.26	14	16	3.4E+00
5624.54	5	5	5.3E-02	2363.86	8	10	5.1E+00	2501.31	2	2	1.4E+00
5633.97	11	13	8.7E-02	2364.83	8	8	6.1E-01	2503.87	10	10	2.4E+00
5638.27	9	7	4.0E-02	2365.77	6	6	2.1E+00	2508.34	8	10	2.7E+00
5650.01	3	5	5.0E-02	2366.59	6	6	9.9E-02	2533.63	12	12	1.3E+00
5655.18	7	9	5.3E-02	2368.60	6	4	5.9E-01	2534.42	8	8	1.2E+00
5658.82	7	7	3.6E-02	2369.95	10	12	5.7E+00	2535.36	6	4	3.3E+00

λ Å	Weights g_i	Weights g_k	A 10^8 s^{-1}	λ Å	Weights g_i	Weights g_k	A 10^8 s^{-1}	λ Å	Weights g_i	Weights g_k	A 10^8 s^{-1}
2535.49	10	8	5.4E-01	2592.78	14	16	2.1E+00	2712.39	10	12	1.3E-01
2536.67	12	12	4.0E-01	2593.72	2	4	1.3E-01	2714.41	8	6	5.5E-01
2537.14	10	10	1.4E+00	2594.96	8	8	1.0E-01	2716.22	6	6	1.1E-01
2538.20	14	12	1.2E+00	2598.37	8	6	1.3E+00	2716.56	14	12	1.6E-01
2538.50	8	6	3.3E-01	2599.40	10	10	2.2E+00	2717.87	16	14	1.4E+00
2538.80	12	10	8.2E-01	2604.05	8	8	1.1E-01	2718.64	10	8	1.3E+00
2538.91	10	8	7.8E-01	2605.04	6	8	2.1E+00	2719.30	6	8	3.7E-01
2538.99	14	12	1.2E+00	2605.34	4	4	1.6E+00	2722.06	8	8	1.1E-01
2540.52	2	2	1.5E+00	2605.42	6	6	2.6E-01	2722.74	6	8	7.8E-01
2541.10	8	6	7.3E-01	2605.90	4	2	1.2E+00	2724.88	6	6	9.7E-01
2541.84	8	6	7.7E-01	2606.51	6	6	1.8E+00	2727.38	12	10	3.2E-01
2542.73	2	2	1.9E+00	2607.09	6	4	1.7E+00	2727.54	6	4	8.5E-01
2543.38	10	12	4.4E-01	2609.13	8	10	3.0E-01	2728.91	8	10	8.8E-02
2543.43	6	4	7.1E-01	2609.87	8	8	1.8E-01	2730.73	4	4	2.5E-01
2544.97	4	6	4.0E-01	2611.87	8	8	1.1E+00	2732.94	8	6	7.8E-01
2545.22	8	10	3.3E-01	2613.82	4	2	2.0E+00	2739.55	8	8	1.9E+00
2545.44	8	10	1.4E-01	2617.62	6	6	4.4E-01	2741.40	6	6	1.7E-01
2546.67	8	8	6.2E-01	2619.07	10	10	2.7E-01	2743.20	2	4	1.8E+00
2547.34	8	8	2.0E-01	2620.17	6	6	1.3E-01	2746.48	4	6	1.9E+00
2548.33	4	6	2.0E-01	2620.70	8	8	3.3E-01	2746.98	6	6	1.6E+00
2548.59	10	10	1.9E-01	2621.67	2	2	4.9E-01	2749.18	4	4	1.1E+00
2548.74	4	2	1.7E+00	2623.11	14	14	1.1E-01	2749.32	6	8	2.1E+00
2548.92	12	10	4.8E-01	2623.73	6	6	2.2E-01	2749.49	2	2	1.1E+00
2549.08	10	8	1.5E+00	2625.49	12	14	2.2E+00	2753.29	10	12	1.2E+00
2549.40	4	4	1.3E+00	2625.67	8	10	3.4E-01	2754.91	8	6	8.4E-01
2549.46	6	6	8.0E-01	2626.50	4	6	3.4E-01	2755.73	8	10	2.1E+00
2549.77	8	6	2.5E-01	2628.29	2	4	8.6E-01	2761.81	2	4	1.1E-01
2550.03	10	10	1.2E+00	2629.59	6	8	6.2E-01	2762.34	6	6	3.7E-01
2550.15	8	10	4.0E-01	2630.07	4	6	5.7E-01	2763.66	14	12	1.3E+00
2550.68	12	12	8.9E-01	2631.05	4	6	7.7E-01	2765.13	10	8	1.2E+00
2551.21	10	8	3.2E-01	2631.32	6	8	6.0E-01	2767.50	12	14	1.9E+00
2555.07	6	8	1.8E-01	2631.61	10	12	5.3E-01	2769.36	12	14	1.6E-01
2555.45	4	6	2.5E-01	2633.20	6	4	1.7E+00	2774.69	2	4	2.4E-01
2557.51	10	8	1.3E-01	2636.69	4	4	1.2E-01	2776.91	8	8	3.0E-01
2559.77	6	8	2.4E-01	2637.50	6	6	5.2E-01	2779.30	10	8	7.6E-01
2559.92	6	8	2.4E-01	2637.64	2	4	8.3E-01	2779.91	2	4	2.3E-01
2560.28	4	4	1.5E+00	2639.56	2	2	1.1E+00	2780.04	2	2	2.9E-01
2562.09	4	2	1.5E+00	2642.01	6	6	3.6E-01	2783.69	12	10	7.0E-01
2562.54	8	6	1.5E+00	2649.47	6	8	1.8E+00	2785.19	12	10	1.0E+00
2563.48	6	4	1.3E+00	2650.48	6	8	1.6E+00	2787.24	8	6	1.3E-01
2566.22	8	10	2.5E+00	2654.63	4	4	7.7E-01	2793.89	10	12	9.6E-02
2566.40	8	6	2.1E+00	2658.25	8	8	3.2E-01	2796.63	10	10	1.0E-01
2566.91	4	2	1.1E+00	2662.56	2	2	9.6E-01	2799.29	10	8	1.1E-01
2568.41	2	4	4.4E-01	2664.66	8	10	1.5E+00	2809.78	8	8	1.6E-01
2569.78	2	4	1.2E+00	2666.64	6	8	1.7E+00	2817.09	6	4	2.1E-01
2570.53	6	8	1.2E+00	2667.22	4	6	9.2E-01	2831.56	4	6	5.8E-01
2570.85	8	6	1.7E+00	2669.93	2	4	4.7E-01	2833.09	6	6	2.7E-01
2573.21	8	10	1.4E-01	2671.40	2	4	5.6E-01	2835.71	4	6	3.1E-01
2574.36	6	4	1.6E+00	2682.51	8	10	7.0E-01	2838.22	4	2	4.2E-01
2576.86	10	12	1.1E+00	2683.00	4	6	6.4E-01	2839.51	10	8	9.9E-01
2577.92	2	2	1.3E+00	2684.75	8	10	1.4E+00	2839.80	8	10	4.1E-01
2582.41	6	8	2.4E-01	2692.60	10	12	1.2E+00	2840.65	2	4	5.3E-01
2582.58	4	4	7.7E-01	2697.33	4	4	2.7E-01	2840.76	10	12	1.1E-01
2585.63	10	10	3.6E-01	2697.46	4	2	1.8E+00	2844.96	2	2	4.5E-01
2585.88	10	8	8.1E-01	2699.20	4	4	6.6E-01	2847.77	4	4	3.3E-01
2587.95	8	10	1.4E+00	2703.99	8	8	1.2E+00	2848.11	6	6	7.0E-01
2588.18	2	2	1.6E-01	2707.13	4	6	8.5E-01	2848.32	6	4	1.1E+00
2590.55	4	6	9.1E-02	2709.05	4	6	3.7E-01	2855.69	8	10	1.0E-01
2591.54	6	6	5.1E-01	2711.84	12	14	3.8E-01	2856.38	6	8	2.7E-01

λ Å	g_i	g_k	A 10^8 s^{-1}	λ Å	g_i	g_k	A 10^8 s^{-1}	λ Å	g_i	g_k	A 10^8 s^{-1}
2856.91	8	8	8.7E-01	**Fe VII**				175.266	2	4	1.72E+03
2857.17	6	8	9.5E-02	150.807	5	7	1.3E+03				
2872.39	10	8	1.5E-01	150.852	7	9	1.3E+03	**Fe XI**			
2873.40	8	10	3.4E-01	151.023	9	11	1.6E+03	72.166	5	7	2.9E+03
2875.35	8	10	9.5E-02	151.046	7	7	2.2E+02	72.310	5	5	1.5E+03
2883.71	12	14	1.0E-01	151.145	9	9	2.1E+02	72.635	5	7	1.6E+03
2884.77	6	8	1.4E-01	151.432	5	7	2.2E+02	91.394	5	7	2.6E+03
2895.22	8	10	8.0E-02	151.512	5	5	5.3E+02	91.472	7	9	2.5E+03
2897.27	6	4	1.4E-01	151.675	7	7	3.9E+02	91.63	3	5	2.3E+03
2944.40	4	2	4.6E-01	151.782	9	9	2.4E+02	91.63	7	9	3.4E+03
2947.66	6	4	2.0E-01	154.307	3	1	8.9E+02	91.63	5	7	2.8E+03
2949.18	10	8	2.0E-01	154.335	5	7	1.2E+03	91.733	9	11	4.1E+03
2959.84	8	6	1.6E-01	154.363	3	3	4.2E+02	92.81	9	11	3.7E+03
2964.63	2	2	9.3E-02	154.565	5	3	3.5E+02	92.87	11	13	3.9E+03
2969.93	8	6	1.8E-01	154.650	5	5	8.8E+02	93.433	9	11	3.2E+03
2982.06	4	6	2.1E-01	154.848	1	3	7.7E+02	179.762	5	7	1.67E+03
2984.82	6	6	3.6E-01	154.921	3	5	9.7E+02				
2985.55	2	4	1.8E-01	154.941	3	3	2.4E+02	**Fe XII**			
2997.30	6	8	8.3E-02	154.949	5	7	1.0E+03	65.905	4	4	2.0E+03
3002.65	4	6	1.4E-01	155.994	9	11	1.8E+03	66.526	6	8	1.7E+03
3036.96	6	6	1.6E-01	158.481	9	9	2.3E+02	66.960	4	6	1.6E+03
3048.99	4	4	2.8E-01	165.087	1	3	6.9E+02	67.164	4	2	1.1E+03
3062.23	12	10	1.2E-01	165.919	7	5	2.8E+03	67.821	4	6	1.4E+03
3071.12	2	4	1.9E-01	166.365	9	7	2.9E+03	68.382	2	4	1.7E+03
3076.44	4	6	2.8E-01	173.441	9	9	3.6E+03	80.541	6	6	8.7E+02
3077.17	14	12	1.1E-01	176.744	9	9	2.7E+03	81.943	6	4	1.4E+03
3078.68	6	8	4.2E-01	176.928	7	7	2.4E+03	82.226	4	2	1.9E+03
3135.36	6	6	8.4E-02	177.172	5	5	1.5E+03	84.48	4	6	4.5E+03
3154.20	10	10	1.5E-01	235.221	5	3	1.7E+02	84.48	8	10	4.9E+03
3167.86	8	8	1.3E-01	240.053	3	1	1.3E+02	84.52	10	12	5.2E+03
3177.54	8	8	8.1E-02	243.379	9	7	2.1E+02	84.52	6	8	4.0E+03
3179.50	6	8	9.9E-02					84.85	6	8	2.3E+03
5247.95	4	6	1.7E+00					85.14	8	10	3.4E+03
5506.20	12	14	1.4E+00	**Fe VIII**				85.477	10	12	4.6E+03
5961.71	10	12	7.7E-01	112.472	4	4	3.6E+02	186.880	6	8	1.0E+03
				112.486	6	6	4.3E+02	192.394	4	2	9.0E+02
Fe III				116.196	4	6	4.5E+02	193.509	4	4	9.1E+02
1843.4	9	7	4.8E+00	117.197	6	8	3.8E+02	195.119	4	6	8.6E+02
1844.3	7	5	4.9E+00	167.486	4	4	3.0E+03				
1846.9	5	3	5.5E+00	168.172	6	6	3.1E+03	**Fe XIII**			
1854.38	3	1	5.7E+00	168.545	6	4	2.0E+03	62.353	1	3	2.0E+03
1865.20	7	7	6.1E+00	168.929	4	2	2.1E+03	62.46	5	7	1.2E+03
1893.98	11	9	5.5E+00	185.213	6	8	1.0E+03	62.699	3	5	2.3E+03
1896.80	13	11	5.0E+00	186.601	4	6	9.4E+02	63.188	5	7	3.9E+03
1904.3	5	5	5.7E+00					64.139	1	3	2.1E+03
1907.58	15	13	5.3E+00	**Fe X**				74.845	5	5	1.0E+03
1915.08	13	15	6.0E+00	76.822	2	2	1.8E+03	75.892	5	3	7.7E+02
1922.79	11	13	5.5E+00	77.865	4	6	1.6E+03	76.117	5	3	2.1E+03
1930.39	9	11	5.1E+00	100.026	8	10	2.6E+03	78.452	9	11	6.3E+03
1931.51	9	11	5.3E+00	101.733	6	8	1.8E+03	84.270	7	9	5.5E+03
1937.35	7	9	5.1E+00	101.846	4	6	1.7E+03	107.384	7	5	1.8E+03
1943.48	5	7	5.0E+00	102.095	10	12	2.9E+03				
1950.33	13	15	5.5E+00	102.192	10	12	2.9E+03	**Fe XIV**			
1951.01	11	11	5.3E+00	102.829	4	6	2.1E+03	58.963	2	4	2.7E+03
1952.65	9	9	4.9E+00	103.319	6	8	2.6E+03	59.579	4	6	3.1E+03
1953.32	7	7	5.1E+00	103.724	6	8	1.7E+03	69.176	4	6	5.6E+02
1987.50	13	13	4.9E+00	104.638	8	10	2.1E+03	69.386	2	4	7.6E+02
				174.534	4	6	1.8E+03				

λ Å	Weights g_i	g_k	A 10^8 s^{-1}	λ Å	Weights g_i	g_k	A 10^8 s^{-1}	λ Å	Weights g_i	g_k	A 10^8 s^{-1}
69.66	2	2	8.9E+02	248	3	1	5.4E+02	12.264	1	3	5.9E+04
69.66	6	6	1.3E+03	284.160	1	3	2.28E+02	12.526	1	3	3.0E+03
70.251	6	4	8.1E+02					12.681	1	3	3.5E+03
70.613	4	2	1.7E+03	**Fe XVI**				13.823	1	3	3.3E+04
72.80	10	12	7.9E+03	31.041	2	4	5.2E+02	13.891	1	3	3.4E+03
76.022	4	6	6.6E+03	31.242	4	6	6.1E+02	15.015	1	3	2.28E+05
76.152	6	8	7.0E+03	32.166	2	4	6.8E+02	15.262	1	3	6.0E+04
91.009	6	4	5.1E+02	32.192	2	2	6.7E+02	16.777	1	3	8.29E+03
91.273	4	2	5.6E+02	32.433	2	4	7.7E+02	17.054	1	3	9.33E+03
188	4	6	2.7E+02	32.652	4	6	9.1E+02	41.37	9	11	4.8E+03
190	6	8	2.8E+02	34.857	2	4	1.23E+03	49.427	3	3	4.0E+03
207	2	2	2.1E+02	35.106	4	6	1.44E+03	50.26	7	9	6.0E+03
211.316	2	4	3.6E+02	35.333	4	6	6.4E+02	58.76	9	11	1.2E+04
213	4	2	2.8E+02	35.368	6	8	6.8E+02				
214	2	2	4.0E+02	36.01	4	2	5.0E+02	**Fe XIX**			
216	6	8	1.7E+02	36.749	2	4	1.1E+03	13.413	5	3	1.3E+04
217	6	8	4.0E+02	36.803	2	2	1.2E+03	13.426	5	7	4.8E+04
217	6	6	2.6E+02	37.096	4	6	1.0E+03	13.47	3	1	1.5E+05
219	2	4	4.8E+02	37.138	6	8	1.07E+03	13.520	5	7	2.0E+05
219	4	6	2.4E+02	39.827	2	4	2.1E+03	13.56	3	5	1.0E+04
219.123	4	6	3.9E+02	40.153	4	6	2.5E+03	13.68	3	1	8.0E+04
220	4	4	3.2E+02	40.161	4	4	4.1E+02	13.69	5	7	2.3E+04
221	4	6	5.9E+02	40.199	4	6	1.7E+03	13.700	1	3	2.7E+05
226	2	4	3.9E+02	40.245	6	8	1.8E+03	13.71	5	5	2.2E+04
234	2	2	2.8E+02	41.91	2	2	4.72E+02	13.738	5	7	1.0E+04
264.787	4	4	3.38E+02	42.30	4	2	9.2E+02	13.796	5	7	7.0E+04
265	4	4	1.5E+02	46.661	4	6	3.46E+03	13.83	5	5	1.4E+04
266	6	4	1.7E+02	46.718	6	8	3.7E+03	13.934	1	3	4.51E+04
268	6	6	2.1E+02	50.350	2	4	1.86E+03	13.961	3	3	2.0E+04
268	4	2	3.3E+02	50.555	2	2	1.98E+03	14.668	5	7	1.1E+04
270.524	4	2	2.1E+02	54.142	2	4	3.41E+03	14.671	5	3	1.1E+04
274.203	2	2	1.8E+02	54.728	4	6	4.16E+03	14.929	3	3	1.2E+04
280	4	6	2.8E+02	54.769	4	4	6.97E+03	14.966	5	3	2.5E+04
283	6	8	2.7E+02	62.879	2	2	1.05E+03	14.995	5	5	2.2E+04
288.45	6	4	1.6E+02	63.719	4	2	2.18E+03	15.015	1	3	1.4E+04
				66.263	4	6	9.39E+03	16.668	3	1	1.1E+04
Fe XV				66.368	6	8	1.00E+04				
38.95	1	3	1.69E+03	66.392	6	6	6.69E+02	**Fe XX**			
52.911	1	3	2.94E+03	76.502	6	4	6.7E+02	12.67	6	6	1.0E+04
59.404	3	5	3.4E+03	76.796	4	2	7.72E+02	12.69	4	6	1.2E+04
63.959	5	7	1.6E+03	80.192	4	6	5.2E+02	12.73	4	2	4.0E+04
65.370	1	3	3.2E+02	80.270	6	8	5.4E+02	12.77	4	4	2.1E+05
65.612	3	3	9.8E+02	85.587	2	4	4.0E+02	12.78	4	2	6.9E+04
66.238	5	3	1.6E+03	86.133	4	6	4.8E+02	12.78	2	4	1.4E+05
68.860	9	11	9.2E+03	96.256	4	6	8.7E+02	12.79	6	4	1.7E+04
69.7	3	1	1.9E+03	96.348	6	8	9.3E+02	12.82	4	4	1.1E+05
69.942	3	5	7.4E+03	117.2	2	4	3.93E+02	12.88	6	4	2.7E+04
69.989	5	7	7.9E+03	117.7	2	2	3.9E+02	12.89	4	4	4.4E+04
70.052	7	9	8.8E+03	123.4	2	4	5.9E+02	12.90	4	2	6.2E+03
70.224	1	3	4.13E+03	124.5	4	6	7.0E+02	12.90	4	6	1.4E+05
70.53	7	5	2.6E+02	144.06	4	6	1.6E+03	12.92	2	4	1.7E+04
70.59	7	7	1.7E+03	144.25	6	8	1.6E+03	12.93	4	6	1.6E+05
73.199	7	9	8.8E+03	148	4	2	6.5E+02	12.93	2	2	1.2E+04
73.473	5	7	6.2E+03	266.7	4	6	3.9E+02	12.98	2	2	6.7E+04
233.857	5	7	2.2E+02	267.0	6	8	4.3E+02	12.99	6	6	5.1E+04
235	1	3	2.5E+02					13.00	6	4	1.1E+04
243	1	3	2.4E+02	**Fe XVII**				13.01	2	4	3.0E+04
243	5	7	2.3E+02	11.023	1	3	2.1E+04	13.03	4	2	8.6E+04
243.790	3	5	4.2E+02	12.123	1	3	8.0E+04	13.07	6	4	8.2E+03

λ Å	Weights g_i	g_k	A 10^8 s^{-1}	λ Å	Weights g_i	g_k	A 10^8 s^{-1}	λ Å	Weights g_i	g_k	A 10^8 s^{-1}
13.13	2	4	8.9E+04	13.14	3	1	2.0E+04	11.325	3	5	1.7E+05
13.24	4	4	1.2E+04	13.41	1	3	7.3E+03	11.338	3	3	9.3E+04
13.28	4	4	6.1E+03					11.429	3	1	1.7E+05
13.70	4	6	1.1E+04	**Fe XXII**				11.433	3	3	1.2E+05
13.71	2	2	9.9E+03	9.002	4	6	5.5E+04	11.441	5	7	2.2E+05
13.78	4	4	1.0E+04	9.006	6	8	5.7E+04	11.445	5	5	5.6E+04
13.79	6	6	1.2E+04	9.006	6	6	5.3E+04	11.485	3	5	1.40E+05
13.83	4	2	9.8E+03	9.163	4	6	6.9E+04	11.491	5	3	5.9E+04
13.90	4	2	1.2E+04	9.183	6	8	8.3E+04	11.519	5	5	1.16E+05
13.98	6	4	1.6E+04	9.241	4	6	5.1E+04	11.520	1	3	2.16E+05
13.99	4	2	2.2E+04	11.748	4	4	1.2E+05	11.524	5	7	2.3E+05
14.05	4	4	1.7E+04	11.748	4	6	1.6E+05	11.593	5	7	3.58E+05
14.23	2	2	6.3E+03	11.748	4	2	1.8E+05	11.613	3	5	1.0E+05
				11.763	2	4	1.6E+05	11.615	3	3	4.4E+04
Fe XXI				11.789	2	2	2.6E+05	11.691	5	7	7.7E+04
8.53	3	1	1.8E+04	11.789	6	8	1.2E+05	11.698	5	5	7.3E+04
8.53	3	5	6.1E+03	11.797	2	4	1.7E+05	11.737	3	5	1.8E+05
8.53	3	3	1.5E+04	11.823	6	4	7.9E+04	11.898	1	3	2.03E+05
8.56	5	7	2.0E+04	11.837	6	8	2.3E+05				
8.56	1	3	2.1E+04	11.837	6	6	1.7E+05	**Fe XXIV**			
8.56	5	3	6.5E+03	11.886	4	6	1.3E+05	1.8523	2	2	1.0E+05
8.64	5	7	1.5E+04	11.898	2	4	8.2E+04	1.8552	2	4	4.82E+06
8.65	5	7	3.9E+04	11.922	4	6	1.8E+05	1.8563	4	2	2.43E+06
8.66	5	5	4.4E+03	11.976	6	8	5.9E+04	1.8572	2	2	3.06E+06
8.74	1	3	2.5E+04	12.027	2	4	6.9E+04	1.858	2	4	1.2E+05
9.42	3	1	4.3E+04	12.045	6	8	2.4E+05	1.8614	4	4	6.24E+06
9.42	3	3	3.3E+04	12.045	4	4	9.7E+04	1.8626	2	4	3.16E+06
9.44	3	5	1.7E+04	12.053	4	6	6.1E+04	1.8627	2	2	5.47E+06
9.45	1	3	5.2E+04	12.077	2	4	1.0E+05	1.8637	2	2	1.91E+06
9.46	5	3	1.5E+04	12.077	4	6	2.4E+05	1.8655	4	6	2.14E+06
9.47	5	7	4.9E+04	12.095	6	6	7.8E+04	1.8672	4	2	1.63E+06
9.47	5	5	6.1E+03	12.193	2	4	7.2E+04	1.8678	4	4	3.5E+05
9.52	3	3	8.1E+03	12.193	4	6	9.9E+04	1.8721	4	6	3.2E+05
9.58	5	5	5.2E+03	12.325	2	2	1.5E+05	1.8721	2	2	2.0E+05
9.59	5	5	1.0E+04					1.8730	2	4	1.5E+05
9.67	1	3	5.7E+04	**Fe XXIII**				1.8739	4	4	8.3E+04
9.68	5	7	4.0E+03	7.733	5	7	3.0E+04	1.891	2	2	9.7E+04
9.74	5	3	5.3E+03	7.849	5	7	4.9E+04	1.897	4	2	9.8E+04
12.02	1	3	1.3E+04	8.307	1	3	4.8E+04	8.231	2	4	6.10E+04
12.13	3	3	1.8E+04	8.529	1	3	4.3E+04	8.316	4	6	7.07E+04
12.18	5	7	2.2E+04	8.550	3	5	6.0E+04	10.619	2	4	7.28E+04
12.19	5	3	6.4E+03	8.552	3	3	3.2E+04	10.663	2	2	7.51E+04
12.21	3	1	1.5E+05	8.614	5	7	7.7E+04	11.030	2	4	1.84E+05
12.21	3	3	1.2E+05	8.664	3	3	4.4E+04	11.171	4	6	2.18E+05
12.25	1	3	2.1E+05	8.669	5	7	6.1E+04				
12.28	5	3	5.2E+04	8.672	1	3	6.8E+04	**Fe XXV**			
12.30	5	7	2.1E+05	8.752	5	7	1.2E+05	1.4607	1	3	2.54E+05
12.36	3	3	3.6E+04	8.764	5	7	4.6E+04	1.4945	1	3	5.05E+05
12.37	5	7	3.1E+05	8.814	3	5	6.2E+04	1.5730	1	3	1.24E+06
12.38	5	3	6.9E+03	10.902	5	5	5.3E+04	1.5749	1	3	1.5E+05
12.47	5	7	5.8E+04	10.910	3	1	6.7E+04	1.778	3	3	8.7E+04
12.47	5	3	1.3E+04	10.927	5	7	6.0E+04	1.782	3	1	4.69E+06
12.49	5	7	1.3E+04	10.934	3	5	5.4E+04	1.787	1	3	2.57E+06
12.53	5	5	1.5E+04	10.979	1	3	7.9E+04	1.787	5	5	1.19E+06
12.57	1	3	7.2E+04	11.018	1	3	4.9E+04	1.788	3	5	2.68E+06
12.73	5	5	8.2E+03	11.086	3	1	6.5E+04	1.788	3	5	1.63E+06
12.95	3	5	6.2E+03	11.165	3	5	6.7E+04	1.789	1	3	1.78E+06
13.00	1	3	7.2E+03	11.255	3	3	3.7E+04	1.790	3	3	1.23E+06
13.03	5	5	1.3E+04	11.298	1	3	1.3E+05	1.791	3	5	4.10E+06

λ Å	Weights g_i	g_k	A 10^8 s^{-1}	λ Å	Weights g_i	g_k	A 10^8 s^{-1}	λ Å	Weights g_i	g_k	A 10^8 s^{-1}
1.791	3	3	2.59E+06	4619.2	4	6	8.1E-01	**Magnesium**			
1.792	3	1	4.92E+06	4633.9	4	6	7.1E-01	**Mg I**			
1.792	5	5	2.81E+06	4658.9	6	4	6.5E-01	2025.8	1	3	8.4E-01
1.793	3	1	2.67E+06	4739.0	6	6	7.6E-01	*2779.8	9	9	5.2E+00
1.794	5	3	2.22E+06	4762.4	2	4	4.2E-01	*2850.0	9	15	2.3E-01
1.797	3	5	8.8E+05	4765.7	4	6	6.7E-01	2852.1	1	3	4.95E+00
1.798	3	3	1.0E+05	4811.8	2	4	1.7E-01	*3094.9	9	15	5.2E-01
1.800	1	3	8.6E+04	4825.2	2	4	1.9E-01	3329.9	1	3	3.3E-02
1.802	3	1	4.1E+05	4832.1	4	2	7.3E-01	3332.2	3	3	9.7E-02
1.810	3	1	5.9E+05	5208.3	4	4	1.4E-01	3336.7	5	3	1.6E-01
1.8502	1	3	4.57E+06	5308.7	4	6	2.4E-02	*3835.3	9	15	1.68E+00
1.8593	1	3	4.42E+05	7407.0	6	6	7.0E-02	4703.0	3	5	2.55E-01
10.038	3	3	8.08E+04					5167.3	1	3	1.16E-01
				Lead				5172.7	3	3	3.46E-01
				Pb I				5183.6	5	3	5.75E-01
Krypton				2022.0	1	3	5.2E-02	5528.4	3	5	1.99E-01
Kr I				2053.3	1	3	1.2E-01				
1164.9	1	3	3.16E+00	2170.0	1	3	1.5E+00	**Mg II**			
1235.8	1	3	3.12E+00	2401.9	3	3	1.9E-01	1239.9	2	4	1.4E-02
4274.0	5	5	2.6E-02	2446.2	3	3	2.5E-01	1240.4	2	2	1.4E-02
4351.4	3	1	3.2E-02	2476.4	3	5	2.8E-01	*2660.8	10	14	3.8E-01
4362.6	5	3	8.4E-03	2577.3	5	3	5.0E-01	2790.8	2	4	4.0E+00
4376.1	3	1	5.6E-02	2613.7	3	3	2.7E-01	2795.5	2	4	2.6E+00
4400.0	3	5	2.0E-02	2614.2	3	5	1.9E+00	2797.9	4	4	7.9E-01
4410.4	3	3	4.4E-03	2628.3	5	3	3.1E-02	2798.1	4	6	4.8E+00
4425.2	3	3	9.7E-03	2657.1	3	5	9.8E-04	2802.7	2	2	2.6E+00
4453.9	3	5	7.8E-03	2663.2	5	5	7.1E-01	2928.8	2	2	1.2E+00
4463.7	3	3	2.3E-02	2802.0	5	7	1.6E+00	2936.5	4	2	2.3E+00
4502.4	3	5	9.2E-03	2823.2	5	5	2.6E-01	*3104.8	10	14	8.1E-01
5562.2	5	5	2.8E-03	2833.1	1	3	5.8E-01	3848.2	6	4	2.8E-02
5570.3	5	3	2.1E-02	2873.3	5	5	3.7E-01	3848.3	4	4	3.0E-03
5649.6	1	3	3.7E-03	3572.7	5	3	9.9E-01	3850.4	4	2	3.0E-02
5870.9	3	5	1.8E-02	3639.6	3	3	3.4E-01	*4481.2	10	14	2.23E+00
6904.7	3	5	1.3E-02	3671.5	5	3	4.4E-01	9218.3	2	4	3.6E-01
7224.1	3	5	1.4E-02	3683.5	3	1	1.5E+00	9244.3	2	2	3.6E-01
7587.4	3	1	5.1E-01	3739.9	5	5	7.3E-01				
7601.5	5	5	3.1E-01	4019.6	5	7	3.5E-02	**Mg IV**			
7685.2	3	1	4.9E-01	4057.8	5	3	8.9E-01	320.99	4	2	1.2E+02
7694.5	5	3	5.6E-02	4062.1	5	3	9.2E-01	323.31	2	2	5.9E+01
7854.8	1	3	2.3E-01	4168.0	5	5	1.2E-02	1219.0	6	6	5.9E+00
8059.5	1	3	1.9E-01	5005.4	1	3	2.7E-01	1375.5	4	4	4.5E+00
8104.4	5	5	1.3E-01	5201.4	1	3	1.9E-01	1459.6	6	4	4.6E+00
8112.9	5	7	3.6E-01	7229.0	5	3	8.9E-03	1495.5	4	6	6.4E+00
8190.1	3	5	1.1E-01					1510.7	4	4	6.7E+00
8263.2	3	5	3.5E-01	**Lithium**				1683.0	6	8	5.8E+00
8281.1	3	3	1.9E-01	**Li I**				1698.8	4	6	3.9E+00
8298.1	3	3	3.2E-01	*2741.2	2	6	1.3E-02	1893.9	6	6	2.8E+00
8508.9	3	3	2.4E-01	*3232.7	2	6	1.17E-02				
8776.7	3	5	2.7E-01	*4602.9	6	10	2.23E-01	**Mg VI**			
8928.7	5	3	3.7E-01	*6103.6	6	10	6.860E-01	*269.92	10	6	3.1E+02
				*6707.8	2	6	3.691E-01	*292.53	6	6	9.0E+01
Kr II								*314.64	6	2	1.8E+02
4250.6	4	4	1.2E-01	**Lutetium**				*349.15	10	10	6.1E+01
4292.9	4	4	9.6E-01	**Lu I**				*387.94	6	10	1.3E+01
4355.5	6	8	1.0E+00	3376.5	4	4	2.23E+00	399.29	4	2	2.8E+01
4431.7	2	2	1.8E+00	3567.8	4	6	5.9E-01	400.68	4	4	2.8E+01
4436.8	2	4	6.6E-01	3620.3	6	4	1.1E-02	403.32	4	6	2.7E+01
4577.2	6	8	9.6E-01	3841.2	6	6	2.5E-01				
4583.0	6	4	7.6E-01	4518.6	4	4	2.1E-01				
4615.3	4	4	5.4E-01								

λ Å	Weights g_i	Weights g_k	A 10^8 s^{-1}	λ Å	Weights g_i	Weights g_k	A 10^8 s^{-1}	λ Å	Weights g_i	Weights g_k	A 10^8 s^{-1}
Mg VII				3007.65	6	8	1.8E-01	3773.86	12	12	2.5E-01
277.01	3	3	9.5E+01	3011.38	8	10	3.1E-01	3800.55	6	8	2.7E-01
278.41	5	3	1.5E+02	3016.45	10	12	2.9E-01	3806.72	10	12	5.9E-01
280.74	5	3	2.0E+02	3043.36	8	8	5.9E-01	3823.51	8	10	5.21E-01
319.02	5	5	8.9E+01	3044.57	10	8	5.7E-01	3823.89	6	6	2.31E-01
*366.42	9	9	4.4E+01	3045.59	10	10	6.7E-01	3833.87	4	4	3.14E-01
*433.04	9	15	1.6E+01	3045.80	8	10	1.7E-01	3834.37	6	8	4.29E-01
1334.3	5	5	5.3E+00	3047.03	12	12	6.1E-01	3839.78	2	2	4.64E-01
1410.0	5	5	2.57E+00	3054.36	8	6	4.6E-01	3841.07	4	6	3.3E-01
1487.0	3	5	3.02E+00	3070.27	6	6	1.9E-01	3843.99	2	4	2.11E-01
1487.9	5	7	3.66E+00	3073.18	4	4	3.7E-01	3889.46	12	14	3.1E-01
				3082.71	14	14	2.9E-01	3898.37	6	8	1.7E-01
Mg VIII				3110.68	6	8	2.7E-01	3899.34	4	6	2.4E-01
*74.976	6	10	4.3E+03	3113.80	12	10	2.6E-01	3924.08	2	4	9.4E-01
315.02	4	4	1.2E+02	3118.10	4	6	1.7E-01	3926.48	6	8	5.4E-01
*342.29	10	6	6.3E+01	3122.88	10	10	1.9E-01	3951.98	2	2	3.1E-01
353.86	4	4	3.89E+01	3126.85	8	6	2.3E-01	3952.84	6	6	4.1E-01
356.00	6	4	5.7E+01	3132.28	10	10	2.1E-01	3975.88	2	4	1.8E-01
*428.52	10	10	3.24E+01	3132.79	8	8	2.7E-01	3982.16	4	2	3.5E-01
*434.62	6	10	1.6E+01	3175.58	8	10	1.8E-01	3982.58	6	4	2.3E-01
*489.33	6	6	3.9E+01	3201.11	4	6	2.2E-01	3982.90	6	4	5.5E-01
*686.92	6	10	9.4E+00	3228.09	10	12	6.4E-01	3991.60	2	2	2.1E-01
				3230.23	10	12	1.9E-01	4011.91	8	8	2.3E-01
Mg IX				3230.72	8	8	3.5E-01	4018.11	10	8	2.54E-01
62.751	1	3	2.87E+03	3240.88	6	4	2.2E-01	4030.76	6	8	1.7E-01
*67.189	9	15	6.20E+03	3243.78	6	6	5.3E-01	4033.07	6	6	1.65E-01
*71.965	9	3	1.22E+03	3251.13	4	2	2.3E-01	4034.49	6	4	1.58E-01
72.312	3	5	4.43E+03	3252.95	4	4	1.8E-01	4041.36	10	10	7.87E-01
77.737	3	1	3.92E+02	3256.14	4	6	5.0E-01	4048.75	6	4	7.5E-01
368.07	1	3	5.27E+01	3258.41	2	2	9.7E-01	4052.48	6	8	3.8E-01
438.69	3	1	7.9E+01	3260.24	2	4	3.8E-01	4055.55	8	8	4.31E-01
*443.74	9	9	4.19E+01	3267.79	14	14	3.5E-01	4058.94	4	2	7.25E-01
749.55	3	5	8.2E+00	3268.72	6	8	3.3E-01	4061.74	8	6	1.9E-01
1639.8	3	5	2.1E+00	3270.35	12	12	2.6E-01	4063.53	6	6	1.69E-01
2814.2	1	3	3.35E-01	3273.02	10	10	2.7E-01	4065.08	12	14	2.5E-01
				3298.23	6	4	2.8E-01	4066.24	10	8	2.2E-01
Mg X				3303.28	4	4	1.9E-01	4070.28	2	2	2.3E-01
57.876	2	4	2.09E+03	3463.66	8	8	3.2E-01	4079.42	2	4	3.8E-01
57.920	2	2	2.09E+03	3470.01	6	8	2.4E-01	4082.95	4	6	2.95E-01
63.152	2	4	5.6E+03	3511.83	12	12	2.7E-01	4083.63	6	8	2.8E-01
63.295	4	6	6.7E+03	3535.30	10	10	1.7E-01	4089.94	8	10	1.7E-01
609.79	2	4	7.53E+00	3559.81	6	6	2.1E-01	4105.37	10	8	1.7E-01
624.94	2	2	7.01E+00	3577.87	10	8	9.4E-01	4135.03	12	12	3.0E-01
2212.5	2	4	9.64E-01	3595.11	6	4	1.8E-01	4141.06	10	10	2.6E-01
2278.7	2	2	8.82E-01	3601.27	12	10	2.3E-01	4148.80	8	8	2.3E-01
5918.7	2	4	3.20E-02	3607.53	8	8	2.3E-01	4176.61	14	12	2.4E-01
6229.6	4	6	3.30E-02	3608.49	6	6	3.6E-01	4189.99	12	10	2.0E-01
				3610.30	4	4	4.2E-01	4201.78	10	8	2.3E-01
Mg XI				3635.70	10	8	2.1E-01	4235.30	8	6	9.17E-01
7.310	1	3	1.15E+04	3660.40	12	14	9.1E-01	4239.74	4	2	3.9E-01
7.473	1	3	2.27E+04	3675.67	6	8	2.2E-01	4257.67	2	2	3.7E-01
7.850	1	3	5.50E+04	3676.96	10	12	7.3E-01	4265.93	4	4	4.92E-01
9.169	1	3	1.97E+05	3680.15	12	10	1.9E-01	4281.10	6	6	2.3E-01
				3682.09	8	10	7.6E-01	4411.87	12	10	2.6E-01
Manganese				3684.87	6	8	2.6E-01	4414.89	8	6	2.93E-01
Mn I				3706.08	12	14	1.4E+00	4419.77	10	8	2.1E-01
2794.82	6	8	3.7E+00	3718.92	10	12	9.6E-01	4436.36	6	4	4.37E-01
2798.27	6	6	3.6E+00	3731.94	8	10	1.0E+00	4451.58	8	8	7.98E-01
2801.08	6	4	3.7E+00	3771.44	14	14	1.9E-01	4453.01	4	2	5.44E-01

λ Å	Weights g_i	g_k	A 10^8 s^{-1}	λ Å	Weights g_i	g_k	A 10^8 s^{-1}	λ Å	Weights g_i	g_k	A 10^8 s^{-1}
4455.82	4	6	1.7E-01	1285.10	5	7	1.1E+01	2751.47	7	9	2.54E-01
4457.04	6	4	2.34E-01	1333.87	7	9	1.0E+01	2756.26	5	3	1.18E-01
4457.55	6	6	4.27E-01					2761.53	9	11	2.06E-01
4458.26	6	8	4.62E-01	**Mercury**				2763.02	3	1	4.44E-01
4461.09	8	8	1.7E-01	**Hg I**				2766.25	3	5	1.17E-01
4462.03	8	10	7.00E-01	2536.52	1	3	8.00E-02	2787.83	9	7	2.85E-01
4464.68	6	6	4.39E-01	2652.04	3	5	3.88E-01	2792.96	5	3	1.53E-01
4470.14	4	4	3.00E-01	2655.13	3	5	1.1E-01	2798.02	7	5	1.22E-01
4472.79	2	2	4.35E-01	2752.78	1	3	6.10E-02	2801.47	5	7	1.24E-01
4479.40	8	10	3.4E-01	2856.94	3	1	1.1E-02	2825.68	5	7	2.53E-01
4490.08	2	4	2.49E-01	2893.60	3	3	1.6E-01	2826.75	7	7	4.23E-01
4498.90	4	6	2.49E-01	2925.4	5	3	7.7E-02	2876.54	9	9	2.84E-01
4502.22	6	8	1.86E-01	2967.3	1	3	4.5E-01	2886.60	11	11	4.74E-01
4605.37	10	12	3.6E-01	3021.50	5	7	5.09E-01	2906.06	3	3	8.04E-01
4626.54	12	14	3.6E-01	3023.48	5	5	9.4E-02	2913.52	5	3	1.38E-01
4709.71	8	8	1.72E-01	3027.49	5	5	2.0E-02	2915.38	5	3	7.31E-01
4727.46	6	6	1.7E-01	3125.66	3	5	6.56E-01	2918.84	5	3	3.79E-01
4739.11	4	4	2.40E-01	3341.48	5	3	1.68E-01	2930.39	1	3	1.91E-01
4754.05	6	8	3.03E-01	3650.15	5	7	1.3E+00	2936.50	11	11	2.33E-01
4761.53	2	4	5.35E-01	3654.83	5	5	1.8E-01	2945.43	7	7	3.66E-01
4762.38	8	10	7.83E-01	4046.56	1	3	2.1E-01	2945.66	3	3	4.08E-01
4765.86	4	6	4.1E-01	4077.81	3	1	4.0E-02	2946.01	5	5	1.68E-01
4766.43	6	8	4.6E-01	4108.1	3	1	3.0E-02	2951.45	9	9	1.43E-01
4783.43	8	8	4.01E-01	4339.22	3	5	2.88E-02	2959.48	9	11	1.75E-01
4823.53	10	8	4.99E-01	4347.50	3	5	8.4E-02	2972.96	5	3	2.69E-01
6013.48	4	6	1.72E-01	4358.34	3	3	5.57E-01	2977.27	9	7	3.28E-01
6021.79	8	6	3.32E-01	4916.07	3	1	5.8E-02	2978.28	7	5	1.50E-01
				5025.64	3	3	2.7E-04	2983.04	1	3	2.82E-01
Mn II				5460.75	5	3	4.87E-01	2987.92	3	5	8.43E-01
2593.72	7	7	2.6E+00	5769.59	3	5	2.36E-01	2988.23	5	7	4.28E-01
2605.68	7	5	2.7E+00	6234.4	1	3	5.3E-03	2988.68	7	9	1.61E-01
2933.05	5	3	2.0E+00	6716.4	1	3	4.3E-03	2989.80	9	7	9.27E-01
2939.31	5	5	1.9E+00	6907.5	3	5	2.8E-02	3000.24	9	9	1.40E-01
2949.20	5	7	1.9E+00	7728.8	1	3	9.7E-03	3000.44	5	5	1.25E-01
3441.99	9	7	4.3E-01	10139.79	3	1	2.71E-01	3000.85	5	7	2.58E-01
3460.32	7	5	3.2E-01					3001.43	5	5	2.31E-01
3474.13	5	3	1.5E-01	**Molybdenum**				3007.71	7	5	1.90E-01
3482.90	5	5	2.0E-01	**Mo I**				3013.39	7	5	6.06E-01
3488.68	3	3	2.5E-01	2616.79	3	5	7.34E-01	3016.78	9	9	2.75E-01
				2621.06	7	7	1.16E-01	3025.00	5	5	8.49E-01
Mn VI				2628.96	3	3	2.81E-01	3036.31	3	5	5.81E-01
307.999	9	9	3.7E+01	2629.85	5	7	7.75E-01	3041.70	13	11	5.94E-01
309.440	9	7	5.7E+01	2631.50	1	3	2.54E-01	3046.80	13	11	1.63E-01
309.579	7	5	4.4E+01	2638.30	5	5	7.57E-01	3047.31	11	9	5.01E-01
310.058	7	7	3.4E+01	2640.98	7	5	1.20E+00	3055.32	9	7	4.29E-01
310.182	5	5	2.8E+01	2644.36	5	7	1.96E-01	3057.56	7	5	2.64E-01
311.748	5	3	5.7E+01	2649.46	7	9	9.84E-01	3061.59	7	5	4.41E-01
320.598	3	5	1.5E+01	2655.02	9	7	4.08E-01	3064.27	13	13	8.46E-01
320.681	1	3	2.2E+01	2658.11	7	7	6.43E-01	3065.04	13	13	3.08E-01
320.874	3	1	7.8E+01	2665.09	7	9	1.32E-01	3069.51	5	5	1.52E-01
320.979	3	3	2.2E+01	2679.85	9	11	1.31E+00	3069.96	11	11	2.72E-01
321.176	5	5	6.0E+01	2684.16	9	9	4.18E-01	3070.89	9	11	1.87E-01
321.541	5	3	2.7E+01	2706.11	3	5	2.03E-01	3074.37	11	11	1.42E+00
325.146	9	7	1.3E+02	2710.74	3	3	1.57E-01	3079.88	9	11	9.55E-01
328.431	5	5	4.4E+01	2725.15	3	5	2.79E-01	3080.40	7	9	3.61E-01
328.558	3	5	1.2E+01	2728.71	3	3	1.26E-01	3081.16	3	5	2.35E-01
329.043	1	3	1.1E+01	2733.39	5	7	2.95E-01	3085.62	9	9	1.63E+00
1236.23	5	3	1.3E+01	2743.71	1	3	2.47E-01	3089.13	11	9	1.53E-01
1255.77	3	1	1.2E+01	2745.38	13	11	1.29E-01	3089.71	5	7	2.34E-01

λ Å	Weights g_i	g_k	A 10^8 s^{-1}	λ Å	Weights g_i	g_k	A 10^8 s^{-1}	λ Å	Weights g_i	g_k	A 10^8 s^{-1}
3094.66	7	7	1.63E+00	3285.35	9	7	4.49E-01	3456.52	3	3	2.96E-01
3099.92	9	7	1.45E-01	3287.38	5	5	1.38E-01	3460.22	5	3	2.77E-01
3100.88	7	9	1.20E+00	3289.01	9	9	5.08E-01	3460.78	9	7	6.03E-01
3101.34	5	5	1.92E+00	3290.82	7	5	5.44E-01	3465.84	3	1	9.99E-01
3106.34	7	5	2.21E-01	3305.56	5	3	1.74E-01	3466.19	9	7	2.11E-01
3117.54	13	13	1.89E-01	3305.91	7	9	3.06E-01	3466.96	7	7	1.52E-01
3123.03	3	3	2.81E-01	3307.13	7	9	1.25E-01	3467.85	5	7	2.63E-01
3125.03	5	3	1.98E-01	3312.33	7	5	1.62E-01	3469.22	5	3	6.96E-01
3132.59	7	9	1.79E+00	3323.95	9	7	2.82E-01	3469.63	13	15	1.51E-01
3135.90	9	11	3.68E-01	3325.13	5	3	2.26E-01	3470.92	3	5	2.91E-01
3136.75	9	11	1.57E-01	3325.67	5	5	1.72E-01	3475.03	3	3	4.68E-01
3142.75	3	5	4.10E-01	3327.30	1	3	2.88E-01	3479.42	7	5	2.26E-01
3147.35	13	11	2.41E-01	3336.56	9	9	1.64E-01	3483.67	7	7	1.13E-01
3155.19	7	7	2.75E-01	3340.16	5	3	1.20E-01	3483.83	7	5	1.41E-01
3158.17	7	7	4.63E-01	3344.73	3	5	6.04E-01	3489.43	7	7	3.27E-01
3170.34	7	7	1.37E+00	3346.83	11	11	1.13E-01	3504.41	7	9	8.06E-01
3171.38	5	7	2.03E-01	3347.00	3	3	2.72E-01	3505.31	7	9	2.25E-01
3175.59	13	11	8.40E-01	3358.12	5	7	7.59E-01	3508.11	9	9	1.59E-01
3179.78	11	13	2.33E-01	3361.37	9	9	1.38E-01	3510.77	13	13	4.75E-01
3183.03	11	9	3.98E-01	3363.78	5	7	2.74E-01	3517.55	11	11	5.41E-01
3184.58	7	5	2.77E-01	3363.87	5	7	1.39E-01	3518.21	3	3	3.64E-01
3185.10	7	7	2.54E-01	3373.81	3	3	2.03E-01	3521.38	9	9	1.39E-01
3185.71	5	3	6.10E-01	3375.22	7	7	1.38E-01	3521.41	9	11	6.06E-01
3188.10	7	9	3.45E-01	3375.65	7	9	1.56E-01	3524.65	5	3	3.10E-01
3188.41	5	7	4.40E-01	3378.19	3	1	1.88E-01	3524.98	7	9	2.25E-01
3192.79	9	11	1.88E-01	3378.46	13	13	3.75E-01	3538.92	11	11	2.24E-01
3193.98	7	5	1.53E+00	3379.96	5	5	4.11E-01	3540.57	5	3	4.46E-01
3194.88	9	11	1.75E-01	3382.48	3	3	2.66E-01	3542.17	7	5	4.93E-01
3195.96	9	7	4.10E-01	3384.61	7	9	7.32E-01	3552.71	9	7	3.64E-01
3197.18	1	3	1.47E-01	3385.87	9	11	3.30E-01	3555.64	3	3	3.46E-01
3198.85	15	13	7.22E-01	3389.79	5	7	1.85E-01	3558.09	5	7	5.43E-01
3200.89	3	5	1.82E-01	3392.17	9	9	1.97E-01	3563.75	1	3	1.53E-01
3205.22	1	3	4.27E-01	3393.65	11	11	2.08E-01	3566.05	9	9	2.67E-01
3205.43	9	11	2.55E-01	3404.33	7	7	2.10E-01	3566.74	7	7	1.43E-01
3205.89	9	9	5.35E-01	3413.37	11	11	1.25E-01	3570.64	15	15	7.18E-01
3208.84	7	5	2.77E-01	3415.27	9	9	1.83E-01	3573.88	3	5	3.58E-01
3210.97	7	5	6.94E-01	3415.61	7	9	1.29E-01	3580.54	13	11	5.49E-01
3214.44	9	7	2.01E-01	3416.14	9	11	2.45E-01	3581.88	11	13	3.81E-01
3215.07	3	5	4.20E-01	3418.52	5	3	1.41E-01	3584.25	3	3	1.73E-01
3216.78	15	13	2.10E-01	3419.69	7	7	1.15E-01	3585.57	7	5	3.95E-01
3221.73	3	1	1.41E+00	3420.04	5	5	3.28E-01	3588.95	7	7	1.18E-01
3228.21	5	7	3.85E-01	3422.31	9	9	2.52E-01	3590.74	7	9	2.23E-01
3229.79	9	11	1.44E-01	3425.13	11	11	2.29E-01	3595.55	5	5	2.32E-01
3233.14	13	13	6.33E-01	3427.90	11	13	4.09E-01	3598.88	13	11	5.67E-01
3237.06	7	9	2.95E-01	3434.79	7	7	1.75E-01	3600.73	9	9	2.07E-01
3244.47	5	3	2.80E-01	3435.45	15	15	1.50E+00	3601.88	7	9	1.15E-01
3247.61	5	5	1.71E-01	3437.21	11	9	8.06E-01	3602.94	5	7	2.96E-01
3249.93	5	3	1.87E-01	3438.87	1	3	2.34E-01	3604.07	9	7	3.25E-01
3251.65	3	5	3.05E-01	3441.87	5	3	1.34E-01	3610.61	5	3	1.78E-01
3256.21	5	3	6.89E-01	3442.66	3	3	2.94E-01	3611.99	7	7	1.16E-01
3256.72	3	3	1.31E-01	3445.03	7	9	1.53E-01	3615.16	7	9	1.96E-01
3259.16	11	13	1.62E-01	3445.26	7	5	2.96E-01	3623.22	11	9	5.58E-01
3262.63	7	9	3.62E-01	3445.80	9	9	1.14E-01	3624.46	9	11	5.27E-01
3264.40	11	9	5.42E-01	3447.12	9	11	8.75E-01	3624.62	5	7	1.37E-01
3265.14	5	7	2.60E-01	3447.29	5	3	1.79E-01	3638.20	5	3	3.51E-01
3266.16	9	11	1.95E-01	3449.07	7	9	1.52E-01	3638.21	5	3	3.33E-01
3270.90	7	7	3.59E-01	3449.85	5	7	1.65E-01	3640.62	7	5	1.94E-01
3276.07	11	9	1.18E-01	3452.60	7	7	2.48E-01	3647.84	7	7	2.11E-01
3285.03	1	3	1.41E-01	3456.15	5	5	3.60E-01	3648.70	7	5	1.15E-01

λ Å	Weights g_i	Weights g_k	A 10^8 s^{-1}	λ Å	Weights g_i	Weights g_k	A 10^8 s^{-1}	λ Å	Weights g_i	Weights g_k	A 10^8 s^{-1}
3654.58	3	3	1.80E-01	3834.64	3	5	1.20E-01	4536.80	13	15	5.03E-01
3657.36	5	7	2.03E-01	3846.18	7	7	1.26E-01	4598.23	1	3	1.47E-01
3658.13	9	9	1.86E-01	3847.25	3	1	2.41E-01	4624.23	9	9	1.32E-01
3659.36	7	9	6.70E-01	3848.30	9	9	1.26E-01	4633.08	3	5	2.35E-01
3660.92	3	5	1.34E-01	3851.99	11	9	1.78E-01	4649.06	3	1	1.25E-01
3662.15	7	9	1.45E-01	3864.10	7	7	6.24E-01	4652.24	5	7	1.55E-01
3662.99	11	11	3.48E-01	3866.69	3	5	1.74E-01	4686.08	3	3	1.72E-01
3663.27	7	5	2.30E-01	3867.67	5	3	2.22E-01	4688.21	13	15	1.54E-01
3664.81	11	13	9.54E-01	3869.08	5	3	1.35E-01	4707.25	7	9	3.63E-01
3664.88	1	3	1.92E-01	3874.15	7	5	1.67E-01	4718.86	5	5	2.17E-01
3669.34	9	7	2.16E-01	3902.95	7	5	6.17E-01	4723.05	9	9	1.23E-01
3672.81	9	11	1.95E-01	3909.54	9	7	1.13E-01	4731.44	9	11	4.49E-01
3672.82	9	9	1.13E-01	3911.94	5	5	1.15E-01	4758.50	11	9	3.01E-01
3676.23	3	1	5.22E-01	3915.43	5	5	1.40E-01	4760.18	11	13	4.67E-01
3680.68	11	11	2.96E-01	3916.43	5	3	1.78E-01	4764.11	9	7	2.16E-01
3681.72	9	7	1.68E-01	3919.55	11	13	2.24E-01	4811.05	13	11	4.36E-01
3683.01	3	5	1.20E-01	3955.48	13	11	1.71E-01	4819.25	11	9	2.71E-01
3687.96	5	7	2.12E-01	3973.76	11	13	4.39E-01	4830.51	9	7	4.07E-01
3688.97	11	9	3.26E-01	3977.90	9	7	1.35E-01	4858.39	13	11	1.24E-01
3690.59	11	9	2.07E-01	3980.20	5	3	2.70E-01	4868.02	7	5	3.11E-01
3694.94	5	7	6.36E-01	3991.85	11	9	1.29E-01	5037.18	9	7	1.14E-01
3696.04	11	11	3.59E-01	4010.13	5	3	4.38E-01	5044.36	7	5	1.31E-01
3698.07	7	5	1.48E-01	4021.01	9	11	2.65E-01	5047.70	3	1	2.61E-01
3708.55	7	9	1.28E-01	4051.18	13	11	1.36E-01	5163.18	9	11	2.03E-01
3715.75	9	7	2.38E-01	4062.08	11	9	1.96E-01	5171.06	5	7	1.84E-01
3718.48	5	7	1.34E-01	4069.88	13	11	3.25E-01	5172.94	5	5	4.11E-01
3720.25	7	9	2.86E-01	4076.19	9	9	1.16E-01	5174.18	5	3	5.83E-01
3725.55	7	7	1.60E-01	4084.37	9	7	1.94E-01	5191.45	7	9	1.62E-01
3727.68	9	11	1.51E-01	4102.15	5	3	1.22E-01	5238.21	7	9	3.74E-01
3728.30	7	5	1.55E-01	4107.46	7	5	2.02E-01	5240.87	7	7	3.89E-01
3728.50	7	9	2.20E-01	4120.09	13	15	6.05E-01	5242.80	7	5	2.01E-01
3733.02	7	7	1.45E-01	4131.92	9	11	1.56E-01	5261.53	5	7	1.13E-01
3733.41	13	13	2.80E-01	4148.98	9	11	1.56E-01	5280.85	5	5	1.28E-01
3735.62	11	11	1.66E-01	4157.40	13	11	2.17E-01	5355.52	9	9	1.21E-01
3742.28	7	7	1.56E-01	4157.90	9	11	1.60E-01	5356.46	11	11	2.11E-01
3747.19	5	7	3.07E-01	4185.82	11	13	3.82E-01	5360.51	9	11	6.19E-01
3748.48	9	11	3.95E-01	4188.32	11	13	3.32E-01	5364.28	9	9	2.26E-01
3755.10	3	5	1.41E-01	4194.56	11	11	2.70E-01	5460.50	5	3	3.46E-01
3755.16	9	9	2.48E-01	4232.59	9	11	3.17E-01	5493.76	7	5	2.13E-01
3758.52	9	9	1.22E-01	4240.83	5	5	1.68E-01	5506.49	5	7	3.61E-01
3759.60	9	7	1.82E-01	4246.02	11	13	2.00E-01	5533.03	5	5	3.72E-01
3760.88	9	9	2.16E-01	4251.88	13	11	1.76E-01	5570.44	5	3	3.30E-01
3768.73	9	9	2.88E-01	4254.95	7	9	2.01E-01	5849.71	3	3	3.02E-01
3769.99	7	9	2.46E-01	4269.28	11	11	1.36E-01	5851.50	3	5	1.55E-01
3777.72	13	11	1.66E-01	4276.91	7	9	2.85E-01	5893.36	5	5	2.60E-01
3788.25	7	9	2.87E-01	4277.24	9	11	1.35E-01	5895.93	5	7	3.12E-01
3794.43	9	9	1.22E-01	4317.92	15	15	1.28E-01	5926.37	7	7	1.63E-01
3797.47	7	5	1.48E-01	4325.80	3	3	1.84E-01	5928.88	7	9	5.32E-01
3798.25	7	9	6.90E-01	4326.14	5	7	2.56E-01	7154.11	9	9	3.45E-01
3801.84	9	7	3.16E-01	4340.74	5	7	1.23E-01				
3805.99	5	5	2.44E-01	4381.63	13	13	2.93E-01	**Neodymium**			
3819.78	9	11	1.47E-01	4382.41	11	13	3.83E-01	**Nd II**			
3824.78	5	7	1.40E-01	4409.94	13	13	1.38E-01	3780.4	16	18	1.4E-01
3827.15	7	7	1.94E-01	4411.69	11	11	2.63E-01	3805.4	14	16	6.9E-01
3828.88	7	7	1.35E-01	4434.95	9	9	2.51E-01	3807.2	10	12	4.9E-02
3830.81	5	5	1.83E-01	4446.42	11	11	1.90E-01	3863.3	8	10	1.5E-01
3831.07	7	9	1.20E-01	4457.35	7	7	1.28E-01	3941.5	10	10	6.1E-01
3832.11	9	9	3.05E-01	4474.57	5	5	2.10E-01	3951.2	12	12	6.0E-01
3833.75	9	9	1.70E-01	4491.65	11	11	2.09E-01	3973.3	18	18	6.3E-01

λ Å	g_i	g_k	A 10^8 s⁻¹	λ Å	g_i	g_k	A 10^8 s⁻¹	λ Å	g_i	g_k	A 10^8 s⁻¹
3979.5	10	12	2.7E-01	3464.3	5	5	6.7E-03	6402.2	5	7	5.14E-01
3990.1	16	16	5.2E-01	3466.6	1	3	1.3E-02	6506.5	3	5	3.00E-01
4012.3	18	20	5.5E-01	3472.6	5	7	1.7E-02	6532.9	1	3	1.08E-01
4061.1	16	18	4.4E-01	3498.1	3	5	5.1E-03	6599.0	3	3	2.32E-01
4106.6	14	16	6.8E-02	3501.2	3	3	1.2E-02	6602.9	3	3	5.9E-03
4109.5	14	16	3.7E-01	3510.7	5	3	2.2E-03	6652.1	3	1	2.9E-03
4133.4	14	12	1.5E-01	3515.2	3	5	6.9E-03	6678.3	3	5	2.33E-01
4156.1	12	14	3.4E-01	3520.5	3	1	9.3E-02	6717.0	3	3	2.17E-01
4205.6	18	16	1.8E-01	3593.5	3	5	9.9E-03	6721.1	3	3	4.9E-04
4284.5	18	18	8.5E-02	3593.6	3	3	6.6E-03	6929.5	3	5	1.74E-01
4303.6	8	10	4.7E-01	3600.2	3	3	4.3E-03	7024.1	3	3	1.89E-02
4325.8	16	16	1.6E-01	3633.7	3	1	1.1E-02	7032.4	5	3	2.53E-01
4358.2	14	14	1.5E-01	3682.2	3	5	1.6E-02	7051.3	3	3	3.0E-02
4382.7	12	10	4.0E-02	3685.7	3	3	3.9E-03	7059.1	3	5	6.8E-02
4400.8	10	12	6.8E-02	3701.2	3	5	2.2E-03	7173.9	3	5	2.87E-02
4451.6	12	14	2.5E-01	4536.3	3	3	5.0E-03	7245.2	3	3	9.35E-02
4456.4	16	18	6.4E-02	4702.5	3	3	2.1E-03	7304.8	1	3	2.55E-03
4463.0	14	16	1.8E-01	4708.9	3	3	4.2E-02	7438.9	1	3	2.31E-02
4958.1	12	10	1.2E-02	4955.4	3	3	3.3E-03	7472.4	3	3	4.0E-02
5130.6	22	20	1.6E-01	5113.7	3	3	1.0E-02	7535.8	3	3	4.3E-01
5192.6	20	18	1.7E-01	5120.5	3	3	5.6E-03	7937.0	5	5	7.8E-03
5249.6	18	16	1.8E-01	5154.4	3	3	1.9E-02	8082.5	3	3	1.2E-03
5276.9	12	10	1.2E-01	5191.3	3	3	1.3E-02	8118.5	3	3	4.9E-02
5293.2	16	14	1.2E-01	5326.4	3	3	6.8E-03	8128.9	3	5	7.2E-03
5302.3	20	18	1.1E-01	5333.3	3	3	5.3E-03	8259.4	5	5	2.03E-02
5311.5	14	12	1.1E-01	5341.1	3	3	1.1E-01	8571.4	3	3	5.5E-02
5319.8	12	10	1.6E-01	5400.6	3	1	9.0E-03	8582.9	3	5	1.00E-02
5357.0	18	16	1.8E-01	5418.6	3	3	5.2E-03	8647.0	5	5	3.91E-02
5371.9	20	20	5.1E-02	5433.7	3	3	2.83E-03	8681.9	3	3	2.1E-01
5485.7	18	18	5.7E-02	5652.6	3	3	8.9E-03	8767.5	3	3	1.1E-03
5594.4	16	16	7.0E-02	5662.5	3	3	6.9E-03	8771.7	3	3	1.6E-01
5620.6	18	18	1.3E-01	5852.5	3	1	6.82E-01	8783.8	3	5	3.13E-01
5688.5	14	14	5.9E-02	5868.4	3	3	1.4E-02	8865.3	3	3	9.4E-03
5718.1	16	16	8.7E-02	5881.9	5	3	1.15E-01	9201.8	3	3	9.1E-02
5726.8	10	10	5.6E-02	5913.6	3	3	4.8E-02	9433.0	3	3	1.1E-03
5740.9	12	12	7.2E-02	5939.3	5	3	2.00E-03	9486.7	3	3	2.5E-02
5804.0	10	10	4.6E-02	5944.8	5	5	1.13E-01	9534.2	3	3	6.3E-02
5865.1	16	18	1.3E-02	5961.6	3	3	3.3E-02	10621	3	3	2.4E-03
6051.9	12	10	1.1E-02	5975.5	5	3	3.51E-02	11409	3	3	4.2E-02
				6030.0	3	3	5.61E-02	11525	3	3	8.4E-02
Neon				6046.1	3	3	2.26E-03	11767	3	3	6.9E-02
Ne I				6074.3	3	1	6.03E-01	12459	3	3	1.5E-02
615.63	1	3	3.8E-01	6096.2	3	5	1.81E-01				
618.67	1	3	9.3E-01	6118.0	5	3	6.09E-03	**Ne II**			
619.10	1	3	3.3E-01	6128.5	3	3	6.7E-03	*357.03	6	10	3.8E+01
626.82	1	3	7.4E-01	6143.1	5	5	2.82E-01	*361.77	6	2	1.6E+01
629.74	1	3	4.8E-01	6150.3	3	3	1.5E-02	*406.28	6	10	1.8E+01
735.90	1	3	6.11E+00	6163.6	1	3	1.46E-01	*446.37	6	6	4.07E+01
743.72	1	3	4.86E-01	6217.3	5	3	6.37E-02	460.73	4	2	4.7E+01
3369.8	5	5	1.0E-03	6266.5	1	3	2.49E-01	462.39	2	2	2.3E+01
3369.9	5	3	7.6E-03	6273.0	3	3	9.7E-03	1907.5	4	2	2.8E-01
3375.6	5	3	2.2E-03	6293.7	3	3	6.39E-03	1916.1	4	4	6.9E-01
3417.9	3	5	9.2E-03	6304.8	3	5	4.16E-02	1930.0	2	2	5.7E-01
3418.0	3	3	2.2E-03	6328.2	5	3	3.39E-02	1938.8	2	4	1.3E-01
3423.9	3	3	1.0E-02	6330.9	3	3	2.3E-02	2858.0	6	6	7.9E-01
3447.7	5	5	2.1E-02	6334.4	5	5	1.61E-01	2870.0	6	6	1.7E-01
3450.8	5	3	4.9E-03	6351.9	1	3	3.45E-03	2873.0	6	4	3.8E-01
3454.2	3	1	3.7E-02	6383.0	3	3	3.21E-01	2876.3	4	6	7.8E-01
3460.5	1	3	7.0E-03	6401.1	3	3	1.39E-02	2876.5	6	4	3.3E-01

λ Å	Weights g_i	g_k	A 10^8 s^{-1}	λ Å	Weights g_i	g_k	A 10^8 s^{-1}	λ Å	Weights g_i	g_k	A 10^8 s^{-1}
2878.1	2	2	6.9E-02	3309.7	4	2	3.1E-01	3568.5	6	8	1.4E+00
2888.4	4	6	7.0E-02	3310.5	4	4	6.9E-02	3571.2	4	4	6.3E-01
2891.5	4	4	6.1E-02	3311.3	4	2	2.6E-01	3574.2	6	6	1.0E-01
2897.0	6	8	5.2E-02	3314.7	6	6	4.4E-02	3574.6	4	6	1.3E+00
2906.8	2	4	5.5E-01	3319.7	4	2	1.6E+00	3590.4	4	6	3.6E-02
2910.1	4	2	1.7E+00	3320.2	8	6	2.1E-01	3594.2	4	2	1.3E+00
2910.4	2	4	5.9E-01	3323.7	4	4	1.6E+00	3612.3	2	4	2.6E-01
2916.2	6	4	9.6E-02	3327.2	4	4	9.1E-01	3628.0	4	4	6.0E-01
2925.6	2	2	5.6E-01	3329.2	8	8	8.8E-01	3632.7	4	4	1.3E-01
2933.7	6	6	6.9E-02	3330.7	6	6	3.9E-02	3643.9	4	4	3.2E-01
2955.7	6	4	1.2E+00	3334.8	6	8	1.8E+00	3644.9	2	4	9.9E-01
3001.7	4	4	8.7E-01	3336.1	4	6	1.1E+00	3659.9	4	6	6.7E-02
3017.3	6	4	3.5E-01	3344.4	2	2	1.5E+00	3664.1	6	4	7.0E-01
3027.0	6	6	1.4E+00	3345.5	6	4	1.4E+00	3679.8	4	2	3.2E-01
3028.7	4	2	8.5E-01	3345.8	4	4	2.2E-01	3694.2	6	6	1.0E+00
3028.9	2	4	4.7E-01	3353.6	4	2	1.2E-01	3697.1	2	2	2.8E-01
3034.5	6	8	3.1E+00	3355.0	4	6	1.3E+00	3701.8	4	6	2.7E-01
3037.7	4	4	2.1E+00	3356.3	6	6	2.0E-01	3709.6	4	2	1.1E+00
3045.6	2	2	2.5E+00	3357.8	6	6	5.0E-01	3713.1	4	6	1.3E+00
3047.6	4	6	1.8E+00	3360.3	2	4	8.6E-01	3721.8	4	6	2.0E-01
3054.7	2	4	9.4E-01	3360.6	2	4	8.2E-01	3726.9	4	4	1.2E-01
3092.9	6	6	1.3E+00	3362.9	4	2	3.5E-01	3727.1	2	4	9.8E-01
3097.1	8	8	1.3E+00	3371.8	4	6	2.2E-01	3734.9	4	4	1.9E-01
3118.0	8	6	4.2E-02	3374.1	4	4	3.0E-01	3744.6	2	4	2.6E-01
3134.1	6	4	2.6E-01	3378.2	2	2	1.7E+00	3751.2	2	2	1.8E-01
3140.4	8	6	2.4E-01	3379.3	2	2	3.0E-01	3753.8	4	6	4.5E-01
3151.1	6	6	4.8E-02	3386.2	4	6	5.5E-02	3766.3	4	6	2.9E-01
3154.8	8	6	1.8E-02	3388.4	4	6	2.2E+00	3777.1	2	4	4.2E-01
3164.4	8	8	1.6E-01	3390.6	2	4	7.7E-02	3800.0	4	4	3.7E-01
3165.7	6	6	1.2E+00	3392.8	2	4	4.4E-01	3818.4	2	4	6.1E-01
3173.6	6	4	4.5E-02	3404.8	4	6	1.9E+00	3829.8	4	6	8.4E-01
3176.1	4	6	6.0E-02	3407.0	6	8	2.3E+00	3942.3	4	6	1.0E-02
3187.6	4	6	1.4E-02	3411.4	4	2	6.1E-01				
3188.7	6	6	3.9E-01	3413.2	4	4	1.8E+00	**Ne V**			
3190.9	4	6	1.5E-01	3414.9	4	6	1.8E-02	*142.61	9	9	6.7E+02
3194.6	4	4	5.2E-01	3416.9	6	6	6.4E-01	*143.32	9	15	1.2E+03
3198.6	6	8	1.7E+00	3417.7	6	8	1.6E+00	147.13	5	7	1.5E+03
3198.9	4	4	2.3E-01	3438.9	2	2	1.4E+00	151.23	5	5	3.38E+02
3209.0	8	8	1.6E-01	3440.7	2	4	3.5E-01	154.50	1	3	7.0E+02
3209.4	2	4	6.0E-01	3453.1	4	4	4.6E-01	*167.69	9	9	1.5E+02
3213.7	2	4	1.7E+00	3454.8	4	4	1.6E+00	*358.93	9	3	2.1E+02
3214.3	4	6	2.2E+00	3456.6	2	4	9.6E-01	365.59	5	3	1.35E+02
3218.2	8	10	3.6E+00	3457.1	4	6	9.9E-02	*482.15	9	9	3.01E+01
3224.8	6	8	3.5E+00	3459.3	6	6	1.6E+00	*571.04	9	15	1.0E+01
3229.5	8	8	1.3E-01	3475.2	4	4	1.2E-02	2259.6	3	5	1.9E+00
3229.6	8	10	3.6E+00	3477.6	4	6	4.3E-01	2265.7	5	7	2.4E+00
3230.1	6	6	1.8E+00	3481.9	4	2	1.4E+00				
3230.4	4	6	1.4E-01	3503.6	2	2	2.0E+00	**Ne VII**			
3232.0	6	4	2.7E-01	3522.7	4	2	2.3E-02	97.502	1	3	1.07E+03
3232.4	4	4	1.6E+00	3538.0	4	2	7.6E-01	*115.46	9	3	4.8E+02
3243.4	6	6	2.3E-01	3539.9	4	4	3.6E-02	116.69	3	5	1.6E+03
3244.1	6	8	1.5E+00	3542.2	6	4	6.0E-01	127.66	3	1	1.9E+02
3248.1	4	4	2.4E-01	3542.9	4	6,	1.2E+00	465.22	1	3	4.09E+01
3255.4	6	4	3.8E-02	3546.2	2	4	6.3E-02	558.61	3	5	8.11E+00
3263.4	2	4	3.9E-01	3551.6	2	4	3.7E-02	559.95	1	3	1.07E+01
3269.9	4	6	5.1E-01	3557.8	2	2	1.9E-01	561.38	3	3	7.99E+00
3270.8	6	4	5.7E-02	3561.2	4	6	2.1E-01	561.73	5	5	2.39E+01
3297.7	6	6	4.3E-01	3565.8	4	4	6.2E-01	562.99	3	1	3.17E+01

λ Å	g_i	g_k	A 10^8 s^{-1}	λ Å	g_i	g_k	A 10^8 s^{-1}	λ Å	g_i	g_k	A 10^8 s^{-1}
564.53	5	3	1.31E+01	2313.98	5	5	5.0E+00	4752.43	3	3	2.0E-01
				2317.16	7	5	3.8E+00	4756.52	9	9	1.5E-01
Ne VIII				2320.03	9	11	6.9E+00	4786.54	11	11	1.8E-01
*88.09	2	6	8.4E+02	2321.38	5	7	5.6E+00	4812.00	3	1	9.5E-02
*98.208	6	10	2.77E+03	2324.65	7	9	1.8E-01	4829.03	5	7	1.9E-01
770.41	2	4	5.90E+00	2325.79	7	9	3.5E+00	4831.18	9	7	1.6E-01
780.32	2	2	5.69E+00	2329.96	5	3	5.3E+00	4838.64	9	7	2.2E-01
2820.7	2	4	7.20E-01	2345.54	9	7	2.2E+00	4855.41	5	5	5.7E-01
2860.1	2	2	6.88E-01	2346.63	7	5	5.5E-01	4904.41	5	3	6.2E-01
				2347.51	9	9	2.2E-01	4912.03	3	3	1.5E-01
Nickel				2348.73	7	7	2.2E-01	4913.97	1	3	2.2E-01
Ni I				2419.31	7	5	2.0E-01	4918.36	9	7	2.3E-01
1963.85	7	7	1.1E-01	2943.91	7	5	1.1E-01	4935.83	7	5	2.4E-01
1976.87	7	9	1.1E+00	2981.65	5	3	2.8E-01	4937.34	9	9	1.2E-01
1981.61	5	5	1.3E-01	3002.48	7	7	8.0E-01	4953.20	5	5	1.2E-01
1990.25	5	7	8.3E-01	3003.62	5	5	6.9E-01	4980.17	9	11	1.9E-01
2007.01	5	5	1.7E-01	3012.00	5	5	1.3E+00	5000.34	7	7	1.4E-01
2007.69	7	7	9.0E-02	3037.93	7	7	2.8E-01	5012.46	7	7	1.1E-01
2014.25	3	5	9.3E-01	3050.82	7	9	6.0E-01	5017.58	11	11	2.0E-01
2025.40	7	5	2.3E-01	3054.31	5	5	4.0E-01	5035.37	7	9	5.7E-01
2026.62	9	7	2.4E-01	3057.64	3	3	1.0E+00	5042.20	3	5	1.4E-01
2047.35	7	5	1.8E-01	3064.62	5	7	1.1E-01	5048.85	7	7	1.6E-01
2052.04	9	9	9.7E-02	3101.56	5	7	6.3E-01	5080.53	9	11	3.2E-01
2055.50	5	3	3.3E-01	3101.88	5	7	4.9E-01	5081.11	7	9	5.7E-01
2059.92	7	5	2.1E-01	3134.11	3	5	7.3E-01	5082.35	3	3	2.5E-01
2060.20	5	3	2.3E-01	3225.02	5	3	9.3E-02	5084.08	7	9	3.1E-01
2064.39	3	1	4.0E-01	3369.56	9	7	1.8E-01	5099.95	7	7	2.9E-01
2069.52	5	5	1.1E-01	3380.57	5	3	1.3E+00	5115.40	11	9	2.2E-01
2085.57	5	5	2.6E+00	3392.98	7	7	2.4E-01	5129.37	7	5	1.2E-01
2089.09	7	5	9.7E-02	3414.76	7	9	5.5E-01	5155.14	5	5	1.1E-01
2095.13	5	7	1.1E-01	3423.71	3	3	3.3E-01	5155.76	5	7	2.9E-01
2114.43	5	5	9.7E-02	3433.56	7	7	1.7E-01	5176.57	5	5	1.8E-01
2121.40	7	5	2.8E-01	3446.26	5	5	4.4E-01	5371.33	7	7	1.6E-01
2124.80	5	3	3.8E-01	3452.88	5	7	9.8E-02	5476.91	1	3	9.5E-02
2147.80	5	3	4.7E-01	3458.46	3	5	6.1E-01	5637.12	3	3	1.1E-01
2157.83	5	3	4.1E-01	3461.66	7	9	2.7E-01	5664.02	5	7	1.1E-01
2158.31	7	5	6.9E-01	3472.55	5	7	1.2E-01	5695.00	3	3	1.7E-01
2161.04	5	5	1.3E-01	3483.77	5	3	1.4E-01	6086.29	3	5	1.1E-01
2173.54	5	3	1.5E-01	3492.96	5	3	9.8E-01	6175.42	3	3	1.7E-01
2174.48	3	1	8.9E-01	3510.33	3	1	1.2E+00	7122.24	5	7	2.1E-01
2182.38	7	5	1.3E-01	3515.05	5	7	4.2E-01	7381.94	9	11	9.7E-02
2183.91	5	5	1.2E-01	3524.54	7	5	1.0E+00	7422.30	7	5	1.8E-01
2190.22	5	5	3.0E-01	3566.37	5	5	5.6E-01	7727.66	7	7	1.1E-01
2197.35	3	3	7.8E-01	3597.70	3	3	1.4E-01				
2201.59	5	3	7.3E-01	3619.39	5	7	6.6E-01	**Ni II**			
2221.94	5	3	2.2E-01	4027.67	5	7	1.3E-01	2165.55	10	10	2.4E+00
2244.46	5	5	3.8E-01	4295.88	9	7	1.7E-01	2169.10	8	8	1.58E+00
2253.57	7	7	1.9E-01	4401.54	9	11	3.8E-01	2174.67	8	10	1.43E+00
2254.81	9	9	9.6E-02	4462.46	3	5	1.7E-01	2175.15	6	6	1.77E+00
2258.15	7	5	1.7E-01	4470.48	5	7	1.9E-01	2184.61	4	4	2.90E+00
2259.56	5	3	2.0E-01	4600.37	5	3	2.6E-01	2201.41	4	6	1.3E+00
2261.42	9	7	9.1E-02	4604.99	9	7	2.3E-01	2206.72	6	8	1.66E+00
2287.32	3	5	1.8E-01	4606.23	5	3	1.0E-01	2216.48	10	12	3.4E+00
2289.98	9	7	2.1E+00	4648.66	11	9	2.4E-01	2220.40	6	8	2.3E+00
2293.11	5	5	3.8E-01	4686.22	5	5	1.4E-01	2222.96	10	10	9.8E-01
2300.77	7	7	7.5E-01	4701.54	9	9	1.4E-01	2224.86	8	8	1.55E+00
2302.97	3	3	4.5E-01	4714.42	13	11	4.6E-01	2226.33	6	6	1.3E+00
2307.35	5	7	1.6E-01	4715.78	7	7	2.0E-01	2253.85	4	6	1.98E+00
2312.34	7	7	5.5E+00	4732.47	7	9	9.3E-02	2264.46	6	8	1.43E+00

λ Å	Weights g_i	g_k	A 10^8 s^{-1}	λ Å	Weights g_i	g_k	A 10^8 s^{-1}	λ Å	Weights g_i	g_k	A 10^8 s^{-1}
2270.21	8	10	1.56E+00	168	6	8	3.2E+02	292	5	7	2.2E+02
2278.77	8	6	2.8E+00	182	2	2	2.5E+02				
2287.09	6	4	2.8E+00	185.23	2	4	4.2E+02	**Ni XVIII**			
2296.55	8	8	1.98E+00	187	4	6	1.2E+02	24.881	2	4	8.6E+02
2297.14	6	4	2.70E+00	187	4	2	3.3E+02	25.070	4	6	9.9E+02
2297.49	4	2	3.0E+00	188	2	2	4.7E+02	26.02	2	4	1.26E+03
2298.27	6	6	2.8E+00	190	6	8	2.0E+02	26.020	2	4	1.1E+03
2303.00	8	6	2.9E+00	192	6	8	4.54E+02	26.046	2	2	1.1E+03
2316.04	10	8	2.88E+00	192	6	6	3.1E+02	26.218	4	6	1.5E+03
2334.58	8	8	8.0E-01	194	4	6	2.8E+02	27.98	4	6	1.0E+03
2375.42	6	8	6.6E-01	194	2	4	5.5E+02	27.982	2	4	2.0E+03
2394.52	8	10	1.70E+00	194	2	2	1.1E+02	28.018	6	8	1.1E+03
2416.13	6	8	2.1E+00	194	4	4	3.5E+02	28.220	4	6	2.33E+03
2437.89	8	10	5.4E-01	194.04	4	6	4.6E+02	29.383	4	6	1.58E+03
2510.87	8	10	5.8E-01	195.27	4	4	9.5E+01	29.422	6	8	1.69E+03
				196	4	6	6.7E+02	29.779	2	4	1.9E+03
Ni III				197	4	6	1.5E+02	29.829	2	2	1.9E+03
1692.51	11	13	7.9E+00	197	4	2	1.2E+02	31.845	4	6	2.7E+03
1709.90	9	11	6.3E+00	199	2	4	4.9E+02	31.890	6	8	3.0E+03
1719.46	5	7	6.0E+00	206	2	2	3.7E+02	32.034	2	4	3.4E+03
1722.28	3	5	5.9E+00	217	4	4	1.1E+02	32.340	4	6	4.0E+03
1724.52	3	1	6.7E+00	218.391	2	4	9.5E+01	36.990	4	6	5.5E+03
1741.96	9	7	5.7E+00	223.119	2	2	1.3E+02	37.049	6	8	5.9E+03
1752.43	7	5	5.5E+00	231	4	4	1.6E+02	41.015	2	4	2.97E+03
1760.56	5	3	6.5E+00	232.475	4	4	4.07E+02	41.218	2	2	3.2E+03
1769.64	11	11	6.2E+00	233	6	4	2.4E+02	43.814	2	4	5.5E+03
1823.06	9	9	5.6E+00	235	4	2	3.8E+02	44.365	4	6	6.8E+03
				235	6	6	2.5E+02	44.405	4	4	1.14E+03
Ni XIV				236	4	4	1.2E+02	52.615	4	6	1.5E+04
164.13	6	8	1.2E+03	237.875	4	2	2.6E+02	52.720	6	8	1.6E+04
168	2	4	2.4E+02	238	6	4	1.3E+02	52.745	6	6	1.06E+03
168.12	4	2	8.5E+02	239.550	2	2	2.6E+02	59.950	6	4	9.6E+02
169.69	4	4	9.8E+02	245	4	4	1.4E+02	60.212	4	2	1.1E+03
170.50	4	4	7.1E+02	245	4	6	3.2E+02	63.512	4	6	7.9E+02
171.37	4	6	9.4E+02	249	6	8	3.3E+02	63.589	6	8	8.5E+02
172.16	6	6	4.7E+02	249	6	4	1.2E+02	69.075	4	6	8.0E+02
172.80	6	4	1.4E+02	250	4	2	1.6E+02	76.254	4	6	1.38E+03
177.28	4	4	5.6E+02	254	6	4	1.8E+02	76.359	6	8	1.47E+03
178	2	4	8.9E+01					99.275	2	4	1.0E+03
181	4	6	7.4E+01	**Ni XVII**				100.4	4	6	1.2E+03
182.14	4	2	1.5E+02	30.919	1	3	2.77E+03	114.46	4	6	2.5E+03
196	4	2	3.8E+01	42.855	1	3	4.75E+03	114.74	6	8	2.7E+03
288.894	4	4	4.6E+01	54.451	9	11	1.5E+04				
292.399	6	6	3.6E+01	55.361	1	3	6.7E+03	**Ni XIX**			
				57.348	7	9	1.4E+04	9.140	1	3	3.1E+04
Ni XV				197.39	1	3	1.6E+02	9.153	1	3	5.2E+03
50.249	5	7	6.8E+03	199.87	3	5	2.1E+02	9.977	1	3	1.1E+05
60.890	9	11	1.0E+04	204	3	3	1.8E+02	10.110	1	3	9.4E+04
64.635	7	9	9.6E+03	205	3	1	2.4E+02	10.283	1	3	4.7E+03
163.64	5	7	5.6E+01	206	1	3	3.0E+02	10.433	1	3	5.1E+03
173.73	5	7	7.6E+02	207.50	5	7	2.5E+02	11.539	1	3	4.8E+04
175	3	1	5.7E+02	215.89	3	5	4.8E+02	11.599	1	3	6.3E+04
179.28	5	7	7.5E+02	216	1	3	2.7E+02	12.435	1	3	3.66E+05
181	1	3	6.8E+02	217	5	7	2.4E+02	12.656	1	3	1.0E+04
269	3	1	5.3E+01	227	5	5	1.6E+02	13.779	1	3	1.23E+04
278.386	5	5	4.3E+01	249.180	1	3	2.75E+02	14.043	1	3	1.31E+04
				281.50	3	1	2.1E+02	40.7	3	3	6.4E+03
Ni XVI				282	3	1	2.4E+02	40.7	3	1	8.4E+03
166	4	6	3.1E+02	284	5	3	1.5E+02	41.132	7	9	9.4E+03

λ Å	Weights g_i	g_k	A 10^8 s^{-1}	λ Å	Weights g_i	g_k	A 10^8 s^{-1}	λ Å	Weights g_i	g_k	A 10^8 s^{-1}
Ni XXI				103.53	4	2	4.17E+02	1.539	1	3	2.6E+06
11.13	3	3	1.7E+04	104.64	2	2	4.7E+02	1.539	3	5	2.6E+06
11.23	5	3	1.7E+04	106.68	4	2	3.67E+02	1.540	3	3	1.7E+06
11.239	5	7	5.7E+04	113.14	4	4	1.65E+02	1.541	3	5	5.5E+06
11.28	3	1	2.2E+05	118.52	2	4	1.5E+02	1.542	3	3	3.6E+06
11.318	5	7	2.8E+05	122.72	6	4	2.17E+02	1.542	5	5	3.5E+06
11.48	3	1	1.1E+05	134.73	6	6	1.44E+02	1.544	5	3	3.2E+06
11.48	1	3	4.0E+05	135.47	4	4	8.0E+01	1.546	3	5	1.6E+06
11.517	5	7	1.4E+05	137.01	4	4	2.6E+02	1.547	3	3	2.1E+05
11.539	5	7	1.2E+05	138.80	4	6	7.2E+01	1.549	1	3	2.0E+05
11.67	1	3	8.0E+04	153.47	2	2	1.27E+02	1.551	3	1	8.2E+05
11.72	3	3	2.3E+04	159.69	2	4	8.9E+01	1.558	3	1	6.5E+05
12.454	5	3	3.3E+04					1.5883	1	3	6.02E+06
12.472	3	3	1.8E+04	**Ni XXV**				1.5963	1	3	7.70E+05
12.502	5	5	2.8E+04	9.30	3	1	9.3E+04				
				9.31	5	7	8.2E+04	**Nitrogen**			
Ni XXII				9.32	3	5	7.8E+04	**N I**			
72.52	4	2	2.84E+02	9.34	1	3	1.1E+05	1163.88	6	6	7.52E-01
84.06	6	4	1.2E+03	9.42	3	1	9.0E+04	1164.00	4	6	1.27E-02
84.24	4	2	5.6E+02	9.49	3	5	8.9E+04	1164.21	6	4	5.17E-02
85.86	4	2	4.9E+02	9.60	1	3	1.8E+05	1164.32	4	4	6.94E-01
88.00	4	2	1.2E+03	9.63	3	5	2.4E+05	1167.45	6	8	1.29E+00
95.95	2	2	4.4E+02	9.64	3	3	1.3E+05	1168.42	6	6	4.24E-02
98.16	4	4	5.2E+02	9.71	3	1	2.3E+05	1168.54	4	6	1.24E+00
98.58	4	4	2.45E+02	9.71	3	3	1.8E+05	1176.51	6	4	9.22E-01
100.60	6	6	3.9E+02	9.74	5	7	3.0E+05	1176.63	4	4	1.02E-01
101.31	6	4	4.83E+02	9.75	3	5	1.3E+05	1177.69	4	2	1.02E+00
103.31	4	2	2.66E+02	9.76	1	3	3.03E+05	1199.55	4	6	4.01E+00
106.04	4	4	2.36E+02	9.76	5	3	7.5E+04	1200.22	4	4	3.99E+00
106.16	4	2	5.1E+02	9.78	5	7	2.9E+05	1200.71	4	2	3.98E+00
124.31	2	2	3.7E+02	9.86	5	7	4.8E+05	1310.54	4	6	8.42E-01
126.32	4	4	3.3E+02	9.87	3	5	2.03E+05	1316.29	4	6	1.42E-02
				9.92	5	5	1.3E+05	1492.63	6	4	3.13E+00
Ni XXIII				9.94	5	7	1.29E+05	1492.82	4	4	3.51E-01
87.66	3	3	2.8E+02	9.97	3	5	2.5E+05	1494.68	4	2	3.72E+00
88.11	5	3	8.3E+02	10.08	1	3	2.80E+05	3822.03	2	2	3.70E-02
90.49	3	3	1.77E+02					3830.43	4	4	4.67E-02
90.96	5	3	2.5E+02	**Ni XXVI**				3834.22	4	2	1.89E-02
91.83	5	3	7.5E+02	1.5930	4	2	3.4E+06	4099.94	2	4	3.48E-02
92.32	3	1	4.39E+02	1.5935	2	2	4.0E+06	4109.95	4	6	3.90E-02
100.42	1	3	2.1E+02	1.5973	4	4	8.1E+06	4113.97	4	4	6.62E-03
102.08	5	5	5.3E+02	1.5977	2	4	4.4E+06	4137.64	2	4	2.80E-03
103.23	3	3	2.4E+02	1.5982	2	2	7.3E+06	4143.43	4	4	6.09E-03
103.67	5	5	1.78E+02	1.5996	2	2	2.7E+06	4151.48	6	4	1.01E-02
104.70	3	1	2.94E+02	1.6005	4	6	2.7E+06	4249.87	4	2	2.59E-02
106.02	5	5	2.87E+02	1.6036	4	2	2.1E+06	4264.00	6	4	2.26E-02
108.27	7	5	3.32E+02	9.390	2	4	2.59E+05	4356.29	6	8	5.10E-02
111.23	3	1	2.26E+02	9.535	4	6	2.96E+05	4385.54	2	2	8.84E-03
111.78	5	3	2.19E+02					4392.41	4	2	1.76E-02
111.86	1	3	1.7E+02	**Ni XXVII**				4435.43	2	4	7.51E-03
112.55	3	1	1.0E+03	1.2534	1	3	3.35E+05	4442.45	4	4	3.81E-02
128.87	5	5	4.02E+02	1.2824	1	3	6.38E+05	4669.89	4	4	7.49E-03
133.54	3	3	1.86E+02	1.3500	1	3	1.63E+06	4914.94	2	2	8.08E-03
137.55	3	1	2.53E+02	1.3516	1	3	2.4E+05	4935.12	4	2	1.76E-02
				1.531	3	3	2.0E+05	5199.84	2	2	1.87E-02
Ni XXIV				1.534	3	1	6.9E+06	5201.61	2	4	1.87E-02
101.13	6	4	1.63E+02	1.537	5	5	2.3E+06	5281.20	6	6	2.45E-03
102.11	4	4	5.4E+02	1.537	1	3	3.7E+06	5344.05	6	6	6.10E-04
103.43	2	4	1.3E+02	1.538	3	5	3.9E+06	5356.62	4	6	1.41E-03

λ Å	g_i	g_k	A 10^8 s⁻¹	λ Å	g_i	g_k	A 10^8 s⁻¹	λ Å	g_i	g_k	A 10^8 s⁻¹
5367.01	4	4	1.07E-03	9810.01	4	2	5.30E-02	1085.55	5	5	9.47E-01
5372.61	2	4	8.34E-04	9814.02	6	8	6.56E-03	1085.70	5	7	3.87E+00
5378.27	2	2	1.66E-03	9822.75	6	6	4.95E-02	3408.13	3	1	2.19E-01
6606.18	4	6	8.87E-04	9834.61	6	4	4.50E-02	3437.14	3	1	2.07E+00
6622.54	6	6	7.93E-03	9863.33	8	8	9.62E-02	3593.60	3	5	1.21E-01
6626.99	2	4	2.20E-03	9872.15	8	6	2.97E-02	3609.10	3	3	1.41E-01
6636.94	4	4	1.40E-02	9883.38	2	2	2.93E-02	3615.86	3	1	1.53E-01
6644.96	8	6	3.49E-02	9905.52	4	2	3.11E-03	3829.80	3	5	2.42E-01
6646.50	2	2	2.18E-02	9909.22	2	4	7.58E-03	3838.37	5	5	6.98E-01
6653.46	6	4	2.74E-02	9931.47	4	4	3.64E-02	3842.19	1	3	3.06E-01
6656.51	4	2	2.17E-02	9947.07	6	8	1.08E-02	3847.40	3	3	2.22E-01
6926.67	4	6	7.75E-03	9965.75	4	6	7.60E-03	3855.10	3	1	8.82E-01
6945.18	6	6	1.83E-02	9968.51	6	4	4.50E-02	3856.06	5	3	3.71E-01
6951.60	2	4	1.03E-02	9980.42	4	6	8.10E-03	3919.00	3	3	6.76E-01
6960.50	4	4	4.67E-03	9997.73	8	8	9.20E-03	3955.85	3	5	1.31E-01
6973.07	2	2	3.83E-03					3995.00	3	5	1.35E+00
6979.18	6	4	9.83E-03	**N II**				4114.33	3	3	1.42E-03
6982.03	4	2	2.04E-02	474.891	5	5	9.66E+00	4124.08	3	5	3.20E-01
7423.64	2	4	5.95E-02	475.647	1	3	1.17E+01	4133.67	5	5	5.30E-01
7442.30	4	4	1.24E-01	475.698	3	5	1.58E+01	4145.77	7	5	7.36E-01
7468.31	6	4	1.93E-01	475.757	3	3	8.75E+00	4374.99	3	5	5.55E-03
7898.98	6	4	2.82E-01	475.803	5	7	2.10E+01	4447.03	3	5	1.14E+00
7899.28	4	4	3.28E-02	475.884	5	5	5.25E+00	4459.94	3	1	1.12E-01
7915.42	4	2	3.13E-01	508.697	5	5	1.91E+00	4465.53	3	3	2.36E-02
8184.86	4	6	8.58E-02	510.758	5	7	1.87E+01	4477.68	5	3	8.85E-02
8188.01	2	4	1.27E-01	513.849	5	5	1.24E+01	4488.09	5	5	1.30E-02
8200.36	2	2	4.95E-02	529.355	1	3	7.23E+00	4507.56	7	5	1.00E-01
8210.72	4	4	4.84E-02	529.413	3	1	2.43E+01	4564.76	3	5	1.41E-02
8216.34	6	6	2.23E-01	529.491	3	3	6.75E+00	4601.48	3	5	2.35E-01
8223.13	4	2	2.64E-01	529.637	3	5	4.92E+00	4607.15	1	3	3.26E-01
8242.39	6	4	1.36E-01	529.722	5	3	1.03E+01	4613.87	3	3	2.26E-01
8567.74	2	4	4.92E-02	529.867	5	5	1.94E+01	4621.39	3	1	9.55E-01
8594.00	2	2	2.09E-01	533.511	1	3	2.39E+01	4630.54	5	5	7.72E-01
8629.24	4	4	2.66E-01	533.581	3	5	3.20E+01	4643.09	5	3	4.51E-01
8655.88	4	2	1.05E-01	533.650	3	3	1.66E+01	4654.53	3	5	2.43E-02
8680.28	6	8	2.46E-01	533.729	5	7	4.13E+01	4667.21	3	3	2.99E-02
8683.40	4	6	1.80E-01	533.815	5	5	9.19E+00	4674.91	3	1	1.05E-01
8686.15	2	4	1.09E-01	547.818	5	3	2.16E+00	4694.27	1	3	1.23E-01
8703.25	2	2	2.10E-01	559.762	1	3	1.14E+01	4695.90	3	5	1.29E-01
8711.70	4	4	1.28E-01	574.650	5	7	3.60E+01	4697.64	3	3	3.06E-02
8718.84	6	6	6.75E-02	582.156	5	5	2.85E+01	4698.55	3	1	3.67E-01
8728.90	4	2	3.76E-02	635.197	1	3	2.33E+01	4700.03	5	7	1.05E-01
8747.37	6	4	1.04E-02	644.634	1	3	1.21E+01	4702.50	5	5	9.15E-02
9028.92	2	2	3.02E-01	644.837	3	3	3.64E+01	4704.25	5	3	2.13E-01
9045.88	6	8	2.80E-01	645.178	5	3	6.07E+01	4706.40	7	9	6.09E-02
9049.49	6	6	1.88E-02	660.286	5	3	3.69E+01	4709.58	7	7	1.82E-01
9049.89	4	6	2.60E-01	671.016	3	5	2.47E+00	4712.07	7	5	1.46E-01
9060.48	2	4	2.95E-01	671.386	5	5	7.40E+00	4718.38	9	9	3.02E-01
9187.45	6	6	2.44E-01	671.411	1	3	3.04E+00	4721.58	9	7	7.75E-02
9187.86	4	6	1.76E-02	671.630	3	3	2.27E+00	4774.24	3	5	3.24E-02
9207.59	6	4	2.70E-02	671.773	3	1	9.85E+00	4779.72	3	3	2.52E-01
9208.00	4	4	2.33E-01	672.001	5	3	3.87E+00	4781.19	5	7	2.05E-02
9386.81	2	4	2.24E-01	745.841	1	3	1.25E+01	4788.14	5	5	2.52E-01
9392.79	4	6	2.63E-01	746.984	5	3	3.85E+00	4793.65	5	3	7.77E-02
9460.68	4	4	3.98E-02	748.369	5	3	3.83E+00	4803.29	7	7	3.18E-01
9776.90	2	4	1.18E-02	775.965	5	5	3.08E+01	4810.30	7	5	4.75E-02
9786.78	4	6	1.13E-02	915.612	1	3	4.38E+00	4860.17	3	5	1.61E-02
9788.29	2	2	2.99E-02	915.962	3	1	1.32E+01	4987.38	3	1	7.48E-01
9798.56	4	4	2.75E-02	1083.99	1	3	2.18E+00	4991.24	3	5	3.54E-01

λ Å	Weights g_i	g_k	A 10^8 s^{-1}	λ Å	Weights g_i	g_k	A 10^8 s^{-1}	λ Å	Weights g_i	g_k	A 10^8 s^{-1}
4994.36	5	7	2.62E-01	5551.92	7	7	2.00E-01	2978.84	2	4	1.66E-01
4994.37	3	3	7.60E-01	5552.68	5	3	1.50E-01	2983.64	4	4	8.24E-01
4997.22	3	3	1.96E-01	5565.26	7	5	3.97E-02	3342.76	2	2	3.80E-01
5001.13	3	5	9.76E-01	5666.63	3	5	3.74E-01	3353.98	2	4	7.66E-01
5001.47	5	7	1.05E+00	5676.02	1	3	2.96E-01	3354.32	4	6	5.51E-01
5002.70	1	3	8.45E-02	5679.56	5	7	5.25E-01	3355.46	4	2	7.51E-01
5005.15	7	9	1.16E+00	5686.21	3	3	1.94E-01	3358.78	2	2	3.05E-01
5005.30	5	5	6.51E-02	5710.77	5	5	1.24E-01	3360.98	4	4	2.44E-01
5007.33	3	5	7.89E-01	5730.66	5	3	1.34E-02	3365.80	4	2	1.52E+00
5010.62	3	3	2.19E-01	5747.30	3	5	3.40E-02	3367.36	6	6	1.27E+00
5011.31	5	3	5.84E-01	5767.45	3	3	2.44E-02	3374.07	6	4	8.13E-01
5012.04	7	7	5.19E-01	5893.15	5	7	2.88E-01	3745.95	2	4	1.90E-01
5016.38	5	5	1.62E-01	5897.25	3	5	2.16E-01	3752.63	2	2	6.67E-02
5023.05	7	5	3.61E-01	5899.83	1	3	1.60E-01	3754.69	4	4	3.78E-01
5025.66	7	7	1.07E-01	5927.81	1	3	3.22E-01	3762.60	4	4	4.24E-02
5040.71	7	5	3.78E-03	5931.78	3	5	4.27E-01	3771.03	6	4	5.59E-01
5045.10	5	3	3.42E-01	5940.24	3	3	2.26E-01	3771.36	6	4	8.28E-02
5073.59	3	3	2.59E-02	5941.65	5	7	5.54E-01	3792.97	8	6	1.03E-01
5168.05	3	5	3.06E-01	5952.39	5	5	1.27E-01	3934.50	2	4	7.49E-01
5170.16	3	3	6.54E-01	5960.91	5	3	1.34E-02	3938.51	4	6	8.96E-01
5171.27	3	1	8.71E-01	6065.00	3	5	2.21E-03	3942.88	4	4	1.49E-01
5171.47	5	7	5.81E-01	6284.32	5	3	7.74E-02	4097.36	2	4	8.70E-01
5172.34	3	5	6.01E-01	6379.62	3	3	6.11E-02	4103.39	2	2	8.67E-01
5172.97	1	3	5.01E-01	6482.05	3	3	3.01E-01	4195.74	2	4	9.37E-01
5173.39	5	7	7.36E-01	6610.56	5	7	6.34E-01	4200.07	4	6	1.12E+00
5174.46	5	5	5.07E-01	6857.03	5	3	2.53E-01	4215.77	4	4	1.85E-01
5175.89	7	9	8.93E-01	6869.58	5	5	2.51E-01	4318.78	2	4	5.40E-02
5176.57	5	3	2.17E-01	6887.83	5	7	2.49E-01	4321.22	2	2	1.08E-01
5177.06	3	3	5.00E-01	7762.24	5	5	8.74E-02	4321.39	4	6	5.03E-02
5179.34	7	9	8.67E-01	8438.74	1	3	2.24E-01	4325.43	4	4	8.60E-02
5179.52	9	11	1.07E+00	8831.75	1	3	8.42E-03	4327.69	6	8	3.06E-02
5180.36	5	5	4.28E-01	8855.30	3	3	2.51E-02	4327.88	4	2	1.07E-01
5183.20	7	7	2.88E-01	8893.29	5	3	4.12E-02	4332.95	6	6	1.23E-01
5184.96	7	7	3.20E-01					4337.01	6	4	7.47E-02
5185.09	5	3	7.11E-02	**N III**				4345.81	8	8	1.82E-01
5186.21	7	5	5.76E-02	374.198	2	4	9.89E+01	4351.11	8	6	4.01E-02
5190.38	9	9	1.77E-01	451.871	2	2	1.03E+01	4510.88	2	4	2.84E-01
5191.96	7	5	4.25E-02	452.227	4	2	2.05E+01	4510.96	4	6	4.77E-01
5199.50	9	7	1.51E-02	684.998	2	4	9.63E+00	4514.85	6	8	6.80E-01
5313.42	3	3	1.41E-01	685.515	2	2	3.83E+01	4518.14	2	2	5.65E-01
5320.20	5	3	4.20E-01	685.817	4	4	4.54E+01	4523.56	4	4	3.61E-01
5320.96	3	5	2.52E-01	686.336	4	2	1.95E+01	4530.86	4	2	1.12E-01
5327.76	5	5	4.65E-02	763.334	2	2	9.58E+00	4534.58	6	6	2.01E-01
5338.73	5	7	1.85E-01	764.351	4	2	1.85E+01	4547.30	6	4	3.33E-02
5340.21	7	5	2.59E-01	771.545	2	4	8.19E+00	4634.13	2	4	6.36E-01
5351.23	7	7	3.67E-01	771.901	4	4	1.64E+01	4640.64	4	6	7.60E-01
5383.72	3	5	3.31E-03	772.384	6	4	2.45E+01	4641.85	4	4	1.26E-01
5452.07	1	3	8.89E-02	772.889	6	4	2.09E+01	4858.70	2	4	4.35E-01
5454.22	3	1	3.34E-01	772.955	4	2	2.34E+01	4858.98	4	6	4.66E-01
5462.58	3	3	1.00E-01	979.832	4	4	8.84E+00	4861.27	6	8	5.32E-01
5478.09	3	5	4.75E-02	979.905	6	6	9.21E+00	4867.12	4	4	1.73E-01
5480.05	5	3	1.30E-01	989.799	2	4	4.18E+00	4867.17	8	10	6.18E-01
5495.65	5	5	2.40E-01	991.511	4	4	8.17E-01	4873.60	6	6	1.50E-01
5526.23	3	5	2.13E-01	991.577	4	6	4.97E+00	4881.78	6	4	1.22E-02
5530.24	5	7	4.04E-01	1747.85	2	4	1.28E+00	4884.14	8	8	8.71E-02
5535.35	7	9	6.04E-01	1751.22	4	4	2.48E-01	4896.58	8	6	5.86E-03
5535.38	3	3	4.53E-01	1751.66	4	6	1.51E+00	5260.86	2	2	2.80E-02
5540.06	3	1	6.03E-01	2972.55	2	2	6.67E-01	5270.57	2	4	6.95E-02
5543.47	5	5	3.51E-01	2977.33	4	2	3.32E-01	5272.68	4	2	1.39E-01

λ Å	g_i	g_k	A 10^8 s⁻¹	λ Å	g_i	g_k	A 10^8 s⁻¹	λ Å	g_i	g_k	A 10^8 s⁻¹
5282.43	4	4	2.21E-02	4769.86	5	3	2.50E-02	935.193	5	5	1.33E+00
5297.75	4	6	4.93E-02	4786.92	7	7	8.79E-02	1028.16	1	3	4.22E-01
5298.95	6	4	7.38E-02	4796.66	7	5	1.53E-02	1152.15	5	5	5.28E+00
5314.36	6	6	1.14E-01	5200.41	3	5	2.67E-01	1217.65	1	3	2.06E+00
5320.87	6	8	5.68E-01	5204.28	5	7	3.55E-01	1302.17	5	3	3.41E+00
5327.19	4	6	5.29E-01	5205.15	1	3	1.97E-01	1304.86	3	3	2.03E+00
5352.46	6	6	3.72E-02	5226.70	3	3	1.46E-01	1306.03	1	3	6.76E-01
6365.84	2	2	2.18E-01	5245.60	5	5	8.66E-02	3823.41	7	7	6.63E-03
6394.75	2	4	2.15E-01	5272.35	5	3	9.48E-03	3823.87	5	3	1.87E-03
6445.34	2	4	8.89E-02	5288.25	5	3	3.22E-01	3824.35	5	5	5.19E-03
6450.79	2	2	1.77E-01	5736.93	3	5	1.84E-01	3825.02	3	3	5.59E-03
6454.08	4	6	1.49E-01	5776.31	1	3	1.85E-01	3825.19	5	7	8.31E-04
6463.09	4	4	1.13E-01	5784.76	3	1	5.51E-02	3855.01	5	5	1.63E-02
6467.02	6	8	2.11E-01	5795.09	3	3	1.37E-02	3947.29	5	7	4.91E-03
6468.57	4	2	3.52E-02	5812.31	3	5	1.36E-02	3947.48	5	5	4.88E-03
6478.76	6	6	6.31E-02	5826.43	5	3	2.25E-02	3947.59	5	3	4.87E-03
6487.84	6	4	1.05E-02	5843.84	5	5	4.01E-02	3951.93	3	1	3.10E-03
7371.51	4	4	3.53E-02	6380.75	1	3	1.42E-01	3952.98	5	3	1.29E-03
7404.54	6	6	3.61E-02	7103.24	1	3	6.28E-02	3953.00	1	3	1.03E-03
8307.51	2	4	1.65E-02	7109.35	3	5	8.46E-02	3954.52	3	5	7.73E-04
8344.95	2	2	6.52E-02	7111.28	3	3	4.70E-02	3954.61	5	5	2.32E-03
8386.39	4	4	8.03E-02	*7116.8	9	15	1.12E-01	3997.95	5	3	2.41E-02
8424.56	4	2	3.17E-02	7122.98	5	7	1.12E-01	4217.09	3	1	5.44E-03
				7127.25	5	5	2.80E-02	4222.77	5	3	2.26E-03
N IV				7127.25	5	3	3.11E-03	4222.82	1	3	1.81E-03
247.205	1	3	1.19E+02	9165.07	3	5	4.23E-02	4233.27	5	5	4.04E-03
*283.52	9	15	3.05E+02	9182.16	5	7	4.45E-02	4368.19	3	1	7.56E-03
*322.64	9	3	8.99E+01	9222.99	7	9	4.95E-02	4368.24	3	5	7.59E-03
335.047	3	5	1.845E+02	9247.04	5	5	7.66E-03	4967.38	3	5	4.43E-03
387.356	3	1	2.55E+01	9311.55	7	7	5.36E-03	4967.88	5	7	8.44E-03
765.147	1	3	2.320E+01					4968.79	7	9	1.27E-02
*923.16	9	9	1.759E+01	**N V**				5019.29	5	5	7.13E-03
955.334	3	1	2.919E+01	*209.29	2	6	1.21E+02	5020.22	7	5	9.98E-03
1718.55	3	5	2.321E+00	*247.66	6	10	4.26E+02	5329.11	3	5	9.48E-03
2649.88	3	3	1.07E+00	1238.82	2	4	3.40E+00	5329.69	5	7	1.81E-02
3052.20	1	3	1.33E-01	1242.80	2	2	3.37E+00	5330.74	7	9	2.71E-02
3059.60	3	3	3.95E-01	4603.74	2	4	4.14E-01	5435.18	3	5	7.74E-03
3075.19	5	3	6.48E-01	4619.97	2	2	4.10E-01	5435.77	5	5	1.29E-02
3443.61	3	5	3.46E-01					5436.86	7	5	1.80E-02
3445.22	1	3	4.60E-01	**N VI**				5512.60	3	5	2.69E-03
3454.65	3	3	3.42E-01	24.8980	1	3	5.158E+03	5512.77	5	7	3.58E-03
3461.36	3	1	1.36E+00	28.7870	1	3	1.809E+04	5554.83	3	3	5.83E-03
3463.36	5	5	1.02E+00	*161.220	3	9	2.859E+02	5555.00	5	3	9.71E-03
3474.53	5	3	5.61E-01	173.275	1	3	2.697E+02	5958.39	3	5	6.80E-03
3478.72	3	5	1.06E+00	*173.93	9	15	8.756E+02	5958.58	5	7	9.06E-03
*3480.8	3	9	1.06E+00	185.192	3	5	8.205E+02	6046.23	3	3	1.05E-02
3483.00	3	3	1.06E+00	*1901	3	9	6.780E-01	6046.44	5	3	1.75E-02
3484.93	3	1	1.06E+00	2896.4	1	3	2.079E-01	6046.49	1	3	3.50E-03
3689.94	3	1	9.10E-02	*6991.1	3	9	8.384E-02	6155.99	3	5	2.67E-02
3694.14	3	3	2.27E-02	9622.0	1	3	3.276E-02	6156.78	5	7	5.08E-02
3707.39	5	3	6.73E-02					6158.19	7	9	7.62E-02
3714.43	5	5	1.34E-02	**Oxygen**				6324.84	7	5	3.76E-05
3735.43	7	5	7.37E-02	**O I**				6453.60	3	5	1.65E-02
3747.54	3	5	9.92E-01	791.973	5	5	4.94E+00	6454.44	5	5	2.75E-02
4057.76	3	5	6.62E-01	792.938	1	3	2.19E+00	6455.98	7	5	3.85E-02
4740.26	3	5	1.53E-02	792.967	3	5	1.64E+00	6726.28	5	5	1.18E-05
4747.96	3	3	7.60E-02	877.798	5	3	2.85E+00	6726.54	5	3	6.44E-06
4752.49	5	7	1.13E-02	877.879	5	5	5.12E+00	7001.92	3	5	2.65E-02
4762.09	5	5	6.99E-02	922.008	5	7	1.23E+00	7002.23	5	7	3.53E-02

λ Å	Weights g_i	g_k	A $10^8\ s^{-1}$	λ Å	Weights g_i	g_k	A $10^8\ s^{-1}$	λ Å	Weights g_i	g_k	A $10^8\ s^{-1}$
7254.15	3	3	2.24E-02	2418.46	6	6	2.30E-01	4104.72	4	6	3.14E-01
7254.45	5	3	3.73E-02	2425.57	6	6	1.77E-01	4104.99	4	4	9.14E-01
7254.53	1	3	7.45E-03	2433.54	2	4	4.21E-01	4106.02	8	6	1.70E-02
7771.94	5	7	3.69E-01	2436.06	4	4	1.69E-01	4109.84	6	6	1.21E-02
7774.17	5	5	3.69E-01	2444.25	4	4	7.56E-02	4110.19	6	4	2.54E-01
7775.39	5	3	3.69E-01	2445.53	4	6	4.98E-01	4110.79	4	2	7.70E-01
7981.94	3	3	2.33E-04	2517.96	4	6	7.72E-02	4112.02	6	6	1.81E-01
7982.40	1	3	3.09E-04	2523.21	2	2	9.63E-02	4113.83	8	6	2.41E-01
7986.98	3	5	4.19E-04	2526.87	4	4	1.20E-01	4119.22	6	8	1.33E+00
7987.33	5	5	1.41E-04	2530.28	6	8	8.16E-02	4120.28	6	6	2.15E-01
7995.07	5	7	5.63E-04	2571.46	2	4	1.15E-01	4120.55	6	4	2.60E-01
8221.82	7	7	2.89E-01	2575.28	4	6	1.37E-01	4121.46	2	2	5.60E-01
8227.65	5	3	8.13E-02	3134.73	8	6	1.23E+00	4129.32	4	2	1.79E-01
8230.00	5	5	2.26E-01	3273.43	8	6	9.99E-01	4132.80	2	4	9.13E-01
8233.00	3	3	2.43E-01	3377.15	2	2	1.27E+00	4140.70	4	4	4.09E-02
8235.35	3	5	4.86E-02	3390.21	2	4	1.22E+00	4153.30	4	6	7.91E-01
8446.25	3	1	3.22E-01	3407.28	6	6	1.02E+00	4156.53	6	4	2.11E-01
8446.36	3	5	3.22E-01	3712.74	2	4	2.84E-01	4169.22	6	6	2.71E-01
8446.76	3	3	3.22E-01	3727.32	4	4	5.81E-01	4185.44	6	8	1.91E+00
8820.42	5	7	2.93E-01	3749.48	6	4	9.31E-01	4189.58	8	8	7.06E-02
9260.81	3	1	4.46E-01	3833.07	6	8	1.02E-02	4189.79	8	10	1.98E+00
9260.85	3	3	3.34E-01	3842.81	2	4	7.45E-02	4192.51	6	4	3.21E-01
9260.94	3	5	1.56E-01	3843.58	4	6	3.55E-02	4196.27	4	4	3.56E-02
9262.58	5	3	1.11E-01	3847.89	2	2	1.95E-01	4196.70	4	2	3.56E-01
9262.67	5	5	2.60E-01	3850.80	4	6	6.00E-03	4317.14	2	4	3.70E-01
9262.78	5	7	2.97E-01	3851.03	4	4	1.59E-01	4319.63	4	6	2.55E-01
9265.83	7	5	2.97E-02	3851.47	8	8	2.72E-02	4319.87	2	2	5.62E-01
9265.93	7	7	1.48E-01	3856.13	4	2	2.28E-01	4325.76	2	2	1.47E-01
9266.01	7	9	4.45E-01	3857.16	6	6	6.59E-02	4327.46	6	6	6.76E-01
9482.89	5	3	2.34E-01	3863.50	6	8	6.49E-02	4327.85	6	4	7.24E-02
9622.11	5	3	5.22E-04	3864.13	2	2	9.12E-02	4328.59	4	2	1.12E+00
9622.16	3	3	1.57E-03	3864.43	6	6	2.15E-01	4331.47	4	6	4.82E-02
9625.26	7	5	3.25E-04	3864.67	6	4	1.80E-01	4331.86	4	4	6.50E-01
9625.30	7	7	1.85E-03	3874.09	2	4	3.26E-02	4336.86	4	4	1.57E-01
9694.66	5	7	4.54E-04	3875.80	8	6	3.38E-02	4345.56	4	2	8.31E-01
9694.91	5	5	4.54E-04	3882.19	8	8	5.50E-01	4347.22	6	4	1.19E-01
9695.06	5	3	4.54E-04	3882.45	4	4	8.94E-02	4347.41	4	4	9.32E-01
				3883.14	8	6	1.13E-01	4349.43	6	6	6.91E-01
O II				3893.52	4	6	1.89E-02	4351.26	6	6	9.89E-01
429.918	4	2	4.25E+01	3907.45	6	6	8.64E-02	4351.46	4	6	5.82E-02
430.041	4	4	4.13E+01	3911.96	6	4	1.09E+00	4359.40	4	6	1.44E-02
430.176	4	6	4.36E+01	3912.12	4	4	1.41E-01	4366.89	6	4	3.98E-01
483.760	4	2	2.05E+01	3919.27	4	2	1.22E+00	4369.27	4	4	3.57E-01
483.980	6	4	1.80E+01	3945.04	2	4	2.05E-01	4395.93	6	6	3.91E-01
484.027	4	4	3.22E+00	3954.36	2	2	8.57E-01	4405.98	6	4	4.30E-02
485.087	6	8	2.60E+01	3973.26	4	4	1.04E+00	4414.90	4	6	8.34E-01
485.470	6	6	1.20E+00	3982.71	4	2	4.27E-01	4416.97	2	4	7.13E-01
485.518	4	6	1.93E+01	4069.62	2	4	1.52E+00	4443.01	6	6	5.05E-01
2290.85	2	4	7.41E-02	4069.88	4	6	1.53E+00	4443.52	6	8	1.89E-02
2293.30	2	2	3.25E-01	4072.15	6	8	1.98E+00	4447.68	8	6	2.52E-02
2300.33	4	4	4.17E-01	4075.86	8	10	2.11E+00	4448.19	8	8	5.10E-01
2302.81	4	2	1.67E-01	4078.84	4	4	5.52E-01	4452.38	4	4	1.37E-01
2365.14	4	2	1.52E-01	4084.65	6	8	7.28E-02	4466.24	2	4	9.00E-01
2375.72	6	4	1.35E-01	4085.11	6	6	4.55E-01	4467.46	2	2	9.00E-01
2406.38	6	4	1.85E-01	4092.93	8	8	2.65E-01	4563.18	4	4	7.18E-03
2407.48	4	4	2.25E-01	4094.14	6	4	4.70E-02	4590.97	6	8	8.85E-01
2411.60	4	2	2.05E-01	4096.53	4	6	1.73E-01	4595.96	6	6	4.87E-02
2411.64	2	2	1.10E-01	4097.22	2	4	3.62E-01	4596.18	4	6	8.34E-01
2415.13	4	2	2.20E-01	4103.00	2	2	5.09E-01	4638.86	2	4	3.71E-01

λ Å	Weights g_i	g_k	A 10^8 s^{-1}	λ Å	Weights g_i	g_k	A 10^8 s^{-1}	λ Å	Weights g_i	g_k	A 10^8 s^{-1}
4641.81	4	6	5.96E-01	263.817	5	7	5.97E+01	3017.62	7	7	5.38E-01
4649.13	6	8	7.81E-01	263.861	5	5	1.49E+01	3023.43	3	5	4.79E-01
4650.84	2	2	6.86E-01	277.386	5	7	9.43E+01	3024.36	7	5	9.39E-02
4661.63	4	4	4.10E-01	279.788	5	5	4.25E+01	3024.54	1	3	6.16E-01
4673.73	4	2	1.35E-01	295.942	1	3	5.56E+01	3035.41	3	3	4.59E-01
4676.23	6	6	2.05E-01	303.413	1	3	4.29E+01	3042.07	3	1	1.94E+00
4690.89	2	4	1.86E-01	303.461	3	1	1.29E+02	3047.10	5	5	1.49E+00
4691.42	2	2	7.43E-01	303.517	3	3	3.21E+01	3059.28	5	3	8.72E-01
4696.35	6	4	3.25E-02	303.622	3	5	3.21E+01	3064.98	1	3	2.17E-01
4698.44	6	6	6.59E-02	303.695	5	3	5.34E+01	3068.13	3	1	6.49E-01
4699.01	6	8	9.88E-01	303.800	5	5	9.61E+01	3068.26	3	3	5.41E-02
4699.22	4	6	9.36E-01	305.596	1	3	1.20E+02	3068.67	5	3	2.27E-01
4701.18	4	4	9.23E-01	305.656	3	5	1.62E+02	3074.14	5	7	1.84E-01
4701.71	4	2	3.69E-01	305.702	3	3	9.01E+01	3074.72	5	3	3.76E-01
4703.16	4	6	9.20E-01	305.767	5	7	2.16E+02	3075.13	5	5	1.61E-01
4705.35	6	8	1.10E+00	305.836	5	5	5.40E+01	3075.95	7	9	1.07E-01
4710.01	4	6	2.98E-01	320.978	5	7	2.17E+02	3083.65	7	7	3.20E-01
4741.70	6	6	4.71E-02	328.448	5	5	1.04E+02	3084.64	7	5	2.55E-01
4751.28	6	8	6.39E-02	345.312	1	3	1.35E+02	3088.04	9	9	5.30E-01
4752.69	6	6	1.45E-02	374.073	5	5	2.85E+01	3095.79	9	7	1.35E-01
4844.92	4	6	1.02E-02	395.557	5	3	2.80E+01	3115.67	3	1	1.39E+00
4856.39	4	6	5.58E-02	507.388	1	3	1.61E+01	3121.63	3	3	1.38E+00
4856.76	4	4	1.00E-01	507.680	3	3	4.82E+01	3132.79	3	5	1.37E+00
4860.97	2	4	4.70E-01	508.178	5	3	8.04E+01	3198.18	3	5	9.57E-02
4864.88	4	2	8.07E-02	525.794	5	3	9.60E+01	3201.14	3	3	4.77E-01
4871.52	4	6	5.60E-01	597.814	1	3	1.49E+01	3202.51	5	7	7.08E-02
4872.02	4	4	9.34E-02	599.590	5	5	5.41E+01	3207.61	5	5	4.40E-01
4890.86	4	2	4.80E-01	702.337	1	3	6.06E+00	3210.58	5	3	1.58E-01
4906.83	4	4	4.54E-01	702.838	3	1	1.83E+01	3216.07	7	7	5.58E-01
4924.53	4	6	5.43E-01	832.929	1	3	3.41E+00	3221.21	7	5	9.75E-02
4941.07	2	4	5.87E-01	835.092	5	5	1.44E+00	3260.86	5	7	1.68E+00
4943.01	4	6	7.78E-01	835.289	5	7	5.99E+00	3265.33	7	9	1.88E+00
4955.71	4	4	1.82E-01	1679.03	3	5	6.57E-01	3267.20	3	5	1.58E+00
5159.94	2	2	3.29E-01	1686.73	3	3	6.48E-01	3281.83	5	5	2.89E-01
5175.90	4	2	1.49E-01	1760.41	3	5	8.38E-01	3284.45	7	7	2.06E-01
5190.50	2	4	1.26E-01	1764.46	5	5	2.50E+00	3299.39	1	3	1.64E-01
5206.65	4	4	3.58E-01	1766.63	1	3	1.11E+00	3312.33	3	3	4.60E-01
5583.22	2	4	2.17E-02	1772.28	3	1	3.29E+00	3326.06	3	3	2.65E-01
5611.07	2	2	2.14E-02	1772.97	5	3	1.37E+00	3330.30	3	5	6.81E-01
6627.37	4	4	1.73E-01	2390.43	3	3	1.62E+00	3330.32	3	5	4.76E-01
6641.03	2	2	9.88E-02	2454.97	3	1	3.43E+00	3332.41	5	3	7.92E-01
6666.66	4	2	6.78E-02	2665.68	3	5	6.75E-01	3332.93	5	7	5.04E-01
6677.87	2	4	3.37E-02	2674.58	5	5	1.11E+00	3336.67	3	3	3.76E-01
6717.75	2	2	1.33E-01	2683.66	3	1	1.85E+00	3336.69	5	5	8.77E-02
6721.39	4	2	1.81E-01	2686.15	7	5	1.54E+00	3340.76	5	3	6.57E-01
6810.48	6	8	1.64E-03	2687.55	3	3	1.84E+00	3344.20	5	5	1.25E-01
6844.10	4	6	2.97E-03	2695.48	3	5	1.82E+00	3344.51	5	7	3.48E-01
6846.80	8	8	3.17E-02	2794.14	3	1	1.82E-01	3347.98	7	5	4.86E-01
6869.48	6	6	5.35E-02	2798.93	3	3	4.52E-02	3350.62	5	3	1.12E+00
6884.88	4	4	6.12E-02	2809.66	5	3	1.34E-01	3350.92	7	7	9.91E-01
6895.10	10	8	2.72E-01	2818.70	5	5	2.66E-02	3355.86	7	7	6.89E-01
6906.44	8	6	2.48E-01	2836.31	7	5	1.46E-01	3362.31	7	5	6.87E-01
6907.87	4	2	3.03E-01	2959.69	3	5	1.83E+00	3376.61	3	1	1.49E+00
6910.56	6	4	2.43E-01	2983.78	3	5	2.15E+00	3376.76	3	3	1.12E+00
				2992.08	3	5	9.32E-02	3377.26	3	5	5.20E-01
O III				2996.48	3	3	4.64E-01	3382.61	5	7	9.86E-01
263.694	1	3	3.32E+01	2997.69	5	7	6.88E-02	3383.31	5	3	3.70E-01
263.727	3	5	4.48E+01	3004.34	5	5	4.27E-01	3383.81	5	5	8.62E-01
263.773	3	3	2.49E+01	3008.78	5	3	1.53E-01	3384.90	7	9	1.48E+00

λ Å	Weights g_i	g_k	A 10^8 s^{-1}
3394.22	7	7	4.88E-01
3395.43	7	5	9.75E-02
3406.88	1	3	1.93E-01
3408.13	3	1	5.79E-01
3415.26	3	3	1.44E-01
3428.63	3	5	1.42E-01
3430.57	5	3	2.37E-01
3444.05	5	5	4.21E-01
3446.68	3	5	9.71E-01
3447.15	1	3	8.09E-01
3447.97	5	7	1.19E+00
3450.91	7	9	1.44E+00
3451.30	3	3	8.06E-01
3454.84	5	5	6.89E-01
3454.99	9	11	1.72E+00
3459.48	5	3	1.14E-01
3459.94	7	7	5.14E-01
3466.13	9	9	2.84E-01
3466.85	7	5	6.82E-02
3475.24	9	7	2.42E-02
3520.94	1	3	1.50E-01
3531.22	3	1	4.45E-01
3533.38	3	3	1.11E-01
3534.90	3	5	1.11E-01
3555.24	5	3	1.82E-01
3556.78	5	5	3.26E-01
3695.38	3	5	4.01E-01
3698.72	5	7	7.62E-01
3703.36	7	9	1.14E+00
3704.75	3	3	8.53E-01
3707.27	3	5	7.34E-01
3709.54	3	1	1.13E+00
3712.49	5	5	6.59E-01
3714.03	3	3	4.06E-01
3715.09	5	7	9.73E-01
3720.89	7	7	3.74E-01
3721.95	5	3	2.80E-01
3725.31	5	5	2.41E-01
3728.51	5	7	1.29E+00
3728.84	7	9	1.45E+00
3729.80	3	5	1.22E+00
3732.13	5	3	2.67E-02
3734.83	7	5	7.40E-02
3742.63	5	5	2.24E-01
3746.90	7	7	1.59E-01
3754.70	3	5	7.53E-01
3757.23	1	3	5.56E-01
3759.88	5	7	9.79E-01
3774.03	3	3	3.91E-01
3791.28	5	5	2.24E-01
3810.98	5	3	2.37E-02
3816.75	5	3	9.63E-02
3961.57	5	7	1.25E+00
4072.64	1	3	3.37E-01
4073.98	3	5	4.54E-01
4081.02	5	7	6.02E-01
4089.30	3	3	2.49E-01
4103.07	5	5	1.48E-01
4118.60	5	3	1.63E-02

λ Å	Weights g_i	g_k	A 10^8 s^{-1}
4440.09	5	3	4.42E-01
4447.69	5	5	4.40E-01
4461.61	5	7	4.36E-01
4524.22	3	1	3.38E-01
4532.78	5	3	1.40E-01
4535.29	3	3	8.40E-02
4555.39	5	5	2.49E-01
4557.91	3	5	8.27E-02
5268.30	1	3	3.50E-01
5508.24	5	5	1.06E-01
5592.25	3	3	3.27E-01

O IV

λ Å	Weights g_i	g_k	A 10^8 s^{-1}
238.360	2	4	2.96E+02
238.570	4	6	3.54E+02
238.579	4	4	5.90E+01
279.631	2	2	2.68E+01
279.933	4	2	5.34E+01
553.329	2	4	1.22E+01
554.076	2	2	4.86E+01
554.513	4	4	6.06E+01
555.263	4	2	2.41E+01
608.397	2	2	1.21E+01
609.829	4	2	2.40E+01
616.952	6	4	2.60E+01
617.005	4	4	2.89E+00
617.036	4	2	2.89E+00
624.619	2	4	1.07E+01
625.127	4	4	2.13E+01
625.853	6	4	3.19E+01
779.736	6	4	1.46E+00
779.820	4	4	1.31E+01
779.912	6	6	1.36E+01
779.997	4	6	9.70E-01
787.710	2	4	5.95E+00
790.112	4	4	1.18E+00
790.199	4	6	7.08E+00
921.296	2	4	2.21E+00
921.365	2	2	8.83E+00
923.367	4	4	1.10E+01
923.436	4	2	4.39E+00
1338.61	2	4	2.17E+00
1342.99	4	4	4.29E-01
1343.51	4	6	2.57E+00
2120.58	2	2	1.05E+00
2132.64	4	4	1.29E+00
2493.39	2	4	1.18E+00
2493.75	4	6	8.48E-01
2493.99	2	2	6.09E-01
2499.27	2	2	4.68E-01
2501.81	4	4	3.73E-01
2507.73	4	2	2.32E+00
2509.22	6	6	1.94E+00
2510.58	4	2	1.19E+00
2517.37	6	4	1.24E+00
2781.22	2	2	1.03E-01
2803.57	6	4	1.26E-01
2805.87	2	4	2.90E-01
2812.50	6	6	3.58E-02

λ Å	Weights g_i	g_k	A 10^8 s^{-1}
2816.53	4	4	5.74E-01
2829.17	8	6	1.56E-01
2836.27	6	4	8.43E-01
2916.31	2	4	1.06E+00
2921.46	4	6	1.27E+00
2926.18	4	4	2.11E-01
3063.43	2	4	1.30E+00
3071.60	2	2	1.29E+00
3177.89	2	4	7.59E-02
3180.77	2	2	1.51E-01
3180.99	4	6	7.06E-02
3185.74	4	4	1.21E-01
3188.22	6	8	4.28E-02
3188.64	4	2	1.50E-01
3194.78	6	6	1.71E-01
3199.58	6	4	1.04E-01
3209.65	8	8	2.53E-01
3216.31	8	6	5.56E-02
3348.06	2	4	8.51E-01
3349.11	4	6	1.02E+00
3354.27	4	2	7.71E-01
3362.55	4	4	7.65E-01
3375.40	4	6	7.56E-01
3378.02	4	4	1.66E-01
3381.21	4	6	7.19E-01
3381.30	2	4	4.28E-01
3385.52	6	8	1.02E+00
3390.19	2	2	8.49E-01
3396.80	4	4	5.40E-01
3405.77	4	2	1.67E-01
3409.70	6	6	3.00E-01
3411.30	4	4	1.69E-01
3411.69	4	6	1.02E+00
3425.55	6	4	4.94E-02
3489.89	4	6	7.29E-01
3492.21	2	4	6.06E-01
3493.43	4	4	1.21E-01
3560.39	4	6	1.03E+00
3563.33	6	8	1.10E+00
3593.08	6	6	7.15E-02
3725.89	2	4	5.61E-01
3725.94	4	6	6.01E-01
3729.03	6	8	6.86E-01
3736.68	4	4	2.23E-01
3736.85	8	10	7.95E-01
3744.89	6	6	1.92E-01
3758.39	8	8	1.11E-01
3930.68	2	2	3.80E-02
3942.06	2	4	9.42E-02
3945.31	4	2	1.88E-01
3956.77	4	4	2.98E-02
3974.58	4	6	6.62E-02
3977.09	6	4	9.91E-02
3995.08	6	6	1.52E-01
4687.03	2	4	2.79E-01
4772.60	2	4	1.23E-01
4779.10	2	2	2.45E-01
4783.42	4	6	2.06E-01
4794.18	4	4	1.56E-01

λ Å	Weights g_i	g_k	A 10^8 s⁻¹	λ Å	Weights g_i	g_k	A 10^8 s⁻¹	λ Å	Weights g_i	g_k	A 10^8 s⁻¹
4798.27	6	8	2.91E-01	4213.35	5	3	1.19E-02	2149.1	4	2	3.18E+00
4813.15	6	6	8.65E-02	4522.66	5	3	1.02E-02	2152.9	2	4	4.85E-01
5305.51	4	4	6.10E-02	4554.53	3	5	2.41E-01	2154.1	4	4	1.73E-01
5362.51	6	6	6.12E-02	5114.06	1	3	1.80E-01	2154.1	4	6	5.8E-01
6876.49	2	4	1.88E-02	5339.94	1	3	1.85E-02	2534.0	2	4	2.00E-01
6931.60	2	2	7.35E-02	5349.74	3	1	7.04E-02	2535.6	4	4	9.5E-01
7004.11	4	4	8.90E-02	5372.71	3	3	1.42E-02	2553.3	2	2	7.1E-01
7061.30	4	2	3.48E-02	5414.59	3	5	9.29E-03	2554.9	4	2	3.00E-01
				5428.38	5	3	2.68E-02				
O V				5471.12	5	5	4.86E-02	**P II**			
172.169	1	3	2.94E+02	5571.81	1	3	8.33E-02	1301.9	1	3	5.0E-01
*192.85	9	15	6.90E+02	5580.12	3	5	1.11E-01	1304.5	3	1	1.5E+00
*215.17	9	3	1.83E+02	5583.23	3	3	6.20E-02	1304.7	3	3	3.7E-01
220.353	3	5	4.292E+02	*5589.9	9	15	1.49E-01	1305.5	3	5	3.8E-01
248.460	3	1	5.59E+01	5597.89	5	7	1.48E-01	1309.9	5	3	6.2E-01
629.732	1	3	2.872E+01	5604.27	5	5	3.68E-02	1310.7	5	5	1.1E+00
758.677	3	5	5.547E+00	5607.41	5	3	4.08E-03	4475.3	5	7	1.3E+00
759.442	1	3	7.373E+00	6330.05	5	7	1.21E-01	4499.2	5	7	1.4E+00
760.227	3	3	5.514E+00	6460.12	3	5	9.37E-02	4530.8	3	5	1.0E+00
760.446	5	5	1.652E+01	6466.14	5	7	1.01E-01	4554.8	3	5	9.6E-01
761.128	3	1	2.197E+01	6500.24	7	9	1.11E-01	4588.0	5	7	1.7E+00
762.004	5	3	9.125E+00	6543.77	5	5	1.64E-02	4589.9	3	5	1.6E+00
774.518	3	1	3.804E+01	6601.28	7	7	1.14E-02	4602.1	7	9	1.9E+00
1371.30	3	5	3.336E+00	6764.72	1	3	4.37E-02	4943.5	7	5	6.3E-01
2729.31	3	5	4.52E-01	6789.62	3	5	5.79E-02	5253.5	3	5	1.0E+00
2731.45	1	3	5.90E-01	6817.40	3	3	3.00E-02	5425.9	5	5	6.9E-01
2743.61	3	3	4.38E-01	6828.95	5	7	7.35E-02	6024.2	3	5	5.1E-01
2752.23	3	1	1.82E+00	6878.76	5	5	1.65E-02	6043.1	5	7	6.8E-01
2755.13	5	5	1.37E+00								
2769.69	5	3	7.88E-01	**O VI**				**P III**			
2781.01	3	5	1.40E+00	*150.10	2	6	2.62E+02	1334.8	2	4	5.5E-01
*2784.0	3	9	1.40E+00	*173.03	6	10	8.78E+02	1344.3	4	6	6.4E-01
2786.99	3	3	1.39E+00	1031.91	2	4	4.16E+00	1344.8	4	4	1.1E-01
2789.85	3	1	1.38E+00	1037.61	2	2	4.09E+00	4057.4	4	4	1.0E-01
3058.68	3	5	1.39E+00	3811.35	2	4	5.14E-01	4059.3	6	4	9.0E-01
3144.66	3	5	8.86E-01	3834.24	2	2	5.05E-01	4080.1	4	2	9.9E-01
3219.24	3	1	1.54E-01								
3222.29	1	3	1.16E-01	**O VII**				**Potassium**			
3227.54	3	3	3.38E-02	18.6270	1	3	9.365E+03	**K I**			
3239.21	3	3	3.28E-01	21.6020	1	3	3.309E+04	4044.1	2	4	1.24E-02
3248.28	5	3	1.18E-01	*120.33	3	9	5.334E+02	4047.2	2	2	1.24E-02
3263.54	5	5	1.86E-02	128.411	1	3	8.982E+02	5084.2	2	2	3.50E-03
3275.64	5	3	4.76E-01	*128.46	9	15	1.615E+03	5099.2	4	2	7.0E-03
3297.62	7	5	1.30E-01	135.820	3	5	1.523E+03	5323.3	2	2	6.3E-03
3690.17	3	5	1.97E-02	*1630.3	3	9	7.935E-01	5339.7	4	2	1.26E-02
3698.36	3	3	1.03E-01	2448.98	1	3	2.514E-01	5343.0	2	4	4.0E-03
3702.72	5	7	1.41E-02	*5933.1	3	9	1.002E-01	5359.6	4	6	4.6E-03
3717.31	5	5	9.63E-02	8241.76	1	3	3.864E-02	5782.4	2	2	1.23E-02
3725.63	5	3	2.91E-02					5801.8	4	2	2.46E-02
3746.64	7	7	1.18E-01	**Phosphorus**				5812.2	2	4	2.8E-03
3761.58	7	5	1.61E-02	**P I**				5831.9	4	6	3.2E-03
4119.37	3	5	3.66E-01	1671.7	4	2	3.9E-01	6911.1	2	2	2.72E-02
4120.49	3	1	3.33E-01	1674.6	4	4	4.0E-01	6938.8	4	2	5.4E-02
4123.96	5	7	4.81E-01	1679.7	4	6	3.9E-01	7664.9	2	4	3.87E-01
4125.49	1	3	2.70E-01	1775.0	4	6	2.17E+00	7699.0	2	2	3.82E-01
4134.11	3	3	3.34E-01	1782.9	4	4	2.14E+00				
4153.27	3	3	1.92E-01	1787.7	4	2	2.13E+00	**K II**			
4158.86	3	5	3.39E-01	2135.5	4	4	2.11E-01	607.93	1	3	1.3E-02
4178.46	5	5	1.12E-01	2136.2	6	4	2.83E+00				

λ Å	Weights g_i	Weights g_k	A 10^8 s^{-1}	λ Å	Weights g_i	Weights g_k	A 10^8 s^{-1}	λ Å	Weights g_i	Weights g_k	A 10^8 s^{-1}
K III				3121.76	6	6	1.1E-01	4056.34	6	4	9.5E-03
2550.0	6	4	2.0E+00	3123.70	10	8	4.6E-02	4082.78	6	4	1.4E-01
2635.1	4	4	1.2E+00	3137.71	4	6	3.3E-02	4097.52	2	4	7.0E-02
2992.4	6	8	2.5E+00	3189.05	6	6	3.03E-01	4121.68	6	6	9.8E-02
3052.1	4	6	1.7E+00	3197.13	6	4	4.35E-02	4128.87	6	8	1.73E-01
3202.0	4	4	1.8E+00	3263.14	6	6	1.3E-01	4135.27	8	8	1.0E-01
3289.1	4	6	2.0E+00	3271.61	6	4	2.0E-01	4196.50	6	8	3.9E-02
3322.4	6	6	1.3E+00	3280.55	8	8	2.36E-01	4211.14	8	10	1.62E-01
3421.8	2	4	1.5E+00	3283.57	6	8	4.4E-01	4244.44	4	4	6.5E-03
				3289.14	4	4	1.0E-01	4278.60	4	6	9.2E-03
K XVI				3323.09	8	10	6.3E-01	4288.71	6	8	6.1E-02
206.27	1	3	9.4E+01	3331.09	4	2	5.40E-02	4373.04	2	4	1.8E-02
				3338.54	8	6	3.5E-02	4374.80	8	10	1.64E-01
K XVII				3360.80	4	4	1.2E-01	4379.92	6	6	2.48E-02
22.020	2	4	4.7E+04	3368.38	6	4	1.1E-01	4492.47	6	6	4.5E-03
22.163	4	6	5.6E+04	3396.82	10	10	6.5E-01	4528.72	6	8	1.35E-02
22.18	4	4	9.3E+03	3399.70	6	8	1.2E-01	4548.73	4	6	5.5E-03
22.60	2	2	2.5E+03	3462.04	6	6	6.2E-01	4551.64	4	4	4.00E-02
22.76	4	2	4.7E+03	3470.66	4	4	8.5E-01	4565.19	4	4	1.1E-02
				3478.91	6	6	3.32E-01	4569.00	6	8	1.0E-02
Praseodymium				3484.04	6	8	9.3E-03	4608.12	2	2	2.1E-02
Pr II				3498.73	4	6	2.12E-01	4675.03	8	8	6.4E-03
3997.0	15	15	1.87E-01	3502.52	10	10	4.3E-01	4721.00	6	4	3.43E-03
4062.8	13	15	1.00E+00	3507.32	6	8	3.4E-01	4745.11	6	6	5.2E-03
4100.7	17	19	8.4E-01	3528.02	8	8	8.5E-01	4755.58	4	4	6.0E-03
4143.1	15	17	5.8E-01	3543.95	4	4	4.65E-01	4842.43	6	8	1.6E-03
4179.4	13	15	5.2E-01	3549.54	6	6	2.22E-01	4963.71	2	2	3.0E-02
4222.9	11	13	3.91E-01	3570.18	4	6	1.82E-01	4977.75	4	4	9.8E-03
4241.0	17	15	2.30E-01	3583.10	8	10	2.6E-01	4979.18	4	6	1.0E-02
4359.8	15	15	1.1E-01	3596.19	6	4	5.5E-01	5090.63	6	6	5.0E-03
4405.8	17	17	9.0E-02	3597.15	6	8	5.9E-01	5120.69	6	8	3.1E-03
4429.3	15	15	2.28E-01	3612.47	4	2	8.90E-01	5130.76	4	4	4.35E-03
4449.8	13	13	1.24E-01	3620.46	6	4	8.5E-02	5155.54	2	4	9.8E-03
4468.7	11	13	1.54E-01	3654.87	8	8	6.0E-02	5184.19	6	8	1.6E-03
4510.2	13	15	1.16E-01	3657.99	8	6	8.8E-01	5212.73	4	2	5.95E-03
4534.2	15	17	4.9E-02	3666.22	6	8	8.4E-02	5292.14	10	10	3.7E-03
4734.2	15	13	2.5E-02	3690.70	6	4	3.23E-01	5390.44	4	6	9.5E-03
4879.1	15	15	1.8E-02	3692.36	10	8	9.1E-01	5424.72	4	4	5.0E-03
4886.0	15	13	1.3E-02	3700.91	8	10	3.9E-01	5599.42	6	8	1.3E-02
4912.6	17	15	5.7E-02	3713.02	4	4	8.3E-02	5983.60	10	10	2.1E-02
5034.4	19	19	1.1E-01	3788.47	4	6	1.4E-01				
5110.8	21	19	2.78E-01	3793.22	8	6	4.2E-01	**Rubidium**			
5135.1	17	17	1.25E-01	3799.31	8	8	5.5E-01	**Rb I**			
5173.9	19	17	3.18E-01	3806.76	6	6	6.2E-02	3022.5	2	4	4.13E-05
5219.1	15	15	9.5E-02	3818.19	6	4	5.8E-02	3032.0	2	4	4.93E-05
5220.1	17	15	2.35E-01	3822.26	6	6	8.5E-01	3044.2	2	4	8.2E-05
5251.7	15	13	1.1E-02	3828.48	6	6	6.2E-01	3060.2	2	4	1.05E-04
5259.7	15	13	2.24E-01	3833.89	6	4	5.8E-01	3082.0	2	4	1.49E-04
5292.6	13	13	9.3E-02	3856.52	8	10	5.9E-01	3112.6	2	4	2.5E-04
5810.6	17	19	2.3E-02	3872.39	4	6	6.7E-03	3113.1	2	2	1.3E-04
5879.3	15	15	7.6E-02	3877.34	8	6	3.7E-02	3157.5	2	4	3.38E-04
6200.8	15	17	1.8E-02	3913.51	8	8	2.5E-03	3158.3	2	2	2.0E-04
6278.7	13	15	2.6E-02	3922.19	4	2	6.25E-02	3228.0	2	4	6.4E-04
6398.0	11	13	1.9E-02	3934.23	8	8	1.58E-01	3229.2	2	2	3.8E-04
				3942.72	4	2	7.15E-01	3348.7	2	4	1.37E-03
Rhodium				3958.86	6	8	5.5E-01	3350.8	2	2	8.9E-04
Rh I				3984.40	4	4	1.1E-01	3587.1	2	4	3.97E-03
3083.96	8	6	4.8E-02	3995.61	4	6	4.7E-02	3591.6	2	2	2.9E-03
3114.91	6	4	4.45E-02	4053.44	2	2	2.8E-02	4201.8	2	4	1.8E-02

λ (Å)	Weights g_i	Weights g_k	A (10^8 s^{-1})	λ (Å)	Weights g_i	Weights g_k	A (10^8 s^{-1})	λ (Å)	Weights g_i	Weights g_k	A (10^8 s^{-1})
4215.5	2	2	1.5E-02	4093.12	4	4	1.23E-01	5339.43	6	6	1.06E-01
7800.3	2	4	3.70E-01	4094.86	6	6	1.44E-01	5341.07	4	2	3.8E-01
7947.6	2	2	3.40E-01	4098.36	8	8	8.7E-02	5349.34	6	4	5.9E-01
				4132.98	4	6	1.19E+00	5350.28	8	8	6.8E-02
Scandium				4140.27	6	8	1.17E+00	5355.79	6	4	3.0E-01
Sc I				4147.38	6	6	1.74E-01	5356.10	8	6	5.7E-01
2116.7	4	4	2.0E-01	4161.85	8	8	1.77E-01	5375.37	8	6	3.4E-01
2120.4	6	6	2.0E-01	4171.53	6	4	1.36E-01	5392.06	10	8	4.2E-01
2262.3	4	4	5.8E-02	4186.42	6	8	8.4E-02	5416.16	4	6	4.4E-02
2266.6	4	2	4.8E-01	4187.61	8	6	1.28E-01	5416.41	6	6	2.0E-02
2270.9	6	4	4.6E-01	4193.53	4	6	6.1E-02	5425.55	6	8	4.5E-02
2280.8	4	6	2.8E-01	4204.52	6	8	3.5E-02	5429.42	2	4	9.0E-02
2289.6	6	6	4.1E-02	4205.20	10	8	1.12E-01	5432.98	4	4	5.4E-02
2311.29	4	6	4.1E-02	4212.32	4	6	1.58E-01	5433.25	6	4	9.7E-02
2315.69	4	4	2.5E-01	4212.48	6	6	8.6E-02	5438.28	4	6	3.4E-02
2320.32	6	6	2.4E-01	4216.08	2	4	2.36E-01	5439.04	2	2	1.74E-01
2324.75	6	4	4.1E-02	4218.23	4	4	2.26E-01	5442.62	4	2	2.15E-01
2328.19	4	6	4.6E-02	4225.54	6	8	9.5E-02	5446.20	8	8	2.8E-01
2334.67	4	2	1.7E-01	4225.69	4	6	7.6E-02	5451.37	6	6	1.50E-01
2346.03	6	4	1.3E-01	4231.64	4	4	1.31E-01	5455.24	4	4	6.6E-02
2429.19	4	4	2.8E-01	4233.59	6	6	4.0E-01	5464.95	4	2	3.2E-01
2438.63	6	6	2.1E-01	4238.05	8	8	7.1E-01	5468.40	6	4	9.7E-02
2468.40	4	2	4.9E-02	4239.55	6	4	2.27E-01	5472.19	8	6	9.7E-02
2692.78	4	2	1.61E-01	4246.14	8	6	1.15E-01	5482.01	8	8	5.2E-01
2699.02	4	6	2.4E-02	4542.55	6	4	1.28E-01	5484.63	6	6	5.2E-01
2706.74	4	4	3.1E-01	4544.67	8	6	1.33E-01	5514.23	6	8	4.1E-01
2707.93	6	4	1.49E-01	4706.94	4	6	2.81E-01	5520.52	8	10	4.3E-01
2711.34	6	6	3.2E-01	4709.31	6	8	4.0E-01	5526.10	4	4	7.1E-02
2965.88	4	6	7.5E-02	4711.72	2	4	1.81E-01	5541.07	6	6	5.5E-02
2974.01	4	4	5.5E-01	4714.30	4	4	2.14E-01	5631.04	2	4	3.0E-02
2980.76	6	6	5.4E-01	4719.31	6	6	1.04E-01	5671.83	10	12	5.4E-01
2988.97	6	4	6.9E-02	4728.77	8	8	1.16E-01	5686.86	8	10	4.9E-01
3015.37	4	6	7.8E-01	4729.20	4	4	2.20E-01	5700.19	6	8	4.6E-01
3019.35	6	8	8.7E-01	4729.24	6	6	1.93E-01	5708.64	10	10	4.7E-02
3030.76	6	6	1.00E-01	4734.11	4	2	1.10E+00	5711.79	4	6	4.5E-01
3255.68	4	4	3.2E-01	4737.65	6	4	8.8E-01	5717.31	8	8	7.5E-02
3269.90	4	2	3.13E+00	4741.02	8	6	9.1E-01	5724.13	6	6	7.4E-02
3273.63	6	4	2.81E+00	4743.82	10	8	9.8E-01	5988.43	6	6	6.6E-02
3907.48	4	6	1.66E+00	4973.67	4	2	8.4E-01	6026.16	4	4	7.2E-02
3911.81	6	8	1.79E+00	4980.36	6	4	5.6E-01	6146.20	6	8	4.2E-02
3933.38	6	6	1.62E-01	4983.43	4	4	2.58E-01	6198.43	4	6	3.5E-02
3996.60	4	6	1.65E-01	4991.91	6	6	3.8E-01	6249.96	6	8	3.2E-02
4020.39	4	4	1.63E+00	4995.00	4	6	5.9E-02	6262.22	4	6	8.4E-02
4023.22	4	4	3.0E-01	5018.41	6	4	2.09E-01	6280.16	2	4	4.0E-02
4023.68	6	6	1.65E+00	5021.52	4	4	2.30E-01	6284.16	6	6	3.9E-02
4031.38	6	6	2.9E-01	5064.31	8	10	7.3E-02	6284.73	4	4	7.1E-02
4036.86	6	4	7.9E-02	5066.38	6	6	3.6E-02	6293.02	2	2	1.04E-01
4043.80	8	8	3.11E-01	5070.17	6	8	1.16E-01	7741.16	10	10	3.8E-02
4047.80	6	4	1.54E-01	5072.71	2	4	2.0E-02	7800.42	8	8	5.1E-02
4051.83	8	6	7.7E-02	5075.82	4	6	1.15E-01				
4054.54	4	2	1.67E-01	5080.22	4	4	4.1E-02	**Sc II**			
4067.00	6	8	1.91E-01	5081.56	10	10	7.6E-01	1880.6	5	3	5.0E+00
4067.63	10	8	4.1E-02	5083.72	8	8·	6.2E-01	2064.3	7	5	2.2E+00
4074.96	4	6	3.7E-01	5085.55	6	6	5.7E-01	2068.0	5	3	2.0E+00
4078.56	2	4	4.3E-01	5086.94	4	4	6.6E-01	2273.1	1	3	7.7E+00
4080.57	4	4	6.6E-02	5096.72	6	4	1.69E-01	2545.20	5	5	4.0E-01
4082.39	6	4	2.73E-01	5099.27	4	6	1.50E-01	2552.35	7	5	2.21E+00
4086.66	6	8	3.7E-01	5101.12	10	8	8.8E-02	2555.79	3	3	6.9E-01
4087.47	4	6	1.12E-01	5331.79	4	4	1.11E-01	2560.23	5	3	2.01E+00

NIST ATOMIC TRANSITION PROBABILITY TABLES (continued)

λ Å	g_i	g_k	A 10^8 s^{-1}	λ Å	g_i	g_k	A 10^8 s^{-1}	λ Å	g_i	g_k	A 10^8 s^{-1}
2563.19	3	1	2.70E+00	5526.79	9	7	3.3E-01	1260.4	2	4	2.0E+01
2611.19	5	5	2.2E+00	5657.91	5	5	1.04E-01	1264.7	4	6	2.3E+01
2667.70	3	5	1.5E+00	5669.06	3	1	1.31E-01	1304.4	2	2	3.6E+00
2746.36	3	1	3.9E+00					1309.3	4	2	7.0E+00
2782.31	5	5	1.3E+00	**Silicon**				1526.7	2	2	3.73E+00
2789.15	7	7	1.3E+00	**Si I**				1533.5	4	2	7.4E+00
2801.31	9	9	1.3E+00	1977.6	1	3	1.8E-01	1808.0	2	4	3.7E-02
2819.49	3	5	2.3E+00	1979.2	3	1	5.1E-01	2904.3	4	6	6.7E-01
2822.12	5	7	2.5E+00	1980.6	3	3	1.3E-01	2905.7	6	8	7.1E-01
2826.64	7	9	2.8E+00	1983.2	3	5	1.4E-01	3210.0	4	6	4.6E-01
2870.85	5	3	1.1E+00	1986.4	5	3	2.1E-01	4128.1	4	6	1.32E+00
2912.98	5	3	1.1E+00	1989.0	5	5	4.1E-01	4130.9	6	8	1.42E+00
2979.68	3	5	1.2E+00	2208.0	1	3	3.11E-01	5041.0	2	4	9.8E-01
2988.92	5	7	2.9E+00	2210.9	3	5	4.16E-01	5056.0	4	6	1.2E+00
3039.92	7	9	3.5E+00	2211.7	3	3	2.32E-01	5957.6	2	2	4.2E-01
3045.73	5	7	3.68E+00	2216.7	5	7	5.5E-01	5978.9	4	2	8.1E-01
3052.92	7	9	3.92E+00	2218.1	5	5	1.38E-01	6347.1	2	4	7.0E+00
3060.54	7	7	3.0E-01	2506.9	3	5	4.66E-01	6371.4	2	2	6.9E-01
3065.12	9	11	4.00E+00	2514.3	1	3	6.1E-01	7848.8	4	6	3.9E-01
3075.36	9	9	2.5E-01	2516.1	5	5	1.21E+00	7849.7	6	8	4.2E-01
3128.27	3	3	1.9E+00	2519.2	3	3	4.56E-01				
3133.07	5	5	1.8E+00	2524.1	3	1	1.81E+00	**Si III**			
3139.72	7	7	2.1E+00	2528.5	5	3	7.7E-01	883.40	5	7	6.3E+01
3190.98	3	3	1.1E+00	2532.4	1	3	2.6E-01	994.79	3	3	7.89E+00
3199.33	5	3	1.9E+00	2631.3	1	3	9.7E-01	997.39	5	3	1.31E+01
3312.72	5	7	1.2E+00	2881.6	5	3	1.89E+00	1141.6	3	5	3.0E+01
3320.40	5	3	1.2E+00	3905.5	1	3	1.18E-01	1144.3	5	7	3.9E+01
3343.23	9	7	1.1E+00	4738.8	3	3	1.0E-02	1161.6	5	5	1.6E+01
3353.72	5	7	1.51E+00	4783.0	5	3	1.7E-02	1206.5	1	3	2.59E+01
3359.67	5	5	2.16E-01	4792.3	5	5	1.7E-02	1206.5	3	5	4.89E+01
3361.26	3	3	3.4E-01	4818.1	5	7	1.1E-02	1207.5	5	5	1.9E+01
3361.93	3	1	1.17E+00	4821.2	3	5	8.0E-03	1294.5	3	5	5.42E+00
3368.94	5	3	8.3E-01	4947.6	3	1	4.2E-02	1296.7	1	3	7.19E+00
3372.15	7	5	9.9E-01	5006.1	3	5	2.8E-02	1298.9	3	3	5.36E+00
3379.16	3	3	2.5E+00	5622.2	3	3	1.6E-02	1299.0	5	5	1.61E+01
3535.71	5	3	6.1E-01	5690.4	3	3	1.2E-02	1301.2	3	1	2.13E+01
3558.53	5	7	3.0E-01	5708.4	5	5	1.4E-02	1303.3	5	3	8.85E+00
3567.70	3	5	3.5E-01	5754.2	5	3	1.5E-02	1328.8	1	3	2.7E+01
3572.53	7	7	1.38E+00	5772.1	3	1	3.6E-02	1417.2	3	1	2.60E+01
3576.34	5	5	1.06E+00	5948.5	3	5	2.2E-02	1435.8	5	7	2.1E+01
3580.93	3	3	1.23E+00	7226.2	3	5	7.9E-03	1589.0	5	3	1.1E+01
3589.63	5	3	4.6E-01	7405.8	3	5	3.7E-02	1778.7	7	9	4.4E+00
3590.47	7	5	2.9E-01	7409.1	5	7	2.3E-02	1783.1	5	7	3.8E+00
3613.83	7	9	1.48E+00	7680.3	3	5	4.6E-02	3241.6	5	3	2.3E+00
3630.74	5	7	1.20E+00	7918.4	3	5	5.2E-02	*3486.9	15	21	1.8E+00
3642.78	3	5	1.13E+00	7932.3	5	7	5.1E-02	3590.5	3	5	3.9E+00
3645.31	7	7	2.74E-01	7944.0	7	9	5.8E-02	4552.6	3	5	1.26E+00
3651.80	5	5	3.0E-01	7970.3	5	5	7.1E-03	4554.0	5	3	7.6E-01
3859.59	7	5	1.1E+00					4567.8	3	3	1.25E+00
4246.82	5	5	1.29E+00	**Si II**				4683.0	5	5	9.5E-01
4314.08	9	7	4.1E-01	989.87	2	4	6.7E+00	4716.7	5	7	2.8E+00
4320.75	7	5	4.0E-01	992.68	4	6	8.0E+00	5451.5	3	5	6.0E-01
4325.00	5	3	4.3E-01	1020.7	2	2	1.3E+00	5473.1	5	7	7.9E-01
4374.46	9	9	1.48E-01	1190.4	2	4	6.9E+00	5716.3	9	7	1.9E-01
4400.39	7	7	1.43E-01	1193.3	2	2	2.8E+01	5739.7	1	3	4.7E-01
4415.54	5	5	1.47E-01	1194.5	4	4	3.6E+01	7462.6	5	3	4.9E-01
4670.41	5	7	1.16E-01	1197.4	4	2	1.4E+01	7466.3	7	5	5.4E-01
5031.01	5	3	3.5E-01	1248.4	4	4	1.3E+01	7612.4	3	5	1.1E+00
5239.81	1	3	1.39E-01	1251.2	6	4	1.9E+01				

λ Å	Weights g_i	g_k	A 10^8 s^{-1}	λ Å	Weights g_i	g_k	A 10^8 s^{-1}	λ Å	Weights g_i	g_k	A 10^8 s^{-1}
Si IV				258.35	4	4	1.4E+02	4747.9	2	2	6.3E-03
457.82	2	4	3.6E+00	261.05	4	2	5.4E+01	4751.8	4	2	1.27E-02
458.16	2	2	3.6E+00	272.00	2	2	3.0E+01	4978.5	2	4	4.1E-02
515.12	2	2	4.1E+00	277.26	4	2	5.7E+01	4982.8	4	4	8.2E-03
516.35	4	2	8.2E+00	287.08	2	4	2.6E+01	4982.8	4	6	4.89E-02
*560.50	6	10	1.0E+00	289.19	4	4	5.0E+01	5148.8	2	2	1.17E-02
*749.94	10	14	1.45E+01	292.22	6	4	7.3E+01	5153.4	4	2	2.33E-02
815.05	2	2	1.23E+01	*347.73	10	10	4.3E+01	5682.6	2	4	1.03E-01
818.13	4	2	2.44E+01	*353.09	6	10	2.1E+01	5688.2	4	6	1.2E-01
*860.74	10	6	1.8E+00					5688.2	4	4	2.1E-02
*1066.6	10	14	3.91E+01	**Si XI**				5890.0	2	4	6.11E-01
1122.5	2	4	2.05E+01	43.763	1	3	6.11E+03	5895.9	2	2	6.10E-01
1128.3	4	4	4.03E+00	*49.116	9	3	2.45E+03	6154.2	2	2	2.6E-02
1128.3	4	6	2.42E+01	49.222	3	5	8.9E+02	6160.8	4	2	5.2E-02
1393.8	2	4	7.73E+00	52.296	3	1	7.6E+02	8183.3	2	4	4.53E-01
1402.8	2	2	7.58E+00	303.30	1	3	6.42E+01	8194.8	4	6	5.4E-01
*1724.1	10	6	5.5E+00	358.29	3	1	1.03E+02	8194.8	4	4	9.0E-02
				358.63	3	5	1.38E+01	11381	2	2	8.9E-02
Si V				361.41	1	3	1.80E+01	11404	4	2	1.76E-01
96.439	1	3	4.8E+02	364.50	3	3	1.32E+01				
97.143	1	3	2.0E+03	365.42	5	5	3.90E+01	**Na II**			
117.86	1	3	3.0E+02	368.28	3	1	5.1E+01	300.15	1	3	3.0E+01
				371.48	5	3	2.07E+01	301.44	1	3	4.9E+01
Si VI				604.14	3	5	1.12E+01	372.08	1	3	3.4E+01
246.00	4	2	1.7E+02	2300.8	1	3	4.34E-01				
249.12	2	2	8.5E+01					**Na III**			
				Si XII				378.14	4	2	7.7E+01
Si VII				*40.924	2	6	4.42E+03	380.10	2	2	3.7E+01
217.83	5	3	4.3E+02	*44.118	6	10	1.4E+04	1991.0	4	6	8.3E+00
272.64	5	3	5.1E+01	499.43	2	4	9.56E+00	2004.2	2	4	4.6E+00
274.18	3	1	1.2E+02	520.72	2	2	8.47E+00	2011.9	6	8	8.4E+00
275.35	5	5	8.9E+01	1862	2	4	1.15E+00	2151.5	2	4	4.4E+00
275.67	3	3	3.0E+01	1949	2	2	1.0E+00	2174.5	4	6	5.3E+00
276.84	1	3	3.9E+01	4620	2	4	4.6E-02	2230.3	6	8	3.7E+00
278.45	3	5	2.9E+01	4942	4	6	4.5E-02	2232.2	4	4	3.3E+00
								2246.7	4	6	2.4E+00
Si VIII				**Silver**				2459.3	4	6	3.0E+00
214.76	4	2	4.1E+02	**Ag I**				2468.9	2	4	2.4E+00
216.92	6	4	3.6E+02	2061.2	2	4	3.1E-02	2497.0	6	6	1.7E+00
232.86	2	2	8.0E+01	2069.9	2	2	1.5E-02				
235.56	4	4	9.7E+01	3280.7	2	4	1.4E+00	**Na V**			
250.45	2	2	7.7E+01	3382.9	2	2	1.3E+00	*307.89	10	6	2.0E+02
250.79	4	2	1.6E+02	5209.1	2	4	7.5E-01	*333.46	6	6	5.6E+01
314.31	4	2	5.2E+01	5465.5	4	6	8.6E-01	*369.01	10	6	1.2E+02
316.20	4	4	5.0E+01	5471.6	4	4	1.4E-01	*400.72	10	10	5.0E+01
319.83	4	6	4.9E+01					*445.14	6	10	7.1E+00
				Sodium				459.90	4	2	2.3E+01
Si IX				**Na I**				461.05	4	4	2.3E+01
223.73	1	3	4.2E+01	3302.4	2	4	2.81E-02	463.26	4	6	2.2E+01
225.03	3	3	1.2E+02	3303.0	2	2	2.81E-02	510.10	2	2	5.6E+01
227.01	5	3	2.0E+02	4390.0	2	4	7.7E-03	511.19	4	4	6.8E+01
227.30	5	3	2.3E+02	4393.3	4	4	1.6E-03				
258.10	5	5	1.04E+02	4393.3	4	6	9.2E-03	**Na VI**			
*294.37	9	9	5.9E+01	4494.2	2	4	1.2E-02	313.75	5	3	1.3E+02
*347.36	9	15	2.2E+01	4497.7	4	6	1.4E-02	361.25	5	5	7.7E+01
				4497.7	4	4	2.4E-03	*416.53	9	9	3.7E+01
Si X				4664.8	2	4	2.33E-02	*492.80	9	15	1.3E+01
253.77	2	4	2.9E+01	4668.6	4	4	4.1E-03	1550.6	5	5	4.35E+00
256.57	2	2	1.1E+02	4668.6	4	6	2.5E-02	1567.8	5	3	2.68E+00

λ Å	Weights g_i	g_k	A 10^8 s^{-1}	λ Å	Weights g_i	g_k	A 10^8 s^{-1}	λ Å	Weights g_i	g_k	A 10^8 s^{-1}
1608.5	3	1	2.6E+00	4607.3	1	3	2.01E+00	4694.1	5	7	6.7E-03
1649.4	5	5	2.05E+00					4695.4	5	5	6.7E-03
1741.5	3	5	2.59E+00	**Sr II**				4696.2	5	3	6.5E-03
1747.5	5	7	3.1E+00	2018.7	2	2	1.2E-01	6403.6	3	5	5.7E-03
				2051.9	4	2	2.4E-01	6408.1	5	5	9.5E-03
Na VII				2282.0	2	4	8.3E-01	6415.5	7	5	1.3E-02
*94.409	6	10	2.7E+03	2322.4	4	6	9.1E-01	*6751.2	15	25	7.9E-02
*105.27	6	2	4.5E+02	2324.5	4	4	1.5E-01	7679.6	3	5	1.2E-02
353.29	4	4	1.0E+02	2423.5	2	2	2.4E-01	7686.1	5	5	2.0E-02
381.30	4	2	4.0E+01	2471.6	2	2	4.8E-01	7696.7	7	5	2.8E-02
397.49	4	4	3.5E+01	3464.5	4	6	3.1E+00				
399.18	6	4	5.2E+01	3474.9	4	4	5.1E-01	**S II**			
*483.28	10	10	2.9E+01	4077.7	2	4	1.42E+00	1124.4	2	4	1.0E+00
486.74	2	4	1.1E+01	4161.8	2	2	6.5E-01	1125.0	4	4	4.6E+00
491.95	4	6	1.3E+01	4215.5	2	2	1.27E+00	1131.0	2	2	3.5E+00
555.80	4	4	2.3E+01	4305.5	4	2	1.4E+00	1131.6	4	2	1.4E+00
777.83	4	6	6.8E+00	4414.8	4	6	1.1E-01	1250.5	4	2	4.6E-01
				4417.5	4	4	1.8E-02	1253.8	4	4	4.2E-01
Na VIII				4585.9	4	2	7.0E-02	1259.5	4	6	3.4E-01
*83.34	9	15	3.94E+03	5303.1	2	4	1.9E-01	4463.6	8	6	5.3E-01
*89.88	9	3	8.09E+02	5379.1	4	6	2.2E-01	4483.4	6	4	3.1E-01
90.536	3	5	2.86E+03	5385.5	4	4	3.7E-02	4486.7	4	2	6.6E-01
411.15	1	3	4.42E+01	5723.7	2	2	7.1E-02	4524.7	4	4	9.3E-02
1239.4	3	3	3.02E+00	5819.0	4	2	1.4E-01	4525.0	6	4	1.2E+00
1802.7	3	1	2.70E+00	8688.9	4	6	5.5E-01	4552.4	4	2	1.2E+00
1867.7	3	5	2.01E+00	8719.6	4	4	9.7E-02	4656.7	2	4	9.0E-02
2059.1	3	5	1.80E+00					4716.2	4	4	2.9E-01
2558.2	5	3	2.26E-02	**Sulfur**				4815.5	6	4	8.8E-01
2772.0	3	5	4.19E-01	**S I**				4885.6	2	4	1.7E-01
3021.0	5	7	4.90E-01	1295.7	5	5	4.9E+00	4917.2	2	2	6.6E-01
3108.9	1	3	2.58E-01	1296.2	5	3	2.7E+00	4924.1	4	6	2.2E-01
3182.3	1	3	2.92E-01	1302.3	3	5	1.8E+00	4925.3	2	4	2.4E-01
				1302.9	3	3	1.6E+00	4942.5	2	2	1.5E-01
Na IX				1303.1	3	1	6.6E+00	4991.9	4	4	1.5E-01
70.615	2	4	1.35E+03	1303.4	5	3	1.9E+00	5009.5	4	2	7.0E-01
70.653	2	2	1.35E+03	1305.9	1	3	2.4E+00	5014.0	4	4	8.4E-01
77.764	2	4	3.6E+03	1401.5	5	3	9.1E-01	5027.2	4	2	2.6E-01
77.911	4	6	4.3E+03	1409.3	3	3	5.0E-01	5032.4	6	6	8.1E-01
681.72	2	4	6.63E+00	1412.9	1	3	1.6E-01	5047.3	4	2	3.6E-01
694.17	2	2	6.30E+00	1425.0	5	7	4.5E+00	5103.3	6	4	5.0E-01
2487.7	2	4	8.32E-01	1425.2	5	5	1.2E+00	5142.3	2	2	1.9E-01
2535.8	2	2	7.89E-01	1433.3	3	5	3.3E+00	5201.0	4	4	7.5E-01
6841.8	2	4	2.59E-02	1433.3	3	3	1.9E+00	5201.3	6	4	6.5E-02
7103.4	4	6	2.78E-02	1437.0	1	3	2.4E+00	5212.6	4	6	9.8E-02
				1448.2	5	3	7.3E+00	5212.6	6	6	8.5E-01
Strontium				1473.0	5	7	4.2E-01	5320.7	6	8	9.2E-01
Sr I				1474.0	5	7	1.6E+00	5345.7	4	6	8.8E-01
2206.2	1	3	6.6E-03	1474.4	5	5	5.0E-01	5345.7	6	6	1.1E-01
2211.3	1	3	8.5E-03	1474.6	5	5	6.2E-02	5428.6	2	4	4.2E-01
2217.8	1	3	1.2E-02	1481.7	3	5	1.7E-01	5432.8	4	6	6.8E-01
2226.3	1	3	1.6E-02	1483.0	3	5	1.2E+00	5453.8	6	8	8.5E-01
2237.7	1	3	2.3E-02	1483.2	3	3	7.5E-01	5473.6	2	2	7.3E-01
2253.3	1	3	3.7E-02	1487.2	1	3	8.7E-01	5509.7	4	4	4.0E-01
2275.3	1	3	6.7E-02	1666.7	5	5	6.3E+00	5526.2	8	8	8.1E-02
2307.3	1	3	1.2E-01	1687.5	1	3	9.4E-01	5536.8	4	6	6.6E-02
2354.3	1	3	1.8E-01	1782.3	1	3	1.9E+00	5556.0	4	2	1.1E-01
2428.1	1	3	1.7E-01	1807.3	5	3	3.8E+00	5564.9	6	6	1.7E-01
2569.5	1	3	5.3E-02	1820.3	3	3	2.2E+00	5578.8	6	6	1.1E-01
2931.8	1	3	1.9E-02	1826.2	1	3	7.2E-01	5606.1	10	8	5.4E-01

λ Å	Weights g_i	Weights g_k	A 10^8 s^{-1}	λ Å	Weights g_i	Weights g_k	A 10^8 s^{-1}	λ Å	Weights g_i	Weights g_k	A 10^8 s^{-1}
5616.6	4	4	1.2E-01	249.27	2	2	3.1E+01	4153	4	6	5.7E-02
5640.0	4	6	6.6E-01	388.94	2	2	4.5E+01				
5645.6	6	4	1.8E-02	390.86	4	2	8.8E+01	**Tantalum**			
5647.0	2	4	5.7E-01	706.48	2	4	4.17E+01	**Ta I**			
5659.9	6	4	4.6E-01	712.68	4	6	4.85E+01	3127.9	4	6	5.7E-03
5664.7	4	2	5.8E-01	712.84	4	4	8.1E+00	3168.3	4	4	6.0E-03
5819.2	4	4	8.5E-02	933.38	2	4	1.7E+01	3170.3	8	10	8.5E-02
6305.5	8	6	1.8E-01	944.52	2	2	1.6E+01	3205.5	6	8	5.6E-03
6312.7	6	4	3.0E-01					3260.2	4	4	5.8E-03
				S VII				3337.8	6	6	1.3E-02
S III				60.161	1	3	9.46E+03	3383.9	6	4	5.3E-03
2496.2	7	5	2.5E+00	60.804	1	3	5.1E+02	3406.9	4	6	6.8E-02
2508.2	5	3	2.3E+00	72.029	1	3	8.61E+02	3419.7	8	8	1.91E-02
2636.9	3	5	4.5E-01					3463.8	4	6	2.62E-02
2665.4	5	5	1.4E+00	**S VIII**				3484.6	4	4	8.5E-03
2680.5	1	3	6.2E-01	198.55	4	2	2.5E+02	3488.8	6	4	7.3E-03
2691.8	3	3	4.6E-01	202.61	2	2	1.2E+02	3497.9	6	8	4.9E-02
2702.8	3	1	1.9E+00					3505.0	8	6	2.72E-02
2718.9	3	3	1.2E+00	**S XI**				3553.4	4	6	3.3E-03
2721.4	5	3	7.7E-01	*189.90	9	3	4.3E+02	3607.4	6	8	4.6E-02
2726.8	3	5	6.0E-01	190.37	5	3	2.8E+02	3625.2	10	8	1.0E-02
2731.1	5	5	1.1E+00	215.95	5	5	1.4E+02	3626.6	8	10	7.1E-02
2756.9	7	7	1.4E+00	217.63	1	3	7.2E+01	3642.1	10	12	5.5E-02
2785.5	3	3	6.1E-01	239.81	1	3	2.6E+01	3657.5	6	6	4.3E-03
2856.0	5	7	5.1E+00	242.57	3	5	1.9E+01	3731.0	4	6	5.3E-03
2863.5	7	9	5.7E+00	242.82	3	3	1.9E+01	3754.5	8	8	6.5E-02
2872.0	3	5	4.7E+00	246.90	5	5	5.4E+01	3784.3	4	6	4.3E-02
2950.2	3	5	3.0E+00	247.12	5	3	3.0E+01	3792.1	4	4	9.0E-03
2964.8	5	7	4.0E+00	*288.49	9	15	2.9E+01	3826.9	6	6	5.2E-03
3662.0	3	3	6.4E-01					3836.6	8	10	4.0E-03
3717.8	5	3	1.0E+00	**S XII**				3848.1	10	8	1.30E-02
3778.9	3	5	4.4E-01	212.14	2	4	3.7E+01	3858.6	10	10	2.5E-03
3831.8	1	3	5.6E-01	215.18	2	2	1.4E+02	3918.5	4	2	2.5E-02
3837.8	3	3	4.2E-01	218.20	4	4	1.7E+02	3922.8	4	4	3.98E-02
3838.3	5	5	1.3E+00	221.44	4	2	6.4E+01	3996.2	2	4	3.35E-02
3860.6	3	1	1.6E+00	227.50	2	2	3.7E+01	3999.3	4	4	1.8E-02
3899.1	5	3	6.7E-01	234.48	4	2	6.8E+01	4003.7	10	8	3.1E-03
4253.6	5	7	1.2E+00					4006.8	6	8	7.6E-03
4285.0	3	5	9.0E-01	**S XIII**				4026.9	4	4	3.60E-02
				32.236	1	3	1.09E+04	4029.9	10	10	2.8E-02
S IV				37.600	3	1	1.3E+03	4030.7	8	10	2.3E-03
551.17	2	2	2.06E+01	256.66	1	3	8.7E+01	4040.9	10	12	7.3E-03
554.07	4	2	4.08E+01	299.89	3	5	1.78E+01	4061.4	2	4	6.5E-02
3097.5	2	4	2.6E+00	303.37	1	3	2.28E+01	4064.6	4	4	3.83E-02
3117.7	2	2	2.5E+00	307.36	3	3	1.64E+01	4067.2	6	4	6.8E-03
				308.91	5	5	4.82E+01	4067.9	6	8	8.4E-03
S V				312.68	3	1	6.3E+01	4097.2	10	10	2.1E-03
437.37	1	3	1.12E+01	316.84	5	3	2.50E+01	4105.0	6	4	1.1E-02
438.19	3	3	3.33E+01	500.42	3	5	1.43E+01	4136.2	8	6	1.82E-02
439.65	5	3	5.5E+01					4147.9	10	8	1.79E-02
*661.52	9	15	6.44E+01	**S XIV**				4175.2	6	8	2.8E-02
*679.01	9	15	8.6E+01	*30.434	2	6	8.28E+03	4205.9	8	10	8.9E-03
*690.75	9	9	5.0E+01	*32.517	6	10	2.6E+04	4303.0	6	6	2.08E-02
786.48	1	3	5.25E+01	417.67	2	4	1.2E+01	4378.8	8	6	4.8E-03
*854.85	9	9	4.18E+01	445.71	2	2	1.0E+01	4386.1	4	6	1.0E-02
				1550	2	4	1.4E+00	4402.5	6	6	2.28E-02
S VI				1663	2	2	1.2E+00	4415.7	2	4	2.53E-02
248.99	2	4	3.1E+01	3967	2	4	5.4E-02	4441.0	4	6	7.5E-03

λ Å	Weights g_i	g_k	A 10^8 s^{-1}	λ Å	Weights g_i	g_k	A 10^8 s^{-1}	λ Å	Weights g_i	g_k	A 10^8 s^{-1}
4441.7	10	8	9.0E-03	5518.9	8	10	3.8E-02	2921.5	4	4	7.6E-02
4473.5	6	8	1.36E-02	5620.7	8	10	6.0E-03	3229.8	4	2	1.73E-01
4511.0	10	12	1.56E-02	5640.2	6	8	4.9E-03	3519.2	4	6	1.24E+00
4511.5	10	8	3.6E-03	5645.9	6	8	1.43E-02	3529.4	4	4	2.20E-01
4514.2	10	10	3.1E-03	5699.2	6	6	4.2E-03	3775.7	2	2	6.25E-01
4521.1	10	10	2.3E-03	5767.9	6	8	2.6E-03	5350.5	4	2	7.05E-01
4530.9	4	6	2.42E-02	5780.7	4	6	3.3E-03				
4547.2	4	6	5.3E-03	5811.1	8	6	5.7E-03	**Thulium**			
4553.7	6	8	9.5E-03	5849.7	10	8	2.8E-03	**Tm I**			
4565.9	8	8	2.5E-02	5877.4	10	12	2.3E-02	2513.8	8	10	6.9E-02
4574.3	4	4	1.2E-02	5939.8	2	4	1.6E-02	2527.0	8	8	1.7E-01
4580.7	8	10	2.1E-03	5944.0	4	6	2.13E-02	2596.5	8	10	1.6E-01
4619.5	6	4	5.3E-02	5997.2	10	10	2.4E-02	2601.1	8	6	1.7E-01
4633.1	4	4	1.2E-02	6020.7	2	4	1.0E-02	2622.5	8	10	6.1E-02
4669.1	6	4	2.85E-02	6045.4	6	8	2.6E-02	2841.1	6	6	2.0E-01
4681.9	6	6	1.5E-02	6047.3	8	10	9.0E-03	2854.2	8	6	2.7E-01
4684.9	10	8	2.8E-03	6249.8	6	6	3.5E-03	2914.8	8	8	7.7E-02
4685.3	6	8	3.4E-03	6258.7	6	8	3.3E-03	2933.0	8	6	1.0E-01
4691.9	2	4	4.08E-02	6309.6	4	6	1.83E-02	2973.2	8	8	2.3E-01
4706.1	6	6	1.4E-02	6360.8	6	8	4.6E-03	3046.9	8	8	1.8E-01
4740.2	4	4	5.0E-02	6428.6	6	6	6.0E-03	3081.1	8	8	1.9E-01
4758.0	4	6	7.5E-03	6430.8	8	8	2.9E-02	3122.5	6	6	5.2E-01
4769.0	8	8	2.8E-02	6450.4	8	10	2.2E-02	3142.4	6	6	8.8E-02
4780.9	10	8	2.16E-02	6485.4	10	10	5.8E-02	3172.7	8	8	1.8E-01
4812.8	4	4	1.2E-02	6514.4	6	4	2.2E-02	3233.7	8	10	5.1E-02
4825.4	6	6	2.63E-02	6516.1	6	8	1.25E-02	3247.0	6	8	3.0E-01
4832.2	4	4	1.7E-02	6612.0	6	4	1.9E-02	3251.8	6	4	5.2E-01
4852.2	4	4	1.7E-02	6673.7	2	4	9.0E-03	3380.7	6	8	2.0E-01
4884.0	6	8	1.1E-02	6771.7	4	4	5.8E-03	3406.0	6	8	1.5E-01
4904.6	12	10	1.95E-02	6866.2	8	6	2.58E-02	3410.1	8	10	1.0E-01
4920.9	8	10	2.1E-03	6927.4	10	12	1.01E-02	3416.6	8	8	5.7E-02
4921.3	2	4	1.2E-02	6928.5	10	8	1.69E-02	3418.6	6	6	1.1E-01
4926.0	4	4	1.5E-02	6951.3	10	10	3.7E-03	3563.9	8	6	9.8E-02
4936.4	8	6	4.5E-02	6953.9	6	8	8.3E-03	3567.4	8	10	4.2E-02
4969.7	4	4	1.0E-02	6966.1	8	8	1.2E-02	3744.1	8	8	9.5E-01
5012.5	4	4	1.9E-02	6969.5	10	10	2.9E-03	3751.8	8	10	1.9E-01
5037.4	10	8	4.4E-02	7407.9	6	4	2.0E-02	3798.5	6	4	1.2E+00
5043.3	6	4	2.73E-02					3807.7	6	6	3.9E-01
5067.9	8	6	2.92E-02	**Thallium**				3883.1	8	6	1.0E+00
5069.9	10	12	1.7E-03	**Tl I**				3887.4	8	8	3.8E-01
5082.3	10	12	1.9E-03	2104.6	2	4	4.0E-02	3916.5	6	8	1.5E+00
5087.4	6	4	1.5E-02	2118.9	2	2	2.0E-02	3949.3	6	6	1.0E+00
5090.7	8	6	9.5E-03	2129.3	2	4	5.8E-02	4022.6	6	8	4.0E-02
5095.3	6	6	5.0E-03	2151.9	2	2	3.1E-02	4044.5	6	4	2.9E-01
5136.5	2	2	4.5E-02	2168.6	2	4	9.8E-02	4094.2	8	6	9.0E-01
5141.6	4	2	1.2E-02	2237.8	2	4	1.9E-01	4105.8	8	10	6.0E-01
5143.7	6	4	1.7E-02	2316.0	2	2	7.8E-02	4138.3	6	4	7.0E-01
5147.6	6	4	9.0E-03	2379.7	2	4	4.4E-01	4158.6	6	8	5.5E-02
5161.8	4	6	6.3E-03	2507.9	4	2	1.1E-01	4187.6	8	8	6.1E-01
5218.7	8	6	8.2E-03	2538.2	4	2	1.6E-02	4203.7	8	10	2.5E-01
5235.4	6	6	4.7E-03	2580.1	2	2	1.8E-01	4222.7	6	8	1.5E-01
5295.0	6	6	7.5E-03	2609.0	4	6	1.0E-01	4271.7	6	6	1.1E-01
5336.1	6	8	5.5E-03	2609.8	4	4	1.9E-02	4359.9	8	6	1.3E-01
5349.6	6	4	2.2E-02	2665.6	4	2	5.7E-02	4386.4	8	8	4.2E-02
5354.7	4	4	6.5E-03	2709.2	4	6	1.7E-01	4394.4	6	4	1.1E-01
5396.0	6	8	2.5E-03	2710.7	4	4	3.7E-02	4643.1	6	6	3.4E-02
5402.5	4	2	1.4E-02	2767.9	2	4	1.26E+00	4681.9	6	8	3.9E-02
5435.3	4	6	1.1E-02	2826.2	4	2	8.0E-02	4691.1	6	6	3.9E-02
5499.4	10	10	6.1E-03	2918.3	4	6	4.2E-01	5307.1	8	10	2.3E-02

λ Å	g_i	g_k	A 10^8 s^{-1}	λ Å	g_i	g_k	A 10^8 s^{-1}	λ Å	g_i	g_k	A 10^8 s^{-1}
5658.3	6	8	1.0E-02	6073.5	3	1	6.3E-02	3642.68	7	9	7.74E-01
5675.8	8	10	1.3E-02	6171.5	3	3	4.9E-02	3653.50	9	11	7.54E-01
5760.2	6	6	1.3E-02					3724.57	9	9	9.1E-01
				Sn II				3725.16	5	3	7.3E-01
Tin				2368.3	4	2	4.4E-03	3729.81	5	5	4.27E-01
Sn I				2449.0	4	6	3.7E-01	3741.06	7	7	4.17E-01
2073.1	1	3	3.6E-02	2487.0	6	8	5.5E-01	3752.86	9	9	5.04E-01
2199.3	3	5	2.9E-01	3283.2	4	6	1.0E+00	3786.04	5	3	1.4E+00
2209.7	5	5	5.6E-01	3352.0	6	8	1.0E+00	3948.67	5	3	4.85E-01
2246.1	1	3	1.6E+00	3472.5	2	4	1.6E-01	3956.34	7	5	3.00E-01
2268.9	5	7	1.2E+00	3575.5	4	6	1.3E-01	3958.21	9	7	4.05E-01
2286.7	5	5	3.1E-01	5332.4	2	4	8.6E-01	3981.76	5	5	3.76E-01
2317.2	5	7	2.0E+00	5562.0	4	6	1.2E+00	3989.76	7	7	3.79E-01
2334.8	3	3	6.6E-01	5588.9	4	6	8.5E-01	3998.64	9	9	4.08E-01
2354.8	3	5	1.7E+00	5596.2	4	4	1.5E-01	4013.24	7	5	2.0E-01
2380.7	3	5	3.1E-02	5797.2	6	6	2.8E-01	4055.01	1	3	2.8E-01
2408.2	5	3	1.8E-01	5799.2	6	8	8.1E-01	4060.26	3	5	2.4E-01
2421.7	5	7	2.5E+00	6453.5	2	4	1.2E+00	4064.20	3	3	2.4E-01
2429.5	5	7	1.5E+00	6761.5	2	2	3.2E-01	4065.09	3	1	7.0E-01
2433.5	5	3	8.0E-03	6844.1	2	2	6.6E-01	4186.12	9	9	2.10E-01
2455.2	5	5	1.1E-02					4266.23	5	5	3.1E-01
2476.4	5	3	1.1E-02	**Titanium**				4284.99	5	5	3.2E-01
2483.4	5	5	2.1E-01	**Ti I**				4289.07	5	5	3.0E-01
2491.8	1	3	1.7E-01	2276.75	7	5	1.3E+00	4290.93	3	3	4.5E-01
2495.7	5	5	6.2E-01	2280.00	9	7	9.4E-01	4295.75	3	1	1.3E+00
2523.9	5	3	7.4E-02	2299.86	5	5	6.9E-01	4393.93	9	11	3.3E-01
2546.6	1	3	2.1E-01	2302.75	7	7	5.7E-01	4417.27	11	9	3.6E-01
2558.0	1	3	3.4E-01	2305.69	9	9	5.2E-01	4449.14	11	11	9.7E-01
2571.6	5	7	4.5E-01	2424.26	9	9	1.7E-01	4450.90	9	9	9.6E-01
2594.4	5	5	3.0E-01	2520.54	5	3	3.8E-01	4453.31	5	5	5.98E-01
2636.9	1	3	1.1E-01	2529.87	7	5	3.8E-01	4453.71	7	7	4.7E-01
2661.2	3	3	1.1E-01	2541.92	9	7	4.3E-01	4455.32	7	7	4.8E-01
2706.5	3	5	6.6E-01	2599.91	5	5	6.7E-01	4457.43	9	9	5.6E-01
2761.8	5	5	3.7E-03	2605.16	7	7	6.4E-01	4465.81	5	7	3.28E-01
2779.8	5	7	1.8E-01	2611.29	9	9	6.4E-01	4481.26	7	7	5.7E-01
2785.0	5	3	1.4E-01	2611.47	7	5	3.3E-01	4496.15	7	5	4.4E-01
2788.0	1	3	1.4E-01	2619.94	9	7	2.1E-01	4518.02	7	9	1.72E-01
2812.6	1	3	2.3E-01	2631.55	7	7	1.7E-01	4522.80	5	7	1.9E-01
2813.6	5	5	1.2E-01	2632.42	5	5	2.7E-01	4527.31	3	5	2.2E-01
2840.0	5	5	1.7E+00	2641.12	5	3	1.8E+00	4533.24	11	11	8.83E-01
2850.6	5	5	3.3E-01	2644.28	7	5	1.4E+00	4534.78	9	9	6.87E-01
2863.3	1	3	5.4E-01	2646.65	9	7	1.5E+00	4544.69	5	3	3.3E-01
2913.5	1	3	8.3E-01	2733.27	5	5	1.9E+00	4548.76	7	5	2.85E-01
3009.1	3	3	3.8E-01	2735.30	3	1	4.1E+00	4552.45	9	7	2.1E-01
3032.8	1	3	6.2E-01	2912.07	5	7	1.3E+00	4563.43	9	11	2.1E-01
3034.1	3	1	2.0E+00	2942.00	5	5	1.0E+00	4617.27	7	9	8.51E-01
3141.8	1	3	1.9E-01	2948.26	7	7	9.3E-01	4623.10	5	7	5.74E-01
3175.1	5	3	1.0E+00	2956.13	9	9	9.7E-01	4639.94	3	3	6.64E-01
3218.7	1	3	4.7E-02	2956.80	7	5	1.8E-01	4640.43	3	1	5.0E-01
3223.6	5	5	1.2E-03	3186.45	5	7	8.0E-01	4645.19	3	1	8.57E-01
3262.3	5	3	2.7E+00	3191.99	7	9	8.5E-01	4650.02	5	3	2.6E-01
3330.6	5	5	2.0E-01	3199.92	9	11	9.4E-01	4742.79	9	9	5.3E-01
3655.8	1	3	4.1E-02	3341.88	5	7	6.5E-01	4758.12	11	11	7.13E-01
3801.0	5	3	2.8E-01	3354.63	7	9	6.9E-01	4759.27	13	13	7.40E-01
4524.7	1	3	2.6E-01	3370.44	5	3	7.6E-01	4778.26	9	9	2.0E-01
5631.7	1	3	2.4E-02	3371.45	9	11	7.2E-01	4805.42	5	7	5.8E-01
5970.3	5	3	9.6E-02	3377.58	7	5	6.9E-01	4840.87	5	5	1.76E-01
6037.7	5	5	5.0E-02	3385.94	9	7	5.0E-01	4856.01	13	15	5.2E-01
6069.0	1	3	4.6E-02	3635.46	5	7	8.04E-01	4885.08	11	13	4.90E-01

λ (Å)	Weights g_i	g_k	A (10^8 s^{-1})	λ (Å)	Weights g_i	g_k	A (10^8 s^{-1})	λ (Å)	Weights g_i	g_k	A (10^8 s^{-1})
4913.62	7	9	4.44E-01	2877.47	8	8	5.7E-01	3278.91	6	4	1.0E+00
4928.34	3	5	6.2E-01	2884.13	10	10	5.2E-01	3282.32	2	2	1.6E+00
4981.73	11	13	6.60E-01	2910.65	8	8	4.6E-01	3287.66	8	10	1.4E+00
4989.14	7	5	3.25E-01	2926.64	10	8	8.9E-01	3315.32	2	4	3.8E-01
4991.07	9	11	5.84E-01	2931.10	6	6	3.2E+00	3321.70	4	4	7.2E-01
4999.50	7	9	5.27E-01	2936.02	4	6	2.7E+00	3322.94	10	10	3.96E-01
5000.99	9	7	3.52E-01	2938.57	6	8	2.4E+00	3329.46	8	8	3.25E-01
5007.21	5	7	4.92E-01	2941.90	8	10	1.8E+00	3332.11	6	4	1.1E+00
5014.28	3	5	6.8E-01	2942.97	8	8	1.1E+00	3340.34	4	4	3.6E-01
5036.47	7	9	3.94E-01	2945.30	10	12	2.7E+00	3361.23	8	10	1.1E+00
5038.40	5	7	3.87E-01	2952.00	8	8	3.0E-01	3372.80	6	8	1.11E+00
5062.11	5	3	2.98E-01	2954.59	10	12	4.0E+00	3383.77	4	6	1.09E+00
5210.39	9	9	3.57E-02	2958.80	8	10	4.0E+00	3452.49	2	2	7.7E-01
5222.69	3	3	1.95E-01	2979.06	4	6	1.2E+00	3456.40	4	4	8.2E-01
5224.30	11	11	3.6E-01	2990.06	6	8	5.6E-01	3465.56	4	2	4.1E-01
5259.98	5	7	2.3E-01	3017.17	12	12	3.6E-01	3483.63	10	8	9.7E-01
5351.07	7	7	3.4E-01	3022.64	10	10	1.2E+00	3492.37	8	6	9.8E-01
5503.90	11	9	2.6E-01	3023.67	8	8	1.0E+00	3504.90	10	10	8.2E-01
5774.04	9	11	5.5E-01	3029.76	10	10	3.5E-01	3510.86	8	8	9.3E-01
5785.98	11	13	6.1E-01	3056.75	2	4	3.2E-01	3520.27	2	4	4.8E-01
5804.27	13	15	6.8E-01	3058.08	6	6	5.0E-01	3535.41	4	6	5.5E-01
6098.66	9	7	2.5E-01	3066.34	4	4	3.3E-01	3641.33	4	2	4.9E-01
6220.46	9	7	1.8E-01	3071.25	6	4	3.6E-01	3706.23	4	4	3.1E-01
				3072.99	4	2	1.6E+00	3741.64	6	6	6.2E-01
Ti II				3075.23	6	4	1.13E+00	3757.70	4	4	4.1E-01
2440.91	4	4	5.1E-01	3078.65	8	6	1.09E+00	3759.30	8	8	9.4E-01
2451.18	6	6	4.5E-01	3081.52	10	8	1.1E+00	3761.33	6	6	9.9E-01
2525.59	10	8	5.6E-01	3088.04	10	8	1.25E+00	4911.18	6	4	3.2E-01
2531.28	8	6	4.9E-01	3089.44	8	6	1.3E+00				
2534.63	6	4	5.4E-01	3097.20	4	6	4.4E-01	**Ti III**			
2535.89	4	2	6.8E-01	3103.81	10	8	1.1E+00	865.79	5	3	6.6E+01
2555.99	6	8	3.2E-01	3105.10	2	4	6.3E-01	1002.37	5	5	7.6E+00
2635.44	4	4	1.9E+00	3106.26	6	6	7.8E-01	1004.67	7	5	4.3E+01
2638.56	6	6	1.7E+00	3117.67	4	2	1.1E+00	1005.80	3	3	1.3E+01
2642.02	8	8	1.9E+00	3119.83	6	4	5.9E-01	1007.16	5	3	3.8E+01
2645.86	10	10	2.7E+00	3127.86	6	6	1.6E+00	1008.12	3	1	5.1E+01
2746.54	6	8	2.6E+00	3128.50	8	8	1.1E+00	1286.37	9	9	2.0E+00
2751.59	8	10	3.7E+00	3161.23	4	2	5.9E-01	1289.30	7	7	2.2E+00
2752.68	8	10	1.1E+00	3161.80	6	4	4.6E-01	1291.62	5	5	2.4E+00
2757.62	6	8	7.2E-01	3162.59	8	6	3.9E-01	1293.23	9	7	1.0E+00
2758.35	4	6	9.9E-01	3168.55	10	8	4.1E-01	1298.97	7	5	4.9E+00
2758.79	2	4	4.4E-01	3181.73	6	6	4.6E-01	1327.59	5	3	3.2E+00
2764.28	4	4	7.4E-01	3182.54	4	4	4.3E-01	1420.44	1	3	1.2E+00
2804.82	6	8	4.6E+00	3189.49	4	4	9.2E-01	1421.63	3	1	4.0E+00
2810.30	8	10	5.1E+00	3190.91	6	8	1.3E+00	1422.41	5	5	3.0E+00
2817.83	10	12	3.8E+00	3202.56	4	6	1.1E+00	1424.14	5	3	1.6E+00
2819.87	8	8	6.5E-01	3224.25	12	10	7.0E-01	1455.19	9	7	6.4E+00
2821.26	6	8	7.9E-01	3228.62	4	2	2.0E+00	1498.70	5	5	2.8E+00
2827.12	8	10	1.0E+00	3232.29	8	6	6.0E-01	2007.36	3	3	3.4E+00
2828.06	12	14	4.4E+00	3234.51	10	10	1.38E+00	2007.60	1	3	1.2E+00
2828.64	6	6	1.2E+00	3236.13	4	4	7.0E-01	2010.80	5	3	5.4E+00
2828.83	10	10	9.1E-01	3236.58	8	8	1.11E+00	2097.30	5	7	3.3E+00
2834.02	10	12	7.9E-01	3239.04	6	6	9.87E-01	2099.86	3	5	2.5E+00
2836.47	8	8	1.2E+00	3239.66	6	4	9.4E-01	2104.86	3	3	1.1E+00
2839.64	12	12	8.3E-01	3241.99	4	4	1.16E+00	2105.09	1	3	1.7E+00
2845.93	10	10	1.2E+00	3251.91	6	4	3.38E-01	2199.22	3	3	5.7E+00
2851.11	2	4	4.1E-01	3252.92	8	6	3.9E-01	2237.77	7	7	2.4E+00
2856.10	12	12	1.5E+00	3272.07	2	4	3.2E-01	2331.35	3	1	4.3E+00
2862.33	4	6	4.0E-01	3278.28	4	4	9.6E-01	2331.66	3	3	1.2E+00

λ Å	Weights g_i	g_k	A 10^8 s^{-1}	λ Å	Weights g_i	g_k	A 10^8 s^{-1}	λ Å	Weights g_i	g_k	A 10^8 s^{-1}
2339.00	5	3	3.0E+00	2862.60	4	2	4.1E+00	308.250	3	5	1.3E+02
2346.79	7	5	3.3E+00	3576.44	4	6	4.6E+00	313.229	5	7	1.6E+02
2374.99	5	3	4.0E+00					318	3	1	1.4E+02
2413.99	5	7	3.8E+00	**Ti VIII**				322.75	5	7	1.99E+02
2516.05	7	9	3.4E+00	249	6	4	1.0E+01	323	1	3	1.8E+02
2567.56	3	3	2.3E+00	258.610	6	8	7.5E+02	327.192	3	5	2.9E+02
2984.75	5	5	1.9E+00	269.533	4	6	6.0E+02	332	3	1	3.25E+02
3066.51	3	3	2.5E+00	272.037	4	4	4.3E+02	386.140	1	3	1.48E+02
3228.89	3	3	1.5E+00	272.843	6	4	6.2E+01	408	7	9	1.37E+02
3278.31	7	9	3.4E+00	276.701	2	4	9.3E+01	425.74	3	1	1.2E+02
3320.94	3	5	2.8E+00	277.813	4	4	3.8E+02	446.69	3	1	1.2E+02
3340.20	7	9	3.7E+00	289.375	2	4	3.6E+01	453	5	7	1.3E+02
3346.18	9	11	3.7E+00	478.971	4	4	1.7E+01				
3354.71	11	13	4.4E+00	480.376	6	6	1.5E+01	**Ti XII**			
3397.24	3	1	1.8E+00					52.896	2	4	1.61E+02
3404.46	3	3	1.8E+00	**Ti IX**				53.140	4	6	1.9E+02
3417.62	3	5	1.9E+00	267.941	5	7	5.1E+02	53.433	2	4	2.1E+02
3915.47	9	11	2.1E+00	278.713	5	7	4.7E+02	53.457	2	2	2.1E+02
4119.14	5	5	9.9E-01	281.446	3	1	3.2E+02	55.181	2	4	2.4E+02
4213.26	9	11	2.2E+00	285.128	1	3	4.1E+02	55.443	4	6	2.81E+02
4215.53	9	11	2.2E+00	433.567	1	3	6.9E+00	59.133	2	4	3.72E+02
4247.62	11	13	1.1E+00	439.513	3	3	7.5E+00	59.435	4	6	4.41E+02
4248.54	5	7	2.3E+00	439.745	3	1	2.1E+01	60.701	2	4	3.4E+02
4250.09	3	5	9.5E-01	447.484	5	5	1.6E+01	60.762	2	2	3.5E+02
4259.01	11	13	9.4E-01	447.701	5	3	6.5E+00	61.286	4	2	1.8E+02
4269.84	9	11	1.7E+00	507.174	3	5	6.5E+00	62.433	4	6	2.08E+02
4285.61	13	15	3.0E+00	516.215	5	7	6.9E+00	62.470	6	8	2.22E+02
4288.66	11	13	1.1E+00					65.540	4	6	3.2E+02
4296.70	11	13	1.6E+00	**Ti X**				65.577	6	8	3.5E+02
4319.56	9	11	1.1E+00	253	4	6	2.1E+02	67.171	2	4	6.2E+02
4343.25	3	1	1.0E+00	254	6	8	2.3E+02	67.555	4	6	7.2E+02
4378.94	3	5	1.6E+00	281	2	2	1.1E+02	70.986	4	6	5.7E+02
4433.91	11	13	1.8E+00	289.579	2	4	2.5E+02	71.031	6	8	6.1E+02
4440.66	1	3	1.2E+00	290.294	4	6	1.1E+02	71.545	2	2	1.8E+02
4533.26	3	5	1.5E+00	291	4	2	1.8E+02	71.987	4	2	3.48E+02
4576.53	9	7	1.3E+00	291	2	2	2.3E+02	82.121	2	4	5.9E+02
4628.07	3	1	1.5E+00	292	6	8	1.1E+02	82.307	4	6	1.13E+03
4652.86	7	9	2.6E+00	293.684	6	8	2.97E+02	82.344	2	2	5.8E+02
4874.00	5	7	1.5E+00	293.798	6	6	1.7E+02	82.368	6	8	1.2E+03
4914.32	3	3	1.1E+00	295.584	4	6	2.9E+02	89.844	2	4	9.9E+02
4971.19	9	11	2.1E+00	296	4	6	1.4E+02	90.512	4	6	1.16E+03
5083.80	5	3	9.7E-01	297	4	6	9.9E+01	90.547	4	4	1.9E+02
5278.33	3	3	9.4E-01	298	4	6	4.3E+02	116.497	4	6	3.0E+03
7506.87	11	13	1.1E+00	302	2	2	1.6E+02	116.597	6	8	3.2E+03
				305	2	4	2.5E+02	116.62	6	6	2.1E+02
Ti IV				317	2	2	1.5E+02	139.884	6	4	2.6E+02
423.49	4	6	4.9E+01	355.815	2	2	1.3E+02	140.361	4	2	2.9E+02
424.16	6	8	5.3E+01	360.133	4	4	2.19E+02	141.6	4	6	1.7E+02
433.63	4	2	5.5E+00	363	4	2	2.1E+02	141.7	6	8	1.7E+02
433.76	6	4	5.0E+00	363	6	6	1.3E+02	169.7	4	6	2.8E+02
729.36	4	2	5.7E+00	365.628	4	2	1.2E+02	169.8	6	8	2.9E+02
1183.64	2	2	6.9E+00	382	4	6	1.8E+02	207.2	2	4	1.5E+02
1195.21	4	2	1.4E+01	385	6	8	1.8E+02	208.5	4	6	1.8E+02
1451.74	2	4	1.8E+01	389.99	6	4	1.1E+02	252.8	4	6	4.8E+02
1467.34	4	6	2.1E+01					253.1	6	8	5.2E+02
2067.56	2	4	5.1E+00	**Ti XI**				257.5	4	2	2.4E+02
2103.16	2	2	5.0E+00	65.403	1	3	5.1E+02				
2541.79	4	6	6.9E+00	87.725	1	3	8.5E+02	**Ti XIII**			
2546.88	6	8	7.4E+00	266	5	7	1.8E+02	23.356	1	3	1.02E+05

λ Å	Weights g_i	g_k	A 10^8 s^{-1}	λ Å	Weights g_i	g_k	A 10^8 s^{-1}	λ Å	Weights g_i	g_k	A 10^8 s^{-1}
23.698	1	3	1.2E+04	145.665	6	6	2.3E+02	16.46	3	3	4.4E+04
23.991	1	3	3.4E+02	157.812	4	2	1.32E+02	16.51	5	7	1.0E+05
26.641	1	3	4.06E+03	161.168	4	4	1.2E+02	16.55	5	5	2.7E+04
26.960	1	3	3.06E+03	163.610	4	2	1.92E+02	16.61	3	1	8.0E+04
117.1	3	3	1.3E+02	169.740	4	6	1.0E+02	16.64	3	3	5.3E+04
117.3	3	1	2.8E+02	176.267	2	2	2.45E+02	16.69	1	3	1.02E+05
120.2	5	3	5.4E+02	178.240	4	4	2.52E+02	16.71	3	5	7.3E+04
120.2	7	5	4.4E+02					16.72	5	3	3.3E+04
128.7	3	3	1.2E+02	**Ti XVII**				16.72	5	5	7.3E+04
				18.05	3	3	4.5E+04	16.74	5	7	1.2E+05
Ti XIV				18.13	5	3	2.4E+04	16.77	3	3	2.6E+04
21.341	4	6	9.8E+03	18.13	1	3	8.1E+04	16.80	5	7	1.81E+05
21.522	2	4	4.5E+04	18.176	5	7	9.2E+04	16.85	3	5	4.4E+04
21.657	4	4	1.3E+04	123.654	3	3	2.3E+02	17.08	3	5	8.3E+04
21.733	4	4	8.8E+04	124.553	5	3	5.2E+02	17.36	1	3	9.5E+04
21.82	4	2	6.4E+04	127.782	5	3	4.6E+02				
21.883	2	4	7.0E+04	135.202	3	1	2.93E+02	**Ti XX**			
21.958	2	4	1.2E+04	136.160	5	3	1.95E+02	2.629	2	4	4.9E+04
22.05	2	2	1.4E+04	136.393	3	3	1.14E+02	2.6295	4	4	3.2E+06
24.592	4	2	6.1E+03	141.948	5	5	3.87E+02	2.631	2	2	6.1E+05
24.891	2	2	7.5E+03	142.589	1	3	1.35E+02	2.6319	2	4	1.5E+06
				144.405	5	5	9.4E+01	2.632	2	2	2.7E+06
Ti XV				146.067	7	5	2.6E+02	2.6355	4	6	1.2E+06
20.19	5	7	6.9E+03	154.133	3	1	1.63E+02	8.621	4	2	1.1E+06
20.234	5	7	1.9E+04	156.54	3	1	1.44E+02	9.788	4	6	5.26E+03
20.234	3	3	4.9E+04	158.469	5	5	1.4E+02	10.046	2	4	7.29E+03
20.246	1	3	4.2E+04	159.62	5	3	1.03E+02	10.109	4	6	8.6E+03
20.250	5	3	6.5E+03	163.049	3	1	6.2E+02	*10.278	2	6	8.4E+03
20.29	3	3	1.1E+04	186.863	5	5	2.66E+02	10.620	2	4	1.34E+04
20.30	1	3	3.4E+04	207.73	3	1	1.07E+02	10.690	4	6	1.58E+04
20.30	1	1	5.8E+04					*11.452	2	6	1.7E+04
20.313	5	3	7.5E+04	**Ti XVIII**				11.872	2	4	2.8E+04
20.418	5	7	8.0E+04	17.22	2	4	7.3E+04	11.958	4	6	3.4E+04
20.538	3	3	3.8E+04	17.365	4	6	8.6E+04	11.958	4	4	5.6E+03
20.54	3	1	4.1E+04	17.39	4	4	1.4E+04	15.211	2	4	3.50E+04
20.551	1	3	1.3E+04	133.852	2	4	5.2E+01	15.253	2	2	3.58E+04
20.689	5	7	4.3E+04	144.759	4	4	3.2E+02	15.907	2	4	8.84E+04
20.698	1	3	1.1E+05	150.15	6	4	1.15E+02	16.049	4	6	1.05E+05
20.771	5	3	1.1E+04	153.15	4	2	1.97E+02	16.067	4	4	1.8E+04
20.897	5	7	2.85E+04	153.23	2	4	6.7E+01	31.586	4	6	5.49E+03
20.928	5	5	8.4E+03	159.00	4	4	1.16E+02	45.650	2	4	9.6E+03
21.065	3	3	1.1E+04	166.225	6	4	1.54E+02	45.996	4	6	1.1E+04
21.079	1	3	1.58E+04	179.902	2	4	6.3E+01				
21.102	3	5	1.3E+04	189.663	6	6	9.6E+01	**Ti XXI**			
22.482	5	3	6.4E+03	191.23	4	4	6.6E+01	2.0633	1	3	1.32E+05
22.936	5	5	1.1E+04	197.838	4	6	4.56E+01	2.1108	1	3	2.60E+05
22.966	5	3	1.1E+04	208.07	4	4	1.2E+02	2.2211	1	3	6.35E+05
23.034	1	3	6.3E+03					2.497	3	1	2.4E+06
				Ti XIX				2.505	5	5	3.5E+05
Ti XVI				15.67	3	1	3.3E+04	2.505	1	3	1.4E+06
110.561	4	2	3.36E+02	15.68	5	5	2.7E+04	2.507	3	5	1.4E+06
116.198	4	4	1.45E+02	15.74	5	7	2.7E+04	2.508	3	5	7.9E+05
118.215	6	4	7.4E+02	15.75	3	5	2.4E+04	2.510	3	3	6.9E+05
121.382	4	2	2.4E+02	15.83	1	3	3.2E+04	2.510	1	3	9.6E+05
124.805	4	2	6.1E+02	15.86	1	3	2.9E+04	2.511	3	3	1.4E+06
129.075	4	2	3.81E+02	16.02	3	1	3.1E+04	2.512	5	5	1.8E+06
134.724	2	2	2.6E+02	16.18	3	5	3.8E+04	2.512	3	1	1.4E+06
138.800	6	4	3.5E+02	16.41	1	3	6.1E+04	2.513	3	1	2.7E+06
143.459	4	4	2.8E+02	16.43	3	5	8.2E+04	2.513	3	5	2.4E+06

λ Å	Weights g_i	g_k	A 10^8 s^{-1}	λ Å	Weights g_i	g_k	A 10^8 s^{-1}	λ Å	Weights g_i	g_k	A 10^8 s^{-1}
2.514	5	3	1.2E+06	3780.8	7	5	4.2E-02	4924.6	13	11	1.75E-03
2.520	3	5	2.6E+05	3809.2	7	5	9.0E-03	4931.6	7	5	1.0E-02
2.527	3	1	1.2E+05	3817.5	7	7	3.1E-02	4948.6	9	11	1.36E-03
2.539	3	1	4.1E+05	3829.1	3	3	3.83E-03	4972.6	9	11	3.9E-03
2.6102	1	3	2.40E+06	3835.1	5	5	5.2E-02	4982.6	1	3	4.17E-03
2.6227	1	3	1.12E+05	3846.3	3	5	2.14E-02	4986.9	11	9	6.3E-03
				3847.5	1	3	8.3E-03	5006.2	9	7	1.2E-02
Tungsten				3864.3	5	5	5.6E-03	5015.3	7	9	5.4E-03
W I				3868.0	7	9	4.6E-02	5040.4	3	5	5.2E-03
2879.4	1	3	2.4E-01	3881.4	7	7	3.6E-02	5053.3	3	3	1.9E-02
2911.0	1	3	7.7E-02	3968.5	1	3	5.07E-03	5071.5	13	11	3.4E-03
2923.5	7	9	1.54E-02	3975.5	9	11	4.1E-03	5117.6	11	11	1.61E-03
2935.0	3	5	1.5E-01	4001.4	9	9	5.6E-03	5124.2	5	5	4.0E-03
3013.8	7	9	6.4E-02	4008.8	7	9	1.63E-01	5141.2	7	9	1.12E-03
3016.5	9	11	9.27E-02	4019.3	5	3	6.7E-03	5224.7	7	5	1.2E-02
3017.4	7	9	1.21E-01	4028.8	1	3	2.0E-02	5243.0	9	7	1.1E-02
3024.9	3	3	1.4E-01	4045.6	7	5	2.88E-02	5254.5	7	5	3.86E-03
3046.4	3	5	5.8E-02	4055.2	7	9	1.79E-03	5268.6	9	9	1.4E-03
3049.7	7	5	1.7E-01	4070.0	7	5	3.60E-02	5500.5	11	9	6.9E-03
3064.9	5	7	1.1E-02	4070.6	3	5	5.6E-03	5514.7	5	3	7.3E-03
3084.9	5	5	1.3E-02	4074.4	7	7	1.0E-01	5537.7	9	11	2.2E-03
3093.5	7	9	4.4E-02	4088.3	5	3	4.13E-03	5617.1	7	7	1.47E-03
3107.2	5	7	2.33E-02	4102.7	9	7	4.9E-02	5631.9	9	7	1.43E-03
3108.0	7	9	1.58E-02	4115.6	11	11	4.8E-03	5660.7	13	11	6.8E-03
3145.5	9	9	4.8E-03	4137.5	5	7	8.4E-03	5675.4	5	5	2.20E-03
3170.2	7	5	6.0E-03	4171.2	7	9	8.6E-03	5796.5	9	7	2.21E-03
3176.6	3	5	2.12E-02	4203.8	9	7	4.9E-03	5891.6	7	7	1.47E-03
3183.5	7	7	2.64E-03	4219.4	9	7	6.1E-03	5947.6	5	7	2.40E-03
3184.4	5	3	2.3E-02	4244.4	9	11	1.38E-02	5965.9	7	5	1.0E-02
3191.6	1	3	3.2E-02	4269.4	7	5	3.04E-02	6021.5	5	3	8.7E-03
3198.8	7	9	4.6E-02	4283.8	9	7	1.69E-03	6081.4	5	3	4.7E-03
3207.3	7	9	3.0E-02	4294.6	7	5	1.2E-01	6203.5	7	7	3.0E-03
3208.3	5	5	4.4E-02	4302.1	7	7	3.6E-02	6285.9	7	5	6.6E-03
3215.6	9	11	2.1E-01	4355.2	9	9	5.1E-03	6292.0	3	5	2.26E-03
3221.9	5	7	1.61E-02	4361.8	9	7	1.64E-03	6303.2	9	9	1.84E-03
3223.1	5	3	3.53E-03	4378.5	7	5	3.48E-03	6404.2	5	7	1.50E-03
3232.5	9	9	2.4E-02	4458.1	3	5	4.2E-03	6439.7	9	9	1.29E-03
3235.1	7	5	2.68E-03	4466.3	7	5	1.5E-02	6445.1	7	5	6.4E-03
3259.7	7	7	1.3E-02	4472.5	13	11	1.55E-03	6532.4	3	5	4.6E-03
3300.8	7	9	8.1E-02	4484.2	3	5	5.6E-03	6538.1	11	9	2.7E-03
3311.4	7	5	5.6E-02	4492.3	9	11	3.6E-03	6563.2	5	5	2.04E-03
3363.3	9	7	6.6E-03	4495.3	11	11	3.3E-03	6814.9	9	9	1.46E-03
3371.0	7	5	1.0E-02	4504.8	9	7	7.0E-03	7285.8	13	11	1.47E-03
3371.4	3	3	6.7E-03	4552.5	9	9	1.42E-03	7569.9	5	3	3.73E-03
3386.1	7	7	2.64E-03	4586.8	1	3	4.20E-03	7664.9	5	3	3.80E-03
3413.0	7	9	9.7E-03	4592.6	7	9	3.4E-03	8017.2	5	7	1.6E-02
3459.5	9	9	2.04E-03	4609.9	7	9	1.42E-02	8358.7	5	7	1.89E-03
3510.0	7	9	5.2E-03	4613.3	9	9	2.9E-03	9381.4	9	7	1.53E-03
3545.2	1	3	3.2E-02	4634.8	9	9	8.8E-03				
3570.6	5	3	6.7E-03	4659.9	1	3	1.0E-02	**Uranium**			
3606.1	3	5	9.6E-03	4680.5	7	7	1.4E-02	**U I**			
3617.5	7	7	1.1E-01	4720.4	3	5	3.22E-03	3553.0	13	13	2.0E-02
3631.9	3	5	1.3E-02	4729.6	7	5	7.8E-03	3553.0	9	7	1.4E-02
3675.6	9	11	1.20E-02	4752.6	3	3	5.20E-03	3553.4	15	13	2.2E-02
3682.1	9	11	2.0E-02	4757.5	7	5	2.72E-03	3554.5	11	9	8.4E-03
3707.9	7	7	2.9E-02	4757.8	11	9	4.1E-03	3554.9	15	17	7.9E-03
3757.9	7	9	1.38E-02	4788.4	9	11	2.6E-03	3555.3	13	15	2.7E-02
3760.1	5	7	1.99E-02	4843.8	5	5	1.9E-02	3555.8	13	11	4.1E-03
3768.5	3	3	3.47E-02	4886.9	9	11	8.1E-03	3556.9	13	11	7.5E-03

λ Å	Weights g_i	g_k	A 10^8 s^{-1}	λ Å	Weights g_i	g_k	A 10^8 s^{-1}	λ Å	Weights g_i	g_k	A 10^8 s^{-1}
3557.8	13	13	2.9E-02	3075.93	4	6	2.8E-01	3818.24	4	2	6.73E-01
3558.0	11	13	1.6E-02	3080.33	2	4	2.7E-01	3828.56	6	4	5.33E-01
3558.6	9	7	3.9E-02	3083.54	6	8	2.5E-01	3840.75	8	6	5.48E-01
3559.4	7	9	1.5E-02	3087.06	2	2	9.2E-01	3855.36	4	4	3.30E-01
3560.3	9	7	6.4E-02	3088.11	4	6	4.9E-01	3855.85	10	8	5.78E-01
3561.4	15	13	5.5E-02	3089.13	4	4	5.3E-01	3863.86	8	6	3.1E-01
3561.5	9	9	2.5E-02	3093.79	6	6	4.1E-01	3864.86	6	6	2.70E-01
3561.8	13	11	5.7E-02	3094.69	2	4	4.3E-01	3871.07	10	8	2.8E-01
3563.7	13	13	2.9E-02	3112.92	4	2	5.0E-01	3875.07	8	8	2.36E-01
3563.8	7	7	1.1E-02	3183.41	6	8	2.4E+00	3902.26	10	10	2.68E-01
3565.0	13	11	2.9E-02	3183.96	8	10	2.5E+00	3921.86	4	2	2.7E-01
3566.0	13	15	1.7E-02	3183.98	4	6	2.4E+00	3922.43	6	6	2.6E-01
3566.6	11	11	2.4E-01	3185.38	10	12	2.7E+00	3930.02	10	10	3.3E-01
3568.8	13	13	3.8E-02	3198.01	6	6	3.9E-01	3934.01	8	8	6.2E-01
3569.1	17	15	1.1E-01	3202.39	8	8	4.0E-01	3992.80	12	10	1.2E+00
3569.4	9	9	1.5E-02	3205.58	8	10	1.3E+00	3998.73	14	12	1.0E+00
3570.1	13	11	1.3E-02	3207.41	10	10	2.6E-01	4050.96	10	10	1.4E+00
3570.2	11	9	5.3E-03	3212.43	10	12	1.4E+00	4051.35	12	12	1.3E+00
3570.6	13	15	2.7E-02	3218.87	8	6	3.5E-01	4090.57	8	10	8.5E-01
3570.7	15	15	1.2E-02	3233.19	10	8	3.2E-01	4092.68	8	10	2.30E-01
3571.2	11	11	6.3E-03	3273.03	8	8	2.7E-01	4095.48	6	8	7.2E-01
3571.6	17	15	1.3E-01	3284.36	10	10	2.8E-01	4099.78	6	8	4.10E-01
3572.9	13	15	1.5E-02	3309.18	4	4	3.2E-01	4102.15	4	6	7.1E-01
3573.9	13	11	4.0E-02	3329.85	6	4	7.7E-01	4104.77	10	8	2.1E+00
3574.1	13	15	3.5E-02	3356.35	4	6	3.1E-01	4105.16	4	6	4.9E-01
3574.8	13	15	1.9E-02	3365.55	2	4	4.8E-01	4109.78	2	4	5.00E-01
3577.1	17	15	4.3E-02	3376.05	4	4	3.2E-01	4111.78	10	10	1.01E+00
3577.5	15	13	7.8E-03	3377.39	4	2	9.0E-01	4115.18	8	8	5.80E-01
3577.8	11	11	8.3E-03	3377.62	6	6	6.0E-01	4116.47	6	6	3.2E-01
3577.9	13	13	2.3E-02	3397.58	6	4	2.3E-01	4116.59	2	2	2.90E-01
3578.3	13	11	2.0E-02	3400.39	8	8	2.5E-01	4123.50	4	2	1.00E+00
3580.0	9	9	1.2E-02	3529.73	4	6	4.1E-01	4128.06	6	4	7.70E-01
3580.2	11	9	2.9E-02	3533.68	6	8	5.2E-01	4131.99	8	6	5.5E-01
3580.4	11	13	7.5E-03	3533.76	2	4	3.7E-01	4134.49	10	8	2.90E-01
3580.9	13	13	2.1E-02	3543.49	2	2	6.7E-01	4232.46	10	10	9.8E-01
3582.6	13	13	2.9E-02	3545.33	4	4	3.7E-01	4232.95	8	8	7.7E-01
3584.6	7	5	2.4E-02	3553.27	6	6	2.2E-01	4268.64	14	14	1.2E+00
3584.9	13	15	1.8E-01	3555.14	4	2	2.6E-01	4271.55	12	12	9.6E-01
3585.4	11	11	1.9E-02	3663.60	4	6	3.1E+00	4276.95	10	10	9.4E-01
3585.8	11	9	2.8E-02	3667.74	6	8	2.7E+00	4284.05	8	8	1.2E+00
3587.8	9	11	1.3E-02	3672.41	12	12	9.2E-01	4291.82	12	14	8.8E-01
3588.3	7	9	1.8E-02	3673.41	8	10	2.7E+00	4296.10	10	12	7.7E-01
3589.7	11	13	2.1E-02	3676.70	14	14	1.3E+00	4297.67	8	10	7.0E-01
3589.8	15	13	5.9E-02	3680.12	10	12	2.2E+00	4298.03	6	8	7.8E-01
3590.7	9	7	2.2E-02	3686.26	10	12	2.3E-01	4379.23	10	12	1.1E+00
3591.7	11	9	5.3E-02	3687.50	12	14	2.9E+00	4384.71	8	10	1.1E+00
3593.0	11	11	1.4E-02	3688.07	8	8	3.5E-01	4389.98	6	8	6.9E-01
3593.2	13	15	4.2E-02	3690.28	2	4	4.5E-01	4395.22	4	6	5.5E-01
3593.7	11	11	7.2E-02	3692.22	6	6	5.4E-01	4400.57	2	4	3.4E-01
				3695.34	14	16	2.8E+00	4406.64	10	10	2.2E-01
Vanadium				3695.86	4	4	6.6E-01	4407.63	8	8	4.4E-01
V I				3703.57	10	8	9.2E-01	4408.20	6	6	6.0E-01
3043.12	6	8	2.3E-01	3704.70	8	6	6.6E-01	4416.47	4	2	2.6E-01
3050.39	10	8	5.3E-01	3705.04	6	4	3.6E-01	4452.01	14	16	9.2E-01
3053.65	4	4	1.3E+00	3706.03	10	10	5.2E-01	4457.75	10	12	2.7E-01
3056.33	6	6	1.3E+00	3708.71	12	12	4.4E-01	4460.33	10	8	3.0E-01
3060.46	8	8	1.4E+00	3790.46	10	8	2.3E-01	4462.36	12	14	7.6E-01
3066.37	10	10	2.1E+00	3794.96	10	10	2.3E-01	4468.00	8	10	2.3E-01
3066.53	6	4	3.2E-01	3806.79	10	10	2.5E-01	4469.71	10	12	6.2E-01

λ Å	Weights g_i	Weights g_k	A 10^8 s^{-1}	λ Å	Weights g_i	Weights g_k	A 10^8 s^{-1}	λ Å	Weights g_i	Weights g_k	A 10^8 s^{-1}
4474.04	10	8	4.7E-01	2893.31	9	7	1.2E+00	2371.06	10	12	5.2E+00
4496.06	8	6	4.0E-01	2903.07	3	5	3.4E-01	2373.06	4	6	2.9E+00
4514.18	6	4	3.3E-01	2906.45	7	7	7.8E-01	2382.46	8	10	5.0E+00
4524.21	12	10	3.0E-01	2908.81	11	9	1.6E+00	2393.58	6	8	4.3E+00
4525.17	4	2	4.1E-01	2910.01	5	5	1.1E+00	2404.18	4	6	2.5E+00
4529.58	10	8	2.4E-01	2910.38	3	3	1.2E+00	2516.14	10	10	3.7E+00
4545.40	10	12	7.6E-01	2911.05	7	9	3.7E-01	2521.55	8	8	3.5E+00
4560.72	8	10	7.0E-01	2912.46	11	9	5.0E-01	2548.21	6	4	2.0E+00
4571.79	6	8	6.0E-01	2915.88	9	7	4.9E-01	2554.22	8	6	1.2E+00
4578.73	4	6	6.8E-01	2924.02	11	11	1.7E+00	2593.05	6	6	2.8E+00
4706.16	6	4	2.4E-01	2924.63	9	9	1.2E+00	2595.10	8	8	2.8E+00
4757.47	4	2	7.6E-01	2930.80	7	7	5.8E-01				
4766.62	6	4	5.6E-01	2941.37	11	9	3.5E-01	**V IV**			
4776.36	8	6	5.1E-01	2944.57	9	7	7.6E-01	677.345	9	9	6.7E+00
4786.50	10	8	4.7E-01	2948.08	9	11	4.0E-01	680.632	9	7	1.2E+01
4796.92	12	10	4.8E-01	2952.07	7	5	7.2E-01	681.145	7	5	1.1E+01
4807.52	14	12	5.8E-01	2955.58	7	9	3.3E-01	682.455	7	7	6.5E+00
5193.00	12	12	4.0E-01	2968.37	7	9	7.0E-01	682.923	5	5	6.9E+00
5195.39	8	8	2.3E-01	2972.26	5	7	5.2E-01	684.450	7	5	7.7E+00
5234.08	10	10	4.9E-01	2973.98	9	11	3.5E-01	691.530	5	3	1.1E+01
5240.87	12	12	4.3E-01	2985.18	7	9	4.4E-01	723.537	3	1	1.5E+01
5415.25	12	14	3.1E-01	3001.20	7	7	7.5E-01	724.068	5	5	1.1E+01
5487.91	12	10	2.9E-01	3014.82	5	3	8.9E-01	724.809	5	3	5.6E+00
5507.75	10	8	3.5E-01	3016.78	7	5	5.0E-01	737.854	9	7	2.4E+01
6090.21	8	6	2.60E-01	3020.21	9	7	5.0E-01	750.110	5	5	1.0E+01
				3048.21	11	13	7.0E-01	884.146	1	3	4.7E+00
V II				3063.25	9	11	1.0E+00	1071.05	5	5	6.1E+00
2527.90	13	13	6.1E-01	3100.94	7	7	5.8E-01	1110.72	3	3	5.0E+00
2528.47	9	9	5.2E-01	3113.56	11	11	5.0E-01	1112.20	7	7	6.3E+00
2528.83	11	11	5.3E-01	3122.89	11	13	7.6E-01	1112.44	5	5	5.0E+00
2554.04	9	9	5.4E-01	3134.93	13	13	5.9E-01	1127.84	7	5	8.9E+00
2589.10	9	9	7.7E-01	3136.50	11	11	5.3E-01	1131.26	9	7	9.4E+00
2640.86	5	7	1.2E+00	3139.73	9	9	5.2E-01	1194.46	7	5	1.0E+01
2677.80	3	5	3.4E-01	3151.32	3	5	4.4E-01	1226.52	5	5	1.5E+01
2679.33	7	7	3.4E-01	3190.69	9	9	3.3E-01	1243.72	3	1	9.4E+00
2683.09	1	3	3.4E-01	3250.78	11	9	5.2E-01	1247.07	5	3	4.7E+00
2687.96	9	9	7.6E-01	3251.87	5	7	3.5E-01	1272.97	3	1	2.7E+01
2689.88	3	1	9.2E-01	3271.12	7	9	6.9E-01	1304.17	3	5	1.5E+01
2690.25	7	5	3.4E-01	3276.12	9	11	5.2E-01	1305.42	5	7	7.0E+00
2690.79	5	3	5.2E-01	3279.84	9	11	5.8E-01	1308.06	7	9	7.9E+00
2700.94	9	11	3.5E-01	3287.71	5	7	7.5E-01	1309.50	5	5	8.7E+00
2706.17	7	9	3.4E-01	3337.85	5	7	5.3E-01	1312.72	7	7	8.6E+00
2734.22	9	7	6.2E-01	3517.30	9	7	3.8E-01	1317.57	5	7	8.7E+00
2753.41	13	11	4.2E-01	3530.77	5	3	4.5E-01	1321.92	7	9	9.9E+00
2784.20	9	9	1.3E+00	3545.19	7	5	4.3E-01	1326.81	3	5	4.0E+00
2787.91	7	9	5.0E-01	3556.80	9	7	5.1E-01	1329.29	5	5	1.5E+01
2825.86	9	7	1.2E+00	3592.01	7	5	4.4E-01	1329.97	3	3	4.8E+00
2843.82	7	5	9.9E-01	3618.92	3	5	3.3E-01	1330.36	1	3	6.0E+00
2847.57	9	7	4.6E-01					1331.67	3	1	1.7E+01
2854.34	11	9	5.0E-01	**V III**				1332.46	5	3	7.5E+00
2862.31	11	11	3.6E-01	2318.06	8	10	4.6E+00	1334.49	9	9	8.3E+00
2868.11	5	3	2.1E+00	2323.82	6	8	3.8E+00	1355.13	7	9	2.5E+01
2869.13	13	11	4.8E-01	2330.42	10	10	3.2E+00	1356.53	5	3	4.9E+00
2882.49	5	5	4.2E-01	2331.75	8	8	2.5E+00	1395.00	5	7	1.4E+01
2884.78	3	3	5.6E-01	2334.21	6	6	2.2E+00	1400.42	5	7	7.5E+00
2889.61	3	1	1.9E+00	2337.13	4	4	2.7E+00	1403.62	7	9	8.4E+00
2891.64	5	3	1.4E+00	2343.10	6	8	3.6E+00	1412.69	3	3	1.1E+01
2892.43	9	9	3.6E-01	2358.73	6	8	4.2E+00	1414.41	5	7	1.2E+01
2892.65	7	5	1.3E+00	2366.31	8	10	4.2E+00	1414.84	5	5	4.6E+00

λ Å	Weights g_i	g_k	A 10^8 s^{-1}	λ Å	Weights g_i	g_k	A 10^8 s^{-1}	λ Å	Weights g_i	g_k	A 10^8 s^{-1}
1418.53	7	7	5.2E+00	**Xe II**				4128.30	6	6	1.6E+00
1419.58	7	9	1.3E+01	4180.1	4	4	2.2E+00	4142.84	4	4	1.6E+00
1423.72	3	5	7.1E+00	4330.5	6	8	1.4E+00	4167.51	6	6	2.38E-01
1426.65	9	11	2.2E+01	4414.8	6	6	1.0E+00	4235.93	6	4	3.0E-01
1429.11	5	5	5.0E+00	4603.0	4	4	8.2E-01	4352.40	4	4	6.7E-03
1434.84	7	7	5.4E+00	4844.3	6	8	1.1E+00	4379.33	6	4	7.83E-01
1451.04	3	3	7.0E+00	4876.5	6	8	6.3E-01	4385.47	4	4	6.9E-02
1454.00	5	3	1.1E+01	5260.4	2	4	2.2E-01	4394.01	8	8	1.9E-02
1520.14	5	7	7.2E+00	5262.0	4	4	8.5E-01	4409.70	4	6	2.7E-03
1522.49	3	5	5.5E+00	5292.2	6	6	8.9E-01	4417.43	10	8	3.2E-02
1601.92	3	3	1.2E+01	5372.4	4	2	7.1E-01	4437.34	6	6	8.64E-02
1611.88	7	7	5.2E+00	5419.2	4	6	6.2E-01	4443.65	10	8	1.1E-01
1806.18	5	3	7.3E+00	5439.0	4	2	7.4E-01	4459.01	4	6	1.8E-02
1809.85	3	1	7.2E+00	5472.6	8	8	9.9E-02	4476.95	8	6	2.8E-01
1817.68	5	3	4.8E+00	5531.1	8	6	8.8E-02	4491.74	10	10	2.3E-02
1825.84	7	5	5.3E+00	5719.6	4	6	6.1E-02	4514.01	4	6	3.34E-01
1861.56	5	7	6.6E+00	5976.5	4	4	2.8E-01	4527.78	8	6	8.33E-01
1939.07	7	9	5.8E+00	6036.2	6	6	7.5E-02	4534.09	6	8	4.4E-02
1951.43	5	7	5.0E+00	6051.2	8	6	1.7E-01	4544.31	6	6	4.10E-01
1963.10	3	5	4.8E+00	6097.6	6	4	2.6E-01	4559.36	2	4	4.0E-01
1997.72	7	7	4.7E+00	6270.8	4	6	1.8E-01	4581.33	6	4	1.5E-01
2084.43	5	5	4.0E+00	6277.5	4	6	3.6E-02	4613.00	6	4	1.8E-01
2120.05	7	9	8.1E+00	6805.7	8	6	6.1E-02	4643.70	4	6	1.8E-01
2141.20	3	5	7.0E+00	6990.9	10	8	2.7E-01	4653.78	4	6	1.6E-01
2146.83	7	9	6.6E+00					4674.85	6	8	1.3E-01
2149.85	5	7	5.1E+00	**Ytterbium**				4725.84	4	4	1.5E-01
2151.09	7	9	4.3E+00	**Yb I**				4762.96	6	4	4.2E-02
2155.34	11	13	1.2E+01	2464.5	1	3	9.1E-01	4780.16	2	4	8.9E-02
2446.80	9	11	5.3E+00	2672.0	1	3	1.18E+00	4781.03	8	10	1.0E-01
2570.72	9	11	7.6E+00	3464.4	1	3	6.2E-01	4799.30	6	8	1.6E-01
3284.56	7	9	5.3E+00	3988.0	1	3	1.76E+00	4804.31	6	4	2.6E-01
3496.42	7	9	4.4E+00	5556.5	1	3	1.14E-02	4804.80	4	4	3.84E-01
3514.25	9	11	4.7E+00					4821.63	6	6	1.0E-01
				Yb II				4845.67	8	8	6.8E-01
Xenon				3289.4	2	4	1.8E+00	4852.68	6	6	6.2E-01
Xe I				3694.2	2	2	1.4E+00	4856.71	6	6	2.0E-01
1043.8	1	3	5.9E-01					4859.84	4	4	7.26E-01
1047.1	1	3	1.3E+00	**Yttrium**				4893.44	6	4	2.2E-01
1050.1	1	3	8.5E-02	**Y I**				4900.08	8	6	2.0E-01
1056.1	1	3	2.45E+00	2948.41	4	4	3.5E-01	4906.11	10	8	1.2E-01
1061.2	1	3	1.9E-01	2974.59	4	6	3.5E-01	4950.01	8	6	2.0E-02
1068.2	1	3	3.99E+00	2984.25	6	8	4.8E-01	4963.49	4	4	1.4E-02
1085.4	1	3	4.10E-01	2995.26	6	4	5.1E-02	4981.97	4	6	4.7E-03
1099.7	1	3	4.34E-01	2996.94	4	6	8.4E-02	5004.44	6	4	1.2E-02
1110.7	1	3	1.5E+00	3005.26	4	4	4.8E-02	5205.01	4	4	8.4E-03
1129.3	1	3	4.4E-02	3022.28	6	6	6.6E-02	5258.47	6	6	2.9E-03
1170.4	1	3	1.6E+00	3045.36	6	6	1.07E-01	5271.82	8	6	1.1E-02
1192.0	1	3	6.2E+00	3053.95	6	4	1.9E-03	5380.63	6	4	3.2E-01
1250.2	1	3	1.4E-01	3155.65	4	6	2.7E-01	5381.24	4	4	9.9E-03
1295.6	1	3	2.46E+00	3172.84	4	4	9.9E-03	5388.39	6	8	1.1E-02
1469.6	1	3	2.81E+00	3185.96	6	8	1.2E-03	5390.81	8	6	2.9E-02
4501.0	5	3	6.2E-03	3209.38	6	6	3.0E-03	5401.88	6	8	6.0E-03
4524.7	5	5	2.1E-03	3227.16	6	4	1.10E-03	5424.36	6	4	3.47E-01
4624.3	5	5	7.2E-03	3484.05	4	6	1.2E-02	5466.24	4	4	1.0E-01
4671.2	5	7	1.0E-02	3549.66	6	6	1.0E-03	5466.47	10	12	6.3E-01
4807.0	3	1	2.4E-02	3552.69	4	4	2.3E-01	5469.10	4	6	3.6E-03
7119.6	7	9	6.6E-02	4077.36	4	6	1.1E+00	5513.65	6	6	2.39E-01
7967.3	1	3	3.0E-03	4083.71	4	4	2.5E-01	5519.88	4	6	1.2E-02
8409.2	5	3	1.0E-02	4102.36	6	8	1.3E+00	5526.43	6	4	3.9E-03

λ Å	g_i	g_k	A $10^8 s^{-1}$	λ Å	g_i	g_k	A $10^8 s^{-1}$	λ Å	g_i	g_k	A $10^8 s^{-1}$
5527.56	8	10	5.4E-01	3448.81	5	5	4.1E-02	5119.11	5	7	1.6E-02
5541.63	8	8	5.2E-02	3467.88	5	3	2.7E-02	5200.41	5	5	1.3E-01
5551.00	4	4	6.9E-02	3496.08	1	3	3.49E-01	5205.73	7	7	1.6E-01
5573.03	6	4	1.8E-02	3549.01	5	7	3.97E-01	5289.82	7	5	6.7E-03
5594.12	6	8	5.0E-02	3584.51	3	5	4.02E-01	5320.78	9	7	3.9E-03
5606.34	10	10	5.84E-02	3600.74	7	7	1.4E+00	5473.39	3	5	4.3E-02
5619.96	6	4	2.0E-02	3601.91	3	3	1.13E+00	5480.73	1	3	7.62E-02
5630.14	4	6	4.9E-01	3611.04	5	5	1.04E+00	5497.41	5	5	1.2E-01
5641.78	2	4	1.9E-02	3628.70	5	3	3.3E-01	5509.90	5	5	4.24E-02
5675.27	6	6	9.3E-02	3664.62	7	5	3.7E-01	5544.61	3	1	1.8E-01
5675.64	4	6	4.3E-02	3710.29	7	9	1.5E+00	5546.01	5	3	5.8E-02
5693.63	4	4	1.1E-01	3747.55	3	5	1.9E-01	5728.89	5	5	3.0E-02
5714.94	8	6	2.0E-02	3774.34	5	7	1.1E+00	6613.74	5	7	1.7E-02
5729.25	6	6	2.2E-03	3776.56	5	3	2.42E-01	6832.48	5	5	3.3E-03
5732.09	6	6	7.5E-02	3788.70	3	5	8.1E-01	7264.16	5	3	1.3E-02
5740.22	8	6	4.0E-02	3818.34	5	5	9.70E-02				
5757.59	4	6	7.6E-03	3832.90	7	7	3.0E-01	**Zinc**			
5788.36	4	4	9.4E-03	3878.29	7	5	2.9E-02	**Zn I**			
5844.13	6	4	5.6E-03	3930.66	5	5	2.1E-02	748.29	1	3	6.0E-02
5879.93	4	2	8.5E-02	3950.36	3	5	2.80E-01	765.60	1	3	7.6E-02
5902.91	6	8	4.0E-02	3951.59	5	3	1.5E-02	792.05	1	3	5.7E-02
6087.94	6	4	1.1E-01	3982.60	5	5	2.7E-01	793.85	1	3	1.8E-01
6191.72	4	4	4.7E-02	4124.91	5	7	1.8E-02	809.92	1	3	2.6E-01
6222.58	4	6	5.9E-03	4177.54	5	5	5.27E-01	1109.1	1	3	3.05E-01
6402.01	6	4	2.7E-03	4199.27	3	5	5.36E-03	2138.6	1	3	7.09E+00
6435.02	6	6	4.0E-02	4204.69	1	3	2.20E-02	3075.9	1	3	3.29E-04
6437.17	10	8	4.8E-02	4235.73	5	5	2.3E-02	3282.3	1	3	9.0E-01
6538.57	10	10	1.5E-01	4309.62	7	5	1.29E-01	3302.6	3	5	1.2E+00
6622.48	8	6	4.5E-03	4358.73	3	3	5.55E-02	3302.9	3	3	6.7E-01
6815.15	2	4	7.18E-02	4374.95	5	5	9.97E-01	3345.0	5	7	1.7E+00
7009.89	2	4	4.4E-02	4398.01	5	3	1.16E-01	3345.6	5	5	4.0E-01
7035.15	4	4	6.3E-02	4422.59	3	1	1.83E-01	3345.9	5	3	4.5E-02
				4682.33	5	5	1.9E-02	6362.3	3	5	4.74E-01
Y II				4786.58	7	7	2.1E-02	11054	3	1	2.43E-01
3112.03	1	3	1.3E-02	4823.31	5	5	4.3E-02				
3179.42	3	5	3.8E-02	4854.87	5	3	3.9E-01	**Zn II**			
3195.62	3	3	8.23E-01	4881.44	5	3	1.5E-03	2025.5	2	4	3.3E+00
3200.27	5	5	4.8E-01	4883.69	9	7	4.7E-01	2064.2	2	4	4.6E+00
3203.32	3	1	2.77E+00	4900.11	7	5	4.51E-01	2099.9	4	6	5.6E+00
3216.69	5	3	2.0E+00	4982.13	7	9	1.5E-02	2102.2	4	4	9.3E-01
3242.28	7	5	2.0E+00	5087.42	9	9	2.0E-01	4911.6	4	6	1.6E+00

ELECTRON AFFINITIES
Thomas M. Miller

Electron affinity is defined as the energy difference between the lowest (ground) state of the neutral and the lowest state of the corresponding negative ion. The accuracy of electron affinity measurements has been greatly improved since the advent of laser photodetachment experiments with negative ions. Electron affinities can be determined with optical precision, though a detailed understanding of atomic and molecular states and splittings is required to specify the photodetachment threshold corresponding to the electron affinity.

Atomic and molecular electron affinities are discussed in two excellent articles reviewing photodetachment studies which appear in *Gas Phase Ion Chemistry*, Vol. 3, Bowers, M. T., Ed., Academic Press, Orlando, 1984: Chapter 21 by Drzaic, P. S., Marks, J., and Brauman, J. I., "Electron Photodetachment from Gas Phase Negative Ions", p. 167, and Chapter 22 by Mead, R. D., Stevens, A. E., and Lineberger, W. C., "Photodetachment in Negative Ion Beams", p. 213. Persons interested in photodetachment details should consult these articles and the critical review of Hotop, H., and Lineberger, W. C., *J. Phys. Chem. Ref. Data*, 14, 731, 1985. For simplicity in the tables below, any electron affinity which was discussed in the articles by Drzaic *et al.* or Hotop and Lineberger is referenced to these sources, where original references are given. A great many additional electron affinities have been provided here by G. B. Ellison, W. C. Lineberger, H. Hotop, D. G. Leopold, and K. H. Bowen. Electron affinities for the lanthanides and actinides have not been measured, but theoretical estimates have been made by Bratch, S. G., *Chem. Phys. Lett.*, 98, 113, 1983, and Bratch, S. G., and Lagowski, J. J., *Chem. Phys. Lett.*, 107, 136, 1984. The development of cluster-ion photodetachment apparatuses has brought an explosion of electron affinity estimates for atomic and molecular clusters. [See Arnold, S. T., Eaton, J. G., Patel-Mistra, D., Sarkas, H. W., and Bowen, K. H., in *Ion and Cluster Ion Spectroscopy and Structure*, Maier, J. P., Ed., Elsevier Science, New York, 1989, p. 417.] The policy in this tabulation is to list the electron affinities for the atoms, diatoms, and triatoms, if adiabatic electron affinities have been determined, but to refer the reader to original sources for higher-order clusters. Additional data on molecular electron affinities may be found in Lias, S. G., Bartmess, J. E., Liebman, J. F., Holmes, J. L., Levin, R. D., and Mallard, W. G., *Gas Phase Ion and Neutral Thermochemistry, J. Phys. Chem. Ref. Data*, 17, (Supplement No. 1), 1988.

For the present tabulation the value $e/hc = 8065.5410 \pm 0.0024$ cm^{-1} eV^{-1} (Cohen, E. R., and Taylor, B. N., *Rev. Mod. Phys.*, 59, 1121, 1987) has been used to convert electron affinities from the units used in spectroscopic work, cm^{-1}, into eV for these tables. The uncertainties in the electron affinities given in the tables do not include the 0.3 ppm uncertainty in *e/hc*; only in a few cases, *e.g.*, atomic oxygen, is this additional uncertainty significant.

Abbreviations used in the tables: calc = calculated value; PT = photodetachment threshold using a lamp as a light source; LPT = laser photodetachment threshold; LPES = laser photoelectron spectroscopy; DA = dissociative attachment; e-scat = electron scattering; CT = charge transfer; and CD = collisional detachment.

Table 1
Atomic Electron Affinities

Atomic number	Atom	Electron affinity in eV	Uncertainty in eV	Method	Ref.
1	D	0.754593	0.000074	LPT	89
	H	0.754195	0.000019	LPT	89
		0.754209	0.000003	calc	1
2	He	not stable	—	calc	1
3	Li	0.6180	0.0005	LPT	1
4	Be	not stable	—	calc	1
5	B	0.277	0.010	LPES	1
6	C	1.2629	0.0003	LPT	1
7	N	not stable	—	DA	1
8	O	1.4611103	0.0000007	LPT	4
9	F	3.401190	0.000004	LPT	74
10	Ne	not stable	—	calc	1
11	Na	0.547926	0.000025	LPT	1
12	Mg	not stable	—	e-scat	1
13	Al	0.441	0.010	LPES	1
14	Si	1.385	0.005	LPES	1
15	P	0.7465	0.0003	LPT	1
16	S	2.077104	0.000001	LPT	1
17	Cl	3.61269	0.00006	LPT	52
18	Ar	not stable	—	calc	1
19	K	0.50147	0.00010	LPT	1
20	Ca	0.02455	0.00010	LPT	44
21	Sc	0.188	0.020	LPES	1
22	Ti	0.079	0.014	LPES	1
23	V	0.525	0.012	LPES	1
24	Cr	0.666	0.012	LPES	1
25	Mn	not stable	—	calc	1

Table 1
Atomic Electron Affinities (continued)

Atomic number	Atom	Electron affinity in eV	Uncertainty in eV	Method	Ref.
26	Fe	0.151	0.003	LPES	27
27	Co	0.662	0.003	LPES	27
28	Ni	1.156	0.010	LPES	1
29	Cu	1.235	0.005	LPES	37
30	Zn	not stable	—	e-scat	1
31	Ga	0.3	0.15	PT	1
32	Ge	1.233	0.003	LPES	28
33	As	0.81	0.03	PT	1
34	Se	2.020670	0.000025	LPT	1
35	Br	3.363590	0.000003	LPT	74
36	Kr	not stable	—	calc	1
37	Rb	0.48592	0.00002	LPT	1
38	Sr	0.048	0.006	LPT	122
39	Y	0.307	0.012	LPES	1
40	Zr	0.426	0.014	LPES	1
41	Nb	0.893	0.025	LPES	1
42	Mo	0.748	0.002	LPES	127
43	Tc	0.55	0.20	calc	1
44	Ru	1.05	0.15	calc	1
45	Rh	1.137	0.008	LPES	1
46	Pd	0.562	0.005	LPES	116
47	Ag	1.302	0.007	LPES	1
48	Cd	not stable	—	e-scat	1
49	In	0.3	0.2	PT	1
50	Sn	1.112	0.004	LPES	28
51	Sb	1.046	0.005	LPES	108
52	Te	1.9708	0.0003	LPT	1
53	I	3.059038	0.000010	LPT	92
54	Xe	not stable	—	calc	1
55	Cs	0.471626	0.000025	LPT	1
56	Ba	0.15	—	calc	57
57	La	0.5	0.3	calc	1
72	Hf	≈0	—	calc	1
73	Ta	0.322	0.012	LPES	1
74	W	0.815	0.002	LPES	37
75	Re	0.15	0.15	calc	1
76	Os	1.1	0.2	calc	1
77	Ir	1.5638	0.0005	LPT	141
78	Pt	2.128	0.002	LPT	1
79	Au	2.30863	0.00003	LPT	1
80	Hg	not stable	—	e-scat	1
81	Tl	0.2	0.2	PT	1
82	Pb	0.364	0.008	LPES	1
83	Bi	0.946	0.010	LPES	1
84	Po	1.9	0.3	calc	1
85	At	2.8	0.2	calc	1
86	Rn	not stable	—	calc	1
87	Fr	0.46	—	calc	82
118	ekaradon	0.056	—	calc	140

Table 2
Electron Affinities for Diatomic Molecules

Molecule	Electron affinity in eV	Uncertainty in eV	Method	Ref.	Molecule	Electron affinity in eV	Uncertainty in eV	Method	Ref.
Ag_2	1.023	0.007	LPES	37	MgH	1.05	0.06	PT	2
Al_2	1.10	0.15	LPES	68	MgI	1.899	0.018	LPES	31
AlO	2.60	0.02	LPES	143	MnD	0.866	0.010	LPES	9
AlS	2.60	0.03	LPES	129	MnH	0.869	0.010	LPES	9
As_2	0	—	PT	2	MoO	1.290	0.006	LPES	127
AsH	1.0	0.1	PT	2	NH	0.370	0.004	LPT	32
Au_2	1.938	0.007	LPES	37	NO	0.026	0.005	LPES	73
BO	3.12	0.09	PT	6	NS	1.194	0.011	LPES	2
BeH	0.7	0.1	PT	2	Na_2	0.430	0.015	LPES	104
Bi_2	1.271	0.008	LPES	119	NaBr	0.788	0.010	LPES	30
Br_2	2.55	0.10	CT	2	NaCl	0.727	0.010	LPES	30
BrO	2.353	0.006	LPES	88	NaF	0.520	0.010	LPES	30
C_2	3.269	0.006	LPES	87	NaI	0.865	0.010	LPES	30
CH	1.238	0.008	LPES	2	NaK	0.465	0.030	LPES	104
CN	3.862	0.004	LPES	111	Ni_2	0.926	0.010	LPES	112
CS	0.205	0.021	LPES	2	NiCu	0.889	0.010	LPES	128
CaH	0.93	0.05	PT	2	NiAg	0.979	0.010	LPES	128
Cl_2	2.38	0.10	CT	2	NiD	0.477	0.007	LPES	29
ClO	2.275	0.006	LPES	88	NiH	0.481	0.007	LPES	29
Co_2	1.110	0.008	LPES	27	NiO	1.470	0.003	LPES	146
CoD	0.680	0.010	LPES	29	O_2	0.451	0.007	LPES	73
CoH	0.671	0.010	LPES	29	OD	1.825534	0.000037	LPT	142
Cr_2	0.505	0.005	LPES	114	OH	1.8276542	0.0000037	LPT	142
CrD	0.568	0.010	LPES	29	P_2	0.589	0.025	LPES	42
CrH	0.563	0.010	LPES	29	PH	1.028	0.010	LPES	2
CrO	1.221	0.006	LPES	5	PO	1.092	0.010	LPES	2
Cs_2	0.469	0.015	LPES	104	Pb_2	1.366	0.010	LPES	117
CsCl	0.455	0.010	LPES	30	PbO	0.722	0.006	LPES	105
CsO	0.273	0.012	LPES	133	Pd_2	1.685	0.008	LPES	112
Cu_2	0.836	0.006	LPES	37	Pt_2	1.898	0.008	LPES	112
CuO	1.777	0.006	LPES	118	PtN	1.240	0.010	LPES	46
F_2	3.08	0.10	CT	2	Rb_2	0.498	0.015	LPES	104
FO	2.272	0.006	LPES	88	RbCl	0.544	0.010	LPES	30
Fe_2	0.902	0.008	LPES	27	RbCs	0.478	0.020	LPES	104
FeD	0.932	0.015	LPES	9	Re_2	1.571	0.008	LPES	33
FeH	0.934	0.011	LPES	9	S_2	1.670	0.015	LPES	53
FeO	1.493	0.005	LPES	45	SD	2.315	0.002	LPES	10
Ge_2	2.035	0.001	LPES	123	SF	2.285	0.006	LPES	93
I_2	2.55	0.05	CT	2	SH	2.314344	0.000004	LPT	47
IBr	2.55	0.10	CT	2	SO	1.125	0.005	LPES	84
IO	2.378	0.006	LPES	88	Sb_2	1.282	0.008	LPES	108
InP	1.95	0.05	LPES	137	Se_2	1.94	0.07	LPES	38
K_2	0.497	0.012	LPES	104	SeH	2.212520	0.000025	LPT	48
KBr	0.642	0.010	LPES	30	SeO	1.456	0.020	LPES	41
KCl	0.582	0.010	LPES	30	Si_2	2.201	0.010	LPES	100
KCs	0.471	0.020	LPES	104	SiH	1.277	0.009	LPES	2
KI	0.728	0.010	LPES	30	Sn_2	1.962	0.010	LPES	117
KRb	0.486	0.020	LPES	104	SnPb	1.569	0.008	LPES	117
LiCl	0.593	0.010	LPES	30	Te_2	1.92	0.07	LPES	38
LiD	0.337	0.012	LPES	102	TeH	2.102	0.015	LPES	39
LiH	0.342	0.012	LPES	102	TeO	1.697	0.022	LPES	40
MgCl	1.589	0.011	LPES	31	ZnH	≤0.95	—	PT	2

Table 3
Electron Affinities for Triatomic Molecules

Molecule	Electron affinity in eV	Uncertainty in eV	Method	Ref.	Molecule	Electron affinity in eV	Uncertainty in eV	Method	Ref.
Ag_3	2.32	0.05	LPES	37	Ge_3	2.23	0.01	LPES	123
Al_3	1.4	0.15	LPES	68	GeH_2	1.097	0.015	LPES	28
AlO_2	4.23	0.02	LPES	143	HCO	0.313	0.005	LPES	35
Al_2S	0.80	0.12	LPES	129	HCl_2	4.896	0.005	LPES	69
AsH_2	1.27	0.03	PT	2	HNO	0.338	0.015	LPES	14
Au_3	3.7	0.3	LPES	37	HO_2	1.078	0.017	LPES	15
BO_2	3.57	0.13	CT	6	InP_2	1.61	0.05	LPES	137
BO_2	4.3	0.2	CT	98	In_2P	2.36	0.05	LPES	137
Bi_3	1.60	0.03	LPES	119	K_3	0.956	0.050	LPES	18
C_3	1.981	0.020	LPES	11	MnD_2	0.465	0.014	LPES	34
CCl_2	1.591	0.010	LPES	95	MnH_2	0.444	0.016	LPES	34
CD_2	0.645	0.006	LPES	12	N_3	2.70	0.12	PT	2
CDF	0.535	0.005	LPES	95	NCN	2.484	0.006	LPES	154
CF_2	0.165	0.010	LPES	95	NCND	2.622	0.005	LPES	154
CH_2	0.652	0.006	LPES	12	NCNH	2.622	0.005	LPES	154
CHBr	1.454	0.005	LPES	95	NCO	3.609	0.005	LPES	111
CHCl	1.210	0.005	LPES	95	NCS	3.537	0.005	LPES	111
CHF	0.542	0.005	LPES	95	NH_2	0.771	0.005	LPES	58
CHI	1.42	0.17	LPES	95	N_2O	0.22	0.10	CT	59
C_2H	2.969	0.006	LPES	87	NO_2	2.273	0.005	LPES	63
C_2O	1.848	0.027	LPES	13	(NO)R	R=Ar,Kr,Xe	—	LPES	90
COS	0.46	0.20	CD	2	Na_3	1.019	0.060	LPES	18
CS_2	0.895	0.020	LPES	11	Ni_3	1.41	0.05	LPES	55
C_2Ti	1.542	0.020	LPES	147	NiCO	0.804	0.012	LPES	2
CoD_2	1.465	0.013	LPES	34	NiD_2	1.926	0.007	LPES	34
CoH_2	1.450	0.014	LPES	34	NiH_2	1.934	0.008	LPES	34
CrH_2	>2.5	—	LPES	34	O_3	2.1028	0.0025	LPT	2
Cr_2D	1.464	0.005	LPES	107	O_2Ar	0.52	0.02	LPES	75
Cr_2H	1.474	0.005	LPES	107	OClO	2.140	0.008	LPES	88
CrO_2	2.413	0.008	LPES	144	OIO	2.577	0.008	LPES	88
Cs_3	0.864	0.030	LPES	18	PH_2	1.271	0.010	LPES	2
Cu_3	2.11	0.05	LPES	37	PO_2	3.42	0.01	LPES	124
DCO	0.301	0.005	LPES	35	Pt_3	1.87	0.02	LPES	55
DNO	0.330	0.015	LPES	14	Pd_3	<1.5	0.1	LPES	55
DO_2	1.089	0.017	LPES	15	Rb_3	0.920	0.030	LPES	18
DS_2	1.912	0.015	LPES	53	S_3	2.093	0.025	LPES	16
HS_2	1.907	0.015	LPES	53	SO_2	1.107	0.008	LPES	16
Fe_3	1.47	0.08	LPES	149	S_2O	1.877	0.008	LPES	16
FeCO	1.157	0.005	LPES	103	Sb_3	1.85	0.03	LPES	108
FeD_2	1.038	0.013	LPES	34	SeO_2	1.823	0.050	LPES	38
FeH_2	1.049	0.014	LPES	34	SiH_2	1.124	0.020	LPES	2
FeO_2	2.358	0.030	LPES	130	VO_2	2.3	0.2	CT	101

Table 4
Electron Affinities for Larger Polyatomic Molecules

Molecule	Electron Affinity in eV	Uncertainty in eV	Method	Ref.	Name
Ag_n	n = 1-10	—	LPES	37	
Al_n	n = 3-32	—	LPES	68	
Al_nO_m	n=1,2	m=1-5	LPES	143	
Al_3O	1.00	0.15	LPES	68	
Al_nS_m	n = 1-5	m = 1-3	LPES	129	
$Ar(H_2O)_n$	n = 2,6,7	—	LPES	77	

Table 4
Electron Affinities for Larger Polyatomic Molecules (continued)

Molecule	Electron Affinity in eV	Uncertainty in eV	Method	Ref.	Name
Au_n	$n = 2\text{-}5$	—	LPES	37	
AuF_6	7.5	estimate	CT	98	
BD_3	0.027	0.014	LPES	62	
BH_3	0.038	0.015	LPES	62	
Bi_4	1.05	0.010	LPES	119	
$Br(CO_2)$	3.582	0.017	LPES	131	
C_n	$n = 2\text{-}84$	—	LPES	70	
$(CO_2)_n$	$n = 1,2$	—	LPES	75	
$(CS)_n$	$n = 2$	—	LPES	75	
$(CS_2)_n$	$n = 1,2$	—	LPES	75	
CF_3Br	0.91	0.2	CD	2	
CF_3I	1.57	0.2	CD	2	
CDO_2	3.510	0.015	LPES	109	
CHO_2	3.498	0.015	LPES	109	
CH_2S	0.465	0.023	LPES	53	
CH_3	0.08	0.03	LPES	2	
CH_3I	0.2	0.1	CT	2	
CH_3NO_2	0.48	0.10	CT	61	
CH_3Si	0.852	0.010	LPES	97	$CH_3\text{-Si}$
CH_3Si	2.010	0.010	LPES	97	$CH_2 = SiH$
CD_3O	1.552	0.022	LPES	2	
CH_3O	1.570	0.022	LPES	2	
CD_3S	1.856	0.006	LPT	2	
CH_3S	1.861	0.004	LPT	2	
CD_3S_2	1.748	0.022	LPES	53	
CH_3S_2	1.757	0.022	LPES	53	
CH_3SiH_2	1.19	0.04	LPT	65	
CFO_2	4.277	0.030	LPES	131	
CO_3	2.69	0.14	LPES	2	
$CO_3(H_2O)$	2.1	0.2	PT	2	
C_2F_2	2.255	0.006	LPES	106	difluorovinylidene
C_2DO	2.350	0.020	LPES	13	
C_2HO	2.350	0.020	LPES	13	
C_2D_2	0.492	0.006	LPES	83	vinylidene-d_2
C_2HD	0.489	0.006	LPES	83	vinylidene-d_1
C_2HF	1.718	0.006	LPES	106	monofluorovinylidene
C_2H_2	0.490	0.006	LPES	83	vinylidene
C_2H_2FO	2.22	0.09	PT	2	acetyl fluoride enolate
C_2D_2N	1.538	0.012	LPES	21	cyanomethyl-d_2 radical
C_2D_2N	1.070	0.024	LPES	21	isocyanomethyl-d_2 radical
C_2H_2N	1.543	0.014	LPES	21	cyanomethyl radical
C_2H_2N	1.059	0.024	LPES	21	isocyanomethyl radical
C_2H_3	0.667	0.024	LPES	90	vinyl
C_2D_3O	1.81897	0.00012	LPT	22	acetaldehyde-d_3 enolate
C_2H_3O	1.82476	0.00012	LPT	22	acetaldehyde enolate
C_2H_5N	0.56	0.01	PT	2	ethyl nitrine
C_2D_5O	1.702	0.033	LPES	23	ethoxide-d_3
C_2H_5O	1.726	0.033	LPES	23	ethoxide
C_2H_5S	1.953	0.006	LPT	2	ethyl sulfide
C_2H_5S	0.868	0.051	LPES	53	CH_3SCH_2
$C_2H_7O_2$	2.26	0.08	PT	50	MeOHOMe
C_3Fe	1.69	0.08	LPES	132	
C_3H	1.858	0.023	LPES	11	
C_3HFe	1.58	0.06	LPES	132	
C_3H_2	1.794	0.008	LPES	153	
$C_3H_2F_3O$	2.625	0.010	LPT	113	1,1,1-trifluoroacetone enolate

Table 4
Electron Affinities for Larger Polyatomic Molecules (continued)

Molecule	Electron Affinity in eV	Uncertainty in eV	Method	Ref.	Name
C_3H_3	0.893	0.025	LPES	24	propargyl radical
C_3H_2D	0.88	0.15	LPES	24	propargyl-d_1 radical
C_3D_2H	0.907	0.023	LPES	24	propargyl-d_2 radical
C_3H_3N	1.247	0.012	LPES	21	CH_3CH-CN
C_3D_5	0.464	0.006	LPES	138	allyl-d_5
C_3H_5	0.481	0.008	LPES	138	allyl
C_3H_4D	0.373	0.019	LPES	25	allyl-d_1
C_3H_5O	1.758	0.019	LPT	113	acetone enolate
C_3H_5O	1.621	0.006	LPT	113	propionaldehyde enolate
$C_3H_5O_2$	1.80	0.06	PT	2	methyl acetate enolate
C_3H_7O	1.789	0.033	LPES	23	n-propyl oxide
C_3H_7O	1.839	0.029	LPES	23	isopropyl oxide
C_3H_7S	2.00	0.02	PT	2	n-propyl sulfide
C_3H_7S	2.02	0.02	PT	2	isopropyl sulfide
C_3O	1.34	0.15	LPES	11	
C_3O_2	0.85	0.15	LPES	11	
C_3Ti	1.561	0.015	LPES	147	
$C_4F_4O_3$	0.5	0.2	CD	2	tetrafluorosuccinic anhydride
C_4Fe	≤2.2	0.2	LPES	132	
C_4HFe	1.67	0.06	LPES	132	
$C_4H_2O_3$	1.44	0.10	CT	61	maleic anhydride
C_4D_4	0.909	0.015	LPES	125	vinylvinylidene-d_4
C_4H_4	0.914	0.015	LPES	125	vinylvinylidene
C_4H_4N	2.39	0.13	PT	2	pyrrolate
C_4H_5O	1.801	0.008	LPT	113	cyclobutanone enolate
C_4H_6	0.431	0.006	LPES	135	trimethylenemethane
$C_4H_6O_2$	0.69	0.10	CT	61	2,3-butanedione
C_4H_6D	0.493	0.008	LPES	138	2-methylallyl-d_7
C_4H_7	0.505	0.006	LPES	138	2-methylallyl
C_4H_7O	1.67	0.05	PT	2	butyraldehyde enolate
C_4H_5DO	1.67	0.05	PT	2	2-butanone-3-d_1 enolate
$C_4H_5D_2O$	1.75	0.06	PT	2	2-butanone-3,3-d_2 enolate
C_4H_9O	1.912	0.054	LPES	23	t-butoxyl
C_4H_9S	2.03	0.02	PT	2	n-butyl sulfide
C_4H_9S	2.07	0.02	PT	2	t-butyl sulfide
C_4O	2.05	0.15	LPES	11	
C_4O_2	2.0	0.2	LPES	11	
C_4Ti	1.494	0.020	LPES	147	
C_5	2.853	0.001	LPT	99	
C_5F_5N	0.68	0.11	CT	67	pentafluoropyridine
$C_5F_6O_3$	1.5	0.2	CD	2	hexafluoroglutaric anhydride
C_5D_5	1.790	0.008	LPES	11	cyclopentadienyl-d_5
C_5H_5	1.804	0.007	LPES	11	cyclopentadienyl
C_5H_7	0.91	0.03	PT	2	pentadienyl
C_5H_7O	1.598	0.007	LPT	113	cyclopentanone enolate
C_5H_9O	1.69	0.05	PT	2	3-penanone enolate
$C_5H_{11}O$	1.93	0.05	LPT	2	neopentoxyl
$C_5H_{11}S$	2.09	0.02	PT	2	n-pentyl sulfide
C_5O_2	1.2	0.2	LPES	11	
C_5Ti	1.748	0.050	LPES	147	
C_6	4.180	0.001	LPT	8	
$C_6Br_4O_2$	2.44	0.20	CT	2	tetrabromo-BQ
$C_6Cl_4O_2$	2.78	0.10	CT	61	tetrachloro-BQ
$C_6F_4O_2$	2.70	0.10	CT	61	tetrafluoro-BQ
C_6F_5Br	1.15	0.11	CT	67	pentafluorobromobenzene

Table 4
Electron Affinities for Larger Polyatomic Molecules (continued)

Molecule	Electron Affinity in eV	Uncertainty in eV	Method	Ref.	Name
C_6F_5Cl	0.82	0.11	CT	67	pentafluorochlorobenzene
C_6F_5I	1.41	0.11	CT	67	pentafluoroiodobenzene
$C_6F_5NO_2$	1.52	0.11	CT	67	pentafluoro-NB
C_6F_6	0.52	0.10	CT	51	hexafluorobenzene
C_6F_{10}	>1.4	0.3	CT	2	perfluorocyclohexene
$C_6H_2Cl_2O_2$	2.48	0.10	CT	61	2,6-dichloro-BQ
$C_6H_3F_2NO_2$	1.17	0.10	CT	61	2,4-difluoro-NB
C_6D_4	0.551	0.010	LPES	36	o-benzyne-d_4
C_6H_4	0.560	0.010	LPES	36	o-benzyne
$C_6H_4BrNO_2$	1.16	0.10	CT	61	o-bromo-NB
$C_6H_4BrNO_2$	1.32	0.10	CT	61	m-bromo-NB
$C_6H_4BrNO_2$	1.29	0.10	CT	61	p-bromo-NB
$C_6H_4ClNO_2$	1.14	0.10	CT	61	o-chloro-NB
$C_6H_4ClNO_2$	1.28	0.10	CT	61	m-chloro-NB
$C_6H_4ClNO_2$	1.26	0.10	CT	61	p-chloro-NB
C_6H_4ClO	≤2.58	0.08	PT	2	o-chloroperoxide
$C_6H_4FNO_2$	1.07	0.10	CT	61	o-fluoro-NB
$C_6H_4FNO_2$	1.23	0.10	CT	61	m-fluoro-NB
$C_6H_4FNO_2$	1.12	0.10	CT	61	p-fluoro-NB
$C_6H_4N_2O_4$	1.65	0.10	CT	61	o-diNB
$C_6H_4N_2O_4$	1.65	0.10	CT	61	m-diNB
$C_6H_4N_2O_4$	2.00	0.10	CT	61	p-diNB
$C_6H_4O_2$	1.91	0.10	CT	61	1,4-benzoquinone (BQ)
C_6D_5	1.092	0.020	LPES	26	phenyl-d_5
C_6D_5N	1.44	0.02	LPES	96	phenylnitrene-d_5
C_6H_5	1.096	0.006	LPES	26	phenyl
C_6H_5N	1.429	0.011	LPT	115	phenylnitrene
$C_6H_5NO_2$	1.01	0.10	CT	61	nitrobenzene (NB)
C_6H_5O	2.253	0.006	LPES	26	phenoxyl
C_6H_5S	≤2.47	0.06	PT	2	thiophenoxide
C_6H_5NH	1.70	0.03	PT	2	anilide
C_6H_7	<1.67	0.04	PT	2	methylchylopentadienyl
C_6H_8Si	1.435	0.004	LPT	65	$C_6H_5SiH_3$
C_6H_9O	1.526	0.010	LPT	113	cyclohexanone enolate
C_6H_{10}	0.645	0.015	LPES	126	t-butyl vinylidene
$C_6H_{11}O$	1.755	+0.05/-0.005	LPT	113	pinacolone enolate
$C_6H_{11}O$	1.82	0.06	PT	2	3,3-dimethylbutananl enolate
C_6N_4	2.3	0.3	PT	2	TCNE
C_7F_5N	1.11	0.11	CT	67	pentafluorobenzonitrile
C_7F_8	0.86	0.11	CT	67	octafluorotoluene
C_7F_{14}	1.08	0.10	CT	61	perfluoromethylcyclohexane
C_7HF_5O	1.10	0.11	CT	67	pentafluorobenzaldehyde
$C_7H_3N_3O_4$	2.16	0.10	CT	61	3,5-$(NO_2)_2$-benzonitrile
$C_7H_4F_3NO_2$	1.41	0.10	CT	61	m-trifluoromethyl-NB
$C_7H_4N_2O_2$	1.61	0.10	CT	61	o-cyano-NB
$C_7H_4N_2O_2$	1.56	0.10	CT	61	m-cyno-NB
$C_7H_4N_2O_2$	1.72	0.10	CT	61	p-cyano-NB
C_7H_6FO	2.218	0.010	LPT	2	m-fluoroacetophenone enolate
C_7H_6FO	2.176	0.010	LPT	2	p-fluoroacetophenone enolate
$C_7H_6FeO_3$	0.990	0.10	CT	120	η^4-1,3-butadiene-$Fe(CO)_3$
$C_7H_6N_2O_4$	1.77	0.05	PT	60	3,4-dintrotoluene
$C_7H_6N_2O_4$	1.77	0.05	PT	60	2,3-dinitrotoluene
$C_7H_6N_2O_4$	1.60	0.05	PT	60	2,4-dinitrotoluene
$C_7H_6N_2O_4$	1.55	0.05	PT	60	2,6-dinitrotoluene
C_7H_7	0.912	0.006	LPES	26	benzyl

Table 4
Electron Affinities for Larger Polyatomic Molecules (continued)

Molecule	Electron Affinity in eV	Uncertainty in eV	Method	Ref.	Name
C_7H_7	0.868	0.006	LPES	136	1-quadricyclanide
C_7H_7	0.962	0.006	LPES	136	2-quadricyclanide
C_7H_7	1.286	0.006	LPES	136	norbornadienide
C_7H_7	0.39	0.04	LPES	136	cycloheptatrienide
C_7H_7	3.046	0.006	LPES	136	1-(1,6-heptadiynide)
C_7H_7	≥1.140	0.006	LPES	136	3-(1,6-heptadiynide)
$C_7H_7NO_2$	0.92	0.10	CT	61	o-methyl-NB
$C_7H_7NO_2$	0.99	0.10	CT	61	m-methyl-NB
$C_7H_7NO_2$	0.95	0.10	CT	61	p-methyl-NB
$C_7H_7NO_3$	1.04	0.10	CT	61	m-OCH_3-NB
$C_7H_7NO_3$	0.91	0.10	CT	61	p-OCH_3-NB
C_7H_7O	≤2.36	0.06	PT	2	o-methyl phenoxide
C_7H_7O	2.14	0.02	PT	50	benzyloxide
$C_7H_7O_2$	1.85	0.10	CT	61	o-CH_3-BQ
C_7H_8FO	<3.05	0.06	PT	50	$PhCH_2OHF$
C_7H_9	1.27	0.03	PT	2	heptatrienyl
C_7H_9O	1.61	0.05	PT	2	2-norbornanone enolate
C_7H_9Si	1.33	0.04	LPT	65	$C_6H_5(CH_3)SiH$
$C_7H_{11}O$	1.598	0.007	LPT	113	cycloheptanone enolate
$C_7H_{11}O$	1.49	0.04	PT	2	2,5-dimethylcyclopentanone enolate
$C_7H_{13}O$	1.72	0.06	PT	2	4-heptanone enolate
$C_7H_{13}O$	1.46	0.04	PT	2	di-isopropyl ketone enolate
$C_8F_{14}N_2$	1.89	0.10	CT	51	1,4-$(CN)_2C_6F_4$
$C_8H_3F_5O$	0.88	0.11	CT	67	pentafluoroacetophenone
$C_8H_3F_6NO_2$	1.79	0.10	CT	61	3,5-$(CF_3)_2$-NB
$C_8H_4O_3$	1.21	0.10	CT	61	phthalic anhydride
C_8H_6	1.044	0.008	LPES	148	
C_8H_7	1.091	0.008	LPES	134	
C_8H_7O	2.057	0.010	PT	2	acetophenone enolate
C_8H_7O	2.10	0.08	LPT	2	phenylacetaldehyde enolate
C_8H_8	0.55	0.02	CT	134	cycooctatetraene
C_8H_8	0.65	0.04	LPES	134	cycooctatetraene
C_8H_8	0.919	0.008	LPES	139	m-xylylene
$C_8H_9NO_2$	1.21	0.05	PT	60	3,5-dimethyl-NB
$C_8H_9NO_2$	2.61	0.05	PT	60	2,6-dimethyl-NB
$C_8H_9NO_2$	0.86	0.10	CT	61	2,3-dimethyl-NB
$C_8H_{13}O$	1.63	0.06	PT	2	cyclooctanone enolate
$C_9H_8FeO_3$	0.76	0.10	CT	120	η^4-1,3-cyclohexadiene-$Fe(CO)_3$
C_9H_9O	2.030	0.010	LPT	2	m-methylacetophenone enolate
C_9H_9SiN	1.43	0.10	PT	2	trimethylsilylnitrene
$C_9H_{11}NO_2$	0.70	0.10	CT	61	2,4,6-trimethyl-NB
$C_9H_{15}O$	1.69	0.06	PT	2	cyclononanone enolate
$C_{10}H_4Cl_2O_2$	2.19	0.10	CT	61	2,3-dichloro-1,4-naphthoquinone
$C_{10}H_6N_2O_4$	1.78	0.10	CT	61	1,3-dinitronaphthalene
$C_{10}H_6N_2O_4$	1.77	0.10	CT	61	1,5-dinitronaphthalene
$C_{10}H_6O_2$	1.81	0.10	CT	61	1,4-naphthoquinone
$C_{10}H_7NO_2$	1.23	0.10	CT	61	1-nitronaphthalene
$C_{10}H_7NO_2$	1.18	0.10	CT	61	2-nitronaphthalene
$C_{10}H_8$	0.69	0.10	CT	61	azulene
$C_{10}H_8CrO_3$	0.93	0.10	CT	120	η^4-1,3,5-cycloheptatriene $Cr(CO)_3$
$C_{10}H_8FeO_3$	0.98	0.10	CT	120	η^4-1,3,5-cycloheptatriene-$Fe(CO)_3$
$C_{10}H_{17}O$	1.83	0.06	PT	2	cyclodecanone enolate
$C_{11}H_8FeO_3$	1.29	0.10	CT	120	η^4-1,3-butadiene-$Fe(CO)_3$
$C_{12}F_{10}$	0.82	0.11	CT	67	decafluorobiphenyl
$C_{12}H_4N_4$	2.8	0.3	CD	2	TCNQ

Table 4
Electron Affinities for Larger Polyatomic Molecules (continued)

Molecule	Electron Affinity in eV	Uncertainty in eV	Method	Ref.	Name
$C_{12}H_9$	1.07	0.10	PT	2	perinaphthenyl
$C_{12}H_{15}O$	2.032	0.010	LPT	2	*t*-butylacetophenone enolate
$C_{12}H_{21}O$	1.90	0.07	PT	2	cyclododecanone enolate
$C_{13}F_{10}O$	1.52	0.11	CT	67	decafluorobenzophenone
$C_{13}H_9FO$	0.64	0.10	CT	61	4-fluorobenzophenone
$C_{13}H_{10}O$	0.62	0.10	CT	61	benzophenone
$C_{14}H_9NO_2$	1.43	0.10	CT	61	9-nitroanthracene
$C_{14}H_{10}$	0.57	0.10	CT	66	anthracene
$C_{18}H_{12}$	1.04	0.10	CT	66	tetracene
$C_{20}H_{12}$	0.79	0.10	CT	66	benz[*a*]pyrene
$C_{20}H_{12}$	0.97	0.10	CT	66	perylene
$C_{22}H_{14}$	1.35	0.10	CT	66	pentacene
CeF_4	3.8	0.4	CT	98	
$Cl(CO_2)$	3.907	0.010	LPES	131	
CoF_4	6.4	0.3	CT	98	
$Cr(CO)_3$	1.349	0.006	LPES	94	
CrO_3	3.6	0.2	CT	98	
Cu_n	$n = 1-41$	—	LPES	37	
Fe_n	n=3-24	—	LPES	149	
$Fe(CO)_2$	1.22	0.02	LPES	2	
$Fe(CO)_3$	1.8	0.2	LPES	2	
$Fe(CO)_4$	2.4	0.3	LPES	2	
FeF_3	3.6	0.1	CT	98	
FeF_4	6.0	estimate	CT	98	
Fe_nO_m	$n=1-4$	$m=1-6$	LPES	152	
Ge_n	$n = 3-15$	—	LPES	71	
Ge_xAs_y	$n = 5-30$	$n = x + y$	LPES	72	
GeH_3	≤ 1.74	0.04	PT	2	
$H(NH_3)_n$	$n = 1,2$	—	LPES	76	
HNO_3	0.57	0.15	CD	2	
$(H_2O)_n$	$n = 2-19$	—	LPES	77	
$I(CO_2)$	3.225	0.001	LPES	131	
In_xP_y	$n = 2-8$	$n = x + y$	LPES	137	
IrF_4	4.7	0.3	CT	98	
IrF_6	6.5	0.4	CT	98	
K_n	$n = 2-7$	—	LPES	18	
MnF_4	5.5	0.2	CT	98	
$Mo(CO)_3$	1.337	0.006	LPES	94	
MoF_5	3.5	0.2	CT	98	
MoF_6	3.8	0.2	CT	98	
MoO_3	2.9	0.2	CT	98	
$(NH_3)_n$	$n = 41-1100$	—	LPES	77	
$NH_2(NH_3)_n$	$n = 1,2$	—	LPES	78	
$NO(H_2O)_n$	$n = 1,2$	—	LPES	75	
NO_3	3.937	0.014	LPES	85	
$NO(N_2O)_n$	$n = 1,2$	—	LPES	79	
$(NO)_2$	≥ 2.1	—	LPES	75	
$(N_2O)_n$	$n = 1,2$	—	LPES	81	
Na_n	$n = 2-5$	—	LPES	18	
$(NaF)_n$	$n = 1-7,12$	—	LPES	64	
$Na(NaF)_n$	$n = 5,7-12$	—	LPES	64	
$Ni(CO)_2$	0.643	0.014	LPES	2	
$Ni(CO)_3$	1.077	0.013	LPES	2	
NiO_2	3.05	0.01	LPES	145	ONiO
NiO_2	0.82	0.03	LPES	145	$Ni(O_2)$

Table 4
Electron Affinities for Larger Polyatomic Molecules (continued)

Molecule	Electron Affinity in eV	Uncertainty in eV	Method	Ref.	Name
$OH(H_2O)$	<2.95	0.15	PT	2	
OsF_4	3.9	0.3	CT	98	
OsF_6	6.0	0.3	CT	98	
PBr_3	1.59	0.15	CD	2	
PBr_2Cl	1.63	0.20	CD	2	
PCl_2Br	1.52	0.20	CD	2	
PCl_3	0.82	0.10	CD	2	
PF_5	0.75	0.15	CT	121	
PO_3	4.5	0.5	CT	98	
$POCl_2$	3.83	0.25	CD	2	
$POCl_3$	1.41	0.20	CD	2	
PtF_4	5.5	0.3	CT	98	
PtF_6	7.0	0.4	CT	98	
ReF_6	4.7	estimate	CT	98	
RhF_4	5.4	0.3	CT	98	
RuF_4	4.8	0.3	CT	98	
RuF_5	5.2	0.4	CT	98	
RuF_6	7.5	0.3	CT	98	
SF_4	1.5	0.2	CT	91	
SF_6	1.05	0.10	CT	56	
SO_3	≥1.70	0.15	CD	2	
$(SO_2)_2$	0.6	0.2	LPES	80	
SeF_6	2.9	0.2	CD	2	
Si_4	2.17	0.01	LPES	110	
Si_n	$n = 3-20$	—	LPES	71	
SiD_3	1.386	0.022	LPES	43	
SiF_3	≤2.95	0.10	PT	17	
SiH_3	1.406	0.014	LPES	43	
TeF_6	3.34	0.17	CD	2	
Ti_n	$n=3-65$	—	LPES	151	
UF_5	3.7	0.2	CT	98	
UF_6	5.1	0.2	CT	98	
UO_3	<2.1	—	CT	98	
V_n	$n=3-65$	—	LPES	150	
VF_4	3.5	0.2	CT	98	
V_4O_{10}	4.2	0.6	CT	101	
$W(CO)_3$	1.859	0.006	LPES	94	
WF_5	1.25	0.3	CD	18	
WF_6	3.36	+0.04/-0.20	CT	19	
WO_3	3.33	+0.04/-0.15	LPT	86	
WO_3	3.9	0.2	CT	98	

REFERENCES

1. Hotop, H., and Lineberger, W. C., *J. Phys. Chem. Ref. Data*, 14, 731, 1985.
2. Drzaic, P. S., Marks, J., and Brauman, J. I., in *Gas Phase Ion Chemistry*, Vol. 3, Bowers, M. T., Ed., Academic Press, Orlando, 1984, p. 167.
3. Schulz, P. A., Mead, R. D., Jones, P. L., and Lineberger, W. C., *J. Chem. Phys.*, 77, 1153, 1982.
4. Neumark, D. M., Lykke, K. R., Anderson, T., and Lineberger, W. C., *Phys. Rev. A*, 32, 1890, 1985.
5. Wenthold, P. G., Gunion, R. F., and Lineberger, W. C., *Chem. Phys. Lett.*, 258, 101, 1996.
6. Srivastava, R. D., Uy, O. M., and Farber, M., *Trans. Faraday Soc.*, 67, 2941, 1971.
7. Klein, R., McGinnis, R. P., and Leone, S. R., *Chem. Phys. Lett.*, 100, 475, 1983.
8. Arnold, C. C., Zhao, Y., Kitsopoulos, T. N., and Neumark, D. M., *J. Chem. Phys.*, 97, 6121, 1992; linear C_6^- to linear C_6.
9. Stevens, A. E., Fiegerle, C. S., and Lineberger, W. C., *J. Chem. Phys.*, 78, 5420, 1983.
10. Breyer, F., Frey, P., and Hotop, H., *Z. Phys.*, A 300, 7, 1981.
11. Oakes, J. M., and Ellison, G. B., *Tetrahedron*, 42, 6263, 1986.
12. Leopold, D. G., Murray, K. K., Miller, A. E. S., and Lineberger, W. C., *J. Chem. Phys.*, 83, 4849, 1985.

13. Oakes, J. M., Jones, M.E., Bierbaum, V. M., and Ellison, G. B., *J. Chem. Phys.*, 87, 4810, 1983.

14. Ellis, H. B., Jr. and Ellison, G. B., *J. Chem. Phys.*, 78, 6541, 1983.

15. Oakes, J. M., Harding, L. B., and Ellison, G. B., *J. Chem. Phys.*, 83, 5400, 1985.

16. Nimlos, M. E., and Ellison, G. B., *J. Chem. Phys.*, 90, 2574, 1986.

17. Richardson, L. M., Stephenson, L. M., and Brauman, J. I., *Chem. Phys. Lett.*, 30, 17, 1975.

18. McHugh, K. M., Eaton, J. G., Lee, G. H., Sarkas, H. W., Kidder, L. H., Snodgrass, J. T., Manaa, M. R., and Bowen, K. H., *J. Chem. Phys.*, 91, 3792, 1989. See also Ref. 104.

19. Viggiano, A. A., Paulson, J. F., Dale, F., Henchman, M., Adams, N. G., and Smith, D., *J. Phys. Chem.*, 89, 2264, 1985.

20. Burnett, S. M., Stevens, A. E., Fiegerle, C. S., and Lineberger, W. C., *Chem. Phys. Lett.*, 100, 124, 1983.

21. Moran, S., Ellis, H. B., DeFrees, D. J., McLean, A. D., and Ellison, G. B., *J. Am. Chem. Soc.*, 109, 5996, 1987; Moran, S., Ellis, H. B., DeFrees, D. J., McLean, A. D., Paulson, S. E., and Ellison, G. B., *J. Am. Chem. Soc.*, 109, 6004, 1987; see also Lykke, K. R., Neumark, D. M., Andersen, T., Trapa, V. J., and Lineberger, W. C., *J. Chem. Phys.*, 87, 6842, 1987.

22. Mead, R. D., Lykke, K. R., Lineberger, W. C., Marks, J., and Brauman, J. I., *J. Chem. Phys.*, 81, 4883, 1984; Lykke, K. R., Mead, R. D., and Lineberger, W. C., *Phys. Rev. Lett.*, 52, 2221, 1984. The EAs are 14717.7 ± 1.0 cm^{-1} for acetaldehyde enolate and 14671.0 ± 1.0 cm^{-1} for acetaldehyde-d$_3$ enolate.

23. Ellison, G. B., Engelking, P. C., and Lineberger, W. C., *J. Chem. Phys.*, 86, 4873, 1982.

24. Oakes, J. M., and Ellison, G. B., *J. Am. Chem. Soc.*, 105, 2969, 1983.

25. Ellison, G. B., and Oakes, J. M., *J. Am. Chem. Soc.*, 106, 7734, 1984. EA(allyl) and EA(allyl-d$_5$) are 0.119 and 0.083 eV too low, respectively, in this work, according to Reference 138. Therefore, EA(allyl-d$_1$) is likely too low by a similar amount.

26. Gunion, R. F., Gilles, M. K., Polak, M. L., and Lineberger, W. C., *Int. J. Mass Spectrom. Ion Processes*, 117, 601, 1992; also, Miller, A. E. S., and Lineberger, W. C., unpublished. See also Reference 136.

27. Leopold, D. G., and Lineberger, W. C., *J. Chem. Phys.*, 85, 51, 1986.

28. Miller, T. M., Miller, A. E. S., and Lineberger, W. C., *Phys. Rev. A*, 33, 3558, 1986.

29. Miller, A. E. S., Fiegerle, C. S., and Lineberger, W. C., *J. Chem. Phys.*, 87, 1549, 1987.

30. Miller, T. M., Leopold, D. G., Murray, K. K., and Lineberger, W. C., *J. Chem. Phys.*, 85, 2368, 1986.

31. Miller, T. M., and Lineberger, W. C., *Chem. Phys. Lett.*, 146, 364, 1988.

32. Neumark, D. M., Lykke, K. R., Andersen, T., and Lineberger, W. C., *J. Chem. Phys.*, 83, 4364, 1985.

33. Leopold, D. G., Miller, T. M., and Lineberger, W. C., *J. Am. Chem. Soc.*, 108, 178, 1986.

34. Miller, A. E. S., Fiegerle, C. S., and Lineberger, W. C., *J. Chem. Phys.*, 84, 4127, 1986.

35. Murray, K. K., Miller, T. M., Leopold, D. G., and Lineberger, W. C., *J. Chem. Phys.*, 84, 2520, 1986.

36. Leopold, D. G., Miller, A. E. S., and Lineberger, W. C., *J. Am. Chem. Soc.*, 108, 1379, 1986.

37. Ho, J., Ervin, K. M., and Lineberger, W. C., *J. Chem. Phys.*, 93, 6987, 1990; Leopold, D. G., Ho, J., and Lineberger, W. C., *J. Chem. Phys.*, 86, 1715, 1987; Pettiette, C. L., Yang, S. H., Craycraft, M. J., Conceicao, J., Laaksonen, R. T., Cheshnovsky, O., and Smalley, R. E., *J. Chem. Phys.*, 88, 5377, 1988.

38. Snodgrass, J. T., Coe, J. V., McHugh, K. M., Friedhoff, C. B., and Bowen, K. H., *J. Phys. Chem.*, 93, 1249, 1989.

39. Friedhoff, C. B., Snodgrass, J. T., Coe, J. V., McHugh, K. M., and Bowen, K. H., *J. Chem. Phys.*, 84, 1051, 1986.

40. Friedhoff, C. B., Coe, J. V., Snodgrass, J. T., McHugh, K. M., and Bowen, K. H., *Chem. Phys. Lett.*, 124, 268, 1986.

41. Coe, J. V., Snodgrass, J. T., Friedhoff, C. B., McHugh, K. M., and Bowen, K. H., *J. Chem. Phys.*, 84, 619, 1986.

42. Snodgrass, J. T., Coe, J. V., Friedhoff, C. B., McHugh, K. M., and Bowen, K. H., *Chem. Phys. Lett.*, 122, 352, 1985.

43. Nimlos, M. R., and Ellison, G. B., *J. Am. Chem. Soc.*, 108, 6522, 1986.

44. Petrunin, V., Andersen, H., Balling, P., and Andersen, T., *Phys. Rev. Lett.*, 76, 744, 1996

45. Andersen, T., Lykke, K. R., Neumark, D. M., and Lineberger, W. C., *J. Chem. Phys.*, 86, 1858, 1987.

46. Murray, K. K., Lykke, K. R., and Lineberger, W. C., *Phys. Rev. A*, 36, 699, 1987.

47. Mansour, N. B., and Larson, D. J., *Abstracts of the XV Int. Conf. on the Phys. of Electronic and Atomic Collisions*, p. 70, 1987.

48. Stonemann, R. C., and Larson, D. J., *Phys. Rev. A*, 35, 2928, 1987.

49. Nimlos, M. R., Harding, L. B., and Ellison, G. B., *J. Chem. Phys.*, 87, 5116, 1987.

50. Moylan, C. R., Dodd, J. A., Han, C.-C., and Braumann, J. I., *J. Chem. Phys.*, 86, 5350, 1987.

51. Chowdhury, S., Grimsrud, E. P., Heinis, T., and Kebarle, P., *J. Am. Chem. Soc.*, 108, 3630, 1986.

52. Trainham, R., Fletcher, G. D., and Larson, D. J., *J. Phys. B: At. Mol. Phys.*, 20, L777, 1987.

53. Moran, S., and Ellison, G. B., *J. Phys. Chem.*, 92, 1794, 1988.

54. Murray, K. K., Leopold, D. G., Miller, T. M., and Lineberger, W. C., *J. Chem. Phys.*, 89, 5442, 1988.

55. Ervin, K. M., Ho, J., and Lineberger, W. C., *J. Chem. Phys.*, 89, 4514, 1988.

56. Grimsrud, E. P., Chowdhury, S., and Kebarle, P., *J. Chem. Phys.*, 85, 4989, 1985.

57. Fischer, C. F., *Phys. Rev. A*, 39, 963, 1989.

58. Wickham-Jones, C. T., Ervin, K. M., Ellision, G. B., and Lineberger, W. C., *J. Chem. Phys.*, 91, 2762, 1989.

59. Hopper, D. G., Wahl, A. C., Wu, R. L. C., and Tiernan, T. O., *J. Chem. Phys.*, 65, 5474, 1976.

60. Mock, R. S., and Grimsrud, E. P., *J. Am. Chem. Soc.*, 111, 2861, 1989.

61. Chowdhury, S., Heinis, T., Grimsrud, E. P., and Kebarle, P., *J. Phys. Chem.*, 90, 2747, 1986. The uncertainty and other results are quoted in Ref. 60.

62. Wickham-Jones, C. T., Moran, S., and Ellison, G. B., *J. Chem. Phys.*, 90, 795, 1989.

63. Ervin, K. M., Ho, J., and Lineberger, W. C., *J. Phys. Chem.*, 92, 5405, 1988.

64. Miller, T. M., and Lineberger, W. C., *Int. J. Mass Spectrom. Ion Processes*, 102, 239, 1990.

65. Wetzel, D. M., Salomon, K. E., Berger, S., and Brauman, J. I., *J. Am. Chem. Soc.*, 111, 3835, 1989.

66. Crocker, L., Wang, T., and Kebarle, P., *J. Am. Chem. Soc.*, 115, 7818, 1993.

67. Dillow, G. W., and Kebarle, P., *J. Am. Chem. Soc.*, 111, 5592, 1989.

68. Gantefor, G., Gausa, M., Meiwes-Broer, K. H., and Lutz, H. O., *Z. Phys. D*, 9, 253, 1988; Taylor, K. J., Petteitte, C. L., Craycraft, M. J., Chesnovsky, O., and Smalley, R. E., *Chem. Phys. Lett.*, 152, 347, 1988.

69. Metz, R. B., Kitsopoulos, T., Weaver, A., and Neumark, D. M., *J. Chem. Phys.*, 88, 1463, 1988.

70. Yang, S., Pettiette, C. L., Conceicao, J., Cheshnovsky, O., and Smalley, R. E., *Chem. Phys. Lett.*, 139, 233, 1987; Yang, S., Taylor, K. J., Craycraft, M. J., Conceicao, J., Pettiette, C. L., Cheshnovsky, O., and Smalley, R. E., *Chem. Phys. Lett.*, 144, 431, 1988; Arnold, D. W., Bradforth, S. E., Kitsopoulos, T. N., and Neumark, D. M., *J. Chem. Phys.*, 95, 5479, 1991.

71. Cheshnovsky, O., Yang, S., Pettiette, C. L., Craycraft, M. J., Liu, Y., and Smalley, R. E., *Chem. Phys. Lett.*, 138, 119, 1987.

73. Travers, M. J., Cowles, D. C., and Ellison, G. B., *Chem. Phys. Lett.*, 164, 449, 1989.
74. Blondel, C., Cacciani, P., Delsart, C., and Trainham, R., *Phys. Rev. A*, 40, 3698, 1989.
75. Bowen, K. H., and Eaton, J. G., in *The Structure of Small Molecules and Ions*, Naaman, R., and Vager, Z., Eds., Plenum, New York, 1988, pp. 147-169; Arnold, S. T., Eaton, J. G., Patel-Mistra, D., Sarkas, H. W., and Bowen, K. H., in *Ion and Cluster Ion Spectroscopy and Structure*, Maier, J. P., Ed., Elsevier Science, New York, 1989, p. 417.
76. Snodgrass, J. T., Coe, J. V., Friedhoff, C. B., McHugh, K. M., and Bowen, K. H., *Faraday Disc. Chem. Soc.*, 88, 1988.
77. Lee, G. H., Arnold, S. T., Eaton, J. G., Sarkas, H. W., Bowen, K. H., Ludewigt, C., and Haberland, H., *Z. Phys. D - Atoms, Mol. and Clusters*, 20, 9, 1991; Coe, J. V., Lee, G. H., Eaton, J. G., Arnold, S. T., Sarkas, H. W., Bowen, K. H., Ludewigt, C., Haberland, H., and Worsnop, D. R., *J. Chem. Phys.*, 92, 3980, 1990.
78. Snodgrass, J. T., Coe, J. V., Friedhoff, C. B., McHugh, K. M., and Bowen, K. H., to be published, quoted in Ref. 75.
79. Coe, J. V., Snodgrass, J. T., Friedhoff, C. B., McHugh, K. M., and Bowen, K. H., *J. Chem. Phys.*, 87, 4302, 1987.
80. Friedhoff, C. B., Snodgrass, J. T., and Bowen, K. H., to be published, quoted in Ref. 75.
81. Coe, J. V., Snodgrass, J. T., Friedhoff, C. B., McHugh, K. M., and Bowen, K. H., *Chem. Phys. Lett.*, 124, 274, 1986.
82. Greene, C. H., *Phys. Rev. A*, 42, 1405, 1990.
83. Ervin, K. M., Ho, J., and Lineberger, W. C., *J. Chem. Phys.*, 91, 5974, 1989.
84. Polak, M. L., Fiala, B. L., Ervin, K. M., and Lineberger, W. C., *J. Chem. Phys.*, 94, 6924, 1991.
85. Weaver, A., Arnold, D. W., Bradforth, S. E., Neumark, D. M., *J. Chem. Phys.*, 94, 1740, 1991.
86. Walter, C. W., Devynck, P., Hertzler, C. F., Bae, Y. K., Smith, G. P., and Peterson, J. R., *Bull. Am. Phys. Soc.*, 35, 1163, 1990.
87. Ervin, K. M., and Lineberger, W. C., *J. Phys. Chem.*, 95, 1167, 1991.
88. Gilles, M. K., Polak, M. L., and Lineberger, W. C., *J. Chem. Phys.*, 96, 8012, 1992.
89. Lykke, K. R., Murray, K. K., and Lineberger, W. C., *Phys. Rev. A*, 43, 6104, 1991.
90. Ervin, K. M., Gronert, S., Barlow, S. E., Gilles, M. K., Harrison, A. G., Bierbaum, V. M., DePuy, C. H., Lineberger, W. C., and Ellison, G. B., *J. Am. Chem. Soc.*, 112, 5750, 1990.
91. Viggiano, A. A., Miller, T. M., Miller, A. E. S., Morris, R. A., Van Doren, J. M., and Paulson, J. F., *Int. J. Mass Spectrom. Ion Processes*, 109, 327, 1991.
92. Hanstorp, D., and Gustafsson, M., *J. Phys. B: At. Mol. Opt. Phys.*, 25, 1773, 1992.
93. Polak, M. L., Gilles, M. K., and Lineberger, W. C., *J. Chem. Phys.*, 96, 7191, 1992.
94. Bengali, A. A., Casey, S. M., Cheng, C.-L., Dick, J. P., Fenn, P. T., Villalta, P. W., and Leopold, D. G., *J. Am. Chem. Soc.*, 114, 5257, 1992.
95. Gilles, M. K., Ervin, K. M., Ho, J., and Lineberger, W. C., *J. Phys. Chem.*, 96, 1130, 1992.
96. Travers, M. J., Cowles, D. C., Clifford, E. P., and Ellison, G. B., *J. Am. Chem. Soc.*, 114, 8699, 1992.
97. Bengali, A. A., and Leopold, D. G., *J. Am. Chem. Soc.*, 114, 9192, 1992.
98. Rudnyi, E. B., Kaibicheva, E. A., and Sidorov, L. N., *Rapid Comm. in Mass Spectrom.*, 6, 356, 1992; Sidorov, L. N., *High Temp. Sci.*, 29, 153, 1990.
99. Kitsopoulos, T. N., Chick, C. J., Zhao, Y., and Neumark, D. M., *J. Chem. Phys.*, 95, 5479, 1991.
100. Arnold, C. C., Kitsopoulos, T. N., and Neumark, D. M., *J. Chem. Phys.*, 99, 766, 1993.
101. Rudnyi, E. B., Kaibicheva, E. A., and Sidorov, L. N., *J. Chem. Thermodynamics*, 25, 929, 1993.
102. Sarkas, H. W., Hendricks, J. H., Arnold, S. T., and Bowen, K. H., *J. Chem. Phys.* 100, 1884, 1994.
103. Villalta, P. W., and Leopold, D. G., *J. Chem. Phys.* 98, 7730, 1993.
104. Eaton, J. G., Sarkas, H. W., Arnold, S. T., McHugh, K. M., and Bowen, K. H., *Chem. Phys. Lett.*, 193, 141, 1992. See also Ref. 18.
105. Polak, M. L., Gilles, M. K., Gunion, R. F., and Lineberger, W. C., *Chem. Phys. Lett.*, 210, 55, 1993.
106. Gilles, M. K., Lineberger, W. C., and Ervin, K. M., *J. Am. Chem. Soc.*, 115, 1031, 1993.
107. Casey, S. M., and Leopold, D. G., *Chem. Phys. Lett.*, 201, 205, 1993.
108. Polak, M. L., Gerber, G., Ho, J., and Lineberger, W. C., *J. Chem. Phys.*, 97, 8990, 1992.
109. Kim, E. H., Bradforth, S. E., Arnold, D. W., Metz, R. B., and Neumark, D. M., *J. Chem. Phys.*, 103, 7801, 1995.
110. Arnold, C. C., and Neumark, D. M., *J. Chem. Phys.*, 99, 3353, 1993.
111. Bradforth, S. E., Kim, E. H., Arnold, D. W., and Neumark, D. M., *J. Chem. Phys.*, 98, 800, 1993.
112. Ho, J., Polak, M. L., Ervin, K. M., and Lineberger, W. C., *J. Chem. Phys.*, 99, 8542, 1993.
113. Brinkman, E. A., Berger, S., Marks, J., and Brauman, J. I., *J. Chem. Phys.*, 99, 7586, 1993.
114. Casey, S. M., and Leopold, D. G., *J. Phys. Chem.*, 97, 816, 1993.
115. McDonald, R. N., and Davidson, S. J., *J. Am. Chem. Soc.*, 115, 10857, 1993.
116. Ho, J., Ervin, K. M., Polak, M. L., Gilles, M. K., and Lineberger, W. C., *J. Chem. Phys.*, 95, 4845, 1991.
117. Ho, J., Polak, M. L., and Lineberger, W. C., *J. Chem. Phys.*, 96, 144, 1992.
118. Polak, M. L., Gilles, M. K., Ho, J., and Lineberger, W. C., *J. Phys. Chem.*, 95, 3460, 1991.
119. Polak, M. L., Ho, J., Gerber, G., and Lineberger, W. C., *J. Chem. Phys.*, 95, 3053, 1991.
120. Sharpe, P., and Kebarle, P., *J. Am. Chem. Soc.*, 115, 782, 1993.
121. Miller, T. M., Miller, A. E. S., Viggiano, A. A., Morris, R. A., and Paulson, J. F., *J. Chem. Phys.*, 100, 7200, 1994.
122. Berkovits, D., Boaretto, E., Gehlberg, S., Heber, O., and Paul, M., *Phys. Rev. Lett.*, 75, 414, 1995.
123. Arnold, C. C., Xu, C., Burton, G. R., and Neumark, D. M., *J. Chem. Phys.*, 102, 6982, 1995. Burton, G. R., Xu, C., Arnold, C. C., and Neumark, D. M., *J. Chem. Phys.*, in press.
124. Xu, C., de Beer, E., and Neumark, D. M., *J. Chem. Phys.*, 104, 2749, 1996.
125. Gunion, R. F., Koppel, H., Leach, G. W., and Lineberger, W. C., *J. Chem. Phys.*, 103, 1250, 1995.
126. Gunion, R. F., and Lineberger, W. C., *J. Phys. Chem.*, 100, 4395, 1996.
127. Gunion, R. F., Dixon-Warren, St. J., Lineberger, W. C., and Morse, M. D., *J. Chem. Phys.*, in press.
128. Dixon-Warren, St. J., Gunion, R. F., and Lineberger, W. C., *J. Chem. Phys.*, 104, 4902, 1996.
129. Nakajima, A., Zhang, N., Kawamata, H., Hayase, T., Nakao, K., and Kaya, K., *Chem. Phys. Lett.*, 241, 295, 1995; Nakajima, A., Taguwa, T., Nakao, K., Hoshino, K., Iwata, S., and Kaya, K., *J. Chem. Phys.*, 102, 660, 1995.
130. Fan, J., and Wang, L.-S., *J. Chem. Phys.*, 102, 8714, 1995.
131. Arnold, D. W., Bradforth, S. E., Kim, E. H., and Neumark, D. M., *J. Chem. Phys.*, 102, 3493, 1995; Zhao, Y., Arnold, C. C., and Neumark, D. M., *J. Chem. Soc. Faraday Trans. 2*, 89, 1449, 1992.
132. Fan, J., Lou, L., and Wang, L.-S., *J. Chem. Phys.*, 102, 2701, 1995.

ELECTRON AFFINITIES (continued)

133. Sarkas, H. W., Hendricks, J. H., Arnold, S. T., Slager, V. L., and Bowen, K. H., *J. Chem. Phys.*, 100, 3358, 1994.
134. Kato, S., Lee, H. S., Gareyev, R., Wenthold, P. G., Lineberger, W. C., DePuy, C. H., and Bierbaum, V. M., *J. Am. Chem. Soc.*, 119, 7863, 1997.
135. Wenthold, P. G., Hu, J., Squires, R. R., and Lineberger, W. C., *J. Am. Chem. Soc.*, 118, 475, 1996.
136. Gunion, R. F., Karney, W., Wenthold, P. G., Borden, W. T., and Lineberger, W. C., *J. Am. Chem. Soc.*, 118, 5074, 1996. The numbers in the abstract for 1,6-heptadiyne were misprinted. EA(cycloheptatrienide) quoted here derives from the LPES data combined with other thermochemical data in Reference 136.
137. Xu, C., de Beer, E., Arnold, D. W., Arnold, C. C., and Neumark, D. M., *J. Chem. Phys.*, 101, 5406, 1996.
138. Wenthold, P. G., Polak, M. L., and Lineberger, W. C., *J. Phys. Chem.*, 100, 6920, 1996.
139. Wenthold, P. G., Kim, J. B., and Lineberger, W. C., *J. Am. Chem. Soc.*, 119, 1354, 1997.
140. Eliav, E., Kaldor, U., Ishikawa, Y., and Pyykkoe, *Phys. Rev. Lett.*, 77, 5350, 1996.
141. Davies, B. J., Ingram, C. W., Larson, D. J., and Ljungblad, U., *J. Chem. Phys.* 106, 5783, 1997.
142. Smith, J. R., Kim, J. B., and Lineberger, W. C., *Phys. Rev. A.*, 55, 2036, 1997. Schulz, P.A., Mead, R. D., Jones, P. L., and Lineberger, W. C., *J. Chem. Phys.*, 77, 1153, 1982. See also Rudmin, J. D., Ratliff, L. P., Yukich, J. N., and Larson, D. J., *J. Phys. B: At. Mol. Opt. Phys.*, 29, L881, 1996.
143. Desai, S. R., Wu, H., Rohlfing, C. M., and Wantg, L.-S., *Int. J. Chem. Phys.*, 106, 1309, 1997.
144. Wenthold, P. G., Joans, K.-L., and Lineberger, W. C., *J. Chem. Phys.*, 106, 9961, 1997.
145. Wu, H. and Wang, L.-S., *J. Chem. Phys.*, 107, 16, 1997.
146. Moravec, V. D. and Jarrold, C. C., *J. Chem. Phys.*, in press.
147. Wang, X.-B., Ding, C.-F., and Wang, L.-S., *J. Phys. Chem. A*, 101, 7699, 1997.
148. Wenthold, P. G. and Lineberger, W. C., *J. Am. Chem. Soc.*, 19, 7772, 1997.
149. Wang, L.-S., Cheng, H.-S., and Fan, J., *J. Chem. Phys.*, 102, 9480, 1995.
150. Wu, H., Desai, S. R., and Wang, L.-S., *Phys. Rev. Lett.*, 77, 2436, 1996.
151. Wu, H., Desai, S. R., and Wang, L.-S., *Phys. Rev. Lett.*, 76, 212, 1996.
152. Wang, L.-S., Wu, H., and Desai, S. R., *Phys. Rev. Lett.*, 76, 4853, 1996.
153. Robinson, M. S., Polak, M. L., Bierbaum, V. M., DePuy, C. H., and Lineberger, W. C., *J. Am. Chem. Soc.*, 117, 6766, 1995.
154. Clifford, E. P., Wenthold, P. G., Lineberger, W. C., Petersson, G. A., and Ellison, G. B., *J. Phys. Chem.*, 101, 4338, 1997.
155. Wenthold, P. G., Kim, J. B., Jonas, K., and Lineberger, W. C., *J. Phys. Chem.*, in press, 1997.

ATOMIC AND MOLECULAR POLARIZABILITIES
Thomas M. Miller

The *polarizability* of an atom or molecule describes the response of the electron cloud to an external field. The atomic or molecular energy shift ΔW due to an external electric field E is proportional to E^2 for external fields which are weak compared to the internal electric fields between the nucleus and electron cloud. The *electric dipole polarizability* α is the constant of proportionality defined by $\Delta W = -\alpha E^2/2$. The induced electric dipole moment is αE. *Hyperpolarizabilities*, coefficients of higher powers of E, are less often required. Technically, the polarizability is a tensor quantity but for spherically symmetric charge distributions reduces to a single number. In any case, an *average polarizability* is usually adequate in calculations. Frequency-dependent or *dynamic polarizabilities* are needed for electric fields which vary in time, except for frequencies which are much lower than electron orbital frequencies, where *static polarizabilities* suffice.

Polarizabilities for atoms and molecules in excited states are found to be larger than for ground states and may be positive or negative. Molecular polarizabilities are very slightly temperature dependent since the size of the molecule depends on its rovibrational state. Only in the case of dihydrogen has this effect been studied enough to warrant consideration in Table 3.

Polarizabilities are normally expressed in cgs units of cm^3. Ground state polarizabilities are in the range of 10^{-24} $cm^3 = 1$ Å^3 and hence are often given in Å^3 units. Theorists tend to use atomic units of a_o^3 where a_o is the Bohr radius. The conversion is $\alpha(cm^3) = 0.148184 \times 10^{-24} \times \alpha$ (a_o^3). Polarizabilities are only recently encountered in SI units, $C \cdot m^2/V = J/(V/m)^2$. The conversion from cgs units to SI units is $\alpha(C \cdot m^2/V) = 4\pi\varepsilon_o \times 10^{-6} \times \alpha(cm^3)$, where ε_o is the permittivity of free space in SI units and the factor 10^{-6} simply converts cm^3 into m^3. Thus, $\alpha(C \cdot m^2/V) = 1.11265 \times 10^{-16} \times \alpha(cm^3)$. Persons measuring excited state polarizabilities by optical methods tend to use units of $MHz/(V/cm)^2$, where the energy shift, ΔW, is expressed in frequency units with a factor of h understood. The polarizability is $-2 \Delta W/E^2$. The conversion into cgs units is $\alpha(cm^3) = 5.95531 \times 10^{-16} \times \alpha[MHz/(V/cm)^2]$.

The polarizability appears in many formulas for low-energy processes involving the valence electrons of atoms or molecules. These formulas are given below in cgs units: the polarizability α is in cm^3; masses m or μ are in grams; energies are in ergs; and electric charges are in esu, where $e = 4.8032 \times 10^{-10}$ esu. The symbol $\alpha(\nu)$ denotes a frequency (ν) dependent polarizability, where $\alpha(\nu)$ reduces to the static polarizability α for $\nu = 0$. For further information and references, see Miller, T. M., and Bederson, B., *Advances in Atomic and Molecular Physics*, 13, 1, 1977. Details on polarizability-related interactions, especially in regard to hyperpolarizabilities and nonlinear optical phenomena, are given by Bogaard, M. P., and Orr, B. J., in *Physical Chemistry*, Series Two, Vol. 2, *Molecular Structure and Properties*, Buckingham, A. D., Ed., Butterworths, London, 1975, pp. 149-194. A tabulation of tensor and hyperpolarizabilities is included. The gas number density, n, in Table 1 is usually taken to be that of 1 atm at 0°C in reporting experimental data.

Table 1
Formulas Involving Polarizability

Description	Formula	Remarks
Lorentz-Lorenz relation	$\alpha(\nu) = \dfrac{3}{4\pi n}\left[\dfrac{\eta^2(\nu)-1}{\eta^2(\nu)+2}\right]$	For a gas of atoms or nonpolar molecules; the index of refraction is $\eta(\nu)$
Refraction by polar molecules	$\alpha(\nu) + \dfrac{d^2}{3kT} = \dfrac{3}{4\pi n}\left[\dfrac{\eta^2(\nu)-1}{\eta^2(\nu)+2}\right]$	The dipole moment is d, in esu·cm $(= 10^{-18}$ D)
Dielectric constant (dimensionless)	$\kappa(\nu) = 1 + 4\pi n\ \alpha(\nu)$	From the Lorentz-Lorenz relation for the usual case of $\kappa(\nu) \approx 1$
Index of refraction (dimensionless)	$\eta(\nu) = 1 + 2\pi n\ \alpha(\nu)$	From $\eta^2(\nu) = \kappa(\nu)$
Diamagnetic susceptibility	$\chi_m = e^2 \left(a_o N\alpha\right)^{1/2} / 4m_e c^2$	From the approximation that the static polarizability is given by the variational formula $\alpha = (4/9a_o)\Sigma(N_i r_i^2)^2$; N is the number of electrons, m_e is the electron mass; a crude approximation is $\chi_m = (E_i/4m_e c^2)\alpha$, where E_i is the ionization energy
Long-range electron- or ion-molecule interaction energy	$V(r) = -e^2\alpha / 2r^4$	The target molecule polarizability is α
Ion mobility in a gas	$\kappa = 13.87/(\alpha\mu)^{1/2}\ cm^2\ / V \cdot s$	This one formula is not in cgs units. Enter α in Å^3 or 10^{-24} cm^3 units and the reduced mass μ of the ion-molecule pair in amu. Classical limit; pure polarization potential

Table 1
Formulas Involving Polarizability (continued)

Description	Formula	Remarks
Langevin capture cross section	$\sigma(v_o) = (2\pi e/v_o)(\alpha/\mu)^{1/2}$	The relative velocity of approach for an ion-molecule pair is v_o; the target molecular polarizability is α and the reduced mass of the ion-molecule pair is μ
Langevin reaction rate coefficient	$k = 2\pi e(\alpha/\mu)^{1/2}$	Collisional rate coefficient for an ion-molecule reaction
Rate coefficient for polar molecules	$k_d = 2\pi e\left[(\alpha/\mu)^{1/2} + cd(2/\mu\pi kT)^{1/2}\right]$	The dipole moment of the neutral is d in esu·cm; the number c is a "locking factor" that depends on α and d, and is between 0 and 1
Modified effective range cross section for electron-neutral scattering	$\sigma(k) = 4\pi A^2$ $+32\pi^4\mu e^2\alpha Ak/3h^2$ $+L$	Here, k is the electron momentum divided by $h/2\pi$, where h is Planck's constant; A is called the "scattering length"; the reduced mass is μ
van der Waals constant between two systems A, B	$C_6 = \dfrac{3}{2}\left[\dfrac{\alpha^A\alpha^B E^A E^B}{E^A + E^B}\right]$	For the interaction potential term $V_6(r) = -C_6 r^{-6}$; $E^{A,B}$ represents average dipole transition energies and $\alpha^{A,B}$ the respective polarizabilities of A, B
Dipole-quadrupole constant between two systems A, B	$C_8 = \dfrac{15}{4}\left[\dfrac{\alpha^A\alpha_q{}^B E^A E_q{}^B}{E^A + E_q{}^B}\right]$ $+ \dfrac{15}{4}\left[\dfrac{\alpha_q{}^A\alpha^B E_q{}^A E^B}{E_q{}^A + E^B}\right]$	For the interaction potential term $V_8(r) = -C_8 r^{-8}$; $E_q{}^{A,B}$ represents average quadrupole transition energies and $\alpha_q{}^{A,B}$ are the respective quadrupole polarizabilities of A, B
van der Waals constant between an atom and a surface	$C_3 = \dfrac{\alpha g E^A E^S}{8(E^A + E^S)}$	For an interaction potential $V_3(r) = -C_3 r^3$; $E^{A,S}$ are characteristic energies of the atom and surface; $g = 1$ for a free-electron metal and $g = (\varepsilon_\infty - 1)/(\varepsilon_\infty + 1)$ for an ionic crystal
Relation between $\alpha(v)$ and oscillator strengths	$\alpha(v) = \dfrac{e^2 h^2}{4\pi^2 m_e}\Sigma\dfrac{f_k}{E_k^2 - (hv)^2}$	Here, f_k is the oscillator strength from the ground state to an excited state k, with excitation energy E_k. This formula is often used to estimate static polarizabilities ($v = 0$)
Dynamic polarizability	$\alpha(v) = \dfrac{\alpha E_r{}^2}{E_r{}^2 - (hv)^2}$	Approximate variation of the frequency-dependent polarizability $\alpha(v)$ from $v = 0$ up to the first dipole-allowed electronic transition, of energy E_r; the static dipole polarizability is $\alpha(0)$; infrared contributions ignored
Rayleigh scattering cross section	$\alpha(v) = \dfrac{8\pi}{9c^4}(2\pi v)^4$ $\times\left[3\alpha^2(v) + 2\gamma^2(v)/3\right]$	The photon frequency is v; the polarizability anisotropy (the difference between polarizabilities parallel and perpendicular to the molecular axis) is $\gamma(v)$
Verdet constant	$V(v) = \dfrac{vn}{2m_e c^2}\left[\dfrac{d\alpha(v)}{dv}\right]$	Defined from $\theta = V(v)B$, where θ is the angle of rotation of linearly polarized light through a medium of number density n, per unit length, for a longitudinal magnetic field strength B (Faraday effect)

Table 2
Static Average Electric Dipole Polarizabilities for Ground State Atoms (in Units of 10^{-24} cm^3)

Atomic number	Atom	Polarizability	Estimated accuracy (%)	Method	Ref.
1	H	0.666793	"exact"	calc	MB77
2	He	0.204956	"exact"	calc	MB77
		0.2050	0.1	index/diel	NB65/OC67
3	Li	24.3	2	beam	MB77
4	Be	5.60	2	calc	MB77
5	B	3.03	2	calc	MB77
6	C	1.76	2	calc	MB77
7	N	1.10	2	calc/index	MB77
8	O	0.802	2	calc/index	MB77
9	F	0.557	2	calc	MB77
10	Ne	0.3956	0.1	diel	OC67
11	Na	24.08	0.4	interferom	ESCHP94
12	Mg	10.6	2	calc	MB77
13	Al	6.8	4.4	beam	MMD90
14	Si	5.38	2	calc	MB77
15	P	3.63	2	calc	MB77
16	S	2.90	2	calc	MB77
17	Cl	2.18	2	calc	MB77
18	Ar	1.6411	0.05	index/diel	NB65/OC67
19	K	43.4	2	beam	MB77
20	Ca	22.8	2	calc	MB77
		25.0	8	beam	MB77
21	Sc	17.8	25	calc	D84
22	Ti	14.6	25	calc	D84
23	V	12.4	25	calc	D84
24	Cr	11.6	25	calc	D84
25	Mn	9.4	25	calc	D84
26	Fe	8.4	25	calc	D84
27	Co	7.5	25	calc	D84
28	Ni	6.8	25	calc	D84
29	Cu	6.1	25	calc	D84
		7.31	25	calc	G84
30	Zn	7.1	25	calc	MB77
		5.6	25	calc	D84
31	Ga	8.12	2	calc	MB77
32	Ge	6.07	2	calc	MB77
33	As	4.31	2	calc	MB77
34	Se	3.77	2	calc	MB77
35	Br	3.05	2	calc	MB77
36	Kr	2.4844	0.05	diel	OC67
37	Rb	47.3	2	beam	MB77
38	Sr	27.6	8	beam	MB77
39	Y	22.7	25	calc	D84
40	Zr	17.9	25	calc	D84
41	Nb	15.7	25	calc	D84
42	Mo	12.8	25	calc	D84
43	Tc	11.4	25	calc	D84
44	Ru	9.6	25	calc	D84
45	Rh	8.6	25	calc	D84
46	Pd	4.8	25	calc	D84
47	Ag	7.2	25	calc	D84
		8.56	25	calc	G84
48	Cd	7.2	25	calc	D84
49	In	10.2	12	beam	GMBSJ84
		9.1	25	calc	D84

Table 2
Static Average Electric Dipole Polarizabiilities for Ground State Atoms (in Units of 10^{-24} cm^3)
(continued)

Atomic number	Atom	Polarizability	Estimated accuracy (%)	Method	Ref.
50	Sn	7.7	25	calc	D84
51	Sb	6.6	25	calc	D84
52	Te	5.5	25	calc	D84
53	I	5.35	25	index	A56
		4.7	25	calc	D84
54	Xe	4.044	0.5	diel	MB77
55	Cs	59.6	2	beam	MB77
56	Ba	39.7	8	beam	MB77
57	La	31.1	25	calc	D84
58	Ce	29.6	25	calc	D84
59	Pr	28.2	25	calc	D84
60	Nd	31.4	25	calc	D84
61	Pm	30.1	25	calc	D84
62	Sm	28.8	25	calc	D84
63	Eu	27.7	25	calc	D84
64	Gd	23.5	25	calc	D84
65	Tb	25.5	25	calc	D84
66	Dy	24.5	25	calc	D84
67	Ho	23.6	25	calc	D84
68	Er	22.7	25	calc	D84
69	Tm	21.8	25	calc	D84
70	Yb	21.0	25	calc	D84
71	Lu	21.9	25	calc	D84
72	Hf	16.2	25	calc	D84
73	Ta	13.1	25	calc	D84
74	W	11.1	25	calc	D84
75	Re	9.7	25	calc	D84
76	Os	8.5	25	calc	D84
77	Ir	7.6	25	calc	D84
78	Pt	6.5	25	calc	D84
79	Au	5.8	25	calc	D84
		6.48	25	calc	G84
80	Hg	5.7	25	calc	D84
		5.1	15	diel	MB77
81	Tl	7.6	15	beam	NYU84
		7.5	25	calc	D84
82	Pb	6.8	25	calc	D84
83	Bi	7.4	25	calc	D84
84	Po	6.8	25	calc	D84
85	At	6.0	25	calc	D84
86	Rn	5.3	25	calc	D84
87	Fr	48.7	25	calc	D84
88	Ra	38.3	25	calc	D84
89	Ac	32.1	25	calc	D84
90	Th	32.1	25	calc	D84
91	Pa	25.4	25	calc	D84
92	U	24.9	6	beam	KB94
93	Np	24.8	25	calc	D84
94	Pu	24.5	25	calc	D84
95	Am	23.3	25	calc	D84
96	Cm	23.0	25	calc	D84
97	Bk	22.7	25	calc	D84
98	Cf	20.5	25	calc	D84
99	Es	19.7	25	calc	D84

Table 2
Static Average Electric Dipole Polarizabiilities for Ground State Atoms (in Units of 10^{-24} cm^3)
(continued)

Atomic number	Atom	Polarizability	Estimated accuracy (%)	Method	Ref.
100	Fm	23.8	25	calc	D84
101	Md	18.2	25	calc	D84
102	No	17.5	25	calc	D84

Note: calc = calculated value; beam = atomic beam deflection technique; interferom = atomic beam interference; index = determination based on the measured index of refraction; diel = determination based on the measured dielectric constant.

REFERENCES

A56.   Atoji, M., *J. Chem. Phys.*, 25, 174, 1956. Semiempirical method based on molecular polarizabilities and atomic radii.

D84.   Doolen, G. D., Los Alamos National Laboratory, unpublished. A relativistic linear response method was used. The method is that described by Zangwill, A., and Soven, P., *Phys. Rev. A*, 21, 1561, 1980. Adjustments of less than 10% across the periodic table have been made to these results to bring them into agreement with accurate experimental values where avaialble, for the purpose of presenting "recommended" polarizabilities in Table 2.

ESCHP94.   Ekstrom, C. R., Schmiedmayer, J., Chapman, M. S., Hammond, T. D., and Pritchard, D. E., *Phys. Rev. Lett.*, in press, 1994.

G84.   Gollisch, H., *J. Phys. B: Atom. Mol. Phys.*, 17, 1463, 1984. Other results and useful references are contained in this paper.

GMBSJ84.   Guella, T. P., Miller, T. M., Bederson, B., Stockdale, J. A. D., and Jaduszliwer, B., *Phys. Rev. A*, 29, 2977, 1984.

KB94.   Kadar-Kallen, M. A., and Bonin, K. D., *Phys. Rev. Lett.*, 72, 828, 1994.

MB77.   Miller, T. M., and Bederson, B., *Adv. At. Mol. Phys.*, 13, 1, 1977. For simplicity, any value in Table 2 which has not changed since this 1977 review is referenced as MB77. Persons interested in original references and further details should consult MB77.

MMD90.   Milani, P., Moullet, I., and de Heer, W. A., *Phys. Rev. A*, 42, 5150, 1990.

NB65.   Newell, A. C., and Baird, R. D., *J. Appl. Phys.*, 36, 3751, 1965.

NYU84.   Preliminary value from the New York University group. See GMBSJ84.

OC67.   Orcutt, R. H., and Cole, R. H., *J. Chem. Phys.*, 46, 697, 1967; see also the later references from this group, given following the tables.

Table 3
Average Electric Dipole Polarizabilities for Ground State Diatomic Molecules (in Units of 10^{-24} cm^3)

Molecule	Polarizability	Ref.	Molecule	Polarizability	Ref.
Al_2	19	23	HI	5.44	3
BH	3.32*	1		5.35	2
Br_2	7.02	2	HgCl	7.4*	9
CO	1.95	3	ICl	12.3	2
Cl_2	4.61	3	K_2	77	22
Cs_2	104	22		72	21
CsK	89	22	Li_2	34	22
D_2 ($v = 0, J = 0$)	0.7921*	5	LiCl	3.46*	10
D_2 (293 K)	0.7954	6	LiF	10.8*	11
DCl	2.84	2	LiH	3.84*	12
F_2	1.38*	7		3.68*	13
H_2 ($v = 0, J = 0$)	0.8023*	5		3.88*	14
H_2 (293 K)	0.8045*	5	N_2	1.7403	6,8
H_2 (293 K)	0.8042	6	NO	1.70	2
H_2 (322 K)	0.8059	8	Na_2	40	22
HBr	3.61	3		38	21
HCl	2.63	3	NaK	51	22
	2.77	2	NaLi	40	4
HD ($v = 0, J = 0$)	0.7976*	5	O_2	1.5812	6
HF	0.80	27	Rb_2	79	22

Table 4
Average Electric Dipole Polarizabilities for Ground State Triatomic Molecules (in Units of 10^{-24} cm^3)

Molecule	Polarizability	Ref.	Molecule	Polarizability	Ref.	Molecule	Polarizability	Ref.
BeH_2	4.34*	14	HCN	2.59	3	O_3	3.21	2
CO_2	2.911	8		2.46	2	OCS	5.71	2
CS_2	8.74	3	$HgBr_2$	14.5	2		5.2	15
	8.86	2	$HgCl_2$	11.6	2	SO_2	3.72	3
D_2O	1.26	2	HgI_2	19.1	2		4.28	2
H_2O	1.45	2	N_2O	3.03	8			
H_2S	3.78	3	NO_2	3.02	2†			
	3.95	2	Na_3	70	21			

Table 5
Average Electric Dipole Polarizabilities for Ground State Inorganic Polyatomic Molecules (Larger than Triatomic) (in Units of 10^{-24} cm^3)

Molecule	Polarizability	Ref.	Molecule	Polarizability	Ref.
$AsCl_3$	14.9	2	$(NaBr)_2$	26.8	16
AsN_3	5.75	2	$(NaCl)_2$	23.4	16
BCl_3	9.38	20	$(NaF)_2$	20.7	16
BF_3	3.31	2	$(NaI)_2$	26.9	16
$(BN_3)_2$	5.73	2	OsO_4	8.17	2
$(BH_2N)_3$	8.0	2†	PCl_3	12.8	2
ClF_3	6.32	2	PF_5	6.10	2
$(CsBr)_2$	54.5	16	PH_3	4.84	2
$(CsCl)_2$	42.4	16	$(RbBr)_2$	48.2	16
$(CsF)_2$	28.4	16	$(RbCl)_2$	43.2	16
$(CsI)_2$	51.8	16	$(RbF)_2$	40.7	16
$GeCl_4$	15.1	2	$(RbI)_2$	46.3	16
GeH_3Cl	6.7	2†	SF_6	6.54	8
$(HgCl)_2$	14.7	9	$(SF_5)_2$	13.2	2
K_n	$n = 2,5,7-9,11,20$	21	SO_3	4.84	2
$(KBr)_2$	42.0	16	SO_2Cl_2	10.5	2
$(KCl)_2$	32.1	16	SeF_6	7.33	2
$(KF)_2$	21.0	16	SiF_4	5.45	2
$(KI)_2$	36.3	16	SiH_4	5.44	2
$(LiBr)_2$	18.9	16	$(SiH_3)_2$	11.1	2
$(LiCl)_2$	13.1	16	$SiHCl_3$	10.7	2
$(LiF)_2$	6.9	16	SiH_2Cl_2	8.92	2
$(LiI)_2$	23.4	16	SiH_3Cl	7.02	2
ND_3	1.70	2	$SnBr_4$	22.0	2
NF_3	3.62	2	$SnCl_4$	18.0	2
NH_3	2.81	20		13.8	15
	2.10	2	SnI_4	32.3	2
	2.26	3	TeF_6	9.00	2
$(NO_2)_2$	6.69	2	$TiCl_4$	16.4	2
Na_n	$n = 1-40$	21	UF_6	12.5	2

Table 6
Average Electric Dipole Polarizabilities for Ground State Hydrocarbon Molecules (in Units of 10^{-24} cm^3)

Molecule	Name	Polarizability	Ref.	Molecule	Name	Polarizability	Ref.
CH_4	methane	2.593	8	C_8H_{10}	ethylbenzene	14.2	2
C_2H_2	acetylene	3.33	3		o-xylene	14.9	2
		3.93	2			14.1	15
C_2H_4	ethylene	4.252	8		p-xylene	13.7	25
C_2H_6	ethane	4.47	3			14.2	15
		4.43	2			14.9	2
C_3H_4	propyne	6.18	2		m-xylene	14.2	15
C_3H_6	propene	6.26	2	C_8H_{16}	ethylcyclohexane	15.9	2
	cyclopropane	5.66	2	C_8H_{18}	n-octane	15.9	2
C_3H_8	propane	6.29	3		3-methylheptane	15.4	27
		6.37	2		2,2,4-trimethylpentane	15.4	27
C_4H_6	1-butyne	7.41	2†	C_9H_{10}	α-methylstyrene	16.05	27
	1,3-butadiene	8.64	2	C_9H_{12}	isopropylbenzene	16.0	2-
C_4H_8	1-butene	7.97	2		mesitylene	15.5	25
		8.52	2			16.1	27
	trans-2-butene	8.49	2	C_9H_{18}	isopropylcyclohexane	17.2	2
	2-methylpropene	8.29	2	C_9H_{20}	nonane	17.4	27
C_4H_{10}	n-butane	8.20	2	$C_{10}H_8$	napthalene	16.5	17
	isobutane	8.14	27			17.5	27
C_5H_6	1,3-cyclopentadiene	8.64	2	$C_{10}H_{14}$	durene	17.3	25
C_5H_8	1-pentyne	9.12	2		t-butylbenzene	17.2	25
	trans-1,3-pentadiene	10.0	2			17.8	2†
	isoprene	9.99	2	$C_{10}H_{20}$	t-butylcyclohexane	19.8	2
C_5H_{10}	cyclopentane	9.15	18	$C_{10}H_{22}$	decane	19.1	27
	1-pentene	9.65	27	$C_{11}H_{10}$	α-methylnaphthalene	19.35	27
	2-pentene	9.84	27		β-methylnaphthalene	19.52	27
C_5H_{12}	pentane	9.99	2	$C_{11}H_{14}$	α,β,β-trimethylstyrene	19.64	27
	neopentane	10.20	18	$C_{11}H_{16}$	pentamethylbenzene	19.1	25
C_6H_6	benzene	10.0	25	$C_{11}H_{24}$	undecane	21.0	27
		10.32	3	$C_{12}H_{10}$	acenaphthene	20.6	27
		10.74	2	$C_{12}H_{12}$	α-ethylnaphthalene	21.19	27
C_6H_{10}	1-hexyne	10.9	2†		β-ethylnaphthalene	21.36	27
	2-ethyl-1,3-butadiene	11.8	2†	$C_{12}H_{18}$	hexamethylbenzene	20.9	25
	3-methyl-1,3-pentadiene	11.8	2†	$C_{12}H_{26}$	dodecane	22.8	27
	2-methyl-1,3-pentadiene	12.1	2†	$C_{13}H_{10}$	fluorene	21.7	27
	2,3-dimethyl-1,3-butadiene	11.8	2†	$C_{14}H_{10}$	anthracene	25.4	17
	cyclohexene	10.7	2†			25.9	27
C_6H_{12}	cyclohexane	11.0	18		phenanthrene	36.8*	17
		10.87	15			24.7	27
	1-hexene	11.7	27	$C_{14}H_{22}$	p-di-t-butylbenzene	24.5	25
C_6H_{14}	n-hexane	11.9	2	$C_{16}H_{10}$	pyrene	28.2	27
C_7H_8	toluene	11.8	25	$C_{17}H_{12}$	2,3-benzfluorene	30.2	27
		12.26	15	$C_{18}H_{12}$	napthacene	32.3	27
		12.3	2		1,2-benzanthracene	32.9	27
C_7H_{12}	1-heptyne	12.8	2†		chrysene	33.1	27
C_7H_{14}	methylcyclohexane	13.1	2		triphenylene	31.1	27
	1-heptene	13.5	27	$C_{18}H_{30}$	1,3,5-tri-t-butylbenzene	31.8	25
C_7H_{16}	n-heptane	13.7	2	$C_{24}H_{12}$	coronene		
C_8H_8	styrene	15.0	2				
		14.4	27				

Table 7
Average Electric Dipole Polarizabilities for Ground State Organic Halides (in Units of 10^{-24} cm^3)

Molecule	Name	Polarizability	Ref.
CBr_2F_2	dibromodifluoromethane	9.0	2[†]
$CClF_3$	chlorotrifluoromethane	5.72	20
		5.59	2
CCl_2F_2	dichlorodifluoromethane	7.93	20
		7.81	2
CCl_2O	phosgene	7.29	2
CCl_2S	thiophosgene	10.2	2
CCl_3F	trichlorofluoromethane	9.47	2
CCl_3NO_2	trichloronitromethane	10.8	2[†]
CCl_4	carbon tetrachloride	11.2	2
		10.5	3
CF_4	carbon tetrafluoride	3.838	8
CF_2O	carbonylfluoride	1.88[*]	17
$CHBr_3$	bromoform	11.8	17
$CHBrF_2$	bromodifluoromethane	5.7	2[†]
$CHClF_2$	chlorodifluoromethane	6.38	20
		5.91	2
$CHCl_2F$	dichlorofluoromethane	6.82	2
$CHCl_3$	chloroform	9.5	8
		8.23	27
CHF_3	fluoroform	3.52	20
		3.57	8
$CHFO$	fluoroformaldehyde	1.76[*]	17
CHI_3	iodoform	18.0	17
CH_2Br_2	dibromomethane	9.32	2
		8.62	27
CH_2ClNO_2	chloronitromethane	6.9	2[†]
CH_2Cl_2	dichloromethane	6.48	3
		7.93	2
CH_2I_2	diiodomethane	12.90	27
CH_3Br	bromomethane	5.87	20
		6.03	2
		5.55	15
CH_3Cl	chloromethane	5.35	20
		4.72	8
CH_3F	fluoromethane	2.97	8
CH_3I	iodomethane	7.97	2
C_2ClF_5	chloropentafluoroethane	6.3	2[†]
$C_2Cl_2F_4$	1,2-dichlorotetrafluoroethane	8.5	2[†]
C_2Cl_3N	trichloroacetonitrile	6.10	18
C_2F_6	hexafluoroethane	6.82	2
C_2HBr	bromoacetylene	7.39	2
C_2HCl	chloroacetylene	6.07	2
C_2HCl_5	pentachloroethane	14.0	2
$C_2H_2Cl_2$	dichloroethylene	10.03	27
	trans-dichloroethylene	9.28	27
	cis-dichloroethylene	9.19	27
$C_2H_2Cl_2F_2$	1,1-dichloro-2,2-difluoroethane	8.4	2[†]
$C_2H_2Cl_2O$	chloroacetyl chloride	8.92	2
$C_2H_2Cl_3F$	1,2,2-trichloro-1-fluoroethane	10.2	2[†]
$C_2H_2Cl_4$	1,1,2,2-tetrachloroethane	12.1	2[†]
C_2H_2ClN	chloroacetonitrile	6.10	18
$C_2H_2F_2$	1,1-difluoroethylene	5.01	20
C_2H_3Br	bromoethylene	7.59	2
C_2H_3Cl	chloroethylene	6.41	2

Table 7
Average Electric Dipole Polarizabilities for Ground State Organic Halides (in Units of 10^{-24} cm^3) (continued)

Molecule	Name	Polarizability	Ref.
$C_2H_3ClF_2$	1-chloro-1,1-difluoroethane	8.05	2
C_2H_3ClO	acetyl chloride	6.62	2
$C_2H_3ClO_2$	ethyl chloroformate	7.1	2[†]
$C_2H_3Cl_3$	1,1,1-trichloroethane	10.7	2
$C_2H_3F_3$	1,1,1-trifluoroethane	4.4	2[†]
C_2H_3I	iodoethylene	9.3	2[†]
C_2H_4BrCl	1-bromo-2-chloroethane	9.5	2[†]
$C_2H_4Br_2$	1,2-dibromomethane	10.7	2[†]
C_2H_4ClF	1-chloro-2-fluoroethane	6.5	2[†]
$C_2H_4ClNO_2$	1-chloro-1-nitroethane	10.9	2
$C_2H_4Cl_2$	1,1-dichloroethane	8.64	2
	1,2-dichloroethane	8.0	2[†]
C_2H_5Br	bromoethane	8.05	2
		7.28	27
C_2H_5Cl	chloroethane	7.27	20
		8.29	2
		6.4	15
C_2H_5ClO	2-chloroethanol	7.1	2[†]
		6.88	27
	chloromethyl methyl ether	7.1	2[†]
C_2H_5F	fluoroethane	4.96	2
C_2H_5I	iodoethane	10.0	2
$C_3H_4Cl_2$	dichloropropene	10.1	2[†]
C_3H_5Cl	chloropropene	8.3	2
C_3H_5ClO	chloroacetone	8.4	2[†]
$C_3H_3ClO_2$	ethyl chloroformate	9.0	2[†]
$C_3H_6ClNO_2$	1-chloro-1-nitropropane	10.4	2[†]
$C_3H_6Cl_2$	dichloropropane	10.9	2[†]
C_3H_7Br	1-bromopropane	9.4	2[†]
		9.07	27
	2-bromopropane	9.6	2[†]
C_3H_7Cl	chloropropane	10.0	2
C_3H_7ClO	β-chloroethyl methyl ether	8.71	27
	2-chloro-1-propanol	8.89	27
	3-chloro-1-propanol	8.84	27
C_3H_7I	1-iodopropane	11.5	2[†]
C_4H_5Cl	4-chloro-1,2-butadiene	10.0	2[†]
C_4H_7Cl	1-chloro-2-methylpropene	10.8	2
$C_4H_7ClO_2$	2-chlorobutyric acid	10.7	27
	3-chlorobutyric acid	10.7	27
	4-chlorobutyric acid	10.6	27
$C_4H_8Cl_2$	1,4-dichlorobutane	12.0	2[†]
C_4H_9Br	bromobutane	13.9	2
		10.86	27
C_4H_9Cl	1-chlorobutane	11.3	2
	1-chloro-2-methylpropane	11.1	2
	2-chloro-2-methylpropane	12.5	2[†]
	2-chlorobutane	12.4	2
C_4H_9ClO	β-chloroethyl ethyl ether	10.56	27
	2-chloro-1-butanol	10.70	27
	3-chloro-1-butanol	10.38	27
C_4H_9I	1-iodobutane	13.3	2[†]
		12.65	27
$C_5H_9ClO_2$	2-chlorobutanoate acid	12.33	27
	3-chlorobutanoate acid	12.31	27

Table 7
Average Electric Dipole Polarizabilities for Ground State Organic
Halides (in Units of 10^{-24} cm^3) (continued)

Molecule	Name	Polarizability	Ref.
	4-chlorobutanoate acid	12.27	27
	2-chloropentanoic acid	12.69	27
	3-chloropentanoic acid	12.57	27
	4-chloropentanoic acid	12.53	27
$C_5H_{11}Br$	1-bromopentane	13.1	2[†]
$C_5H_{11}Cl$	1-chloropentane	12.0	2[†]
$C_5H_{11}F$	fluoropentane	9.95	27
C_6F_6	hexafluorobenzene	9.58	27
C_6HF_5	pentafluorobenzene	9.63	27
$C_6H_2Cl_2O_2$	2,5-dichloro-1,4-benzoquinone	18.4	2
$C_6H_2F_4$	1,2,3,4-tetrafluorobenzene	9.69	27
	1,2,4,5-tetrafluorobenzene	9.69	27
$C_6H_3F_3$	1,3,5-trifluorobenzene	9.74	27
C_6H_4BrF	p-bromofluorobenzene	13.4	2[†]
$C_6H_4ClNO_2$	chloronitrobenzene	14.6	2[†]
$C_6H_4Cl_2$	o-dichlorobenzene	14.17	27
	m-dichlorobenzene	14.23	27
	p-dichlorobenzene	14.20	27
C_6H_4FI	p-fluoroiodobenzene	15.5	2[†]
$C_6H_4FNO_2$	p-fluoronitrobenzene	12.8	2[†]
$C_6H_4F_2$	o-difluorobenzene	9.80	27
	m-difluorobenzene	10.3	2[†]
	p-difluorobenzene	9.80	27
C_6H_5Br	bromobenzene	14.7	2
		13.62	27
C_6H_5Cl	chlorobenzene	14.1	2
		12.3	15
C_6H_5ClO	chlorophenol	13.0	2[†]
C_6H_5F	fluorobenzene	10.3	2
C_6H_5I	iodobenzene	15.5	2[†]
$C_6H_{11}ClO_2$	2-chlorobutanoate acid	14.16	27
	3-chlorobutanoate acid	14.13	27
	4-chlorobutanoate acid	14.11	27
$C_6H_{13}Br$	bromohexane	14.44	27
$C_6H_{13}F$	fluorohexane	11.80	27
C_7H_7Br	p-bromotoluene	14.80	27
C_7H_7Cl	p-chlorotoluene	13.70	27
C_7H_7F	p-fluorotoluene	11.70	27
C_7H_7I	p-iodotoluene	17.10	27
$C_7H_{15}Br$	1-bromoheptane	16.8	2[†]
		16.23	27
$C_7H_{15}F$	fluoroheptane	13.66	27
$C_8H_{17}Br$	bromooctane	18.02	27
$C_8H_{17}F$	fluorooctane	15.46	27
$C_9H_{19}Br$	bromononane	19.81	27
$C_9H_{19}F$	fluorononane	17.34	27
$C_{10}F_8$	octafluoronaphthalene	17.64	27
$C_{10}H_7Br$	α-bromonaphthalene	20.34	27
$C_{10}H_7Cl$	α-chloronaphthalene	19.30	27
	β-chloronaphthalene	19.58	27
$C_{10}H_7I$	α-iodonaphthalene	22.41	27
	β-iodonaphthalene	22.95	27
$C_{10}H_{21}Br$	bromodecane	21.60	27
$C_{10}H_{21}F$	fluorodecane	19.18	27

Table 7
Average Electric Dipole Polarizabilities for Ground State Organic Halides (in Units of 10^{-24} cm^3) (continued)

Molecule	Name	Polarizability	Ref.
$C_{11}H_{23}F$	fluoroundecane	21.00	27
$C_{12}H_{25}Br$	bromododecane	25.18	27
$C_{12}H_{25}F$	fluorododecane	22.83	27
$C_{12}H_8Br_2O$	4,4′–dibromodiphenyl ether	27.8	2[†]
$C_{12}H_9BrO$	4-bromodiphenyl ether	24.2	2[†]
$C_{13}H_{11}BrO$	*p*-bromophenyl-*p*-tolyl ether	26.6	2[†]
$C_{14}H_9Br$	9-bromoanthracene	28.32	27
$C_{14}H_9Cl$	9-chloroanthracene	27.35	27
$C_{14}H_9F$	fluoranthracene	28.34	27
$C_{14}H_{29}F$	fluorotetradecane	26.57	27
$C_{16}H_{33}Br$	bromohexadecane	32.34	27
$C_{18}H_{37}Br$	bromooctadecane	35.92	27

Table 8
Static Average Electric Dipole Polarizabilities for Other Ground State Organic Molecules (in Units of 10^{-24} cm^3)

Molecule	Name	Polarizability	Ref.
CN_4O_8	tetranitromethane	15.3	2
CH_2O	formaldehyde	2.8	2[†]
		2.45	18
CH_2O_2	formic acid	3.4	2[†]
CH_3NO	formamide	4.2	2[†]
		4.08	18
CH_3NO_2	nitromethane	7.37	2
CH_4O	methanol	3.29	2
		3.23	15
		3.32	18
CH_5N	methyl amine	4.7	2
		4.01	19
C_2N_2	cyanogen	7.99	2
C_2H_2O	ketene	4.4	2[†]
C_2H_3N	acetonitrile	4.40	2[†]
		4.48	18
C_2H_4O	acetaldehyde	4.6	2[†]
		4.59	18
	ethylene oxide	4.43	18
$C_2H_4O_2$	acetic acid	5.1	2[†]
	methyl formate	5.05	27
$C_2H_4O_4$	formic acid dimer	12.7	2
C_2H_5NO	acetamide	5.67	18
	N-methyl formamide	5.91	18
$C_2H_5NO_2$	nitroethane	9.63	2
	ethyl nitrite	7.0	15
C_2H_6O	ethanol	5.41	2
		5.11	18
	methyl ether	5.29	20
		5.84	2
		5.16	15
$C_2H_6O_2$	ethylene glycol	5.7	2[†]
		5.61	27
$C_2H_6O_2S$	dimethyl sulfone	7.3	2[†]

Table 8
Static Average Electric Dipole Polarizabilities for Other Ground State Organic Molecules (in Units of 10^{-24} cm^3) (continued)

Molecule	Name	Polarizability	Ref.
C_2H_6S	ethanethiol	7.41	2
C_2H_7N	ethyl amine	7.10	2
	dimethyl amine	6.37	2
$C_2H_8N_2$	ethylene diamine	7.2	2[†]
$C_3H_2N_2$	malononitrile	5.79	18
C_3H_3N	acrylonitrile	8.05	2
$C_3H_4N_2$	pyrazole	7.23	27
C_3H_4O	propenal	6.38	2[†]
C_3H_5N	propionitrile	6.70	2
		6.24	18
C_3H_6O	acetone	6.33	15
		6.4	2[†]
		6.39	18
	allyl alcohol	7.65	2
	propionaldehyde	6.50	2
$C_3H_6O_2$	propionic acid	6.9	2[†]
	ethyl formate	8.01	2
		6.88	27
	methyl acetate	6.94	2
		6.81	27
$C_3H_6O_3$	dimethyl carbonate	7.7	2[†]
C_3H_7NO	N-methyl acetamide	7.82	18
	N,N-dimethyl formamide	7.81	18
$C_3H_7NO_2$	nitropropane	8.5	2[†]
C_3H_8O	2-propanol	7.61	2
		6.97	18
	1-propanol	6.74	2
	ethyl methyl ether	7.93	2
$C_3H_8O_2$	dimethoxymethane	7.7	2[†]
	monomethylether ethylene glycol	7.44	27
C_3H_9N	n-propylamine	7.70	27
		9.20	2
	isopropylamine	7.77	27
	trimethylamine	8.15	2
$C_4H_2N_2$	fumaronitrile	11.8	2
$C_4H_4N_2$	succinonitrile	8.1	2[†]
	pyrimidene	8.53*	17
	pyridazine	9.27*	17
$C_4H_4O_2$	diketene	8.0	2[†]
C_4H_4S	thiophene	9.67	2
C_4H_5N	methacrylonitrile	8.0	2[†]
	trans-crotononitrile	8.2	2[†]
$C_4H_6N_2$	N-methylpyrazole	8.99	27
C_4H_6O	crotonaldehyde	8.5	2[†]
	methacrylaldehyde	8.3	2[†]
$C_4H_6O_2$	biacetyl	8.2	2[†]
$C_4H_6O_3$	acetic anhydride	8.9	2[†]
C_4H_6S	divinyl sulfide	10.9	2[†]
C_4H_7N	butyronitrile	8.4	2[†]
	isobutyronitrile	8.05	18
C_4H_8O	butanal	8.2	2[†]
	methyl ethyl ketone	8.13	15
	trans-2,3-epoxy butane	8.22*	17
$C_4H_8O_2$	ethyl acetate	9.7	2
		8.62	27

Table 8
Static Average Electric Dipole Polarizabilities for Other Ground State
Organic Molecules (in Units of 10^{-24} cm^3) (continued)

Molecule	Name	Polarizability	Ref.
	1,4-dioxane	10.0	2
	p-dioxane	8.60	18
	2-methyl-1,3-dioxolane	9.44	15
	butyric acid	8.38	27
	methyl propionate	8.97	27
$C_4H_9NO_2$	1-nitrobutane	10.4	2[†]
	2-methyl-2-nitropropane	10.3	2[†]
$C_4H_{10}O$	ethyl ether	10.2	2
		8.73	15
	1-butanol	8.88	2
	2-methylpropanol	8.92	2
	n-propyl methyl ether	8.86	27
$C_4H_{10}O_2$	monoethylether ethylene glycol	9.28	27
$C_4H_{10}S$	ethyl sulfide	10.8	2
$C_4H_{11}N$	n-butylamine	13.5	2
	diethylamine	10.2	2
		9.61	27
C_5H_5N	pyridine	9.5	15
		9.18	27
	4-cyano-1,2-butadiene	10.5	2[†]
$C_5H_8N_2$	1,5-dimethylpyrazole	10.72	27
$C_5H_8O_2$	acetyl acetone	10.5	2[†]
C_5H_9N	valeronitrile	10.4	2
	22-DMPN	9.59	18
$C_5H_{10}O$	diethyl ketone	9.93	15
	methyl propyl ketone	9.93	15
$C_5H_{10}O_2$	ethyl propionate	10.41	27
	methyl butanoate	10.41	27
$C_5H_{10}O_3$	diethyl carbonate	11.3	2
$C_5H_{12}O$	ethyl propyl ether	10.68	27
$C_5H_{12}O_4$	tetramethyl orthocarbonate	13.0	2[†]
$C_6H_4N_2O_4$	p-dinitrobenzene	18.4	2
$C_6H_4O_2$	p-benzoquinone	14.5	2
$C_6H_5NO_2$	nitrobenzene	14.7	2
		12.92	15
C_6H_6O	phenol	11.1	2[†]
		9.94*	17
C_6H_7N	aniline	12.1	2[†]
$C_6H_8N_2$	phenylenediamine	13.8	2[†]
	phenylhydrazine	12.91	27
$C_6H_{10}N_2$	1-ethyl-5-methylpyrazole	12.50	27
$C_6H_{10}O_3$	ethyl acetoacetate	12.9	2[†]
$C_6H_{12}N_2$	dimethylketazine	15.6	2
$C_6H_{12}O$	cyclohexanol	11.56	18
$C_6H_{12}O_2$	amyl formate	14.2	2
$C_6H_{12}O_3$	paraldehyde	17.9	2
$C_6H_{14}O$	propyl ether	12.8	2
		12.5	15
$C_6H_{14}O_2$	1,1-diethoxyethane	13.2	2[†]
	1,2-diethoxyethane	11.3	2[†]
$C_6H_{15}N$	triethylamine	13.1	2
		13.38	27
	di-n-propylamine	13.29	27
$C_7H_4N_2O_2$	p-cyanonitrobenzene	19.0	2
C_7H_5N	benzonitrile	12.5	2[†]

Table 8
Static Average Electric Dipole Polarizabilities for Other Ground State Organic Molecules (in Units of 10^{-24} cm^3) (continued)

Molecule	Name	Polarizability	Ref.
$C_7H_7NO_3$	nitroanisole	15.7	2[†]
C_7H_7O	anisole	13.1	2[†]
C_7H_9NO	o-anisidine	14.2	2[†]
$C_7H_{10}N_2$	1,1-methylphenylhydrazine	14.81	27
$C_7H_{14}O$	cyclohexyl methyl ether	13.4	2[†]
	2,4-dimethyl-3-pentanone	13.5	15
$C_7H_{14}O_2$	pentyl acetate	14.9	2
$C_8H_4N_2$	p-dicyanobenzene	19.2	2
$C_8H_6O_2$	quinoxaline	15.13	27
C_8H_8O	acetophenone	15.0	2
$C_8H_8O_2$	2,5-dimethyl-1,4-benzoquinone	18.8	2
$C_8H_{10}O$	phenetole	14.9	2
$C_8H_{11}N$	N-dimethylaniline	16.2	2[†]
$C_8H_{12}N_2$	1,1-ethylphenylhydrazine	16.62	27
$C_8H_{12}O_2$	ethyl sorbate	17.2	2[†]
	tetramethylcyclobutane-1,3-dione	18.6	2
$C_8H_{14}O_4$	diethyl succinate	16.8	2[†]
$C_8H_{18}O$	n-butyl ether	17.2	2
C_9H_7N	quinoline	15.70	27
	isoquinoline	16.43	27
$C_9H_{10}O_2$	ethyl benzoate	16.9	2[†]
$C_9H_{21}N$	tri-n-propylamine	18.87	27
$C_{10}H_9N$	α-naphthylamine	19.50	27
	β-naphthylamine	19.73	27
	1-methylquinoline	18.65	27
	1-methylisoquinoline	18.28	27
$C_{10}H_{10}Fe$	ferrocene	17.1	26
$C_{10}H_{10}N_2$	2,3-dimethylquinoxaline	18.70	27
$C_{10}H_{14}BeO_4$	beryllium acetylacetonate	34.1	2
$C_{11}H_8O$	1-naphthaldehyde	19.75	27
	2-naphthaldehyde	20.06	27
$C_{14}H_8O_2$	anthraquinone	24.46	27
$C_{12}H_8N_2$	phenazine	23.43	27
$C_{12}H_9NO_3$	4-nitrodiphenyl ether	24.7	2[†]
$C_{14}H_{14}O$	di-p-tolyl ether	24.9	2[†]
$C_{15}H_{21}AlO_6$	aluminum acetylacetonate	51.9	2
$C_{15}H_{21}CrO_6$	chromium acetylacetonate	53.7	2
$C_{15}H_{21}FeO_6$	ferric acetylacetonate	58.1	2
$C_{20}H_{28}O_8Th$	thorium acetylacetonate	79.0	2
C_{60}	buckminsterfullerene	~80	24

Note: All polarizabilities in the tables are experimental values except those values marked by an asterisk (*), which indicates a calculated result. The experimental polarizabilities are mostly determined by measurements of a dielectric constant or refractive index which are quite accurate (0.5% or better). However, one should treat many of the results with several percent of caution because of the age of the data and because some of the results refer to optical frequencies rather than static. Comments given with the references are intended to allow one to judge the degree of caution required. Interested persons should consult these references. In many cases, the reference given is to a theoretical paper in which the experimental results are quoted. These papers, noted in the References, contain valuable information on polarizability calculations and experimental data which often includes the tensor components of the polarizability.

ATOMIC AND MOLECULAR POLARIZABILITIES (continued)

REFERENCES

1. McCullough, E. A., Jr., *J. Chem. Phys.*, 63, 5050, 1975. This calculation is for the parallel component, not the average polarizability.
2. Maryott, A. A., and Buckley, F., U. S. National Bureau of Standards Circular No. 537, 1953. A tabulation of dipole moments, dielectric constants, and molar refractions measured between 1910 and 1952, and used here to determine polarizabilities if no more recent result exists. The polarizability is $3/(4\pi N_0)$ times the molar polarization or molar refraction, where N_0 is Avogadro's number. The value $3/(4\pi N_0) = 0.3964308 \times 10^{-24}$ cm^3 was used for this conversion. A dagger (†) following the reference number in the tables indicates that the polarizability was derived from the molar refraction and hence may not include some low-frequency contributions to the static polarizability; these "static" polarizabilities are therefore low by 1 to 30%.
3. Hirschfelder, J. O., Curtis, C. F., and Bird, R. B., *Molecular Theory of Gases and Liquids*, Wiley, New York, 1954, p. 950. Fundamental information on molecular polarizabilities.
4. Miller, T. M., and Bederson, B., *Adv. At. Mol. Phys.*, 13, 1, 1977. Review emphasizing atomic polarizabilities and measurement techniques. The data quoted in Table 3 are accurate to 8 to 12%.
5. Kolos, W., and Wolniewicz, L., *J. Chem. Phys.*, 46, 1426, 1967. Highly accurate molecular hydrogen calculations.
6. Newell, A. C., and Baird, R. C., *J. Appl. Phys.*, 36, 3751, 1965. Highly accurate refractive index measurements at 47.7 GHz (essentially static).
7. Jao, T. C., Beebe, N. H. F., Person, W. B., and Sabin, J. R., *Chem. Phys. Lett.*, 26, 47, 1974. Tensor polarizabilities, derivatives, and other results are reported.
8. Orcutt, R. H., and Cole, R. H., *J. Chem. Phys.*, 46, 697, 1967 (He, Ne, Ar, Kr, H_2, N_2); Sutter, H., and Cole, R. H., *J. Chem. Phys.*, 52, 132, 1970 (CF_3H, CFH_3, $CClF_3$, $CClH_3$); Bose, T. K., and Cole, R. H., *J. Chem. Phys.*, 52, 140, 1970 (CO_2), and 54, 3829, 1971 (C_2H_2, C_2H_4); Nelson, R. D., and Cole, R. H., *J. Chem. Phys.*, 54, 4033, 1971 (SF_6, $CClF_3$); Bose, T. K., Sochanski, J. S., and Cole, R. H., *J. Chem. Phys.*, 57, 3592, 1972 (CH_4, CF_4); Kirouac, S., and Bose, T. K., *J. Chem. Phys.*, 59, 3043, 1973 (N_2O), and 64, 1580, 1976 (He). Highly accurate dielectric constant measurements. These modern data give the most accurate polarizabilities available. A criticism of the interpretation of these data in the case of polar molecules is given in Ref. 20, p. 2905.
9. Huestis, D. L., Technical Report #MP 78-25, SRI International (project PYU 6158), Menlo Park, CA 94025. Molar refractions for mercury-chlorine compounds are analyzed.
10. Bounds, D. G., Clarke, J. H. R., and Hinchliffe, A., *Chem. Phys. Lett.*, 45, 367, 1977. Theoretical tensor polarizability for LiCl.
11. Kolker, H. J., and Karplus, M., *J. Chem. Phys.*, 39, 2011, 1963. Theoretical.
12. Cutschick, V. P., and McKoy, V., *J. Chem. Phys.*, 58, 2397, 1973. Theoretical tensor polarizabilities.
13. Gready, J. E., Bacskay, G. B., and Hush, N. S., *Chem. Phys.*, 22, 141, 1977, and 23, 9, 1977. Theoretical.
14. Amos, A. T., and Yoffe, J. A., *J. Chem. Phys.*, 63, 4723, 1975. Theoretical.
15. Stuart, H. A., *Landolt-Börnstein Zahlenwerte and Funktionen*, Vol. 1, Part 3, Eucken, A., and Hellwege, K. H., Eds., Springer-Verlag, Berlin, 1951, p. 511. Tabulation of molecular polarizabilities. Two misprints in the chemical symbols have been corrected.
16. Guella, T., Miller, T. M., Stockdale, J. A. D., Bederson, B., and Vuskovic, L., *J. Chem. Phys.*, 94, 6857, 1991. Beam measurements with accuracies between 12-24%.
17. Marchese, F. T., and Jaffé, *Theoret. Chim. Acta (Berlin)*, 45, 241, 1977. Theoretical and experimental tensor polarizabilities are tabulated in this paper.
18. Applequist, J., Carl, J. R., and Fung, K.-K., *J. Am. Chem. Soc.*, 94, 2952, 1972. Excellent reference on the calculation of molecular polarizabilities, including extensive tables of tensor polarizabilities, both theoretical and experimental, at 589.3 nm wavelength.
19. Bridge, N. J., and Buckingham, A. D., *Proc. Roy. Soc. (London)*, A295, 334, 1966. Measured tensor polarizabilities at 633 nm wavelength.
20. Barnes, A. N. M., Turner, D. J., and Sutton, L. E., *Trans. Faraday Soc.*, 67, 2902, 1971. Dielectric constants yielding polarizabilities accurate from 0.3-8%.
21. Knight, W. D., Clemenger, K., de Heer, W. A., and Saunders, W. A., *Phys. Rev. B*, 31, 2539, 1985. These data probably correspond to a very low internal temperature.
22. Tarnovsky, V., Bunimovicz, M., Vuskovic, L., Stumpf, B., and Bederson, B., *J. Chem. Phys.*, 98, 3894, 1993. These data correspond to internal temperatures 480-948 K.
23. Milani, P., Moullet, I., and de Heer, W. A., *Phys. Rev. A*, 42, 5150, 1990. Beam measurements accurate to 11%.
24. Bonin, K. D., and Kadar-Kallen, M. A., *Int. J. Mod. Phys.*, 8, 3313, 1994. Review article.
25. Aroney, M. J., and Pratten, S. J., *J. Chem. Soc., Faraday Trans. 1*, 80, 1201, 1984. Uncertainties in the range 1-3%.
26. Le Fevre, R. J. W., Murthy, D. S. N., and Saxby, J. D., *Aust. J. Chem.*, 24, 1057, 1971. Kerr effect.
27. No, K. T., Cho, K. H., Jhon, M. S., and Scheraga, H. A., *J. Am. Chem. Soc.*, 115, 2005, 1993. Theoretical; these results are quoted in numerous valuable papers on calculated polarizabilities, e.g., Miller, K. J., and Savchik, J. A., *J. Am. Chem. Soc.*, 101, 7206, 1979.

IONIZATION POTENTIALS OF ATOMS AND ATOMIC IONS

The ionization potentials of neutral and partially ionized atoms are listed in this table. Data were obtained from the compilations cited below, supplemented by results from the recent research literature. All values have been corrected to the currently recommended value of the conversion factor from wave number to energy, namely 1 eV = 8065.541 cm^{-1} (Reference 5). Values are given in eV.

Following the traditional spectroscopic notation, columns are headed I, II, III, etc. up to XXX, where I indicates the neutral atom, II the singly ionized atom, III the doubly ionized atom, etc. The first section of the table includes spectra I to VIII of all the elements; subsequent sections cover higher spectra (ionization stages) for those elements for which data are available.

REFERENCES

1. Moore, C. E., *Ionization Potentials and Ionization Limits Derived from the Analysis of Optical Spectra*, Natl. Stand. Ref. Data Ser. — Natl. Bur. Stand. (U.S.) No. 34, 1970.
2. Martin, W. C., Zalubas, R., and Hagan, L., *Atomic Energy Levels — The Rare Earth Elements*, Natl. Stand. Ref. Data Ser. — Natl. Bur. Stand. (U.S.), No. 60, 1978.
3. Sugar, J. and Corliss, C., *Atomic Energy Levels of the Iron Period Elements: Potassium through Nickel*, J. Phys. Chem. Ref. Data, Vol.14, Suppl. 2, 1985.
4. References to papers in *J. Phys. Chem. Ref. Data*, in the period 1973—91 covering other elements may be found in the cumulative index to that journal.
5. Cohen, E. R. and Taylor, B. N., *J. Phys. Chem. Ref. Data*, 17, 1795, 1988.
6. Martin, W.C., and Wiese, W.L., in *Atomic, Molecular, and Optical Physics Handbook*, Drake, G.W.F., Ed., AIP Press, New York, 1996.

Neutral Atoms to +7 Ions

Z	Element	I	II	III	IV	V	VI	VII	VIII
1	H	13.59844							
2	He	24.58741	54.41778						
3	Li	5.39172	75.64018	122.45429					
4	Be	9.3227	18.21116	153.89661	217.71865				
5	B	8.29803	25.15484	37.93064	259.37521	340.22580			
6	C	11.26030	24.38332	47.8878	64.4939	392.087	489.99334		
7	N	14.53414	29.6013	47.44924	77.4735	97.8902	552.0718	667.046	
8	O	13.61806	35.11730	54.9355	77.41353	113.8990	138.1197	739.29	871.4101
9	F	17.42282	34.97082	62.7084	87.1398	114.2428	157.1651	185.186	953.9112
10	Ne	21.5646	40.96328	63.45	97.12	126.21	157.93	207.2759	239.0989
11	Na	5.13908	47.2864	71.6200	98.91	138.40	172.18	208.50	264.25
12	Mg	7.64624	15.03528	80.1437	109.2655	141.27	186.76	225.02	265.96
13	Al	5.98577	18.82856	28.44765	119.992	153.825	190.49	241.76	284.66
14	Si	8.15169	16.34585	33.49302	45.14181	166.767	205.27	246.5	303.54
15	P	10.48669	19.7694	30.2027	51.4439	65.0251	220.421	263.57	309.60
16	S	10.36001	23.3379	34.79	47.222	72.5945	88.0530	280.948	328.75
17	Cl	12.96764	23.814	39.61	53.4652	67.8	97.03	114.1958	348.28
18	Ar	15.75962	27.62967	40.74	59.81	75.02	91.009	124.323	143.460
19	K	4.34066	31.63	45.806	60.91	82.66	99.4	117.56	154.88
20	Ca	6.11316	11.87172	50.9131	67.27	84.50	108.78	127.2	147.24
21	Sc	6.5615	12.79967	24.75666	73.4894	91.65	110.68	138.0	158.1
22	Ti	6.8281	13.5755	27.4917	43.2672	99.30	119.53	140.8	170.4
23	V	6.7463	14.66	29.311	46.709	65.2817	128.13	150.6	173.4
24	Cr	6.7665	16.4857	30.96	49.16	69.46	90.6349	160.18	184.7
25	Mn	7.43402	15.63999	33.668	51.2	72.4	95.6	119.203	194.5
26	Fe	7.9024	16.1878	30.652	54.8	75.0	99.1	124.98	151.06
27	Co	7.8810	17.083	33.50	51.3	79.5	102.0	128.9	157.8
28	Ni	7.6398	18.16884	35.19	54.9	76.06	108	133	162
29	Cu	7.72638	20.29240	36.841	57.38	79.8	103	139	166
30	Zn	9.3942	17.96440	39.723	59.4	82.6	108	134	174
31	Ga	5.99930	20.5142	30.71	64				
32	Ge	7.8994	15.93462	34.2241	45.7131	93.5			
33	As	9.7886	18.633	28.351	50.13	62.63	127.6		
34	Se	9.75238	21.19	30.8204	42.9450	68.3	81.7	155.4	
35	Br	11.81381	21.8	36	47.3	59.7	88.6	103.0	192.8
36	Kr	13.99961	24.35985	36.950	52.5	64.7	78.5	111.0	125.802
37	Rb	4.17713	27.285	40	52.6	71.0	84.4	99.2	136
38	Sr	5.6949	11.03013	42.89	57	71.6	90.8	106	122.3
39	Y	6.2171	12.24	20.52	60.597	77.0	93.0	116	129
40	Zr	6.63390	13.13	22.99	34.34	80.348			
41	Nb	6.75885	14.32	25.04	38.3	50.55	102.057	125	

Z	Element	I	II	III	IV	V	VI	VII	VIII
42	Mo	7.09243	16.16	27.13	46.4	54.49	68.8276	125.664	143.6
43	Tc	7.28	15.26	29.54					
44	Ru	7.36050	16.76	28.47					
45	Rh	7.45890	18.08	31.06					
46	Pd	8.3369	19.43	32.93					
47	Ag	7.5763	21.49	34.83					
48	Cd	8.9938	16.90832	37.48					
49	In	5.78636	18.8698	28.03	54				
50	Sn	7.3439	14.63225	30.50260	40.73502	72.28			
51	Sb	8.6084	16.53051	25.3	44.2	56	108		
52	Te	9.0096	18.6	27.96	37.41	58.75	70.7	137	
53	I	10.45126	19.1313	33					
54	Xe	12.1298	21.20979	32.1230					
55	Cs	3.89390	23.15745						
56	Ba	5.21170	10.00390						
57	La	5.5770	11.060	19.1773	49.95	61.6			
58	Ce	5.5387	10.85	20.198	36.758	65.55	77.6		
59	Pr	5.464	10.55	21.624	38.98	57.53			
60	Nd	5.5250	10.73	22.1	40.41				
61	Pm	5.58	10.90	22.3	41.1				
62	Sm	5.6436	11.07	23.4	41.4				
63	Eu	5.6704	11.241	24.92	42.7				
64	Gd	6.1501	12.09	20.63	44.0				
65	Tb	5.8638	11.52	21.91	39.79				
66	Dy	5.9389	11.67	22.8	41.47				
67	Ho	6.0215	11.80	22.84	42.5				
68	Er	6.1077	11.93	22.74	42.7				
69	Tm	6.18431	12.05	23.68	42.7				
70	Yb	6.25416	12.1761	25.05	43.56				
71	Lu	5.4259	13.9	20.9594	45.25	66.8			
72	Hf	6.82507	14.9	23.3	33.33				
73	Ta	7.5496							
74	W	7.8640							
75	Re	7.8335							
76	Os	8.4382							
77	Ir	8.9670							
78	Pt	8.9587	18.563						
79	Au	9.2255	20.5						
80	Hg	10.43750	18.756	34.2					
81	Tl	6.1082	20.428	29.83					
82	Pb	7.41666	15.0322	31.9373	42.32	68.8			
83	Bi	7.2856	16.69	25.56	45.3	56.0	88.3		
84	Po	8.41671							
85	At								
86	Rn	10.74850							
87	Fr	4.0727							
88	Ra	5.2784	10.14716						
89	Ac	5.17	12.1						
90	Th	6.3067	11.5	20.0	28.8				
91	Pa	5.89							
92	U	6.19405							
93	Np	6.2657							
94	Pu	6.0262							
95	Am	5.9738							
96	Cm	6.02							
97	Bk	6.23							
98	Cf	6.30							
99	Es	6.42							
100	Fm	6.50							
101	Md	6.58							
102	No	6.65							

+8 Ions to +15 Ions

Z	Element	IX	X	XI	XII	XIII	XIV	XV	XVI
9	F	1103.1176							
10	Ne	1195.8286	1362.1995						
11	Na	299.864	1465.121	1648.702					
12	Mg	328.06	367.50	1761.805	1962.6650				
13	Al	330.13	398.75	442.00	2085.98	2304.1410			
14	Si	351.12	401.37	476.36	523.42	2437.63	2673.182		
15	P	372.13	424.4	479.46	560.8	611.74	2816.91	3069.842	
16	S	379.55	447.5	504.8	564.44	652.2	707.01	3223.78	3494.1892
17	Cl	400.06	455.63	529.28	591.99	656.71	749.76	809.40	3658.521
18	Ar	422.45	478.69	538.96	618.26	686.10	755.74	854.77	918.03
19	K	175.8174	503.8	564.7	629.4	714.6	786.6	861.1	968
20	Ca	188.54	211.275	591.9	657.2	726.6	817.6	894.5	974
21	Sc	180.03	225.18	249.798	687.36	756.7	830.8	927.5	1009
22	Ti	192.1	215.92	265.07	291.500	787.84	863.1	941.9	1044
23	V	205.8	230.5	255.7	308.1	336.277	896.0	976	1060
24	Cr	209.3	244.4	270.8	298.0	354.8	384.168	1010.6	1097
25	Mn	221.8	248.3	286.0	314.4	343.6	403.0	435.163	1134.7
26	Fe	233.6	262.1	290.2	330.8	361.0	392.2	457	489.256
27	Co	186.13	275.4	305	336	379	411	444	511.96
28	Ni	193	224.6	321.0	352	384	430	464	499
29	Cu	199	232	265.3	369	401	435	484	520
30	Zn	203	238	274	310.8	419.7	454	490	542
36	Kr	230.85	268.2	308	350	391	447	492	541
37	Rb	150	277.1						
38	Sr	162	177	324.1					
39	Y	146.2	191	206	374.0				
42	Mo	164.12	186.4	209.3	230.28	279.1	302.60	544.0	570

+16 Ions to +23 Ions

Z	Element	XVII	XVIII	XIX	XX	XXI	XXII	XXIII	XXIV
17	Cl	3946.2960							
18	Ar	4120.8857	4426.2296						
19	K	1033.4	4610.8	4934.046					
20	Ca	1087	1157.8	5128.8	5469.864				
21	Sc	1094	1213	1287.97	5674.8	6033.712			
22	Ti	1131	1221	1346	1425.4	6249.0	6625.82		
23	V	1168	1260	1355	1486	1569.6	6851.3	7246.12	
24	Cr	1185	1299	1396	1496	1634	1721.4	7481.7	7894.81
25	Mn	1224	1317	1437	1539	1644	1788	1879.9	8140.6
26	Fe	1266	1358	1456	1582	1689	1799	1950	2023
27	Co	546.58	1397.2	1504.6	1603	1735	1846	1962	2119
28	Ni	571.08	607.06	1541	1648	1756	1894	2011	2131
29	Cu	557	633	670.588	1697	1804	1916	2060	2182
30	Zn	579	619	698	738	1856			
36	Kr	592	641	786	833	884	937	998	1051
42	Mo	636	702	767	833	902	968	1020	1082

+24 Ions to +29 Ions

Z	Element	XXV	XXVI	XXVII	XXVIII	XXIX	XXX
25	Mn	8571.94					
26	Fe	8828	9277.69				
27	Co	2219.0	9544.1	10012.12			
28	Ni	2295	2399.2	10288.8	10775.40		
29	Cu	2308	2478	2587.5	11062.38	11567.617	
36	Kr	1151	1205.3	2928	3070	3227	3381
42	Mo	1263	1323	1387	1449	1535	1601

IONIZATION ENERGIES OF GAS-PHASE MOLECULES
Sharon G. Lias

This table presents values for the first ionization energies (IP) of approximately 1000 molecules and atoms. Substances are listed by molecular formula in the modified Hill order (see introduction). Values enclosed in parentheses are considered not to be well established. Data appearing in the 1988 reference, were updated in 1996 for inclusion in the database of ionization energies available at the Internet site of the Standard Reference Data program of the National Institute of Standards and Technology (http://webbook.nist.gov). The list appearing here includes these updates.

The list also includes values for enthalpies of formation of the ions at 298 K, $\Delta_f H_{ion}$, given according to the ion convention used by mass spectrometrists; to convert these values to the electron convention used by thermodynamicists, add 6 kJ/mol. Details on the calculation of $\Delta_f H_{ion}$ as well as data for a much larger number of molecules, may be found in the reference and on the Internet site.

REFERENCE

Lias, S.G., Bartmess, J.E., Liebman, J.F., Holmes, J.L., Levin, R.D., and Mallard, W.G., *Gas-Phase Ion and Neutral Thermochemistry, J. Phys. Chem. Ref. Data*, Vol. 17, Suppl. No. 1, 1988.

Mol. Form.	Name	IP/eV	$\Delta_f H_{ion}$ kJ/mol
	Substances not containing carbon		
Ac	Actinium	5.17	905
Ag	Silver	7.57624	1016
AgCl	Silver(I) chloride	(≤ 10.08)	≤ 1065
AgF	Silver(I) fluoride	(11.0 ± 0.3)	1071
Al	Aluminum	5.98577	905
AlBr	Aluminum monobromide	(9.3)	913
AlBr$_3$	Aluminum tribromide	(10.4)	593
AlCl	Aluminum monochloride	9.4	855
AlCl$_3$	Aluminum trichloride	(12.01)	573
AlF	Aluminum monofluoride	9.73 ± 0.01	673
AlF$_3$	Aluminum trifluoride	≤ 15.45	≤ 282
AlI	Aluminum monoiodide	9.3 ± 0.3	965
AlI$_3$	Aluminum triiodide	(9.1)	673
Am	Americium	5.9738 ± 0.0002	860
Ar	Argon	15.75962	1521
As	Arsenic	9.8152	1250
AsCl$_3$	Arsenic(III) chloride	(10.55 ± 0.025)	754
AsF$_3$	Arsenic(III) fluoride	(12.84 ± 0.05)	452
AsH$_3$	Arsine	(9.89)	1021
Au	Gold	9.22567	1254
B	Boron	8.29803	1363
BBr$_3$	Boron tribromide	(10.51)	809
BCl$_3$	Boron trichloride	11.60 ± 0.02	718
BF	Fluoroborane	11.12 ± 0.01	957
	Difluoroborane	(9.4)	317
BF$_3$	Boron trifluoride	15.7 ± 0.3	365
BH	Boron monohydride	(9.77)	1385
BH$_3$	Borane	12.026 ± 0.024	1261
BI$_3$	Boron triiodide	(9.25 ± 0.03)	964
BO$_2$	Boron dioxide	(13.5 ± 0.3)	1001
B$_2$H$_6$	Diborane	11.38 ± 0.05	1134
B$_2$O$_3$	Boron oxide	13.5 ± 0.15	460
B$_4$H$_{10}$	Tetraborane	10.76 ± 0.04	1105
B$_5$H$_9$	Pentaborane(9)	9.90 ± 0.04	1028
B$_6$H$_{10}$	Hexaborane	(9.0)	965
Ba	Barium	5.21170	683
BaO	Barium oxide	6.91 ± 0.06	543
Be	Beryllium	9.32263	1224
BeO	Beryllium oxide	(10.1 ± 0.4)	1111
Bi	Bismuth	7.2855	908
BiCl$_3$	Bismuth trichloride	(10.4)	736
Bk	Berkelium	6.23	911

Mol. Form.	Name	IP/eV	$\Delta_f H_{ion}$ kJ/mol
Br	Bromine (atomic)	11.81381	1252
BrCl	Bromine chloride	11.01	1079
BrF	Bromine fluoride	11.86	1086
BrF$_5$	Bromine pentafluoride	13.172 ± 0.002	840
BrH	Hydrogen bromide	11.66 ± 0.03	1087
BrH$_3$Si	Bromosilane	10.6	943
BrI	Iodine bromide	9.790 ± 0.004	986
BrK	Potassium bromide	7.85 ± 0.1	578
BrLi	Lithium bromide	(8.7)	685
BrNO	Nitrosyl bromide	10.17 ± 0.03	1065
BrNa	Sodium bromide	8.31 ± 0.1	660
BrO	Bromine monoxide	10.46 ± 0.02	1135
BrRb	Rubidium bromide	7.94 ± 0.03	583
BrTl	Thallium(I) bromide	9.14 ± 0.02	844
Br$_2$	Bromine	10.516 ± 0.005	1046
Br$_2$Hg	Mercury(II) bromide	10.560 ± 0.003	935
Br$_2$Sn	Tin(II) bromide	9.0	839
Br$_3$Ga	Gallium(III) bromide	10.40	711
Br$_3$P	Phosphorus(III) bromide	9.7	798
Br$_4$Hf	Hafnium(IV) bromide	(10.9)	366
Br$_4$Sn	Tin(IV) bromide	10.6	709
Br$_4$Ti	Titanium(IV) bromide	10.3	375
Br$_4$Zr	Zirconium(IV) bromide	(10.7)	388
Ca	Calcium	6.11316	768
CaCl	Calcium monochloride	5.86 ± 0.07	462
CaO	Calcium oxide	6.66 ± 0.18	668
Cd	Cadmium	8.99367	980
Ce	Cerium	5.5387	957
Cf	Californium	6.30	805
Cl	Chlorine (atomic)	12.96764	1373
ClCs	Cesium chloride	(7.84 ± 0.05)	510
ClF	Chlorine fluoride	12.66 ± 0.01	1171
ClFO$_3$	Perchloryl fluoride	(12.945 ± 0.005)	1224
ClF$_2$	Chlorine difluoride	(12.77 ± 0.05)	1128
ClF$_3$	Chlorine trifluoride	(12.65 ± 0.05)	1057
ClF$_5$S	Sulfur chloride pentafluoride	(12.335 ± 0.005)	144
ClH	Hydrogen chloride	12.749 ± 0.009	1137
ClHO	Hypochlorous acid	(11.12 ± 0.01)	993
ClH$_3$Si	Chlorosilane	11.4	899
ClI	Iodine chloride	10.088 ± 0.01	991
ClIn	Indium(I) chloride	(9.51)	842
ClK	Potassium chloride	(8.0 ± 0.4)	557
ClLi	Lithium chloride	9.57	727
ClNO	Nitrosyl chloride	10.87 ± 0.01	1099
ClNO$_2$	Nitryl chloride	(11.84)	1155
ClNa	Sodium chloride	8.92 ± 0.06	681
ClO	Chlorine monoxide	10.95	1159
ClO$_2$	Chlorine dioxide	10.33 ± 0.02	1093
ClRb	Rubidium chloride	(8.50 ± 0.03)	590
ClTl	Thallium(I) chloride	9.70 ± 0.03	869
Cl$_2$	Chlorine	11.480 ± 0.005	1108
Cl$_2$CrO$_2$	Chromyl chloride	11.6	580
Cl$_2$Ge	Germanium(II) chloride	(10.20 ± 0.05)	813
Cl$_2$H$_2$Si	Dichlorosilane	11.4	765
Cl$_2$Hg	Mercury(II) chloride	11.380 ± 0.003	952
Cl$_2$O	Chlorine oxide	10.94	1135
Cl$_2$OS	Thionyl chloride	10.96	844
Cl$_2$O$_2$S	Sulfuryl chloride	12.05	807
Cl$_2$Pb	Lead(II) chloride	(10.2)	791
Cl$_2$S	Sulfur dichloride	9.45 ± 0.03	895

Mol. Form.	Name	IP/eV	$\Delta_f H_{ion}$ kJ/mol
Cl_2Si	Dichlorosilylene	(10.93 ± 0.10)	887
Cl_2Sn	Tin(II) chloride	(10.0)	760
Cl_3Ga	Gallium(III) chloride	11.52	664
Cl_3HSi	Trichlorosilane	(11.7)	648
Cl_3N	Nitrogen trichloride	(10.12 ± 0.1)	1244
Cl_3OP	Phosphorus(V) oxychloride	11.36 ± 0.02	540
Cl_3OV	Vanadyl trichloride	(11.6)	425
Cl_3P	Phosphorus(III) chloride	9.91	668
Cl_3PS	Phosphorus(V) sulfide trichloride	9.71 ± 0.03	573
Cl_3Sb	Antimony(III) chloride	(≤ 10.7)	s719
Cl_4Ge	Germanium(IV) chloride	11.68 ± 0.05	629
Cl_4Hf	Hafnium(IV) chloride	(11.7)	246
Cl_4Si	Tetrachlorosilane	11.79 ± 0.01	527
Cl_4Sn	Tin(IV) chloride	11.7 ± 0.2	656
Cl_4Ti	Titanium(IV) chloride	(11.5)	349
Cl_4V	Vanadium(IV) chloride	(9.2)	361
Cl_4Zr	Zirconium(IV) chloride	(11.2)	210
Cl_5Mo	Molybdenum(V) chloride	(8.7)	392
Cl_5Nb	Niobium(V) chloride	(10.97)	356
Cl_5P	Phosphorus(V) chloride	(10.2)	608
Cl_5Ta	Tantalum(V) chloride	(11.08)	303
Cl_6W	Tungsten(VI) chloride	(9.5)	348
Cm	Curium	6.02	966
Co	Cobalt	7.8810	1187
Cr	Chromium	6.76664	1050
Cs	Cesium	3.89390	452
CsF	Cesium fluoride	(8.80 ± 0.10)	489
CsNa	Cesium sodium	(4.05 ± 0.04)	535
Cu	Copper	7.72638	1084
CuF	Copper(I) fluoride	10.15 ± 0.02	984
Dy	Dysprosium	5.9389	862
Er	Erbium	6.1078	907
Es	Einsteinium	6.42	753
Eu	Europium	5.6704	723
F	Fluorine (atomic)	17.42282	1761
FGa	Gallium monofluoride	(9.6 ± 0.5)	700
FH	Hydrogen fluoride	16.044 ± 0.003	1276
FHO	Hypofluorous acid	12.71 ± 0.01	1130
FH_3Si	Fluorosilane	11.7	752
FI	Iodine fluoride	10.54 ± 0.01	922
FIn	Indium monofluoride	(9.6 ± 0.5)	740
FNO	Nitrosyl fluoride	12.63 ± 0.03	1152
FNO_2	Nitryl fluoride	(13.09)	1154
FNS	Thionitrosyl fluoride (NSF)	11.51 ± 0.04	1090
FO	Fluorine monoxide	12.78 ± 0.03	1342
FO_2	Fluorine superoxide (FOO)	(12.6 ± 0.2)	1228
FS	Sulfur fluoride	10.09	986
FTl	Thallium(I) fluoride	10.52	835
F_2	Fluorine	15.697 ± 0.003	1515
F_2Ge	Germanium(II) fluoride	(≤ 11.65)	551
F_2HN	Difluoramine	(11.53 ± 0.08)	1046
F_2H_2Si	Difluorosilane	(12.2)	386
F_2Mg	Magnesium fluoride	(13.6 ± 0.3)	588
F_2N	Difluoroamidogen	11.628 ± 0.01	1155
F_2N_2	*trans*-Difluorodiazine	(12.8)	1315
F_2O	Fluorine monoxide	13.11 ± 0.01	1290
F_2OS	Thionyl fluoride	12.25	688
F_2O_2S	Sulfuryl fluoride	13.04 ± 0.01	501
F_2Pb	Lead(II) fluoride	(11.5)	679
F_2S	Sulfur difluoride	(10.08)	676

Mol. Form.	Name	IP/eV	$\Delta_f H_{ion}$ kJ/mol
F_2Si	Difluorosilylene	10.78 ± 0.05	450
F_2Sn	Tin(II) fluoride	(11.1)	586
F_2Xe	Xenon difluoride	12.35 ± 0.01	1083
F_3HSi	Trifluorosilane	(14.0)	150
F_3N	Nitrogen trifluoride	13.00 ± 0.02	1125
F_3NO	Trifluoramine oxide	13.31 ± 0.06	1121
F_3OP	Phosphorus(V) oxyfluoride	12.76 ± 0.01	-24
F_3P	Phosphorus(III) fluoride	11.60 ± 0.05	161
F_3PS	Phosphorus(V) sulfide trifluoride	$\leq 11.05 \pm 0.035$	≤ 58
F_3Si	Trifluorosilyl	(9.99)	- 32
F_4Ge	Germanium(IV) fluoride	(15.5)	307
F_4N_2	Tetrafluorohydrazine	11.94 ± 0.03	1119
F_4S	Sulfur tetrafluoride	12.0 ± 0.3	399
F_4Si	Tetrafluorosilane	15.24 ± 0.14	-144
F_4Xe	Xenon tetrafluoride	12.65 ± 0.1	1016
F_5I	Iodine pentafluoride	12.943 ± 0.005	408
F_5P	Phosphorus(V) fluoride	(15.1)	-137
F_5S	Sulfur pentafluoride	9.60 ± 0.05	10
F_6Mo	Molybdenum(VI) fluoride	(14.5 ± 0.1)	-159
F_6S	Sulfur hexafluoride	15.32 ± 0.02	258
F_6U	Uranium(VI) fluoride	14.00 ± 0.10	-796
Fe	Iron	7.9024	1177
Fm	Fermium	6.50	627
Ga	Gallium	5.99930	851
GaI_3	Gallium(III) iodide	9.40	765
Gd	Gadolinium	6.1500	991
Ge	Germanium	7.900	1139
GeH_4	Germane	≤ 10.53	≤ 1108
GeI_4	Germanium(IV) iodide	(9.42)	850
GeO	Germanium(II) oxide	11.25 ± 0.01	1044
GeS	Germanium(II) sulfide	(9.98)	1055
H	Hydrogen (atomic)	13.59844	1530
HI	Hydrogen iodide	10.386 ± 0.001	1028
HLi	Lithium hydride	7.7	882
HN	Imidogen	$\leq 13.49 \pm 0.01$	1678
HNO	Nitrosyl hydride	(10.1)	1075
HNO_2	Nitrous acid	≤ 11.3	≤ 1011
HNO_3	Nitric acid	11.95 ± 0.01	1019
HN_3	Hydrazoic acid	10.72 ± 0.025	1328
HO	Hydroxyl	13.0170 ± 0.0002	1294
HO_2	Hydroperoxy	11.35 ± 0.01	1106
HS	Mercapto	10.4219 ± 0.0004	1145
H_2	Hydrogen	15.42593 ± 0.00005	1488
H_2N	Amidogen	11.14 .01	1264
H_2O	Water	12.6206 ± 0.0020	976
H_2O_2	Hydrogen peroxide	10.58 ± 0.04	885
H_2S	Hydrogen sulfide	10.457 ± 0.012	989
H_2Se	Hydrogen selenide	9.892 ± 0.005	984
H_2Si	Silylene	8.244 ± 0.025	1084
H_3N	Ammonia	10.070 ± 0.020	925
H_3NO	Hydroxylamine	(10.00)	923
H_3P	Phosphine	9.869 ± 0.002	958
H_3Sb	Stibine	9.54 ± 0.03	1067
H_4N_2	Hydrazine	8.1 ± 0.15	877
H_4Si	Silane	11.00 ± 0.02	1095
H_4Sn	Stannane	(10.75)	1200
H_6Si_2	Disilane	9.74 ± 0.02	1019
H_8Si_3	Trisilane	(9.2)	1009
He	Helium	24.58741	2372
Hf	Hafnium	6.82507 ± 0.00004	1278

Mol. Form.	Name	IP/eV	$\Delta_f H_{ion}$ kJ/mol
Hg	Mercury	10.43750	1069
HgI$_2$	Mercury(II) iodide	9.5088 ± 0.0022	900
Ho	Holmium	6.0216	882
I	Iodine (atomic)	10.45126	1115
IK	Potassium iodide	(7.21 ± 0.3)	570
ILi	Lithium iodide	(7.5)	633
INa	Sodium iodide	7.64 ± 0.02	659
ITl	Thallium(I) iodide	8.47 ± 0.02	826
I$_2$	Iodine	9.3074 ± 0.0002	960
I$_4$Ti	Titanium(IV) iodide	(9.1)	602
I$_4$Zr	Zirconium(IV) iodide	(9.3)	500
In	Indium	5.78636	802
Ir	Iridium	9.1	1543
K	Potassium	4.34066	508
KLi	Lithium potassium	4.57 ± 0.04	512
KNa	Potassium sodium	4.41636 ± 0.00017	561
K$_2$	Dipotassium	4.0637 ± 0.0002	519
Kr	Krypton	13.99961	1351
La	Lanthanum	5.5770	969
Li	Lithium	5.39172	680
LiNa	Lithium sodium	5.05 ± 0.04	571
LiO	Lithium monoxide	(8.44)	894
LiRb	Lithium rubidium	4.3 ± 0.1	486
Li$_2$	Dilithium	5.1127 ± 0.0003	709
Lu	Lutetium	5.42585	950
Md	Mendelevium	6.58	635
Mg	Magnesium	7.64624	885
MgO	Magnesium oxide	(8.76 ± 0.22)	901
Mn	Manganese	7.43402	998
Mo	Molybdenum	7.09243	1343
N	Nitrogen (atomic)	14.53414	1875
NO	Nitric oxide	9.26438 ± 0.00005	985
NO$_2$	Nitrogen dioxide	9.586 ± 0.002	958
NP	Phosphorus nitride	11.84 ± 0.04	1247
NS	Nitrogen sulfide	8.87 ± 0.01	1119
N$_2$	Nitrogen	15.5808	1503
N$_2$O	Nitrous oxide	12.886	1325
N$_2$O$_4$	Nitrogen tetroxide	(10.8)	1050
N$_2$O$_5$	Nitrogen pentoxide	(11.9)	1161
Na	Sodium	5.13908	603
NaRb	Rubidium sodium	4.32 ± 0.04	480
Na$_2$	Disodium	4.894 ± 0.003	614
Nb	Niobium	6.75885	1384
Nd	Neodymium	5.5250	859
Ne	Neon	21.56454	2081
Ni	Nickel	7.6398	1167
No	Nobelium	6.65	642
Np	Neptunium	6.2657 ± 0.0003	1069
O	Oxygen (atomic)	13.61806	1563
OPb	Lead(II) oxide	9.08 ± 0.10	939
OS	Sulfur monoxide	10.294 ± 0.004	998
OS$_2$	Sulfur oxide (SSO)	10.584 ± 0.005	971
OSi	Silicon monoxide	11.49 ± 0.20	1008
OSn	Tin(II) oxide	9.60 ± 0.02	944
OSr	Strontium oxide	6.6 ± 0.2	623
O$_2$	Oxygen	12.0697 ± 0.0002	1165
O$_2$S	Sulfur dioxide	12.349 ± 0.001	894
O$_2$Th	Thorium(IV) oxide	(8.7 ± 0.15)	342
O$_2$Ti	Titanium(IV) oxide	(9.54 ± 0.1)	623
O$_2$U	Uranium(IV) oxide	(5.4 ± 0.1)	57

Mol. Form.	Name	IP/eV	$\Delta_f H_{ion}$ kJ/mol
O_3	Ozone	12.43	1342
O_3S	Sulfur trioxide	12.82 ± 0.03	841
O_3U	Uranium(VI) oxide	(10.5 ± 0.5)	214
O_4Os	Osmium(VIII) oxide	(12.32)	850
O_4Ru	Ruthenium(VIII) oxide	12.15 ± 0.03	988
O_7Re_2	Rhenium(VII) oxide	(12.7 ± 0.2)	125
Os	Osmium	8.7	1630
P	Phosphorus	10.48669	1328
P_2	Diphosphorus	10.53	1160
Pa	Protactinium	5.89	1133
Pb	Lead	7.41666	911
PbS	Lead(II) sulfide	(8.5 ± 0.5)	954
Pd	Palladium	8.3367	1181
Pm	Promethium	5.55	536
Pr	Praseodymium	5.464	883
Pt	Platinum	9.0	1433
Pu	Plutonium	6.025	926
Ra	Radium	5.27892	668
Rb	Rubidium	4.17713	484
Re	Rhenium	7.88	1530
Rh	Rhodium	7.45890	1276
Rn	Radon	10.74850	1037
Ru	Ruthenium	7.36050	1355
S	Sulfur	10.36001	1277
SSn	Tin(II) sulfide	(8.8)	966
S_2	Disulfur	9.356 ± 0.002	1031
Sb	Antimony	8.64	1096
Sc	Scandium	6.56144	1010
Se	Selenium	9.75238	1168
Si	Silicon	8.15169	1238
Sm	Samarium	5.6437	751
Sn	Tin	7.34381	1011
Sr	Strontium	5.69484	713
Ta	Tantalum	7.89	1544
Tb	Terbium	5.8639	955
Tc	Technetium	7.28	1380
Te	Tellurium	9.0096	1066
Th	Thorium	6.308 ± 0.003	1207
Ti	Titanium	6.8282	1127
Tl	Thallium	6.10829	771
Tm	Thulium	6.18431	827
U	Uranium	6.19405	1129
V	Vanadium	6.746 ± 0.002	1166
W	Tungsten	7.98	1621
Xe	Xenon	12.12987	1170
Y	Yttrium	6.217	1022
Yb	Ytterbium	6.25416	754
Zn	Zinc	9.39405	1037
Zr	Zirconium	6.63390	1251

Substances containing carbon

Mol. Form.	Name	IP/eV	$\Delta_f H_{ion}$ kJ/mol
C	Carbon	11.26030	1803
$CBrClF_2$	Bromochlorodifluoromethane	(11.21)	642
$CBrCl_3$	Bromotrichloromethane	(10.6)	980
$CBrF_3$	Bromotrifluoromethane	(11.40)	451
CBr_2F_2	Dibromodifluoromethane	11.03 ± 0.04	683
CBr_4	Tetrabromomethane	(10.31 ± 0.02)	1079
CCl	Chloromethylidyne	(8.9 ± 0.2)	1244
$CClF_3$	Chlorotrifluoromethane	12.6 ± 0.2	505

Mol. Form.	Name	IP/eV	$\Delta_f H_{ion}$ kJ/mol
CClN	Cyanogen chloride	12.34 ± 0.01	1329
CCl$_2$	Dichloromethylene	(9.27)	1058
CCl$_2$F$_2$	Dichlorodifluoromethane	12.05 ± 0.24	685
CCl$_2$O	Carbonyl chloride	(11.5)	888
CCl$_3$F	Trichlorofluoromethane	11.77 ± 0.02	868
CCl$_4$	Tetrachloromethane	11.47 ± 0.01	1010
CF	Fluoromethylidyne	9.11 ± 0.01	1134
CFN	Cyanogen fluoride	13.34 ± 0.02	1325
CF$_2$	Difluoromethylene	11.44 ± 0.03	899
CF$_2$O	Carbonyl fluoride	13.035 ± 0.030	617
CF$_3$	Trifluoromethyl	8.7 ± 0.2	379
CF$_3$I	Trifluoroiodomethane	10.23	397
CH	Methylidyne	10.64 ± 0.01	1622
CHBrCl$_2$	Bromodichloromethane	10.6	973
CHBr$_2$Cl	Chlorodibromomethane	10.59 ± 0.01	1030
CHBr$_3$	Tribromomethane	10.48 ± 0.02	1035
CHCl	Chloromethylene	9.84	1247
CHClF$_2$	Chlorodifluoromethane	(12.2)	693
CHCl$_2$F	Dichlorofluoromethane	(11.5)	829
CHCl$_3$	Trichloromethane	11.37 ± 0.02	992
CHF	Fluoromethylene	10.06 ± 0.05	1121
CHF$_3$	Trifluoromethane	(13.86)	643
CHI$_3$	Triiodomethane	9.25 ± 0.02	1010
CHN	Hydrogen cyanide	13.60 ± 0.01	1447
CHN	Hydrogen isocyanide	(12.5 ± 0.1)	1407
CHNO	Isocyanic acid	11.595 ± 0.005	1016
CHNO	Fulminic acid	(10.83)	1263
CHO	Oxomethyl (HCO)	(8.55)	826
CH$_2$	Methylene	10.396 ± .003	1392
CH$_2$BrCl	Bromochloromethane	10.77 ± 0.01	1085
CH$_2$Br$_2$	Dibromomethane	(10.50 ± 0.02)	1013
CH$_2$ClF	Chlorofluoromethane	11.71 ± 0.01	870
CH$_2$Cl$_2$	Dichloromethane	11.32 ± .01	996
CH$_2$F$_2$	Difluoromethane	12.71	774
CH$_2$I$_2$	Diiodomethane	9.46 ± 0.02	1030
CH$_2$N$_2$	Diazomethane	8.999 ± 0.001	1098
CH$_2$N$_2$	Cyanamide	(10.4)	1137
CH$_2$O	Formaldehyde	10.88 ± 0.01	941
CH$_2$O$_2$	Formic acid	11.33 ± 0.01	715
CH$_3$	Methyl	9.843 ± 0.002	1095
CH$_3$BO	Borane carbonyl	11.14 ± 0.02	962
CH$_3$Br	Bromomethane	10.541 ± 0.003	979
CH$_3$Cl	Chloromethane	11.22 ± 0.01	1001
CH$_3$Cl$_3$Si	Methyltrichlorosilane	(11.36 ± 0.03)	548
CH$_3$F	Fluoromethane	12.47 ± 0.02	956
CH$_3$I	Iodomethane	9.538	936
CH$_3$NO	Formamide	10.16 ± 0.06	796
CH$_3$NO$_2$	Nitromethane	11.08 ± 0.07	994
CH$_3$N$_3$	Methyl azide	9.81 ± 0.02	1227
CH$_3$O	Methoxy	(10.72)	1050
CH$_4$	Methane	12.61 ± 0.01	1143
CH$_4$N$_2$O	Urea	9.7	690
CH$_4$O	Methanol	10.85 ± 0.01	845
CH$_4$S	Methanethiol	9.44 ± 0.005	888
CH$_5$N	Methylamine	(8.80)	826
CH$_6$N$_2$	Methylhydrazine	7.7 ± 0.15	835
CH$_6$Si	Methylsilane	(10.7)	1003
CN	Cyanide	13.5984	1748
CNO	Cyanate	11.76 ± 0.01	1290
CO	Carbon monoxide	14.014 ± 0.0003	1242

Mol. Form.	Name	IP/eV	$\Delta_f H_{ion}$ kJ/mol
COS	Carbon oxysulfide	11.18 ± 0.01	936
COSe	Carbon oxyselenide	10.36 ± 0.01	929
CO_2	Carbon dioxide	13.773 ± 0.002	935
CS	Carbon sulfide	11.33 ± 0.01	1361
CS_2	Carbon disulfide	10.0685 ± 0.0020	1089
C_2	Dicarbon	(11.4 ± 0.3)	2000
$C_2Br_2F_4$	1,2-Dibromotetrafluoroethane	(11.1)	280
C_2ClF_3	Chlorotrifluoroethylene	9.81 ± 0.03	373
C_2ClF_5	Chloropentafluoroethane	(12.6)	99
C_2Cl_2	Dichloroacetylene	9.9	1165
$C_2Cl_2F_4$	1,2-Dichlorotetrafluoroethane	12.2	252
$C_2Cl_3F_3$	1,1,1-Trichlorotrifluoroethane	11.5	386
$C_2Cl_3F_3$	1,1,2-Trichlorotrifluoroethane	11.99 ± 0.02	429
C_2Cl_4	Tetrachloroethylene	9.326 ± 0.001	887
$C_2Cl_4F_2$	1,1,2,2-Tetrachloro-1,2-difluoroethane	(11.3)	563
C_2Cl_4O	Trichloroacetyl chloride	(11.0)	827
C_2Cl_6	Hexachloroethane	(11.1)	920
C_2F_3N	Trifluoroacetonitrile	13.93 ± 0.07	845
C_2F_4	Tetrafluoroethylene	10.12 ± 0.02	315
C_2F_6	Hexafluoroethane	(13.6)	-30
C_2H	Ethynyl	(11.61 ± 0.07)	1685
C_2HBr	Bromoacetylene	10.31 ± 0.02	1242
$C_2HBrClF_3$	2-Bromo-2-chloro-1,1,1-trifluoroethane	(11.0)	363
C_2HCl	Chloroacetylene	10.58 ± 0.02	1276
C_2HClF_2	1-Chloro-2,2-difluoroethylene	9.80 ± 0.04	628
C_2HCl_3	Trichloroethylene	9.46 ± 0.02	894
C_2HCl_3O	Dichloroacetyl chloride	(10.9)	809
C_2HCl_5	Pentachloroethane	(11.0)	919
C_2HF	Fluoroacetylene	11.26	1195
C_2HF_3	Trifluoroethylene	10.14	489
$C_2HF_3O_2$	Trifluoroacetic acid	11.46	75
C_2H_2	Acetylene	11.400 ± 0.002	1328
$C_2H_2Cl_2$	1,1-Dichloroethylene	9.81 ± 0.04	949
$C_2H_2Cl_2$	cis-1,2-Dichloroethylene	9.66 ± 0.01	936
$C_2H_2Cl_2$	trans-1,2-Dichloroethylene	9.64 ± 0.02	934
$C_2H_2Cl_2O$	Chloroacetyl chloride	(≤ 10.3)	815
$C_2H_2Cl_4$	1,1,1,2-Tetrachloroethane	(11.1)	920
$C_2H_2Cl_4$	1,1,2,2-Tetrachloroethane	(≤ 11.62)	≤ 971
$C_2H_2F_2$	1,1-Difluoroethylene	10.29 ± 0.01	650
$C_2H_2F_2$	cis-1,2-Difluoroethylene	10.23 ± 0.02	690
C_2H_2O	Ketene	9.617 ± 0.003	880
$C_2H_2O_2$	Glyoxal	10.2	773
$C_2H_2S_2$	Thiirene	8.61	892
C_2H_3Br	Bromoethylene	9.83 ± 0.02	1028
C_2H_3Cl	Chloroethylene	9.99 ± 0.02	985
$C_2H_3ClF_2$	1-Chloro-1,1-difluoroethane	(11.98)	626
C_2H_3ClO	Acetyl chloride	10.82 ± 0.04	801
C_2H_3ClO	Chloroacetaldehyde	(10.48)	815
$C_2H_3ClO_2$	Chloroacetic acid	(10.7)	597
$C_2H_3Cl_3$	1,1,1-Trichloroethane	(11.0)	917
$C_2H_3Cl_3$	1,1,2-Trichloroethane	(11.0)	911
C_2H_3F	Fluoroethylene	10.36 ± 0.01	861
C_2H_3FO	Acetyl fluoride	(11.5)	667
$C_2H_3F_3$	1,1,1-Trifluoroethane	13.3 ± 0.5	536
C_2H_3N	Acetonitrile	12.20 ± 0.01	1253
C_2H_3NO	Methylisocyanate	(10.67)	900
C_2H_4	Ethylene	10.5138 ± 0.0006	1067
$C_2H_4Br_2$	1,2-Dibromoethane	10.35 ± 0.04	961
$C_2H_4Cl_2$	1,1-Dichloroethane	11.04 ± 0.02	935
$C_2H_4Cl_2$	1,2-Dichloroethane	11.04 ± 0.02	931

Mol. Form.	Name	IP/eV	$\Delta_f H_{ion}$ kJ/mol
$C_2H_4F_2$	1,1-Difluoroethane	(11.87)	643
C_2H_4O	Acetaldehyde	10.229 ± 0.0007	821
C_2H_4O	Ethylene oxide	10.56 ± 0.01	966
$C_2H_4O_2$	Acetic acid	10.65 ± 0.02	595
$C_2H_4O_2$	Methyl formate	10.835 ± 0.005	690
C_2H_5Br	Bromoethane	10.29 ± 0.01	931
C_2H_5Cl	Chloroethane	10.98 ± 0.02	947
C_2H_5ClO	2-Chloroethanol	(10.5)	756
C_2H_5F	Fluoroethane	(11.78)	873
C_2H_5I	Iodoethane	9.3492 ± 0.0006	893
C_2H_5N	Ethyleneimine	(9.5 ± 0.3)	1044
C_2H_5NO	Acetamide	9.65 ± 0.03	693
C_2H_5NO	N-Methylformamide	9.83 ± 0.04	760
$C_2H_5NO_2$	Nitroethane	10.88 ± 0.05	948
C_2H_6	Ethane	11.56 ± 0.02	1031
$C_2H_6Cl_2Si$	Dichlorodimethylsilane	(10.7)	576
C_2H_6O	Ethanol	10.43 ± 0.05	772
C_2H_6O	Dimethyl ether	10.025 ± 0.025	783
C_2H_6OS	Dimethyl sulfoxide	9.10 ± 0.03	727
$C_2H_6O_2$	Ethylene glycol	10.16	593
C_2H_6S	Ethanethiol	9.31 ± 0.03	851
C_2H_6S	Dimethyl sulfide	8.69 ± 0.02	801
$C_2H_6S_2$	Dimethyl disulfide	(7.4 ± 0.3)	690
C_2H_7N	Ethylamine	8.86 ± 0.02	808
C_2H_7N	Dimethylamine	8.24 ± 0.08	777
C_2H_7NO	Ethanolamine	8.96	664
$C_2H_8N_2$	1,2-Ethanediamine	(8.6)	812
$C_2H_8N_2$	1,1-Dimethylhydrazine	7.29 ± 0.05	787
C_2N_2	Cyanogen	13.37 ± 0.01	1597
C_3F_6	Perfluoropropene	10.60 ± 0.03	-103
C_3F_6O	Perfluoroacetone	(11.57 ± 0.13)	-282
C_3F_8	Perfluoropropane	(13.38)	-491
C_3HN	Cyanoacetylene	11.64 ± 0.01	1475
C_3H_2O	2-Propynal	(10.7 ± 0.1)	1145
$C_3H_3F_3$	3,3,3-Trifluoropropene	(10.9)	437
C_3H_3N	2-Propenenitrile	10.91 ± 0.01	1237
C_3H_3NO	Oxazole	(9.9)	940
C_3H_3NO	Isoxazole	(9.93)	1038
C_3H_4	Allene	9.692 ± 0.004	1126
C_3H_4	Propyne	10.37 ± 0.01	1187
C_3H_4	Cyclopropene	9.67 ± 0.01	1209
$C_3H_4N_2$	Imidazole	(8.81)	997
C_3H_4O	Propargyl alcohol	10.49 ± 0.02	1060
C_3H_4O	Acrolein	10.103 ± 0.006	900
C_3H_4O	Cyclopropanone	(9.1 ± 0.1)	895
$C_3H_4O_2$	Propenoic acid	10.60	701
$C_3H_4O_2$	2-Oxetanone	(9.70 ± 0.01)	653
C_3H_5Br	3-Bromopropene	(9.96)	1008
C_3H_5Cl	3-Chloropropene	10.04 ± 0.01	965
C_3H_5ClO	Epichlorohydrin	(10.64)	919
$C_3H_5ClO_2$	Methyl chloroacetate	(10.3)	575
C_3H_5F	3-Fluoropropene	(10.11)	821
C_3H_5N	Propanenitrile	11.84 ± 0.02	1194
C_3H_5NO	Acrylamide	(9.5)	720
C_3H_6	Propene	9.73 ± 0.02	959
C_3H_6	Cyclopropane	9.86	1005
$C_3H_6Br_2$	1,2-Dibromopropane	(10.1)	903
$C_3H_6Br_2$	1,3-Dibromopropane	(≤ 10.2)	≤ 919
$C_3H_6Cl_2$	1,2-Dichloropropane	10.8 ± 0.1	886
$C_3H_6Cl_2$	1,3-Dichloropropane	10.89 ± 0.04	892

Mol. Form.	Name	IP/eV	$\Delta_f H_{ion}$ kJ/mol
C_3H_6O	Allyl alcohol	9.67 ± 0.05	808
C_3H_6O	Methyl vinyl ether	8.95 ± 0.01	763
C_3H_6O	Propanal	9.96 ± 0.01	772
C_3H_6O	Acetone	9.703 ± 0.006	719
C_3H_6O	Methyloxirane	(10.22)	892
C_3H_6O	Oxetane	9.65 ± 0.01	851
$C_3H_6O_2$	Propanoic acid	10.525 ± 0.003	568
$C_3H_6O_2$	Ethyl formate	10.61 ± 0.01	639
$C_3H_6O_2$	Methyl acetate	10.25 ± 0.02	579
$C_3H_6O_2$	1,3-Dioxolane	(9.9)	658
$C_3H_6O_3$	1,3,5-Trioxane	(10.3)	528
C_3H_7Br	1-Bromopropane	10.18 ± 0.01	898
C_3H_7Br	2-Bromopropane	10.10 ± 0.03	877
C_3H_7Cl	1-Chloropropane	10.81 ± 0.01	911
C_3H_7Cl	2-Chloropropane	10.79 ± 0.02	896
C_3H_7F	1-Fluoropropane	(11.3)	806
C_3H_7F	2-Fluoropropane	(11.08)	776
C_3H_7I	1-Iodopropane	9.25 ± 0.01	860
C_3H_7I	2-Iodopropane	9.19 ± 0.02	845
C_3H_7N	Allylamine	(8.76)	891
C_3H_7N	Cyclopropylamine	(8.8)	926
C_3H_7N	Propyleneimine	(9.0)	960
C_3H_7NO	N,N-Dimethylformamide	(9.12)	688
$C_3H_7NO_2$	1-Nitropropane	(10.81)	919
$C_3H_7NO_2$	2-Nitropropane	(10.71)	894
C_3H_8	Propane	10.95 ± 0.05	952
C_3H_8O	1-Propanol	10.18 ± 0.06	727
C_3H_8O	2-Propanol	10.17 ± 0.02	709
C_3H_8O	Ethyl methyl ether	9.72 ± 0.07	722
$C_3H_8O_2$	Dimethoxymethane	9.7	588
C_3H_8S	1-Propanethiol	9.20 ± 0.01	819
C_3H_8S	2-Propanethiol	9.145 ± 0.005	806
C_3H_8S	Ethyl methyl sulfide	(8.55)	765
$C_3H_9BO_3$	Trimethyl borate	(10.0)	65
C_3H_9ClSi	Trimethylchlorosilane	(10.15)	624
C_3H_9N	Propylamine	(8.78)	777
C_3H_9N	Isopropylamine	(8.72)	758
C_3H_9N	Trimethylamine	7.82 ± 0.06	731
C_3H_9NO	3-Amino-1-propanol	(9.0)	651
$C_4H_2O_3$	Maleic anhydride	(10.8)	645
C_4H_4	1-Buten-3-yne	9.58 ± 0.02	1230
$C_4H_4N_2$	Succinonitrile	(12.1 ± 0.25)	1377
$C_4H_4N_2$	Pyrimidine	9.23	1087
$C_4H_4N_2$	Pyridazine	8.67 ± 0.03	1112
C_4H_4O	Furan	8.883 ± 0.003	822
$C_4H_4O_2$	Diketene	(9.6 ± 0.02)	736
$C_4H_4O_3$	Succinic anhydride	(10.6)	500
$C_4H_4O_4$	Fumaric acid	(10.7)	355
C_4H_4S	Thiophene	8.86 ± 0.02	970
C_4H_5N	Methylacrylonitrile	10.34	1127
C_4H_5N	Pyrrole	8.207 ± 0.005	900
C_4H_5N	Cyclopropanecarbonitrile	(10.25)	1173
C_4H_6	1,2-Butadiene	(9.03)	1034
C_4H_6	1,3-Butadiene	9.082 ± 0.004	986
C_4H_6	1-Butyne	10.19 ± 0.02	1148
C_4H_6	2-Butyne	9.59 ± 0.03	1071
C_4H_6	Cyclobutene	9.43 ± 0.02	1067
C_4H_6O	Divinyl ether	(8.7)	827
C_4H_6O	trans-2-Butenal	9.73 ± 0.01	835
C_4H_6O	2-Methylpropenal	(9.92)	834

Mol. Form.	Name	IP/eV	$\Delta_f H_{ion}$ kJ/mol
C_4H_6O	Cyclobutanone	(9.35	815
$C_4H_6O_2$	cis-Crotonic acid	(10.08)	625
$C_4H_6O_2$	trans-Crotonic acid	(9.9)	604
$C_4H_6O_2$	Methacrylic acid	(10.15)	611
$C_4H_6O_2$	Vinyl acetate	9.19 ± 0.05	572
$C_4H_6O_2$	Methyl acrylate	(9.9)	641
$C_4H_6O_3$	Acetic anhydride	(10.0)	398
$C_4H_6O_4$	Dimethyl oxalate	(10.0)	287
C_4H_6S	2,5-Dihydrothiophene	(8.4)	898
C_4H_7N	Butanenitrile	(11.2)	1110
C_4H_7N	2-Methylpropanenitrile	(11.3)	1115
C_4H_7NO	2-Pyrrolidone	(9.2)	674
C_4H_8	1-Butene	9.55 ± 0.06	921
C_4H_8	cis-2-Butene	9.11 ± 0.01	871
C_4H_8	trans-2-Butene	9.10 ± 0.01	866
C_4H_8	Isobutene	9.239 ± 0.003	875
C_4H_8	Cyclobutane	(9.82 ± 0.05)	976
C_4H_8	Methylcyclopropane	(9.46)	936
$C_4H_8Br_2$	1,4-Dibromobutane	(10.15)	879
C_4H_8O	Ethyl vinyl ether	(8.98)	709
C_4H_8O	1,2-Epoxybutane	(≤ 10.15)	862
C_4H_8O	Butanal	9.84 ± 0.02	742
C_4H_8O	Isobutanal	9.71 ± 0.01	721
C_4H_8O	2-Butanone	9.52 ± 0.04	678
C_4H_8O	Tetrahydrofuran	9.38 ± 0.05	721
$C_4H_8O_2$	Butanoic acid	10.17 ± 0.05	509
$C_4H_8O_2$	2-Methylpropanoic acid	10.33 ± 0.03	516
$C_4H_8O_2$	Propyl formate	10.52 ± 0.02	555
$C_4H_8O_2$	Ethyl acetate	10.01 ± 0.05	522
$C_4H_8O_2$	Methyl propanoate	10.15 ± 0.03	548
$C_4H_8O_2$	1,3-Dioxane	9.8	607
$C_4H_8O_2$	1,4-Dioxane	9.19 ± 0.01	571
$C_4H_8O_2S$	Sulfolane	(9.8)	577
C_4H_8S	Tetrahydrothiophene	8.38	774
C_4H_9Br	1-Bromobutane	(10.12)	869
C_4H_9Br	2-Bromobutane	10.01 ± 0.02	845
C_4H_9Br	1-Bromo-2-methylpropane	10.09 ± 0.02	861
C_4H_9Br	2-Bromo-2-methylpropane	9.92 ± 0.03	823
C_4H_9Cl	1-Chlorobutane	10.67 ± 0.03	875
C_4H_9Cl	2-Chlorobutane	10.53	857
C_4H_9Cl	1-Chloro-2-methylpropane	10.73 ± 0.07	877
C_4H_9Cl	2-Chloro-2-methylpropane	(10.61)	842
C_4H_9I	1-Iodobutane	9.23 ± 0.01	840
C_4H_9I	2-Iodobutane	9.10 ± 0.02	815
C_4H_9I	1-Iodo-2-methylpropane	9.19 ± 0.01	824
C_4H_9I	2-Iodo-2-methylpropane	(9.02)	798
C_4H_9N	Pyrrolidine	(8.0)	769
C_4H_9NO	N,N-Dimethylacetamide	8.81 ± 0.03	616
C_4H_9NO	Morpholine	(8.2)	841
C_4H_{10}	Butane	10.53 ± 0.10	890
C_4H_{10}	Isobutane	(10.57)	886
$C_4H_{10}O$	1-Butanol	9.99 ± 0.05	689
$C_4H_{10}O$	2-Butanol	9.88 ± 0.03	658
$C_4H_{10}O$	2-Methyl-1-propanol	10.02 ± 0.04	683
$C_4H_{10}O$	2-Methyl-2-propanol	9.90 ± 0.04	642
$C_4H_{10}O$	Diethyl ether	9.51 ± 0.03	666
$C_4H_{10}O$	Methyl propyl ether	9.41 ± 0.07	670
$C_4H_{10}O$	Isopropyl methyl ether	9.45 ± 0.04	661
$C_4H_{10}O_2$	Ethylene glycol monoethyl ether	(9.6)	529
$C_4H_{10}O_2$	Ethylene glycol dimethyl ether	(9.3)	558

Mol. Form.	Name	IP/eV	$\Delta_f H_{ion}$ kJ/mol
$C_4H_{10}S$	1-Butanethiol	9.14 ± 0.01	794
$C_4H_{10}S$	2-Butanethiol	(9.10)	781
$C_4H_{10}S$	2-Methyl-1-propanethiol	(9.12)	783
$C_4H_{10}S$	2-Methyl-2-propanethiol	(9.03)	762
$C_4H_{10}S$	Diethyl sulfide	(8.43)	730
$C_4H_{10}S$	Methyl propyl sulfide	(8.8)	767
$C_4H_{10}S$	Isopropyl methyl sulfide	(8.7)	749
$C_4H_{10}S_2$	Diethyl disulfide	(8.27)	724
$C_4H_{11}N$	Butylamine	8.7 ± 0.1	748
$C_4H_{11}N$	sec-Butylamine	8.46 ± 0.1	711
$C_4H_{11}N$	tert-Butylamine	8.46 ± 0.1	695
$C_4H_{11}N$	Isobutylamine	8.50 ± 0.1	721
$C_4H_{11}N$	Diethylamine	7.85 ± 0.1	684
$C_4H_{12}Si$	Tetramethylsilane	9.80 ± 0.04	713
$C_4H_{12}Sn$	Tetramethylstannane	8.89 ± 0.05	837
C_4NiO_4	Nickel carbonyl	8.27 ± 0.04	200
$C_5H_4O_2$	Furfural	9.22 ± 0.01	739
C_5H_5N	Pyridine	9.25	1031
C_5H_6	1-Penten-3-yne	9.00 ± 0.01	1119
C_5H_6	cis-3-Penten-1-yne	9.14 ± 0.04	1137
C_5H_6	trans-3-Penten-1-yne	9.05 ± 0.01	1128
C_5H_6	2-Methyl-1-buten-3-yne	9.25 ± 0.02	1152
C_5H_6	1,3-Cyclopentadiene	8.55 ± 0.02	955
C_5H_6O	2-Methylfuran	8.38 ± 0.02	729
C_5H_6O	3-Methylfuran	(8.64)	763
C_5H_6S	2-Methylthiophene	(8.14)	867
C_5H_6S	3-Methylthiophene	(8.40)	893
C_5H_8	cis-1,3-Pentadiene	8.63 ± 0.03	914
C_5H_8	trans-1,3-Pentadiene	8.59 ± 0.02	905
C_5H_8	1,4-Pentadiene	9.60 ± 0.02	1032
C_5H_8	2-Methyl-1,3-butadiene	8.84 ± 0.01	928
C_5H_8	1-Pentyne	10.10 ± 0.01	1119
C_5H_8	Cyclopentene	9.01 ± 0.01	905
C_5H_8	Spiropentane	(9.26)	1078
C_5H_8O	Cyclopropyl methyl ketone	(≤ 9.46)	796
C_5H_8O	Cyclopentanone	9.26 ± 0.01	701
C_5H_8O	3,4-Dihydro-2H-pyran	8.35 ± 0.01	681
$C_5H_8O_2$	Ethyl acrylate	(≤ 10.3)	617
$C_5H_8O_2$	Methyl methacrylate	(9.7)	589
$C_5H_8O_2$	2,4-Pentanedione	8.85 ± 0.01	469
C_5H_9NO	N-Methyl-2-pyrrolidone	(≤ 9.17)	≤ 676
C_5H_{10}	1-Pentene	9.51 ± 0.01	896
C_5H_{10}	cis-2-Pentene	9.01 ± 0.03	843
C_5H_{10}	trans-2-Pentene	9.04 ± 0.01	841
C_5H_{10}	2-Methyl-1-butene	9.12 ± 0.01	844
C_5H_{10}	3-Methyl-1-butene	9.52 ± 0.01	891
C_5H_{10}	2-Methyl-2-butene	8.69 ± 0.01	796
C_5H_{10}	Cyclopentane	(10.33 ± 0.15)	918
$C_5H_{10}O$	2,2-Dimethylpropanal	9.51 ± 0.01	675
$C_5H_{10}O$	Cyclopentanol	(9.72)	695
$C_5H_{10}O$	Pentanal	9.74 ± 0.04	709
$C_5H_{10}O$	2-Pentanone	9.38 ± 0.01	646
$C_5H_{10}O$	3-Pentanone	9.31 ± 0.01	640
$C_5H_{10}O$	3-Methyl-2-butanone	9.30 ± 0.01	635
$C_5H_{10}O$	Tetrahydropyran	9.25 ± 0.01	670
$C_5H_{10}O_2$	Pentanoic acid	(≤ 10.53)	≤ 527
$C_5H_{10}O_2$	3-Methylbutanoic acid	(≤ 10.51)	≤ 499
$C_5H_{10}O_2$	Butyl formate	10.52 ± 0.02	584
$C_5H_{10}O_2$	Propyl acetate	(≤ 9.92)	501
$C_5H_{10}O_2$	Isopropyl acetate	9.99 ± 0.03	482

Mol. Form.	Name	IP/eV	$\Delta_f H_{ion}$ kJ/mol
$C_5H_{10}O_2$	Ethyl propanoate	(10.00)	500
$C_5H_{10}O_2$	Methyl butanoate	(10.07)	520
$C_5H_{10}S$	Thiacyclohexane	(8.2)	728
$C_5H_{11}Br$	1-Bromopentane	10.10 ± 0.01	846
$C_5H_{11}I$	1-Iodopentane	9.20 ± 0.01	817
$C_5H_{11}N$	Piperidine	8.03 ± 0.11	726
$C_5H_{11}N$	N-Methylpyrrolidine	$\leq 8.41 \pm 0.02$	≤ 809
C_5H_{12}	Pentane	10.28 ± 0.10	845
C_5H_{12}	Isopentane	10.32 ± 0.05	843
C_5H_{12}	Neopentane	(≤ 10.2)	≤ 818
$C_5H_{12}O$	1-Pentanol	(10.00)	668
$C_5H_{12}O$	2-Pentanol	(9.78)	630
$C_5H_{12}O$	3-Pentanol	9.78	628
$C_5H_{12}O$	2-Methyl-1-butanol	(9.86)	649
$C_5H_{12}O$	2-Methyl-2-butanol	(9.8)	615
$C_5H_{12}O$	3-Methyl-2-butanol	(9.88 ± 0.13)	637
$C_5H_{12}O$	Butyl methyl ether	(9.4 ± 0.1)	648
$C_5H_{12}O$	Methyl tert-butyl ether	(9.24)	608
$C_5H_{12}O$	Ethyl propyl ether	(9.45)	640
$C_5H_{12}S$	tert-Butyl methyl sulfide	(8.38)	687
$C_5H_{12}S$	Ethyl propyl sulfide	(8.50)	716
$C_5H_{12}S$	Ethyl isopropyl sulfide	(8.35)	689
C_6BrF_5	Bromopentafluorobenzene	(9.67)	222
C_6ClF_5	Chloropentafluorobenzene	(9.72)	126
C_6Cl_6	Hexachlorobenzene	(8.98)	822
C_6F_6	Hexafluorobenzene	9.89 ± 0.04	8
C_6F_{12}	Perfluorocyclohexane	(13.2)	-1095
C_6HF_5	Pentafluorobenzene	(9.63)	122
C_6HF_5O	Pentafluorophenol	(9.20)	-71
$C_6H_2F_4$	1,2,3,4-Tetrafluorobenzene	(9.53)	284
$C_6H_2F_4$	1,2,3,5-Tetrafluorobenzene	(9.53)	263
$C_6H_2F_4$	1,2,4,5-Tetrafluorobenzene	(9.35)	254
$C_6H_3Cl_3$	1,2,4-Trichlorobenzene	(9.04)	880
$C_6H_3Cl_3$	1,3,5-Trichlorobenzene	9.32 ± 0.02	899
$C_6H_4ClNO_2$	1-Chloro-3-nitrobenzene	(9.92 ± 0.1)	995
$C_6H_4ClNO_2$	1-Chloro-4-nitrobenzene	(9.96 ± 0.1)	999
$C_6H_4Cl_2$	o-Dichlorobenzene	9.06 ± 0.02	907
$C_6H_4Cl_2$	m-Dichlorobenzene	9.10 ± 0.02	906
$C_6H_4Cl_2$	p-Dichlorobenzene	8.92 ± 0.02	885
$C_6H_4FNO_2$	1-Fluoro-4-nitrobenzene	(9.90)	826
$C_6H_4F_2$	o-Difluorobenzene	9.29 ± 0.01	602
$C_6H_4F_2$	m-Difluorobenzene	9.33 ± 0.01	591
$C_6H_4F_2$	p-Difluorobenzene	9.1589 ± 0.0003	577
$C_6H_4O_2$	p-Benzoquinone	10.01 ± 0.06	844
C_6H_5Br	Bromobenzene	9.00 ± 0.02	971
C_6H_5Cl	Chlorobenzene	9.07 ± 0.02	930
C_6H_5ClO	m-Chlorophenol	8.655 ± 0.001	680
C_6H_5ClO	p-Chlorophenol	(≤ 8.69)	≤ 692
C_6H_5F	Fluorobenzene	9.20 ± 0.01	772
C_6H_5I	Iodobenzene	8.685	1003
$C_6H_5NO_2$	Nitrobenzene	9.86 ± 0.02	1019
$C_6H_5NO_3$	o-Nitrophenol	(9.1)	782
$C_6H_5NO_3$	m-Nitrophenol	(9.0)	755
$C_6H_5NO_3$	p-Nitrophenol	(9.1)	761
C_6H_6	Benzene	9.24378 ± 0.00007	975
C_6H_6	Fulvene	(8.36)	1031
C_6H_6ClN	o-Chloroaniline	(8.50)	883
C_6H_6ClN	m-Chloroaniline	(8.09)	835
C_6H_6ClN	p-Chloroaniline	(≤ 8.18)	≤ 844
$C_6H_6N_2O_2$	o-Nitroaniline	(8.27)	861

Mol. Form.	Name	IP/eV	$\Delta_f H_{ion}$ kJ/mol
$C_6H_6N_2O_2$	*m*-Nitroaniline	(8.31)	865
$C_6H_6N_2O_2$	*p*-Nitroaniline	(8.34)	859
C_6H_6O	Phenol	8.49 ± 0.02	723
$C_6H_6O_2$	*p*-Hydroquinone	7.94 ± 0.01	503
C_6H_6S	Benzenethiol	(8.32)	915
C_6H_7N	Aniline	7.720 ± 0.002	832
C_6H_7N	2-Methylpyridine	(9.02)	970
C_6H_7N	3-Methylpyridine	(9.04)	979
C_6H_7N	4-Methylpyridine	(9.04)	976
$C_6H_8N_2$	*o*-Phenylenediamine	(7.2)	787
$C_6H_8N_2$	*m*-Phenylenediamine	(7.14)	777
$C_6H_8N_2$	*p*-Phenylenediamine	(6.87 ± 0.05)	759
C_6H_{10}	1,5-Hexadiene	9.27 ± 0.05	978
C_6H_{10}	1-Hexyne	10.03 ± 0.05	1089
C_6H_{10}	3,3-Dimethyl-1-butyne	9.90 ± 0.04	1060
C_6H_{10}	Cyclohexene	8.945 ± 0.01	859
$C_6H_{10}O$	Cyclohexanone	9.14 ± 0.01	656
$C_6H_{10}O$	Mesityl oxide	9.10 ± 0.01	694
$C_6H_{10}O_4$	Diethyl oxalate	(9.8)	205
$C_6H_{11}NO$	Caprolactam	(9.07 ± 0.02)	629
C_6H_{12}	1-Hexene	9.44 ± 0.04	869
C_6H_{12}	*cis*-2-Hexene	(8.97 ± 0.01)	818
C_6H_{12}	*trans*-2-Hexene	(8.97 ± 0.01)	814
C_6H_{12}	2-Methyl-1-pentene	(9.08 ± 0.01)	817
C_6H_{12}	4-Methyl-1-pentene	9.45 ± 0.01	862
C_6H_{12}	2-Methyl-2-pentene	(8.58)	761
C_6H_{12}	4-Methyl-*cis*-2-pentene	8.98 ± 0.01	809
C_6H_{12}	4-Methyl-*trans*-2-pentene	(8.97 ± 0.01)	804
C_6H_{12}	2-Ethyl-1-butene	(9.06 ± 0.02)	818
C_6H_{12}	2,3-Dimethyl-1-butene	(9.07 ± 0.01)	812
C_6H_{12}	2,3-Dimethyl-2-butene	8.27 ± 0.01	729
C_6H_{12}	Cyclohexane	9.86 ± 0.03	828
C_6H_{12}	Methylcyclopentane	(9.85)	845
$C_6H_{12}O$	Hexanal	9.72 ± 0.05	691
$C_6H_{12}O$	2-Hexanone	9.3 ± 0.1	626
$C_6H_{12}O$	3-Hexanone	9.12 ± 0.02	600
$C_6H_{12}O$	3-Methyl-2-pentanone	9.21 ± 0.01	600
$C_6H_{12}O$	4-Methyl-2-pentanone	9.30 ± 0.01	609
$C_6H_{12}O$	2-Methyl-3-pentanone	9.10 ± 0.01	592
$C_6H_{12}O$	3,3-Dimethyl-2-butanone	9.12 ± 0.02	589
$C_6H_{12}O$	Cyclohexanol	(9.75)	651
$C_6H_{12}O_2$	Hexanoic acid	≤ 10.12	≤ 463
$C_6H_{12}O_2$	Butyl acetate	(9.92 ± .05)	471
$C_6H_{12}O_2$	*sec*-Butyl acetate	9.90	453
$C_6H_{12}O_2$	Methyl 2,2-dimethylpropanoate	(9.90 ± 0.04)	466
$C_6H_{13}I$	1-Iodohexane	9.179	794
$C_6H_{13}N$	Cyclohexylamine	(8.86)	750
C_6H_{14}	Hexane	10.13	810
C_6H_{14}	2-Methylpentane	(10.12)	802
C_6H_{14}	3-Methylpentane	(10.08)	801
C_6H_{14}	2,2-Dimethylbutane	(10.06)	787
C_6H_{14}	2,3-Dimethylbutane	(10.02)	791
$C_6H_{14}O$	1-Hexanol	(9.89)	639
$C_6H_{14}O$	2-Hexanol	(9.80 ± 0.03)	611
$C_6H_{14}O$	3-Hexanol	(9.63 ± 0.03)	599
$C_6H_{14}O$	Dipropyl ether	(9.27)	602
$C_6H_{14}O$	Diisopropyl ether	9.20 ± 0.05	569
$C_6H_{14}O$	Butyl ethyl ether	(9.36)	610
$C_6H_{14}O$	Methyl pentyl ether	(≤ 9.67)	≤ 657
$C_6H_{14}O_2$	1,1-Diethoxyethane	(9.2)	434

Mol. Form.	Name	IP/eV	$\Delta_f H_{ion}$ kJ/mol
$C_6H_{14}O_3$	Diethylene glycol dimethyl ether	≤ 9.8	≤ 448
$C_6H_{14}S$	Dipropyl sulfide	8.30 ± 0.02	676
$C_6H_{14}S$	Diisopropyl sulfide	(8.2 ± 0.2)	649
$C_6H_{15}N$	Hexylamine	(8.63 ± 0.05)	699
$C_6H_{15}N$	Dipropylamine	(7.84 ± 0.02)	641
$C_6H_{15}N$	Diisopropylamine	(7.73 ± 0.03)	602
$C_6H_{15}N$	Triethylamine	(7.50 ± 0.02)	631
$C_6H_{15}NO_3$	Triethanolamine	(7.9)	206
$C_7H_3F_5$	2,3,4,5,6-Pentafluorotoluene	(9.4)	64
C_7H_5ClO	Benzoyl chloride	(9.53)	815
$C_7H_5Cl_3$	(Trichloromethyl)benzene	(≤ 9.60)	≤ 914
$C_7H_5F_3$	(Trifluoromethyl)benzene	9.685 ± 0.005	335
C_7H_5N	Benzonitrile	9.70 ± 0.01	1154
C_7H_6O	Benzaldehyde	9.49 ± 0.02	878
$C_7H_6O_2$	Benzoic acid	(9.3)	604
C_7H_7Br	p-Bromotoluene	8.67 ± 0.02	908
C_7H_7Cl	o-Chlorotoluene	(8.7 ± 0.1)	856
C_7H_7Cl	m-Chlorotoluene	(8.83)	869
C_7H_7Cl	p-Chlorotoluene	(8.69)	855
C_7H_7Cl	(Chloromethyl)benzene	9.10 ± 0.02	897
C_7H_7F	o-Fluorotoluene	8.91 ± 0.01	709
C_7H_7F	m-Fluorotoluene	8.91 ± 0.01	709
C_7H_7F	p-Fluorotoluene	8.79 ± 0.01	701
C_7H_7NO	Benzamide	(9.25)	792
$C_7H_7NO_2$	o-Nitrotoluene	9.24	946
$C_7H_7NO_2$	m-Nitrotoluene	9.45 ± 0.1	941
$C_7H_7NO_2$	p-Nitrotoluene	9.46 ± 0.05	942
C_7H_8	Toluene	8.8276 ± 0.0006	901
C_7H_8O	o-Cresol	(8.24)	670
C_7H_8O	m-Cresol	8.29 ± 0.07	668
C_7H_8O	p-Cresol	(8.3)	675
C_7H_8O	Benzyl alcohol	(8.3)	701
C_7H_8O	Anisole	8.22 ± 0.03	725
C_7H_9N	Benzylamine	(8.64)	917
C_7H_9N	o-Methylaniline	(7.44 ± 0.02)	772
C_7H_9N	m-Methylaniline	(7.50 ± 0.02)	778
C_7H_9N	p-Methylaniline	(7.24 ± 0.02)	753
C_7H_9N	N-Methylaniline	7.34 ± 0.04	792
C_7H_9N	2,3-Dimethylpyridine	(8.85 ± 0.02)	922
C_7H_9N	2,4-Dimethylpyridine	(8.85 ± 0.03)	918
C_7H_9N	2,5-Dimethylpyridine	$(\leq 8.80 \pm 0.05)$	≤ 916
C_7H_9N	2,6-Dimethylpyridine	8.86 ± 0.03	913
C_7H_9N	3,4-Dimethylpyridine	(≤ 9.15)	≤ 953
C_7H_9N	3,5-Dimethylpyridine	(≤ 9.25)	≤ 965
$C_7H_{10}O$	Dicyclopropyl ketone	(9.1)	1041
C_7H_{14}	1-Heptene	9.34 ± 0.10	839
C_7H_{14}	$trans$-3-Heptene	(8.92)	790
C_7H_{14}	Cycloheptane	9.97	844
C_7H_{14}	Methylcyclohexane	9.64	775
C_7H_{14}	cis-1,2-Dimethylcyclopentane	(9.92 ± 0.05)	828
C_7H_{14}	$trans$-1,2-Dimethylcyclopentane	9.7 ± 0.2	799
$C_7H_{14}O$	1-Heptanal	(9.65)	668
$C_7H_{14}O$	2-Heptanone	9.28 ± 0.10	594
$C_7H_{14}O$	3-Heptanone	9.18 ± 0.08	589
$C_7H_{14}O$	4-Heptanone	9.10 ± 0.06	577
$C_7H_{14}O$	5-Methyl-2-hexanone	(9.28)	586
$C_7H_{14}O$	2,4-Dimethyl-3-pentanone	8.95 ± 0.01	552
$C_7H_{14}O$	1-Methylcyclohexanol	(9.8 ± 0.2)	586
C_7H_{16}	Heptane	9.93 ± 0.10	771
$C_7H_{16}O$	1-Heptanol	(9.84)	614

Mol. Form.	Name	IP/eV	$\Delta_f H_{ion}$ kJ/mol
$C_7H_{16}O$	2-Heptanol	(9.70)	580
$C_7H_{16}O$	3-Heptanol	(9.68)	578
$C_7H_{16}O$	4-Heptanol	(9.61)	572
$C_7H_{16}O$	Ethyl pentyl ether	(≤ 9.49)	≤ 602
$C_8H_4O_3$	Phthalic anhydride	(10.1)	603
$C_8H_6O_4$	Isophthalic acid	(9.98)	268
$C_8H_6O_4$	Terephthalic acid	(9.86)	232
C_8H_7N	2-Methylbenzonitrile	(≤ 9.38)	1085
C_8H_7N	3-Methylbenzonitrile	(≤ 9.34)	1085
C_8H_7N	4-Methylbenzonitrile	9.32 ± 0.02	1083
C_8H_7N	Indole	7.7602 ± 0.0006	908
C_8H_8	Styrene	8.464 ± 0.001	964
C_8H_8O	p-Tolualdehyde	(9.33)	825
C_8H_8O	Acetophenone	9.29 ± 0.03	810
$C_8H_8O_2$	o-Toluic acid	(9.1)	558
$C_8H_8O_2$	m-Toluic acid	(9.43)	579
$C_8H_8O_2$	p-Toluic acid	(9.23)	560
$C_8H_8O_2$	Benzeneacetic acid	(8.26)	479
$C_8H_8O_2$	Methyl benzoate	9.32 ± 0.03	611
C_8H_{10}	Ethylbenzene	8.77 ± 0.01	876
C_8H_{10}	o-Xylene	8.56 ± 0.01	844
C_8H_{10}	m-Xylene	8.56 ± 0.01	843
C_8H_{10}	p-Xylene	8.44 ± 0.01	832
$C_8H_{10}O$	p-Ethylphenol	(7.84)	613
$C_8H_{10}O$	2,3-Xylenol	(8.26)	640
$C_8H_{10}O$	2,4-Xylenol	(8.0)	609
$C_8H_{10}O$	2,6-Xylenol	(8.05)	615
$C_8H_{10}O$	3,4-Xylenol	(8.09)	624
$C_8H_{10}O$	Phenetole	(8.13)	683
$C_8H_{11}N$	2,4,6-Trimethylpyridine	(≤ 8.9)	≤ 880
$C_8H_{11}N$	N-Ethylaniline	(≤ 7.67)	≤ 794
$C_8H_{11}N$	N,N-Dimethylaniline	7.12 ± 0.02	787
C_8H_{14}	1-Octyne	(9.95 ± 0.02)	1040
C_8H_{14}	2-Octyne	9.31 ± 0.01	961
C_8H_{14}	3-Octyne	9.22 ± 0.01	952
C_8H_{14}	4-Octyne	9.20 ± 0.01	946
C_8H_{16}	1-Octene	9.43 ± 0.01	829
C_8H_{16}	Cyclooctane	9.75 ± 0.05	816
C_8H_{16}	Ethylcyclohexane	(9.54)	748
C_8H_{16}	1,1-Dimethylcyclohexane	(9.42)	728
C_8H_{16}	cis-1,2-Dimethylcyclohexane	(<9.78)	772
C_8H_{16}	trans-1,2-Dimethylcyclohexane	9.41	728
C_8H_{16}	cis-1,3-Dimethylcyclohexane	(<9.98)	778
C_8H_{16}	trans-1,3-Dimethylcyclohexane	9.53	743
C_8H_{16}	cis-1,4-Dimethylcyclohexane	(<9.93)	782
C_8H_{16}	trans-1,4-Dimethylcyclohexane	(9.56)	738
C_8H_{16}	Propylcyclopentane	(9.34)	753
$C_8H_{16}O$	2,2,4-Trimethyl-3-pentanone	(8.80)	511
C_8H_{18}	Octane	9.80 ± 0.10	737
C_8H_{18}	2-Methylheptane	(9.84)	734
C_8H_{18}	2,2,4-Trimethylpentane	(9.86)	713
C_8H_{18}	2,2,3,3-Tetramethylbutane	9.8	720
$C_8H_{18}O$	Dibutyl ether	(9.28)	s 560
$C_8H_{18}O$	Di-sec-butyl ether	(9.11)	511
$C_8H_{18}O$	Di-tert-butyl ether	8.88 ± 0.07	493
$C_8H_{18}S$	Dibutyl sulfide	(8.2)	624
$C_8H_{18}S$	Di-tert-butyl sulfide	(8.0)	583
$C_8H_{18}S$	Diisobutyl sulfide	(8.34)	625
$C_8H_{19}N$	Dibutylamine	(7.69)	586
$C_8H_{19}N$	Diisobutylamine	(7.8)	574

Mol. Form.	Name	IP/eV	$\Delta_f H_{ion}$ kJ/mol
$C_8H_{20}Si$	Tetraethylsilane	(8.9)	595
C_9H_7N	Quinoline	8.62 ± 0.01	1041
C_9H_7N	Isoquinoline	8.53 ± 0.03	1032
C_9H_8	Indene	8.14 ± 0.01	949
C_9H_{10}	o-Methylstyrene	(8.20)	908
C_9H_{10}	m-Methylstyrene	(8.15)	899
C_9H_{10}	p-Methylstyrene	(8.1)	895
C_9H_{10}	Cyclopropylbenzene	(8.35)	956
C_9H_{10}	Indan	(8.3)	864
$C_9H_{10}O_2$	Ethyl benzoate	(8.9)	537
C_9H_{12}	Propylbenzene	8.713 ± 0.010	848
C_9H_{12}	Isopropylbenzene	8.73 ± 0.01	847
C_9H_{12}	1,2,3-Trimethylbenzene	8.42 ± 0.02	803
C_9H_{12}	1,2,4-Trimethylbenzene	8.27 ± 0.01	784
C_9H_{12}	1,3,5-Trimethylbenzene	8.41 ± 0.01	796
$C_9H_{13}N$	N,N-Dimethyl-o-toluidine	7.40 ± 0.02	814
$C_9H_{14}O$	Isophorone	(≤ 9.07)	≤ 670
C_9H_{18}	Butylcyclopentane	(9.95)	793
C_9H_{18}	Propylcyclohexane	(9.46)	720
C_9H_{18}	Isopropylcyclohexane	(9.33)	704
$C_9H_{18}O$	2-Nonanone	(9.16)	545
$C_9H_{18}O$	5-Nonanone	(9.07)	530
$C_9H_{18}O$	2,6-Dimethyl-4-heptanone	9.01 ± 0.06	512
C_9H_{20}	Nonane	9.71 ± 0.10	709
$C_{10}F_8$	Perfluoronaphthalene	8.85	-368
$C_{10}H_7Br$	1-Bromonaphthalene	8.08 ± 0.03	955
$C_{10}H_7Cl$	1-Chloronaphthalene	(8.13)	906
$C_{10}H_8$	Naphthalene	8.1442 ± 0.0009	936
$C_{10}H_8$	Azulene	7.38 ± 0.05	1001
$C_{10}H_8O$	1-Naphthol	7.76 ± 0.03	719
$C_{10}H_8O$	2-Naphthol	7.87 ± 0.06	729
$C_{10}H_{10}O_4$	Dimethyl phthalate	(9.64 ± 0.07)	277
$C_{10}H_{12}$	1,2,3,4-Tetrahydronaphthalene	8.46 ± 0.02	841
$C_{10}H_{14}$	Butylbenzene	8.69 ± 0.02	826
$C_{10}H_{14}$	sec-Butylbenzene	8.68 ± 0.02	820
$C_{10}H_{14}$	tert-Butylbenzene	8.68 ± 0.05	816
$C_{10}H_{14}$	Isobutylbenzene	8.69 ± 0.02	817
$C_{10}H_{14}$	p-Cymene	(8.29)	771
$C_{10}H_{14}$	o-Diethylbenzene	(≤ 8.51)	≤ 804
$C_{10}H_{14}$	m-Diethylbenzene	(8.49)	798
$C_{10}H_{14}$	p-Diethylbenzene	(8.40)	790
$C_{10}H_{14}$	1,2,4,5-Tetramethylbenzene	8.04 ± 0.02	730
$C_{10}H_{14}O$	p-tert-Butylphenol	(7.8)	552
$C_{10}H_{16}$	α-Pinene	(8.07)	808
$C_{10}H_{16}O$	Camphor	(8.76)	577
$C_{10}H_{18}$	cis-Decahydronaphthalene	9.36 ± 0.04	734
$C_{10}H_{18}$	trans-Decahydronaphthalene	9.34 ± 0.04	720
$C_{10}H_{20}$	1-Decene	9.42 ± 0.05	786
$C_{10}H_{20}$	Butylcyclohexane	(9.41)	695
$C_{10}H_{22}$	Decane	(9.65)	682
$C_{11}H_{10}$	1-Methylnaphthalene	7.97 ± 0.03	882
$C_{11}H_{10}$	2-Methylnaphthalene	7.91 ± 0.08	877
$C_{11}H_{16}$	p-tert-Butyltoluene	(8.12)	730
$C_{11}H_{24}$	Undecane	(9.56)	650
$C_{11}H_{24}$	2-Methyldecane	(9.7)	658
$C_{12}H_8$	Acenaphthylene	(8.22)	1053
$C_{12}H_9N$	Carbazole	(7.57)	961
$C_{12}H_{10}$	Acenaphthene	7.75 ± 0.07	903
$C_{12}H_{10}$	Biphenyl	8.23 ± 0.10	977
$C_{12}H_{10}N_2O$	trans-Azoxybenzene	(8.1)	1123

Mol. Form.	Name	IP/eV	$\Delta_f H_{ion}$ kJ/mol
$C_{12}H_{10}O$	Diphenyl ether	(8.09)	766
$C_{12}H_{11}N$	Diphenylamine	7.16 ± 0.04	908
$C_{12}H_{18}$	5,7-Dodecadiyne	(8.67)	1079
$C_{12}H_{18}$	Hexamethylbenzene	7.85 ± 0.01	670
$C_{12}H_{22}$	Cyclohexylcyclohexane	(9.41)	690
$C_{12}H_{27}N$	Tributylamine	(7.4)	492
$C_{13}H_{10}$	9H-Fluorene	7.91 ± 0.02	952
$C_{13}H_{10}O$	Benzophenone	9.08 ± 0.05	926
$C_{13}H_{12}$	Diphenylmethane	(8.55)	963
$C_{14}H_{10}$	Anthracene	7.439 ± 0.006	948
$C_{14}H_{10}$	Phenanthrene	7.8914 ± 0.0006	966
$C_{14}H_{10}$	Diphenylacetylene	7.94 ± 0.03	1168
$C_{14}H_{12}$	*cis*-Stilbene	(7.80)	1005
$C_{14}H_{12}$	*trans*-Stilbene	7.656 ± 0.001	973
$C_{14}H_{14}$	1,2-Diphenylethane	8.9 ± 0.1	1002
$C_{16}H_{10}$	Fluoranthene	7.9 ± 0.1	1052
$C_{16}H_{10}$	Pyrene	7.4256 ± 0.0006	935
$C_{18}H_{12}$	Chrysene	7.60 ± 0.01	1017
$C_{18}H_{14}$	*o*-Terphenyl	(7.99)	1056
$C_{18}H_{14}$	*m*-Terphenyl	(8.01)	1057
$C_{18}H_{14}$	*p*-Terphenyl	7.80 ± 0.03	1037
$C_{20}H_{12}$	Perylene	6.960 ± 0.001	981
$C_{24}H_{12}$	Coronene	7.29 ± 0.01	1026

X-RAY ATOMIC ENERGY LEVELS

The energy levels in this tables are the values recommended by Bearden and Burr on the basis of a thorough review of the literature on x-ray wavelengths and related data. All values are in electron volts (eV). Values in parentheses are interpolated, and an asterisk * indicates a level which is not resolved from the level above it. See Reference 1 for uncertainties in the levels and a complete description of how the recommended values were obtained.

REFERENCES

1. Bearden, J. A., and Burr, A. F., *Rev. Mod. Phys.*, 39, 125, 1967; also published as *X-Ray Wavelengths and X-Ray Atomic Energy Levels*, Natl. Stand. Ref. Data Sys.- Natl. Bur. Standards (U.S.), No. 14, 1967.
2. Gray, D. E., Editor, *American Institute of Physics Handbook, Third Edition*, pp. 7-158 to 7-167, McGraw-Hill, New York, 1972.

Level	^{1}H	^{2}He	^{3}Li	^{4}Be	^{5}B	^{6}C	^{7}N	^{8}O
K	13.59811	24.58678	54.75	111.0	188.0	283.8	401.6	532.0
L_I								23.7
$L_{II,III}$					4.7	6.4	9.2	7.1

Level	^{9}F	^{10}Ne	^{11}Na	^{12}Mg	^{13}Al	^{14}Si	^{15}P	^{16}S
K	685.4	866.9	1072.1	1305.0	1559.6	1838.9	2145.5	2472.0
L_I	(31)	(45)	63.3	89.4	117.7	148.7	189.3	229.2
$L_{II,III}$	8.6	18.3	31.1	51.4	73.1	99.2	132.2	164.8

Level	^{17}Cl	^{18}Ar	^{19}K	^{20}Ca	^{21}Sc	^{22}Ti	^{23}V	^{24}Cr
K	2822.4	3202.9	3607.4	4038.1	4492.8	4966.4	5465.1	5989.2
L_I	270.2	320	377.1	437.8	500.4	563.7	628.2	694.6
L_{II}	201.6	247.3	296.3	350.0	406.7	461.5	520.5	583.7
L_{III}	200.0	245.2	293.6	346.4	402.2	455.5	512.9	574.5
M_I	17.5	25.3	33.9	43.7	53.8	60.3	66.5	74.1
$M_{II,III}$	6.8	12.4	17.8	25.4	32.3	34.6	37.8	42.5
$M_{IV,V}$					6.6	3.7	2.2	2.3

Level	^{25}Mn	^{26}Fe	^{27}Co	^{28}Ni	^{29}Cu	^{30}Zn	^{31}Ga	^{32}Ge
K	6539.0	7112.0	7708.9	8332.8	8978.9	9658.6	10367.1	11103.1
L_I	769.0	846.1	925.6	1008.1	1096.1	1193.6	1297.7	1414.3
L_{II}	651.4	721.1	793.6	871.9	951.0	1042.8	1142.3	1247.8
L_{III}	640.3	708.1	778.6	854.7	931.1	1019.7	1115.4	1216.7
M_I	83.9	92.9	100.7	111.8	119.8	135.9	158.1	180.0
M_{II}	48.6	54.0	59.5	68.1	73.6	86.6	106.8	127.9
M_{III}	48.6*	54.0*	59.5*	68.1*	73.6*	86.6*	102.9	120.8
$M_{IV,V}$	3.3	3.6	2.9	3.6	1.6	8.1	17.4	28.7

Level	^{33}As	^{34}Se	^{35}Br	^{36}Kr	^{37}Rb	^{38}Sr	^{39}Y	^{40}Zr
K	11866.7	12657.8	13473.7	14325.6	15199.7	16104.6	17038.4	17997.6
L_I	1526.5	1653.9	1782.0	1921.0	2065.1	2216.3	2372.5	2531.6
L_{II}	1358.6	1476.2	1596.0	1727.2	1863.9	2006.8	2155.5	2306.7
L_{III}	1323.1	1435.8	1549.9	1674.9	1804.4	1939.6	2080.0	2222.3
M_I	203.5	231.5	256.5		322.1	357.5	393.6	430.3
M_{II}	146.4	168.2	189.3	222.7	247.4	279.8	312.4	344.2
M_{III}	140.5	161.9	181.5	213.8	238.5	269.1	300.3	330.5
M_{IV}	41.2	56.7	70.1	88.9	111.8	135.0	159.6	182.4
M_V	41.2*	56.7*	69.0	88.9*	110.3	133.1	157.4	180.0
N_I			27.3	24.0	29.3	37.7	45.4	51.3
N_{II}	2.5	5.6	5.2	10.6	14.8	19.9	25.6	28.7
N_{III}	2.5*	5.6*	4.6	10.6*	14.0	19.9*	25.6*	28.7*

Level	^{41}Nb	^{42}Mo	^{43}Tc	^{44}Ru	^{45}Rh	^{46}Pd	^{47}Ag	^{48}Cd
K	18985.6	19999.5	21044.0	22117.2	23219.9	24350.3	25514.0	26711.2
L_I	2697.7	2865.5	3042.5	3224.0	3411.9	3604.3	3805.8	4018.0
L_{II}	2464.7	2625.1	2793.2	2966.9	3146.1	3330.3	3523.7	3727.0
L_{III}	2370.5	2520.2	2676.9	2837.9	3003.8	3173.3	3351.1	3537.5
M_I	468.4	504.6		585.0	627.1	669.9	717.5	770.2
M_{II}	378.4	409.7	444.9	482.8	521.0	559.1	602.4	650.7
M_{III}	363.0	392.3	425.0	460.6	496.2	531.5	571.4	616.5
M_{IV}	207.4	230.3	256.4	283.6	311.7	340.0	372.8	410.5
M_V	204.6	227.0	252.9	279.4	307.0	334.7	366.7	403.7
N_I	58.1	61.8		74.9	81.0	86.4	95.2	107.6
N_{II}	33.9	34.8	38.9	43.1	47.9	51.1	62.6	66.9
N_{III}	33.9*	34.8*	38.9*	43.1*	47.9*	51.1*	55.9	66.9*
$N_{IV,V}$	3.2	1.8		2.0	2.5	1.5	3.3	9.3

Level	^{49}In	^{50}Sn	^{51}Sb	^{52}Te	^{53}I	^{54}Xe	^{55}Cs	^{56}Ba
K	27939.9	29200.1	30491.2	31813.8	33169.4	34561.4	35984.6	37440.6
L_I	4237.5	4464.7	4698.3	4939.2	5188.1	5452.8	5714.3	5988.8
L_{II}	3938.0	4156.1	4380.4	4612.0	4852.1	5103.7	5359.4	5623.6
L_{III}	3730.1	3928.8	4132.2	4341.4	4557.1	4782.2	5011.9	5247.0
M_I	825.6	883.8	943.7	1006.0	1072.1		1217.1	1292.8
M_{II}	702.2	756.4	811.9	869.7	930.5	999.0	1065.0	1136.7
M_{III}	664.3	714.4	765.6	818.7	874.6	937.0	997.6	1062.2
M_{IV}	450.8	493.3	536.9	582.5	631.3		739.5	796.1
M_V	443.1	484.8	527.5	572.1	619.4	672.3	725.5	780.7
N_I	121.9	136.5	152.0	168.3	186.4		230.8	253.0
N_{II}	77.4	88.6	98.4	110.2	122.7	146.7	172.3	191.8
N_{III}	77.4*	88.6*	98.4*	110.2*	122.7*	146.7*	161.6	179.7
N_{IV}	16.2	23.9	31.4	39.8	49.6		78.8	92.5
N_V	16.2*	23.9*	31.4*	39.8*	49.6*		76.5	89.9
O_I	0.1	0.9	6.7	11.6	13.6		22.7	39.1
O_{II}	0.8	1.1	2.1	2.3	3.3		13.1	16.6
O_{III}	0.8*	1.1*	2.1*	2.3*	3.3*		11.4	14.6

Level	^{57}La	^{58}Ce	^{59}Pr	^{60}Nd	^{61}Pm	^{62}Sm	^{63}Eu	^{64}Gd
K	38924.6	40443.0	41990.6	43568.9	45184.0	46834.2	48519.0	50239.1
L_I	6266.3	6548.8	6834.8	7126.0	7427.9	7736.8	8052.0	8375.6
L_{II}	5890.6	6164.2	6440.4	6721.5	7012.8	7311.8	7617.1	7930.3
L_{III}	5482.7	5723.4	5964.3	6207.9	6459.3	6716.2	6976.9	7242.8
M_I	1361.3	1434.6	1511.0	1575.3		1722.8	1800.0	1880.8
M_{II}	1204.4	1272.8	1337.4	1402.8	1471.4	1540.7	1613.9	1688.3
M_{III}	1123.4	1185.4	1242.2	1297.4	1356.9	1419.8	1480.6	1544.0
M_{IV}	848.5	901.3	951.1	999.9	1051.5	1106.0	1160.6	1217.2
M_V	831.7	883.3	931.0	977.7	1026.9	1080.2	1130.9	1185.2
N_I	270.4	289.6	304.5	315.2		345.7	360.2	375.8
N_{II}	205.8	223.3	236.3	243.3	242	265.6	283.9	288.5
N_{III}	191.4	207.2	217.6	224.6	242*	247.4	256.6	270.9
$N_{IV,V}$	98.9	110.0	113.2	117.5	120.4	129.0	133.2	140.5
$N_{VI,VII}$		0.1	2.0	1.5		5.5	0.0	0.1
O_I	32.3	37.8	37.4	37.5		37.4	31.8	36.1
$O_{II,III}$	14.4	19.8	22.3	21.1		21.3	22.0	20.3

Level	^{65}Tb	^{66}Dy	^{67}Ho	^{68}Er	^{69}Tm	^{70}Yb	^{71}Lu	^{72}Hf
K	51995.7	53788.5	55617.7	57485.5	59389.6	61332.3	63313.8	65350.8
L_I	8708.0	9045.8	9394.2	9751.3	10115.7	10486.4	10870.4	11270.7
L_{II}	8251.6	8580.6	8917.8	9264.3	9616.9	9978.2	10348.6	10739.4
L_{III}	7514.0	7790.1	8071.1	8357.9	8648.0	8943.6	9244.1	9560.7
M_I	1967.5	2046.8	2128.3	2206.5	2306.8	2398.1	2491.2	2600.9

Level	^{65}Tb	^{66}Dy	^{67}Ho	^{68}Er	^{69}Tm	^{70}Yb	^{71}Lu	^{72}Hf
M_{II}	1767.7	1841.8	1922.8	2005.8	2089.8	2173.0	2263.5	2365.4
M_{III}	1611.3	1675.6	1741.2	1811.8	1884.5	1949.8	2023.6	2107.6
M_{IV}	1275.0	1332.5	1391.5	1453.3	1514.6	1576.3	1639.4	1716.4
M_V	1241.2	1294.9	1351.4	1409.3	1467.7	1527.8	1588.5	1661.7
N_I	397.9	416.3	435.7	449.1	471.7	487.2	506.2	538.1
N_{II}	310.2	331.8	343.5	366.2	385.9	396.7	410.1	437.0
N_{III}	385.0	292.9	306.6	320.0	336.6	343.5	359.3	380.4
N_{IV}	147.0	154.2	161.0	176.7	179.6	198.1	204.8	223.8
N_V	147.0*	154.2*	161.0*	167.6	179.6*	184.9	195.0	213.7
$N_{VI,VII}$	2.6	4.2	3.7	4.3	5.3	6.3	6.9	17.1
O_I	39.0	62.9	51.2	59.8	53.2	54.1	56.8	64.9
O_{II}	25.4	26.3	20.3	29.4	32.3	23.4	28.0	38.1
O_{III}	25.4*	26.3*	20.3*	29.4*	32.3*	23.4*	28.0*	30.6

Level	^{73}Ta	^{74}W	^{75}Re	^{76}Os	^{77}Ir	^{78}Pt	^{79}Au	^{80}Hg
K	67416.4	69525.0	71676.4	73870.8	76111.0	78394.8	80724.9	83102.3
L_I	11681.5	12099.8	12526.7	12968.0	13418.5	13879.9	14352.8	14839.3
L_{II}	11136.1	11544.0	11958.7	12385.0	12824.1	13272.6	13733.6	14208.7
L_{III}	9881.1	10206.8	10535.3	10870.9	11215.2	11563.7	11918.7	12283.9
M_I	2708.0	2819.6	2931.7	3048.5	3173.7	3296.0	3424.9	3561.6
M_{II}	2468.7	2574.9	2681.6	2792.2	2908.7	3026.5	3147.8	3278.5
M_{III}	2194.0	2281.0	2367.3	2457.2	2550.7	2645.4	2743.0	2847.1
M_{IV}	1793.2	1871.6	1948.9	2030.8	2116.1	2201.9	2291.1	2384.9
M_V	1735.1	1809.2	1882.9	1960.1	2040.4	2121.6	2205.7	2294.9
N_I	565.5	595.0	625.0	654.3	690.1	722.0	758.8	800.3
N_{II}	464.8	491.6	517.9	546.5	577.1	609.2	643.7	676.9
N_{III}	404.5	425.3	444.4	468.2	494.3	519.0	545.4	571.0
N_{IV}	241.3	258.8	273.7	289.4	311.4	330.8	352.0	378.3
N_V	229.3	245.4	260.2	272.8	294.9	313.3	333.9	359.8
N_{VI}	25.0	36.5	40.6	46.3	63.4	74.3	86.4	102.2
N_{VII}	25.0*	33.6	40.6*	46.3*	60.5	71.1	82.8	98.5
O_I	71.1	77.1	82.8	83.7	95.2	101.7	107.8	120.3
O_{II}	44.9	46.8	45.6	58.0	63.0	65.3	71.7	80.5
O_{III}	36.4	35.6	34.6	45.4	50.5	51.7	53.7	57.6
$O_{IV,V}$	5.7	6.1	3.5		3.8	2.2	2.5	6.4

Level	^{81}Tl	^{82}Pb	^{83}Bi	^{84}Po	^{85}At	^{86}Rn	^{87}Fr	^{88}Ra
K	85530.4	88004.5	90525.9	93105.0	95729.9	98404	101137	103921.9
L_I	15346.7	15860.8	16387.5	16939.3	17493	18049	18639	19236.7
L_{II}	14697.9	15200.0	15711.1	16244.3	16784.7	17337.1	17906.5	18484.3
L_{III}	12657.5	13035.2	13418.6	13813.8	14213.5	14619.4	15031.2	15444.4
M_I	3704.1	3850.7	3999.1	4149.4	(4317)	(4482)	(4652)	4822.0
M_{II}	3415.7	3554.2	3696.3	3854.1	4008	4159	4327	4489.5
M_{III}	2956.6	3066.4	3176.9	3301.9	3426	3538	3663	3791.8
M_{IV}	2485.1	2585.6	2687.6	2798.0	2908.7	3021.5	3136.2	3248.4
M_V	2389.3	2484.0	2579.6	2683.0	2786.7	2892.4	2999.9	3104.9
N_I	845.5	893.6	938.2	995.3	(1042)	(1097)	(1153)	1208.4
N_{II}	721.3	763.9	805.3	851	886	929	980	1057.6
N_{III}	609.0	644.5	678.9	705	740	768	810	879.1
N_{IV}	406.6	435.2	463.6	500.2	533.2	566.6	603.3	635.9
N_V	386.2	412.9	440.0	473.4			577	602.7
N_{VI}	122.8	142.9	161.9					298.9
N_{VII}	118.5	138.1	157.4					298.9*
O_I	136.3	147.3	159.3					254.4
O_{II}	99.6	104.8	116.8					200.4
O_{III}	75.4	86.0	92.8					152.8

X-RAY ATOMIC ENERGY LEVELS (continued)

Level	81Tl	82Pb	83Bi	84Po	85At	86Rn	87Fr	88Ra
O_{IV}	15.3	21.8	26.5	31.4				67.2
O_V	13.1	19.2	24.4	31.4*				67.2*
P_I		3.1						43.5
$P_{II,III}$		0.7	2.7					18.8

Level	89Ac	90Th	91Pa	92U	93Np	94Pu	95Am	96Cm
K	106755.3	109650.9	112601.4	115606.1	118678	121818	125027	128220
L_I	19840	20472.1	21104.6	21757.4	22426.8	23097.2	23772.9	24460
L_{II}	19083.2	19693.2	20313.7	20947.6	21600.5	22266.2	22944.0	23779
L_{III}	15871.0	16300.3	16733.1	17166.3	17610.0	18056.8	18504.1	18930
M_I	(5002)	5182.3	5366.9	5548.0	5723.2	5932.9	6120.5	6288
M_{II}	4656	4830.4	5000.9	5182.2	5366.2	5541.2	5710.2	5895
M_{III}	3909	4046.1	4173.8	4303.4	4434.7	4556.6	4667.0	4797
M_{IV}	3370.2	3490.8	3611.2	3727.6	3850.3	3972.6	4092.1	4227
M_V	3219.0	3332.0	3441.8	3551.7	3665.8	3778.1	3886.9	3971
N_I	(1269)	1329.5	1387.1	1440.8	1500.7	1558.6	1617.1	1643
N_{II}	1080	1168.2	1224.3	1272.6	1327.7	1372.1	1411.8	1440
N_{III}	890	967.3	1006.7	1044.9	1086.8	1114.8	(1135.7)	1154
N_{IV}	674.9	714.1	743.4	780.4	815.9	848.9	878.7	
N_V		676.4	708.2	737.7	770.3	801.4	827.6	
N_{VI}		344.4	371.2	391.3	415.0	445.8		
N_{VII}		335.2	359.5	380.9	404.4	432.4		
O_I		290.2	309.6	323.7		351.9		385
O_{II}		229.4	222.9	259.3	283.4	274.1		
O_{III}		181.8	222.9*	195.1	206.1	206.5		
O_{IV}		94.3	94.1	105.0	109.3	116.0	115.8	
O_V		87.9	94.1*	96.3	101.3	105.4	103.3	
P_I		59.5		70.7				
P_{II}		49.0		42.3				
P_{III}		43.0		32.3				

Level	97Bk	98Cf	99Es	100Fm	101Md	102No	103Lr
K	131590	135960	139490	143090	146780	150540	154380
L_I	25275	26110	26900	27700	28530	29380	30240
L_{II}	24385	25250	26020	26810	27610	28440	29280
L_{III}	19452	19930	20410	20900	21390	21880	22360
M_I	6556	6754	6977	7205	7441	7675	7900
M_{II}	6147	6359	6574	6793	7019	7245	7460
M_{III}	4977	5109	5252	5397	5546	5688	5710
M_{IV}	4366	4497	4630	4766	4903	5037	5150
M_V	4132	4253	4374	4498	4622	4741	4860
N_I	1755	1799	1868	1937	2010	2078	2140
N_{II}	1554	1616	1680	1747	1814	1876	1930
N_{III}	1235	1279	1321	1366	1410	1448	1480
O_I	398	419	435	454	472	484	490

ELECTRON BINDING ENERGIES OF THE ELEMENTS

Gwyn P. Williams

This table gives the binding energies in electron volts (eV) for selected electronic levels of the elements. For metallic elements the binding energy is referred to the Fermi level; for semiconductors, to the valence band maximum; and for gases and insulators, to the vacuum level. The atomic number is listed after the element name.

REFERENCES

1. Fluggle and Martensson, J. *Elect. Spect.*, 21, 275, 1980.
2. Cardona, M. and Ley, L., *Photoemission from Solids*, Springer Verlag, Heidelberg, 1978.
3. Bearden, J. A. and Burr, A. F., *Rev. Mod. Phys.*, 39, 125, 1967.

Actinium (89)

K	1s	106755
L I	2s	19840
L II	$2p_{1/2}$	19083
L III	$2p_{3/2}$	15871
M I	3s	5002
M II	$3p_{1/2}$	4656
M III	$3p_{3/2}$	3909
M IV	$3d_{3/2}$	3370
M V	$3d_{5/2}$	3219
N I	4s	1269[a]
N II	$4p_{1/2}$	1080[a]
N III	$4p_{3/2}$	890[a]
N IV	$4d_{3/2}$	675[a]
N V	$4d_{5/2}$	639[a]
N VI	$4f_{5/2}$	319[a]
N VII	$4f_{7/2}$	319[a]
O I	5s	272[a]
O II	$5p_{1/2}$	215[a]
O III	$5p_{3/2}$	167[a]
O IV	$5d_{3/2}$	80[a]
O V	$5d_{5/2}$	80[a]
P I	6s	—
P II	$6p_{1/2}$	—
P III	$6p_{3/2}$	—

Aluminum (13)

K	1s	1559.0
L I	2s	117.8[a]
L II	$2p_{1/2}$	72.9[a]
L III	$2p_{3/2}$	72.5[a]

Antimony (51)

K	1s	30419
L I	2s	4698
L II	$2p_{1/2}$	4380
L III	$2p_{3/2}$	4132
M I	3s	946[b]
M II	$3p_{1/2}$	812.7[b]
M III	$3p_{3/2}$	766.4[b]
M IV	$3d_{3/2}$	537.5[b]
M V	$3d_{5/2}$	528.2[b]
N I	4s	153.2[b]
N II	$4p_{1/2}$	95.6[b,c]
N III	$4p_{3/2}$	95.6[b]
N IV	$4d_{3/2}$	33.3[b]
N V	$4d_{5/2}$	32.1[b]

Argon (18)

K	1s	3205.9[a]
L I	2s	326.3[a]
L II	$2p_{1/2}$	250.6[a]
L III	$2p_{3/2}$	248.4[a]
M I	3s	29.3[a]
M II	$3p_{1/2}$	15.9[a]
M III	$3p_{3/2}$	15.7[a]

Arsenic (33)

K	1s	11867
L I	2s	1527.0[a,d]
L II	$2p_{1/2}$	1359.1[a,d]
L III	$2p_{3/2}$	1323.6[a,d]
M I	3s	204.7[a]
M II	$3p_{1/2}$	146.2[a]
M III	$3p_{3/2}$	141.2[a]
M IV	$3d_{3/2}$	41.7[a]
M V	$3d_{5/2}$	41.7[a]

Astatine (85)

K	1s	95730
L I	2s	17493
L II	$2p_{1/2}$	16785
L III	$2p_{3/2}$	14214
M I	3s	4317
M II	$3p_{1/2}$	4008
M III	$3p_{3/2}$	3426
M IV	$3d_{3/2}$	2909
M V	$3d_{5/2}$	2787
N I	4s	1042[a]
N II	$4p_{1/2}$	886[a]
N III	$4p_{3/2}$	740[a]
N IV	$4d_{3/2}$	533[a]
N V	$4d_{5/2}$	507[a]
N VI	$4f_{5/2}$	210[a]
N VII	$4f_{7/2}$	210[a]
O I	5s	195[a]
O II	$5p_{1/2}$	148[a]
O III	$5p_{3/2}$	115[a]
O IV	$5d_{3/2}$	40[a]
O V	$5d_{5/2}$	40[a]

Barium (56)

K	1s	37441
L I	2s	5989
L II	$2p_{1/2}$	5624

L III	$2p_{3/2}$	5247
M I	3s	1293[a,d]
M II	$3p_{1/2}$	1137[a,d]
M III	$3p_{3/2}$	1063[a,d]
M IV	$3d_{3/2}$	795.7[a]
M V	$3d_{5/2}$	780.5[a]
N I	4s	253.5[b]
N II	$4p_{1/2}$	192
N III	$4p_{3/2}$	178.6[b]
N IV	$4d_{3/2}$	92.6[b]
N V	$4d_{5/2}$	89.9[b]
N VI	$4f_{5/2}$	—
N VII	$4f_{7/2}$	—
O I	5s	30.3[b]
O II	$5p_{1/2}$	17.0[b]
O III	$5p_{3/2}$	14.8[b]

Beryllium (4)

K	1s	111.5[a]

Bismuth (83)

K	1s	90526
L I	2s	16388
L II	$2p_{1/2}$	15711
L III	$2p_{3/2}$	13419
M I	3s	3999
M II	$3p_{1/2}$	3696
M III	$3p_{3/2}$	3177
M IV	$3d_{3/2}$	2688
M V	$3d_{5/2}$	2580
N I	4s	939[b]
N II	$4p_{1/2}$	805.2[b]
N III	$4p_{3/2}$	678.8[b]
N IV	$4d_{3/2}$	464.0[b]
N V	$4d_{5/2}$	440.1[b]
N VI	$4f_{5/2}$	162.3[b]
N VII	$4f_{7/2}$	157.0[b]
O I	5s	159.3[a,d]
O II	$5p_{1/2}$	119.0[b]
O III	$5p_{3/2}$	92.6[b]
O IV	$5d_{3/2}$	26.9[b]
O V	$5d_{5/2}$	23.8[b]

Boron (5)

K	1s	188[a]

Bromine (35)

K	1s	13474

Cadmium (48)

K	1s	26711
L I	2s	4018
L II	$2p_{1/2}$	3727
L III	$2p_{3/2}$	3538
M I	3s	772.0[b]
M II	$3p_{1/2}$	652.6[b]
M III	$3p_{3/2}$	618.4[b]
M IV	$3d_{3/2}$	411.9[b]
M V	$3d_{5/2}$	405.2[b]
N I	4s	109.8[b]
N II	$4p_{1/2}$	63.9[b,c]
N III	$4p_{3/2}$	63.9[b,c]
N IV	$4d_{3/2}$	11.7[b]
N V	$4d_{5/2}$	10.7[b]

Calcium (20)

K	1s	4038.5[a]
L I	2s	438.4[b]
L II	$2p_{1/2}$	349.7[b]
L III	$2p_{3/2}$	346.2[b]
M I	3s	44.3[b]
M II	$3p_{1/2}$	25.4[b]
M III	$3p_{3/2}$	25.4[b]

Carbon (6)

K	1s	284.2[a]

Cerium (58)

K	1s	40443
L I	2s	6548
L II	$2p_{1/2}$	6164
L III	$2p_{3/2}$	5723
M I	3s	1436[a,d]
M II	$3p_{1/2}$	1274[a,d]
M III	$3p_{3/2}$	1187[a,d]
M IV	$3d_{3/2}$	902.4[a]
M V	$3d_{5/2}$	883.8[a]

N I	4s	291.0[a]
N II	4p$_{1/2}$	223.3
N III	4p$_{3/2}$	206.5[a]
N IV	4d$_{3/2}$	109[a]
N V	4d$_{5/2}$	—
N VI	4f$_{5/2}$	0.1
N VII	4f$_{7/2}$	0.1
O I	5s	37.8
O II	5p$_{1/2}$	19.8[a]
O III	5p$_{3/2}$	17.0[a]

Cesium (55)

K	1s	35985
L I	2s	5714
L II	2p$_{1/2}$	5359
L III	2p$_{3/2}$	5012
M I	3s	1211[a,d]
M II	3p$_{1/2}$	1071[a]
M III	3p$_{3/2}$	1003[a]
M IV	3d$_{3/2}$	740.5[a]
M V	3d$_{5/2}$	726.6[a]
N I	4s	232.3[a]
N II	4p$_{1/2}$	172.4[a]
N III	4p$_{3/2}$	161.3[a]
N IV	4d$_{3/2}$	79.8[a]
N V	4d$_{5/2}$	77.5[a]
N VI	4f$_{5/2}$	—
N VII	4f$_{7/2}$	—
O I	5s	22.7
O II	5p$_{1/2}$	14.2[a]
O III	5p$_{3/2}$	12.1[a]

Chlorine (17)

K	1s	2822.0
L I	2s	270[a]
L II	2p$_{1/2}$	202[a]
L III	2p$_{3/2}$	200[a]

Chromium(24)

K	1s	5989
L I	2s	696.0[b]
L II	2p$_{1/2}$	583.8[b]
L III	2p$_{3/2}$	574.1[b]
M I	3s	74.1[b]
M II	3p$_{1/2}$	42.2[b]
M III	3p$_{3/2}$	42.2[b]

Cobalt (27)

K	1s	7709
L I	2s	925.1[b]
L II	2p$_{1/2}$	793.2[b]
L III	2p$_{3/2}$	778.1[b]
M I	3s	101.0[b]
M II	3p$_{1/2}$	58.9[b]
M III	3p$_{3/2}$	58.9[b]

Copper (29)

K	1s	8979
L I	2s	1096.7[b]

L II	2p$_{1/2}$	952.3[b]
L III	2p$_{3/2}$	932.5[b]
M I	3s	122.5[b]
M II	3p$_{1/2}$	77.3[b]
M III	3p$_{3/2}$	75.1[b]

Dysprosium (66)

K	1s	53789
L I	2s	9046
L II	2p$_{1/2}$	8581
L III	2p$_{3/2}$	7790
M I	3s	2047
M II	3p$_{1/2}$	1842
M III	3p$_{3/2}$	1676
M IV	3d$_{3/2}$	1333
M V	3d$_{5/2}$	1292[a]
N I	4s	414.2[a]
N II	4p$_{1/2}$	333.5[a]
N III	4p$_{3/2}$	293.2[a]
N IV	4d$_{3/2}$	153.6[a]
N V	4d$_{5/2}$	153.6[a]
N VI	4f$_{5/2}$	8.0[a]
N VII	4f$_{7/2}$	4.3[a]
O I	5s	49.9[a]
O II	5p$_{1/2}$	26.3
O III	5p$_{3/2}$	26.3

Erbium (68)

K	1s	57486
L I	2s	9751
L II	2p$_{1/2}$	9264
L III	2p$_{3/2}$	8358
M I	3s	2206
M II	3p$_{1/2}$	2006
M III	3p$_{3/2}$	1812
M IV	3d$_{3/2}$	1453
M V	3d$_{5/2}$	1409
N I	4s	449.8[a]
N II	4p$_{1/2}$	366.2
N III	4p$_{3/2}$	320.2[a]
N IV	4d$_{3/2}$	167.6[a]
N V	4d$_{5/2}$	167.6[a]
N VI	4f$_{5/2}$	—
N VII	4f$_{7/2}$	4.7[a]
O I	5s	50.6[a]
O II	5p$_{1/2}$	31.4[a]
O III	5p$_{3/2}$	24.7[a]

Europium (63)

K	1s	48519
L I	2s	8052
L II	2p$_{1/2}$	7617
L III	2p$_{3/2}$	6977
M I	3s	1800
M II	3p$_{1/2}$	1614
M III	3p$_{3/2}$	1481
M IV	3d$_{3/2}$	1158.6[a]
M V	3d$_{5/2}$	1127.5[a]
N I	4s	360

N II	4p$_{1/2}$	284
N III	4p$_{3/2}$	257
N IV	4d$_{3/2}$	133
N V	4d$_{5/2}$	1227[a]
N VI	4f$_{5/2}$	0
N VII	4f$_{7/2}$	0
O I	5s	32
O II	5p$_{1/2}$	22
O III	5p$_{3/2}$	22

Fluorine (9)

K	1s	696.7[a]

Francium (87)

K	1s	101137
L I	2s	18639
L II	2p$_{1/2}$	17907
L III	2p$_{3/2}$	15031
M I	3s	4652
M II	3p$_{1/2}$	4327
M III	3p$_{3/2}$	3663
M IV	3d$_{3/2}$	3136
M V	3d$_{5/2}$	3000
N I	4s	1153[a]
N II	4p$_{1/2}$	980[a]
N III	4p$_{3/2}$	810[a]
N IV	4d$_{3/2}$	603[a]
N V	4d$_{5/2}$	577[a]
N VI	4f$_{5/2}$	268[a]
N VII	4f$_{7/2}$	268[a]
O I	5s	234[a]
O II	5p$_{1/2}$	182[a]
O III	5p$_{3/2}$	140[a]
O IV	5d$_{3/2}$	58[a]
O V	5d$_{5/2}$	58[a]
P I	6s	34
P II	6p$_{1/2}$	15
P III	6p$_{3/2}$	15

Gadolinium (64)

K	1s	50239
L I	2s	8376
L II	2p$_{1/2}$	7930
L III	2p$_{3/2}$	7243
M I	3s	1881
M II	3p$_{1/2}$	1688
M III	3p$_{3/2}$	1544
M IV	3d$_{3/2}$	1221.9[a]
M V	3d$_{5/2}$	1189.6[a]
N I	4s	378.6[a]
N II	4p$_{1/2}$	286
N III	4p$_{3/2}$	271
N IV	4d$_{3/2}$	—
N V	4d$_{5/2}$	142.6[a]
N VI	4f$_{5/2}$	8.6[a]
N VII	4f$_{7/2}$	8.6[a]
O I	5s	36
O II	5p$_{1/2}$	20
O III	5p$_{3/2}$	20

Gallium (31)

K	1s	10367
L I	2s	1299.0[a,d]
L II	2p$_{1/2}$	1143.2[b]
L III	2p$_{3/2}$	1116.4[b]
M I	3s	159.5[b]
M II	3p$_{1/2}$	103.5[b]
M III	3p$_{3/2}$	100.0[b]
M IV	3d$_{3/2}$	18.7[b]
M V	3d$_{5/2}$	18.7[b]

Germanium (32)

K	1s	11103
L I	2s	1414.6[a,d]
L II	2p$_{1/2}$	1248.1[a,d]
L III	2p$_{3/2}$	1217.0[a,d]
M I	3s	180.1[a]
M II	3p$_{1/2}$	124.9[a]
M III	3p$_{3/2}$	120.8[a]
M IV	3d$_{3/2}$	29.8[a]
M V	3d$_{5/2}$	29.2[a]

Gold (79)

K	1s	80725
L I	2s	14353
L II	2p$_{1/2}$	13734
L III	2p$_{3/2}$	11919
M I	3s	3425
M II	3p$_{1/2}$	3148
M III	3p$_{3/2}$	2743
M IV	3d$_{3/2}$	2291
M V	3d$_{5/2}$	2206
N I	4s	762.1[b]
N II	4p$_{1/2}$	642.7[b]
N III	4p$_{3/2}$	546.3[b]
N IV	4d$_{3/2}$	353.2[b]
N V	4d$_{5/2}$	335.1[b]
N VI	4f$_{5/2}$	87.6[b]
N VII	4f$_{7/2}$	83.9[b]
O I	5s	107.2[a,d]
O II	5p$_{1/2}$	74.2[b]
O III	5p$_{3/2}$	57.2[b]

Hafnium (72)

K	1s	65351
L I	2s	11271
L II	2p$_{1/2}$	10739
L III	2p$_{3/2}$	9561
M I	3s	2601
M II	3p$_{1/2}$	2365
M III	3p$_{3/2}$	2107
M IV	3d$_{3/2}$	1176
M V	3d$_{5/2}$	1662
N I	4s	538[a]
N II	4p$_{1/2}$	438.2[b]
N III	4p$_{3/2}$	380.7[b]
N IV	4d$_{3/2}$	220.0[b]
N V	4d$_{5/2}$	211.5[b]
N VI	4f$_{5/2}$	15.9[b]

N VII	$4f_{7/2}$	14.2[b]
O I	5s	64.2[b]
O II	$5p_{1/2}$	38[a]
O III	$5p_{3/2}$	29.9[b]

Helium (2)

K	1s	24.6[a]

Holmium (67)

K	1s	55618
L I	2s	9394
L II	$2p_{1/2}$	8918
L III	$2p_{3/2}$	8071
M I	3s	2128
M II	$3p_{1/2}$	1923
M III	$3p_{3/2}$	1741
M IV	$3d_{3/2}$	1392
M V	$3d_{5/2}$	1351
N I	4s	432.4[a]
N II	$4p_{1/2}$	343.5
N III	$4p_{3/2}$	308.2[a]
N IV	$4d_{3/2}$	160[a]
N V	$4d_{5/2}$	160[a]
N VI	$4f_{5/2}$	8.6[a]
N VII	$4f_{7/2}$	5.2[a]
O I	5s	49.3[a]
O II	$5p_{1/2}$	30.8[a]
O III	$5p_{3/2}$	24.1[a]

Hydrogen (1)

K	1s	13.6

Indium (49)

K	1s	27940
L I	2s	4238
L II	$2p_{1/2}$	3938
L III	$2p_{3/2}$	3730
M I	3s	827.2[b]
M II	$3p_{1/2}$	703.2[b]
M III	$3p_{3/2}$	665.3[b]
M IV	$3d_{3/2}$	451.4[b]
M V	$3d_{5/2}$	443.9[b]
N I	4s	122.9[b]
N II	$4p_{1/2}$	73.5[b,c]
N III	$4p_{3/2}$	73.5[b,c]
N IV	$4d_{3/2}$	17.7[b]
N V	$4d_{5/2}$	16.9[b]

Iodine (53)

K	1s	33169
L I	2s	5188
L II	$2p_{1/2}$	4852
L III	$2p_{3/2}$	4557
M I	3s	1072[a]
M II	$3p_{1/2}$	931[a]
M III	$3p_{3/2}$	875[a]
M IV	$3d_{3/2}$	631[a]
M V	$3d_{5/2}$	620[a]
N I	4s	186[a]
N II	$4p_{1/2}$	123[a]

N III	$4p_{3/2}$	123[a]
N IV	$4d_{3/2}$	50[a]
N V	$4d_{5/2}$	50[a]

Iridium (77)

K	1s	76111
L I	2s	13419
L II	$2p_{1/2}$	12824
L III	$2p_{3/2}$	11215
M I	3s	3174
M II	$3p_{1/2}$	2909
M III	$3p_{3/2}$	2551
M IV	$3d_{3/2}$	2116
M V	$3d_{5/2}$	2040
N I	4s	691.1[b]
N II	$4p_{1/2}$	577.8[b]
N III	$4p_{3/2}$	495.8[b]
N IV	$4d_{3/2}$	311.9[b]
N V	$4d_{5/2}$	296.3[b]
N VI	$4f_{5/2}$	63.8[b]
N VII	$4f_{7/2}$	60.8[b]
O I	52	95.2[a,d]
O II	$5p_{1/2}$	63.0[a,d]
O III	$5p_{3/2}$	48.0[b]

Iron (26)

K	1s	7112
L I	2s	844.6[b]
L II	$2p_{1/2}$	719.9[b]
L III	$2p_{3/2}$	706.8[b]
M I	3s	91.3[b]
M II	$3p_{1/2}$	52.7[b]
M III	$3p_{3/2}$	52.7[b]

Krypton (36)

K	1s	14326
L I	2s	1921
L II	$2p_{1/2}$	1730.9[a]
L III	$2p_{3/2}$	1678.4[a]
M I	3s	292.8[a]
M II	$3p_{1/2}$	222.2[a]
M III	$3p_{3/2}$	214.4[a]
M IV	$3d_{3/2}$	95.0[a]
M V	$3d_{5/2}$	93.8[a]
N I	4s	27.5[a]
N II	$4p_{1/2}$	14.1[a]
N III	$4p_{3/2}$	14.1[a]

Lanthanum (57)

K	1s	38925
L I	2s	6266
L II	$2p_{1/2}$	5891
L III	$2p_{3/2}$	5483
M I	3s	1362[a,d]
M II	$3p_{1/2}$	1209[a,d]
M III	$3p_{3/2}$	1128[a,d]
M IV	$3d_{3/2}$	853[a]
M V	$3d_{5/2}$	836[a]
N I	4s	247.7[a]
N II	$4p_{1/2}$	205.8

N III	$4p_{3/2}$	196.0[a]
N IV	$4d_{3/2}$	105.3[a]
N V	$4d_{5/2}$	102.5[a]
N VI	$4f_{5/2}$	—
N VII	$4f_{7/2}$	—
O I	5s	34.3[a]
O II	$5p_{1/2}$	19.3[a]
O III	$5p_{3/2}$	16.8[a]

Lead (82)

K	1s	88005
L I	2s	15861
L II	$2p_{1/2}$	15200
L III	$2p_{3/2}$	13055
M I	3s	3851
M II	$3p_{1/2}$	3554
M III	$3p_{3/2}$	3066
M IV	$3d_{3/2}$	2586
M V	$3d_{5/2}$	2484
N I	4s	891.8[b]
N II	$4p_{1/2}$	761.9[b]
N III	$4p_{3/2}$	643.5[b]
N IV	$4d_{3/2}$	434.3[b]
N V	$4d_{5/2}$	412.2[b]
N VI	$4f_{5/2}$	141.7[b]
N VII	$4f_{7/2}$	136.9[b]
O I	5s	147[a,d]
O II	$5p_{1/2}$	106.4[b]
O III	$5p_{3/2}$	83.3[b]
O IV	$5d_{3/2}$	20.7[b]
O V	$5d_{5/2}$	18.1[b]

Lithium (3)

K	1s	54.7[a]

Lutetium

K	1s	63314
L I	2s	10870
L II	$2p_{1/2}$	10349
L III	$2p_{3/2}$	9244
M I	3s	2491
M II	$3p_{1/2}$	2264
M III	$3p_{3/2}$	2024
M IV	$3d_{3/2}$	1639
M V	$3d_{5/2}$	1589
N I	4s	506.8[a]
N II	$4p_{1/2}$	412.4[a]
N III	$4p_{3/2}$	359.2[a]
N IV	$4d_{3/2}$	206.1[a]
N V	$4d_{5/2}$	196.3[a]
N VI	$4f_{5/2}$	8.9[a]
N VII	$4f_{7/2}$	7.5[a]
O I	5s	57.3[a]
O II	$5p_{1/2}$	33.6[a]
O III	$5p_{3/2}$	26.7[a]

Magnesium (12)

K	1s	1303.0[b]
L I	2s	88.6[a]
L II	$2p_{1/2}$	49.6[b]
L III	$2p_{3/2}$	49.2[a]

Manganese (25)

K	1s	6539
L I	2s	769.1[b]
L II	$2p_{1/2}$	649.9[b]
L III	$2p_{3/2}$	638.7[b]
M I	3s	82.3[b]
M II	$3p_{1/2}$	47.2[b]
M III	$3p_{3/2}$	47.2[b]

Mercury (80)

K	1s	83102
L I	2s	14839
L II	$2p_{1/2}$	14209
L III	$2p_{3/2}$	12284
M I	3s	3562
M II	$3p_{1/2}$	3279
M III	$3p_{3/2}$	2847
M IV	$3d_{3/2}$	2385
M V	$3d_{5/2}$	2295
N I	4s	802.2[b]
N II	$4p_{1/2}$	680.2[b]
N III	$4p_{3/2}$	576.6[b]
N IV	$4d_{3/2}$	378.2[b]
N V	$4d_{5/2}$	358.8[b]
N VI	$4f_{5/2}$	104.0[b]
N VII	$4f_{7/2}$	99.9[b]
O I	5s	127[b]
O II	$5p_{3/2}$	83.1[b]
O III	$5p_{3/2}$	64.5[b]
O IV	$5d_{3/2}$	9.6[b]
O V	$5d_{5/2}$	7.8[b]

Molybdenum (42)

K	1s	20000
L I	2s	2866
L II	$2p_{1/2}$	2625
L III	$2p_{3/2}$	2520
M I	3s	506.3[b]
M II	$3p_{1/2}$	411.6[b]
M III	$3p_{3/2}$	394.0[b]
M IV	$3d_{3/2}$	231.1[b]
M V	$3d_{5/2}$	227.9[b]
N I	4s	63.2[b]
N II	$4p_{1/2}$	37.6[b]
N III	$4p_{3/2}$	35.5[b]

Neodymium (60)

K	1s	43569
L I	2s	7126
L II	$2p_{1/2}$	6722
L III	$2p_{3/2}$	6208
M I	3s	1575
M II	$3p_{1/2}$	1403
M III	$3p_{3/2}$	1297
M IV	$3d_{3/2}$	1003.3[a]
M V	$3d_{5/2}$	980.4[a]
N I	4s	319.2[a]
N II	$4p_{1/2}$	243.3
N III	$4p_{3/2}$	224.6
N IV	$4d_{3/2}$	120.5[a]

N V	4d$_{5/2}$	120.5[a]
N VI	4f$_{5/2}$	1.5
N VII	4f$_{7/2}$	1.5
O I	5s	37.5
O II	5p$_{1/2}$	21.1
O III	5p$_{3/2}$	21.1

Neon (10)

K	1s	870.2[a]
L I	2s	48.5[a]
L II	2p$_{1/2}$	21.7[a]
L III	2p$_{3/2}$	21.6[a]

Nickel (28)

K	1s	8333
L I	2s	1008.6[b]
L II	2p$_{1/2}$	870.0[b]
L III	2p$_{3/2}$	852.7[b]
M I	3s	110.8[b]
M II	3p$_{1/2}$	68.0[b]
M III	3p$_{3/2}$	66.2[b]

Niobium (41)

K	1s	18986
L I	2s	2698
L II	2p$_{1/2}$	2465
L III	2p$_{3/2}$	2371
M I	3s	466.6[b]
M II	3p$_{1/2}$	376.1[b]
M III	3p$_{3/2}$	360.6[b]
M IV	3d$_{3/2}$	205.0[b]
M V	3d$_{5/2}$	202.3[b]
N I	4s	56.4[b]
N II	4p$_{1/2}$	32.6[b]
N III	4p$_{3/2}$	30.8[b]

Nitrogen (7)

K	1s	409.9[a]
L I	2s	37.3[a]

Osmium (76)

K	1s	73871
L I	2s	12968
L II	2p$_{1/2}$	12385
L III	2p$_{3/2}$	10871
M I	3s	3049
M II	3p$_{1/2}$	2792
M III	3p$_{3/2}$	2457
M IV	3d$_{3/2}$	2031
M V	3d$_{5/2}$	1960
N I	4s	658.2[b]
N II	4p$_{1/2}$	549.1[b]
N III	4p$_{3/2}$	470.7[b]
N IV	4d$_{3/2}$	293.1[b]
N V	4d$_{5/2}$	278.5[b]
N VI	4f$_{5/2}$	53.4[b]
N VII	4f$_{7/2}$	50.7[b]
O I	5s	84[a]
O II	5p$_{1/2}$	58[a]

O III	5p$_{3/2}$	44.5[b]

Oxygen (8)

K	1s	543.1[a]
L I	2s	41.6[a]

Palladium (46)

K	1s	24350
L I	2s	3604
L II	2p$_{1/2}$	3330
L III	2p$_{3/2}$	3173
M I	3s	671.6[b]
M II	3p$_{1/2}$	559.9[b]
M III	3p$_{3/2}$	532.3[b]
M IV	3d$_{3/2}$	340.5[b]
M V	3d$_{5/2}$	335.2[b]
N I	4s	87.1[a,d]
N II	4p$_{1/2}$	55.7[b,c]
N III	4p$_{3/2}$	50.9[b,c]

Phosphorus (15)

K	1s	2145.5
L I	2s	189[a]
L II	2p$_{1/2}$	136[a]
L III	2p$_{3/2}$	135[a]

Platinum (78)

K	1s	78395
L I	2s	13880
L II	2p$_{1/2}$	13273
L III	2p$_{3/2}$	11564
M I	3s	3296
M II	3p$_{1/2}$	3027
M III	3p$_{3/2}$	2645
M IV	3d$_{3/2}$	2202
M V	3d$_{5/2}$	2122
N I	4s	725.4[b]
N II	4p$_{1/2}$	609.1[b]
N III	4p$_{3/2}$	519.4[b]
N IV	4d$_{3/2}$	331.6[b]
N V	4d$_{5/2}$	314.6[b]
N VI	4f$_{5/2}$	74.5[b]
N VII	4f$_{7/2}$	71.2[b]
O I	5s	101.7[a,d]
O II	5p$_{1/2}$	65.3[a,b]
O III	5p$_{3/2}$	51.7[b]

Polonium (84)

K	1s	93105
L I	2s	16939
L II	2p$_{1/2}$	16244
L III	2p$_{3/2}$	13814
M I	3s	4149
M II	3p$_{1/2}$	3854
M III	3p$_{3/2}$	3302
M IV	3d$_{3/2}$	2798
M V	3d$_{5/2}$	2683
N I	4s	995[a]

N II	4p$_{1/2}$	851[a]
N III	4p$_{3/2}$	705[a]
N IV	4d$_{3/2}$	500[a]
N V	4d$_{5/2}$	473[a]
N VI	4f$_{5/2}$	184[a]
N VII	4f$_{7/2}$	184[a]
O I	5s	177[a]
O II	5p$_{1/2}$	132[a]
O III	5p$_{3/2}$	104[a]
O IV	5d$_{3/2}$	31[a]
O V	5d$_{5/2}$	31[a]

Potassium (19)

K	1s	3608.4[a]
L I	2s	378.6[a]
L II	2p$_{1/2}$	297.3[a]
L III	2p$_{3/2}$	294.6[a]
M I	3s	34.8[a]
M II	3p$_{1/2}$	18.3[a]
M III	3p$_{3/2}$	18.3[a]

Praseodymium (59)

K	1s	41991
L I	2s	6835
L II	2p$_{1/2}$	6440
L III	2p$_{3/2}$	5964
M I	3s	1511
M II	3p$_{1/2}$	1337
M III	3p$_{3/2}$	1242
M IV	3d$_{3/2}$	948.3[a]
M V	3d$_{5/2}$	928.8[a]
N I	4s	304.5
N II	4p$_{1/2}$	236.3
N III	4p$_{3/2}$	217.6
N IV	4d$_{3/2}$	115.1[a]
N V	4d$_{5/2}$	115.1[a]
N VI	4f$_{5/2}$	2.0
N VII	4f$_{7/2}$	2.0
O I	5s	37.4
O II	5p$_{1/2}$	22.3
O III	5p$_{3/2}$	22.3

Promethium (61)

K	1s	45184
L I	2s	7428
L II	2p$_{1/2}$	7013
L III	2p$_{3/2}$	6459
M I	3s	—
M II	3p$_{1/2}$	1471.4
M III	3p$_{3/2}$	1357
M IV	3d$_{3/2}$	1052
M V	3d$_{5/2}$	1027
N I	4s	—
N II	4p$_{1/2}$	242
N III	4p$_{3/2}$	242
N IV	4d$_{3/2}$	120
N V	4d$_{5/2}$	120

Protactinium (91)

K	1s	112601

L I	2s	21105
L II	2p$_{1/2}$	20314
L III	2p$_{3/2}$	16733
M I	3s	5367
M II	3p$_{1/2}$	5001
M III	3p$_{3/2}$	4174
M IV	3d$_{3/2}$	3611
M V	3d$_{5/2}$	3442
N I	4s	1387[a]
N II	4p$_{1/2}$	1224[a]
N III	4p$_{3/2}$	1007[a]
N IV	4d$_{3/2}$	743[a]
N V	4d$_{5/2}$	708[a]
N VI	4f$_{5/2}$	371[a]
N VII	4f$_{7/2}$	360[a]
O I	5s	310[a]
O II	5p$_{1/2}$	232[a]
O III	5p$_{3/2}$	232[a]
O IV	5d$_{3/2}$	94[a]
O V	5d$_{5/2}$	94[a]
P I	6s	—
P II	6p$_{1/2}$	—
P III	6p$_{3/2}$	—

Radium (88)

K	1s	103922
L I	2s	19237
L II	2p$_{1/2}$	18484
L III	2p$_{3/2}$	15444
M I	3s	4822
M II	3p$_{1/2}$	4490
M III	3p$_{3/2}$	3792
M IV	3d$_{3/2}$	3248
M V	3d$_{5/2}$	3105
N I	4s	1208[a]
N II	4p$_{1/2}$	1058
N III	4p$_{3/2}$	879[a]
N IV	4d$_{3/2}$	636[a]
N V	4d$_{5/2}$	603[a]
N VI	4f$_{5/2}$	299[a]
N VII	4f$_{7/2}$	299[a]
O I	5s	254[a]
O II	5p$_{1/2}$	200[a]
O III	5p$_{3/2}$	153[a]
O IV	5d$_{3/2}$	68[a]
O V	5d$_{5/2}$	68[a]
P I	6s	44
P II	6p$_{1/2}$	19
P III	6p$_{3/2}$	19

Radon (86)

K	1s	98404
L I	2s	18049
L II	2p$_{1/2}$	17337
L III	2p$_{3/2}$	14619
M I	3s	4482
M II	3p$_{1/2}$	4159
M III	3p$_{3/2}$	3538
M IV	3d$_{3/2}$	3022
M V	3d$_{5/2}$	2892

N I	4s	1097[a]
N II	4p$_{1/2}$	929[a]
N III	4p$_{3/2}$	768[a]
N IV	4d$_{3/2}$	567[a]
N V	4d$_{5/2}$	541[a]
N VI	4f$_{5/2}$	238[a]
N VII	4f$_{7/2}$	238[a]
O I	5s	214[a]
O II	5p$_{1/2}$	164[a]
O III	5p$_{3/2}$	127[a]
O IV	5d$_{3/2}$	48[a]
O V	5d$_{5/2}$	48[a]
P I	6s	26

Rhenium (75)

K	1s	71676
L I	2s	12527
L II	2p$_{1/2}$	11959
L III	2p$_{3/2}$	10535
M I	3s	2932
M II	3p$_{1/2}$	2682
M III	3p$_{3/2}$	2367
M IV	3d$_{3/2}$	1949
M V	3d$_{5/2}$	1883
N I	4s	625.4[b]
N II	4p$_{1/2}$	518.7[b]
N III	4p$_{3/2}$	446.8[b]
N IV	4d$_{3/2}$	273.9[b]
N V	4d$_{5/2}$	260.5[b]
N VI	4f$_{5/2}$	42.9[a]
N VII	4f$_{7/2}$	40.5[a]
O I	5s	83[b]
O II	5p$_{1/2}$	45.6[b]
O III	5p$_{3/2}$	34.6[a,d]

Rhodium (45)

K	1s	23220
L I	2s	3412
L II	2p$_{1/2}$	3146
L III	2p$_{3/2}$	3004
M I	3s	628.1[b]
M II	3p$_{1/2}$	521.3[b]
M III	3p$_{3/2}$	496.5[b]
M IV	3d$_{3/2}$	311.9[b]
M V	3d$_{5/2}$	307.2[b]
N I	4s	81.4[a,d]
N II	4p$_{1/2}$	50.5[b]
N III	4p$_{3/2}$	47.3[b]

Rubidium (37)

K	1s	15200
L I	2s	2065
L II	2p$_{1/2}$	1864
L III	2p$_{3/2}$	1804
M I	3s	326.7[a]
M II	3p$_{1/2}$	248.7[a]
M III	3p$_{3/2}$	239.1[a]
M IV	3d$_{3/2}$	113.0[a]
M V	3d$_{5/2}$	112[a]

N I	4s	30.5[a]
N II	4p$_{1/2}$	16.3[a]
N III	4p$_{3/2}$	15.3[a]

Ruthenium (44)

K	1s	22117
L I	2s	3224
L II	2p$_{1/2}$	2967
L III	2p$_{3/2}$	2838
M I	3s	586.2[b]
M II	3p$_{1/2}$	483.3[b]
M III	3p$_{3/2}$	461.5[b]
M IV	3d$_{3/2}$	284.2[b]
M V	3d$_{5/2}$	280.0[b]
N I	4s	75.0[b]
N II	4p$_{1/2}$	46.5[b]
N III	4p$_{3/2}$	43.2[b]

Samarium (62)

K	1s	46834
L I	2s	7737
L II	2p$_{1/2}$	7312
L III	2p$_{3/2}$	6716
M I	3s	1723
M II	3p$_{1/2}$	1541
M III	3p$_{3/2}$	1419.8
M IV	3d$_{3/2}$	1110.9[a]
M V	3d$_{5/2}$	1083.4[a]
N I	4s	347.2[a]
N II	4p$_{1/2}$	265.6
N III	4p$_{3/2}$	247.4
N IV	4d$_{3/2}$	129.0
N V	4d$_{5/2}$	129.0
N VI	4f$_{5/2}$	5.2
N VII	4f$_{7/2}$	5.2
O I	5s	37.4
O II	5p$_{1/2}$	21.3
O III	5p$_{3/2}$	21.3

Scandium (21)

K	1s	4492
L I	2s	498.0[a]
L II	2p$_{1/2}$	403.6[a]
L III	2p$_{3/2}$	389.7[a]
M I	3s	51.1[a]
M II	3p$_{1/2}$	28.3[a]
M III	3p$_{3/2}$	28.3[a]

Selenium (34)

K	1s	12658
L I	2s	1652.0[a,d]
L II	2p$_{1/2}$	1474.3[a,d]
L III	2p$_{3/2}$	1433.9[a,d]
M I	3s	229.6[a]
M II	3p$_{1/2}$	166.5[a]
M III	3p$_{3/2}$	160.7[a]
M IV	3d$_{3/2}$	55.5[a]
M V	3d$_{5/2}$	54.6[a]

Silicon (14)

K	1s	1839
L I	2s	149.7[a,d]
L II	2p$_{1/2}$	99.8[a]
L III	2p$_{3/2}$	99.2[a]

Silver (47)

K	1s	25514
L I	2s	3806
L II	2p$_{1/2}$	3524
L III	2p$_{3/2}$	3351
M I	3s	719.0[b]
M II	3p$_{1/2}$	603.8[b]
M III	3p$_{3/2}$	573.0[b]
M IV	3d$_{3/2}$	374.0[b]
M V	3d$_{5/2}$	368.0[b]
N I	4s	97.0[b]
N II	4p$_{1/2}$	63.7[b]
N III	4p$_{3/2}$	58.3[b]

Sodium (11)

K	1s	1070.8[b]
L I	2s	63.5[b]
L II	2p$_{1/2}$	30.4[b]
L III	2p$_{3/2}$	30.5[a]

Strontium (38)

K	1s	16105
L I	2s	2216
L II	2p$_{1/2}$	2007
L III	2p$_{3/2}$	1940
M I	3s	358.7[b]
M II	3p$_{1/2}$	280.3[b]
M III	3p$_{3/2}$	270.0[b]
M IV	3d$_{3/2}$	136.0[b]
M V	3d$_{5/2}$	134.2[b]
N I	4s	38.9[b]
N II	4p$_{1/2}$	21.6[b]
N III	4p$_{3/2}$	20.1[b]

Sulfur (16)

K	1s	2472
L I	2s	230.9[a,d]
L II	2p$_{1/2}$	163.6[a]
L III	2p$_{3/2}$	162.5[a]

Tantalum (73)

K	1s	67416
L I	2s	11682
L II	2p$_{1/2}$	11136
L III	2p$_{3/2}$	9881
M I	3s	2708
M II	3p$_{1/2}$	2469
M III	3p$_{3/2}$	2194
M IV	3d$_{3/2}$	1793
M V	3d$_{5/2}$	1735
N I	4s	563.4[b]
N II	4p$_{1/2}$	463.4[b]
N III	4p$_{3/2}$	400.9[b]
N IV	4d$_{3/2}$	237.9[b]
N V	4d$_{5/2}$	226.4[b]
N VI	4f$_{5/2}$	23.5[b]
N VII	4f$_{7/2}$	21.6[b]
O I	5s	69.7[b]
O II	5p$_{1/2}$	42.2[a]
O III	5p$_{3/2}$	32.7[b]

Technetium (43)

K	1s	21044
L I	2s	3043
L II	2p$_{1/2}$	2793
L III	2p$_{3/2}$	2677
M I	3s	586.1[a]
M II	3p$_{1/2}$	447.6[a]
M III	3p$_{3/2}$	417.7[a]
M IV	3d$_{3/2}$	257.6[a]
M V	3d$_{5/2}$	253.9[a]
N I	4s	69.5[a]
N II	4p$_{1/2}$	42.3[a]
N III	4p$_{3/2}$	39.9[a]

Tellurium (52)

K	1s	31814
L I	2s	4939
L II	2p$_{1/2}$	4612
L III	2p$_{3/2}$	4341
M I	3s	1006[b]
M II	3p$_{1/2}$	870.8[b]
M III	3p$_{3/2}$	820.0[b]
M IV	3d$_{3/2}$	583.4[b]
M V	3d$_{5/2}$	573.0[b]
N I	4s	169.4[b]
N II	4p$_{1/2}$	103.3[b,c]
N III	4p$_{3/2}$	103.3[b,c]
N IV	4d$_{3/2}$	41.9[b]
N V	4d$_{5/2}$	40.4[b]

Terbium (65)

K	1s	51996
L I	2s	8708
L II	2p$_{1/2}$	8252
L III	2p$_{3/2}$	7514
M I	3s	1968
M II	3p$_{1/2}$	1768
M III	3p$_{3/2}$	1611
M IV	3d$_{3/2}$	1267.9[a]
M V	3d$_{5/2}$	1241.1[a]
N I	4s	396.0[a]
N II	4p$_{1/2}$	322.4[a]
N III	4p$_{3/2}$	284.1[a]
N IV	4d$_{3/2}$	150.5[a]
N V	4d$_{5/2}$	150.5[a]
N VI	4f$_{5/2}$	7.7[a]
N VII	4f$_{7/2}$	2.4[a]
O I	5s	45.6[a]
O II	5p$_{1/2}$	28.7[a]
O III	5p$_{3/2}$	22.6[a]

Thallium (81)

K	1s	85530
L I	2s	15347
L II	$2p_{1/2}$	14698
L III	$2p_{3/2}$	12658
M I	3s	3704
M II	$3p_{1/2}$	3416
M III	$3p_{3/2}$	2957
M IV	$3d_{3/2}$	2485
M V	$3d_{5/2}$	2389
N I	4s	846.2[b]
N II	$4p_{1/2}$	720.5[b]
N III	$4p_{3/2}$	609.5[b]
N IV	$4d_{3/2}$	405.7[b]
N V	$4d_{5/2}$	385.0[b]
N VI	$4f_{5/2}$	122.2[b]
N VII	$4f_{7/2}$	117.8[b]
O I	5s	136[a,d]
O II	$5p_{1/2}$	94.6[b]
O III	$5p_{3/2}$	73.5[b]
O IV	$5d_{3/2}$	14.7[b]
O V	$5d_{5/2}$	12.5[b]

Thorium (90)

K	1s	109651
L I	2s	20472
L II	$2p_{1/2}$	19693
L III	$2p_{3/2}$	16300
M I	3s	5182
M II	$3p_{1/2}$	4830
M III	$3p_{3/2}$	4046
M IV	$3d_{3/2}$	3491
M V	$3d_{5/2}$	3332
N I	4s	1330[a]
N II	$4p_{1/2}$	1168[a]
N III	$4p_{3/2}$	966.4[b]
N IV	$4d_{3/2}$	712.1[b]
N V	$4d_{5/2}$	675.2[b]
N VI	$4f_{5/2}$	342.4[b]
N VII	$4f_{7/2}$	333.1[b]
O I	5s	290[a,c]
O II	$5p_{1/2}$	229[a,c]
O III	$5p_{3/2}$	182[a,c]
O IV	$5d_{3/2}$	92.5[b]
O V	$5d_{5/2}$	85.4[b]
P I	6s	41.4[b]
P II	$6p_{1/2}$	24.5[b]
P III	$6p_{3/2}$	16.6[b]

Thulium (69)

K	1s	59390
L I	2s	10116
L II	$2p_{1/2}$	9617
L III	$2p_{3/2}$	8648
M I	3s	2307
M II	$3p_{1/2}$	2090
M III	$3p_{3/2}$	1885
M IV	$3d_{3/2}$	1515

M V	$3d_{5/2}$	1468
N I	4s	470.9[a]
N II	$4p_{1/2}$	385.9[a]
N III	$4p_{3/2}$	332.6[a]
N IV	$4d_{3/2}$	175.5[a]
N V	$4d_{5/2}$	175.5[a]
N VI	$4f_{5/2}$	—
N VII	$4f_{7/2}$	4.6
O I	5s	54.7[a]
O II	$5p_{1/2}$	31.8[a]
O III	$5p_{3/2}$	25.0[a]

Tin (50)

K	1s	29200
L I	2s	4465
L II	$2p_{1/2}$	4156
L III	$2p_{3/2}$	3929
M I	3s	884.7[b]
M II	$3p_{1/2}$	756.5[b]
M III	$3p_{3/2}$	714.6[b]
M IV	$3d_{3/2}$	493.2[b]
M V	$3d_{5/2}$	484.9[b]
N I	4s	137.1[b]
N II	$4p_{1/2}$	83.6[b,c]
N III	$4p_{3/2}$	83.6[b,c]
N IV	$4d_{3/2}$	24.9[b]
N V	$4d_{5/2}$	23.9[b]

Titanium (22)

K	1s	4966
L I	2s	560.9[b]
L II	$2p_{1/2}$	460.2[b]
L III	$2p_{3/2}$	453.8[b]
M I	3s	58.7[b]
M II	$3p_{1/2}$	32.6[b]
M III	$3p_{3/2}$	32.6[b]

Tungsten (74)

K	1s	69525
L I	2s	12100
L II	$2p_{1/2}$	11544
L III	$2p_{3/2}$	10207
M I	3s	2820
M II	$3p_{1/2}$	2575
M III	$3p_{3/2}$	2281
M IV	$3d_{3/2}$	1949
M V	$3d_{5/2}$	1809
N I	4s	594.1[b]
N II	$4p_{1/2}$	490.4[b]
N III	$4p_{3/2}$	423.6[b]
N IV	$4d_{3/2}$	255.9[b]
N V	$4d_{5/2}$	243.5[b]
N VI	$4f_{5/2}$	33.6[a]
N VII	$4f_{7/2}$	31.4[b]
O I	5s	75.6[b]
O II	$5p_{1/2}$	453[a,d]
O III	$5p_{3/2}$	36.8[b]

Uranium (92)

K	1s	115606
L I	2s	21757
L II	$2p_{1/2}$	20948
L III	$2p_{3/2}$	17166
M I	3s	5548
M II	$3p_{1/2}$	5182
M III	$3p_{3/2}$	4303
M IV	$3d_{3/2}$	3728
M V	$3d_{5/2}$	3552
N I	4s	1439[a,d]
N II	$4p_{1/2}$	1271[a,d]
N III	$4p_{3/2}$	1043[b]
N IV	$4d_{3/2}$	778.3[b]
N V	$4d_{5/2}$	736.2[b]
N VI	$4f_{5/2}$	388.2[a]
N VII	$4f_{7/2}$	377.4[b]
O I	5s	321[a,c,d]
O II	$5p_{1/2}$	257[a,c,d]
O III	$5p_{3/2}$	192[a,c,d]
O IV	$5d_{3/2}$	102.8[b]
O V	$5d_{5/2}$	94.2[b]
P I	6s	43.9[b]
P II	$6p_{1/2}$	26.8[b]
P III	$6p_{3/2}$	16.8[b]

Vanadium (23)

K	1s	5465
L I	2s	626.7[b]
L II	$2p_{1/2}$	519.8[b]
L III	$2p_{3/2}$	521.1[b]
M I	3s	66.3[b]
M II	$3p_{1/2}$	37.2[b]
M III	$3p_{3/2}$	37.2[b]

Xenon (54)

K	1s	34561
L I	2s	5453
L II	$2p_{1/2}$	5107
L III	$2p_{3/2}$	4786
M I	3s	1148.7[a]
M II	$3p_{1/2}$	1002.1[a]
M III	$3p_{3/2}$	940.6[a]
M IV	$3d_{3/2}$	689.0[a]
M V	$3d_{5/2}$	676.4[a]
N I	4s	213.2[a]
N II	$4p_{1/2}$	146.7
N III	$4p_{3/2}$	145.5[a]
N IV	$4d_{3/2}$	69.5[a]
N V	$4d_{5/2}$	67.5[a]
N VI	$4f_{5/2}$	—
N VII	$4f_{7/2}$	—
O I	5s	23.3[a]
O II	$5p_{1/2}$	13.4[a]
O III	$5p_{3/2}$	12.1[a]

Ytterbium (70)

K	1s	61332

L I	2s	10486
L II	$2p_{1/2}$	9978
L III	$2p_{3/2}$	8944
M I	3s	2398
M II	$3p_{1/2}$	2173
M III	$3p_{3/2}$	1950
M IV	$3d_{3/2}$	1576
M V	$3d_{5/2}$	1528
N I	4s	480.5[a]
N II	$4p_{1/2}$	388.7[a]
N III	$4p_{3/2}$	339.7[a]
N IV	$4d_{3/2}$	191.2[a]
N V	$4d_{5/2}$	182.4[a]
N VI	$4f_{5/2}$	2.5[a]
N VII	$4f_{7/2}$	1.3[a]
O I	5s	52.0[a]
O II	$5p_{1/2}$	30.3[a]
O III	$5p_{3/2}$	24.1[a]

Yttrium (39)

K	1s	17038
L I	2s	2373
L II	$2p_{1/2}$	2156
L III	$2p_{3/2}$	2080
M I	3s	392.0[a,d]
M II	$3p_{1/2}$	310.6[a]
M III	$3p_{3/2}$	298.8[a]
M IV	$3d_{3/2}$	157.7[b]
M V	$3d_{5/2}$	155.8[b]
N I	4s	43.8[a]
N II	$4p_{1/2}$	24.4[a]
N III	$4p_{3/2}$	23.1[a]

Zinc (30)

K	1s	9659
L I	2s	1196.2[a]
L II	$2p_{1/2}$	1044.9[a]
L III	$2p_{3/2}$	1021.8[a]
M I	3s	139.8[a]
M II	$3p_{1/2}$	91.4[a]
M III	$3p_{3/2}$	88.6[a]
M IV	$3d_{3/2}$	10.2[a]
M V	$3d_{5/2}$	10.1[a]

Zirconium (40)

K	1s	17998
L I	2s	2532
L II	$2p_{1/2}$	2307
L III	$2p_{3/2}$	2223
M I	3s	430.3[b]
M II	$3p_{1/2}$	343.5[b]
M III	$3p_{3/2}$	329.8[b]
M IV	$3d_{3/2}$	181.1[b]
M V	$3d_{5/2}$	178.8[b]
N I	4s	50.6[b]
N II	$4p_{1/2}$	28.5[b]
N III	$4p_{3/2}$	27.1[b]

[a] Reference 1.
[b] Reference 2 (remaining values from Reference 3).
[c] One-particle approximation not valid.
[d] Derived using energy differences from Reference 3.

NATURAL WIDTH OF X-RAY LINES

Natural widths of K X-ray lines in eV:

Element	$K\alpha_1$	$K\alpha_2$	$K\beta_1$	$K\beta_3$	Element	$K\alpha_1$	$K\alpha_2$	$K\beta_1$	$K\beta_3$
Ca	1.00	0.98			Ce	18.60	19.50	20.60	18.60
Ti	1.45	2.13			Nd	21.50	21.50	23.25	21.33
Cr	2.05	2.64			Sm	26.00	24.70	25.65	24.65
Fe	2.45	3.20			Gd	29.50	28.00	29.37	28.00
Ni	3.00	3.70			Dy	33.90	32.20	32.73	32.00
Zn	3.40	3.96			Er	35.00	35.50	36.20	35.70
Ge	3.75	4.18			Yb	38.80	40.60	41.43	41.15
Se	4.10	4.43			Hf	42.70	44.30	46.00	46.10
Kr	4.23	4.62			W	46.80	48.00	51.83	51.50
Sr	5.17	4.97			Os	49.00	49.40	55.90	55.95
Zr	5.70	5.25			Pt	54.10	54.30	59.98	62.13
Mo	6.82	6.80			Hg	64.75	68.20	65.75	68.95
Ru	7.41	7.96			Pb	67.10	72.30	72.20	73.80
Pd	8.80	9.20			Po	73.20	75.10	78.60	80.10
Cd	9.80	10.40			Rn	80.00	81.50	85.50	86.50
Sn	11.20	12.40	11.80	11.00	Ra	87.00	88.20	94.20	95.50
Te	12.80	14.20	13.30	13.10	Th	94.70	95.00	99.70	101.00
Xe	14.20	15.10	15.30	14.50	U	103.00	104.30	105.00	107.30
Ba	16.10	16.80	18.15	16.70					

From Salem, S. I. and Lee, P. L., *At. Data Nucl. Data Tables,* 18, 233, 1976.

Natural widths of L X-ray lines in eV:

Element	$L\alpha_1$	$L\alpha_2$	$L\beta_1$	$L\beta_2$	$L\beta_3$	$L\beta_4$	$L\gamma_1$
Zr	1.68	1.52	1.87	5.13	5.50	5.60	3.34
Mo	1.86	1.80	2.03	5.30	5.90	5.78	3.76
Ru	2.03	1.98	2.18	5.45	6.35	5.96	4.15
Pd	2.21	2.16	2.36	5.63	6.80	6.18	4.50
Gd	2.43	2.40	2.54	5.82	7.23	6.28	4.83
Sn	2.62	2.62	2.75	6.10	7.70	6.60	5.23
Tc	2.88	2.88	2.96	6.25	8.22	6.82	5.60
Xe	3.15	3.15	3.20	6.43	8.70	7.15	5.95
Ba	3.39	3.45	3.45	6.70	9.20	7.42	6.35
Ce	3.70	3.78	3.73	6.86	9.70	7.82	6.75
Nd	3.93	4.08	4.00	7.18	10.30	8.15	7.16
Sm	4.13	4.50	4.33	7.42	10.80	8.60	7.50
Cd	4.46	4.90	4.63	7.70	11.20	9.08	7.83
Dy	4.81	5.35	5.03	7.90	11.50	9.60	8.30
Er	5.17	5.73	5.45	8.28	11.85	10.03	8.75
Yb	5.40	6.22	5.90	8.58	12.20	11.00	9.20
Hf	5.83	6.70	6.36	8.92	12.40	12.80	9.63
W	6.50	7.20	6.90	9.06	13.10	14.60	10.20
Os	7.04	7.70	7.42	9.60	14.60	16.50	10.65
Pt	7.60	8.28	8.00	9.95	16.10	18.00	11.20
Hg	8.10	8.80	8.70	10.40	17.40	19.70	11.80
Pb	8.82	9.35	9.35	10.75	18.65	21.30	12.30
Po	9.50	9.95	10.10	11.25	19.90	22.70	13.05
Rn	10.03	10.50	10.65	11.65	21.00	24.00	13.55
Ra	11.00	11.20	11.60	12.20	22.00	25.20	14.30
Th	11.90	11.80	12.40	12.80	22.85	26.35	15.00
U	12.40	12.40	13.50	13.30	23.70	27.50	15.70
Pu	13.20	13.00	14.10	13.90	24.10	28.30	16.40
Cm	14.80	13.60	15.70	14.60	25.00	29.40	17.10

PHOTON ATTENUATION COEFFICIENTS

Martin J. Berger and John H. Hubbell

This table gives mass attenuation coefficients for photons for all elements at energies between 1 keV (soft x-rays) and 1 GeV (hard gamma rays). The mass attenuation coefficient μ describes the attenuation of radiation as it passes through matter by the relation

$$I(x)/I_o = e^{-\mu\rho x}$$

where I_o is the initial intensity, $I(x)$ the intensity after path length x, and ρ is the mass density of the element in question. To a high approximation the mass attenuation coefficient is additive for the elements present, independent of the way in which they are bound in chemical compounds.

The power of ten is indicated beside each number in the table; i.e., 7.41 + 03 means 7.41×10^3. A vertical line between two columns indicates that an absorption edge lies between those energy values. The various edges are labeled at the bottom of the table.

The attenuation coefficients were calculated with the computer program XCOM (Reference 1), which uses a cross-section database compiled at the Photon and Charged Particle Data Center at the National Institute of Standards and Technology. Their accuracy has been confirmed at all energies by extensive comparisons with experimental attenuation coefficients. Such comparisons for X-ray energies up to 100 keV can be found in Reference 2.

REFERENCES

1. Berger, M. J. and Hubbell, J. H., National Bureau of Standards Report NBSIR-87-3597, 1987.
2. Saloman, E. B., Hubbell, J. H., and Scofield, J. H., *Atomic Data and Nuclear Data Tables*, 38, 1, 1988.

Mass attenuation coefficient, cm²/g

	Atomic no.	Photon energy, MeV								
		0.001	0.002	0.005	0.01	0.02	0.05	0.1	0.2	0.5
H	1	7.21 + 00	1.06 + 00	4.19-01	3.85-01	3.69-01	3.36-01	2.94-01	2.43-01	1.73-01
He	2	6.08 + 01	6.86 + 00	5.77-01	2.48-01	1.96-01	1.70-01	1.49-01	1.22-01	8.71-02
Li	3	2.34 + 02	2.71 + 01	1.62 + 00	3.40-01	1.86-01	1.49-01	1.29-01	1.06-01	7.53-02
Be	4	6.04 + 02	7.47 + 01	4.37 + 00	6.47-01	2.25-01	1.55-01	1.33-01	1.09-01	7.74-02
B	5	1.23 + 03	1.60 + 02	9.68 + 00	1.25 + 00	3.01-01	1.66-01	1.39-01	1.14-01	8.07-02
C	6	2.21 + 03	3.03 + 02	1.91 + 01	2.37 + 00	4.42-01	1.87-01	1.51-01	1.23-01	8.72-02
N	7	3.31 + 03	4.77 + 02	3.14 + 01	3.88 + 00	6.18-01	1.98-01	1.53-01	1.23-01	8.72-02
O	8	4.59 + 03	6.95 + 02	4.79 + 01	5.95 + 00	8.65-01	2.13-01	1.55-01	1.24-01	8.73-02
F	9	5.65 + 03	9.05 + 02	6.51 + 01	8.21 + 00	1.13 + 00	2.21-01	1.50-01	1.18-01	8.27-02
Ne	10	7.41 + 03	1.24 + 03	9.34 + 01	1.20 + 01	1.61 + 00	2.58-01	1.60-01	1.24-01	8.66-02
Na	11	6.54 + 02	1.52 + 03	1.19 + 02	1.56 + 01	2.06 + 00	2.80-01	1.59-01	1.20-01	8.37-02
Mg	12	9.22 + 02	1.93 + 03	1.58 + 02	2.11 + 01	2.76 + 00	3.29-01	1.69-01	1.24-01	8.65-02
Al	13	1.19 + 03	2.26 + 03	1.93 + 02	2.62 + 01	3.44 + 00	3.68-01	1.70-01	1.22-01	8.44-02
Si	14	1.57 + 03	2.78 + 03	2.45 + 02	3.39 + 01	4.46 + 00	4.38-01	1.84-01	1.28-01	8.75-02
P	15	1.91 + 03	3.02 + 02	2.86 + 02	4.04 + 01	5.35 + 00	4.92-01	1.87-01	1.25-01	8.51-02
S	16	2.43 + 03	3.85 + 02	3.49 + 02	5.01 + 01	6.71 + 00	5.85-01	2.02-01	1.30-01	8.78-02
Cl	17	2.83 + 03	4.52 + 02	3.90 + 02	5.73 + 01	7.74 + 00	6.48-01	2.05-01	1.27-01	8.45-02
Ar	18	3.18 + 03	5.12 + 02	4.23 + 02	6.32 + 01	8.63 + 00	7.01-01	2.04-01	1.20-01	7.96-02
K	19	4.06 + 03	6.59 + 02	5.19 + 02	7.91 + 01	1.09 + 01	8.68-01	2.34-01	1.32-01	8.60-02
Ca	20	4.87 + 03	8.00 + 02	6.03 + 02	9.34 + 01	1.31 + 01	1.02 + 00	2.57-01	1.38-01	8.85-02
Sc	21	5.24 + 03	8.70 + 02	6.31 + 02	9.95 + 01	1.41 + 01	1.09 + 00	2.58-01	1.31-01	8.31-02
Ti	22	5.87 + 03	9.86 + 02	6.84 + 02	1.11 + 02	1.59 + 01	1.21 + 00	2.72-01	1.31-01	8.19-02
V	23	6.50 + 03	1.11 + 03	9.29 + 01	1.22 + 02	1.77 + 01	1.35 + 00	2.88-01	1.32-01	8.07-02
Cr	24	7.40 + 03	1.28 + 03	1.08 + 02	1.39 + 02	2.04 + 01	1.55 + 00	3.17-01	1.38-01	8.28-02
Mn	25	8.09 + 03	1.42 + 03	1.21 + 02	1.51 + 02	2.25 + 01	1.71 + 00	3.37-01	1.39-01	8.19-02
Fe	26	9.09 + 03	1.63 + 03	1.40 + 02	1.71 + 02	2.57 + 01	1.96 + 00	3.72-01	1.46-01	8.41-02
Co	27	9.80 + 03	1.78 + 03	1.54 + 02	1.84 + 02	2.80 + 01	2.14 + 00	3.95-01	1.48-01	8.32-02
Ni	28	9.86 + 03	2.05 + 03	1.79 + 02	2.09 + 02	3.22 + 01	2.47 + 00	4.44-01	1.58-01	8.70-02
Cu	29	1.06 + 04	2.15 + 03	1.90 + 02	2.16 + 02	3.38 + 01	2.61 + 00	4.58-01	1.56-01	8.36-02
Zn	30	1.55 + 03	2.37 + 03	2.12 + 02	2.33 + 02	3.72 + 01	2.89 + 00	4.97-01	1.62-01	8.45-02
Ga	31	1.70 + 03	2.52 + 03	2.27 + 02	3.42 + 01	3.93 + 01	3.08 + 00	5.20-01	1.62-01	8.24-02
Ge	32	1.89 + 03	2.71 + 03	2.47 + 02	3.74 + 01	4.22 + 01	3.34 + 00	5.55-01	1.66-01	8.21-02
As	33	2.12 + 03	2.93 + 03	2.71 + 02	4.12 + 01	4.56 + 01	3.63 + 00	5.97-01	1.72-01	8.26-02
Se	34	2.32 + 03	3.10 + 03	2.90 + 02	4.41 + 01	4.82 + 01	3.86 + 00	6.28-01	1.74-01	8.13-02
Br	35	2.62 + 03	3.41 + 03	3.21 + 02	4.91 + 01	5.27 + 01	4.26 + 00	6.86-01	1.84-01	8.33-02
Kr	36	2.85 + 03	3.60 + 03	3.43 + 02	5.26 + 01	5.55 + 01	4.52 + 00	7.22-01	1.87-01	8.23-02
Rb	37	3.17 + 03	3.41 + 03	3.74 + 02	5.77 + 01	5.98 + 01	4.92 + 00	7.80-01	1.96-01	8.36-02
Sr	38	3.49 + 03	2.59 + 03	4.06 + 02	6.27 + 01	6.39 + 01	5.31 + 00	8.37-01	2.04-01	8.44-02
Y	39	3.86 + 03	7.42 + 02	4.42 + 02	6.87 + 01	6.86 + 01	5.76 + 00	9.05-01	2.15-01	8.61-02
Zr	40	4.21 + 03	8.12 + 02	4.76 + 02	7.42 + 01	7.24 + 01	6.17 + 00	9.66-01	2.24-01	8.69-02
Nb	41	4.60 + 03	8.89 + 02	5.13 + 02	8.04 + 01	7.71 + 01	6.64 + 00	1.04 + 00	2.34-01	8.83-02
Mo	42	4.94 + 03	9.60 + 02	5.45 + 02	8.58 + 01	1.31 + 01	7.04 + 00	1.10 + 00	2.42-01	8.85-02
Tc	43	5.36 + 03	1.04 + 03	5.84 + 02	9.23 + 01	1.41 + 01	7.52 + 00	1.17 + 00	2.53-01	8.97-02
Ru	44	5.72 + 03	1.12 + 03	6.17 + 02	9.80 + 01	1.50 + 01	7.92 + 00	1.23 + 00	2.62-01	8.99-02
Rh	45	6.17 + 03	1.21 + 03	6.59 + 02	1.05 + 02	1.61 + 01	8.45 + 00	1.31 + 00	2.74-01	9.13-02
Pd	46	6.54 + 03	1.29 + 03	6.91 + 02	1.11 + 02	1.70 + 01	8.85 + 00	1.38 + 00	2.83-01	9.13-02
Ag	47	7.04 + 03	1.40 + 03	7.39 + 02	1.19 + 02	1.84 + 01	9.45 + 00	1.47 + 00	2.97-01	9.32-02
Cd	48	7.35 + 03	1.47 + 03	7.69 + 02	1.24 + 02	1.92 + 01	9.78 + 00	1.52 + 00	3.04-01	9.25-02
In	49	7.81 + 03	1.58 + 03	8.13 + 02	1.32 + 02	2.04 + 01	1.03 + 01	1.61 + 00	3.17-01	9.37-02
Sn	50	8.16 + 03	1.66 + 03	8.47 + 02	1.38 + 02	2.15 + 01	1.07 + 01	1.68 + 00	3.26-01	9.37-02

L₃ | L₁   K EDGE

L₂

Mass attenuation coefficient, cm²/g

	Atomic no.	Photon energy, MeV								
		1.0	2.0	5.0	10.0	20.0	50.0	100.0	500.0	1000.0
H	1	1.26-01	8.77-02	5.05-02	3.25-02	2.15-02	1.42-02	1.19-02	1.14-02	1.16-02
He	2	6.36-02	4.42-02	2.58-02	1.70-02	1.18-02	8.61-03	7.78-03	7.79-03	7.95-03
Li	3	5.50-02	3.83-02	2.26-02	1.53-02	1.11-02	8.68-03	8.21-03	8.61-03	8.87-03
Be	4	5.65-02	3.94-02	2.35-02	1.63-02	1.23-02	1.02-02	9.94-03	1.08-02	1.12-02
B	5	5.89-02	4.11-02	2.48-02	1.76-02	1.37-02	1.19-02	1.19-02	1.32-02	1.37-02
C	6	6.36-02	4.44-02	2.71-02	1.96-02	1.58-02	1.43-02	1.46-02	1.64-02	1.70-02
N	7	6.36-02	4.45-02	2.74-02	2.02-02	1.67-02	1.57-02	1.63-02	1.85-02	1.92-02
O	8	6.37-02	4.46-02	2.78-02	2.09-02	1.77-02	1.71-02	1.79-02	2.06-02	2.13-02
F	9	6.04-02	4.23-02	2.66-02	2.04-02	1.77-02	1.75-02	1.86-02	2.14-02	2.21-02
Ne	10	6.32-02	4.43-02	2.82-02	2.20-02	1.95-02	1.96-02	2.11-02	2.43-02	2.51-02
Na	11	6.10-02	4.28-02	2.75-02	2.18-02	1.97-02	2.03-02	2.19-02	2.53-02	2.62-02
Mg	12	6.30-02	4.43-02	2.87-02	2.31-02	2.13-02	2.23-02	2.42-02	2.81-02	2.90-02
Al	13	6.15-02	4.32-02	2.84-02	2.32-02	2.17-02	2.31-02	2.52-02	2.93-02	3.03-02
Si	14	6.36-02	4.48-02	2.97-02	2.46-02	2.34-02	2.52-02	2.76-02	3.23-02	3.34-02
P	15	6.18-02	4.36-02	2.91-02	2.45-02	2.36-02	2.58-02	2.84-02	3.33-02	3.45-02
S	16	6.37-02	4.50-02	3.04-02	2.59-02	2.53-02	2.79-02	3.08-02	3.62-02	3.75-02
Cl	17	6.13-02	4.33-02	2.95-02	2.55-02	2.52-02	2.81-02	3.11-02	3.67-02	3.80-02
Ar	18	5.76-02	4.07-02	2.80-02	2.45-02	2.45-02	2.76-02	3.07-02	3.62-02	3.75-02
K	19	6.22-02	4.40-02	3.05-02	2.70-02	2.74-02	3.11-02	3.46-02	4.09-02	4.24-02
Ca	20	6.39-02	4.52-02	3.17-02	2.84-02	2.90-02	3.32-02	3.71-02	4.40-02	4.56-02
Sc	21	5.98-02	4.24-02	3.00-02	2.72-02	2.80-02	3.23-02	3.62-02	4.30-02	4.45-02
Ti	22	5.89-02	4.18-02	2.98-02	2.73-02	2.84-02	3.30-02	3.71-02	4.40-02	4.56-02
V	23	5.79-02	4.11-02	2.96-02	2.74-02	2.88-02	3.36-02	3.78-02	4.49-02	4.65-02
Cr	24	5.93-02	4.21-02	3.06-02	2.86-02	3.03-02	3.56-02	4.01-02	4.76-02	4.93-02
Mn	25	5.85-02	4.16-02	3.04-02	2.87-02	3.07-02	3.63-02	4.09-02	4.86-02	5.04-02
Fe	26	5.99-02	4.26-02	3.15-02	2.99-02	3.22-02	3.83-02	4.33-02	5.15-02	5.33-02
Co	27	5.91-02	4.20-02	3.13-02	3.00-02	3.26-02	3.88-02	4.40-02	5.23-02	5.41-02
Ni	28	6.16-02	4.39-02	3.29-02	3.18-02	3.48-02	4.17-02	4.73-02	5.61-02	5.81-02
Cu	29	5.90-02	4.20-02	3.18-02	3.10-02	3.41-02	4.10-02	4.66-02	5.53-02	5.72-02
Zn	30	5.94-02	4.24-02	3.22-02	3.18-02	3.51-02	4.24-02	4.82-02	5.72-02	5.91-02
Ga	31	5.77-02	4.11-02	3.16-02	3.13-02	3.48-02	4.22-02	4.80-02	5.70-02	5.89-02
Ge	32	5.73-02	4.09-02	3.16-02	3.16-02	3.53-02	4.30-02	4.89-02	5.80-02	6.00-02
As	33	5.73-02	4.09-02	3.19-02	3.21-02	3.60-02	4.40-02	5.01-02	5.95-02	6.15-02
Se	34	5.62-02	4.01-02	3.14-02	3.19-02	3.60-02	4.41-02	5.03-02	5.97-02	6.17-02
Br	35	5.73-02	4.09-02	3.23-02	3.29-02	3.74-02	4.60-02	5.24-02	6.22-02	6.43-02
Kr	36	5.63-02	4.02-02	3.20-02	3.28-02	3.74-02	4.61-02	5.26-02	6.25-02	6.46-02
Rb	37	5.69-02	4.06-02	3.25-02	3.36-02	3.85-02	4.75-02	5.43-02	6.45-02	6.67-02
Sr	38	5.71-02	4.08-02	3.29-02	3.41-02	3.93-02	4.87-02	5.56-02	6.61-02	6.83-02
Y	39	5.80-02	4.14-02	3.35-02	3.50-02	4.05-02	5.03-02	5.75-02	6.83-02	7.06-02
Zr	40	5.81-02	4.15-02	3.38-02	3.55-02	4.12-02	5.13-02	5.87-02	6.98-02	7.22-02
Nb	41	5.87-02	4.18-02	3.44-02	3.63-02	4.22-02	5.27-02	6.03-02	7.17-02	7.42-02
Mo	42	5.84-02	4.16-02	3.44-02	3.65-02	4.26-02	5.33-02	6.10-02	7.26-02	7.51-02
Tc	43	5.88-02	4.19-02	3.48-02	3.71-02	4.35-02	5.45-02	6.24-02	7.43-02	7.68-02
Ru	44	5.85-02	4.16-02	3.48-02	3.73-02	4.39-02	5.50-02	6.30-02	7.51-02	7.77-02
Rh	45	5.89-02	4.20-02	3.53-02	3.80-02	4.48-02	5.63-02	6.45-02	7.69-02	7.94-02
Pd	46	5.85-02	4.16-02	3.52-02	3.80-02	4.50-02	5.66-02	6.49-02	7.73-02	8.00-02
Ag	47	5.92-02	4.21-02	3.58-02	3.88-02	4.61-02	5.81-02	6.67-02	7.93-02	8.20-02
Cd	48	5.83-02	4.14-02	3.54-02	3.85-02	4.59-02	5.79-02	6.64-02	7.91-02	8.18-02
In	49	5.85-02	4.15-02	3.56-02	3.90-02	4.65-02	5.88-02	6.75-02	8.04-02	8.32-02
Sn	50	5.80-02	4.11-02	3.55-02	3.90-02	4.66-02	5.90-02	6.78-02	8.07-02	8.35-02

Mass attenuation coefficients, cm²/g

	Atomic no.	Photon energy, MeV								
		0.001	0.002	0.005	0.01	0.02	0.05	0.1	0.2	0.5
Sb	51	8.58 + 03	1.77 + 03	8.85 + 02	1.46 + 02	2.27 + 01	1.12 + 01	1.76 + 00	3.38-01	9.45-02
Te	52	8.43 + 03	1.83 + 03	9.01 + 02	1.50 + 02	2.34 + 01	1.14 + 01	1.80 + 00	3.43-01	9.33-02
I	53	9.10 + 03	2.00 + 03	8.43 + 02	1.63 + 02	2.54 + 01	1.23 + 01	1.94 + 00	3.66-01	9.70-02
Xe	54	9.41 + 03	2.09 + 03	6.39 + 02	1.69 + 02	2.65 + 01	1.27 + 01	2.01 + 00	3.76-01	9.70-02
Cs	55	9.37 + 03	2.23 + 03	2.30 + 02	1.79 + 02	2.82 + 01	1.34 + 01	2.12 + 00	3.94-01	9.91-02
ZB	56	8.54 + 03	2.32 + 03	2.41 + 02	1.86 + 02	2.94 + 01	1.38 + 01	2.20 + 00	4.05-01	9.92-02
La	57	9.09 + 03	2.46 + 03	2.58 + 02	1.97 + 02	3.12 + 01	1.45 + 01	2.32 + 00	4.24-01	1.01-01
Ce	58	9.71 + 03	2.61 + 03	2.74 + 02	2.08 + 02	3.31 + 01	1.52 + 01	2.45 + 00	4.45-01	1.04-01
Pr	59	1.06 + 04	2.77 + 03	2.92 + 02	2.21 + 02	3.53 + 01	1.60 + 01	2.59 + 00	4.69-01	1.07-01
Nd	60	6.63 + 03	2.88 + 03	3.06 + 02	2.30 + 02	3.68 + 01	1.65 + 01	2.69 + 00	4.84-01	1.08-01
Pm	61	2.06 + 03	3.05 + 03	3.26 + 02	2.44 + 02	3.92 + 01	1.73 + 01	2.84 + 00	5.10-01	1.12-01
Sm	62	2.11 + 03	3.12 + 03	3.36 + 02	2.50 + 02	4.03 + 01	1.77 + 01	2.90 + 00	5.19-01	1.11-01
Eu	63	2.22 + 03	3.28 + 03	3.54 + 02	2.63 + 02	4.24 + 01	1.85 + 01	3.04 + 00	5.43-01	1.14-01
Gd	64	2.29 + 03	3.36 + 03	3.65 + 02	2.69 + 02	4.36 + 01	3.86 + 00	3.11 + 00	5.54-01	1.14-01
Tb	65	2.40 + 03	3.51 + 03	3.84 + 02	2.82 + 02	4.59 + 01	4.06 + 00	3.25 + 00	5.77-01	1.17-01
Dy	66	2.49 + 03	3.47 + 03	3.99 + 02	2.90 + 02	4.76 + 01	4.23 + 00	3.36 + 00	5.95-01	1.18-01
Ho	67	2.62 + 03	3.59 + 03	4.17 + 02	3.01 + 02	4.98 + 01	4.43 + 00	3.49 + 00	6.18-01	1.20-01
Er	68	2.75 + 03	3.52 + 03	4.36 + 02	3.13 + 02	5.20 + 01	4.63 + 00	3.63 + 00	6.41-01	1.23-01
Tm	69	2.90 + 03	3.69 + 03	4.57 + 02	2.83 + 02	5.45 + 01	4.87 + 00	3.78 + 00	6.68-01	1.26-01
Yb	70	3.02 + 03	3.80 + 03	4.72 + 02	2.94 + 02	5.63 + 01	5.04 + 00	3.88 + 00	6.86-01	1.27-01
Lu	71	3.19 + 03	3.45 + 03	4.94 + 02	2.21 + 02	5.88 + 01	5.28 + 00	4.03 + 00	7.13-01	1.30-01
Hf	72	3.34 + 03	3.60 + 03	5.11 + 02	2.30 + 02	6.09 + 01	5.48 + 00	4.15 + 00	7.34-01	1.32-01
Ta	73	3.51 + 03	3.77 + 03	5.33 + 02	2.38 + 02	6.33 + 01	5.72 + 00	4.30 + 00	7.60-01	1.35-01
W	74	3.68 + 03	3.92 + 03	5.53 + 02	9.69 + 01	6.57 + 01	5.95 + 00	4.44 + 00	7.84-01	1.38-01
Re	75	3.87 + 03	3.77 + 03	5.76 + 02	1.01 + 02	6.84 + 01	6.21 + 00	4.59 + 00	8.12-01	1.41-01
Os	76	4.03 + 03	2.22 + 03	5.93 + 02	1.04 + 02	7.04 + 01	6.41 + 00	4.70 + 00	8.33-01	1.43-01
Ir	77	4.24 + 03	1.03 + 03	6.18 + 02	1.09 + 02	7.32 + 01	6.69 + 00	4.86 + 00	8.63-01	1.46-01
Pt	78	4.43 + 03	1.08 + 03	6.40 + 02	1.13 + 02	7.57 + 01	6.95 + 00	4.99 + 00	8.90-01	1.49-01
Au	79	4.65 + 03	1.14 + 03	6.66 + 02	1.18 + 02	7.88 + 01	7.26 + 00	5.16 + 00	9.22-01	1.53-01
Hg	80	4.83 + 03	1.18 + 03	6.87 + 02	1.22 + 02	8.12 + 01	7.50 + 00	5.28 + 00	9.46-01	1.56-01
Tl	81	5.01 + 03	1.23 + 03	7.07 + 02	1.26 + 02	8.36 + 01	7.75 + 00	5.40 + 00	9.69-01	1.58-01
Pb	82	5.21 + 03	1.29 + 03	7.30 + 02	1.31 + 02	8.64 + 01	8.04 + 00	5.55 + 00	9.99-01	1.61-01
Bi	83	5.44 + 03	1.35 + 03	7.58 + 02	1.36 + 02	8.95 + 01	8.38 + 00	5.74 + 00	1.03 + 00	1.66-01
Po	84	5.72 + 03	1.42 + 03	7.93 + 02	1.43 + 02	9.35 + 01	8.80 + 00	5.99 + 00	1.08 + 00	1.71-01
At	85	5.87 + 03	1.49 + 03	8.25 + 02	1.49 + 02	9.70 + 01	9.19 + 00	6.17 + 00	1.12 + 00	1.77-01
Rn	86	5.83 + 03	1.49 + 03	8.16 + 02	1.48 + 02	9.56 + 01	9.12 + 00	6.09 + 00	1.10 + 00	1.73-01
Fr	87	6.08 + 03	1.56 + 03	8.49 + 02	1.54 + 02	9.93 + 01	9.52 + 00	1.66 + 00	1.14 + 00	1.78-01
Ra	88	6.20 + 03	1.62 + 03	8.74 + 02	1.59 + 02	1.02 + 02	9.85 + 00	1.71 + 00	1.17 + 00	1.82-01
Ac	89	6.47 + 03	1.70 + 03	8.69 + 02	1.65 + 02	1.06 + 02	1.03 + 01	1.79 + 00	1.21 + 00	1.87-01
Th	90	6.61 + 03	1.74 + 03	8.88 + 02	1.69 + 02	9.37 + 01	1.05 + 01	1.83 + 00	1.23 + 00	1.90-01
Pa	91	6.53 + 03	1.83 + 03	8.76 + 02	1.77 + 02	7.03 + 01	1.10 + 01	1.92 + 00	1.29 + 00	1.97-01
U	92	6.63 + 03	1.86 + 03	8.89 + 02	1.79 + 02	7.11 + 01	1.12 + 01	1.95 + 00	1.30 + 00	1.98-01
Np	93	6.95 + 03	1.96 + 03	9.32 + 02	1.87 + 02	7.45 + 01	1.18 + 01	2.05 + 00	1.35 + 00	2.05-01
Pu	94	7.19 + 03	2.04 + 03	9.65 + 02	1.94 + 02	7.71 + 01	1.22 + 01	2.13 + 00	1.39 + 00	2.10-01
Am	95	7.37 + 03	2.10 + 03	9.90 + 02	1.98 + 02	7.93 + 01	1.25 + 01	2.19 + 00	1.42 + 00	2.14-01
Cm	96	7.54 + 03	2.15 + 03	1.02 + 03	2.03 + 02	8.14 + 01	1.28 + 01	2.25 + 00	1.44 + 00	2.18-01
Bk	97	7.84 + 03	2.25 + 03	1.06 + 03	2.10 + 02	8.39 + 01	1.34 + 01	2.35 + 00	1.50 + 00	2.25-01
Cf	98	7.89 + 03	2.31 + 03	9.27 + 02	2.15 + 02	8.58 + 01	1.37 + 01	2.41 + 00	1.52 + 00	2.29-01
Es	99	7.79 + 03	2.40 + 03	9.59 + 02	2.22 + 02	4.01 + 01	1.42 + 01	2.51 + 00	1.57 + 00	2.36-01
Fm	100	7.13 + 03	2.46 + 03	9.77 + 02	2.26 + 02	4.09 + 01	1.45 + 01	2.57 + 00	1.59 + 00	2.39-01

N_5 M_5 M_3 L_3

N_4 M_4 M_2 L_2 K EDGE

N_3 M_1 L_1

N_2

N_1

Mass attenuation coefficients, cm^2/g

	Atomic no.	Photon Energy, MeV								
		1.0	2.0	5.0	10.0	20.0	50.0	100.0	500.0	1000.0
Sb	51	5.80-02	4.10-02	3.56-02	3.92-02	4.70-02	5.96-02	6.85-02	8.16-02	8.44-02
Te	52	5.67-02	4.01-02	3.49-02	3.86-02	4.64-02	5.89-02	6.77-02	8.07-02	8.35-02
I	53	5.84-02	4.12-02	3.61-02	4.00-02	4.82-02	6.13-02	7.04-02	8.40-02	8.69-02
Xe	54	5.78-02	4.08-02	3.58-02	3.99-02	4.82-02	6.12-02	7.04-02	8.40-02	8.69-02
Cs	55	5.85-02	4.12-02	3.64-02	4.06-02	4.91-02	6.25-02	7.19-02	8.58-02	8.88-02
ZB	56	5.80-02	4.08-02	3.61-02	4.04-02	4.90-02	6.25-02	7.19-02	8.58-02	8.88-02
La	57	5.88-02	4.12-02	3.66-02	4.11-02	5.00-02	6.37-02	7.34-02	8.76-02	9.06-02
Ce	58	5.96-02	4.18-02	3.73-02	4.19-02	5.10-02	6.52-02	7.50-02	8.96-02	9.27-02
Pr	59	6.07-02	4.24-02	3.80-02	4.29-02	5.23-02	6.68-02	7.69-02	9.19-02	9.50-02
Nd	60	6.07-02	4.24-02	3.81-02	4.30-02	5.26-02	6.72-02	7.74-02	9.25-02	9.56-02
Pm	61	6.19-02	4.31-02	3.88-02	4.40-02	5.38-02	6.89-02	7.94-02	9.48-02	9.81-02
Sm	62	6.11-02	4.24-02	3.83-02	4.35-02	5.34-02	6.84-02	7.88-02	9.41-02	9.73-02
Eu	63	6.19-02	4.28-02	3.88-02	4.42-02	5.42-02	6.96-02	8.02-02	9.57-02	9.90-02
Gd	64	6.12-02	4.23-02	3.84-02	4.38-02	5.38-02	6.91-02	7.97-02	9.51-02	9.83-02
Tb	65	6.20-02	4.27-02	3.89-02	4.45-02	5.47-02	7.03-02	8.11-02	9.67-02	1.00-01
Dy	66	6.20-02	4.26-02	3.90-02	4.46-02	5.49-02	7.06-02	8.15-02	9.72-02	1.00-01
Ho	67	6.26-02	4.29-02	3.93-02	4.50-02	5.55-02	7.14-02	8.24-02	9.83-02	1.02-01
Er	68	6.32-02	4.32-02	3.96-02	4.55-02	5.61-02	7.23-02	8.34-02	9.95-02	1.03-01
Tm	69	6.40-02	4.36-02	4.01-02	4.61-02	5.70-02	7.35-02	8.48-02	1.01-01	1.04-01
Yb	70	6.40-02	4.35-02	4.00-02	4.61-02	5.70-02	7.35-02	8.49-02	1.01-01	1.04-01
Lu	71	6.48-02	4.39-02	4.05-02	4.66-02	5.77-02	7.45-02	8.60-02	1.02-01	1.06-01
Hf	72	6.50-02	4.39-02	4.05-02	4.68-02	5.80-02	7.48-02	8.64-02	1.03-01	1.06-01
Ta	73	6.57-02	4.41-02	4.08-02	4.72-02	5.85-02	7.56-02	8.73-02	1.04-01	1.07-01
W	74	6.62-02	4.43-02	4.10-02	4.75-02	5.89-02	7.62-02	8.80-02	1.05-01	1.08-01
Re	75	6.69-02	4.46-02	4.14-02	4.79-02	5.95-02	7.70-02	8.89-02	1.06-01	1.09-01
Os	76	6.71-02	4.46-02	4.13-02	4.79-02	5.96-02	7.71-02	8.90-02	1.06-01	1.10-01
Ir	77	6.79-02	4.50-02	4.17-02	4.84-02	6.02-02	7.80-02	9.01-02	1.07-01	1.11-01
Pt	78	6.86-02	4.52-02	4.20-02	4.87-02	6.06-02	7.86-02	9.08-02	1.08-01	1.12-01
Au	79	6.95-02	4.57-02	4.24-02	4.93-02	6.14-02	7.95-02	9.19-02	1.09-01	1.13-01
Hg	80	6.99-02	4.57-02	4.25-02	4.94-02	6.15-02	7.98-02	9.22-02	1.10-01	1.13-01
Tl	81	7.03-02	4.58-02	4.25-02	4.94-02	6.16-02	8.00-02	9.24-02	1.10-01	1.14-01
Pb	82	7.10-02	4.61-02	4.27-02	4.97-02	6.21-02	8.06-02	9.31-02	1.11-01	1.15-01
Bi	83	7.21-02	4.66-02	4.32-02	5.03-02	6.28-02	8.15-02	9.42-02	1.12-01	1.16-01
Po	84	7.39-02	4.75-02	4.40-02	5.12-02	6.40-02	8.32-02	9.61-02	1.15-01	1.18-01
At	85	7.54-02	4.82-02	4.46-02	5.20-02	6.49-02	8.44-02	9.76-02	1.16-01	1.20-01
Rn	86	7.30-02	4.65-02	4.30-02	5.01-02	6.26-02	8.14-02	9.42-02	1.12-01	1.16-01
Fr	87	7.45-02	4.72-02	4.36-02	5.08-02	6.35-02	8.26-02	9.56-02	1.14-01	1.18-01
Ra	88	7.53-02	4.75-02	4.38-02	5.10-02	6.38-02	8.31-02	9.61-02	1.15-01	1.19-01
Ac	89	7.69-02	4.82-02	4.44-02	5.17-02	6.47-02	8.43-02	9.75-02	1.16-01	1.20-01
Th	90	7.71-02	4.81-02	4.42-02	5.15-02	6.45-02	8.40-02	9.72-02	1.16-01	1.20-01
Pa	91	7.94-02	4.93-02	4.52-02	5.26-02	6.59-02	8.60-02	9.95-02	1.19-01	1.23-01
U	92	7.90-02	4.88-02	4.46-02	5.19-02	6.51-02	8.49-02	9.83-02	1.17-01	1.21-01
Np	93	8.13-02	4.99-02	4.56-02	5.30-02	6.65-02	8.68-02	1.01-01	1.20-01	1.24-01
Pu	94	8.26-02	5.05-02	4.60-02	5.34-02	6.71-02	8.76-02	1.01-01	1.21-01	1.25-01
Am	95	8.33-02	5.06-02	4.60-02	5.34-02	6.70-02	8.77-02	1.02-01	1.21-01	1.25-01
Cm	96	8.41-02	5.08-02	4.60-02	5.34-02	6.70-02	8.77-02	1.02-01	1.21-01	1.26-01
Bk	97	8.62-02	5.18-02	4.68-02	5.42-02	6.81-02	8.92-02	1.03-01	1.24-01	1.28-01
Cf	98	8.70-02	5.20-02	4.68-02	5.42-02	6.81-02	8.92-02	1.04-01	1.24-01	1.28-01
Es	99	8.89-02	5.28-02	4.74-02	5.48-02	6.89-02	9.04-02	1.05-01	1.25-01	1.29-01
Fm	100	8.94-02	5.28-02	4.72-02	5.45-02	6.86-02	9.00-02	1.05-01	1.25-01	1.29-01

CLASSIFICATION OF ELECTROMAGNETIC RADIATION

Hans Dolezalek

Basic Conversions: $c = \lambda\nu = \nu/k$; $\nu = c/\lambda = ck$; $\lambda = c/\nu = 1/k$; $k = \nu/c = 1/\lambda$
c = speed of light = 2.99792458×10^8 m/s

Frequency (ν)	Wavelength (λ)	Wave number (k)	Names of bands	Approximate photon energies
$3 \times 10^0 - 3 \times 10^1$ Hz 3 — 30 Hz	$10^8 - 10^7$ m 100 — 10 Mm	$10^{-8} - 10^{-7}$ m^{-1} 10 — 100 Gm^{-1}	ELF-(ELF 1), ITU band no. 1	
$3 \times 10^1 - 3 \times 10^2$ Hz 30 — 300 Hz	$10^7 - 10^6$ m 10 — 1 Mm	$10^{-7} - 10^{-6}$ m^{-1} 100 Gm^{-1} — 1 Mm^{-1}	SLF-(ELF 2), ITU band no. 2, mega- meter waves	
$3 \times 10^2 - 3 \times 10^3$ Hz 300 Hz — 3 kHz	$10^6 - 10^5$ m 1 Mm — 100 km	$10^{-6} - 10^{-5}$ m^{-1} 1 — 10 Mm^{-1}	ULF-(ELF 3), ITU band no. 3	
$3 \times 10^3 - 3 \times 10^4$ Hz 3 — 30 kHz	$10^5 - 10^4$ m 100 — 10 km	$10^{-5} - 10^{-4}$ m^{-1} 10 — 100 Mm^{-1}	VLF, ITU band no. 4, myriameter waves	
$3 \times 10^4 - 3 \times 10^5$ Hz 30 — 300 kHz	$10^4 - 10^3$ m 10 — 1 km	$10^{-4} - 10^{-3}$ m^{-1} 100 Mm^{-1} — 1 km^{-1}	LF, ITU band no. 5, kilometer waves	
$3 \times 10^5 - 3 \times 10^6$ Hz 300 kHz — 3 MHz	$10^3 - 10^2$ m 1 km — 100 m	$10^{-3} - 10^{-2}$ m^{-1} 1 — 10 km^{-1}	MF, ITU band no. 6, hectometer waves	
$3 \times 10^6 - 3 \times 10^7$ Hz 3 — 30 MHz	$10^2 - 10^1$ m 100 — 10 m	$10^{-2} - 10^{-1}$ m^{-1} 10 — 100 km^{-1}	HF, ITU band no. 7, decameter waves	
$3 \times 10^7 - 3 \times 10^8$ Hz 30 — 300 MHz	$10^1 - 10^0$ m 10 — 1 m	$10^{-1} - 10^0$ m^{-1} 100 km^{-1} — 1 m^{-1}	VHF, ITU band no. 8, meter waves	
$3 \times 10^8 - 3 \times 10^9$ Hz 300 MHz — 3 GHz	$10^0 - 10^{-1}$ m 1 m — 100 mm	$10^0 - 10^1$ m^{-1} 1 — 10 m^{-1}	UHF, ITU band no. 9, decimeter waves[a]	
$3 \times 10^9 - 3 \times 10^{10}$ Hz 3 — 30 GHz	$10^{-1} - 10^{-2}$ m 100 — 10 mm	$10^1 - 10^2$ m^{-1} 10 — 100 m^{-1}	SHF, ITU band no. 10, centimeter waves[a]	
$3 \times 10^{10} - 3 \times 10^{11}$ Hz 30 — 300 GHz	$10^{-2} - 10^{-3}$ m 10 — 1 mm	$10^2 - 10^3$ m^{-1} 100 m^{-1} — 1 mm^{-1} (1 — 10 cm^{-1})	EHF, ITU band no. 11, millimeter waves	
$3 \times 10^{11} - 3 \times 10^{12}$ Hz 300 GHz — 3 THz	$10^{-3} - 10^{-4}$ m 1 mm — 100 μm	$10^3 - 10^4$ m^{-1} 1 — 10 mm^{-1} (10 — 100 cm^{-1})	Part of micrometer waves, includes part of far or thermal infrared; ITU band no. 12	
$3 \times 10^{12} - 3 \times 10^{13}$ Hz 3 — 30 THz	$10^{-4} - 10^{-5}$ m 100 — 10 μm	$10^4 - 10^5$ m^{-1} 10 — 100 mm^{-1} (100 — 1000 cm^{-1})	Part of micrometer waves includes part of far (thermal) infrared	
$3 \times 10^{13} - 3 \times 10^{14}$ Hz 30 — 300 THz	$10^{-5} - 10^{-6}$ m 10 — 1 μm (100,000 — 10,000 Å)	$10^5 - 10^6$ m^{-1} 100 mm^{-1} — 1 μm^{-1}	Part of μm waves, part of infrared	$(1.6 - 16) \times 10^{-20}$ joule {0.1 — 1 eV}
$3 \times 10^{14} - 3 \times 10^{15}$ Hz 300 THz — 3 PHz	$10^{-6} - 10^{-7}$ m 1 μm — 100 nm (10,000 — 1000 Å)	$10^6 - 10^7$ m^{-1} 1 — 10 μm^{-1}	Near infrared, visible, near ultraviolet	$(1.6 - 16) \times 10^{-19}$ joule {1 — 10 eV}
$3 \times 10^{15} - 3 \times 10^{16}$ Hz 3 — 30 PHz	$10^{-7} - 10^{-8}$ m 100 — 10 nm (1000 — 100 Å)	$10^7 - 10^8$ m^{-1} 10 — 100 μm^{-1}	Part of "vacuum" - ultraviolet	$(1.6 - 16) \times 10^{-18}$ joule {10 — 100 eV}
$3 \times 10^{16} - 3 \times 10^{17}$ Hz 30 — 300 PHz	$10^{-8} - 10^{-9}$ m 10 — 1 nm (100 — 10 Å)	$10^8 - 10^9$ m^{-1} 100 μm^{-1} — 1 nm^{-1}	Part of soft X-rays	$(1.6 - 16) \times 10^{-17}$ joule {100 — 1000 eV)}
$3 \times 10^{17} - 3 \times 10^{18}$ Hz 300 PHz — 3 EHz	$10^{-9} - 10^{-10}$ m 1 nm — 100 pm (10 — 1 Å)	$10^9 - 10^{10}$ m^{-1} 1 — 10 nm^{-1}	Part of soft X-rays	$(1.6 - 16) \times 10^{-16}$ joule {1 — 10 keV}
$3 \times 10^{18} - 3 \times 10^{19}$ Hz 3 — 30 EHz	$10^{-10} - 10^{-11}$ m 100 — 10 pm (1 — 0.1 Å)	$10^{10} - 10^{11}$ m^{-1} 10 — 100 nm^{-1}	Hard X-rays and part of soft γ-rays	$(1.6 - 16) \times 10^{-15}$ joule {10 — 100 keV}
$3 \times 10^{19} - 3 \times 10^{20}$ Hz 30 — 300 EHz	$10^{-11} - 10^{-12}$ m 10 — 1 pm (0.1 — 0.01 Å)	$10^{11} - 10^{12}$ m^{-1} 100 nm^{-1} — 1 pm^{-1}	Part of soft and part of hard γ-rays (limit at 510 keV)	$(1.6 - 16) \times 10^{-14}$ joule {100 keV — 1 MeV}
$3 \times 10^{20} - 3 \times 10^{21}$ Hz 300 — 3000 EHz	$10^{-12} - 10^{-13}$ m 1 pm — 100 fm (0.01 — 0.001 Å)	$10^{12} - 10^{13}$ m^{-1} 1 — 10 pm^{-1}	Part of hard γ-rays and part of "cosmic" γ-rays	$(1.6 - 16) \times 10^{-13}$ joule {1 — 10 MeV}
$3 \times 10^{21} - 3 \times 10^{22}$ Hz 3000 — 30,000 EHz	$10^{-13} - 10^{-14}$ m 100 — 10 fm (0.001 — 0.0001 Å)	$10^{13} - 10^{14}$ m^{-1} 10 — 100 pm^{-1}	γ-rays produced by cosmic rays	$(1.6 - 16) \times 10^{-12}$ joule {10 — 100 MeV}

CLASSIFICATION OF ELECTROMAGNETIC RADIATION (continued)

Note: Abbreviations used in this table: Å—ångstrom (1 Å = 10^{-10} m); EHz—exahertz (10^{18} hertz); EHF—extremely high frequency; ELF—extremely low frequency; eV—electron volt (1 eV = 1.60219×10^{-19} joule); PHz—petahertz (10^{15} hertz); fm—femtometer (10^{-15} m); GHz—gigahertz (10^9 hertz); Gm—gigameter (10^9 m); HF—high frequency; Hz—hertz (s^{-1}); ITU—International Telecommunications Union; keV—kiloelectron volt (10^3 eV); km—kilometer (10^3 m); LF—low frequency; m—meter; MeV—megaelectron volt (10^6 eV); MF—medium frequency; MHz—megahertz (10^6 hertz); Mm—megameter (10^6 meter); mm—millimeter (10^{-3} meter); μm—micrometer (10^{-6} meter); nm—nanometer (10^{-9} meter); pm—picometer (10^{-12} meter); SHF—super high frequency; SLF—super low frequency; THz—terahertz; UHF—ultra high frequency; ULF—ultra low frequency; VHF—very high frequency; VLF—very low frequency.

[a]   Also called "microwaves"; not to be confused with "micrometer waves".

LETTER DESIGNATIONS OF MICROWAVE BANDS

Frequency (GHz)	Wavelength (cm)	Wavenumber (cm^{-1})	Band
1—2	30—15	0.033—0.067	L-Band
1—4	15—7.5	0.067—0.133	S-Band
4—8	7.5—3.7	0.133—0.267	C-Band
8—12	3.7—2.5	0.267—0.4	X-Band
12—18	2.5—1.7	0.4—0.6	Ku-Band
18—27	1.7—1.1	0.6—0.9	K-Band
27—40	1.1—0.75	0.9—1.33	Ka-Band

BLACK BODY RADIATION

The total power radiated from an ideal black body and the wavelength corresponding to maximum power are given here as a function of absolute temperature. Constants used in the calculation are taken from the table "Fundamental Physical Constants" in Section 1. The radiated power in a band $\Delta\lambda$ at λ_{max} may be calculated from:

$$P_{max} = 0.657548 \, (\Delta\lambda/\lambda_{max}) \, P_{tot}$$

T/K	P_{tot}	$\lambda_{max}/\mu m$	T/K	P_{tot}	$\lambda_{max}/\mu m$	T/K	P_{tot}	$\lambda_{max}/\mu m$
50	0.354 W/m^2	57.955	740	17.004	3.916	1520	302.689	1.906
100	5.671	28.978	750	17.942	3.864	1540	318.937	1.882
150	28.707	19.318	760	18.918	3.813	1560	335.831	1.858
200	90.728	14.489	770	19.934	3.763	1580	353.387	1.834
250	221.504	11.591	780	20.989	3.715	1600	371.623	1.811
273	314.973	10.614	790	22.087	3.668	1620	390.555	1.789
280	348.541	10.349	800	23.226	3.622	1640	410.202	1.767
290	401.064	9.992	810	24.410	3.577	1660	430.581	1.746
300	459.311	9.659	820	25.638	3.534	1680	451.710	1.725
310	523.684	9.348	830	26.911	3.491	1700	473.607	1.705
320	594.596	9.055	840	28.232	3.450	1720	496.290	1.685
330	672.478	8.781	850	29.600	3.409	1740	519.779	1.665
340	757.771	8.523	860	31.018	3.369	1760	544.093	1.646
350	850.931	8.279	870	32.486	3.331	1780	569.249	1.628
360	952.428	8.049	880	34.006	3.293	1800	595.267	1.610
370	1.063 kW/m^2	7.832	890	35.578	3.256	1820	622.168	1.592
380	1.182	7.626	900	37.204	3.220	1840	649.970	1.575
390	1.312	7.430	910	38.886	3.184	1860	678.694	1.558
400	1.452	7.244	920	40.623	3.150	1880	708.359	1.541
410	1.602	7.068	930	42.418	3.116	1900	738.987	1.525
420	1.764	6.899	940	44.272	3.083	1920	770.597	1.509
430	1.939	6.739	950	46.187	3.050	1940	803.210	1.494
440	2.125	6.586	960	48.162	3.018	1960	836.848	1.478
450	2.325	6.439	970	50.201	2.987	1980	871.531	1.464
460	2.539	6.299	980	52.303	2.957	2000	907.282	1.449
470	2.767	6.165	990	54.471	2.927	2020	944.121	1.435
480	3.010	6.037	1000	56.705	2.898	2040	982.071	1.420
490	3.269	5.914	1020	61.379	2.841	2060	1.021 MW/m^2	1.407
500	3.544	5.796	1040	66.337	2.786	2080	1.061	1.393
510	3.836	5.682	1060	71.589	2.734	2100	1.103	1.380
520	4.146	5.573	1080	77.147	2.683	2120	1.145	1.367
530	4.474	5.467	1100	83.022	2.634	2140	1.189	1.354
540	4.822	5.366	1120	89.227	2.587	2160	1.234	1.342
550	5.189	5.269	1140	95.773	2.542	2180	1.281	1.329
560	5.577	5.175	1160	102.672	2.498	2200	1.328	1.317
570	5.986	5.084	1180	109.939	2.456	2220	1.377	1.305
580	6.417	4.996	1200	117.584	2.415	2240	1.428	1.294
590	6.871	4.911	1220	125.621	2.375	2260	1.479	1.282
600	7.349	4.830	1240	134.063	2.337	2280	1.532	1.271
610	7.851	4.750	1260	142.924	2.300	2300	1.587	1.260
620	8.379	4.674	1280	152.217	2.264	2320	1.643	1.249
630	8.933	4.600	1300	161.955	2.229	2340	1.700	1.238
640	9.514	4.528	1320	172.154	2.195	2360	1.759	1.228
650	10.122	4.458	1340	182.827	2.163	2380	1.819	1.218
660	10.760	4.391	1360	193.989	2.131	2400	1.881	1.207
670	11.427	4.325	1380	205.655	2.100	2420	1.945	1.197
680	12.124	4.261	1400	217.838	2.070	2440	2.010	1.188
690	12.853	4.200	1420	230.556	2.041	2460	2.077	1.178
700	13.615	4.140	1440	243.822	2.012	2480	2.145	1.168
710	14.410	4.081	1460	257.652	1.985	2500	2.215	1.159
720	15.239	4.025	1480	272.063	1.958	2550	2.398	1.136
730	16.103	3.970	1500	287.070	1.932	2600	2.591	1.115

BLACK BODY RADIATION (continued)

T/K	P_{tot}	$\lambda_{max}/\mu m$	T/K	P_{tot}	$\lambda_{max}/\mu m$	T/K	P_{tot}	$\lambda_{max}/\mu m$
2650	2.796	1.093	3600	9.524	0.805	5100	38.362	0.568
2700	3.014	1.073	3650	10.065	0.794	5200	41.461	0.557
2750	3.243	1.054	3700	10.627	0.783	5300	44.743	0.547
2800	3.485	1.035	3750	11.214	0.773	5400	48.217	0.537
2850	3.741	1.017	3800	11.824	0.763	5500	51.889	0.527
2900	4.011	0.999	3850	12.458	0.753	5600	55.767	0.517
2950	4.294	0.982	3900	13.118	0.743	5700	59.858	0.508
3000	4.593	0.966	3950	13.804	0.734	5800	64.170	0.500
3050	4.907	0.950	4000	14.517	0.724	5900	68.712	0.491
3100	5.237	0.935	4100	16.024	0.707	6000	73.490	0.483
3150	5.583	0.920	4200	17.645	0.690	6500	101.222	0.446
3200	5.946	0.906	4300	19.386	0.674	7000	136.149	0.414
3250	6.326	0.892	4400	21.254	0.659	7500	179.418	0.386
3300	6.725	0.878	4500	23.253	0.644	8000	232.264	0.362
3350	7.142	0.865	4600	25.389	0.630	8500	296.004	0.341
3400	7.578	0.852	4700	27.670	0.617	9000	372.042	0.322
3450	8.033	0.840	4800	30.101	0.604	9500	461.867	0.305
3500	8.509	0.828	4900	32.689	0.591	10000	567.051	0.290
3550	9.006	0.816	5000	35.441	0.580			

The curves below show, for various temperatures, the fraction of radiant power as a function of wavelength. The function plotted is $P_\lambda/\Delta\lambda\, P_{tot}$, where P_λ is the power at wavelength λ in a small interval $\Delta\lambda$ (in μm), and P_{tot} is the total power.

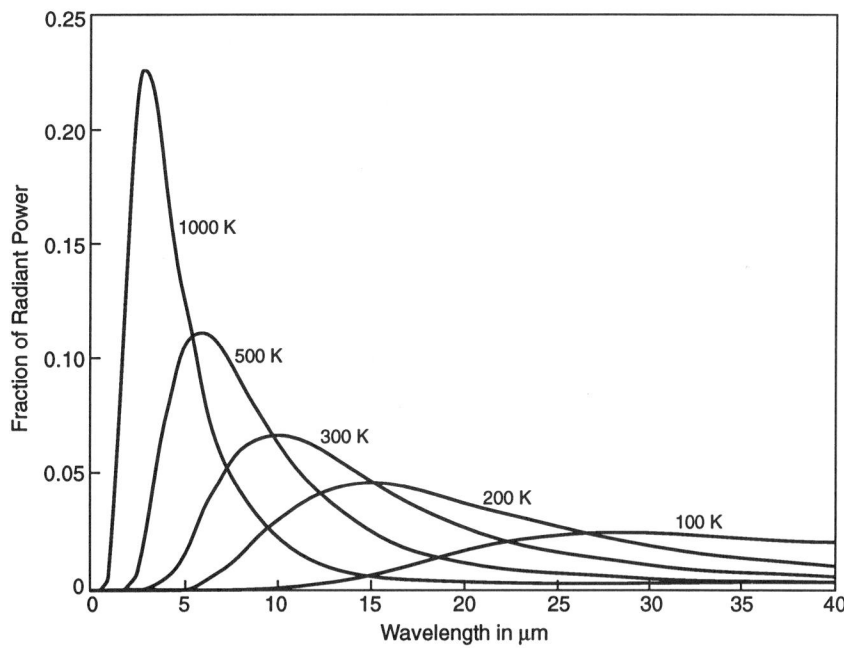

CHARACTERISTICS OF INFRARED DETECTORS

This graph summarizes the wavelength response of some semiconductors used as detectors for infrared radiation. The quantity $D^*(\lambda)$ is the signal to noise ratio for an incident radiant power density of 1 W/cm^2 and a bandwidth of 1 Hz (60° field of view). The Ge, InAs, and InSb detectors are photovoltaics, while the HgCdTe series are photoconductive devices. The cutoff wavelength of the latter can be varied by adjusting the relative amounts of Hg, Cd, and Te (three examples are shown at 77 K). The graph also shows the theoretical background limited sensitivity for ideal detectors which introduce no intrinsic noise.

REFERENCE

Infrared Detectors 1995, EG&G Judson, Montgomeryville, PA.

REFRACTIVE INDEX AND TRANSMITTANCE OF REPRESENTATIVE GLASSES

Typical values of the index of refraction and internal transmittance (fraction of light transmitted through a one centimeter thickness) are tabulated here for selected types of glasses, as well as for synthetic fused (vitreous) silica. Nominal compositions are given in the first part of the table. The second part gives the index of refraction, relative to air, and the internal transmittance for representative samples of each glass at wavelengths in the infrared, visible, and near-ultraviolet regions. It should be emphasized that wide variation of these parameters may be found among subtypes of each glass. More detailed data may be found in Reference 3.

Assuming that the Lambert-Beer Law is followed, the transmittance of a glass plate of thickness d (in centimeters) can be obtained by raising the transmittance value in the table to the power d.

REFERENCES

1. Weber, M.J., *CRC Handbook of Laser Science and Technology*, Vol. IV, Part 2, CRC Press, Boca Raton, FL ,1988.
2. Gray, D.E., Ed., *American Institute of Physics Handbook, Third Edition*, McGraw Hill, New York, 1972.
3. *Schott Optical Glass*, Schott Glass Technologies, Inc., 400 York Ave., Duryea, PA 18642.
4. Kaye, G.W.C., and Laby, T.H., *Tables of Physical and Chemical Constants, 15th Edition*, Longman, London, 1986.

Type	Name	Composition in percent by mass									
		SiO_2	B_2O_3	Al_2O_3	Na_2O	K_2O	CaO	BaO	ZnO	PbO	P_2O_5
PK	Phosphate crown		3	10		12	5				70
PSK	Dense phosphate crown		3	5			4	28			60
BK	Borosilicate crown	70	10		8	8	1	3			
K	Crown	74			9	11	6				
ZK	Zinc crown	71			17				12		
BaK	Barium crown	60	3		3	10		19	5		
SK	Dense crown	39	15	5				41			
KF	Crown flint	67		2	16				3	12	
BaLF	Barium light flint	51			6	5		20	14	4	
SSK	Extra dense crown	35	10	5				42	8		
LLF	Extra light flint	63			5	8				24	
BaF	Barium flint	46				8		16	8	22	
LF	Light flint	53			5	8				34	
F	Flint	47			2	7				44	
BaSF	Dense barium flint	43			1	7		11	5	33	
SF	Dense flint	33				5				62	
KzFS	Short flint										
SiO_2	Fused silica	100									

Type	Index of refraction				Transmittance of 1 cm plate			
	1.060 μm	546.1 nm	365.0 nm	312.6 nm	1.060 μm	546.1 nm	365.0 nm	310 nm
PK	1.51519	1.52736	1.54503	1.5574	0.997	0.998	0.987	0.46
PSK	1.54154	1.55440	1.57342	1.5868	0.996	0.998	0.984	0.46
BK	1.50669	1.51872	1.53627	1.5486	0.999	0.998	0.987	0.35
K	1.50091	1.51314	1.53189	1.5454	0.999	0.998	0.987	0.35
ZK	1.52220	1.53534	1.55588	1.5708	0.998	0.998	0.988	0.40
BaK	1.55695	1.57124	1.59407	1.6108	0.996	0.998	0.976	0.27
SK	1.59490	1.60994	1.63398		0.998	0.997	0.986	0.28
KF	1.50586	1.51978	1.54251	1.5600	0.998	0.998	0.959	0.28
BaLF	1.57579	1.59166	1.61804		0.998	0.996	0.989	0.49
SSK	1.60402	1.61993	1.64595		0.996	0.998	0.933	0.010
LaK	1.69710	1.71616	1.74573		0.999	0.998	0.915	0.010
LLF	1.52775	1.54344	1.57038		0.999	0.998	0.882	0.17
BaF	1.56873	1.58565	1.61524		0.998	0.997	0.990	0.32
LF	1.56594	1.58482	1.61926		0.999	0.997	0.992	0.004
F	1.58636	1.60718	1.64606		0.999	0.998	0.981	0.008
BaSF	1.60889	1.62987	1.66926		0.997	0.998	0.959	
SF	1.71350	1.74620	1.8145		0.999	0.998	0.857	
KzFS	1.59680	1.61639	1.64849	1.6739	0.998	0.997	0.650	
SiO_2	1.44968	1.46008	1.47435[a]	1.53430[b]		0.998	0.672	0.012

[a] At 366.3 nm.

[b] At 213.9 nm.

INDEX OF REFRACTION OF WATER

This table gives the index of refraction of liquid water at atmospheric pressure, relative to a vacuum, at several temperatures and wavelengths. It is generated from the formulation in Reference 1, which covers a wide range of temperature, pressure, and wavelength. The wavelengths listed here correspond to prominent lines of cadmium (226.50 and 361.05 nm), potassium (404.41 nm), sodium (589.00 nm), Ne (632.80 nm, from a helium - neon laser), and mercury (1.01398 μm).

REFERENCES

1. Schiebener, P., Straub, J., Levelt Sengers, J.M.H., and Gallagher, J.S., *J. Phys. Chem. Ref. Data*, 19, 677 (1990); 19, 1617, 1990.
2. Marsh, K.N., Editor, *Recommended Reference Materials for the Realization of Physicochemical Properties*, Blackwell Scientific Publications, Oxford, 1987.

T/°C	226.50 nm	361.05 nm	404.41 nm	589.00 nm	632.80 nm	1.01398 μm
0	1.39450	1.34896	1.34415	1.33432	1.33306	1.32612
10	1.39422	1.34870	1.34389	1.33408	1.33282	1.32591
20	1.39336	1.34795	1.34315	1.33336	1.33211	1.32524
30	1.39208	1.34682	1.34205	1.33230	1.33105	1.32424
40	1.39046	1.34540	1.34065	1.33095	1.32972	1.32296
50	1.38854	1.34373	1.33901	1.32937	1.32814	1.32145
60	1.38636	1.34184	1.33714	1.32757	1.32636	1.31974
70	1.38395	1.33974	1.33508	1.32559	1.32438	1.31784
80	1.38132	1.33746	1.33284	1.32342	1.32223	1.31576
90	1.37849	1.33501	1.33042	1.32109	1.31991	1.31353
100	1.37547	1.33239	1.32784	1.31861	1.31744	1.31114

INDEX OF REFRACTION OF LIQUIDS FOR CALIBRATION PURPOSES

This table gives the index of refraction of six liquids which are available in highly pure form and whose index of refraction has been accurately measured as a function of wavelength and temperature. They are therefore useful for calibration of refractometers. The estimated uncertainty in the values is:

2,2,4-Trimethylpentane	±0.00003
Hexadecane	±0.00008
trans-Bicyclo[4.0.0]decane	±0.00008
1-Methylnaphthalene	±0.00008
Toluene	±0.00003
Methylcyclohexane	±0.00003

Full details are given in the references. This table is reprinted from Reference 1 by permission of the International Union of Pure and Applied Chemistry.

REFERENCES

1. Marsh, K. N., Editor, *Recommended Reference Materials for the Realization of Physicochemical Properties,* Blackwell Scientific Publications, Oxford, 1987.
2. Tilton, L. W., J. *Opt. Soc. Am.*, 32, 71, 1941.

λ	2,2,4-Trimethylpentane			Hexadecane		
nm	20°C	25°C	30°C	20°C	25°C	30°C
667.81	1.38916	1.38670	1.38424	1.43204	1.43001	1.42798
656.28	1.38945	1.38698	1.38452	1.43235	1.43032	1.42829
589.26	1.39145	1.38898	1.38650	1.43453	1.43250	1.43047
546.07	1.39316	1.39068	1.38820	1.43640	1.43436	1.43232
501.57	1.39544	1.39294	1.39044	1.43888	1.43684	1.43480
486.13	1.39639	1.39389	1.39138	1.43993	1.43788	1.43583
435.83	1.40029	1.39776	1.39523	1.44419	1.44213	1.44007

λ	*trans*-Bicyclo[4.4.0]decane			1-Methylnaphthalene		
nm	20°C	25°C	30°C	20°C	25°C	30°C
667.81	1.46654	1.46438	1.46222	1.60828	1.60592	1.60360
656.28	1.46688	1.46472	1.46256	1.60940	1.60703	1.60471
589.26	1.46932	1.46715	1.46498	1.61755	1.61512	1.61278
546.07	1.47141	1.46923	1.46705	1.62488	1.62240	1.62005
501.57	1.47420	1.47200	1.46980	1.63513	1.63259	1.63022
486.13	1.47535	1.47315	1.47095	1.63958	1.63701	1.63463
435.83	1.48011	1.47789	1.47567		1.65627	1.65386

λ	Toluene			Methylcyclohexane		
nm	20°C	25°C	30°C	20°C	25°C	30°C
667.81	1.49180	1.48903	1.48619	1.42064	1.41812	1.41560
656.28	1.49243	1.48966	1.48682	1.42094	1.41842	1.41591
589.26	1.49693	1.49413	1.49126	1.42312	1.42058	1.41806
546.07	1.50086	1.49803	1.49514	1.42497	1.42243	1.41989
501.57	1.50620	1.50334	1.50041	1.42744	1.42488	1.42233
486.13	1.50847	1.50559	1.50265	1.42847	1.42590	1.42334
435.83	1.51800	1.51506	1.51206	1.43269	1.43010	1.42752

INDEX OF REFRACTION OF AIR

This is a table of the index of refraction n of dry air at 15°C and a pressure of 101.325 kPa and containing 0.03% by volume of carbon dioxide ("standard air"). The index of refraction is defined by $n = \lambda_{vac}/\lambda_{air}$, where λ is the wavelength of the radiation. The index is calculated from the expression:

$$(n - 1) \times 10^8 = 8342.13 + 2406030(130 - \sigma^2)^{-1} + 15997(38.9 - \sigma^2)^{-1}$$

where $\sigma = 1/\lambda_{vac}$ and λ_{vac} has units of μm. The equation is valid for λ_{vac} from 200 nm to 2 μm.

The table also gives the correction $(n - 1)\lambda_{air}$ which must be added to the wavelength in air to obtain λ_{vac}.

If the air is at a temperature t in °C and a pressure p in pascals, a value of $(n - 1)$ from this table should be multiplied by

$$\frac{p\left[1 + p(61.3 - t) \times 10^{-10}\right]}{96095.4(1 + 0.003661t)}$$

REFERENCE

Edlen, B., *Metrologia,* 2, 71, 1966.

λ_{vac}	$(n - 1) \times 10^8$	$\lambda_{vac} - \lambda_{air}$	λ_{vac}	$(n - 1) \times 10^8$	$\lambda_{vac} - \lambda_{air}$	λ_{vac}	$(n - 1) \times 10^8$	$\lambda_{vac} - \lambda_{air}$
200 nm	32408	0.0648 nm	540	27803	0.1501	880	27461	0.2416
210	31746	0.0666	550	27782	0.1528	890	27457	0.2443
220	31224	0.0687	560	27763	0.1554	900	27452	0.2470
230	30799	0.0708	570	27745	0.1581	910	27448	0.2497
240	30445	0.0730	580	27728	0.1608	920	27444	0.2524
250	30146	0.0753	590	27712	0.1635	930	27440	0.2551
260	29890	0.0777	600	27697	0.1661	940	27436	0.2578
270	29669	0.0801	610	27682	0.1688	950	27432	0.2605
280	29475	0.0825	620	27669	0.1715	960	27429	0.2632
290	29306	0.0850	630	27656	0.1742	970	27425	0.2660
300	29155	0.0874	640	27643	0.1769	980	27422	0.2687
310	29022	0.0899	650	27631	0.1796	990 nm	27419	0.2714 nm
320	28902	0.0925	660	27620	0.1822			
330	28795	0.0950	670	27609	0.1849	1.00 μm	27416	0.000274 μm
340	28698	0.0975	680	27599	0.1876	1.05	27401	0.000288
350	28611	0.1001	690	27589	0.1903	1.10	27389	0.000301
360	28531	0.1027	700	27579	0.1930	1.15	27378	0.000315
370	28458	0.1053	710	27570	0.1957	1.20	27368	0.000328
380	28392	0.1079	720	27562	0.1984	1.25	27360	0.000342
390	28331	0.1105	730	27553	0.2011	1.30	27352	0.000355
400	28275	0.1131	740	27545	0.2038	1.35	27346	0.000369
410	28223	0.1157	750	27538	0.2065	1.40	27340	0.000383
420	28175	0.1183	760	27530	0.2092	1.45	27334	0.000396
430	28131	0.1209	770	27523	0.2119	1.50	27330	0.0C0410
440	28090	0.1236	780	27516	0.2146	1.55	27325	0.000423
450	28052	0.1262	790	27510	0.2173	1.60	27321	0.000437
460	28016	0.1288	800	27504	0.2200	1.65	27318	0.000451
470	27983	0.1315	810	27498	0.2227	1.70	27314	0.000464
480	27952	0.1341	820	27492	0.2254	1.75	27311	0.000478
490	27923	0.1368	830	27486	0.2281	1.80	27309	0.000491
500	27896	0.1394	840	27481	0.2308	1.85	27306	0.000505
510	27870	0.1421	850	27476	0.2335	1.90	27304	0.000519
520	27846	0.1448	860	27471	0.2362	1.95	27302	0.000532
530	27824	0.1474	870	27466	0.2389	2.00 μm	27300	0.000546 μm

CHARACTERISTICS OF LASER SOURCES
William F. Krupke

Light Amplification by Stimulated Emission of Radiation was first demonstrated by Maiman in 1960, the result of a population inversion produced between energy levels of chromium ions in a ruby crystal when irradiated with a xenon flashlamp. Since then population inversions and coherent emission have been generated in literally thousands of substances (neutral and ionized gases, liquids, and solids) using a variety of incoherent excitation techniques (optical pumping, electrical discharges, gas-dynamic flow, electron-beams, chemical reactions, nuclear decay).

The extrema of laser output parameters which have been demonstrated to date and the laser media used are summarized in Table 1. Note that the extreme power and energy parameters listed in this table were attained with laser *systems* rather than with simple laser oscillators.

Laser sources are commonly classified in terms of the state-of-matter of the active medium: gas, liquid, and solid. Each of these classes is further subdivided into one or more types as shown in Table 2. A well-known representative example of each type of laser is also given in Table 2 together with its nominal operation wavelength and the methods by which it is pumped.

The various lasers together cover a wide spectral range from the far ultraviolet to the far infrared. The particular wavelength of emission (usually a narrow line) is presented for some six dozen lasers in Figures 1A and 1B.

By suitably designing the excitation source and/or by controlling the laser resonator structure, laser systems can provide continuous or pulsed radiation as shown in Table 3.

Besides the method of excitation and the temporal behavior of a laser, there are many other parameters that characterize its operation and efficiency, as shown in Tables 4 and 5.

Although many lasers only emit in one or more narrow spectral "lines", an increasing number of lasers can be tuned by changing the composition or the pressure of the medium, or by varying the wavelength of the pump bands. The spectral regions in which these tunable lasers operate are presented in Figure 2.

REFERENCE

Krupke, W. F., in *Handbook of Laser Science and Technology,* Vol. I, Weber, M. J., Ed., CRC Press, Boca Raton, FL, 1986.

TABLE 1
Extrema of Output Parameters of Laser Devices or Systems

Parameter	Value	Laser medium
Peak power	1×10^{14} W (collimated)	Nd:glass
Peak power density	10^{18} W/cm^2 (focused)	Nd:glass
Pulse energy	$>10^5$ J	CO_2, Nd:glass
Average power	10^5 W	CO_2
Pulse duration	3×10^{-15} s continuous wave (cw)	Rh6G dye; various gases, liquids, solids
Wavelength	60 nm $\leftrightarrow$ 385 μm	Many required
Efficiency (nonlaser pumped)	70%	CO
Beam quality	Diffraction limited	Various gases, liquids, solids
Spectral linewidth	20 Hz (for 10^{-1} s)	Neon-helium
Spatial coherence	10 m	Ruby

TABLE 2
Classes, Types, and Representative Examples of Laser Sources

Class	Type (characteristic)	Representative example	Nominal operating wavelength (nm)	Method(s) of excitation
Gas	Atom, neutral (electronic transition)	Neon-Helium (Ne-He)	633	Glow discharge
	Atom, ionic (electronic transition)	Argon (Ar$^+$)	488	Arc discharge
	Molecule, neutral (electronic transition)	Krypton fluoride (KrF)	248	Glow discharge; e-beam
	Molecule, neutral (vibrational transition)	Carbon dioxide (CO_2)	10600	Glow discharge; gas-dynamic flow
	Molecule, neutral (rotational transition)	Methyl fluoride (CH_3F)	496000	Laser pumping
	Molecule, ionic (electronic transition)	Nitrogen ion (N_2^+)	420	E-beam
Liquid	Organic solvent (dye-chromophore)	Rhodamine dye (Rh6G)	580–610	Flashlamp; laser pumping
Solid	Organic solvent (rare earth chelate)	Europium:TTF	612	Flashlamp
	Inorganic solvent (trivalent rare earth ion)	Neodymium:$POCl_4$	1060	Flashlamp
	Insulator, crystal (impurity)	Neodymium:YAG	1064	Flashlamp, arc lamp
	Insulator, crystal (stoichiometric)	Neodymium:UP(NdP_5O_{14})	1052	Flashlamp
	Insulator, crystal (color center)	F_2^-:LiF	1120	Laser pumping
	Insulator, amorphous (impurity)	Neodymium:glass	1061	Flashlamp
	Semiconductor (p-n junction)	GaAs	820	Injection current
	Semiconductor (electron-hole plasma)	GaAs	890	E-beam, laser pumping

Table 3
Temporal Characteristics of Lasers and Laser Systems

Form	Technique	Pulse width range (s)
Continuous wave	Excitation is continuous; resonator Q is held constant at some moderate value	∞
Pulsed	Excitation is pulsed; resonator Q is held constant at some moderate value	10^{-8}–10^{-3}
Q-Switched	Excitation is continuous or pulsed; resonator Q is switched from a very low value to a moderate value	10^{-8}–10^{-6}
Cavity dumped	Excitation is continuous or pulsed; resonator Q is switched from a very high value to a low value	10^{-7}–10^{-5}
Mode locked	Excitation is continuous or pulsed; phase or loss of the resonator modes is modulated at a rate related to the resonator transit time	10^{-12}–10^{-9}

Table 4
Properties and Performance of Some Continuous Wave (CW) Lasers

Parameter	Unit	Gas			Liquid	Solid	
		Neon helium	Argon ion	Carbon dioxide	Rhodamine 6G dye	Nd:YAG	GaAs
Excitation method		DC discharge	DC discharge	DC discharge	Ar⁺ laser pump	Krypton arc lamp	DC injection
Gain medium composition		Neon:helium	Argon	CO_2:N_2:He	Rh 6G:H_2O	Nd:YAG	p:n:GaAs
Gain medium density	Torr; ions/cm³	0.1:1.0	0.4	0.4:0.8:5.0	2(18):2(22)	1.5(20):2(22)	2(19):3(18):3(22)
Wavelength	nm	633	488	10600	590	1064	810
Laser cross-section	cm^{-2}	3(-13)	1.6(-12)	1.5(-16)	1.8(-16)	7(-19)	~6(-15)
Radiative lifetime (upper level)	s	~1(-7)	7.5(-9)	4(-3)	6.5(-9)	2.6(-4)	~1(-9)
Decay lifetime (upper level)	s	~1(-7)	~5.0(-9)	~4(-3)	6.0(-9)	2.3(-4)	~1(-9)
Gain bandwidth	nm	2(-3)	5(-3)	1.6(-2)	80	0.5	10
Type, gain saturation		Inhomogeneous	Inhomogeneous	Homogeneous	Homogeneous	Homogeneous	Homogeneous
Homogeneous saturation flux	W cm^{-2}			~20	3(5)	2.3(3)	~2(4)
Decay lifetime (lower level)	s	~1(-8)	~4(-10)	~5(-6)[b]	<1(-12)	<1(-7)	<1(-12)
Inversion density	cm^{-3}	~1(9)	2(10)	2(15)	2(16)	6(16)	1(16)
Small signal gain coefficient	cm^{-1}	~1(-3)	~3(-2)	1(-2)	4	5(-2)	40
Pump power density	W cm^{-3}	3	900	0.15	1(6)	150	7(7)
Output power density	W cm^{-3}	2.6(-3)	~1	2(-2)	3(5)	95	5(6)
Laser size (diameter:length)	cm:cm	0.5:100	0.3:100	5.0:600	1(-3):0.3	0.6:10	5(-4):7(-3):2(-2)[a]
Excitation current/voltage	A/V	3(-2):2(3)	30:300	0.1:1.5(4)		90:125	1.0/1.7
Excitation current density	A cm^{-2}	0.15	600	6(-3)		140	4.5(3)
Excitation power	W	60	9(3)	1.5(3)	4	1.1(4)	1.7
Output power	W	0.06	10	240	0.3	300	0.12
Efficiency	%	0.1	0.1	13	7	2.6	7

[a] Junction thickness:width:length.
[b] Pressure dependent.

Table 5
Properties and Performance of Some Pulsed Lasers

Parameter	Unit	Gas				Liquid	Solid	
		Carbon dixoxide		Krypton fluoride		Rhodamine 6G	Nd:YAG	Nd:glass
Excitation method		TEA-discharge	E-beam/sust.	Glow discharge	E-beam	Xenon flashlamp	Xenon flashlamp	Xenon flashlamp
Gain medium composition		CO_2:N_2:He	CO_2:N_2:He	He:Kr:F_2	Ar:Kr:F_2	Rh6G:alcohol	Nd:YAG	Nd:Glass
Gain medium density	torr ions/cm^3	100:50:600	240:240:320	1070:70:3	1235:52:3	1(18):1.5(22)	1.5(20):1(22)	3(20):2(22)
Wavelength	nm	10600	10600	249	249	590	1064	1061
Laser cross-section	cm^{-2}	2(-18)	2(-18)	2(-16)	2(-16)	1.8(-16)	7(-19)	2.8(-20)
Radiative lifetime (upper level)	s	4(-3)	4(-3)	7(-9)	7(-9)	6.5(-9)	2.6(-4)	4.1(-4)
Decay lifetime (upper level)	s	~1(-4)	5(-5)	2(-9)	3(-9)	6.0(-9)	2.3(-4)	3.7(-4)
Gain bandwidth	nm	1	1	2	2	80	0.5	26
Homogeneous saturation fluence	J/cm^2	0.2	0.2	4(-3)	4(-3)	2(-3)	0.6	~5
Decay lifetime (lower level)	s	5(-8)[a]	1(-8)[a]	<1(-12)	<1(-12)	<1(-12)	<1(-7)	<1(-8)
Inversion density	cm^{-3}	3(17)	6(17)	4(14)	2(14)	2(16)	4(17)	3(18)
Small signal gain coefficient	cm^{-1}	2(-2)	4(-2)	8-92)	4(-2)	4	0.3	8(-2)
Medium excitation energy density	J/cm^3	0.1	0.36	0.15	0.13	2.8	0.15	0.6
Output energy density	J/cm^3	2(-2)	1.8(-2)	1.5(-3)	1.2(-2)	0.85	5(-2)	2(-2)
Laser dimensions	cm:cm:cm	4.5:4.5:87	10:10:100	1.5:4.5:100	8.5:10:100	1.2φ25	0.6φ7.5	0.6φ8.3
Excitation current/voltage	A/V	6(4)/3.3(3)	2.4(4)/4(4)	2.5(4)/1.5(5)	1.2(4)/2.5(5)	2(5)/2.5(4)		
Excitation current density	A cm^2	8.5	22	170	11.5	2.6(3)		
Excitation peak power	W	2(8)	9(8)	4(9)	3(9)	5.4(9)	4(4)	9(4)
Output pulse energy	J	35	180	1	102	32	0.1	1.0
Output pulse length	s	1(-6)	4(-6)	2.5(-8)	6(-7)	3.2(-6)	2(-8)	1(-4)
Output pulse power	W	3.5(7)	4(7)	4(7)	2(8)	1(7)	5(6)	1(4)
Efficiency	%	17	5	1	10[b]	0.2	1.5	3.7

a   Pressure dependent.
b   Intrinsic efficiency ≡ energy output/energy deposited in gas.

FIGURE 1A. Wavelengths of lasers operating in the 120 to 1200 nm spectral region.

CHARACTERISTICS OF LASER SOURCES (continued)

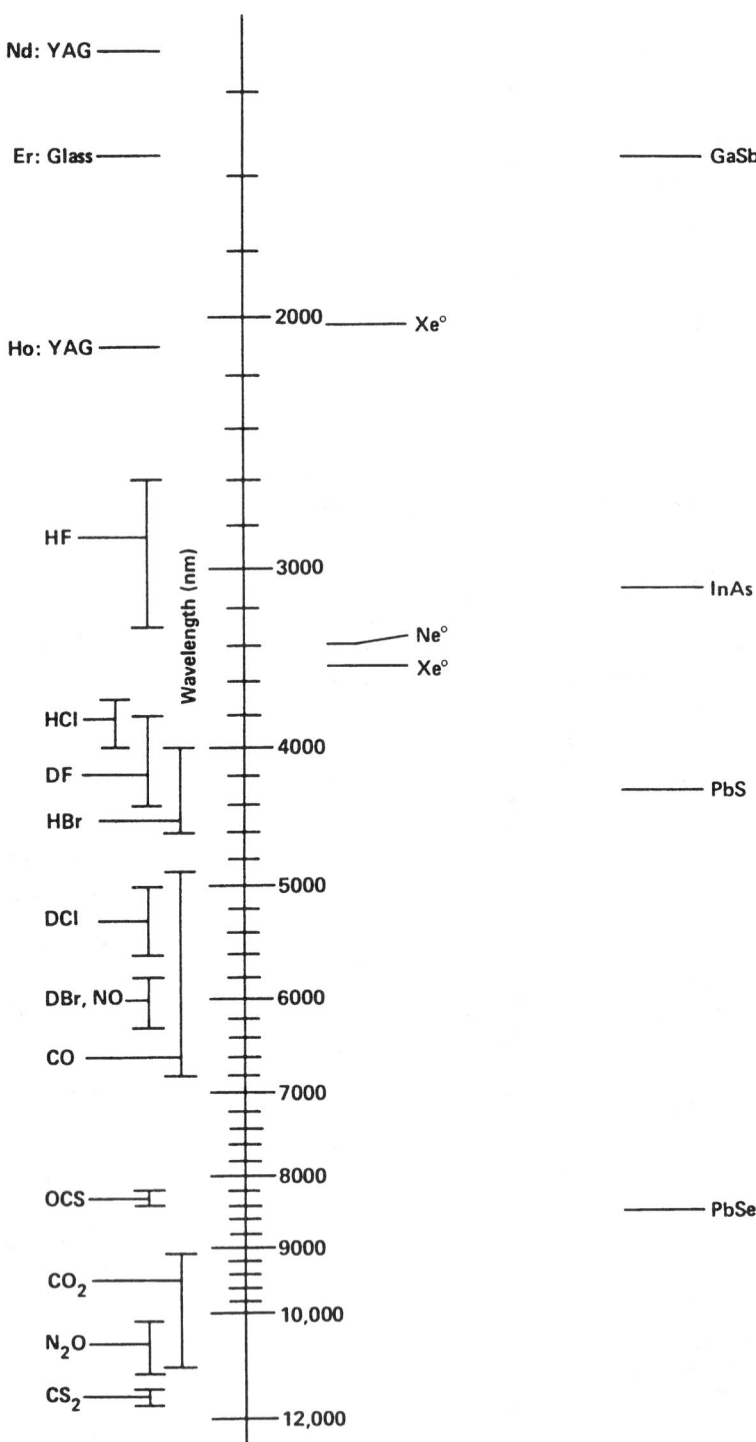

FIGURE 1B. Wavelength of lasers operating in the 1300 to 12,000 nm spectral region.

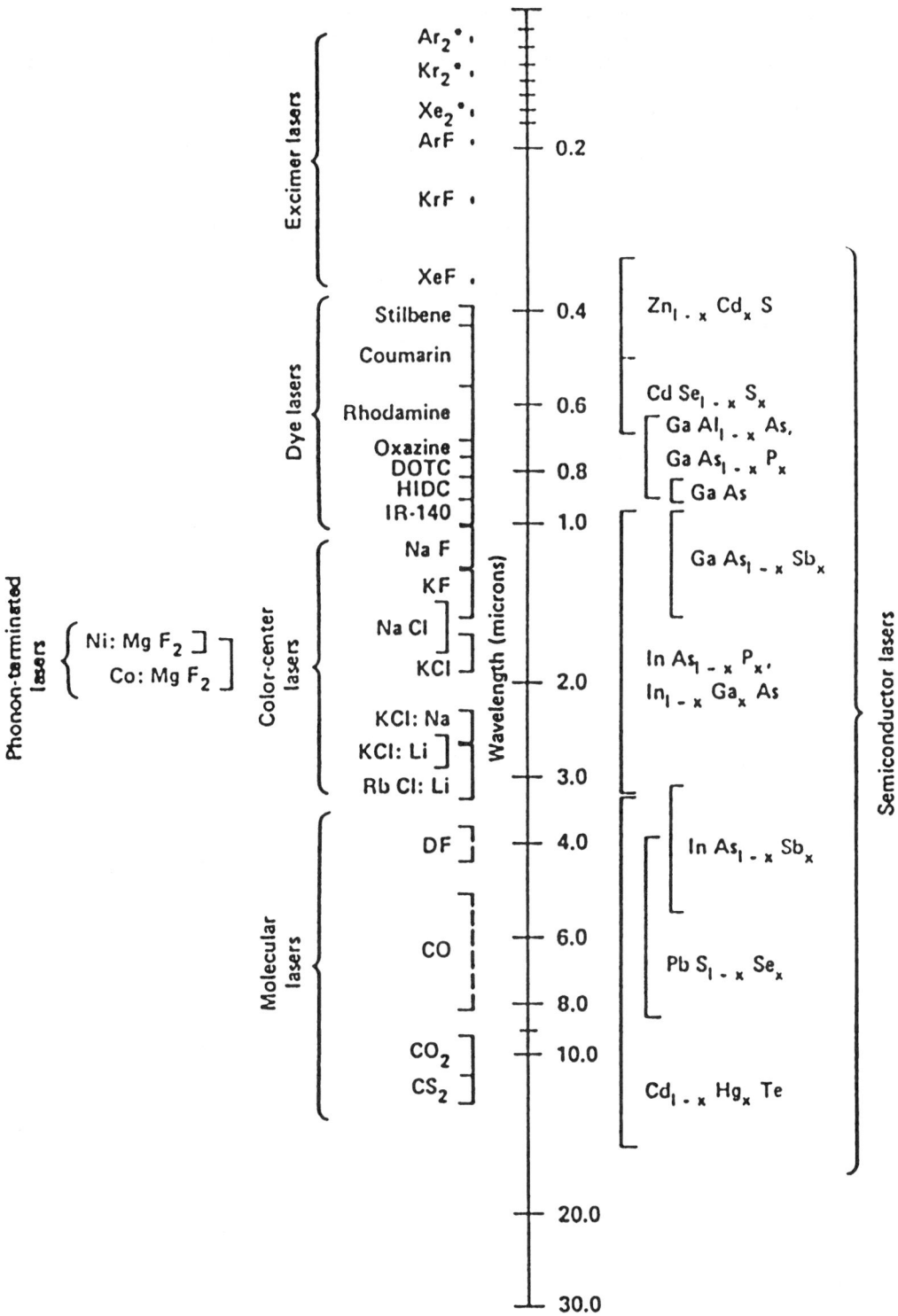

FIGURE 2. Spectral tuning ranges of various types of tunable lasers.

INFRARED LASER FREQUENCIES
Arthur Maki

The CO_2 laser has been the subject of a number of very accurate frequency measurements. Most of the earlier measurements are given by Bradley et al.[1] That analysis was based on a single absolute frequency measurement and many laser frequency differences. New measurements of the methane frequency[2-4] have made it necessary to slightly revise that single absolute frequency measurement. In addition, there have been several other absolute frequency measurements[5-7] that have been used here to improve the accuracy of the present tables. New frequency difference measurements have also been added to the database used for the present tables.[8]

REFERENCES

1. Bradley, L. C., Soohoo, K. L., and Freed, C., *IEEE J. Quantum Electron.*, QE-22, 234-267, 1986.
2. Clairon, A., Dahmani, B., Filimon, A., and Rutman, J., *IEEE Trans. Inst. Meas.*, IM-34, 265-268, 1985.
3. Weiss, C. O., Kramer, G., Lipphardt, B., and Garcia, E., *IEEE J. Quantum Electron.*, QE-24, 1970-1972, 1988.
4. Bagayev, S. N., Baklanov, A. E., Chebotayev, V. P., and Dychkov, A. S., *Appl. Phys.*, B-48, 31-35, 1989.
5. Blaney, T. G., Bradley, C. C., Edwards, G. J., Jolliffe, B. W., Knight, D. J. E., Rowley, W. R. C., Shotten, K. C., and Woods, P. T., *Proc. R. Soc. Lond.*, A-355, 61-88, 1977.
6. Chardonnet, Ch., Van Lerberghe, A., and Bordé, Ch. J., *Opt. Comm.*, 58, 333-337, 1986.
7. Clairon, A., Acef, O., Chardonnet, Ch., and Bordé, Ch. J., *Frequency Standards and Metrology*, De Marchi, A., Ed., Springer-Verlag, Berlin, Heidelberg, 1989, p. 212.
8. Evenson, K., private communication.

Frequencies for the $00°1-(10°0,02°0)_I$ and $00°1-(10°0,02°0)_{II}$ Bands of $^{12}C^{16}O_2$ with the Estimated 2-σ Uncertainties

| | Band I | | | Band II | |
| | Frequency | Uncertainty | | Frequency | Uncertainty |
Line	(MHz)	(MHz)	Line	(MHz)	(MHz)
P(70)	26721305.4647	0.1680	P(70)	29789856.3783	0.0308
P(68)	26794232.6712	0.1217	P(68)	29861850.7690	0.0192
P(66)	26866318.8073	0.0867	P(66)	29933216.1760	0.0122
P(64)	26937571.7234	0.0606	P(64)	30003944.2861	0.0086
P(62)	27007998.9216	0.0415	P(62)	30074026.9127	0.0072
P(60)	27077607.5643	0.0279	P(60)	30143456.0039	0.0066
P(58)	27146404.4834	0.0185	P(58)	30212223.6504	0.0061
P(56)	27214396.1873	0.0121	P(56)	30280322.0930	0.0055
P(54)	27281588.8696	0.0081	P(54)	30347743.7306	0.0049
P(52)	27347988.4161	0.0057	P(52)	30414481.1273	0.0044
P(50)	27413600.4119	0.0043	P(50)	30480527.0196	0.0041
P(48)	27478430.1487	0.0036	P(48)	30545874.3239	0.0039
P(46)	27542482.6310	0.0032	P(46)	30610516.1429	0.0039
P(44)	27605762.5826	0.0030	P(44)	30674445.7724	0.0039
P(42)	27668274.4525	0.0028	P(42)	30737656.7080	0.0039
P(40)	27730022.4206	0.0027	P(40)	30800142.6511	0.0039
P(38)	27791010.4036	0.0026	P(38)	30861897.5150	0.0038
P(36)	27851242.0594	0.0025	P(36)	30922915.4310	0.0037
P(34)	27910720.7927	0.0024	P(34)	30983190.7534	0.0037
P(32)	27969449.7593	0.0023	P(32)	31042718.0652	0.0037
P(30)	28027431.8708	0.0022	P(30)	31101492.1833	0.0036
P(28)	28084669.7981	0.0021	P(28)	31159508.1631	0.0037
P(26)	28141165.9762	0.0020	P(26)	31216761.3029	0.0037
P(24)	28196922.6067	0.0019	P(24)	31273247.1487	0.0037
P(22)	28251941.6622	0.0017	P(22)	31328961.4978	0.0037
P(20)	28306224.8888	0.0016	P(20)	31383900.4028	0.0037
P(18)	28359773.8090	0.0014	P(18)	31438060.1749	0.0037
P(16)	28412589.7245	0.0012	P(16)	31491437.3872	0.0036
P(14)	28464673.7184	0.0011	P(14)	31544028.8776	0.0036
P(12)	28516026.6574	0.0009	P(12)	31595831.7516	0.0036
P(10)	28566649.1935	0.0008	P(10)	31646843.3843	0.0035
P(8)	28616541.7661	0.0008	P(8)	31697061.4225	0.0035
P(6)	28665704.6027	0.0008	P(6)	31746483.7868	0.0035

INFRARED LASER FREQUENCIES (continued)

Line	Band I Frequency (MHz)	Uncertainty (MHz)	Line	Band II Frequency (MHz)	Uncertainty (MHz)
P(4)	28714137.7205	0.0008	P(4)	31795108.6724	0.0035
P(2)	28761840.9272	0.0008	P(2)	31842934.5511	0.0035
R(0)	28832026.2198	0.0008	R(0)	31913172.5691	0.0035
R(2)	28877902.4382	0.0007	R(2)	31958996.0621	0.0034
R(4)	28923046.4303	0.0006	R(4)	32004017.3822	0.0034
R(6)	28967457.0657	0.0005	R(6)	32048236.2498	0.0034
R(8)	29011133.0054	0.0003	R(8)	32091652.6619	0.0034
R(10)	29054072.7010	0.0001	R(10)	32134266.8917	0.0034
R(12)	29096274.3935	0.0003	R(12)	32176079.4878	0.0034
R(14)	29137736.1129	0.0005	R(14)	32217091.2721	0.0035
R(16)	29178455.6759	0.0007	R(16)	32257303.3386	0.0036
R(18)	29218430.6852	0.0009	R(18)	32296717.0510	0.0037
R(20)	29257658.5269	0.0010	R(20)	32335334.0408	0.0038
R(22)	29296136.3689	0.0011	R(22)	32373156.2044	0.0039
R(24)	29333861.1583	0.0012	R(24)	32410185.7003	0.0041
R(26)	29370829.6191	0.0011	R(26)	32446424.9459	0.0042
R(28)	29407038.2491	0.0011	R(28)	32481876.6140	0.0042
R(30)	29442483.3168	0.0011	R(30)	32516543.6293	0.0042
R(32)	29477160.8582	0.0012	R(32)	32550429.1641	0.0042
R(34)	29511066.6733	0.0013	R(34)	32583536.6340	0.0042
R(36)	29544196.3221	0.0015	R(36)	32615869.6937	0.0041
R(38)	29576545.1205	0.0017	R(38)	32647432.2320	0.0040
R(40)	29608108.1360	0.0019	R(40)	32678228.3665	0.0039
R(42)	29638880.1831	0.0022	R(42)	32708262.4386	0.0038
R(44)	29668855.8183	0.0024	R(44)	32737539.0081	0.0039
R(46)	29698029.3350	0.0027	R(46)	32766062.8469	0.0041
R(48)	29726394.7582	0.0032	R(48)	32793838.9334	0.0045
R(50)	29753945.8385	0.0037	R(50)	32820872.4463	0.0055
R(52)	29780676.0464	0.0042	R(52)	32847168.7576	0.0071
R(54)	29806578.5659	0.0047	R(54)	32872733.4269	0.0099
R(56)	29831646.2878	0.0052	R(56)	32897572.1935	0.0141
R(58)	29855871.8032	0.0058	R(58)	32921690.9701	0.0202
R(60)	29879247.3960	0.0074	R(60)	32945095.8355	0.0288
R(62)	29901765.0357	0.0113	R(62)	32967793.0268	0.0407
R(64)	29923416.3695	0.0186	R(64)	32989788.9322	0.0567
R(66)	29944192.7145	0.0302	R(66)	33011090.0831	0.0780
R(68)	29964085.0488	0.0475	R(68)	33031703.1467	0.1060
R(70)	29983084.0036	0.0720	R(70)	33051634.9172	0.1423

Frequencies for the $00^\circ1$-$(10^\circ0,02^\circ0)_I$ and $00^\circ1$-$(10^\circ0,02^\circ0)_{II}$ Bands of $^{13}C^{16}O_2$ with the Estimated 2-σ Uncertainties

Line	Band I Frequency (MHz)	Uncertainty (MHz)	Line	Band II Frequency (MHz)	Uncertainty (MHz)
P(66)	25523832.1808	0.7836	P(66)	28512082.5283	1.2894
P(64)	25590013.4703	0.5415	P(64)	28585121.9396	0.9194
P(62)	25655543.6502	0.3629	P(62)	28657449.4180	0.6420
P(60)	25720428.2487	0.2339	P(60)	28729056.6374	0.4375
P(58)	25784672.4840	0.1430	P(58)	28799935.4147	0.2897
P(56)	25848281.2771	0.0810	P(56)	28870077.7187	0.1853
P(54)	25911259.2627	0.0405	P(54)	28939475.6771	0.1135
P(52)	25973610.8005	0.0157	P(52)	29008121.5846	0.0659
P(50)	26035339.9857	0.0045	P(50)	29076007.9109	0.0357
P(48)	26096450.6582	0.0079	P(48)	29143127.3077	0.0180
P(46)	26156946.4123	0.0101	P(46)	29209472.6164	0.0090

| | Band I | | | Band II | |
Line	Frequency (MHz)	Uncertainty (MHz)	Line	Frequency (MHz)	Uncertainty (MHz)
P(44)	26216830.6053	0.0101	P(44)	29275036.8754	0.0058
P(42)	26276106.3655	0.0090	P(42)	29339813.3270	0.0050
P(40)	26334776.6003	0.0077	P(40)	29403795.4243	0.0044
P(38)	26392844.0030	0.0068	P(38)	29466976.8383	0.0037
P(36)	26450311.0599	0.0063	P(36)	29529351.4635	0.0032
P(34)	26507180.0565	0.0061	P(34)	29590913.4252	0.0029
P(32)	26563453.0836	0.0060	P(32)	29651657.0844	0.0028
P(30)	26619132.0428	0.0058	P(30)	29711577.0447	0.0028
P(28)	26674218.6515	0.0055	P(28)	29770668.1566	0.0031
P(26)	26728714.4479	0.0054	P(26)	29828925.5239	0.0035
P(24)	26782620.7952	0.0054	P(24)	29886344.5074	0.0041
P(22)	26835938.8858	0.0054	P(22)	29942920.7308	0.0046
P(20)	26888669.7451	0.0055	P(20)	29998650.0838	0.0051
P(18)	26940814.2347	0.0055	P(18)	30053528.7271	0.0054
P(16)	26992373.0555	0.0055	P(16)	30107553.0955	0.0055
P(14)	27043346.7508	0.0054	P(14)	30160719.9016	0.0055
P(12)	27093735.7083	0.0052	P(12)	30213026.1388	0.0054
P(10)	27143540.1624	0.0051	P(10)	30264469.0839	0.0054
P(8)	27192760.1962	0.0049	P(8)	30315046.2994	0.0054
P(6)	27241395.7431	0.0048	P(6)	30364755.6359	0.0055
P(4)	27289446.5880	0.0047	P(4)	30413595.2335	0.0056
P(2)	27336912.3682	0.0046	P(2)	30461563.5231	0.0057
R(0)	27407012.8882	0.0045	P(0)	30531879.5415	0.0057
R(2)	27453013.4589	0.0043	P(2)	30577664.6138	0.0056
R(4)	27498426.5430	0.0040	P(4)	30622575.1885	0.0054
R(6)	27543251.1200	0.0037	P(6)	30666611.0128	0.0051
R(8)	27587486.0225	0.0034	P(8)	30709772.1257	0.0047
R(10)	27631129.9356	0.0031	P(10)	30752058.8571	0.0045
R(12)	27674181.3963	0.0029	P(12)	30793471.8269	0.0044
R(14)	27716638.7917	0.0029	P(14)	30834011.9425	0.0043
R(16)	27758500.3577	0.0029	P(16)	30873680.3976	0.0044
R(18)	27799764.1770	0.0029	P(18)	30912478.6694	0.0044
R(20)	27840428.1773	0.0030	P(20)	30950408.5159	0.0044
R(22)	27880490.1283	0.0029	P(22)	30987471.9732	0.0043
R(24)	27919947.6395	0.0029	P(24)	31023671.3517	0.0042
R(26)	27958798.1567	0.0028	P(26)	31059009.2327	0.0042
R(28)	27997038.9591	0.0028	P(28)	31093488.4642	0.0042
R(30)	28034667.1551	0.0027	P(30)	31127112.1569	0.0043
R(32)	28071679.6785	0.0027	P(32)	31159883.6793	0.0045
R(34)	28108073.2842	0.0026	P(34)	31191806.6529	0.0046
R(36)	28143844.5432	0.0026	P(36)	31222884.9469	0.0048
R(38)	28178989.8377	0.0026	P(38)	31253122.6730	0.0053
R(40)	28213505.3554	0.0028	P(40)	31282524.1795	0.0061
R(42)	28247387.0838	0.0033	P(42)	31311094.0452	0.0077
R(44)	28280630.8035	0.0046	P(44)	31338837.0736	0.0108
R(46)	28313232.0818	0.0083	P(46)	31365758.2858	0.0173
R(48)	28345186.2652	0.0161	P(48)	31391862.9147	0.0295
R(50)	28376488.4720	0.0301	P(50)	31417156.3972	0.0505
R(52)	28407133.5839	0.0531	P(52)	31441644.3679	0.0845
R(54)	28437116.2372	0.0887	P(54)	31465332.6516	0.1366
R(56)	28466430.8141	0.1419	P(56)	31488227.2557	0.2138
R(58)	28495071.4324	0.2188	P(58)	31510334.3631	0.3247
R(60)	28523031.9357	0.3271	P(60)	31531660.3243	0.4800
R(62)	28550305.8819	0.4763	P(62)	31552211.6497	0.6932
R(64)	28576886.5323	0.6781	P(64)	31571995.0017	0.9805
R(66)	28602766.8393	0.9467	P(66)	31591017.1868	1.3619

Frequencies for the $00°1$-$(10°0,02°0)_I$ and $00°1$-$(10°0,02°0)_{II}$ Bands of $^{12}C^{18}O_2$ with the Estimated 2-σ Uncertainties

	Band I			Band II	
Line	Frequency (MHz)	Uncertainty (MHz)	Line	Frequency (MHz)	Uncertainty (MHz)
P(70)	27045326.3119	0.4540	P(70)	30695237.5856	0.0858
P(68)	27114914.0922	0.3324	P(68)	30755520.2231	0.0570
P(66)	27183635.7945	0.2392	P(66)	30815311.4928	0.0364
P(64)	27251496.4118	0.1688	P(64)	30874607.2084	0.0223
P(62)	27318500.7361	0.1165	P(62)	30933403.2309	0.0131
P(60)	27384653.3618	0.0783	P(60)	30991695.4724	0.0075
P(58)	27449958.6881	0.0510	P(58)	31049479.9009	0.0049
P(56)	27514420.9224	0.0319	P(56)	31106752.5446	0.0041
P(54)	27578044.0828	0.0191	P(54)	31163509.4964	0.0040
P(52)	27640832.0010	0.0108	P(52)	31219746.9183	0.0040
P(50)	27702788.3248	0.0059	P(50)	31275461.0455	0.0039
P(48)	27763916.5206	0.0035	P(48)	31330648.1908	0.0039
P(46)	27824219.8762	0.0028	P(46)	31385304.7490	0.0039
P(44)	27883701.5029	0.0026	P(44)	31439427.2006	0.0039
P(42)	27942364.3379	0.0025	P(42)	31493012.1163	0.0038
P(40)	28000211.1464	0.0024	P(40)	31546056.1605	0.0038
P(38)	28057244.5242	0.0022	P(38)	31598556.0954	0.0037
P(36)	28113466.8992	0.0021	P(36)	31650508.7847	0.0037
P(34)	28168880.5335	0.0020	P(34)	31701911.1970	0.0037
P(32)	28223487.5256	0.0019	P(32)	31752760.4093	0.0037
P(30)	28277289.8118	0.0017	P(30)	31803053.6105	0.0037
P(28)	28330289.1679	0.0016	P(28)	31852788.1043	0.0038
P(26)	28382487.2111	0.0015	P(26)	31901961.3125	0.0038
P(24)	28433885.4012	0.0013	P(24)	31950570.7773	0.0038
P(22)	28484485.0420	0.0012	P(22)	31998614.1649	0.0038
P(20)	28534287.2828	0.0011	P(20)	32046089.2669	0.0037
P(18)	28583293.1193	0.0010	P(18)	32092994.0036	0.0037
P(16)	28631503.3952	0.0010	P(16)	32139326.4254	0.0036
P(14)	28678918.8025	0.0009	P(14)	32185084.7154	0.0036
P(12)	28725539.8830	0.0010	P(12)	32230267.1907	0.0036
P(10)	28771367.0288	0.0010	P(10)	32274872.3041	0.0037
P(8)	28816400.4829	0.0010	P(8)	32318898.6455	0.0038
P(6)	28860640.3403	0.0011	P(6)	32362344.9434	0.0039
P(4)	28904086.5477	0.0011	P(4)	32405210.0652	0.0041
P(2)	28946738.9048	0.0011	P(2)	32447493.0185	0.0041
R(0)	29009228.1702	0.0010	P(0)	32509824.0580	0.0042
R(2)	29049894.0586	0.0010	P(2)	32550648.1723	0.0042
R(4)	29089764.2368	0.0009	P(4)	32590887.7542	0.0042
R(6)	29128837.8426	0.0008	P(6)	32630542.4457	0.0041
R(8)	29167113.8668	0.0008	P(8)	32669612.0295	0.0041
R(10)	29204591.1529	0.0009	P(10)	32708096.4282	0.0040
R(12)	29241268.3964	0.0010	P(12)	32745995.7040	0.0040
R(14)	29277144.1444	0.0011	P(14)	32783310.0573	0.0040
R(16)	29312216.7955	0.0012	P(16)	32820039.8258	0.0040
R(18)	29346484.5984	0.0012	P(18)	32856185.4827	0.0040
R(20)	29379945.6517	0.0013	P(20)	32891747.6358	0.0040
R(22)	29412597.9024	0.0013	P(22)	32926727.0254	0.0040
R(24)	29444439.1458	0.0013	P(24)	32961124.5220	0.0040
R(26)	29475467.0236	0.0014	P(26)	32994941.1249	0.0040
R(28)	29505679.0230	0.0015	P(28)	33028177.9594	0.0040
R(30)	29535072.4755	0.0016	P(30)	33060836.2743	0.0040
R(32)	29563644.5557	0.0018	P(32)	33092917.4394	0.0041
R(34)	29591392.2794	0.0020	P(34)	33124422.9429	0.0043
R(36)	29618312.5023	0.0023	P(36)	33155354.3878	0.0046
R(38)	29644401.9182	0.0028	P(38)	33185713.4894	0.0049

Line	Band I Frequency (MHz)	Uncertainty (MHz)	Line	Band II Frequency (MHz)	Uncertainty (MHz)
R(40)	29669657.0575	0.0036	P(40)	33215502.0716	0.0056
R(42)	29694074.2853	0.0053	P(42)	33244722.0637	0.0068
R(44)	29717649.7992	0.0082	P(44)	33273375.4969	0.0092
R(46)	29740379.6276	0.0128	P(46)	33301464.5003	0.0134
R(48)	29762259.6274	0.0200	P(48)	33328991.2976	0.0199
R(50)	29783285.4820	0.0307	P(50)	33355958.2027	0.0294
R(52)	29803452.6988	0.0461	P(52)	33382367.6161	0.0427
R(54)	29822756.6072	0.0681	P(54)	33408222.0209	0.0607
R(56)	29841192.3558	0.0985	P(56)	33433523.9780	0.0848
R(58)	29858754.9100	0.1401	P(58)	33458276.1228	0.1165
R(60)	29875439.0495	0.1960	P(60)	33482481.1601	0.1576
R(62)	29891239.3658	0.2702	P(62)	33506141.8605	0.2104
R(64)	29906150.2589	0.3673	P(64)	33529261.0556	0.2775
R(66)	29920165.9352	0.4930	P(66)	33551841.6335	0.3621
R(68)	29933280.4042	0.6540	P(68)	33573886.5352	0.4679
R(70)	29945487.4756	0.8581	P(70)	33595398.7493	0.5992

Frequencies for the $00°1$-$(10°0,02°0)_I$ and $00°1$-$(10°0,02°0)_{II}$ Bands of $^{13}C^{18}O_2$ with the Estimated 2-σ Uncertainties

Line	Band I Frequency (MHz)	Uncertainty (MHz)	Line	Band II Frequency (MHz)	Uncertainty (MHz)
P(70)	25967863.7652	1.1146	P(70)	28960476.2278	0.4069
P(68)	26033448.2798	0.8152	P(68)	29022326.9578	0.2861
P(66)	26098273.9159	0.5860	P(66)	29083661.3546	0.1961
P(64)	26162346.4813	0.4129	P(64)	29144473.5795	0.1303
P(62)	26225671.5466	0.2844	P(62)	29204757.8761	0.0833
P(60)	26288254.4494	0.1906	P(60)	29264508.5768	0.0507
P(58)	26350100.2984	0.1237	P(58)	29323720.1086	0.0290
P(56)	26411213.9778	0.0772	P(56)	29382386.9988	0.0152
P(54)	26471600.1504	0.0459	P(54)	29440503.8809	0.0073
P(52)	26531263.2618	0.0258	P(52)	29498065.4997	0.0038
P(50)	26590207.5442	0.0138	P(50)	29555066.7172	0.0032
P(48)	26648437.0195	0.0077	P(48)	29611502.5178	0.0031
P(46)	26705955.5026	0.0057	P(46)	29667368.0132	0.0031
P(44)	26762766.6051	0.0055	P(44)	29722658.4475	0.0034
P(42)	26818873.7378	0.0055	P(42)	29777369.2022	0.0039
P(40)	26874280.1143	0.0056	P(40)	29831495.8006	0.0044
P(38)	26928988.7531	0.0056	P(38)	29885033.9125	0.0049
P(36)	26983002.4809	0.0056	P(36)	29937979.3584	0.0053
P(34)	27036323.9351	0.0055	P(34)	29990328.1139	0.0054
P(32)	27088955.5657	0.0054	P(32)	30042076.3132	0.0055
P(30)	27140899.6384	0.0051	P(30)	30093220.2534	0.0055
P(28)	27192158.2363	0.0049	P(28)	30143756.3978	0.0054
P(26)	27242733.2620	0.0047	P(26)	30193681.3793	0.0053
P(24)	27292626.4396	0.0044	P(24)	30242992.0038	0.0052
P(22)	27341839.3165	0.0042	P(22)	30291685.2529	0.0051
P(20)	27390373.2651	0.0040	P(20)	30339758.2870	0.0049
P(18)	27438229.4843	0.0037	P(18)	30387208.4477	0.0048
P(16)	27485409.0008	0.0035	P(16)	30434033.2603	0.0046
P(14)	27531912.6704	0.0033	P(14)	30480230.4356	0.0045
P(12)	27577741.1795	0.0031	P(12)	30525797.8725	0.0044
P(10)	27622895.0455	0.0031	P(10)	30570733.6593	0.0043
P(8)	27667374.6182	0.0031	P(8)	30615036.0750	0.0043
P(6)	27711180.0803	0.0033	P(6)	30658703.5912	0.0044
P(4)	27754311.4480	0.0034	P(4)	30701734.8727	0.0045

Line	Band I Frequency (MHz)	Uncertainty (MHz)	Line	Band II Frequency (MHz)	Uncertainty (MHz)
P(2)	27796768.5718	0.0036	P(2)	30744128.7785	0.0045
R(0)	27859189.3155	0.0036	P(0)	30806522.5414	0.0045
R(2)	27899959.0889	0.0035	P(2)	30847319.2956	0.0044
R(4)	27940052.7921	0.0033	P(4)	30887476.2168	0.0043
R(6)	27979469.5315	0.0031	P(6)	30926993.0424	0.0042
R(8)	28018208.2478	0.0028	P(8)	30965869.7046	0.0041
R(10)	28056267.7161	0.0026	P(10)	31004106.3298	0.0040
R(12)	28093646.5448	0.0025	P(12)	31041703.2379	0.0040
R(14)	28130343.1757	0.0025	P(14)	31078660.9408	0.0040
R(16)	28166355.8825	0.0025	P(16)	31114980.1420	0.0040
R(18)	28201682.7706	0.0025	P(18)	31150661.7340	0.0041
R(20)	28236321.7757	0.0025	P(20)	31185706.7976	0.0042
R(22)	28270270.6628	0.0024	P(22)	31220116.5992	0.0043
R(24)	28303527.0249	0.0024	P(24)	31253892.5891	0.0043
R(26)	28336088.2817	0.0023	P(26)	31287036.3991	0.0044
R(28)	28367951.6781	0.0024	P(28)	31319549.8396	0.0043
R(30)	28399114.2823	0.0025	P(30)	31351434.8973	0.0043
R(32)	28429572.9843	0.0026	P(32)	31382693.7318	0.0042
R(34)	28459324.4940	0.0028	P(34)	31413328.6728	0.0042
R(36)	28488365.3390	0.0029	P(36)	31443342.2165	0.0041
R(38)	28516691.8625	0.0029	P(38)	31472737.0219	0.0040
R(40)	28544300.2211	0.0031	P(40)	31501515.9074	0.0039
R(42)	28571186.3823	0.0032	P(42)	31529681.8467	0.0040
R(44)	28597346.1222	0.0032	P(44)	31557237.9646	0.0042
R(46)	28622775.0223	0.0038	P(46)	31584187.5329	0.0046
R(48)	28647468.4672	0.0071	P(48)	31610533.9656	0.0057
R(50)	28671421.6417	0.0148	P(50)	31636280.8146	0.0088
R(52)	28694629.5272	0.0286	P(52)	31661431.7650	0.0151
R(54)	28717086.8993	0.0510	P(54)	31685990.6298	0.0261
R(56)	28738788.3239	0.0852	P(56)	31709961.3449	0.0434
R(58)	28759728.1540	0.1355	P(58)	31733347.9642	0.0693
R(60)	28779900.5263	0.2075	P(60)	31756154.6537	0.1068
R(62)	28799299.3572	0.3078	P(62)	31778385.6867	0.1594
R(64)	28817918.3393	0.4447	P(64)	31800045.4375	0.2317
R(66)	28835750.9374	0.6283	P(66)	31821138.3761	0.3291
R(68)	28852790.3843	0.8707	P(68)	31841669.0622	0.4581
R(70)	28869029.6768	1.1863	P(70)	31861642.1394	0.6268

Frequencies for the $01^{1e}1$-$(11^{1e}0,03^{1e}0)_I$ and $01^{1e}1$-$(11^{1e}0,03^{1e}0)_{II}$ Bands of $^{12}C^{16}O_2$ with the Estimated 2-σ Uncertainties

Line	Band I Frequency (MHz)	Uncertainty (MHz)	Line	Band II Frequency (MHz)	Uncertainty (MHz)
P(59)	26125213.2723	1.6633	P(59)	30427055.2899	0.1962
P(57)	26191576.6703	1.0880	P(57)	30494640.3229	0.1332
P(55)	26257240.7898	0.6844	P(55)	30561557.5929	0.0865
P(53)	26322208.2302	0.4094	P(53)	30627802.0344	0.0530
P(51)	26386481.4313	0.2286	P(51)	30693368.7014	0.0306
P(49)	26450062.6783	0.1155	P(49)	30758252.7710	0.0175
P(47)	26512954.1076	0.0498	P(47)	30822449.5469	0.0123
P(45)	26575157.7109	0.0191	P(45)	30885954.4624	0.0114
P(43)	26636675.3402	0.0160	P(43)	30948763.0834	0.0109
P(41)	26697508.7115	0.0182	P(41)	31010871.1119	0.0100
P(39)	26757659.4084	0.0177	P(39)	31072274.3882	0.0091
P(37)	26817128.8857	0.0160	P(37)	31132968.8940	0.0091
P(35)	26875918.4726	0.0144	P(35)	31192950.7549	0.0102

INFRARED LASER FREQUENCIES (continued)

Line	Band I Frequency (MHz)	Uncertainty (MHz)	Line	Band II Frequency (MHz)	Uncertainty (MHz)
P(33)	26934029.3751	0.0131	P(33)	31252216.2430	0.0118
P(31)	26991462.6787	0.0119	P(31)	31310761.7788	0.0134
P(29)	27048219.3509	0.0106	P(29)	31368583.9339	0.0147
P(27)	27104300.2431	0.0096	P(27)	31425679.4328	0.0155
P(25)	27159706.0925	0.0093	P(25)	31482045.1550	0.0157
P(23)	27214437.5237	0.0097	P(23)	31537678.1367	0.0154
P(21)	27268495.0505	0.0104	P(21)	31592575.5725	0.0147
P(19)	27321879.0769	0.0108	P(19)	31646734.8172	0.0137
P(17)	27374589.8987	0.0108	P(17)	31700153.3868	0.0127
P(15)	27426627.7040	0.0104	P(15)	31752828.9602	0.0119
P(13)	27477992.5747	0.0098	P(13)	31804759.3803	0.0113
P(11)	27528684.4867	0.0096	P(11)	31855942.6551	0.0113
P(9)	27578703.3113	0.0101	P(9)	31906376.9582	0.0116
P(7)	27628048.8151	0.0113	P(7)	31956060.6304	0.0122
P(5)	27676720.6609	0.0127	P(5)	32004992.1796	0.0129
P(3)	27724718.4080	0.0141	P(3)	32053170.2819	0.0136
R(1)	27841759.7696	0.0152	P(1)	32170312.0391	0.0149
R(3)	27887393.2105	0.0146	P(3)	32215845.0845	0.0151
R(5)	27932349.2934	0.0135	P(5)	32260620.8121	0.0152
R(7)	27976627.0108	0.0124	P(7)	32304638.8261	0.0152
R(9)	28020225.2521	0.0115	P(9)	32347898.8990	0.0150
R(11)	28063142.8031	0.0110	P(11)	32390400.9714	0.0148
R(13)	28105378.3457	0.0109	P(13)	32432145.1513	0.0145
R(15)	28146930.4576	0.0109	P(15)	32473131.7137	0.0142
R(17)	28187797.6116	0.0107	P(17)	32513361.0997	0.0140
R(19)	28227978.1750	0.0103	P(19)	32552833.9153	0.0140
R(21)	28267470.4088	0.0099	P(21)	32591550.9309	0.0141
R(23)	28306272.4666	0.0099	P(23)	32629513.0796	0.0143
R(25)	28344382.3939	0.0107	P(25)	32666721.4564	0.0144
R(27)	28381798.1267	0.0122	P(27)	32703177.3164	0.0142
R(29)	28418517.4902	0.0141	P(29)	32738882.0732	0.0136
R(31)	28454538.1976	0.0165	P(31)	32773837.2976	0.0136
R(33)	28489857.8477	0.0213	P(33)	32808044.7156	0.0174
R(35)	28524473.9240	0.0312	P(35)	32841506.2063	0.0279
R(37)	28558383.7917	0.0486	P(37)	32874223.8000	0.0462
R(39)	28591584.6963	0.0754	P(39)	32906199.6761	0.0735
R(41)	28624073.7602	0.1131	P(41)	32937436.1606	0.1114
R(43)	28655847.9806	0.1644	P(43)	32967935.7238	0.1624
R(45)	28686904.2261	0.2328	P(45)	32997700.9775	0.2292
R(47)	28717239.2334	0.3239	P(47)	33026734.6728	0.3151
R(49)	28746849.6038	0.4465	P(49)	33055039.6965	0.4238
R(51)	28775731.7988	0.6142	P(51)	33082619.0689	0.5595
R(53)	28803882.1361	0.8465	P(53)	33109475.9403	0.7272

Frequencies for the $01^{1f}1$-$(11^{1f}0,03^{1f}0)_I$ and $01^{1f}1$-$(11^{1f}0,03^{1f}0)_{II}$ Bands of $^{12}C^{16}O_2$ with the Estimated 2-σ Uncertainties

Line	Band I Frequency (MHz)	Uncertainty (MHz)	Line	Band II Frequency (MHz)	Uncertainty (MHz)
P(60)	26051570.0104	4.4521	P(60)	30355115.0204	0.2752
P(58)	26120964.4932	3.0629	P(58)	30425283.5969	0.1926
P(56)	26189552.8496	2.0516	P(56)	30494732.8293	0.1301
P(54)	26257339.6006	1.3305	P(54)	30563455.6325	0.0840
P(52)	26324329.0344	0.8289	P(52)	30631445.1076	0.0512
P(50)	26390525.2136	0.4901	P(50)	30698694.5456	0.0292
P(48)	26455931.9824	0.2698	P(48)	30765197.4310	0.0163

INFRARED LASER FREQUENCIES (continued)

Line	Band I Frequency (MHz)	Uncertainty (MHz)	Line	Band II Frequency (MHz)	Uncertainty (MHz)
P(46)	26520552.9722	0.1334	P(46)	30830947.4444	0.0111
P(44)	26584391.6075	0.0551	P(44)	30895938.4662	0.0104
P(42)	26647451.1105	0.0181	P(42)	30960164.5794	0.0105
P(40)	26709734.5057	0.0151	P(40)	31023620.0723	0.0105
P(38)	26771244.6242	0.0174	P(38)	31086299.4415	0.0107
P(36)	26831984.1067	0.0157	P(36)	31148197.3941	0.0114
P(34)	26891955.4069	0.0126	P(34)	31209308.8510	0.0126
P(32)	26951160.7945	0.0105	P(32)	31269628.9481	0.0138
P(30)	27009602.3576	0.0096	P(30)	31329153.0395	0.0147
P(28)	27067282.0045	0.0092	P(28)	31387876.6994	0.0151
P(26)	27124201.4662	0.0090	P(26)	31445795.7236	0.0149
P(24)	27180362.2977	0.0089	P(24)	31502906.1318	0.0141
P(22)	27235765.8792	0.0090	P(22)	31559204.1695	0.0128
P(20)	27290413.4182	0.0093	P(20)	31614686.3091	0.0113
P(18)	27344305.9494	0.0096	P(18)	31669349.2515	0.0098
P(16)	27397444.3368	0.0097	P(16)	31723189.9280	0.0086
P(14)	27449829.2733	0.0096	P(14)	31776205.5007	0.0081
P(12)	27501461.2824	0.0096	P(12)	31828393.3642	0.0085
P(10)	27552340.7179	0.0101	P(10)	31879751.1463	0.0095
P(8)	27602467.7649	0.0111	P(8)	31930276.7092	0.0107
P(6)	27651842.4399	0.0125	P(6)	31979968.1497	0.0120
P(4)	27700464.5912	0.0139	P(4)	32028823.8002	0.0131
P(2)	27748333.8988	0.0148	P(2)	32076842.2290	0.0139
R(2)	27864709.8633	0.0146	P(2)	32193218.1935	0.0150
R(4)	27909939.2762	0.0135	P(4)	32238298.4853	0.0151
R(6)	27954412.3294	0.0122	P(6)	32282538.0393	0.0153
R(8)	27998127.7801	0.0112	P(8)	32325936.7244	0.0153
R(10)	28041084.2173	0.0107	P(10)	32368494.6458	0.0154
R(12)	28083280.0620	0.0108	P(12)	32410212.1438	0.0155
R(14)	28124713.5668	0.0110	P(14)	32451089.7941	0.0155
R(16)	28165382.8151	0.0112	P(16)	32491128.4063	0.0154
R(18)	28205285.7213	0.0111	P(18)	32530329.0234	0.0152
R(20)	28244420.0302	0.0110	P(20)	32568692.9211	0.0149
R(22)	28282783.3158	0.0114	P(22)	32606221.6061	0.0147
R(24)	28320372.9812	0.0129	P(24)	32642916.8154	0.0144
R(26)	28357186.2574	0.0149	P(26)	32678780.5147	0.0140
R(28)	28393220.2023	0.0168	P(28)	32713814.8971	0.0133
R(30)	28428471.6994	0.0175	P(30)	32748022.3813	0.0120
R(32)	28462937.4565	0.0165	P(32)	32781405.6101	0.0106
R(34)	28496614.0042	0.0142	P(34)	32813967.4482	0.0122
R(36)	28529497.6934	0.0163	P(36)	32845710.9809	0.0212
R(38)	28561584.6939	0.0309	P(38)	32876639.5111	0.0385
R(40)	28592870.9914	0.0593	P(40)	32906756.5580	0.0646
R(42)	28623352.3850	0.1054	P(42)	32936065.8540	0.1015
R(44)	28653024.4839	0.1779	P(44)	32964571.3426	0.1518
R(46)	28681882.7038	0.2907	P(46)	32992277.1760	0.2184
R(48)	28709922.2632	0.4635	P(48)	33019187.7118	0.3048
R(50)	28737138.1785	0.7231	P(50)	33045307.5105	0.4152

INFRARED AND FAR-INFRARED ABSORPTION FREQUENCY STANDARDS
Arthur Maki

Aside from the CO_2 laser transitions, the absorption spectrum of CO has been more accurately and thoroughly measured than any other spectrum. A bibliography of earlier measurements on CO is given by Maki and Wells,[1] and the present tables were calculated from the measurements referred to in that work. In addition, some new and very accurate frequency measurements[2,3] have been made and were incorporated in the present tables. The frequencies of the rotational transitions of HF and HCl were calculated from constants obtained from fitting the measurements of Evenson et al.[4,5] and Jennings and Wells.[6]

A new report on infrared wavenumber standards from the International Union of Pure and Applied Chemistry, Commission on Molecular Structure and Spectroscopy, may be found in Reference 7.

REFERENCES

1. Maki, A. G. and Wells, J. S., *Wavenumber Calibration Tables from Heterodyne Frequency Measurements*, NIST Special Publication 821, U.S. Dept. of Commerce, Washington, D.C., 1991.
2. Evenson, K. and Stroh, F., private communication.
3. George, T., Urban, W., and co-workers, private communication.
4. Jennings, D. A., Evenson, K. M., Zink, L. R., Demuynck, C., Destombes, J. L., Lemoine, B., and Johns, J. W. C., *J. Mol. Spectrosc.*, 122, 477-480, 1987.
5. Nolt, I. G., Radostitz, J. V., DiLonardo, G., Evenson, K. M., Jennings, D. A., Leopold, K. R., Vanek, M. D., Zink, L. R., Hinz, A., and Chance, K. V., *J. Mol. Spectrosc.*, 125, 274-287, 1987.
6. Jennings, D. A. and Wells, J. S., *J. Mol. Spectrosc.*, 130, 267-268, 1988.
7. *High Resolution Wavenumber Standards for the Infrared, Pure Appl. Chemistry*, 68, 193, 1996.

Wavenumbers for the ν = 1-0 Band of CO

Wavenumber (unc)* cm^{-1}	Transition	Wavenumber (unc) cm^{-1}	Transition
		2147.081132(01)	R(0)
		2150.856006(01)	R(1)
2139.426071(01)	P(1)	2154.595581(01)	R(2)
2135.546178(01)	P(2)	2158.299710(01)	R(3)
2131.631574(01)	P(3)	2161.968245(01)	R(4)
2127.682404(01)	P(4)	2165.601041(01)	R(5)
2123.698816(01)	P(5)	2169.197949(01)	R(6)
2119.680957(01)	P(6)	2172.758824(01)	R(7)
2115.628973(01)	P(7)	2176.283519(01)	R(8)
2111.543012(01)	P(8)	2179.771887(01)	R(9)
2107.423221(01)	P(9)	2183.223782(01)	R(10)
2103.269746(01)	P(10)	2186.639057(01)	R(11)
2099.082734(01)	P(11)	2190.017565(01)	R(12)
2094.862333(01)	P(12)	2193.359161(01)	R(13)
2090.608688(01)	P(13)	2196.663698(01)	R(14)
2086.321947(01)	P(14)	2199.931030(01)	R(15)
2082.002256(01)	P(15)	2203.161010(01)	R(16)
2077.649762(01)	P(16)	2206.353492(01)	R(17)
2073.264612(01)	P(17)	2209.508331(02)	R(18)
2068.846952(01)	P(18)	2212.625379(02)	R(19)
2064.396929(01)	P(19)	2215.704492(02)	R(20)
2059.914688(02)	P(20)	2218.745522(02)	R(21)
2055.400377(02)	P(21)	2221.748326(03)	R(22)
2050.854140(02)	P(22)	2224.712755(03)	R(23)
2046.276126(03)	P(23)	2227.638666(03)	R(24)
2041.666479(03)	P(24)	2230.525912(04)	R(25)
2037.025345(03)	P(25)	2233.374349(04)	R(26)
2032.352870(04)	P(26)	2236.183829(04)	R(27)
2027.649200(04)	P(27)	2238.954210(05)	R(28)
2022.914480(04)	P(28)	2241.685344(05)	R(29)
2018.148857(05)	P(29)	2244.377088(06)	R(30)
2013.352474(05)	P(30)	2247.029296(07)	R(31)
2008.525477(06)	P(31)	2249.641824(08)	R(32)
2003.668012(06)	P(32)	2252.214527(10)	R(33)
1998.780224(07)	P(33)	2254.747262(14)	R(34)
1993.862257(09)	P(34)		

Wavenumber (unc)* cm⁻¹	Transition	Wavenumber (unc) cm⁻¹	Transition
1988.914257(11)	P(35)	2257.239883(18)	R(35)
1983.936367(14)	P(36)	2259.692248(24)	R(36)
1978.928733(18)	P(37)	2262.104213(33)	R(37)
1973.891500(25)	P(38)	2264.475634(45)	R(38)
1968.824811(34)	P(39)	2266.806368(61)	R(39)
1963.728813(46)	P(40)	2269.096273(81)	R(40)
1958.603648(61)	P(41)	2271.345206(106)	R(41)
1953.449462(82)	P(42)	2273.553027(139)	R(42)

* The uncertainty in the last digits (twice the standard error) is given in parentheses.

Wavenumbers for the $v = 2$-0 Band of CO

Wavenumber (unc)* cm⁻¹	Transition	Wavenumber (unc) cm⁻¹	Transition
		4263.837198(02)	R(0)
4256.217140(02)	P(1)	4267.542066(02)	R(1)
4252.302244(02)	P(2)	4271.176630(02)	R(2)
4248.317633(02)	P(3)	4274.740746(02)	R(3)
4244.263453(02)	P(4)	4278.234264(02)	R(4)
4240.139852(02)	P(5)	4281.657039(02)	R(5)
4235.946975(02)	P(6)	4285.008924(02)	R(6)
4231.684972(02)	P(7)	4288.289772(02)	R(7)
4227.353987(02)	P(8)	4291.499437(02)	R(8)
4222.954169(02)	P(9)	4294.637773(02)	R(9)
4218.485665(02)	P(10)	4297.704631(02)	R(10)
4213.948620(02)	P(11)	4300.699868(02)	R(11)
4209.343182(02)	P(12)	4303.623334(02)	R(12)
4204.669499(02)	P(13)	4306.474886(02)	R(13)
4199.927716(02)	P(14)	4309.254375(02)	R(14)
4195.117980(02)	P(15)	4311.961657(02)	R(15)
4190.240439(02)	P(16)	4314.596584(02)	R(16)
4185.295239(02)	P(17)	4317.159011(02)	R(17)
4180.282526(02)	P(18)	4319.648791(02)	R(18)
4175.202447(02)	P(19)	4322.065779(03)	R(19)
4170.055149(03)	P(20)	4324.409829(03)	R(20)
4164.840777(03)	P(21)	4326.680794(03)	R(21)
4159.559478(03)	P(22)	4328.878530(03)	R(22)
4154.211398(03)	P(23)	4331.002889(04)	R(23)
4148.796683(04)	P(24)	4333.053728(04)	R(24)
4143.315479(04)	P(25)	4335.030899(05)	R(25)
4137.767932(04)	P(26)	4336.934259(06)	R(26)
4132.154187(05)	P(27)	4338.763661(07)	R(27)
4126.474391(06)	P(28)	4340.518961(09)	R(28)
4120.728689(07)	P(29)	4342.200014(11)	R(29)
4114.917226(09)	P(30)	4343.806675(16)	R(30)
4109.040148(12)	P(31)	4345.338799(21)	R(31)
4103.097600(16)	P(32)	4346.796243(29)	R(32)
4097.089728(21)	P(33)	4348.178862(40)	R(33)
4091.016676(29)	P(34)	4349.486513(54)	R(34)
4084.878591(40)	P(35)	4350.719052(73)	R(35)
4078.675618(54)	P(36)	4351.876336(96)	R(36)
4072.407901(73)	P(37)	4352.958224(127)	R(37)
4066.075588(97)	P(38)	4353.964572(166)	R(38)
4059.678822(127)	P(39)	4354.895240(214)	R(39)

* The uncertainty in the last digits (twice the standard error) is given in parentheses.

Wavenumbers for the $\nu = 3\text{-}0$ Band of CO

Wavenumber (unc)* cm⁻¹	Transition	Wavenumber (unc) cm⁻¹	Transition
		6354.179057(13)	R(0)
6346.594000(13)	P(1)	6357.813923(13)	R(1)
6342.644103(13)	P(2)	6361.343487(13)	R(2)
6338.589491(13)	P(3)	6364.767599(13)	R(3)
6334.430309(13)	P(4)	6368.086115(13)	R(4)
6330.166705(13)	P(5)	6371.298887(13)	R(5)
6325.798826(13)	P(6)	6374.405768(12)	R(6)
6321.326819(13)	P(7)	6377.406611(12)	R(7)
6316.750831(12)	P(8)	6380.301271(12)	R(8)
6312.071008(12)	P(9)	6383.089600(12)	R(9)
6307.287498(12)	P(10)	6385.771452(12)	R(10)
6302.400447(12)	P(11)	6388.346680(13)	R(11)
6297.410003(12)	P(12)	6390.815139(13)	R(12)
6292.316311(13)	P(13)	6393.176681(13)	R(13)
6287.119520(13)	P(14)	6395.431160(13)	R(14)
6281.819775(13)	P(15)	6397.578430(13)	R(15)
6276.417224(13)	P(16)	6399.618344(13)	R(16)
6270.912012(13)	P(17)	6401.550757(13)	R(17)
6265.304287(13)	P(18)	6403.375523(13)	R(18)
6259.594194(13)	P(19)	6405.092495(14)	R(19)
6253.781880(13)	P(20)	6406.701527(14)	R(20)
6247.867492(14)	P(21)	6408.202474(14)	R(21)
6241.851176(14)	P(22)	6409.595189(15)	R(22)
6235.733077(14)	P(23)	6410.879527(15)	R(23)
6229.513342(15)	P(24)	6412.055343(16)	R(24)
6223.192117(15)	P(25)	6413.122491(17)	R(25)
6216.769547(16)	P(26)	6414.080825(19)	R(26)
6210.245778(17)	P(27)	6414.930201(23)	R(27)
6203.620957(19)	P(28)	6415.670474(28)	R(28)
6196.895229(23)	P(29)	6416.301500(37)	R(29)
6190.068739(28)	P(30)	6416.823133(50)	R(30)
6183.141633(37)	P(31)	6417.235231(67)	R(31)
6176.114058(50)	P(32)	6417.537649(90)	R(32)
6168.986159(67)	P(33)		
6161.758082(90)	P(34)		

* The uncertainty in the last digits (twice the standard error) is given in parentheses.

Frequencies and Wavenumbers for the Rotational Lines of CO

Frequency MHz	Uncertainty* MHz	J'	J''	Wavenumber cm⁻¹	Uncertainty* cm⁻¹
115271.2029	0.0004	1	0	3.84503345	0.00000001
230538.0016	0.0008	2	1	7.68991999	0.00000003
345795.9923	0.0012	3	2	11.53451273	0.00000004
461040.7712	0.0016	4	3	15.37866477	0.00000005
576267.9350	0.0019	5	4	19.22222923	0.00000006
691473.0809	0.0021	6	5	23.06505926	0.00000007
806651.8065	0.0023	7	6	26.90700800	0.00000008
921799.7104	0.0025	8	7	30.74792863	0.00000008
1036912.3919	0.0027	9	8	34.58767438	0.00000009
1151985.4515	0.0029	10	9	38.42609848	0.00000010
1267014.4906	0.0031	11	10	42.26305422	0.00000010
1381995.1119	0.0034	12	11	46.09839491	0.00000011
1496922.9195	0.0038	13	12	49.93197392	0.00000013

Frequency MHz	Uncertainty* MHz	J'	J''	Wavenumber cm^{-1}	Uncertainty* cm^{-1}
1611793.5189	0.0042	14	13	53.76364468	0.00000014
1726602.5173	0.0047	15	14	57.59326065	0.00000016
1841345.5237	0.0052	16	15	61.42067535	0.00000017
1956018.1486	0.0057	17	16	65.24574239	0.00000019
2070616.0050	0.0061	18	17	69.06831542	0.00000020
2185134.7075	0.0065	19	18	72.88824816	0.00000022
2299569.8733	0.0069	20	19	76.70539441	0.00000023
2413917.1217	0.0071	21	20	80.51960806	0.00000024
2528172.0747	0.0073	22	21	84.33074306	0.00000024
2642330.3567	0.0074	23	22	88.13865346	0.00000025
2756387.5949	0.0075	24	23	91.94319341	0.00000025
2870339.4194	0.0077	25	24	95.74421713	0.00000026
2984181.4631	0.0080	26	25	99.54157896	0.00000027
3097909.3621	0.0085	27	26	103.33513334	0.00000028
3211518.7558	0.0090	28	27	107.12473480	0.00000030
3325005.2869	0.0096	29	28	110.91023800	0.00000032
3438364.6013	0.0102	30	29	114.69149772	0.00000034
3551592.3489	0.0107	31	30	118.46836884	0.00000036
3664684.1829	0.0111	32	31	122.24070637	0.00000037
3777635.7608	0.0118	33	32	126.00836545	0.00000039
3890442.7435	0.0137	34	33	129.77120137	0.00000046
4003100.7965	0.0179	35	34	133.52906952	0.00000060
4115605.5892	0.0254	36	35	137.28182546	0.00000085
4227952.7954	0.0370	37	36	141.02932487	0.00000123
4340138.0932	0.0531	38	37	144.77142361	0.00000177
4452157.1657	0.0746	39	38	148.50797766	0.00000249
4564005.7001	0.1025	40	39	152.23884318	0.00000342

* The uncertainty given is twice the standard error.

Frequencies and Wavenumbers for the Rotational Lines of HF

Frequency MHz	Uncertainty* MHz	J'	J''	Wavenumber cm^{-1}	Uncertainty* cm^{-1}
1232476.21	0.12	1	0	41.110981	0.000004
2463428.09	0.19	2	1	82.171116	0.000006
3691334.81	0.25	3	2	123.129676	0.000008
4914682.58	0.51	4	3	163.936165	0.000017
6131968.11	1.10	5	4	204.540439	0.000037
7341702.00	2.00	6	5	244.892818	0.000067
8542412.1	3.21	7	6	284.944197	0.000107
9732646.8	4.72	8	7	324.646153	0.000157
10910978.2	6.51	9	8	363.951056	0.000217
12076004.8	8.55	10	9	402.81216	0.000285
13226355.2	10.81	11	10	441.18372	0.000361
14360689.8	13.25	12	11	479.02105	0.00044
15477704.4	15.86	13	12	516.28065	0.00053
16576131.8	18.61	14	13	552.92024	0.00062
17654744.4	21.48	15	14	588.89888	0.00072
18712356.5	24.44	16	15	624.17703	0.00082
19747825.6	27.43	17	16	658.71656	0.00092
20760054.3	30.32	18	17	692.4809	0.00101
21747991.7	32.91	19	18	725.4349	0.00110
22710634.7	34.94	20	19	757.5452	0.00117
23647028.7	36.08	21	20	788.7800	0.00120
24556268.8	35.93	22	21	819.1090	0.00120
25437499.9	34.12	23	22	848.5037	0.00114

INFRARED AND FAR-INFRARED ABSORPTION FREQUENCY STANDARDS (continued)

Frequency MHz	Uncertainty* MHz	J'	J''	Wavenumber cm⁻¹	Uncertainty* cm⁻¹
26289917.4	30.32	24	23	876.9373	0.00101
27112767.2	24.41	25	24	904.38457	0.00081
27905345.6	16.88	26	25	930.82214	0.00056
28666999.3	10.80	27	26	956.22817	0.00036
29397124.8	14.65	28	27	980.58253	0.00049
30095168.2	24.62	29	28	1003.86676	0.00082
30760624.2	33.36	30	29	1026.0640	0.00111
31393035.7	36.17	31	30	1047.1590	0.00121

* The uncertainty given is twice the standard error.

Frequencies and Wavenumbers for the Rotational Lines of H³⁵Cl

Frequency MHz	Uncertainty* MHz	J'	J''	Wavemumber cm⁻¹	Uncertainty* cm⁻¹
1876226.517	0.065	3	2	62.584180	0.000002
2499864.439	0.066	4	3	83.386502	0.000002
3121986.563	0.064	5	4	104.138262	0.000002
3742216.601	0.076	6	5	124.826909	0.000003
4360180.042	0.098	7	6	145.439951	0.000003
4975504.51	0.11	8	7	165.964966	0.000004
5587820.10	0.12	9	8	186.389615	0.000004
6196759.76	0.22	10	9	206.701656	0.000007
6801959.63	0.50	11	10	226.888951	0.000017
7403059.41	1.02	12	11	246.939481	0.000034
7999702.7	1.8	13	12	266.841359	0.000062
8591537.3	3.1	14	13	286.582837	0.000103
9178215.8	4.8	15	14	306.152324	0.000161

* The uncertainty given is twice the standard error.

Frequencies and Wavenumbers for the Rotational Lines of H³⁷Cl

Frequency MHz	Uncertainty* MHz	J'	J''	Wavenumber cm⁻¹	Uncertainty* cm⁻¹
1873410.72	0.05	3	2	62.490255	0.000002
2496115.33	0.05	4	3	83.261445	0.000002
3117308.69	0.05	5	4	103.982225	0.000002
3736615.64	0.06	6	5	124.640082	0.000002
4353662.84	0.08	7	6	145.222561	0.000003
4968079.04	0.09	8	7	165.717279	0.000003
5579495.53	0.10	9	8	186.111938	0.000003
6187546.42	0.19	10	9	206.394332	0.000006
6791869.04	0.45	11	10	226.552365	0.000015
7392104.3	0.9	12	11	246.574057	0.000030
7987896.9	1.6	13	12	266.447561	0.000054
8578896.1	2.7	14	13	286.161170	0.000089

* The uncertainty given is twice the standard error.

Section 11
Nuclear and Particle Physics

Summary Tables of Particle Properties 11-1
Table of the Isotopes 11-50
Neutron Scattering and Absorption Properties 11-159
Cosmic Radiation 11-176

SUMMARY TABLES OF PARTICLE PROPERTIES

Extracted from the Particle Listings of the
Review of Particle Physics
Published in Eur. Jour. Phys. **C3**, 1 (1998)
Available at http://pdg.lbl.gov

Particle Data Group Authors:

C. Caso, G. Conforto, A. Gurtu, M. Aguilar-Benitez, C. Amsler,
R.M. Barnett, P.R. Burchat, C.D. Carone, O. Dahl, M. Doser,
S. Eidelman, J.L. Feng, M. Goodman, C. Grab, D.E. Groom,
K. Hagiwara, K.G. Hayes, J.J. Hernández, K. Hikasa, K. Honscheid,
F. James, M.L. Mangano, A.V. Manohar, K. Mönig, H. Murayama,
K. Nakamura, K.A. Olive, A. Piepke, M. Roos, R.H. Schindler,
R.E. Shrock, M. Tanabashi, N.A. Törnqvist, T.G. Trippe, P. Vogel,
C.G. Wohl, R.L. Workman, W.-M. Yao

Technical Associates: B. Armstrong, J.L. Casas Serradilla,
B.B. Filimonov, P.S. Gee, S.B. Lugovsky, S. Mankov, F. Nicholson

**Other Authors who have made substantial contributions
to reviews since the 1994 edition:**

K.S. Babu, D. Besson, O. Biebel, R.N. Cahn, R.L. Crawford, R.H. Dalitz,
T. Damour, K. Desler, R.J. Donahue, D.A. Edwards, J. Erler,
V.V. Ezhela, A. Fassò, W. Fetscher, D. Froidevaux, T.K. Gaisser,
L. Garren, S. Geer, H.-J. Gerber, F.J. Gilman, H.E. Haber, C. Hagmann,
I. Hinchliffe, C.J. Hogan, G. Höhler, J.D. Jackson, K.F. Johnson,
D. Karlen, B. Kayser, K. Kleinknecht, I.G. Knowles, C. Kolda, P. Kreitz,
P. Langacker, R. Landua, L. Littenberg, D.M. Manley, J. March-Russell,
T. Nakada, H. Quinn, G. Raffelt, B. Renk, M.T. Ronan, L.J. Rosenberg,
M. Schmitt, D.N. Schramm, D. Scott, T. Sjöstrand, G.F. Smoot,
S. Spanier, M. Srednicki, M. Stanev, M. Suzuki, N.P. Tkachenko,
G. Valencia, K. van Bibber, R. Voss, L. Wolfenstein, S. Youssef

(Approximate closing date for data: January 1, 1998)

GAUGE AND HIGGS BOSONS

$I(J^{PC}) = 0,1(1^{--})$

Mass $m < 2 \times 10^{-16}$ eV
Charge $q < 5 \times 10^{-30}\ e$
Mean life $\tau =$ Stable

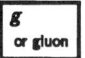
or gluon

$I(J^P) = 0(1^-)$

Mass $m = 0$ [a]
SU(3) color octet

$J = 1$

Charge $= \pm 1\ e$
Mass $m = 80.41 \pm 0.10$ GeV
$m_Z - m_W = 10.78 \pm 0.10$ GeV
$m_{W^+} - m_{W^-} = -0.2 \pm 0.6$ GeV
Full width $\Gamma = 2.06 \pm 0.06$ GeV

W^- modes are charge conjugates of the modes below.

W^+ DECAY MODES		Fraction (Γ_i/Γ)	Confidence level	p (MeV/c)
$\ell^+ \nu$	[b]	(10.74 ± 0.33) %		–
$e^+ \nu$		(10.9 ± 0.4) %		40205
$\mu^+ \nu$		(10.2 ± 0.5) %		40205
$\tau^+ \nu$		(11.3 ± 0.8) %		40185
hadrons		(67.8 ± 1.0) %		–
$\pi^+ \gamma$		$< 2.2 \quad \times 10^{-4}$	95%	40205

$\boxed{Z}$ $\qquad J = 1$

Charge $= 0$
Mass $m = 91.187 \pm 0.007$ GeV [c]
Full width $\Gamma = 2.490 \pm 0.007$ GeV
$\Gamma(\ell^+ \ell^-) = 83.83 \pm 0.27$ MeV [b]
$\Gamma(\text{invisible}) = 498.3 \pm 4.2$ MeV [d]
$\Gamma(\text{hadrons}) = 1740.7 \pm 5.9$ MeV
$\Gamma(\mu^+ \mu^-)/\Gamma(e^+ e^-) = 1.000 \pm 0.005$
$\Gamma(\tau^+ \tau^-)/\Gamma(e^+ e^-) = 0.998 \pm 0.005$ [e]

Average charged multiplicity

$\langle N_{charged} \rangle = 21.00 \pm 0.13$

Couplings to leptons

$g_V^\ell = -0.0377 \pm 0.0007$
$g_A^\ell = -0.5008 \pm 0.0008$
$g_{\nu_e} = 0.53 \pm 0.09$
$g_{\nu_\mu} = 0.502 \pm 0.017$

Asymmetry parameters [f]

$A_e = 0.1519 \pm 0.0034$
$A_\mu = 0.102 \pm 0.034$
$A_\tau = 0.143 \pm 0.008$
$A_c = 0.59 \pm 0.19$
$A_b = 0.89 \pm 0.11$

Charge asymmetry (%) at Z pole

$A_{FB}^{(0\ell)} = 1.59 \pm 0.18$
$A_{FB}^{(0u)} = 4.0 \pm 7.3$
$A_{FB}^{(0s)} = 9.9 \pm 3.1 \quad (S = 1.2)$
$A_{FB}^{(0c)} = 7.32 \pm 0.58$
$A_{FB}^{(0b)} = 10.02 \pm 0.28$

Z DECAY MODES		Fraction (Γ_i/Γ)	Confidence level	p (MeV/c)
$e^+ e^-$		(3.366 ± 0.008) %		45594
$\mu^+ \mu^-$		(3.367 ± 0.013) %		45593
$\tau^+ \tau^-$		(3.360 ± 0.015) %		45559
$\ell^+ \ell^-$	[b]	(3.366 ± 0.006) %		–
invisible		(20.01 ± 0.16) %		–
hadrons		(69.90 ± 0.15) %		–
$(u\bar{u} + c\bar{c})/2$		(10.1 ± 1.1) %		–
$(d\bar{d} + s\bar{s} + b\bar{b})/3$		(16.6 ± 0.6) %		–
$c\bar{c}$		(12.4 ± 0.6) %		–
$b\bar{b}$		(15.16 ± 0.09) %		–
$g g g$		$< 1.1 \quad$ %	95%	–
$\pi^0 \gamma$		$< 5.2 \quad \times 10^{-5}$	95%	45593
$\eta \gamma$		$< 5.1 \quad \times 10^{-5}$	95%	45592
$\omega \gamma$		$< 6.5 \quad \times 10^{-4}$	95%	45590
$\eta'(958) \gamma$		$< 4.2 \quad \times 10^{-5}$	95%	45588
$\gamma \gamma$		$< 5.2 \quad \times 10^{-5}$	95%	45594
$\gamma \gamma \gamma$		$< 1.0 \quad \times 10^{-5}$	95%	45594
$\pi^\pm W^\mp$	[g]	$< 7 \quad \times 10^{-5}$	95%	10139
$\rho^\pm W^\mp$	[g]	$< 8.3 \quad \times 10^{-5}$	95%	10114
$J/\psi(1S)X$		$(3.66 \pm 0.23) \times 10^{-3}$		–
$\psi(2S)X$		$(1.60 \pm 0.29) \times 10^{-3}$		–
$\chi_{c1}(1P)X$		$(2.9 \pm 0.7) \times 10^{-3}$		–
$\chi_{c2}(1P)X$		$< 3.2 \quad \times 10^{-3}$	90%	–
$\Upsilon(1S) X + \Upsilon(2S) X$ $+ \Upsilon(3S) X$		$(1.0 \pm 0.5) \times 10^{-4}$		–
$\Upsilon(1S)X$		$< 5.5 \quad \times 10^{-5}$	95%	–
$\Upsilon(2S)X$		$< 1.39 \quad \times 10^{-4}$	95%	–
$\Upsilon(3S)X$		$< 9.4 \quad \times 10^{-5}$	95%	–
$(D^0/\overline{D}^0) X$		(20.7 ± 2.0) %		–
$D^\pm X$		(12.2 ± 1.7) %		–
$D^*(2010)^\pm X$	[g]	(11.4 ± 1.3) %		–
$B_s^0 X$		seen		–
anomalous γ + hadrons	[h]	$< 3.2 \quad \times 10^{-3}$	95%	–
$e^+ e^- \gamma$	[h]	$< 5.2 \quad \times 10^{-4}$	95%	45594
$\mu^+ \mu^- \gamma$	[h]	$< 5.6 \quad \times 10^{-4}$	95%	45593
$\tau^+ \tau^- \gamma$	[h]	$< 7.3 \quad \times 10^{-4}$	95%	45559
$\ell^+ \ell^- \gamma \gamma$	[i]	$< 6.8 \quad \times 10^{-6}$	95%	–
$q\bar{q} \gamma \gamma$	[i]	$< 5.5 \quad \times 10^{-6}$	95%	–

$\nu\bar{\nu}\gamma\gamma$		[i] < 3.1	$\times 10^{-6}$	95%	45594
$e^{\pm}\mu^{\mp}$	LF	[g] < 1.7	$\times 10^{-6}$	95%	45593
$e^{\pm}\tau^{\mp}$	LF	[g] < 9.8	$\times 10^{-6}$	95%	45576
$\mu^{\pm}\tau^{\mp}$	LF	[g] < 1.2	$\times 10^{-5}$	95%	45576

Higgs Bosons — H^0 and $H^{\pm}$, Searches for

H^0 Mass $m > 77.5$ GeV, CL = 95%

H_1^0 In Supersymmetric Models ($m_{H_1^0} < m_{H_2^0}$)

Mass $m > 62.5$ GeV, CL = 95%

A^0 Pseudoscalar Higgs Boson in Supersymmetric Models [j]

Mass $m > 62.5$ GeV, CL = 95% $\tan\beta > 1$

$H^{\pm}$ Mass $m > 54.5$ GeV, CL = 95%

See the Particle Listings for a Note giving details of Higgs Bosons.

Heavy Bosons Other Than Higgs Bosons, Searches for

Additional W Bosons

W_R — right-handed W
Mass $m > 549$ GeV
(assuming light right-handed neutrino)
W' with standard couplings decaying to $e\nu$, $\mu\nu$
Mass $m > 720$ GeV, CL = 95%

Additional Z Bosons

Z'_{SM} with standard couplings
Mass $m > 690$ GeV, CL = 95% ($p\bar{p}$ direct search)
Mass $m > 779$ GeV, CL = 95% (electroweak fit)
Z_{LR} of SU(2)$_L$×SU(2)$_R$×U(1)
(with $g_L = g_R$)
Mass $m > 630$ GeV, CL = 95% ($p\bar{p}$ direct search)
Mass $m > 389$ GeV, CL = 95% (electroweak fit)
Z_{χ} of SO(10) → SU(5)×U(1)$_{\chi}$
(coupling constant derived from G.U.T.)
Mass $m > 595$ GeV, CL = 95% ($p\bar{p}$ direct search)
Mass $m > 321$ GeV, CL = 95% (electroweak fit)
Z_{ψ} of E_6 → SO(10)×U(1)$_{\psi}$
(coupling constant derived from G.U.T.)
Mass $m > 590$ GeV, CL = 95% ($p\bar{p}$ direct search)
Mass $m > 160$ GeV, CL = 95% (electroweak fit)
Z_{η} of E_6 → SU(3)×SU(2)×U(1)×U(1)$_{\eta}$
(coupling constant derived from G.U.T.);
charges are $Q_{\eta} = \sqrt{3/8}Q_{\chi} - \sqrt{5/8}Q_{\psi}$)
Mass $m > 620$ GeV, CL = 95% ($p\bar{p}$ direct search)
Mass $m > 182$ GeV, CL = 95% (electroweak fit)

Scalar Leptoquarks

Mass $m > 225$ GeV, CL = 95% (1st generation, pair prod.)
Mass $m > 237$ GeV, CL = 95% (1st gener., single prod.)
Mass $m > 119$ GeV, CL = 95% (2nd gener., pair prod.)
Mass $m > 73$ GeV, CL = 95% (2nd gener., single prod.)
Mass $m > 99$ GeV, CL = 95% (3rd gener., pair prod.)
(See the Particle Listings for assumptions on leptoquark quantum numbers and branching fractions.)

Axions (A^0) and Other Very Light Bosons, Searches for

The standard Peccei-Quinn axion is ruled out. Variants with reduced couplings or much smaller masses are constrained by various data. The Particle Listings in the full *Review* contain a Note discussing axion searches.

The best limit for the half-life of neutrinoless double beta decay with Majoron emission is $> 7.2 \times 10^{24}$ years (CL = 90%).

NOTES

In this Summary Table:

When a quantity has "(S = ...)" to its right, the error on the quantity has been enlarged by the "scale factor" S, defined as $S = \sqrt{\chi^2/(N-1)}$, where N is the number of measurements used in calculating the quantity. We do this when S > 1, which often indicates that the measurements are inconsistent. When S > 1.25, we also show in the Particle Listings an ideogram of the measurements. For more about S, see the Introduction.

A decay momentum p is given for each decay mode. For a 2-body decay, p is the momentum of each decay product in the rest frame of the decaying particle. For a 3-or-more-body decay, p is the largest momentum any of the products can have in this frame.

[a] Theoretical value. A mass as large as a few MeV may not be precluded.

[b] ℓ indicates each type of lepton (e, μ, and τ), not sum over them.

[c] The Z-boson mass listed here corresponds to a Breit-Wigner resonance parameter. It lies approximately 34 MeV above the real part of the position of the pole (in the energy-squared plane) in the Z-boson propagator.

[d] This partial width takes into account Z decays into $\nu\bar{\nu}$ and any other possible undetected modes.

[e] This ratio has not been corrected for the τ mass.

[f] Here $A \equiv 2g_V g_A/(g_V^2 + g_A^2)$.

[g] The value is for the sum of the charge states of particle/antiparticle states indicated.

[h] See the Z Particle Listings for the γ energy range used in this measurement.

[i] For $m_{\gamma\gamma} = (60 \pm 5)$ GeV.

[j] The limits assume no invisible decays.

LEPTONS

e

$J = \frac{1}{2}$

Mass $m = 0.51099907 \pm 0.00000015$ MeV [a]
$ = (5.485799111 \pm 0.000000012) \times 10^{-4}$ u
$(m_{e^+} - m_{e^-})/m < 4 \times 10^{-8}$, CL = 90%
$|q_{e^+} + q_{e^-}|/e < 4 \times 10^{-8}$
Magnetic moment $\mu = 1.001159652193 \pm 0.000000000010\ \mu_B$
$(g_{e^+} - g_{e^-}) / g_{average} = (-0.5 \pm 2.1) \times 10^{-12}$
Electric dipole moment $d = (0.18 \pm 0.16) \times 10^{-26}$ ecm
Mean life $\tau > 4.3 \times 10^{23}$ yr, CL = 68% [b]

μ

$J = \frac{1}{2}$

Mass $m = 105.658389 \pm 0.000034$ MeV [c]
$ = 0.113428913 \pm 0.000000017$ u
Mean life $\tau = (2.19703 \pm 0.00004) \times 10^{-6}$ s
$\tau_{\mu^+}/\tau_{\mu^-} = 1.00002 \pm 0.00008$
$c\tau = 658.654$ m
Magnetic moment $\mu = 1.0011659230 \pm 0.0000000084\ e\hbar/2m_\mu$
$(g_{\mu^+} - g_{\mu^-}) / g_{average} = (-2.6 \pm 1.6) \times 10^{-8}$
Electric dipole moment $d = (3.7 \pm 3.4) \times 10^{-19}$ ecm

Decay parameters [d]

$\rho = 0.7518 \pm 0.0026$
$\eta = -0.007 \pm 0.013$
$\delta = 0.749 \pm 0.004$
$\xi P_\mu = 1.003 \pm 0.008$ [e]
$\xi P_\mu \delta / \rho > 0.99682$, CL = 90% [e]
$\xi' = 1.00 \pm 0.04$
$\xi'' = 0.7 \pm 0.4$
$\alpha/A = (0 \pm 4) \times 10^{-3}$
$\alpha'/A = (0 \pm 4) \times 10^{-3}$
$\beta/A = (4 \pm 6) \times 10^{-3}$
$\beta'/A = (2 \pm 6) \times 10^{-3}$
$\bar{\eta} = 0.02 \pm 0.08$

μ^+ modes are charge conjugates of the modes below.

μ^- DECAY MODES		Fraction (Γ_i/Γ)	Confidence level	p (MeV/c)
$e^- \bar{\nu}_e \nu_\mu$		$\approx 100\%$		53
$e^- \bar{\nu}_e \nu_\mu \gamma$	[f]	$(1.4 \pm 0.4)\%$		53
$e^- \bar{\nu}_e \nu_\mu e^+ e^-$	[g]	$(3.4 \pm 0.4) \times 10^{-5}$		53

Lepton Family number (LF) violating modes

			Fraction (Γ_i/Γ)	Confidence level	p (MeV/c)
$e^- \nu_e \bar{\nu}_\mu$	LF	[h]	< 1.2 %	90%	53
$e^- \gamma$	LF		$< 4.9 \times 10^{-11}$	90%	53
$e^- e^+ e^-$	LF		$< 1.0 \times 10^{-12}$	90%	53
$e^- 2\gamma$	LF		$< 7.2 \times 10^{-11}$	90%	53

τ

$J = \frac{1}{2}$

Mass $m = 1777.05\,^{+0.29}_{-0.26}$ MeV
Mean life $\tau = (290.0 \pm 1.2) \times 10^{-15}$ s
$c\tau = 86.93\ \mu$m
Magnetic moment anomaly > -0.052 and < 0.058, CL = 95%
Electric dipole moment $d > -3.1$ and $< 3.1 \times 10^{-16}$ ecm, CL = 95%

Weak dipole moment

$\mathrm{Re}(d_\tau^w) < 0.56 \times 10^{-17}$ ecm, CL = 95%
$\mathrm{Im}(d_\tau^w) < 1.5 \times 10^{-17}$ ecm, CL = 95%

Weak anomalous magnetic dipole moment

$\mathrm{Re}(\alpha_\tau^w) < 4.5 \times 10^{-3}$, CL = 90%
$\mathrm{Im}(\alpha_\tau^w) < 9.9 \times 10^{-3}$, CL = 90%

Decay parameters

See the τ Particle Listings for a note concerning τ-decay parameters.

$\rho^\tau(e\ \text{or}\ \mu) = 0.748 \pm 0.010$
$\rho^\tau(e) = 0.745 \pm 0.012$
$\rho^\tau(\mu) = 0.741 \pm 0.030$
$\xi^\tau(e\ \text{or}\ \mu) = 1.01 \pm 0.04$
$\xi^\tau(e) = 0.98 \pm 0.05$
$\xi^\tau(\mu) = 1.07 \pm 0.08$
$\eta^\tau(e\ \text{or}\ \mu) = 0.01 \pm 0.07$
$\eta^\tau(\mu) = -0.10 \pm 0.18$
$(\delta\xi)^\tau(e\ \text{or}\ \mu) = 0.749 \pm 0.026$
$(\delta\xi)^\tau(e) = 0.733 \pm 0.033$
$(\delta\xi)^\tau(\mu) = 0.78 \pm 0.05$
$\xi^\tau(\pi) = 0.99 \pm 0.05$
$\xi^\tau(\rho) = 0.996 \pm 0.010$
$\xi^\tau(a_1) = 1.02 \pm 0.04$
$\xi^\tau(\text{all hadronic modes}) = 0.997 \pm 0.009$

τ^+ modes are charge conjugates of the modes below. "$h^\pm$" stands for $\pi^\pm$ or $K^\pm$. "ℓ" stands for e or μ. "Neutral" means neutral hadron whose decay products include γ's and/or π^0's.

τ^- DECAY MODES		Fraction (Γ_i/Γ)	Scale factor/ Confidence level	p (MeV/c)
Modes with one charged particle				
particle$^- \geq 0$ neutrals $\geq 0K_L^0\ \nu_\tau$ ("1-prong")		$(84.71 \pm 0.13)\%$	S=1.2	—
particle$^- \geq 0$ neutrals $\geq 0K^0\ \nu_\tau$		$(85.30 \pm 0.13)\%$	S=1.2	—
$\mu^- \bar{\nu}_\mu \nu_\tau$	[i]	$(17.37 \pm 0.09)\%$		885
$\mu^- \bar{\nu}_\mu \nu_\tau \gamma$	[g]	$(3.0 \pm 0.6) \times 10^{-3}$		—
$e^- \bar{\nu}_e \nu_\tau$	[i]	$(17.81 \pm 0.07)\%$		889
$h^- \geq 0$ neutrals $\geq 0K_L^0\ \nu_\tau$		$(49.52 \pm 0.16)\%$	S=1.2	—
$h^- \geq 0K_L^0\ \nu_\tau$		$(12.32 \pm 0.12)\%$	S=1.5	—
$h^- \nu_\tau$		$(11.79 \pm 0.12)\%$	S=1.5	—
$\pi^- \nu_\tau$	[i]	$(11.08 \pm 0.13)\%$	S=1.4	883
$K^- \nu_\tau$	[i]	$(7.1 \pm 0.5) \times 10^{-3}$		820
$h^- \geq 1$ neutrals ν_τ		$(36.91 \pm 0.17)\%$	S=1.2	—
$h^- \pi^0 \nu_\tau$		$(25.84 \pm 0.14)\%$	S=1.1	—
$\pi^- \pi^0 \nu_\tau$	[i]	$(25.32 \pm 0.15)\%$	S=1.1	878
$\pi^- \pi^0$ non-$\rho(770) \nu_\tau$		$(3.0 \pm 3.2) \times 10^{-3}$		878
$K^- \pi^0 \nu_\tau$	[i]	$(5.2 \pm 0.5) \times 10^{-3}$		814
$h^- \geq 2\pi^0 \nu_\tau$		$(10.79 \pm 0.16)\%$	S=1.2	—
$h^- 2\pi^0 \nu_\tau$		$(9.39 \pm 0.14)\%$	S=1.2	—
$h^- 2\pi^0 \nu_\tau$ (ex.K^0)		$(9.23 \pm 0.14)\%$	S=1.2	—
$\pi^- 2\pi^0 \nu_\tau$ (ex.K^0)	[i]	$(9.15 \pm 0.15)\%$	S=1.2	862
$K^- 2\pi^0 \nu_\tau$ (ex.K^0)	[i]	$(8.0 \pm 2.7) \times 10^{-4}$		796
$h^- \geq 3\pi^0 \nu_\tau$		$(1.40 \pm 0.14)\%$	S=1.1	—
$h^- 3\pi^0 \nu_\tau$		$(1.23 \pm 0.10)\%$	S=1.1	—
$\pi^- 3\pi^0 \nu_\tau$ (ex.K^0)	[i]	$(1.11 \pm 0.14)\%$		836
$K^- 3\pi^0 \nu_\tau$ (ex.K^0)	[i]	$(4.3\,^{+10.0}_{-2.9}) \times 10^{-4}$		766
$h^- 4\pi^0 \nu_\tau$ (ex.K^0)		$(1.7 \pm 0.6) \times 10^{-3}$		—
$h^- 4\pi^0 \nu_\tau$ (ex.K^0,η)	[i]	$(1.1 \pm 0.6) \times 10^{-3}$		—
$K^- \geq 0\pi^0 \geq 0K^0\ \nu_\tau$		$(1.66 \pm 0.10)\%$		—
$K^- \geq 1\ (\pi^0\ \text{or}\ K^0)\ \nu_\tau$		$(9.5 \pm 1.0) \times 10^{-3}$		—
Modes with K^0's				
$K^0 (\text{particles})^- \nu_\tau$		$(1.66 \pm 0.09)\%$	S=1.4	—
$h^- \overline{K}^0 \geq 0$ neutrals $\geq 0K_L^0 \nu_\tau$		$(1.62 \pm 0.09)\%$	S=1.4	—
$h^- \overline{K}^0 \nu_\tau$		$(9.9 \pm 0.8) \times 10^{-3}$	S=1.5	—
$\pi^- \overline{K}^0 \nu_\tau$	[i]	$(8.3 \pm 0.8) \times 10^{-3}$	S=1.4	812
$\pi^- \overline{K}^0$ (non-$K^*(892)^-) \nu_\tau$		$< 1.7 \times 10^{-3}$	CL=95%	812
$K^- K^0 \nu_\tau$	[i]	$(1.59 \pm 0.24) \times 10^{-3}$		737
$h^- \overline{K}^0 \pi^0 \nu_\tau$		$(5.5 \pm 0.5) \times 10^{-3}$		—
$\pi^- \overline{K}^0 \pi^0 \nu_\tau$	[i]	$(3.9 \pm 0.5) \times 10^{-3}$		794
$\overline{K}^0 \rho^- \nu_\tau$		$(1.9 \pm 0.7) \times 10^{-3}$		—
$K^- K^0 \pi^0 \nu_\tau$	[i]	$(1.51 \pm 0.29) \times 10^{-3}$		685
$\pi^- \overline{K}^0 \pi^0 \pi^0 \nu_\tau$		$(6 \pm 4) \times 10^{-4}$		—
$K^- K^0 \pi^0 \pi^0 \nu_\tau$		$< 3.9 \times 10^{-4}$	CL=95%	—
$\pi^- K^0 \overline{K}^0 \nu_\tau$	[i]	$(1.21 \pm 0.21) \times 10^{-3}$	S=1.2	682

$\pi^- K^0_S K^0_S \nu_\tau$	$(3.0 \pm 0.5) \times 10^{-4}$	S=1.2	–	
$\pi^- K^0_S K^0_L \nu_\tau$	$(6.0 \pm 1.0) \times 10^{-4}$	S=1.2	–	
$\pi^- K^0_S K^0_S \pi^0 \nu_\tau$	$< 2.0 \quad \times 10^{-4}$	CL=95%	–	
$\pi^- K^0_S K^0_L \pi^0 \nu_\tau$	$(3.1 \pm 1.2) \times 10^{-4}$		–	
$K^- K^0 \geq 0$ neutrals ν_τ	$(3.1 \pm 0.4) \times 10^{-3}$		–	
$K^0 h^+ h^- \geq 0$ neutrals ν_τ	$< 1.7 \quad \times 10^{-3}$	CL=95%	–	
$K^0 h^+ h^- h^- \nu_\tau$	$(2.3 \pm 2.0) \times 10^{-4}$		–	

Modes with three charged particles

$h^- h^- h^+ \geq 0$neut. ν_τ ("3-prong")	(15.18 ± 0.13) %	S=1.2	–	
$h^- h^- h^+ \geq 0$ neutrals ν_τ (ex. $K^0_S \rightarrow \pi^+ \pi^-$)	(14.60 ± 0.13) %	S=1.2	–	
$\pi^- \pi^+ \pi^- \geq 0$ neutrals ν_τ	(14.60 ± 0.14) %		–	
$h^- h^- h^+ \nu_\tau$	(9.96 ± 0.10) %	S=1.1	–	
$h^- h^- h^+ \nu_\tau$ (ex.K^0)	(9.62 ± 0.10) %	S=1.1	–	
$h^- h^- h^+ \nu_\tau$ (ex.K^0,ω)	(9.57 ± 0.10) %	S=1.1	–	
$\pi^- \pi^+ \pi^- \nu_\tau$	(9.56 ± 0.11) %	S=1.1	–	
$\pi^- \pi^+ \pi^- \nu_\tau$ (ex.K^0)	(9.52 ± 0.11) %	S=1.1	–	
$\pi^- \pi^+ \pi^- \nu_\tau$ (ex.K^0,ω)	(9.23 ± 0.11) %	[i]	–	
$h^- h^- h^+ \geq 1$ neutrals ν_τ	(5.18 ± 0.11) %	S=1.2	–	
$h^- h^- h^+ \geq 1$ neutrals ν_τ (ex. $K^0_S \rightarrow \pi^+ \pi^-$)	(4.98 ± 0.11) %	S=1.2	–	
$h^- h^- h^+ \pi^0 \nu_\tau$	(4.50 ± 0.09) %	S=1.1	–	
$h^- h^- h^+ \pi^0 \nu_\tau$ (ex.K^0)	(4.31 ± 0.09) %	S=1.1	–	
$h^- h^- h^+ \pi^0 \nu_\tau$ (ex. K^0, ω)	(2.59 ± 0.09) %		–	
$\pi^- \pi^+ \pi^- \pi^0 \nu_\tau$	(4.35 ± 0.10) %		–	
$\pi^- \pi^+ \pi^- \pi^0 \nu_\tau$ (ex.K^0)	(4.22 ± 0.10) %		–	
$\pi^- \pi^+ \pi^- \pi^0 \nu_\tau$ (ex.K^0,ω)	(2.49 ± 0.10) %	[i]	–	
$h^- (\rho^0)^0 \nu_\tau$	(2.88 ± 0.35) %		–	
$(a_1(1260) h)^- \nu_\tau$	$< 2.0 \quad$ %	CL=95%	–	
$h^- \rho \pi^0 \nu_\tau$	(1.35 ± 0.20) %		–	
$h^- \rho^+ h^- \nu_\tau$	$(4.5 \pm 2.2) \times 10^{-3}$		–	
$h^- \rho^- h^+ \nu_\tau$	(1.17 ± 0.23) %		–	
$h^- h^- h^+ 2\pi^0 \nu_\tau$	$(5.4 \pm 0.4) \times 10^{-3}$		–	
$h^- h^- h^+ 2\pi^0 \nu_\tau$ (ex.K^0)	$(5.3 \pm 0.4) \times 10^{-3}$		–	
$h^- h^- h^+ 2\pi^0 \nu_\tau$ (ex.K^0,ω,η)	$(1.1 \pm 0.4) \times 10^{-3}$	[i]	–	
$h^- h^- h^+ \geq 3\pi^0 \nu_\tau$	$(1.4 \, ^{+0.9}_{-0.7}) \times 10^{-3}$	[i] S=1.5	–	
$h^- h^- h^+ 3\pi^0 \nu_\tau$	$(2.9 \pm 0.8) \times 10^{-4}$		–	
$K^- h^+ h^- \geq 0$ neutrals ν_τ	$(5.4 \pm 0.7) \times 10^{-3}$	S=1.1	–	
$K^- \pi^+ \pi^- \geq 0$ neutrals ν_τ	$(3.1 \pm 0.6) \times 10^{-3}$	S=1.1	–	
$K^- \pi^+ \pi^- \nu_\tau$	$(2.3 \pm 0.4) \times 10^{-3}$		–	
$K^- \pi^+ \pi^- \nu_\tau$ (ex.K^0)	$(1.8 \pm 0.5) \times 10^{-3}$	[i]	–	
$K^- \pi^+ \pi^- \pi^0 \nu_\tau$	$(8 \pm 4) \times 10^{-4}$		–	
$K^- \pi^+ \pi^- \pi^0 \nu_\tau$ (ex.K^0)	$(2.4 \, ^{+4.3}_{-1.6}) \times 10^{-4}$	[i]	–	
$K^- \pi^+ K^- \geq 0$ neut. ν_τ	$< 9 \quad \times 10^{-4}$	CL=95%	–	
$K^- K^+ \pi^- \geq 0$ neut. ν_τ	$(2.3 \pm 0.4) \times 10^{-3}$		–	
$K^- K^+ \pi^- \nu_\tau$	$(1.61 \pm 0.26) \times 10^{-3}$	[i]	685	
$K^- K^+ \pi^- \pi^0 \nu_\tau$	$(6.9 \pm 3.0) \times 10^{-4}$	[i]	–	
$K^- K^+ K^- \geq 0$ neut. ν_τ	$< 2.1 \quad \times 10^{-3}$	CL=95%	–	
$K^- K^+ K^- \nu_\tau$	$< 1.9 \quad \times 10^{-4}$	CL=90%	–	
$\pi^- K^+ \pi^- \geq 0$ neut. ν_τ	$< 2.5 \quad \times 10^{-3}$	CL=95%	–	
$e^- e^- e^+ \overline{\nu}_e \nu_\tau$	$(2.8 \pm 1.5) \times 10^{-5}$		889	
$\mu^- e^- e^+ \overline{\nu}_\mu \nu_\tau$	$< 3.6 \quad \times 10^{-5}$	CL=90%	885	

Modes with five charged particles

$3h^- 2h^+ \geq 0$ neutrals ν_τ (ex. $K^0_S \rightarrow \pi^- \pi^+$) ("5-prong")	$(9.7 \pm 0.7) \times 10^{-4}$		–	
$3h^- 2h^+ \nu_\tau$ (ex.K^0)	$(7.5 \pm 0.7) \times 10^{-4}$	[i]	–	
$3h^- 2h^+ \pi^0 \nu_\tau$ (ex.K^0)	$(2.2 \pm 0.5) \times 10^{-4}$	[i]	–	
$3h^- 2h^+ 2\pi^0 \nu_\tau$	$< 1.1 \quad \times 10^{-4}$	CL=90%	–	

Miscellaneous other allowed modes

$(5\pi)^- \nu_\tau$	$(7.4 \pm 0.7) \times 10^{-3}$		–	
$4h^- 3h^+ \geq 0$ neutrals ν_τ ("7-prong")	$< 2.4 \quad \times 10^{-6}$	CL=90%	–	
$K^*(892)^- \geq 0(h^0 \neq K^0_S) \nu_\tau$	(1.94 ± 0.31) %		–	
$K^*(892)^- \geq 0$ neutrals ν_τ	(1.33 ± 0.13) %		–	
$K^*(892)^- \nu_\tau$	(1.28 ± 0.08) %		665	
$K^*(892)^0 K^- \geq 0$ neutrals ν_τ	$(3.2 \pm 1.4) \times 10^{-3}$		–	
$K^*(892)^0 K^- \nu_\tau$	$(2.1 \pm 0.4) \times 10^{-3}$		539	
$\overline{K}^*(892)^0 \pi^- \geq 0$ neutrals ν_τ	$(3.8 \pm 1.7) \times 10^{-3}$		–	
$\overline{K}^*(892)^0 \pi^- \nu_\tau$	$(2.2 \pm 0.5) \times 10^{-3}$		653	
$(\overline{K}^*(892)\pi)^- \nu_\tau \rightarrow \pi^- \overline{K}^0 \pi^0 \nu_\tau$	$(1.1 \pm 0.5) \times 10^{-3}$		–	
$K_1(1270)^- \nu_\tau$	$(4 \pm 4) \times 10^{-3}$		433	
$K_1(1400)^- \nu_\tau$	$(8 \pm 4) \times 10^{-3}$		335	

$K^*_2(1430)^- \nu_\tau$	$< 3 \quad \times 10^{-3}$	CL=95%	317	
$\eta \pi^- \nu_\tau$	$< 1.4 \quad \times 10^{-4}$	CL=95%	798	
$\eta \pi^- \pi^0 \nu_\tau$	$(1.74 \pm 0.24) \times 10^{-3}$	[i]	778	
$\eta \pi^- \pi^0 \pi^0 \nu_\tau$	$(1.4 \pm 0.7) \times 10^{-4}$		746	
$\eta K^- \nu_\tau$	$(2.7 \pm 0.6) \times 10^{-4}$		720	
$\eta \pi^+ \pi^- \pi^- \geq 0$ neutrals ν_τ	$< 3 \quad \times 10^{-3}$	CL=90%	–	
$\eta \pi^- \pi^+ \pi^- \nu_\tau$	$(3.4 \pm 0.8) \times 10^{-4}$		–	
$\eta a_1(1260)^- \nu_\tau \rightarrow \eta \pi^- \rho^0 \nu_\tau$	$< 3.9 \quad \times 10^{-4}$	CL=90%	–	
$\eta \eta \pi^- \nu_\tau$	$< 1.1 \quad \times 10^{-4}$	CL=95%	637	
$\eta \eta \pi^- \pi^0 \nu_\tau$	$< 2.0 \quad \times 10^{-4}$	CL=95%	559	
$\eta'(958) \pi^- \nu_\tau$	$< 7.4 \quad \times 10^{-5}$	CL=90%	–	
$\eta'(958) \pi^- \pi^0 \nu_\tau$	$< 8.0 \quad \times 10^{-5}$	CL=90%	–	
$\phi \pi^- \nu_\tau$	$< 2.0 \quad \times 10^{-4}$	CL=90%	585	
$\phi K^- \nu_\tau$	$< 6.7 \quad \times 10^{-5}$	CL=90%	–	
$f_1(1285) \pi^- \nu_\tau$	$(5.8 \pm 2.3) \times 10^{-4}$		–	
$f_1(1285) \pi^- \nu_\tau \rightarrow \eta \pi^- \pi^+ \pi^- \nu_\tau$	$(1.9 \pm 0.7) \times 10^{-4}$		–	
$h^- \omega \geq 0$ neutrals ν_τ	(2.36 ± 0.08) %		–	
$h^- \omega \nu_\tau$	(1.93 ± 0.06) %	[i]	–	
$h^- \omega \pi^0 \nu_\tau$	$(4.3 \pm 0.5) \times 10^{-3}$	[i]	–	
$h^- \omega 2\pi^0 \nu_\tau$	$(1.9 \pm 0.8) \times 10^{-4}$		–	

Lepton Family number (LF), Lepton number (L), or Baryon number (B) violating modes
(In the modes below, ℓ means a sum over e and μ modes)

L means lepton number violation (e.g. $\tau^- \rightarrow e^+ \pi^- \pi^-$). Following common usage, LF means lepton family violation *and not* lepton number violation (e.g. $\tau^- \rightarrow e^- \pi^+ \pi^-$). B means baryon number violation.

$e^- \gamma$	LF	< 2.7	$\times 10^{-6}$	CL=90%	888
$\mu^- \gamma$	LF	< 3.0	$\times 10^{-6}$	CL=90%	885
$e^- \pi^0$	LF	< 3.7	$\times 10^{-6}$	CL=90%	883
$\mu^- \pi^0$	LF	< 4.0	$\times 10^{-6}$	CL=90%	880
$e^- K^0$	LF	< 1.3	$\times 10^{-3}$	CL=90%	819
$\mu^- K^0$	LF	< 1.0	$\times 10^{-3}$	CL=90%	815
$e^- \eta$	LF	< 8.2	$\times 10^{-6}$	CL=90%	804
$\mu^- \eta$	LF	< 9.6	$\times 10^{-6}$	CL=90%	800
$e^- \rho^0$	LF	< 2.0	$\times 10^{-6}$	CL=90%	722
$\mu^- \rho^0$	LF	< 6.3	$\times 10^{-6}$	CL=90%	718
$e^- K^*(892)^0$	LF	< 5.1	$\times 10^{-6}$	CL=90%	663
$\mu^- K^*(892)^0$	LF	< 7.5	$\times 10^{-6}$	CL=90%	657
$e^- \overline{K}^*(892)^0$	LF	< 7.4	$\times 10^{-6}$	CL=90%	663
$\mu^- \overline{K}^*(892)^0$	LF	< 7.5	$\times 10^{-6}$	CL=90%	657
$e^- \phi$	LF	< 6.9	$\times 10^{-6}$	CL=90%	596
$\mu^- \phi$	LF	< 7.0	$\times 10^{-6}$	CL=90%	590
$\pi^- \gamma$	L	< 2.8	$\times 10^{-4}$	CL=90%	883
$\pi^- \pi^0$	L	< 3.7	$\times 10^{-4}$	CL=90%	878
$e^- e^+ e^-$	LF	< 2.9	$\times 10^{-6}$	CL=90%	888
$e^- \mu^+ \mu^-$	LF	< 1.8	$\times 10^{-6}$	CL=90%	882
$e^+ \mu^- \mu^-$	L	< 1.5	$\times 10^{-6}$	CL=90%	882
$\mu^- e^+ e^-$	LF	< 1.7	$\times 10^{-6}$	CL=90%	885
$\mu^+ e^- e^-$	L	< 1.5	$\times 10^{-6}$	CL=90%	885
$\mu^- \mu^+ \mu^-$	LF	< 1.9	$\times 10^{-6}$	CL=90%	873
$e^- \pi^+ \pi^-$	LF	< 2.2	$\times 10^{-6}$	CL=90%	877
$e^+ \pi^- \pi^-$	L	< 1.9	$\times 10^{-6}$	CL=90%	877
$\mu^- \pi^+ \pi^-$	LF	< 8.2	$\times 10^{-6}$	CL=90%	866
$\mu^+ \pi^- \pi^-$	L	< 3.4	$\times 10^{-6}$	CL=90%	866
$e^- \pi^+ K^-$	LF	< 6.4	$\times 10^{-6}$	CL=90%	814
$e^- \pi^- K^+$	LF	< 3.8	$\times 10^{-6}$	CL=90%	814
$e^+ \pi^- K^-$	L	< 2.1	$\times 10^{-6}$	CL=90%	814
$e^- K^+ K^-$	LF	< 6.0	$\times 10^{-6}$	CL=90%	739
$e^+ K^- K^-$	L	< 3.8	$\times 10^{-6}$	CL=90%	739
$\mu^- \pi^+ K^-$	LF	< 7.5	$\times 10^{-6}$	CL=90%	800
$\mu^- \pi^- K^+$	LF	< 7.4	$\times 10^{-6}$	CL=90%	800
$\mu^+ \pi^- K^-$	L	< 7.0	$\times 10^{-6}$	CL=90%	800
$\mu^- K^+ K^-$	LF	< 1.5	$\times 10^{-6}$	CL=90%	699
$\mu^+ K^- K^-$	L	< 6.0	$\times 10^{-6}$	CL=90%	699
$e^- \pi^0 \pi^0$	LF	< 6.5	$\times 10^{-6}$	CL=90%	878
$\mu^- \pi^0 \pi^0$	LF	< 1.4	$\times 10^{-5}$	CL=90%	867
$e^- \eta \eta$	LF	< 3.5	$\times 10^{-5}$	CL=90%	700
$\mu^- \eta \eta$	LF	< 6.0	$\times 10^{-5}$	CL=90%	654
$e^- \pi^0 \eta$	LF	< 2.4	$\times 10^{-5}$	CL=90%	798
$\mu^- \pi^0 \eta$	LF	< 2.2	$\times 10^{-5}$	CL=90%	784
$\overline{p} \gamma$	L,B	< 2.9	$\times 10^{-4}$	CL=90%	641
$\overline{p} \pi^0$	L,B	< 6.6	$\times 10^{-4}$	CL=90%	632
$\overline{p} \eta$	L,B	< 1.30	$\times 10^{-3}$	CL=90%	476
e^- light boson	LF	< 2.7	$\times 10^{-3}$	CL=95%	–
μ^- light boson	LF	< 5	$\times 10^{-3}$	CL=95%	–

Heavy Charged Lepton Searches

$L^{\pm}$ – charged lepton

Mass $m > 80.2$ GeV, CL = 95% $m_{\nu} \approx 0$

$L^{\pm}$ – stable charged heavy lepton

Mass $m > 84.2$ GeV, CL = 95%

Neutrinos

See the Particle Listings for a Note "Neutrino Mass" giving details of neutrinos, masses, mixing, and the status of experimental searches.

ν_e

$$J = \tfrac{1}{2}$$

Mass m: Unexplained effects have resulted in significantly negative m^2 in the new, precise tritium beta decay experiments. It is felt that a real neutrino mass as large as 10–15 eV would cause observable spectral distortions even in the presence of the end-point count excesses.

Mean life/mass, $\tau/m_{\nu_e} > 7 \times 10^9$ s/eV   (solar)

Mean life/mass, $\tau/m_{\nu_e} > 300$ s/eV, CL = 90%   (reactor)

Magnetic moment $\mu < 1.8 \times 10^{-10}$ μ_B, CL = 90%

ν_μ

$$J = \tfrac{1}{2}$$

Mass $m < 0.17$ MeV, CL = 90%

Mean life/mass, $\tau/m_{\nu_\mu} > 15.4$ s/eV, CL = 90%

Magnetic moment $\mu < 7.4 \times 10^{-10}$ μ_B, CL = 90%

ν_τ

$$J = \tfrac{1}{2}$$

Mass $m < 18.2$ MeV, CL = 95%

Magnetic moment $\mu < 5.4 \times 10^{-7}$ μ_B, CL = 90%

Electric dipole moment $d < 5.2 \times 10^{-17}$ e cm, CL = 95%

Number of Light Neutrino Types

(including ν_e, ν_μ, and ν_τ)

Number $N = 2.994 \pm 0.012$ (Standard Model fits to LEP data)

Number $N = 3.07 \pm 0.12$ (Direct measurement of invisible Z width)

Massive Neutrinos and Lepton Mixing, Searches for

For excited leptons, see Compositeness Limits below.

See the Particle Listings for a Note "Neutrino Mass" giving details of neutrinos, masses, mixing, and the status of experimental searches.

While no direct, uncontested evidence for massive neutrinos or lepton mixing has been obtained, suggestive evidence has come from solar neutrino observations, from anomalies in the relative fractions of ν_e and ν_μ observed in energetic cosmic-ray air showers, and possibly from a $\overline{\nu}_e$ appearance experiment at Los Alamos. Sample limits are:

Stable Neutral Heavy Lepton Mass Limits

Mass $m > 45.0$ GeV, CL = 95%   (Dirac)

Mass $m > 39.5$ GeV, CL = 95%   (Majorana)

Neutral Heavy Lepton Mass Limits

Mass $m > 69.0$ GeV, CL = 95%   (Dirac ν_L coupling to e, μ, τ with $|U_{\ell j}|^2 > 10^{-12}$)

Mass $m > 58.2$ GeV, CL = 95%   (Majorana ν_L coupling to e, μ, τ with $|U_{\ell j}|^2 > 10^{-12}$)

Solar Neutrinos

Detectors using gallium ($E_\nu \gtrsim 0.2$ MeV), chlorine ($E_\nu \gtrsim 0.8$ MeV), and Čerenkov effect in water ($E_\nu \gtrsim 7$ MeV) measure significantly lower neutrino rates than are predicted from solar models. The deficit in the solar neutrino flux compared with solar model calculations could be explained by oscillations with $\Delta m^2 \leq 10^{-5}$ eV^2 causing the disappearance of ν_e.

Atmospheric Neutrinos

Underground detectors observing neutrinos produced by cosmic rays in the atmosphere have measured a ν_μ/ν_e ratio much less than expected and also a deficiency of upward going ν_μ compared to downward. This could be explained by oscillations leading to the disappearance of ν_μ with $\Delta m^2 \approx 10^{-3}$ to 10^{-2} eV^2.

ν oscillation: $\overline{\nu}_e \nrightarrow \overline{\nu}_e$ ($\theta = $ mixing angle)

$\Delta m^2 < 9 \times 10^{-4}$ eV^2, CL = 90%   (if $\sin^2 2\theta = 1$)

$\sin^2 2\theta < 0.02$, CL = 90%   (if $\Delta(m^2)$ is large)

ν oscillation: ν_μ ($\overline{\nu}_\mu$) $\rightarrow \nu_e$ ($\overline{\nu}_e$) (any combination)

$\Delta m^2 < 0.075$ eV^2, CL = 90%   (if $\sin^2 2\theta = 1$)

$\sin^2 2\theta < 1.8 \times 10^{-3}$, CL = 90%   (if $\Delta(m^2)$ is large)

NOTES

In this Summary Table:

When a quantity has "(S = ...)" to its right, the error on the quantity has been enlarged by the "scale factor" S, defined as $S = \sqrt{\chi^2/(N-1)}$, where N is the number of measurements used in calculating the quantity. We do this when $S > 1$, which often indicates that the measurements are inconsistent. When $S > 1.25$, we also show in the Particle Listings an ideogram of the measurements. For more about S, see the Introduction.

A decay momentum p is given for each decay mode. For a 2-body decay, p is the momentum of each decay product in the rest frame of the decaying particle. For a 3-or-more-body decay, p is the largest momentum any of the products can have in this frame.

[a] The uncertainty in the electron mass in unified atomic mass units (u) is ten times smaller than that given by the 1986 CODATA adjustment, quoted in the Table of Physical Constants (Section 1). The conversion to MeV via the factor 931.49432(28) MeV/u is more uncertain because of the electron charge uncertainty. Our value in MeV differs slightly from the 1986 CODATA result.

[b] This is the best "electron disappearance" limit. The best limit for the mode $e^- \rightarrow \nu\gamma$ is $> 2.35 \times 10^{25}$ yr (CL=68%).

[c] The muon mass is most precisely known in u (unified atomic mass units). The conversion factor to MeV via the factor 931.49432(28) MeV/u is more uncertain because of the electron charge uncertainty.

[d] See the "Note on Muon Decay Parameters" in the μ Particle Listings for definitions and details.

[e] P_μ is the longitudinal polarization of the muon from pion decay. In standard $V-A$ theory, $P_\mu = 1$ and $\rho = \delta = 3/4$.

[f] This only includes events with the γ energy > 10 MeV. Since the $e^- \overline{\nu}_e \nu_\mu$ and $e^- \overline{\nu}_e \nu_\mu \gamma$ modes cannot be clearly separated, we regard the latter mode as a subset of the former.

[g] See the μ Particle Listings for the energy limits used in this measurement.

[h] A test of additive vs. multiplicative lepton family number conservation.

[i] Basis mode for the τ.

QUARKS

The u-, d-, and s-quark masses are estimates of so-called "current-quark masses," in a mass-independent subtraction scheme such as $\overline{MS}$ at a scale $\mu \approx 2$ GeV. The c- and b-quark masses are estimated from charmonium, bottomonium, D, and B masses. They are the "running" masses in the $\overline{MS}$ scheme. These can be different from the heavy quark masses obtained in potential models.

u $I(J^P) = \frac{1}{2}(\frac{1}{2}^+)$

Mass $m = 1.5$ to 5 MeV [a]   Charge $= \frac{2}{3}\,e$ $I_z = +\frac{1}{2}$
$m_u/m_d = 0.20$ to 0.70

d $I(J^P) = \frac{1}{2}(\frac{1}{2}^+)$

Mass $m = 3$ to 9 MeV [a]   Charge $= -\frac{1}{3}\,e$ $I_z = -\frac{1}{2}$
$m_s/m_d = 17$ to 25
$\overline{m} = (m_u + m_d)/2 = 2$ to 6 MeV

s $I(J^P) = 0(\frac{1}{2}^+)$

Mass $m = 60$ to 170 MeV [a]   Charge $= -\frac{1}{3}\,e$ Strangeness $= -1$
$(m_s - (m_u + m_d)/2)/(m_d - m_u) = 34$ to 51

c $I(J^P) = 0(\frac{1}{2}^+)$

Mass $m = 1.1$ to 1.4 GeV   Charge $= \frac{2}{3}\,e$ Charm $= +1$

b $I(J^P) = 0(\frac{1}{2}^+)$

Mass $m = 4.1$ to 4.4 GeV   Charge $= -\frac{1}{3}\,e$ Bottom $= -1$

t $I(J^P) = 0(\frac{1}{2}^+)$

Charge $= \frac{2}{3}\,e$ Top $= +1$

Mass $m = 173.8 \pm 5.2$ GeV   (direct observation of top events)
Mass $m = 170 \pm 7$ (+14) GeV   (Standard Model electroweak
    fit, assuming $M_H = M_Z$. Number in parentheses is shift from
    changing M_H to 300 GeV.

b' (4th Generation) Quark, Searches for

Mass $m > 128$ GeV, CL $= 95\%$ ($p\bar{p}$, charged current decays)
Mass $m > 46.0$ GeV, CL $= 95\%$ ($e^+ e^-$, all decays)

Free Quark Searches

All searches since 1977 have had negative results.

NOTES

[a] The ratios m_u/m_d and m_s/m_d are extracted from pion and kaon masses using chiral symmetry. The estimates of u and d masses are not without controversy and remain under active investigation. Within the literature there are even suggestions that the u quark could be essentially massless. The s-quark mass is estimated from SU(3) splittings in hadron masses.

LIGHT UNFLAVORED MESONS
$(S = C = B = 0)$

For $I = 1$ (π, b, ρ, a): $u\bar{d}$, $(u\bar{u}-d\bar{d})/\sqrt{2}$, $d\bar{u}$;
for $I = 0$ (η, η', h, h', ω, ϕ, f, f'): $c_1(u\bar{u} + d\bar{d}) + c_2(s\bar{s})$

$\pi^{\pm}$ $I^G(J^P) = 1^-(0^-)$

Mass $m = 139.56995 \pm 0.00035$ MeV
Mean life $\tau = (2.6033 \pm 0.0005) \times 10^{-8}$ s   (S = 1.2)
    $c\tau = 7.8045$ m

$\pi^{\pm} \to \ell^{\pm}\nu\gamma$ form factors [a]

$F_V = 0.017 \pm 0.008$
$F_A = 0.0116 \pm 0.0016$ (S = 1.3)
$R = 0.059^{+0.009}_{-0.008}$

π^- modes are charge conjugates of the modes below.

π^+ DECAY MODES		Fraction (Γ_i/Γ)	Confidence level	p (MeV/c)
$\mu^+ \nu_\mu$	[b]	(99.98770 ± 0.00004) %		30
$\mu^+ \nu_\mu \gamma$	[c]	$(1.24 \pm 0.25) \times 10^{-4}$		30
$e^+ \nu_e$	[b]	$(1.230 \pm 0.004) \times 10^{-4}$		70
$e^+ \nu_e \gamma$	[c]	$(1.61 \pm 0.23) \times 10^{-7}$		70
$e^+ \nu_e \pi^0$		$(1.025 \pm 0.034) \times 10^{-8}$		4
$e^+ \nu_e e^+ e^-$		$(3.2 \pm 0.5) \times 10^{-9}$		70
$e^+ \nu_e \nu\bar{\nu}$		< 5	90%	70

Lepton Family number (LF) or Lepton number (L) violating modes

$\mu^+ \bar{\nu}_e$	L	[d]	< 1.5	$\times 10^{-3}$ 90%	30
$\mu^+ \nu_e$	LF	[d]	< 8.0	$\times 10^{-3}$ 90%	30
$\mu^- e^+ e^+ \nu$	LF		< 1.6	$\times 10^{-6}$ 90%	30

π^0 $I^G(J^{PC}) = 1^-(0^{-+})$

Mass $m = 134.9764 \pm 0.0006$ MeV
$m_{\pi^{\pm}} - m_{\pi^0} = 4.5936 \pm 0.0005$ MeV
Mean life $\tau = (8.4 \pm 0.6) \times 10^{-17}$ s   (S = 3.0)
    $c\tau = 25.1$ nm

π^0 DECAY MODES		Fraction (Γ_i/Γ)	Scale factor/ Confidence level	p (MeV/c)
2γ		(98.798 ± 0.032) %	S=1.1	67
$e^+ e^- \gamma$		(1.198 ± 0.032) %	S=1.1	67
γ positronium		$(1.82 \pm 0.29) \times 10^{-9}$		67
$e^+ e^+ e^- e^-$		$(3.14 \pm 0.30) \times 10^{-5}$		67
$e^+ e^-$		$(7.5 \pm 2.0) \times 10^{-8}$		67
4γ		< 2	$\times 10^{-8}$ CL=90%	67
$\nu\bar{\nu}$	[e]	< 8.3	$\times 10^{-7}$ CL=90%	67
$\nu_e \bar{\nu}_e$		< 1.7	$\times 10^{-6}$ CL=90%	67
$\nu_\mu \bar{\nu}_\mu$		< 3.1	$\times 10^{-6}$ CL=90%	67
$\nu_\tau \bar{\nu}_\tau$		< 2.1	$\times 10^{-6}$ CL=90%	67

Charge conjugation (C) or Lepton Family number (LF) violating modes

3γ	C	< 3.1	$\times 10^{-8}$ CL=90%	67
$\mu^+ e^- + e^- \mu^+$	LF	< 1.72	$\times 10^{-8}$ CL=90%	26

η $I^G(J^{PC}) = 0^+(0^{-+})$

Mass $m = 547.30 \pm 0.12$ MeV
Full width $\Gamma = 1.18 \pm 0.11$ keV [f]   (S = 1.8)

C-nonconserving decay parameters

$\pi^+ \pi^- \pi^0$	Left-right asymmetry =	$(0.09 \pm 0.17) \times 10^{-2}$
$\pi^+ \pi^- \pi^0$	Sextant asymmetry =	$(0.18 \pm 0.16) \times 10^{-2}$
$\pi^+ \pi^- \pi^0$	Quadrant asymmetry =	$(-0.17 \pm 0.17) \times 10^{-2}$
$\pi^+ \pi^- \gamma$	Left-right asymmetry =	$(0.9 \pm 0.4) \times 10^{-2}$
$\pi^+ \pi^- \gamma$	β (D-wave) = 0.05 ± 0.06 (S = 1.5)	

Dalitz plot parameter

$\pi^0 \pi^0 \pi^0$ $\alpha = -0.039 \pm 0.015$

η DECAY MODES		Fraction (Γ_i/Γ)	Scale factor/ Confidence level	p (MeV/c)
Neutral modes				
neutral modes		(71.5 ± 0.6) %	S=1.4	–
2γ	[f]	(39.21 ± 0.34) %	S=1.4	274
$3\pi^0$		(32.2 ± 0.4) %	S=1.3	178
$\pi^0 2\gamma$		$(7.1 \pm 1.4) \times 10^{-4}$		257
other neutral modes		< 2.8 %	CL=90%	–
Charged modes				
charged modes		(28.5 ± 0.6) %	S=1.4	–
$\pi^+ \pi^- \pi^0$		(23.1 ± 0.5) %	S=1.4	173
$\pi^+ \pi^- \gamma$		(4.77 ± 0.13) %	S=1.3	235
$e^+ e^- \gamma$		$(4.9 \pm 1.1) \times 10^{-3}$		274
$\mu^+ \mu^- \gamma$		$(3.1 \pm 0.4) \times 10^{-4}$		252
$e^+ e^-$		< 7.7 $\times 10^{-5}$	CL=90%	274
$\mu^+ \mu^-$		$(5.8 \pm 0.8) \times 10^{-6}$		252
$\pi^+ \pi^- e^+ e^-$		$(1.3 ^{+1.2}_{-0.8}) \times 10^{-3}$		235
$\pi^+ \pi^- 2\gamma$		< 2.1 $\times 10^{-3}$		235
$\pi^+ \pi^- \pi^0 \gamma$		< 6 $\times 10^{-4}$	CL=90%	173
$\pi^0 \mu^+ \mu^- \gamma$		< 3 $\times 10^{-6}$	CL=90%	210

Charge conjugation (C), Parity (P),
Charge conjugation × Parity (CP), or
Lepton Family number (LF) violating modes

$\pi^+ \pi^-$	P,CP		< 9	$\times 10^{-4}$ CL=90%	235
3γ	C		< 5	$\times 10^{-4}$ CL=95%	274
$\pi^0 e^+ e^-$	C	[g]	< 4	$\times 10^{-5}$ CL=90%	257
$\pi^0 \mu^+ \mu^-$	C	[g]	< 5	$\times 10^{-6}$ CL=90%	210
$\mu^+ e^- + \mu^- e^+$	LF		< 6	$\times 10^{-6}$ CL=90%	263

$f_0(400{-}1200)$ [h] or σ $I^G(J^{PC}) = 0^+(0^{++})$

Mass $m = (400{-}1200)$ MeV
Full width $\Gamma = (600{-}1000)$ MeV

$f_0(400{-}1200)$ DECAY MODES	Fraction (Γ_i/Γ)	p (MeV/c)
$\pi\pi$	dominant	–
$\gamma\gamma$	seen	–

Meson Summary Table

$\rho(770)$ [l]

$$I^G(J^{PC}) = 1^+(1^{--})$$

Mass $m = 770.0 \pm 0.8$ MeV (S = 1.8)
Full width $\Gamma = 150.7 \pm 1.1$ MeV
$\Gamma_{ee} = 6.77 \pm 0.32$ keV

$\rho(770)$ DECAY MODES	Fraction (Γ_i/Γ)	Scale factor/ Confidence level	p (MeV/c)
$\pi\pi$	~ 100 %		358

$\rho(770)^\pm$ decays

$\pi^\pm\gamma$	$(4.5 \pm 0.5) \times 10^{-4}$	S=2.2	372
$\pi^\pm\eta$	$< 6 \times 10^{-3}$	CL=84%	146
$\pi^\pm\pi^+\pi^-\pi^0$	$< 2.0 \times 10^{-3}$	CL=84%	249

$\rho(770)^0$ decays

$\pi^+\pi^-\gamma$	$(9.9 \pm 1.6) \times 10^{-3}$		358
$\pi^0\gamma$	$(6.8 \pm 1.7) \times 10^{-4}$		372
$\eta\gamma$	$(2.4 ^{+0.8}_{-0.9}) \times 10^{-4}$	S=1.6	189
$\mu^+\mu^-$	[j] $(4.60 \pm 0.28) \times 10^{-5}$		369
e^+e^-	[j] $(4.49 \pm 0.22) \times 10^{-5}$		384
$\pi^+\pi^-\pi^0$	$< 1.2 \times 10^{-4}$	CL=90%	319
$\pi^+\pi^-\pi^+\pi^-$	$< 2 \times 10^{-4}$	CL=90%	246
$\pi^+\pi^-\pi^0\pi^0$	$< 4 \times 10^{-5}$	CL=90%	252

$\omega(782)$

$$I^G(J^{PC}) = 0^-(1^{--})$$

Mass $m = 781.94 \pm 0.12$ MeV (S = 1.5)
Full width $\Gamma = 8.41 \pm 0.09$ MeV
$\Gamma_{ee} = 0.60 \pm 0.02$ keV

$\omega(782)$ DECAY MODES	Fraction (Γ_i/Γ)	Scale factor/ Confidence level	p (MeV/c)
$\pi^+\pi^-\pi^0$	(88.8 ± 0.7) %		327
$\pi^0\gamma$	(8.5 ± 0.5) %		379
$\pi^+\pi^-$	(2.21 ± 0.30) %		365
neutrals (excluding $\pi^0\gamma$)	$(5.3 ^{+8.7}_{-3.5}) \times 10^{-3}$		–
$\eta\gamma$	$(6.5 \pm 1.0) \times 10^{-4}$		199
$\pi^0 e^+e^-$	$(5.9 \pm 1.9) \times 10^{-4}$		379
$\pi^0\mu^+\mu^-$	$(9.6 \pm 2.3) \times 10^{-5}$		349
e^+e^-	$(7.07 \pm 0.19) \times 10^{-5}$	S=1.1	391
$\pi^+\pi^-\pi^0\pi^0$	< 2 %		261
$\pi^+\pi^-\gamma$	$< 3.6 \times 10^{-3}$	CL=95%	365
$\pi^+\pi^-\pi^+\pi^-$	$< 1 \times 10^{-3}$	CL=90%	256
$\pi^0\pi^0\gamma$	$(7.2 \pm 2.5) \times 10^{-5}$		367
$\mu^+\mu^-$	$< 1.8 \times 10^{-4}$	CL=90%	376
3γ	$< 1.9 \times 10^{-4}$	CL=95%	391

Charge conjugation (C) violating modes

$\eta\pi^0$	C	$< 1 \times 10^{-3}$	CL=90%	162
$3\pi^0$	C	$< 3 \times 10^{-4}$	CL=90%	329

$\eta'(958)$

$$I^G(J^{PC}) = 0^+(0^{-+})$$

Mass $m = 957.78 \pm 0.14$ MeV
Full width $\Gamma = 0.203 \pm 0.016$ MeV (S = 1.3)

$\eta'(958)$ DECAY MODES	Fraction (Γ_i/Γ)	Scale factor/ Confidence level	p (MeV/c)
$\pi^+\pi^-\eta$	(43.8 ± 1.5) %	S=1.1	232
$\rho^0\gamma$ (including non-resonant $\pi^+\pi^-\gamma$)	(30.2 ± 1.3) %	S=1.1	169
$\pi^0\pi^0\eta$	(20.7 ± 1.3) %	S=1.2	239
$\omega\gamma$	(3.01 ± 0.30) %		160
$\gamma\gamma$	(2.11 ± 0.13) %	S=1.2	479
$3\pi^0$	$(1.54 \pm 0.26) \times 10^{-3}$		430
$\mu^+\mu^-\gamma$	$(1.03 \pm 0.26) \times 10^{-4}$		467
$\pi^+\pi^-\pi^0$	< 5 %	CL=90%	427
$\pi^0\rho^0$	< 4 %	CL=90%	118
$\pi^+\pi^+\pi^-\pi^-$	< 1 %	CL=90%	372
$\pi^+\pi^+\pi^-\pi^-$ neutrals	< 1 %	CL=95%	–
$\pi^+\pi^+\pi^-\pi^-\pi^0$	< 1 %	CL=90%	298
6π	< 1 %	CL=90%	189
$\pi^+\pi^-e^+e^-$	$< 6 \times 10^{-3}$	CL=90%	458
$\pi^0\gamma\gamma$	$< 8 \times 10^{-4}$	CL=90%	469
$4\pi^0$	$< 5 \times 10^{-4}$	CL=90%	379
e^+e^-	$< 2.1 \times 10^{-7}$	CL=90%	479

Charge conjugation (C) or Parity (P) violating modes

$\pi^+\pi^-$	P,CP	< 2	%	CL=90%	458
$\pi^0\pi^0$	P,CP	< 9	$\times 10^{-4}$	CL=90%	459
$\pi^0 e^+e^-$	C	[g] < 1.3	%	CL=90%	469
ηe^+e^-	C	[g] < 1.1	%	CL=90%	322
3γ	C	< 1.0	$\times 10^{-4}$	CL=90%	479
$\mu^+\mu^-\pi^0$	C	[g] < 6.0	$\times 10^{-5}$	CL=90%	445
$\mu^+\mu^-\eta$	C	[g] < 1.5	$\times 10^{-5}$	CL=90%	274

$f_0(980)$ [k]

$$I^G(J^{PC}) = 0^+(0^{++})$$

Mass $m = 980 \pm 10$ MeV
Full width $\Gamma = 40$ to 100 MeV

$f_0(980)$ DECAY MODES	Fraction (Γ_i/Γ)	Confidence level	p (MeV/c)
$\pi\pi$	dominant		470
$K\overline{K}$	seen		–
$\gamma\gamma$	$(1.19 \pm 0.33) \times 10^{-5}$		490
e^+e^-	$< 3 \times 10^{-7}$	90%	490

$a_0(980)$ [k]

$$I^G(J^{PC}) = 1^-(0^{++})$$

Mass $m = 983.4 \pm 0.9$ MeV
Full width $\Gamma = 50$ to 100 MeV

$a_0(980)$ DECAY MODES	Fraction (Γ_i/Γ)	p (MeV/c)
$\eta\pi$	dominant	321
$K\overline{K}$	seen	–
$\gamma\gamma$	seen	492

$\phi(1020)$

$$I^G(J^{PC}) = 0^-(1^{--})$$

Mass $m = 1019.413 \pm 0.008$ MeV
Full width $\Gamma = 4.43 \pm 0.05$ MeV

$\phi(1020)$ DECAY MODES	Fraction (Γ_i/Γ)	Scale factor/ Confidence level	p (MeV/c)
K^+K^-	(49.1 ± 0.8) %	S=1.3	127
$K^0_L K^0_S$	(34.1 ± 0.6) %	S=1.2	110
$\rho\pi + \pi^+\pi^-\pi^0$	(15.5 ± 0.7) %	S=1.5	–
$\eta\gamma$	(1.26 ± 0.06) %	S=1.1	363
$\pi^0\gamma$	$(1.31 \pm 0.13) \times 10^{-3}$		501
e^+e^-	$(2.99 \pm 0.08) \times 10^{-4}$	S=1.2	510
$\mu^+\mu^-$	$(2.5 \pm 0.4) \times 10^{-4}$		499
ηe^+e^-	$(1.3 ^{+0.8}_{-0.6}) \times 10^{-4}$		363
$\pi^+\pi^-$	$(8 ^{+5}_{-4}) \times 10^{-5}$	S=1.5	490
$\omega\gamma$	< 5 %	CL=84%	210
$\rho\gamma$	$< 7 \times 10^{-4}$	CL=90%	219
$\pi^+\pi^-\gamma$	$< 3 \times 10^{-5}$	CL=90%	490
$f_0(980)\gamma$	$< 1 \times 10^{-4}$	CL=90%	39
$\pi^0\pi^0\gamma$	$< 1 \times 10^{-5}$	CL=90%	492
$\pi^+\pi^-\pi^+\pi^-$	$< 8.7 \times 10^{-4}$	CL=90%	410
$\pi^+\pi^+\pi^-\pi^-\pi^0$	$< 1.5 \times 10^{-4}$	CL=95%	341
$\pi^0 e^+e^-$	$< 1.2 \times 10^{-4}$	CL=90%	501
$\pi^0\eta\gamma$	$< 2.5 \times 10^{-3}$	CL=90%	346
$a_0(980)\gamma$	$< 5 \times 10^{-3}$	CL=90%	36
$\eta'(958)\gamma$	$(1.2 ^{+0.7}_{-0.5}) \times 10^{-4}$		–
$\mu^+\mu^-\gamma$	$(2.3 \pm 1.0) \times 10^{-5}$		–

$h_1(1170)$

$$I^G(J^{PC}) = 0^-(1^{+-})$$

Mass $m = 1170 \pm 20$ MeV
Full width $\Gamma = 360 \pm 40$ MeV

$h_1(1170)$ DECAY MODES	Fraction (Γ_i/Γ)	p (MeV/c)
$\rho\pi$	seen	310

$b_1(1235)$ $I^G(J^{PC}) = 1^+(1^{+-})$

Mass $m = 1229.5 \pm 3.2$ MeV   (S = 1.6)
Full width $\Gamma = 142 \pm 9$ MeV   (S = 1.2)

$b_1(1235)$ DECAY MODES	Fraction (Γ_i/Γ)	Confidence level	p (MeV/c)
$\omega\pi$	dominant		348
[D/S amplitude ratio = 0.29 ± 0.04]			
$\pi^\pm\gamma$	(1.6±0.4) × 10⁻³		608
$\eta\rho$	seen		–
$\pi^+\pi^+\pi^-\pi^0$	< 50 %	84%	536
$(K\overline{K})^\pm\pi^0$	< 8 %	90%	248
$K^0_S K^0_L \pi^\pm$	< 6 %	90%	238
$K^0_S K^0_S \pi^\pm$	< 2 %	90%	238
$\phi\pi$	< 1.5 %	84%	146

$a_1(1260)$ [l] $I^G(J^{PC}) = 1^-(1^{++})$

Mass $m = 1230 \pm 40$ MeV [m]
Full width $\Gamma = 250$ to 600 MeV

$a_1(1260)$ DECAY MODES	Fraction (Γ_i/Γ)	p (MeV/c)
$\rho\pi$	dominant	356
[D/S amplitude ratio = −0.100 ± 0.028]		
$\pi\gamma$	seen	607
$\pi(\pi\pi)_{S}$-wave	possibly seen	575

$f_2(1270)$ $I^G(J^{PC}) = 0^+(2^{++})$

Mass $m = 1275.0 \pm 1.2$ MeV
Full width $\Gamma = 185.5^{+3.8}_{-2.7}$ MeV   (S = 1.5)

$f_2(1270)$ DECAY MODES	Fraction (Γ_i/Γ)	Scale factor/ Confidence level	p (MeV/c)
$\pi\pi$	(84.6 $^{+2.5}_{-1.3}$)%	S=1.3	622
$\pi^+\pi^- 2\pi^0$	(7.2 $^{+1.5}_{-2.7}$)%	S=1.3	562
$K\overline{K}$	(4.6 ±0.4)%	S=2.8	403
$2\pi^+2\pi^-$	(2.8 ±0.4)%	S=1.2	559
$\eta\eta$	(4.5 ±1.0) × 10⁻³	S=2.4	327
$4\pi^0$	(3.0 ±1.0) × 10⁻³		564
$\gamma\gamma$	(1.32 $^{+0.17}_{-0.16}$) × 10⁻⁵		637
$\eta\pi\pi$	< 8 × 10⁻³	CL=95%	475
$K^0 K^-\pi^+ +$ c.c.	< 3.4 × 10⁻³	CL=95%	293
e^+e^-	< 9 × 10⁻⁹	CL=90%	637

$f_1(1285)$ $I^G(J^{PC}) = 0^+(1^{++})$

Mass $m = 1281.9 \pm 0.6$ MeV   (S = 1.7)
Full width $\Gamma = 24.0 \pm 1.2$ MeV   (S = 1.4)

$(4\pi = \rho(\pi\pi)_{Pwave})$

$f_1(1285)$ DECAY MODES	Fraction (Γ_i/Γ)	Scale factor/ Confidence level	p (MeV/c)
4π	(35 ± 4)%	S=1.6	563
$\pi^0\pi^0\pi^+\pi^-$	(23.5± 3.0)%	S=1.6	566
$2\pi^+2\pi^-$	(11.7± 1.5)%	S=1.6	563
$\rho^0\pi^+\pi^-$	(11.7± 1.5)%	S=1.6	340
$4\pi^0$	< 7 × 10⁻⁴	CL=90%	568
$\eta\pi\pi$	(50 ±18)%		479
$a_0(980)\pi$ [ignoring $a_0(980) \to K\overline{K}$]	(34 ± 8)%	S=1.2	234
$\eta\pi\pi$ [excluding $a_0(980)\pi$]	(15 ± 7)%	S=1.1	–
$K\overline{K}\pi$	(9.6± 1.2)%	S=1.5	308
$K\overline{K}^*(892)$	not seen		–
$\gamma\rho^0$	(5.4± 1.2)%	S=2.3	410
$\phi\gamma$	(7.9± 3.0) × 10⁻⁴		236

$\eta(1295)$ $I^G(J^{PC}) = 0^+(0^{-+})$

Mass $m = 1297.0 \pm 2.8$ MeV
Full width $\Gamma = 53 \pm 6$ MeV

$\eta(1295)$ DECAY MODES	Fraction (Γ_i/Γ)	p (MeV/c)
$\eta\pi^+\pi^-$	seen	488
$a_0(980)\pi$	seen	245
$\eta\pi^0\pi^0$	seen	–
$\eta(\pi\pi)_{S}$-wave	seen	–

$\pi(1300)$ $I^G(J^{PC}) = 1^-(0^{-+})$

Mass $m = 1300 \pm 100$ MeV [m]
Full width $\Gamma = 200$ to 600 MeV

$\pi(1300)$ DECAY MODES	Fraction (Γ_i/Γ)	p (MeV/c)
$\rho\pi$	seen	406
$\pi(\pi\pi)_{S}$-wave	seen	–

$a_2(1320)$ $I^G(J^{PC}) = 1^-(2^{++})$

Mass $m = 1318.1 \pm 0.6$ MeV   (S = 1.1)
Full width $\Gamma = 107 \pm 5$ MeV [m]   ($K^\pm K^0_S$ and $\eta\pi$ modes)

$a_2(1320)$ DECAY MODES	Fraction (Γ_i/Γ)	Scale factor/ Confidence level	p (MeV/c)
$\rho\pi$	(70.1±2.7) %	S=1.2	419
$\eta\pi$	(14.5±1.2) %		535
$\omega\pi\pi$	(10.6±3.2) %	S=1.3	362
$K\overline{K}$	(4.9±0.8) %		437
$\eta'(958)\pi$	(5.3±0.9) × 10⁻³		287
$\pi^\pm\gamma$	(2.8±0.6) × 10⁻³		652
$\gamma\gamma$	(9.4±0.7) × 10⁻⁶		659
$\pi^+\pi^-\pi^-$	< 8 %	CL=90%	621
e^+e^-	< 2.3 × 10⁻⁷	CL=90%	659

$f_0(1370)$ [k] $I^G(J^{PC}) = 0^+(0^{++})$

Mass $m = 1200$ to 1500 MeV
Full width $\Gamma = 200$ to 500 MeV

$f_0(1370)$ DECAY MODES	Fraction (Γ_i/Γ)	p (MeV/c)
$\pi\pi$	seen	–
4π	seen	–
$4\pi^0$	seen	–
$2\pi^+2\pi^-$	seen	–
$\pi^+\pi^- 2\pi^0$	seen	–
$2(\pi\pi)_{S}$-wave	seen	–
$\eta\eta$	seen	–
$K\overline{K}$	seen	–
$\gamma\gamma$	seen	–
e^+e^-	not seen	–

$f_1(1420)$ [n] $I^G(J^{PC}) = 0^+(1^{++})$

Mass $m = 1426.2 \pm 1.2$ MeV   (S = 1.3)
Full width $\Gamma = 55.0 \pm 3.0$ MeV

$f_1(1420)$ DECAY MODES	Fraction (Γ_i/Γ)	p (MeV/c)
$K\overline{K}\pi$	dominant	439
$K\overline{K}^*(892) +$ c.c.	dominant	155
$\eta\pi\pi$	possibly seen	571

$\omega(1420)$ [o] $I^G(J^{PC}) = 0^-(1^{--})$

Mass $m = 1419 \pm 31$ MeV
Full width $\Gamma = 174 \pm 60$ MeV

$\omega(1420)$ DECAY MODES	Fraction (Γ_i/Γ)	p (MeV/c)
$\rho\pi$	dominant	488

η(1440) [p]

$$I^G(J^{PC}) = 0^+(0^{-+})$$

Mass $m = 1400 - 1470$ MeV [m]
Full width $\Gamma = 50 - 80$ MeV [m]

η(1440) DECAY MODES	Fraction (Γ_i/Γ)	p (MeV/c)
$K\overline{K}\pi$	seen	–
$K\overline{K}^*(892)$+ c.c.	seen	–
$\eta\pi\pi$	seen	–
$a_0(980)\pi$	seen	–
$\eta(\pi\pi)_{S\text{-wave}}$	seen	–
4π	seen	–

a₀(1450)

$$I^G(J^{PC}) = 1^-(0^{++})$$

Mass $m = 1474 \pm 19$ MeV
Full width $\Gamma = 265 \pm 13$ MeV

a₀(1450) DECAY MODES	Fraction (Γ_i/Γ)	p (MeV/c)
$\pi\eta$	seen	613
$\pi\eta'(958)$	seen	392
$K\overline{K}$	seen	530

ρ(1450) [q]

$$I^G(J^{PC}) = 1^+(1^{--})$$

Mass $m = 1465 \pm 25$ MeV [m]
Full width $\Gamma = 310 \pm 60$ MeV [m]

ρ(1450) DECAY MODES	Fraction (Γ_i/Γ)	Confidence level	p (MeV/c)
$\pi\pi$	seen		719
4π	seen		665
$\omega\pi$	<2.0 %	95%	512
e^+e^-	seen		732
$\eta\rho$	<4 %		317
$\phi\pi$	<1 %		358
$K\overline{K}$	$<1.6 \times 10^{-3}$	95%	541

f₀(1500) [r]

$$I^G(J^{PC}) = 0^+(0^{++})$$

Mass $m = 1500 \pm 10$ MeV (S = 1.3)
Full width $\Gamma = 112 \pm 10$ MeV

f₀(1500) DECAY MODES	Fraction (Γ_i/Γ)	p (MeV/c)
$\eta\eta'(958)$	seen	–
$\eta\eta$	seen	513
4π	seen	–
$4\pi^0$	seen	690
$2\pi^+2\pi^-$	seen	686
2π	seen	–
$\pi^+\pi^-$	seen	737
$2\pi^0$	seen	738
$K\overline{K}$	seen	563

f₂'(1525)

$$I^G(J^{PC}) = 0^+(2^{++})$$

Mass $m = 1525 \pm 5$ MeV [m]
Full width $\Gamma = 76 \pm 10$ MeV [m]

f₂'(1525) DECAY MODES	Fraction (Γ_i/Γ)	p (MeV/c)
$K\overline{K}$	(88.8 ±3.1) %	581
$\eta\eta$	(10.3 ±3.1) %	531
$\pi\pi$	(8.2 ±1.5) $\times 10^{-3}$	750
$\gamma\gamma$	(1.32±0.21) $\times 10^{-6}$	763

ω(1600) [s]

$$I^G(J^{PC}) = 0^-(1^{--})$$

Mass $m = 1649 \pm 24$ MeV (S = 2.3)
Full width $\Gamma = 220 \pm 35$ MeV (S = 1.6)

ω(1600) DECAY MODES	Fraction (Γ_i/Γ)	p (MeV/c)
$\rho\pi$	seen	637
$\omega\pi\pi$	seen	601
e^+e^-	seen	824

ω₃(1670)

$$I^G(J^{PC}) = 0^-(3^{--})$$

Mass $m = 1667 \pm 4$ MeV
Full width $\Gamma = 168 \pm 10$ MeV [m]

ω₃(1670) DECAY MODES	Fraction (Γ_i/Γ)	p (MeV/c)
$\rho\pi$	seen	647
$\omega\pi\pi$	seen	614
$b_1(1235)\pi$	possibly seen	359

π₂(1670)

$$I^G(J^{PC}) = 1^-(2^{-+})$$

Mass $m = 1670 \pm 20$ MeV [m]
Full width $\Gamma = 258 \pm 18$ MeV [m] (S = 1.7)

π₂(1670) DECAY MODES	Fraction (Γ_i/Γ)	p (MeV/c)
3π	(95.8±1.4) %	806
$f_2(1270)\pi$	(56.2±3.2) %	325
$\rho\pi$	(31 ±4) %	649
$f_0(1370)\pi$	(8.7±3.4) %	–
$K\overline{K}^*(892)$+ c.c.	(4.2±1.4) %	453

φ(1680)

$$I^G(J^{PC}) = 0^-(1^{--})$$

Mass $m = 1680 \pm 20$ MeV [m]
Full width $\Gamma = 150 \pm 50$ MeV [m]

φ(1680) DECAY MODES	Fraction (Γ_i/Γ)	p (MeV/c)
$K\overline{K}^*(892)$+ c.c.	dominant	463
$K_S^0 K\pi$	seen	620
$K\overline{K}$	seen	681
e^+e^-	seen	840
$\omega\pi\pi$	not seen	622

ρ₃(1690)

$$I^G(J^{PC}) = 1^+(3^{--})$$

J^P from the 2π and $K\overline{K}$ modes.
Mass $m = 1691 \pm 5$ MeV [m]
Full width $\Gamma = 160 \pm 10$ MeV [m] (S = 1.5)

ρ₃(1690) DECAY MODES	Fraction (Γ_i/Γ)	Scale factor	p (MeV/c)
4π	(71.1 ± 1.9) %		788
$\pi^\pm\pi^+\pi^-\pi^0$	(67 ±22) %		788
$\omega\pi$	(16 ± 6) %		656
$\pi\pi$	(23.6 ± 1.3) %		834
$K\overline{K}\pi$	(3.8 ± 1.2) %		628
$K\overline{K}$	(1.58± 0.26) %	1.2	686
$\eta\pi^+\pi^-$	seen		728

$\rho(1700)$ [q] $I^G(J^{PC}) = 1^+(1^{--})$

Mass $m = 1700 \pm 20$ MeV [m]   ($\eta\rho^0$ and $\pi^+\pi^-$ modes)
Full width $\Gamma = 240 \pm 60$ MeV [m]   ($\eta\rho^0$ and $\pi^+\pi^-$ modes)

$\rho(1700)$ DECAY MODES	Fraction (Γ_i/Γ)	p (MeV/c)
$\rho\pi\pi$	dominant	640
$2(\pi^+\pi^-)$	large	792
$\rho^0\pi^+\pi^-$	large	640
$\rho^\pm\pi^\mp\pi^0$	large	642
$\pi^+\pi^-$	seen	838
$\pi^-\pi^0$	seen	839
$K\overline{K}^*(892) +$ c.c.	seen	479
$\eta\rho$	seen	533
$K\overline{K}$	seen	692
e^+e^-	seen	850
$\pi^0\omega$	seen	662

$f_J(1710)$ [t] $I^G(J^{PC}) = 0^+(\text{even}^{++})$

Mass $m = 1712 \pm 5$ MeV   (S = 1.1)
Full width $\Gamma = 133 \pm 14$ MeV   (S = 1.2)

$f_J(1710)$ DECAY MODES	Fraction (Γ_i/Γ)	p (MeV/c)
$K\overline{K}$	seen	690
$\eta\eta$	seen	648
$\pi\pi$	seen	837

$\pi(1800)$ $I^G(J^{PC}) = 1^-(0^{-+})$

Mass $m = 1801 \pm 13$ MeV   (S = 1.9)
Full width $\Gamma = 210 \pm 15$ MeV

$\pi(1800)$ DECAY MODES	Fraction (Γ_i/Γ)	p (MeV/c)
$\pi^+\pi^-\pi^-$	seen	–
$f_0(980)\pi^-$	seen	623
$f_0(1370)\pi^-$	seen	–
$\rho\pi^-$	not seen	728
$\eta\eta\pi^-$	seen	–
$a_0(980)\eta$	seen	459
$f_0(1500)\pi^-$	seen	240
$\eta\eta'(958)\pi^-$	seen	–
$K_0^*(1430)K^-$	seen	–
$K^*(892)K^-$	not seen	560

$\phi_3(1850)$ $I^G(J^{PC}) = 0^-(3^{--})$

Mass $m = 1854 \pm 7$ MeV
Full width $\Gamma = 87^{+28}_{-23}$ MeV   (S = 1.2)

$\phi_3(1850)$ DECAY MODES	Fraction (Γ_i/Γ)	p (MeV/c)
$K\overline{K}$	seen	785
$K\overline{K}^*(892) +$ c.c.	seen	602

$f_2(2010)$ $I^G(J^{PC}) = 0^+(2^{++})$

Seen by one group only.
Mass $m = 2011^{+60}_{-80}$ MeV
Full width $\Gamma = 202 \pm 60$ MeV

$f_2(2010)$ DECAY MODES	Fraction (Γ_i/Γ)	p (MeV/c)
$\phi\phi$	seen	–

$a_4(2040)$ $I^G(J^{PC}) = 1^-(4^{++})$

Mass $m = 2020 \pm 16$ MeV
Full width $\Gamma = 387 \pm 70$ MeV

$a_4(2040)$ DECAY MODES	Fraction (Γ_i/Γ)	p (MeV/c)
$K\overline{K}$	seen	892
$\pi^+\pi^-\pi^0$	seen	–
$\eta\pi^0$	seen	941

$f_4(2050)$ $I^G(J^{PC}) = 0^+(4^{++})$

Mass $m = 2044 \pm 11$ MeV   (S = 1.4)
Full width $\Gamma = 208 \pm 13$ MeV   (S = 1.2)

$f_4(2050)$ DECAY MODES	Fraction (Γ_i/Γ)	p (MeV/c)
$\omega\omega$	(26 ± 6) %	658
$\pi\pi$	(17.0 ± 1.5) %	1012
$K\overline{K}$	$(6.8^{+3.4}_{-1.8}) \times 10^{-3}$	895
$\eta\eta$	$(2.1 \pm 0.8) \times 10^{-3}$	863
$4\pi^0$	< 1.2 %	977

$f_2(2300)$ $I^G(J^{PC}) = 0^+(2^{++})$

Mass $m = 2297 \pm 28$ MeV
Full width $\Gamma = 149 \pm 40$ MeV

$f_2(2300)$ DECAY MODES	Fraction (Γ_i/Γ)	p (MeV/c)
$\phi\phi$	seen	529

$f_2(2340)$ $I^G(J^{PC}) = 0^+(2^{++})$

Mass $m = 2339 \pm 60$ MeV
Full width $\Gamma = 319^{+80}_{-70}$ MeV

$f_2(2340)$ DECAY MODES	Fraction (Γ_i/Γ)	p (MeV/c)
$\phi\phi$	seen	573

STRANGE MESONS
$(S = \pm 1, C = B = 0)$

$K^+ = u\bar{s}, K^0 = d\bar{s}, \overline{K^0} = \bar{d}s, K^- = \bar{u}s,$ similarly for K^*'s

$\boxed{K^\pm}$ $\qquad I(J^P) = \frac{1}{2}(0^-)$

Mass $m = 493.677 \pm 0.016$ MeV [u] (S = 2.8)
Mean life $\tau = (1.2386 \pm 0.0024) \times 10^{-8}$ s (S = 2.0)
$c\tau = 3.713$ m

Slope parameter g [v]

(See Particle Listings for quadratic coefficients)
$K^+ \to \pi^+\pi^+\pi^- = -0.2154 \pm 0.0035$ (S = 1.4)
$K^- \to \pi^-\pi^-\pi^+ = -0.217 \pm 0.007$ (S = 2.5)
$K^\pm \to \pi^\pm\pi^0\pi^0 = 0.594 \pm 0.019$ (S = 1.3)

$K^\pm$ decay form factors [a,w]

K^+_{e3}	$\lambda_+ = 0.0286 \pm 0.0022$	
$K^+_{\mu3}$	$\lambda_+ = 0.032 \pm 0.008$	(S = 1.6)
$K^+_{\mu3}$	$\lambda_0 = 0.006 \pm 0.007$	(S = 1.6)
K^+_{e3}	$|f_S/f_+| = 0.084 \pm 0.023$	(S = 1.2)
K^+_{e3}	$|f_T/f_+| = 0.38 \pm 0.11$	(S = 1.1)
$K^+_{\mu3}$	$|f_T/f_+| = 0.02 \pm 0.12$	

$K^+ \to e^+\nu_e\gamma$ $|F_A + F_V| = 0.148 \pm 0.010$
$K^+ \to \mu^+\nu_\mu\gamma$ $|F_A + F_V| < 0.23$, CL = 90%
$K^+ \to e^+\nu_e\gamma$ $|F_A - F_V| < 0.49$
$K^+ \to \mu^+\nu_\mu\gamma$ $|F_A - F_V| = -2.2$ to 0.3

K^- modes are charge conjugates of the modes below.

K^+ DECAY MODES	Fraction (Γ_i/Γ)	Scale factor/Confidence level	p (MeV/c)
$\mu^+\nu_\mu$	(63.51 ± 0.18) %	S=1.3	236
$e^+\nu_e$	$(1.55 \pm 0.07) \times 10^{-5}$		247
$\pi^+\pi^0$	(21.16 ± 0.14) %	S=1.1	205
$\pi^+\pi^+\pi^-$	(5.59 ± 0.05) %	S=1.8	125
$\pi^+\pi^0\pi^0$	(1.73 ± 0.04) %	S=1.2	133
$\pi^0\mu^+\nu_\mu$	(3.18 ± 0.08) %	S=1.5	215
Called $K^+_{\mu3}$.			
$\pi^0 e^+\nu_e$	(4.82 ± 0.06) %	S=1.3	228
Called K^+_{e3}.			
$\pi^0\pi^0 e^+\nu_e$	$(2.1 \pm 0.4) \times 10^{-5}$		206
$\pi^+\pi^- e^+\nu_e$	$(3.91 \pm 0.17) \times 10^{-5}$		203
$\pi^+\pi^-\mu^+\nu_\mu$	$(1.4 \pm 0.9) \times 10^{-5}$		151
$\pi^0\pi^0\pi^0 e^+\nu_e$	$< 3.5 \times 10^{-6}$	CL=90%	135
$\pi^+\gamma\gamma$	[x] $(1.10 \pm 0.32) \times 10^{-6}$		227
$\pi^+ 3\gamma$	[x] $< 1.0 \times 10^{-4}$	CL=90%	227
$\mu^+\nu_\mu\nu\bar{\nu}$	$< 6.0 \times 10^{-6}$	CL=90%	236
$e^+\nu_e\nu\bar{\nu}$	$< 6 \times 10^{-5}$	CL=90%	247
$\mu^+\nu_\mu e^+ e^-$	$(1.3 \pm 0.4) \times 10^{-7}$		236
$e^+\nu_e e^+ e^-$	$(3.0 ^{+3.0}_{-1.5}) \times 10^{-8}$		247
$\mu^+\nu_\mu\mu^+\mu^-$	$< 4.1 \times 10^{-7}$	CL=90%	185
$\mu^+\nu_\mu\gamma$	[x,y] $(5.50 \pm 0.28) \times 10^{-3}$		236
$\pi^+\pi^0\gamma$	[x,y] $(2.75 \pm 0.15) \times 10^{-4}$		205
$\pi^+\pi^0\gamma$(DE)	[x,z] $(1.8 \pm 0.4) \times 10^{-5}$		205
$\pi^+\pi^+\pi^-\gamma$	[x,y] $(1.04 \pm 0.31) \times 10^{-4}$		125
$\pi^+\pi^0\pi^0\gamma$	[x,y] $(7.5 ^{+5.5}_{-3.0}) \times 10^{-6}$		133
$\pi^0\mu^+\nu_\mu\gamma$	[x,y] $< 6.1 \times 10^{-5}$	CL=90%	215
$\pi^0 e^+\nu_e\gamma$	[x,y] $(2.62 \pm 0.20) \times 10^{-4}$		228
$\pi^0 e^+\nu_e\gamma$(SD)	[aa] $< 5.3 \times 10^{-5}$	CL=90%	228
$\pi^0\pi^0 e^+\nu_e\gamma$	$< 5 \times 10^{-6}$		206

Lepton Family number (LF), Lepton number (L), $\Delta S = \Delta Q$ (SQ) violating modes, or $\Delta S = 1$ weak neutral current (S1) modes

					p
$\pi^+\pi^+ e^-\bar{\nu}_e$	SQ	< 1.2	$\times 10^{-8}$	CL=90%	203
$\pi^+\pi^+\mu^-\bar{\nu}_\mu$	SQ	< 3.0	$\times 10^{-6}$	CL=95%	151
$\pi^+ e^+ e^-$	S1	$(2.74 \pm 0.23) \times 10^{-7}$			227
$\pi^+\mu^+\mu^-$	S1	$(5.0 \pm 1.0) \times 10^{-8}$			172
$\pi^+\nu\bar{\nu}$	S1	$(4.2 ^{+9.7}_{-3.5}) \times 10^{-10}$			227
$\mu^-\nu e^+ e^+$	LF	< 2.0	$\times 10^{-8}$	CL=90%	236
$\mu^+\nu_e$	LF	[d] < 4	$\times 10^{-3}$	CL=90%	236
$\pi^+\mu^+ e^-$	LF	< 2.1	$\times 10^{-10}$	CL=90%	214
$\pi^+\mu^- e^+$	LF	< 7	$\times 10^{-9}$	CL=90%	214
$\pi^-\mu^+ e^+$	L	< 7	$\times 10^{-9}$	CL=90%	214
$\pi^- e^+ e^+$	L	< 1.0	$\times 10^{-8}$	CL=90%	227
$\pi^-\mu^+\mu^+$	L	[d] < 1.5	$\times 10^{-4}$	CL=90%	172
$\mu^+\bar{\nu}_e$	L	[d] < 3.3	$\times 10^{-3}$	CL=90%	236
$\pi^0 e^+\bar{\nu}_e$	L	< 3	$\times 10^{-3}$	CL=90%	228

 $\boxed{K^0}$ $\qquad I(J^P) = \frac{1}{2}(0^-)$

50% K_S, 50% K_L
Mass $m = 497.672 \pm 0.031$ MeV
$m_{K^0} - m_{K^\pm} = 3.995 \pm 0.034$ MeV (S = 1.1)
$|m_{K^0} - m_{\overline{K^0}}| / m_{average} < 10^{-18}$ [bb]

$\boxed{K^0_S}$ $\qquad I(J^P) = \frac{1}{2}(0^-)$

Mean life $\tau = (0.8934 \pm 0.0008) \times 10^{-10}$ s
$c\tau = 2.6762$ cm

CP-violation parameters [cc]

$Im(\eta_{+-0}) = -0.002 \pm 0.008$
$Im(\eta_{000})^2 < 0.1$, CL = 90%

K^0_S DECAY MODES	Fraction (Γ_i/Γ)	Scale factor/Confidence level	p (MeV/c)
$\pi^+\pi^-$	(68.61 ± 0.28) %	S=1.2	206
$\pi^0\pi^0$	(31.39 ± 0.28) %	S=1.2	209
$\pi^+\pi^-\gamma$	[y,dd] $(1.78 \pm 0.05) \times 10^{-3}$		206
$\gamma\gamma$	$(2.4 \pm 0.9) \times 10^{-6}$		249
$\pi^+\pi^-\pi^0$	$(3.4 ^{+1.1}_{-0.9}) \times 10^{-7}$		133
$3\pi^0$	$< 3.7 \times 10^{-5}$	CL=90%	139
$\pi^\pm e^\mp\nu$	[ee] $(6.70 \pm 0.07) \times 10^{-4}$	S=1.1	229
$\pi^\pm\mu^\mp\nu$	[ee] $(4.69 \pm 0.06) \times 10^{-4}$	S=1.1	216

$\Delta S = 1$ weak neutral current (S1) modes

$\mu^+\mu^-$	S1	< 3.2	$\times 10^{-7}$	CL=90%	225
$e^+ e^-$	S1	< 1.4	$\times 10^{-7}$	CL=90%	249
$\pi^0 e^+ e^-$	S1	< 1.1	$\times 10^{-6}$	CL=90%	231

 $\boxed{K^0_L}$ $\qquad I(J^P) = \frac{1}{2}(0^-)$

$m_{K_L} - m_{K_S} = (0.5301 \pm 0.0014) \times 10^{10}$ $\hbar$ s^{-1}
$\qquad\qquad = (3.489 \pm 0.009) \times 10^{-12}$ MeV
Mean life $\tau = (5.17 \pm 0.04) \times 10^{-8}$ s (S = 1.1)
$c\tau = 15.51$ m

Slope parameter g [v]

(See Particle Listings for quadratic coefficients)
$K^0_L \to \pi^+\pi^-\pi^0 = 0.670 \pm 0.014$ (S = 1.6)

K_L decay form factors [w]

K^0_{e3}	$\lambda_+ = 0.0300 \pm 0.0016$	(S = 1.2)
$K^0_{\mu3}$	$\lambda_+ = 0.034 \pm 0.005$	(S = 2.3)
$K^0_{\mu3}$	$\lambda_0 = 0.025 \pm 0.006$	(S = 2.3)
K^0_{e3}	$|f_S/f_+| < 0.04$, CL = 68%	
K^0_{e3}	$|f_T/f_+| < 0.23$, CL = 68%	
$K^0_{\mu3}$	$|f_T/f_+| = 0.12 \pm 0.12$	

$K_L \to e^+ e^-\gamma$: $\alpha_{K^*} = -0.28 \pm 0.08$

CP-violation parameters [cc]

$\delta = (0.327 \pm 0.012)\%$

$|\eta_{00}| = (2.275 \pm 0.019) \times 10^{-3}$ $(S = 1.1)$

$|\eta_{+-}| = (2.285 \pm 0.019) \times 10^{-3}$

$|\eta_{00}/\eta_{+-}| = 0.9956 \pm 0.0023$ [ff] $(S = 1.8)$

$\epsilon'/\epsilon = (1.5 \pm 0.8) \times 10^{-3}$ [ff] $(S = 1.8)$

$\phi_{+-} = (43.5 \pm 0.6)^\circ$

$\phi_{00} = (43.4 \pm 1.0)^\circ$

$\phi_{00} - \phi_{+-} = (-0.1 \pm 0.8)^\circ$

j for $K_L^0 \to \pi^+\pi^-\pi^0 = 0.0011 \pm 0.0008$

$|\eta_{+-\gamma}| = (2.35 \pm 0.07) \times 10^{-3}$

$\phi_{+-\gamma} = (44 \pm 4)^\circ$

$|\epsilon'_{+-\gamma}|/\epsilon < 0.3$, CL = 90%

$\Delta S = -\Delta Q$ in $K_{\ell 3}^0$ decay

Re $x = 0.006 \pm 0.018$ $(S = 1.3)$

Im $x = -0.003 \pm 0.026$ $(S = 1.2)$

CPT-violation parameters

Re $\Delta = 0.018 \pm 0.020$

Im $\Delta = 0.02 \pm 0.04$

K_L^0 DECAY MODES		Fraction (Γ_i/Γ)	Scale factor/ Confidence level	p (MeV/c)
$3\pi^0$		$(21.12 \pm 0.27)\%$	S=1.1	139
$\pi^+\pi^-\pi^0$		$(12.56 \pm 0.20)\%$	S=1.7	133
$\pi^\pm\mu^\mp\nu$	[gg]	$(27.17 \pm 0.25)\%$	S=1.1	216
Called $K_{\mu 3}^0$.				
$\pi^\pm e^\mp \nu_e$	[gg]	$(38.78 \pm 0.27)\%$	S=1.1	229
Called K_{e3}^0.				
2γ		$(5.92 \pm 0.15) \times 10^{-4}$		249
3γ		$< 2.4 \times 10^{-7}$	CL=90%	249
$\pi^0 2\gamma$	[hh]	$(1.70 \pm 0.28) \times 10^{-6}$		231
$\pi^0\pi^\pm e^\mp\nu$	[gg]	$(5.18 \pm 0.29) \times 10^{-5}$		207
$(\pi\mu\text{atom})\nu$		$(1.06 \pm 0.11) \times 10^{-7}$		–
$\pi^\pm e^\mp \nu_e\gamma$	[y,gg,hh]	$(3.62 ^{+0.26}_{-0.21}) \times 10^{-3}$		229
$\pi^+\pi^-\gamma$	[y,hh]	$(4.61 \pm 0.14) \times 10^{-5}$		206
$\pi^0\pi^0\gamma$		$< 5.6 \times 10^{-6}$		209

Charge conjugation × Parity (CP, CPV) or Lepton Family number (LF) violating modes, or $\Delta S = 1$ weak neutral current ($S1$) modes

				Scale factor/	p
$\pi^+\pi^-$		CP	$(2.067 \pm 0.035) \times 10^{-3}$	S=1.1	206
$\pi^0\pi^0$		CPV	$(9.36 \pm 0.20) \times 10^{-4}$		209
$\mu^+\mu^-$		$S1$	$(7.2 \pm 0.5) \times 10^{-9}$	S=1.4	225
$\mu^+\mu^-\gamma$		$S1$	$(3.25 \pm 0.28) \times 10^{-7}$		225
e^+e^-		$S1$	$< 4.1 \times 10^{-11}$	CL=90%	249
$e^+e^-\gamma$		$S1$	$(9.1 \pm 0.5) \times 10^{-6}$		249
$e^+e^-\gamma\gamma$		$S1$ [hh]	$(6.5 \pm 1.2) \times 10^{-7}$		249
$\pi^+\pi^-e^+e^-$		$S1$ [hh]	$< 4.6 \times 10^{-7}$	CL=90%	206
$\mu^+\mu^-e^+e^-$		$S1$	$(2.9 ^{+6.7}_{-2.4}) \times 10^{-9}$		225
$e^+e^-e^+e^-$		$S1$	$(4.1 \pm 0.8) \times 10^{-8}$	S=1.2	249
$\pi^0\mu^+\mu^-$		$CP,S1$ [ii]	$< 5.1 \times 10^{-9}$	CL=90%	177
$\pi^0 e^+e^-$		$CP,S1$ [ii]	$< 4.3 \times 10^{-9}$	CL=90%	231
$\pi^0\nu\bar\nu$		$CP,S1$ [jj]	$< 5.8 \times 10^{-5}$	CL=90%	231
$e^\pm\mu^\mp$		LF [gg]	$< 3.3 \times 10^{-11}$	CL=90%	238
$e^\pm e^\pm \mu^\mp\mu^\mp$		LF [gg]	$< 6.1 \times 10^{-9}$	CL=90%	–

$K^*(892)$ $\qquad I(J^P) = \frac{1}{2}(1^-)$

$K^*(892)^\pm$ mass $m = 891.66 \pm 0.26$ MeV

$K^*(892)^0$ mass $m = 896.10 \pm 0.28$ MeV $(S = 1.4)$

$K^*(892)^\pm$ full width $\Gamma = 50.8 \pm 0.9$ MeV

$K^*(892)^0$ full width $\Gamma = 50.5 \pm 0.6$ MeV $(S = 1.1)$

$K^*(892)$ DECAY MODES	Fraction (Γ_i/Γ)	Confidence level	p (MeV/c)
$K\pi$	$\sim 100\%$		291
$K^0\gamma$	$(2.30 \pm 0.20) \times 10^{-3}$		310
$K^\pm\gamma$	$(9.9 \pm 0.9) \times 10^{-4}$		309
$K\pi\pi$	$< 7 \times 10^{-4}$	95%	224

$K_1(1270)$ $\qquad I(J^P) = \frac{1}{2}(1^+)$

Mass $m = 1273 \pm 7$ MeV [m]

Full width $\Gamma = 90 \pm 20$ MeV [m]

$K_1(1270)$ DECAY MODES	Fraction (Γ_i/Γ)	p (MeV/c)
$K\rho$	$(42 \pm 6)\%$	76
$K_0^*(1430)\pi$	$(28 \pm 4)\%$	–
$K^*(892)\pi$	$(16 \pm 5)\%$	301
$K\omega$	$(11.0 \pm 2.0)\%$	–
$K f_0(1370)$	$(3.0 \pm 2.0)\%$	–

$K_1(1400)$ $\qquad I(J^P) = \frac{1}{2}(1^+)$

Mass $m = 1402 \pm 7$ MeV

Full width $\Gamma = 174 \pm 13$ MeV $(S = 1.6)$

$K_1(1400)$ DECAY MODES	Fraction (Γ_i/Γ)	p (MeV/c)
$K^*(892)\pi$	$(94 \pm 6)\%$	401
$K\rho$	$(3.0 \pm 3.0)\%$	298
$K f_0(1370)$	$(2.0 \pm 2.0)\%$	–
$K\omega$	$(1.0 \pm 1.0)\%$	285
$K_0^*(1430)\pi$	not seen	–

$K^*(1410)$ $\qquad I(J^P) = \frac{1}{2}(1^-)$

Mass $m = 1414 \pm 15$ MeV $(S = 1.3)$

Full width $\Gamma = 232 \pm 21$ MeV $(S = 1.1)$

$K^*(1410)$ DECAY MODES	Fraction (Γ_i/Γ)	Confidence level	p (MeV/c)
$K^*(892)\pi$	$> 40\%$	95%	408
$K\pi$	$(6.6 \pm 1.3)\%$		611
$K\rho$	$< 7\%$	95%	309

$K_0^*(1430)$ [kk] $\qquad I(J^P) = \frac{1}{2}(0^+)$

Mass $m = 1429 \pm 6$ MeV

Full width $\Gamma = 287 \pm 23$ MeV

$K_0^*(1430)$ DECAY MODES	Fraction (Γ_i/Γ)	p (MeV/c)
$K\pi$	$(93 \pm 10)\%$	621

$K_2^*(1430)$ $\qquad I(J^P) = \frac{1}{2}(2^+)$

$K_2^*(1430)^\pm$ mass $m = 1425.6 \pm 1.5$ MeV $(S = 1.1)$

$K_2^*(1430)^0$ mass $m = 1432.4 \pm 1.3$ MeV

$K_2^*(1430)^\pm$ full width $\Gamma = 98.5 \pm 2.7$ MeV $(S = 1.1)$

$K_2^*(1430)^0$ full width $\Gamma = 109 \pm 5$ MeV $(S = 1.9)$

$K_2^*(1430)$ DECAY MODES	Fraction (Γ_i/Γ)	Scale factor/ Confidence level	p (MeV/c)
$K\pi$	$(49.9 \pm 1.2)\%$		622
$K^*(892)\pi$	$(24.7 \pm 1.5)\%$		423
$K^*(892)\pi\pi$	$(13.4 \pm 2.2)\%$		375
$K\rho$	$(8.7 \pm 0.8)\%$	S=1.2	331
$K\omega$	$(2.9 \pm 0.8)\%$		319
$K^+\gamma$	$(2.4 \pm 0.5) \times 10^{-3}$	S=1.1	627
$K\eta$	$(1.5 ^{+3.4}_{-1.0}) \times 10^{-3}$	S=1.3	492
$K\omega\pi$	$< 7.2 \times 10^{-4}$	CL=95%	110
$K^0\gamma$	$< 9 \times 10^{-4}$	CL=90%	631

$K^*(1680)$ — $I(J^P) = \frac{1}{2}(1^-)$

Mass $m = 1717 \pm 27$ MeV (S = 1.4)
Full width $\Gamma = 322 \pm 110$ MeV (S = 4.2)

$K^*(1680)$ DECAY MODES	Fraction (Γ_i/Γ)	p (MeV/c)
$K\pi$	(38.7 ± 2.5) %	779
$K\rho$	$(31.4^{+4.7}_{-2.1})$ %	571
$K^*(892)\pi$	$(29.9^{+2.2}_{-4.7})$ %	615

$K_2(1770)$ [ll] — $I(J^P) = \frac{1}{2}(2^-)$

Mass $m = 1773 \pm 8$ MeV
Full width $\Gamma = 186 \pm 14$ MeV

$K_2(1770)$ DECAY MODES	Fraction (Γ_i/Γ)	p (MeV/c)
$K\pi\pi$		
$K^*_2(1430)\pi$	dominant	287
$K^*(892)\pi$	seen	653
$K f_2(1270)$	seen	–
$K\phi$	seen	441
$K\omega$	seen	608

$K^*_3(1780)$ — $I(J^P) = \frac{1}{2}(3^-)$

Mass $m = 1776 \pm 7$ MeV (S = 1.1)
Full width $\Gamma = 159 \pm 21$ MeV (S = 1.3)

$K^*_3(1780)$ DECAY MODES	Fraction (Γ_i/Γ)	Confidence level	p (MeV/c)
$K\rho$	(31 ± 9) %		612
$K^*(892)\pi$	(20 ± 5) %		651
$K\pi$	(18.8 ± 1.0) %		810
$K\eta$	(30 ± 13) %		715
$K^*_2(1430)\pi$	< 16 %	95%	284

$K_2(1820)$ [mm] — $I(J^P) = \frac{1}{2}(2^-)$

Mass $m = 1816 \pm 13$ MeV
Full width $\Gamma = 276 \pm 35$ MeV

$K_2(1820)$ DECAY MODES	Fraction (Γ_i/Γ)	p (MeV/c)
$K^*_2(1430)\pi$	seen	325
$K^*(892)\pi$	seen	680
$K f_2(1270)$	seen	186
$K\omega$	seen	638

$K^*_4(2045)$ — $I(J^P) = \frac{1}{2}(4^+)$

Mass $m = 2045 \pm 9$ MeV (S = 1.1)
Full width $\Gamma = 198 \pm 30$ MeV

$K^*_4(2045)$ DECAY MODES	Fraction (Γ_i/Γ)	p (MeV/c)
$K\pi$	(9.9 ± 1.2) %	958
$K^*(892)\pi\pi$	(9 ± 5) %	800
$K^*(892)\pi\pi\pi$	(7 ± 5) %	764
$\rho K\pi$	(5.7 ± 3.2) %	742
$\omega K\pi$	(5.0 ± 3.0) %	736
$\phi K\pi$	(2.8 ± 1.4) %	591
$\phi K^*(892)$	(1.4 ± 0.7) %	363

CHARMED MESONS
$(C = \pm 1)$

$D^+ = c\overline{d}$, $D^0 = c\overline{u}$, $\overline{D}^0 = \overline{c}u$, $D^- = \overline{c}d$, similarly for D^*'s

$D^\pm$ — $I(J^P) = \frac{1}{2}(0^-)$

Mass $m = 1869.3 \pm 0.5$ MeV (S = 1.1)
Mean life $\tau = (1.057 \pm 0.015) \times 10^{-12}$ s
$c\tau = 317$ μm

CP-violation decay-rate asymmetries

$A_{CP}(K^+K^-\pi^\pm) = -0.017 \pm 0.027$
$A_{CP}(K^\pm K^{*0}) = -0.02 \pm 0.05$
$A_{CP}(\phi\pi^\pm) = -0.014 \pm 0.033$
$A_{CP}(\pi^+\pi^-\pi^\pm) = -0.02 \pm 0.04$

$D^+ \to \overline{K}^*(892)^0 \ell^+ \nu_\ell$ form factors

$r_2 = 0.72 \pm 0.09$
$r_v = 1.85 \pm 0.12$
$\Gamma_L/\Gamma_T = 1.23 \pm 0.13$
$\Gamma_+/\Gamma_- = 0.16 \pm 0.04$

D^- modes are charge conjugates of the modes below.

D^+ DECAY MODES		Fraction (Γ_i/Γ)	Scale factor/ Confidence level	p (MeV/c)
Inclusive modes				
e^+ anything		(17.2 ± 1.9) %		–
K^- anything		(24.2 ± 2.8) %	S=1.4	–
$\overline{K}^0$ anything + K^0 anything		(59 ± 7) %		–
K^+ anything		(5.8 ± 1.4) %		–
η anything	[nn]	< 13 %	CL=90%	–
Leptonic and semileptonic modes				
$\mu^+ \nu_\mu$		$< 7.2 \times 10^{-4}$	CL=90%	932
$\overline{K}^0 \ell^+ \nu_\ell$	[oo]	(6.8 ± 0.8) %		868
$\overline{K}^0 e^+ \nu_e$		(6.7 ± 0.9) %		868
$\overline{K}^0 \mu^+ \nu_\mu$		$(7.0^{+3.0}_{-2.0})$ %		865
$K^- \pi^+ e^+ \nu_e$		$(4.1^{+0.9}_{-0.7})$ %		863
$\overline{K}^*(892)^0 e^+ \nu_e$ $\times$ B$(\overline{K}^{*0} \to K^-\pi^+)$		(3.2 ± 0.33) %		720
$K^- \pi^+ e^+ \nu_e$ nonresonant		$< 7 \times 10^{-3}$	CL=90%	863
$K^- \pi^+ \mu^+ \nu_\mu$		(3.2 ± 0.4) %	S=1.1	851
$\overline{K}^*(892)^0 \mu^+ \nu_\mu$ $\times$ B$(\overline{K}^{*0} \to K^-\pi^+)$		(2.9 ± 0.4) %		715
$K^- \pi^+ \mu^+ \nu_\mu$ nonresonant		$(2.7 \pm 1.1) \times 10^{-3}$		851
$(\overline{K}^*(892)\pi)^0 e^+ \nu_e$		< 1.2 %	CL=90%	714
$(\overline{K}\pi\pi)^0 e^+ \nu_e$ non-$\overline{K}^*(892)$		$< 9 \times 10^{-3}$	CL=90%	846
$K^- \pi^+ \pi^0 \mu^+ \nu_\mu$		$< 1.4 \times 10^{-3}$	CL=90%	825
$\pi^0 \ell^+ \nu_\ell$	[pp]	$(3.1 \pm 1.5) \times 10^{-3}$		930

Fractions of some of the following modes with resonances have already appeared above as submodes of particular charged-particle modes.

$\overline{K}^*(892)\ell^+ \nu_\ell$	[oo]	(4.7 ± 0.4) %		720
$\overline{K}^*(892)^0 e^+ \nu_e$		(4.8 ± 0.5) %		720
$\overline{K}^*(892)^0 \mu^+ \nu_\mu$		(4.4 ± 0.6) %	S=1.1	715
$\rho^0 e^+ \nu_e$		$(2.2 \pm 0.8) \times 10^{-3}$		776
$\rho^0 \mu^+ \nu_\mu$		$(2.7 \pm 0.7) \times 10^{-3}$		772
$\phi e^+ \nu_e$		< 2.09 %	CL=90%	657
$\phi \mu^+ \nu_\mu$		< 3.72 %	CL=90%	651
$\eta \ell^+ \nu_\ell$		$< 5 \times 10^{-3}$	CL=90%	–
$\eta'(958) \mu^+ \nu_\mu$		$< 9 \times 10^{-3}$	CL=90%	684
Hadronic modes with a $\overline{K}$ or $\overline{K}K\overline{K}$				
$\overline{K}^0 \pi^+$		(2.89 ± 0.26) %	S=1.1	862
$K^- \pi^+ \pi^+$	[qq]	(9.0 ± 0.6) %		845
$\overline{K}^*(892)^0 \pi^+$ $\times$ B$(\overline{K}^{*0} \to K^-\pi^+)$		(1.27 ± 0.13) %		712
$\overline{K}^*_0(1430)^0 \pi^+$ $\times$ B$(\overline{K}^*_0(1430)^0 \to K^-\pi^+)$		(2.3 ± 0.3) %		368
$\overline{K}^*(1680)^0 \pi^+$ $\times$ B$(\overline{K}^*(1680)^0 \to K^-\pi^+)$		$(3.7 \pm 0.8) \times 10^{-3}$		65
$K^- \pi^+ \pi^+$ nonresonant		(8.5 ± 0.8) %		845
$\overline{K}^0 \pi^+ \pi^0$	[qq]	(9.7 ± 3.0) %	S=1.1	845

$\overline{K}^0 \rho^+$	(6.6 ±2.5) %		680
$\overline{K}^*(892)^0 \pi^+$	(6.3 ±0.4) × 10^{-3}		712
× B($\overline{K}^{*0} \to \overline{K}^0 \pi^0$)			
$\overline{K}^0 \pi^+ \pi^0$ nonresonant	(1.3 ±1.1) %		845
$K^- \pi^+ \pi^+ \pi^0$	(6.4 ±1.1) %	[qq]	816
$\overline{K}^*(892)^0 \rho^+$ total	(1.4 ±0.9) %		423
× B($\overline{K}^{*0} \to K^- \pi^+$)			
$\overline{K}_1(1400)^0 \pi^+$	(2.2 ±0.6) %		390
× B($\overline{K}_1(1400)^0 \to K^- \pi^+ \pi^0$)			
$K^- \rho^+ \pi^+$ total	(3.1 ±1.1) %		616
$K^- \rho^+ \pi^+$ 3-body	(1.1 ±0.4) %		616
$\overline{K}^*(892)^0 \pi^+ \pi^0$ total	(4.5 ±0.9) %		687
× B($\overline{K}^{*0} \to K^- \pi^+$)			
$\overline{K}^*(892)^0 \pi^+ \pi^0$ 3-body	(2.8 ±0.9) %		687
× B($\overline{K}^{*0} \to K^- \pi^+$)			
$K^*(892)^- \pi^+ \pi^+$ 3-body	(7 ±3) × 10^{-3}		688
× B($K^{*-} \to K^- \pi^0$)			
$K^- \pi^+ \pi^+ \pi^0$ nonresonant	(1.2 ±0.6) %	[rr]	816
$\overline{K}^0 \pi^+ \pi^+ \pi^-$	(7.0 ±0.9) %	[qq]	814
$\overline{K}^0 a_1(1260)^+$	(4.0 ±0.9) %		328
× B($a_1(1260)^+ \to \pi^+ \pi^+ \pi^-$)			
$\overline{K}_1(1400)^0 \pi^+$	(2.2 ±0.6) %		390
× B($\overline{K}_1(1400)^0 \to \overline{K}^0 \pi^+ \pi^-$)			
$K^*(892)^- \pi^+ \pi^+$ 3-body	(1.4 ±0.6) %		688
× B($K^{*-} \to \overline{K}^0 \pi^-$)			
$\overline{K}^0 \rho^0 \pi^+$ total	(4.2 ±0.9) %		614
$\overline{K}^0 \rho^0 \pi^+$ 3-body	(5 ±5) × 10^{-3}		614
$\overline{K}^0 \pi^+ \pi^+ \pi^-$ nonresonant	(8 ±4) × 10^{-3}		814
$K^- \pi^+ \pi^+ \pi^+ \pi^-$	(7.2 ±1.0) %	[qq]	772
$\overline{K}^*(892)^0 \pi^+ \pi^+ \pi^-$	(5.4 ±2.3) × 10^{-3}		642
× B($\overline{K}^{*0} \to K^- \pi^+$)			
$\overline{K}^*(892)^0 \rho^0 \pi^+$	(1.9 $^{+1.1}_{-1.0}$) × 10^{-3}		242
× B($\overline{K}^{*0} \to K^- \pi^+$)			
$\overline{K}^*(892)^0 \pi^+ \pi^+ \pi^-$ no-ρ	(2.9 ±1.1) × 10^{-3}		642
× B($\overline{K}^{*0} \to K^- \pi^+$)			
$K^- \rho^0 \pi^+ \pi^+$	(3.1 ±0.9) × 10^{-3}		529
$K^- \pi^+ \pi^+ \pi^+ \pi^-$ nonresonant	< 2.3 × 10^{-3}	CL=90%	772
$K^- \pi^+ \pi^+ \pi^0 \pi^0$	(2.2 $^{+5.0}_{-0.9}$) %		775
$\overline{K}^0 \pi^+ \pi^+ \pi^- \pi^0$	(5.4 $^{+3.0}_{-1.4}$) %		773
$\overline{K}^0 \pi^+ \pi^+ \pi^+ \pi^- \pi^-$	(8 ±7) × 10^{-4}		714
$K^- \pi^+ \pi^+ \pi^+ \pi^- \pi^0$	(2.0 ±1.8) × 10^{-3}		718
$\overline{K}^0 \overline{K}^0 K^+$	(1.8 ±0.8) %		545

Fractions of some of the following modes with resonances have already appeared above as submodes of particular charged-particle modes.

$\overline{K}^0 \rho^+$	(6.6 ±2.5) %		680
$\overline{K}^0 a_1(1260)^+$	(8.0 ±1.7) %		328
$\overline{K}^0 a_2(1320)^+$	< 3 × 10^{-3}	CL=90%	199
$\overline{K}^*(892)^0 \pi^+$	(1.90±0.19) %		712
$\overline{K}^*(892)^0 \rho^+$ total	(2.1 ±1.3) %	[rr]	423
$\overline{K}^*(892)^0 \rho^+$ S-wave	(1.6 ±1.6) %	[rr]	423
$\overline{K}^*(892)^0 \rho^+$ P-wave	< 1 × 10^{-3}	CL=90%	423
$\overline{K}^*(892)^0 \rho^+$ D-wave	(10 ±7) × 10^{-3}		423
$\overline{K}^*(892)^0 \rho^+$ D-wave longitu-dinal	< 7 × 10^{-3}	CL=90%	423
$\overline{K}_1(1270)^0 \pi^+$	< 7 × 10^{-3}	CL=90%	487
$\overline{K}_1(1400)^0 \pi^+$	(4.9 ±1.2) %		390
$\overline{K}^*(1410)^0 \pi^+$	< 7 × 10^{-3}	CL=90%	382
$\overline{K}_0^*(1430)^0 \pi^+$	(3.7 ±0.4) %		368
$\overline{K}^*(1680)^0 \pi^+$	(1.43±0.30) %		65
$\overline{K}^*(892)^0 \pi^+ \pi^0$ total	(6.7 ±1.4) %		687
$\overline{K}^*(892)^0 \pi^+ \pi^0$ 3-body	(4.2 ±1.4) %	[rr]	687
$K^*(892)^- \pi^+ \pi^+$ 3-body	(2.0 ±0.9) %		688
$K^- \rho^+ \pi^+$ total	(3.1 ±1.1) %		616
$K^- \rho^+ \pi^+$ 3-body	(1.1 ±0.4) %		616
$\overline{K}^0 \rho^0 \pi^+$ total	(4.2 ±0.9) %	CL=90%	614
$\overline{K}^0 \rho^0 \pi^+$ 3-body	(5 ±5) × 10^{-3}		614
$\overline{K}^0 f_0(980) \pi^+$	< 5 × 10^{-3}	CL=90%	461
$\overline{K}^*(892)^0 \pi^+ \pi^+ \pi^-$	(8.1 ±3.4) × 10^{-3}	S=1.7	642
$\overline{K}^*(892)^0 \rho^0 \pi^+$	(2.9 $^{+1.7}_{-1.5}$) × 10^{-3}	S=1.8	242
$\overline{K}^*(892)^0 \pi^+ \pi^+ \pi^-$ no-ρ	(4.3 ±1.7) × 10^{-3}		642
$K^- \rho^0 \pi^+ \pi^+$	(3.1 ±0.9) × 10^{-3}		529

Pionic modes

$\pi^+ \pi^0$	(2.5 ±0.7) × 10^{-3}		925
$\pi^+ \pi^+ \pi^-$	(3.6 ±0.4) × 10^{-3}		908
$\rho^0 \pi^+$	(1.05±0.31) × 10^{-3}		769
$\pi^+ \pi^+ \pi^-$ nonresonant	(2.2 ±0.4) × 10^{-3}		908
$\pi^+ \pi^+ \pi^- \pi^0$	(1.9 $^{+1.5}_{-1.2}$) %		882
$\eta \pi^+ $ × B($\eta \to \pi^+ \pi^- \pi^0$)	(1.7 ±0.6) × 10^{-3}		848
$\omega \pi^+ $ × B($\omega \to \pi^+ \pi^- \pi^0$)	< 6 × 10^{-3}	CL=90%	764
$\pi^+ \pi^+ \pi^+ \pi^- \pi^-$	(2.1 ±0.4) × 10^{-3}		845
$\pi^+ \pi^+ \pi^+ \pi^- \pi^- \pi^0$	(2.9 $^{+2.9}_{-2.0}$) × 10^{-3}		799

Fractions of some of the following modes with resonances have already appeared above as submodes of particular charged-particle modes.

$\eta \pi^+$	(7.5 ±2.5) × 10^{-3}		848
$\rho^0 \pi^+$	(1.05±0.31) × 10^{-3}		769
$\omega \pi^+$	< 7 × 10^{-3}	CL=90%	764
$\eta \rho^+$	< 1.2 %	CL=90%	658
$\eta'(958) \pi^+$	< 9 × 10^{-3}	CL=90%	680
$\eta'(958) \rho^+$	< 1.5 %	CL=90%	355

Hadronic modes with a $K\overline{K}$ pair

$K^+ \overline{K}^0$	(7.4 ±1.0) × 10^{-3}		792
$K^+ K^- \pi^+$	(8.8 ±0.8) × 10^{-3}	[qq]	744
$\phi \pi^+ $ × B($\phi \to K^+ K^-$)	(3.0 ±0.3) × 10^{-3}		647
$K^+ \overline{K}^*(892)^0$	(2.8 ±0.4) × 10^{-3}		610
× B($\overline{K}^{*0} \to K^- \pi^+$)			
$K^+ K^- \pi^+$ nonresonant	(4.5 ±0.9) × 10^{-3}		744
$K^0 \overline{K}^0 \pi^+$	—		741
$K^*(892)^+ \overline{K}^0$	(2.1 ±1.0) %		611
× B($K^{*+} \to K^0 \pi^+$)			
$K^+ K^- \pi^+ \pi^0$	—		682
$\phi \pi^+ \pi^0 $ × B($\phi \to K^+ K^-$)	(1.1 ±0.5) %		619
$\phi \rho^+ $ × B($\phi \to K^+ K^-$)	< 7 × 10^{-3}	CL=90%	268
$K^+ K^- \pi^+ \pi^0$ non-ϕ	(1.5 $^{+0.7}_{-0.6}$) %		682
$K^+ \overline{K}^0 \pi^+ \pi^-$	< 2 %	CL=90%	678
$K^0 K^- \pi^+ \pi^+$	(1.0 ±0.6) %		678
$K^*(892)^+ \overline{K}^*(892)^0$	(1.2 ±0.5) %		273
× B^2($K^{*+} \to K^0 \pi^+$)			
$K^0 K^- \pi^+ \pi^+$ non-$K^{*+} \overline{K}^{*0}$	< 7.9 × 10^{-3}	CL=90%	678
$K^+ K^- \pi^+ \pi^+ \pi^-$	—		600
$\phi \pi^+ \pi^+ \pi^-$	< 1 × 10^{-3}	CL=90%	565
× B($\phi \to K^+ K^-$)			
$K^+ K^- \pi^+ \pi^+ \pi^-$ nonresonant	< 3 %	CL=90%	600

Fractions of the following modes with resonances have already appeared above as submodes of particular charged-particle modes.

$\phi \pi^+$	(6.1 ±0.6) × 10^{-3}		647
$\phi \pi^+ \pi^0$	(2.3 ±1.0) %		619
$\phi \rho^+$	< 1.4 %	CL=90%	268
$\phi \pi^+ \pi^+ \pi^-$	< 2 × 10^{-3}	CL=90%	565
$K^+ \overline{K}^*(892)^0$	(4.2 ±0.5) × 10^{-3}		610
$K^*(892)^+ \overline{K}^0$	(3.2 ±1.5) %		611
$K^*(892)^+ \overline{K}^*(892)^0$	(2.6 ±1.1) %		273

Doubly Cabibbo suppressed (DC) modes,
$\Delta C = 1$ weak neutral current (C1) modes, or
Lepton Family number (LF) or Lepton number (L) violating modes

$K^+ \pi^+ \pi^-$	DC	(6.8 ±1.5) × 10^{-4}		845
$K^+ \rho^0$	DC	(2.5 ±1.2) × 10^{-4}		681
$K^*(892)^0 \pi^+$	DC	(3.6 ±1.6) × 10^{-4}		712
$K^+ \pi^+ \pi^-$ nonresonant	DC	(2.4 ±1.2) × 10^{-4}		845
$K^+ K^+ K^-$	DC	< 1.4 × 10^{-4}	CL=90%	550
ϕK^+	DC	< 1.3 × 10^{-4}	CL=90%	527
$\pi^+ e^+ e^-$	C1	< 6.6 × 10^{-5}	CL=90%	929
$\pi^+ \mu^+ \mu^-$	C1	< 1.8 × 10^{-5}	CL=90%	917
$\rho^+ \mu^+ \mu^-$	C1	< 5.6 × 10^{-4}	CL=90%	759
$K^+ e^+ e^-$	[ss]	< 2.0 × 10^{-4}	CL=90%	869
$K^+ \mu^+ \mu^-$	[ss]	< 9.7 × 10^{-5}	CL=90%	856
$\pi^+ e^+ \mu^-$	LF	< 1.1 × 10^{-4}	CL=90%	926
$\pi^+ e^- \mu^+$	LF	< 1.3 × 10^{-4}	CL=90%	926
$K^+ e^+ \mu^-$	LF	< 1.3 × 10^{-4}	CL=90%	866

$K^+ e^- \mu^+$	LF	< 1.2	$\times 10^{-4}$	CL=90%	866
$\pi^- e^+ e^+$	L	< 1.1		CL=90%	929
$\pi^- \mu^+ \mu^+$	L	< 8.7	$\times 10^{-5}$	CL=90%	917
$\pi^- e^+ \mu^+$	L	< 1.1	$\times 10^{-4}$	CL=90%	926
$\rho^- \mu^+ \mu^+$	L	< 5.6	$\times 10^{-4}$	CL=90%	759
$K^- e^+ e^+$	L	< 1.2	$\times 10^{-4}$	CL=90%	869
$K^- \mu^+ \mu^+$	L	< 1.2	$\times 10^{-4}$	CL=90%	856
$K^- e^+ \mu^+$	L	< 1.3	$\times 10^{-4}$	CL=90%	866
$K^*(892)^- \mu^+ \mu^+$	L	< 8.5	$\times 10^{-4}$	CL=90%	703

D^0 $\quad I(J^P) = \frac{1}{2}(0^-)$

Mass $m = 1864.6 \pm 0.5$ MeV (S = 1.1)
$m_{D^\pm} - m_{D^0} = 4.76 \pm 0.10$ MeV (S = 1.1)
Mean life $\tau = (0.415 \pm 0.004) \times 10^{-12}$ s
$\quad c\tau = 124.4$ μm
$|m_{D_1^0} - m_{D_2^0}| < 24 \times 10^{10}$ $\hbar$ s^{-1}, CL = 90% [tt]
$|\Gamma_{D_1^0} - \Gamma_{D_2^0}|/\Gamma_{D^0} < 0.20$, CL = 90% [tt]
$\Gamma(K^+ \ell^- \bar{\nu}_\ell$ (via $\overline{D^0}$))$/\Gamma(K^- \ell^+ \nu_\ell) < 0.005$, CL = 90%
$\dfrac{\Gamma(K^+\pi^- \text{ or } K^+\pi^-\pi^+\pi^- \text{(via} \overline{D^0}\text{))}}{\Gamma(K^-\pi^+ \text{ or } K^-\pi^+\pi^+\pi^-)} < 0.0085$ (or < 0.0037), CL = 90% [uu]

CP-violation decay-rate asymmetries

$A_{CP}(K^+ K^-) = 0.026 \pm 0.035$
$A_{CP}(\pi^+ \pi^-) = -0.05 \pm 0.08$
$A_{CP}(K_S^0 \phi) = -0.03 \pm 0.09$
$A_{CP}(K_S^0 \pi^0) = -0.018 \pm 0.030$

$\overline{D^0}$ modes are charge conjugates of the modes below.

D^0 DECAY MODES		Fraction (Γ_i/Γ)	Scale factor/ Confidence level	p (MeV/c)
Inclusive modes				
e^+ anything		(6.75 ±0.29) %		—
μ^+ anything		(6.6 ±0.8) %		—
K^- anything		(53 ±4) %	S=1.3	—
$\overline{K^0}$ anything + K^0 anything		(42 ±5) %		—
K^+ anything		(3.4 $^{+0.6}_{-0.4}$) %		—
η anything	[nn]	< 13 %	CL=90%	—
Semileptonic modes				
$K^- \ell^+ \nu_\ell$	[oo]	(3.50 ±0.17) %	S=1.3	867
$K^- e^+ \nu_e$		(3.66 ±0.18) %		867
$K^- \mu^+ \nu_\mu$		(3.23 ±0.17) %		863
$K^- \pi^0 e^+ \nu_e$		(1.6 $^{+1.3}_{-0.5}$) %		861
$\overline{K^0} \pi^- e^+ \nu_e$		(2.8 $^{+1.7}_{-0.9}$) %		860
$\overline{K}^*(892)^- e^+ \nu_e$		(1.35 ±0.22) %		719
$\quad \times$ B($K^{*-} \rightarrow \overline{K^0}\pi^-$)				
$K^*(892)^- \ell^+ \nu_\ell$		—		
$\overline{K}^*(892)^0 \pi^- e^+ \nu_e$		—		708
$K^- \pi^+ \pi^- \mu^+ \nu_\mu$		< 1.2 $\times 10^{-3}$	CL=90%	821
$(\overline{K}^*(892)\pi)^- \mu^+ \nu_\mu$		< 1.4 $\times 10^{-3}$	CL=90%	693
$\pi^- e^+ \nu_e$		(3.7 ±0.6) $\times 10^{-3}$		927

A fraction of the following resonance mode has already appeared above as a submode of a charged-particle mode.

$K^*(892)^- e^+ \nu_e$		(2.02 ±0.33) %		719

Hadronic modes with a $\overline{K}$ or $\overline{K}K\overline{K}$				
$K^- \pi^+$		(3.85 ±0.09) %		861
$\overline{K^0} \pi^0$		(2.12 ±0.21) %	S=1.1	860
$\overline{K^0} \pi^+ \pi^-$	[qq]	(5.4 ±0.4) %	S=1.2	842
$\overline{K^0} \rho^0$		(1.21 ±0.17) %		676
$\overline{K^0} f_0(980)$		(3.0 ±0.8) $\times 10^{-3}$		549
$\quad \times$ B($f_0 \rightarrow \pi^+\pi^-$)				
$\overline{K^0} f_2(1270)$		(2.4 ±0.9) $\times 10^{-3}$		263
$\quad \times$ B($f_2 \rightarrow \pi^+\pi^-$)				
$\overline{K^0} f_0(1370)$		(4.3 ±1.3) $\times 10^{-3}$		—
$\quad \times$ B($f_0 \rightarrow \pi^+\pi^-$)				
$K^*(892)^- \pi^+$		(3.4 ±0.3) %		711
$\quad \times$ B($K^{*-} \rightarrow \overline{K^0}\pi^-$)				
$K_0^*(1430)^- \pi^+$		(6.4 ±1.6) $\times 10^{-3}$		364
$\quad \times$ B($K_0^*(1430)^- \rightarrow \overline{K^0}\pi^-$)				
$\overline{K^0} \pi^+ \pi^-$ nonresonant		(1.47 ±0.24) %		842

$K^- \pi^+ \pi^0$	[qq]	(13.9 ±0.9) %	S=1.3	844
$K^- \rho^+$		(10.8 ±1.0) %		678
$K^*(892)^- \pi^+$		(1.7 ±0.2) %		711
$\quad \times$ B($K^{*-} \rightarrow K^-\pi^0$)				
$\overline{K}^*(892)^0 \pi^0$		(2.1 ±0.3) %		709
$\quad \times$ B($\overline{K}^{*0} \rightarrow K^-\pi^+$)				
$K^- \pi^+ \pi^0$ nonresonant		(6.9 ±2.5) $\times 10^{-3}$		844
$\overline{K^0} \pi^0 \pi^0$		—		843
$\overline{K}^*(892)^0 \pi^0$		(1.1 ±0.2) %		709
$\quad \times$ B($\overline{K}^{*0} \rightarrow \overline{K^0}\pi^0$)				
$\overline{K^0} \pi^0 \pi^0$ nonresonant		(7.9 ±2.1) $\times 10^{-3}$		843
$K^- \pi^+ \pi^+ \pi^-$	[qq]	(7.6 ±0.4) %	S=1.1	812
$K^- \pi^+ \rho^0$ total		(6.3 ±0.4) %		612
$K^- \pi^+ \rho^0$ 3-body		(4.8 ±2.1) $\times 10^{-3}$		612
$\overline{K}^*(892)^0 \rho^0$		(9.8 ±2.2) $\times 10^{-3}$		418
$\quad \times$ B($\overline{K}^{*0} \rightarrow K^-\pi^+$)				
$K^- a_1(1260)^+$		(3.6 ±0.6) %		327
$\quad \times$ B($a_1(1260)^+ \rightarrow \pi^+\pi^+\pi^-$)				
$\overline{K}^*(892)^0 \pi^+ \pi^-$ total		(1.5 ±0.4) %		683
$\quad \times$ B($\overline{K}^{*0} \rightarrow K^-\pi^+$)				
$\overline{K}^*(892)^0 \pi^+ \pi^-$ 3-body		(9.5 ±2.1) $\times 10^{-3}$		683
$\quad \times$ B($\overline{K}^{*0} \rightarrow K^-\pi^+$)				
$K_1(1270)^- \pi^+$	[rr]	(3.6 ±1.0) $\times 10^{-3}$		483
$\quad \times$ B($K_1(1270)^- \rightarrow K^-\pi^+\pi^-$)				
$K^- \pi^+ \pi^+ \pi^-$ nonresonant		(1.76 ±0.25) %		812
$\overline{K^0} \pi^+ \pi^- \pi^0$	[qq]	(10.0 ±1.2) %		812
$\overline{K^0} \eta \times$ B($\eta \rightarrow \pi^+\pi^-\pi^0$)		(1.6 ±0.3) $\times 10^{-3}$		772
$\overline{K^0} \omega \times$ B($\omega \rightarrow \pi^+\pi^-\pi^0$)		(1.9 ±0.4) %		670
$K^*(892)^- \rho^+$		(4.1 ±1.6) %		422
$\quad \times$ B($K^{*-} \rightarrow \overline{K^0}\pi^-$)				
$\overline{K}^*(892)^0 \rho^0$		(4.9 ±1.1) $\times 10^{-3}$		418
$\quad \times$ B($\overline{K}^{*0} \rightarrow \overline{K^0}\pi^0$)				
$K_1(1270)^- \pi^+$	[rr]	(5.1 ±1.4) $\times 10^{-3}$		483
$\quad \times$ B($K_1(1270)^- \rightarrow \overline{K^0}\pi^-\pi^+$)				
$\overline{K}^*(892)^0 \pi^+ \pi^-$ 3-body		(4.8 ±1.1) $\times 10^{-3}$		683
$\quad \times$ B($\overline{K}^{*0} \rightarrow \overline{K^0}\pi^0$)				
$\overline{K^0} \pi^+ \pi^- \pi^0$ nonresonant		(2.1 ±2.1) %		812
$K^- \pi^+ \pi^0 \pi^0$		(15 ±5) %		815
$K^- \pi^+ \pi^+ \pi^- \pi^0$		(4.1 ±0.4) %		771
$\overline{K}^*(892)^0 \pi^+ \pi^- \pi^0$		(1.2 ±0.6) %		641
$\quad \times$ B($\overline{K}^{*0} \rightarrow K^-\pi^+$)				
$\overline{K}^*(892)^0 \eta$		(2.9 ±0.8) $\times 10^{-3}$		580
$\quad \times$ B($\overline{K}^{*0} \rightarrow K^-\pi^+$)				
$\quad \times$ B($\eta \rightarrow \pi^+\pi^-\pi^0$)				
$K^- \pi^+ \omega \times$ B($\omega \rightarrow \pi^+\pi^-\pi^0$)		(2.7 ±0.5) %		605
$\overline{K}^*(892)^0 \omega$		(7 ±3) $\times 10^{-3}$		406
$\quad \times$ B($\overline{K}^{*0} \rightarrow K^-\pi^+$)				
$\quad \times$ B($\omega \rightarrow \pi^+\pi^-\pi^0$)				
$\overline{K^0} \pi^+ \pi^+ \pi^- \pi^-$		(5.8 ±1.6) $\times 10^{-3}$		768
$\overline{K^0} \pi^+ \pi^- \pi^0 \pi^0 (\pi^0)$		(10.6 $^{+7.3}_{-3.0}$) %		771
$\overline{K^0} K^+ K^-$		(9.4 ±1.0) $\times 10^{-3}$		544
$\overline{K^0} \phi \times$ B($\phi \rightarrow K^+K^-$)		(4.3 ±0.5) $\times 10^{-3}$		520
$\overline{K^0} K^+ K^-$ non-ϕ		(5.1 ±0.8) $\times 10^{-3}$		544
$K_S^0 K_S^0 K_S^0$		(8.4 ±1.5) $\times 10^{-4}$		538
$K^+ K^- K^- \pi^+$		(2.1 ±0.5) $\times 10^{-4}$		434
$K^+ K^- \overline{K^0} \pi^0$		(7.2 $^{+4.8}_{-3.5}$) $\times 10^{-3}$		435

Fractions of many of the following modes with resonances have already appeared above as submodes of particular charged-particle modes. (Modes for which there are only upper limits and $\overline{K}^*(892)\rho$ submodes only appear below.)

$\overline{K^0} \eta$		(7.1 ±1.0) $\times 10^{-3}$		772
$\overline{K^0} \rho^0$		(1.21 ±0.17) %		676
$K^- \rho^+$		(10.8 ±1.0) %	S=1.2	678
$\overline{K^0} \omega$		(2.1 ±0.4) %		670
$\overline{K^0} \eta'(958)$		(1.72 ±0.26) %		565
$\overline{K^0} f_0(980)$		(5.7 ±1.6) $\times 10^{-3}$		549
$\overline{K^0} \phi$		(8.6 ±1.0) $\times 10^{-3}$		520
$K^- a_1(1260)^+$		(7.3 ±1.1) %		327
$\overline{K^0} a_1(1260)^0$		< 1.9 %	CL=90%	322
$\overline{K^0} f_2(1270)$		(4.2 ±1.5) $\times 10^{-3}$		263
$K^- a_2(1320)^+$		< 2 $\times 10^{-3}$	CL=90%	197
$\overline{K^0} f_0(1370)$		(7.0 ±2.1) $\times 10^{-3}$		—
$K^*(892)^- \pi^+$		(5.1 ±0.4) %	S=1.2	711
$\overline{K}^*(892)^0 \pi^0$		(3.2 ±0.4) %		709
$\overline{K}^*(892)^0 \pi^+ \pi^-$ total		(2.3 ±0.5) %		683
$\overline{K}^*(892)^0 \pi^+ \pi^-$ 3-body		(1.43 ±0.32) %		683

$K^-\pi^+\rho^0$ total	(6.3 ±0.4) %			612
$K^-\pi^+\rho^0$ 3-body	(4.8 ±2.1) × 10⁻³			612
$\overline{K}^*(892)^0\rho^0$	(1.47±0.33) %			418
$\overline{K}^*(892)^0\rho^0$ transverse	(1.5 ±0.5) %			418
$\overline{K}^*(892)^0\rho^0$ S-wave	(2.8 ±0.6) %			418
$\overline{K}^*(892)^0\rho^0$ S-wave long.	< 3 × 10⁻³		CL=90%	418
$\overline{K}^*(892)^0\rho^0$ P-wave	< 3 × 10⁻³		CL=90%	418
$\overline{K}^*(892)^0\rho^0$ D-wave	(1.9 ±0.6) %			418
$K^*(892)^-\rho^+$	(6.1 ±2.4) %			422
$K^*(892)^-\rho^+$ longitudinal	(2.9 ±1.2) %			422
$K^*(892)^-\rho^+$ transverse	(3.2 ±1.8) %			422
$K^*(892)^-\rho^+$ P-wave	< 1.5 %		CL=90%	422
$K^-\pi^+f_0(980)$	< 1.1 %		CL=90%	459
$\overline{K}^*(892)^0 f_0(980)$	< 3 × 10⁻³		CL=90%	–
$K_1(1270)^-\pi^+$	[rr] (1.06±0.29) %			483
$K_1(1400)^-\pi^+$	< 1.2 %		CL=90%	386
$\overline{K}_1(1400)^0\pi^0$	< 3.7 %		CL=90%	387
$K^*(1410)^-\pi^+$	< 1.2 %		CL=90%	378
$K_0^*(1430)^-\pi^+$	(1.04±0.26) %			364
$K_2^*(1430)^-\pi^+$	< 8 × 10⁻³		CL=90%	367
$\overline{K}_2^*(1430)^0\pi^0$	< 4 × 10⁻³		CL=90%	363
$\overline{K}^*(892)^0\pi^+\pi^-\pi^0$	(1.8 ±0.9) %			641
$\overline{K}^*(892)^0\eta$	(1.9 ±0.5) %			580
$K^-\pi^+\omega$	(3.0 ±0.6) %			605
$\overline{K}^*(892)^0\omega$	(1.1 ±0.5) %			406
$K^-\pi^+\eta'(958)$	(7.0 ±1.8) × 10⁻³			479
$\overline{K}^*(892)^0\eta'(958)$	< 1.1 × 10⁻³		CL=90%	99

Pionic modes

$\pi^+\pi^-$	(1.53±0.09) × 10⁻³			922
$\pi^0\pi^0$	(8.5 ±2.2) × 10⁻⁴			922
$\pi^+\pi^-\pi^0$	(1.6 ±1.1) %	S=2.7		907
$\pi^+\pi^-\pi^-\pi^-$	(7.4 ±0.6) × 10⁻³			879
$\pi^+\pi^+\pi^-\pi^-\pi^0$	(1.9 ±0.4) %			844
$\pi^+\pi^+\pi^+\pi^-\pi^-\pi^-$	(4.0 ±3.0) × 10⁻⁴			795

Hadronic modes with a $K\overline{K}$ pair

K^+K^-	(4.27±0.16) × 10⁻³			791
$K^0\overline{K}^0$	(6.5 ±1.8) × 10⁻⁴			788
$K^0K^-\pi^+$	(6.4 ±1.0) × 10⁻³	S=1.1		739
$\overline{K}^*(892)^0K^0$ × B($\overline{K}^{*0}\to K^-\pi^+$)	< 1.1 × 10⁻³	CL=90%		605
$K^*(892)^+K^-$ × B($K^{*+}\to K^0\pi^+$)	(2.3 ±0.5) × 10⁻³			610
$K^0K^-\pi^+$ nonresonant	(2.3 ±2.3) × 10⁻³			739
$\overline{K}^0K^+\pi^-$	(5.0 ±1.0) × 10⁻³			739
$K^*(892)^0\overline{K}^0$ × B($K^{*0}\to K^+\pi^-$)	< 5 × 10⁻⁴	CL=90%		605
$K^*(892)^-K^+$ × B($K^{*-}\to \overline{K}^0\pi^-$)	(1.2 ±0.7) × 10⁻³			610
$\overline{K}^0K^+\pi^-$ nonresonant	(3.9 $^{+2.3}_{-1.9}$) × 10⁻³			739
$K^+K^-\pi^0$	(1.3 ±0.4) × 10⁻³			742
$K_S^0 K_S^0 \pi^0$	< 5.9 × 10⁻⁴			739
$K^+K^-\pi^+\pi^-$	[vv] (2.52±0.24) × 10⁻³			676
$\phi\pi^+\pi^-$ × B($\phi\to K^+K^-$)	(5.3 ±1.4) × 10⁻⁴			614
$\phi\rho^0$ × B($\phi\to K^+K^-$)	(3.0 ±1.6) × 10⁻⁴			260
$K^+K^-\rho^0$ 3-body	(9.1 ±2.3) × 10⁻⁴			309
$K^*(892)^0K^-\pi^+$+c.c. × B($K^{*0}\to K^+\pi^-$)	[ww] < 5 × 10⁻⁴			528
$K^*(892)^0\overline{K}^*(892)^0$ × B²($K^{*0}\to K^+\pi^-$)	(6 ±2) × 10⁻⁴			257
$K^+K^-\pi^+\pi^-$ non-ϕ	—			676
$K^+K^-\pi^+\pi^-$ nonresonant	< 8 × 10⁻⁴	CL=90%		676
$K^0\overline{K}^0\pi^+\pi^-$	(6.9 ±2.7) × 10⁻³			673
$K^+K^-\pi^+\pi^-\pi^0$	(3.1 ±2.0) × 10⁻³			600

Fractions of most of the following modes with resonances have already appeared above as submodes of particular charged-particle modes.

$\overline{K}^*(892)^0K^0$	< 1.6 × 10⁻³		CL=90%	605
$K^*(892)^+K^-$	(3.5 ±0.8) × 10⁻³			610
$K^*(892)^0\overline{K}^0$	< 8 × 10⁻⁴		CL=90%	605
$K^*(892)^-K^+$	(1.8 ±1.0) × 10⁻³			610
$\phi\pi^0$	< 1.4 × 10⁻³		CL=90%	644
$\phi\eta$	< 2.8 × 10⁻³		CL=90%	489

$\phi\omega$	< 2.1 × 10⁻³		CL=90%	239
$\phi\pi^+\pi^-$	(1.08±0.29) × 10⁻³			614
$\phi\rho^0$	(6 ±3) × 10⁻⁴			260
$\phi\pi^+\pi^-$ 3-body	(7 ±5) × 10⁻⁴			614
$K^*(892)^0K^-\pi^+$ + c.c.	[ww] < 8 × 10⁻⁴		CL=90%	–
$K^*(892)^0\overline{K}^*(892)^0$	(1.4 ±0.5) × 10⁻³			257

**Doubly Cabibbo suppressed (DC) modes,
$\Delta C = 2$ forbidden via mixing (C2M) modes,
$\Delta C = 1$ weak neutral current (C1) modes, or
Lepton Family number (LF) violating modes**

$K^+\ell^-\overline{\nu}_\ell$ (via $\overline{D}^0$)	C2M	< 1.7	× 10⁻⁴	CL=90%	–
$K^+\pi^-$ or $K^+\pi^-\pi^+\pi^-$ (via $\overline{D}^0$)	C2M	< 1.0	× 10⁻³	CL=90%	–
$K^+\pi^-$	DC	(2.8 ±0.9)	× 10⁻⁴		861
$K^+\pi^-$ (via $\overline{D}^0$)		< 1.9	× 10⁻⁴	CL=90%	861
$K^+\pi^-\pi^+\pi^-$	DC	(1.9 ±2.7)	× 10⁻⁴		812
$K^+\pi^-\pi^+\pi^-$ (via $\overline{D}^0$)		< 4	× 10⁻⁴	CL=90%	812
μ^- anything (via $\overline{D}^0$)		< 4	× 10⁻⁴	CL=90%	–
e^+e^-	C1	< 1.3	× 10⁻⁵	CL=90%	932
$\mu^+\mu^-$	C1	< 4.1	× 10⁻⁶	CL=90%	926
$\pi^0 e^+e^-$	C1	< 4.5	× 10⁻⁵	CL=90%	927
$\pi^0\mu^+\mu^-$	C1	< 1.8	× 10⁻⁴	CL=90%	915
ηe^+e^-	C1	< 1.1	× 10⁻⁴	CL=90%	852
$\eta\mu^+\mu^-$	C1	< 5.3	× 10⁻⁴	CL=90%	838
$\rho^0 e^+e^-$	C1	< 1.0	× 10⁻⁴	CL=90%	773
$\rho^0\mu^+\mu^-$	C1	< 2.3	× 10⁻⁴	CL=90%	756
ωe^+e^-	C1	< 1.8	× 10⁻⁴	CL=90%	768
$\omega\mu^+\mu^-$	C1	< 8.3	× 10⁻⁴	CL=90%	751
ϕe^+e^-	C1	< 5.2	× 10⁻⁵	CL=90%	654
$\phi\mu^+\mu^-$	C1	< 4.1	× 10⁻⁴	CL=90%	631
$\overline{K}^0 e^+e^-$	[ss] < 1.1	× 10⁻⁴	CL=90%	866
$\overline{K}^0\mu^+\mu^-$	[ss] < 2.6	× 10⁻⁴	CL=90%	852
$\overline{K}^*(892)^0 e^+e^-$	[ss] < 1.4	× 10⁻⁴	CL=90%	717
$\overline{K}^*(892)^0\mu^+\mu^-$	[ss] < 1.18	× 10⁻³	CL=90%	698
$\pi^+\pi^-\pi^0\mu^+\mu^-$	C1	< 8.1	× 10⁻⁴	CL=90%	863
$\mu^\pm e^\mp$	LF	[gg] < 1.9	× 10⁻⁵	CL=90%	929
$\pi^0 e^\pm\mu^\mp$	LF	[gg] < 8.6	× 10⁻⁵	CL=90%	924
$\eta e^\pm\mu^\mp$	LF	[gg] < 1.0	× 10⁻⁴	CL=90%	848
$\rho^0 e^\pm\mu^\mp$	LF	[gg] < 4.9	× 10⁻⁵	CL=90%	769
$\omega e^\pm\mu^\mp$	LF	[gg] < 1.2	× 10⁻⁴	CL=90%	764
$\phi e^\pm\mu^\mp$	LF	[gg] < 3.4	× 10⁻⁵	CL=90%	648
$\overline{K}^0 e^\pm\mu^\mp$	LF	[gg] < 1.0	× 10⁻⁴	CL=90%	862
$\overline{K}^*(892)^0 e^\pm\mu^\mp$	LF	[gg] < 1.0	× 10⁻⁴	CL=90%	712

$D^*(2007)^0$

$I(J^P) = \frac{1}{2}(1^-)$
I, J, P need confirmation.

Mass $m = 2006.7 \pm 0.5$ MeV (S = 1.1)
$m_{D^{*0}} - m_{D^0} = 142.12 \pm 0.07$ MeV
Full width Γ < 2.1 MeV, CL = 90%

$\overline{D}^*(2007)^0$ modes are charge conjugates of modes below.

$D^*(2007)^0$ DECAY MODES	Fraction (Γ_i/Γ)	p (MeV/c)
$D^0\pi^0$	(61.9±2.9) %	43
$D^0\gamma$	(38.1±2.9) %	137

$D^*(2010)^\pm$

$I(J^P) = \frac{1}{2}(1^-)$
I, J, P need confirmation.

Mass $m = 2010.0 \pm 0.5$ MeV (S = 1.1)
$m_{D^*(2010)^+} - m_{D^+} = 140.64 \pm 0.10$ MeV (S = 1.1)
$m_{D^*(2010)^+} - m_{D^0} = 145.397 \pm 0.030$ MeV
Full width Γ < 0.131 MeV, CL = 90%

$D^*(2010)^-$ modes are charge conjugates of the modes below.

$D^*(2010)^\pm$ DECAY MODES	Fraction (Γ_i/Γ)	p (MeV/c)
$D^0\pi^+$	(68.3±1.4) %	39
$D^+\pi^0$	(30.6±2.5) %	38
$D^+\gamma$	(1.1$^{+2.1}_{-0.7}$) %	136

$$D_1(2420)^0 \qquad I(J^P) = \tfrac{1}{2}(1^+)$$

I, J, P need confirmation.

Mass $m = 2422.2 \pm 1.8$ MeV   (S = 1.2)
Full width $\Gamma = 18.9^{+4.6}_{-3.5}$ MeV

$\overline{D}_1(2420)^0$ modes are charge conjugates of modes below.

$D_1(2420)^0$ DECAY MODES	Fraction (Γ_i/Γ)	p (MeV/c)
$D^*(2010)^+\pi^-$	seen	355
$D^+\pi^-$	not seen	474

$$D_2^*(2460)^0 \qquad I(J^P) = \tfrac{1}{2}(2^+)$$

$J^P = 2^+$ assignment strongly favored (ALBRECHT 89B).

Mass $m = 2458.9 \pm 2.0$ MeV   (S = 1.2)
Full width $\Gamma = 23 \pm 5$ MeV

$\overline{D}_2^*(2460)^0$ modes are charge conjugates of modes below.

$D_2^*(2460)^0$ DECAY MODES	Fraction (Γ_i/Γ)	p (MeV/c)
$D^+\pi^-$	seen	503
$D^*(2010)^+\pi^-$	seen	387

$$D_2^*(2460)^\pm \qquad I(J^P) = \tfrac{1}{2}(2^+)$$

$J^P = 2^+$ assignment strongly favored (ALBRECHT 89B).
Mass $m = 2459 \pm 4$ MeV   (S = 1.7)
$m_{D_2^*(2460)^\pm} - m_{D_2^*(2460)^0} = 0.9 \pm 3.3$ MeV   (S = 1.1)
Full width $\Gamma = 25^{+8}_{-7}$ MeV

$D_2^*(2460)^-$ modes are charge conjugates of modes below.

$D_2^*(2460)^\pm$ DECAY MODES	Fraction (Γ_i/Γ)	p (MeV/c)
$D^0\pi^+$	seen	508
$D^{*0}\pi^+$	seen	390

CHARMED, STRANGE MESONS ($C = S = \pm1$)

$D_s^+ = c\bar{s},\ D_s^- = \bar{c}s$,   similarly for D_s^*'s

$$D_s^\pm \qquad I(J^P) = 0(0^-)$$
was $F^\pm$

Mass $m = 1968.5 \pm 0.6$ MeV   (S = 1.1)
$m_{D_s^\pm} - m_{D^\pm} = 99.2 \pm 0.5$ MeV   (S = 1.1)
Mean life $\tau = (0.467 \pm 0.017) \times 10^{-12}$ s
$c\tau = 140\ \mu$m

D_s^+ form factors
$r_2 = 1.6 \pm 0.4$
$r_v = 1.5 \pm 0.5$
$\Gamma_L/\Gamma_T = 0.72 \pm 0.18$

Branching fractions for modes with a resonance in the final state include all the decay modes of the resonance. D_s^- modes are charge conjugates of the modes below.

D_s^+ DECAY MODES		Fraction (Γ_i/Γ)	Scale factor/ Confidence level	p (MeV/c)
Inclusive modes				
K^- anything		(13^{+14}_{-12}) %		–
$\overline{K}^0$ anything + K^0 anything		(39 ± 28) %		–
K^+ anything		(20^{+18}_{-14}) %		–
non-$K\overline{K}$ anything		(64 ± 17) %		–
e^+ anything		(8^{+6}_{-5}) %		–
ϕ anything		(18^{+15}_{-10}) %		–
Leptonic and semileptonic modes				
$\mu^+\nu_\mu$		$(4.0^{+2.2}_{-2.0}) \times 10^{-3}$	S=1.4	981
$\tau^+\nu_\tau$		(7 ± 4) %		182
$\phi\ell^+\nu_\ell$	[xx]	(2.0 ± 0.5) %		–
$\eta\ell^+\nu_\ell + \eta'(958)\ell^+\nu_\ell$	[xx]	(3.4 ± 1.0) %		–
$\eta\ell^+\nu_\ell$		(2.5 ± 0.7) %		–
$\eta'(958)\ell^+\nu_\ell$		$(8.8 \pm 3.4) \times 10^{-3}$		–
Hadronic modes with a $K\overline{K}$ pair (including from a ϕ)				
$K^+\overline{K}^0$		(3.6 ± 1.1) %		850
$K^+K^-\pi^+$	[qq]	(4.4 ± 1.2) %	S=1.1	805
$\phi\pi^+$	[yy]	(3.6 ± 0.9) %		712
$K^+\overline{K}^*(892)^0$	[yy]	(3.3 ± 0.9) %		682
$f_0(980)\pi^+$	[yy]	(1.8 ± 0.8) %	S=1.3	732
$K^+\overline{K}_0^*(1430)^0$	[yy]	$(7 \pm 4) \times 10^{-3}$		186
$f_J(1710)\pi^+ \to K^+K^-\pi^+$	[zz]	$(1.5 \pm 1.9) \times 10^{-3}$		204
$K^+K^-\pi^+$ nonresonant		$(9 \pm 4) \times 10^{-3}$		805
$K^0\overline{K}^0\pi^+$		—		802
$K^*(892)^+\overline{K}^0$	[yy]	(4.3 ± 1.4) %		683
$K^+K^-\pi^+\pi^0$		—		748
$\phi\pi^+\pi^0$	[yy]	(9 ± 5) %		687
$\phi\rho^+$	[yy]	(6.7 ± 2.3) %		407
$\phi\pi^+\pi^0$ 3-body	[yy]	< 2.6 %	CL=90%	687
$K^+K^-\pi^+\pi^0$ non-ϕ		< 9 %	CL=90%	748
$K^+\overline{K}^0\pi^+\pi^-$		< 2.8 %	CL=90%	744
$K^0 K^-\pi^+\pi^+$		(4.3 ± 1.5) %		744
$K^*(892)^+\overline{K}^*(892)^0$	[yy]	(5.8 ± 2.5) %		412
$K^0 K^-\pi^+\pi^+$ non-$K^{*+}\overline{K}^{*0}$		< 2.9 %	CL=90%	744
$K^+K^-\pi^+\pi^+\pi^-$		$(8.3 \pm 3.3) \times 10^{-3}$		673
$\phi\pi^+\pi^+\pi^-$	[yy]	(1.18 ± 0.35) %		640
$K^+K^-\pi^+\pi^+\pi^-$ non-ϕ		$(3.0^{+3.0}_{-2.0}) \times 10^{-3}$		673
Hadronic modes without K's				
$\pi^+\pi^+\pi^-$		(1.0 ± 0.4) %	S=1.2	959
$\rho^0\pi^+$		$< 8 \times 10^{-4}$	CL=90%	827
$f_0(980)\pi^+$	[yy]	(1.8 ± 0.8) %	S=1.7	732
$f_2(1270)\pi^+$	[yy]	$(2.3 \pm 1.3) \times 10^{-3}$		559
$f_0(1500)\pi^+ \to \pi^+\pi^-\pi^+$	[aaa]	$(2.8 \pm 1.6) \times 10^{-3}$		391
$\pi^+\pi^+\pi^-$ nonresonant		$< 2.8 \times 10^{-3}$	CL=90%	959
$\pi^+\pi^+\pi^-\pi^0$		< 12 %	CL=90%	935
$\eta\pi^+$	[yy]	(2.0 ± 0.6) %		902

$\omega\pi^+$	$[yy]$	$(\ 3.1\ \pm\ 1.4\ )\times10^{-3}$	822
$\pi^+\pi^+\pi^+\pi^-\pi^-$		$(\ 6.9\ \pm\ 3.0\ )\times10^{-3}$	899
$\pi^+\pi^+\pi^-\pi^0\pi^0$		—	902
$\eta\rho^+$	$[yy]$	$(10.3\ \pm\ 3.2\ )\ \%$	727
$\eta\pi^+\pi^0$ 3-body	$[yy]$	$<\ 3.0\ \ \ \ \ \ \%$ CL=90%	886
$\pi^+\pi^+\pi^+\pi^-\pi^-\pi^0$		$(\ 4.9\ \pm\ 3.2\ )\ \%$	856
$\eta'(958)\pi^+$	$[yy]$	$(\ 4.9\ \pm\ 1.8\ )\ \%$	743
$\pi^+\pi^+\pi^+\pi^-\pi^-\pi^0\pi^0$		—	803
$\eta'(958)\rho^+$	$[yy]$	$(12\ \ \ \pm\ 4\ \ )\ \%$	470
$\eta'(958)\pi^+\pi^0$ 3-body	$[yy]$	$<\ 3.1\ \ \ \ \ \ \%$ CL=90%	720

Modes with one or three K's

$K^0\pi^+$		$<\ 8\ \ \ \ \ \ \times10^{-3}$ CL=90%	916
$K^+\pi^+\pi^-$		$(\ 1.0\ \pm\ 0.4\ )\ \%$	900
$K^+\rho^0$		$<\ 2.9\ \ \ \ \ \times10^{-3}$ CL=90%	747
$K^*(892)^0\pi^+$	$[yy]$	$(\ 6.5\ \pm\ 2.8\ )\times10^{-3}$	773
$K^+K^+K^-$		$<\ 6\ \ \ \ \ \ \times10^{-4}$ CL=90%	628
ϕK^+	$[yy]$	$<\ 5\ \ \ \ \ \ \times10^{-4}$ CL=90%	607

$\Delta C = 1$ weak neutral current ($C1$) modes, or Lepton number (L) violating modes

$\pi^+\mu^+\mu^-$	$[ss]$	$<\ 4.3\ \ \times10^{-4}$ CL=90%	968
$K^+\mu^+\mu^-$	$C1$	$<\ 5.9\ \ \times10^{-4}$ CL=90%	909
$K^*(892)^+\mu^+\mu^-$	$C1$	$<\ 1.4\ \ \times10^{-3}$ CL=90%	765
$\pi^-\mu^+\mu^+$	L	$<\ 4.3\ \ \times10^{-4}$ CL=90%	968
$K^-\mu^+\mu^+$	L	$<\ 5.9\ \ \times10^{-4}$ CL=90%	909
$K^*(892)^-\mu^+\mu^+$	L	$<\ 1.4\ \ \times10^{-3}$ CL=90%	765

  $I(J^P) = 0(?^?)$

J^P is natural, width and decay modes consistent with 1^-.

Mass $m = 2112.4 \pm 0.7$ MeV (S = 1.1)
$m_{D_s^{*\pm}} - m_{D_s^\pm} = 143.8 \pm 0.4$ MeV
Full width $\Gamma\ <\ 1.9$ MeV, CL = 90%

D_s^{*-} modes are charge conjugates of the modes below.

D_s^{*+} DECAY MODES	Fraction (Γ_i/Γ)	p (MeV/c)
$D_s^+\gamma$	(94.2 ± 2.5) %	139
$D_s^+\pi^0$	$(\ 5.8\pm2.5)$ %	48

$D_{s1}(2536)^\pm$ $I(J^P) = 0(1^+)$

J, P need confirmation.

Mass $m = 2535.35 \pm 0.34 \pm 0.5$ MeV
Full width $\Gamma\ <\ 2.3$ MeV, CL = 90%

$D_{s1}(2536)^-$ modes are charge conjugates of the modes below.

$D_{s1}(2536)^+$ DECAY MODES	Fraction (Γ_i/Γ)	p (MeV/c)
$D^*(2010)^+K^0$	seen	150
$D^*(2007)^0K^+$	seen	169
D^+K^0	not seen	382
D^0K^+	not seen	392
$D_s^{*+}\gamma$	possibly seen	389

$D_{sJ}(2573)^\pm$ $I(J^P) = 0(?^?)$

J^P is natural, width and decay modes consistent with 2^+.

Mass $m = 2573.5 \pm 1.7$ MeV
Full width $\Gamma = 15^{+5}_{-4}$ MeV

$D_{sJ}(2573)^-$ modes are charge conjugates of the modes below.

$D_{sJ}(2573)^+$ DECAY MODES	Fraction (Γ_i/Γ)	p (MeV/c)
D^0K^+	seen	436
$D^*(2007)^0K^+$	not seen	245

BOTTOM MESONS
($B = \pm1$)

$B^+ = u\bar{b}$, $B^0 = d\bar{b}$, $\overline{B^0} = \bar{d}b$, $B^- = \bar{u}b$, similarly for B^*'s

B-particle organization

Many measurements of B decays involve admixtures of B hadrons. Previously we arbitrarily included such admixtures in the $B^\pm$ section, but because of their importance we have created two new sections: "$B^\pm/B^0$ Admixture" for $\Upsilon(4S)$ results and "$B^\pm/B^0/B_s^0/b$-baryon Admixture" for results at higher energies. Most inclusive decay branching fractions are found in the Admixture sections. B^0-$\overline{B^0}$ mixing data are found in the B^0 section, while B_s^0-$\overline{B_s^0}$ mixing data and B-$\overline{B}$ mixing data for a B^0/B_s^0 admixture are found in the B_s^0 section. CP-violation data are found in the B^0 section. b-baryons are found near the end of the Baryon section.

The organization of the B sections is now as follows, where bullets indicate particle sections and brackets indicate reviews.

[Production and Decay of b-flavored Hadrons]
[Semileptonic Decays of B Mesons]
- $B^\pm$
  - mass, mean life
  - branching fractions
- B^0
  - mass, mean life
  - branching fractions
  - polarization in B^0 decay
  - B^0-$\overline{B^0}$ mixing
  - [B^0-$\overline{B^0}$ Mixing and CP Violation in B Decay]
  - CP violation
- $B^\pm B^0$ Admixtures
  - branching fractions
- $B^\pm/B^0/B_s^0/b$-baryon Admixtures
  - mean life
  - production fractions
  - branching fractions
- B^*
  - mass
- B_s^0
  - mass, mean life
  - branching fractions
  - polarization in B_s^0 decay
  - B_s^0-$\overline{B_s^0}$ mixing
  - B-$\overline{B}$ mixing (admixture of B^0, B_s^0)

At end of Baryon Listings:
- Λ_b
  - mass, mean life
  - branching fractions
- b-baryon Admixture
  - mean life
  - branching fractions

$B^\pm$

$$I(J^P) = \tfrac{1}{2}(0^-)$$

I, J, P need confirmation. Quantum numbers shown are quark-model predictions.

Mass $m_{B^\pm} = 5278.9 \pm 1.8$ MeV
Mean life $\tau_{B^\pm} = (1.65 \pm 0.04) \times 10^{-12}$ s
$c\tau = 495$ μm

B^- modes are charge conjugates of the modes below. Modes which do not identify the charge state of the B are listed in the $B^\pm/B^0$ ADMIXTURE section.

The branching fractions listed below assume 50% $B^0 \overline{B}^0$ and 50% $B^+ B^-$ production at the $\Upsilon(4S)$. We have attempted to bring older measurements up to date by rescaling their assumed $\Upsilon(4S)$ production ratio to 50:50 and their assumed D, D_s, D^*, and ψ branching ratios to current values whenever this would affect our averages and best limits significantly.

Indentation is used to indicate a subchannel of a previous reaction. All resonant subchannels have been corrected for resonance branching fractions to the final state so the sum of the subchannel branching fractions can exceed that of the final state.

B^+ DECAY MODES	Fraction (Γ_i/Γ)	Scale factor/ Confidence level	p (MeV/c)
Semileptonic and leptonic modes			
$\ell^+ \nu_\ell$ anything	[pp] (10.3 ± 0.9) %		—
$\overline{D}^0 \ell^+ \nu_\ell$	[pp] (1.86 ± 0.33) %		—
$\overline{D}^*(2007)^0 \ell^+ \nu_\ell$	[pp] (5.3 ± 0.8) %		—
$\pi^0 e^+ \nu_e$	$< 2.2 \times 10^{-3}$	CL=90%	2638
$\omega \ell^+ \nu_\ell$	[pp] $< 2.1 \times 10^{-4}$	CL=90%	—
$\rho^0 \ell^+ \nu_\ell$	[pp] $< 2.1 \times 10^{-4}$	CL=90%	—
$e^+ \nu_e$	$< 1.5 \times 10^{-5}$	CL=90%	2639
$\mu^+ \nu_\mu$	$< 2.1 \times 10^{-5}$	CL=90%	2638
$\tau^+ \nu_\tau$	$< 5.7 \times 10^{-4}$	CL=90%	2340
$e^+ \nu_e \gamma$	$< 2.0 \times 10^{-4}$	CL=90%	—
$\mu^+ \nu_\mu \gamma$	$< 5.2 \times 10^{-5}$	CL=90%	—
D, D^*, or D_s modes			
$\overline{D}^0 \pi^+$	$(5.3 \pm 0.5) \times 10^{-3}$		2308
$\overline{D}^0 \rho^+$	(1.34 ± 0.18) %		2238
$\overline{D}^0 \pi^+ \pi^+ \pi^-$	(1.1 ± 0.4) %		2289
$\overline{D}^0 \pi^+ \pi^+ \pi^-$ nonresonant	$(5 \pm 4) \times 10^{-3}$		2289
$\overline{D}^0 \pi^+ \rho^0$	$(4.2 \pm 3.0) \times 10^{-3}$		2209
$\overline{D}^0 a_1(1260)^+$	$(5 \pm 4) \times 10^{-3}$		2123
$D^*(2010)^- \pi^+ \pi^+$	$(2.1 \pm 0.6) \times 10^{-3}$		2247
$D^- \pi^+ \pi^+$	$< 1.4 \times 10^{-3}$	CL=90%	2299
$\overline{D}^*(2007)^0 \pi^+$	$(4.6 \pm 0.4) \times 10^{-3}$		2256
$D^*(2010)^+ \pi^0$	$< 1.7 \times 10^{-4}$	CL=90%	2254
$\overline{D}^*(2007)^0 \rho^+$	(1.55 ± 0.31) %		2183
$\overline{D}^*(2007)^0 \pi^+ \pi^+ \pi^-$	$(9.4 \pm 2.6) \times 10^{-3}$		2236
$\overline{D}^*(2007)^0 a_1(1260)^+$	(1.9 ± 0.5) %		2062
$D^*(2010)^- \pi^+ \pi^+ \pi^0$	(1.5 ± 0.7) %		2235
$D^*(2010)^- \pi^+ \pi^+ \pi^+ \pi^-$	< 1 %	CL=90%	2217
$\overline{D}_1(2420)^0 \pi^+$	$(1.5 \pm 0.6) \times 10^{-3}$	S=1.3	2081
$\overline{D}_1(2420)^0 \rho^+$	$< 1.4 \times 10^{-3}$		1997
$\overline{D}_2^*(2460)^0 \pi^+$	$< 1.3 \times 10^{-3}$	CL=90%	2064
$\overline{D}_2^*(2460)^0 \rho^+$	$< 4.7 \times 10^{-3}$	CL=90%	1979
$\overline{D}^0 D_s^+$	(1.3 ± 0.4) %		1815
$\overline{D}^0 D_s^{*+}$	$(9 \pm 4) \times 10^{-3}$		1734
$\overline{D}^*(2007)^0 D_s^+$	(1.2 ± 0.5) %		1737
$\overline{D}^*(2007)^0 D_s^{*+}$	(2.7 ± 1.0) %		1650
$D_s^+ \pi^0$	$< 2.0 \times 10^{-4}$	CL=90%	2270
$D_s^{*+} \pi^0$	$< 3.3 \times 10^{-4}$	CL=90%	2214
$D_s^+ \eta$	$< 5 \times 10^{-4}$	CL=90%	2235
$D_s^{*+} \eta$	$< 8 \times 10^{-4}$	CL=90%	2177
$D_s^+ \rho^0$	$< 4 \times 10^{-4}$	CL=90%	2198
$D_s^{*+} \rho^0$	$< 5 \times 10^{-4}$	CL=90%	2139
$D_s^+ \omega$	$< 5 \times 10^{-4}$	CL=90%	2195
$D_s^{*+} \omega$	$< 7 \times 10^{-4}$	CL=90%	2136
$D_s^+ a_1(1260)^0$	$< 2.2 \times 10^{-3}$	CL=90%	2079
$D_s^{*+} a_1(1260)^0$	$< 1.6 \times 10^{-3}$	CL=90%	2014
$D_s^+ \phi$	$< 3.2 \times 10^{-4}$	CL=90%	2141
$D_s^{*+} \phi$	$< 4 \times 10^{-4}$	CL=90%	2079

B^+ DECAY MODES	Fraction (Γ_i/Γ)	Scale factor/ Confidence level	p (MeV/c)
$D_s^+ \overline{K}^0$	$< 1.1 \times 10^{-3}$	CL=90%	2241
$D_s^{*+} \overline{K}^0$	$< 1.1 \times 10^{-3}$	CL=90%	2184
$D_s^+ \overline{K}^*(892)^0$	$< 5 \times 10^{-4}$	CL=90%	2171
$D_s^{*+} \overline{K}^*(892)^0$	$< 4 \times 10^{-4}$	CL=90%	2110
$D_s^- \pi^+ K^+$	$< 8 \times 10^{-4}$	CL=90%	2222
$D_s^{*-} \pi^+ K^+$	$< 1.2 \times 10^{-3}$	CL=90%	2164
$D_s^- \pi^+ K^*(892)^+$	$< 6 \times 10^{-3}$	CL=90%	2137
$D_s^{*-} \pi^+ K^*(892)^+$	$< 8 \times 10^{-3}$	CL=90%	2075
Charmonium modes			
$J/\psi(1S) K^+$	$(9.9 \pm 1.0) \times 10^{-4}$		1683
$J/\psi(1S) K^+ \pi^+ \pi^-$	$(1.4 \pm 0.6) \times 10^{-3}$		1612
$J/\psi(1S) K^*(892)^+$	$(1.47 \pm 0.27) \times 10^{-3}$		1571
$J/\psi(1S) \pi^+$	$(5.0 \pm 1.5) \times 10^{-5}$		1727
$J/\psi(1S) \rho^+$	$< 7.7 \times 10^{-4}$	CL=90%	1613
$J/\psi(1S) a_1(1260)^+$	$< 1.2 \times 10^{-3}$	CL=90%	1414
$\psi(2S) K^+$	$(6.9 \pm 3.1) \times 10^{-4}$	S=1.3	1284
$\psi(2S) K^*(892)^+$	$< 3.0 \times 10^{-3}$	CL=90%	1115
$\psi(2S) K^+ \pi^+ \pi^-$	$(1.9 \pm 1.2) \times 10^{-3}$		909
$\chi_{c1}(1P) K^+$	$(1.0 \pm 0.4) \times 10^{-3}$		1411
$\chi_{c1}(1P) K^*(892)^+$	$< 2.1 \times 10^{-3}$	CL=90%	1265
K or K^* modes			
$K^0 \pi^+$	$(2.3 \pm 1.1) \times 10^{-5}$		2614
$K^+ \pi^0$	$< 1.6 \times 10^{-5}$	CL=90%	2615
$\eta' K^+$	$(6.5 \pm 1.7) \times 10^{-5}$		2528
$\eta' K^*(892)^+$	$< 1.3 \times 10^{-4}$	CL=90%	2472
ηK^+	$< 1.4 \times 10^{-5}$	CL=90%	2587
$\eta K^*(892)^+$	$< 3.0 \times 10^{-5}$	CL=90%	2534
$K^*(892)^0 \pi^+$	$< 4.1 \times 10^{-5}$	CL=90%	2561
$K^*(892)^+ \pi^0$	$< 9.9 \times 10^{-5}$	CL=90%	2562
$K^+ \pi^- \pi^+$ nonresonant	$< 2.8 \times 10^{-5}$	CL=90%	2609
$K^- \pi^+ \pi^+$ nonresonant	$< 5.6 \times 10^{-5}$	CL=90%	—
$K_1(1400)^0 \pi^+$	$< 2.6 \times 10^{-3}$	CL=90%	2451
$K_2^*(1430)^0 \pi^+$	$< 6.8 \times 10^{-4}$	CL=90%	2443
$K^+ \rho^0$	$< 1.9 \times 10^{-5}$	CL=90%	2559
$K^0 \rho^+$	$< 4.8 \times 10^{-5}$	CL=90%	2559
$K^*(892)^+ \pi^+ \pi^-$	$< 1.1 \times 10^{-3}$	CL=90%	2556
$K^*(892)^+ \rho^0$	$< 9.0 \times 10^{-4}$	CL=90%	2505
$K_1(1400)^+ \rho^0$	$< 7.8 \times 10^{-3}$	CL=90%	2389
$K_2^*(1430)^+ \rho^0$	$< 1.5 \times 10^{-3}$	CL=90%	2382
$K^+ \overline{K}^0$	$< 2.1 \times 10^{-5}$	CL=90%	2592
$K^+ K^- \pi^+$ nonresonant	$< 7.5 \times 10^{-5}$	CL=90%	—
$K^+ K^- K^+$	$< 2.0 \times 10^{-5}$	CL=90%	2522
$K^+ \phi$	$< 1.2 \times 10^{-5}$	CL=90%	2516
$K^+ K^- K^+$ nonresonant	$< 3.8 \times 10^{-5}$	CL=90%	2516
$K^*(892)^+ K^+ K^-$	$< 1.6 \times 10^{-3}$	CL=90%	2466
$K^*(892)^+ \phi$	$< 7.0 \times 10^{-5}$	CL=90%	2460
$K_1(1400)^+ \phi$	$< 1.1 \times 10^{-3}$	CL=90%	2339
$K_2^*(1430)^+ \phi$	$< 3.4 \times 10^{-3}$	CL=90%	2332
$K^+ f_0(980)$	$< 8 \times 10^{-5}$	CL=90%	2524
$K^*(892)^+ \gamma$	$(5.7 \pm 3.3) \times 10^{-5}$		2564
$K_1(1270)^+ \gamma$	$< 7.3 \times 10^{-3}$	CL=90%	2486
$K_1(1400)^+ \gamma$	$< 2.2 \times 10^{-3}$	CL=90%	2453
$K_2^*(1430)^+ \gamma$	$< 1.4 \times 10^{-3}$	CL=90%	2447
$K^*(1680)^+ \gamma$	$< 1.9 \times 10^{-3}$	CL=90%	2361
$K_3^*(1780)^+ \gamma$	$< 5.5 \times 10^{-3}$	CL=90%	2343
$K_4^*(2045)^+ \gamma$	$< 9.9 \times 10^{-3}$	CL=90%	2243
Light unflavored meson modes			
$\pi^+ \pi^0$	$< 2.0 \times 10^{-5}$	CL=90%	2636
$\pi^+ \pi^+ \pi^-$	$< 1.3 \times 10^{-4}$	CL=90%	2630
$\rho^0 \pi^+$	$< 4.3 \times 10^{-5}$	CL=90%	2582
$\pi^+ f_0(980)$	$< 1.4 \times 10^{-4}$	CL=90%	2547
$\pi^+ f_2(1270)$	$< 2.4 \times 10^{-4}$	CL=90%	2483
$\pi^+ \pi^- \pi^+$ nonresonant	$< 4.1 \times 10^{-5}$	CL=90%	—
$\pi^+ \pi^0 \pi^0$	$< 8.9 \times 10^{-4}$	CL=90%	2631
$\rho^+ \pi^0$	$< 7.7 \times 10^{-5}$	CL=90%	2582
$\pi^+ \pi^- \pi^+ \pi^0$	$< 4.0 \times 10^{-3}$	CL=90%	2621
$\rho^+ \rho^0$	$< 1.0 \times 10^{-3}$	CL=90%	2525
$a_1(1260)^+ \pi^0$	$< 1.7 \times 10^{-3}$	CL=90%	2494
$a_1(1260)^0 \pi^+$	$< 9.0 \times 10^{-4}$	CL=90%	2494
$\omega \pi^+$	$< 4.0 \times 10^{-4}$	CL=90%	2580

$\eta\pi^+$	< 1.5	$\times 10^{-5}$	CL=90%	2609
$\eta'\pi^+$	< 3.1	$\times 10^{-5}$	CL=90%	2550
$\eta'\rho^+$	< 4.7	$\times 10^{-5}$	CL=90%	2493
$\eta\rho^+$	< 3.2	$\times 10^{-5}$	CL=90%	2554
$\pi^+\pi^+\pi^+\pi^-\pi^-$	< 8.6	$\times 10^{-4}$	CL=90%	2608
$\rho^0 a_1(1260)^+$	< 6.2	$\times 10^{-4}$	CL=90%	2434
$\rho^0 a_2(1320)^+$	< 7.2	$\times 10^{-4}$	CL=90%	2411
$\pi^+\pi^+\pi^+\pi^-\pi^-\pi^0$	< 6.3	$\times 10^{-3}$	CL=90%	2592
$a_1(1260)^+ a_1(1260)^0$	< 1.3	%	CL=90%	2335

Baryon modes

$p\bar{p}\pi^+$	< 1.6	$\times 10^{-4}$	CL=90%	2439
$p\bar{p}\pi^+$ nonresonant	< 5.3	$\times 10^{-5}$	CL=90%	–
$p\bar{p}\pi^+\pi^+\pi^-$	< 5.2	$\times 10^{-4}$	CL=90%	2369
$p\bar{p}K^+$ nonresonant	< 8.9	$\times 10^{-5}$	CL=90%	–
$p\bar{\Lambda}$	< 6	$\times 10^{-5}$	CL=90%	2430
$p\bar{\Lambda}\pi^+\pi^-$	< 2.0	$\times 10^{-4}$	CL=90%	2367
$\bar{\Delta}^0 p$	< 3.8	$\times 10^{-4}$	CL=90%	2402
$\Delta^{++}\bar{p}$	< 1.5	$\times 10^{-4}$	CL=90%	2402
$\Lambda_c^- p\pi^+$	(6.2 ±2.7) $\times 10^{-4}$			–
$\Lambda_c^- p\pi^+\pi^0$	< 3.12	$\times 10^{-3}$	CL=90%	–
$\Lambda_c^- p\pi^+\pi^+\pi^-$	< 1.46	$\times 10^{-3}$	CL=90%	–
$\Lambda_c^- p\pi^+\pi^+\pi^-\pi^0$	< 1.34	%	CL=90%	–

Lepton Family number (LF) or Lepton number (L) violating modes, or $\Delta B = 1$ weak neutral current (B1) modes

$\pi^+ e^+ e^-$	B1	< 3.9	$\times 10^{-3}$	CL=90%	2638
$\pi^+ \mu^+ \mu^-$	B1	< 9.1	$\times 10^{-3}$	CL=90%	2633
$K^+ e^+ e^-$	B1	< 6	$\times 10^{-5}$	CL=90%	2616
$K^+ \mu^+ \mu^-$	B1	< 1.0	$\times 10^{-5}$	CL=90%	2612
$K^*(892)^+ e^+ e^-$	B1	< 6.9	$\times 10^{-4}$	CL=90%	2564
$K^*(892)^+ \mu^+ \mu^-$	B1	< 1.2	$\times 10^{-3}$	CL=90%	2560
$\pi^+ e^+ \mu^-$	LF	< 6.4	$\times 10^{-3}$	CL=90%	2637
$\pi^+ e^- \mu^+$	LF	< 6.4	$\times 10^{-3}$	CL=90%	2637
$K^+ e^+ \mu^-$	LF	< 6.4	$\times 10^{-3}$	CL=90%	2615
$K^+ e^- \mu^+$	LF	< 6.4	$\times 10^{-3}$	CL=90%	2615
$\pi^- e^+ e^+$	L	< 3.9	$\times 10^{-3}$	CL=90%	2638
$\pi^- \mu^+ \mu^+$	L	< 9.1	$\times 10^{-3}$	CL=90%	2633
$\pi^- e^+ \mu^+$	LF	< 6.4	$\times 10^{-3}$	CL=90%	2637
$K^- e^+ e^+$	L	< 3.9	$\times 10^{-3}$	CL=90%	2616
$K^- \mu^+ \mu^+$	L	< 9.1	$\times 10^{-3}$	CL=90%	2612
$K^- e^+ \mu^+$	LF	< 6.4	$\times 10^{-3}$	CL=90%	2615

$$\boxed{B^0} \qquad I(J^P) = \tfrac{1}{2}(0^-)$$

I, J, P need confirmation. Quantum numbers shown are quark-model predictions.

Mass $m_{B^0} = 5279.2 \pm 1.8$ MeV
$m_{B^0} - m_{B^\pm} = 0.35 \pm 0.29$ MeV   (S = 1.1)
Mean life $\tau_{B^0} = (1.56 \pm 0.04) \times 10^{-12}$ s
$\qquad c\tau = 468\ \mu$m
$\tau_{B^+}/\tau_{B^0} = 1.02 \pm 0.04$ (average of direct and inferred)
$\tau_{B^+}/\tau_{B^0} = 1.04 \pm 0.04$ (direct measurements)
$\tau_{B^+}/\tau_{B^0} = 0.95^{+0.15}_{-0.12}$ (inferred from branching fractions)

B^0-$\bar{B}^0$ mixing parameters

$\chi_d = 0.172 \pm 0.010$
$\Delta m_{B^0} = m_{B^0_H} - m_{B^0_L} = (0.464 \pm 0.018) \times 10^{12}\ \hbar\ s^{-1}$
$x_d = \Delta m_{B^0}/\Gamma_{B^0} = 0.723 \pm 0.032$

CP violation parameters

$|\mathrm{Re}(\epsilon_{B^0})| = 0.002 \pm 0.008$

$\bar{B}^0$ modes are charge conjugates of the modes below. Reactions indicate the weak decay vertex and do not include mixing. Modes which do not identify the charge state of the B are listed in the $B^\pm/B^0$ ADMIXTURE section.

The branching fractions listed below assume 50% $B^0\bar{B}^0$ and 50% $B^+ B^-$ production at the $\Upsilon(4S)$. We have attempted to bring older measurements up to date by rescaling their assumed $\Upsilon(4S)$ production ratio to 50:50 and their assumed D, D_s, D^*, and ψ branching ratios to current values whenever this would affect our averages and best limits significantly.

Indentation is used to indicate a subchannel of a previous reaction. All resonant subchannels have been corrected for resonance branching fractions to the final state so the sum of the subchannel branching fractions can exceed that of the final state.

B^0 DECAY MODES	Fraction (Γ_i/Γ)	Scale factor/ Confidence level	p (MeV/c)
$\ell^+ \nu_\ell$ anything	[pp] (10.5 ± 0.8) %		–
$D^- \ell^+ \nu_\ell$	[pp] (2.00± 0.25) %		–
$D^*(2010)^- \ell^+ \nu_\ell$	[pp] (4.60± 0.27) %		–
$\rho^- \ell^+ \nu_\ell$	[pp] (2.5 $^{+0.8}_{-1.0}$) $\times 10^{-4}$		–
$\pi^- \ell^+ \nu_\ell$	(1.8 ± 0.6) $\times 10^{-4}$		–

Inclusive modes

K^+ anything	(78 ±80) %		

D, D^*, or D_s modes

$D^-\pi^+$	(3.0 ± 0.4) $\times 10^{-3}$		2306
$D^-\rho^+$	(7.9 ± 1.4) $\times 10^{-3}$		2236
$\bar{D}^0\pi^+\pi^-$	< 1.6 $\times 10^{-3}$	CL=90%	2301
$D^*(2010)^-\pi^+$	(2.76± 0.21) $\times 10^{-3}$		2254
$D^-\pi^+\pi^+\pi^-$	(8.0 ± 2.5) $\times 10^{-3}$		2287
($D^-\pi^+\pi^+\pi^-$) nonresonant	(3.9 ± 1.9) $\times 10^{-3}$		2287
$D^-\pi^+\rho^0$	(1.1 ± 1.0) $\times 10^{-3}$		2207
$D^- a_1(1260)^+$	(6.0 ± 3.3) $\times 10^{-3}$		2121
$D^*(2010)^-\pi^+\pi^0$	(1.5 ± 0.5) %		2247
$D^*(2010)^-\rho^+$	(6.7 ± 3.3) $\times 10^{-3}$		2181
$D^*(2010)^-\pi^+\pi^+\pi^-$	(7.6 ± 1.7) $\times 10^{-3}$	S=1.3	2235
($D^*(2010)^-\pi^+\pi^+\pi^-$) non-resonant	(0.0 ± 2.5) $\times 10^{-3}$		2235
$D^*(2010)^-\pi^+\rho^0$	(5.7 ± 3.1) $\times 10^{-3}$		2151
$D^*(2010)^- a_1(1260)^+$	(1.30± 0.27) %		2061
$D^*(2010)^-\pi^+\pi^+\pi^-\pi^0$	(3.4 ± 1.8) %		2218
$\bar{D}^*_2(2460)^-\pi^+$	< 2.2 $\times 10^{-3}$	CL=90%	2064
$\bar{D}^*_2(2460)^-\rho^+$	< 4.9 $\times 10^{-3}$	CL=90%	1979
$D^- D_s^+$	(8.0 ± 3.0) $\times 10^{-3}$		1812
$D^*(2010)^- D_s^+$	(9.6 ± 3.4) $\times 10^{-3}$		1735
$D^- D_s^{*+}$	(1.0 ± 0.5) %		1731
$D^*(2010)^- D_s^{*+}$	(2.0 ± 0.7) %		1649
$D_s^+\pi^-$	< 2.8 $\times 10^{-4}$	CL=90%	2270
$D_s^{*+}\pi^-$	< 5 $\times 10^{-4}$	CL=90%	2214
$D_s^+\rho^-$	< 7 $\times 10^{-4}$	CL=90%	2198
$D_s^{*+}\rho^-$	< 8 $\times 10^{-4}$	CL=90%	2139
$D_s^+ a_1(1260)^-$	< 2.6 $\times 10^{-3}$	CL=90%	2079
$D_s^{*+} a_1(1260)^-$	< 2.2 $\times 10^{-3}$	CL=90%	2014
$D_s^- K^+$	< 2.4 $\times 10^{-4}$	CL=90%	2242
$D_s^{*-} K^+$	< 1.7 $\times 10^{-4}$	CL=90%	2185
$D_s^- K^*(892)^+$	< 9.9 $\times 10^{-4}$	CL=90%	2172
$D_s^{*-} K^*(892)^+$	< 1.1 $\times 10^{-3}$	CL=90%	2112
$D_s^-\pi^+ K^0$	< 5 $\times 10^{-3}$	CL=90%	2221
$D_s^{*-}\pi^+ K^0$	< 3.1 $\times 10^{-3}$	CL=90%	2164
$D_s^-\pi^+ K^*(892)^0$	< 4 $\times 10^{-3}$	CL=90%	2136
$D_s^{*-}\pi^+ K^*(892)^0$	< 2.0 $\times 10^{-3}$	CL=90%	2074
$\bar{D}^0\pi^0$	< 1.2 $\times 10^{-4}$	CL=90%	2308
$\bar{D}^0\rho^0$	< 3.9 $\times 10^{-4}$	CL=90%	2238
$\bar{D}^0\eta$	< 1.3 $\times 10^{-4}$	CL=90%	2274
$\bar{D}^0\eta'$	< 9.4 $\times 10^{-4}$	CL=90%	2198
$\bar{D}^0\omega$	< 5.1 $\times 10^{-4}$	CL=90%	2235
$\bar{D}^*(2007)^0\pi^0$	< 4.4 $\times 10^{-4}$	CL=90%	2256
$\bar{D}^*(2007)^0\rho^0$	< 5.6 $\times 10^{-4}$	CL=90%	2183
$\bar{D}^*(2007)^0\eta$	< 2.6 $\times 10^{-4}$	CL=90%	2220
$\bar{D}^*(2007)^0\eta'$	< 1.4 $\times 10^{-4}$	CL=90%	2141
$\bar{D}^*(2007)^0\omega$	< 7.4 $\times 10^{-4}$	CL=90%	2180
$D^*(2010)^+ D^*(2010)^-$	< 2.2 $\times 10^{-3}$	CL=90%	1711
$D^*(2010)^+ D^-$	< 1.8 $\times 10^{-3}$	CL=90%	1790
$D^+ D^*(2010)^-$	< 1.2 $\times 10^{-3}$	CL=90%	1790

Charmonium modes

$J/\psi(1S) K^0$	(8.9 ± 1.2) $\times 10^{-4}$		1683
$J/\psi(1S) K^+\pi^-$	(1.1 ± 0.6) $\times 10^{-3}$		1652
$J/\psi(1S) K^*(892)^0$	(1.35± 0.18) $\times 10^{-3}$		1570
$J/\psi(1S) \pi^0$	< 5.8 $\times 10^{-5}$	CL=90%	1728
$J/\psi(1S) \eta$	< 1.2 $\times 10^{-3}$	CL=90%	1672
$J/\psi(1S) \rho^0$	< 2.5 $\times 10^{-4}$	CL=90%	1614

$J/\psi(1S)\omega$	< 2.7	$\times 10^{-4}$	CL=90%	1609
$\psi(2S)K^0$	< 8	$\times 10^{-4}$	CL=90%	1283
$\psi(2S)K^+\pi^-$	< 1	$\times 10^{-3}$	CL=90%	1238
$\psi(2S)K^*(892)^0$	(1.4 ± 0.9) $\times 10^{-3}$			1113
$\chi_{c1}(1P)K^0$	< 2.7	$\times 10^{-3}$	CL=90%	1411
$\chi_{c1}(1P)K^*(892)^0$	< 2.1	$\times 10^{-3}$	CL=90%	1263

K or K^* modes

$K^+\pi^-$	(1.5 $^{+\,0.5}_{-\,0.4}$) $\times 10^{-5}$			2615
$K^0\pi^0$	< 4.1	$\times 10^{-5}$	CL=90%	2614
$\eta' K^0$	(4.7 $^{+\,2.8}_{-\,2.2}$) $\times 10^{-5}$			2528
$\eta' K^*(892)^0$	< 3.9	$\times 10^{-5}$	CL=90%	2472
$\eta K^*(892)^0$	< 3.0	$\times 10^{-5}$	CL=90%	2534
ηK^0	< 3.3	$\times 10^{-5}$	CL=90%	2593
K^+K^-	< 4.3	$\times 10^{-6}$	CL=90%	2593
$K^0\overline{K^0}$	< 1.7	$\times 10^{-5}$	CL=90%	2592
$K^+\rho^-$	< 3.5	$\times 10^{-5}$	CL=90%	2559
$K^0\rho^0$	< 3.9	$\times 10^{-5}$	CL=90%	2559
$K^0 f_0(980)$	< 3.6	$\times 10^{-4}$	CL=90%	2523
$K^*(892)^+\pi^-$	< 7.2	$\times 10^{-5}$	CL=90%	2562
$K^*(892)^0\pi^0$	< 2.8	$\times 10^{-5}$	CL=90%	2562
$K_2^*(1430)^+\pi^-$	< 2.6	$\times 10^{-3}$	CL=90%	2445
$K^0K^+K^-$	< 1.3	$\times 10^{-3}$	CL=90%	2522
$K^0\phi$	< 8.8	$\times 10^{-5}$	CL=90%	2516
$K^-\pi^+\pi^+\pi^-$	[bbb] < 2.3	$\times 10^{-4}$	CL=90%	2600
$K^*(892)^0\pi^+\pi^-$	< 1.4	$\times 10^{-3}$	CL=90%	2556
$K^*(892)^0\rho^0$	< 4.6	$\times 10^{-4}$	CL=90%	2504
$K^*(892)^0 f_0(980)$	< 1.7	$\times 10^{-4}$	CL=90%	2467
$K_1(1400)^+\pi^-$	< 1.1	$\times 10^{-3}$	CL=90%	2451
$K^- a_1(1260)^+$	[bbb] < 2.3	$\times 10^{-4}$	CL=90%	2471
$K^*(892)^0K^+K^-$	< 6.1	$\times 10^{-4}$	CL=90%	2466
$K^*(892)^0\phi$	< 4.3	$\times 10^{-5}$	CL=90%	2459
$K_1(1400)^0\rho^0$	< 3.0	$\times 10^{-3}$	CL=90%	2389
$K_1(1400)^0\phi$	< 5.0	$\times 10^{-3}$	CL=90%	2339
$K_2^*(1430)^0\rho^0$	< 1.1	$\times 10^{-3}$	CL=90%	2380
$K_2^*(1430)^0\phi$	< 1.4	$\times 10^{-3}$	CL=90%	2330
$K^*(892)^0\gamma$	(4.0 ± 1.9) $\times 10^{-5}$			2564
$K_1(1270)^0\gamma$	< 7.0	$\times 10^{-3}$	CL=90%	2486
$K_1(1400)^0\gamma$	< 4.3	$\times 10^{-3}$	CL=90%	2453
$K_2^*(1430)^0\gamma$	< 4.0	$\times 10^{-4}$	CL=90%	2445
$K^*(1680)^0\gamma$	< 2.0	$\times 10^{-3}$	CL=90%	2361
$K_3^*(1780)^0\gamma$	< 1.0	%	CL=90%	2343
$K_4^*(2045)^0\gamma$	< 4.3	$\times 10^{-3}$	CL=90%	2244
$\phi\phi$	< 3.9	$\times 10^{-5}$	CL=90%	2435

Light unflavored meson modes

$\pi^+\pi^-$	< 1.5	$\times 10^{-5}$	CL=90%	2636
$\pi^0\pi^0$	< 9.3	$\times 10^{-6}$	CL=90%	2636
$\eta\pi^0$	< 8	$\times 10^{-6}$	CL=90%	2609
$\eta\eta$	< 1.8	$\times 10^{-5}$	CL=90%	2582
$\eta'\pi^0$	< 1.1	$\times 10^{-5}$	CL=90%	2551
$\eta'\eta'$	< 4.7	$\times 10^{-5}$	CL=90%	2460
$\eta'\eta$	< 2.7	$\times 10^{-5}$	CL=90%	2522
$\eta'\rho^0$	< 2.3	$\times 10^{-5}$	CL=90%	2493
$\eta\rho^0$	< 1.3	$\times 10^{-5}$	CL=90%	2554
$\pi^+\pi^-\pi^0$	< 7.2	$\times 10^{-4}$	CL=90%	2631
$\rho^0\pi^0$	< 2.4	$\times 10^{-5}$	CL=90%	2582
$\rho^\mp\pi^\pm$	[gg] < 8.8	$\times 10^{-5}$	CL=90%	2582
$\pi^+\pi^-\pi^+\pi^-$	< 2.3	$\times 10^{-4}$	CL=90%	2621
$\rho^0\rho^0$	< 2.8	$\times 10^{-4}$	CL=90%	2525
$a_1(1260)^\mp\pi^\pm$	[gg] < 4.9	$\times 10^{-4}$	CL=90%	2494
$a_2(1320)^\mp\pi^\pm$	[gg] < 3.0	$\times 10^{-4}$	CL=90%	2473
$\pi^+\pi^-\pi^0\pi^0$	< 3.1	$\times 10^{-3}$	CL=90%	2622
$\rho^+\rho^-$	< 2.2	$\times 10^{-3}$	CL=90%	2525
$a_1(1260)^0\pi^0$	< 1.1	$\times 10^{-3}$	CL=90%	2494
$\omega\pi^0$	< 4.6	$\times 10^{-4}$	CL=90%	2580
$\pi^+\pi^+\pi^-\pi^-\pi^0$	< 9.0	$\times 10^{-3}$	CL=90%	2609
$a_1(1260)^+\rho^-$	< 3.4	$\times 10^{-3}$	CL=90%	2434
$a_1(1260)^0\rho^0$	< 2.4	$\times 10^{-3}$	CL=90%	2434
$\pi^+\pi^+\pi^+\pi^-\pi^-\pi^-$	< 3.0	$\times 10^{-3}$	CL=90%	2592
$a_1(1260)^+ a_1(1260)^-$	< 2.8	$\times 10^{-3}$	CL=90%	2336
$\pi^+\pi^+\pi^+\pi^-\pi^-\pi^-\pi^0$	< 1.1	%	CL=90%	2572

Baryon modes

$p\overline{p}$	< 1.8	$\times 10^{-5}$	CL=90%	2467
$p\overline{p}\pi^+\pi^-$	< 2.5	$\times 10^{-4}$	CL=90%	2406
$p\overline{\Lambda}\pi^-$	< 1.8	$\times 10^{-4}$	CL=90%	2401
$\Delta^0\overline{\Delta^0}$	< 1.5	$\times 10^{-3}$	CL=90%	2334
$\Delta^{++}\Delta^{--}$	< 1.1	$\times 10^{-4}$	CL=90%	2334
$\overline{\Sigma}_c^{--}\Delta^{++}$	< 1.0	$\times 10^{-3}$	CL=90%	1839
$\Lambda_c^- p\pi^+\pi^-$	(1.3 ± 0.6) $\times 10^{-3}$			–
$\Lambda_c^- p$	< 2.1	$\times 10^{-4}$	CL=90%	2021
$\Lambda_c^- p\pi^0$	< 5.9	$\times 10^{-4}$	CL=90%	–
$\Lambda_c^- p\pi^+\pi^-\pi^0$	< 5.07	$\times 10^{-3}$	CL=90%	–
$\Lambda_c^- p\pi^+\pi^-\pi^+\pi^-$	< 2.74	$\times 10^{-3}$	CL=90%	–

Lepton Family number (LF) violating modes, or $\Delta B = 1$ weak neutral current (B1) modes

$\gamma\gamma$	B1	< 3.9	$\times 10^{-5}$	CL=90%	2640
e^+e^-	B1	< 5.9	$\times 10^{-6}$	CL=90%	2640
$\mu^+\mu^-$	B1	< 6.8	$\times 10^{-7}$	CL=90%	2637
$K^0 e^+e^-$	B1	< 3.0	$\times 10^{-4}$	CL=90%	2616
$K^0\mu^+\mu^-$	B1	< 3.6	$\times 10^{-4}$	CL=90%	2612
$K^*(892)^0 e^+e^-$	B1	< 2.9	$\times 10^{-4}$	CL=90%	2564
$K^*(892)^0\mu^+\mu^-$	B1	< 2.3	$\times 10^{-5}$	CL=90%	2559
$K^*(892)^0\nu\overline{\nu}$	B1	< 1.0	$\times 10^{-3}$	CL=90%	2244
$e^\pm\mu^\mp$	LF	[gg] < 5.9	$\times 10^{-6}$	CL=90%	2639
$e^\pm\tau^\mp$	LF	[gg] < 5.3	$\times 10^{-4}$	CL=90%	2341
$\mu^\pm\tau^\mp$	LF	[gg] < 8.3	$\times 10^{-4}$	CL=90%	2339

$B^\pm/B^0$ ADMIXTURE

The branching fraction measurements are for an admixture of B mesons at the $\Upsilon(4S)$. The values quoted assume that $B(\Upsilon(4S) \to B\overline{B}) = 100\%$.

For inclusive branching fractions, e.g., $B \to D^\pm$ anything, the treatment of multiple D's in the final state must be defined. One possibility would be to count the number of events with one-or-more D's and divide by the total number of B's. Another possibility would be to count the total number of D's and divide by the total number of B's, which is the definition of average multiplicity. The two definitions are identical when only one of the specified particles is allowed in the final state. Even though the "one-or-more" definition seems sensible, for practical reasons inclusive branching fractions are almost always measured using the multiplicity definition. For heavy final state particles, authors call their results inclusive branching fractions while for light particles some authors call their results multiplicities. In the B sections, we list all results as inclusive branching fractions, adopting a multiplicity definition. This means that inclusive branching fractions can exceed 100% and that inclusive partial widths can exceed total widths, just as inclusive cross sections can exceed total cross sections.

$\overline{B}$ modes are charge conjugates of the modes below. Reactions indicate the weak decay vertex and do not include mixing.

B DECAY MODES		Fraction (Γ_i/Γ)	Scale factor/ Confidence level	p (MeV/c)
Semileptonic and leptonic modes				
$B \to e^+\nu_e$ anything	[ccc]	(10.41±0.29) %	S=1.2	–
$B \to \overline{p}e^+\nu_e$ anything		< 1.6 $\times 10^{-3}$	CL=90%	–
$B \to \mu^+\nu_\mu$ anything	[ccc]	(10.3 ±0.5) %		–
$B \to \ell^+\nu_\ell$ anything	[pp,ccc]	(10.45±0.21) %		–
$B \to D^-\ell^+\nu_\ell$ anything	[pp]	(2.7 ±0.8) %		–
$B \to \overline{D^0}\ell^+\nu_\ell$ anything	[pp]	(7.0 ±1.4) %		–
$B \to \overline{D}^{**}\ell^+\nu_\ell$	[pp,ddd]	(2.7 ±0.7) %		–
$B \to \overline{D}_1(2420)\ell^+\nu_\ell$ anything		(7.4 ±1.6) $\times 10^{-3}$		–
$B \to D\pi\ell^+\nu_\ell$ anything + $D^*\pi\ell^+\nu_\ell$ anything		(2.3 ±0.4) %		–
$B \to \overline{D}_2^*(2460)\ell^+\nu_\ell$ anything		< 6.5 $\times 10^{-3}$	CL=95%	–
$B \to D^{*-}\pi^+\ell^+\nu_\ell$ anything		(1.00±0.34) %		–
$B \to D_s^-\ell^+\nu_\ell$ anything	[pp] < 9	$\times 10^{-3}$	CL=90%	–
$B \to D_s^-\ell^+\nu_\ell K^+$ anything	[pp] < 6	$\times 10^{-3}$	CL=90%	–
$B \to D_s^-\ell^+\nu_\ell K^0$ anything	[pp] < 9	$\times 10^{-3}$	CL=90%	–
$B \to K^+\ell^+\nu_\ell$ anything	[pp]	(6.0 ±0.5) %		–
$B \to K^-\ell^+\nu_\ell$ anything	[pp]	(10 ±4) $\times 10^{-3}$		–
$B \to K^0/\overline{K^0}\ell^+\nu_\ell$ anything	[pp]	(4.4 ±0.5) %		–

D, D^*, or D_s modes

$B \to D^{\pm}$ anything		(24.1 ±1.9) %		–
$B \to D^0 / \overline{D}^0$ anything		(63.1 ±2.9) %	S=1.1	–
$B \to D^*(2010)^{\pm}$ anything		(22.7 ±1.6) %		–
$B \to D^*(2007)^0$ anything		(26.0 ±2.7) %		–
$B \to D_s^{\pm}$ anything	[gg]	(10.0 ±2.5) %		–
$b \to c\overline{c}s$		(22 ±4) %		–
$B \to D_s D, D_s^* D, D_s D^*,$ or $D_s^* D^*$	[gg]	(4.9 ±1.3) %		–
$B \to D^*(2010)\gamma$		< 1.1 $\times 10^{-3}$	CL=90%	–
$B \to D_s^+ \pi^-, D_s^{*+} \pi^-,$ $D_s^+ \rho^-, D_s^{*+} \rho^-, D_s^+ \pi^0,$ $D_s^{*+} \pi^0, D_s^+ \eta, D_s^{*+} \eta,$ $D_s^+ \rho^0, D_s^{*+} \rho^0, D_s^+ \omega,$ $D_s^{*+} \omega$	[gg]	< 5 $\times 10^{-4}$	CL=90%	–
$B \to D_{s1}(2536)^+$ anything		< 9.5 $\times 10^{-3}$	CL=90%	–

Charmonium modes

$B \to J/\psi(1S)$ anything		(1.13±0.06) %		–
$B \to J/\psi(1S)$(direct) anything		(8.0 ±0.8) $\times 10^{-3}$		–
$B \to \psi(2S)$ anything		(3.5 ±0.5) $\times 10^{-3}$		–
$B \to \chi_{c1}(1P)$ anything		(4.2 ±0.7) $\times 10^{-3}$		–
$B \to \chi_{c1}(1P)$(direct) anything		(3.7 ±0.7) $\times 10^{-3}$		–
$B \to \chi_{c2}(1P)$ anything		< 3.8 $\times 10^{-3}$	CL=90%	–
$B \to \eta_c(1S)$ anything		< 9 $\times 10^{-3}$	CL=90%	–

K or K^* modes

$B \to K^{\pm}$ anything	[gg]	(78.9 ±2.5) %		–
$B \to K^+$ anything		(66 ±5) %		–
$B \to K^-$ anything		(13 ±4) %		–
$B \to K^0 / \overline{K}^0$ anything	[gg]	(64 ±4) %		–
$B \to K^*(892)^{\pm}$ anything		(18 ±6) %		–
$B \to K^*(892)^0 / \overline{K}^*(892)^0$ anything	[gg]	(14.6 ±2.6) %		–
$B \to K_1(1400)\gamma$		< 4.1 $\times 10^{-4}$	CL=90%	–
$B \to K_2^*(1430)\gamma$		< 8.3 $\times 10^{-4}$	CL=90%	–
$B \to K_2(1770)\gamma$		< 1.2 $\times 10^{-3}$	CL=90%	–
$B \to K_3^*(1780)\gamma$		< 3.0 $\times 10^{-3}$	CL=90%	–
$B \to K_4^*(2045)\gamma$		< 1.0 $\times 10^{-3}$	CL=90%	–
$B \to \overline{b} \to \overline{s}\gamma$		(2.3 ±0.7) $\times 10^{-4}$		–
$B \to \overline{b} \to \overline{s}$ gluon		< 6.8 %	CL=90%	–

Light unflavored meson modes

$B \to \pi^{\pm}$ anything	[gg,eee]	(359 ±7) %		–
$B \to \eta$ anything		(17.6 ±1.6) %		–
$B \to \rho^0$ anything		(21 ±5) %		–
$B \to \omega$ anything		< 81 %	CL=90%	–
$B \to \phi$ anything		(3.5 ±0.7) %	S=1.8	–

Baryon modes

$B \to \Lambda_c^{\pm}$ anything		(6.4 ±1.1) %		–
$B \to \Lambda_c^- e^+$ anything		< 3.2 $\times 10^{-3}$	CL=90%	–
$B \to \Lambda_c^- p$ anything		(3.6 ±0.7) %		–
$B \to \Lambda_c^- p e^+ \nu_e$		< 1.5 $\times 10^{-3}$	CL=90%	–
$B \to \overline{\Sigma}_c^{--}$ anything		(4.2 ±2.4) $\times 10^{-3}$		–
$B \to \overline{\Sigma}_c^-$ anything		< 9.6 $\times 10^{-3}$	CL=90%	–
$B \to \overline{\Sigma}_c^0$ anything		(4.6 ±2.4) $\times 10^{-3}$		–
$B \to \overline{\Sigma}_c^0 N (N = p$ or $n)$		< 1.5 $\times 10^{-3}$	CL=90%	–
$B \to \Xi_c^0$ anything $\times$ B($\Xi_c^0 \to \Xi^- \pi^+$)		(1.4 ±0.5) $\times 10^{-4}$		–
$B \to \Xi_c^+$ anything $\times$ B($\Xi_c^+ \to \Xi^- \pi^+ \pi^+$)		(4.5 $^{+1.3}_{-1.2}$) $\times 10^{-4}$		–
$B \to p / \overline{p}$ anything	[gg]	(8.0 ±0.4) %		–
$B \to p / \overline{p}$(direct) anything	[gg]	(5.5 ±0.5) %		–
$B \to \Lambda / \overline{\Lambda}$ anything	[gg]	(4.0 ±0.5) %		–
$B \to \Xi^- / \overline{\Xi}^+$ anything	[gg]	(2.7 ±0.6) $\times 10^{-3}$		–
$B \to$ baryons anything		(6.8 ±0.6) %		–
$B \to p\overline{p}$ anything		(2.47 ±0.23) %		–
$B \to \Lambda \overline{p} / \overline{\Lambda} p$ anything	[gg]	(2.5 ±0.4) %		–
$B \to \Lambda \overline{\Lambda}$ anything		< 5 $\times 10^{-3}$	CL=90%	–

Lepton Family number (LF) violating modes or $\Delta B = 1$ weak neutral current (B1) modes

$B \to e^+ e^- s$	B1	< 5.7	$\times 10^{-5}$	CL=90%	–
$B \to \mu^+ \mu^- s$	B1	< 5.8	$\times 10^{-5}$	CL=90%	–
$B \to e^{\pm} \mu^{\mp} s$	LF	< 2.2	$\times 10^{-5}$	CL=90%	–

$B^{\pm}/B^0/B_s^0/b$-baryon ADMIXTURE

These measurements are for an admixture of bottom particles at high energy (LEP, Tevatron, S$p\overline{p}$S).

Mean life $\tau = (1.564 \pm 0.014) \times 10^{-12}$ s

Mean life $\tau = (1.72 \pm 0.10) \times 10^{-12}$ s   Charged b-hadron admixture

Mean life $\tau = (1.58 \pm 0.14) \times 10^{-12}$ s   Neutral b-hadron admixture

$\tau_{\text{charged } b-\text{hadron}} / \tau_{\text{neutral } b-\text{hadron}} = 1.09 \pm 0.13$

The branching fraction measurements are for an admixture of B mesons and baryons at energies above the $\Upsilon(4S)$. Only the highest energy results (LEP, Tevatron, S$p\overline{p}$S) are used in the branching fraction averages. The production fractions give our best current estimate of the admixture at LEP.

For inclusive branching fractions, e.g., $B \to D^{\pm}$ anything, the treatment of multiple D's in the final state must be defined. One possibility would be to count the number of events with one-or-more D's and divide by the total number of B's. Another possibility would be to count the total number of D's and divide by the total number of B's, which is the definition of average multiplicity. The two definitions are identical when only one of the specified particles is allowed in the final state. Even though the "one-or-more" definition seems sensible, for practical reasons inclusive branching fractions are almost always measured using the multiplicity definition. For heavy final state particles, authors call their results inclusive branching fractions while for light particles some authors call their results multiplicities. In the B sections, we list all results as inclusive branching fractions, adopting a multiplicity definition. This means that inclusive branching fractions can exceed 100% and that inclusive partial widths can exceed total widths, just as inclusive cross sections can exceed total cross sections.

The modes below are listed for a $\overline{b}$ initial state. b modes are their charge conjugates. Reactions indicate the weak decay vertex and do not include mixing.

$\overline{b}$ DECAY MODES		Fraction (Γ_i/Γ)	Confidence level	p (MeV/c)

PRODUCTION FRACTIONS

The production fractions for weakly decaying b-hadrons at the Z have been calculated from the best values of mean lives, mixing parameters, and branching fractions in this edition by the LEP B Oscillation Working Group as described in the note "Production and Decay of b-Flavored Hadrons" in the $B^{\pm}$ Particle Listings. Values assume

$$B(\overline{b} \to B^+) = B(\overline{b} \to B^0)$$
$$B(\overline{b} \to B^+) + B(\overline{b} \to B^0) + B(b \to B_s^0) + B(b \to \Lambda_b) = 100 \%.$$

The notation for production fractions varies in the literature (f_{B^0}, $f(b \to \overline{B}^0)$, Br($b \to \overline{B}^0$)). We use our own branching fraction notation here, B($\overline{b} \to B^0$).

B^+		(39.7 $^{+1.8}_{-2.2}$) %	–
B^0		(39.7 $^{+1.8}_{-2.2}$) %	–
B_s^0		(10.5 $^{+1.8}_{-1.7}$) %	–
Λ_b		(10.1 $^{+3.9}_{-3.1}$) %	–

DECAY MODES

Semileptonic and leptonic modes

ν anything		(23.1 ± 1.5) %	–
$\ell^+ \nu_\ell$ anything	[pp,ccc]	(10.99± 0.23) %	–
$e^+ \nu_e$ anything	[ccc]	(10.9 ± 0.5) %	–
$\mu^+ \nu_\mu$ anything	[ccc]	(10.8 ± 0.5) %	–
$D^- \ell^+ \nu_\ell$ anything	[pp]	(2.02± 0.29) %	–
$\overline{D}^0 \ell^+ \nu_\ell$ anything	[pp]	(6.5 ± 0.6) %	–
$D^{*-} \ell^+ \nu_\ell$ anything	[pp]	(2.76± 0.29) %	–
$\overline{D}_j^0 \ell^+ \nu_\ell$ anything	[pp,fff]	seen	–
$D_j^- \ell^+ \nu_\ell$ anything	[pp,fff]	seen	–
$\overline{D}_2^*(2460)^0 \ell^+ \nu_\ell$ anything		seen	–
$D_2^*(2460)^- \ell^+ \nu_\ell$ anything		seen	–
$\tau^+ \nu_\tau$ anything		(2.6 ± 0.4) %	–
$\overline{c} \to \ell^- \overline{\nu}_\ell$ anything	[pp]	(7.8 ± 0.6) %	–

Charmed meson and baryon modes

$\overline{D}^0$ anything	(60.1 ± 3.2) %	–
D^- anything	(23.7 ± 2.3) %	–
$\overline{D}_s$ anything	(18 ± 5) %	–
Λ_c anything	(9.7 ± 2.9) %	–
$\overline{c}/c$ anything	[eee] (117 ± 4) %	–

Charmonium modes

$J/\psi(1S)$ anything	(1.16 ± 0.10) %	–
$\psi(2S)$ anything	(4.8 ± 2.4) × 10^{-3}	–
$\chi_{c1}(1P)$ anything	(1.8 ± 0.5) %	–

K or K* modes

$\overline{s}\gamma$	< 5.4 × 10^{-4}	90%	–
$K^{\pm}$ anything	(88 ±19) %		–
K_S^0 anything	(29.0 ± 2.9) %		–

Pion modes

π^0 anything	[eee] (278 ±60) %	–

Baryon modes

$p/\overline{p}$ anything	(14 ± 6) %	–

Other modes

charged anything	[eee] (497 ± 7) %	–
hadron$^+$ hadron$^-$	(1.7 $^{+1.0}_{-0.7}$) × 10^{-5}	–
charmless	(7 ±21) × 10^{-3}	–

Baryon modes

$\Lambda/\overline{\Lambda}$ anything	(5.9 ± 0.6) %	–

$\Delta B = 1$ weak neutral current (B1) modes

$\mu^+\mu^-$ anything	B1	< 3.2 × 10^{-4}	90%	–

 $\quad I(J^P) = \frac{1}{2}(1^-)$

I, J, P need confirmation. Quantum numbers shown are quark-model predictions.

Mass $m_{B^*} = 5324.9 \pm 1.8$ MeV

$m_{B^*} - m_B = 45.78 \pm 0.35$ MeV

B^* DECAY MODES	Fraction (Γ_i/Γ)	p (MeV/c)
$B\gamma$	dominant	46

BOTTOM, STRANGE MESONS
($B = \pm 1$, $S = \mp 1$)

$B_s^0 = s\overline{b}$, $\overline{B}_s^0 = \overline{s}b$, similarly for B_s^*'s

$\boxed{B_s^0} \qquad\qquad I(J^P) = 0(0^-)$

I, J, P need confirmation. Quantum numbers shown are quark-model predictions.

Mass $m_{B_s^0} = 5369.3 \pm 2.0$ MeV

Mean life $\tau = (1.54 \pm 0.07) \times 10^{-12}$ s

$c\tau = 462~\mu$m

B_s^0-$\overline{B}_s^0$ mixing parameters

χ_B at high energy $= f_d\chi_d + f_s\chi_s = 0.118 \pm 0.006$

$\Delta m_{B_s^0} = m_{B_{sH}^0} - m_{B_{sL}^0} > 9.1 \times 10^{12}~\hbar~\text{s}^{-1}$, CL = 95%

$x_s = \Delta m_{B_s^0}/\Gamma_{B_s^0} > 14.0$, CL = 95%

$\chi_s > 0.4975$, CL = 95%

These branching fractions all scale with B($\overline{b} \rightarrow B_s^0$), the LEP B_s^0 production fraction. The first four were evaluated using B($\overline{b} \rightarrow B_s^0$) = (10.5$^{+1.8}_{-1.7}$)% and the rest assume B($\overline{b} \rightarrow B_s^0$) = 12%.

The branching fraction B($B_s^0 \rightarrow D_s^- \ell^+ \nu_\ell$ anything) is not a pure measurement since the measured product branching fraction B($\overline{b} \rightarrow B_s^0$) × B($B_s^0 \rightarrow D_s^- \ell^+ \nu_\ell$ anything) was used to determine B($\overline{b} \rightarrow B_s^0$), as described in the note on "Production and Decay of b-Flavored Hadrons."

B_s^0 DECAY MODES		Fraction (Γ_i/Γ)	Confidence level	p (MeV/c)
D_s^- anything		(92 ±33) %		–
$D_s^- \ell^+ \nu_\ell$ anything		[ggg] (8.1 ± 2.5) %		–
$D_s^- \pi^+$		< 13 %		2321
$J/\psi(1S)\phi$		(9.3 ± 3.3) × 10^{-4}		1590
$J/\psi(1S)\pi^0$		< 1.2 × 10^{-3}	90%	1788
$J/\psi(1S)\eta$		< 3.8 × 10^{-3}	90%	1735
$\psi(2S)\phi$		seen		1122
$\pi^+\pi^-$		< 1.7 × 10^{-4}	90%	1122
$\pi^0\pi^0$		< 2.1 × 10^{-4}	90%	2861
$\eta\pi^0$		< 1.0 × 10^{-3}	90%	2655
$\eta\eta$		< 1.5 × 10^{-3}	90%	2628
$\pi^+ K^-$		< 2.1 × 10^{-4}	90%	2660
$K^+ K^-$		< 5.9 × 10^{-5}	90%	2639
$p\overline{p}$		< 5.9 × 10^{-5}	90%	2515
$\gamma\gamma$		< 1.48 × 10^{-4}	90%	2685
$\phi\gamma$		< 7 × 10^{-4}	90%	2588

Lepton Family number (LF) violating modes or
$\Delta B = 1$ weak neutral current (B1) modes

$\mu^+\mu^-$	B1	< 2.0 × 10^{-6}	90%	2682
$e^+ e^-$	B1	< 5.4 × 10^{-5}	90%	2864
$e^{\pm}\mu^{\mp}$	LF	[gg] < 4.1 × 10^{-5}	90%	2864
$\phi\nu\overline{\nu}$	B1	< 5.4 × 10^{-3}	90%	–

$c\bar{c}$ MESONS

$\eta_c(1S)$

$I^G(J^{PC}) = 0^+(0^{-+})$

Mass $m = 2979.8 \pm 2.1$ MeV $(S = 2.1)$
Full width $\Gamma = 13.2^{+3.8}_{-3.2}$ MeV

$\eta_c(1S)$ DECAY MODES	Fraction (Γ_i/Γ)	Confidence level	p (MeV/c)
Decays involving hadronic resonances			
$\eta'(958)\pi\pi$	(4.1 ± 1.7) %		1319
$\rho\rho$	(2.6 ± 0.9) %		1275
$K^*(892)^0 K^-\pi^+ +$ c.c.	(2.0 ± 0.7) %		1273
$K^*(892)\overline{K}^*(892)$	$(8.5 \pm 3.1) \times 10^{-3}$		1193
$\phi\phi$	$(7.1 \pm 2.8) \times 10^{-3}$		1086
$a_0(980)\pi$	< 2 %	90%	1323
$a_2(1320)\pi$	< 2 %	90%	1193
$K^*(892)\overline{K} +$ c.c.	< 1.28 %	90%	1307
$f_2(1270)\eta$	< 1.1 %	90%	1142
$\omega\omega$	$< 3.1 \times 10^{-3}$	90%	1268
Decays into stable hadrons			
$K\overline{K}\pi$	(5.5 ± 1.7) %		1378
$\eta\pi\pi$	(4.9 ± 1.8) %		1425
$\pi^+\pi^- K^+ K^-$	$(2.0^{+0.7}_{-0.6})$ %		1342
$2(K^+K^-)$	(2.1 ± 1.2) %		1053
$2(\pi^+\pi^-)$	(1.2 ± 0.4) %		1457
$p\overline{p}$	$(1.2 \pm 0.4) \times 10^{-3}$		1157
$K\overline{K}\eta$	< 3.1 %	90%	1262
$\pi^+\pi^- p\overline{p}$	< 1.2 %	90%	1023
$\Lambda\overline{\Lambda}$	$< 2 \times 10^{-3}$	90%	987
Radiative decays			
$\gamma\gamma$	$(3.0 \pm 1.2) \times 10^{-4}$		1489

$J/\psi(1S)$

$I^G(J^{PC}) = 0^-(1^{--})$

Mass $m = 3096.88 \pm 0.04$ MeV
Full width $\Gamma = 87 \pm 5$ keV
$\Gamma_{ee} = 5.26 \pm 0.37$ keV   (Assuming $\Gamma_{ee} = \Gamma_{\mu\mu}$)

$J/\psi(1S)$ DECAY MODES		Fraction (Γ_i/Γ)	Scale factor/ Confidence level	p (MeV/c)
hadrons		(87.7 ± 0.5) %		–
virtual $\gamma \to$ hadrons		(17.0 ± 2.0) %		–
e^+e^-		(6.02 ± 0.19) %		1548
$\mu^+\mu^-$		(6.01 ± 0.19) %		1545
Decays involving hadronic resonances				
$\rho\pi$		(1.27 ± 0.09) %		1449
$\rho^0\pi^0$		$(4.2 \pm 0.5) \times 10^{-3}$		1449
$a_2(1320)\rho$		(1.09 ± 0.22) %		1125
$\omega\pi^+\pi^+\pi^-\pi^-$		$(8.5 \pm 3.4) \times 10^{-3}$		1392
$\omega\pi^+\pi^-$		$(7.2 \pm 1.0) \times 10^{-3}$		1435
$\omega f_2(1270)$		$(4.3 \pm 0.6) \times 10^{-3}$		1143
$K^*(892)^0\overline{K}_2^*(1430)^0 +$ c.c.		$(6.7 \pm 2.6) \times 10^{-3}$		1005
$\omega K^*(892)\overline{K} +$ c.c.		$(5.3 \pm 2.0) \times 10^{-3}$		1098
$K^+\overline{K}^*(892)^- +$ c.c.		$(5.0 \pm 0.4) \times 10^{-3}$		1373
$K^0\overline{K}^*(892)^0 +$ c.c.		$(4.2 \pm 0.4) \times 10^{-3}$		1371
$\omega\pi^0\pi^0$		$(3.4 \pm 0.8) \times 10^{-3}$		1436
$b_1(1235)^\pm\pi^\mp$	[gg]	$(3.0 \pm 0.5) \times 10^{-3}$		1299
$\omega K^\pm K_S^0\pi^\mp$	[gg]	$(3.0 \pm 0.7) \times 10^{-3}$		1210
$b_1(1235)^0\pi^0$		$(2.3 \pm 0.6) \times 10^{-3}$		1299
$\phi K^*(892)\overline{K} +$ c.c.		$(2.04 \pm 0.28) \times 10^{-3}$		969
$\omega K\overline{K}$		$(1.9 \pm 0.4) \times 10^{-3}$		1268
$\omega f_J(1710) \to \omega K\overline{K}$		$(4.8 \pm 1.1) \times 10^{-4}$		878
$\phi 2(\pi^+\pi^-)$		$(1.60 \pm 0.32) \times 10^{-3}$		1318
$\Delta(1232)^{++}\overline{p}\pi^-$		$(1.6 \pm 0.5) \times 10^{-3}$		1030
$\omega\eta$		$(1.58 \pm 0.16) \times 10^{-3}$		1394
$\phi K\overline{K}$		$(1.48 \pm 0.22) \times 10^{-3}$		1179
$\phi f_J(1710) \to \phi K\overline{K}$		$(3.6 \pm 0.6) \times 10^{-4}$		875
$p\overline{p}\omega$		$(1.30 \pm 0.25) \times 10^{-3}$	S=1.3	769
$\Delta(1232)^{++}\overline{\Delta}(1232)^{--}$		$(1.10 \pm 0.29) \times 10^{-3}$		938
$\Sigma(1385)^-\overline{\Sigma}(1385)^+$ (or c.c.)	[gg]	$(1.03 \pm 0.13) \times 10^{-3}$		692
$p\overline{p}\eta'(958)$		$(9 \pm 4) \times 10^{-4}$	S=1.7	596
$\phi f_2'(1525)$		$(8 \pm 4) \times 10^{-4}$	S=2.7	871

$\phi\pi^+\pi^-$		$(8.0 \pm 1.2) \times 10^{-4}$		1365
$\phi K^\pm K_S^0\pi^\mp$	[gg]	$(7.2 \pm 0.9) \times 10^{-4}$		1114
$\omega f_1(1420)$		$(6.8 \pm 2.4) \times 10^{-4}$		1062
$\phi\eta$		$(6.5 \pm 0.7) \times 10^{-4}$		1320
$\Xi(1530)^-\overline{\Xi}^+$		$(5.9 \pm 1.5) \times 10^{-4}$		597
$pK^-\overline{\Sigma}(1385)^0$		$(5.1 \pm 3.2) \times 10^{-4}$		645
$\omega\pi^0$		$(4.2 \pm 0.6) \times 10^{-4}$	S=1.4	1447
$\phi\eta'(958)$		$(3.3 \pm 0.4) \times 10^{-4}$		1192
$\phi f_0(980)$		$(3.2 \pm 0.9) \times 10^{-4}$	S=1.9	1182
$\Xi(1530)^0\overline{\Xi}^0$		$(3.2 \pm 1.4) \times 10^{-4}$		608
$\Sigma(1385)^-\overline{\Sigma}^+$ (or c.c.)	[gg]	$(3.1 \pm 0.5) \times 10^{-4}$		857
$\phi f_1(1285)$		$(2.6 \pm 0.5) \times 10^{-4}$	S=1.1	1032
$\rho\eta$		$(1.93 \pm 0.23) \times 10^{-4}$		1398
$\omega\eta'(958)$		$(1.67 \pm 0.25) \times 10^{-4}$		1279
$\omega f_0(980)$		$(1.4 \pm 0.5) \times 10^{-4}$		1271
$\rho\eta'(958)$		$(1.05 \pm 0.18) \times 10^{-4}$		1283
$p\overline{p}\phi$		$(4.5 \pm 1.5) \times 10^{-5}$		527
$a_2(1320)^\pm\pi^\mp$	[gg]	$< 4.3 \times 10^{-3}$	CL=90%	1263
$K\overline{K}_2^*(1430) +$ c.c.		$< 4.0 \times 10^{-3}$	CL=90%	1159
$K_2^*(1430)^0\overline{K}_2^*(1430)^0$		$< 2.9 \times 10^{-3}$	CL=90%	588
$K^*(892)^0\overline{K}^*(892)^0$		$< 5 \times 10^{-4}$	CL=90%	1263
$\phi f_2(1270)$		$< 3.7 \times 10^{-4}$	CL=90%	1036
$p\overline{p}\rho$		$< 3.1 \times 10^{-4}$	CL=90%	779
$\phi\eta(1440) \to \phi\eta\pi\pi$		$< 2.5 \times 10^{-4}$	CL=90%	946
$\omega f_2'(1525)$		$< 2.2 \times 10^{-4}$	CL=90%	1003
$\Sigma(1385)^0\overline{\Lambda}$		$< 2 \times 10^{-4}$	CL=90%	911
$\Delta(1232)^+\overline{p}$		$< 1 \times 10^{-4}$	CL=90%	1100
$\Sigma^0\overline{\Lambda}$		$< 9 \times 10^{-5}$	CL=90%	1032
$\phi\pi^0$		$< 6.8 \times 10^{-6}$	CL=90%	1377
Decays into stable hadrons				
$2(\pi^+\pi^-)\pi^0$		(3.37 ± 0.26) %		1496
$3(\pi^+\pi^-)\pi^0$		(2.9 ± 0.6) %		1433
$\pi^+\pi^-\pi^0$		(1.50 ± 0.20) %		1533
$\pi^+\pi^-\pi^0 K^+ K^-$		(1.20 ± 0.30) %		1368
$4(\pi^+\pi^-)\pi^0$		$(9.0 \pm 3.0) \times 10^{-3}$		1345
$\pi^+\pi^- K^+ K^-$		$(7.2 \pm 2.3) \times 10^{-3}$		1407
$K\overline{K}\pi$		$(6.1 \pm 1.0) \times 10^{-3}$		1440
$p\overline{p}\pi^+\pi^-$		$(6.0 \pm 0.5) \times 10^{-3}$	S=1.3	1107
$2(\pi^+\pi^-)$		$(4.0 \pm 1.0) \times 10^{-3}$		1517
$3(\pi^+\pi^-)$		$(4.0 \pm 2.0) \times 10^{-3}$		1466
$n\overline{n}\pi^+\pi^-$		$(4 \pm 4) \times 10^{-3}$		1106
$\Sigma^0\overline{\Sigma}^0$		$(1.27 \pm 0.17) \times 10^{-3}$		992
$2(\pi^+\pi^-) K^+ K^-$		$(3.1 \pm 1.3) \times 10^{-3}$		1320
$p\overline{p}\pi^+\pi^-\pi^0$	[hhh]	$(2.3 \pm 0.9) \times 10^{-3}$	S=1.9	1033
$p\overline{p}$		$(2.14 \pm 0.10) \times 10^{-3}$		1232
$p\overline{p}\eta$		$(2.09 \pm 0.18) \times 10^{-3}$		948
$p\overline{n}\pi^-$		$(2.00 \pm 0.10) \times 10^{-3}$		1174
$n\overline{n}$		$(1.9 \pm 0.5) \times 10^{-3}$		1231
$\Xi^-\overline{\Xi}^-$		$(1.8 \pm 0.4) \times 10^{-3}$	S=1.8	818
$\Lambda\overline{\Lambda}$		$(1.35 \pm 0.14) \times 10^{-3}$	S=1.2	1074
$p\overline{p}\pi^0$		$(1.09 \pm 0.09) \times 10^{-3}$		1176
$\Lambda\overline{\Sigma}^-\pi^+$ (or c.c.)	[gg]	$(1.06 \pm 0.12) \times 10^{-3}$		945
$pK^-\overline{\Lambda}$		$(8.9 \pm 1.6) \times 10^{-4}$		876
$2(K^+K^-)$		$(7.0 \pm 3.0) \times 10^{-4}$		1131
$pK^-\overline{\Sigma}^0$		$(2.9 \pm 0.8) \times 10^{-4}$		820
$K^+ K^-$		$(2.37 \pm 0.31) \times 10^{-4}$		1468
$\Lambda\overline{\Lambda}\pi^0$		$(2.2 \pm 0.7) \times 10^{-4}$		998
$\pi^+\pi^-$		$(1.47 \pm 0.23) \times 10^{-4}$		1542
$K_S^0 K_L^0$		$(1.08 \pm 0.14) \times 10^{-4}$		1466
$\Lambda\overline{\Sigma} +$ c.c.		$< 1.5 \times 10^{-4}$	CL=90%	1032
$K_S^0 K_S^0$		$< 5.2 \times 10^{-6}$	CL=90%	1466
Radiative decays				
$\gamma\eta_c(1S)$		(1.3 ± 0.4) %		116
$\gamma\pi^+\pi^- 2\pi^0$		$(8.3 \pm 3.1) \times 10^{-3}$		1518
$\gamma\eta\pi\pi$		$(6.1 \pm 1.0) \times 10^{-3}$		1487
$\gamma\eta(1440) \to \gamma K\overline{K}\pi$	[p]	$(9.1 \pm 1.8) \times 10^{-4}$		1223
$\gamma\eta(1440) \to \gamma\gamma\rho^0$		$(6.4 \pm 1.4) \times 10^{-5}$		1223
$\gamma\eta(1440) \to \gamma\pi\pi^+\pi^-$		$(3.4 \pm 0.7) \times 10^{-4}$		–
$\gamma\rho\rho$		$(4.5 \pm 0.8) \times 10^{-3}$		1343
$\gamma\eta'(958)$		$(4.31 \pm 0.30) \times 10^{-3}$		1400
$\gamma 2\pi^+ 2\pi^-$		$(2.8 \pm 0.5) \times 10^{-3}$	S=1.9	1517
$\gamma f_4(2050)$		$(2.7 \pm 0.7) \times 10^{-3}$		874
$\gamma\omega\omega$		$(1.59 \pm 0.33) \times 10^{-3}$		1337
$\gamma\eta(1440) \to \gamma\rho^0\rho^0$		$(1.7 \pm 0.4) \times 10^{-3}$	S=1.3	1223
$\gamma f_2(1270)$		$(1.38 \pm 0.14) \times 10^{-3}$		1286
$\gamma f_J(1710) \to \gamma K\overline{K}$		$(8.5^{+1.2}_{-0.9}) \times 10^{-4}$	S=1.2	1075

$\gamma \eta$	(8.6 ±0.8) × 10^{-4}		1500
$\gamma f_1(1420) \to \gamma K \overline{K} \pi$	(8.3 ±1.5) × 10^{-4}		1220
$\gamma f_1(1285)$	(6.5 ±1.0) × 10^{-4}		1283
$\gamma f'_2(1525)$	(4.7 $^{+0.7}_{-0.5}$) × 10^{-4}		1173
$\gamma \phi \phi$	(4.0 ±1.2) × 10^{-4}	S=2.1	1166
$\gamma p \overline{p}$	(3.8 ±1.0) × 10^{-4}		1232
$\gamma \eta(2225)$	(2.9 ±0.6) × 10^{-4}		834
$\gamma \eta(1760) \to \gamma \rho^0 \rho^0$	(1.3 ±0.9) × 10^{-4}		1048
$\gamma \pi^0$	(3.9 ±1.3) × 10^{-5}		1546
$\gamma p \overline{p} \pi^+ \pi^-$	< 7.9 × 10^{-4}	CL=90%	1107
$\gamma \gamma$	< 5 × 10^{-4}	CL=90%	1548
$\gamma \Lambda \overline{\Lambda}$	< 1.3 × 10^{-4}	CL=90%	1074
3γ	< 5.5 × 10^{-5}	CL=90%	1548
$\gamma f_J(2220)$	> 2.50 × 10^{-3}	CL=99.9%	–
$\gamma f_0(1500)$	(5.7 ±0.8) × 10^{-4}		1184
$\gamma e^+ e^-$	(8.8 ±1.4) × 10^{-3}		–

$\chi_{c0}(1P)$

$I^G(J^{PC}) = 0^+(0^{++})$

Mass $m = 3417.3 \pm 2.8$ MeV
Full width $\Gamma = 14 \pm 5$ MeV

$\chi_{c0}(1P)$ DECAY MODES	Fraction (Γ_i/Γ)	Confidence level	p (MeV/c)
Hadronic decays			
$2(\pi^+ \pi^-)$	(3.7±0.7) %		1679
$\pi^+ \pi^- K^+ K^-$	(3.0±0.7) %		1580
$\rho^0 \pi^+ \pi^-$	(1.6±0.5) %		1608
$3(\pi^+ \pi^-)$	(1.5±0.5) %		1633
$K^+ \overline{K}^*(892)^0 \pi^- +$ c.c.	(1.2±0.4) %		1522
$\pi^+ \pi^-$	(7.5±2.1) × 10^{-3}		1702
$K^+ K^-$	(7.1±2.4) × 10^{-3}		1635
$\pi^+ \pi^- p \overline{p}$	(5.0±2.0) × 10^{-3}		1320
$p \overline{p}$	< 9.0 × 10^{-4}	90%	1427
Radiative decays			
$\gamma J/\psi(1S)$	(6.6±1.8) × 10^{-3}		303
$\gamma \gamma$	< 5 × 10^{-4}	95%	1708

$\chi_{c1}(1P)$

$I^G(J^{PC}) = 0^+(1^{++})$

Mass $m = 3510.53 \pm 0.12$ MeV
Full width $\Gamma = 0.88 \pm 0.14$ MeV

$\chi_{c1}(1P)$ DECAY MODES	Fraction (Γ_i/Γ)	p (MeV/c)
Hadronic decays		
$3(\pi^+ \pi^-)$	(2.2±0.8) %	1683
$2(\pi^+ \pi^-)$	(1.6±0.5) %	1727
$\pi^+ \pi^- K^+ K^-$	(9 ±4) × 10^{-3}	1632
$\rho^0 \pi^+ \pi^-$	(3.9±3.5) × 10^{-3}	1659
$K^+ \overline{K}^*(892)^0 \pi^- +$ c.c.	(3.2±2.1) × 10^{-3}	1576
$\pi^+ \pi^- p \overline{p}$	(1.4±0.9) × 10^{-3}	1381
$p \overline{p}$	(8.6±1.2) × 10^{-5}	1483
$\pi^+ \pi^- + K^+ K^-$	< 2.1 × 10^{-3}	–
Radiative decays		
$\gamma J/\psi(1S)$	(27.3±1.6) %	389

$\chi_{c2}(1P)$

$I^G(J^{PC}) = 0^+(2^{++})$

Mass $m = 3556.17 \pm 0.13$ MeV
Full width $\Gamma = 2.00 \pm 0.18$ MeV

$\chi_{c2}(1P)$ DECAY MODES	Fraction (Γ_i/Γ)	Confidence level	p (MeV/c)
Hadronic decays			
$2(\pi^+ \pi^-)$	(2.2±0.5) %		1751
$\pi^+ \pi^- K^+ K^-$	(1.9±0.5) %		1556
$3(\pi^+ \pi^-)$	(1.2±0.8) %		1707
$\rho^0 \pi^+ \pi^-$	(7 ±4) × 10^{-3}		1683
$K^+ \overline{K}^*(892)^0 \pi^- +$ c.c.	(4.8±2.8) × 10^{-3}		1601
$\pi^+ \pi^- p \overline{p}$	(3.3±1.3) × 10^{-3}		1410
$\pi^+ \pi^-$	(1.9±1.0) × 10^{-3}		1773
$K^+ K^-$	(1.5±1.1) × 10^{-3}		1708
$p \overline{p}$	(10.0±1.0) × 10^{-5}		1510
$J/\psi(1S) \pi^+ \pi^- \pi^0$	< 1.5 %	90%	185

Radiative decays

$\gamma J/\psi(1S)$	(13.5±1.1) %		430
$\gamma \gamma$	(1.6±0.5) × 10^{-4}		1778

$\psi(2S)$

$I^G(J^{PC}) = 0^-(1^{--})$

Mass $m = 3686.00 \pm 0.09$ MeV
Full width $\Gamma = 277 \pm 31$ keV (S = 1.1)
$\Gamma_{ee} = 2.14 \pm 0.21$ keV (Assuming $\Gamma_{ee} = \Gamma_{\mu\mu}$)

$\psi(2S)$ DECAY MODES	Fraction (Γ_i/Γ)	Scale factor/ Confidence level	p (MeV/c)
hadrons	(98.10±0.30) %		–
virtual $\gamma \to$ hadrons	(2.9 ±0.4) %		–
$e^+ e^-$	(8.5 ±0.7) × 10^{-3}		1843
$\mu^+ \mu^-$	(7.7 ±1.7) × 10^{-3}		1840
Decays into $J/\psi(1S)$ and anything			
$J/\psi(1S)$ anything	(54.2 ±3.0) %		–
$J/\psi(1S)$ neutrals	(22.8 ±1.7) %		–
$J/\psi(1S) \pi^+ \pi^-$	(30.2 ±1.9) %		477
$J/\psi(1S) \pi^0 \pi^0$	(17.9 ±1.8) %		481
$J/\psi(1S) \eta$	(2.7 ±0.4) %	S=1.7	200
$J/\psi(1S) \pi^0$	(9.7 ±2.1) × 10^{-4}		527
$J/\psi(1S) \mu^+ \mu^-$	(10.0 ±3.3) × 10^{-3}		–
Hadronic decays			
$3(\pi^+ \pi^-) \pi^0$	(3.5 ±1.6) × 10^{-3}		1746
$2(\pi^+ \pi^-) \pi^0$	(3.0 ±1.0) × 10^{-3}		1799
$\pi^+ \pi^- K^+ K^-$	(1.6 ±0.4) × 10^{-3}		1726
$\pi^+ \pi^- p \overline{p}$	(8.0 ±2.0) × 10^{-4}		1491
$K^+ \overline{K}^*(892)^0 \pi^- +$ c.c.	(6.7 ±2.5) × 10^{-4}		1673
$2(\pi^+ \pi^-)$	(4.5 ±1.0) × 10^{-4}		1817
$\rho^0 \pi^+ \pi^-$	(4.2 ±1.5) × 10^{-4}		1751
$\overline{p} p$	(1.9 ±0.5) × 10^{-4}		1586
$3(\pi^+ \pi^-)$	(1.5 ±1.0) × 10^{-4}		1774
$\overline{p} p \pi^0$	(1.4 ±0.5) × 10^{-4}		1543
$K^+ K^-$	(1.0 ±0.7) × 10^{-4}		1776
$\pi^+ \pi^- \pi^0$	(9 ±5) × 10^{-5}		1830
$\rho \pi$	< 8.3 × 10^{-5}	CL=90%	1760
$\pi^+ \pi^-$	(8 ±5) × 10^{-5}		1838
$\Lambda \overline{\Lambda}$	< 4 × 10^{-4}	CL=90%	1467
$\Xi^- \overline{\Xi}^+$	< 2 × 10^{-4}	CL=90%	1285
$K^+ K^- \pi^0$	< 2.96 × 10^{-5}	CL=90%	1754
$K^+ \overline{K}^*(892)^- +$ c.c.	< 5.4 × 10^{-5}	CL=90%	1698
Radiative decays			
$\gamma \chi_{c0}(1P)$	(9.3 ±0.9) %		261
$\gamma \chi_{c1}(1P)$	(8.7 ±0.8) %		171
$\gamma \chi_{c2}(1P)$	(7.8 ±0.8) %		127
$\gamma \eta_c(1S)$	(2.8 ±0.6) × 10^{-3}		639
$\gamma \eta'(958)$	< 1.1 × 10^{-3}	CL=90%	1719
$\gamma \gamma$	< 1.6 × 10^{-4}	CL=90%	1843
$\gamma \eta(1440) \to \gamma K \overline{K} \pi$	< 1.2 × 10^{-4}	CL=90%	1569

$\psi(3770)$

$I^G(J^{PC}) = ?^?(1^{--})$

Mass $m = 3769.9 \pm 2.5$ MeV (S = 1.8)
Full width $\Gamma = 23.6 \pm 2.7$ MeV (S = 1.1)
$\Gamma_{ee} = 0.26 \pm 0.04$ keV (S = 1.2)

$\psi(3770)$ DECAY MODES	Fraction (Γ_i/Γ)	Scale factor	p (MeV/c)
$D \overline{D}$	dominant		242
$e^+ e^-$	(1.12±0.17) × 10^{-5}	1.2	1885

$\psi(4040)$ [III]

$I^G(J^{PC}) = ?^?(1^{--})$

Mass $m = 4040 \pm 10$ MeV
Full width $\Gamma = 52 \pm 10$ MeV
$\Gamma_{ee} = 0.75 \pm 0.15$ keV

$\psi(4040)$ DECAY MODES	Fraction (Γ_i/Γ)	p (MeV/c)
$e^+ e^-$	(1.4±0.4) × 10^{-5}	2020
$D^0 \overline{D}^0$	seen	777
$D^*(2007)^0 \overline{D}^0 +$ c.c.	seen	578
$D^*(2007)^0 \overline{D}^*(2007)^0$	seen	232

$\psi(4160)$ [llll]

$I^G(J^{PC}) = ?^?(1^{--})$

Mass $m = 4159 \pm 20$ MeV
Full width $\Gamma = 78 \pm 20$ MeV
$\Gamma_{ee} = 0.77 \pm 0.23$ keV

$\psi(4160)$ DECAY MODES	Fraction (Γ_i/Γ)	p (MeV/c)
e^+e^-	$(10\pm4) \times 10^{-6}$	2079

$\psi(4415)$ [llll]

$I^G(J^{PC}) = ?^?(1^{--})$

Mass $m = 4415 \pm 6$ MeV
Full width $\Gamma = 43 \pm 15$ MeV   (S = 1.8)
$\Gamma_{ee} = 0.47 \pm 0.10$ keV

$\psi(4415)$ DECAY MODES	Fraction (Γ_i/Γ)	p (MeV/c)
hadrons	dominant	–
e^+e^-	$(1.1\pm0.4) \times 10^{-5}$	2207

$b\bar{b}$ MESONS

$\Upsilon(1S)$

$I^G(J^{PC}) = 0^-(1^{--})$

Mass $m = 9460.37 \pm 0.21$ MeV   (S = 2.7)
Full width $\Gamma = 52.5 \pm 1.8$ keV
$\Gamma_{ee} = 1.32 \pm 0.05$ keV

$\Upsilon(1S)$ DECAY MODES	Fraction (Γ_i/Γ)	Scale factor/ Confidence level	p (MeV/c)
$\tau^+\tau^-$	$(2.67^{+0.14}_{-0.16})$ %		4384
e^+e^-	(2.52 ± 0.17) %		4730
$\mu^+\mu^-$	(2.48 ± 0.07) %	S=1.1	4729
Hadronic decays			
$J/\psi(1S)$ anything	$(1.1 \pm 0.4) \times 10^{-3}$		4223
$\rho\pi$	$< 2 \times 10^{-4}$	CL=90%	4698
$\pi^+\pi^-$	$< 5 \times 10^{-4}$	CL=90%	4728
K^+K^-	$< 5 \times 10^{-4}$	CL=90%	4704
$p\bar{p}$	$< 5 \times 10^{-4}$	CL=90%	4636
Radiative decays			
$\gamma 2h^+2h^-$	$(7.0 \pm 1.5) \times 10^{-4}$		4720
$\gamma 3h^+3h^-$	$(5.4 \pm 2.0) \times 10^{-4}$		4703
$\gamma 4h^+4h^-$	$(7.4 \pm 3.5) \times 10^{-4}$		4679
$\gamma\pi^+\pi^-K^+K^-$	$(2.9 \pm 0.9) \times 10^{-4}$		4686
$\gamma 2\pi^+2\pi^-$	$(2.5 \pm 0.9) \times 10^{-4}$		4720
$\gamma 3\pi^+3\pi^-$	$(2.5 \pm 1.2) \times 10^{-4}$		4703
$\gamma 2\pi^+2\pi^-K^+K^-$	$(2.4 \pm 1.2) \times 10^{-4}$		4658
$\gamma\pi^+\pi^- p\bar{p}$	$(1.5 \pm 0.6) \times 10^{-4}$		4604
$\gamma 2\pi^+2\pi^- p\bar{p}$	$(4 \pm 6) \times 10^{-5}$		4563
$\gamma 2K^+2K^-$	$(2.0 \pm 2.0) \times 10^{-5}$		4601
$\gamma\eta'(958)$	$< 1.3 \times 10^{-3}$	CL=90%	4682
$\gamma\eta$	$< 3.5 \times 10^{-4}$	CL=90%	4714
$\gamma f_2'(1525)$	$< 1.4 \times 10^{-4}$	CL=90%	4607
$\gamma f_2(1270)$	$< 1.3 \times 10^{-4}$	CL=90%	4644
$\gamma\eta(1440)$	$< 8.2 \times 10^{-5}$	CL=90%	4624
$\gamma f_J(1710) \rightarrow \gamma K\bar{K}$	$< 2.6 \times 10^{-4}$	CL=90%	4576
$\gamma f_0(2200) \rightarrow \gamma K^+K^-$	$< 2 \times 10^{-4}$	CL=90%	4475
$\gamma f_J(2220) \rightarrow \gamma K^+K^-$	$< 1.5 \times 10^{-5}$	CL=90%	4469
$\gamma\eta(2225) \rightarrow \gamma\phi\phi$	$< 3 \times 10^{-3}$	CL=90%	4469
γX	$< 3 \times 10^{-5}$	CL=90%	–
X = pseudoscalar with $m< 7.2$ GeV)			
$\gamma X\bar{X}$	$< 1 \times 10^{-3}$	CL=90%	–
$X\bar{X}$ = vectors with $m< 3.1$ GeV)			

$\chi_{b0}(1P)$ [llll]

$I^G(J^{PC}) = 0^+(0^{++})$
J needs confirmation.

Mass $m = 9859.8 \pm 1.3$ MeV

$\chi_{b0}(1P)$ DECAY MODES	Fraction (Γ_i/Γ)	Confidence level	p (MeV/c)
$\gamma\Upsilon(1S)$	<6 %	90%	391

$\chi_{b1}(1P)$ [llll]

$I^G(J^{PC}) = 0^+(1^{++})$
J needs confirmation.

Mass $m = 9891.9 \pm 0.7$ MeV

$\chi_{b1}(1P)$ DECAY MODES	Fraction (Γ_i/Γ)	p (MeV/c)
$\gamma\Upsilon(1S)$	(35 ± 8) %	422

$\chi_{b2}(1P)$ [llll]

$I^G(J^{PC}) = 0^+(2^{++})$
J needs confirmation.

Mass $m = 9913.2 \pm 0.6$ MeV

$\chi_{b2}(1P)$ DECAY MODES	Fraction (Γ_i/Γ)	p (MeV/c)
$\gamma\Upsilon(1S)$	(22 ± 4) %	443

$\Upsilon(2S)$

$I^G(J^{PC}) = 0^-(1^{--})$

Mass $m = 10.02330 \pm 0.00031$ GeV
Full width $\Gamma = 44 \pm 7$ keV
$\Gamma_{ee} = 0.520 \pm 0.032$ keV

$\Upsilon(2S)$ DECAY MODES	Fraction (Γ_i/Γ)	Confidence level	p (MeV/c)
$\Upsilon(1S)\pi^+\pi^-$	(18.5 ± 0.8) %		475
$\Upsilon(1S)\pi^0\pi^0$	(8.8 ± 1.1) %		480
$\tau^+\tau^-$	(1.7 ± 1.6) %		4686
$\mu^+\mu^-$	(1.31 ± 0.21) %		5011
e^+e^-	(1.18 ± 0.20) %		5012
$\Upsilon(1S)\pi^0$	$< 8 \times 10^{-3}$	90%	531
$\Upsilon(1S)\eta$	$< 2 \times 10^{-3}$	90%	127
$J/\psi(1S)$ anything	$< 6 \times 10^{-3}$	90%	4533
Radiative decays			
$\gamma\chi_{b1}(1P)$	(6.7 ± 0.9) %		131
$\gamma\chi_{b2}(1P)$	(6.6 ± 0.9) %		110
$\gamma\chi_{b0}(1P)$	(4.3 ± 1.0) %		162
$\gamma f_J(1710)$	$< 5.9 \times 10^{-4}$	90%	4866
$\gamma f_2'(1525)$	$< 5.3 \times 10^{-4}$	90%	4896
$\gamma f_2(1270)$	$< 2.41 \times 10^{-4}$	90%	4931

$\chi_{b0}(2P)$ [llll]

$I^G(J^{PC}) = 0^+(0^{++})$
J needs confirmation.

Mass $m = 10.2321 \pm 0.0006$ GeV

$\chi_{b0}(2P)$ DECAY MODES	Fraction (Γ_i/Γ)	p (MeV/c)
$\gamma\Upsilon(2S)$	(4.6 ± 2.1) %	210
$\gamma\Upsilon(1S)$	$(9 \pm6) \times 10^{-3}$	746

$\chi_{b1}(2P)$ [llll]

$I^G(J^{PC}) = 0^+(1^{++})$
J needs confirmation.

Mass $m = 10.2552 \pm 0.0005$ GeV
$m_{\chi_{b1}(2P)} - m_{\chi_{b0}(2P)} = 23.5 \pm 1.0$ MeV

$\chi_{b1}(2P)$ DECAY MODES	Fraction (Γ_i/Γ)	Scale factor	p (MeV/c)
$\gamma\Upsilon(2S)$	(21 ± 4) %	1.5	229
$\gamma\Upsilon(1S)$	(8.5 ± 1.3) %	1.3	764

$\chi_{b2}(2P)$ [llll]

$I^G(J^{PC}) = 0^+(2^{++})$
J needs confirmation.

Mass $m = 10.2685 \pm 0.0004$ GeV
$m_{\chi_{b2}(2P)} - m_{\chi_{b1}(2P)} = 13.5 \pm 0.6$ MeV

$\chi_{b2}(2P)$ DECAY MODES	Fraction (Γ_i/Γ)	p (MeV/c)
$\gamma\Upsilon(2S)$	(16.2 ± 2.4) %	242
$\gamma\Upsilon(1S)$	(7.1 ± 1.0) %	776

$\Upsilon(3S)$

$I^G(J^{PC}) = 0^-(1^{--})$

Mass $m = 10.3553 \pm 0.0005$ GeV
Full width $\Gamma = 26.3 \pm 3.5$ keV

$\Upsilon(3S)$ DECAY MODES	Fraction (Γ_i/Γ)	Scale factor/ Confidence level	p (MeV/c)
$\Upsilon(2S)$ anything	(10.6 ± 0.8) %		296
$\Upsilon(2S)\pi^+\pi^-$	(2.8 ± 0.6) %	S=2.2	177
$\Upsilon(2S)\pi^0\pi^0$	(2.00 ± 0.32) %		190
$\Upsilon(2S)\gamma\gamma$	(5.0 ± 0.7) %		327
$\Upsilon(1S)\pi^+\pi^-$	(4.48 ± 0.21) %		814
$\Upsilon(1S)\pi^0\pi^0$	(2.06 ± 0.28) %		816
$\Upsilon(1S)\eta$	$< 2.2 \times 10^{-3}$	CL=90%	–
$\mu^+\mu^-$	(1.81 ± 0.17) %		5177
e^+e^-	seen		5177
Radiative decays			
$\gamma\chi_{b2}(2P)$	(11.4 ± 0.8) %	S=1.3	87
$\gamma\chi_{b1}(2P)$	(11.3 ± 0.6) %		100
$\gamma\chi_{b0}(2P)$	(5.4 ± 0.6) %	S=1.1	123

$\Upsilon(4S)$ 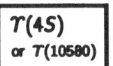 or $\Upsilon(10580)$

$I^G(J^{PC}) = ?^?(1^{--})$

Mass $m = 10.5800 \pm 0.0035$ GeV
Full width $\Gamma = 10 \pm 4$ MeV
$\Gamma_{ee} = 0.248 \pm 0.031$ keV   (S = 1.3)

$\Upsilon(4S)$ DECAY MODES	Fraction (Γ_i/Γ)	Confidence level	p (MeV/c)
$B\bar{B}$	> 96 %	95%	–
non-$B\bar{B}$	< 4 %	95%	–
e^+e^-	$(2.8 \pm 0.7) \times 10^{-5}$		5290
$J/\psi(3097)$ anything	$(2.2 \pm 0.7) \times 10^{-3}$		–
D^{*+} anything + c.c.	< 7.4 %	90%	5099
ϕ anything	$< 2.3 \times 10^{-3}$	90%	5240
$\Upsilon(1S)$ anything	$< 4 \times 10^{-3}$	90%	1053

$\Upsilon(10860)$

$I^G(J^{PC}) = ?^?(1^{--})$

Mass $m = 10.865 \pm 0.008$ GeV   (S = 1.1)
Full width $\Gamma = 110 \pm 13$ MeV
$\Gamma_{ee} = 0.31 \pm 0.07$ keV   (S = 1.3)

$\Upsilon(10860)$ DECAY MODES	Fraction (Γ_i/Γ)	p (MeV/c)
e^+e^-	$(2.8 \pm 0.7) \times 10^{-6}$	5432

$\Upsilon(11020)$

$I^G(J^{PC}) = ?^?(1^{--})$

Mass $m = 11.019 \pm 0.008$ GeV
Full width $\Gamma = 79 \pm 16$ MeV
$\Gamma_{ee} = 0.130 \pm 0.030$ keV

$\Upsilon(11020)$ DECAY MODES	Fraction (Γ_i/Γ)	p (MeV/c)
e^+e^-	$(1.6 \pm 0.5) \times 10^{-6}$	5509

NOTES

In this Summary Table:

When a quantity has "(S = ...)" to its right, the error on the quantity has been enlarged by the "scale factor" S, defined as $S = \sqrt{\chi^2/(N-1)}$, where N is the number of measurements used in calculating the quantity. We do this when S > 1, which often indicates that the measurements are inconsistent. When S > 1.25, we also show in the Particle Listings an ideogram of the measurements. For more about S, see the Introduction.

A decay momentum p is given for each decay mode. For a 2-body decay, p is the momentum of each decay product in the rest frame of the decaying particle. For a 3-or-more-body decay, p is the largest momentum any of the products can have in this frame.

[a] See the "Note on $\pi^\pm \to \ell^\pm \nu\gamma$ and $K^\pm \to \ell^\pm \nu\gamma$ Form Factors" in the $\pi^\pm$ Particle Listings for definitions and details.

[b] Measurements of $\Gamma(e^+\nu_e)/\Gamma(\mu^+\nu_\mu)$ always include decays with γ's, and measurements of $\Gamma(e^+\nu_e\gamma)$ and $\Gamma(\mu^+\nu_\mu\gamma)$ never include low-energy γ's. Therefore, since no clean separation is possible, we consider the modes with γ's to be subreactions of the modes without them, and let $[\Gamma(e^+\nu_e) + \Gamma(\mu^+\nu_\mu)]/\Gamma_{total} = 100\%$.

[c] See the $\pi^\pm$ Particle Listings for the energy limits used in this measurement; low-energy γ's are not included.

[d] Derived from an analysis of neutrino-oscillation experiments.

[e] Astrophysical and cosmological arguments give limits of order 10^{-13}; see the π^0 Particle Listings.

[f] See the "Note on the Decay Width $\Gamma(\eta \to \gamma\gamma)$" in our 1994 edition, Phys. Rev. **D50**, 1 August 1994, Part I, p. 1451.

[g] C parity forbids this to occur as a single-photon process.

[h] See the "Note on scalar mesons" in the $f_0(1370)$ Particle Listings. The interpretation of this entry as a particle is controversial.

[i] See the "Note on $\rho(770)$" in the $\rho(770)$ Particle Listings.

[j] The e^+e^- branching fraction is from $e^+e^- \to \pi^+\pi^-$ experiments only. The $\omega\rho$ interference is then due to $\omega\rho$ mixing only, and is expected to be small. If $e\mu$ universality holds, $\Gamma(\rho^0 \to \mu^+\mu^-) = \Gamma(\rho^0 \to e^+e^-) \times 0.99785$.

[k] See the "Note on scalar mesons" in the $f_0(1370)$ Particle Listings.

[l] See the "Note on $a_1(1260)$" in the $a_1(1260)$ Particle Listings.

[m] This is only an educated guess; the error given is larger than the error on the average of the published values. See the Particle Listings for details.

[n] See the "Note on the $f_1(1420)$" in the $\eta(1440)$ Particle Listings.

[o] See also the $\omega(1600)$ Particle Listings.

[p] See the "Note on the $\eta(1440)$" in the $\eta(1440)$ Particle Listings.

[q] See the "Note on the $\rho(1450)$ and the $\rho(1700)$" in the $\rho(1700)$ Particle Listings.

[r] See the "Note on non-$q\bar{q}$ mesons" in the Particle Listings (see the index for the page number).

[s] See also the $\omega(1420)$ Particle Listings.

[t] See the "Note on $f_J(1710)$" in the $f_J(1710)$ Particle Listings.

[u] See the note in the $K^\pm$ Particle Listings.

[v] The definition of the slope parameter g of the $K \to 3\pi$ Dalitz plot is as follows (see also "Note on Dalitz Plot Parameters for $K \to 3\pi$ Decays" in the $K^\pm$ Particle Listings):
$$|M|^2 = 1 + g(s_3 - s_0)/m_{\pi^+}^2 + \cdots.$$

[w] For more details and definitions of parameters see the Particle Listings.

[x] See the $K^\pm$ Particle Listings for the energy limits used in this measurement.

[y] Most of this radiative mode, the low-momentum γ part, is also included in the parent mode listed without γ's.

[z] Direct-emission branching fraction.

[aa] Structure-dependent part.

[bb] Derived from measured values of ϕ_{+-}, ϕ_{00}, $|\eta|$, $|m_{K_L^0} - m_{K_S^0}|$, and $\tau_{K_S^0}$, as described in the introduction to "Tests of Conservation Laws."

[cc] The CP-violation parameters are defined as follows (see also "Note on CP Violation in $K_S \to 3\pi$" and "Note on CP Violation in K_L^0 Decay" in the Particle Listings):

$$\eta_{+-} = |\eta_{+-}|e^{i\phi_{+-}} = \frac{A(K_L^0 \to \pi^+\pi^-)}{A(K_S^0 \to \pi^+\pi^-)} = \epsilon + \epsilon'$$

$$\eta_{00} = |\eta_{00}|e^{i\phi_{00}} = \frac{A(K_L^0 \to \pi^0\pi^0)}{A(K_S^0 \to \pi^0\pi^0)} = \epsilon - 2\epsilon'$$

$$\delta = \frac{\Gamma(K_L^0 \to \pi^-\ell^+\nu) - \Gamma(K_L^0 \to \pi^+\ell^-\bar\nu)}{\Gamma(K_L^0 \to \pi^-\ell^+\nu) + \Gamma(K_L^0 \to \pi^+\ell^-\bar\nu)},$$

$$\mathrm{Im}(\eta_{+-0})^2 = \frac{\Gamma(K_S^0 \to \pi^+\pi^-\pi^0)^{CP \text{ viol.}}}{\Gamma(K_L^0 \to \pi^+\pi^-\pi^0)},$$

$$\mathrm{Im}(\eta_{000})^2 = \frac{\Gamma(K_S^0 \to \pi^0\pi^0\pi^0)}{\Gamma(K_L^0 \to \pi^0\pi^0\pi^0)}.$$

where for the last two relations CPT is assumed valid, i.e., $\mathrm{Re}(\eta_{+-0}) \simeq 0$ and $\mathrm{Re}(\eta_{000}) \simeq 0$.

[dd] See the K_S^0 Particle Listings for the energy limits used in this measurement.

[ee] Calculated from K_L^0 semileptonic rates and the K_S^0 lifetime assuming $\Delta S = \Delta Q$.

[ff] ϵ'/ϵ is derived from $|\eta_{00}/\eta_{+-}|$ measurements using theoretical input on phases.

[gg] The value is for the sum of the charge states of particle/antiparticle states indicated.

[hh] See the K_L^0 Particle Listings for the energy limits used in this measurement.

[ii] Allowed by higher-order electroweak interactions.

[jj] Violates CP in leading order. Test of direct CP violation since the indirect CP-violating and CP-conserving contributions are expected to be suppressed.

[kk] See the "Note on $f_0(1370)$" in the $f_0(1370)$ Particle Listings and in the 1994 edition.

[ll] See the note in the $L(1770)$ Particle Listings in Reviews of Modern Physics **56** No. 2 Pt. II (1984), p. S200. See also the "Note on $K_2(1770)$ and the $K_2(1820)$" in the $K_2(1770)$ Particle Listings .

[mm] See the "Note on $K_2(1770)$ and the $K_2(1820)$" in the $K_2(1770)$ Particle Listings .

[nn] This is a weighted average of $D^\pm$ (44%) and D^0 (56%) branching fractions. See "D^+ and $D^0 \to$ (η anything) / (total D^+ and D^0)" under "D^+ Branching Ratios" in the Particle Listings.

[oo] This value averages the e^+ and μ^+ branching fractions, after making a small phase-space adjustment to the μ^+ fraction to be able to use it as an e^+ fraction; hence our ℓ^+ here is really an e^+.

[pp] An ℓ indicates an e or a μ mode, not a sum over these modes.

[qq] The branching fraction for this mode may differ from the sum of the submodes that contribute to it, due to interference effects. See the relevant papers in the Particle Listings.

[rr] The two experiments measuring this fraction are in serious disagreement. See the Particle Listings.

[ss] This mode is not a useful test for a $\Delta C = 1$ weak neutral current because both quarks must change flavor in this decay.

[tt] The D_1^0-D_2^0 limits are inferred from the D^0-$\overline{D}^0$ mixing ratio $\Gamma(K^+\ell^-\bar\nu_\ell(\text{via } \overline{D}^0)) / \Gamma(K^-\ell^+\nu_\ell)$.

[uu] The larger limit (from E791) allows interference between the doubly Cabibbo-suppressed and mixing amplitudes; the smaller limit (from E691) doesn't. See the papers for details.

[vv] The experiments on the division of this charge mode amongst its submodes disagree, and the submode branching fractions here add up to considerably more than the charged-mode fraction.

[ww] However, these upper limits are in serious disagreement with values obtained in another experiment.

[xx] For now, we average together measurements of the $X e^+\nu_e$ and $X\mu^+\nu_\mu$ branching fractions. This is the average, not the sum.

[yy] This branching fraction includes all the decay modes of the final-state resonance.

[zz] This value includes only K^+K^- decays of the $f_J(1710)$, because branching fractions of this resonance are not known.

[aaa] This value includes only $\pi^+\pi^-$ decays of the $f_0(1500)$, because branching fractions of this resonance are not known.

[bbb] B^0 and B_s^0 contributions not separated. Limit is on weighted average of the two decay rates.

[ccc] These values are model dependent. See 'Note on Semileptonic Decays' in the B^+ Particle Listings.

[ddd] D^{**} stands for the sum of the $D(1\,^1P_1)$, $D(1\,^3P_0)$, $D(1\,^3P_1)$, $D(1\,^3P_2)$, $D(2\,^1S_0)$, and $D(2\,^1S_1)$ resonances.

[eee] Inclusive branching fractions have a multiplicity definition and can be greater than 100%.

[fff] D_j represents an unresolved mixture of pseudoscalar and tensor D^{**} (P-wave) states.

[ggg] Not a pure measurement. See note at head of B_s^0 Decay Modes.

[hhh] Includes $p\overline{p}\pi^+\pi^-\gamma$ and excludes $p\overline{p}\eta$, $p\overline{p}\omega$, $p\overline{p}\eta'$.

[iii] J^{PC} known by production in e^+e^- via single photon annihilation. I^G is not known; interpretation of this state as a single resonance is unclear because of the expectation of substantial threshold effects in this energy region.

[jjj] Spectroscopic labeling for these states is theoretical, pending experimental information.

Meson Summary Table

See also the table of suggested $q\bar{q}$ quark-model assignments in the Quark Model section.

- Indicates particles that appear in the preceding Meson Summary Table. We do not regard the other entries as being established.

† Indicates that the value of J given is preferred, but needs confirmation.

LIGHT UNFLAVORED ($S = C = B = 0$)

$I^G(J^{PC})$

Particle	$I^G(J^{PC})$	Particle	$I^G(J^{PC})$
• $\pi^\pm$	$1^-(0^-)$	$X(1650)$	$0^+(?^{?-})$
• π^0	$1^-(0^{-+})$	• $\omega_3(1670)$	$0^-(3^{--})$
• η	$0^+(0^{-+})$	• $\pi_2(1670)$	$1^-(2^{-+})$
• $f_0(400\text{--}1200)$	$0^+(0^{++})$	• $\phi(1680)$	$0^-(1^{--})$
• $\rho(770)$	$1^+(1^{--})$	• $\rho_3(1690)$	$1^+(3^{--})$
• $\omega(782)$	$0^-(1^{--})$	• $\rho(1700)$	$1^+(1^{--})$
• $\eta'(958)$	$0^+(0^{-+})$	• $f_J(1710)$	$0^+(\text{even}^{++})$
• $f_0(980)$	$0^+(0^{++})$	$\eta(1760)$	$0^+(0^{-+})$
• $a_0(980)$	$1^-(0^{++})$	$X(1775)$	$1^-(?^{-+})$
• $\phi(1020)$	$0^-(1^{--})$	• $\pi(1800)$	$1^-(0^{-+})$
• $h_1(1170)$	$0^-(1^{+-})$	$f_2(1810)$	$0^+(2^{++})$
• $b_1(1235)$	$1^+(1^{+-})$	• $\phi_3(1850)$	$0^-(3^{--})$
• $a_1(1260)$	$1^-(1^{++})$	$\eta_2(1870)$	$0^+(2^{-+})$
• $f_2(1270)$	$0^+(2^{++})$	$X(1910)$	$0^+(?^{?+})$
• $f_1(1285)$	$0^+(1^{++})$	$f_2(1950)$	$0^+(2^{++})$
• $\eta(1295)$	$0^+(0^{-+})$	$X(2000)$	$1^-(?^{?+})$
• $\pi(1300)$	$1^-(0^{-+})$	• $f_2(2010)$	$0^+(2^{++})$
• $a_2(1320)$	$1^-(2^{++})$	$f_0(2020)$	$0^+(0^{++})$
• $f_0(1370)$	$0^+(0^{++})$	• $a_4(2040)$	$1^-(4^{++})$
$h_1(1380)$	$?^-(1^{+-})$	• $f_4(2050)$	$0^+(4^{++})$
$\hat{\rho}(1405)$	$1^-(1^{-+})$	$f_0(2060)$	$0^+(0^{++})$
• $f_1(1420)$	$0^+(1^{++})$	$\pi_2(2100)$	$1^-(2^{-+})$
• $\omega(1420)$	$0^-(1^{--})$	$f_2(2150)$	$0^+(2^{++})$
$f_2(1430)$	$0^+(2^{++})$	$\rho(2150)$	$1^+(1^{--})$
• $\eta(1440)$	$0^+(0^{-+})$	$f_0(2200)$	$0^+(0^{++})$
• $a_0(1450)$	$1^-(0^{++})$	$f_J(2220)$	$0^+(2^{++}\text{ or }4^{++})$
• $\rho(1450)$	$1^+(1^{--})$	$\eta(2225)$	$0^+(0^{-+})$
• $f_0(1500)$	$0^+(0^{++})$	$\rho_3(2250)$	$1^+(3^{--})$
$f_1(1510)$	$0^+(1^{++})$	• $f_2(2300)$	$0^+(2^{++})$
• $f_2'(1525)$	$0^+(2^{++})$	$f_4(2300)$	$0^+(4^{++})$
$f_2(1565)$	$0^+(2^{++})$	• $f_2(2340)$	$0^+(2^{++})$
• $\omega(1600)$	$0^-(1^{--})$	$\rho_5(2350)$	$1^+(5^{--})$
$X(1600)$	$2^+(2^{++})$	$a_6(2450)$	$1^-(6^{++})$
$f_2(1640)$	$0^+(2^{++})$	$f_6(2510)$	$0^+(6^{++})$
$\eta_2(1645)$	$0^+(2^{-+})$	$X(3250)$	$?^?(?^{??})$

OTHER LIGHT UNFLAVORED ($S = C = B = 0$)

Particle	$I^G(J^{PC})$
$e^+e^-(1100\text{--}2200)$	$?^?(1^{--})$
$\bar{N}N(1100\text{--}3600)$	
$X(1900\text{--}3600)$	

STRANGE ($S = \pm1$, $C = B = 0$)

$I(J^P)$

Particle	$I(J^P)$
• $K^\pm$	$1/2(0^-)$
• K^0	$1/2(0^-)$
• K_S^0	$1/2(0^-)$
• K_L^0	$1/2(0^-)$
• $K^*(892)$	$1/2(1^-)$
• $K_1(1270)$	$1/2(1^+)$
• $K_1(1400)$	$1/2(1^+)$
• $K^*(1410)$	$1/2(1^-)$
• $K_0^*(1430)$	$1/2(0^+)$
• $K_2^*(1430)$	$1/2(2^+)$
$K(1460)$	$1/2(0^-)$
$K_2(1580)$	$1/2(2^-)$
$K_1(1650)$	$1/2(1^+)$
• $K^*(1680)$	$1/2(1^-)$
• $K_2(1770)$	$1/2(2^-)$
• $K_3^*(1780)$	$1/2(3^-)$
• $K_2(1820)$	$1/2(2^-)$
$K(1830)$	$1/2(0^-)$
$K_0^*(1950)$	$1/2(0^+)$
$K_2^*(1980)$	$1/2(2^+)$
• $K_4^*(2045)$	$1/2(4^+)$
$K_2(2250)$	$1/2(2^-)$
$K_3(2320)$	$1/2(3^+)$
$K_5^*(2380)$	$1/2(5^-)$
$K_4(2500)$	$1/2(4^-)$
$K(3100)$	$?^?(?^{??})$

CHARMED ($C = \pm1$)

Particle	$I(J^P)$
• $D^\pm$	$1/2(0^-)$
• D^0	$1/2(0^-)$
• $D^*(2007)^0$	$1/2(1^-)$
• $D^*(2010)^\pm$	$1/2(1^-)$
• $D_1(2420)^0$	$1/2(1^+)$
$D_1(2420)^\pm$	$1/2(?^?)$
• $D_2^*(2460)^0$	$1/2(2^+)$
• $D_2^*(2460)^+$	$1/2(2^+)$

CHARMED, STRANGE ($C = S = \pm1$)

Particle	$I(J^P)$
• $D_s^\pm$	$0(0^-)$
• $D_s^{*\pm}$	$0(?^?)$
• $D_{s1}(2536)^\pm$	$0(1^+)$
• $D_{sJ}(2573)^\pm$	$0(?^?)$

BOTTOM ($B = \pm1$)

Particle	$I(J^P)$
• $B^\pm$	$1/2(0^-)$
• B^0	$1/2(0^-)$
• B^*	$1/2(1^-)$
$B_J^*(5732)$	$?(?^?)$

BOTTOM, STRANGE ($B = \pm1$, $S = \mp1$)

$I^G(J^{PC})$

Particle	$I^G(J^{PC})$
• B_s^0	$0(0^-)$
B_s^*	$0(1^-)$
$B_{sJ}^*(5850)$	$?(?^?)$

BOTTOM, CHARMED ($B = C = \pm1$)

Particle	$I^G(J^{PC})$
$B_c^\pm$	$0(0^-)$

$c\bar{c}$

Particle	$I^G(J^{PC})$
• $\eta_c(1S)$	$0^+(0^{-+})$
• $J/\psi(1S)$	$0^-(1^{--})$
• $\chi_{c0}(1P)$	$0^+(0^{++})$
• $\chi_{c1}(1P)$	$0^+(1^{++})$
$h_c(1P)$	$?^?(?^{??})$
• $\chi_{c2}(1P)$	$0^+(2^{++})$
$\eta_c(2S)$	$?^?(?^{?+})$
• $\psi(2S)$	$0^-(1^{--})$
• $\psi(3770)$	$?^?(1^{--})$
• $\psi(4040)$	$?^?(1^{--})$
• $\psi(4160)$	$?^?(1^{--})$
• $\psi(4415)$	$?^?(1^{--})$

$b\bar{b}$

Particle	$I^G(J^{PC})$
• $\Upsilon(1S)$	$0^-(1^{--})$
• $\chi_{b0}(1P)$	$0^+(0^{++})$
• $\chi_{b1}(1P)$	$0^+(1^{++})$
• $\chi_{b2}(1P)$	$0^+(2^{++})$
• $\Upsilon(2S)$	$0^-(1^{--})$
• $\chi_{b0}(2P)$	$0^+(0^{++})$
• $\chi_{b1}(2P)$	$0^+(1^{++})$
• $\chi_{b2}(2P)$	$0^+(2^{++})$
• $\Upsilon(3S)$	$0^-(1^{--})$
• $\Upsilon(4S)$	$?^?(1^{--})$
• $\Upsilon(10860)$	$?^?(1^{--})$
• $\Upsilon(11020)$	$?^?(1^{--})$

NON-$q\bar{q}$ CANDIDATES

Non-$q\bar{q}$ Candidates

This short table gives the name, the quantum numbers (where known), and the status of baryons in the Review. Only the baryons with 3- or 4-star status are included in the main Baryon Summary Table. Due to insufficient data or uncertain interpretation, the other entries in the short table are not established as baryons. The names with masses are of baryons that decay strongly. See our 1986 edition (Physics Letters **170B**) for listings of evidence for Z baryons (KN resonances).

p	P_{11}	****	$\Delta(1232)$	P_{33}	****	Λ	P_{01}	****	Σ^+	P_{11}	****	Ξ^0		P_{11}	****
n	P_{11}	****	$\Delta(1600)$	P_{33}	***	$\Lambda(1405)$	S_{01}	****	Σ^0	P_{11}	****	Ξ^-		P_{11}	****
$N(1440)$	P_{11}	****	$\Delta(1620)$	S_{31}	****	$\Lambda(1520)$	D_{03}	****	Σ^-	P_{11}	****	$\Xi(1530)$		P_{13}	****
$N(1520)$	D_{13}	****	$\Delta(1700)$	D_{33}	****	$\Lambda(1600)$	P_{01}	***	$\Sigma(1385)$	P_{13}	****	$\Xi(1620)$			*
$N(1535)$	S_{11}	****	$\Delta(1750)$	P_{31}	*	$\Lambda(1670)$	S_{01}	****	$\Sigma(1480)$		*	$\Xi(1690)$			***
$N(1650)$	S_{11}	****	$\Delta(1900)$	S_{31}	**	$\Lambda(1690)$	D_{03}	****	$\Sigma(1560)$		**	$\Xi(1820)$		D_{13}	***
$N(1675)$	D_{15}	****	$\Delta(1905)$	F_{35}	****	$\Lambda(1800)$	S_{01}	***	$\Sigma(1580)$	D_{13}	**	$\Xi(1950)$			***
$N(1680)$	F_{15}	****	$\Delta(1910)$	P_{31}	****	$\Lambda(1810)$	P_{01}	***	$\Sigma(1620)$	S_{11}	**	$\Xi(2030)$			***
$N(1700)$	D_{13}	***	$\Delta(1920)$	P_{33}	***	$\Lambda(1820)$	F_{05}	****	$\Sigma(1660)$	P_{11}	***	$\Xi(2120)$			*
$N(1710)$	P_{11}	***	$\Delta(1930)$	D_{35}	***	$\Lambda(1830)$	D_{05}	****	$\Sigma(1670)$	D_{13}	****	$\Xi(2250)$			**
$N(1720)$	P_{13}	****	$\Delta(1940)$	D_{33}	*	$\Lambda(1890)$	P_{03}	****	$\Sigma(1690)$		**	$\Xi(2370)$			**
$N(1900)$	P_{13}	**	$\Delta(1950)$	F_{37}	****	$\Lambda(2000)$		*	$\Sigma(1750)$	S_{11}	***	$\Xi(2500)$			*
$N(1990)$	F_{17}	**	$\Delta(2000)$	F_{35}	**	$\Lambda(2020)$	F_{07}	*	$\Sigma(1770)$	P_{11}	*				
$N(2000)$	F_{15}	**	$\Delta(2150)$	S_{31}	*	$\Lambda(2100)$	G_{07}	****	$\Sigma(1775)$	D_{15}	****	Ω^-			****
$N(2080)$	D_{13}	**	$\Delta(2200)$	G_{37}	*	$\Lambda(2110)$	F_{05}	***	$\Sigma(1840)$	P_{13}	*	$\Omega(2250)^-$			***
$N(2090)$	S_{11}	*	$\Delta(2300)$	H_{39}	**	$\Lambda(2325)$	D_{03}	*	$\Sigma(1880)$	P_{11}	**	$\Omega(2380)^-$			**
$N(2100)$	P_{11}	*	$\Delta(2350)$	D_{35}	*	$\Lambda(2350)$	H_{09}	***	$\Sigma(1915)$	F_{15}	****	$\Omega(2470)^-$			**
$N(2190)$	G_{17}	****	$\Delta(2390)$	F_{37}	*	$\Lambda(2585)$		**	$\Sigma(1940)$	D_{13}	***				
$N(2200)$	D_{15}	**	$\Delta(2400)$	G_{39}	**				$\Sigma(2000)$	S_{11}	*	Λ_c^+			****
$N(2220)$	H_{19}	****	$\Delta(2420)$	$H_{3,11}$	****				$\Sigma(2030)$	F_{17}	****	$\Lambda_c(2593)^+$			***
$N(2250)$	G_{19}	****	$\Delta(2750)$	$I_{3,13}$	**				$\Sigma(2070)$	F_{15}	*	$\Lambda_c(2625)^+$			***
$N(2600)$	$I_{1,11}$	***	$\Delta(2950)$	$K_{3,15}$	**				$\Sigma(2080)$	P_{13}	**	$\Sigma_c(2455)$			****
$N(2700)$	$K_{1,13}$	**							$\Sigma(2100)$	G_{17}	*	$\Sigma_c(2520)$			***
									$\Sigma(2250)$		***	Ξ_c^+			***
									$\Sigma(2455)$		**	Ξ_c^0			***
									$\Sigma(2620)$		**	$\Xi_c(2645)$			***
									$\Sigma(3000)$		*	Ω_c^0			***
									$\Sigma(3170)$		*				
												Λ_b^0			***
												Ξ_b^0, Ξ_b^-			*

**** Existence is certain, and properties are at least fairly well explored.

*** Existence ranges from very likely to certain, but further confirmation is desirable and/or quantum numbers, branching fractions, etc. are not well determined.

** Evidence of existence is only fair.

* Evidence of existence is poor.

N BARYONS
$(S = 0, I = 1/2)$

$p, N^+ = uud; \quad n, N^0 = udd$

p

$I(J^P) = \frac{1}{2}(\frac{1}{2}^+)$

Mass $m = 938.27231 \pm 0.00028$ MeV [a]
$= 1.007276470 \pm 0.000000012$ u
$|\frac{q_{\bar{p}}}{m_{\bar{p}}}|/(\frac{q_p}{m_p}) = 1.0000000015 \pm 0.0000000011$
$|q_p + q_{\bar{p}}|/e < 2 \times 10^{-5}$
$|q_p + q_e|/e < 1.0 \times 10^{-21}$ [b]
Magnetic moment $\mu = 2.79284739 \pm 0.00000006 \, \mu_N$
Electric dipole moment $d = (-4 \pm 6) \times 10^{-23}$ ecm
Electric polarizability $\bar{\alpha} = (12.1 \pm 0.9) \times 10^{-4}$ fm^3
Magnetic polarizability $\bar{\beta} = (2.1 \pm 0.9) \times 10^{-4}$ fm^3
Mean life $\tau > 1.6 \times 10^{25}$ years (independent of mode)
$> 10^{31}$ to 5×10^{32} years [c] (mode dependent)

Below, for N decays, p and n distinguish proton and neutron partial lifetimes. See also the "Note on Nucleon Decay" in our 1994 edition (Phys. Rev. **D50**, 1673) for a short review.

The "partial mean life" limits tabulated here are the limits on τ/B_i, where τ is the total mean life and B_i is the branching fraction for the mode in question.

p DECAY MODES	Partial mean life $(10^{30}$ years$)$	Confidence level	p (MeV/c)
Antilepton + meson			
$N \to e^+ \pi$	$> 130 \, (n), > 550 \, (p)$	90%	459
$N \to \mu^+ \pi$	$> 100 \, (n), > 270 \, (p)$	90%	453
$N \to \nu \pi$	$> 100 \, (n), > 25 \, (p)$	90%	459
$p \to e^+ \eta$	> 140	90%	309
$p \to \mu^+ \eta$	> 69	90%	296
$n \to \nu \eta$	> 54	90%	310
$N \to e^+ \rho$	$> 58 \, (n), > 75 \, (p)$	90%	153
$N \to \mu^+ \rho$	$> 23 \, (n), > 110 \, (p)$	90%	119
$N \to \nu \rho$	$> 19 \, (n), > 27 \, (p)$	90%	153
$p \to e^+ \omega$	> 45	90%	142
$p \to \mu^+ \omega$	> 57	90%	104
$n \to \nu \omega$	> 43	90%	144
$N \to e^+ K$	$> 1.3 \, (n), > 150 \, (p)$	90%	337
$p \to e^+ K^0_S$	> 76	90%	337
$p \to e^+ K^0_L$	> 44	90%	337
$N \to \mu^+ K$	$> 1.1 \, (n), > 120 \, (p)$	90%	326
$p \to \mu^+ K^0_S$	> 64	90%	326
$p \to \mu^+ K^0_L$	> 44	90%	326
$N \to \nu K$	$> 86 \, (n), > 100 \, (p)$	90%	339
$p \to e^+ K^*(892)^0$	> 52	90%	45
$N \to \nu K^*(892)$	$> 22 \, (n), > 20 \, (p)$	90%	45
Antilepton + mesons			
$p \to e^+ \pi^+ \pi^-$	> 21	90%	448
$p \to e^+ \pi^0 \pi^0$	> 38	90%	449
$n \to e^+ \pi^- \pi^0$	> 32	90%	449
$p \to \mu^+ \pi^+ \pi^-$	> 17	90%	425
$p \to \mu^+ \pi^0 \pi^0$	> 33	90%	427
$n \to \mu^+ \pi^- \pi^0$	> 33	90%	427
$n \to e^+ K^0 \pi^-$	> 18	90%	319
Lepton + meson			
$n \to e^- \pi^+$	> 65	90%	459
$n \to \mu^- \pi^+$	> 49	90%	453
$n \to e^- \rho^+$	> 62	90%	154
$n \to \mu^- \rho^+$	> 7	90%	120
$n \to e^- K^+$	> 32	90%	340
$n \to \mu^- K^+$	> 57	90%	330
Lepton + mesons			
$p \to e^- \pi^+ \pi^+$	> 30	90%	448
$n \to e^- \pi^+ \pi^0$	> 29	90%	449
$p \to \mu^- \pi^+ \pi^+$	> 17	90%	425
$n \to \mu^- \pi^+ \pi^0$	> 34	90%	427
$p \to e^- \pi^+ K^+$	> 20	90%	320
$p \to \mu^- \pi^+ K^+$	> 5	90%	279

	Partial mean life	Confidence level	p (MeV/c)
Antilepton + photon(s)			
$p \to e^+ \gamma$	> 460	90%	469
$p \to \mu^+ \gamma$	> 380	90%	463
$n \to \nu \gamma$	> 24	90%	470
$p \to e^+ \gamma \gamma$	> 100	90%	469
Three (or more) leptons			
$p \to e^+ e^+ e^-$	> 510	90%	469
$p \to e^+ \mu^+ \mu^-$	> 81	90%	457
$p \to e^+ \nu \nu$	> 11	90%	469
$n \to e^+ e^- \nu$	> 74	90%	470
$n \to \mu^+ e^- \nu$	> 47	90%	464
$n \to \mu^+ \mu^- \nu$	> 42	90%	458
$p \to \mu^+ e^+ e^-$	> 91	90%	464
$p \to \mu^+ \mu^+ \mu^-$	> 190	90%	439
$p \to \mu^+ \nu \nu$	> 21	90%	463
$p \to e^- \mu^+ \mu^+$	> 6	90%	457
$n \to 3\nu$	> 0.0005	90%	470
Inclusive modes			
$N \to e^+$ anything	$> 0.6 \, (n, p)$	90%	–
$N \to \mu^+$ anything	$> 12 \, (n, p)$	90%	–
$N \to e^+ \pi^0$ anything	$> 0.6 \, (n, p)$	90%	–
$\Delta B = 2$ dinucleon modes			

The following are lifetime limits per iron nucleus.

	Partial mean life	Confidence level	p (MeV/c)
$pp \to \pi^+ \pi^+$	> 0.7	90%	–
$pn \to \pi^+ \pi^0$	> 2	90%	–
$nn \to \pi^+ \pi^-$	> 0.7	90%	–
$nn \to \pi^0 \pi^0$	> 3.4	90%	–
$pp \to e^+ e^+$	> 5.8	90%	–
$pp \to e^+ \mu^+$	> 3.6	90%	–
$pp \to \mu^+ \mu^+$	> 1.7	90%	–
$pn \to e^+ \bar{\nu}$	> 2.8	90%	–
$pn \to \mu^+ \bar{\nu}$	> 1.6	90%	–
$nn \to \nu_e \bar{\nu}_e$	> 0.000012	90%	–
$nn \to \nu_\mu \bar{\nu}_\mu$	> 0.000006	90%	–

$\bar{p}$ DECAY MODES

$\bar{p}$ DECAY MODES	Partial mean life (years)	Confidence level	p (MeV/c)
$\bar{p} \to e^- \gamma$	> 1848	95%	469
$\bar{p} \to e^- \pi^0$	> 554	95%	459
$\bar{p} \to e^- \eta$	> 171	95%	309
$\bar{p} \to e^- K^0_S$	> 29	95%	337
$\bar{p} \to e^- K^0_L$	> 9	95%	337

n

$I(J^P) = \frac{1}{2}(\frac{1}{2}^+)$

Mass $m = 939.56563 \pm 0.00028$ MeV [a]
$= 1.008664904 \pm 0.000000014$ u
$m_n - m_p = 1.293318 \pm 0.000009$ MeV
$= 0.001388434 \pm 0.000000009$ u
Mean life $\tau = 886.7 \pm 1.9$ s $(S = 1.2)$
$c\tau = 2.658 \times 10^8$ km
Magnetic moment $\mu = -1.9130428 \pm 0.0000005 \, \mu_N$
Electric dipole moment $d < 0.97 \times 10^{-25}$ ecm, CL = 90%
Electric polarizability $\alpha = (0.98^{+0.19}_{-0.23}) \times 10^{-3}$ fm^3 $(S = 1.1)$
Charge $q = (-0.4 \pm 1.1) \times 10^{-21} \, e$
Mean $n\bar{n}$-oscillation time $> 1.2 \times 10^8$ s, CL = 90% [d] (bound n)
$> 0.86 \times 10^8$ s, CL = 90% (free n)

Decay parameters [e]

$pe^- \bar{\nu}_e$	$g_A/g_V = -1.2670 \pm 0.0035$	$(S = 1.9)$
"	$A = -0.1162 \pm 0.0013$	$(S = 1.8)$
"	$B = 0.990 \pm 0.008$	
"	$a = -0.102 \pm 0.005$	
"	$\phi_{AV} = (180.07 \pm 0.18)°$ [f]	
"	$D = (-0.5 \pm 1.4) \times 10^{-3}$	

n DECAY MODES	Fraction (Γ_i/Γ)	Confidence level	p (MeV/c)
$pe^- \bar{\nu}_e$	100 %		1.19
Charge conservation (Q) violating mode			
$p\nu_e \bar{\nu}_e$	$Q \quad < 8 \times 10^{-27}$	68%	1.29

$N(1440)$ P_{11} $I(J^P) = \frac{1}{2}(\frac{1}{2}^+)$

Breit-Wigner mass = 1430 to 1470 (≈ 1440) MeV
Breit-Wigner full width = 250 to 450 (≈ 350) MeV
$p_{beam} = 0.61$ GeV/c $4\pi \lambda^2 = 31.0$ mb
Re(pole position) = 1345 to 1385 (≈ 1365) MeV
-2Im(pole position) = 160 to 260 (≈ 210) MeV

$N(1440)$ DECAY MODES	Fraction (Γ_i/Γ)	p (MeV/c)
$N\pi$	60–70 %	397
$N\pi\pi$	30–40 %	342
$\quad \Delta\pi$	20–30 %	143
$\quad N\rho$	<8 %	†
$\quad N(\pi\pi)_{S\text{-wave}}^{I=0}$	5–10 %	–
$p\gamma$	0.035–0.048 %	414
$\quad p\gamma$, helicity=1/2	0.035–0.048 %	414
$n\gamma$	0.009–0.032 %	413
$\quad n\gamma$, helicity=1/2	0.009–0.032 %	413

$N(1520)$ D_{13} $I(J^P) = \frac{1}{2}(\frac{3}{2}^-)$

Breit-Wigner mass = 1515 to 1530 (≈ 1520) MeV
Breit-Wigner full width = 110 to 135 (≈ 120) MeV
$p_{beam} = 0.74$ GeV/c $4\pi \lambda^2 = 23.5$ mb
Re(pole position) = 1505 to 1515 (≈ 1510) MeV
-2Im(pole position) = 110 to 120 (≈ 115) MeV

$N(1520)$ DECAY MODES	Fraction (Γ_i/Γ)	p (MeV/c)
$N\pi$	50–60 %	456
$N\pi\pi$	40–50 %	410
$\quad \Delta\pi$	15–25 %	228
$\quad N\rho$	15–25 %	†
$\quad N(\pi\pi)_{S\text{-wave}}^{I=0}$	<8 %	–
$p\gamma$	0.46–0.56 %	470
$\quad p\gamma$, helicity=1/2	0.001–0.034 %	470
$\quad p\gamma$, helicity=3/2	0.44–0.53 %	470
$n\gamma$	0.30–0.53 %	470
$\quad n\gamma$, helicity=1/2	0.04–0.10 %	470
$\quad n\gamma$, helicity=3/2	0.25–0.45 %	470

$N(1535)$ S_{11} $I(J^P) = \frac{1}{2}(\frac{1}{2}^-)$

Breit-Wigner mass = 1520 to 1555 (≈ 1535) MeV
Breit-Wigner full width = 100 to 250 (≈ 150) MeV
$p_{beam} = 0.76$ GeV/c $4\pi \lambda^2 = 22.5$ mb
Re(pole position) = 1495 to 1515 (≈ 1505) MeV
-2Im(pole position) = 90 to 250 (≈ 170) MeV

$N(1535)$ DECAY MODES	Fraction (Γ_i/Γ)	p (MeV/c)
$N\pi$	35–55 %	467
$N\eta$	30–55 %	182
$N\pi\pi$	1–10 %	422
$\quad \Delta\pi$	<1 %	242
$\quad N\rho$	<4 %	†
$\quad N(\pi\pi)_{S\text{-wave}}^{I=0}$	<3 %	–
$\quad N(1440)\pi$	<7 %	†
$p\gamma$	0.15–0.35 %	481
$\quad p\gamma$, helicity=1/2	0.15–0.35 %	481
$n\gamma$	0.004–0.29 %	480
$\quad n\gamma$, helicity=1/2	0.004–0.29 %	480

$N(1650)$ S_{11} $I(J^P) = \frac{1}{2}(\frac{1}{2}^-)$

Breit-Wigner mass = 1640 to 1680 (≈ 1650) MeV
Breit-Wigner full width = 145 to 190 (≈ 150) MeV
$p_{beam} = 0.96$ GeV/c $4\pi \lambda^2 = 16.4$ mb
Re(pole position) = 1640 to 1680 (≈ 1660) MeV
-2Im(pole position) = 150 to 170 (≈ 160) MeV

$N(1650)$ DECAY MODES	Fraction (Γ_i/Γ)	p (MeV/c)
$N\pi$	55–90 %	547
$N\eta$	3–10 %	346
ΛK	3–11 %	161
$N\pi\pi$	10–20 %	511
$\quad \Delta\pi$	1–7 %	344
$\quad N\rho$	4–12 %	†
$\quad N(\pi\pi)_{S\text{-wave}}^{I=0}$	<4 %	–
$\quad N(1440)\pi$	<5 %	147
$p\gamma$	0.04–0.18 %	558
$\quad p\gamma$, helicity=1/2	0.04–0.18 %	558
$n\gamma$	0.003–0.17 %	557
$\quad n\gamma$, helicity=1/2	0.003–0.17 %	557

$N(1675)$ D_{15} $I(J^P) = \frac{1}{2}(\frac{5}{2}^-)$

Breit-Wigner mass = 1670 to 1685 (≈ 1675) MeV
Breit-Wigner full width = 140 to 180 (≈ 150) MeV
$p_{beam} = 1.01$ GeV/c $4\pi \lambda^2 = 15.4$ mb
Re(pole position) = 1655 to 1665 (≈ 1660) MeV
-2Im(pole position) = 125 to 155 (≈ 140) MeV

$N(1675)$ DECAY MODES	Fraction (Γ_i/Γ)	p (MeV/c)
$N\pi$	40–50 %	563
ΛK	<1 %	209
$N\pi\pi$	50–60 %	529
$\quad \Delta\pi$	50–60 %	364
$\quad N\rho$	< 1–3 %	†
$p\gamma$	0.004–0.023 %	575
$\quad p\gamma$, helicity=1/2	0.0–0.015 %	575
$\quad p\gamma$, helicity=3/2	0.0–0.011 %	575
$n\gamma$	0.02–0.12 %	574
$\quad n\gamma$, helicity=1/2	0.006–0.046 %	574
$\quad n\gamma$, helicity=3/2	0.01–0.08 %	574

$N(1680)$ F_{15} $I(J^P) = \frac{1}{2}(\frac{5}{2}^+)$

Breit-Wigner mass = 1675 to 1690 (≈ 1680) MeV
Breit-Wigner full width = 120 to 140 (≈ 130) MeV
$p_{beam} = 1.01$ GeV/c $4\pi \lambda^2 = 15.2$ mb
Re(pole position) = 1665 to 1675 (≈ 1670) MeV
-2Im(pole position) = 105 to 135 (≈ 120) MeV

$N(1680)$ DECAY MODES	Fraction (Γ_i/Γ)	p (MeV/c)
$N\pi$	60–70 %	567
$N\pi\pi$	30–40 %	532
$\quad \Delta\pi$	5–15 %	369
$\quad N\rho$	3–15 %	†
$\quad N(\pi\pi)_{S\text{-wave}}^{I=0}$	5–20 %	–
$p\gamma$	0.21–0.32 %	578
$\quad p\gamma$, helicity=1/2	0.001–0.011 %	578
$\quad p\gamma$, helicity=3/2	0.20–0.32 %	578
$n\gamma$	0.021–0.046 %	577
$\quad n\gamma$, helicity=1/2	0.004–0.029 %	577
$\quad n\gamma$, helicity=3/2	0.01–0.024 %	577

$N(1700)$ D_{13} — $I(J^P) = \frac{1}{2}(\frac{3}{2}^-)$

Breit-Wigner mass = 1650 to 1750 ($\approx$ 1700) MeV
Breit-Wigner full width = 50 to 150 ($\approx$ 100) MeV
p_{beam} = 1.05 GeV/c $4\pi\lambda^2$ = 14.5 mb
Re(pole position) = 1630 to 1730 ($\approx$ 1680) MeV
$-$2Im(pole position) = 50 to 150 ($\approx$ 100) MeV

$N(1700)$ DECAY MODES	Fraction (Γ_i/Γ)	p (MeV/c)
$N\pi$	5–15 %	580
ΛK	<3 %	250
$N\pi\pi$	85–95 %	547
$N\rho$	<35 %	†
$p\gamma$	0.01–0.05 %	591
$p\gamma$, helicity=1/2	0.0–0.024 %	591
$p\gamma$, helicity=3/2	0.002–0.026 %	591
$n\gamma$	0.01–0.13 %	590
$n\gamma$, helicity=1/2	0.0–0.09 %	590
$n\gamma$, helicity=3/2	0.01–0.05 %	590

$N(1710)$ P_{11} — $I(J^P) = \frac{1}{2}(\frac{1}{2}^+)$

Breit-Wigner mass = 1680 to 1740 ($\approx$ 1710) MeV
Breit-Wigner full width = 50 to 250 ($\approx$ 100) MeV
p_{beam} = 1.07 GeV/c $4\pi\lambda^2$ = 14.2 mb
Re(pole position) = 1670 to 1770 ($\approx$ 1720) MeV
$-$2Im(pole position) = 80 to 380 ($\approx$ 230) MeV

$N(1710)$ DECAY MODES	Fraction (Γ_i/Γ)	p (MeV/c)
$N\pi$	10–20 %	587
ΛK	5–25 %	264
$N\pi\pi$	40–90 %	554
$\Delta\pi$	15–40 %	393
$N\rho$	5–25 %	48
$N(\pi\pi)_{S\text{-wave}}^{I=0}$	10–40 %	–
$p\gamma$	0.002–0.05%	598
$p\gamma$, helicity=1/2	0.002–0.05%	598
$n\gamma$	0.0–0.02%	597
$n\gamma$, helicity=1/2	0.0–0.02%	597

$N(1720)$ P_{13} — $I(J^P) = \frac{1}{2}(\frac{3}{2}^+)$

Breit-Wigner mass = 1650 to 1750 ($\approx$ 1720) MeV
Breit-Wigner full width = 100 to 200 ($\approx$ 150) MeV
p_{beam} = 1.09 GeV/c $4\pi\lambda^2$ = 13.9 mb
Re(pole position) = 1650 to 1750 ($\approx$ 1700) MeV
$-$2Im(pole position) = 110 to 390 ($\approx$ 250) MeV

$N(1720)$ DECAY MODES	Fraction (Γ_i/Γ)	p (MeV/c)
$N\pi$	10–20 %	594
ΛK	1–15 %	278
$N\pi\pi$	>70 %	561
$N\rho$	70–85 %	104
$p\gamma$	0.003–0.10 %	604
$p\gamma$, helicity=1/2	0.003–0.08 %	604
$p\gamma$, helicity=3/2	0.001–0.03 %	604
$n\gamma$	0.002–0.39 %	603
$n\gamma$, helicity=1/2	0.0–0.002 %	603
$n\gamma$, helicity=3/2	0.001–0.39 %	603

$N(2190)$ G_{17} — $I(J^P) = \frac{1}{2}(\frac{7}{2}^-)$

Breit-Wigner mass = 2100 to 2200 ($\approx$ 2190) MeV
Breit-Wigner full width = 350 to 550 ($\approx$ 450) MeV
p_{beam} = 2.07 GeV/c $4\pi\lambda^2$ = 6.21 mb
Re(pole position) = 1950 to 2150 ($\approx$ 2050) MeV
$-$2Im(pole position) = 350 to 550 ($\approx$ 450) MeV

$N(2190)$ DECAY MODES	Fraction (Γ_i/Γ)	p (MeV/c)
$N\pi$	10–20 %	888

$N(2220)$ H_{19} — $I(J^P) = \frac{1}{2}(\frac{9}{2}^+)$

Breit-Wigner mass = 2180 to 2310 ($\approx$ 2220) MeV
Breit-Wigner full width = 320 to 550 ($\approx$ 400) MeV
p_{beam} = 2.14 GeV/c $4\pi\lambda^2$ = 5.97 mb
Re(pole position) = 2100 to 2240 ($\approx$ 2170) MeV
$-$2Im(pole position) = 370 to 570 ($\approx$ 470) MeV

$N(2220)$ DECAY MODES	Fraction (Γ_i/Γ)	p (MeV/c)
$N\pi$	10–20 %	905

$N(2250)$ G_{19} — $I(J^P) = \frac{1}{2}(\frac{9}{2}^-)$

Breit-Wigner mass = 2170 to 2310 ($\approx$ 2250) MeV
Breit-Wigner full width = 290 to 470 ($\approx$ 400) MeV
p_{beam} = 2.21 GeV/c $4\pi\lambda^2$ = 5.74 mb
Re(pole position) = 2080 to 2200 ($\approx$ 2140) MeV
$-$2Im(pole position) = 280 to 680 ($\approx$ 480) MeV

$N(2250)$ DECAY MODES	Fraction (Γ_i/Γ)	p (MeV/c)
$N\pi$	5–15 %	923

$N(2600)$ $I_{1,11}$ — $I(J^P) = \frac{1}{2}(\frac{11}{2}^-)$

Breit-Wigner mass = 2550 to 2750 ($\approx$ 2600) MeV
Breit-Wigner full width = 500 to 800 ($\approx$ 650) MeV
p_{beam} = 3.12 GeV/c $4\pi\lambda^2$ = 3.86 mb

$N(2600)$ DECAY MODES	Fraction (Γ_i/Γ)	p (MeV/c)
$N\pi$	5–10 %	1126

Δ BARYONS
($S = 0$, $I = 3/2$)

$\Delta^{++} = uuu$, $\Delta^+ = uud$, $\Delta^0 = udd$, $\Delta^- = ddd$

$\Delta(1232)$ P_{33} — $I(J^P) = \frac{3}{2}(\frac{3}{2}^+)$

Breit-Wigner mass (mixed charges) = 1230 to 1234 ($\approx$ 1232) MeV
Breit-Wigner full width (mixed charges) = 115 to 125 ($\approx$ 120) MeV
p_{beam} = 0.30 GeV/c $4\pi\lambda^2$ = 94.8 mb
Re(pole position) = 1209 to 1211 ($\approx$ 1210) MeV
$-$2Im(pole position) = 98 to 102 ($\approx$ 100) MeV

$\Delta(1232)$ DECAY MODES	Fraction (Γ_i/Γ)	p (MeV/c)
$N\pi$	>99 %	227
$N\gamma$	0.52–0.60 %	259
$N\gamma$, helicity=1/2	0.11–0.13 %	259
$N\gamma$, helicity=3/2	0.41–0.47 %	259

$\Delta(1600)$ P_{33} — $I(J^P) = \frac{3}{2}(\frac{3}{2}^+)$

Breit-Wigner mass = 1550 to 1700 ($\approx$ 1600) MeV
Breit-Wigner full width = 250 to 450 ($\approx$ 350) MeV
p_{beam} = 0.87 GeV/c $4\pi\lambda^2$ = 18.6 mb
Re(pole position) = 1500 to 1700 ($\approx$ 1600) MeV
$-$2Im(pole position) = 200 to 400 ($\approx$ 300) MeV

$\Delta(1600)$ DECAY MODES	Fraction (Γ_i/Γ)	p (MeV/c)
$N\pi$	10–25 %	512
$N\pi\pi$	75–90 %	473
$\Delta\pi$	40–70 %	301
$N\rho$	<25 %	†
$N(1440)\pi$	10–35 %	74
$N\gamma$	0.001–0.02 %	525
$N\gamma$, helicity=1/2	0.0–0.02 %	525
$N\gamma$, helicity=3/2	0.001–0.005 %	525

$\Delta(1620)$ S_{31} $I(J^P) = \frac{3}{2}(\frac{1}{2}^-)$

Breit-Wigner mass = 1615 to 1675 (≈ 1620) MeV
Breit-Wigner full width = 120 to 180 (≈ 150) MeV
$p_{beam} = 0.91$ GeV/c $4\pi\lambda^2 = 17.7$ mb
Re(pole position) = 1580 to 1620 (≈ 1600) MeV
-2Im(pole position) = 100 to 130 (≈ 115) MeV

$\Delta(1620)$ DECAY MODES	Fraction (Γ_i/Γ)	p (MeV/c)
$N\pi$	20–30 %	526
$N\pi\pi$	70–80 %	488
$\Delta\pi$	30–60 %	318
$N\rho$	7–25 %	†
$N\gamma$	0.004–0.044 %	538
$N\gamma$, helicity=1/2	0.004–0.044 %	538

$\Delta(1700)$ D_{33} $I(J^P) = \frac{3}{2}(\frac{3}{2}^-)$

Breit-Wigner mass = 1670 to 1770 (≈ 1700) MeV
Breit-Wigner full width = 200 to 400 (≈ 300) MeV
$p_{beam} = 1.05$ GeV/c $4\pi\lambda^2 = 14.5$ mb
Re(pole position) = 1620 to 1700 (≈ 1660) MeV
-2Im(pole position) = 150 to 250 (≈ 200) MeV

$\Delta(1700)$ DECAY MODES	Fraction (Γ_i/Γ)	p (MeV/c)
$N\pi$	10–20 %	580
$N\pi\pi$	80–90 %	547
$\Delta\pi$	30–60 %	385
$N\rho$	30–55 %	†
$N\gamma$	0.12–0.26 %	591
$N\gamma$, helicity=1/2	0.08–0.16 %	591
$N\gamma$, helicity=3/2	0.025–0.12 %	591

$\Delta(1905)$ F_{35} $I(J^P) = \frac{3}{2}(\frac{5}{2}^+)$

Breit-Wigner mass = 1870 to 1920 (≈ 1905) MeV
Breit-Wigner full width = 280 to 440 (≈ 350) MeV
$p_{beam} = 1.45$ GeV/c $4\pi\lambda^2 = 9.62$ mb
Re(pole position) = 1800 to 1860 (≈ 1830) MeV
-2Im(pole position) = 230 to 330 (≈ 280) MeV

$\Delta(1905)$ DECAY MODES	Fraction (Γ_i/Γ)	p (MeV/c)
$N\pi$	5–15 %	713
$N\pi\pi$	85–95 %	687
$\Delta\pi$	<25 %	542
$N\rho$	>60 %	421
$N\gamma$	0.01–0.03 %	721
$N\gamma$, helicity=1/2	0.0–0.1 %	721
$N\gamma$, helicity=3/2	0.004–0.03 %	721

$\Delta(1910)$ P_{31} $I(J^P) = \frac{3}{2}(\frac{1}{2}^+)$

Breit-Wigner mass = 1870 to 1920 (≈ 1910) MeV
Breit-Wigner full width = 190 to 270 (≈ 250) MeV
$p_{beam} = 1.46$ GeV/c $4\pi\lambda^2 = 9.54$ mb
Re(pole position) = 1830 to 1880 (≈ 1855) MeV
-2Im(pole position) = 200 to 500 (≈ 350) MeV

$\Delta(1910)$ DECAY MODES	Fraction (Γ_i/Γ)	p (MeV/c)
$N\pi$	15–30 %	716
$N\gamma$	0.0–0.2 %	725
$N\gamma$, helicity=1/2	0.0–0.2 %	725

$\Delta(1920)$ P_{33} $I(J^P) = \frac{3}{2}(\frac{3}{2}^+)$

Breit-Wigner mass = 1900 to 1970 (≈ 1920) MeV
Breit-Wigner full width = 150 to 300 (≈ 200) MeV
$p_{beam} = 1.48$ GeV/c $4\pi\lambda^2 = 9.37$ mb
Re(pole position) = 1850 to 1950 (≈ 1900) MeV
-2Im(pole position) = 200 to 400 (≈ 300) MeV

$\Delta(1920)$ DECAY MODES	Fraction (Γ_i/Γ)	p (MeV/c)
$N\pi$	5–20 %	722

$\Delta(1930)$ D_{35} $I(J^P) = \frac{3}{2}(\frac{5}{2}^-)$

Breit-Wigner mass = 1920 to 1970 (≈ 1930) MeV
Breit-Wigner full width = 250 to 450 (≈ 350) MeV
$p_{beam} = 1.50$ GeV/c $4\pi\lambda^2 = 9.21$ mb
Re(pole position) = 1840 to 1940 (≈ 1890) MeV
-2Im(pole position) = 200 to 300 (≈ 250) MeV

$\Delta(1930)$ DECAY MODES	Fraction (Γ_i/Γ)	p (MeV/c)
$N\pi$	10–20 %	729
$N\gamma$	0.0–0.02 %	737
$N\gamma$, helicity=1/2	0.0–0.01 %	737
$N\gamma$, helicity=3/2	0.0–0.01 %	737

$\Delta(1950)$ F_{37} $I(J^P) = \frac{3}{2}(\frac{7}{2}^+)$

Breit-Wigner mass = 1940 to 1960 (≈ 1950) MeV
Breit-Wigner full width = 290 to 350 (≈ 300) MeV
$p_{beam} = 1.54$ GeV/c $4\pi\lambda^2 = 8.91$ mb
Re(pole position) = 1880 to 1890 (≈ 1885) MeV
-2Im(pole position) = 210 to 270 (≈ 240) MeV

$\Delta(1950)$ DECAY MODES	Fraction (Γ_i/Γ)	p (MeV/c)
$N\pi$	35–40 %	741
$N\pi\pi$		716
$\Delta\pi$	20–30 %	574
$N\rho$	<10 %	469
$N\gamma$	0.08–0.13 %	749
$N\gamma$, helicity=1/2	0.03–0.055 %	749
$N\gamma$, helicity=3/2	0.05–0.075 %	749

$\Delta(2420)$ $H_{3,11}$ $I(J^P) = \frac{3}{2}(\frac{11}{2}^+)$

Breit-Wigner mass = 2300 to 2500 (≈ 2420) MeV
Breit-Wigner full width = 300 to 500 (≈ 400) MeV
$p_{beam} = 2.64$ GeV/c $4\pi\lambda^2 = 4.68$ mb
Re(pole position) = 2260 to 2400 (≈ 2330) MeV
-2Im(pole position) = 350 to 750 (≈ 550) MeV

$\Delta(2420)$ DECAY MODES	Fraction (Γ_i/Γ)	p (MeV/c)
$N\pi$	5–15 %	1023

Λ BARYONS
$(S = -1, I = 0)$
$\Lambda^0 = uds$

Λ

$I(J^P) = 0(\frac{1}{2}^+)$

Mass $m = 1115.683 \pm 0.006$ MeV
Mean life $\tau = (2.632 \pm 0.020) \times 10^{-10}$ s (S = 1.6)
$c\tau = 7.89$ cm
Magnetic moment $\mu = -0.613 \pm 0.004$ μ_N
Electric dipole moment $d < 1.5 \times 10^{-16}$ $e$cm, CL = 95%

Decay parameters

$p\pi^-$	$\alpha_- = 0.642 \pm 0.013$
"	$\phi_- = (-6.5 \pm 3.5)°$
"	$\gamma_- = 0.76$ [g]
"	$\Delta_- = (8 \pm 4)°$ [g]
$n\pi^0$	$\alpha_0 = +0.65 \pm 0.05$
$pe^-\bar{\nu}_e$	$g_A/g_V = -0.718 \pm 0.015$ [e]

Λ DECAY MODES	Fraction (Γ_i/Γ)	p (MeV/c)
$p\pi^-$	(63.9 ± 0.5) %	101
$n\pi^0$	(35.8 ± 0.5) %	104
$n\gamma$	$(1.75 \pm 0.15) \times 10^{-3}$	162
$p\pi^-\gamma$	[h] $(8.4 \pm 1.4) \times 10^{-4}$	101
$pe^-\bar{\nu}_e$	$(8.32 \pm 0.14) \times 10^{-4}$	163
$p\mu^-\bar{\nu}_\mu$	$(1.57 \pm 0.35) \times 10^{-4}$	131

Λ(1405) S_{01}

$I(J^P) = 0(\frac{1}{2}^-)$

Mass $m = 1407 \pm 4$ MeV
Full width $\Gamma = 50.0 \pm 2.0$ MeV
Below $\overline{K}N$ threshold

Λ(1405) DECAY MODES	Fraction (Γ_i/Γ)	p (MeV/c)
$\Sigma\pi$	100 %	152

Λ(1520) D_{03}

$I(J^P) = 0(\frac{3}{2}^-)$

Mass $m = 1519.5 \pm 1.0$ MeV [i]
Full width $\Gamma = 15.6 \pm 1.0$ MeV [i]
$p_{beam} = 0.39$ GeV/c $4\pi\lambda^2 = 82.8$ mb

Λ(1520) DECAY MODES	Fraction (Γ_i/Γ)	p (MeV/c)
$N\overline{K}$	45 ± 1%	244
$\Sigma\pi$	42 ± 1%	267
$\Lambda\pi\pi$	10 ± 1%	252
$\Sigma\pi\pi$	0.9 ± 0.1%	152
$\Lambda\gamma$	0.8 ± 0.2%	351

Λ(1600) P_{01}

$I(J^P) = 0(\frac{1}{2}^+)$

Mass $m = 1560$ to 1700 (≈ 1600) MeV
Full width $\Gamma = 50$ to 250 (≈ 150) MeV
$p_{beam} = 0.58$ GeV/c $4\pi\lambda^2 = 41.6$ mb

Λ(1600) DECAY MODES	Fraction (Γ_i/Γ)	p (MeV/c)
$N\overline{K}$	15–30 %	343
$\Sigma\pi$	10–60 %	336

Λ(1670) S_{01}

$I(J^P) = 0(\frac{1}{2}^-)$

Mass $m = 1660$ to 1680 (≈ 1670) MeV
Full width $\Gamma = 25$ to 50 (≈ 35) MeV
$p_{beam} = 0.74$ GeV/c $4\pi\lambda^2 = 28.5$ mb

Λ(1670) DECAY MODES	Fraction (Γ_i/Γ)	p (MeV/c)
$N\overline{K}$	15–25 %	414
$\Sigma\pi$	20–60 %	393
$\Lambda\eta$	15–35 %	64

Λ(1690) D_{03}

$I(J^P) = 0(\frac{3}{2}^-)$

Mass $m = 1685$ to 1695 (≈ 1690) MeV
Full width $\Gamma = 50$ to 70 (≈ 60) MeV
$p_{beam} = 0.78$ GeV/c $4\pi\lambda^2 = 26.1$ mb

Λ(1690) DECAY MODES	Fraction (Γ_i/Γ)	p (MeV/c)
$N\overline{K}$	20–30 %	433
$\Sigma\pi$	20–40 %	409
$\Lambda\pi\pi$	~ 25 %	415
$\Sigma\pi\pi$	~ 20 %	350

Λ(1800) S_{01}

$I(J^P) = 0(\frac{1}{2}^-)$

Mass $m = 1720$ to 1850 (≈ 1800) MeV
Full width $\Gamma = 200$ to 400 (≈ 300) MeV
$p_{beam} = 1.01$ GeV/c $4\pi\lambda^2 = 17.5$ mb

Λ(1800) DECAY MODES	Fraction (Γ_i/Γ)	p (MeV/c)
$N\overline{K}$	25–40 %	528
$\Sigma\pi$	seen	493
$\Sigma(1385)\pi$	seen	345
$N\overline{K}^*(892)$	seen	†

Λ(1810) P_{01}

$I(J^P) = 0(\frac{1}{2}^+)$

Mass $m = 1750$ to 1850 (≈ 1810) MeV
Full width $\Gamma = 50$ to 250 (≈ 150) MeV
$p_{beam} = 1.04$ GeV/c $4\pi\lambda^2 = 17.0$ mb

Λ(1810) DECAY MODES	Fraction (Γ_i/Γ)	p (MeV/c)
$N\overline{K}$	20–50 %	537
$\Sigma\pi$	10–40 %	501
$\Sigma(1385)\pi$	seen	356
$N\overline{K}^*(892)$	30–60 %	†

Λ(1820) F_{05}

$I(J^P) = 0(\frac{5}{2}^+)$

Mass $m = 1815$ to 1825 (≈ 1820) MeV
Full width $\Gamma = 70$ to 90 (≈ 80) MeV
$p_{beam} = 1.06$ GeV/c $4\pi\lambda^2 = 16.5$ mb

Λ(1820) DECAY MODES	Fraction (Γ_i/Γ)	p (MeV/c)
$N\overline{K}$	55–65 %	545
$\Sigma\pi$	8–14 %	508
$\Sigma(1385)\pi$	5–10 %	362

Λ(1830) D_{05}

$I(J^P) = 0(\frac{5}{2}^-)$

Mass $m = 1810$ to 1830 (≈ 1830) MeV
Full width $\Gamma = 60$ to 110 (≈ 95) MeV
$p_{beam} = 1.08$ GeV/c $4\pi\lambda^2 = 16.0$ mb

Λ(1830) DECAY MODES	Fraction (Γ_i/Γ)	p (MeV/c)
$N\overline{K}$	3–10 %	553
$\Sigma\pi$	35–75 %	515
$\Sigma(1385)\pi$	>15 %	371

$\Lambda(1890)\ P_{03}$ $\qquad I(J^P) = 0(\frac{3}{2}^+)$

Mass m = 1850 to 1910 ($\approx$ 1890) MeV
Full width Γ = 60 to 200 ($\approx$ 100) MeV
p_{beam} = 1.21 GeV/c $\qquad 4\pi\lambda^2$ = 13.6 mb

$\Lambda(1890)$ DECAY MODES	Fraction (Γ_i/Γ)	p (MeV/c)
$N\overline{K}$	20–35 %	599
$\Sigma\pi$	3–10 %	559
$\Sigma(1385)\pi$	seen	420
$N\overline{K}^*(892)$	seen	233

$\Lambda(2100)\ G_{07}$ $\qquad I(J^P) = 0(\frac{7}{2}^-)$

Mass m = 2090 to 2110 ($\approx$ 2100) MeV
Full width Γ = 100 to 250 ($\approx$ 200) MeV
p_{beam} = 1.68 GeV/c $\qquad 4\pi\lambda^2$ = 8.68 mb

$\Lambda(2100)$ DECAY MODES	Fraction (Γ_i/Γ)	p (MeV/c)
$N\overline{K}$	25–35 %	751
$\Sigma\pi$	$\sim$ 5 %	704
$\Lambda\eta$	<3 %	617
ΞK	<3 %	483
$\Lambda\omega$	<8 %	443
$N\overline{K}^*(892)$	10–20 %	514

$\Lambda(2110)\ F_{05}$ $\qquad I(J^P) = 0(\frac{5}{2}^+)$

Mass m = 2090 to 2140 ($\approx$ 2110) MeV
Full width Γ = 150 to 250 ($\approx$ 200) MeV
p_{beam} = 1.70 GeV/c $\qquad 4\pi\lambda^2$ = 8.53 mb

$\Lambda(2110)$ DECAY MODES	Fraction (Γ_i/Γ)	p (MeV/c)
$N\overline{K}$	5–25 %	757
$\Sigma\pi$	10–40 %	711
$\Lambda\omega$	seen	455
$\Sigma(1385)\pi$	seen	589
$N\overline{K}^*(892)$	10–60 %	524

$\Lambda(2350)\ H_{09}$ $\qquad I(J^P) = 0(\frac{9}{2}^+)$

Mass m = 2340 to 2370 ($\approx$ 2350) MeV
Full width Γ = 100 to 250 ($\approx$ 150) MeV
p_{beam} = 2.29 GeV/c $\qquad 4\pi\lambda^2$ = 5.85 mb

$\Lambda(2350)$ DECAY MODES	Fraction (Γ_i/Γ)	p (MeV/c)
$N\overline{K}$	$\sim$ 12 %	915
$\Sigma\pi$	$\sim$ 10 %	867

Σ BARYONS $(S = -1,\ I = 1)$

$\Sigma^+ = uus,\quad \Sigma^0 = uds,\quad \Sigma^- = dds$

Σ^+ $\qquad I(J^P) = 1(\frac{1}{2}^+)$

Mass m = 1189.37 $\pm$ 0.07 MeV (S = 2.2)
Mean life τ = (0.799 $\pm$ 0.004) $\times 10^{-10}$ s
$\quad c\tau$ = 2.396 cm
Magnetic moment μ = 2.458 $\pm$ 0.010 μ_N (S = 2.1)
$\Gamma(\Sigma^+ \to n\ell^+\nu)/\Gamma(\Sigma^- \to n\ell^-\bar{\nu})$ < 0.043

Decay parameters

$p\pi^0$	$\alpha_0 = -0.980^{+0.017}_{-0.015}$
"	$\phi_0 = (36 \pm 34)^\circ$
"	$\gamma_0 = 0.16$ [g]
"	$\Delta_0 = (187 \pm 6)^\circ$ [g]
$n\pi^+$	$\alpha_+ = 0.068 \pm 0.013$
"	$\phi_+ = (167 \pm 20)^\circ$ (S = 1.1)
"	$\gamma_+ = -0.97$ [g]
"	$\Delta_+ = (-73^{+133}_{-10})^\circ$ [g]
$p\gamma$	$\alpha_\gamma = -0.76 \pm 0.08$

Σ^+ DECAY MODES	Fraction (Γ_i/Γ)		Confidence level	p (MeV/c)
$p\pi^0$	(51.57$\pm$0.30) %			189
$n\pi^+$	(48.31$\pm$0.30) %			185
$p\gamma$	(1.23$\pm$0.05) $\times 10^{-3}$			225
$n\pi^+\gamma$	[h] (4.5 $\pm$0.5) $\times 10^{-4}$			185
$\Lambda e^+\nu_e$	(2.0 $\pm$0.5) $\times 10^{-5}$			71

$\Delta S = \Delta Q$ (SQ) violating modes or
$\Delta S = 1$ weak neutral current ($S1$) modes

$ne^+\nu_e$	SQ	< 5	$\times 10^{-6}$	90%	224
$n\mu^+\nu_\mu$	SQ	< 3.0	$\times 10^{-5}$	90%	202
pe^+e^-	$S1$	< 7	$\times 10^{-6}$	90%	225

Σ^0 $\qquad I(J^P) = 1(\frac{1}{2}^+)$

Mass m = 1192.642 $\pm$ 0.024 MeV
$m_{\Sigma^-} - m_{\Sigma^0}$ = 4.807 $\pm$ 0.035 MeV (S = 1.1)
$m_{\Sigma^0} - m_\Lambda$ = 76.959 $\pm$ 0.023 MeV
Mean life τ = (7.4 $\pm$ 0.7) $\times 10^{-20}$ s
$\quad c\tau$ = 2.22 $\times 10^{-11}$ m
Transition magnetic moment $|\mu_{\Sigma\Lambda}|$ = 1.61 $\pm$ 0.08 μ_N

Σ^0 DECAY MODES	Fraction (Γ_i/Γ)	Confidence level	p (MeV/c)
$\Lambda\gamma$	100 %		74
$\Lambda\gamma\gamma$	< 3 %	90%	74
$\Lambda e^+ e^-$	[j] 5 $\times 10^{-3}$		74

Σ^- $\qquad I(J^P) = 1(\frac{1}{2}^+)$

Mass $m = 1197.449 \pm 0.030$ MeV (S = 1.2)
$m_{\Sigma^-} - m_{\Sigma^+} = 8.08 \pm 0.08$ MeV (S = 1.9)
$m_{\Sigma^-} - m_\Lambda = 81.766 \pm 0.030$ MeV (S = 1.2)
Mean life $\tau = (1.479 \pm 0.011) \times 10^{-10}$ s (S = 1.3)
$\quad c\tau = 4.434$ cm
Magnetic moment $\mu = -1.160 \pm 0.025\ \mu_N$ (S = 1.7)

Decay parameters

$n\pi^-$	$\alpha_- = -0.068 \pm 0.008$	
"	$\phi_- = (10 \pm 15)°$	
"	$\gamma_- = 0.98$	[g]
"	$\Delta_- = (249^{+12}_{-120})°$	[g]
$ne^-\bar{\nu}_e$	$g_A/g_V = 0.340 \pm 0.017$	[e]
"	$f_2(0)/f_1(0) = 0.97 \pm 0.14$	
"	$D = 0.11 \pm 0.10$	
$\Lambda e^-\bar{\nu}_e$	$g_V/g_A = 0.01 \pm 0.10$ [e] (S = 1.5)	
"	$g_{WM}/g_A = 2.4 \pm 1.7$	[e]

Σ^- DECAY MODES	Fraction (Γ_i/Γ)	p (MeV/c)
$n\pi^-$	(99.848 ± 0.005) %	193
$n\pi^-\gamma$	[h] $(4.6 \pm 0.6) \times 10^{-4}$	193
$ne^-\bar{\nu}_e$	$(1.017 \pm 0.034) \times 10^{-3}$	230
$n\mu^-\bar{\nu}_\mu$	$(4.5 \pm 0.4) \times 10^{-4}$	210
$\Lambda e^-\bar{\nu}_e$	$(5.73 \pm 0.27) \times 10^{-5}$	79

$\Sigma(1385)\ P_{13}$ $\qquad I(J^P) = 1(\frac{3}{2}^+)$

$\Sigma(1385)^+$mass $m = 1382.8 \pm 0.4$ MeV (S = 2.0)
$\Sigma(1385)^0$mass $m = 1383.7 \pm 1.0$ MeV (S = 1.4)
$\Sigma(1385)^-$mass $m = 1387.2 \pm 0.5$ MeV (S = 2.2)
$\Sigma(1385)^+$full width $\Gamma = 35.8 \pm 0.8$ MeV
$\Sigma(1385)^0$full width $\Gamma = 36 \pm 5$ MeV
$\Sigma(1385)^-$full width $\Gamma = 39.4 \pm 2.1$ MeV (S = 1.7)
Below $\overline{K}N$ threshold

$\Sigma(1385)$ DECAY MODES	Fraction (Γ_i/Γ)	p (MeV/c)
$\Lambda\pi$	88 ± 2 %	208
$\Sigma\pi$	12 ± 2 %	127

$\Sigma(1660)\ P_{11}$ $\qquad I(J^P) = 1(\frac{1}{2}^+)$

Mass $m = 1630$ to $1690\ (\approx 1660)$ MeV
Full width $\Gamma = 40$ to $200\ (\approx 100)$ MeV
$\quad p_{beam} = 0.72$ GeV/c $\qquad 4\pi\lambda^2 = 29.9$ mb

$\Sigma(1660)$ DECAY MODES	Fraction (Γ_i/Γ)	p (MeV/c)
$N\overline{K}$	10–30 %	405
$\Lambda\pi$	seen	439
$\Sigma\pi$	seen	385

$\Sigma(1670)\ D_{13}$ $\qquad I(J^P) = 1(\frac{3}{2}^-)$

Mass $m = 1665$ to $1685\ (\approx 1670)$ MeV
Full width $\Gamma = 40$ to $80\ (\approx 60)$ MeV
$\quad p_{beam} = 0.74$ GeV/c $\qquad 4\pi\lambda^2 = 28.5$ mb

$\Sigma(1670)$ DECAY MODES	Fraction (Γ_i/Γ)	p (MeV/c)
$N\overline{K}$	7–13 %	414
$\Lambda\pi$	5–15 %	447
$\Sigma\pi$	30–60 %	393

$\Sigma(1750)\ S_{11}$ $\qquad I(J^P) = 1(\frac{1}{2}^-)$

Mass $m = 1730$ to $1800\ (\approx 1750)$ MeV
Full width $\Gamma = 60$ to $160\ (\approx 90)$ MeV
$\quad p_{beam} = 0.91$ GeV/c $\qquad 4\pi\lambda^2 = 20.7$ mb

$\Sigma(1750)$ DECAY MODES	Fraction (Γ_i/Γ)	p (MeV/c)
$N\overline{K}$	10–40 %	486
$\Lambda\pi$	seen	507
$\Sigma\pi$	<8 %	455
$\Sigma\eta$	15–55 %	81

$\Sigma(1775)\ D_{15}$ $\qquad I(J^P) = 1(\frac{5}{2}^-)$

Mass $m = 1770$ to $1780\ (\approx 1775)$ MeV
Full width $\Gamma = 105$ to $135\ (\approx 120)$ MeV
$\quad p_{beam} = 0.96$ GeV/c $\qquad 4\pi\lambda^2 = 19.0$ mb

$\Sigma(1775)$ DECAY MODES	Fraction (Γ_i/Γ)	p (MeV/c)
$N\overline{K}$	37–43%	508
$\Lambda\pi$	14–20%	525
$\Sigma\pi$	2–5%	474
$\Sigma(1385)\pi$	8–12%	324
$\Lambda(1520)\pi$	17–23%	198

$\Sigma(1915)\ F_{15}$ $\qquad I(J^P) = 1(\frac{5}{2}^+)$

Mass $m = 1900$ to $1935\ (\approx 1915)$ MeV
Full width $\Gamma = 80$ to $160\ (\approx 120)$ MeV
$\quad p_{beam} = 1.26$ GeV/c $\qquad 4\pi\lambda^2 = 12.8$ mb

$\Sigma(1915)$ DECAY MODES	Fraction (Γ_i/Γ)	p (MeV/c)
$N\overline{K}$	5–15 %	618
$\Lambda\pi$	seen	622
$\Sigma\pi$	seen	577
$\Sigma(1385)\pi$	<5 %	440

$\Sigma(1940)\ D_{13}$ $\qquad I(J^P) = 1(\frac{3}{2}^-)$

Mass $m = 1900$ to $1950\ (\approx 1940)$ MeV
Full width $\Gamma = 150$ to $300\ (\approx 220)$ MeV
$\quad p_{beam} = 1.32$ GeV/c $\qquad 4\pi\lambda^2 = 12.1$ mb

$\Sigma(1940)$ DECAY MODES	Fraction (Γ_i/Γ)	p (MeV/c)
$N\overline{K}$	<20 %	637
$\Lambda\pi$	seen	639
$\Sigma\pi$	seen	594
$\Sigma(1385)\pi$	seen	460
$\Lambda(1520)\pi$	seen	354
$\Delta(1232)\overline{K}$	seen	410
$N\overline{K}^*(892)$	seen	320

$\Sigma(2030)\ F_{17}$ $\qquad I(J^P) = 1(\frac{7}{2}^+)$

Mass $m = 2025$ to $2040\ (\approx 2030)$ MeV
Full width $\Gamma = 150$ to $200\ (\approx 180)$ MeV
$\quad p_{beam} = 1.52$ GeV/c $\qquad 4\pi\lambda^2 = 9.93$ mb

$\Sigma(2030)$ DECAY MODES	Fraction (Γ_i/Γ)	p (MeV/c)
$N\overline{K}$	17–23 %	702
$\Lambda\pi$	17–23 %	700
$\Sigma\pi$	5–10 %	657
ΞK	<2 %	412
$\Sigma(1385)\pi$	5–15 %	529
$\Lambda(1520)\pi$	10–20 %	430
$\Delta(1232)\overline{K}$	10–20 %	498
$N\overline{K}^*(892)$	<5 %	438

Σ(2250)

$$I(J^P) = 1(?^?)$$

Mass $m = 2210$ to 2280 (≈ 2250) MeV
Full width $\Gamma = 60$ to 150 (≈ 100) MeV
$p_{beam} = 2.04$ GeV/c $4\pi\lambda^2 = 6.76$ mb

Σ(2250) DECAY MODES	Fraction (Γ_i/Γ)	p (MeV/c)
$N\overline{K}$	<10 %	851
$\Lambda\pi$	seen	842
$\Sigma\pi$	seen	803

Ξ BARYONS
($S = -2, I = 1/2$)
$\Xi^0 = uss$, $\Xi^- = dss$

Ξ⁰

$$I(J^P) = \tfrac{1}{2}(\tfrac{1}{2}^+)$$

P is not yet measured; $+$ is the quark model prediction.

Mass $m = 1314.9 \pm 0.6$ MeV
$m_{\Xi^-} - m_{\Xi^0} = 6.4 \pm 0.6$ MeV
Mean life $\tau = (2.90 \pm 0.09) \times 10^{-10}$ s
$c\tau = 8.71$ cm
Magnetic moment $\mu = -1.250 \pm 0.014\ \mu_N$

Decay parameters

$\Lambda\pi^0$	$\alpha = -0.411 \pm 0.022$	(S = 2.1)
"	$\phi = (21 \pm 12)^\circ$	
"	$\gamma = 0.85$ [g]	
"	$\Delta = (218^{+12}_{-19})^\circ$ [g]	
$\Lambda\gamma$	$\alpha = 0.4 \pm 0.4$	
$\Sigma^0\gamma$	$\alpha = 0.20 \pm 0.32$	

Ξ⁰ DECAY MODES	Fraction (Γ_i/Γ)	Confidence level	p (MeV/c)
$\Lambda\pi^0$	(99.54 ± 0.05) %		135
$\Lambda\gamma$	$(1.06 \pm 0.16) \times 10^{-3}$		184
$\Sigma^0\gamma$	$(3.5 \pm 0.4) \times 10^{-3}$		117
$\Sigma^+ e^- \bar{\nu}_e$	$< 1.1 \times 10^{-3}$	90%	120
$\Sigma^+ \mu^- \bar{\nu}_\mu$	$< 1.1 \times 10^{-3}$	90%	64

$\Delta S = \Delta Q$ (SQ) violating modes or
$\Delta S = 2$ forbidden (S2) modes

$\Sigma^- e^+ \nu_e$	SQ	< 9	$\times 10^{-4}$	90%	112
$\Sigma^- \mu^+ \nu_\mu$	SQ	< 9	$\times 10^{-4}$	90%	49
$p\pi^-$	S2	< 4	$\times 10^{-5}$	90%	299
$p e^- \bar{\nu}_e$	S2	< 1.3	$\times 10^{-3}$		323
$p\mu^- \bar{\nu}_\mu$	S2	< 1.3	$\times 10^{-3}$		309

Ξ⁻

$$I(J^P) = \tfrac{1}{2}(\tfrac{1}{2}^+)$$

P is not yet measured; $+$ is the quark model prediction.

Mass $m = 1321.32 \pm 0.13$ MeV
Mean life $\tau = (1.639 \pm 0.015) \times 10^{-10}$ s
$c\tau = 4.91$ cm
Magnetic moment $\mu = -0.6507 \pm 0.0025\ \mu_N$

Decay parameters

$\Lambda\pi^-$	$\alpha = -0.456 \pm 0.014$	(S = 1.8)
"	$\phi = (4 \pm 4)^\circ$	
"	$\gamma = 0.89$ [g]	
"	$\Delta = (188 \pm 8)^\circ$ [g]	
$\Lambda e^- \bar{\nu}_e$	$g_A/g_V = -0.25 \pm 0.05$ [e]	

Ξ⁻ DECAY MODES	Fraction (Γ_i/Γ)	Confidence level	p (MeV/c)
$\Lambda\pi^-$	(99.887 ± 0.035) %		139
$\Sigma^-\gamma$	$(1.27 \pm 0.23) \times 10^{-4}$		118
$\Lambda e^- \bar{\nu}_e$	$(5.63 \pm 0.31) \times 10^{-4}$		190
$\Lambda \mu^- \bar{\nu}_\mu$	$(3.5 ^{+3.5}_{-2.2}) \times 10^{-4}$		163
$\Sigma^0 e^- \bar{\nu}_e$	$(8.7 \pm 1.7) \times 10^{-5}$		122
$\Sigma^0 \mu^- \bar{\nu}_\mu$	$< 8 \times 10^{-4}$	90%	70
$\Xi^0 e^- \bar{\nu}_e$	$< 2.3 \times 10^{-3}$	90%	6

$\Delta S = 2$ forbidden (S2) modes

$n\pi^-$	S2	< 1.9	$\times 10^{-5}$	90%	303
$n e^- \bar{\nu}_e$	S2	< 3.2	$\times 10^{-3}$	90%	327
$n \mu^- \bar{\nu}_\mu$	S2	< 1.5	%	90%	314
$p\pi^- \pi^-$	S2	< 4	$\times 10^{-4}$	90%	223
$p\pi^- e^- \bar{\nu}_e$	S2	< 4	$\times 10^{-4}$	90%	304
$p\pi^- \mu^- \bar{\nu}_\mu$	S2	< 4	$\times 10^{-4}$	90%	250
$p\mu^- \mu^-$	L	< 4	$\times 10^{-4}$	90%	272

Ξ(1530) P_{13}

$$I(J^P) = \tfrac{1}{2}(\tfrac{3}{2}^+)$$

$\Xi(1530)^0$ mass $m = 1531.80 \pm 0.32$ MeV   (S = 1.3)
$\Xi(1530)^-$ mass $m = 1535.0 \pm 0.6$ MeV
$\Xi(1530)^0$ full width $\Gamma = 9.1 \pm 0.5$ MeV
$\Xi(1530)^-$ full width $\Gamma = 9.9^{+1.7}_{-1.9}$ MeV

Ξ(1530) DECAY MODES	Fraction (Γ_i/Γ)	Confidence level	p (MeV/c)
$\Xi\pi$	100 %		152
$\Xi\gamma$	<4 %	90%	200

Ξ(1690)

$$I(J^P) = \tfrac{1}{2}(?^?)$$

Mass $m = 1690 \pm 10$ MeV [i]
Full width $\Gamma < 50$ MeV

Ξ(1690) DECAY MODES	Fraction (Γ_i/Γ)	p (MeV/c)
$\Lambda\overline{K}$	seen	240
$\Sigma\overline{K}$	seen	51
$\Xi^- \pi^+ \pi^-$	possibly seen	214

Ξ(1820) D_{13}

$$I(J^P) = \tfrac{1}{2}(\tfrac{3}{2}^-)$$

Mass $m = 1823 \pm 5$ MeV [i]
Full width $\Gamma = 24^{+15}_{-10}$ MeV [i]

Ξ(1820) DECAY MODES	Fraction (Γ_i/Γ)	p (MeV/c)
$\Lambda\overline{K}$	large	400
$\Sigma\overline{K}$	small	320
$\Xi\pi$	small	413
$\Xi(1530)\pi$	small	234

Ξ(1950)

$$I(J^P) = \tfrac{1}{2}(?^?)$$

Mass $m = 1950 \pm 15$ MeV [i]
Full width $\Gamma = 60 \pm 20$ MeV [i]

Ξ(1950) DECAY MODES	Fraction (Γ_i/Γ)	p (MeV/c)
$\Lambda\overline{K}$	seen	522
$\Sigma\overline{K}$	possibly seen	460
$\Xi\pi$	seen	518

Ξ(2030)

$$I(J^P) = \tfrac{1}{2}(\geq \tfrac{5}{2}^?)$$

Mass $m = 2025 \pm 5$ MeV [i]
Full width $\Gamma = 20^{+15}_{-5}$ MeV [i]

Ξ(2030) DECAY MODES	Fraction (Γ_i/Γ)	p (MeV/c)
$\Lambda\overline{K}$	~ 20 %	589
$\Sigma\overline{K}$	~ 80 %	533
$\Xi\pi$	small	573
$\Xi(1530)\pi$	small	421
$\Lambda\overline{K}\pi$	small	501
$\Sigma\overline{K}\pi$	small	430

Ω BARYONS
$(S = -3, I = 0)$

$\Omega^- = sss$

$I(J^P) = 0(\frac{3}{2}+)$

J^P is not yet measured; $\frac{3}{2}+$ is the quark model prediction.

Mass $m = 1672.45 \pm 0.29$ MeV
Mean life $\tau = (0.822 \pm 0.012) \times 10^{-10}$ s
$c\tau = 2.46$ cm
Magnetic moment $\mu = -2.02 \pm 0.05 \mu_N$

Decay parameters

ΛK^- $\alpha = -0.026 \pm 0.026$
$\Xi^0 \pi^-$ $\alpha = 0.09 \pm 0.14$
$\Xi^- \pi^0$ $\alpha = 0.05 \pm 0.21$

Ω^- DECAY MODES	Fraction (Γ_i/Γ)	Confidence level	p (MeV/c)
ΛK^-	(67.8 ± 0.7) %		211
$\Xi^0 \pi^-$	(23.6 ± 0.7) %		294
$\Xi^- \pi^0$	(8.6 ± 0.4) %		290
$\Xi^- \pi^+ \pi^-$	$(4.3^{+3.4}_{-1.3}) \times 10^{-4}$		190
$\Xi(1530)^0 \pi^-$	$(6.4^{+5.1}_{-2.0}) \times 10^{-4}$		17
$\Xi^0 e^- \bar{\nu}_e$	$(5.6 \pm 2.8) \times 10^{-3}$		319
$\Xi^- \gamma$	$< 4.6 \times 10^{-4}$	90%	314

$\Delta S = 2$ forbidden (S2) modes

$\Lambda \pi^-$	S2	$< 1.9 \times 10^{-4}$	90%	449

$\Omega(2250)^-$

$I(J^P) = 0(?^?)$

Mass $m = 2252 \pm 9$ MeV
Full width $\Gamma = 55 \pm 18$ MeV

$\Omega(2250)^-$ DECAY MODES	Fraction (Γ_i/Γ)	p (MeV/c)
$\Xi^- \pi^+ K^-$	seen	531
$\Xi(1530)^0 K^-$	seen	437

CHARMED BARYONS
$(C = +1)$

$\Lambda_c^+ = udc,\quad \Sigma_c^{++} = uuc,\quad \Sigma_c^+ = udc,\quad \Sigma_c^0 = ddc,$
$\Xi_c^+ = usc,\quad \Xi_c^0 = dsc,\quad \Omega_c^0 = ssc$

Λ_c^+

$I(J^P) = 0(\frac{1}{2}+)$

J not confirmed; $\frac{1}{2}$ is the quark model prediction.

Mass $m = 2284.9 \pm 0.6$ MeV
Mean life $\tau = (0.206 \pm 0.012) \times 10^{-12}$ s
$c\tau = 61.8\ \mu m$

Decay asymmetry parameters

$\Lambda \pi^+$ $\alpha = -0.98 \pm 0.19$
$\Sigma^+ \pi^0$ $\alpha = -0.45 \pm 0.32$
$\Lambda \ell^+ \nu_\ell$ $\alpha = -0.82^{+0.11}_{-0.07}$

Nearly all branching fractions of the Λ_c^+ are measured relative to the $pK^-\pi^+$ mode, but there are no model-independent measurements of this branching fraction. We explain how we arrive at our value of B($\Lambda_c^+ \to pK^-\pi^+$) in a Note at the beginning of the branching-ratio measurements, in the Listings. When this branching fraction is eventually well determined, all the other branching fractions will slide up or down proportionally as the true value differs from the value we use here.

Λ_c^+ DECAY MODES		Fraction (Γ_i/Γ)	Scale factor/ Confidence level	p (MeV/c)
Hadronic modes with a p and one $\overline{K}$				
$p\overline{K}^0$		(2.5 ± 0.7) %		872
$pK^-\pi^+$	[k]	(5.0 ± 1.3) %		822
$p\overline{K}^*(892)^0$	[l]	(1.8 ± 0.6) %		681
$\Delta(1232)^{++} K^-$		$(8 \pm 5) \times 10^{-3}$		709
$\Lambda(1520) \pi^+$	[l]	$(4.5 ^{+2.5}_{-2.1}) \times 10^{-3}$		626
$pK^-\pi^+$ nonresonant		(2.8 ± 0.9) %		822
$p\overline{K}^0 \eta$		(1.3 ± 0.4) %		567
$p\overline{K}^0 \pi^+ \pi^-$		(2.4 ± 1.1) %		753
$pK^-\pi^+\pi^0$		seen		758
$pK^*(892)^- \pi^+$	[l]	(1.1 ± 0.6) %		579
$p(K^-\pi^+)_{\text{nonresonant}} \pi^0$		(3.6 ± 1.2) %		758
$\Delta(1232)\overline{K}^*(892)$		seen		416
$pK^-\pi^+\pi^+\pi^-$		$(1.1 \pm 0.8) \times 10^{-3}$		670
$pK^-\pi^+\pi^0\pi^0$		$(8 \pm 4) \times 10^{-3}$		676
$pK^-\pi^+\pi^0\pi^0\pi^0$		$(5.0 \pm 3.4) \times 10^{-3}$		573
Hadronic modes with a p and zero or two K's				
$p\pi^+\pi^-$		$(3.5 \pm 2.0) \times 10^{-3}$		926
$p f_0(980)$	[l]	$(2.8 \pm 1.9) \times 10^{-3}$		621
$p\pi^+\pi^+\pi^-\pi^-$		$(1.8 \pm 1.2) \times 10^{-3}$		851
pK^+K^-		$(2.3 \pm 0.9) \times 10^{-3}$		615
$p\phi$	[l]	$(1.2 \pm 0.5) \times 10^{-3}$		589
Hadronic modes with a hyperon				
$\Lambda \pi^+$		$(9.0 \pm 2.8) \times 10^{-3}$		863
$\Lambda \pi^+ \pi^0$		(3.6 ± 1.3) %		843
$\Lambda \rho^+$		< 5 %	CL=95%	638
$\Lambda \pi^+\pi^+\pi^-$		(3.3 ± 1.0) %		806
$\Lambda \pi^+ \eta$		(1.7 ± 0.6) %		690
$\Sigma(1385)^+ \eta$	[l]	$(8.5 \pm 3.3) \times 10^{-3}$		569
$\Lambda K^+ \overline{K}^0$		$(6.0 \pm 2.1) \times 10^{-3}$		441
$\Sigma^0 \pi^+$		$(9.9 \pm 3.2) \times 10^{-3}$		824
$\Sigma^+ \pi^0$		(1.00 ± 0.34) %		826
$\Sigma^+ \eta$		$(5.5 \pm 2.3) \times 10^{-3}$		712
$\Sigma^+ \pi^+\pi^-$		(3.4 ± 1.0) %		803
$\Sigma^+ \rho^0$		< 1.4 %	CL=95%	578
$\Sigma^- \pi^+\pi^+$		(1.8 ± 0.8) %		798
$\Sigma^0 \pi^+\pi^0$		(1.8 ± 0.8) %		802
$\Sigma^0 \pi^+\pi^+\pi^-$		(1.1 ± 0.4) %		762
$\Sigma^+ \pi^+\pi^-\pi^0$		—		766
$\Sigma^+ \omega$	[l]	(2.7 ± 1.0) %		568
$\Sigma^+ \pi^+\pi^+\pi^-\pi^-$		$(3.0 ^{+4.1}_{-2.1}) \times 10^{-3}$		707
$\Sigma^+ K^+ K^-$		$(3.5 \pm 1.2) \times 10^{-3}$		346
$\Sigma^+ \phi$	[l]	$(3.5 \pm 1.7) \times 10^{-3}$		292
$\Sigma^+ K^+ \pi^-$		$(7 ^{+6}_{-4}) \times 10^{-3}$		668
$\Xi^0 K^+$		$(3.9 \pm 1.4) \times 10^{-3}$		652
$\Xi^- K^+ \pi^+$		$(4.9 \pm 1.7) \times 10^{-3}$		564
$\Xi(1530)^0 K^+$	[l]	$(2.6 \pm 1.0) \times 10^{-3}$		471
Semileptonic modes				
$\Lambda \ell^+ \nu_\ell$	[m]	(2.0 ± 0.6) %		—
$\Lambda e^+ \nu_e$		(2.1 ± 0.6) %		—
$\Lambda \mu^+ \nu_\mu$		(2.0 ± 0.7) %		—
e^+ anything		(4.5 ± 1.7) %		—
$p e^+$ anything		(1.8 ± 0.9) %		—
Λe^+ anything		—		—
$\Lambda \mu^+$ anything		—		—
$\Lambda \ell^+ \nu_\ell$ anything		—		—
Inclusive modes				
p anything		(50 ± 16) %		—
p anything (no Λ)		(12 ± 19) %		—
p hadrons		—		—
n anything		(50 ± 16) %		—
n anything (no Λ)		(29 ± 17) %		—
Λ anything		(35 ± 11) %	S=1.4	—
$\Sigma^\pm$ anything	[n]	(10 ± 5) %		—

$\Delta C = 1$ weak neutral current ($C1$) modes, or
Lepton number (L) violating modes

$p\mu^+\mu^-$	$C1$	< 3.4	$\times 10^{-4}$	CL=90%	936
$\Sigma^-\mu^+\mu^+$	L	< 7.0	$\times 10^{-4}$	CL=90%	811

$\Lambda_c(2593)^+$ $\qquad I(J^P) = 0(\frac{1}{2}^-)$

The spin-parity follows from the fact that $\Sigma_c(2455)\pi$ decays, with little available phase space, are dominant.

Mass $m = 2593.9 \pm 0.8$ MeV
$m - m_{\Lambda_c^+} = 308.9 \pm 0.6$ MeV   (S = 1.1)
Full width $\Gamma = 3.6^{+2.0}_{-1.3}$ MeV

$\Lambda_c^+\pi\pi$ and its submode $\Sigma_c(2455)\pi$ — the latter just barely — are the only strong decays allowed to an excited Λ_c^+ having this mass; and the $\Lambda_c^+\pi^+\pi^-$ mode seems to be largely via $\Sigma_c^{++}\pi^-$ or $\Sigma_c^0\pi^+$.

$\Lambda_c(2593)^+$ DECAY MODES	Fraction (Γ_i/Γ)	p (MeV/c)
$\Lambda_c^+\pi^+\pi^-$	$[o] \approx 67\%$	124
$\Sigma_c(2455)^{++}\pi^-$	$24 \pm 7\%$	17
$\Sigma_c(2455)^0\pi^+$	$24 \pm 7\%$	23
$\Lambda_c^+\pi^+\pi^-$ 3-body	$18 \pm 10\%$	124
$\Lambda_c^+\pi^0$	not seen	261
$\Lambda_c^+\gamma$	not seen	290

$\Lambda_c(2625)^+$ $\qquad I(J^P) = 0(?^?)$

J^P is expected to be $3/2^-$.

Mass $m = 2626.6 \pm 0.8$ MeV   (S = 1.2)
$m - m_{\Lambda_c^+} = 341.7 \pm 0.6$ MeV   (S = 1.6)
Full width $\Gamma < 1.9$ MeV, CL = 90%

$\Lambda_c^+\pi\pi$ and its submode $\Sigma(2455)\pi$ are the only strong decays allowed to an excited Λ_c^+ having this mass.

$\Lambda_c(2625)^+$ DECAY MODES	Fraction (Γ_i/Γ)	p (MeV/c)
$\Lambda_c^+\pi^+\pi^-$	seen	184
$\Sigma_c(2455)^{++}\pi^-$	small	100
$\Sigma_c(2455)^0\pi^+$	small	101
$\Lambda_c^+\pi^+\pi^-$ 3-body	large	184
$\Lambda_c^+\pi^0$	not seen	293
$\Lambda_c^+\gamma$	not seen	319

$\Sigma_c(2455)$ $\qquad I(J^P) = 1(\frac{1}{2}^+)$

J^P not confirmed; $\frac{1}{2}^+$ is the quark model prediction.

$\Sigma_c(2455)^{++}$mass $m = 2452.8 \pm 0.6$ MeV
$\Sigma_c(2455)^+$ mass $m = 2453.6 \pm 0.9$ MeV
$\Sigma_c(2455)^0$ mass $m = 2452.2 \pm 0.6$ MeV
$m_{\Sigma_c^{++}} - m_{\Lambda_c^+} = 167.87 \pm 0.19$ MeV
$m_{\Sigma_c^+} - m_{\Lambda_c^+} = 168.7 \pm 0.6$ MeV
$m_{\Sigma_c^0} - m_{\Lambda_c^+} = 167.30 \pm 0.20$ MeV
$m_{\Sigma_c^{++}} - m_{\Sigma_c^0} = 0.57 \pm 0.23$ MeV
$m_{\Sigma_c^+} - m_{\Sigma_c^0} = 1.4 \pm 0.6$ MeV

$\Lambda_c^+\pi$ is the only strong decay allowed to a Σ_c having this mass.

$\Sigma_c(2455)$ DECAY MODES	Fraction (Γ_i/Γ)	p (MeV/c)
$\Lambda_c^+\pi$	$\approx 100\%$	90

$\Sigma_c(2520)$ $\qquad I(J^P) = 1(?^?)$

$\Sigma_c(2520)^{++}$mass $m = 2519.4 \pm 1.5$ MeV
$\Sigma_c(2520)^0$ mass $m = 2517.5 \pm 1.4$ MeV
$m_{\Sigma_c(2520)^{++}} - m_{\Lambda_c^+} = 234.5 \pm 1.4$ MeV
$m_{\Sigma_c(2520)^0} - m_{\Lambda_c^+} = 232.6 \pm 1.3$ MeV
$m_{\Sigma_c(2520)^{++}} - m_{\Sigma_c(2520)^0} = 1.9 \pm 1.9$ MeV
$\Sigma_c(2520)^{++}$full width $\Gamma = 18 \pm 5$ MeV
$\Sigma_c(2520)^0$ full width $\Gamma = 13 \pm 5$ MeV

Ξ_c^+ $\qquad I(J^P) = \frac{1}{2}(\frac{1}{2}^+)$

$I(J^P)$ not confirmed; $\frac{1}{2}(\frac{1}{2}^+)$ is the quark model prediction.

Mass $m = 2465.6 \pm 1.4$ MeV
Mean life $\tau = (0.35^{+0.07}_{-0.04}) \times 10^{-12}$ s
$c\tau = 106$ μm

Ξ_c^+ DECAY MODES	Fraction (Γ_i/Γ)	p (MeV/c)
$\Lambda K^-\pi^+\pi^+$	seen	784
$\Lambda \overline{K}^*(892)^0\pi^+$	not seen	601
$\Sigma(1385)^+ K^-\pi^+$	not seen	676
$\Sigma^+ K^-\pi^+$	seen	808
$\Sigma^+ \overline{K}^*(892)^0$	seen	653
$\Sigma^0 K^-\pi^+\pi^+$	seen	733
$\Xi^-\pi^+$	seen	875
$\Xi^-\pi^+\pi^+$	seen	850
$\Xi(1530)^0\pi^+$	not seen	748
$\Xi^0\pi^+\pi^0$	seen	854
$\Xi^0\pi^+\pi^+\pi^-$	seen	817
$\Xi^0 e^+\nu_e$	seen	882

Ξ_c^0 $\qquad I(J^P) = \frac{1}{2}(\frac{1}{2}^+)$

$I(J^P)$ not confirmed; $\frac{1}{2}(\frac{1}{2}^+)$ is the quark model prediction.

Mass $m = 2470.3 \pm 1.8$ MeV   (S = 1.3)
$m_{\Xi_c^0} - m_{\Xi_c^+} = 4.7 \pm 2.1$ MeV   (S = 1.2)
Mean life $\tau = (0.098^{+0.023}_{-0.015}) \times 10^{-12}$ s
$c\tau = 29$ μm

Ξ_c^0 DECAY MODES	Fraction (Γ_i/Γ)	p (MeV/c)
$\Lambda \overline{K}^0$	seen	864
$\Xi^-\pi^+$	seen	875
$\Xi^-\pi^+\pi^+\pi^-$	seen	816
$p K^- \overline{K}^*(892)^0$	seen	406
$\Omega^- K^+$	seen	522
$\Xi^- e^+\nu_e$	seen	882
$\Xi^- \ell^+$ anything	seen	–

$\Xi_c(2645)$ $\qquad I(J^P) = ?(?^?)$

$\Xi_c(2645)^+$ mass $m = 2644.6 \pm 2.1$ MeV   (S = 1.2)
$\Xi_c(2645)^0$ mass $m = 2643.8 \pm 1.8$ MeV
$m_{\Xi_c(2645)^+} - m_{\Xi_c^0} = 174.3 \pm 1.1$ MeV
$m_{\Xi_c(2645)^0} - m_{\Xi_c^+} = 178.2 \pm 1.1$ MeV
$\Xi_c(2645)^+$ full width $\Gamma < 3.1$ MeV, CL = 90%
$\Xi_c(2645)^0$ full width $\Gamma < 5.5$ MeV, CL = 90%

$\Xi_c\pi$ is the only strong decay allowed to a Ξ_c resonance having this mass.

$\Xi_c(2645)$ DECAY MODES	Fraction (Γ_i/Γ)	p (MeV/c)
$\Xi_c^0\pi^+$	seen	101
$\Xi_c^+\pi^-$	seen	107

$$I(J^P) = 0(\tfrac{1}{2}^+)$$

$I(J^P)$ not confirmed; $0(\tfrac{1}{2}^+)$ is the quark model prediction.

Mass $m = 2704 \pm 4$ MeV   (S = 1.8)
Mean life $\tau = (0.064 \pm 0.020) \times 10^{-12}$ s
$c\tau = 19$ μm

Ω_c^0 DECAY MODES	Fraction (Γ_i/Γ)	p (MeV/c)
$\Sigma^+ K^- K^- \pi^+$	seen	697
$\Xi^- K^- \pi^+ \pi^+$	seen	838
$\Omega^- \pi^+$	seen	827
$\Omega^- \pi^- \pi^+ \pi^+$	seen	759

BOTTOM BARYONS
$(B = -1)$
$\Lambda_b^0 = u d b, \ \Xi_b^0 = u s b, \ \Xi_b^- = d s b$

Λ_b^0

$$I(J^P) = 0(\tfrac{1}{2}^+)$$

$I(J^P)$ not yet measured; $0(\tfrac{1}{2}^+)$ is the quark model prediction.
Mass $m = 5624 \pm 9$ MeV   (S = 1.8)
Mean life $\tau = (1.24 \pm 0.08) \times 10^{-12}$ s
$c\tau = 372$ μm

These branching fractions are actually an average over weakly decaying b-baryons weighted by their production rates in Z decay (or high-energy $p\overline{p}$), branching ratios, and detection efficiencies. They scale with the LEP Λ_b production fraction $B(b \to \Lambda_b)$ and are evaluated for our value $B(b \to \Lambda_b) = (10.1^{+3.9}_{-3.1})\%$.

The branching fractions $B(b\text{-baryon} \to \Lambda \ell^- \overline{\nu}_\ell \text{anything})$ and $B(\Lambda_b^0 \to \Lambda_c^+ \ell^- \overline{\nu}_\ell \text{anything})$ are not pure measurements because the underlying measured products of these with $B(b \to \Lambda_b)$ were used to determine $B(b \to \Lambda_b)$, as described in the note "Production and Decay of b-Flavored Hadrons."

Λ_b^0 DECAY MODES	Fraction (Γ_i/Γ)	Confidence level	p (MeV/c)
$J/\psi(1S)\Lambda$	$(4.7 \pm 2.8) \times 10^{-4}$		1744
$\Lambda_c^+ \pi^-$	seen		2345
$\Lambda_c^+ a_1(1260)^-$	seen		2156
$\Lambda_c^+ \ell^- \overline{\nu}_\ell \text{anything}$	[p] $(9.0^{+3.1}_{-3.8})\%$		–
$p\pi^-$	$< 5.0 \times 10^{-5}$	90%	2732
pK^-	$< 5.0 \times 10^{-5}$	90%	2711

b-baryon ADMIXTURE ($\Lambda_b, \Xi_b, \Sigma_b, \Omega_b$)

Mean life $\tau = (1.20 \pm 0.07) \times 10^{-12}$ s

These branching fractions are actually an average over weakly decaying b-baryons weighted by their production rates in Z decay (or high-energy $p\overline{p}$), branching ratios, and detection efficiencies. They scale with the LEP Λ_b production fraction $B(b \to \Lambda_b)$ and are evaluated for our value $B(b \to \Lambda_b) = (10.1^{+3.9}_{-3.1})\%$.

The branching fractions $B(b\text{-baryon} \to \Lambda \ell^- \overline{\nu}_\ell \text{anything})$ and $B(\Lambda_b^0 \to \Lambda_c^+ \ell^- \overline{\nu}_\ell \text{anything})$ are not pure measurements because the underlying measured products of these with $B(b \to \Lambda_b)$ were used to determine $B(b \to \Lambda_b)$, as described in the note "Production and Decay of b-Flavored Hadrons."

b-baryon ADMIXTURE ($\Lambda_b, \Xi_b, \Sigma_b, \Omega_b$)	Fraction (Γ_i/Γ)	p (MeV/c)
$p\mu^- \overline{\nu} \text{anything}$	$(4.9 \pm 2.4)\%$	–
$\Lambda \ell^- \overline{\nu}_\ell \text{anything}$	$(3.1^{+1.0}_{-1.2})\%$	–
$\Lambda / \overline{\Lambda} \text{anything}$	$(35^{+12}_{-14})\%$	–
$\Xi^- \ell^- \overline{\nu}_\ell \text{anything}$	$(5.5^{+2.0}_{-2.4}) \times 10^{-3}$	–

NOTES

This Summary Table only includes established baryons. The Particle Listings include evidence for other baryons. The masses, widths, and branching fractions for the resonances in this Table are Breit-Wigner parameters, but pole positions are also given for most of the N and Δ resonances.

For most of the resonances, the parameters come from various partial-wave analyses of more or less the same sets of data, and it is not appropriate to treat the results of the analyses as independent or to average them together. Furthermore, the systematic errors on the results are not well understood. Thus, we usually only give ranges for the parameters. We then also give a best guess for the mass (as part of the name of the resonance) and for the width. The *Note on N and Δ Resonances* and the *Note on Λ and Σ Resonances* in the Particle Listings review the partial-wave analyses.

When a quantity has "(S = ...)" to its right, the error on the quantity has been enlarged by the "scale factor" S, defined as $S = \sqrt{\chi^2/(N-1)}$, where N is the number of measurements used in calculating the quantity. We do this when $S > 1$, which often indicates that the measurements are inconsistent. When $S > 1.25$, we also show in the Particle Listings an ideogram of the measurements. For more about S, see the Introduction.

A decay momentum p is given for each decay mode. For a 2-body decay, p is the momentum of each decay product in the rest frame of the decaying particle. For a 3-or-more-body decay, p is the largest momentum any of the products can have in this frame. For any resonance, the *nominal* mass is used in calculating p. A dagger ("†") in this column indicates that the mode is forbidden when the nominal masses of resonances are used, but is in fact allowed due to the nonzero widths of the resonances.

[a] The masses of the p and n are most precisely known in u (unified atomic mass units). The conversion factor to MeV, 1 u = 931.49432 ± 0.00028 MeV, is less well known than are the masses in u.

[b] The limit is from neutrality-of-matter experiments; it assumes $q_n = q_p + q_e$. See also the charge of the neutron.

[c] The first limit is geochemical and independent of decay mode. The second entry, a range of limits, assumes the dominant decay modes are among those investigated. For antiprotons the best limit, inferred from the observation of cosmic ray $\overline{p}$'s is $\tau_{\overline{p}} > 10^7$ yr, the cosmic-ray storage time, but this limit depends on a number of assumptions. The best direct observation of stored antiprotons gives $\tau_{\overline{p}}/B(\overline{p} \to e^- \gamma) > 1848$ yr.

[d] There is some controversy about whether nuclear physics and model dependence complicate the analysis for bound neutrons (from which the best limit comes). The second limit here is from reactor experiments with free neutrons.

[e] The parameters g_A, g_V, and g_{WM} for semileptonic modes are defined by $\overline{B}_f[\gamma_\lambda(g_V + g_A \gamma_5) + i(g_{WM}/m_{B_i}) \sigma_{\lambda\nu} q^\nu]B_i$, and ϕ_{AV} is defined by $g_A/g_V = |g_A/g_V|e^{i\phi_{AV}}$. See the "Note on Baryon Decay Parameters" in the neutron Particle Listings.

[f] Time-reversal invariance requires this to be $0°$ or $180°$.

[g] The decay parameters γ and Δ are calculated from α and ϕ using

$$\gamma = \sqrt{1-\alpha^2}\cos\phi, \qquad \tan\Delta = -\frac{1}{\alpha}\sqrt{1-\alpha^2}\sin\phi.$$

See the "Note on Baryon Decay Parameters" in the neutron Particle Listings.

[h] See the Particle Listings for the pion momentum range used in this measurement.

[i] The error given here is only an educated guess. It is larger than the error on the weighted average of the published values.

[j] A theoretical value using QED.

[k] See the "Note on Λ_c^+ Branching Fractions" in the Branching Fractions of the Λ_c^+ Particle Listings.

[l] This branching fraction includes all the decay modes of the final-state resonance.

[m] An ℓ indicates an e or a μ mode, not a sum over these modes.

[n] The value is for the sum of the charge states of particle/antiparticle states indicated.

[o] Assuming isospin conservation, so that the other third is $\Lambda_c^+ \pi^0 \pi^0$.

[p] Not a pure measurement. See note at head of Λ_b^0 Decay Modes.

MONOPOLES, SUPERSYMMETRY, COMPOSITENESS, etc., SEARCHES FOR

Magnetic Monopole Searches

Isolated supermassive monopole candidate events have not been confirmed. The most sensitive experiments obtain negative results.
Best cosmic-ray supermassive monopole flux limit:

$$< 1.0 \times 10^{-15} \text{ cm}^{-2}\text{sr}^{-1}\text{s}^{-1} \quad \text{for } 1.1 \times 10^{-4} < \beta < 0.1$$

Supersymmetric Particle Searches

Limits are based on the Minimal Supersymmetric Standard Model.
Assumptions include: 1) $\tilde{\chi}_1^0$ (or $\tilde{\gamma}$) is lightest supersymmetric particle;
2) R-parity is conserved; 3) All scalar quarks (except $\tilde{t}_L$ and $\tilde{t}_R$) are degenerate in mass, and $m_{\tilde{q}_R} = m_{\tilde{q}_L}$. 4) Limits for selectrons and smuons refer to the $\tilde{\ell}_R$ states.
See the Particle Listings for a Note giving details of supersymmetry.

$\tilde{\chi}_i^0$ — neutralinos (mixtures of $\tilde{\gamma}$, $\tilde{Z}^0$, and $\tilde{H}_i^0$)

Mass $m_{\tilde{\chi}_1^0} > 10.9$ GeV, CL = 95%

Mass $m_{\tilde{\chi}_2^0} > 45.3$ GeV, CL = 95%   [$\tan\beta > 1$]

Mass $m_{\tilde{\chi}_3^0} > 75.8$ GeV, CL = 95%   [$\tan\beta > 1$]

Mass $m_{\tilde{\chi}_4^0} > 127$ GeV, CL = 95%   [$\tan\beta > 3$]

$\tilde{\chi}_i^\pm$ — charginos (mixtures of $\tilde{W}^\pm$ and $\tilde{H}_i^\pm$)

Mass $m_{\tilde{\chi}_1^\pm} > 65.7$ GeV, CL = 95%   [$m_{\tilde{\chi}_1^\pm} - m_{\tilde{\chi}_1^0} \geq 2$ GeV]

Mass $m_{\tilde{\chi}_2^\pm} > 99$ GeV, CL = 95%   [GUT relations assumed]

$\tilde{\nu}$ — scalar neutrino (sneutrino)

Mass $m > 37.1$ GeV, CL = 95%   [one flavor]

Mass $m > 43.1$ GeV, CL = 95%   [three degenerate flavors]

$\tilde{e}$ — scalar electron (selectron)

Mass $m > 58$ GeV, CL = 95%   [$m_{\tilde{e}_R} - m_{\tilde{\chi}_1^0} \geq 4$ GeV]

$\tilde{\mu}$ — scalar muon (smuon)

Mass $m > 55.6$ GeV, CL = 95%   [$m_{\tilde{\mu}_R} - m_{\tilde{\chi}_1^0} \geq 4$ GeV]

$\tilde{\tau}$ — scalar tau (stau)

Mass $m > 45$ GeV, CL = 95%   [if $m_{\tilde{\chi}_1^0} < 38$ GeV]

$\tilde{q}$ — scalar quark (squark)

These limits include the effects of cascade decays, evaluated assuming a fixed value of the parameters μ and $\tan\beta$. The limits are weakly sensitive to these parameters over much of parameter space. Limits assume GUT relations between gaugino masses and the gauge coupling; in particular that for $|\mu|$ not small, $m_{\tilde{\chi}_1^0} \approx m_{\tilde{g}}/6$.

Mass $m > 176$ GeV, CL = 95%   [any $m_{\tilde{g}} < 300$ GeV, $\mu = -250$ GeV, $\tan\beta = 2$]

Mass $m > 224$ GeV, CL = 95%   [$m_{\tilde{g}} \leq m_{\tilde{q}}$, $\mu = -400$ GeV, $\tan\beta = 4$]

$\tilde{g}$ — gluino

There is some controversy on whether gluinos in a low-mass window ($1 \lesssim m_{\tilde{g}} \lesssim 5$ GeV) are excluded or not. See the Supersymmetry Listings for details.

The limits summarised here refer to the high-mass region ($m_{\tilde{g}} \gtrsim 5$ GeV), and include the effects of cascade decays, evaluated assuming a fixed value of the parameters μ and $\tan\beta$. The limits are weakly sensitive to these parameters over much of parameter space. Limits assume GUT relations between gaugino masses and the gauge coupling; in particular that for $|\mu|$ not small, $m_{\tilde{\chi}_1^0} \approx m_{\tilde{g}}/6$.

Mass $m > 173$ GeV, CL = 95%   [any $m_{\tilde{q}}$, $\mu = -200$ GeV, $\tan\beta = 2$]

Mass $m > 212$ GeV, CL = 95%   [$m_{\tilde{g}} \geq m_{\tilde{q}}$, $\mu = -250$ GeV, $\tan\beta = 2$]

Quark and Lepton Compositeness, Searches for

Scale Limits Λ for Contact Interactions
(the lowest dimensional interactions with four fermions)

If the Lagrangian has the form

$$\pm \frac{g^2}{2\Lambda^2} \overline{\psi}_L \gamma_\mu \psi_L \overline{\psi}_L \gamma^\mu \psi_L$$

(with $g^2/4\pi$ set equal to 1), then we define $\Lambda \equiv \Lambda_{LL}^\pm$. For the full definitions and for other forms, see the Note in the Listings on Searches for Quark and Lepton Compositeness in the full *Review* and the original literature.

$\Lambda_{LL}^+(eeee) > 2.4$ TeV, CL = 95%

$\Lambda_{LL}^-(eeee) > 3.6$ TeV, CL = 95%

$\Lambda_{LL}^+(ee\mu\mu) > 2.6$ TeV, CL = 95%

$\Lambda_{LL}^-(ee\mu\mu) > 2.9$ TeV, CL = 95%

$\Lambda_{LL}^+(ee\tau\tau) > 1.9$ TeV, CL = 95%

$\Lambda_{LL}^-(ee\tau\tau) > 3.0$ TeV, CL = 95%

$\Lambda_{LL}^+(\ell\ell\ell\ell) > 3.5$ TeV, CL = 95%

$\Lambda_{LL}^-(\ell\ell\ell\ell) > 3.8$ TeV, CL = 95%

$\Lambda_{LL}^+(eeqq) > 2.5$ TeV, CL = 95%

$\Lambda_{LL}^-(eeqq) > 3.7$ TeV, CL = 95%

$\Lambda_{LL}^+(eebb) > 3.1$ TeV, CL = 95%

$\Lambda_{LL}^-(eebb) > 2.9$ TeV, CL = 95%

$\Lambda_{LL}^+(\mu\mu qq) > 2.9$ TeV, CL = 95%

$\Lambda_{LL}^-(\mu\mu qq) > 4.2$ TeV, CL = 95%

$\Lambda_{LR}^\pm(\nu_\mu\nu_e\mu e) > 3.1$ TeV, CL = 90%

$\Lambda_{LL}^\pm(qqqq) > 1.6$ TeV, CL = 95%

Excited Leptons

The limits from $\ell^{*+}\ell^{*-}$ do not depend on λ (where λ is the $\ell\ell^*$ transition coupling). The λ-dependent limits assume chiral coupling, except for the third limit for e^* which is for nonchiral coupling. For chiral coupling, this limit corresponds to $\lambda_\gamma = \sqrt{2}$.

$e^{*\pm}$ — excited electron

Mass $m > 85.0$ GeV, CL = 95%   (from $e^{*+}e^{*-}$)

Mass $m > 91$ GeV, CL = 95%   (if $\lambda_Z > 1$)

Mass $m > 194$ GeV, CL = 95%   (if $\lambda_\gamma = 1$)

$\mu^{*\pm}$ — excited muon

Mass $m > 85.3$ GeV, CL = 95%   (from $\mu^{*+}\mu^{*-}$)

Mass $m > 91$ GeV, CL = 95%   (if $\lambda_Z > 1$)

$\tau^{*\pm}$ — excited tau

Mass $m > 84.6$ GeV, CL = 95%   (from $\tau^{*+}\tau^{*-}$)

Mass $m > 90$ GeV, CL = 95%   (if $\lambda_Z > 0.18$)

ν^* — excited neutrino

Mass $m > 84.9$ GeV, CL = 95%   (from $\nu^*\overline{\nu}^*$)

Mass $m > 91$ GeV, CL = 95%   (if $\lambda_Z > 1$)

Mass $m = $ none 40–96 GeV, CL = 95%   (from $ep \to \nu^* X$)

q^* — excited quark

Mass $m > 45.6$ GeV, CL = 95%   (from $q^*\overline{q}^*$)

Mass $m > 88$ GeV, CL = 95%   (if $\lambda_Z > 1$)

Mass $m > 570$ GeV, CL = 95%   ($p\overline{p} \to q^* X$)

Color Sextet and Octet Particles

Color Sextet Quarks (q_6)

Mass $m > 84$ GeV, CL = 95%   (Stable q_6)

Color Octet Charged Leptons (ℓ_8)

Mass $m > 86$ GeV, CL = 95%   (Stable ℓ_8)

Color Octet Neutrinos (ν_8)

Mass $m > 110$ GeV, CL = 90%   ($\nu_8 \to \nu g$)

TESTS OF CONSERVATION LAWS

Revised by L. Wolfenstein and T.G. Trippe, May 1998.

In keeping with the current interest in tests of conservation laws, we collect together a Table of experimental limits on all weak and electromagnetic decays, mass differences, and moments, and on a few reactions, whose observation would violate conservation laws. The Table is given only in the full *Review of Particle Physics*, not in the Particle Physics Booklet. For the benefit of Booklet readers, we include the best limits from the Table in the following text. Limits in this text are for CL=90% unless otherwise specified. The Table is in two parts: "Discrete Space-Time Symmetries," *i.e.*, C, P, T, CP, and CPT; and "Number Conservation Laws," *i.e.*, lepton, baryon, hadronic flavor, and charge conservation. The references for these data can be found in the the Particle Listings in the *Review*. A discussion of these tests follows.

CPT INVARIANCE

General principles of relativistic field theory require invariance under the combined transformation CPT. The simplest tests of CPT invariance are the equality of the masses and lifetimes of a particle and its antiparticle. The best test comes from the limit on the mass difference between K^0 and $\overline{K}^0$. Any such difference contributes to the CP-violating parameter ϵ. Assuming CPT invariance, ϕ_ϵ, the phase of ϵ should be very close to 44°. (See the "Note on CP Violation in K_L^0 Decay" in the Particle Listings.) In contrast, if the entire source of CP violation in K^0 decays were a $K^0 - \overline{K}^0$ mass difference, ϕ_ϵ would be $44° + 90°$. Assuming that there is no other source of CPT violation than this mass difference, it is possible to deduce that [1]

$$ m_{\overline{K}^0} - m_{K^0} \approx \frac{2(m_{K_L^0} - m_{K_S^0})\,|\eta|\,(\frac{2}{3}\phi_{+-} + \frac{1}{3}\phi_{00} - \phi_0)}{\sin\phi_0}\, , $$

where $\phi_0 = 43.5°$ with an uncertainty of less than 0.1°. Using our best values of the CP-violation parameters, we get $|(m_{\overline{K}^0} - m_{K^0})/m_{K^0}| \leq 10^{-18}$. Limits can also be placed on specific CPT-violating decay amplitudes. Given the small value of $(1 - |\eta_{00}/\eta_{+-}|)$, the value of $\phi_{00} - \phi_{+-}$ provides a measure of CPT violation in $K_L^0 \rightarrow 2\pi$ decay. Results from CERN [1] and Fermilab [2] indicate no CPT-violating effect.

CP AND T INVARIANCE

Given CPT invariance, CP violation and T violation are equivalent. So far the only evidence for CP or T violation comes from the measurements of η_{+-}, η_{00}, and the semileptonic decay charge asymmetry for K_L, *e.g.*, $|\eta_{+-}| = |A(K_L^0 \rightarrow \pi^+\pi^-)/A(K_S^0 \rightarrow \pi^+\pi^-)| = (2.285 \pm 0.019) \times 10^{-3}$ and $[\Gamma(K_L^0 \rightarrow \pi^-e^+\nu) - \Gamma(K_L^0 \rightarrow \pi^+e^-\overline{\nu})]/[\text{sum}] = (0.333 \pm 0.014)\%$. Other searches for CP or T violation divide into (a) those that involve weak interactions or parity violation, and (b) those that involve processes otherwise allowed by the strong or electromagnetic interactions. In class (a) the most sensitive are probably the searches for an electric dipole moment of the neutron, measured to be $< 1.0 \times 10^{-25}$ e cm, and the electron $(-0.18 \pm 0.16) \times 10^{-26}$ e cm. A nonzero value requires both P and T violation. Class (b) includes the search for C violation in η decay, believed to be an electromagnetic process, *e.g.*, as measured by $\Gamma(\eta \rightarrow \mu^+\mu^-\pi^0)/\Gamma(\eta \rightarrow \text{all}) < 5 \times 10^{-6}$, and searches for T violation in a number of nuclear and electromagnetic reactions.

CONSERVATION OF LEPTON NUMBERS

Present experimental evidence and the standard electroweak theory are consistent with the absolute conservation of three separate lepton numbers: electron number L_e, muon number L_μ, and tau number L_τ. Searches for violations are of the following types:

a) $\Delta L = 2$ **for one type of lepton.** The best limit comes from the search for neutrinoless double beta decay $(Z, A) \rightarrow (Z + 2, A) + e^- + e^-$. The best laboratory limit is $t_{1/2} > 1.1 \times 10^{25}$ yr (CL=90%) for ^{76}Ge.

b) **Conversion of one lepton type to another.** For purely leptonic processes, the best limits are on $\mu \rightarrow e\gamma$ and $\mu \rightarrow 3e$, measured as $\Gamma(\mu \rightarrow e\gamma)/\Gamma(\mu \rightarrow \text{all}) < 5 \times 10^{-11}$ and $\Gamma(\mu \rightarrow 3e)/\Gamma(\mu \rightarrow \text{all}) < 1.0 \times 10^{-12}$. For semileptonic processes, the best limit comes from the coherent conversion process in a muonic atom, $\mu^- + (Z, A) \rightarrow e^- + (Z, A)$, measured as $\Gamma(\mu^-\text{Ti} \rightarrow e^-\text{Ti})/\Gamma(\mu^-\text{Ti} \rightarrow \text{all}) < 4 \times 10^{-12}$. Of special interest is the case in which the hadronic flavor also changes, as in $K_L \rightarrow e\mu$ and $K^+ \rightarrow \pi^+e^-\mu^+$, measured as $\Gamma(K_L \rightarrow e\mu)/\Gamma(K_L \rightarrow \text{all}) < 3.3 \times 10^{-11}$ and $\Gamma(K^+ \rightarrow \pi^+e^-\mu^+)/\Gamma(K^+ \rightarrow \text{all}) < 2.1 \times 10^{-10}$. Limits on the conversion of τ into e or μ are found in τ decay and are much less stringent than those for $\mu \rightarrow e$ conversion, *e.g.*, $\Gamma(\tau \rightarrow \mu\gamma)/\Gamma(\tau \rightarrow \text{all}) < 3.0 \times 10^{-6}$ and $\Gamma(\tau \rightarrow e\gamma)/\Gamma(\tau \rightarrow \text{all}) < 2.7 \times 10^{-6}$.

c) **Conversion of one type of lepton into another type of antilepton.** The case most studied is $\mu^- + (Z, A) \rightarrow e^+ + (Z - 2, A)$, the strongest limit being $\Gamma(\mu^-\text{Ti} \rightarrow e^+\text{Ca})/\Gamma(\mu^-\text{Ti} \rightarrow \text{all}) < 9 \times 10^{-11}$.

d) **Relation to neutrino mass.** If neutrinos have mass, then it is expected even in the standard electroweak theory that the lepton numbers are not separately conserved, as a consequence of lepton mixing analogous to Cabibbo quark mixing. However, in this case lepton-number-violating processes such as $\mu \rightarrow e\gamma$ are expected to have extremely small probability. For small neutrino masses, the lepton-number violation would be observed first in neutrino oscillations, which have been the subject of extensive experimental searches. For example, searches for $\overline{\nu}_e$ disappearance, which we label as $\overline{\nu}_e \not\rightarrow \overline{\nu}_e$, give measured limits $\Delta(m^2) < 9 \times 10^{-4}$ eV^2 for $\sin^2(2\theta) = 1$, and $\sin^2(2\theta) < 0.02$ for large $\Delta(m^2)$, where θ is the neutrino mixing angle. Possible evidence for mixing has come from two sources. The deficit in the solar neutrino flux compared with solar model calculations could be explained by oscillations with $\Delta(m^2) \leq 10^{-5}$ eV^2 causing the disappearance of ν_e. In addition underground detectors observing neutrinos produced by cosmic rays in the atmosphere have measured a ν_μ/ν_e ratio much less than expected and also a deficiency of upward going ν_μ compared to downward. This could be explained by oscillations leading to the disappearance of ν_μ with $\Delta(m^2)$ of the order 10^{-2}–10^{-3} eV^2.

CONSERVATION OF HADRONIC FLAVORS

In strong and electromagnetic interactions, hadronic flavor is conserved, *i.e.* the conversion of a quark of one flavor (d, u, s, c, b, t) into a quark of another flavor is forbidden. In the Standard Model, the weak interactions violate these conservation laws in a manner described by the Cabibbo-Kobayashi-Maskawa mixing (see the section "Cabibbo-Kobayashi-Maskawa Mixing Matrix"). The way in which these conservation laws are violated is tested as follows:

a) $\Delta S = \Delta Q$ **rule.** In the semileptonic decay of strange particles, the strangeness change equals the change in charge of the hadrons. Tests come from limits on decay rates such as

$\Gamma(\Sigma^+ \to n e^+ \nu)/\Gamma(\Sigma^+ \to \text{all}) < 5 \times 10^{-6}$, and from a detailed analysis of $K_L \to \pi e \nu$, which yields the parameter x, measured to be $(\text{Re}\,x, \text{Im}\,x) = (0.006 \pm 0.018, -0.003 \pm 0.026)$. Corresponding rules are $\Delta C = \Delta Q$ and $\Delta B = \Delta Q$.

b) Change of flavor by two units. In the Standard Model this occurs only in second-order weak interactions. The classic example is $\Delta S = 2$ via $K^0 - \overline{K}^0$ mixing, which is directly measured by $m(K_S) - m(K_L) = (3.489 \pm 0.009) \times 10^{-12}$ MeV. There is now evidence for $B^0 - \overline{B}^0$ mixing ($\Delta B = 2$), with the corresponding mass difference between the eigenstates $(m_{B_H^0} - m_{B_L^0})$ $= (0.723 \pm 0.032)\Gamma_{B^0} = (3.05 \pm 0.12) \times 10^{-10}$ MeV, and for $B_s^0 - \overline{B}_s^0$ mixing, with $(m_{B_{sH}^0} - m_{B_{sL}^0}) > 14\Gamma_{B_s^0}$ or $> 6 \times 10^{-9}$ MeV (CL=95%). No evidence exists for $D^0 - \overline{D}^0$ mixing, which is expected to be much smaller in the Standard Model.

c) Flavor-changing neutral currents. In the Standard Model the neutral-current interactions do not change flavor. The low rate $\Gamma(K_L \to \mu^+ \mu^-)/\Gamma(K_L \to \text{all}) = (7.2 \pm 0.5) \times 10^{-9}$ puts limits on such interactions; the nonzero value for this rate is attributed to a combination of the weak and electromagnetic interactions. The best test should come from $K^+ \to \pi^+ \nu \bar{\nu}$, which occurs in the Standard Model only as a second-order weak process with a branching fraction of (1 to 8)$\times 10^{-10}$. Observation of one event has been reported [4], yielding $\Gamma(K^+ \to \pi^+ \nu \bar{\nu})/\Gamma(K^+ \to \text{all}) = (4.2^{+9.7}_{-3.5}) \times 10^{-10}$. Limits for charm-changing or bottom-changing neutral currents are much less stringent: $\Gamma(D^0 \to \mu^+ \mu^-)/\Gamma(D^0 \to \text{all}) < 4 \times 10^{-6}$ and $\Gamma(B^0 \to \mu^+ \mu^-)/\Gamma(B^0 \to \text{all}) < 7 \times 10^{-7}$. One cannot isolate flavor-changing neutral current (FCNC) effects in non leptonic decays. For example, the FCNC transition $s \to d + (\overline{u} + u)$ is equivalent to the charged-current transition $s \to u + (\overline{u} + d)$. Tests for FCNC are therefore limited to hadron decays into lepton pairs. Such decays are expected only in second-order in the electroweak coupling in the Standard Model.

References

1. R. Carosi *et al.*, Phys. Lett. **B237**, 303 (1990).
2. M. Karlsson *et al.*, Phys. Rev. Lett. **64**, 2976 (1990); L.K. Gibbons *et al.*, Phys. Rev. Lett. **70**, 1199 (1993).
3. B. Schwingenheuer *et al.*, Phys. Rev. Lett. **74**, 4376 (1995).
4. S. Adler *et al.*, Phys. Rev. Lett. **79**, 2204 (1997).

TESTS OF DISCRETE SPACE-TIME SYMMETRIES

CHARGE CONJUGATION (C) INVARIANCE

$\Gamma(\pi^0 \to 3\gamma)/\Gamma_{\text{total}}$	$<3.1 \times 10^{-8}$, CL = 90%
η C-nonconserving decay parameters	
$\pi^+ \pi^- \pi^0$ left-right asymmetry parameter	$(0.09 \pm 0.17) \times 10^{-2}$
$\pi^+ \pi^- \pi^0$ sextant asymmetry parameter	$(0.18 \pm 0.16) \times 10^{-2}$
$\pi^+ \pi^- \pi^0$ quadrant asymmetry parameter	$(-0.17 \pm 0.17) \times 10^{-2}$
$\pi^+ \pi^- \gamma$ left-right asymmetry parameter	$(0.9 \pm 0.4) \times 10^{-2}$
$\pi^+ \pi^- \gamma$ parameter β (D-wave)	0.05 ± 0.06 (S = 1.5)
$\Gamma(\eta \to 3\gamma)/\Gamma_{\text{total}}$	$<5 \times 10^{-4}$, CL = 95%
$\Gamma(\eta \to \pi^0 e^+ e^-)/\Gamma_{\text{total}}$	[a] $<4 \times 10^{-5}$, CL = 90%
$\Gamma(\eta \to \pi^0 \mu^+ \mu^-)/\Gamma_{\text{total}}$	[a] $<5 \times 10^{-6}$, CL = 90%
$\Gamma(\omega(782) \to \pi \pi^0)/\Gamma_{\text{total}}$	$<1 \times 10^{-3}$, CL = 90%
$\Gamma(\omega(782) \to 3\pi^0)/\Gamma_{\text{total}}$	$<3 \times 10^{-4}$, CL = 90%
$\Gamma(\eta'(958) \to \pi^0 e^+ e^-)/\Gamma_{\text{total}}$	[a] $<1.3 \times 10^{-2}$, CL = 90%
$\Gamma(\eta'(958) \to \eta e^+ e^-)/\Gamma_{\text{total}}$	[a] $<1.1 \times 10^{-2}$, CL = 90%
$\Gamma(\eta'(958) \to 3\gamma)/\Gamma_{\text{total}}$	$<1.0 \times 10^{-4}$, CL = 90%
$\Gamma(\eta'(958) \to \mu^+ \mu^- \pi^0)/\Gamma_{\text{total}}$	[a] $<6.0 \times 10^{-5}$, CL = 90%
$\Gamma(\eta'(958) \to \mu^+ \mu^- \eta)/\Gamma_{\text{total}}$	[a] $<1.5 \times 10^{-5}$, CL = 90%

PARITY (P) INVARIANCE

e electric dipole moment	$(0.18 \pm 0.16) \times 10^{-26}$ ecm
μ electric dipole moment	$(3.7 \pm 3.4) \times 10^{-19}$ ecm
τ electric dipole moment (d_τ)	> -3.1 and $< 3.1 \times 10^{-16}$ ecm, CL = 95%
$\Gamma(\eta \to \pi^+ \pi^-)/\Gamma_{\text{total}}$	$<9 \times 10^{-4}$, CL = 90%
$\Gamma(\eta'(958) \to \pi^+ \pi^-)/\Gamma_{\text{total}}$	$<2 \times 10^{-2}$, CL = 90%
$\Gamma(\eta'(958) \to \pi^0 \pi^0)/\Gamma_{\text{total}}$	$<9 \times 10^{-4}$, CL = 90%
p electric dipole moment	$(-4 \pm 6) \times 10^{-23}$ ecm
n electric dipole moment	$<0.97 \times 10^{-25}$ ecm, CL = 90%
Λ electric dipole moment	$<1.5 \times 10^{-16}$ ecm, CL = 95%

TIME REVERSAL (T) INVARIANCE

Limits on e, μ, τ, p, n, and Λ electric dipole moments under Parity Invariance above are also tests of Time Reversal Invariance.

μ decay parameters		
transverse e^+ polarization normal to plane of μ spin, e^+ momentum		0.007 ± 0.023
α'/A		$(0 \pm 4) \times 10^{-3}$
β'/A		$(2 \pm 6) \times 10^{-3}$
τ electric dipole moment (d_τ)		> -3.1 and $< 3.1 \times 10^{-16}$ ecm, CL = 95%
Im(ξ) in $K_{\mu3}^\pm$ decay (from transverse μ pol.)		-0.017 ± 0.025
Im(ξ) in $K_{\mu3}^0$ decay (from transverse μ pol.)		-0.007 ± 0.026
$n \to p e^- \nu$ decay parameters		
ϕ_{AV}, phase of g_A relative to g_V	[b]	$(180.07 \pm 0.18)^\circ$
triple correlation coefficient D		$(-0.5 \pm 1.4) \times 10^{-3}$
triple correlation coefficient D for $\Sigma^- \to n e^- \overline{\nu}_e$		0.11 ± 0.10

CP INVARIANCE

Re(d_τ^w)		$<0.56 \times 10^{-17}$ ecm, CL = 95%
Im(d_τ^w)		$<1.5 \times 10^{-17}$ ecm, CL = 95%
$\Gamma(\eta \to \pi^+ \pi^-)/\Gamma_{\text{total}}$		$<9 \times 10^{-4}$, CL = 90%
$\Gamma(\eta'(958) \to \pi^+ \pi^-)/\Gamma_{\text{total}}$		$<2 \times 10^{-2}$, CL = 90%
$\Gamma(\eta'(958) \to \pi^0 \pi^0)/\Gamma_{\text{total}}$		$<9 \times 10^{-4}$, CL = 90%
$K^\pm \to \pi^\pm \pi^+ \pi^-$ rate difference/average		$(0.07 \pm 0.12)\%$
$K^\pm \to \pi^\pm \pi^0 \pi^0$ rate difference/average		$(0.0 \pm 0.6)\%$
$K^\pm \to \pi^\pm \pi^0 \gamma$ rate difference/average		$(0.9 \pm 3.3)\%$
$(g_{\tau^+} - g_{\tau^-}) / (g_{\tau^+} + g_{\tau^-})$ for $K^\pm \to \pi^\pm \pi^+ \pi^-$		$(-0.7 \pm 0.5)\%$
CP-violation parameters in K_S^0 decay		
Im(η_{+-0}) = Im($A(K_S^0 \to \pi^+ \pi^- \pi^0$, CP-violating) / $A(K_L^0 \to \pi^+ \pi^- \pi^0)$)		-0.002 ± 0.008
Im(η_{000})^2 = $\Gamma(K_S^0 \to 3\pi^0)$ / $\Gamma(K_L^0 \to 3\pi^0)$		<0.1, CL = 90%
charge asymmetry j for $K_L^0 \to \pi^+ \pi^- \pi^0$		0.0011 ± 0.0008
$|\epsilon'_{+-\gamma}|/\epsilon$ for $K_L^0 \to \pi^+ \pi^- \gamma$		<0.3, CL = 90%
$\Gamma(K_L^0 \to \pi^0 \mu^+ \mu^-)/\Gamma_{\text{total}}$	[c]	$<5.1 \times 10^{-9}$, CL = 90%
$\Gamma(K_L^0 \to \pi^0 e^+ e^-)/\Gamma_{\text{total}}$	[c]	$<4.3 \times 10^{-9}$, CL = 90%
$\Gamma(K_L^0 \to \pi^0 \nu \bar{\nu})/\Gamma_{\text{total}}$	[d]	$<5.8 \times 10^{-5}$, CL = 90%
$A_{CP}(K^+ K^- \pi^\pm)$ in $D^\pm \to K^+ K^- \pi^\pm$		-0.017 ± 0.027
$A_{CP}(K^\pm K^{*0})$ in $D^+ \to K^+ \overline{K}^{*0}$ and $D^- \to K^- K^{*0}$		-0.02 ± 0.05
$A_{CP}(\phi \pi^\pm)$ in $D^\pm \to \phi \pi^\pm$		-0.014 ± 0.033
$A_{CP}(\pi^+ \pi^- \pi^\pm)$ in $D^\pm \to \pi^+ \pi^- \pi^\pm$		-0.02 ± 0.04
$A_{CP}(K^+ K^-)$ in $D^0, \overline{D}^0 \to K^+ K^-$		0.026 ± 0.035
$A_{CP}(\pi^+ \pi^-)$ in $D^0, \overline{D}^0 \to \pi^+ \pi^-$		-0.05 ± 0.08
$A_{CP}(K_S^0 \phi)$ in $D^0, \overline{D}^0 \to K_S^0 \phi$		-0.03 ± 0.09
$A_{CP}(K_S^0 \pi^0)$ in $D^0, \overline{D}^0 \to K_S^0 \pi^0$		-0.018 ± 0.030
$|\text{Re}(\epsilon_{B^0})|$		0.002 ± 0.008
$[\alpha_-(\Lambda) + \alpha_+(\overline{\Lambda})] / [\alpha_-(\Lambda) - \alpha_+(\overline{\Lambda})]$		-0.03 ± 0.06

Limits are given at the 90% confidence level, while errors are given as ± 1 standard deviation.

Tests of Conservation Laws

CP VIOLATION OBSERVED

K_L^0 branching ratios

charge asymmetry in $K_{\ell 3}^0$ decays

$\delta(\mu) = [\Gamma(\pi^- \mu^+ \nu_\mu)$ $- \Gamma(\pi^+ \mu^- \bar{\nu}_\mu)]/\text{sum}$	$(0.304 \pm 0.025)\%$
$\delta(e) = [\Gamma(\pi^- e^+ \nu_e)$ $- \Gamma(\pi^+ e^- \bar{\nu}_e)]/\text{sum}$	$(0.333 \pm 0.014)\%$

parameters for $K_L^0 \to 2\pi$ decay

$|\eta_{00}| = |A(K_L^0 \to 2\pi^0) /$ $A(K_S^0 \to 2\pi^0)|$	$(2.275 \pm 0.019) \times 10^{-3}$ (S = 1.1)
$|\eta_{+-}| = |A(K_L^0 \to \pi^+ \pi^-) /$ $A(K_S^0 \to \pi^+ \pi^-)|$	$(2.285 \pm 0.019) \times 10^{-3}$
$\epsilon'/\epsilon \approx \text{Re}(\epsilon'/\epsilon) = (1-|\eta_{00}/\eta_{+-}|)/3$	[e] $(1.5 \pm 0.8) \times 10^{-3}$ (S = 1.8)
ϕ_{+-}, phase of η_{+-}	$(43.5 \pm 0.6)^\circ$
ϕ_{00}, phase of η_{00}	$(43.4 \pm 1.0)^\circ$

parameters for $K_L^0 \to \pi^+ \pi^- \gamma$ decay

$|\eta_{+-\gamma}| = |A(K_L^0 \to \pi^+ \pi^- \gamma, CP$ violating)/$A(K_S^0 \to \pi^+ \pi^- \gamma)|$	$(2.35 \pm 0.07) \times 10^{-3}$
$\phi_{+-\gamma}$ = phase of $\eta_{+-\gamma}$	$(44 \pm 4)^\circ$
$\Gamma(K_L^0 \to \pi^+ \pi^-)/\Gamma_{\text{total}}$	$(2.067 \pm 0.035) \times 10^{-3}$ (S = 1.1)
$\Gamma(K_L^0 \to \pi^0 \pi^0)/\Gamma_{\text{total}}$	$(9.36 \pm 0.20) \times 10^{-4}$

CPT INVARIANCE

$(m_{W^+} - m_{W^-}) / m_{\text{average}}$	-0.002 ± 0.007
$(m_{e^+} - m_{e^-}) / m_{\text{average}}$	$<4 \times 10^{-8}$, CL = 90%
$|q_{e^+} + q_{e^-}|/e$	$<2 \times 10^{-18}$
$(g_{e^+} - g_{e^-}) / g_{\text{average}}$	$(-0.5 \pm 2.1) \times 10^{-12}$
$(\tau_{\mu^+} - \tau_{\mu^-}) / \tau_{\text{average}}$	$(2 \pm 8) \times 10^{-5}$
$(g_{\mu^+} - g_{\mu^-}) / g_{\text{average}}$	$(-2.6 \pm 1.6) \times 10^{-8}$
$(m_{\pi^+} - m_{\pi^-}) / m_{\text{average}}$	$(2 \pm 5) \times 10^{-4}$
$(\tau_{\pi^+} - \tau_{\pi^-}) / \tau_{\text{average}}$	$(6 \pm 7) \times 10^{-4}$
$(m_{K^+} - m_{K^-}) / m_{\text{average}}$	$(-0.6 \pm 1.8) \times 10^{-4}$
$(\tau_{K^+} - \tau_{K^-}) / \tau_{\text{average}}$	$(0.11 \pm 0.09)\%$ (S = 1.2)
$K^\pm \to \mu^\pm \nu_\mu$ rate difference/average	$(-0.5 \pm 0.4)\%$
$K^\pm \to \pi^\pm \pi^0$ rate difference/average	[f] $(0.8 \pm 1.2)\%$
$|m_{K^0} - m_{\bar{K}^0}| / m_{\text{average}}$	[g] $<10^{-18}$
phase difference $\phi_{00} - \phi_{+-}$	$(-0.1 \pm 0.8)^\circ$

CPT-violation parameters in K^0 decay

real part of Δ	0.018 ± 0.020
imaginary part of Δ	0.02 ± 0.04
$(|\frac{q_{\bar{p}}}{m_{\bar{p}}}| - |\frac{q_p}{m_p}|)/|\frac{q}{m}|_{\text{average}}$	$(1.5 \pm 1.1) \times 10^{-9}$
$|q_p + q_{\bar{p}}|/e$	$<2 \times 10^{-5}$
$(\mu_p + \mu_{\bar{p}}) / |\mu|_{\text{average}}$	$(-2.6 \pm 2.9) \times 10^{-3}$
$(m_n - m_{\bar{n}}) / m_{\text{average}}$	$(9 \pm 5) \times 10^{-5}$
$(m_\Lambda - m_{\bar{\Lambda}}) / m_\Lambda$	$(-1.0 \pm 0.9) \times 10^{-5}$
$(\tau_\Lambda - \tau_{\bar{\Lambda}}) / \tau_{\text{average}}$	0.04 ± 0.09
$(\mu_{\Sigma^+} + \mu_{\bar{\Sigma}^-}) / |\mu|_{\text{average}}$	0.014 ± 0.015
$(m_{\Xi^-} - m_{\bar{\Xi}^+}) / m_{\text{average}}$	$(1.1 \pm 2.7) \times 10^{-4}$
$(\tau_{\Xi^-} - \tau_{\bar{\Xi}^+}) / \tau_{\text{average}}$	0.02 ± 0.18
$(m_{\Omega^-} - m_{\bar{\Omega}^+}) / m_{\text{average}}$	$(0 \pm 5) \times 10^{-4}$

TESTS OF NUMBER CONSERVATION LAWS

LEPTON FAMILY NUMBER

Lepton family number conservation means separate conservation of each of L_e, L_μ, L_τ.

$\Gamma(Z \to e^\pm \mu^\mp)/\Gamma_{\text{total}}$	[h] $<1.7 \times 10^{-6}$, CL = 95%
$\Gamma(Z \to e^\pm \tau^\mp)/\Gamma_{\text{total}}$	[h] $<9.8 \times 10^{-6}$, CL = 95%
$\Gamma(Z \to \mu^\pm \tau^\mp)/\Gamma_{\text{total}}$	[h] $<1.2 \times 10^{-5}$, CL = 95%
limit on $\mu^- \to e^-$ conversion	
$\sigma(\mu^- {}^{32}\text{S} \to e^- {}^{32}\text{S}) /$ $\sigma(\mu^- {}^{32}\text{S} \to \nu_\mu {}^{32}\text{P}^*)$	$<7 \times 10^{-11}$, CL = 90%
$\sigma(\mu^- \text{Ti} \to e^- \text{Ti}) /$ $\sigma(\mu^- \text{Ti} \to \text{capture})$	$<4.3 \times 10^{-12}$, CL = 90%
$\sigma(\mu^- \text{Pb} \to e^- \text{Pb}) /$ $\sigma(\mu^- \text{Pb} \to \text{capture})$	$<4.6 \times 10^{-11}$, CL = 90%
limit on muonium $\to$ antimuonium conversion $R_g = G_C / G_F$	<0.018, CL = 90%
$\Gamma(\mu^- \to e^- \nu_e \bar{\nu}_\mu)/\Gamma_{\text{total}}$	[i] $<1.2 \times 10^{-2}$, CL = 90%
$\Gamma(\mu^- \to e^- \gamma)/\Gamma_{\text{total}}$	$<4.9 \times 10^{-11}$, CL = 90%
$\Gamma(\mu^- \to e^- e^+ e^-)/\Gamma_{\text{total}}$	$<1.0 \times 10^{-12}$, CL = 90%
$\Gamma(\mu^- \to e^- 2\gamma)/\Gamma_{\text{total}}$	$<7.2 \times 10^{-11}$, CL = 90%
$\Gamma(\tau^- \to e^- \gamma)/\Gamma_{\text{total}}$	$<2.7 \times 10^{-6}$, CL = 90%
$\Gamma(\tau^- \to \mu^- \gamma)/\Gamma_{\text{total}}$	$<3.0 \times 10^{-6}$, CL = 90%
$\Gamma(\tau^- \to e^- \pi^0)/\Gamma_{\text{total}}$	$<3.7 \times 10^{-6}$, CL = 90%
$\Gamma(\tau^- \to \mu^- \pi^0)/\Gamma_{\text{total}}$	$<4.0 \times 10^{-6}$, CL = 90%
$\Gamma(\tau^- \to e^- K^0)/\Gamma_{\text{total}}$	$<1.3 \times 10^{-3}$, CL = 90%
$\Gamma(\tau^- \to \mu^- K^0)/\Gamma_{\text{total}}$	$<1.0 \times 10^{-3}$, CL = 90%
$\Gamma(\tau^- \to e^- \eta)/\Gamma_{\text{total}}$	$<8.2 \times 10^{-6}$, CL = 90%
$\Gamma(\tau^- \to \mu^- \eta)/\Gamma_{\text{total}}$	$<9.6 \times 10^{-6}$, CL = 90%
$\Gamma(\tau^- \to e^- \rho^0)/\Gamma_{\text{total}}$	$<2.0 \times 10^{-6}$, CL = 90%
$\Gamma(\tau^- \to \mu^- \rho^0)/\Gamma_{\text{total}}$	$<6.3 \times 10^{-6}$, CL = 90%
$\Gamma(\tau^- \to e^- K^*(892)^0)/\Gamma_{\text{total}}$	$<5.1 \times 10^{-6}$, CL = 90%
$\Gamma(\tau^- \to \mu^- K^*(892)^0)/\Gamma_{\text{total}}$	$<7.5 \times 10^{-6}$, CL = 90%
$\Gamma(\tau^- \to e^- \bar{K}^*(892)^0)/\Gamma_{\text{total}}$	$<7.4 \times 10^{-6}$, CL = 90%
$\Gamma(\tau^- \to \mu^- \bar{K}^*(892)^0)/\Gamma_{\text{total}}$	$<7.5 \times 10^{-6}$, CL = 90%
$\Gamma(\tau^- \to e^- \phi)/\Gamma_{\text{total}}$	$<6.9 \times 10^{-6}$, CL = 90%
$\Gamma(\tau^- \to \mu^- \phi)/\Gamma_{\text{total}}$	$<7.0 \times 10^{-6}$, CL = 90%
$\Gamma(\tau^- \to e^- e^+ e^-)/\Gamma_{\text{total}}$	$<2.9 \times 10^{-6}$, CL = 90%
$\Gamma(\tau^- \to e^- \mu^+ \mu^-)/\Gamma_{\text{total}}$	$<1.8 \times 10^{-6}$, CL = 90%
$\Gamma(\tau^- \to e^+ \mu^- \mu^-)/\Gamma_{\text{total}}$	$<1.5 \times 10^{-6}$, CL = 90%
$\Gamma(\tau^- \to \mu^- e^+ e^-)/\Gamma_{\text{total}}$	$<1.7 \times 10^{-6}$, CL = 90%
$\Gamma(\tau^- \to \mu^+ e^- e^-)/\Gamma_{\text{total}}$	$<1.5 \times 10^{-6}$, CL = 90%
$\Gamma(\tau^- \to \mu^- \mu^+ \mu^-)/\Gamma_{\text{total}}$	$<1.9 \times 10^{-6}$, CL = 90%
$\Gamma(\tau^- \to e^- \pi^+ \pi^-)/\Gamma_{\text{total}}$	$<2.2 \times 10^{-6}$, CL = 90%
$\Gamma(\tau^- \to \mu^- \pi^+ \pi^-)/\Gamma_{\text{total}}$	$<8.2 \times 10^{-6}$, CL = 90%
$\Gamma(\tau^- \to e^- \pi^+ K^-)/\Gamma_{\text{total}}$	$<6.4 \times 10^{-6}$, CL = 90%
$\Gamma(\tau^- \to e^- \pi^- K^+)/\Gamma_{\text{total}}$	$<3.8 \times 10^{-6}$, CL = 90%
$\Gamma(\tau^- \to e^- K^+ K^-)/\Gamma_{\text{total}}$	$<6.0 \times 10^{-6}$, CL = 90%
$\Gamma(\tau^- \to \mu^- \pi^+ K^-)/\Gamma_{\text{total}}$	$<7.5 \times 10^{-6}$, CL = 90%
$\Gamma(\tau^- \to \mu^- \pi^- K^+)/\Gamma_{\text{total}}$	$<7.4 \times 10^{-6}$, CL = 90%
$\Gamma(\tau^- \to \mu^- K^+ K^-)/\Gamma_{\text{total}}$	$<1.5 \times 10^{-5}$, CL = 90%
$\Gamma(\tau^- \to e^- \pi^0 \pi^0)/\Gamma_{\text{total}}$	$<6.5 \times 10^{-6}$, CL = 90%
$\Gamma(\tau^- \to \mu^- \pi^0 \pi^0)/\Gamma_{\text{total}}$	$<1.4 \times 10^{-5}$, CL = 90%
$\Gamma(\tau^- \to e^- \eta\eta)/\Gamma_{\text{total}}$	$<3.5 \times 10^{-5}$, CL = 90%
$\Gamma(\tau^- \to \mu^- \eta\eta)/\Gamma_{\text{total}}$	$<6.0 \times 10^{-5}$, CL = 90%
$\Gamma(\tau^- \to e^- \pi^0 \eta)/\Gamma_{\text{total}}$	$<2.4 \times 10^{-5}$, CL = 90%
$\Gamma(\tau^- \to \mu^- \pi^0 \eta)/\Gamma_{\text{total}}$	$<2.2 \times 10^{-5}$, CL = 90%
$\Gamma(\tau^- \to e^- \text{light boson})/\Gamma_{\text{total}}$	$<2.7 \times 10^{-3}$, CL = 95%
$\Gamma(\tau^- \to \mu^- \text{light boson})/\Gamma_{\text{total}}$	$<5 \times 10^{-3}$, CL = 95%

ν oscillations. (For other lepton mixing effects in particle decays, see the Particle Listings.)

$\bar{\nu}_e \not\to \bar{\nu}_e$

$\Delta(m^2)$ for $\sin^2(2\theta) = 1$	$<9 \times 10^{-4}$ eV^2, CL = 90%
$\sin^2(2\theta)$ for "Large" $\Delta(m^2)$	<0.02, CL = 90%

$\nu_e \to \nu_\tau$

$\Delta(m^2)$ for $\sin^2(2\theta) = 1$	<9 eV^2, CL = 90%
$\sin^2(2\theta)$ for "Large" $\Delta(m^2)$	<0.25, CL = 90%

$\bar{\nu}_e \to \bar{\nu}_\tau$

$\sin^2(2\theta)$ for "Large" $\Delta(m^2)$	<0.7, CL = 90%

$\nu_\mu \to \nu_e$

Limits are given at the 90% confidence level, while errors are given as ± 1 standard deviation.

$\Delta(m^2)$ for $\sin^2(2\theta) = 1$ — <0.09 eV^2, CL = 90%
$\sin^2(2\theta)$ for "Large" $\Delta(m^2)$ — $<3.0 \times 10^{-3}$, CL = 90%

$\bar{\nu}_\mu \to \bar{\nu}_e$
  $\Delta(m^2)$ for $\sin^2(2\theta) = 1$ — <0.14 eV^2, CL = 90%
  $\sin^2(2\theta)$ for "Large" $\Delta(m^2)$ — <0.004, CL = 95%

$\nu_\mu(\bar{\nu}_\mu) \to \nu_e(\bar{\nu}_e)$
  $\Delta(m^2)$ for $\sin^2(2\theta) = 1$ — <0.075 eV^2, CL = 90%
  $\sin^2(2\theta)$ for "Large" $\Delta(m^2)$ — $<1.8 \times 10^{-3}$, CL = 90%

$\nu_\mu \to \nu_\tau$
  $\Delta(m^2)$ for $\sin^2(2\theta) = 1$ — <0.9 eV^2, CL = 90%
  $\sin^2(2\theta)$ for "Large" $\Delta(m^2)$ — <0.004, CL = 90%

$\bar{\nu}_\mu \to \bar{\nu}_\tau$
  $\Delta(m^2)$ for $\sin^2(2\theta) = 1$ — <2.2 eV^2, CL = 90%
  $\sin^2(2\theta)$ for "Large" $\Delta(m^2)$ — $<4.4 \times 10^{-2}$, CL = 90%

$\nu_\mu(\bar{\nu}_\mu) \to \nu_\tau(\bar{\nu}_\tau)$
  $\Delta(m^2)$ for $\sin^2(2\theta) = 1$ — <1.5 eV^2, CL = 90%
  $\sin^2(2\theta)$ for "Large" $\Delta(m^2)$ — $<8 \times 10^{-3}$, CL = 90%

$\nu_e \not\to \nu_e$
  $\Delta(m^2)$ for $\sin^2(2\theta) = 1$ — <0.17 eV^2, CL = 90%
  $\sin^2(2\theta)$ for "Large" $\Delta(m^2)$ — $<7 \times 10^{-2}$, CL = 90%

$\nu_\mu \not\to \nu_\mu$
  $\Delta(m^2)$ for $\sin^2(2\theta) = 1$ — <0.23 or >1500 eV^2
  $\sin^2(2\theta)$ for $\Delta(m^2) = 100$eV^2 — [j] <0.02, CL = 90%

$\bar{\nu}_\mu \not\to \bar{\nu}_\mu$
  $\Delta(m^2)$ for $\sin^2(2\theta) = 1$ — <7 or >1200 eV^2
  $\sin^2(2\theta)$ for 190 eV$^2 < \Delta(m^2) <$ 320 eV^2 — [k] <0.02, CL = 90%

$\Gamma(\pi^+ \to \mu^+ \nu_e)/\Gamma_{total}$	[l]	$<8.0 \times 10^{-3}$, CL = 90%
$\Gamma(\pi^+ \to \mu^- e^+ e^+ \nu)/\Gamma_{total}$		$<1.6 \times 10^{-6}$, CL = 90%
$\Gamma(\pi^0 \to \mu^+ e^- + e^- \mu^+)/\Gamma_{total}$		$<1.72 \times 10^{-8}$, CL = 90%
$\Gamma(\eta \to \mu^+ e^- + \mu^- e^+)/\Gamma_{total}$		$<6 \times 10^{-6}$, CL = 90%
$\Gamma(K^+ \to \mu^- \nu e^+ e^+)/\Gamma_{total}$		$<2.0 \times 10^{-8}$, CL = 90%
$\Gamma(K^+ \to \mu^+ \nu_e)/\Gamma_{total}$	[l]	$<4 \times 10^{-3}$, CL = 90%
$\Gamma(K^+ \to \pi^+ \mu^+ e^-)/\Gamma_{total}$		$<2.1 \times 10^{-10}$, CL = 90%
$\Gamma(K^+ \to \pi^+ \mu^- e^+)/\Gamma_{total}$		$<7 \times 10^{-9}$, CL = 90%
$\Gamma(K_L^0 \to e^\pm \mu^\mp)/\Gamma_{total}$	[h]	$<3.3 \times 10^{-11}$, CL = 90%
$\Gamma(K_L^0 \to e^\pm e^\pm \mu^\mp)/\Gamma_{total}$	[h]	$<6.1 \times 10^{-9}$, CL = 90%
$\Gamma(D^+ \to \pi^+ e^+ \mu^-)/\Gamma_{total}$		$<1.1 \times 10^{-4}$, CL = 90%
$\Gamma(D^+ \to \pi^+ e^- \mu^+)/\Gamma_{total}$		$<1.3 \times 10^{-4}$, CL = 90%
$\Gamma(D^+ \to K^+ e^+ \mu^-)/\Gamma_{total}$		$<1.3 \times 10^{-4}$, CL = 90%
$\Gamma(D^+ \to K^+ e^- \mu^+)/\Gamma_{total}$		$<1.2 \times 10^{-4}$, CL = 90%
$\Gamma(D^0 \to \mu^\pm e^\mp)/\Gamma_{total}$	[h]	$<1.9 \times 10^{-5}$, CL = 90%
$\Gamma(D^0 \to \pi^0 e^\pm \mu^\mp)/\Gamma_{total}$	[h]	$<8.6 \times 10^{-5}$, CL = 90%
$\Gamma(D^0 \to \eta e^\pm \mu^\mp)/\Gamma_{total}$	[h]	$<1.0 \times 10^{-4}$, CL = 90%
$\Gamma(D^0 \to \rho^0 e^\pm \mu^\mp)/\Gamma_{total}$	[h]	$<4.9 \times 10^{-5}$, CL = 90%
$\Gamma(D^0 \to \omega e^\pm \mu^\mp)/\Gamma_{total}$	[h]	$<1.2 \times 10^{-4}$, CL = 90%
$\Gamma(D^0 \to \phi e^\pm \mu^\mp)/\Gamma_{total}$	[h]	$<3.4 \times 10^{-5}$, CL = 90%
$\Gamma(D^0 \to \bar{K}^0 e^\pm \mu^\mp)/\Gamma_{total}$	[h]	$<1.0 \times 10^{-4}$, CL = 90%
$\Gamma(D^0 \to \bar{K}^*(892)^0 e^\pm \mu^\mp)/\Gamma_{total}$	[h]	$<1.0 \times 10^{-4}$, CL = 90%
$\Gamma(B^+ \to \pi^+ e^+ \mu^-)/\Gamma_{total}$		$<6.4 \times 10^{-3}$, CL = 90%
$\Gamma(B^+ \to \pi^+ e^- \mu^+)/\Gamma_{total}$		$<6.4 \times 10^{-3}$, CL = 90%
$\Gamma(B^+ \to K^+ e^+ \mu^-)/\Gamma_{total}$		$<6.4 \times 10^{-3}$, CL = 90%
$\Gamma(B^+ \to K^+ e^- \mu^+)/\Gamma_{total}$		$<6.4 \times 10^{-3}$, CL = 90%
$\Gamma(B^+ \to \pi^- e^+ \mu^+)/\Gamma_{total}$		$<6.4 \times 10^{-3}$, CL = 90%
$\Gamma(B^+ \to K^- e^+ \mu^+)/\Gamma_{total}$		$<6.4 \times 10^{-3}$, CL = 90%
$\Gamma(B^0 \to e^\pm \mu^\mp)/\Gamma_{total}$	[h]	$<5.9 \times 10^{-6}$, CL = 90%
$\Gamma(B^0 \to e^\pm \tau^\mp)/\Gamma_{total}$	[h]	$<5.3 \times 10^{-4}$, CL = 90%
$\Gamma(B^0 \to \mu^\pm \tau^\mp)/\Gamma_{total}$	[h]	$<8.3 \times 10^{-4}$, CL = 90%
$\Gamma(B \to e^\pm \mu^\mp s)/\Gamma_{total}$		$<2.2 \times 10^{-5}$, CL = 90%
$\Gamma(B_s^0 \to e^\pm \mu^\mp)/\Gamma_{total}$	[h]	$<4.1 \times 10^{-5}$, CL = 90%

TOTAL LEPTON NUMBER

Violation of total lepton number conservation also implies violation of lepton family number conservation.

limit on $\mu^- \to e^+$ conversion		
$\sigma(\mu^- {}^{32}S \to e^+ {}^{32}Si^*)$ / $\sigma(\mu^- {}^{32}S \to \nu_\mu {}^{32}P^*)$		$<9 \times 10^{-10}$, CL = 90%
$\sigma(\mu^- {}^{127}I \to e^+ {}^{127}Sb^*)$ / $\sigma(\mu^- {}^{127}I \to$ anything$)$		$<3 \times 10^{-10}$, CL = 90%
$\sigma(\mu^- Ti \to e^+ Ca)$ / $\sigma(\mu^- Ti \to$ capture$)$		$<8.9 \times 10^{-11}$, CL = 90%
$\Gamma(\tau^- \to \pi^- \gamma)/\Gamma_{total}$		$<2.8 \times 10^{-4}$, CL = 90%
$\Gamma(\tau^- \to \pi^- \pi^0)/\Gamma_{total}$		$<3.7 \times 10^{-4}$, CL = 90%
$\Gamma(\tau^- \to e^+ \pi^- \pi^-)/\Gamma_{total}$		$<1.9 \times 10^{-6}$, CL = 90%
$\Gamma(\tau^- \to \mu^+ \pi^- \pi^-)/\Gamma_{total}$		$<3.4 \times 10^{-6}$, CL = 90%
$\Gamma(\tau^- \to e^+ \pi^- K^-)/\Gamma_{total}$		$<2.1 \times 10^{-6}$, CL = 90%
$\Gamma(\tau^- \to e^+ K^- K^-)/\Gamma_{total}$		$<3.8 \times 10^{-6}$, CL = 90%
$\Gamma(\tau^- \to \mu^+ \pi^- K^-)/\Gamma_{total}$		$<7.0 \times 10^{-6}$, CL = 90%
$\Gamma(\tau^- \to \mu^+ K^- K^-)/\Gamma_{total}$		$<6.0 \times 10^{-6}$, CL = 90%
$\Gamma(\tau^- \to \bar{p}\gamma)/\Gamma_{total}$		$<2.9 \times 10^{-4}$, CL = 90%
$\Gamma(\tau^- \to \bar{p}\pi^0)/\Gamma_{total}$		$<6.6 \times 10^{-4}$, CL = 90%
$\Gamma(\tau^- \to \bar{p}\eta)/\Gamma_{total}$		$<1.30 \times 10^{-3}$, CL = 90%

$\nu_e \to (\bar{\nu}_e)_L$
  $\alpha\Delta(m^2)$ for $\sin^2(2\theta) = 1$ — <0.14 eV^2, CL = 90%
  $\alpha^2 \sin^2(2\theta)$ for "Large" $\Delta(m^2)$ — <0.032, CL = 90%

$\nu_\mu \to (\bar{\nu}_e)_L$
  $\alpha\Delta(m^2)$ for $\sin^2(2\theta) = 1$ — <0.16 eV^2, CL = 90%
  $\alpha^2 \sin^2(2\theta)$ for "Large" $\Delta(m^2)$ — <0.001, CL = 90%

$\Gamma(\pi^+ \to \mu^+ \bar{\nu}_e)/\Gamma_{total}$	[l]	$<1.5 \times 10^{-3}$, CL = 90%
$\Gamma(K^+ \to \pi^- \mu^+ e^+)/\Gamma_{total}$		$<7 \times 10^{-9}$, CL = 90%
$\Gamma(K^+ \to \pi^- e^+ e^+)/\Gamma_{total}$		$<1.0 \times 10^{-8}$, CL = 90%
$\Gamma(K^+ \to \pi^- \mu^+ \mu^+)/\Gamma_{total}$	[l]	$<1.5 \times 10^{-4}$, CL = 90%
$\Gamma(K^+ \to \mu^+ \bar{\nu}_e)/\Gamma_{total}$	[l]	$<3.3 \times 10^{-3}$, CL = 90%
$\Gamma(K^+ \to \pi^0 e^+ \bar{\nu}_e)/\Gamma_{total}$		$<3 \times 10^{-3}$, CL = 90%
$\Gamma(D^+ \to \pi^- e^+ e^+)/\Gamma_{total}$		$<1.1 \times 10^{-4}$, CL = 90%
$\Gamma(D^+ \to \pi^- \mu^+ \mu^+)/\Gamma_{total}$		$<8.7 \times 10^{-5}$, CL = 90%
$\Gamma(D^+ \to \pi^- e^+ \mu^+)/\Gamma_{total}$		$<1.1 \times 10^{-4}$, CL = 90%
$\Gamma(D^+ \to \rho^- \mu^+ \mu^+)/\Gamma_{total}$		$<5.6 \times 10^{-4}$, CL = 90%
$\Gamma(D^+ \to K^- e^+ e^+)/\Gamma_{total}$		$<1.2 \times 10^{-4}$, CL = 90%
$\Gamma(D^+ \to K^- \mu^+ \mu^+)/\Gamma_{total}$		$<1.2 \times 10^{-4}$, CL = 90%
$\Gamma(D^+ \to K^- e^+ \mu^+)/\Gamma_{total}$		$<1.3 \times 10^{-4}$, CL = 90%
$\Gamma(D^+ \to K^*(892)^- \mu^+ \mu^+)/\Gamma_{total}$		$<8.5 \times 10^{-4}$, CL = 90%
$\Gamma(D_s^+ \to \pi^- \mu^+ \mu^+)/\Gamma_{total}$		$<4.3 \times 10^{-4}$, CL = 90%
$\Gamma(D_s^+ \to K^- \mu^+ \mu^+)/\Gamma_{total}$		$<5.9 \times 10^{-4}$, CL = 90%
$\Gamma(D_s^+ \to K^*(892)^- \mu^+ \mu^+)/\Gamma_{total}$		$<1.4 \times 10^{-3}$, CL = 90%
$\Gamma(B^+ \to \pi^- e^+ e^+)/\Gamma_{total}$		$<3.9 \times 10^{-3}$, CL = 90%
$\Gamma(B^+ \to \pi^- \mu^+ \mu^+)/\Gamma_{total}$		$<9.1 \times 10^{-3}$, CL = 90%
$\Gamma(B^+ \to K^- e^+ e^+)/\Gamma_{total}$		$<3.9 \times 10^{-3}$, CL = 90%
$\Gamma(B^+ \to K^- \mu^+ \mu^+)/\Gamma_{total}$		$<9.1 \times 10^{-3}$, CL = 90%
$\Gamma(\Xi^- \to p\mu^- \mu^-)/\Gamma_{total}$		$<4 \times 10^{-4}$, CL = 90%
$\Gamma(\Lambda_c^+ \to \Sigma^- \mu^+ \mu^+)/\Gamma_{total}$		$<7.0 \times 10^{-4}$, CL = 90%

BARYON NUMBER

$\Gamma(\tau^- \to \bar{p}\gamma)/\Gamma_{total}$	$<2.9 \times 10^{-4}$, CL = 90%
$\Gamma(\tau^- \to \bar{p}\pi^0)/\Gamma_{total}$	$<6.6 \times 10^{-4}$, CL = 90%
$\Gamma(\tau^- \to \bar{p}\eta)/\Gamma_{total}$	$<1.30 \times 10^{-3}$, CL = 90%
p mean life	$>1.6 \times 10^{25}$ years

A few examples of proton or bound neutron decay follow. For limits on many other nucleon decay channels, see the Baryon Summary Table.

$\tau(N \to e^+ \pi)$	> 130 (n), > 550 (p) $\times 10^{30}$ years, CL = 90%
$\tau(N \to \mu^+ \pi)$	> 100 (n), > 270 (p) $\times 10^{30}$ years, CL = 90%
$\tau(N \to e^+ K)$	> 1.3 (n), > 150 (p) $\times 10^{30}$ years, CL = 90%
$\tau(N \to \mu^+ K)$	> 1.1 (n), > 120 (p) $\times 10^{30}$ years, CL = 90%
limit on $n\bar{n}$ oscillations (bound n)	[m] $>1.2 \times 10^8$ s, CL = 90%
limit on $n\bar{n}$ oscillations (free n)	$>0.86 \times 10^8$ s, CL = 90%

Limits are given at the 90% confidence level, while errors are given as ± 1 standard deviation.

ELECTRIC CHARGE (Q)

e mean life / branching fraction	[n]	$>4.3 \times 10^{23}$ yr, CL = 68%
$\Gamma(n \to p\nu_e \bar{\nu}_e)/\Gamma_{total}$		$<8 \times 10^{-27}$, CL = 68%

$\Delta S = \Delta Q$ RULE

Allowed in second-order weak interactions.

$\Gamma(K^+ \to \pi^+ \pi^+ e^- \bar{\nu}_e)/\Gamma_{total}$	$<1.2 \times 10^{-8}$, CL = 90%
$\Gamma(K^+ \to \pi^+ \pi^+ \mu^- \bar{\nu}_\mu)/\Gamma_{total}$	$<3.0 \times 10^{-6}$, CL = 95%
$x = A(\overline{K}^0 \to \pi^- \ell^+ \nu)/A(K^0 \to \pi^- \ell^+ \nu) = A(\Delta S = -\Delta Q)/A(\Delta S = \Delta Q)$	
real part of x	0.006 ± 0.018 (S = 1.3)
imaginary part of x	-0.003 ± 0.026 (S = 1.2)
$\Gamma(\Sigma^+ \to n\ell^+\nu)/\Gamma(\Sigma^- \to n\ell^-\bar{\nu})$	<0.043
$\Gamma(\Sigma^+ \to ne^+\nu_e)/\Gamma_{total}$	$<5 \times 10^{-6}$, CL = 90%
$\Gamma(\Sigma^+ \to n\mu^+\nu_\mu)/\Gamma_{total}$	$<3.0 \times 10^{-5}$, CL = 90%
$\Gamma(\Xi^0 \to \Sigma^- e^+\nu_e)/\Gamma_{total}$	$<9 \times 10^{-4}$, CL = 90%
$\Gamma(\Xi^0 \to \Sigma^- \mu^+\nu_\mu)/\Gamma_{total}$	$<9 \times 10^{-4}$, CL = 90%

$\Delta S = 2$ FORBIDDEN

Allowed in second-order weak interactions.

$\Gamma(\Xi^0 \to p\pi^-)/\Gamma_{total}$	$<4 \times 10^{-5}$, CL = 90%
$\Gamma(\Xi^0 \to pe^-\bar{\nu}_e)/\Gamma_{total}$	$<1.3 \times 10^{-3}$
$\Gamma(\Xi^0 \to p\mu^-\bar{\nu}_\mu)/\Gamma_{total}$	$<1.3 \times 10^{-3}$
$\Gamma(\Xi^- \to n\pi^-)/\Gamma_{total}$	$<1.9 \times 10^{-5}$, CL = 90%
$\Gamma(\Xi^- \to ne^-\bar{\nu}_e)/\Gamma_{total}$	$<3.2 \times 10^{-3}$, CL = 90%
$\Gamma(\Xi^- \to n\mu^-\bar{\nu}_\mu)/\Gamma_{total}$	$<1.5 \times 10^{-2}$, CL = 90%
$\Gamma(\Xi^- \to p\pi^-\pi^-)/\Gamma_{total}$	$<4 \times 10^{-4}$, CL = 90%
$\Gamma(\Xi^- \to p\pi^- e^-\bar{\nu}_e)/\Gamma_{total}$	$<4 \times 10^{-4}$, CL = 90%
$\Gamma(\Xi^- \to p\pi^- \mu^-\bar{\nu}_\mu)/\Gamma_{total}$	$<4 \times 10^{-4}$, CL = 90%
$\Gamma(\Omega^- \to \Lambda\pi^-)/\Gamma_{total}$	$<1.9 \times 10^{-4}$, CL = 90%

$\Delta S = 2$ VIA MIXING

Allowed in second-order weak interactions, e.g. mixing.

$m_{K_L^0} - m_{K_S^0}$	$(0.5301 \pm 0.0014) \times 10^{10} \, \hbar \, s^{-1}$
$m_{K_L^0} - m_{K_S^0}$	$(3.489 \pm 0.009) \times 10^{-12}$ MeV

$\Delta C = 2$ VIA MIXING

Allowed in second-order weak interactions, e.g. mixing.

$\|m_{D_1^0} - m_{D_2^0}\|$	[o]	$<24 \times 10^{10} \, \hbar \, s^{-1}$, CL = 90%
$\|\Gamma_{D_1^0} - \Gamma_{D_2^0}\|/\Gamma_{D^0}$ mean life difference/average	[o]	<0.20, CL = 90%
$\Gamma(K^+ \ell^- \bar{\nu}_\ell \text{(via } \overline{D}^0))/\Gamma(K^- \ell^+ \nu_\ell)$		<0.005, CL = 90%
$\Gamma(K^+ \pi^- \text{ or } K^+ \pi^- \pi^+ \pi^- \text{(via } \overline{D}^0))/\Gamma(K^- \pi^+ \text{ or } K^- \pi^+ \pi^+ \pi^-)$	[p]	<0.0085 (or <0.0037), CL = 90%
$\Gamma(D^0 \to K^+ \ell^- \bar{\nu}_\ell \text{(via } \overline{D}^0))/\Gamma_{total}$		$<1.7 \times 10^{-4}$, CL = 90%
$\Gamma(D^0 \to K^+ \pi^- \text{ or } K^+ \pi^- \pi^+ \pi^- \text{(via } \overline{D}^0))/\Gamma_{total}$		$<1.0 \times 10^{-3}$, CL = 90%

$\Delta B = 2$ VIA MIXING

Allowed in second-order weak interactions, e.g. mixing.

χ_d	0.172 ± 0.010
$\Delta m_{B^0} = m_{B_H^0} - m_{B_L^0}$	$(0.464 \pm 0.018) \times 10^{12} \, \hbar \, s^{-1}$
$x_d = \Delta m_{B^0}/\Gamma_{B^0}$	0.723 ± 0.032
χ_B at high energy	0.118 ± 0.006
$\Delta m_{B_s^0} = m_{B_{sH}^0} - m_{B_{sL}^0}$	$>9.1 \times 10^{12} \, \hbar \, s^{-1}$, CL = 95%
$x_s = \Delta m_{B_s^0}/\Gamma_{B_s^0}$	>14.0, CL = 95%
χ_s	>0.4975, CL = 95%

$\Delta S = 1$ WEAK NEUTRAL CURRENT FORBIDDEN

Allowed by higher-order electroweak interactions.

$\Gamma(K^+ \to \pi^+ e^+ e^-)/\Gamma_{total}$		$(2.74 \pm 0.23) \times 10^{-7}$
$\Gamma(K^+ \to \pi^+ \mu^+ \mu^-)/\Gamma_{total}$		$(5.0 \pm 1.0) \times 10^{-8}$
$\Gamma(K^+ \to \pi^+ \nu\bar{\nu})/\Gamma_{total}$		$(4.2^{+9.7}_{-3.5}) \times 10^{-10}$
$\Gamma(K_S^0 \to \mu^+ \mu^-)/\Gamma_{total}$		$<3.2 \times 10^{-7}$, CL = 90%
$\Gamma(K_S^0 \to e^+ e^-)/\Gamma_{total}$		$<1.4 \times 10^{-7}$, CL = 90%
$\Gamma(K_S^0 \to \pi^0 e^+ e^-)/\Gamma_{total}$		$<1.1 \times 10^{-6}$, CL = 90%
$\Gamma(K_L^0 \to \mu^+ \mu^-)/\Gamma_{total}$		$(7.2 \pm 0.5) \times 10^{-9}$ (S = 1.4)
$\Gamma(K_L^0 \to \mu^+ \mu^- \gamma)/\Gamma_{total}$		$(3.25 \pm 0.28) \times 10^{-7}$
$\Gamma(K_L^0 \to e^+ e^-)/\Gamma_{total}$		$<4.1 \times 10^{-11}$, CL = 90%
$\Gamma(K_L^0 \to e^+ e^- \gamma)/\Gamma_{total}$		$(9.1 \pm 0.5) \times 10^{-6}$
$\Gamma(K_L^0 \to e^+ e^- \gamma\gamma)/\Gamma_{total}$	[q]	$(6.5 \pm 1.2) \times 10^{-7}$
$\Gamma(K_L^0 \to \pi^+ \pi^- e^+ e^-)/\Gamma_{total}$	[q]	$<4.6 \times 10^{-7}$, CL = 90%
$\Gamma(K_L^0 \to \mu^+ \mu^- e^+ e^-)/\Gamma_{total}$		$(2.9^{+6.7}_{-2.4}) \times 10^{-9}$
$\Gamma(K_L^0 \to e^+ e^- e^+ e^-)/\Gamma_{total}$		$(4.1 \pm 0.8) \times 10^{-8}$ (S = 1.2)
$\Gamma(K_L^0 \to \pi^0 \mu^+ \mu^-)/\Gamma_{total}$		$<5.1 \times 10^{-9}$, CL = 90%
$\Gamma(K_L^0 \to \pi^0 e^+ e^-)/\Gamma_{total}$		$<4.3 \times 10^{-9}$, CL = 90%
$\Gamma(K_L^0 \to \pi^0 \nu\bar{\nu})/\Gamma_{total}$		$<5.8 \times 10^{-5}$, CL = 90%
$\Gamma(\Sigma^+ \to pe^+ e^-)/\Gamma_{total}$		$<7 \times 10^{-6}$

$\Delta C = 1$ WEAK NEUTRAL CURRENT FORBIDDEN

Allowed by higher-order electroweak interactions.

$\Gamma(D^+ \to \pi^+ e^+ e^-)/\Gamma_{total}$	$<6.6 \times 10^{-5}$, CL = 90%
$\Gamma(D^+ \to \pi^+ \mu^+ \mu^-)/\Gamma_{total}$	$<1.8 \times 10^{-5}$, CL = 90%
$\Gamma(D^+ \to \rho^+ \mu^+ \mu^-)/\Gamma_{total}$	$<5.6 \times 10^{-4}$, CL = 90%
$\Gamma(D^0 \to e^+ e^-)/\Gamma_{total}$	$<1.3 \times 10^{-5}$, CL = 90%
$\Gamma(D^0 \to \mu^+ \mu^-)/\Gamma_{total}$	$<4.1 \times 10^{-6}$, CL = 90%
$\Gamma(D^0 \to \pi^0 e^+ e^-)/\Gamma_{total}$	$<4.5 \times 10^{-5}$, CL = 90%
$\Gamma(D^0 \to \pi^0 \mu^+ \mu^-)/\Gamma_{total}$	$<1.8 \times 10^{-4}$, CL = 90%
$\Gamma(D^0 \to \eta e^+ e^-)/\Gamma_{total}$	$<1.1 \times 10^{-4}$, CL = 90%
$\Gamma(D^0 \to \eta \mu^+ \mu^-)/\Gamma_{total}$	$<5.3 \times 10^{-4}$, CL = 90%
$\Gamma(D^0 \to \rho^0 e^+ e^-)/\Gamma_{total}$	$<1.0 \times 10^{-4}$, CL = 90%
$\Gamma(D^0 \to \rho^0 \mu^+ \mu^-)/\Gamma_{total}$	$<2.3 \times 10^{-4}$, CL = 90%
$\Gamma(D^0 \to \omega e^+ e^-)/\Gamma_{total}$	$<1.8 \times 10^{-4}$, CL = 90%
$\Gamma(D^0 \to \omega \mu^+ \mu^-)/\Gamma_{total}$	$<8.3 \times 10^{-4}$, CL = 90%
$\Gamma(D^0 \to \phi e^+ e^-)/\Gamma_{total}$	$<5.2 \times 10^{-5}$, CL = 90%
$\Gamma(D^0 \to \phi \mu^+ \mu^-)/\Gamma_{total}$	$<4.1 \times 10^{-4}$, CL = 90%
$\Gamma(D^0 \to \pi^+ \pi^- \pi^0 \mu^+ \mu^-)/\Gamma_{total}$	$<8.1 \times 10^{-4}$, CL = 90%
$\Gamma(D_s^+ \to K^+ \mu^+ \mu^-)/\Gamma_{total}$	$<5.9 \times 10^{-4}$, CL = 90%
$\Gamma(D_s^+ \to K^*(892)^+ \mu^+ \mu^-)/\Gamma_{total}$	$<1.4 \times 10^{-3}$, CL = 90%
$\Gamma(\Lambda_c^+ \to p\mu^+ \mu^-)/\Gamma_{total}$	$<3.4 \times 10^{-4}$, CL = 90%

Limits are given at the 90% confidence level, while errors are given as ± 1 standard deviation.

$\Delta B = 1$ WEAK NEUTRAL CURRENT FORBIDDEN

Allowed by higher-order electroweak interactions.

$\Gamma(B^+ \to \pi^+ e^+ e^-)/\Gamma_{total}$	$<3.9 \times 10^{-3}$, CL = 90%
$\Gamma(B^+ \to \pi^+ \mu^+ \mu^-)/\Gamma_{total}$	$<9.1 \times 10^{-3}$, CL = 90%
$\Gamma(B^+ \to K^+ e^+ e^-)/\Gamma_{total}$	$<6 \times 10^{-5}$, CL = 90%
$\Gamma(B^+ \to K^+ \mu^+ \mu^-)/\Gamma_{total}$	$<1.0 \times 10^{-5}$, CL = 90%
$\Gamma(B^+ \to K^*(892)^+ e^+ e^-)/\Gamma_{total}$	$<6.9 \times 10^{-4}$, CL = 90%
$\Gamma(B^+ \to K^*(892)^+ \mu^+ \mu^-)/\Gamma_{total}$	$<1.2 \times 10^{-3}$, CL = 90%
$\Gamma(B^0 \to \gamma\gamma)/\Gamma_{total}$	$<3.9 \times 10^{-5}$, CL = 90%
$\Gamma(B^0 \to e^+ e^-)/\Gamma_{total}$	$<5.9 \times 10^{-6}$, CL = 90%
$\Gamma(B^0 \to \mu^+ \mu^-)/\Gamma_{total}$	$<6.8 \times 10^{-7}$, CL = 90%
$\Gamma(B^0 \to K^0 e^+ e^-)/\Gamma_{total}$	$<3.0 \times 10^{-4}$, CL = 90%
$\Gamma(B^0 \to K^0 \mu^+ \mu^-)/\Gamma_{total}$	$<3.6 \times 10^{-4}$, CL = 90%
$\Gamma(B^0 \to K^*(892)^0 e^+ e^-)/\Gamma_{total}$	$<2.9 \times 10^{-4}$, CL = 90%
$\Gamma(B^0 \to K^*(892)^0 \mu^+ \mu^-)/\Gamma_{total}$	$<2.3 \times 10^{-5}$, CL = 90%
$\Gamma(B^0 \to K^*(892)^0 \nu\bar{\nu})/\Gamma_{total}$	$<1.0 \times 10^{-3}$, CL = 90%
$\Gamma(B \to e^+ e^- s)/\Gamma_{total}$	$<5.7 \times 10^{-5}$, CL = 90%
$\Gamma(B \to \mu^+ \mu^- s)/\Gamma_{total}$	$<5.8 \times 10^{-5}$, CL = 90%
$\Gamma(\bar{b} \to \mu^+ \mu^- \text{ anything})/\Gamma_{total}$	$<3.2 \times 10^{-4}$, CL = 90%
$\Gamma(B_s^0 \to \mu^+ \mu^-)/\Gamma_{total}$	$<2.0 \times 10^{-6}$, CL = 90%
$\Gamma(B_s^0 \to e^+ e^-)/\Gamma_{total}$	$<5.4 \times 10^{-5}$, CL = 90%
$\Gamma(B_s^0 \to \phi\nu\bar{\nu})/\Gamma_{total}$	$<5.4 \times 10^{-3}$, CL = 90%

NOTES

In this Summary Table:

When a quantity has "(S = ...)" to its right, the error on the quantity has been enlarged by the "scale factor" S, defined as $S = \sqrt{\chi^2/(N-1)}$, where N is the number of measurements used in calculating the quantity. We do this when $S > 1$, which often indicates that the measurements are inconsistent. When $S > 1.25$, we also show in the Particle Listings an ideogram of the measurements. For more about S, see the Introduction.

[a] C parity forbids this to occur as a single-photon process.

[b] Time-reversal invariance requires this to be 0° or 180°.

[c] Allowed by higher-order electroweak interactions.

[d] Violates CP in leading order. Test of direct CP violation since the indirect CP-violating and CP-conserving contributions are expected to be suppressed.

[e] ϵ'/ϵ is derived from $|\eta_{00}/\eta_{+-}|$ measurements using theoretical input on phases.

[f] Neglecting photon channels. See, e.g., A. Pais and S.B. Treiman, Phys. Rev. **D12**, 2744 (1975).

[g] Derived from measured values of ϕ_{+-}, ϕ_{00}, $|\eta|$, $|m_{K_L^0} - m_{K_S^0}|$, and $\tau_{K_S^0}$, as described in the introduction to "Tests of Conservation Laws."

[h] The value is for the sum of the charge states of particle/antiparticle states indicated.

[i] A test of additive vs. multiplicative lepton family number conservation.

[j] $\Delta(m^2) = 100 \text{ eV}^2$.

[k] $190 \text{ eV}^2 < \Delta(m^2) < 320 \text{ eV}^2$.

[l] Derived from an analysis of neutrino-oscillation experiments.

[m] There is some controversy about whether nuclear physics and model dependence complicate the analysis for bound neutrons (from which the best limit comes). The second limit here is from reactor experiments with free neutrons.

[n] This is the best "electron disappearance" limit. The best limit for the mode $e^- \to \nu\gamma$ is $> 2.35 \times 10^{25}$ yr (CL=68%).

[o] The D_1^0-D_2^0 limits are inferred from the D^0-$\overline{D}^0$ mixing ratio $\Gamma(K^+ \ell^- \bar{\nu}_\ell(\text{via } \overline{D}^0)) / \Gamma(K^- \ell^+ \nu_\ell)$.

[p] The larger limit (from E791) allows interference between the doubly Cabibbo-suppressed and mixing amplitudes; the smaller limit (from E691) doesn't. See the papers for details.

[q] See the K_L^0 Particle Listings for the energy limits used in this measurement.

Limits are given at the 90% confidence level, while errors are given as ±1 standard deviation.

TABLE OF THE ISOTOPES
(Revised 1998)

Norman E. Holden

Reactor Division

Brookhaven National Laboratory

Upton, New York 11973

This table presents an evaluated set of values for the experimental quantities which characterize the decay of radioactive nuclides. A list of the major references used in this evaluation is given below. When uncertainties are not listed, they are assumed to be five or less in the last digit quoted. If they exceed five in the last digit, the value is prefaced by an approximate sign. The effective literature cutoff date for data in this edition of the Table is January, 1998. The assistance of Felicia Kramer in assembling this file is gratefully acknowledged.

TABLE LAYOUT

Column No.	Column Title	Description
1	Isotope or Element	For elements, the atomic number and chemical symbol are listed. For nuclides, the mass number and chemical symbol are listed. Isomers are indicated by the addition of m, m1, or m2.
2	Isotopic Abundance	in atom percent.
3	Atomic Mass or Atomic Weight	Atomic mass relative to $^{12}C = 12$. Atomic weight is given on the same scale.
4	Half-life	Half-life in decimal notation. μs = microseconds; ms = milliseconds; s = seconds; m = minutes; h = hours; d = days; and y = years.
5	Decay Mode/Energy	Decay modes are α = alpha particle emission; β^- = negative beta emission; β^+ = positron emission; EC = orbital electron capture; IT = isomeric transition from upper to lower isomeric state; n = neutron emission; SF = spontaneous fission. Total disintegration energy in MeV units.
6	Particle Energy/Intensity	End point energies of beta transitions and discrete energies of alpha particles in MeV and their intensities in percent.
7	Spin and Parity	Nuclear spin or angular momentum of the nuclides in units of $h/2\pi$; parity is positive or negative.
8	Magnetic Dipole Moment	Magnetic dipole moments in nuclear magneton units.
9	Electric Quadrupole Moment	Electric quadrupole moments in barn units (10^{-24} cm^2).
10	Gamma Ray Energy/Intensity	Gamma ray energies in MeV and intensities in percent. Ann. rad. refers to the 511.006 keV photons emitted in the annihilation of positrons in matter.

GENERAL NUCLEAR DATA REFERENCES

The following references represent the major sources of the nuclear data presented, along with subsequent published journals and reports:

1. G. Audi, O. Bersillon, J. Blachot, A.H. Wapstra, *The Nubase Evaluation of Nuclear and Decay Properties*, Nuclear Physics A624, 1 (1997).
2. International Commission on Atomic Weights, *Atomic Weights of the Elements - 1995*, Pure & Applied Chemistry 68, 2339 (1996).
3. J.R. Parrington, H.D. Knox, S. Breneman, E.M. Baum, F. Feiner, *Chart of the Nuclides, 15th Edition*, Knolls Atomic Power Lab. (1996)
4. N.E. Holden, *Total and Spontaneous Fission Half-lives for Uranium, Plutonium, Americium and Curium Nuclides*, Pure & Applied Chemistry 61, 1483 (1989).
5. N.E. Holden, *Half-lives of Selected Nuclides*, Pure & Applied Chemistry 62, 941 (1990).
6. N.E. Holden, *Review of Thermal Neutron Cross Sections and Isotopic Composition of the Elements*, BNL-NCS-42224 (March 1989).
7. P. Raghavan, *Table of Nuclear Moments*, Atomic Data Nuclear Data Tables 42, 189 (1989).
8. E. Brown, R. Firestone, *Radioactivity Handbook*, Wiley Interscience Press (1986).
9. J.K. Tuli, *Nuclear Wallet Cards*, Brookhaven National Laboratory (July 1995).

*This research was carried out under the auspices of the US Department of Engery Contract No. DE-AC02-76CH00016.

T A B L E   O F   T H E   I S O T O P E S

Elem. or Isot.	Natural Abundance (%)	Atomic Mass or Weight	Half-Life	Decay Mode/Energy (/MeV)	Particle Energy /Intensity (MeV / %)	Spin (h/2π)	Nuclear Magnetic Mom.(nm)	Elect. Quadr. Mom. (b)	γ-ray Energy /Intensity (MeV / %)
₀n		1.008664924	10.3 m	β⁻/0.78235	0.782/100.	1/2+	-1.913043		
₁H		1.00794(7)							
¹H	99.985(1)	1.007825032				1/2+	+2.79285		
²H	0.015(1)	2.014101778				1+	+0.85744	+2.86 mb	
³H		3.016049268	12.32 y	β⁻/0.01859	0.01860/100.	1/2+	+2.97896		
₂He		4.002602(2)							
³He	1.37x10⁻⁴	3.016029309				1/2+	-2.12762		
⁴He	≈100.	4.002603250				0+			
⁵He		5.01222	7.6x10⁻²² s	n,α		3/2-			
⁶He		6.018888	0.807 s	β⁻/3.508,d	3.510/100.	0+			
⁷He		7.02803	3.x10⁻²¹ s	n		(3/2)-			
⁸He		8.03392	0.119 s	β⁻/10.65, t	13/88.	0+			0.9807/84.
				n/	/12.				0.4776/5.
⁹He		9.0438		n					
¹⁰He		10.0524		2n					
₃Li		6.941(2)							
⁵Li		5.01254	≈3.x10⁻²² s			3/2-			
⁶Li	7.5(2)	6.0151223				1+	+0.82205	-0.8 mb	
⁷Li	92.5(2)	7.0160041				3/2-	+3.25644	-0.041	
⁸Li		8.022486	0.84 s	β⁻/16.004	12.5/100.	2+	+1.6536	+0.032	
				α/	α(1.6)				
⁹Li		9.026789	0.178 s	β⁻/13.606	13.5/75.	3/2-	3.439	0.028	
				β⁻/	11/25.				
¹⁰Li		10.03590	4. x 10⁻²² s	β⁻ /20.84					
¹¹Li		11.04380	8.4 ms	β⁻/20.6		3/2(-)	3.668	0.031	3.367/35.
				n,2n,3n,α					(0.22-2.81)
₄Be		9.012182(3)							
⁶Be		6.01973	5.0x10⁻²¹ s	2p,α		0+			
⁷Be		7.0169293	53.28 d	EC/0.8618		3/2-			0.4776/10.4
⁸Be		8.00530509	≈ 7.x10⁻¹⁷ s	2α/0.046		0+			
⁹Be	100.	9.0121822				3/2-	-1.1776	+0.053	
¹⁰Be		10.0135338	1.52x10⁶ y	β⁻/0.5559	0.555/100.	0+			
¹¹Be		11.02166	13.8 s	β⁻,β⁻α/11.51	11.48/61.	1/2+			2.125/35.5
									(0.478-7.97)
¹²Be		12.02692	24. ms	β⁻,(n)/11.71		0+			(0.95 - 4.4)
¹³Be		13.0361							
¹⁴Be		14.0428	4.3 ms	β⁻,(n)/16.2		0+			
₅B		10.811(7)							
⁷B		7.0299	4.x10⁻²² s	p					
⁸B		8.024607	0.770 s	β⁺, 2α/17.979	13.7(β⁺)/93.	2+	1.0355	0.068	ann.rad.
⁹B		9.013329	8x10⁻¹⁹ s	p2α/		3/2-			
¹⁰B	19.9(2)	10.0129371				3+	+1.8006	+0.085	
¹¹B	80.1(2)	11.0093055				3/2-	+2.6886	+0.0406	
¹²B		12.014352	0.0202 s	β⁻/13.369		1+	+1.0031	0.0132	4.438/1.3
				β⁻ α/1.6/					3.215/0.00065
¹³B		13.017780	0.0174 s	β⁻ /13.437	13.4	3/2-	+3.17778	0.037	3.68/7.6
				β⁻ n/0.25/	2.43(n)/0.09				
					3.55(n)/0.16				
¹⁴B		14.02540	14. ms	β⁻ /20.64		2-	1.185	0.0298	6.094/90.
¹⁵B		15.03110	10.4 ms	β⁻ ,(n)/19.09		(3/2-)	2.66	0.038	
¹⁶B		16.0398	<1.9x10⁻¹⁰ s						
¹⁷B		17.0469	5.1 ms	β⁻,(n)/22.7			2.55		
¹⁹B		19.0637		β⁻/26.5					
₆C		12.0107(8)							
⁸C		8.03768	2.0x10⁻²¹ s	p					
⁹C		9.031040	127. ms	β⁺ ,p, 2α/16.498		(3/2-)	-1.391		ann.rad.
¹⁰C		10.0168532	19.3 s	β⁺ /3.648	1.865	0+			ann.rad.
									0.71829/100.
¹¹C		11.011433	20.3 m	β⁺ ,EC/1.982	0.9608/99.	3/2-	-0.964	0.0333	ann.rad.
¹²C	98.93(8)	12.000000000				0+			
¹³C	1.07(8)	13.003354838				1/2-	+0.70241		
¹⁴C		14.003241991	5715. y	β⁻ /0.15648	0.1565/100.	0+			
¹⁵C		15.010599	2.45 s	β⁻ /9.772	4.51/68.	1/2+	1.32		5.298/68.
					9.82/32.				(7.30-9.05)

TABLE OF THE ISOTOPES

Elem. or Isot.	Natural Abundance (%)	Atomic Mass or Weight	Half-Life	Decay Mode/Energy (/MeV)	Particle Energy /Intensity (MeV / %)	Spin (h/2π)	Nuclear Magnetic Mom.(nm)	Elect. Quadr. Mom. (b)	γ-ray Energy /Intensity (MeV / %)
^{16}C		16.014701	0.75 s	β^- ,n/8.012					
^{17}C		17.02258	0.19 s	β^- ,n/13.17					1.375
									1.849
									1.906
^{18}C		18.02676	0.09 s	β^- ,n/11.81					
^{19}C		19.0353	0.05 s	n					
^{20}C		20.0403	0.01 s						
^{21}C		21.0493							
^{22}C		22.056							
,N		**14.00674(7)**							
^{12}N		12.018613	11.00 ms	β^+ ,β^+ α/17.338	16.38/95.	1+	+0.457	+0.026	ann.rad.
									4.438/2.
^{13}N		13.0057386	9.97 m	β^+ /2.2204	1.190/100.	1/2-	0.3222		
^{14}N	99.632(7)	14.003074007				1+	+0.40376	+0.020	
^{15}N	0.368(7)	15.00010897				1/2-	-0.28319		
^{16}N		16.006100	7.13 s	β^- /10.419	4.27/68.	2-			6.129/68.8
					10.44/26.				7.115/4.7
				β^-, α	1.85/.0012				(0.99-8.87)
^{17}N		17.00845	4.17 s	β^- ,β^- n/8.68	3.7/100.	1/2-	0.352		0.871/3.
				0.4-1.7n/95.					2.1842/0.3
				β^- α/	8.0, 8.2				
^{18}N		18.01408	0.62 s	β^- /13.90	9.4/100.	1-			0.822/61.
									1.65/60.5
									1.982/98.
									(0.535-7.13)
^{19}N		19.01703	0.32 s	β^- /12.53					(0.096-3.14)
^{20}N		20.02337	0.14 s	β^- /17.97					
^{21}N		21.0271	0.08 s						
^{22}N		22.0344	0.02s						
^{23}N		23.0405							
,O		**15.9994(3)**							
^{12}O		12.03440	$\approx 1. \times 10^{21}$ s	2p					
^{13}O		13.02481	8.9 ms	β^+ ,p/17.77	1.56 (p)/	(3/2-)	1.389		ann.rad.
									4.438/0.56
^{14}O		14.0085953	70.60 s	β^+ /5.1430	1.81/99.	0+			ann.rad.
									2.312/99.4
^{15}O		15.0030655	122.2 s	β^+ /2.754	1.723/100.	1/2-	0.719		ann.rad.
^{16}O	99.757(16)	15.994914622				0+			
^{17}O	0.038(1)	16.9991315				5/2+	-1.8938	-0.026	
^{18}O	0.205(14)	17.999160				0+			
^{19}O		19.003579	26.9 s	β^- /4.820	3.25/60.	5/2+			0.197/95.9
					4.60/40.				1.3569/50.4
									(0.11-4.18)
^{20}O		20.004076	13.5 s	β^- /3.814		0+			1.057/100.
^{21}O		21.00866	3.4 s	β^- /8.11					(0.28-4.6)
^{22}O		22.00997	2.2 s	β^- /6.5					(0.64-1.86)
^{23}O		23.0157	0.08 s						
^{24}O		24.0204	0.06 s						
,F		**18.9984032(5)**							
^{15}F		15.0180	$5. \times 10^{-22}$ s	p		(1/2+)			
^{16}F		16.01147	$\approx 1. \times 10^{-20}$ s	p		0-			
^{17}F		17.0020952	64.5 s	β^+ /2.761	1.75/	5/2+	+4.722	0.058	ann.rad.
^{18}F		18.000938	1.830 h	β^+ ,EC/1.656	0.635/97.	1+			
^{19}F	100.	18.9984032				1/2+	+2.62887	0.072	
^{20}F		19.9999813	11.00 s	β^- /7.0245	5.398/100.	2+	+2.094	0.042	1.634/100.
									3.33/0.009
^{21}F		20.999949	4.16 s	β^- /5.684	3.7/8.	5/2+			0.3507/90.
					5.0/63.				1.395/15.
					5.4/29.				(1.746-4.684)
^{22}F		22.00300	4.23 s	β^- /10.82	3.48/15.	4+			1.2746/100.
					4.67/7.				2.0826/82.
					5.50/62.				(0.82-4.37)
^{23}F		23.00357	2.2 s	β^- /8.5		5/2+			1.701/48.
									2.129/34.
									(0.493-3.83)
^{24}F		24.0081	0.3 s	β^- /13.5					1.9816/
^{25}F		25.0121	0.06 s	β^- ,(n)					
^{26}F		26.0196	0.03 s						
^{27}F		27.0269	5. ms						
^{29}F		29.043	2. ms						

T A B L E   O F   T H E   I S O T O P E S

Elem. or Isot.	Natural Abundance (%)	Atomic Mass or Weight	Half-Life	Decay Mode/Energy (/MeV)	Particle Energy /Intensity (MeV / %)	Spin (h/2π)	Nuclear Magnetic Mom.(nm)	Elect. Quadr. Mom. (b)	γ-ray Energy /Intensity (MeV / %)
₁₀Ne		20.1797(6)							
¹⁶Ne		16.02575	4.x10²¹ s	2p					
¹⁷Ne		17.01770	109. ms	β⁺ .p/14.53	1.4-10.6/	1/2-			ann.rad./ 0.495
¹⁸Ne		18.005697	1.67 s	β⁺ /4.446	3.416/92.	0+			ann.rad./ 1.0413/7.8 (0.658-1.70)
¹⁹Ne		19.001880	17.22 s	β⁺ /3.238	2.24/99.	1/2+	-1.885		ann.rad./ (0.11-1.55)
²⁰Ne	90.48(3)	19.992440176				0+			
²¹Ne	0.27(1)	20.99384674				3/2+	-0.66180	+0.103	
²²Ne	9.25(3)	21.9913855				0+		-0.19	
²³Ne		22.9944673	37.2 s	β⁻ /4.376	3.95/32. 4.39/67.	5/2+	-1.08		0.440/33. (1.64-2.98)
²⁴Ne		23.99362	3.38 m	β⁻ /2.47	1.10/8. 1.98/92.	0+			0.4723/100. 0.874/7.9
²⁵Ne		24.99779	0.61 s	β⁻ /7.30	6.3/ 7.3/	1/2+			0.0895/96. (0.98-3.69)
²⁶Ne		26.00046	197 ms	β⁻ /7.3					0.233/
²⁷Ne		27.0076	32 ms	β⁻, n/12.7					
²⁸Ne		28.0121	18. ms	β⁻, n/12.3					
²⁹Ne		29.0194	15. ms	β⁻ /15.4					
³⁰Ne		30.024	7. ms						
³¹Ne		31.033							
³²Ne		32.040							
₁₁Na		22.989770(2)							
¹⁹Na		19.01388	0.03 s	β⁺ ,p/11.18					
²⁰Na		20.00735	0.446 s	β⁺ /13.89 α	2.15/	2+	+0.3694		ann.rad./ 1.634/79.
²¹Na		20.997655	22.48 s	β⁺ /3.547	2.50/95.	3/2+	+2.3863	+0.05	ann.rad./ 0.351/5.
²²Na		21.9944366	2.605 y	β⁺ /90/2.842 EC/10/	0.545/90.	3+	+1.746		ann.rad./ 1.2745/99.9
²³Na	100.	22.9897697				3/2+	+2.21752	+0.101	
²⁴ᵐNa			20.2 ms	I.T.,β⁻		1+			0.4723/100.
²⁴Na		23.9909633	14.96 h	β⁻ /5.5158	1.389/>99.	4+	+1.690		1.3686/100. 2.754/100. (0.997-4.238)
²⁵Na		24.989954	59.3 s	β⁻ /3.835	2.6/7. 3.15/25. 4.0/65.	5/2+	+3.683	-0.10	0.3897/12.7 0.5850/13. 0.9747/14.9 (0.836-2.80)
²⁶Na		25.99259	1.07 s	β⁻ /9.31		3+	+2.851	-0.08	1.809/98.9
²⁷Na		26.99401	0.290 s	β⁻ /9.01 β⁻ ,n/	7.95/	5/2+	+3.90	-0.06	0.9847/87.4 1.698/11.9
²⁸Na		27.9989	31. ms	β⁻ /14.0 β⁻ ,n/	12.3/	1+	+2.43	-0.02	1.473/37. 2.389/18.6
²⁹Na		29.0028	44. ms	β⁻ ,n/13.3	11.5/	3/2+	+2.45	-0.03	2.560/36. (1.04-3.99)
³⁰Na		30.0092	50. ms	β⁻ /17.5		2	+2.08		1.483/46.
³¹Na		31.0136	17.2 ms	/15.9 β⁻ ,n		(3/2-)	+2.31		1.483/14. (0.05-3.54)
³²Na		32.0197	13.5 ms	β⁻ /19.1					0.886/60.
³³Na		33.027	8.1 ms	β⁻ /20.					0.886/16.
³⁴Na		34.035	5. ms	β⁻ /24.					0.886/60.
³⁵Na		35.044	1.5 ms	β⁻ /24					
₁₂Mg		24.30505(6)							
²⁰Mg		20.01886	96. ms	β⁺ ,p/10.73					
²¹Mg		21.01171	122. ms	β⁺ ,p/13.10		5/2+			0.332/51.
²²Mg		21.999574	3.86 s	β⁺ /4.786	3.05/	0+			0.0729/60. 0.5820/100. (1.28-1.93)
²³Mg		22.994125	11.32 s	β⁺ /4.057	3.09/92.	3/2+	0.536		0.440/8.2
²⁴Mg	78.99(4)	23.9850419				0+			
²⁵Mg	10.00(1)	24.9858370				5/2+	-0.85545	+0.199	
²⁶Mg	11.01(3)	25.9825930				0+			
²⁷Mg		26.9843407	9.45 m	β⁻ /2.6103	1.59/41. 1.75/58. 2.65/0.3	1/2+			0.17068/0.8 0.84376/73. 1.01443/29.
²⁸Mg		27.983877	20.9 h	β⁻ /1.832	0.459/95.	0+			0.0306/95. 0.4006/36. 0.9418/36. 1.342/54.

Elem. or Isot.	Natural Abundance (%)	Atomic Mass or Weight	Half-Life	Decay Mode/Energy (/MeV)	Particle Energy /Intensity (MeV / %)	Spin (h/2π)	Nuclear Magnetic Mom.(nm)	Elect. Quadr. Mom. (b)	γ-ray Energy /Intensity (MeV / %)
^{29}Mg		28.98855	1.3 s	β^- /7.55	5.4/	3/2 +			0.960/15. 1.398/16. 2.224/36.
^{30}Mg		29.9905	0.32 s	β^- /7.0		0+			0.224/85.
^{31}Mg		30.9966	0.24 s	β^- /11.7		(3/2 +)			1.61/26.
^{32}Mg		31.9992	0.12 s	β^- /10.3					2.765/25.
^{33}Mg		33.0056	0.09 s	β^- /13.7					1.848/
^{34}Mg		34.0091	0.02 s	β^- /11.3					
^{35}Mg		35.0175							
^{36}Mg		36.022							
^{37}Mg		37.031							
^{38}Mg									
$_{13}$Al		26.981538(2)							
^{22}Al		22.0195	59. ms	β^+ /18.6 β^+,p,2p,α/		4 +			ann.rad./
^{23}Al		23.00727	0.47 s	β^+ /12.24 β^+ ,p/					ann.rad./
^{24m}Al			0.129 s	I.T./0.4259 β^+	13.3	1 +			1.3686/5.3
^{24}Al		23.999941	2.07 s	β^+ /13.878,p	3.40/48. 4.42/41. 6.80/3. 8.74/8.	4 +			1.078(2)/16. 1.368(2)/96. 2.753(2)/43. 4.315(3)/15. 5.392(3)/20. 7.0662(2)/41.
^{25}Al		24.990429	7.17 s	β^+ /4.277	3.27/	5/2 +	3.646		ann.rad./ 1.6115(2)/100. 0.975(2)/5.
^{26m}Al			6.345 s	β^+ /	3.2/	0+			ann.rad./
^{26}Al		25.9868917	7.1x10^5 y	β^+ /82/4.0042 EC/18/	1.16/	5 +	+2.804		ann.rad./ 1.80865(7)/99.8
^{27}Al	100.	26.9815384				5/2 +	+3.64151	+0.15	
^{28}Al		27.9819102	2.25 m	β^- /4.6422	2.865/100.	3 +	3.24	0.18	1.7778(6)/100.
^{29}Al		28.980445	6.5 m	β^- /3.680	1.4/30. 2.5/70.	5/2 +			1.2732(8)/89. 2.0282(8)/4. 2.4262(8)/7.
^{30}Al		29.98296	3.68 s	β^- /8.56	5.05/	3 +			1.26313(3)/35. 2.23525(5)/65.
^{31}Al		30.98395	0.64 s	β^- /8.00	6.25/				0.75223(3)/18. 1.69473(3)/59. 2.31664(4)/73.
^{32}Al		31.9881	33. ms	β^- /13.0		1 +			
^{33}Al		32.9909	41. ms						
^{34}Al		33.9969	≈42. ms	β^- /17.1					
^{35}Al		34.9999	30 ms						
^{36}Al		36.0064	0.09 s						
^{37}Al		37.010							
^{38}Al		38.0169							
^{39}Al		39.022							
^{40}Al									
^{41}Al									
$_{14}$Si		28.0855(3)							
^{22}Si		22.0345	29. ms	β^+,p	1.99/20				
^{23}Si		23.0255	40.7 ms	β^+,p/5.9	1.32,2.40,2.83				
^{24}Si		24.01155	0.14 s	β^+ ,p/10.81					ann.rad./
^{25}Si		25.00411	221 ms	β^+ ,p/12.74		5/2 +			ann.rad./
^{26}Si		25.992330	2.23 s	β^+ /5.066	3.282/	0+			ann.rad./ 0.8294(8)/22.
^{27}Si		26.9867048	4.14 s	β^+ /4.8118	3.85/100.	5/2 +	-0.8554		ann.rad./ 2.211(5)/0.2
^{28}Si	92.22(2)	27.97692653				0+			
^{29}Si	4.69(1)	28.97649472				1/2 +	-0.5553		
^{30}Si	3.09(1)	29.97377022				0+			
^{31}Si		30.9753633	2.62 h	β^- /1.4920	1.471/99.9	3/2 +			1.2662(5)/0.05
^{32}Si		31.974148	1.6x10^2 y	β^- /0.224	0.213/100.	0+			
^{33}Si		32.97800	6.1 s	β^- /5.85	3.92	(3/2 +)			1.4313(5)/13. 1.84769(5)/100. 2.538(2)/10.
^{34}Si		33.97858	2.8 s	β^- /4.60	3.09/	0+			0.42907(5)/60. 1.17852(2)/64. 1.60756(5)/36.
^{35}Si		34.98458	0.9 s	β^- /10.50					
^{36}Si		35.9867	0.5 s	β^- /7.9					
^{37}Si		36.9930	≈0.09 s						
^{38}Si		37.9960							

Elem. or Isot.	Natural Abundance (%)	Atomic Mass or Weight	Half-Life	Decay Mode/Energy (/MeV)	Particle Energy /Intensity (MeV / %)	Spin (h/2π)	Nuclear Magnetic Mom.(nm)	Elect. Quadr. Mom. (b)	γ-ray Energy /Intensity (MeV / %)
³⁹Si		39.0023							
⁴⁰Si		40.0058							
⁴¹Si		41.013							
⁴²Si		42.016							
₁₅P		30.973761(2)							
²⁶P		26.0118	≈20. ms	β⁺ ,p/18.1		3+			
²⁷P		26.99919	0.3 s	β⁺,p /11.63		1/2+			
²⁸P		27.992312	270. ms	β⁺ /14.332	3.94/13. 5.25/13. 6.96/16. 8.8/7. 11.49/52.	3+			ann.rad./ 1.779(2)/98. 2.839(2)/2.8 3.040(2)/3.2 4.498(2)/12. 7.537(2)/9.
²⁹P		28.981801	4.14 s	β⁺ /4.9431	3.945/98.	1/2+	1.2349		ann.rad./ 1.273/1.32 2.426/0.39
³⁰P		29.9783138	2.50 m	β⁺ /4.2323	3.245/99.9	1+			ann.rad./ 2.230(3)/0.07
³¹P	100.	30.9737615				1/2+	+1.13160		
³²P		31.9739071	14.28 d	β⁻ /1.7106	1.710/100.	1+	-0.2524		
³³P		32.971725	25.3 d	β⁻ /0.249	0.249/100.	1/2+			
³⁴P		33.973636	12.4 s	β⁻ /5.374	3.2/15. 5.1/85.	1+			1.78-4.1/ 2.127(5)/15.
³⁵P		34.973314	47. s	β⁻ /3.989	2.34/100.	1/2+			1.572(1)/100.
³⁶P		35.97826	5.7 s	β⁻ /10.41					0.902/77. 3.291/100.
³⁷P		36.97961	2.3 s	β⁻ /7.90					0.6462/ 1.5829/
³⁸P		37.9845	0.6 s	β⁻ /12.4					1.2923/ 2.224/
³⁹P		38.9864	≈ 0.16 s						
⁴⁰P		39.9911							
⁴¹P		40.9948	0.12 s						
⁴²P		42.0001	0.11 s						
⁴³P		43.0033	33. ms	β⁻,(n)/					
⁴⁴P		44.010							
⁴⁵P		45.015							
⁴⁶P		46.024							
₁₆S		32.066(6)							
²⁷S		27.0188	21. ms	β⁺, 2p /18.3					
²⁹S		28.99661	0.188 s	β⁺ /13.79 β⁺ ,p/		5/2+			ann.rad./
³⁰S		29.984903	1.18 s	β⁺ /6.138	4.42/78. 5.08/20.	0+			ann.rad./ 0.678/79.
³¹S		30.979555	2.56 s	β⁺ /5.396	4.39/99.	1/2+	0.48793		ann.rad./ 1.2662(5)/1.2
³²S	94.93(31)	31.9720707				0+			
³³S	0.76(2)	32.9714585				3/2+	+0.64382	-0.07	
³⁴S	4.29(28)	33.9678668				0+			
³⁵S		34.9690321	87.2 d	β⁻ /0.1672	0.1674/100.	3/2+	+1.00	+0.047	
³⁶S	0.02(1)	35.9670809				0+			
³⁷S		36.9711257	5.05 m	β⁻ /4.8653	1.64/94. 4.75/5.6				0.9083(4)/0.06 3.1033(2)/94.2
³⁸S		37.97116	2.84 h	β⁻ /2.94	1.00/	0+			0.1962(4)/0.2 1.9421(3)/84.
³⁹S		38.97514	11.5 s	β⁻ /6.64					1.301/52. 1.697/44.
⁴⁰S		39.9755	9. s	β⁻ /4.7					
⁴¹S		40.9800							
⁴²S		41.9815	≈0.56 s	β,(n)/					
⁴³S		42.987							
⁴⁴S		43.9883	0.12 s	β⁻, n/9.					
⁴⁵S		44.9948	0.08 s	β⁻,n/					
⁴⁶S		45.9996							
⁴⁷S		47.008							
⁴⁸S		48.013							
₁₇Cl		35.4527(9)							
³¹Cl		30.99242	0.15 s	β⁺ ,p/11.98	1.52	3/2+			ann.rad./

Elem. or Isot.	Natural Abundance (%)	Atomic Mass or Weight	Half-Life	Decay Mode/Energy (/MeV)	Particle Energy /Intensity (MeV / %)	Spin (h/2π)	Nuclear Magnetic Mom.(nm)	Elect. Quadr. Mom. (b)	γ-ray Energy /Intensity (MeV / %)
³²Cl		31.98569	297. ms	β⁺ /12.69	4.75/25. 6.18/10. 7.48/14. 9.47/50. 11.6/1.	1+			ann.rad./ 1.548(2)/3.5 2.2305(1)/92. 2.4638(1)/4. 2.885(1)/1. 4.770(1)/20.
³³Cl		32.977452	2.511 s	β⁺ /5.583	4.51/98.	3/2 +	+0.752		ann.rad./ 0.8409/0.52 1.966/0.45 2.866/0.44
³⁴ᵐCl			32.2 m	β⁺ /	1.35/24. 2.47/28.	3 +			ann.rad./
				I.T./					0.1457(8)/42. 2.1276(5)/42.
³⁴Cl		33.9737620	1.528 s	β⁺ /5.4922	4.50/100.	0+			ann.rad./
³⁵Cl	75.78(4)	34.96885271				3/2 +	+0.82187	-0.08	
³⁶Cl		35.9683069	3.01x10⁵ y	β⁻ /0.7086 β⁺,EC/1.1421	0.7093/98. 0.115/0.002	0+	+1.28547	-0.018	ann.rad./
³⁷Cl	24.22(4)	36.96590260				3/2 +	+0.68412	-0.065	
³⁸ᵐCl			0.715 s	I.T./		5-			0.67138(2)/100. 1.64216(1)/31.
³⁸Cl		37.9680106	37.2 m	β⁻ /4.9168	1.11/31. 2.77/11. 4.91/58.	2-	2.05		2.16760(2)/42.
³⁹Cl		38.968008	55.6 m	β⁻ /3.442	1.91/85. 2.18/8. 3.45/7.	3/2 +			0.25026(1)/47. 1.26720(5)/54. 0.986-1.517
⁴⁰Cl		39.97042	1.38 m	β⁻ /7.48		2-			0.6431(3)/6. 1.4608(1)/77. 2.8402(2)/17. (0.167-1.359)
⁴¹Cl		40.9707	34. s	β⁻ /5.7	3.8/				
⁴²Cl		41.9732	6.8 s	β⁻ /9.4					
⁴³Cl		42.9742	3.3 s	β⁻ /8.0					
⁴⁴Cl		43.9785	≈0.43 s	β⁻,n/12.3					
⁴⁵Cl		44.980	0.40 s	β⁻, n/11.					
⁴⁶Cl		45.984	0.22 s	β⁻, n/14.9					
⁴⁷Cl		46.988		β⁻, n/15.					
⁴⁸Cl		47.995							
⁴⁹Cl		48.9999							
⁵¹Cl		51.014							
₁₈Ar		39.948(1)							
³¹Ar		31.0121	15. ms	β⁺ /18.4 β⁺, 2p β⁺, 3p					
³²Ar		31.99766	98. ms	β⁺ ,p/11.2					ann.rad./
³³Ar		32.98993	174. ms	β⁺ /11.62 β⁺ ,p/	3.12/	1/2 +	-0.72		ann.rad./ 0.810(2)/48.
³⁴Ar		33.980270	0.844 s	β⁺/6.061	5.0/95.	0+			ann.rad./ 0.6658(1)/2.5 3.1290(1)/1.3
³⁵Ar		34.975257	1.77 s	β⁺/5.965	4.94/93.	3/2 +	+0.633	-0.08	ann.rad./ 1.2185(5)/1.22 1.763(1)/0.25 2.964(1)/0.2
³⁶Ar	0.3365(30)	35.9675463				0+			
³⁷Ar		36.9667759	35.0 d	EC/.813		3/2 +	+1.15	+0.076	
³⁸Ar	0.0632(5)	37.9627322				0+			
³⁹Ar		38.964313	268. y	β⁻/0.565	0.565/100.	7/2-	-1.3	-0.12	
⁴⁰Ar	99.6003(30)	39.962383123				0+			
⁴¹Ar		40.964501	1.82 h	β⁻/2.492	1.198/	7/2-			1.29364(5)/99. 1.6770(3)/0.05
⁴²Ar		41.96305	33. y	β⁻/0.60	0.60/100.	0+			
⁴³Ar		42.9657	5.4 m	β⁻/4.6					0.4791(2)/10. 0.7380(1)/43. 0.9752(1)/100. 1.4400(3)/39. 0.182-1.866
⁴⁴Ar		43.96537	11.87 m	β⁻/3.55		0+			0.0610/25.
⁴⁵Ar		44.96809	21.5 s	β⁻/6.9		7/2-			1.020/35. 3.707/34. 1.944/
⁴⁶Ar		45.96809	8.4 s	β⁻/5.70		0+			
⁴⁷Ar		46.9722	≈0.7 s	β⁻					
⁴⁹Ar		48.9822							
⁵⁰Ar		49.986							
⁵¹Ar		50.993							

Elem. or Isot.	Natural Abundance (%)	Atomic Mass or Weight	Half-Life	Decay Mode/Energy (/MeV)	Particle Energy /Intensity (MeV / %)	Spin (h/2π)	Nuclear Magnetic Mom.(nm)	Elect. Quadr. Mom. (b)	γ-ray Energy /Intensity (MeV / %)
₁₉K		39.0983(1)							
³⁵K		34.98801	0.19 s	β⁺ /11.88 β⁺ ,p/		3/2+			ann.rad./ 1.751/14. 2.5698/26. 2.9827/51.
³⁶K		35.98129	0.342 s	β⁺ /12.81	5.3/42. 9.9/44.	2+	+0.548		ann.rad./ 1.97044(5)/82. 2.20783(5)/30. 2.43343(2)/32.
³⁷K		36.9733769	1.23 s	β⁺ /6.149	5.13/	3/2+	+0.2032		ann.rad./ 2.7944(8)/2. 3.602(2)/0.05
³⁸ᵐK			0.924 s	β⁺ /6.742	5.02/100.	0+			ann.rad./
³⁸K		37.969080	7.63 m	β⁺ /5.913	2.60/99.8	3+	+1.37		ann.rad./ 2.1675(3)/99.8 3.9356(5)/0.2
³⁹K	93.2581(44)	38.9637069				3/2+	+0.39146	+0.049	
⁴⁰K	0.0117(1)	39.9639987	1.26x10⁹ y	β⁻ /1.3111 β⁺ ,EC/1.505	1.312/89. 1.50/10.7	4-	-1.298	-0.061	ann.rad./ 1.46081(5)/10.5
⁴¹K	6.7302(44)	40.9618260				3/2+	+0.21487	+0.060	
⁴²K		41.9624031	12.36 h	β⁻ /3.525	1.97/19. 3.523/81.	2-	-1.1425		0.31260(2)/0.3 1.5246(3)/18.
⁴³K		42.96072	22.3 h	β⁻ /1.82	0.465/8. 0.825/87. 1.24/3.5 1.814/1.3	3/2+	+0.163		0.2211(2)/4. 0.3729(2)/88. 0.3971(2)/11. 0.6178(2)/81.
⁴⁴K		43.96156	22.1 m	β⁻ /5.66	5.66/34.	2-	-0.856		0.368207(1)/2.2 1.15700(1)/58. 2.15079(2)/22.
⁴⁵K		44.96070	17.8 m	β⁻ /4.20	1.1/23. 2.1/69. 4.0/8.	3/2+	+0.1734		0.1743(5)/80. 1.2607(8)/7. 1.7056(6)/69. 2.3542(5)/14.
⁴⁶K		45.96198	1.8 m	β⁻ /7.72	6.3/	2-	-1.05		1.347(1)/91. 3.700(5)/28.
⁴⁷K		46.96168	17.5 s	β⁻ /6.64	4.1/99. 6.0/1.	1/2+	+1.93		0.56474(3)/15. 0.58575(3)/85. 2.01313(3)/100.
⁴⁸K		47.96551	6.8 s	β⁻ /12.09	5.0/	(2-)			0.67122(1)/4. 0.6723(5)/20. 0.78016(1)/32. 3.83153(7)/80.
⁴⁹K		48.9675	1.26 s	β⁻ /11.0					2.025/ 2.252/
⁵⁰K		49.9728	0.472 s	β⁻ /14.2					
⁵¹K		50.9764	0.365 s	β⁻ /					
⁵²K		51.983	0.105 s	β⁻					
⁵³K		52.987	30. ms	β⁻		3/2+			
⁵⁴K		53.994	10. ms	β⁻					
₂₀Ca		40.078(4)							
³⁵Ca		35.0048	0.05 s	β⁺ /15.6					
³⁶Ca		35.99309	0.10 s	β⁺,(p)/10.99 β⁺ ,n/	2.52				ann.rad./
³⁷Ca		36.98587	0.18 s	β⁺ /11.64 β⁺ ,n/	3.103	3/2+			ann.rad./ 1.369
³⁸Ca		37.976319	0.44 s	β⁺ /6.74		0+			ann.rad./ 1.5677(5)/25. 3.210(2)/1.
³⁹Ca		38.970718	0.861 s	β⁺ /6.531	5.49/100.	3/2+	1.02168		ann.rad./
⁴⁰Ca	96.941(156)	39.9625912				0+			
⁴¹Ca		40.9622783	1.02x10⁵ y	EC/0.4214		7/2-	-1.595	-0.08	
⁴²Ca	0.647(23)	41.9586183				0+			
⁴³Ca	0.135(10)	42.9587668				7/2-	-1.3173	-0.05	
⁴⁴Ca	2.086(110)	43.955481				0+			
⁴⁵Ca		44.956186	162.7 d	β⁻ /0.257	0.257/100.	7/2-	-1.327	+0.05	
⁴⁶Ca	0.004(3)	45.953693				0+			
⁴⁷Ca		46.954546	4.536 d	β⁻ /1.992	0.684/84. 1.98/16.	7/2-	-1.38	+0.02	1.297/75 (0.041-1.88)
⁴⁸Ca	0.187(21)	47.952533				0+			
⁴⁹Ca		48.955673	8.72 m	β⁻ /5.262	0.89/7. 1.95/92.	3/2-			3.0844(1)/92. 4.0719(1)/7.
⁵⁰Ca		49.95752	14. s	β⁻ /4.97	3.12/	0+			0.2569/98. (0.0715 - 1.59)
⁵¹Ca		50.9615	10. s	β⁻ /7.3		(3/2-)			
⁵²Ca		51.9651	4.6 s	β⁻ /8.0					
⁵³Ca		52.9701	0.09 s	β⁻ /10.9					

Elem. or Isot.	Natural Abundance (%)	Atomic Mass or Weight	Half-Life	Decay Mode/Energy (/MeV)	Particle Energy /Intensity (MeV / %)	Spin (h/2π)	Nuclear Magnetic Mom.(nm)	Elect. Quadr. Mom. (b)	γ-ray Energy /Intensity (MeV / %)
^{54}Ca		53.975							
^{55}Ca		54.981							
^{56}Ca		55.986							
$_{21}$Sc		**44.955910(8)**							
^{39}Sc		38.98479		p					
^{40}Sc		39.977964	0.182 s	β⁺ /14.320	5.73/50. 7.53/15. 8.76/15. 9.58/20.	4-			ann.rad./ 0.752/41. 3.732/99.5 (1.12-3.92)
^{41}Sc		40.9692513	0.596 s	β⁺ /6.4953	5.61/100.	7/2-	5.431	0.166	ann.rad./
^{42m}Sc			61.6 s	β⁺ /	2.82/	7+			ann.rad./ 0.4375(5)/100. 1.2270(5)/100. 1.5245(5)/100.
^{42}Sc		41.9655168	0.682 s	β⁺ /6.4259	5.32/100.	0+			ann.rad./
^{43}Sc		42.961151	3.89 h	β⁺ ,EC/2.221	0.82/22. 1.22/78.	7/2-	+4.62	-0.26	ann.rad./ 0.3729(1)/22.
^{44m}Sc			2.44 d	I.T./0.27 EC/3.926		6+	+3.88		0.27124(1)/87. (1.00-1.16)
^{44}Sc		43.959403	3.93 h	β⁺, EC/3.653	1.47/	2+	+2.56	+0.10	ann.rad./ 1.15700(1)/100.
^{45}Sc	100.	44.955910				7/2-	+4.75649	-0.22	
^{46m}Sc			18.7 s	I.T./0.14253		1-			0.14253(2)/62.
^{46}Sc		45.955170	83.81 d	β⁻ /2.367	0.357/100.	4+	+3.03	+0.12	0.88925(3)/100. 1.12051(1)/100.
^{47}Sc		46.952408	3.349 d	β⁻ /0.600	0.439/69. 0.601/31.	7/2-	+5.34	-0.22	0.15938(1)/68.
^{48}Sc		47.95224	43.7 h	β⁻ /3.99	0.655/	6+			0.98350(1)/100. 1.03750(1)/97. 1.31209(3)/100.
^{49}Sc		48.950024	57.3 m	β⁻ /2.006	2.00/99.9.	7/2-			1.7619(3)/0.05
^{50}Sc		49.95219	1.71 m	β⁻ /6.89	3.05/76. 3.60/24.	(5+)			0.5235(1)/88. 1.1210(1)/100. 1.5537(2)/100.
^{51}Sc		50.95360	12.4 s	β⁻ /6.51	4.4/ 5.0/	7/2-			1.4373(4)/52. 0.718-2.144
^{52}Sc		51.9566	8.2 s	β⁻ /9.0		(3+)			
^{53}Sc		52.9592	< 1.6 s	β⁻ /8.1					
^{54}Sc		53.9630	≈0.23 s	β⁻ /11.6					
^{55}Sc		54.967	0.13 s	β⁻ /13					
^{56}Sc		55.973							
^{57}Sc		56.977							
^{58}Sc		57.983							
$_{22}$Ti		**47.867(1)**							
^{39}Ti		39.0013	28. ms	β⁺ /15.4					
^{40}Ti		39.9905	52. ms	β⁺ /11.7					
^{41}Ti		40.98313	80. ms	β⁺ ,p/12.93		3/2+			ann.rad./
^{42}Ti		41.97303	0.20 s	β⁺ /7.000	6.0/				ann.rad./ 0.6107(5)/56.
^{43}Ti		42.96852	0.50 s	β⁺ /6.87	5.80/	7/2-	0.85		ann.rad./
^{44}Ti		43.959690	66. y	EC/0.268		0+			0.067868(1)/91. 0.078322(1)/97.
^{45}Ti		44.958124	3.078 h	β⁺ /86/2.062 EC/14/	1.04	7/2-	0.095	0.015	ann.rad./ (0.36-1.66)/weak
^{46}Ti	8.25(3)	45.952630				0+			
^{47}Ti	7.44(2)	46.951764				5/2-	-0.78848	+0.30	
^{48}Ti	73.72(3)	47.947947				0+			
^{49}Ti	5.41(2)	48.947871				7/2-	-1.10417	+0.24	
^{50}Ti	5.18(2)	49.944792				0+			
^{51}Ti		50.946616	5.76 m	β⁻ /2.471	1.50/92. 2.13/	3/2-			0.3197(2)/93. 0.6094-0.9291
^{52}Ti		51.94690	1.7 m	β⁻ /1.97	1.8/100.	0+			0.0170(5)/100. 0.12445(5)/100.
^{53}Ti		52.9497	33. s	β⁻ /5.0	(2.2-3)/	3/2-			0.1008(1)/20. 0.1276(1)/45. 0.2284(1)/39. 1.6755(5)/45. (1.72-2.8)/
^{54}Ti		53.9509	1.5 s	β⁻ /4.3					
^{55}Ti		54.9551	0.60 s	β⁻ /7.4					
^{56}Ti		55.9580	0.15 s	β⁻ /7.0					
^{57}Ti		56.963	≈ 56. ms	β⁻ /11.					
^{58}Ti		57.966							
^{59}Ti		58.972							
^{60}Ti		59.976							

Elem. or Isot.	Natural Abundance (%)	Atomic Mass or Weight	Half-Life	Decay Mode/Energy (/MeV)	Particle Energy /Intensity (MeV / %)	Spin (h/2π)	Nuclear Magnetic Mom.(nm)	Elect. Quadr. Mom. (b)	γ-ray Energy /Intensity (MeV / %)
^{61}Ti		60.982							
$_{23}$V		**50.9415(1)**							
^{43}V		42.9807	>0.8 s	β^+ /11.3					
^{44}V		43.9744	0.09 s	β^+ ,α/13.7					ann.rad./
^{45}V		44.96578	0.54 s	β^+ /7.13		7/2-			
^{46}V		45.960200	0.4223 s	β^+ /7.051	6.03/100.	0+			ann.rad./
^{47}V		46.954907	32.6 m	β^+ ,EC/2.928	1.90/99.+	3/2-			ann.rad./ 1.7949(8)/0.19 (0.2-2.16)
^{48}V		47.952254	15.98 d	β^+ /4.012	0.698/50.	4+	2.01		ann.rad./ 0.98350(2)/100. (1.3-2.4)
^{49}V		48.948517	337. d	EC/0.602		7/2-	4.47		
^{50}V	0.250(4)	49.9947163	>1.4x10^{17}y	EC, β^-		6+	+3.34569	+0.21	
^{51}V	99.750(4)	50.943964				7/2-	+5.148706	-0.04	
^{52}V		51.944780	3.76 m	β^- /3.976	2.47/	3+			1.4341(1)/100.
^{53}V		52.944342	1.61 m	β^- /3.436	2.52/	7/2-			1.0060(5)/90. 1.2891(3)/10.
^{54}V		53.94644	49.8 s	β^- /7.04	1.00/5. 2.00/12. 2.95/45. 5.20/11.	3+			0.8348/97. 0.9887/80. 2.259/46. (0.56-3.38)
^{55}V		54.9472	6.5 s	β^- /6.0	6.0/	(7/2-)			0.5177/73. (0.224-1.21)
^{56}V		55.9504	0.23 s	β^- /9.1					
^{57}V		56.9524	0.33 s	β^- /8.1					
^{58}V		57.9567	0.20 s	β^- /11.6					
^{59}V		58.9593	0.07 s	β^- /9.9					
^{60}V		59.965		β^- /14.					
^{61}V		60.967							
^{62}V		61.973							
^{63}V		62.977							
^{64}V									
$_{24}$Cr		**51.9961(6)**							
^{43}Cr		42.9977	21. ms						
^{44}Cr		43.9855	53. ms	β^+ ,(p) /10.3	p/0.95-3.1				
^{45}Cr		44.9792	0.05 s	β^+ ,p/12.5		7/2-			ann.rad./
^{46}Cr		45.96836	0.3 s	β^+ /7.60					ann.rad./
^{47}Cr		46.96291	0.51 s	β^+ /7.45		3/2-			ann.rad./
^{48}Cr		47.95404	21.6 h	EC/1.66					ann.rad./ 0.116(2)/95. 0.305(10)/100.
^{49}Cr		48.951341	42.3 m	β^+,EC/2.631	1.39/ 1.45/ 1.54/	5/2-	0.476		ann.rad./ 0.09064(1)/51. 0.15293(1)/27. (0.062-1.6)
^{50}Cr	4.345(13)	49.946050				0+			
^{51}Cr		50.944772	27.70 d	EC/0.7527		7/2-	-0.934		0.320076(1)/10.2
^{52}Cr	83.789(18)	51.940512				0+			
^{53}Cr	9.501(17)	52.940653				3/2-	-0.47454	-0.15	
^{54}Cr	2.365(7)	53.938885				0+		-0.2	
^{55}Cr		54.940844	3.497 m	β^- /2.603	2.5/	3/2-			1.5282(2)/0.04 (0.13-2.37)
^{56}Cr		55.94065	5.9 m	β^- /1.62	1.50/100.	0+			0.026(2)/100. 0.083(3)/100.
^{57}Cr		56.9438	21. s	β^- /5.1	3.3/ 3.5/	3/2-	0.0834		0.850/8. (0.083-2.62)
^{58}Cr		57.9443	7.0 s	β^- /4.0					(0.131-0.683)
^{59}Cr		58.9487	1.0 s	β^- /7.7					1.236
^{60}Cr		59.9497	0.6 s	β^- /6.0					
^{61}Cr		60.9541		β^- /8.8					
^{62}Cr		61.9558	0.20 s	β^- /7.3					
^{63}Cr		62.962							
^{65}Cr		64.970							
^{66}Cr									
^{67}Cr									
$_{25}$Mn		**54.938049(9)**							
^{46}Mn		45.9867	≈41. ms	β^+ /17.1					
^{47}Mn		46.9761	≈0.1 s	β^+ /12.3					
^{48}Mn		47.9689	0.15 s	β^+ /13.5	5.79/58. 4.43/10.	4+			
^{49}Mn		48.95962	0.38 s	β^+ /7.72	6.69/	5/2-			ann.rad./

Elem. or Isot.	Natural Abundance (%)	Atomic Mass or Weight	Half-Life	Decay Mode/Energy (/MeV)	Particle Energy /Intensity (MeV / %)	Spin (h/2π)	Nuclear Magnetic Mom.(nm)	Elect. Quadr. Mom. (b)	γ-ray Energy /Intensity (MeV / %)
50mMn			1.74 m	β⁺ /7.887	3.54/	5+			ann.rad./ 1.0980/94. 0.783/91. (0.66-3.11)
50Mn		49.954244	0.283 s	β⁺ /7.6330	6.61/	0+			ann.rad./
51Mn		50.948215	46.2 m	β⁺ ,EC/3.208	2.2/	5/2-	3.568	0.4	ann.rad./ 0.7491(1)/0.26 (1.148-1.164)
52mMn			21.1 m	β⁺ /98/5.09 I.T./2/0.378	2.631/	2+	0.0076		ann.rad./ 0.3778 (I.T.) 1.43406(1)/98. (0.7-4.8)
52Mn		51.945570	5.591 d	β⁺ /4.712 EC/	0.575/	6+	+3.063	+0.5	ann.rad./ 0.74421(1)/90. 1.43406(1)/100.
53Mn		52.941294	3.7x10⁶ y	EC/0.5970		7/2-	5.024		
54Mn		53.940363	312.1 d	EC/1.377		3+	+3.282	+0.33	0.83403(5)/100.
55Mn	100.	54.938049				5/2-	+3.4687	+0.32	
56Mn		55.938909	2.579 h	β⁻ /3.6954	0.718/18. 1.028/34.	3+	+3.2266		0.84675(2)/98.9 1.81072(4)/27. 2.11305(5)/14.5
57Mn		56.938287	1.45 m	β⁻ /2.691		5/2-			
58Mn		57.93999	65 s	β⁻ /6.25	3.8/ 5.1/	3+			0.45916(2)/20. 0.81076(1)/82. 1.32309(5)/53.
59Mn		58.94045	4.6 s	β⁻ /5.19	4.5/				0.471/ 0.531-0.726
60mMn			1.77 s	β⁻ /IT	5.7/	3+			0.824/
60Mn		59.9433	50. s	β⁻ /8.6		0+			1.969/
61Mn		60.9446	0.71 s	β⁻ /7.4		(5/2)-			
62Mn		61.9480	0.9 s	β⁻ /10.4		(3+)			0.877/ 0.942-1.299
63Mn		62.9498	0.25 s	β⁻ /8.8					0.356
64Mn		63.9537	0.17 s	β⁻ /11.8					
65Mn		64.9561		β⁻ /10.					
66Mn		65.961							
67Mn		66.964							
68Mn									
69Mn									
26Fe		55.845(2)							
45Fe		45.0146							
46Fe		46.0008	≈0.02 s	β⁺ /13.1					
47Fe		46.9929	≈0.03 s	β⁺ /15.6					
48Fe		47.9806	≈ 44. ms	β⁺ /11.2					
49Fe		48.9763	70. ms	β⁺ /13.0		(7/2-)			ann.rad./ 0.651
50Fe		49.9630	0.15 s	β⁺ /8.2		0+			
51Fe		50.95683	0.31 s	β⁺ /8.02		(5/2-)			ann.rad./
52mFe			46. s	β⁺ /4.4		(12+)			ann.rad./ (0.622-2.286)/
52Fe		51.94812	8.28 h	β⁺ /57/2.37 EC/43/ I.T./	0.804/	0+			ann.rad./ 0.16868(1)/99. 0.377 (I.T.)/
53mFe			2.6 m	I.T./3.0407		19/2-			0.7011(1)/99. 1.0115(1)/87. 1.3281(1)/87. 2.3396(1)/13.
53Fe		52.945312	8.51 m	β⁺ /3.743	2.40/42. 2.80/57.	7/2-			ann.rad./ 0.3779(1)/42. (1.2 - 3.2)
54Fe	5.845(35)	53.939615				0+			
55Fe		54.938298	2.73 y	EC/0.2314		3/2-			
56Fe	91.754(36)	55.934942				0+			
57Fe	2.119(10)	56.935398				1/2-	+0.0906	0.16	
58Fe	0.282(4)	57.933280				0+			
59Fe		58.934880	44.51 d	β⁻ /1.565	0.273/48. 0.475/51.	3/2-	- 0.336		1.099/57 1.292/43. (0.14-1.48)
60Fe		59.934077	1.5x10⁶ y	β⁻ /0.237	0.184/100.	0+			0.0586/100.(IT)
61Fe		60.93675	6.0 m	β⁻ /3.98	2.5/13. 2.63/54. 2.80/31.				1.205/44. 1.028/43. (0.12-3.37)
62Fe		61.93677	68. s	β⁻ /2.53	2.5/100.	0+			0.5061(1)/100.
63Fe		62.9404	6. s	β⁻ /6.3		5/2-			0.995/ (1.365-1.427)
64Fe		63.9411	2.0 s	β⁻ /4.9					
65Fe		64.9449	≈0.45 s	β⁻ /7.9					
66Fe		65.9460		β⁻ /5.7					

T A B L E   O F   T H E   I S O T O P E S

Elem. or Isot.	Natural Abundance (%)	Atomic Mass or Weight	Half-Life	Decay Mode/Energy (/MeV)	Particle Energy /Intensity (MeV / %)	Spin (h/2π)	Nuclear Magnetic Mom.(nm)	Elect. Quadr. Mom. (b)	γ-ray Energy /Intensity (MeV / %)
⁴⁷Fe		66.9500		β⁻ /8.8					
⁶⁸Fe		67.953	≈0.10 s	β⁻ / ≈7.6					
⁶⁹Fe		68.958							
⁷⁰Fe									
⁷¹Fe									
⁷²Fe									
₂₇Co		58.933200(9)							
⁵⁰Co		49.9812	44. ms	β⁺ /17.0	2.03-2.79				
⁵¹Co		50.9705		β⁺ /12.8					
⁵²Co		51.9632	0.12 s	β⁺ /14.0					0.849-1.942
⁵²ᵐCo			0.25 s	β⁺ ,p/		19/2-			ann.rad./
⁵³Co		52.95423	0.26 s	β⁺ /8.30		7/2-			ann.rad./
⁵⁴ᵐCo			1.46 m	β⁺ /8.44	4.25/100.	7+			ann.rad./ 0.411(1)/99. 1.130(1)/100. 1.408(1)/100.
⁵⁴Co		53.948464	0.1932 s	β⁺ /8.2430	7.34/100.	0+			ann.rad./
⁵⁵Co		54.942003	17.53 h	β⁺ /3.4513 EC/	0.53/ 1.03/ 1.50/	7/2-	+4.822		ann.rad./ 0.9312/75. 0.4772/20. (0.092-3.11)
⁵⁶Co		55.939844	77.3 d	β⁺ /4.566 EC/	1.459/18.	4+	3.85	+0.25	ann.rad./ 0.84678(6)/99.9 1.2383/68. (0.26-3.61)
⁵⁷Co		56.936296	271.8 d	EC/0.8361		7/2-	+4.72	+0.5	0.12206(2)/85.6 (0.014-0.706)
⁵⁸ᵐCo			9.1 h	I.T./		5+			0.02489/0.035
⁵⁸Co		57.935757	70.88 d	β⁺ /2.307 EC/		2+	+4.04	+0.22	ann.rad./ 0.810755(3)/99.
⁵⁹Co	100.	58.933200				7/2-	+4.63	+0.41	
⁶⁰ᵐCo			10.47 m	I.T./99.8/0.059 β⁻ /0.2/1.56		2+	+4.40	+0.3	0.058603(1)/2.0
⁶⁰Co		59.933822	5.271 y	β⁻ /2.824	0.315/99.7	5+	+3.799	+0.44	1.173210(2)/100. 1.332470(2)/100.
⁶¹Co		60.932479	1.650 h	β⁻ /1.322	1.22/95.	7/2-			0.067415(1)/86. (0.8417-0.9092)
⁶²ᵐCo			13.9 m	β⁻ /	0.88/25. 2.88/75.	5+			1.1635(3)/70. 1.1730(3)/98. 2.0039(3)/19.
⁶²Co		61.93405	1.50 m	β⁻ /5.32	1.03/10. 1.76/5. 2.9/20. 4.05/60.	2+			1.1292(3)/13. 1.1730(3)/83. 1.9851(1)/3. 2.3020(1)/19.
⁶³Co		62.93362	27.5 s	β⁻ /3.67	3.6/	7/2-			0.08713(1)/49. 0.9817(3)/2.6 0.156-2.17
⁶⁴Co		63.93581	0.30 s	β⁻ /7.31	7.0/	1+			
⁶⁵Co		64.93648	1.14 s	β⁻ /5.96		(7/2)-			
⁶⁶Co		65.9398	0.25 s	β⁻ /10.0					(1.245-1.425)
⁶⁷Co		66.9406	0.32 s	β⁻ /8.4					0.694
⁶⁸Co		67.9444	≈ 0.18 s	β⁻ /11.7					
⁶⁹Co		68.9452	0.27 s	β⁻ /9.3					
⁷⁰Co		69.950	0.17 s	β⁻ 13.					
⁷¹Co		70.952	0.20 s	β					
⁷²Co		71.956	0.12 s	β					
⁷³Co									
⁷⁴Co									
⁷⁵Co									
₂₈Ni		58.6934(2)							
⁴⁹Ni									
⁵⁰Ni		49.9959							
⁵¹Ni		50.9877		β⁺ /16.0					
⁵²Ni		51.9757	38. ms	β⁺ /11.7					
⁵³Ni		52.9685	0.05 s	β⁺ ,p/13.3		7/2-			ann.rad./
⁵⁴Ni		53.95791	0.14 s	β⁺ /8.80					0.937
⁵⁵Ni		54.95134	0.20 s	β⁺ /8.70	7.66/	7/2-			ann.rad./
⁵⁶Ni		55.94214	6.08 d	EC/2.14		0+			0.15838(3)/98.8 0.81185(3)/87. 0.2695-0.7500
⁵⁷Ni		56.939800	35.6 h	β⁺ /3.264 EC/	0.712/10. 0.849/76.	3/2-	- 0.798		ann.rad./ 1.3776/78. (0.127-3.177)
⁵⁸Ni	68.0769(89)	57.935348				0+			
⁵⁹Ni		58.934351	≈7.6x10⁴ y	EC/		3/2-			

Elem. or Isot.	Natural Abundance (%)	Atomic Mass or Weight	Half-Life	Decay Mode/Energy (/MeV)	Particle Energy /Intensity (MeV / %)	Spin (h/2π)	Nuclear Magnetic Mom.(nm)	Elect. Quadr. Mom. (b)	γ-ray Energy /Intensity (MeV / %)
^{60}Ni	26.2231(77)	59.930790				0+			
^{61}Ni	1.1399(6)	60.931060				3/2-	-0.75002	+0.16	
^{62}Ni	3.6345(17)	61.928348				0+			
^{63}Ni		62.929673	100. y	β^- /0.066945	0.065/	1/2-			
^{64}Ni	0.9256(9)	63.927969				0+			
^{65}Ni		64.930088	2.517 h	β^- /2.137	0.65/30. 1.020/11. 2.140/58.	5/2-	0.69		0.36627(3)/5. 1.11553(4)/16. 1.48184(5)/23.
^{66}Ni		65.92912	54.6 h	β^- /0.23		0+			
^{67}Ni		66.93157	21. s	β^- /3.56	3.8/	1/2-			1.0722/100. 1.6539/100. (0.10-1.98)
^{68}Ni		67.93185	19. s	β^- /2.06		0+			
^{69}Ni		68.9352	10. s	β^- /5.4					0.6807(3)/100. (0.207-1.213)
^{70}Ni		69.9361		β^- /3.5					
^{71}Ni		70.9400	1.9 s	β^- /6.9					
^{72}Ni		71.9413	2.1 s	β^- /5.2					
^{73}Ni		72.946	0.9 s	β^- /9.					
^{74}Ni		73.948	1.1 s	β^- /7.					
^{75}Ni		74.953	≈ 0.47 s						
^{76}Ni		75.955	≈ 0.24 s						
^{77}Ni		76.961							
^{78}Ni		77.964							
$_{29}$Cu		63.546(3)							
^{55}Cu		54.9655		β^+ /13.2					
^{56}Cu		55.9586		β^+ /15.3					
^{57}Cu		56.94922	196. ms	β^+ /8.77		3/2-			0.77-3.01
^{58}Cu		57.944541	3.21 s	β^+ /8.563 EC/	4.5/15. 7.439/83.	1+			ann.rad./ 0.0403(4)/5. 1.4483(2)/11. 1.4546(2)/16.
^{59}Cu		58.939504	1.36 m	β^+ /4.800	1.9/ 3.75/	3/2-			ann.rad./ 0.3393(1)/8. 0.8780(1)/12. 1.3015(1)/15. (0.4 - 2.6)
^{60}Cu		59.937368	23.7 m	β^+ /6.127 EC/	2.00/69. 3.00/18. 3.92/6.	2+	+1.219		ann.rad./ 1.3325/88. 1.7915/45. (0.12-5.048)
^{61}Cu		60.933462	3.35 h	β^+ /2.237	0.56/3. 0.94/5. 1.15/2. 1.220/51.	3/2-	+2.14		ann.rad./ 0.2830/13. 0.6560/11. (0.067-2.123)
^{62}Cu		61.932587	9.67 m	β^+ /98/3.948 EC/	2.93/98.	1+	-0.380		ann.rad./ 1.17302(1)/0.6 (0.87-3.37)
^{63}Cu	69.17(3)	62.929601				3/2-	+2.2233	-0.211	
^{64}Cu		63.929768	12.701 h	β^- /39/0.579 β^+ /19/1.6751 EC/41/	0.578/ 0.65/	1+	-0.217		ann.rad./ 1.3459(3)/0.6
^{65}Cu	30.83(3)	64.927794				3/2-	+2.3817	-0.195	
^{66}Cu		65.928873	5.09 m	β^- /2.642	1.65/6. 2.7/94.	1+	-0.282		0.8330(1)/0.22 1.0392(2)/9.2
^{67}Cu		66.92775	2.580 d	β^- /0.58	0.395/56. 0.484/23. 0.577/20.	3/2-			0.09125(1)/7. 0.09325(1)/17. 0.18453(1)/47.
^{68m}Cu			3.79 m	I.T./86/ β^- /14/1.8		6-			0.0843(5)/70. 0.1112(5)/18. 0.5259(5)/74. (0.64-1.34)
^{68}Cu		67.92964	31. s	β^- /4.46	3.5/40. 4.6/31.	1+			1.0774(5)/58. 1.2613(5)/17. (0.15-2.34)
^{69}Cu		68.92943	2.8 m	β^- /2.68	2.48/80.	3/2-			0.5307(3)/3. 0.8340(5)/6. 1.0065(8)/10.
^{70m}Cu			47. s	β^- /	2.52/10.	5-			0.8848(2)/100. 0.9017(2)/90. 1.2517(5)/60. (0.39-3.06)
^{70}Cu		69.93241	5. s	β^- /6.60	5.42/54. 6.09/46.	1+			0.8848(2)/54.
^{71}Cu		70.93262	20. s	β^- /4.56		3/2-			0.490/
^{72}Cu		71.9357	6.6 s	β^- /8.2		(1+)			0.652/
^{73}Cu		72.9365	3.9 s	β^- /6.3					0.450/

Elem. or Isot.	Natural Abundance (%)	Atomic Mass or Weight	Half-Life	Decay Mode/Energy (/MeV)	Particle Energy /Intensity (MeV / %)	Spin (h/2π)	Nuclear Magnetic Mom.(nm)	Elect. Quadr. Mom. (b)	γ-ray Energy /Intensity (MeV / %)
^{74}Cu		73.9401	1.6 s	β⁻ /9.9					
^{75}Cu		74.9414	1.2 s	β⁻ /7.9					
^{76}Cu		75.9455	0.64 s	β⁻ /11.					
^{77}Cu		76.947	0.47 s	β⁻ /≈10.					
^{78}Cu		77.952	0.34 s	β⁻ /12.					
^{79}Cu		78.954	0.19 s	β⁻ /11.					
^{80}Cu		79.962							
$_{30}$Zn		65.39(2)							
^{57}Zn		56.9649	0.04 s	β⁺,p/14.6		(7/2-)			ann.rad./
^{59}Zn		58.94927	183. ms	β⁺,p/9.09	8.1/	3/2-			ann.rad./
^{60}Zn		59.94183	2.40 m	β⁺ /97/4.16 EC/3/		0+			(0.491-0.914) ann.rad./ 0.669/47.
^{61}Zn		60.93951	1.485 m	β⁺ /5.64	4.38/68.	3/2-			(0.062-0.947) ann.rad./ 0.4748/17.
^{62}Zn		61.93433	9.22 h	β⁺ /3/1.63 EC/93/	0.66/7.	0+			(0.15-3.52) ann.rad./ 0.0408/25 0.5967/26.
^{63}Zn		62.933215	38.5 m	β⁺ /93/3.367 EC/7/	1.02/ 1.40/ 1.71/ 2.36/84.	3/2-	-0.28164	+0.29	(0.20-1.526)/ ann.rad./ 0.66962(5)/8.4 0.96206(5)/6.6 (0.24-3.1)
^{64}Zn	48.63(60)	63.929146				0+			
^{65}Zn		64.929245	243.8 d	β⁺ /98/1.3514 EC/1.5/	0.325/	5/2-	+0.7690	-0.023	ann.rad./ 1.11552(2)/50.8
^{66}Zn	27.90(27)	65.926036				0+			
^{67}Zn	4.10(13)	66.927131				5/2-	+0.8755	+0.15	
^{68}Zn	18.75(51)	67.924847				0+			
^{69m}Zn			13.76 h	I.T./99+/0.439		9/2+			0.4390(2)/95.
^{69}Zn		68.926553	56. m	β⁻ /0.906	0.905/99.9	1/2-			0.318/
^{70}Zn	0.62(3)	69.925325				0+			
^{71m}Zn			3.97 h	β⁻ /	1.45/	9/2+			0.3864/93. 0.4874/62. 0.6203/57. (0.099-2.489) 0.5116(1)/30. 0.9103(1)/7.5
^{71}Zn		70.92773	2.4 m	β⁻ /2.81		1/2-			(0.12-2.29) 0.0164(3)/8. 0.1447(1)/83.
^{72}Zn		71.92686	46.5 h	β⁻ /0.46	0.25/14. 0.30/86.	0+			0.1915(2)/9.4
^{73m}Zn			6. s	I.T./0.196		(7/2+)			0.042
^{73}Zn		72.92978	24. s	β⁻ /4.29	4.7/	(1/2-)			0.216(1)/100.
^{74}Zn		73.92946	1.60 m	β⁻ /2.3	2.1/				0.496-0.911 0.0565/ 0.1401/
^{75}Zn		74.9329	10.2 s	β⁻ /6.0					(0.05-0.35) 0.229/
^{76}Zn		75.9334	5.7 s	β⁻ /4.2	3.6/				0.119/
^{77m}Zn			1.0 s	β⁻ /		(1/2-)			0.772
^{77}Zn		76.9371	2.1 s	β⁻ /7.3	4.8/				0.189/
^{78}Zn		77.9386	1.5 s	β⁻ /6.4					0.225/
^{79}Zn		78.9421	1.0 s	β⁻ /8.6					0.702/
^{80}Zn		79.9444	0.54 s	β⁻ /7.3					0.713/
^{81}Zn		80.9505	0.29 s	β⁻ /11.9					0.2248/
^{82}Zn		81.9548							
^{83}Zn									
$_{31}$Ga		69.723(1)							
^{60}Ga		59.9571							
^{61}Ga		60.9492	0.15 s	β⁺ /9.0					
^{62}Ga		61.94418	0.116 s	β⁺ /9.17 EC/	8.3/	0+			ann.rad./
^{63}Ga		62.9391	32. s	β⁺ /5.5 EC/	4.5/				ann.rad./ 0.6271(2)/10. 0.6370(2)/11. 1.0652(4)/45.
^{64}Ga		63.936838	2.63 m	β⁺ /7.165	2.79/ 6.05/	0+			ann.rad./ 0.80785(1)/14. 0.99152(1)/43. 1.38727(1)/12. 3.3659(1)/13.

Elem. or Isot.	Natural Abundance (%)	Atomic Mass or Weight	Half-Life	Decay Mode/Energy (/MeV)	Particle Energy /Intensity (MeV / %)	Spin (h/2π)	Nuclear Magnetic Mom.(nm)	Elect. Quadr. Mom. (b)	γ-ray Energy /Intensity (MeV / %)
⁶⁵Ga		64.9394	15.2 m	β⁺ /86/3.255 EC/	0.82/10. 1.39/19. 2.113/56. 2.237/15.	3/2-			ann.rad./ 0.1151(2)/55. 0.1530(2)/96. 0.2069(2)/39. (0.06-2.4)
⁶⁶Ga		65.931592	9.5 h	β⁺ /56/5.175 EC/43/	0.74/1. 1.84/54. 4.153/51.	0+			ann.rad./ 1.03935(8)/38. 2.7523(1)/23. (0.28-5.01)
⁶⁷Ga		66.928205	3.260 d	EC/1.001		3/2-	+1.8507	0.20	0.09332/37. 0.18459/20. 0.30024/17. (0.091-0.89)
⁶⁸Ga		67.927983	1.130 h	β⁺ /90/2.921 EC/10/	1.83/	1+	0.01175	0.028	ann.rad./ 1.0774(1)/3. (0.57-2.33)/
⁶⁹Ga	60.108(9)	68.925581				3/2-	+2.01659	+0.17	
⁷⁰Ga		69.926027	21.1 m	EC/0.2/0.655 β⁻ /99.8/1.656	1.65/99.	1+			0.1755(5)/0.15 1.042(5)/0.48
⁷¹Ga	39.892(9)	70.924707				3/2-	+2.56227	+0.11	
⁷²Ga		71.926372	14.10 h	β⁻ /4.001	0.64/40. 1.51/9. 2.52/8. 3.15/11.	3-	-0.13224	+0.5	0.62986(5)/24. 2.2016(2)/26. 2.5077(2)/12.8 (0.11-3.3)/
⁷³Ga		72.92517	74.87 h	β⁻ /1.59		3/2-			0.05344(5)/10. 0.29732(5)/47. (0.01-1.00)/
⁷⁴ᵐGa			10. s	I.T./		1+			0.0565(1)/75.
⁷⁴Ga		73.92694	8.1 m	β⁻ /5.4	2.6/	3-			0.5959/92. 2.354/45. (0.23-3.99)
⁷⁵Ga		74.92650	2.10 m	β⁻ /3.39	3.3/	3/2-			0.2529/ 0.5746/ (0.12-2.10)
⁷⁶Ga		75.9289	29. s	β⁻ /7.0		3-			0.5629/66. 0.5455/26. (0.34-4.25)
⁷⁷Ga		76.9293	13.0 s	β⁻ /5.3	5.2/				0.469/ 0.459/
⁷⁸Ga		77.9317	5.09 s	β⁻ /8.2		3+			0.619/77. 1.187/20.
⁷⁹Ga		78.9329	2.85 s	β⁻ /7.0	4.6/				0.465/
⁸⁰Ga		79.9366	1.68 s	β⁻ /10.4	10./				0.659/
⁸¹Ga		80.9377	1.22 s	β⁻ /8.3	5.1/				0.217/
⁸²Ga		81.9432	0.599 s	β⁻ /12.6					1.348/
⁸³Ga		82.9469	0.308 s	β⁻ /≈ 11.5					
⁸⁴Ga		83.952	≈0.085 s	β⁻ /14					
⁸⁵Ga									
⁸⁶Ga									
₃₂Ge		72.61(2)							
⁶¹Ge		60.9638	0.04 s	β⁺ /13.6					
⁶³Ge		62.9496	0.10 s	β⁻ /9.8					
⁶⁴Ge		63.9416	1.06 m	β⁺ /4.4 EC/	3.0/	0+			ann.rad./ 0.1282(2)/11. 0.4270(3)/37. 0.6671(3)/17.
⁶⁵Ge		64.9394	31. s	β⁺ /6.2 EC/	0.82/10. 1.39/19. 2.113/56. 2.237/15.				ann.rad./ 0.0620/27. 0.6497/33. 0.8091/21. (0.19-3.28)
⁶⁶Ge		65.93385	2.26 h	β⁺ /27/2.10 EC/73/		0+			ann.rad./ 0.0438/29. 0.3819/28. (0.022-1.77)
⁶⁷Ge		66.932738	19.0 m	β⁺ /96/4.225 EC/4/	1.6/ 2.3/ 3.15/	1/2-			ann.rad./ 0.1670/84. (0.25-3.73)
⁶⁸Ge		67.92810	270.8 d	EC/0.11		0+			Ga k x-ray/39.
⁶⁹Ge		68.927973	1.63 d	β⁺ /36/2.2273 EC/64/	0.70/ 1.2/	5/2-	0.735	0.02	ann.rad./ 0.574/13. 1.1068/36. (0.2-2.04)
⁷⁰Ge	20.84(87)	69.924250				0+			
⁷¹ᵐGe			20.4 ms	I.T./0.0234		9/2+			0.1749
⁷¹Ge		70.924954	11.2 d	EC/0.229		1/2-	+0.547		
⁷²Ge	27.54(34)	71.922076				0+			

Elem. or Isot.	Natural Abundance (%)	Atomic Mass or Weight	Half-Life	Decay Mode/Energy (/MeV)	Particle Energy /Intensity (MeV / %)	Spin (h/2π)	Nuclear Magnetic Mom.(nm)	Elect. Quadr. Mom. (b)	γ-ray Energy /Intensity (MeV / %)
73Ge	7.73(5)	72.923460				9/2+	-0.879467	-0.17	
74Ge	36.28(73)	73.921178				0+			
75mGe			48. s	I.T./		7/2+			0.13968(3)/39.
75Ge		74.922860	1.380 h	β⁻ /1.177	1.19/	1/2-	+0.510		0.26461(5)/11.
									0.41931(5)/0.2
76Ge	7.61(38)	75.921403				0+			
77mGe			53. s	I.T./20/		1/2-			0.1597(1)/11.
				β⁻ /80/2.861	2.9/				0.21551/21.
									(0.19-0.61)
77Ge		76.923549	11.25 h	β⁻ /2.702	0.71/23.	7/2+			0.2110/29.
					1.38/35.				0.2155/27.
					2.19/42.				0.2644/51.
									(0.15-2.35)
78Ge		77.922853	1.45 h	β⁻ /0.95	0.70/	0+			0.2773(5)/96.
									0.2939(5)/4.
79mGe			39. s	β⁻ /IT		7/2+			
79Ge		78.9254	19.1 s	β⁻ /4.2	4.0/20.	1/2-			0.1096/21.
					4.3/80.				(0.10-2.59)
80Ge		79.92545	29.5 s	β⁻ /2.67	2.4/	0+			0.5427(4)/15.
									0.1104(4)/6.
81mGe			≈7.6 s	β⁻ /	3.75/	1/2+			0.2656(4)/25.
									0.3362(4)/
81Ge		80.9288	≈7.6 s	β⁻ /6.2	3.44/	9/2+			0.7935(4)/
									0.1976(4)/21.
									0.3362(4)/100.
82Ge		81.9296	4.6 s	β⁻ /4.7		0+			1.093/
83Ge		82.9345	1.9 s	β⁻ /8.9					
84Ge		83.9373	0.98 s	β⁻ /7.7					
85Ge		84.943	0.54 s	β⁻ /10.					
86Ge		85.946							
87Ge									
88Ge									
89Ge									
33As		74.92160(2)							
64As		63.9576							
65As		64.9495	0.19 s	β⁺ /9.4					
66As		65.94410	95.8 ms	β⁺ /9.55					
67As		66.9392	42. s	β⁺ /6.0	5.0/	5/2-			0.121/
				EC/					0.123/
68As		67.9368	2.53 m	β⁺ /8.1		3+			0.244/
									ann.rad./
									0.652/32.
									0.762/33.
									1.016/77.
									(0.61-3.55)
69As		68.93228	15.2 m	β⁺ /98/4.01	2.95/	5/2-	1.2		ann.rad./
				EC/2/					0.0868(5)/1.5
70As		69.93093	52.6 m	β⁺ /84/6.22	1.44/	4+	+2.1061	+0.09	0.1458(3)/2.4
				EC/16/2.14					ann.rad./
				/2.89					1.0395(7)/82.
71As		70.927114	2.72 d	β⁺ /32/2.013		5/2-	+1.6735	-0.02	(0.17-4.4)/
				EC/68/					ann.rad./
									0.1749(2)/84.
72As		71.926753	26.0 h	β⁺ /77/4.356	0.669/5.	2-	-2.1566	-0.08	1.0957(2)/4.2
					1.884/12.				ann.rad./
					2.498/62.				0.83395(5)/80.
					3.339/19.				1.0507(1)/9.6
73As		72.923825	80.3 d	EC/0.341		3/2-			(0.1-4.0)
									0.013263(1)/0.1
									0.053437(1)/10.5
									Se k x-ray/90.
74As		73.923829	17.78 d	β⁺ /31/2.562	0.94/26.	2-	-1.597		ann.rad./
				EC/37/	1.53/3.				0.59588(1)/60.
				β⁻ /1.353	0.71/16.				0.6084(1)/0.6
					1.35/16.				0.6348(1)/15.
75As	100.	74.921597				3/2-	+1.43947	+0.31	0.5591(1)/45.
76As		75.922394	26.3 h	β⁻ /2.962	0.54/3.	2-	-0.906		0.65703(5)/6.2
					1.785/8.				1.21602(1)/3.4
					2.410/36.				(0.3-2.67)
					2.97/51.				0.2391(2)/1.6
77As		76.920648	38.8 h	β⁻ /0.683	0.70/98.	3/2-			0.2500(3)/0.4
									0.52078(1)/0.43
78As		77.92183	1.512 h	β⁻ /4.21	3.00/12.	2-			0.6136(3)/54.
					3.70/17.				0.6954(3)/18.
					4.42/37.				1.3088(3)/10.

Elem. or Isot.	Natural Abundance (%)	Atomic Mass or Weight	Half-Life	Decay Mode/Energy (/MeV)	Particle Energy /Intensity (MeV / %)	Spin (h/2π)	Nuclear Magnetic Mom.(nm)	Elect. Quadr. Mom. (b)	γ-ray Energy /Intensity (MeV / %)
⁷⁹As		78.92095	9.0 m	β⁻ /2.28	1.80/95.	3/2-			0.0955(5)/16.
									0.3645(5)/1.9
⁸⁰As		79.92258	16. s	β⁻ /5.64	3.38/	1+			0.6662(2)/42.
									(2.5-3.0)
⁸¹As		80.92213	33. s	β⁻ /3.856		3/2-			0.4676(2)/20.
									0.4911(2)/8.
⁸²ᵐAs			13.7 s	β⁻ /	3.6/	5-			0.6544(1)/72.
									0.8186(4)/27.
									1.7313(2)/27.
									1.8954(2)/38.
⁸²As		81.9246	19. s	β⁻ /7.4	7.2/80.	1+			0.6544(1)/15.
⁸³As		82.9250	13.4 s	β⁻ /5.5					0.7345/100.
									1.1131/34.
									2.0767/28.
⁸⁴ᵐAs			0.6 s	β⁻					
⁸⁴As		83.9291	4. s	β⁻, n/7.2		1-			0.6671(2)/21.
									1.4439(5)/49.
									(0.325-5.150)
⁸⁵As		84.9318	2.03 s	β⁻, n/8.9		3/2-			0.667(1)/42.
									1.4551(2)/100.
⁸⁶As		85.9362	0.95 s	β⁻, n/11.4					0.704/
⁸⁷As		86.9396	0.49 s	β⁻, n/10.					0.704/
⁸⁸As		87.945							
⁸⁹As		88.949							
⁹⁰As									
⁹¹As									
⁹²As									
₃₄Se		78.96(3)							
⁶⁵Se		64.965	0.011 s	β⁺ /60/14.					
				β⁺, p	3.55/				
⁶⁶Se		65.9552							
⁶⁷Se		66.9501	0.06 s	β⁺ /10.2					ann.rad./
				β⁺, (p)/					0.352
⁶⁸Se		67.9419	36. s	β⁺ /4.7					ann.rad./
									(0.050-0.426)
⁶⁹Se		68.93956	27.4 s	β⁺ /6.78	5.006/				ann.rad./
				EC/					0.0664(4)/27.
									0.0982(4)/63.
⁷⁰Se		69.9335	41.1 m	β⁺ /2.4		0+			ann.rad
									0.04951(5)/35.
									0.4262(2)/29.
⁷¹Se		70.9319	4.7 m	β⁺ /4.4	3.4/36.	5/2-			ann.rad
				EC/					0.1472(3)/47.
									0.8309(3)/13.
									1.0960(3)/10.
⁷²Se		71.92711	8.5 d	EC/0.34		0+			0.0460(2)/57.
⁷³ᵐSe			40. m	I.T./73/0.0257	0.85	3/2-			ann.rad.
				β⁺ /27/2.77	1.45/				0.0257(2)/27.
					1.70/				0.2538(1)/2.5
⁷³Se		72.92678	7.1 h	β⁺ /65/2.74	0.80/	9/2+			ann.rad
				EC/35/	1.32/95.				0.0670(1)/72.
					1.68/1.				0.3609(1)/97.
									(0.6-1.5)
⁷⁴Se	0.89(4)	73.922477				0+			
⁷⁵Se		74.922524	119.78 d	EC/0.864		5/2+	0.67	1.0	0.136000(1)/55.
									0.264651(1)/58.
									(0.024-0.821)
⁷⁶Se	9.37(29)	75.919214				0+			
⁷⁷ᵐSe			17.4 s	I.T./		7/2+			0.1619(2)/52.
⁷⁷Se	7.63(16)	76.919915				1/2-	+0.53506		
⁷⁸Se	23.77(28)	77.917310				0+			
⁷⁹ᵐSe			3.92 m	I.T./					0.09573(3)/9.5
⁷⁹Se		78.918500	1.1x10⁶ y	β⁻/0.151		7/2+	-1.018	+0.8	
⁸⁰Se	49.61(41)	79.916522				0+			
⁸¹ᵐSe			57.3 m	I.T./99/0.1031		7/2+			0.1031(3)/9.7
									0.2602(2)/0.06
									0.27599(1)/0.06
⁸¹Se		80.917993	18.5 m	β⁻/1.585	1.6/98.	1/2-			0.27594(5)/0.85
									0.29008(5)/0.75
									0.82827(5)/0.32
⁸²Se	8.73(22)	81.916700				0+			
⁸³ᵐSe			1.17 m	β⁻ /3.96	2.88/	1/2-			0.35666(6)/17.
					3.92/				0.9879(1)/15.
									1.0305(1)/21.
									2.0514(2)/11.
									(0.19-3.1)

Elem. or Isot.	Natural Abundance (%)	Atomic Mass or Weight	Half-Life	Decay Mode/Energy (/MeV)	Particle Energy /Intensity (MeV / %)	Spin (h/2π)	Nuclear Magnetic Mom.(nm)	Elect. Quadr. Mom. (b)	γ-ray Energy /Intensity (MeV / %)
^{83}Se		82.919119	22.3 m	β^- /3.668	0.93/ 1.51/	9/2+			0.22516(6)/33. 0.35666(6)/69. 0.51004(8)/45. (0.21-2.42)
^{84}Se		83.91847	3.3 m	β^- /1.83	1.41/100.	0+			0.4088(5)/100.
^{85}Se		84.92225	32. s	β^- /6.18	5.9/	5/2+			0.3450(1)/22. 0.6094(1)/41.
^{86}Se		85.92428	15. s	β^- /5.10		5/2+			2.0124(1)/24. 2.4433(8)/100. 2.6619(1)/49.
^{87}Se		86.92853	5.4 s	β^- /7.28 n/					0.468(1)/100. 1.4979(1)/23.
^{88}Se		87.93143	1.5 s	β^- , n/6.85					0.5346/
^{89}Se		88.9360	0.41s	β^- , n/9.0					
^{90}Se		89.9394							
^{91}Se		90.945	0.27 s	β^- , n/8.					
^{92}Se		91.949							
^{93}Se									
^{94}Se									
$_{35}$Br		79.904(1)							
^{69}Br		68.9502		β^+ /9.6					
^{70}Br		69.9446	79. ms	β^+ /10.0					
^{71}Br		70.9392	21. s	β^+ /6.9					
^{72}Br		71.9365	1.31 m	β^+ /8.7		3			
^{73}Br		72.9318	3.4 m	β^+ /4.7	3.7/	3/2-	≈0.55		0.4547-1.3167 ann.rad
^{74m}Br			46. m	β^+ /	4.5/	4-	1.8		(0.0649-0.6995) ann.rad 0.6348 0.7285
^{74}Br		73.92989	25.4 m	β^+ /6.91					(0.2 - 4.38) ann.rad 0.6341 0.6348
^{75}Br		74.92578	1.62	β^+ /76/3.03		3/2-	0.75(11)		(0.2-4.7) ann.rad 0.28650
^{76m}Br			1.4 s	I.T./5.05		4+			(0.1-1.56) 0.104548
^{76}Br		75.92454	16.0 h	β^+ /57 /4.96	1.9/ 3.68/	1-	0.54821	0.270	0.05711 ann.rad 0.55911 1.85368
^{77m}Br			4.3 m	I.T./0.1059		9/2+	0.97		(0.4-4.6) 0.1059
^{77}Br		76.921380	2.376 d	EC/99 /1.365		3/2-			ann.rad. 0.23898 0.52069
^{78}Br		77.921146	6.45 m	β^+ /92/3.574 EC/8 /	1.2/ 2.5/	1+	0.1		(0.08-1.2) ann.rad 0.61363
^{79m}Br			4.86 s	I.T./0.207		9/2+	+2.106		(0.7-3.0) 0.2072
^{79}Br	50.69(7)	78.918338				3/2-	+2.106400	+0.331	
^{80m}Br			4.42 h	I.T./0.04885		5-	+1.3177		Br k x-ray 0.03705/39.1 0.04885/0.3
^{80}Br		79.918530	17.66 m	β^- /92 /2.004 EC/5.7/1.8706 β^+ /2.6/	1.38 β^- /7.6 1.99 β^- /82 0.85 β^+ /2.8	1+	0.5140	0.196	ann.rad. 0.6169/6.7 (0.64-1.45)
^{81}Br	49.31(7)	80.916291				3/2-	+2.270562	+0.276	
^{82m}Br			6.1 m	I.T./98/0.046 β^- /2 /3.139		2-			0.046/0.24 (0.62-2.66)
^{82}Br		81.916805	1.471 d	β^- /3.093	0.444/	5-	+1.6270	0.751	0.5544/71 0.61905/43 0.77649/84
^{83}Br		82.915181	2.40 h	β^- /0.972	,0.395/1 0.925/99	3/2-			(0.013-1.96) 0.52964
^{84m}Br			6.0 m	β^- /4.97	2.2/100	(6-)			(0.12-0.68) 0.4240/100 0.8817/98
^{84}Br		83.91651	31.8 m	β^- /4.65	2.70/11 3.81/20 4.63/34	2-	2.		1.4637/101 0.8816/41 1.8976/13
^{85}Br		84.91561	2.87 m	β^- /2.87	2.57	3/2-			(0.23-4.12) 0.80241/2.56 0.92463/1.6 (0.09-2.4)

Elem. or Isot.	Natural Abundance (%)	Atomic Mass or Weight	Half-Life	Decay Mode/Energy (/MeV)	Particle Energy /Intensity (MeV / %)	Spin (h/2π)	Nuclear Magnetic Mom.(nm)	Elect. Quadr. Mom. (b)	γ-ray Energy /Intensity (MeV / %)
⁸⁶Br		85.91880	55.5 s	β⁻ /7.63	3.3 7.4	(2-)			1.56460/64 2.75106/21 (0.5-6.8)
⁸⁷Br		86.92072	55.6 s	β⁻ /6.85 n/	6.1/	3/2-			1.41983 1.4762 (0.2-6.1)
⁸⁸Br		87.92407	16.3 s	β⁻ /8.96 n/		1-			0.7649 0.7753 0.8021 (0.1-6.99)
⁸⁹Br		88.92640	4.35 s	β⁻ /8.16 n/		3/2-			0.7753 1.0978
⁹⁰Br		89.9306	1.91 s	β⁻ /10.4 n/	8.3/ 9.8/	2-			0.6555 0.7071 1.3626
⁹¹Br		90.9339	0.54 s	β⁻ /90 /9.80 β⁻ n/10 /					0.263 0.803
⁹²Br		91.9392	0.31 s	β⁻ /12.20 β⁻ n/					0.740
⁹³Br		92.9431	0.10 s	β⁻ n/11.1					
⁹⁴Br		93.9487	0.07 s	β⁻ n/					
⁹⁵Br									
⁹⁶Br									
⁹⁷Br									
₃₆Kr		83.80(1)							
⁶⁹Kr		68.9653	0.03 s	β⁺,(p)	4.07/				
⁷⁰Kr		69.9560							
⁷¹Kr		70.9505	100. ms	β⁺,EC/10.1					(0.198-0.207)
⁷²Kr		71.9419	17. s	β⁺ /5.0 EC/		0+			ann.rad. 0.3100/29 0.4150/36 (0.12-0.58)
⁷³Kr		72.9389	27. s	β⁺ /6.7 EC/ β⁺,p/		5/2-			ann.rad. 0.1781/66 (0.06-0.86)
⁷⁴Kr		73.9333	11.5 m	β⁺ /3.1 EC/		0+			ann.rad. 0.08970/31 0.2030/20 (0.010-1.06)
⁷⁵Kr		74.93104	4.3 m	β⁺ /4.90 EC/	3.2/		- 0.531	+ 1.1	ann.rad. 0.1325/68 0.1547/21 (0.02-1.7)
⁷⁶Kr		75.92595	14.8 h	EC/1.31		0+			Br k x-ray 0.270/21 0.3158/39 (0.03-1.07)
⁷⁷Kr		76.92467	1.24 h	β⁺ /80 /3.06 EC/20 /	1.55/ 1.70/ 1.87/	5/2+	- 0.583	+ 0.9	ann.rad. 0.1297/80 0.1465/38 (0.02-2.3)
⁷⁸Kr	0.35(1)	77.92039				0+			
⁷⁹ᵐKr			53. s	I.T./0.1299		7/2+	- 0.786	+ 0.40	Kr x-ray
⁷⁹Kr		78.920083	1.455 d	β⁺ /7 /1.626 EC/93 /		1/2-	+ 0.536		ann.rad. 0.2613/13 0.39756/19 0.6061/8 (0.04-1.3)
⁸⁰Kr	2.28(6)	79.916379				0+			
⁸¹ᵐKr			13.1 s	I.T./0.1904		1/2-	+ 0.586		0.1904
⁸¹Kr		80.916593	2.1x10⁵ y	EC/0.2807		7/2+	- 0.908	+ 0.64	Br k x-ray 0.2760
⁸²Kr	11.58(14)	81.913485				0+			Kr k x-ray
⁸³ᵐKr			1.86 h	I.T./0.0416		1/2-	+ 0.591		0.00940 0.03216
⁸³Kr	11.49(6)	82.914137				9/2+	-0.970699	+0.253	
⁸⁴Kr	57.00(4)	83.911508				0+			
⁸⁵ᵐKr			4.48 h	β⁻ /79 / I.T./21 /0.305	0.83/79	1/2-	+ 0.633		0.30487 0.15118
⁸⁵Kr		84.912530	10.73 y	β⁻ /0.687	0.15/0.4	9/2+	1.005	+0.43	0.51399
⁸⁶Kr	17.30(22)	85.910615				0+			
⁸⁷Kr		86.913359	1.27 h	β⁻ /3.887	1.33/8 3.49/43 3.89/30	5/2+	-1.023	- 0.30	0.40258/49.6 2.5548/9.2 (0.13-3.31)

Elem. or Isot.	Natural Abundance (%)	Atomic Mass or Weight	Half-Life	Decay Mode/Energy (/MeV)	Particle Energy /Intensity (MeV / %)	Spin (h/2π)	Nuclear Magnetic Mom.(nm)	Elect. Quadr. Mom. (b)	γ-ray Energy /Intensity (MeV / %)
^{88}Kr		87.91445	2.84 h	β⁻ /2.91		0+			0.19632/26. 2.392/34.6 (0.03-2.8)
^{89}Kr		88.91764	3.15 m	β⁻ /4.99	3.8/ 4.6/ 4.9/	5/2 +	- 0.330	+ 0.16	0.19746 0.2209/19.9 0.5858/16.4 1.4728/6.8 (0.2-4.7)
^{90}Kr		89.91953	32.3 s	β⁻ /4.39	2.6/77 2.8/6	0+			0.12182/32.9 0.5395/28.6 1.1187/36.2 (0.1 - 4.2)
^{91}Kr		90.9234	8.6 s	β⁻ /6.4	4.33/ 4.59/	5/2 +	- 0.583	+ 0.30	0.10878/43.5 0.50658/19. (0.2-4.4)
^{92}Kr		91.92611	1.84 s	β⁻ /5.99 n/					0.1424/66. (0.14 - 3.7)
^{93}Kr		92.9312	1.29 s	β⁻ /8.6 n/	7.1/	1/2 +	- 0.413		0.1820 0.2534/42. 0.32309/24.6 (0.057-4.03)
^{94}Kr		93.9343	0.21 s	β⁻ /7.3					0.2196/67 0.6293/100. (0.098-0.985)
^{95}Kr		94.9397	0.78 s	β⁻ /9.7			- 0.410		
^{96}Kr		95.9431							
^{97}Kr		96.9486	< 0.1 s	β⁻					
^{98}Kr									
^{99}Kr									
^{100}Kr									
$_{37}$Rb		85.4678(3)							
^{74}Rb		73.9445	65. ms	β⁺ /10.4					
^{75}Rb		74.93857	19. s	β⁺ /7.02	2.31/				ann. rad. 0.179
^{76}Rb		75.93508	39. s	β⁺ /8.50	4.7/	1-	-0.372623	+0.4	ann.rad. 0.4240/92. (0.064-1.68)
^{77}Rb		76.93041	3.8 m	β⁺ /5.34	3.86/	3/2-	+0.654468	+0.70	ann.rad. 0.0665/59 (0.04 - 2.82)
^{78m}Rb			5.7 m	I.T./0.1034 β⁺ / EC/	3.4	4-			ann.rad. 0.4553/81. (0.103-4.01)
^{78}Rb		77.92814	17.7 m	β⁺ /7.22 EC/		0+			ann.rad. 0.4553/63. (0.42-5.57)
^{79}Rb		78.92400	23. m	β⁺ /84 /3.65 EC/16 /		5/2 +	+ 0.3358	-0.10	ann.rad. 0.68812/23. (0.017-3.02)
^{80}Rb		79.92252	34. s	β⁺ /5.72	4.1/22 4.7/74	1+	-0.0836	+0.35	ann.rad. 0.6167/25.
^{81m}Rb			30.5 m	I.T./0.85 β⁺,EC/	1.4	9/2 +			ann. rad. (0.085-1.9)
^{81}Rb		80.91900	4.57 h	β⁺/27 /2.24 EC/73	1.05/	3/2-	+2.060	+0.40	ann.rad./ 0.19030/64. (0.05 - 1.9)
^{82m}Rb			6.47 h	β⁺/26 / EC/74 /	0.80/	5-	+1.5		ann.rad./ 0.5544/63. 0.7765/85. (0.092 - 2.3)
^{82}Rb		81.91821	1.258 m	β⁺/96 /4.40 EC/4 /	3.3/	1+	+0.554508	+0.19	ann.rad./ 0.7665/13. (0.47 - 3.96)
^{83}Rb		82.91511	86.2 d	EC/0.91		5/2-	+1.425	+0.20	Kr x-ray 0.5205/46. (0.03-0.80)
^{84m}Rb			20.3 m	I.T./0.216		6-			0.2163/34. 0.2482/63. 0.4645/32.
^{84}Rb		83.914387	32.9 d	β⁺/22 /2.681 EC/75 / β⁻/3 /0.894	0.780/11 1.658/11 0.893/	2-	-1.32412	-0.015	ann.rad./ 0.8817/68. (1.02-1.9)
^{85}Rb	72.17(2)	84.911792				5/2-	+1.353	+0.23	
^{86m}Rb			1.018 m	I.T./0.5560		6-			0.556/98.
^{86}Rb		85.911170	18.65 d	β⁻/1.775	1.774/8.8	2-	-1.6920		1.0768/8.8
^{87}Rb	27.83(2)	86.909186	4.88x10^{10}y	β⁻/0.283	0.273/100	3/2-	+2.7512		

Elem. or Isot.	Natural Abundance (%)	Atomic Mass or Weight	Half-Life	Decay Mode/Energy (/MeV)	Particle Energy /Intensity (MeV / %)	Spin (h/2π)	Nuclear Magnetic Mom.(nm)	Elect. Quadr. Mom. (b)	γ-ray Energy /Intensity (MeV / %)
⁸⁸Rb		87.911323	17.7 m	β⁻/5.316	5.31	2-	0.508		0.8980/14. 1.8360/21. (0.34-4.85)
⁸⁹Rb		88.91229	15.4 m	β⁻/4.50	1.26/38 1.9/5 2.2/34 4.49/18	3/2-	+2.304	+0.14	1.032/58. 1.248/42. 2.1960/13 (0.12-4.09)
⁹⁰ᵐRb			4.3 m	β⁻/4.50	1.7/ 6.5/	4-			0.1069(IT) 0.8317/94 (0.20-5.00)
⁹⁰Rb		89.91481	2.6 m	β⁻/6.59	6.6	1-			0.8317/28. (0.31-5.60)
⁹¹Rb		90.91649	58.0 s	β⁻/5.861	5.9	3/2-	+2.182	+0.15	0.0936/34. (0.35-4.70)
⁹²Rb		91.91968	4.48 s	β⁻/8.11	8.1/94	1-			0.8148/8. (0.1-6.1)
⁹³Rb		92.92195	5.85 s	β⁻/7.46 n/1	7.4/	5/2	+1.410	+0.18	0.2134/4.8 0.4326/12.5 0.9861/4.9 (0.16-5.41)
⁹⁴Rb		93.92643	2.71 s	β⁻/10.31 n/10	9.5/	3	+1.498	+0.16	0.8369/87. 1.5775/32. (0.12-6.35)
⁹⁵Rb		94.92929	0.377 s	β⁻/9.30 n/8	8.6/	5/2	+1.334	+0.21	0.352/65. 0.680/22. (0.20-2.27)
⁹⁶Rb		95.93427	0.199 s	β⁻/11.76 n/13/	10.8/	2+	+1.466	+0.25	0.815/76. (0.20-5.42)
⁹⁷Rb		96.93733	0.169 s	β⁻/10.42 n/27/	10.0	3/2	+1.841	+0.58	0.167/100. 0.585/79. 0.599/56. 1.258/52. (0.14-2.08)
⁹⁸Rb		97.94174	0.107 s	β⁻/12.34 n/13	0.144/				(0.07-3.68)
⁹⁹Rb		98.9453	59. ms	β⁻/11.3					
¹⁰⁰Rb		99.9499	53. ms	β⁻ /13.5					
¹⁰¹Rb		100.9532	0.03 s	β⁻ /11.8					
¹⁰²Rb		101.9592	0.09 s	β⁻					
₃₈Sr		87.62(1)							
⁷³Sr		72.966							
⁷⁴Sr		73.9563							
⁷⁵Sr		74.9499	≈ 0.07 s						
⁷⁶Sr		75.9416	8.9 s	β⁺ /6.1					
⁷⁷Sr		76.9378	9.0 s	β⁺ /6.9	5.6		-0.35	1.4	0.147 (0.047-0.793)
⁷⁸Sr		77.93218	2.7 m	β⁺ /3.76					
⁷⁹Sr		78.92971	2.1 m	β⁺ /5.32	4.1	3/2-	-0.474	+0.74	ann.rad./ 0.039/28. 0.105/22. (0.135-0.612)
⁸⁰Sr		79.92453	1.77 h	β⁺ /1.87		0+			ann.rad./ 0.174/10. 0.589/39. (0.24-0.55)
⁸¹Sr		80.92322	22.3 m	β⁺ /87 /3.93 EC/13 /	2.43/ 2.68/	1/2-	+0.544		ann.rad./ 0.148/31. 0.1534/35 (0.06-1.7)
⁸²Sr		81.91840	25.36 d	EC/0.18					Rb x-ray
⁸³ᵐSr			5.0 s	I.T./0.2591		1/2-			0.2591/87.5
⁸³Sr		82.91756	1.350 d	β⁺/24 /2.28 EC/76/	0.465/ 0.803/ 1.227/	7/2+	-0.898	+0.8	ann.rad./ 0.3816/12. 0.3816 0.7627/30. (0.094-2.15)
⁸⁴Sr	0.56(1)	83.913426				0+			
⁸⁵ᵐSr			1.127 h	I.T./87 /0.2387 EC/13		1/2-			0.2318/84. (0.15-0.24)
⁸⁵Sr		84.912936	64.85 d	EC/1.065		9/2+	-1.001	+0.32	0.51399/99.3
⁸⁶Sr	9.86(1)	85.909265				0+			
⁸⁷ᵐSr			2.81 h	I.T./0.3884		1/2-			0.3884(IT)
⁸⁷Sr	7.00(1)	86.908882				9/2+	-1.09360	+0.34	
⁸⁸Sr	82.58(1)	87.905617				0+			
⁸⁹Sr		88.907455	50.52 d	β⁻/1.497	1.492/100	5/2+	-1.149	-0.3	0.9092
⁹⁰Sr		89.907738	29.1 y	β⁻/0.546	0.546/100	0+			

Elem. or Isot.	Natural Abundance (%)	Atomic Mass or Weight	Half-Life	Decay Mode/Energy (/MeV)	Particle Energy /Intensity (MeV / %)	Spin (h/2π)	Nuclear Magnetic Mom.(nm)	Elect. Quadr. Mom. (b)	γ-ray Energy /Intensity (MeV / %)
⁹¹Sr		90.91020	9.5 h	β⁻/2.70	0.61/7 1.09/33 1.36/29 2.66/26	5/2+	-0.887	+0.044	0.5556/61. 0.7498/24. 1.0243/33. (0.12-2.4)
⁹²Sr		91.91098	2.71 h	β⁻/1.91	0.55/96 1.5/3	0+			1.3831/90. (0.24-1.1)
⁹³Sr		92.91394	7.4 m	β⁻/4.08	2.2/10 2.6/25 3.2/65	5/2+	-0.794	+0.27	0.5903/ 0.7104 0.87573 0.8883/ (0.17-3.97)
⁹⁴Sr		93.91537	1.25 m	β⁻/3.511	2.1/ 3.3/	0+			0.6219 0.7043 0.7241 0.8064 1.4283
⁹⁵Sr		94.91931	25.1 s	β⁻/6.08	6.1/50	1/2+	-0.5379		0.6859 0.8269 2.7173 2.9332
⁹⁶Sr		95.92165	1.06 s	β⁻/5.37	4.2/	0+			0.1222 0.5305 0.8094 0.9318
⁹⁷Sr		96.92615	0.42 s	β⁻/7.47	5.3	(1/2+)	-0.500		0.2164 0.3071 0.6522 0.9538 1.2580 1.9050
⁹⁸Sr		97.92845	0.65 s	β⁻/5.83	5.1				0.0365 0.1190 0.4286 0.4447 0.5636
⁹⁹Sr		98.9333	0.27 s	β⁻/8.0					
¹⁰⁰Sr		99.9354	0.201 s	β⁻/7.1					
¹⁰¹Sr		100.9405	0.115 s	β⁻/9.5					
¹⁰²Sr		101.9430	68. ms	β⁻/8.8					
¹⁰³Sr		102.9490							
¹⁰⁴Sr		103.952							
¹⁰⁵Sr									
₃₉Y		88.90585(2)							
⁷⁸Y		77.9435		β⁺/10.5					
⁷⁹Y		78.9374	15. s	β⁺/7.1					
⁸⁰Y		79.9320	36. s	β⁺/7.0	5.5 5.0/	(4)			(0.152-1.106) ann.rad./ 0.3858 0.5951 1.1852
⁸¹Y		80.9291	1.21 m	β⁺/5.5	3.7/ 4.2/				ann.rad./ 0.428 0.469
⁸²Y		81.9268	9.5 s	β⁺/7.8	6.3/	1+			ann.rad./ 0.5736 0.6017 0.7375
⁸³ᵐY			2.85 m	β⁺/95/4.6 EC/5 /	2.9	1/2-			ann.rad./ 0.2591 0.4218 0.4945
⁸³Y		82.92235	7.1 m	β⁺/4.47 EC/	3.3	9/2+			ann.rad./ 0.0355 0.4899 0.8821 (0.03 - 3.4)
⁸⁴ᵐY			4.6 s	β⁺/ EC/		1+			ann.rad./ 0.7930
⁸⁴Y		83.9203	40. m	β⁺/6.4 EC/	1.64/47 2.24/25 2.64/21 3.15/7	5-			ann.rad./ 0.4628 0.6606 0.7931 0.9744 1.0398 (0.2 - 3.3)

Elem. or Isot.	Natural Abundance (%)	Atomic Mass or Weight	Half-Life	Decay Mode/Energy (/MeV)	Particle Energy /Intensity (MeV / %)	Spin (h/2π)	Nuclear Magnetic Mom.(nm)	Elect. Quadr. Mom. (b)	γ-ray Energy /Intensity (MeV / %)
⁸³ᵐY			4.9 h	β⁺/70 / EC/30 /		9/2+			ann.rad./ 0.2317 0.5356 0.7673 2.1238 (0.1 - 3.1)
⁸³Y		84.91643	2.6 h	β⁺/55 /3.26 EC/45 /	1.54/	1/2-			ann.rad./ 0.2317 0.5045 0.9140 (0.07 - 1.4)
⁸⁴ᵐY			48. m	I.T./99 / β⁺/ EC/		8+			ann.rad./ 0.0102(IT) 0.2080 (0.09 - 1.1)
⁸⁵Y		85.91489	14.74 h	β⁺/5.24 EC/		4-	<0.6		ann.rad./ 0.3070 0.6277 1.0766 1.1531 1.9207 (0.1 - 3.8)
⁸⁷ᵐY			13. h	I.T./98 / β⁺/0.7 / EC/	1.15/0.7	9/2+			0.3807
⁸⁷Y		86.910880	3.35 d	EC/99 +/1.862	0.78/	1/2-			0.3880 0.4870
⁸⁸Y		87.909506	106.6 d	EC/99 + /3.623 β⁺/0.2 /	0.76/	4-			ann.rad./ 0.89802 1.83601 2.73404 3.2190
⁸⁹ᵐY			15.7 s	I.T./0.909		9/2+			0.9092(IT)
⁸⁹Y	100.	88.905849				1/2-	-0.13742		
⁹⁰ᵐY			3.24 h	I.T./99 + /0.68204 β⁻/0.002/		7+			0.2025 0.4794 0.6820
⁹⁰Y		89.907152	2.67 d	β⁻/2.282	2.28/	2-	-1.630	-0.155	
⁹¹ᵐY			49.7 m	I.T./0.555		9/2+	6.0		0.5556(IT)
⁹¹Y		90.907301	58.5 d	β⁻/1.544	1.545/	1/2-	0.1641		1.208
⁹²Y		91.90893	3.54 h	β⁻/3.63	3.64/	2-			0.4485 0.5611 0.9345 1.4054 (0.4 - 3.3)
⁹³ᵐY			0.82 s	I.T./0.759		9/2+			0.1686(IT) 0.5902
⁹³Y		92.90956	10.2 h	β⁻/2.87	2.88/90	1/2-			0.2669 0.9471 1.9178
⁹⁴Y		93.91160	18.7 m	β⁻/4.919	4.92/	2-			0.3816 0.9188 1.1389 (0.3 - 4.1)
⁹⁵Y		94.91279	10.3 m	β⁻/4.42		1/2-			0.4324 0.9542 2.1760 3.5770
⁹⁶ᵐY			9.6 s	β⁻/		(3+)			0.1467 0.6174 0.9150 1.1071 1.7507
⁹⁶Y		95.91588	6.2 s	β⁻ /7.09	7.12/	0-			1.594
⁹⁷ᵐY			1.21 s	β⁻ /7.4	4.8/ 6.0/	9/2+			0.1614 0.9700 1.1030
⁹⁷Y		96.91813	3.76 s	β⁻ /6.69	6.7	1/2-			0.2969 1.9960 3.2876 3.4013
⁹⁸ᵐY			2.1 s	β⁻ /9.8	5.5/	(4-)			0.2415 0.6205 0.6473 1.2228 1.8016

Elem. or Isot.	Natural Abundance (%)	Atomic Mass or Weight	Half-Life	Decay Mode/Energy (/MeV)	Particle Energy /Intensity (MeV / %)	Spin (h/2π)	Nuclear Magnetic Mom.(nm)	Elect. Quadr. Mom. (b)	γ-ray Energy /Intensity (MeV / %)
⁹⁸Y		97.92224	0.59 s	β⁻ /8.83	8.7/	1+			0.2131 1.2228 1.5907 2.9413 4.4501
⁹⁹Y		98.92463	1.47 s	β⁻ /7.57 n	/2.5/	1/2-			0.1218/43.8 0.5362 0.7242 1.0130
¹⁰⁰ᵐY			0.94 s	β⁻,n /		3+			
¹⁰⁰Y		99.9278	0.73 s	β⁻,n /9.3	n/1.8/	1+			
¹⁰¹Y		100.9303	0.43 s	β⁻,n /8.6	n/1.5/	(5/2)			
¹⁰²Y		101.9336	0.36 s	β⁻,n /9.9	n/4.0/				
¹⁰³Y		102.9369	0.23 s	β⁻,n	n/8.3/				
¹⁰⁴Y		103.9414							
¹⁰⁵Y		104.9451							
¹⁰⁶Y		105.950							
¹⁰⁷Y									
¹⁰⁸Y									
₄₀Zr		91.224(2)							
⁸⁰Zr		79.9406		β⁺ /8.0					0.290 0.538
⁸¹Zr		80.9368	15. s	β⁺ /7.2	6.1				ann.rad./
⁸²Zr		81.9311	32. s	β⁺ /4.0	3.				ann.rad./
⁸³ᵐZr			7. s	β⁺ /7.0		(7/2+)			ann.rad./
⁸³Zr		82.9287	44. s	β⁺ /5.9 EC	4.8	(1/2-)			ann.rad./ 0.0556 0.1050 0.2560 0.474 1.525
⁸⁴Zr		83.9233	26. m	β⁺ /2.7 EC/		0+			ann.rad./ 0.0449 0.1125 0.3729 0.667
⁸⁵ᵐZr			10.9 s	I.T./0.2922 β⁺ ,EC/		1/2-			ann.rad./ 0.2922(IT) 0.4165
⁸⁵Zr		84.9215	7.9 m	β⁺ /4.7 EC/	3.1	7/2+			ann.rad./ 0.2663 0.4163 0.4543
⁸⁶Zr		85.91647	16.5 h	EC/1.47		0+			0.0280 0.243 0.612
⁸⁷ᵐZr			14.0 s	I.T./0.3362		1/2-			0.1352(IT) 0.2010
⁸⁷Zr		86.91482	1.73 h	β⁺ /3.67 EC/	2.26	9/2+			ann.rad./ 0.3811 1.228
⁸⁸Zr		87.91023	83.4 d	EC/0.67		0+			0.3929
⁸⁹ᵐZr			4.18 m	I.T./94 /0.5877 β⁺ /1.5 / EC/4.7 /		1/2-			ann.rad./ 0.5877(IT) 1.507
⁸⁹Zr		88.908889	3.27 d	β⁺ /23 /2.832 EC/77 /	0.9/	9/2+	-1.07		ann.rad./ 0.9092
⁹⁰ᵐZr			0.809 s	I.T./		5-			0.1326 2.1862 2.3189(IT)
⁹⁰Zr	51.45(40)	89.904702				0+			
⁹¹Zr	11.22(5)	90.905643				5/2+	-1.30362	-0.21	
⁹²Zr	17.15(8)	91.905039				0+			
⁹³Zr		92.906474	1.5x10⁶y	β⁻ /0.091		5/2+			0.0304
⁹⁴Zr	17.38(28)	93.906314				0+			
⁹⁵Zr		94.908041	64.02 d	β⁻ /1.125	0.366/55 0.400/44	5/2+	1.10	+0.29	0.7242 0.7567
⁹⁶Zr	2.80(9)	95.908275				0+			
⁹⁷Zr		96.910950	16.8 h	β⁻ /2.658	1.91/	1/2-			0.7434
⁹⁸Zr		97.91276	30.7 s	β⁻ /2.26	2.2/100	0+			
⁹⁹Zr		98.91651	2.2 s	β⁻ /4.56	3.9/ 3.5/	1/2+			0.4692/55.2 0.5459 0.5940
¹⁰⁰Zr		99.91776	7.1 s	β⁻ /3.34		0+			0.4006 0.5043

Elem. or Isot.	Natural Abundance (%)	Atomic Mass or Weight	Half-Life	Decay Mode/Energy (/MeV)	Particle Energy /Intensity (MeV / %)	Spin (h/2π)	Nuclear Magnetic Mom.(nm)	Elect. Quadr. Mom. (b)	γ-ray Energy /Intensity (MeV / %)
¹⁰¹Zr		100.92114	2.1 s	β⁻ /5.49	6.2/	3/2-			0.1194 0.2057 0.2089
¹⁰²Zr		101.92298	2.9 s	β⁻ /4.61					
¹⁰³Zr		102.9266	1.3 s	β⁻ /7.0					
¹⁰⁴Zr		103.9288	1.2 s	β⁻ /5.9					
¹⁰⁵Zr		104.9331	≈1. s	β⁻ /8.5					
¹⁰⁶Zr		105.9359							
¹⁰⁷Zr		106.941							
¹⁰⁸Zr		107.944							
¹⁰⁹Zr									
¹¹⁰Zr									
₄₁Nb		92.90638(2)							
⁸²Nb		81.9431		β⁺ /11.					
⁸³Nb		82.9367		β⁺ /7.5					
⁸⁴Nb		83.9336	12. s	β⁺ ,EC/9.6		(3+)			
⁸⁵Nb		84.9279	2.3 m	β⁺ /6.0					
⁸⁶ᵐNb			56. s	β⁺					
⁸⁶Nb		85.9250	1.46 m	β⁺ /8.0					ann.rad./ 0.751 1.003
⁸⁷ᵐNb			3.7 m	β⁺ / EC/		1/2-			ann.rad./ 0.1352 0.2010
⁸⁷Nb		86.92036	2.6 m	β⁺ 5.2/ EC/		(9/2+)			ann.rad./ 0.2010 0.4706 0.6165 1.0665 1.8842
⁸⁸ᵐNb			7.7 m	β⁺ / EC/		4-			ann.rad./ 0.2625 0.3996 1.0569 1.0825
⁸⁸Nb		87.9183	14.3 m	β⁺ /7.6 EC/	3.2/	8+			ann.rad./ 1.0570 1.0828 (0.07 - 2.5)
⁸⁹ᵐNb			2.0 h	β⁺ / EC/	3.3/	9/2+			0.5880/10(D) (0.17 - 4.0)
⁸⁹Nb		88.91349	1.10 h	β⁺ /74 /4.29 EC/26 /	2.8/	1/2-	+6.216		ann.rad./ 0.5074 0.5880 0.7696 1.2775
⁹⁰ᵐNb			18.8 s	I.T./0.1246		4-			0.002 0.1225
⁹⁰Nb		89.911263	14.6 h	β⁺ /53 /6.111 EC/47 /	0.86/5 1.5/92	8+	4.961	+0.005	ann.rad./ 0.1412 1.1292 2.1862 2.3189 (0.1 - 3.3)
⁹¹ᵐNb			62. d	I.T./97 / EC/3 /		1/2-			0.1045(IT) 1.2050
⁹¹Nb		90.906989	7x10²y	EC/1.253		9/2+			Mo k x-ray
⁹²ᵐNb			10.13 d	EC/99+ /		2+	6.114		0.9126 0.9345 1.8475
⁹²Nb		91.907192	3.7x10⁷y	EC/2.006		7+			0.5611 0.9345
⁹³ᵐNb			16.1 y	I.T./0.0304		1/2-			Nb x-ray 0.0304
⁹³Nb	100.	92.906376				9/2+	+6.1705	-0.32	Nb k x-ray
⁹⁴ᵐNb			6.26 m	I.T./99+ /2.086 β⁻ /0.5 /		3+			0.0409 0.87109
⁹⁴Nb		93.907282	2.4x10⁴y	β⁻ /2.045	0.47/	6+			0.70263 0.87109
⁹⁵ᵐNb			3.61 d	I.T./97.5 /0.2357 β⁻ /2.5 /		1/2-			0.2040 0.2356
⁹⁵Nb		94.906834	34.97 d	β⁻ /0.926	0.160/	9/2+	6.141		0.76578
⁹⁶Nb		95.908099	23.4 h	β⁻ /3.187	0.5/10 0.75/90	6+	4.976		0.7782 0.2191-1.498
⁹⁷ᵐNb			58.1 s	I.T./0.7434	0.734/98	1/2-			0.7434

Elem. or Isot.	Natural Abundance (%)	Atomic Mass or Weight	Half-Life	Decay Mode/Energy (/MeV)	Particle Energy /Intensity (MeV / %)	Spin (h/2π)	Nuclear Magnetic Mom.(nm)	Elect. Quadr. Mom. (b)	γ-ray Energy /Intensity (MeV / %)
^{97}Nb		96.908096	1.23 h	β^- /1.934	1.27/98	9/2+	6.15		0.4809
^{98m}Nb			51. m	β^- /4.67		5+			0.6579
									0.7874
^{98}Nb		97.91033	2.9 s	β^- /4.59	4.6/	1+			0.1726-1.89
									0.6451
									0.7874
^{99m}Nb			2.6 m	β^- /	3.2/	1/2-			1.0243
									0.0978/100
^{99}Nb		98.91162	15.0 s	β^- /3.64	3.5/100	9/2+			(0.138-3.010)
									0.0977
^{100m}Nb			3.0 s	β^- /6.74	5.8				0.1378/3.1
									Nb k x-ray
									0.159
									0.6364
									1.0637
^{100}Nb		99.91418	1.5 s	β^- /6.25	6.2/ 5.3/				0.5354
^{101}Nb		100.91525	7.1 s	β^- /4.57	4.3/				0.6001-1.566
^{102m}Nb			4.3 s	β^- /					0.1105-0.810
^{102}Nb		101.91804	1.3 s	β^- /7.21	7.2/				
^{103}Nb		102.91914	1.5 s	β^- /5.53	5.3/	5/2+			0.2960-2.184
^{104m}Nb			0.9 s	β^-,n/	n/0.06				
^{104}Nb		103.9225	4.8 s	β^-,n/8.1	n/0.05				
^{105}Nb		104.9239	3.0 s	β^-,n/6.5	n/1.7				
^{106}Nb		105.9282	1.0 s	β^-,n/9.3	n/4.5				
^{107}Nb		106.9303	0.30 s	β^-,n/7.9	n/6.0				
^{108}Nb		107.9350	0.19 s	β ,n/	n/6.2				(0.193-0.590)
^{109}Nb		108.9376	0.19 s	β ,n/	n/31				
^{110}Nb		109.943	0.17 s	β ,n/	n/40				
^{111}Nb									
^{112}Nb									
^{113}Nb									
$_{42}$Mo		95.94(1)							
^{84}Mo		83.9401		β^+ /6.					
^{85}Mo		84.9366	3.2 s	β^+/8.1					
^{86}Mo		85.9302	20. s	β^+ /4.8					
^{87}Mo		86.9273	14. s	EC, β^+/6.5					(0.752-1.004)
^{88}Mo		87.92195	8.0 m	β^+ /3.4 EC		0+			ann.rad./
									0.0800
									0.1399
^{89m}Mo			0.19 s	I.T./0.118		1/2-			0.1707
									0.118(IT)
^{89}Mo		88.91948	2.2 m	β^+ /5.58 EC/		9/2+			0.268
									ann.rad./
									0.659
									0.803
									1.155
^{90}Mo		89.91394	5.7 h	β^+ /25 /2.489 EC/75 /	1.085/	0+			1.272
									ann.rad./
									0.04274
									0.12237
^{91m}Mo			1.08 m	I.T./50 /0.653 β^+ ,EC/50 /	2.5/ 2.8/ 4.0/	1/2-			0.25734 ann.rad./ 0.6529 1.2081 1.5080
^{91}Mo		90.91175	15.5 m	β^+ /94 /4.43 EC/6 /	3.44/94	9/2-			2.2407 ann.rad./ 1.6373 2.6321 3.0286
^{92}Mo	14.84(35)	91.906810				0+			(0.1 - 4.2)
^{93m}Mo			6.9 h	I.T./99+ /2.425	21/2+	+9.21			0.26306(IT)
									0.68461
									1.47711
^{93}Mo		92.906811	3.5x10^{3}y	EC/0.405		5/2+			0.0304
^{94}Mo	9.25(12)	93.905087				0+			
^{95}Mo	15.92(13)	94.905841				5/2+	-0.9142	-0.02	
^{96}Mo	16.68(2)	95.904678				0+			
^{97}Mo	9.55(8)	96.906020				5/2+	-0.9335	+0.26	
^{98}Mo	24.13(31)	97.905407				0+			
^{99}Mo		98.907711	2.7476 d	β^- /1.357	0.45/14 0.84/2 1.21/84	1/2+	0.375		0.144048 0.18109 0.36644 0.73947
^{100}Mo	9.63(23)	99.90748				0+			

Elem. or Isot.	Natural Abundance (%)	Atomic Mass or Weight	Half-Life	Decay Mode/Energy (/MeV)	Particle Energy /Intensity (MeV / %)	Spin (h/2π)	Nuclear Magnetic Mom.(nm)	Elect. Quadr. Mom. (b)	γ-ray Energy /Intensity (MeV / %)
[101]Mo		100.91035	14.6 m	β⁻ /2.82	2.23/ 0.7/	1/2 +			0.0063 0.19193 0.5909 (0.0809-2.405)
[102]Mo		101.91030	11.3 m	β⁻ /1.01	1.2/	0+			0.1493/89. 0.2116/100. 0.2243/32.
[103]Mo		102.91320	1.13 m	β⁻ /3.8		3/2 +			0.1028(2)/ 0.1440(2) 0.2511(2)
[104]Mo		103.91376	1.00 m	β⁻ /2.16		0+			0.0686(1)/100. 0.4239(4)/21.
[105]Mo		104.9170	36. s	β⁻ /4.95		3/2 +			0.0642/ 0.0856/ 0.2495/
[106]Mo		105.91814	8.4 s	β⁻ /3.52		0+			0.1894(2)/22. 0.3644(2)/6. 0.3723(2)/12.
[107]Mo		106.9217	3.5 s	β⁻ /6.2					
[108]Mo		107.9236	1.1 s	β⁻ /5.1					(0.028-0.636)
[109]Mo		108.9278	0.5 s	β⁻ /7.2					
[110]Mo		109.9297	0.30 s	β⁻ /5.7					Tc k x-ray 0.142 (0.039-0.599)
[111]Mo		110.9345							
[112]Mo		111.937							
[113]Mo		112.942							
[114]Mo									
[115]Mo									
[116]Mo									
[117]Mo									
₄₃Tc									
[86]Tc		85.9430		β⁺ /11.9					
[87]Tc		86.9365		β⁺ /8.6					
[88]Tc		87.9328		β⁺ /10.1					
[89m]Tc			13. s						
[89]Tc		88.9275	13. s	β⁺ /7.5					
[90m]Tc			49.2 s	β⁺	5.3/	6+			ann.rad./ 0.9479/ 1.0542/
[90]Tc		89.9235	8.3 s	β⁺ /8.9	7.0/15 7.9/95.	1+			ann.rad./ 0.9479/
[91m]Tc			3.3 m	β⁺ EC		1/2 +			ann.rad./170. 0.3375(2)/1.1 0.8110(5)/5. 1.6052(1)/7.8 1.6339(1)/9.1 1.9023(1)/6. 2.4509(1)/13.5
[91]Tc		90.9184	3.14 m	β⁺ /6.2	5.2	9/2 +			ann.rad./200.
[92]Tc		91.91526	4.4 m	β⁺ /7.87 EC	4.1	8+			ann.rad./200. 0.0850/ 0.1475 0.3293 0.7731 1.5096
[93m]Tc			43. m	I.T./13 EC/20		1/2-			0.3924(IT) 0.9437 2.6445
[93]Tc		92.910248	2.73 h	β⁺ /13/3.201 EC/87/	0.81	9/2 +	6.26		ann.rad./ 1.3629 1.4771 1.5203 (0.1 - 3.0)
[94m]Tc			52. m	β⁺ /72/4.33 EC/28/		2+			ann.rad./ 0.8710 1.8686
[94]Tc		93.909655	4.88 h	β⁺ /11/4.256 EC/89/		7+	5.08		ann.rad./ 0.4491 0.7026 0.8496 0.8710

Elem. or Isot.	Natural Abundance (%)	Atomic Mass or Weight	Half-Life	Decay Mode/Energy (/MeV)	Particle Energy /Intensity (MeV / %)	Spin (h/2π)	Nuclear Magnetic Mom.(nm)	Elect. Quadr. Mom. (b)	γ-ray Energy /Intensity (MeV / %)
^{93m}Tc			61. d	I.T./4/ β⁺/0.3 EC/96	0.5/ 0.7/	1/2-			ann.rad./ 0.0389(IT) 0.2041 0.5821 0.5821 0.8351
^{93}Tc		94.90766	20.0 h	EC/100/1.691		9/2+	5.89		0.7657 1.0738
^{94m}Tc			52. m	I.T./90/ β⁺,EC/2/		4+			0.0342(IT) 0.7782 1.2002
^{94}Tc		95.90787	4.3 d	EC/2.973		7+	+5.04		Mo k x-ray 0.7782 0.8125 0.8498 1.12168
^{97m}Tc			90. d	I.T./0.0965		1/2-			Tc k x-ray 0.0965
^{97}Tc		96.906364	2.6x10⁶ y	EC/100/0.320		9/2+			Mo k x-ray
^{98}Tc		97.907215	4.2x10⁶ y	β⁻/1.80	0.40/100	6+			0.65241 0.74535
^{99m}Tc			6.01 h	I.T./100/0.142		1/2-			Tc k x-ray 0.14049 0.14261
^{99}Tc		98.906254	2.13x10⁵ y	β⁻/0.294	0.293/100	9/2+	+5.6847	-0.129	
^{100}Tc		99.907657	15.8 s	β⁻/3.202 EC /1.8(10)⁻³/0.17	2.2/ 2.9/ 3.3	1+			0.5396 0.5908 1.5122 (0.3 - 2.6)
^{101}Tc		100.90731	14.2 m	β⁻/1.61	1.32/	9/2+			0.1272 0.1841 0.3068 0.5451 (0.073-0.969)
^{102m}Tc			4.4 m	I.T./2/4.8 β⁻/98/	1.8/				0.4184 0.4752 0.6281 0.6302 1.0464 1.1033 1.6163 2.2447
^{102}Tc		101.90921	5.3 s	β⁻/4.53	3.4/ 4.2 2.2/	1+			0.4686 0.4751 1.1055
^{103}Tc		102.90918	54. s	β⁻/2.66	2.0/ 2.2/	5/2+			0.1361 0.1743 0.2104 0.3464 0.5629 (0.13 - 1.0)
^{104}Tc		103.91144	18.2 m	β⁻/5.60	5.3/	(3+)			0.3483 0.3580 0.5305 0.5351 0.8844 0.8931 1.6768 (0.3 - 3.7)
^{105}Tc		104.91166	7.6 m	β⁻/3.6	3.4/	5/2+			0.1079 0.1432 0.3215
^{106}Tc		105.91436	36. s	β⁻/6.55		2+			0.2703 0.5222 1.9694 2.2393 2.7893
^{107}Tc		106.9151	21.2 s	β⁻/4.8					0.1027 0.1063 0.1770
^{108}Tc		107.9185	5.1 s	β⁻/7.72		(3)			0.4587 0.2422 0.4656 0.7078 0.7326 1.5835
^{109}Tc		108.9200	1.4 s	β⁻/6.3	p/0.08				
^{110}Tc		109.9234	0.83 s	β⁻/8.8	p/0.04				0.2407
^{111}Tc		110.9250	0.30 s	β⁻.n /7.0	n/0.85				

Elem. or Isot.	Natural Abundance (%)	Atomic Mass or Weight	Half-Life	Decay Mode/Energy (/MeV)	Particle Energy /Intensity (MeV / %)	Spin (h/2π)	Nuclear Magnetic Mom.(nm)	Elect. Quadr. Mom. (b)	γ-ray Energy /Intensity (MeV / %)
^{112}Tc		111.9292	0.21 s	β,n	n/2.6				
^{113}Tc		112.931	0.13 s	$β^-$ /8.					
^{114}Tc		113.936							
^{115}Tc		114.938							
^{116}Tc									
^{117}Tc									
^{119}Tc									
$_{44}$Ru		**101.07(2)**							
^{87}Ru		86.949							
^{88}Ru		87.9404							
^{89}Ru		88.936		$β^+$ /8.					
^{90}Ru		89.9298	11. s	$β^+$ /5.9					ann.rad./ 0.155 - 1.551
^{91}Ru		90.9264	9. s	$β^+$,EC/7.4		9/2+			ann.rad./
^{92}Ru		91.9201	3.7 m	$β^+$ /53/4.5 EC/47/		0+			ann.rad./ 0.1346 0.2138 0.2593
^{93m}Ru			10.8 s	I.T./21/ $β^+$,EC/79/	5.3/	1/2-			ann.rad./ 0.7344 1.1112 1.3962 2.0931
^{93}Ru		92.9171	1.0 m	$β^+$ /6.3 EC/		9/2+			ann.rad./ 0.6807 1.4349 (0.5- 4.2)weak
^{94}Ru		93.91137	52. m	EC/100/1.59					0.3672 0.5247 0.8922
^{95}Ru		94.91042	1.64 h	EC/85/2.57 $β^+$ /15/	1.20/ 0.91/	5/2+	0.86		ann.rad./ 0.2904 0.3364 0.6268
^{96}Ru	5.52(20)	95.90760				0+			Tc k x-ray
^{97}Ru		96.90756	2.89 d	EC/1.12		5/2+	-0.78		0.2157 0.3245 0.4606
^{98}Ru	1.88(9)	97.90529				0+			
^{99}Ru	12.74(26)	98.905939				5/2+	-0.6413	+0.079	
^{100}Ru	12.60(19)	99.904219				0+			
^{101}Ru	17.05(7)	100.905582				5/2+	-0.7188	+0.46	
^{102}Ru	31.57(31)	101.904349				0+			
^{103}Ru		102.906323	39.27 d	$β^-$ /0.763	0.223	3/2+	0.20	+0.62	0.05329 0.29498 0.4438 0.49708 0.55704 0.61033 (0.04 - 1.6)
^{104}Ru	18.66(44)	103.905430				0+			
^{105}Ru		104.907750	4.44 h	$β^-$ /1.917	1.11/22 1.134/13 1.187/49	3/2+	-0.3		0.12968 / 0.0 0.1491 0.2629 0.31664 0.46943 0.67634 0.72420 (0.1 - 1.8)
^{106}Ru		105.90733	1.020 y	$β^-$ /0.0394	0.0394/100	0+			
^{107}Ru		106.9099	3.8 m	$β^-$ /2.9	2.1/ 3.2/				0.1939 0.3741 0.4625 0.8488
^{108}Ru		107.9102	4.5 m	$β^-$ /1.4	1.2/				0.0923 0.1651 0.4339 0.4975 0.6189
^{109}Ru		108.91320	34.5 s	$β^-$ /4.2					0.1164 0.3584
^{110}Ru		109.9140	15. s	$β^-$ /2.81					0.1121 0.3737 0.4397 0.7967
^{111}Ru		110.9176	1.5 s	$β^-$ /5.5					

Elem. or Isot.	Natural Abundance (%)	Atomic Mass or Weight	Half-Life	Decay Mode/Energy (/MeV)	Particle Energy /Intensity (MeV / %)	Spin (h/2π)	Nuclear Magnetic Mom.(nm)	Elect. Quadr. Mom. (b)	γ-ray Energy /Intensity (MeV / %)
^{112}Ru		111.9188	4.5 s	β^- /4.5					
^{113}Ru		112.9225	2.7 s	β^- /7.					
^{114}Ru		113.9239	0.57 s	β^- /6.1					0.127/24
^{115}Ru		114.928	≈0.74 s	β^- /8.					(0.053-0.180)
^{116}Ru		115.930							
^{117}Ru		116.935							
^{118}Ru		117.937							
^{119}Ru									
^{120}Ru									
$_{45}$Rh		102.90550(2)							
^{89}Rh		88.9494							
^{90}Rh		89.9429							
^{91}Rh		90.9366							
^{92}Rh		91.9320		β^+ /11.1					
^{93}Rh		92.9257		β^+ /8.1					(0.138-1.493)
^{94m}Rh			25.8 s	β^+ /		8+			ann.rad./ 0.1264 0.3117 0.7562 1.0752 1.4307
^{94}Rh		93.9217	1.18 m	β^+ /9.6	6.4/	3+			ann.rad./ 0.1461 0.3117 0.7562 1.4307
^{95m}Rh			1.96 m	I.T./88/ β^+,EC/12/		1/2+			ann.rad./ 0.5433(TT) 0.7837
^{95}Rh		94.9159	5.0 m	β^+ /5.1	3.2	9/2+			ann.rad./ 0.2293 0.4103 0.6610 0.9416 1.3520
^{96m}Rh			1.51 m	I.T./60 /0.052 β^+,EC/40/	4.70/	2+			(0.2 - 3.8) ann.rad./ Tc,Ru x-rays 0.8326 1.0985 1.6921
^{96}Rh		95.91452	9.6 m	β^+/6.45 EC/	3.3/	5+			(0.4 - 3.3) ann.rad./ 0.4299 0.6315 0.6853 0.7418 0.8326
^{97m}Rh			46.m	I.T./5 / β^+,EC/95 /	2.6/	1/2-			(0.2 - 3.4) ann.rad./ 0.1886 0.4215 2.2452
^{97}Rh		96.91134	31.0m	β^+ /3.52	2.1/	9/2+			ann.rad./ 0.1886 0.3892 0.4515 0.8398 0.8788
^{98m}Rh			3.5 m	β^+ /		5+			(0.2 - 3.5) ann.rad./ 0.6154 0.6524 0.7452
^{98}Rh		97.91072	8.7 m	β^+ /90 /5.06	3.4/	2+			ann.rad./ 0.6524 0.7623
^{99m}Rh			4.7 h	β^+ /8 / EC/92 /	.74/	9/2+			ann.rad./ 0.2766/ 0.3408 0.6178 1.2612

Elem. or Isot.	Natural Abundance (%)	Atomic Mass or Weight	Half-Life	Decay Mode/Energy (/MeV)	Particle Energy /Intensity (MeV / %)	Spin (h/2π)	Nuclear Magnetic Mom.(nm)	Elect. Quadr. Mom. (b)	γ-ray Energy /Intensity (MeV / %)
^{99}Rh		98.90820	16. d	β⁺/4 /2.10 EC/97 /	0.54/ 0.68/	1/2-			ann.rad./ 0.0894/ 0.3530 0.5277 (0.1 - 2.0)
^{100m}Rh			4.7 m	I.T./99 / β⁺ /0.4 /		5+			ann.rad./ 0.0748/ 0.2647(IT)
^{100}Rh		99.90812	20.8 h	β⁺ /3.63 EC/	2.62/ 2.07/	1-			0.4462 0.5396 0.5882 0.8225 1.5534 2.3761
^{101m}Rh			4.35 d	EC/92 / I.T./8 /0.1573		9/2+	+5.51		Rh k x-ray 0.1272/ 0.3069 0.5451
^{101}Rh		100.90616	3.3 y	EC/0.54		1/2-			Ru k x-ray 0.1272 0.1980 0.3252
^{102m}Rh			207. d	I.T./5 / β⁻ /19/ β⁺ /14/ EC/62/					ann.rad./ 0.4686 0.4751 0.5566 0.6280 1.1032 (0.4 - 1.6)
^{102}Rh		101.906842	2.9 y	EC/2.323		6+	4.04		Ru k x-ray 0.4751 0.6313 0.6975 0.7668 1.0466 1.1032
^{103m}Rh			56.12 m	IT		7/2+			
^{103}Rh	100.	102.905504				1/2-	-0.0884		
^{104m}Rh			4.36 m	I.T./99+ / β⁻	1.3/	5+			Rh k x-ray 0.0514 0.0971 0.5558
^{104}Rh		103.906655	42.3 s	β⁻ /99+ /2.441 EC/0.4/1.141	1.88/2 2.44/98	1+			0.3581 0.5558 1.2370 (0.35 - 1.8)
^{105m}Rh			40. s	I.T./1.296		1/2-			Rh k x-ray 0.1296
^{105}Rh		104.905692	35.4 h	β⁻ /0.567	0.247/30 0.567/70	7/2+	+4.45		0.2801 0.3061 0.3189
^{106m}Rh			2.18 h	β⁻ /	0.92/	6+			0.2217 0.4510 0.5119 0.6162 0.7173 0.7484 1.0458 1.5277
^{106}Rh		105.90729	29.9 s	β⁻ /3.54	2.4/2 3.0/12 3.54/79	1+	2.58		0.51186/ 0.61612 0.62187 (0.05 - 3.04)
^{107}Rh		106.90675	21.7 m	β⁻ /1.51	1.20/65 1.5/17	7/2+			0.2776 0.3028 0.3925
^{108m}Rh			6.0 m	β⁻ /	1.57/				0.4339 0.4973 0.6189
^{108}Rh		107.9087	17. s	β⁻ /4.5		1+			0.4046 0.4339 0.4973 0.5811 0.6146 0.9014 0.9471

Elem. or Isot.	Natural Abundance (%)	Atomic Mass or Weight	Half-Life	Decay Mode/Energy (/MeV)	Particle Energy /Intensity (MeV / %)	Spin (h/2π)	Nuclear Magnetic Mom.(nm)	Elect. Quadr. Mom. (b)	γ-ray Energy /Intensity (MeV / %)
¹⁰⁹Rh		108.90874	1.34 m	β⁻ /2.59	2.25/	7/2+			0.1134 0.1780 0.2914 0.3254 0.3268 0.4261 (0.1 - 1.6)
¹¹⁰ᵐRh			29. s	β⁻ /	[.6/				0.3737 0.4397 0.7967
¹¹⁰Rh		109.9110	3.1 s	β⁻ /5.4	5.5/	1+			0.3737 0.4400 0.5463 0.6877 0.8381 0.9045
¹¹¹Rh		110.9117	11. s	β⁻ /3.7					0.275
¹¹²ᵐRh			6.8 s	β⁻ /					
¹¹²Rh		111.9140	4. s	β⁻ /6.2		1+			
¹¹³Rh		112.9154	0.9 s	β⁻ /4.9					0.3489 0.1285
¹¹⁴ᵐRh			1.8 s	β⁻ /					
¹¹⁴Rh		113.9173	1.8 s	β⁻ /6.5		1+			
¹¹⁵Rh		114.9201	0.99 s	β⁻ /6.0					
¹¹⁶ᵐRh			0.9 s	β⁻ /					0.3405
¹¹⁶Rh		115.9228	0.7 s	β⁻ /8.0		1+			
¹¹⁷Rh		116.925	0.44 s	β⁻ /7.					0.0346 0.1317
¹¹⁸Rh		117.929							
¹¹⁹Rh		118.931							
¹²⁰Rh		119.936							
¹²¹Rh		120.938							
¹²²Rh									
₄₆Pd		106.42(1)							
⁹¹Pd		90.949							
⁹²Pd		91.9404							
⁹³Pd		92.9359							
⁹⁴Pd		93.9288	9. s	EC,β⁺ /≈ 6.6					0.5582 (0.0546-0.798)
⁹⁵ᵐPd		94.92684	13.4 s	EC,β⁺ /10.2		21/2+			
⁹⁵Pd									
⁹⁶Pd		95.9182	2.03 m	EC,β⁺ /3.5	1.15/				0.1248 0.4995
⁹⁷Pd		96.9165	3.1 m	β⁺ ,EC/4.8	3.5/	5/2+			ann.rad./ 0.2653 0.4752 0.7927 (0.2 - 3.4)
⁹⁸Pd		97.91273	17.7 m	β⁺ /1.87 EC/		0+			ann.rad./ 0.0677 0.1125 0.6630 0.8379
⁹⁹Pd		98.91181	21.4 m	β⁺ /49 /3.37 EC/51 /	2.18/	5/2+			ann.rad./ 0.1360 0.2636 0.6734 (0.2 - 2.85)
¹⁰⁰Pd		99.90851	3.7 d	EC/0.36		0+			0.03271 0.0748 0.0840
¹⁰¹Pd		100.90829	8.4 h	β⁺ /5 /1.980 EC/95 /	0.776/	5/2+	-0.66		ann.rad./ 0.0244 0.2963 0.5904
¹⁰²Pd	1.02(1)	101.905607				0+			
¹⁰³Pd		102.906087	16.99 d	EC/0.543		5/2+			Rh k x-ray 0.03975 0.3575 0.4971
¹⁰⁴Pd	11.14(8)	103.904034				0+			
¹⁰⁵Pd	22.33(8)	104.905083				5/2+	-0.642	+0.66	
¹⁰⁶Pd	27.33(3)	105.903484				0+			
¹⁰⁷ᵐPd			20.9 s	I.T./0.2149		11/2-			Pd k x-ray 0.2149(IT)
¹⁰⁷Pd		106.90513	6.5x10⁶y	β⁻ /0.033	0.03/	5/2+			
¹⁰⁸Pd	26.46(9)	107.903895				0+			

Elem. or Isot.	Natural Abundance (%)	Atomic Mass or Weight	Half-Life	Decay Mode/Energy (/MeV)	Particle Energy /Intensity (MeV / %)	Spin (h/2π)	Nuclear Magnetic Mom.(nm)	Elect. Quadr. Mom. (b)	γ-ray Energy /Intensity (MeV / %)
^{109m}Pd			4.75 m	I.T./0.1889		11/2-			Pd x-ray 0.1889(IT)
^{109}Pd		108.905954	13.5 h	β^- /1.116	1.028	5/2+			0.0880 (0.08 - 1.0)
^{110}Pd	11.72(9)	109.905153				0+			
^{111m}Pd			5.5 h	I.T./73 /0.172 β^- /27 /	0.35 0.77	11/2-			0.0704 0.1722 0.3912 (0.1 - 1.97)
^{111}Pd		110.90764	23.4 m	β^- /2.19	2.2/95	5/2+			0.0598 0.2454 0.5800 0.6504 1.3885 1.4590
^{112}Pd		111.90731	21.04 h	β^- /0.29	0.28/	0+			0.018 0.0959
^{113m}Pd			1.48 m	β^- /		5/2+			0.0958
^{113}Pd		112.91015	1.64 m	β^- /3.34					0.4824 0.6436 0.7394
^{114}Pd		113.91037	2.48 m	β^- /1.45		0+			0.1266 0.2320 0.5582 0.5760
^{115}Pd		114.9137	47. s	β^- /4.58					0.1255 0.2554 0.3428
^{116}Pd		115.9142	12.7 s	β^- /2.61					0.1015 0.1147 0.1778
^{117}Pd		116.9178	4.4 s	β^- /5.7					0.2473 (0.0766-0.4025)
^{118}Pd		117.9189	2.4 s	β^- /4.1					0.1254 (0.0284-0.5957)
^{119}Pd		118.9227	0.9 s	β^- /6.5					0.2566 (0.0699-0.3261)
^{120}Pd		119.9240	0.5 s	β^- /5.0					0.1581 (0.0525-0.5952)
^{121}Pd		120.9282							
^{122}Pd		121.9298							
^{123}Pd									
$_{47}$Ag		107.8682(2)							
^{94}Ag		93.9428	0.42 s	β^+, p/					
^{95}Ag		94.9355	2.0 s	β^+, p/					ann.rad./
^{96}Ag		95.9307	5.1 s	β^+ /11.6 EC/					0.1248 0.4995 (0.1066-1.416)
^{97}Ag		96.9240	19. s	β^+ /7.0 EC/					ann.rad./ 0.6862 1.2941 (0.352-3.294)
^{98}Ag		97.9218	47. s	β^+ /8.4 EC/		5+			ann.rad./ 0.5711 0.6786 0.8631 (0.153-1.185)
^{99m}Ag			11. s	I.T./100/		1/2-			Ag k x-ray 0.1636(IT) 0.3426
^{99}Ag		98.9176	2.07 m	β^+ /87 /5.4 EC/13 /		9/2+			ann.rad./ 0.2199 0.2645 0.8056 0.8323 (0.2 - 3.5)
^{100m}Ag			2.3 m	β^+ / EC/		2+			ann.rad./ 0.6657 1.6941
^{100}Ag		99.9161	2.0m	β^+/7.1 EC/	4.7/	5+			ann.rad./ 0.2807 0.4503 0.6657 0.7508 0.7732

Elem. or Isot.	Natural Abundance (%)	Atomic Mass or Weight	Half-Life	Decay Mode/Energy (/MeV)	Particle Energy /Intensity (MeV / %)	Spin (h/2π)	Nuclear Magnetic Mom.(nm)	Elect. Quadr. Mom. (b)	γ-ray Energy /Intensity (MeV / %)
^{101m}Ag			3.1 s	I.T./0.23		1/2-			Ag k x-ray 0.0981 0.176(IT)
^{101}Ag		100.9128	11.1 m	β⁺ /69 /4.2 EC/31 /	2.7/ 2.18/ 2.73/ 3.38/	9/2+	5.7		ann.rad./ 0.2610 0.2747 0.3269 0.4392 0.6673 1.1739 (0.2 - 3.1)
^{102m}Ag			7.8 m	β⁺ /38 / EC/13 / I.T./49 /	3.4	2+	+4.14		ann.rad./ 0.5567 0.9777 1.8347 2.0545 2.1594 3.2386
^{102}Ag		101.91197	13.0 m	β⁺ /78 /5.92 EC/22 /	2.26/	5+	4.6		ann.rad./ 0.5567 0.7194 0.8354 1.2571 1.5816 1.7446
^{103m}Ag			5.7 s	I.T./0.134		1/2-			Ag k x-ray 0.1344
^{103}Ag		102.90897	1.10 h	β⁺ /28 /2.69 EC/72 /	1.7 1.3	7/2+	+4.47		ann.rad./ 0.1187
^{104m}Ag			33. m	β⁺ /64/ EC/36/ I.T./0.07/	2.71/	2+	+3.7		0.1482 ann.rad./ 0.5558 0.7657 (0.5 - 3.4)
^{104}Ag		103.90863	69. m	β⁺ /16 /4.28 EC/84 /	0.99/	5+	3.92		ann.rad./ 0.5558 0.9259 0.9416 (0.18 - 2.27)
^{105m}Ag			7.2 m	I.T./98 /0.0255 EC/2 /		7/2+			Ag x-ray 0.3063 0.3192 (0.1 - 1.0)
^{105}Ag		104.90653	41.3 d	EC/1.35		1/2-	0.1014		0.0640 0.2804 0.3445 0.4434 Pd k x-ray
^{106m}Ag			8.4 d	EC/		6+	3.71		0.4510 0.5118 0.7173 1.0458 ann.rad./
^{106}Ag		105.90667	24.0 m	β⁺ /59 /2.965 EC/41 /	/1.96	1+	+2.85		0.5119
^{107m}Ag			44.2 s	I.T./0.093		7/2+			Ag x-ray 0.0931
^{107}Ag	51.839(8)	106.905093				1/2-	-0.11357		Ag k x-ray
^{108m}Ag			1.3x10²y	EC/92 / I.T./8 /0.079		6+	3.580		Pd k x-ray 0.43392 0.61427 0.72290 ann.rad./
^{108}Ag		107.905954	2.39 m	β⁻ /97/1.65 EC/2/ β⁻ /1/1.92	1.02/1.7 1.65/96 0.88/0.3	1+	+2.6884		0.43392 0.61885 0.63298
^{109m}Ag			39.8 s	I.T./0.088		7/2+			Ag k x-ray 0.0880
^{109}Ag	48.161(8)	108.904756				1/2-	-0.13056		
^{110m}Ag			249.8 d	β⁻ /99 / I.T./1 /0.1164	0.087 0.530	6+	+3.607		0.65774 0.76393 0.88467 0.93748 1.38427 (0.447-1.56)
^{110}Ag		109.906111	24.6 s	β⁻ /2.892	2.22/5 2.89/95	1+	+2.7271		0.65774 0.8154 1.1257

Elem. or Isot.	Natural Abundance (%)	Atomic Mass or Weight	Half-Life	Decay Mode/Energy (/MeV)	Particle Energy /Intensity (MeV / %)	Spin (h/2π)	Nuclear Magnetic Mom.(nm)	Elect. Quadr. Mom. (b)	γ-ray Energy /Intensity (MeV / %)
^{111m}Ag			1.08 m	I.T./99 /0.0598 β⁻ /1/		7/2+			Ag k x-ray 0.0598 0.2454
^{111}Ag		110.905295	7.47 d	β⁻ /1.037	1.035/	1/2-	-0.146		0.2454 0.3421
^{112}Ag		111.90701	3.13 h	β⁻ /3.96	3.94/ 3.4	2-	0.0547		0.6067 0.6174 1.3877 (0.4 - 2.9)
^{113m}Ag			1.14 m	I.T./80 /0.043 β⁻ /20 /	1.5	7/2+			0.1422 0.2983 0.3161 0.3923
^{113}Ag		112.90657	5.3 h	β⁻ /2.02	2.01/	1/2-	0.159		0.2588 0.2986
^{114}Ag		113.90881	4.6 s	β⁻ /5.08	4.9/	1+			0.5582 0.5760 1.9946
^{115m}Ag			18.7 s	β⁻ /		7/2+			0.1134 0.1315 0.2288 0.3887
^{115}Ag		114.90876	20. m	β⁻ /3.10		1/2-			0.1316 0.2128 0.2291 0.4727 (0.13 - 2.49)
^{116m}Ag			10.5 s	I.T./2 / β⁻ /98 /	3.2/ 2.9	5+			0.1027 0.2549 0.5134 0.7055 1.0289
^{116}Ag		115.91137	2.68 m	β⁻ /6.16	5.3	2-			0.5134 0.6993 2.4779
^{117m}Ag			5.3 s	β⁻ /	3.2/	7/2+			0.1354 0.2981 0.3868
^{117}Ag		116.91171	1.22 m	β⁻ /4.18	2.3	1/2-			0.1354 0.1571 0.3377
^{118m}Ag			2.8 s	β⁻ /59 / I.T./41 /0.1277					0.1277 0.4878 0.6771 0.7709 1.0586
^{118}Ag		117.9145	4.0 s	β⁻ /7.1					0.4878 0.6771 3.2259
^{119}Ag		118.9157	2.1 s	β⁻ /5.35		7/2+			0.0674 0.3662 0.3991 0.6264
^{120m}Ag			0.32 s	β⁻ / I.T./					0.2030 0.5059 0.6978 0.8300 0.9258
^{120}Ag		119.9188	1.23 s	β⁻ /8.2					0.5059 0.6978 0.8171 1.3231
^{121}Ag		120.9198	0.78 s	β⁻ /6.4					0.1150 0.3148 0.3537 0.3696 0.5007 1.5105 (0.11 - 2.5)
^{122m}Ag			1. s	β⁻ /					
^{122}Ag		121.9233	0.44 s	β⁻ /9.2					
^{123}Ag		122.9249	0.31 s	β⁻ /7.4					
^{124}Ag		123.9285	0.22 s	β⁻ /10.1					
^{125}Ag		124.9305	0.17 s						
^{126}Ag		125.9345	0.11 s						
^{127}Ag		126.9369	0.11 s						

Elem. or Isot.	Natural Abundance (%)	Atomic Mass or Weight	Half-Life	Decay Mode/Energy (/MeV)	Particle Energy /Intensity (MeV / %)	Spin (h/2π)	Nuclear Magnetic Mom.(nm)	Elect. Quadr. Mom. (b)	γ-ray Energy /Intensity (MeV / %)
$_{48}$Cd		**112.411(8)**							
^{97}Cd		96.9349	3. s	β^+,(p)					
^{98}Cd		97.9276	9.2 s	β^+ /5.4 (p)	/0.025				
^{99}Cd		98.9250	16. s	β^+ ,EC/6.9					
^{100}Cd		99.9203	1.1 m	β^+ ,EC/3.9					ann.rad./
^{101}Cd		100.9187	1.2 m	β^+ /83 /5.5 EC/17 /	4.5	5/2+			ann.rad./ (0.090-1.043) In k x-ray 0.0985 1.7225
^{102}Cd		101.91474	5.8 m	β^+ /27 /2.59 EC/73		0+			(0.31 - 2.84) ann.rad./ 0.0974 0.4810 1.0366 1.3598
^{103}Cd		102.91342	7.5 m	β^+ /33 /4.14 EC/67 /		5/2+	-0.81	-0.8	ann.rad./ Ag k x-ray 1.0799 1.4487 1.4618 (0.1 - 2.8)
^{104}Cd		103.90985	58. m	EC/1.14		0+			Ag k x-ray 0.0835 0.7093
^{105}Cd		104.90947	55.5 m	β^+ /26 /2.739 EC/74 /	1.69/	5/2+	-0.7393	+0.43	Ag k x-ray 0.3469 0.6072 0.9618 1.3025 (0.25 - 2.4)
^{106}Cd	1.25(6)	105.90646	>2.6x10^{17} y	β^+,EC		0+			
^{107}Cd		106.90661	6.52 h	EC/99+ /1.417 β^+ /		5/2+	-0.615055	+0.68	Ag k x-ray 0.0931 0.8289
^{108}Cd	0.89(3)	107.90418				0+			
^{109}Cd		108.904985	462.0 d	EC/0.214		5/2+	-0.827846	+0.69	Ag k x-ray 0.08804
^{110}Cd	12.49(18)	109.903006				0+			Cd k x-ray
^{111m}Cd			48.5 m	I.T./		11/2-			0.1508(IT) 0.2454
^{111}Cd	12.80(12)	110.904182				1/2+	-0.594886		
^{112}Cd	24.13(21)	111.902758				0+			
^{113m}Cd			14.1 y	β^- /99.9 /0.59	0.59/99.9	11/2-	-1.087		0.2637
^{113}Cd	12.22(12)	112.904401	7.7x10^{15}y	β^-		1/2+	-0.622301		
^{114}Cd	28.73(42)	113.903359				0+			
^{115m}Cd			44.6 d	β^- /1.629	0.68/1.6 1.62/97	11/2-	-1.042		0.48450 0.93381 1.29064
^{115}Cd		114.905431	2.228 d	β^- /1.446	0.593/42 1.11/58	1/2+	-0.648426		0.23141 0.26085 0.33624 0.49227 0.52780
^{116}Cd	7.49(18)	115.904756				0+			
^{117m}Cd			3.4 h	β^- /2.66	0.72/	11/2-			0.1586 0.5529 0.37 - 2.42
^{117}Cd		116.907219	2.49 h	β^- /2.52	0.67/51 2.2/10	1/2+			0.2209 0.2733 0.3445 1.3033
^{118}Cd		117.90692	50.3 m	β^- /0.52		0+			
^{119m}Cd			2.20 m	β^- /		11/2-			0.1056 0.7208 1.0250 2.0213
^{119}Cd		118.90992	2.69 m	β^- /3.8	≈ 3.5/	1/2+			0.1340 0.2929 0.3429
^{120}Cd		119.90985	50.8 s	β^- /1.76	1.5/	0+			
^{121m}Cd			8. s	β^- /		11/2-			0.1008 0.9878 1.0209 1.1815 2.0594

Elem. or Isot.	Natural Abundance (%)	Atomic Mass or Weight	Half-Life	Decay Mode/Energy (/MeV)	Particle Energy /Intensity (MeV / %)	Spin (h/2π)	Nuclear Magnetic Mom.(nm)	Elect. Quadr. Mom. (b)	γ-ray Energy /Intensity (MeV / %)
¹²¹Cd		120.9131	13.5 s	β⁻ /4.9		(3/2+)			0.2102 0.3242 0.3492 1.0403
¹²²Cd		121.9135	5.3 s	β⁻ /3.0		0+			
¹²³ᵐCd			1.9 s	β⁻ /					
¹²³Cd		122.91770	2.09 s	β⁻ /6.12		3+			
¹²⁴Cd		123.9177	1.24 s	β⁻ /4.17		0+			0.0365 0.0628 0.1799
¹²⁵ᵐCd			0.66 s	β⁻ /					
¹²⁵Cd		124.92129	0.68 s	β⁻ /7.16		3/+			
¹²⁶Cd		125.9224	0.52 s	β⁻ /5.49		0+			0.2601
¹²⁷Cd		126.9264	0.4 s	β⁻ /8.5		3/+			
¹²⁸Cd		127.9278	0.28 s	β⁻ /7.1		0+			0.247
¹²⁹Cd		128.9323	0.27 s	β⁻ /5.9					0.281
¹³⁰Cd		129.9340	0.20 s	β⁻ /		0+			
₄₉In		**114.818(3)**							
⁹⁸In		97.9422							
⁹⁹In		98.9346		β⁺ /8.9					
¹⁰⁰In		99.9316	6. s	β⁺,(p)/10.5					
¹⁰¹In		100.9266	15. s	β⁺ /7.3					
¹⁰²In		101.9243	22. s	EC/8.9		(5)			0.1566 0.7767 (0.397-0.923)
¹⁰³ᵐIn			34. s						
¹⁰³In		102.91991	1.1 m	β⁺,EC/6.05	4.2	9/2+			ann.rad./ 0.1879 (0.157-3.98)
¹⁰⁴ᵐIn			16. s	IT/0.0935					
¹⁰⁴In		103.9183	1.84 m	β⁺,EC/7.9	4.8	5+	+4.44	+0.7	ann.rad./ 0.6580 0.8341 0.8781
¹⁰⁵ᵐIn			43. s	I.T.		1/2-			In k x-ray 0.6740
¹⁰⁵In		104.91467	5.1 m	β⁺,EC/4.85	3.7	9/2+	+5.675	+0.83	0.1310 0.2600 0.6038
¹⁰⁶ᵐIn			5.3 m	β⁺ /85 / EC/15 /	4.90	3+			ann.rad./ 0.6326 0.8611 1.7164
¹⁰⁶In		105.91346	6.2 m	β⁺ /65 /6.52 EC/35 /	2.6	7+	+4.92	+0.97	ann.rad./ 0.2259 0.6327 0.8611 0.9978 1.0091
¹⁰⁷ᵐIn			51. s	I.T./0.6786		1/2-			In k x-ray 0.6785
¹⁰⁷In		106.91029	32.4 m	β⁺ /35 /3.43 E.C/65 /	2.20/	9/2+	+5.59	+0.81	ann.rad./ Cd k x-ray 0.2050 0.3209 0.5055 (0.2 - 2.99)
¹⁰⁸ᵐIn			57. m	β⁺ /53 / EC/47 /	1.3	6+			ann.rad./ Cd k x-ray 0.6329 1.9863 3.4522
¹⁰⁸In		107.90971	40. m	β⁺ /33 /5.15 EC/67 /	3.49/	3+	+4.56	+1.01	ann.rad./ Cd k x-ray 0.2429 0.6331 0.8756
¹⁰⁹ᵐIn			1.3 m	I.T./0.650		1/2-			In k x-ray 0.6498
¹⁰⁹In		108.90715	4.2 h	β⁺ /8 /2.02 EC/92 /	0.79/	9/2+	+5.54	+0.84	ann.rad./ Cd k x-ray 0.2035 0.6235

Elem. or Isot.	Natural Abundance (%)	Atomic Mass or Weight	Half-Life	Decay Mode/Energy (/MeV)	Particle Energy /Intensity (MeV / %)	Spin (h/2π)	Nuclear Magnetic Mom.(nm)	Elect. Quadr. Mom. (b)	γ-ray Energy /Intensity (MeV / %)
^{110m}In			4.9 h	EC/		7+	+4.72	+1.00	Cd k x-ray 0.6577 0.8847 0.9375 (0.1 - 1.98)
^{110}In		109.90717	1.15 h	β⁺ /62 /3.88 EC/38 /	2.22/	2+	+4.37	+0.35	ann.rad./ Cd k x-ray 0.6577 (0.6 - 3.6)
^{111m}In			7.7 m	I.T./0.537		1/2-	+5.53		In k x-ray 0.537
^{111}In		110.90511	2.8049 d	EC/0.866		9/2+	+5.50	+0.80	Cd k x-ray 0.1712 0.2453
^{112m}In			20.8 m	I.T./0.155		4+			In k x-ray 0.1555
^{112}In		111.90553	14.4 m	β⁺ /22 /2.586 EC/34 / β⁻ /0.663		1+	+2.82	+0.09	ann.rad./ Cd k x-ray 0.6171
^{113m}In			1.658 h	I.T./0.3917		1/2-	-0.210		In k x-ray 0.3917
^{113}In	4.29(5)	112.904062				9/2+	+5.529	+0.80	
^{114m}In			49.51 d	I.T./97 /0.190 EC/3 /		5+	+4.7		In k x-ray 0.19027
^{114}In		113.904918	1.198 m	β⁻ /97 /1.989 EC/3 /1.453	1.984/	1+	+2.82		Cd k x-ray 0.5584 0.5727 1.2998
^{115m}In			4.486 h	I.T./95 /0.336 β⁻ /5 /0.83		1/2-	-0.255		In k x-ray 0.3362 0.4974
^{115}In	95.71(5)	114.903879	4.4x10¹⁴ y	β⁻ /0.495		9/2+	+5.541	+0.81	
^{116m2}In			2.16 s	I.T./0.162		8-			In k x-ray 0.1624
^{116m1}In			54.1 m	β⁻ /	1.0	5+	+4.3		0.13792 0.41688/27 1.09723/58.5 1.29349/85
^{116}In		115.905261	14.1 s	β⁻ /3.274	3.3/99	1+	2.788	0.11	0.46313 1.2526 1.29349
^{117m}In			1.94 h	β⁻ /53 /1.769 I.T./47 /	1.77/	1/2-	0.25		In k x-ray 0.15855 0.31531 0.55294
^{117}In		116.90452	44. m	β⁻ /1.455	0.74/	9/2+	+5.52	+0.83	0.15855 0.3966 0.55294
^{118m2}In			8.5 s	I.T./98 / β⁻ /2 /		(8-)			In k x-ray 0.1382
^{118m1}In			4.40 m	β⁻ /	1.3 2.0	5+			0.2086 0.6833 1.2295
^{118}In		117.90636	5.0 s	β⁻ /4.42	4.2/	1+			0.5282 1.1734 1.2295 2.0432
^{119m}In			17.9 m	β⁻ /97 / I.T./3 /0.311	2.7/	1/2-			0.3114 0.7631
^{119}In		118.90585	2.3 m	β⁻ /2.36	1.6/	9/2+	+5.52	+0.85	0.0239 0.6495 0.7631 1.2149
^{120m}In			47. s	β⁻ /	2.2/	8-			0.7042 1.1725 2.0398 2.3902
^{120}In		119.90796	3.1 s	β⁻ /5.37	5.6/ 3.1/	(1+)	+4.30	+0.81	0.4146 0.5924 0.8637 1.0232 1.1714 (0.4 - 2.7)
^{121m1}In			3.8 m	β⁻ /99 / I.T./1 /0.313	3.7/	1/2-			0.0601 0.3136 0.9256 1.0412 1.1022 1.1204

Elem. or Isot.	Natural Abundance (%)	Atomic Mass or Weight	Half-Life	Decay Mode/Energy (/MeV)	Particle Energy /Intensity (MeV / %)	Spin (h/2π)	Nuclear Magnetic Mom.(nm)	Elect. Quadr. Mom. (b)	γ-ray Energy /Intensity (MeV / %)
^{121}In		120.90785	23. s	β^- /3.36	2.5	9/2+	+5.50	+0.81	0.2620
									0.6573
									0.9256
^{122m}In			10. s	β^- /	4.4/	8-			1.0014
									1.1403
^{122}In		121.91028	1.5 s	β^- /6.37	5.3/	(1+)			0.2391
									1.0014
									1.1403
									1.164
									1.1903
^{123m}In			47. s	β^- /	4.6/	(1/2-)	-0.40		0.1258
									1.170
									3.234
^{123}In		122.91044	6.0 s	β^- /4.39	3.3/	(9/2+)	+5.49	+0.76	0.6188
									1.0197
									1.1305
^{124m}In			3.4 s	β^-		8-	+3.89	+0.66	0.1029
									0.9699
									1.0729
									1.1316
^{124}In		123.91318	3.18 s	β^- /7.36	5/	3+	+4.04	+0.61	0.7070
									0.9978
									1.1316
									3.2142
									(0.3 - 4.6)
^{125m}In			12.2 s	β^- /	5.5/	1/2-	-0.43		0.1876
^{125}In		124.91360	2.33 s	β^- /5.42	4.1/	9/2+	+5.50	+0.71	0.4260
									1.0318
									1.3350
^{126m}In			1.53 s		4.9/	3+	+4.03	+0.49	0.9086
									0.9696
									1.1411
^{126}In		125.91646	1.63 s	β^- /8.21	4.2/	8-	+4.06		0.1118
									0.9086
									1.1411
^{127m}In			3.73 s	β^- /	6.4/	(1/2-)			0.2523
									3.074
^{127}In		126.91734	1.14 s	β^- /6.51	4.9/	(9/2+)	+5.52	+0.59	0.4680
									0.6461
									0.8051
									1.5977
^{128m}In			0.7 s	β^- /	5.4/	(8-)			1.8670
									1.9739
									(0.1205-2.12)
^{128}In		127.92017	0.80 s	β^- /8.98	5.0/	3+			0.9352
									1.1688
									3.5198
									4.2970
^{129m}In			1.23 s	β^- /98 / n/2 /	≈ 7.5/	1/2-			0.3153
									0.9067
									1.2220
^{129}In		128.9217	0.63 s	β^- /7.66	5.5/	9/2+			0.2853
									0.7693
									1.8650
									2.1180
^{130m2}In			0.53 s	β^- /	8.8/	5+			0.0892
									0.7744
									1.2212
^{130m1}In			0.51 s	β^- /	6.1/	10-			0.0892
									0.1298
									0.7744
									1.2212
									1.9052
^{130}In		129.92486	0.29 s	β^- /10.25	10.0/	1-			
^{131m2}In			0.3 s	β^- /		(21/2+)			
^{131m1}In			0.35 s	β^- /		(1/2-)			
^{131}In		130.9268	0.28 s	β^- /9.18	6.4/	(9/2+)			0.3328
									2.433
^{132}In		131.9323	0.20 s	β^- /13.6	6.0/ 8.8/	(7-)			0.1320
									0.2992
									0.3747
									4.0406
^{133}In		132.9383	0.18 s	β^- ,(n)					
^{134}In		133.9447	0.14 s						(0.354-2.005)
$_{50}$Sn		118.710(7)							
^{100}Sn		99.9394	≈ 0.9 s	β^+ /7.3	3.4/				
^{101}Sn		100.9361	3. s	β^+ /9.					

Elem. or Isot.	Natural Abundance (%)	Atomic Mass or Weight	Half-Life	Decay Mode/Energy (/MeV)	Particle Energy /Intensity (MeV / %)	Spin (h/2π)	Nuclear Magnetic Mom.(nm)	Elect. Quadr. Mom. (b)	γ-ray Energy /Intensity (MeV / %)
¹⁰²Sn		101.9243	≈ 5 s	β⁺ /5.8					
¹⁰³Sn		102.9281	7. s	β⁺ /7.7					
¹⁰⁴Sn		103.9232	21. s	β⁺, EC/4.5					
¹⁰⁵Sn		104.9214	28. s	β⁺ /6.3					In-x-ray (0.2879-3.819)
¹⁰⁶Sn		105.91688	2.0 m	β⁺ /20 /3.18 EC/80 /					ann.rad./ In k x-ray 0.3865 0.4772
¹⁰⁷Sn		106.9157	2.92 m	EC/5.0 β⁺ /	1.2/				0.4218 0.6105 0.6785 1.0013 1.1290 1.542
¹⁰⁸Sn		107.91196	10.3 m	β⁺ /1 /2.09 EC/99 /	0.36/	0+			In k x-ray 0.2724 0.3965 (0.105-1.68)
¹⁰⁹Sn		108.91129	18.0 m	β⁺ /9 /3.85 EC/91 /	1.52/	7/2+	-1.08	+0.3	ann.rad./ In k x-ray 0.6498 1.0992
¹¹⁰Sn		109.90785	4.1 h	EC/0.64		0+			In k x-ray 0.283
¹¹¹Sn		110.90774	35. m	β⁺ /31 /2.45 EC/69 /	1.5/	7/2+	+0.61	+0.2	In k x-ray 0.7620 1.1530 1.9147
¹¹²Sn	0.97(1)	111.904822				0+			
¹¹³ᵐSn			21.4 m	I.T./92 /0.077 EC/8 /		7/2+			Sn k x-ray In x-ray 0.0774
¹¹³Sn		112.905174	115.1 d	EC/1.036		1/2+	-0.879		In k x-ray 0.25511 0.39169
¹¹⁴Sn	0.65(1)	113.902783				0+			
¹¹⁵Sn	0.34(1)	114.903347				1/2+	-0.9188		
¹¹⁶Sn	14.54(9)	115.901745				0+			
¹¹⁷ᵐSn			13.60 d	I.T./0.3146		11/2-	-1.40	-0.4	Sn k x-ray 0.15856
¹¹⁷Sn	7.68(7)	116.902955				1/2+	-1.0010		
¹¹⁸Sn	24.22(9)	117.901608				0+			
¹¹⁹ᵐSn			293. d	I.T./0.0896		11/2-	-1.4	0.21	Sn k x-ray 0.02387
¹¹⁹Sn	8.59(4)	118.903311				1/2+	-1.0473		
¹²⁰Sn	32.59(9)	119.902199				0+			
¹²¹ᵐSn			≈ 55. y	I.T./78 /0.006 β⁻ /22 /	0.354/	11/2-			Sn k x-ray 0.03715
¹²¹Sn		120.904239	1.128 d	β⁻ /0.388	0.383/100	3/2+	0.698	-0.02	
¹²²Sn	4.63(3)	121.903441				0+			
¹²³ᵐSn			40.1 m	β⁻ /1.428	1.26/99	3/2+			0.1603 0.3814
¹²³Sn		122.905723	129.2 d	β⁻ /1.404	1.42/99.4	11/2-	-1.370	+0.03	0.1603 1.0302 1.0886
¹²⁴Sn	5.79(5)	123.905275				0+			
¹²⁵ᵐSn			9.51 m	β⁻ /2.387	2.03/98	3/2+			0.3321 1.4040
¹²⁵Sn		124.907785	9.63 d	β⁻ /2.364	2.35/82	11/2-	-1.35	+0.1	1.0671 (0.2-2.3)
¹²⁶Sn		125.90765	2.1x10⁵y	β⁻ /0.38	0.25/100	0+			0.0643 0.0876 0.4148 0.6663 0.6950
¹²⁷ᵐSn			4.15 m	β⁻ /3.21	2.72/	3/2+			0.4909 1.3480 1.5640
¹²⁷Sn		126.91035	2.12 h	β⁻ /3.20	2.42/ 3.2/	11/2-			0.8231 1.0956 (0.120-2.84)
¹²⁸ᵐSn			6.5 s	IT/0.091		(7-)			
¹²⁸Sn		127.91054	59.1 m	β⁻ /1.27	0.48/ 0.63/	0+			0.4823 0.5573 0.6805 1.1611
¹²⁹ᵐSn			6.9 m	β⁻ /		11/2-			
¹²⁹Sn		128.9134	2.4 m	β⁻ /4.0		3/2+			0.6456

Elem. or Isot.	Natural Abundance (%)	Atomic Mass or Weight	Half-Life	Decay Mode/Energy (/MeV)	Particle Energy /Intensity (MeV / %)	Spin (h/2π)	Nuclear Magnetic Mom.(nm)	Elect. Quadr. Mom. (b)	γ-ray Energy /Intensity (MeV / %)
¹³⁰ᵐSn			1.7 m	β⁻ /		(7-)			0.1449 0.8992
¹³⁰Sn		129.91386	3.7 m	β⁻ /2.15	1.10/	0+			0.0700 0.1925 0.7798
¹³¹ᵐSn			1.02m	β⁻/	3.4/	11/2-			0.3043 0.4500 0.7985 1.2260 (0.08 - 3.21)
¹³¹Sn		130.9169	39. s	β⁻ /4.64	3.8/	3/2+			see ¹³¹ᵐ Sn
¹³²Sn		131.91775	40. s	β⁻ /3.3	1.8/	0+			0.0855 0.2467 0.3402 0.8985
¹³³Sn		132.9236	1.44 s	β⁻ /7.8	7.5/	7/2-			
¹³⁴Sn		133.9278	1.04 s	β⁻ /6.8					
₅₁Sb		121.760(1)							
¹⁰³Sb		102.9401							
¹⁰⁴Sb		103.9363	0.5 s						
¹⁰⁵Sb		104.9315	1.1 s						
¹⁰⁶Sb		105.9288	0.6 s	β⁺ /10.5					
¹⁰⁷Sb		106.9242	≈ 4.6 s	β⁺ /7.9					(0.253-2.154)
¹⁰⁸Sb		107.9222	7.0 s	β⁺ /9.5					(0.151-1.280)
¹⁰⁹Sb		108.91814	17. s	β⁺ /6.38 EC/	4.42/ 4.67/ 4.33/	5/2+			ann.rad./ 0.2467 0.6645 0.9254 1.0617 1.4958
¹¹⁰Sb		109.9175	24. s	β⁺ /9.0 EC/	6.8/	3+			ann.rad./ 0.6365 0.9847 1.2117 1.2433
¹¹¹Sb		110.91254	1.25 m	β⁺ /87 /4.47 EC/13 /	3.3/	5/2+			ann.rad./ 0.1002 0.1545 0.4891 1.0326
¹¹²Sb		111.91240	51.4 s	β⁺ /90 /7.06 EC/10 /	4.75/	3+			ann.rad./ 0.6700 0.9909 1.2571 (0.3 - 3.6)
¹¹³Sb		112.90937	6.7 m	β⁺ /65 /3.91 EC/35 /	2.42/	5/2+			ann.rad./ Sn k x-ray 0.3324 0.4980
¹¹⁴Sb		113.9091	3.49 m	β⁺ /78 /5.9 EC/22 /	3.4/	3+	1.7		ann.rad./ Sn k x-ray 0.8876 1.2999
¹¹⁵Sb		114.90660	32.1 m	β⁺ /67 /3.03 EC/33 /	1.51/	5/2+	+3.46	-0.4	ann.rad./ Sn k x-ray 0.4973
¹¹⁶ᵐSb			1.00 h	β⁺ /78 / EC/22 /	1.16/	8-			ann.rad./ Sn k x-ray 0.4073 0.5429 0.9725 1.2935 (0.0998-1.501)
¹¹⁶Sb		115.90680	16. m	β⁺ /50 /4.707 EC/50 /	1.3/ 2.3/	3+	2.72		ann.rad./ Sn k x-ray 0.93180 1.29354 (0.138-3.903)
¹¹⁷Sb		116.90484	2.80 h	β⁺ /2 /1.76 EC/98 /	0.57/	5/2+	+3.4		Sn k x-ray 0.1586
¹¹⁸ᵐSb			5.00 h	EC/99 /		8-	2.3		Sn k x-ray 0.25368 1.05069 1.22964
¹¹⁸Sb		117.905533	3.6 m	β⁺ /74 /3.657 EC/26 /	2.65/	1+	2.5		ann.rad./ Sn k x-ray 1.22964

Elem. or Isot.	Natural Abundance (%)	Atomic Mass or Weight	Half-Life	Decay Mode/Energy (/MeV)	Particle Energy /Intensity (MeV / %)	Spin (h/2π)	Nuclear Magnetic Mom.(nm)	Elect. Quadr. Mom. (b)	γ-ray Energy /Intensity (MeV / %)
^{119}Sb		118.90395	38.1 h	EC/0.59		5/2+	+3.45	-0.4	Sn k x-ray 0.0239
^{120m}Sb			5.76 d	EC/		8-	2.34		Sn k x-ray 0.0898, 0.19730, 1.02301, 1.17121
^{120}Sb		119.90508	15.89 m	β^+ /41 /2.68 EC/59 /	1.72/	1+	+2.3		ann.rad./ Sn k x-ray 0.7038, 1.17121
^{121}Sb	57.21(5)	120.903822				5/2+	+3.363	-0.4	
^{122m}Sb			4.19 m	I.T./0.162		8-			Sb x-ray 0.0614, 0.0761, 0.56409, 0.69277, 1.14050
^{122}Sb		121.90518	2.72 d	β^- /98 /1.979 β^+ /2 /1.620	1.414/65 1.980/26	2-	-1.90	+0.9	1.2569
^{123}Sb	42.79(5)	122.904216				7/2+	+2.550	-0.5	
^{124m2}Sb			20.3 m	I.T./0.035		8-			
^{124m1}Sb			1.6 m	I.T./80 / β^- /20 /	1.2/ 1.7/	5+			0.4984, 0.6027, 0.6458, 1.1010
^{124}Sb		123.905938	60.20 d	β^- /2.905	0.61/52 2.301/23	3-	1.2	+1.9	0.60271/97.8, 0.64583/7.4, 0.72277/10.5, 1.69094/48.2
^{125}Sb		124.905247	2.758 y	β^- /0.767	0.13/30 0.302/45 0.62/13	7/2+	+2.63		(0.0274-2.808), 0.0355, 0.17632, 0.38044, 0.42786, 0.46336, 0.60060, 0.63595
^{126m2}Sb			11. s	I.T./		3-			L x-ray 0.0227
^{126m1}Sb			19.0 m	β^- /86 / I.T./14 /	1.9	5+			0.4148, 0.6663, 0.6950
^{126}Sb		125.90725	12.4 d	β^- /3.67	1.9	8-	1.3		0.2786, 0.4148, 0.6663, 0.6950, 0.7205
^{127}Sb		126.906914	3.84 d	β^- /1.581	0.89/ 1.10/ 1.50/	7/2+	2.70		0.2524, 0.2908, 0.4121, 0.4370, 0.6857, 0.7837
^{128m}Sb			10.1 m	β^- /96 / I.T./4 /	2.6/	5+			0.3140, 0.5941, 0.7432
^{128}Sb		127.90917	9.1 h	β^- /4.38	2.3/	8-	1.3		0.7539, 0.2148, 0.3141, 0.5265, 0.7433, 0.7540
^{129m}Sb			17.7 m	β^- /					0.4338, 0.6578, 0.7598
^{129}Sb		128.90915	4.40 h	β^- /2.38	0.65/	7/2-	2.82		0.0278, 0.1808, 0.3594, 0.4596, 0.5447, 0.8128, 0.9146, 1.0301
^{130m}Sb			6.5 m	β^- /2.6	2.12/				0.1023, 0.7934, 0.8394

Elem. or Isot.	Natural Abundance (%)	Atomic Mass or Weight	Half-Life	Decay Mode/Energy (/MeV)	Particle Energy /Intensity (MeV / %)	Spin (h/2π)	Nuclear Magnetic Mom.(nm)	Elect. Quadr. Mom. (b)	γ-ray Energy /Intensity (MeV / %)
¹³⁰Sb		129.91155 ·	38.4 m	β⁻ /4.96	2.9/	8-			0.1823 0.3309 0.4680 0.7394 0.8394
¹³¹Sb		130.9120	23.0 m	β⁻ /3.2	1.31/ 3.0/	7/2+			0.6423 0.6579 0.9331 0.9434
¹³²ᵐSb			2.8 m	β⁻ /	3.9/	4+			0.1034 0.3538 0.6968 0.9739 0.9896
¹³²Sb		131.91420	4.2 m	β⁻ /5.29		8-			0.1034 0.1506 0.6968 0.9739
¹³³Sb		132.9152	2.5 m	β⁻ /4.00	1.20/	7/2+	3.00		0.4235 0.6318 0.8165 1.0764
¹³⁴ᵐSb			10.4 s	β⁻ /	6.1	7-			0.1152
¹³⁴Sb		133.9206	0.8 s	β⁻ /8.4	8.4	0-			0.2970 0.7063 1.2791
¹³⁵Sb		134.9252	1.71 s	β⁻ /8.12		7/2+			1.127 1.279
¹³⁶Sb		135.9301	0.82 s	β⁻ /9.3					
₅₂Te		127.60(3)							
¹⁰⁶Te		105.9377	0.06 ms	α/4.32					
¹⁰⁷Te		106.9350	3.1 ms	α/ 70/ β⁺ ,EC/10.1	3.86(1)/				
¹⁰⁸Te		107.9295	2.1 s	α/68 / β⁺ ,EC/32 /6.8	3.314(4)/	0+			
¹⁰⁹Te		108.9275	4.6 s	β⁺ EC/96 /8.7 α/4 /	3.107(4)/				
¹¹⁰Te		109.9224	19. s	β⁺ ,EC/4.5		0+			ann.rad./ 0.2191 0.6059
¹¹¹Te		110.9211	19.3 s	β⁺ ,EC/8.0		(7/2+)			ann.rad./ 0.267 0.322 0.341
¹¹²Te		111.9171	2.0 m	β⁺ ,EC/4.3		0+			ann.rad./ 0.2962 0.3727 0.4187
¹¹³Te		112.9154	1.7 s	β⁺ /85 /5.7 EC/15 /	4.5/	(7/2+)			ann.rad./ Sb k x-ray 0.8144 1.0181 1.1812
¹¹⁴Te		113.9125	15. m	β⁺ /40 /3.2 EC/60 /		0+			ann.rad./ Sb k x-ray 0.0838 0.0903
¹¹⁵ᵐTe			6.7 m	β⁺ /45 / EC/55 /		(1/2+)			ann.rad./ Sb k x-ray 0.7236 0.7704
¹¹⁵Te		114.9116	5.8 m	β⁺ /45 /4.6 EC/55 /	2.7/	7/2+			ann.rad./ Sb k x-ray 0.7236 1.3268 1.3806 (0.22 - 2.7)
¹¹⁶Te		115.9084	2.49 h	EC/1.5		0+			Sb k x-ray 0.0937
¹¹⁷Te		116.90864	1.03 h	EC/75 /3.54 β⁺ /25 /	1.78/	1/2+			ann.rad./ Sb k x-ray 0.9197 1.7164 2.3000
¹¹⁸Te		117.90583	6.00 d	EC/0.28		0+			Sb k x-ray

Elem. or Isot.	Natural Abundance (%)	Atomic Mass or Weight	Half-Life	Decay Mode/Energy (/MeV)	Particle Energy /Intensity (MeV / %)	Spin (h/2π)	Nuclear Magnetic Mom.(nm)	Elect. Quadr. Mom. (b)	γ-ray Energy /Intensity (MeV / %)
^{118m}Te			4.69 d	EC/		11/2-	0.89		Sb k x-ray 0.15360 0.2705 1.21271
^{119}Te		118.90641	16.0 h	β^+ /2 /2.293 EC/98 /	0.627/	1/2+	0.25		ann.rad./ Sb k x-ray 0.6440 0.6998
^{120}Te	0.09(1)	119.90403				0+			
^{121m}Te			≈ 154. d	I.T.(89%) EC(11%)		11/2-			Te k x-ray 0.2122
^{121}Te		120.90494	16.8 d	EC/1.04		1/2+			Sb k x-ray 0.5076 0.5731
^{122}Te	2.55(12)	121.903056				0+			
^{123m}Te			119.7 d	I.T./0.247		11/2-	-0.93		Te k x-ray 0.1590/84.1
^{123}Te	0.89(3)	122.904271	2.4×10^{19} y	EC/0.051		1/2+	-0.73695		
^{124}Te	4.74(14)	123.902819				0+			
^{125m}Te			58. d	I.T./0.145		11/2-	-0.99	-0.06	Te k x-ray 0.0355
^{125}Te	7.07(15)	124.904424				1/2+	-0.8885		
^{126}Te	18.84(25)	125.903305				0+			
^{127m}Te			109. d	I.T./98 /0.088 β^- /2 /0.77		11/2-	-1.04		Te k x-ray 0.0883 0.3603
^{127}Te		126.905217	9.4 h	β^- /0.698	0.696/	3/2+	0.64		
^{128}Te	31.74(8)	127.904462				0+			
^{129m}Te			33.6 d	I.T./63 /0.105 β^- /37 /	1.60/	11/2-			Te k x-ray 0.45984 0.6959
^{129}Te		128.906596	1.16 h	β^- /1.498	0.99/9 1.45/89	3/2+	0.70	0.06	0.0278 0.45984 0.48728
^{130}Te	34.08(62)	129.906223	2.5×10^{21} y			0+			
^{131m}Te			1.35 d	β^- /78 /2.4 I.T./22 /0.18	0.42/	11/2-	-1.04		0.0811 0.1021 0.14973 0.77369 0.79375 0.85225
^{131}Te		130.908522	25.0 m	β^- /2.233	1.35/12 1.69/22 2.14/60	3/2+	0.70		0.14973 0.45327 0.49269
^{132}Te		131.90852	3.26 d	β^- /0.493	0.215	0+			0.049725 0.11198 0.22830
^{133m}Te			55.4 m	β^- /82 / I.T./18 /0.334	2.4/30	11/2-			Te k x-ray 0.0949 0.1689 0.3121 0.3341
^{133}Te		132.9109	12.4 m	β^- /2.9	2.25/25 2.65	3/2+			0.3121 0.4079 1.3334
^{134}Te		133.9116	42. m	β^- /1.6	0.6/ 0.7/	0+			0.7672/29 0.0794-0.9255
^{135}Te		134.9165	19.0 s	β^- /6.0	5.4/ 6.0				0.267 0.603 0.870
^{136}Te		135.92010	17.5 s	β^- /5.1	2.5/	0+			2.0779/25 0.0873-3.235
^{137}Te		136.9253	2.5 s	β^- /98 /6.9 n/2 /	6.8	7/2-			0.2436
^{138}Te		137.9292	1.4 s	β^- /6.4					
$_{53}$I		126.90447(3)							
^{108}I		107.9436	0.04 s	α/91 /4.	3.95				
^{109}I		108.9382	0.11 ms	p					
^{110}I		109.9346	0.65 s	β^+,EC/83 /11.4 α/17 /≈3.6 p /11 /	3.457(10)/				ann.rad./
^{111}I		110.9303	2.5 s	β^+,E../8.5					ann.rad./ 0.2665 0.3215 0.3412
^{112}I		111.9280	3.4 s	β^+,EC/10.2					ann.rad./ 0.6889 0.7869

Elem. or Isot.	Natural Abundance (%)	Atomic Mass or Weight	Half-Life	Decay Mode/Energy (/MeV)	Particle Energy /Intensity (MeV / %)	Spin (h/2π)	Nuclear Magnetic Mom.(nm)	Elect. Quadr. Mom. (b)	γ-ray Energy /Intensity (MeV / %)
^{113}I		112.9237	5.9 s	β^+ ,EC/7.6					ann.rad./ 0.4625/100 0.6224/74 0.0550-1.422
^{114}I		113.9219	2.1 s	β^+ ,EC/8.7					ann.rad./ 0.6826 0.7088
^{115}I		114.9188	1.3 m	β^+ ,EC/6.7		5/2+			ann.rad./ 0.275 0.284 0.460 0.709
^{116}I		115.9167	2.9 s	β^+ /97 /7.8 EC/3 /	6.7/	1+			ann.rad./ 0.5402 0.6789
^{117}I		116.9136	2.22 m	β^+ ,EC/4.7	3.2/	(5/2+)	3.1		ann.rad./ 0.2744 0.3259
^{118m}I			8.5 m	β^+ ,EC/ I.T.	4.9/	7-	4.2		ann.rad./ 0.104 0.5998 0.6052 0.6138
^{118}I		117.9134	14. m	β^+ ,EC/7.0		2-	2.0		ann.rad./ 0.5448 0.6052 1.3384
^{119}I		118.9102	19. m	β^+ /54 /3.5 EC/46 /	2.4/	(5/2+)	+2.9		ann.rad./ Te k x-ray 0.2575
^{120m}I			53. m	β^+ /80 / EC/20 /	3.8		4.2		ann.rad. Te k x-ray 0.4257 0.5604 0.6147 1.3459
^{120}I		119.91005	1.35 h	β^+ /81 /5.62 EC/19 /	4.03 4.60	2-	1.23		ann.rad./ Tek x-ray 0.5604 0.6411 1.5230 (0.43 - 3.1)
^{121}I		120.90737	2.12 h	β^+ /13 /2.27 EC/87 /	1.2/	5/2+	2.3		ann.rad./ Te k x-ray 0.2122 (0.14 - 1.1)
^{122}I		121.90760	3.6 m	β^+ /4.234 EC/	3.1/	1+	0.94		ann.rad./ Te k x-ray 0.5641
^{123}I		122.905605	13.2 h	EC/1.242		5/2+	2.82		Te k x-ray 0.1590
^{124}I		123.906211	4.18 d	β^+ /23 /3.160 EC/77 /	1.54/ 2.14/ 0.75/	2-	1.44		ann.rad./ Te k x-ray 0.6027/62.9 0.7228/10.3 1.6910/11.2 (0.31-1.73)
^{125}I		124.904624	59.4 d	EC/0.1861		5/2+	2.82	-0.89	Te k x-ray 0.0355
^{126}I		125.905619	13.0 d	EC/ β^+ /2.155 β^- /1.258	1.13/ 0.87/ 1.25/	2-	1.44		ann.rad./ Te k x-ray 0.3887 0.6622
^{127}I	100.	126.904468				5/2+	+2.8133	-0.79	
^{128}I		127.905805	25.00 m	β^- /2.118 EC/1.251	2.13/	1+			Te k x-ray 0.44287 0.52658
^{129}I		128.904988	1.7×10^7 y	β^- /0.194	0.15/	7/2+	+2.621	-0.55	Xe k x-ray 0.0396
^{130m}I			9.0 m	I.T./83 /0.048 β^- /17 /		2+			I k x-ray 0.5361
^{130}I		129.906674	12.36 h	β^- /2.949	1.04/ 0.62	5+	3.35		0.4180 0.5361 0.6685 0.7395
^{131}I		130.906125	8.040 d	β^- /0.971	0.606/	7/2+	+2.742	-0.40	0.08017 0.28431 0.36446 0.63699

Elem. or Isot.	Natural Abundance (%)	Atomic Mass or Weight	Half-Life	Decay Mode/Energy (/MeV)	Particle Energy /Intensity (MeV / %)	Spin (h/2π)	Nuclear Magnetic Mom.(nm)	Elect. Quadr. Mom. (b)	γ-ray Energy /Intensity (MeV / %)
^{132m}I			1.39 h	IT		8-			
^{132}I		131.90800	2.28 h	β^- /14 /3.58 I.T./86 /	0.80/ 1.03/ 1.2/ 1.6/ 2.16/	4+	3.09	0.09	I k x-ray 0.0980 0.5059 0.52264 0.63019 0.6506 0.66768 0.77260 0.95457
^{133m}I			9. s	I.T./1.63		19/2-			I kx-ray 0.0730 0.6474 0.9126
^{133}I		132.90781	20.8 h	β^- /1.77	1.24/85	7/2+	+2.86	-0.27	0.51056 0.52989 0.87537
^{134m}I			3.7 m	I.T./98 /0.316 β^- /2 /		8-			I k x-ray 0.0444 0.2719
^{134}I		133.9099	52.6 m	β^- /4.2	1.2/	4+			0.1354 0.84702 0.88409
^{135}I		134.91005	6.57 h	β^- /2.65	0.9/ 1.3/	7/2+			0.2884 0.41768 0.52658 1.13156 1.26046
^{136m}I			47. s	β^- /	4.7/ 5.2/	6-			0.1973 0.3468 0.3701 0.3814 1.3130 (0.16 - 2.36)
^{136}I		135.91466	1.39 m	β^- /6.93	4.3/ 5.6/	2-			0.3447 1.3130 1.3211 2.2896 (0.3 - 6.1)
^{137}I		136.91787	24.5 s	β^- /5.88	5.0/	(7/2+)			0.6010 1.2180 1.2201 1.3026 1.5343 (0.25 - 4.4)
^{138}I		137.9224	6.5 s	β^- /7.8	6.9/ 7.4/	2-			0.4836 0.5888 0.8752 (0.4 - 5.3)
^{139}I		138.92609	2.30 s	β^- /6.81 n/					0.192 0.198 0.273 0.382 0.386 0.468 0.683 1.313
^{140}I		139.9310	0.86 s	β^- /8.8 n/		(3)			0.372 0.377 0.457
^{141}I		140.9351	0.45 s	β^- /7.8					
^{142}I		141.9402	≈ 0.2 s	β^-					
$_{54}$Xe		131.29(2)							
^{110}Xe		109.9445	0.2 s	β^+ /9.2					
^{111m}Xe			0.9 s	EC,β^+					
^{111}Xe		110.9416	0.7 s	EC,β^+ /10.6 α/	3.58(1)/				
^{112}Xe		111.9357	3. s	EC,β^+ /7.2 α/0.8 /					
^{113}Xe		112.9334	2.8 s	EC,β^+ /9.1					
^{114}Xe		113.9281	10.0 s	β^+ ,EC/5.9		0+			ann.rad./ 0.1031 0.1616 0.3085 0.6826 0.7088
^{115}Xe		114.9270	18. s	β^+ ,EC/7.6		(5/2+)			ann.rad./

Elem. or Isot.	Natural Abundance (%)	Atomic Mass or Weight	Half-Life	Decay Mode/Energy (/MeV)	Particle Energy /Intensity (MeV / %)	Spin (h/2π)	Nuclear Magnetic Mom.(nm)	Elect. Quadr. Mom. (b)	γ-ray Energy /Intensity (MeV / %)
^{116}Xe		115.9214	56. s	β^+ ,EC/4.3	3.3/	0+			ann.rad./ 0.1042 0.1916 0.2477 0.3107 0.4127
^{117}Xe		116.9206	1.02 m	β^+ ,EC/6.5		(5/2+)			ann.rad./ 0.2214 0.5190 0.6389 0.6613
^{118}Xe		117.917	≈ 4. m	β^+ ,EC/3.	2.7/	0+			ann.rad./ 0.0535 0.0600 0.1199
^{119}Xe		118.9156	5.8 m	β^+ ,EC/5.0	3.5/	7/2+	0.6		0.0873 0.1000 0.2318 0.4615
^{120}Xe		119.91216	40. m	β^+ ,EC/97 /1.96 β^+ /3 /		0+			I k x-ray 0.0251 0.0726 0.1781 (0.1 - 1.03)
^{121}Xe		120.91138	39. m	β^+ /44 /3.73 EC/56 /	2.8/	5/2+	0.65		ann.rad./ I k x-ray 0.1328 0.2527 0.4452 (0.1 - 3.1)
^{122}Xe		121.9086	20.1 h	EC/0.9		0+			I k x-ray 0.3501
^{123}Xe		122.90848	2.00 h	β^+ /23 /2.68 EC/77 /	1.51/	1/2+			ann.rad./ I k x-ray 0.1489 0.1781 (0.1 - 2.1)
^{124}Xe	0.09(1)	123.905895							
^{125m}Xe			57. s	I.T./0.252		(9/2-)			Xe k x-ray 0.1111 0.141
^{125}Xe		124.906398	17.1 h	EC/1.653	0.47/	1/2+			I k x-ray 0.1884 0.2434
^{126}Xe	0.09(1)	125.90427				0+			
^{127m}Xe			1.15 m	I.T./0.297		(9/2-)			Xe k x-ray 0.1246 0.1725
^{127}Xe		126.905179	36.4 d	EC/0.662		1/2+	-0.504		I k x-ray 0.1721 0.2029 0.3750
^{128}Xe	1.92(3)	127.903531				0+			
^{129m}Xe			8.89 d	I.T./0.236		11/2-			Xe k x-ray 0.0396 0.1966
^{129}Xe	26.44(24)	128.904780				1/2+	-0.7780		
^{130}Xe	4.08(2)	129.903509				0+			
^{131m}Xe			11.9 d	I.T./0.164		11/2-	+0.6908		Xe k x-ray 0.16398
^{131}Xe	21.18(3)	130.905083				3/2+	+0.69186	-0.12	
^{132}Xe	26.89(6)	131.904155				0+			
^{133m}Xe			2.19 d	I.T./0.233		11/2-	-1.082		Xe k x-ray 0.23325
^{133}Xe		132.905906	5.243 d	β^- /0.427	0.346/99	3/2+	+0.813	+0.14	Cs k x-ray 0.080998 0.1606
^{134}Xe	10.44(10)	133.905395				0+			
^{135m}Xe			15.3 m	I.T./		11/2-	1.103		Xe k x-ray 0.52658
^{135}Xe		134.90721	9.10 h	β^- /1.15	0.91/	3/2+	0.903		0.24975 0.60807
^{136}Xe	8.87(16)	135.90722				0+			
^{137}Xe		136.91156	3.82 m	β^- /4.17	4.1/ 3.6/	7/2-			0.45549 0.8489 0.9822 1.2732 1.7834 2.8498

T A B L E   O F   T H E   I S O T O P E S

Elem. or Isot.	Natural Abundance (%)	Atomic Mass or Weight	Half-Life	Decay Mode/Energy (/MeV)	Particle Energy /Intensity (MeV / %)	Spin (h/2π)	Nuclear Magnetic Mom.(nm)	Elect. Quadr. Mom. (b)	γ-ray Energy /Intensity (MeV / %)
^{138}Xe		137.91399	14.1 m	β^- /2.77	0.8/ 2.4/	0+			0.1538 0.2426 0.2583 0.4345 1.76826 2.0158
^{139}Xe		138.91879	39.7 s	β^- /5.06	4.5/ 5.0/				0.1750 0.2186 0.2965 (0.1 - 3.37)
^{140}Xe		139.9216	13.6 s	β^- /4.1	2.6	0+			0.0801 0.6220 0.8055 1.4137 (0.04 - 2.3)
^{141}Xe		140.9267	1.72 s	β^- /6.2	6.2/	5/2+			0.1187 0.9095 (0.05 - 2.55)
^{142}Xe		141.9297	1.22 s	β^- /5.0	3.7/ 4.2/	0+			0.0338 0.0729 0.2038 0.3091 0.4145 0.5382 0.5718 0.6181 0.6448
^{143m}Xe			0.96 s	β^-					
^{143}Xe		142.9352	0.30 s	β^- /7.3					
^{144}Xe		143.9385	1.2 s	β^- /6.1					
^{145}Xe		144.9437	0.9 s	β^- ,(n)					
$_{55}$Cs		132.90545(2)							
^{112}Cs		111.9503	0.5 ms	p	0.81				
^{113}Cs		112.9445	17. μs	p	0.96				
^{114}Cs		113.9408	0.58 s	β^+ ,EC/11.8		1+			ann.rad./ 0.6826 0.7088
^{115}Cs		114.9359	≈ 1.4 s	β^+ ,EC/8.4					ann.rad./
^{116m}Cs			0.7 s	β ,EC/					ann.rad./ 0.3935
^{116}Cs		115.9330	3.8 s	β^+ ,EC/10.8					ann.rad./ 0.3935 0.5243 0.6151 0.6223
^{117m}Cs			6.5 s	β^+ ,EC/					
^{117}Cs		116.9286	≈ 8.4 s	β^+ ,EC/7.5					ann.rad./
^{118m}Cs			17. s	β^+ ,EC/					
^{118}Cs		117.92654	14. s	β^+ ,EC/9.		2			ann.rad./ 0.3372 0.4727 0.5865 0.5906
^{119m}Cs			28. s			3/2			
^{119}Cs		118.92234	38. s	β^+ ,EC/6.3		9/2+	+5.5	+2.8	ann.rad./ 0.169 0.176 0.224 0.257
^{120m}Cs			60. s	β^+ ,EC/					
^{120}Cs		119.92066	64. s	β^+ ,EC/7.92		2+	+3.87	+1.45	ann.rad./ 0.3224 0.4735 0.5534 (0.3 - 3.28)
^{121m}Cs			2.0 m	I.T./60 / β^+ /40 /	4.4	(9/2+)	+5.41	+2.7	ann.rad./ 0.1794 0.1961
^{121}Cs		120.91718	2.3 m	β^+ ,EC/5.40	4.38/	3/2+	+0.77	+0.84	ann.rad./ 0.1537 (0.08 - 0.56)
^{122m2}Cs			4.4 m	β^+ ,EC		8-	+4.77	+3.3	ann.rad./
^{122m1}Cs			0.36 s	IT					0.3311 0.4971 0.6385 (0.27 - 2.22)

Elem. or Isot.	Natural Abundance (%)	Atomic Mass or Weight	Half-Life	Decay Mode/Energy (/MeV)	Particle Energy /Intensity (MeV / %)	Spin (h/2π)	Nuclear Magnetic Mom.(nm)	Elect. Quadr. Mom. (b)	γ-ray Energy /Intensity (MeV / %)
^{122}Cs		121.91614	21. s	β^+ ,EC/7.1	5.8/	(1+)	-0.133	-0.19	ann.rad./ 0.3311 0.5120 0.8179
^{123m}Cs			1.6 s	I.T./		11/2-			Cs k x-ray 0.0946
^{123}Cs		122.91299	5.87 m	β^+ /75 /4.20 EC/25 /	3.0/	1/2+	+1.38		ann.rad./ Xe k x-ray 0.0974 0.5964
^{124m}Cs ^{124}Cs		123.91225	6.3 s 30. s	IT β^+ /92 /5.92 EC/8 /	≈ 5.	7+ 1+	+0.673	-0.74	ann.rad./ Xe k x-ray 0.3539 0.4925 0.9418
^{125}Cs		124.90972	45. m	β^+ /40 /3.09 EC/60 /	2.06/	1/2+	+1.41		ann.rad./ Xe k x-ray 0.112 0.526
^{126}Cs		125.90945	1.64 m.	β^+ /81 /4.83 EC/19 /	3.4 3.7/	1+	+0.78	-0.68	ann.rad./ Xe k x-ray 0.3886 0.4912 0.9252
^{127}Cs		126.90741	6.2 h	β^+ /96 /2.08 EC/4 /	0.65/ 1.06	1/2+	+1.46		Xe k x-ray 0.1247 0.4119
^{128}Cs		127.90775	3.62 m	β^+ /68 /3.930 EC/32 /	2.44/ 2.88/	1+	+0.97	-0.57	ann.rad./ Xe k x-ray 0.4429
^{129}Cs		128.90606	1.336 d	EC/1.195		1/2+	+1.49		Xe k x-ray 0.3719 0.4115
^{130m}Cs ^{130}Cs		129.90671	3.5 m 29.21 m	IT,β^+ ,EC β^+ /55 /2.98 EC/43 / β^- /1.6 /0.37	1.98/ 0.44/1.6	5- 1+	+1.46	-0.06	ann.rad./ Xe k x-ray 0.5361
^{131}Cs ^{132}Cs		130.90546 131.906430	9.69 d 6.48 d	EC/0.352 EC/98 / β^+ /0.3 /2.120 β^- / /1.280		5/2+ 2-	+3.54 +2.22	-0.58 +0.51	Xe k x-ray Xe k x-ray 0.4646 0.6302 0.66769
^{133}Cs ^{134m}Cs	100.	132.905447	2.91 h	I.T./0.139		7/2+ 8-	+2.582 +1.098	-0.0037 +1.0	Cs k x-ray 0.12749
^{134}Cs		133.906714	2.065 y	β^- /2.059 EC/1.22	0.089/27 0.658/70	4+	+2.994	+0.39	0.56327 0.56935 0.60473 0.79584
^{135m}Cs			53. m	I.T./1.627		19/2-	+2.18	+0.9	0.7869 0.8402
^{135}Cs ^{136m}Cs ^{136}Cs		134.905972 135.907307	2.3x10^6 y 19. s 13.16 d	β^- /0.269 I.T./ β^- /2.548	0.205/100 0.341/	7/2+ 8 5+	+2.732 +3.71	+0.05 +0.2	0.06691 0.34057 0.81850 1.04807
^{137}Cs		136.907085	30.2 y	β^- /1.176	0.514/95	7/2+	+2.84	+0.05	Ba k x-ray 0.66164
^{138m}Cs			2.9 m	I.T./75 /0.080 β^- /25 /	3.3	6-	+1.71	-0.40	Cs k x-ray 0.0799 0.1917 0.4628 1.43579
^{138}Cs		137.91101	32.2 m	β^- /5.37	2.9/	3-	+0.700	+0.12	0.1381 0.46269 1.00969 1.43579 2.21788
^{139}Cs		138.913359	9.3 m	β^- /4.213	4.21	7/2+	+2.70	-0.07	0.6272 1.2832 (0.4 - 3.66)
^{140}Cs		139.91727	1.06 m	β^- /6.22	5.7/ 6.21/	1-	+0.13390	-0.11	0.5283 0.6023 0.9084 (0.41 - 3.94)

Elem. or Isot.	Natural Abundance (%)	Atomic Mass or Weight	Half-Life	Decay Mode/Energy (/MeV)	Particle Energy /Intensity (MeV / %)	Spin (h/2π)	Nuclear Magnetic Mom.(nm)	Elect. Quadr. Mom. (b)	γ-ray Energy /Intensity (MeV / %)
^{141}Cs		140.92005	24.9 s	β⁻ /5.26	5.20/	7/2+	+2.44	-0.4	Ba k x-ray 0.0485 0.5616 0.5887 1.1940 (0.05 - 3.33)
^{142}Cs		141.92430	1.8 s	β⁻ /7.31	6.9/ 7.28				0.3596 0.9668 1.1759 1.3265
^{143}Cs		142.92732	1.78 s	β⁻ /6.24	6.1	(3/2+)	+0.87	+0.47	0.1955 0.2324 0.3064 (0.17 - 1.98)
^{144}Cs		143.93203	1.01 s	β⁻ /8.47	8.46/ 7.9/	1	-0.546	+0.30	0.1993 0.5598 0.6392 0.7587
^{145}Cs		144.93541	0.59 s	β⁻ /7.89	7.4/ 7.9/	3/2+	+0.784	+0.6	0.1126 0.1755 0.1990
^{146}Cs		145.94024	0.322 s	β⁻ ,(n)/9.38	≈ 9.0	2-			
^{147}Cs		146.9439	0.227 s	β⁻ ,(n)/9.3					
^{148}Cs		147.9490	0.15 s	β⁻ ,(n)/10.5					
$_{56}$Ba		137.327(7)							
^{114}Ba		113.9509	0.43 s	β⁺,(p)	p/20				
^{115}Ba		114.948	0.45 s	β⁺,(p)	p/ < 15				
^{116}Ba		115.9417	1.3 s	β⁺,(p)	p/3				
^{117}Ba		116.9377	1.8 s	β⁺,(p),EC/8.4	p/13	(3/2-)			(0.0457-0.364)
^{118}Ba		117.9466	5.2 s	β⁺,					(0.040-0.156)
^{119}Ba		118.931	5.4 s	β⁺ ,EC/8.					
^{120}Ba		119.9260	24. s	β⁺ ,EC/5.0		0+			ann.rad./ 0.140 (0.075-0.146)
^{121}Ba		120.9245	30. s	β⁺ ,EC/6.8		5/2	+0.660	+1.8	ann.rad./
^{122}Ba		121.9203	2.0 m	β⁺ ,EC/3.8		0+			ann.rad./
^{123}Ba		122.9189	2.7 m	β⁺ ,EC/5.5			-0.68	+1.5	ann.rad./ 0.0306 0.0927 0.1161 0.1235
^{124}Ba		123.91509	12. m	β⁺ ,EC/2.65					ann.rad./ 0.1695 0.1888 1.2160
^{125m}Ba			8. m	β⁺ ,EC/	4.5				
^{125}Ba		124.9146	3.5 m	β⁺ ,EC/4.6	3.4	1/2+	+0.18		ann.rad./ 0.0550 0.0776 0.0854 0.1409
^{126}Ba		125.91124	1.65 h	β⁺ /2 /1.67 EC/98 /		0+			Cs k x-ray 0.2179 0.2336 0.2576
^{127}Ba		126.9111	12.9 m	β⁺ /54 /3.5 EC/46 /		1/2+	+0.09		ann.rad./ Cs k x-ray 0.1148 0.1808 (0.07 - 2.5)
^{128}Ba		127.90831	2.43 d	EC/0.52		0+			Cs k x-ray 0.27344
^{129m}Ba			2.17 h	EC/98 / β⁺ /2 /		7/2+			Cs k x-ray 0.1769 0.1823 0.2023 1.4593
^{129}Ba		128.90868	2.2 h	β⁺ /20 /2.43 EC/80 /	1.42/	1/2+	-0.40		ann.rad./ Cs k x-ray 0.1291 0.2143 0.2208
^{130}Ba	0.106(1)	129.90631				0+			
^{131m}Ba			14.6 m	I.T./0.187		9/2-	-0.87	+1.5	Ba k x-ray 0.1085

Elem. or Isot.	Natural Abundance (%)	Atomic Mass or Weight	Half-Life	Decay Mode/Energy (/MeV)	Particle Energy /Intensity (MeV / %)	Spin (h/2π)	Nuclear Magnetic Mom.(nm)	Elect. Quadr. Mom. (b)	γ-ray Energy /Intensity (MeV / %)
^{131}Ba		130.90693	11.7 d	EC/1.37		1/2+	0.7081		Cs k x-ray 0.12381/28.4 0.21608/21.3 0.49636/42.9 (0.0549-1.171)
^{132}Ba	0.101(1)	131.905056				0+			
^{133m}Ba			1.621 d	I.T./0.288		11/2-	-0.91	+0.9	Ba k x-ray 0.2761
^{133}Ba		132.906003	10.53 y	EC/0.517		1/2+	0.7717		Cs k x-ray 0.08099 0.35600
^{134}Ba	2.417(18)	133.904504				0+			
^{135m}Ba			1.20 d	I.T./0.2682		11/2-	-1.00	+1.0	Ba k x-ray 0.2682
^{135}Ba	6.592(12)	134.905684				3/2+	+0.838	+0.16	
^{136m}Ba			0.308 s	I.T./2.0305		7-			Ba k x-ray 0.8185 1.0481
^{136}Ba	7.854(24)	135.904571				0+			
^{137m}Ba			2.552 m	I.T./0.6617		11/2-	-0.99	+0.8	Ba k x-ray 0.66164
^{137}Ba	11.232(24)	136.905822				3/2+	+0.9374	+0.245	
^{138}Ba	71.698(42)	137.905242				0+			
^{139}Ba		138.908836	1.396 h	β^-/2.317	2.14/27 2.27/72	7/2-	-0.97	-0.57	0.16585 1.2544 1.42033
^{140}Ba		139.91060	12.75 d	β^-/1.05	0.48 1.0/ 1.02/	0+			0.16268 0.30485 0.53727
^{141}Ba		140.91441	18.3 m	β^-/3.22	2.59/ 2.73/	3/2-	-0.34	+0.45	0.1903 0.2770 0.3042 (0.1 - 2.5)
^{142}Ba		141.91645	10.7 m	β^-/2.212	1.0/ 1.10/	0+			0.23152 0.25512 0.3090 1.2040
^{143}Ba		142.92061	14.3 s	β^-/4.24	4.2/	5/2+	+0.44	-0.88	0.1786 0.21148 0.7988 (0.17 - 2.4)
^{144}Ba		143.92294	11.4 s	β^-/3.1	2.4/ 2.9/	0+			La k x-ray 0.10386 0.1566 0.1728 0.3882 0.43048
^{145}Ba		144.9269	4.0 s	β^-/4.9	4.9/	(5/2-)	-0.28	+1.22	La k x-ray 0.0918 0.09709
^{146}Ba		145.9302	2.20 s	β^-/4.12	3.9/	0+			0.0644 0.2513 0.3270 0.3329 0.3622
^{147}Ba		146.9340	0.892 s	β^-/5.75	5.5/				
^{148}Ba		147.9377	0.64 s	β^-,n/5.11					
^{149}Ba		148.9421	0.36 s	β^-,(n)/7.3					
$_{57}$La		138.9055(2)							
^{120}La		119.938	2.8 s	EC,β^+/11.					
^{121}La		120.9330	5.3 s						
^{122}La		121.9307	9. s	EC,β^+/ ≈ 9.7					
^{123}La		122.9262	17. s	EC /7.					
^{124}La		123.9245	30. s	EC/ ≈ 8.8		(7+)			
^{125}La		124.9207	1.2 m	β^+,EC/5.6		11/2-			ann.rad./ 0.0436 0.0676
^{126}La		125.9194	1.0 m	β^+,EC/7.6					ann.rad./ 0.2561 0.340 0.4555 0.6214
^{127}La		126.9162	3.8 m	β^+,EC/4.7		3/2+			ann.rad./ 0.025 0.0562

Elem. or Isot.	Natural Abundance (%)	Atomic Mass or Weight	Half-Life	Decay Mode/Energy (/MeV)	Particle Energy /Intensity (MeV / %)	Spin (h/2π)	Nuclear Magnetic Mom.(nm)	Elect. Quadr. Mom. (b)	γ-ray Energy /Intensity (MeV / %)
^{128}La		127.9155	5.0 m	β⁺ /80 /6.7 EC/20 /		(5-)			ann.rad./ Ba k x-ray 0.2841/87 0.4793/54 (0.315-2.212)
^{129m}La			0.56 s	IT		(11/2-)			
^{129}La		128.91267	11.6 m	β⁺ /58 /3.72 EC/42 /	2.42/	3/2 +			ann.rad./ Ba k x-ray 0.1105 0.2786 (0.1 - 1.8)
^{130}La		129.9123	8.7 m	β⁺ /78 /5.6 EC/22 /		3+			ann.rad./ Ba k x-ray 0.3573/81 0.5506/27 (0.1965-1.989)
^{131}La		130.9101	59. m	β⁺ /76 /3.0 EC/24 /	1.42/ 1.94/	3/2 +			ann.rad./ Ba k x-ray 0.1085 0.3658 0.5263
^{132m}La			24. m	I.T./76 / β⁺ ,EC/24 /		6-			La k x-ray 0.1352 0.4645
^{132}La		131.91011	4.8 h	β⁺ /40 /4.71 EC/60 /	2.6/ 3.2 3.7/	2-			ann.rad./ Ba k x-ray 0.4645 0.5671
^{133}La		132.9084	3.91 h	β⁺ /4 /2.2 EC/96 /	1.2/	5/2 +			Ba k x-ray 0.2788 0.2901 0.3024
^{134}La		133.90849	6.5 m	β⁺ /63 /3.71 EC/37/	2.67/	1+			ann.rad./ Ba k x-ray 0.6047 (0.5 - 1.9)
^{135}La		134.90697	19.5 h	EC/1.20		5/2 +			Ba k x-ray 0.4805
^{136}La		135.9077	9.87 m	β⁺ /36 /2.9 EC/64 /	1.8/	1+			ann.rad./ Ba k x-ray 0.8185
^{137}La		136.90647	6x10⁴ y	EC/0.60		7/2 +	+2.70	+0.2	0.2836
^{138}La	0.090(1)	137.907107	1.06x10¹¹ y			5 +			1.4358/65 0.7887/35
^{139}La	99.910(1)	138.906349				7/2 +			
^{140}La		139.909473	1.678 d	β⁻ /3.762	1.35 1.24/ 1.67/	3-			
^{141}La		140.910958	3.90 h	β⁻ /2.502	2.43/	7/2 +			
^{142}La		141.91408	1.54 h	β⁻ /4.505	2.11/ 2.98/ 4.52/	2-			
^{143}La		142.91606	14.1 m	β⁻ /3.43	3.3/	7/2-			
^{144}La		143.9196	40.7 s	β⁻ /5.5	4.1/				
^{145}La		144.9217	24. s	β⁻ /4.1	4.1/	3/2 +			
^{146m}La			10.0 s	β⁻ /6.7	5.5/	(6)			
^{146}La		145.9258	6.3 s	β⁻ /6.6	6.2/	(2-)			
^{147}La		146.9278	4.02 s	β⁻ /5.0	4.6/				
^{148}La		147.9322	1.1 s	β⁻ /7.26		2-			
^{149}La		148.9342	1.10 s	β⁻ /5.5					
^{150}La		149.9386	0.51 s						x-ray (0.097-0.209)
$_{58}$Ce		140.116(1)							
^{121}Ce		121.944	1.1 s	β⁺,p					
^{123}Ce		122.936	3.8 s	β⁺ ,EC/ ≈ 8.6					ann.rad./
^{124}Ce		123.931	6. s	EC / ≈ 5.6					
^{125}Ce		124.929	10. s	β⁺ ,EC/7.		(5/2 +)			ann.rad./
^{126}Ce		125.9241	50. s	EC/4.					
^{127}Ce		126.9228	32. s	β⁺ ,EC/6.1					ann.rad./ (0.058-1.148)
^{128}Ce		127.9189	4.1 m	β⁺ ,EC/3.2					ann.rad./ (0.023-0.880)
^{129}Ce		128.9187	3.5 m	β⁺ ,EC/5.6					ann.rad./ (0.0675-1.015)
^{130}Ce		129.9147	26. m	β⁺ ,EC/2.2		0+			ann.rad./ La k x-ray (0.047-1.431)

Elem. or Isot.	Natural Abundance (%)	Atomic Mass or Weight	Half-Life	Decay Mode/Energy (/MeV)	Particle Energy /Intensity (MeV / %)	Spin (h/2π)	Nuclear Magnetic Mom.(nm)	Elect. Quadr. Mom. (b)	γ-ray Energy /Intensity (MeV / %)
131mCe			5. m	β+ ,EC/					ann.rad./ 0.2304 0.3955 0.4213
131Ce		130.9144	10. m	β+ ,EC/4.0	2.8/				ann.rad./ 0.119 0.169 0.414
132Ce		131.9115	3.5 h	EC/1.3		0+			La k x-ray 0.1554 0.1821
133mCe			1.6 h	β+ ,EC/		1/2+			ann.rad./ 0.0769 0.0973 0.5577
133Ce		132.9116	5.4 h	β+ /8 /2.9 EC/92 /	1.3/	9/2-			ann.rad./ La k x-ray 0.0584 0.1308 0.4722 0.5104
134Ce		133.9090	3.16 d	EC/0.5		0+			La k x-ray 0.1304 0.1623 0.6047
135mCe			20. s	I.T./0.446		11/2-			Ce k x-ray 0.0826 0.1497 0.2134
135Ce		134.90915	17.7 h	β+ /1 /2.026 EC/99 /	0.8/	1/2+			La k x-ray 0.0345 0.2656 0.3001 0.6068
136Ce	0.19(1)	135.90714				0+			Ce k x-ray
137mCe			1.43 d	I.T./99 /0.254 EC/0.8 /		11/2-	0.70		0.1693 0.2543
137Ce		136.90778	9.0 h	β+ /1.222		3/2+	0.9		La k x-ray 0.4472
138Ce	0.25(1)	137.90599				0+			
139mCe			56.4 s	I.T./0.7542		11/2-			Ce k x-ray 0.7542
139Ce		138.90665	137.6 d	EC/0.28		3/2+	0.9		La k x-ray 0.16585
140Ce	88.48(10)	139.905435				0+			
141Ce		140.908272	32.50 d	β- /0.581	0.436/69 0.581/31	7/2-	1.1		Pr k x-ray 0.14544/48.0
142Ce	11.08(10)	141.909241				0+			
143Ce		142.912382	1.38 d	β- /1.462	1.404/ 1.110/47	3/2-	≈ 1.		Pr k x-ray 0.0574 0.2933
144Ce		143.913643	284.6 d	β- /0.319	0.185/20 0.318/	0+			Pr k x-ray 0.0801 0.1335
145Ce		144.91723	3.00 m	β- /2.54	1.7/24 1.3	3/2-			Pr k x-ray 0.0627 0.7245
146Ce		145.9187	13.5 m	β- /1.04	0.7/90	0+			Pr k x-ray 0.0986 0.2182 0.3167
147Ce		146.9225	56. s	β- /3.29	3.3/				0.0930 0.2687
148Ce		147.9244	56. s	β- /2.1	1.66/	0+			0.0904 0.0985 0.1212 0.2918
149Ce		148.9283	5.2 s	β- /4.2					0.0577 0.0864 0.3800
150Ce		149.9302	4.4 s	β- /3.0					0.1099
151Ce		150.9340	1.0 s	β- /5.3					0.0526
152Ce		151.9366	1.4 s	β- /4.4					Pr k x-ray 0.098 0.115

Elem. or Isot.	Natural Abundance (%)	Atomic Mass or Weight	Half-Life	Decay Mode/Energy (/MeV)	Particle Energy /Intensity (MeV / %)	Spin (h/2π)	Nuclear Magnetic Mom.(nm)	Elect. Quadr. Mom. (b)	γ-ray Energy /Intensity (MeV / %)
59Pr		140.90765(2)							
124Pr		123.943	1.2 s	β+,EC/12.					ann.rad./
125Pr		124.9378	≈3.3 s	β+					ann.rad./ 0.1358
126Pr		125.9353	3.1 s	β+,EC/≈10.4					ann.rad./ (0.170-0.985)
127Pr		126.9308	4.2 s	β+/≈7.5					ann.rad./ (0.0288-0.8949)
128Pr		127.9288	3.2 s	β+,EC/≈9.3					ann.rad./
129Pr		128.9249	32 s	β+,EC/5.8					ann.rad./ (0.0395-1.865)
130Pr		129.9234	40. s	β+,EC/8.1					ann.rad./
131mPr			5.7 s						(0.06 - 0.16)
131Pr		130.9201	1.7 m	β+,EC/5.3				≈5.5	ann.rad./ (0.059-0.980)
132Pr		131.9191	1.6 m	β+,EC/7.1					ann.rad./ 0.325 0.496 0.533
133Pr		132.9162	6.5 m	β+,EC/4.3		5/2+			ann.rad./ 0.074 0.1343 0.2419 0.3156 0.3308 0.4650
134mPr			≈11. m	β+,EC/					ann.rad./ 0.294 0.460 0.495 0.632
134Pr		133.9157	17. m	β+,EC/6.2		2+			ann.rad./ 0.294 0.495
135Pr		134.9131	24. m	β+,EC/3.7	2.5/	3/2+			ann.rad./ 0.0826 0.2135 0.2961 0.5832
136Pr		135.91265	13.1 m	β+/57/5.13 EC/43	2.98/	2+			ann.rad./ Ce k x-ray 0.5398 0.5522
137Pr		136.91068	1.28 h	β+/26/2.70 EC/74/	1.68/	5/2+			ann.rad./ Ce k x-ray 0.4339 0.5140 0.8367 (0.16 - 1.8)
138mPr			2.1 h	β+/24/ EC/76/	1.65/	7-			ann.rad./ Ce k x-ray 0.3027 0.7887 1.0378 (0.07 - 2.0)
138Pr		137.91075	1.45 m	β+/75/4.44 EC/25/	3.42/	1+			ann.rad./ Ce k x-ray 0.7887
139Pr		138.90893	4.41 h	β+/8/2.129 EC/92/	1.09/	5/2+			ann.rad./ Ce k x-ray 0.2551 1.3473 1.6307
140Pr		139.90907	3.39 m	β+/51/3.39 EC/49/	2.37/	1+			ann.rad./ Ce k x-ray 0.3069 1.5965
141Pr	100.	140.907648				5/2+	+4.275	-0.059	
142mPr			14.6 m	I.T./0.004	c.e./	5-	2.2		
142Pr		141.910041	19.12 h	β-/2.162 EC/0.744	0.58/4 2.16/96	2-	+0.234	+0.030	0.5088 1.57580
143Pr		142.910813	13.57 d	β-/0.934	0.933/	7/2+	+2.7	+0.8	0.7420
144mPr			7.2 m	I.T./99+/0.059 β-/		3-			Pr k x-ray 0.0590 0.6965 0.8142

Elem. or Isot.	Natural Abundance (%)	Atomic Mass or Weight	Half-Life	Decay Mode/Energy (/MeV)	Particle Energy /Intensity (MeV / %)	Spin (h/2π)	Nuclear Magnetic Mom.(nm)	Elect. Quadr. Mom. (b)	γ-ray Energy /Intensity (MeV / %)
^{144}Pr		143.913301	17.28 m	β^- /2.998	0.807/1 2.30/ 2.996/98	0-			0.69649 1.48912 2.18562
^{145}Pr		144.91451	5.98 h	β^- /1.81	1.80/97	7/2+			0.0725 0.6758 0.7483
^{146}Pr		145.9176	24.2 m	β^- /4.2	2.2/30 3.7/10 4.2/40	2-			0.4539/48 1.5247
^{147}Pr		146.91898	13.4 m	β^- /2.69	1.5/ 2.1/	3/2+			0.3146/24. 0.5779/16 0.6413/19.
^{148m}Pr			2.0 m	β^- /	4.0/ 3.8/	(4)			0.3016 0.4506 0.6975
^{148}Pr		147.9222	2.27 m	β^- /4.9	4.8/ 4.5/	1-			0.3017
^{149}Pr		148.92379	2.3 m	β^- /3.40	3.0	(5/2+)			0.1085 0.1385 0.1651
^{150}Pr		149.9270	6.2 s	β^- /5.7	≈ 5.5	1-			0.1302 0.8044 0.8527
^{151}Pr		150.9283	22.4 s	β^- /4.2					
^{152}Pr		151.9319	3.2 s	β^- /6.7		(3)			0.0726 0.164 0.285
^{153}Pr		152.9339	4.3 s	β^- /5.5					
^{154}Pr		153.9381	2.3 s	β^- /7.9					
$_{60}$Nd		144.24(3)							
^{127}Nd		126.941	1.8 s	β^+ ,EC/9.		(5/2)			ann.rad./
^{128}Nd		127.935	4. s	β^+ ,EC/6.					ann.rad./
^{129}Nd		128.933	4.9 s	β^+ ,EC/8.		5/2(-)			ann.rad./ (0.091-0.875)
^{130}Nd		129.929	28. s	β^+ ,EC/5.					ann.rad./
^{131}Nd		130.9271	0.5 m	β^+ ,EC/6.6					ann.rad./ (0.09-0.36)
^{132}Nd		131.9231	1.5 m	β^+ ,EC/3.7					ann.rad./ (0.099-0.567)
^{133}Nd		132.9222	1.2 m	β^+ ,EC/5.6					ann.rad./ (0.06-0.37)
^{134}Nd		133.9187	≈ 8.5 m	β^+ /17 /2.8 EC/83 /		0+			ann.rad./ Pr k x-ray 0.1631/58 (0.09-1.00)
^{135m}Nd			5.5 m	β^+ /					
^{135}Nd		134.9182	12. m	β^+ /65 /4.8 EC/35 /		9/2-	0.78	+2.1	ann.rad./ Pr k x-ray 0.0415/23. 0.204/51. (0.11-1.8)
^{136}Nd		135.9150	50.6 m	EC/94 /2.21 β^+ /6 /	1.04/	0+			Pr kx-ray 0.0401/21. 0.1091/35. (0.10-0.97)
^{137m}Nd			1.6 s	I.T./0.5196		11/2-			Nd k x-ray 0.1084 0.1775 0.2337
^{137}Nd		136.9146	38. m	β^+ /40 /3.69 EC/60 /	1.7/20 2.40/20	1/2+	-0.63		ann.rad./ Pr k x-ray 0.0755 0.5806
^{138}Nd		137.9119	5.1 h	EC/1.1		0+			Pr k x-ray 0.1995 0.3258
^{139m}Nd			5.5 h	I.T./12 /0.231 β^+ /88 /	1.17/	11/2-			Nd k x-ray Pr k x-ray 0.1139/34. 0.7382/30.
^{139}Nd		138.91192	30. m	β^+ /25 /2.79 EC/75 /	1.77/	3/2+	0.91	+0.3	ann.rad./ Pr k x-ray 0.4050
^{140}Nd		139.90931	3.37 d	EC /0.22		0+			Pr k x-ray
^{141m}Nd			1.04 m	I.T./99+ /0.756		11/2-			Nd k x-ray 0.7565

Elem. or Isot.	Natural Abundance (%)	Atomic Mass or Weight	Half-Life	Decay Mode/Energy (/MeV)	Particle Energy /Intensity (MeV / %)	Spin (h/2π)	Nuclear Magnetic Mom.(nm)	Elect. Quadr. Mom. (b)	γ-ray Energy /Intensity (MeV / %)
^{141}Nd		140.909605	2.49 h	EC/98 /1.823 β⁺ /2 /	0.802/	3/2+	1.01	+0.3	Pr k x-ray (0.15-1.7)
^{142}Nd	27.13(12)	141.907719				0+			
^{143}Nd	12.18(6)	142.909810				7/2-	-1.07	-0.6	
^{144}Nd	23.80(12)	143.910083	2.1x10¹⁵ y			0+			
^{145}Nd	8.30(6)	144.912569				7/2-	-0.66	-0.33	
^{146}Nd	17.19(9)	145.913113				0+			
^{147}Nd		146.916096	10.98 d	β⁻ /0.896	0.805/	5/2-	0.58	0.9	Pr k x-ray 0.09111 0.53102
^{148}Nd	5.76(3)	147.916889				0+			
^{149}Nd		148.920145	1.73 h	β⁻ /1.691	1.03/25 1.13/26 1.42/	5/2-	0.35	1.3	Pr k x-ray 0.1143/19. 0.2113/27. (0.06 - 1.6)
^{150}Nd	5.64(3)	149.920887				0+			
^{151}Nd		150.923825	12.4 m	β⁻ /2.442	1.2/	(3/2+)			Pm k x-ray 0.1168 0.2557 1.1806 (0.10 - 1.9)m
^{152}Nd		151.92468	11.4 m	β⁻ /1.1		0+			0.2785/29. 0.2501/18. (0.016 - 0.66)
^{153}Nd		152.9280	28.9 s	β⁻ /3.6					0.418
^{154}Nd		153.9296	25.9 s	β⁻ /2.8					0.1519 0.7998
^{155}Nd		154.9334	8.9 s	β⁻ /5.0					0.1807
^{156}Nd		155.9355	5.5 s	β⁻ /4.1					0.0848
$_{61}$Pm									
^{130}Pm		129.940	2.2 s	β⁺ ,EC/11.					ann.rad./
^{132}Pm		131.934	6. s	β⁺ ,EC/10.					ann.rad./
^{133}Pm		132.930	12. s	β⁺ ,EC/≈ 7.0					ann.rad./
^{134}Pm		133.9282	24. s	β⁺ ,EC/≈ 8.9		(5+)			ann.rad./ 0.294 0.495
^{135}Pm		134.9247	0.8 m	β⁺ ,EC/6.0		11/2-			(0.13-0.47)
^{136}Pm		135.9235	1.8 m	β⁺ /89 /7.9 EC/11 /		(3+)			ann.rad./ Nd k x-ray 0.3735 0.6027
^{137}Pm		136.9206	2.4 m	β⁺ ,EC/5.6		(11/2-)			ann.rad./ 0.1086 0.1775
^{138m}Pm			3.2 m	β⁺ /50 /≈ 7.0 EC/50 /	3.9/	3+			ann.rad./ Nd k x-ray 0.5209 0.7290
^{138}Pm		137.9193	10. s	β⁺ /6.9	6.1/	1+			ann.rad./
^{139m}Pm			0.18 s	IT/		(11/2-)			0.1887
^{139}Pm		138.91678	4.14 m	β⁺ /68 /4.52 EC/32 /	3.52/	(5/2+)			ann.rad./ Nd k x-ray 0.4028 (0.27 - 2.4)
^{140m}Pm			5.87 m	β⁺ /70 / EC/30 /	3.2	7/2-			ann.rad./ Nd k x-ray 0.4199 0.7738 1.0283
^{140}Pm		139.91585	9.2 s	β⁺ /89 /6.09 EC/11 /	5.07/74	1+			ann.rad./ Nd k x-ray 0.7738 1.4898
^{141}Pm		140.91359	20.9 m	β⁺ /52 /3.72 EC/48 /	2.71	5/2+			ann.rad./ Nd k x-ray 0.8862 1.2233
^{142}Pm		141.91295	40.5 s	β⁺ /86 /4.87 EC/20 /	3.8/	1+			ann.rad./ Nd k x-ray 0.6414 1.5758
^{143}Pm		142.910928	265. d	EC/1.041 β⁺ /<6 x 10⁶/		5/2+	3.8		Nd k x-ray 0.7420
^{144}Pm		143.912586	360. d	EC/2.332 β⁺ /<8 x 10⁵/		5-	1.7		Nd k x-ray 0.6180 0.6965

Elem. or Isot.	Natural Abundance (%)	Atomic Mass or Weight	Half-Life	Decay Mode/Energy (/MeV)	Particle Energy /Intensity (MeV / %)	Spin (h/2π)	Nuclear Magnetic Mom.(nm)	Elect. Quadr. Mom. (b)	γ-ray Energy /Intensity (MeV / %)
^{145}Pm		144.912745	17.7 y	EC/0.163		5/2+			Nd k x-ray 0.0723
^{146}Pm		145.914693	5.53 y	EC/63 /1.472 β^- /37 /1.542	0.795/	3-			Nd k x-ray 0.4538 0.7362 0.7474
^{147}Pm		146.915134	2.6234 y	β^- /0.224	0.224/	7/2+	+2.6	+0.7	0.1213 0.1974
^{148m}Pm			41.3 d	β^- /95 /2.6 I.T./5 /0.137	0.4/60 0.5/17 0.7/21	6-	1.8		0.5503/94. 0.6300/89. 0.7257/33
^{148}Pm		147.91747	5.37 d	β^- /2.47	1.02/ 2.47/	1-	+2.0	+0.2	0.5503 0.9149 1.4651
^{149}Pm		148.918330	2.212 d	β^- /1.071	0.78/9 1.072/90	7/2+	3.3		0.2859 0.5909 0.8594
^{150}Pm		149.92098	2.68 h	β^- /3.45	1.6/ 2.3/ 1.8/	(1-)			0.3339/69. 1.1658/16. 1.3245/17. (0.25 - 2.9)
^{151}Pm		150.92120	1.183 d	β^- /1.187	0.84/	5/2+	+1.8	1.9	0.1677/8 0.2751/7 0.3401/22
^{152m2}Pm			15. m	β^- ,I.T./		(>6)			(0.14-1.4)
^{152m1}Pm			7.5 m	β^- /		(4-)			0.1218 0.2447 0.3404 1.0971 1.4375
^{152}Pm		151.9235	4.1 m	β^- /3.5	3.5/20 3.50/60	1+			0.1218 (0.12 - 2.1)
^{153}Pm		152.92414	5.4 m	β^- /1.90	1.7/	(5/2-)			0.0910 0.1198 0.1273
^{154m}Pm			2.7 m	β^- /	2.0/				0.0820 0.1848 1.4403
^{154}Pm		153.9266	1.7 m	β^- /4.1	1.9/				0.0820 0.8396 1.3940 2.0589 (0.08 - 2.8)
^{155}Pm		154.9280	48. s	β^- /3.2		(5/2-)			(0.05-0.78)
^{156}Pm		155.93106	26.7 s	β^- /5.16					
^{157}Pm		156.9332	10.9 s	β^- /4.6					
^{158}Pm		157.9367	5. s	β^- /6.3					
$_{62}$Sm		150.36(3)							
^{131}Sm		130.946	1.2 s	β^+ ,EC/					ann.rad./
^{132}Sm		131.941	4.0 s	β^+					
^{133}Sm		132.939	2.9 s	β^+ ,EC/$\approx$8.4		5/2+			ann.rad./
^{134}Sm		133.934	11. s	β^+ ,EC/5.		0+			ann.rad./
^{135}Sm		134.932	10. s	β^+ ,EC/7.		7/2+			ann.rad./
^{136}Sm		135.9283	42. s	β^+ ,EC/4.5		0+			ann.rad./
^{137}Sm		136.9271	45. s	β^+ ,EC/6.1					ann.rad./
^{138}Sm		137.9235	3.0 m	β^+ ,EC/3.9		0+			ann.rad./ 0.0536 0.0747
^{139m}Sm			10. s	I.T./94 /0.457 β^+ /6 /	4.7	(11/2-)			Sm k x-ray 0.1118 0.1553 0.1901 0.2673
^{139}Sm		138.9226	2.6 m	β^+ /75 /5.5 EC/25 /	4.1/	1/2+	-0.53		Pm k x-ray 0.3678 0.4028 (0.27 - 2.4)
^{140}Sm		139.9195	14.8 m	β^+ ,EC/3.4	1.9/	0+			ann.rad./ Pm k x-ray 0.1396 0.2255 (0.07 - 1.7)
^{141m}Sm			22.6 m	β^+ /32 / EC/68 / I.T./0.3 /0.1758	1.6/ 2.19/	11/2-	-0.83	+1.6	ann.rad./ Pm k x-ray 0.1966 0.4318 0.7774

Elem. or Isot.	Natural Abundance (%)	Atomic Mass or Weight	Half-Life	Decay Mode/Energy (/MeV)	Particle Energy /Intensity (MeV / %)	Spin (h/2π)	Nuclear Magnetic Mom.(nm)	Elect. Quadr. Mom. (b)	γ-ray Energy /Intensity (MeV / %)
^{141}Sm		140.91847	10.2 m	β^+ /52 /4.54 EC/48 /	3.2/	1/2+	-0.74		ann.rad./ Pm k x-ray 0.4382
^{142}Sm		141.91520	1.208 h	β^+ /6 /2.10 EC/94 /	1.0/	0+			ann.rad./ Pm k x-ray 0.7540
^{143m}Sm			1.10 m	I.T./99 /0.7540		11/2-			Sm k x-ray 0.7540
^{143}Sm		142.914624	8.83 m	β^+ /46 /3.443 EC/54 /	2.47/	3/2+	+1.01	+0.4	ann.rad./ Pm k x-ray 1.0565
^{144}Sm	3.1(1)	143.911996				0+			
^{145}Sm		144.913407	340. d	EC/0.617		7/2-	-1.1	-0.6	Pm k x-ray 0.0613 0.4924
^{146}Sm		145.913038	1.03x10^{8}y	α/	2.50/	0+			
^{147}Sm	15.0(2)	146.914894	1.06x10^{11}y	α/	2.23/	7/2-	-0.815	-0.26	
^{148}Sm	11.3(1)	147.914818	7x10^{15}y	α/	1.96/	0+			
^{149}Sm	13.8(1)	148.917180	10^{16}y	α/		7/2-	-0.672	+0.075	
^{150}Sm	7.4(1)	149.917272				0+			
^{151}Sm		150.919929	90. y	β^- /0.0768	0.076/	5/2-	-0.363	+0.7	0.02154
^{152}Sm	26.7(2)	151.919729				0+			
^{153}Sm		152.922094	1.929 d	β^- /0.808	0.64/ 0.69/	3/2+	-0.0216	+1.3	Eu k x-ray 0.069676 0.10318
^{154}Sm	22.7(2)	153.922206				0+			
^{155}Sm		154.924636	22.2 m	β^- /1.627	1.52	3/2-		1.1	Eu k x-ray 0.1043/75.
^{156}Sm		155.92553	9.4 h	β^- /0.72	0.43/ 0.71/	0+			0.0872 0.1657 0.2038
^{157}Sm		156.9283	8.0 m	β^- /2.7	2.4/	3/2-			Eu k x-ray 0.1964 0.1978 0.3942
^{158}Sm		157.9299	5.5 m	β^- /2.0		0+			0.1894/100. 0.3636/82.
^{159}Sm		158.9332	11.3 s	β^- /3.8					0.1898
^{160}Sm		159.9353	9.6 s	β^- /3.6		0+			0.110
$_{63}$Eu		151.964(1)							
^{134}Eu		133.946	0.5 s	EC,β^+					ann.rad./
^{135}Eu		134.942	1.5 s	EC,β^+ / ≈ 8.7					ann.rad./ 0.255
^{136m}Eu			≈ 3.2 s			7+			
^{136}Eu		135.940	≈ 3.9 s	EC,β^+ /10.		1+			ann.rad./
^{137}Eu		136.935	11. s	EC/ ≈7.5		11/2-			ann.rad./
^{138}Eu		137.9335	12. s	EC,β^+ / ≈ 9.2		7+			ann.rad./
^{139}Eu		138.9298	18. s	EC,β^+ /6.7					ann.rad./
^{140m}Eu			0.125 s	EC,β^+					ann.rad./
^{140}Eu		139.9285	1.51 s	EC,β^+ /8.4		1-			ann.rad./
^{141m}Eu			3.0 s	β^+ /58 / EC/9 / I.T./33 /0.0964		11/2-			ann.rad./ Eu k x-ray (0.09 - 1.6)
^{141}Eu		140.9244	40. s	β^+/81 /5.6 EC/15 /		5/2+	+3.49	+0.85	ann.rad./ Sm k x-ray 0.3845 0.3940
^{142m}Eu			1.22 m	β^+/83 / EC/17 /	4.8/	8-	+2.98	+1.4	ann.rad./ Sm k x-ray 0.5566 0.7680 1.0233
^{142}Eu		141.9231	2.4 s	β^- /94/7.4 EC/6 /	7.0/	1+	+1.54	+0.12	ann.rad./ 0.7680
^{143}Eu		142.92017	2.62 m	β^+ /72/5.17 EC/28/	4.1/ 5.1/	5/2+	+3.67	+0.51	ann.rad./ Sm k x-ray 0.1107/7 1.5368/3. 1.9127/2.
^{144}Eu		143.91879	10.2 s	β^+ /86 /6.33 EC/13 /	5.31/	1+	+1.89	+0.10	ann.rad./ Sm k x-ray 1.6601
^{145}Eu		144.916263	5.93 d	β^+ /2 /2.660 EC/98 /1.71	0.79/	5/2+	+3.99	+0.29	ann.rad./ Sm k x-ray 0.6535 0.8937 1.6587

Elem. or Isot.	Natural Abundance (%)	Atomic Mass or Weight	Half-Life	Decay Mode/Energy (/MeV)	Particle Energy /Intensity (MeV / %)	Spin (h/2π)	Nuclear Magnetic Mom.(nm)	Elect. Quadr. Mom. (b)	γ-ray Energy /Intensity (MeV / %)
¹⁴⁶Eu		145.91720	4.57 d	β⁺ /5 /3.88 EC/95 /	1.47/	4-	+1.43	-0.18	ann.rad./ Sm k x-ray 0.6336 0.6341 0.7470 (0.27 - 2.64)
¹⁴⁷Eu		146.916742	24.4 d	EC/99. /1.722 β⁺ /0.4 /		5/2+	+3.72	+0.55	Sm k x-ray 0.12113 0.19725 0.6776
¹⁴⁸Eu		147.91815	54.5 d	EC/3.11	0.92	5-	+2.34	+0.35	Sm k x-ray 0.5503/99. 0.6299/71. (0.067-2.17)
¹⁴⁹Eu		148.91792	93.1 d	EC/0.692		5/2+	+3.57	+0.75	Sm k x-ray 0.2770 0.3275
¹⁵⁰Eu		149.91970	36. y	EC/2.26		5-	+2.71	+1.13	Sm k x-ray 0.3340 0.4394 0.5843 (0.25 - 1.8)
¹⁵⁰ᵐEu			12.8 h	β⁻ /92 / β⁺ /0.4 / EC/8 /	1.013/ 1.24/	0-			Sm k x-ray 0.3339 0.4065
¹⁵¹Eu ¹⁵²ᵐ²Eu	47.8(1.5)	150.919846	1.60 h	I.T./0.1478		5/2+ 8-	+3.472	+0.90	Eu k x-ray 0.0898
¹⁵²ᵐ¹Eu			9.30 h	β⁻ /72 / EC/28 /	1.85/ 0.89/	0-			Sm k x-ray 0.12178 0.84153 0.96334
¹⁵²Eu		151.921741	13.5 y	EC/72 /1.874 β⁻ /28 /1.818	0.69/ 1.47/	3-	-1.941	+2.71	Sm k x-ray Gd k x-ray 0.12178 0.34427 1.40802 (0.252-1.528)
¹⁵³Eu ¹⁵⁴ᵐEu	52.2(15)	152.921227	46.1 m	I.T./≈ 0.16		5/2+ 8-	+1.533	+2.41	Eu k x-ray 0.0682 0.1009
¹⁵⁴Eu		153.922976	8.59 y	β⁻ /99.9 /1.969 EC./0.02 /0.717	0.27/29 0.58/38 0.84/17 0.98/4 1.87/11	3-	-2.01	+2.8	Gd k x-ray 0.12299/40. 0.72331/20. 1.2745/36 (0.059-1.90)
¹⁵⁵Eu		154.922890	4.76 y	β⁻ /0.252	0.15/	5/2+	+1.6	+2.3	Gd k x-ray 0.0865/30 0.1053/20
¹⁵⁶Eu		155.92475	15.2 d	β⁻ /2.451	0.30/11 0.49/30 1.2/12 2.45/31	1+	≈ 1.1		0.08899/9. 0.64623/7. 0.723441/6. 0.8118/10.
¹⁵⁷Eu		156.92542	15.13 h	β⁻ /1.36	0.98/ 1.30/41	(5/2+)			Gd k x-ray 0.0639/100. 0.3705/48. 0.4107/76.
¹⁵⁸Eu		157.9278	45.9 m	β⁻ /3.5	2.5/	(1-)			0.0795 0.8976 0.9442 0.9771
¹⁵⁹Eu		158.92909	18.1 m	β⁻ /2.51	2.4/ 2.57/	(5/2+)			0.0678 0.0786 0.0957
¹⁶⁰Eu		159.9315	38. s	β⁻ /4.1	2.7/ 4.1/	(0-)			0.0753 0.1735 0.4131 0.5155 0.8217 0.9110 0.9246 0.0719
¹⁶¹Eu ¹⁶²Eu		160.9337 161.9370	27. s 11. s	β⁻ /3.7 β⁻ /5.6					
₆₄Gd		157.25(3)							
¹³⁵Gd ¹³⁷Gd		136.945	1.1 s 7. s	β⁺ EC,β⁺ / ≈ 8.8					(0.163-0.360) ann.rad./

Elem. or Isot.	Natural Abundance (%)	Atomic Mass or Weight	Half-Life	Decay Mode/Energy (/MeV)	Particle Energy /Intensity (MeV / %)	Spin (h/2π)	Nuclear Magnetic Mom.(nm)	Elect. Quadr. Mom. (b)	γ-ray Energy /Intensity (MeV / %)
^{139}Gd		138.9381	5. s	EC,β^+ / ≈7.7				ann.rad./	
^{140}Gd		139.934	16. s	EC/4.8		0+			0.1748
^{141m}Gd			25. s	EC,β^+ /		11/2-			ann.rad./
^{141}Gd		140.9322	21. s	β^+ /7.3		0+			ann.rad./
^{142}Gd		141.9276	1.17 m	EC,β^+ /4.2		1/2+			ann.rad./
^{143m}Gd			1.84 m	β^+ /67 /		11/2-			ann.rad./
				EC/33 /					Eu k x-ray
				I.T./					0.1176
									0.2719
									0.5880
									0.6681
									0.7999
^{143}Gd		142.9266	39. s	β^+ /82 /6.0		1/2+			ann.rad./
				EC/18 /					Eu k x-ray
									0.2048
									0.2588
^{144}Gd		143.9234	4.5 m	β^+ /45 /4.3	3.3/	0+			ann.rad./
				EC/55 /					Eu k x-ray
									0.3332
^{145m}Gd			1.44 m	I.T./95 /0.749		11/2-			0.0273
				β^+ /4 /5.7					0.3295
									0.3866
									0.7214
^{145}Gd		144.92169	23.4 m	β^+ /33 /5.05	2.5/	1/2+			ann.rad./
				EC/67 /					Eu k x-ray
									1.7579
									1.8806
									(0.32 - 3.69)
^{146}Gd		145.91831	48.3 d	EC/99.9 /1.03	0.35/	0+			Eu k x-ray
				β^+ /0.2					0.1147
									0.1155
									0.1546
^{147}Gd		146.919090	1.588 d	EC/99.8 /2.188	0.93/	7/2-	1.0		Eu k x-ray
				EC/0.2 /					0.2293
									0.3699
									0.3960
									0.9289
									(0.1 - 1.8)
^{148}Gd		147.918111	75. y	α/3.27	3.1828/	0+			
^{149}Gd		148.919339	9.3 d	EC/1.32		7/2-	0.9		Eu k x-ray
									0.1496
									0.2985
									0.3465
^{150}Gd		149.91866	1.8×10^6 y	α/2.80	2.73/	0+			
^{151}Gd		150.920345	124. d	EC/0.464		7/2-	0.8		Eu k x-ray
									0.1536
									0.2432
^{152}Gd	0.20(1)	151.919789				0+			
^{153}Gd		152.921747	241.6 d	EC/0.485		3/2-	0.4		Eu k x-ray
									0.09743
									0.10318
^{154}Gd	2.18(3)	153.920862				0+			
^{155}Gd	14.80(5)	154.922619				3/2-	-2.59	+1.30	
^{156}Gd	20.47(4)	155.922120				0+			
^{157}Gd	15.65(3)	156.923957				3/2-	-3.40	+1.36	
^{158}Gd	24.84(12)	157.924101				0+			
^{159}Gd		158.926385	18.6 h	β^- 0.971	0.60/11	3/2-	-0.44		Tb k x-ray
					0.89/26				0.36351
					0.96/63				(0.05845-0.8550)
^{160}Gd	21.86(4)	159.927051				0+			
^{161}Gd		160.929666	3.66 m	β^- /1.956	1.56/85	5/2-			Tb k x-ray
									0.1023
									0.3149
									0.3609
									0.4030
^{162}Gd		161.930981	8.4 m	β^- /1.39	1.0/	0+			0.4421
									0.2868
^{163}Gd		162.9340	1.13 m	β^- /3.1					0.214
									1.685
^{164}Gd		163.9359	45. s	β^- /2.3					
$_{65}$Tb		158.92534(2)							
^{140}Tb		139.946	2.4 s	β^+ ,EC/11					
^{141}Tb		140.941	3.5 s	β^+ ,EC/ ≈ 8.3					
^{142m}Tb			0.30 s	β^+ ,EC/		4-			
^{142}Tb		141.939	0.60 s	β^+ ,EC/10.		0+			
^{143}Tb		142.9346	12. s	β^+ ,EC/7.4		11/2-			
^{144m}Tb			4.1 s	IT		5-			

Elem. or Isot.	Natural Abundance (%)	Atomic Mass or Weight	Half-Life	Decay Mode/Energy (/MeV)	Particle Energy /Intensity (MeV / %)	Spin (h/2π)	Nuclear Magnetic Mom.(nm)	Elect. Quadr. Mom. (b)	γ-ray Energy /Intensity (MeV / %)
^{144}Tb		143.9324	< 1.5 s	β⁺ ,EC/8.4		1+			
^{145m}Tb			30. s	β⁺ ,EC/ ≈ 6.6		11/2-			ann.rad./ 0.2577 0.5370 0.9876
^{145}Tb		144.9287		β⁺ ,EC/6.5		1/2+			
^{146m}Tb			23. s	β⁺ /76 / EC/24 /		(5-)			ann.rad./ Gd k x-ray 1.0789 1.5795
^{146}Tb		145.9270	≈ 8. s	β⁺ /8.1		1+			
^{147m}Tb			1.8 m	β⁺ /35 / EC/65 /		11/2-			ann.rad./ Gd k x-ray 1.3977 1.7978
^{147}Tb		146.92404	1.6 h	β⁺ /42 /4.61 EC/58 /		5/2+			ann.rad./ Gd k x-ray 0.6944 1.1522 (0.120-3.318)
^{148m}Tb			2.3 m	β⁺ /25 / EC/75 /		9+			ann.rad./ Gd k x-ray 0.3945 0.6319 0.7845 0.8824
^{148}Tb		147.92422	1.00 h	β⁺ ,EC/5.69		2-			ann.rad./ Gd k x-ray 0.4888 0.7845 (0.14 - 3.8)
^{149m}Tb			4.16 m	EC/88 / β⁺ /12 /		11/2-			ann.rad./ Gd k x-ray 0.1650 0.7960
^{149}Tb		148.923243	4.13 h	β⁺ /4 /3.636 α/16/	1.8/ 3.97/	1/2+	< 0.9		Gd k x-ray 0.1650 0.3522 0.3886 (0.1 - 3.2)
^{150m}Tb			6.0 m	β⁺ /17 / EC/83 /					ann.rad./ Gd k x-ray 0.4384 0.6380 0.6504 0.8275
^{150}Tb		149.92366	3.3 h	β⁺ ,EC/4.66		2-			ann.rad./ 0.4963 0.6380 (0.3 - 4.29)
^{151m}Tb			25. s	I.T./95 / β⁺ ,EC/7 /		11/2-			0.0229 0.0495 0.3797 0.8305
^{151}Tb		150.923099	17.61 h	β⁺ /1 /2.565 EC/99 /	0.70/	1/2+			Gd k x-ray 0.1083 0.2517 0.2870 (0.1 - 1.8)
^{152m}Tb			4.3 m	I.T./79 /0.5018 EC/21 /4.35		(8+)			Tb k x-ray Gd k x-ray 0.2833 0.3443 0.4111
^{152}Tb		151.92407	17.5 h	β⁺ /20 /3.99 EC/80 /	2.5/ 2.8/	2-	0.9		ann.rad./ Gd k x-ray 0.3443 (0.2 - 2.88)
^{153}Tb		152.923433	2.34 d	EC/1.570		5/2+	3.5		Gd k x-ray 0.2119 (0.05 - 1.1)
^{154m2}Tb			23.1 h	EC/98 / I.T./2 /		(7-)			Gd k x-ray 0.1231 0.2479 0.3467 1.4199

Elem. or Isot.	Natural Abundance (%)	Atomic Mass or Weight	Half-Life	Decay Mode/Energy (/MeV)	Particle Energy /Intensity (MeV / %)	Spin (h/2π)	Nuclear Magnetic Mom.(nm)	Elect. Quadr. Mom. (b)	γ-ray Energy /Intensity (MeV / %)
¹⁵⁴ᵐ¹Tb			9. h	β⁺ /78 / I.T./22 /		(3-)	1.8	+3.	Gd k x-ray 0.1231 0.2479 0.5401 (0.12 - 2.57)
¹⁵⁴Tb		153.92469	21.5 h	EC/99 /3.56 β⁺ /1 /	1.86/ 2.45	0-	0.9		Gd k x-ray 0.1231 1.2744 2.1872 (0.12 - 3.14)
¹⁵⁵Tb		154.92350	5.3 d	EC/0.82		3/2 +	2.0		Gd k x-ray 0.08654 0.10530
¹⁵⁶ᵐ²Tb			1.02 d	I.T./		(7-)			Tb k x-ray 0.0496
¹⁵⁶ᵐ¹Tb			5.3 h	I.T./0.0884		(0+)			Tb k x-ray 0.0884
¹⁵⁶Tb		155.924744	5.3 d	EC/2.444		3-	1.4	+2.3	Gd k x-ray 0.08896 0.19921 0.53435 1.22245
¹⁵⁷Tb		156.924021	1.1x10² y	EC/0.0601		3/2 +	2.0		Gd k x-ray 0.0545
¹⁵⁸ᵐTb			10.5 s	I.T./0.11		0-			Gd k x-ray 0.0110
¹⁵⁸Tb		157.925410	1.8x10² y	EC/80 /1.220 β⁻ /20 /0.937		3-	+1.76	+2.7	Gd k x-ray 0.0795 0.9442 0.9621
¹⁵⁹Tb	100.	158.925343				3/2 +	+2.014	+1.43	
¹⁶⁰Tb		159.927164	72.3 d	β⁻ /1.835	0.57/47 0.86/27	3-	+1.79	3.8	Dy k x-ray 0.08678 0.29857 0.87936 0.96615
¹⁶¹Tb		160.927566	6.91 d	β⁻ /0.593	0.46/23 0.52/66 0.6/10	3/2 +	2.2	+1.2	Dy k x-ray 0.02565 0.04892 0.07458
¹⁶²Tb		161.92948	7.6 m	β⁻ /2.51	1.4	(1/2-)			Dy k x-ray 0.2600 0.8075 0.8882
¹⁶³Tb		162.930644	19.5 m	β⁻ /1.785	0.80/	3/2 +			Dy k x-ray 0.3511 0.3897 0.4945
¹⁶⁴Tb		163.9334	3.0 m	β⁻ /3.9	1.7/	(5+)			Dy k x-ray 0.1689 0.2157 0.6110 0.6885 0.7548
¹⁶⁵Tb		164.9349	2.1 m	β⁻ /3.0		3/2 +			0.5389 1.1785 1.2920 1.6648
¹⁶⁶Tb		165.9380	≈ 21 s	β/					
₆₆Dy		162.50(3)							
¹⁴¹Dy		140.951	0.9 s	EC,β⁺ /9.					
¹⁴²Dy		141.946	2.3 s	EC,β⁺ /7.1					
¹⁴³Dy		142.9440	3.9 s	EC,β⁺ /≈ 8.8					
¹⁴⁴Dy		143.9391	9.1 s	EC,β⁺ /≈ 6.2					
¹⁴⁵ᵐDy		144.9365	14. s	EC,β⁺		11/2-			
¹⁴⁶ᵐDy			0.15 s	I.T.		10+			
¹⁴⁶Dy		145.9325	30. s	EC,β⁺ /5.2					
¹⁴⁷ᵐDy			56. s	I.T./40 / β⁺ ,EC/60 /		(11/2-)	-0.66	+0.7	Dy k x-ray / 0.072 0.6787
¹⁴⁷Dy		146.9309	75. s	EC,β⁺ /6.37		1/2+	-0.92	0.1007	ann.rad./ 0.2534 0.3653
¹⁴⁸Dy		147.92710	3.1 m	β⁺ /4 /2.68 EC/96 /	1.2/	0+			ann.rad./ Tb k x-ray 0.6202

Elem. or Isot.	Natural Abundance (%)	Atomic Mass or Weight	Half-Life	Decay Mode/Energy (/MeV)	Particle Energy /Intensity (MeV / %)	Spin (h/2π)	Nuclear Magnetic Mom.(nm)	Elect. Quadr. Mom. (b)	γ-ray Energy /Intensity (MeV / %)
¹⁴⁹Dy		148.92734	4.2 m	β⁺ ,EC/3.81		(7/2-)	-0.12	-0.62	ann.rad./ 0.1008 0.1063 0.2534 0.6536 0.7894 1.7765 1.8062
¹⁵⁰Dy		149.92558	7.18 m	β⁺ ,EC/67 /1.79 α /33 /	4.233/	0+			Tb k x-ray 0.3967
¹⁵¹Dy		150.926181	17. m	β⁺ /5 /2.871 EC/89 / α /6 /	4.067/	7/2-	-0.95	-0.30	Tb k x-ray 0.1764 0.3030 0.3861 0.5463 (0.16 - 2.09)
¹⁵²Dy		151.92472	2.37 h	EC/0.60 α /	3.63/	0+			Tb k x-ray 0.2569
¹⁵³Dy		152.925763	6.3 h	β⁺ /1 /2.171 EC/99 / α /0.01 /	0.89/ 3.46/	(7/2-)	-0.78	-0.02	Tb k x-ray 0.0807 0.0997 0.2137 (0.08 - 1.66)
¹⁵⁴Dy		153.92442	3.x10⁶ y	α/2.95	2.87/	0+			
¹⁵⁵Dy		154.92575	9.9 h	β⁺ /2 /2.095 EC/98 /	0.845/	3/2-	-0.385	+1.04	Tb k x-ray 0.0655 0.2269
¹⁵⁶Dy	0.06(1)	155.92428				0+			
¹⁵⁷Dy		156.92546	8.1 h	EC/1.34		3/2-	-0.301	+1.30	Tb k x-ray (0.0609-1.319)
¹⁵⁸Dy	0.10(1)	157.924405				0+			
¹⁵⁹Dy		158.925736	144. d	EC/0.366		3/2-	-0.354	+1.37	Tb k x-ray 0.3262
¹⁶⁰Dy	2.34(6)	159.925194				0+			
¹⁶¹Dy	18.9(2)	160.926930				5/2+	-0.480	+2.51	
¹⁶²Dy	25.5(2)	161.926795				0+			
¹⁶³Dy	24.9(2)	162.928728				5/2-	+0.673	+2.65	
¹⁶⁴Dy	28.2(2)	163.929171				0+			
¹⁶⁵ᵐDy			1.26 m	I.T./98 /0.108 β⁻ /2 /		1/2-			Dy k x-ray 0.1082 0.5155
¹⁶⁵Dy		164.931700	2.33 h	β⁻ /1.286	1.29/	7/2+	-0.52	-3.5	Ho k x-ray 0.09468
¹⁶⁶Dy		165.932803	3.400 d	β⁻ /0.486	0.40/	0+			Ho k x-ray 0.0282 0.0825
¹⁶⁷Dy		166.9357	6.2 m	β⁻ / ≈ 2.35	1.78	(1/2-)			Ho k x-ray 0.2593 0.3103 0.5697 (0.06 - 1.4)
¹⁶⁸Dy		167.9372	8.5 m	β⁻ /1.6		0+			Ho k x-ray 0.1925 0.4867
¹⁶⁹Dy		168.9403	≈ 39. s	β⁻ /3.2					
₆₇Ho		164.93032(2)							
¹⁴⁴Ho		143.952	0.7 s	β⁺ ,EC/12					
¹⁴⁵Ho		144.947	2.4 s	β⁺					
¹⁴⁶Ho		145.9440	3.3 s	β⁺ ,EC/10.7		(10+)			ann.rad./
¹⁴⁷Ho		146.9396	5.8 s	β⁺ ,EC/8.2		11/2-			ann.rad./
¹⁴⁸ᵐHo			9. s	β⁺ ,EC/		4-			ann.rad./
¹⁴⁸Ho		147.9372	2. s	β⁺ ,EC/9.4		1+			ann.rad./ 0.6615 1.6883
¹⁴⁹ᵐHo			21. s	β⁺ ,EC/		11/2-			ann.rad./ 1.0733 1.0911
¹⁴⁹Ho		148.93379	> 30. s	β⁺ ,EC/6.01		1/2+			
¹⁵⁰ᵐHo			25. s	β⁺ ,EC/		(9+)			ann.rad./ 0.3939 0.5511 0.6534 0.8034
¹⁵⁰Ho		149.9326	1.3 m	β⁺ ,EC/6.6					ann.rad./ 0.5913 0.6534 0.8034

Elem. or Isot.	Natural Abundance (%)	Atomic Mass or Weight	Half-Life	Decay Mode/Energy (/MeV)	Particle Energy /Intensity (MeV / %)	Spin (h/2π)	Nuclear Magnetic Mom.(nm)	Elect. Quadr. Mom. (b)	γ-ray Energy /Intensity (MeV / %)
^{151m}Ho			47. s	β^+ ,EC/87 / α/13	 4.605/				ann.rad./ 0.2102 0.4889 0.6948 0.7762
^{151}Ho		150.93169	35.2 s	β^+ ,EC/80/5.13 α/20 /	 4.519/				ann.rad./ 0.3522 0.5274 0.9676 1.0471
^{152m}Ho			50. s	β^+ ,EC/90 / α/10 /	 4.453/	(9+)	+5.9	-1.	ann.rad./ 0.4929 0.6138 0.6474 0.6835
^{152}Ho		151.93166	2.4 m	β^+ ,EC/88 /6.47 α/12 /	 4.387/	(3+)	-1.02	+0.1	ann.rad./ 0.6140 0.6476
^{153m}Ho			9.3 m	β^+ ,EC/99+ /4.12 α/	 4.01/	5/2	+1.19		ann.rad./ 0.0905 0.1089 0.1618 0.2302 0.2707 0.3659 0.4565
^{153}Ho		152.93020	2.0 m	β^+ ,EC/99+ /4.13 α/	 3.91/	11/2-	+6.8	-1.	ann.rad./ 0.2958 0.3346 0.4381 0.6383
^{154m}Ho			3.3 m	β^+ ,EC/		(8+)	5.7	-.1.	ann.rad./ 0.3346 0.4124 0.4771
^{154}Ho		153.93060	12. m	β^+ ,EC/5.75		1-	-0.64	+0.2	ann.rad./ Dy k x-ray 0.3346 0.5700 0.8734
^{155}Ho		154.92908	48. m	β^+ /6 /3.10 EC/94 /		(5/2+)	+3.51	+1.5	ann.rad./ Dy k x-ray 0.0474 0.1363 0.3254 (0.06 - 2.24)
^{156m}Ho			5.8 m	I.T./0.0352 β^+ /25 / EC/75 /	 1.8/ 2.9/		+2.99	+2.3	ann.rad./ Dy k x-ray 0.1378 0.2666 (0.28 - 2.9)
^{156}Ho		155.9290	56. m	β^+ ,EC/4.4		(5+)			ann.rad./ 0.1378 0.2665
^{157}Ho		156.92819	12.6 m	β^+ /5 /2.54 EC/95 /	1.18/	7/2-	+4.35	+3.0	ann.rad./ Dy k x-ray 0.2800 0.3411
^{158m2}Ho			28. m	I.T./44 / EC/56 /		2-	+2.44	+1.6	ann.rad./ Dy k x-ray 0.0989 0.2182
^{158m1}Ho			21. m	β^+ ,EC/		(9+)			ann.rad./ 0.0981 0.1664 0.2182 0.3205 0.4062 0.9774 1.0532 0.4846
^{158}Ho		157.92895	11.3 m	β^+ /8 /4.24 EC/92 /	1.30/	5+	+3.77	+4.1	ann.rad./ Dy k x-ray 0.0989 0.2182 0.9488
^{159m}Ho			8.3 s	I.T./0.206		1/2+			Ho k x-ray 0.1660 0.2059

T A B L E   O F   T H E   I S O T O P E S

Elem. or Isot.	Natural Abundance (%)	Atomic Mass or Weight	Half-Life	Decay Mode/Energy (/MeV)	Particle Energy /Intensity (MeV / %)	Spin (h/2π)	Nuclear Magnetic Mom.(nm)	Elect. Quadr. Mom. (b)	γ-ray Energy /Intensity (MeV / %)
¹⁵⁹Ho		158.927708	33.0 m	EC/1.838		7/2-	+4.28	+3.2	Dy k x-ray 0.1210 0.1320 0.2529 0.3096 (0.06 - 1.2)
¹⁶⁰ᵐ²Ho ¹⁶⁰ᵐHo			3. s 5.0 h	I.T./67 /0.060 EC/33 /3.35		1+ 2-	+2.52	+1.8	0.0868 0.1970 0.6464 0.7281 0.8791 0.9619 0.9658
¹⁶⁰Ho		159.92873	25.6 m	β⁺ ,EC/3.29	0.57/	5+	+3.71	+4.0	See Ho[166m] 0.7282 0.8794
¹⁶¹ᵐHo			6.8 s	I.T./0.211					Ho k x-ray 0.2112
¹⁶¹Ho		160.927852	2.48 h	EC/0.859		7/2-	+4.25	+3.2	Dy k x-ray 0.0256 0.0592 0.0774 0.1031
¹⁶²ᵐHo			1.12 h	I.T./61 / EC/39 /		6-	+3.60	+4.	Dy k x-ray Ho k x-ray 0.0807 0.1850 0.2828 0.9372 1.2200
¹⁶²Ho		161.929092	15. m	EC/96 /0.295 β⁺ /4 /		1+			Dy k x-ray 0.0807 1.3196 1.3728
¹⁶³ᵐHo			1.09 s	I.T./0.298		(1/2+)			Ho k x-ray 0.2798
¹⁶³Ho ¹⁶⁴ᵐHo		162.928730	4.57x10³ y 38. m	EC/0.00258 I.T./0.140		7/2- (6-)	+4.23	+3.6	Dy M x-rays Ho k x-ray 0.0373 0.0566 0.0940
¹⁶⁴Ho		163.930231	29. m	EC/58 /0.987 β⁻ /42 /0.963		1+			Dy k x-ray 0.0734 0.0914
¹⁶⁵Ho ¹⁶⁶ᵐHo	100.	164.930319	1.2x10³ y	β⁻ /		7/2- 7-	+4.127 3.6	+3.49 -3.	Er k x-ray 0.18407 0.71169 0.81031
¹⁶⁶Ho		165.932281	1.117 d	β⁻ /1.855	1.776/48 1.855/51	0-			Er k x-ray 0.08057 1.37943
¹⁶⁷Ho		166.933127	3.1 h	β⁻ /1.007	0.31/43 0.61/21 0.96/15 0.97/15	(7/2-)			Er k x-ray 0.0793 0.0835 0.2379 0.3213 0.3465
¹⁶⁸ᵐHo ¹⁶⁸Ho		167.93550	2.2 m 3.0 m	I.T./ β⁻ /2.91	2.0/	3+			Er k x-ray 0.7413 0.8159 0.8211 (0.08 - 2.34)
¹⁶⁹Ho		168.93687	4.7 m	β⁻ /2.12	1.2/ 2.0/	(7/2-)			0.1496 0.7610 0.7784 0.7884 0.8529
¹⁷⁰ᵐHo			43. s	β⁻ /		1+			0.0787 0.8123 1.8940 1.9726

Elem. or Isot.	Natural Abundance (%)	Atomic Mass or Weight	Half-Life	Decay Mode/Energy (/MeV)	Particle Energy /Intensity (MeV / %)	Spin (h/2π)	Nuclear Magnetic Mom.(nm)	Elect. Quadr. Mom. (b)	γ-ray Energy /Intensity (MeV / %)
^{170}Ho		169.93962	2.8 m	β⁻ /3.87		6+			Er k x-ray 0.1816 0.2582 0.8902 0.9321 0.9414 1.1387
^{171}Ho		170.941	53 s	β/					
^{172}Ho		171.9448	25. s	β⁻ /					Er k x-ray (0.077-1.186)
$_{68}$Er		167.26(3)							
^{145}Er		144.957	0.9 s	β⁺					
^{146}Er		145.952	≈ 1.7 s	β⁺					
^{147}Er		146.9494	2.5 s	E.C,β⁺ / ≈ 9.1					
^{148}Er		147.9444	4.5 s	β⁺ ,EC/6.8					
^{149m}Er			10. s	IT		11/2-			
^{149}Er		148.9425	10.7 s	ECβ⁺ /8.1		1/2+			
^{150}Er		149.9370	18. s	β⁺ /36 /4.11 EC/64 /		0+			ann.rad./ Ho k x-ray 0.4758
^{151}Er		150.9373	23. s	β⁺ ,EC/5.2		7/2-			ann.rad./
^{152}Er		151.93500	10.2 s	β⁺ ,EC/10 /3.11 α/90 /	4.804/	0+			ann.rad./
^{153}Er		152.93509	37.1 s	α/ β⁺ ,EC/47 /4.56	4.674 4.35/				0.351 (0.0945-1.700)
^{154}Er		153.93278	3.7 m	β⁺ ,EC/99+ /2.03 α/0.5 /	4.166/	0+			ann.rad./
^{155}Er		154.93321	5.3 m	β⁺ ,EC/47 /3.84 EC/53 /		(7/2-)	-0.669	-0.27	ann.rad./ Ho k x-ray 0.1101 0.2415
^{156}Er		155.9308	20. m	β⁺ ,EC/1.7		0+			ann.rad./ 0.0298 0.0352 0.0522 0.1336
^{157}Er		156.9319	25. m	β⁺ ,EC/3.5		3/2-	-0.412	+0.92	ann.rad./ 0.117 0.385 1.320 1.660 1.820 2.000
^{158}Er		157.93087	2.2 h	EC/99.5 /1.78 β⁺ /0.5 /	0.74/	0+			Ho k x-ray 0.0719 0.2486 0.3868
^{159}Er		158.930681	36. m	β⁺ /7 /2.769 EC/93 /		3/2-	-0.304	+1.17	ann.rad./ Ho k x-ray 0.6245 0.6493 (0.07 - 2.5)
^{160}Er		159.92908	1.191 d	EC/0.33		0+			Ho k x-ray (0.05 - 0.96)
^{161}Er		160.93000	3.21 h	EC/2.00		3/2-	-0.37	+1.36	Ho k x-ray 0.8265 (0.07 - 1.74)
^{162}Er	0.14(1)	161.928775				0+			
^{163}Er		162.93003	1.25 h	EC/1.210		5/2-	+0.557	+2.55	Ho k x-ray 0.4361 0.4399 1.1135
^{164}Er	1.61(2)	163.929197				0+			
^{165}Er		164.930723	10.36 h	EC/0.376		5/2-	+0.643	+2.71	Ho k x-ray
^{166}Er	33.6(2)	165.930290				0+			
^{167m}Er			2.27 s	I.T./0.208		1/2-			Er k x-ray 0.2078
^{167}Er	22.95(15)	166.932046				7/2+	-0.5639	+3.57	
^{168}Er	26.8(2)	167.932368				0+			
^{169}Er		168.934588	9.40 d	β⁻ /0.351	0.35/ ≈ 100	1/2-	+0.515		Tm k x-ray 0.1098 0.1182
^{170}Er	14.9(2)	169.935461				0+			

Elem. or Isot.	Natural Abundance (%)	Atomic Mass or Weight	Half-Life	Decay Mode/Energy (/MeV)	Particle Energy /Intensity (MeV / %)	Spin (h/2π)	Nuclear Magnetic Mom.(nm)	Elect. Quadr. Mom. (b)	γ-ray Energy /Intensity (MeV / %)
¹⁷¹Er		170.938026	7.52 h	β⁻ /1.491		5/2-	0.66	2.9	Tm k x-ray 0.11160 0.29591 0.30832 (0.08 - 1.4)
¹⁷²Er		171.939352	2.05 d	β⁻ /0.891	0.28/48 0.36/46				Tm k x-ray 0.0597 0.4073 0.6101
¹⁷³Er		172.9424	1.4 m	β⁻ /2.6		(7/2-)			Tm k x-ray 0.1928 0.1992 0.8952
¹⁷⁴Er		173.9441	3.1 m	β⁻ /1.8					Tm k x-ray (0.100-0.152)
¹⁷⁵Er		174.9479	1.2 m	β⁻					(0.0765-1.168)
₆₉Tm		168.93421(2)							
¹⁴⁶ᵐTm			0.21 s	β⁺,p	p/1.118				
¹⁴⁶Tm		145.967	0.06 s	β⁺ /14. p	1.119/ 1.189/				
¹⁴⁷ᵐTm			0.4 ms	β⁺,p	p/1.115				
¹⁴⁷Tm		146.961	0.56 s	EC, β⁺ /85/≈10.7 p/15/	1.052/				
¹⁴⁸ᵐTm		147.9573	0.7 s	β⁺ ,EC/12.					ann.rad./
¹⁴⁸Tm									
¹⁴⁹Tm		148.9524	0.9 s	β⁺ ,EC/ ≈9.2		11/2-			
¹⁵⁰Tm		149.9494	2.3 s	β⁺ ,EC/ ≈11.5		6-			(0.1007-2.177)
¹⁵¹Tm		150.9454	4. s	β⁺ ,EC/7.5					ann.rad./
¹⁵²ᵐTm			8. s	β⁺ ,EC/		9+			
¹⁵²Tm		151.9443	5. s	β⁺ ,EC/8.8					ann.rad./
¹⁵³Tm		152.94203	1.6 s	β⁺ ,EC/10 /6.46 α/90 /	5.109/				ann.rad./
¹⁵⁴ᵐTm			3.3 s	β⁺ ,EC/15 / α /	α/5.031/100 4.84/0.24				ann.rad./ (0.4605-0.7960)
¹⁵⁴Tm		153.9407	8.1 s	β⁺ ,EC/56 /7.4 α /44 /	α/4.956/100 4.83/0.45				ann.rad./
¹⁵⁵Tm		154.93919	30. s	β⁺ ,EC/5.58 α/	4.46/				0.0315 0.0638 0.0881 0.2268 0.5320 0.6067
¹⁵⁶ᵐTm			19. s	α/	4.46/				
¹⁵⁶Tm		155.9389	1.40 m	β⁺ ,EC/7.6 α/	4.23/	2-	+0.40	-0.5	ann.rad./ 0.3446 0.4529 0.5860
¹⁵⁷Tm		156.9367	3.6 m	β⁺ ,EC/4.5 α/	2.6 3.97/	1/2	+0.48		ann.rad./ 0.1104 0.3484 0.3855 0.4550 (0.1 - 1.58)
¹⁵⁸Tm		157.9379	4.0 m	β⁺ ,EC/74 /6.5 EC/26 /		(2-)	+0.04	+0.7	ann.rad./ Er k x-ray 0.1921 0.3351 0.6280 1.1498 (0.18 - 2.81)
¹⁵⁹Tm		158.9348	9.1 m	β⁺ /23 /3.9 EC/77 /		5/2 +	+3.42	+1.9	ann.rad./ Er k x-ray 0.0591 0.0848 0.2713 (0.05 - 1.27)
¹⁶⁰ᵐTm			1.24 m	IT		(5)			
¹⁶⁰Tm		159.9354	9.4 m	β⁺ /15 /5.9 EC/85 /		1-	+0.16	+0.58	ann.rad./ Er k x-ray 0.1264 0.2642 0.7285 0.8544 0.8614 1.3685

Elem. or Isot.	Natural Abundance (%)	Atomic Mass or Weight	Half-Life	Decay Mode/Energy (/MeV)	Particle Energy /Intensity (MeV / %)	Spin (h/2π)	Nuclear Magnetic Mom.(nm)	Elect. Quadr. Mom. (b)	γ-ray Energy /Intensity (MeV / %)
¹⁶¹Tm		160.9334	31. m	β⁺ ,EC/3.2		7/2+	+2.40	+2.9	ann.rad./ Er k x-ray 0.0595 0.0844 1.6481 (0.04 - 2.15)
¹⁶²ᵐTm			24. s	I.T./90 / β⁺ ,EC/10 /		5+			Tm k x-ray Er k x-ray 0.0669 0.8115 0.9003
¹⁶²Tm		161.93394	21.7 m	β⁺ /8 /4.81 EC/92 /		1-	+0.07	+0.69	ann.rad./ Er k x-ray 0.1020 0.7987 (0.1 - 3.75)m
¹⁶³Tm		162.93265	1.81 h	EC/98 /2.439 β⁺ /1 /		1/2+	0.082		Er k x-ray 0.0692 0.1043 0.2414
¹⁶⁴ᵐTm			5.1 m	I.T./80 / β⁺ ,EC/20 /		6-			0.0914 0.1394 0.2081 0.2405 0.3149
¹⁶⁴Tm		163.93345	2.0 m	β⁺ /36 /3.96 EC/64 /	2.94/	1+	+2.38	+0.71	ann.rad./ Er k x-ray 0.0914
¹⁶⁵Tm		164.932433	1.253 d	EC/1.593		1/2+	0.139		Er k x-ray 0.0472 0.0544 0.29728 0.80636
¹⁶⁶Tm		165.93355	7.70 h	EC/98 /3.04 β⁺ /2 /		2+	+0.092	+2.14	Er k x-ray 0.0806 0.1844 0.7789 1.2734 2.0524
¹⁶⁷Tm		166.932849	9.24 d	EC/0.748		1/2+	-0.197		Er k x-ray 0.0571 0.20778
¹⁶⁸Tm		167.934171	93.1 d	EC/1.679		3+	0.23	+3.2	Er k x-ray 0.19825 0.4475 0.81595
¹⁶⁹Tm	100	168.934211				1/2+	-0.2316	-1.2	
¹⁷⁰Tm		169.935798	128.6 d	β⁻ /99.8 /0.968 EC/0.2 /0.314	0.883/24 0.968/76	1-	0.2476	+0.74	Yb k x-ray 0.08425
¹⁷¹Tm		170.936426	1.92 y	β⁻ /0.096	0.03/2 0.096/98	1/2+	-0.2303		0.06674
¹⁷²Tm		171.93840	2.65 d	β⁻ /1.88	1.79/36 1.88/29	2-			Yb k x-ray 0.07879 1.38722 1.46601 1.52982 1.60861
¹⁷³Tm		172.93960	8.2 h	β⁻ /1.298	0.80/21 0.86/71	1/2+			Yb k x-ray 0.3988 0.4613
¹⁷⁴Tm		173.94216	5.4 m	β⁻ /3.08	0.70/14 1.20/83	(4-)			Yb k x-ray 0.07664 0.17669 0.27332 0.3666 0.99205 (0.08 - 1.6)
¹⁷⁵Tm		174.94383	15.2 m	β⁻ /2.39	0.9/36 1.9/23	(1/2+)			Yb k x-ray 0.36396 0.51487 0.94125 0.98247
¹⁷⁶Tm		175.9471	1.9 m	β⁻ /4.2	2.0/ 1.2/	(4+)			Yb k x-ray 0.1898 0.3819 1.0691
¹⁷⁷Tm		176.9490	1.4 m	β⁻					

Elem. or Isot.	Natural Abundance (%)	Atomic Mass or Weight	Half-Life	Decay Mode/Energy (/MeV)	Particle Energy /Intensity (MeV / %)	Spin (h/2π)	Nuclear Magnetic Mom.(nm)	Elect. Quadr. Mom. (b)	γ-ray Energy /Intensity (MeV / %)
₇₀Yb		173.04(3)							
¹⁵¹Yb		150.9545	1.6 s	β⁺ /8.5					
¹⁵²Yb		151.9502	3.2 s	β⁺ EC/5.5					
¹⁵³Yb		152.9492	4. s	β⁺ EC/6.7					
¹⁵⁴Yb		153.9455	0.40 s	β⁺ EC/7 /4.49 α/93 /	5.32/				ann.rad./
¹⁵⁵Yb		154.9456	1.7 s	β⁺ ,EC/16 /6.0 α/84 /	5.19/		-0.8		ann.rad./
¹⁵⁶Yb		155.94277	26. s	β⁺ ,EC/21 /3.57 α/79 /	4.69/	0+			ann.rad./
¹⁵⁷Yb		156.9427	39. s	β⁺ ,EC/99+ /5.5 α/0.5 /	4.69/				ann.rad./ 0.231 (0.035-0.670)
¹⁵⁸Yb		157.93986	1.5 m	β⁺ ,EC/1.9		0+			ann.rad./ 0.0741 0.2526
¹⁵⁹Yb		158.9402	1.4 m	EC,β⁺ /5.1					Tm k x-ray 0.1661 0.1772 0.3297 0.3903
¹⁶⁰Yb		159.9376	4.8 m	β⁺ ,EC/2.0		0+			ann.rad./ 0.1404 0.1737 0.2158
¹⁶¹Yb		160.9375	4.2 m	β⁺ ,EC/3.9		3/2-	-0.33	+1.03	ann.rad./ Tm k x-ray 0.0782 0.5999 0.6315
¹⁶²Yb		161.9358	18.9 m	β⁺ ,EC/1.7		0+			ann.rad./ Tm k x-ray 0.1188 0.1635
¹⁶³Yb		162.9363	11.1 m	β⁺ /26 /3.4	1.4/	3/2-	-0.37	+1.2	ann.rad./ Tm k x-ray 0.0636 0.8603 (0.06 - 1.9)
¹⁶⁴Yb		163.9345	1.26 h	EC/1.0		0+			Tm k x-ray 0.0914 0.6752
¹⁶⁵Yb		164.93540	9.9 m	β⁺ /10 /2.76 EC/90 /	1.58/	(5/2-)	+0.48	+2.48	ann.rad./ Tm k x-ray 0.0801 1.0903
¹⁶⁶Yb		165.93388	2.363 d	EC/0.30		0+			Tm k x-ray 0.0828 0.1844 0.7789 1.2734 2.0524
¹⁶⁷Yb		166.934947	17.5 m	β⁺ /0.5 /1.954 EC/99.5 /	0.639/	5/2-	+0.62	+2.70	Tm k x-ray 0.06296 0.10616 0.11337 0.17633
¹⁶⁸Yb	0.13(1)	167.933895				0+			
¹⁶⁹ᵐYb			46. s	I.T./0.0242		1/2-			Yb L x-ray 0.0242
¹⁶⁹Yb		168.935187	32.03 d	EC/0.909		7/2+	-0.63	+3.5	Tm k x-ray (0.0498-0.3078)
¹⁷⁰Yb	3.05(6)	169.934759				0+			
¹⁷¹Yb	14.3(2)	170.936323				1/2-	+0.49367		
¹⁷²Yb	21.9(3)	171.936378				0+			
¹⁷³Yb	16.12(21)	172.938207				5/2-	-0.67989	+2.80	
¹⁷⁴Yb	31.8(4)	173.938858				0+			
¹⁷⁵Yb		174.941273	4.19 d	β⁻ /0.470	0.466/73 0.071/21 0.353/6.2	7/2-	0.6		Lu k x-ray 0.3963/13 (0.114 - 0.28)
¹⁷⁶ᵐYb			11.4 s	I.T./1.051		(8-)			Yb k x-ray 0.0961 0.1901 0.2929 0.3897
¹⁷⁶Yb	12.7(2)	175.942569				0+			

Elem. or Isot.	Natural Abundance (%)	Atomic Mass or Weight	Half-Life	Decay Mode/Energy (/MeV)	Particle Energy /Intensity (MeV / %)	Spin (h/2π)	Nuclear Magnetic Mom.(nm)	Elect. Quadr. Mom. (b)	γ-ray Energy /Intensity (MeV / %)
^{177m}Yb			6.41 s	I.T./0.3315		1/2-			Yb k x-ray 0.1131 0.2084
^{177}Yb		176.945257	1.9 h	β^- /1.399	1.40	9/2+			Lu k x-ray 0.1504
^{178}Yb		177.94664	1.23 h	β^- /0.65	0.25/	0+			0.1415 0.3246 0.3516 0.3815 0.6125
^{179}Yb		178.9499	8. m	β^- /2.4					
^{180}Yb		179.9523	2. m	β^-					(0.1028-0.4423)
$_{71}$Lu		174.967(1)							
^{150}Lu		149.973	≈35. ms	p					
^{151}Lu		150.967	0.08 s	p/1.231					
^{152}Lu		151.963	0.7 s						
^{153}Lu		152.959							
^{154}Lu		153.9571	1.0 s	β^+ ,EC/10.8					
^{155m}Lu			2.6 ms	α/7.41					
^{155}Lu		154.9542	0.07 s	EC/8.0 α/	5.66/90				
^{156m}Lu			0.20 s	α/	5.57/				
^{156}Lu		155.9529	≈ 0.5 s	β^+ ,EC/9.5 α/	5.45/				ann.rad./
^{157m}Lu			≈9.6 s	α	4.925/				
^{157}Lu		156.95010	4.8 s	β^+ ,EC/94 /6.93 α/	5.00/				ann.rad./
^{158}Lu		157.94984	10.4 s	β^+ ,EC/99 /8.0 α/	4.67/				ann.rad./ 0.3682 0.4770
^{159}Lu		158.9467	12.3 s	β^+ ,EC/6.0					ann.rad./ 0.1505 0.1875 0.3693
^{160}Lu		159.94654	36.1 s	β^+ ,EC/7.3					ann.rad./ 0.2434 0.3957 0.5773
^{161}Lu		160.9432	1.2 m	β^+ ,EC/5.3					ann.rad./ 0.0437 0.0671 0.1003 0.1108 0.1562 0.2562
^{162m}Lu			≈ 1.5 m	EC/		4-			
^{162}Lu		161.9432	1.37 m	β^+ ,EC/6.9		1-			ann.rad./ 0.1666 0.6314
^{163}Lu		162.9412	4.1 m	β^+ ,EC/4.6					ann.rad./ 0.0539 0.0581 0.1504 0.1631 0.3717
^{164}Lu		163.9412	3.14 m	β^+ ,EC/6.3	1.6/ 3.8/				0.1238 0.2621 0.7404 0.8639 0.8804
^{165}Lu		164.9396	10.7 m	β^+ ,EC/3.9	2.06/	1/2+			ann.rad./ 0.1206 0.1324 0.1742 0.2036 (0.04 - 2.0)
^{166m2}Lu			2.1 m	β^+ /35 / EC/65 /		(0-)			ann.rad./ Yb k x-ray 1.0673 1.2566 2.0986
^{166m1}Lu			1.4 m	β^+ ,EC/58 / I.T./42 /0.0344		(3-)			ann.rad./ 0.1024 0.2281 0.2861 0.8119 0.8301

Elem. or Isot.	Natural Abundance (%)	Atomic Mass or Weight	Half-Life	Decay Mode/Energy (/MeV)	Particle Energy /Intensity (MeV / %)	Spin (h/2π)	Nuclear Magnetic Mom.(nm)	Elect. Quadr. Mom. (b)	γ-ray Energy /Intensity (MeV / %)
¹⁶⁶Lu		165.9398	2.8 m	β⁺ /25 /5.5 EC/75 /		(6-)			ann.rad./ Yb k x-ray 0.1024 0.2281 0.3375 0.3679
¹⁶⁷Lu		166.9383	52. m	β⁺ /2 /3.1 EC/98 /	2.1/	7/2+			Yb k x-ray 0.0297 0.2392 (0.03 - 2.0)
¹⁶⁸ᵐLu			6.7 m	β⁺ /12 / EC/88 /		3+			ann.rad./ Yb k x-ray 0.1988 0.8960 0.9792
¹⁶⁸Lu		167.9387	5.5 m	β⁺ /6 /4.5 EC/94 /	1.2/	(6-)			ann.rad./ Yb k x-ray 0.1114 0.1124 0.2286 0.3483 1.4836
¹⁶⁹ᵐLu			2.7 m	I.T./0.0290		1/2-			Lu L x-ray 0.0290
¹⁶⁹Lu		168.93765	1.419 d	EC/2.293	1.271/	7/2+	2.30	3.4	Yb k x-ray 0.19121 0.9606 (0.08 - 2.1)
¹⁷⁰ᵐLu			0.7 s	I.T./0.0929		4-			Lu L x-ray 0.04449 0.0484
¹⁷⁰Lu		169.93847	2.01 d	EC/3.46	2.44/	0+			Yb k x-ray 0.58711 0.5908 1.28029 (0.1 - 3.38)
¹⁷¹ᵐLu			1.31 m	I.T./0.0711		1/2-			Lu k x-ray 0.07119
¹⁷¹Lu		170.937910	8.24 d	EC/1.479	0.362/	7/2+	2.31	3.38	Yb k x-ray 0.01939 0.66744 (0.02 - 1.3)
¹⁷²ᵐLu			3.7 m	I.T./0.0419		1-			Lu L x-rays 0.04186
¹⁷²Lu		171.939082	6.70 d	EC/2.519		4-	2.89	3.79	Yb k x-ray 0.18156 1.09367 (0.07 - 2.2)
¹⁷³Lu		172.938927	1.37 y	EC/0.671		7/2+	2.28	3.56	Yb k x-ray 0.07860 0.27198
¹⁷⁴ᵐLu			142. d	I.T./99.3 /0.17086 EC/0.7 /		6-	1.5		Lu k x-ray 0.067055
¹⁷⁴Lu		173.940334	3.3 y	EC/1.374		1-	1.9		Yb k x-ray 0.07664 1.2419
¹⁷⁵Lu	97.41(2)	174.940768				7/2+	+2.2327	+3.49	
¹⁷⁶ᵐLu			3.66 h	β⁻ /1.315	1.229/ 1.317/	1-	+0.318	-1.47	Hf k x-ray 0.088372
¹⁷⁶Lu	2.59(2)	175.942683	3.8x10¹⁰ y	β⁻ /1.192		7-	+3.169	+4.92	Hf k x-ray 0.20187 0.30691
¹⁷⁷ᵐLu			160.7 d	I.T./22 /0.9702 β⁻ /78		23/2-	2.34	5.2	Lu k x-ray Hf k x-ray 0.11295 0.20836 0.37850 0.41853
¹⁷⁷Lu		176.943755	6.75 d	β⁻ /0.498	0.497/	7/2+	+2.239	+3.39	0.11295 0.20836
¹⁷⁸ᵐLu			23.1 m	β⁻ /		(9-)			0.2166 0.3317
¹⁷⁸Lu		177.945952	28.5 m	β⁻ /2.099	2.03/	1+			Hf k x-ray 0.0932 1.3099 1.3408 (0.09 - 1.7)
¹⁷⁹Lu		178.94732	4.6 h	β⁻ /1.405	1.35/	7/2+			0.2143 0.3377

Elem. or Isot.	Natural Abundance (%)	Atomic Mass or Weight	Half-Life	Decay Mode/Energy (/MeV)	Particle Energy /Intensity (MeV / %)	Spin (h/2π)	Nuclear Magnetic Mom.(nm)	Elect. Quadr. Mom. (b)	γ-ray Energy /Intensity (MeV / %)
^{180}Lu		179.9499	5.7 m	β^- /3.1	1.49/				
^{181}Lu		180.9518	3.5 m	β^- /2.5		(7/2+)			0.40795/50. (0.07-1.9) 0.0458 0.2059
^{182}Lu		181.9552	2.0 m	β^- / ≈ 4.1					0.5749 0.0978 0.7208
^{183}Lu		182.9576	58. s	β^- /		7/2+			0.8182
^{184}Lu		183.9612	20 s	β^-					
$_{72}$Hf		178.49(2)							
^{154}Hf		153.964	2. s	EC,β^+ / ≈ 6.7					
^{155}Hf		154.963	0.9 s	EC,β^+ /8.					
^{156}Hf		155.9593	25. ms	α/					
^{157}Hf		156.9581	0.11 s	α/					
^{158}Hf		157.9539	2.9 s	EC/54 /5.1 α/46 /	5.27/	0+			
^{159}Hf		158.9538	5.6 s	β^+ ,EC/88 /6.9 α/12 /	5.09/				ann.rad./
^{160}Hf		159.95063	≈ 12. s	β^+ ,EC/97 /4.9 α /4.78		0+			ann.rad./
^{161}Hf		160.9503	17. s	α/	4.60/				
^{162}Hf		161.94720	38. s	β^+ ,EC/3.7		0+			ann.rad./ 0.1739 0.1963 0.4101
^{163}Hf		162.9471	40. s	β^+ ,EC/5.5					ann.rad./ 0.0454 0.0621 0.0710 0.6882
^{164}Hf		163.9536	2.8 m	EC,β^+ /3.0					
^{165}Hf		164.9445	1.32 m	EC/4.6		11/2-			
^{166}Hf		165.9423	6.8 m	EC/93 /2.3 β^+ /7 /					ann.rad./ Lu k x-ray 0.0788
^{167}Hf		166.9426	2.0 m	β^+ /40 /4.0 EC/60 /		(5/2-)			ann.rad./ Lu k x-ray 0.1754 0.3152
^{168}Hf		167.9406	25.9 m	β^+ ,EC/1.8		0+			ann.rad./ (0.0144-1.311)
^{169}Hf		168.9412	3.25 m	EC/85 /3.3 β^+ /15 /		(5/2-)			ann.rad./ Lu k x-ray 0.3695 0.4929
^{170}Hf		169.9397	16.0 h	EC/1.1		0+			Lu k x-ray 0.0985 0.1202 0.1647 0.5729 0.6207
^{171m}Hf			30. s						
^{171}Hf		170.9405	12.2 h	EC,β^+ /2.4		7/2+			ann.rad./ Lu k x-ray 0.1221 0.6620 1.0714
^{172}Hf		171.93946	1.87 y	EC/0.35		0+			Lu k x-ray 0.02399 0.12582 (0.0818-0.123)
^{173}Hf		172.9407	23.6 h	EC/1.6		1/2-			Lu k x-ray 0.12367 0.13963 0.29697 0.31124 (0.1 - 2.1)
^{174}Hf	0.162(3)	173.940042	2.0x10^{15} y			0+			
^{175}Hf		174.941504	70. d	EC/0.686		5/2-	-0.62	+2.6	Lu k x-ray 0.08936 0.34340
^{176}Hf	5.206(5)	175.941403				0+			

T A B L E   O F   T H E   I S O T O P E S

Elem. or Isot.	Natural Abundance (%)	Atomic Mass or Weight	Half-Life	Decay Mode/Energy (/MeV)	Particle Energy /Intensity (MeV / %)	Spin (h/2π)	Nuclear Magnetic Mom.(nm)	Elect. Quadr. Mom. (b)	γ-ray Energy /Intensity (MeV / %)
¹⁷⁷ᵐ²Hf			51.4 m	I.T./2.740		37/2-			Hf k x-ray 0.2140 0.2951 0.3115 0.3267
¹⁷⁷ᵐ¹Hf			1.1 s	I.T./		23/2+			Hf k x-ray 0.20836 0.22847 0.37851
¹⁷⁷Hf	18.606(4)	176.943220				7/2-	+0.7935	+0.337	
¹⁷⁸ᵐ²Hf			31. y	I.T./		16+	+8.2	+6.0	Hf k x-ray 0.32555 0.42635 (0.0889-0.5742)
¹⁷⁸ᵐ¹Hf			4.0 s	I.T./		8-			Hf k x-ray 0.21342 0.32555 0.42635
¹⁷⁸Hf	27.297(4)	177.943698				0+			
¹⁷⁹ᵐ²Hf			25.1 d	I.T./1.1057		25/2-	7.4		Hf k x-ray 0.1227 0.1461 0.3626 0.4537
¹⁷⁹ᵐ¹Hf			18.7 s	I.T./0.375		1/2-			Hf k x-ray 0.1607 0.2141
¹⁷⁹Hf	13.629(6)	178.945815				9/2+	-0.6409	+3.79	
¹⁸⁰ᵐHf			5.52 h	I.T./1.1416		8-	+8.7	+4.6	Hf k x-ray 0.2152 0.3323 0.4432
¹⁸⁰Hf	35.100(7)	179.946549				0+			
¹⁸¹Hf		180.949099	42.4 d	β⁻ /1.027	0.408/	1/2-			Ta k x-ray 0.13294 0.48200
¹⁸²ᵐHf			62. m	β⁻ /54 /1.60 I.T./46 /1.1729	0.49/43 0.95/10	8-			Hf k x-ray 0.0509 0.2244 0.3441 0.4558 0.5066 0.9428
¹⁸²Hf		181.95055	9.x10⁶ y	β⁻ /0.37		0+			Ta k x-ray 0.2704
¹⁸³Hf		182.95353	1.07 h	β⁻ /2.01	1.18/68 1.54/25	3/2-			Ta k x-ray 0.0732 0.4591 0.7837
¹⁸⁴Hf		183.95545	4.1 h	β⁻ /1.34	0.74/38 0.85/16 1.10/46	0+			Ta k x-ray 0.0414 0.1391 0.3449
¹⁸⁵Hf		184.9588	≈3.5m	β⁻ /					0.165
₇₃Ta		180.9479(1)							
¹⁵⁶Ta		155.972	0.11 s	β⁺ / ≈11.6 p/	1.02/ ≈100				
¹⁵⁷Ta		156.968	10 ms	α/ p/	6.117 0.927/3.4				
¹⁵⁸Ta		157.9664	37. ms	α/	6.05/100 5.97/100				
¹⁵⁹Ta		158.9629	0.6 s	β⁺ ,EC/20 /8.5 α/80 /	α/5.52/34 5.60/55				ann.rad./
¹⁶⁰Ta		159.9615	1.4 s	β⁺ ,EC/10.1 α	5.41/				ann.rad./
¹⁶¹Ta		160.9584	2.9 s	β⁺ ,EC/7.5 α/	5.15				ann.rad./
¹⁶²Ta		161.9564	4. s	EC/8.6					
¹⁶³Ta		162.9544	10.6 s	EC/6.8					
¹⁶⁴Ta		163.9536	14.2 s	β⁺ /8.5 α/	4.62/	3+			ann.rad./ 0.2110 0.3768
¹⁶⁵Ta		164.9508	31. s	ECβ⁺ /5.9					

Elem. or Isot.	Natural Abundance (%)	Atomic Mass or Weight	Half-Life	Decay Mode/Energy (/MeV)	Particle Energy /Intensity (MeV / %)	Spin (h/2π)	Nuclear Magnetic Mom.(nm)	Elect. Quadr. Mom. (b)	γ-ray Energy /Intensity (MeV / %)
^{166}Ta		165.9505	34. s	β⁺ /82 /7.7; EC/18 /					ann.rad./; Hf k x-ray; 0.1587; 0.3117; 0.8101
^{167}Ta		166.9486	1.4 m	β⁺ ,EC/5.6					ann.rad./
^{168}Ta		167.9478	2.4 m	β⁺ /77 /6.7; EC/23 /		3+			ann.rad./; Hf k x-ray; 0.1239; 0.2615; 0.7502
^{169}Ta		168.9459	4.9 m	β⁺ ,EC/4.4					ann.rad./; 0.0288; 0.1535; 0.1924
^{170}Ta		169.9461	6.8 m	β⁺ /70 /6.0; EC/35 /		(3+)			ann.rad./; Hf k x-ray; 0.1008; 0.2212
^{171}Ta		170.9445	23.3 m	β⁺ ,EC/3.7		(5/2-)			0.0496; 0.5018; 0.5064; (0.05 - 1.02)
^{172}Ta		171.9447	36.8 m	β⁺ /25 /4.9; EC/75 /		(3-)			ann.rad./; Hf k x-ray; 0.21396; 1.10923; (0.09 - 3.8)
^{173}Ta		172.9446	3.6 h	β⁺ /24 /3.7; EC/76 /		(5/2-)	1.7	-1.9	ann.rad./; Hf k x-ray; 0.06972; 0.17219; (0.06 - 2.7)
^{174}Ta		.173.9442	1.12 h	β⁺ /27 /3.8; EC/73 /		(3+)			ann.rad./; Hf k x-ray; 0.09089; 0.20638; (0.09 - 3.64)
^{175}Ta		174.9437	10.5 h	EC/2.0		7/2+	2.27	+3.7	Hf k x-ray; 0.2077; 0.2671; 0.3487
^{176}Ta		175.9447	8.1 h	EC/3.1		1-			Hf k x-ray; 0.08837; 1.15735
^{177}Ta		176.944472	2.356 d	EC/1.166		7/2+	2.25		Hf k x-ray; 0.11295; (0.07 - 1.06)
^{177m}Ta			2.4 h	EC/		(7-)			Hf k x-ray; 0.08886; 0.21342; 0.32555; 0.42635
^{178}Ta		177.9458	9.29 m	EC/99 /1.9; β⁺ /1 /		1+	+2.74	+0.65	ann.rad./; Hf k x-ray; 0.09316
^{179}Ta		178.94593	1.8 y	EC/0.110		7/2+	2.29	3.37	Hf k x-ray
^{180m}Ta	0.012(2)		> 1.2x10¹⁵ y			(9-)	4.77		
^{180}Ta		179.947466	8.15 h	EC/87 /0.854; β⁻ /13 /0.708	0.61/3; 0.71/10	1+			Hf k x-ray; W k x-ray; 0.09333; 0.10340
^{181}Ta	99.988(2)	180.947996				7/2+	+2.370	+3.3	
^{182m}Ta			15.8 m	I.T./0.5198		10-			Ta k x-ray; 0.14678; 0.17157
^{182}Ta		181.950152	114.43 d	β⁻ /1.814	0.25/30; 0.44/20; 0.52/40	3-	+3.02		W k x-ray; 0.06775; 0.10010; 1.12127; 1.22138
^{183}Ta		182.951373	5.1 d	β⁻ /1.070	0.45/5; 0.62/91	7/2+	+2.36		W k x-ray; 0.0847; 0.0991; 0.1079; 0.2461; 0.3540

Elem. or Isot.	Natural Abundance (%)	Atomic Mass or Weight	Half-Life	Decay Mode/Energy (/MeV)	Particle Energy /Intensity (MeV / %)	Spin (h/2π)	Nuclear Magnetic Mom.(nm)	Elect. Quadr. Mom. (b)	γ-ray Energy /Intensity (MeV / %)
¹⁸⁴Ta		183.95401	8.7 h	β⁻ /2.87	1.11/15 1.17/81	(5-)			W k x-ray 0.2528/44. 0.4140/74. (0.09-1.4)
¹⁸⁵Ta		184.95556	49. m	β⁻ /1.99	1.21/5 1.77/81	(7/2+)			W k x-ray 0.0697 0.1739 0.1776
¹⁸⁶Ta		185.9586	10.5 m	β⁻ /3.9	2.2/	(3-)			W k x-ray 0.1979 0.2149 0.5106 (0.09 - 1.5)
₇₄W		183.84(1)							
¹⁵⁸ᵐW			<1. ms	α	8.28(3)/				
¹⁵⁸W		157.974	0.9 ms	α/	6.433				
¹⁵⁹W		158.972	7. ms	α/					
¹⁶⁰W		159.9684	0.08 s	α/	5.92/	0+			
¹⁶¹W		160.9671	0.41 s	β⁺ ,EC/18 /8.1 α/82 /	5.78/				
¹⁶²W		161.9626	1.39 s	β⁺ ,EC/54 /5.8 α/46 /	5.54/	0+			
¹⁶³W		162.9624	2.8 s	β⁺ ,EC/59 /7.5 α/41 /	5.38/				
¹⁶⁴W		163.95890	6. s	β⁺ ,EC/97 /5.0 α/3 /	5.15/	0+			ann.rad./
¹⁶⁵W		164.9583	5.1 s	β⁺ ,EC/99 /7.0 α/1 /	4.91/				ann.rad./
¹⁶⁶W		165.95502	16. s	β⁺ ,EC/99 /4.2 α/1 /	4.74/	0+			ann.rad./
¹⁶⁷W		166.9547	20. s	EC/5.6					ann.rad./
¹⁶⁸W		167.9519	53. s	EC/3.8 α/10⁻³/	4.40(1)				Ta k x-ray 0.1755 (0.037-0.573)
¹⁶⁹W		168.9518	1.3 m	EC/5.4					ann.rad./ Ta k x-ray 0.123 (0.097-0.699)
¹⁷⁰W		169.9485	2.4 m	EC/2.2					ann.rad./ Ta k x-ray 0.3162 (0.060-0.144)
¹⁷¹W		170.9494	2.4 m	EC/4.6					ann.rad./ Ta k x-ray 0.1842 (0.052-0.479)
¹⁷²W		171.9474	6.6 m	β⁺ ,EC/2.5					ann.rad./ Ta k x-ray 0.0389 (0.034-0.674)
¹⁷³W		172.9489	6.3 m	EC/4.0					ann.rad./ Ta k x-ray 0.4576 (0.035-0.623)
¹⁷⁴W		173.9462	35. m	EC/1.9		0+			ann.rad./ Ta k x-ray 0.3287 0.4288 (0.056-0.429)
¹⁷⁵W		174.9468	35. m	EC/2.9		1/2-			(0.015-0.27)
¹⁷⁶W		175.9456	2.5 h	β⁺ ,EC/0.8		0+			0.03358 0.06129 0.09487 0.10020
¹⁷⁷W		176.9466	2.21 h	EC/2.0		(1/2-)			Ta k x-ray 0.15505 0.18569 0.42694
¹⁷⁸W		177.9459	21.6 d	EC/0.091		0+			Ta k x-ray
¹⁷⁹ᵐW			6.4 m	I.T./99.7 /0.222 EC/0.3 /		(1/2-)			W k x-ray 0.2220
¹⁷⁹W		178.94707	38. m	EC/1.06		(7/2-)			Ta k x-ray 0.0307
¹⁸⁰W	0.120(1)	179.946706	7.4x10¹⁶ y	α/		0+			

Elem. or Isot.	Natural Abundance (%)	Atomic Mass or Weight	Half-Life	Decay Mode/Energy (/MeV)	Particle Energy /Intensity (MeV / %)	Spin (h/2π)	Nuclear Magnetic Mom.(nm)	Elect. Quadr. Mom. (b)	γ-ray Energy /Intensity (MeV / %)
^{181}W		180.94820	121.2 d	EC/0.188		9/2+			Ta k x-ray 0.13617 0.15221
^{182}W	26.498(29)	181.948205	8.3x10^{18} y	α/		0+			
^{183m}W			5.15 s	I.T./		(11/2+)			W k x-ray 0.0465 0.0526 0.0991 0.1605
^{183}W	14.314(4)	182.950224	1.9x10^{18} y	α/		1/2-	+0.1177848		
^{184}W	30.642(8)	183.950932	4.0x10^{18} y	α/		0+			
^{185m}W			1.6 m	I.T./0.1974		11/2+			W k x-ray 0.0659 0.1315 0.1737
^{185}W		184.953420	74.8 d	β$^-$ /0.433	0.433/99.9	3/2-			0.12536
^{186}W	28.426(37)	185.954362	6.5x10^{18} y	α/		0+			
^{187}W		186.957158	23.9 h	β$^-$ /1.311	0.624/ 1.315/	3/2-	0.62		Re k x-ray 0.0725 0.47951 0.68572
^{188}W		187.958487	69.4 d	β$^-$ /0.349	0.349/99	0+			0.0636 0.2271 0.2907
^{189}W		188.9619	11.5 m	β$^-$ /2.5	1.4/ 2.5/	(3/2-)			(0.1262-1.466)
^{190}W		189.9632	30. m	β$^-$ /1.3	0.95/	0+			Re k x-ray 0.1576 0.1621
$_{75}$Re		186.207(1)							
^{160}Re		159.981	0.7 ms	p/ α/	1.261(6)/91 6.54/				
^{161}Re		160.978	14 ms	α/ p	6.24 1.35				
^{162}Re		161.9757	0.10 s	α/	6.12/94 6.09/94				
^{163}Re		162.9721	0.26 s	β$^+$,EC/9.0 α/	α/5.87/32 5.92/66				
^{164}Re		163.9704	0.9 s	β$^+$,EC/10.7 α/	5.78/				
^{165}Re		164.9671	2. s	β$^+$,EC/87 /8.1 α/	5.51/				
^{166}Re		165.9651	2.5 s	β$^+$,EC/9.4 α/	5.50/				
^{167m}Re			6.2 s	α, EC/					
^{167}Re		166.9626	3.4 s	β$^+$,EC/7.4 α/	5.015/				
^{168}Re		167.9616	4.4 s	β$^+$,EC/9.1 α/	4.833/				0.1117
^{169m}Re			8.1 s	α	4.70/ 4.87/				
^{169}Re		168.9588	16. s						
^{170}Re		169.9582	9.2 s	β$^+$, EC/9.0					0.1560 0.3055 0.4125
^{171}Re		170.9555	15.2 s	EC/ ≈ 5.7					
^{172m}Re			55. s	β$^+$,EC/		(2)			ann.rad./ 0.1234 0.2537 0.3504
^{172}Re		171.9553	15. s	β$^+$,EC/7.3					ann.rad./ 0.1234 0.2537
^{173}Re		172.9531	2.0 m	EC/ ≈ 3.9					ann.rad./
^{174}Re		173.9521	2.4 m	β$^+$,EC/5.6					ann.rad./ 0.1119 0.2430
^{175}Re		174.9514	5.8 m	β$^+$,EC/4.3					ann.rad./
^{176}Re		175.9516	5.3 m	β$^+$,EC/5.6		(3+)			ann.rad./ 0.1089 0.2406
^{177}Re		176.9503	14. m	EC/78 /3.4 β$^+$ /22 /		(5/2-)			ann.rad./ W k x-ray 0.0797 0.0843 0.1968

Elem. or Isot.	Natural Abundance (%)	Atomic Mass or Weight	Half-Life	Decay Mode/Energy (/MeV)	Particle Energy /Intensity (MeV / %)	Spin (h/2π)	Nuclear Magnetic Mom.(nm)	Elect. Quadr. Mom. (b)	γ-ray Energy /Intensity (MeV / %)
¹⁷⁸Re		177.9509	13.2 m	β⁺ /11 /4.7 EC/89 /	3.3/	(3)			ann.rad./ W k x-ray 0.1059 0.2373 0.9391
¹⁷⁹Re		178.9500	19.7 m	EC/99 /2.71 β⁺ /1 /	0.95/	(5/2+)			W k x-ray 0.1199 0.2900 0.4154 0.4302 1.6803
¹⁸⁰Re		179.95079	2.45 m	EC/92 /3.80 β⁺ /8 /	1.76/	1-			ann.rad./ W k x-ray 0.1036 0.9028 (0.07 - 2.2)
¹⁸¹Re		180.95006	20. h	EC /1.74		5/2+	3.19		W k x-ray 0.3607 0.3655 0.6390
¹⁸²ᵐRe			12.7 h	EC/	0.55/ 1.74/	2+	3.3	+1.8	W k x-ray 0.0677 1.1214 1.2215 (0.06 - 2.2)
¹⁸²Re		181.9512	2.67 d	EC/2.8		(7+)	2.8	+4.1	W k x-ray 0.0678 0.2293 1.1213 1.2214
¹⁸³Re		182.95082	70. d	EC/0.56		(5/2+)	+3.17	+2.3	W k x-ray 0.16232
¹⁸⁴ᵐRe			165. d	I.T./75 /0.188 EC/25 /		8+	+2.9		Re k x-ray 0.1047 0.2165 0.92093 (0.10 - 1.1)
¹⁸⁴Re		183.95252	38. d	EC/1.48		3-	+2.53	+2.8	W k x-ray 0.79207 0.90328 (0.1 - 1.4)
¹⁸⁵Re	37.40(2)	184.952955				5/2+	+3.1871	+2.18	
¹⁸⁶ᵐRe			2.0x10⁵ y	I.T./0.150		8+			Re k x-ray 0.0590
¹⁸⁶Re		185.954986	3.718 d	β⁻ /92 /1.070 EC/8 /0.582	0.973/21 1.07/71	1-	+1.739	+0.62	W k x-ray 0.1227/0.6 0.1372/9.5 (0.63-0.77)
¹⁸⁷Re	62.60(2)	186.955751	4.4x10¹⁰ y	β⁻ /0.00266	0.0025/	5/2+	+3.2197	+2.07	
¹⁸⁸ᵐRe			18.6 m	I.T./0.172		(6-)			Re k x-ray 0.0925 0.1059
¹⁸⁸Re		187.958112	16.94 h	β⁻ /2.120	1.962/20 2.118/79	1-	+1.788	+0.57	Os k x-ray 0.15502
¹⁸⁹Re		188.959228	24. h	β⁻ /1.01	1.01/	(5/2+)			0.1471 0.2167 0.2194 0.2451
¹⁹⁰ᵐRe			3.0 h	β⁻ /51 / I.T./49 /		(6-)			Re k x-ray 0.1191 0.2238 0.6731 (0.1 - 1.79)
¹⁹⁰Re		189.9618	3.0 m	β⁻ /3.2	1.8/	(2-)			Os k x-ray 0.1867 0.5580 0.6051
¹⁹¹Re		190.96312	9.7 m	β⁻ /2.05	1.8/				
¹⁹²Re		191.9660	16. s	β⁻ /4.2	≈ 2.5/				(0.2-0.75)
₇₆Os		190.23(3)							
¹⁶²Os		161.984	1.5 ms	α/	6.62				
¹⁶³Os		162.982	5.5 ms	α/	6.51				
¹⁶⁴Os		163.9779	0.04 s	α					
¹⁶⁵Os		164.9765	0.07 s	α					
¹⁶⁶Os		165.9718	0.18 s	β⁺ ,EC/28 /6.3 α/72 /	6.27/ 5.98/	0+			ann. rad./

Elem. or Isot.	Natural Abundance (%)	Atomic Mass or Weight	Half-Life	Decay Mode/Energy (/MeV)	Particle Energy /Intensity (MeV / %)	Spin (h/2π)	Nuclear Magnetic Mom.(nm)	Elect. Quadr. Mom. (b)	γ-ray Energy /Intensity (MeV / %)
^{147}Os		166.9714	0.7 s	β$^+$,EC/76 /8.2 α/24 /	 5.84/				ann.rad./
^{168}Os		167.96775	2.2 s	β$^+$,EC/51 /5.7 α/49 /		0+			ann. rad./
^{169}Os		168.9671	3.3 s	β$^+$,EC/89 /7.7 α/13 /	5.57/80 5.51/12 5.54/8				ann.rad./
^{170}Os		169.96357	7.1 s	β$^+$,EC/5.0 α/	 5.40/	0+			ann.rad./ (0.162-0.216)
^{171}Os		170.9630	8.4 s	β$^+$,EC/98 /7.1 α/19 /	α/5.24/93.5 5.17/6.5				ann.rad./ (0.1898-0.7050)
^{172}Os		171.9601	19. s	β$^+$,EC/99 /4.5 α/1.1/	 5.10/	0+			ann.rad./ (0.063-1.120)
^{173}Os		172.9598	16. s	β$^+$,EC/6.3 α/0.4 /	 4.94/				ann.rad./ (0.1416-0.2993)
^{174}Os		173.9563	44. s	β$^+$,EC/3.9 α/0.02 /	 4.76/	0+			0.118 0.138 / 0.001 0.158 0.325
^{175}Os		174.9570	1.4 m	β$^+$,EC/5.3					0.125 0.181 0.248
^{176}Os		175.9550	3.6 m	β$^+$,EC/3.2		0+			0.8155 0.7758 0.8573 1.2093 1.2909
^{177}Os		176.9551	2.8 m	β$^+$,EC/4.5		(1/2-)			0.0848 0.1958 0.3002 1.2686
^{178}Os		177.9534	5.0 m	β$^+$,EC/2.3		0+			ann.rad./ 0.5946 0.6850 0.9687 1.3311
^{179}Os		178.9539	7. m	β$^+$,EC/3.7					ann.rad./ 0.0654 0.2186 0.5938
^{180}Os		179.9524	21.5 m	β$^+$,EC/1.5		0+			Re k x-ray 0.0202-0.7174
^{181m}Os			1.75 h	EC/		(1/2-)			ann.rad./ 0.0489
^{181}Os		180.9532	2.7 m	EC/2.9		(7/2-)			ann.rad./ 0.11794 0.23868 0.8267 (0.07 - 2.64)
^{182}Os		181.95219	21.5 h	EC/0.9		0+			Re k x-ray 0.1802 0.5100
^{183m}Os			9.9 h	EC/84 / I.T./16 /		1/2-			Os k x-ray Re k x-ray 1.1020 1.1080
^{183}Os		182.9531	13. h	EC/2.1		9/2+	-0.79	+3.1	Re k x-ray 0.1144 0.3818
^{184}Os	0.020(3)	183.952491				0+			
^{185}Os		184.954043	93.6 d	EC/1.013		1/2-			Re k x-ray 0.6461 0.8748 0.8805
^{186}Os	1.58(10)	185.953838	2.x10^{15} y	α/	≈ 2.75/	0+			
^{187}Os	1.6(1)	186.955748				1/2-	+0.0646519		
^{188}Os	13.3(2)	187.955836				0+			
^{189m}Os			5.8 h	I.T./0.0308		9/2-			Os L x-ray 0.0308
^{189}Os	16.1(3)	188.958145				3/2+	+0.65993	+0.86	
^{190m}Os			9.9 m	I.T./1.705		10-	-0.6		Os k x-ray 0.1867 0.3611 0.5026 0.6161
^{190}Os	26.4(4)	189.958445				0+			
^{191m}Os			13.1 h	I.T./0.0744		3/2-			Os k x-ray 0.0744

Elem. or Isot.	Natural Abundance (%)	Atomic Mass or Weight	Half-Life	Decay Mode/Energy (/MeV)	Particle Energy /Intensity (MeV / %)	Spin (h/2π)	Nuclear Magnetic Mom.(nm)	Elect. Quadr. Mom. (b)	γ-ray Energy /Intensity (MeV / %)
^{191}Os		190.960928	15.4 d	β⁻ /0.314	0.140/100	9/2-		+2.5	Ir k x-ray 0.1294
^{192m}Os			6.0 s	I.T./2.0154		(10-)			Os k x-ray 0.2058/65.9 0.5692/70 (0.201-1.000)
^{192}Os	41.0(3)	191.961479				0+			
^{193}Os		192.964148	30.5 h	β⁻ /1.141	1.04/20	3/2-	+0.730	+0.47	Ir k x-ray 0.1389 0.4605
^{194}Os		193.965179	6.0 y	β⁻ /0.097	0.054/33 0.096/67	0+			Ir L x-ray 0.0429
^{195}Os		194.9681	6.5 m	β⁻ /2.0	2.0/				
^{196}Os		195.96962	34.9 m	β⁻ /1.16	0.84/	0+			0.1262/5 0.4079/5.9
$_{77}$Ir		192.217(3)							
^{165}Ir		164.9876	0.3 ms	p/87 α/13	1.71 6.72				
^{166m}Ir			15 ms	α/98.2 p/1.8	6.56 1.32				
^{166}Ir		165.9855	≈ 11 ms	α/93 p/6.9	6.56 1.15				
^{167m}Ir			30 ms	α/48,β⁺ p/32	6.35 1.06				
^{167}Ir		166.9817	35. ms	α/80,β⁺ p/0.4	6.41 1.24				
^{168}Ir		167.9799		α/					
^{169}Ir		168.9764	0.4 s	α/					
^{170}Ir		169.9743	1.0 s	α/	6.03/				
^{171}Ir		170.9718	1.5 s	α/	5.91/				
^{172}Ir		171.9706	2.1 s	α/	5.811/				0.228 (0.379-0.475)
^{173}Ir		172.9677	3.0 s	α/	5.665/				0.0493 (0.092-0.296)
^{174}Ir		173.9668	4. s	α/	5.478/				0.1587 (0.276-1.33)
^{175}Ir		174.9641	≈ 4.5 s	α/	5.393/				0.1056
^{176}Ir		175.9635	8. s	EC, β⁺/80 α/3.2/	5.118/				0.260 (0.135-0.415)
^{177}Ir		176.9612	30. s	EC, β⁺/5.7 α/0.06/	5.011/				0.184 (0.062-0.194)
^{178}Ir		177.9601	12. s	β⁺ ,EC/6.3					0.1320 0.2667 0.3633
^{179}Ir		178.9592	4. m	EC/4.9					0.0975 (0.045-0.220)
^{180}Ir		179.9593	1.5 m	EC/6.4					0.2765 ((0.132-1.106)
^{181}Ir		180.9576	4.9 m	β⁺ ,EC/4.1		(7/2+)			ann.rad./ 0.1076 (0.0196-1.715)
^{182}Ir		181.9582	15. m	β⁺ /44 /5.6 EC/56 /					ann.rad./ Os k x-ray 0.1273 0.2370
^{183}Ir		182.9568	57. m	β⁺ ,EC/3.5					ann.rad./ 0.0877 0.2285 0.2824
^{184}Ir		183.9574	3.0 h	β⁺ /12 /4.6 EC/88 /	2.3/ 2.9/	5-	0.70	+2.0	ann.rad./ Os k x-ray 0.11968 0.2640 0.3904
^{185}Ir		184.9566	14. h	β⁺ /3 /2.4 EC/97 /		(5/2-)	2.60	-2.1	ann.rad./ Os k x-ray 0.2543 1.8288
^{186m}Ir			1.7 h	EC /		(2-)			Os k x-ray 0.1371 0.7675
^{186}Ir		185.95795	15.7 h	EC/98 /3.83 β⁺ /2 /		(5+)	3.9	-2.5	Os k x-ray 0.1372 0.2968 0.4348 (0.13 - 3.0)

Elem. or Isot.	Natural Abundance (%)	Atomic Mass or Weight	Half-Life	Decay Mode/Energy (/MeV)	Particle Energy /Intensity (MeV / %)	Spin (h/2π)	Nuclear Magnetic Mom.(nm)	Elect. Quadr. Mom. (b)	γ-ray Energy /Intensity (MeV / %)
187Ir		186.95736	10.5 h	EC/1.50		3/2+			Os k x-ray 0.0743 0.4009 0.4271 0.6109 0.9128
188Ir		187.95885	1.72 d	β+ /2.81 EC/99+ /	1.13/ 1.64/	(2-)	0.30	+5.4	Os k x-ray 0.1550 0.4780 0.6330 2.2146
189Ir		188.95872	13.2 d	EC/0.53		3/2+	0.13	+1.0	Os k x-ray 0.2449 0.376
190m2Ir			3.09 h	β+ ,EC/95 / I.T./5 /		(11-)			Ir L x-ray
190m1Ir 190Ir		189.9606	1.12 h 11.8 d	I.T. /0.0263 EC/2.0		7+ (4+)	0.04	+2.8	Os k x-ray 0.1867 0.4072 0.5186 0.5580 0.6051 (0.2 - 1.4)
191mIr			4.93 s	I.T./0.1714		11/2-	0.603		Ir k x-ray 0.1294
191Ir	37.3(5)	190.960591				3/2+	+0.151	+0.82	
192m2Ir			241. y	I.T./0.161		(9+)			Ir k x-ray
192m1Ir			1.44 m	I.T./0.0580		(1+)			Ir L x-ray 0.0580 0.3165
192Ir		191.962602	73.83 d	β- /1.460		(4-)	+1.92	+2.3	Pt k x-ray 0.31649/83. 0.46806/48.
193mIr			10.53 d	I.T./0.0802		11/2-			Ir L x-ray 0.0803
193Ir 194mIr	62.7(5)	192.962923	170. d	β- /		3/2+ 11	+0.164	+0.75	Pt k x-ray 0.3284 0.4829 0.5624
194Ir		193.965075	19.3 h	β- /2.247	1.92/9 2.25/86	1-	0.39	+0.34	0.2935 0.3284 0.6451 (0.1 - 2.2)
195mIr			3.9 h	β- /	0.41/ 0.97/	(11/2-)			Pt k x-ray 0.3199/9.6 0.3649/9.5 0.4329/9.6 0.6849/9.6
195Ir		194.965976	2.8 h	β- /1.120	1.0/80 1.11/13 1.16/	(3/2+)			Pt k x-ray 0.0989/9.7
196mIr			1.40 h	β- /					Pt k x-ray 0.3557 0.3935 0.4471 0.5214 0.6473
196Ir		195.96838	52.s	β- /3.21	2.1/15 3.2/80	0-			0.3329 0.3557 0.7796
197mIr			8.9 m	β- / I.T./		(11/2-)			0.3465 See Ir[197]
197Ir		196.96964	5.8 m	β- /2.16	1.5/ 2.0/	(3/2+)			0.0531 0.1351 0.4306 0.4697 0.4074 0.5070
198Ir		197.9723	8. s	β- /4.1					
78Pt		195.078(2)							
166Pt			0.3 ms	α/	7.11/				
167Pt			0.7 ms	α/	6.99/				
168Pt		167.9880	2.0 ms	α	6.83				
169Pt		168.9864	3. ms	α					
170Pt		169.9816	15 ms	α	6.55				
171Pt		170.9811	0.03 s	α					
172Pt		171.97730	0.10 s	α /	6.31/94	0+			

Elem. or Isot.	Natural Abundance (%)	Atomic Mass or Weight	Half-Life	Decay Mode/Energy (/MeV)	Particle Energy /Intensity (MeV / %)	Spin (h/2π)	Nuclear Magnetic Mom.(nm)	Elect. Quadr. Mom. (b)	γ-ray Energy /Intensity (MeV / %)
¹⁷³Pt		172.9765	0.34 s	β⁺ ,EC/8.2 α/	6.20/				
¹⁷⁴Pt		173.97281	0.89 s	β⁺ ,EC/17 /5.6 α/83 /	6.040/	0+			
¹⁷⁵Pt		174.9723	2.5 s	β⁺ ,EC/65 /7.6 α/35 /	5.831/5 5.96/54 6.038/				0.0774 0.1354 0.2128
¹⁷⁶Pt		175.9690	6.3 s	β⁺ ,EC/60 /5.1 α/40 /	5.528/0.6 5.750/41	0+			ann.rad./ 0.2277
¹⁷⁷Pt		176.9685	11. s	EC/91 /6.8 α/9 /	5.53/ 5.485/3 5.525/6				0.0908
¹⁷⁸Pt		177.9649	21. s	EC/93 /4.5 α/7 /	5.286/0.2 5.442/7	0+			
¹⁷⁹Pt		178.9653	33. s	β⁺ ,EC/5.7 α /	5.16/				
¹⁸⁰Pt		179.9632	52. s	β⁺ ,EC/99.7 /3.7 α/0.3 /	5.140/	0+			
¹⁸¹Pt		180.9632	51. s	β⁺ ,EC/5.2					
¹⁸²Pt		181.9613	2.7 m	β⁺ ,EC/2.9		0+			ann.rad./ 0.1360 0.1460 0.2100
¹⁸³ᵐPt			43. s	β⁺ ,EC/ I.T./		(7/2-)			ann.rad./ 0.3132 0.3164 0.6296
¹⁸³Pt		182.9617	7. m	β⁺ ,EC/4.6					ann.rad./
¹⁸⁴Pt		183.9599	17.3 m	β⁺ ,EC/2.3					ann.rad./ 0.1549 0.1919 0.5484
¹⁸⁵ᵐPt			33. m	β⁺ ,EC/		1/2-			
¹⁸⁵Pt		184.9607	1.18 h	β⁺ ,EC/3.8		(9/2+)			ann.rad./ 0.1353 0.1974 0.2296 0.2551
¹⁸⁶Pt		185.95943	2.0 h	β⁺ ,EC/1.38		0+			ann.rad./ 0.6115 0.6892
¹⁸⁷Pt		186.9607	2.35 h	β⁺ ,EC/3.1		3/2			ann.rad./ Ir k x-ray 0.1064 0.1100 0.2015 0.2849 0.7092
¹⁸⁸Pt		187.95940	10.2 d	EC/0.51		0+			Ir k x-ray 0.1876 0.1951
¹⁸⁹Pt		188.96083	10.9 h	β⁺ ,EC/1.97		3/2-	0.43	-0.6	Ir k x-ray 0.0943 0.6076 0.7214 (0.09 - 1.47)
¹⁹⁰Pt	0.01(1)	189.95993	4.5x10¹¹ y			0+			
¹⁹¹Pt		190.961684	2.96 d	EC/1.02		(3/2-)	0.50	-0.6	Ir k x-ray 0.3599 0.4094 0.5389
¹⁹²Pt	0.79(6)	191.961035				0+			
¹⁹³ᵐPt			4.33 d	I.T./0.1498		13/2+	-0.75		Pt k x-ray 0.1355
¹⁹³Pt		192.962984	60. y	EC/0.0566		(1/2-)			Ir k x-rays
¹⁹⁴Pt	32.9(6)	193.962663				0+			
¹⁹⁵ᵐPt			4.02 d	I.T./0.2952		13/2+	0.61	+1.4	Pt k x-ray 0.0989
¹⁹⁵Pt	33.8(6)	194.964774				1/2-	+0.6095		
¹⁹⁶Pt	25.3(6)	195.964934				0+			
¹⁹⁷ᵐPt			1.590 h	I.T./97 / β⁻ /3 /		13/2+			Pt k x-ray 0.0530 0.3465
¹⁹⁷Pt		196.967323	18.3 h	β⁻ /0.719		1/2-	0.51		Au k x-ray 0.1914 0.2688

Elem. or Isot.	Natural Abundance (%)	Atomic Mass or Weight	Half-Life	Decay Mode/Energy (/MeV)	Particle Energy /Intensity (MeV / %)	Spin (h/2π)	Nuclear Magnetic Mom.(nm)	Elect. Quadr. Mom. (b)	γ-ray Energy /Intensity (MeV / %)
^{198}Pt	7.2(2)	197.967875				0+			
^{199m}Pt			13.6 s	I.T./0.424		13/2+			Pt k x-ray 0.3919
^{199}Pt		198.970576	30.8 m	β⁻ /1.70	0.90/18 1.14/14	(5/2-)			0.3170/4.9 0.49375/5.7 0.5430/14.8 (0.055-1.293)
^{200}Pt		199.97142	12.5 h	β⁻ / ≈0.66		0+			Au k x-ray 0.13590 0.22747 0.24371
^{201}Pt		200.9745	2.5 m	β⁻ /2.66		(5/2-)			0.070 0.152 0.222 1.760
^{202}Pt		201.9757	1.8 d						0.440
$_{79}$Au		196.96655(2)							
^{171}Au		170.9918	1.0 ms	p/46 α/54	1.69 7.00				
^{172}Au		171.9901	4 ms	α/7.02	6.86				
^{173}Au		172.9864	0.06 s	α					
^{174}Au		173.9842	0.12 s	α					
^{175}Au		174.9817	0.20 s	α					
^{176}Au		175.9803	1.2 s	β⁺ ,EC/10.5 α /	6.260/80 6.290/20				
^{177}Au		176.9772	1.2 s	α /	6.115/ 6.150/				
^{178}Au		177.9760	2.6 s	α /	5.920/				
^{179}Au		178.9732	7.5 s	α /	5.85/				
^{180}Au		179.9724	8.1 s	EC/8.6 α/	5.65 5.61 5.50				0.1522 0.2564 0.5242 0.6765 0.8084 0.8597
^{181}Au		180.9700	11.4 s	EC/97.5/6.3 α/2.7/	5.482/				
^{182}Au		181.9686	21. s	β⁺ ,EC/6.9 α/0.13/					ann.rad./ 0.1549 0.2649 (0.13 - 1.4)
^{183}Au		182.9676	42. s	EC/5.5 α/0.8/			+1.97		0.1630 0.2730 0.3625
^{184m}Au			48 s			(2+)	+1.44	+1.9	0.069(IT)
^{184}Au		183.9675	21. s	EC,β⁺ /7.1 α/0.013/		(5+)	+2.07	+4.7	
^{185m}Au			6.8 m	β⁺ ,EC/ I.T./0.145					
^{185}Au		184.9657	4.3 m	β⁺ ,EC/4.71 α/0.26/		(5/2-)	+2.17	-1.1	ann.rad./ 0.1915
^{186m}Au			< 2. m	β⁺ ,EC/					
^{186}Au		185.9659	10.7 m	β⁺ ,EC/6.0 α/8(10)⁻⁴/		3-	-1.26	+3.1	ann.rad./ 0.1915 0.2988
^{187m}Au			2.3 s	IT		9/2-			
^{187}Au		186.9646	8.3 m	β⁺ ,EC/3.60		1/2+	+0.54		ann.rad./ 0.9152 1.2668 1.3321 1.4081
^{188}Au		187.9651	8.8 m	β⁺ ,EC/5.3		(1-)	-0.07		ann.rad./ 0.2660 0.3404 0.6061
^{189m}Au			4.6 m	β⁺ ,EC/		11/2-	+6.19		0.1667
^{189}Au		188.9642	28.7 m	EC/96 /3.2 β⁺ /4 /		1/2+	+0.49		ann.rad./ Pt k x-ray 0.4478 0.7133 0.8128
^{190}Au		189.96470	43. m	β⁺ /2 /4.44 EC/98 /		1-	-0.07		ann.rad./ Pt k x-ray 0.2958 0.3018 0.5977

Elem. or Isot.	Natural Abundance (%)	Atomic Mass or Weight	Half-Life	Decay Mode/Energy (/MeV)	Particle Energy /Intensity (MeV / %)	Spin (h/2π)	Nuclear Magnetic Mom.(nm)	Elect. Quadr. Mom. (b)	γ-ray Energy /Intensity (MeV / %)
¹⁹¹ᵐAu			0.9 s	I.T./0.2663		(11/2-	6.6		Au k x-ray 0.2414 0.2526
¹⁹¹Au		190.96365	3.2 h	EC/1.83		3/2 +	+0.137	+0.72	Pt k x-ray 0.5864/16 (0.088-1.30)
¹⁹²Au		191.96481	4.9 h	β⁺ /5 /3.52 EC/95 /	2.19/ 2.49/	1-	-0.011	-0.23	ann.rad./ Pt k x-ray 0.2959 0.3165
¹⁹³ᵐAu			3.9 s	I.T./0.2901		11/2-	6.2	+1.98	Au k x-ray 0.2580
¹⁹³Au		192.96413	17.6 h	EC/1.07		3/2 +	+0.140	+0.66	Pt k x-ray 0.1862 0.2556
¹⁹⁴Au		193.96534	1.64 d	β⁺ /3 /2.49 EC/97 /	1.49/	1-	+0.075	-0.24	ann.rad./ Pt k x-ray 0.2935 0.3284/61
¹⁹⁵ᵐAu			30.5 s	I.T./0.3186		11/2-	6.2	+1.9	Au k x-ray 0.2617
¹⁹⁵Au		194.965017	186.12 d	EC/0.227		3/2 +	+0.149	+0.61	Pt k x-ray
¹⁹⁶ᵐ²Au			9.7 h	I.T./0.5954		12-	5.7		Au k x-ray 0.1478 0.1883
¹⁹⁶ᵐ¹Au			8.1 s	I.T./0.0846		8 +			0.0847
¹⁹⁶Au		195.966551	6.18 d	EC/92 /1.506		2-	+0.591	0.81	Pt k x-ray
¹⁹⁷ᵐAu			7.8 s	I.T./0.4094 β⁻ /8 /0.686		11/2-	+6.0	+1.7	Au k x-ray 0.1302 0.2790
¹⁹⁷Au	100.	196.966551				3/2 +	+0.14575	+0.55	
¹⁹⁸ᵐAu			2.30 d	I.T./0.812		(12-)			Au k x-ray 0.0972 0.1803 0.2419
¹⁹⁸Au		197.968225	2.694 d	β⁻ /1.372	0.290/1 0.961/99	2-	+0.5934	+0.68	Hg k x-ray 0.411794
¹⁹⁹Au		198.968748	3.14 d	β⁻ /0.453	0.25/22 0.292/72 0.462/6	3/2 +	+0.2715	0.55	Hg k x-ray 0.15837 0.20820
²⁰⁰ᵐAu			18.7 h	β⁻ /84 /1.0 I.T./16 /	0.56/	12-	5.9		Au k x-ray 0.2559/71 0.3680/77 0.4978/73 0.5793/72 0.084-0.904)
²⁰⁰Au		199.97072	48.4 m	β⁻ /2.24	0.7/15 2.2/77	1-			0.3679/19 1.2254/10.6 (0.077-1.570)
²⁰¹Au		200.97165	26. m	β⁻ /1.28	1.27/82	3/2 +			(0.027-0.732)
²⁰²Au		201.9738	29. s	β⁻ /3.0		(1-)			0.4396
²⁰³Au		202.97515	1.0 m	β⁻ /2.14	≈ 1.9/	3/2 +			(0.04-0.37)
²⁰⁴Au		203.9783	40. s	β⁻ /4.5		(2-)			0.4366 1.5113
²⁰⁵Au		204.9796	31. s	β⁻ /					(0.38 - 1.33)
₈₀Hg		200.59(2)							
¹⁷⁴Hg			≈ 2.7 ms	α	7.07				
¹⁷⁵Hg		174.9912	0.02 s	α					
¹⁷⁶Hg		175.98733	0.03 s	α	6.77				
¹⁷⁷Hg		176.9863	0.13 s	α					
¹⁷⁸Hg		177.98248	0.26 s	EC/50 /6.1 α/50 /	6.43/	0+			
¹⁷⁹Hg		178.9818	1.09 s	EC/8.0 α /	6.29/				
¹⁸⁰Hg		179.9783	2.6 s	EC/5.5 α /	6.12/33 5.69/.03	0+			0.1250 0.3005 0.3812
¹⁸¹Hg		180.9778	3.6 s	β⁺ EC/74 / ≈7.3 α/26 /		(1/2-)	+0.5071		0.0663 0.0811 0.0924 0.1474 0.1587 0.2142 0.2398

Elem. or Isot.	Natural Abundance (%)	Atomic Mass or Weight	Half-Life	Decay Mode/Energy (/MeV)	Particle Energy /Intensity (MeV / %)	Spin (h/2π)	Nuclear Magnetic Mom.(nm)	Elect. Quadr. Mom. (b)	γ-ray Energy /Intensity (MeV / %)
^{182}Hg		181.9739	10.8 s	β^+ ,EC/85 /5.0 α/15 /	 5.87/8.6 5.45/0.03	0+			0.1289 0.2168 0.4126
^{183}Hg		182.9744	9. s	β^+ ,EC/77 /6.3 α /	 5.83/ 5.91/	1/2-	+0.524		0.0714 0.0874 0.1538
^{184}Hg		183.9719	30.9 s	β^+ ,EC/99 /4.1 α/1 /	 5.54/1.3 5.07/2 x 10^3	0+			0.0915 0.1265 0.1560 0.2362
^{185m}Hg			21. s	β^+ , EC,IT,α /	5.37/	13/2 +	-1.02	+0.2	0.211 0.292
^{185}Hg		184.9720	51. s	β^+ ,EC/95 /5.8		1/2-	+0.509		(0.02 - 0.55)
^{186}Hg		185.9695	1.4 m	β^+ ,EC/3.3 α	 5.09/0.02	0+			0.1119 0.2518
^{187m}Hg			1.7 m	β^+ ,EC/		13/2 +	-1.04	+0.5	See Hg[187]
^{187}Hg		186.9698	2.4 m	β^+ ,EC/4.9		3/2-	-0.594		0.1034/32. 0.2334/100. 0.2403/33. 0.27151/31. 0.3763/38. 0.5254/30. (0.10-2.18)
^{188}Hg		187.9676	3.2 m	β^+ ,EC/2.3 α	 4.61	0+			0.0988 0.1148 0.1424 0.1900
^{189m}Hg			8.6 m	EC/		13/2 +	-1.06	+0.7	0.0780 0.3210 0.4345 0.5655 (0.08 - 2.170)
^{189}Hg		188.9687	7.6 m	EC/4.2		3/2-	-0.6086	-0.8	0.2005 0.2038 0.2386 0.2485
^{190}Hg		189.9663	20.0 m	EC/1.5		0+			0.1296 0.1426
^{191m}Hg			51. m	β^+ /6 / EC/94 /		13/2 +	-1.07	+0.6	ann.rad./ Au k x-ray 0.2741 0.4203 0.5787 (0.07 - 1.9)
^{191}Hg		190.9671	50. m	β^+ ,EC/3.2		(3/2-)	-0.62	-0.8	0.1963 0.2247 0.2524
^{192}Hg		191.9653	5.0 h	EC/ ≈0.5		0+			Au k x-ray 0.1572 0.2748 0.3065
^{193m}Hg			11.8 h	β^+ ,EC/91 / I.T./9 /0.2901		13/2 +	-1.05843	+0.92	Hg k x-ray 0.1866 0.2580 0.4076 0.5733 0.9324 (0.1 - 1.96)
^{193}Hg		192.96664	3.8 h	EC,B$^+$/2.34		3/2-	-0.6276		0.1866 0.2580 0.8611
^{194}Hg		193.96538	520. y	EC/0.04		0+			Au L x-rays
^{195m}Hg			1.67 d	I.T./(54)/0.3186 EC/(46)/		13/2 +	-1.04465	+1.1	Hg k x-ray Au k x-ray 0.2617 0.5603 0.7798
^{195}Hg		194.96664	9.5 h	EC/1.51		1/2-	+0.541475		Au k x-ray 0.0614 0.7798
^{196}Hg	0.15(1)	195.965814				0+			
^{197m}Hg			23.8 h	I.T./(93)/0.2989		13/2 +	-1.02768	+1.2	Hg k x-ray Au k x-ray 0.13398
^{197}Hg		196.967195	2.672 d	EC/0.600		1/2-	+0.527374		Au k x-ray 0.07735
^{198}Hg	9.97(8)	197.966752				0+			
^{199m}Hg			42.6 m	I.T./0.532		13/2 +	-1.014703	+1.2	Hg k x-ray 0.15841

Elem. or Isot.	Natural Abundance (%)	Atomic Mass or Weight	Half-Life	Decay Mode/Energy (/MeV)	Particle Energy /Intensity (MeV / %)	Spin (h/2π)	Nuclear Magnetic Mom.(nm)	Elect. Quadr. Mom. (b)	γ-ray Energy /Intensity (MeV / %)
^{199}Hg	16.87(10)	198.968262				1/2-	+0.505885		
^{200}Hg	23.10(16)	199.968309				0+			
^{201}Hg	13.18(8)	200.970285				3/2-	-0.560226	+0.39	
^{202}Hg	29.86(20)	201.970625				0+			
^{203}Hg		202.972857	46.61 d	β⁻ /0.492	0.213/100	5/2-	+0.8489	+0.34	Tl k x-ray 0.279188
^{204}Hg	6.87(4)	203.973475				0+			
^{205}Hg		204.976056	5.2 m	β⁻ /1.531	1.33/4	1/2-	+0.6010		0.20378 (0.2 - 1.4)
^{206}Hg		205.97750	8.2 m	β⁻ /1.31	0.935/34 1.3/63	0+			Tl k x-ray 0.3052 0.6502
^{207}Hg		206.9825	2.9 m	β⁻ /4.8		(9/2+)			
^{208}Hg		207.9859	0.7 h	β⁻					0.474
$_{81}$Tl		204.3833(2)							
^{179}Tl		178.9917	0.2 s	α					
^{182}Tl		181.9856	3. s	β⁺, EC/10.9					0.351 (0.26 - 0.41)
^{183m}Tl			0.06 s	α		9/2-			
^{183}Tl		182.9826	5. s	β⁺, EC/7.7		1/2+			0.208
^{184}Tl		183.9818	11. s	β⁺, EC/(98)/9.2 α/(2)/	6.16/				0.2868 0.3399 0.3667
^{185m}Tl			1.8 s	I.T./0.453 α/5.97	6.01/	(9/2-)			0.1688 0.2840
^{185}Tl		184.9791	20. s	EC/β⁺/6.6					
^{186m}Tl			4. s	I.T./0.374					0.3738
^{186}Tl		185.9776	28. s	β⁺, EC/7.5					0.3567 0.4026 0.4053
^{187m}Tl			15.6 s	I.T./≈ 0.33		(9/2+)	3.8	-2.4	0.2995
^{187}Tl		186.9762	50. s	β⁺, EC/6.0		1/2+	1.6		
^{188m}Tl			1.18 m	β⁺, EC/		(7+)			Hg k x-ray 0.4129 0.5043 0.5921
^{188}Tl	See Tl[188m]	187.9759	1.2 m	β⁺, EC/7.8		(2-)	0.48	0.13	See Tl[188m] 0.4129
^{189m}Tl			1.4 m	β⁺, EC/		(9/2-)	+3.878	-2.29	0.2156 0.2284 0.3175 0.4452
^{189}Tl		188.9743	2.3 m	β⁺, EC/5.2		(1/2+)			0.3337 0.4510 0.5223 0.9422
^{190m}Tl			3.7 m	β⁺, EC/	4.2/	(7+)	+0.495	0.29	0.1968 0.4164 0.7311
^{190}Tl		189.9738	2.6 m	β⁺, EC/7.0	5.7/	(2-)	+0.25	-0.33	0.4164 0.6254 0.6838 1.0999
^{191m}Tl			5.2 m	β⁺, EC/(98)/		(9/2+)	+3.903	-2.3	0.2157 0.2647 0.3256 0.3359
^{191}Tl		190.9723				(1/2)	1.59		
^{192m}Tl			10.8 m	β⁺, EC/		(7+)	+0.518	0.46	0.1740 0.4228 0.6348 0.7863 0.7455
^{192}Tl		191.972	9.6 m	β⁺, EC/6.4		(2-)	+0.20	-0.33	0.3975 0.4228 0.6908
^{193m}Tl			2.1 m	I.T./(75)/		(9/2-)	+3.948	-2.2	0.3650
^{193}Tl		192.9706	22. m	β⁺, EC/3.6		(1/2+)	+1.591		0.2077 0.3244 0.3440 0.6761 1.0447 1.5793

Elem. or Isot.	Natural Abundance (%)	Atomic Mass or Weight	Half-Life	Decay Mode/Energy (/MeV)	Particle Energy /Intensity (MeV / %)	Spin (h/2π)	Nuclear Magnetic Mom.(nm)	Elect. Quadr. Mom. (b)	γ-ray Energy /Intensity (MeV / %)
¹⁹⁴ᵐTl			32.8 m	β⁺ /(20)/≈ 0.30 EC/(80)/		(7+)	+0.540	0.62	ann.rad./ Hg k x-ray 0.4282 0.6363 0.7490
¹⁹⁴Tl		193.9711	34. m	β⁺ ,EC/5.3		2-	0.14	-0.28	0.3955 0.4282 0.6363
¹⁹⁵ᵐTl			3.6 s	I.T./0.483		9/2-			Tl k x-ray 0.0990 0.3836
¹⁹⁵Tl		194.9697	1.16 h	EC/(97)/2.8 β⁺ /(3)/		1/2+	+1.58		ann.rad./ Hg k x-ray 0.2422 0.5635 0.8845 1.3639 (0.13 - 2.5)
¹⁹⁶ᵐTl			1.41 h	β⁺ ,EC/(95)/4.9		(7+)	0.55	+0.76	0.0840 0.4261 0.6353 0.6954 (0.08 - 1.0)
¹⁹⁶Tl		195.9705	1.84 h	β⁺ /(15)/4.4 EC/(85)/		2-	0.07	-0.18	ann.rad./ Hg k x-ray 0.4257 0.6105 (0.03 - 2.4)
¹⁹⁷ᵐTl			0.54 s	I.T./(53)/0.608 β⁺ ,EC/(47)/		9/2-			Tl k x-ray 0.2262 0.4118 0.5872 0.6367
¹⁹⁷Tl		196.96954	2.83 h	β⁺ /(1)/2.18 EC/(99)/		1/2+	+1.58		Hg k x-ray 0.1522/8.2 0.4258
¹⁹⁸ᵐTl			1.87 h	β⁺ ,EC/(53)/ I.T./(47)/0.5347		7+	+0.64		Hg k x-ray Tl k x-ray 0.4118 0.5872 0.6367
¹⁹⁸Tl		197.9405	5.3 h	EC,β⁺ /(1)/3.5	1.4/ 2.1/ 2.4/	2-	0.00(1)		Hg k x-ray 0.4118 0.6367 0.6759 (0.23 - 2.8)
¹⁹⁹Tl		198.9698	7.4 h	EC/1.4		1/2-	+1.60		Hg k x-ray 0.2082 0.2473 0.4555
²⁰⁰Tl		199.97095	1.087 d	EC/2.46	1.07/ 1.44/	2-	0.04		Hg k x-ray 0.36799 1.2057 (0.11 - 2.3)
²⁰¹Tl		200.97080	3.040 d	EC/0.48		1/2+	+1.605		Hg k x-ray 0.13528 0.16740/10.0
²⁰²Tl		201.97209	12.23 d	EC/1.36		2-	0.06		Hg k x-ray 0.43957
²⁰³Tl	29.524(14)	202.972329				1/2+	+1.622258		
²⁰⁴Tl		203.973848	3.78 y	β⁻ /(97)/0.7637 EC/(3)/0.347	0.763/97	2-	0.09		Hg k x-ray
²⁰⁵Tl	70.476(14)	204.974412				1/2+	+1.638215		
²⁰⁶ᵐTl			3.76 m	I.T./2.644		12-			Tl k x-ray 0.2166 0.2661 0.4534 0.6866 1.0219
²⁰⁶Tl		205.976095	4.20 m	β⁻ /1.533	1.53/99.9	0-			Pb k x-ray 0.80313
²⁰⁷ᵐTl			1.3 s	I.T./1.350		11/2-			Tl k x-ray 0.3501 1.0000
²⁰⁷Tl		206.97741	4.77 m	β⁻ /1.423	1.43/99.8	1/2+	+1.88		0.89723

Elem. or Isot.	Natural Abundance (%)	Atomic Mass or Weight	Half-Life	Decay Mode/Energy (/MeV)	Particle Energy /Intensity (MeV / %)	Spin (h/2π)	Nuclear Magnetic Mom.(nm)	Elect. Quadr. Mom. (b)	γ-ray Energy /Intensity (MeV / %)
^{208}Tl		207.982004	3.053 m	β^- /5.001	1.28/23 1.52/22 1.796/51	(5+)	+0.29		Pb k x-ray 0.27728 0.51061 0.58302 2.61448
^{209}Tl		208.98535	2.16 m	β^- /3.98	1.8 /100	(1/2+)			Pb k x-ray 1.5670/100 0.4651/95 (0.12 - 1.33)
^{210}Tl		209.99006	1.30 m	β^- /5.48	1.3/25 1.9/56	(5+)			Pb k x-ray 0.081 0.2981 0.79788
$_{82}$Pb		207.2(1)							
^{180}Pb			5 ms	α/	7.23				
^{181}Pb		180.9967	0.05 s	α/	7.07				
^{182}Pb		181.99268	0.06 s	α	6.92				
^{183}Pb		182.9919	0.3 s	α/		1/2+			
^{184}Pb		183.9882	0.6 s	α/	6.63/	0+			
^{185}Pb		184.9876	4.1 s	α/	6.34/ 6.40/ 6.48/				
^{186}Pb		185.9835	5. s	β^+ ,EC/(95)/5.5 α/(5)/	6.32/ 6.34/<100 6.01/<0.2	0+			
^{187m}Pb			15.2 s	β^+ ,EC/	5.99/ 6.19/	(1/2-)			0.0674 0.2080 0.2755 0.2995 0.4487 0.7477
^{187}Pb		186.9839	18.3 s	EC/7.2 α/	6.08/	13/2+			0.1930 0.3314 0.3435 0.3934
^{188}Pb		187.9811	23. s	EC/(78)/4.8 α/(22)/	5.98/<10 5.61/<0.1	0+			0.1850 0.7582
^{189}Pb		188.9809	51. s	EC/6.1 α/	5.58/				
^{190}Pb		189.9782	1.2 m	β^+ (13)/4.1 EC/(86)/ α/(0.9)/	5.58/	0+			ann.rad./ Tl k x-ray 0.1415 0.1512 0.9422
^{191m}Pb			2.2 m	β^+ ,EC/		13/2+			ann.rad./ 0.3871 0.6135 0.7122
^{191}Pb		190.9782	1.3 m	β^+ ,EC/5.5					ann.rad./ 0.9368
^{192}Pb		191.9758	3.5 m	β^+ ,EC/$\approx$3.4 α/.006/	5.11	0+			ann.rad./ 0.1675 0.6082 1.1954
^{193m}Pb			5.8 m	β^+ ,EC/		13/2+	-1.16		ann.rad./ 0.3650 0.3922
^{193}Pb		192.9761	$\approx$ 2. m	EC/5.2		3/2			
^{194}Pb		193.9740	11. m	β^+ ,EC/2.7 α	4.64	0+			ann.rad./ 0.2036
^{195m}Pb			15. m	β^+ /(8)/ EC/(92)/		13/2+	-1.132	+0.29	ann.rad./ Tl k x-ray 0.3836 0.3942 0.8784
^{195}Pb		194.976	$\approx$ 15. m	β^+ ,EC/5.8					ann.rad./ 0.3836 0.3937 0.7776
^{196}Pb		195.9727	37. m	β^+ ,EC/2.1		0+			Tl k x-ray 0.2531 0.5021

Elem. or Isot.	Natural Abundance (%)	Atomic Mass or Weight	Half-Life	Decay Mode/Energy (/MeV)	Particle Energy /Intensity (MeV / %)	Spin (h/2π)	Nuclear Magnetic Mom.(nm)	Elect. Quadr. Mom. (b)	γ-ray Energy /Intensity (MeV / %)
197mPb			43. m	EC/(79)/ β⁺/(2)/ I.T./(19)/0.3193		13/2+	-1.105	+0.5	Tl k x-ray 0.3079 0.3877 0.7743 (0.2 - 2.2)
197Pb		196.9734	≈ 8. m	EC/(97)/3.6 β⁺/(3)/		(3/2-)	-1.075	-0.08	Tl k x-ray 0.3755 0.3858 0.7611
198Pb		197.9720	2.4 h	EC/1.4		0+			Tl k x-ray 0.1734 0.2903 0.3654
199mPb			12.2 m	I.T./(93)/0.4248 β⁺,EC/(7)/		13/2+	-1.074	+0.08	Pb k x-ray 0.4255
199Pb		198.9729	1.5 h	EC/(99)/2.9 β⁺/(1)/		5/2-			Tl k x-ray 0.3534 0.7202 1.1350 (0.22 - 2.4)
200Pb		199.97182	21.5 h	EC/0.81		0+			Tl k x-ray 0.14763
201mPb			1.02 m	I.T./0.6291		13/2+			Pb k x-ray 0.6288
201Pb		200.97285	9.33 h	EC/1.90		5/2-	+0.675	-0.009	Tl k x-ray 0.33120 0.36131 (0.11 - 1.8)
202mPb			3.53 h	I.T./(90)/2.170 β⁺/(10)/		9-	-0.228	+0.58	Pb k x-ray Tl k x-ray 0.42219 0.78700 0.96271
202Pb		201.97214	5.3x10⁴ y	EC/0.05		0+			Tl L x-ray
203mPb			6.2 s	I.T./0.8252		13/2+			Pb k x-ray 0.8203 0.8252
203Pb		202.97338	2.1615 d	EC/0.98		5/2-	+0.686	+0.10	Tl k x-ray 0.279188
204mPb			1.12 h	I.T./2.185		9-			Pb k x-ray 0.37481 0.89922 0.91175
204Pb	1.4(1)	203.973028				0+			
205Pb		204.974467	1.51x10⁷ y	EC/0.0512		5/2-	+0.712	+0.23	Tl L x-ray
206Pb	24.1(1)	205.974449				0+			
207mPb			0.80 s	I.T./1.632		13/2+			Pb k x-ray 0.56915 1.06310
207Pb	22.1(1)	206.975880				1/2-	+0.59258		
208Pb	52.4(1)	207.976636	>2x10¹⁹ y	SF		0+			
209Pb		208.981075	3.25 h	β⁻/0.644	0.645/100	9/2+	-1.474	-0.3	
210Pb		209.984174	22.6 y	β⁻/0.0635	0.017/81 0.061/19 α 3.72	0+			
211Pb		210.988732	36.1 m	β⁻/1.37	0.57/5 1.36/92	(9/2+)	-1.404	+0.09	0.40486 0.42700 0.83186 (0.09 - 1.27)
212Pb		211.991887	10.64 h	β⁻/0.574	0.28/83 0.57/12	0+			Bi k x-ray 0.23858
213Pb		212.9966	10.2 m	β⁻/2.1					
214Pb		213.999797	26.9 m	β⁻/1.0	0.67/48 0.73/42	0+			Bi k x-ray 0.24192 0.29509 0.35187
215Pb			36 s						
₈₃Bi		208.98038(2)							
185Bi		184.9977	0.04 ms	p/86 α/14	1.59				
186Bi		185.9965	10 ms	α	7.16 7.26				
187mBi			≈ 8. ms	α					
187Bi		186.9935	35. ms	α					
188Bi		187.9922		α					
189mBi			7.0 ms	α	7.30				
189Bi		188.9895	0.68 s	α					

Elem. or Isot.	Natural Abundance (%)	Atomic Mass or Weight	Half-Life	Decay Mode/Energy (/MeV)	Particle Energy /Intensity (MeV / %)	Spin (h/2π)	Nuclear Magnetic Mom.(nm)	Elect. Quadr. Mom. (b)	γ-ray Energy /Intensity (MeV / %)
¹⁹⁰Bi		189.9875	5. s	β⁺,EC/(10)/8.7 α/(90)/	6.45/				
¹⁹¹Bi		190.9861	12. s	β⁺,EC/(60)/7.3 α/(40)/	6.32/				
¹⁹²Bi		191.9854	40. s	β⁺,EC/(80)/9.0 α/(20)/	6.06/				
¹⁹³ᵐBi			3.2 s	β⁺,EC/ α/	6.48/	1/2+			
¹⁹³Bi		192.9837	1.11 m	β⁺,EC/(40)/7.1 α/(60)/	5.91/	9/2+			
¹⁹⁴Bi		193.9828	1.8 m	β⁺,EC/(99.9)/8.2 α/0.1/		(10-)			0.1661 0.1740 0.2802 0.421 0.5754 0.9650
¹⁹⁵ᵐBi			1.45 m	β⁺,EC/(94)/ α/(6)/	6.11/				
¹⁹⁵Bi		194.9811	2.9 m	β⁺,EC/(99.8)/5.8 α/(0.2)	5.45/	3/2-			
¹⁹⁶Bi		195.9806	5. m	EC/≈7.4					0.1376 0.3720 0.6880 1.0486
¹⁹⁷Bi		196.9789	5. m	β⁺,EC/5.2		1/2+			
¹⁹⁸ᵐBi			7.7 s	I.T./0.2485		(10-)			0.2485
¹⁹⁸Bi		197.9790	11.8 m	β⁺,EC/6.6		(7+)			0.0900 0.1976 0.5624 1.0635
¹⁹⁹ᵐBi			24.7 m	β⁺,EC/					ann.rad./
¹⁹⁹Bi		198.9776	27. m	β⁺,EC/4.3		9/2-			0.7203 0.8374 0.8417 0.9460 1.0528 1.3056 (0.12 - 3.2)
²⁰⁰ᵐBi			31. m	β⁺,EC/		(2+)			0.2453 0.4198 0.4624 1.0265
²⁰⁰Bi		199.9781	36. m	EC/(90)/5.9 β⁺/(10)/		7+			ann.rad./ Pb k x-ray 0.4198 0.4623 1.0265
²⁰¹ᵐBi			59.1 m	I.T./0.846 β⁺,EC/		(1/2+)			Bi k x-ray 0.8464
²⁰¹Bi		200.97697	1.8 h	EC/3.84		9/2-			Pb k x-ray 0.6288 0.9357 1.0138 (0.13 - 2.4)
²⁰²Bi		201.97768	1.72 h	β⁺/(3)/5.16 EC/(97)/		5+	+4.26	-0.72	ann.rad./ Pb k x-ray 0.57860 0.92734 (0.08 - 3.5)
²⁰³Bi		202.97687	11.8 h	EC/(99.8)/3.25 β⁺/(0.2)/	1.35/	9/2-	+4.02	-0.69	Pb k x-ray 0.1865 0.8203 0.8969 1.8475 (0.1 - 2.9)
²⁰⁴Bi		203.97779	11.2 h	EC/4.44		6+	+4.32	-0.43	Pb k x-ray 0.37481 0.89922 0.98409
²⁰⁵Bi		204.97737	15.31 d	EC/2.71		9/2-	+4.07	-0.59	Pb k x-ray 0.70347 1.76435
²⁰⁶Bi		205.97848	6.243 d	EC/3.76		6+	+4.36	-0.39	Pb k x-ray 0.51619 0.80313 0.88100
²⁰⁷Bi		206.978456	35. y	EC/2.399		9/2-	4.08	-0.6	Pb k x-ray 0.56915 1.06310

Elem. or Isot.	Natural Abundance (%)	Atomic Mass or Weight	Half-Life	Decay Mode/Energy (/MeV)	Particle Energy /Intensity (MeV / %)	Spin (h/2π)	Nuclear Magnetic Mom.(nm)	Elect. Quadr. Mom. (b)	γ-ray Energy /Intensity (MeV / %)
^{208}Bi		207.979727	3.68x10^5 y	EC/2.880		5+			Pb k x-ray 2.61435
^{209}Bi	100.	208.980384				9/2-	+4.111	-0.37	
^{210m}Bi			3.0x10^6 y	α/	4.420(3)/0.29 4.569(3)/3.9 4.584(3)/1.4 4.908(4)/39 4.946(3)/55	9-	+2.73	-0.47	Tl k x-ray 0.2661 0.3052 0.6502
^{210}Bi		209.984105	5.01 d	β$^-$/1.163	1.16/99	1-	-0.0445	+0.136	0.2661 0.3.52
^{211}Bi		210.98726	2.14 m	α/(99.7)/ β$^-$/(0.3)/0.58	6.279/16 6.623/84	9/2-			Tl k x-ray 0.3501
^{212m2}Bi			7. m	β$^-$/		(15-)			
^{212m1}Bi			25.0 m	α/(93)/ β$^-$/(7)/	6.300/40 6.340/53	(9-)			0.120 0.233 0.275 0.404 0.727
^{212}Bi		211.991271	1.009 h	β$^-$/(64)/2.254 α/(36)/	6.051/25 6.090/9.6	(1-)	+0.32	+0.1	Tl k x-ray Po k x-ray 0.2881 0.72725 0.78551 1.62066
^{213}Bi		212.99437	45.6 m	β$^-$/(98)/1.43 α/(2)/	1.02/31 1.42/66 5.549/0.16 5.869/2.0	9/2-	+3.72	-0.60	Po k x-ray 0.4404 (0.15 - 1.328)
^{214}Bi		213.99870	19.7 m	β$^-$/3.27					1.10006 0.60931 1.12027 1.76449 (0.19 - 3.2)
^{215}Bi		215.0018	7.7 m	β$^-$/2.3					0.2937 (0.27 - 0.835)
^{216}Bi		216.0062	3.6 m	β$^-$/4.0					0.5498 0.4192
^{217}Bi			97 s	β/					
$_{84}$Po									
^{190}Po		189.9951	2.4 ms	α/	7.53				
^{191}Po		190.9947	17 ms	α/	7.33				
^{192}Po		191.9915	34. ms	α/8.5	7.17				
^{193m}Po			0.24 s	α/	7.00				
^{193}Po		192.9911	0.45 s	α/	6.95				
^{194}Po		193.9883	0.39 s	α/	6.84/93 6.19/0.22	0+			
^{195m}Po			1.9 s	α/	6.70/				
^{195}Po		194.9881	4.6 s	α/	6.61/				
^{196}Po		195.9855	5.8 s	α/(95)/ β$^+$,EC/(5)/≈4.6	6.52/94 5.77/0.02	0+			
^{197m}Po			25.8 s	α/(84)/ β$^+$,EC/(16)/	6.385(3)/55	13/2+			
^{197}Po		196.9856	53. s	α/(44)/ β$^+$,EC/(56)/6.2	6.282(4)/76	(3/2-)			
^{198}Po		197.9834	1.76 m	α/(70)/ β$^+$,EC/(30)/4.0	6.18/57 5.27/7.6 x 10^{-4}	0+			
^{199m}Po			4.2 m	β$^+$,EC/(51)/ α/(39)/	6.059/24	13/2+	0.99		ann.rad./ 0.2745 0.4998 1.0020
^{199}Po		198.985	5.2 m	β$^+$,EC/(88)/7. α/(12)/	5.952/7.5	(3/2-)			Bi k x-ray 0.1877 0.3616 1.0214 1.0344
^{200}Po		199.9817	11.5 m	β$^+$,EC/(85)/3.4 α/(15)/	5.863/11.1	0+			0.14748 0.32792 0.6176 0.6709
^{201m}Po			8.9 m	β$^+$,EC/(57)/ I.T./(40)/0.418 α/(3)/	5.786/≈3.	13/2+	1.0		Bi k x-ray Po k x-ray 0.2726 0.4123 0.4179 0.9670

Elem. or Isot.	Natural Abundance (%)	Atomic Mass or Weight	Half-Life	Decay Mode/Energy (/MeV)	Particle Energy /Intensity (MeV / %)	Spin (h/2π)	Nuclear Magnetic Mom.(nm)	Elect. Quadr. Mom. (b)	γ-ray Energy /Intensity (MeV / %)
²⁰¹Po		200.9822	15.3 m	β⁺ ,EC/(98)/4.9 α/(2)/	5.683(3)/1.1	3/2-	0.94		Bi k x-ray 0.2056 0.2250 0.8483 0.9048
²⁰²Po		201.9807	45. m	β⁺ ,EC/(98)/2.8 α/(2)/	5.588/1.9	0+			0.0410 0.1656 0.3158 0.6884
²⁰³ᵐPo			1.2 m	I.T./(96)/0.6414 β⁻ EC/(4)/		13/2+			Bi k x-ray Po k x-ray 0.6414
²⁰³Po		202.9814	35. m	β⁺ ,EC/4.2		5/2-	+0.74		0.17516 0.21477 0.89350 0.90863 1.09095
²⁰⁴Po		203.98031	3.53 h	EC/2.34 α	5.377/0.66	0+			Bi k x-ray 0.2702 0.8844 1.0162 (0.11 - 1.9)
²⁰⁵Po		204.98117	1.7 h	β⁺ ,EC/3.53		5/2-	+0.76	+0.17	Bi k x-ray 0.83681 0.84983 0.87241 1.00124 (0.12 - 2.7)
²⁰⁶Po		205.98047	8.8 d	EC/(95)/1.85 α/(5)/	5.223/5.5	0+			Bi k x-ray 0.28644 0.31156 0.51134 0.80737 1.03228 (0.11 - 1.5)
²⁰⁷ᵐPo			2.8 s	I.T./1.383		19/2-			Po k x-ray 0.2682 0.30074 0.81448
²⁰⁷Po		206.98158	5.80 h	EC,β⁺ /2.91		5/2-	+0.79	+0.28	Bi k x-ray 0.74263 0.91176 0.99225
²⁰⁸Po		207.981231	2.898 y	α/5.213	4.233/0.0002 5.1158/100	0+			
²⁰⁹Po		208.982415	102. y	α/4.976	4.624/0.56 4.879/99.2	1/2-	+0.77		0.26049 0.8964
²¹⁰Po		209.982857	138.38 d	α/5.407	4.516/0.001 5.304/100	0+			0.80313
²¹¹ᵐPo			25.2 s	α/	7.273/91 7.994/1.7 8.316/0.25 8.875/7.0	25/2+			Pb k x-ray 0.32808 0.56915 0.89723 1.06310
²¹¹Po		210.986637	0.516 s	α/7.594	6.570/0.54 6.892/0.55 7.450/98.9	9/2+			0.56915 0.89723
²¹²ᵐPo			45. s	α/	8.514/2.0 9.086/1.0 11.650/97	16+			
²¹²Po		211.988852	0.298 μs	α/8.953	8.784/100	0+			
²¹³Po		212.992843	4. μs	α/8.537	7.614/0.003 8.375/100	9/2+			
²¹⁴Po		213.995186	163.7 μs	α/7.833	6.904/0.01 7.686/99.99	0+			0.7995 0.298
²¹⁵Po		214.999415	1.780 ms	α/7.526	6.950/0.02 6.957/0.03 7.386/100	(9/2+)			
²¹⁶Po		216.001905	0.145 s	α/6.906	5.895/0.002 6.778/99.99	0+			
²¹⁷Po		217.0064	< 10. s	α/6.662	6.539/				
²¹⁸Po		218.008965	3.04 m	α/6.114	5.181/1.00	0+			
₈₅At									
¹⁹³At		192.9998	40 ms	α/					
¹⁹⁴At		193.9990	40 ms	α/					
¹⁹⁵At		194.9965	140 ms	α/					
¹⁹⁶At		195.9957	0.3 s	α/	7.05/				

Elem. or Isot.	Natural Abundance (%)	Atomic Mass or Weight	Half-Life	Decay Mode/Energy (/MeV)	Particle Energy /Intensity (MeV / %)	Spin (h/2π)	Nuclear Magnetic Mom.(nm)	Elect. Quadr. Mom. (b)	γ-ray Energy /Intensity (MeV / %)
^{197m}At			4. s	α		(1/2+)			
^{197}At		196.9939	0.35 s	$β^+$,EC/7.8 α/	6.96/	(9/2-)			
^{198m}At			1.5 s	$β^+$,EC/(75)/ α/(25)/	6.85/86				
^{198}At		197.9928	5. s	α/	6.75/94				
^{199}At		198.9910	7.1 s	$β^+$,EC/(8)/5.6 α/(92)/	6.64/	9/2-			
^{200m}At			4.3 s	$β^+$,EC/(80) α/(20)/	6.536/12	10-			
^{200}At		199.990	43. s	$β^+$,EC/(65)/≈8.0 α/(35)/	6.412/44 6.465/57	5+			
^{201}At		200.9885	1.48 s	$β^+$,EC/(29)/5.9 α/(71)/6.474	6.344/	9/2-			
^{202m}At			≤ 1.5 s	I.T./0.391					
^{202}At		201.9885	3.02 m	$β^+$,EC/(88)/7.2 α/(12)/	6.135/7.7 6.225/4.3	5+			ann.rad./ 0.4413 0.5697 0.6753
^{203}At		202.9868	7.4 m	$β^+$,EC/(69)/5.1 α/(31)/6.210	6.088/	9/2-			0.1458 0.2459 0.6414 1.0020 1.0340
^{204}At		203.9873	9.1 m	$β^+$,EC/(95)/6.5 α/(5)/	5.951/	(5+)			Po k x-ray 0.3271 0.4254 0.5156 0.6837
^{205}At		204.98604	26. m	$β^+$,EC/(90)/4.54 α/(10)/6.020	5.902/	(9/2-)			Po k x-ray 0.1543 0.6696 0.7194
^{206}At		205.98660	29.4 m	$β^+$,EC/(99)/5.72 α/(1)/5.881	5.703/	5+			Po k x-ray 0.20186 0.39561 0.47716 0.70071
^{207}At		206.98578	1.81 h	$β^+$,EC/(90)/3.91 α/(10)/5.873	5.758/	9/2-			Po k x-ray 0.16801 0.58842 0.81448
^{208}At		207.98657	1.63 h	$β^+$,EC/(99)/4.97 α/(1)/5.752	5.626/0.01 5.641/0.53	(6+)			Po k x-ray 0.1770 0.2060 0.6601 0.6852 0.8450 1.0281
^{209}At		208.98616	5.4 h	$β^+$,EC/(96)/3.49 α/(4)/5.757	5.647/4.1	(6+)			Po k x-ray 0.10422 0.54503 0.78189 0.79020 (0.1 - 2.6)
^{210}At		209.98713	8.1 h	EC/99.8/3.98 α/(0.2)/5.632	5.361/0.05 5.442/0.05	5+			Po k x-ray 0.24535 0.52758 1.18143 1.43678 1.48335 (0.04 - 2.4)
^{211}At		210.987481	7.21 h	EC/(58)/0.787 α/(42)/5.980	5.211/0.004 5.868/42	9/2-			Po k x-ray 0.66956 0.6870 0.74263
^{212m}At			0.119 s	α/	7.837/65 7.897/33	(9-)			
^{212}At		211.990735	0.314 s	α/7.828	7.058/0.4 7.088/0.6 7.618/15 7.681/84	(1-)			
^{213}At		212.992922	0.11 μs	α/9.254	9.080/	9/2-			
^{214m}At			0.76 μs	α/8.762		(9-)			
^{214}At		213.996357	0.56 μs	α/8.987	8.819/100	(1-)			
^{215}At		214.99864	0.10 ms	α/8.178	7.626/0.045 8.023/99.9	(9/2-)			0.40486

Elem. or Isot.	Natural Abundance (%)	Atomic Mass or Weight	Half-Life	Decay Mode/Energy (/MeV)	Particle Energy /Intensity (MeV / %)	Spin (h/2π)	Nuclear Magnetic Mom.(nm)	Elect. Quadr. Mom. (b)	γ-ray Energy /Intensity (MeV / %)
²¹⁶At		216.002408	0.30 ms	α/7.947	7.595/0.2 7.697/2.1 7.800/97	(1-)			
²¹⁷At		217.00471	32. ms	α/7.202	6.812/0.06 7.067/99.9	(9/2-)			0.2595 0.3345 0.5940
²¹⁸At		218.00868	1.6 s	α/6.883	6.654/6 6.695/90 6.748/4				
²¹⁹At		219.0113	50. s	α/6.390	6.275/				
²²⁰At		220.0153	3.71 m	β⁻ /3.7					(0.24-0.70)
²²¹At		221.0181	2.3 m	β					
²²²At		222.0223	0.9 m	β					
²²³At		223.0253	50. s	β					
₈₆Rn									
¹⁹⁶Rn		195.9977	≈ 3 ms	α/	7.49				
¹⁹⁷ᵐRn			0.02 s	α	7.36				
¹⁹⁷Rn		196.9983	0.07 s	α/	7.26				
¹⁹⁸Rn		197.9988	0.05 s	α					
¹⁹⁹ᵐRn			0.3 s	α		(13/2+)			
¹⁹⁹Rn		198.9983	0.62 s	α/		3/2-			
²⁰⁰Rn		199.9957	1.06 s	α/(98)/ EC/(2)/5.	6.901/	0+			0.4329 0.5043
²⁰⁰ᵐRn			3.8 s	EC/(10)/ α/(90)/	6.773/	13/2+			
²⁰¹Rn		200.9955	7.0 s	α/(80)/ EC/(20)/	6.725/ α/6.778	(3/2-)			
²⁰²Rn		201.9932	9.9 s	α/(12)/ EC/(88)/	6.641/	0+			0.5695 (0.2876-0.6255)
²⁰³ᵐRn			28. s	α/	6.551	13/2+	-0.96	+1.3	
²⁰³Rn		202.9948	45. s	α/(66)/6.629 EC/(34)/ ≈7.4	6.499/	0			
²⁰⁴Rn		203.9914	1.24 m	α/(68)/ EC/(32)/3.8	6.420/	0+			
²⁰⁵Rn		204.9917	2.8 m	α/(23)/6.390 EC/(77)/5.2	6.123(3)/0.02 6.262(3)/23	(5/2-)	+0.80	+0.06	0.2652 0.3553 0.4648 0.6205 0.6753 0.7300
²⁰⁶Rn		205.9902	5.7 m	α/(68)/6.384 EC/(32)/3.3	6.258(3)/	0+			0.06170 0.0968 0.3245 0.3862 0.4822 0.4973 0.7728
²⁰⁷Rn		206.9907	9.3 m	β⁺ ,EC/(77)/4.6 α/(23)/6.252	5.995(4)/0.02 6.068(3)/0.15 6.126(3)/22.8	5/2-	+0.82	+0.22	At k x-ray 0.32947 0.34455 0.36767 0.40267 0.74723 (0.18 - 1.4)
²⁰⁸Rn		207.98963	24.3 m	α/(60)/6.260 EC/(40)/2.85	5.469(2)/0.003 6.140(2)/60	0+			
²⁰⁹Rn		208.99038	29. m	β⁺ /(83)/3.93 α/(17)/	2.16/2.3 5.887(3)/0.04 5.898(3)/0.02 6.039(2)/16.9	5/2-	+0.8388	+0.31	At k x-ray 0.27933 0.33753 0.40841 0.68942 0.74594 (0.18 - 3.2)
²¹⁰Rn		209.98968	2.4 h	α/(96)/6.157 EC/(4)/2.37	5.351(2)/0.005 6.039(2)/96	0+			At k x-ray 0.19625 0.45824 0.57104 0.64868 (0.14 - 1.7)
²¹¹Rn		210.99059	14.6 h	β⁺ ,EC/(74)/2.89 α/(26)/5.964	5.619(1)/0.7 5.784(1)/16.4 5.851(1)/8.8	1/2-	+0.60		At k x-ray 0.16877 0.25022 0.37049 0.67412 0.67839 1.36298 (0.11 - 2.7)

Elem. or Isot.	Natural Abundance (%)	Atomic Mass or Weight	Half-Life	Decay Mode/Energy (/MeV)	Particle Energy /Intensity (MeV / %)	Spin (h/2π)	Nuclear Magnetic Mom.(nm)	Elect. Quadr. Mom. (b)	γ-ray Energy /Intensity (MeV / %)
²¹²Rn		211.990689	24. m	α/6.385	5.587(4)/0.05 6.260(4)/99.95	0+			
²¹³Rn		212.99387	25.0 ms	α/8.243	7.552(8)/1.0 8.087(8)/99	9/2+			
²¹⁴Rn		213.99535	0.27 μs	α/9.209	9.037(9)/	0+			
²¹⁵Rn		214.99873	2.3 μs	α/8.840	8.674(8)/	(9/2+)			
²¹⁶Rn		216.00026	45. μs	α					
²¹⁷Rn		217.003915	0.6 ms	α/7.885	7.500/0.1 7.742(4)/100	9/2+			
²¹⁸Rn		218.005586	35. ms	α/7.267	6.534(1)/0.16 7.133(1)/99.8	0+			0.6093 0.6653
²¹⁹Rn		219.009475	3.96 s	α/6.946(1)	6.3130(5)/0.05 6.425(3)/7.5 6.5309(4)/0.12 6.5531(3)/12.2 6.8193(3)/81	(5/2+)	-0.44	+0.93	Po k x-ray 0.13057 0.27113 0.40170 (0.1 - 1.05)
²²⁰Rn		220.011384	55.6 s	α/6.404	5.7486(5)/0.07 6.2883(1)/99.9	0+			
²²¹Rn		221.0156	25. m	α/(22)/6.148 β⁻/(78)/1.2	5.778(3)/1.8 5.788(3)/2.2 6.037(3)/18	7/2+	-0.020	-0.38	Fr L x-ray 0.07384 0.08323 0.0610 0.18639
²²²Rn		222.017570	3.8235 d	α/5.590	4.987(1)/0.08 5.4897(3)/99.9	0+			0.510
²²³Rn		223.0218	23. m	β⁻/					
²²⁴Rn		224.0241	1.8 h	β⁻/		0+			0.1085 0.2601 0.2655
²²⁵Rn		225.0284	4.5 m	β⁻/		7/2	-0.70	+0.84	
²²⁶Rn		226.0309	7.4 m	β⁻/					
²²⁷Rn		227.0354	2. s	β⁻/					
²²⁸Rn		228.0381	65. s	β⁻/					
₈₇Fr									
²⁰⁰Fr		200.0065	≈ 20 ms	α	7.47				
²⁰¹Fr		201.0046	0.05 s	α/	7.36/	(9/2-)			
²⁰²Fr		202.0033	0.34 s	α/7.590	7.237(8)/100				
²⁰³Fr		203.0014	0.55 s	α/7.280	7.132(5)/	(9/2-)			
²⁰⁴Fr		204.001	2.1 s	α/	7.03/96 6.97/90 7.01/74				
²⁰⁵Fr		204.9987	3.9 s	α/7.050	6.914(5)/	(9/2-)			
²⁰⁶ᵐFr			0.7 s	α/	6.93				0.531(TT)
²⁰⁶Fr		205.9985	16.0 s	α/7.416	6.792(5)/84				
²⁰⁷Fr		206.9969	14.8 s	α/6.900	6.766(5)/	9/2-	+3.9	-0.16	
²⁰⁸Fr		207.99713	59.1 s	α/(77)/6.770 EC/(23)/6.99	6.636(5)/	7+	-4.8	+0.004	
²⁰⁹Fr		208.99592	50.0 s	α/(89)/5.1 EC/(11)/5.16	6.646(3)/	9/2-	+3.9	-0.24	0.7978 (0.1103-1.384)
²¹⁰Fr		209.99640	3.2 m	α/6.670 EC/6.26	6.543(5)/	6+	+4.4	+0.19	0.2030 0.6438 0.8175 0.9008
²¹¹Fr		210.99553	3.10 m	α/6.660 EC/4.61	6.534(5)/	9/2-	+4.0	-0.19	0.220 0.2799 0.5389 0.9169
²¹²Fr		211.99618	20. m	EC/(57)/5.12 α/(43)/6.529	6.261(1)/16 6.335(1)/4 6.335(1)/4 6.343(1)/1.3 6.383(1)/10 6.406(1)/9.5 6.08-6.18	(5+)	+4.6	-0.10	Rn x-ray 0.08107 0.08378 0.2277 1.1856 1.2748 0.014-1.178
²¹³Fr		212.99617	34.6 s	α/6.905	8.476(4)/51	9/2-	+4.0	-0.14	
²¹⁴ᵐFr			3.4 ms	α/	8.547(4)/46 6.775-8.046	9-			
²¹⁴Fr		213.99895	5.1 ms	α/8.587	ʼ7.409(3)/0.3 7.605(8)/1.0 7.940(3)/1.0 8.355(3)/4.7 8.427(3)/93	(1-)			
²¹⁵Fr		215.00033	0.12 μs	α/9.537	9.360(8)/	(9/2-)			
²¹⁶Fr		216.00319	0.70 μs	α/9.175	9.005(10)/95				(0.045-0.160)
²¹⁷Fr		217.00462	0.016 ms	α/8.471	8.315(8)/	(9/2-)			
²¹⁸ᵐFr			22. ms	α					

Elem. or Isot.	Natural Abundance (%)	Atomic Mass or Weight	Half-Life	Decay Mode/Energy (/MeV)	Particle Energy /Intensity (MeV / %)	Spin (h/2π)	Nuclear Magnetic Mom.(nm)	Elect. Quadr. Mom. (b)	γ-ray Energy /Intensity (MeV / %)
²¹⁸Fr		218.00756	1. ms	α/8.014	7.384(10)/0.5 7.542(15)/1.0 7.572(10)/5 7.732(10)/0.5 7.867(2)/93	(1-)			
²¹⁹Fr		219.00924	21. ms	α/8.132	6.802(2)/0.25 6.967(2)/0.6 7.146(2)/0.25 7.313(2)/99	(9/2-)			
²²⁰Fr		220.012313	27.4 s	α/6.800	6.582(1)/10 6.630(2)/6 6.641(1)/12 6.686(1)/61 6.39-6.58	1+	-0.67	+0.47	0.0450 0.061 0.1060 0.1539 0.1617
²²¹Fr		221.01425	4.8 m	α/6.457	5.9393(7)/0.17 5.9797(7)/0.49 6.0751(7)/0.15 6.1270(7)/ 6.2433(3)/1.3 6.3410(7)/83.4	(5/2-)	+1.58	-1.0	At k x-ray 0.0995 0.21798 0.4091
²²²Fr		222.01754	14.3 m	β⁻/2.03 α/5.850	1.78/	2-	+0.63	+0.51	
²²³Fr		223.019731	22.0 m	β⁻/1.149	1.17/65	(3/2+)	+1.17	+1.17	0.05014 0.07972 (0.13 - 0.9)
²²⁴Fr		224.02323	3.0 m	β⁻/2.82		1-	+0.40	+0.517	0.13150 0.21575 0.8367 (0.1 - 2.21)
²²⁵Fr		225.02561	3.9 m	β⁻/1.87		3/2	+1.07	+1.3	
²²⁶Fr		226.0293	49. s	β⁻/3.6		1	+0.071	-1.35	0.18606 0.25373
²²⁷Fr		227.0318	2.48 m	β⁻/2.5		1/2	+1.50		
²²⁸Fr		228.0357	39. s	β⁻/≈3.5		2-	-0.76	+2.4	
²²⁹Fr		229.0384	50. s	β⁻/					
²³⁰Fr		230.0425	19. s	β⁻/		(3)			
²³¹Fr		231.0454	17. s	β⁻/					
²³²Fr		232.0500	5. s	β⁻/					
₈₈Ra									
²⁰²Ra			≈ 3 ms	α	7.86				
²⁰³ᵐRa			0.03 s	α	7.62				
²⁰³Ra		203.0092	≈ 4 ms	α	7.58				
²⁰⁴Ra		204.0065	0.06 s	α	7.48				
²⁰⁵ᵐRa			≈ 0.17 s						
²⁰⁵Ra		205.0062	0.22 s	α	7.34				
²⁰⁶Ra		206.0038	0.4 s	α/7.416	7.272(5)/	0+			
²⁰⁷Ra		207.0037	1.3 s	α/7.270	7.133(5)/				
²⁰⁸Ra		208.0018	1.4 s	α/7.273	7.133(5)/	0+			
²⁰⁹Ra		209.0019	4.6 s	α/7.150	7.008(5)/	5/2	+0.87	+0.38	
²¹⁰Ra		210.0005	3.7 s	α/7.610	7.020(5)/	0+			
²¹¹Ra		211.0009	13. s	α/7.046 EC/5.0	6.912(5)/	(5/2-)	+0.878	+0.46	
²¹²Ra		211.99978	13.0 s	α/7.033	6.901(2)/	0+			
²¹³ᵐRa			2.1 ms	IT					
²¹³Ra		213.00034	2.7 m	EC/(20)/3.88 α/(80)/6.860	6.521(3)/4.8 6.622(3)/39 6.730(3)/36	(1/2-)	+0.613		0.1024 0.11010 0.2125
²¹⁴Ra		214.00009	2.46 s	α/7.272	7.136(4)/	0+			
²¹⁵Ra		215.00270	1.6 ms	α/8.864	7.883(6)/2.8 8.171(3)/1.4 8.700(3)/95.9	(9/2+)			
²¹⁶Ra		216.00352	0.18 μs	α/9.526	9.349(8)/	0+			
²¹⁷Ra		217.00631	1.6 μs	α/9.161	8.992(8)/	9/2-			
²¹⁸Ra		218.00712	26. μs	α/8.547	8.390(8)/	0+			
²¹⁹Ra		219.01006	0.010 s	α/8.132	7.680(10)/65 7.982(9)/35				
²²⁰Ra		220.01101	25. ms	α/7.593	6.998(7)/1.0 7.455(7)/99	0+			0.465
²²¹Ra		221.01391	29. s	α/6.879	6.254(10)/0.7 6.578(5)/3 6.585(3)/8 6.608(3)/35 6.669(3)/21 6.758(3)/31	5/2	-0.180	+1.9	
²²²Ra		222.015361	36.2 s	α/5.590	6.237(2)/3.0 6.556(2)/97	0+			0.324 (0.1448-0.8402)

Elem. or Isot.	Natural Abundance (%)	Atomic Mass or Weight	Half-Life	Decay Mode/Energy (/MeV)	Particle Energy /Intensity (MeV / %)	Spin (h/2π)	Nuclear Magnetic Mom.(nm)	Elect. Quadr. Mom. (b)	γ-ray Energy /Intensity (MeV / %)
^{223}Ra		223.018497	11.435 d	α/5.979	5.287(1)/0.15 5.338(1)/0.13 5.365(1)/0.13 5.433(5)/2.3 5.502(1)/1.0 5.540(1)/9.2 5.607(3)/24 5.716(3)/52 5.747(1)/9 5.857(1)/0.32 5.872(1)/0.85	(3/2+)	+0.271	+1.2	Rn k x-ray 0.12231 0.14418 0.15418 0.15859 0.26939 0.32388 0.33328 0.44494 (0.10 - 0.7)
^{224}Ra		224.020202	3.66 d	α/5.789	5.034(10)/0.003 5.047(1)/0.007 5.164(5)/0.007 5.449(2)/4.9 5.685(2)/95	0+			Rn k x-ray 0.2407 0.4093 0.6501
^{225}Ra		225.023603	14.9 d	β⁻ /0.36	0.32/100	(3/2+)	-0.734		Ac k x-ray 0.0434
^{226}Ra		226.025402	1599. y	α/4.870	4.194(1)/0.001 4.343(1)/0.006 4.601(1)/5.5 4.784(1)/94	0+			Rn k x-ray 0.1861 0.2624
^{227}Ra		227.029170	42. m	β⁻ /1.325	1.03/ 1.30/	(3/2+)	-0.404	+1.5	Ac L x-ray Ac k x-ray 0.02739
^{228}Ra		228.031063	5.76 y	β⁻ /0.046	0.039/50 0.014/30 0.026/20	0+			0.0135 (0.006-0.0306)
^{229}Ra		229.0348	4.0 m	β⁻ /1.76	1.76/	(3/2+)	+0.503	+3.0	(0.0145-0.1715)
^{230}Ra		230.03708	1.5 h	β⁻ /1.0	0.7/	0+			0.0631 0.0720 0.2028 0.4698 0.4787
^{231}Ra		231.0412	1.7 m	β⁻					
^{232}Ra		232.0437	4. m	β⁻					
^{233}Ra		233.0480	30. s	β⁻					
^{234}Ra		234.051	≈ 30. s	β⁻/					
$_{89}$Ac									
^{206m}Ac			0.04 s	α	7.79				
^{206}Ac			≈ 26 ms	α	7.75				
^{207}Ac		207.0121	27 ms	α/	7.69				
^{208m}Ac			≈25. ms	α/	7.72				
^{208}Ac		208.0115	≈0.1 s	α/	7.62				
^{209}Ac		209.0096	≈0.09 s	α/	7.58				
^{210}Ac		210.0093	0.4 s	α/7.610	7.462(8)/				
^{211}Ac		211.0076	0.3 s	α/7.620	7.480(8)/				
^{212}Ac		212.0078	0.9 s	α/7.520	7.379(8)/				
^{213}Ac		213.0066	0.8 s	α/7.500	7.364(8)/	(9/2-)			
^{214}Ac		214.0069	8.2 s	α/(86)/7.350 EC/(14)/6.34	7.007(8)/3 7.082(5)/38 7.214(5)/45	(5+)			
^{215}Ac		215.0065	0.17 s	α/7.750	7.604(5)/	(9/2-)			
^{216m}Ac			0.33 ms	α/	8.198(8)/1.7 8.283(8)/2.5 9.028(5)/49 9.106(5)/46	(9-)			
^{216}Ac		216.00871	≈ 0.3 ms	α/9.241	8.990(2)/10 9.070(8)/90	(1)			
^{217m}Ac			0.7 μs	α/	10.540/100				
^{217}Ac		217.00933	0.07 μs	α/9.832	9.650(10)/100	9/2-			
^{218}Ac		218.01162	1.1 μs	α/9.380	9.205(15)/				
^{219}Ac		219.01241	0.012 ms	α/8.830	8.664(10)/	(9/2-)			
^{220}Ac		220.0148	26. ms	α/8.350	7.610(20)/23 4.680(20)/21 7.790(10)/13 7.850(10)/24 7.985(10)/4 8.005(10)/5 8.060(10)/6 8.195(10)/3				
^{221}Ac		221.01558	52. ms	α/7.790	7.170(10)/2 7.375(10)/10 7.440(15)/20 7.645(10)/70				

Elem. or Isot.	Natural Abundance (%)	Atomic Mass or Weight	Half-Life	Decay Mode/Energy (/MeV)	Particle Energy /Intensity (MeV / %)	Spin (h/2π)	Nuclear Magnetic Mom.(nm)	Elect. Quadr. Mom. (b)	γ-ray Energy /Intensity (MeV / %)
^{222m}Ac			63. s	α/(>89)/ EC/(1)/ I.T./(<10)/	6.710(20)/7 6.750(20)/13 6.810(20)/24 6.840(20)/9 6.890(20)/13 6.970(20)/7 7.000(20)/13				
^{222}Ac		222.01782	5. s	α/7.141	6.967(10)/6 7.013(2)/94	1-			
^{223}Ac		223.01913	2.1 m	α/(>:)/6.783 EC/(1)/0.59	6.131(2)/0.12 6.177(2)/0.94 6.293(1)/0.47 6.326(1)/0.3 6.332(2)/0.14 6.360(1)/0.22 6.397(1)/0.13 6.448(1)/0.2 6.473(1)/3.1 6.523(2)/0.6 6.528(1)/3.1 6.563(1)/13.6 6.582(3)/0.3 6.646(1)/44 6.661(1)/31	(5/2-)			0.0725 0.0839 0.0927 0.0990 0.1917 0.2158 0.3588 0.4768
^{224}Ac		224.021708	2.7 h	EC/(90)/1.403 α/(10)/6.323	5.841(1)/0.5 5.860(1)/0.75 5.875(1)/1.7 5.941(1)/4.4 6.000(1)/6.7 6.013(1)/1.4 6.056(1)/22 6.138(1)/26 6.154(1)/1.0 6.204(1)/12 6.210(1)/20	0-			Ra L kx-ray Ra k x-ray 0.08426 0.13150 0.1571 0.21575 0.2619 (0.03 - 0.3)
^{225}Ac		225.02322	10.0 d	α/5.935	5.286(1)/0.2 5.444(3)/0.1 5.554(1)/0.1 5.608(1)/1.1 5.636(1)/4.5 5.681(1)/1.4 5.722(1)/2.9 5.731(1)/10 5.791(1)/9 5.793(1)/18	3/2			Fr k x-ray 0.9958 0.9982 0.1084 0.1116 0.1451 0.1539 0.15724 0.18799 0.19575 0.2162 0.21686 (0.025 - 0.52)
^{226}Ac		226.026089	1.224 d	EC/(17)/0.640 β⁻/(83)/1.116 α/(0.006)/5.510	5.399(5)/0.006	(1-)			Ra k x-ray Th k x-ray 0.07218 0.15816 0.23034
^{227}Ac		227.027747	21.77 y	β⁻/(98.6)/0.045 α/(1.4)/5.043	0.045/54 4.869(1)/0.09 4.938(1)/0.52 4.951(1)/0.65	(3/2-)	+1.1	+1.7	0.0838/23. 0.0811/14. 0.2696/13. (0.044 - 1.27)
^{228}Ac		228.031014	6.15 h	β⁻/2.127	1.11/32 1.85/12 2.18/11	(3+)			Th L x-ray Th k x-ray 0.12903 0.33842 0.91116 0.96897 (0.2 - 1.96)
^{229}Ac		229.03293	1.04 h	β⁻/1.10	1.1/	(3/2+)			0.07450 0.16451 0.26188 0.5085 0.56916
^{230}Ac		230.0360	2.03 m	β⁻/2.7	1.4/	1+			Th k x-ray 0.45497 0.50820 (0.12 - 2.5)
^{231}Ac		231.0386	7.5 m	β⁻/2.1	2.1/100	(1/2+)			0.14379 0.18574 0.22140 0.28250 0.3070
^{232}Ac		232.0420	2.0 m	β⁻/3.7		(2-)			
^{233}Ac		233.0446	2.4 m	β⁻/		(1/2+)			

Elem. or Isot.	Natural Abundance (%)	Atomic Mass or Weight	Half-Life	Decay Mode/Energy (/MeV)	Particle Energy /Intensity (MeV / %)	Spin (h/2π)	Nuclear Magnetic Mom.(nm)	Elect. Quadr. Mom. (b)	γ-ray Energy /Intensity (MeV / %)
²³⁴Ac		234.0484	40. s	β⁻ /		(1+)			
₉₀Th		232.0381(1)							
²⁰⁰Th			≈ 0.01 s	α	8.08				
²¹⁰Th		210.0150	≈ 12 ms	α	7.90				
²¹¹Th		211.0149	0.04 s	α	7.79				
²¹²Th		212.0129	≈ 30. ms	α/	7.80/	0+			
²¹³Th		213.0130	0.14 s	α/7.840	7.692(10)/				
²¹⁴Th		214.0115	0.09 s	α/7.825	7.677(10)/	0+			
²¹⁵Th		215.0117	1.2 s	α/7.660	7.33(10)/8 7.395(8)/52	(1/2-)			7.524(8)/40
²¹⁶ᵐTh			0.18 ms	α					
²¹⁶Th		216.01105	28. ms	α/8.071	7.921(8)/	0+			
²¹⁷Th		217.01306	0.25 ms	α/9.424	9.250(10)/				
²¹⁸Th		218.01327	0.11 μs	α/9.847	9.665(10)/	0+			
²¹⁹Th		219.01552	1.05 μs	α/9.510	9.340(20)/				
²²⁰Th		220.01573	10. μs	α/8.953	8.790(20)/	0+			
²²¹Th		221.01817	1.7 ms	α/8.628	7.743(8)/6 8.146(5)/56 8.4272(5)/39				
²²²Th		222.01845	2.8 ms	α/8.129	7.982(8)/9.7 7.600(15)/3	0+			
²²³Th		223.02079	0.65 s	α/7.454	7.29(1)/41(5) 7.32(1)/29(5) 7.350(15)/20(5) 7.390(15)/10(4)				
²²⁴Th		224.02146	1.05 s	α/7.305	6.768(5)/1.2 6.997(5)/19 7.170(5)/79				
²²⁵Th		225.02394	8.72 m	EC/(10)/0.68 α/(90)/6.920	6.441(2)/15 6.479(2)/43 6.501(3)/14 6.627(3)/3 6.650(5)/3 6.700(5)/2 6.743(3)/7 6.796(2)/9	(3/2+)			
²²⁶Th		226.024891	30.83 m	α/6.454	6.026(1)/0.2 6.041(1)/0.19 6.098(1)/1.3 6.2283(4)/23 6.3375(4)/75	0+			Ra k x-ray 0.1112 0.2421 0.1310 (0.1733-0.9295)
²²⁷Th		227.027699	18.72 d	α/6.146		(3/2+)			Ra L x-ray Ra k x-ray 0.05014 0.23597 0.25624 (0.02 - 1.0)
²²⁸Th		228.028731	1.913 y	α/5.520	5.1770(2)/0.18 5.2114(1)/0.4 5.3405(1)/26.7 5.4233(1)/73	0+			
²²⁹Th		229.031754	7.9x10³ y	α/5.168	4.814/9.3 4.845(5)/56 4.9008(5)/10.2 4.689-5.077	5/2+	+0.46	+4.	
²³⁰Th		230.033126	7.54x10⁴ y	α/4.771	4.4383(6)/0.03 4.4798(6)/0.12 4.6211(6)/23.4 4.6876(6)/76.3	0+			0.0677/0.46 0.1439/0.078
²³¹Th		231.036296	1.063 d	β⁻ /0.390	0.138/22 0.218/20 0.305/52	5/2+			Pa L x-ray Pa k x-ray 0.02564 0.084203/ (0.02 - 0.3)
²³²Th	100.	232.038050	1.4x10¹⁰ y	α/4.081 SF/9x10¹⁸	3.830(10)/0.2 3.952(5)/23 4.010(5)/77	0+			0.0590 0.124
²³³Th		233.041576	22.3 m	β⁻ /1.245	1.245/	1/2+			Pa L x-ray Pa k x-ray 0.02938 0.08653 0.45930 (0.02 - 1.2)
²³⁴Th		234.043596	24.10 d	β⁻ /0.273	0.102/20 0.198/72	0+			Pa L x-ray 0.06329/4.1 0.09235/2.4

Elem. or Isot.	Natural Abundance (%)	Atomic Mass or Weight	Half-Life	Decay Mode/Energy (/MeV)	Particle Energy /Intensity (MeV / %)	Spin (h/2π)	Nuclear Magnetic Mom.(nm)	Elect. Quadr. Mom. (b)	γ-ray Energy /Intensity (MeV / %)
²³⁵Th		235.04751	7.2 m	β⁻ /1.9					0.09278/2.4 0.4162 0.6594 0.7272 0.747 0.9318
²³⁶Th		236.0497	37.5 m	β⁻ / ≈ 1.0					Pa k x-ray 0.1107
²³⁷Th		237.0539	5.0 m	β⁻					
₉₁Pa		231.03588(2)							
²¹²Pa			≈ 5 ms	α	8.27				
²¹³Pa		213.0212	7 ms	α	8.24				
²¹⁴Pa		214.0207	17 ms	α	8.12				
²¹⁵Pa		215.0190	15. ms	α	8.08/100				
²¹⁶Pa		216.0190	0.19 s	α/	7.95/57 7.86/18 7.84/14				
²¹⁷ᵐPa			1.5 ms	α/	10.16/100				
²¹⁷Pa		217.0183	3.4 ms	α/8.490	8.340(10)/100				
²¹⁸Pa		218.0200	0.12 ms	α/	9.54/31 9.61/69				
`		221.0219	6. μs	α	9.08(3)				
		222.0237	≈ 4.3 ms	α/8.700	8.180/50 8.330/20 8.540/30				
		223.0240	≈ 6.5 ms	α/8.340	8.006(10)/55 8.196(10)/45				
		224.0256	0.84 s	α/7.630	7.555(10)/75(3) 7.46(1)/25(3)				0.1945 (0.028-0.412)
		225.0261	1.8 s	α/7.380	7.195(10)/30 7.245(10)/70				
		226.02792	1.8 m	α/(74)/6.987 EC/(26)/2.83	6.728(10)/0.7 6.823(10)/35 6.863(10)/39				
²²⁷Pa		227.02879	38.3 m	α/(85)/6.582 EC/(15)/1.02	6.357(4)/7 6.376(10)/2.2 6.401(4)/8 6.416(4)/13 6.423(10)/10 6.465(4)/43	(5/2-)			0.0649 0.0669 0.1100
²²⁸Pa		228.03100	22. h	EC/(98)/2.111 α/(2)	5.779/0.23 5.805/0.15 6.078/0.4 6.105/0.25 6.118/0.22	(3+)	+3.5		Th k x-ray 0.20939 0.27026 0.28202 0.32767 0.32807 0.46310 0.91116 0.96464 0.96897 (0.1 - 1.96)
²²⁹Pa		229.03209	1.5 d	EC/(99.8)/0.32 α/(0.2)/5.836	5.536(2)/0.02 5.579(2)/0.09 5.668(2)/0.05	(5/2)			0.04244 (0.024 - 0.18)
²³⁰Pa		230.034532	17.4 d	EC/(90)/1.310 β⁻ /(10)/0.563	0.51/	(2-)	2.0		Th L x-ray Th k x-ray 0.4437 0.45477 0.89876 0.91856 0.95199 (0.053-1.07)
²³¹Pa		231.035878	3.25x10⁴ y	α/5.148	4.6781(5)/1.5 4.7102(5)/1.0 4.7343(5)/8.4 4.8513(5)/1.4 4.9339(5)/3 4.9505(5)/22.8 4.9858(5)/1.4 5.0131(5)/25.4 5.0292(5)/20 5.0318(5)/2.5 5.0587(5)/11	3/2-	2.01	-1.	Ac L x-ray Ac k x-ray 0.01899 0.027396 0.03823 0.04639 0.25586 0.26029 0.28367 0.30007 0.30264 0.33007 (0.02 - 0.61)

Elem. or Isot.	Natural Abundance (%)	Atomic Mass or Weight	Half-Life	Decay Mode/Energy (/MeV)	Particle Energy /Intensity (MeV / %)	Spin (h/2π)	Nuclear Magnetic Mom.(nm)	Elect. Quadr. Mom. (b)	γ-ray Energy /Intensity (MeV / %)
²³²Pa		232.03858	1.31 d	β⁻ /1.34		(2-)			U k x-ray 0.10900 0.15009 0.89439 0.96934 (0.10 - 1.17)
²³³Pa		233.040239	27.0 d	β⁻ /0.571	0.15/40 0.256/60	3/2-	+4.0	-3.0	U L x-ray U k x-ray 0.30017 0.31201 0.34059
²³⁴ᵐPa			1.17 m	β⁻ /(99.9)/2.29 IT/(0.13)/		(0-)			U k x-ray 0.25818/0.07 0.76641/0.32 1.0009/0.85 (0.06 - 1.96)
²³⁴Pa		234.043303	6.69 h	β⁻ /2.197	0.51/	(4+)			U L x-ray U k x-ray 0.1312/0.03 0.5695/0.02 0.9256/0.02 (0.02 - 1.99)
²³⁵Pa		235.04544	24.4 m	β⁻ /1.41	1.4/97	(3/2-)			0.0308-0.65893
²³⁶Pa		236.0487	9.1 m	β⁻ /2.9	1.1/40 2.0/50 3.1/10	(1-)			U k x-ray 0.64235 0.68759 1.7630 (0.04 - 2.18)
²³⁷Pa		237.0511	8.7 m	β⁻ /2.3	1.1/60 1.6/30 2.3/10	(1/2+)			0.4986 0.5293 0.5407 0.8536 0.8650 (0.04 - 1.4)
²³⁸Pa		238.0545	2.3 m	β⁻ /3.5	1.2/ 1.7/	(3-)			0.10350 0.1785 0.4484 0.6350 0.6800 1.01446 (0.04 - 2.5)
²³⁹Pa		239.0571	1.8 h						
₉₂U		238.0289(1)							
²¹⁸U		218.0235	≈ 0.002 s	α	8.63(3)/				
²¹⁹U		219.0249	0.04 ms	α	9.68(4)/				
²²²U		222.0261	≈ 1.µs	α					
²²³U		223.0277	0.02 s	α/	8.78(4)/				
²²⁴U		224.02759	≈ 1. ms	α/	8.46/100				
²²⁵U		225.02938	0.09 s	α/	7.88/85 7.82/15				
²²⁶U		226.02933	0.5 s	α/7.560	7.430/	0+			
²²⁷U		227.03113	1.1 m	α/7.200	6.870/				
²²⁸U		228.03137	9.1 m	α/6.803	6.404(6)/0.6 6.440(5)/0.7 6.589(5)/29 6.681(6)/70	0+			0.095 0.152 0.187 0.246
²²⁹U		229.03350	58. m	EC/(80)/1.31 α/(20)/6.473	6.223/3 6.297(3)/11 6.332(3)/20 6.360(3)/64	(3/2+)			
²³⁰U		230.033927	20.8 d	α/5.992	5.5866(3)/0.01 5.6624(3)/0.26 5.6663(3)/0.38 5.8178(3)/32 5.8887(3)/67	0+			Th L x-ray 0.07218 0.15421 0.23034 (0.081-0.8565)
²³¹U		231.03626	4.2 d	EC/0.36 α/(10⁻³)	5.46/1.6 x 10⁻³ 5.47/1.4 x 10⁻³ 5.40/1. x 10⁻³	(5/2-)			Pa L x-ray Pa k x-ray 0.02564 0.08420
²³²U		232.037146	68.9 y	α/5.414	4.9979(1)/0.003 5.1367(1)/0.3 5.2635(1)/31 5.3203(1)/69	0+			
²³³U		233.039627	1.59x10⁵ y	α/4.909	4.7830(8)/13.2 4.8247(8)/84.4 4.510-4.804	5/2+	+0.59	3.66	Th L x-ray 0.04244 0.09714 (0.0252-1.119)

Elem. or Isot.	Natural Abundance (%)	Atomic Mass or Weight	Half-Life	Decay Mode/Energy (/MeV)	Particle Energy /Intensity (MeV / %)	Spin (h/2π)	Nuclear Magnetic Mom.(nm)	Elect. Quadr. Mom. (b)	γ-ray Energy /Intensity (MeV / %)
²³⁴U	0.0055(5)	234.040945	2.45x10⁵ y	α/4.856	4.604(1)/0.24 4.7231(1)/27.5 4.776(1)/72.5	0+			0.05323/0.156 0.12091
²³⁵ᵐU			26. m	IT/0.0007		1/2+			
²³⁵U	0.720(1)	235.043922	7.04x10⁸ y	α/4.6793	4.1525(9)/0.9 4.2157(9)/5.7 4.3237(9)/4.6 4.3641(9)/11 4.370(4)/6 4.3952(9)/55 4.4144(9)/2.1 4.5025(9)/1.7 4.5558(9)/4.2 4.5970(9)/5.0	7/2-	-0.38	4.6	Th L x-ray Th k x-ray 0.10917 0.14378 0.16338 0.18574 0.20213 0.20533 0.22140 (0.03 - 0.79)
²³⁶U		236.045561	2.34x10⁷ y	α/4.569	4.332(8)/0.26 4.445(5)/26 4.494(3)/74	0+			Th L x-ray 0.04937 0.11275
²³⁷U		237.048723	6.75 d	β⁻/0.519	0.24/ 0.25/	1/2+			Np L x-ray Np k x-ray 0.05953 0.20801
²³⁸U	99.2745(15)	238.050784	4.46x10⁹ y	α	4.039(5)/0.23 4.147(5)/23 4.196(5)/77	0+			Th L x-ray 0.04955/.06 0.1135/.01
²³⁹U		239.054289	23.5 m	β⁻/1.265	1.2/ 1.3/	5/2+			(0.522-0.681)
²⁴⁰U		240.056585	14.1 h	β⁻/0.39	0.36/	0+			Np L x-ray 0.04410 0.05558 0.06760
²⁴²U		242.0629	16.8 m	β⁻ / ≈ 1.2					
₉₃Np									
²²⁵Np		225.0339							
²²⁶Np		226.0351	0.03 s	α/	8.04(2)/				
²²⁷Np		227.0350	0.51 s	α/	7.65(2)/ 7.68(1)/				
²²⁸Np		228.0362	61. s	EC/60(7)/ α/40(7)/,SF					
²²⁹Np		229.0363	4.0 m	α/7.010	6.890(20)				
²³⁰Np		230.0378	4.6 m	EC/97 /3.6 α/3	6.660(20)				
²³¹Np		231.03823	48.8 m	EC/98 /1.8 α/2 /6.368	6.280/2	5/2			0.2629 0.3475 0.3703
²³²Np		232.0400	14.7 m	EC/99 /2.7		(4-)			U L x-ray U k x-ray 0.3268 0.81925 0.86683
²³³Np		233.0410	36.2 m	EC/1.2		(5/2+)			U L x-ray U k x-ray 0.29887 0.31201
²³⁴Np		234.04289	4.4 d	β⁺,EC/1.81	0.79/	(0+)			U L x-ray U k x-ray 1.5272 1.5587 1.6022
²³⁵Np		235.044055	1.085 y	EC/99.9 /0.124 α/0.001/5.191		5/2+			U k x-ray
²³⁶ᵐNp			22.5 h	EC/52 / β⁻ /48 /		(1-)			U L x-ray Pu L x-ray U k x-ray 0.64235 0.68759
²³⁶Np		236.04657	1.55x10⁵ y	EC/91 /0.94 β⁻ /9 /0.49		(6-)			U L x-ray U k x-ray 0.10423 0.16031
²³⁷Np		237.048166	2.14x10⁶ y	α/4.957	4.6395(5)/6.5 4.766(5)/9.7 4.7715(5)/22.7 4.7884(5)/47.8 4.558-4.873	5/2+	+3.14	+3.89	Pa L x-ray Pa k x-ray 0.029378/15 0.08653/12 (0.03-0.28)

Elem. or Isot.	Natural Abundance (%)	Atomic Mass or Weight	Half-Life	Decay Mode/Energy (/MeV)	Particle Energy /Intensity (MeV / %)	Spin (h/2π)	Nuclear Magnetic Mom.(nm)	Elect. Quadr. Mom. (b)	γ-ray Energy /Intensity (MeV / %)
²³⁸Np		238.050940	2.117 d	β⁻ /1.292	1.2/	2+			Pu L x-ray Pu k x-ray 0.98447/25.2 1.02855/18.3 (.044-1.026)
²³⁹Np		239.052931	2.355 d	β⁻ /0.722	0.341/30 0.438/48	5/2+			Pu L x-ray Pu k x-ray 0.10613 0.228186/11 0.27760/15 (0.04-0.50)
²⁴⁰ᵐNp			7.22 m	β⁻ /99.9 / IT/0.1 /	2.18/	(1+)			0.25143 0.26333 0.55454 0.59735
²⁴⁰Np		240.05617	1.032 h	β⁻ /2.20	0.89/	5+			0.1471/ 0.5664 0.6008
²⁴¹Np		241.0583	13.9 m	β⁻ /1.3	1.3/	5/2+			0.1330/ 0.1740 0.280
²⁴²ᵐNp			2.2 m	β⁻ /		(1+)			0.15910 0.2651/ 0.78570 0.9448/
²⁴²Np		242.0616	5.5 m	β⁻ /2.7	2.7/	6+			0.6209 0.73620 0.78074 1.47340 (0.04-2.37)
²⁴³Np		243.0643	1.9 m						
²⁴⁴Np		244.0678	2.3 m						
₉₄Pu									
²²⁸Pu		228.0387		α/	7.81(2)/				
²²⁹Pu		229.0362	4.0 m	α/	7.46(3)/				
²³⁰Pu		230.03964	4.6 m	α/	7.05/				
²³²Pu		232.04118	34. m	EC/ > 80/1.1 α/ < 20/6.716	6.542(10)/38 6.600(10)/62	0+			
²³³Pu		233.04299	20.9 m	EC(99.9)/1.9 α/0.1 /6.416	6.300(20)/0.1				0.1503 0.1804 0.2353 0.5002 0.5346/ 1.0352/
²³⁴Pu		234.04331	8.8 h	EC/94 /0.39 α/6 /6.310	6.035(3)/0.024 6.149(3)/1.9 6.200(3)/4.0	0+			
²³⁵Pu		235.0453	25.3 m	EC/99+ /1.2 α/0.003/5.957	5.850(20)/0.003	(5/2+)			
²³⁶Pu		236.046048	2.87 y	α/5.867	5.611/0.21 5.7210/30.5 5.7677(1)/69.3	0+			0.0476/0.07 0.109/0.02 (0.17 - 0.97)
²³⁷Pu		237.048403	45.7 d	EC/99.9 /0.220 α/0.003 /5.747	5.334(4)/0.0015 5.356(4)/0.0006 5.650(4)/0.0007	7/2-			Np L x-ray Np k x-ray 0.026344 0.03319 0.05954 (0.03-0.5)
²³⁸Pu		238.049553	87.74 y	α/5.593	5.3583(1)/0.10 5.465(1)/28.3 5.4992(1)/71.6	0+			U k x-ray 0.04347 (0.04-1.1)
²³⁹Pu		239.052156	2.411x10⁴ y	α/5.244	5.055/0.047 5.076/0.078 5.106/11.9 5.144/17.1 5.157/70.8 (4.74 -5.03)	1/2+	+0.203		U k x-ray 0.05162 0.05682 0.12928 0.37502 0.41369
²⁴⁰Pu		240.053807	6537. y	α/5.255	5.0212(1)/0.07 5.1237(1)/26.4 5.1681(1)/73.5	0+			U L x-ray 0.04524 0.10423 (0.04-0.97)
²⁴¹Pu		241.056844	14.4 y	β⁻ /99+ /0.0208 α/0.002 /5.139	4.853(7)/3x10⁻⁴ 4.8966(7)/0.002	5/2+	-0.683	+6.	0.14854 0.1600

Elem. or Isot.	Natural Abundance (%)	Atomic Mass or Weight	Half-Life	Decay Mode/Energy (/MeV)	Particle Energy /Intensity (MeV / %)	Spin (h/2π)	Nuclear Magnetic Mom.(nm)	Elect. Quadr. Mom. (b)	γ-ray Energy /Intensity (MeV / %)
²⁴²Pu		242.058736	3.76x10⁵ y	α/4.983	4.7546(7)/0.098 4.8564(7)/22.4 4.9006(7)/78	0+			U L x-ray 0.04491 0.10350
²⁴³Pu		243.061996	4.956 h	β⁻ /0.582	0.49/21 0.58/60	7/2+			Am L x-ray 0.0417 0.0839
²⁴⁴Pu		244.064197	8.2x10⁷ y	α/99.9/4.665 SF/0.1 /	4.546(1)/19.4 4.589(1)/80.5	0+		U L x-ray	0.0439
²⁴⁵Pu		245.06774	10.5 h	β⁻ /1.21	0.93/57 1.21/11	(9/2-)			Am L x-ray Am k x-ray 0.2804 / 0.30832 0.32752 0.56014 (0.03-1.2)
²⁴⁶Pu		246.07020	10.85 d	β⁻ /0.40	0.150/85 0.35/10	0+			Am L x-ray Am k x-ray 0.04379 0.22371
²⁴⁷Pu		247.0741	2.3 d						
₉₅Am									
²³²Am		232.0466	0.9 m	EC/ ≈ 5.0					
²³⁴Am		234.0478	2.3 m	EC/4.2					
²³⁵Am		235.0480	≈ 15 m	EC					Pu K x-ray
²³⁷Am		237.0503	1.22 h	EC/99.98 /1.7 α/0.02 /6.20	6.042(5)/0.02	(5/2-)			Pu k x-ray 0.14559 0.28026 0.43845
²³⁸Am		238.05198	1.63 h	EC/2.26 α/0.0001 /6.04	5.940/0.0001	1+			Pu L x-ray Pu k x-ray 0.91870 0.96278
²³⁹Am		239.053018	11.9 h	EC/99.99 /0.803 a/0.01 /5.924	5.734(2)/0.001 5.776(2)/0.008	5/2-			Pu L x-ray Pu k x-ray 0.18172 0.22818 0.27760
²⁴⁰Am		240.05529	2.12 d	EC/1.38 α/5.592	5.378(1)/16x10⁻⁴	(3-)			Pu L x-ray Pu k x-ray 0.88878 0.98764 (0.1-1.3)
²⁴¹Am		241.056822	432.2 y	α/5.637	5.2443(1)/0.002 5.3221(1)/0.015 5.3884(1)/1.4 5.4431(1)/12.8 5.4857(1)/85.2 5.5116(1)/0.20 5.5442(1)/0.34	5/2-	+1.61	+4.	Np L x-ray 0.02634 0.033192 0.059536 (0.03-1.128)
²⁴²ᵐAm			141. y	IT/99.5 /0.048 α/0.5 /5.62	5.141(4)/0.026 5.2070(2)/0.4	5-	+1.0	+6.5	Am L x-ray 0.04863 0.08648 0.10944 0.16304
²⁴²Am		242.059542	16.02 h	β⁻ /83 /0.665 EC/17 /0.750	0.63/46 0.67/37	1-	+0.388	-2.4	Pu L x-ray Cm L x-ray Pu k x-ray 0.0422 0.04453
²⁴³Am		243.061372	7.37x10³ y	α/5.438	5.1798(5)/1.1 5.2343(5)/11 5.2766(5)/88 5.394(5)/0.12 5.3500(5)/0.16	5/2-	+1.61	+4.	0.04354 0.07467 0.08657 0.11770 0.14197 0.0429
²⁴⁴ᵐAm			≈ 26. m	β⁻ /1.498		(1-)			
²⁴⁴Am		244.064279	10.1 h	β⁻ /1.428					Am L x-ray Cm k x-ray 0.7460 0.9000
²⁴⁵Am		245.066444	2.05 h	β⁻ /0.894	0.65/19 0.90/77	(5/2+)			Cm L x-ray Cm k x-ray 0.25299

Elem. or Isot.	Natural Abundance (%)	Atomic Mass or Weight	Half-Life	Decay Mode/Energy (/MeV)	Particle Energy /Intensity (MeV / %)	Spin (h/2π)	Nuclear Magnetic Mom.(nm)	Elect. Quadr. Mom. (b)	γ-ray Energy /Intensity (MeV / %)
²⁴⁶ᵐAm			25.0 m	β⁻ /	1.3/79. 1.60/14 2.1/7	2-			Cm L x-ray Cm k x-ray 0.27002 0.79881 1.06201 1.07885 (0.04-2.29)
²⁴⁶Am		246.06977	39. m	β⁻ /2.38	1.2/	(7-)			Cm L x-ray Cm k x-ray 0.1529 0.2046 0.6786
²⁴⁷Am		247.0722	22. m	β⁻ /1.7					Cm L x-ray Cm k x-ray 0.2267 / 0.2853 /
₉₆Cm									
²³⁶Cm		236.0514		EC/1.7					
²³⁷Cm		237.0529		EC/2.5					
²³⁸Cm		238.05302	2.4 h	EC/ > 90 /0.97 α/ < 10 /6.632	6.520(50)/ < 10	0+			
²³⁹Cm		239.0548	≈ 3. h	EC/1.7					0.0407 0.1466 0.1874
²⁴⁰Cm		240.055519	27. d	α/6.397	5.989/0.014 6.147/0.05 6.2478(6) /28.8 6.2906(6) /70.6	0+			
²⁴¹Cm		241.057646	32.8 d	EC/99 /0.768 α/1 /6.184	5.8842(4)/0.12 5.9291(4)/0.18 5.9389(4)/0.69	1/2+			Am k x-ray 0.13241 0.16505 0.18028 0.43063 0.47181
²⁴²Cm		242.058828	162.8 d	α/6.216	5.9694(1)/0.035 6.069(1)/25 6.1129(1)/74	0+			Pu L x-ray 0.04408 0.10189 (0.04-1.2)
²⁴³Cm		243.061381	28.5 y	α/6.167	5.6815(5) /0.2 5.6856(5)/1.6 5.7420(5)/10.6 5.7859(5)/73.3 5.9922(5)/6.5 6.0103(5)/1.0 6.0589(5)/5 6.0666(5)/1.5	5/2+	0.41		Pu L x-ray Pu k x-ray 0.10612 0.20975 0.22819 0.27760 0.28546 0.33431 (0.04-0.7)
²⁴⁴Cm		244.062745	18.11 y	α/5.902	5.6656(1)/0.02 5.7528(1)/24 5.8050(1)/76	0+			Pu L x-ray 0.04282 0.09885 0.15262
²⁴⁵Cm		245.065485	8.5x10³ y	α /5.623	5.235(10)/0.3 5.3038(10)/5.0 5.3620(7)/93 5.4927(11)/0.8 5.5331(11)/0.6	7/2+	0.5		Pu L x-ray Pu k x-ray 0.04195 0.13299 0.13606 0.17494
²⁴⁶Cm		246.067217	4.78x10³ y	α/5.476	5.343(3)/21 5.386(3)/79	0+			Pu L x-ray 0.04453
²⁴⁷Cm		247.070346	1.56x10⁷ y	α/5.352	4.818(4)/4.7 4.8690(20)/71 4.941(4)/1.6 4.9820(20)/2.0 5.1436(20)/1.2 5.2104(20)/5.7 5.2659(20)/13.8	9/2- 9/2-	0.37		Pu k x-ray 0.2792 0.2886 0.3471 0.4035
²⁴⁸Cm		248.072341	3.4x10⁵ y	α/92 /5.162 SF/8/	4.931(5)/0.07 5.0349(2)/16.5 5.0784(2)/(75)/1	0+			
²⁴⁹Cm		249.075946	64.15 m	β⁻ /0.900	0.9/	1/2+			Bk k x-ray 0.56039 0.63431
²⁵⁰Cm		250.07835	≈ 9.7x10³ y	SF/ α/5.27		0+			

Elem. or Isot.	Natural Abundance (%)	Atomic Mass or Weight	Half-Life	Decay Mode/Energy (/MeV)	Particle Energy /Intensity (MeV / %)	Spin (h/2π)	Nuclear Magnetic Mom.(nm)	Elect. Quadr. Mom. (b)	γ-ray Energy /Intensity (MeV / %)
^{251}Cm		251.08228	16.8 m	β⁻ /1.42	0.90/16	(1/2+)			0.3896 / 0.5299 0.5425
^{252}Cm		252.0849	< 2 d						
$_{97}$Bk									
^{238}Bk		238.0583	2.4 m	EC/5.0					
^{240}Bk		240.0598	≈ 4.8 m						
^{242}Bk		242.0621	7.0 m	EC/3.0					
^{243}Bk		243.063001	4.5 h	EC/99.8 /1.508 α/0.15 /6.871	6.542(4)/0.03 6.5738(2)/0.04 6.7180(22)/0.02 6.7581(20)/0.02	(3/2-)			0.1466 0.1874 0.755 0.840 0.946
^{244}Bk		244.0652	4.4 h	EC/99.99 /2.26 α/0.01 /6.778	6.625(4)/0.003 6.667(4)/0.003	(4-)			0.1445 0.1876 0.2176 0.9815 0.9215/
^{245}Bk		245.066355	4.94 d	EC/99.9 /0.810 α/0.1 /6.453	5.8851(5)/0.03 6.1176(9)/0.01 6.1467(5)/0.02 6.3087(5)/0.014 6.3492(5)/0.018	3/2-			Cm L x-ray Cm k x-ray 0.25299 0.3809 0.3851
^{246}Bk		246.0687	1.80 d	EC/1.35		(2-)			Cm L x-ray Cm k x-ray 0.79881 1.08142
^{247}Bk		247.07030	1.4x10³ y	α/5.889	5.465(5)/1.5 5.501(5)/7 5.532(5)/45 5.6535(20)/5.5 5.678(2)/13 5.712(2)/17 5.753(2)/4.3 5.794(2)/5.5	(3/2-)			0.04175 0.0839 0.268
^{248}Bk		248.07311	23.7 h	β⁻ /70 /0.87 EC/30 /0.72	0.86/	(1-)			Cm L x-ray Cf L x-ray Cm k x-ray Cf k x-ray 0.5507
^{249}Bk		249.074980	320. d	β⁻ /0.125 α/0.001 /5.525	0.125/100 5.390(1)/0.0002 5.4174(6)/0.001	7/2+	2.0		0.327/10⁻⁵ 0.308/10⁻⁶
^{250}Bk		250.078309	3.217 h	β⁻ /1.780	0.74/	2-			Cf L x-ray Cf k x-ray 0.98912 1.03184 (0.04-1.6)
^{251}Bk		251.08075	56. m	β⁻ /1.09		(3/2-)			0.02481 0.1528 0.1776
^{252}Bk		252.0843	1.8 m						
$_{98}$Cf									
^{237}Cf		237.0621	2.1 s	α,SF/10					
^{238}Cf		238.0614	21 ms	SF					
^{239}Cf		239.0626	≈ 0.7 m	α					
^{240}Cf		240.0623	1.1 m	α/7.719,SF/2	7.590(10)/	0+			
^{241}Cf		241.0637	4. m	EC/3.3 α/7.60	7.335(5)/				
^{242}Cf		242.06369	3.5 m	α/7.509 SF/0.014	7.351(6)/20 7.385(4)/80	0+			
^{243}Cf		243.0654	11. m	EC/86 /2.2 α/14 /7.40	7.060(6)/20 7.170/4	(1/2+)			
^{244}Cf		244.065990	20. m	α/7.328	7.168(5)/25 7.210(5)/75	0+			
^{245}Cf		245.068038	44. m	α/30 /7.255 EC/70 /1.569	7.15/91.7 6.983/0.31 7.09/7 7.065/0.68				Cm K x-ray 0.5709 0.6014 0.6163
^{246}Cf		246.068798	1.49 h	α/6.869	6.6156(10)/0.18 6.7086(7)/21.8 6.7501(7)/78.0	0+			Cm L x-ray 0.04221 0.0945 0.147

Elem. or Isot.	Natural Abundance (%)	Atomic Mass or Weight	Half-Life	Decay Mode/Energy (/MeV)	Particle Energy /Intensity (MeV / %)	Spin (h/2π)	Nuclear Magnetic Mom.(nm)	Elect. Quadr. Mom. (b)	γ-ray Energy /Intensity (MeV / %)
²⁴⁷Cf		247.07099	3.11 h	EC/99.96 /0.65 α/0.04 /6.55	6.301(5)/	7/2+			Bk k x-ray 0.2941 0.4778
²⁴⁸Cf		248.07218	334. d	α/6.369 SF/0.0029	6.220(5)/17 6.262(5)/83	0+			
²⁴⁹Cf		249.074846	351. y	α/6.295	5.7582(2)/3.7 5.8119(2)/84 5.8488(2)/1.0 5.9029(2)/2.8 5.9451(2)/4.0 6.1401(2)/1.1 6.1940(2)/2.2	9/2-			Cm L x-ray Cm k x-ray 0.25299 0.33351 0.38832 (0.0376-1.103)
²⁵⁰Cf		250.076399	13.1 y	α/6.129 SF/0.077	5.8913(4)/0.3 5.9889(4)/15 6.0310(4)/84.5	0+			Cm L x-ray 0.04285
²⁵¹Cf		251.079579	9.0x10² y	α/6.172	5.56448(7)/1.5 5.632(1)/4.5 5.648(1)/3.5 5.6773(6)/35 5.762(3)/3.8 5.7937(7)/2.0 5.8124(8)/4.2 5.8514(6)/27 6.0140(7)/11.6 6.0744(7)/2.7	1/2+			
²⁵²Cf		252.081619	2.65 y	α/96.9 /6.217 SF/3.1/	5.7977(1)/0.23 6.0756(4)/15.2 6.1184(4)/81.6	0+			Cm L x-ray 0.04339 0.1002
²⁵³Cf		253.08512	17.8 d	β⁻ /99.7 /0.29 α/0.3 /6.126	0.27/100 5.921(5)/0.02	(7/2+)			
²⁵⁴Cf		254.08732	60.5 d	SF/99.7/ α/0.3/5.930	5.792(5)/0.05 5.834(5)/0.26	0+			
²⁵⁵Cf		255.0910	1.4 h	β⁻ /0.7					
²⁵⁶Cf		256.0934	12. m	SF					
₉₉Es									
²⁴¹Es		241.0687	≈ 8 s	α	8.11				
²⁴²Es		242.0697	16 s	α	7.92				
²⁴³Es		243.0696	21. s	α/ >30 / EC/ <70 /4.0	7.89/ >30				
²⁴⁴Es		244.0709	37. s	EC/76 /4.6 α/4 /	7.57/4				
²⁴⁵Es		245.0713	1.3 m	α/40 /7.858 EC/60 /3.1	7.74				
²⁴⁶Es		246.0730	7.7 m	EC/90 /3.9 α/10 /	7.35				
²⁴⁷Es		247.07365	4.8 m	EC/93 /2.48 α/7 /	7.32				
²⁴⁸Es		248.0755	26. m	EC/99.7 /3.1 α/0.3 /	6.87				
²⁴⁹Es		249.07640	1.70 h	EC/99.4 /1.45 α/0.6 /	6.77	(7/2+)			0.3795 0.8132
²⁵⁰ᵐEs			2.2 h	EC/ β⁺		(1-)			Cf L x-ray Cf k x-ray 0.9891 1.0319
²⁵⁰Es		250.0787	8.6 h	EC/2.1		(6+)			Cf L x-ray Cf k x-ray 0.30339 0.34948 0.82883
²⁵¹Es		251.07998	1.38 d	EC/99.5 /0.38 α/0.5 /	6.462/0.05 6.492/0.4	(3/2-)			
²⁵²Es		252.08297	1.29 y	α/76 / EC/24 /1.26	6.632/61.0 6.562/10.3	(5-)			
²⁵³Es		253.084818	20.47 d	α/	6.633/89.8 6.5916/6.6	7/2+	+4.10	7.	0.04180 0.3892
²⁵⁴ᵐEs			1.64 d	β⁻ /99.6 / α/0.3 /6.67	0.475 6.382	2+ 2+	2.9	3.7	Fm L x-ray Fm k x-ray 0.6488 0.6938
²⁵⁴Es		254.088017	276. d	α/	6.429	(7+)			0.064
²⁵⁵Es		255.09027	40. d	β⁻ /92 /0.29 α/8 / SF/0.004/	6.26 6.300	(7/2+)			

Elem. or Isot.	Natural Abundance (%)	Atomic Mass or Weight	Half-Life	Decay Mode/Energy (/MeV)	Particle Energy /Intensity (MeV / %)	Spin (h/2π)	Nuclear Magnetic Mom.(nm)	Elect. Quadr. Mom. (b)	γ-ray Energy /Intensity (MeV / %)
^{256m}Es			7.6 h	β⁻ /		(8+)			0.218 0.232 0.862
^{256}Es		256.0936	25. m	β⁻ /1.7		(1+)			
$_{100}$Fm									
^{242}Fm		242.0734	0.8 ms	SF					
^{243}Fm		243.0745	0.2 s	α/	8.55				
^{244}Fm		244.0741	3.7 ms	SF/		0+			
^{245}Fm		245.0754	4. s	α/	8.15/				
^{246}Fm		246.07528	1.2 s	α/92 / SF/8/	8.24/	0+			
^{247m}Fm			9. s	α/	8.18/				
^{247}Fm		247.0768	35. s	α/8.20 EC/2.9	7.87/70 7.93/30				
^{248}Fm		248.07718	36. s	α/99.9 /8.001 SF/0.1/	7.83/20 7.87/80	0+			
^{249}Fm		249.0790	3. m	EC/2.4 α/	7.53	(7/2+)			
^{250m}Fm			1.8 s	IT/					
^{250}Fm		250.07951	30. m	α/ EC/0.8	7.43/	0+			
^{251}Fm		251.08157	5.3 h	EC/98 /1.47 α/2 /	6.833	(9/2-)			
^{252}Fm		252.08246	1.058 d	α/7.154 SF	6.998/15 7.039/85	0+			
^{253}Fm		253.085175	3.0 d	EC(88%)/0.333 α/12 /	6.676/ 6.943/	1/2+			Es k x-ray 0.2719
^{254}Fm		254.086847	3.240 h	α/ SF/0.06/	7.150 7.192	0+			
^{255}Fm		255.089955	20.1 h	α/	6.9635(5)/5.0 7.0225(5)/93.4	7/2+			
^{256}Fm		256.09177	2.63 h	SF/92/ α/18 /	6.92/	0+			
^{257}Fm		257.09510	100.5 d	α/99.8 / SF/0.2	6.519	(9/2+)			0.1794 0.2410
^{258}Fm		258.0971	0.37 ms	SF/					
^{259}Fm		259.1006	1.5 s	SF/					
$_{101}$Md									
^{245m}Md			≈ 0.4 s	α	8.64, 8.68				
^{245}Md		245.0810	0.9 ms	SF					
^{246}Md		246.0819	1.0 s	α	8.74 8.50-8.56				
^{247}Md		247.0818	3. s	α	8.43				
^{248}Md		248.0828	7. s	EC/80 /5.3 α/20 /	8.32/15 8.36/5				
^{249}Md		249.0830	24. s	EC > / < 80 /3.7 α/ > 20 /8.46	8.030(20)/				
^{250}Md		250.0845	50. s	EC/94 /4.6 α/6 /8.25	7.75/4 7.83/2				
^{251}Md		251.0849	4.0 m	EC/ > 94 /3.1 α/ < 6 /	7.55/				
^{252}Md		252.0866	2. m	EC/ > 50 /3.9 α/ < 50 /	7.73/				
^{253}Md		253.0873	≈ 6 m	EC/2.0					
^{254m}Md			30. m	EC/					
^{254}Md		254.0897	10. m	EC/2.7					
^{255}Md		255.09108	27. m	EC/92 /1.04 α/8 /	7.33	(7/2-)			0.4531 0.4056
^{256}Md		256.0941	1.30 h	EC/89 /2.13 α/11 /	7.22/47 7.155/21				Fm k x-ray (0.112-0.834)
^{257}Md		257.095535	5.5 h	EC/85 /0.41 α/15 /,SF	7.074 7.014	(7/2-)			Fm k x-ray (0.181-0.389)
^{258m}Md			57. m	EC/		(1-)			Fm k x-ray
^{258}Md		258.098427	51.5 d	α/7.40	6.718(2)/ 6.763(4)/	(8-)			0.3678 (0.057 - 0.448)
^{259}Md		259.1005	1.64 h	SF/ α/ < 1.3		7/2+			
^{260}Md		260.104	≈ 27.8 d	SF/					
$_{102}$No									
^{250}No		250.0875	0.25 ms	SF/		0+			
^{251}No		251.0889	0.8 s	α/	8.60/80 8.68/20				

Elem. or Isot.	Natural Abundance (%)	Atomic Mass or Weight	Half-Life	Decay Mode/Energy (/MeV)	Particle Energy /Intensity (MeV / %)	Spin (h/2π)	Nuclear Magnetic Mom.(nm)	Elect. Quadr. Mom. (b)	γ-ray Energy /Intensity (MeV / %)
²⁵²No		252.08897	2.3 s	α/73 /8.551 SF/27/	8.42 8.37	0+			
²⁵³No		253.0907	1.7 m	α/ EC/3.2	8.010(20)	(9/2-)			
²⁵⁴ᵐNo			0.28 s	I.T./					
²⁵⁴No		254.09095	55. s	α/ EC/1.1	8.09	0+			
²⁵⁵No		255.09323	3.1 m	α/62 / EC/38/2.01	8.12/ 7.93 8.08	1/2+			0.187
²⁵⁶No		256.09428	2.9 s	α/	8.43	0+			
²⁵⁷No		257.09685	25. s	α/	8.22 8.27 8.32	(7/2+)			
²⁵⁸No		258.0983	≈ 1.2 ms	SF/		0+			
²⁵⁹No		259.1011	58. m	α/78 /7.794 EC/22/0.5	7.52 7.55	(9/2+)			
²⁶²No		262.108	≈ 8. ms	SF/					
₁₀₃Lr									
²⁵³Lr		253.0953	1.3 s	α/	8.80 8.72				
²⁵⁴Lr		254.0965	13. s	α/ EC/5.2	8.45				
²⁵⁵Lr		255.0967	22. s	α/ EC/3.2	8.37/60 8.43/40				
²⁵⁶Lr		256.0988	28. s	α/99.7 /8.554 EC/4.2	8.43/ 8.39				
²⁵⁷Lr		257.0996	0.65 s	α/ EC/2.5	8.80 8.80	7/2+			
²⁵⁸Lr		258.1019	3.9 s	α/ EC/3.4	8.60 8.62 8.56				
²⁵⁹Lr		259.1030	6.1 s	α/80 SF/20	8.44(1)				
²⁶⁰Lr		260.1056	3. m	α/	8.03				
²⁶¹Lr		261.1069	40. m	SF					
²⁶²Lr		262.110	3.6 h	EC/2.					
₁₀₄Rf									
²⁵³Rf		253.1007	≈ 48. μs	SF α/ < 10					
²⁵⁴Rf		254.1002	23. μs	SF α/ < 1.5					
²⁵⁵Rf		255.1015	1.7 s	α SF/45	8.72 8.77				
²⁵⁶Rf		256.10118	6.2 ms	SF α/0.32	8.81				
²⁵⁷Rf		257.1031	4.7 s	α/9.22 EC/11 SF/ < 1.4	8.77 9.01 8.95 8.62				0.117
²⁵⁸Rf		258.1036	12. ms	SF,α					
²⁵⁹Rf		259.1056	3.4 s.	α/9.09/93 SF/7	8.77(2)/ 8.86/				
²⁶⁰Rf		260.1064	20. ms	SF/					
²⁶¹Rf		261.1088	1.1 m	α/8.78,SF/ < 10	8.28/				
²⁶²Rf		262.1101	2.1 s	SF/					
²⁶³Rf		263.1125	< 20. m	SF					
₁₀₅Db									
²⁵⁵Db		255.1074	≈ 1.5 s	α,SF					
²⁵⁶Db		256.1081	≈ 2.6 s	α,SF,E.C.					
²⁵⁷Db		257.1079	1.5 s	α/ SF	8.97/ 9.07/ 9.16/				
²⁵⁸Db		258.1094	4.2 s	α/ E.C/5.3	9.17/ 9.08/				
²⁵⁹Db		259.1097	≈ 1.2 s	SF/					
²⁶⁰Db		260.1114	1.5 s	α/ SF/	9.05/ 9.08/ 9.13/				
²⁶¹Db		261.1121	1.8 s	α/ SF/	8.93/				

Elem. or Isot.	Natural Abundance (%)	Atomic Mass or Weight	Half-Life	Decay Mode/Energy (/MeV)	Particle Energy /Intensity (MeV / %)	Spin (h/2π)	Nuclear Magnetic Mom.(nm)	Elect. Quadr. Mom. (b)	γ-ray Energy /Intensity (MeV / %)
²⁶²Db		262.1141	34. s	SF/					
				α/	8.45/				
					8.53/				
					8.67/				
²⁶³Db		263.1151	≈0.45 m	SF/57/, α/43/	8.35/43				
₁₀₆Sg									
²⁵⁸Sg		258.1132	≈ 2.9 ms	SF					
				α/ < 20					
²⁵⁹Sg		259.1147	0.5 s	α,SF/	9.62				
²⁶⁰Sg		260.11444	4. ms	α,SF/	9.77				
					9.72				
²⁶¹Sg		261.1162	0.3 s	α,SF/	9.56				
²⁶³Sg		263.1183	0.8 s	SF,α	9.06				
					9.25/				
²⁶⁵Sg		265.1211	<30. s	α/ > 50					
				SF/ < 50					
²⁶⁶Sg		266.1219	<30. s	α/ > 50	8.63(5)				
				SF/ < 50					
₁₀₇Bh									
²⁶⁰Bh		260.122		α					
²⁶¹Bh		261.1218	12. ms	α/,SF	10.40				
					10.10				
					10.03				
²⁶²ᵐBh			8. ms	α/	10.37				
					10.24				
²⁶²Bh		262.1230	0.10 s	α/	10.06				
					9.91				
					9.74				
²⁶⁴Bh		264.1247	0.44 s	α	9.47				
					9.62				
₁₀₈Hs									
²⁶³Hs		263.1287		α/					
²⁶⁴Hs		264.1284	≈ 0.08 ms	α/,SF/ ≈ 50	11.0				
²⁶⁵ᵐHs			0.9 ms	α	10.57				
²⁶⁵Hs		265.1300	≈1.6 ms	α/	10.31				
²⁶⁷Hs		267.1371	60 ms	α/ > 88	9.74-9.87				
²⁶⁹Hs		269.1341	19.7 s	α	9.23				
₁₀₉Mt									
²⁶⁶ᵐMt			≈1.2 ms	α	10.46-10.81				
²⁶⁶Mt		266.1379	≈0.7 ms	α	10.48-11.31				
²⁶⁷Mt		267.138	19 ms	α					
²⁶⁸Mt		268.1388	0.07 s	α/ > 68	10.10,10.24				
₁₁₀110									
²⁶⁷110		267.1440	≈ 3 μs	α/ > 32	11.6				
²⁶⁹110		269.1451	0.17 ms	α/ > 75	11.11				
²⁷¹ᵐ110			≈ 56 ms	α	10.71				
²⁷¹110		271.1461	≈ 1.1 ms	α	10.68,10.74				
²⁷³110		273.1492	0.3 ms	α/ > 32	11.35				
₁₁₁111									
²⁷²111		272.1535	≈ 1.5 ms	α/ > 68	10.8				
₁₁₂112									
²⁷⁷112			≈ 0.24 ms	α	11.45				

NEUTRON SCATTERING AND ABSORPTION PROPERTIES*
(Revised 1996)

Norman E. Holden
Reactor Division
Brookhaven National Laboratory
Upton, New York 11973

This Table presents an evaluated set of values for the experimental quantities, which characterize the properties for scattering and absorption of neutrons. The neutron cross section is given for room temperature neutrons, 20.43° C, corresponds to a thermal neutron energy of 0.0253 electron volts (eV) or a neutron velocity of 2200 meters/second. The neutron resonance integral is defined over the energy range from 0.5 eV to 0.1 x 10^6 eV, or 0.1 MeV. Bound neutron scattering lengths and neutron cross sections averaged over a maxwellian spectrum at 30 keV for astrophysical applications are also presented. A list of the major references used is given below. The literature cutoff date is December 1996. Uncertainties are given in parentheses. Parentheses with two or more numbers indicate values to the excited state(s) and to the ground state of the product nucleus. Discussions with Frank Feiner and Harold Knox of KAPL on thermal cross sections and resonance integrals and the assistance of Felicia Kramer in assembling this file are gratefully acknowledged.

TABLE LAYOUT

Column No.	Column Title	Description
1	Isotope or Element	For elements, the atomic number and chemical symbol are listed. For nuclides, the mass number and chemical symbol are listed. Isomers are indicated by the addition of m, m1, or m2.
2	Isotopic Abundance	in atom percent.
3	Half-life	Half-life in decimal notation. μs = microseconds; ms = milliseconds; s = seconds; m = minute; h = hours; d = days; and y = years.
4	Thermal Neutron Cross Section	Cross sections for neutron capture reactions in barns (10^{24} cm²) or in millibarns (mb). Proton, alpha production and fission reactions are designated by σ_p, σ_a, σ_f, respectively. Separate values are listed for isomeric production.
5	Neutron Resonance Integral	Resonance integrals for neutron capture reactions in barns (10^{24} cm²) or millibarns (mb). Proton, alpha production and fission reactions are designated by RI_p, RI_a, RI_f, respectively. Separate values are listed for isomeric production.
6	Neutron Scattering Length	Bound Coherent scattering length for neutron scattering reactions in units of femtometer (fm), which is equal to fermis (10^{-13} cm).
7	Maxwellian Averaged Cross Section	Astrophysical Cross Sections, averaged over a stellar neutron maxwellian spectrum characterized by a thermal energy of 30 keV, expressed in barns (10^{24} cm²), millibarns (mb) or microbarns (μb).

GENERAL NUCLEAR DATA REFERENCES

The following references represent the major sources of the neutron data presented:

1. Mughabghab, S.F.; Divadeenam, M.; Holden, N.E.; Neutron Cross Sections, Vol. I *Neutron Resonance Parameters and Thermal Cross Sections* Part A Z=1-60. Academic Press Inc., New York, New York (1981); Mughabghab, S.F.; Part B, Z-61-100. Academic Press Inc., Orlando, Florida (1984).

2. Parrington, J.R.; Knox, H.D.; Breneman, S.L.; Baum, E.M.; Feiner, F.: *Chart of the Nuclides, 15th Edition*, Knolls Atomic Power Laboratory (Oct. 1996).

3. Holden, N.E.; *Fifty Years with Nuclear Fission* Conference, Wash., D.C., Gaithersburg, Md. April 26-29, 1989, p. 946. American Nuclear Society, LaGrange Park, Illinois (1989).

4. Holden, N.E.; *Review of Thermal Neutron Cross Sections and Isotopic Composition of the Elements*, BNL-NCS-42224 (March 1989) and Holden, N.E.; *Thermal Neutron Cross Sections for the 1991 Table of the Isotopes*, BNL-45255 (Feb. 1991).

5. Tuli, J.K.; *Nuclear Wallet Cards*, Brookhaven National Laboratory (July 1995).

6. Holden, N.E.; *Half-lives of Selected Nuclides*, Pure & Applied Chemistry 62, 941 (1990).

7. Koester, L; Rauch, H.; Seymann, E.; *Neutron Scattering Lengths: A Survey of Experimental Data and Methods*, Atomic Data Nuclear Data Tables 49, 65 (1991).

8. Sears, V.F.; *Neutron Scattering Lengths and Cross Sections*, Neutron News 3, (3), 26 (1992).

*Research carried out under auspices of the US Department of Energy, Contract No. DE-AC02-76CH00636

Elem. or Isot.	Natural Abundance (%)	Half-life	Thermal Neut. Cross Section (barns)	Resonance Integral (barns)	Coh. Scat. Length (fm)	σ(30KEV) Maxw. Avg. (barns)
$_1$H			0.332(2)	0.149(1)	- 3.739(1)	
^{1}H	99.985(1)		0.332(2)	0.149(1)	- 3.741(1)	0.25(2) mb*
^{2}H	0.015(1)		0.51(1)mb	0.23(2) mb	6.671(4)	
^{3}H		12.32 y	< 6. μb		4.79(3)	
$_2$He			< 0.05		3.26(3)	
^{3}He	1.37x10^{-6}		σ_p=5.33(1)x10^3 0.05(1) mb	RI$_p$=2.39(1)x10^3	5.74(7)	8.(1) μb*
^{4}He	≈ 100.				3.26(3)	
$_3$Li			71.(2)	32.(1)	- 1.90(2)	
^{6}Li	7.5(2)		σ_α=9.4(1)x10^2 39.(5) mb	RI$_\alpha$=422.(4) 17.(2) mb	2.0(1)	
^{7}Li	92.5(2)		45.(5) mb	20.(2) mb	- 2.22(2)	39(6) μb
$_4$Be			8.8(4) mb	3.9(2) mb	7.79(1)	
^{7}Be		53.28 d	σ_p=3.9(1)x10^4	RI$_p$=1.75(5)x10^4		σ_p 16(4)
^{9}Be	100.		σ_α ≈ 0.1 8.8(4) mb	3.9(2) mb	7.79(1)	
^{10}Be		1.52x10^6 y	<1. mb			
$_5$B			7.6(1)x10^2	343.(5)	5.30(4)	
^{10}B	19.9(2)		σ_α=38.4(1)x10^2 0.3(1) σ_p=7.(1) mb σ_t=8.(2) mb	RI$_\alpha$=17.3(1)x10^2 0.13(4)	- 0.1(3)	
^{11}B	80.1(2)		5.(3) mb	2(1) mb	6.65(4)	
$_6$C			3.5(1) mb	1.6(1) mb	6.646(1)	
^{12}C	98.89(1)		3.5(1) mb	1.6(1) mb	6.651(2)	16(1) μb*
^{13}C	1.11(1)		1.4(1) mb	1.7(2) mb	6.19(9)	0.2(1) mb
^{14}C		5715. y	<1. μb			1.5(4) μb*
$_7$N			1.9(1)	0.85(5)	9.36(2)	
^{14}N	99.634(9)		σ_p=1.83(7) 0.080(1)	RI$_p$=0.82(3) 0.034(1)	9.37(2)	σ_p 1.8(2) mb* 0.04(7) mb
^{15}N	0.366(9)		0.04(1) mb	0.11(3) mb	6.44(3)	4.9(5) μb*
$_8$O			0.29(1) mb	0.40(4) mb	5.805(4)	
^{16}O	99.76(1)		0.19(1) mb	0.36(4) mb	5.805(5)	34(4) μb
^{17}O	0.04		σ_α=0.24(1) 0.54(7) mb	0.11(1) 0.39(5) mb	5.8(2)	σ_α 3.9(5) mb*
^{18}O	0.20(1)		0.16(1) mb	0.81(4) mb	5.84(7)	7.2(6) μb*

Elem. or Isot.	Natural Abundance (%)	Half-life	Thermal Neut. Cross Section (barns)	Resonance Integral (barns)	Coh. Scat. Length (fm)	σ(30KEV) Maxw. Avg. (barns)
$_9$F			9.5(7) mb	21.(3) mb	5.65(1)	
^{19}F	100.		9.5(7) mb	21.(3) mb	5.65(1)	6.(1) mb
$_{10}$Ne			42(5) mb	19(3) mb	4.566(6)	
^{20}Ne	90.48(3)		39(5) mb	18(3) mb	4.631(6)	0.12(1) mb
^{21}Ne	0.27(1)		0.7(1),$\sigma_\alpha < 1.5$	0.31(5)	6.7(2)	$\approx$ 1.6 mb
^{22}Ne	9.25(3)		51(5) mb	23(3) mb	3.87(1)	60.(5) μb*
$_{11}$Na			0.53(2)	0.32(2)	3.63(2)	2.3(2) mb
^{22}Na		2.605 y	$\sigma_p = 2.8(3) \times 10^4$ $\sigma_\alpha = 2.6(4) \times 10^2$ $\sigma_m = 0.43(3)$	< 2x10^5 1.2(2) x 10^2 RI$_m$=0.30(6)		
^{23}Na	100.				3.63(2)	
$_{12}$Mg			66(6) mb	38.(5) mb	5.375(4)	
^{24}Mg	78.99(3)		0.053(6)	32.(4) mb	5.7(2)	4.2(2) mb
^{25}Mg	10.00(1)		.20(1)	98.(15) mb	3.6(2)	6.5(3) mb
^{26}Mg	11.01(2)		.038(1)	25.(2) mb	4.9(2)	66.(3) μb
^{27}Mg		9.45 m	0.07(2)	0.03(1)		
$_{13}$Al			0.230(2)	0.17(1)	3.45(1)	
^{26}Al			$\sigma_p \approx 2.1$			0.14(2)
^{27}Al	100.		0.230(2)	0.17(1)	3.45(1)	3.8(3) mb
$_{14}$Si			0.166(9)	0.12(2)	4.15(1)	
^{28}Si	92.21(2)		0.17(1)	0.11(2)	4.11(1)	1.8(2) mb
^{29}Si	4.69(2)		0.12(1)	0.08(2)	4.7(1)	7.9(9) mb
^{30}Si	3.10(1)		0.107(4)	0.62(6)	4.61(1)	6.4(6) mb
^{31}Si		2.62 h	73.(6) mb	33.(3) mb		
$_{15}$P			0.17(1)	0.08(1)	5.13(1)	
^{31}P	100.		0.17(1)	0.08(1)	5.13(1)	1.7(1) mb
$_{16}$S			0.54(2)	0.24(2)	2.847(1)	
^{32}S	95.02(9)		0.55(5) σ_α < 0.5 mb	0.25(2)	2.804(2)	4.7(2) mb
^{33}S	0.75(4)		0.46(3) $\sigma_\alpha = 0.12(1)$ $\sigma_p = 2$ mb	0.21(2) RI$_\alpha$ = 0.05(1)	4.7(2)	7.4(15) mb $\sigma_\alpha = 0.18(1)$
^{34}S	4.21(8)		0.30(2)	0.13	3.48(3)	2.9(9) mb
^{36}S	0.02(1)		0.23(2)	0.26(3)		0.17(1) mb
$_{17}$Cl			33.6(3)	15.(2)	9.58(1)	
^{35}Cl	75.77(7)		43.7(4);σ_p=0.44(1) $\sigma_\alpha \approx 0.08$ mb	20.(2);RI$_p$=0.2	11.7(1)	10.(3) mb σ_p 1.7(2) mb

Elem. or Isot.	Natural Abundance (%)	Half-life	Thermal Neut. Cross Section (barns)	Resonance Integral (barns)	Coh. Scat. Length (fm)	σ(30KEV) Maxw. Avg. (barns)
^{36}Cl		3.01×10^5 y	$\sigma_p=46.(2)$ mb $<10.$ $\sigma_\alpha = 0.59(7)$ mb	$RI_p=0.02$		σ_p 91.(8) mb σ_α 0.9(2) mb
^{37}Cl	24.23(7)		$(0.05 + 0.38)$	$(0.04+0.26)$	3.1(1)	2.15(8) mb
$_{18}$Ar			0.66(3)	0.42(5)	1.91(1)	
^{36}Ar	0.337(3)		$5(1); \sigma_p<1.5$ mb σ_α 6(1) mb	2.(1)	24.9(1)	
^{37}Ar		35.0 d	$\sigma_\alpha=2.0(3) \times 10^3$ $\sigma_p = 69(14)$	$RI_\alpha=900.$ $RI_p=31.$		
^{38}Ar	0.063(1)		0.8(2)	0.4(1)	3.5(35)	
^{39}Ar		268. y	$6.(2) \times 10^2$			
^{40}Ar	99.600(3)		0.64(3)	0.41(5)	1.83(1)	2.3(2)mb
^{41}Ar		1.82 h	0.5(1)	0.2(1)		
$_{19}$K			2.1(1)	1.0(1)	3.67(2)	
^{39}K	93.2581(44)		2.1(2) $\sigma_\alpha = 4.3(5)$ mb $\sigma_p < 0.05$ mb	0.9(1)	3.74(2)	11.8(4)mb
^{40}K	0.0117(1)	1.26×10^9 y	30.(8) $\sigma_p=4.4(4)$ $\sigma_\alpha = 0.42(8)$	13.(4) 2.0(2)		σ_p .7(1)mb σ_α 40(6)mb
^{41}K	6.7302(44)		1.46(3)	1.4(2)	2.69(8)	22.0(7)mb
$_{20}$Ca			0.43(2)	0.23(2)	4.70(2)	
^{40}Ca	96.941(18)		0.41(3) σ_α 0.13(4) mb	0.22(4)	4.80(2)	6.7(7)mb
^{41}Ca		1.02×10^5 y	$\approx 4.$ $\sigma_\alpha = 0.19$			
^{42}Ca	0.647(9)		0.65(10)	0.39(4)	3.4(1)	16.(2)mb
^{43}Ca	0.135(6)		6.(1)	3.9(2)	- 1.56(9)	51.(6)mb
^{44}Ca	2.086(12)		0.8(2)	0.56(1)	1.42(6)	9.(1)mb
^{45}Ca		162.7 d	$\approx 15.$			
^{46}Ca	0.004(3)		0.70(3)	0.9(1)	3.6(2)	5.7(5)mb
^{48}Ca	0.187(4)		1.1(1)	0.5(1)	0.39(9)	1.0(2)mb
$_{21}$Sc			27.2(2)	12.(1)	12.3(1)	
^{45}Sc	100.		$(10.+17.)$	$(5.6+6.4)$	12.3(1)	90.(6)mb
^{46}Sc		83.81 d	8.(1)	3.6(5)		
$_{22}$Ti			6.1(1)	2.8(2)	- 3.438(2)	
^{46}Ti	8.25(3)		0.6(2)	0.4(1)	4.93(6)	27.(3)mb
^{47}Ti	7.44(2)		1.6(2)	1.6(2)	3.63(1)	64.(8)mb
^{48}Ti	73.72(3)		7.9(9)	3.6(2)	- 6.09(2)	31.(5)mb
^{49}Ti	5.41(2)		1.9(5)	1.2(2)	1.04(5)	22.(2)mb
^{50}Ti	5.18(2)		0.179(3)	0.12(2)	6.18(8)	4.0(5)mb

Elem. or Isot.	Natural Abundance (%)	Half-life	Thermal Neut. Cross Section (barns)	Resonance Integral (barns)	Coh. Scat. Length (fm)	σ(30KEV) Maxw. Avg. (barns)
$_{23}$V			5.0(2)	2.8(1)	- 0.382(1)	
^{50}V	0.250(4)	>1.4x10^{17} y	21(4) $\sigma_p = 0.7(4)$ mb	50.(20)	7.6(6)	
^{51}V	99.750(4)		4.9(1)	2.7(2)	- 0.402(2)	42.(3)mb
$_{24}$Cr			3.0(2)	1.7(1)	3.635(7)	
^{50}Cr	4.345(13)		15.(1)	8.(1)	- 4.5(1)	52.(1)mb
^{51}Cr		27.70 d	< 10.			
^{52}Cr	83.79(2)		0.8(1)	0.6(2)	4.91(2)	8.8(4)mb
^{53}Cr	9.50(2)		18.(2)	9(1)	- 4.2(1)	0.06(1)
^{54}Cr	2.365(7)		0.36(4)	0.25(5)	4.6(1)	7.(2)mb
$_{25}$Mn			13.3(1)	14.0(3)	- 3.75(2)	
^{53}Mn		3.7x10^6 y	70.(10)	32.(5)		
^{54}Mn		312.2 d	< 10.			
^{55}Mn	100.		13.3(1)	14.0(3)	- 3.75(2)	40.(2)mb
$_{26}$Fe			2.6(1)	1.4(2)	9.45(2)	
^{54}Fe	5.85(4)		2.7(5) $\sigma_\alpha = 10 \mu$b	1.3(2) $RI_\alpha = 1.1(1)$ mb	4.2(1)	29.(2)mb
^{55}Fe		2.73 y	13.(2) $\sigma_\alpha = 0.03$	6.(1)		
^{56}Fe	91.75(4)		2.6(2)	1.4(2)	9.93(3)	13.5(5)mb
^{57}Fe	2.12(1)		2.5(5)	1.6(2)	2.3(1)	35.(4)mb
^{58}Fe	0.28(1)		1.3(1)	1.3(2)	15.(7)	13.(2)mb
^{59}Fe		44.51 d	13.(3)	6.(1)		
$_{27}$Co			37.19(8)	74.(2)	2.49(2)	
^{58m}Co		9.1 h	1.4(1)x10^5	2.5(10)x10^5		
^{58}Co		70.88 d	1.9(2)x10^3	7.(1)x10^3		
^{59}Co	100.		(20.7+16.5)	(39.+35.)	2.49(2)	38.(4)mb
^{60m}Co		10.47 m	58.(3)	230.(50)		
^{60}Co		5.271 y	2.0(2)	4.3(10)		
$_{28}$Ni			4.5(2)	2.3(2)	10.3(1)	
^{58}Ni	68.077(9)		4.6(4) $\sigma_\alpha < 0.03$ mb	2.3(2)	14.4(1)	42.(3)mb
^{59}Ni		≈7.6x10^4 y	σabs=92.(4) $\sigma_\alpha = 14(2)$ $\sigma_p = 2(1)$	RIabs=1.4(1)x10^2		
^{60}Ni	26.223(8)		2.9(3)	1.5(2)	2.8(1)	27.(2)mb
^{61}Ni	1.140(1)		2.5(5) $\sigma_\alpha = 0.03$ mb	1.5(4)	7.60(6)	82.(1)mb
^{62}Ni	3.634(2)		15.(1)	6.8(3)	- 8.7(2)	36.(4)mb
^{63}Ni		100. y	20(5)	9.(2)	26.(8)mb	
^{64}Ni	0.926(1)		1.6(1)	1.(2)	- 0.37(7)	10.(1)mb
^{65}Ni		2.517 h	22.(2)	10.(1)		

Elem. or Isot.	Natural Abundance (%)	Half-life	Thermal Neut. Cross Section (barns)	Resonance Integral (barns)	Coh. Scat. Length (fm)	σ(30KEV) Maxw. Avg. (barns)
$_{29}$Cu			3.8(1)	4.1(4)	7.718(4)	
^{63}Cu	69.17(3)		4.5(2)	5.(1)	6.43(15)	0.09(1)
^{64}Cu		12.701 h	$\approx$ 270.			
^{65}Cu	30.83(3)		2.17(3)	2.2(1)	10.61(19)	53(5)mb
^{66}Cu		5.09 m	1.4(1)x10^2	60.(20)		
$_{30}$Zn			1.1(2)	2.8(4)	5.680(5)	
^{64}Zn	48.6(6)		0.74(5) $\sigma_p < 12\ \mu b$ $\sigma_\alpha = 11(3)\ \mu b$	1.4(3)	5.23(4)	59(5)mb
^{65}Zn		243.8 d	66.(8) $\sigma_\alpha = 2.0(2)$	30.(4)		
^{66}Zn	27.9(3)		0.9(3) $\sigma_\alpha < 0.02$ mb	1.8(2)	5.98(5)	35(3)mb
^{67}Zn	4.1(1)		6.9(1.4) $\sigma_\alpha = 0.4$ mb	25.(5)	7.58(8)	0.15(2)
^{68}Zn	18.8(5)		(0.072 + 0.8) $\sigma_\alpha < 0.02$ mb	(0.2 + 2.9)	6.04(3)	19(2)mb
^{70}Zn	0.6(3)		(8.1+83.) mb	0.9(2)		
$_{31}$Ga			2.9(1)	22.(3)	7.288(2)	
^{69}Ga	60.11(3)		1.68(7)	16.(2)	7.88(4)	0.15(1)
^{71}Ga	39.89(3)		4.7(2);σ_m=0.15(5)	31.(3)	6.40(3)	0.12(1)
$_{32}$Ge			2.2(1)	6.(2)	8.19(2)	
^{70}Ge	21.23(4)		(0.3 + 2.7)	2.3(1)	10.0(1)	86(5)mb
^{72}Ge	27.66(3)		0.9(2)	0.8(3)	8.5(1)	0.06(2)
^{73}Ge	7.73(1)		15.(1)	66.(20)	5.02(4)	0.28(8)
^{74}Ge	35.94(2)		(0.14 + 0.28)	(0.4+0.5)	7.6(1)	54(6)mb
^{76}Ge	7.44(2)		(0.09 + 0.06)	(1.3+0.6)	8.2(15)	16(1)mb
$_{33}$As			4.0(4)	61.(5)	6.58(1)	
^{75}As	100.		4.0(4)	61.(5)	6.58(1)	0.46(2)
$_{34}$Se			12.(1)	14.(3)	7.970(9)	
^{74}Se	0.89(4)		50.(2)	520(50)	0.8(3)	
^{75}Se		119.78 d	3.3(10)x10^2			
^{76}Se	9.37(29)		(22. + 63.)	(9.+31.)	12.2(1)	164(8)mb
^{77}Se	7.63(16)		42.(4),σ_α=0.97(3)μb	30.(5)	8.25(8)	
^{78}Se	23.77(28)		σ_m=0.38(2)	RI$_m$=4.3(4)	8.24(9)	0.08
^{80}Se	49.61(41)		(0.06+0.35)	(0.36+1.0)	7.48(3)	44.(3)mb
^{82}Se	8.73(22)		(39.+ 5.2) mb	39.(4)mb	6.34(8)	
$_{35}$Br			6.8(2)	92.(8)	6.79(2)	
^{76}Br		16.0 h	224.(42)			
^{79}Br	50.69(7)		(2.5+8.3)	(36.+96)	6.79(7)	0.74(3)
^{81}Br	49.31(7)		(2.4+0.24)	51.(5)	6.78(7)	0.24(1)

Elem. or Isot.	Natural Abundance (%)	Half-life	Thermal Neut. Cross Section (barns)	Resonance Integral (barns)	Coh. Scat. Length (fm)	σ(30KEV) Maxw. Avg. (barns)
$_{36}$Kr			24.(1)	39.(6)	7.81(2)	
^{78}Kr	0.35(1)		(0.17+6.)	20.(1)		0.34(4)
^{80}Kr	2.28(6)		(4.6+7.)	57.(6)		0.24(1)
^{82}Kr	11.58(14)		(14.+7.)	130.(13)		79(6)mb
^{83}Kr	11.50(6)		183.(30)	183.(20)	0.24(2)	
^{84}Kr	56.99(4)		σ_m=0.09;σ_m+g=0.11	2.4(3)		(17+17)mb
^{85}Kr		10.73 y	1.7(2)	1.8(10)		0.07(2)
^{86}Kr	17.30(22)		3.(2)mb	$\approx$1. mb	8.1(3)	3.5(3)mb
$_{37}$Rb			0.39(4)	6.(3)	7.08(2)	
^{84}Rb		32.9 d	σ_p=12.(2)			
^{85}Rb	72.17(2)		(0.06+0.38)	(0.7+7.)	7.0(1)	240(9)mb
^{86}Rb		18.65 d	<20.			
^{87}Rb	27.83(2)	4.88x10^{10} y	0.10(1)	2.3(4)	7.3(1)	18.0(5)mb
^{88}Rb		17.7 m	1.2(3)	0.5(1)		
$_{38}$Sr			1.2(1)	10.(1)	7.02(2)	
^{84}Sr	0.56(1)		(0.6+0.2)	(9.+1.)		
^{86}Sr	9.86(1)		σ_m=0.81(4)	RI$_m$=4.(1)	5.68(5)	(48+22)mb
^{87}Sr	7.00(1)		16.(3)	118.(30)	7.41(7)	97.(5)mb
^{88}Sr	82.58(1)		5.8(4)mb	0.07(3)	7.16(6)	6.1(2)mb
^{89}Sr		50.52 d	0.42(4)	0.2		
^{90}Sr		29.1 y	9.7(7)mb	4. mb		
$_{39}$Y			1.25(5)	1.0(1)	7.75(2)	
^{89}Y	100.		(0.001+1.25)	(0.006+1.0)	7.75(2)	19.(1)mb
^{90}Y		2.67 d	<6.5			
^{91}Y		58.5 d	1.4(3)	0.6(1)		
$_{40}$Zr			0.19(1);σ_α<0.1mb	0.95(9)	7.16(3)	
^{90}Zr	51.45(40)		$\approx$0.014	0.2(1)	6.4(1)	21.(2)mb
^{91}Zr	11.22(5)		1.2(3)	5.(2)	8.8(1)	60.(8)mb
^{92}Zr	17.15(8)		0.2(1)	0.6(2)	7.5(2)	33.(4)mb
^{93}Zr		1.5x10^6 y	<4.	16.(5)		0.10(1)
^{94}Zr	17.38(28)		0.049(6)	0.25(3)	8.3(2)	26.(1)mb
^{96}Zr	2.80(9)		0.020(3)	5.0(5)	5.5(1)	10.0(3)mb
$_{41}$Nb			1.11(1);σ_α<0.1mb	8.5(6)	7.14(3)	
^{93}Nb	100.		1.1;σ_m=0.86	(6.3+2.2)	7.14(3)	266.(5)mb
^{94}Nb		2.4x10^4 y	σ_m+g=15.(1) σ_m=0.6(1)	126.(13)		
^{95}Nb		34.97 d	<7.	<200.		
$_{42}$Mo			2.5(1),σ_α<0.1mb	26.(5)	6.72(2)	
^{92}Mo	14.84(35)		0.06,σ_m=0.2μb	$\approx$0.8	6.93(8)	
^{94}Mo	9.25(12)		0.02	$\approx$0.8	6.82(7)	
^{95}Mo	15.92(13)		13.4(3),σ_α=30.(4)μb	109.(5)	6.93(6)	0.29(1)

Elem. or Isot.	Natural Abundance (%)	Half-life	Thermal Neut. Cross Section (barns)	Resonance Integral (barns)	Coh. Scat. Length (fm)	σ(30KEV) Maxw. Avg. (barns)
^{96}Mo	16.68(2)		0.5	17.(3)	6.22(6)	112.(5)mb
^{97}Mo	9.55(8)		2.5(2),σ_α=0.4(2)μb	14.(3)	7.26(8)	0.34(1)
^{98}Mo	24.13(31)		0.14(1)	7.2(7)	6.60(7)	99.(5)mb
^{100}Mo	9.63(23)		0.19(1)	3.6(3)	6.75(7)	
$_{43}$Tc						
^{98}Tc		4.2x10^6 y	σ_m = 0.9(2)			
^{99}Tc		2.13x10^5 y	23.(2)	4.0(4)x10^2	6.8(3)	0.78(4)
$_{44}$Ru			2.6 (1)	48.(5)	7.03(3)	
^{96}Ru	5.52(20)		0.23(4)	7.(2)		
^{98}Ru	1.86(9)		< 8.			
^{99}Ru	12.74(26)		4.(1)	195.(20)		
^{100}Ru	12.60(19)		5.8(6)	11.(2)		0.21(1)
^{101}Ru	17.05(7)		5.(1),σ_α<0.15μb	1.1(3)x10^2		1.00(4)
^{102}Ru	31.57(31)		1.2(1)	4.3(5)		0.19(1)
^{103}Ru		39.27 d	<20.	$\approx$30.		
^{104}Ru	18.66(44)		0.49(2)	6.(2)		0.16(1)
^{105}Ru		4.44 h	0.29(3)	0.13(1)		
^{106}Ru		1.020 y	0.15(4)	2.0(6)		
$_{45}$Rh			145.(2)	1.2(1)x10^3	5.88(4)	
^{103}Rh	100.		(11.+ 134.)	(0.08+1.1)x10^3	5.88(4)	0.81(1)
^{104m}Rh		4.36 m	800.(100)			
^{104}Rh		42.3 s	40.(30)			
^{105}Rh		35.4 h	1.1(3)x10^4	1.7(4)x10^4		
$_{46}$Pd			7.(1)	82.(8)	5.91(6)	
^{102}Pd	1.02(1)		3.2(10)	10.(2)		
^{104}Pd	11.14(11)			16.(2)		
^{105}Pd	22.33(13)		22.(2),σ_α=0.5(2)μb	60.(20)	5.5(3)	1.20(6)
^{106}Pd	27.33(6)		(0.013+0.28)	(0.2+5.5)	6.4(4)	0.25(3)
^{107}Pd		6.5x10^6 y	1.8(2)	108.(4)		1.34(6)
^{108}Pd	26.46(15)		(0.19+8.5)	(2.+240.)	4.1(3)	0.20(2)
^{110}Pd	11.72(16)		(0.033+0.7)	(0.7+8.)		0.15(2)
$_{47}$Ag			62.(1)	767.(60)	5.922(7)	
^{107}Ag	51.839(8)		(0.37+35.)	(1.2+105.)	7.56(1)	0.80(2)
^{109}Ag	48.161(8)		(4.2 + 87.)	(0.7+14.1)x10^2	4.17(1)	0.78(2)
^{110m}Ag		249.8 d	82.(11)	20.(4))		
^{111}Ag		7.47 d	3.(2)	105.(20)		
$_{48}$Cd			2.52(5)x10^3	73.(8)	4.87(5)	
^{106}Cd	1.25(6)		0.20(3)	4.(1)		0.56(6)
^{108}Cd	0.89(3)		1.	14.(3)	5.4(1)	0.41(7)
^{109}Cd		462.0 d	$\approx$180.,σ_α<0.05	6.7(12)x10^3		
^{110}Cd	12.49(18)		(0.06+11.)	(6.+34.)	5.9(1)	0.25(3)
^{111}Cd	12.80(12)		3.5(20)	51.(6)	6.5(1)	1.06(13)

Elem. or Isot.	Natural Abundance (%)	Half-life	Thermal Neut. Cross Section (barns)	Resonance Integral (barns)	Coh. Scat. Length (fm)	σ(30KEV) Maxw. Avg. (barns)
^{112}Cd	24.13(21)		(0.012+2.2)	15.	6.4(1)	0.23(3)
^{113}Cd	12.22(12)	9×10^{15} y	$2.06(4) \times 10^4, \sigma_\alpha < 1. \mu b$	390.(40)	- 8.0(2)	0.73(8)
^{114}Cd	28.73(42)		(0.04+0.29)	16.(7)	7.5(1)	0.16(3)
^{116}Cd	7.49(18)		(26.+52.)mb	1.2	6.3(1)	0.09(1)
$_{49}$In			197.(4)	$3.3(2) \times 10^3$	4.07(2)	
^{113}In	4.29(5)		(3.1+5.0+3.9)	(220.+90)	5.39(6)	0.79(7)
^{115}In	95.71(5)	4.4×10^{14} y	(88.+73.+44.)	$(1.5+1.2+0.7) \times 10^3$	4.01(2)	0.74(7)
$_{50}$Sn			0.61(3)	8.(2)	6.225(2)	
^{112}Sn	0.97(1)		(0.15+0.40)	(8.+19.)		0.20(5)
^{113}Sn		115.1 d	$\approx 9.$	210.(50)		
^{114}Sn	0.66(1)		≈ 0.12	5.(1)	6.2(3)	134.(3) mb
^{115}Sn	0.34(1)		$\sigma_\alpha = 0.06$ mb	29.(6)		0.34(1)
^{116}Sn	14.54(9)		(0.006+0.14)	(0.5+11.)	5.93(5)	91.(2)mb
^{117}Sn	7.68(7)		1.1(1)	16.(5)	6.48(5)	319.(7)mb
^{118}Sn	24.22(9)		$\sigma_m = 4.$ mb	4.7(5)	6.07(5)	62.(1)mb
^{119}Sn	8.58(4)		2.(1)	2.9(5)	6.12(5)	0.25(5)
^{120}Sn	32.59(9)		(0.001+0.13)	1.2(3)	6.49(5)	(0.5+36)mb
^{122}Sn	4.63(3)		(0.15+0.001)	0.81(4)	5.74(5)	
^{124}Sn	5.79(5)		(0.13+0.005)	$RI_m = 8.0(2)$	5.97(5)	
$_{51}$Sb			5.2(2)	169.(20)	5.57(3)	
^{121}Sb	57.21(5)		(0.4+5.6)	(13.+192.)	5.71(6)	0.53(2)
^{123}Sb	42.79(5)		(0.02+0.04+4.0)	(1.+119.)	5.38(7)	0.30(1)
^{124}Sb		60.20 d	17.(3)	$\approx 8.$		
$_{52}$Te			4.2(1)	47.(3)	5.80(3)	
^{120}Te	0.096(2)		(0.25+2.0)	$\approx 1.$	5.3(5)	
^{122}Te	2.603(4)		2.4(5)	80.(20)	3.8(2)	295.(3)mb
^{123}Te	0.908(2)	1.3×10^{13} y	370.(40), $\sigma_\alpha = 0.05$mb	$4.5(3) \times 10^3$	- 0.05	0.83(1)
^{124}Te	4.816(6)		(0.05+7.)	5.2(7)	8.0(1)	155.(2)mb
^{125}Te	7.139(6)		1.6(2)	21.(4)	5.02(8)	431.(4)mb
^{126}Te	18.952(11)		(0.12+0.8)	8.2(6)	5.56(7)	81.(1)mb
^{128}Te	31.687(11)		(0.016+0.20)	(0.08+1.5)	5.89(7)	
^{130}Te	33.799(10)	2.5×10^{21} y	(0.03+0.20)	(0.08+0.34)	6.02(7)	
$_{53}$I			6.15(10)	148.(7)	5.28(2)	
^{125}I		59.4 d	900.(100)	$1.4(2) \times 10^4$		
^{127}I	100.		6.15(10)	148.(7)	5.28(2)	0.64(3)
^{128}I		25.00 m	22.(4)	$\approx 10.$		
^{129}I		1.7×10^7 y	(20.7+10.3),	36.(4)		0.44(2)
^{130}I		12.36 h	18.(3)	$\approx 8.$		
^{131}I		8.040 d	≈ 0.7	8.(4)		
$_{54}$Xe			25.(1)	263.(50)	4.92(3)	
^{124}Xe	0.09(1)		(28.+137.)	$(0.6+3.0) \times 10^3$		
^{125}Xe		17.1 h	$\sigma_\alpha \leq < 0.03$			

Elem. or Isot.	Natural Abundance (%)	Half-life	Thermal Neut. Cross Section (barns)	Resonance Integral (barns)	Coh. Scat. Length (fm)	σ(30KEV) Maxw. Avg. (barns)
^{126}Xe	0.09(1)		(0.45+3.)	(8.+52.)		
^{127}Xe		36.4 d	$\sigma_\alpha \leq 0.01$			
^{128}Xe	1.92(3)		$\sigma_m = 0.48$	$RI_m = 38.(10)$		0.35(11)
^{129}Xe	26.44(24)		22.(5)	250.(50)		0.76(7)
^{130}Xe	4.08(2)		$\sigma_m = 0.45$	$RI_m = 16.(4)$		0.18(5)
^{131}Xe	21.18(3)		90.(10)	$9.(1) \times 10^2$		0.45(8)
^{132}Xe	26.89(6)		(0.05+0.4)	(0.9+3.7)		65.(5)mb
^{133}Xe		5.243 d	190.(90)			
^{134}Xe	10.44(10)		(0.003 + 0.26)	0.40(4)		20.(2)mb
^{135}Xe		9.10 h	$2.65(11) \times 10^6$	$7.6(5) \times 10^3$		
^{136}Xe	8.87(16)		0.26(2)	0.7(2)		0.9(1)mb
$_{55}$Cs			30.4(8)	422.(50)	5.42(2)	
^{132}Cs		6.48 d	$\sigma_\alpha < 0.15$			
^{133}Cs	100.		(2.7+27.7)	(32.+390.)	5.42(2)	0.51(2)
^{134}Cs		2.065 y	140.(10)	54.(9)		
^{135}Cs		2.3×10^6 y	8.9(5)	90.(20)		
^{137}Cs		30.2 y	0.25(1)	0.36(7)		
$_{56}$Ba			1.3(2)	10.(2)	5.07(3)	
^{130}Ba	0.106(1)		(1.+8.)	(25.+200.)	- 3.6(6)	
^{132}Ba	0.101(1)		(0.84+9.7)	(4.7+24.)	7.8(3)	
^{133}Ba		10.53 y	4.(1)	85.(30)		
^{134}Ba	2.417(18)		(0.1+1.3)	(5.6+18.)	5.7(1)	0.18(6)
^{135}Ba	6.592(12)		(0.014+5.8)	(0.47+131.)	4.7(1)	0.30(6)
^{136}Ba	7.854(24)		(0.010+0.44)	(0.1+1.5)	4.91(8)	62.(2)mb
^{137}Ba	11.232(24)		5.(1)	4.(1)	6.8(1)	0.05(1)
^{138}Ba	71.698(42)		0.41(2)	0.4(1)	4.84(8)	4.1(2)mb
^{139}Ba		1.396 h	5.(1)	2.2(5)		
^{140}Ba		12.75 d	1.6(3)	14.(1)		
$_{57}$La			9.2(2)	12.(1)	8.24(4)	
^{138}La	0.090(1)	1.06×10^{11} y	57.(6)	$4.1(9) \times 10^2$		
^{139}La	99.910(1)		9.2(2)	12.(1)	8.24(4)	38.(3)mb
^{140}La		1.678 d	2.7(3)	69.(4)		
$_{58}$Ce			0.64(4)	0.71(6)	4.84(2)	
^{136}Ce	0.185(2)		(1.0+6.5)	58.(12)	5.80(9)	(28+272)mb
^{138}Ce	0.251(2)		(0.018+1.)	(1.5+5.2)	6.70(9)	179.(5)mb
^{140}Ce	88.450(18)		0.58(4)	0.50(5)	4.84(9)	11.0(4)mb
^{141}Ce		32.50 d	29.(3)	13.(2)		
^{142}Ce	11.114(17)		0.97(3)	1.3(3)	4.75(9)	28.(1)mb
^{143}Ce		1.38 d	6.1(7)	2.7(3)		
^{144}Ce		284.6 d	1.0(1)	2.6(3)		
$_{59}$Pr			11.5(4)	14.(3)	4.58(5)	
^{141}Pr	100.		(4.+7.5)	14.(3)	4.58(5)	0.12(2)
^{142}Pr		19.12 h	20.(3)	9.(1)		
^{143}Pr		13.57 d	90.(10)	190.(25)		

Elem. or Isot.	Natural Abundance (%)	Half-life	Thermal Neut. Cross Section (barns)	Resonance Integral (barns)	Coh. Scat. Length (fm)	σ(30KEV) Maxw. Avg. (barns)
$_{60}$Nd			51.(2)	49.(5)	7.69(5)	
^{142}Nd	27.2(5)		19.(1)	34.(11)	7.7(3)	46.(4)mb
^{143}Nd	12.2(2)		330.(10),σ_α=17.mb	128.(30)		0.24(1)
^{144}Nd	23.8(3)	2.1×10^{15} y	3.6(3)	3.9(5)	2.8(3)	0.11(1)
^{145}Nd	8.30(1)		47.(6),σ_α=12.μb	260.(40)		0.49(10)
^{146}Nd	17.2(3)		1.5(2)	3.0(4)	8.7(2)	0.16(4)
^{147}Nd		10.98 d	440.(150)	200.		
^{148}Nd	5.7(1)		2.4(1)	13.(2)	5.7(3)	0.19(4)
^{150}Nd	5.6(2)		1.0(1)	14.(2)	5.3(2)	
$_{61}$Pm						
^{146}Pm		5.53 y	8.4(1.7)x10^3			
^{147}Pm		2.6234 y	(84.+96.)	(1000.+1280.)	12.6(4)	
^{148m}Pm		41.3 d	10600.(800)			
^{148}Pm		5.37 d	$\approx 10^3$	2.6(2.4)x10^3		
^{149}Pm		2.212 d	1400.(200)			
^{151}Pm		1.183 d	$\approx$ 150.			
$_{62}$Sm			5.6(1)x10^3	1.4(2)x10^3		
^{144}Sm	3.08(7)		1.6(1)	2.4(3)		
^{145}Sm		340. d	280.(20)	600.(90)		
^{147}Sm	14.99(17)	1.06×10^{11} y	56.(4),σ_α=0.6mb	710.(50)	14.(3)	0.97(1)
^{148}Sm	11.24(10)	7×10^{15} y	2.4(6)	27.(14)		241.(2)mb
^{149}Sm	13.82(7)	10^{16} y	40100.(600),σ_α=31.mb	3100.(500)		1.82(2)
^{150}Sm	7.38(1)		102.(5)	290.(30)	14.(3)	422.(4)mb
^{151}Sm		90. y	15200.(300)	3520.(60)		
^{152}Sm	26.74(14)		206.(15)	3000.(300)	- 5.0(6)	473.(4)mb
^{153}Sm		1.929 d	420.(180)			
^{154}Sm	22.75(27)		7.5(3)	32.(6)	9.(1)	
$_{63}$Eu			4570.(100)	3.8(5)x10^3	5.3(3)	
^{151}Eu	47.81(6)		(4.+3150.+6000.) σ_α=8.7(3)μb	(2000.+4000.)		4.4(2)
^{152m1}Eu		9.30 h	68000.(15000)	< 100000.		
^{152}Eu		13.5 y	11000.(2000)	1600.(200)		
^{153}Eu	52.19(6)		300.(20),σ_α<1.μb	1800.(400)	8.2(1)	3.2(3)
^{154}Eu		8.59 y	1500.(300)	1600.(200)		4.4(7)
^{155}Eu		4.76 y	3900.(200)	16000.(2000)		
$_{64}$Gd			48.8(6)x10^3	400.(10)	9.5(2)	
^{152}Gd	0.20(1)		700.(200),σ_α<7.mb	700.(200)		1.05(7)
^{153}Gd		241.6 d	20000.(10000),σ_α=0.03			
^{154}Gd	2.18(3)		(0.06+60.)	230.(50)		0.88(3)
^{155}Gd	14.80(12)		61.(1)x10^3,σ_α=.08mb	1540.(100)		2.72(9)
^{156}Gd	20.47(9)		$\approx$2.0	104.(15)	6.3(4)	0.64(6)
^{157}Gd	15.65(2)		2.54(3)x10^5,σ_α<0.05	800.(100)		1.36(4)
^{158}Gd	24.84(7)		2.3(3)	73.(7)	9.(2)	0.22(2)
^{160}Gd	21.86(19)		1.5(7)	6.(1)	9.15(5)	
^{161}Gd		3.66 m	2.0(6)x10^4			

Elem. or Isot.	Natural Abundance (%)	Half-life	Thermal Neut. Cross Section (barns)	Resonance Integral (barns)	Coh. Scat. Length (fm)	σ(30KEV) Maxw. Avg. (barns)
$_{65}$Tb			23.2(5)	420.(50)	7.38(3)	
^{159}Tb	100.		23.2(5)	420.(50)	7.38(3)	1.75(10)
^{160}Tb		72.3 d	570.(110)			
$_{66}$Dy			9.5(2)x10^2	1.5(2)x10^3	16.9(3)	
^{156}Dy	0.06(1)		33.(3),σ_α<9mb	1000.(100)		
^{158}Dy	0.10(1)		43.(6),σ_α<6mb	120.(10)	6.1(5)	
^{159}Dy		144. d	8.(2)x10^3			
^{160}Dy	2.34(8)		60.(10),σ_α<0.3mb	1100.(200)	6.7(4)	0.77(4)
^{161}Dy	18.91(24)		600.(50),σ_α<1μb	1100.(100)	10.3(4)	2.01(6)
^{162}Dy	25.51(26)		170.(20)	2755.(300)	- 1.4(5)	0.47(5)
^{163}Dy	24.90(16)		120.(10),σ_α<20μb	1600.(400)	5.0(4)	1.14(4)
^{164}Dy	28.18(36)		(1.7+1.0)x10^3	(420.+100.)	49.4(2)	0.27(3)
^{165m}Dy		1.26 m	2.0(6)x10^3			
^{165}Dy		2.33 h	3.5(3)x10^3	2.2(3)x10^4		
$_{67}$Ho			61.(2)	670.(40)	8.01(8)	
^{165}Ho	100.		(3.1+58.),σ_α<20μb	(?+670.)	8.01(8)	1.32(7)
^{166}Ho		1.2x10^3 y	9140.(650)	1140.(90)		
$_{68}$Er			169.(20)	730.(10)	7.79(2)	
^{162}Er	0.14(1)		19.(3),σ_α<11mb	480.(50)	8.8(2)	
^{164}Er	1.61(3)		13.(3),σ_α<1.2mb	105.(10)	8.2(2)	0.71(6)
^{166}Er	33.61(35)		(15.+5.),σ_α<70μb	96.(12)	10.6(2)	0.61(6)
^{167}Er	22.93(17)		700.(200),σ_α=3μb	2970.(70)	3.0(3)	1.5(2)
^{168}Er	26.78(26)		2.0(6),σ_α=0.09mb	37.(5)	7.4(4)	0.37(4)
^{170}Er	14.93(27)		6.(1)	26.(4)	9.6(5)	0.22(3)
^{171}Er		7.52 h	370.(40)	170.(20)		
$_{69}$Tm			106.(5)	1.5(2)x10^3	7.07(3)	
^{169}Tm	100		(8.+98.)	1.5(2)x10^3	7.07(3)	1.13(6)
^{170}Tm		128.6 d	100.(20)	460.(50)		
^{171}Tm		1.92 y	$\approx$ 160.	118.(6)		
$_{70}$Yb				52.(10)	1.7(2)x10^2	12.43(3)
^{168}Yb	0.13(1)		2.4(2)x103,σ_α<0.1mb	2.0(5)x10^4	-4.07(2)	
^{169}Yb		32.03 d	3.6(3)x10^3	5200.(500)		
^{170}Yb	3.04(15)		12.(2),σ_α<10μb	320.(30)	6.8(1)	0.74(3)
^{171}Yb	14.3(6)		53.(5),σ_α<1.5μb	315.(30)	9.7(1)	1.41(5)
^{172}Yb	21.8(7)		$\approx$ 1.3,σ_α<1μb	25.(3)	9.4(1)	0.41(3)
^{173}Yb	16.13(3)		16.(2),σ_α<1μb	380.(30)	9.56(7)	0.87(7)
^{174}Yb	31.8(9)		(46.+74.),σ_α<0.02mb	(13.+47.)	19.3(1)	0.18(2)
^{176}Yb	12.8(4)		3.1(2),σ_α<1μb	8.(2)	8.7(1)	
$_{71}$Lu			78.(7)	8.3(7)x10^2	7.21(3)	
^{175}Lu	97.41(2)		(16.+8.)	(550.+270.)	7.24(3)	1.18(4)
^{176}Lu	2.59(2)	3.8x10^{10} y	(2.+2100.)	(3.+1100.)	6.1(2)	1.54(6)

Elem. or Isot.	Natural Abundance (%)	Half-life	Thermal Neut. Cross Section (barns)	Resonance Integral (barns)	Coh. Scat. Length (fm)	$\sigma(30KEV)$ Maxw. Avg. (barns)
^{177m}Lu		160.7 d	3.2(3)	1.4(2)		
^{177}Lu		6.75 d	1000.(300)			
$_{72}$Hf			106.(3)	$19.7(5)\times10^2$	7.8(1)	
^{174}Hf	0.162(3)	2.0×10^{15} y	600.(50)	400.(50)	11.(1)	
^{176}Hf	5.206(5)		23.(4)	700.(100)	6.6(2)	0.46(2)
^{177}Hf	18.606(13)		$(1.+375.),\sigma_\alpha<20\mu b$	7170.(200)		1.37(6)
^{178}Hf	27.297(4)		(53.+32.)	(340.+1500.)	5.9(2)	0.31(1)
^{179}Hf	13.629(6)		(0.43+46.)	(6.8+620.)	7.5(2)	0.99(3)
^{180}Hf	35.100(7)		$13.(1),\sigma_\alpha<13\mu b$	32.(2)	13.2(3)	175.(5)mb
^{181}Hf		42.4 d	30.(25)			
$_{73}$Ta			20.(1)	650(20.)	6.91(7)	
^{180m}Ta	0.012(2)	$> 1.2\times10^{15}$ y	$\approx$ 560.	1350.(100)		
^{181}Ta	99.998(2)		$(0.012 + 20),\sigma_\alpha<1\mu b$	(0.4+650.)	6.91(7)	0.77(2)
^{182}Ta		114.43 d	8200.(600)	900.(90)		
$_{74}$W			18.(1)	$3.6(3)\times10^2$	4.86(2)	
^{180}W	0.12(1)		$\approx$ 4.	210.(30)		
^{182}W	26.50(16)		20.(1)	600.(90)	6.97(4)	274.(8)mb
^{183}W	14.31(4)		10.5(3)	340.(50)	6.53(4)	0.52(2)
^{184}W	30.64(2)		(0.002 + 2.0)	15.(2)	7.48(6)	0.23(1)
^{185}W		74.8 d	$\approx$ 3.3	300.(50)		
^{186}W	28.43(19)		37.(2)	510.(50)	- 0.72(4)	176.(5)mb
^{187}W		23.9 h	70.(10)	2760.(550)		
$_{75}$Re			90.(4)	$8.4(2)\times10^2$	9.2(3)	
^{185}Re	37.40(2)		(0.33+110.)	1700.(50)	9.0(3)	1.54(6)
^{187}Re	62.60(2)	4.4×10^{10} y	(2.+72.)	(9.+310.)	9.3(3)	1.16(6)
$_{76}$Os			17.(1)	$1.5(1)\times10^2$	10.7(2)	
^{184}Os	0.020(3)		$3.3(3)\times10^3,\sigma_\alpha<10mb$	$1.4(1)\times10^3$		
^{186}Os	1.58(2)	$2.\times10^{15}$ y	$\approx$ $80.,\sigma_\alpha<0.1mb$	280.(40)	12(2)	0.42(2)
^{187}Os	1.6(4)		$2.(1)\times10^2,\sigma_\alpha<0.1mb$	500.(70)		0.87(3)
^{188}Os	13.3(1)		$\approx$ $5.,\sigma_\alpha<30\mu b$	150.(20)	7.6(3)	0.40(2)
^{189}Os	16.1(1)		$(0.00026+40),\sigma_\alpha<10\mu b$	(0.013+670.)	10.7(3)	1.17(5)
^{190}Os	26.4(2)		$(9.+4.),\sigma_\alpha<20\mu b$	(22.+8.)	11.0(3)	0.30(5)
^{191}Os		15.4 d	$3.8(6)\times10^2$			
^{192}Os	41.0(3)		$3.(1),\sigma_\alpha<10\mu b$	6.(1)	11.5(4)	0.31(5)
^{193}Os		30.5 h	40.(10)			
$_{77}$Ir			$4.2(1)\times10^2$	$2.8(4)\times10^3$	10.6(3)	
^{191}Ir	37.3(2)		(660.+260.)	(1000.+4200.)		1.35(7)
^{192}Ir		73.83 d	1450.(250)	4800.(700)		
^{193}Ir	62.7(2)		(6.+110.)	1400.(200)		0.86(4)
^{194}Ir		19.3 h	$1.5(3)\times10^3$			
$_{78}$Pt			10.(1)	$1.3(1)\times10^2$	9.60(1)	

Elem. or Isot.	Natural Abundance (%)	Half-life	Thermal Neut. Cross Section (barns)	Resonance Integral (barns)	Coh. Scat. Length (fm)	σ(30KEV) Maxw. Avg. (barns)
^{190}Pt	0.014(1)	6.5x10^{11} y	1.5(1)x10^2,σ_α<8mb	70.(10)	9(1)	
^{192}Pt	0.782(7)		(2.0+6.),σ_α<0.2mb	115.(20)	9.9(5)	0.20(6)
^{194}Pt	32.97(10)		(0.1+1.1),σ_α< 5μb	(4.+?)	10.55(8)	0.37(8)
^{195}Pt	33.832(10)		28.(1),σ_α< 5μb	365.(50)	8.8(1)	0.9(2)
^{196}Pt	25.242(41)		(0.045+0.55)	7.(2)	9.89(8)	0.20(2)
^{198}Pt	7.16(6)		(0.027+3.6)	(0.5+56.)	7.8(1)	
^{199}Pt		30.8 m	$\approx$ 16.			
$_{79}$Au			98.7(1)	1.55(3)x10^3	7.63(6)	
^{197}Au	100.		98.7(1)	1.55(3)x10^3	7.63(6)	582.(9)mb
^{198}Au		2.694 d	26.5(15)x10^3	$\approx$ 4.x10^4		
^{199}Au		3.14 d	$\approx$ 30.			
$_{80}$Hg			3.7(1)x10^2	87.(5)	12.69(2)	
^{196}Hg	0.15(1)		(105.+3000.)	(53.+410.)	30.(1)	
^{198}Hg	9.97(20)		(0.017+2.)	(1.7+70.)		0.17(2)
^{199}Hg	16.87(22)		2.1(2)x10^3	435(20)	16.9(4)	0.37(2)
^{200}Hg	23.10(19)		< 60.	2.1(5)		0.12(1)
^{201}Hg	13.18(9)		< 60.	30.(3)		0.26(1)
^{202}Hg	29.86(26)		4.9(5)	4.5(2)		74.(6)mb
^{204}Hg	6.87(15)		0.4(1)	0.8(2)		42.(4)mb
$_{81}$Tl			3.3(1)	12.5(8)	8.776(5)	
^{203}Tl	29.52(1)		11.(1),σ_α<0.3mb	41.(2)	7.0(2)	124.(8)mb
^{204}Tl		3.78 y	22.(2)	90.(20)		
^{205}Tl	70.48(1)		0.11(2)	0.6(2)	9.52(7)	54.(4)mb
$_{82}$Pb			0.172(2)	0.14(4)	9.402(2)	
^{204}Pb	1.4(1)		0.68(7)	2.0(2)	10.6(20)	90.(6)mb
^{205}Pb		1.51x10^7 y	$\approx$ 5.			
^{206}Pb	24.1(1)		0.030(1)	0.10(1)	9.23(5)	14.(1)mb
^{207}Pb	22.1(1)		0.70(1)	0.38(1)	9.28(4)	8.4(4)mb
^{208}Pb	52.4(1)		0.49(3)mb,σ_α<8μb	2.0(2) mb	9.50(6)	0.36(2)mb
^{210}Pb		22.6 y	< 0.5			
$_{83}$Bi			0.034(1)	0.19(2)	8.532(2)	
^{209}Bi	100.		(11+23)mb,σ_α<0.3μb	0.19(2)	8.532(2)	3.1(3)mb
^{210m}Bi		3.0x10^6 y	54(4)mb	0.20(3)		
$_{84}$Po						
^{210}Po		138.38 d	σ_m<0.5mb,σ_α< 2.mb σ_g<30.mb,σ_f< 0.1			
$_{85}$At						
$_{86}$Rn						
^{220}Rn		55.6 s	<0.2			

Elem. or Isot.	Natural Abundance (%)	Half-life	Thermal Neut. Cross Section (barns)	Resonance Integral (barns)	Coh. Scat. Length (fm)	σ(30KEV) Maxw. Avg. (barns)
^{222}Rn		3.8235 d	0.74(5)			
$_{88}$Ra						
^{223}Ra		11.435d	$1.3(2) \times 10^2, \sigma_f < 0.7$			
^{224}Ra		3.66 d	12.0(5)			
^{226}Ra		1599. y	$\approx 13., \sigma_f < 7.\mu b$	280.(50)	10.(1)	
^{228}Ra		5.76 y	$36.(5), \sigma_f < 2.$			
$_{89}$Ac						
^{227}Ac		21.77 y	$8.8(7) \times 10^2, \sigma_f < .35$ mb	$1.5(4) \times 10^3$		
$_{90}$Th			7.4	85.(3)	10.31(3)	
^{227}Th		18.72 d	$\sigma_f = 2.0(2) \times 10^2$			
^{228}Th		1.913 y	$1.2(2) \times 10^2, \sigma_f < 0.3$	1014.(400)		
^{229}Th		7.9×10^3 y	$\approx 60.$ $\sigma_f = 30(3)$	$1.0(2) \times 10^3$ $RI_f = 466.(75)$		
^{230}Th		7.54×10^4 y	23.4(5) $\sigma_f < 0.5$mb	$1.0(1) \times 10^3$		
^{232}Th	100.	1.4×10^{10}y	7.37(4) $\sigma_f = 3.(1) \mu b$ $\sigma_\alpha < 1. \mu b$	85.(3)	10.31(3)	
^{233}Th		22.3 m	$1.5(1) \times 10^3$ $\sigma_f = 15.(2)$	$4.(1) \times 10^2$		
^{234}Th		24.10 d	1.8(5) $\sigma_f < 0.01$			
$_{91}$Pa						
^{230}Pa		17.4 d	$1.5(3) \times 10^3$			
^{231}Pa		3.25×10^4 y	$2.0(1) \times 10^2$ $\sigma_f = 20.(1)$ mb	750.(80) $RI_f = 0.05(1)$	9.1(3)	
^{232}Pa		1.31 d	$4.6(10) \times 10^2$ $\sigma_f = 7.(1) \times 10^2$	300.(70)		
^{233}Pa		27.0 d	39.(2) $\sigma_m = 20.(4)$ $\sigma_g = 19.(3)$ $\sigma_f < 0.1$	$(460. + 440)$		
$_{92}$U			$3.4(3); \sigma_f = 4.2(1)$	$280.(20), RI_f = 2.0$	8.417(5)	
^{230}U		20.8 d	$\sigma_f \approx 25.$			
^{231}U		4.2 d	$\sigma_f \approx 250.$			
^{232}U		68.9 y	73.(2) $\sigma_f = 74.(8)$	280.(15) $RI_f = 350.(30)$		
^{233}U		1.59×10^5 y	47.(2) $\sigma_f = 5.3(1) \times 10^2$ $\sigma_\alpha < 0.2$ mb	137.(6) $RI_f = 760.(17)$	10.1(2)	
^{234}U	0.0055(5)	2.45×10^5 y	96.(2) $\sigma_f < 5.$ mb	660.(70) $RI_f = 6.5$	12.(4)	

Elem. or Isot.	Natural Abundance (%)	Half-life	Thermal Neut. Cross Section (barns)	Resonance Integral (barns)	Coh. Scat. Length (fm)	σ(30KEV) Maxw. Avg. (barns)
^{235}U	0.720(1)	7.04×10^8 y	95.(5) $\sigma_f=586.(2)$ $\sigma_\alpha< 0.1$ mb	144.(6) $RI_f=275(5)$	10.47(4)	
^{236}U		2.34×10^7 y	5.1(3) $\sigma_f=0.04$	360.(15) $RI_f=4.38(50)$		
^{237}U		6.75 d	$\approx 10^2$ $\sigma_f< 0.35$	1200.(200)		
^{238}U	99.2745(15)	4.46×10^9 y	2.7(1) $\sigma_f\approx 3.\mu b$ $\sigma_\alpha=1.4(5)$ μb	277.(3) 1.54(15) mb	8.402(5)	
^{239}U		23.5 m	22.(2) $\sigma_f=15.(3)$			
$_{93}$Np						
^{234}Np		4.4 d	$\sigma_f=9.(3)\times 10^2$			
^{235}Np		1.085 y	$1.6(1)\times 10^2$			
^{236m}Np		22.5 h	$\sigma_f=2.7(2)\times 10^3$	700.(400)		
^{236}Np		1.55×10^5 y	$\sigma_f=2.6(5)\times 10^3$	1030.(100)		
^{237}Np		2.14×10^6 y	160.(10) $\sigma_f=20.(1)$mb	$8.(2)\times 10^2$ $RI_f=0.32$	10.6(1)	
^{238}Np		2.117 d	$\sigma_f=2.1(1)\times 10^3$	$9.(1)\times 10^2$		
^{239}Np		2.355 d	(32.+19.) $\sigma_f<1.$			
$_{94}$Pu						
^{236}Pu		2.87 y	$\sigma_f=1.6(3)\times 10^2$	1000.(60)		
^{237}Pu		45.7 d	$\sigma_f=2.3(3)\times 10^3$			
^{238}Pu		87.74 y	$5.1(2)\times 10^2$ $\sigma_f=17.(1)$	$1.6(2)\times 10^2$ $RI_f=33.(5)$	14.1(5)	
^{239}Pu		2.411×10^4y	$2.7(1)\times 10^2$ $\sigma_f=752.(3)$ $\sigma_\alpha\leq 0.3$ mb	$2.0(2)\times 10^2$ $3.0(1)\times 10^2$	7.7(1)	
^{240}Pu		6537. y	$2.9(1)\times 10^2$ $\sigma_f\approx 44.$mb	$8.4(3)\times 10^3$ $RI_f=2.4$	3.5(1)	
^{241}Pu		14.4 y	$3.7(1)\times 10^2, \sigma_\alpha<0.2$mb $\sigma_f=1.01(1)\times 10^3$	$1.6(1)\times 10^2$ $5.7(4)\times 10^2$		
^{242}Pu		3.76×10^5 y	19.(1) $\sigma_f<0.2$	$1.1(1)\times 10^3$ $RI_f=0.23$	8.1(1)	
^{243}Pu		4.956 h	<100. $\sigma_f=2.0(2)\times 10^2$			
^{244}Pu		8.2×10^7 y	1.7(1)	41.(3)		
^{245}Pu		10.5 h	$1.5(3)\times 10^2$	220(40)		
$_{95}$Am						
^{241}Am		432.2 y	$(0.5+5.7)\times 10^2$ $\sigma_f=3.1(2)$	$(2.0+12.3)\times 10^2$ 14.(1)		
^{242m}Am		141. y	$1.7(4)\times 10^3$ $\sigma_f=7.0(3)\times 10^3$	$\approx 200.$ $RI_f=1.8(1)\times 10^3$		
^{242}Am		16.02 h	$\sigma_f=2.1(2)\times 10^3$	$RI_f= < 300.$		
^{243}Am		7.37×10^3y	(75.+5.) $\sigma_f=74.(4)$ mb	$(17.1+1.0)\times 10^2$ $RI_f=0.056$	8.3(2)	

Elem. or Isot.	Natural Abundance (%)	Half-life	Thermal Neut. Cross Section (barns)	Resonance Integral (barns)	Coh. Scat. Length (fm)	σ(30KEV) Maxw. Avg. (barns)
^{244m}Am		≈ 26. m	σ_f=1.6(3)x10^3			
^{244}Am		10.1 h	σ_f=2.2(3)x10^3			
$_{96}$Cm						
^{242}Cm		162.8 d	≈ 20. σ_f≈ 5.	120.(50)		
^{243}Cm		28.5 y	1.3(1)x10^2 σ_f=6.2(2)x10^2	214.(20) RI$_f$=1.6(1)x10^3		
^{244}Cm		18.11 y	15.(1) σ_f=1.1(2)	640.(50) RI$_f$=10.8(8)	9.5(3)	
^{245}Cm		8.5x10^3 y	3.5(2)x10^2 σ_f=2.1(1)x10^3	110(10) RI$_f$=8.(1)x10^2		
^{246}Cm		4.78x10^{3}y	1.2(2) σ_f=0.16(7)	120.(10) 13.(2)	9.3(2)	
^{247}Cm		1.56x10^{7}y	60.(30) σ_f=82.(5)	5.(1)x10^2 7.3(7)x10^2		
^{248}Cm		3.4x10^5 y	2.6(3) σ_f=0.36(7)	270.(30) 13.(2)	7.7(2)	
^{249}Cm		64.15 m	≈1.6			
^{250}Cm		≈ 9.7x10^3 y	≈80.			
$_{97}$Bk						
^{249}Bk		320. d	7.(1)x10^2 σ_f≈0.1	9.(1)x10^2		
^{250}Bk		3.217 h	σ_f=1.0(2)x10^3			
$_{98}$Cf						
^{249}Cf		351. y	5.0(3)x10^2 σ_f=1.7(1)x10^3	7.7(4)x10^2 RI$_f$=2.1(3)x10^3		
^{250}Cf		13.1 y	2.0(2)x10^3 σ_f=110.(90)	12.(2)x10^3 RI$_f$=160.(40)		
^{251}Cf		9.0x10^2 y	2.9(2)x10^3 σ_f=4.5(5)x10^3	1.6(1)x10^3 RI$_f$=5.5(3)x10^3		
^{252}Cf		2.65 y	20.(2) σ_f=32.(4)	43.(3) RI$_f$=1.1(3)x10^2		
^{253}Cf		17.8 d	18.(2) σ_f=1.3(2)x10^3	8.(1)		
^{254}Cf		60.5 d	4.5(10)	2.		
$_{99}$Es						
^{253}Es		20.47 d	(180.+5.8)	(37.5+1.1)x10^2		
^{254m}Es		1.64 d	σ_f=1.8(1)x10^3			
^{254}Es		276. d	28.(3)	18.(2)		
^{255}Es		40. d	σ_f=1.8(2)x10^3 ≈ 55.	RI$_f$=1.2(3)x10^3		
$_{100}$Fm						
^{255}Fm		20.1 h	26.(3) σ_f=3.3(2)x10^3	14.(2)		
^{257}Fm		100.5 d	σ_f=3.0(2)x10^3			

COSMIC RADIATION

A.G. Gregory and R.W. Clay

The Nature of Cosmic Rays

Primary cosmic radiation, in the form of high energy nuclear particles, electrons and photons from outside the solar system and from the Sun, continually bombards our atmosphere. Secondary radiation, resulting from the interaction of the primary cosmic rays with atmospheric gas, is present at sea-level and throughout the atmosphere.

The secondary radiation is collimated by absorption and scattering in the atmosphere and consists of a number of components associated with different particle species. High energy primary particles can produce large numbers of secondary particles forming an extensive air shower. Thus, a number of particles may then be detected simultaneously at sea-level.

Primary particle energies accessible in the vicinity of the earth range from $\sim 10^8$ eV to $\sim 10^{20}$ eV. At the lower energies, the limit is determined by the inability of charged particles to traverse the heliosphere to us through the outward-moving solar wind. The upper energy limit is set by the practicality of building detectors to record particles with the extremely low fluxes found at those energies (J.G. Wilson, 1976; O.C. Allkofer, 1975a).

Primary Cosmic Rays

Primary Particle Energy Spectrum

Figure 1 shows the spectrum of primary particle energies. This includes all particle species. In differential form it is roughly a power law of intensity versus energy with an index of ~ -3. There appears to be a knee (a steepening) at a little above 10^{15} eV and an ankle (a flattening) above $\sim 10^{18}$ eV. Figure 2 emphasizes the features in the spectrum at the highest energies through multiplying the flux with a strongly rising power law of energy. This figure should be used with caution as errors for the two axes are not now independent.

Data on the high energy cosmic ray spectrum are uncertain largely because of limited event statistics due to the very low flux which might best be measured in particles per square kilometer per century. The highest energy event recorded to 1995 had an energy of 3×10^{20} eV (D.J. Bird et al., 1993).

It is expected that the highest energy cosmic rays will interact with the 2.7 K cosmic microwave background through photoproduction or photodisintegration. These interactions will appreciably reduce the observed flux of cosmic rays with energies above 5×10^{19} eV if they travel further than ~ 150 million light years. This process is known as the Greisen-Zatsepin-Kuz'min (GZK) cut off (P. Sokolsky, 1989).

At energies below $\sim 10^{13}$ eV, solar system magnetic fields and plasma can modulate the primary component and Figure 3 shows the extent of this modulation between solar maximum and minimum (E. Juliusson, 1975; J. Linsley, 1981).

Primary Particle Energy Density

If the above spectrum is corrected for solar effects, the energy density above a particle energy of 10^9 eV outside the solar system is found to be $\sim 5 \times 10^5$ eV m^{-3}. As the threshold energy is increased, the energy density decreases rapidly, being 2×10^4 eV m^{-3} above 10^{12} eV and 10^2 eV m^{-3} above 10^{15} eV. The energy density at lower energies outside the heliosphere is unknown but may be substantially greater if the particle rest mass energy is included together with the kinetic energy (A. W. Wolfendale, 1979).

Primary Particle Isotropy

This is measured as an anisotropy $(I_{max} - I_{min})/(I_{max} + I_{min}) \times 100\%$, where I, the intensity (m^{-2}s^{-1}sr^{-1}), is usually measured with an angular resolution of a few degrees.

The measured anisotropy is small and energy dependent. It is roughly constant in amplitude at between 0.05 and 0.1% (with a phase of 0 to 6 hours in right ascension) for energies between 10^{11} eV and 10^{14} eV and appears to increase at higher energies roughly as $0.4 \times (\text{Energy(eV)}/10^{16})^{0.5}\%$ up to $\sim 10^{18}$ eV. The latter rise may well be an artifact of the progressively more limited statistics as the flux drops rapidly with energy. It appears possible that a real anisotropy has been observed at the highest energies (above a few times 10^{19} eV) with a directional preference for the supergalactic plane (this plane reflects the directions of galaxies within about 100 million light years) (A.W. Wolfendale, 1979; R.W. Clay, 1987; T. Stanev et al., 1995).

Primary Particle Composition

The composition of low energy cosmic rays is close to universal nuclear abundances except where propagation effects are present. For example, Li, Be, and B which are spallation products, are over-abundant by about six orders of magnitude.

Composition at 10^{11} eV per nucleus

Charge	1	2	(3–5)	(6–8)	(10–14)	(16–24)	(26–28)	≥30
% Composition	50	25	1	12	7	4	4	0.1
(10% uncertainty)								

Measurements at higher energies indicate that there is an increase in the relative abundances of nuclei with charge greater than 6 at energies above 50 TeV/nucleus (K. Asakimori et al., 1993) (1 TeV = 10^{12} eV).

Cosmic ray composition at low energies is often quoted at a fixed energy per nucleon. When presented in this way, protons constitute roughly 90% of the flux, helium nuclei about 10% and the remainder sum to a total of about 1%.

Certain radioactive isotopic ratios show lifetime effects. The ratio of Be^{10}/B^9 abundances is used to measure an "age" of cosmic rays since Be^{10} is unstable with a half life of about 1.6×10^6 years. A ratio of 0.6 is expected in the absence of Be^{10} decay and a ratio of about 0.2 is found experimentally (E. Juliusson, 1975; P. Meyer, 1981).

FIGURE 1. The energy spectrum of cosmic ray particles. This spectrum is of a differential form and can be converted to an integral spectrum by integration over all energies above a required threshold (E). Insofar as the spectrum approximates a power law of index –3, a simple conversion to the integral at an energy $E/1.8$ is obtained by multiplying the differential flux by the energy and dividing by 0.62.

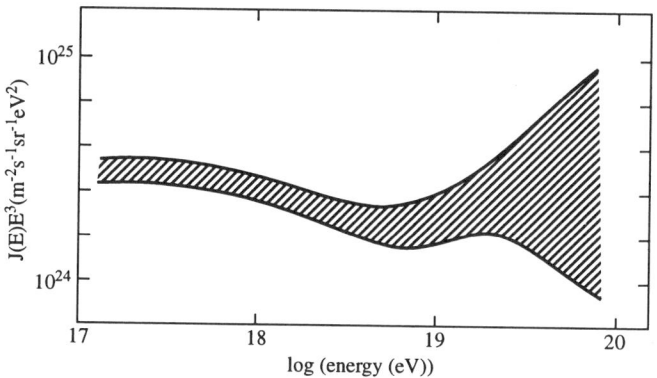

FIGURE 2. Energy spectrum at the highest energies. This spectrum (after Yoshida et al., 1995) has the differential spectrum multiplied by energy cubed. It is from a compilation of a number of measurements and indicates the good general agreement at the lower energies and a spread due to inadequate statistics at the highest energies.

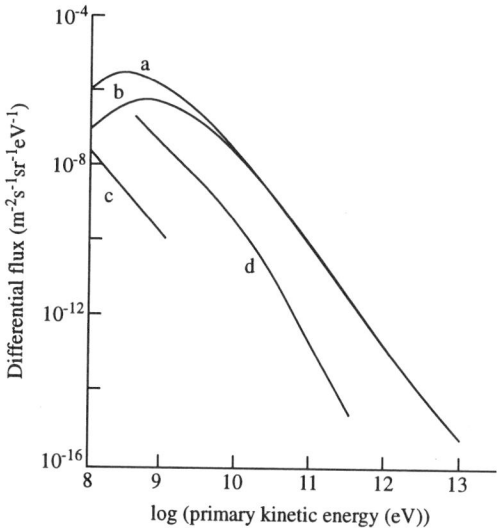

FIGURE 3. Energy spectrum of particles at lower energies. (a) Solar minimum proton energy spectrum. (b) Solar maximum proton energy spectrum. (c) Gamma-ray energy spectrum. (d) Local interstellar electron spectrum.

At higher energies, composition determinations are indirect and are rather contradictory and controversial. Experiments aim to differentiate between broad composition models. The measurement technique is based on studies of cosmic ray shower development. A rather direct technique for such studies is to use fluorescence observations of the shower development to determine the atmospheric depth of maximum development of the shower. Such observations suggest a heavy composition (large atomic number) at energies ~10^{17} eV which changes with increasing energy to a light composition (perhaps protonic) above ~10^{19} eV (T. K. Gaisser et al., 1993).

Primary Electrons

Primary electrons constitute about 1% of the cosmic ray beam. The positron to negative electron ratio is about 10% (J. M. Clem et al., 1995).

Antimatter in the Primary eam

The ratio of antiprotons to protons in the primary cosmic ray beam (at about 400 MeV) is about 10^{-5}. At about 10 GeV the ratio is about 10^{-3}. At the highest measured energies (10 TeV), the upper limit to the ratio is about 20% (S. Orito et al., 1995; M. Amenomori et al., 1995).

Primary Gamma-Rays

The flux of primary gamma-rays is low at high energies. At 1 GeV the ratio of gamma-rays to protons is about 10^{-6}. The arrival directions of these gamma-rays are strongly concentrated in the plane of the Milky Way although there is a diffuse, near isotropic background flux and some point sources have been detected.

Since the absorption cross section for gamma-rays above 100 MeV is approximately 20 mbarn/electron, less than 10% of gamma-rays reach mountain altitudes (A. W. Wolfendale, 1979; P. F. Michelson, 1994).

Sea Level Cosmic Radiation

The sea level cosmic ray dose is 300 millirad·yr^{-1} and the sea level ionization is 2.2×10^6 ion pairs m^{-3}s^{-1}. The sea level flux has a soft component, which can be absorbed in about 100 mm of lead (about 100 g·cm^{-2} of absorber) and a more penetrating (largely muon) hard component. The sea level radiation is largely produced in the atmosphere and is a secondary component from interactions of the primary particles. The steep primary energy spectrum means that most secondaries at sea level are from rather low energy primaries. Thus the secondary flux is dependent on the solar cycle and the geomagnetic latitude of the observer.

Absolute Flux of the Hard Component

Vertical Integral Intensity I(0) ~100 m^{-2}s^{-1}sr^{-1}
Angular dependence I(θ) ~ I(0) cos^2(θ)
Integrated Intensity ~200 m^{-2}s^{-1}
(O.C. Allkofer, 1975b).

Flux of the Soft Component

In free air, the soft component comprises about one third of the total cosmic ray flux.

Latitude Effect

The geomagnetic field influences the trajectories of lower energy cosmic rays approaching the Earth. As a result, the background flux is reduced by about 7% at the geomagnetic equator. The effect decreases towards the poles and is negligible at latitudes above about 40°.

Flux of Protons

The proton component is strongly attenuated by the atmosphere with an attenuation length (reduction by a factor of e) of about 120 g·cm^{-2}. It constitutes about 1% of the total vertical sea level flux.

Absorption

The soft component is absorbed in about 100 g·cm^{-2} of matter. The hard component is absorbed much more slowly:
    Absorption in lead, 6% per 100 g·cm^{-2}
    Absorption in rock, 8.5% per 100 g·cm^{-2}
    Absorption in water, 10% per 100 g·cm^{-2}
    (Absorption for depths less than 100 g·pd cm^{-2} is given by K. Greisen, 1943.)

Altitude Dependence

The cosmic ray background in the atmosphere has a maximum intensity of about 15 times that at sea level at a depth of about 150 g·cm^{-2} (15 km altitude). At maximum intensity, the soft and hard components contribute roughly equally but the hard component is then attenuated more slowly (S. Hayakawa, 1969).

Cosmic Ray Showers

High energy cosmic rays produce particle cascades in the atmosphere which can be detected at sea level provided that their energy exceeds about 100 GeV (such low energy cascades may be detected by using the most sensitive atmospheric Cerenkov detectors). The primary particle progressively loses energy which is transferred through the production of successive generations of secondary particles to a cascade of hadrons, an electromagnetic shower component (both positively and negatively charged electrons and gamma-rays) and muons. The secondary particles are relativistic and all travel effectively at the speed of light. As a result, they reach sea level at approximately the same time but, due to Coulomb scattering (for the electrons) and production angles (for the pions producing the muons), are spread laterally into a disk-like shower front with a characteristic lateral width of several tens of meters and thickness (near the central shower core) of 2 to 3 m. The number of particles at sea level is roughly proportional to the primary particle energy:

Number of particles at sea level ~10^{-10} × energy (eV).

At altitudes below a few kilometers, the number of particles in a shower attenuates with an *attenuation length* of about 200 g·cm^{-2}.

i.e., particle number = original number × exp(–(depth increase)/200)

The above applies to an individual shower. The rate of observation of showers of a given size (particle number at the detector) at different depths of absorber attenuates with an *absorption length* of about 100 g·cm^{-2} (J.G. Wilson, 1976).

Atmospheric Background Light from Cosmic Rays

Cosmic ray particles produce Cerenkov light in the atmosphere and produce fluorescent light through the excitation of atmospheric molecules.

Cerenkov Light

High energy charged particles will cause the emission of Cerenkov light in air if their energies are above about 30 MeV (electrons). This threshold is pressure (and hence altitude) dependent. A typical Cerenkov light pulse (at sea level, 100 m from the central shower core) has a time spread of a few nanoseconds. Over this time, the photon flux between 430 and 530 nm would be ~10^{14} m^{-2}s^{-1} for a primary particle energy of 10^{16} eV. For comparison, the night sky background flux is ~6×10^{11} photons m^{-2}s^{-1}sr^{-1} in the same wavelength band (J.V. Jelley, 1967).

Fluorescence Light

Cosmic ray particles in the atmosphere excite atmospheric molecules which then emit fluorescence light. This is weak compared to the highly collimated Cerenkov component when viewed in the direction of the incident cosmic ray particle but is emitted isotropically. Typical pulse widths are longer than 50 ns and may be up to several microseconds for the total pulse from distant large showers (R.M. Baltrusaitis et al., 1985).

Effects of Cosmic Rays

Cerenkov Effects in Transparent Media

Background cosmic ray particles will produce Cerenkov light in transparent material with a photon yield between wavelengths λ_1 and λ_2

$$\sim (2\pi/137)\sin^2(\theta_c)\int_{\lambda_1}^{\lambda_2} d\lambda/\lambda^2 \text{ photons (unit length)}^{-1}$$

where θ_c (the Cerenkov angle) = $\cos^{-1}$ (1/refractive index).

This background light is known to affect light detectors, e.g., photomultipliers, and can be a major source of background noise (R.W. Clay and A.G. Gregory, 1977).

Effects on Electronic Components

If background cosmic ray particles pass through electronic components, they may deposit sufficient energy to affect the state of, e.g., a transistor flip-flop. This effect may be significant where reliability is of great importance or the background flux is high. For instance, it has been estimated that, in communication satellite operation, an error rate of about 2×10^{-3} per transistor per year may be found. Permanent damage may also result. A significant error rate may be found even at sea level in large electronic memories. This error rate is dependent on the sensitivity of the component devices to the deposition of electrons in their sensitive volumes (J.F. Ziegler, 1981).

Biophysical Significance

When cosmic rays interact with living tissue, they produce radiation damage. The amount of the damage depends on the total dose of radiation. At sea level, this dose is small compared with doses from other sources but both the quantity and quality of the radiation change rapidly with altitude. Approximate dose rates under various conditions are:

Dose rates (mrem·yr^{-1})
Sea level cosmic rays, 30
Cosmic rays at 10 km (subsonic jets), 2000
Cosmic rays at 18 km (supersonic transports), 10,000
(c.f., mean total sea level dose, 300)

Astronauts would be subject to radiation from galactic (0.05 rads per day) and solar (a few hundred rads per solar flare) cosmic rays as well as large fluxes of low energy radiation when passing through the Van Allen belts (about 0.3 rads per traverse).
Both astronauts and SST travellers would be subject to a small flux of low energy heavy nuclei stopping in the body. Such particles are capable of destroying cell nuclei and could be particularly harmful in the early stages of the development of an embryo. The rates of heavy nuclei stopping in tissue in supersonic transports and spacecraft are approximately as follows:

Stopping nuclei ((cm^3 tissue)$^{-1}$ hr^{-1})
Supersonic transport (16 km), 0.0005
Supersonic transport (20 km), 0.005
Spacecraft, 0.15
(O. C. Allkofer, 1975a; O. C. Allkofer et al., 1974).

Carbon Dating

Radiocarbon is produced in the atmosphere due to the action of cosmic ray slow neutrons. Solar cycle modulation of the very low energy cosmic rays causes an anticorrelation of the atmospheric ^{14}C activity with sunspot number with a mean amplitude of about 0.5%. In the long term, modulation of cosmic rays by a varying magnetic field may be important (A.A. Burchuladze et al., 1979).

Practical Uses of Cosmic Rays

There are few direct practical uses of cosmic rays. Their attenuation in water and snow have, however, enabled automatic monitors of water and snow depth to be constructed. A search for hidden cavities in pyramids has been carried out using a muon "telescope".

Other Effects

Stellar X-rays have been observed to affect the transmission times of radio signals between distant stations by altering the depth of the ionospheric reflecting layer. It has also been suggested that variations in ionization of the atmosphere due to solar modulation may have observable effects on climatic conditions.

REFERENCES

O.C. Allkofer, (1975a) *Introduction to Cosmic Radiation,* Verlag Karl Thiemig, Munchen, Germany.

O.O. Allkofer, (1975b) *J. Phys. G: Nucl. Phys.,* 1, L51.

O.C. Allkofer and W. Heinrich, (1974) *Health Phys.,* 27, 543.

M. Amenomori et al., (1995) Proc. 24th Int. Cosmic Ray Conf. Rome, 3, 85. Universita La Sapienza, Roma.

K. Asakimori et al., (1993) Proc. 23rd Int. Cosmic Ray Conf. Calgary, 2, 25, University of Calgary, Canada.

R.M. Baltrusaitis et al., (1985) *Nucl. Inst. Meth.,* A420, 410.

D.J. Bird et al., (1993) *Phys. Rev. Lett.,* 71, 3401.

A.A. Burchuladze, S.V. Pagava, P. Povinec, G. I. Togonidze, S. Usacev, (1979) Proc. 16th Int. Cosmic Ray Conf. Kyoto, 3, 201, Univ. of Tokyo, Japan.

R.W. Clay, (1987) *Aust. J. Phys.,* 40, 423.

R.W. Clay and A.G. Gregory, (1977) *J. Phys. A: Math. Gen.,* 10, 135.

J.M. Clem et al., (1995) Proc. 24th Int. Cosmic Ray Conf. Rome, 3, 5, Universita La Sapienza, Roma.

T.K. Gaisser et al., (1993) *Phys. Rev. D,* 47, 1919.

K. Greisen, (1943) *Phys. Rev.,* 63, 323.

S. Hayakawa, (1969) *Cosmic Ray Physics,* Wiley-Interscience, New York.

J.V. Jelley, (1967) *Prog. in Elementary Particle and Cosmic Ray Physics,* 9, 41.

E. Juliusson, (1975) Proc. 14th Int. Cosmic Ray Conf. Munich, 8, 2689, Max Planck Institute fur Extraterrestriche Physik, Munchen, Germany.

J. Linsley, (1981) *Origin of Cosmic Rays,* I.A.U. Symposium 94, 53, D. Reidel Publishing Co Dordrecht, Holland.

P. Meyer, (1981) *Origin of Cosmic Rays,* I.A.U. Symposium 94, 7, D. Reidel Publishing Co. Dordrecht, Holland.

P.F. Michelson (1994) in *Towards a Major Atmospheric Cerenkov Detector III,* 257, Ed. T. Kifune, Universal Academy Press Inc., Tokyo, Japan.

P. Sokolsky, (1989) *Introduction to Ultrahigh Energy Cosmic Ray Physics,* Addison Wesley Publishing Company.

T. Stanev et al., (1995) *Phys. Rev. Lett.,* 75, 3056.

S. Orito et al., (1995) Proc. 24th Int. Cosmic Ray Conf. Rome, 3, 76. Universita La Sapienza, Roma.

J.G. Wilson, (1976) *Cosmic Rays,* Wykeham Pub. (London) Lt., U.K.

A.W. Wolfendale, (1979) *Pramana,* 12, 631.

S.Yoshida et al., (1995) *Astroparticle Phys.,* 3, 105.

J.F. Ziegler, (1981) IEEE Trans. Electron Devices, ED-28, 560.

Section 12
Properties of Solids

Techniques for Materials Characterization 12-1
Symmetry of Crystals 12-7
Ionic Radii in Crystals 12-14
Polarizability of Atoms and Ions in Solids 12-17
Crystal Structures and Lattice Parameters of Allotropes of the Elements 12-19
Lattice Energies 12-22
The Madelung Constant and Crystal Lattice Energy 12-36
Elastic Constants of Single Crystals 12-37
Electrical Resistivity of Pure Metals 12-45
Electrical Resistivity of Selected Alloys 12-48
Permittivity (Dielectric Constant) of Inorganic Solids 12-52
Curie Temperature of Selected Ferroelectric Crystals 12-61
Properties of Antiferroelectric Crystals 12-62
Dielectric Constants of Glasses 12-63
Properties of Superconductors 12-64
High Temperature Superconductors 12-91
Organic Superconductors 12-94
Properties of Semiconductors 12-97
Diffusion Data for Semiconductors 12-108
Properties of Magnetic Materials 12-117
Organic Magnets 12-126
Electron Work Function of the Elements 12-130
Secondary Electron Emission 12-131
Optical Properties of Metals and Semiconductors 12-133
Elasto-optic, Electro-optic, and Magneto-optic Constants 12-157
Nonlinear Optical Constants 12-174
Phase Diagrams 12-178
Heat Capacity of Selected Solids 12-196
Thermal and Physical Properties of Pure Metals 12-197
Thermal Conductivity of Metals and Semiconductors as a Function of Temperature 12-199
Thermal Conductivity of Alloys as a Function of Temperature 12-201
Thermal Conductivity of Crystalline Dielectrics 12-202
Thermal Conductivity of Ceramics and Other Insulating Materials 12-204
Thermal Conductivity of Glasses 12-206
Commercial Metals and Alloys 12-210
Hardness of Minerals and Ceramics 12-211

TECHNIQUES FOR MATERIALS CHARACTERIZATION

EXPERIMENTAL TECHNIQUES USED TO DETERMINE THE COMPOSITION, STRUCTURE, AND ENERGY STATES OF SOLIDS AND LIQUIDS

H.P.R.Frederikse

The many experimental methods, originally designed to study the chemical and physical behavior of solids and liquids, have grown into a new field known as Materials Characterization (or Materials Analysis). During the past 30 years a host of techniques aimed at the study of surfaces and thin films has been added to the many tools for the analysis of bulk samples. The field has benefitted particularly from the development of computers and microprocessors, which have vastly increased the speed and accuracy of the measuring devices and the recording of their output. Materials characterization was and is a very important tool in the search for new physical and chemical phenomena. It plays an essential role in new applications of solids and liquids in industry, communications, and medicine. Many of its techniques are used in quality control, in safety regulations, and in the fight against pollution.

In most Materials Characterization experiments the sample is subjected to some kind of radiation: electromagnetic, acoustic, thermal, or particles (electrons, ions, neutrons, etc.). The surface analysis techniques usually require a high vacuum. As a result of interactions between the solid (or liquid) and the incoming radiation a beam of a similar (or a different) nature will emerge from the sample. Measurement of the physical and/or chemical attributes of this emerging radiation will yield qualitative, and often quantitative, information about the composition and the properties of the material being probed.

The modern tendency of describing practically everything in this world by a combination of a few letters (acronyms) has also penetrated the field of Materials Characterization. The table below gives the meaning of the acronym for every technique listed, the form and size of the required sample (bulk, surface, film, liquid, powder, etc.), the nature of the incoming radiation, the nature of the emerging radiation, the depth and the lateral spatial resolution that can be probed, and the information obtained from the experiment. The last column lists one or two major references to the technique described.

OPTICAL AND MASS SPECTROSCOPIES FOR CHEMICAL ANALYSIS

	Technique	Sample	In	Out	Depth	Lateral resolution	Information obtained	Ref.
1.	AAS Atomic Absorption Spectroscopy	Atomize (flame, electro, thermal, etc.)	Light e.g. glow discharge	Absorption spectrum	—	—	Concentration of atomic species (quantitative, using standards)	1,2
2.	ICP-AES Induct. Coupled Plasma — Atomic Emission Spectroscopy	Atomize (flame, electro, thermal, ICP, etc.)	—	Emission spectrum	—	—	Concentration of atomic species (quantitative, using standards)	3
3.	Dynamic SIMS Dynamic Secondary Ion Mass Spectroscopy	Surface	Ion beam (1–20 keV)	Secondary ions; analysis with mass spectrometer	2 nm–1 µm (or deeper: ion milling)	0.50 nm	Elemental and isotopic analysis; depth profile (all elements); detection limits: ppb-ppm	4
4.	Static SIMS Static Secondary Ion Mass Spectroscopy	Surface	Ion beam (0.5–20 keV)	Secondary ions, analysis with mass spectrometer	0.1–0.5 nm	10 µm	Elemental analysis of surface layers; molecular analysis; detection limits: ppb-ppm	4
5.	SNMS Sputtered Neutral Mass Spectroscopy	Surface, bulk	Plasma discharge; noble gases: 0.5–20 keV	Sputtered atoms ionized by atoms or electrons; then mass analyzed	0.1–0.5 nm (or deeper: ion milling)	1 cm	Elemental analysis Z ≥ 3; depth profile; detection limit: ppm	4,6
6.	SALI Surface Analysis by Laser Ionization	Surface	e-beam, ion-beam, or laser for sputtering	Sputtered atoms ionized by laser; then mass analyzed	0.1–0.5 nm up to 3 µm in milling mode	60 nm	Surface analysis; depth profiling	7
7.	LIMS Laser Ionization Mass Spectroscopy	Surface, bulk	u.v. laser (ns pulses)	Ionized species; analyzed with mass spectrometer	50–150 nm	5 µm–1 mm	Elemental (micro)analysis; detection limits: 1–100 ppm	8
8.	SSMS Spark Source Mass Spectroscopy	Sample in the form of two electrodes	High voltage R.F. spark produces ions	Ions — analyzed in mass spectrometer	1–5 µm	—	Survey of trace elements; detection limit: 0.01–0.05 ppm	9
9.	GDMS Glow Discharge Mass Spectroscopy	Sample forms the cathode for a D.C. glow discharge	Sputtered atoms ionized in plasma	Ions — analyzed in mass spectrometer	0.1–100 µm	3–4 mm	(Bulk) trace element analysis; detection limit: sub-ppb	9,10
10.	ICPMS Induct. Coupled Plasma Mass Spectroscopy	Liquid-dissolved sample carried by gas stream into R.F. induction coil	Ions produced in argon plasma	Ions — analyzed in quadrupole mass spectrometer	—	—	High sensitivity analysis of trace elements	11

TECHNIQUES FOR MATERIALS CHARACTERIZATION (continued)

PHOTONS — ABSORPTION, REFLECTION AND ELECTRON EMISSION

Technique	Sample	In	Out	Depth	Lateral resolution	Information obtained	Ref.
11. IRS Infrared Spectroscopy	Thin crystal, glass, liquid	I.R. light (W-filament, globar, Hg-arc)	I.R. spectrum	—	—	Electronic transitions (mainly in semiconductors and superconductors); vibrational modes (in crystals and molecules)	12,13, 14
12. FTIR Fourier Transform I.R. Spectroscopy	Solid, liquid; transmission or reflection	White light (all frequencies)	Fourier Transform of spectrum (interferometer)	—	—	Spectra obtained at higher speed and resolution	15
13. ATR Attenuated Total Reflection	Surface or thin crystal	—	—	μm's	—	Atomic or molecular spectra of surfaces and films	16
14. (μ)-RS (Micro-) Raman Spectroscopy	Solid, liquid (1 μm–1 cm)	Laser beam, e.g., Ar-line, YAG-line	Raman spectra	0.5 μm	0.5 μm	Molecular and crystal vibrations	12,14, 17
15. CARS Coherent Anti-Stokes Raman Spectroscopy	Solid, liquid (50 μm–3 cm)	Pump beam (ω_0)+ probe beam (ω_a)	Anti-Stokes spectrum	—	—	High resolution Raman spectra	14
16. Ellipsometry	Transparent films, crystals, adsorbed layers	Polarized light	Change in polarization	0.05 nm–5 μm	25 μm (or sample thickness)	Refractive index and absorption	18,19
17. UPS Ultraviolet Photo-electron Spectroscopy	Surfaces, adsorbed layers	u.v. light, 10–100 eV; 200 eV (synchrotron)	Electrons	0.2–10 nm	0.1–10 nm	Energies of electronic states of surfaces and free molecules	20,21
18. PSD Photon Stimulated Desorption	Surfaces with adsorbed species	Far u.v. light E > 10 eV	Ions — analyzed with mass spectrometer	0.1–2 nm	—	Structure and desorption kinetics of adsorbed atoms and molecules	22

X-RAYS

Technique	Sample	In	Out	Depth	Lateral resolution	Information obtained	Ref.
19. XRD X-Ray Diffraction	Single crystals, powders films	X-rays: λ = 0.05–0.2 nm (6–17 keV)	Diffracted X-ray beam	1–1000 μm	0.1–10 mm	Identification of crystallographic structures; all elements (low Z difficult)	23,24
20. XRF/EDS X-Ray Fluorescence/Energy Dispersive Spectroscopy	Thin films, single layer	Prim. X-ray beam λ = 0.02–0.1 nm 12–80 keV	Fluorescent X-rays	1–100 μm	10 mm	Elemental analysis; all elements except H, He, Li — (EDS also used in XRD, SEM, TEM and EPMA)	25,26
21. EXAFS Extended X-Ray Absorption Fine Structure	Films, foils	High intensity X-rays (synchrotron)	Spectrum near absorption edge	nm–μm	—	Local atomic structure: order/disorder in vicinity of absorbing atom	27
22. XPS/ESCA X-Ray Photo-electron Spectroscopy/Electron Spect. for Chemical Analysis	Surfaces, thin films (≈20 atomic layers)	Soft X-rays (1–20 keV)	Core electrons; valence electrons	0.5–10 nm	5 nm–50 μm	(Quantitative) identification of all elements in surface layer or film	28,29

ELECTRONS

Technique	Sample	In	Out	Depth	Lateral resolution	Information obtained	Ref.
23. CL Cathode Luminescence	Insulators, semiconductors	Electrons 5–50 keV	Photons 0.1–5 eV	1 nm–2 μn	1 or 2 μm	Energy levels of impurities and point defects	30
24. APS Appearance Potential Spectroscopy	Surface (≈20 atomic layers)	Electrons (energy scan) 50–2000 eV	X-rays to pinpoint electron energy threshold	—	—	Identification of surface species	21, see also C

TECHNIQUES FOR MATERIALS CHARACTERIZATION (continued)

	Technique	Sample	In	Out	Depth	Lateral resolution	Information obtained	Ref.
25.	AES Auger Electron Spectroscopy	Thin films, surfaces	Electrons 3–10 keV	Auger electrons 20–2000 eV	0.3–3 nm	≈30 nm	Elemental composition of surface (except H, He); detection limit 0.1–1%	28,29
26.	EELS Electron Energy Loss Spectroscopy	Very thin samples (<200 nm)	Electrons (100–400 keV)	(Retarded) electrons; minus 1–1000 eV	<200 nm	1–100 nm	Local elemental concentration; electronic structure, chem. bonding; interatomic distances	31
27.	EXELFS Extended Electron Energy Loss Fine Structure	Thin films	Electrons (100–400 keV)	Electrons energies 0–30 eV above edge	<200 nm	1–100 nm	Density of states of valence electrons (above Fermi level)	27,32
28.	ESD Electron Stimulated Desorption	Adsorbed species	Electrons E > 10 eV	Ions — analyzed with mass spectrometer	—	—	Structure and desorption properties of adsorbed atoms and molecules	22
29.	ESDIAD ESD-Ion Angular Distribution	(See ESD)	(See ESD)	Directional dependence of emitted ions	—	—	Geometries of adsorbed species (atoms or molecules)	22
30.	EPMA Electron Probe (X-Ray) Micro Analysis	Solid conductors and insulators <1 cm thick	Electrons 5–30 keV	Characteristic X-ray 0.1–15 keV	100 nm–5 μm	1 μm	Elemental analysis, $Z \leq 4$, major, minor and trace amounts	33,34
31.	LEED Low Energy Electron Diffraction	Surface	Mono-energetic electron beam 10–1000 eV	Diffracted electrons	0.4–2 nm	<5 μm	Crystallographic structure of surface; resolution: 0.01 nm	35
32.	RHEED Reflection High Energy Electron Diffraction	Surface	Electron beam at grazing angle 5–50 keV	Reflected electrons	0.2–10 nm	<5 μm	Surface symmetry	36,37
33.	SEM Scanning Electron Microscopy	Bulk, films (conducting)	High energy electrons usually ~30 keV	Secondary and backscattered electrons	1 nm–5 μm	1–20 nm	Surface image, defect structure; resolution 5–15 nm; magnification 300,000×	33,34
34.	(S)TEM (Scanning) Transmission Electron Microscopy	Thin specimen — <200 nm	High energy electrons typically 300 keV	Transmitted and diffracted electrons	(Sample thickness)	2–20 nm	(Defect) structure of cryst. solids; microchemistry; high resol.: 0.2 nm	33
35.	FEM Field Emission Microscopy	Metals, alloys (sharp point)	—	Electron emission (with appl. electric field — 50 kV)	≈0.5 nm	10–100 nm	Surface image, crystallographic structure	34
36.	STM Scanning Tunneling Microscopy	Polished or cleaved surface (conducting)	Tunneling current controls distance between sample and very sharp tip		1–5 nm	2–10 nm	Atomic-scale relief map of surface; resolution: vert. 0.002 nm, hor. 0.2 nm	39
37.	SPM Scanned Probe Microscopy	Very flat surface	Any field: e.g. mechan. vibration recorded with laser probe; same with magnetic, electric or thermal field		1–100 nm	1–100 nm	Surface-magnetic field, surface-thermal conductivity, etc.	39a
38.	AFM Atomic Force Microscopy	Very flat surface	Similar to STM; force measured with cantilever spring		0.5–5 nm	0.2–130 nm	Surface topography with atomic resolution; interatomic force	40
	IONS AND NEUTRONS							
39.	ISS (or LEIS) Ion Scattering Spectroscopy (Low Energy Ion Scattering)	Surface	Ion beam He+ or Ne+ <3 keV	Sputtered ions (energy analysis)	0.1–0.5 nm	1–100 μm	Elemental analysis (better for low Z) detection limits: 0.01–1%	41
40.	FIM Field Ion Microscopy	Surface: metals, alloys; very sharp tip	(He gas above sample)	He ions + high electric field produce image	≈0.1 nm	0.1–2 nm	Atomic structure of surface	34,42
41.	RBS Rutherford Back Scattering	Solids, thin films	Mono-energetic ions (H+ or He++) 0.5–3 MeV	Backscattered ions	10 nm–1 μm	1 mm	Element identification (Li to U) detection limit: 0.01–1%	46

TECHNIQUES FOR MATERIALS CHARACTERIZATION (continued)

Technique	Sample	In	Out	Depth	Lateral resolution	Information obtained	Ref.
42. NRA Nuclear Reaction Analysis	Solids, thin films	Mono-energetic ions (Li, Be, B, etc.) 200 keV–6 MeV	Protons, deuterons ^{3}He, α-particles, γ-rays	0.1–5 μm	10 μm–10 mm	Element identification (all) detection limit: 10^{-12}–10^{-2}	47
43. PIXE Particle Induced X-ray Emission	Thin films, surface layers	High energy ions (H^+ or He^{++})	Characteristic X-rays	<10 μm	1 μm–2 mm	Trace impurities: Z >3 detection limit: 0.1–100 ppm (depending on sample thickness)	48
44. INS Ion Neutralization Spectroscopy	Surface	He-ions (≈5 eV)	Electrons	—	—	Energies of valence electrons	49
45. NAA Neutron Activation Analysis	Bulk, >0.5 g	Thermal neutrons	Characteristic γ-rays, (≈1 MeV)	Bulk	—	Trace concentrations (of isotopes) of elements: trans. metals, Pt-group; detection limit: 10^8–10^{14} atoms/cm^3	43
46. N(P)D Neutron (Powder) Diffraction	Crystalline solids	Thermal neutrons E =0.0025 eV	Diffracted neutrons	Bulk	—	Crystallographic structure; porosity, particle size	44
47. SANS Small Angle Neutron Scattering	Inhomogeneous solids; powders; porous samples	Thermal neutrons 2θ = 10^{-2}–10^{-4}	Scattered neutrons	1–25 mm	—	Average size of inhomogeneities; range: 1 nm–1 mm	45
ACOUSTIC							
48. SLAM Scanning Laser Acoustic Microscopy	Bulk, film	Acoustic wave produced by laser 1 MHz–1 GHz	Reflected acoustic wave	μm–cm	0.1–20 mm	Defect structure; thickness measurement	50
THERMAL							
49. DTA Differential Thermal Analysis	Specimen and reference sample	Uniform heating	Temperature difference	Bulk	—	Phase transitions, crystallization	51
50. DSC Differential Scanning Calorimetry	Specimen and ref. sample	Controlled heating	Measure heat required for equal temperature	Bulk	—	Phase transitions, crystallization; activation energies	51
51. TGA Thermo Gravimetric Analysis	Bulk, 1–100 g	Controlled heating	Weight as function of temperature (and time)	Bulk	—	Decomposition, non-stoichiometry, kinetics of reaction	52
RESONANCE							
52. EPR (ESR) Electron Paramagnetic (Spin) Resonance	Paramagnetic solids or liquids	Microwave radiation in magnetic field 3–300 GHz; 1–100 kG	Microwave absorption (at resonance)	Bulk	—	Local environment of paramagnetic ion; concentration of paramagnetic species; detection limit: 10^{11} spins/cm^3	53,54
53. ECR Electron Cyclotron Resonance	Semiconductors, metals; free electrons (low temperature)	Microwave radiation in magnetic field 10–30 GHz; 5–10 kG	Microwave absorption (at resonance)	Bulk	—	Electronic energy bands, effective masses	55
54. Mössbauer Effect	Source and absorber	Mono-energetic γ-rays: 5–100 keV	Mössbauer spectrum (Doppler shifted lines)	50 m	1 cm	Interaction between nucleus and its environment (local electric, magnetic fields; bonds; valency; diffusion, etc.)	56
55. NMR (MRI) Nuclear Magnetic Resonance (Magnetic Resonance Imaging)	Solids, liquids	R.F. radiation + magnetic field; e.g. for protons: 60 MHz, 14 kG	R.F. absorption	<1 cm	1 cm	Quant. analysis; local magnetic environment; diffusion; imaging	58
56. ENDOR Electron Nuclear Double Resonance	Solids, liquids	R.F. + microwave radiation in magn. field.	Microwave absorption	—	—	Hyperfine interaction → local atomic structure	54

TECHNIQUES FOR MATERIALS CHARACTERIZATION (continued)

	Technique	Sample	In	Out	Depth	Lateral resolution	Information obtained	Ref.
57.	NQR Nuclear Quadrupole Resonance	Solids	R.F. radiation 0.5–1000 MHz	R.F. absorption	—	—	Asymmetry of the charge distribution at the nucleus	55,59
				OTHER				
58.	BET Brunauer-Emmett-Teller	(Large) surface area 1–20 m²/g	Adsorbed gas (e.g., N_2 at low temp.) as function of pressure (monolayer coverage)	—	—	—	Surface area measurement	60

REFERENCES

General References

A. Wachtman, J. B., *Characterization of Materials*, Butterworth-Heinemann, Boston, 1993.

B. Brundle, C. R., Evans, C. A., and Wilson, S., Eds., *Encyclopedia of Materials*, Butterworth-Heinemann, Boston, 1992.

C. Woodruff, D. P. and Delchar, T. A., *Modern Techniques of Surface Science*, Cambridge University Press, Cambridge, 1986.

D. *Metals Handbook*, 9th Edition, Vol. 10, Materials Characterization, Whan, R. E., Coordinator, American Society for Metals, Metals Park, OH, 1986.

Specific References

1. Slavin, M., *Atomic Absorption Spectroscopy*, 2nd Edition, John Wiley & Sons, New York, 1978.
2. Schrenk, W. G., *Analytical Atomic Spectroscopy*, Plenum Press, New York, 1975.
3. Dean, J. A. and Rains, T. E., *Flame Emission and Atomic Absorption Spectroscopy*, Vols. 1—3, Marcel Dekker, New York, 1969.
4. Benninghoven, A., Rudenauer, F. G., and Werner, H. W., *Secondary Ion Mass Spectroscopy*, John Wiley & Sons, New York, 1987.
5. Bird, J. R. and Williams, J. S., Eds., in *Ion Beams for Materials Analysis*, Academic Press, New York, 1989, pp. 515—537.
6. Smith, G. C., *Quantitative Surface Analysis for Materials Science*, The Institute of Metals, London, 1991.
7. Becker, E. H., in *Ion Spectroscopies for Surface Analysis*, Czanderna, A. W. and Hercules, D. M., Eds., Plenum Press, New York, 1991, p. 273.
8. Simons, D. S., *Int. J. Mass Spectrometry and Ion Processes*, 55, 15, 1983.
9. White, F. A. and Wood, G. M., *Mass Spectrometry: Applications in Science and Engineering*, John Wiley & Sons, New York, 1986.
10. Harrison, W. W. and Bentz, B. L., *Prog. Anal. Spectrometry*, 11, 53, 1988.
11. Bowmans, P. W. J. M., *Inductively Coupled Plasma Emission Spectroscopy*, Parts I and II, John Wiley & Sons, New York, 1987.
12. Brame, Jr., E. G. and Grasselli, J., *Infrared and Raman Spectroscopy*, Practical Spectroscopy Series, Vol. I, Marcel Dekker, New York, 1976.
13. Hollas, J. M., *Modern Spectroscopy*, John Wiley & Sons, New York, 1987.
14. Turrell, G., *Infrared and Raman Spectroscopy of Crystals*, Academic Press, New York and London, 1972.
15. Griffith, P. R. and Haseth, J. A., *Fourier Transform Infrared Spectroscopy*, John Wiley & Sons, New York, 1986.
16. Barnowski, M. K., *Fundamentals of Optical Fiber Communications*, Academic Press, New York, 1976.
17. Long, D. A., *Raman Spectroscopy*, McGraw-Hill, New York, 1977.
18. Azzam, R. M. A., *Ellipsometry and Polarized Light*, Elsevier-North Holland, Amsterdam, 1977.
19. Hecht, E., *Optics*, 2nd Edition, Addison-Wesley, Reading MA, 1987.
20. Brundle, C. R., in *Molecular Spectroscopy*, West, A. R., Ed., Heyden, London, 1976.
21. Park, R. L., in *Experimental Methods in Catalytic Research*, Vol. III, Anderson, R. B. and Dawson, P. T., Academic Press, New York, 1976, pp. 1—39.
22. Madey, T. E. and Stockbauer, R., in *Solid State Physics: Surfaces*, Vol. 22 of Methods of Experimental Physics, Park, R.L. and Lagally, M. G., Eds., Academic Press, New York, 1985.
23. Cullity, B. D., *Elements of X-Ray Diffraction*, 2nd Edition, Addison-Wesley, Reading, MA, 1978.
24. Schwartz, L. H. and Cohen, J. B., *Diffraction from Materials*, Springer Verlag, Berlin, 1987.

12-5

TECHNIQUES FOR MATERIALS CHARACTERIZATION (continued)

25. deBoer, D. K. G., in *Advances in X-Ray Analysis*, Vol. 34, Barrett, C. S. et. al., Eds., Plenum Press, New York, 1991.
26. Birks, L. S., *X-Ray Spectrochemical Analysis*, 2nd Edition, John Wiley & Sons, New York, 1969.
27. Bonnelle, C. and Mande, C., *Advances in X-Ray Spectroscopy*, Pergamon Press, Oxford, 1982.
28. *Practical Surface Analysis by Auger and X-Ray Photo-Electric Spectroscopy*, Briggs, D. and Seah, M. P., Eds., John Wiley & Sons, New York, 1983.
29. Powell, C. J. and Seah, M. P., *J. Vac. Sci. Technol. A*, Vol. 8, 735, 1990.
30. Yacobi, G. G. and Holt, D. B., *Cathodeluminescence Microscopy of Inorganic Solids*, Plenum Press, New York, 1990.
31. Egerton, R. F., *Electron Energy Loss Spectroscopy in the Electron Microscope*, Plenum Press, New York, 1986.
32. Disko, M. M., Krivanek, O. L., and Rez, P., *Phys. Rev.*, B25, 4252, 1982.
33. Goldstein, J. I., et. al., *Scanning Electron Microscopy and X-Ray Microanalysis*, 2nd Edition, Plenum Press, New York, 1986.
34. Murr, L. E. *Electron and Ion Microscopy and Microanalysis*, Marcel Dekker, New York, 1982.
35. Armstrong, R. A., in *Experimental Methods in Catalytic Research*, Vol. III, Anderson, R. B., and Dawson, P. T.,Eds., Academic Press, New York, 1976.
36. Dobson, P. J. et. al., *Vacuum*, 33, 593, 1983.
37. Ryner, T. B., *Electron Diffraction*, Methuen, London, 1970.
38. Reimer, L., *Transmission Election Microscopy*, Springer-Verlag, Berlin, 1984.
39. *Scanning Tunneling Microscopy and Related Methods*, Behm, R. J., Garcia, N., and Rohrer, H., Kluwer, Eds., Academic Publishers, Norwell, MA, 1990.
39a. Wikramasinghe, H.K., *Scientific American*, Vol. 261, No. 4, pp. 98—105, Oct. 1989.
40. Rugar, D. and Hansma, P., *Physics Today*, 43(10), pp. 23—30, 1990.
41. Feldman, C. C. and Mayer, J. W., *Fundamentals of Surface and Thin Film Analysis*, North-Holland, Amsterdam, 1986.
42. Muller, E. W. and Tsong, T. T., *Field Ion Microscopy*, Elsevier, Amsterdam, 1969.
43. Amiel, S., *Nondestructive Activation Analysis*, Elsevier, Amsterdam, 1981.
44. Bacon, G. E., *Neutron Diffraction*, 3rd Edition, Clarendon Press, Oxford 1975.
45. Neutron Scattering, Part A.,,in *Methods of Experimental Physics*, Vol. 23, Skold, K. and Price, D. L., Eds., Academic Press, New York, 1986.
46. Chu, W. K., Mayer, J. W., and Nicolet, M. A., *Backscattering Spectroscopy*, Academic Press, New York, 1987.
47. Rickey, F. A., in *High Energy and Heavy Ion Beams in Materials Analysis*, Tesmer, J. R., et. al., Eds., MRS, 1990, pp. 3—26.
48. Johansson, S. A. E. and Campbell, J. L., *PIXE: A Novel Technique for Elemental Analysis*, John Wiley & Sons, New York, 1988.
49. Hagstrum, H. D., in *Inelastic Ion-Surface Collisions*, Tolk, N. H. et. al., Eds., Academic Press, New York, 1977, pp. 1—46.
50. Nikoonahad, M., in *Research Techniques in Nondestructive Testing*, Vol. VI, Sharpe, R.S., Ed., Academic Press, New York, 1984, pp. 217—257.
51. Gallagher, P. K., *Characterization of Materials by Thermoanalytical Techniques*, MRS - Bulletin, Vol. 13, No. 7, pp. 23—27, 1988.
52. Earnest, C. M., *Compositional Analysis of Materials by Thermogravimetry*, ASTM Special Technical Publication 997, 1988.
53. Poole, C. P., *Electron Spin Resonance — A Comprehensive Treatise on Experimental Techniques*, 2nd Edition, John Wiley & Sons, New York, 1983.
54. Atherton, N. M., *Principles of Electron Spin Resonance*, Ellis Horwood Ltd., Chichester, U.K., 1993.
55. Kittel, C., *Introduction to Solid State Physics*, 6th Edition, John Wiley & Sons, New York, 1986, p. 196.
56. Gibb, T. C., *Principles of Mössbauer Spectroscopy*, Chapman & Hall, London, 1976.
57. Slichter, C. P., *Principles of Magnetic Resonance*, 3rd Edition, Springer-Verlag, Berlin, 1990.
58. *NMR Spectroscopy Techniques*, Dybrowski, C. and Lichter, R. L., Eds., Marcel Dekker, New York, 1987.
59. Das, T. P. and Hahn, E. L., *Nuclear Quadrupole Resonance Spectroscopy*, Academic Press, New York, 1958.
60. Somorjai, G. A., *Principles of Surface Chemistry*, Prentice-Hall, Englewood Cliffs, NJ, 1972, p. 216.

SYMMETRY OF CRYSTALS

L. I. Berger

The ability of a body to coincide with itself in its different positions regarding a coordinate system is called its symmetry. This property reveals itself in iteration of the parts of the body in space. The iteration may be done by reflection in mirror planes, rotation about certain axes, inversions and translations. These actions are called the symmetry operations. The planes, axes, points, etc., are known as symmetry elements. Essentially, mirror reflection is the only truly primitive symmetry operation. All other operations may be done by a sequence of reflections in certain mirror planes. Hence, the mirror plane is the only true basic symmetry element. But for clarity, it is convenient to use the other symmetry operations, and accordingly, the other aforementioned symmetry elements. The symmetry elements and operations are presented in Table 1.

The entire set of symmetry elements of a body is called its symmetry class. There are thirty-two symmetry classes that describe all crystals which have ever been noted in mineralogy or been synthesized (more than 150,000). The denominations and symbols of the symmetry classes are presented in Table 2.

There are several known approaches to classification of individual crystals in accordance with their symmetry and crystallochemistry. The particles which form a crystal are distributed in certain points in space. These points are separated by certain distances (translations) equal to each other in any chosen direction in the crystal. Crystal lattice is a diagram that describes the location of particles (individual or groups) in a crystal. The lattice parameters are three non-coplanar translations that form the crystal lattice. Three basic translations form the unit cell of a crystal. August Bravais (1848) has shown that all possible crystal lattice structures belong to one or another of fourteen lattice types (Bravais lattices). The Bravais lattices, both primitive and non-primitive, are the contents of Table 3.

Among the three-dimensional figures, there is a group of polyhedrons that are called regular, which have all faces of the same shape and all edges of the same size (regular polygons). It has been shown that there are only five regular polyhedrons. Because of their importance in crystallography and solid state physics, a brief description of these polyhedrons is included in Table 4.

The systematic description of crystal structures is presented primarily in the well known *Structurbericht*. The classification of crystals by the Structurbericht does not reflect their crystal class, the Bravais lattice, but is based on the crystallochemical type. This makes it inconvenient to use the Structurbericht categories for comparison of some individual crystals. Thus, there have been several attempts to provide a more convenient classification of crystals. Table 5 presents a compilation of different classifications which allows the reader to correlate the Structurbericht type with the international and Schoenflies point and space groups and with Pearson's symbols, based on the Bravais lattice and chemical composition of the class prototype. The information included in Table 5 has been chosen as an introduction to a more detailed crystallophysical and crystallochemical description of solids.

TABLE 1
Symmetry Operations and Elements

Symmetry operation	Name	Symmetry element — Symbol International (Hermann-Mauguin)	Schoenflies	Presentation on the stereographic projection — Parallel	Perpendicular
Reflection in a plane	Plane	m	C_s		
Rotation by angle $\alpha = 360°/n$ about an axis	Axis	n = 1, 2, 3, 4 or 6	C_n		
		n = 2	C_2		
		n = 3	C_3		
		n = 4	C_4		
		n = 6	C_6		
Rotation about an axis and inversion in a symmetry center lying on the axis	Inversion (improper) axis	$\bar{n} = \bar{3}, \bar{4}, \bar{6}$	C_{ni}		
		$\bar{n} = \bar{3}$	C_{3i}		
		$\bar{n} = \bar{4}$	C_{4i}		

TABLE 1
Symmetry Operations and Elements (continued)

Symmetry operation	Name	International (Hermann-Mauguin)	Schoenflies	Parallel	Perpendicular
		$\bar{n} = \bar{6}$	C_{6i}		
Inversion in a point	Center	$\bar{1}$	C_i		
Parallel translation	Translation vector $\vec{a}, \vec{b}, \vec{c}$				
Reflection in a plane and translation parallel to the plane	Glide–plane	a, b, c, n, d			
Rotation about an axis and translation parallel to the axis	Screw axis	n_m $(m = 1, 2, .., n - 1)$			
Rotation about an axis and reflection in a plane perpendicular to the axis	Rotatory-reflection axis	$\tilde{n}$ $\tilde{n} = \tilde{1}, \tilde{2}, \tilde{3}, \tilde{4}, \tilde{6}$	S_n		

(Column headers: **Symmetry element** — Name, Symbol [International (Hermann-Mauguin), Schoenflies]; **Presentation on the stereographic projection** — Parallel, Perpendicular)

TABLE 2
The Thirty-Two Symmetry Classes

Crystal symbol	Primitive Int	Primitive Sch	Central Int	Central Sch	Planal Int	Planal Sch	Axial Int	Axial Sch	Plane-axial Int	Plane-axial Sch	Inversion primitive Int	Inversion primitive Sch	Inversion-planal Int	Inversion-planal Sch
Triclinic	1	C_1	$\bar{1}$	C_i										
Monoclinic					m	C_s	2	C_2	2/m	C_{2h}				
Ortho-rhombic					mm2	C_{2v}	222	D_2	mmm	D_{2h}				
Trigonal	3	C_3	$\bar{3}$	C_{3i}	3m	C_{3v}	32	D_3	$\bar{3}$m	C_{3d}				
Tetragonal	4	C_4	4/m	C_{4h}	4mm	C_{4v}	422	D_4	4/mmm	D_{4h}	$\bar{4}$	S_4	$\bar{4}$2m	D_{2d}
Hexagonal	6	C_6	6/m	C_{6h}	6mm	C_{6v}	622	D_6	6/mmm	D_{6h}	$\bar{6}$	C_{3h}	$\bar{6}$m2	D_{3h}
Cubic	23	T	m3	T_h	$\bar{4}$3m	T_d	432	O	m3m	O_h				

[a] Per Fedorov Institute of Crystallography, USSR Academy of Sciences, nomenclature.

SYMMETRY OF CRYSTALS (continued)

TABLE 3
The Fourteen Possible Space Lattices (Bravais Lattices)

Crystal system	Metric category of the system	No. of different lattices in the system	Lattice type[a] (marked by +)					No. of identi-points per unit cell	Characteristic parameters (marked by +)						Description of characteristic parameters $a \subset X$, $b \subset Y$, $c \subset Z$; $\alpha \equiv (b,c)$, $\beta \equiv (a,c)$, $\gamma \equiv (a,b)$	Symmetry of the lattice	
			P	C	I	F	R		a	b	c	α	β	γ		Int	Sch
Triclinic	Trimetric	1	+					1	+	+	+	+	+	+	$a \neq b \neq c$, $\alpha \neq \beta \neq \gamma$	1	C
Monoclinic	Trimetric	2	+	+				1 or 2	+	+	+		+		$a \neq b \neq c$, $\alpha = \gamma = 90° \neq \beta$	2/m	C_{2h}
Orthorhombic	Trimetric	4	+	+	+	+		1, 2 or 4	+	+	+				$a \neq b \neq c$, $\alpha = \beta = \gamma = 90°$	mmm	D_{2h}
Trigonal (rhombohedral)	Dimetric	1					+	1	+			+			$a = b = c$, $120° > \alpha = \beta = \gamma \neq 90°$	3m	D_{3d}
Tetragonal	Dimetric	2	+		+			1 or 2	+		+				$a = b \neq c$, $\alpha = \beta = \gamma = 90°$	4/mmm	D_{4h}
Hexagonal	Dimetric	1	+					1	+		+				$a = b \neq c$, $\alpha = \beta = 90°$, $\gamma = 120°$	6/mmm	D_{6h}
Isometric (cubic)	Monometric	3	+		+	+		1, 2 or 4	+						$a = b = c$, $\alpha = \beta = \gamma = 90°$	m3m	O_h

a Designations of the space-lattice types: P — primitive, C — side-centered (base-centered), I — body-centered, F — face-centered, R — rhombohedral.

TABLE 4
The Five Possible Regular Polyhedrons

Polyhedron	Symmetry (Schoenflies)		Form of faces	Number of[a]		
	Class	Elements		Faces (F)	Edges (E)	Vertices (V)
Tetrahedron	T	$4C_3 3C_2$	Equilateral triangle	4	6	4
Cube (hexahedron)	O	$3C_4 4C_3 6C_2$	Square	6	12	8
Octahedron	O	$3C_4 4C_3 6C_2$	Equilateral triangle	8	12	6
Pentagonal dodecahedron	J	$6C_5 10C_3 15C_2$	Regular pentagon	12	30	20
Icosahedron	J	$6C_5 10C_3 15C_2$	Equilateral triangle	20	30	12

[a] Per formula by Leonhard Euler: $F + V - E = 2$

TABLE 5
Classification of Crystals

Strukturbericht symbol	Structure name	Symmetry group		Pearson symbol[a]	Standard ASTM E157-82a symbol[b]
1	2	International	Schoenflies	5	6
		3	4		
A1	Cu	Fm3m	O^4_h	cF4	F
A2	W	Im3m	O^9_h	cI2	B
A3	Mg	P6₃/mmc	D^4_{6h}	hP2	H
A4	C	Fd3m	O^7_h	cF8	F
A5	Sn	If₁/amd	D^{19}_{4h}	tI4	U
A6	In	I4/mmm	D^{17}_{4h}	tI2	U
A7	As	$R\bar{3}m$	D^5_{3d}	hR2	R
A8	Se	P3₁21 or P3₂21	$D^4_3 (D^6_3)$	hP3	H
A10	Hg	$R\bar{3}m$	D^5_{3d}	hR1	R
A11	Ga	Cmca	D^{18}_{2h}	oC8	Q
A12	α-Mn	$I\bar{4}3m$	T^3_d	cI58	B
A13	β-Mn	P4₁32	O^7	cP20	C
A15	OW₃	Pm3n	O^3_h	cP8	C
A20	α-U	Cmcm	D^{17}_{2h}	oC4	Q
B1	ClNa	Fm3m	O^5_h	cF8	F
B2	ClCs	Pm3m	O^1_h	cP2	C
B3	SZn	$F\bar{4}3m$	T^2_d	cF8	F
B4	SZn	P6₃mc	C^4_{6v}	hP4	H
B8₁	AsNi	P6₃/mmc	D^4_{6h}	hP4	H
B8₂	InNi₂	P6₃/mmc	D^4_{6h}	hP6	H
B9	HgS	P3₁21 or P3₂21	D^4_3 or D^6_3	hP6	H
B10	OPb	P4/nmm	D^7_{4h}	tP4	T
B11	γ-CuTi	P4/nmm	D^7_{4h}	tP4	T
B13	NiS	$R\bar{3}m$	D^5_{3d}	hR6	R
B16	GeS	Pnma	D^{16}_{2h}	oP8	O
B17	PtS	P4₂/mmc	D^9_{4h}	tP4	T
B18	CuS	P6₃/mmc	D^4_{6h}	hP12	H
B19	AuCd	Pmma	D^5_{2h}	oP4	O
B20	FeSi	P2₁3	T^4	cP8	C
B27	BFe	Pnma	D^{16}_{2h}	oP8	O
B31	MnP	Pnma	D^{16}_{2h}	oP8	O
B32	NaTl	Fd3m	O^7_h	cF16	F
B34	Pds	P4₂/m	C^2_{4h}	tP16	T

TABLE 5
Classification of Crystals (continued)

Strukturbericht symbol	Structure name	Symmetry group International	Symmetry group Schoenflies	Pearson symbol[a]	Standard ASTM E157-82a symbol[b]
1	2	3	4	5	6
B35	CoSn	P6/mmm	D^1_{6h}	hP6	H
B37	SeTl	I4/mcm	D^{18}_{4h}	tI16	U
B_e	CdSb	Pbca	D^{15}_{2h}	oP16	O
B_f (B33)	ξ-BCr	Cmcm	D^{17}_{2h}	oC8	Q
B_g	BMo	I4$_1$/amd	D^{19}_{4h}	tI4	U
B_h	CW	P6m2	D^1_{3h}	hP2	H
B_i	γ-CMo (AsTi)	P6$_3$/mmc	D^4_{6h}	hP8	H
C1	CaF$_2$	Fm3m	O^5_h	cF12	F
C1$_b$	AgAsMg	F43m	T^2_d	cF12	F
C2	FeS$_2$	Pa3	T^6_h	cP12	C
C3	Cu$_2$O	Pn3m	O^4_h	cP6	C
C4	O$_2$Ti	P4$_2$/mnm	D^{14}_{4h}	tP6	T
C6	CdI$_2$	P3m1	D^3_{3d}	hP3	H
C7	MoS$_2$	P6$_3$/mmc	D^4_{6h}	hP6	H
C11$_a$	C$_2$Ca	I4/mmm	D^{17}_{4h}	tI6	U
C11$_b$	MoSi$_2$	I4/mmm	D^{17}_{4h}	tI6	U
C12	CaSi$_2$	R3m	D^5_{3d}	hR6	R
C14	MgZn$_2$	P6$_3$/mmc	D^4_{6h}	hP12	H
C15	Cu$_2$Mg	Fd3m	O^7_h	cF24	F
C15$_b$	AuBe$_5$	F43m or F23	T^2_d or T^2	cF24	F
C16	Al$_2$Cu	I4/mcm	D^{18}_{4h}	tI12	U
C18	FeS$_2$	Pnnm	D^{12}_{2h}	oP6	O
C19	CdCl$_2$	R3m	D^5_{3d}	hR3	R
C22	Fe$_2$P	P26m	D^1_{3h}	hP9	H
C23	Cl$_2$Pb	Pnma	D^{16}_{2h}	oP12	O
C32	AlB$_2$	P6/mmm	D^1_{6h}	hP3	H
C33	Bi$_2$STe$_2$	R3m	D^5_{3d}	hR5	R
C34	AuTe$_2$	C2/m (P2/m)	C^3_{2h} (C^1_{2h})	mC6	N
C36	MgNi$_2$	P6$_3$/mmc	D^4_{6h}	hP24	H
C38	Cu$_2$Sb	P4/nmm	D^7_{4h}	tP6	T
C40	CrSi$_2$	P6$_2$22	D^4_6	hP9	H
C42	SiS$_2$	Ibam	D^{26}_{2h}	oI12	P
C44	GeS$_2$	Fdd2	C^{19}_{2v}	oF72	S
C46	AuTe$_2$	Pma2	C^4_{2v}	oP24	O
C49	Si$_2$Zr	Cmcm	D^{17}_{2h}	oC12	Q
C54	Si$_2$Ti	Fddd	D^{24}_{2h}	oF24	S
C_c	Si$_2$Th	I4$_1$/amd	D^{19}_{4h}	tI12	U
C_e	CoGe$_2$	Aba2	C^{17}_{2v}	oC23	Q
DO$_2$	As$_3$Co	Im3	T^5_h	cI32	B
DO$_3$	BiF$_3$	Fm3m	O^5_h	cF16	F
DO$_9$	O$_3$Re	Pm3m	O^1_h	cP4	C
DO$_{11}$	CFe$_3$	Pnma	D^{16}_{2h}	oP16	O
DO$_{18}$	AsNa$_3$	P6$_3$/mmc	D^4_{6h}	hP8	H
DO$_{19}$	Ni$_3$Sn	P6$_3$/mmc	D^4_{6h}	hP8	H
DO$_{20}$	Al$_3$Ni	Pnma	D^{16}_{2h}	oP16	O
DO$_{21}$	Cu$_3$P	P3c1	D^4_{3d}	hP24	H
DO$_{22}$	Cu$_3$P	I4/mmm	D^{17}_{4h}	tI8	U
DO$_{23}$	Al$_3$Zr	I4/mmm	D^{17}_{4h}	tI16	U
DO$_{24}$	Ni$_3$Ti	P6$_3$/mmc	D^4_{6h}	hP16	H
DO$_c$	SiU$_3$	I4/mcm	D^{18}_{4h}	tI16	U
DO$_e$	Ni$_3$P	I4	S^2_4	tI32	U
D1$_3$	Al$_4$Ba	I4/mmm	D^{17}_{4h}	tI10	U
D1$_a$	MoNi$_4$	I4/m	C^5_{4h}	tI10	U

TABLE 5
Classification of Crystals (continued)

Strukturbericht symbol	Structure name	Symmetry group		Pearson symbol[a]	Standard ASTM E157-82a symbol[b]
		International	Schoenflies		
1	**2**	**3**	**4**	**5**	**6**
$D1_b$	Al_4U	Imma	D^{28}_{2h}	oI20	P
$D1_c$	$PtSn_4$	Aba2	C^{17}_{2v}	oC20	Q
$D1_e$	B_4Th	P4/mbm	D^5_{4h}	tP20	T
$D1_f$	BMn_4	Fddd	D^{24}_{2h}	oF40	S
$D2_1$	B_6Ca	Pm3m	O^1_h	cP7	C
$D2_3$	$NaZn_{13}$	Fm3m	O^5_h	cF112	F
$D2_b$	$Mn_{12}Th$	I4/mmm	D^{17}_{4h}	tI26	U
$D2_c$	MnU_6	I4/mcm	D^{18}_{4h}	tI28	U
$D2_d$	$CaCu_5$	P6/mmm	D^1_{6h}	hP6	H
$D2_f$	$B_{12}U$	Fm3m	O^5_h	cF52	F
$D2_h$	Al_6Mn	Cmcm	D^{17}_{2h}	oC28	Q
$D5_1$	$\alpha\text{-}Al_2O_3$	R$\bar{3}$c	D^6_{3d}	hR10	R
$D5_2$	La_2O_3	P$\bar{3}$m1	D^3_{3d}	hP5	H
$D5_3$	Mn_2O_3	Ia3	T^7_h	cI80	B
$D5_8$	S_3Sb_2	Pnma	D^{16}_{2h}	oP20	O
$D5_9$	P_2Zn_3	P4$_2$/mmc	D^9_{4h}	tP40	T
$D5_{10}$	C_2C_3	Pnma	D^{16}_{2h}	oP20	O
$D5_{13}$	Al_3Ni_2	P$\bar{3}$m1	D^3_{3d}	hP5	H
$D5_a$	Si_2U_3	P4/mbm	D^5_{4h}	tP10	T
$D5_c$	C_3Pu_2	I$\bar{4}$3d	T^6_d	cI40	B
$D7_1$	Al_4C_3	R$\bar{3}$m	D^5_{3d}	hR7	R
$D7_3$	P_4Th_3	I$\bar{4}$3d	T^6_d	cI28	B
$D7_b$	B_4Ta_3	Immm	D^{25}_{2h}	oI14	P
$D8_1$	Fe_3Zn_{10}	Im3m	O^9_h	cI52	B
$D8_2$	Cu_5Zn_8	I$\bar{4}$3m	T^3_d	cI52	B
$D8_3$	Al_4Cu_9	P$\bar{4}$3m	T^1_d	cP52	C
$D8_4$	C_6Cr23	Fm3m	O^5_h	cF116	F
$D8_5$	Fe_7W_6	R$\bar{3}$m	D^5_{3d}	hR13	R
$D8_6$	$Cu_{15}Si_4$	I$\bar{4}$3m	T^3_d	cI76	B
$D8_8$	Mn_5Si_3	P6$_3$/mcm	D^3_{6h}	hP16	H
$D8_9$	Co_9S_8	Fm3m	O^5_h	cF68	F
$D8_{10}$	Al_8Cr_5	R3m	C^5_{3v}	hR26	R
$D8_{11}$	Al_5Co_2	P6$_3$/mcm	D^3_{6h}	hP28	H
$D8_a$	$Mn_{23}Th_6$	Fm3m	O^5_h	cF116	F
$D8_b$	σ-phase of Cr-Fe	p$\bar{4}_2$/mnm	D^{14}_{4h}	tP30	T
$D8_e$	$(Al,Zn)_{49}Mg_{32}$	Im3	T^5_h	cI162	B
$D8_f$	Ge_7Ir_3	Im3m	O^9_h	cI40	B
$D8_h$	B_5W_2	P6$_3$/mmc	D^4_{6h}	hP14	H
$D8_i$	B_5Mo_2	R$\bar{3}$m	D^5_{3d}	hR7	R
$D8_l$	B_3Cr_5	I4/mcm	D^{18}_{4h}	tI32	U
$D8_m$	Si_3W_5	I4/mcm	D^{18}_{4h}	tI32	U
$D10_1$	C_3Cr_7	P31c	C^4_{3v}	hP80	H
$D10_2$	Fe_3Th_7	P6$_3$mc	C^4_{6v}	hP20	H
$E0_1$	$ClFPb$	P4/nmm	D^7_{4h}	tP6	T
$E1_1$	$CuFeS_2$	I$\bar{4}$2d	D^{12}_{2d}	tI16	U
$E2_1$	CaO_3Ti	Pm3m	O^1_h	cP5	C
$E2_4$	S_3Sn_2	Pnma	D^{16}_{2h}	oP20	O
$E3$	Al_2CdS_4	I$\bar{4}$	S^2_4	tI14	U
$E9_3$	$SiFe_3W_3$	Fd3m	O^7_h	cF112	F
$E9_a$	Al_7Cu_2Fe	P4/mnc	D^6_{4h}	tP40	T
$E9_b$	$AlLi_3N_2$	Ia3	T^7_h	cI96	B
$F0_1$	$NiSSb$	P2$_1$3	T^4	cP12	C
$F5_1$	$CrNaS_2$	R3m or R32	D^5_{3d} or D^7_3	hR4	R
$F5_6$	CuS_2Sb	Pnma	D^{16}_{2h}	oP16	O

TABLE 5
Classification of Crystals (continued)

Strukturbericht symbol	Structure name	Symmetry group		Pearson symbol[a]	Standard ASTM E157-82a symbol[b]
		International	Schoenflies		
1	2	3	4	5	6
$H1_1$	Al_2MgO_4	Fd3m	O^7_h	cF56	F
$H2_4$	Cu_3S_4V	P$\bar{4}$3m	T^1_d	cP8	C
$H2_5$	$AsCu_3S_4$	Pmn2$_1$	C^7_{2v}	oP16	O
$L1_0$	AuCu	P4/mmm	D^1_{4h}	tP4	T
$L1_2$	$AlCu_3$	Pm3m	O^1_h	cP4	C
$L2_1$	$AlCu_2Mn$	Fm3m	O^5_h	cF16	F
$L2_2$	Sb_2Tl_7	Im3m	O^9_h	cI54	B
$L'2_b$	H_2Th	I4/mmm	D^{17}_{4h}	tI6	U
$L'3$	Fe_2N	P6$_3$/mmc	D^4_{6h}	hP3	H
$L6_0$	$CuTi_3$	P4/mmm	D^1_{4h}	tP4	T

[a]   The first letter denotes the crystal system: triclinic (a), monoclinic (m), orthorhombic (o), tetragonal (t), hexagonal (h) and cubic (c). Trigonal (rhombohedral) system is presented by combination hR. The second letter of Pearson's symbol denotes lattice type: primitive (P), edge-(base-) centered (C), body-centered (I) or face-centered (F). The following number denotes amount of atoms in the crystal unit cell.

[b]   Standard ASTM E157-82a has the Bravais lattices designations as following: C — primitive cubic; B — body-centered cubic; F — face-centered cubic; T — primitive tetragonal; U — body-centered tetragonal; R — rhombohedral; H — hexagonal; O — primitive orthorhombic; P — body-centered orthorhombic; Q — base-centered orthorhombic; S — face-centered orthorhombic; M — primitive monoclinic; N — centered monoclinic; A — triclinic.

REFERENCES

1.   A. Schoenflies, *Kristallsysteme und Kristallstructur*, Leipzig, 1891.
2.   E. S. Fedorow, Zusammenstellung der kristallographischen Resultate, *Zs. Krist.*, 20, 1892.
3.   P. Groth, *Elemente der physikalischen und chemischen Krystallographie*, R. Oldenbourg, München/Berlin, 1921.
4.   N. V. Belov, *Class Method of Deriving Space Groups of Symmetry*, Trudy Instituta Kristallodraffi imeni Fedorova (Transactions of the Fedorov Inst. of Crystallography), 5, 25, 1951, in Russian.
5.   W. B. Pearson, *Handbook of Lattice Spacings and Structures of Metals and Alloys*, Vol. 1, Pergamon Press, 1958; Vol. 2, 1967.
6.   Ch. Kittel, *Introduction to Solid State Physics*, John Wiley & Sons, 1956.
7.   G. S. Zhdanov, *Fizika Tverdogo Tela (Solid State Physics)*, Moscow University Press, 1962, in Russian.
8.   M. J. Buerger, *Elementary Crystallography*, John Wiley & Sons, 1963.
9.   F. D. Bloss, *Crystallography & Crystal Chemistry*, Holt, Rinehart & Winston, 1971.
10.   T. Janssen, *Crystallographic Groups*, North-Holland/American Elsevier, 1973.
11.   M. P. Shaskolskaya, *Kristallografiya (Crystallography)*, Vysshaya Shkola, Moscow, 1976, in Russian.
12.   T. Hahn, Ed., Internat. *Tables for Crystallography*, Vol. A, D. Reidel Publishing, Boston, 1983.
13.   Crystal Data. Determinative Tables, Volumes 1—6, 1966—1983, JCPDS-Intern Centre for Diffraction Data and U.S. Dept. of Commerce.
14.   R. W. G. Wyckoff, *Crystal Structures,* 2nd ed., Volumes 1–6, Interscience, New York, 1963.
15.   C.J. Bradley and A.P. Cracknell, *The Mathematical Theory of Symmetry in Solids,* Clarendon Press, Oxford, 1972.
16.   International Tables for Crystallography. Volume A, *Space–Group Symmetry,* T. Hahn, Ed., 1989; Volume B, *Reciprocal Space,* U. Schmueli, Ed.; Volume C, *Mathematical, Physical and Chemical Tables,* A. J. C. Wilson, Ed., Kluwer Academic Publishers, Dordrecht, 1989.
17.   G. R. Desiraju, *Crystal Engineering: The Design of Organic Solids,* Elsevier, Amsterdam, 1989.
18.   M. Senechal, *Crystalline Symmetries: An Informal Mathematical Introduction,* Adam Hilger Publ., Bristol, 1990.
19.   C. Hammond, *Introduction to Crystallography,* Oxford University Press, 1990.
20.   N.W. Alcock, *Bonding and Structure: Structural Principles in Inorganic and Organic Chemistry,* Ellis Norwood Publ., 1990.
21.   T. C. W. Mak and G. D. Zhou. *Crystallography in Modern Chemistry: A Resource Book of Crystal Structures,* Wiley–Interscience, New York, 1992.
22.   S. C. Abrahams, K. Mirsky, and R. M. Nielson, *Acta Cryst,* B52, 806 (1996); B52, 1057 (1996).
23.   C. Marcos, A. Panalague, D. B. Morciras, S. Garcia–Granda and M. R. Dias. *Acta Cryst,* B52, 899 (1996).

Crystallographic Computing

24.   A. C. Larson, *Crystallographic Computing,* Manksgaard, Copenhagen, 1970.
25.   G. M. Sheldrick, SHELXS86. Crystallographic Computing 3, Clarendon Press, Oxford, 1986; SHELXL93. Program for the Refinement of Crystal Structures, University of Göttingen Press, 1993.
26.   Inorganic Crystal Structure Database, CD–ROM. Sci. Inf. Service. E-mail: SISI@Delphi.com.

IONIC RADII IN CRYSTALS

Ionic radii are a useful tool for predicting and visualizing crystal structures. This table lists a set of ionic radii R_i in Å units for the most common coordination numbers CN of positive and negative ions. The values are based on experimental crystal structure determinations, supplemented by empirical relationships, and theoretical calculations. The notation sq after the coordination number indicates a square configuration, while py indicates pyramidal.

The advice of Howard T. Evans and Marvin J. Weber in preparing this table is appreciated.

REFERENCES

1. Shannon, R. D., *Acta Crystallogr.* A32, 751, 1976.
2. Jia, Y.Q., *J. Solid State Chem.* 95, 184, 1991.

Ion	CN	R_i/Å	Ion	CN	R_i/Å	Ion	CN	R_i/Å
Anions			C^{+4}	4	0.15	Er^{+3}	6	0.89
F^{-1}	6	1.33		6	0.16		8	1.00
Cl^{-1}	6	1.81	Ca^{+2}	6	1.00	Eu^{+2}	6	1.17
Br^{-1}	6	1.96		8	1.12		8	1.25
I^{-1}	6	2.20		10	1.23		10	1.35
OH^{-1}	4	1.35		12	1.34	Eu^{+3}	6	0.95
	6	1.37	Cd^{+2}	4	0.78		8	1.07
O^{-2}	2	1.21		6	0.95	F^{+7}	6	0.08
	6	1.40		8	1.10	Fe^{+2}	4	0.63
	8	1.42		12	1.31		6	0.61
S^{-2}	6	1.84	Ce^{+3}	6	1.01		8	0.92
Se^{-2}	6	1.98		8	1.14	Fe^{+3}	4	0.49
Te^{-2}	6	2.21		10	1.25		6	0.55
				12	1.34		8	0.78
Cations			Ce^{+4}	6	0.87	Fr^{+1}	6	1.80
Ac^{+3}	6	1.12		8	0.97	Ga^{+3}	4	0.47
Ag^{+1}	4	1.00		10	1.07		6	0.62
	6	1.15		12	1.14	Gd^{+3}	6	0.94
	8	1.28	Cf^{+3}	6	0.95		8	1.05
Ag^{+2}	4sq	0.79	Cf^{+4}	6	0.82	Ge^{+2}	6	0.73
	6	0.94		8	0.92	Ge^{+4}	4	0.39
Al^{+3}	4	0.39	Cl^{+5}	3py	0.12		6	0.53
	5	0.48	Cl^{+7}	4	0.08	Hf^{+4}	4	0.58
	6	0.54		6	0.27		6	0.71
Am^{+3}	6	0.98	Cm^{+3}	6	0.97		8	0.83
	8	1.09	Cm^{+4}	6	0.85	Hg^{+1}	6	1.19
Am^{+4}	6	0.85		8	0.95	Hg^{+2}	2	0.69
	8	0.95	Co^{+2}	4	0.56		4	0.96
As^{+3}	6	0.58		6	0.65		6	1.02
As^{+5}	4	0.34		8	0.90		8	1.14
	6	0.46	Co^{+3}	6	0.55	I^{+5}	3py	0.44
Au^{+1}	6	1.37	Cr^{+2}	6	0.73		6	0.95
Au^{+3}	4sq	0.64	Cr^{+3}	6	0.62	I^{+7}	4	0.42
	6	0.85	Cr^{+4}	4	0.41		6	0.53
Ba^{+2}	6	1.35		6	0.55	In^{+3}	4	0.62
	8	1.42	Cr^{+6}	4	0.26		6	0.80
	12	1.61		6	0.44	Ir^{+3}	6	0.68
Be^{+2}	4	0.27	Cs^{+1}	6	1.67	Ir^{+4}	6	0.63
	6	0.45		8	1.74	Ir^{+5}	6	0.57
Bi^{+3}	5	0.96		10	1.81	K^{+1}	4	1.37
	6	1.03		12	1.88		6	1.38
	8	1.17	Cu^{+1}	2	0.46		8	1.51
Bi^{+5}	6	0.76		4	0.60		12	1.64
Bk^{+3}	6	0.96		6	0.77	La^{+3}	6	1.03
Bk^{+4}	6	0.83	Cu^{+2}	4sq	0.57		8	1.16
	8	0.93		6	0.73		10	1.27
Br^{+5}	3py	0.31	Dy^{+2}	6	1.07		12	1.36
Br^{+7}	4	0.25		8	1.19	Li^{+1}	4	0.59
	6	0.39	Dy^{+3}	6	0.91		6	0.76
				8	1.03			

Ion	CN	R_i/Å	Ion	CN	R_i/Å	Ion	CN	R_i/Å
	8	0.92		6	0.78	Sr^{+2}	6	1.18
Lu^{+3}	6	0.86		8	0.94		8	1.26
	8	0.97	Pd^{+2}	4sq	0.64		10	1.36
Mg^{+2}	4	0.57		6	0.86		12	1.44
	6	0.72	Pd^{+3}	6	0.76	Ta^{+3}	6	0.72
	8	0.89	Pd^{+4}	6	0.62	Ta^{+4}	6	0.68
Mn^{+2}	4	0.66	Pm^{+3}	6	0.97	Ta^{+5}	6	0.64
	6	0.83		8	1.09	Tb^{+3}	6	0.92
	8	0.96	Po^{+4}	6	0.97		8	1.04
Mn^{+3}	6	0.58	Pr^{+3}	6	0.99	Tb^{+4}	6	0.76
Mn^{+4}	4	0.39		8	1.13		8	0.88
	6	0.53	Pr^{+4}	6	0.85	Tc^{+4}	6	0.65
Mn^{+5}	4	0.33		8	0.96	Te^{+4}	4	0.66
Mn^{+6}	4	0.26	Pt^{+2}	4sq	0.60		6	0.97
Mn^{+7}	4	0.25		6	0.80	Te^{+6}	4	0.43
Mo^{+3}	6	0.69	Pt^{+4}	6	0.63		6	0.56
Mo^{+4}	6	0.65	Pu^{+3}	6	1.00	Th^{+4}	6	0.94
Mo^{+5}	4	0.46	Pu^{+4}	6	0.86		8	1.05
	6	0.61	Pu^{+5}	6	0.74		10	1.13
Mo^{+6}	4	0.41	Pu^{+6}	6	0.71		12	1.21
	6	0.59	Ra^{+2}	8	1.48	Ti^{+2}	6	0.86
	7	0.73		12	1.70	Ti^{+3}	6	0.67
N^{+3}	6	0.16	Rb^{+1}	6	1.52	Ti^{+4}	4	0.42
N^{+5}	6	0.13		8	1.61		6	0.61
Na^{+1}	4	0.99		10	1.66		8	0.74
	6	1.02		12	1.72	Tl^{+1}	6	1.50
	8	1.18	Re^{+4}	6	0.63		8	1.59
	9	1.24	Re^{+5}	6	0.58		12	1.70
	12	1.39	Re^{+6}	6	0.55	Tl^{+3}	4	0.75
Nb^{+3}	6	0.72	Re^{+7}	4	0.38		6	0.89
	8	0.79		6	0.53		8	0.98
Nb^{+4}	6	0.68	Rh^{+3}	6	0.67	Tm^{+2}	6	1.01
Nb^{+5}	4	0.48	Rh^{+4}	6	0.60		7	1.09
	6	0.64	Rh^{+5}	6	0.55	Tm^{+3}	6	0.88
	8	0.74	Ru^{+3}	6	0.68		8	0.99
Nd^{+3}	6	0.98	Ru^{+4}	6	0.62	U^{+3}	6	1.03
	8	1.12	Ru^{+5}	6	0.57	U^{+4}	6	0.89
	9	1.16	Ru^{+7}	4	0.38		8	1.00
	12	1.27	Ru^{+8}	4	0.36		12	1.17
Ni^{+2}	4sq	0.49	S^{+4}	6	0.37	U^{+5}	6	0.76
	6	0.69	S^{+6}	4	0.12	U^{+6}	2	0.45
Ni^{+3}	6	0.56		6	0.29		4	0.52
Np^{+3}	6	1.01	Sb^{+3}	4py	0.76		6	0.73
Np^{+4}	6	0.87		6	0.76		8	0.86
Np^{+5}	6	0.75	Sb^{+5}	6	0.60	V^{+2}	6	0.79
Np^{+6}	6	0.72	Sc^{+3}	6	0.75	V^{+3}	6	0.64
Os^{+4}	6	0.63		8	0.87	V^{+4}	5	0.53
Os^{+5}	6	0.58	Se^{+4}	6	0.50		6	0.58
Os^{+6}	6	0.55	Se^{+6}	4	0.28		8	0.72
Os^{+8}	4	0.39		6	0.42	V^{+5}	4	0.36
P^{+5}	4	0.17	Si^{+4}	4	0.26		5	0.46
	6	0.38		6	0.40		6	0.54
Pa^{+3}	6	1.04	Sm^{+2}	6	1.19	W^{+4}	6	0.66
Pa^{+4}	6	0.90		8	1.27	W^{+5}	6	0.62
Pa^{+5}	6	0.78	Sm^{+3}	6	0.96	W^{+6}	4	0.42
Pb^{+2}	6	1.19		8	1.08		5	0.51
	8	1.29		12	1.24		6	0.60
	10	1.40	Sn^{+4}	4	0.55	Y^{+3}	6	0.90
	12	1.49		6	0.69		8	1.02
Pb^{+4}	4	0.65		8	0.81		9	1.08

Ion	CN	R_i/Å	Ion	CN	R_i/Å	Ion	CN	R_i/Å
Yb^{+2}	6	1.02	Zn^{+2}	4	0.60		6	0.72
	8	1.14		6	0.74		8	0.84
Yb^{+3}	8	0.99		8	0.90		9	0.89
	9	1.04	Zr^{+4}	4	0.59			

POLARIZABILITIES OF ATOMS AND IONS IN SOLIDS
H. P. R. Frederikse

The polarization of a solid dielectric medium, P, is defined as the dipole moment per unit volume averaged over the volume of a crystal cell. A component of P can be expanded as a function of the electric field E:

$$P_i = \sum_j a_j E_j + \sum_{jk} b_{jk} E_j E_k$$

For relatively small electric fields in isotropic substances $P = \chi_e E$, where χ_e is the electric susceptibility. If the medium is made up of N atoms (or ions) per unit volume, the polarization is $P = N p_m$ where p_m is the average dipole moment per atom. The polarizability α can be defined as $p_m = \alpha E_0$, where E_0 is the local field at the position of the atom. Using the Lorentz method to calculate the local field one finds:

$$P = N\alpha(E + 4\pi P) = \chi_e E$$

Together with the definition of the dielectric constant (relative permittivity) , $\varepsilon = 1 + 4\pi\chi_e$, this leads to:

$$\alpha = \frac{3}{4\pi N}\left(\frac{\varepsilon - 1}{\varepsilon + 2}\right)$$

This expression is known as the Clausius-Mossotti equation.

The total polarization associated with atoms, ions, or molecules is due to three different sources:

1. Electronic polarization arises because the center of the local electronic charge cloud around the nucleus is displaced under the action of the field: $P_e = N\alpha_e E_0$ where α_e is the *electronic polarizability*.
2. Ionic polarization occurs in ionic materials because the electric field displaces cations and anions in opposite directions: $P_i = N\alpha_i E_0$, where α_i is the *ionic polarizability*.
3. Orientational polarization can occur in substances composed of molecules that have permanent electric dipoles. The alignment of these dipoles depends on temperature and leads to an *orientational polarizability* per molecule: $\alpha_{or} = p^2/3kT$, where p is the permanent dipole moment per molecule, k is the Boltzmann constant, and T is the temperature.

Because of the different nature of these three polarization processes the response of a dielectric solid to an applied electric field will strongly depend on the frequency of the field. The resonance of the electronic excitation in insulators (dielectrics) takes place in the ultraviolet part of the spectrum; the characteristic frequency of the lattice vibrations is located in the infrared, while the orientation of dipoles requires fields of much lower frequencies (below 10^{10} Hz). This response to electric fields of different frequencies is shown in Figure 1. Values of the electronic polarizabilities for selected atoms and ions are given in Table 1.

REFERENCES

1. Kittel, C., *Introduction to Solid State Physics*, Fourth Edition, John Wiley & Sons, New York, 1971.
2. Lerner, R.G., and Trigg, G.L., Editors, *Encyclopedia of Physics, Second Edition*, VCH Publishers, New York, 1990.
3. Ralls, K.M., Courtney, T.H., and Wulff, J., *An Introduction to Materials Science and Engineering*, John Wiley & Sons, New York, 1976.

POLARIZABILITIES OF ATOMS AND IONS IN SOLIDS (continued)

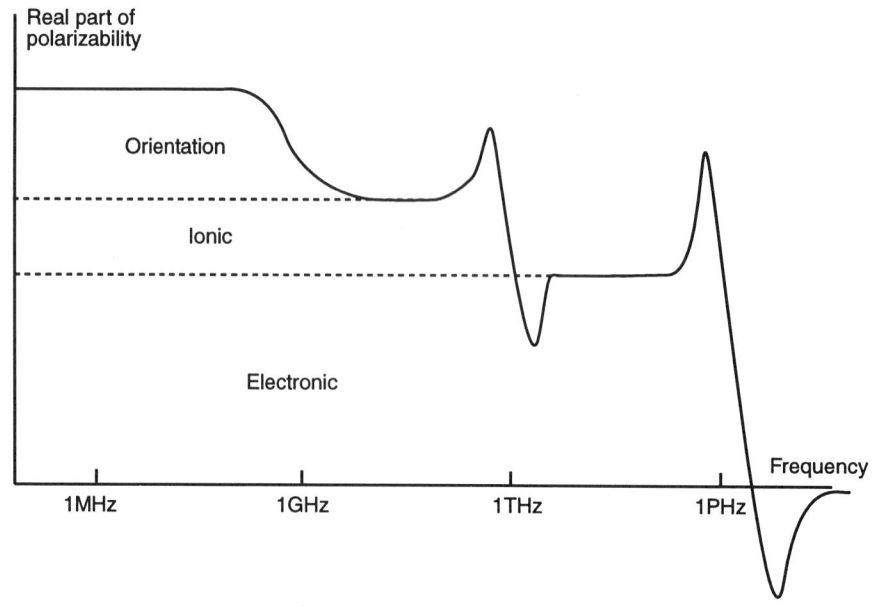

Figure 1. Schematic graph of the frequency dependence of the different contributions to polarizability.

TABLE 1
Electronic Polarizabilities in Units of 10^{-24} cm^3

							He 0.201
Li$^+$ 0.029	Be^{2+} 0.008	B^{3+} 0.003	C^{4+} 0.0013	O^{2-} 3.88	F$^-$ 1.04	Ne 0.39	
Na$^+$ 0.179	Mg^{2+} 0.094	Al^{3+} 0.052	Si^{4+} 0.0165	S^{2-} 10.2	Cl$^-$ 3.66	Ar 1.62	
K$^+$ 0.83	Ca^{2+} 0.47	Sc^{3+} 0.286	Ti^{4+} 0.185	Se^{2-} 10.5	Br$^-$ 4.77	Kr 2.46	
Rb$^+$ 1.40	Sr^{2+} 0.86	Y^{3+} 0.55	Zr^{4+} 0.37	Te^{2-} 14.0	I$^-$ 7.1	Xe 3.99	
Cs$^+$ 2.42	Ba^{2+} 1.55	La^{3+} 1.04	Ce^{4+} 0.73				

Data from Pauling, L., *Proc. R. Soc. London*, A114, 181, 1927. See also Jaswal, S.S. and Sharma, T.P., *J. Phys. Chem. Solids*, 34, 509, 1973.
Values are appropriate for cgs units. To convert to SI, use the relation
$$\alpha(SI)/C\ m^2V^{-1} = 1.11265 \cdot 10^{-16}\ \alpha(cgs)/cm^3$$

CRYSTAL STRUCTURES AND LATTICE PARAMETERS OF ALLOTROPES OF THE ELEMENTS

H. W. King

The crystal structures of the allotropic forms of the elements are presented in terms of the Pearson symbol, the Strukturbericht designation, and the prototype of the structure. The temperatures of the phase transformations are listed in degrees Celsius and the pressures are in GPa. A consistent nomenclature is used, whereby all allotropes are labeled by Greek letters. The lattice parameters of the unit cells are given in nanometers (nm) and are considered to be accurate to ±2 in the last reported digit.

This compilation is restricted to changes in crystal structure that occur as a result of a change in temperature or pressure. Low-temperature structures are included for the diatomic and rare gases, which show many similarities with respect to the metallic elements. The elements identified with an asterisk (*) have polymorphic structures based on different molecular configurations. The crystal data given for these elements refer to the most stable structure at room temperature.

Reprinted with the permission of ASM International from T. B. Massalski, Ed., Binary Alloy Phase Diagrams, ASM International, Metals Park, Ohio, 1986; certain data on rare earth elements were provided by K. A. Gschneidner.

Element	Temperature, °C	Pressure, GPa	Pearson symbol	Space group	Strukturbericht designation	Prototype	a	b	c	Comment, c/a or α or β
Ac	25	atm	cF4	Fm3m	A1	Cu	0.5311	...	...	...
Ag	25	atm	cF4	Fm3m	A1	Cu	0.40857	...	...	...
αAl	25	atm	cF4	Fm3m	A1	Cu	0.40496	...	...	...
βAl	25	>20.5	hP2	P6₃/mmc	A3	Mg	0.2693	...	0.4398	1.6331
α'Am	25	atm	hP4	P6₃/mmc	A3'	αLa	0.34681	...	1.1241	2*1.621
αAm	>769	atm	cF4	Fm3m	A1	Cu	0.4894	...	...	...
βAm	>1074	atm	cI2	Im3m	A2	W	?	...	...	...
γAm	25	>15	oC4	Cmcm	A20	αU	0.3063	0.5968	0.5169	...
αAr	<−189.35	atm	cF4	Fm3m	A1	Cu	0.5316	...	...	...
(βAr)	<−189.40	atm	hP2	P6₃/mmc	A3	Mg	0.3760	...	0.6141	1.633
αAs	25	atm	hR2	R3m	A7	αAs	0.41319	...	...	α = 54.12°
εAs	>448	atm	oC8	Cmca	...	P (black)	0.362	1.085	0.448	...
Au	25	atm	cF4	Fm3m	A1	Cu	0.40782	...	...	...
βB	25	atm	hR105	R3m	...	βB	1.017	...	...	α = 65.12°
αBa	25	atm	cI2	Im3m	A2	W	0.50227	...	...	...
βBa	25	>5.33	hP2	P6₃/mmc	A3	Mg	0.3901	...	0.6154	1.5775
γBa	25	>23	?	?	...	...	...	...	...	...
αBe	25	atm	hP2	P6₃/mmc	A3	Mg	0.22859	...	0.35845	1.5681
βBe	>1270	atm	cI2	Im3m	A2	W	0.25515	...	...	...
γBe	25	>9.3	?	...	...	...	...	...	...	...
αBi	25	atm	hR2	R3m	A7	αAs	0.47460	...	...	α = 57.23°
βBi	25	>2.6	mC4	C2/m	...	βBi	0.6674	0.6117	0.3304	β = 110.33°
γBi	25	>3.0	mP3	?	...	...	0.605	0.42	0.465	β = 85.33°
σBi	25	>4.3	?	?	...	...	...	...	...	...
εBi	25	>6.5	?	?	...	...	...	...	...	...
ζBi	25	>9.0	cI2	Im3m	A2	W	0.3800	...	...	...
αBk	25	atm	hP4	P6₃/mmc	A3'	αLa	0.3416	...	1.1069	2*1.620
βBk	>977	atm	cF4	Fm3m	A1	Cu	0.4997	...	...	...
Br	<7.25	atm	oC8	Cmca	...	Cl	0.668	0.449	0.874	...
C (graphite)	25	atm	hP4	P6₃/mmc	A9	C (graphite)	0.24612	...	0.6709	2.7258
C (diamond)	25	>60	cF8	Fd3m	A4	C (diamond)	0.35669	...	...	...
C (hd)	25	HP	hP4	P6₃/mmc	...	C (hd)	0.2522	...	0.4119	1.633
αCa	25	atm	cF4	Fm3m	A1	Cu	0.55884	...	...	...
βCa	>443	atm	cI2	Im3m	A2	W	0.4480	...	...	...
γCa	25	>1.5	?	...	...	...	...	...	...	...
Cd	25	atm	hP2	P6₃/mmc	A3	Mg	0.29793	...	0.56196	1.8862
αCe	<−177	atm	cF4	Fm3m	A1	Cu	0.485	...	...	...
βCe	25	atm	hP4	P6₃/mmc	A3'	αLa	0.36810	...	1.1857	2*1.611
γCe	25	atm	cF4	Fm3m	A1	Cu	0.51610	...	...	...
δ-Ce	>726	atm	cI2	Im3m	A2	W	0.412	...	...	...
α'Ce	25	>5.4	oC4	Cmcm	A20	αU	0.3049	0.5998	0.5215	...
αCf	25	atm	hP4	P6₃/mmc	A3'	αLa	0.339	...	1.1015	2*1.625
βCf	>590	atm	cF4	Fm3m	A1	Cu	?	...	...	...
Cl	25	atm	oC8	Cmca	...	Cl	0.624	0.448	0.826	...
αCm	25	atm	hP4	P6₃/mmc	A3'	αLa	0.3496	...	1.1331	2*1.621
βCm	>1277	atm	cF4	Fm3m	A1	Cu	0.4382	...	...	...
εCo	25	atm	hP2	P6₃/mmc	A3	Mg	0.25071	...	0.40686	1.6228
αCo	>422	atm	cF4	Fm3m	A1	Cu	0.35447	...	...	...
αCr	25	atm	cI2	Im3m	A2	W	0.38848	...	...	...
α'Cr	25	HP	tI2	I4/mmm	...	α'Cr	0.2882	...	0.2887	1.002
αCs	25	atm	cI2	Im3m	A2	W	0.6141	...	...	...
βCs	25	>2.37	cF4	Fm3m	A1	Cu	0.6465	...	...	...
β'Cs	25	>4.22	cF4	Fm3m	A1	Cu	0.5800	...	...	...
γCs	25	>4.27	?	...	...	...	...	...	...	...
Cu	25	atm	cF4	Fm3m	A1	Cu	0.36146	...	...	...
α'Dy	<−187	atm	oC4	Cmcm	...	α'Dy	0.3595	0.6184	0.5678	...
αDy	25	atm	hP2	P6₃/mmc	A3	Mg	0.35915	...	0.56501	1.5732

CRYSTAL STRUCTURES AND LATTICE PARAMETERS OF ALLOTROPES OF THE ELEMENTS
(continued)

Element	Temperature, °C	Pressure, GPa	Pearson symbol	Space group	Strukturbericht designation	Prototype	Lattice parameters, nm a	b	c	Comment, c/a or α or β
βDy	>1381	atm	cI2	Im3m	A2	W	0.403	...	...	...
γDy	25	>7.5	hR3	R3m	...	αSm	0.3436	...	2.483	4.5*1.606
Er	25	atm	hP2	P6₃/mmc	A3	Mg	0.35592	...	0.55850	1.5692
αEs	25	atm	hP4	P6₃/mmc	A3′	αLa	?	...	...	...
βEs	?	atm	cF4	Fm3m	A1	Cu	?	...	...	...
Eu	25	atm	cI2	Im3m	A2	W	0.45827	...	...	...
αF	<−227.6	atm	mC8	C2/c	...	αF	0.550	0.338	0.728	β = 102.17°
βF	<−219.67	atm	cP16	Pm3n	...	γO	0.667	...	...	...
αFe	25	atm	cI2	Im3m	A2	W	0.28665	...	...	...
γFe	>912	atm	cF4	Fm3m	A1	Cu	0.36467	...	...	...
σFe	>1394	atm	cI2	Im3m	A2	W	0.29315	...	...	...
εFe	25	>13	hP2	P6₃/mmc	A3	Mg	0.2468	...	0.396	1.603
αGa	25	atm	oC8	Cmca	A11	αGa	0.45186	0.76570	0.45258	...
βGa	25	>1.2	tI2	I4/mmm	A6	In	0.2808	...	0.4458	1.588
γGa	−53	>3.0	oC40	Cmcm	...	γGa	1.0593	1.3523	0.5203	...
αGd	25	atm	hP2	P6₃/mmc	A3	Mg	0.36336	...	0.57810	1.5910
βGd	>1235	atm	cI2	Im3m	A2	W	0.406	...	...	...
γGd	25	>3.0	hR3	R3m	...	αSm	0.361	...	2.603	4*1.60
αGe	25	atm	cF8	Fd3m	A4	C (diamond)	0.56574	...	...	...
βGe	25	>12	tI4	I4₁/amd	A5	βSn	0.4884	...	0.2692	0.551
γGe	25	>12 → atm	tP12	P4₃2₁2	...	σGe	0.593	...	0.698	1.18
σGe	LT	>12	cI16	Im3m	...	γSi	0.692	...	...	...
αH	<−271.9	atm	cF4	Fm3m	A1	Cu	0.5338	...	...	...
βH	<−259.34	atm	hP2	P6₃/mmc	A3	Mg	0.3776	...	0.6162	1.632
αHe	<−268.94	atm	hP2	P6₃/mmc	A3	Mg	0.3555	...	0.5798	1.631
βHe	>−258	0.125	cF4	Fm3m	A1	Cu	0.4240	...	...	...
γHe	<−271.47	0.03	cI2	Im3m	A2	W	0.4110	...	...	...
αHf	25	atm	hP2	P6₃/mmc	A3	Mg	0.31946	...	0.50510	1.5811
βHf	>1995	atm	cI2	Im3m	A2	W	0.3610	...	...	...
αHg	<−38.84	atm	hR1	R3m	A10	αHg	0.3005	...	...	α = 70.53°
βHg	<−194	HP	tI2	I4/mmm	...	βHg	0.3995	...	0.2825	0.707
γHg	<−194	c.w.	hR1	?	...	...	...	...	...	...
αHo	25	atm	hP2	P6₃/mmc	A3	Mg	0.35778	...	0.56178	1.5702
βHo	25	>7.5	hR3	R3m	...	αSm	0.334	...	2.45	4.5*1.63
I	25	atm	oC8	Cmca	...	Cl	0.72697	0.47903	0.97942	...
In	25	atm	tI2	I4/mmm	A6	In	0.3253	...	0.49470	1.5210
Ir	25	atm	cF4	Fm3m	A1	Cu	0.38392	...	...	...
K	25	atm	cI2	Im3m	A2	W	0.5321	...	...	...
Kr	<−157.39	atm	cF4	Fm3m	A1	Cu	0.5810	...	...	...
αLa	25	atm	hP4	P6₃/mmc	A3′	αLa	0.37740	...	1.2171	2*1.6125
βLa	>310	atm	cF4	Fm3m	A1	Cu	0.5303	...	...	...
γLa	>865	atm	cI2	Im3m	A2	W	0.426	...	...	...
β′La	25	>2.0	cF4	Fm3m	A1	Cu	0.517	...	...	...
αLi	<−193	atm	hP2	P6₃/mmc	A3	Mg	0.3111	...	0.5093	1.637
βLi	25	atm	cI2	Im3m	A2	W	0.35093	...	...	...
γLi	<−201	c.w.	cF4	Fm3m	A1	Cu	0.4388	...	...	...
Lu	25	atm	hP2	P6₃/mmc	A3	Mg	0.35052	...	0.55494	1.5832
Mg	25	atm	hP2	P6₃/mmc	A3	Mg	0.32094	...	0.52107	1.6236
αMn	25	atm	cI58	I43m	A12	αMn	0.89126	...	...	...
βMn	>710	atm	cP20	P4₁32	A13	βMn	0.63152	...	...	...
γMn	>1079	atm	cF4	Fm3m	A1	Cu	0.3860	...	...	...
σMn	>1143	atm	cI2	Im3m	A2	W	0.3080	...	...	...
Mo	25	atm	cI2	Im3m	A2	W	0.31470	...	...	...
αN	<−237.6	atm	cP8	Pa3	...	αN	0.5661	...	...	...
βN	<−210.00	atm	hP4	P6₃/mmc	...	βN	0.4050	...	0.6604	1.631
γN	<−253	>3.3	tP4	P4₂/mnm	...	γN	0.3957	...	0.5109	1.291
αNa	<−233	atm	hP2	P6₃/mmc	A3	Mg	0.3767	...	0.6154	1.634
βNa	25	atm	cI2	Im3m	A2	W	0.42906	...	...	...
Nb	25	atm	cI2	Im3m	A2	W	0.33004	...	...	...
αNd	25	atm	hP4	P6₃/mmc	A3′	αLa	0.36582	...	1.17966	2*1.6124
βNd	>863	atm	cI2	Im3m	A2	W	0.413	...	...	...
γNd	25	>5.0	cF4	Fm3m	A1	Cu	0.480	...	...	...
Ne	<−243.59	atm	cF4	Fm3m	A1	Cu	0.4462	...	...	...
Ni	25	atm	cF4	Fm3m	A1	Cu	0.35240	...	...	...
αNp	25	atm	oP8	Pnma	A_c	αNp	0.6663	0.4723	0.4887	...
βNp	>280	atm	tP4	P42₁2	A_d	βNp	0.4883	...	0.3389	0.694
γNp	>576	atm	cI2	Im3m	A2	W	0.352	...	...	...
αO	<−243.3	atm	mC4	C2m	...	αO	0.5403	0.3429	0.5086	β = 132.53°
βO	<−229.6	atm	hR2	R3m	...	βO	0.4210	...	...	α = 46.27°
γO	<−218.79	atm	cP16	Pm3n	...	γO	0.683	...	...	...
Os	25	atm	hP2	P6₃/mmc	A3	Mg	0.27341	...	0.43918	1.6063
P (black)	25	atm	oC8	Cmca	...	P (black)	0.33136	1.0478	0.43763	...
αPa	25	atm	tI2	I4/mmm	A_a	αPa	0.3921	...	0.3235	0.825
βPa	>1170	atm	cI2	Im3m	A2	W	0.381	...	...	...
αPb	25	atm	cF4	Fm3m	A1	Cu	0.49502	...	...	...
βPb	25	>10.3	hP2	P6₃/mmc	A3	Mg	0.3265	...	0.5387	1.650

Element	Temperature, °C	Pressure, GPa	Pearson symbol	Space group	Struktur-bericht designation	Proto-type	Lattice parameters, nm			Comment, c/a or α or β
							a	b	c	
Pd	25	atm	cF4	Fm3m	A1	Cu	0.38903	…	…	…
αPm	25	atm	hP4	P6₃/mmc	A3'	αLa	0.365	…	1.165	2*1.60
βPm	>890	atm	cI2	Im3m	A2	W	(0.410)	…	…	…
αPo	25	atm	cP1	Pm3m	Aₕ	αPo	0.3366	…	…	…
βPo	>54	atm	hR1	R3m	…	βPo	0.3373	…	…	α = 98.08°
αPr	25	atm	hP4	P6₃/mmc	A3'	αLa	0.36721	…	1.18326	2*1.6111
βPr	>795	atm	cI2	Im3m	A2	W	0.413	…	…	…
γPr	25	>4.0	cF4	Fm3m	A1	Cu	0.488	…	…	…
Pt	25	atm	cF4	Fm3m	A1	Cu	0.39236	…	…	…
αPu	25	atm	mP16	P2₁/m	…	αPu	0.6183	0.4822	1.0963	β = 101.97°
βPu	>125	atm	mI34	I2/m	…	βPu	0.9284	1.0463	0.7859	β = 92.13°
γPu	>215	atm	oF8	Fddd	…	γPu	0.31587	0.57682	1.0162	…
σPu	>320	atm	cF4	Fm3m	A1	Cu	0.46371	…	…	…
σ'Pu	>463	atm	tI2	I4/mmm	A6	In	0.33261	…	0.44630	1.3418
εPu	>483	atm	cI2	Im3m	A2	W	0.36343	…	…	…
Ra	25	atm	cI2	Im3m	A2	W	0.5148	…	…	…
αRb	25	atm	cI2	Im3m	A2	W	0.5705	…	…	…
βRb	25	>1.08	?	…	…	…	…	…	…	…
γRb	25	>2.05	?	…	…	…	…	…	…	…
Re	25	atm	hP2	P6₃/mmc	A3	Mg	0.27609	…	0.4458	1.6145
Rh	25	atm	cF4	Fm3m	A1	Cu	0.38032	…	…	…
Ru	25	atm	hP2	P6₃/mmc	A3	Mg	0.27058	…	0.42816	1.5824
αS	25	atm	oF128	Fddd	A16	αS	1.0464	1.28660	2.44860	…
αSb	25	atm	hR2	R3m	A7	αAs	0.45067	…	…	α = 57.11°
βSb	25	>5.0	cP1	Pm3m	Aₕ	αPo	0.2992	…	…	…
γSb	25	>7.5	hP2	P6₃/mmc	A3	Mg	0.3376	…	0.5341	1.582
σSb	25	>14.0	mP3	?	…	…	0.556	0.404	0.422	β = 86.0°
αSc	25	atm	hP2	P6₃/mmc	A3	Mg	0.33088	…	0.52680	1.5921
βSc	>1337	atm	cI2	Im3m	A2	W	(0.373)	…	…	…
γSe	25	atm	hP3	P3₁21	A8	γSe	0.43659	…	0.49537	1.1346
αSi	25	atm	cF8	Fd3m	A4	C (diamond)	0.54306	…	…	…
βSi	25	>9.5	tI4	I4₁/amd	A5	βSn	0.4686	…	0.2585	0.552
γSi	25	>16.0	cI16	Im3m	…	γSi	0.6636	…	…	…
σSi	25	>16 → atm	hP4	P6₃/mmc	A3'	αLa	0.380	…	0.628	1.653
αSm	25	atm	hR3	R3m	…	αSm	0.36290	…	2.6207	4*1.6048
βSm	>734	atm	hP2	P6₃/mmc	A3	Mg	0.36630	…	0.58448	1.5956
γ'Sm	>922	atm	cI2	Im3m	A2	W	(0.410)	…	…	…
σSm	25	>4.0	hP4	P6₃/mmc	A3'	αLa	0.3618	…	1.166	2*1.611
αSn	<13	atm	cF8	Fd3m	A4	C (diamond)	0.64892	…	…	…
βSn	25	atm	tI4	I4₁/amd	A5	βSn	0.58318	…	0.31818	0.5456
γSn	25	>9.0	tI2	?	…	γSn	0.370	…	0.337	0.91
αSr	25	atm	cF4	Fm3m	A1	Cu	0.6084	…	…	…
βSr	>547	atm	cI2	Im3m	A2	W	0.487	…	…	…
β'Sr	25	>3.5	cI2	Im3m	A2	W	0.4437	…	…	…
Ta	25	atm	cI2	Im3m	A2	W	0.33030	…	…	…
α'Tb	<−53	atm	oC4	Cmcm	…	α'Dy	0.3605	0.6244	0.5706	…
αTb	25	atm	hP2	P6₃/mmc	A3	Mg	0.36055	…	0.56966	1.5800
βTb	>1289	atm	cI2	Im3m	A2	W	(0.407)	…	…	…
γTb	25	>6.0	hR3	R3m	…	αSm	0.341	…	2.45	4*1.60
Tc	25	atm	hP2	P6₃/mmc	A3	Mg	0.2738	…	0.4393	1.604
αTe	25	atm	hP3	P3₁21	A8	γSe	0.44566	…	0.59264	1.3298
βTe	25	>2.0	hR2	R3m	A7	αAs	0.469	…	…	α = 53.30°
γTe	25	>7.0	hR1	R3m	…	βPo	0.3002	…	…	α = 103.3°
αTh	25	atm	cF4	Fm3m	A1	Cu	0.50842	…	…	…
βTh	>1360	atm	cI2	Im3m	A2	W	0.411	…	…	…
αTi	25	atm	hP2	P6₃/mmc	A3	Mg	0.29506	…	0.46835	1.5873
βTi	>882	atm	cI2	Im3m	A2	W	0.33065	…	…	…
ωTi	25	HP → atm	hP3	P6/mmm	…	ωTi	0.4625	…	0.2813	0.6082
αTl	25	atm	hP2	P6₃/mmc	A3	Mg	0.34566	…	0.55248	1.5983
βTl	>230	atm	cI2	Im3m	A2	W	0.3879	…	…	…
γTl	25	HP	cF4	Fm3m	A1	Cu	?	…	…	…
Tm	25	atm	hP2	P6₃/mmc	A3	Mg	0.35375	…	0.55540	1.5700
αU	25	atm	oC4	Cmcm	A20	αU	0.28537	0.58695	0.49548	…
βU	>668	atm	tP30	P4₂/mnm	Aₕ	βU	1.0759	…	0.5656	0.526
γU	>776	atm	cI2	Im3m	A2	W	0.3524	…	…	…
V	25	atm	cI2	Im3m	A2	W	0.30240	…	…	…
W	25	atm	cI2	Im3m	A2	W	0.31652	…	…	…
Xe	<−111.76	atm	cF4	Fm3m	A1	Cu	0.6350	…	…	…
αY	25	atm	hP2	P6₃/mmc	A3	Mg	0.36482	…	0.57318	1.5711
βY	>1478	atm	cI2	Im3m	A2	W	(0.410)	…	…	…
αYb	<−3	atm	hP2	P6₃/mmc	A3	Mg	0.38799	…	0.63859	1.6459
βYb	25	atm	cF4	Fm3m	A1	Cu	0.54848	…	…	…
γYb	>795	atm	cI2	Im3m	A2	W	0.444	…	…	…
Zn	25	atm	hP2	P6₃/mmc	A3	Mg	0.26650	…	0.49470	1.8563
αZr	25	atm	hP2	P6₃/mmc	A3	Mg	0.32316	…	0.51475	1.5929
βZr	>863	atm	cI2	Im3m	A2	W	0.36090	…	…	…
ωZr	25	HP → atm	hP2	P6/mmm	…	ωTi	0.5036	…	0.3109	0.617

LATTICE ENERGIES

H. D. B. Jenkins and H. K. Roobottom

THERMOCHEMICAL CYCLE AND CALCULATED VALUES

Table 1 contains calculated values of the lattice energies (total lattice potential energies), U_{POT}, of crystalline salts, M_aX_b,. U_{POT} is expressed in units of kilojoules per mole, kJ mol^{-1}. M and X can be either simple or complex ions. Substances are arranged by chemical class.

Also listed in the table is the lattice energy, U_{POT}^{BFHC}, obtained from the application of the Born - Fajans - Haber cycle (BHFC) described below, using the "Standard Thermochemical Properties of Chemical Substances" table in Section 5 of this *Handbook*, References 1 through 4, and certain other data which are given in Table 3 below.

The lattice enthalpy, ΔH_L, is given by the cycle:

where (ss) is the standard state of the element concerned.

The lattice enthalpy, ΔH_L, is obtained using the equation:

$$\Delta H_L = a\Delta_f H^o(M^{b+}, g) + b\Delta_f H^o(X^{a-}, g) - \Delta_f H^o(M_aX_b, c)$$

and is futher related to the total lattice potential energy, U_{POT}, by the relationship:

$$\Delta H_L = U_{POT} + \left[a\left(\frac{n_M}{2} - 2\right) + b\left(\frac{n_X}{2} - 2\right) \right] RT$$

where n_M and n_X equal 3 for monatomic ions, 5 for linear polyatomic ions and 6 for polyatomic non-linear ions.

METHOD OF ESTIMATION OF VALUES NOT TABULATED

In cases where the lattice energy is not tabulated and we want to furnish an estimate, then the Kapustinskii equation[5] can be used to obtain a value (in kJ mol^{-1}):

$$U_{POT} = \frac{121.4 z_a z_b v}{(r_a + r_b)} \left(1 - \frac{0.0345}{(r_a + r_b)} \right)$$

where z_a and z_b are the moduli of the charges on the v ions in hte lattice and r_a and r_b (in nm) are the thermochemical radii given in Table 2. The r_a for metal ions is taken to be the Goldschmidt[6] radius.

To cite an example, if we wish to estimate the lattice energy of the salt $[NH_4^+][HF_2^-]$ using the above procedure, we see that Table 2 gives the thermochemical radius (r_a) for NH_4^+ to be 0.136 nm and that for HF_2^- (r_b) to be 0.172 nm. The lattice potential energy is then estimated to be 700 kJ mol^{-1} compared with the calculated value of 705 kJ mol^{-1} and the Born - Fajans - Haber cycle value of 658 kJ mol^{-1}.

REFERENCES

1. Wagman, D. D., Evans, W. H., Parker, V. B., Schumm, R. H., Halow, I., Bailey, S. M., Churney, K. L., and Nuttall, R. L., *The NBS Tables of Chemical Thermodynamic Properties, J. Phys. Chem. Ref. Data*, Vol. 11, Suppl. 2, 1982.
2. Chase, M. W., Davies, C. A., Downey, J. R., Frurip, D. J., McDonald, R. A., and Syverud, A. N., *JANAF Thermochemical Tables, Third Edition, J. Phys. Chem. Ref. Data*, Vol. 14, Suppl. 1, 1985.
3. Lias, S. G., Bartmess, J. E., Liebman, J. F., Holmes, J. L., Levin, R. D., and Mallard, W. G., *Gas-Phase Ion and Neutral Thermochemistry, J. Phys. Chem. Ref. Data*, Vol. 17, Suppl. 1, 1988.
4. Jenkins, H. D. B., and Pratt, K. F., *Adv. Inorg. Chem. Radiochem.*, 22, 1, 1978.
5. Kapustinskii, A. F., *Quart. Rev.*, 10, 283-294., 1956.
6. Goldschmidt, V. M., *Skrifter Norske Videnskaps-Akad.* Oslo, I, Mat.-Naturn. Kl, 1926. See also Dasent, W. E., *Inorganic Energetics*, 2nd ed., Cambridge University Press, 1982.
7. Jenkins, H. D. B., Roobottom, H. K., Passmore, J., and Glasser, L., *J. Chem. Education,* in press.

Table 1
LATTICE ENERGIES (kJ mol^{-1})

Substance	Calc. U_{POT}	U_{POT}^{BHFC}	Substance	Calc. U_{POT}	U_{POT}^{BHFC}
Acetates			TbB_6	7489	-
$Li(CH_3COO)$	-	843	DyB_6	7489	-
$Na(CH_3COO)$	828	807	HoB_6	7489	-
$K(CH_3COO)$	749	726	ErB_6	7489	-
$Rb(CH_3COO)$	715	-	TmB_6	7489	-
$Cs(CH_3COO)$	682	-	YbB_6	5146	-
Acetylides			LuB_6	7489	-
CaC_2	2911	2902	ThB_6	10167	-
SrC_2	2788	2782	**Borohydrides**		
BaC_2	2647	2652	$LiBH_4$	778	-
Azides			$NaBH_4$	703	694
LiN_3	861	875	KBH_4	655	638
NaN_3	770	784	$RbBH_4$	648	-
KN_3	697	-	$CsBH_4$	628	-
RbN_3	674	691	**Borohalides**		
CsN_3	665	674	$LiBF_4$	699	749
AgN_3	854	910	$NaBF_4$	657	674
TlN_3	689	742	KBF_4	611	616
$Ca(N_3)_2$	2186	2316	$RbBF_4$	577	590
$Sr(N_3)_2$	2056	2187	$CsBF_4$	556	565
$Ba(N_3)_2$	2021	-	NH_4BF_4	582	-
$Mn(N_3)_2$	2408	2348	$KBCl_4$	506	497
$Cu(N_3)_2$	2730	2738	$RbBCl_4$	489	486
$Zn(N_3)_2$	2840	2970	$CsBCl_4$	473	-
$Cd(N_3)_2$	2446	2576	**Carbonates**		
$Pb(N_3)_2$	-	2300	Li_2CO_3	2523	2254
Bihalide Salts			Na_2CO_3	2301	2016
$LiHF_2$	821	847	K_2CO_3	2084	1846
$NaHF_2$	755	748	Rb_2CO_3	2000	1783
KHF_2	657	660	Cs_2CO_3	1920	1722
$RbHF_2$	627	631	$MgCO_3$	3138	3122
$CsHF_2$	607	-	$CaCO_3$	2804	2811
NH_4HF_2	705	658	$SrCO_3$	2720	2688
$CsHCl_2$	601	-	$BaCO_3$	2615	2554
Me_4NHCl_2	427	-	$MnCO_3$	3046	3092
Et_4NHCl_2	346	-	$FeCO_3$	3121	3169
Bu_4NHCl_2	290	-	$CoCO_3$	3443	3235
Bicarbonates			$CuCO_3$	3494	-
$NaHCO_3$	820	656	$ZnCO_3$	3121	3273
$KHCO_3$	741	573	$CdCO_3$	2929	3052
$RbHCO_3$	707	522	$SnCO_3$	2904	-
$CsHCO_3$	678	520	$PbCO_3$	2728	2750
NH_4HCO_3	-	577	**Cyanates**		
Borides			$LiNCO$	849	-
CaB_6	5146	-	$NaNCO$	807	816
SrB_6	5104	-	$KNCO$	726	734
BaB_6	5021	-	$RbNCO$	692	-
YB_6	7447	-	$CsNCO$	661	-
LaB_6	7406	-	NH_4NCO	724	-
CeB_6	10083	-	**Cyanides**		
PrB_6	7447	—	$LiCN$	874	
NdB_6	7447	-	$NaCN$	766	759
PmB_6	7406	-	KCN	692	686
SmB_6	7447	-	$RbCN$	638	-
EuB_6	5104	-	$CsCN$	601	-
GdB_6	7489	-	$Ca(CN)_2$	2268	2240

Table 1
LATTICE ENERGIES (kJ mol^{-1}) (continued)

Substance	Calc. U_{POT}	U_{POT}^{BHFC}	Substance	Calc. U_{POT}	U_{POT}^{BHFC}
Sr(CN)$_2$	2138	-	FrBr	611	-
Ba(CN)$_2$	2001	2009	FrI	582	-
NH$_4$CN	617	691	CuCl	992	996
AgCN	(741)	935	CuBr	969	978
Zn(CN)$_2$	2809	2817	CuI	948	966
Cd(CN)$_2$	2583	2591	AgF	953	974
Formates			AgCl	910	918
Li(HCO$_2$)	865	-	AgBr	897	905
Na(HCO$_2$)	791	804	AgI	881	892
K(HCO$_2$)	713	722	AuCl	1013	1066
Rb(HCO$_2$)	685	-	AuBr	1029	1059
Cs(HCO$_2$)	651	-	AuI	1027	1070
NH$_4$(HCO$_2$)	715	-	InCl	-	764
Germanates			InBr	-	767
Mg$_2$GeO$_4$	7991	-	InI	-	733
Ca$_2$GeO$_4$	7301	7306	TlF	-	850
Sr$_2$GeO$_4$	6987	-	TlCl	738	751
Ba$_2$GeO$_4$	6653	6643	TlBr	720	734
Halates			TlI	692	710
LiBrO$_3$	883	880	Me$_4$NCl	566	-
NaBrO$_3$	803	791	Me$_4$NBr	553	-
KBrO$_3$	740	722	Me$_4$NI	544	-
RbBrO$_3$	720	705	PH$_4$Br	616	-
CsBrO$_3$	694	681	PH$_4$I	590	-
NaClO$_3$	770	785	BeF$_2$	3464	3526
KClO$_3$	711	721	BeCl$_2$	3004	3033
RbClO$_3$	690	703	BeBr$_2$	2950	2914
CsClO$_3$	-	679	BeI$_2$	2780	2813
LiIO$_3$	975	974	MgF$_2$	2926	2978
NaIO$_3$	883	876	MgCl$_2$	2477	2540
KIO$_3$	820	780	MgBr$_2$	2406	2451
RbIO$_3$	791	-	MgI$_2$	2293	2340
CsIO$_3$	761	-	CaF$_2$	2640	2651
Halides			CaCl$_2$	2268	2271
LiF	1030	1049	CaBr$_2$	2132	-
LiCl	834	864	CaI$_2$	1971	2087
LiBr	788	820	SrF$_2$	2476	2513
LiI	730	764	SrCl$_2$	2142	2170
NaF	910	930	SrI$_2$	1984	1976
NaCl	769	790	BaF$_2$	2347	2373
NaBr	732	754	BaCl$_2$	2046	2069
NaI	682	705	BaBr$_2$	1971	1995
KF	808	829	BaI$_2$	1862	1890
KCl	701	720	RaF$_2$	2284	-
KBr	671	691	RaCl$_2$	2004	-
KI	632	650	RaBr$_2$	1929	-
RbF	774	795	RaI$_2$	1803	-
RbCl	680	695	ScCl$_2$	2380	-
RbBr	651	668	ScBr$_2$	2291	-
RbI	617	632	ScI$_2$	2201	-
CsF	744	759	TiF$_2$	2724	-
CsCl	657	670	TiCl$_2$	2439	2514
CsBr	632	647	TiBr$_2$	2360	2430
CsI	600	613	TiI$_2$	2259	2342
FrF	715	-	VCl$_2$	2607	2593
FrCl	632	-	VBr$_2$	-	2534

Table 1
LATTICE ENERGIES (kJ mol⁻¹) (continued)

Substance	Calc. U_{POT}	U_{POT}^{BHFC}	Substance	Calc. U_{POT}	U_{POT}^{BHFC}
VI_2	-	2470	YF_3	4983	-
CrF_2	2778	2939	YCl_3	4506	4524
$CrCl_2$	2540	2601	YI_3	4240	4258
$CrBr_2$	2377	2536	TiF_3	5644	-
CrI_2	2269	2440	$TiCl_3$	5134	5153
$MoCl_2$	2737	2746	$TiBr_3$	5012	5023
$MoBr_2$	2742	2753	TiI_3	4845	-
MoI_2	2630	-	$ZrCl_3$	-	4791
MnF_2	2644	-	$ZrBr_3$	-	4758
$MnCl_2$	2510	2551	ZrI_3	-	4591
$MnBr_2$	2448	2482	VF_3	5895	-
MnI_2	2212	-	VCl_3	5322	5329
FeF_2	2849	2967	VBr_3	5214	5224
$FeCl_2$	2569	2641	VI_3	5121	5136
$FeBr_2$	2515	2577	$NbCl_3$	5062	-
FeI_2	2439	2491	$NbBr_3$	4980	-
CoF_2	3004	3042	NbI_3	4860	-
$CoCl_2$	2707	2706	CrF_3	6033	6065
$CoBr_2$	2640	2643	$CrCl_3$	5518	5529
CoI_2	2569	2561	$CrBr_3$	5355	-
NiF_2	3098	3089	CrI_3	5275	5294
$NiCl_2$	2753	2786	MoF_3	6459	-
$NiBr_2$	2729	2721	$MoCl_3$	5246	5253
NiI_2	2607	2637	$MoBr_3$	5156	-
$PdCl_2$	2778	2818	MoI_3	5073	-
$PdBr_2$	2741	2751	MnF_3	6017	-
PdI_2	2748	2760	$MnCl_3$	5544	-
CuF_2	3046	3102	$MnBr_3$	5448	-
$CuCl_2$	2774	2824	MnI_3	5330	-
$CuBr_2$	2715	2774	$TcCl_3$	5270	-
CuI_2	2640	-	$TcBr_3$	5215	-
AgF_2	2942	2967	TcI_3	5188	-
ZnF_2	3021	3053	FeF_3	5870	-
$ZnCl_2$	2703	2748	$FeCl_3$	5364	5436
$ZnBr_2$	2648	2689	$FeBr_3$	5333	5347
ZnI_2	2581	2619	FeI_3	5117	-
CdF_2	2809	2830	$RuCl_3$	5245	5257
$CdCl_2$	2552	2565	$RuBr_3$	5223	5232
$CdBr_2$	2507	2517	RuI_3	5222	5235
CdI_2	2441	2455	CoF_3	5991	-
HgF_2	2757	-	$RhCl_3$	5641	5665
$HgCl_2$	2657	2664	IrF_3	(6112)	-
$HgBr_2$	2628	2639	$IrBr_3$	(4794)	-
HgI_2	2628	2624	NiF_3	(6111)	-
SnF_2	2551	-	AuF_3	(5777)	-
$SnCl_2$	2297	2310	$AuCl_3$	(4605)	-
$SnBr_2$	2251	2256	$ZnCl_3$	5832	-
SnI_2	2193	2206	$ZnBr_3$	5732	-
PbF_2	2535	2543	ZnI_3	5636	-
$PbCl_2$	2270	2282	AlF_3	5924	6252
$PbBr_2$	2219	2230	$AlCl_3$	5376	5513
PbI_2	2163	2177	$AlBr_3$	5247	5360
ScF_3	5492	5540	AlI_3	5070	5227
$ScCl_3$	4874	4901	GaF_3	5829	6238
$ScBr_3$	4729	4761	$GaCl_3$	5217	5665
ScI_3	4640	-	$GaBr_3$	4966	5569

Table 1
LATTICE ENERGIES (kJ mol^{-1}) (continued)

Substance	Calc. U_{POT}	U_{POT}^{BHFC}	Substance	Calc. U_{POT}	U_{POT}^{BHFC}
GaI$_3$	4611	5496	CrF$_2$Br	5753	-
InCl$_3$	4736	5183	CrF$_2$I	5669	-
InBr$_3$	4535	5117	CrCl$_2$Br	5448	-
InI$_3$	4234	5001	CrCl$_2$I	5381	5429
TlF$_3$	5493	-	CrBr$_2$I	5330	5370
TlCl$_3$	5258	5278	CuFCl	2891	-
TlBr$_3$	5171	-	CuFBr	2853	-
TlI$_3$	5088	-	CuFI	2803	-
AsBr$_3$	5497	5365	CuClBr	2753	-
AsI$_3$	4824	5295	CuClI	2694	-
SbF$_3$	5295	5324	CuBrI	2669	-
SbCl$_3$	5032	4857	FeF$_2$Cl	5711	-
SbBr$_3$	4954	4776	FeF$_2$Br	5653	-
SbI$_3$	4867	4692	FeF$_2$I	5569	-
BiCl$_3$	4689	4707	FeCl$_2$Br	5339	-
BiI$_3$	3774	-	FeCl$_2$I	5272	-
LaF$_3$	4682	-	FeBr$_2$I	5209	-
LaCl$_3$	4263	4242	LiIO$_2$F$_2$	845	-
LaBr$_3$	4209	-	NaIO$_2$F$_2$	766	756
LaI$_3$	3916	3986	KIO$_2$F$_2$	699	689
CeCl$_3$	4394	4348	RbIO$_2$F$_2$	674	-
CeI$_3$	-	4061	CsIO$_2$F$_2$	636	-
PrCl$_3$	4322	4387	NH$_4$IO$_2$F$_2$	678	-
PrI$_3$	-	4101	AgIO$_2$F$_2$	736	685
NdCl$_3$	4343	4415	**Hydrides**		
SmCl$_3$	4376	4450	LiH	916	918
EuCl$_3$	4393	4490	NaH	807	807
GdCl$_3$	4406	4495	KH	711	713
DyCl$_3$	4481	4529	RbH	686	684
HoCl$_3$	4501	4572	CsH	648	653
ErCl$_3$	4527	4591	VH	1184	(1344)
TmCl$_3$	4548	4608	NbH	1163	(1633)
TmI$_3$	-	4340	PdH	979	1368
YbCl$_3$	-	4651	CuH	828	1254
AcCl$_3$	4096	-	TiH	996	1407
UCl$_3$	4243	-	ZrH	916	1590
NpCl$_3$	4268	-	HfH	904	-
PuCl$_3$	4289	-	LaH	828	-
PuBr$_3$	(3959)	-	TaH	1021	-
AmCl$_3$	4293	-	CrH	1050	-
TiF$_4$	10012	9908	NiH	929	-
TiCl$_4$	9431	-	PtH	937	-
TiBr$_4$	9288	9059	AgH	941	-
TiI$_4$	9108	8918	AuH	1033	1108
ZrF$_4$	8853	8971	TlH	745	-
ZrCl$_4$	8021	8144	GeH	950	-
ZrBr$_4$	7661	7984	PbH	778	-
ZrI$_4$	7155	7801	BeH$_2$	3205	3306
MoF$_4$	8795	-	MgH$_2$	2791	2718
MoCl$_4$	8556	9603	CaH$_2$	2410	2406
MoBr$_4$	8510	9500	SrH$_2$	2250	2265
MoI$_4$	8427	-	BaH$_2$	2121	2133
SnCl$_4$	8355	8930	ScH$_2$	2711	2744
SnBr$_4$	7970	8852	YH$_2$	(2598)	2733
PbF$_4$	9519	-	LaH$_2$	2380	2522
CrF$_2$Cl	5795	-	CeH$_2$	2414	2509

Table 1
LATTICE ENERGIES (kJ mol^{-1}) (continued)

Substance	Calc. U_{POT}	U_{POT}^{BHFC}	Substance	Calc. U_{POT}	U_{POT}^{BHFC}
PrH_2	2448	2405	AgOH	918	845
NdH_2	2464	2394	AuOH	1033	-
PmH_2	2519	-	TlOH	705	874
SmH_2	2510	2389	$Zn(OH)_2$	2795	3151
GdH_2	2494	2651	$Cd(OH)_2$	2607	2909
AcH_2	2372	-	$Hg(OH)_2$	2669	-
ThH_2	2711	2738	$Sn(OH)_2$	2489	2721
PuH_2	2519	-	$Pb(OH)_2$	2376	-
AmH_2	2544	-	$Sc(OH)_3$	5063	5602
TiH_2	2866	2864	$Y(OH)_3$	4707	-
ZrH_2	2711	2999	$La(OH)_3$	4443	-
CuH_2	2941	-	$Cr(OH)_3$	5556	6299
ZnH_2	2870	-	$Mn(OH)_3$	6213	-
HgH_2	2707	-	$Al(OH)_3$	5627	-
AlH_3	5924	5969	$Ga(OH)_3$	5732	6368
FeH_3	5724	-	$In(OH)_3$	5280	-
ScH_3	5439	-	$Tl(OH)_3$	5314	-
YH_3	5063	4910	$Ti(OH)_4$	9456	-
LaH_3	4895	4493	$Zr(OH)_4$	8619	-
FeH_3	5724	-	$Mn(OH)_4$	10933	-
GaH_3	5690	-	$Sn(OH)_4$	9188	9879
InH_3	5092	-	**Imides**		
TlH_3	5092	-	CaNH	3293	-
Hydroselenides			SrNH	3146	-
NaHSe	703	732	BaNH	2975	-
KHSe	644	712	**Metavanadates**		
RbHSe	623	689	Li_3VO_4	3945	-
CsHse	598	669	Na_3VO_4	3766	-
Hydrosulphides			K_3VO_4	3376	-
LiHS	768	862	Rb_3VO_4	3243	-
NaHS	723	771	Cs_3VO_4	3137	-
RbHS	655	682	**Nitrates**		
CsHS	628	657	$LiNO_3$	848	854
NH_4HS	661	718	$NaNO_3$	755	763
$Ca(HS)_2$	2184	(2171)	KNO_3	685	694
$Sr(HS)_2$	2063	-	$RbNO_3$	662	671
$Ba(HS)_2$	1979	(1956)	$CsNO_3$	648	650
Hydroxides			$AgNO_3$	820	832
LiOH	1021	1028	$TlNO_3$	690	707
NaOH	887	892	$Mg(NO_3)_2$	2481	2521
KOH	789	796	$Ca(NO_3)_2$	2268	2247
RbOH	766	765	$Sr(NO_3)_2$	2176	2151
CsOH	721	732	$Ba(NO_3)_2$	2062	2035
$Be(OH)_2$	3477	3620	$Mn(NO_3)_2$	2318	2478
$Mg(OH)_2$	2870	2998	$Fe(NO_3)_2$	-	(2580)
$Ca(OH)_2$	2506	2637	$Co(NO_3)_2$	2560	2647
$Sr(OH)_2$	2330	2474	$Ni(NO_3)_2$	-	2729
$Ba(OH)_2$	2142	2330	$Cu(NO_3)_2$	-	2739
$Ti(OH)_2$	-	2953	$Zn(NO_3)_2$	2376	2649
$Mn(OH)_2$	2909	3008	$Cd(NO_3)_2$	2238	2462
$Fe(OH)_2$	2653	3044	$Sn(NO_3)_2$	2155	2254
$Co(OH)_2$	2786	3109	$Pb(NO_3)_2$	2067	2208
$Ni(OH)_2$	2832	3186	**Nitrides**		
$Pd(OH)_2$	-	3189	ScN	7547	7506
$Cu(OH)_2$	2870	3229	LaN	6876	6793
CuOH	1006	-	TiN	8130	8033

Table 1
LATTICE ENERGIES (kJ mol^{-1}) (continued)

Substance	Calc. U_{POT}	U_{POT}^{BHFC}	Substance	Calc. U_{POT}	U_{POT}^{BHFC}
ZrN	7633	7723	Nd_2O_3	12736	-
VN	8283	8233	Pm_2O_3	12811	-
NbN	7939	8022	Sm_2O_3	12878	-
CrN	8269	8358	Eu_2O_3	12945	-
Nitrites			Gd_2O_3	12996	-
$NaNO_2$	774	772	Tb_2O_3	13071	-
KNO_2	699	687	Dy_2O_3	13138	-
$RbNO_2$	724	765	Ho_2O_3	13180	-
$CsNO_2$	690	-	Er_2O_3	13263	-
Oxides			Tm_2O_3	13322	-
Li_2O	2799	2814	Yb_2O_3	13380	-
Na_2O	2481	2478	Lu_2O_3	13665	-
K_2O	2238	2232	Ac_2O_3	12573	-
Rb_2O	2163	2161	Ti_2O_3	-	14149
Cs_2O	2131	2063	V_2O_3	15096	14520
Cu_2O	3273	3189	Cr_2O_3	15276	14957
Ag_2O	3002	2910	Mn_2O_3	15146	15035
Tl_2O	2659	2575	Fe_2O_3	14309	14774
LiO_2	(878)	(872)	Al_2O_3	15916	-
NaO_2	799	821	Ga_2O_3	15590	15220
KO_2	741	751	In_2O_3	13928	-
RbO_2	706	721	Pb_2O_3	(14841)	-
CsO_2	679	696	CeO_2	9627	-
Li_2O_2	2592	2557	ThO_2	10397	-
Na_2O_2	2309	22717	PaO_2	10573	-
K_2O_2	2114	2064	$VO_2(g)$	10644	-
Rb_2O_2	2025	1994	NpO_2	10707	-
Cs_2O_2	1948	1512	PuO_2	10786	-
MgO_2	3356	3526	AmO_2	10799	-
CaO_2	3144	3132	CmO_2	10832	-
SrO_2	3037	2977	TiO_2	12150	-
KO_3	697	707	ZrO_2	11188	-
BeO	4514	4443	MoO_2	11648	-
MgO	3795	3791	MnO_2	12970	-
CaO	3414	3401	SiO_2	13125	-
SrO	3217	3223	GeO_2	12828	-
BaO	3029	3054	SnO_2	11807	-
TiO	3832	3811	PbO_2	11217	-
VO	3932	3863	**Perchlorates**		
MnO	3724	3745	$LiClO_4$	709	715
FeO	3795	3865	$NaClO_4$	643	641
CoO	3837	3910	$KClO_4$	599	595
NiO	3908	4010	$RbClO_4$	564	576
PdO	3736	-	$CsClO_4$	636	550
CuO	4135	4050	NH_4ClO_4	583	580
ZnO	4142	3971	$Ca(ClO_4)_2$	1958	1971
CdO	3806	-	$Sr(ClO_4)_2$	1862	1862
HgO	3907	-	$Ba(ClO_4)_2$	1795	1769
GeO	3919	-	**Permanganates**		
SnO	3652	-	$NaMnO_4$	661	-
PbO	3520	-	$KMnO_4$	607	-
Sc_2O_3	13557	13708	$RbMnO_4$	586	-
Y_2O_3	12705	-	$CsMnO_4$	565	-
La_2O_3	12452	-	$Ca(MnO_4)_2$	1937	-
Ce_2O_3	12661	-	$Sr(MnO_4)_2$	1845	-
Pr_2O_3	12703	-	$Ba(MnO_4)_2$	1778	-

Table 1
LATTICE ENERGIES (kJ mol^{-1}) (continued)

Substance	Calc. U_{POT}	U_{POT}^{BHFC}	Substance	Calc. U_{POT}	U_{POT}^{BHFC}
Phosphates			Rb_2S	1929	1949
$Mg_3(PO_4)_2$	11632	11407	Cs_2S	1892	1850
$Ca_3(PO_4)_2$	10602	10479	$(NH_4)_2S$	2008	(2026)
$Sr_3(PO_4)_2$	10125	10075	Cu_2S	2786	2865
$Ba_3(PO_4)_2$	9652	9654	Ag_2S	2606	2677
$MnPO_4$	7397	-	Au_2S	2908	-
$FePO_4$	7251	7300	Tl_2S	2298	2258
BPO_4	8201	-	**Sulphates**		
$AlPO_4$	7427	7507	Li_2SO_4	2229	2142
$GaPO_4$	7381	-	Na_2SO_4	1827	1938
Selenides			K_2SO_4	1700	1796
Li_2Se	2364		Rb_2SO_4	1636	1748
Na_2Se	2130		Cs_2SO_4	1596	1658
K_2Se	1933		$(NH_4)_2SO_4$	1766	1777
Rb_2Se	1837		Cu_2SO_4	2276	2166
Cs_2Se	1745		Ag_2SO_4	2104	1989
Ag_2Se	2686	-	Tl_2SO_4	1828	1722
Tl_2Se	2209	-	Hg_2SO_4	-	2127
$BeSe$	3431		$CaSO_4$	2489	2480
$MgSe$	3071	-	$SrSO_4$	2577	2484
$CaSe$	2858	2862	$BaSO_4$	2469	2374
$SrSe$	2736	-	$MnSO_4$	2920	2825
$BaSe$	2611	-	**Ternary Salts**		
$MnSe$	3176	-	Cs_2CuCl_4	1393	
Selenites			Rb_2ZnCl_4	1529	-
Li_2SeO_3	2171	-	Cs_2ZnCl_4	1492	-
Na_2SeO_3	1950	1916	Rb_2ZnBr_4	1498	-
K_2SeO_3	1774	1749	Cs_2ZnBr_4	1454	-
Rb_2SeO_3	1715	1675	Cs_2ZnI_4	1386	-
Cs_2SeO_3	1640	-	$CsGaCl_4$	494	-
Tl_2SeO_3	1879	-	$NaAlCl_4$	556	-
Ag_2SeO_3	2113	2148	$CsAlCl_4$	486	-
$BeSeO_3$	3322	-	$NaFeCl_4$	492	-
$MgSeO_3$	3012	2998	Rb_2CoCl_4	1447	-
$CaSeO_3$	2732	-	Cs_2CoCl_4	1391	-
$SrSeO_3$	2586	2588	K_2PtCl_4	1574	1550
$BaSeO_3$	2460	2451	Cs_2GeF_6	1573	-
Selenates			$(NH_4)_2GeF_6$	1657	-
Li_2SeO_4	2054	-	Cs_2GeCl_6	1404	1419
Na_2SeO_4	1879	-	K_2HfCl_6	1345	1461
K_2SeO_4	1732	-	K_2IrCl_6	1442	1440
Rb_2SeO_4	1686	-	Na_2MoCl_6	1526	1504
Cs_2SeO_4	1615	-	K_2MoCl_6	1418	1412
Cu_2SeO_4	2201	-	Rb_2MoCl_6	1399	1399
Ag_2SeO_4	2033	-	Cs_2MoCl_6	1347	1347
Tl_2SeO_4	1766	-	K_2NbCl_6	1375	1398
Hg_2SeO_4	2163	-	Rb_2NbCl_6	1371	1385
$BeSeO_4$	3448	-	Cs_2NbCl_6	1381	1344
$MgSeO_4$	2895	-	K_2OsCl_6	1447	1447
$CaSeO_4$	2632	-	Cs_2OsCl_6	1409	-
$SrSeO_4$	2489	-	K_2OsBr_6	1396	-
Sulphides			K_2PdCl_6	1481	1493
Li_2S	2464	2472	Rb_2PdCl_6	1449	-
Na_2S	2192	2203	Cs_2PdCl_6	1426	-
K_2S	1979	(2052)	Rb_2PbCl_6	1343	1343

Table 1
LATTICE ENERGIES (kJ mol^{-1}) (continued)

Substance	Calc. U_{POT}	U_{POT}^{BHFC}	Substance	Calc. U_{POT}	U_{POT}^{BHFC}
Cs_2PbCl_6	1344	-	Rb_2TiCl_6	1415	1416
$(NH_4)_2PbCl_6$	1355	-	Cs_2TiCl_6	1402	1384
K_2PtCl_6	1468	1471	Tl_2TiCl_6	1560	1553
Rb_2PtCl_6	1464	-	K_2TiBr_6	1379	1379
Cs_2PtCl_6	1444	-	Rb_2TiBr_6	1341	1331
$(NH_4)_2PtCl_6$	1468	-	Cs_2TiBr_6	1339	1306
Tl_2PtCl_6	1546	-	Na_2UBr_6	1504	-
Ag_2PtCl_6	1773	1881	K_2UBr_6	1484	-
$BaPtCl_6$	2047	2070	Rb_2UBr_6	1473	-
K_2PtBr_6	1423	1392	Cs_2UBr_6	1459	-
Ag_2PtBr_6	1791	2276	K_2WCl_6	1398	1423
K_2PtI_6	1421	-	Rb_2WCl_6	1397	1434
K_2ReCl_6	1416	1442	Cs_2WCl_6	1392	1366
Rb_2ReCl_6	1414	-	K_2WBr_6	1408	1408
Cs_2ReCl_6	1398	-	Rb_2WBr_6	1361	1391
K_2ReBr_6	1375	1375	Cs_2WBr_6	1362	1332
K_2SiF_6	1670	1765	K_2ZrCl_6	1339	1371
Rb_2SiF_6	1639	1673	Rb_2ZrCl_6	1341	-
Cs_2SiF_6	1604	1498	Cs_2ZrCl_6	1339	1307
Tl_2SiF_6	1675	-	**Tellurides**		
K_2SnCl_6	1363	1390	Li_2Te	2212	-
Rb_2SnCl_6	1361	1363	Na_2Te	1997	2095
Cs_2SnCl_6	1358	-	K_2Te	1830	-
Tl_2SnCl_6	1437	-	Rb_2Te	1837	-
$(NH_4)_2SnCl_6$	1370	1344	Cs_2Te	1745	-
Rb_2SnBr_6	1309	-	Cu_2Te	2706	2683
Cs_2SnBr_6	1306	-	Ag_2Te	2607	2600
Rb_2SnI_6	1226	-	Tl_2Te	2084	2172
Cs_2SnBr_6	1243	-	$BeTe$	3319	-
K_2TeCl_6	1318	1320	$MgTe$	2878	3081
Rb_2TeCl_6	1321	-	$CaTe$	2721	-
Cs_2TeCl_6	1323	-	**Thiocyanates**		
Tl_2TeCl_6	1392	-	$LiCNS$	764	(765)
$(NH_4)_2TeCl_6$	1318	-	$NaCNS$	682	682
K_2RuCl_6	1451	-	$KCNS$	623	616
Rb_2CoF_6	1688	-	$RbCNS$	623	619
Cs_2CoF_6	1632	-	$CsCNS$	623	568
K_2NiF_6	1721	-	NH_4CNS	605	611
Rb_2NiF_6	1688	-	$Ca(CNS)_2$	2184	2118
Rb_2SbCl_6	1357	-	$Sr(CNS)_2$	2063	1957
Rb_2SeCl_6	1409	-	$Ba(CNS)_2$	1979	1852
Cs_2SeCl_6	1397	-	$Mn(CNS)_2$	2280	2351
$(NH_4)_2SeCl_6$	1420	-	$Zn(CNS)_2$	2335	2560
$(NH_4)_2PoCl_6$	1338	-	$Cd(CNS)_2$	2201	2374
Cs_2PoBr_6	1286	-	$Hg(CNS)_2$	2146	2492
Cs_2CrF_6	1603	-	$Sn(CNS)_2$	2117	2142
Rb_2MnF_6	1688	-	$Pb(CNS)_2$	2058	-
Cs_2MnF_6	1620	-	**Vanadates**		
K_2MnCl_6	1462	-	$LiVO_3$	810	-
Rb_2MnCl_6	1451	-	$NaVO_3$	761	-
$(NH_4)_2MnCl_6$	1464	-	KVO_3	686	-
Cs_2TeBr_6	1306	-	$RbVO_3$	657	-
Cs_2TeI_6	1246	-	$CsVO_3$	628	-
K_2TiCl_6	1412	1447			

Table 2
THERMOCHEMICAL RADII (nm)

Ion	Radius		Ion	Radius	
Singly Charged Anions			NbF_6^-	0.254	± 0.019
AgF_4^-	0.231	± 0.019	$Nb_2F_{11}^-$	0.311	± 0.038
$AlBr_4^-$	0.321	± 0.023	NbO_3^-	0.194	± 0.019
$AlCl_4^-$	0.317	± 0.019	NH_2^-	0.168	± 0.019
AlF_4^-	0.214	± 0.023	$NH_2CH_2COO^-$	0.252	± 0.019
AlH_4^-	0.226	± 0.019	NO_2^-	0.187	± 0.019
AlI_4^-	0.374	± 0.019	NO_3^-	0.200	± 0.019
AsF_6^-	0.243	± 0.019	O_2^-	0.165	± 0.019
AsO_2^-	0.211	± 0.019	O_3^-	0.199	± 0.034
$Au(CN)_2^-$	0.266	± 0.019	OH^-	0.152	± 0.019
$AuCl_4^-$	0.288	± 0.019	OsF_6^-	0.252	± 0.020
AuF_4^-	0.240	± 0.019	PaF_6^-	0.249	± 0.019
AuF_6^-	0.235	± 0.038	PdF_6^-	0.252	± 0.019
$B(OH)_4^-$	0.229	± 0.019	PF_6^-	0.242	± 0.019
BF_4^-	0.205	± 0.019	PO_3^-	0.204	± 0.019
BH_4^-	0.205	± 0.019	PtF_6^-	0.247	± 0.019
Br^-	0.190	± 0.019	PuF_5^-	0.239	± 0.019
BrF_4^-	0.231	± 0.019	ReF_6^-	0.240	± 0.019
BrO_3^-	0.214	± 0.019	ReO_4^-	0.227	± 0.019
$CF_3SO_3^-$	0.230	± 0.049	RuF_6^-	0.242	± 0.019
$CH_3CO_2^-$	0.194	± 0.019	S_6^-	0.305	± 0.019
Cl^-	0.168	± 0.019	SCN^-	0.209	± 0.019
ClO_2^-	0.195	± 0.019	$SbCl_6^-$	0.320	± 0.019
ClO_3^-	0.208	± 0.019	SbF_6^-	0.252	± 0.019
ClO_4^-	0.225	± 0.019	$Sb_2F_{11}^-$	0.312	± 0.038
$ClS_2O_6^-$	0.260	± 0.049	$Sb_3F_{14}^-$	0.374	± 0.038
CN^-	0.187	± 0.023	$SeCl_5^-$	0.258	± 0.038
$Cr_3O_8^-$	0.276	± 0.019	$SeCN^-$	0.230	± 0.019
$CuBr_4^-$	0.315	± 0.019	SeH^-	0.195	± 0.019
F^-	0.126	± 0.019	SH^-	0.191	± 0.019
$FeCl_4^-$	0.317	± 0.019	SO_3F^-	0.214	± 0.019
$GaCl_4^-$	0.328	± 0.019	$S_3N_3^-$	0.231	± 0.038
H^-	0.148	± 0.019	$S_3N_3O_4^-$	0.252	± 0.038
$H_2AsO_4^-$	0.227	± 0.019	$TaCl_6^-$	0.352	± 0.019
$H_2PO_4^-$	0.213	± 0.019	TaF_6^-	0.250	± 0.019
HCO_2^-	0.200	± 0.019	TaO_3^-	0.192	± 0.019
HCO_3^-	0.207	± 0.019	UF_6^-	0.301	± 0.019
HF_2^-	0.172	± 0.019	VF_6^-	0.235	± 0.019
HSO_4^-	0.221	± 0.019	VO_3^-	0.201	± 0.019
I^-	0.211	± 0.019	WCl_6^-	0.337	± 0.019
I_2Br^-	0.261	± 0.019	WF_6^-	0.246	± 0.019
I_3^-	0.272	± 0.019	WOF_5^-	0.241	± 0.019
I_4^-	0.300	± 0.019	**Doubly Charged Anions**		
IBr_2^-	0.251	± 0.019	AmF_6^{2-}	0.255	± 0.019
ICl_2^-	0.235	± 0.019	$Bi_2Br_8^{2-}$	0.392	± 0.055
ICl_4^-	0.307	± 0.019	$Bi_6Cl_{20}^{2-}$	0.501	± 0.073
$IO_2F_2^-$	0.233	± 0.019	$CdCl_4^{2-}$	0.307	± 0.019
IO_3^-	0.218	± 0.019	$CeCl_6^{2-}$	0.352	± 0.019
IO_4^-	0.231	± 0.019	CeF_6^{2-}	0.249	± 0.019
IrF_6^-	0.242	± 0.019	CO_3^{2-}	0.189	± 0.019
MnO_4^-	0.220	± 0.019	$CoCl_4^{2-}$	0.306	± 0.019
MoF_6^-	0.241	± 0.019	CoF_4^{2-}	0.209	± 0.019
$MoOF_5^-$	0.241	± 0.019	CoF_6^{2-}	0.256	± 0.019
N_3^-	0.180	± 0.019	$Cr_2O_7^{2-}$	0.292	± 0.019
NCO^-	0.193	± 0.019	CrF_6^{2-}	0.253	± 0.019
$NbCl_6^-$	0.338	± 0.049	CrO_4^{2-}	0.229	± 0.019

Table 2
THERMOCHEMICAL RADII (nm) (continued)

Ion	Radius		Ion	Radius	
$CuCl_4^{2-}$	0.304	± 0.019	$S_2O_6^{2-}$	0.283	± 0.019
CuF_4^{2-}	0.213	± 0.019	$S_2O_7^{2-}$	0.275	± 0.019
$GeCl_6^{2-}$	0.335	± 0.019	$S_2O_8^{2-}$	0.291	± 0.019
GeF_6^{2-}	0.244	± 0.019	$S_3O_6^{2-}$	0.302	± 0.019
HfF_6^{2-}	0.248	± 0.019	$S_4O_6^{2-}$	0.325	± 0.019
HgI_4^{2-}	0.377	± 0.019	$S_6O_6^{2-}$	0.382	± 0.019
$IrCl_6^{2-}$	0.332	± 0.019	ScF_6^{2-}	0.276	± 0.019
$MnCl_6^{2-}$	0.314	± 0.031	Se^{2-}	0.181	± 0.019
MnF_4^{2-}	0.219	± 0.019	$SeBr_6^{2-}$	0.363	± 0.019
MnF_6^{2-}	0.241	± 0.019	$SeCl_6^{2-}$	0.336	± 0.019
$MoBr_6^{2-}$	0.364	± 0.019	SeO_4^{2-}	0.229	± 0.019
$MoCl_6^{2-}$	0.338	± 0.019	SiF_6^{2-}	0.248	± 0.019
MoF_6^{2-}	0.274	± 0.019	SiO_3^{2-}	0.195	± 0.019
MoO_4^{2-}	0.231	± 0.019	SmF_4^{2-}	0.218	± 0.019
$NbCl_6^{2-}$	0.343	± 0.019	$Sn(OH)_6^{2-}$	0.279	± 0.020
NH^{2-}	0.128	± 0.019	$SnBr_6^{2-}$	0.374	± 0.019
$Ni(CN)_4^{2-}$	0.322	± 0.019	$SnCl_6^{2-}$	0.345	± 0.019
NiF_4^{2-}	0.211	± 0.019	SnF_6^{2-}	0.265	± 0.019
NiF_6^{2-}	0.249	± 0.019	SnI_6^{2-}	0.427	± 0.019
O^{2-}	0.141	± 0.019	SO_3^{2-}	0.204	± 0.019
O_2^{2-}	0.167	± 0.019	SO_4^{2-}	0.218	± 0.019
$OsBr_6^{2-}$	0.365	± 0.019	$TcBr_6^{2-}$	0.363	± 0.019
$OsCl_6^{2-}$	0.336	± 0.019	$TcCl_6^{2-}$	0.337	± 0.019
OsF_6^{2-}	0.276	± 0.019	TcF_6^{2-}	0.244	± 0.019
$PbCl_4^{2-}$	0.279	± 0.019	TcH_9^{2-}	0.260	± 0.019
$PbCl_6^{2-}$	0.347	± 0.019	TcI_6^{2-}	0.419	± 0.019
PbF_6^{2-}	0.268	± 0.019	Te^{2-}	0.220	± 0.019
$PdBr_6^{2-}$	0.354	± 0.019	$TeBr_6^{2-}$	0.383	± 0.019
$PdCl_4^{2-}$	0.313	± 0.019	$TeCl_6^{2-}$	0.353	± 0.019
$PdCl_6^{2-}$	0.333	± 0.019	TeI_6^{2-}	0.430	± 0.019
PdF_6^{2-}	0.252	± 0.019	TeO_4^{2-}	0.238	± 0.019
$PoBr_6^{2-}$	0.380	± 0.019	$Th(NO_3)_6^{2-}$	0.424	± 0.019
PoI_6^{2-}	0.428	± 0.019	$ThCl_6^{2-}$	0.360	± 0.019
$Pt(NO_2)_3Cl_3^{2-}$	0.364	± 0.019	ThF_6^{2-}	0.263	± 0.019
$Pt(NO_2)_4Cl_2^{2-}$	0.383	± 0.019	$TiBr_6^{2-}$	0.356	± 0.019
$Pt(OH)_2^{2-}$	0.333	± 0.019	$TiCl_6^{2-}$	0.335	± 0.019
$Pt(SCN)_6^{2-}$	0.451	± 0.019	TiF_6^{2-}	0.252	± 0.019
$PtBr_4^{2-}$	0.324	± 0.019	UCl_6^{2-}	0.354	± 0.019
$PtBr_6^{2-}$	0.363	± 0.019	UF_6^{2-}	0.256	± 0.019
$PtCl_4^{2-}$	0.307	± 0.019	VO_3^{2-}	0.204	± 0.019
$PtCl_6^{2-}$	0.333	± 0.019	WBr_6^{2-}	0.363	± 0.019
PtF_6^{2-}	0.245	± 0.019	WCl_6^{2-}	0.339	± 0.019
$PuCl_6^{2-}$	0.349	± 0.019	WO_4^{2-}	0.237	± 0.019
$ReBr_6^{2-}$	0.371	± 0.019	$WOCl_5^{2-}$	0.334	± 0.019
$ReCl_6^{2-}$	0.337	± 0.019	$ZnBr_4^{2-}$	0.335	± 0.019
ReF_6^{2-}	0.256	± 0.019	$ZnCl_4^{2-}$	0.306	± 0.019
ReF_8^{2-}	0.276	± 0.019	ZnF_4^{2-}	0.219	± 0.019
ReH_9^{2-}	0.257	± 0.019	ZnI_4^{2-}	0.384	± 0.019
ReI_6^{2-}	0.421	± 0.026	$ZrBr_4^{2-}$	0.334	± 0.019
RhF_6^{2-}	0.240	± 0.019	$ZrCl_4^{2-}$	0.306	± 0.019
$RuCl_6^{2-}$	0.336	± 0.019	$ZrCl_6^{2-}$	0.348	± 0.019
RuF_6^{2-}	0.248	± 0.019	ZrF_6^{2-}	0.258	± 0.019
S^{2-}	0.189	± 0.019	**Multi-Charged Anions**		
$S_2O_3^{2-}$	0.251	± 0.019	AlH_6^{3-}	0.256	± 0.042
$S_2O_4^{2-}$	0.262	± 0.019	AsO_4^{3-}	0.237	± 0.042
$S_2O_5^{2-}$	0.270	± 0.019	$CdBr_6^{4-}$	0.374	± 0.038

Table 2
THERMOCHEMICAL RADII (nm) (continued)

Ion	Radius		Ion	Radius	
$CdCl_6^{4-}$	0.352	± 0.038	Br_5^+	0.229	± 0.027
CeF_6^{3-}	0.278	± 0.038	$BrClCNH_2^+$	0.175	± 0.027
CeF_7^{3-}	0.282	± 0.038	BrF_2^+	0.183	± 0.027
$Co(CN)_6^{3-}$	0.349	± 0.038	BrF_4^+	0.172	± 0.027
$Co(NO_2)_6^{3-}$	0.343	± 0.038	$C_{10}F_8^+$	0.265	± 0.027
$CoCl_5^{3-}$	0.320	± 0.038	$C_6F_6^+$	0.228	± 0.027
CoF_6^{3-}	0.258	± 0.042	$Cl(SNSCN)_2^+$	0.347	± 0.027
$Cr(CN)_6^{3-}$	0.351	± 0.038	$Cl_2C=NH_2^+$	0.173	± 0.027
CrF_6^{3-}	0.254	± 0.042	Cl_2F^+	0.165	± 0.027
$Cu(CN)_4^{3-}$	0.312	± 0.038	Cl_3^+	0.182	± 0.027
$Fe(CN)_6^{3-}$	0.347	± 0.038	ClF_2^+	0.147	± 0.027
FeF_6^{3-}	0.298	± 0.042	ClO_2^+	0.118	± 0.027
HfF_7^{3-}	0.277	± 0.042	$GaBr_4^-$	0.317	± 0.038
InF_6^{3-}	0.268	± 0.038	I_2^+	0.185	± 0.027
$Ir(CN)_6^{3-}$	0.347	± 0.038	I_3^+	0.225	± 0.027
$Ir(NO_2)_6^{3-}$	0.338	± 0.038	I_5^+	0.263	± 0.027
$Mn(CN)_6^{3-}$	0.350	± 0.038	IBr_2^+	0.196	± 0.027
$Mn(CN)_6^{5-}$	0.401	± 0.042	ICl_2^+	0.175	± 0.036
$MnCl_6^{4-}$	0.349	± 0.038	IF_6^+	0.209	± 0.027
N^{3-}	0.180	± 0.042	$N(S_3N_2)_2^+$	0.258	± 0.027
$Ni(NO_2)_6^{3-}$	0.342	± 0.038	$N(SCl)_2^+$	0.186	± 0.027
$Ni(NO_2)_6^{4-}$	0.383	± 0.038	$N(SeCl)_2^+$	0.246	± 0.027
NiF_6^{3-}	0.250	± 0.042	$N(SF_2)_2^+$	0.214	± 0.027
O^{3-}	0.288	± 0.038	N_2F^+	0.156	± 0.027
P^{3-}	0.224	± 0.042	NO^+	0.145	± 0.027
PaF_8^{3-}	0.299	± 0.042	NO_2^+	0.153	± 0.027
PO_4^{3-}	0.230	± 0.042	O_2^+	0.140	± 0.027
PrF_6^{3-}	0.281	± 0.038	$O_2(SCCF_3Cl)_2^+$	0.275	± 0.027
$Rh(NO_2)_6^{3-}$	0.345	± 0.038	$ONCH_3CF_3^+$	0.200	± 0.027
$Rh(SCN)_6^{3-}$	0.428	± 0.042	$OsOF_5^-$	0.246	± 0.038
TaF_8^{3-}	0.284	± 0.042	$P(CH_3)_3Cl^+$	0.197	± 0.027
TbF_7^{3-}	0.290	± 0.038	$P(CH_3)_3D^+$	0.196	± 0.027
$Tc(CN)_6^{5-}$	0.410	± 0.042	PCl_4^+	0.235	± 0.027
ThF_7^{3-}	0.282	± 0.042	$ReOF_5^-$	0.245	± 0.038
$TiBr_6^{3-}$	0.315	± 0.038	$S(CH_3)_3Cl^+$	0.207	± 0.027
TlF_6^{3-}	0.271	± 0.038	$S(N(C_2H_5)_3)_3^+$	0.439	± 0.027
UF_7^{3-}	0.285	± 0.042	$S_2(CH_3)_2Cl^+$	0.265	± 0.027
YF_6^{3-}	0.275	± 0.038	$S_2(CH_3)_2CN^+$	0.223	± 0.027
ZrF_7^{3-}	0.273	± 0.038	$S_2(CH_3)_3^+$	0.233	± 0.027
Singly Charged Cations			$S_2Br_5^+$	0.267	± 0.027
$N(CH_3)_4^+$	0.234	± 0.019	S_2N^+	0.159	± 0.034
$N_2H_5^+$	0.158	± 0.019	$S_2N_2C_2H_3^+$	0.211	± 0.027
$N_2H_6^{2+}$	0.158	± 0.029	$S_2NC_2(PhCH_3)_2^+$	0.310	± 0.027
$NH(C_2H_5)_3^+$	0.274	± 0.019	$S_2NC_3H_4^+$	0.218	± 0.027
$NH_3C_2H_5^+$	0.193	± 0.019	$S_2NC_4H_8^+$	0.225	± 0.027
$NH_3C_3H_7^+$	0.225	± 0.019	$S_3(CH_3)_3^+$	0.239	± 0.027
$NH_3CH_3^+$	0.177	± 0.019	$S_3Br_3^+$	0.245	± 0.027
NH_3OH^+	0.147	± 0.019	$S_3C_3H_7^+$	0.199	± 0.027
NH_4^+	0.136	± 0.019	$S_3C_4F_6^+$	0.261	± 0.027
$NH_3C_2H_4OH^+$	0.203	± 0.019	$S_3CF_3CN^+$	0.263	± 0.027
$As_3S_4^+$	0.244	± 0.027	$S_3Cl_3^+$	0.233	± 0.027
$As_3Se_4^+$	0.253	± 0.027	$S_3N_2^+$	0.201	± 0.027
$AsCl_4^+$	0.221	± 0.027	$S_3N_2Cl^+$	0.232	± 0.027
Br_2^+	0.155	± 0.027	$S_4N_3^+$	0.231	± 0.027
Br_3^+	0.204	± 0.027	$S_4N_3(Ph)_2^+$	0.316	± 0.027
Br_3^-	0.238	± 0.027	$S_4N_4H^+$	0.178	± 0.027

Table 2
THERMOCHEMICAL RADII (nm) (continued)

Ion	Radius		Ion	Radius	
$S_5N_5^+$	0.257	± 0.027	XeF^+	0.174	± 0.027
S_7I^+	0.262	± 0.027	XeF_3^+	0.183	± 0.027
$Sb(NPPh_3)_4^+$	0.518	± 0.027	XeF_5^+	0.186	± 0.027
SBr_3^+	0.220	± 0.027	$XeOF_3^+$	0.186	± 0.027
$SCH_3O_2^+$	0.183	± 0.027	**Doubly Charged Cations**		
$SCH_3P(CH_3)_3^+$	0.248	± 0.027	$Co_2S_2(CO)_6^{2+}$	0.263	± 0.035
$SCH_3PCH_3Cl_2^+$	0.205	± 0.027	$FeW(Se)_2(CO)^{2+}$	0.260	± 0.035
$SCl(C_2H_5)_2^+$	0.207	± 0.027	I_4^{2+}	0.207	± 0.035
$SCl_2CF_3^+$	0.207	± 0.027	$Mo(Te_3)(CO)_4^{2+}$	0.234	± 0.035
$SCl_2CH_3^+$	0.204	± 0.027	S_{19}^{2+}	0.292	± 0.035
SCl_3^+	0.185	± 0.027	$S_2(S(CH_3)_2)_2^{2+}$	0.230	± 0.035
$Se_3Br_3^+$	0.253	± 0.027	$S_2I_4^{2+}$	0.231	± 0.035
$Se_3Cl_3^+$	0.245	± 0.027	$S_3N_2^{2+}$	0.184	± 0.035
$Se_3N_2^+$	0.288	± 0.042	$S_3NCCNS_3^{2+}$	0.220	± 0.035
$Se_3NC_{12}^+$	0.163	± 0.027	S_3Se^{2+}	0.326	± 0.035
Se_6I^+	0.260	± 0.027	$S_4N_4^{2+}$	0.186	± 0.035
$SeBr_3^+$	0.182	± 0.027	$S_6N_4^{2+}$	0.232	± 0.035
$SeCl_3^+$	0.192	± 0.027	S_8^{2+}	0.182	± 0.035
SeF_3^+	0.179	± 0.027	Se_{10}^{2+}	0.253	± 0.035
SeI_3^+	0.238	± 0.027	Se_{17}^{2+}	0.236	± 0.035
SeN_2Cl^+	0.196	± 0.027	Se_{19}^{2+}	0.296	± 0.035
$SeNCl_2^+$	0.157	± 0.027	$Se_2I_4^{2+}$	0.218	± 0.035
$(SeNMe_3)_3^+$	0.406	± 0.027	$Se_3N_2^{2+}$	0.182	± 0.035
$SeS_2N_2^+$	0.282	± 0.042	Se_4^{2+}	0.152	± 0.035
$SF(C_6F_5)_2^+$	0.294	± 0.027	$Se_4S_2N_4^{2+}$	0.224	± 0.035
$SF_2CF_3^+$	0.198	± 0.027	Se_8^{2+}	0.186	± 0.035
$SF_2N(CH_3)_2^+$	0.210	± 0.027	$SeN_2S_2^{2+}$	0.182	± 0.035
SF_3^+	0.172	± 0.027	$(SNP(C_2H_5)_3)_2^{2+}$	0.312	± 0.035
$SFS(C(CF_3)_2)_2^+$	0.275	± 0.027	$TaBr_6^-$	0.351	± 0.049
$SH_2C_3H_7^+$	0.210	± 0.027	$Te(trtu)_4^{2+}$	0.328	± 0.035
SN^+	0.158	± 0.027	$Te(tu)_4^{2+}$	0.296	± 0.035
$SNCl_5(CH_3CN)^-$	0.290	± 0.038	$Te_2(esu)_4Br_2^{2+}$	0.356	± 0.035
$(SNPMe_3)_3^+$	0.308	± 0.027	$Te_2(esu)_4Cl_2^{2+}$	0.361	± 0.035
$SNSC(CH_3)N^+$	0.225	± 0.027	$Te_2(esu)_4I_2^{2+}$	0.342	± 0.035
$SNSC(CN)CH^+$	0.209	± 0.027	$Te_2Se_2^{2+}$	0.192	± 0.035
$SNSC(Ph)N^+$	0.251	± 0.027	$Te_2Se_4^{2+}$	0.222	± 0.035
$SNSC(Ph)NS_3N_2^+$	0.327	± 0.027	$Te_2Se_8^{2+}$	0.252	± 0.035
$SNSC(PhCH_3)N^+$	0.264	± 0.027	$Te_3S_3^{2+}$	0.217	± 0.035
$(Te(N(SiMe_3)_2)_2^+$	0.371	± 0.027	Te_3Se^{2+}	0.193	± 0.035
$Te(N_3)_3^+$	0.226	± 0.027	Te_4^{2+}	0.169	± 0.035
$Te_4Nb_3OTe_2I_6^+$	0.407	± 0.027	Te_8^{2+}	0.187	± 0.035
$TeBr_3^+$	0.235	± 0.027	$W(CO)_4(h3-Te)^{2+}$	0.234	± 0.035
$TeCl_3^+$	0.216	± 0.027	$W_2(CO)_{10}Se_4^{2+}$	0.290	± 0.035
$TeCl_3(15\text{-crown-}5)^+$	0.282	± 0.027	**Multi-Charged Cations**		
TeI_3^+	0.243	± 0.027	I_{15}^{3+}	0.442	± 0.051
$Xe_2F_{11}^+$	0.266	± 0.027	$Te_2(su)_6^{4+}$	0.453	± 0.034
$Xe_2F_3^+$	0.221	± 0.027			

Ligand abbreviations: su = selenourea; esu = ethyleneselenourea; tu = thiourea; ph = phenyl.

Table 3
ANCILLARY THERMOCHEMICAL DATA (kJ mol^{-1})

Species	State	$\Delta_f H^o$
AsO_4^{3-}	g	(289)
BrO_3^-	g	-145
ClO_4^-	g	-344
CN^-	g	66
CO_3^{2-}	g	-321
$Fe(NO_3)_2$	c	(-448)
HF_2^-	g	-774
$HfCl_6^{2-}$	g	-1640
$IO_2F_2^-$	g	-693
IO_3^-	g	-208
$IrCl_6^{2-}$	g	-785
$LiCH_3O_2$	c	(-745)
$NbCl_6^{2-}$	g	-1224
$NH_2CH_2CO_2^-$	g	-564
O_2^{2-}	g	553
$PdCl_6^{2-}$	g	-749
PO_4^-	g	291
$PtCl_6^{2-}$	g	-774
$ReBr_6^{2-}$	g	-689
$ReCl_6^{2-}$	g	-919
$Ti(OH)_2$	c	-778

THE MADELUNG CONSTANT AND CRYSTAL LATTICE ENERGY

If U is the crystal lattice energy and M is the Madelung constant, then[a]

$$U = \frac{N M z_i z_j e^2}{r}(1 - 1/n)$$

Substance	Ion type	Crystal form[b]	M
Sodium chloride, NaCl	M^+, X^-	FCC	1.74756
Cesium chloride, CsCl	M^+, X^-	BCC	1.76267
Calcium chloride, $CaCl_2$	M^{++}, $2X^-$	Cubic	2.365
Calcium fluoride (fluorite), CaF_2	M^{++}, $2X^-$	Cubic	2.51939
Cadmium chloride, $CdCl_2$	M^{++}, $2X^-$	Hexagonal	2.244[c]
Cadmium iodide (α), CdI_2	M^{++}, $2X^-$	Hexagonal	2.355[c]
Magnesium fluoride, MgF_2	M^{++}, $2X^-$	Tetragonal	2.381[c]
Cuprous oxide (cuprite), Cu_2O	$2M^+$, X^{--}	Cubic	2.22124
Zinc oxide, ZnO	M^{++}, X^{--}	Hexagonal	1.4985[c]
Sphalerite (zinc blende), ZnS	M^{++}, X^{--}	FCC	1.63806
Wurtzite, ZnS	M^{++}, X^{--}	Hexagonal	1.64132[c]
Titanium dioxide (anatase), TiO_2	M^{4+}, $2X^{--}$	Tetragonal	2.400[c]
Titanium dioxide (rutile), TiO_2	M^{4+}, $2X^{--}$	Tetragonal	2.408[c]
β-Quartz, SiO_2	M^{4+}, $2X^{--}$	Hexagonal	2.2197[c]
Corundum, Al_2O_3	$2M^{3+}$, $3X^{--}$	Rhombohedral	4.1719

[a] N is Avogadro's number, z_i and z_j are the integral charges on the ions(in units of e), and e is the charge on the electron in electrostatic units ($e = 4.803 \times 10^{-10}$ esu). r is the shortest distance between cation-anion pairs in centimeters. Then U is in ergs (1 erg = 10^{-7} J).

[b] FCC = face centered cubic; BCC = body centered cubic.

[c] For tetragonal and hexagonal crystals the value of M depends on the details of the lattice parameters.

The Born Exponent, n is:

Ion type	n
He, Li^+	5
Ne, Na^+, F^-	7
Ar, K^+, Cu^+, Cl^-	9
Kr, Rb^+, Ag^+, Br^-	10
Xe, Cs^+, Au^+, I^-	12

For a crystal with a mixed-ion type, an average of the values of n in this table is to be used (6 for LiF, for example).

ELASTIC CONSTANTS OF SINGLE CRYSTALS
H. P. R. Frederikse

This table gives selected values of elastic constants for single crystals. The values believed most reliable were selected from the original literature. The substances are arranged by crystal system and, within each system, alphabetically by name. A reference to the original literature is given for each value; a useful compilation of published values from many sources may be found in Reference 1 below.

Data are given for the single-crystal density and for the elastic constants c_{ij}, in units of 10^{11} N/m^2, which is equivalent to 10^{12} dyn/cm^2.

GENERAL REFERENCES

1. Simmons, G., and Wang, H., *Single Crystal Elastic Constants and Calculated Aggregate Properties: A Handbook, Second Edition,* The MIT Press, Cambridge, MA, 1971.
2. Gray, D.E., Ed., *American Institute of Physics Handbook, Third Edition*, McGraw-Hill, New York, 1972.

CUBIC CRYSTALS

Name	Formula	ρ/g cm^{-3}	T/K	Ref.	C_{11}	C_{12}	C_{44}
Aluminum	Al	2.6970	298	1	1.0675	0.6041	0.2834
Aluminum antimonide	AlSb	4.3600	300	2	0.8939	0.4427	0.4155
Ammonium bromide	NH$_4$Br	2.4314	300	3	0.3414	0.0782	0.0722
Ammonium chloride	NH$_4$Cl	1.5279	290	4	0.3814	0.0866	0.0903
Argon	Ar	1.7710	4.2	5	0.0529	0.0135	0.0159
Barium fluoride	BaF$_2$	4.8860	298	6	0.9199	0.4157	0.2568
Barium nitrate	Ba(NO$_3$)$_2$	3.2560	293	7	0.2925	0.2065	0.1277
Calcium fluoride	CaF$_2$	3.810	298	8	1.6420	0.4398	0.8406
Calcium telluride	CaTe	5.8544	298	9	0.5351	0.3681	0.1994
Cesium	Cs	1.9800	78	10	0.0247	0.0206	0.0148
Cesium bromide	CsBr	4.4560	298	11	0.3063	0.0807	0.0750
Cesium chloride	CsCl	3.9880	298	11	0.3644	0.0882	0.0804
Cesium iodide	CsI	4.5250	298	11	0.2446	0.0661	0.0629
Chromite	FeCr$_2$O$_4$	4.4500	RT	12	3.2250	1.4370	1.1670
Chromium	Cr	7.20	298	13	3.398	0.586	0.990
Cobalt oxide	CoO	6.44	298	14	2.6123	1.4699	0.8300
Cobalt zinc ferrite	CoZnFeO$_2$	5.43	303	12	2.660	1.530	0.780
Copper	Cu	8.932	298	15	1.683	1.221	0.757
Gallium antimonide	GaSb	5.6137	298	16	0.8839	0.4033	0.4316
Gallium arsenide	GaAs	5.3169	298	17	1.1877	0.5372	0.5944
Gallium phosphide	GaP	4.1297	300	18	1.4120	0.6253	0.7047
Garnet (yttrium-iron)	Y$_3$Fe$_2$(FeO$_4$)$_3$	5.17	298	19	2.680	1.106	0.766
Germanium	Ge	5.313	298	20	1.2835	0.4823	0.6666
Gold	Au	19.283	296.5	21	1.9244	1.6298	0.4200
Indium antimonide	InSb	5.7890	298	22	0.6720	0.3670	0.3020
Indium arsenide	InAs	5.6720	293	23	0.8329	0.4526	0.3959
Indium phosphide	InP	4.78	RT	24	1.0220	0.5760	0.4600
Iridium	Ir	22.52	300	25	5.80	2.42	2.56
Iron	Fe	7.8672	298	26	2.26	1.40	1.16
Lead	Pb	11.34	296	27	0.4966	0.4231	0.1498
Lead fluoride	PbF$_2$	7.79	300	28	0.8880	0.4720	0.2454
Lead nitrate	Pb(NO$_3$)$_2$	4.547	293	29	0.3729	0.2765	0.1347
Lead telluride	PbTe	8.2379	303.2	30	1.0795	0.0764	0.1343
Lithium	Li	0.5326	298	31	0.1350	0.1144	0.0878
Lithium bromide	LiBr	3.47	RT	32	0.3940	0.1880	0.1910
Lithium chloride	LiCl	2.068	295	33	0.4927	0.2310	0.2495
Lithium fluoride	LiF	2.638	RT	34	1.1397	0.4767	0.6364
Lithium iodide	LiI	4.061	RT	32	0.2850	0.1400	0.1350
Magnesium oxide	MgO	3.579	298	20	2.9708	0.9536	1.5613
Magnetite	Fe$_3$O$_4$	5.18	RT	32	2.730	1.060	0.971
Manganese oxide	MnO	5.39	298	35	2.23	1.20	0.79
Mercury telluride	HgTe	8.079	290	36	0.548	0.381	0.204
Molybdenum	Mo	10.2284	273	37	4.637	1.578	1.092

CUBIC CRYSTALS (continued)

Name	Formula	ρ/g cm^{-3}	T/K	Ref.	C_{11}	C_{12}	C_{44}
Nickel	Ni	8.91	298	15	2.481	1.549	1.242
Niobium	Nb	8.578	300	38	2.4650	1.3450	0.2873
Palladium	Pd	12.038	300	39	2.2710	1.7604	0.7173
Platinum	Pt	21.50	300	40	3.4670	2.5070	0.7650
Potassium	K	0.851	295	41	0.0370	0.0314	0.0188
Potassium bromide	KBr	2.740	298	11	0.3468	0.0580	0.0507
Potassium chloride	KCl	1.984	298	11	0.4069	0.0711	0.0631
Potassium cyanide	KCN	1.553	RT	32	0.1940	0.1180	0.0150
Potassium fluoride	KF	2.480	295	33	0.6490	0.1520	0.1232
Potassium iodide	KI	3.128	300	42	0.2710	0.0450	0.0364
Pyrite	FeS$_2$	5.016	RT	43	3.818	0.310	1.094
Rubidium	Rb	1.58	170	44	0.0296	0.0250	0.0171
Rubidium bromide	RbBr	3.350	300	45	0.3152	0.0500	0.0380
Rubidium chloride	RbCl	2.797	300	45	0.3624	0.0612	0.0468
Rubidium iodide	RbI	3.551	300	45	0.2556	0.0382	0.0278
Silicon	Si	2.331	298	46	1.6578	0.6394	0.7962
Silver	Ag	10.50	300	47	1.2399	0.9367	0.4612
Silver bromide	AgBr	5.585	300	48	0.5920	0.3640	0.0616
Sodium	Na	0.971	299	49	0.0739	0.0622	0.0419
Sodium bromate	NaBrO$_3$	3.339	RT	32	0.5450	0.1910	0.1500
Sodium bromide	NaBr	3.202	300	33	0.3970	0.1001	0.0998
Sodium chlorate	NaClO$_3$	2.485	RT	50	0.4920	0.1420	0.1160
Sodium chloride	NaCl	2.163	298	11	0.4947	0.1288	0.1287
Sodium fluoride	NaF	2.804	300	51	0.9700	0.2380	0.2822
Sodium iodide	NaI	3.6689	300	52	0.3007	0.0912	0.0733
Spinel	MgAl$_2$O$_4$	3.6193	298	53	2.9857	1.5372	1.5758
Strontium fluoride	SrF$_2$·	4.277	300	54	1.2350	0.4305	0.3128
Strontium nitrate	Sr(NO$_3$)$_2$	2.989	293	29	0.4255	0.2921	0.1590
Strontium oxide	SrO	4.99	300	55	1.601	0.435	0.590
Strontium titanate	SrTiO$_3$	5.123	RT	56	3.4817	1.0064	4.5455
Tantalum	Ta	16.626	298	57	2.6023	1.5446	0.8255
Tantalum carbide	TaC	14.65	RT	58	5.05	0.73	0.79
Thallium bromide	TlBr	7.4529	298	59	0.3760	0.1458	0.0757
Thorium	Th	11.694	300	60	0.7530	0.4890	0.4780
Thorium oxide	ThO$_2$	9.991	298	61	3.670	1.060	0.797
Tin telluride	SnTe	6.445	300	62	1.1250	0.0750	0.1172
Titanium carbide	TiC	4.940	RT	107	5.00	1.13	1.75
Tungsten	W	19.257	297	64	5.2239	2.0437	1.6083
Uranium carbide	UC	13.63	300	65	3.200	0.850	0.647
Uranium dioxide	UO$_2$	10.97	298	66	3.960	1.210	0.641
Vanadium	V	6.022	300	67	2.287	1.190	0.432
Zinc selenide	ZnSe	5.262	298	68	0.8096	0.4881	0.4405
Zinc sulfide	ZnS	4.088	298	68	1.0462	0.6534	0.4613
Zinc telluride	ZnTe	5.636	298	68	0.7134	0.4078	0.3115
Zirconium carbide	ZrC	6.606	298	63	4.720	0.987	1.593

TETRAGONAL CRYSTALS

Name	Formula	ρ/g cm^{-3}	T/K	Ref.	C_{11}	C_{12}	C_{13}	C_{16}	C_{33}	C_{44}	C_{66}
Ammonium dihydrogen arsenate (ADA)	$NH_4H_2AsO_4$	2.3110	298	69	0.6747	-0.106	0.1652		0.3022	0.0685	0.0639
Ammonium dihydrogen phosphate (ADP)	$NH_4H_2PO_4$	1.8030	293	69	0.6200	-0.050	0.1400		0.3000	0.0910	0.0610
Barium titanate	$BaTiO_3$	5.9988	298	70	2.7512	1.7897	1.5156		1.6486	0.5435	1.1312
Calcium molybdate	$CaMoO_4$	4.255	298	79	1.447	0.664	0.466	0.134	1.265	0.369	0.451
Indium	In	7.300	RT	71	0.4450	0.3950	0.4050		0.4440	0.0655	0.1220
Magnesium fluoride	MgF_2	3.177	RT	72	1.237	0.732	0.536		1.770	0.552	0.978
Nickel sulfate hexahydrate	$NiSO_4 \cdot 6H_2O$	2.070	RT	73	0.3209	0.2315	0.0209		0.2931	0.1156	0.1779
Potassium dihydrogen arsenate (KDA)	KH_2AsO_4	2.867	RT	12	0.530	-0.060	-0.020		0.370	0.120	0.070
Potassium dihydrogen phosphate (KDP)	KH_2PO_4	2.388	RT	71	0.7140	-0.049	0.1290		0.5620	0.1270	0.0628
Rubidium dihydrogen phosphate (RDP)	RbH_2PO_4	2.800	298	74	0.5562	-0.064	0.0279		0.4398	0.1142	0.0350
Rutile	TiO_2	4.260	298	75	2.7143	1.7796	1.4957		4.8395	1.2443	1.9477
Tellurium oxide	TeO_2	5.99	RT	76	0.5320	0.4860	0.2120		1.0850	0.2440	0.5520
Tin (white)	Sn	7.29	288	77	0.7529	0.6156	0.4400		0.9552	0.2193	0.2336
Zircon	$ZrSiO_4$	4.70	RT	78	2.585	1.791	1.542		3.805	0.733	1.113

ORTHORHOMBIC CRYSTALS

Name	Formula	ρ/g cm^{-3}	T/K	Ref.	C_{11}	C_{12}	C_{13}	C_{22}	C_{23}	C_{33}	C_{44}	C_{55}	C_{66}
Acenaphthene	$C_{12}H_{10}$	1.220	293	80	0.1380	0.0210	0.0410	0.1262	0.0460	0.1117	0.0265	0.0290	0.0185
Ammonium sulfate	$(NH_4)_2SO_4$	1.774	293	81	0.3607	0.1651	0.1580	0.2981	0.1456	0.3534	0.1025	0.0717	0.0974
Aragonite	$CaCO_3$	2.93	RT	82	1.5958	0.3663	0.0197	0.8697	0.1597	0.8503	0.4132	0.2564	0.4274
Barite	$BaSO_4$	4.40	RT	82	0.8941	0.4614	0.2691	0.7842	0.2676	1.0548	0.1190	0.2874	0.2778
Benzene	C_6H_6	1.061	250	83	0.0614	0.0352	0.0401	0.0656	0.0390	0.0583	0.0197	0.0378	0.0153
Benzophenone	$(C_6H_5)_2CO$	1.219	RT	32	0.1070	0.0550	0.0169	0.1000	0.0321	0.0710	0.0203	0.0155	0.0353
Bronzite	$(MgFe)SiO_3$	3.38	RT	78	1.876	0.686	0.605	1.578	0.561	2.085	0.700	0.592	0.544
Calcium sulfate	$CaSO_4$	2.962	RT	84	0.9382	0.1650	0.1520	1.845	0.3173	1.1180	0.3247	0.2653	0.0926
Celestite	$SrSO_3$	3.96	RT	12	1.044	0.773	0.605	1.061	0.619	1.286	0.135	0.279	0.266
Cesium sulfate	Cs_2SO_4	4.243	293	81	0.4490	0.1958	0.1815	0.4283	0.1800	0.3785	0.1326	0.1319	0.1323
Fosterite	Mg_2SiO_4	3.224	298	85	3.2848	0.6390	0.6880	1.9980	0.7380	2.3530	0.6515	0.8120	0.8088
Iodic acid	HIO_3	4.630	RT	73	0.3030	0.1194	0.1169	0.5448	0.0548	0.4359	0.1835	0.2193	0.1736
Lithium ammonium tartrate	$LiNH_4C_4H_4O_6 \cdot 4H_2O$	1.71	RT	12	0.3864	0.1655	0.0875	0.5393	0.2007	0.3624	0.1190	0.0667	0.2326
Magnesium sulfate heptahydrate	$MgSO_4 \cdot 7H_2O$	1.68	RT	86	0.325	0.174	0.182	0.288	0.182	0.315	0.078	0.156	0.090
Natrolite	$(Na,Al)SiO_3$	2.25	RT	78	0.716	0.261	0.297	0.632	0.297	1.378	0.196	0.248	0.423
Nickel sulfate heptahydrate	$NiSO_4 \cdot 7H_2O$	1.948	RT	86	0.353	0.198	0.201	0.311	0.201	0.335	0.091	0.172	0.099
Olivine	$(MgFe)SiO_4$	3.324	RT	87	3.240	0.590	0.790	1.980	0.780	2.490	0.667	0.810	0.793
Potassium pentaborate	$KB_5O_8 \cdot 4H_2O$	1.74	RT	71	0.582	0.229	0.174	0.359	0.231	0.255	0.164	0.046	0.057
Potassium sulfate	K_2SO_4	2.665	293	81	0.5357	0.1999	0.2095	0.5653	0.1990	0.5523	0.195	0.1879	0.1424
Rochelle salt	$NaK(C_4H_4O_6) \cdot 4H_2O$	1.79	RT	71	0.255	0.141	0.116	0.381	0.146	0.371	0.134	0.032	0.098
Rubidium sulfate	Rb_2SO_4	3.621	293	81	0.5029	0.1965	0.1999	0.5098	0.1925	0.4761	0.1626	0.1589	0.1407
Sodium ammonium tartrate	$NaNH_4C_4H_4O_6 \cdot 4H_2O$	1.587	RT	12	0.3685	0.2725	0.3083	0.5092	0.3472	0.5541	0.1058	0.0303	0.0870
Sodium tartrate	$Na_2C_4H_4O_6 \cdot 2H_2O$	1.794	RT	12	0.461	0.286	0.320	0.547	0.352	0.665	0.124	0.031	0.098
Strontium formate dihydrate	$Sr(CHO_2)_2 \cdot 2H_2O$	2.25	RT	12	0.4391	0.1037	-0.149	0.3484	-0.014	0.3746	0.1538	0.1075	0.1724
Sulfur	S	2.07	RT	12	0.240	0.133	0.171	0.205	0.159	0.483	0.043	0.087	0.076
Thallium sulfate	$TlSO_4$	6.776	293	81	0.4106	0.2573	0.2288	0.3885	0.2174	0.4268	0.1125	0.1068	0.0751
Topaz	$Al_2SiO_3(OH,F)_2$	3.52	RT	82	2.8136	1.2582	0.8464	3.8495	0.8815	2.9452	1.0811	1.3298	1.3089
Uranium (alpha)	U	19.0453	293	88	2.1486	0.4622	0.2176	1.9983	1.0764	2.6763	1.2479	0.7379	0.7454
Zinc sulfate heptahydrate	$ZnSO_4 \cdot 7H_2O$	1.970	RT	86	0.3320	0.1720	0.2000	0.2930	0.1980	0.3200	0.0780	0.1530	0.0830

MONOCLINIC CRYSTALS

Name	Formula	ρ/g cm^{-3}	T/K	Ref.	C_{11}	C_{12}	C_{13}	C_{15}	C_{22}
Aegirine	$(NaFe)Si_2O_6$	3.50	RT	89	1.858	0.685	0.707	0.098	1.813
Anthracene	$C_{14}H_{10}$	1.258	RT	90	0.0852	0.0672	0.0590	-0.0192	0.1170
Cobalt sulfate heptahydrate	$CoSO_4 \cdot 7H_2O$	1.948	RT	86	0.335	0.205	0.158	0.016	0.378
Diopside	$(CaMg)Si_2O_6$	3.31	RT	91	2.040	0.884	0.0883	-0.193	1.750
Dipotassium tartrate	$KHC_4H_4O_6$	1.97	RT	12	0.4294	0.1399	0.3129	-0.0105	0.3460
Feldspar (microceine)	$KAlSi_3O_8$	2.56	RT	92	0.664	0.438	0.259	-0.033	1.710
Ferrous sulfate heptahydrate	$FeSO_4 \cdot 7H_2O$	1.898	RT	86	0.349	0.208	0.174	-0.020	0.376
Lithium sulfate monohydrate	$Li_2SO_4 \cdot H_2O$	2.221	RT	32	0.5250	0.1715	0.1730	-0.0196	0.5060
Naphthalene	$C_{10}H_8$	1.127	RT	93	0.0780	0.0445	0.0340	-0.006	0.0990
Potassium tartrate	$K_2C_4H_4O_6$	1.987	RT	32	0.3110	0.1720	0.1690	0.0287	0.3900
Sodium thiosulfate	$Na_2S_2O_3$	1.7499	RT	12	0.3323	0.1814	0.1875	0.0225	0.2953
Stilbene	$(C_6H_5CH)_2$	1.60	RT	94	0.0930	0.0570	0.0670	-0.003	0.0920
Triglycine sulfate (TGS)	$(NH_2CH_2COOH)_3 \cdot$ H_2SO_4	1.68	RT	32	0.4550	0.1720	0.1980	-0.030	0.3210

Name	C_{23}	C_{25}	C_{33}	C_{35}	C_{44}	C_{46}	C_{55}	C_{66}
Aegirine	0.626	0.094	2.344	0.214	0.692	0.077	0.510	0.474
Anthracene	0.0375	-0.0170	0.1522	-0.0187	0.0272	0.0138	0.0242	0.0399
Cobalt sulfate heptahydrate	0.158	-0.018	0.371	-0.047	0.060	0.016	0.058	0.101
Diopside	0.482	-0.196	2.380	-0.336	0.675	-0.113	0.588	0.705
Dipotassium tartrate	0.1173	0.0176	0.6816	0.0294	0.0961	-0.0044	0.1270	0.0841
Feldspar (microceine)	0.192	-0.148	1.215	-0.131	0.143	-0.015	0.238	0.361
Ferrous sulfate heptahydrate	0.172	-0.019	0.360	-0.014	0.064	0.001	0.056	0.096
Lithium sulfate monohydrate	0.0368	0.0571	0.5400	-0.0254	0.1400	-0.0054	0.1565	0.2770
Naphthalene	0.0230	-0.0270	0.1190	0.0290	0.0330	-0.0050	0.0210	0.0415
Potassium tartrate	0.1330	0.0182	0.5540	0.0710	0.0870	0.0072	0.1040	0.0826
Sodium thiosulfate	0.1713	0.0983	0.4590	-0.0678	0.0569	-0.0268	0.1070	0.0598
Stilbene	0.0485	-0.005	0.0790	-0.005	0.0325	0.0050	0.0640	0.0245
Triglycine sulfate (TGS)	0.2080	-0.0036	0.2630	-0.0500	0.0950	-0.0026	0.1110	0.0620

HEXAGONAL CRYSTALS

Name	Formula	ρ/g cm^{-3}	T/K	Ref.	C_{11}	C_{12}	C_{13}	C_{33}	C_{55}
Apatite	$Ca_5(PO_4)_3(OH,F,Cl)$	3.218	RT	12	1.667	0.131	0.655	1.396	0.663
Beryl	$Be_3Al_2Si_6O_{18}$	2.68	RT	12	2.800	0.990	0.670	2.480	0.658
Beryllium	Be	1.8477	300	95	2.923	0.267	0.140	3.364	1.625
Beryllium oxide	BeO	3.01	RT	96	4.70	1.68	1.19	4.94	1.53
Cadmium	Cd	8.652	300	97	1.1450	0.3950	0.3990	0.5085	0.1985
Cadmium selenide	CdSe	5.655	298	68	0.7046	0.4516	0.3930	0.8355	0.1317
Cadmium sulfide	CdS	4.824	298	98	0.8431	0.5208	0.4567	0.9183	0.1458
Cobalt	Co	8.836	298	99	3.071	1.650	1.027	3.581	0.755
Dysprosium	Dy	8.560	298	100	0.7466	0.2616	0.2233	0.7871	0.2427
Erbium	Er	9.064	298	100	0.8634	0.3050	0.2270	0.8554	0.2809
Gadolinium	Gd	7.888	298	101	0.6667	0.2499	0.2132	0.7191	0.2089
Hafnium	Hf	12.727	298	102	1.881	0.772	0.661	1.969	0.557
Ice	H_2O(solid)	0.920	250	103	0.1410	0.0660	0.0624	0.1515	0.0288
Indium	In	7.2788	300	104	0.4535	0.4006	0.4151	0.4515	0.0651
Magnesium	Mg	1.7364	298	105	0.5950	0.2612	0.2180	0.6155	0.1635
Rhenium	Re	21.024	298	100	6.1820	2.7530	2.0780	6.8350	1.6060
Ruthenium	Ru	12.3615	298	100	5.6260	1.8780	1.6820	6.2420	1.8060
Thallium	Tl	11.560	300	106	0.4080	0.3540	0.2900	0.5280	0.0726
Titanium	Ti	4.5063	298	102	1.6240	0.9200	0.6900	1.8070	0.4670
Titanium diboride	TiB_2	4.95	RT	107	6.90	4.10	3.20	4.40	2.50
Yttrium	Y	4.472	300	108	0.7790	0.2850	0.2100	0.7690	0.2431
Zinc	Zn	7.134	295	109	1.6368	0.3640	0.5300	0.6347	0.3879
Zinc oxide	ZnO	5.6760	298	110	2.0970	1.2110	1.0510	2.1090	0.4247
Zinc sulfide	ZnS	4.089	298	96	1.2420	0.6015	0.4554	1.4000	0.2864
Zirconium	Zr	6.505	298	102	1.434	0.728	0.653	1.648	0.320

TRIGONAL CRYSTALS

Name	Formula	ρ/g cm^{-3}	T/K	Ref.	C_{11}	C_{12}	C_{13}	C_{14}	C_{33}	C_{44}
Aluminum oxide	Al_2O_3	3.986	300	111	4.9735	1.6397	1.1220	-0.2358	4.9911	1.4739
Aluminum phosphate	$AlPO_4$	2.556	RT	73	1.0503	0.2934	0.6927	-0.1271	1.3353	0.2314
Antimony	Sb	6.70	295	112	1.0130	0.3450	0.2920	0.2090	0.4500	0.3930
Bismuth	Bi	9.80	295	112	0.6370	0.2490	0.2470	0.0717	0.3820	0.1123
Calcite	$CaCO_3$	2.712	300	113	1.4806	0.5578	0.5464	-0.2058	0.8557	0.3269
Hematite	Fe_2O_3	5.240	RT	82	2.4243	0.5464	0.1542	-0.1247	2.2734	0.8569
Lithium niobate	$LiNbO_3$	4.70	RT	114	2.030	0.530	0.750	0.090	2.450	0.600
Lithium tantalate	$LiTaO_3$	7.45	RT	114	2.330	0.470	0.800	-0.110	2.750	0.940
Quartz	SiO_2	2.6485	298	115	0.8680	0.0704	0.1191	-0.1804	1.0575	0.5820
Selenium	Se	4.838	300	116	0.1870	0.0710	0.2620	0.0620	0.7410	0.1490
Sodium nitrate	$NaNO_3$	2.27	RT	12	0.8670	0.1630	0.1600	0.0820	0.3740	0.2130
Tourmaline		3.05	RT	82	2.7066	0.6927	0.0872	-0.0774	1.6070	0.6682

REFERENCES

1. Thomas, J. F., *Phys. Rev.*, 175, 955-962, 1968.
2. Bolef, D. I. and M. Menes, *J. Appl. Phys.*, 31, 1426-1427, 1960.
3. Garland, C. W. and C. F. Yarnell, *J. Chem. Phys.*, 44, 1112-1120, 1966.
4. Garland, C. W. and R. Renard, *J. Chem. Phys.*, 44, 1130-1139, 1966.

ELASTIC CONSTANTS OF SINGLE CRYSTALS (continued)

5. Gsänger, M., H. Egger and E. Lüscher, *Phys. Letters*, 27A, 695-696, 1968.
6. Wong, C. and D. E. Schuele, *J. Phys. Chem. Solids*, 29, 1309-1330, 1968.
7. Haussühl, S., *Phys. Stat. Sol.*, 3, 1072-1076, 1963.
8. Wong, C. and D. E. Schuele, *J. Phys. Chem. Solids*, 28, 1225-1231, 1967.
9. McSkimin, H. J. and D. G. Thomas, *J. Appl. Phys.*, 33, 56-59, 1962.
10. Kollarits, F. J. and J. Trivisonno, *J. Phys. Chem. Solids*, 29, 2133-2139, 1968.
11. Slagle, D. D. and H. A. McKinstry, *J. Appl. Phys.*, 38, 446-458, 1967.
12. Hearmon, R. F. S., *Adv. Phys.*, 5, 323-382, 1956.
13. Sumer, A. and J. F. Smith, *J. Appl. Phys.*, 34, 2691-2694, 1963.
14. Alexandrov, K. S. et. al., *Sov. Phys. Sol. State*, 10, 1316-1321, 1968.
15. Epstein, S. G. and O. N. Carlson, *Acta Metal.*, 13, 487-491, 1965.
16. McSkimin, H. J., et. al., *J. Appl. Phys.*, 39, 4127-4128, 1968.
17. McSkimin, H. J., et. al., *J. Appl. Phys.*, 38, 2362-2364, 1967.
18. Weil, R. and W. O. Groves, *J. Appl. Phys.*, 39, 4049-4051, 1968.
19. Bateman, T. B., *J. Appl. Phys.*, 37, 2194-2195, 1966.
20. Bogardus, E. H., *J. Appl. Phys.*, 36, 2504-2513, 1965.
21. Golding, B., S. C. Moss and B. L. Averbach, *Phys. Rev.*, 158, 637-645, 1967.
22. Bateman, T. B., H. J. McSkimin and J. M. Whelan, *J. Appl. Phys.*, 30, 544-545, 1959.
23. Gerlich, D., *J. Appl. Phys.*, 35, 3062, 1964.
24. Hickernell, F. S. and W. R. Gayton, *J. Appl. Phys.*, 37, 462, 1966.
25. MacFarlane, R. E., et. al., *Phys. Letters*, 20, 234-235, 1966.
26. Leese, J. and A. E. Lord Jr., *J. Appl. Phys.*, 39, 3986-3988, 1968.
27. Miller, R. A. and D. E. Schuele, *J. Phys. Chem. Solids*, 30, 589-600, 1969.
28. Wasilik, J. H. and M. L. Wheat, *J. Appl. Phys.*, 36, 791-793, 1965.
29. Haussühl, S., *Phys. Stat. Sol.*, 3, 1072-1076, 1963.
30. Houston, B., et. al., *J. Appl. Phys.*, 39, 3913-3916, 1968.
31. Trivisonno, J. and C. S. Smith, *Acta Metal.*, 9, 1064-1071, 1961.
32. Alexandrov, K. S. and T. V. Ryzhova, *Sov. Phys. Cryst.*, 6, 228-252, 1961.
33. Lewis, J. T., A. Lehoczky and C. V. Briscoe, *Phys. Rev.*, 161, 877-887, 1967.
34. Drabble, J. R. and R. E. B. Strathen, *Proc. Phys. Soc.*, 92, 1090-1995, 1967.
35. Oliver, D. W., *J. Appl. Phys.*, 40, 893, 1969.
36. Alper, T., and G. A. Saunders, *J. Phys. Chem. Solids*, 28, 1637-1642, 1967.
37. Dickinson, J. M. and P. E. Armstrong, *J. Appl. Phys.*, 38, 602-606, 1967.
38. Bolef, D. I., *J. Appl. Phys.*, 32, 100-105, 1961.
39. Rayne, J. A., *Phys. Rev.*, 112, 1125-1130, 1958.
40. MacFarlane, R. E., et. al., *Phys. Letters*, 18, 91-92, 1965.
41. Smith, P. A. and C. S. Smith, *J. Phys. Chem. Solids*, 26, 279-289, 1965.
42. Norwood, M. H. and C. V. Briscoe, *Phys. Rev.*, 112, 45-48, 1958.
43. Simmons, G. and F. Birch, *J. Appl. Phys.*, 34, 2736-2738, 1963.
44. Gutman, E. J. and J. Trivisonno, *J. Phys. Chem. Sol.*, 28, 805-809, 1967.
45. Ghafelehbashi, M., et. al., *J. Appl. Phys.*, 41, 652-666, 1970.
46. McSkimin, H. J. and P. Andreatch Jr., *J. Appl. Phys.*, 35, 2161-2165, 1964.
47. Neighbours, J. R. and G. A. Alers, *Phys. Rev.*, 111, 707-712, 1958.
48. Hidshaw, W., J. T. Lewis, and C. V. Briscoe, *Phys. Rev.*, 163, 876-881, 1967.
49. Daniels, W. B., *Phys. Rev.*, 119, 1246-1252, 1960.
50. Viswanathan, R., *J. Appl. Phys.*, 37, 884-886, 1966.
51. Miller, R. A. and C. S. Smith, *J. Phys. Chem. Sol.*, 25, 1279-1292, 1964.
52. Claytor, R. N. and B. J. Marshall, *Phys. Rev.*, 120, 332-334, 1960.
53. Schreiber, E., *J. Appl. Phys.*, 38, 2508-2511, 1967.
54. Gerlich, D., *Phys. Rev.*, 136, A1366-A1368, 1964.
55. Johnston, D. L., P. H. Thrasher and R. J. Kearney, *J. Appl. Phys.*, 41, 427-428, 1970.
56. Poindexter, E. and A. A. Giardini, *Phys. Rev.*, 110, 1069, 1958.
57. Soga, N., *J. Appl. Phys.*, 37, 3416-3420, 1966.
58. Bartlett, R. W. and C. W. Smith, *J. Appl. Phys.*, 38, 5428-5429, 1967.
59. Morse, G. E. and A. W. Lawson, *J. Phys. Chem. Sol.*, 28, 939-950, 1967.
60. Armstrong, P. E., O. N. Carlson and J. F. Smith, *J. Appl. Phys.*, 30, 36-41, 1959.
61. Macedo, P. M., W. Capps and J. B. Wachtman, *J. Am. Cer. Soc.*, 47, 651, 1964.
62. Beattie, A. G., *J. Appl. Phys.*, 40, 4818-4821, 1969.
63. Chang, R. and L. J. Graham, *J. Appl Phys.*, 37, 3778-3783, 1966.
64. Lowrie, R. and A. M. Gonas, *J. Appl. Phys.*, 38, 4505-4509, 1967.
65. Graham, L. J., H. Nadler and R. Chang, *J. Appl. Phys.*, 34, 1572-1573, 1963.

66. Wachtman, J. B. Jr., et. al., *J. Nucl. Mat.*, 16, 39-41, 1965.
67. Bolef, D. I., *J. Appl. Phys.*, 32, 100-105, 1961.
68. Berlincourt, D., H. Jaffe and L. R. Shiozawa, *Phys. Rev.*, 129, 1009-1017, 1963.
69. Adhav. R. S. J. *Acoust. Soc. Am.*, 43, 835-838, 1968.
70. Berlincourt, D. and H. Jaffe, *Phys. Rev.*, 111, 143-148, 1958.
71. Huntington, H. B., in *Solid State Pysics, Vol. 7*, Seitz, F., and Turnbull, D., Ed., pp. 213-285; Academic Press, New York 1958.
72. Cutler, H. R., J. J. Gibson and K. A. McCarthy, *Sol. State Comm.*, 6, 431-433, 1968.
73. Mason, W. P., *Piezoelectric Crystals and Their Application to Ultrasonics*, D. Van Nostrand Co., Inc., New York, 1950.
74. Adhav, R. S., *J. Appl. Phys.*, 40, 2725-2727, 1969.
75. Manghnani, M. H., *J. Geophys. Res.*, 74, 4317-4328, 1969.
76. Uchida, N. and Y. Ohmachi, *J. Appl Phys.*, 40, 4692-4695, 1969.
77. House, D. G. and E. Y. Vernon, *Br. J. Appl. Phys.*, 11, 254-259, 1960.
78. Ryzhova, T. V., et. al., *Bull. Acad. Sci. USSR, Earth Phys. Ser.*, English Transl., no. 2, 111-113, 1966.
79. Alton, W. J. and A. J. Barlow, *J. Appl. Phys.*, 38, 3817-3820, 1967.
80. Michard, F., et. al., *C. R. Acad. Sci., Paris*, 265, 565-567, 1967.
81. Haussühl, S., *Acta Cryst.*, 18, 839-842, 1965.
82. Hearmon, R. F. S., *Rev. Mod. Phys.*, 18, 409-440, 1946.
83. Heseltine, J. C. W., D. W. Elliott and O. B. Wilson, *J. Chem. Phys.*, 40, 2584-2587, 1964.
84. Schwerdtner, W. M., et. al., *Canad. J. Earth Sci.*, 2, 673-683, 1965.
85. Kumazawa, M. and O. L. Anderson, *J. Geophys. Res.*, 74, 5961-5972, 1969 .
86. Alexandrov, K. S., et. al., *Sov. Phys. Cryst.*, 7, 753-755, 1963.
87. Verma, R. K., *J. Geophys. Soc.*, 65, 757-766, 1960.
88. McSkimin, H. J. and E. S. Fisher, *J. Appl. Phys.*, 31, 1627-1639, 1960.
89. Alexandrov, K. S. and T.V. Ryzhova, *Bull. Acad. Sci. USSR, Geophys. Ser.*, English Transl., no.8, 871-875, 1961.
90. Afanaseva, G. K., et. al, *Phys. Stat. Sol.*, 24, K61-K63, 1967.
91. Alexandrov, K. S., et. al., *Sov. Phys. Cryst.*, 8, 589-591, 1964.
92. Alexandrov, K. S. and T. V Ryzhova, *Bull Acad. Sci. USSR, Geophys. Ser.*, English Transl., no.2, 129-131, 1962.
93. Alexandrov, K. S., et. al., *Sov. Phys. Cryst.*, 8, 164-166, 1963.
94. Teslenko, V. F., et. al., *Sov. Phys. Cryst.*, 10, 744-747, 1966.
95. Smith, J. F. and C. L. Arbogast, *J. Appl. Phys.*, 31, 99-102, 1960.
96. Cline, C. F., H. L. Dunegan and G. M. Henderson, *J. Appl. Phys.*, 38, 1944-1948, 1967.
97. Chang, Y. A. and L. Himmel, *J. Appl. Phys.*, 37, 3787-3790, 1966.
98. Gerlich, D., *J. Phys. Chem. Solids*, 28, 2575-2579, 1967.
99. McSkimin, H. J., *J. Appl. Phys.*, 26, 406-409, 1955.
100. Fisher, E. S. and D. Dever, *Trans. Met. Soc. AIME*, 239, 48-57, 1967.
101. Fisher, E. S. and D. Dever, *Proc. Conf. Rare Earth Res.*, 6th, Gatlinburg, Tenn., 522-533, 1967.
102. Fisher, E. S. and C. J. Renken, Phys. Rev., 135, A482-A494, 1964.
103. Proctor, T. M. Jr., *J. Acoust. Soc. Am.*, 39, 972-977, 1966.
104. Chandrasekhar, B. S. and J. A. Rayne, *Phys. Rev.*, 124, 1011-1041, 1961.
105. Wazzan, A. R. and L. B. Robinson, *Phys. Rev.*, 155, 586-594, 1967.
106. Ferris, R. W., et. al., *J. Appl. Phys.*, 34, 768-770, 1963.
107. Gilman, J. J. and B. W. Roberts, *J. Appl. Phys.*, 32, 1405, 1961.
108. Smith, J. F. and J. A. Gjevre, *J. Appl. Phys.*, 31, 645-647, 1960.
109. Alers, G. A. and J. R. Neighbours, *J. Phys. Chem. Solids*, 7, 58-64, 1908.
110. Bateman, T. B., *J. Appl. Phys.*, 33, 3309-3312, 1962.
111. Tefft, W. E., *J. Res. Natl. Bur. Stand.*, 70A, 277-280, 1966.
112. DeBretteville, Jr., A. et. al., *Phys. Rev.*, 148, 575-579, 1966.
113. Dandekar, D. P. and A. L. Ruoff, *J. Appl. Phys.*, 39, 6004-6009, 1968.
114. Warner, A. W., M. Onoe and G. A. Coquin, *J. Acoust. Soc. Am.*, 42, 1223-1231, 1967.
115. McSkimin, H. J., P. Andreatch and R. N. Thurston, *J. Appl. Phys.*, 36, 1624-1632, 1965.
116. Mort, J., *J. Appl. Phys.*, 38, 3414-3415, 1967.

ELECTRICAL RESISTIVITY OF PURE METALS

The first part of this table gives the electrical resistivity, in units of 10^{-8} Ω m, for 28 common metallic elements as a function of temperature. The data refer to polycrystalline samples. The number of significant figures indicates the accuracy of the values. However, at low temperatures (especially below 50 K) the electrical resistivity is extremely sensitive to sample purity. Thus the low-temperature values refer to samples of specified purity and treatment. The references should be consulted for further information on this point, as well as for values at additional temperatures.

The second part of the table gives resistivity values in the neighborhood of room temperature for other metallic elements that have not been studied over an extended temperature range.

REFERENCES

1. C. Y. Ho, et al., *J. Phys. Chem. Ref. Data*, 12, 183—322, 1983; 13, 1069—1096, 1984; 13, 1097—1130, 1984, 13, 1131—1172, 1984.
2. R. A. Matula, *J. Phys Chem. Ref. Data*, 8, 1147—1298, 1979.
3. T. C. Chi, *J. Phys. Chem. Ref. Data*, 8, 339—438, 1979; 8, 439—498, 1979.
4. K. H. Hellwege, Ed., *Landolt-Börnstein Numerical Data and Functional Relationships in Science and Technology*, Group III, Vol. 15, Subvolume a, Springer-Verlag, Heidelberg, 1982.
5. L. A. Hall, *Survey of Electrical Resistivity Measurements on 16 Pure Metals in the Temperature Range 0 to 273 K*, NBS Technical Note 365, U.S. Superintendent of Documents, 1968.

ELECTRICAL RESISTIVITY IN 10^{-8} Ω m

T/K	Aluminum	Barium	Beryllium	Calcium	Cesium	Chromium	Copper
1	0.000100	0.081	0.0332	0.045	0.0026		0.00200
10	0.000193	0.189	0.0332	0.047	0.243		0.00202
20	0.000755	0.94	0.0336	0.060	0.86		0.00280
40	0.0181	2.91	0.0367	0.175	1.99		0.0239
60	0.0959	4.86	0.067	0.40	3.07		0.0971
80	0.245	6.83	0.075	0.65	4.16		0.215
100	0.442	8.85	0.133	0.91	5.28	1.6	0.348
150	1.006	14.3	0.510	1.56	8.43	4.5	0.699
200	1.587	20.2	1.29	2.19	12.2	7.7	1.046
273	2.417	30.2	3.02	3.11	18.7	11.8	1.543
293	2.650	33.2	3.56	3.36	20.5	12.5	1.678
298	2.709	34.0	3.70	3.42	20.8	12.6	1.712
300	2.733	34.3	3.76	3.45	21.0	12.7	1.725
400	3.87	51.4	6.76	4.7		15.8	2.402
500	4.99	72.4	9.9	6.0		20.1	3.090
600	6.13	98.2	13.2	7.3		24.7	3.792
700	7.35	130	16.5	8.7		29.5	4.514
800	8.70	168	20.0	10.0		34.6	5.262
900	10.18	216	23.7	11.4		39.9	6.041

T/K	Gold	Hafnium	Iron	Lead	Lithium	Magnesium	Manganese
1	0.0220	1.00	0.0225		0.007	0.0062	7.02
10	0.0226	1.00	0.0238		0.008	0.0069	18.9
20	0.035	1.11	0.0287		0.012	0.0123	54
40	0.141	2.52	0.0758		0.074	0.074	116
60	0.308	4.53	0.271		0.345	0.261	131
80	0.481	6.75	0.693	4.9	1.00	0.557	132
100	0.650	9.12	1.28	6.4	1.73	0.91	132
150	1.061	15.0	3.15	9.9	3.72	1.84	136
200	1.462	21.0	5.20	13.6	5.71	2.75	139
273	2.051	30.4	8.57	19.2	8.53	4.05	143
293	2.214	33.1	9.61	20.8	9.28	4.39	144
298	2.255	33.7	9.87	21.1	9.47	4.48	144
300	2.271	34.0	9.98	21.3	9.55	4.51	144
400	3.107	48.1	16.1	29.6	13.4	6.19	147
500	3.97	63.1	23.7	38.3		7.86	149

T/K	Gold	Hafnium	Iron	Lead	Lithium	Magnesium	Manganese
600	4.87	78.5	32.9			9.52	151
700	5.82		44.0			11.2	152
800	6.81		57.1			12.8	
900	7.86					14.4	

T/K	Molybdenum	Nickel	Palladium	Platinum	Potassium	Rubidium	Silver
1	0.00070	0.0032	0.0200	0.002	0.0008	0.0131	0.00100
10	0.00089	0.0057	0.0242	0.0154	0.0160	0.109	0.00115
20	0.00261	0.0140	0.0563	0.0484	0.117	0.444	0.0042
40	0.0457	0.068	0.334	0.409	0.480	1.21	0.0539
60	0.206	0.242	0.938	1.107	0.90	1.94	0.162
80	0.482	0.545	1.75	1.922	1.34	2.65	0.289
100	0.858	0.96	2.62	2.755	1.79	3.36	0.418
150	1.99	2.21	4.80	4.76	2.99	5.27	0.726
200	3.13	3.67	6.88	6.77	4.26	7.49	1.029
273	4.85	6.16	9.78	9.6	6.49	11.5	1.467
293	5.34	6.93	10.54	10.5	7.20	12.8	1.587
298	5.47	7.12	10.73	10.7	7.39	13.1	1.617
300	5.52	7.20	10.80	10.8	7.47	13.3	1.629
400	8.02	11.8	14.48	14.6			2.241
500	10.6	17.7	17.94	18.3			2.87
600	13.1	25.5	21.2	21.9			3.53
700	15.8	32.1	24.2	25.4			4.21
800	18.4	35.5	27.1	28.7			4.91
900	21.2	38.6	29.4	32.0			5.64

T/K	Sodium	Strontium	Tantalum	Tungsten	Vanadium	Zinc	Zirconium
1	0.0009	0.80	0.10	0.000016		0.0100	0.250
10	0.0015	0.80	0.102	0.000137	0.0145	0.0112	0.253
20	0.016	0.92	0.146	0.00196	0.039	0.0387	0.357
40	0.172	1.70	0.751	0.0544	0.304	0.306	1.44
60	0.447	2.68	1.65	0.266	1.11	0.715	3.75
80	0.80	3.64	2.62	0.606	2.41	1.15	6.64
100	1.16	4.58	3.64	1.02	4.01	1.60	9.79
150	2.03	6.84	6.19	2.09	8.2	2.71	17.8
200	2.89	9.04	8.66	3.18	12.4	3.83	26.3
273	4.33	12.3	12.2	4.82	18.1	5.46	38.8
293	4.77	13.2	13.1	5.28	19.7	5.90	42.1
298	4.88	13.4	13.4	5.39	20.1	6.01	42.9
300	4.93	13.5	13.5	5.44	20.2	6.06	43.3
400		17.8	18.2	7.83	28.0	8.37	60.3
500		22.2	22.9	10.3	34.8	10.82	76.5
600		26.7	27.4	13.0	41.1	13.49	91.5
700		31.2	31.8	15.7	47.2		104.2
800		35.6	35.9	18.6	53.1		114.9
900			40.1	21.5	58.7		123.1

Element	T/K	Electrical resistivity $10^{-8}\,\Omega$ m
Antimony	273	39
Bismuth	273	107
Cadmium	273	6.8
Cerium (β, hex)	290—300	82.8
Cerium (γ, cub)	298	74.4
Cobalt	273	5.6
Dysprosium	290—300	92.6
Erbium	290—300	86.0
Europium	290—300	90.0
Gadolinium	290—300	131
Gallium	273	13.6
Holmium	290—300	81.4
Indium	273	8.0
Iridium	273	4.7
Lanthanum	290—300	61.5
Lutetium	290—300	58.2
Mercury	298	96.1
Neodymium	290—300	64.3
Niobium	273	15.2
Osmium	273	8.1
Polonium	273	40
Praseodymium	290—300	70.0
Promethium	290—300	75 est.
Protactinium	273	17.7
Rhenium	273	17.2
Rhodium	273	4.3
Ruthenium	273	7.1
Samarium	290—300	94.0
Scandium	290—300	56.2
Terbium	290—300	115
Thallium	273	15
Thorium	273	14.7
Thulium	290—300	67.6
Tin	273	11.5
Titanium	273	39
Uranium	273	28
Ytterbium	290—300	25.0
Yttrium	290—300	59.6

ELECTRICAL RESISTIVITY OF SELECTED ALLOYS

These values were obtained by fitting all available measurements to a theoretical formulation describing the temperature and composition dependence of the electrical resistivity of metals. Some of the values listed here fall in regions of temperature and composition where no actual measurements exist. Details of the procedure may be found in the reference.

Values of the resistivity are given in units of 10^{-8} Ω m. General comments in the preceding table for pure metals also apply here.

REFERENCE

C.Y. Ho, et al., *J. Phys. Chem. Ref. Data*, 12, 183-322, 1983.

Aluminum-Copper

Wt % Al	100 K	273 K	293 K	300 K	350 K	400 K
99[a]	0.531	2.51	2.74	2.82	3.38	3.95
95[a]	0.895	2.88	3.10	3.18	3.75	4.33
90[b]	1.38	3.36	3.59	3.67	4.25	4.86
85[b]	1.88	3.87	4.10	4.19	4.79	5.42
·80[b]	2.34	4.33	4.58	4.67	5.31	5.99
70[b]	3.02	5.03	5.31	5.41	6.16	6.94
60[b]	3.49	5.56	5.88	5.99	6.77	7.63
50[b]	4.00	6.22	6.55	6.67	7.55	8.52
40[c]		7.57	7.96	8.10	9.12	10.2
30[c]		11.2	11.8	12.0	13.5	15.2
25[f]		16.3	17.2	17.6	19.8	22.2
15[h]			12.3			
10[g]	8.71	10.8	11.0	11.1	11.7	12.3
5[c]	7.92	9.43	9.61	9.68	10.2	10.7
1[b]	3.22	4.46	4.60	4.65	5.00	5.37

Aluminum-Magnesium

Wt % Al	100 K	273 K	293 K	300 K	350 K	400 K
99[c]	0.958	2.96	3.18	3.26	3.82	4.39
95[c]	3.01	5.05	5.28	5.36	5.93	6.51
90[c]	5.42	7.52	7.76	7.85	8.43	9.02
10[b]	14.0	17.1	17.4	17.6	18.4	19.2
5[b]	9.93	13.1	13.4	13.5	14.3	15.2
1[a]	2.78	5.92	6.25	6.37	7.20	8.03

Copper-Gold

Wt % Cu	100 K	273 K	293 K	300 K	350 K	400 K
99[c]	0.520	1.73	1.86	1.91	2.24	2.58
95[c]	1.21	2.41	2.54	2.59	2.92	3.26
90[c]	2.11	3.29	4.42	3.46	3.79	4.12
85[c]	3.01	4.20	4.33	4.38	4.71	5.05
80[c]	3.95	5.15	5.28	5.32	5.65	5.99
70[c]	5.91	7.12	7.25	7.30	7.64	7.99
60[c]	8.04	9.18	9.13	9.36	9.70	10.05
50[c]	9.88	11.07	11.20	11.25	11.60	11.94
40[c]	11.44	12.70	12.85	12.90	13.27	13.65
30[c]	12.43	13.77	13.93	13.99	14.38	14.78
25[c]	12.59	13.93	14.09	14.14	14.54	14.94
15[c]	11.38	12.75	12.91	12.96	13.36	13.77
10[c]	9.33	10.70	10.86	10.91	11.31	11.72
5[c]	5.91	7.25	7.41	7.46	7.87	8.28
1[c]	2.00	3.40	3.57	3.62	4.03	4.45

Copper-Nickel

Wt % Cu	100 K	273 K	293 K	300 K	350 K	400 K
99[c]	1.45	2.71	2.85	2.91	3.27	3.62
95[c]	6.19	7.60	7.71	7.82	8.22	8.62
90[c]	12.08	13.69	13.89	13.96	14.40	14.81
85[c]	18.01	19.63	19.83	19.90	20.32	20.70
80[c]	23.89	25.46	25.66	25.72	26.12	26.44
70[i]	35.73	36.67	36.72	36.76	36.85	36.89
60[i]	45.76	45.43	45.38	43.35	45.20	45.01
50[i]	50.22	50.19	50.05	50.01	49.73	49.50
40[c]	36.77	47.42	47.73	47.82	48.28	48.49
30[i]	26.73	40.19	41.79	42.34	44.51	45.40
25[c]	22.22	33.46	35.11	35.69	39.67	42.81
15[c]	13.49	22.00	23.35	23.85	27.60	31.38
10[c]	9.28	16.65	17.82	18.26	21.51	25.19
5[c]	5.20	11.49	12.50	12.90	15.69	18.78
1[c]	1.81	7.23	8.08	8.37	10.63	13.18

Copper-Palladium

Wt % Cu	100 K	273 K	293 K	300 K	350 K	400 K
99[c]	0.91	2.10	2.23	2.27	2.59	2.92
95[c]	2.99	4.21	4.35	4.40	4.74	5.08
90[c]	5.69	6.89	7.03	7.08	7.41	7.74
85[c]	8.30	9.48	9.61	9.66	10.01	10.36
80[c]	10.74	11.99	12.12	12.16	12.51	12.87
70[c]	15.67	16.87	17.01	17.06	17.41	17.78
60[c]	20.45	21.73	21.87	21.92	22.30	22.69
50[c]	26.07	27.62	27.79	27.86	28.25	28.64
40[c]	33.53	35.31	35.51	35.57	36.03	36.47
30[c]	45.03	46.50	46.66	46.71	47.11	47.47
25[c]	44.12	46.25	46.45	46.52	46.99	47.43
15[c]	31.79	36.52	36.99	37.16	38.28	39.35
10[c]	23.00	28.90	29.51	29.73	31.19	32.56
5[c]	13.09	20.00	20.75	21.02	22.84	24.54
1[c]	8.97	11.90	12.67	12.93	14.82	16.68

Copper-Zinc

Wt % Cu	100 K	273 K	293 K	300 K	350 K	400 K
99[b]	0.671	1.84	1.97	2.02	2.36	2.71
95[b]	1.54	2.78	2.92	2.97	3.33	3.69
90[b]	2.33	3.66	3.81	3.86	4.25	4.63
85[b]	2.93	4.37	4.54	4.60	5.02	5.44
80[b]	3.44	5.01	5.19	5.26	5.71	6.17
70[b]	4.08	5.87	6.08	6.15	6.67	7.19

Gold-Palladium

Wt % Au	100 K	273 K	293 K	300 K	350 K	400 K
99[c]	1.31	2.69	2.86	2.91	3.32	3.73
95[c]	3.88	5.21	5.35	5.41	5.79	6.17
90[i]	6.70	8.01	8.17	8.22	8.56	8.93
85[b]	9.14	10.50	10.66	10.72	11.10	11.48
80[b]	11.23	12.75	12.93	12.99	13.45	13.93
70[c]	16.44	18.23	18.46	18.54	19.10	19.67
60[b]	24.64	26.70	26.94	27.02	27.63	28.23

Gold-Palladium (continued)

Wt % Au	100 K	273 K	293 K	300 K	350 K	400 K
50[a]	23.09	27.23	27.63	27.76	28.64	29.42
40[a]	19.40	24.65	25.23	25.42	26.74	27.95
30[b]	14.94	20.82	21.49	21.72	23.35	24.92
25[b]	12.72	18.86	19.53	19.77	21.51	23.19
15[a]	8.54	15.08	15.77	16.01	17.80	19.61
10[a]	6.54	13.25	13.95	14.20	16.00	17.81
5[a]	4.58	11.49	12.21	12.46	14.26	16.07
1[a]	3.01	10.07	10.85	11.12	12.99	14.80

Gold-Silver

Wt % Au	100 K	273 K	293 K	300 K	350 K	400 K
99[b]	1.20	2.58	2.75	2.80	3.22	3.63
95[a]	3.16	4.58	4.74	4.79	5.19	5.59
90[j]	5.16	6.57	6.73	6.78	7.19	7.58
85[j]	6.75	8.14	8.30	8.36	8.75	9.15
80[j]	7.96	9.34	9.50	9.55	9.94	10.33
70[j]	9.36	10.70	10.86	10.91	11.29	11.68
60[j]	9.61	10.92	11.07	11.12	11.50	11.87
50[j]	8.96	10.23	10.37	10.42	10.78	11.14
40[j]	7.69	8.92	9.06	9.11	9.46	9.81
30[a]	6.15	7.34	7.47	7.52	7.85	8.19
25[a]	5.29	6.46	6.59	6.63	6.96	7.30
15[a]	3.42	4.55	4.67	4.72	5.03	5.34
10[a]	2.44	3.54	3.66	3.71	4.00	4.31
5[i]	1.44	2.52	2.64	2.68	2.96	3.25
1[b]	0.627	1.69	1.80	1.84	2.12	2.42

Iron-Nickel

Wt % Fe	100 K	273 K	293 K	300 K	400 K
99[a]	3.32	10.9	12.0	12.4	18.7
95[c]	10.0	18.7	19.9	20.2	26.8
90[c]	14.5	24.2	25.5	25.9	33.2
85[c]	17.5	27.8	29.2	29.7	37.3
80[c]	19.3	30.1	31.6	32.2	40.0
70[b]	20.9	32.3	33.9	34.4	42.4
60[c]	28.6	53.8	57.1	58.2	73.9
50[d]	12.3	28.4	30.6	31.4	43.7
40[d]	7.73	19.6	21.6	22.5	34.0
30[c]	5.97	15.3	17.1	17.7	27.4
25[b]	5.62	14.3	15.9	16.4	25.1
15[c]	4.97	12.6	13.8	14.2	21.1
10[c]	4.20	11.4	12.5	12.9	18.9
5[c]	3.34	9.66	10.6	10.9	16.1
1[b]	1.66	7.17	7.94	8.12	12.8

Silver-Palladium

Wt % Ag	100 K	273 K	293 K	300 K	350 K	400 K
99[b]	0.839	1.891	2.007	2.049	2.35	2.66
95[b]	2.528	3.58	3.70	3.74	4.04	4.34
90[b]	4.72	5.82	5.94	5.98	6.28	6.59
85[k]	6.82	7.92	8.04	8.08	8.38	8.68
80[k]	8.91	10.01	10.13	10.17	10.47	10.78

Silver-Palladium (continued)

Wt % Ag	100 K	273 K	293 K	300 K	350 K	400 K
70[k]	13.43	14.53	14.65	14.69	14.99	15.30
60[i]	19.4	20.9	21.1	21.2	21.6	22.0
50[k]	29.3	31.2	31.4	31.5	32.0	32.4
40[m]	40.8	42.2	42.2	42.2	42.3	42.3
30[b]	37.1	40.4	40.6	40.7	41.3	41.7
25[k]	32.4	36.67	37.06	37.19	38.1	38.8
15[i]	21.0	27.08	26.68	27.89	29.3	30.6
10[i]	14.95	21.69	22.39	22.63	24.3	25.9
5[b]	8.91	15.98	16.72	16.98	18.8	20.5
1[a]	3.97	11.06	11.82	12.08	13.92	15.70

[a] Uncertainty in resistivity is ± 2%.

[b] Uncertainty in resistivity is ± 3%.

[c] Uncertainty in resistivity is ± 5%.

[d] Uncertainty in resistivity is ± 7% below 300 K and ± 5% at 300 and 400 K.

[e] Uncertainty in resistivity is ± 7%.

[f] Uncertainty in resistivity is ± 8%.

[g] Uncertainty in resistivity is ± 10%.

[h] Uncertainty in resistivity is ± 12%.

[i] Uncertainty in resistivity is ± 4%.

[j] Uncertainty in resistivity is ± 1%.

[k] Uncertainty in resistivity is ± 3% up to 300 K and ± 4% above 300 K.

[m] Uncertainty in resistivity is ± 2% up to 300 K and ± 4% above 300 K.

PERMITTIVITY (DIELECTRIC CONSTANT) OF INORGANIC SOLIDS

H. P. R. Frederikse

This table lists the permittivity ϵ, frequently called the dielectric constant, of a number of inorganic solids. When the material is not isotropic, the individual components of the permittivity are given. A superscript S indicates a measurement made under constant strain ("clamped") dielectric constant. If the constraint is removed, the measurement yields ϵ^T, the "unclamped" or free dielectric constant.

The temperature of the measurement is given when available; the symbol r.t. indicates a value at nominal room temperature. The frequency of the measurement is given in the last column (i.r. indicates a measurement in the infrared).

Substances are listed in alphabetical order by chemical formula.

REFERENCE

Young, K. F. and Frederikse, H. P. R., *J. Phys. Chem. Ref. Data*, 2, 313, 1973.

Formula	Name	ϵ_{ijk}	T/K	ν/Hz
Ag_3AsS_3	Silver thioarsenate (Proustite)	$\epsilon_{11}^T = 16.5$, $\epsilon_{11}^S = 14.5$	r.t.	2×10^7
		$\epsilon_{33}^T = 20.0$, $\epsilon_{33}^S = 18.0$	r.t.	2×10^7
$AgBr$	Silver bromide	12.50	r.t.	
$AgCN$	Silver cyanide	5.6	r.t.	10^6
$AgCl$	Silver chloride	11.15	r.t.	
$AgNO_3$	Silver nitrate	9.0	293	5×10^5
$AgNa(NO_2)_2$	Silver sodium nitrite	4.5 ± 0.5	r.t.	9.4×10^9
Ag_2O	Silver oxide	8.8	r.t.	
$(AlF)_2SiO_4$	Aluminum fluosilicate (topaz)	$\epsilon_{11} = 6.62$	297	7×10^3
		$\epsilon_{22} = 6.58$	297	7×10^3
		$\epsilon_{33} = 6.95$	297	7×10^3
Al_2O_3	Aluminum oxide (alumina)	$\epsilon_{11} = \epsilon_{22} = 9.34$	298	$10^2 - 8 \times 10^9$
		$\epsilon_{33} = 11.54$	298	$10^2 - 8 \times 10^9$
$AlPO_4$	Aluminum phosphate	$\epsilon_{11}^T = 6.05$	r.t.	10^3
$AlSb$	Aluminum antimonide	11.21	300	i.r.
AsF_3	Arsenic trifluoride	5.7	r.t.	
BN	Boron nitride	7.1	r.t.	i.r.
$BaCO_3$	Barium carbonate	8.53	291	2×10^5
$Ba(COOH)_2$	Barium formate	$\epsilon_{11} = 7.9$	r.t.	10^3
		$\epsilon_{22} = 5.9$	r.t.	10^3
		$\epsilon_{33} = 7.5$	r.t.	10^3
$BaCl_2$	Barium chloride	9.81	r.t.	
$BaCl_2 \cdot 2H_2O$	Barium chloride dihydrate	9.00	r.t.	10^3
BaF_2	Barium fluoride	7.32	292	$5 \times 10^2 - 10^{11}$
$Ba(NO_3)_2$	Barium nitrate	4.95	292	2×10^5
$Ba_2NaNb_5O_{15}$	Barium sodium niobate ("Bananas")	$\epsilon_{11}^S = 222$, $\epsilon_{11}^T = 235$	296	10^4
		$\epsilon_{22}^S = 227$, $\epsilon_{22}^T = 247$	296	
		$\epsilon_{33}^S = 32$, $\epsilon_{33}^T = 51$	296	
BaO	Barium oxide (baria)	34 ± 1	248, 333	60×10^7
BaO_2	Barium peroxide	10.7	r.t.	2×10^6
BaS	Barium sulfide	19.23	r.t.	7.25×10^6
$BaSO_4$	Barium sulfate	11.4	288	10^8
$BaSnO_3$	Barium stannate	18	298	25×10^5
$BaTiO_3$	Barium titanate	$\epsilon_{11}^T = 3600$	298	10^5
		$\epsilon_{11}^S = 2300$	298	2.5×10^8
		$\epsilon_{33}^T = 150$	298	10^5
		$\epsilon_{33}^S = 80$	298	2.5×10^8
$Ba_6Ti_2Nb_8O_{30}$	Barium titanium niobate	$\epsilon_{11} = \epsilon_{22} \approx 190$	298	
		$\epsilon_{33} \approx 220$	298	
$BaWO_4$	Barium tungstate	$\epsilon_{11} = \epsilon_{22} = 35.5 \pm 0.2$	297.5	1.6×10^3
		$\epsilon_{33} = 37.2 \pm 0.2$	297.5	1.6×10^3
$BaZrO_3$	Barium zirconate	43	r.t.	

Formula	Name	ϵ_{ijk}	T/K	ν/Hz
$Be_3Al_2Si_6O_{18}$	Beryllium aluminum silicate (Beryl)	$\epsilon_{33} = 5.95$	297	7×10^3
		$\epsilon_{11} = \epsilon_{22} = 6.86$	297	7×10^3
$BeCO_3$	Beryllium carbonate	9.7	291	2×10^5
BeO	Beryllium oxide (beryllia)	7.35 ± 0.2	293	2×10^6
$BiFeO_3$	Bismuth iron oxide	40 ± 3	300	9.4×10^9
$Bi_{12}GeO_{20}$	Bismuth germanite	$\epsilon_{11}^S = 38$	r.t.	
$Bi(GeO_4)_3$	Bismuth germanate	16	293	
Bi_2O_3	Bismuth sesquioxide	18.2	r.t.	2×10^6
$Bi_4Ti_3O_{12}$	Bismuth titanate	112	r.t.	10^3
C	Diamond			
	Type I	5.87 ± 0.19	300	10^3
	Type IIa	5.66 ± 0.04	300	10^3
$C_4H_4O_6$	Tartaric acid	$\epsilon_{11} = \epsilon_{22} = 4.3$	298	
		$\epsilon_{33} = 4.5$	298	
		$\epsilon_{13} = 0.55$	298	
$C_6H_{14}N_2O_6$	Ethylene diamine tartrate (EDT)	$\epsilon_{11}^T = 5.0$	293	
		$\epsilon_{22}^T = 8.3$	293	
		$\epsilon_{33}^T = 6.0$	293	
		$\epsilon_{13}^T = 0.7$	293	
$C_6H_{12}O_6NaBr$	Dextrose sodium bromide	$\epsilon_{11}^T = 4.0$	r.t.	10^3
$(CH_3NH_3)Al(SO_4)_2 \cdot 2H_2O$	Methyl ammonium alum (MASD)	19	197	
$Ca_2B_6O_{11} \cdot 5H_2O$	Colemanite	$\epsilon_{11} = 20$	293	10^3
		$\epsilon_{33} = 25$	293	10^3
$CaCO_3$	Calcium carbonate	$\epsilon_{11} = 8.67$	r.t.	9.4×10^{10}
		$\epsilon_{22} = 8.69$	r.t.	9.4×10^{10}
		$\epsilon_{33} = 8.31$	r.t.	9.4×10^{10}
$CaCeO_3$	Calcium cerate	21	r.t.	
CaF_2	Calcium fluoride	6.81	300	$5 \times 10^2 - 10^{11}$
$CaMoO_4$	Calcium molybdate	$\epsilon_{11} = \epsilon_{22} = 24.0 \pm 0.2$	297.5	<10
		$\epsilon_{33} = 20.0 \pm 0.2$	297.5	<10
$Ca(NO_3)_2$	Calcium nitrate	6.54	292	2×10^5
$CaNb_2O_6$	Calcium niobate	$\epsilon_{11} = 22.8 \pm 1.9$	r.t.	$(5-500) \times 10^3$
$Ca_2Nb_2O_7$	Calcium pyroniobate	~ 45	r.t.	5×10^7
CaO	Calcium oxide	11.8 ± 0.3	283	2×10^6
CaS	Calcium sulfide	6.699	r.t.	7.25×10^6
$CaSO_4 \cdot 2H_2O$	Calcium sulfate dihydrate	$\epsilon_{11} = 5.10$	r.t.	
		$\epsilon_{22} = 5.24$	r.t.	
		$\epsilon_{33} = 10.30$	r.t.	
$CaTiO_3$	Calcium titanate	165	r.t.	
$CaWO_4$	Calcium tungstate	$\epsilon_{11} = \epsilon_{22} = 11.7 \pm 0.1$	297.5	1.59×10^3
		$\epsilon_{33} = 9.5 \pm 0.2$	297.5	1.59×10^3
Cd_3As_2	Cadmium arsenide	$\epsilon_{33} = 18.5$	4	
$CdBr_2$	Cadmium bromide	8.6	293	5×10^5
CdF_2	Cadmium fluoride	8.33 ± 0.08	300	$10^5 - 10^7$
CdS	Cadmium sulfide	$\epsilon_{11} = \epsilon_{22} = 8.7$	300	i.r.
		$\epsilon_{33} = 9.25$	300	i.r.
		$\epsilon_{11} = \epsilon_{22} = 8.37$	8	i.r.
		$\epsilon_{33} = 9.00$	8	i.r.
		$\epsilon_{11}^T = 8.48$	77	10^4
		$\epsilon_{33}^T = 9.48$	77	10^4
		$\epsilon_{11}^S = 9.02, \epsilon_{11}^T = 9.35$	298	10^4
		$\epsilon_{33}^S = 9.53, \epsilon_{33}^T = 10.33$	298	10^4
$CdSe$	Cadmium selenide	$\epsilon_{11}^S = 9.53, \epsilon_{11}^T = 9.70$	298	10^4
		$\epsilon_{33}^S = 10.2, \epsilon_{33}^T = 10.65$	298	10^4
$CdTe$	Cadmium telluride	$\epsilon_{11} = \epsilon_{22} = 10.60 \pm 0.15$	297	i.r.
		$\epsilon_{33} = 7.05 \pm 0.05$	297	i.r.
$Cd_2Nb_2O_7$	Cadmium pyroniobate	$500-580$	293	10^3
CeO_2	Cerium oxide	7.0	r.t.	2×10^6

Formula	Name	ϵ_{ijk}	T/K	ν/Hz
$CoNb_2O_6$	Cobalt niobate	$\epsilon_{11} = 18.4 \pm 0.6$	r.t.	$(5\text{---}500) \times 10^3$
		$\epsilon_{22} = 21.4 \pm 1.1$	r.t.	$(5\text{---}500) \times 10^3$
		$\epsilon_{33} = 33.0 \pm 0.7$	r.t.	$(5\text{---}500) \times 10^3$
CoO	Cobalt oxide	12.9	298	$10^2\text{---}10^{10}$
Cr_2O_3	Chromic sesquioxide	$\epsilon_{11} = \epsilon_{22} = 13.3$	298.5	10^3
		$\epsilon_{33} = 11.9$	298.5	10^3
		8	315 (T_N)	6×10^{10}
$CsAl(SO_4)_2 \cdot 12H_2O$	Cesium alum	5.0	r.t.	$20\text{---}20 \times 10^3$
$CsBr$	Cesium bromide	6.38	298	1.6×10^3
Cs_2CO_3	Cesium carbonate	6.53	291	2×10^5
$CsCl$	Cesium chloride	7.2	298	
$Cs_2H_2AsO_4$	Cesium dihydrogen arsenate (CDA)	4.8	273	9.5×10^9
$Cs_2H_2PO_4$	Cesium dihydrogen phosphate (CDP)	6.15	285	9.5×10^9
$CsH_3(SeO_3)_2$	Cesium trihydrogen selenite	$\epsilon_{11} = 80$	273	10^5
		$\epsilon_{22} = 63$	273	10^5
		$\epsilon_{33} = 12$	273	10^5
CsI	Cesium iodide	6.31	298	1.6×10^3
$CsNO_3$	Cesium nitrate	$\epsilon_{11} = \epsilon_{22} = 9.4$	r.t.	5×10^5
		$\epsilon_{33} = 8.3$	r.t.	5×10^5
$CsPbCl_3$	Cesium lead chloride	14.37	300	$10^5\text{---}10^6$
$CuBr$	Cuprous bromide	8.0	293	5×10^5
$CuCl$	Cuprous chloride	9.8 ± 0.5	r.t.	10^3
CuO	Cupric oxide	18.1	r.t.	2×10^6
Cu_2O	Cuprous oxide (Cuprite)	7.60 ± 0.06	r.t.	10^5
$CuSO_4 \cdot 5H_2O$	Cupric sulfate pentahydrate	6.60	r.t.	
EuF_2	Europium fluoride	7.7 ± 0.2	298	$(1\text{---}300) \times 10^3$
$Eu_2(MoO_4)_3$	Europium molybdate	9.5	298	
EuS	Europium sulfide	13.10 ± 0.04	80	$5 \times 10^2\text{---}10^5$
FeO	Ferrous oxide	14.2	r.t.	2×10^6
Fe_2O_3	Ferric sesquioxide	4.5	r.t.	$10^5\text{---}10^7$
$Fe_2O_3\text{-}\alpha$	Ferric sesquioxide (hematite)	12		6×10^{10}
Fe_3O_4	Ferrosoferric oxide (magnetite)	20	r.t.	$10^5\text{---}10^7$
$GaAs$	Gallium arsenide	13.13	300	
		12.90	4	i.r.
GaP	Gallium phosphide	11.1	r.t.	
		10.75 ± 0.1	1.6	i.r.
$GaSb$	Gallium antimonide	15.69	r.t.	
		15.7	4	i.r.
$Gd_2(MoO_4)_3$	Gadolinium molybdate	$\epsilon^T = 10$	298	
		$\epsilon^S = 9.5$	298	10^3
Ge	Germanium	16.0 ± 0.3	4	9.2×10^9
		15.8 ± 0.2	r.t.	$500\text{---}3 \times 10^{10}$
GeO_2	Germanium dioxide	$\epsilon_{11} = \epsilon_{22} = 7.44$	r.t.	i.r.
HIO_3	Iodic acid	$\epsilon_{11} = 7.5$	r.t.	10^3
		$\epsilon_{22} = 12.4$	r.t.	10^3
		$\epsilon_{33} = 8.1$	r.t.	10^3
$HNH_4(ClCH_2COO)_2$	Hydrogen ammonium dichloroacetate	$\epsilon_{[102]} = 5.9$	r.t.	10^5
H_2O	Ice I (P = 0 kbar)	99	243	
	Ice III (P = 3 kbar)	117	243	
	Ice V(P = 5 kbar)	114	243	
	Ice VI (P = 8 kbar)	193	243	
$HgCl$	Mercurous chloride (Calumel)	$\epsilon_{11} = \epsilon_{22} = 14.0$	r.t.	10^{12}
$HgCl_2$	Mercuric chloride	6.5	r.t.	10^{12}
HgS	Mercurous sulfide (Cinnabar)	$\epsilon_{11} = \epsilon_{22} = 18.0$	r.t.	i.r.
		$\epsilon_{33} = 32.5$	r.t.	i.r.
$HgSe$	Mercurous selenide	25.6	r.t.	$10^4\text{---}10^6$
I_2	Iodine	$\epsilon_{11} = 6$	r.t.	$5 \times 10^4\text{---}10^7$
		$\epsilon_{22} = 3$	r.t.	$5 \times 10^4\text{---}10^7$
		$\epsilon_{33} = 40$	r.t.	$5 \times 10^4\text{---}10^7$

Formula	Name	ϵ_{ijk}	T/K	ν/Hz
InAs	Indium arsenide	14.55 ± 0.3	r.t.	i.r.
		15.15	4	i.r.
InP	Indium phosphide	12.61	r.t.	i.r.
InSb	Indium antimonide	17.88	4	i.r.
$KAl(SO_4)_2 \cdot 12H_2O$	Potassium alum	6.5	r.t.	$20—20 \times 10^3$
KBr	Potassium bromide	4.88	300	
		4.53	4.2	
$KBrO_3$	Potassium bromate	7.3	r.t.	2×10^6
KCN	Potassium cyanide	6.15	r.t.	2×10^6
K_2CO_3	Potassium carbonate	4.96	291	2×10^5
$K_2C_4H_4O_6 \cdot \frac{1}{2} H_2O$	Dipotassium tartrate (DKT)	$\epsilon_{11} = 6.44$	r.t.	
		$\epsilon_{22} = 5.80$	r.t.	
		$\epsilon_{33} = 6.49$	r.t.	
		$\epsilon_{13} = 0.005$	r.t.	
KCl	Potassium chloride	4.86 ± 0.02	r.t.	5×10^3
		4.50	4.2	
$KClO_3$	Potassium chlorate	5.1	r.t.	2×10^6
$KClO_4$	Potassium perchlorate	5.9	r.t.	2×10^6
K_2CrO_4	Potassium chromate	7.3	r.t.	6×10^7
$KCr(SO_4)_2 \cdot 12H_2O$	Potassium chrome alum	6.5	100—240	175×10^3
KD_2AsO_4	Potassium dideuterium arsenate (KDDA)	$\epsilon_{11} = 70$	298	
		$\epsilon_{33} = 31$	298	
KD_2PO_4	Potassium dideuterium phosphate (KDDP)	50 ± 2	297	10^3
KF	Potassium fluoride	6.05		2×10^6
KH_2AsO_4	Potassium dihydrogen arsenate (KDA)	$\epsilon_{11} = 60$	298	
		$\epsilon_{33} = 24$	298	
KH_2PO_4	Potassium dihydrogen phosphate (KDP)	46	298	10^3
		$\epsilon_{11} = 42$	r.t.	
		$\epsilon_{33} = 21$	r.t.	
K_2HPO_4	Dipotassium monohydrogen orthophosphate	9.05	r.t.	2×10^6
KI	Potassium iodide	5.00	r.t.	9.4×10^{10}
KIO_3	Potassium iodate	170	255	10^5
		10	293	10^5
		$\epsilon_{[101]} \approx 40,70$	r.t.	10^5
		16.85	r.t.	2×10^6
$(K,H)Al_3(SiO_4)_3$	Mica (muscovite)	5.4	299	$10^2—3 \times 10^9$
$(K,H)Mg_3Al(SiO_4)_3$	Mica (Canadian)	$\epsilon_{11} = \epsilon_{22} = 6.9$	298	$10^2—10^4$
		$\epsilon_{33} = 7.3$	298	10^4
KNO_2	Potassium nitrite	25	305	
KNO_3	Potassium nitrate	4.37	293	2×10^5
$KNbO_3$	Potassium niobate	700	r.t.	
K_3PO_4	Potassium orthophosphate	7.75	r.t.	2×10^6
KSCN	Potassium thiocyanate	7.9	r.t.	2×10^6
K_2SO_4	Potassium sulfate	6.4	r.t.	2×10^6
$K_2S_3O_6$	Potassium trithionate	5.7	293	1.8×10^6
$K_2S_4O_6$	Potassium tetrathionate	5.5	293	1.8×10^6
$K_2S_5O_6 \cdot H_2O$	Potassium pentathionate	7.8	293	1.8×10^6
$K_2S_6O_6$	Potassium hexathionate	7.8	293	1.8×10^6
K_2SeO_4	Potassium selenate	$\epsilon_{11} = 5.9$	r.t.	10^3
		$\epsilon_{22} = 7.7$	r.t.	10^3
$KSr_2Nb_5O_{15}$	Potassium strontium niobate	$\epsilon_{11} = \epsilon_{11} \approx 1200$	298	
		$\epsilon_{33} \approx 800$	298	
$KTaNbO_3$	Potassium tantalate niobate (KTN)	34,000	273	10^4
		6,000	293	10^4
$KTaO_3$	Potassium tantalate	242	298	2×10^5
$LaScO_3$	Lanthanum scandate	30	r.t.	
LiBr	Lithium bromide	12.1	r.t.	2×10^6
Li_2CO_3	Lithium carbonate	4.9	291	2×10^5

Formula	Name	ϵ_{ijk}	T/K	ν/Hz
LiCl	Lithium chloride	11.05	r.t.	2×10^6
LiD	Lithium deuteride	14.0 ± 0.5	r.t.	i.r.
LiF	Lithium fluoride	9.00	298	10^2—10^7
		9.11	353	10^2—10^7
LiGaO$_2$	Lithium metagallate	$\epsilon_{11}^T = 7.0$, $\epsilon_{22}^T = 6.0$	r.t.	
		$\epsilon_{33}^T = 9.5$	r.t.	
		$\epsilon_{11}^S = 6.8$, $\epsilon_{22}^S = 5.8$	r.t.	
Li^{6}H	Lithium-6 hydride	13.2 ± 0.5	r.t.	
Li^{7}H	Lithium-7 hydride	12.9 ± 0.5	r.t.	
LiH$_3$(SeO$_3$)$_2$	Lithium trihydrogen selenite	29	298	10^4
		$\epsilon_{11} = 13.0$	r.t.	
		$\epsilon_{22} = 12.9$	r.t.	
		$\epsilon_{33} = 46$	r.t.	
LiI	Lithium iodide	11.03	r.t.	2×10^6
LiIO$_3$	Lithium iodate	$\epsilon_{11} = \epsilon_{22} = 65$	294.5	10^3
		$\epsilon_{33} = 554$	298	
LiNH$_4$C$_4$H$_4$O$_6 \cdot$ H$_2$O	Lithium ammonium tartrate (LAT)	$\epsilon_{11}^T = 7.2$	298	
		$\epsilon_{22}^T = 8.0$	298	
		$\epsilon_{33}^T = 6.9$	298	
LiNa$_3$CrO$_4 \cdot$ 6H$_2$O	Lithium trisodium chromate	8.0	r.t.	10^3
LiNa$_3$MoO$_4 \cdot$ 6H$_2$O	Lithium trisodium molybdate	$\epsilon_{11} = 6.7$	r.t.	10^3
		$\epsilon_{33} = 5.3$	r.t.	10^3
LiNbO$_3$	Lithium niobate	$\epsilon_{11} = \epsilon_{22} = 82$	298	10^5
		$\epsilon_{33} = 30$	298	10^5
Li$_2$SO$_4 \cdot$ H$_2$O	Lithium sulfate monohydrate	$\epsilon_{11} = 5.6$	298	
		$\epsilon_{22} = 10.3$	298	
		$\epsilon_{33} = 6.5$	298	
		$\epsilon_{13} = 0.07$	298	
LiTaO$_3$	Lithium tantalate	$\epsilon_{11} = \epsilon_{22} = 53$	r.t.	10^5
		$\epsilon_{33} = 46$	r.t.	10^5
		$\epsilon_{11}^S = \epsilon_{22}^S = 41$	r.t.	
		$\epsilon_{33}^S = 43$	r.t.	
		$\epsilon_{11}^T = \epsilon_{22}^T = 51$	r.t.	
		$\epsilon_{33}^T = 45$	r.t.	
LiTlC$_4$H$_4$O$_6 \cdot$ H$_2$O	Lithium thallium tartrate (LTT)	$\epsilon_{11} \approx 20$	80	
Mg$_3$B$_7$O$_{13}$Cl	Magnesium borate monochloride (boracite)	$\epsilon_{11} = 14.1$	r.t.	5×10^5
MgCO$_3$	Magnesium carbonate	8.1	291	2×10^5
MgNb$_2$O$_6$	Magnesium niobate	$\epsilon_{11} = 16.4 \pm 0.5$	r.t.	$(5$—$500) \times 10^3$
		$\epsilon_{22} = 20.9 \pm 0.5$	r.t.	$(5$—$500) \times 10^3$
		$\epsilon_{33} = 32.4 \pm 0.5$	r.t.	$(5$—$500) \times 10^3$
MgO	Magnesium oxide (Periclase)	9.65	298	10^2—10^8
(MgO)$_x$Al$_2$O$_3$	Spinel	8.6	r.t.	
MgSO$_4$	Magnesium sulfate	8.2	r.t.	
MgSO$_4 \cdot$ 7H$_2$O	Magnesium sulfate septa hydrate	5.46	r.t.	
MgTiO$_3$	Magnesium titanate	13.5	r.t.	
MgWO$_4$	Magnesium tungstate	$\epsilon_{11} = 18.0 \pm 1$	r.t.	$(5$—$500) \times 10^3$
		$\epsilon_{22} = 18.0 \pm 1$	r.t.	$(5$—$500) \times 10^3$
MnNb$_2$O$_6$	Manganese niobate	$\epsilon_{11} = 17.4 \pm 2$	r.t.	$(5$—$500) \times 10^3$
		$\epsilon_{22} = 16.1 \pm 0.5$	r.t.	$(5$—$500) \times 10^3$
		$\epsilon_{33} = 30.7 \pm 1$	r.t.	$(5$—$500) \times 10^3$
MnO	Manganese oxide (Pyrolusite)	12.8	r.t.	6×10^{10}
MnO$_2$	Manganese dioxide	$\sim 10^4$	298	10^4
Mn$_2$O$_3$	Manganese sesquioxide	8	r.t.	6×10^{10}
MnWO$_4$	Manganese tungstate	$\epsilon_{11} = 19.3 \pm 1.3$	r.t.	$(5$—$500) \times 10^3$
		$\epsilon_{22} = 14.3 \pm 0.5$	r.t.	$(5$—$500) \times 10^3$
		$\epsilon_{33} = 16.5 \pm 1.1$	r.t.	$(5$—$500) \times 10^3$
N(CH$_3$)$_4$HgBr$_3$	Tetramethylammonium tribromo mercurate (TTM)	~ 10	233—373	

Formula	Name	ϵ_{ijk}	T/K	ν/Hz
$N(CH_3)_4HgI_3$	Tetramethylammonium triiodo mercurate (TTM)	~10	233—373	
$N_4(CH_2)_6$	Hexamethylene tetramine (HMTA)	2.6 ± 0.2	r.t.	10^9—10^{10}
$(ND_4)_2BeF_4$	Deuteroammonium fluoberyllate	$\epsilon_{11} = 10$	r.t.	
		$\epsilon_{22} = 9$	r.t.	
		$\epsilon_{33} = 9$	r.t.	
$(ND_4)_2SO_4$	Deuteroammonium sulfate	$\epsilon_{11} = 9$	r.t.	
		$\epsilon_{22} = 10$	r.t.	
		$\epsilon_{33} = 9$	r.t.	
$(NH_2 \cdot CH_2COOH)_3 \cdot H_2SO_4$	Triglycine sulfate (TGS)	$\epsilon_{11} = 9$	273	10^4
		$\epsilon_{22} = 30$	273	10^4
		$\epsilon_{33} = 6.5$	273	10^4
$(NH_2 \cdot CH_2COOH)_3 \cdot H_2SeO_4$	Triglycine selenate (TGSe)	200	293	1.6×10^3
$(NH_2 \cdot CH_2 COOH)_3 \cdot H_2BeF_4$	Triglycine fluorberyllate (TGFB)	$\epsilon_{22} = 12$	273	10^4
$NH_4Al(SO_4)_2 \cdot 12H_2O$	Ammonium alum	6	r.t.	10^{12}
$(NH_4)_2BeF_4$	Ammonium fluorberyllate	$\epsilon_{11} = \epsilon_{22} = 7.8$	123	10^5
		$\epsilon_{33} = 7.1$	123	10^5
		$\epsilon_{11} = \epsilon_{22} = 8.8$	293	10^5
		$\epsilon_{33} = 9.2$	293	10^5
NH_4Br	Ammonium bromide	7.1	r.t.	7×10^5
NH_4I	Ammonium iodide	9.8	r.t.	
$(NH_4)_2C_2H_6O_6$	Ammonium tartrate	$\epsilon_{11} = 6.45$	r.t.	10^3
		$\epsilon_{22} = 6.8$	r.t.	10^3
		$\epsilon_{33} = 6.0$	r.t.	10^3
$(NH_4)_2Cd_2(SO_4)_3$	Ammonium cadmium sulfate	10.0	r.t.	10^4
NH_4Cl	Ammonium chloride	6.9	r.t.	7×10^5
$NH_4(ClCH_2COO)$	Ammonium monochloroacetate	5	r.t.	2×10^6
$NH_4Cr(SO_4)_2 \cdot 12H_2O$	Ammonium chrome alum	6.5	r.t.	175×10^3
NH_4HSO_4	Ammonium bisulfate	165	273	5×10^4
$NH_4H_2AsO_4$	Ammonium dihydrogen arsenate (ADA)	5.1	265	9.5×10^9
		$\epsilon_{11} = \epsilon_{22} = 85$	298	10^3
		$\epsilon_{33} = 22$	298	
$NH_4H_2PO_4$	Ammonium dihydrogen phosphate (ADP)	$\epsilon_{11} = \epsilon_{22} = 57.1 \pm 0.6$	294.5	10^5—35×10^9
		$\epsilon_{33} = 14.0 \pm 0.3$	294	10^5—36×10^9
$ND_4D_2PO_4$	Ammonium dideuterium phosphate (ADDP)	$\epsilon_{11} = \epsilon_{22} = 74, \epsilon_{33} = 24$	300	
NH_4NO_3	Ammonium nitrate	10.7	322	$(5—50) \times 10^3$
$(NH_4)_2SO_4$	Ammonium sulfate	$\epsilon_{11} = \epsilon_{22} = 8.0$	123	10^5
		$\epsilon_{33} = 6.3$	123	10^5
		$\epsilon_{11} = \epsilon_{22} = 10.0$	293	10^5
		$\epsilon_{33} = 9.3$	293	10^5
$(NH_4)_2UO_2(C_2O_4)_2$	Ammonium uranyl oxalate	8.03	r.t.	10^4—3.3×10^9
$(NH_4)_2UO_2(C_2O_4)_2 \cdot 3H_2O$	Ammonium uranyl oxalate trihydrate	6.06	r.t.	10^4—3.3×10^9
$NaBr$	Sodium bromide	6.44	298	1.6×10^3
$NaBrO_3$	Sodium bromate	$\epsilon_{11}^T = 5.70$	298	10^3
$NaCN$	Sodium cyanide	7.55	293	10^5
$NaCO_3$	Sodium carbonate	8.75	291	2×10^5
$NaCO_3 \cdot 10H_2O$	Sodium carbonate decahydrate	5.3	r.t.	6×10^7
$NaCl$	Sodium chloride	5.9	298	10^2—10^7
		5.45	4.2	
$NaClO_3$	Sodium chlorate	$\epsilon_{11}^T = 5.76$	301	10^3
		5.28	r.t.	10^3
$NaClO_4$	Sodium perchlorate	5.76	r.t.	10^3
NaF	Sodium fluoride	5.08 ± 0.02	r.t.	5×10^3
$NaH_3(SeO_3)_2$	Sodium trihydrogen selenite	$\epsilon_{11} \approx 75$	273	2×10^5
$NaD_3(SeO_3)_2$	Sodium trideuterium selenite	$\epsilon_{11} \approx 220$	273	2×10^5
NaI	Sodium iodide	7.28 ± 0.03	r.t.	
$NaK(C_4H_2D_2O_6) \cdot 4D_2O$	Sodium potassium tartrate tetradeutrate (double deuterated Rochelle salt)	$\epsilon_{11} = 70$	273	10^3
		$\epsilon_{22} = 8.9$	273	10^3

Formula	Name	ϵ_{ijk}	T/K	ν/Hz
$NaK(C_4H_4O_6) \cdot 4H_2O$	Sodium potassium tartrate tetrahydrate (Rochelle salt)	$\epsilon_{11} = 170$	273	10^3
		$\epsilon_{22} = 9.1$	273	10^3
$NaNH_4(C_4H_4O_6) \cdot 4H_2O$	Sodium ammonium tartrate (Ammonium Rochelle salt)	$\epsilon_{11} = 8.4$	298	
		$\epsilon_{22} = 9.2$	298	
		$\epsilon_{33} = 9.5$	298	
$NaNbO_3$	Sodium niobate	$\epsilon_{33} = 670 \pm 13$	r.t.	
		$\epsilon_{11} = \epsilon_{22} = 76 \pm 2$	r.t.	
$NaNO_2$	Sodium nitrite	$\epsilon_{11} = 7.4$	r.t.	5×10^5
		$\epsilon_{22} = 5.5$	r.t.	5×10^5
		$\epsilon_{33} = 5.0$	r.t.	5×10^5
$NaNO_3$	Sodium nitrate	6.85	292	2×10^5
$NaSO_4$	Sodium sulfate	7.90	r.t.	
$NaSO_4 \cdot 10H_2O$	Sodium sulfate decahydrate	5.0	r.t.	
$Na_2S_2O_3 \cdot 5H_2O$	Sodium sulfate pentahydrate	7	250—290	$300—10^4$
$Na_2UO_2(C_2O_4)_2$	Sodium uranyl oxalate	5.18	r.t.	
$NdAlO_3$	Neodymium aluminate	17.5	r.t.	
$NdScO_3$	Neodymium scandate	27	r.t.	
$Ni_3B_7O_{13}I$	Nickel iodine boracite	$\epsilon_{11} = 14$	260	
$NiNb_2O_6$	Nickel niobate	$\epsilon_{11} = 16.0 \pm 0.5$	r.t.	$(5—500) \times 10^3$
		$\epsilon_{22} = 23.8 \pm 1.8$	r.t.	$(5—500) \times 10^3$
		$\epsilon_{33} = 31.3 \pm 2.5$	r.t.	$(5—500) \times 10^3$
NiO	Nickel oxide	11.9	298	10^5
$NiSO_4 \cdot 6H_2O$	Nickel sulfate hexahydrate	$\epsilon_{11} = 6.2$	r.t.	
		$\epsilon_{33} = 6.8$	r.t.	
$NiWO_4$	Nickel tungstate	$\epsilon_{11} = 17.4 \pm 2.4$	r.t.	$(5—500) \times 10^3$
		$\epsilon_{22} = 13.6 \pm 1.0$	r.t.	$(5—500) \times 10^3$
		$\epsilon_{33} = 19.7 \pm 0.6$	r.t.	$(5—500) \times 10^3$
P	Phosphorous (red)	4.1	r.t.	10^8
	Phosphorous (yellow)	3.6	r.t.	10^8
$[P(CH_3)_4]HgBr_3$	Tetramethylphosphonium tribromo mercurate (TTM)	~ 10	233—373	
$PbBr_2$	Lead bromide	>30	293	$(0.5—3) \times 10^6$
$PbCO_3$	Lead carbonate	18.6	288	10.8
$Pb(C_2H_3O_2)_2$	Lead acetate	2.6	290—295	10^6
$PbCl_2$	Lead chloride	33.5	273	$(0.5—3) \times 10^6$
Pb_2CoWO_6	Lead cobalt tungstate	~ 250	r.t.	
PbF_2	Lead fluoride	26.3	r.t.	
$PbHfO_3$	Lead hafnate	390	300	10^5
		185	400	
PbI_2	Lead iodide	20.8	293	$(0.5—3) \times 10^6$
$Pb_3MgNb_2O_9$	Lead magnesium niobate	10,000	297	
$PbMoO_4$	Lead molybdate	$\epsilon_{11} = 34.0 \pm 0.4$	297.5	1.6×10^3
		$\epsilon_{33} = 40.6 \pm 0.2$	297.5	1.6×10^3
$Pb(NO_3)_2$	Lead nitrate	16.8	r.t.	$(0.5—3) \times 10^6$
$PbNb_2O_6$	Lead niobate	$\epsilon_{33}^T = 180$	298	
PbO	Lead oxide	25.9	r.t.	2×10^6
PbS	Lead sulfide (Galena)	190	77	i.r.
		200 ± 35	r.t.	i.r.
$PbSO_4$	Lead sulfate	14.3	290—295	10^6
$PbSe$	Lead selenide	280	r.t.	i.r.
$PbTa_2O_6$	Lead metatantalate	$\epsilon_{11} = \epsilon_{22} \approx 300$	r.t.	10^4
		$\epsilon_{33} = 150$	r.t.	10^4
$PbTe$	Lead telluride	450	r.t.	i.r.
		40	77	$10^4—15 \times 10^4$
		430	4.2	$10^4—15 \times 10^4$
$PbTiO_3$	Lead titanate	~ 200	r.t.	10^3
$PbWO_4$	Lead tungstate	$\epsilon_{11} = \epsilon_{22} = 23.6 \pm 0.3$	297.5	1.59×10^3
		$\epsilon_{33} = 31.0 \pm 0.4$	297.5	1.59×10^3
$Pb(Zn_{1/3}Nb_{2/3})O_3$	Lead zinc niobate	7	300	$10^3, 300 \times 10^3$
$PbZrO_3$	Lead zirconate	200	400	

Formula	Name	ϵ_{ijk}	T/K	ν/Hz
$RbAl(SO_4)_2 \cdot 12H_2O$	Rubidium alum	5.1	r.t.	10^{12}
$RbBr$	Rubidium bromide	4.83	300	
Rb_2CO_3	Rubidium carbonate	4.87 ± 0.02	r.t.	5×10^3
$RbCl$	Rubidium chloride	4.91 ± 0.02	r.t.	5×10^3
$RbCr(SO_4)_2 \cdot 12H_2O$	Rubidium chrome alum	5.0	r.t.	10^{12}
RbF	Rubidium fluoride	5.91	r.t.	2×10^6
$RbHSO_4$	Rubidium bisulfate	$\epsilon_{11} = 7$	r.t.	10^5
		$\epsilon_{22} = 8$	r.t.	10^5
		$\epsilon_{33} = 10$	r.t.	10^5
RbH_2AsO_4	Rubidium dihydrogen arsenate (RDA)	3.90	273	9.5×10^9
RbH_2PO_4	Rubidium dihydrogen phosphate (RDP)	6.15	285	9.5×10^9
RbI	Rubidium iodide	4.94 ± 0.02	r.t.	5×10^3
$RbInSO_4$	Rubidium indium sulfate	6.85	r.t.	
$RbNO_3$	Rubidium nitrate	20—380	433—488	10^6
		30	488—538	10^6
S	Sulfur	$\epsilon_{11} = 3.75$	298	10^2—10^3
		$\epsilon_{22} = 3.95$	298	10^2—10^3
		$\epsilon_{33} = 4.44$	298	10^2—10^3
	sublimed	3.69	298	10^2—10^3
$SC(NH_2)_2$	Thiourea	$\epsilon_{11} = \epsilon_{33} \approx 3$	77—300	10^3
		$\epsilon_{22} = 35$	300	10^3
Sb_2O_3	Antimonous sesquioxide	12.8	r.t.	$(1.5—2) \times 10^3$
Sb_2S_3	Antimonous sulfide (stibnite)	$\epsilon_{11} = \epsilon_{22} = 15$	r.t.	10^3
		$\epsilon_{33} = 180$	r.t.	10^3
Sb_2Se_3	Antimonous selenide	~110	r.t.	$(10—16.5) \times 10^9$
$SbSI$	Antimonous sulfide iodide	2000	273	10^5
		$\epsilon_{11} = \epsilon_{22} \approx 25$	r.t.	10^3—10^5
		$\epsilon_{33} \approx 5 \times 10^4$	295	10^3—10^5
Se	Selenium	$\epsilon_{11} = \epsilon_{22} = 11$	300	24×10^9
	(monocrystal)			
		$\epsilon_{33} = 21$	300	24×10^9
	(amorphous)	6.0	298	10^2—10^{10}
Si	Silicon	12.1	4.2	10^7—10^9
SiC	Silicon carbide			
	cubic	9.72	r.t.	i.r.
	6H	$\epsilon_{11} = \epsilon_{22} = 9.66$	r.t.	i.r.
		$\epsilon_{33} = 10.03$	r.t.	i.r.
		9.7 ± 0.1	1.8	i.r.
Si_3N_4	Silicon nitride	4.2 (film)	r.t.	10^3
SiO	Silicon monoxide	5.8	r.t.	10^3
SiO_2	Silicon dioxide	$\epsilon_{11} = 4.42$	r.t.	9.4×10^{10}
		$\epsilon_{22} = 4.41$	r.t.	9.4×10^{10}
		$\epsilon_{33} = 4.60$	r.t.	9.4×10^{10}
$Sm_2(MoO_4)_3$	Samarium molybdate	12	298	
SnO_2	Stannic dioxide	$\epsilon_{11} = \epsilon_{22} = 14 \pm 2$	r.t.	10^4—10^{10}
		$\epsilon_{33} = 9.0 \pm 0.5$	r.t.	10^4—10^{10}
$SnSb$	Tin antimonide	147	r.t.	10^4—10^6
$SnTe$	Tin telluride	1770 ± 300	r.t.	i.r.
$Sr(COOH)_2 \cdot 2H_2O$	Strontium formate dihydrate	6.1	r.t.	10^3
$SrCO_3$	Strontium carbonate	8.85	298	2×10^5
$SrCl_2$	Strontium chloride	9.19	r.t.	
$Sr_4Cl_2 \cdot 6H_2O$	Strontium chloride hexahydrate	8.52	r.t.	
SrF_2	Strontium fluoride	6.50	300	5×10^2—10^{11}
$SrMoO_4$	Strontium molybdate	$\epsilon_{11} = \epsilon_{22} = 31.7 \pm 0.2$	297.5	1.59×10^3
		$\epsilon_{33} = 41.7 \pm 0.2$	297.5	1.59×10^3
$Sr(NO_3)_2$	Strontium nitrate	5.33	292	2×10^5
$Sr_2Nb_2O_7$	Strontium niobate	$\epsilon_{11} = 75$	r.t.	10^3
		$\epsilon_{22} = 46$	r.t.	10^3
		$\epsilon_{33} = 43$	r.t.	10^3

Formula	Name	ϵ_{ijk}	T/K	ν/Hz
SrO	Strontium oxide	13.3 ± 0.3	273	2×10^6
SrS	Strontium sulfide	11.3	r.t.	7.25×10^6
SrSO$_4$	Strontium sulfate	11.5	r.t.	
SrTiO$_3$	Strontium titanate	332	298	10^3
		2080	78	10^3
SrWO$_4$	Strontium tungstate	$\epsilon_{11} = \epsilon_{22} = 25.7 \pm 0.2$	297.5	1.6×10^3
		$\epsilon_{33} = 34.1 \pm 0.2$	297.5	1.6×10^3
Ta$_2$O$_5$	Tantalum pentoxide (tantala)			
	α phase	$\epsilon_{11} = \epsilon_{22} = 30$	77	10^3
		$\epsilon_{33} = 65$	77	10^3
	β phase	24	292	10^3
Tb(MoO$_4$)$_3$	Terbium molybdate	11	298	
		$\epsilon_{11} = \epsilon_{22} = 33$	100—200	9.4×10^9
		$\epsilon_{33} = 53$	100—200	9.4×10^9
Te	Tellurium	$\epsilon_{11} = \epsilon_{22} = 33$	r.t.	
		$\epsilon_{33} = 54$	r.t.	
	polycrystalline	27.5	r.t.	i.r.
	monocrystalline	28.0	r.t.	i.r.
ThO$_2$	Thorium dioxide	18.9 ± 0.4	r.t.	3×10^5
TiO$_2$	Titanium dioxide (rutile)	$\epsilon_{11} = \epsilon_{22} = 86$	300	$10^4–10^6$
		$\epsilon_{33} = 170$	300	$10^4–10^6$
Ti$_2$O$_3$	Titanium sesquioxide	30	77	6×10^{10}
TlBr	Thallium bromide	30	293	$10^3–10^7$
TlCl	Thallous chloride	32.2 ± 0.2	293	$10^3–10^5$
TlI	Thallous iodide (orthorhombic)	20.7 ± 0.2	293	10^4
		37.3	193	10^7
TlNO$_3$	Thallous nitrate	16.5	293	5×10^5
TlSO$_4$	Thallous sulfate	25.5	293	5×10^5
UO$_2$	Uranium dioxide	24	r.t.	3×10^5
WO$_3$	Tungsten trioxide	300		
YMnO$_3$	Yttrium manganate	20	r.t.	2×10^7
Y$_2$O$_3$	Yttrium sesquioxide	10	r.t.	10^6
YbMnO$_3$	Ytterbium manganate	20	r.t.	2×10^7
Yb$_2$O$_3$	Ytterbium sesquioxide	5.0 (film)	r.t.	10^3
ZnO	Zinc monoxide	$\epsilon_{11}^S = 8.33$	r.t.	
		$\epsilon_{33}^S = 8.84$	r.t.	
		$\epsilon_{11}^T = 9.26$	r.t.	
		$\epsilon_{33}^T = 11.0$	r.t.	
		$\epsilon_{11} = 9.26$	r.t.	
		$\epsilon_{33} = 8.2$	r.t.	
		8.15	r.t.	i.r.
ZnS	Zinc sulfide	$\epsilon_{11}^S = 8.08 \pm 2\%$	77	10^4
		$\epsilon_{11}^S = 8.32 \pm 2\%$	298	10^4
		$\epsilon_{11}^T = 8.14 \pm 2\%$	77	10^4
		$\epsilon_{11}^T = 8.37 \pm 2\%$	298	10^4
ZnSe	Zinc selenide	$\epsilon_{11}^T = \epsilon_{11}^S = 9.12 \pm 2\%$	298	10^4
ZnTe	Zinc telluride	$\epsilon_{11}^T = \epsilon_{11}^S = 10.10 \pm 2\%$	r.t.	
ZnWO$_4$	Zinc tungstate	$\epsilon_{22} = 16.1 \pm 0.5$	r.t.	$(5–500) \times 10^3$
ZrO$_2$	Zirconium dioxide (zirconia)	12.5	r.t.	2×10^6

CURIE TEMPERATURE OF SELECTED FERROELECTRIC CRYSTALS

H. P. R. Frederikse

The following table lists the major ferroelectric crystals and their Curie temperatures, T_C.

REFERENCE

Young, K. F. and Frederikse, H. P. R., *J. Phys. Chem. Ref. Data*, 2, 313, 1973.

Name or acronym	Formula	T_C/K
Potassium dihydrogen phosphate group		
KDP	KH_2PO_4	123
KDA	KH_2AsO_4	97
KDDP	KD_2PO_4	213
KDDA	KD_2AsO_4	162
RDP	RbH_2PO_4	146
RDA	RbH_2AsO_4	111
RDDP	RbD_2PO_4	218
RDDA	RbD_2AsO_4	178
CDP	CsH_2PO_4	159
CDA	CsH_2AsO_4	143
CDDA	CsD_2AsO_4	212
Rochelle salt group		
Rochelle salt	$NaKC_4H_4O_6 \cdot 4H_2O$	255-297
Deuterated Rochelle salt	$NaKC_4H_2D_2O_6 \cdot 4H_2O$	251-308
Ammonium Rochelle salt	$NaNH_4C_4H_4O_6 \cdot 4H_2O$	109
LAT	$LiNH_4C_4H_4O_6 \cdot H_2O$	106
Triglycine sulfate group		
TGS	$(NH_2CH_2COOH)_3 \cdot H_2SO_4$	322
TGSe	$(NH_2CH_2COOH)_3 \cdot H_2SeO_4$	295
TGFB	$(NH_2CH_2COOH)_3 \cdot H_2BeF_4$	346
AFB	$(NH_4)_2BeF_4$	176
HADA	$HNH_4(ClCH_2COO)_2$	128
Perovskites and related compounds		
Barium titanate	$BaTiO_3$	406, 278, 193
Lead titanate	$PbTiO_3$	765
Potassium niobate	$KNbO_3$	712
Potassium tantalate niobate	$KTa_{2/3}Nb_{1/3}O_3$	241, 220, 170
Lithium niobate	$LiNBO_3$	1483
Lithium tantalate	$LiTaO_3$	891
Barium titanium niobate	$Ba_6Ti_2Nb_8O_{30}$	521
Ba-Na niobate ("Bananas")	$Ba_2NaNb_5O_{15}$	833
Potassium iodate	KIO_3	485, 343, 257-263, 83
Lithium iodate	$LiIO_3$	529
Potassium nitrate	KNO_3	397
Sodium nitrate	$NaNO_3$	548
Rubidium nitrate	$RbNO_3$	437-487

Name or acronym	Formula	T_C/K
Miscellaneous compounds		
Cesium trihydrogen selenite	$CsH_3(SeO_3)_2$	143
Lithium trihydrogen selenite	$LiH_3(SeO_3)_2$	$T_C > T_{mp}$
Potassium selenate	K_2SeO_4	93
Methyl ammonium alum (MASD)	$CH_3NH_3Al(SO_4)_2 \lozenge 12H_2O$	177
Ammonium cadmium sulfate	$(NH_4)_2Cd_2(SO_4)_3$	95
Ammonium bisulfate	$(NH_4)HSO_4$	271
Ammonium sulfate	$(NH_4)_2SO_4$	224
Ammonium nitrate	NH_4NO_3	398, 357, 305, 255
Colemanite	$CaB_3O_4(OH)_3 \lozenge H_2O$	266
Cadmium pyroniobite	$Cd_2Nb_2O_7$	185
Gadolinium molybdate	$Gd_2(MoO_4)_3$	432

PROPERTIES OF ANTIFERROELECTRIC CRYSTALS

H. P. R. Frederikse

Some important antiferroelectric crystals are listed here with their Curie Temperatures T_C. The last column gives the constant T_0 which appears in the Curie-Weiss law describing the dielectric constant of these materials above the Curie Temperature:

$$\varepsilon = const./(T - T_0)$$

Name or acronym	Formula	T_C/K	T_0/K
ADP	$NH_4H_2PO_4$	148	
ADA	$NH_4H_2AsO_4$	216	
ADDP	$NH_4D_2PO_4$	242, 245	
ADDA	$NH_4D_2AsO_4$	299	
A_dDDP	$ND_4D_2PO_4$	243	
A_dDDA	$ND_4D_2AsO_4$	304	
Sodium niobate	$NaNbO_3$	911, 793	
Lead hafnate	$PbHfO_3$	476	378
Lead zirconate	$PbZrO_3$	503	475
Lead metaniobate	$PbNb_2O_6$	843	530
Lead metatantalate	$PbTa_2O_6$	543	533
Tungsten trioxide	WO_3	1010	
Potassium strontium niobate	$KSr_2Nb_5O_{15}$	427	413
Sodium nitrite	$NaNO_2$	437	437
Sodium trihydrogen selenite	$NaH_3(SeO_3)_2$	193	192
Sodium trideuterium selenite	$NaD_3(SeO_3)_2$	271	245
Ammonium trihydrogen periodate	$(NH_4)_2H_3IO_6$	245	

DIELECTRIC CONSTANTS OF GLASSES

Type	Dielectric constant at 100 MHz (20°C)	Volume resistivity (350°C megohm-cm)	Loss factor[a]
Corning 0010	6.32	10	0.015
Corning 0080	6.75	0.13	0.058
Corning 0120	6.65	100	0.012
Pyrex 1710	6.00	2,500	0.025
Pyrex 3320	4.71	—	0.019
Pyrex 7040	4.65	80	0.013
Pyrex 7050	4.77	16	0.017
Pyrex 7052	5.07	25	0.019
Pyrex 7060	4.70	13	0.018
Pyrex 7070	4.00	1,300	0.0048
Vycor 7230	3.83	—	0.0061
Pyrex 7720	4.50	16	0.014
Pyrex 7740	5.00	4	0.040
Pyrex 7750	4.28	50	0.011
Pyrex 7760	4.50	50	0.0081
Vycor 7900	3.9	130	0.0023
Vycor 7910	3.8	1,600	0.00091
Vycor 7911	3.8	4,000	0.00072
Corning 8870	9.5	5,000	0.0085
G. E. Clear (silica glass)	3.81	4,000—30,000	0.00038
Quartz (fused)	3.75 4.1 (1 MHz)	—	0.0002 (1 MHz)

[a]   Power factor × dielectric constant equals loss factor.

PROPERTIES OF SUPERCONDUCTORS

L. I. Berger and B. W. Roberts

The following tables include superconductive properties of selected elements, compounds, and alloys. Individual tables are given for thin films, elements at high pressures, superconductors with high critical magnetic fields, and high critical temperature superconductors.

The historically first observed and most distinctive property of a superconductive body is the near total loss of resistance at a critical temperature (T_c) that is characteristic of each material. Figure 1(a) below illustrates schematically two types of possible transitions. The sharp vertical discontinuity in resistance is indicative of that found for a single crystal of a very pure element or one of a few well annealed alloy compositions. The broad transition, illustrated by broken lines, suggests the transition shape seen for materials that are not homogeneous and contain unusual strain distributions. Careful testing of the resistivity limit for superconductors shows that it is less than 4×10^{-23} ohm cm, while the lowest resistivity observed in metals is of the order of 10^{-13} ohm cm. If one compares the resistivity of a superconductive body to that of copper at room temperature, the superconductive body is at least 10^{17} times less resistive.

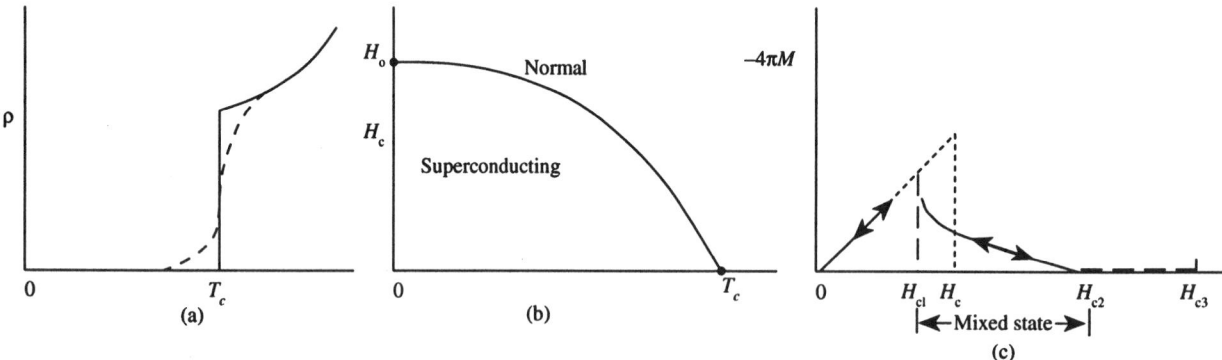

FIGURE 1. Physical properties of superconductors. (a) Resistivity vs. temperature for a pure and perfect lattice (solid line); impure and/or imperfect lattice (broken line). (b) Magnetic-field temperature dependence for Type-I or "soft" superconductors. (c) Schematic magnetization curve for "hard" or Type-II superconductors.

The temperature interval ΔT_c, over which the transition between the normal and superconductive states takes place, may be of the order of as little as 2×10^{-5} K or several K in width, depending on the material state. The narrow transition width was attained in 99.9999% pure gallium single crystals.

A Type-I superconductor below T_c, as exemplified by a pure metal, exhibits perfect diamagnetism and excludes a magnetic field up to some critical field H_c, whereupon it reverts to the normal state as shown in the H-T diagram of Figure 1(b).

The magnetization of a typical high-field superconductor is shown in Figure 1(c). The discovery of the large current-carrying capability of Nb_3Sn and other similar alloys has led to an extensive study of the physical properties of these alloys. In brief, a high-field superconductor, or Type-II superconductor, passes from the perfect diamagnetic state at low magnetic fields to a mixed state and finally to a sheathed state before attaining the normal resistive state of the metal. The magnetic field values separating the four stages are given as H_{c1}, H_{c2}, and H_{c3}. The superconductive state below H_{c1} is perfectly diamagnetic, identical to the state of most pure metals of the "soft" or Type-I superconductor. Between H_{c1} and H_{c2} a "mixed superconductive state" is found in which fluxons (a minimal unit of magnetic flux) create lines of normal flux in a superconductive matrix. The volume of the normal state is proportional to $-4\pi M$ in the "mixed state" region. Thus at H_{c2} the fluxon density has become so great as to drive the interior volume of the superconductive body completely normal. Between H_{c2} and H_{c3} the superconductor has a sheath of current-carrying superconductive material at the body surface, and above H_{c3} the normal state exists. With several types of careful measurement, it is possible to determine H_{c1}, H_{c2}, and H_{c3}. Table 6 contains some of the available data on high-field superconductive materials.

High-field superconductive phenomena are also related to specimen dimension and configuration. For example, the Type-I superconductor, Hg, has entirely different magnetization behavior in high magnetic fields when contained in the very fine sets of filamentary tunnels found in an unprocessed Vycor glass. The great majority of superconductive materials are Type-II. The elements in very pure form and a very few precisely stoichiometric and well annealed compounds are Type I with the possible exceptions of vanadium and niobium.

Metallurgical Aspects. The sensitivity of superconductive properties to the material state is most pronounced and has been used in a reverse sense to study and specify the detailed state of alloys. The mechanical state, the homogeneity, and the presence of impurity atoms and other electron-scattering centers are all capable of controlling the critical temperature and the current-carrying capabilities in high-magnetic fields. Well annealed specimens tend to show sharper transitions than those that are strained or inhomogeneous. This sensitivity to mechanical state underlines a general problem in the tabulation of properties for superconductive materials. The occasional divergent values of the critical temperature and of the critical fields quoted for a Type-II superconductor may lie in the variation in sample preparation. Critical temperatures of materials studied early in the history of superconductivity must be evaluated in light of the probable metallurgical state of the material, as well as the availability of less pure starting elements. It has been noted that recent work has given extended consideration to the metallurgical aspects of sample preparation.

Symbols in tables: T_c: Critical temperature; H_o: Critical magnetic field in the $T = 0$ limit; θ_D: Debye temperature; and γ: Electronic specific heat.

TABLE 1
Selective Properties of Superconductive Elements

Element	T_c(K)	H_o(oersted)	θ_D(K)	γ(mJ mol^{-1}K^{-1})
Al	1.175 ± 0.002	104.9 ± 0.3	420	1.35
Am* (α,?)	0.6			
Am* (β,?)	1.0			
Be	0.026			0.21
Cd	0.517 ± 0.002	28 ± 1	209	0.69
Ga	1.083 ± 0.001	58.3 ± 0.2	325	0.60
Ga (β)	5.9, 6.2	560		
Ga (γ)	7	950, HF^a		
Ga (Δ)	7.85	815, HF		
Hf	0.128	12.7		2.21
Hg (α)	4.154 ± 0.001	411 ± 2	87, 71.9	1.81
Hg (β)	3.949	339	93	1.37
In	3.408 ± 0.001	281.5 ± 2	109	1.672
Ir	0.1125 ± 0.001	16 ± 0.05	425	3.19
La (α)	4.88 ± 0.02	800 ± 10	151	9.8
La (β)	6.00 ± 0.1	1096, 1600	139	11.3
Lu	0.1 ± 0.03	350 ± 50		
Mo	0.915 ± 0.005	96 ± 3	460	1.83
Nb	9.25 ± 0.02	2060 ± 50, HF	276	7.80
Os	0.66 ± 0.03	70	500	2.35
Pa	1.4			
Pb	7.196 ± 0.006	803 ± 1	96	3.1
Re	1.697 ± 0.006	200 ± 5	4.5	2.35
Ru	0.49 ± 0.015	69 ± 2	580	2.8
Sn	3.722 ± 0.001	305 ± 2	195	1.78
Ta	4.47 ± 0.04	829 ± 6	258	6.15
Tc	7.8 ± 0.1	1410, HF	411	6.28
Th	1.38 ± 0.02	1.60 ± 3	165	4.32
Ti	0.40 ± 0.04	56	415	3.3
Tl	2.38 ± 0.02	178 ± 2	78.5	1.47
U	0.2			
V	5.40 ± 0.05	1408	383	9.82
W	0.0154 ± 0.0005	1.15 ± 0.03	383	0.90
Zn	0.85 ± 0.01	54 ± 0.3	310	0.66
Zr	0.61 ± 0.15	47	290	2.77
Zr (ω)	0.65, 0.95			

TABLE 2
Range of Critical Temperatures Observed for Superconductive Elements in Thin Films Condensed Usually at Low Temperatures

Element	T_c Range (K)	Comments	Element	T_c Range (K)	Comments
Al	1.15—5.7	HF^a	Nb	2.0—10.1	
Be	5—9.75	HF	Pb	1.8—7.5	
Bi	6.17—6.6		Re	1.7—7	
Cd			Sn	3.5—6	
(Disordered)	0.79—0.91		Ta	<1.7—4.51	HF^a
(Ordered)	0.53—0.59		Tc	4.6—7.7	
Ga	2.5—8.5	HF	Ti	1.3 Max	
Hg	3.87—4.5		Tl	2.33—2.96	
In	2.2—5.6	HF	V	1.8—6.02	
La	3.55—6.74		W	<1.0—4.1	
Mo	3.3—8.0		Zn	0.77—1.9	

^{a} HF denotes high magnetic field superconductive properties.

TABLE 3
Elements Exhibiting Superconductivity Under or After Application of High Pressure

Element	T_c Range (K)	Pressure (kbar)	Element	T_c Range (K)	Pressure (kbar)
Al	1.98—0.075	0—62	Pb II	3.55	160
As	0.31—0.5	220—140	Re II	2.3 Max.	"Plastic" compression
	0.2—0.25	140—100			
Ba II	1—1.8	55—85	Sb (prepared 120	2.6—2.7	
III	1.8—5	85—144	kbar, held below		
IV	4.5—5.4	144—190	77K)		
Bi II	3.9	25—27	Sb II	3.55—3.40	85—150
III	6.55—7.25	28—38	Se II	6.75, 6.95	130
IV	7.0, 8.7—6.0	43, 43—62	Si	6.7—7.1	120—130
V	6.7, 8.3	48—80	Sn II	5.2—4.85	125—160
VI	8.55	90, 92—101	III	5.30	113
VII(?)	8.2	30	Te II	2.4—5.1	38—55
Ce (α)	0.020—0.045	20—35		4.1—4.2	53—62
Ce (α')	1.9—1.3	45—125	IV	4.72—4	63—80
Cs V	1.5	>125	( )	3.3—2.8	100—260
Ga II	6.38	≥35	Tl (cubic form)	1.45	35
II'	7.5	≥35 then P removed	(hexagonal form)	1.95	35
			U	2.4—0.4	10—85
Ge	5.35	115	Y	1.7—2.5	110—160
La	5.5—12.9	0—210	Zr (omega form, metastable)	1—1.7	60—130
Lu	0.022—1.0	45—190			
P	5.8	170			

TABLE 4
Superconductive Compounds and Alloys

All compositions are denoted on an atomic basis, i.e., AB, AB$_2$, or AB$_3$ for compounds, unless noted. Solid solutions or odd compositions may be denoted as A$_z$B$_{1-z}$ or A$_z$B. A series of three or more alloys is indicated as A$_x$B$_{1-x}$ or by actual indication of the atomic fraction range, such as A$_{0-0.6}$B$_{1-0.4}$. The critical temperature of such a series of alloys is denoted by a range of values or possibly the maximum value.

The selection of the critical temperature from a transition in the effective permeability, or the change in resistance, or possibly the incremental changes in frequency observed by certain techniques is not often obvious from the literature. Most authors choose the mid-point of such curves as the probable critical temperature of the idealized material, while others will choose the highest temperature at which a deviation from the normal state property is observed. In view of the previous discussion concerning the variability of the superconductive properties as a function of purity and other metallurgical aspects, it is recommended that appropriate literature be checked to determine the most probable critical temperature or critical field of a given alloy.

A very limited amount of data on critical fields, H_o, is available for these compounds and alloys; these values are given at the end of the table.

A. SUPERCONDUCTORS WITH $T_c < 10$ K

Substance	T_c, K	Crystal structure type
Ag$_{3.3}$Al	0.34	A12-cI58 (Mn)
Ag$_x$Al$_y$Zn$_{1-x-y}$	0.15 .	Cubic
AgBi$_2$	2.87—3.0	
Ag$_7$F$_{0.25}$N$_{0.75}$O$_{10.25}$	0.85—0.90	
Ag$_2$F	0.0.066	
Ag$_7$FO$_8$	0.3	Cubic
Ag$_{0.8-0.3}$Ga$_{0.2-0.7}$	6.5—8	
Ag$_4$Ge	0.85	Hex., c.p.
Ag$_{0.438}$Hg$_{0.562}$	0.64	D8$_2$
AgIn$_2$	~2.4	C16

TABLE 4
Superconductive Compounds and Alloys (continued)

Substance	T_c, K	Crystal structure type
$Ag_{0.1}In_{0.9}Te$ ($n = 1.4 \times 10^{22}$)*	1.2—1.89	B1
$Ag_{0.2}In_{0.8}Te$ ($n = 1.07 \times 10^{22}$)	0.77—1.00	B1
AgLa	0.94	B2-cP2 (CsCl)
AgLa (9.5 kbar)	1.2	B2
AgLu	0.33	B2-cP2
$AgMo_4S_5$	9.1	hR15 (Mo_6PbS_8)
$Ag_{1.2}Mo_6Se_8$	5.9	Same
Ag_7NO_{11}	1.04	Cubic
Ag_xPb_{1-x}	7.2 max.	
Ag_4Sn	0.1	h**
Ag_xSn_{1-x}	1.5—3.7	
Ag_xSn_{1-x} (film)	2.0—3.8	
$AgTe_3$	2.6	Cubic
AgTh	2.2	C16-tI12 (Al_2Cu)
$AgTh_2$	2.26	C16
$Ag_{0.03}Tl_{0.97}$	2.67	
$Ag_{0.94}Tl_{0.06}$	2.32	
AgY	0.33	B2-cP2 (CsCl)
Ag_xZn_{1-x}	0.5—0.845	
$AlAu_4$	0.4—0.7	Like A13
Al_2Au	0.1	C1-cF12 (CaF_2)
Al_2CMo_3	9.8—10.2	A13+trace 2nd. phase)
Al_2CaSi	5.8	
$Al_{0.131}Cr_{0.088}V_{0.781}$	1.46	Cubic
$AlGe_2$	1.75	
Al_2Ge_2U	1.6	LI_2-cP4 (Cu_3Au)
$AlLa_3$	5.57	DO_{19}
Al_2La	3.23	C15
Al_2Lu	1.02	C15-cF24 (Cu_2Mg)
Al_3Mg_2	0.84	F.C.C.
$AlMo_3$	0.58	A15
$AlMo_6Pd$	2.1	
AlN	1.55	B4
Al_2NNb_3	1.3	A13
Al_3Nb	0.64	tI8 (Al_3Ti)
AlOs	0.39	B2
Al_3Os	5.90	
AlPb (film)	1.2—7	
Al_2Pt	0.48—0.55	C1
Al_5Re_{24}	3.35	A12
AlSb	2.8	B4-tI4 (Sn)
Al_2Sc	1.02	C15-cF24 (Cu_2Mg)
Al_2Si_2U	1.34	LI_2-cP4 (Cu_3Au)
$AlTh_2$	0.1	C16-tI12 (Al_2Cu)
Al_3Th	0.75	DO_{19}
$Al_xTi_yV_{1-x-y}$	2.05—3.62	Cubic
$Al_{0.108}V_{0.892}$	1.82	Cubic
Al_2Y	0.35	C15-cF24 (Cu_2Mg)
Al_3Yb	0.94	LI_2-cP4 (Cu_3Au)
Al_xZn_{1-x}	0.5—0.845	
$AlZr_3$	0.73	LI_2
AsBiPb	9.0	
AsBiPbSb	9.0	
AsHfOs	3.2	C22-hP9 (Fe_2P)
AsHfRu	4.9	same

TABLE 4
Superconductive Compounds and Alloys (continued)

Substance	T_c, K	Crystal structure type
$As_{0.33}InTe_{0.67}$ ($n = 1.24 \times 10^{22}$)	0.85—1.15	B1
$As_{0.5}InTe_{0.5}$ ($n = 0.97 \times 10^{22}$)	0.44—0.62	B1
As_4La_3	0.6	cI28 (Th_3P_4)
$AsNb_3$	0.3	$L1_2$-tP32
$As_{0.50}Ni_{0.06}Pd_{0.44}$	1.39	C2
$AsNi_{0.25}Pd_{0.75}$	1.6	$B8_1$-hP4 (NiAs)
AsOsZr	8.0	C22-hP9 (Fe_2P)
AsPb	8.4	
$AsPd_2$ (low-temp. phase)	0.60	Hexagonal
$AsPd_2$ (high-temp. phase)	1.70	C22
$AsPd_5$	0.46	Complex
As_3Pd_5	1.9	
AsRh	0.58	B31
$AsRh_{1.4—1.6}$	< 0.03—0.56	Hexagonal
AsSn	4.10	
AsSn ($n = 2.14 \times 10^{22}$)	3.41—3.65	B1
$As_{-2}Sn_{-3}$	3.5—3.6; 1.21—1.17	
As_3Sn_4 ($n = 0.56 \times 10^{22}$)	1.16—1.19	Rhombohedral
AsV_3	0.20	A15-cP8 (Cr_3Si)
Au_5Ba	0.4—0.7	$D2_d$
AuBe	2.64	B20
Au_2Bi	1.80	C15
Au_5Ca	0.34—0.38	$C15_b$
$AuGa_2$	1.6	C1-cF12 (CaF_2)
AuGa	1.2	B31
$Au_{0.40—0.92}Ge_{0.60—0.08}$	<0.32—1.63	Complex
$AuIn_2$	0.2	C1-cF12
AuIn	0.4—0.6	Complex
AuLu	<0.35	B2
$AuNb_3$	1.2	A2
$AuPb_2$	3.15	
$AuPb_2$ (film)	4.3	
$AuPb_3$	4.40	
$AuPb_3$ (film)	4.25	
Au_2Pb	1.18; 6—7	C15
$AuSb_2$	0.58	C2
AuSn	1.25	$B8_1$
Au_xSn_{1-x} (film)	2.0—3.8	
Au_5Sn	0.7—1.1	A3
$AuTa_{4.3}$	0.55	A15-cP8 (Cr_3Si)
Au_3Te_5	1.62	Cubic
$AuTh_2$	3.08	C16
AuTl	1.92	
AuV_3	0.74	A15
Au_xZn_{1-x}	0.50—0.845	
$AuZn_3$	1.21	Cubic
Au_xZr_y	1.7—2.8	A3
$AuZr_3$	0.92	A15
$B_2Ba_{0.67}Pt_3$	5.60	hP12 (B_2BaPt_3)
$BCMo_2$	5.4	Orthorhombic
$BCMo_2$	5.3—7.0	Same
$B_2Ca_{0.67}Pt_3$	1.57	hP12
B_4ErIr_4	2.1	tP18 (B_4CeCo_4)
B_4ErRh_4	4.3	oC108 (B_4LuRh_4)
B_4ErRh_4	8.7	tP18 (B_4CeCo_4)

TABLE 4
Superconductive Compounds and Alloys (continued)

Substance	T_c, K	Crystal structure type
BHf	3.1	Cubic
B_4HoIr_4	2.0	tP18
B_4HoRh_4	1.4	oC108
B_2Ir_3La	1.65	hP6 ($CaCu_5$)
B_2Ir_3Th	2.09	Same
B_4Ir_4Tm	1.6	tP18
B_6La	5.7	
B_2LaRh_3	2.82	hP6
$B_{12}Lu$	0.48	
B_2LuOs	2.66	oP16 (B_2LuRu)
B_2LuOs_3	4.62	hP6
B_4LuRh_4	6.2	oC108
B_2LuRu	9.86	oP16
B_4LuRu_4	2.0	tI72 (B_4LuRu_4)
BMo	0.5 (extrapol.)	
BMo_2	4.74	C16
BNb	8.25	B_f
B_4NdRh_4	5.3	tP18
B_2OsSc	1.34	oP16
B_2OsY	2.22	oP16
$B_2Pt_3Sr_{0.67}$	2.78	hP12 (B_2BaPt_3)
BRe_2	2.80; 4.6	
$B_4Rh_{3.4}Ru_{0.6}$	8.38	tI72
B_4Rh_4Sm	2.7	tP18
B_4Rh_4Th	4.3	Same
B_4Rh_4Tm	9.8	Same
B_4Rh_4Tm	5.4	oC108
$B_{0.3}Ru_{0.7}$	2.58	$D10_2$
B_4Ru_4Sc	7.2	tI72
B_2Ru_3Th	1.79	hP6
B_2Ru_3Y	2.85	Same
$B_2Ru\,Y$	7.80	oP16
B_4Ru_4Y	1.4	tI72
$B_{12}Sc$	0.39	
BTa	4.0	B_f
BTa_2	3.12	C16-tI12 (Al_2Cu)
B_6Th	0.74	
BW_2	3.1	C16
B_6Y	6.5—7.1	
$B_{12}Y$	4.7	
BZr	3.4	Cubic
$B_{12}Zr$	5.82	
$BaBi_3$	5.69	Tetragonal
$Ba_2Mo_{15}Se_{19}$	2.75	hP15 (Mo_6PbS_8)
$Ba_xO_3Sr_{1-x}Ti$ ($n = 4.2 \times 10^{19}$)	<0.1—0.55	
$Ba_{0.13}O_3W$	1.9	Tetragonal
$Ba_{0.14}O_3W$	<1.25—2.2	Hexagonal
$BaRh_2$	6.0	C15
$Be_{22}Mo$	2.51	Cubic ($Be_{22}Re$)
$Be_8Nb_5Zr_2$	5.2	
$Be_{0.98—0.92}Re_{0.02—0.08}$ (quenched)	9.5—9.75	Cubic
$Be_{0.957}Re_{0.043}$	9.62	Cubic ($Be_{22}Re$)
BeTc	5.21	Cubic
$Be_{22}W$	4.12	Cubic ($Be_{22}Re$)

TABLE 4
Superconductive Compounds and Alloys (continued)

Substance	T_c, K	Crystal structure type
$Be_{13}W$	4.1	Tetragonal
Bi_3Ca	2.0	
$Bi_{0.5}Cd_{0.13}Pb_{0.25}Sn_{0.12}$ (weight fractions)	8.2	
BiCo	0.42—0.49	
Bi_2Cs	4.75	C15
Bi_xCu_{1-x} (electrodeposited)	2.2	
BiCu	1.33—1.40	
Bi_3Fe	1.0	m**
$Bi_{0.019}In_{0.981}$	3.86	
$Bi_{0.05}In_{0.95}$	4.65	α-phase
$Bi_{0.10}In_{0.90}$	5.05	Same
$Bi_{0.15—0.30}In_{0.85—0.70}$	5.3—5.4	α- and β-phases
$Bi_{0.34—0.48}In_{0.66—0.52}$	4.0—4.1	
Bi_3In_5	4.1	
$BiIn_2$	5.65	β-phase
Bi_2Ir	1.7—2.3	
Bi_2Ir (quenched)	3.0—3.96	
BiK	3.6	
Bi_2K	3.58	C15
BiLi	2.47	$L1_0$, α-phase
$Bi_{4—9}Mg$	0.7—~1.0	
Bi_3Mo	3—3.7	
BiNa	2.25	$L1_0$
$BiNb_3$	4.5	A15-cP8 (Cr_3Si)
$BiNb_3$ (high pressure and temperature)	3.05	A15
BiNi	4.25	$B8_1$
Bi_3Ni	4.06	Orthorhombic
$BiNi_{0.5}Rh_{0.5}$	3.0	$B8_1$-hP4 (AsNi)
$Bi_{0.5}NiSb_{0.5}$	2.0	Same
$Bi_{1.0}Pb_{0-1}$	7.26—9.14	
$Bi_{1.0}Pb_{0-1}$ (film)	7.25—8.67	
$Bi_{0.05—0.40}Pb_{0.95—0.60}$	7.35—8.4	H.C.P. to ε-phase
Bi_2Pb	4.25	t**
BiPbSb	8.9	
$Bi_{0.5}Pb_{0.31}Sn_{0.19}$ (weight fractions)	8.5	
$Bi_{0.5}Pb_{0.25}Sn_{0.25}$	8.5	
$BiPd_2$	4.0	
$Bi_{0.4}Pd_{0.6}$	3.7—4	Hexagonal, ordered
BiPd	3.7	Orthorhombic
Bi_2Pd	1.70	Monoclinic, α-phase
Bi_2Pd	4.25	Tetragonal, β-phase
$BiPd_{0.45}Pt_{0.55}$	3.7	$B8_1$-hP4 (NiAs)
BiPdSe	1.0	C2
BiPdTe	1.2	C2
BiPt	1.21	$B8_1$
$Bi_{0.1}PtSb_{0.9}$	2.05; 1.5	$B8_1$-hP4 (NiAs)
BiPtSe	1.45	C2
BiPtTe	1.15	C2
Bi_2Pt	0.155	Hexagonal
Bi_2Rb	4.25	C15
$BiRe_2$	1.9—2.2	
BiRh	2.06	$B8_1$

TABLE 4
Superconductive Compounds and Alloys (continued)

Substance	T_c, K	Crystal structure type
Bi_3Rh	3.2	Orthorhombic (NiB_3)
Bi_4Rh	2.7	Hexagonal
$BiRu$	5.7	m**
Bi_3Sn	3.6—3.8	
$BiSn$	3.8	
Bi_xSn_y	3.85—4.18	
Bi_3Sr	5.62	$L1_2$
Bi_3Te	0.75—1.0	
Bi_5Tl_3	6.4	
$Bi_{0.26}Tl_{0.74}$	4.4	Cubic, disordered
$Bi_{0.26}Tl_{0.74}$	4.15	$L1_2$, ordered (?)
Bi_2Y_3	2.25	
Bi_3Zn	0.8—0.9	
$Bi_{0.3}Zr_{0.7}$	1.51	
$BiZr_3$	2.4—2.8	
$BrMo_6Se_7$	7.1	hP15 (Mo_6PbS_8)
$Br_3Mo_6Se_5$	7.1	Same
CCs_x	0.020—0.135	Hexagonal
CFe_3	1.30	DO_{11}-oP16 (Fe_3C)
$CGaMo_2$	3.7—4.1	Hexagonal
$CHf_{0.5}Mo_{0.5}$	3.4	B1
$CHf_{0.3}Mo_{0.7}$	5.5	B1
$CHf_{0.25}Mo_{0.75}$	6.6	B1
$CHf_{0.7}Nb_{0.3}$	6.1	B1
$CHf_{0.6}Nb_{0.4}$	4.5	B1
$CHf_{0.5}Nb_{0.5}$	4.8	B1
$CHf_{0.4}Nb_{0.6}$	5.6	B1
$CHf_{0.25}Nb_{0.75}$	7.0	B1
$CHf_{0.2}Nb_{0.8}$	7.8	B1
$CHf_{0.9—0.1}Ta_{0.1—0.9}$	5.0—9.0	B1
CK (excess K)	0.55	Hexagonal
C_8K	0.39	Hexagonal
C_2La	1.66	tI6 (CaC_2)
C_2Lu	3.33	Same
$C_{0.40—0.44}Mo_{0.60—0.56}$	9—13	
C_3MoRe	3.8	B1-cF8
$C_{0.6}Mo_{4.8}Si_3$	7.6	$D8_8$
$CMo_{0.2}Ta_{0.8}$	7.5	B1
$CMo_{0.5}Ta_{0.5}$	7.7	B1
$CMo_{0.75}Ta_{0.25}$	8.5	B1
$CMo_{0.8}Ta_{0.2}$	8.7	B1
$CMo_{0.85}Ta_{0.15}$	8.9	B1
CMo_xV_{1-x}	2.9—9.3	B1
CMo_xZr_{1-x}	9.8	B1
$C_{0.984}Nb$	9.8	B1
CNb_2	9.1	
CNb_xTi_{1-x}	<4.2—8.8	B1
$CNb_{0.1—0.9}Zr_{0.9—0.1}$	4.2—8.4	B1
CRb_x (Au)	0.023—0.151	Hexagonal
$CRe_{0.06}W$	5.0	
CRu	2.00	hP2 (CW)
$C_{0.98}7Ta$	9.7	
$C_{0.848—0.987}$	2.04—9.7	
CTa (film)	5.09	B1
CTa_2	3.26	L'_3

TABLE 4
Superconductive Compounds and Alloys (continued)

Substance	T_c, K	Crystal structure type
$CTa_{0.4}Ti_{0.6}$	4.8	B1
$Cta_{1-0.4}W_{0-0.6}$	8.5—10.5	B1
$CTa_{0.2-0.9}Zr_{0.8-0.1}$	4.6—8.3	B1
CTc (excess C)	3.85	Cubic
$CTi_{0.5-0.7}W_{0.5-0.3}$	6.7—2.1	B1
CW	1.0	
CW_2	2.74	L'_3
CW_2	5.2	F.C.C.
C_2Y	3.88	tI6 (CaC_2)
$Ca_3Co_4Sn_{13}$	5.9	cP40 ($Pr_3Rh_2Sn_{13}$)
$Ca_3Ge_{13}Rh_4$	2.1	Same
CaHg	1.6	B2-cP2 (CsCl)
$CaHg_3$	1.6	hP8 (Ni_3Sn)
$CaIr_2$	6.15	C15
$Ca_3Ir_4Sn_{13}$	7.1	cP40
$Ca_xO_3Sr_{1-x}Ti$ ($n = 3.7—11 \times 10^{19}$)	<0.1—0.55	
$Ca_{0.1}O_3W$	1.4—3.4	Hexagonal
CaPb	7.0	
$CaRh_2$	6.40	C15
$CaRh_{1.2}Sn_{4.5}$	8.7	cP40
$CaTl_3$	2.0	B2-cP2
$Cd_{0.3-0.5}Hg_{0.7-0.5}$	1.70—1.92	
CdHg	1.77; 2.15	Tetragonal
$Cd_{0.0075-0.05}In_{0.9925-0.95}$	3.24—3.36	Tetragonal
$Cd_{0.97}Pb_{0.03}$	4.2	
CdSn	3.65	
$Cd_{0.17}Tl_{0.83}$	2.3	
$Cd_{0.18}Tl_{0.82}$	2.54	
$CeCo_2$	0.84	C15
$CeCo_{1.67}Ni_{0.33}$	0.46	C15
$CeCo_{1.67}Rh_{0.33}$	0.47	C15
$Ce_xGd_{1-x}Ru_2$	3.2—5.2	C15
$CeIr_3$	3.34	
$CeIr_5$	1.82	
$Ce_{0.005}La_{0.995}$	4.6	
Ce_xLa_{1-x}	1.3—6.3	
$Ce_xPr_{1-x}Ru_2$	1.4—5.3	C15
Ce_xPt_{1-x}	0.7—1.55	
$CeRu_2$	6.0	C15
$Ce_3Mo_6Se_5$	5.7	hR15 (Mo_6PbS_8)
$Ce_2Mo_6Te_6$	1.7	Same
$Co_xFe_{1-x}Si_2$	1.4 (max.)	C1
$CoHf_2$	0.56	$E9_3$
$CoLa_3$	4.28	
$Co_4La_3Sn_{13}$	2.8	cP40
$CoLu_3$	~0.35	
Co_xLuSn_y	1.5	cP40
$Co_{0-0.01}Mo_{0.8}Re_{0.2}$	2—10	
$Co_{0.02-0.10}Nb_3Rh_{0.98-0.90}$	2.28—1.90	A15
$Co_xNi_{1-x}Si_2$	1.4 (max.)	C1
$Co_{0.5}Rh_{0.5}Si_2$	2.5	
$Co_xRh_{1-x}Si_2$	3.65 (max.)	
$Co_{-0.3}So_{-0.7}$	~0.35	
$Co_4Sc_5Si_{10}$	5.0	tP38 ($Co_4Sc_5Si_{10}$)
$CoSi_2$	1.40; 1.22	C1

TABLE 4
Superconductive Compounds and Alloys (continued)

Substance	T_c, K	Crystal structure type
Co_xSn_yYb	2.5	cP40
Co_3Th_7	1.83	$D10_2$
Co_xTi_{1-x}	2.8 (max.)	Co in α-Ti
Co_xTi_{1-x}	3.8 (max.)	Co in β-Ti
$CoTi_2$	3.44	$E9_3$
CoTi	0.71	A2
CoU	1.7	B2, distorted
CoU_6	2.29	$D2_c$
$Co_{0.28}Y_{0.72}$	0.34	
CoY_3	<0.34	
$CoZr_2$	6.3	C16
$Co_{0.1}Zr_{0.9}$	3.9	A3
$Cr_{0.6}Ir_{0.4}$	0.4	H.C.P.
$Cr_{0.65}Ir_{0.35}$	0.59	H.C.P.
$Cr_{0.7}Ir_{0.3}$	0.76	H.C.P.
$Cr_{0.72}Ir_{0.28}$	0.83	
Cr_3Ir	0.45	A15
$Cr_{0-0.1}Nb_{1-0.9}$	4.6—9.2	A2
$Cr_{0.80}Os_{0.20}$	2.5	Cubic
Cr_3Os	4.68	A15-cP8 (Cr_3Si)
Cr_xRe_{1-x}	1.2—5.2	
$Cr_{0.4}Re_{0.6}$	2.15	$D8_b$
$Cr_{0.8-0.6}Rh_{0.2-0.4}$	0.5—1.10	A3
Cr_3Rh	0.3	A15-cP8
Cr_3Ru (annealed)	3.3	A15
Cr_2Ru	2.02	$D8_b$
Cr_3Ru_2	2.10	$D8_b$-tP30 (CrFe)
$Cr_{0.1-0.5}Ru_{0.9-0.5}$	0.34—1.65	A3
Cr_xTi_{1-x}	3.6 (max.)	Cr in α-Ti
Cr_xTi_{1-x}	4.2 (max.)	Cr in β-Ti
$Cr_{0.1}Ti_{0.3}V_{0.6}$	5.6	
$Cr_{0.0175}U_{0.9825}$	0.75	β-phase
$Cs_{0.32}O_3W$	1.12	Hexagonal
$Cu_{0.15}In_{0.85}$ (film)	3.75	
$Cu_{0.04-0.08}In_{0.94-0.92}$	4.4	
CuLa	5.85	
$Cu_2Mo_6O_2S_6$	9	hR15 (Mo_6PbS_8)
$Cu_2Mo_6Se_8$	5.9	Same
Cu_xPb_{1-x}	5.7—7.7	
CuS	1.62	B18
CuS_2	1.48—1.53	C18
CuSSe	1.5—2.0	C18
$CuSe_2$	2.3—2.43	C18
CuSeTe	1.6—2.0	C18
Cu_xSn_{1-x}	3.2—3.7	
Cu_xSn_{1-x} (film, made at 10K)	3.6—7	
Cu_xSn_{1-x} (film, made at 300K)	2.8—3.7	
$CuTe_2$	<1.25—1.3	C18
$CuTh_2$	3.49	C16
$Cu_{0-0.027}V$	3.9—5.3	A2
CuY	0.33	B2-cP2 (CsCl)
Cu_xZn_{1-x}	0.5—0.845	
$DyMo_6S_8$	2.1	hR15
Er_xLa_{1-x}	1.4—6.3	
$ErMo_6S_8$	2.2	hR15

TABLE 4
Superconductive Compounds and Alloys (continued)

Substance	T_c, K	Crystal structure type
$ErMo_6Se_8$	6.2	hR15
$Fe_3Lu_2Si_5$	6.1	tP40 ($Fe_3Sc_2Si_5$)
$Fe_{0-0.04}Mo_{0.8}Re_{0.2}$	1—10	
$Fe_{0.05}Ni_{0.05}Zr_{0.90}$	~3.9	
Fe_3Re_2	6.55	$D8_b$-tP30 (FeCr)
$Fe_3Sc_2Si_5$	4.52	tP40
Fe_3Si_5Tm	1.3	Same
$Fe_3Si_5Y_2$	2.4	Same
Fe_3Th_7	1.86	D10
Fe_xTi_{1-x}	3.2 (max.)	Fe in α-Ti
Fe_xTi_{1-x}	3.7 (max.)	Fe in β-Ti
$Fe_xTi_{0.6}V_{1-x}$	6.8 (max.)	
FeU_6	3.86	$D2_c$
$Fe_{0.1}Zr_{0.9}$	1.0	A3
$Ga_{0.5}Ge_{0.5}Nb_3$	7.3	A15
Ga_2Ge_2U	0.87	B2-cP2
$GaHf_2$	0.21	C16-tI12 (Al_2Cu)
$GaLa_3$	5.84	
Ga_3Lu	2.3	B2-cP2
Ga_2Mo	9.5	
$GaMo_3$	0.76	A15
GaN (black)	5.85	B4
$Ga_{0.7}Pt_{0.3}$	2.9	C1
GaPt	1.74	B20
GaSb (120kbar, 77K, annealed)	4.24	A5
GaSb (unannealed)	~5.9	
$Ga_{0-1}Sn_{1-0}$ (quenched)	3.47—4.18	
$Ga_{0-1}Sn_{1-0}$ (annealed)	2.6—3.85	
GaTe	0.17	mC24 (GaTe)
Ga_5V_2	3.55	Tetragonal (Mn_2Hg_5)
$GaV_{4.5}$	9.15	
Ga_3Zr	1.38	
Ga_3Zr_5	3.8	$D8_b$-hP16 (Mn_5Si_3)
Gd_xLa_{1-x}	< 1.0—5.5	
$GdMo_6S_8$	3.5	hR15
$GdMo_6Se_8$	5.6	hR15
$Gd_xOs_2Y_{1-x}$	1.4—4.7	
$Gd_xRu_2Th_{1-x}$	3.6 (max.)	C15
$Ge_{10}As_4Y_5$	9.06	tP38 ($C0_4Sc_5Si_{10}$)
GeIr	4.7	B31
GeIrLa	1.64	tI12 (LaPtSi)
$Ge_{10}Ir_4Lu_5$	2.60	tP38
$Ge_{10}Ir_4Y_5$	2.62	tP38
Ge_2La	1.49; 2.2	Orthorhombic, distorted (Mn_2Hg_5)
GeLaPt	3.53	tI12
$Ge_{13}Lu_3Os_4$	3.6	cP40 ($Pr_3Rh_2Sn_{13}$)
$Ge_{10}Lu_5Rh_4$	2.79	tP38
$Ge_{13}Lu_3Ru_4$	2.3	cP40
$GeMo_3$	1.43	A15
$GeNb_2$	1.9	
$Ge_{0.29}Nb_{0.71}$	6	A15
GePt	0.40	B31
Ge_3Rh_5	2.12	Orthorhombic, related to $InNi_2$

TABLE 4
Superconductive Compounds and Alloys (continued)

Substance	T_c, K	Crystal structure type
GeRh	0.96	B31-oP8 (MnP)
$Ge_{13}Rh_4Sc_3$	1.9	c P40
$Ge_{10}Rh_4Y_5$	1.35	tP38
$Ge_{13}Ru_4Y_3$	1.7	cP40
Ge_2So	1.3	
$GeTa_3$	8.0	A15-cP8 (Cr_3Si)
Ge_3Te_4 ($n = 1.06 \times 10^{22}$)	1.55—1.80	Rhombohedral
Ge_xTe_{1-x} ($n = 8.5$—64×10^{20})	0.07—0.41	R1
GeV_3	6.01	A15
Ge_2Y	3.80	C_c
$Ge_{1.62}Y$	2.4	
Ge_2Zr	0.30	oC12 ($ZrSi_2$)
$GeZr_3$	0.4	$L1_2$-tP32 (Ti_3P)
$H_{0.33}Nb_{0.67}$	7.28	B.C.C.
$H_{0.1}Nb_{0.9}$	7.38	Same
$H_{0.05}Nb_{0.95}$	7.83	Same
$H_{0.12}Ta_{0.88}$	2.81	B.C.C.
$H_{0.08}Ta_{0.92}$	3.26	Same
$H_{0.04}Ta_{0.96}$	3.62	Same
HfIrSi	3.50	C37-cP12 (Co_2Si)
$HfMo_2$	0.05	hP24 (Ni_2Mn)
$HfN_{0.989}$	6.6	B1
$Hf_{0—0.5}Nb_{1—0.5}$	8.3—9.5	A2
$Hf_{0.75}Nb_{0.25}$	>4.2	
$HfOs_2$	2.69	C14
HfOsP	6.1	C22-hP9 (Fe_2P)
HfPRu	9.9	Same
$HfRe_2$	4.80	C14
$Hf_{0.14}Re_{0.86}$	5.86	A12
$Hf_{0.99—0.96}Rh_{0.01—0.04}$	0.85—1.51	
$Hf_{0—0.55}Ta_{1—0.45}$	4.4—6.5	A2
HfV_2	8.9—9.6	C15
Hg_xIn_{1-x}	3.14—4.55	
HgIn	3.81	
Hg_2K	1.20	Orthorhombic
Hg_3K	3.18	
Hg_4K	3.27	
Hg_8K	3.42	
Hg_3Li	1.7	Hexagonal
$HgMg_3$	0.17	hP8 (Na_3As)
Hg_2Mg	4.0	tI6 ($MoSi_2$)
Hg_3Mg_5	0.48	$D8_b$-hP16 (Mn_5Si_3)
Hg_2Na	1.62	Hexagonal
Hg_4Na	3.05	
Hg_xPb_{1-x}	4.14—7.26	
HgSn	4.2	
Hg_xTl_{1-x}	2.30—4.19	
Hg_5Tl_2	3.86	
Ho_xLa_{1-x}	1.3—6.3	
$Ho_{1.2}Mo_6Se_8$	6.1	$D10_2$-hR12 (Be_3Nb)
$In_{1—0.86}Mg_{0—0.14}$	3.395—3.363	
$In_2Mo_6Te_6$	2.6	hR15 (Mo_6PbS_8)
$InNb_3$ (high pressure and temp.)	4—8; 9.2	A15
$In_{0.5}Nb_3Zr_{0.5}$	6.4	
$In_{0.11}O_3W$	< 1.25—2.8	Hexagonal

TABLE 4
Superconductive Compounds and Alloys (continued)

Substance	T_c, K	Crystal structure type
$In_{0.95—0.85}Pb_{0.05—0.15}$	3.6—5.05	
$In_{0.98—0.91}Pb_{0.02—0.09}$	3.45—4.2	
InPb	6.65	
InPd	0.7	B2
InSb (quenched from 170 kbar into liquid N_2)	4.8	Like A5
InSb	2.1	B4
$(InSb)_{0.95—0.10}Sn_{0.05—0.90}$ (various heat treatments)	3.8—5.1	
$(InSb)_{0—0.07}Sn_{1—0.93}$	3.67—3.74	
In_3Sn	~5.5	
In_xSn_{1-x}	3.4—7.3	
$In_{0.82—1}Te$ ($n = 0.83—1.71 \times 10^{22}$)	1.02—3.45	B1
$In_{1.000}Te_{1.002}$	3.5—3.7	B1
In_3Te_4 ($n = 4.7 \times 10^{21}$)	1.15—1.25	Rhombohedral
In_xTl_{1-x}	2.7—3.374	
$In_{0.8}Tl_{0.2}$	3.223	
$In_{0.62}Tl_{0.38}$	2.760	
$In_{0.78—0.69}Tl_{0.22—0.31}$	3.18—3.32	Tetragonal
$In_{0.69—0.62}Tl_{0.31—0.38}$	2.98—3.3	F.C.C.
Ir_2La	0.48	C15
Ir_3La	2.32	$D10_2$
Ir_3La_7	2.24	$D10_2$
Ir_5La	2.13	
$IrLaSi_2$	2.03	oC16 ($CeNiSi_2$)
$IrLaSi_3$	2.7	tI10 ($BaNiSn_3$)
Ir_2Lu	2.47	C15
Ir_3Lu	2.89	C15
$Ir_4Lu_5Si_{10}$	3.9	tP38 ($Co_4Sc_5Si_{10}$)
IrMo	< 1.0	A3
$IrMo_3$	9.6	A15
$IrMo_3$	6.8	$D8_b$
$IrNb_3$	1.9	A15
$Ir_{0.4}Nb_{0.6}$	9.8	$D8_b$
$Ir_{0.37}Nb_{0.63}$	2.32	$D8_b$
IrNb	7.9	$D8_b$
$Ir_{1.15}Nb_{0.85}$	4.6	oP12 (IrTa)
$Ir_{0.02}Nb_3Rh_{0.98}$	2.43	A15
$Ir_{0.05}Nb_3Rh_{0.95}$	2.38	A15
$Ir_{0.287}O_{0.14}Ti_{0.573}$	5.5	$E9_3$
$Ir_{0.265}O_{0.035}Ti_{0.65}$	2.30	$E9_3$
Ir_xOs_{1-x}	0.3—0.98	
$Ir_{1.5}Os_{0.5}$	2.4	C14
IrOsY	2.6	C15
IrSiY	2.70	C37-oP12 (Co_2Si)
IrSiZr	2.04	Same
Ir_2Sc	2.07	C15
$Ir_{2.5}Sc$	2.46	C15
$Ir_4Sc_5Si_{10}$	8.46	tP38
Ir_2Si_2Th	2.14	tI10
$IrSi_3Th$	1.75	tI10
IrSiTh	6.50	tI12 (LaPtSi)
Ir_2Si_2Y	2.60	tI10 (Al4Ba)
$Ir_4Si_{10}Y_5$	3.10	tP38
$Ir_3Si_5Y_2$	2.83	oI40

TABLE 4
Superconductive Compounds and Alloys (continued)

Substance	T_c, K	Crystal structure type
$IrSn_2$	0.65—0.78	C1
Ir_2Sr	5.70	C15
Ir_7Ta_{13}	1.2	$D8_b$-tP30 (FeCr)
$Ir_{0.5}Te_{0.5}$	~3	
$IrTe_3$	1.18	C2
IrTh	< 0.37	B_f
Ir_2Th	6.50	C15
Ir_3Th	4.71	
Ir_3Th_7	1.52	$D10_2$
Ir_5Th	3.93	$D2_d$
$IrTi_3$	5.40	A15
IrV_2	1.39	A15
IrW_3	3.82	
$Ir_{0.28}W_{0.72}$	4.49	
Ir_2Y	2.18; 1.38	C15
$Ir_{0.69}Y_{0.31}$	1.98; 1.44	C15
$Ir_{0.70}Y_{0.30}$	2.16	C15
Ir_2Y_3	1.61	
Ir_3Y	3.50	$D10_2$-hR13 (Be_3Nb)
Ir_xY_{1-x}	0.3—3.7	
Ir_2Zr	4.10	C15
$Ir_{0.1}Zr_{0.9}$	5.5	A3
$K_2Mo_{15}S_{19}$	3.32	hR15
$K_{0.27—0.31}O_3W$	0.50	Hexagonal
$K_{0.40—0.57}O_3W$	1.5	Tetragonal
$La_{0.55}Lu_{0.45}$	2.2	Hexagonal, La type
$La_{0.8}Lu_{0.2}$	3.4	Same
$LaMg_2$	1.05	C15
$LaMo_6S_8$	7.1	hR15
LaN	1.35	
$LaOs_2$	6.5	C15
$LaPt_2$	0.46	C15
$La_{0.28}Pt_{0.72}$	0.54	C15
LaPtSi	3.48	tI12
$LaRh_3$	2.60	
$LaRh_5$	1.62	
La_7Rh_3	2.58	$D10_2$
$LaRhSi_2$	3.42	oC16 ($CeNiSi_2$)
$La_2Rh_3Si_5$	4.45	oI40 ($Co_3Si_5U_2$)
$LaRhSi_3$	2.7	tI10 ($BaNiSn_3$)
$LaRh_2Si_2$	3.90	tI10 (Al_4Ba)
$LaRu_2$	1.63	C15
La_3S_4	6.5	$D7_3$
La_3Se_4	8.6	$D7_3$
$LaSi_2$	2.3	C_c
La_xY_{1-x}	1.7—5.4	
LaZn	1.04	B2
$Li_2Mo_6S_8$	4.2	hR15
LiPb	7.2	
$LuOs_2$	3.49	C14
$Lu_{0.275}Rh_{0.725}$	1.27	C15
$LuRh_5$	0.49	
$Lu_5Rh_4Si_{10}$	3.95	tP38 ($Co_4So_5Si_{10}$)
$LuRu_2$	0.86	C14
$Mg_{1.14}Mo_{6.6}S_8$	3.5	hR15

TABLE 4
Superconductive Compounds and Alloys (continued)

Substance	T_c, K	Crystal structure type
Mg2Nb	5.6	
$Mg_{-0.47}Tl_{-0.53}$	2.75	B2
MgZn	0.9	A3-oP4 (AuCd)
Mn_xTi_{1-x}	2.3 (max.)	Mn in -Ti
Mn_xTi_{1-x}	1.1—3.0	Mn in -Ti
MnU_6	2.32	$D2_c$
Mo_2N	5.0	F.C.C.
$Mo_6Na_2S_8$	8.6	hR15
Mo_xNb_{1-x}	0.016—9.2	
$Mo_{5.25}Nb_{0.75}Se_8$	6.2	hR15
Mo_6NdSa_8	8.2	hR15
Mo_3Os	7.2	A15
$Mo_{0.62}Cs_{0.38}$	5.65	$D8_b$
Mo_3P	5.31	DO_e
$Mo_6Pb_{1.2}Se_8$	6.75	hR15
$Mo_{0.5}Pd_{0.5}$	3.52	A3
Mo_6PrSe_8	9.2	hR15
MoRe	7.8	$D8_b$-tP30
$MoRe_3$	9.25; 9.89	A12
Mo_xRe_{1-x}	1.2—12.2	
$Mo_{0.42}Re_{0.58}$	6.35	$D8_b$
MoRh	1.97	A3
Mo_xRh_{1-x}	1.5—8.2	B.C.C.
MoRu	9.5—10.5	A3
$Mo_{0.61}Ru_{0.39}$	7.18	$D8_b$
$Mo_{0.2}Ru_{0.8}$	1.66	A3
Mo_3Ru_2	7.0	$D8_b$-tP30
$Mo_4Ru_2Te_8$	1.7	hR15
Mo_6S_8	1.85	hR15
Mo_6S_8Sc	3.6	hR15
$Mo_6S_8Sm_{1.2}$	2.9	hR15
Mo_6S_8Tb	2.0	hR15
Mo_6S_8Tl	8.7	hR15
$Mo_6S_8Tm_{1.2}$	2.1	hR15
$Mo_6S_8Y_{1.2}$	3.0	hR15
Mo_6S_8Yb	9.2	hR15
$Mo_{6.6}S_8Zn_{11}$	3.6	hR15
Mo_3Sb_4	2.1	
Mo_6Se_8	6.3	hR15
$Mo_6Se_8Sm_{1.2}$	6.8	hR15
$Mo_6Se_8Sn_{1.2}$	6.8	hR15
Mo_6Se_8Tb	5.7	hR15
Mo_3Se_3Tl	4.0	hP14
$Mo_6Se_8Tm_{1.2}$	6.3	hR15
Mo_6Se_8Yb	6.2	hR15
Mo_3Si	1.30	A15
$MoSi_{0.7}$	1.34	
Mo_xSiV_{3-x}	4.54—16.0	A15
$Mo_{5.25}Ta_{0.75}Te_8$	1.7	hR15
Mo_6Te_8	1.7	hR15
$Mo_{0.16}Ti_{0.84}$	4.18; 4.25	
$Mo_{0.913}Ti_{0.087}$	2.95	
$Mo_{0.04}Ti_{0.96}$	2.0	Cubic
$Mo_{0.025}Ti_{0.975}$	1.8	
Mo_xU_{1-x}	0.7—2.1	

TABLE 4
Superconductive Compounds and Alloys (continued)

Substance	T_c, K	Crystal structure type
Mo_xV_{1-x}	0—~5.3	
Mo_2Zr	4.25—4.75	C15
NNb (film)	6—9	B1
$N_xO_yTi_z$	2.9—5.6	Cubic
$N_xO_yV_z$	5.8—8.2	Cubic
$N_{0.34}Re$	4—5	F.C.C.
NTa (film)	4.84	B1
$N_{0.6—0.987}Ti$	<1.17—5.8	B1
$N_{0.82—0.99}V$	2.9—7.9	B1
NZr	9.8	B1
$N_{0.906—0.984}Zr$	3.0—9.5	B1
$Na_{0.28—0.35}O_3W$	0.56	Tetragonal
$Na_{0.28}Pb_{0.72}$	7.2	
NbO	1.25	
$NbOs_2$	2.52	A12
Nb_3Os	1.05	A15
$Nb_{0.6}Os_{0.4}$	1.89; 1.78	$D8_b$
$Nb_3Os_{0.02—0.10}Rh_{0.98—0.90}$	2.42—2.30	A15
Nb_3P	1.8	$L1_2$tP32 (Ti_3P)
NbPRh	4.08	C37-oP12 (Co_2Si)
$Nb_{0.6}Pd_{0.4}$	1.60	$D8_f$ plus cubic
$Nb_3Pd_{0.02—0.10}Rh_{0.92—0.90}$	2.49—2.55	A15
$Nb_{0.62}Pt_{0.38}$	4.21	$D8_b$
Nb_5Pt_3	3.73	$D8_b$
$Nb_3Pt_{0.02—0.98}Rh_{0.98—0.02}$	2.52—9.6	A15
$NbRe_3$	5.27	$D8_b$-tP30 (FeCr)
$Nb_{0.38—0.18}Re_{0.62—0.82}$	2.43—9.70	A15
NbRe	3.8	$D8_b$-tP30
NbReSi	5.1	oI36 (FeTiSi)
Nb_3Rh	2.64	A15
$Nb_{0.6}Rh_{0.40}$	4.21	$D8_b$ plus other
$Nb_{0.9}Rh_{1.1}$	3.07	A3-oP4 (AuCd)
$Nb_3Rh_{0.98—0.90}Ru_{0.02—0.10}$	2.42—2.44	A15
Nb_xRu_{1-x}	1.2—4.8	
NbRuSi	2.65	oI36
NbS_2	6.1—6.3	Hexagonal, $NbSe_2$ type
NbS_2	5.0—5.5	Hexagonal, three-layer type
Nb_3Sb	0.2	$L1_2$-tP32 (Ti_3P)
$Nb_3Sb_{0—0.7}Sn_{1—0.3}$	6.8—18	A15
$NbSe_2$	5.15—5.62	Hexagonal
$Nb_{1—1.05}Se_2$	2.2—7.0	Same
Nb_3Se_4	2.0	hP14
Nb_3Si	1.5	$L1_2$
Nb_3SiSnV_3	4.0	
$NbSn_2$	2.60	Orthorhombic
Nb_6Sn_5	2.8	oI44 (Sn_5Ti_6)
NbSnTaV	6.2	A15
$NbSnV_2$	5.5	A15
Nb_2SnV	9.8	A15
Nb_xTa_{1-x}	4.4—9.2	A2
Nb_3Te_4	1.8	hP14
Nb_xTi_{1-x}	0.6—9.8	
$Nb_{0.6}Ti_{0.4}$	9.8	
Nb_xU_{1-x}	1.95 (max.)	
$Nb_{0.88}V_{0.12}$	5.7	A2

TABLE 4
Superconductive Compounds and Alloys (continued)

Substance	T_c, K	Crystal structure type
$Nb_{0.5}V_{1.5}Zr$	4.3	C15-hP12 ($MgZn_2$)
$Ni_{0.3}Th_{0.7}$	1.98	$D10_2$
$NiZr_2$	1.52	
$Ni_{0.1}Zr_{0.9}$	1.5	A3
$O_3Rb_{0.27-0.29}W$	1.98	Hexagonal
OSn	3.81	tP4 (PbO)
O_3SrTi ($n = 1.7-12.0 \times 10^{19}$)	0.12—0.37	
O_3SrTi ($n = 10^{18}-10^{21}$)	0.05—0.47	
O_3SrTi ($n = 10^{20}$)	0.47	
$O_3Sr_{0.08}W$	2—4	Hexagonal
OTi	0.58	
$O_3Tl_{0.30}W$	2.0—2.14	Hexagonal
OV_3Zr_3	7.5	$E9_3$
OW_3 (film)	3.35; 1.1	A15
OsPti	1.2	C22-hP9 (Fe_2P)
OsPZr	7.4	Same
OsReY	2.0	C14
Os_2Sc	4.6	C14
OsTa	1.95	A12
Os_3Th_7	1.51	$D10_2$
Os_xW_{1-x}	0.9—4.1	
OsW_3	~3	
Os_2Y	4.7	C14
Os_2Zr	3.0	C14
Os_xZr_{1-x}	1.5—5.6	
PPb	7.8	
OsW_2	3.81	$D8_b$-tP30 (FeCr)
$PPd_{3.0-3.2}$	<0.35—0.7	DO_{11}
P_3Pd_7 (high temperature)	1.0	Rhombohedral
P_3Pd_7 (low temperature)	0.70	Complex
PRh	1.22	
PRh_2	1.3	C1
P_4Rh_5	1.22	oP28 ($CaFe_2O_4$)
PRhTa	4.41	C37-oP12 (Co_2Si)
PRhZr	1.55	Same
PRuTi	1.3	C22-hP9 (Fe_2P)
PRuZr	3.46	C37-oP12
PW_3	2.26	DO_c
Pb_2Pd	2.95	C16
Pb_4Pt	2.80	Related to C16
Pb_2Rh	2.66	C16
PbSb	6.6	
PbTe (plus 0.1 w/o Pb)	5.19	
PbTe (plus 0.1 w/o Te)	5.24—5.27	
$PbTl_{0.27}$	6.43	
$PbTl_{0.17}$	6.73	
$PbTl_{0.12}$	6.88	
$PbTl_{0.075}$	6.98	
$PbTl_{0.04}$	7.06	
$Pb_{1-0.26}Tl_{0-0.74}$	7.20—3.68	
$PbTl_2$	3.75—4.1	
Pb_3Zr_5	4.60	$D8_8$
$PbZr_3$	0.76	A15
$Pd_{0.9}Pt_{0.1}Te_2$	1.65	C6
$Pd_{0.05}Ru_{0.05}Zr_{0.9}$	~9	

TABLE 4
Superconductive Compounds and Alloys (continued)

Substance	T_c, K	Crystal structure type
$Pd_{2.2}S$ (quenched)	1.63	Cubic
$PdSb_2$	1.25	C2
PdSb	1.5	$B8_1$
PdSbSe	1.0	C2
PdSbTe	1.2	C2
Pd_4Se	0.42	Tetragonal
$Pd_{6-7}Se$	0.66	Like Pd_4Te
$Pd_{2.8}Se$	2.3	
Pd_xSe_{1-x}	2.5 (max.)	
PdSi	0.93	B31
PdSn	0.41	B31
$PdSn_2$	3.34	
Pd_2Sn	0.41	C37
Pd_3Sn	0.47—0.64	$B8_2$
Pd_2SnTm	1.77	DO_3-cF16 (BiF_3)
Pd_2SnY	4.92	Same
Pd_2SnYb	1.79	Same
PdTe	2.3; 3.85	$B8_1$
$PdTe_{1.02-1.08}$	2.56—1.88	$B8_1$
$PdTe_2$	1.69	C6
$PdTe_{2.1}$	1.89	C6
$PdTe_{2.3}$	1.85	C6
$Pd_{1.1}Te$	4.07	$B8_1$
Pd_3Te	0.76	cI2 (W)
$PdTh_2$	0.85	C16
$Pd_{0.1}Zr_{0.9}$	7.5	A3
PtSb	2.1	$B8_1$
PtSi	0.88	B31
PtSn	0.37	$B8_1$
$PtSn_4$	2.38	C16-oC20 ($PdSn_4$)
Pt_3Ta_7	1.5	$D8_b$-tP30
$PtTa_3$	0.4	A15-cP8 (Cr_3Si)
PtTe	0.59	Orthorhombic
PtTh	0.44	B_f
Pt_3Th_7	0.98	$D10_2$
Pt_5Th	3.13	
$PtTi_3$	0.58	A15
$Pt_{0.02}U_{0.98}$	0.87	β-phase
$PtV_{2.5}$	1.36	A15
PtV_3	2.87—3.20	A15
$PtV_{3.5}$	1.26	A15
$Pt_{0.5}W_{0.5}$	1.45	A1
Pt_xW_{1-x}	0.4—2.7	
Pt_2Y_3	0.90	
Pt_2Y	1.57; 1.70	C15
Pt_3Y_7	0.82	$D10_2$
PtZr	3.0	A3
Re_2Sc	4.2	C15-hP12 ($MgZn_2$)
$Re_{24}Sc_5$	2.2	A12-cI58 (Mg)
ReSiTa	4.4	oI36 (FeTiSi)
$Re_3Si_5Y_2$	1.76	tP40 ($Fe_3Sc_2Si_5$)
Re_3Ta_2	1.4	$D8_b$-tP30 (FeCr)
$Re_{0.64}Ta_{0.36}$	1.46	A12
Re_3Ta	6.78	A12-cI58 (Mn)
$Re_{24}Ti_5$	6.60	A12

TABLE 4
Superconductive Compounds and Alloys (continued)

Substance	T_c, K	Crystal structue type
Re_xTi_{1-x}	6.6 (max.)	
$Re_{0.76}V_{0.24}$	4.52	$D8_b$
Re_3V	6.26	$D8_b$-tP30
$Re_{0.92}V_{0.08}$	6.8	A3
$Re_{0.6}W_{0.4}$	6.0	
$Re_{0.5}W_{0.5}$	5.12	$D8_b$
$Re_{13}W_{12}$	5.2	$D8_b$-tP30
Re_3W	9.0	A12-cI58
Re_2Y	1.83	C14
Re_2Zr	5.9	C14
Re_3Zr	7.40	A12-cI58
Re_6Zr	7.40	Same
$Rh_{17}S_{15}$	5.8	Cubic
$Rh_{-0.24}Sc_{-0.76}$	0.88; 0.92	
$Rh_4Sc_5Si_{10}$	8.54	tP38
$Rh_4Sc_3Sn_{13}$	4.5	cP40
Rh_xSe_{1-x}	6.0 (max.)	
$RhSi_3Th$	1.76	tI10
$Rh_{0.86}Sc_{1.04}Th$	6.45	tI12
Rh_2Si_2Y	3.11	tI10
$Rh_3Si_5Y_2$	2.70	oI40
$Rh_4Sn_{13}Sr_3$	4.3	cP40
Rh_xSn_yTh	1.9	cI2 (W)
Rh_xSn_yTm	2.3	cP40
$Rh_4Sn_{13}Y_3$	3.2	cP40
Rh_2Sr	6.2	C15
$Rh_{0.4}Ta_{0.6}$	2.35	$D8_b$
$RhTe_2$	1.51	C2
$Rh_{0.67}Te_{0.33}$	0.49	
Rh_xTe_{1-x}	1.51 (max.)	
$RhTh$	0.36	B_f
Rh_3Th_7	2.15	$D10_2$
Rh_5Th	1.07	
Rh_xTi_{1-x}	2.25—3.95	
$Rh_{0.02}U_{0.98}$	0.96	
RhV_3	0.38	A15
RhW	~3.4	A3
RhY_3	0.65	
Rh_2Y_3	1.48	
Rh_3Y	1.07	C15
Rh_5Y	0.56	
Rh_3Y_7	0.32	hP20 (Fe_3Th_7)
$Rh_{0.005}Zr$ (annealed)	5.8	
$Rh_{0-0.45}Zr_{1-0.55}$	2.1—10.8	
$Rh_{0.1}Zr_{0.9}$	9.0	H.C.P.
Ru_2Sc	1.67	C14
$RuSiTa$	3.15·	oI36
Ru_3Si_2Th	3.98	hP12
Ru_3Si_2Y	3.51	hP12
$Ru_{1.1}Sn_{3.1}Y$	1.3	cP40
Ru_2Th	3.56	C15
$RuTi$	1.07	B2
$Ru_{0.05}Ti_{0.95}$	2.5	
$Ru_{0.1}Ti_{0.9}$	3.5	
$Ru_xTi_{0.6}V_y$	6.6 (max.)	

TABLE 4
Superconductive Compounds and Alloys (continued)

Substance	T_c, K	Crystal structue type
Ru_3U	0.15	$L1_2$-cP4
$Ru_{0.45}V_{0.55}$	4.0	B2
RuW	7.5	A3
Ru_2Y	1.52	C14
Ru_2Zr	1.84	C14
$Ru_{0.1}Zr_{0.9}$	5.7	A3
STh	0.5	B1-cF8 (NaCl)
SbSn	1.30—1.42	B1 or distorted
$SbTa_3$	0.72	A15-cP8 (Cr_3Si)
$SbTi_3$	5.8	Same
Sb_2Ti_7	5.2	
$Sb_{0.01—0.03}V_{0.99—0.97}$	3.76—2.63	A2
SbV_3	0.80	A15
SeTh	1.7	B1-cF8
$SiMo_3$	1.4	A15-cP8
Si_2Th	3.2	C_c, α-phase
Si_2Th	2.4	C32, β-phase
$SiV_{2.7}Ru_{0.3}$	2.9	A15
Si_2W_3	2.8; 2.84	A15
$SiZr_3$	0.5	$L1_2$-tP32 (Ti_3P)
$Sn_{0.174—0.104}Ta_{0.826—0.896}$	6.5—< 4.2	A15
$SnTa_3$	8.35	A15, highly ordered
$SnTa_3$	6.2	A15, partially ordered
$SnTaV_2$	2.8	A15
$SnTa_2V$	3.7	A15
Sn_xTe_{1-x} (n = 10.5—20 × 10^{20})	0.07—0.22	B1
Sn_3Th	3.33	$L1_2$-cP4
$SnTi_3$	5.80	A15-cP8
Sn_xTl_{1-x}	2.37—5.2	
SnV_3	3.8	A15
$Sn_{0.02—0.057}V_{0.98—0.943}$	2.87—~1.6	A2
$SnZr_3$	0.92	A15-cP8
$Ta_{0.025}Ti_{0.975}$	1.3	Hexagonal
$Ta_{0.05}Ti_{0.95}$	2.9	Hexagonal
$Ta_{0.05—0.75}V_{0.95—0.25}$	4.30—2.65	A2
$Ta_{0.8—1}W_{0.2—0}$	1.2—4.4	A2
$Tc_{0.1—0.4}W_{0.9—0.6}$	1.25—7.18	Cubic
$Tc_{0.50}W_{0.50}$	7.52	α plus
$Tc_{0.60}W_{0.40}$	7.88	plus α
Tc_6Zr	9.7	A12
TeY	1.02	B1-cF8
$ThTl_3$	0.87	$L1_2$-cP4
$Th_{0—0.55}Y_{1—0.45}$	1.2—1.8	
$Ti_{0.70}V_{0.30}$	6.14	Cubic
Ti_xV_{1-x}	0.2—7.5	
$Ti_{0.5}Zr_{0.5}$ (annealed)	1.23	
$Ti_{0.5}Zr_{0.5}$ (quenched)	2.0	
Tl_3Y	1.52	$L1_2$-cP4
V_2Zr	8.80	C15
$V_{0.26}Zr_{0.74}$	5.9	
W_2Zr	2.16	C15
YZn	0.33	B2-cP2 (CsCl)

* n denotes current carriers concentration in cm^{-3}.

TABLE 4
Superconductive Compounds and Alloys (continued)

B. SUPERCONDUCTORS WITH $T_c > 10K$

Substance	T_c, K	Crystal structure type	
Al_2CMo_3	10.0	A13	
$Al_{0.5}Ge_{0.5}Nb$	12.6	A15	
$Al_{-0.8}Ge_{-0.2}Nb_3$	20.7	A15	
$AlNb_3$	18.0	A15	(Cr_3Si)
$AlNb_3$	12.0		$(FeCr)$
Al_xNb_{1-x}	<4.2—13.5	$D8_b$	
Al_xNb_{1-x}	12—17.5	A15	
$Al_{0.27}Nb_{0.73—0.48}V_{0—0.25}$	14.5—17.5	A15	
$Al\ Nb_xV_{1-x}$	4.4—13.5		
$Al_{0.1}Si_{0.9}V_3$	14.05		
AlV_3	11.8	A15	(Cr_3Si)
$AuNb_3$	11.5	A15	
$Au_{0—0.3}Nb_{1—0.7}$	1.1—11.0		
$Au_{0.02—0.98}Nb_3Rh_{0.98—0.02}$	2.53—10.9	A15	
$AuNb_{3(1-x)}V_{3x}$	1.5—11.0	A15	
$B_{0.03}C_{0.51}Mo_{0.47}$	12.5		
B_4LuRh_4	11.7		(B_4CeCo_4)
B_2LuRu	10		
B_4Rh_4Y	11.3		(B_4CeCo_4)
$B_{0.1}Si_{0.9}V_3$	15.8	A15	
$BaBi_{0.2}O_3Pb_{0.8}$	13.2		
$Ba_2CaCu_2O_8Tl_2$	120		
$Ba_2Cu_3LaO_6$	80		
$Ba_2Cu_3O_7Tm$	101		
$Ba_2Cu_3O_7Y$	90		
$(Ba,La)_2CuO_4$	36	A15	(K_2NiF_4)
$Bi_2CaCu_2O_8Sr_2$	110		
$Br_2Mo_6S_6$	13.8		(Mo_6PbS_8)
C_3La	11.0		(C_3Pu_2)
CMo	14.3	B1	$(NaCl)$
CMo_2	12.2		o**
$C_{0.5}Mo_xNb_{1-x}$	10.8—12.5	B1	
CMo_xTi_{1-x}	10.2(max)	B1	
$CMo_{0.83}Ti_{0.17}$	10.2	B1	
$C_{0—0.38}N_{1—0.62}Ta$	10.0—11.3		
CNb (whiskers)	7.5—10.5		
CNb	11.5	B1	
$C_{0.7—1.0}Nb_{0.3—0}$	6—11	B1	
CNb_xTa_{1-x}	8.2—13.9		
$CNb_{0.6—0.9}W_{0.4—0.1}$	12.5—11.6	B1	
$C_{0.1}Si_{0.9}V_3$	16.4	A15	
CTa	10.3	B1	
$CTa_{1—0.4}W_{0—0.6}$	8.5—10.5	B1	
$C_{0.66}Th_{0.13}Y_{0.21}$	17		(C_3Pu_2)
C_3Y_2	11.5		(C_3Pu_2)
CW	10	B1	
$(Ca,La)_2CuO_4$	18		(K_2NiF_4)
$Cu(La,Sr)_2O_4$	39		
$Cu_{1.8}Mo_6S_8$	10.8		(Mo_6PbS_8)
$Cr_{0.3}SiV_{2.7}$	11.3	A15	
$GaNb_3$	14.5	A15	(Cr_3Si)
$Ga_xNb_3Sn_{1-x}$	14—18.37	A15	

TABLE 4
Superconductive Compounds and Alloys (continued)

Substance	T_c, K	Crystal structure type	
GaV_3	16.8	A15	
$GaV_{2.1-3.5}$	6.3—14.45	A15	
$GeNb_3$	23.2	A15	
$GeNb_3$ (quenched)	6—17	A15	
$Ge_xNb_3Sn_{1-x}$	17.6—18.0	A15	
$Ge_{0.5}Nb_3Sn_{0.5}$	11.3		
$Ge_{0.1}Si_{0.9}V_3$	14.0	A15	
GeV_3	11	A15	
$InLa_3$	9.83; 10.4	LI_2	$(AuCu_3)$
$InLa_3$ (0—35 kbar)	9.75—10.55		
$In_{0-0.3}Nb_3Sn_{1-0.7}$	18.0—18.19	A15	
InV_3	13.9	A15	
$Ir_{0.4}Nb_{0.6}$	10		(FeCr)
$LaMo_6Se_8$	11.4		(Mo_6PbS_8)
LiO_4Ti_2	13.7		(Al_2MgO_4)
MoN	12; 14.8	h*	
Mo_3Os	12.7	A15	
$Mo_6Pb_{0.9}S_{7.5}$	15.2		(Mo_6PbS_8)
Mo_3Re	10.0; 15	A15	
Mo_xRe_{1-x}	1.2—12.2		
$Mo_{0.52}Re_{0.48}$	11.1		
$Mo_{0.57}Re_{0.43}$	14.0		
$Mo_{-0.60}Re_{0.395}$	10.6		
$MoRu$	9.5—10.5	A3	
Mo_3Ru	10.6	A15	
Mo_6Se_8Tl	12.2		(Mo_6PbS_8)
$Mo_{0.3}SiV_{2.7}$	11.7	A15	
Mn_3Si	12.5	A15	
Mo_3Tc	15	A15	
$Mo_{0.3}Tc_{0.7}$	12.0	A15	
Mo_xTc_{1-x}	10.8—15.8		
$MoTc_3$	15.8		
NNb (whiskers)	10—14.5		
NNb (diffusion wires)	16.10		
$N_{0.988}Nb$	14.9; 17.3	B1	
$N_{0.824-0.988}Nb$	14.4—15.3	B1	
$N_{0.7-0.795}Nb$	11.3—12.9		
NNb_xO_y	13.5—17.0	B1	
NNb_xO_y	6.0—11		
$N_{100-42w/o}Nb_{0-58w/o}Ti$	15—16.8		
$N_{100-75w/o}Nb_{0-25w/o}Zr$	12.5—16.35		
NNb_xZr_{1-x}	9.8—13.8	B1	
$N_{0.93}Nb_{0.85}Zr_{0.15}$	13.8	B1	
NTa	12—14	B1	
NZr	10.7	B1	
Nb_3Pt	10.9	A15	
$Nb_{0.18}Re_{0.82}$	10		(Mn)
Nb_3Si	19	A15	
$Nb_{0.3}SiV_{2.7}$	12.8	A15	
Nb_3Sn	18.05	A15	
$Nb_{0.8}Sn_{0.2}$	18.18; 18.5	A15	
Nb_xSn_{1-x} (film)	2.6—18.5	o*	
Nb_3Sn_2	16.6	t*	
$NbSnTa_2$	10.8	A15	
Nb_2SnTa	16.4	A15	

TABLE 4
Superconductive Compounds and Alloys (continued)

Substance	T_c,K	Crystal structure type
$Nb_{2.5}SnTa_{0.5}$	17.6	A15
$Nb_{2.75}SnTa_{0.25}$	17.8	A15
$Nb_{3x}SnTa_{3(1-x)}$	6.0—18.0	
$Nb_2SnTa_{0.5}V_{0.5}$	12.2	A15
$NbTc_3$	10.5	A12
$Nb_{0.75}Zr_{0.25}$	10.8	
$Nb_{0.66}Zr_{0.33}$	10.8	
$PbTa_3$	17	A15
$RhTa_3$	10	A15
$RhZr_2$	10.8; 11.3	C16 (Al$_2$Cu)
$Rh_{0—0.45}Zr_{1—0.55}$	2.1—10.8	
$SiTi_{0.3}V_{2.7}$	10.9	A15
SiV_3	17.1	A15
$SiV_{2.7}Zr_{0.3}$	13.2	A15

TABLE 5
Critical Field Data

Substance	H_o oersteds	Substance	H_o oersteds
Ag_2F	2.5	InSb	1100
Ag_7NO_{11}	57	In_xTl_{1-x}	252—284
Al_2CMo_3	1700	$In_{0.8}Tl_{0.2}$	252
$BaBi_3$	740	$Mg_{0.47}Tl_{0.53}$	220
Bi_2Pt	10	$Mo_{0.16}Ti_{0.84}$	<985
Bi_3Sr	530	$NbSn_2$	620
Bi_5Tl_3	>400	$PbTl_{0.27}$	756
CdSn	>266	$PbTl_{0.17}$	796
$CoSi_2$	105	$PbTl_{0.12}$	849
$Cr_{0.1}Ti_{0.3}V_{0.6}$	1360	$PbTl_{0.075}$	880
$In_{1-0.86}Mg_{0-0.14}$	272.4—259.2	$PbTl_{0.04}$	864

TABLE 6
High Critical Magnetic-Field Superconductive Compounds and Alloys

Substance	T_c, K	H_{c1}, kOe	H_{c2}, kOe	H_{c3}, kOe	T_{obs}, K[a]
Al_2CMo_3	9.8—10.2	0.091	156		1.2
$AlNb_3$		0.375			
$Ba_xO_3Sr_{1-x}Ti$	<0.1—0.55	0.0039 max.			
$Bi_{0.5}Cd_{0.1}Pb_{0.27}Sn_{0.13}$			>24		3.06
Bi_xPb_{1-x}	7.35—8.4	0.122 max.	30 max.		4.2
$Bi_{0.56}Pb_{0.44}$	8.8		15		4.2
$Bi_{7.5w/o}Pb_{92.5w/o}$[b]			2.32		
$Bi_{0.099}Pb_{0.901}$		0.29	2.8		
$Bi_{0.02}Pb_{0.98}$		0.46	0.73		
$Bi_{0.53}Pb_{0.32}Sn_{0.16}$			>25		3.06
$Bi_{1-0.93}Sn_{0-0.07}$			0—0.032		3.7
Bi_5Tl_3	6.4		>5.6		3.35
C_8K (excess K)	0.55		0.160 (H⊥c)		0.32
			0.730 (H∥c)		0.32

TABLE 6
High Critical Magnetic-Field Superconductive Compounds and Alloys (continued)

Substance	T_c, K	H_{c1}, kOe	H_{c2}, kOe	H_{c3}, kOe	T_{obs}, K[a]
C_8K	0.39		0.025 (H⊥c)		0.32
			0.250 (H∥c)		0.32
$C_{0.44}Mo_{0.56}$	12.5—13.5	0.087	98.5		1.2
CNb	8—10	0.12	16.9		4.2
$CNb_{0.4}Ta_{0.6}$	10—13.6	0.19	14.1		1.2
CTa	9—11.4	0.22	4.6		1.2
$Ca_xO_3Sr_{1-x}Ti$	<0.1—0.55	0.002—0.004			
$Cd_{0.1}Hg_{0.9}$		0.23	0.34		2.04
(by weight)					
$Cd_{0.05}Hg_{0.95}$		0.28	0.31		2.16
$Cr_{0.10}Ti_{0.30}V_{0.60}$	5.6	0.071	84.4		0
GaN	5.85	0.725			4.2
Ga_xNb_{1-x}			>28		4.2
GaSb (annealed)	4.24		2.64		3.5
$GaV_{1.95}$	5.3		73[e]		
$GaV_{2.1-3.5}$	6.3—14.45		230—300[d]		0
GaV_3		0.4	350[e]		0
			500[d]		
$GaV_{4.5}$	9.15		121[e]		0
Hf_xNb_y			>52—>102		1.2
Hf_xTa_y			>28—>86		1.2
$Hg_{0.05}Pb_{0.95}$		0.235	2.3		
$Hg_{0.101}Pb_{0.899}$		0.23	4.3		4.2
$Hg_{0.15}Pb_{0.85}$	6.75		>13		2.93
$In_{0.98}Pb_{0.02}$	3.45	0.1		0.12	2.76
$In_{0.96}Pb_{0.04}$	3.68	0.1	0.12	0.25	2.94
$In_{0.94}Pb_{0.06}$	3.90	0.095	0.18	0.35	3.12
$In_{0.913}Pb_{0.087}$	4.2	~10.17	0.55	2.65	
$In_{0.316}Pb_{0.684}$		0.155	3.7		4.2
$In_{0.17}Pb_{0.83}$			2.8	5.5	4.2
$In_{1.000}Te_{1.002}$	3.5—3.7		1.2[e]		0
$In_{0.95}Tl_{0.05}$		0.263	0.263		3.3
$In_{0.90}Tl_{0.10}$		0.257	0.257		3.25
$In_{0.83}Tl_{0.17}$		0.242	0.39		3.21
$In_{0.75}Tl_{0.25}$		0.216	0.50		3.16
LaN	1.35	0.45			0.76
La_3S_4	6.5	≈0.15	>25		1.3
La_3Se_4	8.6	≈0.2	>25		1.25
$Mo_{0.52}Re_{0.48}$	11.1		14—21	22—33	4.2
			18—28	37—43	1.3
$Mo_{0.6}Re_{0.395}$	10.6		14—20	20—37	4.2
			19—26	26—37	1.3
$Mo_{0.5}Ti_{0.5}$			75[e]		0
$Mo_{0.16}Ti_{0.84}$	4.18	0.028	98.7[e]		0
			36—38		3.0
$Mo_{0.913}Ti_{0.087}$	2.95	0.060	15		4.2
$Mo_{0.1-0.3}U_{0.9-0.7}$	1.85—2.06		>25		
$Mo_{0.17}Zr_{0.83}$			30		
$N_{(12.8\ w/o)}Nb$	15.2		>9.5		13.2
NNb (wires)	16.1		153[e]		0
			132		4.2
			95		8
			53		12
NNb_xO_{1-x}	13.5—17.0		38		

TABLE 6
High Critical Magnetic-Field Superconductive Compounds and Alloys (continued)

Substance	T_c, K	H_{c1}, kOe	H_{c2}, kOe	H_{c3}, kOe	T_{obs}, K[a]
NNb_xZr_{1-x}	9.8—13.8		4- >130		4.2
$N_{0.93}Nb_{0.85}Zr_{0.15}$	13.8		>130		4.2
$Na_{0.086}Pb_{0.914}$		0.19	6.0		
$Na_{0.016}Pb_{0.984}$		0.28	2.05		
Nb	9.15		2.020		1.4
			1.710		4.2
Nb		0.4—1.1	3—5.5		4.2
Nb (unstrained)		1.1—1.8	3.40	6—9.1	4.2
Nb (strained)		1.25—1.92	3.44	6.0—8.7	4.2
Nb (cold-drawn wire)		2.48	4.10	≈10	4.2
Nb (film)			>25		4.2
NbSc			>30		
Nb_3Sn		0.170	221		4.2
			70		14.15
			54		15
			34		16
			17		17
$Nb_{0.1}Ta_{0.9}$		0.084	0.154		4.195
$Nb_{0.2}Ta_{0.8}$			10		4.2
$Nb_{0.65-0.73}Ta_{0.02-0.10}Zr_{0.25}$			>70—>90		4.2
Nb_xTi_{1-x}			148 max.		1.2
			120 max.		4.2
$Nb_{0.222}U_{0.778}$		1.98	23		1.2
Nb_xZr_{1-x}			127 max.		1.2
			94 max.		4.2
O_3SrTi	0.43	0.0049[c]	0.504[c]		0
O_3SrTi	0.33	0.00195[c]	0.420[c]		0
$PbSb_{1\,w/o}$(quenched)			>1.5		4.2
$PbSb_{1\,w/o}$(annealed)			>0.7		4.2
$PbSb_{2.8\,w/o}$(quenched)			>2.3		4.2
$PbSb_{2.8\,w/o}$(annealed)			>0.7		4.2
$Pb_{0.871}Sn_{0.129}$		0.45	1.1		
$Pb_{0.965}Sn_{0.035}$		0.53	0.56		
$Pb_{1-0.26}Tl_{0-0.74}$	7.20—3.68		2—6.9[c]		0
$PbTl_{0.17}$	6.73		4.5[c]		0
$Re_{0.26}W_{0.74}$			>30		
$Sb_{0.93}Sn_{0.07}$			0.12		3.7
SiV_3	17.0	0.55	156[e]		
Sn_xTe_{1-x}		0.00043—0.00236	0.005—0.0775		0.012—0.079
Ta (99.95%)		0.425	1.850		1.3
		0.325	1.425		2.27
		0.275	1.175		2.66
		0.090	0.375		3.72
$Ta_{0.5}Nb_{0.5}$			3.55		4.2
$Ta_{0.65-0}Ti_{0.35-1}$	4.4—7.8		>14—138		1.2
$Ta_{0.5}Ti_{0.5}$			138		1.2
Te	3.3	0.25[c]			0
Tc_xW_{1-x}	5.75—7.88		8—44		4.2
Ti				2.7	4.2
$Ti_{0.75}V_{0.25}$	5.3	0.029[c]	199[c]		0
$Ti_{0.775}V_{0.225}$	4.7	0.024[c]	172[c]		0
$Ti_{0.615}V_{0.385}$	7.07	0.050	34		4.2
$Ti_{0.516}V_{0.484}$	7.20	0.062	28		4.2
$Ti_{0.415}V_{0.585}$	7.49	0.078	25		4.2

TABLE 6
High Critical Magnetic-Field Superconductive Compounds and Alloys (continued)

Substance	T_c, K	H_{c1}, kOe	H_{c2}, kOe	H_{c3}, kOe	T_{obs}, K[a]
$Ti_{0.12}V_{0.88}$			17.3	28.1	4.2
$Ti_{0.09}V_{0.91}$			14.3	16.4	4.2
$Ti_{0.06}V_{0.94}$			8.2	12.7	4.2
$Ti_{0.03}V_{0.97}$			3.8	6.8	4.2
Ti_xV_{1-x}			108 max.		1.2
V	5.31	0.8	3.4		1.79
		0.75	3.15		2
		0.45	2.2		3
		0.30	1.2		4
$V_{0.26}Zr_{0.74}$	≈5.9	0.238			1.05
		0.227			1.78
		0.185			3.04
		0.165			3.5
W (film)	1.7—4.1		>34		1

[a] Temperature of critical field measurement.
[b] w/o denotes weight percent.
[c] Extrapolated.
[d] Linear extrapolation.
[e] Parabolic extrapolation.

REFERENCES

1. B. W. Roberts, in *Superconductive Materials and Some of Their Properties. Progress in Cryogenics*, Vol. IV, 1964, pp. 160—231.
2. B. W. Roberts, Superconductive Materials and Some of Their Properties, NBS Technical Notes 408 and 482, U.S. Government Printing Office, 1966 and 1969; B. W. Roberts, *J. Phys. Chem. Ref. Data*, 5, 581, 1976.
3. B. W. Roberts, Properties of Selected Superconductive Materials, 1978 Supplement, NBS Technical Note 983, 1978.
4. T. Claeson, *Phys. Rev.*, 147, 340, 1966.
5. C. J. Raub, W. H. Zachariasen, T. H. Geballe, and B. T. Matthias, *J. Phys. Chem. Solids*, 24, 1093, 1963.
6. T. H. Geballe, B. T. Matthias, V. B. Compton, E. Corenzwit, G. W. Hull, Jr., and L. D. Longinotti, *Phys. Rev.*, 1A, 119, 1965.
7. C. J. Raub, V. B. Compton, T. H. Geballe, B. T. Matthias, J. P. Maita, and G. W. Hull, Jr., *J. Phys. Chem. Solids*, 26, 2051, 1965.
8. R. D. Blaugher, J. K. Hulm, and P. N. Yocom, *J. Phys. Chem. Solids*, 26, 2037, 1965.
9. T. Claeson and H. L. Luo, *J. Phys. Chem. Solids*, 27, 1081, 1966.
10. S. C. Ng and B. N. Brockhouse, *Solid State Comm.*, 5, 79, 1967.
11. O. I. Shulishova and I. A. Shcherbak, *Izv. AN SSSR, Neorg. Materials*, 3, 1495, 1967.
12. T. F. Smith and H. L. Luo, *J. Phys. Chem. Solids*, 28, 569, 1967.
13. A. C. Lawson, *J. Less–Common Metals*, 23, 103, 1971.
14. R. Chevrel, M. Sergent, and J. Prigent, *J. Solid State Chem.*, 3, 515, 1971.
15. M. Marezio, P. D. Dernier, J. P. Remeika, and B. T. Matthias, *Mat. Res. Bull.*, 8, 657, 1973.
16. J. K. Hulm and R. D., *Blaugher in Superconductivity in d– and f–Band Metals*, D. H. Douglass, Ed., American Institute of Physics, 4, 1, 1972.
17. R. N. Shelton, A. C. Lawson, and D. C. Johnston, *Mat. Res. Bull.*, 10, 297, 1975.
18. H. D. Wiesinger, *Phys. Status Sol.*, 41A, 465, 1977.
19. O. Fisher, *Applied Phys.*, 16, 1, 1978.
20. D. C. Johnston, *Solid State Comm.*, 24, 699, 1977.
21. H. C. Ku and R. H. Shelton, *Mat. Res. Bull.*, 15, 1441, 1980.
22. H. Barz, *Mat. Res. Bull.*, 15, 1489, 1980.
23. G. P. Espinosa, A. S. Cooper, H. Barz, and J. P. Remeika, *Mat. Res. Bull.*, 15, 1635, 1980.
24. E. M. Savitskii, V. V. Baron, Yu. V. Efimov, M. I. Bychkova, and L. F. Myzenkova, in *Superconducting Materials*, Plenum Press, 1981, p. 107.
25. R. Fluckiger and R. Baillif, in Topics in *Current Physics*, O. Fischer and M. B. Maple, Eds., Springer Verlag, 34, 113, 1982.
26. R. N. Shelton, in *Superconductivity in d– and f–Band Metals*, W. Buckel and W. Weber, Eds., Kernforschungszentrum, Karlsruhe, 1982, p. 123.
27. D. C. Johnston and H. F. Braun, *Topics in Current Phys.*, 32, 11, 1982.
28. R. Chevrel and M. Sergent, *Topics in Current Phys.*, 32, 25, 1982.

29. G. P. Espinosa, A. S. Cooper, and H. Barz, *Mat. Res. Bull.*, 17, 963, 1982.
30. R. Muller, R. N. Shelton, J. W. Richardson, Jr., and R. A. Jacobson, *J. Less–Comm. Met.*, 92, 177, 1983.
31. You–Xian Zhao and Shou–An He, in *High Pressure in Science and Technology*, North Holland, 22, 51, 1983.
32. You–Xian Zhao and Shou–An He, *Solid State Comm.*, 24, 699, 1983.
33. G. P. Meisner and H. C. Ku, *Appl. Phys.*, A31, 201, 1983.
34. R. J. Cava, D. W. Murphy, and S. M. Zahurak, *J. Electrochem. Soc.*, 130, 2345, 1983.
35. R. N. Shelton, *J. Less–Comm. Met.*, 94, 69, 1983.
36. B. Chevalier, P. Lejay, B. Lloret, Wang Xian–Zhong, J. Etourneau, and P. Hagenmuller, *Annales de Chemie*, 9, 191, 1984.
37. G. Venturini, M. Meot–Meyer, E. McRae, J. F. Mareche, and B. Rogues, *Mat. Res. Bull.*, 19, 1647, 1984.
38. J. M. Tarascon, F. G. DiSalvo, D. W. Murphy, G. Hull, and J. V. Waszczak, *Phys. Rev.*, 29B, 172, 1984.
39. G. V. Subba and G. Balakrishnan, *Bull. Mat. Sci.*, 6, 283, 1984.
40. B. Batlog, *Physica*, 126B, 275, 1984.
41. M. J. Johnson, Ames Lab (USA) Report IS-T-1140, 1984.
42. I. M. Chapnik, *J. Mat. Sci. Lett.*, 4, 370, 1985.
43. W. Rong–Yao, L. Qi–Guang, and Z. Xiao, *Phys. Status Sol.*, 90A, 763, 1985.
44. W. Xian–Zhong, B. Chevalier, J. Etourneau, and P. Hagenmuller, *Mat. Res. Bull.*, 20, 517, 1985.
45. H. R. Ott, F. Hulliger, H. Rudigier, and Z. Fisk, *Phys. Rev.*, 31B, 1329, 1985.
46. P. Villars and L. D. Calver, *Pearson's Handbook of Crystallographic Data for Intermetallic Phases*, Vol. 1—3, ASM, 1985.
47. G. V. Subba Rao, K. Wagner, G. Balakhrishnan, J. Jakani, W. Paulus, and R. Scollhorn, *Bull. Mat. Sci.*, 7, 215, 1985.
48. J. G. Bednorz and K. A. Muller, *Zs. Physik*, B64, 189, 1986.
49. W. Rong–Yao, *Phys. Status Sol.*, 94A, 445, 1986.
50. H. D. Yang, R. N. Shelton, and H. F. Braun, *Phys. Rev.*, 33B, 5062, 1986.
51. G. Venturini, M. Kanta, E. McRae, J. F. Mareche, B. Malaman, and B. Roques, *Mat. Res. Bull.*, 21, 1203, 1986.
52. W. Rong–Yao, *J. Mat. Sci. Lett.*, 5, 87, 1986.
53. M. K. Wu, J. R. Ashburn, C. J. Torng, P. H. Hor, R. L. Meng, L. Gao, Z. J. Huang, Y. Q. Wang, and C. W. Chu, *Phys. Rev. Lett.*, 58, 908, 1987.
54. R. J. Cava, R. B. Van Dover, B. Batlog, and E. A. Rietman, *Phys. Rev. Lett.*, 58, 408, 1987.
55. L. C. Porter, T. J. Thorn, U. Geiser, A. Umezawa, H. H. Wang, W. K. Kwok, H–C. I. Kao, M. R. Monaghan, G. W. Crabtree, K. D. Carlson, and J. M. Williams, *Inorg. Chem.*, 26, 1645, 1987.
56. A. M. Kini, U. Geiser, H–C. I. Kao, K. D. Carlson, H. H. Wang, M. R. Monaghan, and K. M. Williams, *Inorg. Chem.*, 26, 1834, 1987.
57. T. Penney, S. von Molnar, D. Kaiser, F. Holtzberg, and A. W. Kleinsasser, *Phys. Rev.*, B38, 2918, 1988.
58. Y. K. Tao, J. S. Swinnea, A. Manthiram, J. S. Kim, J. B. Goodenoug, and H. Steinfink, *J. Mat. Res.*, 3, 248, 1988.
59. G. G. Peterson, B. R. Weinberger, L. Lynds, and H. A. Krasinski, *J. Mat. Res.*, 3, 605, 1988.
60. J. B. Torrance, Y. Tokura, A. Nazzai, and S. S. P. Parkin, *Phys. Rev. Lett.*, 60, 542, 1988.
61. K. Kourtakis, M. Robbins, P. K. Gallagher, and T. Teifel, *J. Mat. Res.*, 4, 1289, 1989.
62. J. C. Phillips, *Physics of High-T$_c$ Superconductors*, Academic Press, 1989, p. 336.
63. Shui Wai Lin and L. I. Berger, *Rev. Sci. Instrum.*, 60, 507, 1989.
64. M. Tinkham, *Introduction to Superconductivity*, McGraw–Hill, New York, 1975.
65. O. Fischer and M.B. Maple, Eds., *Topics in Current Physics*, Volume 32: Superconductivity in Ternary Compounds I; Volume 34: Superconductivity in Ternary Compounds II, Springer–Verlag, Berlin, 1982.
66. K. J. Dunn and F. P. Bundy, *Phys. Rev.*, B25, 194, 1982.
67. A. Barone and G. Paterno, *Physics and Applications of the Josephson Effect*, Wiley, New York, 1982.
68. D. H. Douglass, Ed., *Superconductivity in d- and f-band Metals*, Plenum Press, New York, 1976.
69. D. M. Ginsberg, Ed., *Physical Properties of High Temperature Superconductors*, (Volume II, 1990; Volume III, 1992; Volume V, 1996), World Scientific, Singapore.
70. T. Ishiguro and K. Yamji, *Organic Superconductors*, Springer Verlag, Berlin, 1990.
71. Sh. Okada, K. Shimizu, T. C. Kobayashi, K. Amaya, and Sh. Endo., *J. Phys. Soc. Jpn.*, 65, 1924, 1996.
72. A. Bourdillon and N. X. Tan Bourdillon, *High Temperature Superconductors: Processing and Science*, Academic Press, 1994.
73. J. M. Williams, J. R. Ferraro, R. J. Thorn, K. Carlson, U. Geiser, H. H. Wang, A. M. Kini, and M.-H. Whangbo, *Organic Superconductors (Including Fullerenes): Synthesis, structure, Properties, and Theory*, Prentice–Hall, 1992.

HIGH TEMPERATURE SUPERCONDUCTORS
C. N. R. Rao and A. K. Raychaudhuri

The following tables give properties of a number of high temperature superconductors. Table 1 lists the crystal structure (space group and lattice constants) and the critical transition temperature T_c for the more important high temperature superconductors so far studied. Table 2 gives energy gap, critical current density, and penetration depth in the superconducting state. Table 3 gives electrical and thermal properties of some of these materials in the normal state. The tables were prepared in November 1992 and updated in November 1994.

REFERENCES

1. Ginsburg, D.M., Ed., *Physical Properties of High-Temperature Superconductors*, Vols. I—III, World Scientific, Singapore, 1989—1992.
2. Rao, C.N.R., Ed., *Chemistry of High-Temperature Superconductors*, World Scientific, Singapore, 1991.
3. Shackelford, J.F., *The CRC Materials Science and Engineering Handbook*, CRC Press, Boca Raton, 1992, 98—99 and 122—123.
4. Kaldis, E., Ed., *Materials and Crystallographic Aspects of HT_c-Superconductivity*, Kluwer Academic Publ., Dordrecht, The Netherlands, 1992.
5. Malik, S.K. and Shah, S.S., Ed., *Physical and Material Properties of High Temperature Superconductors*, Nova Science Publ., Commack, N.Y., 1994.
6. Chmaissem, O. et. al., *Physica*, C230, 231—238, 1994.
7. Antipov, E.V. et. al., *Physica*, C215, 1—10, 1993.

Table 1
Structural Parameters and Approximate T_c Values of High-Temperature Superconductors

Material	Structure	T_c/K (maximum value)
$La_2CuO_{4+\delta}$	Bmab; $a = 5.355$, $b = 5.401$, $c = 13.15$ Å	39
$La_{2-x}Sr_x(Ba_x)CuO_4$	I4/mmm; $a = 3.779$, $c = 13.23$ Å	35
$La_2Ca_{1-x}Sr_xCu_2O_6$	I4/mmm; $a = 3.825$, $c = 19.42$ Å	60
$YBa_2Cu_3O_7$	Pmmm; $a = 3.821$, $b = 3.885$, $c = 11.676$ Å	93
$YBa_2Cu_4O_8$	Ammm; $a = 3.84$, $b = 3.87$, $c = 27.24$ Å	80
$Y_2Ba_4Cu_7O_{15}$	Ammm; $a = 3.851$, $b = 3.869$, $c = 50.29$ Å	93
$Bi_2Sr_2CuO_6$	Amaa; $a = 5.362$, $b = 5.374$, $c = 24.622$ Å	10
$Bi_2CaSr_2Cu_2O_8$	A_2aa; $a = 5.409$, $b = 5.420$, $c = 30.93$ Å	92
$Bi_2Ca_2Sr_2Cu_3O_{10}$	A_2aa; $a = 5.39$, $b = 5.40$, $c = 37$ Å	110
$Bi_2Sr_2(Ln_{1-x}Ce_x)_2Cu_2O_{10}$	P4/mmm; $a = 3.888$, $c = 17.28$ Å	25
$Tl_2Ba_2CuO_6$	A_2aa; $a = 5.468$, $b = 5.472$, $c = 23.238$ Å; I4/mmm; $a = 3.866$, $c = 23.239$ Å	92
$Tl_2CaBa_2Cu_2O_8$	I4/mmm; $a = 3.855$, $c = 29.318$ Å	119
$Tl_2Ca_2Ba_2Cu_3O_{10}$	I4/mmm; $a = 3.85$, $c = 35.9$ Å	128
$Tl(BaLa)CuO_5$	P4/mmm; $a = 3.83$, $c = 9.55$ Å	40
$Tl(SrLa)CuO_5$	P4/mmm; $a = 3.7$, $c = 9$ Å	40
$(Tl_{0.5}Pb_{0.5})Sr_2CuO_5$	P4/mmm; $a = 3.738$, $c = 9.01$ Å	40
$TlCaBa_2Cu_2O_7$	P4/mmm; $a = 3.856$, $c = 12.754$ Å	103
$(Tl_{0.5}Pb_{0.5})CaSr_2Cu_2O_7$	P4/mmm; $a = 3.80$, $c = 12.05$ Å	90
$TlSr_2Y_{0.5}Ca_{0.5}Cu_2O_7$	P4/mmm; $a = 3.80$, $c = 12.10$ Å	90
$TlCa_2Ba_2Cu_3O_8$	P4/mmm; $a = 3.853$, $c = 15.913$ Å	110
$(Tl_{0.5}Pb_{0.5})Sr_2Ca_2Cu_3O_9$	P4/mmm; $a = 3.81$, $c = 15.23$ Å	120
$TlBa_2(La_{1-x}Ce_x)_2Cu_2O_9$	I4/mmm; $a = 3.8$, $c = 29.5$ Å	40
$Pb_2Sr_2La_{0.5}Ca_{0.5}Cu_3O_8$	Cmmm; $a = 5.435$, $b = 5.463$, $c = 15.817$ Å	70
$Pb_2(Sr,La)_2Cu_2O_6$	$P22_12$; $a = 5.333$, $b = 5.421$, $c = 12.609$ Å	32
$(Pb,Cu)Sr_2(La,Ca)Cu_2O_7$	P4/mmm; $a = 3.820$, $c = 11.826$ Å	50
$(Pb,Cu)(Sr,Eu)(Eu,Ce)Cu_2O_x$	I4/mmm; $a = 3.837$, $c = 29.01$ Å	25
$Nd_{2-x}Ce_xCuO_4$	I4/mmm; $a = 3.95$, $c = 12.07$ Å	30
$Ca_{1-x}Sr_xCuO_2$	P4/mmm; $a = 3.902$, $c = 3.35$ Å	110
$Sr_{1-x}Nd_xCuO_2$	P4/mmm; $a = 3.942$, $c = 3.393$ Å	40
$Ba_{0.6}K_{0.4}BiO_3$	Pm3m; $a = 4.287$ Å	31
Rb_2CsC_{60}	$a = 14.493$ Å	31
$NdBa_2Cu_3O_7$	Pmmm; $a = 3.878$, $b = 3.913$, $c = 11.753$	58

Table 1
Structural Parameters and Approximate T_c Values of High-Temperature Superconductors
(continued)

Material	Structure	T_c/K (maximum value)
$SmBaSrCu_3O_7$	I4/mmm; $a = 3.854$, $c = 11.62$	84
$EuBaSrCu_3O_7$	I4/mmm; $a = 3.845$, $c = 11.59$	88
$GdBaSrCu_3O_7$	I4/mmm; $a = 3.849$, $c = 11.53$	86
$DyBaSrCu_3O_7$	Pmmm; $a = 3.802$, $b = 3.850$, $c = 11.56$	90
$HoBaSrCu_3O_7$	Pmmm; $a = 3.794$, $b = 3.849$, $c = 11.55$	87
$ErBaSrCu_3O_7$ (multiphase)	Pmmm; $a = 3.787$, $b = 3.846$, $c = 11.54$	82
$TmBaSrCu_3O_7$ (multiphase)	Pmmm; $a = 3.784$, $b = 3.849$, $c = 11.55$	88
$YBaSrCu_3O_7$	Pmmm; $a = 3.803$, $b = 3.842$, $c = 11.54$	84
$HgBa_2CuO_4$	I4/mmm; $a = 3.878$, $c = 9.507$	94
$HgBa_2CaCu_2O_6$ (annealed in O_2)	I4/mmm; $a = 3.862$, $c = 12.705$	127
$HgBa_2Ca_2Cu_3O_8$	Pmmm; $a = 3.85$, $c = 15.85$	133
$HgBa_2Ca_3Cu_4O_{10}$	Pmmm; $a = 3.854$, $c = 19.008$	126

Table 2
Superconducting Properties

J_c (0): Critical current density extrapolated to 0 K
λ_{ab}: Penetration depth in a-b plane
k_B: Boltzmann constant

Material	Form	Energy gap (Δ)		$10^{-6} \times J_c$ (0)/A cm^{-2}	λ_{ab}/Å
		$2\Delta_{pp}/k_BT_c$*	$2\Delta_{fit}/k_BT_c$†		
Y $Ba_2Cu_3O_7$	Single Crystal	5–6	4–5	30 (film)	1400
$Bi_2Sr_2CaCu_2O_8$	Single Crystal	8–9	5.5–6.5	2	2700
$Tl_2Ba_3CaCu_2O_8$	Ceramic	6–7	4–6	10 (film, 80 K)	2000
$La_{2-x}Sr_xCuO_4$, $x = 0.15$	Ceramic	7–9	4–6		
$Nd_{2-x}Ce_xCuO_4$	Ceramic	8	4–5	0.2 (film)	

*   Obtained from peak to peak value.
†   Obtained from fit to BCS-type relation.

Table 3
Normal State Properties

ρ$_{ab}$:   Resistivity in the a-b plane
ρ$_c$:   Resistivity along the c axis
+ve: ρ$_c$ has positive temperature coefficient of resistivity
−ve: ρ$_c$ has negative temperature coefficient of resistivity
n$_H$:   Hall density
k: Thermal conductivity
in plane: Along a-b plane
out of plane: Perpendicular to a-b plane

Material	Form	ρ$_{ab}$/μΩ cm		ρ$_c$/mΩ cm	dρ$_c$/dT	$10^{-21} \times n_H$/cm^{-3}		k/(mW/cm K) at 300 K	
		300 K	100 K	300 K		300 K	100 K	in plane	out of plane
YBa$_2$Cu$_3$O$_7$	Single Crystal	110	35	5	+ve	11–16	4–6	120	3
	film	200–300	60–100			5–9	2–3		
YBa$_2$Cu$_4$O$_8$	Single Crystal	75	20	10	−ve	14	17		
	film	100–200	20–50			22			
Bi$_2$Sr$_2$CuO$_6$	Single Crystal	300	150	5000	−ve	6	5		
Bi$_2$Sr$_2$CaCu$_2$O$_8$	Single Crystal	150	50	>1000	−ve	4	3	60	8
Tl$_2$Ba$_2$CuO$_6$	Single Crystal	300–400	50–75	200–300	+ve	3.1	2.5		
Tl$_2$Ba$_2$Ca$_2$Cu$_3$O$_{10}$	Ceramic	***	**		+ve for T >225 K	2.5	≈2*		
La$_{2-x}$Sr$_x$CuO$_4$, x = 0.12	Single Crystal	900	350	200	+ve for T >150 K	2.5		50 (for x = 0.04)	
La$_{2-x}$Sr$_x$CuO$_4$, x = 0.20	Single Crystal	400	200	80		10	6.3		20
	film	400	160			8.4		250 (for x = 0.15)	
Nd$_{2-x}$Ce$_x$CuO$_4$, x = 0.17	Single Crystal	500	275			53	17		
x = 0.15	film	140–180	35			32	11		

*   At 200 K
**   ρ ~0.4 mΩ cm at 120 K
***   ρ ~1.5 mΩ cm at 300 K

ORGANIC SUPERCONDUCTORS
H.P.R. Frederikse

Although the vast majority of organic compounds are insulators, a small number of organic solids show considerable electrical conductivity. Some of these materials appear to be superconductors. The superconducting organics fall primarily into two groups: those containing fulvalenes (pentagonal rings containing sulfur or selenium) and those based on fullerenes, involving the nearly spherical cluster C_{60}.

The transition temperatures T_c of the fulvalene derivatives are shown in Table 1. The abbreviations of the various molecular groups are listed in Table 2 and their chemical structures are depicted in Figure 1. Most of the T_c's are between 1 and 12 K. Several of the compounds only show superconductivity under pressure.

The fullerenes are A_3C_{60} compounds, where A represents a single or a combination of alkali atoms. The C_{60} cluster is shown in Figure 2a, while Figure 2b illustrates how the alkali atoms fit into the A_3C_{60} molecule to form the A15 crystallographic structure. Their superconducting transition temperatures range from 8 to 31.3 K (see Table 3).

REFERENCES

1. Ishigura, T. and Yamaji, K., *Organic Superconductors*, Springer-Verlag, Berlin, 1990.
2. Williams, Jack M. et al., *Organic Superconductors (Including Fullerenes)*, Prentice Hall, Englewood Cliffs, N.J., 1992.
3. *The Fullerenes*, Ed.: Krato, H.W., Fisher, J.E., and Cox, D.E., Pergamon Press, Oxford, 1993.
4. Schluter, M. et al., in *The Fullerenes* (Ref. 3), p. 303.

Table 1
Critical Pressure and Maximum Critical Temperature of Organic Superconductors

Material	P_c/kbar	T_c/K	Material	P_c/kbar	T_c/K
$(TMTSF)_2PF_6$	6.5	1.2	β-$(ET)_2IBr_2$	0	2.8
$(TMTSF)_2AsF_6$	9	1.3	β-$(ET)_2AuI_2$	0	4.8
$(TMTSF)_2SbF_6$	11	0.4	$(ET)_4Hg_{2.89}Cl_8$	0	4.2
$(TMTSF)_2TaF_6$	12	1.4	$(ET)_4Hg_{2.89}Br_8$	12	1.8
$(TMTSF)_2ClO_4$	0	1.4	$(ET)_3Cl_2(H_2O)_2$	16	2
$(TMTSF)_2ReO_4$	9.5	1.3	κ-$(ET)_2Cu(NCS)_2$	0	10.4
$(TMTSF)_2FSO_3$	5	3	κ-$(d$-$ET)_2Cu(NCS)_2$	0	11.4
$(ET)_4(ReO_4)_2$	4.5	2	$(DMET)_2Au(CN)_2$	1.5	0.9
β_L-$(ET)_2I_3$	0	1.4	$(DMET)_2AuI_2$	5	0.6
β_H-$(ET)_2I_3$	0	8.1	$(DMET)_2AuBr_2$	0	1.9
γ-$(ET)_3I_{2.5}$	0	2.5	$(DMET)_2AuCl_2$	0	0.9
ε-$(ET)_2I_3(I_8)_{0.5}$	0	2.5	$(DMET)_2I_3$	0	0.6
α-$(ET)_2I_3I_2$-doped	0	3.3	$(DMET)_2IBr_2$	0	0.7
α_t-$(ET)_2I_3$	0	8	$(MDT$-$TTF)_2AuI_2$	0	3.5
$\varepsilon \rightarrow \beta$-$(ET)_2I_3$[a]	0	6	$TTF[Ni(dmit)_2]_2$	2	1.6[b]
θ-$(ET)_2I_3$	0	3.6	$TTF[Pd(dmit)_2]_2$	20	6.5
κ-$(ET)_2I_3$	0	3.6	$(CH_3)_4N[Ni(dmit)_2]_2$	7	5

[a]   Converted form ε-type to β-type by thermal treatment.

[b]   For 7 kbar.

From Ishigura, T. and Yamaji, K., *Organic Superconductors*, Springer-Verlag, Berlin, 1990. With permission.

Table 2
List of Symbols and Abbreviations

TTF	tetrathiafulvalene
TMTSF	tetramethyltetraselenafulvalene
BEDT-TTF or "ET"	bis(ethylenedithio)tetrathiafulvalene
MDT-TTF	methylenedithiotetrathiafulvalene
DMET	[dimethyl(ethylenedithio)diselenadithiafulvalene]
dmit	4,5-dimercapto-1,3-dithiole-2-thione
T_c	transition temperature to superconducting state
P_c	minimum pressure required for superconducting transition

TMTSF

Tetramethyltetraselenafulvalene

TTF

Tetrathiafulvalene

BEDT – TTF or ET

Bis(ethylenedithio)tetrathiafulvalene

DMET

Dimethyl(ethylenedithio)diselenadithiafulvalene

MDT – TTF

Methylenedithiotetrathiafulvalene

$M = Ni, Pd, Pt$

$M(dmit)_2^{2-}$

Ligand is 4,5-dimercapto-1.3-dithiole-2-thione

FIGURE 1. Structures of various donor molecules and acceptor species.

a

b

FIGURE 2. (a) C_{60} cluster placed in a fcc lattice. Each crystal axis crosses a double bond shared by two hexagons. (b) A hypothetical A_3C_{60} with the A15 structure. The structure can be seen to be an ordered defect structure of A_6C_{60}.

Table 3
Unit Cell and T_c for FCC-$A_3 C_{60}$

	Lattice parameter(s) (Å)	T_c/K
$Na_2Rb_{0.5}Cs_{0.5}C_{60}$	14.148(3)	8.0
Na_2CsC_{60} No. 1[a]	14.132(2)	10.5
Na_2CsC_{60} No. 2[a]	14.176(9)	14.0
K_3C_{60}	14.253(3)	19.3
K_2RbC_{60}	14.299(2)	21.8
Rb_2KC_{60} No. 1[a]	14.336(1)	24.4
Rb_2KC_{60} No. 2[a]	14.364(5)	26.4
Rb_3C_{60}	14.436(2)	29.4
Rb_2CsC_{60}	14.493(2)	31.3

[a]   Samples labeled No. 1 and No. 2 have the same nominal composition.

From Schluter, M et. al., *The Fullerenes*, Ed.: Krato, H.W., Fisher, J.E., and Cox, D.E., Pergamon Press, Oxford, 1993. With permission.

PROPERTIES OF SEMICONDUCTORS

L. I. Berger and B. R. Pamplin

The term "semiconductor" is applied to a material in which electric current is carried by electrons or holes and whose electrical conductivity, when extremely pure, rises exponentially with temperature and may be increased from its low "intrinsic" value by many orders of magnitude by "doping" with electrically active impurities.

Semiconductors are characterized by an energy gap in the allowed energies of electrons in the material which separates the normally filled energy levels of the *valence band* (where "missing" electrons behave like positively charged current carriers "holes") and the *conduction band* (where electrons behave rather like a gas of free negatively charged carriers with an effective mass dependent on the material and the direction of the electrons' motion). This energy gap depends on the nature of the material and varies with direction in anisotropic crystals. It is slightly dependent on temperature and pressure, and this dependence is usually almost linear at normal temperatures and pressures.

Data are presented in three tables. Table I "General Properties of Semiconductors" lists the main crystallographic and semiconducting properties of a large number of semiconducting materials in three main categories: "Tetrahedral Semiconductors" in which every atom is tetrahedrally co-ordinated to four nearest neighbor atoms (or atomic sites) as for example in the diamond structure; "Octahedral Semiconductors" in which every atom is octahedrally co-ordinated to six nearest neighbor atoms — as for examples the halite structure; and "Other Semiconductors."

Table II gives more detailed information about some better known semiconductors, while Table III gives some information about the electronic energy band structure parameters of the best known materials.

Table I
PHYSICO-CHEMICAL PROPERTIES OF SEMICONDUCTORS
(LISTED BY CRYSTAL STRUCTURE)

Substance	Molecular mass	Average atomic mass	Lattice parameters (Å, room temp.)	Density (g/cm³)	Melting point (K)	Microhardness, N/mm² (M-Mohs Scale)	Specific heat, J/kg·K (300 K)	Debye temp. (K)	Coefficient of thermal linear expansion [10^{-6} K^{-1} (300K)]	Thermal conductivity [mW/cm·K (300K)]

PART A. ADAMANTINE SEMICONDUCTORS

§A1. Diamond Structure Elements (Strukturbericht symbol A4, Space Group Fd3m-O_h^7)

Substance	Molecular mass	Average atomic mass	Lattice parameters (Å, room temp.)	Density (g/cm³)	Melting point (K)	Microhardness, N/mm² (M-Mohs Scale)	Specific heat, J/kg·K (300 K)	Debye temp. (K)	Coefficient of thermal linear expansion [10^{-6} K^{-1} (300K)]	Thermal conductivity [mW/cm·K (300K)]
C	12.01	3.56683	3.51		≈3850 Transition to graphite > 980	10(M)	471.5	2340	1.18	9900(I) 23200(IIA) 13600(IIB)
Si	28.09	5.43072	2.3283	1685 ± 2	11270	702	645	2.49	1240
Ge	72.59	5.65754	5.3234	1231	7644	321.9	374	6.1	640
α-Sn	118.69	6.4912	5.765 (281 K)	505.2 (Tr. 286.4)			213	230	5.4 (220 K)	

§A2. Sphalerite (Zinc Blende) Structure Compounds (Strukturbericht symbol B3 Space Group F $\overline{4}$ 3m-T_d^2)

Substance	Molecular mass	Average atomic mass	Lattice parameters (Å, room temp.)	Density (g/cm³)	Melting point (K)	Microhardness, N/mm² (M-Mohs Scale)	Specific heat, J/kg·K (300 K)	Debye temp. (K)	Coefficient of thermal linear expansion [10^{-6} K^{-1} (300K)]	Thermal conductivity [mW/cm·K (300K)]
I VII Compounds										
CuF	82.54	41.27	4.255		1181					
CuCl	98.99	49.49	5.4057	3.53	695	2.3 (M)	490	240	12.1	8.4
CuBr	143.36	71.73	5.6905	4.98	770	2.5(M)	381	207	15.4	12.5
CuI	190.46	95.23	6.60427	5.63	878	192	276	181	19.2	16.8
AgBr	187.78	93.89		6.473	>1570 (Tr. 410)	2.5(M)	270			
AgI	234.77	117.39	6.502	5.67	831	2.5(M)	232	134	−2.5	4.2
II VI Compounds										
BeS	41.08	20.54	4.865	2.36						
BeSe	87.97	43.99	5.139	4.315						
BeTe	136.61	68.31	5.626	5.090						
BePo	(218)	(109)	5.838	7.3						
ZnO	81.37	40.69	4.63	5.675	2248	5.0 (M)	494	416	2.9	234
ZnS	97.43	48.72	5.4093	4.079	2100 (Tr. 1295)	1780	472	530	6.36	251
ZnSe	144.34	72.17	5.6676	5.42	1790	1350	339	400	7.2	140
ZnTe	192.99	96.5	6.101	6.34	1568	900	264	223	8.19	108
ZnPo	(274)	(137)	6.309							
CdS	144.46	72.23	5.832	4.826	1750	1250	330	219	4.7	200
CdSe	191.36	95.68	6.05	5.674	1512	1300	255	181	3.8	90
CdTe	240.00	120.00	6.477	5.86	1365	600	205	200	4.9	58.5
CdPo	(321)	(161)	6.665							
HgS	232.65	116.33	5.8517	7.73	1820	3(M)	210			
HgSe	279.55	139.78	6.084	8.25	1070	2.5(M)	178	151	5.46	10
HgTe	328.19	164.10	6.4623	8.17	943	300	164	242	4.6	20
III V Compounds										
BN	24.82	12.41	3.615	3.49	≈3300	10(M)	793	≈1900		200
BP(L.T.)	41.78	20.87	4.538	2.9	≈2800	37000		≈980		
BAs	85.73	42.87	4.777		≈2300	19000		≈625		

Substance	Molecular mass	Average atomic mass	Lattice parameters (Å, room temp.)	Density (g/cm³)	Melting point (K)	Microhardness, N/mm² (M-Mohs Scale)	Specific heat, J/kg·K (300 K)	Debye temp. (K)	Coefficient of thermal linear expansion [10^{-6} K^{-1} (300K)]	Thermal conductivity [mW/cm·K (300K)]
AlP	57.95	28.98	5.451	2.42	≈2100	5.5(M)		588		920
AlAs	101.90	50.95	5.6622	3.81	2013	5000		417	3.5	840
AlSb	148.73	74.37	6.1355	4.218	1330	4000		292	4.2	600
GaP	100.69	50.35	5.4505	4.13	1750	9450		446	5.3	752
GaAs	144.64	72.32	5.65315	5.316	1510	7500		344	5.4	560
GaSb	191.47	95.74	6.0954	5.619	980	4480	320	265	6.1	270
InP	145.79	72.90	5.86875	4.787	1330	4100		321	4.6	800
InAs	189.74	94.87	6.05838	5.66	1215	3300	268	249	4.7	290
InSb	236.57	118.29	6.47877	5.775	798	2200	144	202	4.7	160
Other sphalerite structure compounds										
MnS	87.0	43.5	5.011							
MnSe	133.9	66.95	5.82							
β-SiC	40.1	20.1	4.348	3.21	3070					
Ga₂Se₃	376.32	75.26	5.429	4.92	1020	3160			8.9	50
Ga₂Te₃	522.24	104.45	5.899	5.75	1063	2370				47
In₂Te₃(H.T.)	608.44	121.7	6.150	5.8	940	1660				69
MgGeP₂	158.84	39.71	5.652							
ZnSnP₂	246.00	61.5	5.65		1200					
ZnSnAs₂(H.T.)	333.90	82.38	5.851	5.53	1050					76
ZnSnSb₂	427.56	106.89	6.281	5.67	870	2500				76

§A3. Wurtzite (Zincite) Structure Compounds (Strukturbericht symbol B4, Space Group P 6₃mc-C⁴₆ᵥ)

Substance	Molecular mass	Average atomic mass	Lattice parameters (Å, room temp.)	Density (g/cm³)	Melting point (K)	Microhardness, N/mm² (M-Mohs Scale)	Specific heat, J/kg·K (300 K)	Debye temp. (K)	Coefficient of thermal linear expansion [10^{-6} K^{-1} (300K)]	Thermal conductivity [mW/cm·K (300K)]
I VII Compounds										
CuCl	99.0	49.5	3.91	6.42		T_c680K				
CuBr	143.46	71.73	4.06	6.66		T_c658K				
CuI	190.46	95.23	4.31	7.09						
AgI	234.80	117.40	4.580	7.494						
II VI Compounds										
BeO	25.01	12.51	2.698	4.380		2800				
MgTe	151.9	76.0	4.54	7.39	3.85	≈2800				
ZnO	81.37	40.69	3.24950	5.2069	5.66	2250				600
ZnS	97.43	48.72	3.8140	6.2576	4.1	2100				460
ZnTe	192.99	46.50	4.27	6.99		1568				
CdS	144.46	72.23	4.1348	6.7490	4.82	1748				401
CdSe	191.36	95.68	4.299	7.010	5.66	1512				316
CdTe	240.00	120.00	4.57	7.47						
III V Compounds										
BP(H.T.)	41.79	20.90	3.562	5.900						
AlN	40.99	20.50	3.111	4.978	3.26	≈2500				823
GaN	83.73	41.87	3.190	5.189	6.10	1500				656
InN	128.83	64.42	3.533	5.693	6.88	1200				556
Other wurtzite structure compounds										
MnS	87.0	43.5	3.985	6.45	3.248					
MnSe	133.9	66.95	4.12	6.72						
SiC	40.1	20.1	3.076	5.048						
MnTe	182.54	91.27	4.078	6.701						
Al₂S₃	150.14	30.03	3.579	5.829	2.55	1400				
Al₂Se₃	290.84	58.17	3.890	6.30	3.91	1250				

§A4. Chalcopyrite Structure Compounds (Strukturbericht symbol E1₁, Space Group I 4̄ 2d-D¹²₂ᵈ)

Substance	Molecular mass	Average atomic mass	Lattice parameters (Å, room temp.)	Density (g/cm³)	Melting point (K)	Microhardness, N/mm² (M-Mohs Scale)	Specific heat, J/kg·K (300 K)	Debye temp. (K)	Coefficient of thermal linear expansion [10^{-6} K^{-1} (300K)]	Thermal conductivity [mW/cm·K (300K)]
I III VI₂ Compounds										
CuAlS₂	154.65	38.66	5.323	10.44	3.47	2500				
CuAlSe₂	248.45	62.11	5.617	10.92	4.70	2260				
CuAlTe₂	345.73	86.43	5.976	11.80	5.50	2550				
CuGaS₂	197.39	49.35	5.360	10.49	4.35	2500				
CuGaSe₂	291.19	72.80	5.618	11.01	5.56	1970	4200	275	5.4	42
CuGaTe₂	388.47	97.12	6.013	11.93	5.99	2400	3500		6.9	27
CuInS₂	242.49	60.62	5.528	11.08	4.75	1400	2550			
CuInSe₂	336.29	84.07	5.785	11.56	5.77	1600	2050		6.6	37
CuInTe₂	433.57	108.39	6.179	12.365	6.10	1660	400	195	7.1	49
CuTlS₂	332.05	83.01	5.580	11.17	6.32					
CuTlSe₂(L.T.)	425.85	106.46	5.844	11.65	7.11	900				
CuFeS₂	183.51	45.88	5.25	10.32	4.088					
CuFeSe₂	277.31	69.33				850				

Substance	Molecular mass	Average atomic mass	Lattice parameters (Å, room temp.)		Density (g/cm³)	Melting point (K)	Microhardness, N/mm² (M-Mohs Scale)	Specific heat, J/kg·K (300 K)	Debye temp. (K)	Coefficient of thermal linear expansion [10⁻⁶ K⁻¹ (300K)]	Thermal conductivity [mW/cm·K (300K)]
CuLaS₂	266.58	66.65	5.65	10.86							
AgAlS₂	198.97	49.74	5.707	10.28	3.94						
AgAlSe₂	292.77	73.19	5.968	10.77	5.07	1220					
AgAlTe₂	390.05	97.51	6.309	11.85	6.18	1000					
AgGaS₂	241.71	60.43	5.755	10.28	4.72						
AgGaSe₂	335.51	83.88	5.985	10.90	5.84	1120	4400				
AgGaTe₂	432.79	108.2	6.301	11.96	6.05	990	1800		212		10
AgInS₂(L.T.)	286.87	71.70	5.828	11.19	5.00		2250				
AgInSe₂	380.61	95.15	6.102	11.69	5.81	1053	1850				30
AgInTe₂	477.89	119.47	6.42	12.59	6.12	965				9.49, 0.69	
AgFeS₂	227.83	56.96	5.66	10.30	4.53						
II IV V₂ Compounds											
ZnSiP₂	155.40	38.85	5.400	10.441	3.39	1640	11000				
ZnGeP₂	199.90	49.98	5.465	10.771	4.17	1295	8100				180
ZnSnP₂	246.00	61.5					6500				
CdSiP₂	202.43	50.61	5.678	10.431	4.00	≈1470	10500		282		
CdGeP₂	246.94	61.74	5.741	10.775	4.48	1049	5650				110
CdSnP₂	243.03	73.26	5.900	11.518			5000		195		140
ZnSiAs₂	242.20	60.55	5.61	10.88	4.70	1311	9200				
ZnGeAs₂	287.80	71.95	5.672	11.153	5.32	1150	6800		263		110
ZnSnAs₂	333.90	83.48	5.8515	11.704	5.53	1048	4550		271		150
CdSiAs₂	290.34	72.58	5.884	10.882			6850				
CdGeAs₂	334.83	83.71	5.9427	11.2172	5.60	938	4700				48
CdSnAs₂	380.93	95.23	6.0944	11.9182	5.72	880	3450				40

§A5. Other Ternary Semiconductors with Tetrahedral Coordination

Substance	Molecular mass	Average atomic mass	Lattice parameters (Å, room temp.)		Density (g/cm³)	Melting point (K)	Microhardness, N/mm² (M-Mohs Scale)	Specific heat, J/kg·K (300 K)	Debye temp. (K)	Coefficient of thermal linear expansion [10⁻⁶ K⁻¹ (300K)]	Thermal conductivity [mW/cm·K (300K)]
I₂ IV VI₃ Compounds											
Cu₂SiS₃(H.T.)	251.36	41.89	3.684	6.004	3.81	1200					23
Cu₂SiS₃(L.T.)			5.290	10.156	3.63						
Cu₂SiTe₃	537.98	89.66	5.93		5.47						
Cu₂GeS₃(H.T.)	295.88	49.31	5.317		4.45	1210	4550	510	254	7.2	12
Cu₂GeS₃(L.T.)			5.327	5.215	4.46						
Cu₂GeSe₃	436.56	72.76	5.589	5.485	5.57	1030	3840	340	168	8.4	24
Cu₂GeTe₃	582.51	97.09	5.958	5.935	5.92		2890				130
Cu₂SnS₃	341.98	57.00	5.436		5.02	1110	2770	440	214	7.8	28
CuSnSe₃	482.66	80.44	5.687		5.94	960	2510	310	148	8.9	35
Cu₂SnTe₃	628.61	104.77	6.048		6.51	680	1970				144
Ag₂GeSe₃	525.21	87.54									
Ag₂SnSe₃	571.31	95.22									
Ag₂GeTe₃	671.13	111.86									
Ag₂SnTe₃	717.23	119.54									
I₃ V VI₄ Compounds											
Cu₃PS₄	349.85	40.73	7.44	6.19							
Cu₃AsS₄	393.79	49.22	6.43	6.14	4.37					3.2	30.2
Cu₃AsSe₄	581.37	72.67	5.570	10.957	5.61				169	9.5	19
Cu₃SbS₄	440.64	55.08	5.38	16.76	4.90						
Cu₃SbSe₄	628.22	78.53	5.654	11.256	6.0				131	12.4	14.6
I IV₂ V₃ Compounds											
CuSi₂P₃	212.64	35.44	5.25								
CuGe₂P₃	301.65	50.28	5.375		4.318	1113	8500	429		8.21	37.6
AgGe₂P₃	345.97	57.66				1015	6150				

§A6. "Defect Chalcopyrite" Structure Compounds (Strukturbericht symbol E3, Space Group I $\overline{4}$-S$_4^2$)

Substance	Molecular mass	Average atomic mass	Lattice parameters (Å, room temp.)		Density (g/cm³)	Melting point (K)					
ZnAl₂Se₄	435.18	62.17	5.503	10.90	4.37						
ZnAl₂Te₄(?)	629.74	84.96	5.904	12.05	4.95						
ZnGa₂S₄(?)	333.06	47.58	5.274	10.44	3.80						
ZnGa₂Se₄(?)	520.66	74.38	5.496	10.99	5.21						
ZnGa₂Te₄(?)	715.22	102.17	5.937	11.87	5.67						
ZnIn₂Se₄	610.86	87.27	5.711	11.42	5.44	1250					
ZnIn₂Te₄	805.42	115.06	6.122	12.24	5.83	1075					
CdAl₂S₄	294.61	42.09	5.564	10.32	3.06						
CdAl₂Se₄	482.21	68.89	5.747	10.68	4.54						
CdAl₂Te₄(?)	676.77	97.68	6.011	12.21	5.10						

Substance	Molecular mass	Average atomic mass	Lattice parameters (Å, room temp.)		Density (g/cm³)	Melting point (K)	Microhardness, N/mm² (M-Mohs Scale)	Specific heat, J/kg·K (300 K)	Debye temp. (K)	Coefficient of thermal linear expansion [10⁻⁶ K⁻¹ (300K)]	Thermal conductivity [mW/cm·K (300K)]
CdGa₂S₄	380.09	54.30	5.577	10.08	4.03						
CdGa₂Se₄	567.69	81.10	5.743	10.73	5.32						
CdGa₂Te₄	762.25	108.89	6.093	11.81	5.77						
CdIn₂Te₄	852.45	121.78	6.205	12.41	5.9	1060					
HgAl₂S₄	382.79	54.68	5.488	10.26	4.11						
HgAl₂Se₄	570.39	82.48	5.708	10.74	5.05						
HgAl₂Te₄(?)	764.48	109.28	6.004	12.11	5.81						
HgGa₂S₄	468.27	66.90	5.507	10.23	5.00						
HgGa₂Se₄	655.87	93.70	5.715	10.78	6.18						
HgIn₂Se₄	746.07	106.58	5.764	11.80	6.3	1100					
HgIn₂Te₄(?)	940.63	134.38	6.186	12.37	6.3	980					

§A7. Other Adamantine Compounds

Substance	Molecular mass	Average atomic mass	Lattice parameters (Å, room temp.)		Density (g/cm³)	Melting point (K)	Microhardness	Specific heat	Debye temp.	Coeff. thermal	Thermal cond.
αSiC	40.1	20.1	3.0817		3.21	3070					
			15.1183								
Hg₅Ga₂Te₈	2163.19	144.21	6.235								
Hg₅In₂Te₈	2253.39	150.23	6.328								
CdIn₂Se₄	657.89	93.98	a = c = 5.823								

PART B. OCTAHEDRAL SEMICONDUCTORS
§B1. HALITE STRUCTURE SEMICONDUCTORS (Strukturbericht symbol B1, Space Group Fm3m-O$_h^5$)

Substance	Molecular mass	Average atomic mass	Lattice param.	Density (g/cm³)	Melting point (K)	Microhardness	Specific heat	Debye temp.	Coeff. thermal	Thermal conductivity
GeTe	200.19	100.1	5.98	6.14						
SnSe	197.65	98.83	6.020		1133					
SnTe	246.29	123.15	6.313	6.45	1080 (max)					91
PbS	239.26	119.63	5.9362	7.61	1390					23
PbSe	286.16	143.08	6.1243	8.15	1340					17
PbTe	334.8	167.4	6.454	8.16	1180					23

Selected other binary halites

Substance	Molecular mass	Average atomic mass	Lattice param.	Density (g/cm³)	Melting point (K)	Microhardness	Specific heat	Debye temp.	Coeff. thermal	Thermal conductivity
BiSe	287.94	143.97	5.99	7.98	880					
BiTe	336.58	168.29	6.47							
EuSe	230.92	115.46	6.191		2300					2.4
GdSe	236.21	118.11	5.771		2400					
NiD	60.71	30.35	4.1684	6.6	2260					
CdO	128.41	64.21	4.6953		1700					7
SrS	119.68	59.84	6.0199	3.643	3000					

PART C. OTHER SEMICONDUCTORS
§C1. Antifluorite Structure Compounds (Fm3m - O$_h^5$)

Substance	Molecular mass	Average atomic mass	Lattice param.	Density (g/cm³)	Melting point (K)	Microhardness	Specific heat	Debye temp.	Coeff. thermal	Thermal conductivity
Mg₂Si	76.70	25.57	6.338	1.88	1375				11.5	
Mg₂Ge	121.20	40.4	6.380	3.08	1388				15.0	
Mg₂Sn	167.3	55.77	6.765	3.53	1051				9.9	92
Mg₂Pb	225.81	85.27	6.836	5.1	823				10.0	

§C2. Tetradymite Structure Compounds ($\bar{R}3m - D_{3d}^5$)

Substance	Molecular mass	Average atomic mass	Lattice parameters		Density (g/cm³)	Melting point (K)	Microhardness	Specific heat	Coeff. thermal	Thermal conductivity
Sb₂Te₃	626.3	125.26	4.25	30.3	6.44	895				24
Bi₂Se₃	654.84	130.97	4.14	28.7	7.51	979	167			24
Bi₂Te₃	800.76	160.15	4.38	30.45	7.73	858	155	16		30

§C3. Skutterudite Structure Compounds (Im3 - T$_h^5$)

Substance	Molecular mass	Average atomic mass	Lattice param.	Density (g/cm³)	Melting point (K)	Microhardness	Specific heat	Debye temp.	Coeff. thermal	Thermal conductivity
CoP₃	151.85	37.96	7.7073		>1270					
CoAs₃	286.70	71.65	8.2060	6.73	1230					
CoSb₃	424.18	106.05	9.0385		1123			307		50
NiAs₃	283.45	70.86	8.330	6.43						
RhP₃	195.83	48.96	7.9951		>1470					
RhAs₃	327.67	81.92	8.4427		>1270					
RhSb₃	468.16	117.04	9.2322		1170					100
IrP₃	285.14	71.29	8.0151	7.36	>1470					
IrAs₃	416.98	104.25	8.4673	9.12	>1470					
IrSb₃	557.47	139.37	9.2533	9.35	1170			303		90

§C4. Selected Multinary Compounds

Substance	Molecular mass	Average atomic mass	Lattice param.	Density (g/cm³)	Melting point (K)	Microhardness	Specific heat	Debye temp.	Coeff. thermal	Thermal conductivity
AgSbSe₂	387.54	96.88	5.786	6.60	910					10.5
AgSbTe₂ (or Ag₁₉Sb₂₉Te₅₂)	484.82	121.2	6.078	7.12	830					86, 0.3
AgBiS₂(H.T.)	380.97	95.24	5.648							
AgBiSe₂(H.T.)	474.77	118.69	5.82							
AgBiTe₂(H.T.)	572.05	143.01	6.155							
Cu₇CdSnS₄	486.43	60.80	5.586	10.83						

Table I
PHYSICO-CHEMICAL PROPERTIES OF SEMICONDUCTORS
(LISTED BY CRYSTAL STRUCTURE) (continued)

Substance	Molecular mass	Average atomic mass	Lattice parameters (Å, room temp.)	Density (g/cm³)	Melting point (K)	Microhardness, N/mm² (M-Mohs Scale)	Specific heat, J/kg·K (300 K)	Debye temp. (K)	Coefficient of thermal linear expansion [10⁻⁶ K⁻¹ (300K)]	Thermal conductivity [mW/cm·K (300K)]
§C5. Some Elemental Semiconductors										
B	10.81	4.91	12.6	2.34	2348	9.5(M)	1277	1370	8.3	600
Se(gray)	78.96	4.36	4.95	4.81	493	350	292.6		(∥C) 17.89 (⊥C) 74.09	(∥C) 45.2 (⊥C) 13.1
Te	127.6	4.45	5.91	6.23	723		196.5		16.8	(∥C) 33.8 (⊥C) 19.7

Table II
BASIC THERMODYNAMIC, ELECTRICAL, AND MAGNETIC PROPERTIES OF SEMICONDUCTORS (LISTED BY CRYSTAL STRUCTURE)

Substance	Heat of formation [kJ/mole (300K)]	Volume compressibility (10⁻¹⁰ m²/N)	Static dielectric constant	Atomic magnetic susceptibility (10⁻⁶ CGS)	Index of refraction	Minimum room temperature energy gap (eV)	Mobility (Room temp.) (cm²/V·s) Electrons	Holes	Optical transition	Remarks
PART A. ADAMANTINE SEMICONDUCTORS											

§A1. Diamond Structure Elements (Strukturbericht symbol A4, Space Group Fd3m-O_h^7)

Substance	Heat of formation	Volume compressibility	Static dielectric constant	Atomic magnetic susceptibility	Index of refraction	Energy gap (eV)	Electrons	Holes	Optical transition	Remarks
C	714.4	18	5.7	− 5.88	2.419 (589 nm)	5.4	1800	1400	i*	
Si	324	0.306	11.8	− 3.9	3.49 (589 nm)	1.107	1900	500	i	
Ge	291	0.768	16	− 0.12	3.99 (589 nm)	0.67	3800	1820	i	
α-Sn	267.5		24		2.75 (589 nm)	0.0; 0.8	2500	2400		

§A2. Sphalerite (Zinc Blende) Structure Compounds (Strukturbericht symbol B3 Space Group F $\overline{4}$ 3m-T_d^2)

Substance	Heat of formation	Volume compressibility	Static dielectric constant	Atomic magnetic susceptibility	Index of refraction	Energy gap (eV)	Electrons	Holes	Optical transition	Remarks
I VII Compounds										
CuF										
CuCl	481	0.26	7.9		1.93	3.17			d	Nantokite
CuBr	481	0.26	7.9		2.12	2.91			d	
CuI	439	0.27	6.5		2.346	2.95			d	Marshite
AgBr	486		12.4		2.253	2.50	4000		i	Bromirite
AgI	389	0.41	10		2.22	2.22	30		d	Miersite
II VI Compounds										
BeS					4.17				i	
BeSe					3.61				i	
BeTe					1.45		20		d	
BePo										
ZnO										See A3
ZnS	477		8.9	− 9.9	2.356	3.54	180	5(400°C)	d	See also A3
ZnSe	422		9.2		2.89	2.58	540	28	d	
ZnTe	376		10.4		3.56	2.26	340	100	d	
ZnPo										
CdS										See A3
CdSe										See A3
CdTe	339		7.2		2.50	1.44	1200	50	d	
CdPo										
HgS					2.85		250		d	Metacinnabarite
HgSe	247				2.10 (α)		20000	≈1.5	s	Tiemannite
HgTe	242					−0.06	25000	350	s	Coloradoite
III V Compounds										
BN	815					4.6				Borazone
BP(L.T.)						≈2.1	500	70		Ignites 470K
BAs						≈1.5				
AlP						2.45	80		i	
AlAs	627		10.9			2.16	1200	420	i	
AlSb	585	0.571	11		3.2	1.60	200—400	550	i	
GaP	635	0.110	11.1	− 13.8	3.2	2.24	300	150	i	
GaAs	535	0.771	13.2	− 16.2	3.30	1.35	8800	400	d	
GaSb	493	0.457	15.7	− 14.2	3.8	0.67	4000	1400	d	
InP	560	0.735	12.4	− 22.8	3.1	1.27	4600	150	d	
InAs	477	0.549	14.6	− 27.7	3.5	0.36	33000	460	d	
InSb	447	0.442	17.7	− 32.9	3.96	0.163	78000	750	d	

* i = indirect, d = direct, s = semimetal.

Substance	Heat of formation [kJ/mole] (300K)]	Volume compressibility (10^{-10} m²/N)	Static dielectric constant	Atomic magnetic susceptibility (10^{-6} CGS)	Index of refraction	Minimum room temperature energy gap (eV)	Mobility (Room temp.) (cm²/V·s) Electrons	Holes	Optical transition	Remarks
Other sphalerite structure compounds										
MnS										See also §A3
MnSe										See also §A3
β-SiC					2.697	2.3	4000			
Ga_2Te_3	271			−13.5		1.35	50			
In_2Te_3(H.T.)	198			−13.6		1.04	50			
$MgGeP_2$										$E1$—T^{d12}
$ZnSnP_2$						2.1				Same
$ZnSnAs_2$(H.T.)						≈0.7				Same
$ZnSnSb_2$						0.4				Same

§A3. Wurtzite (Zincite) Structure Compounds (Strukturbericht symbol B4, Space Group P 6_3mc-C_{6v}^4)

Substance	Heat of formation [kJ/mole] (300K)]	Volume compressibility (10^{-10} m²/N)	Static dielectric constant	Atomic magnetic susceptibility (10^{-6} CGS)	Index of refraction	Minimum room temperature energy gap (eV)	Mobility (Room temp.) (cm²/V·s) Electrons	Holes	Optical transition	Remarks
I VII Compounds										
CuCl										
CuBr										
CuI										
AgI						2.63				Iodargirite
II VI Compounds										
BeO										
MgTe										
ZnO	−350					3.2	180			
ZnS	−206					3.67				
ZnTe	−163									
CdS			8.45; 9.12		2.32	2.42	350	40	d	Greenockide
CdSe						1.74	900	50	d	Cadmoselite
CdTe						1.50	650			
III V Compounds										
BP(H.T.)										
AlN						6.02				
GaN						3.34				
InN						2.0				
Other wurtzite structure compounds										
MnS										
MnSe										
SiC					2.654					
MnTe						≈1.0				
Al_2S_3	426					4.1				
Al_2Se_3	367					3.1				

§A4. Chalcopyrite Structure Compounds (Strukturbericht symbol E1₁, Space Group I $\bar{4}$ 2d-D_{2d}^{12})

Substance	Heat of formation [kJ/mole] (300K)]	Volume compressibility (10^{-10} m²/N)	Static dielectric constant	Atomic magnetic susceptibility (10^{-6} CGS)	Index of refraction	Minimum room temperature energy gap (eV)	Mobility (Room temp.) (cm²/V·s) Electrons	Holes	Optical transition	Remarks
I III VI₂ Compounds										
$CuAlS_2$	0.106					2.5				
$CuAlSe_2$						1.1				
$CuAlTe_2$						0.88				
$CuGaS_2$	0.106					2.38				
$CuGaSe_2$	0.141					0.96, 1.63				
$CuGaTe_2$	0.227					0.82, 1.0				
$CuInS_2$	0.141					1.2				
$CuInSe_2$	0.187					0.86, 0.92				
$CuInTe_2$	0.278					0.95				
$CuTlS_2$										
$CuTlSe_2$(L.T.)						1.07				
$CuFeS_2$						0.53				Chalcopyrite
$CuFeSe_2$						0.16				
$CuLaS_2$										
$AgAlS_2$										
$AgAlSe_2$						0.7				
$AgAlTe_2$						0.56				
$AgGaS_2$	0.150					1.66				
$AgGaSe_2$	0.182					1.1				
$AgGaTe_2$	0.280					1.9				
$AgInS_2$(L.T.)	0.185					1.18				
$AgInSe_2$	0.238					0.96, 0.52				
$AgInTe_2$	0.338									
$AgFeS_2$										

Substance	Heat of formation [kJ/mole (300K)]	Volume compressibility $(10^{-10}\ m^2/N)$	Static dielectric constant	Atomic magnetic susceptibility $(10^{-6}\ CGS)$	Index of refraction	Minimum room temperature energy gap (eV)	Mobility (Room temp.) $(cm^2/V \cdot s)$ Electrons	Holes	Optical transition	Remarks
II IV V_2 Compounds										
$ZnSiP_2$	312					2.3	1000			
$ZnGeP_2$	293					2.2				
$ZnSnP_2$	275					1.45				
$CdSiP_2$		0.103				2.2	1000			
$CdGeP_2$	289					1.8				
$CdSnP_2$	270					1.5				
$ZnSiAs_2$	290					1.7		50		
$ZnGeAs_2$	271			-14.4		0.85				
$ZnSnAs_2$	252			-18.4		0.65		300		Disorders at 910K
$CdSiAs_2$		0.143				1.6				
$CdGeAs_2$	266			-23.4		0.53	70	25		Disorders at 903
$CdSnAs_2$	247		13.7	-21.5		0.26	22000	250		

§A5. Other Ternary Semiconductors with Tetrahedral Coordination

Substance	Heat of formation [kJ/mole (300K)]	Volume compressibility $(10^{-10}\ m^2/N)$	Static dielectric constant	Atomic magnetic susceptibility $(10^{-6}\ CGS)$	Index of refraction	Minimum room temperature energy gap (eV)	Electrons	Holes	Optical transition	Remarks
I_2 IV VI_3 Compounds										
$Cu_2SiS_3(H.T.)$										Wurtzite
$Cu_2SiS_3(L.T.)$										Tetragonal
Cu_2SiTe_3										Cubic
$Cu_2GeS_3(H.T.)$				-18.7						Cubic
$Cu_2GeS_3(L.T.)$							360			Tetragonal
Cu_2GeSe_3	211.5			-21.3		0.94	238			Same
Cu_2GeTe_3	190.2			-23.4						Same
Cu_2SnS_3				-18.2		0.91	405			Cubic
$CuSnSe_3$				-21.0		0.66	870			Cubic
Cu_2SnTe_3				-28.4						Cubic
Ag_2GeSe_3				-29.6		0.91 (77K)				
Ag_2SnSe_3				-29.5		0.81				
Ag_2GeTe_3				-31.4		0.25				
Ag_2SnTe_3				-31.0		0.08				
I_3 V VI_4 Compounds										
Cu_3PS_4										Enargite
Cu_3AsS_4	269.6			-15.8		1.24				
Cu_3AsSe_4	161.3			-13.1		0.88				Famatinite
Cu_3SbS_4				-8.3		0.74				Famatinite
Cu_3SbSe_4	127.1			-20.5		0.31				
I IV_2 V_3 Compounds										
$CuSi_2P_3$										E1
$CuGe_2P_3$		0.12				0.9				E1
$AgGe_2P_3$										

§A6. "Defect Chalcopyrite" Structure Compounds (Strukturbericht symbol E3, Space Group I $\bar{4}$-S_4^2)

Substance	Heat of formation [kJ/mole (300K)]	Volume compressibility $(10^{-10}\ m^2/N)$	Static dielectric constant	Atomic magnetic susceptibility $(10^{-6}\ CGS)$	Index of refraction	Minimum room temperature energy gap (eV)	Electrons	Holes	Optical transition	Remarks
$ZnAl_2Se_4$										
$ZnAl_2Te_4(?)$										
$ZnGa_2S_4(?)$						≈ 3.4				
$ZnGa_2Se_4(?)$						≈ 2.2				
$ZnGa_2Te_4(?)$						1.35				
$ZnIn_2Se_4$	206					1.82	35			
$ZnIn_2Te_4$	198					1.2				
$CdAl_2S_4$										
$CdAl_2Se_4$										
$CdAl_2Te_4(?)$										
$CdGa_2S_4$	256					3.44	60			
$CdGa_2Se_4$	216					2.43	33			
$CdGa_2Te_4$										
$CdIn_2Te_4$	195					(1.26 or 0.9)	4000			
$HgAl_2S_4$										
$HgAl_2Se_4$										
$HgAl_2Te_4(?)$										
$HgGa_2S_4$	249					2.84				
$HgGa_2Se_4$	204					1.95	400			
$HgIn_2Se_4$	196					0.6	290			
$HgIn_2Te_4(?)$	188					0.86	200			

Table II

BASIC THERMODYNAMIC, ELECTRICAL, AND MAGNETIC PROPERTIES OF SEMICONDUCTORS (LISTED BY CRYSTAL STRUCTURE) (continued)

Substance	Heat of formation [kJ/mole (300K)]	Volume compressibility (10^{-10} m²/N)	Static dielectric constant	Atomic magnetic susceptibility (10^{-6} CGS)	Index of refraction	Minimum room temperature energy gap (eV)	Mobility (Room temp.) (cm²/V·s)		Optical transition	Remarks
							Electrons	Holes		
§A7. Other Adamantine Compounds										
αSiC			10.2	−6.4	2.67	2.86	400			6H structure
$Hg_5Ga_2Te_8$										B3 with superlattice
$Hg_5In_2Te_8$						0.7	2000			B3 with superlattice
$CdIn_2Se_4$						1.55				

PART B. OCTAHEDRAL SEMICONDUCTORS

§B1. HALITE STRUCTURE SEMICONDUCTORS (Strukturbericht symbol B1, Space Group Fm3m-O_h^5)

Substance	Heat of formation [kJ/mole (300K)]	Volume compressibility (10^{-10} m²/N)	Static dielectric constant	Atomic magnetic susceptibility (10^{-6} CGS)	Index of refraction	Minimum room temperature energy gap (eV)	Electrons	Holes	Optical transition	Remarks
GeTe										
SnSe										
SnTe										
PbS	435					0.5	600	600		
PbSe	393		161			0.37	1000	900		
PbTe	393		280			0.26	1600	600		Altaite
			360			0.25				
Selected other binary halites										
BiSe										
BiTe						0.4				
EuSe										
GdSe						1.8	4			
NiD						2.0 or 3.7	100			
CdO	531					2.5				
SrSW						4.1				

PART C. OTHER SEMICONDUCTORS

§C1. Antifluorite Structure Compounds (Fm3m - O_h^5)

Substance	Heat of formation [kJ/mole (300K)]	Volume compressibility (10^{-10} m²/N)	Static dielectric constant	Atomic magnetic susceptibility (10^{-6} CGS)	Index of refraction	Minimum room temperature energy gap (eV)	Electrons	Holes	Optical transition	Remarks
Mg_2Si	79.08					0.77	405	70		
Mg_2Ge						0.74	520	110		
Mg_2Sn	76.57					0.36	320	260		
Mg_2Pb	52.72					0.1				

§C2. Tetradymite Structure Compounds (R$\overline{3}$m − D_{3d}^5)

Substance						Minimum room temperature energy gap (eV)	Electrons	Holes	Optical transition	Remarks
Sb_2Te_3						0.3		360		
Bi_2Se_3						0.35	600			
Bi_2Te_3						0.21	1140	680		R3m (166)

§C3. Skutterudite Structure Compounds (Im3-T_h^5)

Substance						Minimum room temperature energy gap (eV)	Electrons	Holes	Optical transition	Remarks
CoP_3						0.43				
$CoAs_3$						0.69		~4000		
$CoSb_3$						0.63	70	~3000		
RhP_3								700		
$RhAs_3$						0.85		~3000		
$RhSb_3$						0.80		~7000		
$IrSb_3$						1.18		1500		

§C4. Selected Multinary Compounds

Substance						Minimum room temperature energy gap (eV)	Electrons	Holes	Optical transition	Remarks
$AgSbSe_2$						0.58				
$AgSbTe_2$ (or $Ag_{19}Sb_{29}Te_{52}$)						0.7, 0.27				
$AgBiS_2$(H.T.)										
$AgBiSe_2$(H.T.)										
$AgBiTe_2$(H.T.)										
Cu_2CdSnS_4						1.16	<2			

§C5. Some Elemental Semiconductors

Substance	Heat of formation [kJ/mole (300K)]	Volume compressibility (10^{-10} m²/N)	Static dielectric constant	Atomic magnetic susceptibility (10^{-6} CGS)	Index of refraction	Minimum room temperature energy gap (eV)	Electrons	Holes	Optical transition	Remarks
B	397.1			−6.7	3.4	1.55	10			
Se(gray)			6.6 (0.1 GHz)	−22.1	2.5	1.5		5		$P3_121(152)$
Te				−39.5	3.3	0.33	1700	1200		Same

Table III
SEMICONDUCTING PROPERTIES OF SELECTED MATERIALS

Substance	Minimum Energy Gap (eV) R.T.	0 K	dEg/dT x10⁴ eV/°C	dEg/dP x10⁶ eV·cm²/kg	Density of States Electron Effective Mass m_{d_a} (m_n)	Electron Mobility and Temperature Dependence μ_n cm²/V·s	−x	Density of States Hole Effective Mass m_{d_p} (m_n)	Hole Mobility and Temperature Dependence μ_p cm²/V·s	−x
Si	1.107	1.153	−2.3	−2.0	1.1	1,900	2.6	0.56	500	2.3
Ge	0.67	0.744	−3.7	+7.3	0.55	3,800	1.66	0.3	1,820	2.33
αSn	0.08	0.094	−0.5		0.02	2,500	1.65	0.3	2,400	2.0
Te	0.33				0.68	1,100		0.19	560	
III–V Compounds										
AlAs	2.2	2.3			0.09	1,200			420	
AlSb	1.6	1.7	−3.5	−1.6	0.09	200	1.5	0.4	500	1.8
GaP	2.24	2.40	−5.4	−1.7	0.35	300	1.5	0.5	150	1.5
GaAs	1.35	1.53	−5.0	+9.4	0.068	9,000	1.0	0.5	500	2.1
GaSb	0.67	0.78	−3.5	+12	0.050	5,000	2.0	0.23	1,400	0.9
InP	1.27	1.41	−4.6	+4.6	0.067	5,000	2.0		200	2.4
InAs	0.36	0.43	−2.8	+8	0.022	33,000	1.2	0.41	460	2.3
InSb	0.165	0.23	−2.8	+15	0.014	78,000	1.6	0.4	750	2.1
II–VI Compounds										
ZnO	3.2		−9.5	+0.6	0.38	180	1.5			
ZnS	3.54		−5.3	+5.7		180			5(400°C)	
ZnSe	2.58	2.80	−7.2	+6		540			28	
ZnTe	2.26			+6		340			100	
CdO	2.5 ± .1		−6		0.1	120				
CdS	2.42		−5	+3.3	0.165	400		0.8		
CdSe	1.74	1.85	−4.6		0.13	650	1.0	0.6		
CdTe	1.44	1.56	−4.1	+8	0.14	1,200		0.35	50	
HgSe	0.30				0.030	20,000	2.0			
HgTe	0.15		−1		0.017	25,000		0.5	350	
Halite Structure Compounds										
PbS	0.37	0.28	+4		0.16	800		0.1	1,000	2.2
PbSe	0.26	0.16	+4		0.3	1,500		0.34	1,500	2.2
PbTe	0.25	0.19	+4	−7	0.21	1,600		0.14	750	2.2
Others										
ZnSb	0.50	0.56			0.15	10				1.5
CdSb	0.45	0.57	−5.4		0.15	300			2,000	1.5
Bi₂S₃	1.3					200			1,100	
Bi₂Se₃	0.27					600			675	
Bi₂Te₃	0.13		−0.95		0.58	1,200	1.68	1.07	510	1.95
Mg₂Si		0.77	−6.4		0.46	400	2.5		70	
Mg₂Ge		0.74	−9			280	2		110	
Mg₂Sn	0.21	0.33	−3.5		0.37	320			260	
Mg₃Sb₂		0.32				20			82	
Zn₃As₂	0.93					10	1.1		10	
Cd₃As₂	0.55				0.046	100,000	0.88			
GaSe	2.05		3.8						20	
GaTe	1.66	1.80	−3.6			14	−5			
InSe	1.8					900				
TlSe	0.57		−3.9		0.3	30		0.6	20	1.5
CdSnAs₂	0.23				0.05	25,000	1.7			
Ga₂Te₃	1.1	1.55	−4.8							
α-In₂Te₃	1.1	1.2			0.7				50	1.1
β-In₂Te₃	1.0								5	
Hg₃In₂Te₆	0.5								11,000	
SnO₂									78	

Table IV
BAND PROPERTIES OF SEMICONDUCTORS

PART A. DATA ON VALENCE BANDS OF SEMICONDUCTORS (ROOM TEMPERATURES)

Substance	Band Curvature Effective Mass Heavy Holes	Light Holes	"Split-off" Band Holes	Energy Separation of "Split-off" Band (eV)	Measured (Light) Hole Mobility cm²/V·s
	(Expressed as fraction of free electron mass)				
Semiconductors with Valence Band Maximum at the Center of the Brillouin Zone ('Γ')					
Si	0.52	0.16	0.25	0.044	500
Ge	0.34	0.043	0.08	0.3	1,820
Sn	0.3				2,400
AlAs					
AlSb	0.4			0.7	550
GaP				0.13	100
GaAs	0 8	0.12	0.20	0.34	400
GaSb	0.23	0.06		0.7	1,400
InP				0.21	150
InAs	0.41	0.025	0.083	0.43	460
InSb	0.4	0.015		0.85	750
CdTe	0.35				50
HgTe	0.5				350

Table IV
BAND PROPERTIES OF SEMICONDUCTORS (continued)

Semiconductors with Multiple Valence Band Maxima

Substance	Number of Equivalent Valleys and Direction	Band Curvature Effective Masses		Anisotropy $K = m_L/m_T$	Measured (Light) Hole Mobility cm²/V·s
		Longitudinal m_L	Transverse m_T		
PbSe	4 "L" [111]	0.095	0.047	2.0	1,500
PbTe	4 "L" [111]	0.27	0.02	10	750
Bi_2Te_3	6	0.207	~0.045	4.5	515

PART B. DATA ON CONDUCTION BANDS OF SEMICONDUCTORS (Room Temperature Data)

Single Valley Semiconductors

Substance	Energy Gap (eV)	Effective Mass (m_o)	Mobility (cm²/V.s)	Comments
GaAs	1.35	0.067	8,500	3(or 6?) equivalent [100] valleys 0.36 eV above this maximum with a mobility of ~50
InP	1.27	0.067	5,000	3(or 6?) equivalent [100] valleys 0.4 eV above this minimum.
InAs	0.36	0.022	33,000	equivalent valleys ~1.0 eV above this minimum
InSb	0.165	0.014	78,000	
CdTe	1.44	0.11	1,000	4(or 8?) equivalent [111] valleys 0.51 eV above this minimum.

Multivalley Semiconductors

Substance	Energy Gap	Number of Equivalent Valleys and Direction	Band Curvature Effective Mass		Anisotropy $K = m_L/m_T$	Comments
			Longitudinal m_L	Transverse m_T		
Si	1.107	6 in [100] "Δ"	0.90	0.192	4.7	
Ge	0.67	4 in [111] at "L"	1.588	0.0815	19.5	
GaSb	0.67	as Ge (?)	~1.0	~0.2	~5	
PbSe	0.26	4 in [111] at "L"	0.085	0.05	1.7	
PbTe	0.25	4 in [111] at "L"	0.21	0.029	5.5	
Bi_2Te_3	0.13	6			~0.05	

Table V
RESISTIVITY OF SEMICONDUCTING MINERALS

Mineral	ρ (ohm · m)	Mineral	ρ (ohm · m)
Diamond (C)	2.7	Gersdorffite, NiAsS	1 to 160 × 10⁻⁶
Sulfides		Glaucodote, (Co, Fe)AsS	5 to 100 × 10⁻⁶
Argentite, Ag_2S	1.5 to 2.0 × 10⁻³	Antimonide	
Bismuthinite, Bi_2S_3	3 to 570	Dyscrasite, Ag_3Sb	0.12 to 1.2 × 10⁻⁶
Bornite, $Fe_2S_3 \cdot nCu_2S$	1.6 to 6000 × 10⁻⁶	Arsenides	
Chalcocite, Cu_2S	80 to 100 × 10⁻⁶	Allemonite, $SbAs_2$	70 to 60,000
Chalcopyrite, $Fe_2S_3 \cdot Cu_2S$	150 to 9000 × 10⁻⁶	Lollingite, $FeAs_2$	2 to 270 × 10⁻⁶
Covellite, CuS	0.30 to 83 × 10⁻⁶	Nicollite, NiAs	0.1 to 2 × 10⁻⁶
Galena, PbS	6.8 × 10⁻⁶ to 9.0 × 10⁻²	Skutterudite, $CoAs_3$	1 to 400 × 10⁻⁶
Haverite, MnS_2	10 to 20	Smaltite, $CoAs_2$	1 to 12 × 10⁻⁶
Marcasite, FeS_2	1 to 150 × 10⁻³	Tellurides	
Metacinnabarite, 4HgS	2 × 10⁻⁶ to 1 × 10⁻³	Altaite, PbTe	20 to 200 × 10⁻⁶
Millerite, NiS	2 to 4 × 10⁻⁷	Calavarite, $AuTe_2$	6 to 12 × 10⁻⁶
Molybdenite, MoS_2	0.12 to 7.5	Coloradoite, HgTe	4 to 100 × 10⁻⁶
Pentlandite, $(Fe, Ni)_9S_8$	1 to 11 × 10⁻⁶	Hessite, Ag_2Te	4 to 100 × 10⁻⁶
Pyrrhotite, Fe_7S_8	2 to 160 × 10⁻⁶	Nagyagite, $Pb_6Au(S, Te)_{14}$	20 to 80 × 10⁻⁶
Pyrite, FeS_2	1.2 to 600 × 10⁻³	Sylvanite, $AgAuTe_4$	4 to 20 × 10⁻⁶
Sphalerite, ZnS	2.7 × 10⁻³ to 1.2 × 10⁴	Oxides	
Antimony-sulfur compounds		Braunite, Mn_2O_3	0.16 to 1.0
Berthierite, $FeSb_2S_4$	0.0083 to 2.0	Cassiterite, SnO_2	4.5 × 10⁻⁴ to 10,000
Boulangerite, $Pb_5Sb_4S_{11}$	2 × 10⁵ to 4 × 10⁶	Cuprite, Cu_2O	10 to 50
Cylindrite, $Pb_3Sn_4Sb_2S_{14}$	2.5 to 60	Hollandite, $(Ba, Na, K)Mn_8O_{16}$	2 to 100 × 10⁻³
Franckeite, $Pb_5Sn_3Sb_2S_{14}$	1.2 to 4	Ilmenite, $FeTiO_3$	0.001 to 4
Hauchecornite, $Ni_9(Bi, Sb)_2S_8$	1 to 83 × 10⁻⁶	Magnetite, Fe_3O_4	52 × 10⁻⁶
Jamesonite, $Pb_4FeSb_6S_{14}$	0.020 to 0.15	Manganite, MnO·OH	0.018 to 0.5
Tetrahedrite, Cu_3SbS_3	0.30 to 30,000	Melaconite, CuO	6000
Arsenic-sulfur compounds		Psilomelane, $KMnO \cdot MnO_2 \cdot nH_2O$	0.04 to 6000
Arsenopyrite, FeAsS	20 to 300 × 10⁻⁶	Pyrolusite, MnO_2	0.007 to 30
Cobaltite, CoAsS	6.5 to 130 × 10⁻³	Rutile, TiO_2	29 to 910
Enargite, Cu_3AsS_4	0.2 to 40 × 10⁻³	Uraninite, UO	1.5 to 200

From Carmichael, R. S., ed., *Handbook of Physical Properties of Rocks*, Vol. I, CRC Press, 1982.

REFERENCES

1. Beer, A. C., *Galvanomagnetic Effects in Semiconductors*, Academic Press, 1963.
2. Goryunova, N. A., *The Chemistry of Diamond-Like Semiconductors*, The MIT Press, 1965.
3. Abrikosov, N. Kh., Bankina, V. F., Poretskaya, L. E., Shelimova, L. E., and Skudnova, E. V., *Semiconducting II-VI, IV-VI, and V-VI Compounds*, Plenum Press, 1969.
4. Berger, L. I. and Prochukhan, V. D., *Ternary Diamond-Like Semiconductors*, Cons. Bureau/Plenum Press, 1969.
5. Shay, J. L. and Wernick, J. H., *Ternary Chalcopyrite Semiconductors: Growth, Electronic Properties, and Applications*, Pergamon Press, 1975.
6. Bergman, R., *Thermal Conductivity in Solids*, Clarendon, Oxford, 1976.
7. Handbook of Semiconductors, Vol. 1, Moss, T. S. and Paul, W., Eds., *Band Theory and Transport Properties*; Vol. 2, Moss, T. S. and Balkanski, M., Eds., *Optical Properties of Solids*; Vol. 3, Moss, T. S. and Keller, S. P., Eds., *Materials Properties and Preparation*, North Holland Publ. Co., 1980.
8. Böer, K. W., *Survey of Semiconductor Physics*, Van Nostrand Reinhold, 1990.
9. Rowe, D. M., Ed., *CRC Handbook of Thermoelectrics*, CRC Press, Boca Raton, FL, 1995.
10. Berger, L. I., *Semiconductor Materials*, CRC Press, Boca Raton, FL, 1997.
11. Glazov, V. M., Chizhevskaya, S. N., and Glagoleva, N. N., *Liquid Semiconductors*, Plenum Press, New York, 1969.
12. Phillips, J. C., *Bonds and Bands in Semiconductors*, Academic Press, New York, 1973.
13. Harrison, W. A., *Electronic Structure and the Properties of Solids*, Freeman Publ. House, San Francisco, 1980.
14. Balkanski, M., Ed., *Optical Properties of Solids*, North-Holland, Amsterdam, 1980.
15. *Landolt-Börnstein. Numerical Data and Functional Relationships in Science and Technology, New Series, Group III: Crystal and Solid State Physics*, Hellwege, K.-H., and Madelung, O., Eds., Volumes 17 and 22, Springer Verlag, Berlin, 1984 (and further).
16. Shklovskii, B. L., and Efros, A. L., *Electronic Processes in Doped Semiconductors*, Springer Verlag, Berlin, 1984.
17. Cohen, M. L., and Chelikowsky, J. R., *Electronic Structure and Optical Properties of Semiconductors*, Springer Verlag, New York, 1988.
18. Glass, J. T., Messier, R. F., and Fujimori, N., Eds., *Diamond, Silicon Carbide, and Related Wide Bandgap Semiconductors*, MRS Symposia Proc. 1652, Mater. Res. Soc., Pittsburgh, 1990.
19. Palik, E., Ed., *Handbook of Optical Constants of Solids II*, Academic Press, New York, 1991.
20. Reed, M., Ed., *Semiconductors and Semimetals*, Volume 35, Academic Press, Boston, 1992.
21. Haug, H., and Koch, S. W., *Quantum Theory of the Optical and Electronic Properties of Semiconductors*, 2nd Edition, World Scientific, Singapore, 1993.
22. Lockwood, D. J., Ed., *Proc. 22nd Intl. Conf. on the Physics of Semiconductors, Vancouver, 1994*, World Scientific, Singapore, 1994.
23. Morelli, D. T., Caillat, T., Fleurial, J.-P., Borshchevsky, A., Vandersande, J., Chen, B., and Uher, C., *Phys. Rev.*, B51, 9622, 1995.
24. Caillat, T., Borshchevsky, A., and Fleurial, J.-P., *J. Appl. Phys.*, 80, 4442, 1996.
25. Fleurial, J.-P., Caillat, T., and Borshchevsky, A., *Proc. XVI Intl. Conf. Thermoelectrics*, Dresden, Germany, August 26–29, 1997 (in print).
26. Borshchevsky A. et al., U.S. Patents 5,610,366 (March 1997) and 5,831,286 (March 1998).

DIFFUSION DATA FOR SEMICONDUCTORS
B. L. Sharma

The diffusion coefficient D in many semiconductors may be expressed by an Arrhenius-type relation

$$D = D_0 \exp(-Q/kT)$$

where D_0 is a frequency factor, Q is the activation energy for diffusion, k is the Boltzmann constant, and T is the absolute temperature. This table lists D_0 and Q for various diffusants in common semiconductors.

Abbreviations used in the table are

AES — Auger Electron Spectroscopy
DLTS — Deep Level Transient Spectroscopy
SEM — Scanning Electron Microscopy
SIMS — Secondary Ion Mass Spectrometry
D(c) — Concentration Dependent Diffusion Coefficient
D_{max} — Maximum Diffusion Coefficient

(f) — Fast Diffusion Component
(i) — Interstitial Diffusion Component
(s) — Slow Diffusion Component
($\parallel$) — Parallel to c Direction
($\perp$) — Perpendicular to c Direction

Semiconductor	Diffusant	Frequency factor, D_0 (cm^2/s)	Activation energy, Q (eV)	Temperature range (°C)	Method of measurement	Ref.
Si	H	6×10^{-1}	1.03	120—1207	Electrical and SIMS	1
	Li	2.5×10^{-3}	0.65	25—1350	Electrical	2
	Na	1.65×10^{-3}	0.72	530—800	Electrical and flame photometry	3
	K	1.1×10^{-3}	0.76	740—800	Electrical and flame photometry	3
	Cu	4×10^{-2}	1.0	800—1100	Radioactive	4
		4.7×10^{-3}	0.43 (i)	300—700	Radioactive	5
	Ag	2×10^{-3}	1.6	1100—1350	Radioactive	6
	Au	2.4×10^{-4}	0.39 (i)	700—1300	Radioactive	7
		2.75×10^{-3}	2.05 (s)			
	Be	($D \sim 10^{-7}$)	—	1050	Electrical	8
	Ca	($D \sim 6 \times 10^{-14}$)	—	1100	Electrical and SIMS	1
	Zn	1×10^{-1}	1.4	980—1270	Electrical	9
	B	2.46	3.59	1100—1250	Electrical	10
		2.4×10^{1}	3.87	840—1250	Electrical	11
	Al	1.38	3.41	1119—1390	Electrical	12
		1.8	3.2	1025—1175	Electrical	13
	Ga	3.74×10^{-1}	3.39	1143—1393	Electrical	12
		6×10^{1}	3.89	900—1050	Radioactive	14
	In	7.85×10^{-1}	3.63	1180—1389	Electrical	12
		1.94×10^{1}	3.86	1150—1242	Radioactive	15
	Tl	1.37	3.7	1244—1338	Electrical	12
		1.65×10^{1}	3.9	1105—1360	Electrical	16
	Sc	8×10^{-2}	3.2	1100—1250	Radioactive	1
	Ce	($D \sim 3.9 \times 10^{-13}$)	—	1050	SIMS	1
	Pr	2.5×10^{-7}	1.74	1100—1280	Electrical	1
	Pm	7.5×10^{-9}	1.2 (s)	730—1270	Radioactive	1
		4.2×10^{-12}	0.13 (f)			
	Er	2×10^{-3}	2.9	1100—1250	Radioactive	1
	Tm	8×10^{-3}	3.0	1100—1280	Radioactive	1
	Yb	2.8×10^{-5}	0.95	947—1097	Neutron activation	1
	Ti	1.45×10^{-2}	1.79	950—1200	DLTS	17
	C	3.3×10^{-1}	2.92	1070—1400	Radioactive	18
	Si (self)	1.54×10^{2}	4.65	855—1175	SIMS	19
		1.6×10^{3}	4.77	1200—1400	Radioactive	20
	Ge	3.5×10^{-1}	3.92	855—1000	Radioactive	21
		2.5×10^{3}	4.97	1030—1302	Radioactive	21
		7.55×10^{3}	5.08	1100—1300	SIMS	22
	Sn	3.2×10^{1}	4.25	1050—1294	Neutron activation	23

Semiconductor	Diffusant	Frequency factor, D_o (cm²/s)	Activation energy, Q (eV)	Temperature range (°C)	Method of measurement	Ref.
	N	2.7×10^{-3}	2.8	800—1200	Out Diffusion; SIMS	1
	P	2.02×10^1	3.87	1100—1250	Electrical	10
		1.1	3.4	900—1200	Radioactive	24
		7.4×10^{-2}	3.3	1130—1405	Electrical	25
	As	6.0×10^1	4.2	950—1350	Radioactive	26
		6.55×10^{-2}	3.44	1167—1394	Electrical	27
		2.29×10^1	4.1	900—1250	Electrical	28
	Sb	1.29×10^1	3.98	1190—1398	Radioactive	29
		2.14×10^{-1}	3.65	1190—1405	Electrical	27
	Bi	1.03×10^3	4.64	1220—1380	Electrical	16
		1.08	3.85	1190—1394	Electrical	27
	Cr	1×10^{-2}	1	1100—1250	Radioactive	30
	Mo	$(D \sim 2 \times 10^{-10})$	—	1000	DLTS	1
	W	$(D \sim 10^{-12})$	—	1100	DLTS	1
	O	7×10^{-2}	2.44	700—1250	SIMS	31
		1.4×10^{-1}	2.53	700—1160	SIMS	32
	S	5.95×10^{-3}	1.83	975—1200	Radioactive	33
	Se	9.5×10^{-1}	2.6	1050—1250	Electrical	34
	Te	5×10^{-1}	3.34	900—1250	SIMS	1
	Mn	6.9×10^{-4}	0.63	900—1200	Radioactive	35
	Fe	1.3×10^{-3}	0.68	30—1250	Radioactive	36
	Co	2×10^{-3}	0.69	700—1300	Radioactive	37
	Ni	2×10^{-3}	0.47	800—1300	Radioactive	38
	Ru	$(D \sim 5 \times 10^{-7}$ — $5 \times 10^{-6})$	—	1000—1280	Electrical	1
	Rh	$(D \sim 10^{-6}$—$10^{-4})$	—	1000—1200	Electrical	39
	Pd	2.95×10^{-4}	0.22 (i)	702—1320	Nuclear Activation	1
	Pt	1.5×10^2	2.22	800—1000	Electrical	1
	Os	$(D \sim 2 \times 10^{-6})$	—	1280	Electrical	40
	Ir	4.2×10^{-2}	1.3	950—1250	Electrical	41
Ge	Li	1.3×10^{-3}	0.46	350—800	Electrical	42
		9.1×10^{-3}	0.57	800—500	Electrical	43
	Na	3.95×10^{-1}	2.03	700—850	Radioactive	44
	Cu	1.9×10^{-4}	0.18 (i)	750—900	Radioactive	45
		4×10^{-2}	0.99 (s)	600—700		
		4×10^{-3}	0.33 (i)	350—750	Radioactive	5
	Ag	4.4×10^{-2}	1.0 (i)	700—900	Radioactive	46, 47
		4×10^{-2}	2.23 (s)	800—900	Radioactive	48
	Au	2.25×10^2	2.5	600—900	Radioactive	49
	Be	5×10^{-1}	2.5	720—900	Electrical	50
	Mg	$(D \sim 8 \times 10^{-9})$	—	900	Electrical	1
	Zn	5	2.7	600—900	Radioactive and electrical	51
	Cd	1.75×10^9	4.4	760—915	Radioactive	52
	B	1.8×10^9	4.55	600—900	Electrical	51
	Al	1.0×10^3	3.45	554—905	SIMS	53
		$\sim 1.6 \times 10^2$	~3.24	750—850	Electrical	54
	Ga	1.4×10^2	3.35	554—916	SIMS	55
		3.4×10^1	3.1	600—900	Electrical	51
	In	1.8×10^4	3.67	554—919	SIMS	56
		3.3×10^1	3.02	700—855	Radioactive	57
	Tl	1.7×10^3	3.4	800—930	Radioactive	58
	Si	2.4×10^{-1}	2.9	650—900	(γ) resonance	59
	Ge (self)	2.48×10^1	3.14	549—891	Radioactive	60
		7.8	2.95	766—928	Radioactive	61
	Sn	1.7×10^{-2}	1.9	—	Radioactive	45
	P	3.3	2.5	600—900	Electrical	51

Semiconductor	Diffusant	Frequency factor, D_0 (cm²/s)	Activation energy, Q (eV)	Temperature range (°C)	Method of measurement	Ref.
	As	2.1	2.39	700—900	Electrical	62
	Sb	3.2	2.41	700—855	Radioactive	57
		1.0×10^1	2.5	600—900	Radioactive and electrical	51
	Bi	3.3	2.57	650—850	—	63
	O	4×10^{-1}	2.08	—	Optical	64
	S	$(D \sim 10^{-9})$	—	920	—	65
	Se	$(D \sim 10^{-10})$	—	920	—	65
	Te	5.6	2.43	750—900	Radioactive	66
	Fe	1.3×10^{-1}	1.08	750—900	Radioactive	67
	Co	1.6×10^{-1}	1.12	750—850	Radioactive	47
	Ni	8×10^{-1}	0.9	670—900	Electrical	68
GaAs	Li	5.3×10^{-1}	1.0	250—500	Electrical and chemical	69
	Cu	3×10^{-2}	0.53	100—500	Radioactive	69
		6×10^{-2}	0.98	450—750	Ultrasonic	69
		1.5×10^{-3}	0.6	800—1000	Radioactive	69
	Ag	4×10^{-4}	0.8	500—1150	Radioactive	69
	Au	1×10^{-3}	1.0	740—1025	Radioactive	69
	Be	7.3×10^{-6}	1.2	800—990	Electrical	69
	Mg	4×10^{-5}	1.22	800—1200	Electrical	69
	Zn	1.5×10^1	2.49	600—980	Radioactive	69
		2.5×10^{-1}	3.0	750—1000	Radioactive	69
	Cd	1.3×10^{-3}	2.2	800—1100	Radioactive	69
		5×10^{-2}	2.43	868—1149	Radioactive	69
	Hg	$(D \sim 5 \times 10^{-14})$	—	1100	Radioactive	69
	Al	$(D \sim 4 \times 10^{-18}$— $10^{-14})$	4.3	850—1100	AES	70
	Ga (self)	4×10^{-5}	2.6	1025—1100	Radioactive	69
		1×10^7	5.6	1125—1230	Radioactive	69
	In	$(D \sim 7 \times 10^{-11})$	—	1000	Radioactive	69
	C	$(D \sim 1.04 \times 10^{-16})$	—	825	SIMS	69
	Si	1.1×10^{-1}	2.5	850—1050	SIMS	69
	Ge	1.6×10^{-5}	2.06	650—850	SIMS	69
	Sn	6×10^{-4}	2.5	1060—1200	Radioactive	69
		1×10^{-5}	2	800—1000	Radioactive	69
	P	$(D \sim 10^{-12}$—$10^{-10})$	2.9	800—1150	Reflectance measurements	69
	As (self)	7×10^{-1}	3.2	—	Radioactive	69
	Cr	2.04×10^{-6}	0.83 (f)	750—1000	SIMS	69
			1.7 (s)	700—900		
		7.9×10^{-3}	2.2	800—1100	Chemical analysis	69
	O	2×10^{-3}	1.1	700—900	Mass spectroscopy	69
	S	1.85×10^{-2}	2.6	1000—1300	Radioactive	69
		1.1×10^1	2.95	750—900	Electrical	69
	Se	3×10^3	4.16	1025—1200	Radioactive	69
	Te	1.5×10^{-1}	3.5	1000—1150	Radioactive	69
	Mn	6.5×10^{-1}	2.49	850—1100	Radioactive	69
	Fe	4.2×10^{-2}	1.8	850—1150	Radioactive	69
		2.2×10^{-3}	2.32	750—1050	Radioactive	69
	Co	5×10^2	2.5	800—1000	Radioactive	69
		1.2×10^{-1}	2.64	750—1050	Radioactive	69
	Tm	2.3×10^{-16}	1.0	800—1000	Radioactive	69
GaSb	Li	2.3×10^{-4}	1.9 (s)	527—657	Electrical and flame photometry	69
		1.2×10^{-1}	0.7 (f)	277—657		
	Cu	4.7×10^{-3}	0.9	470—650	Radioactive	69

Semiconductor	Diffusant	Frequency factor, D_o (cm²/s)	Activation energy, Q (eV)	Temperature range (°C)	Method of measurement	Ref.
	Zn	$(D \sim 2 \times 10^{-13}$— $1 \times 10^{-11})$	2	510—600	Radioactive	69
	Cd	1.5×10^{-6}	0.72	640—800	Electrical	69
	Ga (self)	3.2×10^3	3.15	658—700	Radioactive	69
	In	1.2×10^{-7}	0.53	320—650	Radioactive	69
	Sn	2.4×10^{-5}	0.8	320—650	Radioactive	69
		1.3×10^{-5}	1.1	500—650	Radioactive	69
	Sb (self)	3.4×10^4	3.45	658—700	Radioactive	69
	Se	$(D \sim 2.4 \times 10^{-13}$— $1.37 \times 10^{-11})$	—	400—500	Radioactive	69
	Te	3.8×10^{-4}	1.20	320—650	Radioactive	69
	Fe	5×10^{-2}	1.9 (I)	500—650	Radioactive	69
		5×10^2	2.3 (II)	500—650		
GaP	Ag	—	—	1000—1300	Radioactive	69
	Au	8	2.5 (I)	1050—1250	Radioactive	69
		20	2.4 (II)	1100—1250	Diffusion (I) A face and (II) B face	
	Be	$(D_{max} \sim 2.4 \times 10^{-9}$— $8.5 \times 10^{-8})$	—	900—1000	Atomic absorption analysis	69
	Mg	5×10^{-5}	1.4	700—1050	Electrical	69
	Zn	1.0	2.1	700—1300	Radioactive	69
	Ge	—	—	900—1000	Radioactive	69
	Cr	6.2×10^{-4}	1.2	900—1130	Radioactive; ESR	69
	S	3.2×10^3	4.7	1120—1305	Radioactive	69
	Mn	2.1×10^9	4.7	$T < 950$	Radioactive; ESR	69
		1.1×10^{-6}	0.9	950—1130		
	Fe	1.6×10^{-1}	2.3	980—1180	Radioactive	69
	Co	2.8×10^{-3}	2.9	850—1100	Radioactive	69
InP	Cu	3.8×10^{-3}	0.69	600—900	Radioactive	69
	Ag	3.6×10^{-4}	0.59	500—900	Radioactive	69
	Au	1.32×10^{-5}	0.48	600—820	Radioactive	69
		1.37×10^{-4}	0.73	600—900	Radioactive	69
	Zn	1.6×10^{-8}	0.3	750—900	Electrical	69
		$(D \sim 2 \times 10^{-9}$— $4 \times 10^{-8})$	—	700—900	Radioactive	69
	Cd	1.8	1.9	700—900	Radioactive	69
		1.1×10^{-7}	0.72	700—900	Electrical	69
		$(D \sim 7 \times 10^{-13}$— $2 \times 10^{-10})$	—	450—650	Electrical	69
	In (self)	1×10^5	3.85	830—990	Radioactive	69
	Sn	$(D \sim 3 \times 10^{-8})$	—	550	Etching and cathodo- luminescence	69
	P (self)	7×10^{10}	5.65	900—1000	Radioactive	69
	Cr	—	—	600—900	Radioactive	69
	S	3.6×10^{-4}	1.94	585—708	Electrical	69
	Se	$(D \sim 2 \times 10^{-8})$	—	550	Cathodoluminescence	69
	Mn	—	2.9	650—750	SIMS	69
	Fe	3	2	600—950	Radioactive	69
		6.8×10^5	3.4	600—700	SIMS	69
	Co	9×10^{-1}	1.8	600—950	Radioactive	69
InAs	Cu	3.6×10^{-3}	0.52	342—875	Radioactive	69
		2.2×10^{-2}	0.54	525—890	Radioactive	69
	Ag	7.3×10^{-4}	0.26	450—900	Radioactive	69
	Au	5.8×10^{-3}	0.65	600—900	Radioactive	69
	Mg	1.98×10^{-6}	1.17	600—900	Electrical	69
	Zn	4.2×10^{-3}	0.96	600—900	Radioactive	69
		3.11×10^{-3}	1.17	600—900	Electrical	69

Semiconductor	Diffusant	Frequency factor, D_0 (cm²/s)	Activation energy, Q (eV)	Temperature range (°C)	Method of measurement	Ref.
	Cd	7.4×10^{-4}	1.15	650—900	Radioactive	69
	Hg	1.45×10^{-5}	1.32	650—850	Radioactive	69
	In (self)	6×10^5	4.0	740—900	Radioactive	69
	Ge	3.74×10^{-6}	1.17	600—900	Electrical	69
	Sn	1.49×10^{-6}	1.17	600—900	Electrical	69
	As (self)	3×10^7	4.45	740—900	Radioactive	69
	S	6.78	2.2	600—900	Electrical	69
	Se	12.6	2.2	600—900	Electrical	69
	Te	3.43×10^{-5}	1.28	600—900	Electrical	69
InSb	Li	7×10^{-4}	0.28	0—210	Electrical	69
	Cu	9×10^{-4}	1.08	200—500	Radioactive	69
		3×10^{-5}	0.37	230—490	Radioactive	69
	Ag	1×10^{-7}	0.25	440—510	Radioactive	69
	Au	7×10^{-4}	0.32	140—510	Radioactive	69
	Zn	5×10^{-1}	1.35	362—508	Radioactive	69
		—	1.5	355—455	SIMS	69
	Cd	1×10^{-5}	1.1	250—500	Radioactive	69
		1.3×10^{-4}	1.2	360—500	Electrical	69
	Hg	4×10^{-6}	1.17	425—500	Radioactive	69
	In (self)	6×10^{-7}	1.45	400—500	Radioactive	69
		1.8×10^{13}	4.3	475—517	Radioactive	69
	Sn	5.5×10^{-8}	0.75	390—512	Radioactive	69
	Pb	$(D \sim 2.7 \times 10^{-15})$	—	500	Radioactive	71
	Sb (self)	5.35×10^{-4}	1.91	400—500	Radioactive	69
		3.1×10^{13}	4.3	475—517	Radioactive	69
	S	9×10^{-2}	1.4	360—500	Electrical	69
	Se	1.6	1.87	380—500	Electrical	69
	Te	1.7×10^{-7}	0.57	300—500	Radioactive	69
	Fe	1×10^{-7}	0.25	440—510	Radioactive	69
	Co	2.7×10^{-11}	0.39	420—500	Radioactive	69
AlAs	Ga	$(D \sim 2 \times 10^{-18} - 10^{-15})$	3.6	850—1100	AES	70
	Zn	$(D \sim 9 \times 10^{-11})$	—	557	SEM	69
AlSb	Cu	3.5×10^{-3}	0.36	150—500	Radioactive	69
	Zn	3.3×10^{-1}	1.93	660—860	Radioactive	69
	Cd	$D(c) \sim 4 \times 10^{-12} - 3 \times 10^{-10}$	—	900	Radioactive	69
	Al (self)	2	1.88	570—620	X-ray	69
	Sb (self)	1	1.7	570—620	X-ray	69
ZnS	Cu	2.6×10^{-3}	0.79	470—750	Radioactive	69
		4.3×10^{-4}	0.64	250—1200	Electroluminescence	69
		9.75×10^{-3}	1.04	400—800	Luminescence	69
	Au	1.75×10^{-4}	1.16	500—800	Radioactive	69
	Zn (self)	3×10^{-4}	1.5	925<T<940	Radioactive	69
		1.5×10^4	3.26	940<T<1030		
		1×10^{16}	6.5	1030<T<1075		
	Cd	$(D \sim 10^{-10})$	—	1100	Luminescence	72
	Al	5.69×10^{-4}	1.28	800—1000	Luminescence	69
	In	3×10^1	2.2	750—1000	Radioactive	69
	S (self)	2.16×10^4	3.15	600—800	Radioactive	69
		8×10^{-5}	2.2	740—1100	Radioactive	69
	Se	$(D \sim 5 \times 10^{-13})$	—	1070	X-ray microprobe	69
	Mn	2.3×10^3	2.46	500—800	Radioactive	69
ZnSe	Li	2.66×10^{-6}	0.49	950—980	Electrical	69
	Cu	1×10^{-4}	0.66	400—800	Luminescence	69
		1.7×10^{-5}	0.56	200—570	Radioactive	69
	Ag	2.2×10^{-2}	1.18	400—800	Luminescence	69

Semiconductor	Diffusant	Frequency factor, D_o (cm²/s)	Activation energy, Q (eV)	Temperature range (°C)	Method of measurement	Ref.
	Zn (self)	9.8	3.0	760—1150	Radioactive	69
	Cd	6.39×10^{-4}	1.87	700—950	Photoluminescence	69
	Al	2.3×10^{-2}	1.8	800—1100	Luminescence	69
	Ga	1.81×10^2	3.0	900—1100	Luminescence	69
		—	1.3	700—850	Electron probe	69
	In	$(D \sim 2 \times 10^{-12})$	—	940	—	69
	S	$(D \sim 8 \times 10^{-12})$	—	1060	X-ray microprobe	69
	Se (self)	1.3×10^1	2.5	860—1020	Radioactive	69
		2.3×10^{-1}	2.7	1000—1050	Radioactive	69
	Ni	$(D \sim 1.5 \times 10^{-8}$—				
		$1.7 \times 10^{-7})$	—	740—910	Luminescence	69
ZnTe	Li	2.9×10^{-2}	1.22 (s)	400—700	Nuclear and chemical analysis	69
		1.7×10^{-4}	0.78 (f)			
	Zn (self)	2.34	2.56	760—860	Radioactive	69
		1.4×10^1	2.69	667—1077	Radioactive	69
	Al	—	2.0	700—1000	Electrical and optical	69
	In	4	1.96	1100—1300	Radioactive	69
	Te (self)	2×10^4	3.8	727—977	Radioactive	69
CdS	Li	3×10^{-6}	0.68	610—960	Microhardness	69
	Na	$(D \sim 3 \times 10^{-7})$	—	800	Radioactive	69
	Cu	1.5×10^{-3}	0.76	400—700	Radioactive	69
		1.2×10^{-2}	1.05	300—700	Ultrasonic	69
		8×10^{-5}	0.72	20—200	Electrical	69
	Ag	2.5×10^1	1.2 (s)	300—500	Radioactive	69
		2.4×10^{-1}	0.8 (f)			
	Au	2×10^2	1.8	500—800	Radioactive	69
	Zn	1.27×10^{-9}	0.86 (s)	720—1000	Radioactive	69
		1.22×10^{-8}	0.66 (f)			
	Cd (self)	3.4	2.0	700—1100	Radioactive	69
	Ga	—	—	667—967	Optical and microprobe	69
	In	6×10^1	2.3 (∥)	650—930	Radioactive, optical and microprobe	69
		1×10^1	2.03 (⊥)			
	P	6.5×10^{-4}	1.6	800—1100	Radioactive	69
	S (self)	1.6×10^{-2}	2.05	800—900	Radioactive	69
		—	2.4	750—1050	Radioactive	69
	Se	$(D \sim 1.2 \times 10^{-9})$	—	900	Radioactive	69
	Te	1.3×10^{-7}	10.4	700—1000	Radioactive	69
	Cl	$(D \sim 3 \times 10^{-10})$	—	800	Electrical	69
	I	$(D \sim 5 \times 10^{-12})$	—	1000	Radioactive	69
	Ni	6.75×10^{-3}	10.9	570—900	Luminescence	69
	Yb	$(D \sim 1.3 \times 10^{-9})$	—	960	Photoluminescence	69
CdSe	Ag	2×10^{-4}	0.53	22—400	Ultrasonic	69
	Cd (self)	1.6×10^{-3}	1.5	700—1000	Radioactive	69
		6.3×10^{-2}	1.25 (I)	600—900	Radioactive;	69
		4.12×10^{-2}	2.18 (II)	600—900	(I) saturated Cd and (II) saturated Se pressure	
	P	$(D \sim 5.3 \times 10^{-12}$—				
		$6 \times 10^{-11})$	—	900—1000	Radioactive	69
	Se (self)	2.6×10^3	1.55	700—1000	Radioactive; saturated Se pressure	69
CdTe	Li	$(D \sim 1.5 \times 10^{-10})$	—	300	Ion microprobe	69
	Cu	3.7×10^{-4}	0.67	97—300	Radioactive	69
		8.2×10^{-8}	0.64	290—350	Ion backscattering	69
	Ag	—	—	700—800	Electrical and photo-	

Semiconductor	Diffusant	Frequency factor, D_o (cm^2/s)	Activation energy, Q (eV)	Temperature range (°C)	Method of measurement	Ref.
					luminescence	69
	Au	6.7×10^1	2.0	600—1000	Radioactive	69
	Cd (self)	1.26	2.07	700—1000	Radioactive	69
		3.26×10^2	2.67 (I)	650—900	Radioactive;	69
		1.58×10^1	2.44 (II)		(I) saturated Cd and (II) saturated Te pressure	
	In	8×10^{-2}	1.61	650—1000	Radioactive	69
		$1.17 \times^2$	2.21 (I)	500—850	Radioactive; (I) saturated Cd and (II) saturated Te pressure	
		6.48×10^{-4}	1.15 (II)			69
	Sn	8.3×10^{-2}	2.2	700—925	Radioactive	69
	P	($D \sim 1.2 \times 10^{-10}$)	—	900	Radioactive	69
	As	—	—	850	—	69
	O	5.6×10^{-9}	1.22	200—650	Mass spectrometry	69
		6.0×10^{-10}	0.29	650—900		
	Se	1.7×10^{-4}	1.35	700—1000	Radioactive	69
	Te (self)	8.54×10^{-7}	1.42 (I)	600—900	Radioactive; (I) saturated Cd and (II) saturated Te pressure	69
		1.66×10^{-4}	1.38 (II)	500—800		
	Cl	7.1×10^{-2}	1.6	520—800	Radioactive	69
	Fe	($D \sim 4 \times 10^{-8}$)	0.77	900	Radioactive	69
HgSe	Sb	6.3×10^{-5}	0.85	540—630	Radioactive	69
	Se (self)	—	—	200—400	Radioactive	69
HgTe	Ag	6×10^{-4}	0.8	250—350	Radioactive	69
	Zn	5×10^{-8}	0.6	250—350	Radioactive	69
	Cd	3.1×10^{-4}	0.66	250—350	Radioactive	69
	Hg (self)	2×10^{-8}	0.6	200—350	Radioactive	69
	In	6×10^{-6}	0.9	200—300	Radioactive	69
	Sn	1.72×10^{-6}	0.66 (s)	200—300	Radioactive	69
		1.8×10^{-3}	0.80 (f)			
	Te (self)	10^{-6}	1.4	200—400	Radioactive	69
	Mn	1.5×10^{-4}	1.3	250—350	Radioactive	69
PbS	Cu	4.6×10^{-4}	0.36	150—450	Electrical	69
		5×10^{-3}	0.31	100—400	Electrical	69
	Pb (self)	8.6×10^{-5}	1.52	500—800	Radioactive	69
	S (self)	6.8×10^{-5}	1.38	500—750	Radioactive	69
	Ni	1.78×10^1	0.95	200—500	Electrical	69
PbSe	Na	1.5×10^1	1.74 (s)	400—850	Radioactive	69
		5.6×10^{-6}	0.4 (f)			
	Cu	2×10^{-5}	0.31	93—520	Radioactive	69
	Ag	7.4×10^{-4}	0.35	400—850	Radioactive	69
	Pb (self)	4.98×10^{-6}	0.83	400—800	Radioactive	69
	Sb	3.4×10^{-1}	2.0	650—850	Radioactive	69
	Se (self)	2.1×10^{-5}	1.2	650—850	Radioactive	69
	Cl	1.6×10^{-8}	0.45	400—850	Radioactive	69
	Ni	($D \sim 1 \times 10^{-10}$)	—	700	Radioactive	69
PbTe	Na	1.7×10^{-1}	1.91	600—850	Radioactive	69
	Sn	3.1×10^{-2}	1.56	500—800	Radioactive	69
	Pb (self)	2.9×10^{-5}	0.6	250—500	Radioactive	69
	Sb	4.9×10^{-2}	1.54	500—800	Radioactive	69
	Te	2.7×10^{-6}	0.75	500—800	Radioactive	69
	Cl	($D > 2.3 \times 10^{-10}$)	—	700	Radioactive	69
	Ni	($D > 1 \times 10^{-6}$)	—	700	Radioactive	69

REFERENCES

1. N. A. Stolwijk and H. Bracht, in *Diffusion in Semiconductors and Non-Metallic Solids,* D. L. Beke, Ed., Springer-Verlag, Berlin, 1998, 2-1.
2. E. M. Pell, *Phys. Rev.,* 119, 1960; 119, 1014, 1960.
3. L. Svob, *Solid State Electron,* 10, 991, 1967.
4. B. I. Boltaks and I. I. Sosinov, *Zh. Tekh. Fiz.,* 28, 3, 1958.
5. R. N. Hall and J. N. Racette, *J. Appl. Phys.,* 35, 379, 1964.
6. B. I. Boltaks and Hsueh Shih-Yin, *Sov. Phys. Solid State,* 2, 2383, 1961.
7. W. R. Wilcox and T. J. LaChapelle, *J. Appl. Phys.,* 35, 240, 1964.
8. E. A. Taft and R. O. Carlson, *J. Electrochem. Soc.,* 117, 711, 1970.
9. R. Sh. Malkovich and N. A. Alimbarashvili, *Sov. Phys. Solid State,* 4, 1725, 1963.
10. R. N. Ghoshtagore, *Solid State Electron,* 15, 1113, 1972.
11. C. Hill, *Semiconductor Silicon 1981,* H. R. Huff, R. J. Kreiger, and Y. Takeishi, Eds., p. 988, *Electrochem. Soc.,* 1981.
12. R. N. Ghoshtagore, *Phys. Rev. B,* 3, 2507, 1971.
13. W. Rosnowski, *J. Electrochem. Soc.,* 125, 957, 1978.
14. J. S. Makris and B. J. Masters, *J. Appl. Phys.,* 42, 3750, 1971.
15. M. F. Millea, *J. Phys. Chem. Solids,* 27, 315, 1965 (refer Reference 2).
16. C. S. Fuller and J. A. Ditzenberger, *J. Appl. Phys.,* 27, 544, 1956.
17. S. Hocine and D. Mathiot, *Appl. Phys. Lett.,* 53, 1269, 1988.
18. R. C. Newman and J. Wakefield, *J. Phys. Chem. Solids,* 19, 230, 1961.
19. L. Kalinowski and R. Seguin, *Appl. Phys. Lett.,* 35, 211, 1979; *Appl. Phys. Lett.,* 36, 171, 1980.
20. R. F. Peart, *Phys. Stat. Sol.,* 15, K 119, 1966.
21. G. Hettich, H. Mehrer and K. Maler, *Inst. Phys. Conf. Ser.,* 46, 500, 1979.
22. M. Ogina, Y. Oana and M. Watanabe, *Phys. Stat. Sol. (a),* 72, 535, 1982.
23. T. H. Yeh, S. M. Hu, and R. H. Kastl, *Appl. Phys.,* 39, 4266, 1968.
24. I. Franz and W. Langheinrich, *Solid State Electron,* 14, 835, 1971.
25. R. N. Ghoshtagore, *Hys. Rev. B,* 3, 389, 1971.
26. B. J. Masters and J. M. Fairfield, *J. Appl. Phys.,* 40, 2390, 1969.
27. R. N. Goshtagore, *Phys. Rev. B,* 3, 397, 1971.
28. R. S. Fair and J. C. C. Tsai, *J. Electrochem. Soc.,* 122, 1689, 1975.
29. J. J. Rohan, N. E. Pickering and J. Kennedy, *J. Electrochem. Soc.,* 106, 705, 1969.
30. W. Wuerker, K. Roy, and J. Hesse, *Matsr. Res. Bull.,* 9, 971, 1974.
31. J. C. Mikkelsen, Jr., *Appl. Phys. Lett.,* 40, 336, 1982.
32. S. Tang Lee and D. Nicols, *Appl. Phys. Lett.,* 47, 1001, 1985.
33. P. L. Gruzin, S. V. Zemskii, A. D. Bullkin, and N. M. Makarov, *Sov. Phys. Sem.,* 7, 1241, 1974.
34. N. S. Zhdanovich and Yu. I. Kozlov, *Svoistva Legir, Poluprovodn.,* V. S. Zemskov, Ed., Nauka, Moscow, 1977, 115-120; *Fiz Tekh. Poluprovod.,* 9, 1594, 1975.
35. D. Gilles, W. Bergholze, and W. Schroeter, *J. Appl. Phys.,* 59, 3590, 1986.
36. E. R. Weber, *Appl. Phys. A,* 30, 1, 1983.
37. E. R. Weber, Properties of Silicon, EMIS Datareviews Ser. No. 4, INSPEC Publications, 1988, 409-451.
38. M. K. Bakhadyrkhanov, S. Zainabidinov, and A. Khamidov, *Sov. Phys. Sem.,* 14, 243, 1980.
39. S. A. Azimov, M. S. Yunosov, F. K. Khatamkulov, and G. Nasyrov, *Poluprovod.,* N. Kh. Abrikosov and V. S. Zemskov, Eds., Nauka, Moscow, 1975, 21-23.
40. S. A. Azimov, M. S. Yunosov, G. Nurkuziev, and F. R. Karimov, *Sov. Phys. Sem.,* 12, 981, 1978.
41. S. A. Azimov, B. V. Umarov, and M. S. Yunusov, *Sov. Phys. Sem.,* 10, 842, 1976.
42. C. S. Fuller and J. A. Ditzenberger, *Phys. Rev.,* 91, 193, 1953.
43. B. Pratt and F. Friedman, *J. Appl. Phys.,* 37, 1893, 1966.
44. M. Stojic, V. Spiric, and D. Kostoski, *Inst. Phys. Conf. Ser.,* 31, 304, 1976.
45. B. I. Boltaks, *Diffusion in Semiconductors,* Inforsearch, London, 1963, 162.
46. A. A. Bugai, V. E. Kosenko, and E. G. Miselyuk, *Zh. Tekh. Fiz.,* 27, 67, 1957.
47. L. Y. Wei, *J. Phys. Chem. Solids,* 18, 162, 1961.
48. V. E. Kosenko, *Sov. Phys. Solid State,* 4, 42, 1962.
49. W. C. Dunlap, Jr., *Phys. Rev.,* 97, 614, 1955
50. Yu. I. Belyaev and V. A. Zhidkov, *Sov. Phys. Solid State,* 3, 133, 1961.
51. W. C. Dunlap, Jr. *Phys. Rev.,* 94, 1531, 1954.
52. V. E. Kosenko, *Sov. Phys. Solid State,* 1, 1481, 1960.
53. P. Dorner, W. Gust, A. Lodding, H. Odelius, B. Predel, and U. Roll, *Acta Metall.,* 30, 941, 1982.
54. W. Meer and D. Pommerrening, *Z. Agnew. Phys.,* 23, 369, 1967.
55. U. Sodervall, H. Odelius, A. Lodding, U. Roll, B. Predel, W. Gust, and P. Dorner, *Phil. Mag. A,* 54, 539, 1986.
56. P. Dorner, W. Gust, A. Lodding, H. Odelius, B. Predel, and U. Roll, *Z. Metalkd.,* 73, 325, 1982.

57. P. V. Pavlov, *Sov. Phys. Solid State,* 8, 2377, 1967.
58. V. I. Tagirov and A. A. Kuliev, *Sov. Phys. Solid State,* 4, 196, 1962.
59. J. Raisanen, J. Hirvonen, and A. Anttila, *Solid State Electron.,* 24, 333, 1981.
60. C. Vogel, G. Hettich, and H. Mehrer, *J. Phys. C.,* 16, 6197, 1983.
61. H. Letaw, Jr., W. M. Portnoy, and L. Slifkin, *Phys. Rev.,* 102, 363, 1956.
62. W. Bosenberg, *Z. Naturforsch.,* 10a, 285, 1955.
63. V. M. Glazov and V. S. Zemskov, Physicochemical Principles of Semiconductor Doping, Israel Program for Scientific Translation, Jerusalem, 1968.
64. J. W. Corbett, R. S. McDonald, and G. D. Watkins, *J. Phys. Chem. Solids,* 25, 873, 1964.
65. W. W. Tyler, *J. Phys. Chem. Solids,* 8, 59, 1959.
66. V. D. Ignatkov and V. E. Kosenko, *Sov. Phys. Solid State,* 4, 1193, 1962.
67. A. A. Bugal, V. E. Kosenko, and E. G. Miseluk, *Zh. Tekh. Fiz.,* 27, 210, 1957.
68. F. van der Maesen and J. A. Brenkman, *Phillips Res. Rep.,* 9, 255, 1954.
69. M. B. Dutt and B. L. Sharma, in *Diffusion in Semiconductors and Non-Metallic Solids,* D. L. Beke, Ed., Springer-Verlag, Berlin, 1998, 3-1.
70. L. L. Chang and A. Koma, *Appl. Physics Lett.,* 29, 138, 1976.
71. D. L. Kendall, *Semiconductors and Semimetals,* Vol. 4, R. K. Willardson and A. C. Beer, Eds., Academic, 1968, 255.
72. H. J. Biter and F. Williams, *J. Luminescence,* 3, 395, 1971.

PROPERTIES OF MAGNETIC MATERIALS

H. P. R. Frederikse

Glossary of Symbols

Quantity	Symbol	Units	
		SI	emu
Magnetic field	H	A m^{-1}	Oe (oersted)
Magnetic induction	B	T (tesla)	G (gauss)
Magnetization	M	A m^{-1}	emu cm^{-3}
Spontaneous magnetization	M_s	A m^{-1}	emu cm^{-3}
Saturation magnetization	M_0	A m^{-1}	emu cm^{-3}
Magnetic flux	Φ	Wb (weber)	maxwell
Magnetic moment	m, μ	A m^2	erg/G
Coercive field	H_c	A m^{-1}	Oe
Remanence	B_r	T	G
Saturation magnetic polarization	J_s	T	G
Magnetic susceptibility	χ		
Magnetic permeability	μ	H m^{-1} (henry/meter)	
Magnetic permeability of free space	μ_0	H m^{-1}	
Saturation magnetostriction	λ ($\Delta l/l$)		
Curie temperature	T_C	K	K
Néel temperature	T_N	K	K

Magnetic moment $\mu = \gamma \hbar J = g\, \mu_B\, J$

where

γ = gyromagnetic ratio; J = angular momentum; g = spectroscopic splitting factor (~2)
μ_B = bohr magneton = $9.2741 \cdot 10^{-24}$ J/T = $9.2741 \cdot 10^{-21}$ erg/G

Earth's magnetic field H = 56 A m^{-1} = 0.7 Oe
For iron: $M_0 = 1.7 \cdot 10^6$ A m^{-1}; $B_r = 0.8 \cdot 10^6$ A m^{-1}
1 Oe = $(1000/4\pi)$ A m^{-1}; 1 G = 10^{-4} T; 1 emu cm^{-3} = 10^3 A m^{-1}
1 maxwell = 10^{-8} Wb
$\mu_0 = 4\pi \cdot 10^{-7}$ H m^{-1}

Relation Between Magnetic Induction and Magnetic Field

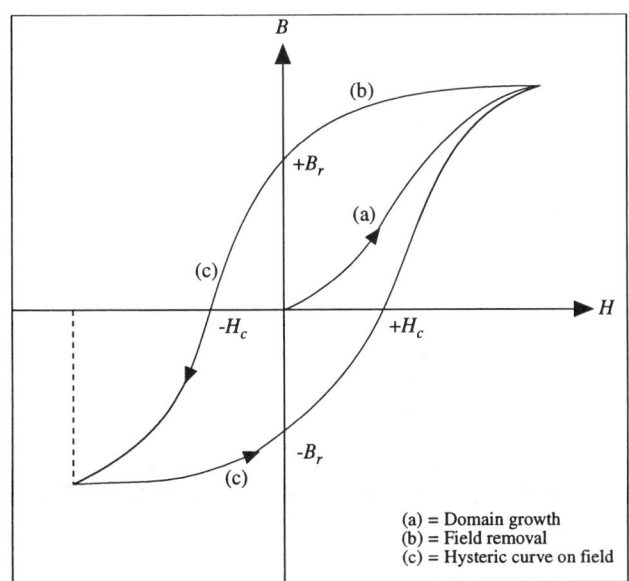

(a) = Domain growth
(b) = Field removal
(c) = Hysteric curve on field

Figure 1. Typical curve representing the dependence of magnetic induction B on magnetic field H for a ferromagnetic material. When H is first applied, B follows curve a as the favorably oriented magnetic domains grow. This curve flattens as saturation is approached. When H is then reduced, B follows curve b, but retains a finite value (the remanence B_r) at $H = 0$. In order to demagnetize the material, a negative field -H_c (where H_c is called the coercive field or coercivity) must be applied. As H is further decreased and then increased to complete the cycle (curve c), a hysteresis loop is obtained. The area within this loop is a measure of the energy loss per cycle for a unit volume of the material.

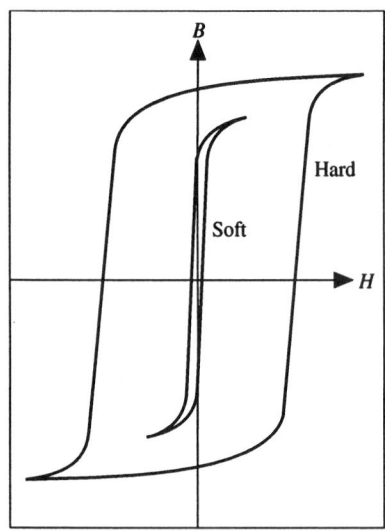

Figure 2. Schematic curve illustrating the *B* vs. *H* dependence for hard and soft magnetic materials. Hard materials have a larger remanence and coercive field, and a correspondingly large hysteresis loss.

REFERENCE

Ralls, K.M., Courtney, T.H., and Wulff, J., *Introduction to Materials Science and Engineering,* J. Wiley & Sons, New York, 1976, p. 577, 582. With permission.

Magnetic Susceptibility of the Elements

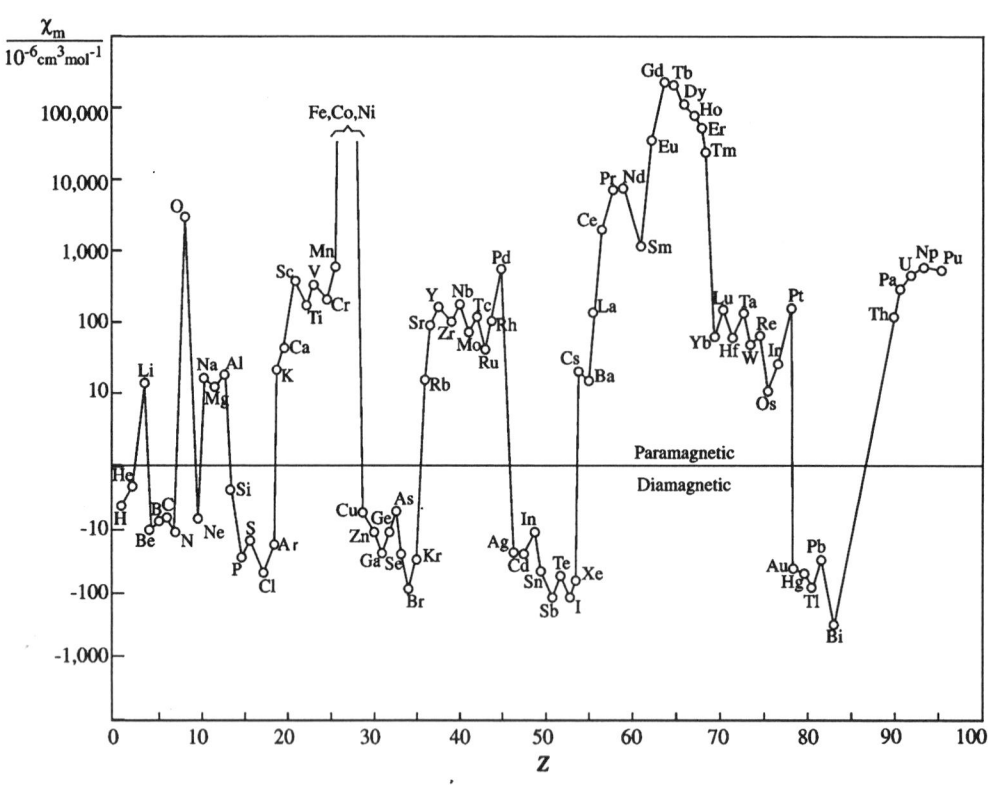

Figure 3. Molar susceptibility of the elements at room temperature (cgs units of 10^{-6} cm^3/mol). Values are not available for Z=9, 61, and 84–89; Fe, Co, and Ni (Z = 26–28) are ferromagnetic. Data taken from the table "Magnetic Susceptibility of the Elements and Inorganic Compounds" in Section 4.

REFERENCE

Gray, D.E., Ed., *American Institute of Physics Handbook, Third Edition*, McGraw Hill, New York, 1972, p. 5-224. With permission.

Ground State of Ions with Partly Filled d or f Shells

Z	Element	n	S	L	J	Gr. state	p_{calc}[a]	p_{calc}[b]	p_{meas}
22	Ti^{3+}	1	1/2	2	3/2	$^2D_{3/2}$	1.73	1.55	1.8
23	V^{4+}	1	1/2	2	3/2	$^2D_{3/2}$	1.73	1.55	1.8
23	V^{3+}	2	1	3	2	$^3F_2$	2.83	1.63	2.8
23	V^{2+}	3	3/2	3	3/2	$^4F_{3/2}$	3.87	0.77	3.8
24	Cr^{3+}	3	3/2	3	3/2	$^4F_{3/2}$	3.87	0.77	3.7
25	Mn^{4+}	3	3/2	3	3/2	$^4F_{3/2}$	3.87	0.77	4.0
24	Cr^{2+}	4	2	2	0	$^5D_0$	4.90	0	4.9
25	Mn^{3+}	4	2	2	0	$^5D_0$	4.90	0	5.0
25	Mn^{2+}	5	5/2	0	5/2	$^6S_{5/2}$	5.92	5.92	5.9
26	Fe^{3+}	5	5/2	0	5/2	$^6S_{5/2}$	5.92	5.92	5.9
26	Fe^{2+}	6	2	2	4	$^5D_4$	4.90	6.70	5.4
27	Co^{2+}	7	3/2	3	9/2	$^4F_{9/2}$	3.87	6.54	4.8
28	Ni^{2+}	8	1	3	4	$^3F_4$	2.83	5.59	3.2
29	Cu^{2+}	9	1/2	2	5/2	$^2D_{5/2}$	1.73	3.55	1.9

Z	Element	n	S	L	J	Gr. state	p_{calc}[c]		p_{meas}
58	Ce^{3+}	1	1/2	3	5/2	$^2F_{5/2}$	2.54		2.4
59	Pr^{3+}	2	1	5	4	$^3H_4$	3.58		3.5
60	Nd^{3+}	3	3/2	6	9/2	$^4I_{9/2}$	3.62		3.5
61	Pm^{3+}	4	2	6	4	$^5I_4$	2.68		
62	Sm^{3+}	5	5/2	5	5/2	$^6H_{5/2}$	0.84		1.5
63	Eu^{3+}	6	3	3	0	$^7F_0$	0.0		3.4
64	Gd^{3+}	7	7/2	0	7/2	$^8S_{7/2}$	7.94		8.0
65	Tb^{3+}	8	3	3	6	$^7F_6$	9.72		9.5
66	Dy^{3+}	9	5/2	5	15/2	$^6H_{15/2}$	10.63		10.6
67	Ho^{3+}	10	2	6	8	$^5I_8$	10.60		10.4
68	Er^{3+}	11	3/2	6	15/2	$^4I_{15/2}$	9.59		9.5
69	Tm^{3+}	12	1	5	6	$^3H_6$	7.57		7.3
70	Yb^{3+}	13	1/2	3	7/2	$^2F_{7/2}$	4.54		4.5

[a] $p_{calc} = 2[S(S + 1)]^{1/2}$
[b] $p_{calc} = 2[J(J + 1)]^{1/2}$
[c] $p_{calc} = g[J(J + 1)]^{1/2}$

REFERENCES

1. Jiles, D., *Magnetism and Magnetic Materials*, Chapman & Hall, London, 1991, p. 243.
2. Kittel, C., *Introduction to Solid State Physics, 6th Edition*, J. Wiley & Sons, New York, 1986, pp. 405—406.
3. Ashcroft, N.W. and Mermin, N.D., *Solid State Physics,* Holt, Rinehart, and Winston, New York, 1976, p. 652.

Ferro- and Antiferromagnetic Elements

M_0 is the saturation magnetization at $T = 0$ K
n_B is the number of Bohr magnetons per atom
T_C is the Curie temperature
T_N is the Néel temperature

	M_0/gauss	n_B	T_C/K	T_N/K	Comments
Fe	1752	2.22	1043		
Co	1446	1.72	1388		
Ni	510	0.62	627		
Cr				311	
Mn				100	
Ce				12.5	c-Axis antiferromagnetic
Nd				19.2	Basal plane modulation on hexagonal sites

Ferro- and Antiferromagnetic Elements (continued)

	M_0/gauss	n_B	T_C/K	T_N/K	Comments
				7.8	Cubic sites order (periodicity different from high-T phase)
Sm				106	Ordering on hexagonal sites
				13.8	Cubic site order
Eu				90.5	Spiral along cube axis
Gd	1980	7	293		
Tb		9	220		Basal plane ferromagnet
				230.2	Basal plane spiral
Dy		10	87		Basal plane ferromagnet
				176	Basal plane spiral
Ho		10	20		Bunched cone structure
				133	Basal plane spiral
Er		9	32		c-Axis ferrimagnetic cone structure
				80	c-Axis modulated structure
Tm		7	32		c-Axis ferrimagnetic cone structure
				56	c-Axis modulated structure

REFERENCES

1. Ashcroft, N.W., and Mermin, N.D., *Solid State Physics,* Holt, Rinehart, and Winston, New York, 1976, p.652.
2. Gschneidner, K.A., and Eyring, L., *Handbook on the Physics and Chemistry of Rare Earths*, North Holland Publishing Co., Amsterdam, 1978.

Selected Ferromagnetic Compounds

M_0 is the saturation magnetization at $T = 293$ K
T_C is the Curie temperature

Compound	M_0/gauss	T_C/K	Crystal system
MnB	152	578	orthorh(FeB)
MnAs	670	318	hex(FeB)
MnBi	620	630	hex(FeB)
MnSb	710	587	hex(FeB)
M_4N	183	743	
MnSi		34	cub(FeSi)
CrTe	247	339	hex(NiAs)
$CrBr_3$	270	37	hex(BiI_3)
CrI_3		68	hex(BiI_3)
CrO_2	515	386	tetr(TiO_2)
EuO	1910*	77	cub
EuS	1184*	16.5	cub
$GdCl_3$	550*	2.2	orthorh
FeB		598	orthorh
Fe_2B		1043	tetr ($CuAl_2$)
$FeBe_5$		75	cub($MgCu_2$)
Fe_3C		483	orthorh
FeP		215	orthorh (MnP)

* At $T = 0$ K

REFERENCES

1. Kittel, C., *Introduction to Solid State Physics, 6th Edition*, J. Wiley & Sons, New York, 1986.
2. Ashcroft, N.W., and Mermin, N.D., *Solid State Physics*, Holt, Rinehart, and Winston, New York, 1976.

Magnetic Properties of High-Permeability Metals and Alloys (Soft)

μ_i is the initial permeability
μ_m is the maximum permeability
H_c is the coercive force
J_s is the saturation polarization
W_H is the hysteresis loss per cycle
T_C is the Curie temperature

Material	Composition (mass %)	μ_i/μ_0	μ_m/μ_0	H_c/A m^{-1}	J_s/T	W_H/J m^{-3}	T_C/K
Iron	Commercial 99Fe	200	6000	70	2.16	500	1043
Iron	Pure 99.9Fe	25000	350000	0.8	2.16	60	1043
Silicon-iron	96Fe-4Si	500	7000	40	1.95	50-150	1008
Silicon-iron (110) [001]	97Fe-3Si	9000	40000	12	2.01	35-140	1015
Silicon-iron {100} <100>	97Fe-3Si		100000	6	2.01		1015
Mild steel	Fe-0.1C-0.1Si-0.4Mn	800	1100	200			
Hypernik	50Fe-50Ni	4000	70000	4	1.60	22	753
Deltamax {100} <100>	50Fe-50Ni	500	200000	16	1.55		773
Isoperm {100} <100>	50Fe-50Ni	90	100	480	1.60		
78 Permalloy	78Ni-22Fe	4000	100000	4	1.05	50	651
Supermalloy	79Ni-16Fe-5Mo	100000	1000000	0.15	0.79	2	673
Mumetal	77Ni-16Fe-5Cu-2Cr	20000	100000	4	0.75	20	673
Hyperco	64Fe-35Co-0.5Cr	650	10000	80	2.42	300	1243
Permendur	50Fe-50Co	500	6000	160	2.46	1200	1253
2V-Permendur	49Fe-49Co-2V	800	4000	160	2.45	600	1253
Supermendur	49Fe-49Co-2V		60000	16	2.40	1150	1253
25Perminvar	45Ni-30Fe-25Co	400	2000	100	1.55		
7Perminvar	70Ni-23Fe-7Co	850	4000	50	1.25		
Perminvar (magnet. annealed)	43Ni-34Fe-23Co		400000	2.4	1.50		
Alfenol (or Alperm)	84Fe-16Al	3000	55000	3.2	0.8		723
Alfer	87Fe-13Al	700	3700	53	1.20		673
Aluminum-Iron	96.5Fe-3.5Al	500	19000	24	1.90		
Sendust	85Fe-10Si-5Al	36000	120000	1.6	0.89		753

REFERENCES

1. McCurrie, R.A., *Structure and Properties of Ferromagnetic Materials*, Academic Press, London, 1994, p. 42.
2. Gray, D.E., Ed., *American Institute of Physics Handbook, Third Edition*, McGraw Hill, New York, 1972, p. 5-224.

Applications of High-Permeability Materials

Applications	Requirements
POWER APPLICATIONS	
Distribution and power transformers	Low core losses, high permeability, high saturation magnetic polarization
High-quality motors and generators, stators and armatures, switched-mode power supplies	
INSTRUMENT TRANSFORMERS	
Audiofrequency transformers	Low core losses, high permeability, high magnetic polarization
Pulse transformers	High permeability

Applications	Requirements

CORES FOR INDUCTOR COILS

Applications	Requirements
Audiofrequency	Low hysteresis, high permeability
Carrier frequency	Very low hysteresis and eddy current loss
Radiofrequency	High permeability at low fields

MISCELLANEOUS

Applications	Requirements
Relays, switches Earth leakage circuit }	High permeability, low remanence, low coercivity
Magnetic shielding	Low core loss for AC applications
Magnetic recording heads	High initial permeability, low or zero remanence
Magnetic amplifiers Saturable reactors Saturable transformers Transformer cores }	Rectangular hysteresis loops, low hysteresis loss
Magnetic shunts for temperature compensation in magnetic circuits	Low Curie temperature, appropriate decrease in permeability with increase in temperature
Electromagnets in indicating instruments, fire detection, quartz watches, electromechanical devices	High permeability, high saturation magnetic polarization
Magnetic yokes in permanent magnet devices, such as lifting and holding magnets, loudspeakers	High permeability, high saturation magnetic polarization

REFERENCE

McCurrie, R.A., *Structure and Properties of Ferromagnetic Materials*, Academic Press, London, 1994. With permission.

Saturation Magnetostriction of Selected Materials

The tabulated parameter λ_s is related to the fractional change in length $\Delta l/l$ by $\Delta l/l = (3/2)\lambda_s(\cos^2\theta - 1/3)$, where θ is the angle of rotation.

Material	$\lambda_s \times 10^6$
Iron	-7
Fe - 3.2% Si	+9
Nickel	-33
Cobalt	-62
45 Permalloy, 45% Ni - 55% Fe	+27
Permalloy, 82% Ni - 18% Fe	0
Permendur, 49% Co - 49% Fe - 2% V	+70
Alfer, 87% Fe - 13% Al	+30
Magnetite, Fe_3O_4	+40
Cobalt ferrite, $CoFe_2O_4$	-110
$SmFe_2$	-1560
$TbFe_2$	+1753
$Tb_{0.3}Dy_{0.7}Fe_{1.93}$ (Terfenol D)	+2000
$Fe_{66}Co_{18}B_{15}Si$ (amorphous)	+35
$Co_{72}Fe_3B_6A_{13}$ (amorphous)	0

REFERENCE

McCurrie, R.A., *Structure and Properties of Ferromagnetic Materials*, Academic Press, London, 1994, p. 91; additional data provided by A.E. Clark, Adelphi, MD.

Properties of Various Permanent Magnetic Materials (Hard)

B_r is the remanence
$_BH_c$ is the flux coercivity
$_iH_c$ is the intrinsic coercivity
$(BH)_{max}$ is the maximum energy product
T_C is the Curie temperature
T_{max} is the maximum operating temperature

Composition	B_r/T	$_BH_c/10^3$ A m^{-1}	$_iH_c/10^3$ A m^{-1}	$(BH)_{max}/$ kJ m^{-3}	$T_C/°C$	$T_{max}/°C$
Alnico1 20Ni;12Al;5Co	0.72		35	25		
Alnico2 17Ni;10Al;12.5Co;6Cu	0.72		40-50	13-14		
Alnico3 24-30Ni;12-14Al;0-3Cu	0.5-0.6		40-54	10		
Alnico4 21-28Ni;11-13Al;3-5Co;2-4Cu	0.55-0.75		36-56	11-12		
Alnico5 14Ni;8Al;24Co;3Cu	1.25	53	54	40	850	520
Alnico6 16Ni;8Al;24Co;3Cu;2Ti	1.05		75	52		
Alnico8 15Ni;7Al;35Co;4Cu;5Ti	0.83	1.6	160	45		
Alnico9 15Ni;7Al;35Co;4Cu;5Ti	1.10	1.45	1.45	75	850	520
Alnico12 13.5Ni;8Al;24.5Co;2Nb	1.20		64	76.8		
BaFe$_{12}$O$_{19}$ (Ferroxdur)	0.4	1.6	192	29	450	400
SrFe$_{12}$O$_{19}$	0.4	2.95	3.3	30	450	400
LaCo$_5$	0.91			164	567	
CeCo$_5$	0.77			117	380	
PrCo$_5$	1.20			286	620	
NdCo$_5$	1.22			295	637	
SmCo$_5$	1.00	7.9	696	196	700	250
Sm(Co$_{0.76}$Fe$_{0.10}$Cu$_{0.14}$)$_{6.8}$	1.04	4.8	5	212	800	300
Sm(Co$_{0.65}$Fe$_{0.28}$Cu$_{0.05}$Zr$_{0.02}$)$_{7.7}$	1.2	10	16	264	800	300
Nd$_2$Fe$_{14}$B sintered	1.22	8.4	1120	280	300	100
Fe;52Co;14V (Vicalloy II)	1.0	42		28	700	500
Fe;24Cr;15Co;3Mo (anisotropic)	1.54	67		76	630	500
Fe;28Cr;10.5Co (Chromindur II)	0.98	32		16	630	500
Fe;23Cr;15Co;3V;2Ti	1.35	4		44	630	500
Cu;20Ni;20Fe (Cunife)	0.55	4		12	410	350
Cu;21Ni;29Fe (Cunico)	0.34	0.5		8		
Pt;23Co	0.64	4		76	480	350
Mn;29.5Al;0.5C (anisotropic)	0.61	2.16	2.4	56	300	120

REFERENCES

1. McCurrie, R.A., *Structure and Properties of Ferromagnetic Materials*, Academic Press, London, 1994, p. 204.
2. Gray, D.E., Ed., *American Institute of Physics Handbook, Third Edition*, McGraw Hill, New York, 1972, p. 5-165.
3. Jiles, D., *Magnetism and Magnetic Materials,* Chapman & Hall, London, 1991.

Selected Ferrites

J_s is the saturation magnetic polarization
T_C is the Curie temperature
ΔH is the line width

Material	J_s/T	$T_C/°C$	$\Delta H/$ kA m^{-1}	Applications
Spinels				
γ-Fe$_2$O$_3$	0.52	575		
Fe$_3$O$_4$	0.60	585		
NiFe$_2$O$_4$	0.34	575	350	Microwave devices
MgFe$_2$O$_4$	0.14	440	70	
NiZnFe$_2$O$_4$	0.50	375	120	Transformer cores
MnFe$_2$O$_4$	0.50	300	50	Microwave devices

Selected Ferrites (continued)

Material	J_s/T	T_C/°C	ΔH/ kA m^{-1}	Applications
$NiCoFe_2O_4$	0.31	590	140	Microwave devices
$NiCoAlFe_2O_4$	0.15	450	330	Microwave devices
$NiAl_{0.35}Fe_{1.65}O_4$	0.12	430	67	Microwave devices
$NiAlFe_2O_4$	0.05	1860	32	Microwave devices
$Mg_{0.9}Mn_{0.1}Fe_2O_4$	0.25	290	56	Microwave devices
$Ni_{0.5}Zn_{0.5}Al_{0.8}Fe_{1.2}O_4$	0.14		17	Microwave devices
$CuFe_2O_4$	0.17	455		Electromechanical transducers
$CoFe_2O_4$	0.53	520		
$LiFe_5O_8$	0.39	670		Microwave devices
Garnets				
$Y_3Fe_5O_{12}$	0.178	280	55	Microwave devices
$Y_3Fe_5O_{12}$ (single crys.)	0.178	292	0.5	Microwave devices
$(Y,Al)_3Fe_5O_{12}$	0.12	250	80	Microwave devices
$(Y,Gd)_3Fe_5O_{12}$	0.06	250	150	Microwave devices
$Sm_3Fe_5O_{12}$	0.170	305		Microwave devices
$Eu_3Fe_5O_{12}$	0.116	293		Microwave devices
$GdFe_5O_{12}$	0.017	291		Microwave devices
Hexagonal crystals				
$BaFe_{12}O_{19}$	0.45	430	1.5	Permanent magnets
$Ba_3Co_2Fe_{24}O_{41}$	0.34	470	12	Microwave devices
$Ba_2Zn_2Fe_{12}O_{22}$	0.28	130	25	Microwave devices
$Ba_3Co_{1.35}Zn_{0.65}Fe_{24}O_{41}$		390	16	Microwave devices
$Ba_2Ni_2Fe_{12}O_{22}$	0.16	500	8	Microwave devices
$SrFe_{12}O_{19}$	0.4	450		Permanent magnets

REFERENCE

McCurrie, R.A., *Structure and Properties of Ferromagnetic Materials*, Academic Press, London, 1994.

Spinel Structure (AB$_2$O$_4$)

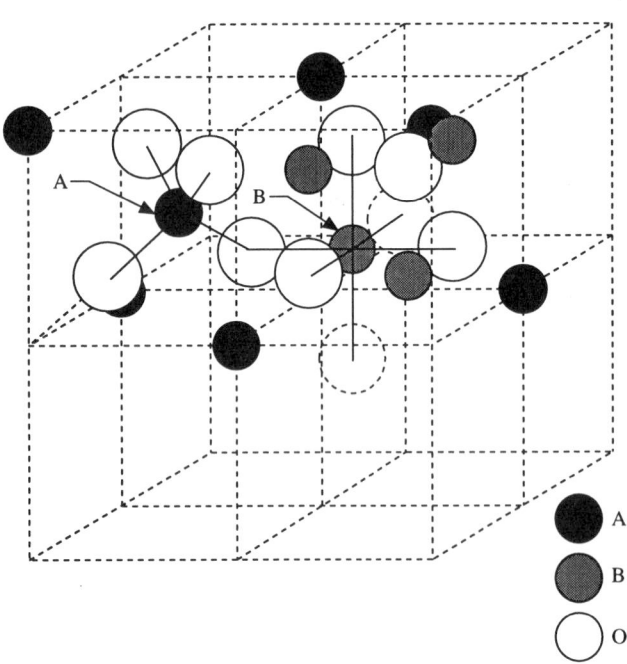

Figure 4. Arrangement of metal ions in the two octants A and B, showing tetrahedrally (A) and octahedrally (B) coordinated sites. (Reprinted from McCurrie, R.A., *Ferromagnetic Materials*, Academic Press, London, 1994. With permission.)

A
B
O^{2-}

Selected Antiferromagnetic Solids

T_N is the Néel temperature

Material	Structure	T_N/K	Material	Structure	T_N/K
Binary oxides			**NiAs and related structures**		
MnO	cub(fcc)	122	CrAs	orth	300
FeO	cub(fcc)	198	CrSb	hex	705-723
CoO	cub(fcc)	291	CrSe	hex	300
NiO	cub(fcc)	525	MnTe	hex	320-323
α-Mn_2O_3	cub	90	NiS	hex	263
CuO	monocl	230	CrS	monocl	460
UO_2	cub	30.8	**Rutile and related structures**		
Er_2O_3	cub	3.4	CoF_2	tetr	38
Gd_2O_3	cub	1.6	CrF_2	monocl	53
Perovskites			FeF_2	tetr	79
$LaCrO_3$	orth	282	MnF_2	tetr	67
$LaMnO_3$	orth	100	NiF_2	tetr	83
$LaFeO_3$	orth	750	$CrCl_2$	orth	20
$NdCrO_3$	orth	224	MnO_2	tetr	84
$NdFeO_3$	orth	760	FeOF	tetr	315
$YbCrO_3$	orth	118	**Corundum and related structures**		
$CaMnO_3$	cub	110	Cr_2O_3	rhomb	318
$EuTiO_3$	cub	5.3	α-Fe_2O_3	rhomb	948
$YCrO_3$	orth	141	$FeTiO_3$	rhomb	68
$BiFeO_3$	cub*	673	$MnTiO_3$	rhomb	41
$KCoF_3$	cub	125	$CoTiO_3$	rhomb	38
$KMnF_3$	cub*	88.3	**VF_3 and related structures**		
$KFeF_3$	cub	115	CoF_3	rhomb	460
$KNiF_3$	cub	275	CrF_3	rhomb	80
$NaMnF_3$	cub*	60	FeF_3	rhomb	394
$NaNiF_3$	orth	149	MnF_3	monocl	43
$RbMnF_3$	cub	82	MoF_3	rhomb	185
Spinels			**Miscellaneous**		
Co_3O_4	cub	40	K_2NiF_4	tetr	97
$NiCr_2O_4$	tetr	65	MnI_2	hex	3.4
$ZnCr_2O_4$	cub	15	$CoUO_4$	orth	12
$ZnFe_2O_4$	cub	9	$CaMn_2O_4$	orth	225
$GeFe_2O_4$	cub	10	CrN	cub*	273
MgV_2O_4	cub	45	CeC_2	tetr	33
$MnGa_2O_4$	cub	33	FeSn	hex	373
			Mn_2P	hex	103

* Distorted.

REFERENCES

1. Gray, D.E., Ed., *American Institute of Physics Handbook, Third Edition*, McGraw Hill, New York, 1972, p. 5-168 to 183.
2. Kittel, C., *Introduction to Solid State Physics, 6th Edition*, J. Wiley & Sons, New York, 1986.
3. Ashcroft, N.W., and Mermin, N.D., *Solid State Physics*, Holt, Rinehart, and Winston, New York, 1976, p. 697.

ORGANIC MAGNETS
J.S. Miller

Magnetic ordering, e.g., ferromagnetism, like superconductivity, is a property of a solid, not of an individual molecule or ion, and very rarely occurs for organic compounds. In contrast to superconductivity, where all electron spins pair to form a perfect diamagnetic material, magnetic ordering requires unpaired electron spins; hence, superconductivity and ferromagnetism are mutually exclusive.

The vast majority of organic compounds are diamagnetic (i.e., all electron spins are paired), and a relative few possess unpaired electrons (designated by an arrow, ↑) and are paramagnetic (PM), i.e., they are oriented in random directions. A few organic solids, however, exhibit strong magnetic behavior and magnetically order as ferromagnets (FO) with all spins aligned in the same direction. In some cases the spins align in the opposite direction and compensate to form an antiferromagnet (AF). In some cases these spins are not opposed to each other and do not compensate and lead to a canted antiferromagnet or weak ferromagnet (WF). If the number of spins that align in one direction differs from the number of spins that align in the opposite direction, the spins cannot compensate and a ferrimagnet (FI) results. Metamagnets (MM) are antiferromagnets in which all the spins become aligned like a ferromagnet in an applied magnetic field. Above the ordering or critical temperature, T_c, all magnets are paramagnets (PM). Organic magnets all possess electron spins in p-orbitals, but these may be in conjunction with metal ion-based spins.

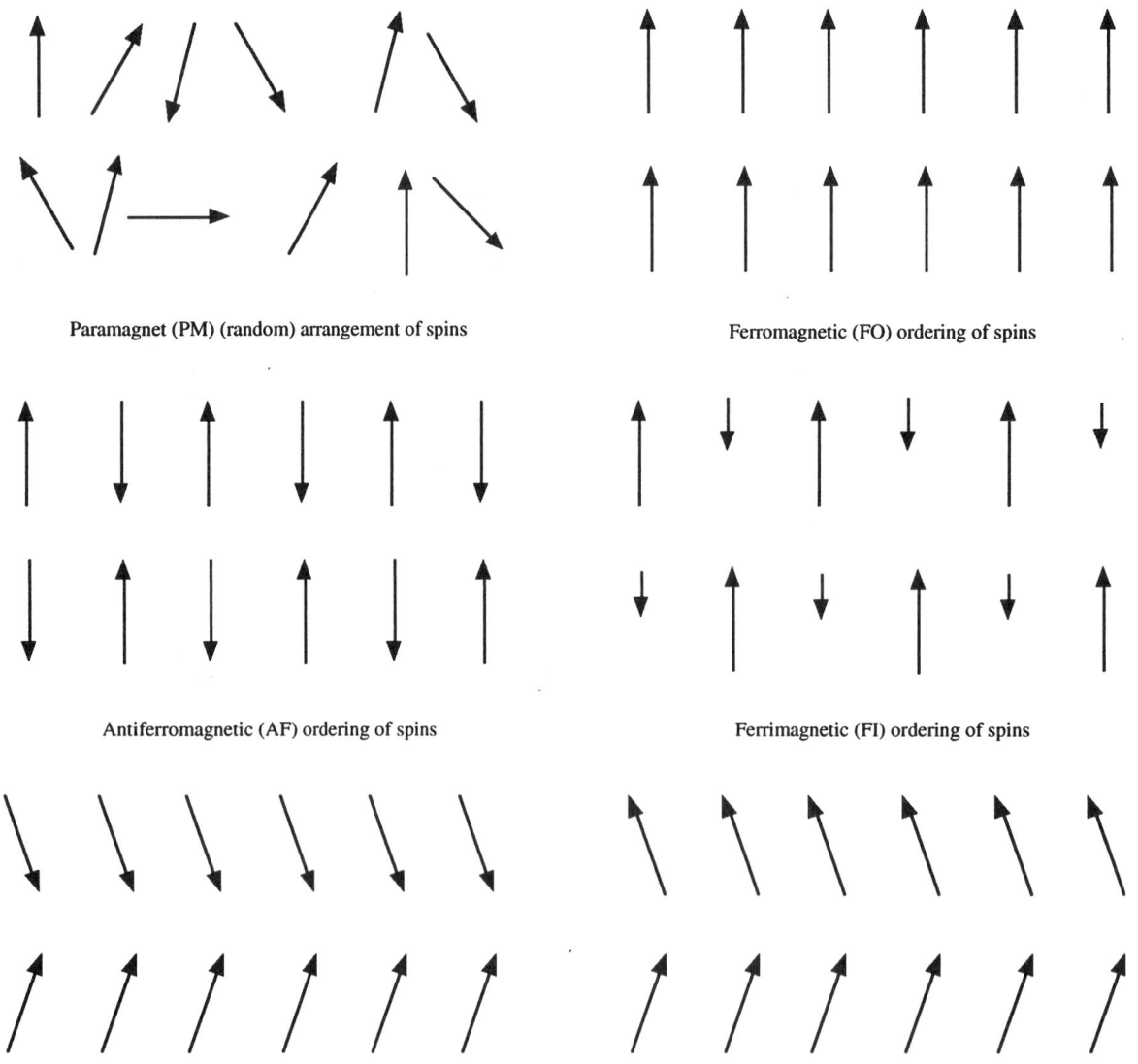

Paramagnet (PM) (random) arrangement of spins

Ferromagnetic (FO) ordering of spins

Antiferromagnetic (AF) ordering of spins

Ferrimagnetic (FI) ordering of spins

Canted antiferromagnet or weak ferromagnet (WF) ordering of spins

Figure 1. Schematic illustration of the different types of magnetic behavior.

Summary of the Critical Temperature, T_c, Saturation Magnetization, M_s, Coercive Field, H_{cr}, and Remanent Magnetization, M_r, for Selected Organic-Based Magnets

Magnet	Type	T_c/K	M_s/A m^{-1}	H_{cr}/T	M_r/A m^{-1}
α-1,3,5,7-Tetramethyl-2,6-diazaadamantane-N,N'-doxyl	FO	1.48	48,300	<0.00001	—
β-2-(4'-Nitrophenyl)-4,4,5,5-tetramethyl-4,5-dihydro-1H-imidazol-1-oxyl-3-N-oxide	FO	0.6	22,300	0.00008	<200
{Fe^{III}[C$_5$(CH$_3$)$_5$]$_2$}[TCNE]	FO	4.8	37,600	0.10	2,300
{Mn^{III}[C$_5$(CH$_3$)$_5$]$_2$}[TCNE]	FO	8.8	58,200	0.12	3,700
{Cr^{III}[C$_5$(CH$_3$)$_5$]$_2$}[TCNE]	FO	3.65	46,300	—	—
α-{Fe^{III}[C$_5$(CH$_3$)$_5$]$_2$}[TCNQ]	MM	2.55	34,200	—	—
β-{Fe^{III}[C$_5$(CH$_3$)$_5$]$_2$}[TCNQ]	FO	3.0	21,600	—	—
Tanol subarate	MM	0.38	20,700	—	—
NCC$_6$F$_4$CN$_2$S$_2$	WF	35.5	45	0.00009	—
Mn^{II}(hfac)$_2$NITC$_2$H$_5$	FI	7.8	39,400	0.03	27,600
Mn^{II}(hfac)$_2$NIT(i-C$_3$H$_8$)	FI	7.6	42,400	<0.0005	<420
[Mn(hfac)$_2$]$_3$[{ON[C$_6$H$_3$(t-C(CH$_3$)$_3$)$_2$]$_2$NO]$_2$}	FI	46	24,400	—	—
[MnTPP][TCNE]·2C$_6$H$_5$CH$_3$	FI	13	18,400	2.4	10,300
V[TCNE]$_x$·yCH$_2$Cl$_2$ ($x \sim 2$; $y \sim 0.5$)	FI	~400	28,200	0.0015 - 0.006	1,650
Mn[TCNE]$_x$·yCH$_2$Cl$_2$ ($x \sim 2$; $y \sim 0.5$)	FI	75	52,000	0.002	270
Fe[TCNE]$_x$·yCH$_2$Cl$_2$ ($x \sim 2$; $y \sim 0.5$)	FI	97	46,300	0.23	3
Co[TCNE]$_x$·yCH$_2$Cl$_2$ ($x \sim 2$; $y \sim 0.5$)	FI	44	22,000	0.65	—

List of Symbols and Abbreviations

M_s	Saturation magnetization at 2 K
H_{cr}	Coercive Field
T_c	Critical Temperature
M_r	Remanent magnetization at 2 K
TCNE	Tetracyanoethylene
TCNQ	7,7,8,8-Tetracyano-p-quinodimethane
hfac	Hexafluoroacetonate
NIT	Nitronyl nitroxide
FO	Ferromagnet
FI	Ferrimagnet
MM	Metamagnet
WF	Weak ferromagnet

1,3,5,7-Tetramethyl-2,6-diazaadamantane-N,N'-doxyl

2-(4'-Nitrophyenyl)-4,4,5,5-tetramethyl-4,5-dihydro-1H-imidazol-1-oxyl-3-N-oxide

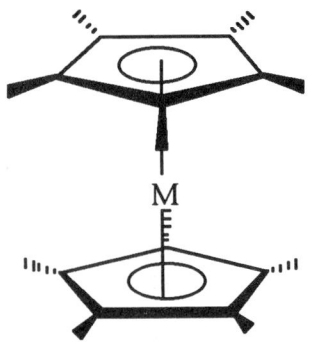

$M[C_5(CH_3)_5]_2(M = Cr, Mn, Fe)$

TCNE

TCNQ

Mn(hfac)₂

Tanol subarate

NITR (R = C_2H_5, i-C_3H_8, n-C_3H_8)

{ON[C_6H_3(t-C(CH₃)₃)₂NO]₂}

ORGANIC MAGNETS (continued)

MnTPP

NCC$_6$F$_4$CN$_2$S$_2$

REFERENCES

1. Miller, J. S. and Epstein, A. J., *Angew. Chem. Internat. Ed.*, 33, 385, 1994.
2. Chiarelli, R., Rassat, A., Dromzee, Y., Jeannin, Y., Novak, M. A., and Tholence, J. L., *Phys. Scrip.*, T49, 706, 1993.
3. Kinoshita, M., *Jap. J. Appl. Phys.*, 33, 5718, 1994.
4. Gatteschi, D., *Adv. Mat.*, 6, 635, 1994.
5. Miller, J. S. and Epstein, A. J., *J. Chem. Soc., Chem. Commun.*, 1319, 1998.
6. Broderick, W. E., Eichorn, D. M., Lu, X., Toscano, P. J., Owens, S. M. and Hoffman, B. M., *J. Am. Chem. Soc.*, 117, 3641, 1995.
7. Banister, A. J., Bricklebank, N., Lavander, I., Rawson, J., Gregory, C. I., Tanner, B. K., Clegg, W. J., Elsegood, M. R., and Palacio, F., *Angew. Chem. Internat. Ed.*, 35, 2533, 1996.

ELECTRON WORK FUNCTION OF THE ELEMENTS

The electron work function Φ is a measure of the minimum energy required to extract an electron from the surface of a solid. It is defined more precisely as the energy difference between the state in which an electron has been removed to a distance from the surface of a single crystal face that is large enough that the image force is negligible but small compared to the distance to any other face (typically about 10^{-4} cm) and the state in which the electron is in the bulk solid. In general, Φ differs for each face of a monocrystalline sample.

Since Φ is dependent on the cleanliness of the surface, measurements reported in the literature often cover a considerable range. This table contains selected values for the electron work function of the elements which may be regarded as typical values for a reasonably clean surface. The method of measurement is indicated for each value. The following abbreviations appear:

TE — Thermionic emission
PE — Photoelectric effect
FE — Field emission
CPD — Contact potential difference
polycr — Polycrystalline sample
amorp — Amorphous sample

Values in parentheses are only approximate.

REFERENCES

1. Hölzl, J., and Schulte, F. K., Work Functions of Metals, in *Solid Surface Physics*, Höhler, G., Editor, Springer-Verlag, Berlin, 1979.
2. Riviere, J. C., Work Function: Measurements and Results, in *Solid State Surface Science, Vol.1*, Green, M., Editor, Decker, New York, 1969.
3. Michaelson, H. B., *J. Appl. Phys.*, 48, 4729, 1977.

Element	Plane	Φ/eV	Method	Element	Plane	Φ/eV	Method	Element	Plane	Φ/eV	Method
Ag	100	4.64	PE		210	5.00	PE	Ru	polycr	4.71	PE
	110	4.52	PE	K	polycr	2.29	PE	Sb	amorp	4.55	
	111	4.74	PE	La	polycr	3.5	PE		100	4.7	
Al	100	4.20	PE	Li	polycr	2.93	FE	Sc	polycr	3.5	PE
	110	4.06	PE	Lu	polycr	(3.3)	CPD	Se	polycr	5.9	PE
	111	4.26	PE	Mg	polycr	3.66	PE	Si	n	4.85	CPD
As	polycr	(3.75)	PE	Mn	polycr	4.1	PE		p 100	(4.91)	CPD
Au	100	5.47	PE	Mo	100	4.53	PE		p 111	4.60	PE
	110	5.37	PE		110	4.95	PE	Sm	polycr	2.7	PE
	111	5.31	PE		111	4.55	PE	Sn	polycr	4.42	CPD
B	polycr	(4.45)	TH		112	4.36	PE	Sr	polycr	(2.59)	TH
Ba	polycr	2.52	TH		114	4.50	PE	Ta	polycr	4.25	TH
Be	polycr	4.98	PE		332	4.55	PE		100	4.15	TH
Bi	polycr	4.34	PE	Na	polycr	2.36	PE		110	4.80	TH
C	polycr	(5.0)	CPD	Nb	001	4.02	TH		111	4.00	TH
Ca	polycr	2.87	PE		110	4.87	TH	Tb	polycr	3.0	PE
Cd	polycr	4.08	CPD		111	4.36	TH	Te	polycr	4.95	PE
Ce	polycr	2.9	PE		112	4.63	TH	Th	polycr	3.4	TH
Co	polycr	5.0	PE		113	4.29	TH	Ti	polycr	4.33	PE
Cr	polycr	4.5	PE		116	3.95	TH	Tl	polycr	(3.84)	CPD
Cs	polycr	1.95	PE		310	4.18	TH	U	polycr	3.63	PE
Cu	100	5.10	FE	Nd	polycr	3.2	PE		100	3.73	PE
	110	4.48	PE	Ni	100	5.22	PE		110	3.90	PE
	111	4.94	PE		110	5.04	PE		113	3.67	PE
	112	4.53	PE		111	5.35	PE	V	polycr	4.3	PE
Eu	polycr	2.5	PE	Os	polycr	5.93	PE	W	polycr	4.55	CPD
Fe	100	4.67	PE	Pb	polycr	4.25	PE		100	4.63	FE
	111	4.81	PE	Pd	polycr	5.22	PE		110	5.22	FE
Ga	polycr	4.32	PE		111	5.6	PE		111	4.45	FE
Gd	polycr	2.90	CPD	Pt	polycr	5.64	PE		113	4.46	FE
Ge	polycr	5.0	CPD		110	5.84	FE		116	4.32	TH
Hf	polycr	3.9	PE		111	5.93	FE	Y	polycr	3.1	PE
Hg	liquid	4.475	PE		320	5.22	FE	Zn	polycr	3.63	PE
In	polycr	4.09	PE		331	5.12	FE		polycr	(4.9)	CPD
Ir	100	5.67	PE	Rb	polycr	2.261	PE	Zr	polycr	4.05	PE
	110	5.42	PE	Re	polycr	4.72	TE				
	111	5.76	PE	Rh	polycr	4.98	PE				

SECONDARY ELECTRON EMISSION

The secondary emission yield, or secondary emission ratio, δ, is the average number of secondary electrons emitted from a bombarded material for every incident primary electron. It is a function of the primary electron energy E_p. The maximum yield δ_{max} corresponds to a primary electron energy E_{pmax} (see figure). The two primary electron energies corresponding to a yield of unity are denoted the first and second crossovers (E_I and E_{II}). An insulating target, or a conducting target that is electrically floating, will charge positively or negatively depending on the primary electron energy. For $E_I < E_p < E_{II}$, $\delta > 1$ and the surface charges positively provided there is a collector present that is positive with respect to the target. For $E_p < E_I$ or $E_p > E_{II}$, $\delta < 1$, and the surface charges negatively with respect to the potential of the source of primary electrons.

Primary Electron Energy (Ep)

Element	δ_{max}	E_{pmax} (eV)	E_I (eV)	E_{II} (eV)
Ag	1.5	800	200	>2000
Al	1.0	300	300	300
Au	1.4	800	150	>2000
B	1.2	150	50	600
Ba	0.8	400	None	None
Bi	1.2	550	None	None
Be	0.5	200	None	None
C (diamond)	2.8	750	None	>5000
C (graphite)	1.0	300	300	300
C (soot)	0.45	500	None	None
Cd	1.1	450	300	700
Co	1.2	600	200	None
Cs	0.7	400	None	None
Cu	1.3	600	200	1500
Fe	1.3	400	120	1400
Ga	1.55	500	75	None
Ge	1.15	500	150	900
Hg	1.3	600	350	>1200
K	0.7	200	None	None

Element	δ_{max}	E_{pmax} (eV)	E_I (eV)	E_{II} (eV)
Li	0.5	85	None	None
Mg	0.95	300	None	None
Mo	1.25	375	150	1200
Na	0.82	300	None	None
Nb	1.2	375	150	1050
Ni	1.3	550	150	>1500
Pb	1.1	500	250	1000
Pd	>1.3	>250	120	None
Pt	1.8	700	350	3000
Rb	0.9	350	None	None
Sb	1.3	600	250	2000
Si	1.1	250	125	500
Sn	1.35	500	None	None
Ta	1.3	600	250	>2000
Th	1.1	800	None	None
Ti	0.9	280	None	None
Tl	1.7	650	70	>1500
W	1.4	650	250	>1500
Zr	1.1	350	None	None

Compound	δ_{max}	E_{pmax} (eV)
Alkali halides		
CsCl	6.5	
KBr (crystal)	14	1800
KCl (crystal)	12	1600
KCl (layer)	7.5	1200
KI (crystal)	10	1600
KI (layer)	5.6	
LiF (crystal)	8.5	
LiF (layer)	5.6	700
NaBr (crystal)	24	1800
NaBr (layer)	6.3	
NaCl (crystal)	14	1200

Compound	δ_{max}	E_{pmax} (eV)
NaCl (layer)	6.8	600
NaF (crystal)	14	1200
NaF (layer)	5.7	
NaI (crystal)	19	1300
NaI (layer)	5.5	
RbCl (layer)	5.8	
Oxides		
Ag_2O	1.0	
Al_2O_3 (layer)	2—9	
BaO (layer)	2.3—4.8	400
BeO	3.4	2000
CaO	2.2	500

Compound	δ_{max}	E_{pmax} (eV)	Compound	δ_{max}	E_{pmax} (eV)
Cu_2O	1.2	400	Others		
MgO (crystal)	20—25	1500	BaF_2 (layer)	4.5	
MgO (layer)	3—15	400—1500	CaF_2 (layer)	3.2	
MoO_2	1.2		$BiCs_3$	6	1000
SiO_2 (quartz)	2.1—4	400	BiCs	1.9	1000
SnO_2	3.2	640	GeCs	7	700
Sulfides			Rb_3Sb	7.1	450
MoS_2	1.1		$SbCs_3$	6	700
PbS	1.2	500	Mica	2.4	350
WS_2	1.0		Glasses	2—3	300—450
ZnS	1.8	350			

OPTICAL PROPERTIES OF METALS AND SEMICONDUCTORS
J. H. Weaver and H. P. R. Frederikse

These tables list the index of refraction n, the extinction coefficient k, and the normal incidence reflection $R\,(\phi = 0)$ as a function of photon energy E, which is expressed in electron volts (eV). To convert the energy in eV to wavelength in μm, use $\lambda = 1.2398/E$. To compute the dielectric function $\tilde{\varepsilon} = \varepsilon_1 + i\varepsilon_2$ from the complex index of refraction $\tilde{N} = n + ik$, use $\varepsilon_1 = n^2 - k^2$ and $\varepsilon_2 = 2nk$.

The optical constants in these tables are abridged from three more extensive tabulations:

- *Optical Properties of Metals* (OPM), Volumes I and II, *Physics Data, Nr.* 18-1 and 18-2, J. H. Weaver, C. Krafka, D. W. Lynch, and E. E. Koch, Fachinformationzentrum, Karlsruhe, Germany.
- *Handbook of Optical Constants* (HOC), Vol. I, 1985, and Vol. II, 1991. Edited by E. D. Palik, published by Academic Press, Inc.
- *American Institute of Physics Handbook* (AIPH), 3rd Edition, Coord. Editor D. E. Gray, published by McGraw-Hill Book Co., New York, 1972.

The first two of these major sources provide detailed comparisons of all optical data available in the literature at the time of the compilation. For critical applications the reader should refer to the original work. References for individual metals and semiconductors are listed at the end of the tables. Generally, tabulated values for the optical properties are accurate to better than 10%. Data in parentheses are extrapolated or interpolated values. For most elements the spectral range covered is from the far infrared (0.010 or 0.10 eV) to the far ultraviolet (10, 30 or 300 eV). The intervals between successive energies in the tables are chosen in such a way that the major spectral features are preserved.

Very small values of k are expressed in exponential notation, e.g., 1.23E-5 means 1.23×10^{-5}.

The following table is convenient for identifying the energy entries in these tables with the corresponding wavelengths:

λ	E/eV	λ	E/eV
1 mm	0.00124	6000 Å	2.066
500 μm	0.00248	5000 Å	2.480
100 μm	0.01240	4000 Å	3.100
50 μm	0.02480	3000 Å	4.133
10 μm	0.12398	2000 Å	6.199
5 μm	0.24797	1000 Å	12.398
1 μm	1.240	400 Å	30.996

Energy (eV)	n	k	$R(\phi = 0)$	Energy (eV)	n	k	$R(\phi = 0)$	Energy (eV)	n	k	$R(\phi = 0)$
Aluminium[1]				2.200	1.018	6.846	0.9200	14.200	0.053	0.373	0.8312
0.040	98.595	203.701	0.9923	2.400	0.826	6.283	0.9228	14.400	0.058	0.327	0.8102
0.050	74.997	172.199	0.9915	2.600	0.695	5.800	0.9238	14.600	0.067	0.273	0.7802
0.060	62.852	150.799	0.9906	2.800	0.598	5.385	0.9242	14.800	0.086	0.211	0.7202
0.070	53.790	135.500	0.9899	3.000	0.523	5.024	0.9241	15.000	0.125	0.153	0.6119
0.080	45.784	123.734	0.9895	3.200	0.460	4.708	0.9243	15.200	0.178	0.108	0.4903
0.090	39.651	114.102	0.9892	3.400	0.407	4.426	0.9245	15.400	0.234	0.184	0.3881
0.100	34.464	105.600	0.9889	3.600	0.363	4.174	0.9246	15.600	0.280	0.073	0.3182
0.125	24.965	89.250	0.9884	3.800	0.326	3.946	0.9247	15.800	0.318	0.065	0.2694
0.150	18.572	76.960	0.9882	4.000	0.294	3.740	0.9248	16.000	0.351	0.060	0.2326
0.175	14.274	66.930	0.9879	4.200	0.267	3.552	0.9248	16.200	0.380	0.055	0.2031
0.200	11.733	59.370	0.9873	4.400	0.244	3.380	0.9249	16.400	0.407	0.050	0.1789
0.250	8.586	48.235	0.9858	4.600	0.223	3.222	0.9249	16.750	0.448	0.045	0.1460
0.300	6.759	40.960	0.9844	4.800	0.205	3.076	0.9249	17.000	0.474	0.042	0.1278
0.350	5.438	35.599	0.9834	5.000	0.190	2.942	0.9244	17.250	0.498	0.040	0.1129
0.400	4.454	31.485	0.9826	6.000	0.130	2.391	0.9257	17.500	0.520	0.038	0.1005
0.500	3.072	25.581	0.9817	6.500	0.110	2.173	0.9260	17.750	0.540	0.036	0.0899
0.600	2.273	21.403	0.9806	7.000	0.095	1.983	0.9262	18.000	0.558	0.035	0.0809
0.700	1.770	18.328	0.9794	7.500	0.082	1.814	0.9265	18.500	0.591	0.032	0.0664
0.800	1.444	15.955	0.9778	8.000	0.072	1.663	0.9269	19.000	0.620	0.030	0.0554
0.900	1.264	14.021	0.9749	8.500	0.063	1.527	0.9272	19.500	0.646	0.028	0.0467
1.000	1.212	12.464	0.9697	9.000	0.056	1.402	0.9277	20.000	0.668	0.027	0.0398
1.100	1.201	11.181	0.9630	9.500	0.049	1.286	0.9282	20.500	0.689	0.025	0.0342
1.200	1.260	10.010	0.9521	10.000	0.044	1.178	0.9286	21.000	0.707	0.024	0.0296
1.300	1.468	8.949	0.9318	10.500	0.040	1.076	0.9293	21.500	0.724	0.023	0.0258
1.400	2.237	8.212	0.8852	11.000	0.036	0.979	0.9298	22.000	0.739	0.022	0.0226
1.500	2.745	8.309	0.8678	11.500	0.033	0.883	0.9283	22.500	0.753	0.021	0.0199
1.600	2.625	8.597	0.8794	12.000	0.033	0.791	0.9224	23.000	0.766	0.021	0.0177
1.700	2.143	8.573	0.8972	12.500	0.034	0.700	0.9118	23.500	0.778	0.020	0.0157
1.800	1.741	8.205	0.9069	13.000	0.038	0.609	0.8960	24.000	0.789	0.019	0.0140
1.900	1.488	7.821	0.9116	13.500	0.041	0.517	0.8789	24.500	0.799	0.018	0.0126
2.000	1.304	7.479	0.9148	14.000	0.048	0.417	0.8486	25.000	0.809	0.018	0.0113

Energy (eV)	n	k	R(φ = 0)
25.500	0.817	0.017	0.0102
26.000	0.826	0.016	0.0092
27.000	0.840	0.015	0.0076
28.000	0.854	0.014	0.0063
29.000	0.865	0.014	0.0053
30.000	0.876	0.013	0.0044
35.000	0.915	0.010	0.0020
40.000	0.940	0.008	0.0010
45.000	0.957	0.007	0.0005
50.000	0.969	0.006	0.0003
55.000	0.979	0.005	0.0001
60.000	0.987	0.004	0.0000
65.000	0.995	0.004	0.0000
70.000	1.006	0.004	0.0000
72.500	1.025	0.004	0.0002
75.000	1.011	0.024	0.0002
77.500	1.008	0.025	0.0002
80.000	1.007	0.024	0.0002
85.000	1.007	0.028	0.0002
90.000	1.005	0.031	0.0002
95.000	0.999	0.036	0.0003
100.000	0.991	0.030	0.0002
110.000	0.994	0.025	0.0002
120.000	0.991	0.024	0.0002
130.000	0.987	0.021	0.0001
140.000	0.989	0.016	0.0001
150.000	0.990	0.015	0.0001
160.000	0.989	0.014	0.0001
170.000	0.989	0.011	0.0001
180.000	0.990	0.010	0.0000
190.000	0.990	0.009	0.0000
200.000	0.991	0.007	0.0000
220.000	0.992	0.006	0.0000
240.000	0.993	0.005	0.0000
260.000	0.993	0.004	0.0000
280.000	0.994	0.003	0.0000
300.000	0.995	0.002	0.0000

Carbon (diamond)[2]

Energy (eV)	n	k	R(φ = 0)
0.06199	2.3741		0.166
0.06888	2.3741		0.166
0.07749	2.3745		0.166
0.08856	2.3750		0.166
0.1033	2.3757		0.166
0.1240	2.3765		0.166
0.1550	2.3772		0.166
0.1907		3.1 E-05	
0.2066	2.3779	5.7 E-05	0.166
0.22		1.21E-04	
0.23		2.36E-04	
0.24		3.82E-04	
0.25		5.21E-04	
0.26		2.96E-04	
0.27		4.39E-04	
0.28		2.75E-04	
0.29		7.82E-05	
0.30		1.32E-04	
0.31	2.3787	1.30E-04	0.167
0.32		1.11E-04	
0.33		2.99E-05	
0.34		1.89E-05	
0.35		2.11E-05	
0.36		2.47E-05	
0.37		2.80E-05	
0.38		3.11E-05	
0.39		3.67E-05	
0.40		3.58E-05	
0.41		3.25E-05	
0.4133	2.3795		0.167
0.42		2.94E-05	
0.43		2.87E-05	
0.44		3.14E-05	
0.45		3.62E-05	
0.46		3.22E-05	
0.47		1.57E-05	
0.48		6.17E-06	
0.4959	2.3801		0.167
0.6199	2.3813		0.167
0.8266	2.3837		0.167
1.240	2.3905		0.168
1.378	2.3934		0.169
1.459	2.3953		0.169
1.550	2.3975		0.169
1.653	2.4003		0.170
1.771	2.4036		0.170
1.889	2.4073		0.171
1.926	2.4084		0.171
2.066	2.4133		0.171
2.105	2.4147		0.172
2.271	2.4210		0.173
2.480	2.4299		0.174
2.650	2.4380		0.175
2.845		3.82E-07	0.178
3.100	2.4627		0.182
3.434	2.4849		0.183
3.576	2.4955		
3.961		8.97E-07	
4.160	2.5465		0.190
4.511		1.29E-06	
4.8187	2.6205	1.47E-06	0.200
5.00	2.6383		0.203
5.30		2.98E-06	
5.35		6.45E-06	
5.40		1.04E-05	
5.50		3.41E-05	
5.55		5.48E-04	
5.60	2.740	1.48E-03	0.216
5.80	2.780	5.02E-03	0.222
6.00	2.826	7.99E-03	0.228
6.10	2.852	8.62E-03	0.231
6.20	2.879	9.30E-03	0.235
6.30	2.910	9.74E-03	0.239
6.40	2.944	9.87E-03	0.243
6.50	2.985	1.10E-02	0.248
6.60	3.031	1.47E-02	0.254
6.70	3.085	2.20E-02	0.261
6.80	3.146	3.44E-02	0.268
6.90	3.220	5.24E-02	0.277
7.00	3.322	9.35E-02	0.289
7.10	3.444	0.210	0.304
7.15	3.464	0.307	0.308
7.20	3.437	0.388	0.307
7.30	3.376	0.473	0.303
7.40	3.335	0.515	0.300
7.50	3.321	0.533	0.299
7.60	3.306	0.592	0.300
7.80	3.276	0.659	0.300
8.00	3.251	0.712	0.300
8.25	3.232	0.765	0.301
8.50	3.228	0.806	0.303
8.75	3.247	0.855	0.308
9.00	3.272	0.910	0.314
9.25	3.308	0.978	0.322
9.50	3.348	1.055	0.331
9.75	3.398	1.147	0.342
10.00	3.453	1.258	0.355
10.25	3.514	1.403	0.371
10.50	3.565	1.581	0.389
10.75	3.600	1.813	0.411
11.00	3.582	2.078	0.434
11.25	3.507	2.380	0.460
11.50	3.346	2.693	0.488
11.75	3.090	2.986	0.518
12.00	2.736	3.228	0.551
12.20	2.383	3.354	0.580
12.40	1.983	3.382	0.610
12.60	1.532	3.265	0.641
12.80	1.312	2.953	0.627
13.00	1.223	2.722	0.604
13.50	1.129	2.379	0.557
14.00	1.070	2.178	0.526
14.50	1.018	2.034	0.504
15.00	0.972	1.929	0.489
15.50	0.917	1.845	0.482
16.00	0.861	1.767	0.477
16.50	0.805	1.692	0.474
17.00	0.753	1.619	0.471
17.50	0.707	1.546	0.467
18.00	0.665	1.476	0.463
18.50	0.626	1.408	0.459
19.00	0.589	1.341	0.455
19.50	0.557	1.273	0.449
20.00	0.527	1.203	0.442
21.00	0.487	1.052	0.413
22.00	0.518	0.888	0.330
23.00	0.597	0.850	0.270
24.00	0.586	0.829	0.268
25.00	0.562	0.787	0.265
26.00	0.538	0.736	0.260
27.00	0.516	0.679	0.252
28.00	0.501	0.616	0.239
29.00	0.494	0.552	0.221
30.00	0.493	0.490	0.201

Cesium (evaporated)[3]

Energy (eV)	n	k	R(φ = 0)
2.145	0.264	1.123	0.631
2.271	0.278	0.950	0.561
2.845	0.425	0.438	0.235
3.064	0.540	0.320	0.127
3.397	0.671	0.233	0.057
3.966	0.827	0.174	0.018
4.889	0.916	0.143	0.007

Chromium[4]

Energy (eV)	n	k	R(φ = 0)
0.06	21.19	42.00	0.962
0.10	11.81	29.76	0.955
0.14	15.31	26.36	0.936
0.18	8.73	25.37	0.53
0.22	5.30	20.62	0.954
0.26	3.91	17.12	0.951
0.30	3.15	14.28	0.943
0.42	3.47	8.97	0.862
0.54	3.92	7.06	0.788
0.66	3.96	5.95	0.736
0.78	4.13	5.03	0.680

OPTICAL PROPERTIES OF METALS AND SEMICONDUCTORS (continued)

Energy (eV)	n	k	$R(\phi = 0)$	Energy (eV)	n	k	$R(\phi = 0)$	Energy (eV)	n	k	$R(\phi = 0)$
0.90	4.43	4.60	0.650	20.00	0.77	0.64	0.130	5.60	1.16	1.75	0.400
1.00	4.47	4.43	0.639	20.5	0.76	0.63	0.129	5.80	1.10	1.73	0.406
1.12	4.53	4.31	0.631	21.0	0.74	0.58	0.121	6.00	1.03	1.68	0.407
1.24	4.50	4.28	0.629	21.5	0.72	0.55	0.116	6.20	0.97	1.62	0.401
1.36	4.42	4.30	0.631	22.0	0.71	0.52	0.112	6.40	0.94	1.53	0.386
1.46	4.31	4.32	0.632	22.5	0.70	0.50	0.109	6.60	0.91	1.46	0.368
1.77	3.84	4.37	0.639	23.0	0.69	0.48	0.105	6.80	0.91	1.38	0.345
2.00	3.48	4.36	0.644	23.5	0.68	0.45	0.101	7.00	0.91	1.32	0.326
2.20	3.18	4.41	0.656	24.0	0.68	0.43	0.096	7.00	0.91	1.26	0.305
2.40	2.75	4.46	0.677	24.5	0.67	0.39	0.089	7.40	0.92	1.21	0.286
2.60	2.22	4.36	0.698	25.0	0.68	0.36	0.080	7.60	0.93	1.17	0.269
2.80	1.80	4.06	0.703	25.5	0.68	0.33	0.072	7.80	0.94	1.13	0.253
3.00	1.54	3.71	0.695	26.0	0.70	0.31	0.063	8.00	0.95	1.09	0.239
3.20	1.44	3.40	0.670	26.5	0.71	0.28	0.055				
3.40	1.39	3.24	0.657	27.0	0.72	0.26	0.048	**Cobalt, single crystal, $\vec{E} \perp \hat{c}$** [5]			
3.60	1.26	3.12	0.661	27.5	0.73	0.25	0.043	0.10	5.83	32.36	0.979
3.80	1.12	2.95	0.660	28.0	0.75	0.23	0.037	0.15	4.24	21.37	0.965
4.00	1.02	2.76	0.651	29.0	0.77	0.22	0.032	0.20	3.87	15.53	0.042
4.20	0.94	2.58	0.639	30.0	0.78	0.21	0.030	0.30	4.34	10.01	0.865
4.40	0.90	2.42	0.620					0.40	4.66	7.39	0.785
4.50	0.89	2.35	0.607	**Cobalt, single crystal, $\vec{E} \parallel \hat{c}$** [5]				0.50	5.17	5.75	0.709
4.60	0.88	2.28	0.598	0.10	6.71	37.87	0.982	0.60	5.77	5.17	0.682
4.70	0.86	2.21	0.586	0.15	4.66	25.47	0.973	0.70	6.15	5.20	0.685
4.80	0.86	2.13	0.572	0.20	3.55	18.78	0.962	0.80	6.08	5.61	0.702
4.90	0.86	2.07	0.557	0.25	3.98	14.59	0.933	0.90	5.57	5.93	0.715
5.00	0.85	2.01	0.542	0.30	4.04	12.16	0.907	1.00	4.83	5.94	0.721
5.10	0.86	1.94	0.523	0.40	4.24	9.13	0.847	1.10	4.31	5.60	0.711
5.20	0.87	1.87	0.503	0.50	4.41	7.19	0.782	1.20	4.02	5.34	0.701
5.40	0.93	1.80	0.466	0.60	4.91	6.13	0.729	1.30	3.78	5.16	0.694
5.60	0.95	1.74	0.443	0.70	5.24	5.85	0.713	1.40	3.55	5.05	0.692
5.80	0.97	1.74	0.437	0.80	5.17	5.89	0.716	1.50	3.26	4.93	0.692
6.00	0.94	1.73	0.444	0.90	4.94	5.95	0.720	1.60	3.03	4.74	0.687
6.20	0.89	1.69	0.446	1.00	4.46	5.86	0.722	1.70	2.83	4.60	0.684
6.40	0.85	1.66	0.447	1.10	4.07	5.61	0.715	1.80	2.61	4.45	0.683
6.60	0.80	1.59	0.444	1.20	3.81	5.36	0.706	1.90	2.41	4.27	0.677
6.80	0.75	1.51	0.439	1.30	3.60	5.20	0.701	2.00	2.25	4.09	0.670
7.00	0.74	1.45	0.425	1.40	3.37	5.09	0.701	2.10	2.13	3.89	0.659
7.20	0.71	1.39	0.414	1.50	3.10	4.96	0.701	2.20	2.04	3.72	0.646
7.40	0.69	1.33	0.404	1.60	2.84	4.77	0.697	2.30	1.99	3.56	0.632
7.60	0.66	1.23	0.378	1.70	2.66	4.57	0.690	2.40	1.95	3.44	0.620
7.80	0.67	1.15	0.347	1.80	2.45	4.41	0.687	2.50	1.90	3.34	0.611
8.00	0.68	1.07	0.315	1.90	2.31	4.18	0.675	2.60	1.86	3.26	0.605
8.20	0.71	1.00	0.278	2.00	2.21	4.00	0.664	2.70	1.79	3.19	0.602
8.50	0.74	0.92	0.235	2.10	2.13	3.85	0.654	2.80	1.72	3.11	0.596
9.0	0.83	0.81	0.170	2.20	2.07	3.70	0.642	2.90	1.66	3.03	0.591
9.50	0.92	0.74	0.132	2.30	2.01	3.59	0.634	3.00	1.60	2.94	0.586
10.00	0.98	0.73	0.120	2.40	1.95	3.49	0.627	3.20	1.50	2.78	0.571
10.50	1.01	0.72	0.112	2.50	1.88	3.40	0.622	3.40	1.42	2.62	0.553
11.00	1.05	0.69	0.103	2.60	1.81	3.32	0.618	3.60	1.36	2.47	0.533
11.50	1.09	0.69	0.100	2.70	1.73	3.24	0.615	3.80	1.33	2.33	0.511
12.00	1.13	0.70	0.101	2.80	1.66	3.13	0.607	4.00	1.31	2.21	0.488
12.50	1.15	0.73	0.108	2.90	1.61	3.05	0.600	4.20	1.28	2.12	0.471
13.00	1.15	0.77	0.119	3.00	1.55	2.96	0.594	4.40	1.27	2.03	0.452
13.50	1.12	0.80	0.128	3.20	1.46	2.80	0.579	4.60	1.26	1.95	0.435
14.00	1.09	0.82	0.135	3.40	1.38	2.64	0.563	4.80	1.25	1.90	0.423
14.50	1.03	0.82	0.142	3.60	1.31	2.48	0.544	5.00	1.24	1.84	0.411
15.00	1.00	0.82	0.143	3.80	1.28	2.33	0.519	5.20	1.22	1.80	0.403
15.50	0.96	0.80	0.141	4.00	1.26	2.20	0.495	5.40	1.21	1.78	0.399
16.00	0.92	0.77	0.139	4.20	1.25	2.10	0.471	5.60	1.17	1.76	0.400
16.50	0.31	0.75	0.134	4.40	1.24	2.01	0.452	5.80	1.11	1.74	0.406
17.00	0.90	0.73	0.132	4.60	1.24	1.94	0.435	6.00	1.04	1.69	0.407
17.50	0.88	0.72	0.130	4.80	1.23	1.88	0.423	6.20	0.98	1.62	0.401
18.00	0.87	0.70	0.129	5.00	1.22	1.83	0.411	6.40	0.94	1.54	0.386
18.50	0.84	0.69	0.130	5.20	1.21	1.79	0.403	6.60	0.92	1.46	0.368
19.00	0.82	0.68	0.131	5.40	1.19	1.77	0.399	6.80	0.91	1.38	0.345

Energy (eV)	n	k	R(φ = 0)	Energy (eV)	n	k	R(φ = 0)	Energy (eV)	n	k	R(φ = 0)
7.00	0.91	1.32	0.326	27.00	0.88	0.38	0.043	0.02108		1.60E-03	
7.20	0.91	1.26	0.305	28.00	0.86	0.35	0.039	0.02232		1.55E-03	
7.40	0.92	1.21	0.285	29.00	0.85	0.30	0.032	0.02356		1.53E-03	
7.60	0.93	1.17	0.269	30.00	0.86	0.26	0.025	0.02480		1.50E-03	
7.80	0.94	1.13	0.253	31.00	0.88	0.24	0.020	0.02604		1.25E-03	
				32.00	0.89	0.22	0.017	0.02728		8.50E-04	
Copper[6]				33.00	0.90	0.21	0.015	0.02852		6.50E-04	
0.10	29.69	71.57	0.980	34.00	0.91	0.20	0.014	0.02976		7.00E-04	
0.50	1.71	17.63	0.979	35.00	0.92	0.20	0.013	0.03100	3.9827	8.50E-04	0.358
1.00	0.44	8.48	0.976	36.00	0.92	0.19	0.012	0.03224		1.55E-03	
1.50	0.26	5.26	0.965	37.00	0.92	0.19	0.011	0.03348		2.75E-03	
1.70	0.22	4.43	0.958	38.00	0.93	0.18	0.010	0.03472		3.55E-03	
1.75	0.21	4.25	0.956	39.00	0.93	0.17	0.009	0.03596	(3.9900)	3.05E-03	0.359
1.80	0.21	4.04	0.952	40.00	0.93	0.17	0.009	0.03720		2.75E-03	
1.85	0.22	3.85	0.947	41.00	0.94	0.16	0.008	0.03844		2.70E-03	
1.90	0.21	3.67	0.943	42.00	0.94	0.16	0.007	0.03968	(3.9930)	2.90E-03	0.359
2.00	0.27	3.24	0.910	43.00	0.94	0.15	0.007	0.04092		2.95E-03	
2.10	0.47	2.81	0.814	44.00	0.95	0.15	0.007	0.04215		3.20E-03	
2.20	0.83	2.60	0.673	45.00	0.95	0.15	0.006	0.04339		6.30E-03	
2.30	1.04	2.59	0.618	46.00	0.95	0.15	0.006	0.04463		3.40E-03	
2.40	1.12	2.60	0.602	47.00	0.95	0.14	0.006	0.04587	(3.9955)	2.50E-03	0.360
2.60	1.15	2.50	0.577	48.00	0.95	0.14	0.006	0.04711		2.10E-03	
2.80	1.17	2.36	0.545	49.00	0.95	0.14	0.005	0.04835		2.00E-03	
3.00	1.18	2.21	0.509	50.00	0.95	0.13	0.005	0.04959		8.00E-04	
3.20	1.23	2.07	0.468	51.00	0.95	0.13	0.005	0.05083		1.40E-03	
3.40	1.27	1.95	0.434	52.00	0.95	0.13	0.005	0.05207		1.35E-03	
3.60	1.31	1.87	0.407	53.00	0.96	0.12	0.004	0.05331		1.10E-03	
3.80	1.34	1.81	0.387	54.00	0.96	0.12	0.004	0.05455		8.00E-04	
4.00	1.34	1.72	0.364	55.00	0.96	0.12	0.004	0.05579		6.00E-04	
4.20	1.42	1.64	0.336	56.00	0.96	0.11	0.004	0.05703		9.0 E-04	
4.40	1.49	1.64	0.329	57.00	0.96	0.11	0.004	0.05827		6.5 E-04	
4.60	1.52	1.67	0.334	58.00	0.96	0.11	0.004	0.05951		4.6 E-04	
4.80	1.53	1.71	0.345	59.00	0.97	0.11	0.003	0.06075		4.0 E-04	
5.00	1.47	1.78	0.366	60.00	0.97	0.11	0.003	0.06199	3.9992	3.98E-04	0.360
5.20	1.38	1.80	0.380	61.00	0.97	0.11	0.003	0.06323		4.0 E-04	
5.40	1.28	1.78	0.389	62.00	0.97	0.11	0.003	0.06447		4.3 E-04	
5.60	1.18	1.74	0.391	63.00	0.96	0.10	0.003	0.06571		4.4 E-04	
5.80	1.10	1.67	0.389	64.00	0.96	0.10	0.003	0.06695	(4.0000)	4.3 E-04	0.360
6.00	1.04	1.59	0.380	65.00	0.97	0.10	0.003	0.06819		3.1 E-04	
6.50	0.96	1.37	0.329	66.00	0.97	0.10	0.003	0.06943		3.3 E-04	
7.00	0.97	1.20	0.271	67.00	0.97	0.09	0.003	0.07067		3.8 E-04	
7.50	1.00	1.09	0.230	68.00	0.97	0.09	0.002	0.07191		3.3 E-04	
8.00	1.03	1.03	0.206	69.00	0.97	0.09	0.002	0.07315		2.5 E-04	
8.50	1.03	0.98	0.189	70.00	0.97	0.09	0.002	0.07439		1.9 E-04	
9.00	1.03	0.92	0.171	75.00	0.98	0.09	0.002	0.07514		1.58E-04	
9.50	1.03	0.87	0.154	80.00	0.98	0.09	0.002	0.07749	4.0009	9.55E-05	0.360
10.00	1.04	0.82	0.139	85.00	0.97	0.09	0.002	0.07999	4.0011	1.71E-04	0.360
11.00	1.07	0.75	0.118	90.00	0.96	0.08	0.002	0.08266	4.0013	9.78E-05	0.360
12.00	1.09	0.73	0.111					0.08551	4.0015	5.77E-05	0.360
13.00	1.08	0.72	0.109	**Gallium (liquid)[7]**				0.08920		3.98E-05	
14.00	1.06	0.72	0.111	1.425	2.40	9.20	0.900	0.09460		4.59E-05	
14.50	1.03	0.72	0.111	1.550	2.09	8.50	0.898	0.09840		3.51E-05	
15.00	1.01	0.71	0.111	1.771	1.65	7.60	0.898	0.1	4.0063	3.70E-05	0.361
15.50	0.98	0.69	0.109	2.066	1.25	6.60	0.897	0.2	4.0108		0.361
16.00	0.95	0.67	0.106	2.480	0.89	5.60	0.898	0.3	4.0246		0.362
17.00	0.91	0.62	0.097	3.100	0.59	4.50	0.896	0.4	4.0429		0.364
18.00	0.89	0.56	0.084					0.5	(4.074)		0.367
19.00	0.88	0.51	0.071	**Germanium, single crystal[8]**				0.6	(4.104)	6.58E-07	0.370
20.00	0.88	0.45	0.059	0.01240	(4.0065)	3.00E-03	0.361	0.7	4.180	1.27E-04	0.377
21.00	0.90	0.41	0.048	0.01364	4.0063	2.40E-03	0.361	0.8	4.275	5.67E-03	0.385
22.00	0.92	0.38	0.040	0.01488	(4.0060)	1.70E-03	0.361	0.9	4.285	7.45E-02	0.386
23.00	0.94	0.37	0.035	0.01612	(4.0060)	1.55E-03	0.361	1.0	4.325	8.09E-02	0.390
24.00	0.96	0.37	0.035	0.01736	(4.0060)	1.50E-03	0.361	1.1	4.385	0.103	0.395
25.00	0.96	0.40	0.040	0.01860		1.50E-03		1.2	4.420	0.123	0.398
26.00	0.92	0.40	0.044	0.01984		1.60E-03		1.3	4.495	0.167	0.405

Energy (eV)	n	k	$R(\phi = 0)$
1.4	4.560	0.190	0.411
1.5	4.635	0.298	0.418
1.6	4.763	0.345	0.428
1.7	4.897	0.401	0.439
1.8	5.067	0.500	0.453
1.9	5.380	0.540	0.475
2.0	5.588	0.933	0.495
2.1	5.748	1.634	0.523
2.2	5.283	2.049	0.516
2.3	5.062	2.318	0.519
2.4	4.610	2.455	0.508
2.5	4.340	2.384	0.492
2.6	4.180	2.309	0.480
2.7	4.082	2.240	0.471
2.8	4.035	2.181	0.464
2.9	4.037	2.140	0.461
3.0	4.082	2.145	0.463
3.1	4.141	2.215	0.471
3.2	4.157	2.340	0.482
3.3	4.128	2.469	0.490
3.4	4.070	2.579	0.497
3.5	4.020	2.667	0.502
3.6	3.985	2.759	0.509
3.7	3.958	2.863	0.517
3.8	3.936	2.986	0.527
3.9	3.920	3.137	0.539
4.0	3.905	3.336	0.556
4.1	3.869	3.614	0.579
4.2	3.745	4.009	0.612
4.3	3.338	4.507	0.659
4.4	2.516	4.669	0.705
4.5	1.953	4.297	0.713
4.6	1.720	3.960	0.702
4.7	1.586	3.709	0.690
4.8	1.498	3.509	0.677
4.9	1.435	3.342	0.664
5.0	1.394	3.197	0.650
5.1	1.370	3.073	0.636
5.2	1.364	2.973	0.622
5.3	1.371	2.897	0.609
5.4	1.383	2.854	0.600
5.5	1.380	2.842	0.598
5.6	1.360	2.846	0.602
5.7	1.293	2.163	0.479
5.8	1.209	2.873	0.632
5.9	1.108	2.813	0.641
6.0	1.30	2.34	0.517
6.5	1.10	2.05	0.489
7.0	1.00	1.80	0.448
7.5		1.60	
8.0	0.92	1.40	0.348
8.5	0.92	1.20	0.282
9.0	0.92	1.14	0.262
9.5		1.00	
10.0	0.93	0.86	0.167
20.0		0.237	
22.0		0.179	
24.0		0.144	
26.0		0.110	
28.0		0.0747	
30.0		0.1020	
32.0		0.0999	
34.0		0.0856	
36.0		0.0740	
38.0		0.0651	

Energy (eV)	n	k	$R(\phi = 0)$
40.0		0.0604	

Gold, electropolished, Au (110)[9]

Energy (eV)	n	k	$R(\phi = 0)$
0.10	8.17	82.83	0.995
0.20	2.13	41.73	0.995
0.30	0.99	27.82	0.995
0.40	0.59	20.83	0.995
0.50	0.39	16.61	0.994
0.60	0.28	13.78	0.994
0.70	0.22	11.75	0.994
0.80	0.18	10.21	0.993
0.90	0.15	9.01	0.993
1.00	0.13	8.03	0.992
1.20	0.10	6.54	0.991
1.40	0.08	5.44	0.989
1.60	0.08	4.56	0.986
1.80	0.09	3.82	0.979
2.00	0.13	3.16	0.953
2.10	0.18	2.84	0.925
2.20	0.24	2.54	0.880
2.40	0.50	1.86	0.647
2.50	0.82	1.59	0.438
2.60	1.24	1.54	0.331
2.70	1.43	1.72	0.356
2.80	1.46	1.77	0.368
2.90	1.50	1.79	0.368
3.00	1.54	1.80	0.369
3.10	1.54	1.81	0.371
3.20	1.54	1.80	0.368
3.30	1.55	1.78	0.362
3.40	1.56	1.76	0.356
3.50	1.58	1.73	0.349
3.60	1.62	1.73	0.346
3.70	1.64	1.75	0.351
3.80	1.63	1.79	0.360
3.90	1.59	1.81	0.366
4.00	1.55	1.81	0.369
4.10	1.51	1.79	0.368
4.20	1.48	1.78	0.367
4.30	1.45	1.77	0.368
4.40	1.41	1.76	0.370
4.50	1.35	1.74	0.370
4.60	1.30	1.69	0.364
4.70	1.27	1.64	0.354
4.80	1.25	1.59	0.344
4.90	1.23	1.54	0.332
5.00	1.22	1.49	0.319
5.20	1.21	1.40	0.295
5.40	1.21	1.33	0.275
5.60	1.21	1.27	0.256
5.80	1.21	1.20	0.236
6.00	1.22	1.14	0.218
6.20	1.24	1.09	0.203
6.40	1.25	1.05	0.190
6.60	1.27	1.01	0.177
6.80	1.30	0.97	0.167
7.00	1.34	0.95	0.162
7.20	1.36	0.95	0.161
7.40	1.38	0.96	0.164
7.60	1.38	0.98	0.169
7.80	1.35	0.99	0.171
8.00	1.31	0.96	0.165
8.20	1.30	0.92	0.155
8.40	1.30	0.89	0.147
8.60	1.31	0.88	0.144

Energy (eV)	n	k	$R(\phi = 0)$
8.80	1.31	0.86	0.140
9.00	1.30	0.83	0.133
9.20	1.31	0.81	0.126
9.40	1.33	0.78	0.122
9.60	1.36	0.78	0.121
9.80	1.37	0.79	0.124
10.00	1.37	0.80	0.126
10.20	1.36	0.80	0.127
10.40	1.35	0.80	0.125
10.60	1.34	0.79	0.123
10.80	1.34	0.77	0.120
11.00	1.34	0.76	0.116
11.20	1.34	0.74	0.113
11.40	1.35	0.73	0.111
11.60	1.36	0.72	0.109
11.80	1.38	0.71	0.108
12.00	1.39	0.71	0.109
12.40	1.44	0.73	0.115
12.80	1.45	0.79	0.127
13.20	1.42	0.84	0.137
13.60	1.37	0.86	0.140
14.00	1.33	0.86	0.140
14.40	1.29	0.86	0.139
14.80	1.26	0.84	0.135
15.20	1.24	0.83	0.132
15.60	1.22	0.81	0.127
16.00	1.21	0.79	0.123
16.40	1.20	0.78	0.119
16.80	1.19	0.76	0.116
17.20	1.19	0.75	0.114
17.60	1.19	0.74	0.111
18.00	1.19	0.74	0.109
18.40	1.19	0.73	0.109
18.80	1.20	0.74	0.110
19.20	1.21	0.76	0.116
19.60	1.21	0.80	0.125
20.00	1.18	0.83	0.133
20.40	1.14	0.85	0.141
20.80	1.10	0.87	0.149
21.20	1.05	0.88	0.156
21.60	1.00	0.88	0.162
22.00	0.94	0.86	0.164
22.40	0.89	0.83	0.163
22.80	0.85	0.79	0.157
23.20	0.82	0.75	0.149
23.60	0.80	0.70	0.138
24.00	0.80	0.66	0.125
24.40	0.80	0.62	0.113
24.80	0.80	0.58	0.101
25.20	0.82	0.56	0.090
25.60	0.83	0.54	0.084
26.00	0.84	0.52	0.079
26.40	0.85	0.51	0.074
26.80	0.85	0.50	0.071
27.20	0.86	0.49	0.068
27.60	0.86	0.49	0.065
28.00	0.87	0.48	0.063
28.40	0.88	0.48	0.062
28.80	0.88	0.48	0.062
29.20	0.88	0.48	0.062
29.60	0.87	0.48	0.064
30.00	0.86	0.48	0.064

Hafnium, single crystal, $\vec{E} \parallel \hat{c}$ [10]

Energy (eV)	n	k	$R(\phi = 0)$
0.52	1.48	4.11	0.747

Energy (eV)	n	k	$R(\phi = 0)$
0.56	1.84	3.29	0.615
0.60	2.34	2.62	0.486
0.66	3.21	2.13	0.428
0.70	3.70	2.03	0.441
0.76	4.31	2.10	0.476
0.80	4.61	2.31	0.504
0.86	4.71	2.70	0.533
0.90	4.64	2.85	0.541
0.95	4.54	2.96	0.545
1.00	4.45	3.00	0.545
1.10	4.28	3.08	0.547
1.20	4.08	3.10	0.544
1.30	3.87	3.04	0.536
1.40	3.72	2.95	0.525
1.50	3.60	2.85	0.514
1.60	3.52	2.73	0.500
1.70	3.52	2.61	0.488
1.80	3.57	2.56	0.485
1.90	3.63	2.59	0.489
2.00	3.65	2.67	0.498
2.10	3.64	2.81	0.511
2.20	3.53	2.99	0.526
2.30	3.34	3.09	0.534
2.40	3.15	3.11	0.537
2.50	2.99	3.13	0.540
2.60	2.83	3.12	0.542
2.70	2.68	3.10	0.542
2.80	2.54	3.08	0.543
2.90	2.40	3.04	0.544
3.00	2.27	3.00	0.544
3.10	2.14	2.95	0.544
3.20	2.00	2.89	0.544
3.30	1.87	2.79	0.538
3.40	1.78	2.68	0.528
3.50	1.71	2.58	0.517
3.60	1.66	2.48	0.503
3.70	1.63	2.40	0.491
3.80	1.60	2.33	0.481
3.90	1.56	2.27	0.473
4.00	1.52	2.21	0.466
4.10	1.48	2.14	0.455
4.20	1.45	2.07	0.442
4.30	1.43	2.01	0.431
4.40	1.41	1.95	0.420
4.50	1.39	1.89	0.407
4.60	1.39	1.83	0.394
4.70	1.39	1.79	0.382
4.80	1.38	1.75	0.373
4.90	1.38	1.71	0.364
5.00	1.37	1.68	0.356
5.20	1.36	1.61	0.341
5.40	1.35	1.55	0.324
5.60	1.35	1.51	0.314
5.80	1.32	1.48	0.308
6.00	1.28	1.41	0.295
6.20	1.26	1.35	0.278
6.40	1.26	1.28	0.258
6.60	1.27	1.22	0.240
6.80	1.28	1.16	0.224
7.00	1.31	1.13	0.212
7.20	1.33	1.10	0.204
7.40	1.34	1.07	0.197
7.60	1.36	1.05	0.191
7.80	1.37	1.02	0.183
8.00	1.40	1.01	0.179

Energy (eV)	n	k	$R(\phi = 0)$
8.20	1.43	1.01	0.178
8.40	1.45	1.01	0.180
8.60	1.47	1.02	0.183
8.80	1.48	1.04	0.186
9.00	1.49	1.07	0.193
9.20	1.50	1.10	0.201
9.40	1.48	1.14	0.211
9.60	1.46	1.18	0.222
9.80	1.41	1.21	0.230
10.00	1.36	1.22	0.235
10.20	1.32	1.22	0.238
10.40	1.28	1.22	0.240
10.60	1.24	1.21	0.241
10.80	1.20	1.20	0.242
11.00	1.16	1.19	0.242
11.20	1.13	1.17	0.241
11.40	1.10	1.16	0.241
11.60	1.07	1.14	0.239
11.80	1.04	1.12	0.238
12.00	1.02	1.10	0.236
12.40	0.96	1.06	0.232
12.80	0.92	1.01	0.225
13.20	0.88	0.96	0.218
13.60	0.84	0.90	0.205
14.00	0.83	0.83	0.186
14.40	0.83	0.80	0.172
14.80	0.81	0.76	0.167
15.20	0.79	0.70	0.153
15.60	0.79	0.64	0.132
16.00	0.83	0.60	0.111
16.40	0.81	0.60	0.114
16.80	0.79	0.55	0.105
17.20	0.79	0.50	0.089
17.60	0.80	0.46	0.077
18.00	0.81	0.42	0.064
18.40	0.84	0.38	0.051
18.80	0.87	0.34	0.040
19.00	0.89	0.33	0.036
19.60	0.93	0.32	0.030
20.00	0.94	0.31	0.027
20.60	0.97	0.30	0.023
21.00	0.99	0.29	0.022
21.60	1.01	0.28	0.020
22.00	1.03	0.28	0.020
22.60	1.06	0.28	0.020
23.00	1:07	0.28	0.021
23.60	1.09	0.29	0.022
24.00	1.09	0.30	0.023
24.60	1.10	0.31	0.024

Hafnium, single crystal, $\vec{E} \perp \hat{c}$ [10]

Energy (eV)	n	k	$R(\phi = 0)$
0.52	2.25	4.65	0.723
0.56	2.34	3.66	0.623
0.60	2.84	2.89	0.512
0.66	3.71	2.35	0.469
0.70	4.26	2.21	0.482
0.76	4.97	2.33	0.521
0.80	5.41	2.62	0.554
0.86	5.46	3.36	0.593
0.90	5.22	3.62	0.601
0.95	4.95	3.72	0.602
1.00	4.76	3.76	0.602
1.10	4.43	3.80	0.601
1.20	4.07	3.74	0.594
1.30	3.79	3.55	0.578
1.40	3.61	3.36	0.561
1.50	3.55	3.13	0.540
1.60	3.58	3.01	0.529
1.70	3.63	2.98	0.526
1.80	3.66	3.02	0.530
1.90	3.63	3.14	0.541
2.00	3.51	3.26	0.551
2.10	3.35	3.33	0.558
2.20	3.18	3.36	0.563
2.30	2.99	3.39	0.568
2.40	2.78	3.35	0.569
2.50	2.65	3.26	0.562
2.60	2.54	3.22	0.560
2.70	2.42	3.17	0.559
2.80	2.31	3.13	0.558
2.90	2.20	3.08	0.558
3.00	2.08	3.05	0.561
3.10	1.94	2.98	0.560
3.20	1.83	2.88	0.555
3.30	1.74	2.78	0.547
3.40	1.68	2.69	0.538
3.50	1.62	2.61	0.529
3.60	1.57	2.52	0.519
3.70	1.53	2.45	0.510
3.80	1.49	2.38	0.501
3.90	1.45	2.32	0.493
4.00	1.41	2.25	0.484
4.10	1.38	2.18	0.474
4.20	1.35	2.11	0.462
4.30	1.33	2.05	0.451
4.40	1.31	1.99	0.438
4.50	1.30	1.93	0.427
4.60	1.29	1.88	0.415
4.70	1.28	1.82	0.402
4.80	1.28	1.77	0.389
4.90	1.27	1.73	0.379
5.00	1.27	1.69	0.367
5.20	1.27	1.62	0.349
5.40	1.27	1.57	0.335
5.60	1.26	1.52	0.322
5.80	1.24	1.48	0.313
6.00	1.21	1.42	0.302
6.20	1.19	1.36	0.285
6.40	1.18	1.29	0.265
6.60	1.19	1.22	0.244
6.80	1.21	1.18	0.230
7.00	1.22	1.14	0.217
7.20	1.23	1.10	0.206
7.40	1.26	1.06	0.194
7.60	1.28	1.04	0.187
7.80	1.30	1.02	0.180
8.00	1.33	1.00	0.174
8.20	1.35	0.99	0.173
8.40	1.38	0.99	0.173
8.60	1.40	1.00	0.174
8.80	1.42	1.02	0.178
9.00	1.43	1.04	0.184
9.20	1.45	1.08	0.193
9.40	1.43	1.12	0.204
9.60	1.40	1.16	0.214
9.80	1.37	1.19	0.223
10.00	1.32	1.21	0.230
10.20	1.27	1.21	0.234
10.40	1.23	1.20	0.235
10.60	1.19	1.20	0.237

Energy (eV)	n	k	$R(\phi = 0)$	Energy (eV)	n	k	$R(\phi = 0)$	Energy (eV)	n	k	$R(\phi = 0)$
10.80	1.15	1.19	0.237	2.40	2.07	4.14	0.689	17.60	1.30	0.87	0.140
11.00	1.12	1.17	0.237	2.50	1.98	4.00	0.682	18.00	1.30	0.93	0.154
11.20	1.08	1.16	0.237	2.60	1.91	3.86	0.673	18.40	1.27	0.97	0.166
11.40	1.05	1.14	0.236	2.70	1.85	3.73	0.665	18.80	1.24	1.00	0.176
11.60	1.03	1.12	0.235	2.80	1.81	3.61	0.655	19.20	1.20	1.03	0.187
11.80	1.00	1.10	0.233	2.90	1.77	3.51	0.646	19.60	1.15	1.05	0.197
12.00	0.97	1.08	0.231	3.00	1.73	3.43	0.640	20.00	1.10	1.06	0.205
12.40	0.92	1.04	0.226	3.20	1.62	3.26	0.629	20.50	1.04	1.05	0.210
12.80	0.88	0.99	0.219	3.40	1.53	3.05	0.610	21.00	0.99	1.04	0.215
13.20	0.83	0.94	0.211	3.60	1.52	2.81	0.573	21.50	0.94	1.02	0.220
13.60	0.80	0.88	0.196	3.80	1.61	2.69	0.541	22.00	0.89	1.00	0.222
14.00	0.79	0.81	0.177	4.00	1.64	2.68	0.535	22.50	0.84	0.99	0.228
14.40	0.80	0.77	0.160	4.20	1.58	2.71	0.549	23.00	0.79	0.96	0.232
14.80	0.77	0.73	0.154	4.40	1.45	2.68	0.561	23.50	0.76	0.92	0.228
15.20	0.76	0.68	0.140	4.60	1.31	2.60	0.567	24.00	0.73	0.87	0.223
15.60	0.76	0.61	0.119	4.80	1.18	2.49	0.570	24.50	0.70	0.83	0.218
16.00	0.81	0.58	0.099	5.00	1.10	2.35	0.559	25.00	0.69	0.79	0.209
16.40	0.78	0.57	0.102	5.20	1.04	2.22	0.543	25.50	0.68	0.76	0.200
16.80	0.77	0.53	0.092	5.40	1.00	2.09	0.522	26.00	0.67	0.72	0.192
17.20	0.77	0.48	0.077	5.60	0.98	1.98	0.499	26.50	0.67	0.69	0.181
17.60	0.79	0.44	0.065	5.80	0.96	1.86	0.474	27.00	0.66	0.66	0.174
18.00	0.80	0.39	0.053	6.00	0.95	1.78	0.454	27.50	0.66	0.63	0.166
18.40	0.82	0.36	0.041	6.20	0.94	1.68	0.427	28.00	0.66	0.61	0.158
18.80	0.86	0.33	0.032	6.40	0.94	1.59	0.401	28.50	0.66	0.59	0.151
19.00	0.88	0.32	0.030	6.60	0.94	1.50	0.375	29.00	0.65	0.57	0.148
19.60	0.91	0.31	0.025	6.80	0.95	1.42	0.345	29.50	0.64	0.55	0.145
20.00	0.93	0.30	0.023	7.00	0.97	1.34	0.318	30.00	0.64	0.53	0.140
20.60	0.96	0.29	0.021	7.20	0.99	1.27	0.290	32.00	0.62	0.44	0.119
21.00	0.97	0.29	0.020	7.40	1.02	1.20	0.262	34.00	0.64	0.35	0.091
21.60	1.00	0.28	0.019	7.60	1.03	1.14	0.241	36.00	0.69	0.27	0.059
22.00	1.01	0.28	0.019	7.80	1.08	1.06	0.208	38.00	0.73	0.24	0.044
22.60	1.03	0.27	0.018	8.00	1.13	1.03	0.191	40.00	0.76	0.22	0.034
23.00	1.05	0.28	0.019	8.20	1.18	1.00	0.179				
23.60	1.06	0.28	0.020	8.40	1.22	0.98	0.171	**Iron[5]**			
24.00	1.07	0.29	0.021	8.60	1.26	0.96	0.164	0.10	6.41	33.07	0.978
24.60	1.09	0.30	0.022	8.80	1.29	0.95	0.160	0.15	6.26	22.82	0.956
				9.00	1.33	0.94	0.157	0.20	3.68	18.23	0.958
Iridium[11]				9.20	1.36	0.95	0.159	0.26	4.98	13.68	0.911
0.10	28.49	60.62	0.975	9.40	1.39	0.95	0.161	0.30	4.87	12.05	0.892
0.15	15.32	45.15	0.973	9.60	1.42	0.97	0.163	0.36	4.68	10.44	0.867
0.20	9.69	35.34	0.972	9.80	1.44	0.99	0.169	0.40	4.42	9.75	0.858
0.25	6.86	28.84	0.969	10.00	1.45	1.01	0.175	0.50	4.14	8.02	0.817
0.30	5.16	24.25	0.967	10.20	1.45	1.04	0.182	0.60	3.93	6.95	0.783
0.35	4.11	20.79	0.964	10.40	1.44	1.07	0.187	0.70	3.78	6.17	0.752
0.40	3.42	18.06	0.960	10.60	1.43	1.09	0.193	0.80	3.65	5.60	0.725
0.45	3.05	15.82	0.954	10.80	1.41	1.12	0.200	0.90	3.52	5.16	0.700
0.50	2.98	14.06	0.944	11.00	1.38	1.13	0.206	1.00	3.43	4.79	0.678
0.60	2.79	11.58	0.925	11.20	1.34	1.14	0.208	1.10	3.33	4.52	0.660
0.70	2.93	9.78	0.895	11.40	1.31	1.13	0.208	1.20	3.24	4.26	0.641
0.80	3.14	8.61	0.862	11.60	1.28	1.12	0.206	1.30	3.16	4.07	0.626
0.90	3.19	7.88	0.840	11.80	1.25	1.10	0.203	1.40	3.12	3.87	0.609
1.00	3.15	7.31	0.822	12.00	1.24	1.08	0.199	1.50	3.05	3.77	0.601
1.10	3.04	6.84	0.808	12.40	1.21	1.05	0.191	1.60	3.00	3.60	0.585
1.20	2.96	6.41	0.791	12.80	1.19	1.01	0.181	1.70	2.98	3.52	0.577
1.30	2.85	6.07	0.779	13.20	1.18	0.98	0.173	1.80	2.92	3.46	0.573
1.40	2.72	5.74	0.767	13.60	1.17	0.95	0.165	1.90	2.89	3.37	0.563
1.50	2.65	5.39	0.750	14.00	1.16	0.91	0.155	2.00	2.85	3.36	0.563
1.60	2.68	5.08	0.728	14.40	1.17	0.88	0.147	2.10	2.80	3.34	0.562
1.70	2.69	4.92	0.716	14.80	1.18	0.87	0.142	2.20	2.74	3.33	0.563
1.80	2.64	4.81	0.710	15.20	1.19	0.84	0.136	2.30	2.65	3.34	0.567
1.90	2.57	4.68	0.704	15.60	1.20	0.83	0.133	2.40	2.56	3.31	0.567
2.00	2.50	4.57	0.699	16.00	1.21	0.83	0.131	2.50	2.46	3.31	0.570
2.10	2.40	4.48	0.697	16.40	1.23	0.82	0.129	2.60	2.34	3.30	0.576
2.20	2.29	4.38	0.695	16.80	1.25	0.82	0.127	2.70	2.23	3.25	0.575
2.30	2.18	4.26	0.692	17.20	1.28	0.83	0.131	2.80	2.12	3.23	0.580

OPTICAL PROPERTIES OF METALS AND SEMICONDUCTORS (continued)

Energy (eV)	n	k	$R(\phi = 0)$	Energy (eV)	n	k	$R(\phi = 0)$	Energy (eV)	n	k	$R(\phi = 0)$
2.90	2.01	3.17	0.580	13.17	0.84	0.79	0.161	24.00	0.74	0.27	0.045
3.00	1.88	3.12	0.583	13.33	0.84	0.78	0.160	24.17	0.74	0.26	0.044
3.10	1.78	3.04	0.580	13.50	0.83	0.77	0.159	24.33	0.74	0.26	0.043
3.20	1.70	2.96	0.576	13.67	0.82	0.76	0.157	24.50	0.74	0.25	0.042
3.30	1.62	2.87	0.572	13.83	0.81	0.75	0.154	24.67	0.75	0.25	0.040
3.40	1.55	2.79	0.565	14.00	0.81	0.73	0.151	24.83	0.75	0.24	0.039
3.50	1.50	2.70	0.556	14.17	0.80	0.72	0.149	25.00	0.75	0.24	0.038
3.60	1.47	2.63	0.548	14.33	0.80	0.71	0.146	26.00	0.76	0.21	0.031
3.70	1.43	2.56	0.542	14.50	0.79	0.79	0.144	27.00	0.78	0.18	0.026
3.83	1.38	2.49	0.534	14.67	0.79	0.69	0.141	28.00	0.79	0.16	0.021
4.00	1.30	2.39	0.527	14.83	0.78	0.67	0.138	29.00	0.81	0.14	0.017
4.17	1.26	2.27	0.510	15.00	0.78	0.66	0.135	30.00	0.82	0.13	0.014
4.33	1.23	2.18	0.494	15.17	0.78	0.65	0.131				

Lithium[12]

Energy (eV)	n	k	$R(\phi = 0)$
0.14	0.659	38.0	0.998
0.54	0.661	12.6	0.984
0.75	0.561	7.68	0.963
1.05	0.448	5.58	0.946
1.35	0.338	4.36	0.935
1.65	0.265	3.55	0.925
1.95	0.221	2.94	0.913
2.25	0.206	2.48	0.892
2.55	0.217	2.11	0.854
2.85	0.247	1.82	0.797
3.15	0.304	1.60	0.715
3.45	0.334	1.45	0.656
3.75	0.345	1.32	0.611
4.05	0.346	1.21	0.578
4.35	0.333	1.11	0.557
4.65	0.317	1.01	0.540
4.95	0.302	0.906	0.520
5.25	0.299	0.795	0.484
5.55	0.310	0.688	0.434
5.85	0.342	0.594	0.365
6.15	0.376	0.522	0.306
6.45	0.408	0.460	0.256
6.75	0.440	0.407	0.214
7.05	0.466	0.364	0.183
7.35	0.492	0.320	0.155
7.65	0.517	0.282	0.131
7.95	0.545	0.246	0.109
8.25	0.572	0.214	0.091
8.55	0.601	0.189	0.075
8.85	0.624	0.163	0.063
9.15	0.657	0.144	0.050
9.45	0.680	0.130	0.042
9.75	0.708	0.119	0.034
10.1	0.726	0.108	0.029
10.4	0.743	0.102	0.025
10.6	0.753	0.080	0.022

Magnesium (evaporated)[13]

Energy (eV)	n	k	$R(\phi = 0)$
2.145	0.48	3.71	0.880
2.270	0.57	3.47	0.843
2.522	0.53	2.92	0.805
2.845	0.52	2.65	0.777
3.064	0.52	2.05	0.681
5.167	0.10	1.60	0.894
5.636	0.15	1.50	0.832
6.200	0.20	1.40	0.765
6.889	0.25	1.30	0.693
7.750	0.20	1.20	0.722
8.857	0.15	0.95	0.730
10.335	0.25	0.40	0.419

Continuation of first material (columns 1 and 2)

Energy (eV)	n	k	$R(\phi = 0)$	Energy (eV)	n	k	$R(\phi = 0)$
4.50	1.20	2.10	0.482	15.33	0.78	0.64	0.238
4.67	1.16	2.02	0.470	15.50	0.77	0.63	0.126
4.83	1.14	1.93	0.451	15.67	0.77	0.62	0.123
5.00	1.14	1.87	0.435	15.83	0.77	0.61	0.119
5.17	1.12	1.81	0.425	16.00	0.77	0.60	0.116
5.33	1.11	1.75	0.408	16.17	0.78	0.58	0.112
5.50	1.09	1.17	0.401	16.33	0.78	0.58	0.110
5.67	1.09	1.65	0.383	16.50	0.78	0.57	0.107
5.83	1.10	1.61	0.373	16.67	0.77	0.56	0.106
6.00	1.09	1.59	0.366	16.83	0.78	0.55	0.103
6.17	1.08	1.57	0.365	17.00	0.78	0.55	0.102
6.33	1.04	1.55	0.365	17.17	0.78	0.54	0.100
6.50	1.02	1.51	0.358	17.33	0.78	0.54	0.098
6.67	1.00	1.47	0.351	17.50	0.77	0.53	0.097
6.83	0.97	1.43	0.346	17.67	0.77	0.52	0.095
7.00	0.96	1.39	0.333	17.83	0.78	0.51	0.092
7.17	0.94	1.35	0.327	18.00	0.78	0.51	0.091
7.33	0.94	1.30	0.311	18.17	0.78	0.51	0.090
7.50	0.94	1.26	0.298	18.33	0.78	0.50	0.089
7.67	0.94	1.23	0.288	18.50	0.77	0.50	0.089
7.83	0.94	1.21	0.279	18.67	0.77	0.50	0.088
8.00	0.94	1.18	0.272	18.83	0.77	0.49	0.087
8.17	0.94	1.16	0.265	19.00	0.77	0.49	0.087
8.33	0.94	1.14	0.258	19.17	0.76	0.49	0.088
8.50	0.94	1.12	0.251	19.33	0.76	0.48	0.087
8.67	0.94	1.10	0.246	19.50	0.75	0.47	0.086
8.83	0.92	1.08	0.240	19.67	0.75	0.47	0.085
9.00	0.93	1.07	0.236	19.83	0.75	0.46	0.084
9.17	0.92	1.06	0.233	20.00	0.74	0.45	0.083
9.33	0.91	1.04	0.231	20.17	0.74	0.44	0.081
9.50	0.90	1.02	0.226	20.33	0.74	0.44	0.081
9.67	0.90	1.00	0.221	20.50	0.74	0.42	0.080
9.83	0.89	0.99	0.218	20.67	0.73	0.43	0.079
10.00	0.88	0.97	0.213	20.83	0.73	0.42	0.078
10.17	0.87	0.94	0.203	21.00	0.73	0.41	0.077
10.33	0.87	0.91	0.196	21.17	0.72	0.40	0.076
10.50	0.87	0.89	0.189	21.33	0.72	0.39	0.074
10.67	0.88	0.87	0.179	21.50	0.72	0.38	0.073
10.83	0.89	0.85	0.170	21.67	0.72	0.38	0.071
11.00	0.91	0.83	0.162	21.83	0.72	0.37	0.070
11.17	0.92	0.83	0.159	22.00	0.72	0.36	0.068
11.33	0.93	0.84	0.159	22.17	0.71	0.35	0.067
11.50	0.93	0.84	0.160	22.33	0.72	0.34	0.064
11.67	0.93	0.84	0.162	22.50	0.72	0.34	0.063
11.83	0.92	0.84	0.163	22.67	0.72	0.33	0.062
12.00	0.91	0.84	0.163	22.83	0.72	0.32	0.059
12.17	0.90	0.84	0.165	23.00	0.72	0.31	0.058
12.33	0.89	0.83	0.164	23.17	0.72	0.30	0.056
12.50	0.98	0.83	0.165	23.33	0.72	0.29	0.054
12.67	0.87	0.82	0.166	23.50	0.73	0.28	0.050
12.83	0.86	0.81	0.166	23.67	0.73	0.28	0.049
13.00	0.85	0.80	0.162	23.83	0.74	0.27	0.047

Energy (eV)	n	k	$R(\phi = 0)$	Energy (eV)	n	k	$R(\phi = 0)$	Energy (eV)	n	k	$R(\phi = 0)$
Manganese[14]				2.4	1.384	4.407	0.779	17.0	1.177	0.367	0.034
0.64	3.89	5.95	0.738	2.6	1.186	4.090	0.779	17.5	1.184	0.366	0.034
0.77	3.78	5.41	0.710	2.8	1.027	3.802	0.779	18.0	1.191	0.367	0.035
0.89	3.65	5.02	0.688	3.0	0.898	3.538	0.777	18.5	1.195	0.367	0.035
1.02	3.48	4.74	0.673	3.2	0.798	3.294	0.773	19.0	1.200	0.366	0.035
1.14	3.30	4.53	0.662	3.4	0.713	3.074	0.770	19.5	1.208	0.364	0.035
1.26	3.10	4.35	0.653	3.6	0.644	2.860	0.763				
1.39	2.97	4.18	0.643	3.8	0.589	2.665	0.755	**Molybdenum[16]**			
1.51	2.83	4.03	0.634	4.0	0.542	2.502	0.749	0.10	18.53	68.51	0.985
1.64	2.70	3.91	0.627	4.2	0.507	2.341	0.738	0.15	8.78	47.54	0.985
1.76	2.62	3.78	0.617	4.4	0.477	2.195	0.727	0.20	5.10	35.99	0.985
1.88	2.56	3.65	0.606	4.6	0.452	2.058	0.715	0.25	3.36	28.75	0.984
2.01	2.51	3.54	0.596	4.8	0.431	1.929	0.701	0.30	2.44	23.80	0.983
2.13	2.47	3.43	0.585	5.0	0.414	1.806	0.685	0.34	2.00	20.84	0.982
2.26	2.39	3.33	0.577	5.2	0.401	1.687	0.666	0.38	1.70	18.44	0.980
2.38	2.32	3.23	0.567	5.4	0.394	1.569	0.642	0.42	1.57	16.50	0.978
2.50	2.25	3.14	0.559	5.6	0.386	1.454	0.617	0.46	1.46	14.91	0.975
2.63	2.19	3.06	0.552	5.7	0.386	1.396	0.601	0.50	1.37	13.55	0.971
2.75	2.11	2.98	0.545	5.8	0.386	1.341	0.585	0.54	1.35	12.36	0.966
2.88	2.06	2.90	0.536	5.9	0.385	1.287	0.569	0.58	1.34	11.34	0.960
3.00	2.00	2.82	0.528	6.0	0.386	1.232	0.551	0.62	1.38	10.44	0.952
3.12	1.96	2.74	0.518	6.1	0.388	1.176	0.531	0.66	1.43	9.67	0.942
3.25	1.92	2.67	0.509	6.2	0.390	1.118	0.510	0.70	1.48	8.99	0.932
3.37	1.89	2.59	0.498	6.3	0.399	1.058	0.481	0.74	1.51	8.38	0.921
3.50	1.89	2.51	0.484	6.4	0.412	1.002	0.450	0.78	1.60	7.83	0.906
3.62	1.87	2.45	0.475	6.5	0.428	0.949	0.418	0.82	1.64	7.35	0.892
3.74	1.86	2.38	0.463	6.6	0.436	0.898	0.392	0.86	1.70	6.89	0.876
3.87	1.86	2.32	0.451	6.7	0.438	0.836	0.367	9.90	1.74	6.48	0.859
3.99	1.86	2.25	0.438	6.8	0.459	0.756	0.320	1.00	1.94	5.58	0.805
4.12	1.86	2.19	0.427	6.9	0.510	0.676	0.255	1.10	2.15	4.85	0.743
4.24	1.85	2.14	0.417	7.0	0.585	0.617	0.191	1.20	2.44	4.22	0.671
4.36	1.85	2.08	0.406	7.1	0.663	0.589	0.148	1.30	2.77	3.74	0.608
4.49	1.86	2.03	0.395	7.2	0.717	0.584	0.128	1.40	3.15	3.40	0.562
4.61	1.85	1.99	0.388	7.3	0.769	0.575	0.111	1.50	3.53	3.30	0.550
4.74	1.84	1.94	0.378	7.4	0.817	0.574	0.100	1.60	3.77	3.41	0.562
4.86	1.83	1.91	0.372	7.5	0.860	0.580	0.094	1.70	3.84	3.51	0.570
4.98	1.82	1.86	0.362	7.6	0.893	0.597	0.093	1.80	3.81	3.58	0.576
5.11	1.82	1.82	0.354	7.8	0.929	0.623	0.096	1.90	3.74	3.58	0.576
5.23	1.81	1.79	0.348	8.0	0.946	0.639	0.098	2.00	3.68	3.52	0.571
5.36	1.78	1.76	0.342	8.2	0.952	0.645	0.099	2.10	3.68	3.45	0.565
5.48	1.74	1.73	0.337	8.4	0.953	0.638	0.097	2.20	3.76	3.41	0.562
5.60	1.73	1.70	0.331	8.6	0.956	0.624	0.093	2.30	3.79	3.61	0.578
5.73	1.72	1.67	0.325	8.8	0.965	0.607	0.087	2.40	3.59	3.78	0.594
5.85	1.70	1.64	0.319	9.0	0.975	0.588	0.082	2.50	3.36	3.73	0.591
5.98	1.67	1.61	0.313	9.2	0.988	0.568	0.076	2.60	3.22	3.61	0.582
6.10	1.63	1.58	0.307	9.4	1.009	0.548	0.069	2.70	3.13	3.51	0.573
6.22	1.62	1.55	0.301	9.6	1.044	0.541	0.066	2.80	3.08	3.42	0.565
6.35	1.59	1.52	0.295	9.8	1.061	0.557	0.069	2.90	3.05	3.33	0.566
6.47	1.55	1.50	0.292	10.0	1.062	0.567	0.071	3.00	3.04	3.27	0.550
6.60	1.48	1.47	0.288	10.2	1.054	0.569	0.072	3.10	3.03	3.21	0.544
				10.4	1.045	0.561	0.070	3.20	3.05	3.18	0.540
Mercury (liquid)[15]				10.6	1.041	0.550	0.068	3.30	3.06	3.18	0.540
0.2	13.99	14.27	0.869	10.8	1.039	0.537	0.065	3.40	3.06	3.19	0.541
0.3	11.37	11.95	0.846	11.0	1.039	0.523	0.062	3.50	3.06	3.21	0.543
0.4	9.741	10.65	0.830	11.5	1.050	0.491	0.055	3.60	3.05	3.23	0.546
0.5	8.528	9.805	0.818	12.0	1.064	0.467	0.050	3.70	3.04	3.27	0.550
0.6	7.574	9.195	0.808	12.5	1.078	0.445	0.045	3.80	3.04	3.31	0.554
0.8	6.086	8.312	0.796	13.0	1.092	0.430	0.042	3.90	3.04	3.40	0.564
1.0	4.962	7.643	0.789	13.5	1.104	0.416	0.040	4.00	3.01	3.51	0.576
1.2	4.050	7.082	0.786	14.0	1.115	0.404	0.038	4.20	2.77	3.77	0.610
1.4	3.324	6.558	0.785	14.5	1.125	0.394	0.037	4.40	2.39	3.88	0.640
1.6	2.746	6.054	0.783	15.0	1.135	0.383	0.035	4.60	2.06	3.84	0.658
1.8	2.284	5.582	0.782	15.5	1.146	0.374	0.034	4.80	1.75	3.76	0.678
2.0	1.910	5.150	0.782	16.0	1.159	0.368	0.034	5.00	1.46	3.62	0.695
2.2	1.620	4.751	0.780	16.5	1.170	0.367	0.034	5.20	1.22	3.42	0.706

Energy (eV)	n	k	$R(\phi = 0)$	Energy (eV)	n	k	$R(\phi = 0)$	Energy (eV)	n	k	$R(\phi = 0)$
5.40	1.07	3.20	0.706	27.00	0.73	0.29	0.050	4.80	1.53	2.11	0.435
5.60	0.96	2.99	0.700	27.50	0.76	0.28	0.041	5.00	1.40	2.10	0.449
5.80	0.89	2.80	0.688	28.00	0.79	0.27	0.036	5.20	1.27	2.04	0.454
6.00	0.85	2.64	0.674	28.50	0.81	0.26	0.031	5.40	1.16	1.94	0.449
6.20	0.81	2.50	0.660	29.00	0.83	0.26	0.028	5.60	1.09	1.83	0.435
6.40	0.79	2.36	0.641	29.50	0.86	0.26	0.025	5.80	1.04	1.73	0.417
6.60	0.78	2.24	0.619	30.00	0.88	0.26	0.023	6.20	1.00	1.54	0.371
6.80	0.78	2.13	0.592	31.00	0.92	0.29	0.024	6.40	1.01	1.46	0.345
7.00	0.80	2.04	0.568	32.00	0.92	0.32	0.030	6.60	1.01	1.40	0.325
7.20	0.81	1.98	0.548	33.00	0.90	0.33	0.032	6.80	1.02	1.35	0.308
7.40	0.81	1.95	0.542	34.00	0.91	0.34	0.034	7.00	1.03	1.30	0.291
7.60	0.75	1.90	0.552	35.00	0.87	0.37	0.043	7.20	1.03	1.27	0.282
7.80	0.71	1.81	0.542	36.00	0.82	0.34	0.043	7.40	1.03	1.24	0.273
8.00	0.69	1.73	0.530	37.00	0.81	0.30	0.038	7.60	1.02	1.22	0.265
8.20	0.67	1.65	0.512	38.00	0.81	0.27	0.033	7.80	1.01	1.18	0.256
8.40	0.66	1.57	0.495	39.00	0.82	0.25	0.029	8.00	1.01	1.15	0.248
8.60	0.65	1.49	0.475	40.00	0.83	0.23	0.025	8.20	1.00	1.13	0.242
8.80	0.65	1.41	0.450					8.40	0.99	1.11	0.235
9.00	0.65	1.33	0.420	**Nickel**[17]				8.60	0.98	1.08	0.228
9.20	0.67	1.25	0.385	0.10	9.54	45.82	0.983	8.80	0.97	1.05	0.220
9.40	0.69	1.19	0.355	0.15	5.45	30.56	0.978	9.00	0.97	1.01	0.211
9.60	0.71	1.12	0.320	0.20	4.12	22.48	0.969	9.20	0.96	0.99	0.203
9.80	0.74	1.05	0.285	0.25	4.25	17.68	0.950	9.40	0.95	0.96	0.194
10.00	0.77	0.99	0.250	0.30	4.19	15.05	0.934	9.60	0.95	0.93	0.185
10.20	0.81	0.93	0.217	0.35	4.03	13.05	0.918	9.80	0.95	0.89	0.175
10.40	0.86	0.88	0.188	0.40	3.84	11.43	0.900	10.00	0.95	0.87	0.166
10.60	0.91	0.83	0.162	0.50	4.03	9.64	0.864	10.20	0.95	0.83	0.155
10.80	0.98	0.79	0.138	0.60	3.84	8.35	0.835	10.40	0.95	0.80	0.145
11.00	1.05	0.77	0.125	0.70	3.59	7.48	0.813	10.60	0.97	0.76	0.129
11.20	1.12	0.78	0.123	0.80	3.38	6.82	0.794	10.80	0.99	0.75	0.123
11.40	1.18	0.80	0.125	0.90	3.18	6.23	0.774	11.00	1.01	0.73	0.115
11.60	1.23	0.85	0.135	1.00	3.06	5.74	0.753	11.25	1.04	0.72	0.111
11.80	1.25	0.89	0.145	1.10	2.97	5.38	0.734	11.50	1.05	0.71	0.109
12.00	1.26	0.92	0.154	1.20	2.85	5.10	0.721	11.75	1.07	0.71	0.108
12.40	1.25	0.98	0.168	1.30	2.74	4.85	0.708	12.00	1.07	0.71	0.108
12.80	1.23	1.00	0.178	1.40	2.65	4.63	0.695	12.25	1.07	0.71	0.107
13.20	1.20	1.02	0.185	1.50	2.53	4.47	0.688	12.50	1.08	0.71	0.106
13.60	1.17	1.02	0.187	1.60	2.43	4.31	0.679	12.75	1.08	0.71	0.106
14.00	1.15	1.01	0.185	1.70	2.28	4.18	0.677	13.00	1.08	0.71	0.105
14.40	1.13	1.00	0.182	1.80	2.14	4.01	0.670	13.25	1.08	0.71	0.105
14.80	1.13	0.99	0.179	1.90	2.02	3.82	0.659	13.50	1.07	0.70	0.105
15.00	1.14	0.99	0.179	2.00	1.92	3.65	0.649	13.75	1.07	0.70	0.105
15.60	1.15	1.01	0.184	2.10	1.85	3.48	0.634	14.00	1.07	0.71	0.106
16.00	1.14	1.04	0.194	2.20	1.80	3.33	0.620	14.25	1.06	0.70	0.106
16.60	1.10	1.10	0.216	2.30	1.75	3.19	0.605	14.50	1.05	0.70	0.106
17.00	1.04	1.12	0.233	2.40	1.71	3.06	0.590	14.75	1.04	0.70	0.107
17.60	0.94	1.14	0.257	2.50	1.67	2.93	0.575	15.00	1.03	0.70	0.107
18.00	0.87	1.12	0.270	2.60	1.65	2.81	0.557	15.25	1.02	0.69	0.106
18.60	0.77	1.08	0.283	2.70	1.64	2.71	0.542	15.50	1.01	0.69	0.105
19.00	0.71	1.02	0.284	2.80	1.63	2.61	0.525	15.75	1.00	0.68	0.104
19.60	0.66	0.94	0.275	2.90	1.62	2.52	0.509	16.00	0.99	0.67	0.103
20.00	0.64	0.89	0.264	3.00	1.61	2.44	0.495	16.50	0.98	0.66	0.101
20.60	0.62	0.81	0.245	3.10	1.61	2.36	0.480	17.00	0.96	0.64	0.098
21.00	0.61	0.77	0.234	3.20	1.61	2.30	0.467	17.50	0.94	0.63	0.096
21.60	0.61	0.71	0.215	3.30	1.61	2.23	0.454	18.00	0.92	0.61	0.092
22.00	0.60	0.69	0.207	3.40	1.62	2.17	0.441	18.50	0.91	0.58	0.087
22.60	0.59	0.63	0.195	3.50	1.63	2.11	0.428	19.00	0.90	0.56	0.082
23.00	0.58	0.60	0.185	3.60	1.64	2.07	0.416	19.50	0.90	0.54	0.077
23.60	0.58	0.53	0.166	3.70	1.66	2.02	0.405	20.00	0.89	0.51	0.071
24.00	0.58	0.49	0.151	3.80	1.69	1.99	0.397	20.50	0.89	0.49	0.066
24.60	0.60	0.43	0.124	3.90	1.72	1.98	0.393	21.00	0.90	0.47	0.061
25.00	0.62	0.39	0.106	4.00	1.73	1.98	0.392	21.50	0.91	0.46	0.057
25.60	0.66	0.35	0.085	4.20	1.74	2.01	0.396	22.00	0.91	0.45	0.055
26.00	0.68	0.33	0.072	4.40	1.71	2.06	0.409	22.50	0.91	0.44	0.053
26.50	0.71	0.31	0.060	4.60	1.63	2.09	0.421	23.00	0.92	0.44	0.051

Energy (eV)	n	k	R(φ = 0)	Energy (eV)	n	k	R(φ = 0)	Energy (eV)	n	k	R(φ = 0)
23.50	0.91	0.44	0.052	4.60	2.39	2.56	0.470	23.60	0.88	0.30	0.029
24.00	0.90	0.43	0.051	4.80	2.32	2.52	0.465	24.00	0.91	0.29	0.025
24.50	0.90	0.43	0.051	5.00	2.26	2.57	0.475	24.60	0.94	0.28	0.022
25.00	0.89	0.42	0.050	5.20	2.16	2.62	0.487	25.00	0.96	0.27	0.020
26.00	0.88	0.39	0.046	5.40	2.00	2.68	0.505	25.60	0.99	0.26	0.018
27.00	0.87	0.37	0.042	5.60	1.81	2.67	0.518	26.00	1.00	0.26	0.017
28.00	0.87	0.35	0.040	5.80	1.63	2.60	0.522	26.60	1.03	0.25	0.016
29.00	0.86	0.34	0.037	6.00	1.49	2.49	0.520	27.00	1.04	0.25	0.015
30.00	0.86	0.32	0.034	6.20	1.38	2.38	0.512	27.60	1.06	0.25	0.015
35.00	0.86	0.24	0.022	6.40	1.31	2.25	0.496	28.00	1.08	0.24	0.015
40.00	0.87	0.18	0.014	6.60	1.26	2.14	0.480	28.60	1.11	0.24	0.016
45.00	0.88	0.13	0.008	6.80	1.24	2.04	0.460	29.00	1.13	0.25	0.017
50.00	0.92	0.10	0.004	7.00	1.23	1.96	0.441	29.60	1.16	0.26	0.020
60.00	0.96	0.08	0.002	7.20	1.22	1.91	0.430	30.00	1.18	0.28	0.023
65.00	0.98	0.09	0.002	7.40	1.20	1.88	0.427	31.00	1.18	0.31	0.026
68.00	0.96	0.12	0.004	7.60	1.14	1.85	0.430	32.00	1.20	0.34	0.031
70.00	0.94	0.11	0.004	7.80	1.07	1.78	0.428	33.00	1.21	0.38	0.038
75.00	0.94	0.09	0.003	8.00	1.02	1.69	0.412	34.00	1.20	0.42	0.044
80.00	0.94	0.07	0.002	8.20	1.00	1.60	0.390	35.20	1.17	0.47	0.051
90.00	0.94	0.06	0.002	8.40	0.99	1.51	0.365	36.00	1.15	0.50	0.056
				8.60	0.99	1.43	0.340	37.50	1.07	0.53	0.064
Niobium[18]				8.70	0.99	1.39	0.328	39.50	0.95	0.50	0.063
0.12	15.99	53.20	0.979	8.80	1.00	1.36	0.315	40.50	0.92	0.47	0.059
0.20	7.25	34.14	0.976	9.00	1.01	1.29	0.290				
0.24	5.47	28.88	0.975	9.20	1.04	1.22	0.265	**Osmium (Polycrystalline)[9]**			
0.28	4.26	24.95	0.974	9.40	1.07	1.18	0.245	0.10	4.08	50.23	0.994
0.35	3.11	20.03	0.970	9.60	1.10	1.13	0.227	0.15	2.90	33.60	0.990
0.45	2.28	15.58	0.964	9.80	1.13	1.09	0.209	0.20	2.44	25.11	0.985
0.55	1.83	12.67	0.956	10.00	1.18	1.05	0.194	0.25	2.35	19.99	0.977
0.65	1.57	10.59	0.947	10.20	1.23	1.04	0.187	0.30	2.23	16.54	0.969
0.75	1.41	9.00	0.935	10.40	1.27	1.04	0.185	0.35	2.33	14.06	0.955
0.85	1.35	7.74	0.918	10.60	1.30	1.06	0.190	0.40	2.45	12.32	0.940
0.95	1.35	6.70	0.893	10.80	1.32	1.08	0.195	0.45	2.43	11.02	0.927
1.05	1.44	5.86	0.857	11.00	1.32	1.10	0.200	0.50	2.41	9.97	0.913
1.15	1.55	5.18	0.814	11.20	1.31	1.12	0.204	0.55	2.33	9.12	0.901
1.25	1.65	4.63	0.768	11.40	1.30	1.13	0.207	0.60	2.21	8.37	0.890
1.35	1.76	4.13	0.715	11.60	1.28	1.13	0.209	0.65	2.11	7.68	0.877
1.45	1.95	3.68	0.650	11.80	1.27	1.13	0.210	0.70	2.02	7.04	0.862
1.55	2.15	3.37	0.595	12.00	1.25	1.12	0.209	0.75	2.00	6.46	0.842
1.65	2.36	3.13	0.552	12.40	1.24	1.10	0.204	0.80	2.00	5.95	0.820
1.75	2.54	2.99	0.527	12.80	1.24	1.09	0.200	0.85	2.01	5.51	0.796
1.85	2.69	2.89	0.510	13.20	1.24	1.09	0.201	0.90	2.03	5.10	0.769
1.95	2.82	2.86	0.505	13.60	1.23	1.12	0.208	0.95	2.05	4.74	0.742
2.05	2.89	2.87	0.505	14.00	1.20	1.13	0.216	1.00	2.09	4.41	0.712
2.15	2.92	2.87	0.505	14.40	1.16	1.15	0.225	1.10	2.15	3.84	0.651
2.25	2.93	2.87	0.505	14.80	1.11	1.16	0.234	1.20	2.16	3.35	0.592
2.35	2.92	2.88	0.506	15.00	1.08	1.16	0.238	1.30	2.25	2.77	0.506
2.45	2.89	2.90	0.509	15.60	0.99	1.14	0.247	1.40	2.49	2.23	0.419
2.55	2.83	2.92	0.512	16.00	0.92	1.11	0.250	1.50	2.84	1.80	0.369
2.65	2.74	2.90	0.511	16.60	0.85	1.04	0.245	1.60	3.36	1.62	0.379
2.75	2.66	2.86	0.507	17.00	0.80	0.99	0.240	1.70	3.70	1.75	0.411
2.85	2.58	2.80	0.500	17.20	0.79	0.96	0.236	1.80	3.78	1.83	0.423
3.00	2.51	2.68	0.485	17.40	0.77	0.93	0.230	1.90	3.81	1.75	0.418
3.10	2.48	2.60	0.475	17.80	0.75	0.87	0.217	2.00	3.98	1.60	0.418
3.20	2.45	2.53	0.465	18.00	0.74	0.85	0.209	2.10	4.26	1.54	0.432
3.30	2.44	2.45	0.453	18.60	0.73	0.77	0.185	2.20	4.58	1.62	0.457
3.40	2.46	2.38	0.442	19.00	0.72	0.72	0.170	2.30	4.84	1.76	0.479
3.50	2.48	2.33	0.435	19.60	0.72	0.66	0.150	2.40	5.10	2.01	0.506
3.60	2.52	2.29	0.428	20.00	0.72	0.62	0.137	2.50	5.28	2.38	0.532
3.70	2.56	2.27	0.426	20.60	0.71	0.55	0.119	2.60	5.36	2.82	0.557
3.80	2.59	2.28	0.427	21.00	0.72	0.50	0.100	2.70	5.30	3.29	0.580
3.90	2.62	2.29	0.429	21.60	0.75	0.43	0.075	2.80	5.07	3.78	0.603
4.00	2.64	2.33	0.434	22.00	0.78	0.40	0.063	2.90	4.65	4.18	0.624
4.20	2.64	2.42	0.447	22.60	0.82	0.35	0.045	3.00	4.05	4.40	0.639
4.40	2.53	2.56	0.467	23.00	0.85	0.33	0.038	3.20	3.29	3.96	0.614

Energy (eV)	n	k	$R(\phi = 0)$	Energy (eV)	n	k	$R(\phi = 0)$	Energy (eV)	n	k	$R(\phi = 0)$
3.40	2.93	3.79	0.607	20.00	0.96	1.10	0.239	2.50	1.41	3.48	0.685
3.60	2.75	3.45	0.577	20.40	0.93	1.09	0.240	2.60	1.37	3.36	0.676
3.80	2.73	3.32	0.562	20.80	0.89	1.05	0.240	2.70	1.32	3.25	0.668
4.00	2.71	3.34	0.565	21.20	0.86	1.02	0.237	2.80	1.29	3.13	0.658
4.20	2.53	3.44	0.584	21.60	0.83	0.99	0.235	2.90	1.26	3.03	0.648
4.40	2.24	3.44	0.599	22.00	0.80	0.96	0.230	3.00	1.23	2.94	0.639
4.60	2.01	3.31	0.598	22.40	0.78	0.93	0.226	3.10	1.20	2.85	0.630
4.80	1.88	3.19	0.592	22.80	0.77	0.90	0.220	3.20	1.17	2.77	0.622
5.00	1.74	3.12	0.596	23.20	0.75	0.88	0.217	3.30	1.14	2.68	0.613
5.20	1.58	3.00	0.597	23.60	0.75	0.86	0.211	3.40	1.12	2.60	0.602
5.40	1.46	2.88	0.593	24.00	0.73	0.84	0.209	3.50	1.10	2.52	0.591
5.60	1.36	2.77	0.589	24.40	0.72	0.82	0.207	3.60	1.08	2.45	0.581
5.80	1.27	2.65	0.582	24.80	0.70	0.80	0.205	3.70	1.07	2.38	0.570
6.00	1.20	2.54	0.575	25.20	0.69	0.77	0.202	3.80	1.06	2.31	0.558
6.20	1.13	2.44	0.571	25.60	0.67	0.75	0.199	3.90	1.05	2.25	0.547
6.40	1.06	2.33	0.562	26.00	0.66	0.72	0.195	4.00	1.03	2.19	0.537
6.60	1.01	2.21	0.548	26.40	0.65	0.69	0.189	4.20	1.04	2.09	0.510
6.80	0.97	2.11	0.532	26.80	0.63	0.66	0.183	4.40	1.03	2.01	0.493
7.00	0.95	2.00	0.514	27.20	0.65	0.62	0.165	4.60	1.03	1.94	0.476
7.20	0.92	1.91	0.497	28.00	0.64	0.59	0.156	4.80	1.01	1.90	0.470
7.40	0.91	1.81	0.476	28.40	0.64	0.57	0.148	5.00	0.96	1.86	0.472
7.60	0.90	1.72	0.451	28.80	0.65	0.55	0.140	5.20	0.90	1.79	0.474
7.80	0.90	1.63	0.426	29.20	0.65	0.53	0.134	5.40	0.85	1.70	0.463
8.00	0.91	1.55	0.400	29.60	0.65	0.51	0.128	5.60	0.81	1.62	0.449
8.20	0.91	1.48	0.375	30.00	0.65	0.49	0.121	5.80	0.78	1.54	0.437
8.40	0.94	1.40	0.344	31.00	0.65	0.45	0.111	6.00	0.76	1.45	0.418
8.60	0.96	1.34	0.319	32.00	0.66	0.41	0.095	6.20	0.74	1.37	0.397
8.80	0.98	1.29	0.296	33.00	0.68	0.37	0.079	6.40	0.73	1.29	0.375
9.00	1.01	1.24	0.274	34.00	0.70	0.34	0.068	6.60	0.72	1.21	0.350
9.20	1.04	1.19	0.255	35.00	0.72	0.31	0.057	6.80	0.73	1.13	0.316
9.40	1.08	1.16	0.238	36.00	0.74	0.29	0.048	7.00	0.73	1.05	0.287
9.60	1.10	1.14	0.229	37.00	0.77	0.27	0.040	7.20	0.75	0.98	0.255
9.80	1.13	1.11	0.217	38.00	0.79	0.26	0.035	7.40	0.77	0.91	0.223
10.00	1.16	1.10	0.209	39.00	0.81	0.26	0.031	7.60	0.79	0.85	0.195
10.20	1.19	1.08	0.203	40.00	0.84	0.26	0.026	7.80	0.83	0.78	0.163
10.30	1.20	1.08	0.201					8.00	0.88	0.73	0.133
10.40	1.22	1.08	0.200	Palladium[19]				8.20	0.94	0.70	0.117
10.50	1.23	1.09	0.201	0.10	4.13	54.15	0.994	8.40	0.96	0.70	0.114
10.60	1.24	1.10	0.203	0.15	3.13	35.82	0.990	8.60	1.00	0.65	0.097
10.80	1.25	1.11	0.206	0.20	3.07	26.59	0.983	8.80	1.04	0.65	0.094
11.00	1.24	1.13	0.213	0.26	3.11	20.15	0.971	9.00	1.07	0.64	0.090
11.20	1.23	1.14	0.217	0.30	3.56	17.27	0.955	9.50	1.12	0.65	0.089
11.40	1.19	1.15	0.223	0.36	3.98	14.41	0.932	10.00	1.14	0.65	0.088
11.60	1.17	1.12	0.216	0.40	4.27	13.27	0.916	10.50	1.16	0.65	0.087
11.80	1.16	1.10	0.211	0.46	4.27	12.11	0.902	11.00	1.18	0.64	0.086
12.00	1.15	1.08	0.205	0.50	4.10	11.44	0.896	11.50	1.19	0.65	0.087
12.40	1.14	1.03	0.191	0.56	3.92	10.49	0.883	12.00	1.20	0.66	0.089
12.80	1.15	1.01	0.183	0.60	3.80	9.96	0.876	12.50	1.19	0.67	0.091
13.20	1.16	0.98	0.174	0.72	3.51	8.70	0.854	13.00	1.18	0.67	0.091
13.60	1.17	0.97	0.170	0.80	3.35	8.06	0.840	13.50	1.18	0.67	0.092
14.00	1.17	0.96	0.169	1.00	2.99	6.89	0.811	14.00	1.17	0.67	0.093
14.40	1.16	0.94	0.165	1.10	2.81	6.46	0.800	14.50	1.15	0.68	0.095
14.80	1.16	0.91	0.156	1.20	2.65	6.10	0.790	15.00	1.13	0.69	0.098
15.20	1.17	0.89	0.148	1.30	2.50	5.78	0.781	15.50	1.10	0.68	0.096
15.60	1.20	0.86	0.140	1.40	2.34	5.50	0.774	16.00	1.08	0.66	0.092
16.00	1.25	0.87	0.140	1.50	2.17	5.22	0.767	16.50	1.06	0.63	0.086
16.40	1.28	0.90	0.147	1.60	2.08	4.95	0.755	17.00	1.07	0.61	0.081
16.80	1.28	0.94	0.157	1.70	2.00	4.72	0.745	17.50	1.06	0.61	0.080
17.20	1.27	0.97	0.167	1.80	1.92	4.54	0.737	18.00	1.07	0.59	0.077
17.60	1.26	1.01	0.178	1.90	1.82	4.35	0.729	18.50	1.07	0.59	0.077
18.00	1.23	1.04	0.189	2.00	1.75	4.18	0.721	19.00	1.08	0.59	0.077
18.40	1.19	1.08	0.200	2.10	1.67	4.03	0.714	19.50	1.08	0.61	0.080
18.80	1.14	1.10	0.210	2.20	1.60	3.88	0.707	20.00	1.07	0.65	0.090
19.20	1.10	1.10	0.219	2.30	1.53	3.75	0.700	20.50	1.03	0.67	0.098
19.60	1.05	1.11	0.227	2.40	1.47	3.61	0.693	21.00	0.99	0.67	0.103

Energy (eV)	n	k	R(φ = 0)	Energy (eV)	n	k	R(φ = 0)	Energy (eV)	n	k	R(φ = 0)
21.50	0.95	0.66	0.103	6.00	1.38	1.40	0.276	28.50	0.75	0.59	0.121
22.00	0.91	0.64	0.103	6.20	1.39	1.35	0.261	29.00	0.75	0.58	0.118
22.50	0.88	0.62	0.101	6.40	1.42	1.29	0.246	29.50	0.74	0.58	0.120
23.00	0.86	0.59	0.097	6.60	1.45	1.26	0.236	30.00	0.73	0.58	0.124
23.50	0.85	0.56	0.091	6.80	1.48	1.24	0.231				
24.00	0.84	0.54	0.086	7.00	1.50	1.24	0.230	**Potassium[21]**			
25.00	0.81	0.51	0.084	7.20	1.50	1.25	0.231	0.55	0.139	7.10	0.989
26.40	0.80	0.43	0.066	7.40	1.49	1.23	0.228	0.58	0.119	6.72	0.990
27.80	0.81	0.38	0.052	7.60	1.48	1.22	0.225	0.63	0.106	6.32	0.990
29.20	0.82	0.35	0.046	7.80	1.48	1.20	0.221	0.67	0.091	5.79	0.990
				8.00	1.47	1.18	0.216	0.73	0.079	5.30	0.989
Platinum[20]				8.20	1.47	1.17	0.212	0.81	0.066	4.75	0.989
0.10	13.21	44.72	0.976	8.40	1.47	1.15	0.209	0.92	0.056	4.19	0.988
0.15	8.18	31.16	0.969	8.60	1.47	1.14	0.205	1.05	0.044	3.58	0.987
0.20	5.90	23.95	0.962	8.80	1.47	1.13	0.202	1.23	0.040	3.04	0.985
0.25	4.70	19.40	0.954	9.00	1.48	1.12	0.200	1.44	0.040	2.56	0.979
0.30	3.92	16.16	0.945	9.20	1.49	1.11	0.198	1.65	0.044	2.19	0.970
0.35	3.28	13.66	0.936	9.40	1.49	1.12	0.200	1.87	0.050	1.84	0.955
0.40	2.81	11.38	0.922	9.60	1.49	1.13	0.203	2.07	0.053	1.62	0.943
0.45	3.03	9.31	0.882	9.80	1.48	1.15	0.207	2.27	0.049	1.43	0.938
0.50	3.91	7.71	0.813	10.00	1.46	1.15	0.209	2.45	0.046	1.28	0.933
0.55	4.58	7.14	0.777	10.20	1.43	1.16	0.211	2.64	0.043	1.14	0.928
0.60	5.13	6.75	0.753	10.40	1.40	1.15	0.210	2.82	0.043	1.02	0.919
0.65	5.52	6.66	0.746	10.60	1.37	1.14	0.207	2.95	0.041	0.898	0.913
0.70	5.71	6.83	0.751	10.80	1.35	1.12	0.203	3.06	0.041	0.799	0.905
0.75	5.57	7.02	0.759	11.00	1.33	1.10	0.199	3.40	0.052	0.549	0.852
0.80	5.31	7.04	0.762	11.20	1.31	1.08	0.194	3.71	0.089	0.288	0.719
0.85	5.05	6.98	0.763	11.40	1.30	1.06	0.188	3.97	0.287	0.091	0.310
0.90	4.77	6.91	0.765	11.60	1.29	1.04	0.183	4.00	0.34	0.08	0.245
0.95	4.50	6.77	0.763	11.80	1.29	1.01	0.177	4.065	0.38	0.07	0.204
1.00	4.25	6.62	0.762	12.00	1.29	1.00	0.173	4.133	0.41	0.07	0.177
1.10	3.86	6.24	0.753	12.40	1.29	0.97	0.165	4.203	0.45	0.06	0.145
1.20	3.55	5.92	0.746	12.80	1.29	0.94	0.158	4.275	0.48	0.06	0.125
1.30	3.29	5.61	0.736	13.20	1.31	0.93	0.155	4.350	0.52	0.05	0.101
1.40	3.10	5.32	0.725	13.60	1.31	0.93	0.155	4.428	0.55	0.05	0.085
1.50	2.92	5.07	0.716	14.00	1.31	0.93	0.155	4.509	0.58	0.05	0.072
1.60	2.76	4.84	0.706	14.40	1.30	0.93	0.156	4.592	0.61	0.05	0.060
1.70	2.63	4.64	0.697	14.80	1.27	0.93	0.157	4.679	0.64	0.04	0.049
1.80	2.51	4.43	0.686	15.20	1.27	0.93	0.155	4.769	0.66	0.04	0.043
1.90	2.38	4.26	0.678	15.60	1.25	0.92	0.151	4.862	0.68	0.04	0.037
2.00	2.30	4.07	0.664	16.00	1.24	0.89	0.146	4.959	0.70	0.04	0.032
2.10	2.23	3.92	0.654	16.50	1.24	0.87	0.142	5.061	0.72	0.04	0.027
2.20	2.17	3.77	0.642	17.00	1.25	0.86	0.138	5.166	0.74	0.04	0.023
2.30	2.10	3.67	0.636	17.50	1.27	0.85	0.135	5.276	0.76	0.04	0.019
2.40	2.03	3.54	0.626	18.00	1.31	0.88	0.142	5.391	0.78	0.04	0.016
2.50	1.96	3.42	0.616	18.50	1.30	0.94	0.157	5.510	0.79	0.05	0.015
2.60	1.91	3.30	0.605	19.00	1.28	0.99	0.171	5.637	0.81	0.05	0.012
2.70	1.87	3.20	0.595	19.50	1.23	1.03	0.184	5.767	0.83	0.05	0.009
2.80	1.83	3.10	0.585	20.00	1.18	1.06	0.197	6.048	0.85	0.05	0.007
2.90	1.79	3.01	0.575	20.50	1.11	1.09	0.212	6.199	0.87	0.05	0.006
3.00	1.75	2.92	0.565	21.00	1.03	1.10	0.226	6.358	0.88	0.05	0.005
3.20	1.68	2.76	0.546	21.50	0.94	1.08	0.238	6.526	0.90	0.06	0.004
3.40	1.63	2.62	0.527	22.00	0.87	1.04	0.240	6.702	0.91	0.06	0.003
3.60	1.58	2.48	0.507	22.50	0.81	0.98	0.235	6.888	0.92	0.06	0.003
3.80	1.53	2.37	0.491	23.00	0.77	0.92	0.226	7.085	0.92	0.06	0.003
4.00	1.49	2.25	0.472	23.50	0.75	0.87	0.213	7.293	0.93	0.06	0.002
4.20	1.45	2.14	0.452	24.00	0.74	0.82	0.201	7.514	0.93	0.06	0.002
4.40	1.43	2.04	0.432	24.50	0.73	0.77	0.187	7.749	0.94	0.06	0.002
4.60	1.39	1.95	0.415	25.00	0.73	0.73	0.174	7.999	0.94	0.06	0.002
4.80	1.38	1.85	0.392	25.50	0.73	0.70	0.162	8.260	0.94	0.06	0.002
5.00	1.36	1.76	0.372	26.00	0.74	0.67	0.150	8.551	0.94	0.06	0.002
5.20	1.36	1.67	0.350	26.50	0.74	0.65	0.142	8.856	0.94	0.05	0.002
5.40	1.36	1.61	0.332	27.00	0.74	0.63	0.136	9.184	0.94	0.05	0.002
5.60	1.36	1.54	0.315	27.50	0.74	0.62	0.130	9.537	0.94	0.04	0.001
5.80	1.36	1.47	0.295	28.00	0.75	0.60	0.125	9.919	0.94	0.04	0.001

OPTICAL PROPERTIES OF METALS AND SEMICONDUCTORS (continued)

Energy (eV)	n	k	R(φ = 0)	Energy (eV)	n	k	R(φ = 0)	Energy (eV)	n	k	R(φ = 0)
10.33	0.94	0.03	0.001	7.60	1.12	1.99	0.470	31.00	0.62	0.29	0.086
11.0		0.03		7.80	1.08	1.89	0.454	32.00	0.66	0.26	0.065
12.0		0.028		8.00	1.05	1.80	0.435	33.00	0.68	0.24	0.054
				8.20	1.05	1.71	0.411	34.00	0.72	0.21	0.041
Rhenium, single crystal, $\vec{E} \parallel \hat{c}$[9]				8.40	1.05	1.62	0.386	35.00	0.76	0.20	0.031
0.10	6.06	51.03	0.991	8.60	1.06	1.55	0.360	36.00	0.79	0.20	0.025
0.15	4.66	33.96	0.984	8.80	1.09	1.48	0.336	37.00	0.82	0.19	0.021
0.20	4.16	25.36	0.975	9.00	1.11	1.43	0.317	38.00	0.85	0.20	0.018
0.25	4.03	20.10	0.962	9.20	1.13	1.39	0.301	39.00	0.89	0.21	0.016
0.30	4.37	16.69	0.943	9.40	1.16	1.34	0.281	40.00	0.88	0.26	0.022
0.35	4.50	14.53	0.925	9.60	1.18	1.32	0.274	42.00	0.88	0.26	0.022
0.40	4.53	12.96	0.909	9.80	1.20	1.29	0.264	44.00	0.89	0.29	0.026
0.45	4.53	11.78	0.893	10.00	1.23	1.26	0.252	46.00	0.85	0.32	0.035
0.50	4.53	10.88	0.878	10.20	1.25	1.25	0.246	48.00	0.82	0.30	0.036
0.55	4.50	10.26	0.867	10.40	1.28	1.25	0.242	50.00	0.80	0.30	0.038
0.60	4.29	9.75	0.861	10.60	1.29	1.25	0.242	52.00	0.78	0.30	0.044
0.65	4.07	9.35	0.856	10.80	1.30	1.26	0.244	54.00	0.72	0.30	0.055
0.70	3.80	8.94	0.853	11.00	1.30	1.27	0.247	56.00	0.66	0.24	0.061
0.75	3.48	8.55	0.850	11.20	1.29	1.28	0.249	58.00	0.65	0.16	0.055
0.80	3.21	8.10	0.846	11.40	1.28	1.28	0.252				
0.85	2.96	7.68	0.841	11.60	1.26	1.28	0.252	**Rhenium, single crystal, $\vec{E} \perp \hat{c}$[9]**			
0.90	2.73	7.24	0.835	11.80	1.24	1.26	0.249	0.10	4.25	42.83	0.991
0.95	2.56	6.79	0.826	12.00	1.23	1.24	0.244	0.15	3.28	28.08	0.984
1.00	2.45	6.36	0.813	12.40	1.22	1.21	0.237	0.20	3.28	20.66	0.971
1.10	2.38	5.61	0.778	12.80	1.21	1.18	0.230	0.25	3.47	16.27	0.951
1.20	2.35	5.02	0.742	13.20	1.22	1.16	0.222	0.30	3.73	13.44	0.926
1.30	2.39	4.54	0.702	13.60	1.22	1.13	0.215	0.35	3.93	11.54	0.900
1.40	2.44	4.13	0.662	14.00	1.24	1.12	0.209	0.40	3.99	10.15	0.875
1.50	2.50	3.79	0.624	14.40	1.27	1.11	0.204	0.45	4.17	9.03	0.846
1.60	2.59	3.49	0.587	14.80	1.29	1.15	0.213	0.50	4.34	8.26	0.821
1.70	2.70	3.27	0.557	15.20	1.29	1.19	0.225	0.55	4.45	7.73	0.801
1.80	2.82	3.10	0.535	15.60	1.26	1.22	0.236	0.60	4.53	7.40	0.788
1.90	2.90	3.00	0.520	16.00	1.23	1.25	0.248	0.65	4.44	7.26	0.784
2.00	2.97	2.91	0.510	16.40	1.19	1.27	0.259	0.70	4.13	7.09	0.784
2.10	3.03	2.86	0.504	16.80	1.14	1.29	0.269	0.75	3.77	6.75	0.779
2.20	3.06	2.84	0.501	17.00	1.12	1.30	0.275	0.80	3.55	6.32	0.766
2.30	3.07	2.82	0.499	17.40	1.07	1.30	0.286	0.85	3.39	5.95	0.752
2.40	3.06	2.81	0.498	18.00	0.99	1.30	0.300	0.90	3.26	5.61	0.737
2.50	3.02	2.80	0.497	18.40	0.93	1.29	0.311	0.95	3.17	5.27	0.719
2.60	2.96	2.77	0.493	18.80	0.87	1.28	0.321	1.00	3.09	4.96	0.701
2.70	2.89	2.68	0.482	19.20	0.81	1.25	0.330	1.10	3.05	4.39	0.658
2.80	2.89	2.57	0.468	19.60	0.77	1.21	0.332	1.20	3.08	3.89	0.613
2.90	2.99	2.47	0.457	20.00	0.73	1.18	0.333	1.30	3.20	3.56	0.578
3.00	3.11	2.57	0.470	20.40	0.70	1.14	0.332	1.40	3.23	3.38	0.559
3.20	2.90	2.68	0.482	20.80	0.67	1.11	0.332	1.50	3.23	3.12	0.532
3.40	2.83	2.50	0.459	21.20	0.64	1.08	0.334	1.60	3.29	2.88	0.507
3.60	2.93	2.48	0.457	21.60	0.61	1.04	0.335	1.70	3.38	2.72	0.491
3.80	2.86	2.56	0.467	22.00	0.58	1.01	0.340	1.80	3.47	2.59	0.480
4.00	2.81	2.51	0.460	22.40	0.55	0.97	0.341	1.90	3.54	2.50	0.473
4.20	2.86	2.55	0.466	22.80	0.53	0.93	0.338	2.00	3.63	2.43	0.469
4.40	2.81	2.74	0.489	23.20	0.51	0.89	0.334	2.10	3.74	2.40	0.470
4.60	2.56	2.83	0.504	23.60	0.50	0.85	0.329	2.20	3.83	2.38	0.472
4.80	2.41	2.71	0.493	24.00	0.48	0.80	0.319	2.30	3.93	2.44	0.481
5.00	2.39	2.68	0.488	24.40	0.48	0.76	0.207	2.40	4.00	2.55	0.492
5.20	2.34	2.75	0.500	24.80	0.47	0.72	0.296	2.50	4.01	2.70	0.505
5.40	2.20	2.81	0.515	25.20	0.47	0.68	0.282	2.60	3.90	2.84	0.514
5.60	2.02	2.84	0.530	25.60	0.47	0.65	0.270	2.70	3.74	2.92	0.517
5.80	1.83	2.80	0.538	26.00	0.47	0.61	0.255	2.80	3.57	2.88	0.511
6.00	1.65	2.71	0.541	26.40	0.48	0.57	0.240	2.90	3.49	2.75	0.497
6.20	1.54	2.59	0.532	26.80	0.48	0.54	0.225	3.00	3.53	2.71	0.493
6.40	1.45	2.50	0.526	27.20	0.49	0.51	0.208	3.20	3.55	2.84	0.506
6.80	1.32	2.31	0.508	27.60	0.50	0.48	0.193	3.40	3.34	2.88	0.508
7.00	1.26	2.23	0.500	28.00	0.51	0.45	0.176	3.60	3.25	2.83	0.501
7.20	1.20	2.15	0.493	29.00	0.54	0.39	0.145	3.80	3.24	2.84	0.502
7.40	1.16	2.06	0.480	30.00	0.57	0.33	0.114	4.00	3.19	2.94	0.513

Energy (eV)	n	k	$R(\phi = 0)$
4.20	3.05	3.06	0.526
4.40	2.88	3.15	0.539
4.60	2.67	3.18	0.548
4.80	2.44	3.17	0.554
5.00	2.25	3.12	0.556
5.20	2.10	3.04	0.555
5.40	1.96	2.96	0.553
5.60	1.84	2.88	0.551
5.80	1.73	2.81	0.549
6.00	1.61	2.74	0.549
6.20	1.51	2.64	0.545
6.40	1.42	2.56	0.541
6.80	1.28	2.37	0.526
7.00	1.22	2.28	0.517
7.20	1.16	2.19	0.508
7.40	1.12	2.08	0.493
7.60	1.12	1.98	0.468
7.80	1.08	1.93	0.463
8.00	1.05	1.83	0.443
8.20	1.05	1.74	0.418
8.40	1.05	1.66	0.397
8.60	1.06	1.58	0.372
8.80	1.07	1.52	0.351
9.00	1.09	1.46	0.327
9.20	1.11	1.41	0.309
9.40	1.14	1.36	0.290
9.60	1.17	1.31	0.273
9.80	1.20	1.27	0.258
10.00	1.24	1.24	0.244
10.20	1.29	1.22	0.234
10.40	1.33	1.23	0.233
10.60	1.36	1.25	0.238
10.80	1.38	1.28	0.245
11.00	1.37	1.31	0.253
11.20	1.36	1.33	0.259
11.40	1.33	1.34	0.264
11.60	1.31	1.34	0.266
11.80	1.28	1.33	0.266
12.00	1.26	1.32	0.264
12.40	1.23	1.29	0.257
12.80	1.22	1.26	0.251
13.20	1.20	1.23	0.245
13.60	1.19	1.20	0.236
14.00	1.20	1.16	0.225
14.40	1.22	1.13	0.214
14.80	1.27	1.12	0.207
15.20	1.31	1.17	0.218
15.60	1.31	1.23	0.234
16.00	1.28	1.28	0.251
16.40	1.24	1.33	0.270
16.80	1.17	1.37	0.288
17.00	1.14	1.38	0.297
17.40	1.06	1.39	0.314
18.00	0.95	1.38	0.334
18.40	0.88	1.36	0.346
18.80	0.82	1.33	0.355
19.20	0.76	1.29	0.360
19.60	0.72	1.25	0.363
20.00	0.67	1.21	0.369
20.40	0.64	1.15	0.364
20.80	0.61	1.10	0.357
21.20	0.60	1.06	0.349
21.60	0.58	1.02	0.342
22.00	0.57	0.98	0.336
22.40	0.56	0.95	0.328

Energy (eV)	n	k	$R(\phi = 0)$
22.80	0.55	0.92	0.325
23.20	0.53	0.89	0.322
23.60	0.52	0.85	0.317
24.00	0.50	0.82	0.314
24.40	0.49	0.79	0.309
24.80	0.48	0.75	0.303
25.20	0.47	0.72	0.295
25.60	0.47	0.68	0.286
26.00	0.46	0.64	0.276
26.40	0.46	0.61	0.263
26.80	0.46	0.57	0.249
27.20	0.47	0.53	0.231
27.60	0.48	0.50	0.216
28.00	0.49	0.47	0.198
29.00	0.51	0.41	0.164
30.00	0.55	0.34	0.129
31.00	0.59	0.29	0.097
32.00	0.64	0.26	0.072
33.00	0.67	0.24	0.060
34.00	0.70	0.22	0.047
35.00	0.74	0.20	0.036
36.00	0.77	0.19	0.029
37.00	0.80	0.19	0.023
38.00	0.84	0.19	0.018
39.00	0.88	0.21	0.016
40.00	0.87	0.25	0.023
42.00	0.87	0.25	0.023
44.00	0.88	0.28	0.026
46.00	0.84	0.31	0.035
48.00	0.82	0.30	0.036
50.00	0.80	0.30	0.039
52.00	0.77	0.30	0.044
54.00	0.71	0.29	0.055
56.00	0.66	0.23	0.061
58.00	0.64	0.16	0.055

Rhodium[11]

Energy (eV)	n	k	$R(\phi = 0)$
0.10	18.48	69.43	0.986
0.20	8.66	37.46	0.977
0.30	5.85	25.94	0.967
0.40	4.74	19.80	0.955
0.50	4.20	16.07	0.941
0.60	3.87	13.51	0.925
0.70	3.67	11.72	0.908
0.80	3.63	10.34	0.887
0.90	3.62	9.36	0.867
1.00	3.71	8.67	0.848
1.10	3.67	8.26	0.837
1.20	3.51	7.94	0.832
1.30	3.26	7.63	0.829
1.40	3.01	7.31	0.827
1.50	2.78	6.97	0.823
1.60	2.60	6.64	0.818
1.70	2.42	6.33	0.813
1.80	2.30	6.02	0.805
1.90	2.20	5.76	0.798
2.00	2.12	5.51	0.789
2.10	2.05	5.30	0.780
2.20	2.00	5.11	0.772
2.30	1.94	4.94	0.765
2.40	1.90	4.78	0.756
2.50	1.88	4.65	0.748
2.60	1.85	4.55	0.743
2.70	1.80	4.49	0.742
2.90	1.63	4.36	0.748

Energy (eV)	n	k	$R(\phi = 0)$
3.00	1.53	4.29	0.753
3.10	1.41	4.20	0.760
3.20	1.30	4.09	0.764
3.30	1.20	3.97	0.767
3.40	1.11	3.84	0.769
3.50	1.04	3.71	0.768
3.60	0.99	3.58	0.764
3.70	0.95	3.45	0.759
3.80	0.91	3.34	0.753
3.90	0.88	3.23	0.747
4.00	0.86	3.12	0.739
4.20	0.83	2.94	0.722
4.40	0.80	2.76	0.706
4.60	0.78	2.60	0.684
4.80	0.79	2.46	0.659
5.00	0.79	2.34	0.635
5.20	0.79	2.23	0.613
5.40	0.80	2.14	0.591
5.60	0.80	2.06	0.573
5.80	0.79	2.00	0.561
6.00	0.76	1.93	0.556
6.20	0.73	1.85	0.544
6.40	0.70	1.77	0.534
6.60	0.68	1.69	0.518
6.80	0.67	1.60	0.498
7.00	0.66	1.52	0.476
7.20	0.66	1.43	0.452
7.40	0.66	1.35	0.423
7.60	0.67	1.27	0.394
7.80	0.68	1.20	0.363
8.00	0.69	1.12	0.329
8.20	0.71	1.04	0.288
8.40	0.74	0.97	0.252
8.60	0.78	0.89	0.212
8.80	0.83	0.83	0.179
9.00	0.88	0.77	0.148
9.20	0.95	0.73	0.125
9.40	1.01	0.71	0.110
9.60	1.07	0.69	0.102
9.80	1.12	0.69	0.098
10.00	1.17	0.69	0.098
10.60	1.26	0.73	0.106
11.00	1.29	0.76	0.113
11.60	1.32	0.80	0.124
12.00	1.32	0.82	0.127
12.60	1.32	0.82	0.129
13.00	1.32	0.83	0.131
13.60	1.32	0.85	0.134
14.00	1.32	0.86	0.138
14.60	1.30	0.89	0.144
15.00	1.28	0.90	0.147
15.60	1.25	0.90	0.147
16.00	1.24	0.89	0.147
16.50	1.23	0.88	0.145
17.00	1.22	0.88	0.144
17.50	1.22	0.87	0.143
18.00	1.23	0.88	0.145
18.50	1.25	0.92	0.155
19.00	1.24	0.98	0.172
19.50	1.18	1.05	0.193
20.00	1.10	1.09	0.213
20.50	1.00	1.09	0.230
21.00	0.91	1.05	0.234
21.50	0.86	1.00	0.228
22.00	0.83	0.95	0.219

Energy (eV)	n	k	$R(\phi=0)$
22.50	0.81	0.92	0.214
23.00	0.79	0.90	0.213
23.50	0.75	0.87	0.214
24.00	0.73	0.84	0.210
24.50	0.70	0.81	0.208
25.00	0.69	0.77	0.202
25.50	0.67	0.74	0.195
26.00	0.66	0.70	0.188
26.50	0.65	0.66	0.176
27.00	0.65	0.64	0.168
27.50	0.65	0.61	0.159
28.00	0.65	0.59	0.152
29.00	0.65	0.54	0.137
30.00	0.66	0.51	0.127
31.00	0.64	0.49	0.127
32.00	0.61	0.44	0.126
33.00	0.60	0.37	0.110
34.00	0.65	0.30	0.074
35.00	0.69	0.28	0.058
36.00	0.73	0.27	0.049
37.00	0.74	0.28	0.047
38.00	0.74	0.27	0.045
39.00	0.75	0.25	0.041

Ruthenium, single crystal, $\vec{E} \parallel \hat{c}$ [9]

Energy (eV)	n	k	$R(\phi=0)$
0.10	11.50	51.38	0.984
0.20	5.93	27.14	0.970
0.30	4.33	18.50	0.953
0.40	3.60	13.97	0.933
0.50	3.18	11.04	0.909
0.60	3.28	8.89	0.865
0.70	3.62	7.73	0.822
0.80	3.42	7.02	0.801
0.90	3.25	6.12	0.766
1.00	3.39	5.33	0.715
1.10	3.66	4.83	0.675
1.20	3.84	4.57	0.654
1.30	3.94	4.38	0.638
1.40	4.02	4.19	0.624
1.50	4.16	4.07	0.614
1.60	4.33	4.08	0.615
1.70	4.42	4.21	0.624
1.80	4.40	4.38	0.636
1.90	4.29	4.61	0.651
2.00	4.04	4.81	0.667
2.10	3.69	4.90	0.679
2.20	3.35	4.82	0.683
2.30	3.09	4.70	0.681
2.40	2.89	4.55	0.677
2.50	2.74	4.40	0.671
2.60	2.64	4.25	0.663
2.70	2.58	4.14	0.656
2.80	2.54	4.05	0.650
2.90	2.48	4.03	0.650
3.00	2.38	4.03	0.656
3.10	2.26	4.00	0.661
3.20	2.13	3.96	0.666
3.30	2.00	3.91	0.671
3.40	1.87	3.83	0.673
3.50	1.76	3.74	0.674
3.60	1.66	3.65	0.675
3.70	1.57	3.55	0.673
3.80	1.49	3.45	0.672
3.90	1.42	3.35	0.668
4.00	1.37	3.24	0.661

Energy (eV)	n	k	$R(\phi=0)$
4.20	1.29	3.08	0.649
4.40	1.22	2.93	0.639
4.60	1.16	2.79	0.628
4.80	1.11	2.67	0.617
5.00	1.06	2.56	0.607
5.20	1.01	2.46	0.600
5.40	0.95	2.35	0.593
5.60	0.92	2.23	0.576
5.80	0.90	2.14	0.559
6.00	0.88	2.05	0.545
6.20	0.87	1.98	0.531
6.40	0.84	1.91	0.521
6.60	0.82	1.84	0.510
6.80	0.79	1.77	0.500
7.00	0.76	1.69	0.489
7.20	0.75	1.61	0.472
7.40	0.73	1.54	0.455
7.60	0.73	1.46	0.433
7.80	0.73	1.39	0.411
8.00	0.72	1.33	0.391
8.20	0.72	1.26	0.366
8.40	0.73	1.20	0.342
8.60	0.74	1.14	0.318
8.80	0.74	1.08	0.295
9.00	0.75	1.02	0.267
9.20	0.77	0.97	0.243
9.40	0.79	0.91	0.217
9.60	0.82	0.86	0.190
9.80	0.85	0.81	0.167
10.00	0.88	0.76	0.144
10.20	0.92	0.72	0.125
10.40	0.96	0.69	0.110
10.60	1.01	0.67	0.100
10.80	1.05	0.66	0.094
11.00	1.09	0.65	0.090
11.20	1.12	0.65	0.088
11.40	1.15	0.65	0.087
11.60	1.18	0.65	0.088
11.80	1.21	0.66	0.090
12.00	1.23	0.67	0.092
12.40	1.26	0.69	0.098
12.80	1.27	0.72	0.104
13.20	1.28	0.74	0.108
13.60	1.28	0.75	0.111
14.00	1.28	0.76	0.114
14.40	1.27	0.76	0.114
14.80	1.27	0.76	0.114
15.00	1.27	0.76	0.114
15.60	1.28	0.77	0.115
16.00	1.30	0.78	0.118
16.50	1.32	0.80	0.123
17.00	1.34	0.85	0.136
17.50	1.32	0.93	0.155
18.00	1.26	0.99	0.173
18.50	1.18	1.02	0.185
19.00	1.11	1.02	0.192
19.50	1.05	1.02	0.199
20.00	0.99	1.02	0.208
20.50	0.92	0.99	0.212
21.00	0.86	0.94	0.209
21.50	0.83	0.90	0.203
22.00	0.81	0.86	0.193
23.00	0.77	0.79	0.182
24.00	0.74	0.74	0.171
25.00	0.71	0.69	0.163

Energy (eV)	n	k	$R(\phi=0)$
26.00	0.68	0.63	0.154
27.00	0.67	0.57	0.140
28.00	0.66	0.51	0.124
29.00	0.67	0.46	0.107
30.00	0.67	0.43	0.097
31.00	0.67	0.37	0.084
32.00	0.69	0.33	0.070
33.00	0.71	0.30	0.058
34.00	0.73	0.27	0.048
35.00	0.75	0.25	0.039
36.00	0.77	0.24	0.035
37.00	0.79	0.23	0.039
38.00	0.80	0.22	0.027
39.00	0.82	0.22	0.024
40.00	0.83	0.22	0.022

Ruthenium, single crystal, $\vec{E} \perp \hat{c}$ [5]

Energy (eV)	n	k	$R(\phi=0)$
0.10	11.85	50.81	0.983
0.20	6.68	27.18	0.966
0.30	4.94	18.92	0.950
0.40	3.90	14.51	0.933
0.50	3.27	11.63	0.915
0.60	2.98	9.54	0.888
0.70	2.82	7.99	0.856
0.80	2.73	6.71	0.815
0.90	2.82	5.54	0.751
1.00	3.17	4.59	0.670
1.10	3.69	3.91	0.604
1.20	4.28	3.66	0.585
1.30	4.66	3.72	0.593
1.40	4.86	3.79	0.601
1.50	4.99	3.89	0.609
1.60	5.08	4.03	0.618
1.70	5.12	4.22	0.629
1.80	5.10	4.45	0.642
1.90	4.96	4.78	0.660
2.00	4.61	5.06	0.677
2.10	4.21	5.09	0.682
2.20	3.94	5.00	0.681
2.30	3.69	4.97	0.684
2.40	3.44	4.88	0.684
2.50	3.27	4.77	0.681
2.60	3.14	4.66	0.677
2.70	3.06	4.59	0.674
2.80	2.99	4.59	0.676
2.90	2.87	4.64	0.686
3.00	2.64	4.69	0.701
3.10	2.40	4.64	0.710
3.20	2.18	4.55	0.717
3.30	2.00	4.43	0.721
3.40	1.84	4.30	0.723
3.50	1.71	4.16	0.723
3.60	1.60	4.03	0.722
3.70	1.50	3.90	0.721
3.80	1.41	3.77	0.718
3.90	1.35	3.64	0.713
4.00	1.29	3.53	0.707
4.20	1.21	3.31	0.694
4.40	1.16	3.13	0.679
4.60	1.13	2.97	0.662
4.80	1.09	2.86	0.652
5.00	1.03	2.75	0.648
5.20	0.97	2.64	0.643
5.40	0.91	2.52	0.635
5.60	0.88	2.40	0.622

Energy (eV)	n	k	$R(\phi = 0)$
5.80	0.86	2.29	0.605
6.00	0.84	2.20	0.591
6.20	0.82	2.11	0.576
6.40	0.81	2.04	0.564
6.60	0.78	1.97	0.556
6.80	0.76	1.89	0.545
7.00	0.73	1.82	0.538
7.20	0.70	1.75	0.527
7.40	0.68	1.67	0.513
7.60	0.67	1.59	0.496
7.80	0.66	1.51	0.476
8.00	0.66	1.44	0.454
8.20	0.65	1.36	0.430
8.40	0.66	1.29	0.403
8.60	0.66	1.22	0.378
8.80	0.68	1.15	0.346
9.00	0.69	1.09	0.317
9.20	0.70	1.02	0.286
9.40	0.73	0.95	0.251
9.60	0.77	0.89	0.216
9.80	0.82	0.84	0.185
10.00	0.86	0.81	0.163
10.20	0.90	0.77	0.143
10.40	0.94	0.74	0.127
10.60	0.99	0.72	0.115
10.80	1.04	0.71	0.108
11.00	1.08	0.70	0.104
11.20	1.11	0.70	0.102
11.40	1.14	0.70	0.101
11.60	1.17	0.71	0.102
11.80	1.20	0.72	0.104
12.00	1.22	0.73	0.107
12.40	1.25	0.76	0.113
12.80	1.26	0.78	0.118
13.20	1.27	0.81	0.124
13.60	1.27	0.83	0.129
14.00	1.26	0.84	0.132
14.40	1.25	0.84	0.132
14.80	1.25	0.84	0.133
15.00	1.25	0.84	0.133
15.60	1.25	0.85	0.134
16.00	1.27	0.85	0.134
16.50	1.28	0.89	0.145
17.00	1.28	0.94	0.158
17.50	1.25	1.00	0.175
18.00	1.19	1.04	0.190
18.50	1.12	1.05	0.200
19.00	1.07	1.05	0.205
19.50	1.02	1.04	0.212
20.00	0.97	1.04	0.219
20.50	0.91	1.03	0.228
21.00	0.85	1.01	0.234
21.50	0.80	0.97	0.234
22.00	0.77	0.94	0.233
23.00	0.71	0.87	0.229
24.00	0.67	0.79	0.218
25.00	0.64	0.73	0.205
26.00	0.61	0.66	0.194
27.00	0.60	0.59	0.177
28.00	0.60	0.53	0.155
29.00	0.61	0.48	0.134
30.00	0.62	0.45	0.123
31.00	0.61	0.40	0.114
32.00	0.63	0.34	0.093
33.00	0.65	0.31	0.077

Energy (eV)	n	k	$R(\phi = 0)$
34.00	0.67	0.28	0.065
35.00	0.70	0.26	0.054
36.00	0.72	0.25	0.047
37.00	0.73	0.23	0.041
38.00	0.75	0.22	0.035
39.00	0.77	0.22	0.031
40.00	0.79	0.22	0.028

Selenium, single crystal, $\vec{E} \parallel \hat{c}$ [22]

Energy (eV)	n	k	$R(\phi = 0)$
0.01364	2.914	0.248	0.242
0.01488	3.175	9.95E-02	0.272
0.01612	3.263	2.13E-03	0.282
0.01736	3.306	3.81E-02	0.287
0.01860	3.330	7.04E-03	0.290
0.01984	3.346	4.23E-02	0.291
0.02108	3.358	3.40E-03	0.293
0.02232	3.366	5.31E-02	0.294
0.02356	3.372	1.96E-03	0.294
0.02480	3.377	2.39E-02	0.295
0.02604	3.380		0.295
0.02728		1.16E-02	
0.02976		7.96E-03	
0.03224		8.57E-03	
0.03472		2.70E-02	
0.03720	3.397	1.72E-02	0.297
0.04463		1.13E-02	
0.04959	3.403	2.79E-03	0.298
0.05703		1.56E-03	
0.06199	3.405	1.35E-03	0.298
0.06819		5.79E-04	
0.07439	3.407	4.44E-04	0.298
0.08059		4.41E-04	
0.08679	3.408	4.32E-04	0.298
0.09299		2.44E-04	
0.09919	3.409	3.23E-04	0.299
0.1116	3.409	2.87E-04	0.299
0.1240	3.410	2.71E-04	0.299
0.2480	3.417	2.67E-04	0.299
0.3720	3.427	1.90E-04	0.301
0.4959	3.442	1.41E-04	0.302
0.6199	3.462	1.12E-04	0.304
0.7439	3.486	9.42E-05	0.307
0.8679	3.516	8.07E-05	0.310
0.9919	3.551	7.11E-05	0.314
1.116	3.592	6.37E-05	0.319
1.240	3.640	5.81E-05	0.324
1.50		1.33E-04	
1.60		1.59E-04	
1.70		6.27E-04	
1.80	4.46	2.20E-02	0.402
2.0	4.79	0.76	0.438
2.2	4.49	1.19	0.431
2.4	4.28	1.21	0.417
2.6	4.40	1.32	0.430
2.8	4.59	1.70	0.462
3.0	4.44	2.29	0.490
3.2	3.92	2.59	0.493
3.4	3.69	2.76	0.502
3.6	3.39	3.01	0.521
3.8	(3.00)		
4.0	(2.65)		
4.2	(2.30)		
4.5	1.92	2.78	0.528
5.0	1.50	2.31	0.482
6.0	1.57	1.49	0.288

Energy (eV)	n	k	$R(\phi = 0)$
7.0	1.84	1.45	0.276
8.0	1.35	1.68	0.353
9.0	1.35	1.64	0.342
10.0	0.92	1.07	0.238
12.0	1.00	1.10	0.232
14.0	0.81	0.91	0.211
16.0	0.65	0.61	0.160
18.0	0.65	0.48	0.120
20.0	0.69	0.36	0.076
22.0	0.81	0.25	0.030
24.0	0.91	0.18	0.011
26.0	0.86	0.15	0.012
28.0	0.85	0.13	0.011
30.0	0.87	0.11	0.008

Selenium, single crystal, $\vec{E} \perp \hat{c}$ [22]

Energy (eV)	n	k	$R(\phi = 0)$
0.01364	2.854	0.0239	0.231
0.01488	2.932	0.0325	0.241
0.01612	3.140	0.1750	0.269
0.01736	2.959	1.3300	0.321
0.01860	2.111	0.2550	0.133
0.01984	2.356	0.0746	0.164
0.02108	2.462	0.0276	0.178
0.02232	2.502	0.0442	0.184
0.02356	2.543	0.0097	0.190
0.02480	2.550	0.0239	0.191
0.02604	2.582		0.195
0.02728	2.600	0.0101	0.198
0.02976	2.576	9.95E-03	0.194
0.03224	2.598	1.16E-02	0.197
0.03472	2.607	1.68E-02	0.199
0.03720	2.613	1.54E-02	0.199
0.04463		1.17E-02	
0.04959	2.627	3.58E-03	0.201
0.05703		8.65E-04	
0.06199	2.632	2.07E-03	0.202
0.06819		2.89E-04	
0.07439	2.635	1.59E-04	0.202
0.08059		1.35E-04	
0.08679	2.636	1.42E-04	0.202
0.09299		1.04E-04	
0.09919	2.637	8.95E-05	0.203
0.1116	2.638	8.84E-05	0.203
0.1240	2.639	8.51E-05	0.203
0.2480	2.645	5.97E-05	0.204
0.3720	2.652	5.44E-05	0.205
0.4959	2.654	4.58E-05	0.205
0.6199	2.675	3.82E-05	0.208
0.7439	2.692	3.32E-05	0.210
0.8679	2.713	2.96E-05	0.213
0.9919	2.739	2.69E-05	0.216
1.116	2.772	2.48E-05	0.221
1.240	2.816	2.31E-05	0.226
1.50		7.37E-05	
1.60		8.63E-05	
1.70		3.60E-04	
1.80	3.32	0.11	0.289
2.00	3.38	0.65	0.310
2.20	3.07	0.73	0.282
2.40	2.93	0.61	0.259
2.60	3.00	0.53	0.263
2.80	3.12	0.58	0.279
3.00	3.30	0.70	0.305
3.20	3.35	1.01	0.328
3.40	3.22	1.24	0.334

Energy (eV)	n	k	$R(\phi = 0)$
3.60	3.06	1.47	0.344
3.80	2.84	1.66	0.351
4.00	2.51	1.81	0.356
4.20	2.18	1.83	0.352
4.50	1.75	1.94	0.382
5.00	1.25	1.50	0.316
6.00	1.32	0.73	0.107
7.00	1.62	0.61	0.105
8.00	1.81	0.69	0.135
9.00	1.66	1.02	0.182
10.00	1.72	0.95	0.171
12.00	1.25	1.02	0.181
14.00	0.98	0.92	0.178
16.00	0.68	0.96	0.274
18.00	0.61	0.65	0.191
20.00	0.73	0.48	0.094
22.00	0.78	0.39	0.060
24.00	0.78	0.32	0.046
26.00	0.78	0.26	0.036
28.00	0.80	0.19	0.023
30.00	0.79	0.14	0.020

Silicon, single crystal[23]

Energy (eV)	n	k	$R(\phi = 0)$
0.01240	3.4185	2.90E-04	0.300
0.01488	3.4190	2.30E-04	0.300
0.01736	3.4192	1.90E-04	0.300
0.01984	3.4195	1.70E-04	0.300
0.02480	3.4197		0.300
0.03100	3.4199		0.300
0.04092	3.4200		0.300
0.04463		1.08E-04	
0.04959	3.4201	9.15E-05	0.300
0.05703		1.56E-04	
0.06199	3.4204	2.86E-04	0.300
0.06943		3.84E-04	
0.07439		7.16E-04	
0.08059	(3.4207)	1.52E-04	0.300
0.08679		1.02E-04	
0.09299		2.59E-04	
0.09919		1.77E-04	
0.1054		1.53E-04	
0.1116		2.02E-04	
0.1178		1.22E-04	
0.1240	3.4215	6.76E-05	0.300
0.1364		5.49E-05	
0.1488		2.41E-05	
0.1612		2.49E-05	
0.1736	(3.4230)	1.68E-05	0.300
0.1798		2.45E-05	
0.1860		2.66E-06	
0.1922		1.74E-06	
0.1984		8.46E-07	
0.2046		5.64E-07	
0.2108	(3.4244)	4.17E-07	0.300
0.2170		4.05E-07	
0.2232		3.94E-07	
0.2294		3.26E-07	
0.2356		2.97E-07	
0.2418		2.82E-07	
0.2480	3.4261	1.99E-07	0.300
0.3100	3.4294		0.301
0.3626	3.4327		0.301
0.4568	3.4393	2.50E-09	0.302
0.6199	3.4490		0.303
0.8093	3.4784		0.306
1.033	3.5193		0.311
1.1	(3.5341)	1.30E-05	0.312
1.2		1.80E-04	
1.3		2.26E-03	
1.4		7.75E-03	
1.5	3.673	5.00E-03	0.327
1.6	3.714	8.00E-03	0.331
1.7	3.752	1.00E-02	0.335
1.8	3.796	0.013	0.340
1.9	3.847	0.016	0.345
2.0	3.906	0.022	0.351
2.1	3.969	0.030	0.357
2.2	4.042	0.032	0.364
2.3	4.123	0.048	0.372
2.4	4.215	0.060	0.380
2.5	4.320	0.073	0.390
2.6	4.442	0.090	0.400
2.7	4.583	0.130	0.412
2.8	4.753	0.163	0.426
2.9	4.961	0.203	0.442
3.0	5.222	0.269	0.461
3.1	5.570	0.387	0.486
3.2	6.062	0.630	0.518
3.3	6.709	1.321	0.561
3.4	6.522	2.705	0.592
3.5	5.610	3.014	0.575
3.6	5.296	2.987	0.564
3.7	5.156	3.058	0.563
3.8	5.065	3.182	0.568
3.9	5.016	3.346	0.577
4.0	5.010	3.587	0.591
4.1	5.020	3.979	0.614
4.2	4.888	4.639	0.652
4.3	4.086	5.395	0.703
4.4	3.120	5.344	0.726
4.5	2.451	5.082	0.740
4.6	1.988	4.678	0.742
4.7	1.764	4.278	0.728
4.8	1.658	3.979	0.710
4.9	1.597	3.749	0.693
5.0	1.570	3.565	0.675
5.1	1.571	3.429	0.658
5.2	1.589	3.354	0.646
5.3	1.579	3.353	0.647
5.4	1.471	3.366	0.663
5.5	1.340	3.302	0.673
5.6	1.247	3.206	0.675
5.7	1.180	3.112	0.673
5.8	1.133	3.045	0.672
5.9	1.083	2.982	0.673
6.0	1.010	2.909	0.677
6.5	0.847	2.73	0.688
7.0	0.682	2.45	0.691
7.5	0.563	2.21	0.693
8.0	0.478	2.00	0.691
8.5	0.414	1.82	0.688
9.0	0.367	1.66	0.683
9.5	0.332	1.51	0.672
10.0	0.306	1.38	0.661
12.0	0.257	0.963	0.590
14.0	0.275	0.641	0.460
16.0	0.345	0.394	0.297
18.0	0.455	0.219	0.159
20.0	0.567	0.0835	0.079
22.14	0.675	0.0405	0.038
24.31	0.752	0.0243	0.020
26.38	0.803	0.0178	0.012
28.18	0.834	0.0152	0.008
30.24	0.860	0.0138	0.006
31.79	0.877	0.0132	0.004
34.44	0.899	0.0121	0.003
36.47	0.913	0.0113	0.002
38.75	0.925	0.0104	0.002
40.00	0.930	0.0100	0.001

Silver[6]

Energy (eV)	n	k	$R(\phi = 0)$
0.10	9.91	90.27	0.995
0.20	2.84	45.70	0.995
0.30	1.41	30.51	0.994
0.40	0.91	22.89	0.993
0.50	0.67	18.32	0.992
1.00	0.28	9.03	0.987
1.50	0.27	5.79	0.969
2.00	0.27	4.18	0.944
2.50	0.24	3.09	0.914
3.00	0.23	2.27	0.864
3.25	0.23	1.86	0.816
3.50	0.21	1.42	0.756
3.60	0.23	1.13	0.671
3.70	0.30	0.77	0.475
3.77	0.53	0.40	0.154
3.80	0.73	0.30	0.053
3.90	1.30	0.36	0.040
4.00	1.61	0.60	0.103
4.10	1.73	0.85	0.153
4.20	1.75	1.06	0.194
4.30	1.73	1.13	0.208
4.50	1.69	1.28	0.238
4.75	1.61	1.34	0.252
5.00	1.55	1.36	0.257
5.50	1.45	1.34	0.257
6.00	1.34	1.28	0.246
6.50	1.25	1.18	0.225
7.00	1.18	1.06	0.196
7.50	1.14	0.91	0.157
8.00	1.16	0.75	0.114
9.00	1.33	0.56	0.074
10.00	1.46	0.56	0.082
11.00	1.52	0.56	0.088
12.00	1.61	0.59	0.100
13.00	1.66	0.64	0.112
14.00	1.72	0.78	0.141
14.50	1.64	0.88	0.152
15.00	1.56	0.92	0.156
16.00	1.42	0.91	0.151
17.00	1.33	0.86	0.139
18.00	1.28	0.80	0.124
19.00	1.27	0.75	0.111
20.00	1.29	0.71	0.103
21.00	1.35	0.75	0.112
21.50	1.37	0.80	0.124
22.00	1.34	0.87	0.141
22.50	1.26	0.93	0.157
23.00	1.17	0.94	0.163
23.50	1.10	0.93	0.165
24.00	1.04	0.90	0.165
24.50	0.99	0.87	0.160
25.00	0.95	0.83	0.154
25.50	0.91	0.78	0.144
26.00	0.90	0.74	0.133

Energy (eV)	n	k	$R(\phi = 0)$	Energy (eV)	n	k	$R(\phi = 0)$	Energy (eV)	n	k	$R(\phi = 0)$
26.50	0.89	0.69	0.121	6.358	0.454		0.141	3.20	2.73	2.31	0.432
27.00	0.89	0.65	0.109	6.526	0.485		0.120	3.40	2.61	2.33	0.435
27.50	0.89	0.62	0.099	6.702	0.533		0.093	3.60	2.49	2.30	0.430
28.00	0.90	0.59	0.090	6.888	0.574		0.073	3.80	2.40	2.22	0.418
28.50	0.91	0.57	0.084	7.130	0.616		0.056	4.00	2.36	2.14	0.406
29.00	0.92	0.56	0.079	7.328	0.641		0.048	4.20	2.35	2.06	0.392
30.00	0.93	0.54	0.074	7.583	0.674		0.038	4.40	2.39	2.01	0.384
31.00	0.93	0.53	0.072	7.847	0.700		0.031	4.60	2.45	2.00	0.384
32.00	0.92	0.53	0.072	8.015	0.710		0.029	4.80	2.53	2.06	0.394
33.00	0.90	0.51	0.071	8.634	0.762		0.018	5.00	2.58	2.20	0.416
34.00	0.88	0.49	0.067	9.143	0.800		0.012	5.20	2.52	2.44	0.450
35.00	0.86	0.45	0.061	9.709	0.819		0.010	5.40	2.31	2.61	0.480
36.00	0.89	0.44	0.055	10.20	0.843		0.007	5.60	2.06	2.67	0.501
38.00	0.89	0.39	0.043	11.08	0.870		0.005	5.80	1.83	2.63	0.510
40.00	0.90	0.37	0.039	11.83	0.887		0.004	6.00	1.63	2.56	0.515
42.00	0.90	0.35	0.036	12.73	0.907		0.002	6.20	1.48	2.45	0.512
44.00	0.90	0.33	0.033	13.05	0.913		0.002	6.40	1.37	2.33	0.504
46.00	0.90	0.32	0.031	13.42	0.914		0.002	6.60	1.29	2.22	0.492
48.00	0.89	0.31	0.030	13.73	0.917		0.002	6.80	1.23	2.11	0.478
50.00	0.88	0.29	0.027	14.07	0.922		0.002	7.00	1.18	2.01	0.462
52.00	0.89	0.28	0.024	14.83	0.934		0.001	7.20	1.15	1.91	0.445
54.00	0.88	0.17	0.024	15.05	0.936		0.001	7.40	1.13	1.82	0.425
56.00	0.87	0.26	0.024	15.46	0.942		0.001	7.60	1.12	1.75	0.406
58.00	0.87	0.24	0.021	16.21	0.948		0.001	7.80	1.11	1.68	0.390
60.00	0.87	0.22	0.018	18.10	0.964		0.000	8.00	1.11	1.61	0.370
62.00	0.88	0.21	0.016	21.12	0.979		0.000	8.20	1.12	1.55	0.350
64.00	0.88	0.21	0.016	25.51	0.993		0.000	8.40	1.13	1.50	0.332
66.00	0.88	0.21	0.016	26.95	1.00		0.000	8.60	1.14	1.45	0.317
68.00	0.87	0.21	0.017	27.68	1.01		0.000	8.80	1.17	1.41	0.301
70.00	0.83	0.20	0.021	28.37	1.01		0.000	9.00	1.19	1.40	0.294
72.00	0.85	0.18	0.016	29.52	1.02		0.000	9.20	1.21	1.38	0.289
74.00	0.85	0.17	0.014					9.40	1.21	1.38	0.287
76.00	0.85	0.16	0.013	**Tantalum[16]**				9.60	1.21	1.38	0.285
78.00	0.85	0.15	0.013	0.10	10.14	66.39	0.984	9.80	1.21	1.37	0.285
80.00	0.85	0.14	0.012	0.15	9.45	46.41	0.9834	10.00	1.20	1.37	0.286
85.00	0.85	0.11	0.011	0.20	5.77	35.46	0.982	10.20	1.19	1.37	0.286
90.00	0.85	0.08	0.009	0.26	3.67	27.53	0.981	10.40	1.18	1.37	0.287
95.00	0.86	0.06	0.007	0.30	2.87	23.90	0.980	10.60	1.16	1.36	0.288
100.00	0.87	0.04	0.005	0.38	2.03	18.87	0.978	10.80	1.15	1.36	0.289
				0.50	1.37	14.26	0.974	11.00	1.13	1.35	0.290
Sodium[24]				0.58	1.15	12.19	0.970	11.20	1.11	1.35	0.292
0.55	0.262	9.97	0.990	0.70	0.96	9.92	0.962	11.40	1.09	1.34	0.293
0.58	0.241	9.45	0.989	0.78	0.89	8.77	0.956	11.60	1.07	1.33	0.294
0.63	0.207	8.80	0.990	0.90	0.84	7.38	0.942	11.80	1.05	1.32	0.295
0.67	0.175	8.09	0.990	1.00	0.89	6.47	0.992	12.00	1.02	1.31	0.296
0.73	0.147	7.42	0.990	1.10	0.93	5.75	0.899	12.20	1.00	1.29	0.295
0.81	0.123	6.67	0.989	1.20	0.98	5.14	0.872	12.40	0.98	1.28	0.294
0.92	0.099	5.82	0.989	1.30	1.00	4.62	0.842	12.60	0.96	1.26	0.292
1.05	0.078	5.11	0.989	1.40	1.04	4.15	0.805	12.80	0.94	1.24	0.289
1.23	0.064	4.35	0.987	1.50	1.09	3.73	0.762	13.00	0.93	1.22	0.286
1.44	0.053	3.72	0.986	1.60	1.15	3.33	0.707	13.60	0.91	1.16	0.272
1.65	0.050	3.22	0.983	1.70	1.24	2.95	0.640	14.00	0.90	1.15	0.272
1.87	0.049	2.76	0.978	1.80	1.35	2.60	0.560	14.60	0.85	1.15	0.285
2.07	0.053	2.48	0.971	1.90	1.57	2.24	0.460	15.00	0.80	1.13	0.293
2.27	0.059	2.23	0.961	2.00	1.83	1.99	0.388	15.60	0.72	1.08	0.301
2.45	0.063	2.07	0.953	2.10	2.10	1.84	0.354	16.00	0.68	1.04	0.304
2.64	0.066	1.88	0.943	2.20	2.36	1.81	0.351	16.60	0.63	0.97	0.301
2.82	0.068	1.76	0.936	2.30	2.56	1.86	0.365	17.00	0.60	0.92	0.296
2.95	0.068	1.63	0.928	2.40	2.68	1.92	0.378	17.60	0.60	0.92	0.296
3.06	0.069	1.54	0.921	2.50	2.75	1.98	0.388	18.00	0.55	0.79	0.274
3.20	0.065	1.47	0.921	2.60	2.80	2.02	0.395	18.60	0.53	0.71	0.254
3.40	0.061	1.33	0.916	2.70	2.84	2.08	0.405	19.00	0.53	0.65	0.236
3.71	0.055	1.13	0.908	2.80	2.85	2.14	0.412	19.60	0.53	0.57	0.207
3.97	0.049	1.01	0.908	2.90	2.84	2.20	0.420	20.00	0.54	0.52	0.185
6.199	0.390		0.193	3.00	2.81	2.24	0.425	20.60	0.55	0.44	0.153

Column group 1

Energy (eV)	n	k	R(φ = 0)
21.00	0.57	0.39	0.127
21.60	0.64	0.34	0.089
22.00	0.64	0.32	0.081
22.60	0.69	0.27	0.058
23.00	0.73	0.24	0.043
23.60	0.80	0.26	0.033
24.00	0.80	0.26	0.034
24.60	0.82	0.25	0.029
25.00	0.83	0.25	0.026
25.60	0.86	0.24	0.022
26.00	0.88	0.25	0.022
26.60	0.87	0.26	0.023
27.00	0.87	0.25	0.022
27.60	0.89	0.23	0.019
28.00	0.90	0.23	0.017
28.60	0.91	0.22	0.015
29.00	0.92	0.22	0.014
29.60	0.94	0.22	0.014
30.00	0.95	0.22	0.014
31.00	0.97	0.23	0.014
32.00	0.98	0.24	0.015
33.00	0.98	0.25	0.015
34.00	0.99	0.25	0.016
35.00	0.99	0.26	0.017
36.00	0.99	0.27	0.018
37.00	0.99	0.28	0.019
38.00	0.98	0.28	0.021
39.00	0.97	0.29	0.022
40.00	0.95	0.29	0.023

Tellurium, E $\parallel$ $\hat{c}$ [25]

Energy (eV)	n	k	R(φ = 0)
0.01364	4.82	0.118	0.431
0.01488	5.26	0.0505	0.463
0.01612	5.47	0.0278	0.477
0.01736	5.59	0.0174	0.485
0.01860		0.0796	
0.01984		0.0696	
0.02108		0.0749	
0.02232		0.1900	
0.02356		0.2220	
0.02480		0.0716	
0.02604		0.0682	
0.02728		0.0832	
0.02976		0.0149	
0.03224		2.14E-03	
0.03472		1.71E-02	
0.03720	5.94	3.71E-03	0.507
0.03968		2.44E-03	
0.04339	5.96	1.59E-03	0.508
0.04711		7.85E-04	
0.05083		7.38E-04	
0.05579		3.89E-04	
0.06199	5.98	3.09E-04	0.509
0.07439		2.52E-04	
0.08679		2.96E-04	
0.09919		3.68E-04	
0.12400	6.246	3.34E-04	0.524
0.15500	6.253		0.525
0.20660	6.286		0.526
0.24800	6.316	7.48E-05	0.528
0.31	6.372	1.18E-05	0.531
0.35		4.93E-04	
0.41		6.74E-03	
0.5	6.53	2.30E-02	0.539
0.6	6.71	7.50E-02	0.549

Column group 2

Energy (eV)	n	k	R(φ = 0)
0.7	7.00	0.24	0.563
0.8	7.23	0.48	0.574
0.9	7.48	0.94	0.589
1.0	7.70	1.56	0.606
1.2	6.99	2.22	0.593
1.4	7.11	2.46	0.604
1.6	6.75	2.91	0.606
1.8	6.89	3.70	0.637
2.0	4.67	4.67	0.654
2.2	4.94	5.16	0.681
2.4	3.94	5.08	0.686
2.6	3.25	4.77	0.681
2.8	2.73	4.42	0.674
3.0	2.30	4.16	0.674
3.5	1.69	3.44	0.646
4.0	1.33	2.64	0.571
4.5	1.32	1.96	0.428
5.0	1.63	1.60	0.312
5.5	1.72	1.57	0.302
6.0	1.73	1.45	0.276
6.5	1.78	1.36	0.257
7.0	1.83	1.36	0.257
7.5	1.72	1.51	0.289
8.0	1.54	1.37	0.260
8.5	1.55	1.23	0.226
9.0	0.99	0.93	0.179
9.5	1.47	1.25	0.233
10.0	0.86	0.86	0.181
11.0	0.80	0.77	0.165
12.0	0.79	0.76	0.164
14.0	0.67	0.59	0.146
16.0	0.59	0.49	0.147
18.0	0.48	0.31	0.160
20.0	0.74	0.20	0.035
22.0	0.83	0.18	0.018
24.0	0.85	0.15	0.013
26.0	0.87	0.12	0.009
28.0	0.89	0.090	0.006
30.0	0.90	0.045	0.003

Tellurium, E $\perp$ $\hat{c}$ [25]

Energy (eV)	n	k	R(φ = 0)
0.01364	2.61	0.2980	0.204
0.01488	3.65	0.0894	0.325
0.01612	4.10	0.0535	0.370
0.01736	4.63	0.4990	0.420
0.01860		0.1170	
0.01984		0.0343	
0.02108	(4.42)	0.0421	0.398
0.02232		0.1060	
0.02356		0.0880	
0.02480		0.0458	
0.02604		0.0928	
0.02728		0.0886	
0.02976		0.0232	
0.03224		3.06E-03	
0.03472		1.25E-02	
0.03720	4.71	2.65E-03	0.422
0.03968		1.89E-03	
0.04339	4.74	1.41E-03	0.425
0.04711		8.38E-04	
0.05083		6.79E-04	
0.05579		1.59E-04	
0.06199	4.77	1.16E-04	0.427
0.07439		7.23E-05	
0.08679		5.34E-05	

Column group 3

Energy (eV)	n	k	R(φ = 0)
0.09919		4.28E-05	
0.1240	4.796	3.18E-05	0.429
0.1550	4.809		0.430
0.2066	4.838		0.432
0.2480	4.864	2.19E-05	0.434
0.31	4.929	3.18E-05	0.439
0.35		7.89E-02	
0.41		0.149	
0.5	4.90		0.437
0.6	4.93		0.439
0.7	4.95	0.11	0.441
0.8	5.10	0.13	0.452
0.9	5.22	0.22	0.461
1.0	5.35	0.45	0.472
1.2	5.17	0.63	0.462
1.4	5.56	0.63	0.488
1.6	5.88	1.15	0.517
1.8	6.10	1.80	0.545
2.0	5.94	2.69	0.571
2.2	5.10	3.61	0.594
2.4	4.24	3.77	0.593
2.6	3.57	3.75	0.591
2.8	3.03	3.63	0.588
3.0	2.51	3.39	0.578
3.5	1.72	2.70	0.532
4.0	1.32	2.01	0.440
4.5	1.28	1.28	0.251
5.0	1.47	0.82	0.132
5.5	1.74	0.51	0.104
6.0	1.94	0.39	0.118
6.5	2.19	0.32	0.148
7.0	2.48	0.40	0.192
7.5	2.60	0.69	0.226
8.0	2.59	0.91	0.245
8.5	2.39	1.00	0.235
9.0	1.11	1.24	0.259
9.5	2.08	1.11	0.224
10.0	0.99	1.04	0.215
11.0	0.84	1.01	0.237
12.0	0.87	0.87	0.182
14.0	0.59	0.87	0.282
16.0	0.64	0.55	0.144
18.0	0.52	0.41	0.161
20.0	0.50	0.38	0.165
22.0	0.56	0.29	0.110
24.0	0.54	0.25	0.113
26.0	0.50	0.20	0.127
28.0	0.48	0.17	0.135
30.0	0.46	0.088	0.140

Titanium (Polycrystalline) [14]

Energy (eV)	n	k	R(φ = 0)
0.10	5.03	23.38	0.965
0.15	3.00	15.72	0.954
0.20	2.12	11.34	0.939
0.25	2.05	8.10	0.890
0.30	6.39	9.94	0.833
0.35	2.74	6.21	0.792
0.40	2.49	4.68	0.708
0.45	3.35	3.25	0.545
0.50	4.43	3.22	0.555
0.60	4.71	3.77	0.597
0.70	4.38	3.89	0.603
0.80	4.04	3.82	0.596
0.90	3.80	3.65	0.582
1.00	3.62	3.52	0.570

Energy (eV)	n	k	$R(\phi = 0)$	Energy (eV)	n	k	$R(\phi = 0)$	Energy (eV)	n	k	$R(\phi = 0)$
1.10	3.47	3.40	0.560	11.00	0.79	0.72	0.152	0.86	2.92	4.37	0.661
1.20	3.35	3.30	0.550	11.20	0.81	0.69	0.139	0.90	3.11	4.44	0.660
1.30	3.28	3.25	0.546	11.40	0.81	0.69	0.139	0.94	3.15	4.43	0.658
1.40	3.17	3.28	0.549	11.60	0.79	0.68	0.139	0.98	3.15	4.36	0.653
1.50	2.98	3.32	0.557	11.80	0.78	0.67	0.137	1.00	3.14	4.32	0.649
1.60	2.74	3.30	0.559	12.00	0.77	0.65	0.132	1.10	3.05	4.04	0.627
1.70	2.54	3.23	0.557	12.80	0.76	0.55	0.106	1.20	3.00	3.64	0.590
1.80	2.36	3.11	0.550	13.20	0.76	0.52	0.097	1.30	3.12	3.24	0.545
1.90	2.22	2.99	0.540	13.60	0.76	0.48	0.087	1.40	3.29	2.96	0.515
2.00	2.11	2.88	0.530	14.00	0.77	0.45	0.077	1.50	3.48	2.79	0.500
2.10	2.01	2.77	0.520	14.40	0.77	0.42	0.069	1.60	3.67	2.68	0.494
2.20	1.92	2.67	0.509	14.80	0.79	0.38	0.058	1.70	3.84	2.79	0.507
2.30	1.86	2.56	0.495	15.20	0.79	0.36	0.052	1.80	3.82	2.91	0.518
2.40	1.81	2.47	0.483	15.60	0.79	0.32	0.045	1.90	3.70	2.94	0.518
2.50	1.78	2.39	0.471	16.00	0.83	0.31	0.037	2.00	3.60	2.89	0.512
2.60	1.75	2.34	0.462	16.40	0.84	0.28	0.030	2.10	3.54	2.84	0.506
2.70	1.71	2.29	0.456	16.80	0.87	0.27	0.025	2.20	3.49	2.76	0.497
2.80	1.68	2.25	0.451	17.20	0.90	0.25	0.020	2.30	3.49	2.72	0.494
2.90	1.63	2.21	0.447	17.60	0.93	0.25	0.017	2.40	3.45	2.72	0.493
3.00	1.59	2.17	0.444	18.00	0.94	0.24	0.165	2.50	3.38	2.68	0.487
3.10	1.55	2.15	0.442	18.40	0.94	0.23	0.017	2.60	3.34	2.62	0.480
3.20	1.50	2.12	0.442	18.80	0.95	0.24	0.016	2.70	3.31	2.55	0.472
3.30	1.44	2.09	0.442	19.20	0.96	0.25	0.016	2.80	3.31	2.49	0.466
3.40	1.37	2.06	0.443	19.60	0.97	0.25	0.017	2.90	3.32	2.45	0.461
3.50	1.30	2.01	0.443	20.00	0.98	0.27	0.018	3.00	3.35	2.42	0.459
3.60	1.24	1.96	0.441	20.40	0.98	0.27	0.019	3.10	3.39	2.41	0.460
3.70	1.17	1.90	0.436	20.60	1.00	0.29	0.020	3.20	3.43	2.45	0.465
3.80	1.11	1.83	0.430	21.20	0.99	0.31	0.023	3.30	3.45	2.55	0.476
3.85	1.08	1.78	0.423	21.60	0.99	0.31	0.024	3.40	3.39	2.66	0.485
3.90	1.06	1.73	0.413	22.00	0.98	0.32	0.025	3.50	3.24	2.70	0.488
4.00	1.04	1.62	0.389	22.40	0.98	0.33	0.027	3.60	3.13	2.67	0.482
4.20	1.05	1.45	0.333	22.80	0.97	0.33	0.028	3.70	3.05	2.62	0.476
4.40	1.13	1.33	0.284	23.20	0.96	0.34	0.030	3.80	2.99	2.56	0.468
4.60	1.17	1.29	0.265	23.60	0.95	0.35	0.031	3.90	2.96	2.50	0.460
4.80	1.21	1.23	0.244	24.00	0.92	0.35	0.033	4.00	2.95	2.43	0.451
5.00	1.24	1.21	0.236	24.5	0.91	0.34	0.032	4.20	3.02	2.33	0.440
5.20	1.27	1.20	0.228	25.0	0.91	0.33	0.032	4.40	3.13	2.32	0.442
5.40	1.17	1.16	0.228	25.5	0.89	0.33	0.032	4.60	3.24	2.41	0.455
5.60	1.24	1.21	0.234	26.0	0.89	0.33	0.032	4.80	3.33	2.57	0.475
5.80	1.21	1.22	0.241	26.5	0.88	0.32	0.032	5.00	3.40	2.85	0.505
6.00	1.15	1.21	0.244	27.0	0.86	0.31	0.032	5.20	3.27	3.27	0.548
6.20	1.11	1.18	0.240	27.5	0.85	0.30	0.033	5.40	2.92	3.58	0.586
6.40	1.08	1.14	0.232	28.0	0.84	0.29	0.033	5.60	2.43	3.70	0.618
6.60	1.04	1.06	0.212	28.5	0.82	0.26	0.029	5.80	2.00	3.61	0.637
6.80	1.05	1.02	0.198	29.0	0.83	0.25	0.027	6.00	1.70	3.42	0.643
7.00	1.06	0.97	0.182	30.0	0.84	0.22	0.022	6.20	1.47	3.24	0.646
7.20	1.07	0.95	0.175					6.40	1.32	3.04	0.640
7.40	1.11	0.94	0.167	**Tungsten[27]**				6.60	1.21	2.87	0.631
7.60	1.09	0.92	0.165	0.10	14.06	54.71	0.983	6.80	1.12	2.70	0.619
7.80	1.11	0.93	0.165	0.20	3.87	28.30	0.981	7.00	1.06	2.56	0.607
8.00	1.10	0.94	0.169	0.25	2.56	22.44	0.980	7.20	1.01	2.43	0.593
8.20	1.10	0.95	0.171	0.30	1.83	18.32	0.979	7.40	0.98	2.30	0.573
8.40	1.08	0.95	0.175	0.34	1.71	15.71	0.973	7.60	0.95	2.18	0.556
8.60	1.04	0.96	0.181	0.38	1.86	13.88	0.963	7.80	0.93	2.06	0.533
8.80	1.02	0.95	0.181	0.42	1.92	12.63	0.954	8.00	0.94	1.95	0.505
9.00	1.00	0.94	0.182	0.46	1.69	11.59	0.952	8.20	0.94	1.86	0.481
9.20	0.97	0.93	0.182	0.50	1.40	10.52	0.952	8.40	0.96	1.76	0.449
9.40	0.95	0.91	0.181	0.54	1.23	9.45	0.948	8.60	0.99	1.70	0.422
9.60	0.94	0.90	0.179	0.58	1.17	8.44	0.938	8.80	1.01	1.65	0.401
9.80	0.91	0.88	0.179	0.62	1.28	7.52	0.917	9.00	1.01	1.60	0.388
10.00	0.89	0.88	0.180	0.66	1.45	6.78	0.888	9.20	1.02	1.55	0.369
10.20	0.86	0.85	0.178	0.70	1.59	6.13	0.856	9.40	1.03	1.50	0.352
10.40	0.85	0.83	0.175	0.74	1.83	5.52	0.810	9.60	1.05	1.44	0.329
10.60	0.81	0.79	0.167	0.78	2.12	5.00	0.759	9.80	1.09	1.38	0.307
10.80	0.80	0.76	0.162	0.82	2.36	4.61	0.710	10.00	1.13	1.34	0.287

Energy (eV)	n	k	$R(\phi = 0)$	Energy (eV)	n	k	$R(\phi = 0)$	Energy (eV)	n	k	$R(\phi = 0)$
10.20	1.19	1.33	0.274	**Vanadium**[9]				13.50	0.87	0.53	0.079
10.40	1.24	1.34	0.270	0.10	12.83	45.89	0.978	14.00	0.86	0.51	0.075
10.60	1.27	1.36	0.274	0.20	3.90	24.30	0.975	14.50	0.86	0.49	0.070
10.80	1.29	1.39	0.282	0.28	2.13	17.35	0.973	15.00	0.86	0.47	0.065
11.00	1.28	1.42	0.290	0.36	1.54	13.32	0.966	15.50	0.86	0.46	0.062
11.20	1.27	1.44	0.297	0.44	1.28	10.74	0.957	16.00	0.85	0.45	0.061
11.40	1.25	1.46	0.305	0.52	1.16	8.93	0.945	16.50	0.84	0.43	0.059
11.60	1.22	1.48	0.313	0.60	1.10	7.59	0.929	17.00	0.84	0.41	0.056
11.80	1.20	1.48	0.318	0.68	1.07	6.54	0.909	17.50	0.83	0.40	0.054
12.00	1.16	1.48	0.323	0.76	1.08	5.67	0.882	18.00	0.82	0.38	0.051
12.40	1.10	1.47	0.329	0.80	1.10	5.30	0.864	18.50	0.82	0.37	0.048
12.80	1.04	1.44	0.333	0.90	1.18	4.50	0.811	19.00	0.82	0.35	0.045
13.20	0.98	1.40	0.332	1.00	1.34	3.80	0.730	19.50	0.82	0.34	0.043
13.60	0.94	1.35	0.325	1.10	1.60	3.26	0.632	20.00	0.81	0.32	0.041
14.00	0.91	1.28	0.312	1.20	1.93	2.88	0.543	20.50	0.81	0.31	0.038
14.40	0.90	1.23	0.296	1.30	2.25	2.71	0.498	21.00	0.81	0.29	0.036
14.80	0.90	1.17	0.276	1.40	2.48	2.72	0.491	21.50	0.81	0.28	0.033
15.20	0.93	1.13	0.255	1.50	2.57	2.79	0.499	22.00	0.81	0.27	0.032
15.60	0.97	1.12	0.246	1.60	2.57	2.84	0.507	22.50	0.81	0.25	0.029
16.00	0.98	1.14	0.249	1.70	2.52	2.88	0.512	23.00	0.82	0.24	0.027
16.40	0.97	1.17	0.260	1.80	2.45	2.88	0.515	23.50	0.82	0.23	0.025
16.80	0.94	1.19	0.273	1.90	2.36	2.85	0.514	24.00	0.82	0.22	0.024
17.20	0.90	1.21	0.289	2.00	2.34	2.81	0.509	24.50	0.83	0.21	0.022
17.60	0.85	1.21	0.304	2.10	2.31	2.78	0.506	25.00	0.83	0.20	0.020
18.00	0.80	1.20	0.317	2.20	2.28	2.80	0.510	25.50	0.83	0.19	0.019
18.40	0.74	1.18	0.330	2.30	2.23	2.83	0.516	26.00	0.83	0.18	0.018
18.80	0.69	1.15	0.340	2.40	2.15	2.88	0.528	26.50	0.84	0.17	0.016
19.20	0.64	1.11	0.347	2.50	2.02	2.91	0.540	27.00	0.84	0.16	0.015
19.60	0.60	1.07	0.353	2.60	1.89	2.92	0.552	27.50	0.85	0.16	0.014
20.00	0.56	1.02	0.354	2.70	1.74	2.89	0.561	28.00	0.85	0.15	0.013
20.40	0.54	0.97	0.350	2.80	1.61	2.85	0.569	28.50	0.86	0.14	0.012
20.80	0.52	0.92	0.342	2.90	1.48	2.80	0.577	29.00	0.86	0.14	0.011
21.20	0.50	0.87	0.331	3.00	1.36	2.73	0.582	29.50	0.86	0.13	0.010
21.60	0.50	0.82	0.318	3.20	1.16	2.55	0.585	30.00	0.87	0.13	0.009
22.00	0.49	0.77	0.303	3.40	0.99	2.37	0.586	31.00	0.88	0.12	0.008
22.40	0.49	0.73	0.287	3.60	0.87	2.17	0.575	32.00	0.90	0.11	0.007
22.80	0.49	0.69	0.272	3.80	0.80	1.96	0.547	33.00	0.90	0.10	0.005
23.20	0.49	0.66	0.263	4.00	0.78	1.76	0.503	34.00	0.91	0.10	0.005
23.60	0.48	0.62	0.252	4.20	0.80	1.60	0.449	35.00	0.92	0.09	0.004
24.00	0.49	0.57	0.234	4.40	0.83	1.47	0.400	36.00	0.94	0.10	0.004
24.40	0.50	0.53	0.213	4.60	0.87	1.38	0.355	37.00	0.94	0.10	0.004
24.80	0.51	0.49	0.191	4.80	0.90	1.31	0.326	38.00	0.95	0.11	0.004
25.20	0.53	0.46	0.171	5.00	0.91	1.26	0.304	39.00	0.95	0.12	0.004
25.60	0.55	0.43	0.150	5.25	0.93	1.18	0.271	40.00	0.95	0.13	0.005
26.00	0.57	0.40	0.132	5.50	0.94	1.14	0.258				
26.40	0.59	0.38	0.117	5.75	0.96	·1.09	0.235	**Zinc, E $\parallel \hat{c}$**[28]			
26.80	0.61	0.37	0.105	6.00	0.98	1.06	0.223	0.7514	1.9241	7.5619	0.883
27.00	0.62	0.36	0.099	6.25	0.97	1.02	0.212	0.827	1.7921	6.9973	0.874
27.50	0.64	0.34	0.085	6.50	0.97	0.98	0.199	0.866	1.5571	6.7753	0.881
28.00	0.67	0.32	0.073	6.75	0.97	0.94	0.185	0.952	1.4824	6.2296	0.868
28.50	0.69	0.31	0.065	7.00	0.98	0.91	0.175	0.992	1.5762	5.8843	0.847
29.00	0.71	0.30	0.057	7.33	0.97	0.89	0.170	1.033	1.5407	5.3192	0.823
29.50	0.73	0.30	0.052	7.66	0.98	0.87	0.162	1.078	1.5853	4.9013	0.793
30.00	0.75	0.29	0.047	8.00	0.98	0.85	0.155	1.127	1.7768	4.5307	0.748
31.00	0.78	0.29	0.042	8.33	0.98	0.81	0.146	1.181	1.9808	4.2004	0.701
32.00	0.79	0.29	0.040	8.66	0.98	0.81	0.145	1.240	2.8821	3.4766	0.575
33.00	0.82	0.28	0.033	9.00	0.96	'0.79	0.142	1.305	3.2039	3.0042	0.520
34.00	0.84	0.29	0.032	9.50	0.94	0.77	0.136	1.377	2.9459	3.5761	0.584
35.00	0.85	0.31	0.033	10.00	0.91	0.74	0.133	1.459	3.2523	4.2447	0.640
36.00	0.85	0.32	0.036	10.50	0.89	0.71	0.126	1.550	3.8086	4.6212	0.657
37.00	0.84	0.33	0.039	11.00	0.87	0.65	0.112	1.653	3.7577	4.6239	0.659
38.00	0.83	0.33	0.040	11.50	0.88	0.58	0.091	1.722	3.5908	4.4614	0.650
39.00	0.81	0.33	0.042	12.00	0.90	0.58	0.089	1.823	3.4234	4.3232	0.642
40.00	0.80	0.33	0.045	12.50	0.89	0.57	0.086	1.937	3.0132	3.9974	0.624
				13.00	0.88	0.55	0.082	1.984	1.8562	3.9706	0.690

Energy (eV)	n	k	$R(\phi = 0)$
2.066	1.4856	4.0555	0.737
2.094	1.2525	3.9961	0.762
2.119	1.0017	3.8683	0.789
2.275	0.7737	3.9129	0.832
2.445	0.6395	3.4013	0.821
2.666	0.4430	3.1379	0.851
2.917	0.3589	2.8140	0.853
3.220	0.3069	2.5088	0.847
3.594	0.2737	2.1737	0.828
4.065	0.2510	1.8528	0.799
4.678	0.2354	1.6357	0.776

Zinc, $\vec{E} \perp \hat{c}$ [28]

Energy (eV)	n	k	$R(\phi = 0)$
0.751	1.4469	7.4158	0.905
0.827	1.4744	6.9688	0.892
0.866	1.3628	6.6886	0.892
0.952	1.3165	6.2212	0.881
0.992	1.3835	5.8910	0.863
1.033	1.2889	5.4001	0.850
1.078	1.3095	4.9025	0.822
1.127	1.6897	4.4062	0.746
1.181	1.9701	4.0176	0.684
1.240	2.8717	3.2873	0.555
1.305	3.3991	2.7684	0.497
1.377	3.1807	3.4709	0.569
1.459	3.5064	4.1994	0.630
1.550	4.1241	4.7768	0.664
1.653	4.0269	4.8027	0.667
1.722	3.9369	4.6356	0.657
1.823	3.7549	4.3042	0.635
1.937	3.4512	4.1942	0.631
1.984	3.2515	4.2980	0.644
2.066	2.0802	4.7231	0.738
2.094	1.7084	4.7923	0.774
2.119	1.3329	4.4751	0.791
2.275	0.9725	4.2879	0.825
2.455	0.7568	3.7627	0.824
2.666	0.5470	3.4277	0.845
2.917	0.4774	3.0476	0.834
3.220	0.3911	2.7463	0.835
3.594	0.3147	2.3041	0.821
4.065	0.3013	2.0077	0.789
4.678	0.2806	1.7997	0.770

Zirconium (Polycrystalline) [28]

Energy (eV)	n	k	$R(\phi = 0)$
0.10	6.18	1.76	0.300
0.15	3.37	1.30	0.123
0.20	2.34	1.08	0.058
0.26	2.24	1.06	0.052
0.30	2.59	1.14	0.073
0.36	3.17	1.26	0.110
0.40	3.09	1.24	0.105
0.46	3.36	1.30	0.123
0.50	4.13	1.44	0.175
0.56	5.01	1.58	0.231
0.60	5.18	1.61	0.242
0.70	4.54	1.51	0.202
0.80	4.03	1.42	0.168
0.90	3.74	1.37	0.149
0.96	3.69	1.36	0.145
1.00	3.66	1.35	0.143
1.10	3.65	1.35	0.142
1.20	3.53	1.33	0.134
1.30	3.25	1.27	0.116
1.40	3.10	1.25	0.106
1.50	3.02	1.23	0.100
1.60	2.88	1.20	0.091
1.70	2.68	1.16	0.078
1.80	2.49	1.12	0.067
2.00	2.14	1.03	0.047
2.10	1.99	1.00	0.040
2.20	1.87	0.97	0.034
2.30	1.78	0.94	0.030
2.40	1.71	0.92	0.027
2.50	1.62	0.90	0.024
2.60	1.54	0.88	0.022
2.70	1.46	0.86	0.019
2.80	1.40	0.84	0.018
2.90	1.34	0.82	0.016
3.00	0.30	0.81	0.016
3.10	1.26	0.80	0.015
3.30	1.19	0.77	0.014
3.40	1.16	0.76	0.013
3.50	1.13	0.75	0.013
3.60	1.10	0.74	0.013
3.70	1.07	0.73	0.013
3.80	1.04	0.72	0.012
3.90	1.01	0.71	0.012
4.00	0.98	0.70	0.012
4.20	0.94	0.68	0.013
4.40	0.89	0.67	0.013
4.60	0.85	0.65	0.014
4.80	0.81	0.64	0.014
5.00	0.78	0.63	0.015
5.20	0.77	0.62	0.016
5.40	0.77	0.62	0.016
5.60	0.80	0.63	0.014
5.80	0.87	0.66	0.013
6.00	1.00	0.71	0.012
6.20	1.11	0.75	0.013
6.40	1.23	0.78	0.014
6.60	1.33	0.81	0.016
6.80	1.42	0.84	0.018
7.00	1.49	0.86	0.020
7.20	1.54	0.88	0.022
7.40	1.58	0.89	0.023
7.60	1.61	0.90	0.024
7.80	1.63	0.90	0.025
8.00	1.66	0.91	0.026
8.20	0.67	0.91	0.026
8.40	1.68	0.92	0.026
8.60	1.68	0.92	0.026
8.80	1.66	0.91	0.026
9.00	1.65	0.91	0.025
9.20	1.63	0.90	0.025
9.40	1.60	0.89	0.024
9.60	1.57	0.89	0.023
9.80	1.52	0.87	0.021
10.00	1.47	0.86	0.020
10.20	1.42	0.84	0.018
10.40	1.35	0.82	0.016
10.50	1.32	0.81	0.016
10.60	1.28	0.80	0.015
10.80	1.23	0.78	0.014
11.00	1.19	0.77	0.014
11.20	1.16	0.76	0.013
11.40	1.13	0.75	0.013
11.60	1.11	0.74	0.013
11.80	1.09	0.74	0.013
12.00	1.08	0.73	0.013
12.40	1.05	0.72	0.012
12.80	1.01	0.71	0.012
13.20	0.98	0.70	0.012
13.60	0.95	0.69	0.013
14.00	0.92	0.68	0.013
14.40	0.89	0.67	0.013
14.80	0.90	0.67	0.013
15.20	0.92	0.68	0.013
15.60	0.95	0.69	0.013
16.00	0.98	0.70	0.012
16.40	1.01	0.71	0.012
16.80	1.04	0.72	0.012
17.20	1.09	0.74	0.013
17.60	1.13	0.75	0.013
18.00	1.17	0.76	0.014
18.40	1.21	0.78	0.014
18.80	1.24	0.79	0.014
19.20	1.27	0.80	0.015
19.60	1.29	0.80	0.015
20.00	1.30	0.81	0.015
20.60	1.29	0.80	0.015
21.00	1.27	0.80	0.015
21.60	1.23	0.78	0.014
22.00	1.20	0.77	0.014
22.60	1.15	0.76	0.013
23.00	1.12	0.75	0.013
23.60	1.08	0.73	0.013
24.00	1.05	0.73	0.013
24.60	1.02	0.71	0.012
25.00	1.00	0.71	0.012
25.60	0.97	0.69	0.012
26.00	0.95	0.69	0.013
26.60	0.91	0.67	0.013
27.00	0.88	0.66	0.013
27.60	0.84	0.65	0.014
28.00	0.83	0.64	0.014
28.60	0.82	0.64	0.014
29.00	0.81	0.64	0.014
29.60	0.82	0.64	0.014
30.00	0.82	0.64	0.014

REFERENCES

1. Shiles, E., Sasaki, T., Inokuti, M., and Smith, D. Y., *Phys. Rev. Sect. B*, 22, 1612, 1980.
2. Edwards, D. F., and Philipp, H. R., in *HOC-I*, p.665.
3. Ives, H. E., and Briggs, N. B., *J. Opt. Soc. Am.*, 27, 395, 1937.
4. Bos, L. W., and Lynch, D. W., *Phys. Rev. Sect. B*, 2, 4567, 1970.
5. Weaver, J. H., Colavita, E., Lynch, D. W., and Rosei, R., *Phys. Rev. Sect. B*, 19, 3850, 1979.
6. Hagemann, H. J., Gudat, W., and Kunz, C., *J. Opt. Soc. Am.*, 65, 742, 1975.
7. Schulz, L. G., *J. Opt. Soc. Am.*, 47, 64, 1957.
8. Potter, R. F., in *HOC-I*, p.465.
9. Olson, C. G., Lynch, D. W., and Weaver, J. H., unpublished.
10. Lynch, D. W., Olson, C. G., and Weaver, J. H., unpublished.
11. Weaver, J. H., Olson, C. G., and Lynch, D. W., *Phys. Rev. Sect. B*, 15, 4115, 1977.
12. Lynch, D. W., and Hunter, W. R., in *HOC-II*, p.345.
13. Priol, M. A., Daudé, A., and Robin, S., *Compt. Rend.*, 264, 935, 1967.
14. Johnson, P. B., and Christy, R. W., *Phys. Rev. Sect. B*, 9, 5056, 1974.
15. Arakawn, E. T., and Inagaki, T., in *HOC-II*, p.461.
16. Weaver, J. H., Lynch, D. W., and Olson, D. G., *Phys. Rev. Sect. B*, 10, 501, 1973.
17. Lynch, D. W., Rosei, R., and Weaver, J. H., *Solid State Commun.*, 9, 2195, 1971.
18. Weaver, J. H., Lynch, D. W., and Olson, C. G., *Phys. Rev. Sect. B*, 7, 4311, 1973.
19. Weaver, J. H., and Benbow, R. L., *Phys. Rev. Sect. B*, 12, 3509, 1975.
20. Weaver, J. H., *Phys. Rev.,Sect. B*, 11, 1416, 1975.
21. Lynch, D. W., and Hunter, W. R., in *HOC-II*, p.364.
22. Palik, E. D., in *HOC-II*, p. 691.
23. Edwards, D. F., in *HOC-I*, p. 547.
24. Lynch, D. W., and Hunter, W. R., in *HOC-II*, p.354.
25. Palik, E. D., in *HOC-II*, p. 709.
26. Lynch, D. W., Olson, C. G., and Weaver, J. H., *Phys. Rev. Sect. B*, 11, 3671, 1975.
27. Weaver, J. H., Lynch, D. W., and Olson, C. G., *Phys. Rev. Sect. B*, 12, 1293, 1975.
28. Lanham, A. P., and Terherne, D. M., *Proc. Phys. Soc.*, 83, 1059, 1964.

ELASTO-OPTIC, ELECTRO-OPTIC, AND MAGNETO-OPTIC CONSTANTS

When a crystal is subjected to a stress field, an electric field, or a magnetic field, the resulting optical effects are in general dependent on the orientation of these fields with respect to the crystal axes. It is useful, therefore, to express the optical properties in terms of the refractive index ellipsoid (or indicatrix):

$$\frac{x^2}{n_x^2} + \frac{y^2}{n_y^2} + \frac{z^2}{n_z^2} = 1$$

or

$$\sum_{ij} B_{ij} x_i y_j = 1 \quad (i, j = 1, 2, 3)$$

where

$$B_{ij} = \left[\frac{1}{\varepsilon}\right]_{ij} = \left[\frac{1}{n^2}\right]_{ij}$$

ε is the dielectric constant or permeability; the quantity B_{ij} has the name impermeability.

A crystal exposed to a *stress* S will show a change of its impermeability. The photo-elastic (or elasto-optic) constants, P_{ijkl}, are defined by

$$\Delta\left[\frac{1}{\varepsilon}\right]_{ij} = \Delta\left[\frac{1}{n^2}\right]_{ij} = \sum_{kl} P_{ijkl} S_{kl}$$

where n is the refractive index and S_{kl} are the strain tensor elements; the P_{ijkl} are the elements of a 4th rank tensor.

When a crystal is subjected to an *electric field E* two possible changes of the refractive index may occur depending on the symmetry of the crystal.

1. All materials, including isotropic solids and polar liquids, show an electro-optic birefringence (Kerr effect) which is proportional to the square of the electric field, E:

$$\left[\frac{1}{n^2}\right]_{ij} = \sum_{k,l=1,2,3} K_{ijkl} E_k E_l = \sum_{k,l=1,2,3} g_{ijkl} P_k P_l$$

where E_k and E_l are the components of the electric field and P_k and P_l the electric polarizations. The coefficients, K_{ijkl}, are the quadratic electro-optic coefficients, while the constants g_{ijkl} are known as the Kerr constants.

2. The other electro-optic effect only occurs in the 20 piezo-electric crystal classes (no center of symmetry). This effect is known as the Pockels effect. The optical impermeability changes linearly with the static field

$$\Delta\left[\frac{1}{n^2}\right]_{ij} = \sum_k r_{ij,k} E_k$$

The coefficients $r_{ij,k}$ have the name (linear) electro-optic coefficients.

The values of the electro-optic coefficients depend on the boundary conditions. If the superscripts T and S denote respectively the conditions of zero stress (free) and zero strain (clamped) one finds:

$$r_{ij}^T = r_{ij}^S + q_{ik}^E e_{jk} = r_{ij}^S + P_{ik}^E d_{jk}$$

where $e_{jk} = (\partial T_k/\partial E_j)_S$ and $d_{jk} = (\partial S_k/\partial E_j)_T$ are the appropriate piezo-electric coefficients.

The interaction between a *magnetic field* and a light wave propagating in a solid or in a liquid gives rise to a rotation of the plane of polarization. This effect is known as *Faraday rotation*. It results from a difference in propagation velocity for left and right circular polarized light.

The Faraday rotation, θ_F, is linearly proportional to the magnetic field H:

$$\theta_F = VlH$$

where l is the light path length and V is the *Verdet* constant (minutes/oersted·cm).

For ferromagnetic, ferrimagnetic, and antiferromagnetic materials the magnetic field in the above expression is replaced by the magnetization M and the magneto-optic coefficient in this case is known as the Kund constant K:

$$\text{Specific Faraday rotation } F = KM$$

In the tables below the *Faraday rotation* is listed at the saturation magnetization per unit length, together with the absorption coefficient α, the temperature T, the critical temperature T_C (or T_N), and the wavelength of the measurement.

ELASTO-OPTIC, ELECTRO-OPTIC, AND MAGNETO-OPTIC CONSTANTS (continued)

In the tables which follow, the properties are presented in groups:
- Elasto-optic coefficients (photoelastic constants)
- Linear electro-optic coefficients (Pockels constants)
- Quadratic electro-optic coefficients (Kerr constants)
- Magneto-optic coefficients:
  - Verdet constants
  - Faraday rotation parameters

Within each group, materials are classified by crystal system or physical state. References are given at the end of each group of tables.

ELASTO-OPTIC COEFFICIENTS (PHOTOELASTIC CONSTANTS)

Name

Cubic (43m, 432, m3m)	Formula	$\lambda/\mu m$	p_{11}	p_{12}	p_{44}	p_{11}-p_{12}	Ref.
Sodium fluoride	NaF	0.633	0.08	0.20	-0.03	-0.12	1
Sodium chloride	NaCl	0.589	0.115	0.159	-0.011	-0.042	2
Sodium bromide	NaBr	0.589	0.148	0.184	-0.0036	-0.035	1
Sodium iodide	NaI	0.589	—	—	0.0048	-0.0141	3
Potassium fluoride	KF	0.546	0.26	0.20	-0.029	0.06	1
Potassium chloride	KCl	0.633	0.22	0.16	-0.025	0.06	4
Potassium bromide	KBr	0.589	0.212	0.165	-0.022	0.047	5
Potassium iodide	KI	0.590	0.212	0.171	—	0.041	6
Rubidium chloride	RbCl	0.589	0.288	0.172	-0.041	0.116	7,8
Rubidium bromide	RbBr	0.589	0.293	0.185	-0.034	0.108	7,8
Rubidium iodide	RbI	0.589	0.262	0.167	-0.023	0.095	7,8
Lithium fluoride	LiF	0.589	0.02	0.13	-0.045	-0.11	5
Lithium chloride	LiCl	0.589	—	—	-0.0177	-0.0407	3
Ammonium chloride	NH$_4$Cl	0.589	0.142	0.245	0.042	-0.103	9
Cadmium telluride	CdTe	1.06	-0.152	-0.017	-0.057	-0.135	10
Calcium fluoride	CaF$_2$	0.55-0.65	0.038	0.226	0.0254	-0.183	11
Copper chloride	CuCl	0.633	0.120	0.250	-0.082	-0.130	12
Copper bromide	CuBr	0.633	0.072	0.195	-0.083	-0.123	12
Copper iodide	CuI	0.633	0.032	0.151	-0.068	-0.119	12
Diamond	C	0.540-0.589	-0.278	0.123	-0.161	-0.385	13
Germanium	Ge	3.39	-0.151	-0.128	-0.072	-0.023	14
Gallium arsenide	GaAs	1.15	-0.165	-0.140	-0.072	-0.025	15
Gallium phosphide	GaP	0.633	-0.151	-0.082	-0.074	-0.069	15
Strontium fluoride	SrF$_2$	0.633	0.080	0.269	0.0185	-0.189	16
Strontium titanate	SrTiO$_3$	0.633	0.15	0.095	0.072	—	17
KRS-5	Tl(Br,I)	0.633	-0.140	0.149	-0.0725	-0.289	18,20
KRS-6	Tl(Br,Cl)	0.633	-0.451	-0.337	-0.164	-0.114	19,20
Zinc sulfide	Zn	0.633	0.091	-0.01	0.075	0.101	15

Rare Gases	Formula	$\lambda/\mu m$	p_{11}	p_{12}	p_{44}	p_{11}-p_{12}	Ref.
Neon ($T = 24.3$ K)	Ne	0.488	0.157	0.168	0.004	-0.011	21
Argon ($T = 82.3$ K)	Ar	0.488	0.256	0.302	0.015	-0.046	22
Krypton ($T = 115.6$ K)	Kr	0.488	0.34	0.34	0.037	0	21
Xenon ($T = 160.5$ K)	Xe	0.488	0.284	0.370	0.029	-0.086	22

Garnets	Formula	$\lambda/\mu m$	p_{11}	p_{12}	p_{44}	p_{11}-p_{12}	Ref.
GGG	Gd$_3$Ga$_5$O$_{12}$	0.514	-0.086	-0.027	-0.078	-0.059	23
YIG	Y$_3$Fe$_5$O$_{12}$	1.15	0.025	0.073	0.041	—	15
YGG	Y$_3$Ga$_5$O$_{12}$	0.633	0.091	0.019	0.079	—	17
YAG	Y$_3$Al$_5$O$_{12}$	0.633	-0.029	0.0091	-0.0615	-0.038	15

Name

Cubic (23, m3)	Formula	$\lambda/\mu m$	p_{11}	p_{12}	p_{44}	p_{13}	Ref.
Barium nitrate	Ba(NO3)2	0.589	—	p_{11}-p_{22} = 0.992	-0.0205	p_{11}-p_{13} = 0.713	13
Lead nitrate	$Pb(NO_3)_2$	0.589	0.162	0.24	-0.0198	0.20	24,25
Sodium bromate	$NaBrO_3$	0.589	0.185	0.218	-0.0139	0.213	26
Sodium chlorate	$NaClO_3$	0.589	0.162	0.24	-0.0198	0.20	26
Strontium nitrate	$Sr(NO_3)_2$	0.41	0.178	0.362	-0.014	0.316	27

Hexagonal (mmc,6mm)	Formula	$\lambda/\mu m$	p_{11}	p_{12}	p_{13}	p_{31}	p_{33}	p_{44}	Ref.
Beryl	$Be_3Al_2Si_6O_{18}$	0.589	0.0099	0.175	0.191	0.313	0.023	-0.152	28
Cadmium sulfide	CdS	0.633	-0.142	-0.066	-0.057	-0.041	-0.20	-0.099	15,2
Zinc oxide	ZnO	0.633	±0.222	±0.099	-0.111	±0.088	-0.235	0.0585	30
Zinc sulfide	ZnS	0.633	-0.115	0.017	0.025	0.0271	-0.13	-0.0627	31

Trigonal (3m,32,$\bar{3}$m)	Formula	$\lambda/\mu m$	p_{11}	p_{12}	p_{13}	p_{14}	p_{31}
Sapphire	Al_2O_3	0.644	-0.23	-0.03	0.02	0.00	-0.04
Calcite	$CaCO_3$	0.514	0.062	0.147	0.186	-0.011	0.241
Lithium niobate	$LiNbO_3$	0.633	±0.034	±0.072	±0.139	±0.066	±0.178
Lithium tantalate	$LiTaO_3$	0.633	-0.081	0.081	0.093	-0.026	0.089
Cinnabar	HgS	0.633			±0.445		
Quartz	SiO_2	0.589	0.16	0.27	0.27	-0.030	0.29
Proustite	Ag_3AsS_3	0.633	±0.10	±0.19	±0.22		±0.24
Sodium nitrite	$NaNO_3$	0.633		±0.21	±0.215	±0.027	±0.25
Tellurium	Te	10.6	0.155	0.130	—	—	—

Trigonal (3m,32,$\bar{3}$m) (continued)	p_{33}	p_{41}	p_{44}	Ref.
Sapphire	-0.20	0.01	-0.10	15,32
Calcite	0.139	-0.036	-0.058	33
Lithium niobate	+-0.060	±0.154	±0.300	15,34
Lithium tantalate	-0.044	-0.085	0.028	15,35
Cinnabar	+-0.115	—	—	36
Quartz	0.10	-0.047	-0.079	37
Proustite	+-0.20	—	—	38
Sodium nitrite		0.055	-0.06	39
Tellurium		—	—	15

Tetragonal (4/mmm,$\bar{4}$2m,422)	Formula	$\lambda/\mu m$	p_{11}	p_{12}	p_{13}	p_{31}
Ammonium dihydrogen phosphate	ADP	0.589	0.319	0.277	0.169	0.197
Barium titanate	$BaTiO_3$	0.633	0.425	—	—	—
Cesium dihydrogen arsenate	CDA	0.633	0.267	0.225	0.200	0.195
Magnesium fluoride	MgF_2	0.546	—	—	—	—
Calomel	Hg_2Cl_2	0.633	±0.551	±0.440	±0.256	±0.137
Potassium dihydrogen phosphate	KDP	0.589	0.287	0.282	0.174	0.241
Rubidium dihydrogen arsenate	RDA	0.633	0.227	0.239	0.200	0.205
Rubidium dihydrogen phosphate	RDP	0.633	0.273	0.240	0.218	0.210
Strontium barium niobate	$Sr_{0.75}Ba_{0.25}Nb_2O_6$	0.633	0.16	0.10	0.08	0.11
Strontium barium niobate	$Sr_{0.5}Ba_{0.5}Nb_2O_6$	0.633	0.06	0.08	0.17	0.09
Tellurium oxide	TeO_2	0.633	0.0074	0.187	0.340	0.090
Rutile	TiO_2	0.633	0.017	0.143	-0.139	-0.080

Name Tetragonal (4/mmm,$\bar{4}$2m,422) (continued)	p_{33}	p_{44}	p_{66}	Ref.
Ammonium dihydrogen phosphate	0.167	-0.058	-0.091	40
Barium titanate	—	—	—	41
Cesium dihydrogen arsenate	0.227	—	—	42
Magnesium fluoride	—	±0.0776	±0.0488	43
Calomel	±0.010	—	±0.047	44
Potassium dihydrogen phosphate	0.122	-0.019	-0.064	45
Rubidium dihydrogen arsenate	0.182	—	—	41
Rubidium dihydrogen phosphate	0.208	—	—	41
Strontium barium niobate	0.47	—	—	46
Strontium barium niobate	0.23	—	—	46
Tellurium oxide	0.240	-0.17	-0.046	47
Rutile	-0.057	-0.009	-0.060	48

Tetragonal (4,$\bar{4}$,4/m)	Formula	$\lambda/\mu m$	p_{11}	p_{12}	p_{13}	p_{16}	p_{31}
Cadmium molybdate	$CdMoO_4$	0.633	0.12	0.10	0.13	—	0.11
Lead molybdate	$PbMoO_4$	0.633	0.24	0.24	0.255	0.017	0.175
Sodium bismuth molybdate	$NaBi(MoO_4)_2$	0.633	0.243	0.205	0.25	—	0.21

Tetragonal (4,$\bar{4}$,4/m) (continued)	p_{33}	p_{44}	p_{45}	p_{61}	p_{66}	Ref.
Cadmium molybdate	0.18	—	—	—	—	49
Lead molybdate	0.300	0.067	-0.01	0.013	0.05	52
Sodium bismuth molybdate	0.29	—	—	—	—	—

Orthorhombic (222,m22,mmm)	Formula	$\lambda/\mu m$	p_{11}	p_{12}	p_{13}	p_{21}	p_{22}	p_{23}
Ammonium chlorate	NH_4ClO_4	0.633	—	0.24	0.18	0.23	—	0.20
Ammonium sulfate	$(NH_4)_2SO_4$	0.633	0.26	0.19	±0.260	±0.230	±0.27	±0.254
Rochelle salt	$NaKC_4H_4O_6$	0.589	0.35	0.41	0.42	0.37	0.28	0.34
Iodic acid (α)	HIO_3	0.633	0.302	0.496	0.339	0.263	0.412	0.304
Sulfur (α)	S	0.633	0.324	0.307	0.268	0.272	0.301	0.310
Barite	$BaSO_4$	0.589	0.21	0.25	0.16	0.34	0.24	0.19
Topaz	$Al_2SiO_4(OH,F)_2$	—	-0.085	0.069	0.052	0.095	-0.120	0.065

Orthorhombic (222,m22,mmm) (continued)	p_{31}	p_{32}	p_{33}	p_{44}	p_{55}	p_{66}	Ref.
Ammonium chlorate	0.19	0.18	±0.02	<±0.02	—	±0.04	51
Ammonium sulfate	0.20	±0.26	0.26	0.015	±0.0015	0.012	52
Rochelle salt	0.36	0.35	0.36	-0.030	0.0046	-0.025	53
Iodic Acid (α)	0.251	0.345	0.336	0.084	-0.030	0.098	54
Sulfur (α)	0.203	0.232	0.270	0.143	0.019	0.118	54
Barite	0.28	0.22	0.31	0.002	-0.012	0.037	55
Topaz	0.095	0.085	-0.083	-0.095	-0.031	0.098	28

Monoclinic (2,m,2/m)	Formula	$\lambda/\mu m$			
Taurine	$C_2H_7NO_3S$	0.589	$p_{11} = 0.313$	$p_{25} = -0.0025$	$p_{51} = -0.014$
			$p_{12} = 0.251$	$p_{31} = 0.362$	$p_{52} = 0.006$
			$p_{13} = 0.270$	$p_{32} = 0.275$	$p_{53} = 0.0048$
			$p_{15} = -0.10$	$p_{33} = 0.308$	$p_{55} = 0.047$

Name

Monoclinic (2,m,2/m)	Formula	$\lambda/\mu m$			
Taurine (continued)			$p_{21} = 0.281$	$p_{35} = -0.003$	$p_{64} = 0.0024$
			$p_{22} = 0.252$	$p_{44} = 0.0025$	$p_{66} = 0.0028$
			$p_{23} = 0.272$	$p_{46} = -0.0056$	

Isotropic	Formula	$\lambda/\mu m$	p_{11}	p_{12}	p_{44}	Ref.
Fused silica	SiO_2	0.633	0.121	0.270	-0.075	15
Water	H_2O	0.633	±0.31	±0.31		15
Polystyrene		0.633	±0.30	±0.31		25
Lucite		0.633	±0.30	0.28		25
Orpiment	As_2S_3-glass	1.15	0.308	0.299	0.0045	15
Tellurium oxide	TeO_2-glass	0.633	0.257	0.241	0.0079	56
Laser glasses	LGS-247-2	0.488	±0.168	±0.230		57
	LGS-250-3		±0.135	±0.198		
	LGS-1		±0.214	±0.250		
	KGSS-1621		±0.205	±0.239		
Dense flint glasses	LaSF	0.633	0.088	0.147	-0.030	58
(examples)	SF_4		0.215	0.243	-0.014	
	U10502		0.172	0.179	-0.004	
	$TaFd_7$		0.099	0.138	-0.020	

REFERENCES

A. Narasimhamurty, T. S., *Photoelastic and Electro-Optic Properties of Crystals*, Plenum Press, New York, 1981; pp. 290-293.

B. Weber, M. J., Ed., *CRC Handbook of Laser Science and Technology*, Volume IV, Part 2, CRC Press, Boca Raton, FL, 1986; pp. 324-331.

1. Petterson, H. E., *J. Opt. Soc. Am.*, 63, 1243, 1973.
2. Burstein, E. and Smith, P. L., *Phys. Rev.*, 74, 229, 1948.
3. Pakhnev, A. V., et al., *Sov. Phys. J. (transl.)*, 18, 1662, 1975.
4. Feldman, A., Horovitz, D., and Waxler, R. M., *Appl. Opt.*, 16, 2925, 1977.
5. Iyengar, K. S., *Nature (London)*, 176, 1119, 1955.
6. Bansigir, K. G. and Iyengar, K. S., *Acta Crystallogr.*, 14, 727, 1961.
7. Pakhev, A. V., et al., *Sov. Phys. J. (transl.)*, 20, 648, 1975.
8. Bansigir, K. G., *Acta Crystallogr.*, 23, 505, 1967.
9. Krishna Rao, K. V. and Krishna Murty, V. G., *Ind. J. Phys.*, 41, 150, 1967.
10. Weil, R. and Sun, M. J., *Proc. Int. Symp. CdTe (Detectors)*, Strasbourg Centre de Rech. Nucl., 1971, XIX-1 to 6, 1972.
11. Schmidt, E. D. D. and Vedam, K., *J. Phys. Chem. Solids*, 27, 1563, 1966.
12. Biegelsen, D. K., et al., *Phys. Rev. B*, 14, 3578, 1976.
13. Helwege, K. H., *Landolt-Börnstein, New Series Group III*, Vol. II, Springer-Verlag Berlin, 1979.
14. Feldman, A., Waxler, R. M., and Horovitz, D., *J. Appl. Phys.*, 49, 2589, 1978.
15. Dixon, R. W., *J. Appl. Phys.*, 38, 5149, 1967.
16. Shabin, O. V., et al., *Sov. Phys. Sol. State (transl.)*, 13, 3141, 1972.
17. Reintjes, J. and Schultz, M. B., *J. Appl. Phys.*, 39, 5254, 1968.
18. Rivoallan, L. and Favre, F., *Opt. Commun.*, 8, 404, 1973.
19. Rivoallan, L. and Favre, F., *Opt. Commun.*, 11, 296, 1974.
20. Afanasev, I. I., et al., *Sov. J. Opt. Technol.*, 46, 663, 1979.
21. Rand, S. C., et al., *Phys. Rev. B*, 19, 4205, 1979.
22. Sipe, J. E., *Can J. Phys.*, 56, 199, 1978.
23. Christyi, I. L., et al., *Sov. Phys. Sol. State (transl.)*, 17, 922, 1975.
24. Narasimhamurty, T. S., *Curr. Sci. (India)*, 23, 149, 1954.
25. Smith, T. M. and Korpel, A., *IEEE J. Quant. Electron.*, QE-1, 283, 1965.
26. Narasimhamurty, T. S., *Proc. Ind. Acad. Sci.*, A40, 164, 1954.
27. Rabman, A., *Bhagarantam Commem. Vol.*, Bangalore Print. and Publ., 173, 1969.
28. Eppendahl, R., *Ann. Phys. (IV)*, 61, 591, 1920.
29. Laurenti, J. P. and Rouzeyre, M., *J. Appl. Phys.*, 52, 6484, 1981.
30. Sasaki, H., et al., *J. Appl. Phys.*, 47, 2046, 1976.
31. Uchida, N. and Saito, S., *J. Appl. Phys.*, 43, 971, 1972.

32. Waxler, R. M. and Farabaugh, E. M., *J. Res. Natl. Bur. Stand.*, A74, 215, 1970.
33. Nelson, D. F., Lazay, P. D., and Lax, M., *Phys. Rev.*, B6, 3109, 1972.
34. O'Brien, R. J., Rosasco, G. J., and Weber, A., *J. Opt. Soc. Am.*, 60, 716, 1970.
35. Avakyants, L. P., et al., *Sov. Phys.*, 18, 1242, 1976.
36. Sapriel, J., *Appl. Phys. Litt.*, 19, 533, 1971.
37. Narasimhamurty, T. S., *J. Opt. Soc. Am.*, 59, 682, 1969.
38. Zubrinov, I. I., et al., *Sov. Phys. Sol. State (transl.)*, 15, 1921, 1974.
39. Kachalov, O. V. and Shpilko, I. O., *Sov. Phys. JETP (transl.)*, 35, 957, 1972.
40. Narasimhamurty, T. S., et al., *J. Mater. Sci.*, 8, 577, 1973.
41. Tada, K. and Kikuchi, K., *Jpn. J. Appl. Phys.*, 19, 1311, 1980.
42. Aleksandrov, K. S., et al., *Sov. Phys. Sol. State (transl.)*, 19, 1090, 1977.
43. Afanasev, I. I., et al., *Sov. Phys. Sol. State (transl.)*, 17, 2006, 1975.
44. Silvestrova, I. M., et al., *Sov. Phys. Cryst. (transl.)*, 20, 649, 1975.
45. Veerabhadra Rao, K. and Narasimhamurty, T. S., *J. Mater. Sci.*, 10, 1019, 1975.
46. Venturini, E. L., et al., *J. Appl. Phys.*, 40, 1622, 1969.
47. Vehida, N. and Ohmachi, Y., *J. Appl. Phys.*, 40, 4692, 1969.
48. Grimsditch, M. H. and Ramdus, A. K., *Phys. Rev. B*, 22, 4094, 1980.
49. Schinke, D. P. and Viehman, W., unpublished Data.
50. Coquin, G. A., et al., *J. Appl. Phys.*, 42, 2162, 1971.
51. Vasquez, F., et al., *J. Phys. Chem. Solids*, 37, 451, 1976.
52. Luspin, Y. and Hauret, G., *C.R.Ac. Sci. Paris*, B274, 995 1972.
53. Narasimhamurty, T. S., *Phys. Rev.*, 186, 945, 1969.
54. Haussühl, S. and Weber, H. J., *Z. Kristall.*, 132, 266, 1970.
55. Vedam, K., *Proc. Ind. Ac. Sci.*, A34, 161, 1951.
56. Yano, T., Fukumoto, A., and Watanabe, A., *J. Appl. Phys.*, 42, 3674, 1971.
57. Manenkov, A. A. and Ritus, A. I., *Sov. J. Quant. Electr.*, 8, 78, 1978.
58. Eschler, H. and Weidinger, F., *J. Appl. Phys.*, 46, 65, 1975.

LINEAR ELECTRO-OPTIC COEFFICIENTS

Name

Cubic ($\bar{4}3m$)	Formula	$\lambda/\mu m$	r_{41} pm/V
Cuprous bromide	CuBr	0.525	0.85
Cuprous chloride	CuCl	0.633	3.6
Cuprous iodide	CuI	0.55	-5.0
Eulytite (BSO)	$Bi_4Si_3O_{12}$	0.63	0.54
Germanium eulytite (BGO)	$Bi_4Ge_3O_{12}$	0.63	1.0
Gallium arsenide	GaAs	10.6	1.6
Gallium phosphide	GaP	0.56	-1.07
Hexamethylenetetramine	$C_6H_{12}N_4$	0.633	0.78
Sphalerite	ZnS	0.65	2.1
Zinc selenide	ZnSe	0.546	2.0
Zinc telluride	ZnTe	3.41	4.2
Cadmium telluride	CdTe	3.39	6.8

Cubic (23)	Formula	$\lambda/\mu m$	r_{41} pm/V
Ammonium chloride (77 K)	NH_4Cl	—	1.5
Ammonium cadmium langbeinite	$(NH_4)_2Cd_2(SO_4)_3$	0.546	0.70
Ammonium manganese langbeinite	$(NH_4)_2Mn_2(SO_4)_3$	0.546	0.53
Thallium cadmium langbeinite	$Tl_2Cd_2(SO_4)_3$	0.546	0.37
Potassium magnesium langbeinite	$K_2Mg_2(SO_4)_3$	0.546	0.40
Bismuth monogermanate	$Bi_{12}GeO_{20}$	—	3.3
Bismuth monosilicate	$Bi_{12}SiO_{20}$	—	3.3
Sodium chlorate	$NaClO_3$	0.589	0.4
Sodium uranyl acetate	$NaUO_2(CH_3COO)_3$	0.546	0.87

Name

Cubic (23)	Formula	$\lambda/\mu m$	r_{41} pm/V
Trenhydrobromide	$N(CH_2CH_2NH_2)_3 \, 3HBr$	—	1.5
Trenhydrochloride	$N(CH_2CH_2NH_2)_3 \, 3HCl$	—	1.7

Tetragonal ($\bar{4}2m$)	Formula	T_{tran} K	r_{41} pm/V	r_{63} pm/V
Ammonium dihydrogen phosphate (ADP)	$NH_4H_2PO_4$	148	24.5	-8.5
Ammonium dideuterium phosphate (AD*P)	$NH_4D_2PO_4$	242	—	11.9
Ammonium dihydrogen arsenate (ADA)	$NH_4H_2AsO_4$	—	—	9.2
Cesium dihydrogen arsenate (CsDA)	CsH_2AsO_4	143	—	18.6
Cesium dideuterium arsenate (CsD*A)	CsD_2AsO_4	212	—	36.6
Potassium dihydrogen phosphate (KDP)	KH_2PO_4	123	8.6	-10.5
Potassium dideuterium phosphate (KD*P)	KD_2PO_4	222	8.8	23.8
Potassium dihydrogen arsenate (KDA)	KH_2AsO_4	97	12.5	10.9
Potassium dideuterium arsenate (KD*A)	KD_2AsO_4	162	—	18.2
Rubidium dihydrogen phosphate (RDP)	RbH_2PO_4	147	—	15.5
Rubidium dihydrogen arsenate (RDA)	RbH_2AsO_4	110	—	13.0
Rubidium dideuterium arsenate (RD*A)	RbD_2AsO_4	178	—	21.4

Tetragonal (4mm)	Formula	T_{tran} K	r_{13} pm/V	r_{33} pm/V	r_{51} pm/V
Barium titanate	$BaTiO_3$	406	8	28	—
Potassium lithium niobate	$K_3Li_2Nb_5O_{15}$	693	8.9	5.9	—
Lead titanate	$PbTiO_3$	765	13.8	5.9	—
Strontium barium niobate (SBN75)	$Sr_{0.75}Ba_{0.25}Nb_2O_6$	330	6.7	1340	42
Strontium barium niobate (SBN46)	$Sr_{0.46}Ba_{0.54}Nb_2O_6$	602	~180	35	—

Hexagonal (6mm)	Formula	r_{13} pm/V	r_{33} pm/V	r_{42} pm/V	r_{51} pm/V
Greenockite	CdS	3.1	2.9	2.0	3.7
Greenockite (const. strain)	CdS	1.1	2.4	—	—
Wurzite	ZnS	0.9	1.8	—	—
Zincite	ZnO	-1.4	+2.6	—	—

Hexagonal (6)	Formula	r_{13} pm/V	r_{33} pm/V	r_{42} pm/V	r_{51} pm/V
Lithium iodate	$LiIO_3$	4.1	6.4	1.4	3.3
Lithium potassium sulfate	$LiKSO_4$	r_{13}-r_{33} = 1.6	—	—	—

Trigonal (3m)	Formula	T_{tran} K	r_{13} pm/V	r_{22} pm/V	r_{33} pm/V	r_{42} pm/V
Cesium nitrate	$CsNO_3$	425	—	0.43	—	—
Lithium niobate	$LiNbO_3$	1483	8.6	7.0	30.8	28
Lithium tantalate	$LiTaO_3$	890	8.4	—	30.5	—
Lithium sodium sulfate	$LiNaSO_4$	—	—	<0.02	—	—
Tourmaline	—	—	—	0.3	—	—

Trigonal (32)	Formula	T_{tran} K	r_{11} pm/V	r_{41} pm/V
Cesium tartrate	$Cs_2C_4H_4O_6$	—	1.0	—
Cinnabar	HgS	659	3.1	1.5

LINEAR ELECTRO-OPTIC COEFFICIENTS (continued)

Name

Trigonal (32)	Formula	T_{tran} K	r_{11} pm/V	r_{41} pm/V
Potassium dithionate	$K_2S_2O_6$	—	0.26	—
Strontium dithionate	$SrS_2O_6 \cdot 4H_2O$	—	0.1	—
Quartz	SiO_2	1140	-0.47	0.2
Selenium	Se	398	2.5	

Orthorhombic (222)	Formula	T_{tran} K	r_{41} pm/V	r_{52} pm/V	r_{63} pm/V
Ammonium oxalate	$(NH_4)_2C_2O_4 \cdot 4H_2O$	—	230	330	250
Rochelle salt	$KNaC_4H_4O_6 \cdot 4H_2O$	$T_u = 297$ $T_l = 255$	-2.0	-1.7	+0.32

Orthorhombic (mm2)	Formula	T_{trans} K	r_{13} pm/V	r_{23} pm/V	r_{33} pm/V	r_{42} pm/V	r_{51} pm/V
Barium sodium niobate (BSN)	Ba_2NaNbO_{15}	833	15	13	48	92	90
Potassium niobate	$KNbO_3$	476	28	1.3	64	380	105

Monoclinic (2)	Formula	T_{trans} K	r_{22} pm/V	r_{32} pm/V
Calcium pyroniobate	$Ca_2Nb_2O_7$	—	0.33	13.7
Triglycine sulfate (TGS)	$(NH_2CH_2COOH)_3 \cdot H_2SO_4$	322	7.2	13.6

REFERENCES

1. Narasimhamurty, T. S., *Photoelastic and Electro-Optic Properties of Crystals*, Plenum Press, New York, 1981, pp. 405-407.
2. Weber, M. J., Ed., *CRC Handbook of Laser Science and Technology*, Vol. IV, CRC Press, Boca Raton, FL, 1986, pp. 258-278.

QUADRATIC ELECTRO-OPTIC COEFFICIENTS
Kerr Constants of Ferroelectric Crystals[1,2]

Name	Formula	T_{tran} K	λ μm	g_{11} 10^{10} esu	g_{12} 10^{10} esu	$g_{11}-g_{12}$ 10^{10} esu	g_{44} 10^{10} esu
Barium titanate	$BaTiO_3$	406	0.633	1.33	-0.11	1.44	
Strontium titanate	$SrTiO_3$	—	0.633	—	—	1.56	—
Potassium tantalate niobate	$KTa_{0.65}Nb_{0.35}O_3$	330	0.633	1.50	-0.42	1.92	1.63
Potassium tantalate	$KTaO_3$	13	0.633	—	—	1.77	1.33
Lithium niobate	$LiNbO_3$	1483	—	0.94	0.25	0.7	0.6
Lithium tantalate	$LiTaO_3$	938	—	1.0	0.17	0.8	0.7
Barium sodium niobate (BSN)	$Ba_{0.8}Na_{0.4}Nb_2O_6$	833	—	1.55	0.44	1.11	

Kerr Constants of Selected Liquids[2]
K is the Kerr constant at a wavelength of 589 nm and at room temperature; ε is the static dielectric constant;
T_m is the melting point; and T_b is the normal boiling point

Name	Molecular formula	K 10^{-7} esu	ε	T_m °C	T_b °C
Carbon disulfide	CS_2	+3.23	2.63	-111.5	+46.3
Acetone	C_3H_6O	+16.3	21.0	-94.8	+56.1
Methyl ethyl ketone	C_4H_8O	+13.6	18.56	-86.67	+79.6

QUADRATIC ELECTRO-OPTIC COEFFICIENTS (continued)
Kerr Constants of Selected Liquids (continued)[2]

Name	Molecular formula	K 10^{-7} esu	ε	T_m °C	T_b °C
Pyridine	C_5H_5N	+20.4	13.26	-42	+115.23
Ethyl cyanoacetate	$C_5H_7NO_2$	+38.8	31.6	-22.5	205
o-Dichlorobenzene	$C_6H_4Cl_2$	+42.6	10.12	-16.7	180
Benzenesulfonyl chloride	$C_6H_5ClO_2S$	+89.9	28.90	+14.5	247
Nitrobenzene	$C_6H_5NO_2$	+326	35.6	+5.7	210.8
Ethyl 3-aminocrotonate	$C_6H_{11}NO_2$	+31.0	—	+33.9	210
Paraldehyde	$C_6H_{12}O_3$	-23.0	14.7 / 12.0[a]	+12.6	124
Benzaldehyde	C_7H_6O	+80.8	17.85 / 14.1[a]	-26	179.05
p-Chlorotoluene	C_7H_7Cl	+23.0	6.25	+7.5	162.4
o-Nitrotoluene	$C_7H_7NO_2$	+174	26.26	-10	222.3
m-Nitrotoluene	$C_7H_7NO_2$	+177	24.95	+15.5	232
p-Nitrotoluene	$C_7H_7NO_2$	+222	22.2	+51.6	238.3
Benzyl alcohol	C_7H_8O	-15.4	11.92 / 10.8[a]	-15.3	205.8
m-Cresol	C_7H_8O	+21.2	12.44 / 5.0[a]	+11.8	202.27
m-Chloroacetophenone	C_8H_7ClO	+69.1			
Acetophenone	C_8H_8O	+66.6	17.44 / 15.8[a]	+19.7	202.3
Quinoline	C_9H_7N	+15.0	9.16	-14.78	237.16
Ethyl salicylate	$C_9H_{10}O_3$	+19.6	8.48	+1.3	231.5
Carvone	$C_{10}H_{14}O$	+23.6	11.2	<0	230
Ethyl benzoylacetate	$C_{11}H_{12}O_3$	+16.0	13.50	<0	270
Water	H_2O	+4.0	80.10	0.00	100.0

[a] Dielectric constant at radiofrequencies (10^8-10^9 Hz).

REFERENCES

1. Narasimhamurty, T. S., *Photoelastic and Electro-Optic Properties of Crystals*, Plenum Press, New York, 1981, p. 408.
2. Gray, D. E., Ed., *AIP Handbook of Physics*, McGraw Hill, New York, 1972, p. 6-241.

MAGNETO-OPTIC CONSTANTS
Verdet Constants of Non-Magnetic Crystals[1]
V is the Verdet constant; *n* is the refractive index; and λ is the wavelength

Material	T K	λ nm	n	V min/Oe cm
Al_2O_3	300	546.1	1.771	0.0240
	300	589.3	1.768	0.0210
$BaTaO_3$	403	427		0.95
	403	496		0.38
	403	620		0.18
	403	826		0.072
$Bi_4Ge_3O_{12}$	300	442	2.077	0.289
	300	632.8	2.048	0.099
	300	1064	2.031	0.026
C (diamond)	300	589.3	2.417	0.0233
$CaCO_3$	300	589.3	1.658	0.019
CaF_2	300	589.3	1.434	0.0088
$Cd_{0.55}Mn_{0.45}Te$	300	632.8		6.87

MAGNETO-OPTIC CONSTANTS (continued)
Verdet Constants of Non-Magnetic Crystals (continued)[1]

Material	T K	λ nm	n	V min/Oe cm
CuCl	300	546.1	1.93	0.20
GaSe	298	632.8		0.80
$KAl(SO_4)_2 \cdot 12H_2O$	300	589.3	1.456	0.0124
KBr	300	546.1	1.564	0.0500
	300	589.3	1.560	0.0425
KCl	300	589.3	1.490	0.0275
KI	300	546.1	1.673	0.083
	300	589.3	1.666	0.070
$KTaO_3$	296	352		0.44
	296	413		0.19
	296	496		0.096
	296	620		0.051
	296	826		0.022
LaF_3	300	325	1.639	0.054
(H‖c)	300	442	1.615	0.028
	300	632.8	1.601	0.012
	300	1064	1.592	0.006
$MgAl_2O_4$	300	589.3	1.718	0.021
$NH_4AlSO_4 \cdot 12H_2O$	300	589.3	1.459	0.0128
NH_4Br	300	589.3	1.711	0.0504
NH_4Cl	300	546.1		0.0410
	300	589.3	1.643	0.0362
NaBr	300	546.1		0.0621
NaCl	300	546.1		0.0410
	300	589.3	1.544	0.0345
$NaClO_3$	300	546.1		0.0105
	300	589.3	1.515	0.0081
$NiSO_4 \cdot 6H_2O$	297	546.1		0.0256
	297	589.3	1.511	0.0221
SiO_2	300	546.1	1.546	0.0195
	300	589.3	1.544	0.0166
$SrTiO_3$	298	413	2.627	0.78
	298	496		0.31
	298	620		0.14
	298	826		0.066
ZnS	300	546.1		0.287
	300	589.3	2.368	0.226
ZnSe	300	476	2.826	1.50
	300	496	2.759	1.04
	300	514	2.721	0.839
	300	587	2.627	0.529
	300	632.8	2.592	0.406

Verdet Constants of Rare-Earth Aluminum Garnets at Various Wavelengths[1]
The absorption coefficient α for these materials ranges from 0.2 to 0.6 cm^{-1} at 300 K

Material	T/K	V in min/Oe cm							
		λ = 405 nm	450 nm	480 nm	520 nm	546 nm	578 nm	635 nm	670 nm
$Tb_2Al_5O_{12}$	300	-2.266	-1.565	-1.290	-1.039	-0.912	-0.787	-0.620	-0.542
	77		-102.16	-83.45	-3.425	-3.051	-2.603	-2.008	-1.815
	4.2				-64.80	-58.35	-53.77	48.39	-45.15
	1.45		-200.95	-172.52	-139.28	-125.07	-111.27	97.47	-93.42
$Dy_3Al_5O_{12}$	300	-1.241	-0.942	-0.803	-0.667	-0.592	-0.518	-0.411	-0.359
$Ho_3Al_5O_{12}$	300	-0.709	-0.320	-0.260	-0.335	-0.304	-0.299		-0.206

MAGNETO-OPTIC CONSTANTS (continued)
Verdet Constants of Rare-Earth Aluminum Garnets at Various Wavelengths (continued)[1]

Material	T/K	λ = 405 nm	450 nm	480 nm	520 nm	546 nm	578 nm	635 nm	670 nm
					V in min/Oe cm				
$Er_3Al_5O_{12}$	300	-0.189	-0.240	-0.154	-0.162	-0.157	-0.145	-0.105	-0.089
$Tm_3Al_5O_{12}$	300	+0.151	+0.103	+0.093	0.076	0.069	+0.059	+0.048	
$Yb_3Al_5O_{12}$	298	0.287	0.215	0.186	0.140	0.133	0.116	0.094	
	77	0.718	0.540	0.481	0.393	0.342	0.302	0.239	

Verdet Constants for KDP-Type Crystals[1]
Measurements refer to T = 298 K and
λ = 632.8 nm, with $k \parallel$ [001]

Material	V min/Oe cm
KH_2PO_4 (KDP)	0.0124
$KH_{0.3}D_{1.7}PO_4$ (KD*P)	0.145
$NH_4H_2PO_4$ (ADP)	0.138
KH_2AsO_4 (KDA)	0.238
$KH_{0.1}D_{1.9}AsO_4$ (KD*A)	0.245
$NH_4H_2AsO_4$ (ADH)	0.244

Verdet Constants of Gases[2]
Values refer to T = 0°C and P = 101.325 kPa (760 mmHg); n_D is the refractive index at a wavelength of 589 nm

Gas	$(n_D - 1) \times 10^3$	$10^6 \times V$ min/Oe cm
He	0.036	+0.40
Ar	2.81	+9.36
H_2		+6.29
N_2	0.297	+6.46
O_2	0.272	+5.69
Air	0.293	+6.27
Cl_2	0.773	+31.9
HCl	0.447	+21.5
H_2S	0.63	+41.5
NH_3	0.376	+19.0
CO	0.34	+11.0
CO_2	0.45	+9.39
NO	0.297	-58
CH4	0.444	+17.4
n-C4H10		+44.0

Verdet Constants of Liquids[2]
n_D is the refractive index at a wavelength of 589 nm and a temperature of 20°C, unless otherwise indicated. V is the Verdet constant

Liquid	λ/nm	T/°C	$10^2 \times V$ min/Oe cm	n_D
P	589	33	+13.3	
S	589	114	+8.1	1.929 (110°C)
H_2O	589	20	+1.309	1.3328
D_2O	589	19.7	+1.257	1.3384
H_3PO_4	578	97.4	+1.35	
CS_2	589	20	+4.255	1.6255
CCl_4	578-589	25.1	+1.60	1.463 (15°C)
$SbCl_5$	578	18	+7.45	1.601 (14°C)
$TiCl_4$	578	17	-1.65	1.61
$TiBr_4$	578	46	-5.3	
Methanol	589	18.7	+0.958	1.3289
Acetone	578-589	20.0	+1.116	1.3585
Toluene	578-589	15.0	+2.71	1.4950
Benzene	578-589	15.0	+3.00	1.5005
Chlorobenzene	589	15	+2.92	1.5246
Nitrobenzene	589	15	+2.17	1.5523
Bromoform	589	17.9	+3.13	1.5960

MAGNETO-OPTIC CONSTANTS (continued)

Verdet Constants of Rare Earth Paramagnetic Crystals[1]

n is the refractive index, and V is the Verdet constant at the wavelength and temperature indicated

Rare Earth	Host	T/K	λ/nm	n	V min/Oe cm
Ce^{3+}(30%)	CaF_2	300	325	1.516	-0.956
		300	442	1.502	-0.297
		300	633	1.494	-0.111
		300	1064	1.489	-0.035
Ce^{3+}	CeF_3	300	442	1.613	-1.05
		300	633	1.598	-0.406
		77	633		-1.418
		300	1064	1.589	-0.113
Pr^{3+}(5%)	CaF_2	300	266	1.471	-0.172
		300	325	1.461	-0.0818
		300	442	1.451	-0.0089
		300	633	1.445	-0.0168
		300	1064	1.441	-0.0045
Nd^{3+}(2.9%)	CaF_2	4.2	426		-0.19
Nd^{3+}	NdF_3	300	442	1.60	-0.553
		290	633	1.59	-0.209
		77	633		-0.755
		300	1064	1.58	-0.097
Eu^{3+}(3%)	CaF_2	4.2	430		29
		4.2	440		22
Eu^{2+}	EuF_2	300	450		-4.5
		300	500		-2.6
		300	550		-1.6
		300	600		-1.1
		300	650		-0.8
		300	1064		-0.19
Tb^{3+}	KTb_3F_{10}	300	325	1.531	-2.174
		300	442	1.518	-0.933
		300	633	1.510	-0.386
		77	633		-1.94
		300	1064	1.505	-0.114
Tb^{3+}	$LiTbF_4$	300	325	1.493	-1.9
		300	442	1.481	-0.98
		300	633	1.473	-0.44
		300	1064	1.469	-0.13
Tb^{3+}	$Tb_3Ga_5O_{12}$	300	500	1.989	-0.749
		300	570	1.981	-0.581
		300	633	1.976	-0.461
		300	830	1.967	-0.21
		300	1060	1.954	-0.12

MAGNETO-OPTIC CONSTANTS (continued)

Verdet Constants of Paramagnetic Glasses[1]

The Verdet constant V is given at room temperature for the wavelengths indicated

Rare earth phosphate glasses of composition $R_2O_3 \cdot xP_2O_5$, where x is given in the second column

R	x	λ = 405 nm	λ = 436 nm	λ = 480 nm	λ = 500 nm	λ = 520 nm	λ = 546 nm	λ = 578 nm	λ = 600 nm	λ = 635 nm	λ = 670 nm
						Verdet constant V in min/Oe cm					
La		0.037	0.030	0.024	0.022	0.020	0.018	0.015	-0.014	0.013	—
Ce	2.67	-0.672	0.510	-0.366	-0.326	-0.287	-0.253	-0.217	-0.197	-0.173	-0.150
Pr	3.09	-0.447	-0.332	-0.283	-0.261	-0.236	-0.208	-0.182	-0.170	-0.150	-0.132
Nd	2.92	-0.250	-0.209	-0.167	-0.155	-0.136	-0.134	-0.094	-0.080	-0.080	-0.071
Sm	2.87	0.026	0.024	0.020	0.020	0.017	0.015	0.014	0.012	0.011	0.010
Eu	2.93	-0.025	-0.017	-0.010	-0.006	-0.006	-0.005	-0.004	-0.003	-0.002	-0.002
Gd	3.01	0.018	0.015	0.014	0.012	0.012	0.011	0.011	0.010	0.009	0.009
Tb	2.94	-0.560	-0.458	-0.357	-0.323	-0.295	-0.261	-0.226	-0.206	-0.190	-0.164
Dy	2.51	-0.540	-0.453	-0.359	-0.331	-0.301	0.268	-0.237	-0.217	-0.197	-0.173
Ho	2.94	-0.299	-0.313	-0.156	-0.153	-0.138	-0.138	-0.119	-0.110	-0.098	-0.084
Er	3.01	-0.139	-0.121	-0.100	-0.111	-0.095	-0.062	-0.060	-0.057	-0.051	-0.044
Tm	2.79	0.019	0.013	0.012	0.009	0.008	0.006	0.005	0.004	0.004	0.007
Yb	3.01	0.087	0.072	0.056	0.050	0.045	0.041	0.036	0.032	0.029	0.024

The following are rare earth borate glasses with composition:
for La and Pr: $R_2O_3 \cdot xP_2O_5$; for Tb-Pr and Dy-Pr: $R_2O_3 \cdot xB_2O_3$; and for other elements: $R_2O_3 \cdot 0.85La_3O_3 \cdot xB_2O_3$.

R	x	λ = 405 nm	λ = 436 nm	λ = 480 nm	λ = 500 nm	λ = 520 nm	λ = 546 nm	λ = 578 nm	λ = 600 nm	λ = 635 nm	λ = 670 nm
La	3.04	0.043	0.036	0.029	0.026	0.023	0.022	0.019	0.018	0.016	0.014
Pr-La	5.44	-0.380	-0.307	-0.230	-0.220	-0.201	-0.178	-0.153	-0.146	-0.128	-0.110
Nd-La	5.41	-0.180	-0.147	-0.120	-0.111	-0.096	-0.094	-0.100	-0.059	-0.056	-0.046
Sm-La	4.97	0.032	0.030	0.025	0.024	0.022	0.019	0.017	0.016	0.014	0.012
Eu-La	4.69	-0.081	-0.060	-0.038	-0.033	-0.029	-0.024	0.019	-0.016	0.014	-0.012
Gd-La	4.71	0.032	0.026	0.024	0.022	0.021	0.020	0.018	0.017	0.015	0.013
Tb-La	4.73	-0.512	-0.419	-0.319	-0.288	-0.262	-0.234	-0.205	-0.186	-0.167	-0.142
Dy-La	4.88	-0.436	-0.361	-0.299	-0.273	-0.246	-0.220	-0.193	-0.177	-0.159	-0.138
Ho-La	4.36	-0.269	-0.252	-0.123	-0.131	-0.112	-0.128	-0.104	-0.096	—	-0.074
Er-La	4.50	-0.093	-0.078	-0.068	-0.082	—	-0.045	-0.042	-0.040	-0.035	-0.034
Tm-La	4.75	0.060	0.046	0.039	0.034	0.031	0.026	0.023	0.021	0.018	0.016
Yb-La	8.58	0.115	0.094	0.073	0.066	0.060	0.054	0.046	0.043	0.037	0.033
Tb-Pr	4.99	-0.940	-0.786	-0.560	-0.536	-0.489	-0.436	-0.380	-0.348	-0.306	-0.265
Dy-Pr	4.63	-0.850	—	—	-0.497	-0.465	-0.413	-0.358	-0.332	-0.290	-0.252
Pr	2.56	-0.843	-0.646	-0.471	-0.480	-0.432	-0.390	-0.334	-0.317	-0.271	-0.243

MAGNETO-OPTIC CONSTANTS (continued)

Verdet Constants of Diamagnetic Glasses[1]

The Verdet constant V is given at room temperature for the wavelengths indicated

Glass type	Composition (wt. %)	Verdet constant V in min/Oe cm			
		$\lambda = 325$ nm	$\lambda = 442$ nm	$\lambda = 633$ nm	$\lambda = 1064$ nm
SiO_2	100% SiO_2			0.013	
B_2O_3	100% B_2O_3			0.010	
CdO	47.5% CdO, 52.5% P_2O_5	0.079	0.033	0.022	
ZnO	36.4% ZnO, 63.6% P_2O_5	0.072	0.044	0.020	
TeO_2	88.9% TeO_2, 11.1% P_2O_5		0.196	0.076	0.022
ZrF_4	63.1% ZrF_4, 14.9% BaF_2, 7.2% LaF_3, 1.9% AlF_3, 9.1% PbF_2, 3.8% LiF			0.011	
		$\lambda = 700$ nm	$\lambda = 853$ nm	$\lambda = 1060$ nm	
Bi_2O_3	95% Bi_2O_3, 5% B_2O_3	0.086	0.051	0.033	
PbO	95% PbO, 5% B_2O_3	0.093	0.061	0.031	
	82% PbO, 18% SiO_2	0.077	0.045	0.027	
	50% PbO, 15% K_2O, 35% SiO_2	0.032	0.020	0.011	
Tl_2O	95% Tl_2O, 5% B_2O_3	0.092	0.061	0.032	
	82% Tl_2O, 18% SiO_2	0.100	0.067	0.043	
	50% Tl_2O, 15% K_2O, 35% SiO_2	0.036	0.022	0.012	
SnO	76% SnO, 13% B_2O_3, 11% SiO_2	0.071	0.046	0.026	
TeO_3	75% TeO_2, 25% Sb_2O_3	0.076	0.052	0.032	
	80% TeO_2, 20% $ZnCl_2$	0.073	0.046	0.025	
	84% TeO_2, 16% BaO	0.056	0.041	0.029	
	70% TeO_2, 30% WO_3	0.052	0.035	0.022	
	20% TeO_2, 80% PbO	0.128	0.075	0.048	
Sb_2O_3	25% Sb_2O_3, 75% TeO_2	0.076	0.050	0.032	
	75% Sb_2O_3, 75% Cs_2O, 5% Al_2O_3	0.074	0.044	0.025	
	75% Sb_2O_3, 10% Cs_2O, 10% Rb_2O, 5% Al_2O_3	0.078	0.052	0.030	

MAGNETO-OPTIC CONSTANTS (continued)
Verdet Constants of Commercial Glasses[1]

This table gives the density, ρ, refractive index at 589 nm, n_D, and Verdet constant, V, for the wavelengths indicated; the data refer to room temperature

Glass type	ρ g/cm^3	n_D	V in min/Oe cm				
			λ = 365.0 nm	λ = 404.7 nm	λ = 435.8 nm	λ = 546.1 nm	λ = 578.0 nm
BSC	2.49	1.5096	0.0499	0.0392	0.0333	0.02034	0.01798
HC	2.53	1.5189	0.0561	0.0440	0.0372	0.0225	0.01995
LBC	2.87	1.5406	0.0609	0.0477	0.0403	0.0245	0.0216
LF	3.23	1.5785	0.1143	0.0850	0.0693	0.0394	0.0344
BLF	3.48	1.6047	0.1112	0.0832	0.0685	0.0393	0.0344
DBC	3.56	1.6122	0.0662	0.0517	0.0435	0.0261	0.0231
DF	3.63	1.6203	0.1473	0.1076	0.0872	0.0485	0.0423
EDF	3.9	1.6533	0.1725	0.1248	0.1007	0.0556	0.0483

The composition of the glasses in weight percent is:

Glass type	SiO_2	B_2O_3	K_2O	CaO	Al_2O_3	As_2O_3	Na_2O	BaO	ZnO	PbO
BSC	69.6	6.7	20.5	2.9	0.3	0.1	—	—	—	—
HC	72.0	—	10.1	11.4	0.3	0.2	6.1	—	—	—
LBC	57.1	1.8	13.7	0.3	0.2	0.1	—	26.9	—	—
LF	52.5	—	9.5	0.3	0.2	0.1	—	—	—	37.6
BLF	45.2	—	7.8	—	—	0.4	—	16.0	8.3	22.2
DBC	36.2	7.7	0.2	0.2	3.5	0.7	—	44.6	6.7	—
DF	46.3	—	1.1	0.3	0.2	0.1	5.0	—	—	47.0
EDF	40.6	—	7.5	0.2	0.2	0.2	0.1	—	—	51.5

REFERENCES

1. Weber, M. J., *CRC Handbook of Laser Science and Technology*, Vol. IV, Part 2, CRC Press, Boca Raton, FL, 1988, p. 299-310.
2. Gray, D. E., Ed., *American Institute of Physics Handbook*, Third Edition, McGraw Hill, New York, 1972, p. 6-230.

FARADAY ROTATION
Ferro-, Ferri-, and Antiferromagnetic Solids

Material	T_c K	$4\pi M_s$ gauss	F deg/cm	α cm^{-1}	$2 F/\alpha$	T K	λ nm
Fe	1043	21,800	4.4×10^5	6.5×10^5	1.4	300	500
			6.5×10^5	5.0×10^5	2.6	300	1000
			7×10^5	4.2×10^5	3.3	300	1500
			7×10^5	3.5×10^5	4.0	300	2000
Co	1390	18,200	2.9×10^5	—	—	300	500
			5.5×10^5	6.1×10^5	1.8	300	1000
			5.5×10^5	4.5×10^5	2.4	300	1500
			5.5×10^5	3.6×10^5	2.7	300	2000
Ni	633	6,400	0.8×10^5	—	—	300	500
			2.6×10^5	5.8×10^5	0.9	300	1000
			1.5×10^5	4.8×10^5	0.6	300	1500
			1×10^5	4.1×10^5	0.25	300	2000
Permalloy (Ni/Fe = 82/18)	803	10,700	1.2×10^5	6×10^5	0.4	300	500

Material	T_c K	$4\pi M_s$ gauss	F deg/cm	α cm^{-1}	$2F/\alpha$	T K	λ nm
Ni/Fe = 100/0		6,000	1.2×10^5	7.05×10^5	0.34	300	632.8
Ni/Fe = 80/20		10,800	2.2×10^5	7.10×10^5	0.62	300	632.8
Ni/Fe = 60/40		14,900	2.9×10^5	7.54×10^5	0.77	300	632.8
Ni/Fe = 40/60		14,400	2.2×10^5	8.17×10^5	0.54	300	632.8
Ni/Fe = 20/80		19,400	3.3×10^5	8.10×10^5	0.81	300	632.8
Ni/Fe = 0/100	639	21,600	3.5×10^5	8.13×10^5	0.86	300	632.8
MnBi		7,700	4.2×10^5	6.1×10^5	1.4	300	450
			7.5×10^5	4.2×10^5	3.6	300	900
MnAs	313	—	0.44×10^5	5.0×10^5	0.174	300	500
			0.62×10^5	4.4×10^5	0.28	300	900
CrTe	334	1015	0.5×10^5	2.0×10^5	0.5	300	550
			0.4×10^5	1.2×10^5	0.7	300	900
FeRh	333	—	0.9×10^5	3.3×10^5	0.56	348	700
$Y_3Fe_5O_{12}$ (YIG)	560	2500	2400	1500	3.2	300	555
			1250	1400	1.8	300	625
			750	450	3.3	300	770
			175	<0.06	$>3 \times 10^3$	300	5000 to 1500
$Gd_3Fe_5O_{12}$ (GdIG)	Tn = 564	7300	-2000	6000	0.6	300	500
	T = 286		-1050	900	2.3	300	600
			-300	100	6.0	300	800
			-80	70	2.3	300	1000
$NiFe_2O_4$	858	3350	2.0×10^4	5.9×10^4	0.7	300	286
			-1.0×10^4	10×10^4	0.2	300	500
			-120	38	6	300	1500
			+75	15	10	300	3000
			+110	32	7	300	5000
$CoFe_2O_4$	793	4930	2.75×10^4	12×10^4	0.5	300	286
			3.6×10^4	17×10^4	0.4	300	400
			-2.5×10^4	6×10^4	0.8	300	660
$MgFe_2O_4$	593-713[e]	1450[e]	-60	100	1	300	2500
			0	12	0	300	4000
			+35	6	11	300	6000
$Li_{0.5}Fe_{2.5}O_4$	863-953[e]	3240[e] to 3900	-440	150	6	300	1500
			+10	85	0.2	300	3000
			+110	44	5	300	5000
			+135	80	3	300	7000
$BaFe_{12}O_{19}$	723	—	-50	-38	3	300	2000
			+75	20	7.5	300	3000
			+150	20	15	300	5000
			+165	22	15	300	7000
$Ba_2Zn_2Fe_{12}O_{19}$	—	—	90	120	1.5	300	5000
			75	65	2.0	300	7000
$RbNiF_3$	220	1250	360	35	20	77	450[a]
			70	10	14	77	600[a]
			310	70	9	77	800[a]
			75	25	6	77	1000[a]
$RbNi_{0.75}Co_{0.25}F_3$	109	—	180	9	40	77	600[b]
$RbFeF_3$	102	—	3400	7	900	82	300[c]
			1600	3	1100	82	400[c]
			620	1.5	830	82	600[c]
			300	2.5	240	82	800[c]
FeF_3	365	40 at 300 K	670	14	95	300	349[d]
			180	4.4	82	300	522.5[d]
$CrCl_3$	16.8	3880	2000	200	20	1.5	410

FARADAY ROTATION (continued)
Ferro-, Ferri-, and Antiferromagnetic Solids (continued)

Material	T_c K	$4\pi M_s$ gauss	F deg/cm	α cm^{-1}	$2F/\alpha$	T K	λ nm
			-500	300	3	1.5	450
			-1000	70	30	1.5	590
$CrBr_3$	32.5	3390	3×10^5	3×10^3	200	1.5	478
			1.6×10^5	1.4×10^4	23	1.5	500
CrI_3	68	2690	1.1×10^5	6.3×10^3	35	1.5	970
			0.8×10^5	3×10^3	53	1.5	1000
$FeBO_3$	348	115	3200	140	45	300	500
		at 300 K	450	38	24	300	700
EuO	69	23700	-1.0×10^5	0.5×10^4	40	5	1100
			5×10^5	9.7×10^4	10	5	700
			0.5×10^5	7.8×10^4	1.3	5	500
			3×10^4	>0.5	~105	20	2500
			660	>1.0	1300	20	10600
EuS	16.3	—	-1.6×10^5	0	—	6	825
			-9.6×10^5	3.3×10^4	58	6	690
			$+5.5 \times 10^5$	1.2×10^5	9.2	6	563
$EuSe$	7.0	13,200	1.45×10^5	80	3600	4.2	750
			0.95×10^5	60	3170	4.2	800

[a] Measured along the C-axis (magnetic hard axis).
[b] Measured along the C-axis (magnetic easy axis).
[c] Measured along the C-axis ([100]-direction at room temperature).
[d] Strong natural birefringence interferes with the Faraday effect.
[e] Depends on heat treatment.

REFERENCE

1. Weber, M. J., Ed., *CRC Handbook of Laser Science and Technology,* Vol. IV, Part 2, CRC Press, Boca Raton, FL, 1988, pp. 288-296.

NONLINEAR OPTICAL CONSTANTS

H. P. R. Frederikse

The relation between the polarization density P of a dielectric medium and the electric field E is linear when E is small, but becomes nonlinear as E acquires values comparable with interatomic electric fields (10^5 to 10^8 V/cm). Under these conditions the relation between P and E can be expanded in a Taylor's series

$$P = \varepsilon_0 \chi^{(1)} E + 2\chi^{(2)} E^2 + 4\chi^{(3)} E^3 + \cdots \tag{1}$$

where ε_0 is the permittivity of free space, while $\chi^{(1)}$ is the linear and $\chi^{(2)}$, $\chi^{(3)}$ etc. the nonlinear optical susceptibilities.

If we consider two optical fields, the first $E_j^{\omega_1}$ (along the j-direction at frequency ω_1) and the second $E_k^{\omega_2}$ (along the k-direction at frequency ω_2) one can write the second term of the Taylor's series as follows

$$P_i(\omega_1\omega_2) = 2\chi_{ijk}^{\omega_3=\omega_1\pm\omega_2} E_j^{\omega_1} E_k^{\omega_2}$$

When $\omega_1 \neq \omega_2$ the (parametric) mixing of the two fields gives rise to two new polarizations at the frequencies $\omega_3 = \omega_1 + \omega_2$ and $\omega_3' = \omega_1 - \omega_2$. When the two frequencies are equal, $\omega_1 = \omega_2 = \omega$, the result is Second Harmonic Generation (SHG) $\chi_{ijk}(2\omega, \omega, \omega)$, while equal and opposite frequencies, $\omega_1 = \omega$ and $\omega_2 = -\omega$ leads to Optical Rectification (OR): $\chi_{ijk}(0,\omega,-\omega)$. In the SHG case the following convention is adopted: the second order nonlinear coefficient d is equal to one half of the second order nonlinear susceptibility

$$d_{ijk} = 1/2\,\chi^{(2)}$$

Because of the symmetry of the indices j and k one can replace these two by a single index (subscript) m. Consequently the notation for the SHG nonlinear coefficient in reduced form is d_{im} where m takes the values 1 to 6. Only noncentrosymmetric crystals can possess a nonvanishing d_{ijk} tensor (third rank). The unit of the SHG coefficients is m/V (in the MKSQ/SI system).

In centrosymmetric media the dominant nonlinearity is of the third order. This effect is represented by the third term in the Taylor's series (Equation 1); it is the result of the interaction of a number of optical fields (one to three) producing a new frequency $\omega_4 = \omega_1 + \omega_2 + \omega_3$. The third order polarization is given by

$$P_j(\omega_1\omega_2\omega_3) = g_4 \chi_{jklm} E_k^{\omega_1} E_l^{\omega_2} E_m^{\omega_3}$$

Third Harmonic Generation (THG) is achieved when $\omega_1 = \omega_2 = \omega_3 = \omega$. In this case the constant $g_4 = 1/4$. The third order nonlinear coefficient C is related to the third order susceptibility as follows

$$C_{jklm} = 1/4\,\chi_{jklm}$$

This coefficient is a fourth rank tensor. In the THG case the matrices must be invariant under permutation of the indices k, l, and m; as a result the notation for the third order nonlinear coefficient can be simplified to C_{jn}. The unit of C_{jn} is $m^2 \cdot V^{-2}$ (in the MKSQ/SI system).

Applications of second order nonlinear optical materials include the generation of higher (up to sixth) optical harmonics, the mixing of monochromatic waves to generate sum or difference frequencies (frequency conversion), the use of two monochromatic waves to amplify a third wave (parametric amplification) and the addition of feedback to such an amplifier to create an oscillation (parametric oscillation).

Third order nonlinear optical materials are used for THG, self-focusing, four wave mixing, optical amplification, and optical conjugation. Many of these effects — as well as the variation and modulation of optical propagation caused by mechanical, electric, and magnetic fields (see the preceeding table on "Elasto-Optic, Electro-Optic, and Magneto-Optic Constants") are used in the areas of optical communication, optical computing, and optical imaging.

REFERENCES (NONLINEAR OPTICS)

1. *Handbook of Laser Science and Technology*, Vol. 111, Part 1; Ed.: Marvin J. Weber, Publ.: CRC Press, Inc., Boca Raton, FL, 1986.
2. Dmitriev, V.G., Gurzadyan, G.G., and Nikogosyan, D., *Handbook of Nonlinear Optical Crystals*, Springer-Verlag, Berlin, 1991.
3. Shen, Y.R., *The Principles of Nonlinear Optics*, John Wiley, New York, 1984.
4. Yariv, A., *Quantum Electronics*, 3rd edition, John Wiley, New York, 1988.
5. Bloembergen, N., *Nonlinear Optics*, W.A. Benjamin, New York, 1965.
6. Zernike F. and Midwinter, J.E., *Applied Nonlinear Optics*, John Wiley, New York, 1973.
7. Hopf, F.A. and Stegeman, G.I., *Applied Classical Electrodynamics*, Volume 2: Nonlinear Optics, John Wiley, New York, 1986.
8. *Nonlinear Optical Properties of Organic Molecules and Crystals*, Eds.: D.S. Chemla and J. Zyss, Publ.: Academic Press, Orlando, FL, 1987.
9. *Optical Phase Conjugation*, Ed.: R.A. Fisher, Publ.: Academic Press, New York, 1983.
10. Zyss, J., *Molecular Nonlinear Optics: Materials, Devices and Physics*, Academic Press, Boston, 1994.
11. Nonlinear Optics, 5 articles in *Physics Today, (Am. Inst. of Phys.)*, Vol. 47, No. 5, May, 1994.

Selected SHG Coefficients of NLO Crystals*

Material	Symmetry class	$d_{im} \times 10^{12}$ m/V	λ μm
GaAs	$\bar{4}3$ m	$d_{14} = 134.1 \pm 42$	10.6
GaP	$\bar{4}3$ m	$d_{14} = 71.8 \pm 12.3$	1.058
InAs	$\bar{4}3$ m	$d_{14} = 364 \pm 47$	1.058
		$d_{14} = 210$	10.6
ZnSe	$\bar{4}3$ m	$d_{14} = 78.4 \pm 29.3$	10.6
		$d_{36} = 26.6 \pm 1.7$	1.058
β-ZnS	$\bar{4}3$ m	$d_{14} = 30.6 \pm 8.4$	10.6
		$d_{36} = 20.7 \pm 1.3$	1.058
ZnTe	$\bar{4}3$ m	$d_{14} = 92.2 \pm 33.5$	10.6
		$d_{14} = 83.2 \pm 8.4$	1.058
		$d_{36} = 89.6 \pm 5.7$	1.058
CdTe	$\bar{4}3$ m	$d_{14} = 167.6 \pm 63$	10.6
Bi_4GeO_{12}	$\bar{4}3$ m	$d_{14} = 1.28$	1.064
$N_4(CH_2)_6$ (hexamine)	$\bar{4}3$ m	$d_{14} = 4.1$	1.06
$LiIO_3$	6	$d_{33} = -7.02$	1.06
		$d_{31} = -5.53 \pm 0.3$	1.064
ZnO	6 mm	$d_{33} = -5.86 \pm 0.16$	1.058
		$d_{31} = 1.76 \pm 0.16$	1.058
		$d_{15} = 1.93 \pm 0.16$	1.058
α-ZnS	6 mm	$d_{33} = 11.37 \pm 0.07$	1.058
		$d_{33} = 37.3 \pm 12.6$	10.6
		$d_{31} = -18.9 \pm 6.3$	10.6
		$d_{15} = 21.37 \pm 8.4$	10.6
CdS	6 mm	$d_{33} = 25.8 \pm 1.6$	1.058
		$d_{31} = -13.1 \pm 0.8$	1.058
		$d_{15} = 14.4 \pm 0.8$	1.058
CdSe	6 mm	$d_{33} = 54.5 \pm 12.6$	10.6
		$d_{31} = -26.8 \pm 2.7$	10.6
$BaTiO_3$	4 mm	$d_{33} = 6.8 \pm 1.0$	1.064
		$d_{31} = 15.7 \pm 1.8$	1.064
		$d_{15} = 17.0 \pm 1.8$	1.064
$PbTiO_3$	4 mm	$d_{33} = 7.5 \pm 1.2$	1.064
		$d_{31} = 37.6 \pm 5.6$	1.064
		$d_{15} = 33.3 \pm 5$	1.064
$K_3Li_2Nb_5O_{15}$	4 mm	$d_{33} = 11.2 \pm 1.6$	1.064
		$d_{31} = 6.18 \pm 1.28$	1.064
		$d_{15} = 5.45 \pm 0.54$	1.064
$K_{0.8}Na_{0.2}Ba_2Nb_5O_{15}$	4 mm	$d_{31} = 13.6 \pm 1.6$	1.064
$SrBaNb_5O_{15}$	4 mm	$d_{33} = 11.3 \pm 3.3$	1.064
		$d_{31} = 4.31 \pm 1.32$	1.064
		$d_{15} = 5.98 \pm 2$	1.064
$NH_4H_2PO_4$ (ADP)	$\bar{4}2$ m	$d_{36} = 0.53$	1.064
		$d_{36} = 0.85$	0.694
KH_2PO_4 (KDP)	$\bar{4}2$ m	$d_{36} = 0.44$	1.064
		$d_{36} = 0.47 \pm 0.07$	0.694
KD_2PO_4 (KD*P)	$\bar{4}2$ m	$d_{36} = 0.38 \pm 0.016$	1.058
		$d_{36} = 0.34 \pm 0.06$	0.694
		$d_{14} = 0.37$	1.058
KH_2AsO_4 (KDA)	$\bar{4}2$ m	$d_{36} = 0.43 \pm 0.025$	1.06
		$d_{36} = 0.39 \pm 0.4$	0.694
$CdGeAs_2$	$\bar{4}2$ m	$d_{36} = 351 \pm 105$	10.6
$AgGaS_2$	$\bar{4}2$ m	$d_{36} = 18 \pm 2.7$	10.6
$AgGaSe_2$	$\bar{4}2$ m	$d_{36} = 37.4 \pm 6.0$	10.6
$(NH_2)_2CO$ (urea)	$\bar{4}2$ m	$d_{36} = 1.3$	1.06
$AlPO_4$	32	$d_{11} = 0.35 \pm 0.03$	1.058
Se	32	$d_{11} = 97 \pm 25$	10.6

Material	Symmetry class	$d_{im} \times 10^{12}$ m/V	λ μm
Te	32	$d_{11} = 650 \pm 30$	10.6
SiO_2 (quartz)	32	$d_{11} = 0.335$	1.064
HgS	32	$d_{11} = 50.3 \pm 17$	10.6
$(C_6H_5CO)_2$ [benzil]	32	$d_{11} = 3.6 \pm 0.5$	1.064
β-BaB_2O_4 [BBO]	3 m	$d_{22} = 2.22 \pm 0.09$	1.06
		$d_{31} = 0.16 \pm 0.08$	1.06
$LiNbO_3$	3 m	$d_{33} = 34.4$	1.06
		$d_{31} = -5.95$	1.06
		$d_{22} = 2.76$	1.06
$LiTaO_3$	3 m	$d_{33} = -16.4 \pm 2$	1.058
		$d_{31} = -1.07 \pm 0.2$	1.058
		$d_{22} = +1.76 \pm 0.2$	1.058
Ag_3AsS_3 [proustite]	3 m	$d_{31} = 11.3 \pm 2.5$	10.6
		$d_{22} = 18.0 \pm 2.5$	10.6
Ag_3SbS_3 [pyrargerite]	3m	$d_{31} = 12.6 \pm 4$	10.6
		$d_{22} = 13.4 \pm 4$	10.6
α-HIO_3	222	$d_{36} = 5.15 \pm 0.16$	1.064
$NO_2 \cdot CH_3NOC_5H_4 \cdot$ (POM)	222	$d_{36} = 6.4 \pm 1.0$	1.064
$Ba_2NaNb_5O_{15}$ [Banana]	mm 2	$d_{33} = -17.6 \pm 1.28$	1.064
		$d_{31} = -12.8 \pm 1.28$	1.064
$C_6H_4(NO_2)_2$ [MDB]	mm 2	$d_{33} = 0.74$	1.064
		$d_{32} = 2.7$	1.064
		$d_{31} = 1.78$	1.064
$Gd_2(MoO_4)_3$	mm 2	$d_{33} = -0.044 \pm 0.008$	1.064
		$d_{32} = +2.42 \pm 0.36$	1.064
		$d_{31} = -2.49 \pm 0.37$	1.064
$KNbO_3$	mm 2	$d_{33} = -19.58 \pm 1.03$	1.064
		$d_{32} = +11.34 \pm 1.03$	1.064
		$d_{31} = -12.88 \pm 1.03$	1.064
$KTiOPO_4$ [KTP]	mm 2	$d_{33} = 13.7$	1.06
		$d_{32} = \pm 5.0$	1.06
		$d_{31} = \pm 6.5$	1.06
$NO_2C_6H_4 \cdot NH_2$ [mNA]	mm 2	$d_{33} = 13.12 \pm 1.28$	1.064
		$d_{32} = 1.02 \pm 0.22$	1.064
		$d_{31} = 12.48 \pm 1.28$	1.064
$C_{10}H_{12}N_3O_6$ [MAP]	2	$d_{23} = 10.67 \pm 1.3$	1.064
		$d_{22} = 11.7 \pm 1.3$	1.064
		$d_{21} = 2.35 \pm 0.5$	1.064
		$d_{25} = -0.35 \pm 0.3$	1.064
$(NH_2CH_2COOH)_3H_2SO_4$ [TGS]	2	$d_{23} = 0.32$	0.694

* These data are taken from References 1 and 2.

Selected THG Coefficients of Some NLO Materials*

Material	NLO process	$C_{jn} \times 10^{20}$ m^2/V^{-2}	λ μm
$NH_4H_2PO_4$ [ADP]	$(-3\omega,\omega,\omega,\omega)$	$C_{11} = 0.0104$	1.06
		$C_{18} = 0.0098$	1.06
C_6H_6 [benzene]	$(-3\omega,\omega,\omega,\omega)$	$C_{11} = 0.0184 \pm 0.0042$	1.89
$CdGeAs_2$	$(-3\omega,\omega,\omega,\omega)$	$C_{11} = 182 \pm 84$	10.6
p-type: 5×10^{16} cm^{-3}		$C_{16} = 175$	10.6
		$C_{18} = -35$	10.6
$C_{40}H_{56}$ [β-carotene]	$(-3\omega,\omega,\omega,\omega)$	$C_{11}\ 0.263 \pm 0.08$	1.89
GaAs	$(-3\omega,\omega,\omega,-\omega)$	$C_{11} = 62 \pm 31$	1.06
high-resistivity			
Ge	$(-3\omega,\omega,\omega,-\omega)$	$C_{11} = 23.5 \pm 12$	1.06
$LiIO_3$	$(-3\omega,\omega,\omega,-\omega)$	$C_{12} = 0.2285$	1.06
		$C_{35} = 6.66 \pm 1$	1.06
KBr	$(-3\omega,\omega,\omega,-\omega)$	$C_{11} = 0.0392$	1.06
		$C_{18}/C_{11} = 0.3667$	1.06
KCl	$(-3\omega,\omega,\omega,-\omega)$	$C_{11} = 0.0168$	1.06
		$C_{18}/C_{11} = 0.28$	1.06
KH_2PO_4 [KDP]	$(-3\omega,\omega,\omega,-\omega)$	$C_{11}-3C_{18} = 0.04$	1.06
Si	$(-3\omega,\omega,\omega,-\omega)$	$C_{11} = 82.8 \pm 25$	1.06
p-type: 10^{14} cm^{-3}			
NaCl	$(-3,\omega,\omega,\omega,-\omega)$	$C_{11} = 0.0168$	1.06
		$C_{18}/C_{11} = 0.4133$	1.06
NaF	$(-3\omega,\omega,\omega,-\omega)$	$C_{11} = 0.0035$	1.06

* These data are taken from Reference 1.

PHASE DIAGRAMS
H. P. R. Frederikse

A phase is a structurally homogeneous portion of matter. Regardless of the number of chemical constituents of a gas, there is only one vapor phase. This is true also for the liquid form of a pure substance, although a mixture of several liquid substances may exist as one or several phases, depending on the interactions among the substances. On the other hand a pure solid may exist in several phases at different temperatures and pressures because of differences in crystal structure (Reference 1). At the phase transition temperature, T_{tr}, the chemical composition of the solid remains the same, but often a change in the physical properties will take place. Such changes are found in ferroelectric crystals (example $BaTiO_3$) which develop a spontaneous polarization below T_{tr}, in superconductors (example Pb) which loose all electrical resistance below the transition point, and in many other classes of solids.

In quite a few cases it is difficult to bring about the phase transition, and the high- (or low-) temperature phase persists in its metastable form. Many liquids remain in the liquid state for shorter or longer periods of time when cooled below the melting point (supercooling). However, often the slightest disturbance will cause solidification. Persistence of the high temperature phase in solid-solid transitions is usually of much longer duration. An example of this behavior is found in white tin; although gray tin is the thermodynamically stable form below T_{tr} (286.4 K), the metal remains in its undercooled, white tin state all the way to $T = 0$ K, and crystals of gray tin are very difficult to produce.

A *phase diagram* is a map which indicates the areas of stability of the various phases as a function of external conditions (temperature and pressure). Pure materials, such as mercury, helium, water, and methyl alcohol are considered one-component systems and they have *unary* phase diagrams. The equilibrium phases in two-component systems are presented in *binary* phase diagrams. Because many important materials consist of three, four, and more components, many attempts have been made to deduce their multicomponent phase diagrams. However, the vast majority of systems with three or more components are very complex, and no overall maps of the phase relationships have been worked out.

It has been shown during the last 20 to 25 years that very useful partial phase diagrams of complex systems can be obtained by means of thermodynamic modeling (References 2, 3). Especially for complicated, multicomponent alloy systems the CALPHAD method has proved to be a successful approach for producing valuable portions of very intricate phase diagrams (Reference 4). With this method thermodynamic descriptions of the free energy functions of various phases are obtained which are consistent with existing (binary) phase diagram information and other thermodynamic data. Extrapolation methods are then used to extend the thermodynamic functions into a ternary system. Comparison of the results of this procedure with available experimental data is then used to fine-tune the phase diagram and add ternary interaction functions if necessary. In principle this approximation strategy can be extended to four, five, and more component systems.

The nearly two dozen phase diagrams shown below present the reader with examples of some important types of single and multicomponent systems, especially for ceramics and metal alloys. This makes it possible to draw attention to certain features like the kinetic aspects of phase transitions (see Figure 22, which presents a time-temperature-transformation, or TTT, diagram for the precipitation of α-phase particles from the β-phase in a Ti-Mo alloy; Reference 1, pp.358-360). The general references listed below and the references to individual figures contain phase diagrams for many additional systems.

GENERAL REFERENCES

1. Ralls, K.M., Courtney, T.H., and Wulff, J., *Introduction to Materials Science and Engineering*, Chapters 16 and 17, John Wiley & Sons, New York, 1976.
2. Kaufman, L., and Bernstein, H., *Computer Calculation of Phase Diagrams*, Academic Press, New York, 1970.
3. Kattner, U.R., Boettinger, W.J.B., and Coriell, S.R., *Z. Metallkd.*, 87, 9, 1996.
4. Dinsdale, A.T., Editor, *CALPHAD*, Vol. 1–20, Pergamon Press, Oxford, 1977–1996 and continuing.
5. Baker, H., Editor, *ASM Handbook, Volume 3: Alloy Phase Diagrams*, ASM International, Materials Park, OH, 1992.
6. Massalski, T.B., Editor, *Binary Alloy Phase Diagrams, Second Edition*, ASM International, Materials Park, OH, 1990.
7. Roth. R.S., Editor, *Phase Diagrams for Ceramists*, Vol. I (1964) to Volume XI (1995), American Ceramic Society, Waterville, OH.

REFERENCES TO INDIVIDUAL PHASE DIAGRAMS

Figure 1. Carbon: Reference 7, Vol. X (1994), Figure 8930. Reprinted with permission.
Figure 2. Si-Ge : Ref.5, p. 2.231. Reprinted with permission.
Figure 3. H_2O (ice): See figure.
Figure 4. SiO_2: Reference 7, Vol. XI (1995), Figure 9174. Reprinted with permission.
Figure 5. Fe-O: Darken, L.S., and Gurry, R.W., *J. Am. Chem. Soc.*, 68, 798, 1946. Reprinted with permission.
Figure 6. Ti-O: Reference 5, p. 2.324. Reprinted with permission.
Figure 7. $BaO-TiO_2$: Reference 7, Vol. III (1975), Figure 4302. Reprinted with permission.
Figure 8. $MgO-Al_2O_3$: Reference 7, Vol. XI (1995), Figure 9239. Reprinted with permission.
Figure 9. $Y_2O_3-ZrO_2$: Reference 7, Vol. XI (1995), Figure 9348. Reprinted with permission.
Figure 10. Si-N-Al-O (Sialon): Reference 7, Vol. X (1994), Figure 8759. Reprinted with permission.
Figure 11. $PbO-ZrO_2-TiO_2$ (PZT): Reference 7, Vol. III (1975), Figure 4587. Reprinted with permission.
Figure 12. Al-Si-Ca-O: Reference 7 (1964), Vol. I, Figure 630. Reprinted with permission.
Figure 13. Y-Ba-Cu-O: Whitler, J.D., and Roth, R.S., *Phase Diagrams for High T_c Superconductors*, Figure S-082, American Ceramic Society, Waterville, OH, 1990. Reprinted with permission.
Figure 14. Al-Cu: Reference 5, p. 2.44. Reprinted with permission.
Figure 15. Fe-C: Ralls, K.M., Courtney, T.H., and Wulff, J., *Introduction to Materials Science and Engineering*, Figure 16.13 , John Wiley & Sons, New York, 1976. Reprinted with permission.
Figure 16. Fe-Cr: Reference 5, p. 2.152. Reprinted with permission.

Figure 17. Cu-Sn: Reference 5, p. 2.178. Reprinted with permission.
Figure 18. Cu-Ni: Reference 5, p. 2.173. Reprinted with permission.
Figure 19. Pb-Sn (solder): Reference 5, p. 2.335. Reprinted with permission.
Figure 20. Cu-Zn (brass): Subramanian, P.R., Chakrabarti, D.J., and Laughlin, D.E., Editors, *Phase Diagrams of Binary Copper Alloys*, p. 487, ASM
    International, Materials Park, OH, 1994. Reprinted with permission.
Figure 21. Co-Sm: Reference 5, p. 2.148. Reprinted with permission.
Figure 22. Ti-Mo: Reference 5, p. 2.296; Reference 1, p. 359. Reprinted with permission.
Figure 23: Fe-Cr-Ni: Reference 5, Figure 48. Reprinted with permission.

Figure 1

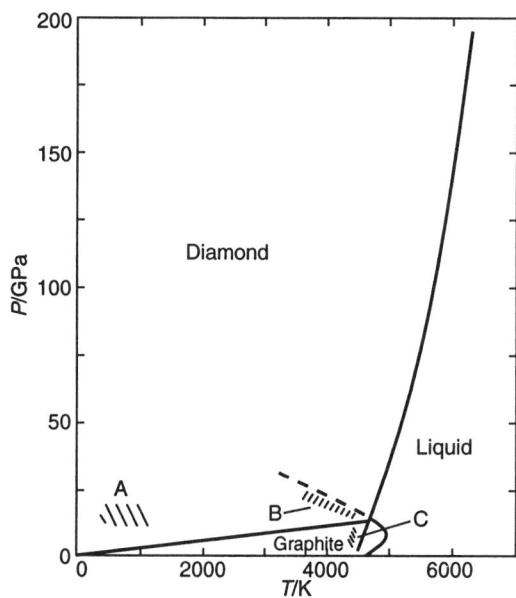

Figure 1. Phase diagram of carbon. (A) Martensitic transition: hex graphite →
hex diamond. (B) Fast graphite-to-diamond transition. (C) Fast diamond-to-
graphite transition.

Figure 2

Figure 2. Si-Ge system.

Phase	Composition, mass % Si	Pearson symbol	Space group
(Ge,Si)	0 to 100	*cF*8	$Fd\bar{3}m$
High-pressure phases			
GeII	—	*tI*4	$I4_1/amd$
SiII	—	*tI*4	$I4_1/amd$

FIGURE 3

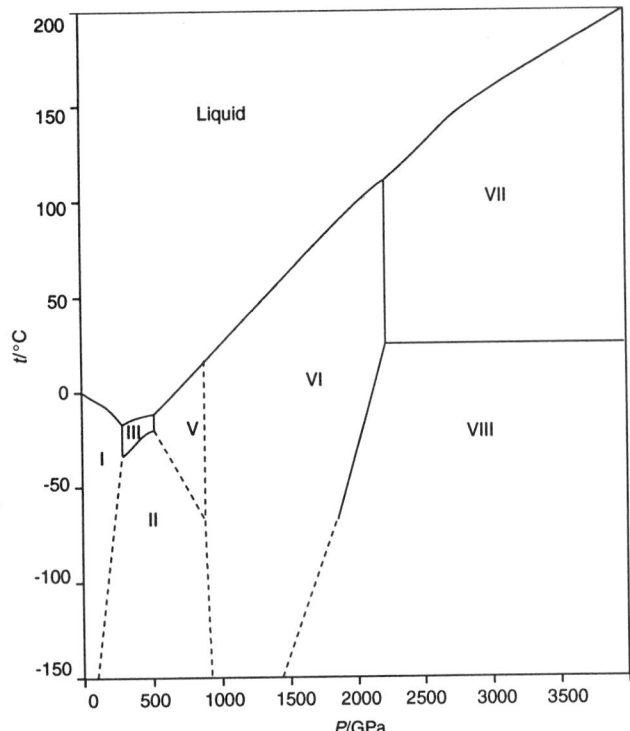

Figure 3. Diagram of the principal phases of ice. Solid lines are measured boundaries between stable phases; dotted lines are extrapolated. Ice IV is a metastable phase which exists in the region of ice V. Ice IX exists in the region below -100°C and pressures in the range 200–400 MPa. Ice X exists at pressures above 44 GPa. See Table 1 for the coordinates of the triple points, where liquid water is in equilibrium with two adjacent solid phases.

Table 1. Crystal Structure, Density, and Transition Temperatures for the Phases of Ice

Phase	Crystal system	Cell parameters	Z	n	$\rho/g\ cm^{-3}$	Triple ponts
Ih	Hexagonal	$a = 4.513; c = 7352$	4	4	0.93	I-III: -21.99°C, 209.9 MPa
Ic	Cubic	$a = 6.35$	8	4	0.94	
II	Rhombohedral	$a = 7.78; \alpha = 113.1°$	12	4	1.18	
III	Tetragonal	$a = 6.73; c = 6.83$	12	4	1.15	III-V: -16.99°C, 350.1 MPa
IV	Rhombohedral	$a = 7.60; \alpha = 70.1°$	16	4	1.27	
V	Monoclinic	$a = 9.22; b = 7.54,$ $c = 10.35; \beta = 109.2°$	28	4	1.24	V-VI: 0.16°C, 632.4 MPa
VI	Tetragonal	$a = 6.27; c = 5.79$	10	4	1.31	VI-VII: 82°C, 2216 MPa
VII	Cubic	$a = 3.41$	2	8	1.56	
VIII	Tetragonal	$a = 4.80; c = 6.99$	8	8	1.56	
IX	Tetragonal	$a = 6.73; c = 6.83$	12	4	1.16	
X	Cubic	$a = 2.83$	2	8	2.51	

REFERENCES

1. Wagner, W., Saul, A., and Pruss, A., *J. Phys. Chem. Ref. Data*, 23, 515, 1994.
2. Lerner, R.G. and Trigg, G.L., Editors, *Encyclopedia of Physics*, VCH Publishers, New York, 1990.
3. Donnay, J.D.H. and Ondik, H.M, *Crystal Data Determinative Tables, Third Edition, Volume 2, Inorganic Compounds*, Joint Committee on Powder Diffraction Standards, Swarthmore, PA, 1973.
4. Hobbs, P.V., *Ice Physics*, Oxford University Press, Oxford, 1974.

Figure 4

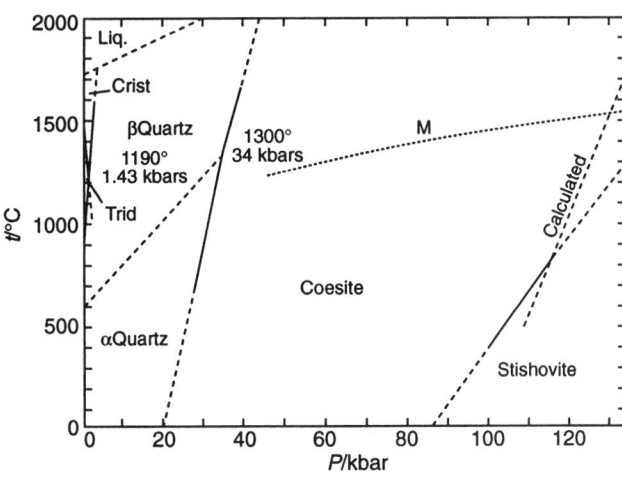

Figure 4. SiO$_2$ system. Crist = cristobalite; Trid = tridymite.

Figure 5

Figure 5. Fe-O system.

Figure 5 (continued)

Point	$t/°C$	% O	p_{CO_2}/p_{CO}	Point	$t/°C$	% O	p_{CO_2}/p_{CO}	p_{O_2}/atm
A	1539			Q	560	23.26	1.05	
B	1528	0.16	0.209	R	1583	28.30		1
C	1528	22.60	0.209	R'	1583	28.07		1
G	1400[a]	22.84	0.263	S	1424	27.64	16.2	
H	1424	25.60	16.2	V	1597	27.64		0.0575
I	1424	25.31	16.2	Y	1457	28.36		1
J	1371	23.16	0.282	Z	1457	30.04		1
L	911[a]	23.10	0.447	Z'		30.6		
N	1371	22.91	0.282					

[a] Values for pure iron.

Figure 6

Figure 6. Ti-O system.

Phase	Composition, mass % O	Pearson symbol	Space group
(βTi)	0 to 3	$cI2$	$Im\bar{3}m$
(αTi)	0 to 13.5	$hP2$	$P63/mmc$
Ti_3O	~8 to ~13	$hP\sim16$	$P\bar{3}c$
Ti_2O	~10 to 14.4	$hP3$	$P\bar{3}m1$
γTiO	15.2 to 29.4	$cF8$	$Fm\bar{3}m$
Ti_3O_2	~18	$hP\sim5$	$P6/mmm$
βTiO	~24 to ~29.4	c^{**}	—
αTiO	~25.0	$mC16$	$A2/m$ or $B*/*$
$\beta Ti_{1-x}O$	~29.5	$oI12$	$I222$
$\alpha Ti_{1-x}O$	~29.5	$tI18$	$I4/m$
βTi_2O_3	33.2 to 33.6	$hR30$	$R\bar{3}c$
αTi_2O_3	33.2 to 33.6	$hR30$	$R\bar{3}c$
βTi_3O_5	35.8	m^{**}	—
αTi_3O_5	35.8	$mC32$	$C2/m$
$\alpha' Ti_3O_5$	35.8	$mC32$	Cc

Figure 6 (continued)

Phase	Composition, mass % O	Pearson symbol	Space group
γTi_4O_7	36.9	$aP44$	$P\bar{1}$
βTi_4O_7	36.9	$aP44$	$P\bar{1}$
$\alpha Ti4O_7$	36.9	$aP44$	$P\bar{1}$
γTi_5O_9	37.6	$aP28$	$P\bar{1}$
βTi_6O_{11}	38.0	$aC68$	$A\bar{1}$
Ti_7O_{13}	38.3	$aP40$	$P\bar{1}$
Ti_8O_{15}	38.5	$aC92$	$A\bar{1}$
Ti_9O_{17}	38.7	$aP52$	$P\bar{1}$
Rutile TiO_2	40.1	$tP6$	$P4_2/mnm$
Metastable phases			
Anatase	—	$tI12$	$I4_1/amd$
Brookite	—	$oP24$	$Pbca$
High-pressure phases			
TiO_2-II	—	$oP12$	$Pbcn$
TiO_2-III	—	$hP{\sim}48$	—

FIGURE 7

Figure 7. BaO-TiO$_2$ system.

Figure 8

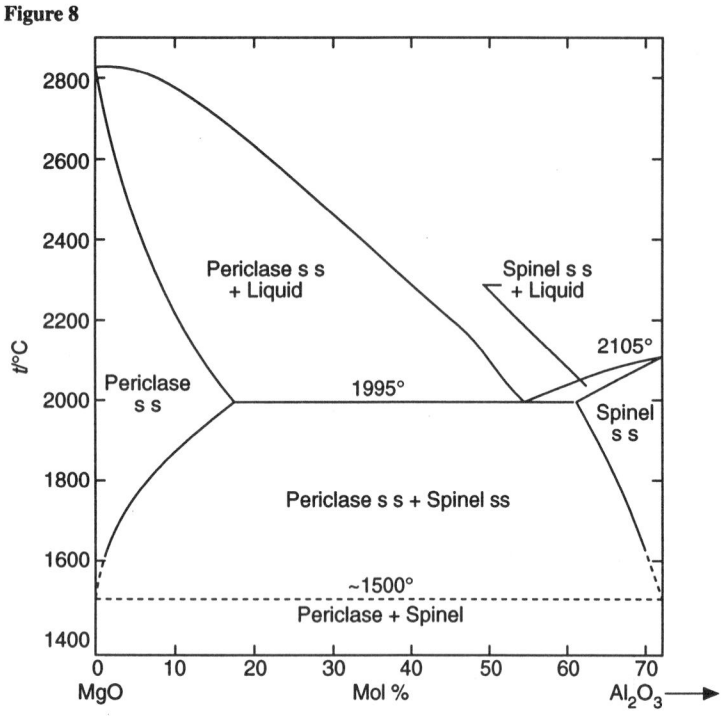

Figure 8. MgO-Al$_2$O$_3$ system.

Figure 9

Figure 9. Y$_2$O$_3$-ZrO$_2$ system. C$_{ss}$ = cubic ZrO$_2$ ss (fluorite-type ss); Y$_{ss}$ = cubic Y$_2$O$_3$ ss; Tet$_{ss}$ = tetragonal ZrO$_2$ ss; Mon$_{ss}$ = monoclinic ZrO$_2$ ss; H$_{ss}$ = hexagonal Y$_2$O$_3$ ss; 3:4 = Zr$_3$Y$_4$O$_{12}$; 1:6 = ZrY$_6$O$_{11}$ ss.

Figure 10

Figure 10. 3(SiO$_2$)-Si$_3$N$_4$-4(AlN)-2(Al$_2$O$_3$) system. "Behavior" diagram at 1700°C. The labels 8H, 15R, 12H, 21R, 27R, 2H$^\delta$ indicate defect AlN polytypes. β' = 3-sialon (Si$_{6-x}$Al$_x$O$_x$N$_{8-x}$); O' = sialon of Si$_2$ON$_2$ type; X = SiAlO$_2$N ("nitrogen mullite"). ALON ss = aluminum oxynitride ss extending from approximately Al$_7$O$_9$N to Al$_3$O$_3$N.

Figure 11

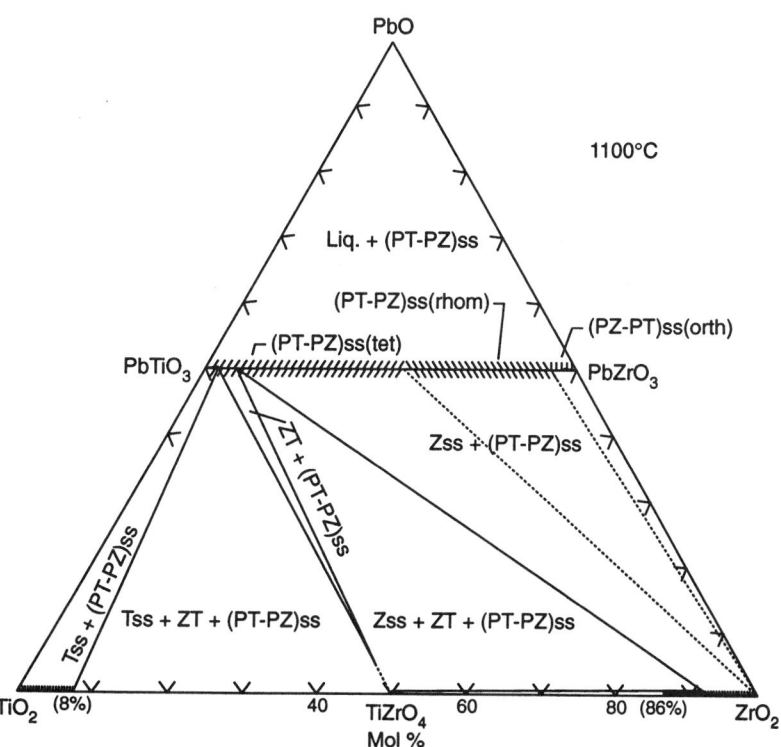

Figure 11. PbO-ZrO$_2$-TiO$_2$ (PZT) system, subsolidus at 1100°C. P = PbO; T = TiO$_2$; Z = ZrO$_2$.

Figure 12

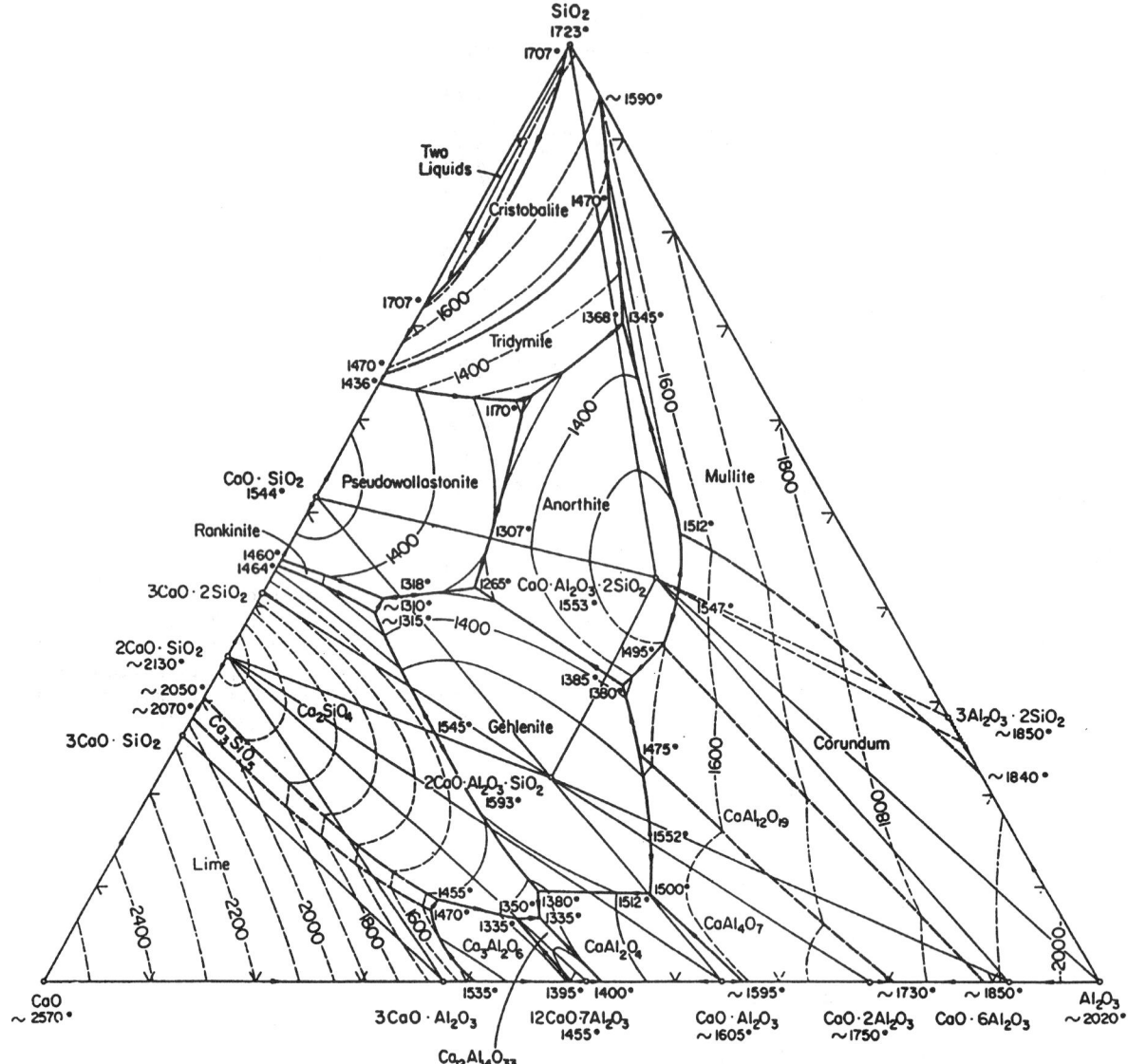

Figure 12. CaO-Al$_2$O$_3$-SiO$_2$ system (temperatures in °C).

Crystalline Phases

Notation	Oxide formula
Cristobalite } Tridymite	SiO$_2$
Pseudowollastonite	CaO·SiO$_2$
Rankinite	3CaO·2SiO$_2$
Lime	CaO
Corundum	Al$_2$O$_3$
Mullite	3Al$_2$O$_3$·2SiO$_2$
Anorthite	CaO·Al$_2$O$_3$·2SiO$_2$
Gehlenite	2CaO·Al$_2$O$_3$·SiO$_2$

Temperatures up to approximately 1550°C are on the Geophysical Laboratory Scale; those above 1550°C are on the 1948 International Scale.

Figure 13

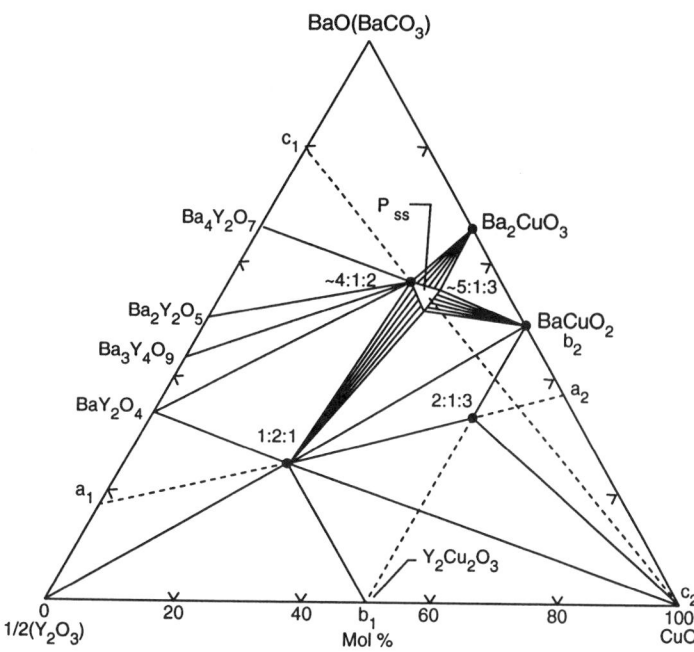

Figure 13. BaO-Y_2O_3-CuO system. 2:1:3 = $Ba_2YCu_3O_{7-x}$; 1:2:1 = BaY_2CuO_5; 4:1:2 = $Ba_4YCu_2O_{7.5+x}$; and 5:1:3 = $Ba_5YCu_3O_{9.5+x}$. The superconducting 2:1:3 phase was prepared using barium peroxide.

Figure 14

Figure 14. Al-Cu system.

Figure 14 (continued)

Phase	Composition, wt % Cu	Pearson symbol	Space group
(Al)	0 to 5.65	$cF4$	$Fm\bar{3}m$
θ	52.5 to 53.7	$tI12$	$I4/mcm$
η_1	70.0 to 72.2	$oP16$ or $oC16$	$Pban$ or $Cmmm$
η_2	70.0 to 72.1	$mC20$	$C2/m$
ζ_1	74.4 to 77.8	$hP42$	$P6/mmm$
ζ_2	74.4 to 75.2	(a)	—
ε_1	77.5 to 79.4	(b)	—
ε_2	72.2 to 78.7	$hP4$	$P63/mmc$
δ	77.4 to 78.3	(c)	$R\bar{3}m$
γ_0	77.8 to 84	(d)	—
γ_1	79.7 to 84	$cP52$	$P\bar{4}3m$
β_0	83.1 to 84.7	(d)	—
β	85.0 to 91.5	$cI2$	$Im\bar{3}m$
α_2	88.5 to 89	(e)	—
(Cu)	90.6 to 100	$cF4$	$Fm\bar{3}m$
Metastable phases			
θ'	—	$tP6$	—
β'	—	$cF16$	$Fm\bar{3}m$
Al_3Cu_2	61 to 70	$hp5$	$P3m1$

(a) Monoclinic? (b) Cubic? (c) Rhombohedral. (d) Unknown. (e) $D0_{22}$-type long-period superlattice.

Figure 15

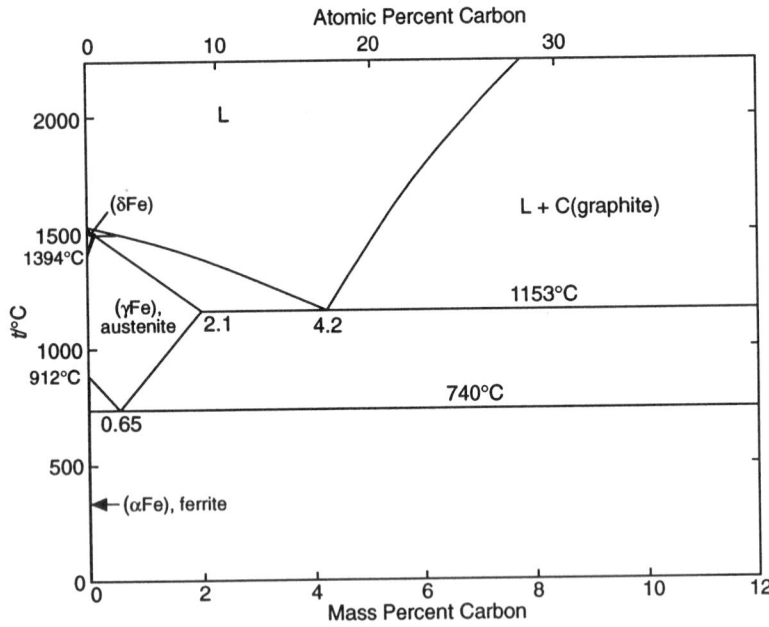

Figure 15. Fe-C system.

Phase	Composition, mass % C	Pearson symbol	Space group
(δFe)	0 to 0.09	$cI2$	$Im\bar{3}m$
(γFe)	0 to 2.1	$cF4$	$Fm\bar{3}m$
(αFe)	0 to 0.021	$cI2$	$Im\bar{3}m$
(C)	100	$hP4$	$P6_3/mmc$
Metastable/high-pressure phases			
(εFe)	0	$hP2$	$P6_3/mmc$
Martensite	< 2.1	$tI4$	$I4/mmm$

Figure 15 (continued)

Phase	Composition, mass % C	Pearson symbol	Space group
Fe_4C	5.1	$cP5$	$P\bar{4}3m$
Fe_3C (θ)	6.7	$oP16$	$Pnma$
Fe_5C_2 (χ)	7.9	$mC28$	$C2/c$
Fe_7C_3	8.4	$hP20$	$P6_3mc$
Fe_7C_3	8.4	$oP40$	$Pnma$
Fe_2C (η)	9.7	$oP6$	$Pnnm$
Fe_2C (ε)	9.7	$hP*$	$P6_322$
Fe_2C	9.7	$hP*$	$P\bar{3}m1$
(C)	100	$cF8$	$Fd\bar{3}m$

Figure 16

Figure 16. Fe-Cr system.

Phase	Composition, mass % Cr	Pearson symbol	Space group
(aFe, Cr)	0 to 100	$cI2$	$Im\bar{3}m$
(γFe)	0 to 11.2	$cF4$	$Fm\bar{3}m$
σ	42.7 to 48.2	$tP30$	$P4_2/mnm$

Figure 17

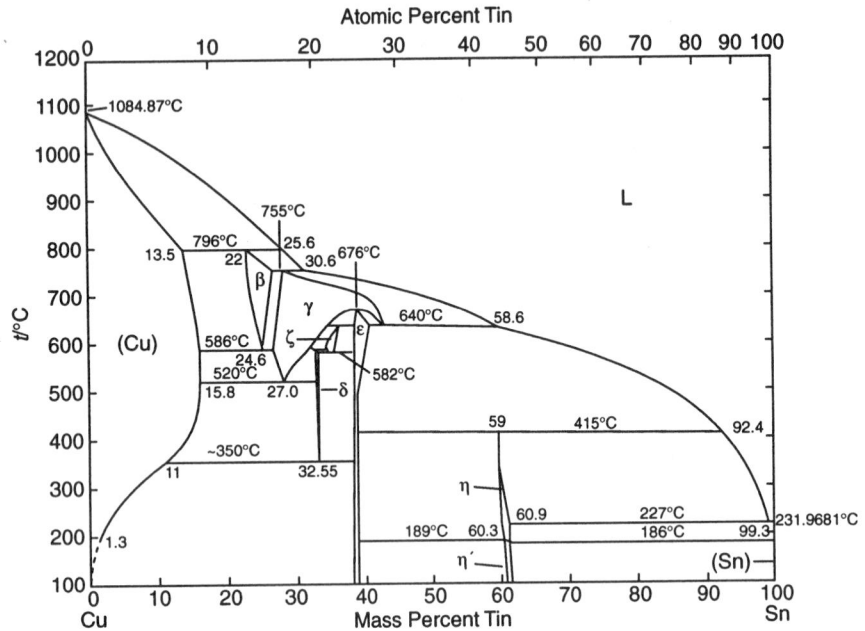

Figure 17. Cu-Sn system.

Phase	Composition, mass % Sn	Pearson symbol	Space group
α	0 to 15.8	$cF4$	$Fm\bar{3}m$
β	22.0 to 27.0	$cI2$	$Im\bar{3}m$
γ	25.5 to 41.5	$cF16$	$Fm\bar{3}m$
δ	32 to 33	$cF416$	$F\bar{4}3m$
ζ	32.2 to 35.2	$hP26$	$P6_3$
ε	27.7 to 39.5	$oC80$	$Cmcm$
η	59.0 to 60.9	$hP4$	$P6_3/mmc$
η′	44.8 to 60.9	(a)	—
(βSn)	~100	$tI4$	$I4_1/amd$
(αSn)	100	$cF8$	$Fd\bar{3}m$

(a) Hexagonal; superlattice based on NiAs-type structure.

Figure 18

Figure 18. Cu-Ni system.

Phase	Composition, mass % Ni	Pearson symbol	Space group
(Cu, Ni) (above 354.5°C)	0 to 100	$cF4$	$Fm\bar{3}m$

Figure 19

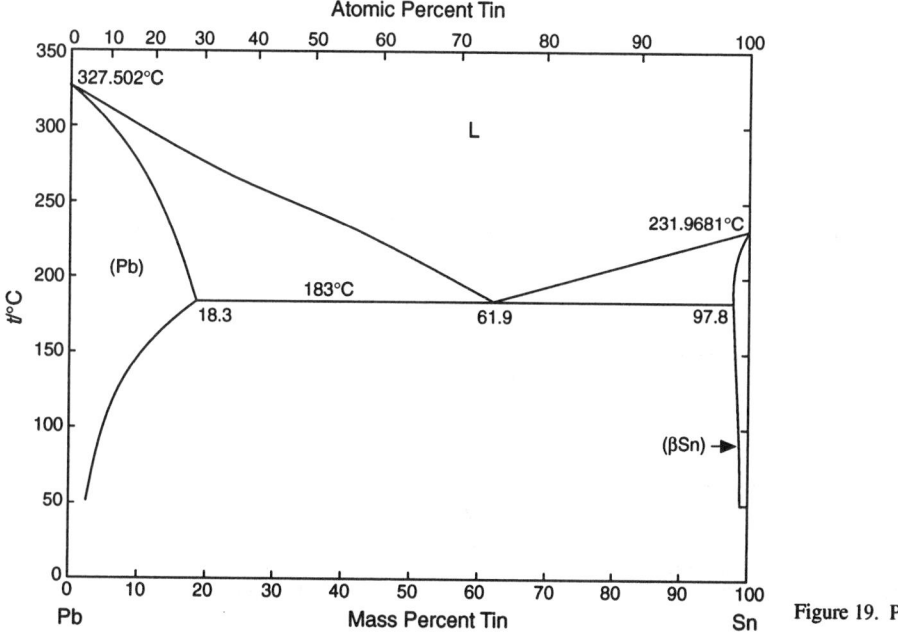

Figure 19. Pb-Sn system.

Figure 19 (continued)

Phase	Composition, mass % Sn	Pearson symbol	Space group
(Pb)	0 to 18.3	cF4	$Fm\bar{3}m$
(βSn)	97.8 to 100	tI4	$I4_1/amd$
(αSn)	100	cF8	$Fd\bar{3}m$
High-pressure phases			
ε(a)	52 to 74	hP1	$P6/mmm$
ε'(b)	52	hP2	$P6_3/mmc$

(a) From phase diagram calculated at 2500 MPa. (b) This phase was claimed for alloys at 350°C and 5500 MPa.

Figure 20

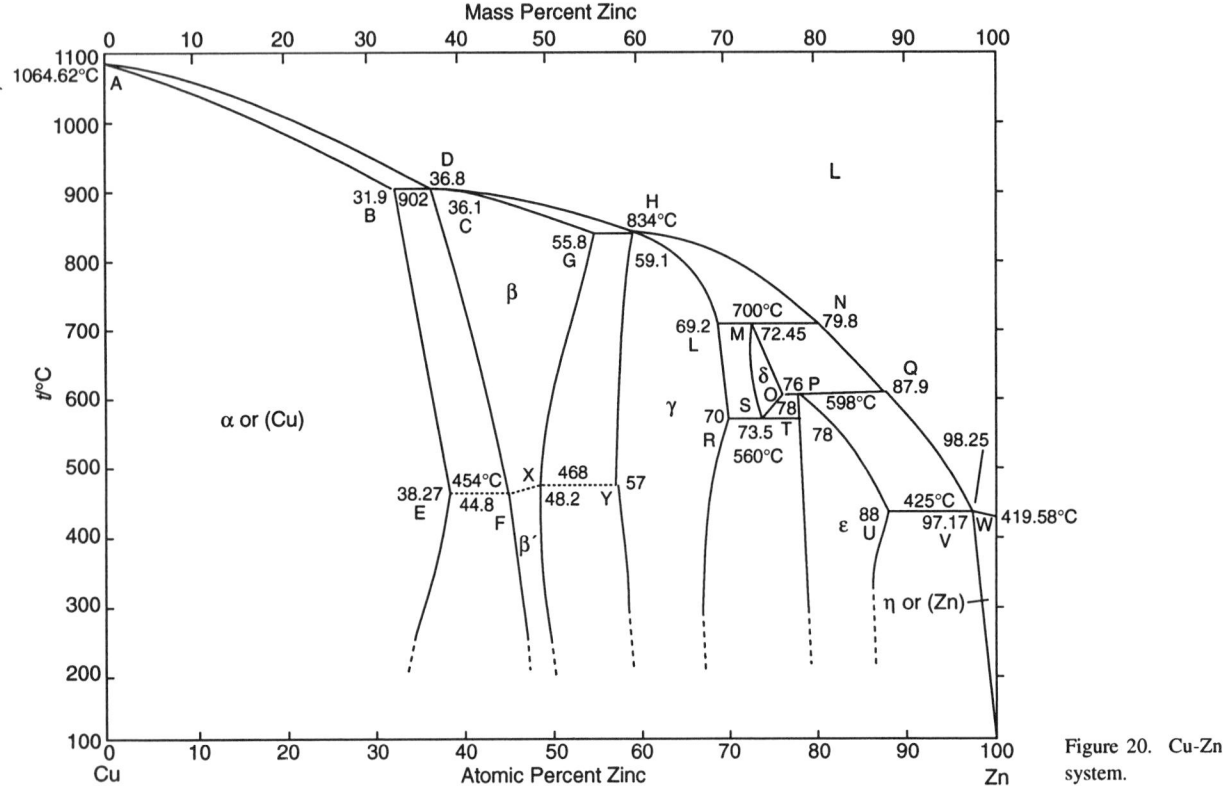

Figure 20. Cu-Zn system.

Phase	Composition, mass % Zn	Pearson symbol	Space group
α or (Cu)	0 to 38.95	cF4	$Fm\bar{3}m$
β	36.8 to 56.5	cI2	$Im\bar{3}m$
β'	45.5 to 50.7	cP2	$Pm\bar{3}m$
γ	57.7 to 70.6	cI52	$I\bar{4}3m$
δ	73.02 to 76.5	hP3	$P\bar{6}$
ε	78.5 to 88.3	hP2	$P6_3/mmc$
η or (Zn)	97.25 to 100	hP2	$P6_3/mmc$

Figure 21

Figure 21. Co-Sm system.

Phase	Composition, mass % Sm	Pearson symbol	Space group
(αCo)	0 to ~3.7	cF4	$Fm\bar{3}m$
(εCo)	~0	hP2	$P6_3/mmc$
βCo₁₇Sm₂	~23.0	hP38	$P6_3/mmc$
αCo₁₇Sm₂	~23.0	hR19	$R\bar{3}m$
		hP8	$P6/mmm$
Co₅₊ₓSm	~33 to 34	—	—
Co₅₋ₓSm	~34 to 35	—	—
Co₁₉Sm₅	~40.1	hR24	$R\bar{3}m$
		hP48	$P6_3/mmc$
αCo₇Sm₂	~42.1	hR18	$R\bar{3}m$
βCo₇Sm₂	~42.1	hP36	$P6_3/mmc$
Co₃Sm	46	hR12	$R\bar{3}m$
Co₂Sm	56.0	hR4	$R\bar{3}m$
		cF24	$Fd\bar{3}m$
Co₄Sm₉	~85.1	o**	—
CoSm₃	88	oP16	$Pnma$
(γSm)	~100	cI2	$Im\bar{3}m$
(βSm)	~100	hP2	$P6_3/mmc$
(αSm)	~100	hR3	$R\bar{3}m$
Other reported phases			
Co₅Sm	~33.8	hP6	$P6/mmm$
Co₂Sm₅	~86.4	mC28	$C2/c$

Figure 22

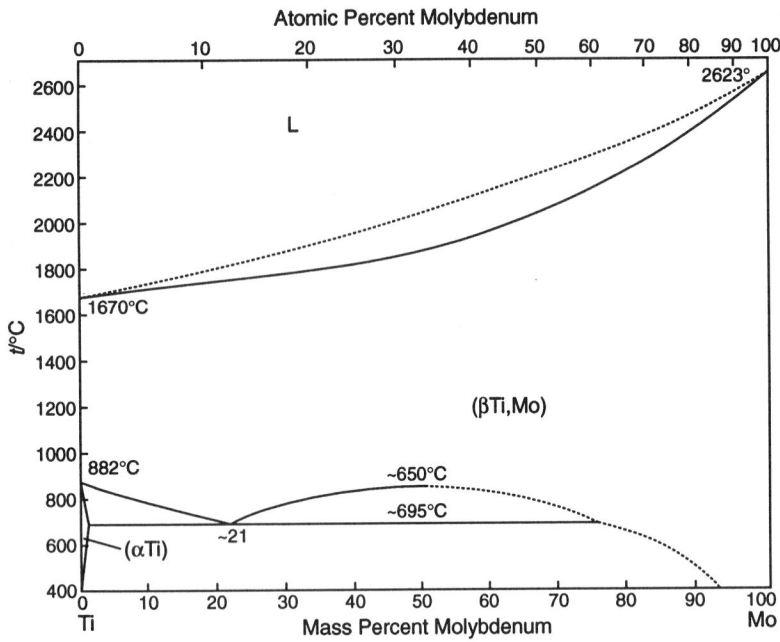

Figure 22. Ti-Mo system.

Phase	Composition, mass % Mo	Pearson symbol	Space group
(βTi, Mo)	0 to 100	$cI2$	$Im\bar{3}m$
(αTi)	0 to 0.8	$hP2$	$P6_3/mmc$
α′	(a)	$hP2$	$P6_3/mmc$
α″	(a)	$oC4$	$Cmcm$
ω	(a)	$hP3$	$P6/mmm$

(a) Metastable.

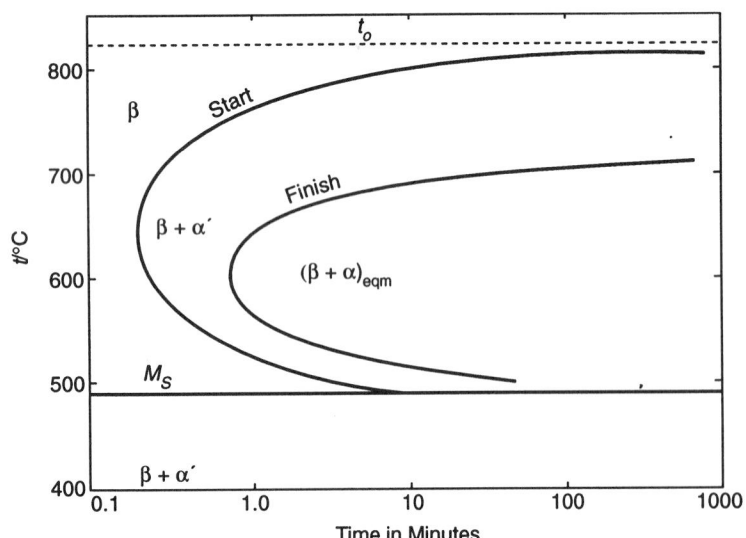

Experimental time-temperature-transformation (TTT) diagram for Ti-Mo. The start and finish times of the isothermal precipitation reaction vary with temperature as a result of the temperature dependence of the nucleation and growth processes. Precipitation is complete, at any temperature, when the equilibrium fraction of α is established in accordance with the lever rule. The solid horizontal line represents the athermal (or nonthermally activated) martensitic transformation that occurs when the β phase is quenched.

Figure 23

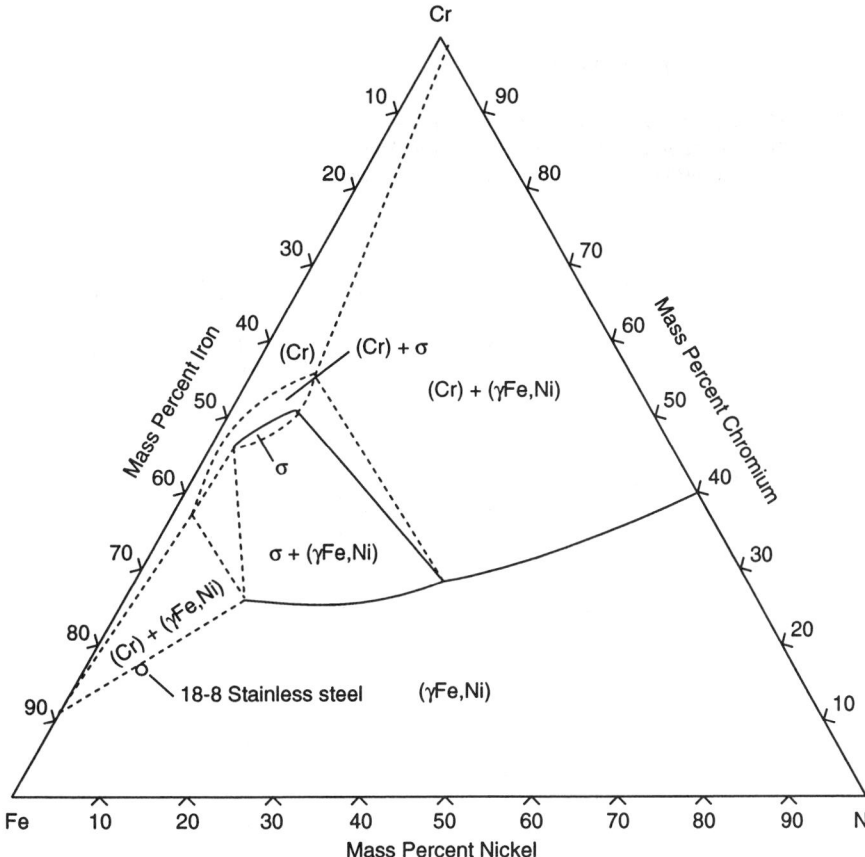

Figure 23. The isothermal section at 900°C (1652°F) of the iron-chromium-nickel ternary phase diagram, showing the nominal composition of 18-8 stainless steel.

HEAT CAPACITY OF SELECTED SOLIDS

This table gives the molar heat capacity at constant pressure of representative metals, semiconductors, and other crystalline solids as a function of temperature in the range 200 to 600 K.

REFERENCES

1. Chase, M. W., et al., *JANAF Thermochemical Tables, 3rd ed., J. Phys. Chem. Ref. Data,* 14, Suppl. 1, 1985.
2. Garvin, D., Parker, V. B., and White, H. J., *CODATA Thermodynamic Tables,* Hemisphere Press, New York, 1987.
3. DIPPR Database of Pure Compound Properties, Design Institute for Physical Properties Data, American Institute of Chemical Engineers, New York, 1987.

Name	C_p in J/mol K						
	200 K	250 K	300 K	350 K	400 K	500 K	600 K
Aluminum	21.33	23.08	24.25	25.11	25.78	26.84	27.89
Aluminum oxide	51.12	67.05	79.45	88.91	96.14	106.17	112.55
Anthracene	138.6	173.9	210.7	248.8	288.4		
Benzoic acid	102.7	123.5	147.4	172.0			
Beryllium	9.98	13.58	16.46	18.53	19.95	21.94	23.34
Biphenyl	131.0	162.5	197.2				
Boron	5.99	8.82	11.40	13.65	15.69	18.72	20.78
Calcium	24.54	25.41	25.94	26.32	26.87	28.49	30.38
Calcium carbonate	66.50	75.66	83.82	91.51	96.97	104.52	109.86
Calcium oxide	33.64	38.59	42.18	45.07	46.98	49.33	50.72
Cesium chloride	50.13	51.34	52.48	53.58	54.68	56.90	59.10
Chromium	19.86	22.30	23.47	24.39	25.23	26.63	27.72
Cobalt	22.23	23.98	24.83	25.68	26.53	28.20	29.66
Copper	22.63	23.77	24.48	24.95	25.33	25.91	26.48
Copper oxide	34.80		42.41	44.95	46.78	49.19	50.83
Copper sulfate	77.01	89.25	99.25	107.65	114.93	127.19	136.31
Germanium			23.25	23.85	24.31	24.96	25.45
Gold			25.41	25.37	25.51	26.06	26.65
Graphite	5.01	6.82	8.58	10.24	11.81	14.62	16.84
Hexachlorobenzene	162.7	183.6	202.4				
Iodine	51.57	53.24	54.51	58.60			
Iron	21.59	23.74	25.15	26.28	27.39	29.70	32.05
Lead	25.87	26.36	26.85	27.30	27.72	28.55	29.40
Lithium	21.57	23.42	24.64	25.96	27.60	29.28	
Lithium chloride	43.35	46.08	48.10	49.66	50.97	53.34	55.59
Magnesium	22.72	24.02	24.90	25.57	26.14	27.17	28.18
Magnesium oxide			37.38	40.59	42.77	45.56	47.30
Manganese	23.05	24.95	26.35	27.52	28.53	30.29	31.90
Naphthalene	105.8	134.1	167.8	204.1			
Potassium	27.00	28.01	29.60				
Potassium chloride	48.44	50.10	51.37	52.31	53.08	54.71	56.35
Silicon	15.64	18.22	20.04	21.28	22.14	23.33	24.15
Silicon dioxide	32.64	39.21	44.77	49.47	53.43	59.64	64.42
Silver			25.36	25.55	25.79	26.36	26.99
Sodium	22.45	27.01	28.20	30.14			
Sodium chloride	46.89	48.85	50.21	51.25	52.14	53.96	55.81
Tantalum	24.08	24.86	25.31	25.60	25.84	26.35	26.84
Titanium	22.37	24.07	25.28	26.17	26.86	27.88	28.60
Tungsten	22.49	23.69	24.30	24.65	24.92	25.36	25.79
Vanadium	21.88	23.70	24.93	25.68	26.23	26.94	27.49
Zinc	24.05	25.02	25.45	25.88	26.35	27.39	28.59
Zirconium	23.87	24.69	25.22	25.61	25.93	26.56	27.28

THERMAL AND PHYSICAL PROPERTIES OF PURE METALS

This table gives the following properties for the metallic elements:

t_m: Melting point in °C
t_b: Normal boiling point in °C, at a pressure of 101.325 kPa (760 Torr)
$\Delta_{fus} H$: Enthalpy of fusion at the melting point in J/g
ρ_{25}: Density at 25°C in g/cm³
α: Coefficient of linear expansion at 25°C in K⁻¹ (the quantity listed is $10^6 \times \alpha$)
c_p: Specific heat capacity at constant pressure at 25°C in J/g K
λ: Thermal conductivity at 27°C in W/cm K

REFERENCES

1. Dinsdale, A. T., *CALPHAD*, 15, 317, 1991 (melting points, enthalpy of fusion).
2. Touloukian, Y. S., *Thermophysical Properties of Matter*, Vol. 12, Thermal Expansion, IFI/Plenum, New York, 1975 (coefficient of expansion, density).
3. Ho, C. Y., Powell, R. W., and Liley, P. E., *J. Phys. Chem. Ref. Data*, 3, Suppl. 1, 1974 (thermal conductivity).
4. Cox, J. D., Wagman, D. D., and Medvedev, V. A., *CODATA Key Values for Thermodynamics*, Hemisphere Publishing Corp., New York, 1989 (heat capacity).
5. Glushko, V. P., Ed., *Thermal Constants of Substances*, VINITI, Moscow, (enthalpy of fusion, heat capacity).
6. Wagman, D. D., et. al., *The NBS Tables of Chemical Thermodynamic Properties, J. Phys. Chem. Ref. Data*, 11, Suppl. 2, 1982 (heat capacity).
7. Chase, M. W., et. al., *JANAF Thermochemical Tables*, 3rd ed., J. Phys. Chem. Ref. Data, 14, Suppl. 1, 1985 (heat capacity, enthalpy of fusion).
8. Gschneidner, K. A., *Bull. Alloy Phase Diagrams*, 11, 216—224, 1990 (various properties of the rare earth metals).
9. Hellwege, K. H., Ed., *Landolt Börnstein, Numerical Values and Functions in Physics, Chemistry, Astronomy, Geophysics, and Technology*, Vol. 2, Part 1, Mechanical-Thermal Properties of State, 1971 (density).
10. *Physical Encyclopedic Dictionary*, Vol. 1–5, Encyclopedia Publishing House, Moscow, 1960–66.

Metal (symbol)	Atomic weight	t_m °C	t_b °C	$\Delta_{fus} H$ J/g	ρ_{25} g/cm³	$\alpha \times 10^6$ K⁻¹	c_p J/g K	λ W/cm K
Actinium (Ac)		1051	3198		10			
Aluminum (Al)	26.98	660.32	2519	399.9	2.70	23.1	0.897	2.37
Antimony (Sb)	121.76	630.63	1587	162.5	6.68	11.0	0.207	0.243
Barium (Ba)	137.33	727	1897	51.8	3.62	20.6	0.204	0.184
Beryllium (Be)	9.01	1287	2471	876.0	1.85	11.3	1.825	2.00
Bismuth (Bi)	208.98	271.40	1564	53.3	9.79	13.4	0.122	0.0787
Cadmium (Cd)	112.41	321.07	767	55.2	8.69	30.8	0.232	0.968
Calcium (Ca)	40.08	842	1484	213.1	1.54	22.3	0.647	2.00
Cerium (Ce)	140.11	798	3443	39.0	6.77	6.3	0.192	0.113
Cesium (Cs)	132.91	28.44	671	15.7	1.93	97	0.242	0.359
Chromium (Cr)	52.00	1907	2671	404	7.15	4.9	0.449	0.937
Cobalt (Co)	58.93	1495	2927	272.5	8.86	13.0	0.421	1.00
Copper (Cu)	63.55	1084.62	2562	203.5	8.96	16.5	0.385	4.01
Dysprosium (Dy)	162.50	1412	2567	68.1	8.55	9.9	0.170	0.107
Erbium (Er)	167.26	1529	2868	119	9.07	12.2	0.168	0.145
Europium (Eu)	151.96	822	1529	60.6	5.24	35.0	0.182	0.139[a]
Gadolinium (Gd)	157.25	1313	3273	63.6	7.90	9.4[b]	0.236	0.105
Gallium (Ga)	69.72	29.76	2204	80.0	5.91	18	0.371	0.406
Gold (Au)	196.97	1064.18	2856	64.6	19.3	14.2	0.129	3.17
Hafnium (Hf)	178.49	2233	4603	152.4	13.3	5.9	0.144	0.230
Holmium (Ho)	164.93	1474	2700	103[a]	8.80	11.2	0.165	0.162
Indium (In)	114.82	156.60	2072	28.6	7.31	32.1	0.233	0.816
Iridium (Ir)	192.22	2446	4428	213.9	22.5	6.4	0.131	1.47
Iron (Fe)	55.85	1538	2861	247.3	7.87	11.8	0.449	0.802
Lanthanum (La)	138.91	918	3464	44.6	6.15	12.1	0.195	0.134
Lead (Pb)	207.20	327.46	1749	23.1	11.3	28.9	0.129	0.353
Lithium (Li)	6.94	180.5	1342	432	0.534	46	3.582	0.847
Lutetium (Lu)	174.97	1663	3402	126[a]	9.84	9.9	0.154	0.164
Magnesium (Mg)	24.30	650	1090	348.9	1.74	24.8	1.023	1.56

THERMAL AND PHYSICAL PROPERTIES OF PURE METALS (continued)

Metal (symbol)	Atomic weight	t_m °C	t_b °C	$\Delta_{fus} H$ J/g	ρ_{25} g/cm³	$\alpha \times 10^6$ K⁻¹	c_p J/g K	λ W/cm K
Manganese (Mn)	54.94	1246	2061	235.0	7.3	21.7	0.479	0.0782
Mercury (Hg)	200.59	-38.83	356.73	11.4	13.5336	60.4	0.140	0.0834
Molybdenum (Mo)	95.94	2623	4639	390.7	10.2	4.8	0.251	1.38
Neodymium (Nd)	144.24	1021	3074	49.5	7.01	9.6	0.190	0.165
Neptunium (Np)		644			13.5	20.2		0.063
Nickel (Ni)	58.69	1455	2913	290.3	8.90	13.4	0.444	0.907
Niobium (Nb)	92.91	2477	4744	323	8.57	7.3	0.265	0.537
Osmium (Os)	190.23	3033	5012	304.1	22.59	5.1	0.130	0.876
Palladium (Pd)	106.42	1554.9	2963	157.3	12.0	11.8	0.246	0.718
Platinum (Pt)	195.08	1768.4	3825	113.6	21.5	8.8	0.133	0.716
Plutonium (Pu)		640	3228	11.6	19.7	46.7		0.0674
Polonium (Po)		254	962		9.20	23.5		0.20
Potassium (K)	39.10	63.38	759	59.6	0.89	83.3	0.757	1.024
Praseodymium (Pr)	140.91	931	3520	48.9	6.77	6.7	0.193	0.125
Promethium (Pm)		1042	3000ᵃ		7.26	11ᵃ	0.19ᵃ	0.15ᵃ
Protactinium (Pa)	231.04	1572		53.4	15.4			
Radium (Ra)		700			5			
Rhenium (Re)	186.21	3186	5596	324.5	20.8	6.2	0.137	0.479
Rhodium (Rh)	102.91	1964	3695	258.4	12.4	8.2	0.243	1.50
Rubidium (Rb)	85.47	39.30	688	25.6	1.53		0.363	0.582
Ruthenium (Ru)	101.07	2334	4150	381.8	12.1	6.4	0.238	1.17
Samarium (Sm)	150.36	1074	1794	57.3	7.52	12.7	0.197	0.133
Scandium (Sc)	44.96	1541	2836	314	2.99	10.2	0.568	0.158
Silver (Ag)	107.87	961.78	2162	104.6	10.5	18.9	0.235	4.29
Sodium (Na)	22.99	97.72	883	113.1	0.97	71	1.228	1.41
Strontium (Sr)	87.62	777	1382	84.8	2.64	22.5	0.301	0.353
Tantalum (Ta)	180.95	3017	5458	202.1	16.4	6.3	0.140	0.575
Technetium (Tc)		2157	4265	339.7	11			0.506
Terbium (Tb)	158.93	1356	3230	67.9	8.23	10.3	0.182	0.111
Thallium (Tl)	204.38	304	1473	20.3	11.8	29.9	0.129	0.461
Thorium (Th)	232.04	1750	4788	59.5	11.7	11.0	0.113	0.540
Thulium (Tm)	168.93	1545	1950	99.7	9.32	13.3	0.160	0.169
Tin (Sn)	118.71	231.93	2602	60.4	7.26	22.0	0.228	0.666
Titanium (Ti)	47.88	1668	3287	295.6	4.51	8.6	0.523	0.219
Tungsten (W)	183.84	3422	5555	284.5	19.3	4.5	0.132	1.74
Uranium (U)	238.03	1135	4131	38.4	19.1	13.9	0.116	0.276
Vanadium (V)	50.94	1910	3407	422	6.0	8.4	0.489	0.307
Ytterbium (Yb)	173.04	819	1196	44.3	6.90	26.3	0.155	0.385
Yttrium (Y)	88.91	1522	3345	128	4.47	10.6	0.298	0.172
Zinc (Zn)	65.39	419.53	907	108.1	7.14	30.2	0.388	1.16
Zirconium (Zr)	91.22	1855	4409	230.2	6.52	5.7	0.278	0.227

ᵃ Estimated.
ᵇ At 100°C.

THERMAL CONDUCTIVITY OF METALS AND SEMICONDUCTORS AS A FUNCTION OF TEMPERATURE

This table gives the temperature dependence of the thermal conductivity of several metals and of carbon, germanium, and silicon. For graphite, separate entries are given for the thermal conductivity parallel (∥) and perpendicular (⊥) to the layer planes. The thermal conductivity of all these materials is very sensitive to impurities at low temperatures, especially below 100 K. Therefore, the values given here should be regarded as typical values for a highly purified specimen; the thermal conductivity of different specimens can vary by more than an order of magnitude in the low-temperature range. See Reference 2 for details.

REFERENCES

1. Ho, C. Y., Powell, R. W., and Liley, P. E., *J. Phys. Chem. Ref. Data*, 1, 279, 1972.
2. White, G. K., and Minges, M. L., *Thermophysical Properties of Some Key Solids*, CODATA Bulletin No. 59, 1985.

Thermal Conductivity in W/cm K

T/K	Ag	Al	Au	Carbon (C) Diamond (type) I	IIa	IIb	Pyrolytic graphite ∥	⊥	Cr	Cu
1	39.4	41.1	5.46						0.402*	42.2
2	78.3	81.8	10.9	0.0138*	0.033*	0.0200*			0.803	84.0
3	115	121	16.1	0.0461	0.111	0.0676			1.20	125
4	147	157	20.9	0.108	0.261	0.160			1.60	162
5	172	188	25.2	0.206	0.494	0.307			2.00	195
6	187	213	28.5	0.344	0.820	0.510			2.39	222
7	193	229	30.9	0.523	1.24	0.778			2.27	239
8	190	237	32.3	0.762	1.77	1.12			3.14	248
9	181	239	32.7	1.05	2.41	1.53			3.50	249
10	168	235	32.4	1.40	3.17	2.03	0.811	0.0116	3.85	243
15	96.0	176	24.6	3.96	8.65	5.66			5.24	171
20	51.0	117	15.8	7.87	16.8	11.2	4.20	0.0397	5.93	108
30	19.3	49.5	7.55	18.8	38.9	26.5	9.86	0.0786	5.49	44.5
40	10.5	24.0	5.15	29.4	65.9	44.0	16.4	0.120	4.25	21.7
50	7.0	13.5	4.21	35.3	92.1	59.1	23.1	0.152	3.17	12.5
60	5.5	8.5	3.74	37.4	112	67.5	29.8	0.173	2.48	8.29
70	4.97	5.85	3.48	36.9	119	69.1	36.6	0.181	2.07	6.47
80	4.71	4.32	3.32	35.1	117	65.7	42.8	0.181	1.84	5.57
90	4.60	3.42	3.28	32.7	109	60.0	47.5	0.176	1.69	5.08
100	4.50	3.02	3.27	30.0	100	54.2	49.7	0.168	1.59	4.82
150	4.32	2.48	3.25	19.5	60.2	32.5	45.1	0.125	1.29	4.29
200	4.30	2.37	3.23	14.1	40.3	22.6	32.3	0.0923	1.11	4.13
250	4.29	2.35	3.21	11.0	29.7	17.0	24.4	0.0711	1.00	4.06
300	4.29	2.37	3.17	8.95	23.0	13.5	19.5	0.0570	0.937	4.01
350	4.27	2.40	3.14	7.55*	18.5*	11.1*	16.2	0.0477	0.929	3.96
400	4.25	2.40	3.11	6.5*	15.4*	9.32*	13.9	0.0409	0.909	3.93
500	4.19	2.36	3.04				10.8	0.0322	0.860	3.86
600	4.12	2.31	2.98				8.92	0.0268	0.807	3.79
800	3.96	2.18	2.84				6.67	0.0201	0.713	3.66
1000	3.79		2.70				5.34	0.0160	0.654	3.52
1200	3.61*		2.55				4.48	0.0134	0.619	3.39
1400							3.84	0.0116	0.588	
1600							3.33	0.0100	0.556	
1800							2.93	0.00895	0.526*	
2000							2.62	0.00807	0.494*	

THERMAL CONDUCTIVITY OF METALS AND SEMICONDUCTORS AS A FUNCTION OF TEMPERATURE (continued)

T/K	Fe	Ge[a]	Mg	Ni	Pb	Pt	Si[a]	Sn	Ti	W
1	1.71	0.274	9.86	2.17	27.9	2.31	0.0693*	183	0.0144*	14.4
2	3.42	2.06	19.6	4.34	44.6	4.60	0.454	323	0.0288*	28.7
3	5.11	5.35	29.0	6.49	35.8	6.79	1.38	297	0.0432	42.8
4	6.77	8.77	37.6	8.59	22.2	8.8	2.97	181	0.0575	56.3
5	8.39	11.6	45.0	10.6	13.8	10.5	5.27	117	0.0719	68.7
6	9.93	13.9	50.8	12.5	8.10	11.8	8.23	76	0.0863	79.5
7	11.4	15.5	54.7	14.2	4.86	12.6	11.7	52	0.101	88.0
8	12.7	16.6	56.7	15.8	3.20	12.9	15.5	36	0.115	93.8
9	13.9	17.3	57.0	17.1	2.30	12.8	19.5	26	0.129	96.8
10	14.8	17.7	55.8	18.1	1.78	12.3	23.3	19.3	0.143	97.1
15	17.0	17.3	41.1	19.5	0.845	8.41	41.6	6.3	0.212	72.0
20	15.4	14.9	27.2	16.5	0.591	4.95	49.8	3.2	0.275	40.5
30	10.0	10.8	12.9	9.56	0.477	2.15	48.1	1.79	0.365	14.4
40	6.23	7.98	7.19	5.82	0.451	1.39	35.3	1.33	0.390	6.92
50	4.05	6.15	4.65	4.00	0.436	1.09	26.8	1.15	0.374	4.27
60	2.85	4.87	3.27	3.08	0.425	0.947	21.1	1.04	0.355	3.14
70	2.16	3.93	2.49	2.50	0.416	0.862	16.8	0.96	0.340	2.58
80	1.75	3.25	2.02	2.10	0.409	0.815	13.4	0.915	0.326	2.29
90	1.50	2.70	1.78	1.83	0.403	0.789	10.8	0.880	0.315	2.17
100	1.34	2.32	1.69	1.64	0.397	0.775	8.84	0.853	0.305	2.08
150	1.04	1.32	1.61	1.22	0.379	0.740	4.09	0.779	0.270	1.92
200	0.94	0.968	1.59	1.07	0.367	0.726	2.64	0.733	0.245	1.85
250	0.865	0.749	1.57	0.975	0.360	0.718	1.91	0.696	0.229	1.80
300	0.802	0.599	1.56	0.907	0.353	0.716	1.48	0.666	0.219	1.74
350	0.744	0.495	1.55	0.850	0.347	0.717	1.19	0.642	0.210	1.67
400	0.695	0.432	1.53	0.802	0.340	0.718	0.989	0.622	0.204	1.59
500	0.613	0.338	1.51	0.722	0.328	0.723	0.762	0.596	0.197	1.46
600	0.547	0.273	1.49	0.656	0.314	0.732	0.619		0.194	1.37
800	0.433	0.198	1.46*	0.676		0.756	0.422		0.197	1.25
1000	0.323	0.174		0.718		0.787	0.312		0.207	1.18
1200	0.283	0.174		0.762		0.826	0.257		0.220	1.12
1400	0.312			0.804		0.871	0.235		0.236	1.08
1600	0.330					0.919	0.221		0.253	1.04
1800	0.345*					0.961			0.270*	1.01
2000						0.994*				0.98

[a] Values below 300 K are typical values.
* Extrapolated.

THERMAL CONDUCTIVITY OF ALLOYS AS A FUNCTION OF TEMPERATURE

This table lists the thermal conductivity of selected alloys at various temperatures. The indicated compositions refer to weight percent. Since the thermal conductivity is sensitive to exact composition and processing history, especially at low temperatures, these values should be considered approximate.

REFERENCES

1. Powell, R. L., and Childs, G. E., in *American Institute of Physics Handbook, 3rd Edition*, Gray, D. E., Ed., McGraw-Hill, New York, 1972.
2. Ho, C. Y., et al., *J. Phys. Chem. Ref. Data*, 7, 959, 1978.

Thermal conductivity in W/m K

	Alloy	4 K	20 K	77 K	194 K	273 K	373 K	573 K	973 K
Aluminum:	1100	50	240	270	220	220			
	2024	3.2	17	56	95	130			
	3003	11	58	140	150	160			
	5052	4.8	25	77	120	140			
	5083, 5086	3	17	55	95	120			
	Duralumin	5.5	30	91	140	160	180		
Bismuth:	Rose metal		5.5	8.3	14	16			
	Wood's metal	4	17	23					
Copper:	electrolytic tough pitch	330	1300	550	400	390	380	370	350
	free cutting, leaded	200	800	460	380	380			
	phosphorus, deoxidized	7.5	42	120	190	220			
	brass, leaded	2.3	12	39	70	120			
	bronze, 68% Cu; 32% Zn	2.3	16	48	92	110			
	beryllium	2	17	36	70	90	113	172	
	german silver	0.75	7.5	17	20	23	25	30	40
	silicon bronze A		3.4	11	23	30			
	manganin	0.48	3.2	14	17	22			
	constantan	0.9	8.6	17	19	22			
Ferrous:	commercial pure iron	15	72	106	82	76	66	54	34
	plain carbon steel(AISI 1020)	13	20	58	65	65			
	plain carbon steel(AISI 1095)		8.5	31	41	45			
	3% Ni; 0.7% Cr; 0.6% Mo		6	22		33	35	36	30
	4% Si					20	24	28	26
	stainless steel	0.3	2	8	13	14	16	19	25
	27% Ni; 15% Cr		1.7	55		11	12	16	21
Gold:	colbalt thermocouple	1.2	8.6	20					
	65% Au; 35% Ag		12	24		61	89		
Indium:	85.5% In; 14.5% Pb	1.9	7.8	24	41				
Lead:	60% Pb; 40% Sn (soft solder)		28	44					
	64.35% Pb; 35.65% In	0.8	3.26	9.1		20.2			
Nickel:	80% Ni; 20% Cr					12	14	17	23
	contracid	0.2	2	7.3	9.5	13			
	inconel	0.5	4.2	12.5	13	15	16	19	26
	monel	0.9	7.1	15	20	21	24	30	43
Platinum:	90% Pt; 10% Ir					31	31.4		
	90% Pt; 10% Rh					30.1	30.5		
Silver:	silver solder		12	34	58				
	normal Ag thermocouple	48	230	310					
Tin:	60% Sn; 40% Pb	16	55	51					
Titanium:	5.5% Al; 2.5% Sn;0.2% Fe		1.8	4.3	6.4	7.8	8.4	10.8	
	4.7% Mn; 3.99% Al; 0.14% C		1.7	4.5	6.5	8.5			

THERMAL CONDUCTIVITY OF CRYSTALLINE DIELECTRICS

This table lists the thermal conductivity of a number of crystalline dielectrics, including some which find use as optical materials. Values are given at temperatures for which data are available.

REFERENCE

Powell, R. L., and Childs, G. E., in *American Institute of Physics Handbook, 3rd Edition*, Gray, D. E., Ed., McGraw-Hill, New York, 1972.

Material	T/K	Ther. cond. W/m K	Material	T/K	Ther. cond. W/m K
AgCl	223	1.3	BeO	4.2	0.3
	273	1.2		20	16
	323	1.1		77	270
	373	1.1		373	210
Al,B silicate (tourmaline)	398	2.9		573	120
∥ to c axis	540	3.2		1273	29
	723	3.5	Bi_2Te_3	80	6.4
Al,Be silicate (beryl)	315	6.4		204	2.8
Al,F silicate (topaz)	315	17.7		303	3.6
∥ to c axis	358	15.6		370	4.6
	417	13.3	C (diamond)	4.2	13
Al,Fe silicate (garnet)	315	35.8	type I	20	800
	358	35.4		77	3550
	377	35.6		194	1450
Al_2O_3 (sapphire):				273	1000
36° to c axis	4.2	110	$CaCO_3$		
	20	3500	∥ to c axis	83	25
	35	6000		273	5.5
	77	1100	⊥ to c axis	83	17
⊥ to c axis	373	2.6		194	6.5
	523	3.9		273	4.6
	773	5.8		373	3.6
Al_2O_3 (sintered)	4.2	0.5	CaF_2	83	39
	20	23		223	18
	77	150		273	10
	194	48		323	9.2
	273	35		373	9
	373	26	$CaWO_4$ (scheelite)	422	11.3
	973	8	CdTe	160	7.0
Ar	8	6.0		297	3.6
	10	3.7		422	2.9
	20	1.4	CsBr	223	1.2
	77	0.31		273	0.94
As_2S_3 (glass)	283	0.16		323	0.81
	323	0.21		373	0.77
	373	0.27	CsI	223	1.4
BN	1047	36.2		273	1.2
	1475	22.7		323	1
	1928	21.9		373	0.95
	2111	18.5	Cu_2O (cuprite)	102	3.74
BaF_2	225	20		163	7.76
	260	13.4		299	5.58
	305	10.9		360	4.86
	370	10.5	Fe_3O_4 (magnetite)	4.5	27.4
$BaTiO_3$	5	4.2		20.5	293.0
	30	24.0		126.5	7.4
	40	25.0		304	7.0
	100	12.0	Glass:		
	250	4.8	phoenix	4.2	0.095
	300	6.2		20	0.13
				77	0.37

Material	T/K	Ther. cond. W/m K	Material	T/K	Ther. cond. W/m K
plastic perspex	4.2	0.058	NaCl	4.2	440
	20	0.074		20	300
pyrex	77	0.44		77	30
	194	0.88		273	6.4
	273	1		323	5.6
H_2 (para + 0.5% ortho)	2.5	100		373	5.4
	3	150	NaF	5	1100
	4	200		50	250
	6	30		100	90
	10	3	Ne	2	3.0
H_2O (ice)	173	3.5		3	4.6
	223	2.8		4.2	4.2
	273	2.2		10	0.8
He^3 (high pressure)	0.6	25		20	0.3
	1	2	NH_4Cl	77	17
	1.5	0.57		194	23
	2	0.21		230	38
He^4 (high pressure)	0.5	42		273	27
	0.8	120	$NH_4H_2PO_4$		
	1	24	‖ to optic axis	315	0.71
	2	0.18		339	0.71
I_2	300	0.45	⊥ to optic axis	313	1.26
	325	0.42		342	1.34
	350	0.4	NiO	4.2	5.9
KBr	2	150		40	400
	4.2	360		194	82
	100	12	SiO_2 (quartz)		
	273	5	‖ to c axis	20	720
	323	4.8		194	20
	373	4.8		273	12
KCl	4.2	500	⊥ to c axis	20	370
	25	140		194	10
	80	35		273	6.8
	194	10	SiO_2 (fused silica)	4.2	0.25
	273	7.0		20	0.7
	323	6.5		77	0.8
	373	6.3		194	1.2
KI	4.2	700		273	1.4
	80	13		373	1.6
	194	4.6		673	1.8
	273	3.1	$SrTiO_3$	5	2.4
Kr	4.2	0.48		30	21.0
	10	1.7		40	19.2
	20	1.2		100	18.5
	77	0.36		250	12.5
LaF_3	78	7.8		300	11.2
	197	5.0	TlBr	316	0.59
	274	5.4	TlCl	311	0.75
LiF	4.2	620	TiO_2 (rutile)		
	20	1800	‖ to optic axis	4.2	200
	77	150		20	1000
$MgO \cdot Al_2O_3$ (spinel)	373	13		273	13
	773	8.5	⊥ to optic axis	4.2	160
MnO	4.2	0.25		20	690
	40	55		273	9
	120	8			
	573	3.5			

THERMAL CONDUCTIVITY OF CERAMICS AND OTHER INSULATING MATERIALS

Thermal conductivity values for ceramics, refractory oxides, and miscellaneous insulating materials are given here. The thermal conductivity refers to samples with density indicated in the second column. Since most of these materials are highly variable, the values should only be considered as a rough guide.

REFERENCES

1. Powell, R. L., and Childs, G. E., in *American Institute of Physics Handbook, 3rd Edition*, Gray, D. E., Ed., McGraw-Hill, New York, 1972.
2. Perry, R. H., and Green, D., *Perry's Chemical Engineers' Handbook, Sixth Edition*, McGraw-Hill, New York, 1984.

Material	Dens. g/cm^3	t °C	Ther. cond. W/m K	Material	Dens. g/cm^3	t °C	Ther. cond. W/m K
Alumina (Al$_2$O$_3$)	3.8	100	30	Diatomite	0.2	0	0.05
		400	13			400	0.09
		1300	6		0.5	0	0.09
		1800	7.4			400	0.16
	3.5	100	17	Ebonite	1.2	0	0.16
		800	7.6	Felt, flax	0.2	30	0.05
Al$_2$O$_3$ + MgO		100	15		0.3	30	0.04
		400	10	Fuller's earth	0.53	30	0.1
		1000	5.6	Glass wool	0.2	-200 to 20	0.005
Asbestos	0.4	-100	0.07			50	0.04
		100	1			100	0.05
		0	0.09			300	0.08
Asbestos + 85% MgO	0.3	30	0.08	Graphite			
Asphalt	2.1	20	0.06	100 mesh	0.48	40	0.18
Beryllia (BeO)	2.8	100	210	20-40 mesh	0.7	40	1.29
		400	90	Linoleum cork	0.54	20	0.08
		1000	20	Magnesia (MgO)		100	36
		1800	15			400	18
	1.85	50	64			1200	5.8
		200	40			1700	9.2
		600	23	MgO + SiO$_2$		100	5.3
Brick, dry	1.54	0	0.04			400	3.5
Brick, refractory:						1500	2.3
alosite		1000	1.3	Mica:			
aluminous	1.99	400	1.2	muscovite		100	0.72
		1000	1.3			300	0.65
diatomaceous	0.77	100	0.2			600	0.69
		500	0.24	phlogopite		100	0.66
	0.4	100	0.08	Canadian		300	0.19
		500	0.1			600	0.2
fireclay	2	400	1	Micanite		30	0.3
		1000	1.2	Mineral wool	0.15	30	0.04
silicon carbide	2	200	2	Perlite, expanded	0.1	-200 to 20	0.002
		600	2.4	Plastics:			
vermiculite	0.77	200	0.26	bakelite	1.3	20	1.4
		600	0.31	celluloid	1.4	30	0.02
Calcium oxide		100	16	polystyrene foam	0.05	-200 to 20	0.033
		400	9	mylar foil	0.05	-200 to 20	0.0001
		1000	7.5	nylon		-253	0.10
Cement mortar	2	90	0.55			-193	0.23
Charcoal	0.2	20	0.055			25	0.30
Coal	1.35	20	0.26	polytetrafluoroethylene		-253	0.13
Concrete	1.6	0	0.8			-193	0.16
Cork	0.05	0	0.03			25	0.26
		100	0.04			230	2.5
	0.35	0	0.06	urethane foam	0.07	20	0.06
		100	0.08	Porcelain		90	1
Cotton wool	0.08	30	0.04				

THERMAL CONDUCTIVITY OF CERAMICS AND OTHER INSULATING MATERIALS (continued)

Material	Dens. g/cm^3	t °C	Ther. cond. W/m K	Material	Dens. g/cm^3	t °C	Ther. cond. W/m K
Rock:				Uranium dioxide		100	9.8
basalt		20	2			400	5.5
chalk		20	0.92			1000	3.4
granite	2.8	20	2.2	Wood:			
limestone	2	20	1	balsa, ⊥	0.11	30	0.04
sandstone	2.2	20	1.3	fir, ⊥	0.54	20	0.14
slate, ⊥		95	1.4	fir, ‖	0.54	20	0.35
slate, ‖		95	2.5	oak		20	0.16
Rubber:				plywood		20	0.11
sponge	0.2	20	0.05	pine, ⊥	0.45	60	0.11
92 percent		25	0.16	pine, ‖	0.45	60	0.26
Sand, dry	1.5	20	0.33	walnut, ⊥	0.65	20	0.14
Sawdust	0.2	30	0.06	Wool	0.09	30	0.04
Shellac		20	0.23	Zinc oxide		200	17
Silica aerogel	0.1	-200 to 20	0.003			800	5.3
Snow	0.25	0	0.16	Zirconia (ZrO$_2$)		100	2
Steel wool	0.1	55	0.09			400	2
Thoria (ThO$_2$)		100	10			1500	2.5
		400	5.8	Zirconia + silica		200	5.6
		1500	2.4			600	4.6
Titanium dioxide		100	6.5			1500	3.7
		400	3.8				
		1200	3.3				

THERMAL CONDUCTIVITY OF GLASSES

This table gives the composition of various types of glasses and the thermal conductivity k as a function of temperature. Because of the variability of glasses, the data should be regarded as only approximate.

Type of glass	SiO$_2$ (wt%)	Other oxides (wt%)		t °C	k W/m K
Vitreous silica	100			−150	0.85
				−100	1.05
				−50	1.20
				0	1.30
				50	1.40
				100	1.50
Vycor glass	96	B$_2$O$_3$	3	−100	1.00
				0	1.25
				100	1.40
Pyrex type chemically-resistant borosilicate glasses	80–81	B$_2$O$_3$	12–13	−100	0.90
		Na$_2$O	4	0	1.10
		Al	2	100	1.25
Borosilicate crown glasses	60–65	B$_2$O$_3$	15–20	−100	0.65–0.75
				0	0.90–0.95
				100	1.00–1.05
	65–70	B$_2$O$_3$	10–15	−100	0.75–0.80
				0	0.95–1.00
				100	1.05–1.15
	70–75	B$_3$O$_3$	5–10	−100	0.80–0.85
				0	1.05–1.10
				100	1.15–1.20
Zinc crown glasses (i)	55–65	ZnO Remainder: B$_2$O$_3$, Al$_2$O$_3$	5–15	−100 0 100	0.88–0.92 1.10–1.15 1.15–1.25
		ZnO Remainder: Na$_2$O, K$_2$O	5–15	−100 0 100	0.60–0.70 0.70–0.90 0.85–0.95
		ZnO Remainder: B$_2$O$_3$, Al$_2$O$_3$	15–25	−100 0 100	0.88–0.92 1.10–1.15 1.15–1.20
		ZnO Remainder: Na$_2$O, K$_2$O	15–25	−100 0 100	0.65–0.80 0.85–0.95 0.90–1.05
Zinc crown glasses (ii)	65–75	ZnO Remainder: B$_2$O$_3$, Al$_2$O$_3$	5–15	−100 0 100	0.88–0.92 1.15–1.15 1.20–1.30
		ZnO Remainder: Na$_2$O, K$_2$O	5–15	−100 0 100	0.70–0.85 0.90–1.05 1.00–1.15

Type of glass	Composition			t	k
	SiO_2 (wt%)	Other oxides (wt%)		°C	W/m K
		ZnO	15–25	−100	0.90–0.95
		Remainder:		0	1.15–1.15
		B_2O_3, Al_2O_3		100	1.20–1.25
		ZnO	15–25	−100	0.65–0.85
		Remainder:		0	0.85–1.00
		Na_2O, K_2O		100	1.05–1.20
Barium crown glasses	31	B_2O_3	12	−100	0.55
		Al_2O_3	8	0	0.70
		BaO	48	100	0.80
	41	B_2O_3	6	−100	0.60
		Al_2O_3	2	0	0.75
		ZnO	8	100	0.85
		BaO	43		
	47	B_2O_3	4	−100	0.65
		Na_2O	1	0	0.75
		K_2O	7	100	0.90
		ZnO	8		
		BaO	32		
	65	B_2O_3	2	−100	0.70
		Na_2O	5	0	0.90
		K_2O	15	100	1.00
		ZnO	2		
		BaO	10		
Borate glasses					
Borate flint glass	9	B_2O_3	36	−100	0.55
		Na_2O	1	0	0.65
		K_2O	2	100	0.80
		PbO	36		
		Al_2O_3	10		
		ZnO	6		
Borate flint glass	0	B_2O_3	56	−100	0.50
		Al_2O_3	12	0	0.65
		PbO	32	100	0.85
Borate flint glass	0	B_2O_3	43	−100	0.40
		Al_2O_3	5	0	0.55
		PbO	52	100	0.70
Borate glass	4	B_2O_3	55	−100	0.65
		Al_2O_3	14	0	0.80
		PbO	11	100	0.90
		K_2O	4		
		ZnO	12		
Borate crown glass	0	B_2O_3	64	−100	0.50
		Na_2O	8	0	0.65
		K_2O	3	100	0.85
		BaO	4		
		PbO	3		
		Al_2O_3	18		

Type of glass	Composition SiO$_2$ (wt%)	Other oxides (wt%)		t °C	k W/m K
Light borate crown glass	0	B$_2$O$_3$	69	−100	0.55
		Na$_2$O	8	0	0.70
		BaO	5	100	0.90
		Al$_2$O$_3$	18		
Zinc borate glass	0	B$_2$O$_3$	40	−100	0.65
		ZnO	60	0	0.75
				100	0.85
Phosphate crown glasses Potash phosphate glass	0	P$_2$O$_5$	70	0	0.75
		B$_2$O$_3$	3	100	0.85
		K$_2$O	12		
		Al$_2$O$_3$	10		
		MgO	4		
Baryta phosphate glass	0	P$_2$O$_5$	60	45	0.75
		B$_2$O$_3$	3		
		Al$_2$O$_3$	8		
		BaO	28		
Soda-lime glasses	75	Na$_2$O	17	−100	0.75
		CaO	8	0	0.95
				100	1.10
	75	Na$_2$O	12	−100	0.90
		CaO	13	0	1.10
				100	1.15
	72	Na$_2$O	15	−100	0.80
		CaO	11	0	1.00
		Al$_2$O$_3$	2	100	1.15
	65	Na$_2$O	25	−100	0.65
		CaO	10	0	0.85
				100	0.95
	65	Na$_2$O	15	−100	0.85
		CaO	20	0	1.00
				100	1.10
	60	Na$_2$O	20	−100	0.75
		CaO	20	0	0.90
				100	1.00
Other crown glasses Crown glass	75	Na$_2$O	9	−100	0.80
		K$_2$O	11	0	1.00
		CaO	5	100	1.10
High dispersion crown glass	68	Na$_2$O	16	−100	0.65
		ZnO	3	0	0.85
		PbO	13	100	1.00

| Type of glass | Composition | | | t | k |
	SiO$_2$ (wt%)	Other oxides (wt%)		°C	W/m K
Miscellaneous flint glasses					
(i) Silicate flint glasses					
Light flint glasses	65	PbO	25	−100	0.65–0.70
		Others	10	0	0.88–0.92
				100	1.00–1.05
	55	PbO	35	−100	0.60–0.65
		Others	10	0	0.75–0.85
				100	0.88–0.92
Ordinary flint glass	45	PbO	45	−100	0.50–0.60
		Others	10	0	0.65–0.75
				100	0.80–0.85
Heavy flint glass	35	PbO	60	−100	0.45–0.50
		Others	5	0	0.60–0.65
				100	0.70–0.75
Very heavy flint glasses	25	PbO	73	−100	0.40–0.45
		Others	2	0	0.55–0.60
				100	0.63–0.67
	20	PbO	80	−100	0.40
				0	0.50
				100	0.60
(ii) Borosilicate flint glass	33	B$_2$O$_3$	31	−100	0.65
		PbO	25	0	0.85
		Al$_2$O$_3$	7	100	0.95
		K$_2$O	3		
		Na$_2$O	1		
(iii) Barium flint glass	50	BaO	24	−100	0.60
		PbO	6	0	0.70
		K$_2$O	8	100	0.85
		Na$_2$O	3		
		ZnO	8		
		Sb$_2$O$_3$	1		
Other glasses					
Potassium glass	59	K$_2$O	33	50	0.88–0.92
		CaO	8		
Iron glasses	63	Fe$_2$O$_3$	10	−100	0.80
		Na$_2$O	17	0	0.95
		MgO	4	100	1.05
		CaO	3		
		Al$_2$O$_3$	2		
	67	Fe$_2$O$_3$	15	0	0.88—0.92
		Na$_2$O$_3$	18	100	1.00—1.05
	62	Fe$_2$O$_3$	20	0	0.85—0.90
		Na$_2$O	18	100	0.95—1.00
Rock glasses					
Obsidian				0	1.35
Artificial diabase				100	1.25

COMMERCIAL METALS AND ALLOYS

This table gives typical values of mechanical, thermal, and electrical properties of several common commercial metals and alloys. Values refer to ambient temperature (0 to 25°C). All values should be regarded as typical, since these properties are dependent on the particular type of alloy, heat treatment, and other factors. Values for individual specimens can vary widely.

REFERENCES

1. *ASM Metals Reference Book, Second Edition*, American Society for Metals, Metals Park, OH, 1983.
2. Lynch, C. T., *CRC Practical Handbook of Materials Science*, CRC Press, Boca Raton, FL, 1989.
3. Shackelford, J. F., and Alexander, W., *CRC Materials Science and Engineering Handbook*, CRC Press, Boca Raton, FL, 1991.

Common name	Thermal conductivity W/cm K	Density g/cm^3	Coeff. of linear expansion 10^{-6}/°C	Electrical resistivity $\mu\Omega$ cm	Modulus of elasticity GPa	Tensile strength MPa	Approx. melting point °C
Ingot iron	0.7	7.86	11.7	9.7	205	-	1540
Plain carbon steel AISI-SAE 1020	0.52	7.86	11.7	18	205	450	1515
Stainless steel type 304	0.15	7.9	17.3	72	195	550	1425
Cast gray iron	0.47	7.2	10.5	67	90	180	1175
Malleable iron		7.3	12	30	170	345	1230
Hastelloy C	0.12	8.94	11.3	125	200	780	1350
Inconel	0.15	8.25	11.5	103	200	800	1370
Aluminum alloy 3003, rolled	1.9	2.73	23.2	3.7	70	110	650
Aluminum alloy 2014, annealed	1.9	2.8	23.0	3.4	70	185	650
Aluminum alloy 360	1.5	2.64	21.0	7.5	70	325	565
Copper, electrolytic (ETP)	3.9	8.94	16.5	1.7	120	300	1080
Yellow brass (high brass)	1.2	8.47	20.3	6.4	100	300-800	930
Aluminum bronze	0.7	7.8	16.4	12	120	400-600	1050
Beryllium copper 25	0.8	8.23	17.8	7	130	500-1400	925
Cupronickel 30%	0.3	8.94	16.2		150	400-600	1200
Red brass, 85%	1.6	8.75	18.7	11	90	300-700	1000
Chemical lead	0.35	11.34	29.3	21	13	17	327
Antimonial lead (hard lead)	0.3	10.9	26.5	23	20	47	290
Solder 50-50	0.5	8.89	23.4	15	-	42	215
Magnesium alloy AZ31B	1.0	1.77	26	9	45	260	620
Monel	0.3	8.84	14.0	58	180	545	1330
Nickel (commercial)	0.9	8.89	13.3	10	200	460	1440
Cupronickel 55-45 (constantan)	0.2	8.9	18.8	49	160	-	1260
Titanium (commercial)	1.8	4.5	8.5	43	110	330-500	1670
Zinc (commercial)	1.1	7.14	32.5	6	-	130	419
Zirconium (commercial)	0.2	6.5	5.85	41	95	450	1855

HARDNESS OF MINERALS AND CERAMICS

There are several hardness scales for describing the resistance of a material to indentation or scratching. This table lists a number of common materials in order of increasing hardness. Values are given, when available, on three different hardness scales: the original Mohs Scale (range 1 to 10); the modified Mohs Scale (range 1 to 15), and the Knoop Hardness Scale. In the last case, a load of 100 g is assumed.

REFERENCE

Shackelford, J. F. and Alexander, W., *CRC Materials Science and Engineering Handbook*, CRC Press, Boca Raton, FL, 1991.

Material	Formula	Mohs	Modified mohs	Knoop
Graphite	C	0.5		
Talc	$3MgO \cdot 4SiO_2 \cdot H_2O$	1	1	
Alabaster	$CaSO_4 \cdot 2H_2O$	1.7		
Gypsum	$CaSO_4 \cdot 2H_2O$	2	2	32
Halite (rock salt)	NaCl	2		
Stibnite (antimonite)	Sb_2S_3	2.0		
Galena	PbS	2.5		
Mica		2.8		
Calcite	$CaCO_3$	3	3	135
Barite	$BaSO_4$	3.3		
Marble		3.5		
Aragonite	$CaCO_3$	3.5		
Dolomite	$CaMg(CO_3)_2$	3.5		
Fluorite	CaF_2	4	4	163
Magnesia	MgO	5		370
Apatite	$CaF_2 \cdot 3Ca_3(PO_4)_2$	5	5	430
Opal		5		
Feldspar (orthoclase)	$K_2O \cdot Al_2O \cdot 6SiO_2$	6	6	560
Augite		6		
Hematite	Fe_2O_3	6		750
Magnetite	Fe_3O_4	6		
Rutile	TiO_2	6.2		
Pyrite	FeS_2	6.3		
Agate	SiO_2	6.5		
Uranium dioxide	UO_2	6.7		600
Silica (fused)	SiO_2		7	
Quartz	SiO_2	7	8	820
Flint		7		
Silicon	Si	7		
Andalusite	Al_2OSiO_4	7.5		
Zircon	$ZrSiO_4$	7.5		
Zirconia	ZrO_2			1200
Aluminum nitride	AlN			1225
Beryl	$Be_3Al_2Si_6O_{18}$	7.8		
Beryllia	BeO			1300
Topaz	$Al_2SiO_4(OH,F)_2$	8	9	1340
Garnet	$Al_2O_3 \cdot 3FeO \cdot 3SiO_2$		10	1360
Emery	Al_2O_3 (impure)	8		
Zirconium nitride	ZrN	8+		1510
Zirconium boride	ZrB_2			1560
Titanium nitride	TiN	9		1770
Zirconia (fused)	ZrO_2		11	
Tantalum carbide	TaC			1800
Tungsten carbide	WC			1880
Corundum (alumina)	Al_2O_3	9		2025
Zirconium carbide	ZrC			2150
Alumina (fused)	Al_2O_3		12	
Beryllium carbide	Be_2C			2400

HARDNESS OF MINERALS AND CERAMICS (continued)

Material	Formula	Mohs	Modified mohs	Knoop
Titanium carbide	TiC			2470
Carborundum (silicon carbide)	SiC	9.3	13	2500
Aluminum boride	AlB			2500
Tantalum boride	TaB$_2$			2600
Boron carbide	B$_4$C		14	2800
Boron	B	9.5		
Titanium boride	TiB$_2$			2850
Diamond	C	10	15	7000

Section 13
Polymer Properties

Naming Organic Polymers **13**-1
Solvents for Common Polymers **13**-3
Glass Transition Temperature for Selected Polymers **13**-4
Dielectric Constant of Selected Polymers **13**-12
Pressure-Volume-Temperature Relationship for Polymer Melts **13**-13

NAMING ORGANIC POLYMERS

Robert B. Fox

Organic polymers have been traditionally named on the basis of the monomer used, a hypothetical monomer, or a semisystematic structure. Alternatively, they may be named in the same way as organic compounds, i.e., on the basis of a structure as written. While the former methods, often called "source-based nomenclature", frequently result in ambiguity and multiple names for a single material, the latter method, termed "structure-based nomenclature", generates a sometimes cumbersome unique name for a given polymer, independent of its source. Within their limitations, both types of names are acceptable and well-documented.[1] The use of stereochemical descriptors with both types of polymer nomenclature has been published.[1]

TRADITIONAL POLYMER NAMES

Monomer-Based Names

"Polystyrene" is the name of a polymer made from the monomer styrene. Generally, the name of the monomer should be enclosed in parentheses, as in "poly(methyl methacrylate)" or "poly(4-bromostyrene)" to more clearly identify the monomer. Many names are often possible for a given polymer with this method: "poly(ethylene glycol)", "poly(ethylene oxide)", and "poly(oxirane)" describe the same polymer. Sometimes, the name of a hypothetical monomer is used, as in "poly(vinyl alcohol)". Even though a name like "polyethylene" covers a multitude of materials, the system does provide understandable names when a single monomer is involved in the synthesis of a single polymer. When one monomer can yield more than one polymer, as with 1,3-butadiene, some sort of structural notation must be used to identify the product, and one is not far from a formal structure-based name.

Copolymers, Block Polymers, and Graft Polymers. Monomer-based names are more complex when more than one monomer is involved. Some common polymers have been given names based on an apparent structure, as with "poly(ethylene terephthalate)". A better system has been approved by the IUPAC.[1] With this method, the arrangement of the monomeric units is introduced through use of an italicized connective placed between the names of the monomers. For monomer names represented by A, B, and C, the various types of arrangements are

Type	Connective	Example
unspecified	*-co-*	poly(A-*co*-B)
statistical	*-stat-*	poly(A-*stat*-B)
random	*-ran-*	poly(A-*ran*-B)
alternating	*-alt-*	poly(A-*alt*-B)
periodic	*-per-*	poly(A-*per*-B-*per*-C)
block	*-block-*	polyA-*block*-polyB
graft	*-graft-*	polyA-*graft*-polyB

A long dash may be used in place of the word "block". These connectives can be used in combination and with small joining units. Names such as "poly[ethylene-*ran*-(vinyl chloride)]" and "polystyrene-*block*-dimethylsilylene-*block*-polybutadiene" result.

STRUCTURE-BASED POLYMER NOMENCLATURE

Regular Single-Strand Polymers

Structure-based nomenclature has been approved by the IUPAC[4] and is currently being updated; it is used by *Chemical Abstracts*.[5] Monomer names are not involved. To the extent that a polymer chain can be described by a repeating segment of the chain, it can be named "poly(repeating unit)", where "repeating unit" is, in most cases, a bivalent group.

Since there are usually many possible repeating units in a given chain, it is necessary to select systematically one, called the "constitutional repeating unit" (CRU), to provide a unique and unambiguous name, "poly(CRU)", where "CRU" is a recitation of the names of successive units as one proceeds through the CRU. For this purpose, a portion of the main chain structure that includes at least two repeating sequences is written out. These sequences will be composed of bivalent subunits such as -CH_2-, -O-, and groups from ring systems, each of which can be named by the usual rules of nomenclature.[6,7]

Where a chain is simply one long sequence of a single subunit, that subunit is itself the CRU, as in "poly(methylene)" or "poly(1,4-phenylene)". In chains having more than one kind of subunit, a seniority system is used to determine the beginning of the CRU and the direction to move along the main chain atoms (following the shortest path in rings) to complete the CRU. For citation of the first, most senior, subunit, the descending order of seniority is (1) heterocyclic rings, (2) hetero atoms, (3) carbocyclic rings, and, lowest, (4) acyclic groups containing only carbon atoms.

Within each of these classes, there is a further order of seniority that follows the usual rules of nomenclature. Thus, among hetero atoms, the descending order of seniority is O, S, Se, N, P, As, Si, Ge, B, with other atoms fitted in according to their place in the Periodic Table. The descending order of seniority among heterocyclic rings is (a) a ring system containing N; (b) a ring system containing a hetero atom (other than N) as high as possible in the order of hetero atoms given above; (c) a ring system having the greatest number of rings; (d) a ring system with the largest individual ring; (e) a ring system with the largest number of hetero atoms. Among equivalent ring systems, preference is given to the one having lowest locants for the free valences in the subunit, and among otherwise identical ring systems, the one having least hydrogenation is senior. Lowest locants in unsaturated chains are also given preference. Lowest locants for substituents, independent of kind except for alphabetical order, are the final determinant of seniority.

Direction within the repeating unit depends upon the shortest path, in main chain atoms, from the most senior subunit to another subunit of the same kind or to a subunit next lower in seniority. Once identification of the CRU is established, the name of the CRU is formed from the names of largest possible subunits within the CRU. For example, the main chain of the polymer traditionally named "poly(ethylene terephthalate)" has the structure

...CH$_2$CH$_2$OC(O) —⬡— C(O) [OCH$_2$CH$_2$OC(O)—⬡— C(O)]OCH$_2$CH$_2$OC(O)—⬡— C(O) O...

The CRU is enclosed in brackets and read from left to right. It is selected because -O- is the most senior subunit, and two carbon atoms is the shortest path to the next -O-. The structure-based name of this polymer is then "poly(oxyethyleneoxyterephthaloyl)", not much longer than the traditional name and much more adaptable to the complexities of substitution. As organic nomenclature evolves, more systematic names can be applied to the subunits, such as "ethane-1,2-diyl" instead of "ethylene".

Structure-based nomenclature can also be used when the CRU has no carbon atoms. An example is the polymer traditionally named "poly(dimethylsiloxane)", which on the basis of structure would be named "poly(oxydimethylsilylene)" or "poly(oxydimethylsilanediyl)". This nomenclature method has also been applied to inorganic and coordination polymers[8] and to double-strand (ladder and spiro) organic polymers.[9]

Additional examples of both structure-based and traditional names will be found in the tables of glass transition temperatures (**13**-4 to 11) in this *Handbook* as well as in the *Chemical Abstracts* indexes and in the IUPAC documents.

Irregular Single-Strand Polymers

Polymers that cannot be described by the repetition of a single CRU or comprise units not all connected identically in a directional sense can also be named on a structure basis.[10] These include copolymers, block and graft polymers, star polymers, and chemically modified polymers. They are given names of the type "poly(A/B/C...)", where A, B, C, etc. are the names of the component constitutional units, the number of which are minimized. The constitutional units may include regular or irregular blocks as well as atoms or atomic groupings, and each is named by the method described above or by the rules of organic nomenclature.

The oblique stroke denotes an unspecified arrangement of the units within the main chain. For example, a statistical copolymer derived from styrene and vinyl chloride with the monomeric units joined head-to-tail is named "poly(1-chloroethylene/1-phenylethylene)". A polymer obtained by 1,4 polymerization and both head-to-head and head-to-tail 1,2 polymerization of 1,3-butadiene would be named "poly(but-2-ene-1,4-diyl/1-vinylethylene/2-vinyl-ethylene)". A long dash is used separate components of block polymers, as in "poly(A)—poly(B)—poly(C)" or "poly(A)—X—poly(B)". Graft polymers are named in the same way as a substituted polymer but without the ending "yl" for the grafted chain: "poly[Z/poly(A)Z]" is the name of a regular polymer consisting of Z units in which some have grafts of "poly(A)". Star polymers are treated as a central unit with substituent blocks, as in "tetrakis(polymethylene)silane".

REFERENCES

1. International Union of Pure and Applied Chemistry, *Compendium of Macromolecular Nomenclature*, Blackwell Scientific Publications, Oxford, 1991.
2. International Union of Pure and Applied Chemistry, Stereochemical Definitions and Notations Relating to Polymers (Recommendations 1980), *Pure & Appl. Chem.*, 53, 733—752, 1981.
3. International Union of Pure and Applied Chemistry, Source-Based Nomenclature for Copolymers (Recommendations 1985), *Pure & Appl. Chem.*, 57, 1427—1440, 1985)
4. International Union of Pure and Applied Chemistry, Nomenclature of Regular Single-Strand Organic Polymers (1975), *Pure & Appl. Chem.*, 48, 373—385, 1976.
5. Chemical Abstracts Service, Naming and Indexing of Chemical Substances for Chemical Abstracts, Appendix IV, *Chemical Abstracts 1994 Index Guide.*
6. International Union of Pure and Applied Chemistry, *A Guide to IUPAC Nomenclature of Organic Compounds (1993)*, Blackwell Scientific Publications, Oxford, 1993.
7. International Union of Pure and Applied Chemistry, *Nomenclature of Organic Chemistry, Sections A, B, C, D, E, F, and H*, Pergamon Press, Oxford, 1979.
8. International Union of Pure and Applied Chemistry, Nomenclature of Regular Double-Strand and Quasi-Single-Strand Inorganic and Coordination Polymers (Recommendations 1984), *Pure & Appl. Chem.*, 57, 149—168, 1985.
9. International Union of Pure and Applied Chemistry, Nomenclature of Regular Double-Strand (Ladder and Spiro) Organic Polymers (Recommendations 1993), *Pure & Appl. Chem.*, 65, 1561—1580, 1993.
10. International Union of Pure and Applied Chemistry, Structure-Based Nomenclature for Irregular Single-Strand Organic Polymers (Recommendations 1994), *Pure & Appl. Chem.*, 66, 873—889, 1994.

SOLVENTS FOR COMMON POLYMERS

Abbreviations: HC: hydrocarbons; MEK: methyl ethyl ketone; THF: tetrahydrofuran; DMF: dimethylformamide; DMSO: dimethylsulfoxide

Polyethylene (HDPE)	HC and halogenated HC
Polypropylene (atactic)	HC and halogenated HC
Polybutadiene	HC, THF, ketones
Polystyrene	ethylbenzene, $CHCl_3$, CCl_4, THF, MEK
Polyacrylates	aromatic HC, chlorinated HC, THF, esters, ketones
Polymethacrylates	aromatic HC, chlorinated HC, THF, esters, MEK
Polyacrylamide	water
Poly(vinyl ethers)	halogenated HC, MEK, butanol
Poly(vinyl alcohol)	glycols (hot), DMF
Poly(vinyl acetate)	aromatic HC, chlorinated HC, THF, esters, DMF
Poly(vinyl chloride)	THF, DMF, DMSO
Poly(vinylidene chloride)	THF (hot), dioxane, DMF
Poly(vinyl fluoride)	DMF, DMSO (hot)
Polyacrylonitrile	DMF, DMSO
Poly(oxyethylene)	aromatic HC, $CHCl_3$, alcohols, esters, DMF
Poly(2,6-dimethylphenylene oxide)	aromatic HC, halogenated HC
Poly(ethylene terephthalate)	phenol, DMSO (hot)
Polyurethanes (linear)	aromatic HC, THF, DMF
Polyureas	phenol, formic acid
Polysiloxanes	HC, THF, DMF
Poly[bis(2,2,2-trifluoroethoxy)-phosphazene]	THF, ketones, ethyl acetate

GLASS TRANSITION TEMPERATURE FOR SELECTED POLYMERS

Robert B. Fox

Polymer names are based on the IUPAC structure-based nomenclature system described in the table "Naming Organic Polymers". Within each category, names are listed in alphabetical order. Source-based and trivial names are also given (in italics) for the most common polymers. The table does not include polymers for which T_g is not clearly defined because of variability of structure or because of reactions taking place near the glass transition.

All values of T_g cited in this table have been determined by differential scanning calorimetry (DSC) except those values indicated by:

(D)   dynamic method
(Dil) dilatometry
(M)   mechanical method

Polymer name	Glass transition temperature (T_g/K)
ACYCLIC CARBON CHAINS	
Polyalkadienes	
Poly(alkenylene) *Polyalkadiene* –[CH=CHCH$_2$CH$_2$]–	
Poly(*cis*-1-butenylene)	171
cis-1,3-polybutadiene [PBD]	
Poly(*trans*-1-butenylene)	215
trans-1,3-polybutadiene [PBD]	
Poly(1-chloro-*cis*-1-butenylene)	253
cis-1,3-polychloroprene	
Poly(1-chloro-*trans*-1-butenylene)	233
trans-1,3-polychloroprene	
Poly(1-methyl-*cis*-1-butenylene)	200
cis-1,3-polyisoprene	
Poly(1-methyl-*trans*-1-butenylene)	207
trans-1,3-polyisoprene	
Poly(1,4,4-trifluoro-1-butenylene)	238
Polyalkenes	
Poly(alkylethylene) *Poly(alkylethylene)* -[RCHCH$_2$]-	
Poly(1-benzylethylene)	333
Poly(1-butylethylene)	223
Poly(1-cyclohexylethylene) (atactic)	393
Poly(1-cyclohexylethylene) (isotactic)	406 (D)
Poly(1,1-dimethylethylene)	200
Polyisobutylene [PIB]	
Poly(ethylene)	148
Poly(methylene)	155
Poly(1-phenethylethylene)	283
Poly(propylene) (isotactic)	272
Poly(propylene) (syndiotactic)	ca. 265
Poly[1-(2-pyridyl)ethylene]	377
Poly[1-(4-pyridyl)ethylene]	415
Poly(1-vinylethylene)	273
Polyacrylics	
Poly[1-(alkoxycarbonyl)ethylene] *Poly(alkyl acrylate)* –[(ROCO)CHCH$_2$]–	
Poly[1-(benzyloxycarbonyl)ethylene]	279
Poly[1-(butoxycarbonyl)ethylene]	219 (M)
Poly(butyl acrylate) [PBA]	

Polymer name	Glass transition temperature (T_g/K)
Poly[1-(*sec*-butoxycarbonyl)ethylene]	251
Poly[1-(butoxycarbonyl)-1-cyanoethylene]	358
Poly[1-(butylcarbamoyl)ethylene]	319 (M)
Poly(1-carbamoylethylene)	438
Polyacrylamide [PAM]	
Poly(1-carboxyethylene)	379
Poly(acrylic acid) [PAA]	
Poly[1-(2-chlorophenoxycarbonyl)ethylene]	326
Poly[1-(4-chlorophenoxycarbonyl)ethylene]	331
Poly[1-(4-cyanobenzyloxycarbonyl)ethylene]	317
Poly[1-(2-cyanoethoxycarbonyl)ethylene]	277
Poly[1-(cyanomethoxycarbonyl)ethylene)]	433 Dil
Poly[1-(4-cyanophenoxycarbonyl)ethylene]	363
Poly[1-(cyclohexyloxycarbonyl)ethylene]	292
Poly[1-(2,4-dichlorophenoxycarbonyl)ethylene]	333
Poly[1-(dimethylcarbamoyl)ethylene]	362
Poly[1-(ethoxycarbonyl)ethylene]	249
Poly(ethyl acrylate) [PEA]	
Poly[1-(ethoxycarbonyl)-1-fluoroethylene]	316
Poly[1-(2-ethoxycarbonylphenoxycarbonyl)ethylene]	303
Poly[1-(3-ethoxycarbonylphenoxycarbonyl)ethylene]	297
Poly[1-(4-ethoxycarbonylphenoxycarbonyl)ethylene]	310
Poly[1-(2-ethoxyethoxycarbonyl)ethylene]	223
Poly[1-(3-ethoxypropoxycarbonyl)ethylene]	218
Poly[1-(isopropoxycarbonyl)ethylene]	267-270
Poly[1-(methoxycarbonyl)ethylene]	283
Poly(methyl acrylate) [PMA]	
Poly[1-(2-methoxycarbonylphenoxycarbonyl)ethylene]	319
Poly[1-(3-methoxycarbonylphenoxycarbonyl)ethylene]	311
Poly[1-(4-methoxycarbonylphenoxycarbonyl)ethylene]	340
Poly[1-(2-methoxyethoxycarbonyl)ethylene]	223
Poly[1-(4-methoxyphenoxycarbonyl)ethylene]	324
Poly[1-(3-methoxypropoxycarbonyl)ethylene]	198
Poly[1-(2-naphthyloxycarbonyl)ethylene]	358
Poly[1-(pentachlorophenoxycarbonyl)ethylene]	420
Poly[1-(phenethoxycarbonyl)ethylene]	270
Poly[1-(phenoxycarbonyl)ethylene]	330
Poly[1-(*m*-tolyloxycarbonyl)ethylene]	298
Poly[1-(*o*-tolyloxycarbonyl)ethylene]	325
Poly[1-(*p*-tolyloxycarbonyl)ethylene]	316
Poly[1-(2,2,2-trifluorethoxycarbonyl)ethylene]	263

Polymethacrylics

   Poly[1-(alkoxycarbonyl)-1-methylethylene] *Poly(alkyl methacrylate)* –[(ROCO)(Me)CCH$_2$]–

Poly[1-(benzyloxycarbonyl)-1-methylethylene]	327
Poly[1-(2-bromoethoxycarbonyl)-1-methylethylene]	325
Poly[(1-(butoxycarbonyl)-1-methylethylene]	293
Poly(butyl methacrylate) [PBMA]	
Poly[1-(*sec*-butoxycarbonyl)-1-methylethylene]	333
Poly[1-(*tert*-butoxycarbonyl)-1-methylethylene)]	391
Poly[1-(2-chloroethoxycarbonyl)-1-methylethylene]	ca 315
Poly[1-(2-cyanoethoxycarbonyl)-1-methylethylene]	364
Poly[1-(4-cyanophenoxycarbonyl)-1-methylethylene]	428
Poly[1-(cyclohexyloxycarbonyl)-1-methylethylene] (atactic)	356
Poly[1-(cyclohexyloxycarbonyl)-1-methylethylene)] (isotactic)	324

Polymer name	Glass transition temperature (T_g/K)
Poly[1-(dimethylaminoethoxycarbonyl)-1-methylethylene]	292
Poly[1-(ethoxycarbonyl)-1-ethylethylene]	300
Poly[1-(ethoxycarbonyl)-1-methylethylene] (atactic) *Poly(ethyl methacrylate)* [PEMA]	338
Poly[1-(ethoxycarbonyl)-1-methylethylene] (isotactic)	285
Poly[1-(ethoxycarbonyl)-1-methylethylene)] (syndiotactic)	339
Poly[1-(hexyloxycarbonyl)-1-methylethylene]	268
Poly[1-(isobutoxycarbonyl)-1-methylethylene]	326
Poly[1-(isopropoxycarbonyl)-1-methylethylene]	354
Poly[1-(methoxycarbonyl)-1-methylethylene] (atactic) *Poly(methyl methacrylate)* [PMMA]	378
Poly[1-(methoxycarbonyl)-1-methylethylene)] (isotactic)	311
Poly[1-(methoxycarbonyl)-1-methylethylene)] (syndiotactic)	378
Poly[1-(4-methoxycarbonylphenoxy)-1-methylethylene]	379
Poly[1-(methoxycarbonyl)-1-phenylethylene)] (atactic)	391
Poly[1-(methoxycarbonyl)-1-phenylethylene)] (isotactic)	397
Poly[1-methyl-1-(phenethoxycarbonyl)ethylene]	299
Poly[1-methyl-1-(phenoxycarbonyl)ethylene]	383

Polyvinyl ethers, alcohols, and ketones
    Poly(1-alkoxyethylene) *Poly(alkyl vinyl ether)* –[ROCHCH$_2$]–
    Poly(1-hydroxyethylene) *Poly(vinyl alcohol)* –[HOCHCH$_2$]–
    Poly(1-alkanoylethylene) *Poly(alkyl vinyl ketone)* –[RCOCHCH$_2$]–

Poly(1-butoxyethylene)	218
Poly(1-*sec*-butoxyethylene)	253
Poly(1-*tert*-butoxyethylene)	361
Poly[1-(butylthio)ethylene]	253
Poly(1-ethoxyethylene)	230
Poly[1-(4-ethylbenzoyl)ethylene]	325
Poly(1-hydroxyethylene)	358 (D)
Poly(vinyl alcohol) [PVA]	
Poly(hydroxymethylene)	407
Poly(1-isopropoxyethylene)	270
Poly[1-(4-methoxybenzoyl)ethylene]	319 (M)
Poly(1-methoxyethylene)	242
Poly(methyl vinyl ether) [PMVE]	
Poly[1-(methylthio)ethylene]	272
Poly(1-propoxyethylene)	224
Poly[1-(trifluoromethoxy)trifluoroethylene]	268

Polyvinyl halides and nitriles
    Poly(1-haloethylene) *Poly(vinyl halide)* –[XCHCH$_2$]–
    Poly(1-cyanoethylene) *Poly(acrylonitrile)* –[NCCHCH$_2$]–

Poly(1-chloroethylene)	354
Poly(vinyl chloride) [PVC]	
Poly(chlorotrifluoroethylene)	373
Poly(1-cyanoethylene)	370
Polyacrylonitrile [PAN]	
Poly(1-cyano-1-methylethylene)	393
Polymethacrylonitrile	
Poly(1,1-dichloroethylene)	255
Poly(vinylidene chloride)	
Poly(1,1-difluoroethylene)	ca 233
Poly(vinylidene fluoride)	
Poly(1-fluoroethylene)	314 (M)
Poly(vinyl fluoride)	

Polymer name	Glass transition temperature (T_g/K)
Poly(1-hexafluoropropylene)	425
Poly[1-(2-iodoethyl)ethylene]	343
Poly(tetrafluoroethylene)	(160)
Poly[1-(trifluoromethyl)ethylene]	300

Polyvinyl esters

Poly[1-(alkanoyloxy)ethylene] *Poly(vinyl alkanoate)* –[RCOOCHCH₂]–

Poly(1-acetoxyethylene)	305
Poly(vinyl acetate) [PVAc]	
Poly[1-(benzoyloxy)ethylene]	344
Poly[1-(4-bromobenzoyloxy)ethylene]	365
Poly[1-(2-chlorobenzoyloxy)ethylene]	335
Poly[1-(3-chlorobenzoyloxy)ethylene]	338
Poly[1-(4-chlorobenzoyloxy)ethylene]	357
Poly[1-(cyclohexanoyloxy)ethylene]	349 (M)
Poly[1-(4-ethoxybenzoyloxy)ethylene]	343
Poly[1-(4-ethylbenzoyloxy)ethylene]	326
Poly[1-(4-isopropylbenzoyloxy)ethylene]	342
Poly[1-(2-methoxybenzoyloxy)ethylene]	338
Poly[1-(3-methoxybenzoyloxy)ethylene]	ca 317
Poly[1-(4-methoxybenzoyloxy)ethylene]	360
Poly[1-(4-methylbenzoyloxy)ethylene]	343
Poly[1-(4-nitrobenzoyloxy)ethylene]	395
Poly[1-(propionoyloxy)ethylene]	283 (M)

Polystyrenes

Poly(1-phenylethylene) *Polystyrene* –[C₆H₅CHCH₂]–

Poly[1-(4-acetylphenyl)ethylene]	389 (M)
Poly[1-(4-benzoylphenyl)ethylene]	371 (M)
Poly[1-(4-bromophenyl)ethylene]	391
Poly[1-(4-butoxyphenyl)ethylene]	ca 320 (M)
Poly[1-(4-butoxycarbonylphenyl)ethylene]	349 (M)
Pol[(1-(4-butylphenyl)ethylene]	279
Poly[1-(4-carboxyphenyl)ethylene]	386 (M)
Poly[1-(2-chlorophenyl)ethylene]	392
Poly[1-(3-chlorophenyl)ethylene]	363
Poly[1-(4-chlorophenyl)ethylene]	383
Poly[1-(2,4-dichlorophenyl)ethylene]	406
Poly[1-(2,5-dichlorophenyl)ethylene]	379
Poly[1-(2,6-dichlorophenyl)ethylene]	440
Poly[1-(3,4-dichlorophenyl)ethylene]	401
Poly[1-(2,4-dimethylphenyl)ethylene]	385
Poly[1-(4-(dimethylamino)phenyl)ethylene]	398 (M)
Poly[1-(4-ethoxyphenyl)ethylene]	ca 359 (M)
Poly[1-(4-ethoxycarbonylphenyl)ethylene]	367 (M)
Poly[1-(4-fluorophenyl)ethylene]	368
Poly[1-(4-iodophenyl)ethylene]	429
Poly[1-(4-methoxyphenyl)ethylene]	386
Poly[1-(4-methoxycarbonylphenyl)ethylene]	386 (M)
Poly(1-methyl-1-phenylethylene)	373
Poly(α-methylstyrene)	
Poly[1-(2-(methylamino)phenyl)ethylene]	462 (M)
Poly(1-phenylethylene)	373
Polystyrene [PS]	

Polymer name	Glass transition temperature (T_g/K)
Poly[1-(4-propoxyphenyl)ethylene]	343 (M)
Poly[1-(4-propoxycarbonylphenyl)ethylene]	365 (M)
Poly(1-*o*-tolylethylene)	409

CHAINS WITH CARBOCYCLIC UNITS

Poly(arylenealkylene) $-[-Ar-(CH_2)_n]-$

Poly[1-(2-bromo-1,4-phenylene)ethylene]	353 (M)
Poly[1-(2-chloro-1,4-phenylene)ethylene]	343 (M)
Poly[1-(2-cyano-1,4-phenylene)ethylene]	363 (M)
Poly[1-(2,5-dimethyl-1,4-phenylene)ethylene]	373 (M)
Poly[1-(2-ethyl-1,4-phenylene)ethylene]	298 (M)
Poly[1-(1,4-naphthylene)ethylene]	433 (M)
Poly[1-(1,4-phenylene)ethylene]	ca 353 (M)

CHAINS WITH HETEROATOM UNITS

Main chain oxide units
   Poly(oxyalkylene) *Poly(alkylene oxide)* $-[O(CH_2)_n]-$

Poly[oxy(1,1-bis(chloromethyl)trimethylene)]	265
Poly[oxy(1-(bromomethyl)ethylene)]	259
Poly[oxy(1-(butoxymethyl)ethylene)]	194
Poly[oxy(1-butylethylene)]	203
Poly[oxy(1-*tert*-butylethylene)]	308
Poly[oxy(1-(chloromethyl)ethylene)]	251
Poly(epichlorohydrin)	
Poly[oxy(2,6-dimethoxy-1,4-phenylene)]	440
Poly[oxy(1,1-dimethylethylene)]	264
Poly[oxy(2,6-dimethyl-1,4-phenylene)]	482
Poly[oxy(2,6-diphenyl-1,4-phenylene)]	493
Poly[oxy(1-ethylethylene)]	203
Poly(oxyethylidene)	243
Polyacetaldehyde	
Poly[oxy(1-(methoxymethyl)ethylene)]	211
Poly[oxy(2-methyl-6-phenyl-1,4-phenylene)]	428
Poly[oxy(1-methyltrimethylene)]	223 (D)
Poly[oxy(2-methyltrimethylene)]	218
Poly(oxy-1,4-phenylene)	358
Poly(phenylene oxide) [PPO]	
Poly[oxy(1-phenylethylene)]	313
Poly(oxytetramethylene)	189
Poly(tetrahydrofuran) [PTMO]	
Poly(oxytrimethylene)	195

Main-chain ester or anhydride units
   Poly(oxyalkyleneoxyalkanedioyl) *Poly(alkylene alkanedioate)*--$[O(CH_2)_mOCO(CH_2)_nCO]-$

Poly(oxyadipoyloxydecamethylene)	217
Poly(oxyadipoyloxy-1,4-phenyleneisopropylidene-1,4-phenylene)	341
Poly(oxycarbonyloxy-1,4-phenylene-isopropylidene-1,4-phenylene)	422
Bisphenol A polycarbonate	
Poly(oxycarbonylpentamethylene)	213
Poly(oxycarbonyl-1,4-phenylenemethylene-1,4-phenylene)	395
Poly(oxycarbonyl-1,4-phenyleneisopropylidene-1,4-phenylene)	333

Polymer name	Glass transition temperature (T_g/K)
Poly[oxy(2,6-dimethyl-1,4-phenyleneisopropylidene-3,5-dimethyl-1,4-phenylene)oxysebacoyl]	318
Poly(oxyethylenecarbonyl-1,4-cyclohexylenecarbonyl) (trans)	291
Poly(oxyethyleneoxycarbonyl-1,4-naphthylenecarbonyl)	337
Poly(oxyethyleneoxycarbonyl-1,5-naphthylenecarbonyl)	344
Poly(oxyethyleneoxycarbonyl-2,6-naphthylenecarbonyl)	386
Poly(oxyethyleneoxycarbonyl-2,7-naphthylenecarbonyl)	392
Poly(oxyethyleneoxyterephthaloyl)	342
Poly(ethylene terephthalate) [PET]	
Poly(oxyisophthaloyl)	403 (D)
Poly(oxy(1-oxo-2,2-dimethyltrimethylene))	263
Poly(pivalolactone)	
Poly(oxy-1,4-phenyleneisopropylidene-1,4-phenyleneoxysebacoyl)	280
Poly(oxy-1,4-phenyleneoxy-1,4-phenyleneoxy-carbonyl-1,4-phenylene) [PEEK]	416
Poly(oxypropyleneoxyterephthaloyl)	341
Poly[oxyterephthaloyloxy(2,6-dimethyl-1,4-phenyleneisopropylidene-3,5-dimethyl-1,4-(D)phenylene)]	498
Poly(oxyterephthaloyloxyoctamethylene)	318 (D)
Poly(oxyterephthaloyloxy-1,4-phenyleneisopropylidene-1,4-phenylene)	478
Poly(bisphenol A terephthalate)	
Poly(oxytetramethyleneoxyterephthaloyl)	323
Poly(butylene terephthalate) [PBT]	

Main-chain amide units

     Poly(iminoalkyleneiminoalkanedioyl) *Poly(alkylene alkanediamide)*–[NH(CH$_2$)$_m$NHCO(CH$_2$)$_n$CO]–

Poly(iminoadipoyliminodecamethylene)	313
Nylon 10,6	
Poly(iminoadipoyliminohexamethylene)	ca 323
Nylon 6,6	
Poly(iminoadipoyliminooctamethylene)	318
Nylon 8,6	
Poly[iminoadipoyliminotrimethylene(methylimino)trimethylene]	278
Poly(iminocarbonyl-1,4-cyclohexylenemethylene)	466
Poly[iminocarbonyl-1,4-phenylene(2-oxoethylene)iminohexamethylene]	377
Poly(iminoethylene-1,4-phenyleneethyleneiminosebacoyl)	378 (D)
Poly(iminohexamethyleneiminoazelaoyl)	331
Nylon 6,9	
Poly(iminohexamethyleneiminododecanedioyl)	319
Nylon 6, 12	
Poly(iminohexamethyleneiminopimeloyl)	331
Nylon 6,7	
Poly(iminohexamethyleneiminosebacoyl)	323
Nylon 6,10	
Poly(iminohexamethyleneiminosuberoyl)	330
Nylon 6,8	
Poly(iminoisophthaloylimino-4,4'-biphenylylene)	558
Poly(iminoisophthaloyliminohexamethylene)	390
Poly(iminoisophthaloyliminomethylene-1,4-cyclohexylenemethylene)	481
Poly(iminoisophthaloyliminomethylene-1,3-phenylenemethylene)	438 (M)
Poly[iminomethylene(2,5-dimethyl-1,4-phenylene)methyleneiminosuberoyl]	351
Poly(imino-1,5-naphthyleneiminoisophthaloyl)	598
Poly(imino-1,5-naphthyleneiminoterephthaloyl)	578
Poly(iminooctamethyleneiminodecanedioyl)	333
Nylon 8,10	
Poly(iminooxalyliminohexamethylene)	430
Nylon 6,2	
Poly[imino(1-oxohexamethylene)]	326
Nylon 6	

Polymer name	Glass transition temperature (T_g/K)
Poly[imino(1-oxodecamethylene)]	315
Nylon 10	
Poly[imino(1-oxoheptamethylene)]	325
Nylon 7	
Poly[imino(1-oxo-3-methyltrimethylene]	369
Poly[imino(1-oxononamethylene)]	319
Nylon 9	
Poly[imino(1-oxooctamethylene)]	323
Nylon 8	
Poly[imino(1-oxotrimethylene)]	384
Nylon 3	
Poly(iminopentamethyleneiminoadipoyl)	318
Nylon 5,6	
Poly[iminopentamethyleneiminocarbonyl-1,4-phenylene(2-oxoethylene)]	376
Poly(imino-1,3-phenyleneiminoisophthaloyl)	553 (M)
Poly(imino-1,4-phenyleneiminoterephthaloyl)	618
Poly(iminopimeloyliminoheptamethylene)	328
Nylon 7,7	
Poly(iminoterephthaloylimino-4,4′-biphenylylene)	613
Poly(iminotetramethyleneiminoadipoyl)	316
Nylon 4,6	
Poly[iminotetramethyleneiminocarbonyl-1,4-phenylene(2-oxoethylene)]	357
Poly(iminotrimethyleneiminoadipoyliminotrimethylene)	307
Poly[iminotrimethyleneiminocarbonyl-1,4-phenylene(2-oxoethylene)]	382
Poly(oxy-1,4-phenyleneiminoterephthaloyl-imino-1,4-phenylene)	613
Poly(sulfonylimino-1,4-phenyleneiminoadipoylimino-1,4-phenylene)	467

Main-chain urethane units
   Poly(oxyalkyleneoxycarbonyliminoalkyleneiminocarbonyl)–[O(CH$_2$)$_m$OCONH(CH$_2$)$_n$NHCO]–

Poly(oxyethyleneoxycarbonyliminohexamethyleneiminocarbonyl)	329
Poly[oxyethyleneoxycarbonylimino(6-methyl-1,3-phenylene)iminocarbonyl]	325
Poly(oxyethyleneoxycarbonylimino-1,4-phenylenemethylene-1,4-phenyleneiminocarbonyl)	412
Poly(oxyhexamethyleneoxycarbonyliminohexamethyleneiminocarbonyl)	332
Poly[oxyhexamethyleneoxycarbonylimino(6-methyl-1,3-phenylene)iminocarbonyl]	305
Poly(oxyhexamethyleneoxycarbonylimino-1,4-phenylenemethylene-1,4-phenyleneiminocarbonyl)	364
Poly(oxyoctamethyleneoxycarbonyliminohexamethyleneiminocarbonyl)	331
Poly[oxyoctamethyleneoxycarbonylimino(6-methyl-1,3-phenylene)iminocarbonyl]	337
Poly(oxyoctamethyleneoxycarbonylimino-1,4-phenylenemethylene-1,4-phenyleneiminocarbonyl)	352
Poly(oxytetramethyleneoxycarbonyliminohexamethyleneiminocarbonyl)	332
Poly[oxytetramethyleneoxycarbonylimino(6-methyl-1,3-phenylene)iminocarbonyl]	315
Poly(oxytetramethyleneoxycarbonylimino-1,4-phenylenemethylene-1,4-phenyleneiminocarbonyl)	382

Main-chain siloxanes
   Poly[oxy(dialkylsilylene)] *Poly(dialkylsiloxane)* –[O(R$_2$Si)]–

Poly[oxy(dimethylsilylene)]	148
Poly(dimethylsiloxane) [PDMS]	
Poly[oxy(dimethylsilylene)oxy-1,4-phenylene]	363 (M)
Poly[oxy(dimethylsilylene)oxy-1,4-phenyleneisopropylidene-1,4-phenylene]	318 (M)
Poly[oxy(diphenylsilylene)]	238
Poly(diphenylsiloxane)	
Poly[oxy(diphenylsilylene)-1,3-phenylene]	ca 331
Poly[oxy((methyl)phenylsilylene)]	187
Poly[oxy((methyl)-3,3,3-trifluoropropylsilylene]	<193

Polymer name	Glass transition temperature (T_g/K)
Main-chain sulfur-containing units	
Poly(dithioethylene)	223
Poly(dithiomethylene-1,4-phenylenemethylene)	296
Poly(oxy-4,4′-biphenylylene-1,4-phenylenesulfonyl-1,4-phenylene)	503 (M)
Poly(oxycarbonyloxy-1,4-phenylenethio-1,4-phenylene)	ca 383
Poly(oxyethylenedithioethylene)	220 (M)
Poly[oxy(2-hydroxytrimethylene)oxy-1,4-phenylenesulfonyl-1,4-phenylene]	428
Poly(oxymethyleneoxyethylenedithioethylene)	214
Poly(oxy-1,4-phenylenesulfinyl-1,4-phenyleneoxy-1,4-phenylenecarbonyl-1,4-phenylene)	478 (M)
Poly(oxy-1,4-phenylenesulfinyl-1,4-phenyleneoxy-1,4-phenyleneisopropylidene-1,4-phenylene)	438 (M)
Poly(oxy-1,4-phenylenesulfonyl-1,4-phenylene)	487
Poly(oxy-1,4-phenylenesulfonyl-4,4′-biphenylylenesulfonyl-1,4-phenylene)	533
Poly[oxy-1,4-phenylenesulfonyl-1,4-phenyleneoxy(2,6-dimethyl-1,4-phenylene)isopropylidene (3,5-dimethyl-1,4-phenylene)]	508 (M)
Poly(oxy-1,4-phenylenesulfonyl-1,4-phenyleneoxy-1,4-phenylenecarbonyl-1,4-phenylene)	478 (M)
Poly[oxy-1,4-phenylenesulfonyl-1,4-phenyleneoxy-1,4-phenylene(hexafluoroisopropylidene)1,4-phenylene]	478 (M)
Poly(oxy-1,4-phenylenesulfonyl-1,4-phenyleneoxy-1,4-phenyleneisopropylidene-1,4-phenylene)	449
Poly(oxy-1,4-phenylenesulfonyl-1,4-phenyleneoxy-1.4-phenylenemethylene-1,4-phenylene)	453 (M)
Poly(oxy-1,4-phenylenesulfonyl-1,4-phenyleneoxy-1.4-phenylenethio-1,4-phenylene)	448 (M)
Poly(oxy-1,4-phenylenesulfonyl-1,4-phenyleneoxyterephthaloyl)	522
Poly(oxytetramethylenedithiotetramethylene)	197
Poly(sulfonyl-1,2-cyclohexylene)	401
Poly(sulfonyl-1,3-cyclohexylene)	381
Poly(sulfonyl-1,4-phenylenemethylene-1,4-phenylene)	497
Poly(thio-1,3-cyclohexylene)	221
Poly[thio(difluoromethylene)]	155
Poly(thioethylene)	223
Poly[thio(1-ethylethylene]	218
Poly[thio(1-methyl-3-oxotrimethylene)]	285
Poly[thio(1-methyltrimethylene)]	214
Pol[(thio(1-oxohexamethylene)]	292
Poly(thio-1,4-phenylene)	370
Poly(thiopropylene)	226
Main-chain heterocyclic units	
Poly(1,3-dioxa-4,6-cyclohexylenemethylene)	378
Poly(vinyl formal)	
Poly[(2,6-dioxopiperidine-1,4-diyl)trimethylene]	363
Poly[(2-methyl-1,3-dioxa-4,6-cyclohexylene)methylene]	355
Poly(vinyl acetal)	
Poly(1,4-piperazinediylcarbonyloxyethyleneoxycarbonyl)	333
Poly(1,4-piperazinediylisophthaloyl)	465 (M)
Poly[(2-propyl-1,3-dioxa-4,6-cyclohexylene)methylene]	322
Poly(vinyl butyral)	
Poly(3,6-pyridazinediyloxy-1,4-phenyleneisopropylidene-1,4-phenyleneoxy)	453 (M)
Poly(2,5-pyridinediylcarbonyliminohexamethyleneiminocarbonyl)	322

DIELECTRIC CONSTANT OF SELECTED POLYMERS

This table lists typical values of the dielectric constant (more properly called relative permittivity) of some important polymers. Values are given for frequencies of 1 kHz, 1 MHz, and 1 GHz; in most cases the dielectric constant at frequencies below 1 kHz does not differ significantly from the value at 1 kHz. Since the dielectric constant of a polymeric material can vary with density, degree of crystallinity, and other details of a particular sample, the values given here should be regarded only as typical or average values.

REFERENCES

1. Gray, D.E., Ed., *American Institute of Physics Handbook, Third Edition*, p. 5-132, McGraw Hill, New York, 1972.
2. Anderson, H.L., Editor, *A Physicist's Desk Reference*, American Institute of Physics, New York, 1989.
3. Brandrup, J., and Immergut, E.H., *Polymer Handbook, Third Edition*, John Wiley & Sons, New York, 1989.

Name	$t/°C$	1 kHz	1 MHz	1 GHz
Polyacrylonitrile	25	5.5	4.2	
Polyamides (nylons)	25	3.50	3.14	2.8
	84	11	4.4	2.8
Polybutadiene	25	2.5		
Polycarbonate	23	2.92	2.8	
Polychloroprene (neoprene)	25	6.6	6.3	4.2
Polychlorotrifluoroethylene	23	2.65	2.46	2.39
Polyethylene	23	2.3		
Poly(ethylene terephthalate) (Mylar)	23	3.25	3.0	2.8
Polyisoprene (natural rubber)	27	2.6	2.5	2.4
Poly(methyl methacrylate)	27	3.12	2.76	2.6
	80	3.80	2.7	2.6
Polyoxymethylene (polyformaldehyde)	25	3.8		
Poly(phenylene oxide)	23	2.59	2.59	
Polypropylene	25	2.3	2.3	2.3
Polystyrene	25	2.6	2.6	2.6
Polysulfones	25	3.13	2.10	
Polytetrafluoroethylene (teflon)	25	2.1	2.1	2.1
Poly(vinyl acetate)	50		3.5	
	150		8.3	
Poly(vinyl chloride)	25	3.39	2.9	2.8
	100	7.8	3.3	2.7
Poly(vinylidene chloride)	23	4.6	3.2	2.7
Poly(vinylidene fluoride)	23	12.2	8.9	4.7

PRESSURE-VOLUME-TEMPERATURE RELATIONSHIP FOR POLYMER MELTS

Christian Wohlfarth

Numerous theoretical equations of state for polymer liquids have been developed. These, at the minimum, have to provide accurate fitting functions to experimental data. However, for the purpose of this table, the empirical Tait equation along with a polynomial expression for the zero pressure isobar is used. This equation is able to represent the experimental data for the melt state within the limits of experimental errors, i.e., the maximum deviations between measured and calculated specific volumes are about 0.001-0.002 cm^3/g.

The general form of the Tait equation is:

$$V(P,T) = V(0,T)\{1 - C \ln[1 + P/B(T)]\} \tag{1}$$

where the coefficient C is usually taken to be a universal constant equal to 0.0894. T is the absolute temperature in K and P the pressure in MPa. The volume V is the specific volume in cm^3/g. The Tait parameter $B(T)$ has the very simple meaning that it is inversely proportional to the compressibility κ at constant temperature and zero pressure:

$$\kappa(0,T) = -[1/V(0,T)](dV/dP) = C/B(T) \tag{2}$$

The $B(T)$ function is usually given by:

$$B(T) = B_0 \exp[-B_1(T-273.15)] \tag{3}$$

but, sometimes a polynomial expression is used:

$$B(T) = b_0 + b_1(T-273.15) + b_2(T-273.15)^2 \tag{4}$$

The zero-pressure isobar $V(0,T)$ is usually given by:

$$V(0,T) = A_0 + A_1(T-273.15) + A_2(T-273.15)^2 \tag{5}$$

where A_0, A_1, A_2 are specific constants for a given polymer (the expression T-273.15 is used because fitting to the zero-pressure isobar is usually done in terms of Celsius temperature). Other forms for $V(0,T)$ are also found in the literature, such as

$$V(0,T) = A_3 \exp[A_4(T-273.15)] \tag{6}$$

or

$$V(0,T) = A_5 \exp(A_6 T^{1.5}) \tag{7}$$

where A_3 and A_4 or A_5 and A_6 are again specific constants for a given polymer.

The Tait equation is particularly useful to calculate derivative quantities, such as the isothermal compressibility and the thermal expansivity coefficients. The isothermal compressibility $\kappa(P,T)$ is derived from equation (1) as:

$$\kappa(P,T) = -(1/V)(dV/dP) = 1/\{[P + B(T)][1/C - \ln(1 + P/B(T))]\} \tag{8}$$

and the thermal expansivity $\alpha(P,T)$ as:

$$\alpha(P,T) = (1/V)(dV/dT) = \alpha(0,T) - PB_1\kappa(P,T) \tag{9}$$

where $\alpha(0,T)$ represents the thermal expansivity at zero (atmospheric) pressure and is calculated from any suitable fit for the zero-pressure volume, such as equations (5) through (7) above.

Because polymer melt PVT-behavior depends only slightly on polymer molar mass above the oligomeric region, usually no information is given in the original literature for the average molar mass of the polymers.

Table 1 summarizes the polymers or copolymers considered here and the experimental ranges of pressure and temperature over which data are available. In Table 2 the Tait-equation functions, with parameters obtained from the fit, are given for 90 polymer or copolymer melts.

REFERENCES

1. Zoller, P., *J. Appl. Polym. Sci.* 23, 1051-1056, 1979.
2. Starkweather, H. W., Jones, G. A., and Zoller, P., *J. Polym. Sci., Pt. B Polym. Phys.* 26, 257-266, 1988.
3. Fakhreddine, Y. A., and Zoller, P., *J. Polym. Sci., Pt. B Polym. Phys.* 29, 1141-1146, 1991.
4. Rodgers, P. A., *J. Appl. Polym. Sci.* 48, 1061-1080, 1993.
5. Rodgers, P. A., *J. Appl. Polym. Sci.* 48, 2075-2083, 1993.
6. Yi, Y. X., and Zoller, P., *J. Polym. Sci., Pt. B Polym. Phys.* 31, 779-788, 1993.
7. Callaghan, T. A., and Paul, D. R., *Macromolecules* 26, 2439–2450, 1993.
8. Wang, Y. Z., Hsieh, K. H., Chen, L. W.,and Tseng, H. C., *J. Appl. Polym. Sci.* 53, 1191-1201, 1994.
9. Privalko, V. P., Arbuzova, A. P., Korskanov, V. V., and Zagdanskaya, N. E., *Polym. Intern.* 35, 161-169, 1994.
10. Sachdev, V. K., Yashi, U., and Jain, R. K., *J. Polym. Sci., Pt. B Polym. Phys.* 36, 841-850, 1998.

Table 1

Names of the Polymers, Abbreviation Used, and Range of Experimental Data Applied in the Determination of the Equation Constants

Polymer	Symbol	T/K	P/MPa	Ref.
Ethylene/propylene copolymer (50 wt%)	EP50	413-523	0.1-63	4
Ethylene/vinyl acetate copolymer				
18 wt% vinyl acetate	EVA18	385-491	0.1-177	4
25 wt% vinyl acetate	EVA25	367-506	0.1-177	4
28 wt% vinyl acetate	EVA28	367-508	0.1-177	4
40 wt% vinyl acetate	EVA40	348-508	0.1-177	4
Polyamide-6	PA6	509-569	0.1-196	4
Polyamide-11	PA11	478-542	0.1-200	5
Polyamide-66	PA66	519-571	0.1-196	4
cis-1,4-Polybutadiene	cPBD	277-328	0.1-284	4
Polybutadiene, 8% 1,2-content	PBD-8	298-473	0.1-200	6
Polybutadiene, 24% 1,2-content	PBD-24	298-473	0.1-200	6
Polybutadiene, 40% 1,2-content	PBD-40	298-473	0.1-200	6
Polybutadiene, 50% 1,2-content	PBD-50	298-473	0.1-200	6
Polybutadiene, 87% 1,2-content	PBD-87	298-473	0.1-200	6
Poly(1-butene), isotactic	iPB	406-519	0.1-196	4
Poly(butyl methacrylate)	PnBMA	307-473	0.1-200	4
Poly(butylene terephthalate)	PBT	508-576	0.1-200	3
Poly(ε-caprolactone)	PCL	373-421	0.1-200	4
Polycarbonate-bisphenol-A	PC	424-613	0.1-177	4
Polycarbonate-bisphenol-chloral	BCPC	428-557	0.1-200	4
Polycarbonate-hexafluorobisphenol-A	HFPC	432-553	0.1-200	4
Polycarbonate-tetramethylbisphenol-A	TMPC	491-563	0.1-160	4
Poly(cyclohexyl methacrylate)	PcHMA	396-471	0.1-200	4
Poly(2,5-dimethylphenylene oxide)	PPO	473-593	0.1-177	4
Poly(dimethyl siloxane)	PDMS	298-343	0.1-100	4
Poly(dimethyl siloxane) M_n = 1000	PDMS-10	304-420	0.1-250	10
Poly(dimethyl siloxane) M_n = 4000	PDMS-40	298-418	0.1-250	10
Poly(dimethyl siloxane) M_n = 6000	PDMS-60	291-423	0.1-250	10
Poly(epichlorohydrin)	PECH	333-413	0.1-200	4
Poly(ether ether ketone)	PEEK	619-671	0.1-200	4
Poly(ethyl acrylate)	PEA	310-490	0.1-196	4
Poly(ethyl methacrylate)	PEMA	386-434	0.1-196	4
Polyethylene, high density	HDPE	413-476	0.1-196	4
Polyethylene, linear	LPE	415-473	0.1-200	4
Polyethylene, linear, high MW	HMLPE	410-473	0.1-200	4
Polyethylene, branched	BPE	398-471	0.1-200	4
Polyethylene, low density	LDPE	394-448	0.1-196	4
Polyethylene, low density, type A	LDPE-A	385-498	0.1-196	1
Polyethylene, low density, type B	LDPE-B	385-498	0.1-196	1
Polyethylene, low density, type C	LDPE-C	385-498	0.1-196	1
Poly(ethylene oxide)	PEO	361-497	0.1- 68	4
Poly(ethylene terephthalate)	PET	547-615	0.1-196	4
Poly(4-hexylstyrene)	P4HS	303-403	30-100	4
Polyisobutylene	PIB	326-383	0.1-100	4
Polyisoprene, 8% 3,4-content	PI-8	298-473	0.1-200	6
Polyisoprene, 14% 3,4-content	PI-14	298-473	0.1-200	6
Polyisoprene, 41% 3,4-content	PI-41	298-473	0.1-200	6
Polyisoprene, 56% 3,4-content	PI-56	298-473	0.1-200	6
Poly(methyl acrylate)	PMA	310-493	0.1-196	4

Table 1

Names of the Polymers, Abbreviation Used, and Range of Experimental Data Applied in the Determination of the Equation Constants

Polymer	Symbol	T/K	P/MPa	Ref.
Poly(methyl methacrylate)	PMMA	387-432	0.1-200	4
Poly(4-methyl-1-pentene)	P4MP	514-592	0.1-196	4
Poly(α-methylstyrene)	PαMS	473-533	0.1-170	7
Poly(o-methylstyrene)	PoMS	412-471	0.1-180	4
Polyoxymethylene	POM	463-493	0.1-196	2
Phenoxy[a]	PH	341-573	0.1-177	4
Polysulfone[b]	PSF	475-644	0.1-196	4
Polyarylate[c]	PAr	450-583	0.1-177	4
Polypropylene, atactic	aPP	353-393	0.1-100	4
Polypropylene, isotactic	iPP	443-570	0.1-196	4
Polystyrene	PS	388-469	0.1-200	4
Poly(tetrafluoroethylene)	PTFE	603-645	0.1- 39	4
Poly(tetrahydrofuran)	PTHF	335-439	0.1- 78	4
Poly(vinyl acetate)	PVAc	308-373	0.1- 80	4
Poly(vinyl chloride)	PVC	373-423	0.1-200	4
Poly(vinyl methyl ether)	PVME	303-471	0.1-200	4
Poly(vinylidene fluoride)	PVdF	451-521	0.1-200	5
Styrene/acrylonitrile copolymer				
2.7 wt% acrylonitrile	SAN3	378-539	0.1-200	4
5.7 wt% acrylonitrile	SAN6	370-540	0.1-200	4
15.3 wt% acrylonitrile	SAN15	405-531	0.1-200	4
18.0 wt% acrylonitrile	SAN18	377-528	0.1-200	4
40 wt% acrylonitrile	SAN40	373-543	0.1-200	4
70 wt% acrylonitrile	SAN70	373-544	0.1-200	4
Styrene/butadiene copolymer				
10 wt% styrene	SBR10	393-533	0.1-196	8
23.5 wt% styrene	SBR23	393-533	0.1-196	8
60 wt% styrene	SBR60	393-533	0.1-196	8
85 wt% styrene	SBR85	393-533	0.1-196	8
Styrene/methyl methacrylate copolymer				
20 wt% methyl methacrylate	SMMA20	383-543	0.1-200	4
60 wt% methyl methacrylate	SMMA60	383-543	0.1-200	4
N-Vinylcarbazole/4-ethylstyrene copolymer				
50 mol% ethylstyrene	VCES50	393-443	30-100	9
N-Vinylcarbazole/4-hexylstyrene copolymer				
80 mol% hexylstyrene	VCHS80	313-423	30-100	9
67 mol% hexylstyrene	VCHS67	333-423	30-100	9
60 mol% hexylstyrene	VCHS60	383-453	30-100	9
50 mol% hexylstyrene	VCHS50	373-443	30-100	9
40 mol% hexylstyrene	VCHS40	423-493	30-100	9
33 mol% hexylstyrene	VCHS33	463-523	30-100	9
20 mol% hexylstyrene	VCHS20	473-523	30-100	9
N-Vinylcarbazole/4-octylstyrene copolymer				
50 mol% octylstyrene	VCOS50	403-453	30-100	9
N-Vinylcarbazole/4-pentylstyrene copolymer				
50 mol% pentylstyrene	VCPS50	383-443	30-100	9

[a]Phenoxy = Poly(oxy-2-hydroxytrimethyleneoxy-1,4-phenyleneisopropylidene-1,4-phenylene)

[b]Polysulfone = Poly(oxy-1,4-phenylenesulfonyl-1,4-phenyleneoxy-1,4-phenyleneisopropylidene-1,4-phenylene)

[c]Polyarylate = Poly(oxyterephthaloyl/isophthaloyl T/I=50/50)oxy-1,4-phenyleneisopropylidene-1,4-phenylene

Table 2
Tait Equation Parameter Functions for Polymer Melts

Polymer	$V(0,T)/\text{cm}^3\text{g}^{-1}$	$B(T)/\text{MPa}$
EP50	$1.2291 + 5.799 \cdot 10^{-5}(T-273.15) + 1.964 \cdot 10^{-6}(T-273.15)^2$	$487.0 \exp[-8.103 \cdot 10^{-3}(T-273.15)]$
EVA18	$1.02391 \exp(2.173 \cdot 10^{-5}T^{1.5})$	$188.2 \exp[-4.537 \cdot 10^{-3}(T-273.15)]$
EVA25	$1.00416 \exp(2.244 \cdot 10^{-5}T^{1.5})$	$184.4 \exp[-4.734 \cdot 10^{-3}(T-273.15)]$
EVA28	$1.00832 \exp(2.241 \cdot 10^{-5}T^{1.5})$	$183.5 \exp[-4.457 \cdot 10^{-3}(T-273.15)]$
EVA40	$1.06332 \exp(2.288 \cdot 10^{-5}T^{1.5})$	$205.1 \exp[-4.989 \cdot 10^{-3}(T-273.15)]$
PA6	$0.7597 \exp[4.701 \cdot 10^{-4}(T-273.15)]$	$376.7 \exp[-4.660 \cdot 10^{-3}(T-273.15)]$
PA11	$0.9581 \exp[6.664 \cdot 10^{-4}(T-273.15)]$	$254.7 \exp[-4.178 \cdot 10^{-3}(T-273.15)]$
PA66	$0.7657 \exp[6.600 \cdot 10^{-4}(T-273.15)]$	$316.4 \exp[-5.040 \cdot 10^{-3}(T-273.15)]$
cPBD	$1.0970 \exp[6.600 \cdot 10^{-4}(T-273.15)]$	$177.1 \exp[-3.593 \cdot 10^{-3}(T-273.15)]$
PBD-8	$1.1004 + 6.718 \cdot 10^{-4}(T-273.15) + 6.584 \cdot 10^{-7}(T-273.15)^2$	$200.0 \exp[-4.606 \cdot 10^{-3}(T-273.15)]$
PBD-24	$1.1049 + 6.489 \cdot 10^{-4}(T-273.15) + 7.099 \cdot 10^{-7}(T-273.15)^2$	$193.0 \exp[-4.519 \cdot 10^{-3}(T-273.15)]$
PBD-40	$1.1013 + 6.593 \cdot 10^{-4}(T-273.15) + 5.776 \cdot 10^{-7}(T-273.15)^2$	$188.0 \exp[-4.437 \cdot 10^{-3}(T-273.15)]$
PBD-50	$1.1037 + 5.955 \cdot 10^{-4}(T-273.15) + 7.789 \cdot 10^{-7}(T-273.15)^2$	$183.0 \exp[-4.425 \cdot 10^{-3}(T-273.15)]$
PBD-87	$1.1094 + 6.729 \cdot 10^{-4}(T-273.15) + 4.470 \cdot 10^{-7}(T-273.15)^2$	$175.0 \exp[-4.538 \cdot 10^{-3}(T-273.15)]$
iPB	$1.1417 \exp[6.751 \cdot 10^{-4}(T-273.15)]$	$167.5 \exp[-4.533 \cdot 10^{-3}(T-273.15)]$
PnBMA	$0.9341 + 5.5254 \cdot 10^{-4}(T-273.15) + 6.5803 \cdot 10^{-6}(T-273.15)^2 + 1.5691 \cdot 10^{-10}(T-273.15)^3$	$226.7 \exp[-5.344 \cdot 10^{-3}(T-273.15)]$
PBT	$0.9640 - 1.017 \cdot 10^{-3}(T-273.15) + 3.065 \cdot 10^{-6}(T-273.15)^2$	$263.0 \exp[-3.444 \cdot 10^{-3}(T-273.15)]$
PCL	$0.9049 \exp[6.392 \cdot 10^{-4}(T-273.15)]$	$189.0 \exp[-3.931 \cdot 10^{-3}(T-273.15)]$
PC	$0.73565 \exp(1.859 \cdot 10^{-5}T^{1.5})$	$310.0 \exp[-4.078 \cdot 10^{-3}(T-273.15)]$
BCPC	$0.6737 + 3.634 \cdot 10^{-4}(T-273.15) + 2.370 \cdot 10^{-7}(T-273.15)^2$	$363.4 \exp[-4.921 \cdot 10^{-3}(T-273.15)]$
HFPC	$0.6111 + 4.898 \cdot 10^{-4}(T-273.15) + 1.730 \cdot 10^{-7}(T-273.15)^2$	$236.6 \exp[-5.156 \cdot 10^{-3}(T-273.15)]$
TMPC	$0.8497 + 5.073 \cdot 10^{-4}(T-273.15) + 3.832 \cdot 10^{-7}(T-273.15)^2$	$231.4 \exp[-4.242 \cdot 10^{-3}(T-273.15)]$
PcHMA	$0.8793 + 4.0504 \cdot 10^{-4}(T-273.15) + 7.774 \cdot 10^{-7}(T-273.15)^2 - 7.7534 \cdot 10^{-10}(T-273.15)^3$	$295.2 \exp[-5.220 \cdot 10^{-3}(T-273.15)]$
PPO	$0.78075 \exp(2.151 \cdot 10^{-5}T^{1.5})$	$227.8 \exp[-4.290 \cdot 10^{-3}(T-273.15)]$
PDMS	$1.0079 \exp[9.121 \cdot 10^{-4}(T-273.15)]$	$89.4 \exp[-5.701 \cdot 10^{-3}(T-273.15)]$
PDMS-10	$0.8343 + 5.991 \cdot 10^{-4}(T-273.15) + 5.734 \cdot 10^{-7}(T-273.15)^2$	$542.63 \exp[-6.69 \cdot 10^{-3}(T-273.15)]$
PDMS-40	$0.8018 + 7.072 \cdot 10^{-4}(T-273.15) + 3.635 \cdot 10^{-7}(T-273.15)^2$	$482.73 \exp[-6.09 \cdot 10^{-3}(T-273.15)]$
PDMS-60	$0.8146 + 5.578 \cdot 10^{-4}(T-273.15) + 5.774 \cdot 10^{-7}(T-273.15)^2$	$482.73 \exp[-6.09 \cdot 10^{-3}(T-273.15)]$
PECH	$0.7216 \exp[5.825 \cdot 10^{-4}(T-273.15)]$	$238.3 \exp[-4.171 \cdot 10^{-3}(T-273.15)]$
PEEK	$0.7158 \exp[6.690 \cdot 10^{-4}(T-273.15)]$	$388.0 \exp[-4.124 \cdot 10^{-3}(T-273.15)]$
PEA	$0.8756 \exp[7.241 \cdot 10^{-4}(T-273.15)]$	$193.2 \exp[-4.839 \cdot 10^{-3}(T-273.15)]$
PEMA	$0.8614 \exp[7.468 \cdot 10^{-4}(T-273.15)]$	$260.9 \exp[-5.356 \cdot 10^{-3}(T-273.15)]$
HDPE	$1.1595 + 8.0394 \cdot 10^{-4}(T-273.15)$	$179.9 \exp[-4.739 \cdot 10^{-3}(T-273.15)]$
LPE	$0.9172 \exp[7.806 \cdot 10^{-4}(T-273.15)]$	$176.7 \exp[-4.661 \cdot 10^{-3}(T-273.15)]$
HMLPE	$0.8992 \exp[8.502 \cdot 10^{-4}(T-273.15)]$	$168.3 \exp[-4.292 \cdot 10^{-3}(T-273.15)]$
BPE	$0.9399 \exp[7.341 \cdot 10^{-4}(T-273.15)]$	$177.1 \exp[-4.699 \cdot 10^{-3}(T-273.15)]$
LDPE	$1.1944 + 2.841 \cdot 10^{-4}(T-273.15) + 1.872 \cdot 10^{-6}(T-273.15)^2$	$202.2 \exp[-5.243 \cdot 10^{-3}(T-273.15)]$
LDPE-A	$1.1484 \exp[6.950 \cdot 10^{-4}(T-273.15)]$	$192.9 \exp[-4.701 \cdot 10^{-3}(T-273.15)]$
LDPE-B	$1.1524 \exp[6.700 \cdot 10^{-4}(T-273.15)]$	$196.6 \exp[-4.601 \cdot 10^{-3}(T-273.15)]$
LDPE-C	$1.1516 \exp[6.730 \cdot 10^{-4}(T-273.15)]$	$186.7 \exp[-4.391 \cdot 10^{-3}(T-273.15)]$
PEO	$0.8766 \exp[7.087 \cdot 10^{-4}(T-273.15)]$	$207.7 \exp[-3.947 \cdot 10^{-3}(T-273.15)]$
PET	$0.6883 + 5.90 \cdot 10^{-4}(T-273.15)$	$369.7 \exp[-4.150 \cdot 10^{-3}(T-273.15)]$
P4HS	$0.8251 + 6.77 \cdot 10^{-4}T$	$103.1 \exp[-2.417 \cdot 10^{-3}(T-273.15)]$
PIB	$1.0750 \exp[5.651 \cdot 10^{-4}(T-273.15)]$	$200.3 \exp[-4.329 \cdot 10^{-3}(T-273.15)]$
PI-8	$1.1030 + 6.488 \cdot 10^{-4}(T-273.15) + 5.125 \cdot 10^{-7}(T-273.15)^2$	$188.0 \exp[-4.541 \cdot 10^{-3}(T-273.15)]$
PI-14	$1.0943 + 6.293 \cdot 10^{-4}(T-273.15) + 6.231 \cdot 10^{-7}(T-273.15)^2$	$202.0 \exp[-4.653 \cdot 10^{-3}(T-273.15)]$
PI-41	$1.0951 + 6.188 \cdot 10^{-4}(T-273.15) + 6.629 \cdot 10^{-7}(T-273.15)^2$	$199.0 \exp[-4.622 \cdot 10^{-3}(T-273.15)]$
PI-56	$1.0957 + 6.655 \cdot 10^{-4}(T-273.15) + 5.661 \cdot 10^{-7}(T-273.15)^2$	$200.0 \exp[-4.644 \cdot 10^{-3}(T-273.15)]$
PMA	$0.8365 \exp[6.795 \cdot 10^{-4}(T-273.15)]$	$235.8 \exp[-4.493 \cdot 10^{-3}(T-273.15)]$
PMMA	$0.8254 + 2.8383 \cdot 10^{-4}(T-273.15) + 7.792 \cdot 10^{-7}(T-273.15)^2$	$287.5 \exp[-4.146 \cdot 10^{-3}(T-273.15)]$
P4MP	$1.4075 - 9.095 \cdot 10^{-4}(T-273.15) + 3.497 \cdot 10^{-6}(T-273.15)^2$	$37.67 + 0.2134(T-273.15)] - 7.0445 \cdot 10^{-4}(T-273.15)^2$
PαMS	$0.89365 + 3.4864 \cdot 10^{-4}(T-273.15) + 5.0184 \cdot 10^{-7}(T-273.15)^2$	$297.7 \exp[-4.074 \cdot 10^{-3}(T-273.15)]$
PoMS	$0.9396 \exp[5.306 \cdot 10^{-4}(T-273.15)]$	$261.9 \exp[-4.114 \cdot 10^{-3}(T-273.15)]$
POM	$0.7484 \exp[6.770 \cdot 10^{-4}(T-273.15)]$	$305.6 \exp[-4.326 \cdot 10^{-3}(T-273.15)]$
PH	$0.76644 \exp(1.921 \cdot 10^{-5}T^{1.5})$	$359.9 \exp[-4.378 \cdot 10^{-3}(T-273.15)]$

Table 2
Tait Equation Parameter Functions for Polymer Melts

Polymer	$V(0,T)/cm^3g^{-1}$	$B(T)/MPa$
PSF	$0.7644 + 3.419 \cdot 10^{-4}(T-273.15) + 3.126 \cdot 10^{-7}(T-273.15)^2$	$365.9 \exp[-3.757 \cdot 10^{-3}(T-273.15)]$
PAr	$0.73381 \exp(1.626 \cdot 10^{-5}T^{1.5})$	$296.9 \exp[-3.375 \cdot 10^{-3}(T-273.15)]$
aPP	$1.1841 - 1.091 \cdot 10^{-4}(T-273.15) + 5.286 \cdot 10^{-6}(T-273.15)^2$	$162.1 \exp[-6.604 \cdot 10^{-3}(T-273.15)]$
iPP	$1.1606 \exp[6.700 \cdot 10^{-4}(T-273.15)]$	$149.1 \exp[-4.177 \cdot 10^{-3}(T-273.15)]$
PS	$0.9287 \exp[5.131 \cdot 10^{-4}(T-273.15)]$	$216.9 \exp[-3.319 \cdot 10^{-3}(T-273.15)]$
PTFE	$0.3200 + 9.5862 \cdot 10^{-4}(T-273.15)$	$425.2 \exp[-9.380 \cdot 10^{-3}(T-273.15)]$
PTHF	$1.0043 \exp[6.691 \cdot 10^{-4}(T-273.15)]$	$178.6 \exp[-4.223 \cdot 10^{-3}(T-273.15)]$
PVAc	$0.82496 + 5.820 \cdot 10^{-4}(T-273.15) + 2.940 \cdot 10^{-7}(T-273.15)^2$	$204.9 \exp[-4.346 \cdot 10^{-3}(T-273.15)]$
PVC	$0.7196 + 5.581 \cdot 10^{-5}(T-273.15) + 1.468 \cdot 10^{-6}(T-273.15)^2$	$294.2 \exp[-5.321 \cdot 10^{-3}(T-273.15)]$
PVME	$0.9585 \exp[6.653 \cdot 10^{-4}(T-273.15)]$	$215.8 \exp[-4.588 \cdot 10^{-3}(T-273.15)]$
PVdF	$0.5790 \exp[8.051 \cdot 10^{-4}(T-273.15)]$	$244.0 \exp[-5.210 \cdot 10^{-3}(T-273.15)]$
SAN3	$0.9233 + 3.936 \cdot 10^{-4}(T-273.15) + 5.685 \cdot 10^{-7}(T-273.15)^2$	$239.8 \exp[-4.376 \cdot 10^{-3}(T-273.15)]$
SAN6	$0.9211 + 4.370 \cdot 10^{-4}(T-273.15) + 5.846 \cdot 10^{-7}(T-273.15)^2$	$226.9 \exp[-4.286 \cdot 10^{-3}(T-273.15)]$
SAN15	$0.9044 + 4.207 \cdot 10^{-4}(T-273.15) + 4.077 \cdot 10^{-7}(T-273.15)^2$	$238.4 \exp[-3.943 \cdot 10^{-3}(T-273.15)]$
SAN18	$0.9016 + 4.036 \cdot 10^{-4}(T-273.15) + 4.206 \cdot 10^{-7}(T-273.15)^2$	$240.4 \exp[-3.858 \cdot 10^{-3}(T-273.15)]$
SAN40	$0.8871 + 3.406 \cdot 10^{-4}(T-273.15) + 4.938 \cdot 10^{-7}(T-273.15)^2$	$289.3 \exp[-4.431 \cdot 10^{-3}(T-273.15)]$
SAN70	$0.8528 + 3.616 \cdot 10^{-4}(T-273.15) + 2.634 \cdot 10^{-7}(T-273.15)^2$	$335.4 \exp[-3.923 \cdot 10^{-3}(T-273.15)]$
SBR10	$0.9053 \exp(2.437 \cdot 10^{-5}T^{1.5})$	$530.3 \exp[-3.99 \cdot 10^{-3}(T-273.15)]$
SBR23	$0.8986 \exp(2.317 \cdot 10^{-5}T^{1.5})$	$551.6 \exp[-4.17 \cdot 10^{-3}(T-273.15)]$
SBR60	$0.8812 \exp(2.031 \cdot 10^{-5}T^{1.5})$	$486.0 \exp[-4.34 \cdot 10^{-3}(T-273.15)]$
SBR85	$0.8704 \exp(1.846 \cdot 10^{-5}T^{1.5})$	$356.7 \exp[-4.24 \cdot 10^{-3}(T-273.15)]$
SMMA20	$0.9063 + 3.570 \cdot 10^{-4}(T-273.15) + 6.532 \cdot 10^{-7}(T-273.15)^2$	$232.0 \exp[-4.143 \cdot 10^{-3}(T-273.15)]$
SMMA60	$0.8610 + 3.350 \cdot 10^{-4}(T-273.15) + 6.980 \cdot 10^{-7}(T-273.15)^2$	$261.0 \exp[-4.611 \cdot 10^{-3}(T-273.15)]$
VCES50	$0.6676 + 6.63 \cdot 10^{-4}T$	$5281.7 \exp[-9.264 \cdot 10^{-3}(T-273.15)]$
VCHS80	$0.7753 + 6.17 \cdot 10^{-4}T$	$247.6 \exp[-2.604 \cdot 10^{-3}(T-273.15)]$
VCHS67	$0.8028 + 6.50 \cdot 10^{-4}T$	$581.7 \exp[-4.553 \cdot 10^{-3}(T-273.15)]$
VCHS60	$0.8213 + 6.23 \cdot 10^{-4}T$	$229.1 \exp[-2.133 \cdot 10^{-3}(T-273.15)]$
VCHS50	$0.7827 + 5.05 \cdot 10^{-4}T$	$136.0 \exp[-1.083 \cdot 10^{-3}(T-273.15)]$
VCHS40	$0.7805 + 4.92 \cdot 10^{-4}T$	$155.0 \exp[-1.605 \cdot 10^{-3}(T-273.15)]$
VCHS33	$0.7710 + 4.86 \cdot 10^{-4}T$	$460.4 \exp[-3.453 \cdot 10^{-3}(T-273.15)]$
VCHS20	$0.6416 + 5.42 \cdot 10^{-4}T$	$489.8 \exp[-3.193 \cdot 10^{-3}(T-273.15)]$
VCOS50	$0.7081 + 7.40 \cdot 10^{-4}T$	$666.5 \exp[-4.503 \cdot 10^{-3}(T-273.15)]$
VCPS50	$0.7814 + 4.36 \cdot 10^{-4}T$	$880.1 \exp[-4.393 \cdot 10^{-3}(T-273.15)]$

Section 14
Geophysics, Astronomy, and Acoustics

Astronomical Constants 14-1
Properties of the Solar System 14-2
Satellites of the Planets 14-4
Mass, Dimensions, and other Parameters of the Earth 14-6
Geological Time Scale 14-8
Acceleration Due to Gravity 14-9
Density, Pressure, and Gravity as a Function of Depth within the Earth 14-10
Ocean Pressure as a Function of Depth and Latitude 14-11
Properties of Seawater 14-12
Abundance of Elements in the Earth's Crust and in the Sea 14-14
Solar Spectral Irradiance 14-15
U.S. Standard Atmosphere (1976) 14-16
Geographical and Seasonal Variation in Solar Radiation 14-23
Infrared Absorption by the Earth's Atmosphere 14-24
Atmospheric Concentration of Carbon Dioxide, 1958-1996 14-25
Mean Temperatures in the United States, 1900-1992 14-27
Global Temperature Trend, 1866-1996 14-29
Atmospheric Electricity 14-30
Speed of Sound in Various Media 14-38
Attenuation and Speed of Sound in Air as a Function of Humidity and Frequency 14-40
Speed of Sound in Dry Air 14-41
Musical Scales 14-42
Characteristics of Human Hearing 14-44

ASTRONOMICAL CONSTANTS
Victor Abalakin

The constants in this table are based primarilarly on the set of constants adopted by the International Astronomical Union (IAU) in 1976. Updates have been made when new data were available. All values are given in SI Units; thus masses are expressed in kilograms and distances in meters. The astronomical unit of time is a time interval of one day (1 d) equal to 86400 s. An interval of 36525 d is one Julian century (1 cy).

REFERENCES

1. Seidelmann, P. K., *Explanatory Supplement to the Astronomical Almanac*, University Science Books, Mill Valley, CA, 1990.
2. Lang, K. R., *Astrophysical Data: Planets and Stars*, Springer-Verlag, New York, 1992.

Defining constants

Gaussian gravitational constant	$k = 0.01720209895$ m^3 kg^{-1} s^{-2}
Speed of light	$c = 299792458$ m s^{-1}

Primary constants

Light-time for unit distance (1 AU)	$\tau_A = 499.004782$ s
Equatorial radius of earth	$a_e = 6378140$ m
Equatorial radius of earth (IUGG value)	$a_e = 6378136$ m
Dynamical form-factor for earth	$J_2 = 0.001082626$
Geocentric gravitational constant	$GE = 3.986005 \times 10^{14}$ m^{3}s^{-2}
Constant of gravitation	$G = 6.672 \times 10^{-11}$ m^{3}kg^{-1}s^{-2}
Ratio of mass of moon to that of earth	$\mu = 0.01230002$
	$1/\mu = 81.300587$
General precession in longitude, per Julian century, at standard epoch J2000	$\rho = 5029''.0966$
Obliquity of the ecliptic at standard epoch J2000	$\varepsilon = 23°26'21''.448$

Derived constants

Constant of nutation at standard epoch J2000	$N = 9''.2025$
Unit distance (AU $= c\tau_A$)	$AU = 1.49597870 \times 10^{11}$ m
Solar parallax ($\pi_0 = \arcsin(a_e/AU)$)	$\pi_0 = 8''.794148$
Constant of aberration for standard epoch J2000	$\kappa = 20''.49552$
Flattening factor for the earth	$f = 1/298.257 = 0.00335281$
Heliocentric gravitational constant ($GS = A^3k^2/D^2$)	$GS = 1.32712438 \times 10^{20}$ m^3 s^{-2}
Ratio of mass of sun to that of the earth ($(S/E) = (GS)/(GE)$)	$S/E = 332946.0$
Ratio of mass of sun to that of earth + moon	$(S/E)/(1 + \mu) = 328900.5$
Mass of the sun ($S = (GS)/G$)	$S = 1.9891 \times 10^{30}$ kg

Ratios of mass of sun to masses of the planets

Mercury	6023600
Venus	408523.5
Earth + moon	328900.5
Mars	3098710
Jupiter	1047.355
Saturn	3498.5
Uranus	22869
Neptune	19314
Pluto	3000000

PROPERTIES OF THE SOLAR SYSTEM

The following tables give various properties of the planets and characteristics of their orbits in the solar system. Certain properties of the sun and of the earth's moon are also included.

Explanations of the column headings:

- *Den.*: mean density in g/cm^3
- *Radius*: radius at the equator in km
- *Flattening*: degree of oblateness, defined as $(r_e - r_p)/r_e$, where r_e and r_p are the equatorial and polar radii, respectively
- *Potential coefficients*: coefficients in the spherical harmonic representation of the gravitational potential U by the equation

$$U(r,\phi) = (GM/r) \left[1 - \sum J_n (a/r)^n P_n (\sin \phi) \right]$$

  where G is the gravitational constant, r the distance from the center of the planet, a the radius of the planet, M the mass, ϕ the latitude, and P_n the Legendre polynomial of degree n.
- *Gravity*: acceleration due to gravity at the surface
- *Escape velocity*: velocity needed at the surface of the planet to escape the gravitational pull
- *Dist. to sun*: semi-major axis of the elliptical orbit (1 AU = 1.496×10^8 km)
- ε: eccentricity of the orbit
- *Ecliptic angle*: angle between the planetary orbit and the plane of the earth's orbit around the sun
- *Inclin.*: angle between the equatorial plane and the plane of the planetary orbit
- *Rot. period*: period of rotation of the planet measured in earth days
- *Albedo*: ratio of the light reflected from the planet to the light incident on it
- T_{sur}: mean temperature at the surface
- P_{sur}: pressure of the atmosphere at the surface

The following general information on the solar system is of interest:

Mass of the earth = M_e = 5.9742×10^{24} kg
Total mass of planetary system = 2.669×10^{27} kg = 447 M_e
Total angular momentum of planetary system = 3.148×10^{43} kg m^2/s
Total kinetic energy of the planets = 1.99×10^{35} J
Total rotational energy of planets = 0.7×10^{35} J

Properties of the sun:

Mass = 1.9891×10^{30} kg = 332946.0 M_e
Radius = 6.9599×10^8 m
Surface area = 6.087×10^{18} m^2
Volume = 1.412×10^{27} m^3
Mean density = 1.409 g/cm^3
Gravity at surface = 27398 cm/s^2
Escape velocity at surface = 6.177×10^5 m/s
Effective temperature = 5780 K
Total radiant power emitted (luminosity) = 3.86×10^{26} W
Surface flux of radiant energy = 6.340×10^7 W/m^2
Flux of radiant energy at the earth (Solar Constant) = 1373 W/m^2

REFERENCES

1. Seidelmann, P. K., Editor, *Explanatory Supplement to the Astronomical Almanac*, University Science Books, Mill Valley, CA, 1992.
2. Lang, K. R., *Astrophysical Data: Planets and Stars*, Springer-Verlag, New York, 1992.
3. Allen, C. W., *Astrophysical Quantities, Third Edition*, Athlone Press, London, 1977.

Planet	Mass 10^{24} kg	Den. g/cm^3	Radius km	Flattening	$10^3 J_2$	$10^6 J_3$	$10^6 J_4$	Gravity cm/s^2	Escape vel. km/s
Mercury	0.33022	5.43	2439.7	0				370	4.25
Venus	4.8690	5.24	6051.9	0	0.027			887	10.4
Earth	5.9742	5.515	6378.140	0.00335364	1.08263	-2.54	-1.61	980	11.2
(Moon)	0.073483	3.34	1738	0	0.2027			162	2.37
Mars	0.64191	3.94	3397	0.00647630	1.964	36		371	5.02
Jupiter	1898.8	1.33	71492	0.0648744	14.75	-580		2312	59.6
Saturn	568.50	0.70	60268	0.0979624	16.45	-1000		896	35.5
Uranus	86.625	1.30	25559	0.0229273	12			777	21.3
Neptune	102.78	1.76	24764	0.0171	4			1100	23.3
Pluto	0.015	1.1	1151	0				72	1.1

The spanning header over $10^3 J_2$, $10^6 J_3$, $10^6 J_4$ is **Potential coeffients**.

Planet	Dist. to sun AU	ε	Ecliptic angle	Inclin.	Rot. period d	Albedo	No. of satellites
Mercury	0.38710	0.2056	7.00°	0°	58.6462	0.106	0
Venus	0.72333	0.0068	3.39°	177.3°	-243.01	0.65	0
Earth	1.00000	0.0167		23.45°	0.99726968	0.367	1
(Moon)				6.68°	27.321661	0.12	
Mars	1.52369	0.0933	1.85°	25.19°	1.02595675	0.150	2
Jupiter	5.20283	0.048	1.31°	3.12°	0.41354	0.52	16
Saturn	9.53876	0.056	2.49°	26.73°	0.4375	0.47	18
Uranus	19.19139	0.046	0.77°	97.86°	-0.65	0.51	15
Neptune	30.06107	0.010	1.77°	29.56°	0.768	0.41	8
Pluto	39.52940	0.248	17.15°	118°	-6.3867	0.3	1

Planet	T_{sur} K	P_{sur} bar	CO_2	N_2	O_2	H_2O	H_2	He	Ar	Ne	CO
Mercury	440	2×10^{-15}					2%	98%			
Venus	730	90	96.4%	3.4%	69 ppm	0.1%			4 ppm		20 ppm
Earth	288	1	0.03%	78.08%	20.95%	0 to 3%			0.93%	18 ppm	1 ppm
Mars	218	0.007	95.32%	2.7%	0.13%	0.03%			1.6%	3 ppm	0.07%
Jupiter	129						86.1%	13.8%			
Saturn	97						92.4%	7.4%			
Uranus	58						89%	11%			
Neptune	56						89%	11%			
Pluto	50	1×10^{-5}									

The spanning header over the last nine columns is **Atmospheric composition**.

SATELLITES OF THE PLANETS

This table gives characteristics of the known satellites of the planets. The parameters covered are:

- Orbital period in units of earth days. An R following the value indicates a retrograde motion.
- Distance from the planet, as measured by the semi-major axis of the orbit.
- Eccentricity of the orbit.
- Inclination of the satellite orbit with respect to the equator of planet.
- Mass of the satellite relative to the planet.
- Radius of the satellite in km.
- Mean density of the satellite.
- Geometric albedo, which is a measure of the fraction of incident sunlight reflected by the satellite.

REFERENCES

1. Seidelmann, P. K., Editor, *Explanatory Supplement to the Astronomical Almanac*, University Science Books, Mill Valley, CA, 1992.
2. Lang, K. R., *Astrophysical Data: Planets and Stars*, Springer-Verlag, New York, 1992.
3. Burns, J. A., and Matthews, M. S., Eds., *Satellites*, University of Arizona Press, Tucson, 1986.

Planet		Satellite	Orb. Period d	Distance 10³ km	Eccentricity	Inclination	Rel. mass	Radius km	Den. g/cm³	Albedo
Earth		Moon	27.321661	384.400	0.054900489	18.28-28.58°	0.01230002	1738	3.34	0.12
Mars	I	Phobos	0.31891023	9.378	0.015	1.0°	1.5×10^{-8}	$13.5 \times 10.8 \times 9.4$	<2	0.06
	II	Deimos	1.2624407	23.459	0.0005	0.9-2.7°	3×10^{-9}	$7.5 \times 6.1 \times 5.5$	<2	0.07
Jupiter	I	Io	1.769137786	422	0.004	0.04°	4.68×10^{-5}	1815	3.55	0.61
	II	Europa	3.551181041	671	0.009	0.47°	2.52×10^{-5}	1569	3.04	0.64
	III	Ganymede	7.15455296	1070	0.002	0.21°	7.80×10^{-5}	2631	1.93	0.42
	IV	Callisto	16.6890184	1883	0.007	0.51°	5.66×10^{-5}	2400	1.83	0.20
	V	Amalthea	0.49817905	181	0.003	0.40°	3.8×10^{-9}	$135 \times 83 \times 75$		0.05
	VI	Himalia	250.5662	11480	0.15798	27.63°	5.0×10^{-9}	93		0.03
	VII	Elara	259.6528	11737	0.20719	24.77°	4×10^{-10}	38		0.03
	VIII	Pasiphae	735 R	23500	0.378	145°	1×10^{-10}	25		
	IX	Sinope	758 R	23700	0.275	153°	0.4×10^{-10}	18		
	X	Lysithea	259.22	11720	0.107	29.02°	0.4×10^{-10}	18		
	XI	Carme	692 R	22600	0.20678	164°	0.5×10^{-10}	20		
	XII	Ananke	631 R	21200	0.16870	147°	0.2×10^{-10}	15		
	XIII	Leda	238.72	11094	0.14762	26.07°	0.03×10^{-10}	8		
	XIV	Thebe	0.6745	222	0.015	0.8°	4×10^{-10}	55×45		0.05
	XV	Adrastea	0.29826	129			0.1×10^{-10}	$12.5 \times 10 \times 7.5$		0.05
	XVI	Metis	0.294780	128			0.5×10^{-10}	20		0.05
Saturn	I	Mimas	0.942421813	185.52	0.0202	1.53°	8.0×10^{-8}	196	1.44	0.5
	II	Enceladus	1.370217855	238.02	0.00452	1.86°	1.3×10^{-7}	250	1.13	1.0
	III	Tethys	1.887802160	294.66	0.000000	1.86°	1.3×10^{-6}	530	1.20	0.9
	IV	Dione	2.736914742	377.40	0.002230	0.02°	1.85×10^{-6}	560	1.41	0.7
	V	Rhea	4.517500436	527.04	0.00100	0.35°	4.4×10^{-6}	765	1.33	0.7
	VI	Titan	15.94542068	1221.83	0.029192	0.33°	2.38×10^{-4}	2575	1.88	0.21
	VII	Hyperion	21.2766088	1481.1	0.104	0.43°	3×10^{-8}	$205 \times 130 \times 110$		0.3
	VIII	Iapetus	79.3301825	3561.3	0.02828	14.72°	3.3×10^{-6}	730	1.15	0.2
	IX	Phoebe	550.48 R	12952	0.16326	177°	7×10^{-10}	110		0.06

SATELLITES OF THE PLANETS (continued)

Planet		Satellite	Orb. Period d	Distance 10³ km	Eccentricity	Inclination	Rel. mass	Radius km	Den. g/cm³	Albedo
	X	Janus	0.6945	151.472	0.007	0.14°		$110 \times 100 \times 80$		0.8
	XI	Epimetheus	0.6942	151.422	0.009	0.34°		$70 \times 60 \times 50$		0.8
	XII	Helene	2.7369	377.40	0.005	0.0°		$18 \times 16 \times 15$		0.7
	XIII	Telesto	1.8878	294.66				$17 \times 14 \times 13$		0.5
	XIV	Calypso	1.8878	294.66				$17 \times 11 \times 11$		0.6
	XV	Atlas	0.6019	137.670	0.000	0.3°		20×10		0.9
	XVI	Prometheus	0.6130	139.353	0.003	0.0°		$70 \times 50 \times 40$		0.6
	XVII	Pandora	0.6285	141.700	0.004	0.0°		$55 \times 45 \times 35$		0.9
	XVIII	Pan	0.5750	133.583				10		0.5
Uranus	I	Ariel	2.52037935	191.02	0.0034	0.3°	1.56×10^{-5}	579	1.55	0.34
	II	Umbriel	4.1441772	266.30	0.0050	0.36°	1.35×10^{-5}	586	1.58	0.18
	III	Titania	8.7058717	435.91	0.0022	0.14°	4.06×10^{-5}	790	1.69	0.27
	IV	Oberon	13.4632389	583.52	0.0008	0.10°	3.47×10^{-5}	762	1.64	0.24
	V	Miranda	1.41347925	129.39	0.0027	4.2°	0.08×10^{-5}	240	1.25	0.27
	VI	Cordelia	0.335033	49.77	<0.001	0.1°		13		0.07
	VII	Ophelia	0.376409	53.79	0.010	0.1°		15		0.07
	VIII	Bianca	0.434577	59.17	<0.001	0.2°		21		0.07
	IX	Cressida	0.463570	61.78	<0.001	0.0°		31		0.07
	X	Desdemona	0.473651	62.68	<0.001	0.2°		27		0.07
	XI	Juliet	0.493066	64.35	<0.001	0.1°		42		0.07
	XII	Portia	0.513196	66.09	<0.001	0.1°		54		0.07
	XIII	Rosalind	0.558459	69.94	<0.001	0.3°		27		0.07
	XIV	Belinda	0.623525	75.26	<0.001	0.0°		33		0.07
	XV	Puck	0.761832	86.01	<0.001	0.31°		77		0.07
Neptune	I	Triton	5.8768541 R	354.76	0.000016	157.345°	2.09×10^{-4}	1353	2.05	0.7
	II	Nereid	360.13619	5513.4	0.7512	27.6°	2×10^{-7}	170		0.4
	III	Naiad	0.294396	117.6	<0.001	4.74°		29		0.06
	IV	Thalassa	0.311485	73.6	<0.001	0.21°		40		0.06
	V	Despina	0.334655	52.6	<0.001	0.07°		74		0.06
	VI	Galatea	0.428745	62.0	<0.001	0.05°		79		0.06
	VII	Larissa	0.554654	50.0	<0.0014	0.20°		104×89		0.06
	VIII	Proteus	1.122315	48.2	<0.001	0.55°		$218 \times 208 \times 201$		0.06
Pluto	I	Charon	6.38725	19.6	<0.001	99°	0.22	593		0.5

MASS, DIMENSIONS, AND OTHER PARAMETERS OF THE EARTH

This table is a collection of data on various properties of the Earth. Most of the values are given in SI units. Note that 1 AU (astronomical unit) = 149,597,870 km.

REFERENCES

1. Seidelmann, P. K., Editor, *Explanatory Supplement to the Astronomical Almanac*, University Science Books, Mill Valley, CA, 1992.
2. Lang, K. R., *Astrophysical Data: Planets and Stars*, Springer-Verlag, New York, 1992.

Quantity	Symbol	Value	Unit
Mass	M	$5.9742 \cdot 10^{27}$	g
Major orbital semi-axis	a_{orb}	1.000000	AU
		$1.4959787 \cdot 10^8$	km
Distance from sun at perihelion	r_π	0.9833	AU
Distance from sun at aphelion	r_α	1.0167	AU
Moment of perihelion passage	T_π	Jan. 2, 4 h 52 min	
Moment of aphelion passage	T_α	July 4, 5 h 05 min	
Siderial rotation period around sun	P_{orb}	$31.5581 \cdot 10^6$	s
		365.25636	d
Mean rotational velocity	U_{orb}	29.78	km/s
Mean equatorial radius	$\overline{a}$	6378.140	km
Mean polar compression (flattening factor)	α	1/298.257	
Difference in equatorial and polar semi-axes	$a - c$	21.385	km
Compression of meridian of major equatorial axis	α_a	1/295.2	
Compression of meridian of minor equatorial axis	α_b	1/298.0	
Equatorial compression	ε	1/30 000	
Difference in equatorial semi-axes	$a - b$	213	m
Difference in polar semi-axes	$c_N - c_S$	~70	m
Polar asymmetry	η	$\sim 1 \cdot 10^{-5}$	
Mean acceleration of gravity at equator	g_e	9.78036	m/s^2
Mean acceleration of gravity at poles	g_p	9.83208	m/s^2
Difference in acceleration of gravity at pole and at equator	$g_p - g_e$	5.172	cm/s^2
Mean acceleration of gravity for entire surface of terrestrial ellipsoid	g	9.7978	m/s^2
Mean radius	R	6371.0	km
Area of surface	S	$5.10 \cdot 10^8$	km^2
Volume	V	$1.0832 \cdot 10^{12}$	km^3
Mean density	ρ	5.515	g/cm^3
Siderial rotational period	P	86,164.09	s
Rotational angular velocity	ω	$7.292116 \cdot 10^{-5}$	rad/s
Mean equatorial rotational velocity	v	0.46512	km/s
Rotational angular momentum	L	$5.861 \cdot 10^{33}$	J s
Rotational energy	E	$2.137 \cdot 10^{29}$	J
Ratio of centrifugal force to force of gravity at equator	q_c	0.0034677 = 1/288	
Moment of inertia	I	$8.070 \cdot 10^{37}$	kg m^2
Relative braking of earth's rotation due to tidal friction	$\Delta \omega_e/\omega$	$-4.2 \cdot 10^{-8}$	century^{-1}
Relative secular acceleration of earth's rotation	$\Delta \omega_i/\omega$	$+1.4 \cdot 10^{-8}$	century^{-1}
Not secular braking of earth's rotation	$\Delta \omega/\omega$	$-2.8 \cdot 10^{-8}$	century^{-1}
Probable value of total energy of tectonic deformation of earth	E_t	$\sim 1 \cdot 10^{23}$	J/century
Secular loss of heat of earth through radiation into space	$\Delta' E_k$	$1 \cdot 10^{23}$	J/century
Portion of earth's kinetic energy transformed into heat as a result of lunar and solar tides in the hydrosphere	$\Delta'' E_k$	$1.3 \cdot 10^{23}$	J/century

Quantity	Symbol	Value	Unit
Differences in duration of days in March and August	ΔP	0.0025 (March-August)	s
Corresponding relative annual variation in earth's rotational velocity	$\Delta^*\omega/\omega$	$2.9 \cdot 10^{-8}$ (Aug.-March)	
Presumed variation in earth's radius between August and March	Δ^*R	-9.2 (Aug.-March)	cm
Annual variation in level of world ocean	Δh_o	~10 (Sept.-March)	cm
Area of continents	S_C	$1.49 \cdot 10^8$	km^2
		29.2	% of surface
Area of world ocean	S_o	$3.61 \cdot 10^8$	km^2
		70.8	% of surface
Mean height of continents above sea level	h_C	875	m
Mean depth of world ocean	h_o	3794	m
Mean thickness of lithosphere within the limits of the continents	$h_{c.l.}$	35	km
Mean thickness of lithosphere within the limits of the ocean	$h_{o.l.}$	4.7	km
Mean rate of thickening of continental lithosphere	$\Delta h/\Delta t$	10 - 40	$m/10^6$ y
Mean rate of horizontal extension of continental lithosphere	$\Delta l/\Delta t$	0.75 - 20	$km/10^6$ y
Mass of crust	m_l	$2.36 \cdot 10^{22}$	kg
Mass of mantle		$4.05 \cdot 10^{24}$	kg
Amount of water released from the mantle and core in the course of geological time		$3.40 \cdot 10^{21}$	kg
Total reserve of water in the mantle		$2 \cdot 10^{23}$	kg
Present content of free and bound water in the earth's lithosphere		$2.4 \cdot 10^{21}$	kg
Mass of hydrosphere	m_h	$1.664 \cdot 10^{21}$	kg
Amount of oxygen bound in the earth's crust		$1.300 \cdot 10^{21}$	kg
Amount of free oxygen		$1.5 \cdot 10^{18}$	kg
Mass of atmosphere	m_a	$5.136 \cdot 10^{18}$	kg
Mass of biosphere	m_b	$1.148 \cdot 10^{16}$	kg
Mass of living matter in the biosphere		$3.6 \cdot 10^{14}$	kg
Density of living matter on dry land		0.1	g/cm^2
Density of living matter in ocean		$15 \cdot 10^{-8}$	g/cm^3
Age of the earth		$4.55 \cdot 10^9$	y
Age of oldest rocks		$4.0 \cdot 10^9$	y
Age of most ancient fossils		$3.4 \cdot 10^9$	y

GEOLOGICAL TIME SCALE

Period or Epoch	Beginning and end, in 10^6 years	Key events
Cenozoic era		
Quaternian		
Contemporary	0–10,000 y ± 2,000 y	
Pleistocene	10,000–1,000,000 y ± 50,000 y	Homo Erectus breakout
Tertiary		
Pliocene	1.8–5.3	Ape man fossils
Miocene	5–25	Origin of grass
Oligocene	25–37	Rise of cats, dogs, pigs
Eocene	37–55	Debut of hoofed mammals
Paleocene	55–67	Earliest primates
Mesozoic era		
Cretaceous	67–138	Demise of dinosaurs
Jurassic	138–208	First birds
Triassic	208–245	Appearance of dinosaurs
Paleozoic era		
Permian	245–290	Flowers, insect pollination
Carboniferous	290–360	First conifers
Devonian	360–410	First vertebrates ashore
Silurian	410–435	Spore-bearing plants
Ordovician	435–520	First animals ashore
Cambrian	520–570	Vertebrates appear
Pre-Cambrian		
Pre-Cambrian III (Proterozoic)	570–2500	First plants, jellyfish
Pre-Cambrian II (Archean)	2500–3800	Photosynthetic bacteria
Pre-Cambrian I (Hadean)	3800–4450	Earth formed 4600 million years ago

Reference: Calder, N., *Timescale - An Atlas of the Fourth Dimension,* Viking Press, New York, 1983.

ACCELERATION DUE TO GRAVITY

The acceleration due to gravity is tabulated here as a function of latitude and height above the earth's surface. Values were calculated from the expression

$$g/(m/s^2) = 9.780356 \, (1 + 0.0052885 \sin^2 \phi - 0.0000059 \sin^2 2\phi) - 0.003086 \, H$$

where ϕ is the latitude and H is the height in kilometers.

REFERENCE

Jursa, A. S., Ed., *Handbook of Geophysics and the Space Environment,* 4th ed., Air Force Geophysics Laboratory, 1985, p. 14-17.

ϕ	$H = 0$	$H = 1$ km	$H = 5$ km	$H = 10$ km
0	9.78036	9.77727	9.76493	9.74950
5	9.78075	9.77766	9.76532	9.74989
10	9.78191	9.77882	9.76648	9.75105
15	9.78381	9.78072	9.76838	9.75295
20	9.78638	9.78330	9.77095	9.75552
25	9.78956	9.78647	9.77413	9.75870
30	9.79324	9.79016	9.77781	9.76238
35	9.79732	9.79424	9.78189	9.76646
40	9.80167	9.79858	9.78624	9.77081
45	9.80616	9.80307	9.79073	9.77530
50	9.81065	9.80757	9.79522	9.77979
55	9.81501	9.81193	9.79958	9.78415
60	9.81911	9.81602	9.80368	9.78825
65	9.82281	9.81972	9.80738	9.79195
70	9.82601	9.82292	9.81058	9.79515
75	9.82860	9.82551	9.81317	9.79774
80	9.83051	9.82743	9.81508	9.79965
85	9.83168	9.82860	9.81625	9.80082
90	9.83208	9.82899	9.81665	9.80122

DENSITY, PRESSURE, AND GRAVITY AS A FUNCTION OF DEPTH WITHIN THE EARTH

This table gives the density ρ, pressure p, and acceleration due to gravity g as a function of depth below the earth's surface, as calculated from the model of the structure of the earth in Reference 1. The model assumes a radius of 6371 km for the earth. The boundary between the crust and mantle (the Mohorovicic discontinuity) is taken as 21 km, while in reality it varies considerable with location.

REFERENCES

1. Anderson, D. L., and Hart, R. S., *J. Geophys. Res.*, 81, 1461, 1976.
2. Carmichael, R. S., *CRC Practical Handbook of Physical Properties of Rocks and Minerals*, p.467, CRC Press, Boca Raton, FL, 1989.

Depth km	ρ g/cm³	p kbar	g cm/s²	Depth km	ρ g/cm³	p kbar	g cm/s²
				1771	4.96	752	994
				2071	5.12	903	1002
Crust				2371	5.31	1061	1017
				2671	5.45	1227	1042
0	1.02	0	981	2886	5.53	1352	1069
3	1.02	3	982				
3	2.80	3	982	**Outer core (liquid)**			
21	2.80	5	983				
				2886	9.96	1352	1069
Mantle (solid)				2971	10.09	1442	1050
				3371	10.63	1858	953
21	3.49	5	983	3671	11.00	2154	874
41	3.51	12	983	4071	11.36	2520	760
61	3.52	19	984	4471	11.69	2844	641
81	3.48	26	984	4871	11.99	3116	517
101	3.44	33	984	5156	12.12	3281	427
121	3.40	39	985				
171	3.37	56	987	**Inner core (solid)**			
221	3.34	73	989				
271	3.37	89	991	5156	12.30	3281	427
321	3.47	106	993	5371	12.48	3385	355
371	3.59	124	994	5771	12.52	3529	218
571	3.95	199	999	6071	12.53	3592	122
871	4.54	328	997	6371	12.58	3617	0
1171	4.67	466	992				
1471	4.81	607	991				

OCEAN PRESSURE AS A FUNCTION OF DEPTH AND LATITUDE

The following table is based upon an ocean model which takes into account the equation of state of standard seawater and the dependence on latitude of the acceleration of gravity. The tabulated pressure value is the excess pressure over the ambient atmospheric pressure at the surface.

REFERENCES

1. *International Oceanographic Tables, Volume 4*, Unesco Technical Papers in Marine Science No. 40, Unesco, Paris, 1987.
2. Saunders, P.M., and Fofonoff, N.P., *Deep-Sea Res.* 23, 109-111, 1976.

Pressure in MPa at the Specified Latitude

Depth (meters)	0°	15°	30°	45°	60°	75°	90°
0	0.0000	0.0000	0.0000	0.0000	0.0000	0.0000	0.0000
500	5.0338	5.0355	5.0404	5.0471	5.0537	5.0586	5.0605
1000	10.0796	10.0832	10.0930	10.1064	10.1198	10.1296	10.1333
1500	15.1376	15.1431	15.1577	15.1778	15.1980	15.2127	15.2182
2000	20.2076	20.2148	20.2344	20.2613	20.2882	20.3080	20.3153
2500	25.2895	25.2985	25.3231	25.3568	25.3905	25.4153	25.4244
3000	30.3831	30.3940	30.4236	30.4641	30.5047	30.5345	30.5453
3500	35.4886	35.5012	35.5358	35.5832	35.6307	35.6654	35.6782
4000	40.6056	40.6201	40.6598	40.7140	40.7683	40.8082	40.8229
4500	45.7342	45.7505	45.7952	45.8564	45.9176	45.9626	45.9791
5000	50.8742	50.8924	50.9421	51.0102	51.0785	51.1285	51.1469
5500	56.0255	56.0456	56.1004	56.1755	56.2508	56.3059	56.3262
6000	61.1882	61.2100	61.2700	61.3521	61.4344	61.4947	61.5168
6500	66.3619	66.3857	66.4508	66.5399	66.6292	66.6947	66.7187
7000	71.5467	71.5724	71.6427	71.7388	71.8352	71.9059	71.9318
7500	76.7426	76.7701	76.8456	76.9488	77.0523	77.1282	77.1560
8000	81.9493	81.9788	82.0594	82.1697	82.2804	82.3614	82.3911
8500	87.1669	87.1983	87.2841	87.4016	87.5193	87.6057	87.6373
9000	92.3950	92.4284	92.5194	92.6440	92.7689	92.8606	92.8941
9500	97.6346	97.6698	97.7661	97.8978	98.0300	98.1269	98.1624
10000	102.8800	102.9170	103.0185	103.1572	103.2961	103.3981	103.4355

PROPERTIES OF SEAWATER

In addition to the dependence on temperature and pressure, the physical properties of seawater vary with the concentration of the dissolved constituents. A convenient parameter for describing the composition is the salinity, S, which is defined in terms of the electrical conductivity of the seawater sample. The defining equation for the practical salinity is:

$$S = a_0 + a_1 K^{1/2} + a_2 K + a_3 K^{3/2} + a_4 K^2 + a_5 K^{5/2} ,$$

where K is the ratio of the conductivity of the seawater sample at 15°C and atmospheric pressure to the conductivity of a potassium chloride solution in which the mass fraction of KCl is 0.0324356, at the same temperature and pressure. The values of the coefficients are:

$$a_0 = 0.080 \qquad a_3 = 14.0941$$
$$a_1 = -0.1692 \qquad a_4 = -7.0261$$
$$a_2 = 25.3851 \qquad a_5 = 2.7081$$
$$\Sigma\, a_i = 35.0000$$

Thus when $K = 1$, $S = 35$ exactly (S is normally quoted in units of ‰, i.e., parts per thousand). The value of S can be roughly equated with the mass of dissolved material in grams per kilogram of seawater. Salinity values in the oceans at mid latitudes typically fall between 34 and 36.

The freezing point of seawater at normal atmospheric pressure varies with salinity as follows:

S	0	5	10	15	20	25	30	35	40
t_f /°C	0.000	-0.274	-0.542	-0.812	-1.083	-1.358	-1.636	-1.922	-2.212

The first table below gives several properties of seawater as a function of temperature for a salinity of 35. The second gives electrical conductivity as a function of salinity at several temperatures, and the third lists typical concentrations of the main constituents of seawater as a function of salinity.

REFERENCES

1. *The Practical Salinity Scale 1978 and the International Equation of State of Seawater 1980*, Unesco Technical Papers in Marine Science No. 36, Unesco, Paris, 1981; sections No. 37, 38, 39, and 40 in this series give background papers and detailed tables.
2. Kennish, M.J., *CRC Practical Handbook of Marine Science*, CRC Press, Boca Raton, FL, 1989.
3. Poisson, A. *IEEE J. Ocean. Eng*. OE-5, 50, 1981.

Properties of Seawater as a Function of Temperature at Salinity $S=35$ and Normal Atmospheric Pressure
(100 kPa)

ρ = density in g/cm³
$\beta = (1/\rho)\,(d\rho/dS)$ = fractional change in density per unit change in salinity
$\alpha = (1/\rho)\,(d\rho/dt)$ = fractional change in density per unit change in temperature (°C⁻¹)
κ = electrical conductivity in S/cm
η = viscosity in mPa s (equal to cP)
c_p = specific heat in J/kg
v = speed of sound in m/s

t/°C	ρ/g cm⁻³	$10^7 \cdot \beta$	$10^7 \cdot \alpha$/°C⁻¹	κ/S cm⁻¹	η/mPa s	c_p/J kg⁻¹ °C⁻¹	v/m s⁻¹
0	1.028106	7854	526	0.029048	1.892	3986.5	1449.1
5	1.027675	7717	1136	0.033468	1.610		
10	1.026952	7606	1668	0.038103	1.388	3986.3	1489.8
15	1.025973	7516	2141	0.042933	1.221		
20	1.024763	7444	2572	0.047934	1.085	3993.9	1521.5
25	1.023343	7385	2970	0.053088	0.966		
30	1.021729	7338	3341	0.058373	0.871	4000.7	1545.6
35	1.019934	7300	3687				
40		7270	4004			4003.5	1563.2

Electrical Conductivity of Seawater in S/cm as a Function of Temperature and Salinity

t/°C	$S = 5$	$S = 10$	$S = 15$	$S = 20$	$S = 25$	$S = 30$	$S = 35$	$S = 40$
0	0.004808	0.009171	0.013357	0.017421	0.021385	0.025257	0.029048	0.032775
5	0.005570	0.010616	0.015441	0.020118	0.024674	0.029120	0.033468	0.037734
10	0.006370	0.012131	0.017627	0.022947	0.028123	0.033171	0.038103	0.042935
15	0.007204	0.013709	0.019905	0.025894	0.031716	0.037391	0.042933	0.048355
20	0.008068	0.015346	0.022267	0.028948	0.035438	0.041762	0.047934	0.053968
25	0.008960	0.017035	0.024703	0.032097	0.039276	0.046267	0.053088	0.059751
30	0.009877	0.018771	0.027204	0.035330	0.043213	0.050888	0.058373	0.065683

Composition of Seawater and Ionic Strength at Various Salinities
Concentration of major constituents expressed as molality (moles per kilogram of H_2O)

Constituent	$S = 30$	$S = 35$	$S = 40$
Cl^-	0.482	0.562	0.650
Br^-	0.00074	0.00087	0.00100
F^-		0.00007	
SO_4^{2-}	0.0104	0.0114	0.0122
HCO_3^-	0.00131	0.00143	0.00100
$NaSO_4^-$	0.0085	0.0108	0.0139
KSO_4^-	0.00010	0.00012	0.00015
Na^+	0.405	0.472	0.544
K^+	0.00892	0.01039	0.01200
Mg^{2+}	0.0413	0.0483	0.0561
Ca^{2+}	0.00131	0.00143	0.00154
Sr^+	0.00008	0.00009	0.00011
$MgHCO_3^+$	0.00028	0.00036	0.00045
$MgSO_4$	0.00498	0.00561	0.00614
$CaSO_4$	0.00102	0.00115	0.00126
$NaHCO_3$	0.00015	0.00020	0.00024
H_3BO_3	0.00032	0.00037	0.00042
Ionic strength (mol/kg)	0.5736	0.6675	0.7701

ABUNDANCE OF ELEMENTS IN THE EARTH'S CRUST AND IN THE SEA

This table gives the estimated abundance of the elements in the continental crust (in mg/kg, equivalent to parts per million by mass) and in seawater near the surface (in mg/L). Values represent the median of reported measurements. The concentrations of the less abundant elements may vary with location by several orders of magnitude.

REFERENCES

1. Carmichael, R. S., Ed., *CRC Practical Handbook of Physical Properties of Rocks and Minerals,* CRC Press, Boca Raton, FL, 1989.
2. Bodek, I., et al., *Environmental Inorganic Chemistry*, Pergamon Press, New York, 1988.
3. Ronov, A. B., and Yaroshevsky, A. A., "Earth's Crust Geochemistry", in *Encyclopedia of Geochemistry and Environmental Sciences,* Fairbridge, R. W., Ed., Van Nostrand, New York, 1969.

Element	Abundance Crust mg/kg	Abundance Sea mg/L	Element	Abundance Crust mg/kg	Abundance Sea mg/L
Ac	5.5×10^{-10}		N	1.9×10^1	5×10^{-1}
Ag	7.5×10^{-2}	4×10^{-5}	Na	2.36×10^4	1.08×10^4
Al	8.23×10^4	2×10^{-3}	Nb	2.0×10^1	1×10^{-5}
Ar	3.5	4.5×10^{-1}	Nd	4.15×10^1	2.8×10^{-6}
As	1.8	3.7×10^{-3}	Ne	5×10^{-3}	1.2×10^{-4}
Au	4×10^{-3}	4×10^{-6}	Ni	8.4×10^1	5.6×10^{-4}
B	1.0×10^1	4.44	O	4.61×10^5	8.57×10^5
Ba	4.25×10^2	1.3×10^{-2}	Os	1.5×10^{-3}	
Be	2.8	5.6×10^{-6}	P	1.05×10^3	6×10^{-2}
Bi	8.5×10^{-3}	2×10^{-5}	Pa	1.4×10^{-6}	5×10^{-11}
Br	2.4	6.73×10^1	Pb	1.4×10^1	3×10^{-5}
C	2.00×10^2	2.8×10^1	Pd	1.5×10^{-2}	
Ca	4.15×10^4	4.12×10^2	Po	2×10^{-10}	1.5×10^{-14}
Cd	1.5×10^{-1}	1.1×10^{-4}	Pr	9.2	6.4×10^{-7}
Ce	6.65×10^1	1.2×10^{-6}	Pt	5×10^{-3}	
Cl	1.45×10^2	1.94×10^4	Ra	9×10^{-7}	8.9×10^{-11}
Co	2.5×10^1	2×10^{-5}	Rb	9.0×10^1	1.2×10^{-1}
Cr	1.02×10^2	3×10^{-4}	Re	7×10^{-4}	4×10^{-6}
Cs	3	3×10^{-4}	Rh	1×10^{-3}	
Cu	6.0×10^1	2.5×10^{-4}	Rn	4×10^{-13}	6×10^{-16}
Dy	5.2	9.1×10^{-7}	Ru	1×10^{-3}	7×10^{-7}
Er	3.5	8.7×10^{-7}	S	3.50×10^2	9.05×10^2
Eu	2.0	1.3×10^{-7}	Sb	2×10^{-1}	2.4×10^{-4}
F	5.85×10^2	1.3	Sc	2.2×10^1	6×10^{-7}
Fe	5.63×10^4	2×10^{-3}	Se	5×10^{-2}	2×10^{-4}
Ga	1.9×10^1	3×10^{-5}	Si	2.82×10^5	2.2
Gd	6.2	7×10^{-7}	Sm	7.05	4.5×10^{-7}
Ge	1.5	5×10^{-5}	Sn	2.3	4×10^{-6}
H	1.40×10^3	1.08×10^5	Sr	3.70×10^2	7.9
He	8×10^{-3}	7×10^{-6}	Ta	2.0	2×10^{-6}
Hf	3.0	7×10^{-6}	Tb	1.2	1.4×10^{-7}
Hg	8.5×10^{-2}	3×10^{-5}	Te	1×10^{-3}	
Ho	1.3	2.2×10^{-7}	Th	9.6	1×10^{-6}
I	4.5×10^{-1}	6×10^{-2}	Ti	5.65×10^3	1×10^{-3}
In	2.5×10^{-1}	2×10^{-2}	Tl	8.5×10^{-1}	1.9×10^{-5}
Ir	1×10^{-3}		Tm	5.2×10^{-1}	1.7×10^{-7}
K	2.09×10^4	3.99×10^2	U	2.7	3.2×10^{-3}
Kr	1×10^{-4}	2.1×10^{-4}	V	1.20×10^2	2.5×10^{-3}
La	3.9×10^1	3.4×10^{-6}	W	1.25	1×10^{-4}
Li	2.0×10^1	1.8×10^{-1}	Xe	3×10^{-5}	5×10^{-5}
Lu	8×10^{-1}	1.5×10^{-7}	Y	3.3×10^1	1.3×10^{-5}
Mg	2.33×10^4	1.29×10^3	Yb	3.2	8.2×10^{-7}
Mn	9.50×10^2	2×10^{-4}	Zn	7.0×10^1	4.9×10^{-3}
Mo	1.2	1×10^{-2}	Zr	1.65×10^2	3×10^{-5}

SOLAR SPECTRAL IRRADIANCE

The solar luminosity (total radiant power emitted) is $3.86 \cdot 10^{26}$ W, of which 1373 W/m^2 reaches the top of the earth's atmosphere. To a zeroth approximation the sun can be considered a black body with an effective temperature of 5780 K, which implies a peak in the radiation at around 0.520 μm (5200 Å). The actual solar spectral emission is more complex, especially at ultraviolet and shorter wavelengths. The graph below, which was taken from Reference 1, summarizes the solar irradiance at the top of the atmosphere in the range 0.3 to 10 μm.

REFERENCES

1. Jursa, A.S., ed., *Handbook of Geophysics and the Space Environment*, Air Force Geophysics Laboratory, 1985.
2. Pierce, A.K., and Allen, R.G., "The Solar Spectrum between 0.3 and 10 μm", in *The Solar Output and its Variation*, White, O.R., Ed., Colorado Associated University Press, Boulder, CO, 1977.
3. Lang, K.R., *Astrophysical Data. Planets and Stars*, Springer-Verlag, New York, 1992.

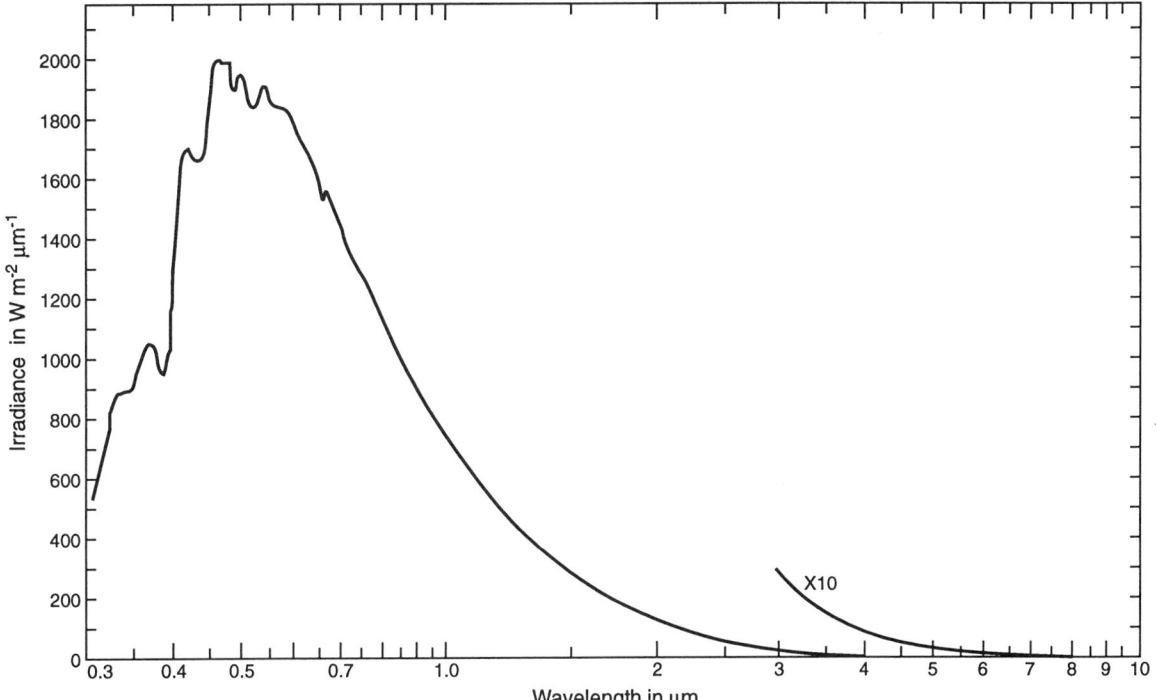

U.S. STANDARD ATMOSPHERE (1976)

A Standard Atmosphere is a hypothetical vertical distribution of atmospheric temperature, pressure, and density which is roughly representative of year-round, midlatitude conditions. Typical uses are to serve as a basis for pressure altimeter calibrations, aircraft performance calculations, aircraft and rocket design, ballistic tables, meteorological diagrams, and various types of atmospheric modeling. The air is assumed to be dry and to obey the perfect gas law and the hydrostatic equation which, taken together, relate temperature, pressure, and density with vertical position. The atmosphere is considered to rotate with the earth and to be an average over the diurnal cycle, the semiannual variation, and the range from active to quiet geomagnetic and sunspot conditions.

The U.S. Standard Atmosphere, 1976 is an idealized, steady-state representation of mean annual conditions of the earth's atmosphere from the surface to 1000 km at latitude 45°N, as it is assumed to exist during a period with moderate solar activity. The defining meteorological elements are sea-level temperature and pressure and a temperature-height profile to 1000 km. The 1976 Standard Atmosphere uses the following sea-level values which have been standard for many decades:

Temperature — 288.15 K (15°C)
Pressure — 101325 Pa (1013.25 mbar, 760 mm of Hg, or 29.92 in. of Hg)
Density — 1225 g/m^3 (1.225 g/L)
Mean molar mass — 28.964 g/mol

The parameters included in this condensed version of the U.S. Standard Atmosphere are:

Z — Height (geometric) above mean sea level in meters
T — Temperature in kelvins
P — Pressure in pascals (1 Pa = 0.01 millibars)
ρ — Density in kilograms per cubic meter (1 kg/m^3 = 1 g/L)
n — Number density in molecules per cubic meter
ν — Mean collision frequency in collisions per second
l — Mean free path in meters
η — Absolute viscosity in pascal seconds (1 Pa s = 1000 cP)
k — Thermal conductivity in joules per meter second kilogram (W/m K)
v_s — Speed of sound in meters per second
g — Acceleration of gravity in meters per second square

The sea-level composition (percent by volume) is taken to be:

N_2 — 78.084%   He — 0.000524
O_2 — 20.9476   Kr — 0.000114
Ar — 0.934   Xe — 0.0000087
CO_2 — 0.0314 CH_4 — 0.0002
Ne — 0.001818 H_2 — 0.00005

The T and P columns for the troposphere and lower stratosphere were generated from the following formulas:

	T/K	P/Pa
$H \leq 1100$ m	288.15 - 0.0065 H	$101325(288.15/T)^{-5.25577}$
11000 m $< H \leq 20000$ m	216.65	$22632\ e^{-0.00015768832(H-11000)}$
20000 m $< H \leq 32000$ m	216.65 + 0.0010(H-20000)	$5474.87(216.65/T)^{34.16319}$

where $H = rZ/(r + Z)$ is the geopotential height in meters and r is the mean earth radius at 45° N latitude, taken as 6356766 m. For altitudes up to 32 km, $\rho = 0.003483677(P/T)$ in the units used here. Formulas for the other quantities may be found in the references.

REFERENCES

1. COESA, U.S. Standard Atmosphere, 1976, U.S. Government Printing Office, Washington, D.C., 1976.
2. Jursa, A.S., ed., *Handbook of Geophysics and the Space Environment*, Air Force Geophysics Laboratory, 1985.

Z/m	T/K	P/Pa	ρ/kg m⁻³	n/m⁻³	ν/s⁻¹	l/m	η/Pa s	k/J m⁻¹s⁻¹K⁻¹	v_s/m s⁻¹	g/m s⁻²
-5000	320.68	1.778E+05	1.931	4.015E+25	1.151E+10	4.208E-08	1.942E-05	0.02788	359.0	9.822
-4500	317.42	1.685E+05	1.849	3.845E+25	1.096E+10	4.395E-08	1.927E-05	0.02763	357.2	9.830
-4000	314.17	1.596E+05	1.770	3.680E+25	1.044E+10	4.592E-08	1.912E-05	0.02738	355.3	9.819
-3500	310.91	1.511E+05	1.693	3.520E+25	9.933E+09	4.800E-08	1.897E-05	0.02713	353.5	9.818
-3000	307.66	1.430E+05	1.619	3.366E+25	9.448E+09	5.019E-08	1.882E-05	0.02688	351.6	9.816
-2500	304.41	1.352E+05	1.547	3.217E+25	8.982E+09	5.252E-08	1.867E-05	0.02663	349.8	9.814
-2000	301.15	1.278E+05	1.478	3.102E+25	8.623E+09	5.447E-08	1.852E-05	0.02638	347.9	9.813
-1500	297.90	1.207E+05	1.411	2.935E+25	8.106E+09	5.757E-08	1.836E-05	0.02613	346.0	9.811
-1000	294.65	1.139E+05	1.347	2.801E+25	7.693E+09	6.032E-08	1.821E-05	0.02587	344.1	9.810
-500	291.40	1.075E+05	1.285	2.672E+25	7.298E+09	6.324E-08	1.805E-05	0.02562	342.2	9.808
0	288.15	1.013E+05	1.225	2.547E+25	6.919E+09	6.633E-08	1.789E-05	0.02533	340.3	9.807
500	284.90	9.546E+04	1.167	2.427E+25	6.556E+09	6.961E-08	1.774E-05	0.02511	338.4	9.805
1000	281.65	8.988E+04	1.112	2.311E+25	6.208E+09	7.310E-08	1.758E-05	0.02485	336.4	9.804
1500	278.40	8.456E+04	1.058	2.200E+25	5.874E+09	7.680E-08	1.742E-05	0.02459	334.5	9.802
2000	275.15	7.950E+04	1.007	2.093E+25	5.555E+09	8.073E-08	1.726E-05	0.02433	332.5	9.801
2500	271.91	7.469E+04	0.957	1.990E+25	5.250E+09	8.491E-08	1.710E-05	0.02407	330.6	9.799
3000	268.66	7.012E+04	0.909	1.891E+25	4.959E+09	8.937E-08	1.694E-05	0.02381	328.6	9.797
3500	265.41	6.579E+04	0.863	1.795E+25	4.680E+09	9.411E-08	1.678E-05	0.02355	326.6	9.796
4000	262.17	6.166E+04	0.819	1.704E+25	4.414E+09	9.917E-08	1.661E-05	0.02329	324.6	9.794
4500	258.92	5.775E+04	0.777	1.616E+25	4.160E+09	1.046E-07	1.645E-05	0.02303	322.6	9.793
5000	255.68	5.405E+04	0.736	1.531E+25	3.918E+09	1.103E-07	1.628E-05	0.02277	320.6	9.791
5500	252.43	5.054E+04	0.697	1.450E+25	3.687E+09	1.165E-07	1.612E-05	0.02250	318.5	9.790
6000	249.19	4.722E+04	0.660	1.373E+25	3.467E+09	1.231E-07	1.595E-05	0.02224	316.5	9.788
6500	245.94	4.408E+04	0.664	1.299E+25	3.258E+09	1.302E-07	1.578E-05	0.02197	314.4	9.787
7000	242.70	4.111E+04	0.590	1.227E+25	3.058E+09	1.377E-07	1.561E-05	0.02170	312.3	9.785
7500	239.46	3.830E+04	0.557	1.159E+25	2.869E+09	1.458E-07	1.544E-05	0.02144	310.2	9.784
8000	236.22	3.565E+04	0.526	1.093E+25	2.689E+09	1.545E-07	1.527E-05	0.02117	308.1	9.782
8500	232.97	3.315E+04	0.496	1.031E+25	2.518E+09	1.639E-07	1.510E-05	0.02090	306.0	9.781
9000	229.73	3.080E+04	0.467	9.711E+24	2.356E+09	1.740E-07	1.493E-05	0.02063	303.9	9.779
9500	226.49	2.858E+04	0.440	9.141E+24	2.202E+09	1.848E-07	1.475E-05	0.02036	301.7	9.777
10000	223.25	2.650E+04	0.414	8.598E+24	2.056E+09	1.965E-07	1.458E-05	0.02009	299.5	9.776
10500	220.01	2.454E+04	0.389	8.079E+24	1.918E+09	2.091E-07	1.440E-05	0.01982	297.4	9.774
11000	216.77	2.270E+04	0.365	7.585E+24	1.787E+09	2.227E-07	1.422E-05	0.01954	295.2	9.773
11500	216.65	2.098E+04	0.337	7.016E+24	1.653E+09	2.408E-07	1.422E-05	0.01953	295.1	9.771
12000	216.65	1.940E+04	0.312	6.486E+24	1.528E+09	2.605E-07	1.422E-05	0.01953	295.1	9.770
12500	216.65	1.793E+04	0.288	5.996E+24	1.412E+09	2.818E-07	1.422E-05	0.01953	295.1	9.768
13000	216.65	1.658E+04	0.267	5.543E+24	1.306E+09	3.048E-07	1.422E-05	0.01953	295.1	9.767
13500	216.65	1.533E+04	0.246	5.124E+24	1.207E+09	3.297E-07	1.422E-05	0.01953	295.1	9.765
14000	216.65	1.417E+04	0.228	4.738E+24	1.116E+09	3.566E-07	1.422E-05	0.01953	295.1	9.764
14500	216.65	1.310E+04	0.211	4.380E+24	1.032E+09	3.857E-07	1.422E-05	0.01953	295.1	9.762
15000	216.65	1.211E+04	0.195	4.049E+24	9.538E+08	4.172E-07	1.422E-05	0.01953	295.1	9.761
16000	216.65	1.035E+04	0.166	3.461E+24	8.153E+08	4.881E-07	1.422E-05	0.01953	295.1	9.758
17000	216.65	8.850E+03	0.142	2.959E+24	6.969E+08	5.710E-07	1.422E-05	0.01953	295.1	9.754
18000	216.65	7.565E+03	0.122	2.529E+24	5.958E+08	6.680E-07	1.422E-05	0.01953	295.1	9.751
19000	216.65	6.467E+03	0.104	2.162E+24	5.093E+08	7.814E-07	1.422E-05	0.01953	295.1	9.748

Z/m	T/K	P/Pa	ρ/kg m⁻³	n/m⁻³	v/s⁻¹	l/m	η/Pa s	k/J m⁻¹s⁻¹K⁻¹	v_s/m s⁻¹	g/m s⁻²
20000	216.65	5.529E+03	8.891E-02	1.849E+24	4.354E+08	9.139E-07	1.422E-05	0.01953	295.1	9.745
21000	217.58	4.729E+03	7.572E-02	1.574E+24	3.716E+08	1.073E-06	1.427E-05	0.01961	295.1	9.742
22000	218.57	4.048E+03	6.451E-02	1.341E+24	3.173E+08	1.260E-06	1.432E-05	0.01970	296.4	9.739
23000	219.57	3.467E+03	5.501E-02	1.144E+24	2.712E+08	1.477E-06	1.438E-05	0.01978	297.1	9.736
24000	220.56	2.972E+03	4.694E-02	9.759E+23	2.319E+08	1.731E-06	1.443E-05	0.01986	297.7	9.733
25000	221.55	2.549E+03	4.008E-02	8.334E+23	1.985E+08	2.027E-06	1.448E-05	0.01995	298.4	9.730
26000	222.54	2.188E+03	3.426E-02	7.123E+23	1.700E+08	2.372E-06	1.454E-05	0.02003	299.1	9.727
27000	223.54	1.880E+03	2.930E-02	6.092E+23	1.458E+08	2.773E-06	1.459E-05	0.02011	299.7	9.724
28000	224.53	1.610E+03	2.508E-02	5.214E+23	1.250E+08	3.240E-06	1.465E-05	0.02020	300.4	9.721
29000	225.52	1.390E+03	2.148E-02	4.466E+23	1.073E+08	3.783E-06	1.470E-05	0.02028	301.1	9.718
30000	226.51	1.197E+03	1.841E-02	3.828E+23	9.219E+07	4.414E-06	1.475E-05	0.02036	301.7	9.715
31000	227.50	1.031E+03	1.579E-02	3.283E+23	7.925E+07	5.146E-06	1.481E-05	0.02044	302.4	9.712
32000	228.49	8.891E+02	1.356E-02	2.813E+23	6.818E+07	5.995E-06	1.486E-05	0.02053	303.0	9.709
33000	230.97	7.673E+02	1.157E-02	2.406E+23	5.852E+07	7.021E-06	1.499E-05	0.02073	304.7	9.706
34000	233.74	6.634E+02	9.887E-03	2.056E+23	5.030E+07	8.218E-06	1.514E-05	0.02096	306.5	9.703
35000	236.51	5.746E+02	8.463E-03	1.760E+23	4.331E+07	9.601E-06	1.529E-05	0.02119	308.3	9.700
36000	239.28	4.985E+02	7.258E-03	1.509E+23	3.736E+07	1.120E-05	1.543E-05	0.02142	310.1	9.697
38000	244.82	3.771E+02	5.367E-03	1.116E+23	2.794E+07	1.514E-05	1.572E-05	0.02188	313.7	9.690
40000	250.35	2.871E+02	3.996E-03	8.308E+22	2.104E+07	2.034E-05	1.601E-05	0.02233	317.2	9.684
42000	255.88	2.200E+02	2.995E-03	6.227E+22	1.594E+07	2.713E-05	1.629E-05	0.02278	320.7	9.678
44000	261.40	1.695E+02	2.259E-03	4.697E+22	1.215E+07	3.597E-05	1.657E-05	0.02323	324.1	9.672
46000	266.93	1.313E+02	1.714E-03	3.564E+22	9.318E+06	4.740E-05	1.685E-05	0.02376	327.5	9.666
48000	270.65	1.023E+02	1.317E-03	2.738E+22	7.208E+06	6.171E-05	1.704E-05	0.02397	329.8	9.660
50000	270.65	7.978E+01	1.027E-03	2.135E+22	5.620E+06	7.913E-05	1.703E-05	0.02397	329.8	9.654
52000	269.03	6.221E+01	8.056E-04	1.675E+22	4.397E+06	1.009E-04	1.696E-05	0.02384	328.8	9.648
54000	263.52	4.834E+01	6.390E-04	1.329E+22	3.452E+06	1.272E-04	1.660E-05	0.02340	325.4	9.642
56000	258.02	3.736E+01	5.045E-04	1.049E+22	2.696E+06	1.611E-04	1.640E-05	0.02296	322.0	9.636
58000	252.52	2.872E+01	3.963E-04	8.239E+21	2.095E+06	2.051E-04	1.612E-05	0.02251	318.6	9.632
60000	247.02	2.196E+01	3.097E-04	6.439E+21	1.620E+06	2.624E-04	1.584E-05	0.02206	315.1	9.624
65000	233.29	1.093E+01	1.632E-04	3.393E+21	8.294E+05	4.979E-04	1.512E-05	0.02093	306.2	9.609
70000	219.59	5.221	8.283E-05	1.722E+21	4.084E+05	9.810E-04	1.438E-05	0.01978	297.1	9.594
75000	208.40	2.388	3.992E-05	8.300E+20	1.918E+05	2.035E-03	1.376E-05	0.01883	289.4	9.579
80000	198.64	1.052	1.846E-05	3.838E+20	8.656E+04	4.402E-03	1.321E-05	0.01800	282.5	9.564
85000	188.89	4.457E-01	8.220E-06	1.709E+20	3.766E+04	9.886E-03	1.265E-05	0.01716	275.5	9.550
90000	186.87	1.836E-01	3.416E-06	7.116E+19	1.560E+04	2.370E-02				9.535
95000	188.42	7.597E-02	1.393E-06	2.920E+19	6.440E+03	5.790E-02				9.520
100000	195.08	3.201E-02	5.604E-07	1.189E+19	2.680E+03	1.420E-01				9.505
110000	240.00	7.104E-03	9.708E-08	2.144E+18	5.480E+02	7.880E-01				9.476
120000	360.00	2.538E-03	2.222E-08	5.107E+17	1.630E+02	3.310				9.447
130000	469.27	1.251E-03	8.152E-09	1.930E+17	7.100E+01	8.800				9.418
140000	559.63	7.203E-04	3.831E-09	9.322E+16	3.800E+01	1.800E+01				9.389
150000	634.39	4.542E-04	2.076E-09	5.186E+16	2.300E+01	3.300E+01				9.360
160000	696.29	3.040E-04	1.233E-09	3.162E+16	1.500E+01	5.300E+01				9.331
170000	747.57	2.121E-04	7.815E-10	2.055E+16	1.000E+01	8.200E+01				9.302
180000	790.07	1.527E-04	5.194E-10	1.400E+16	7.200	1.200E+02				9.274

Z/m	T/K	P/Pa	ρ/kg m^{-3}	n/m^{-3}	v/s^{-1}	l/m	η/Pa s	k/J m^{-1}s^{-1}K^{-1}	v_s/m s^{-1}	g/m s^{-2}
190000	825.16	1.127E-04	3.581E-10	9.887E+15	5.200	1.700E+02				9.246
200000	854.56	8.474E-05	2.541E-10	7.182E+15	3.900	2.400E+02				9.218
220000	899.01	5.015E-05	1.367E-10	4.040E+15	2.300	4.200E+02				9.162
240000	929.73	3.106E-05	7.858E-11	2.420E+15	1.400	7.000E+02				9.106
260000	950.99	1.989E-05	4.742E-11	1.515E+15	9.300E-01	1.100E+03				9.051
280000	965.75	1.308E-05	2.971E-11	9.807E+14	6.100E-01	1.700E+03				8.997
300000	976.01	8.770E-06	1.916E-11	6.509E+14	4.200E-01	2.600E+03				8.943
320000	983.16	5.980E-06	1.264E-11	4.405E+14	2.900E-01	3.800E+03				8.889
340000	988.15	4.132E-06	8.503E-12	3.029E+14	2.000E-01	5.600E+03				8.836
360000	991.65	2.888E-06	5.805E-12	2.109E+14	1.400E-01	8.000E+03				8.784
380000	994.10	2.038E-06	4.013E-12	1.485E+14	1.000E-01	1.100E+04				8.732
400000	995.83	1.452E-06	2.803E-12	1.056E+14	7.200E-02	1.600E+04				8.680
450000	998.22	6.447E-07	1.184E-12	4.678E+13	3.300E-02	3.600E+04				8.553
500000	999.24	3.024E-07	5.215E-13	2.192E+13	1.600E-02	7.700E+04				8.429
550000	999.67	1.514E-07	2.384E-13	1.097E+13	8.400E-03	1.500E+05				8.307
600000	999.85	8.213E-08	1.137E-13	5.950E+12	4.800E-03	2.800E+05				8.188
650000	999.93	4.887E-08	5.712E-14	3.540E+12	3.100E-03	4.800E+05				8.072
700000	999.97	3.191E-08	3.070E-14	2.311E+12	2.200E-03	7.300E+05				7.958
750000	999.98	2.260E-08	1.788E-14	1.637E+12	1.700E-03	1.000E+06				7.846
800000	999.99	1.704E-08	1.136E-14	1.234E+12	1.400E-03	1.400E+06				7.737
850000	1000.00	1.342E-08	7.824E-15	9.717E+11	1.200E-03	1.700E+06				7.630
900000	1000.00	1.087E-08	5.759E-15	7.876E+11	1.000E-03	2.100E+06				7.525
950000	1000.00	8.982E-09	4.453E-15	6.505E+11	8.700E-04	2.600E+06				7.422
1000000	1000.00	7.514E-09	3.561E-15	5.442E+11	7.500E-04	3.100E+06				7.322

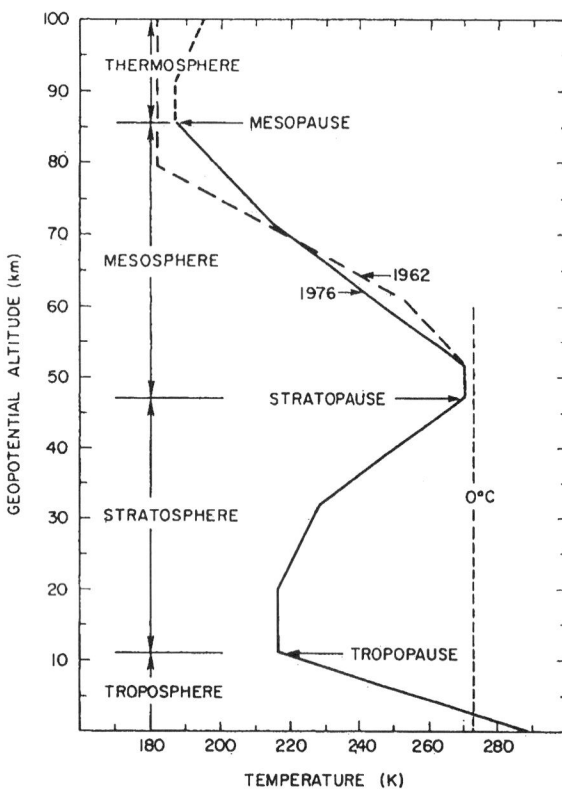

FIGURE 1. Temperature-height profile for U.S. Standard Atmosphere.

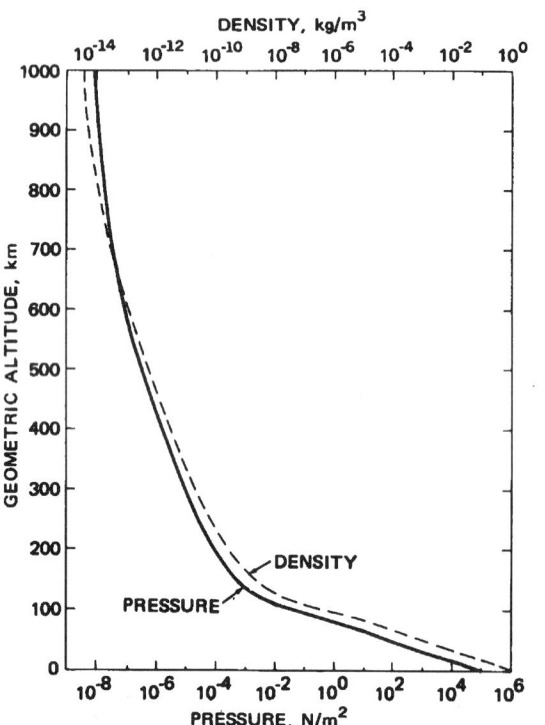

FIGURE 2. Total pressure and mass density as a function of geometric altitude.

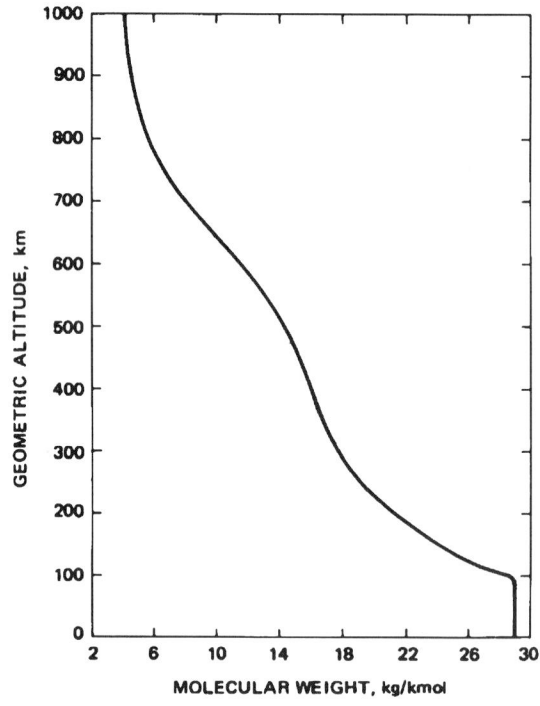

FIGURE 3. Mean molecular weight as a function of geometric altitude.

FIGURE 4. Number density of individual species and total number density as a function of geometric altitude.

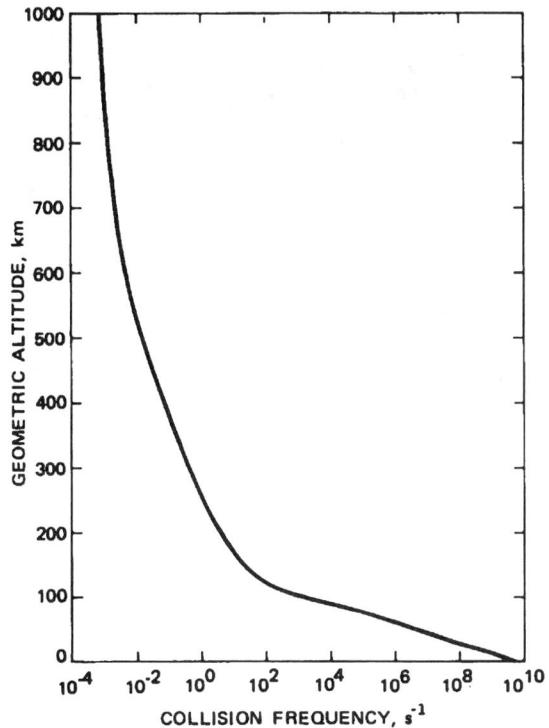

FIGURE 5. Collision frequency as a function of geometric altitude.

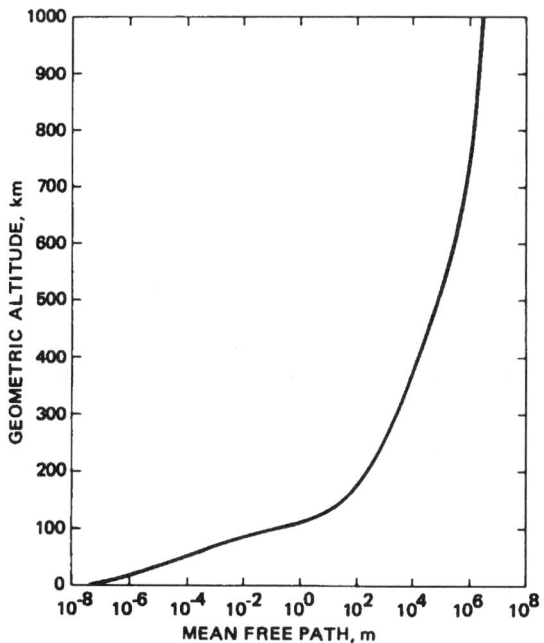

FIGURE 6. Mean free path as a function of geometric altitude.

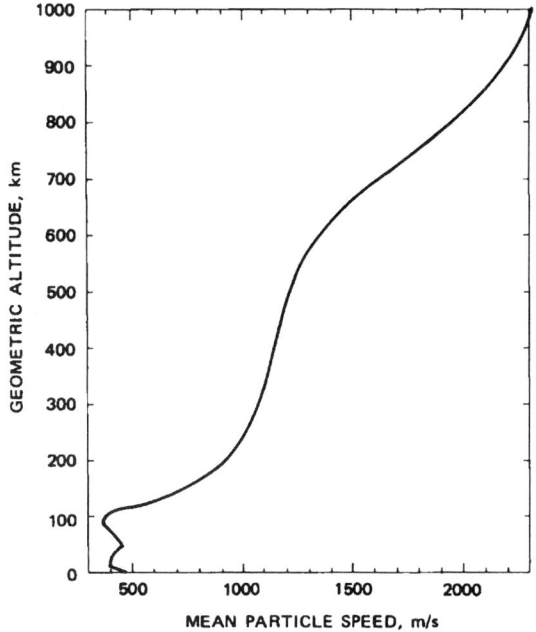

FIGURE 7. Mean air-particle speed as a function of geometric altitude.

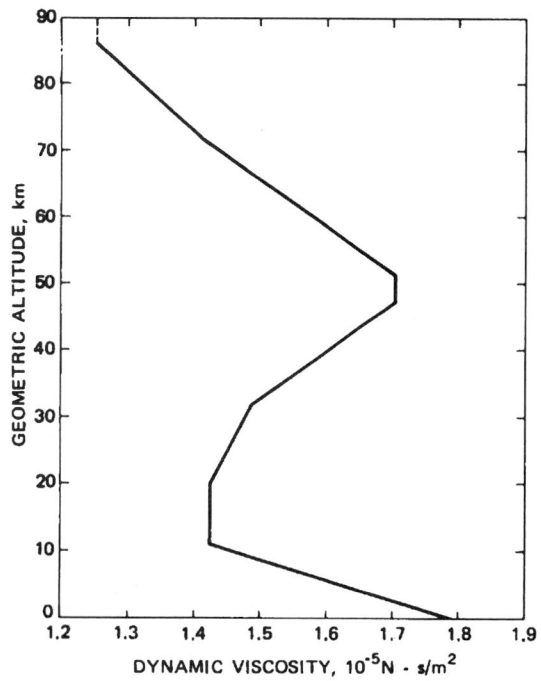

FIGURE 8. Dynamic viscosity as a function of geometric altitude.

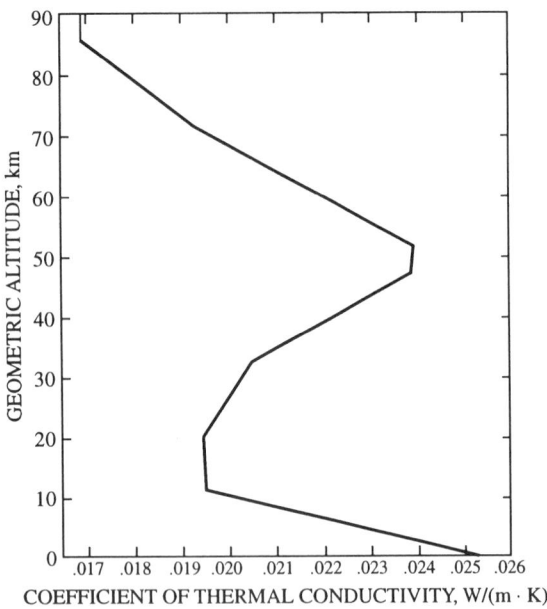

FIGURE 9. Coefficient of thermal conductivity as a function of geometric altitude.

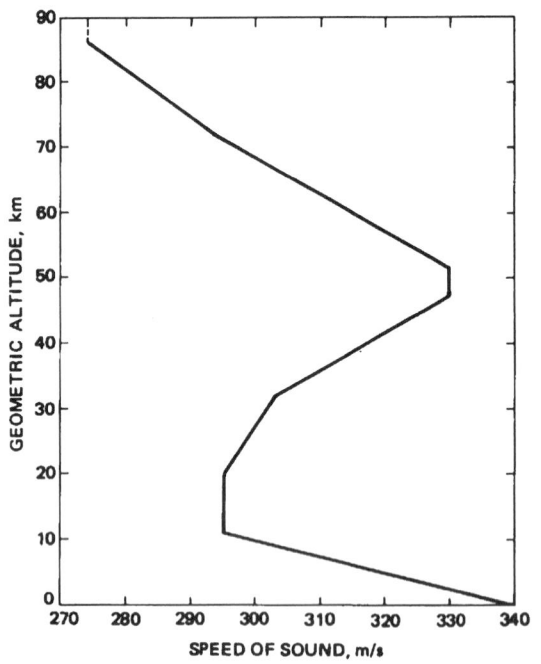

FIGURE 10. Speed of sound as a function of geometric altitude.

FIGURE 11. Molecular-diffusion and eddy-diffusion coefficients as a function of geometric altitude.

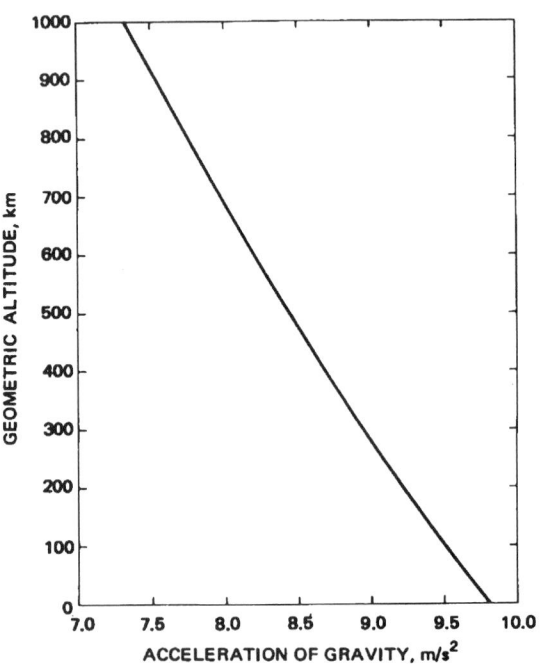

FIGURE 12. Acceleration of gravity as a function of geometric altitude.

GEOGRAPHICAL AND SEASONAL VARIATION IN SOLAR RADIATION

This table gives the amount of solar radiation reaching a unit area at the top of the earth's atmosphere per day as a function of latitude and approximate date. It is based upon a solar constant (total energy per unit area at the earth's average orbital distance) of 1373 W/m^2. Absorption of radiation by the atmosphere is not taken into consideration.

REFERENCE

List, R.J., *Smithsonian Meteorological Tables, Seventh Edition*, Smithsonian Institution Press, Washington, D.C., 1962.

Daily Solar Radiation in MJ/m^2

Lat.	Mar. 21	Apr. 13	May 6	May 29	Jun. 2	Jul. 15	Aug. 8	Aug. 31
90°		18.0	32.8	42.4	45.7	42.2	32.5	17.7
80	6.6	18.0	32.3	41.8	45.0	41.6	32.0	17.7
70	13.0	22.3	31.8	39.9	43.0	39.7	31.5	22.0
60	19.0	27.0	34.4	39.7	41.6	39.4	34.0	26.7
50	24.4	31.1	36.8	40.7	42.0	40.5	36.5	30.8
40	29.1	34.3	38.6	41.3	42.1	41.1	38.3	33.9
30	32.9	36.7	39.4	41.1	41.4	40.8	39.1	36.3
20	35.7	38.0	39.2	39.7	39.7	39.5	38.9	37.5
10	37.4	38.1	37.9	37.4	37.1	37.2	37.6	37.7
0	38.0	37.1	35.5	34.1	33.5	34.0	35.2	36.6
−10	37.4	35.0	32.3	30.0	29.2	29.9	32.0	34.6
−20	35.7	31.8	28.0	25.2	24.1	25.1	27.8	31.5
−30	32.9	27.8	23.1	19.7	18.5	19.7	22.8	27.4
−40	29.1	22.8	17.5	14.0	12.6	13.9	17.4	22.6
−50	24.4	17.3	11.7	8.2	7.0	8.2	11.6	17.2
−60	19.0	11.4	5.9	2.9	2.0	2.9	5.9	11.3
−70	13.0	5.4	1.0				1.0	5.3
−80	6.6	0.3						0.3
−90								

Lat.	Sep. 23	Oct. 16	Nov. 8	Nov. 30	Dec. 22	Jan. 13	Feb. 4	Feb. 26
90°								
80	6.5	0.3						0.3
70	12.9	5.5	1.0				1.0	5.6
60	18.8	11.6	6.2	3.1	2.1	3.1	6.2	11.7
50	24.1	17.6	12.1	8.7	7.5	8.7	12.3	17.8
40	28.7	23.1	18.2	14.8	13.5	14.9	18.4	23.5
30	32.5	28.2	23.9	20.9	19.8	21.0	24.1	28.4
20	35.3	32.3	29.1	26.6	25.7	26.7	29.3	32.7
10	37.0	35.5	33.5	31.8	31.1	31.9	33.8	35.9
0	37.6	37.6	36.9	36.1	35.8	36.3	37.3	38.0
−10	37.0	38.6	39.4	39.5	39.6	39.7	39.7	39.1
−20	35.3	38.5	40.7	42.0	42.4	42.2	41.1	39.0
−30	32.5	37.2	40.9	43.3	44.2	43.5	41.3	37.7
−40	28.7	34.8	40.1	43.6	45.0	43.8	40.5	35.2
−50	24.1	31.5	38.3	43.1	44.8	43.2	38.6	31.9
−60	18.8	27.3	35.7	41.9	44.4	42.1	36.0	27.7
−70	12.9	22.6	33.0	42.2	45.9	42.4	33.3	22.9
−80	6.5	18.2	33.5	44.2	48.1	44.4	33.8	18.4
−90		18.2	34.0	44.8	48.8	45.1	34.4	18.4

INFRARED ABSORPTION BY THE EARTH'S ATMOSPHERE

This graph summarizes the absorption by various atmospheric constituents in the wavelength range from 1 to 13 μm (wavenumber range 10,000 to 770 cm^{-1}). The vertical scale is in arbitrary units and does not take into account the wavelength variation of either the solar background radiation or the infrared detector response. Thus the intensities of the absorption bands have only qualitative significance.

WAVELENGTH IN MICROMETERS

14-24

ATMOSPHERIC CONCENTRATION OF CARBON DIOXIDE, 1958-1996

The data in this table were taken at Mauna Loa Observatory in Hawaii and represent averages adjusted to the 15th of each month. The concentration of CO_2 is given in parts per million by volume. Data from other measurement sites may be found in the references.

REFERENCES

1. Keeling, C. D., and Whorf, T. P., Carbon Dioxide Information Analysis Center WWW site <http://cdiac.esd.ornl.gov/trends_html/trends/co2>, August 1997.
2. Keeling, C. D., and Whorf, T. P., in *Trends '93: A Compendium of Data on Global Change*, p.16, Boden, T. A., Kaiser, D. P., Sepanski, R. J., and Stoss, F. W., Editors, ORNL/CDIAC-65, Oak Ridge National Laboratory, Oak Ridge, TN, 1994.

Year	Jan.	Feb.	Mar.	Apr.	May	June	July	Aug.	Sept.	Oct.	Nov.	Dec.	Annual
1958	N/A	N/A	315.56	317.29	317.34	N/A	315.69	314.78	313.05	N/A	313.18	314.50	N/A
1959	315.42	316.31	316.50	317.56	318.13	318.00	316.39	314.65	313.68	313.18	314.66	315.43	315.83
1960	316.27	316.81	317.42	318.87	319.87	319.43	318.01	315.74	314.00	313.68	314.84	316.03	316.75
1961	316.73	317.54	318.38	319.31	320.42	319.61	318.42	316.63	314.84	315.16	315.94	316.85	317.49
1962	317.78	318.40	319.53	320.42	320.85	320.45	319.45	317.25	316.11	315.27	316.53	317.53	318.30
1963	318.58	318.93	319.70	321.22	322.08	321.31	319.58	317.61	316.05	315.83	316.91	318.20	318.83
1964	319.41	N/A	N/A	N/A	322.05	321.73	320.27	318.53	316.54	316.72	317.53	318.55	N/A
1965	319.27	320.28	320.73	321.97	322.00	321.71	321.05	318.71	317.66	317.14	318.70	319.25	319.87
1966	320.46	321.43	322.23	323.54	323.91	323.59	322.24	320.20	318.48	317.94	319.63	320.87	321.21
1967	322.17	322.34	322.88	324.25	324.83	323.93	322.38	320.76	319.10	319.24	320.56	321.80	322.02
1968	322.40	322.99	323.73	324.86	325.40	325.20	323.98	321.95	320.18	320.09	321.16	322.74	322.89
1969	323.83	324.27	325.47	326.50	327.21	326.54	325.72	323.50	322.22	321.62	322.69	323.95	324.46
1970	324.89	325.82	326.77	327.97	327.90	327.50	326.18	324.53	322.93	322.90	323.85	324.96	325.52
1971	326.01	326.51	327.02	327.62	328.76	328.40	327.20	325.27	323.20	323.40	324.63	325.85	326.16
1972	326.60	327.47	327.58	329.56	329.90	328.92	327.88	326.16	324.68	325.04	326.34	327.39	327.29
1973	328.37	329.40	330.14	331.33	332.31	331.90	330.70	329.15	327.35	327.02	327.99	328.48	329.51
1974	329.18	330.55	331.32	332.48	332.92	332.08	331.01	329.23	327.27	327.21	328.29	329.41	330.08
1975	330.23	331.25	331.87	333.14	333.80	333.43	331.73	329.90	328.40	328.17	329.32	330.59	330.99
1976	331.58	332.39	333.33	334.41	334.71	334.17	332.89	330.77	329.14	328.78	330.14	331.52	331.98
1977	332.75	333.24	334.53	335.90	336.57	336.10	334.76	332.59	331.42	330.98	332.24	333.68	333.73
1978	334.80	335.22	336.47	337.59	337.84	337.72	336.37	334.51	332.60	332.38	333.75	334.78	335.34
1979	336.05	336.59	337.79	338.71	339.30	339.12	337.56	335.92	333.75	333.70	335.12	336.56	336.68
1980	337.84	338.19	339.91	340.60	341.29	341.00	339.39	337.43	335.72	335.84	336.93	338.04	338.52
1981	339.06	340.30	341.21	342.33	342.74	342.08	340.32	338.26	336.52	336.68	338.19	339.44	339.76
1982	340.57	341.44	342.53	343.39	343.96	343.18	341.88	339.65	337.81	337.69	339.09	340.32	340.96
1983	341.20	342.35	342.93	344.77	345.58	345.14	343.81	342.21	339.69	339.82	340.98	342.82	342.61
1984	343.52	344.33	345.11	346.89	347.25	346.62	345.22	343.11	340.90	341.18	342.80	344.04	344.25
1985	344.79	345.82	347.25	348.17	348.74	348.07	346.38	344.52	342.92	342.62	344.06	345.38	345.73
1986	346.11	346.78	347.68	349.37	350.03	349.37	347.76	345.73	344.68	343.99	345.48	346.72	346.97
1987	347.84	348.29	349.24	350.80	351.66	351.08	349.33	347.92	346.27	346.18	347.64	348.78	348.75
1988	350.25	351.54	352.05	353.41	354.04	353.63	352.22	350.27	348.55	348.72	349.91	351.18	351.31
1989	352.60	352.92	353.53	355.26	355.52	354.97	353.75	351.52	349.64	349.83	351.14	352.37	352.75
1990	353.50	354.55	355.23	356.04	357.00	356.07	354.67	352.76	350.82	351.04	352.70	354.07	354.04
1991	354.59	355.63	357.03	358.48	359.22	358.12	356.06	353.92	352.05	352.11	353.64	354.89	355.48
1992	355.88	356.63	357.72	359.07	359.58	359.17	356.94	354.92	352.94	353.23	354.09	355.33	356.29
1993	356.63	357.10	358.32	359.41	360.23	359.55	357.53	355.48	353.67	353.95	355.30	356.78	356.99
1994	358.34	358.89	359.95	361.25	361.67	360.94	359.55	357.49	355.84	356.00	357.59	359.05	358.88
1995	359.98	361.03	361.66	363.48	363.82	363.30	361.93	359.49	358.08	357.77	359.57	360.69	360.90
1996	362.03	363.22	363.99	364.66	365.36	364.91	363.58	361.41	359.37	359.50	360.65	362.21	362.57

CO$_2$ Concentration in Parts per Million (Annual Average)

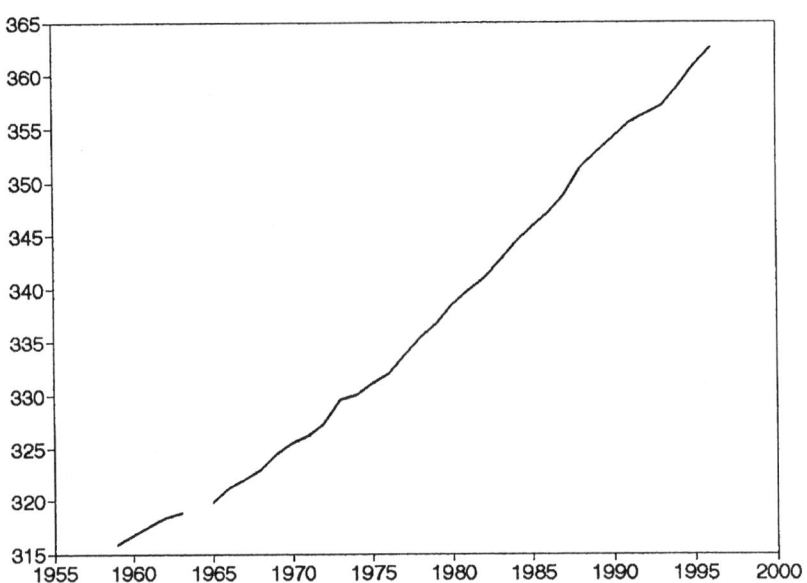

MEAN TEMPERATURES IN THE UNITED STATES, 1900-1992

Historical records of atmospheric temperatures have been analyzed to obtain mean temperatures in °C for 23 climatically distinct regions of the United States. The table below gives the average over these 23 regions, which cover completely the contiguous 48 states. Data for the individual regions and for other parts of the world may be found in the references.

The data are presented as temperature anomalies, i.e., as deviations (in °C) from the average temperature at each individual recording station over a 1961-1990 reference period. The trend in the temperature anomaly thus gives an indication of the long-term variation in average temperatures.

CY Mean: Calendar year mean (January-December)
Winter: December-February
Spring: March-May
Summer: June-August
Fall: September-November

REFERENCES

1. Karl, T. R., Easterling, D. R., Knight, R. W., and Hughes, P. Y., in *Trends '93: A Compendium of Data on Global Change*, p. 686, Boden, T. A., Kaiser, D. P., Sepanski, R. J., and Stoss, F. W., Editors, ORNL/CDIAC-65, Oak Ridge National Laboratory, Oak Ridge, TN, 1994.
2. Carbon Dioxide Information Analysis Center, WWW site <http://cdiac.esd.ornl.gov/ftp/trends93>.

Year	CY Mean	Winter	Spring	Summer	Fall
1900	0.46		0.21	0.27	0.85
1901	-0.21	0.07	-0.46	0.48	-0.28
1902	-0.14	-0.69	0.48	-0.58	0.12
1903	-0.77	-0.84	0.10	-0.83	-1.02
1904	-0.72	-1.86	-0.39	-0.92	-0.21
1905	-0.45	-1.86	0.67	-0.32	-0.25
1906	-0.04	0.23	-0.69	-0.42	-0.10
1907	-0.23	1.09	-0.61	-0.85	-0.39
1908	-0.11	0.73	0.36	-0.64	-0.59
1909	-0.36	0.82	-1.06	0.06	0.01
1910	-0.14	-2.08	0.95	-0.55	0.13
1911	0.02	0.52	0.20	-0.19	-0.62
1912	-0.88	-1.50	-0.75	-0.76	-0.51
1913	-0.23	-0.74	-0.65	-0.07	0.22
1914	-0.05	0.48	0.04	0.14	0.32
1915	-0.11	-0.37	-0.18	-1.16	0.31
1916	-0.77	-0.29	-0.36	-0.30	-1.23
1917	-1.34	-1.93	-1.75	-0.73	-1.04
1918	-0.14	-2.02	0.30	0.03	-0.09
1919	-0.16	0.69	0.00	0.12	-0.13
1920	-0.37	-0.83	-0.96	-0.67	0.02
1921	0.87	1.56	0.83	0.52	0.32
1922	0.01	-0.44	0.15	0.16	0.41
1923	-0.10	0.23	-1.02	-0.07	-0.09
1924	-1.01	0.13	-1.27	-0.64	-0.53
1925	0.20	-0.44	0.57	0.05	-0.41
1926	-0.01	0.97	-0.58	-0.27	0.02
1927	0.20	1.11	0.41	-0.83	0.83
1928	-0.08	-0.40	-0.28	-0.43	-0.08
1929	-0.68	-1.94	0.30	-0.39	-0.78
1930	-0.12	0.07	0.09	-0.07	-0.25
1931	0.81	1.16	-0.71	0.51	1.41
1932	-0.14	1.75	-0.58	0.26	-0.78
1933	0.35	-0.60	-0.08	0.45	0.54
1934	0.86	1.45	0.77	0.86	0.82
1935	0.12	0.84	0.12	0.31	-0.47
1936	-0.10	-2.23	0.48	1.00	-0.32
1937	-0.13	-0.65	-0.24	0.66	-0.03
1938	0.71	1.31	0.98	0.39	0.08
1939	0.38	0.36	0.34	0.18	0.15
1940	0.06	0.03	-0.10	0.18	0.12

MEAN TEMPERATURES IN THE UNITED STATES, 1901-1992 (continued)

Year	CY Mean	Winter	Spring	Summer	Fall
1941	0.79	1.56	0.44	0.28	0.85
1942	0.01	0.35	0.22	0.08	0.06
1943	-0.17	0.20	-0.46	0.54	-0.79
1944	0.04	0.61	-0.38	-0.20	0.43
1945	-0.02	0.30	0.25	-0.47	0.08
1946	0.53	-0.26	1.21	-0.21	0.18
1947	0.10	0.47	-0.42	0.03	0.85
1948	-0.22	-0.67	-0.01	0.02	-0.15
1949	0.05	-0.77	0.39	0.33	0.23
1950	-0.37	0.39	-1.02	-0.86	0.13
1951	-0.45	0.05	-0.69	-0.30	-0.70
1952	0.03	0.68	-0.40	0.49	-1.22
1953	0.61	1.91	0.03	0.31	0.31
1954	0.57	1.47	-0.22	0.36	0.70
1955	-0.25	-0.33	0.12	0.19	-0.60
1956	-0.05	-0.16	-0.42	0.01	-0.37
1957	0.45	1.02	0.44	0.25	-0.13
1958	0.14	0.93	0.06	0.11	0.45
1959	0.12	-0.60	0.16	0.50	-0.52
1960	-0.27	0.43	-0.98	0.01	0.56
1961	-0.01	-0.02	-0.17	0.18	-0.29
1962	-0.04	-0.52	-0.09	-0.41	0.65
1963	-0.06	-1.35	0.61	0.12	1.38
1964	-0.27	-1.30	-0.24	-0.08	-0.40
1965	-0.03	-0.07	-0.56	-0.42	0.28
1966	-0.34	-0.31	-0.25	-0.07	-0.13
1967	-0.13	0.23	-0.05	-0.37	-0.32
1968	-0.28	-0.31	-0.16	-0.12	0.01
1969	-0.06	-0.36	-0.54	0.12	-0.16
1970	-0.12	-0.17	-0.44	0.32	-0.07
1971	-0.04	-0.08	-0.99	0.01	0.56
1972	-0.13	0.20	0.26	-0.19	0.05
1973	0.48	-0.23	0.47	0.22	0.90
1974	0.16	0.52	0.74	-0.32	-0.39
1975	-0.09	0.63	-0.79	0.03	-0.20
1976	-0.62	0.88	-0.09	-0.54	-1.72
1977	0.32	-1.95	1.07	0.60	0.68
1978	-0.37	-1.31	0.23	0.09	0.21
1979	-0.53	-2.92	0.09	-0.21	-0.07
1980	0.18	0.72	-0.25	0.43	-0.04
1981	0.64	0.90	0.80	0.57	0.36
1982	-0.08	-0.86	0.03	-0.11	0.06
1983	0.40	2.33	-0.36	0.58	0.94
1984	0.21	-0.78	-0.30	0.31	0.07
1985	-0.26	-0.78	1.24	-0.23	0.05
1986	0.93	0.22	1.22	0.45	0.49
1987	0.67	1.52	0.97	0.33	-0.17
1988	-0.07	-0.26	-0.06	0.57	0.04
1989	-0.30	-0.28	0.36	0.12	-0.27
1990	0.72	0.41	0.72	0.41	0.66
1991	0.77	0.32	1.36	0.56	-0.31
1992		2.48	0.82	-0.70	

GLOBAL TEMPERATURE TREND, 1866-1996

This table and graph summarize the trend in annual mean global temperature from 1886 to 1996. The temperatures were calculated from mean temperature anomalies by assuming an absolute global mean of 14.00 °C (which is the best estimate for the 1951-1980 period). The estimated uncertainty in the year-to-year change in the recent years is about 0.07 °C.

REFERENCES

1. NASA Goddard Institute for Space Studies, World Wide Web site, www.giss.nasa.gov/data/gistemp/
2. Hansen, J., and Wilson, H., *Climatic Change*, 25, 285, 1993.

Year	t/°C	Year	t/°C	Year	t/°C	Year	t/°C	Year	t/°C
1866	13.52	1897	13.85	1928	14.03	1959	14.05	1990	14.47
1867	13.42	1898	13.63	1929	13.82	1960	14.00	1991	14.42
1868	13.52	1899	13.78	1930	13.93	1961	14.08	1992	14.13
1869	13.64	1900	14.00	1931	14.03	1962	14.03	1993	14.21
1870	13.57	1901	13.92	1932	14.02	1963	14.02	1994	14.31
1871	13.60	1902	13.76	1933	13.86	1964	13.74	1995	14.47
1872	13.76	1903	13.64	1934	14.02	1965	13.85	1996	14.36
1873	13.81	1904	13.51	1935	13.93	1966	13.91		
1874	13.51	1905	13.73	1936	14.02	1967	13.98		
1875	13.41	1906	13.81	1937	14.12	1968	13.88		
1876	13.58	1907	13.57	1938	14.13	1969	14.03		
1877	13.89	1908	13.70	1939	13.97	1970	14.04		
1878	13.98	1909	13.70	1940	14.14	1971	13.89		
1879	13.50	1910	13.71	1941	14.12	1972	13.93		
1880	13.68	1911	13.71	1942	14.09	1973	14.19		
1881	13.67	1912	13.72	1943	14.03	1974	13.93		
1882	13.61	1913	13.77	1944	14.11	1975	13.95		
1883	13.61	1914	13.96	1945	13.96	1976	13.78		
1884	13.35	1915	13.98	1946	14.01	1977	14.16		
1885	13.58	1916	13.76	1947	14.12	1978	14.09		
1886	13.60	1917	13.58	1948	13.99	1979	14.14		
1887	13.48	1918	13.65	1949	13.93	1980	14.28		
1888	13.70	1919	13.83	1950	13.85	1981	14.38		
1889	13.88	1920	13.83	1951	13.98	1982	14.07		
1890	13.60	1921	13.90	1952	14.04	1983	14.29		
1891	13.58	1922	13.87	1953	14.13	1984	14.11		
1892	13.61	1923	13.84	1954	13.93	1985	14.11		
1893	13.55	1924	13.85	1955	13.94	1986	14.16		
1894	13.65	1925	13.85	1956	13.83	1987	14.32		
1895	13.64	1926	14.10	1957	14.10	1988	14.35		
1896	13.81	1927	13.96	1958	14.10	1989	14.25		

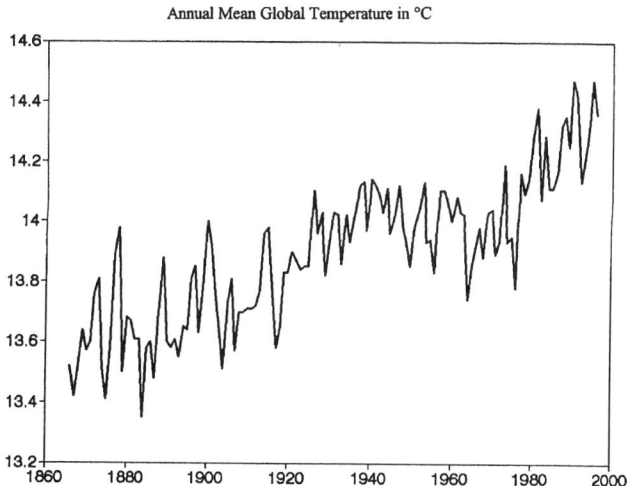

Annual Mean Global Temperature in °C

ATMOSPHERIC ELECTRICITY
Hans Dolezalek, Hannes Tammet, John Latham, and Martin A. Uman

I. SURVEY AND GLOBAL CIRCUIT
Hans Dolezalek

The science of atmospheric electricity originated in 1752 by an experimental proof of a related earlier hypothesis (that lightning is an electrical event). In spite of a large effort, in part by such eminent physicists as Coulomb, Lord Kelvin, and many others, an overall, proven theory able to generate models with sufficient resolution is not yet available. Generally accepted and encompassing text books are now more than 20 years old. The voluminous proceedings of the, so far, nine international atmospheric electricity conferences (1954 to 1992) give much valuable detail and demonstrate impressive progress, as do a number of less comprehensive textbooks published in the last 20 years, but a general theory as indicated above is not yet created. Only now, certain related measuring techniques and mathematical possibilities are emerging.

Applications to practical purposes do exist in the field of lightning research (including the electromagnetic radiation emanating from lightning) by the establishment of lightning-location networks and by the now developing possibility to detect electrified clouds which pose hazards to aircraft. Application of atmospheric electricity to other parts of meteorology seems to be promising but so far has seldom been instituted. Because some atmospheric electric signals propagate around the earth and because of the existence of a global circuit, applications for the monitoring of global change processes and conditions are now being proposed. Significant secular changes in the global circuit would indicate a change in the global climate; the availability of many old data (about a span of 100 years) could help detect a long term trend.

The concept of the "global circuit" is based on the theory of the global spherical capacitor: both, the solid (and liquid) earth as one electrode, and the high atmospheric layers (about the ionosphere) as the other, are by orders of magnitude more electrically conductive than the atmosphere between them. According to the "classical picture of atmospheric electricity", this capacitor is continuously charged by the common action of all thunderstorms to a d.c. voltage difference of several hundred kilovolts, the earth being negative. The much smaller but still existing conductivity of the atmosphere allows a current flowing from the ionosphere to the ground, integrated for all sink areas of the whole earth, of the order of 1.5 kA. In this way, a global circuit is created with many generators and sink-areas both interspaced and distributed over the whole globe, all connected to two nodes: ionosphere and ground. Within the scope of the global circuit, for each location, the current density (order of several pA/m^2) is determined by the voltage difference between ionosphere and ground (which is the same for all locations but varying in time) and the columnar resistance reaching from the ground up to the ionosphere (in the order of $10^{17} \Omega\, m^2$).

Natural processes, especially meteorological processes and some human activity, which produce or move electric charges ("space charges") or affect the ion distribution, constitute local generators and thereby "local circuits", horziontally and/or in parallel or antiparallel to the local part of the global circuit. In many cases, the local currents are much stronger than the global ones, making the measurement of the global current at a given location and/or during a period of time very difficult or, often, impossible. The strongest local circuits usually occur with certain weather conditions (precipitation, fog, high wind, blown-up dust or snow, heavy cloudiness) which make measurement of the global circuit impossible everywhere; but even in their absence local generators exist in varying magnitudes and of different characters. The separation of the local and global shares in the measured values of current density is a central problem of the science of atmospheric electricity. Aerological measurements are of high value in this regard.

The above description is within the "classical picture" of atmospheric electricity, a group of hypotheses to explain the electricification of the atmosphere. It is probably fundamentally correct but certainly not complete; it has not been confirmed by systems of measurements resulting in no inner contradictions. In particular, extraterrestrial influences must be permitted; their general significance is still under debate.

Within this "classical picture" a kind of electric standard atmosphere may be constructed as shown in Table 1.

Values with a star, *, are rough average values from measurement. A star in parentheses, (*), points to a typical value from one or a few measurements. All other values have been calculated from starred values, under the assumption that at 2 km 50% and at 12 km 90% of the columnar resistance is reached. Voltage drop along one of the partial columns can be calculated by subtracting the value for the lower column from that of the upper one. Columnar resistances, conductances, and capacitances are valid for that particular part of the column which is indicated at the left. Capacitances are calculated with the formula for plate capacitors, and this fact must be considered also for the time constants for columns.

According to measurements, U, the potential difference between 0 m and 65 km may vary by a factor of approximately 2. The total columnar resistance, R_c, is estimated to vary up to a factor of 3, the variation being due to either reduction of conductivity in the exchange layer (about lowest 2 km of this table) or to the presence of high mountains; in both cases the variation is caused in the troposphere. Smaller variations in the stratosphere and mesosphere are being discussed because of aerosols there. The air-earth current density in fair weather varies by a factor of 3 to 6 accordingly. Conductivity near the ground varies by a factor of about 3 but only decreasing; increase of conductivity due to extraordinary radioactivity is a singular event. The field strength near the ground varies as a consequence of variations of air-earth current density and conductivity from about 1/3 to about 10 times of the value quoted in the table. Conductivity near the ground shows a diurnal and an annual variation which depends strongly on the locality; air-earth current density shows a diurnal and annual variation because the earth-ionosphere potential difference undergoes such variations, and also because the columnar resistance is supposed to have a diurnal and probably an annual variation.

Conductivities and air-earth current densities on high mountains are greater than at sea level by factors of up to 10. Conductivity decreases when atmospheric humidity increases. Values for space charges are not quoted because measurements are too few to allow calculation of average values. Values of parameters over the oceans are still rather uncertain.

Theoretically, in fair-weather conditions, Ohm's law must be fulfilled for the electric field, the conduction current density, and the electrical conductivity of the atmosphere. Deviations point to shortcomings in the applied measuring techniques. Data which are representative for a large area (in the extreme, "globally representative data", i.e. data on the global circuit), can on the ground be obtained only by stations on an open plane and only if local generators are either small or constant or are independently measured. Certain measurements with instrumented aircraft provide globally representative information valid for the period of the actual measurement.

TABLE 1
Electrical Parameters of the Clear (Fair-Weather) Atmosphere, Pertinent to the Classical Picture of Atm. Electricity (Electric Standard Atmosphere)

Part of atmosphere for which the values are calculated (elements are in free, cloudless atmosphere)	Currents, I, in A; and current densities, i, in A/m²	Potential differences, U, in V; field strength E in V/m; $U = 0$ at sea level	Resistances, R, in Ω; columnar resistances, R_c, in Ω m² and resistivities, ρ, in Ω m	Conductances, G, in Ω⁻¹; columnar conductances G_c, in Ω⁻¹ m⁻²; total conductivities, γ, in Ω⁻¹ m⁻¹	Capacitances, C, in F; columnar capacitances, C_c, in F m⁻² and capacitivities, ε, in F m⁻¹	Time constants τ, in seconds
Volume element at about sea level, 1 m³	$i = 3 \times 10^{-12}$*	$E_0 = 1.2 \times 10^2$*	$\rho_0 = 4 \times 10^{13}$	$\gamma_0 = 2.5 \times 10^{-14}$	$\varepsilon_0 = 8.9 \times 10^{-12}$*	$\tau_0 = 3.6 \times 10^2$
Lower column of 1 m² cross section from sea level to 2 km height	$i = 3 \times 10^{-12}$	At upper end: $U_1 = 1.8 \times 10^5$	$R_{c1} = 6 \times 10^{16}$	$G_{c1} = 1.7 \times 10^{-17}$	$C_{c1} = 4.4 \times 10^{-15}$	$\tau_{c1} = 2.6 \times 10^2$
Volume element at about 2 km height, 1 m³	$i = 3 \times 10^{-12}$	$E_2 = 6.6 \times 10^1$	$\rho_2 = 2.2 \times 10^{13(*)}$	$\gamma_2 = 4.5 \times 10^{-14}$	$\varepsilon_2 = 8.9 \times 10^{-12}$	$\tau_2 = = 2 \times 10^2$
Center column of 1 m² cross section from 2 to 12 km	$i = 3 \times 10^{-12}$	At upper end: $U_m = 3.15 \times 10^5$	$R_{cm} = 4.5 \times 10^{16}$	$G_{cm} = 5 \times 10^{-17}$	$C_{cm} = 8.8 \times 10^{-16}$	$\tau_{cm} = 1.8 \times 10^1$
Volume element at about 12 km height, 1 m³	$i = 3 \times 10^{-12}$	$E_{12} = 3.9 \times 10^0$	$\rho_{12} = 1.3 \times 10^{12(*)}$	$\gamma_{12} = 7.7 \times 10^{-13}$	$\varepsilon_{12} = 8.9 \times 10^{-12}$	$\tau_{12} = 1.2 \times 10^1$
Upper column of 1 m² cross section from 12 to 65 km height	$i = 3 \times 10^{-12}$	At upper end: $U_u = 3.5 \times 10^5$	$R_{cu} = 1.5 \times 10^{16}$	$G_{cu} = 2.5 \times 10^{-17}$	$C_{cu} = 1.67 \times 10^{-16}$	$\tau_{cm} = 6.7 \times 10^0$
Whole column of 1 m² cross section from 0 to 65 km height	$i = 3 \times 10^{-12}$	At upper end: $U = 3.5 \times 10^5$	$R_c = 1.2 \times 10^{17}$	$G_c = 8.3 \times 10^{-18}$	$C_c = 1.36 \times 10^{-16}$	$\tau_c = 1.64 \times 10^1$
Total spherical capacitor area: 5×10^{14} m²	$I = 1.5 \times 10^3$	$U = 3.5 \times 10^5$*	$R = 2.4 \times 10^2$	$G = 4.2 \times 10^{-3}$	$C = 6.8 \times 10^{-2}$	$\tau = 1.64 \times 10^1$

Note: All currents and fields listed are part of the global circuit, i.e., circuits of local generators are not included. Values are subject to variations due to latitude and longitude, locality with respect to sources of disturbances, meteorological and climatological factors, and man-made changes. For more explanations, see text.

II. AIR IONS
Hannes Tammet

The term "air ions" signifies all airborne particles which are the carriers of the electrical current in the air and have drift velocities determined by the electric field.

The probability of electrical dissociation of molecules in the atmospheric air under thermodynamic equilibrium is near to zero. The average ionization at the ground level over the ocean is $2 \cdot 10^6$ ion pairs $m^{-3}s^{-1}$. This ionization is produced mainly by cosmic rays. Over the continents the ionizing radiation from soil and from radioactive substances in the air each add about $4 \cdot 10^6$ $m^{-3}s^{-1}$. The total average ionization rate of 10^7 $m^{-3}s^{-1}$ is equivalent to 17 $\mu R/h$ which is a customary expression of the background level of the ionizing radiations. The ionization rate over the ground varies in space due to the radioactivity of soil, and in time depending on the exchange of air between the atmosphere and radon-containing soil. Radioactive pollution increases the ionization rate. A temporary increase of about 10 times was registered in Sweden after the Chernobyl accident in 1986. The emission of Kr^{85} from nuclear power plants can noticeably increase the global ionization rate in the next century. The ionization rate decreases with altitude near the ground and increases at higher altitudes up to 15 km, where it has a maximum about $5 \cdot 10^7$ $m^{-3}s^{-1}$. Solar X-ray and extreme UV radiation cause a new increase at altitudes over 60 km.

Local sources of air ions are point discharges in strong electric fields, fluidization of charged drops from waves, etc.

The enhanced chemical activity of an ion results in a chain of ion-molecule reactions with the colliding neutrals, and, in the first microsecond of the life of an air ion, a charged molecular cluster called the *cluster ion* is formed. According to theoretical calculations in the air free from exotic trace gases the following cluster ions should be dominant:

$$NO_3^- \cdot (HNO_3) \cdot H_2O, \ NO_2^- \cdot (H_2O)_2, \ NO_3^- \cdot H_2O, \ O_2^- \cdot (H_2O)_4, \ O_2^- \cdot (H_2O)_5,$$
$$H_3O^+ \cdot (H_2O)_6, \ NH_4^+ \cdot (H_2O)_2, \ NH_4^+ \cdot (H_2O), \ H_3O^+ \cdot (H_2O)_5, \ NH_4^+ \cdot NH_3$$

A measurable parameter of air ions is the electrical mobility k, characterizing the drift velocity in the unit electric field. The mobility is inversely proportional to the density of air, and the results of measurements are as a rule reduced to normal conditions. According to mobility the air ions are called: fast or small or light ions with mobility $k > 5 \cdot 10^{-5}$ $m^2V^{-1}s^{-1}$, intermediate ions, and slow or large or heavy ions with mobility $k < 10^{-6}$ m^2V^{-1} s^{-1}. The boundary between intermediate and slow ions is conventional.

Cluster ions are fast ions. The masses of cluster ions may be measured with mass spectrometers, but the possible ion-molecule reactions during the passage of the air through nozzles to the vacuum chamber complicate the measurement. Mass and mobility of cluster ions are highly correlated. The experimental results[5] can be expressed by the empirical formula

$$m \approx \frac{850 \ u}{\left[0.3 + k / \left(10^{-4} \, m^2 V^{-1} s^{-1}\right)\right]^3}$$

where u is the unified atomic mass unit.

The value of the transport cross-section of a cluster ion is needed to calculate its mobility according to the kinetic theory of Chapman and Enskog. The theoretical estimation of transport cross-sections is rough and cannot be used to identify the chemical structures of cluster ions. Mass spectrometry is the main technique of identification of cluster ions.[2]

Märk and Castleman (1986) presented an overview of over 1000 publications on the experimental studies of cluster ions. Most of them present information about ions of millisecond age range. The low concentration makes it difficult to get detailed information about masses and mobilities of the natural atmospheric ions at ground level. The results of a 1-year continuous measurement[6] are as follows:

	+ ions	- ions	unit
Average mobility	1.36	1.56	10^{-4} $m^2V^{-1}s^{-1}$
The corresponding mass	190	130	u
The corresponding diameter	0.69	0.61	nm
The average concentration	400	360	10^6 m^{-3}
The corresponding conductivity	8.7	9.0	fS

The distribution of tropospheric cluster ions according to the mobility and estimated mass is depicted in Figure 1.

The problems and results of direct mass spectrometry of natural cluster ions are analysed by Eisele (1986) for ground level and by Meyerott, Reagan and Joiner (1980) for stratospheric measurements. Air ions in the high atmosphere are a subject of ionospheric physics.

During its lifetime (about 1 min), a cluster ion at ground level collides with nearly 10^{12} molecules. Thus the cluster ions are able to concentrate trace gases of very low concentration if they have an extra high electron or proton affinity. For example, Eisele (1986) demonstrated that a considerable fraction of positive atmospheric cluster ions in the unpolluted atmosphere at ground level probably consist of a molecule derived from pyridine. The concentration of these constituents is estimated to be about 10^{-12}. Therefore, air-ion mass and mobility spectrometry is considered as a promising technique for trace analysis in the air. Mass and mobility spectrometry of millisecond-age air ions has been developed as a technique of chemical analysis known as "plasma chromatography" (Carr, 1984). The sensitivity of the detection grows with the age of the cluster ions measured.

The mechanisms of annihilation of cluster ions are ion-ion recombination (on the average 3%) and sedimentation on aerosol particles (on the average 97% of cluster ions at ground level). The result of the combination of a cluster ion and neutral particle is a charged particle called an *aerosol ion*. In conditions of detailed thermodynamic equilibrium the probability that a spherical particle of diameter d carries q elementary charges is calculated from the Boltzmann distribution:

$$p_q(d) = (2 \ \pi d/d_0)^{1/2} \exp(-q^2 \ d_0/2d)$$

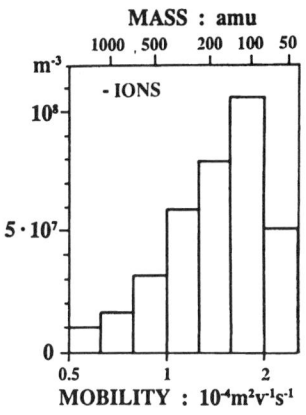

Figure 1. Average mobility and mass spectra of natural tropospheric cluster ions. Concentrations of the mobility fractions were measured in a rural site every 5 min over 1 year.[6] Ion mass is estimated according to the above empirical formula.

where $d_0 = 115$ nm (at 18°C). The supposition about the detailed equilibrium is an approximation and the formula is not valid for particles less than d_0. On the basis of numerical calculations by Hoppel and Frick (1990) the following charge probabilities can be derived:

d	3	10	30	100	300	1000	3000	nm
p_0	98	90	70	42	24	14	8	%
$p_{-1} + p_1$	2	10	30	48	41	25	15	%
$p_{-2} + p_2$	0	0	0	10	23	21	14	%
$p_{q>2}$	0	0	0	0	12	40	63	%
k_1	15000	1900	250	28	5.1	1.11	0.33	$10^{-9} m^2 V^{-1} s^{-1}$

The last line of the table presents the mobility of a particle carrying one elementary charge. The distribution of the atmospheric aerosol ions over mobility is demonstrated in Figure 2.

Although the concentration of aerosol in continental air at ground level is an order of magnitude higher than the concentration of cluster ions, the mobilities of aerosol ions are so small that their percentage in air conductivity is less than 1%.

A specific class of aerosol ions are condensed aerosol ions produced as a result of the condensation of gaseous matter on the cluster ions. In aerosol physics the process is called ion-induced nucleation; it is considered as one among the processes of gas-to-particle conversion. The condensed aerosol ions have an inherent charge. Their sizes and mobilities are between the sizes and mobilities of cluster ions and of ordinary aerosol ions. Water and standard constituents of atmospheric air are not able to condense on the cluster ions in the real atmosphere. Thus the concentration of condensed aerosol ions depends on the trace constituents in the air and is very low in unpolluted air. Knowledge about condensed aerosol ions is poor because of measurement difficulties.

Figure 2. Mobility and size spectra of tropospheric aerosol ions.[6] The wide bars mark the fraction concentrations theoretically estimated on the basis of the standard size distribution of tropospheric aerosol. The pin bars with head + and - mark average values of positive and negative aerosol ion fraction concentrations measured in a rural site every 5 min during 4 months.

REFERENCES

1. Carr, T. W., Ed., *Plasma Chromatography*, Plenum Press, New York and London, XII + 259 pp., 1984.
2. Eisele, F. L., Identification of Tropospheric Ions, *J. Geophys. Res.*, vol. 91, no. D7, pp. 7897-7906, 1986.

3. Hoppel, W. A., and Frick, G. M., The Nonequilibrium Character of the Aerosol Charge Distributions Produced by Neutralizers, *Aerosol Sci. Technol.*, vol. 12, no. 3, pp. 471-496, 1990.

4. Mark, T. D., and Castleman, A. W., Experimental Studies on Cluster Ions, in *Advances in Atomic and Molecular Physics*, vol. 20, pp. 65.-172, Academic Press, 1985.

5. Meyerott, R. E., Reagan, J. B., and Joiner, R. G., The Mobility and Concentration of Ions and the Ionic Conductivity in the Lower Stratosphere, *J. Geophys. Res.*, vol. 85, no. A3, pp. 1273-1278, 1980.

6. Salm, J., Tammet, H., Iher, H., and Hörrak, U., Atmospheric Electrical Measurements in Tahkuse, Estonia (in Russian), in *Voprosy Atmosfernogo Elektrichestva*, pp. 168-175, Gidrometeoizdat, Leningrad, 1990.

III. THUNDERSTORM ELECTRICITY
John Latham

The development of improved radar techniques and instruments for in-cloud electrical and physical measurements, coupled with a much clearer recognition by the research community that establishment of the mechanism or mechanisms responsible for electric field development in thunderclouds, culminating in lightning, is inextricably linked to the concomitant dynamical and microphysical evolution of the clouds, has led to significant progress over the past decade.

Field studies indicate that in most thunderclouds the electrical development is associated with the process of glaciation, which can occur in a variety of incompletely understood ways. In the absence of ice, field growth is slow, individual hydrometeor charges are low, and lightning is produced only rarely. Precipitation — in the solid form, as graupel — also appears to be a necessary ingredient for significant electrification, as does significant convective activity and mixing between the clouds and their environments, via entrainment.

Increasingly, the view is being accepted that charge transfer leading to field-growth is largely a consequence of rebounding collisions between graupel pellets and smaller vapor-grown ice crystals, followed by the separation under gravity of these two types of hydrometeor. These collisions occur predominantly within the temperature range -15 to -30°C, and for significant charge transfer need to occur in the presence of supercooled cloud droplets.

The field evidence is inconsistent with an inductive mechanism, and extensive laboratory studies indicate that the principal charging mechanism is non-inductive and associated — in ways yet to be identified — with differences in surface characteristics of the interacting hydrometeors.

Laboratory studies indicate that the two most favored sites for corona emission leading to the lightning discharge are the tips of ephemeral liquid filaments, produced during the glancing collisions of supercooled raindrops, and protuberances on large ice crystals or graupel pellets. The relative importance of these alternatives will depend on the hydrometeor characteristics and the temperature in the regions of strongest fields; these features are themselves dependent on air-mass characteristics and climatological considerations.

A recently identified but unresolved question is why, in continental Northern Hemisphere thunderclouds at least, the sign of the charge brought to ground by lightning is predominantly negative in summer but more evenly balanced in winter.

IV. LIGHTNING
Martin A. Uman

From both ground-based weather-station data and satellite measurements, it has been estimated that there are about 100 lightning discharges, both cloud and ground flashes, over the whole earth each second; representing an average global lightning flash density of about 6 $km^{-2}yr^{-1}$. Most of this lightning occurs over the earth's land masses. For example, in central Florida, where thunderstorms occur about 90 days/yr, the flash density for discharges to earth is about 15 $km^{-2}yr^{-1}$. Some tropical areas of the earth have thunderstorms up to 300 days/yr.

Lightning can be defined as a transient, high-current electric discharge whose path length is measured in kilometers and whose most common source is the electric charge separated in the ordinary thunderstorm or cumulonimbus cloud. Well over half of all lightning discharges occur totally within individual thunderstorm clouds and are referred to as intracloud discharges. Cloud-to-ground lightning, however, has been studied more extensively than any other lightning form because of its visibility and its more practical interest. Cloud-to-cloud and cloud-to-air discharges are less common than intracloud or cloud-to-ground lightning.

Lightning between the cloud and earth can be categorized in terms of the direction of motion, upward or downward, and the sign of the charge, positive or negative, of the developing discharge (called a *leader*) which initiates the overall event. Over 90% of the worldwide cloud-to-ground discharges is initiated in the thundercloud by downward-moving negatively-charged leaders and subsequently results in the lowering of negative charge to earth. Cloud-to-ground lightning can also be initiated by downward-moving positive leaders, less than 10% of the worldwide cloud-to-ground lightning being of this type although the exact percentage is a function of season and latitude. Lightning between cloud and ground can also be initiated by leaders which develop upward from the earth. These upward-initiated discharges are relatively rare, may be of either polarity, and generally occur from mountaintops and tall man-made structures.

We discuss next the most common type of cloud-to-ground lightning. A negative cloud-to-ground discharge or *flash* has an overall duration of some tenths of a second and is made up of various components, among which are typically three or four high-current pulses called *strokes*. Each stroke lasts about a millisecond, the separation time between strokes being typically several tens of milliseconds. Such lightning often appears to "flicker" because the human eye can just resolve the individual light pulse associated with each stroke. A drawing of the components of a negative cloud-to-ground flash is found in Figure 3. Some values for salient parameters are found in Table 1. The negatively-charged *stepped leader* initiates the first stroke in a flash by propagating from cloud to ground through virgin air in a series of discrete steps. Photographically observed leader steps in clear

Figure 3. Sequence of steps in cloud-to-ground lightning.

air are typically 1 µs in duration and tens of meters in length, with a pause time between steps of about 50 µs. A fully developed stepped leader lowers up to 10 or more coulombs of negative cloud charge toward ground in tens of milliseconds with an average downward speed of about 2×10^5 m/s. The average leader current is in the 100 to 1000 A range. The steps have pulse currents of at least 1 kA. Associated with these currents are electric- and magnetic-field pulses with widths of about 1 µs or less and risetimes of about 0.1 µs or less. The stepped leader, during its trip toward ground, branches in a downward direction, resulting in the characteristic downward-branched geometrical structure commonly observed. The electric potential of the bottom of the negatively-charged leader channel with respect to ground has a magnitude in excess of 10^7 V. As the leader tip nears ground, the electric field at sharp objects on the ground or at irregularities of the ground itself exceeds the breakdown value of air, and one or more upward-moving discharges (often called upward leaders) are initiated from those points, thus beginning the *attachment process*. An understanding of the physics of the attachment process is central to an understanding of the operation of lightning protection of ground-based objects and the effects of lightning on humans and animals, since it is the attachment process that determines where the lightning connects to objects on the ground and the value of the early currents which flow. When one of the upward-moving discharges from the ground (or from a lightning rod or an individual) contacts the tip of the downward-moving stepped leader, typically some tens of meters above the ground, the leader tip is effectively connected to ground potential. The negatively-charged leader channel is then discharged to earth when a ground potential wave, referred to as the first *return stroke,* propagates continuously up the leader path. The upward speed of a return stroke near the ground is typically near one third the speed of light, and the speed decreases with height. The first return stroke produces a peak current near ground of typically 30 kA, with a time from zero to peak of a few microseconds. Currents measured at the ground fall to half of the peak value in about 50 µs, and currents of the order of hundreds of amperes may flow for times of a few milliseconds up to several hundred milliseconds. The longer-lasting currents are known as *continuing currents.* The rapid release of return stroke energy heats the leader channel to a temperature near 30,000 K and creates a high-pressure channel which expands and generates the shock waves that eventually become thunder, as further discussed later. The return stroke effectively lowers to ground the charge originally deposited onto the stepped-leader channel and additionally initiates the lowering of other charge which may be available to the top of its channel. First return-stroke electric fields exhibit a microsecond scale rise to peak with a typical peak value of 5 V/m, normalized to a distance of 100 km by an inverse distance relationship. Roughly half of the field rise to peak, the so-called "fast transition", takes place in tenths of a microsecond, an observation that can only be made if the field propagation is over a highly conducting surface such as salt water.

After the first return-stroke current has ceased to flow, the flash, including charge motion in the cloud, may end. The lightning is then called a single-stroke flash. On the other hand, if additional charge is made available to the top of the channel, a continuous or *dart leader* may propagate down the residual first-stroke channel at a typical speed of about 1×10^7 m/s. The dart leader lowers a charge of the order of 1 C by virtue of a current of about 1 kA. The dart leader then initiates the second (or any subsequent) return stroke. Subsequent return-stroke currents generally have faster zero-to-peak rise times than do first-stroke currents, but similar maximum rates of change, about 100 kA/μs. Some leaders begin as dart leaders, but toward the end of their trip toward ground become stepped leaders. These leaders are known as *dart-stepped leaders* and may have different ground termination points (and separate upward leaders) from the first stroke. Most often the dart-stepped leaders are associated with the second stroke of the flash. Nearly half of all flashes exhibit more than one termination point on ground with the distance between separate terminations being up to several kilometers. Subsequent return-stroke radiated electric and magnetic fields are similar to, but usually a factor of two or so smaller, than first return-stroke fields. About one third of all multiple-stroke flashes has at least one subsequent stroke which is larger than the first stroke.

Cloud-to-ground flashes that lower positive charge, though not common, are of considerable practical interest because their peak currents and total charge transfer can be much larger than for the more common negative ground flash. The largest recorded peak currents, those in the 200- to 300-kA range, are due to the return strokes of positive lightning. Such positive flashes to ground are initiated by downward-moving leaders which do not exhibit the distinct steps of their negative counterparts. Rather, they show a luminosity which is more or less continuous but modulated in intensity. Positive flashes are generally composed of a single stroke followed by a period of continuing current. Positive flashes are probably initiated from the upper positive charge in the thundercloud charge dipole when that cloud charge is horizontally separated from the negative charge beneath it, the source of the usual negative cloud-to-ground lightning. Positive flashes are relatively common in winter thunderstorms (snow storms), which produce few flashes overall, and are relatively uncommon in summer thunderstorms. The fraction of positive lightning in summer thunderstorms apparently increases with increasing latitude and with increasing height of the ground above sea level.

Distant lightning return stroke fields are often referred to as sferics (called "atmospherics" in the older literature). The peak in the sferics frequency spectrum is near 5 kHz due to the bipolar or ringing nature of the distant return-stroke electromagnetic signal and to the effects of propagation.

Thunder, the acoustic radiation associated with lightning, is sometimes divided into the categories "audible", sounds that one can hear, and "infrasonic", below a few tens of hertz, a frequency range that is inaudible. This division is made because it is thought that the mechanisms that produce audible and infrasonic thunder are different. Audible thunder is thought to be due to the expansion of a rapidly heated return stroke channel, as noted earlier, whereas infrasonic thunder is thought to be associated with the conversion to sound of the energy stored in the electrostatic field of the thundercloud when lightning rapidly reduces that cloud field.

The technology of artificially initiating lightning by firing upward small rocket trailing grounded wire of a few hundred meters length has been well-developed during the past decade. Such "triggered" flashes are similar to natural upward-initiated discharges from tall structure. They often contain subsequent strokes which, when they occur, are similar to the subsequent strokes in natural lightning. These triggered subsequent strokes have been the subject of considerable recent research.

Also in the past 10 years or so sophisticated lightning locating equipment has been installed throughout the world. For example, all ground flashes in the U.S. are now centrally monitored for research, for better overall weather prediction, and for hazard warning for aviation, electric utilities and other lightning-sensitive facilities.

Information on lightning physics can be found in M. A. Uman, *The Lightning Discharge*, Academic Press, San Diego, 1987; on lightning death and injury in *Medical Aspects of Lightning Injury*, editors C. Andrews, M. A. Cooper, M. Darveniza, and D. Mackerras, CRC Press, 1992. Ground flash location information for the U.S., in real time or archived, is available from Geomet Data Service of Tucson, AZ, which is also a source of the names of providers of those data in other countries.

Table 2 has data for cloud-to-ground lightning discharges bringing negative charge to earth. The values listed are intended to convey a rough feeling for the various physical parameters of lightning. No great accuracy is claimed since the results of different investigators are often not in good agreement. These values may, in fact, depend on the particular environment in which the lightning discharge is generated. The choice of some of the entries in the table is arbitrary.

TABLE 2
Data for Cloud-to-Ground Lightning Discharges

	Minimum[a]	Representative values	Maximum[a]
Stepped leader			
Length of step, m	3	50	200
Time interval between steps, μs	30	50	125
Average speed of propagation of stepped leader, m/s[b]	1.0×10^5	2.0×10^5	3.0×10^6
Charged deposited on stepped-leader channel, coulombs	3	5	20
Dart leader			
Speed of propagation, m/s[b]	1.0×10^6	1.0×10^7	2.4×10^7
Charged deposited on dart-leader channel, coulombs	0.2	1	6
Return stroke[c]			
Speed of propagation, m/s[b]	2.0×10^7	1.0×10^8	2.0×10^8
Maximum current rate of increase, kA/μs	<1	100	400
Time to peak current, μs	<1	2	30
Peak current, kA	2	30	200
Time to half of peak current, μs	10	50	250
Charge transferred excluding continuing current, coulombs	0.02	3	20
Channel length, km	2	5	15
Lightning flash			
Number of strokes per flash	1	4	26
Time interval between strokes in absence of continuing current, ms	3	60	100
Time duration of flash, s	10^{-2}	0.5	2
Charge transferred including continuing current, coulombs	3	30	200

[a]   The words maximum and minimum are used in the sense that most measured values fall between these limits.

[b]   Speeds of propagation are generally determined from photographic data and are "two-dimensional". Since many lightning flashes are not vertical, values stated are probably slight underestimates of actual values.

[c]   First return strokes have longer times to current peak and generally larger charge transfer than do subsequent return strokes.

Adapted from Uman, M. A., *Lightning*, Dover Paperbook, New York, 1986, and Uman, M. A., *The Lightning Discharge*, Academic Press, San Diego, 1987.

SPEED OF SOUND IN VARIOUS MEDIA

The speed of sound in various solids, liquids, and gases is given in these tables. While only a single parameter v is needed for liquids and gases, sound propagation in isotropic solids is characterized by three velocity parameters. For a solid of infinite extent (or of finite extent if all dimensions are much larger than a wavelength), there are two relevant quantities,

v_l: velocity of longitudinal waves
v_s: velocity of shear waves.

For a cylindrical rod with diameter much smaller than a wavelength,

v_{ext}: velocity of extensional waves along the rod. (Torsional waves in the rod are propagated at the same speed as sheer waves in an infinite solid.)

Table 1 lists values for a variety of solid materials. Table 2 covers gases liquids and gases; values for cryogenic liquids are given at the normal boiling point. Table 3 gives the speed of sound in pure water and in seawater of salinity $S = 3.5\%$ as a function of temperature. All values are in meters per second and are given for normal atmospheric pressure.

REFERENCES

1. Gray, D.E., Ed., *American Institute of Physics Handbook, Third Edition*, McGraw Hill, New York, 1972.
2. Anderson, H.L., Ed., *A Physicist's Desk Reference*, American Institute of Physics, New York, 1989.
3. Younglove, B.A., *Thermophysical Proeprties of Fluids. Part I, J. Phys. Chem. Ref. Data*, 11, Suppl. 1, 1982.
4. Younglove, B.A., and Ely, J.F., *Thermophysical Properties of Fluids. Part II, J. Phys. Chem. Ref. Data*, 16, 577, 1987.
5. Haar, L., Gallagher, J.S., and Kell, G.S., *NBS/NRC Steam Tables*, Hemisphere Publishing Corp., New York, 1984.
6. Mason, W.P., *Physical Acoustics and the Properties of Solids*, D. Van Nostrand Co., Princeton, N.J., 1958.
7. Landolt-Börnstein, *Numerical Data and Functional Relationships in Science and Technology, New Series, II/5, Molecular Acoustics*, Springer-Verlag, Heidelberg, 1967.

TABLE 1
Speed of Sound in Solids at Room Temperature

Name	v_l/m s^{-1}	v_s/m s^{-1}	v_{ext}/m s^{-1}	Name	v_l/m s^{-1}	v_s/m s^{-1}	v_{ext}/m s^{-1}
Metals				Steel, 347 Stainless	5790	3100	5000
Aluminum, rolled	6420	3040	5000	Steel, K9	5940	3250	5250
Berylium	12890	8880	12870	Tin, rolled	3320	1670	2730
Brass (70 Cu, 30 Zn)	4700	2110	3480	Titanium	6070	3125	5090
Constantan	5177	2625	4270	Tungsten, annealed	5220	2890	4620
Copper, annealed	4760	2325	3810	Tungsten, drawn	5410	2640	4320
Copper, rolled	5010	2270	3750	Zinc, rolled	4210	2440	3850
Duralumin 17S	6320	3130	5150				
Gold, hard-drawn	3240	1200	2030	**Other materials**			
Iron, cast	4994	2809	4480	Fused silica	5968	3764	5760
Iron, electrolytic	5950	3240	5120	Glass, heavy silicate flint	3980	2380	3720
Iron, Armco	5960	3240	5200	Glass, light borate crown	5100	2840	4540
Lead, annealed	2160	700	1190	Glass, pyrex	5640	3280	5170
Lead, rolled	1960	690	1210	Lucite	2680	1100	1840
Magnesium, annealed	5770	3050	4940	Nylon 6-6	2620	1070	1800
Molybdenum	6250	3350	5400	Polyethylene	1950	540	920
Monel metal	5350	2720	4400	Polystyrene	2350	1120	1840
Nickel	6040	3000	4900	Rubber, butyl	1830		
Platinum	3260	1730	2800	Rubber, gum	1550		
Silver	3650	1610	2680	Rubber, neoprene	1600		
Steel (1% C)	5940	3220	5180	Tungsten carbide	6655	3980	6220

TABLE 2
Speed of Sound in Liquids and Gases

Name	$t/°C$	$v/\text{m s}^{-1}$	Name	$t/°C$	$v/\text{m s}^{-1}$
Liquids			Pentane	20	1008
Acetone	20	1203	Propane	-42.1	1158
Argon	-185.9	813	1-Propanol	20	1223
Benzene	25	1310	Tetrachloromethane	25	930
Bromobenzene	20	1169	Trichloromethane	25	987
Butane	-0.5	1034	1-Undecene	20	1275
1-Butanol	20	1258	Water	25	1497
Carbon disulphide	25	1140	Water (sea, $S = 3.5\%$)	25	1535
Chlorobenzene	20	1311			
Cyclohexane	19	1280	**Gases at 1 atm**		
1-Decene	20	1250	Air, dry	25	346
Diethyl ether	25	976	Ammonia	0	415
Ethane	-88.6	1326	Argon	27	323
Ethanol	20	1162	Carbon monoxide	0	338
Ethylene	-103.8	1309	Carbon dioxide	0	259
Ethylene glycol	25	1658	Chlorine	0	206
Fluorobenzene	20	1183	Deuterium	0	890
Glycerol	25	1904	Ethane	27	312
Helium	-268.9	180	Ethylene	27	331
Heptane	20	1162	Helium	0	965
1-Heptene	20	1128	Hydrogen	27	1310
Hexane	20	1083	Hydrogen bromide	0	200
Hydrogen	-252.9	1101	Hydrogen chloride	0	296
Iodobenzene	20	1114	Hydrogen iodide	0	157
Mercury	25	1450	Hydrogen sulfide	0	289
Methane	-161.5	1337	Methane	27	450
Methanol	20	1121	Neon	0	435
Nitrobenzene	25	1463	Nitric oxide	10	325
Nitrogen	-195.8	939	Nitrogen	27	353
1-Nonene	20	1218	Nitrous oxide	0	263
Octane	20	1197	Oxygen	27	330
1-Octene	20	1184	Sulfur dioxide	0	213
Oxygen	-183.0	906	Water (steam)	100	473
1-Pentadecene	20	1351			

TABLE 3
Speed of Sound in Water and Seawater (S = 3.5%) at Different Temperatures

	$v/\text{m s}^{-1}$	
$t/°C$	Water	Seawater
0	1401.0	1449.4
10	1447.8	1490.4
20	1483.2	1522.2
25	1497.4	1535.1
30	1509.5	1546.2
40	1528.4	
50	1541.4	
60	1549.5	
70	1553.2	
80	1552.8	

ATTENUATION AND SPEED OF SOUND IN AIR AS A FUNCTION OF HUMIDITY AND FREQUENCY

This table gives the attenuation and speed of sound as a function of frequency at various values of relative humidity. All values refer to still air at 20°C.

REFERENCES

1. Tables of Absorption and Velocity of Sound in Still Air at 68°F (20°C), AD-738576, National Technical Information Service, Springfield, VA.
2. Evans, L. B., Bass, H. E., and Sutherland, L. C., *J. Acoust. Soc. Am.,* 51, 1565, 1972.

Frequency (Hz)	Attenuation (dB/km)	Speed (m/s)	Frequency (Hz)	Attenuation (dB/km)	Speed (m/s)
Relative humidity 0%			**Relative humidity 60%**		
20	0.51	343.477	20	0.02	344.182
40	1.07	343.514	40	0.06	344.183
50	1.26	343.525	50	0.09	344.183
63	1.43	343.536	63	0.15	344.184
100	1.67	343.550	100	0.34	344.185
200	1.84	343.559	200	0.99	344.190
400	1.96	343.561	400	1.94	344.197
630	2.11	343.562	630	2.57	344.200
800	2.27	343.562	800	2.94	344.201
1250	2.82	343.562	1250	4.01	344.202
2000	4.14	343.562	2000	6.55	344.203
4000	8.84	343.564	4000	18.73	344.204
6300	14.89	343.565	6300	42.51	344.204
10000	26.28	343.566	10000	101.84	344.206
12500	35.81	343.566	12500	155.67	344.208
16000	52.15	343.567	16000	247.78	344.211
20000	75.37	343.567	20000	373.78	344.215
40000	267.01	343.567	40000	1195.37	344.238
63000	644.66	343.567	63000	2220.64	344.262
80000	1032.14	343.567	80000	2951.71	344.274
Relative humidity 30%			**Relative humidity 100%**		
20	0.03	343.807	20	0.01	344.685
40	0.11	343.808	40	0.04	344.685
50	0.17	343.810	50	0.06	344.685
63	0.25	343.810	63	0.09	344.685
100	0.50	343.814	100	0.22	344.686
200	1.01	343.821	200	0.77	344.689
400	1.59	343.826	400	2.02	344.695
630	2.24	343.827	630	3.05	344.699
800	2.85	343.828	800	3.57	344.701
1250	5.09	343.828	1250	4.59	344.704
2000	10.93	343.829	2000	6.29	344.705
4000	38.89	343.831	4000	13.58	344.706
6300	90.61	343.836	6300	27.72	344.706
10000	204.98	343.846	10000	63.49	344.706
12500	294.08	343.854	12500	96.63	344.707
16000	422.51	343.865	16000	154.90	344.708
20000	563.66	343.877	20000	237.93	344.709
40000	1110.97	343.911	40000	884.28	344.718
63000	1639.47	343.924	63000	1973.62	344.731
80000	2083.08	343.929	80000	2913.01	344.742

SPEED OF SOUND IN DRY AIR

The values in this table were calculated from the equation of state for dry air (average molecular weight 28.96) treated as a real gas. Values refer to standard atmospheric pressure. The speed of sound varies only slightly with pressure; at two atmospheres and -100°C the value decreases by 0.13%, while at two atmospheres and 80°C the speed increases by 0.04%.

REFERENCE

Sytchev, V.V., Vasserman, A.A., Kozlov, A.D., Spiridonov, G.A., and Tsymarny, V.A., *Thermodynamic Properties of Air*, Hemisphere Publishing Corp., New York, 1987.

$t/°C$	$v_s/m\ s^{-1}$	$t/°C$	$v_s/m\ s^{-1}$	$t/°C$	$v_s/m\ s^{-1}$
-100	263.5	-35	309.5	30	349.1
-95	267.3	-30	312.7	35	352.0
-90	271.1	-25	315.9	40	354.8
-85	274.8	-20	319.1	45	357.6
-80	278.5	-15	322.3	50	360.4
-75	282.1	-10	325.4	55	363.2
-70	285.7	-5	328.4	60	365.9
-65	289.2	0	331.5	65	368.6
-60	292.7	5	334.5	70	371.3
-55	296.1	10	337.5	75	374.0
-50	299.5	15	340.4	80	376.7
-45	302.9	20	343.4		
-40	306.2	25	346.3		

MUSICAL SCALES

EQUAL TEMPERED CHROMATIC SCALE
$A_4 = 440$ Hz

American Standard pitch. Adopted by the American Standards Association in 1936

Note	Frequency	Note	Frequency	Note	Frequency	Note	Frequency
C_0	16.35	C_2	65.41	C_4	261.63	C_6	1046.50
$C\#_0$	17.32	$C\#_2$	69.30	$C\#_4$	277.18	$C\#_6$	1108.73
D_0	18.35	D_2	73.42	D_4	293.66	D_6	1174.66
$D\#_0$	19.45	$D\#_2$	77.78	$D\#_4$	311.13	$D\#_6$	1244.51
E_0	20.60	E_2	82.41	E_4	329.63	E_6	1318.51
F_0	21.83	F_2	87.31	F_4	349.23	F_6	1396.91
$F\#_0$	23.12	$F\#_2$	92.50	$F\#_4$	369.99	$F\#_6$	1479.98
G_0	24.50	G_2	98.00	G_4	392.00	G_6	1567.98
$G\#_0$	25.96	$G\#_2$	103.83	$G\#_4$	415.30	$G\#_6$	1661.22
A_0	27.50	A_2	110.00	A_4	440.00	A_6	1760.00
$A\#_0$	29.14	$A\#_2$	116.54	$A\#_4$	466.16	$A\#_6$	1864.66
B_0	30.87	B_2	123.47	B_4	493.88	B_6	1975.53
C_1	32.70	C_3	130.81	C_5	523.25	C_7	2093.00
$C\#_1$	34.65	$C\#_3$	138.59	$C\#_5$	554.37	$C\#_7$	2217.46
D_1	36.71	D_3	146.83	D_5	587.33	D_7	2349.32
$D\#_1$	38.89	$D\#_3$	155.56	$D\#_5$	622.25	$D\#_7$	2489.02
E_1	41.20	E_3	164.81	E_5	659.26	E_7	2637.02
F_1	43.65	F_3	174.61	F_5	698.46	F_7	2793.83
$F\#_1$	46.25	$F\#_3$	185.00	$F\#_5$	739.99	$F\#_7$	2959.96
G_1	49.00	G_3	196.00	G_5	783.99	G_7	3135.96
$G\#_1$	51.91	$G\#_3$	207.65	$G\#_5$	830.61	$G\#_7$	3322.44
A_1	55.00	A_3	220.00	A_5	880.00	A_7	3520.00
$A\#_1$	58.27	$A\#_3$	233.08	$A\#_5$	932.33	$A\#_7$	3729.31
B_1	61.74	B_3	246.94	B_5	987.77	B_7	3951.07
						C_8	4186.01

EQUAL TEMPERED CHROMATIC SCALE
$A_4 = 435$ Hz

International Pitch, adopted 1891

Note	Frequency	Note	Frequency	Note	Frequency	Note	Frequency
C_0	16.17	C_2	64.66	C_4	258.65	C_6	1034.61
$C\#_0$	17.13	$C\#_2$	68.51	$C\#_4$	274.03	$C\#_6$	1096.13
D_0	18.15	D_2	72.58	D_4	290.33	D_6	1161.31
$D\#_0$	19.22	$D\#_2$	76.90	$D\#_4$	307.59	$D\#_6$	1230.37
E_0	20.37	E_2	81.47	E_4	325.88	E_6	1303.53
F_0	21.58	F_2	86.31	F_4	345.26	F_6	1381.04
$F\#_0$	22.86	$F\#_2$	91.45	$F\#_4$	365.79	$F\#_6$	1463.16
G_0	24.22	G_2	96.89	G_4	387.54	G_6	1550.16
$G\#_0$	25.66	$G\#_2$	102.65	$G\#_4$	410.59	$G\#_6$	1642.34
A_0	27.19	A_2	108.75	A_4	435.00	A_6	1740.00
$A\#_0$	28.80	$A\#_2$	115.22	$A\#_4$	460.87	$A\#_6$	1843.47
B_0	30.52	B_2	122.07	B_4	488.27	B_6	1953.08
C_1	32.33	C_3	129.33	C_5	517.31	C_7	2069.22
$C\#_1$	34.25	$C\#_3$	137.02	$C\#_5$	548.07	$C\#_7$	2192.26
D_1	36.29	D_3	145.16	D_5	580.66	D_7	2322.62
$D\#_1$	38.45	$D\#_3$	153.80	$D\#_5$	615.18	$D\#_7$	2460.73
E_1	40.74	E_3	162.94	E_5	651.76	E_7	2607.05
F_1	43.16	F_3	172.63	F_5	690.52	F_7	2762.08
$F\#_1$	45.72	$F\#_3$	182.89	$F\#_5$	731.58	$F\#_7$	2926.32
G_1	48.44	G_3	193.77	G_5	775.08	G_7	3100.33

MUSICAL SCALES (continued)

EQUAL TEMPERED CHROMATIC SCALE
A_4 = 435 Hz

International Pitch, adopted 1891

Note	Frequency	Note	Frequency	Note	Frequency	Note	Frequency
$G\#_1$	51.32	$G\#_3$	205.29	$G\#_5$	821.17	$G\#_7$	3284.68
A_1	54.38	A_3	217.50	A_5	870.00	A_7	3480.00
$A\#_1$	57.61	$A\#_3$	230.43	$A\#_5$	921.73	$A\#_7$	3686.93
B_1	61.03	B_3	244.14	B_5	976.54	B_7	3906.17
						C_8	4138.44

SCIENTIFIC OR JUST SCALE
C_4 = 256 Hz

Note	Frequency	Note	Frequency	Note	Frequency	Note	Frequency
C_0	16	C_2	64	C_4	256	C_6	1024
D_0	18	D_2	72	D_4	288	D_6	1152
E_0	20	E_2	80	E_4	320	E_6	1280
F_0	21.33	F_2	85.33	F_4	341.33	F_6	1365.33
G_0	24	G_2	96	G_4	384	G_6	1536
A_0	26.67	A_2	106.67	A_4	426.67	A_6	1706.67
B_0	30	B_2	120	B_4	480	B_6	1920
C_1	32	C_3	128	C_5	512	C_7	2048
D_1	36	D_3	144	D_5	576	D_7	2304
E_1	40	E_3	160	E_5	640	E_7	2560
F_1	42.67	F_3	170.67	F_5	682.67	F_7	2730.67
G_1	48	G_3	192	G_5	768	G_7	3072
A_1	53.33	A_3	213.33	A_5	853.33	A_7	3413.33
B_1	60	B_3	240	B_5	960	B_7	3840
						C_8	4096

CHARACTERISTICS OF HUMAN HEARING

The human ear is sensitive to sound waves with frequencies in the range from a few hertz to almost 20 kHz. Auditory response is usually expressed in terms of the *loudness level* of a sound, which is a measure of the sound pressure. The reference level, which is given in the unit *phon,* is a pure tone of frequency 1000 Hz with sound pressure of 20 mPa (in cgs units, $2 \cdot 10^{-4}$ dyn/cm^2); loudness level is usually expressed in decibels (dB) relative to this reference level. If a normal observer perceives an arbitrary sound to be equally loud as this reference sound, the sound is said to have the loudness level of the reference. The sensitivity of the typical human ear ranges from about 0 dB, the threshold loudness level, to about 140 dB, the level at which pain sets in. The minimum detectable level thus represents a sound wave of pressure 20 mPa and intensity (power density) 10^{-16} W/cm^2.

The following figure illustrates the frequency dependence of the threshold for an average young adult.

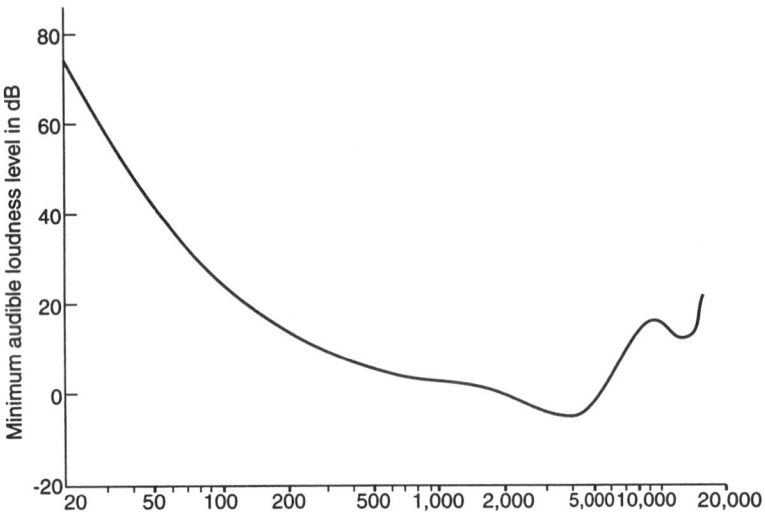

The relation between loudness level and frequency for a typical person is expressed by the following table:

Sound pressure level in dB relative to 20 mPa	Frequency in Hz					
	125	500	1000	4000	8000	10000
10			10	18		
20		16	20	28	11	
30	4	27	30	37	21	17
40	17	39	40	45	30	26
50	34	52	50	54	38	35
60	52	65	60	64	47	44
70	70	76	70	73	56	54
80	86	86	80	83	66	64
90	98	96	90	94	77	74
100	108	105	100	106	88	86

Thus, a 10,000 Hz tone at a pressure level of 50 dB seems equally loud as a 1000 Hz tone at a pressure of 35 dB.

The term *noise* refers to any unwanted sound, either a pure tone or a mixture of frequencies. Since the sensitivity of the ear is frequency dependent, as illustrated by the above table, noise level is expressed in a frequency-weighted scale, known as A-weighting. Decibel readings on this scale are designated as dBa. Typical noise levels from various sources are illustrated in this table:

Source	Noise level in dBa
Rocket engine	200
Jet aircraft engine	160
Light aircraft, cruising	140
Tractor, 150 hp	115
Electric motor, 100 hp at 2600 rpm	105
Pneumatic drill	100
Subway train	90
Vacuum cleaner	85
Heavy automobile traffic	75
Conversational speech	65
Whispered speech	40
Background noise, recording studio	25-30

Recommended noise thresholds in the workplace have been established by the American Conference of Government Industrial Hygenists. Some examples of the maximum safe levels for different daily exposure times are given below.

Duration of exposure	Max. level in dBa
24 h	80
8 h	85
4 h	88
1 h	94
30 min	97
15 min	100
2 min	109
28 s	115
0.11 s	139

No exposure greater than 140 dBa is permitted. Further details may be found in Reference 3.

REFERENCES

1. Anderson, H. L., Editor, *A Physicist's Desk Reference*, American Institute of Physics, New York, 1989, chap. 2.
2. Gray, D. E., Ed., *American Institute of Physics Handbook, Third Edition*, McGraw Hill, New York, 1972, chap. 3.
3. *Threshold Limit Values for Chemical Substances and Physical Agents; Biological Exposure Indices,* 1999 Edition, American Conference of Governmental Industrial Hygienists, 1330 Kemper Meadow Drive, Cincinnati, OH 45240-1634.

Section 15
Practical Laboratory Data

Standard ITS-90 Thermocouple Tables **15**-1
Properties of Common Laboratory Solvents **15**-14
Dependence of Boiling Point on Pressure **15**-19
Ebullioscopic Constants for Calculation of Boiling Point Elevation **15**-20
Cryogenic Constants for Calculation of Freezing Point Depression **15**-21
Freezing Point Lowering by Electrolytes in Aqueous Solution **15**-22
Determination of Relative Humidity from Dew Point **15**-23
Determination of Relative Humidity from Wet and Dry Bulb Temperatures **15**-24
Constant Humidity Solutions **15**-25
Standard Salt Solutions for Humidity Calibration **15**-26
Low Temperature Baths for Maintaining Constant Temperature **15**-26
Wire Tables **15**-27
Characteristics of Particles and Particle Dispersoids **15**-28
Density of Various Solids **15**-29
Dielectric Strength of Insulating Materials **15**-30

STANDARD ITS-90 THERMOCOUPLE TABLES

The Instrument Society of America (ISA) has assigned standard letter designations to a number of thermocouple types having specified emf-temperature relations. These designations and the approximate metal compositions which meet the required relations, as well as the useful temperature ranges, are given below:

Type B	(Pt + 30% Rh) vs. (Pt + 6% Rh)	0 to 1820°C
Type E	(Ni + 10% Cr) vs. (Cu + 43% Ni)	-270 to 1000°C
Type J	Fe vs. (Cu + 43%Ni)	-210 to 1200°C
Type K	(Ni + 10% Cr) vs. (Ni + 2% Al + 2% Mn + 1% Si)	-270 to 1372°C
Type N	(Ni + 14% Cr + 1.5% Si) vs. (Ni + 4.5% Si + 0.1% Mg)	-270 to 1300°C
Type R	(Pt + 13% Rh) vs. Pt	-50 to 1768°C
Type S	(Pt + 10% Rh) vs. Pt	-50 to 1768°C
Type T	Cu vs. (Cu + 43% Ni)	-270 to 400°C

The compositions are given in weight percent, and the positive leg is listed first. It should be emphasized that the standard letter designations do not imply a precise composition but rather that the specified emf-temperature relation is satisfied.

The first set of tables below lists, for each thermocouple type, the emf as a function of temperature on the International Temperature Scale of 1990 (ITS-90). The coefficients in the equation used to generate the table are also given. The second set of tables gives the inverse relationships, i.e., the coefficients in the polynomial equation which expresses the temperature as a function of thermocouple emf. The accuracy of these equations is also stated.

Further details and tables at closer intervals may be found in Reference 1.

REFERENCES

1. Burns, G. W., Seroger, M. G., Strouse, G. F., Croarkin, M. C., and Guthrie, W. F., *Temperature-Electromotive Force Reference Functions and Tables for the Letter-Designated Thermocouple Types Based on the ITS-90*, Natl. Inst. Stand. Tech. (U.S.) Monogr. 175, 1993.
2. Schooley, J. F., *Thermometry*, CRC Press, Boca Raton, FL, 1986.

Type B thermocouples: emf-temperature (°C) reference table and equations.

Thermocouple emf as a Function of Temperature in Degrees Celsius (ITS-90)

emf in Millivolts   Reference Junctions at 0 °C

°C	0	10	20	30	40	50	60	70	80	90	100
0	0.000	−0.002	−0.003	−0.002	−0.000	0.002	0.006	0.011	0.017	0.025	0.033
100	0.033	0.043	0.053	0.065	0.078	0.092	0.107	0.123	0.141	0.159	0.178
200	0.178	0.199	0.220	0.243	0.267	0.291	0.317	0.344	0.372	0.401	0.431
300	0.431	0.462	0.494	0.527	0.561	0.596	0.632	0.669	0.707	0.746	0.787
400	0.787	0.828	0.870	0.913	0.957	1.002	1.048	1.095	1.143	1.192	1.242
500	1.242	1.293	1.344	1.397	1.451	1.505	1.561	1.617	1.675	1.733	1.792
600	1.792	1.852	1.913	1.975	2.037	2.101	2.165	2.230	2.296	2.363	2.431
700	2.431	2.499	2.569	2.639	2.710	2.782	2.854	2.928	3.002	3.078	3.154
800	3.154	3.230	3.308	3.386	3.466	3.546	3.626	3.708	3.790	3.873	3.957
900	3.957	4.041	4.127	4.213	4.299	4.387	4.475	4.564	4.653	4.743	4.834
1000	4.834	4.926	5.018	5.111	5.205	5.299	5.394	5.489	5.585	5.682	5.780
1100	5.780	5.878	5.976	6.075	6.175	6.276	6.377	6.478	6.580	6.683	6.786
1200	6.786	6.890	6.995	7.100	7.205	7.311	7.417	7.524	7.632	7.740	7.848
1300	7.848	7.957	8.066	8.176	8.286	8.397	8.508	8.620	8.731	8.844	8.956
1400	8.956	9.069	9.182	9.296	9.410	9.524	9.639	9.753	9.868	9.984	10.099
1500	10.099	10.215	10.331	10.447	10.563	10.679	10.796	10.913	11.029	11.146	11.263
1600	11.263	11.380	11.497	11.614	11.731	11.848	11.965	12.082	12.199	12.316	12.433
1700	12.433	12.549	12.666	12.782	12.898	13.014	13.130	13.246	13.361	13.476	13.591
1800	13.591	13.706	13.820								

Temperature Ranges and Coefficients of Equations Used to Compute the Above Table

The equations are of the form: $E = c_0 + c_1 t + c_2 t^2 + c_3 t^3 + \ldots c_n t^n$, where E is the emf in millivolts, t is the temperature in degrees Celsius (ITS-90), and c_0, c_1, c_2, c_3, *etc.* are the coefficients. These coefficients are extracted from NIST Monograph 175.

		0 °C to 630.615 °C	630.615 °C to 1820 °C
c_0	=	0.000 000 000 0 ...	−3.893 816 862 1 ...
c_1	=	−2.465 081 834 6 X 10^{-4}	2.857 174 747 0 X 10^{-2}
c_2	=	5.904 042 117 1 X 10^{-6}	−8.488 510 478 5 X 10^{-5}
c_3	=	−1.325 793 163 6 X 10^{-9}	1.578 528 016 4 X 10^{-7}
c_4	=	1.566 829 190 1 X 10^{-12}	−1.683 534 486 4 X 10^{-10}
c_5	=	−1.694 452 924 0 X 10^{-15}	1.110 979 401 3 X 10^{-13}
c_6	=	6.299 034 709 4 X 10^{-19}	−4.451 543 103 3 X 10^{-17}
c_7	=	. . . .	9.897 564 082 1 X 10^{-21}
c_8	=	. . . .	−9.379 133 028 9 X 10^{-25}

Type E thermocouples: emf-temperature (°C) reference table and equations.

Thermocouple emf as a Function of Temperature in Degrees Celsius (ITS-90)

emf in Millivolts   Reference Junctions at 0 °C

°C	0	−10	−20	−30	−40	−50	−60	−70	−80	−90	−100
−200	−8.825	−9.063	−9.274	−9.455	−9.604	−9.718	−9.797	−9.835			
−100	−5.237	−5.681	−6.107	−6.516	−6.907	−7.279	−7.632	−7.963	−8.273	−8.561	−8.825
0	0.000	−0.582	−1.152	−1.709	−2.255	−2.787	−3.306	−3.811	−4.302	−4.777	−5.237

°C	0	10	20	30	40	50	60	70	80	90	100
0	0.000	0.591	1.192	1.801	2.420	3.048	3.685	4.330	4.985	5.648	6.319
100	6.319	6.998	7.685	8.379	9.081	9.789	10.503	11.224	11.951	12.684	13.421
200	13.421	14.164	14.912	15.664	16.420	17.181	17.945	18.713	19.484	20.259	21.036
300	21.036	21.817	22.600	23.386	24.174	24.964	25.757	26.552	27.348	28.146	28.946
400	28.946	29.747	30.550	31.354	32.159	32.965	33.772	34.579	35.387	36.196	37.005
500	37.005	37.815	38.624	39.434	40.243	41.053	41.862	42.671	43.479	44.286	45.093
600	45.093	45.900	46.705	47.509	48.313	49.116	49.917	50.718	51.517	52.315	53.112
700	53.112	53.908	54.703	55.497	56.289	57.080	57.870	58.659	59.446	60.232	61.017
800	61.017	61.801	62.583	63.364	64.144	64.922	65.698	66.473	67.246	68.017	68.787
900	68.787	69.554	70.319	71.082	71.844	72.603	73.360	74.115	74.869	75.621	76.373
1000	76.373										

Temperature Ranges and Coefficients of Equations Used to Compute the Above Table
The equations are of the form: $E = c_0 + c_1 t + c_2 t^2 + c_3 t^3 + \ldots c_n t^n$, where E is the emf in millivolts, t is the temperature in degrees Celsius (ITS-90), and c_0, c_1, c_2, c_3, etc. are the coefficients. These coefficients are extracted from NIST Monograph 175.

		−270 °C to 0 °C	0 °C to 1000 °C
c_0	=	0.000 000 000 0 . . .	0.000 000 000 0 . . .
c_1	=	5.866 550 870 8 X 10^{-2}	5.866 550 871 0 X 10^{-2}
c_2	=	4.541 097 712 4 X 10^{-5}	4.503 227 558 2 X 10^{-5}
c_3	=	−7.799 804 868 6 X 10^{-7}	2.890 840 721 2 X 10^{-8}
c_4	=	−2.580 016 084 3 X 10^{-8}	−3.305 689 665 2 X 10^{-10}
c_5	=	−5.945 258 305 7 X 10^{-10}	6.502 440 327 0 X 10^{-13}
c_6	=	−9.321 405 866 7 X 10^{-12}	−1.919 749 550 4 X 10^{-16}
c_7	=	−1.028 760 553 4 X 10^{-13}	−1.253 660 049 7 X 10^{-18}
c_8	=	−8.037 012 362 1 X 10^{-16}	2.148 921 756 9 X 10^{-21}
c_9	=	−4.397 949 739 1 X 10^{-18}	−1.438 804 178 2 X 10^{-24}
c_{10}	=	−1.641 477 635 5 X 10^{-20}	3.596 089 948 1 X 10^{-28}
c_{11}	=	−3.967 361 951 6 X 10^{-23}	. . . .
c_{12}	=	−5.582 732 872 1 X 10^{-26}	. . . .
c_{13}	=	−3.465 784 201 3 X 10^{-29}	. . . .

Type J thermocouples: emf-temperature (°C) reference table and equations.

Thermocouple emf as a Function of Temperature in Degrees Celsius (ITS-90)

emf in Millivolts   Reference Junctions at 0 °C

°C	0	−10	−20	−30	−40	−50	−60	−70	−80	−90	−100
−200	−7.890	−8.095									
−100	−4.633	−5.037	−5.426	−5.801	−6.159	−6.500	−6.821	−7.123	−7.403	−7.659	−7.890
0	0.000	−0.501	−0.995	−1.482	−1.961	−2.431	−2.893	−3.344	−3.786	−4.215	−4.633

°C	0	10	20	30	40	50	60	70	80	90	100
0	0.000	0.507	1.019	1.537	2.059	2.585	3.116	3.650	4.187	4.726	5.269
100	5.269	5.814	6.360	6.909	7.459	8.010	8.562	9.115	9.669	10.224	10.779
200	10.779	11.334	11.889	12.445	13.000	13.555	14.110	14.665	15.219	15.773	16.327
300	16.327	16.881	17.434	17.986	18.538	19.090	19.642	20.194	20.745	21.297	21.848
400	21.848	22.400	22.952	23.504	24.057	24.610	25.164	25.720	26.276	26.834	27.393
500	27.393	27.953	28.516	29.080	29.647	30.216	30.788	31.362	31.939	32.519	33.102
600	33.102	33.689	34.279	34.873	35.470	36.071	36.675	37.284	37.896	38.512	39.132
700	39.132	39.755	40.382	41.012	41.645	42.281	42.919	43.559	44.203	44.848	45.494
800	45.494	46.141	46.786	47.431	48.074	48.715	49.353	49.989	50.622	51.251	51.877
900	51.877	52.500	53.119	53.735	54.347	54.956	55.561	56.164	56.763	57.360	57.953
1000	57.953	58.545	59.134	59.721	60.307	60.890	61.473	62.054	62.634	63.214	63.792
1100	63.792	64.370	64.948	65.525	66.102	66.679	67.255	67.831	68.406	68.980	69.553
1200	69.553										

Temperature Ranges and Coefficients of Equations Used to Compute the Above Table

The equations are of the form: $E = c_0 + c_1 t + c_2 t^2 + c_3 t^3 + \ldots c_n t^n$, where E is the emf in millivolts, t is the temperature in degrees Celsius (ITS-90), and c_0, c_1, c_2, c_3, etc. are the coefficients. These coefficients are extracted from NIST Monograph 175.

		−210 °C to 760 °C	760 °C to 1200 °C
c_0	=	0.000 000 000 0 ...	$2.964\ 562\ 568\ 1 \times 10^2$
c_1	=	$5.038\ 118\ 781\ 5 \times 10^{-2}$	−1.497 612 778 6 ...
c_2	=	$3.047\ 583\ 693\ 0 \times 10^{-5}$	$3.178\ 710\ 392\ 4 \times 10^{-3}$
c_3	=	$-8.568\ 106\ 572\ 0 \times 10^{-8}$	$-3.184\ 768\ 670\ 1 \times 10^{-6}$
c_4	=	$1.322\ 819\ 529\ 5 \times 10^{-10}$	$1.572\ 081\ 900\ 4 \times 10^{-9}$
c_5	=	$-1.705\ 295\ 833\ 7 \times 10^{-13}$	$-3.069\ 136\ 905\ 6 \times 10^{-13}$
c_6	=	$2.094\ 809\ 069\ 7 \times 10^{-16}$	. . . .
c_7	=	$-1.253\ 839\ 533\ 6 \times 10^{-19}$	. . . .
c_8	=	$1.563\ 172\ 569\ 7 \times 10^{-23}$	. . . .

Type K thermocouples: emf-temperature (°C) reference table and equations.

Thermocouple emf as a Function of Temperature in Degrees Celsius (ITS-90)

emf in Millivolts   Reference Junctions at 0 °C

°C	0	−10	−20	−30	−40	−50	−60	−70	−80	−90	−100
−200	−5.891	−6.035	−6.158	−6.262	−6.344	−6.404	−6.441	−6.458			
−100	−3.554	−3.852	−4.138	−4.411	−4.669	−4.913	−5.141	−5.354	−5.550	−5.730	−5.891
0	0.000	−0.392	−0.778	−1.156	−1.527	−1.889	−2.243	−2.587	−2.920	−3.243	−3.554

°C	0	10	20	30	40	50	60	70	80	90	100
0	0.000	0.397	0.798	1.203	1.612	2.023	2.436	2.851	3.267	3.682	4.096
100	4.096	4.509	4.920	5.328	5.735	6.138	6.540	6.941	7.340	7.739	8.138
200	8.138	8.539	8.940	9.343	9.747	10.153	10.561	10.971	11.382	11.795	12.209
300	12.209	12.624	13.040	13.457	13.874	14.293	14.713	15.133	15.554	15.975	16.397
400	16.397	16.820	17.243	17.667	18.091	18.516	18.941	19.366	19.792	20.218	20.644
500	20.644	21.071	21.497	21.924	22.350	22.776	23.203	23.629	24.055	24.480	24.905
600	24.905	25.330	25.755	26.179	26.602	27.025	27.447	27.869	28.289	28.710	29.129
700	29.129	29.548	29.965	30.382	30.798	31.213	31.628	32.041	32.453	32.865	33.275
800	33.275	33.685	34.093	34.501	34.908	35.313	35.718	36.121	36.524	36.925	37.326
900	37.326	37.725	38.124	38.522	38.918	39.314	39.708	40.101	40.494	40.885	41.276
1000	41.276	41.665	42.053	42.440	42.826	43.211	43.595	43.978	44.359	44.740	45.119
1100	45.119	45.497	45.873	46.249	46.623	46.995	47.367	47.737	48.105	48.473	48.838
1200	48.838	49.202	49.565	49.926	50.286	50.644	51.000	51.355	51.708	52.060	52.410
1300	52.410	52.759	53.106	53.451	53.795	54.138	54.479	54.819			

Temperature Ranges and Coefficients of Equations Used to Compute the Above Table
The equations are of the form: $E = c_0 + c_1 t + c_2 t^2 + c_3 t^3 + \ldots c_n t^n$, where E is the emf in millivolts, t is the temperature in degrees Celsius (ITS-90), and c_0, c_1, c_2, c_3, *etc.* are the coefficients. In the 0 °C to 1372 °C range there is also an exponential term that must be evaluated and added to the equation. The exponential term is of the form: $c_0 e^{c_1 (t - 126.9686)^2}$, where t is the temperature in °C, e is the natural logarithm base, and c_0 and c_i are the coefficients. These coefficients are extracted from NIST Monograph 175.

		−270 °C to 0 °C	0 °C to 1372 °C	0 °C to 1372 °C (exponential term)
c_0	=	0.000 000 000 0 ...	−1.760 041 368 6 X 10^{-2}	1.185 976 X 10^{-1}
c_1	=	3.945 012 802 5 X 10^{-2}	3.892 120 497 5 X 10^{-2}	−1.183 432 X 10^{-4}
c_2	=	2.362 237 359 8 X 10^{-5}	1.855 877 003 2 X 10^{-5}	. . . .
c_3	=	−3.285 890 678 4 X 10^{-7}	−9.945 759 287 4 X 10^{-8}	. . . .
c_4	=	−4.990 482 877 7 X 10^{-9}	3.184 094 571 9 X 10^{-10}	. . . .
c_5	=	−6.750 905 917 3 X 10^{-11}	−5.607 284 488 9 X 10^{-13}	. . . .
c_6	=	−5.741 032 742 8 X 10^{-13}	5.607 505 905 9 X 10^{-16}	. . . .
c_7	=	−3.108 887 289 4 X 10^{-15}	−3.202 072 000 3 X 10^{-19}	. . . .
c_8	=	−1.045 160 936 5 X 10^{-17}	9.715 114 715 2 X 10^{-23}	. . . .
c_9	=	−1.988 926 687 8 X 10^{-20}	−1.210 472 127 5 X 10^{-26}	. . . .
c_{10}	=	−1.632 269 748 6 X 10^{-23}	. . . .	. . . .

Type N thermocouples: emf-temperature (°C) reference table and equations.

Thermocouple emf as a Function of Temperature in Degrees Celsius (ITS-90)

emf in Millivolts   Reference Junctions at 0 °C

°C	0	−10	−20	−30	−40	−50	−60	−70	−80	−90	−100
−200	−3.990	−4.083	−4.162	−4.226	−4.277	−4.313	−4.336	−4.345			
−100	−2.407	−2.612	−2.808	−2.994	−3.171	−3.336	−3.491	−3.634	−3.766	−3.884	−3.990
0	0.000	−0.260	−0.518	−0.772	−1.023	−1.269	−1.509	−1.744	−1.972	−2.193	−2.407

°C	0	10	20	30	40	50	60	70	80	90	100
0	0.000	0.261	0.525	0.793	1.065	1.340	1.619	1.902	2.189	2.480	2.774
100	2.774	3.072	3.374	3.680	3.989	4.302	4.618	4.937	5.259	5.585	5.913
200	5.913	6.245	6.579	6.916	7.255	7.597	7.941	8.288	8.637	8.988	9.341
300	9.341	9.696	10.054	10.413	10.774	11.136	11.501	11.867	12.234	12.603	12.974
400	12.974	13.346	13.719	14.094	14.469	14.846	15.225	15.604	15.984	16.366	16.748
500	16.748	17.131	17.515	17.900	18.286	18.672	19.059	19.447	19.835	20.224	20.613
600	20.613	21.003	21.393	21.784	22.175	22.566	22.958	23.350	23.742	24.134	24.527
700	24.527	24.919	25.312	25.705	26.098	26.491	26.883	27.276	27.669	28.062	28.455
800	28.455	28.847	29.239	29.632	30.024	30.416	30.807	31.199	31.590	31.981	32.371
900	32.371	32.761	33.151	33.541	33.930	34.319	34.707	35.095	35.482	35.869	36.256
1000	36.256	36.641	37.027	37.411	37.795	38.179	38.562	38.944	39.326	39.706	40.087
1100	40.087	40.466	40.845	41.223	41.600	41.976	42.352	42.727	43.101	43.474	43.846
1200	43.846	44.218	44.588	44.958	45.326	45.694	46.060	46.425	46.789	47.152	47.513
1300	47.513										

Temperature Ranges and Coefficients of Equations Used to Compute the Above Table

The equations are of the form: $E = c_0 + c_1 t + c_2 t^2 + c_3 t^3 + \ldots c_n t^n$, where E is the emf in millivolts, t is the temperature in degrees Celsius (ITS-90), and c_0, c_1, c_2, c_3, *etc.* are the coefficients. These coefficients are extracted from NIST Monograph 175.

		−270 °C to 0 °C	0 °C to 1300 °C
c_0	=	0.000 000 000 0 . . .	0.000 000 000 0 . . .
c_1	=	$2.615\ 910\ 596\ 2 \times 10^{-2}$	$2.592\ 939\ 460\ 1 \times 10^{-2}$
c_2	=	$1.095\ 748\ 422\ 8 \times 10^{-5}$	$1.571\ 014\ 188\ 0 \times 10^{-5}$
c_3	=	$-9.384\ 111\ 155\ 4 \times 10^{-8}$	$4.382\ 562\ 723\ 7 \times 10^{-8}$
c_4	=	$-4.641\ 203\ 975\ 9 \times 10^{-11}$	$-2.526\ 116\ 979\ 4 \times 10^{-10}$
c_5	=	$-2.630\ 335\ 771\ 6 \times 10^{-12}$	$6.431\ 181\ 933\ 9 \times 10^{-13}$
c_6	=	$-2.265\ 343\ 800\ 3 \times 10^{-14}$	$-1.006\ 347\ 151\ 9 \times 10^{-15}$
c_7	=	$-7.608\ 930\ 079\ 1 \times 10^{-17}$	$9.974\ 533\ 899\ 2 \times 10^{-19}$
c_8	=	$-9.341\ 966\ 783\ 5 \times 10^{-20}$	$-6.086\ 324\ 560\ 7 \times 10^{-22}$
c_9	=	. . . .	$2.084\ 922\ 933\ 9 \times 10^{-26}$
c_{10}	=	. . . .	$-3.068\ 219\ 615\ 1 \times 10^{-29}$

Type R thermocouples: emf-temperature (°C) reference table and equations.

Thermocouple emf as a Function of Temperature in Degrees Celsius (ITS-90)

emf in Millivolts   Reference Junctions at 0 °C

°C	0	−10	−20	−30	−40	−50	−60	−70	−80	−90	−100
0	0.000	−0.051	−0.100	−0.145	−0.188	−0.226					

°C	0	10	20	30	40	50	60	70	80	90	100
0	0.000	0.054	0.111	0.171	0.232	0.296	0.363	0.431	0.501	0.573	0.647
100	0.647	0.723	0.800	0.879	0.959	1.041	1.124	1.208	1.294	1.381	1.469
200	1.469	1.558	1.648	1.739	1.831	1.923	2.017	2.112	2.207	2.304	2.401
300	2.401	2.498	2.597	2.696	2.796	2.896	2.997	3.099	3.201	3.304	3.408
400	3.408	3.512	3.616	3.721	3.827	3.933	4.040	4.147	4.255	4.363	4.471
500	4.471	4.580	4.690	4.800	4.910	5.021	5.133	5.245	5.357	5.470	5.583
600	5.583	5.697	5.812	5.926	6.041	6.157	6.273	6.390	6.507	6.625	6.743
700	6.743	6.861	6.980	7.100	7.220	7.340	7.461	7.583	7.705	7.827	7.950
800	7.950	8.073	8.197	8.321	8.446	8.571	8.697	8.823	8.950	9.077	9.205
900	9.205	9.333	9.461	9.590	9.720	9.850	9.980	10.111	10.242	10.374	10.506
1000	10.506	10.638	10.771	10.905	11.039	11.173	11.307	11.442	11.578	11.714	11.850
1100	11.850	11.986	12.123	12.260	12.397	12.535	12.673	12.812	12.950	13.089	13.228
1200	13.228	13.367	13.507	13.646	13.786	13.926	14.066	14.207	14.347	14.488	14.629
1300	14.629	14.770	14.911	15.052	15.193	15.334	15.475	15.616	15.758	15.899	16.040
1400	16.040	16.181	16.323	16.464	16.605	16.746	16.887	17.028	17.169	17.310	17.451
1500	17.451	17.591	17.732	17.872	18.012	18.152	18.292	18.431	18.571	18.710	18.849
1600	18.849	18.988	19.126	19.264	19.402	19.540	19.677	19.814	19.951	20.087	20.222
1700	20.222	20.356	20.488	20.620	20.749	20.877	21.003				

Temperature Ranges and Coefficients of Equations Used to Compute the Above Table

The equations are of the form: $E = c_0 + c_1 t + c_2 t^2 + c_3 t^3 + \ldots c_n t^n$, where E is the emf in millivolts, t is the temperature in degrees Celsius (ITS-90), and c_0, c_1, c_2, c_3, *etc.* are the coefficients. These coefficients are extracted from NIST Monograph 175.

		−50 °C to 1064.18 °C	1064.18 °C to 1664.5 °C	1664.5 °C to 1768.1 °C
c_0	=	0.000 000 000 00 ...	2.951 579 253 16 ...	1.522 321 182 09 X 10^2
c_1	=	5.289 617 297 65 X 10^{-3}	−2.520 612 513 32 X 10^{-3}	−2.688 198 885 45 X 10^{-1}
c_2	=	1.391 665 897 82 X 10^{-5}	1.595 645 018 65 X 10^{-5}	1.712 802 804 71 X 10^{-4}
c_3	=	−2.388 556 930 17 X 10^{-8}	−7.640 859 475 76 X 10^{-9}	−3.458 957 064 53 X 10^{-8}
c_4	=	3.569 160 010 63 X 10^{-11}	2.053 052 910 24 X 10^{-12}	−9.346 339 710 46 X 10^{-15}
c_5	=	−4.623 476 662 98 X 10^{-14}	−2.933 596 681 73 X 10^{-16}	. . . .
c_6	=	5.007 774 410 34 X 10^{-17}	. . . .	. . . .
c_7	=	−3.731 058 861 91 X 10^{-20}	. . . .	. . . .
c_8	=	1.577 164 823 67 X 10^{-23}	. . . .	. . . .
c_9	=	−2.810 386 252 51 X 10^{-27}	. . . .	. . . .

STANDARD ITS-90 THERMOCOUPLE TABLES (continued)

Type S thermocouples: emf-temperature (°C) reference table and equations.

Thermocouple emf as a Function of Temperature in Degrees Celsius (ITS-90)

emf in Millivolts   Reference Junctions at 0 °C

°C	0	−10	−20	−30	−40	−50	−60	−70	−80	−90	−100
0	0.000	−0.053	−0.103	−0.150	−0.194	−0.236					

°C	0	10	20	30	40	50	60	70	80	90	100
0	0.000	0.055	0.113	0.173	0.235	0.299	0.365	0.433	0.502	0.573	0.646
100	0.646	0.720	0.795	0.872	0.950	1.029	1.110	1.191	1.273	1.357	1.441
200	1.441	1.526	1.612	1.698	1.786	1.874	1.962	2.052	2.141	2.232	2.323
300	2.323	2.415	2.507	2.599	2.692	2.786	2.880	2.974	3.069	3.164	3.259
400	3.259	3.355	3.451	3.548	3.645	3.742	3.840	3.938	4.036	4.134	4.233
500	4.233	4.332	4.432	4.532	4.632	4.732	4.833	4.934	5.035	5.137	5.239
600	5.239	5.341	5.443	5.546	5.649	5.753	5.857	5.961	6.065	6.170	6.275
700	6.275	6.381	6.486	6.593	6.699	6.806	6.913	7.020	7.128	7.236	7.345
800	7.345	7.454	7.563	7.673	7.783	7.893	8.003	8.114	8.226	8.337	8.449
900	8.449	8.562	8.674	8.787	8.900	9.014	9.128	9.242	9.357	9.472	9.587
1000	9.587	9.703	9.819	9.935	10.051	10.168	10.285	10.403	10.520	10.638	10.757
1100	10.757	10.875	10.994	11.113	11.232	11.351	11.471	11.590	11.710	11.830	11.951
1200	11.951	12.071	12.191	12.312	12.433	12.554	12.675	12.796	12.917	13.038	13.159
1300	13.159	13.280	13.402	13.523	13.644	13.766	13.887	14.009	14.130	14.251	14.373
1400	14.373	14.494	14.615	14.736	14.857	14.978	15.099	15.220	15.341	15.461	15.582
1500	15.582	15.702	15.822	15.942	16.062	16.182	16.301	16.420	16.539	16.658	16.777
1600	16.777	16.895	17.013	17.131	17.249	17.366	17.483	17.600	17.717	17.832	17.947
1700	17.947	18.061	18.174	18.285	18.395	18.503	18.609				

Temperature Ranges and Coefficients of Equations Used to Compute the Above Table
The equations are of the form: $E = c_0 + c_1 t + c_2 t^2 + c_3 t^3 + \ldots c_n t^n$, where E is the emf in millivolts, t is the temperature in degrees Celsius (ITS-90), and c_0, c_1, c_2, c_3, etc. are the coefficients. These coefficients are extracted from NIST Monograph 175.

		−50 °C to 1064.18 °C	1064.18 °C to 1664.5 °C	1664.5 °C to 1768.1 °C
c_0	=	0.000 000 000 00 ...	1.329 004 440 85 ...	$1.466\ 282\ 326\ 36 \times 10^2$
c_1	=	$5.403\ 133\ 086\ 31 \times 10^{-3}$	$3.345\ 093\ 113\ 44 \times 10^{-3}$	$-2.584\ 305\ 167\ 52 \times 10^{-1}$
c_2	=	$1.259\ 342\ 897\ 40 \times 10^{-6}$	$6.548\ 051\ 928\ 18 \times 10^{-6}$	$1.636\ 935\ 746\ 41 \times 10^{-4}$
c_3	=	$-2.324\ 779\ 686\ 89 \times 10^{-9}$	$-1.648\ 562\ 592\ 09 \times 10^{-9}$	$-3.304\ 390\ 469\ 87 \times 10^{-8}$
c_4	=	$3.220\ 288\ 230\ 36 \times 10^{-11}$	$1.299\ 896\ 051\ 74 \times 10^{-14}$	$-9.432\ 236\ 906\ 12 \times 10^{-15}$
c_5	=	$-3.314\ 651\ 963\ 89 \times 10^{-14}$	. . . .	. . . .
c_6	=	$2.557\ 442\ 517\ 86 \times 10^{-17}$	. . . .	. . . .
c_7	=	$-1.250\ 688\ 713\ 93 \times 10^{-20}$	. . . .	. . . .
c_8	=	$2.714\ 431\ 761\ 45 \times 10^{-24}$	. . . .	. . . .

Type T thermocouples: emf-temperature (°C) reference table and equations.

Thermocouple emf as a Function of Temperature in Degrees Celsius (ITS-90)

emf in Millivolts   Reference Junctions at 0 °C

°C	0	−10	−20	−30	−40	−50	−60	−70	−80	−90	−100
−200	−5.603	−5.753	−5.888	−6.007	−6.105	−6.180	−6.232	−6.258			
−100	−3.379	−3.657	−3.923	−4.177	−4.419	−4.648	−4.865	−5.070	−5.261	−5.439	−5.603
0	0.000	−0.383	−0.757	−1.121	−1.475	−1.819	−2.153	−2.476	−2.788	−3.089	−3.379

°C	0	10	20	30	40	50	60	70	80	90	100
0	0.000	0.391	0.790	1.196	1.612	2.036	2.468	2.909	3.358	3.814	4.279
100	4.279	4.750	5.228	5.714	6.206	6.704	7.209	7.720	8.237	8.759	9.288
200	9.288	9.822	10.362	10.907	11.458	12.013	12.574	13.139	13.709	14.283	14.862
300	14.862	15.445	16.032	16.624	17.219	17.819	18.422	19.030	19.641	20.255	20.872
400	20.872										

Temperature Ranges and Coefficients of Equations Used to Compute the Above Table
The equations are of the form: $E = c_0 + c_1 t + c_2 t^2 + c_3 t^3 + \ldots c_n t^n$, where E is the emf in millivolts, t is the temperature in degrees Celsius (ITS-90), and c_0, c_1, c_2, c_3, *etc.* are the coefficients. These coefficients are extracted from NIST Monograph 175.

		−270 °C to 0 °C	0 °C to 400 °C
c_0	=	0.000 000 000 0 . . .	0.000 000 000 0 . . .
c_1	=	3.874 810 636 4 X 10^{-2}	3.874 810 636 4 X 10^{-2}
c_2	=	4.419 443 434 7 X 10^{-5}	3.329 222 788 0 X 10^{-5}
c_3	=	1.184 432 310 5 X 10^{-7}	2.061 824 340 4 X 10^{-7}
c_4	=	2.003 297 355 4 X 10^{-8}	−2.188 225 684 6 X 10^{-9}
c_5	=	9.013 801 955 9 X 10^{-10}	1.099 688 092 8 X 10^{-11}
c_6	=	2.265 115 659 3 X 10^{-11}	−3.081 575 877 2 X 10^{-14}
c_7	=	3.607 115 420 5 X 10^{-13}	4.547 913 529 0 X 10^{-17}
c_8	=	3.849 393 988 3 X 10^{-15}	−2.751 290 167 3 X 10^{-20}
c_9	=	2.821 352 192 5 X 10^{-17}	. . . .
c_{10}	=	1.425 159 477 9 X 10^{-19}	. . . .
c_{11}	=	4.876 866 228 6 X 10^{-22}	. . . .
c_{12}	=	1.079 553 927 0 X 10^{-24}	. . . .
c_{13}	=	1.394 502 706 2 X 10^{-27}	. . . .
c_{14}	=	7.979 515 392 7 X 10^{-31}	. . . .

Type B thermocouples: coefficients (c_i) of polynomials for the computation of temperatures in °C as a function of the thermocouple emf in various temperature and emf ranges.

Temperature Range:	250 °C to 700 °C	700 °C to 1820 °C
emf Range:	0.291 mV to 2.431 mV	2.431 mV to 13.820 mV
$c_0 =$	$9.842\ 332\ 1 \times 10^1$	$2.131\ 507\ 1 \times 10^2$
$c_1 =$	$6.997\ 150\ 0 \times 10^2$	$2.851\ 050\ 4 \times 10^2$
$c_2 =$	$-8.476\ 530\ 4 \times 10^2$	$-5.274\ 288\ 7 \times 10^1$
$c_3 =$	$1.005\ 264\ 4 \times 10^3.$	$9.916\ 080\ 4 \dots$
$c_4 =$	$-8.334\ 595\ 2 \times 10^2$	$-1.296\ 530\ 3 \dots$
$c_5 =$	$4.550\ 854\ 2 \times 10^2$	$1.119\ 587\ 0 \times 10^{-1}$
$c_6 =$	$-1.552\ 303\ 7 \times 10^2$	$-6.062\ 519\ 9 \times 10^{-3}$
$c_7 =$	$2.988\ 675\ 0 \times 10^1$	$1.866\ 169\ 6 \times 10^{-4}$
$c_8 =$	$-2.474\ 286\ 0 \dots$	$-2.487\ 858\ 5 \times 10^{-6}$

NOTE—The above coefficients are extracted from NIST Monograph 175 and are for an expression of the form shown in Section 10.3.2. They yield approximate values of temperature that agree within ± 0.03 °C with the values given in Table 10.2.

Type E thermocouples: coefficients (c_i) of polynomials for the computation of temperatures in °C as a function of the thermocouple emf in various temperature and emf ranges.

Temperature Range:	−200 °C to 0 °C	0 °C to 1000 °C
emf Range:	−8.825 mV to 0.0 mV	0.0 mV to 76.373 mV
$c_0 =$	$0.000\ 000\ 0 \dots$	$0.000\ 000\ 0 \dots$
$c_1 =$	$1.697\ 728\ 8 \times 10^1$	$1.705\ 703\ 5 \times 10^1$
$c_2 =$	$-4.351\ 497\ 0 \times 10^{-1}$	$-2.330\ 175\ 9 \times 10^{-1}$
$c_3 =$	$-1.585\ 969\ 7 \times 10^{-1}$	$6.543\ 558\ 5 \times 10^{-3}$
$c_4 =$	$-9.250\ 287\ 1 \times 10^{-2}$	$-7.356\ 274\ 9 \times 10^{-5}$
$c_5 =$	$-2.608\ 431\ 4 \times 10^{-2}$	$-1.789\ 600\ 1 \times 10^{-6}$
$c_6 =$	$-4.136\ 019\ 9 \times 10^{-3}$	$8.403\ 616\ 5 \times 10^{-8}$
$c_7 =$	$-3.403\ 403\ 0 \times 10^{-4}$	$-1.373\ 587\ 9 \times 10^{-9}$
$c_8 =$	$-1.156\ 489\ 0 \times 10^{-5}$	$1.062\ 982\ 3 \times 10^{-11}$
$c_9 =$	$\dots\dots$	$-3.244\ 708\ 7 \times 10^{-14}$

NOTE—The above coefficients are extracted from NIST Monograph 175 and are for an expression of the form shown in Section 10.3.2. They yield approximate values of temperature that agree within ± 0.02 °C with the values given in Table 10.4.

STANDARD ITS-90 THERMOCOUPLE TABLES (continued)

Type J thermocouples: coefficients (c_i) of polynomials for the computation of temperatures in °C as a function of the thermocouple emf in various temperature and emf ranges.

Temperature Range:	−210 °C to 0 °C	0 °C to 760 °C	760 °C to 1200 °C
emf Range:	−8.095 mV to 0.0 mV	0.0 mV to 42.919 mV	42.919 mV to 69.553 mV
$c_0 =$	0.000 000 0 . . .	0.000 000 . . .	−3.113 581 87 X 10^3
$c_1 =$	1.952 826 8 X 10^1	1.978 425 X 10^1	3.005 436 84 X 10^2
$c_2 =$	−1.228 618 5 . . .	−2.001 204 X 10^{-1}	−9.947 732 30 . . .
$c_3 =$	−1.075 217 8 . . .	1.036 969 X 10^{-2}	1.702 766 30 X 10^{-1}
$c_4 =$	−5.908 693 3 X 10^{-1}	−2.549 687 X 10^{-4}	−1.430 334 68 X 10^{-3}
$c_5 =$	−1.725 671 3 X 10^{-1}	3.585 153 X 10^{-6}	4.738 860 84 X 10^{-6}
$c_6 =$	−2.813 151 3 X 10^{-2}	−5.344 285 X 10^{-8}	. . . .
$c_7 =$	−2.396 337 0 X 10^{-3}	5.099 890 X 10^{-10}	. . . .
$c_8 =$	−8.382 332 1 X 10^{-5}	. . . .	. . . .

NOTE—The above coefficients are extracted from NIST Monograph 175 and are for an expression of the form shown in Section 10.3.2. They yield approximate values of temperature that agree within ± 0.05 °C with the values given in Table 10.6.

Type K thermocouples: coefficients (c_i) of polynomials for the computation of temperatures in °C as a function of the thermocouple emf in various temperature and emf ranges.

Temperature Range:	−200 °C to 0 °C	0 °C to 500 °C	500 °C to 1372 °C
emf Range:	−5.891 mV to 0.0 mV	0.0 mV to 20.644 mV	20.644 mV to 54.886 mV
$c_0 =$	0.000 000 0 . . .	0.000 000 0 . . .	−1.318 058 X 10^2
$c_1 =$	2.517 346 2 X 10^1	2.508 355 X 10^1	4.830 222 X 10^1
$c_2 =$	−1.166 287 8 . . .	7.860 106 X 10^{-2}	−1.646 031 . . .
$c_3 =$	−1.083 363 8 . . .	−2.503 131 X 10^{-1}	5.464 731 X 10^{-2}
$c_4 =$	−8.977 354 0 X 10^{-1}	8.315 270 X 10^{-2}	−9.650 715 X 10^{-4}
$c_5 =$	−3.734 237 7 X 10^{-1}	−1.228 034 X 10^{-2}	8.802 193 X 10^{-6}
$c_6 =$	−8.663 264 3 X 10^{-2}	9.804 036 X 10^{-4}	−3.110 810 X 10^{-8}
$c_7 =$	−1.045 059 8 X 10^{-2}	−4.413 030 X 10^{-5}	. . . .
$c_8 =$	−5.192 057 7 X 10^{-4}	1.057 734 X 10^{-6}	. . . .
$c_9 =$	. . . .	−1.052 755 X 10^{-8}	. . . .

NOTE—The above coefficients are extracted from NIST Monograph 175 and are for an expression of the form shown in Section 10.3.2. They yield approximate values of temperature that agree within ± 0.05 °C with the values given in Table 10.8.

Type N thermocouples: coefficients (c_i) of polynomials for the computation of temperatures in °C as a function of the thermocouple emf in various temperature and emf ranges.

Temperature Range:	−200 °C to 0 °C	0 °C to 600 °C	600 °C to 1300 °C
emf Range:	−3.990 mV to 0.0 mV	0.0 mV to 20.613 mV	20.613 mV to 47.513 mV
$c_0 =$	0.000 000 0 . . .	0.000 00 . . .	1.972 485 X 10^1
$c_1 =$	3.843 684 7 X 10^1	3.868 96 X 10^1	3.300 943 X 10^1
$c_2 =$	1.101 048 5 . . .	−1.082 67 . . .	−3.915 159 X 10^{-1}
$c_3 =$	5.222 931 2 . . .	4.702 05 X 10^{-2}	9.855 391 X 10^{-3}
$c_4 =$	7.206 052 5 . . .	−2.121 69 X 10^{-6}	−1.274 371 X 10^{-4}
$c_5 =$	5.848 858 6 . . .	−1.172 72 X 10^{-4}	7.767 022 X 10^{-7}
$c_6 =$	2.775 491 6 . . .	5.392 80 X 10^{-6}	. . . .
$c_7 =$	7.707 516 6 X 10^{-1}	−7.981 56 X 10^{-8}	. . . .
$c_8 =$	1.158 266 5 X 10^{-1}	. . . .	. . . .
$c_9 =$	7.313 886 8 X 10^{-3}	. . . .	. . . .

NOTE—The above coefficients are extracted from NIST Monograph 175 and are for an expression of the form shown in Section 10.3.2. They yield approximate values of temperature that agree within ± 0.04 °C with the values given in Table 10.10.

Type R thermocouples: coefficients (c_i) of polynomials for the computation of temperatures in °C as a function of the thermocouple emf in various temperature and emf ranges.

	−50 °C to 250 °C	250 °C to 1200 °C	1064 °C to 1664.5 °C	1664.5 °C to 1768.1 °C
Temperature Range:				
emf Range:	−0.226 mV to 1.923 mV	1.923 mV to 13.228 mV	11.361 mV to 19.739 mV	19.739 mV to 21.103 mV
$c_0 =$	0.000 000 0 . . .	1.334 584 505 X 10^1	−8.199 599 416 X 10^1	3.406 177 836 X 10^4
$c_1 =$	1.889 138 0 X 10^2	1.472 644 573 X 10^2	1.553 962 042 X 10^2	−7.023 729 171 X 10^3
$c_2 =$	−9.383 529 0 X 10^1	−1.844 024 844 X 10^1	−8.342 197 663 . . .	5.582 903 813 X 10^2
$c_3 =$	1.306 861 9 X 10^2	4.031 129 726 . . .	4.279 433 549 X 10^{-1}	−1.952 394 635 X 10^1
$c_4 =$	−2.270 358 0 X 10^2	−6.249 428 360 X 10^{-1}	−1.191 577 910 X 10^{-2}	2.560 740 231 X 10^{-1}
$c_5 =$	3.514 565 9 X 10^2	6.468 412 046 X 10^{-2}	1.492 290 091 X 10^{-4}	. . . .
$c_6 =$	−3.895 390 0 X 10^2	−4.458 750 426 X 10^{-3}	. . . .	. . . .
$c_7 =$	2.823 947 1 X 10^2	1.994 710 149 X 10^{-4}	. . . .	. . . .
$c_8 =$	−1.260 728 1 X 10^2	−5.313 401 790 X 10^{-6}	. . . .	. . . .
$c_9 =$	3.135 361 1 X 10^1	6.481 976 217 X 10^{-6}	. . . .	. . . .
$c_{10} =$	−3.318 776 9 . . .	. . . .	. . . .	. . . .

NOTE—The above coefficients are extracted from NIST Monograph 175 and are for an expression of the form shown in Section 10.3.2. They yield approximate values of temperature that agree within ± 0.02 °C with the values given in Table 10.12.

Type S thermocouples: coefficients (c_i) of polynomials for the computation of temperatures in °C as a function of the thermocouple emf in various temperature and emf ranges.

Temperature Range:	−50 °C to 250 °C	250 °C to 1200 °C	1064 °C to 1664.5 °C	1664.5 °C to 1768.1 °C
emf Range:	−0.235 mV to 1.874 mV	1.874 mV to 11.950 mV	10.332 mV to 17.536 mV	17.536 mV to 18.693 mV
$c_0 =$	$0.000\,000\,00\,\ldots$	$1.291\,507\,177 \times 10^1$	$-8.087\,801\,117 \times 10^1$	$5.333\,875\,126 \times 10^4$
$c_1 =$	$1.849\,494\,60 \times 10^2$	$1.466\,298\,863 \times 10^2$	$1.621\,573\,104 \times 10^2$	$-1.235\,892\,298 \times 10^4$
$c_2 =$	$-8.005\,040\,62 \times 10^1$	$-1.534\,713\,402 \times 10^1$	$-8.536\,869\,453\,\ldots$	$1.092\,657\,613 \times 10^3$
$c_3 =$	$1.022\,374\,30 \times 10^2$	$3.145\,945\,973\,\ldots$	$4.719\,686\,976 \times 10^{-1}$	$-4.265\,693\,686 \times 10^1$
$c_4 =$	$-1.522\,485\,92 \times 10^2$	$-4.163\,257\,839 \times 10^{-1}$	$-1.441\,693\,666 \times 10^{-2}$	$6.247\,205\,420 \times 10^{-1}$
$c_5 =$	$1.888\,213\,43 \times 10^2$	$3.187\,963\,771 \times 10^{-2}$	$2.081\,618\,890 \times 10^{-4}$	$\ldots\ldots\ldots$
$c_6 =$	$-1.590\,859\,41 \times 10^2$	$-1.291\,637\,500 \times 10^{-3}$	$\ldots\ldots\ldots$	$\ldots\ldots\ldots$
$c_7 =$	$8.230\,278\,80 \times 10^1$	$2.183\,475\,087 \times 10^{-5}$	$\ldots\ldots\ldots$	$\ldots\ldots\ldots$
$c_8 =$	$-2.341\,819\,44 \times 10^1$	$-1.447\,379\,511 \times 10^{-7}$	$\ldots\ldots\ldots$	$\ldots\ldots\ldots$
$c_9 =$	$2.797\,862\,60\,\ldots$	$8.211\,272\,125 \times 10^{-9}$	$\ldots\ldots\ldots$	$\ldots\ldots\ldots$

NOTE—The above coefficients are extracted from NIST Monograph 175 and are for an expression of the form shown in Section 10.3.2. They yield approximate values of temperature that agree within ± 0.02 °C with the values given in Table 10.14.

Type T thermocouples: coefficients (c_i) of polynomials for the computation of temperatures in °C as a function of the thermocouple emf in various temperature and emf ranges.

Temperature Range:	−200 °C to 0 °C	0 °C to 400 °C
emf Range:	−5.603 mV to 0.0 mV	0.0 mV to 20.872 mV
$c_0 =$	$0.000\,000\,0\,\ldots$	$0.000\,000\,\ldots$
$c_1 =$	$2.594\,919\,2 \times 10^1$	$2.592\,800 \times 10^1$
$c_2 =$	$-2.131\,696\,7 \times 10^{-1}$	$-7.602\,961 \times 10^{-1}$
$c_3 =$	$7.901\,869\,2 \times 10^{-1}$	$4.637\,791 \times 10^{-2}$
$c_4 =$	$4.252\,777\,7 \times 10^{-1}$	$-2.165\,394 \times 10^{-3}$
$c_5 =$	$1.330\,447\,3 \times 10^{-1}$	$6.048\,144 \times 10^{-5}$
$c_6 =$	$2.024\,144\,6 \times 10^{-2}$	$-7.293\,422 \times 10^{-7}$
$c_7 =$	$1.266\,817\,1 \times 10^{-3}$	$\ldots\ldots\ldots$

NOTE—The above coefficients are extracted from NIST Monograph 175 and are for an expression of the form shown in Section 10.3.2. They yield approximate values of temperature that agree within ± 0.04 °C with the values given in Table 10.16.

PROPERTIES OF COMMON LABORATORY SOLVENTS

This table give properties of 200 organic solvents which are frequently used in laboratory and industrial applications. Compounds are listed in alphabetical order by the most common name; synonyms are given in some cases. The properties tabulated are:

MF: Molecular formula
CAS RN: Chemical Abstracts Service Registry Number
M_r: Molecular weight
t_m: Melting point in °C
t_b: Normal boiling point in °C
ρ: Density in g/cm^3 at the temperature in °C indicated by the superscript
c_p: Specific heat capacity of the liquid at constant pressure at 25°C in J/g K
vp: Vapor pressure at 25°C in kPa (1 kPa = 7.50 mmHg)
μ: Electric dipole moment in debye units. Values in parentheses are measurements on the pure liquid or in solution; these are less reliable than the other values, which were obtained in the gas phase.
FP: Flash point temperature in °C. The fact that no flash point is listed does not necessarily mean that the substance is nonflammable, because some liquids will burn if the quantity is large or impurities are present.
Fl. Lim.: Flammable (explosive) range in air in percent by volume
Ign. Temp.: Autoignition temperature in °C
TLV: Threshold limit for allowable airborne concentration, given in parts per million by volume at 25°C and atmospheric pressure (see table "Threshold Limit Values for Airborne Contaminants" in Section 16)

REFERENCES

1. Lide, D.R., *Handbook of Organic Solvents*, CRC Press, Boca Raton, FL, 1994.
2. Lide, D.R., and Kehiaian, H.V., *CRC Handbook of Thermophysical and Thermochemical Data*, CRC Press, Boca Raton, FL, 1994.
3. Riddick, J.A., Bunger, W.B., and Sakano, T.K., *Organic Solvents, Fourth Edition*, John Wiley & Sons, New York, 1986.
4. *Fire Protection Guide to Hazardous Materials*, 10th Edition, National Fire Protection Association, Quincy, MA, 1991.
5. Urben, P.G., Ed., *Bretherick's Handbook of Reactive Chemical Hazards, 5th Edition*, Butterworth-Heinemann, Oxford, 1995.

Name	MF	CAS RN	M_r	t_m/°C	t_b/°C	ρ/g cm^{-3}	c_p/J g^{-1}K^{-1}	vp/kPa	μ/D	FP/°C	Fl. Lim.	Ign. Temp./°C	TLV
Acetal (1,1-Diethoxyethane)	$C_6H_{14}O_2$	105-57-7	118.18	-100	102.2	0.8254[20]	2.01	3.68	(1.4)	-21	2-10%	230	
Acetic acid	$C_2H_4O_2$	64-19-7	60.05	17	118	1.0492[20]	2.06	2.07	1.70	39	4-20%	463	10
Acetone	C_3H_6O	67-64-1	58.08	-95	56	0.7899[20]	2.18	30.8	2.88	-20	3-13%	465	750
Acetonitrile	C_2H_3N	75-05-8	41.05	-44	82	0.7857[20]	2.23	11.8	3.92	6	3-16%	524	40
Acetylacetone	$C_5H_8O_2$	123-54-6	100.12	-23	138	0.9721[25]	2.08	1.02	(2.8)	34		340	
Acrylonitrile	C_3H_3N	107-13-1	53.06	-83.5	77.3	0.8060[20]	2.05	14.1	3.87	0	3-17%	481	2
Adiponitrile	$C_6H_8N_2$	111-69-3	108.14	1	295	0.9676[20]	1.19	<0.01		93	2-5%	550	2
Allyl alcohol	C_3H_6O	107-18-6	58.08	-129	97.0	0.8540[20]	2.39	3.14	1.60	21	3-18%	378	2
Allylamine	C_3H_7N	107-11-9	57.10	-88.2	53.3	0.758[20]		33.1	1.2	-29	2-22%	374	
2-Aminoisobutanol	$C_4H_{11}NO$	124-68-5	89.14	25.5	165.5	0.934[20]				67			
Benzal chloride	$C_7H_6Cl_2$	98-87-3	161.03	-17	205	1.26[25]	1.62	0.06	(2.1)				
Benzaldehyde	C_7H_6O	100-52-7	106.12	-26	179.0	1.0415[10]	1.74	0.17	(3.0)	63		192	
Benzene	C_6H_6	71-43-2	78.11	6	80	0.8765[20]	1.60	12.7	0	-11	1-8%	498	10
Benzonitrile	C_7H_5N	100-47-0	103.12	-12.7	191.1	1.0093[15]	1.44	0.11	4.18				
Benzyl chloride	C_7H_7Cl	100-44-7	126.59	-45	179	1.1004[20]	0.41	0.16	(1.8)	67	1%-	585	1
Bromochloromethane	CH_2BrCl	74-97-5	129.38	-87.9	68.0	1.9344[20]	0.52	19.5	(1.7)				200
Bromoform (Tribromomethane)	$CHBr_3$	75-25-2	252.73	8.0	149	2.899[15]		0.73	0.99	83			0.5

PROPERTIES OF COMMON LABORATORY SOLVENTS (continued)

Name	MF	CAS RN	M_r	t_m/°C	t_b/°C	ρ/g cm^{-3}	c_p/J g^{-1}K^{-1}	vp/kPa	μ/D	FP/°C	Fl. Lim.	Ign. Temp./°C	TLV
Butyl acetate	$C_6H_{12}O_2$	123-86-4	116.16	-78	126	0.8825[20]	1.96	1.66	(1.9)	22	2-8%	425	150
Butyl alcohol	$C_4H_{10}O$	71-36-3	74.12	-90	118	0.8098[20]	2.39	0.86	1.66	37	1-11%	343	50
sec-Butyl alcohol	$C_4H_{10}O$	78-92-2	74.12	-114.7	99.5	0.8063[20]	2.66	2.32	(1.8)	24	2-10%	405	100
tert-Butyl alcohol	$C_4H_{10}O$	75-65-0	74.12	26	82	0.7887[20]	2.97	5.52	(1.7)	11	2-8%	478	100
Butylamine	$C_4H_{11}N$	109-73-9	73.14	-49	77	0.7414[20]	2.45	12.2	1.0	-12	2-10%	312	5
tert-Butylamine	$C_4H_{11}N$	75-64-9	73.14	-67	44	0.6958[20]	2.63	48.4	(1.3)	-9	2-9%	380	
Butyl methyl ketone	$C_6H_{12}O$	591-78-6	100.16	-56	128	0.8113[20]	2.13	1.54	(2.7)	25	1-8%	423	5
p-tert-Butyltoluene	$C_{11}H_{16}$	98-51-1	148.25	-52	190	0.8612[20]		0.09	=0	68			10
γ-Butyrolactone	$C_4H_6O_2$	96-48-0	86.09	-43.3	204	1.1284[16]	1.64	0.43	4.27	98			
Caprolactam	$C_6H_{11}NO$	105-60-2	113.16	69	270		1.38	<0.01	(3.9)	125			5
Carbon disulfide	CS_2	75-15-0	76.14	-112	46	1.2632[20]	1.00	48.2	0	-30	1-50%	90	10
Carbon tetrachloride	CCl_4	56-23-5	153.82	-23	77	1.5940[20]	0.85	15.2	0				5
1-Chloro-1,1-difluoroethane	$C_2H_3ClF_2$	75-68-3	100.50	-131	-10	1.107[25]	1.30	351	2.14				
Chlorobenzene	C_6H_5Cl	108-90-7	112.56	-45	132	1.1058[20]	1.33	1.6	1.69	28	1-10%	593	10
Chloroform	$CHCl_3$	67-66-3	119.38	-64	61	1.4832[20]	0.96	26.2	1.04				10
Chloropentafluoroethane	C_2ClF_5	76-15-3	154.47	-99	-38	1.5678[-42]	1.19	912	0.52				1000
Cumene (Isopropylbenzene)	C_9H_{12}	98-82-8	120.19	-96.0	152	0.8618[20]	1.75	0.61	0.79	36	1-7%	424	50
Cyclohexane	C_6H_{12}	110-82-7	84.16	7	81	0.7785[20]	1.84	13.0	=0	-20	1-8%	245	300
Cyclohexanol	$C_6H_{12}O$	108-93-0	100.16	25	161	0.9624[20]	2.08	0.10		68	1-9%	300	50
Cyclohexanone	$C_6H_{10}O$	108-94-1	98.14	-31	155	0.9478[20]	1.86	0.53	2.87	44	1-9%	420	25
Cyclohexylamine	$C_6H_{13}N$	108-91-8	99.18	-18	134	0.8191[20]		1.20	(1.3)	31	1-9%	293	10
Cyclopentane	C_5H_{10}	287-92-3	70.13	-93.8	49.3	0.7457[20]	1.84	42.3	=0	<-7	2%-	361	600
Cyclopentanone	C_5H_8O	120-92-3	84.12	-51.3	130.5	0.9487[20]	1.84	1.55	3.3	26			
p-Cymene	$C_{10}H_{14}$	99-87-6	134.22	-69	177	0.8573[20]	1.76	0.19	=0	47	1-6%	436	
cis-Decalin	$C_{10}H_{18}$	493-01-6	138.25	-42.9	195.8	0.8965[20]	1.68	0.10	=0				
trans-Decalin	$C_{10}H_{18}$	493-02-7	138.25	-30.3	187.3	0.8699[20]	1.65	0.16	=0	54	1-5%	255	
Diacetone alcohol	$C_6H_{12}O_2$	123-42-2	116.16	-44	168	0.9387[20]	1.91	0.22	(3.2)	58	2-7%	643	50
1,2-Dibromoethane	$C_2H_4Br_2$	106-93-4	187.86	9.9	131.6	2.1791[20]	0.72	1.55	(1.2)				
Dibromofluoromethane	$CHBr_2F$	1868-53-7	191.83	-78	64.9	2.421[20]							
Dibromomethane	CH_2Br_2	74-95-3	173.83	-52.5	97	2.4969[20]	0.61	6.12	1.43				
1,2-Dibromotetrafluoroethane	$C_2Br_2F_4$	124-73-2	259.82	-110.4	47.3	2.149[25]	0.69	43.4					
Dibutylamine	$C_8H_{19}N$	111-92-2	129.25	-62	160	0.7670[20]	2.27	0.34	(1.0)	47	1-6%		
o-Dichlorobenzene	$C_6H_4Cl_2$	95-50-1	147.00	-17	180	1.3059[20]	1.10	0.18	2.50	66	2-9%	648	25
1,1-Dichloroethane	$C_2H_4Cl_2$	75-34-3	98.96	-97	57	1.1757[20]	1.28	30.5	2.06	-17	5-11%	458	100
1,2-Dichloroethane	$C_2H_4Cl_2$	107-06-2	98.96	-36	84	1.2351[20]	1.30	10.6	(1.8)	13	6-16%	413	10
1,1-Dichloroethylene	$C_2H_2Cl_2$	75-35-4	96.94	-122.5	31.6	1.213[20]	1.15	80.0	1.34	-15	7-16%	570	5
cis-1,2-Dichloroethylene	$C_2H_2Cl_2$	156-59-2	96.94	-80	60	1.2837[20]	1.20	26.8	1.90	6	3-15%	460	200
trans-1,2-Dichloroethylene	$C_2H_2Cl_2$	156-60-5	96.94	-50	49	1.2565[20]	1.20	44.2	0	2	6-13%	460	200
Dichloroethyl ether	$C_4H_8Cl_2O$	111-44-4	143.01	-52	179	1.22[20]	1.54	0.14	(2.6)	55	3%-	369	5
Dichloromethane	CH_2Cl_2	75-09-2	84.93	-95	40	1.3266[20]	1.19	58.2	1.60		13-23%	556	50
1,2-Dichloropropane	$C_3H_6Cl_2$	78-87-5	112.99	-100	96	1.1560[20]	1.32	6.62	(1.8)	16	3-15%	557	75
1,2-Dichlorotetrafluoroethane	$C_2Cl_2F_4$	76-14-2	170.92	-94	4	1.518[4]	0.96	215	0.5				1000
Diethanolamine	$C_4H_{11}NO_2$	111-42-2	105.14	28	269	1.0966[20]	2.22	<0.01	(2.8)	172	2-13%	662	0.46
Diethylamine	$C_4H_{11}N$	109-89-7	73.14	-50	55	0.7056[20]	2.31	30.1	0.92	-23	2-10%	312	5
Diethyl carbonate	$C_5H_{10}O_3$	105-58-8	118.13	-43	126	0.9752[20]	1.80	1.63	1.10	25			
Diethylene glycol	$C_4H_{10}O_3$	111-46-6	106.12	-10	246	1.1197[15]	2.31	<0.01	(2.3)	124	2-17%	224	
Diethylene glycol dimethyl ether	$C_6H_{14}O_3$	111-96-6	134.18	-68	162	0.9434[20]	2.04	0.31	(2.0)	67			
Diethylene glycol monoethyl ether	$C_6H_{14}O_3$	111-90-0	134.18		196	0.9885[20]	2.24	0.02	(1.6)	96			

Name	MF	CAS RN	M_r	$t_m/°C$	$t_b/°C$	$\rho/\text{g cm}^{-3}$	$c_p/\text{J g}^{-1}\text{K}^{-1}$	vp/kPa	μ/D	FP/°C	Fl. Lim.	Ign. Temp./°C	TLV
Diethylene glycol monoethyl ether acetate	$C_8H_{16}O_4$	112-15-2	176.21	-25	218.5	1.0096^{20}		0.03	(1.8)	110		425	
Diethylene glycol monomethyl ether	$C_5H_{12}O_3$	111-77-3	120.15		193	1.035^{20}	2.26	0.02	(1.6)	96	1-23%	240	
Diethylenetriamine	$C_4H_{13}N_3$	111-40-0	103.17	-39	207	0.9569^{20}	2.46	0.03	(1.9)	98	2-7%	358	1
Diethyl ether	$C_4H_{10}O$	60-29-7	74.12	-116	34	0.7138^{20}	2.33	71.7	1.15	-45	2-36%	180	400
Diisobutyl ketone	$C_9H_{18}O$	108-83-8	142.24	-42	169	0.8062^{20}	2.09	0.23	(2.7)	49	1-7%	396	25
Diisopropyl ether	$C_6H_{14}O$	108-20-3	102.18	-87	69	0.7241^{20}	2.12	19.9	1.13	-28	1-8%	443	250
N,N-Dimethylacetamide	C_4H_9NO	127-19-5	87.12	-20	165	0.9366^{25}	2.02	0.07	(3.7)	70	2-12%	490	10
Dimethylamine	C_2H_7N	124-40-3	45.08	-92	7	0.6804^{0}	3.05	203	1.01	20	3-14%	400	5
Dimethyl disulfide	$C_2H_6S_2$	624-92-0	94.20	-85	109.8	1.0625^{20}	1.55	3.82	(1.8)	24			
N,N-Dimethylformamide	C_3H_7NO	68-12-2	73.09	-60	153	0.944^{25}	2.06	0.44	3.82	58	2-15%	445	10
Dimethyl sulfoxide	C_2H_6OS	67-68-5	78.14	19	189	1.1014^{20}	1.96	0.08	3.96	95	3-42%	215	10
1,4-Dioxane	$C_4H_8O_2$	123-91-1	88.11	12	101	1.0337^{20}	1.74	4.95	0	12	2-22%	180	25
1,3-Dioxolane	$C_3H_6O_2$	646-06-0	74.08	-95	78	1.060^{20}	1.59	14.6	1.19	2			
Dipentene	$C_{10}H_{16}$	7705-14-8	136.24	-95.5	178	0.8402^{21}	1.83	0.26		45		237	
Epichlorohydrin	C_3H_5ClO	106-89-8	92.52	-26	116	1.1812^{20}	1.42	2.2	(1.8)	31	4-21%	411	2
Ethanolamine (Glycinol)	C_2H_7NO	141-43-5	61.08	11	171	1.0180^{20}	3.20	0.05	(2.3)	86	3-24%	410	3
Ethyl acetate	$C_4H_8O_2$	141-78-6	88.11	-84	77	0.9003^{20}	1.94	12.6	1.78	-4	2-12%	426	400
Ethyl acetoacetate	$C_6H_{10}O_3$	141-97-9	130.14	-45	180.8	1.0368^{10}	1.91	0.09		57	1-10%	295	
Ethyl alcohol	C_2H_6O	64-17-5	46.07	-114	78	0.7893^{20}	2.44	7.87	1.69	13	3-19%	363	1000
Ethylamine	C_2H_7N	75-04-7	45.08	-81	17	0.686^{17}	2.88	142	1.22	<-18	4-14%	385	5
Ethylbenzene	C_8H_{10}	100-41-4	106.17	-95	136	0.8670^{20}	1.73	1.28	0.59	21	1-7%	432	100
Ethyl bromide	C_2H_5Br	74-96-4	108.97	-118.6	38.5	1.4604^{20}	0.93	62.5	2.03		7-8%	511	5
Ethyl chloride	C_2H_5Cl	75-00-3	64.51	-139	12	0.909^{12}	1.62	160	2.05	-50	4-15%	519	1000
Ethylene carbonate	$C_3H_4O_3$	96-49-1	88.06	36.4	248	1.3214^{39}	1.52	<0.01	(4.9)	143			
Ethylenediamine	$C_2H_8N_2$	107-15-3	60.10	11	117	0.8979^{20}	2.87	1.62	1.99	40	3-12%	385	10
Ethylene glycol	$C_2H_6O_2$	107-21-1	62.07	-13	197	1.1088^{20}	2.41	0.01	2.28	111	3-22%	398	50
Ethylene glycol diethyl ether	$C_6H_{14}O_2$	629-14-1	118.18	-74	119.4	0.8484^{20}	2.19	4.33		35			
Ethylene glycol dimethyl ether	$C_4H_{10}O_2$	110-71-4	90.12	-58	85	0.8691^{20}	2.14	9.93		-2		202	
Ethylene glycol monobutyl ether	$C_6H_{14}O_2$	111-76-2	118.18	-75	168	0.9015^{20}	2.38	0.15	(2.1)	69	4-13%	238	25
Ethylene glycol monoethyl ether	$C_4H_{10}O_2$	110-80-5	90.12	-70	135	0.9297^{20}	2.34	0.71	(2.1)	43	3-18%	235	5
Ethylene glycol ethyl ether acetate	$C_6H_{12}O_3$	111-15-9	132.16	-62	156	0.9740^{20}	2.85	0.24	(2.2)	56	2-8%	379	5
Ethylene glycol monomethyl ether	$C_3H_8O_2$	109-86-4	76.10	-85	124	0.9647^{20}	2.25	1.31	2.36	39	2-14%	285	5
Ethylene glycol monomethyl ether acetate	$C_5H_{10}O_3$	110-49-6	118.13	-70	143	1.0074^{19}	2.62	0.67	(2.1)	49	2-12%	392	5
Ethyl formate	$C_3H_6O_2$	109-94-4	74.08	-80	54	0.9168^{20}	2.02	32.3	1.9	-20	3-16%	455	100
Furan	C_4H_4O	110-00-9	68.08	-86	31	0.9514^{20}	1.69	80.0	0.66	<0	2-14%	316	2
Furfural	$C_5H_4O_2$	98-01-1	96.09	-37	162	1.1594^{20}	1.70	0.29	(3.5)	60	2-19%	491	2
Furfuryl alcohol	$C_5H_6O_2$	98-00-0	98.10	-31	171	1.1296^{20}	2.08	0.10	(1.9)	75	2-16%	370	10
Glycerol	$C_3H_8O_3$	56-81-5	92.09	18	290	1.2613^{20}	2.38	<0.01	(2.6)	199	3-19%	204	400
Heptane	C_7H_{16}	142-82-5	100.20	-91	98	0.6837^{20}	2.24	6.09	≈0	-4	1-7%	204	400
1-Heptanol	$C_7H_{16}O$	111-70-6	116.20	-34	176.4	0.8219^{20}	2.34						
Hexane	C_6H_{14}	110-54-3	86.18	-95	69	0.6548^{25}	2.27	20.2	≈0	-22	1-8%	225	50
1-Hexanol (Caproyl alcohol)	$C_6H_{14}O$	111-27-3	102.18	-44.6	157.6	0.8136^{20}	2.35	0.11		63			
Hexylene glycol	$C_6H_{14}O_2$	107-41-5	118.18	-50	197	0.923^{15}	2.84	<0.01	(2.9)	102	1-9%	306	25
Hexyl methyl ketone	$C_8H_{16}O$	111-13-7	128.21	-16	172.5	0.820^{20}	2.13		(2.7)	52			
Isobutyl acetate	$C_6H_{12}O_2$	110-19-0	116.16	-99	117	0.8712^{20}	2.01	2.39	(1.9)	18	1-11%	421	150
Isobutyl alcohol	$C_4H_{10}O$	78-83-1	74.12	-108	108	0.8018^{20}	2.44	1.39	1.64	28	2-11%	415	50

PROPERTIES OF COMMON LABORATORY SOLVENTS (continued)

Name	MF	CAS RN	M_r	t_m/°C	t_b/°C	ρ/g cm^{-3}	c_p/J g^{-1}K^{-1}	vp/kPa	μ/D	FP/°C	Fl. Lim.	Ign. Temp./°C	TLV
Isobutylamine	$C_4H_{11}N$	78-81-9	73.14	-87	68	0.724^{25}	2.50	19.0	(1.3)	-9	2-12%	378	
Isopentyl acetate	$C_7H_{14}O_2$	123-92-2	130.19	-79	143	0.876^{15}	1.91	0.73	(1.9)	25	1-8%	360	100
Isophorone	$C_9H_{14}O$	78-59-1	138.21	-8	215	0.9255^{20}	1.83	0.06		84	1-4%	460	5
Isopropyl acetate	$C_5H_{10}O_2$	108-21-4	102.13	-73	89	0.8718^{20}	1.95	8.1	1.56	2	2-8%	460	250
Isopropyl alcohol	C_3H_8O	67-63-0	60.10	-90	82	0.7855^{20}	2.58	6.02		12	2-13%	399	400
Isoquinoline	C_9H_7N	119-65-3	129.16	26.47	243.2	1.0910^{30}	1.52		2.73	49			
d-Limonene (Citrene)	$C_{10}H_{16}$	5989-27-5	136.24	-97	178	0.8411^{20}	1.83	0.28	(1.7)				
2,6-Lutidine	C_7H_9N	108-48-5	107.16	-6.1	144.1	0.9226^{20}	1.73	0.75					
Mesitylene	C_9H_{12}	108-67-8	120.19	-45	165	0.8652^{20}	1.74	0.33	0	50	1-5%	559	25
Mesityl oxide	$C_6H_{10}O$	141-79-7	98.14	-59	130	0.8653^{20}	2.17	1.47	(2.8)	31	1-7%	344	15
Methyl acetate	$C_3H_6O_2$	79-20-9	74.08	-98	57	0.9342^{20}	1.92	28.8	1.72	-10	3-16%	454	200
Methylal	$C_3H_8O_2$	109-87-5	76.10	-105	42	0.8593^{20}	2.12	53.1	(0.7)	-32	2-14%	237	1000
Methyl alcohol	CH_4O	67-56-1	32.04	-98	65	0.7914^{20}	2.53	16.9	1.70	11	6-36%	464	200
Methylamine	CH_5N	74-89-5	31.06	-93	-6	0.656^{25}	3.29	353	1.31	0	5-21%	430	5
Methyl benzoate	$C_8H_8O_2$	93-58-3	136.15	-15	199	1.0933^{15}	1.63	0.05	(1.9)	83			
Methylcyclohexane	C_7H_{14}	108-87-2	98.19	-127	101	0.7694^{20}	1.88	6.18	=0	-4	1-7%	250	400
Methyl ethyl ketone	C_4H_8O	78-93-3	72.11	-87	80	0.8054^{20}	2.20	12.6	2.78	-9	1-11%	404	200
N-Methylformamide	C_2H_5NO	123-39-7	59.07	-3.8	199.5	1.011^{19}	2.10		3.83				
Methyl formate	$C_2H_4O_2$	107-31-3	60.05	-99	32	0.9742^{20}	1.98	78.1	1.77	-19	5-23%	449	100
Methyl iodide	CH_3I	74-88-4	141.94	-66.4	42.5	2.279^{20}	0.89	53.9	1.62				2
Methyl isobutyl ketone	$C_6H_{12}O$	108-10-1	100.16	-84	116	0.7978^{20}	2.13	2.64		18	1-8%	448	50
Methyl isopentyl ketone	$C_7H_{14}O$	110-12-3	114.19		144	0.888^{20}		0.69		36	1-8%	191	50
2-Methylpentane	C_6H_{14}	107-83-5	86.18	-153.7	60.2	0.650^{25}	2.25	28.2	=0	<-29	1-7%	264	
4-Methyl-2-pentanol	$C_6H_{14}O$	108-11-2	102.18	-90	132	0.8075^{20}	2.67	0.70		41	1-6%		25
Methyl pentyl ketone	$C_7H_{14}O$	110-43-0	114.19	-35	151	0.8111^{20}	2.04	0.49	(2.6)	39	1-8%	393	50
Methyl propyl ketone	$C_5H_{10}O$	107-87-9	86.13	-77	102	0.809^{20}	2.14	4.97	(2.7)	7	2-8%	452	200
N-Methyl-2-pyrrolidone	C_5H_9NO	872-50-4	99.13	-24	202	1.0230^{25}	3.11	0.04	(4.1)	96	1-10%	346	
Morpholine	C_4H_9NO	110-91-8	87.12	-5	128	1.0005^{20}	1.89	1.34	1.55	37	1-11%	290	20
Nitrobenzene	$C_6H_5NO_2$	98-95-3	123.11	6	211	1.2037^{20}	1.51	0.03	4.22	88	2-9%	482	1
Nitroethane	$C_2H_5NO_2$	79-24-3	75.07	-90	114	1.0448^{25}	1.79	2.79	3.23	28	3-17%	414	100
Nitromethane	CH_3NO_2	75-52-5	61.04	-29	101	1.1371^{20}	1.75	4.79	3.46	35	7-22%	418	20
1-Nitropropane	$C_3H_7NO_2$	108-03-2	89.09	-108	131.1	0.9961^{20}	1.97	1.36	3.66	36	2%-	421	25
2-Nitropropane	$C_3H_7NO_2$	79-46-9	89.09	-91	120	0.9821^{25}	1.91	2.3	3.73	24	3-11%	428	10
Octane	C_8H_{18}	111-65-9	114.23	-57	126	0.6986^{20}	2.23	1.86	=0	13	1-7%	206	300
1-Octanol	$C_8H_{18}O$	111-87-5	130.23	-15.5	195.1	0.8262^{25}	2.34	0.01	(1.8)	81			
Pentachloroethane	C_2HCl_5	76-01-7	202.29	-29	160	1.6796^{20}	0.86	0.48	0.92				
Pentamethylene glycol	$C_5H_{12}O_2$	111-29-5	104.15	-18	239	0.9914^{20}	3.08		(2.5)	129		335	
Pentane	C_5H_{12}	109-66-0	72.15	-130	36	0.6262^{20}	2.32	68.3	=0	<-40	2-8%	260	600
1-Pentanol	$C_5H_{12}O$	71-41-0	88.15	-79	138	0.8144^{20}	2.36	0.26	(1.7)	33	1-10%	300	
Pentyl acetate	$C_7H_{14}O_2$	628-63-7	130.19	-71	149	0.8756^{20}	2.00	0.60	1.75	16	1-8%	360	100
2-Picoline	C_6H_7N	109-06-8	93.13	-67	129	0.9443^{20}	1.70	1.5	1.85	39		538	
α-Pinene	$C_{10}H_{16}$	80-56-8	136.24	-64	156	0.8539^{25}		0.64		35		275	
β-Pinene	$C_{10}H_{16}$	127-91-3	136.24	-61.5	166	0.8606^{20}		0.61		38		275	
Piperidine	$C_5H_{11}N$	110-89-4	85.15	-11	106	0.860^{25}	2.11	4.28	(1.2)	16	1-10%		
Propanenitrile	C_3H_5N	107-12-0	55.08	-93	97	0.7818^{20}	2.17	6.14	4.05	2	3-14%	512	
Propyl acetate	$C_5H_{10}O_2$	109-60-4	102.13	-93	102	0.8878^{20}	1.92	4.49	(1.8)	13	2-8%	450	200
Propyl alcohol	C_3H_8O	71-23-8	60.10	-126	97	0.8035^{20}	2.39	2.76	1.55	23	2-14%	412	200
Propylamine	C_3H_9N	107-10-8	59.11	-83	47	0.7173^{20}	2.75	42.1	1.17	-37	2-10%	318	
Propylbenzene	C_9H_{12}	103-65-1	120.19	-99.5	159.2	0.8620^{20}	1.79		=0	30	1-6%	450	

PROPERTIES OF COMMON LABORATORY SOLVENTS (continued)

Name	MF	CAS RN	M_r	t_m/°C	t_b/°C	ρ/g cm^{-3}	c_p/J g^{-1} K^{-1}	vp/kPa	μ/D	FP/°C	Fl. Lim.	Ign. Temp./°C	TLV
Propylene glycol	C$_3$H$_8$O$_2$	57-55-6	76.10	-60	188	1.0361^{20}	2.51	0.02	(2.2)	99	3-13%	371	
Pseudocumene	C$_9$H$_{12}$	95-63-6	120.19	-44	169	0.8758^{20}	1.79	0.30	=0	44	1-6%	500	25
Pyridine	C$_5$H$_5$N	110-86-1	79.10	-42	115	0.9819^{20}	1.68	2.76	2.21	20	2-12%	482	5
Pyrrole	C$_4$H$_5$N	109-97-7	67.09	-23.4	129.7	0.9698^{20}	1.90	1.10	1.74	39			
Pyrrolidine	C$_4$H$_9$N	123-75-1	71.12	-57.8	86.5	0.8586^{20}	2.20	8.40	(1.6)	3			
2-Pyrrolidone	C$_4$H$_7$NO	616-45-5	85.11	25	251	1.120^{20}	1.99		(3.5)	129		480	
Quinoline	C$_9$H$_7$N	91-22-5	129.16	-14.78	237.1	1.0977^{15}	1.51		2.29			490	50
Styrene	C$_8$H$_8$	100-42-5	104.15	-31	145	0.9060^{20}	1.75	0.81		31	1-7%		50
Sulfolane	C$_4$H$_8$O$_2$S	126-33-0	120.17	28	287	1.2723^{18}	1.50	<0.01	(4.8)	177			
α-Terpinene	C$_{10}$H$_{16}$	99-86-5	136.24		174	0.8375^{19}							
1,1,1,2-Tetrachloro-2,2-difluoroethane	C$_2$Cl$_4$F$_2$	76-11-9	203.83	40.6	91.5	1.649^{25}		7.36					500
1,1,2,2-Tetrachloro-1,2-difluoroethane	C$_2$Cl$_4$F$_2$	76-12-0	203.83	26	93	1.6447^{25}	0.85	7.51					500
1,1,1,2-Tetrachloroethane	C$_2$H$_2$Cl$_4$	630-20-6	167.85	-70	131	1.5406^{20}	0.92	1.6			5-12%		
1,1,2,2-Tetrachloroethane	C$_2$H$_2$Cl$_4$	79-34-5	167.85	-44	146	1.5953^{20}	0.97	0.62	1.32		20-54%		1
Tetrachloroethylene	C$_2$Cl$_4$	127-18-4	165.83	-22	121	1.6227^{20}	0.86	2.42	0				50
Tetraethylene glycol	C$_8$H$_{18}$O$_5$	112-60-7	194.23	-6.2	328	1.1285^{15}	2.21			182		321	
Tetrahydrofuran	C$_4$H$_8$O	109-99-9	72.11	-108	65	0.8892^{20}	1.72	21.6	1.75	-14	2-12%	385	200
1,2,3,4-Tetrahydronaphthalene	C$_{10}$H$_{12}$	119-64-2	132.21	-36	208	0.9660^{25}	1.65	0.05	=0	71	1-5%		
Tetrahydropyran	C$_5$H$_{10}$O	142-68-7	86.13	-45	88	0.8814^{20}	1.82	9.54	1.74	-20			
Tetramethylsilane	C$_4$H$_{12}$Si	75-76-3	88.22	-99.0	26.6	0.648^{19}	2.31	94.2	0				
Toluene	C$_7$H$_8$	108-88-3	92.14	-95	111	0.8669^{20}	1.70	3.79	0.37	4	1-7%	480	50
o-Toluidine	C$_7$H$_9$N	95-53-4	107.16	-16.3	200.3	0.9984^{20}	1.96	0.04	(1.6)	85	1-%	482	2
Triacetin	C$_9$H$_{14}$O$_6$	102-76-1	218.21	-78	259	1.1583^{20}	1.76	<0.01		138	1-5%	433	
Tributylamine	C$_{12}$H$_{27}$N	102-82-9	185.35	-70	217	0.7770^{20}		0.01	(0.8)	86			
1,1,1-Trichloroethane	C$_2$H$_3$Cl$_3$	71-55-6	133.40	-30	74	1.3390^{20}	1.08	16.5	1.76		8-13%	537	350
1,1,2-Trichloroethane	C$_2$H$_3$Cl$_3$	79-00-5	133.40	-37	114	1.4397^{20}	1.13	3.1	(1.4)		6-28%	460	10
Trichloroethylene	C$_2$HCl$_3$	79-01-6	131.39	-85	87	1.4642^{20}	0.95	9.91	(0.8)		8-11%	420	50
Trichlorofluoromethane	CCl$_3$F	75-69-4	137.37	-111	24	1.478^{24}	0.89	106	0.46				1000
1,1,2-Trichlorotrifluoroethane	C$_2$Cl$_3$F$_3$	76-13-1	187.38	-35	48	1.5635^{25}	0.91	44.8			1-%		1000
Triethanolamine	C$_6$H$_{15}$NO$_3$	102-71-6	149.19	21	335	1.1242^{20}	2.61	<0.01	(3.6)	179	1-10%		0.5
Triethylamine	C$_6$H$_{15}$N	121-44-8	101.19	-115	89	0.7275^{20}	2.17	7.70	0.66	-7	1-8%	249	1
Triethylene glycol	C$_6$H$_{14}$O$_4$	112-27-6	150.17	-7	285	1.1274^{15}	2.18			177	1-9%	371	
Triethyl phosphate	C$_6$H$_{15}$O$_4$P	78-40-0	182.16	-56.4	215.5	1.0695^{20}			(3.1)	115		454	
Trimethylamine	C$_3$H$_9$N	75-50-3	59.11	-117	3	0.627^{25}	2.33	215	0.61	-7	2-12%	190	5
Trimethylene glycol	C$_3$H$_8$O$_2$	504-63-2	76.10	-26.7	214.4	1.0538^{20}			(2.5)				
Trimethyl phosphate	C$_3$H$_9$O$_4$P	512-56-1	140.08	-46	197.2	1.2144^{20}		0.11	(3.2)	107		400	
Veratrole	C$_8$H$_{10}$O$_2$	91-16-7	138.17	22.5	206	1.0810^{25}			(1.3)				
o-Xylene	C$_8$H$_{10}$	95-47-6	106.17	-25	144	0.8802^{10}	1.75	0.88	0.64	32	1-7%	463	100
m-Xylene	C$_8$H$_{10}$	108-38-3	106.17	-48	139	0.8642^{20}	1.72	1.13	=0	27	1-7%	527	100
p-Xylene	C$_8$H$_{10}$	106-42-3	106.17	13	138	0.8611^{20}	1.71	1.19	0	27	1-7%	528	100

DEPENDENCE OF BOILING POINT ON PRESSURE

The normal boiling point of a liquid is defined as the temperature at which the vapor pressure reaches standard atmospheric pressure, 101.325 kPa. The change in boiling point with pressure may be calculated from the representation of the vapor pressure by the Antoine Equation,

$$\ln p = A_1 - A_2/(T + A_3)$$

where p is the vapor pressure, T the absolute temperature, and A_1, A_2, and A_3 are constants. This table, which has been calculated using the Antoine constants in Reference 1, gives values of $\Delta t/\Delta p$ for a number of liquids, in units of both °C/kPa and °C/mmHg. The correction to the boiling point is generally accurate to 0.1 to 0.2 °C as long as the pressure is within 10% of standard atmospheric pressure.

A slightly less accurate estimate of $\Delta t/\Delta p$ may be obtained from the Claussius-Clapeyron equation, with the assumption that the change in volume upon vaporization equals the ideal-gas volume of the vapor. This leads to the equation

$$\Delta t/\Delta p = RT_b^2/p_0 \Delta_{vap}H(T_b)$$

where R is the molar gas constant, p_0 is 101.325 kPa, T_b is the normal boiling point temperature (absolute), and $\Delta_{vap}H(T_b)$ is the molar enthalpy of vaporization at the normal boiling point. Values of the last quantity may be obtained from the table "Enthalpy of Vaporization" in Section 6.

REFERENCE

1. Lide, D.R., and Kehiaian, H.V., *CRC Handbook of Thermophysical and Thermochemical Data*, CRC Press, Boca Raton, FL, 1994, pp. 49-59.

Compound	t_b °C	$\Delta t/\Delta p$ °C/kPa	$\Delta t/\Delta p$ °C/mmHg	Compound	t_b °C	$\Delta t/\Delta p$ °C/kPa	$\Delta t/\Delta p$ °C/mmHg
Acetaldehyde	20.1	0.261	0.0348	1-Hexanol	157.6	0.318	0.0424
Acetic acid	117.9	0.324	0.0432	Hydrogen fluoride	20.1	0.276	0.0368
Acetone	56.0	0.289	0.0385	Iodomethane	42.5	0.291	0.0388
Acetonitrile	81.6	0.316	0.0421	Isobutane	-11.7	0.254	0.0339
Ammonia	-33.33	0.198	0.0264	Methanol	64.6	0.251	0.0335
Aniline	184.1	0.378	0.0504	Methyl acetate	56.8	0.282	0.0376
Anisole	153.7	0.367	0.0489	Methyl formate	31.7	0.582	0.0776
Benzaldehyde	179.0	0.392	0.0523	*N*-Methylaniline	196.2	0.396	0.0528
Benzene	80.0	0.321	0.0428	*N*-Methylformamide	199.5	0.371	0.0495
Bromine	58.8	0.300	0.0400	Nitrobenzene	210.8	0.418	0.0557
Butane	-0.5	0.267	0.0356	Nitromethane	101.1	0.320	0.0427
1-Butanol	117.7	0.278	0.0371	1-Octanol	195.1	0.360	0.0480
Carbon disulfide	46.2	0.304	0.0405	Pentane	36.0	0.289	0.0385
Chlorine	-34.04	0.224	0.0299	1-Pentanol	137.9	0.296	0.0395
Chlorobenzene	131.7	0.365	0.0487	Phenol	181.8	0.349	0.0465
1-Chlorobutane	78.6	0.321	0.0428	Propane	-42.1	0.224	0.0299
Chloroethane	12.3	0.262	0.0349	1-Propanol	97.2	0.261	0.0348
Chloroethylene	-13.3	0.241	0.0321	2-Propanol	82.3	0.247	0.0329
Cyclohexane	80.7	0.328	0.0437	Pyridine	115.2	0.340	0.0453
Cyclohexanol	160.8	0.344	0.0459	Pyrrole	129.7	0.330	0.0440
Cyclohexanone	155.4	0.382	0.0509	Pyrrolidine	86.5	0.309	0.0412
Decane	174.1	0.388	0.0517	Styrene	145.1	0.369	0.0492
Dibutyl ether	140.2	0.363	0.0484	Sulfur dioxide	-10.05	0.221	0.0295
Dichloromethane	39.6	0.276	0.0368	Tetrachloroethylene	121.3	0.354	0.0472
Diethyl ether	34.5	0.278	0.0371	Tetrachloromethane	76.8	0.325	0.0433
Dimethyl sulfoxide	189.0	0.379	0.0505	Toluene	110.6	0.353	0.0471
1,4-Dioxane	101.5	0.321	0.0428	Trichloroethylene	87.2	0.330	0.0440
Dipropyl ether	90.0	0.326	0.0435	Trichloromethane	61.1	0.302	0.0403
Ethanol	78.2	0.249	0.0332	Trimethylamine	2.8	0.248	0.0331
Ethyl acetate	77.1	0.300	0.0400	Water	100.0	0.276	0.0368
Ethylene glycol	197.3	0.331	0.0441	*o*-Xylene	144.5	0.373	0.0497
Heptane	98.5	0.336	0.0448	*m*-Xylene	139.1	0.368	0.0491
Hexafluorobenzene	80.2	0.305	0.0407	*p*-Xylene	138.3	0.369	0.0492
Hexane	68.7	0.314	0.0419				

EBULLIOSCOPIC CONSTANTS FOR CALCULATION OF BOILING POINT ELEVATION

The boiling point T_b of a dilute solution of a non-volatile, non-dissociating solute is elevated relative to that of the pure solvent. If the solution is ideal (i.e., follows Raoult's Law), the amount of elevation depends only on the number of particles of solute present. Hence the change in boiling point ΔT_b can be expressed as

$$\Delta T_b = E_b\, m_2$$

where m_2 is the molality (moles of solute per kilogram of solvent) and E_b is the Ebullioscopic Constant, a characteristic property of the solvent. The Ebullioscopic Constant may be calculated from the relation

$$E_b = R\, T_b^2\, M/\Delta_{vap}H$$

where R is the molar gas constant, T_b is the normal boiling point temperature (absolute) of the solvent, M the molar mass of the solvent, and $\Delta_{vap}H$ the molar enthalpy (heat) of vaporization of the solvent at its normal boiling point.

This table lists E_b values for some common solvents, as calculated from data in the table "Enthalpy of Vaporization" in Section 6.

Compound	E_b/K kg mol^{-1}	Compound	E_b/K kg mol^{-1}
Acetic acid	3.22	Hexane	2.90
Acetone	1.80	Iodomethane	4.31
Acetonitrile	1.44	Methanol	0.86
Aniline	3.82	Methyl acetate	2.21
Anisole	4.20	N-Methylaniline	4.3
Benzaldehyde	4.24	N-Methylformamide	2.2
Benzene	2.64	Nitrobenzene	5.2
1-Butanol	2.17	Nitromethane	2.09
Carbon disulfide	2.42	1-Octanol	5.06
Chlorobenzene	4.36	Phenol	3.54
1-Chlorobutane	3.13	1-Propanol	1.66
Cyclohexane	2.92	2-Propanol	1.58
Cyclohexanol	3.5	Pyridine	2.83
Decane	6.10	Pyrrole	2.33
Dichloromethane	2.42	Pyrrolidine	2.32
Diethyl ether	2.20	Tetrachloroethylene	6.18
Dimethyl sulfoxide	3.22	Tetrachloromethane	5.26
1,4-Dioxane	3.01	Toluene	3.40
Ethanol	1.23	Trichloroethylene	4.52
Ethyl acetate	2.82	Trichloromethane	3.80
Ethylene glycol	2.26	Water	0.513
Heptane	3.62	o-Xylene	4.25

CRYOSCOPIC CONSTANTS FOR CALCULATION OF FREEZING POINT DEPRESSION

The freezing point T_f of a dilute solution of a non-volatile, non-dissociating solute is depressed relative to that of the pure solvent. If the solution is ideal (i.e., follows Raoult's Law), this lowering is a function only of the number of particles of solute present. Thus the absolute value of the lowering of freezing point ΔT_f can be expressed as

$$\Delta T_f = E_f m_2$$

where m_2 is the molality (moles of solute per kilogram of solvent) and E_f is the Cryoscopic Constant, a characteristic property of the solvent. The Cryoscopic Constant may be calculated from the relation

$$E_f = R\, T_f^2\, M/\Delta_{fus}H$$

where R is the molar gas constant, T_b is the freezing point temperature (absolute) of the solvent, M the molar mass of the solvent, and $\Delta_{fus}H$ the molar enthalpy (heat) of fusion of the solvent.

This table lists cryscopic constants for selected substances, as calculated from data in the table "Enthalpy of Fusion" in Section 6.

Compound	E_f/K kg mol^{-1}	Compound	E_f/K kg mol^{-1}
Acetamide	3.92	1,4-Dioxane	4.63
Acetic acid	3.63	Diphenylamine	8.38
Acetophenone	5.16	Ethylene glycol	3.11
Aniline	5.23	Formamide	4.25
Benzene	5.07	Formic acid	2.38
Benzonitrile	5.35	Glycerol	3.56
Benzophenone	8.58	Methylcyclohexane	2.60
(+)-Camphor	37.8	Naphthalene	7.45
1-Chloronaphthalene	7.68	Nitrobenzene	6.87
o-Cresol	5.92	Phenol	6.84
m-Cresol	7.76	Pyridine	4.26
p-Cresol	7.20	Quinoline	6.73
Cyclohexane	20.8	Succinonitrile	19.3
Cyclohexanol	42.2	1,1,2,2-Tetrabromoethane	21.4
cis-Decahydronaphthalene	6.42	1,1,2,2-Tetrachloro-1,2-difluoroethane	41.0
trans-Decahydronaphthalene	4.70	Toluene	3.55
Dibenzyl ether	6.17	p-Toluidine	4.91
p-Dichlorobenzene	7.57	Tribromomethane	15.0
Diethanolamine	3.16	Water	1.86
Dimethyl sulfoxide	3.85	p-Xylene	4.31

FREEZING POINT LOWERING BY ELECTROLYTES IN AQUEOUS SOLUTION

REFERENCE

Forsythe, W. E., *Smithsonian Physical Tables, Ninth Edition*, Smithsonian Institution, Washington, 1956.

Compound	Lowering of freezing point of water (in °C) as function of molality (mol/kg)									
	0.05	0.10	0.25	0.50	0.75	1.00	1.50	2.00	2.50	3.00
$CaCl_2$	0.25	0.49	1.27	2.66	4.28	6.35	10.78	15.27	20.42	28.08
$CuSO_4$	0.13	0.23	0.47	0.96						
HCl	0.18	0.36	0.90	1.86	2.90	4.02	6.63	9.94		
HNO_3	0.18	0.35	0.88	1.80	2.78	3.80	5.98	8.34	10.95	13.92
H_2SO_4	0.20	0.39	0.96	1.95	3.04	4.28	7.35	11.35	16.32	
KBr	0.18	0.36	0.92	1.78						
KCl	0.17	0.35	0.86	1.68	2.49	3.29	4.88	6.50	8.14	9.77
KNO_3	0.17	0.33	0.78	1.47	2.11	2.66				
K_2SO_4	0.23	0.43	1.01	1.87						
LiCl	0.18	0.35	0.88	1.80	2.78					
$MgSO_4$	0.13	0.24	0.55	1.01	1.50	2.08	3.41			
NH_4Cl	0.17	0.34	0.85	1.70	2.55					
NaCl	0.18	0.35	0.85	1.68	2.60					
$NaNO_3$	0.18	0.36	0.80	1.62	2.63	3.10				

DETERMINATION OF RELATIVE HUMIDITY FROM DEW POINT

The relative humidity of a water vapor-air mixture is defined as 100 times the partial pressure of water divided by the saturation vapor pressure of water at the same temperature. The relative humidity may be determined from the dew point t_{dew}, which is the temperature at which liquid water first condenses when the mixture is cooled from an initial temperature t. This table gives relative humidity as a function of the dew point depression $t - t_{dew}$ for several values of the dew point. Values are calculated from the vapor pressure table in Section 6.

$t - t_{dew}$	t_{dew}/°C					$t - t_{dew}$	t_{dew}/°C				
	-10	0	10	20	30		-10	0	10	20	30
0.0	100	100	100	100	100	8.2	54	56	59	61	63
0.2	99	99	99	99	99	8.4	53	56	58	60	63
0.4	97	97	97	98	98	8.6	52	55	57	60	62
0.6	95	96	96	96	97	8.8	51	54	57	59	61
0.9	94	94	95	95	96	9.0	51	53	56	58	61
1.0	92	93	94	94	94	9.2	50	53	55	58	60
1.2	91	92	92	93	93	9.4	49	52	55	57	59
1.4	90	90	91	92	92	9.6	48	51	54	56	59
1.6	88	89	90	91	91	9.8	48	51	53	56	58
1.8	87	88	89	90	90	10.0	47	50	53	55	57
2.0	86	87	88	88	89	10.5	45	48	51	54	56
2.2	84	85	86	87	89	11.0	44	47	49	52	55
2.4	83	84	85	86	87	11.5	42	45	48	51	53
2.6	82	83	84	85	86	12.0	41	44	47	49	52
2.8	80	82	83	84	85	12.5	39	42	45	48	50
3.0	79	81	82	83	84	13.0	38	41	44	46	49
3.2	78	80	81	82	83	13.5	37	40	43	45	48
3.4	77	79	80	81	82	14.0	35	38	41	44	47
3.6	76	77	79	80	82	14.5	34	37	40	43	45
3.8	75	76	78	79	81	15.0	33	36	39	42	44
4.0	73	75	77	78	80	15.5	32	35	38	40	
4.2	72	74	76	77	79	16.0	31	34	37	39	
4.4	71	73	75	77	78	16.5	30	33	36	38	
4.6	70	72	74	76	77	17.0	29	32	35	37	
4.8	69	71	73	75	76	17.5	28	31	34	36	
5.0	69	70	72	74	75	18.0	27	30	33	35	
5.2	67	69	71	73	75	18.5	26	29	32	34	
5.4	66	68	70	72	74	19.0	25	28	31	33	
5.6	65	67	69	71	73	19.5	24	27	30	33	
5.9	64	66	69	70	72	20.0	24	26	29	32	
6.0	63	66	68	70	71	21.0	22	25	27	30	
6.2	62	65	67	69	71	22.0	21	23	26	29	
6.4	61	64	66	68	70	23.0	19	22	24	27	
6.6	60	63	65	67	69	24.0	18	21	23	26	
6.8	60	62	64	66	68	25.0	17	19	22	24	
7.0	59	61	63	66	68	26.0	16	18	21	23	
7.2	58	60	63	65	67	27.0	15	17	20	22	
7.4	57	60	62	64	66	28.0	14	16	19	21	
7.6	56	59	61	63	65	29.0	13	15	18	20	
7.8	55	58	60	63	65	30.0	12	14	17	19	
8.0	54	57	60	62	64						

DETERMINATION OF RELATIVE HUMIDITY FROM WET AND DRY BULB TEMPERATURES

Relative humidity may be determined by comparing temperature readings of wet and dry bulb thermometers. The following table, extracted from more extensive U.S. National Weather Service tables, gives the relative humidity as a function of air temperature t_d (dry bulb) and the difference $t_d - t_w$ between dry and wet bulb temperatures. The data assume a pressure near normal atmospheric pressure and an instrumental configuration with forced ventilation.

t_d/°C	0.5	1.0	1.5	2.0	2.5	3.0	3.5	4.0	4.5	5.0	5.5	6.0
-10	83	67	51	35	19							
-8	86	71	57	43	29	15						
-6	88	74	61	49	37	25	8					
-4	89	77	66	55	44	33	23	12				
-2	90	79	69	60	50	40	31	22	12			
0	91	81	72	64	55	46	38	29	21	13	5	
2	91	84	76	68	60	52	44	37	29	22	14	7
4	92	85	78	71	63	57	49	43	36	29	22	16
6	93	86	79	73	66	60	54	48	41	35	29	24
8	93	87	81	75	69	63	57	51	46	40	35	29
10	94	88	82	77	71	66	60	55	50	44	39	34
12	94	89	83	78	73	68	63	58	53	48	43	39
14	95	90	85	79	75	70	65	60	56	51	47	42
16	95	90	85	81	76	71	67	63	58	54	50	46
18	95	91	86	82	77	73	69	65	61	57	53	49
20	96	91	87	83	78	74	70	66	63	59	55	51
22	96	92	87	83	80	76	72	68	64	61	57	54
24	96	92	88	84	80	77	73	69	66	62	59	56
26	96	92	88	85	81	78	74	71	67	64	61	58
28	96	93	89	85	82	78	75	72	69	65	62	59
30	96	93	89	86	83	79	76	73	70	67	64	61
35	97	94	90	87	84	81	78	75	72	69	67	64
40	97	94	91	88	85	82	80	77	74	72	69	67

t_d/°C	6.5	7.0	7.5	8.0	8.5	9.0	10.0	11.0	12.0	13.0	14.0	15.0
4	9											
6	17	11	5									
8	24	19	14	8								
10	29	24	20	15	10	6						
12	34	29	25	21	16	12	5					
14	38	34	30	26	22	18	10					
16	42	38	34	30	26	23	15	8				
18	45	41	38	34	30	27	20	14	7			
20	48	44	41	37	34	31	24	18	12	6		
22	50	47	44	40	37	34	28	22	17	11	6	
24	53	49	46	43	40	37	31	26	20	15	10	5
26	54	51	49	46	43	40	34	29	24	19	14	10
28	56	53	51	48	45	42	37	32	27	22	18	13
30	58	55	52	50	47	44	39	35	30	25	21	17
32	60	57	54	51	49	46	41	37	32	28	24	20
34	61	58	56	53	51	48	43	39	35	30	26	23
36	62	59	57	54	52	50	45	41	37	33	29	25
38	63	61	58	56	54	51	47	43	39	35	31	27
40	64	62	59	57	54	53	48	44	40	36	33	29

CONSTANT HUMIDITY SOLUTIONS

Anthony Wexler

An excess of a water soluble salt in contact with its saturated solution and contained within an enclosed space produces a constant relative humidity and water vapor pressure according to

$$RH = A \exp(B/T)$$

where RH is the percent relative humidity (generally accurate to ± 2 %), T is the temperature in kelvin, and the constants A and B and the range of valid temperatures are given in the table below. The vapor pressure, p, can be calculated from

$$p = (RH/100) \times p_0$$

where p_0 is the vapor pressure of pure water at temperature T as given in the table in Section 6 titled "Vapor Pressure of Water from 0 to 370°C".

REFERENCES

1. Wexler, A. S. and Seinfeld, J. H., *Atmospheric Environment*, 25A, 2731, 1991.
2. Greenspan, L., *J. Res. National Bureau of Standards*, 81A, 89, 1977.
3. Broul, et al., *Solubility of Inorganic Two-Component Systems*, Elsevier, New York, 1981.
4. Wagman, D. D. et al., *J. Phys. Chem. Ref. Data*, Vol. 11, Suppl. 2, 1982.

Compound	Temperature range (°C)	RH 25°C	A	B
$NaOH \cdot H_2O$	15—60	6	5.48	27
$LiBr \cdot 2H_2O$	10—30	6	0.23	996
$ZnBr_2 \cdot 2H_2O$	5—30	8	1.69	455
$KOH \cdot 2H_2O$	5—30	9	0.014	1924
$LiCl \cdot H_2O$	20—65	11	14.53	−75
$CaBr_2 \cdot 6H_2O$	11—22	16	0.17	1360
$LiI \cdot 3H_2O$	15—65	18	0.15	1424
$CaCl_2 \cdot 6H_2O$	15—25	29	0.11	1653
$MgCl_2 \cdot 6H_2O$	5—45	33	29.26	34
$NaI \cdot 2H_2O$	5—45	38	3.62	702
$Ca(NO_3)_2 \cdot 4H_2O$	10—30	51	1.89	981
$Mg(NO_3)_2 \cdot 6H_2O$	5—35	53	25.28	220
$NaBr \cdot 2H_2O$	0—35	58	20.49	308
NH_4NO_3	10—40	62	3.54	853
KI	5—30	69	29.35	254
$SrCl_2 \cdot 6H_2O$	5—30	71	31.58	241
$NaNO_3$	10—40	74	26.94	302
$NaCl$	10—40	75	69.20	25
NH_4Cl	10—40	79	35.67	235
KBr	5—25	81	40.98	203
$(NH_4)_2SO_4$	10—40	81	62.06	79
KCl	5—25	84	49.38	159
$Sr(NO_3)_2 \cdot 4H_2O$	5—25	85	28.34	328
$BaCl_2 \cdot 2H_2O$	5—25	90	69.99	75
CsI	5—25	91	70.77	75
KNO_3	0—50	92	43.22	225
K_2SO_4	10—50	97	86.75	34

STANDARD SALT SOLUTIONS FOR HUMIDITY CALIBRATION

Saturated aqueous solutions of inorganic salts are convenient secondary standards for calibration of instruments for measurement of relative humidity. The International Union of Pure and Applied Chemistry has recommended salt solutions for calibrations in the range of 10% to 90% relative humidity, and the American Society for Testing and Materials has published similar standards. The data in this table are taken from the IUPAC recommendations, except for K_2CO_3 and K_2SO_4, which are ASTM recommendations.

Details on the preparation and use of these standards may be found in References 1 and 2. Data for other salts are given in Reference 3.

REFERENCES

1. Marsh, K. N., Editor, *Recommended Reference Materials for the Realization of Physicochemical Properties*, Blackwell Scientific Publications, Oxford, 1987, pp.157-162.
2. *Standard Practice for Maintaining Constant Relative Humidity by Means of Aqueous Solutions*, ASTM Standard E 104-85, Reapproved 1991.
3. Greenspan, L., *J. Res. Nat. Bur. Stand.*, 81A, 89, 1977.

Relative Humidity in %

$t/°C$	LiCl	$MgCl_2$	K_2CO_3	$Mg(NO_3)_2$	NaCl	KCl	K_2SO_4
0		33.66±0.33	43.1±0.7	60.35±0.55	75.51±0.34	88.61±0.53	98.8±2.1
5		33.60±0.28	43.1±0.5	58.86±0.43	75.65±0.27	87.67±0.45	98.5±0.9
10		33.47±0.24	43.1±0.4	57.36±0.33	75.67±0.22	86.77±0.39	98.2±0.8
15		33.30±0.21	43.2±0.3	55.87±0.27	75.61±0.18	85.92±0.33	97.9±0.6
20	11.31±0.31	33.07±0.18	43.2±0.3	54.38±0.23	75.47±0.14	85.11±0.29	97.6±0.5
25	11.30±0.27	32.78±0.16	43.2±0.4	52.89±0.22	75.29±0.12	84.34±0.26	97.3±0.5
30	11.28±0.24	32.44±0.14	43.2±0.5	51.40±0.24	75.09±0.11	83.62±0.25	97.0±0.4
35	11.25±0.22	32.05±0.13		49.91±0.29	74.87±0.12	82.95±0.25	96.7±0.4
40	11.21±0.21	31.60±0.13		48.42±0.37		82.32±0.25	96.4±0.4
45	11.16±0.21	31.10±0.13		46.93±0.47		81.74±0.28	96.1±0.4
50	11.10±0.22	30.54±0.14		45.44±0.60		81.20±0.31	95.8±0.5
55	11.03±0.23	29.93±0.16				80.70±0.35	
60	10.95±0.26	29.26±0.18				80.25±0.41	
65	10.86±0.29	28.54±0.21				79.85±0.48	
70	10.75±0.33	27.77±0.25				79.49±0.57	
75	10.64±0.38	26.94±0.29				79.17±0.66	
80	10.51±0.44	26.05±0.34				78.90±0.77	

LOW TEMPERATURE BATHS FOR MAINTAINING CONSTANT TEMPERATURE

A liquid-solid slurry is a convenient means of maintaining a constant temperature environment below room temperature. The following is a list of readily available organic liquids suitable for this purpose, arranged in order of their melting (freezing) points t_m. The normal boiling points t_b are also given.

Compound	$t_m/°C$	$t_b/°C$
Isopentane (2-Methylbutane)	-159.9	27.8
Methylcyclopentane	-142.5	71.8
3-Chloropropene (Allyl chloride)	-134.5	45.1
Pentane	-129.7	36.0
Allyl alcohol	-129	97.0
Ethanol	-114.1	78.2
Carbon disulfide	-111.5	46
Isobutyl alcohol	-108	107.8
Toluene	-94.9	110.6
Acetone	-94.8	56.0
Ethyl acetate	-83.6	77.1
Dry ice + acetone	-78	
p-Cymene	-68.9	177.1
Trichloromethane (Chloroform)	-63.6	61.1
N-Methylaniline	-57	196.2
Chlorobenzene	-45.2	131.7
Anisole	-37.5	153.7
Bromobenzene	-30.6	156.0
Tetrachloromethane (Carbon tetrachloride)	-23	76.8
Benzonitrile	-12.7	191.1

WIRE TABLES

The resistance per unit length of wires of various metals is tabulated here. Values were calculated from resistivity values in the tables "Electrical Resistivity of Pure Metals" and "Electrical Resistivity of Selected Alloys", which appear in Section 12. In practice, resistance may vary because of differing heat treatments and metal composition. The values in the table refer to 20°C, but values at other temperatures may be calculated from the following resistivity data:

Metal	Resistivity in 10^{-8} Ω m at temperature			
	0°C	20°C	25°C	100°C
Aluminum	2.417	2.650	2.709	3.56
Brass (70% Cu, 30% Zn)	5.87	6.08	6.13	6.91
Constantan (60% Cu, 40% Ni)	45.43	45.38	45.35	45.11
Copper	1.543	1.678	1.712	2.22
Nichrome (79% Ni, 21% Cr)	107.3	107.5	107.6	108.3
Platinum	9.6	10.5	10.7	13.6
Silver	1.467	1.587	1.617	2.07
Tungsten	4.82	5.28	5.39	7.18

Resistance per unit length at 20°C in Ω/m

B & S Gauge	Diameter (mm)	Aluminum	Brass	Constantan	Copper	Nichrome	Platinum	Silver	Tungsten
0	8.252	0.000495	0.00114	0.00848	0.000314	0.0201	0.00196	0.000297	0.00099
2	6.543	0.000788	0.00181	0.0135	0.000499	0.0320	0.00312	0.000472	0.00157
4	5.189	0.00125	0.00287	0.0214	0.000793	0.0508	0.00496	0.000750	0.00250
6	4.115	0.00199	0.00457	0.0341	0.00126	0.0808	0.00789	0.00119	0.00397
8	3.264	0.00317	0.00727	0.0542	0.00200	0.128	0.0125	0.00190	0.00631
10	2.588	0.00504	0.0115	0.0863	0.00319	0.204	0.0200	0.00302	0.0100
12	2.053	0.00800	0.0184	0.137	0.00507	0.325	0.0317	0.00479	0.0159
14	1.628	0.0127	0.0292	0.218	0.00806	0.516	0.0504	0.00762	0.0254
16	1.291	0.0202	0.0464	0.347	0.0128	0.821	0.0802	0.0121	0.0403
18	1.024	0.0322	0.0738	0.551	0.0204	1.30	0.127	0.0193	0.0641
20	0.8118	0.0512	0.117	0.877	0.0324	2.08	0.203	0.0307	0.102
22	0.6439	0.0814	0.187	1.39	0.0515	3.30	0.322	0.0487	0.162
24	0.5105	0.129	0.297	2.22	0.0820	5.25	0.513	0.0775	0.258
26	0.4049	0.206	0.472	3.52	0.130	8.35	0.815	0.123	0.410
28	0.3211	0.327	0.751	5.60	0.207	13.3	1.30	0.196	0.652
30	0.2548	0.520	1.19	8.90	0.329	21.1	2.06	0.311	1.03
32	0.2019	0.828	1.90	14.2	0.524	33.6	3.28	0.496	1.65
34	0.1601	1.32	3.02	22.5	0.833	53.4	5.22	0.788	2.62
36	0.1270	2.09	4.80	35.8	1.32	84.9	8.29	1.25	4.17
38	0.1007	3.33	7.63	57.0	2.11	135	13.2	1.99	6.63
40	0.07988	5.29	12.1	90.5	3.35	214	20.9	3.17	10.5

CHARACTERISTICS OF PARTICLES AND PARTICLE DISPERSOIDS

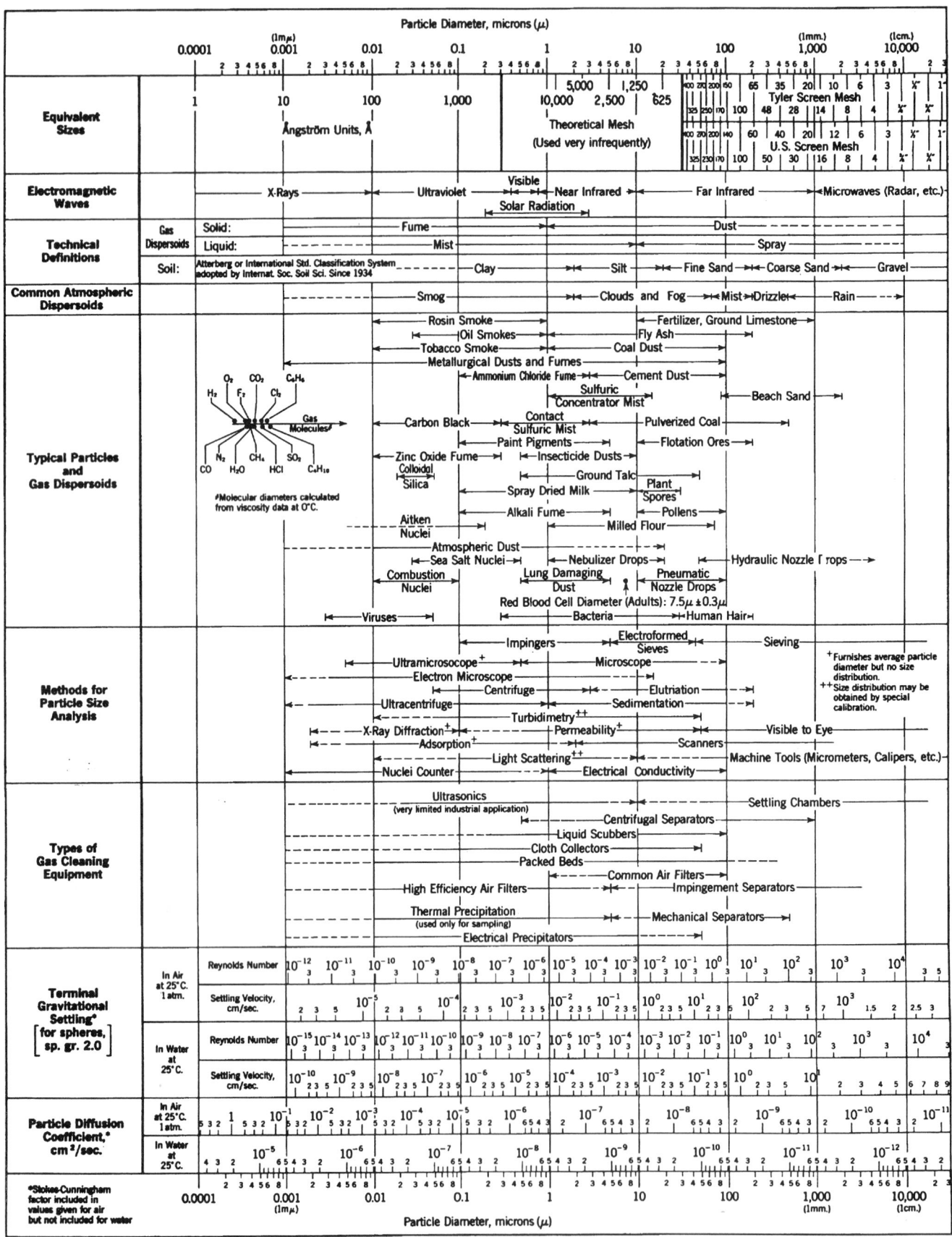

DENSITY OF VARIOUS SOLIDS

This table gives the range of density for miscellaneous solid materials whose characteristics depend on the source or method of preparation.

REFERENCES

1. Forsythe, W. E., *Smithsonian Physical Tables, Ninth Edition*, Smithsonian Institution, Washington, 1956.
2. Kaye, G. W. C., and Laby, T. H., *Tables of Physical and Chemical Constants, 16th Edition*, Longman, London, 1995.
3. Brandrup, J., and Immergut, E. H., *Polymer Handbook, Third Edition*, John Wiley & Sons, New York, 1989.

Material	$\rho/\ g\ cm^{-3}$	Material	$\rho/\ g\ cm^{-3}$	Material	$\rho/\ g\ cm^{-3}$
Agate	2.5-2.7	Pyrex	2.23	Soapstone	2.6-2.8
Alabaster,		Granite	2.64-2.76	Solder	8.7-9.4
carbonate	2.69-2.78	Graphite	2.30-2.72	Starch	1.53
sulfate	2.26-2.32	Gum arabic	1.3-1.4	Steel, stainless	7.8
Albite	2.62-2.65	Gypsum	2.31-2.33	Sugar	1.59
Amber	1.06-1.11	Hematite	4.9-5.3	Talc	2.7-2.8
Amphiboles	2.9-3.2	Hornblende	3.0	Tallow, beef	0.94
Anorthite	2.74-2.76	Ice	0.917	Tar	1.02
Asbestos	2.0-2.8	Iron, cast	7.0-7.4	Topaz	3.5-3.6
Asbestos slate	1.8	Ivory	1.83-1.92	Tourmaline	3.0-3.2
Asphalt	1.1-1.5	Kaolin	2.6	Tungsten carbide	14.0-15.0
Basalt	2.4-3.1	Leather, dry	0.86	Wax, sealing	1.8
Beeswax	0.96-0.97	Lime, slaked	1.3-1.4	Wood (seasoned)	
Beryl	2.69-2.70	Limestone	2.68-2.76	alder	0.42-0.68
Biotite	2.7-3.1	Linoleum	1.18	apple	0.66-0.84
Bone	1.7-2.0	Magnetite	4.9-5.2	ash	0.65-0.85
Brasses	8.44-8.75	Malachite	3.7-4.1	balsa	0.11-0.14
Brick	1.4-2.2	Marble	2.6-2.84	bamboo	0.31-0.40
Bronzes	8.74-8.89	Meerschaum	0.99-1.28	basswood	0.32-0.59
Butter	0.86-0.87	Mica	2.6-3.2	beech	0.70-0.90
Calamine	4.1-4.5	Muscovite	2.76-3.00	birch	0.51-0.77
Calcspar	2.6-2.8	Ochre	3.5	blue gum	1.00
Camphor	0.99	Opal	2.2	box	0.95-1.16
Cardboard	0.69	Paper	0.7-1.15	butternut	0.38
Celluloid	1.4	Paraffin	0.87-0.91	cedar	0.49-0.57
Cement, set	2.7-3.0	Peat blocks	0.84	cherry	0.70-0.90
Chalk	1.9-2.8	Pitch	1.07	dogwood	0.76
Charcoal,		Polyamides	1.15-1.25	ebony	1.11-1.33
oak	0.57	Polyethylene	0.92-0.97	elm	0.54-0.60
pine	0.28-0.44	Poly(methyl methacrylate)	1.19	hickory	0.60-0.93
Cinnabar	8.12	Polypropylene	0.91-0.94	holly	0.76
Clay	1.8-2.6	Polystyrene	1.06-1.12	juniper	0.56
Coal,		Polytetrafluoroethylene	2.28-2.30	larch	0.50-0.56
anthracite	1.4-1.8	Poly(vinyl acetate)	1.19	locust	0.67-0.71
bituminous	1.2-1.5	Poly(vinyl chloride)	1.39-1.42	logwood	0.91
Coke	1.0-1.7	Porcelain	2.3-2.5	mahogany	0.66-0.85
Copal	1.04-1.14	Porphyry	2.6-2.9	maple	0.62-0.75
Cork	0.22-0.26	Pyrite	4.95-5.10	oak	0.60-0.90
Corundum	3.9-4.0	Quartz	2.65	pear	0.61-0.73
Diamond	3.51	Resin	1.07	pine, pitch	0.83-0.85
Dolomite	2.84	Rock salt	2.18	white	0.35-0.50
Ebonite	1.15	Rubber,		yellow	0.37-0.60
Emery	4.0	hard	1.19	plum	0.66-0.78
Epidote	3.25-3.50	soft	1.1	poplar	0.35-0.50
Feldspar	2.55-2.75	pure gum	0.91-0.93	satinwood	0.95
Flint	2.63	Neoprene	1.23-1.25	spruce	0.48-0.70
Fluorite	3.18	Sandstone	2.14-2.36	sycamore	0.40-0.60
Galena	7.3-7.6	Serpentine	2.50-2.65	teak, Indian	0.66-0.98
Garnet	3.15-4.3	Silica, fused,	2.21	walnut	0.64-0.70
Gelatin	1.27	Silicon carbide	3.16	water gum	1.00
Glass,		Slag	2.0-3.9	willow	0.40-0.60
common	2.4-2.8	Slate	2.6-3.3	Wood's metal	9.70
lead	3-4				

DIELECTRIC STRENGTH OF INSULATING MATERIALS

L. I. Berger

The loss of the dielectric properties by a sample of a gaseous, liquid, or solid insulator as a result of application to the sample of an electric field* greater than a certain critical magnitude is called *dielectric breakdown*. The critical magnitude of electric field at which the breakdown of a material takes place is called the *dielectric strength* of the material (or *breakdown voltage*). The dielectric strength of a material depends on the specimen thickness (as a rule, thin films have greater dielectric strength than that of thicker samples of a material), the electrode shape**, the rate of the applied voltage increase, the shape of the voltage vs. time curve, and the medium surrounding the sample, e.g., air or other gas (or a liquid — for solid materials only).

Breakdown in Gases

The current carriers in gases are free electrons and ions generated by external radiation. The equilibrium concentration of these particles at normal pressure is about 10^3 cm^{-3}, and hence the electrical conductivity is very small, of the order of 10^{-16} - 10^{-15} S/cm. But in a strong electric field, these particles acquire kinetic energy along their free pass, large enough to ionize the gas molecules. The new charged particles ionize more molecules; this avalanche-like process leads to formation between the electrodes of channels of conducting plasma (streamers), and the electrical resistance of the space between the electrodes decreases virtually to zero.

Because the dielectric strength (breakdown voltage) of gases strongly depends on the electrode geometry and surface condition and the gas pressure, it is generally accepted to present the data for a particular gas as a fraction of the dielectric strength of either nitrogen or sulfur hexafluoride measured at the same conditions. In Table 1, the data are presented in comparison with the dielectric strength of nitrogen, which is considered equal to 1.00. For convenience to the reader, a few average magnitudes of the dielectric strength of some gases are expressed in kilovolts per millimeter. The data in the table relate to the standard conditions, unless indicated otherwise.

Breakdown in Liquids

If a liquid is pure, the breakdown mechanism in it is similar to that in gases. If a liquid contains liquid impurities in the form of small drops with greater dielectric constant than that of the main liquid, the breakdown is the result of formation of ellipsoids from these drops by the electric field. In a strong enough electric field, these ellipsoids merge and form a high-conductivity channel between the electrodes. The current increases the temperature in the channel, liquid boils, and the current along the steam canal leads to breakdown. Formation of a conductive channel (bridge) between the electrodes is observed also in liquids with solid impurities. If a liquid contains gas impurities in the form of small bubbles, breakdown is the result of heating of the liquid in strong electric fields. In the locations with the highest current density, the liquid boils, the size of the gas bubbles increases, they merge and form gaseous channels between the electrodes, and the breakdown medium is again the gas plasma.

Breakdown in Solids

It is known that the current in solid insulators does not obey Ohm's law in strong electric fields. The current density increases almost exponentially with the electric field, and at a certain field magnitude it jumps to very high magnitudes at which a specimen of a material is destroyed. The two known kinds of electric breakdown are thermal and electrical breakdowns. The former is the result of material heating by the electric current. Destruction of a sample of a material happens when, at a certain voltage, the amount of heat produced by the current exceeds the heat release through the sample surface; the breakdown voltage in this case is proportional to the square root of the ratio of the thermal conductivity and electrical conductivity of the material. The electrical breakdown results from the tunneling of the charge carriers from electrodes or from the valence band or from the impurity levels into the conduction band, or by the impact ionization. The tunnel effect breakdown happens mainly in thin layers, e.g., in thin p-n junctions. Otherwise, the impact ionization mechanism dominates. For this mechanism, the dielectric strength of an insulator can be estimated using Boltzmann's kinetic equation for electrons in a crystal.

In the following tables, the dielectric strength values are for room temperature and normal atmospheric pressure, unless indicated otherwise.

---

*   The unit of electric field in the SI system is newton per coulomb or volt per meter.
** For example, the U.S. standard ASTM D149 is based on use of symmetrical electrodes, while per U.K. standard BS2918 one electrode is a plane and the other is a rod with the axis normal to the plane.

Table 1
Dielectric Strength of Gases

Material	Dielectric* Strength	Ref.	Material	Dielectric* Strength	Ref.
Nitrogen, N_2	1.00		Trichlorofluoromethane, CCl_3F	3.50	1
Hydrogen, H_2	0.50	1,2		4.53	2
Helium, He	0.15	1	Trichloromethane, $CHCl_3$	4.2	1
Oxygen, O_2	0.92	2		4.39	2
Air	0.97	6	Methylamine, CH_3NH_2	0.81	1
Air (flat electrodes), kV/mm	3.0	3	Difluoromethane, CH_2F_2	0.79	2
Air, kV/mm	0.4-0.7	4	Trifluoromethane, CHF_3	0.71	2
Air, kV/mm	1.40	5	Bromochlorodifluoromethane, CF_2ClBr	3.84	2
Neon, Ne	0.25	1	Chlorodifluoromethane, $CHClF_2$	1.40	1
	0.16	2		1.11	2
Argon, Ar	0.18	2	Dichlorofluoromethane, $CHCl_2F$	1.33	1
Chlorine, Cl_2	1.55	1		2.61	2
Carbon monoxide, CO	1.02	1	Chlorofluoromethane, CH_2ClF	1.03	1
	1.05	2	Hexafluoroethane, C_2F_6	1.82	1
Carbon dioxide, CO_2	0.88	1		2.55	2
	0.82	2	Ethyne (Acetylene), C_2H_2	1.10	1
	0.84	6		1.11	2
Nitrous oxide, N_2O	1.24	2	Chloropentafluoroethane, C_2ClF_5	2.3	1
Sulfur dioxide, SO_2	2.63	2		3.0	6
	2.68	6	Dichlorotetrafluoroethane, $C_2Cl_2F_4$	2.52	1
Sulfur monochloride, S_2Cl_2	1.02	1	Chlorotrifluoroethylene, C_2ClF_3	1.82	2
(at 12.5 Torr)			1,1,1-Trichloro-2,2,2-trifluoroethane	6.55	2
Thionyl fluoride, SOF_2	2.50	1	1,1,2-Trichloro-1,2,2-trifluoroethane	6.05	2
Sulfur hexafluoride, SF_6	2.50	1	Chloroethane, C_2H_5Cl	1.00	1
	2.63	2	1,1-Dichloroethane	2.66	2
Sulfur hexafluoride, SF_6, kV/mm	8.50	7	Trifluoroacetonitrile, CF_3CN	3.5	1
	9.8	8	Acetonitrile, CH_3CN	2.11	2
Perchloryl fluoride, ClO_3F	2.73	1	Dimethylamine, $(CH_3)_2NH$	1.04	1
Tetrachloromethane, CCl_4	6.33	1	Ethylamine, $C_2H_5NH_2$	1.01	1
	6.21	2	Ethylene oxide (oxirane), CH_3CHO	1.01	1
Tetrafluoromethane, CF_4	1.01	1	Perfluoropropene, C_3F_6	2.55	2
Methane, CH_4	1.00	1	Octafluoropropane, C_3F_8	2.19	1
	1.13	2		2.47	2
Bromotrifluoromethane, CF_3Br	1.35	1	3,3,3-Trifluoro-1-propene, CH_2CHCF_3	2.11	2
	1.97	2	Pentafluoroisocyanoethane, C_2F_5NC	4.5	1
Bromomethane, CH_3Br	0.71	2	1,1,1,4,4,4-Hexafluoro-2-butyne, CF_3CCCF_3	5.84	2
Chloromethane, CH_3Cl	1.29	2	Octafluorocyclobutane, C_4F_8	3.34	2
Iodomethane, CH_3I	3.02	2	1,1,1,2,3,4,4,4-Octafluoro-2-butene	2.8	1
Iodomethane, CH_3I, at 370 Torr	2.20	7	Decafluorobutane, C_4F_{10}	3.08	1
Dichloromethane, CH_2Cl_2	1.92	2	Perfluorobutanenitrile, C_3F_7CN	5.5	1
Dichlorodifluoromethane, CCl_2F_2	2.42	1	Perfluoro-2-methyl-1,3-butadiene, C_5F_8	5.5	1
	2.63	2,6	Hexafluorobenzene, C_6F_6	2.11	2
Chlorotrifluoromethane, $CClF_3$	1.43	1	Perfluorocyclohexane, C_6F_{12}, (saturated vapor)	6.18	2
	1.53	2			

*Relative to nitrogen, unless units of kV/mm are indicated.

Table 2
Dielectric Strength of Liquids

Material	Dielectric strength kV/mm	Ref.	Material	Dielectric strength kV/mm	Ref.
Helium, He, liquid, 4.2 K	10	9		20.4	15
Static	10	11		179	17,18
Dynamic	5	11	Ethylbenzene, C_8H_{10}	226	17,18
	23	12	Propylbenzene, C_9H_{12}	250	17,18
Nitrogen, N_2, liquid, 77K			Isopropylbenzene, C_9H_{12}	238	17,18
Coaxial cylinder electrodes	20	10	Decane, $C_{10}H_{22}$	192	17,18
Sphere to plane electrodes	60	10	Synthetic Paraffin Mixture		
Water, H_2O, distilled	65-70	13	Synfluid 2cSt PAO	29.5	37
Carbon tetrachloride, CCl_4	5.5	14	Butylbenzene, $C_{10}H_{14}$	275	17,18
	16.0	15	Isobutylbenzene, $C_{10}H_{14}$	222	17,18
Hexane, C_6H_{14}	42.0	16	Silicone oils—polydimethylsiloxanes,		
Two 2.54 cm diameter spherical			$(CH_3)_3Si\text{-}O\text{-}[Si(CH_3)_2]_x\text{-}O\text{-}Si(CH_3)_3$		
electrodes, 50.8 µm space	156	17,18	Polydimethylsiloxane silicone fluid	15.4	20
Cyclohexane, C_6H_{12}	42-48	16	Dimethyl silicone	24.0	21,22
2-Methylpentane, C_6H_{14}	149	17,18	Phenylmethyl silicone	23.2	22
2,2-Dimethylbutane, C_6H_{14}	133	17,18	Silicone oil, Basilone M50	10-15	23
2,3-Dimethylbutane, C_6H_{14}	138	17,18	Mineral insulating oils	11.8	6
Benzene, C_6H_6	163	17,18	Polybutene oil for capacitors	13.8	6
Chlorobenzene, C_6H_5Cl	7.1	14	Transformer dielectric liquid	28-30	6
	18.8	15	Isopropylbiphenyl capacitor oil	23.6	6
2,2,4-Trimethylpentane, C_8H_{18}	140	17,18	Transformer oil	110.7	24
Phenylxylylethane	23.6	19	Transformer oil Agip ITE 360	9-12.6	23
Heptane, C_7H_{16}	166	17,18	Perfluorinated hydrocarbons		
2,4-Dimethylpentane, C_7H_{16}	133	17,18	Fluorinert FC 6001	8.0	23
Toluene, $C_6H_5CH_3$	199	17,18	Fluorinert FC 77	10.7	23
	46	16	Perfluorinated polyethers		
	12.0	14	Galden XAD (Mol. wt. 800)	10.5	23
	20.4	15	Galden D40 (Mol. wt. 2000)	10.2	23
Octane, C_8H_{18}	16.6	14	Castor oil	65	25

Table 3
Dielectric Strength of Solids

Material	Dielectric strength kV/mm	Ref	Material	Dielectric strength kV/mm	Ref
Sodium chloride, NaCl, crystalline	150	26	Phlogopite, amber, natural	118	6
Potassium bromide, KBr, crystalline	80	26	Fluorophlogopite, synthetic	118	6
Ceramics			Glass-bonded mica	14.0-15.7	6
Alumina (99.9% Al_2O_3)	13.4	6,27a	Thermoplastic Polymers		
Aluminum silicate, Al_2SiO_5	5.9	6	Polypropylene	23.6	6
Berillia (99% BeO)	13.8	6,27b	Amide polymer nylon 6/6, dry	23.6	6
Boron nitride, BN	37.4	6	Polyamide-imide copolymer	22.8	6
Cordierite, $Mg_2Al_4Si_5O_{18}$	7.9	6,27c	Modified polyphenylene oxide	21.7	6
Forsterite, Mg_2SiO_4	9.8	28	Polystyrene	19.7	6
Porcelain	35-160	26	Polymethyl methacrylate	19.7	6
Steatite, $Mg_3Si_4O_{11} \cdot H_2O$	9.1-15.4	6	Polyetherimide	18.9	6
Titanates of Mg, Ca, Sr, Ba, and Pb	20-120	3	Amide polymer nylon 11(dry)	16.7	6
Barium titanate, glass bonded	>30	36	Polysulfone	16.7	6
Zirconia, ZrO_2	11.4	29	Styrene-acrylonitrile copolymer	16.7	6
Glasses			Acrylonitrile-butadiene-styrene	16.7	6
Fused silica, SiO_2	470-670	26	Polyethersulfone	15.7	6
Alkali-silicate glass	200	26	Polybutylene terephthalate	15.7	6
Standard window glass	9.8-13.8	28	Polystyrene-butadiene copolymer	15.7	6
Micas			Acetal homopolymer	15.0	6
Muscovite, ruby, natural	118	6	Acetal copolymer	15.0	6

Table 3
Dielectric Strength of Solids (continued)

Material	Dielectric strength kV/mm	Ref.	Material	Dielectric strength kV/mm	Ref.
Polyphenylene sulfide	15.0	6	Varnishes		
Polycarbonate	15.0	6	Vacuum-pressure impregnated baking type solventless polyester varnish		
Acetal homopolymer resin (molding resin)	15.0	6	Rigid, two-part	70.9	6
Acetal copolymer resin	15.0	6	Semiflexible high-bond thixotropic	78.7	6
Thermosetting Molding Compounds			Rigid high-bond high-flash freon-resistant	68.9	6
Glass-filled allyl (Type GDI-30 per MIL-M-14G)	15.7	6	Baking type epoxy varnish		
Glass-filled epoxy, electrical grade	15.4	6	Solventless, rigid, low viscosity, one-part	90.6	6
Glass-filled phenolic (Type GPI-100 per MIL-M-14G)	15.0	6	Solventless, semiflexible, one-part	82.7	6
Glass-filled alkyd/polyester (Type MAI-60 per MIL-M-14G)	14.8	6	Solventless, semirigid, chemical resistant, low dielectric constant	106.3	6
Glass-filled melamine (Type MMI-30 per MIL-M-14G)	13.4	6	Solvable, for hermetic electric motors	181.1	6
Extrusion Compounds for High-Temperature Insulation			Polyurethane coating		
Polytetrafluoroethylene	19.7	6	Clear conformal, fast cure		
Perfluoroalkoxy polymer	21.7	6	Standard conditions	78.7	6
Fluorinated ethylene-propylene copolymer	19.7	6	Immersion conditions	47.2	6
Ethylene-tetrafluoroethylene copolymer	15.7	6	Insulating Films and Tapes		
Polyvinylidene fluoride	10.2	6	Low-density polyethylene film (40 μm thick)	300	31
Ethylene-chlorotrifluoroethylene copolymer	19.3	6	Poly-p-xylylene film	410-590	32
Polychlorotrifluoroethylene	19.7	6	Aromatic polymer films		
Extrusion Compounds for Low-Temperature Insulation			Kapton H (Du Pont)	389-430	33
Polyvinyl chloride			Ultem (GE Plastic and Roem AG)	437-565	33
Flexible	11.8-15.7	30	Hostaphan (Hoechst AG)	338-447	33
Rigid	13.8-19.7	30	Amorphous Stabar K2000 (ICI film)	404-422	33
Polyethylene	18.9	28	Stabar S100 (ICI film)	353-452	33
Polyethylene, low-density	21.7	6	Polyetherimide film (26 μm)	486	34
	300	31	Parylene N/D (poly-p-xylylene/poly-dichloro-p-xylylene) 25 μm film	275	6
Polyethylene, high-density	19.7	6	Cellulose acetate film	157	6
Polypropylene/polyethylene copolymer	23.6	6	Cellulose triacetate film	157	6
Embedding Compounds			Polytetrafluoroethylene film	87-173	6
Basic epoxy resin: bisphenol-A/epichlorohydrin polycondensate	19.7	6	Perfluoroalkoxy film	157-197	6
Cycloaliphatic epoxy: alicyclic diepoxy carboxylate	19.7	6	Fluorinated ethylene-propylene copolymer film	197	6
Polyetherketone	18.9	30	Ethylene-tetrafluoroethylene film	197	6
Polyurethanes			Ethylene-chlorotrifluoroethylene copolymer film	197	6
Two-component, polyol-cured	25.4	6	Polychlorotrifluoroethylene film	118-153.5	6
Two-part solventless, polybutylene-based	24.0	6	High-voltage rubber insulating tape	28	6
Silicones			Composites		
Clear two-part heat curing eletrical grade silicone embedding resin	21.7	6	Isophthalic polyester (vinyl toluene monomer) filled with		
Red insulating enamel (MIL-E-22118)			Calcium carbonate, $CaCO_3$	15.0	38
Dry	47.2	6	Gypsum, $CaSO_4$	14.4	38
Wet	11.8	6	Alumina trihydrate	15.4	38
Enamels			Clay	14.4	38
Red enamel, fast cure			BPA fumarate polyester (vinyl toluene monomer) filled with		
Standard conditions	78.7	6	Calcium carbonate	6.1	38
Immersion conditions	47.2	6	Gypsum	5.9	38
Black enamel			Alumina trihydrate	11.8	38
Standard conditions	70.9	6	Clay	12.6	38
Immersion conditions	47.2	6			

Table 3
Dielectric Strength of Solids (continued)

Material	Dielectric strength kV/mm	Ref.	Material	Dielectric strength kV/mm	Ref.
Polysulfone resin—30% glass fiber	16.5-18.7	38	Butyl rubber	23.6	6
Polyamid resin (Nylon 66)—			Neoprene	15.7-27.6	6
30% carbon fiber	13.0	38	Silicone rubber	26-36	6
Polyimide thermoset resin,			Room-temperature vulcanized	9.2-10.9	35
glass reinforced	12.0	39	silicone rubber		
Polyester resin (thermoplastic)—			Ureas (from carbamide	11.8-15.7	28
40% glass fiber	20.0	38	to tetraphenylurea)		
Epoxy resin (diglycidyl ether of			Dielectric papers		
bisphenol A), glass reinforced	16.0	40	Aramid paper, calendered	28.7	6
Various Insulators			Aramid paper, uncalendered	12.2	6
Rubber, natural	100-215	26	Aramid with Mica	39.4	6

REFERENCES

1. Vijh, A. K. *IEEE Trans.,* EI-12, 313, 1997.
2. Brand, K. P., *IEEE Trans.,* EI-17, 451, 1982.
3. *Encyclopedic Dictionary in Physics,* Vedensky, B. A. and Vul, B. M., Eds., Vol. 4, Soviet Encyclopedia Publishing House, Moscow, 1965.
4. Kubuki, M., Yoshimoto, R., Yoshizumi, K., Tsuru, S., and Hara, M., *IEEE Trans.,* DEI-4, 92, 1997.
5. Al-Arainy, A. A. Malik, N. H., and Cureshi, M. I., *IEEE Trans.,* DEI-1, 305, 1994.
6. Shugg, W. T., *Handbook of Electrical and Electronic Insulating Materials,* Van Nostrand Reinhold, New York, 1986.
7. Devins, J. C., *IEEE Trans.,* EI-15, 81, 1980.
8. Xu, X., Jayaram, S., and Boggs, S. A., *IEEE Trans.,* DEI-3, 836, 1996.
9. Okubo, H., Wakita, M., Chigusa, S., Nayakawa, N., and Hikita, M., *IEEE Trans.,* DEI-4, 120, 1997.
10. Hayakawa, H., Sakakibara, H., Goshima, H., Hikita, M., and Okubo, H., *IEEE Trans.,* DEI-4, 127, 1997.
11. Okubo, H., Wakita, M., Chigusa, S., Hayakawa, N., and Hikita, M., *IEEE Trans.,* DEI-4, 220, 1997.
12. Von Hippel, A. R., *Dielectric Materials and Applications,* MIT Press, Cambridge, MA, 1954.
13. Jones, H. M. and Kunhards, E. E., *IEEE Trans.,* DEI-1, 1016, 1994.
14. Nitta, Y. and Ayhara, Y., *IEEE Trans.,* EI-11, 91, 1976.
15. Gallagher, T. J., *IEEE Trans.,* EI-12, 249, 1977.
16. Wong, P. P. and Forster, E. O., in *Dielectric Materials. Measurements and Applications,* IEE Conf. Publ. 177, 1, 1979.
17. Kao, K. C. *IEEE Trans.,* EI-11, 121, 1976.
18. Sharbaugh, A. H., Crowe, R. W., and Cox, E. B., *J. Appl. Phys.,* 27, 806, 1956.
19. Miller, R. L., Mandelcorn, L., and Mercier, G. E., in *Proc. Intl. Conf. on Properties and Applications of Dielectric Materials,* Xian, China, June 24-28, 1985; cited in Ref. 6, p. 492.
20. Hakim, R. M., Oliver, R. G., and St-Onge, H., *IEEE Trans.,* EI-12, 360, 1977.
21. Hosticka, C., *IEEE Trans.,* 389, 1977.
22. Yasufuku, S., Umemura, T., and Ishioka, Y., *IEEE Trans.,* EI-12, 402, 1977.
23. Forster, E. O., Yamashita, H., Mazzetti, C., Pompini, M., Caroli, L., and Patrissi, S., *IEEE Trans.,* DEI-1, 440, 1994.
24. Bell, W. R., *IEEE Trans.,* 281, 1977.
25. Ramu, T. C. and Narayana Rao, Y., in *Dielectric Materials. Measurements and Applications,* IEE Conf. Publ. 177, 37.
26. Skanavi, G. I., *Fizika Dielektrikov; Oblast Silnykh Polei* (Physics of Dielectrics; Strong Fields). Gos. Izd. Fiz. Mat. Nauk (State Publ. House for Phys. and Math. Scis.), Moscow, 1958.
27. Kleiner, R. N., in *Practical Handbook of Materials Science,* Lynch, C. T., Ed., CRC Press, 1989; 27a: p. 304; 27b: p.300; 27c: p. 316.
28. *Materials Selector Guide. Materials and Methods,* Reinhold Publ., New York, 1973.
29. Flinn, R. A. and Trojan, P. K., *Engineering Materials and Their Applications,* 2nd ed., Houghton Mifflin, 1981, p. 614.
30. Lynch, C. T., Ed., *Practical Handbook of Materials Science,* CRC Press, Boca Raton, FL, 1989.
31. Suzuki, H., Mukai, S., Ohki, Y., Nakamichi, Y., and Ajiki, K., *IEEE Trans.,* DEI-4, 238, 1997.
32. Mori, T., Matsuoka, T., and Muzitani, T., *IEEE Trans.,* DEI-1, 71, 1994.
33. Bjellheim, P. and Helgee, B., *IEEE Trans.,* DEI-1, 89, 1994.
34. Zheng, J. P., Cygan, P. J., and Jow, T. R., *IEEE Trans.,* DEI-3, 144, 1996.
35. Danukas, M. G., *IEEE Trans.,* DEI-1, 1196, 1994.
36. Burn, I. and Smithe, D. H., *J. Mater. Sci.,* 7, 339, 1972.
37. Hope, K.D., Chevron Chemical, Private Communication.

38. *Engineering Materials Handbook*, vol. 1, Composites, C.A. Dostal, Ed., ASM Intl., 1987.
39. 1985 Materials Selector, *Mater. Eng.*, (12) 1984.
40. *Modern Plastics Encyclopedia*, McGraw-Hill, v. 62 (No. 10A) 1985–1986.

Review Literature on the Subject

R1. Kuffel, E. and Zaengl, W. S., *HV Engineering Fundamentals*, Pergamon, 1989.
R2. Kok, J. A., *Electrical Breakdown of Insulating Liquids*, Phillips Tech. Library, Cleaver-Hum, Longon, 1961.
R3. Gallagher, T. J., *Simple Dielectric Liquids*, Clarendon, Oxford, 1975.
R4. Meek, J. M. and Craggs, J. D., Eds., *Electric Breakdown in Gases*, John Wiley & Sons, 1976.
R5. Von Hippel, A. R., *Dielectric Materials and Applications*, MIT Press, Cambridge, MA, 1954.

Section 16
Health and Safety Information

Handling and Disposal of Chemicals in Laboratories **16**-1
Flammability of Chemical Substances **16**-16
Threshold Limits for Airborne Contaminants **16**-32
Octanol-Water Partition Coefficients **16**-43
Protection Against Ionizing Radiation **16**-48
Annual Limits on Intakes of Radionuclides **16**-49
Chemical Carcinogens **16**-53

HANDLING AND DISPOSAL OF CHEMICALS IN LABORATORIES

Robert Joyce and Blaine C. McKusick

The following material has been extracted from two books prepared under the auspices of the Committee on Hazardous Substances in the Laboratory of the National Academy of Sciences — National Research Council. Readers are referred to these books for full details:

Prudent Practices for Handling Hazardous Chemicals in Laboratories, National Academy Press, Washington, 1981.
Prudent Practices for Disposal of Chemicals from Laboratories, National Academy Press, Washington, 1983.

The permission of the National Academy Press to use these extracts is gratefully acknowledged.

INCOMPATIBLE CHEMICALS

The term "incompatible chemicals" refers to chemicals that can react with each other

- Violently
- With evolution of substantial heat
- To produce flammable products
- To produce toxic products

Good laboratory safety practice requires that incompatible chemicals be stored, transported, and disposed of in ways that will prevent their coming together in the event of an accident. Tables 1 and 2 give some basic guidelines for the safe handling of acids, bases, reactive metals, and other chemicals. Neither of these tables is exhaustive, and additional information on incompatible chemicals can be found in the following references.

1. L. Bretherick, *Handbook of Reactive Chemical Hazards*, 3rd ed., Butterworths, London–Boston, 1985.
2. L. Bretherick, Ed., *Hazards in the Chemical Laboratory*, 3rd ed., Royal Society of Chemistry, London, 1981.
3. *Manual of Hazardous Chemical Reactions, A Compilation of Chemical Reactions Reported to be Potentially Hazardous*, National Fire Protection Association, NFPA 491M, 1975, NFPA, 470 Atlantic Avenue, Boston, MA 02210.

TABLE 1
General Classes of Incompatible Chemicals

A	B
Acids	Bases, reactive metals
Oxidizing agents[a]	Reducing agents[a]
Chlorates	Ammonia, anhydrous and aqueous
Chromates	Carbon
Chromium trioxide	Metals
Dichromates	Metal hydrides
Halogens	Nitrites
Halogenating agents	Organic compounds
Hydrogen peroxide	Phosphorus
Nitric acid	Silicon
Nitrates	Sulfur
Perchlorates	
Peroxides	
Permanganates	
Persulfates	

[a] The examples of oxidizing and reducing agents are illustrative of common laboratory chemicals; they are not intended to be exhaustive.

TABLE 2
Examples of Incompatible Chemicals

Chemical	Is incompatible with
Acetic acid	Chromic acid, nitric acid, hydroxyl compounds, ethylene glycol, perchloric acid, peroxides, permanaganates
Acetylene	Chlorine, bromine, copper, fluorine, silver, mercury
Acetone	Concentrated nitric and sulfuric acid mixtures
Alkali and alkaline earth metals (such as powdered aluminum or magnesium, calcium, lithium, sodium, potassium)	Water, carbon tetrachloride or other chlorinated hydrocarbons, carbon dioxide, halogens
Ammonia (anhydrous)	Mercury (in manometers, for example), chlorine, calcium hypochlorite, iodine, bromine, hydrofluoric acid (anhydrous)
Ammonium nitrate	Acids, powdered metals, flammable liquids, chlorates, nitrites, sulfur, finely divided organic or combustible materials
Aniline	Nitric acid, hydrogen peroxide
Arsenical materials	Any reducing agent
Azides	Acids
Bromine	See Chlorine
Calcium oxide	Water
Carbon (activated)	Calcium hypochlorite, all oxidizing agents
Carbon tetrachloride	Sodium
Chlorates	Ammonium salts, acids, powdered metals, sulfur, finely divided organic or combustible materials
Chromic acid and chromium troixide	Acetic acid, naphthalene, camphor, glycerol, alcohol, flammable liquids in general
Chlorine	Ammonia, acetylene, butadiene, butane, methane, propane (or other petroleum gases), hydrogen, sodium carbide, benzene, finely divided metals, turpentine
Chlorine dioxide	Ammonia, methane, phosphine, hydrogen sulfide
Copper	Acetylene, hydrogen peroxide
Cumene hydroperoxide	Acids (organic or inorganic)
Cyanides	Acids
Flammable liquids	Ammonium nitrate, chromic acid, hydrogen peroxide, nitric acid, sodium peroxide, halogens
Fluorine	Everything
Hydrocarbons (such as butane, propane, benzene)	Fluorine, chlorine, bromine, chromic acid, sodium peroxide
Hydrocyanic acid	Nitric acid, alkali
Hydrofluoric acid (anhydrous)	Ammonia (aqueous or anhydrous)
Hydrogen peroxide	Copper, chromium, iron, most metals or their salts, alcohols, acetone, organic materials, aniline, nitromethane, combustible matierals
Hydrogen sulfide	Fuming nitric acid, oxidizing gases
Hypochlorites	Acids, activated carbon
Iodine	Acetylene, ammonia (aqueous or anhydrous), hydrogen
Mercury	Acetylene, fulminic acid, ammonia
Nitrates	Sulfuric acid
Nitric acid (concentrated)	Acetic acid, aniline, chromic acid, hydrocyanic acid, hydrogen sulfide, flammable liquids, flammable gases, copper, brass, any heavy metals
Nitrites	Acids
Nitroparaffins	Inorganic bases, amines
Oxalic acid	Silver, mercury

TABLE 2
Examples of Incompatible Chemicals (continued)

Chemical	Is incompatible with
Oxygen	Oils, grease, hydrogen, flammable liquids, solids, or gases
Perchloric acid	Acetic anhydride, bismuth and its alloys, alcohol, paper, wood, grease, oils
Peroxides, organic	Acids (organic or mineral), avoid friction, store cold
Phosphorus (white)	Air, oxygen, alkalis, reducing agents
Potassium	Carbon tetrachloride, carbon dioxide, water
Potassium chlorate	Sulfuric and other acids
Potassium perchlorate (see also chlorates)	Sulfuric and other acids
Potassium permanganate	Glycerol, ethylene glycol, benzaldehyde, surfuric acid
Selenides	Reducing agents
Silver	Acetylene, oxalic acid, tartartic acid, ammonium compounds, fulminic acid
Sodium	Carbon tetrachloride, carbon dioxide, water
Sodium nitrite	Ammonium nitrate and other ammonium salts
Sodium peroxide	Ethyl or methyl alcohol, glacial acetic acid, acetic anhydride, benzaldehyde, carbon disulfide, glycerin, ethylene glycol, ethyl acetate, methyl acetate, furfural
Sulfides	Acids
Sulfuric acid	Potassium chlorate, potassium perchlorate, potassium permanganate (similar compounds of light metals, such as sodium, lithium)
Tellurides	Reducing agents

EXPLOSION HAZARDS

Table 3 lists some common classes of laboratory chemicals that have potential for producing a violent explosion when subjected to shock or friction. These chemicals should never be disposed of as such, but should be handled by procedures given in *Prudent Practices for Disposal of Chemicals from Laboratories*, National Academy Press, 1983, chapters 6 and 7. Additional information on these, as well as on some less common classes of explosives, can be found in L. Bretherick, *Handbook of Reactive Chemical Hazards,* 3rd ed., Butterworths, London–Boston, 1985.

Table 4 lists some illustrative combinations of common laboratory reagents that can produce explosions when they are brought together or that form reaction products that can explode without any apparent external initiating action. This list is not exhaustive, and additional information on potentially explosive reagent combinations can be found in Manual of Hazardous Chemical Reactions, A Compilation of Chemical Reactions Reported to be Potentially Hazardous, National Fire Protection Association, NFPA 491M, 1975, NFPA, 470 Atlantic Avenue, Boston, MA 02210.

WATER-REACTIVE CHEMICALS

Table 5 lists some common laboratory chemicals that react violently with water and that should always be stored and handled so that they do not come into contact with liquid water or water vapor. Procedures for decomposing laboratory quantities are given in Prudent Practices for Disposal of Chemicals from Laboratories, chapter 6; the pertinent section of that chapter is given in parentheses.

PYROPHORIC CHEMICALS

Many members of the classes of readily oxidized, common laboratory chemicals listed in Table 6 ignite spontaneously in air. A more extensive list can be found in L. Bretherick, *Handbook of Reactive Chemical Hazards*, 3rd ed., Butterworths, London-Boston, 1985. Pyrophoric chemicals should be stored in tightly closed containers under an inert atmosphere (or, for some, an inert liquid), and all transfers and manipulations of them must be carried out under an inert atmosphere or liquid. Suggested procedures for decomposing them are given in Prudent Practices for Disposal of Chemicals from Laboratories, chapter 6; the pertinent section of that chapter is given in parentheses.

TABLE 3
Shock–Sensitive Compounds

Acetylenic compounds, especially polyacetylenes, haloacetylenes, and heavy metal salts of acetylenes (copper, silver, and mercury salts are particularly sensitive)

Acyl nitrates

Alkyl nitrates, particularly polyol nitrates such as nitrocellulose and nitroglycerine

Alkyl and acyl nitrites

Alkyl perchlorates

Amminemetal oxosalts: metal compounds with coordinated ammonia, hydrazine, or similar nitrogenous donors and ionic perchlorate, nitrate, per manganate, or other oxidizing group

Azides, including metal, nonmetal, and organic azides

Chlorite salts of metals, such as $AgClO_2$ and $Hg(ClO_2)_2$

Diazo compounds such as CH_2N_2

Diazonium slats, when dry

Fulminates (silver fulminate, AgCNO, can form in the reaction mixture from the Tollens' test for aldehydes if it is allowed to stand for some time; this can be prevented by adding dilute nitric acid to the test mixture as soon as the test has been completed)

Hydrogen peroxide becomes increasingly treacherous as the concentration rises above 30%, forming explosive mixtures with organic materials and decomposing violently in the presence of traces of transition metals

N–Halogen compounds such as difluoroamino compounds and halogen azides

N–Nitro compounds such as N–nitromethylamine, nitrourea, nitroguanidine, and nitric amide

Oxo salts of nitrogenous bases: perchlorates, dichromates, nitrates, iodates, chlorites, chlorates, and permanganates of ammonia, amines, hydroxylamine, guanidine, etc.

Perchlorate salts. Most metal, nonmetal, and amine perchlorates can be detonated and may undergo violent reaction in contact with combustible materials

Peroxides and hydroperoxides, organic (see Chapter 6, Section II.P)

Peroxides (solid) that crystallize from or are left from evaporation of peroxidizable solvents (see Chapter 6 and Appendix I)

Peroxides, transition–metal salts

Picrates, especially salts of transition and heavy metals, such as Ni, Pb, Hg, Cu, and Zn; picric acid is explosive but is less sensitive to shock or friction than its metal salts and is relatively safe as a water–wet paste (see Chapter 7)

Polynitroalkyl compounds such as tetranitromethane and dinitroacetonitrile

Polynitroaromatic compounds, especially polynitro hydrocarbons, phenols, and amines

TABLE 4
Potentially Explosive Combinations of Some Common Reagents

Acetone + chloroform in the presence of base

Acetylene + copper, silver, mercury, or their salts

Ammonia (including aqueous solutions) + Cl_2, Br_2, or I_2

Carbon disulfide + sodium azide

Chlorine + an alcohol

Chloroform or carbon tetrachloride + powdered Al or Mg

Decolorizing carbon + an oxidizing agent

Diethyl ether + chlorine (including a chlorine atmosphere)

Dimethyl sulfoxide + an acyl halide, $SOCl_2$ or $POCl_3$

Dimethyl sulfoxide + CrO_3

Ethanol + calcium hypochlorite

Ethanol + silver nitrate

Nitric acid + acetic anhydride or acetic acid

Picric acid + a heavy–metal salt, such as of Pb, Hg, or Ag

Silver oxide + ammonia + ethanol

Sodium + a cholrinated hydrocarbon

Sodium hypochlorite + an amine

TABLE 5
Water–Reactive Chemicals

Alkali metals (III.D)
Alkali metal hydrides (III.C.2)
Alkali metal amides (III.C.7)
Metal alkyls, such as lithium alkyls and aluminum alkyls (IV.A)
Grignard reagents (IV.A)
Halides of nonmetals, such as BCl_3, BF_3, PCl_3, PCl_5, $SiCl_4$, S_2Cl_2 (III.F)
Inorganic acid halides, such as $POCl_3$, $SOCl_2$, SO_2Cl_2 (III.F)
Anhydrous metal halides, such as $AlCl_3$, $TiCl_4$, $ZrCl_4$, $SnCl_4$ (III.E)
Phosphorus pentoxide (III.I)
Calcium carbide (IV.E)
Organic acid halides and anhydrides of low molecular weight (II.J)

TABLE 6
Classes of Pyrophoric Chemicals

Grignard reagents, RMgX (IV.A)
Metal alkyls and aryls, such as RLi, RNa, R_3Al, R_2Zn (IV.A)
Metal carbonyls, such as $Ni(CO)_4$, $Fe(CO)_5$, $Co_2(CO)_8$ (IV.B)
Alkali metals such as Na, K (III.D.l)
Metal powders, such as Al, Co, Fe, Mg, Mn, Pd, Pt, Ti, Sn, Zn, Zr (III.D.2)
Metal hydrides, such as NaH, $LiAlH_4$ (IV.C.2)
Nonmetal hydrides, such as B_2H_6 and other boranes, PH_3, AsH_3 (III.G)
Nonmetal alkyls, such as R_3B, R_3P, R_3As (IV.C)
Phosphorus (white) (III.H)

HAZARDS FROM PEROXIDE FORMATION

Many common laboratory chemicals can form peroxides when allowed access to air over a period of time. A single opening of a container to remove some of the contents can introduce enough air for peroxide formation to occur. Some types of compounds form peroxides that are treacherously and violently explosive in concentrated solution or as solids. Accordingly, peroxide–containing liquids should never be evaporated near to or to dryness. Peroxide formation can also occur in many polymerizable unsaturated compounds, and these peroxides can initiate a runaway, sometimes explosive, polymerization reaction. Procedures for testing for peroxides and for removing small amounts from laboratory chemicals are given in Prudent Practices for Disposal of Chemicals from Laboratories, chapter 6, Section II.P.

Table 7 provides a list of structural characteristics in organic compounds that can peroxidize. These structures are listed in approximate order of decreasing hazard. Reports of serious incidents involving the last five structural types are extremely rare, but these structures are listed because laboratory workers should be aware that they can form peroxides that can influence the course of experiments in which they are used.

Table 8 gives examples of common laboratory chemicals that are prone to form peroxides on exposure to air. The lists are not exhaustive, and analogous organic compounds that have any of the structural features given in Table 7 should be tested for peroxides before being used as solvents or reagents, or before being distilled. The recommended retention times begin with the date of synthesis or of opening the original container.

DISPOSAL OF TOXIC CHEMICALS

It is often desirable to precipitate toxic cations or hazardous anions from solution to facilitate recovery or disposal. Table 9 lists precipitants for many common cations, and Table 10 gives precipitants for some hazardous anions. Many cations can be precipitated as sulfides by adding sodium sulfide solution (preferable to the highly toxic hydrogen sulfide) to a neutral solution of the cation (Table 11). Control of pH is important because some sulfides will redissolve in excess sulfide ion. After precipitation, excess sulfide can be destroyed by addition of hypochlorite.

Most metal cations are precipitated as hydroxides or oxides at high pH. Since many of these precipitates will redissolve in excess base, it is often necessary to control pH. Table 12 shows the recommended pH range for precipitating many cations in their most common oxidation state. The notation "1 N" in the right–hand column indicates that the precipitate will not dissolve in 1 N sodium hydroxide (pH 14).

The distinctions between high and low toxicity or hazard are based on toxicological and other data, and are relative. There is no implication of a sharp distinction between high and low, or that any cations or anions are totally without hazard.

TABLE 7
Types of Chemicals That Are Prone to Form Peroxides

A. Organic structures (in approximate order of decreasing hazard)

1.   Ethers and acetals with a hydrogen atoms

2.   Olefins with allylic hydrogen atoms

3.   Chloroolefins and fluoroolefins

4.   Vinyl halides, esters, and ethers

5.   Dienes

6.   Vinylacetylenes with α hydrogen atoms

7.   Alkylacetylenes with α hydrogen atoms

8.   Alkylarenes that contain tertiary hydrogen atoms

9.   Alkanes and cycloalkanes that contain tertiary hydrogen atoms

10.   Acrylates and methacrylates

11.   Secondary alcohols

12.   Ketones that contain a hydrogen atoms

13.   Aldehydes

14.   Ureas, amides, and lactams that have a hydrogen atom on a carbon atom attached to nitrogen

16-6

TABLE 7
Types of Chemicals That Are Prone to Form Peroxides (continued)

B. Inorganic substances

1. Alkali metals, especially potassium, rubidium, and cesium (see Chapter 6, Section III.D)
2. Metal amides (see Chapter 6, Section III.C.7)
3. Organometallic compounds with a metal atom bonded to carbon (see Chapter 6, Section IV)
4. Metal alkoxides

TABLE 8
Common Peroxide-Forming Chemicals

LIST A

Severe Peroxide Hazard on Storage with Exposure to Air

Discard within 3 months

- Diisopropyl ether (isopropyl ether)
- Divinylacetylene (DVA)[a]
- Potassium metal
- Potassium amide

- Sodium amide (sodamide)
- Vinylidene chloride (1,1–dichloro-ethylene)[a]

LIST B

Peroxide Hazard on Concentration; Do Not Distill or Evaporate Without First Testing for the Presence of Peroxides

Discard or test for peroxides after 6 months

- Acetaldehyde diethyl acetal (acetal)
- Cumene (isopropylbenzene)
- Cyclohexene
- Cyclopentene
- Decalin (decahydronaphthalene)
- Diacetylene
- Dicyclopentadiene
- Diethyl ether (ether)
- Diethylene glycol dimethyl ether (diglyme)

- Dioxane
- Ethylene glycol dimethyl ether (glyme)
- Ethylene glycol ether acetates
- Ethylene glycol monoethers (cello–solves)
- Furan
- Methylacetylene
- Methylcyclopentane
- Methyl isobutyl ketone
- Tetrahydrofuran (THF)
- Tetralin (tetrahydronaphthalene)
- Vinyl ethers[a]

LIST C

Hazard of Rapid Polymerization Initiated by Internally Formed Peroxides[a]

a. Normal Liquids; Discard or test for peroxides after 6 months[b]

- Chloroprene (2–chloro–1,3–buta–diene)[c]
- Styrene

- Vinyl acetate
- Vinylpyridine

b. Normal Gases; Discard after 12 months[d]

- Butadiene[c]
- Tetrafluoroethylene (TFE)[c]

- Vinylacetylene (MVA)[c]
- Vinyl chloride

[a] Polymerizable monomers should be stored with a polymerization inhibitor from which the monomer can be separated by distillation just before use.

[b] Although common acrylic monomers such as acrylonitrile, acrylic acid, ethyl acrylate, and methyl methacrylate can form peroxides, they have not been reported to develop hazardous levels in normal use and storage.

[c] The hazard from peroxides in these compounds is substantially greater when they are stored in the liquid phase, and if so stored without an inhibitor they should be considered as in LIST A.

TABLE 8
Common Peroxide-Forming Chemicals (continued)

ᵈ   Although air will not enter a gas cylinder in which gases are stored under pressure, these gases are sometimes transferred from the original cylinder to another in the laboratory, and it is difficult to be sure that there is no residual air in the receiving cylinder. An inhibitor should be put into any such secondary cylinder before one of these gases is transferred into it; the supplier can suggest inhibitors to be used. The hazard posed by these gases is much greater if there is a liquid phase in such a secondary container, and even inhibited gases that have been put into a secondary container under conditions that create a liquid phase should be discarded within 12 months.

Note: Laboratory workers should label all containers of peroxidizable solvents or reagents with one of the following:

[LIST A]

                      Peroxidizable compound
                             Received   Opened
Date   _____   _____
                      Discard 3 months after opening

[LISTS B AND C]

                      Peroxidizable compound
                             Received   Opened
Date   _____   _____
                      Discard or test for peroxides
                      6 months after opening

TABLE 9
Relative Toxicity of Cations

High toxic hazard	Precipitant[a]	Low toxic hazard	Precipitant[a]
Antimony	OH^-, S^{2-}	Aluminum	OH^-
Arsenic	S^{2-}	Bismuth	OH^-, S^{2-}
Barium	SO_4^{2-}, CO_3^{2-}	Calcium	SO_4^{2-}, CO_3^{2-}
Beryllium	OH^-	Cerium	OH^-
Cadmium	OH^-, S^{2-}	Cesium	
Chromium (III)[b]	OH^-	Copper[c]	OH^-, S^{2-}
Cobalt (II)[b]	OH^-, S^{2-}	Gold	OH^-, S^{2-}
Gallium	OH^-	Iron[c]	OH^-, S^{2-}
Germanium	OH^-, S^{2-}	Lanthanides	OH^-
Hafnium	OH^-	Lithium	
Indium	OH^-, S^{2-}	Magnesium	OH^-
Iridium	OH^-, S^{2-}	Molybdenum (VI)[b,d]	
Lead	OH^-, S^{2-}	Niobium (V)	OH^-
Manganese (II)[b]	OH^-, S^{2-}	Palladium	OH^-, S^{2-}
Mercury	OH^-, S^{2-}	Potassium	
Nickel	OH^-, S^{2-}	Rubidium	
Osmium (IV)[b,e]	OH^-, S^{2-}	Scandium	OH^-
Platinum (II)[b]	OH^-, S^{2-}	Sodium	
Rhenium (VII)[b]	S^{2-}	Strontium	SO_4^{2-}, CO_3^{2-}
Rhodium (III)[b]	OH^-, S^{2-}	Tantalum	OH^-
Ruthenium (III)[b]	OH^-, S^{2-}	Tin	OH^-, S^{2-}
Selenium	S^{2-}	Titanium	OH^-
Silver	Cl^-, OH^-, S^{2-}	Yttrium	OH^-
Tellurium	S^{2-}	Zinc[c]	OH^-, S^2
Thallium	OH^-, S^{2-}	Zirconium	OH^-
Tungsten (VI)[b,d]			
Vanadium	OH^-, S^{2-}		

TABLE 9
Relative Toxicity of Cations (continued)

[a]   Precipitants are listed in order of preference:
    OH^- = base (sodium hydroxide or sodium carbonate)
    S^{2-} = sulfide
    Cl^- = chloride
    SO_4^{2-} = sulfate
    CO_3^{2-} = carbonate

[b]   The precipitant is for the indicated valence state.

[c]   Maximum tolerance levels have been set for these low–toxicity ions by the U.S. Public Health Service, and large amounts should not be put into public sewer systems. The small amounts typically used in laboratories will not normally affect water supplies.

[d]   These ions are best precipitated as calcium molybdate or calcium tungstate.

[e]   CAUTION: OsO_4, a volatile, extremely poisonous substance, is formed from almost any osmium compound under acid conditions in the presence of air.

TABLE 10
Relative Hazard of Anions

High–hazard anions			
Ion	Hazard type[a]	Precipitant	Low-hazard anions
Aluminum hydride, AlH_4	F	—	Bisulfite, HSO_3^-
Amide, NH_2^-	F, E[b]	—	Borate, BO_3^{3-}, $B_4O_7^{2-}$
Arsenate, AsO_3^-, AsO_4^{3-}	T	Cu^{2+}, Fe^{2+}	Bromide, Br^-
Arsenite, AsO_2^-, AsO_3^{3-}	T	Pb^{2+}	Carbonate, CO_3^{2-}
Azide, N_3^-	E, T	—	Chloride, Cl^-
Borohydride, BH_4^-	F	—	Cyanate, OCN^-
Bromate, BrO_3^-	O, E	—	Hydroxide, OH^-
Chlorate, ClO_3^-	O, E	—	Iodide, I^-
Chromate, CrO_4^{2-}, $Cr_2O_7^{2-}$	T, O	[c]	Oxide, O^{2-}
Cyanide, CN^-	T	—	Phosphate, PO_4^{3-}
Ferricyanide, $Fe(CN)_6^{3-}$	T	Fe^{2+}	Sulfate, SO_4^{2-}
Ferrocyanide, $Fe(CN)_6^{4-}$	T	Fe^{3+}	Sulfite, SO_3^{2-}
Fluoride, F^-	T	Ca^{2+}	Thiocyanate, SCN^-
Hydride, H^-	F	—	
Hydroperoxide, O_2H^-	O, E	—	
Hydrosulfide, SH^-	T	—	
Hypochlorite, OCl^-	O	—	
Iodate, IO_3^-	O, E	—	
Nitrate, NO_3^-	O	—	
Nitrite, NO_2^-	T, O	—	
Perchlorate, ClO_4^-	O, E	—	
Permanganate, MnO_4^-	T, O	[d]	
Peroxide, O_2^{2-}	O, E	—	
Persulfate, $S_2O_8^{2-}$	O	—	
Selenate, SeO_4^{2-}	T	Pb^{2+}	
Selenide, Se^{2-}	T	Cu^{2+}	
Sulfide, S^{2-}	T	[e]	

[a]   Toxic, T: oxidant, O; flammable, F; explosive, E.
[b]   Metal amides readily form explosive peroxides on exposure to air.
[c]   Reduce and precipitate as Cr(III); see Table 9.
[d]   Reduce and precipitate as Mn(II); see Table 9.
[e]   See Table 11.

TABLE 11
Precipitation of Sulfides

Precipitated at pH 7	Not precipitated at low pH	Forms a soluble complex at high pH
Ag^+		
As^{3+a}		X
Au^{+a}		X
Bi^{3+}		
Cd^{2+}		
Co^{2+}	X	
Cr^{3+a}		
Cu^{2+}		
Fe^{2+a}	X	
Ge^{2+}		X
Hg^{2+}		X
In^{3+}	X	
Ir^{4+}		X
Mn^{2+a}	X	
Mo^{3+}		X
Ni^{2+}	X	
Os^{4+}		
Pb^{2+}		
Pd^{2+a}		
Pt^{2+a}		X
Re^{4+}		
Rh^{2+a}		
Ru^{4+}		
Sb^{3+a}		X
Se^{2+}		X
Sn^{2+}		X
Te^{4+}		X
Tl^{+a}	X	
V^{4+a}		
Zn^{2+}	X	

[a] Higher oxidation states of this ion are reduced by sulfide ion and precipitated as this sulfide.

TABLE 12
pH Range for Precipitation of Metal Hydroxides and Oxides

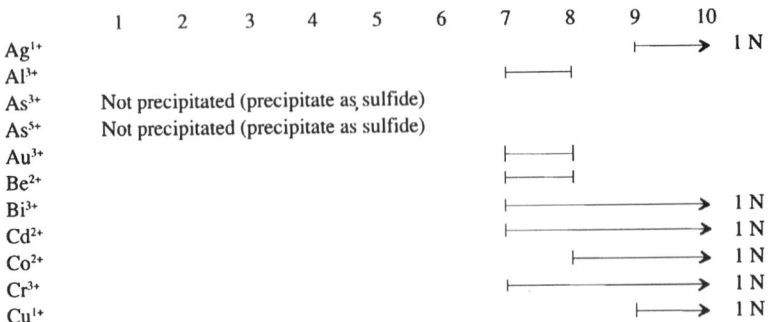

TABLE 12
pH Range for Precipitation of Metal Hydroxides and Oxides (continued)

REFERENCES

L. Erdey, *Gravimetric Analysis*, Part II, Pergamon Press, New York, 1965.

D. T. Burns, A. Towsend, and A. H. Carter, *Inorganic Reaction Chemistry*, Vol. 2, Ellis Horwood, New York, 1981.

FIRE HAZARDS

Flammable solvents are a common source of laboratory fires. The relative ease with which some common laboratory solvents can be ignited is indicated by the following properties.

Flash Point — The lowest temperature, as determined by standard tests, at which a liquid emits vapor in sufficient concentration to form an ignitable mixture with air near the surface of the liquid in a test vessel. Note that many of these common chemicals have flash points below room temperature.

Ignition Temperature — The minimum temperature required to initiate self–sustained combustion, regardless of the heat source.

Flammable Limits — The lower flammable limit is the minimum concentration (percent by volume) of a vapor in air below which a flame is not propagated when an ignition source is present. Below this concentration the mixture is too lean to burn. The upper flammable limit is the maximum concentration (percent by volume) of the vapor in air above which a flame is not propagated. Above this concentration the mixture is too rich to burn. The flammable range comprises all concentrations between these two limits. This range becomes wider with increasing temperature and in oxygen–rich atmospheres. Table 13 lists these properties for a few common laboratory chemicals.

GLOVE MATERIALS

It is good safety practice (and mandated in some laboratories) to wear rubber gloves while handling chemicals that can cause injury when in contact with, or absorbed through, the skin. The various common rubbers are not equally resistant to all chemicals. Table 14 provides guidelines for selecting the best, and avoiding the poorest, glove material for handling a given chemical.

RESPIRATORS

In the event of a laboratory accident or spill, it will be necessary for someone to enter the contaminated area for cleanup. If significant quantities of a chemical are spilled, or even minor quantities of a known toxic material, it is essential to wear the correct kind of respirator equipment when entering the area. If it is not known whether the contamination is of a chemical "Immediately dangerous to life or health", the prudent course is to assume that it is, and to use the corresponding type of respirator. Guidelines are presented in Table 15.

TABLE 13
Flash Points, Boiling Points, Ignition Temperatures, and Flammable Limits of Some Common Laboratory Chemicals

Chemical	Flash point (°C)	Boiling point (°C)	Ignition temp. (°C)	Flammable limit (percent by volume in air) Lower	Upper
Acetaldehyde	−37.8	21.1	175.0	4.0	60.0
Acetone	−19.0	56.0	538.0	2.6	12.8
Benzene	−11.1	80.1	560.0	1.4	8.0
Carbon disulfide	−30.0	45.8	90.0	1.0	44.0
Cyclohexane	−18.0	80.7	260.0	1.3	8.0
Diethyl ether	−45.0	34.4	160.0	1.8	48.0
Ethanol	12.0	78.3	363.0	3.3	19.0
n–Heptane	−3.9	98.4	204.0	1.0	6.7
n–Hexane	−21.7	68.7	223.0	1.2	7.5
Isopropyl alcohol	11.7	82.2	398.9	2.0	12.0
Methanol	11.1	64.5	385.0	6.0	36.5
Methyl ethyl ketone	−6.1	79.6	515.6	1.9	11.0
Pentane	−40.0	36.1	260.0	1.4	7.8
Styrene	31.0	145.0	490.0	1.1	6.1
Toluene	4.4	110.6	530.0	1.3	7.0
p–Xylene	25.0	132.4	529.0	1.1	7.0

Note: For a more extensive listing, see the table "Properties of Common Solvents" in Section 15.

TABLE 14
Resistance to Chemicals of Common Glove Materials
(E = Excellent, G = Good, F = Fair, P = Poor)

Chemical	Natural rubber	Neoprene	Nitrile	Vinyl
Acetaldehyde	G	G	E	G
Acetic acid	E	E	E	E
Acetone	G	G	G	F
Acrylonitrile	P	G	—	F
Ammonium hydroxide (sat)	G	E	E	E
Aniline	F	G	E	G
Benzaldehyde	F	F	E	G
Benzene[a]	P	F	G	F
Benzyl chloride[a]	F	P	G	P
Bromine	G	G	—	G
Butane	P	E	—	P
Butyraldehyde	P	G	—	G
Calcium hypochlorite	P	G	G	G
Carbon disulfide	P	P	G	F
Carbon tetrachloride[a]	P	F	G	F
Chlorine	G	G	—	G
Chloroacetone	F	E	—	P
Chloroform[a]	P	F	G	P
Chromic acid	P	F	F	E
Cyclohexane	F	E	—	P
Dibenzyl ether	F	G	—	P
Dibutyl phthalate	F	G	—	P
Diethanolamine	F	E	—	E
Diethyl ether	F	G	E	P
Dimethyl sulfoxide[b]	—	—	—	—
Ethyl acetate	F	G	G	F
Ethylene dichloride[a]	P	F	G	
Ethylene glycol	G	G	E	E
Ethylene trichloride[a]	P	P	—	P
Fluorine	G	G	—	G
Formaldehyde	G	E	E	E
Formic acid	G	E	E	E
Glycerol	G	G	E	E
Hexane	P	E	—	P
Hydrobromic acid (40%)	G	E	—	E
Hydrochloric acid (conc)	G	G	G	E
Hydrofluoric acid (30%)	G	G	G	E
Hydrogen peroxide	G	G	G	E
Iodine	G	G	—	G
Methylamine	G	G	E	E
Methyl cellosolve	F	E	—	P
Methyl chloride[a]	P	E	—	P
Methyl ethyl ketone	F	G	G	P
Methylene chloride[a]	F	F	G	F
Monoethanolamine	F	E	—	E
Morpholine	F	E	—	E
Naphthalene[a]	G	G	E	G
Nitric acid (conc)	P	P	P	G
Perchloric acid	F	G	F	E
Phenol	G	E	—	E
Phosphoric acid	G	E	—	E
Potassium hydroxide (sat)	G	G	G	E

TABLE 14
Resistance to Chemicals of Common Glove Materials
(E = Excellent, G = Good, F = Fair, P = Poor) (continued)

Chemical	Natural rubber	Neoprene	Nitrile	Vinyl
Propylene dichloride[a]	P	F	—	P
Sodium hydroxide	G	G	G	E
Sodium hypochlorite	G	P	F	G
Sulfuric acid (conc)	G	G	F	G
Toluene[a]	P	F	G	F
Trichloroethylene[a]	P	F	G	F
Tricresyl phosphate	P	F	—	F
Triethanolamine	F	E	E	E
Trinitrotoluene	P	E	—	P

[a] Aromatic and halogenated hydrocarbons will attack all types of natural and synthetic glove materials. Should swelling occur, the user should change to fresh gloves and allow the swollen gloves to dry and return to normal.

[b] No data on the resistance to dimethyl sulfoxide of natural rubber, neoprene, nitrile rubber, or vinyl materials are available; the manufacturer of the substance recommends the use of butyl rubber gloves.

TABLE 15
Guide for Selection of Respirators

Type of hazard	Type of respirator
Oxygen deficiency	Self–contained breathing apparatus Hose mask with blower Combination of air–line respirator and auxiliary self–contained air supply or air–storage receiver with alarm

Gas and vapor contaminants
  Immediately dangerous to
    life or health
	Self–contained breathing apparatus
	Hose mask with blower
	Air–purifying full–facepiece respirator with chemical canister (gas mask)
	Self–rescue mouthpiece respirator (for escape only)
	Combination of air–line respirator and auxiliary self–contained air supply or air–storage receiver with alarm

Not immediately dangerous to life or health	Air–line respirator Hose mask with blower Air–purifying half–mask or mouthpiece respirator with chemical cartridge

Particulate Contaminants
  Immediately dangerous to
    life or health
	Self–contained breathing apparatus
	Hose mask with blower
	Air–purifying full–facepiece respirator with appropriate filter
	Self-rescue mouthpiece respirator (for escape only)
	Combination of air–line respirator and auxiliary self–contained air supply or air–storage receiver with alarm

TABLE 15
Guide for Selection of Respirators (continued)

Type of hazard	Type of respirator
Not immediately dangerous to life or health	Air–purifying half–mask or mouthpiece respirator with filter pad or cartridge Air–line respirator Air–line abrasive–blasting respirator Hose mask with blower
Combination of gas, vapor, and particulate contaminants Immediately dangerous to life or health	Self–contained breathing apparatus Hose mask with blower Air–purifying full–facepiece respirator with chemical canister and appropriate filter (gas mask with filter) Self–rescue mouthpiece respirator (for escape only) Combination of air–line respirator and auxiliary self–contained air supply or air–storage receiver with alarm
Not immediately dangerous to life or health	Air–line respirator Hose mask without blower Air–purifying half–mask or mouthpiece respirator with chemical cartridge and appropriate filter

Source: ANSI Standard Z88.2 (1969).

FLAMMABILITY OF CHEMICAL SUBSTANCES

This table gives properties related to the flammability of about 900 chemical substances. The properties listed are:

t_B : Normal boiling point in °C (at 101.325 kPa pressure).

FP: Flash point, which is the minimum temperature at which the vapor pressure of a liquid is sufficient to form an ignitable mixture with air near the surface of the liquid. Flash point is not an intrinsic physical property but depends on the conditions of measurement (see Reference 1).

Fl. Limits: Flammable limits (often called explosive limits), which specify the range of concentration of the vapor in air (in percent by volume) for which a flame can propagate. Below the lower flammable limit, the gas mixture is too lean to burn; above the upper flammable limit, the mixture is too rich. Values refer to ambient temperature and pressure and are dependent on the precise test conditions. A ? indicates that one of the limits is not known.

IT: Ignition temperature (sometimes called autoignition temperature), which is the minimum temperature required for self-sustained combustion in the absence of an external ignition source. As in the case of flash point, the value depends on specified test conditions.

Even in cases where very careful measurements of flash point have been replicated in several laboratories, observed values can differ by 3 to 6°C (Reference 4). For more typical measurements, larger uncertainties should be assumed in both flash points and autoignition temperatures. The absence of a flash point entry in this table does not mean that the substance is nonflammable, but only that no reliable value is available.

Compounds are listed by molecular formula following the Hill convention. Substances not containing carbon are listed first, followed by those that contain carbon. To locate an organic compound by name or CAS Registry Number when the molecular formula is not known, use the table "Physical Constants of Organic Compounds" in Section 3 and its indexes to determine the molecular formula.

REFERENCES

1. *Fire Protection Guide to Hazardous Materials, 11th Edition*, National Fire Protection Association, Quincy, MA, 1994.
2. Urben, P.G., Editor, *Bretherick's Handbook of Reactive Chemical Hazards, 5th Edition*, Butterworth-Heinemann, Oxford, 1995.
3. Daubert, T.E., Danner, R.P., Sibul, H.M., and Stebbins, C.C., *Physical and Thermodynamic Properties of Pure Compounds: Data Compilation*, extant 1994 (core with 4 supplements), Taylor & Francis, Bristol, PA.
4. *Report of Investigation: Flash Point Reference Materials*, National Institute of Standards and Technology, Standard Reference Materials Program, Gaithersburg, MD, 1995.

Mol. Form.	Name	t_B/°C	FP/°C	Fl. Limits	IT/°C
Compounds not containing carbon					
B_2H_6	Diborane	-92.4	-90	1-98%	≈40
B_5H_9	Pentaborane(9)	60	30	0.4-?	35
BrH_3Si	Bromosilane	1.9	<0		≈20
Br_3HSi	Tribromosilane	109			≈20
Cl_2H_2Si	Dichlorosilane	8.3		4.1-99%	36
Cl_3HSi	Trichlorosilane	33	-50		104
GeH_4	Germane	-88.1			≈20
Ge_2H_6	Digermane	30.8			≈50
H_2	Hydrogen	-252.8		4-74%	
H_2S	Hydrogen sulfide	-59.55		4-44%	260
H_2S_2	Hydrogen disulfide	70.7	<22		
H_2Te	Hydrogen telluride	-2			-50
H_3N	Ammonia	-33.33		16-25%	
H_3P	Phosphine	-87.75		1.8-?	
H_4N_2	Hydrazine	113.55	38	5-100%	
H_4P_2	Diphosphine	63.5			≈20
H_4Si	Silane	-111.9	-112	1.4-?	≈20
H_6Si_2	Disilane	-14.3	-14		≈20
H_8Si_3	Trisilane	52.9	<0		≈20
P	Phosphorus (white)	280.5			38
Compounds containing carbon					
CHN	Hydrogen cyanide	26	-18	6-40%	538
CH_2Cl_2	Dichloromethane	40		13-23%	556
CH_2N_2	Cyanamide		141		

Mol. Form	Name	$t_B/°C$	FP/°C	Fl. Limits	IT/°C
CH_2O	Formaldehyde	-19.1	85	7.0-73%	424
$(CH_2O)_x$	Paraformaldehyde		70	7.0-73%	300
CH_2O_2	Formic acid	101	50	18-57%	434
CH_3Br	Bromomethane	3.5		10-16%	537
CH_3Cl	Chloromethane	-24.0		8.1-17.4%	632
CH_3Cl_3Si	Methyltrichlorosilane	65.6	-9	7.6->20%	>404
CH_3NO	Formamide	220	154		
CH_3NO_2	Nitromethane	101.1	35	7.3-?	418
CH_4	Methane	-161.5		5.0-15.0%	537
CH_4Cl_2Si	Dichloromethylsilane	41	-9	6.0-55%	316
CH_4O	Methanol	64.6	11	6.0-36%	464
CH_4S	Methanethiol	5.9	-18	3.9-21.8%	
CH_5N	Methylamine	-6.3	0	4.9-20.7%	430
CH_6N_2	Methylhydrazine	87.5	-8	2.5-92%	194
CO	Carbon monoxide	-191.5		12.5-74%	609
COS	Carbon oxysulfide	-50		12-29%	
CS_2	Carbon disulfide	46	-30	1.3-50.0%	90
C_2ClF_3	Chlorotrifluoroethylene	-27.8		8.4-16.0%	
C_2F_4	Tetrafluoroethylene	-75.9		10.0-50.0%	200
C_2HCl_3	Trichloroethylene	87.2		8-10.5%	420
C_2HCl_3O	Dichloroacetyl chloride	108	66		
C_2H_2	Acetylene	-84.7		2.5-100%	305
$C_2H_2Cl_2$	1,1-Dichloroethylene	31.6	-28	6.5-15.5%	570
$C_2H_2Cl_2$	cis-1,2-Dichloroethylene	60.1	6	3-15%	460
$C_2H_2Cl_2$	trans-1,2-Dichloroethylene	48.7	2	6-13%	460
$C_2H_2F_2$	1,1-Difluoroethylene	-85.7		5.5-21.3%	
C_2H_3Br	Bromoethylene	15.8		9-15%	530
C_2H_3Cl	Chloroethylene	-13.3	-78	3.6-33.0%	472
$C_2H_3ClF_2$	1-Chloro-1,1-difluoroethane	-9.7		6-18%	632
C_2H_3ClO	Acetyl chloride	50.7	4		390
$C_2H_3Cl_2NO_2$	1,1-Dichloro-1-nitroethane	123.5	76		
$C_2H_3Cl_3$	1,1,1-Trichloroethane	74.0		8-10.5%	500
$C_2H_3Cl_3$	1,1,2-Trichloroethane	113.8	32	6-28%	460
$C_2H_3Cl_3Si$	Trichlorovinylsilane	91.5	21		
C_2H_3F	Fluoroethylene	-72		2.6-21.7%	
C_2H_3N	Acetonitrile	81.6	6	3.0-16.0%	524
C_2H_3NO	Methyl isocyanate	39.5	-7	5.3-26%	534
C_2H_4	Ethylene	-103.7		2.7-36%	450
$C_2H_4ClNO_2$	1-Chloro-1-nitroethane	124.5	56		
$C_2H_4Cl_2$	1,1-Dichloroethane	57.4	-17	5.4-11.4%	458
$C_2H_4Cl_2$	1,2-Dichloroethane	83.5	13	6.2-16%	413
C_2H_4O	Acetaldehyde	20.1	-39	4.0-60%	175
C_2H_4O	Ethylene oxide	10.6	-20	3.0-100%	429
$C_2H_4O_2$	Acetic acid	117.9	39	4.0-19.9%	463
$C_2H_4O_2$	Methyl formate	31.7	-19	4.5-23%	449
$C_2H_4O_3$	Ethaneperoxoic acid	110	41		
C_2H_5Br	Bromoethane	38.5		6.8-8.0%	511
C_2H_5Cl	Chloroethane	12.3	-50	3.8-15.4%	519
C_2H_5ClO	Ethylene chlorohydrin	128.6	60	4.9-15.9%	425
$C_2H_5Cl_3Si$	Trichloroethylsilane	100.5	22		
C_2H_5N	Ethyleneimine	56	-11	3.3-54.8%	320
$C_2H_5NO_2$	Nitroethane	114.0	28	3.4-17%	414
$C_2H_5NO_2$	Ethyl nitrite	18	-35	4.0-50%	90
$C_2H_5NO_3$	Ethyl nitrate	87.2	10	4-?	
C_2H_6	Ethane	-88.6		3.0-12.5%	472
$C_2H_6Cl_2Si$	Dichlorodimethylsilane	70.3	<21	3.4-9.5%	
C_2H_6O	Ethanol	78.2	13	3.3-19%	363
C_2H_6O	Dimethyl ether	-24.8	-41	3.4-27.0%	350
C_2H_6OS	2-Mercaptoethanol	158	74		
C_2H_6OS	Dimethyl sulfoxide	189	95	2.6-42%	215
$C_2H_6O_2$	Ethylene glycol	197.3	111	3.2-22%	398

Mol. Form.	Name	$t_B/°C$	FP/°C	Fl. Limits	IT/°C
$C_2H_6O_4S$	Dimethyl sulfate		83		188
C_2H_6S	Ethanethiol	35.1	-17	2.8-18.0%	300
C_2H_6S	Dimethyl sulfide	37.3	-37	2.2-19.7%	206
$C_2H_6S_2$	Dimethyl disulfide	109.8	24		
C_2H_7N	Ethylamine	16.5	-16	3.5-14%	385
C_2H_7N	Dimethylamine	6.8	20	2.8-14.4%	400
C_2H_7NO	Ethanolamine	171	86	3.0-23.5%	410
$C_2H_8N_2$	1,2-Ethanediamine	117	40	2.5-12.0%	385
$C_2H_8N_2$	1,1-Dimethylhydrazine	63.9	-15	2-95%	249
C_2N_2	Cyanogen	-21.1		6.6-32%	
C_3H_3Br	3-Bromo-1-propyne	89	10	3.0-?	324
C_3H_3N	2-Propenenitrile	77.3	0	3.0-17.0%	481
C_3H_4	Propyne	-23.2		2.1-12.5%	
C_3H_4ClN	3-Chloropropanenitrile	175.5	76		
$C_3H_4Cl_2$	2,3-Dichloropropene	94	15	2.6-7.8%	
C_3H_4O	Propargyl alcohol	113.6	36		
C_3H_4O	Acrolein	52.6	-26	2.8-31%	220
$C_3H_4O_2$	Propenoic acid	141	50	2.4-8.0%	438
$C_3H_4O_2$	2-Oxetanone	162	74	2.9-?	
$C_3H_4O_3$	Ethylene carbonate	248	143		
C_3H_5Br	3-Bromopropene	70.1	-1	4.4-7.3%	295
C_3H_5Cl	2-Chloropropene	22.6	-37	4.5-16%	
C_3H_5Cl	3-Chloropropene	45.1	-32	2.9-11.1%	485
C_3H_5ClO	Epichlorohydrin	118	31	3.8-21.0%	411
C_3H_5ClO	Propanoyl chloride	80	12		
$C_3H_5ClO_2$	2-Chloropropanoic acid	185	107		500
$C_3H_5ClO_2$	Ethyl chloroformate	95	16		500
$C_3H_5ClO_2$	Methyl chloroacetate	129.5	57	7.5-18.5%	
$C_3H_5Cl_2NO_2$	1,1-Dichloro-1-nitropropane	145	66		
$C_3H_5Cl_3$	1,2,3-Trichloropropane	157	71	3.2-12.6%	
$C_3H_5Cl_3Si$	Trichloro-2-propenylsilane	117.5	35		
C_3H_5N	Propanenitrile	97.1	2	3.1-14%	512
C_3H_5NO	3-Hydroxypropanenitrile	221	129		
$C_3H_5N_3O_9$	Trinitroglycerol				270
C_3H_6	Propene	-47.6		2.0-11.1%	455
C_3H_6	Cyclopropane	-32.8		2.4-10.4%	498
$C_3H_6ClNO_2$	1-Chloro-1-nitropropane	142	62		
$C_3H_6ClNO_2$	2-Chloro-2-nitropropane		57		
$C_3H_6Cl_2$	1,2-Dichloropropane	96.4	21	3.4-14.5%	557
$C_3H_6Cl_2O$	1,3-Dichloro-2-propanol	176	74		
$C_3H_6N_2$	Dimethylcyanamide	163.5	71		
C_3H_6O	Allyl alcohol	97.0	21	2.5-18.0%	378
C_3H_6O	Methyl vinyl ether	5.5			287
C_3H_6O	Propanal	48	-30	2.6-17%	207
C_3H_6O	Acetone	56.0	-20	2.5-12.8%	465
C_3H_6O	Methyloxirane	35	-37	3.1-27.5%	449
$C_3H_6O_2$	Propanoic acid	141.1	52	2.9-12.1%	465
$C_3H_6O_2$	Ethyl formate	54.4	-20	2.8-16.0%	455
$C_3H_6O_2$	Methyl acetate	56.8	-10	3.1-16%	454
$C_3H_6O_2$	1,3-Dioxolane	78	2		
$C_3H_6O_3$	Dimethyl carbonate	90.5	19		
$C_3H_6O_3$	1,3,5-Trioxane	114.5	45	3.6-29%	414
C_3H_7Br	1-Bromopropane	71.1			490
C_3H_7Cl	1-Chloropropane	46.5	<-18	2.6-11.1%	520
C_3H_7Cl	2-Chloropropane	35.7	-32	2.8-10.7%	593
C_3H_7ClO	2-Chloro-1-propanol	133.5	52		
C_3H_7ClO	1-Chloro-2-propanol	127	52		
$C_3H_7Cl_3Si$	Trichloropropylsilane	123.5	37		
C_3H_7N	Allylamine	53.3	-29	2.2-22%	374
C_3H_7NO	*N*,*N*-Dimethylformamide	153	58	2.2-15.2%	445
$C_3H_7NO_2$	1-Nitropropane	131.1	36	2.2-?	421

Mol. Form.	Name	t_B/°C	FP/°C	Fl. Limits	IT/°C
C$_3$H$_7$NO$_2$	2-Nitropropane	120.2	24	2.6-11.0%	428
C$_3$H$_7$NO$_3$	Propyl nitrate	110	20	2-100%	175
C$_3$H$_8$	Propane	-42.1	-104	2.1-9.5%	450
C$_3$H$_8$O	1-Propanol	97.2	23	2.2-13.7%	412
C$_3$H$_8$O	2-Propanol	82.3	12	2.0-12.7%	399
C$_3$H$_8$O	Ethyl methyl ether	7.4	-37	2.0-10.1%	190
C$_3$H$_8$O$_2$	1,2-Propylene glycol	187.6	99	2.6-12.5%	371
C$_3$H$_8$O$_2$	1,3-Propylene glycol	214.4			400
C$_3$H$_8$O$_2$	Ethylene glycol monomethyl ether	124.1	39	1.8-14%	285
C$_3$H$_8$O$_2$	Dimethoxymethane	42	-32	2.2-13.8%	237
C$_3$H$_8$O$_3$	Glycerol	290	199	3-19%	370
C$_3$H$_9$BO$_3$	Trimethyl borate	67.5	-8		
C$_3$H$_9$ClSi	Trimethylchlorosilane	60	-28		395
C$_3$H$_9$N	Propylamine	47.2	-37	2.0-10.4%	318
C$_3$H$_9$N	Isopropylamine	31.7	-37		402
C$_3$H$_9$N	Trimethylamine	2.8	-5	2.0-11.6%	190
C$_3$H$_9$NO	3-Amino-1-propanol	187.5	80		
C$_3$H$_9$NO	1-Amino-2-propanol	159.4	77		374
C$_3$H$_9$NO	N-Methyl-2-ethanolamine	158	74		
C$_3$H$_9$O$_3$P	Trimethyl phosphite	111.5	54		
C$_3$H$_9$O$_4$P	Trimethyl phosphate	197.2	107		
C$_3$H$_{10}$N$_2$	1,3-Propanediamine	139.8	24		
C$_4$Cl$_6$	Hexachloro-1,3-butadiene	215			610
C$_4$H$_2$O$_3$	Maleic anhydride	202	102	1.4-7.1%	477
C$_4$H$_4$	1-Buten-3-yne	5.1		21-100%	
C$_4$H$_4$N$_2$	Succinonitrile	266	132		
C$_4$H$_4$O	Furan	31.5	-36	2.3-14.3%	
C$_4$H$_4$O$_2$	Diketene	126.1	34		
C$_4$H$_4$S	Thiophene	84.0	-1		
C$_4$H$_5$Cl	2-Chloro-1,3-butadiene	59.4	-20	4.0-20.0%	
C$_4$H$_5$N	2-Butenenitrile	120.5	16		
C$_4$H$_5$N	Methylacrylonitrile	90.3	1	2-6.8%	
C$_4$H$_5$N	Pyrrole	129.7	39		
C$_4$H$_6$	1,3-Butadiene	-4.4		2.0-12.0%	420
C$_4$H$_6$	2-Butyne	26.9	-31	1.4-?	
C$_4$H$_6$O	Divinyl ether	28.3	<-30	1.7-27%	360
C$_4$H$_6$O	Ethoxyacetylene	50	<-7		
C$_4$H$_6$O	trans-2-Butenal	102.2	13	2.1-15.5%	232
C$_4$H$_6$O	3-Buten-2-one	81.4	-7	2.1-15.6%	491
C$_4$H$_6$O	Vinyloxirane	68	<-50		
C$_4$H$_6$O$_2$	Methacrylic acid	162.5	77	1.6-8.8%	68
C$_4$H$_6$O$_2$	Vinyl acetate	72.5	-8	2.6-13.4%	402
C$_4$H$_6$O$_2$	Methyl acrylate	80.7	-3	2.8-25%	468
C$_4$H$_6$O$_2$	2,3-Butanedione	88	27		
C$_4$H$_6$O$_2$	gamma-Butyrolactone	204	98		
C$_4$H$_6$O$_3$	Acetic anhydride	139.5	49	2.7-10.3%	316
C$_4$H$_6$O$_3$	Propylene carbonate	242	135		
C$_4$H$_6$O$_6$	L-Tartaric acid		210		425
C$_4$H$_7$Br	1-Bromo-2-butene	104.5		4.6-12.0%	
C$_4$H$_7$BrO$_2$	Ethyl bromoacetate	168.5	48		
C$_4$H$_7$Cl	2-Chloro-1-butene	58.5	-19	2.3-9.3%	
C$_4$H$_7$Cl	3-Chloro-2-methylpropene	71.5	-12	3.2-8.1%	
C$_4$H$_7$ClO	2-Chloroethyl vinyl ether	108	27		
C$_4$H$_7$ClO$_2$	Ethyl chloroacetate	144.3	64		
C$_4$H$_7$N	Butanenitrile	117.6	24	1.6-?	501
C$_4$H$_7$N	2-Methylpropanenitrile	103.9	8		482
C$_4$H$_7$NO	Acetone cyanohydrin		74	2.2-12.0%	688
C$_4$H$_7$NO	2-Pyrrolidone	251	129		
C$_4$H$_8$	1-Butene	-6.2		1.6-10.0%	385
C$_4$H$_8$	cis-2-Butene	3.7		1.7-9.0%	325
C$_4$H$_8$	trans-2-Butene	0.8		1.8-9.7%	324

Mol. Form.	Name	$t_B/°C$	FP/°C	Fl. Limits	IT/°C
C_4H_8	Isobutene	-6.9		1.8-9.6%	465
C_4H_8	Cyclobutane	12.6	<10	1.8-?	
$C_4H_8Cl_2$	1,2-Dichlorobutane	124.1			275
$C_4H_8Cl_2$	1,4-Dichlorobutane	161	52		
$C_4H_8Cl_2O$	Bis(2-chloroethyl) ether	178.5	55	2.7-?	369
C_4H_8O	2-Buten-1-ol	121.5	27	4.2-35.3%	349
C_4H_8O	2-Methyl-2-propenol	114.5	33		
C_4H_8O	Ethyl vinyl ether	35.5	<-46	1.7-28%	202
C_4H_8O	1,2-Epoxybutane	63.4	-22	1.7-19%	439
C_4H_8O	Butanal	74.8	-22	1.9-12.5%	218
C_4H_8O	Isobutanal	64.5	-18	1.6-10.6%	196
C_4H_8O	2-Butanone	79.5	-9	1.4-11.4%	404
C_4H_8O	Tetrahydrofuran	65	-14	2-11.8%	321
C_4H_8OS	1,4-Oxathiane	147	42		
$C_4H_8O_2$	Butanoic acid	163.7	72	2.0-10.0%	443
$C_4H_8O_2$	2-Methylpropanoic acid	154.4	56	2.0-9.2%	481
$C_4H_8O_2$	Propyl formate	80.9	-3		455
$C_4H_8O_2$	Isopropyl formate	68.2	-6		485
$C_4H_8O_2$	Ethyl acetate	77.1	-4	2.0-11.5%	426
$C_4H_8O_2$	Methyl propanoate	79.8	-2	2.5-13%	469
$C_4H_8O_2$	3-Hydroxybutanal		66		250
$C_4H_8O_2$	1,4-Dioxane	101.5	12	2.0-22%	180
$C_4H_8O_2S$	Sulfolane	287.3	177		
$C_4H_8O_3$	Methyl lactate	144.8	49	2.2-?	385
$C_4H_8O_3$	Ethylene glycol monoacetate	188	102		
C_4H_9Br	1-Bromobutane	101.6	18	2.6-6.6%	265
C_4H_9Br	2-Bromobutane	91.2	21		
C_4H_9Cl	1-Chlorobutane	78.6	-12	1.9-10.1%	240
C_4H_9Cl	2-Chlorobutane	68.2	-10		
C_4H_9Cl	1-Chloro-2-methylpropane	68.5	-6	2.0-8.7%	
C_4H_9Cl	2-Chloro-2-methylpropane	50.9	0		
$C_4H_9Cl_3Si$	Butyltrichlorosilane	148.5	54		
C_4H_9N	Pyrrolidine	86.5	3		
C_4H_9NO	N-Ethylacetamide	205	110		
C_4H_9NO	N,N-Dimethylacetamide	165	70	1.8-11.5%	490
C_4H_9NO	Butanal oxime	154	58		
C_4H_9NO	2-Butanone oxime	152.5	≈70		
C_4H_9NO	Morpholine	128	37	1.4-11.2%	290
$C_4H_9NO_2$	N-Acetylethanolamine		179		460
$C_4H_9NO_3$	Butyl nitrate	133	36		
C_4H_{10}	Butane	-0.5	-60	1.9-8.5%	287
C_4H_{10}	Isobutane	-11.7	-87	1.8-8.4%	460
$C_4H_{10}N_2$	Piperazine	146	81		
$C_4H_{10}O$	1-Butanol	117.7	37	1.4-11.2%	343
$C_4H_{10}O$	2-Butanol	99.5	24	1.7-9.8%	405
$C_4H_{10}O$	2-Methyl-1-propanol	107.8	28	1.7-10.6%	415
$C_4H_{10}O$	2-Methyl-2-propanol	82.4	11	2.4-8.0%	478
$C_4H_{10}O$	Diethyl ether	34.5	-45	1.9-36.0%	180
$C_4H_{10}O$	Methyl propyl ether	39.1	-20	2.0-14.8%	
$C_4H_{10}O_2$	1,2-Butanediol	190.5	40		
$C_4H_{10}O_2$	1,3-Butanediol	207.5	121		395
$C_4H_{10}O_2$	1,4-Butanediol	235	121		
$C_4H_{10}O_2$	2,3-Butanediol	182.5			402
$C_4H_{10}O_2$	Ethylene glycol monoethyl ether	135	43	3-18%	235
$C_4H_{10}O_2$	Ethylene glycol dimethyl ether	85	-2		202
$C_4H_{10}O_2$	tert-Butyl hydroperoxide		27		
$C_4H_{10}O_2S$	2,2'-Thiodiethanol	282	160		298
$C_4H_{10}O_3$	Diethylene glycol	245.8	124	2-17%	224
$C_4H_{10}O_4S$	Diethyl sulfate	208	104		436
$C_4H_{10}S$	1-Butanethiol	98.5	2		
$C_4H_{10}S$	2-Butanethiol	85	-23		

Mol. Form.	Name	$t_B/°C$	FP/°C	Fl. Limits	IT/°C
$C_4H_{10}S$	2-Methyl-1-propanethiol	88.5	2		
$C_4H_{10}S$	2-Methyl-2-propanethiol	64.3	<-29		
$C_4H_{10}Se$	Diethyl selenide	108		2.5-?	
$C_4H_{11}N$	Butylamine	77.0	-12	1.7-9.8%	312
$C_4H_{11}N$	sec-Butylamine	63.5	-9		
$C_4H_{11}N$	tert-Butylamine	44.0	-9	1.7-8.9%	380
$C_4H_{11}N$	Isobutylamine	67.7	-9	2-12%	378
$C_4H_{11}N$	Diethylamine	55.5	-23	1.8-10.1%	312
$C_4H_{11}NO$	2-Amino-1-butanol	178	74		
$C_4H_{11}NO$	2-Amino-2-methyl-1-propanol	165.5	67		
$C_4H_{11}NO_2$	Diethanolamine	268.8	172	2-13%	662
$C_4H_{12}Sn$	Tetramethylstannane	78	-12	1.9-?	
$C_4H_{13}N_3$	Diethylenetriamine	207	98	2-6.7%	358
$C_5H_4O_2$	Furfural	161.7	60	2.1-19.3%	316
C_5H_5N	Pyridine	115.2	20	1.8-12.4%	482
C_5H_6	2-Methyl-1-buten-3-yne	32	<-7		
$C_5H_6N_2$	2-Methylpyrazine	137	50		
C_5H_6O	3-Methylfuran	66	-30		
$C_5H_6O_2$	Furfuryl alcohol	171	75	1.8-16.3%	491
C_5H_7N	1-Methylpyrrole	115	16		
C_5H_7NO	2-Furanmethanamine	145.5	37		
$C_5H_7NO_2$	Ethyl cyanoacetate	205	110		
C_5H_8	2-Methyl-1,3-butadiene	34.0	-54	1.5-8.9%	395
C_5H_8	1-Pentyne	40.1	<-20		
C_5H_8	Cyclopentene	44.2	-29		395
C_5H_8O	3-Methyl-3-buten-2-one	98		1.8-9.0%	
C_5H_8O	Cyclopentanone	130.5	26		
C_5H_8O	3,4-Dihydro-2H-pyran	86	-18		
$C_5H_8O_2$	Allyl acetate	103.5	22		374
$C_5H_8O_2$	Isopropenyl acetate	94	26		432
$C_5H_8O_2$	Vinyl propanoate	91.2	1		
$C_5H_8O_2$	Ethyl acrylate	99.4	10	1.4-14%	372
$C_5H_8O_2$	Methyl methacrylate	100.5	10	1.7-8.2%	
$C_5H_8O_2$	2,4-Pentanedione	138	34		340
$C_5H_8O_3$	Methyl acetoacetate	171.7	77		280
C_5H_9NO	N-Methyl-2-pyrrolidone	202	96	1-10%	346
C_5H_{10}	1-Pentene	29.9	-18	1.5-8.7%	275
C_5H_{10}	cis-2-Pentene	36.9	<-20		
C_5H_{10}	trans-2-Pentene	36.3	<-20		
C_5H_{10}	2-Methyl-1-butene	31.2	-20		
C_5H_{10}	3-Methyl-1-butene	20.1	-7	1.5-9.1%	365
C_5H_{10}	2-Methyl-2-butene	38.5	-20		
C_5H_{10}	Cyclopentane	49.3	-25	1.5-?	361
$C_5H_{10}Cl_2$	1,5-Dichloropentane	179	>27		
$C_5H_{10}N_2$	3-(Dimethylamino)propanenitrile	173	65		
$C_5H_{10}O$	Cyclopentanol	140.4	51		
$C_5H_{10}O$	Pentanal	103	12		222
$C_5H_{10}O$	2-Pentanone	102.2	7	1.5-8.2%	452
$C_5H_{10}O$	3-Pentanone	101.9	13	1.6-?	450
$C_5H_{10}O$	Tetrahydropyran	88	-20		
$C_5H_{10}O$	2-Methyltetrahydrofuran	78	-11		
$C_5H_{10}O_2$	Pentanoic acid	186.1	96		400
$C_5H_{10}O_2$	3-Methylbutanoic acid	176.5			416
$C_5H_{10}O_2$	Butyl formate	106.1	18	1.7-8.2%	322
$C_5H_{10}O_2$	Isobutyl formate	98.2	5	2-9%	320
$C_5H_{10}O_2$	Propyl acetate	101.5	13	1.7-8%	450
$C_5H_{10}O_2$	Isopropyl acetate	88.6	2	1.8-8%	460
$C_5H_{10}O_2$	Ethyl propanoate	99.1	12	1.9-11%	440
$C_5H_{10}O_2$	Methyl butanoate	102.8	14		
$C_5H_{10}O_2$	3-Ethoxypropanal	135.2	38		
$C_5H_{10}O_2$	Tetrahydrofurfuryl alcohol	178	75	1.5-9.7%	282

Mol. Form.	Name	t_B/°C	FP/°C	Fl. Limits	IT/°C
$C_5H_{10}O_3$	Diethyl carbonate	126	25		
$C_5H_{10}O_3$	Ethylene glycol monomethyl ether acetate	143	49	1.5-12.3%	392
$C_5H_{10}O_3$	Ethyl lactate	154.5	46	1.5-?	400
$C_5H_{11}Br$	1-Bromopentane	129.8	32		
$C_5H_{11}Cl$	1-Chloropentane	107.8	13	1.6-8.6%	260
$C_5H_{11}Cl$	2-Chloro-2-methylbutane	85.6		1.5-7.4%	345
$C_5H_{11}Cl$	1-Chloro-3-methylbutane	98.9	<21	1.5-7.4%	
$C_5H_{11}Cl_3Si$	Trichloropentylsilane	172	63		
$C_5H_{11}N$	Piperidine	106.2	16		
$C_5H_{11}N$	N-Methylpyrrolidine	81	-14		
$C_5H_{11}NO$	4-Methylmorpholine	116	24		
$C_5H_{11}NO_2$	Isopentyl nitrite	99.2			210
C_5H_{12}	Pentane	36.0	-40	1.4-8.0%	260
C_5H_{12}	Isopentane	27.8	-51	1.4-7.6%	420
C_5H_{12}	Neopentane	9.4	-65	1.4-7.5%	450
$C_5H_{12}N_2$	1-Methylpiperazine	138	42		
$C_5H_{12}N_2O$	Tetramethylurea	176.5	77		
$C_5H_{12}O$	1-Pentanol	137.9	33	1.2-10.0%	300
$C_5H_{12}O$	2-Pentanol	119.3	34	1.2-9.0%	343
$C_5H_{12}O$	3-Pentanol	116.2	41	1.2-9.0%	435
$C_5H_{12}O$	2-Methyl-1-butanol	128	50		385
$C_5H_{12}O$	3-Methyl-1-butanol	131.1	43	1.2-9.0%	350
$C_5H_{12}O$	2-Methyl-2-butanol	102.4	19	1.2-9.0%	437
$C_5H_{12}O$	3-Methyl-2-butanol	112.9	38		
$C_5H_{12}O$	2,2-Dimethyl-1-propanol	113.5	37		
$C_5H_{12}O$	Ethyl propyl ether	63.2	<-20	1.7-9.0%	
$C_5H_{12}O_2$	1,5-Pentanediol	239	129		335
$C_5H_{12}O_2$	2-Isopropoxyethanol	145	33		
$C_5H_{12}O_2$	2,2-Dimethyl-1,3-propanediol	208	129		399
$C_5H_{12}O_3$	Diethylene glycol monomethyl ether	193	96	1.38-22.7%	240
$C_5H_{12}S$	1-Pentanethiol	126.6	18		
$C_5H_{12}S$	3-Methyl-2-butanethiol		3		
$C_5H_{13}N$	Pentylamine	104.3	-1	2.2-22%	
$C_5H_{13}N$	Butylmethylamine	91	13		
$C_6H_2Cl_4$	1,2,4,5-Tetrachlorobenzene	244.5	155		
$C_6H_3ClN_2O_4$	1-Chloro-2,4-dinitrobenzene	315	194	2.0-22%	
$C_6H_3Cl_3$	1,2,4-Trichlorobenzene	213.5	105	2.5-6.6%	571
$C_6H_4ClNO_2$	1-Chloro-4-nitrobenzene	242	127		
$C_6H_4Cl_2$	o-Dichlorobenzene	180	66	2.2-9.2%	648
$C_6H_4Cl_2$	m-Dichlorobenzene	173	72		
$C_6H_4Cl_2$	p-Dichlorobenzene	174	66		
$C_6H_4Cl_2O$	2,4-Dichlorophenol	210	114		
C_6H_5Br	Bromobenzene	156.0	51		565
C_6H_5Cl	Chlorobenzene	131.7	28	1.3-9.6%	593
C_6H_5ClO	o-Chlorophenol	174.9	64		
C_6H_5ClO	p-Chlorophenol	220	121		
$C_6H_5Cl_2N$	3,4-Dichloroaniline	272	166		
$C_6H_5Cl_3Si$	Trichlorophenylsilane	201	91		
C_6H_5F	Fluorobenzene	84.7	-15		
$C_6H_5NO_2$	Nitrobenzene	210.8	88	1.8-?	482
$C_6H_5N_3O_4$	2,4-Dinitroaniline		224		
C_6H_6	1,5-Hexadien-3-yne	85	<-20	1.5-?	
C_6H_6	Benzene	80.0	-11	1.2-7.8%	498
$C_6H_6N_2O_2$	p-Nitroaniline	332	199		
C_6H_6O	Phenol	181.8	79	1.8-8.6%	715
$C_6H_6O_2$	1,2-Benzenediol	245	127		
$C_6H_6O_2$	Resorcinol		127	1.4-?	608
$C_6H_6O_2$	p-Hydroquinone	287	165		516
C_6H_7N	Aniline	184.1	70	1.3-11%	615
C_6H_7N	2-Methylpyridine	129.3	39		538
C_6H_7N	4-Methylpyridine	145.3	57		

Mol. Form.	Name	t_B/°C	FP/°C	Fl. Limits	IT/°C
C₆H₈ClN	Aniline, hydrochloride		193		
C₆H₈Cl₂O₂	Hexanedioyl dichloride		72		
C₆H₈N₂	Adiponitrile	295	93	1.0-?	550
C₆H₈N₂	o-Phenylenediamine	257	156	1.5-?	
C₆H₈N₂	Phenylhydrazine	243.5	88		
C₆H₈N₂	2,5-Dimethylpyrazine	155	64		
C₆H₈O	2,5-Dimethylfuran	93.5	7		
C₆H₈O₄	Dimethyl maleate	202	113		
C₆H₁₀	1,4-Hexadiene	65	-21	2.0-6.1%	
C₆H₁₀	2-Methyl-1,3-pentadiene	75.8	-12		
C₆H₁₀	4-Methyl-1,3-pentadiene	76.5	-34		
C₆H₁₀	2-Hexyne	84.5	-10		
C₆H₁₀	Cyclohexene	82.9	-12	1.2-?	310
C₆H₁₀O	Diallyl ether	94	-7		
C₆H₁₀O	Cyclohexanone	155.4	44	1.1-9.4%	420
C₆H₁₀O	Mesityl oxide	130	31	1.4-7.2%	344
C₆H₁₀O₂	Vinyl butanoate	116.7	20	1.4-8.8%	
C₆H₁₀O₂	Ethyl 2-butenoate	136.5	2		
C₆H₁₀O₂	Ethyl methacrylate	117	20		
C₆H₁₀O₂	2,5-Hexanedione	194	79		499
C₆H₁₀O₃	Ethyl acetoacetate	180.8	57	1.4-9.5%	295
C₆H₁₀O₃	Propanoic anhydride	170	63	1.3-9.5%	285
C₆H₁₀O₄	Adipic acid	337.5	196		420
C₆H₁₀O₄	Diethyl oxalate	185.7	76		
C₆H₁₀O₄	Ethylene glycol diacetate	190	88	1.6-8.4%	482
C₆H₁₁Cl	Chlorocyclohexane	142	32		
C₆H₁₁NO	Caprolactam	270	125		
C₆H₁₁NO₂	Nitrocyclohexane	205	88		
C₆H₁₁NO₂	4-Acetylmorpholine		113		
C₆H₁₂	1-Hexene	63.4	-26	1.2-6.9%	253
C₆H₁₂	cis-2-Hexene	68.8	-21		
C₆H₁₂	2-Methyl-1-pentene	62.1	-28		300
C₆H₁₂	4-Methyl-1-pentene	53.9	-7		300
C₆H₁₂	4-Methyl-cis-2-pentene	56.3	-32		
C₆H₁₂	4-Methyl-trans-2-pentene	58.6	-29		
C₆H₁₂	2-Ethyl-1-butene	64.7	<-20		315
C₆H₁₂	2,3-Dimethyl-1-butene	55.6	<-20		360
C₆H₁₂	2,3-Dimethyl-2-butene	73.3	<-20		401
C₆H₁₂	Cyclohexane	80.7	-20	1.3-8%	245
C₆H₁₂	Methylcyclopentane	71.8	-29	1.0-8.35%	258
C₆H₁₂	Ethylcyclobutane	70.8	-15	1.2-7.7%	210
C₆H₁₂	2-Methyl-2-pentene	67.3	<-7		
C₆H₁₂Cl₂O₂	1,2-Bis(2-chloroethoxy)ethane	232	121		
C₆H₁₂O	cis-3-Hexen-1-ol	156.5	54		
C₆H₁₂O	Butyl vinyl ether	94	-9		255
C₆H₁₂O	Isobutyl vinyl ether	83	-9		
C₆H₁₂O	Hexanal	131	32		
C₆H₁₂O	2-Ethylbutanal		21	1.2-7.7%	
C₆H₁₂O	2-Methylpentanal	117	17		199
C₆H₁₂O	2-Hexanone	127.6	25	1-8%	423
C₆H₁₂O	3-Hexanone	123.5	35	1-8%	
C₆H₁₂O	4-Methyl-2-pentanone	116.5	18	1.2-8.0%	448
C₆H₁₂O	Cyclohexanol	160.8	68	1-9%	300
C₆H₁₂O₂	Hexanoic acid	205.2	102		380
C₆H₁₂O₂	2-Methylpentanoic acid	195.6	107		378
C₆H₁₂O₂	Diethylacetic acid	194	99		400
C₆H₁₂O₂	Pentyl formate	130.4	26		
C₆H₁₂O₂	Butyl acetate	126.1	22	1.7-7.6%	425
C₆H₁₂O₂	sec-Butyl acetate	112	31	1.7-9.8%	
C₆H₁₂O₂	Isobutyl acetate	116.5	18	1.3-10.5%	421
C₆H₁₂O₂	Propyl propanoate	122.5	79		

Mol. Form.	Name	$t_B/°C$	FP/°C	Fl. Limits	IT/°C
$C_6H_{12}O_2$	Ethyl butanoate	121.5	24		463
$C_6H_{12}O_2$	Ethyl 2-methylpropanoate	110.1	13		
$C_6H_{12}O_2$	Diacetone alcohol	167.9	58	1.8-6.9%	643
$C_6H_{12}O_3$	Ethylene glycol monoethyl ether acetate	156.4	56	2-8%	379
$C_6H_{12}O_3$	Paraldehyde	124.3	36	1.3-?	238
$C_6H_{12}S$	Cyclohexanethiol	158.9	43		
$C_6H_{13}Cl$	1-Chlorohexane	135	35		
$C_6H_{13}N$	Cyclohexylamine	134	31	1.9-9.4%	293
$C_6H_{13}NO$	N-Butylacetamide	229	116		
$C_6H_{13}NO$	2,6-Dimethylmorpholine	146.6	44		
$C_6H_{13}NO$	N-Ethylmorpholine	138.5	32		
$C_6H_{13}NO_2$	4-Morpholineethanol	227	99		
C_6H_{14}	Hexane	68.7	-22	1.1-7.5%	225
C_6H_{14}	2-Methylpentane	60.2	<-29	1.0-7.0%	264
C_6H_{14}	3-Methylpentane	63.2	-7	1.2-7.0%	278
C_6H_{14}	2,2-Dimethylbutane	49.7	-48	1.2-7.0%	405
C_6H_{14}	2,3-Dimethylbutane	57.9	-29	1.2-7.0%	405
$C_6H_{14}N_2O$	1-Piperazineethanol	246	124		
$C_6H_{14}O$	1-Hexanol	157.6	63		
$C_6H_{14}O$	2-Methyl-1-pentanol	149	54	1.1-9.65%	310
$C_6H_{14}O$	4-Methyl-2-pentanol	131.6	41	1.0-5.5%	
$C_6H_{14}O$	2-Ethyl-1-butanol	147	57		
$C_6H_{14}O$	Dipropyl ether	90.0	21	1.3-7.0%	188
$C_6H_{14}O$	Diisopropyl ether	68.5	-28	1.4-7.9%	443
$C_6H_{14}O$	Butyl ethyl ether	92.3	4		
$C_6H_{14}O_2$	2,5-Hexanediol	218	110		
$C_6H_{14}O_2$	2-Methyl-2,4-pentanediol	197.1	102	1-9%	306
$C_6H_{14}O_2$	Ethylene glycol monobutyl ether	168.4	69	4-13%	238
$C_6H_{14}O_2$	1,1-Diethoxyethane	102.2	-21	1.6-10.4%	230
$C_6H_{14}O_2$	Ethylene glycol diethyl ether	119.4	27		205
$C_6H_{14}O_3$	1,2,6-Hexanetriol		191		
$C_6H_{14}O_3$	Diethylene glycol monoethyl ether	196	96		
$C_6H_{14}O_3$	Diethylene glycol dimethyl ether	162	67		
$C_6H_{14}O_3$	Trimethylolpropane		149		
$C_6H_{14}O_4$	Triethylene glycol	285	177	0.9-9.2%	371
$C_6H_{15}N$	Hexylamine	132.8	29		
$C_6H_{15}N$	Butylethylamine	107.5	18		
$C_6H_{15}N$	Dipropylamine	109.3	17		299
$C_6H_{15}N$	Diisopropylamine	83.9	-1	1.1-7.1%	316
$C_6H_{15}N$	Triethylamine	89	-7	1.2-8.0%	249
$C_6H_{15}NO_2$	Diisopropanolamine	250	127		374
$C_6H_{15}NO_3$	Triethanolamine	335.4	179	1-10%	
$C_6H_{15}N_3$	1-Piperazineethanamine	220	93		
$C_6H_{15}O_4P$	Triethyl phosphate	215.5	115		454
$C_6H_{16}N_2$	N,N-Diethylethylenediamine	144	46		
$C_7H_3ClF_3NO_2$	1-Chloro-4-nitro-2-(trifluoromethyl)benzene	232	135		
$C_7H_4ClF_3$	1-Chloro-2-(trifluoromethyl)benzene	152.2	59		
$C_7H_4F_3NO_2$	1-Nitro-3-(trifluoromethyl)benzene	202.8	103		
C_7H_5ClO	Benzoyl chloride	197.2	72		
C_7H_5ClO	4-Chlorobenzaldehyde	213.5	88		
$C_7H_5Cl_3$	(Trichloromethyl)benzene	221	127		211
$C_7H_5F_3$	(Trifluoromethyl)benzene	102.1	12		
$C_7H_6N_2O_4$	1-Methyl-2,4-dinitrobenzene		207		
C_7H_6O	Benzaldehyde	179.0	63		192
$C_7H_6O_2$	Benzoic acid	249.2	121		570
$C_7H_6O_2$	Salicylaldehyde	197	78		
$C_7H_6O_3$	Salicylic acid		157	1.1-?	540
C_7H_7Br	o-Bromotoluene	181.7	79		
C_7H_7Br	p-Bromotoluene	184.3	85		
C_7H_7Cl	(Chloromethyl)benzene	179	67	1.1-?	585
$C_7H_7NO_2$	o-Nitrotoluene	222	106		

Mol. Form.	Name	$t_B/°C$	FP/°C	Fl. Limits	IT/°C
$C_7H_7NO_2$	m-Nitrotoluene	232	106		
$C_7H_7NO_2$	p-Nitrotoluene	238.3	106		
C_7H_8	Toluene	110.6	4	1.1-7.1%	480
C_7H_8	Bicyclo[2.2.1]hepta-2,5-diene	89.5	-21		
C_7H_8O	o-Cresol	191.0	81	1.4-?	599
C_7H_8O	m-Cresol	202.2	86	1.1-?	558
C_7H_8O	p-Cresol	201.9	86	1.1-?	558
C_7H_8O	Benzyl alcohol	205.3	93		436
C_7H_8O	Anisole	153.7	52		475
$C_7H_8O_2$	4-Methoxyphenol	243	132		421
$C_7H_8O_3S$	p-Toluenesulfonic acid		184		
C_7H_9N	o-Methylaniline	200.3	85		482
C_7H_9N	p-Methylaniline	200.4	87		482
C_7H_9NO	o-Anisidine	224	118		
$C_7H_{10}O$	3-Cyclohexene-1-carboxaldehyde	105	57		
$C_7H_{10}O_4$	3,3-Diacetoxy-1-propene	180	82		
C_7H_{12}	4-Methylcyclohexene	102.7	-1		
$C_7H_{12}O_2$	Butyl acrylate	145	29	1.7-9.9%	292
$C_7H_{12}O_2$	Isobutyl acrylate	132	30		427
$C_7H_{12}O_2$	Cyclohexyl formate	162	51		
$C_7H_{12}O_4$	Diethyl malonate	200	93		
C_7H_{14}	1-Heptene	93.6	-1		260
C_7H_{14}	trans-2-Heptene	98	<0		
C_7H_{14}	Cycloheptane	118.4	<21	1.1-6.7%	
C_7H_{14}	Methylcyclohexane	100.9	-4	1.2-6.7%	250
C_7H_{14}	Ethylcyclopentane	103.5	<21	1.1-6.7%	260
$C_7H_{14}O$	2-Heptanone	151.0	39	1.1-7.9%	393
$C_7H_{14}O$	3-Heptanone	147	46		
$C_7H_{14}O$	4-Heptanone	144	49		
$C_7H_{14}O$	5-Methyl-2-hexanone	144	36	1.0-8.2%	191
$C_7H_{14}O$	cis-2-Methylcyclohexanol	165	65		296
$C_7H_{14}O$	trans-2-Methylcyclohexanol	167.5	65		296
$C_7H_{14}O$	cis-3-Methylcyclohexanol	174.5	70		295
$C_7H_{14}O$	trans-3-Methylcyclohexanol	174.5	70		295
$C_7H_{14}O$	cis-4-Methylcyclohexanol	173	70		295
$C_7H_{14}O$	trans-4-Methylcyclohexanol	174	70		295
$C_7H_{14}O_2$	Pentyl acetate	149.2	16	1.1-7.5%	360
$C_7H_{14}O_2$	Isopentyl acetate	142.5	25	1.0-7.5%	360
$C_7H_{14}O_2$	sec-Pentyl acetate	130.5	32		
$C_7H_{14}O_2$	Butyl propanoate	146.8	32		426
$C_7H_{14}O_2$	Propyl butanoate	143.0	37		
$C_7H_{15}NO_2$	Ethyl N-butylcarbamate	202	92		
C_7H_{16}	Heptane	98.5	-4	1.05-6.7%	204
C_7H_{16}	2-Methylhexane	90.0	-1	1.0-6.0%	280
C_7H_{16}	3-Methylhexane	92	-4		280
C_7H_{16}	2,3-Dimethylpentane	89.7	-56	1.1-6.7%	335
C_7H_{16}	2,4-Dimethylpentane	80.4	-12		
C_7H_{16}	2,2,3-Trimethylbutane	80.8	<0		412
$C_7H_{16}N_2O$	4-Morpholinepropanamine	220	104		
$C_7H_{16}O$	2-Heptanol	159	71		
$C_7H_{16}O$	3-Heptanol	157	60		
$C_7H_{16}O$	2,4-Dimethyl-3-pentanol	138.7	49		
$C_7H_{16}O$	2,3,3-Trimethyl-2-butanol	131	<0		375
$C_7H_{17}N$	Heptylamine	156	54		
$C_7H_{18}N_2$	N,N-Diethyl-1,3-propanediamine	168.5	59		
$C_8H_4O_3$	Phthalic anhydride	295	152	1.7-10.5%	570
$C_8H_6O_4$	Phthalic acid		168		
$C_8H_6O_4$	Terephthalic acid		260		496
C_8H_7ClO	α-Chloroacetophenone	247	118		
C_8H_7N	Benzeneacetonitrile	233.5	113		
C_8H_8	Styrene	145	31	0.9-6.8%	490

Mol. Form.	Name	t_B/°C	FP/°C	Fl. Limits	IT/°C
C_8H_8O	Phenyloxirane	194.1	74		498
C_8H_8O	Benzeneacetaldehyde	195	71		
C_8H_8O	Acetophenone	202	77		570
$C_8H_8O_2$	Benzeneacetic acid	265.5	>100		
$C_8H_8O_2$	Phenyl acetate	196	80		
$C_8H_8O_2$	Methyl benzoate	199	83		
$C_8H_8O_2$	2-Methoxybenzaldehyde	243.5	118		
$C_8H_8O_3$	Methyl salicylate	222.9	96		454
C_8H_9Cl	1-Chloro-4-ethylbenzene	184.4	64		
C_8H_9NO	Acetanilide	304	169		530
$C_8H_9NO_2$	Methyl 2-aminobenzoate	256	>100		
C_8H_{10}	Ethylbenzene	136.1	21	0.8-6.7%	432
C_8H_{10}	o-Xylene	144.5	32	0.9-6.7%	463
C_8H_{10}	m-Xylene	139.1	27	1.1-7.0%	527
C_8H_{10}	p-Xylene	138.3	27	1.1-7.0%	528
$C_8H_{10}O$	p-Ethylphenol	217.9	104		
$C_8H_{10}O$	Benzeneethanol	218.2	96		
$C_8H_{10}O$	α-Methylbenzyl alcohol	205	93		
$C_8H_{10}O$	Phenetole	169.8	63		
$C_8H_{10}O$	Benzyl methyl ether	170	135		
$C_8H_{10}O$	4-Methylanisole	175.5	60		
$C_8H_{10}O_2$	2-Phenoxyethanol	245	121		
$C_8H_{11}N$	N-Ethylaniline	203.0	85		
$C_8H_{11}N$	N,N-Dimethylaniline	194.1	63		371
$C_8H_{11}N$	2,3-Xylidine	221.5	97	1.0-?	
$C_8H_{11}N$	2,6-Xylidine	215	96		
$C_8H_{11}N$	α-Methylbenzylamine	187	79		
$C_8H_{11}N$	5-Ethyl-2-picoline	178.3	68	1.1-6.6%	
$C_8H_{11}NO$	N-Phenylethanolamine	279.5	152		
$C_8H_{11}NO$	o-Phenetidine	232.5	115		
$C_8H_{11}NO$	p-Phenetidine	254	116		
C_8H_{12}	1,5-Cyclooctadiene	150.8	35		
C_8H_{12}	4-Vinylcyclohexene	128	16		269
$C_8H_{12}O_4$	Diethyl maleate	223	121		350
$C_8H_{12}O_4$	Diethyl fumarate	214	104		
$C_8H_{14}O_2$	Cyclohexyl acetate	173	58		335
$C_8H_{14}O_2$	Butyl methacrylate	160	52		
$C_8H_{14}O_3$	Butanoic anhydride	200	54	0.9-5.8%	279
$C_8H_{14}O_3$	2-Methylpropanoic anhydride	183	59	1.0-6.2%	329
$C_8H_{14}O_3$	Butyl acetoacetate		85		
$C_8H_{14}O_4$	Ethyl succinate	217.7	90		
$C_8H_{14}O_5$	Diethylene glycol diacetate	200	135		
$C_8H_{14}O_6$	Diethyl tartrate	281	93		
$C_8H_{15}ClO$	Octanoyl chloride	195.6	82		
C_8H_{16}	1-Octene	121.2	21		230
C_8H_{16}	2,4,4-Trimethyl-1-pentene	101.4	-5	0.8-4.8%	391
C_8H_{16}	2,4,4-Trimethyl-2-pentene	104.9	2		305
C_8H_{16}	Ethylcyclohexane	131.9	35	0.9-6.6%	238
C_8H_{16}	cis-1,2-Dimethylcyclohexane	129.8	16		304
C_8H_{16}	trans-1,2-Dimethylcyclohexane	123.5	11		304
C_8H_{16}	cis-1,4-Dimethylcyclohexane	124.4	16		
C_8H_{16}	Propylcyclopentane	131			269
$C_8H_{16}O$	Octanal	171	52		
$C_8H_{16}O$	2-Ethylhexanal	163	44	0.85-7.2%	190
$C_8H_{16}O$	2-Octanone	172.5	52		
$C_8H_{16}O_2$	Hexyl acetate	171.5	45		
$C_8H_{16}O_2$	sec-Hexyl acetate	147.5	45		
$C_8H_{16}O_2$	2-Ethylbutyl acetate	162.5	54		
$C_8H_{16}O_2$	Pentyl propanoate	168.6	41		378
$C_8H_{16}O_2$	Butyl butanoate	166	53		

Mol. Form.	Name	t_B/°C	FP/°C	Fl. Limits	IT/°C
$C_8H_{16}O_2$	Isobutyl butanoate	156.9	50		
$C_8H_{16}O_2$	Isobutyl isobutanoate	148.6	38	0.96-7.59%	432
$C_8H_{16}O_2$	Ethyl hexanoate	167	49		
$C_8H_{16}O_2$	1,4-Cyclohexanedimethanol	283	167		316
$C_8H_{16}O_3$	Pentyl lactate		79		
$C_8H_{16}O_4$	Diethylene glycol monoethyl ether acetate	218.5	110		425
$C_8H_{17}Cl$	1-Chlorooctane	181.5	70		
$C_8H_{17}Cl$	3-(Chloromethyl)heptane	172	60		
C_8H_{18}	Octane	125.6	13	1.0-6.5%	206
C_8H_{18}	2,3-Dimethylhexane	115.6	7		438
C_8H_{18}	2,4-Dimethylhexane	109.5	10		
C_8H_{18}	3-Ethyl-2-methylpentane	115.6	<21		460
C_8H_{18}	2,2,3-Trimethylpentane	110	<21		346
C_8H_{18}	2,2,4-Trimethylpentane	99.2	-12		418
C_8H_{18}	2,3,3-Trimethylpentane	114.8	<21		425
$C_8H_{18}O$	1-Octanol	195.1	81		
$C_8H_{18}O$	2-Octanol	180	88		
$C_8H_{18}O$	2-Ethyl-1-hexanol	184.6	73	0.88-9.7%	231
$C_8H_{18}O$	Dibutyl ether	140.2	25	1.5-7.6%	194
$C_8H_{18}O_2$	2-Ethyl-1,3-hexanediol	244	127		360
$C_8H_{18}O_2$	2,2,4-Trimethyl-1,3-pentanediol	235	113		346
$C_8H_{18}O_2$	Di-tert-butyl peroxide	111	18		
$C_8H_{18}O_3$	Diethylene glycol diethyl ether	188	82		
$C_8H_{18}O_4$	2,5,8,11-Tetraoxadodecane	216	111		
$C_8H_{18}O_5$	Tetraethylene glycol	328	182		
$C_8H_{18}S$	1-Octanethiol	199.1	69		
$C_8H_{18}S$	Dibutyl sulfide	185	76		
$C_8H_{19}N$	Octylamine	179.6	60		
$C_8H_{19}N$	Dibutylamine	159.6	47	1.1-6%	
$C_8H_{19}N$	Diisobutylamine	139.6	29		
$C_8H_{19}N$	2-Ethylhexylamine	169.2	60		
$C_8H_{20}O_4Si$	Ethyl silicate	168.8	52		
$C_8H_{23}N_5$	Tetraethylenepentamine	341.5	163		321
$C_9H_6N_2O_2$	Toluene-2,4-diisocyanate	251	127	0.9-9.5%	
C_9H_7N	Quinoline	237.1			480
C_9H_{10}	o-Methylstyrene	169.8	53	0.8-11.0%	538
C_9H_{10}	m-Methylstyrene	164	53	0.8-11.0%	538
C_9H_{10}	p-Methylstyrene	172.8	53	0.8-11.0%	538
C_9H_{10}	Isopropenylbenzene	165.4	54	1.9-6.1%	574
$C_9H_{10}O$	1-Phenyl-1-propanone	217.5	99		
$C_9H_{10}O$	4-Methylacetophenone	226	96		
$C_9H_{10}O_2$	Ethyl benzoate	212	88		490
$C_9H_{10}O_2$	Benzyl acetate	213	90		460
$C_9H_{10}O_2$	Methyl 2-phenylacetate	216.5	91		
$C_9H_{11}NO$	4-Methylacetanilide	307	168		
C_9H_{12}	Propylbenzene	159.2	30	0.8-6.0%	450
C_9H_{12}	Isopropylbenzene	152.4	36	0.9-6.5%	424
C_9H_{12}	o-Ethyltoluene	165.2			440
C_9H_{12}	m-Ethyltoluene	161.3			480
C_9H_{12}	p-Ethyltoluene	162			475
C_9H_{12}	1,2,3-Trimethylbenzene	176.1	44	0.8-6.6%	470
C_9H_{12}	1,2,4-Trimethylbenzene	169.3	44	0.9-6.4%	500
C_9H_{12}	1,3,5-Trimethylbenzene	164.7	50	1-5%	559
$C_9H_{12}O$	α–Ethylbenzyl alcohol	219	100		
$C_9H_{12}O_2$	Ethylene glycol monobenzyl ether	256	129		352
$C_9H_{12}O_3S$	Ethyl p-toluenesulfonate		158		
$C_9H_{13}N$	Amphetamine	203	<100		
$C_9H_{14}O$	Phorone	197.5	85		
$C_9H_{14}O$	Isophorone	215.2	84	0.8-3.8%	460
$C_9H_{14}O_6$	Triacetin	259	138	1.0-?	433
C_9H_{16}	Octahydroindene	167			296

Mol. Form.	Name	t_B/°C	FP/°C	Fl. Limits	IT/°C
$C_9H_{16}O_2$	Allyl hexanoate	186	66		
C_9H_{18}	1-Nonene	146.9	26		
C_9H_{18}	Propylcyclohexane	156.7			248
C_9H_{18}	Isopropylcyclohexane	154.8			283
C_9H_{18}	Butylcyclopentane	156.6			250
$C_9H_{18}O$	2-Nonanone	195.3	60	0.9-5.9%	360
$C_9H_{18}O$	Diisobutyl ketone	169.4	49	0.8-7.1%	396
$C_9H_{18}O_2$	Pentyl butanoate	186.4	57		
$C_9H_{18}O_2$	Isopentyl butanoate	179	59		
$C_9H_{18}O_2$	Butyl 3-methylbutanoate		53		
C_9H_{20}	Nonane	150.8	31	0.8-2.9%	205
C_9H_{20}	3-Ethyl-4-methylhexane	140	24		
C_9H_{20}	4-Ethyl-2-methylhexane	133.8	<21	0.7-?	280
C_9H_{20}	2,2,5-Trimethylhexane	124.0	13		
C_9H_{20}	3,3-Diethylpentane	146.3		0.7-5.7%	290
C_9H_{20}	3-Ethyl-2,4-dimethylpentane	136.7	390		
C_9H_{20}	2,2,3,3-Tetramethylpentane	140.2	<21	0.8-4.9%	430
C_9H_{20}	2,2,3,4-Tetramethylpentane	133.0	<21		
$C_9H_{21}BO_3$	Triisopropyl borate	140	28		
$C_9H_{21}N$	Tripropylamine	156	41		
$C_9H_{21}NO_3$	Triisopropanolamine		160		320
$C_{10}H_7Cl$	1-Chloronaphthalene	259	121		>558
$C_{10}H_8$	Naphthalene	217.9	79	0.9-5.9%	526
$C_{10}H_8O$	2-Naphthol	285	153		
$C_{10}H_9N$	1-Naphthalenamine	300.8	157		
$C_{10}H_{10}O_2$	Safrole	234.5	100		
$C_{10}H_{10}O_4$	Dimethyl phthalate	283.7	146	0.9-?	490
$C_{10}H_{10}O_4$	Dimethyl isophthalate	282	138		
$C_{10}H_{10}O_4$	Dimethyl terephthalate	288	153		518
$C_{10}H_{11}NO_2$	Acetoacetanilide		185		
$C_{10}H_{12}$	1,2,3,4-Tetrahydronaphthalene	207.6	71	0.8-5.0%	385
$C_{10}H_{12}O_2$	Isopropyl benzoate	216	99		
$C_{10}H_{12}O_2$	Ethyl phenylacetate	227	99		
$C_{10}H_{14}$	Butylbenzene	183.3	71	0.8-5.8%	410
$C_{10}H_{14}$	sec-Butylbenzene	173.3	52	0.8-6.9%	418
$C_{10}H_{14}$	tert-Butylbenzene	169.1	60	0.7-5.7%	450
$C_{10}H_{14}$	Isobutylbenzene	172.7	55	0.8-6.0%	427
$C_{10}H_{14}$	p-Cymene	177.1	47	0.7-5.6%	436
$C_{10}H_{14}$	1,2,3,4-Tetramethylbenzene	205	74		427
$C_{10}H_{14}$	1,2,3,5-Tetramethylbenzene	198	71		427
$C_{10}H_{14}$	1,2,4,5-Tetramethylbenzene	196.8	54		
$C_{10}H_{14}$	o-Diethylbenzene	184	57		395
$C_{10}H_{14}$	m-Diethylbenzene	181.1	56		450
$C_{10}H_{14}$	p-Diethylbenzene	183.7	55	0.7-6.0%	430
$C_{10}H_{14}O$	Butyl phenyl ether	210	82		
$C_{10}H_{14}O_2$	4-tert-Butyl-1,2-benzenediol	285	130		
$C_{10}H_{15}N$	N-Butylaniline	243.5	107		
$C_{10}H_{15}N$	N,N-Diethylaniline	216.3	85		630
$C_{10}H_{15}NO_2$	N-Phenyl-N,N-diethanolamine		196	0.7-?	387
$C_{10}H_{16}$	Dipentene	178	45		237
$C_{10}H_{16}$	d-Limonene	178	45	0.7-6.1%	237
$C_{10}H_{16}$	α-Pinene	156.2	33		255
$C_{10}H_{16}$	β-Pinene	166	38		275
$C_{10}H_{16}$	β-Phellandrene	171.5	49		
$C_{10}H_{16}O$	Camphor	207.4	66	0.6-3.5%	466
$C_{10}H_{18}$	trans-Decahydronaphthalene	187.3	54	0.7-5.4%	255
$C_{10}H_{18}O$	Borneol		66		
$C_{10}H_{18}O$	Linalol	198	71		
$C_{10}H_{18}O$	α-Terpineol	220	90		
$C_{10}H_{18}O$	Cineole	176.4	48		

Mol. Form.	Name	t_B/°C	FP/°C	Fl. Limits	IT/°C
$C_{10}H_{18}O$	*trans*-Geraniol	230	>100		
$C_{10}H_{18}O_4$	Dibutyl oxalate	241	104		
$C_{10}H_{19}NO_2$	*N-tert*-Butylaminoethyl methacrylate		96		
$C_{10}H_{20}$	1-Decene	170.5	<55		235
$C_{10}H_{20}$	Butylcyclohexane	180.9			246
$C_{10}H_{20}$	Isobutylcyclohexane	171.3			274
$C_{10}H_{20}$	*tert*-Butylcyclohexane	171.5			342
$C_{10}H_{20}O$	Citronellol	224	96		
$C_{10}H_{20}O_2$	2-Ethylhexyl acetate	199	71	0.76-8.14%	268
$C_{10}H_{20}O_2$	Ethyl octanoate	208.5	79		
$C_{10}H_{21}N$	*N*-Butylcyclohexanamine		93		
$C_{10}H_{22}$	Decane	174.1	51	0.8-5.4%	210
$C_{10}H_{22}$	2-Methylnonane	167.1			210
$C_{10}H_{22}$	3-Ethyloctane	166.5			230
$C_{10}H_{22}$	4-Ethyloctane	163.7			229
$C_{10}H_{22}O$	1-Decanol	231.1	82		288
$C_{10}H_{22}O$	Dipentyl ether	190	57		170
$C_{10}H_{22}O_2$	Ethylene glycol dibutyl ether	203.3	85		
$C_{10}H_{22}O_5$	Tetraethylene glycol dimethyl ether	275.3	141		
$C_{10}H_{22}S$	Dipentyl sulfide		85		
$C_{10}H_{23}N$	Decylamine	220.5	99		
$C_{10}H_{23}N$	Dipentylamine	202.5	51		
$C_{11}H_{10}$	1-Methylnaphthalene	244.7			529
$C_{11}H_{12}O_3$	Ethyl benzoylacetate		141		
$C_{11}H_{14}O_2$	Butyl benzoate	250.3	107		
$C_{11}H_{16}$	*p-tert*-Butyltoluene	190	68		
$C_{11}H_{16}$	Pentylbenzene	205.4	66		
$C_{11}H_{16}$	1,3-Diethyl-5-methylbenzene	205			455
$C_{11}H_{16}$	Pentamethylbenzene	232	93		427
$C_{11}H_{16}O$	4-*tert*-Butyl-2-methylphenol	237	118		
$C_{11}H_{17}N$	*p-tert*-Pentylaniline	260.5	102		
$C_{11}H_{20}O_2$	2-Ethylhexyl acrylate		82		252
$C_{11}H_{22}$	Pentylcyclohexane	203.7			239
$C_{11}H_{22}O$	2-Undecanone	231.5	89		
$C_{11}H_{22}O_2$	Nonyl acetate	210	68		
$C_{11}H_{24}$	Undecane	195.9	69		
$C_{11}H_{24}$	2-Methyldecane	189.3			225
$C_{11}H_{24}O$	2-Undecanol	228	113		
$C_{12}H_9Br$	4-Bromo-1,1'-Biphenyl	310	144		
$C_{12}H_{10}$	Biphenyl	256.1	113	0.6-5.8%	540
$C_{12}H_{10}Cl_2Si$	Dichlorodiphenylsilane	305	142		
$C_{12}H_{10}O$	*o*-Phenylphenol	286	124		530
$C_{12}H_{10}O$	Diphenyl ether	258.0	112	0.8-1.5%	618
$C_{12}H_{11}N$	2-Aminobiphenyl	299			450
$C_{12}H_{11}N$	Diphenylamine	302	153		634
$C_{12}H_{12}$	1-Ethylnaphthalene	258.6			480
$C_{12}H_{14}O_4$	Diethyl phthalate	295	161	0.7-?	457
$C_{12}H_{14}O_4$	Diethyl terephthalate	302	117		
$C_{12}H_{16}$	Cyclohexylbenzene	240.1	99		
$C_{12}H_{16}O_3$	Pentyl salicylate	270	132		
$C_{12}H_{17}NO$	*N*-Butyl-*N*-phenylacetamide	281	141		
$C_{12}H_{18}$	1,5,9-Cyclododecatriene	240	71		
$C_{12}H_{20}O_4$	Dibutyl maleate	280	141		
$C_{12}H_{22}O_4$	Dimethyl sebacate		145		
$C_{12}H_{22}O_6$	Dibutyl tartrate	320	91		284
$C_{12}H_{23}N$	Dicyclohexylamine		>99		
$C_{12}H_{24}$	1-Dodecene	213.8	79		
$C_{12}H_{24}O_2$	Ethyl decanoate	241.5	>100		
$C_{12}H_{25}Br$	1-Bromododecane	276	144		
$C_{12}H_{26}$	Dodecane	216.3	74	0.6-?	203
$C_{12}H_{26}O$	1-Dodecanol	259	127		275

Mol. Form.	Name	$t_B/°C$	FP/°C	Fl. Limits	IT/°C
$C_{12}H_{26}O$	2-Butyl-1-octanol	246.5	110		
$C_{12}H_{26}O_3$	Diethylene glycol dibutyl ether	256	118		310
$C_{12}H_{26}S$	1-Dodecanethiol	277	128		
$C_{12}H_{27}BO_3$	Tributyl borate	234	93		
$C_{12}H_{27}N$	Tributylamine	216.5	63		
$C_{12}H_{27}O_4P$	Tributyl phosphate	289	146		
$C_{13}H_{12}$	2-Methylbiphenyl	255.5	137		502
$C_{13}H_{12}$	Diphenylmethane	265.0	130		485
$C_{13}H_{14}N_2$	p,p'-Diaminodiphenylmethane	398	220		
$C_{13}H_{26}$	1-Tridecene	232.8	79		
$C_{13}H_{26}O$	2-Tridecanone	263	107		
$C_{13}H_{28}$	Tridecane	235.4	79		
$C_{13}H_{28}O$	1-Tridecanol		121		
$C_{14}H_8O_2$	9,10-Anthracenedione	377	185		
$C_{14}H_{10}$	Anthracene	339.9	121	0.6-?	540
$C_{14}H_{10}$	Phenanthrene	340	171		
$C_{14}H_{12}O_2$	Benzyl benzoate	323.5	148		480
$C_{14}H_{12}O_3$	Benzyl salicylate	320	>100		
$C_{14}H_{14}$	1,1-Diphenylethane	272.6	>100		440
$C_{14}H_{14}O$	Dibenzyl ether	298	135		
$C_{14}H_{16}$	1-Butylnaphthalene	289.3	360		
$C_{14}H_{16}N_2O_2$	o-Dianisidine		206		
$C_{14}H_{23}N$	N,N-Dibutylaniline	274.8	110		
$C_{14}H_{28}$	1-Tetradecene	233	110		235
$C_{14}H_{30}$	Tetradecane	253.5	112	0.5-?	200
$C_{14}H_{30}O$	1-Tetradecanol	289	141		
$C_{15}H_{18}$	1-Pentylnaphthalene	307	124		
$C_{15}H_{24}$	Nonylbenzene	280.5	99		
$C_{15}H_{24}O$	2,6-Di-tert-butyl-4-methylphenol	265	127		
$C_{15}H_{26}O_6$	Tributyrin	307.5	180	0.5-?	407
$C_{15}H_{33}N$	Tripentylamine	242.5	102		
$C_{16}H_{14}O$	1,3-Diphenyl-2-buten-1-one	342.5	177		
$C_{16}H_{18}$	2-Butyl-1,1'-biphenyl		>100		430
$C_{16}H_{22}O_4$	Dibutyl phthalate	340	157	0.5-?	402
$C_{16}H_{26}$	Decylbenzene	298	107		
$C_{16}H_{34}$	Hexadecane	286.8	136		202
$C_{16}H_{34}O$	Dioctyl ether	283	>100		205
$C_{16}H_{35}N$	Bis(2-ethylhexyl)amine		132		
$C_{17}H_{20}N_2O$	N,N'-Diethylcarbanilide		150		
$C_{17}H_{34}O$	2-Heptadecanone	320	120		
$C_{17}H_{36}O$	1-Heptadecanol	333	154		
$C_{18}H_{14}$	o-Terphenyl	332	163		
$C_{18}H_{14}$	m-Terphenyl	363	191		
$C_{18}H_{15}O_3P$	Triphenyl phosphite	360	218		
$C_{18}H_{15}O_4P$	Triphenyl phosphate		220		
$C_{18}H_{15}P$	Triphenylphosphine		180		
$C_{18}H_{30}$	Dodecylbenzene	328	140		
$C_{18}H_{32}O_7$	Butyl citrate		157		368
$C_{18}H_{34}O_2$	Oleic acid	360	189		363
$C_{18}H_{34}O_4$	Dibutyl sebacate	344.5	178	0.4-?	365
$C_{18}H_{36}O_2$	Stearic acid		196		395
$C_{18}H_{37}Cl_3Si$	Trichlorooctadecylsilane		89		
$C_{18}H_{38}$	Octadecane	316.3	>100		227
$C_{18}H_{38}O$	1-Octadecanol				450
$C_{19}H_{16}$	Triphenylmethane	359	>100		
$C_{19}H_{38}O$	2-Nonadecanone		124		
$C_{19}H_{38}O_2$	Methyl stearate	443	153		
$C_{19}H_{40}$	Nonadecane	329.9	>100		230
$C_{20}H_{14}O_4$	Diphenyl phthalate		224		
$C_{20}H_{28}$	1-Decylnaphthalene	379	177		
$C_{20}H_{42}$	Eicosane	343	>100		232

Mol. Form.	Name	t_B/°C	FP/°C	Fl. Limits	IT/°C
$C_{21}H_{21}O_4P$	Tri-*o*-cresyl phosphate	410	225		385
$C_{21}H_{26}O_3$	4-Octylphenyl salicylate		216		416
$C_{21}H_{32}O_2$	Methyl abietate		180		
$C_{22}H_{42}O_2$	Butyl oleate		180		
$C_{22}H_{42}O_4$	Bis(2-ethylhexyl) adipate		206	0.4-?	377
$C_{22}H_{44}O_2$	Butyl stearate	343	160		355
$C_{23}H_{46}O_2$	Pentyl stearate		185		
$C_{24}H_{20}Sn$	Tetraphenylstannane	420	232		
$C_{24}H_{38}O_4$	Bis(2-ethylhexyl) phthalate	384	218		
$C_{25}H_{48}O_4$	Bis(2-ethylhexyl) azelate		227	0.3-?	374

THRESHOLD LIMITS FOR AIRBORNE CONTAMINANTS

Several organizations recommend limits of exposure to airborne contaminants in the workplace. These include the Occupational Safety and Health Administration (OSHA), the National Institute for Occupational Safety and Health (NIOSH), and the non-governmental organization, American Conference of Governmental Industrial Hygienists (ACGIH). The threshold limit value (TLV) for a substance is defined as the concentration level under which the majority of workers may be repeatedly exposed, day after day, without adverse effects. The TLV recommendations are given in two forms:

- Time-weighted average (TWA) concentration for a normal 8-h workday and 40-h workweek.
- Short-term exposure limit (STEL), which should not be exceeded for more than 15 min.

Both kinds of limits are specified for some substances.

The following table gives threshold limit values for a number of substances that may be encountered in the atmosphere of a chemical laboratory or industrial facility. All values refer to the concentration in air at 25°C and normal atmospheric pressure. Data for gases are given both in parts per million by volume (ppm) and in mass concentration (mg/m³). Values for liquids refer to mists or aerosols, and those for solids to dusts or fumes; both are stated in mg/m³. A "C" following a value indicates a ceiling limit which should not be exceeded even for very brief periods because of acute toxic effects of the substance.

Substances are listed by systematic name, which is followed by molecular formula in the Hill format and Chemical Abstracts Service Registry Number. Common synonyms are given in brackets [ ] for some compounds.

REFERENCES

1. *1999 TLV's and BEI's,* American Conference of Governmental Industrial Hygienists, 1330 Kemper Meadow Drive, Cincinnati, OH 45240-1634, 1999.
2. *NIOSH Pocket Guide to Chemical Hazards*, U.S. Department of Health and Human Services, National Institute for Occupational Health and Safety, U.S. Government Printing Office, Washington, DC, 1994.
3. *Chemical Information Manual*, U.S. Department of Labor, Occupational Safety and Health Administration, Washington, DC, 1991.

Substance	Molecular Formula	CAS Reg. No.	Time-Weighted Average		Short-Term Exposure Limit	
			ppm	mg/m³	ppm	mg/m³
Abate [Temephos]	$C_{16}H_{20}O_6P_2S_3$	3383-96-8		10		
Acetaldehyde	C_2H_4O	75-07-0			25 C	45 C
Acetic acid	$C_2H_4O_2$	64-19-7	10	25	15	37
Acetic anhydride	$C_4H_6O_3$	108-24-7	5	21		
Acetone	C_3H_6O	67-64-1	500	1188	750	1780
Acetone cyanohydrin	C_4H_7NO	75-86-5			4.7 C	5 C
Acetonitrile	C_2H_3N	75-05-8	40	67	60	101
Acetophenone	C_8H_8O	98-86-2	10	49		
2-(Acetyloxy)benzoic acid [Aspirin]	$C_9H_8O_4$	50-78-2		5		
Acrolein [2-Propenal]	C_3H_4O	107-02-8			0.1 C	0.23 C
Acrylamide	C_3H_5NO	79-06-1		0.03		
Acrylic acid [2-Propenoic acid]	$C_3H_4O_2$	79-10-7	2	5.9		
Acrylonitrile [Propenenitrile]	C_3H_3N	107-13-1	2	4.3		
Adipic acid	$C_6H_{10}O_4$	124-04-9		5		
Adiponitrile	$C_6H_8N_2$	111-69-3	2	9		
Aldrin	$C_{12}H_8Cl_6$	309-00-2		0.25		
Allyl alcohol [2-Propen-1-ol]	C_3H_6O	107-18-6	0.5	1.2		
Allyl glycidyl ether	$C_6H_{10}O_2$	106-92-3	1	5		
Allyl propyl disulfide	$C_6H_{12}S_2$	2179-59-1	2	12	3	18
Aluminum (metal dust)	Al	7429-90-5		10		
Aluminum oxide	Al_2O_3	1344-28-1		10		
4-Amino-3,5,6-trichloropyridinecarboxlic acid [Picloram]	$C_6H_3Cl_3N_2O_2$	1918-02-1		10		
Ammonia	H_3N	7664-41-7	25	17	35	24
Ammonium chloride	ClH_4N	12125-02-9		10		20
Ammonium perfluorooctanoate	$C_8H_4F_{15}NO_2$	3825-26-1		0.01		
Ammonium sulfamate	$H_6N_2O_3S$	7773-06-0		10		
Aniline	C_6H_7N	62-53-3	2	7.6		
Antimony	Sb	7440-36-0		0.5		
Arsenic	As	7440-38-2		0.01		
Arsine	AsH_3	7784-42-1	0.05	0.16		

Substance	Molecular Formula	CAS Reg. No.	Time-Weighted Average ppm	Time-Weighted Average mg/m^3	Short-Term Exposure Limit ppm	Short-Term Exposure Limit mg/m^3
Atrazine	$C_8H_{14}ClN_5$	1912-24-9		5		
Azinphos-methyl	$C_{10}H_{12}N_3O_3PS_2$	86-50-0		0.2		
Barium	Ba	7440-39-3		0.5		
Barium sulfate	BaO_4S	7727-43-7		10		
Benomyl	$C_{14}H_{18}N_4O_3$	17804-35-2	0.84	10		
Benzene	C_6H_6	71-43-2	0.5	1.6	2.5	8
1,3-Benzenedimethanamine [*m*-Xylene diamine]	$C_8H_{12}N_2$	1477-55-0				0.1 C
Benzenethiol [Phenyl mercaptan]	C_6H_6S	108-98-5	0.5	2.3		
p-Benzoquinone [Quinone]	$C_6H_4O_2$	106-51-4	0.1	0.44		
Benzoyl chloride	C_7H_5ClO	98-88-4			0.5 C	2.8 C
Benzoyl peroxide	$C_{14}H_{10}O_4$	94-36-0		5		
Benzyl acetate	$C_9H_{10}O_2$	140-11-4	10	61		
Beryllium	Be	7440-41-7		0.002		0.01
Biphenyl	$C_{12}H_{10}$	92-52-4	0.2	1.3		
Bis(4-amino-3-chlorophenyl)methane [4,4-Methylene bis(2-chloroaniline)]	$C_{13}H_{12}Cl_2N_2$	101-14-4	0.01	0.11		
Bis(2-chloroethyl) ether [2,2'-Dichlorethyl ether]	$C_4H_8Cl_2O$	111-44-4	5	29	10	58
Bis(chloromethyl) ether	$C_2H_4Cl_2O$	542-88-1	0.001	0.0047		
Bis(2-ethylhexyl) phthalate [Di-*sec*-octyl phthalate]	$C_{24}H_{38}O_4$	117-81-7		5		10
Bismuth telluride	Bi_2Te_3	1304-82-1		10		
Boron oxide	B_2O_3	1303-86-2		10		
Boron tribromide	BBr_3	10294-33-4			1 C	10 C
Boron trifluoride	BF_3	7637-07-2			1 C	2.8 C
Bromacil	$C_9H_{13}BrN_2O_2$	314-40-9		10		
Bromine	Br_2	7726-95-6	0.1	0.66	0.2	1.3
Bromine pentafluoride	BrF_5	7789-30-2	0.1	0.72		
Bromochloromethane [Halon 1011]	CH_2BrCl	74-97-5	200	1060		
2-Bromo-2-chloro-1,1,1-trifluoroethane [Halothane]	$C_2HBrClF_3$	151-67-7	50	404		
Bromoethane [Ethyl bromide]	C_2H_5Br	74-96-4	5	22		
Bromoethene [Vinyl bromide]	C_2H_3Br	593-60-2	0.5	2.2		
Bromomethane [Methyl bromide]	CH_3Br	74-83-9	1	3.9		
Bromotrifluoromethane	$CBrF_3$	75-63-8	1000	6090		
1,3-Butadiene	C_4H_6	106-99-0	2	4.4		
Butane	C_4H_{10}	106-97-8	800	1900		
1-Butanethiol [Butyl mercaptan]	$C_4H_{10}S$	109-79-5	0.5	1.8		
1-Butanol	$C_4H_{10}O$	71-36-3			50 C	152 C
2-Butanol [*sec*-Butyl alcohol]	$C_4H_{10}O$	78-92-2	100	303		
2-Butanone [Methyl ethyl ketone]	C_4H_8O	78-93-3	200	590	300	885
trans-2-Butenal [Crotonaldehyde]	C_4H_6O	4170-30-3			0.3 C	0.9 C
3-Buten-2-one	C_4H_6O	78-94-4			0.2 C	0.6 C
Butyl acetate	$C_6H_{12}O_2$	123-86-4	150	713	200	950
sec-Butyl acetate	$C_6H_{12}O_2$	105-46-4	200	950		
tert-Butyl acetate	$C_6H_{12}O_2$	540-88-5	200	950		
Butyl acrylate	$C_7H_{12}O_2$	141-32-2	2	10		
Butylamine	$C_4H_{11}N$	109-73-9			5 C	15 C
tert-Butyl chromate	$C_8H_{18}CrO_4$	1189-85-1				0.1 C
Butyl glycidyl ether	$C_7H_{14}O_2$	2426-08-6	25	133		
Butyl lactate	$C_7H_{14}O_3$	138-22-7	5	30		
o-sec-Butylphenol	$C_{10}H_{14}O$	89-72-5	5	31		
p-tert-Butyltoluene	$C_{11}H_{16}$	98-51-1	1	6.1		
Cadmium	Cd	7440-43-9		0.01		
Calcium carbonate	$CCaO_3$	1317-65-3		10		

Substance	Molecular Formula	CAS Reg. No.	Time-Weighted Average		Short-Term Exposure Limit	
			ppm	mg/m^3	ppm	mg/m^3
Calcium chromate	$CaCrO_4$	13765-19-0		0.003		
Calcium cyanamide	$CCaN_2$	156-62-7		0.5		
Calcium hydroxide	CaH_2O_2	1305-62-0		5		
Calcium metasilicate	CaO_3Si	1344-95-2		10		
Calcium oxide	CaO	1305-78-8		2		
Calcium sulfate	CaO_4S	7778-18-9		10		
Camphor	$C_{10}H_{16}O$	76-22-2	2	12	4	24
Caprolactam	$C_6H_{11}NO$	105-60-2	5 (gas)	1 (solid)	10 (gas)	3 (solid)
Captafol	$C_{10}H_9Cl_4NO_2S$	2425-06-1		0.1		
Captan	$C_9H_8Cl_3NO_2S$	133-06-2		5		
Carbaryl	$C_{12}H_{11}NO_2$	63-25-2		5		
Carbofuran	$C_{12}H_{15}NO_3$	1563-66-2		0.1		
Carbon black	C	1333-86-4		3.5		
Carbon dioxide	CO_2	124-38-9	5000	9000	30,000	54,000
Carbon disulfide	CS_2	75-15-0	10	31		
Carbon monoxide	CO	630-08-0	25	29		
Carbonyl chloride [Phosgene]	CCl_2O	75-44-5	0.1	0.40		
Carbonyl fluoride	CF_2O	353-50-4	2	5.4	5	13
Cesium hydroxide	$CsHO$	21351-79-1		2		
Chlordane	$C_{10}H_6Cl_8$	57-74-9		0.5		
Chlorine	Cl_2	7782-50-5	0.5	1.5	1	2.9
Chlorine dioxide	ClO_2	10049-04-4	0.1	0.28	0.3	0.83
Chlorine trifluoride	ClF_3	7790-91-2			0.1 C	0.38 C
Chloroacetaldehyde	C_2H_3ClO	107-20-0			1 C	3.2 C
Chloroacetone	C_3H_5ClO	78-95-5			1 C	3.8 C
α-Chloroacetophenone	C_8H_7ClO	532-27-4	0.05	0.32		
Chloroacetyl chloride	$C_2H_2Cl_2O$	79-04-9	0.05	0.23	0.15	0.69
Chlorobenzene	C_6H_5Cl	108-90-7	10	46		
o-Chlorobenzylidene malononitrile	$C_{10}H_5ClN_2$	2698-41-1			0.05 C	0.39 C
2-Chloro-1,3-butadiene [Chloroprene]	C_4H_5Cl	126-99-8	10	36		
Chlorodifluoromethane	$CHClF_2$	75-45-6	1000	3540		
Chloroethane [Ethyl chloride]	C_2H_5Cl	75-00-3	100	264		
2-Chloroethanol [Ethylene chlorohydrin]	C_2H_5ClO	107-07-3			1 C	3.3 C
Chloroethene [Vinyl chloride]	C_2H_3Cl	75-01-4	1	2.5		
Chloromethane [Methyl chloride]	CH_3Cl	74-87-3	50	103	100	207
(Chloromethyl)benzene [Benzyl chloride]	C_7H_7Cl	100-44-7	1	5.2		
1-Chloro-4-nitrobenzene	$C_6H_4ClNO_2$	100-00-5	0.1	0.64		
1-Chloro-1-nitropropane	$C_3H_6ClNO_2$	600-25-9	2	10		
Chloropentafluoroethane	C_2ClF_5	76-15-3	1000	6320		
2-Chloropropanoic acid	$C_3H_5ClO_2$	598-78-7	0.1	0.44		
3-Chloropropene [Allyl chloride]	C_3H_5Cl	107-05-1	1	3	2	6
2-Chlorostyrene	C_8H_7Cl	2039-87-4	50	283	75	425
o-Chlorotoluene	C_7H_7Cl	95-49-8	50	259		
Chlorpyrifos	$C_9H_{11}Cl_3NO_3PS$	2921-88-2		0.2		
Chromium	Cr	7440-47-3		0.5		
Chromyl chloride	Cl_2CrO_2	14977-61-8	0.025	0.16		
Clopidol	$C_7H_7Cl_2NO$	2971-90-6		10		
Cobalt	Co	7440-48-4		0.02		
Cobalt carbonyl	$C_8Co_2O_8$	10210-68-1		0.1		
Cobalt hydrocarbonyl	C_4HCoO_4	16842-03-8		0.1		
Copper	Cu	7440-50-8		0.2		
Cresol (all isomers)	C_7H_8O	1319-77-3	5	22		
Crufomate	$C_{12}H_{19}ClNO_3P$	299-86-5		5		
Cyanamide	CH_2N_2	420-04-2		2		
Cyanogen	C_2N_2	460-19-5	10	21		
Cyanogen chloride	$CClN$	506-77-4			0.3 C	0.75 C
Cyclohexane	C_6H_{12}	110-82-7	300	1030		

Substance	Molecular Formula	CAS Reg. No.	Time-Weighted Average		Short-Term Exposure Limit	
			ppm	mg/m^3	ppm	mg/m^3
Cyclohexanol	$C_6H_{12}O$	108-93-0	50	206		
Cyclohexanone	$C_6H_{10}O$	108-94-1	25	100		
Cyclohexene	C_6H_{10}	110-83-8	300	1010		
Cyclohexylamine	$C_6H_{13}N$	108-91-8	10	41		
Cyclonite [Hexahydro-1,3,5-trinitro-1,3,5-triazine]	$C_3H_6N_6O_6$	121-82-4		0.5		
1,3-Cyclopentadiene	C_5H_6	542-92-7	75	203		
Cyclopentane	C_5H_{10}	287-92-3	600	1720		
Cyhexatin	$C_{18}H_{34}OSn$	13121-70-5		5		
Decaborane(14)	$B_{10}H_{14}$	17702-41-9	0.05	0.25	0.15	0.75
Diacetone alcohol	$C_6H_{12}O_2$	123-42-2	50	238		
4,4'-Diaminodiphenylmethane [4,4-Methylene dianiline]	$C_{13}H_{14}N_2$	101-77-9	0.1	0.81		
Diazinon	$C_{12}H_{21}N_2O_3PS$	333-41-5		0.1		
Diazomethane	CH_2N_2	334-88-3	0.2	0.34		
Diborane	B_2H_6	19287-45-7	0.1	0.11		
Dibromodifluoromethane	CBr_2F_2	75-61-6	100	858		
2-Dibutylaminoethanol	$C_{10}H_{23}NO$	102-81-8	0.5	3.5		
2,6-Di-*tert*-butyl-4-methylphenol	$C_{15}H_{24}O$	128-37-0		10		
Dibutylphenyl phosphate	$C_{14}H_{23}O_4P$	2528-36-1	0.3	3.5		
Dibutyl phosphate	$C_8H_{19}O_4P$	107-66-4	1	8.6	2	17
Dibutyl phthalate	$C_{16}H_{22}O_4$	84-74-2		5		
Dichloroacetylene	C_2Cl_2	7572-29-4			0.1 C	0.39 C
o-Dichlorobenzene	$C_6H_4Cl_2$	95-50-1	25	150	50	301
p-Dichlorobenzene	$C_6H_4Cl_2$	106-46-7	10	60		
1,4-Dichloro-2-butene (unspecified isomer)	$C_4H_6Cl_2$	764-41-0	0.005	0.026		
Dichlorodifluoromethane	CCl_2F_2	75-71-8	1000	4950		
1,3-Dichloro-5,5-dimethyl hydantoin	$C_5H_6Cl_2N_2O_2$	118-52-5		0.2		0.4
Dichlorodiphenyltrichloroethane [DDT]	$C_{14}H_9Cl_5$	50-29-3		1		
1,1-Dichloroethane [Ethylidene dichloride]	$C_2H_4Cl_2$	75-34-3	100	405		
1,2-Dichloroethane [Ethylene dichloride]	$C_2H_4Cl_2$	107-06-2	10	40		
1,1-Dichloroethene [Vinylidene chloride]	$C_2H_2Cl_2$	75-35-4	5	20		
1,2-Dichloroethylene (both isomers)	$C_2H_2Cl_2$	540-59-0	200	793		
Dichlorofluoromethane	$CHCl_2F$	75-43-4	10	42		
Dichloromethane [Methylene chloride]	CH_2Cl_2	75-09-2	50	174		
1,1-Dichloro-1-nitroethane	$C_2H_3Cl_2NO_2$	594-72-9	2	12		
(2,4-Dichlorophenoxy)acetic acid	$C_8H_6Cl_2O_3$	94-75-7		10		
1,2-Dichloropropane	$C_3H_6Cl_2$	78-87-5	75	347	110	508
2,2-Dichloropropanoic acid	$C_3H_4Cl_2O_2$	75-99-0	1	5.8		
1,3-Dichloropropene (both isomers)	$C_3H_4Cl_2$	542-75-6	1	4.5		
1,2-Dichloro-1,1,2,2-tetrafluoroethane	$C_2Cl_2F_4$	76-14-2	1000	7000		
Dichlorvos	$C_4H_7Cl_2O_4P$	62-73-7	0.1	0.90		
Dicrotophos	$C_8H_{16}NO_5P$	141-66-2		0.25		
m-Dicyanobenzene [*m*-Phthalodinitrile]	$C_8H_4N_2$	626-17-5		5		
Dicyclopentadiene	$C_{10}H_{12}$	77-73-6	5	27		
Dieldrin	$C_{12}H_8Cl_6O$	60-57-1		0.25		
Diethanolamine	$C_4H_{11}NO_2$	111-42-2	0.46	2		
Diethylamine	$C_4H_{11}N$	109-89-7	5	15	15	45
2-Diethylaminoethanol	$C_6H_{15}NO$	100-37-8	2	9.6		
Diethylenetriamine [Bis(2-amimoethyl)amine]	$C_4H_{13}N_3$	111-40-0	1	4.2		
Diethyl ether	$C_4H_{10}O$	60-29-7	400	1210	500	1520
Diethyl phthalate	$C_{12}H_{14}O_4$	84-66-2		5		
1,1-Difluoroethene	$C_2H_2F_2$	75-38-7	500	1310		
Diglycidyl ether	$C_6H_{10}O_3$	2238-07-5	0.1	0.53		
Diisopropylamine	$C_6H_{15}N$	108-18-9	5	21		
Diisopropyl ether	$C_6H_{14}O$	108-20-3	250	1040	310	1300
Dimethoxymethane [Methylal]	$C_3H_8O_2$	109-87-5	1000	3110		

Substance	Molecular Formula	CAS Reg. No.	Time-Weighted Average		Short-Term Exposure Limit	
			ppm	mg/m^3	ppm	mg/m^3
Dimethyl mercury	C_2H_6Hg	593-74-8		0.01		0.03
N,N-Dimethylacetamide	C_4H_9NO	127-19-5	10	36		
Dimethylamine	C_2H_7N	124-40-3	5	9.2	15	27.6
N,N-Dimethylaniline	$C_8H_{11}N$	121-69-7	5	25	10	50
2,2-Dimethylbutane	C_6H_{14}	75-83-2	500	1760	1000	3500
2,3-Dimethylbutane	C_6H_{14}	79-29-8	500	1760	1000	3500
N,N-Dimethylformamide	C_3H_7NO	68-12-2	10	30		
2,6-Dimethyl-4-heptanone [Diisobutyl ketone]	$C_9H_{18}O$	108-83-8	25	145		
1,1-Dimethylhydrazine	$C_2H_8N_2$	57-14-7	0.01	0.025		
Dimethyl phthalate	$C_{10}H_{10}O_4$	131-11-3		5		
Dimethyl sulfate	$C_2H_6O_4S$	77-78-1	0.1	0.52		
Dinitrobenzene (all isomers)	$C_6H_4N_2O_4$	25154-54-5	0.15	1.0		
Dinitrotoluene (all isomers)	$C_7H_6N_2O_4$	25321-14-6		0.2		
1,4-Dioxane	$C_4H_8O_2$	123-91-1	20	72		
Dioxathion	$C_{12}H_{26}O_6P_2S_4$	78-34-2		0.2		
Diphenylamine	$C_{12}H_{11}N$	122-39-4		10		
Diphenyl ether	$C_{12}H_{10}O$	101-84-8	1	7	2	14
4,4'-Diphenylmethane diisocyanate	$C_{15}H_{10}N_2O_2$	101-68-8	0.005	0.051		
Dipropylene glycol monomethyl ether	$C_7H_{16}O_3$	34590-94-8	100	600	150	900
Diquat	$C_{12}H_{12}N_2$	231-36-7		0.5		
Disulfiram	$C_{10}H_{20}N_2S_4$	97-77-8		2		
Disulfoton	$C_8H_{19}O_2PS_3$	298-04-4		0.1		
Diuron	$C_9H_{10}Cl_2N_2O$	330-54-1		10		
Divinyl benzene (all isomers)	$C_{10}H_{10}$	1321-74-0	10	53		
Endosulfan	$C_9H_6Cl_6O_3S$	115-29-7		0.1		
Endrin	$C_{12}H_8Cl_6O$	72-20-8		0.1		
Enflurane	$C_3H_2ClF_5O$	13838-16-9	75	566		
Epichlorohydrin [(Chloromethyl)oxirane]	C_3H_5ClO	106-89-8	0.5	1.9		
1,2-Epoxy-4-(epoxyethyl)cyclohexane [Vinylcyclohexene dioxide]	$C_8H_{12}O_2$	106-87-6	0.1	0.57		
1,2-Ethanediamine [Ethylenediamine]	$C_2H_8N_2$	107-15-3	10	25		
Ethanethiol [Ethyl mercaptan]	C_2H_6S	75-08-1	0.5	1.3		
Ethanol	C_2H_6O	64-17-5	1000	1880		
Ethanolamine	C_2H_7NO	141-43-5	3	7.5	6	15
Ethion	$C_9H_{22}O_4P_2S_4$	563-12-2		0.4		
Ethoxydimethylsilane	$C_4H_{12}OSi$	14857-34-2	0.5	2.1	1.5	6.4
Ethyl acetate	$C_4H_8O_2$	141-78-6	400	1440		
Ethyl acrylate	$C_5H_8O_2$	140-88-5	5	20	15	61
Ethylamine	C_2H_7N	75-04-7	5	9.2	15	27.6
Ethylbenzene	C_8H_{10}	100-41-4	100	434	125	543
Ethylene glycol	$C_2H_6O_2$	107-21-1				100 C
Ethylene glycol dinitrate	$C_2H_4N_2O_6$	628-96-6	0.05	0.31		
Ethylene glycol monobutyl ether [2-Butoxyethanol]	$C_6H_{14}O_2$	111-76-2	20	97		
Ethylene glycol monoethyl ether [2-Ethoxyethanol]	$C_4H_{10}O_2$	110-80-5	5	18		
Ethylene glycol monoethyl ether acetate [2-Ethoxyethyl acetate]	$C_6H_{12}O_3$	111-15-9	5	27		
Ethylene glycol monomethyl ether [2-Methoxyethanol]	$C_3H_8O_2$	109-86-4	5	16		
Ethylene glycol monomethyl ether acetate [2-Methoxyethyl acetate]	$C_5H_{10}O_3$	110-49-6	5	24		
Ethyleneimine	C_2H_5N	151-56-4	0.5	0.88		
Ethylene oxide [Oxirane]	C_2H_4O	75-21-8	1	1.8		
Ethyl formate	$C_3H_6O_2$	109-94-4	100	303		
Ethylidene norbornene	C_9H_{12}	16219-75-3			5 C	25 C
N-Ethylmorpholine	$C_6H_{13}NO$	100-74-3	5	24		

Substance	Molecular Formula	CAS Reg. No.	Time-Weighted Average		Short-Term Exposure Limit	
			ppm	mg/m^3	ppm	mg/m^3
Ethyl p-nitrophenyl benzenethiophosphate [EPN]	$C_{14}H_{14}NO_4PS$	2104-64-5		0.1		
Ethyl silicate	$C_8H_{20}O_4Si$	78-10-4	10	85		
Fenamiphos	$C_{13}H_{22}NO_3PS$	22224-92-6		0.1		
Fensulfothion	$C_{11}H_{17}O_4PS_2$	115-90-2		0.1		
Fenthion	$C_{10}H_{15}O_3PS_2$	55-38-9		0.2		
Ferbam	$C_9H_{18}FeN_3S_6$	14484-64-1		10		
Ferrocene [Dicyclopentadienyl iron]	$C_{10}H_{10}Fe$	102-54-5		10		
Fluorine	F_2	7782-41-4	1	1.6	2	3.1
Fluorine monoxide [Oxygen difluoride]	F_2O	7783-41-7			0.05 C	0.11 C
Fonofos	$C_{10}H_{15}OPS_2$	944-22-9		0.1		
Formaldehyde	CH_2O	50-00-0			0.3 C	0.37 C
Formamide	CH_3NO	75-12-7	10	18		
Formic acid	CH_2O_2	64-18-6	5	9.4	10	19
Furfural [2-Furaldehyde]	$C_5H_4O_2$	98-01-1	2	7.9		
Furfuryl alcohol [2-Furanmethanol]	$C_5H_6O_2$	98-00-0	10	40	15	60
Germane [Germanium tetrahydride]	GeH_4	7782-65-2	0.2	0.63		
Glycerol	$C_3H_8O_3$	56-81-5		10		
Graphite	C	7440-44-0		2		
Hafnium	Hf	7440-58-6		0.5		
Heptachlor	$C_{10}H_5Cl_7$	76-44-8		0.05		
Heptane	C_7H_{16}	142-82-5	400	1640	500	2050
2-Heptanone [Methyl pentyl ketone]	$C_7H_{14}O$	110-43-0	50	233		
3-Heptanone [Ethyl butyl ketone]	$C_7H_{14}O$	106-35-4	50	233	75	350
4-Heptanone [Dipropyl ketone]	$C_7H_{14}O$	123-19-3	50	233		
Hexachlorobenzene	C_6Cl_6	118-74-1		0.002		
Hexachloro-1,3-butadiene	C_4Cl_6	87-68-3	0.02	0.21		
1,2,3,4,5,6-Hexachlorocyclohexane [Lindane]	$C_6H_6Cl_6$	58-89-9		0.5		
Hexachloro-1,3-cyclopentadiene	C_5Cl_6	77-47-4	0.01	0.11		
Hexachloroethane [Perchloroethane]	C_2Cl_6	67-72-1	1	9.7		
Hexachloronaphthalene (all isomers)	$C_{10}H_2Cl_6$	1335-87-1		0.2		
Hexamethylene diisocyanate	$C_8H_{12}N_2O_2$	822-06-0	0.005	0.034		
Hexane	C_6H_{14}	110-54-3	50	176		
1,6-Hexanediamine [Hexamethylenediamine]	$C_6H_{16}N_2$	124-09-4	0.5	2.3		
2-Hexanone [Butyl methyl ketone]	$C_6H_{12}O$	591-78-6	5	20	10	40
1-Hexene	C_6H_{12}	592-41-6	30	103		
sec-Hexyl acetate	$C_8H_{16}O_2$	108-84-9	50	295		
Hydrazine	H_4N_2	302-01-2	0.01	0.013		
Hydrazoic acid	HN_3	7782-79-8			0.11 C	0.19 C
Hydrogen bromide	BrH	10035-10-6			3 C	9.9 C
Hydrogen chloride	ClH	7647-01-0			5 C	7.5 C
Hydrogen cyanide	CHN	74-90-8			4.7 C	5 C
Hydrogen fluoride	FH	7664-39-3			3 C	2.3 C
Hydrogen peroxide	H_2O_2	7722-84-1	1	1.4		
Hydrogen selenide	H_2Se	7783-07-5	0.05	0.16		
Hydrogen sulfide	H_2S	7783-06-4	10	14	15	21
p-Hydroquinone [1,4-Benzenediol]	$C_6H_6O_2$	123-31-9		2		
2-Hydroxypropyl acrylate	$C_6H_{10}O_3$	999-61-1	0.5	2.8		
Indene	C_9H_8	95-13-6	10	48		
Indium	In	7440-74-6		0.1		
Iodine	I_2	7553-56-2			0.1 C	1.0 C
Iodomethane [Methyl iodide]	CH_3I	74-88-4	2	12		
Iron(III) oxide	Fe_2O_3	1309-37-1		5		
Iron pentacarbonyl	C_5FeO_5	13463-40-6	0.1	0.23	0.2	0.45
Isobutyl acetate	$C_6H_{12}O_2$	110-19-0	150	713		
Isopentane	C_5H_{12}	78-78-4	600	1770		
Isopentyl acetate [Isoamyl acetate]	$C_7H_{14}O_2$	123-92-2	100	532		

Substance	Molecular Formula	CAS Reg. No.	Time-Weighted Average		Short-Term Exposure Limit	
			ppm	mg/m^3	ppm	mg/m^3
Isophorone	$C_9H_{14}O$	78-59-1			5 C	28 C
Isophorone diisocyanate	$C_{12}H_{18}N_2O_2$	4098-71-9	0.005	0.045		
Isopropenylbenzene [α-Methyl styrene]	C_9H_{10}	98-83-9	50	242	100	483
2-Isopropoxyethanol	$C_5H_{12}O_2$	109-59-1	25	106		
Isopropyl acetate	$C_5H_{10}O_2$	108-21-4	250	1040	310	1290
Isopropylamine	C_3H_9N	75-31-0	5	12	10	24
N-Isopropylaniline	$C_9H_{13}N$	768-52-5	2	11		
Isopropylbenzene [Cumene]	C_9H_{12}	98-82-8	50	246		
Isopropyl glycidyl ether	$C_6H_{12}O_2$	4016-14-2	50	238	75	356
Kaolin		1332-58-7		2		
Ketene	C_2H_2O	463-51-4	0.5	0.86	1.5	2.6
Lead	Pb	7439-92-1		0.05		
Lead(II) arsenate	$As_2O_8Pb_3$	7784-40-9		0.15		
Lead(II) chromate	CrO_4Pb	7758-97-6		0.075		
Lithium hydride	HLi	7580-67-8		0.025		
Magnesium carbonate [Magnesite]	$CMgO_3$	546-93-0		10		
Magnesium oxide	MgO	1309-48-4		10		
Malathion	$C_{10}H_{19}O_6PS_2$	121-75-5		10		
Maleic anhydride	$C_4H_2O_3$	108-31-6	0.25	1.0		
Manganese	Mn	7439-96-5		0.2		
Manganese cyclopentadienyl tricarbonyl	$C_8H_5MnO_3$	12079-65-1		0.4		
Mercury	Hg	7439-97-6		0.025		
Mesityl oxide	$C_6H_{10}O$	141-79-7	15	60	25	100
Methacrylic acid [2-Methylpropenoic acid]	$C_4H_6O_2$	79-41-4	20	70		
Methanethiol [Methyl mercaptan]	CH_4S	74-93-1	0.5	0.98		
Methanol	CH_4O	67-56-1	200	262	250	328
Methomyl	$C_5H_{10}N_2O_2S$	16752-77-5		2.5		
o-Methoxyaniline [o-Anisidine]	C_7H_9NO	90-04-0	0.1	0.5		
p-Methoxyaniline [p-Anisidine]	C_7H_9NO	104-94-9	0.1	0.5		
Methoxychlor	$C_{16}H_{15}Cl_3O_2$	72-43-5		10		
4-Methoxyphenol	$C_7H_8O_2$	150-76-5		5		
Methyl acetate	$C_3H_6O_2$	79-20-9	200	606	250	757
Methyl acrylate	$C_4H_6O_2$	96-33-3	2	7		
2-Methylacrylonitrile	C_4H_5N	126-98-7	1	2.7		
Methylamine	CH_5N	74-89-5	5	6.4	15	19
o-Methylaniline [o-Toluidine]	C_7H_9N	95-53-4	2	8.8		
m-Methylaniline [m-Toluidine]	C_7H_9N	108-44-1	2	8.8		
p-Methylaniline [p-Toluidine]	C_7H_9N	106-49-0	2	8.8		
N-Methylaniline	C_7H_9N	100-61-8	0.5	2.2		
3-Methyl-1-butanol [Isoamyl alcohol]	$C_5H_{12}O$	123-51-3	100	361	125	452
3-Methyl-2-butanone [Methyl isopropyl ketone]	$C_5H_{10}O$	563-80-4	200	705		
Methyl tert-butyl ether [MTBE]	$C_5H_{12}O$	1634-04-4	40	144		
Methyl 2-cyanoacrylate	$C_5H_5NO_2$	137-05-3	0.2	0.9		
Methylcyclohexane	C_7H_{14}	108-87-2	400	1610		
Methylcyclohexanol (all isomers)	$C_7H_{14}O$	25639-42-3	50	234		
2-Methylcyclohexanone	$C_7H_{12}O$	583-60-8	50	229	75	344
2-Methylcyclopentadienyl manganese tricarbonyl	$C_9H_7MnO_3$	12108-13-3		0.8		
Methyl demeton	$C_6H_{15}O_3PS_2$	8022-00-2		0.5		
2-Methyl-3,5-dinitrobenzamide [Dinitolmide]	$C_8H_7N_3O_5$	148-01-6		5		
2-Methyl-4,6-dinitrophenol [Dinitro-o-cresol]	$C_7H_6N_2O_5$	534-52-1		0.2		
Methylene bis(4-cyclohexylisocyanate)	$C_{15}H_{22}N_2O_2$	5124-30-1	0.005	0.054		
Methyl ethyl ketone peroxide	$C_8H_{18}O_2$	1338-23-4			0.2 C	1.5 C
Methyl formate	$C_2H_4O_2$	107-31-3	100	246	150	368
6-Methyl-1-heptanol [Isooctyl alcohol]	$C_8H_{18}O$	26952-21-6	50	266		
5-Methyl-3-heptanone	$C_8H_{16}O$	541-85-5	25	131		
5-Methyl-2-hexanone [Methyl isopentyl ketone]	$C_7H_{14}O$	110-12-3	50	234		

Substance	Molecular Formula	CAS Reg. No.	Time-Weighted Average		Short-Term Exposure Limit	
			ppm	mg/m^3	ppm	mg/m^3
Methylhydrazine	CH_6N_2	60-34-4	0.01	0.019		
Methyl isocyanate	C_2H_3NO	624-83-9	0.02	0.047		
Methyl methacrylate	$C_5H_8O_2$	80-62-6	100	410		
Methyloxirane [1,2-Propylene oxide]	C_3H_6O	75-56-9	20	48		
Methyl parathion	$C_8H_{10}NO_5PS$	298-00-0		0.2		
2-Methylpentane	C_6H_{14}	107-83-5	500	1760	1000	3500
3-Methylpentane	C_6H_{14}	96-14-0	500	1760	1000	3500
2-Methyl-2,4-pentanediol [Hexylene glycol]	$C_6H_{14}O_2$	107-41-5			25 C	121 C
4-Methyl-2-pentanol [Methyl isobutyl carbinol]	$C_6H_{14}O$	108-11-2	25	104	40	167
4-Methyl-2-pentanone [Isobutyl methyl ketone]	$C_6H_{12}O$	108-10-1	50	205	75	307
2-Methyl-1-propanol [Isobutyl alcohol]	$C_4H_{10}O$	78-83-1	50	152		
2-Methyl-2-propanol [*tert*-Butyl alcohol]	$C_4H_{10}O$	75-65-0	100	303		
Methylstyrene (all isomers)	C_9H_{10}	25013-15-4	50	242	100	483
N-Methyl-*N*,2,4,6-tetranitroaniline [Tetryl]	$C_7H_5N_5O_8$	479-45-8		1.5		
Metribuzin	$C_8H_{14}N_4OS$	21087-64-9		5		
Mevinphos	$C_7H_{13}O_6P$	7786-34-7	0.01	0.092	0.03	0.27
Mica		12001-26-2		3		
Molybdenum	Mo	7439-98-7		10		
Monocrotophos	$C_7H_{14}NO_5P$	6923-22-4		0.25		
Morpholine	C_4H_9NO	110-91-8	20	71		
Naled	$C_4H_7Br_2Cl_2O_4P$	300-76-5		3		
Naphthalene	$C_{10}H_8$	91-20-3	10	52	15	79
1-Naphthalenylthiourea [ANTU]	$C_{11}H_{10}N_2S$	86-88-4		0.3		
Neopentane	C_5H_{12}	463-82-1	600	1770		
Nickel	Ni	7440-02-0		1.5		
Nickel carbonyl	C_4NiO_4	13463-39-3	0.05	0.12		
Nickel(III) sulfide	Ni_3S_2	12035-72-2		0.14		
Nicotine	$C_{10}H_{14}N_2$	54-11-5		0.5		
Nitrapyrin	$C_6H_3Cl_4N$	1929-82-4		10		20
Nitric acid	HNO_3	7697-37-2	2	5.2	4	10
Nitric oxide	NO	10102-43-9	25	31		
p-Nitroaniline	$C_6H_6N_2O_2$	100-01-6		3		
Nitrobenzene	$C_6H_5NO_2$	98-95-3	1	5		
Nitroethane	$C_2H_5NO_2$	79-24-3	100	307		
Nitrogen dioxide	NO_2	10102-44-0	3	5.6	5	9.4
Nitrogen trifluoride	F_3N	7783-54-2	10	29		
Nitromethane	CH_3NO_2	75-52-5	20	50		
1-Nitropropane	$C_3H_7NO_2$	108-03-2	25	91		
2-Nitropropane	$C_3H_7NO_2$	79-46-9	10	36		
Nitrotoluene (all isomers)	$C_7H_7NO_2$	1321-12-6	2	11		
Nitrous oxide	N_2O	10024-97-2	50	90		
Nonane (all isomers)	C_9H_{20}	111-84-2	200	1050		
Octachloronaphthalene	$C_{10}Cl_8$	2234-13-1		0.1		0.3
Octane (all isomers)	C_8H_{18}	111-65-9	300	1400	375	1750
Osmium(VIII) oxide [Osmium tetroxide]	O_4Os	20816-12-0	0.0002	0.0016	0.0006	0.0047
Oxalic acid	$C_2H_2O_4$	144-62-7		1		2
2-Oxetanone [β-Propiolactone]	$C_3H_4O_2$	57-57-8	0.5	1.5		
Oxiranemethanol [Glycidol]	$C_3H_6O_2$	556-52-5	2	6.1		
Ozone	O_3	10028-15-6	0.1	0.2		
Paraquat	$C_{12}H_{14}N_2$	4685-14-7		0.5		
Parathion	$C_{10}H_{14}NO_5PS$	56-38-2		0.1		
Pentaborane(9)	B_5H_9	19624-22-7	0.005	0.013	0.015	0.039
Pentachloronaphthalene (unspecified isomer)	$C_{10}H_3Cl_5$	1321-64-8		0.5		
Pentachloronitrobenzene	$C_6Cl_5NO_2$	82-68-8		0.5		
Pentachlorophenol	C_6HCl_5O	87-86-5		0.5		
Pentaerythritol	$C_5H_{12}O_4$	115-77-5		10		
Pentanal [Valeraldehyde]	$C_5H_{10}O$	110-62-3	50	176		

Substance	Molecular Formula	CAS Reg. No.	Time-Weighted Average		Short-Term Exposure Limit	
			ppm	mg/m³	ppm	mg/m³
Pentane	C_5H_{12}	109-66-0	600	1770	750	2210
Pentanedial [Glutaraldehyde]	$C_5H_8O_2$	111-30-8			0.05 C	0.2 C
2-Pentanone [Methyl propyl ketone]	$C_5H_{10}O$	107-87-9	200	705	250	881
3-Pentanone [Diethyl ketone]	$C_5H_{10}O$	96-22-0	200	700	300	1050
Pentyl acetate [Amyl acetate]	$C_7H_{14}O_2$	628-63-7	100	532		
sec-Pentyl acetate	$C_7H_{14}O_2$	626-38-0	125	665		
Perchloromethyl mercaptan	CCl_4S	594-42-3	0.1	0.76		
Perchloryl fluoride	$ClFO_3$	7616-94-6	3	13	6	25
Perfluoroacetone [Hexafluoroacetone]	C_3F_6O	684-16-2	0.1	0.68		
Perfluoroisobutene	C_4F_8	382-21-8			0.01 C	0.082 C
Phenol	C_6H_6O	108-95-2	5	19		
10H-Phenothiazine	$C_{12}H_9NS$	92-84-2		5		
Phenylenediamine (all isomers)	$C_6H_8N_2$	25265-76-3		0.1		
Phenyl glycidyl ether	$C_9H_{10}O_2$	122-60-1	0.1	0.6		
Phenylhydrazine	$C_6H_8N_2$	100-63-0	0.1	0.44		
Phenylphosphine	C_6H_7P	638-21-1			0.05 C	0.23 C
Phorate	$C_7H_{17}O_2PS_3$	298-02-2		0.05		0.2
Phosphine	H_3P	7803-51-2	0.3	0.42	1	1.4
Phosphoric acid	H_3O_4P	7664-38-2		1		3
Phosphorus (white)	P	7723-14-0	0.02	0.1		
Phosphorus(III) chloride [Phosphorus trichloride]	Cl_3P	7719-12-2	0.2	1.1	0.5	2.8
Phosphorus(V) chloride [Phosphorus pentachloride]	Cl_5P	10026-13-8	0.1	0.85		
Phosphorus(V) oxychloride [Phosphoryl chloride]	Cl_3OP	10025-87-3	0.1	0.63		
Phosphorus(V) sulfide	P_2S_5	1314-80-3		1		3
Phthalic anhydride	$C_8H_4O_3$	85-44-9	1	6.1		
Piperazine dihydrochloride	$C_4H_{12}Cl_2N_2$	142-64-3		5		
2-Pivaloyl-1,3-indandione [Pindone]	$C_{14}H_{14}O_3$	83-26-1		0.1		
Platinum	Pt	7440-06-4		1		
Potassium hydroxide	HKO	1310-58-3		2 C		
Propane	C_3H_8	74-98-6	2500	4500		
Propanoic acid	$C_3H_6O_2$	79-09-4	10	30		
1-Propanol	C_3H_8O	71-23-8	200	492	250	614
2-Propanol [Isopropyl alcohol]	C_3H_8O	67-63-0	400	983	500	1230
Propargyl alcohol [2-Propyn-1-ol]	C_3H_4O	107-19-7	1	2.3		
Propoxur	$C_{11}H_{15}NO_3$	114-26-1		0.5		
Propyl acetate	$C_5H_{10}O_2$	109-60-4	200	835	250	1040
1,2-Propylene glycol dinitrate	$C_3H_6N_2O_6$	6423-43-4	0.05	0.34		
Propylene glycol monomethyl ether	$C_4H_{10}O_2$	107-98-2	100	369	150	553
Propyleneimine	C_3H_7N	75-55-8	2	4.7		
Propyl nitrate	$C_3H_7NO_3$	627-13-4	25	107	40	172
Propyne [Methylacetylene]	C_3H_4	74-99-7	1000	1640		
2-Pyridinamine [2-Aminopyridine]	$C_5H_6N_2$	504-29-0	0.5	1.9		
Pyridine	C_5H_5N	110-86-1	5	16		
Pyrocatechol [Catechol]	$C_6H_6O_2$	120-80-9	5	23		
Resorcinol	$C_6H_6O_2$	108-46-3	10	45	20	90
Rhodium	Rh	7440-16-6		1		
Ronnel	$C_8H_8Cl_3O_3PS$	299-84-3		10		
Rotenone	$C_{23}H_{22}O_6$	83-79-4		5		
Selenium	Se	7782-49-2		0.2		
Selenium hexafluoride	F_6Se	7783-79-1	0.12	0.95		
Sesone	$C_8H_7Cl_2NaO_5S$	136-78-7		10		
Silane	H_4Si	7803-62-5	5	6.6		
Silicon	Si	7440-21-3		10		
Silicon carbide	CSi	409-21-2		10		

Substance	Molecular Formula	CAS Reg. No.	Time-Weighted Average		Short-Term Exposure Limit	
			ppm	mg/m^3	ppm	mg/m^3
Silicon dioxide (α-quartz)	O_2Si	14808-60-7		0.1		
Silicon dioxide (tridymite)	O_2Si	15468-32-3		0.05		
Silicon dioxide (cristobalite)	O_2Si	14464-46-1		0.05		
Silicon dioxide (vitreous)	O_2Si	60676-86-0		0.1		
Silver	Ag	7440-22-4		0.1		
Sodium azide	N_3Na	26628-22-8				0.29 C
Sodium fluoroacetate	$C_2H_2FNaO_2$	62-74-8		0.05		
Sodium hydrogen sulfite	$HNaO_3S$	7631-90-5		5		
Sodium hydroxide	HNaO	1310-73-2				2 C
Sodium metabisulfite	$Na_2O_5S_2$	7681-57-4		5		
Sodium pyrophosphate	$Na_4O_7P_2$	7722-88-5		5		
Sodium tetraborate decahydrate	$B_4H_{20}Na_2O_{17}$	1303-96-4		5		
Stibine	H_3Sb	7803-52-3	0.1	0.51		
Strontium chromate	CrO_4Sr	7789-06-2		0.002		
Strychnine	$C_{21}H_{22}N_2O_2$	57-24-9		0.15		
Styrene	C_8H_8	100-42-5	20	85	40	170
Sucrose	$C_{12}H_{22}O_{11}$	57-50-1		10		
Sulfotep	$C_8H_{20}O_5P_2S_2$	3689-24-5		0.2		
Sulfur chloride	Cl_2S_2	10025-67-9			1 C	5.5 C
Sulfur decafluoride	$F_{10}S_2$	5714-22-7			0.01 C	0.10 C
Sulfur dioxide	O_2S	7446-09-5	2	5.2	5	13
Sulfur hexafluoride	F_6S	2551-62-4	1000	6000		
Sulfur tetrafluoride	F_4S	7783-60-0			0.1 C	0.44 C
Sulfuric acid	H_2O_4S	7664-93-9		1		3
Sulfuryl fluoride	F_2O_2S	2699-79-8	5	21	10	42
Sulprofos	$C_{12}H_{19}O_2PS_3$	35400-43-2		1		
Talc		14807-96-6		2		
Tantalum	Ta	7440-25-7		5		
Tantalum(V) oxide	O_5Ta_2	1314-61-0		5		
Tellurium	Te	13494-80-9		0.1		
Tellurium hexafluoride	F_6Te	7783-80-4	0.02	0.10		
Terephthalic acid	$C_8H_6O_4$	100-21-0		10		
Terphenyl (all isomers)	$C_{18}H_{14}$	26140-60-3			0.53 C	5 C
1,1,2,2-Tetrabromoethane [Acetylene tetrabromide]	$C_2H_2Br_4$	79-27-6	1	14		
Tetrabromomethane [Carbon tetrabromide]	CBr_4	558-13-4	0.1	1.4	0.3	4.1
1,1,1,2-Tetrachloro-2,2-difluoroethane	$C_2Cl_4F_2$	76-11-9	500	4170		
1,1,2,2-Tetrachloro-1,2-difluoroethane	$C_2Cl_4F_2$	76-12-0	500	4170		
1,1,2,2-Tetrachloroethane	$C_2H_2Cl_4$	79-34-5	1	6.9		
Tetrachloroethene [Perchloroethylene]	C_2Cl_4	127-18-4	25	170	100	685
Tetrachloromethane [Carbon tetrachloride]	CCl_4	56-23-5	5	31	10	63
Tetrachloronaphthalene (all isomers)	$C_{10}H_4Cl_4$	1335-88-2		2		
Tetraethyl lead	$C_8H_{20}Pb$	78-00-2		0.1		
Tetraethyl pyrophosphate [TEPP]	$C_8H_{20}O_7P_2$	107-49-3		0.05		
Tetrahydrofuran [Oxolane]	C_4H_8O	109-99-9	200	590	250	737
Tetramethyl lead	$C_4H_{12}Pb$	75-74-1		0.15		
Tetramethyl silicate	$C_4H_{12}O_4Si$	681-84-5	1	6		
Tetramethyl succinonitrile	$C_8H_{12}N_2$	3333-52-6	0.5	2.8		
Tetranitromethane	CN_4O_8	509-14-8	0.005	0.04		
Thallium	Tl	7440-28-0		0.1		
4,4'-Thiobis(6-*tert*-butyl-*m*-cresol)	$C_{22}H_{30}O_2S$	96-69-5		10		
Thioglycolic acid	$C_2H_4O_2S$	68-11-1	1	3.8		
Thionyl chloride	Cl_2OS	7719-09-7			1 C	4.9 C
Thiram	$C_6H_{12}N_2S_4$	137-26-8		1		
Tin	Sn	7440-31-5		2		
Titanium(IV) oxide [Titanium dioxide]	O_2Ti	13463-67-7		10		
Toluene	C_7H_8	108-88-3	50	188		

THRESHOLD LIMITS FOR AIRBORNE CONTAMINANTS (continued)

Substance	Molecular Formula	CAS Reg. No.	Time-Weighted Average		Short-Term Exposure Limit	
			ppm	mg/m^3	ppm	mg/m^3
Toluene-2,4-diisocyanate	$C_9H_6N_2O_2$	584-84-9	0.005	0.036	0.02	0.14
1H-1,2,4-Triazol-3-amine	$C_2H_4N_4$	61-82-5		0.2		
Tribromomethane [Bromoform]	$CHBr_3$	75-25-2	0.5	5.2		
Tributyl phosphate	$C_{12}H_{27}O_4P$	126-73-8	0.2	2.2		
Trichloroacetic acid	$C_2HCl_3O_2$	76-03-9	1	6.7		
1,2,4-Trichlorobenzene	$C_6H_3Cl_3$	120-82-1			5 C	37 C
1,1,1-Trichloroethane [Methyl chloroform]	$C_2H_3Cl_3$	71-55-6	350	1910	450	2460
1,1,2-Trichloroethane	$C_2H_3Cl_3$	79-00-5	10	55		
Trichloroethene	C_2HCl_3	79-01-6	50	269	100	537
Trichlorofluoromethane	CCl_3F	75-69-4			1000 C	5620 C
Trichloromethane [Chloroform]	$CHCl_3$	67-66-3	10	49		
(Trichloromethyl)benzene [Benzotrichloride]	$C_7H_5Cl_3$	98-07-7			0.01 C	0.08 C
Trichloronaphthalene (all isomers)	$C_{10}H_5Cl_3$	1321-65-9		5		
Trichloronitromethane [Chloropicrin]	CCl_3NO_2	76-06-2	0.1	0.67		
2,4,5-Trichlorophenoxyacetic acid	$C_8H_5Cl_3O_3$	93-76-5		10		
1,2,3-Trichloropropane	$C_3H_5Cl_3$	96-18-4	10	60		
1,1,2-Trichloro-1,2,2-trifluoroethane	$C_2Cl_3F_3$	76-13-1	1000	7670	1250	9590
Tri-o-cresyl phosphate	$C_{21}H_{21}O_4P$	78-30-8		0.1		
Triethanolamine	$C_6H_{15}NO_3$	102-71-6		5		
Triethylamine	$C_6H_{15}N$	121-44-8	1	4.1	3	12
Triiodomethane [Iodoform]	CHI_3	75-47-8	0.6	10		
Trimellitic anhydride [1,2,4-Benzenetricarboxylic anhydride]	$C_9H_4O_5$	552-30-7				0.04 C
Trimethylamine	C_3H_9N	75-50-3	5	12	15	36
Trimethylbenzene (all isomers)	C_9H_{12}	25551-13-7	25	123		
Trimethyl phosphite	$C_3H_9O_3P$	121-45-9	2	10		
Trinitroglycerol [Nitroglycerin]	$C_3H_5N_3O_9$	55-63-0	0.05	0.46		
2,4,6-Trinitrophenol [Picric acid]	$C_6H_3N_3O_7$	88-89-1		0.1		
2,4,6-Trinitrotoluene [TNT]	$C_7H_5N_3O_6$	118-96-7		0.1		
Triphenylamine	$C_{18}H_{15}N$	603-34-9		5		
Triphenyl phosphate	$C_{18}H_{15}O_4P$	115-86-6		3		
Tungsten	W	7440-33-7		5		10
Uranium	U	7440-61-1		0.2		0.6
Vanadium(V) oxide	O_5V_2	1314-62-1		0.05		
Vinyl acetate	$C_4H_6O_2$	108-05-4	10	35	15	53
4-Vinylcyclohexene	C_8H_{12}	100-40-3	0.1	0.44		
Warfarin	$C_{19}H_{16}O_4$	81-81-2		0.1		
Xylene (all isomers)	C_8H_{10}	1330-20-7	100	434	150	651
Xylidine (all isomers)	$C_8H_{11}N$	1300-73-8	0.5	2.5		
Yttrium	Y	7440-65-5		1		
Zinc chloride	Cl_2Zn	7646-85-7		1		2
Zinc chromate, basic	CrH_2O_4Zn	13530-65-9		0.045		
Zinc oxide	OZn	1314-13-2		5		10
Zirconium	Zr	7440-67-7		5		10

OCTANOL-WATER PARTITION COEFFICIENTS

The octanol-water partition coefficient, P, is a widely used parameter for correlating biological effects of organic substances. It is a property of the two-phase system in which water and 1-octanol are in equilibrium at a fixed temperature and the substance is distributed between the water-rich and octanol-rich phases. P is defined as the ratio of the equilibrium concentration of the substance in the octanol-rich phase to that in the water-rich phase, in the limit of zero concentration. In general, P tends to be large for compounds with extended non-polar structures (such as long chain or multi-ring hydrocarbons) and small for compounds with highly polar groups. Thus P (or, in its more common form of expression, log P) provides a measure of the lipophilic vs. hydrophilic nature of a compound, which is an important consideration in assessing the potential toxicity. A discussion of methods of measurement and accuracy considerations for log P may be found in Reference 1.

This table gives selected values of log P for about 450 organic compounds, including many of environmental importance. All values refer to a nominal temperature of 25°C. The source of each value is indicated in the last column. These references contain data on many more compounds than are included here.

Compounds are listed by molecular formula following the Hill convention. To locate a compound by name or CAS Registry Number when the molecular formula is not known, use the table "Physical Constants of Organic Compounds" in Section 3 and its indexes to determine the molecular formula.

REFERENCES

1. Sangster, J., *J. Phys. Chem. Ref. Data*, 18, 1111, 1989.
2. Mackay, D., Shiu, W.Y., and Ma, K.C., *Illustrated Handbook of Physical-Chemical Properties and Environmental Fate for Organic Chemicals*, Lewis Publishers/CRC Press, Boca Raton, FL, 1992.
3. Shiu, W.Y., and Mackay, D., *J. Phys. Chem. Ref. Data*, 15, 911, 1986.
4. Pinsuwan, S., Li, L., and Yalkowsky, S.H., *J. Chem. Eng. Data*, 40, 623, 1995.
5. *Solubility Data Series, International Union of Pure and Applied Chemistry, Vol. 20*, Pergamon Press, Oxford, 1985.
6. *Solubility Data Series, International Union of Pure and Applied Chemistry, Vol. 38*, Pergamon Press, Oxford, 1985.
7. Miller, M.M., Ghodbane, S., Wasik, S.P., Tewari, Y.B., and Martire, D.E., *J. Chem. Eng. Data*, 29, 184, 1984.

Mol. Form.	Name	log P	Ref.	Mol. Form.	Name	log P	Ref.
CCl_2F_2	Dichlorodifluoromethane	2.16	2	C_2H_4O	Acetaldehyde	0.45	1
CCl_3F	Trichlorofluoromethane	2.53	2	C_2H_4O	Ethylene oxide	-0.30	1
CCl_4	Tetrachloromethane	2.64	2	$C_2H_4O_2$	Acetic acid	-0.17	1
$CHBr_3$	Tribromomethane	2.38	2	C_2H_5Br	Bromoethane	1.6	2
$CHCl_3$	Trichloromethane	1.97	2	C_2H_5Cl	Chloroethane	1.43	2
CH_2BrCl	Bromochloromethane	1.41	2	C_2H_5I	Iodoethane	2	2
CH_2Br_2	Dibromomethane	2.3	2	C_2H_5NO	Acetamide	-1.26	1
CH_2Cl_2	Dichloromethane	1.25	2	$C_2H_5NO_2$	Nitroethane	0.18	1
CH_2F_2	Difluoromethane	0.20	1	C_2H_6O	Ethanol	-0.30	1
CH_2I_2	Diiodomethane	2.5	2	C_2H_6O	Dimethyl ether	0.10	1
CH_2O	Formaldehyde	0.35	1	C_2H_6OS	Dimethyl sulfoxide	-1.35	1
CH_2O_2	Formic acid	-0.54	1	$C_2H_6O_2S$	Dimethyl sulfone	-1.41	1
CH_3Br	Bromomethane	1.19	2	C_2H_7N	Ethylamine	-0.13	1
CH_3Cl	Chloromethane	0.91	2	C_2H_7N	Dimethylamine	-0.38	1
CH_3F	Fluoromethane	0.51	1	C_3H_3N	2-Propenenitrile	0.25	1
CH_3I	Iodomethane	1.5	2	$C_3H_4Cl_2$	*cis*-1,3-Dichloropropene	2.03	2
CH_3NO	Formamide	-1.51	1	C_3H_4O	Propargyl alcohol	-0.38	1
CH_3NO_2	Nitromethane	-0.33	1	C_3H_4O	Acrolein	-0.01	1
CH_4O	Methanol	-0.74	1	C_3H_5Br	3-Bromopropene	1.79	1
CH_5N	Methylamine	-0.57	1	C_3H_5ClO	Epichlorohydrin	0.30	2
$C_2Cl_3F_3$	1,1,2-Trichlorotrifluoroethane	3.16	2	$C_3H_5Cl_3$	1,2,3-Trichloropropane	2.63	2
C_2Cl_4	Tetrachloroethylene	2.88	2	C_3H_5N	Propanenitrile	0.16	1
C_2Cl_6	Hexachloroethane	4.00	4	C_3H_5NO	Acrylamide	-0.78	1
C_2HCl_3	Trichloroethylene	2.53	2	$C_3H_6Cl_2$	1,2-Dichloropropane	2.0	2
C_2HCl_5	Pentachloroethane	2.89	2	C_3H_6O	Allyl alcohol	0.17	1
$C_2H_2Cl_2$	1,1-Dichloroethylene	2.13	2	C_3H_6O	Propanal	0.59	1
$C_2H_2Cl_2$	*cis*-1,2-Dichloroethylene	1.86	2	C_3H_6O	Acetone	-0.24	1
$C_2H_2Cl_2$	*trans*-1,2-Dichloroethylene	1.93	2	C_3H_6O	Methyloxirane	0.03	1
$C_2H_2Cl_4$	1,1,2,2-Tetrachloroethane	2.39	2	$C_3H_6O_2$	Propanoic acid	0.33	1
C_2H_3Cl	Chloroethylene	1.38	2	$C_3H_6O_2$	Methyl acetate	0.18	1
$C_2H_3Cl_3$	1,1,1-Trichloroethane	2.49	2	C_3H_7Br	1-Bromopropane	2.1	2
$C_2H_3Cl_3$	1,1,2-Trichloroethane	2.38	2	C_3H_7Br	2-Bromopropane	1.9	2
C_2H_3N	Acetonitrile	-0.34	1	C_3H_7Cl	1-Chloropropane	2.04	1
$C_2H_4Cl_2$	1,1-Dichloroethane	1.79	2	C_3H_7Cl	2-Chloropropane	1.90	1
$C_2H_4Cl_2$	1,2-Dichloroethane	1.48	2	C_3H_7I	1-Iodopropane	2.5	2

Mol. Form.	Name	log P	Ref.	Mol. Form.	Name	log P	Ref.
C_3H_7N	Allylamine	0.03	1	C_5H_{10}	Cyclopentane	3.00	1
C_3H_7NO	*N,N*-Dimethylformamide	-1.01	1	$C_5H_{10}O$	2-Pentanone	0.84	1
C_3H_7NO	*N*-Methylacetamide	-1.05	1	$C_5H_{10}O$	3-Pentanone	0.82	1
$C_3H_7NO_2$	1-Nitropropane	0.87	1	$C_5H_{10}O$	3-Methyl-2-butanone	0.56	1
C_3H_8O	1-Propanol	0.25	1	$C_5H_{10}O$	Tetrahydropyran	0.82	1
C_3H_8O	2-Propanol	0.05	1	$C_5H_{10}O$	2-Methyltetrahydrofuran	1.85	2
C_3H_8S	1-Propanethiol	1.81	1	$C_5H_{10}O_2$	Pentanoic acid	1.39	1
C_3H_9N	Propylamine	0.48	1	$C_5H_{10}O_2$	Propyl acetate	1.24	1
C_3H_9N	Isopropylamine	0.26	1	$C_5H_{10}O_2$	Ethyl propanoate	1.21	1
C_3H_9N	Ethylmethylamine	0.15	1	$C_5H_{10}O_3$	Diethyl carbonate	1.21	1
C_3H_9N	Trimethylamine	0.16	1	$C_5H_{11}Br$	1-Bromopentane	3.37	1
C_4H_4O	Furan	1.34	1	$C_5H_{11}F$	1-Fluoropentane	2.33	1
C_4H_4S	Thiophene	1.81	1	$C_5H_{11}N$	Piperidine	0.84	1
C_4H_5N	Pyrrole	0.75	1	$C_5H_{11}NO_2$	1-Nitropentane	2.01	1
C_4H_6	1,3-Butadiene	1.99	1	C_5H_{12}	Pentane	3.45	1
C_4H_6	2-Butyne	1.46	1	C_5H_{12}	Neopentane	3.11	1
C_4H_6O	2,5-Dihydrofuran	0.46	1	$C_5H_{12}O$	1-Pentanol	1.51	1
$C_4H_6O_2$	Methacrylic acid	0.93	1	$C_5H_{12}O$	2-Pentanol	1.25	1
$C_4H_6O_2$	Vinyl acetate	0.73	1	$C_5H_{12}O$	3-Pentanol	1.21	1
$C_4H_6O_2$	Methyl acrylate	0.80	1	$C_5H_{12}O$	3-Methyl-1-butanol	1.28	1
C_4H_7N	Butanenitrile	0.60	1	$C_5H_{12}O$	2-Methyl-2-butanol	0.89	1
C_4H_8	*cis*-2-Butene	2.33	1	$C_5H_{12}O$	3-Methyl-2-butanol	1.28	1
C_4H_8	*trans*-2-Butene	2.31	1	$C_5H_{12}O$	2,2-Dimethyl-1-propanol	1.31	1
C_4H_8	Isobutene	2.35	1	$C_5H_{12}O$	Methyl *tert*-butyl ether	0.94	1
$C_4H_8Cl_2O$	Bis(2-chloroethyl) ether	1.12	2	$C_5H_{13}N$	Pentylamine	1.49	1
C_4H_8O	Ethyl vinyl ether	1.04	1	C_6Cl_6	Hexachlorobenzene	5.47	5
C_4H_8O	Butanal	0.88	1	C_6HCl_5	Pentachlorobenzene	5.03	5
C_4H_8O	2-Butanone	0.29	1	C_6HCl_5O	Pentachlorophenol	5.07	4
C_4H_8O	Tetrahydrofuran	0.46	1	$C_6H_2Cl_4$	1,2,3,4-Tetrachlorobenzene	4.55	5
$C_4H_8O_2$	Butanoic acid	0.79	1	$C_6H_2Cl_4$	1,2,3,5-Tetrachlorobenzene	4.65	5
$C_4H_8O_2$	Propyl formate	0.83	1	$C_6H_2Cl_4$	1,2,4,5-Tetrachlorobenzene	4.51	5
$C_4H_8O_2$	Ethyl acetate	0.73	1	$C_6H_3Cl_3$	1,2,3-Trichlorobenzene	4.04	5
C_4H_9Br	1-Bromobutane	2.75	1	$C_6H_3Cl_3$	1,2,4-Trichlorobenzene	3.98	5
C_4H_9Cl	1-Chlorobutane	2.64	2	$C_6H_3Cl_3$	1,3,5-Trichlorobenzene	4.02	5
C_4H_9F	1-Fluorobutane	2.58	1	$C_6H_4Cl_2$	*o*-Dichlorobenzene	3.38	5
C_4H_9I	1-Iodobutane	3	2	$C_6H_4Cl_2$	*m*-Dichlorobenzene	3.48	5
C_4H_9N	Pyrrolidine	0.46	1	$C_6H_4Cl_2$	*p*-Dichlorobenzene	3.38	5
C_4H_9NO	Butanamide	-0.21	1	$C_6H_4Cl_2O$	2,4-Dichlorophenol	3.23	4
C_4H_9NO	*N,N*-Dimethylacetamide	-0.77	1	C_6H_5Br	Bromobenzene	2.99	2
$C_4H_9NO_2$	1-Nitrobutane	1.47	1	C_6H_5Cl	Chlorobenzene	2.84	1
C_4H_{10}	Isobutane	2.8	2	C_6H_5F	Fluorobenzene	2.27	2
$C_4H_{10}O$	1-Butanol	0.84	1	C_6H_5I	Iodobenzene	3.28	2
$C_4H_{10}O$	2-Butanol	0.65	1	$C_6H_5NO_2$	Nitrobenzene	1.85	1
$C_4H_{10}O$	2-Methyl-1-propanol	0.76	1	C_6H_6	Benzene	2.13	1
$C_4H_{10}O$	2-Methyl-2-propanol	0.35	1	C_6H_6O	Phenol	1.48	4
$C_4H_{10}O$	Diethyl ether	0.89	1	C_6H_6S	Benzenethiol	2.52	1
$C_4H_{10}S$	1-Butanethiol	2.28	1	C_6H_7N	Aniline	0.90	1
$C_4H_{10}S$	Diethyl sulfide	1.95	1	C_6H_7N	2-Methylpyridine	1.11	1
$C_4H_{11}N$	Butylamine	0.86	1	C_6H_7N	3-Methylpyridine	1.20	1
$C_4H_{11}N$	*tert*-Butylamine	0.40	1	C_6H_7N	4-Methylpyridine	1.22	1
$C_4H_{11}N$	Diethylamine	0.58	1	C_6H_8	1,4-Cyclohexadiene	2.3	2
C_5H_5N	Pyridine	0.65	1	C_6H_8O	5-Hexyn-2-one	0.58	1
C_5H_6O	2-Methylfuran	1.85	1	C_6H_8O	2-Cyclohexen-1-one	0.61	1
C_5H_7N	1-Methylpyrrole	1.21	1	C_6H_8O	2-Ethylfuran	2.40	1
C_5H_8	1,4-Pentadiene	2.48	1	C_6H_{10}	1,5-Hexadiene	2.8	2
C_5H_8	1-Pentyne	1.98	1	C_6H_{10}	1-Hexyne	2.73	2
$C_5H_8O_2$	Methyl methacrylate	1.38	1	C_6H_{10}	Cyclohexene	2.86	1
$C_5H_8O_2$	Ethyl acrylate	1.32	1	$C_6H_{10}O$	5-Hexen-2-one	1.02	1
C_5H_9N	Pentanenitrile	0.94	1	$C_6H_{10}O$	Cyclohexanone	0.81	1
C_5H_{10}	1-Pentene	2.2	2	$C_6H_{10}O_2$	Ethyl methacrylate	1.94	1

Mol. Form.	Name	log P	Ref.	Mol. Form.	Name	log P	Ref.
$C_6H_{11}Br$	Bromocyclohexane	3.20	1	C_7H_{16}	Heptane	4.50	1
$C_6H_{11}N$	Hexanenitrile	1.66	1	$C_7H_{16}O$	1-Heptanol	2.62	1
C_6H_{12}	1-Hexene	3.40	1	$C_7H_{16}O$	2-Heptanol	2.31	1
C_6H_{12}	4-Methyl-1-pentene	2.5	2	$C_7H_{16}O$	3-Heptanol	2.24	1
C_6H_{12}	Cyclohexane	3.44	1	$C_7H_{16}O$	4-Heptanol	2.22	1
C_6H_{12}	Methylcyclopentane	3.37	2	$C_7H_{17}N$	Heptylamine	2.57	1
$C_6H_{12}O$	Cyclohexanol	1.23	1	C_8H_6	Phenylacetylene	2.40	1
$C_6H_{12}O$	Hexanal	1.78	1	C_8H_6O	Benzofuran	2.67	1
$C_6H_{12}O$	2-Hexanone	1.38	1	C_8H_6S	Benzo[b]thiophene	3.12	1
$C_6H_{12}O$	4-Methyl-2-pentanone	1.31	1	C_8H_7N	Benzeneacetonitrile	1.56	1
$C_6H_{12}O_2$	Hexanoic acid	1.92	1	C_8H_7N	Indole	2.14	1
$C_6H_{12}O_2$	Butyl acetate	1.82	1	C_8H_8	Styrene	3.05	1
$C_6H_{13}Br$	1-Bromohexane	3.80	1	C_8H_8O	Acetophenone	1.63	1
$C_6H_{13}N$	Cyclohexylamine	1.49	1	C_8H_8O	2-Methylbenzaldehyde	2.26	1
C_6H_{14}	Hexane	4.00	1	C_8H_8O	Benzeneacetaldehyde	1.78	1
C_6H_{14}	3-Methylpentane	3.60	2	C_8H_8O	2,3-Dihydrobenzofuran	2.14	1
C_6H_{14}	2,2-Dimethylbutane	3.82	1	C_8H_8O	Phenyloxirane	1.61	1
C_6H_{14}	2,3-Dimethylbutane	3.85	2	$C_8H_8O_2$	o-Toluic acid	2.32	4
$C_6H_{14}O$	1-Hexanol	2.03	1	$C_8H_8O_2$	m-Toluic acid	2.37	1
$C_6H_{14}O$	2-Hexanol	1.76	1	$C_8H_8O_2$	p-Toluic acid	2.34	1
$C_6H_{14}O$	3-Hexanol	1.65	1	$C_8H_8O_2$	Benzeneacetic acid	1.41	1
$C_6H_{14}O$	3,3-Dimethyl-2-butanol	1.48	1	$C_8H_8O_2$	Phenyl acetate	1.49	1
$C_6H_{14}O$	Dipropyl ether	2.03	1	$C_8H_8O_2$	Methyl benzoate	2.20	1
$C_6H_{14}O$	Diisopropyl ether	1.52	1	C_8H_{10}	Ethylbenzene	3.15	1
$C_6H_{15}N$	Hexylamine	2.06	1	C_8H_{10}	o-Xylene	3.12	1
$C_6H_{15}N$	Dipropylamine	1.67	1	C_8H_{10}	m-Xylene	3.20	1
$C_6H_{15}N$	Triethylamine	1.45	1	C_8H_{10}	p-Xylene	3.15	1
$C_7H_5BrO_2$	2-Bromobenzoic acid	2.20	4	$C_8H_{10}O$	o-Ethylphenol	2.47	1
$C_7H_5BrO_2$	3-Bromobenzoic acid	2.87	4	$C_8H_{10}O$	m-Ethylphenol	2.50	1
$C_7H_5BrO_2$	4-Bromobenzoic acid	2.86	4	$C_8H_{10}O$	p-Ethylphenol	2.50	1
C_7H_5N	Benzonitrile	1.56	1	$C_8H_{10}O$	2,4-Xylenol	2.35	1
C_7H_6O	Benzaldehyde	1.48	1	$C_8H_{10}O$	2,5-Xylenol	2.34	1
$C_7H_6O_2$	Benzoic acid	1.88	4	$C_8H_{10}O$	2,6-Xylenol	2.36	1
$C_7H_6O_2$	Phenyl formate	1.26	1	$C_8H_{10}O$	3,4-Xylenol	3.23	1
$C_7H_6O_3$	Salicylic acid	2.20	4	$C_8H_{10}O$	3,5-Xylenol	2.35	1
C_7H_7Br	(Bromomethyl)benzene	2.92	1	$C_8H_{10}O$	Benzeneethanol	1.36	1
C_7H_7Cl	o-Chlorotoluene	3.42	1	$C_8H_{10}O$	α-Methylbenzyl alcohol	1.42	1
C_7H_7Cl	m-Chlorotoluene	3.28	1	$C_8H_{10}O$	3-Methylbenzenemethanol	1.60	1
C_7H_7Cl	p-Chlorotoluene	3.33	1	$C_8H_{10}O$	4-Methylbenzenemethanol	1.58	1
C_7H_7Cl	(Chloromethyl)benzene	2.30	1	$C_8H_{10}O$	Phenetole	2.51	1
$C_7H_7NO_2$	p-Nitrotoluene	2.42	1	$C_8H_{10}O$	Benzyl methyl ether	1.35	1
C_7H_8	Toluene	2.73	1	$C_8H_{10}O$	2-Methylanisole	2.74	1
C_7H_8	1,3,5-Cycloheptatriene	2.63	2	$C_8H_{10}O$	3-Methylanisole	2.66	1
C_7H_8O	o-Cresol	1.98	1	$C_8H_{10}O$	4-Methylanisole	2.81	1
C_7H_8O	m-Cresol	1.98	1	$C_8H_{11}N$	p-Ethylaniline	1.96	1
C_7H_8O	p-Cresol	1.97	1	$C_8H_{11}N$	N,N-Dimethylaniline	2.31	1
C_7H_8O	Benzyl alcohol	1.05	1	$C_8H_{11}N$	Benzeneethanamine	1.41	1
C_7H_8O	Anisole	2.11	1	$C_8H_{14}O_2$	Butyl methacrylate	2.88	1
C_7H_9N	Benzylamine	1.09	1	$C_8H_{15}N$	Octanenitrile	2.75	1
C_7H_9N	o-Methylaniline	1.32	1	C_8H_{16}	1-Octene	4.57	1
C_7H_9N	m-Methylaniline	1.40	1	C_8H_{16}	Cyclooctane	4.45	2
C_7H_9N	p-Methylaniline	1.39	1	$C_8H_{16}O$	2-Octanone	2.37	1
C_7H_9N	N-Methylaniline	1.66	1	$C_8H_{16}O_2$	Octanoic acid	3.05	1
C_7H_{14}	1-Heptene	3.99	1	$C_8H_{17}Br$	1-Bromooctane	4.89	1
C_7H_{14}	Methylcyclohexane	3.88	1	C_8H_{18}	Octane	5.15	1
$C_7H_{14}O$	2-Heptanone	1.98	1	$C_8H_{18}O$	1-Octanol	3.07	1
$C_7H_{14}O$	5-Methyl-2-hexanone	1.88	1	$C_8H_{18}O$	2-Octanol	2.90	1
$C_7H_{15}Br$	1-Bromoheptane	4.36	1	$C_8H_{18}O$	4-Octanol	2.68	1
$C_7H_{15}Cl$	1-Chloroheptane	4.15	1	$C_8H_{18}O$	Dibutyl ether	3.21	1
$C_7H_{15}I$	1-Iodoheptane	4.70	1	C_9H_7N	Quinoline	2.03	1

Mol. Form.	Name	log P	Ref.	Mol. Form.	Name	log P	Ref.
C_9H_7N	Isoquinoline	2.08	1	$C_{12}H_2Cl_8$	2,2′,3,3′,5,5′,6,6′-		
C_9H_8	Indene	2.92	1		Octachlorobiphenyl	7.10	3
$C_9H_8O_2$	*trans*-Cinnamic acid	2.13	1	$C_{12}H_3Cl_7$	2,2′,3,3′,4,4′,6-Heptachlorobiphenyl	6.70	3
C_9H_9N	Benzenepropanenitrile	1.72	1	$C_{12}H_4Cl_6$	2,2′,3,3′,4,4′-Hexachlorobiphenyl	7.00	3
C_9H_{10}	Indan	3.33	1	$C_{12}H_4Cl_6$	2,2′,4,4′,6,6′-Hexachlorobiphenyl	7.00	3
$C_9H_{10}O$	1-Phenyl-1-propanone	2.19	1	$C_{12}H_4Cl_6$	2,2′,3,3′,6,6′-Hexachlorobiphenyl	6.70	3
$C_9H_{10}O$	1-Phenyl-2-propanone	1.44	1	$C_{12}H_5Cl_5$	2,3,4,5,6-Pentachlorobiphenyl	6.30	3
$C_9H_{10}O$	4-Methylacetophenone	2.19	1	$C_{12}H_5Cl_5$	2,2′,4,5,5′-Pentachlorobiphenyl	6.40	3
$C_9H_{10}O_2$	2-Phenylpropanoic acid	1.80	1	$C_{12}H_6Cl_4$	2,3,4,5-Tetrachlorobiphenyl	5.72	3
$C_9H_{10}O_2$	Benzyl acetate	1.96	1	$C_{12}H_6Cl_4$	2,2′,4′,5-Tetrachlorobiphenyl	5.73	7
$C_9H_{10}O_2$	4-Methylphenyl acetate	2.11	1	$C_{12}H_7Cl_3$	2,4,5-Trichlorobiphenyl	5.60	3
$C_9H_{10}O_2$	Ethyl benzoate	2.64	1	$C_{12}H_7Cl_3$	2,4,6-Trichlorobiphenyl	5.47	3
C_9H_{12}	Propylbenzene	3.69	1	$C_{12}H_8Cl_2$	2,5-Dichlorobiphenyl	5.10	3
C_9H_{12}	Isopropylbenzene	3.66	1	$C_{12}H_8Cl_2$	2,6-Dichlorobiphenyl	5.00	3
C_9H_{12}	*o*-Ethyltoluene	3.53	1	$C_{12}H_8O$	Dibenzofuran	4.12	1
C_9H_{12}	*p*-Ethyltoluene	3.63	2	$C_{12}H_9Cl$	2-Chlorobiphenyl	4.52	1
C_9H_{12}	1,2,3-Trimethylbenzene	3.60	1	$C_{12}H_9Cl$	3-Chlorobiphenyl	4.58	1
C_9H_{12}	1,2,4-Trimethylbenzene	3.63	1	$C_{12}H_9Cl$	4-Chlorobiphenyl	4.61	1
C_9H_{12}	1,3,5-Trimethylbenzene	3.42	1	$C_{12}H_9N$	Carbazole	3.72	1
$C_9H_{12}O$	2-Propylphenol	2.93	1	$C_{12}H_{10}$	Acenaphthene	3.96	4
$C_9H_{12}O$	4-Propylphenol	3.20	1	$C_{12}H_{10}$	Biphenyl	3.76	6
$C_9H_{12}O$	2,3,6-Trimethylphenol	2.67	1	$C_{12}H_{10}N_2$	Azobenzene	3.82	1
$C_9H_{12}O$	2,4,6-Trimethylphenol	2.46	1	$C_{12}H_{10}O$	Diphenyl ether	4.21	1
$C_9H_{12}O$	Benzenepropanol	1.88	1	$C_{12}H_{10}S$	Diphenyl sulfide	4.45	1
$C_9H_{13}N$	*N,N*-Dimethylbenzylamine	1.98	1	$C_{12}H_{11}N$	Diphenylamine	3.44	4
$C_9H_{13}N$	Amphetamine	1.76	1	$C_{12}H_{12}$	1-Ethylnaphthalene	4.40	1
C_9H_{18}	1-Nonene	5.15	1	$C_{12}H_{12}$	1,2-Dimethylnaphthalene	4.31	1
$C_9H_{18}O$	2-Nonanone	3.16	1	$C_{12}H_{12}$	1,4-Dimethylnaphthalene	4.37	1
$C_9H_{18}O$	5-Methyl-2-octanone	2.92	1	$C_{12}H_{14}O$	4-Phenylcyclohexanone	2.45	1
C_9H_{20}	Nonane	5.65	1	$C_{12}H_{18}$	Hexylbenzene	5.52	1
$C_9H_{20}O$	1-Nonanol	4.02	1	$C_{12}H_{18}$	Hexamethylbenzene	4.69	4
$C_9H_{21}N$	Tripropylamine	2.79	1	$C_{12}H_{22}O$	Cyclododecanone	4.10	1
$C_{10}H_7Cl$	1-Chloronaphthalene	3.90	1	$C_{12}H_{24}O_2$	Dodecanoic acid	4.6	1
$C_{10}H_7Cl$	2-Chloronaphthalene	3.98	1	$C_{12}H_{26}O$	1-Dodecanol	5.13	1
$C_{10}H_8$	Naphthalene	3.34	4	$C_{13}H_8O$	9H-Fluoren-9-one	3.58	1
$C_{10}H_8$	Azulene	3.22	1	$C_{13}H_9N$	Acridine	3.40	1
$C_{10}H_8O$	1-Naphthol	2.84	1	$C_{13}H_{10}$	9H-Fluorene	4.20	4
$C_{10}H_8O$	2-Naphthol	2.70	1	$C_{13}H_{10}O$	Benzophenone	3.18	1
$C_{10}H_{12}O_2$	Isopropyl benzoate	3.18	1	$C_{13}H_{10}O_2$	Phenyl benzoate	3.59	1
$C_{10}H_{14}$	Butylbenzene	4.26	1	$C_{13}H_{11}NO$	*N*-Phenylbenzamide	2.62	1
$C_{10}H_{14}$	*tert*-Butylbenzene	4.11	1	$C_{13}H_{12}$	Diphenylmethane	4.14	1
$C_{10}H_{14}$	Isobutylbenzene	4.01	2	$C_{13}H_{12}$	4-Methylbiphenyl	4.63	1
$C_{10}H_{14}$	*p*-Cymene	4.10	1	$C_{13}H_{12}O$	Diphenylmethanol	2.67	1
$C_{10}H_{14}$	1,2,4,5-Tetramethylbenzene	4.10	2	$C_{13}H_{12}O$	Benzyl phenyl ether	3.79	1
$C_{10}H_{14}$	1,2,3,4-Tetramethylbenzene	4.00	1	$C_{14}H_{10}$	Anthracene	4.56	4
$C_{10}H_{14}$	1,2,3,5-Tetramethylbenzene	4.10	1	$C_{14}H_{10}$	Phenanthrene	4.52	4
$C_{10}H_{14}O$	4-Butylphenol	3.65	1	$C_{14}H_{12}$	*trans*-Stilbene	4.81	1
$C_{10}H_{20}O$	2-Decanone	3.77	1	$C_{14}H_{12}$	1-Methylfluorene	4.97	1
$C_{10}H_{20}O_2$	Decanoic acid	4.09	1	$C_{14}H_{12}O$	2-Phenylacetophenone	3.18	1
$C_{10}H_{22}$	Decane	6.25	1	$C_{14}H_{12}O_2$	Benzyl benzoate	3.97	1
$C_{10}H_{22}O$	1-Decanol	4.57	1	$C_{14}H_{14}$	1,2-Diphenylethane	4.70	1
$C_{11}H_9N$	4-Phenylpyridine	2.59	1	$C_{14}H_{14}$	4,4′-Dimethylbiphenyl	5.09	1
$C_{11}H_{10}$	1-Methylnaphthalene	3.87	1	$C_{14}H_{22}$	Octylbenzene	6.30	1
$C_{11}H_{10}$	2-Methylnaphthalene	4.00	1	$C_{14}H_{28}O_2$	Tetradecanoic acid	6.1	1
$C_{11}H_{16}$	Pentylbenzene	4.90	1	$C_{15}H_{12}$	2-Methylanthracene	5.15	2
$C_{11}H_{16}$	Pentamethylbenzene	4.56	1	$C_{15}H_{12}$	9-Methylanthracene	5.07	1
$C_{11}H_{22}O$	2-Undecanone	4.09	1	$C_{15}H_{12}$	1-Methylphenanthrene	5.14	2
$C_{11}H_{22}O_2$	Methyl decanoate	4.41	1	$C_{16}H_{10}$	Fluoranthene	5.07	4
$C_{12}Cl_{10}$	Decachlorobiphenyl	8.26	3	$C_{16}H_{10}$	Pyrene	5.08	4
$C_{12}HCl_9$	2,2′,3,3′,4,5,5′,6,6′-			$C_{16}H_{14}$	9,10-Dimethylanthracene	5.69	1
	Nonachlorobiphenyl	8.16	3	$C_{16}H_{32}O_2$	Hexadecanoic acid	7.17	1

Mol. Form.	Name	log P	Ref.	Mol. Form.	Name	log P	Ref.
$C_{17}H_{12}$	11H-Benzo[a]fluorene	5.40	1	$C_{18}H_{36}O_2$	Stearic acid	8.23	1
$C_{17}H_{12}$	11H-Benzo[b]fluorene	5.75	1	$C_{19}H_{16}O$	Triphenylmethanol	3.68	1
$C_{18}H_{12}$	Benz[a]anthracene	5.91	1	$C_{20}H_{12}$	Perylene	6.25	1
$C_{18}H_{12}$	Chrysene	5.73	4	$C_{20}H_{12}$	Benzo[a]pyrene	6.20	4
$C_{18}H_{12}$	Naphthacene	5.76	1	$C_{20}H_{32}O_2$	Arachidonic acid	6.98	1
$C_{18}H_{12}$	Triphenylene	5.49	4	$C_{20}H_{40}O_2$	Arachidic acid	9.29	1
$C_{18}H_{15}N$	Triphenylamine	5.74	1	$C_{21}H_{16}$	1,2-Dihydro-3-methylbenz[j]		
$C_{18}H_{30}O_2$	Linolenic acid	6.46	1		aceanthrylene	6.75	1
$C_{18}H_{32}O_2$	Linoleic acid	7.05	1	$C_{22}H_{12}$	Benzo[ghi]perylene	6.90	1
$C_{18}H_{34}O_2$	Oleic acid	7.64	1	$C_{24}H_{12}$	Coronene	6.05	4

PROTECTION AGAINST IONIZING RADIATION

The following data and rules of thumb are helpful in estimating the penetrating capability of and danger of exposure to various types of ionizing radiation. More precise data should be used for critical applications.

Alpha Particles

Alpha particles of at least 7.5 MeV are required to penetrate the epidermis, the protective layer of skin, 0.07 mm thick.

Electrons

Electrons of at least 70 keV are required to penetrate the epidermis, the protective layer of skin, 0.07 mm thick.

The range of electrons in g/cm^2 is approximately equal to the maximum energy (E) in MeV divided by 2.

The range of electrons in air is about 3.65 m per MeV; for example, a 3 MeV electron has a range of about 11 m in air.

A chamber wall thickness of 30 mg/cm^2 will transmit 70% of the initial fluence of 1 MeV electrons and 20% of that of 0.4 MeV electrons.

When electrons of 1 to 2 MeV pass through light materials such as water, aluminum, or glass, less than 1% of their energy is dissipated as bremsstrahlung.

The bremsstrahlung from 1 Ci of ^{32}P aqueous solution in a glass bottle is about 1 mR/h at 1 meter distance.

When electrons from a 1 Ci source of ^{90}Sr - ^{90}Y are absorbed, the bremsstrahlung hazard is approximately equal to that presented by the gamma radiation from 12 mg of radium. The average energy of the bremsstrahlung is about 300 keV.

Gamma Rays

The air-scattered radiation (sky-shine) from a 100 Ci ^{60}Co source placed 1 ft behind a 4 ft high shield is about 100 mrad/h at 6 ft from the outside of the shield.

Within ±20% for point source gamma emitters with energies between 0.07 and 4 MeV, the exposure rate (R/h) at 1 ft is $6C \cdot E \cdot n$ where C is the activity in curies, E is the energy in MeV, and n is the number of gammas per disintegration.

Neutrons

An approximate HVL (thickness of absorber for which the neutron flux falls to half its initial value) for 1 MeV neutrons is 3.2 cm of paraffin; that for 5 MeV neutrons is 6.9 cm of paraffin).

Miscellaneous

The activity of any radionuclide is reduced to less than 1% after 7 half-lives (i.e., $2^{-7} = 0.8\%$).

For nuclides with a half-life greater than 6 days, the change in activity in 24 hours will be less than 10%.

10 HVL (half-value layers) attenuates approximately by 10^{-3}.

There is 0.64 mm^3 of radon gas at STP in transient equilibrium with 1 Ci of radium.

The natural background from all sources in most parts of the world leads to an equivalent dose rate of about 0.04 to 4 mSv per year for the average person. About 84% of this comes from terrestrial sources, the remainder from cosmic rays. The U. S. average is about 3.6 mSv/yr but can range up to 50 mSv/yr in some areas. A passenger in a plane flying at 12,000 meters receives 5 μSv/hr from cosmic rays (as compared to about 0.03 μSv/hr at sea level).

The ICRP recommended exposure limit to man-made sources of ionizing radiation (Reference 2) is 20 mSv/yr averaged over 5 years, with the dose in any one year not to exceed 50 mSv.

A whole-body dose of about 3 Gy over a short time interval will typically lead to 50% mortality in 30 days assuming no medical treatment.

Units

The gray (Gy) is the SI unit of absorbed dose; it is a measure of the mean energy imparted to a sample of irradiated matter, divided by the mass of the sample. Gy is a special name for the SI unit J/kg.

The sievert (Sv) is the SI unit of equivalent dose, which is defined as the absorbed dose multiplied by a weighting factor that expresses the long-term biological risk from low-level chronic exposure to a specified type of radiation. The Sv is another special name for J/kg.

1 curie (Ci) = $3.7 \cdot 10^{10}$ becquerel (Bq); i.e., $3.7 \cdot 10^{10}$ disintegrations per second.

1 roentgen (R) = $2.58 \cdot 10^{-4}$ coulomb per kilogram (C/kg); a measure of the charge (positive or negative) liberated by x-ray or gamma radiation in air, divided by the mass of air.

1 rad = 0.01 Gy

1 rem = 0.01 Sv

REFERENCES

1. Padikal, T.N., and Fivozinsky, S.P., *Medical Physics Data Book, National Bureau of Standards Handbook 138*, U. S. Government Printing Office, Washington, D.C., 1981.
2. *1990 Recommendations of the International Commission on Radiological Protection*, ICRP Publication 60, *Annals of the ICRP*, Pergamon Press, Oxford, 1991.
3. *Radiation: Doses, Effects, Risks*, United Nations Sales No. E.86.III.D.4, 1985.
4. *Review of Particle Properties, Phys. Rev. D*, 50, 1173, 1994 (p. 1268).

ANNUAL LIMITS ON INTAKES OF RADIONUCLIDES

K. F. Eckerman

The following table lists, for workers, the annual limits on oral and inhalation intakes (ALI) for selected radionuclides based on the occupational radiation protection guidance of the International Commission on Radiological Protection (References 1 and 2). An intake of one ALI corresponds to an annual whole body dose of 0.02 Sv (2 rem).

The ALI is expressed in the SI unit of activity, the becquerel (Bq), and in the conventional unit, the microcurie (μCi); 1 μCi = 3.7·10⁴ Bq. The chemical form of inhaled radionuclides is, in most instances, stated in terms of the rate of absorption to blood from the lungs and the fractional absorption from the small intestine. Type F, M, and S denote chemical forms which are absorbed from the lungs at rates characterized as fast, moderate, and slow, respectively. The time to absorb 90% of the deposited radionuclide, in the absence of radioactive decay, corresponds to about 10 minutes, 150 days, and 7000 days for Type F, M, and S compounds, respectively. Type F compounds can be considered to be more soluble than M or S, S being the most insoluble. Chemical form consideration for ingestion is specified by the fractional absorption from the small intestine, denoted as f_1. The f_1 values range from 10^{-5} to 1. Higher fractional absorption is associated with greater solubility of the compound.

REFERENCES

1. *1990 Recommendations of the International Commission on Radiological Protection, ICRP Publication 60, Annals of the ICRP 21, (1—3)*, Pergamon Press, Oxford, 1991.
2. *Dose Coefficients for Intakes of Radionuclides by Workers, ICRP Publication 68, Annals of the ICRP*, 24(4), Pergamon Press, Oxford, 1995.

	Physical half-life	Inhalation intakes			Oral intakes		
		Chemical form Type/f_1	ALI Bq	ALI μCi	Chemical form f_1	ALI Bq	ALI μCi
^{3}H	12.3 y	HT gas	1.1E+13	3.0E+08	1.000	1.1E+13	3.0E+08
		HTO vapor	1.1E+09	3.0E+04			
^{11}C	0.340 h	CO	1.7E+10	4.5E+05	1.000	8.3E+08	2.3E+04
		CO_2	9.1E+09	2.5E+05			
		Organic compounds	6.2E+09	1.7E+05			
^{14}C	5730 y	CO	2.5E+10	6.8E+05	1.000	3.4E+07	9.3E+02
		CO_2	3.1E+09	8.3E+04			
		Organic compounds	3.4E+07	9.3E+02			
^{18}F	1.83 h	F 1.000	3.7E+08	1.0E+04	1.000	4.1E+08	1.1E+04
		M 1.000	2.2E+08	6.1E+03			
		S 1.000	2.2E+08	5.8E+03			
^{22}Na	2.60 y	F 1.000	1.0E+07	2.7E+02	1.000	6.3E+06	1.7E+02
^{24}Na	15.0 h	F 1.000	3.8E+07	1.0E+03	1.000	4.7E+07	1.3E+03
^{32}P	14.3 d	F 0.800	1.8E+07	4.9E+02	0.800	8.3E+06	2.3E+02
		M 0.800	6.9E+06	1.9E+02			
^{35}S	87.4 d	Inorganic compounds					
		F 0.800	2.5E+08	6.8E+03	0.800	1.4E+08	3.9E+03
		M 0.800	1.8E+07	4.9E+02	0.100	1.1E+08	2.8E+03
		Vapor	1.7E+08	4.5E+03			
		Organic compounds			1.000	2.6E+07	7.0E+02
^{42}K	12.4 h	F 1.000	1.0E+08	2.7E+03	1.000	4.7E+07	1.3E+03
^{43}K	22.6 h	F 1.000	7.7E+07	2.1E+03	1.000	8.0E+07	2.2E+03
^{45}Ca	163 d	M 0.300	8.7E+06	2.4E+02	0.300	2.6E+07	7.1E+02
^{47}Ca	4.53 d	M 0.300	9.5E+06	2.6E+02	0.300	1.3E+07	3.4E+02
^{51}Cr	27.7 d	F 0.100	6.7E+08	1.8E+04	0.100	5.3E+08	1.4E+04
		M 0.100	5.9E+08	1.6E+04	0.010	5.4E+08	1.5E+04
		S 0.100	5.6E+08	1.5E+04			
^{54}Mn	312 d	F 0.100	1.8E+07	4.9E+02	0.100	2.8E+07	7.6E+02
		M 0.100	1.7E+07	4.5E+02			
^{52}Fe	8.28 h	F 0.100	2.9E+07	7.8E+02	0.100	1.4E+07	3.9E+02
		M 0.100	2.1E+07	5.7E+02			
^{55}Fe	2.70 y	F 0.100	2.2E+07	5.9E+02	0.100	6.1E+07	1.6E+03
		M 0.100	6.1E+07	1.6E+03			

	Physical half-life	Inhalation intakes			Oral intakes		
		Chemical form Type/f_1	ALI		Chemical form f_1	ALI	
			Bq	µCi		Bq	µCi
^{59}Fe	44.5 d	F 0.100	6.7E+06	1.8E+02	0.100	1.1E+07	3.0E+02
		M 0.100	6.3E+06	1.7E+02			
^{57}Co	271 d	M 0.100	5.1E+07	1.4E+03	0.100	9.5E+07	2.6E+03
		S 0.050	3.3E+07	9.0E+02	0.050	1.1E+08	2.8E+03
^{58}Co	70.8 d	M 0.100	1.4E+07	3.9E+02	0.100	2.7E+07	7.3E+02
		S 0.050	1.2E+07	3.2E+02	0.050	2.9E+07	7.7E+02
^{60}Co	5.27 y	M 0.100	2.8E+06	7.6E+01	0.100	5.9E+06	1.6E+02
		S 0.050	1.2E+06	3.2E+01	0.050	8.0E+06	2.2E+02
^{64}Cu	12.7 h	F 0.500	2.9E+08	7.9E+03	0.500	1.7E+08	4.5E+03
		M 0.500	1.3E+08	3.6E+03			
		S 0.500	1.3E+08	3.6E+03			
^{59}Ni	75000 y	F 0.050	9.1E+07	2.5E+03	0.050	3.2E+08	8.6E+03
		M 0.050	2.1E+08	5.8E+03			
		Vapor	2.4E+07	6.5E+02			
^{63}Ni	96.0 y	F 0.050	3.8E+07	1.0E+03	0.050	1.3E+08	3.6E+03
		M 0.050	6.5E+07	1.7E+03			
		Vapor	1.0E+07	2.7E+02			
^{65}Zn	244 d	S 0.500	7.1E+06	1.9E+02	0.500	5.1E+06	1.4E+02
^{67}Ga	3.26 d	F 0.001	1.8E+08	4.9E+03	0.001	1.1E+08	2.8E+03
		M 0.001	7.1E+07	1.9E+03			
^{68}Ga	1.13 h	F 0.001	4.1E+08	1.1E+04	0.001	2.0E+08	5.4E+03
		M 0.001	2.5E+08	6.7E+03			
^{68}Ge	288 d	F 1.000	2.4E+07	6.5E+02	1.000	1.5E+07	4.2E+02
		M 1.000	2.5E+06	6.8E+01			
^{75}Se	120 d	F 0.800	1.4E+07	3.9E+02	0.800	7.7E+06	2.1E+02
		M 0.800	1.2E+07	3.2E+02	0.050	4.9E+07	1.3E+03
^{79}Se	65000 y	F 0.800	1.3E+07	3.4E+02	0.800	6.9E+06	1.9E+02
		M 0.800	6.5E+06	1.7E+02	0.050	5.1E+07	1.4E+03
^{86}Rb	18.6 d	F 1.000	1.5E+07	4.2E+02	1.000	7.1E+06	1.9E+02
^{85}Sr	64.8 d	F 0.300	3.6E+07	9.7E+02	0.300	3.6E+07	9.7E+02
		S 0.010	3.1E+07	8.4E+02	0.010	6.1E+07	1.6E+03
^{87m}Sr	2.80 h	F 0.300	9.1E+08	2.5E+04	0.300	6.7E+08	1.8E+04
		S 0.010	5.7E+08	1.5E+04	0.010	6.1E+08	1.6E+04
^{89}Sr	50.5 d	F 0.300	1.4E+07	3.9E+02	0.300	7.7E+06	2.1E+02
		S 0.010	3.6E+06	9.7E+01	0.010	8.7E+06	2.4E+02
^{90}Sr	29.1 y	F 0.300	6.7E+05	1.8E+01	0.300	7.1E+05	1.9E+01
		S 0.010	2.6E+05	7.0E+00	0.010	7.4E+06	2.0E+02
^{99}Mo	2.75 d	F 0.800	5.6E+07	1.5E+03	0.800	2.7E+07	7.3E+02
		S 0.050	1.8E+07	4.9E+02	0.050	1.7E+07	4.5E+02
^{99m}Tc	6.02 h	F 0.800	1.0E+09	2.7E+04	0.800	9.1E+08	2.5E+04
		M 0.800	6.9E+08	1.9E+04			
^{99}Tc	213000 y	F 0.800	5.0E+07	1.4E+03	0.800	2.6E+07	6.9E+02
		M 0.800	6.3E+06	1.7E+02			
^{106}Ru	1.01 y	F 0.050	2.0E+06	5.5E+01	0.050	2.9E+06	7.7E+01
		M 0.050	1.2E+06	3.2E+01			
		S 0.050	5.7E+05	1.5E+01			
^{111}In	2.83 d	F 0.020	9.1E+07	2.5E+03	0.020	6.9E+07	1.9E+03
		M 0.020	6.5E+07	1.7E+03			
^{113m}In	1.66 h	F 0.020	1.1E+09	2.8E+04	0.020	7.1E+08	1.9E+04
		M 0.020	6.3E+08	1.7E+04			
^{113}Sn	115 d	F 0.020	2.5E+07	6.8E+02	0.020	2.7E+07	7.4E+02
		M 0.020	1.1E+07	2.8E+02			
^{123}I	13.2 h	F 1.000	1.8E+08	4.9E+03	1.000	9.5E+07	2.6E+03
		Vapor	9.5E+07	2.6E+03			
^{125}I	60.1 d	F 1.000	2.7E+06	7.4E+01	1.000	1.3E+06	3.6E+01
		Vapor	1.4E+06	3.9E+01			
^{129}I	1.57·10^7 y	F 1.000	3.9E+05	1.1E+01	1.000	1.8E+05	4.9E+00

	Physical half-life	Inhalation intakes			Oral intakes		
		Chemical form Type/f_1	ALI		Chemical form f_1	ALI	
			Bq	µCi		Bq	µCi
^{131}I	8.04 d	Vapor	2.1E+05	5.6E+00			
		F 1.000	1.8E+06	4.9E+01	1.000	9.1E+05	2.5E+01
^{129}Cs	1.34 d	Vapor	1.0E+06	2.7E+01			
		F 1.000	2.5E+08	6.7E+03	1.000	3.3E+08	9.0E+03
^{134}Cs	2.06 y	F 1.000	2.1E+06	5.6E+01	1.000	1.1E+06	2.8E+01
^{136}Cs	13.1 d	F 1.000	1.1E+07	2.8E+02	1.000	6.7E+06	1.8E+02
^{137}Cs	30.0 y	F 1.000	3.0E+06	8.1E+01	1.000	1.5E+06	4.2E+01
^{141}Ce	32.5 d	M 5.0E-04	7.4E+06	2.0E+02	5.0E-04	2.8E+07	7.6E+02
		S 5.0E-04	6.5E+06	1.7E+02			
^{144}Ce	284 d	M 5.0E-04	8.7E+05	2.4E+01	5.0E-04	3.8E+06	1.0E+02
		S 5.0E-04	6.9E+05	1.9E+01			
^{133}Ba	10.7 y	F 0.100	1.1E+07	3.0E+02	0.100	2.0E+07	5.4E+02
^{140}Ba	12.7 d	F 0.100	1.3E+07	3.4E+02	0.100	8.0E+06	2.2E+02
^{169}Yb	32.0 d	M 5.0E-04	9.5E+06	2.6E+02	5.0E-04	2.8E+07	7.6E+02
		S 5.0E-04	8.3E+06	2.3E+02			
^{198}Au	2.69 d	F 0.100	5.1E+07	1.4E+03	0.100	2.0E+07	5.4E+02
		M 0.100	2.0E+07	5.5E+02			
		S 0.100	1.8E+07	4.9E+02			
^{198m}Au	2.30 d	F 0.100	3.4E+07	9.2E+02	0.100	1.5E+07	4.2E+02
		M 0.100	1.0E+07	2.7E+02			
		S 0.100	1.1E+07	2.8E+02			
^{197}Hg	2.67 d	Inorganic compounds					
		F 0.400	2.4E+08	6.4E+03	1.000	2.0E+08	5.5E+03
					0.400	1.2E+08	3.2E+03
		Vapor Organic compounds	4.5E+06	1.2E+02			
		F 0.020	2.0E+08	5.4E+03	0.020	8.7E+07	2.4E+03
		M 0.020	7.1E+07	1.9E+03			
^{203}Hg	46.6 d	Inorganic compounds					
		F 0.400	2.7E+07	7.2E+02	1.000	1.1E+07	2.8E+02
					0.400	1.8E+07	4.9E+02
		Vapor Organic compounds	2.9E+06	7.7E+01			
		F 0.020	3.4E+07	9.2E+02	0.020	3.7E+07	1.0E+03
		M 0.020	1.1E+07	2.8E+02			
^{201}Tl	3.04 d	F 1.000	2.6E+08	7.1E+03	1.000	2.1E+08	5.7E+03
^{210}Pb	22.3 y	F 0.200	1.8E+04	4.9E-01	0.200	2.9E+04	7.9E-01
^{207}Bi	38.0 y	F 0.050	2.4E+07	6.4E+02	0.050	1.5E+07	4.2E+02
		M 0.050	6.3E+06	1.7E+02			
^{210}Po	138 d	F 0.100	2.8E+04	7.6E-01	0.100	8.3E+04	2.3E+00
		M 0.100	9.1E+03	2.5E-01			
^{224}Ra	3.66 d	M 0.200	8.3E+03	2.3E-01	0.200	3.1E+05	8.3E+00
^{226}Ra	1600 y	M 0.200	1.7E+03	4.5E-02	0.200	7.1E+04	1.9E+00
^{228}Ra	5.75 y	M 0.200	1.2E+04	3.2E-01	0.200	3.0E+04	8.1E-01
^{228}Th	1.91 y	M 5.0E-04	8.7E+02	2.4E-02	5.0E-04	2.9E+05	7.7E+00
		S 2.0E-04	6.3E+02	1.7E-02	2.0E-04	5.7E+05	1.5E+01
^{230}Th	77000 y	M 5.0E-04	7.1E+02	1.9E-02	5.0E-04	9.5E+04	2.6E+00
		S 2.0E-04	2.8E+03	7.5E-02	2.0E-04	2.3E+05	6.2E+00
^{232}Th	1.40·10^{10} y	M 5.0E-04	6.9E+02	1.9E-02	5.0E-04	9.1E+04	2.5E+00
		S 2.0E-04	1.7E+03	4.5E-02	2.0E-04	2.2E+05	5.9E+00
^{234}U	2.44·10^5 y	F 0.020	3.1E+04	8.4E-01	0.020	4.1E+05	1.1E+01
		M 0.020	9.5E+03	2.6E-01	0.002	2.4E+06	6.5E+01
		S 0.002	2.9E+03	7.9E-02			

| | Physical half-life | Inhalation intakes | | | Oral intakes | | |
| | | Chemical form Type/f_1 | ALI | | Chemical form f_1 | ALI | |
			Bq	µCi		Bq	µCi
[235]U	7.04·10[8] y	F 0.020	3.3E+04	9.0E-01	0.020	4.3E+05	1.2E+01
		M 0.020	1.1E+04	3.0E-01	0.002	2.4E+06	6.5E+01
		S 0.002	3.3E+03	8.9E-02			
[238]U	4.47·10[9] y	F 0.020	3.4E+04	9.3E-01	0.020	4.5E+05	1.2E+01
		M 0.020	1.3E+04	3.4E-01	0.002	2.6E+06	7.1E+01
		S 0.002	3.5E+03	9.5E-02			
[237]Np	2.14·10[6] y	M 5.0E-04	1.3E+03	3.6E-02	5.0E-04	1.8E+05	4.9E+00
[239]Np	2.36 d	M 5.0E-04	1.8E+07	4.9E+02	5.0E-04	2.5E+07	6.8E+02
[238]Pu	87.7 y	M 5.0E-04	6.7E+02	1.8E-02	5.0E-04	8.7E+04	2.4E+00
		S 1.0E-05	1.8E+03	4.9E-02	1.0E-05	2.3E+06	6.1E+01
					1.0E-04	4.1E+05	1.1E+01
[239]Pu	24100 y	M 5.0E-04	6.3E+02	1.7E-02	5.0E-04	8.0E+04	2.2E+00
		S 1.0E-05	2.4E+03	6.5E-02	1.0E-05	2.2E+06	6.0E+01
					1.0E-04	3.8E+05	1.0E+01
[241]Pu	14.4 y	M 5.0E-04	3.4E+04	9.3E-01	5.0E-04	4.3E+06	1.2E+02
		S 1.0E-05	2.4E+05	6.4E+00	1.0E-05	1.8E+08	4.9E+03
					1.0E-04	2.1E+07	5.6E+02
[241]Am	432 y	M 5.0E-04	7.4E+02	2.0E-02	5.0E-04	1.0E+05	2.7E+00
[244]Cm	18.1 y	M 5.0E-04	1.2E+03	3.2E-02	5.0E-04	1.7E+05	4.5E+00
[252]Cf	2.64 y	M 5.0E-04	1.5E+03	4.2E-02	5.0E-04	2.2E+05	6.0E+00

CHEMICAL CARCINOGENS

The following substances are listed in the *Eighth Report on Carcinogens*, released by the National Institute of Environmental Health Sciences (NIEHS) in 1997 under the National Toxicology Program (NTP). Substances are grouped in two classes:

1. Known to be human carcinogens — substances for which there is sufficient evidence of carcinogenicity from studies in humans that indicate a causal relationship between exposure to the substance and human cancer.
2. Reasonably anticipated to be human carcinogens — substances for which there is limited evidence of carcinogenicity in humans and/ or sufficient evidence of carcinogenicity in experimental animals.

The NTP report also lists many poorly defined materials such as soots, tars, mineral oils, coke oven emissions, etc. These materials are not included here.

The table lists the name normally used in this *Handbook*, followed by additional names by which the substance is known. In many cases the primary name given here is different from that used in the NTP report; however names used in the NTP report appear in the *Other names* column. The Chemical Abstracts Service Registry Number (CAS RN), given in the last column, is taken from the NTP report. Extensive details on each substance are given in the reference.

REFERENCE

Public Health Service, National Toxicology Program, *Eighth Report on Carcinogens*, OCR Services, Inc., P.O. Box 12510, Research Triangle Park, NC 27709-2510. Also available on the Internet at http://ehis.niehs.nih.gov/roc/

Substance	Other names	CAS RN
Known to be Human Carcinogens		
Aflatoxins		1402-68-2
4-Aminobiphenyl	*p*-Biphenylamine	92-67-1
Arsenic		7440-38-2
Arsenic(III) oxide	Arsenic trioxide	1327-53-3
Arsenic(V) oxide	Arsenic pentoxide	1303-28-2
Asbestos		1332-21-4
Azathioprine	1H-Purine, 6-[(1-methyl-4-nitro-1H-imidazol-5-yl)thio]-	446-86-6
Barium chromate(VI)		10294-40-3
Benzene		71-43-2
p-Benzidine	[1,1'-Biphenyl]-4,4'-diamine	92-87-5
Bis(2-chloroethyl) sulfide	Mustard gas	505-60-2
Bis(chloromethyl) ether		542-88-1
1,4-Butanediol dimethylsulfonate	Myleran; Busulfan	55-98-1
Calcium arsenate	Arsenic acid, calcium salt (2:3)	10103-62-5
Calcium arsenite (2:1)	Arsenenous acid, calcium salt (2:1); Protars	15194-98-6
Calcium arsenite (1:1)	Arsonic acid, calcium salt (1:1)	52740-16-6
Calcium arsenite (2:3)	Tricalcium diarsenite; Arsenous acid, calcium salt (2:3)	27152-57-4
Calcium chromate		13765-19-0
Chlorambucil		305-03-3
Chloroethylene	Vinyl chloride	75-01-4
1-(2-Chloroethyl)-3-(4-methylcyclohexyl)-1-nitrosourea	MeCCNU; Urea, 1-(2-chloroethyl)-3-(4-methylcyclohexyl)-1-nitroso-	13909-09-6
Chloromethyl methyl ether		107-30-2
Chromium		7440-47-3
Chromium(III) acetate		1066-30-4
Chromium carbonate		29689-14-3
Chromium(III) oxide	Chromic oxide	1308-38-9
Chromium(VI) oxide	Chromium trioxide	1333-82-0
Chromium(III) phosphate	Chromic phosphate	7789-04-0
Cobalt chromium alloy		11114-92-4
Cyclophosphamide	2H-1,3,2-oxazaphosphorin-2-amine, *N,N*-bis(2-chloroethyl)tetrahydro-, 2-oxide	50-18-0
Cyclosporin A	Cyclosporine	59865-13-3
Diethylstilbestrol		56-53-1

Substance	Other names	CAS RN
Erionite		66733-21-9
Lead(II) chromate		7758-97-6
Lead(II) chromate(VI) oxide		18454-12-1
Lead(II) hydrogen arsenate		7784-40-9
Melphalan	L-Phenylalanine, 4-[bis(2-chloroethyl)amino]-	148-82-3
Methoxsalen (with UV therapy)	PUVA; 9-Methoxy-7H-furo[3,2-g][1]benzopyran-7-one	298-81-7
2-Naphthylamine	2-Aminonaphthalene; β-Naphthylamine	91-59-8
Piperazine estrone sulfate	Estrone, hydrogen sulfate, compd. with piperazine	7280-37-7
Potassium arsenite		13464-35-2
Potassium chromate		7789-00-6
Potassium dichromate		7778-50-9
Potassium dihydrogen arsenate		7784-41-0
Radon		10043-92-2
Sodium arsenate		7631-89-2
Sodium arsenite		7784-46-5
Sodium chromate		7775-11-3
Sodium dichromate		10588-01-9
Sodium equilin sulfate	Estra-1,3,5(10),7-tetraen-17-one, 3-hydroxy, hydrogen sulfate, sodium salt	16680-47-0
Sodium estrone sulfate	Estrone, hydrogen sulfate, sodium salt	438-67-5
Sodium hydrogen arsenate heptahydrate	Disodium hydrogen arsenate heptahydrate	10048-95-0
Strontium chromate		7789-06-2
Thorium(IV) oxide	Thorium dioxide	1314-20-1
Triethylenethiophosphoramide	Thiotepa; Tris(1-aziridinyl)phosphine, sulfide	52-24-4
Zinc chromate, basic		13530-65-9
Reasonably Anticipated to be Human Carcinogens		
Acetaldehyde	Ethanal	75-07-0
2-(Acetylamino)fluorene		53-96-3
Acrylamide		79-06-1
Acrylonitrile	Propenenitrile	107-13-1
Adriamycin, hydrochloride	Doxorubicin hydrochloride	25316-40-9
2-Amino-9,10-anthracenedione	2-Aminoanthraquinone	117-79-3
1-Amino-2-methyl-9,10-anthracenedione	1-Amino-2-methylanthraquinone	82-28-0
Arochlor 1254	Polychlorinated biphenyls (54% Cl)	11097-69-1
Arochlor 1260	Polychlorinated biphenyls (60% Cl)	11096-82-5
Azacitidine	5-Azacytidine; 1,3,5-Triazine-2(1H)-one, 4-amino-1-beta-D-ribofuranosyl-	320-67-2
Benz[a]anthracene		56-55-3
Benzo[b]fluoranthene	Benz[e]acephenanthrylene	205-99-2
Benzo[j]fluoranthene		205-82-3
Benzo[k]fluoranthene	2,3,1',8'-Binaphthylene	207-08-9
Benzo[a]pyrene		50-32-8
Beryllium		7440-41-7
Beryllium aluminum alloy		12770-50-2
Beryllium aluminum metasilicate	Beryl	1302-52-9
Beryllium carbonate, basic	Beryllium, bis[carbonato(2-)]dihydroxytri-	66104-24-3
Beryllium chloride		7787-47-5
Beryllium fluoride		7787-49-7
Beryllium hydroxide		13327-32-7
Beryllium oxide	Beryllia	1304-56-9
Beryllium phosphate		13598-15-7
Beryllium sulfate		13510-49-1
Beryllium sulfate tetrahydrate		7787-56-6
Beryllium zinc silicate		39413-47-3
2,2'-Bioxirane	Diepoxybutane	1464-53-5
Bis(4-amino-3-chlorophenyl)methane	4,4-Methylene-bis(2-chloraniline); MBOCA	101-14-4
Bis(2-chloroethyl)methylamine	Nitrogen mustard hydrochloride	55-86-7
N,N'-Bis(2-chloroethyl)-N-nitrosourea	BCNU; Carmustine	154-93-8
Bis[4-(dimethylamino)phenyl]methane	Michler's Base; 4,4-Methylenebis(N,N-dimethylbenzenamine)	101-61-1
1,3-Bis(2,3-epoxypropoxy)benzene	Diglycidyl resorcinol ether	101-90-6

Substance	Other names	CAS RN
Bis(2-ethylhexyl) phthalate	DEHP; Di(2-ethylhexyl) phthalate	117-81-7
Bromodichloromethane		75-27-4
1,3-Butadiene		106-99-0
tert-Butyl-4-hydroxyanisole	BHA; Butylated hydroxyanisole	25013-16-5
N-Butyl-*N*-(4-hydroxybutyl)nitrosamine	*N*-Nitrosobutyl-*N*-(4-hydroxybutyl)amine	3817-11-6
Cadmium		7440-43-9
Cadmium carbonate		513-78-0
Cadmium chloride		10108-64-2
Cadmium nitrate		10325-94-7
Cadmium oxide		1306-19-0
Cadmium sulfate		10124-36-4
Cadmium sulfide		1306-23-6
Cadmium tetrafluoroborate	Cadmium fluoborate	14486-19-2
Chlorendic acid	5-Norbornene-2,3-dicarboxylic acid, 1,4,5,6,7,7-hexachloro	115-28-6
Chlorinated paraffins (C_{12}, 60% Cl)		108171-26-2
4-Chloro-1,2-benzenediamine	4-Chloro-*o*-phenylenediamine	95-83-0
1-(2-Chloroethyl)-3-cyclohexyl-1-nitrosourea	CCNU; Lomustine; Belustine	13010-47-4
4-Chloro-2-methylaniline	*p*-Chloro-*o*-toluidine	95-69-2
4-Chloro-2-methylaniline hydrochloride	*p*-Chloro-*o*-toluidine hydrochloride	3165-93-3
1-Chloro-2-methylpropene	Dimethylvinyl chloride	513-37-1
3-Chloro-2-methylpropene		563-47-3
Chlorozotocin	*D*-Glucose, 2-[[[(2-chloroethyl)nitrosoamino] carbonyl]amino]-2-deoxy-	54749-90-5
C.I. Basic Red 9, monohydrochloride		569-61-9
C.I. Direct Black 38, disodium salt	Direct Black 38	1937-37-7
C.I. Direct Blue 6, tetrasodium salt	Direct Blue 6	2602-46-2
Cupferron		135-20-6
Dacarbazine	1H-Imidazole-4-carboxamide, 5-(3,3-dimethyl-1-triazenyl)-	4342-03-4
Decabromobiphenyl		13654-09-6
cis-Diaminedichloroplatinum	Cisplatin	15663-27-1
2,4-Diaminoanisole sulfate	1,3-Benzenediamine, 4-methoxy, sulfate	39156-41-7
4,4'-Diaminodiphenyl ether	4,4-Oxydianiline	101-80-4
4,4'-Diaminodiphenylmethane	4,4'-Methylenedianiline	101-77-9
Dibenz[a,h]acridine		226-36-8
Dibenz[a,j]acridine		224-42-0
Dibenz[a,h]anthracene		53-70-3
7H-Dibenzo[c,g]carbazole		194-59-2
Dibenzo[a,e]pyrene	Naphtho[1,2,3,4-def]chrysene	192-65-4
Dibenzo[a,h]pyrene	Dibenzo[b,def]chrysene	189-64-0
Dibenzo[a,i]pyrene	Benzo[rst]pentaphene	189-55-9
Dibenzo[a,l]pyrene	Dibenzo[def,p]chrysene	191-30-0
1,2-Dibromo-3-chloropropane		96-12-8
1,2-Dibromoethane	Ethylene dibromide	106-93-4
2,3-Dibromo-1-propanol, phosphate (3:1)	Tris(2,3-dibromopropyl) phosphate	126-72-7
p-Dichlorobenzene	1,4-Dichlorobenzene	106-46-7
3,3'-Dichloro-*p*-benzidine	[1,1'-Biphenyl]-4,4'-diamine, 3,3'-dichloro-	91-94-1
3,3'-Dichloro-*p*-benzidine dihydrochloride	3,3'-Dichloro-[1,1'-biphenyl]-4,4'-diamine dihydrochloride	612-83-9
1,2-Dichloroethane	Ethylene dichloride	107-06-2
Dichloromethane	Methylene chloride	75-09-2
1,3-Dichloropropene (unspecified isomer)		542-75-6
Diethyl sulfate		64-67-5
1,8-Dihydroxy-9,10-anthracenedione	Danthron; 1,8-Dihydroxyanthraquinone	117-10-2
3,3'-Dimethoxybenzidine	Dianisidine	119-90-4
p-(Dimethylamino)azobenzene		60-11-7
2',3-Dimethyl-4-aminoazobenzene	*o*-Aminoazotoluene; 4-*o*-Tolylazo-*o*-toluidine	97-56-3
Dimethylcarbamic chloride	Dimethylcarbamoyl chloride	79-44-7
1,1-Dimethylhydrazine	UDMH	57-14-7
Dimethyl sulfate		77-78-1
1,6-Dinitropyrene		42397-64-8

Substance	Other names	CAS RN
1,8-Dinitropyrene		42397-65-9
1,4-Dioxane		123-91-1
1,2-Diphenylhydrazine	Hydrazobenzene	122-66-7
Disperse Blue No. 1	9,10-Anthracenedione, 1,4,5,8-tetraamino-	2475-45-8
Epichlorohydrin	(Chloromethyl)oxirane	106-89-8
1,2-Epoxy-4-(epoxyethyl)cyclohexane	4-Vinyl-1-cyclohexene dioxide	106-87-6
Estra-1,3,5(10)-triene-3,17-diol (17β)-	Estradiol-17β	50-28-2
Estrone		53-16-7
N-(4-Ethoxyphenyl)acetamide	Phenacetin	62-44-2
Ethyl acrylate	Ethyl propenoate	140-88-5
Ethyl carbamate	Urethane	51-79-6
Ethylene oxide	Oxirane	75-21-8
Ethyl methanesulfonate		62-50-0
N-Ethyl-N-nitrosourea	ENU; N-Nitroso-N-ethylurea	759-73-9
Formaldehyde		50-00-0
Furan		110-00-9
Hexabromobiphenyl (unspecified isomer)	Firemaster FF-1	67774-32-7
Hexachlorobenzene	Perchlorobenzene	118-74-1
1,2,3,4,5,6-Hexachlorocyclohexane, (1α,2α,3β,4α,5α,6β)	Lindane	58-89-9
1,2,3,4,5,6-Hexachlorocyclohexane, (1α,2α,3β,4α,5β,6β)	α-Hexachlorocyclohexane	319-84-6
1,2,3,4,5,6-Hexachlorocyclohexane, (1α,2β,3α,4β,5α,6β)-	β-Hexachlorocyclohexane	319-85-7
Hexachlorocyclohexane (unspecified isomer)		608-73-1
Hexachloroethane	Perchloroethane	67-72-1
Hexamethylphosphoric triamide	Hexamethylphosphoramide; Tris(dimethylamino)phosphine oxide	680-31-9
Hydrazine		302-01-2
Hydrazine sulfate		10034-93-2
2-Imidazolidinethione	Ethylene thiourea	96-45-7
Indeno[1,2,3-cd]pyrene	1,10-(1,2-Phenylene)pyrene	193-39-5
Kanechlor 500	Polychlorinated biphenyls	37317-41-2
Kepone	Chlordecone	143-50-0
Lead(II) acetate		301-04-2
Lead(II) phosphate		7446-27-7
Mestranol	19-Norpregna-1,3,5(10)-trien-20-yn-17-ol, 3-methoxy-, (17α)-	72-33-3
o-Methoxyaniline hydrochloride	o-Anisidine hydrochloride	134-29-2
2-Methoxy-5-methylaniline	p-Cresidine; 5-Methyl-o-anisidine	120-71-8
o-Methylaniline	o-Toluidine	95-53-4
o-Methylaniline hydrochloride	o-Toluidine hydrochloride	636-21-5
5-Methylchrysene		3697-24-3
4,4-Methylenedianiline dihydrochloride	Benzenamine, 4,4'-methylenedi-, dihydrochloride	13552-44-8
Methyl methanesulfonate		66-27-3
N-Methyl-N'-nitro-N-nitrosoguanidine		70-25-7
N-methyl-N-nitrosourea	N-Nitroso-N-methylurea	684-93-5
Methyloxirane	1,2-Propylene oxide	75-56-9
Metronidazole	2-Methyl-5-nitro-1H-imidazole-1-ethanol	443-48-1
Mirex	1,3,4-Metheno-1H-cyclobuta[cd]pentalene, 1,1a,2,2,3,3a,4,5,5,5a,5b,6-dodecachlorooctahydro-	2385-85-5
Nickel		7440-02-0
Nickel(II) acetate		373-02-4
Nickel(II) carbonate		3333-67-3
Nickel carbonyl		13463-39-3
Nickel(II) hydroxide		12054-48-7
Nickel hydroxide (unspecified oxidation state)		11113-74-9
Nickelocene	Bis(η 5-2,4-cyclopentadien-1-yl)nickel	1271-28-9
Nickel(II) oxide		1313-99-1
Nickel(III) sulfide	Nickel subsulfide	12035-72-2
Nitrilotriacetic acid	N,N-Bis(carboxymethyl)glycine	139-13-9
2-Nitroanisole	1-Methoxy-2-nitrobenzene	91-23-6
6-Nitrochrysene		7496-02-8
Nitrofen	Benzene, 2,4-dichloro-1-(4-nitrophenoxy)-	1836-75-5

Substance	Other names	CAS RN
2-Nitropropane		79-46-9
1-Nitropyrene		5522-43-0
4-Nitropyrene		57835-92-4
N-Nitrosobutyl-N-(3-carboxypropyl)amine	Butyl(3-carboxypropyl)nitrosoamine	38252-74-3
N-Nitrosodibutylamine		924-16-3
N-Nitrosodiethanolamine	Ethanol, 2,2'-(nitrosoimino)-	1116-54-7
N-Nitrosodiethylamine	DEN; Diethylnitrosamine	55-18-5
N-Nitrosodimethylamine	DMN; Dimethylnitrosamine	62-75-9
4-(N-Nitrosomethylamino)-1-(3-pyridyl)-1-butanone	NNK; Ketone, 3-pyridyl-3-(N-methyl-N-nitrosamino)propyl	64091-91-4
N-Nitroso-N-methylvinylamine	Ethenamine, N-methyl-N-nitroso-	4549-40-0
4-Nitrosomorpholine	N-Nitrosomorpholine	59-89-2
N-Nitrosonornicotine		16543-55-8
N-Nitrosopiperidine	1-Nitrosopiperidine	100-75-4
N-Nitroso-N-propyl-1-propanamine	N-Nitrosodipropylamine	621-64-7
N-Nitrosopyrrolidine		930-55-2
N-Nitrososarcosine	Glycine, N-methyl-N-nitroso-	13256-22-9
Norethisterone	19-Norpregn-4-en-20-yn-3-one, 17-hydroxy-, (17 α)-	68-22-4
19-Norpregna-1,3,5(10)-trien-20-yne-3,17-diol, (17α)-	Ethinylestradiol	57-63-6
Ochratoxin A		303-47-9
Octabromobiphenyl (unspecified isomer)		61288-13-9
2-Oxetanone	β-Propiolactone	57-57-8
Oxiranemethanol	Glycidol	556-52-5
Oxymetholone	Androstan-3-one, 17-hydroxy-2-(hydroxymethylene)-17-methyl-	434-07-1
Phenazopyridine hydrochloride	2,6-Pyridinediamine, 3-(phenylazo)-, monohydrochloride	136-40-3
Phenoxybenzamine hydrochloride	Benzenemethanamine, N-(2-chloroethyl)-N-(1-methyl-2-phenoxyethyl)-, hydrochloride	63-92-3
Phenytoin	5,5-Diphenyl-2,4-imidazolidinedione	57-41-0
Polychlorinated biphenyls	PCBs	1336-36-3
Procarbazine hydrochloride		366-70-1
Progesterone	Pregn-4-ene-3,20-dione	57-83-0
1,3-Propane sultone	1,2-Oxathiolane, 2,2-dioxide	1120-71-4
Propyleneimine	2-Methylaziridine	75-55-8
4(1H)-Pyrimidinone, 2,3-dihydro-6-propyl-2-thioxo-	Propylthiouracil	51-52-5
Reserpine		50-55-5
Saccharin sodium	1,2-Benzisothiazolin-3-one, 1,1-dioxide, sodium salt	128-44-9
Safrole	5-(2-Propenyl)-1,3-benzodioxole	94-59-7
Selenium sulfide		7446-34-6
Silicon dioxide (β-quartz)		14808-60-7
Silicon dioxide (cristobalite)		14464-46-1
Silicon dioxide (tridymite)		15468-32-3
Streptozotocin	D-Glucopyranose, 2-deoxy-2-[[(methylnitrosoamino)carbonyl]amino]-	18883-66-4
Sulfallate	N,N-Diethyldithiocarbamic acid, 2-chloroallyl ester	95-06-7
2,3,7,8-Tetrachlorodibenzo-p-dioxin	TCDD; Dioxin	1746-01-6
Tetrachloroethylene	Perchloroethylene	127-18-4
Tetrachloromethane	Carbon tetrachloride	56-23-5
N,N,N',N'-Tetramethyl-4,4'-diaminobenzophenone	Bis(dimethylamino)benzophenone; Michler's Ketone	90-94-8
Tetranitromethane		509-14-8
Thioacetamide		62-55-5
Thiourea		62-56-6
o-Tolidine	3,3-Dimethylbenzidine	119-93-7
Toluene-2,4-diamine	2,4-Diaminotoluene	95-80-7
Toluene diisocyanate (unspecified isomer)		26471-62-5
Toxaphene	Polychlorocamphene	8001-35-2
1H-1,2,4-Triazol-3-amine	Amitrole	61-82-5
1,1,1-Trichloro-2,2-bis(4-chlorophenyl)ethane	DDT; Dichlorodiphenyltrichloroethane	50-29-3
Trichloromethane	Chloroform	67-66-3
(Trichloromethyl)benzene	Benzotrichloride	98-07-7
2,4,6-Trichlorophenol		88-06-2
1,2,3-Trichloropropane		96-18-4

Appendix A
Mathematical Tables

Miscellaneous Mathematical Constants A-1
Exponential and Hyperbolic Functions and their Common Logarithms A-2
Natural Trigonomic Functions to Four Places A-7
Relation of Angular Functions in Terms of One Another A-11
Derivatives A-12
Integration A-15
Integrals A-20
Differential Equations A-65
Fourier Series A-75
The Fourier Transforms A-80
Series Expansion A-84
Vector Analysis A-87
Moment of Inertial for Various Bodies of Mass A-96
Bessel Functions A-96
The Gamma Function A-99
The Beta Function A-101
The Error Function A-101
Orthogonal Polynomials A-102
Normal Probability Function A-104
Percentage Points, Student's *t*-Distribution A-105
Percentage Points, Chi-Square Distribution A-105
Percentage Points, F-Distribution A-106

MISCELLANEOUS MATHEMATICAL CONSTANTS

π CONSTANTS

$$\pi = 3.14159\ 26535\ 89793\ 23846\ 26433\ 83279\ 50288\ 41971\ 69399\ 37511$$
$$1/\pi = 0.31830\ 98861\ 83790\ 67153\ 77675\ 26745\ 02872\ 40689\ 19291\ 48091$$
$$\pi^2 = 9.86960\ 44010\ 89358\ 61883\ 44909\ 99876\ 15113\ 53136\ 99407\ 24079$$
$$\log_e \pi = 1.14472\ 98858\ 49400\ 17414\ 34273\ 51353\ 05871\ 16472\ 94812\ 91531$$
$$\log_{10} \pi = 0.49714\ 98726\ 94133\ 85435\ 12682\ 88290\ 89887\ 36516\ 78324\ 38044$$
$$\log_{10} \sqrt{2\pi} = 0.39908\ 99341\ 79057\ 52478\ 25035\ 91507\ 69595\ 02099\ 34102\ 92128$$

CONSTANTS INVOLVING e

$$e = 2.71828\ 18284\ 59045\ 23536\ 02874\ 71352\ 66249\ 77572\ 47093\ 69996$$
$$1/e = 0.36787\ 94411\ 71442\ 32159\ 55237\ 70161\ 46086\ 74458\ 11131\ 03177$$
$$e^2 = 7.38905\ 60989\ 30650\ 22723\ 04274\ 60575\ 00781\ 31803\ 15570\ 55185$$
$$M = \log_{10} e = 0.43429\ 44819\ 03251\ 82765\ 11289\ 18916\ 60508\ 22943\ 97005\ 80367$$
$$1/M = \log_e 10 = 2.30258\ 50929\ 94045\ 68401\ 79914\ 54684\ 36420\ 76011\ 01488\ 62877$$
$$\log_{10} M = 9.63778\ 43113\ 00536\ 78912\ 29674\ 98645\ -10$$

π^e AND e^π CONSTANTS

$$\pi^e = 22.45915\ 77183\ 61045\ 47342\ 71522$$
$$e^\pi = 23.14069\ 26327\ 79269\ 00572\ 90864$$
$$e^{-\pi} = 0.04321\ 39182\ 63772\ 24977\ 44177$$
$$e^{1/2\pi} = 4.81047\ 73809\ 65351\ 65547\ 30357$$
$$i^i = e^{-1/2\pi} = 0.20787\ 95763\ 50761\ 90854\ 69556$$

NUMERICAL CONSTANTS

$$\sqrt{2} = 1.41421\ 35623\ 73095\ 04880\ 16887\ 24209\ 69807\ 85696\ 71875\ 37695$$
$$\sqrt[3]{2} = 1.25992\ 10498\ 94873\ 16476\ 72106\ 07278\ 22835\ 05702\ 51464\ 70151$$
$$\log_e 2 = 0.69314\ 71805\ 59945\ 30941\ 72321\ 21458\ 17656\ 80755\ 00134\ 36026$$
$$\log_{10} 2 = 0.30102\ 99956\ 63981\ 19521\ 37388\ 94724\ 49302\ 67881\ 89881\ 46211$$
$$\sqrt{3} = 1.73205\ 08075\ 68877\ 29352\ 74463\ 41505\ 87236\ 69428\ 05253\ 81039$$
$$\sqrt[3]{3} = 1.44224\ 95703\ 07408\ 38232\ 16383\ 10780\ 10958\ 83918\ 69253\ 49935$$
$$\log_e 3 = 1.09861\ 22886\ 68109\ 69139\ 52452\ 36922\ 52570\ 46474\ 90557\ 82275$$
$$\log_{10} 3 = 0.47712\ 12547\ 19662\ 43729\ 50279\ 03255\ 11530\ 92001\ 28864\ 19070$$

OTHER CONSTANTS

$$\text{Euler's Constant } \gamma = 0.57721\ 56649\ 01532\ 86061$$
$$\log_e \gamma = -0.54953\ 93129\ 81644\ 82234$$
$$\text{Golden Ratio } \phi = 1.61803\ 39887\ 49894\ 84820\ 45868\ 34365\ 63811\ 77203\ 09180$$

EXPONENTIAL AND HYPERBOLIC FUNCTIONS AND THEIR COMMON LOGARITHMS

x	e^x Value	e^x log^{10}	e^{-x} (value)	sinh x Value	sinh x log^{10}	cosh x Value	cosh x log^{10}	tanh x (value)
0.00	1.0000	0.00000	1.00000	0.0000	— ∞	1.0000	0.00000	0.00000
0.01	1.0101	.00434	0.99005	.0100	2̲.00001	1.0001	.00002	.01000
0.02	1.0202	.00869	.98020	.0200	2̲.30106	1.0002	.00009	.02000
0.03	1.0305	.01303	.97045	.0300	2̲.47719	1.0005	.00020	.02999
0.04	1.0408	.01737	.96079	.0400	2̲.60218	1.0008	.00035	.03998
0.05	1.0513	.02171	.95123	.0500	2̲.69915	1.0013	.00054	.04996
0.06	1.0618	.02606	.94176	.0600	2̲.77841	1.0018	.00078	.05993
0.07	1.0725	.03040	.93239	.0701	2̲.84545	1.0025	.00106	.06989
0.08	1.0833	.03474	.92312	.0801	2̲.90355	1.0032	.00139	.07983
0.09	1.0942	.03909	.91393	.0901	2̲.95483	1.0041	.00176	.08976
0.10	1.1052	.04343	.90484	.1002	1̲.00072	1.0050	.00217	.09967
0.11	1.1163	.04777	.89583	1102	1̲.04227	1.0061	.00262	.10956
0.12	1.1275	.05212	.88692	.1203	1̲.08022	1.0072	.00312	.11943
0.13	1.1388	.05646	.87809	.1304	1̲.11517	1.0085	.00366	.12927
0.14	1.1503	.06080	.86936	.1405	1̲.14755	1.0098	.00424	.13909
0.15	1.1618	.06514	.86071	.1506	1̲.17772	1.0113	.00487	.14889
0.16	1.1735	.06949	.85214	.1607	1̲.20597	1.0128	.00554	.15865
0.17	1.1853	.07383	.84366	.1708	1̲.23254	1.0145	.00625	.16838
0.18	1.1972	.07817	.83527	.1810	1̲.25762	1.0162	.00700	.17808
0.19	1.2092	.08252	.82696	.1911	1̲.28136	1.0181	.00779	.18775
0.20	1.2214	.08686	.81873	.2013	1̲.30392	1.0201	.00863	.19738
0.21	1.2337	.09120	.81058	.2115	1̲.32541	1.0221	.00951	.20697
0.22	1.2461	.09554	.80252	.2218	1̲.34592	1.0243	.01043	.21652
0.23	1.2586	.09989	.79453	.2320	1̲.36555	1.0266	.01139	.22603
0.24	1.2712	.10423	.78663	.2423	1̲.38437	1.0289	.01239	.23550
0.25	1.2840	.10857	.77880	.2526	1̲.40245	1.0314	.01343	.24492
0.26	1.2969	.11292	.77105	.2629	1̲.41986	1.0340	.01452	.25430
0.27	1.3100	.11726	.76338	.2733	1̲.43663	1.0367	.01564	.26362
0.28	1.3231	12160	.75578	.2837	1̲.45282	1.0395	.01681	.27291
0.29	1.3364	.12595	.74826	.2941	1̲.46847	1.0423	.01801	.28213
0.30	1.3499	.13029	.74082	.3045	1̲.48362	1.0453	.01926	.29131
0.31	1.3634	.13463	.73345	.3150	1̲.49830	1.0484	.02054	.30044
0.32	1.3771	.13897	.72615	.3255	1̲.51254	1.0516	.02187	.30951
0.33	1.3910	.14332	.71892	.3360	1̲.52637	1.0549	.02323	.31852
0.34	1.4049	.14766	.71177	.3466	1̲.53981	1.0584	.02463	.32748
0.35	1.4191	.15200	.70469	.3572	1̲.55290	1.0619	.02607	.33638
0.36	1.4333	.15635	.69768	.3678	1̲.56564	1.0655	.02755	.34521
0.37	1.4477	.16069	.69073	.3785	1̲.57807	1.0692	.02907	.35399
0.38	1.4623	.16503	.68386	.3892	1̲.59019	1.0731	.03063	.36271
0.39	1.4770	.16937	.67706	.4000	1̲.60202	1.0770	.03222	.37136
0.40	1.4918	.17372	.67032	.4108	1̲.61358	1.0811	.03385	.37995
0.41	1.5063	.17806	.66365	.4216	1̲.62488	1.0852	.03552	.33847
0.42	1.5220	.18240	.65705	.4325	1̲.63594	1.0895	.03723	.39693
0.43	1.5373	.18675	.65051	.4434	1̲.64677	1.0939	.03897	.40532
0.44	1.5527	.19109	.64404	.4543	1̲.65738	1.0984	.04075	.41364
0.45	1.5683	.19543	.63763	.4653	1̲.66777	1.1030	.04256	.42190
0.46	1.5841	.19978	.63128	.4764	1̲.67797	1.1077	.04441	.43008
0.47	1.6000	.20412	.62500	.4875	1̲.68797	1.1125	.04630	.43820
0.48	1.6161	.20846	.61878	.4986	1̲.69779	1.1174	.04822	.44624
0.49	1.6323	.21280	.61263	.5098	1̲.70744	1.1225	.05018	.45422
0.50	1.6487	.21715	.60653	.5211	1̲.71692	1.1276	.05217	.46212
0.51	1.6653	.22149	.60050	.5324	1̲.72624	1.1329	.05419	.46995
0.52	1.6820	.22583	.59452	.5438	1̲.73540	1.1383	.05625	.47770
0.53	1.6989	.23018	.58860	.5552	1̲.74442	1.1438	.05834	.48538
0.54	1.7160	.23452	.58275	.5666	1̲.75330	1.1494	.06046	.49299
0.55	1.7333	.23886	.57695	.5782	1̲.76204	1.1551	.06262	.50052
0.56	1.7507	.24320	.57121	.5897	1̲.77065	1.1609	.06481	.50798
0.57	1.7683	.24755	.56553	.6014	1̲.77914	1.1669	.06703	.51536
0.58	1.7860	.25189	.55990	.6131	1̲.78751	1.1730	.06929	.52267
0.59	1.8040	.25623	.55433	.6248	1̲.79576	1.1792	.07157	.52990
0.60	1.8221	.26058	.54881	.6367	1̲.80390	1.1855	.07389	.53705
0.61	1.8404	.26492	.54335	.6485	1̲.81194	1.1919	.07624	.54413
0.62	1.8589	.26926	.53794	.6605	1̲.81987	1.1984	.07861	.55113
0.63	1.8776	.27361	.53259	.6725	1̲.82770	1.2051	.08102	.55805
0.64	1.8965	.27795	.52729	.6846	1̲.83543	1.2119	.08346	.56490
0.65	1.9155	.28229	.52205	.6967	1̲.84308	1.2188	.08593	.57167
0.66	1.9348	.28664	.51685	.7090	1̲.85063	1.2258	.08843	.57836
0.67	1.9542	.29098	.51171	.7213	1̲.85809	1.2330	.09095	.58498
0.68	1.9739	.29532	.50662	.7336	1̲.86548	1.2402	.09351	.59152
0.69	1.9937	.29966	.50158	.7461	1̲.87278	1.2476	.09609	.59798
0.70	2.0138	.30401	.49659	.7586	1̲.88000	1.2552	.09870	.60437
0.71	2.0340	.30835	.49164	.7712	1̲.88715	1.2628	.10134	.61068
0.72	2.0544	.31269	.48675	.7838	1̲.89423	1.2706	.10401	.61691

x	e^x Value	e^x $\log^{10}$	e^{-x} (value)	sinh x Value	sinh x $\log^{10}$	cosh x Value	cosh x $\log^{10}$	tanh x (value)
0.73	2.0751	.31703	.48191	.7966	$\bar{1}$.90123	1.2785	.10670	.62307
0.74	2.0959	.32138	.47711	.8094	$\bar{1}$.90817	1.2865	.10942	.62915
0.75	2.1170	.32572	.47237	.8223	$\bar{1}$.91504	1.2947	.11216	.63515
0.76	2.1383	.33006	.46767	.8353	$\bar{1}$.92185	1.3030	.11493	.64108
0.77	2.1598	.33441	.46301	.8484	$\bar{1}$.92859	1.3114	.11773	.64693
0.78	2.1815	.33875	.45841	.8615	$\bar{1}$.93527	1.3199	.12055	.65721
0.79	2.2034	.34309	.45384	.8748	$\bar{1}$.94190	1.3286	.12340	.65841
0.80	2.2255	.34744	.44933	.8881	$\bar{1}$.94846	1.3374	.12627	.66404
0.81	2.2479	.35178	.44486	.9015	$\bar{1}$.95498	1.3464	.12917	.66959
0.82	2.2705	.35612	.44043	.9150	$\bar{1}$.96144	1.3555	.13209	.67507
0.83	2.2933	.36046	.43605	.9286	$\bar{1}$.96784	1.3647	.13503	.68048
0.84	2.3164	.36481	.43171	.9423	$\bar{1}$.97420	1.3740	.13800	.68581
0.85	2.3396	.36915	.42741	.9561	$\bar{1}$.98051	1.3835	.14099	.69107
0.86	2.3632	.37349	.42316	.9700	$\bar{1}$.98677	1.3932	.14400	.69626
0.87	2.3869	.37784	.41895	.9840	$\bar{1}$.99299	1.4029	.14704	.70137
0.88	2.4100	.38218	.41478	.9981	$\bar{1}$.99916	1.4128	.15009	.70642
0.89	2.4351	.38652	.41066	1.0122	0.00528	1.4229	.15317	.71139
0.90	2.4596	.39087	.40657	1.0265	.01137	1.4331	.15627	.21630
0.91	2.4843	39521	.40242	1.0409	.01741	1.4434	.15939	.72113
0.92	2.5093	.39955	.39852	1.0554	.02341	1.4539	.16254	.72590
0.93	2.5345	.40389	.39455	1.0700	.02937	1.4645	.16570	.73059
0.94	2.5600	.40824	.39063	1.0847	.03530	1.4753	.16888	.73522
0.95	2.5857	.41258	.38674	1.0995	.04119	1.4862	.17208	.73978
0.96	2.6117	.41692	.38289	1.1144	.04704	1.4973	.17531	.74428
0.97	2.6379	.42127	.37908	1.1294	.05286	1.5085	.17855	.74870
0.98	2.6645	.42561	.37531	1.1446	.05864	1.5199	.18181	.75307
0.99	2.6912	.42995	.37158	1.1598	.06439	1.5314	.18509	.75736
1.00	2.7183	.43429	.36788	1.1752	.07011	1.5431	.18839	.76159
1.01	2.7456	.43864	.36422	1.1907	.07580	1.5549	.19171	.76576
1.02	2.7732	.44298	.36060	1.2063	.06146	1.5669	.19504	.76987
1.03	2.8011	.44732	.35701	1.2220	.08708	1.5790	.19839	.77391
1.04	2.8292	.45167	.35345	1.2379	.09268	1.5913	.20176	.77789
1.05	2.8577	.45601	.34994	1.2539	.09825	1.6038	.20515	.78181
1.06	2.8864	.46035	.34646	1.2700	.10379	1.6164	.20855	.78566
1.07	2.9154	.46470	.34301	1.2862	.10930	1.6292	.21197	.78946
1.08	2.9447	.46904	.33960	1.3025	.11479	1.6421	.21541	.79320
1.09	2.9743	.47338	.33622	1.3190	.12025	1.6552	.21886	.79688
1.10	3.0042	.47772	.33287	1.3356	.12569	1.6685	.22233	.80050
1.11	3.0344	.48207	.32956	1.3524	.13111	1.6820	.22582	.80406
1.12	3.0659	.48641	.32628	1.3693	.13649	1.6956	.22931	.80757
1.13	3.0957	.49075	.32303	1.3863	.14186	1.7083	.23283	.81102
1.14	3.1268	.49510	.31982	1.4035	.14720	1.7233	.23636	.81441
1.15	3.1582	.49944	.31644	1.4208	.15253	1.7374	.23990	.81775
1.16	3.1899	.50378	.31349	1.4382	.15783	1.7517	.24346	.82104
1.17	3.2220	.50812	.31037	1.4558	.16311	1.7662	.24703	.82427
1.18	3.2544	.51247	.30728	1.4735	.16836	1.7808	.25062	.82745
1.19	3.2871	.51681	.30422	1.4914	.17360	1.7957	.25422	.83058
1.20	3.3201	.52115	.30119	1.5095	.17882	1.8107	.25784	.83365
1.21	3.3535	.52550	.29820	1.5276	.18402	1.8258	.26146	.83668
1.22	3.3872	.52984	.29523	1.5460	.18920	1.8412	.26510	.83965
1.23	3.4212	.53418	.29229	1.5645	.19437	1.8568	.26876	.84258
1.24	3.4556	.53853	.28938	1.5831	.19951	1.8725	.27242	.83546
1.25	3.4903	.54287	.28650	1.6019	.20464	1.8884	.27610	.84828
1.26	3.5254	.54721	.28365	1.6209	.20975	1.9045	.27979	.85106
1.27	3.5609	.55155	.28083	1.6400	.21485	1.9208	.28349	.85380
1.28	3.5996	.55590	.27804	1.6593	.21993	1.9373	.28721	.85648
1.29	3.6328	.56024	.27527	1.6788	.22499	1.9540	.29093	.85913
1.30	3.6693	.56458	.27253	1.6984	.23004	1.9709	.29467	.86172
1.31	3.7062	.56893	.26982	1.7182	.23507	1.9880	.29842	.86428
1.32	3.7434	.57327	.26714	1.7381	.24009	2.0053	.30217	.86678
1.33	3.7810	.57761	.26448	1.7583	.24509	2.0228	.30594	.86925
1.34	3.8190	.58195	.26185	1.7786	.25008	2.0404	.30972	.87167
1.35	3.8574	.58630	.25924	1.7991	.25505	2.0583	.31352	.87405
1.36	3.8962	.59064	.25666	1.8198	.26002	2.0764	.31732	.87639
1.37	3.9354	.59498	.25411	1.8406	.26496	2.0947	.32113	.87869
1.38	3.9749	.59933	.25158	1.8617	.26990	2.1132	.32495	.88095
1.39	4.0149	.60367	.24908	1.8829	.27482	2.1320	.32878	.88317
1.40	4.0552	.60801	.24660	1.9043	.27974	2.1509	.33262	.88535
1.41	4.0960	.61236	.24414	1.9259	.28464	2.1700	.33647	.88749
1.42	4.1371	.61670	.24171	1.9477	.28952	2.1894	.34033	.88960
1.43	4.1787	.62104	.23931	1.9697	.29440	2.2090	.34420	.89167

EXPONENTIAL AND HYPERBOLIC FUNCTIONS AND THEIR COMMON LOGARITHMS
(continued)

	e^x		e^{-x}	sinh x		cosh x		tanh x
x	Value	$\log^{10}$	(value)	Value	$\log^{10}$	Value	$\log^{10}$	(value)
1.44	4.2207	.62538	.23693	1.9919	.29926	2.2288	.34807	.89370
1.45	4.2631	.62973	.23457	2.0143	.30412	2.2488	.35196	.89569
1.46	4.3060	.63407	.23224	2.0369	.30896	2.2691	.35585	.89765
1.47	4.3492	.63841	.22993	2.0597	.31379	2.2896	.35976	.89958
1.48	4.3929	.64276	.22764	2.0827	.31862	2.3103	.36367	.90147
1.49	4.4371	.64710	.22537	2.1059	.32343	2.3312	.36759	.90332
1.50	4.4817	.65144	.22313	2.1293	.32823	2.3524	.37151	.90515
1.51	4.5267	.65578	.22091	2.1529	.33303	2.3738	.37545	.90694
1.52	4.5722	.66013	.21871	2.1768	.33781	2.3955	.37939	.90870
1.53	4.6182	.66447	.21654	2.2008	.34258	2.4174	.38334	.91042
1.54	4.6646	.66881	.21438	2.2251	.34735	2.4395	.38730	.91212
1.55	4.7115	.67316	.21225	2.2496	.35211	2.4619	.39126	.91379
1.56	4.7588	.67750	.21014	2.2743	.35686	2.4845	.39524	.91542
1.57	4.8066	.68184	.20805	2.2993	.36160	2.5073	.39921	.91703
1.58	4.8550	.68619	.20598	2.3245	.36633	2.5305	.40320	.91860
1.59	4.9037	.69053	.20393	2.3499	.37105	2.5538	.40719	.92015
1.60	4.9530	.69487	.20190	2.3756	.37577	2.5775	.41119	.92167
1.61	5.0028	.69921	.19989	2.4015	.38048	2.6013	.41520	.92316
1.62	5.0531	.70356	.19790	2.4276	.38518	2.6255	.41921	.92462
1.63	5.1039	.70790	.19593	2.4540	.38987	2.6499	.42323	.92606
1.64	5.1552	.71224	.19398	2.4806	.39456	2.6746	.42725	.92747
1.65	5.2070	.71659	.19205	2.5075	.39923	2.6995	.43129	.92886
1.66	5.2593	.72093	.19014	2.5346	.40391	2.7247	.43532	.93022
1.67	5.3122	.72527	.18825	2.5620	.40857	2.7502	.43937	.93155
1.68	5.3656	.72961	.18637	2.5896	.41323	2.7760	.44341	.93286
1.69	5.4195	.73396	.18452	2.6175	.41788	2.8020	.44747	.93415
1.70	5.4739	.73830	.18268	2.6456	.42253	2.8283	.45153	.93541
1.71	5.5290	.74264	.18087	2.6740	.42717	2.8549	.45559	.93665
1.72	5.5845	.74699	.17907	2.7027	.43180	2.8818	.45966	.93786
1.73	5.6407	.75133	.17728	2.7317	.43643	2.9090	.46374	.93906
1.74	5.6973	.75567	.17552	2.7609	.44105	2.9364	.46782	.94023
1.75	5.7546	.76002	.17377	2.7904	.44567	2.9642	.47191	.94138
1.76	5.8124	.76436	.17204	2.8202	.45028	2.9922	.47600	.94250
1.77	5.8709	.76870	.17033	2.8503	.45488	3.0206	.48009	.94361
1.78	5.9299	.77304	.16864	2.8806	.45948	3.0492	.48419	.94470
1.79	5.9895	.77739	.16696	2.9112	.46408	3.0782	.48830	.94576
1.80	6.0496	.78173	.16530	2.9422	.46867	3.1075	.49241	.94681
1.81	6.1104	.78607	.16365	2.9734	.47325	3.1371	.49652	.94783
1.82	6.1719	.79042	.16203	3.0049	.47783	3.1669	.50064	.94884
1.83	6.2339	.79476	.16041	3.0367	.48241	3.1972	.50476	.94983
1.84	6.2965	.79910	.15882	3.0689	.48698	3.2277	.50889	.95080
1.85	6.3598	.80344	.15724	3.1013	.49154	3.2585	.51302	.95175
1.86	6.4237	.80779	.15567	3.1340	.49610	3.2897	.51716	.95268
1.87	6.4383	.81213	.15412	3.1671	.50066	3.3212	.52130	.95359
1.88	6.5535	.81647	.15259	3.2005	.50521	3.3530	.52544	.95449
1.89	6.6194	.82082	.15107	3.2341	.50976	3.3852	.52959	.95537
1.90	6.6859	.82516	.14957	3.2682	.51430	3.4177	.53374	.95624
1.91	6.7531	.82950	.14808	3.3025	.51884	3.4506	.53789	.95709
1.92	6.8210	.83385	.14661	3.3372	.52338	3.4838	.54205	.95792
1.93	6.8895	.83819	.14515	3.3722	.52791	3.5173	.54621	.95873
1.94	6.9588	.84253	.14370	3.4075	.53244	3.5512	.55038	.95953
1.95	7.0287	.84687	.14227	3.4432	.53696	3.5855	.55455	.96032
1.96	7.0993	.85122	.14086	3.4792	.54148	3.6201	.55872	.96109
1.97	7.1707	.85556	.13946	3.5156	.54600	3.6551	.56290	96185
1.98	7.2427	.85990	.13807	3.5923	.55051	3.6904	.56707	.96259
1.99	7.3155	.86425	.13670	3.5894	.55502	3.7261	.57126	.96331
2.00	7.3891	.86859	.13534	3.6269	.55953	3.7622	.57544	.96403
2.01	7.4633	.87293	.13399	3.6647	.56403	3.7987	.57963	.96473
2.02	7.5383	.87727	.13266	3.7028	.56853	3.8335	.58382	.96541
2.03	7.6141	.88162	.13134	3.7414	.57303	3.8727	.58802	.96609
2.04	7.6906	.88596	.13003	3.7803	.57753	3.9103	.59221	.96675
2.05	7.7679	.89030	.12873	3.8196	.58202	3.9483	.59641	.96740
2.06	7.8460	.89465	.12745	3.8593	.58650	3.9867	.60061	.96803
2.07	7.9248	.89899	.12619	3.8993	.59099	4.0255	.60482	.96865
2.08	8.0045	.90333	.12493	3.9398	.59547	4.0647	.60903	.96926
2.09	8.0849	.90768	.12369	3.9806	.59995	4.1043	.61324	.96986
2.10	8.1662	.91202	.12246	4.0219	.60443	4.1443	.61745	.97045
2.11	8.2482	.91636	.12124	4.0635	.60890	4.1847	.62167	.97103
2.12	8.3311	.92070	.12003	4.1056	.61337	4.2256	.62589	.97159
2.13	8.4149	.92505	.11884	4.1480	.61784	4.2669	.63011	.97215
2.14	8.4994	.92939	.11765	4.1909	.62231	4.3085	.63433	.97269
2.15	8.5849	.93373	.11648	4.2342	62677	4.3507	.63856	.97323

EXPONENTIAL AND HYPERBOLIC FUNCTIONS AND THEIR COMMON LOGARITHMS
(continued)

x	e^x Value	e^x log^{10}	(value)	sinh x Value	sinh x log^{10}	cosh x Value	cosh x log^{10}	tanh x (value)
2.16	8.6711	.93808	.11533	4.2779	.63123	4.3932	.64278	.97375
2.17	8.7583	.94242	.11418	4.3221	.63569	4.4362	.64701	.97426
2.18	8.8463	.94676	.11304	4.3666	.64015	4.4797	.65125	.97477
2.19	8.9352	.95110	.11192	4.4116	.64460	4.5236	.65548	.97526
2.20	9.0250	.95545	.11080	4.4571	.64905	4.5679	.65972	.97574
2.21	9.1157	.95979	.10970	4.5030	.65350	4.6127	.66396	.97622
2.22	9.2073	.96413	.10861	4.5494	.65795	4.6580	.66820	.97668
2.23	9.2999	.96848	.10753	4.5962	.66240	4.7037	.67244	.97714
2.24	9.3933	.97282	.10646	4.6434	.66684	4.7499	.67668	.97759
2.25	9.4877	.97716	.10540	4.6912	.67128	4.7966	.68093	.97803
2.26	9.5831	.98151	.10435	4.7394	.67572	4.8437	.68518	.97846
2.27	9.6794	.98585	.10331	4.7880	.68016	4.8914	.68943	.97888
2.28	9.7767	.99019	.10228	4.8372	.68459	4.9395	.69368	.97929
7.29	9.8749	.99453	.10127	4.8868	.68903	4.9881	.69794	97970
2.30	9.9742	.99888	.10026	4.9370	.69346	5.0372	.70219	.98010
2.31	10.074	1.00322	.09926	4.9876	.69789	5.0868	.70645	.98049
2.32	10.176	1.00756	.09827	5.0387	.70232	5.1370	.71071	.98087
2.33	10.278	1.01191	.09730	5.0903	.70675	5.1876	.71497	.98124
2.34	10.381	1.01625	.09633	5.1425	.71117	5.2388	.71923	.98161
2.35	10.486	1.02059	.09537	5.1951	.71559	5.2905	.72349	.98197
2.36	10.591	1.02493	.09442	5.2483	.72002	5.3427	.72776	.98233
2.37	10.697	1.02928	.09348	5.3020	.72444	5.3954	.73203	.98267
2.38	10.805	1.03362	.09255	5.3562	.72885	5.4487	.73630	.98301
2.39	10.913	1.03796	.09163	5.4109	.73327	5.5026	.74056	.98335
2.40	11.023	1.04231	.09072	5.4662	.73769	5.5569	.74484	.98367
2.41	11.134	1.04665	.08982	5.5221	.74210	5.6119	.74911	.98400
2.42	11.246	1.05099	.08892	5.5785	.74652	5.6674	.75338	.98431
2.43	11.359	1.05534	08804	5.6354	.75093	5.7235	.75766	.98462
2.44	11.473	1.05968	.08716	5.6929	.75534	5.7801	.76194	.98492
2.45	11.588	1.06402	.08629	5.7510	.75975	5.8373	.76621	.98522
2.46	11.705	1.06836	.08543	5.8097	.76415	5.8951	.77049	.98551
2.47	11.822	1.07271	.08458	5.8689	.76856	5.9535	.77477	.98579
2.48	11.941	1.07705	.08374	5.9288	.77296	6.0125	.77906	.98607
2.49	12.061	1.08139	.08291	5.9892	.77737	6.0721	.78334	.98635
2.50	12.182	1.08574	.08208	6.0502	.78177	6.1323	.78762	.98661
2.51	12.305	1.09008	.08127	6.1118	.78617	6.1931	.79191	.98688
2.52	12.429	1.09442	.08046	6.1741	.79057	6.2545	.79619	.98714
2.53	12.554	1.09877	.07966	6.2369	.79497	6.3166	.80048	.98739
2.54	12.680	1.10311	.07887	6.3004	.79937	6.3793	.80477	.98764
2.55	12.807	1.10745	.07808	6.3645	.80377	6.4426	.80906	.98788
2.56	12.936	1.11179	.07730	6.4293	.80816	6.5066	.81335	.98812
2.57	13.066	1.11614	.07654	6.4946	.81256	6.5712	.81764	.98835
2.58	13.197	1.12048	.07577	6.5607	.81695	6.6365	.82194	.98858
2.59	13.330	1.12482	.07502	6.6274	.82134	6.7024	.82623	.98881
2.60	13.464	1.12917	.07427	6.6947	.82573	6.7690	.83052	.98903
2.61	13.599	1.13351	.07353	6.7628	.83012	6.8363	.83482	.98924
2.62	13.736	1.13785	.07280	6.8315	.83451	6.9043	.83912	.98946
2.63	13.874	1.14219	.07208	6.9008	.83890	6.9729	.84341	.98966
2.64	14.013	1.14654	.07136	6.9709	.84329	7.0423	.84771	98987
2.65	14.154	1.15008	.07065	7.0417	.84768	7.1123	.85201	.99007
2.66	14.296	1.15522	.06995	7.1132	.85206	7.1831	.85631	.99026
2.67	14.440	1.15957	.06925	7.1854	.85645	7.2546	.86061	.99045
2.68	14.585	1.16391	.06856	7.2583	.86083	7.3268	.86492	.99064
2.69	14.732	1.16825	.06788	7.3319	.86522	7.3998	.86922	.99083
2.70	14.880	1.17260	.06721	7.4063	.86960	7.4735	.87352	.99101
2.71	15.029	1.17694	.06654	7.4814	.87398	7.5479	.87783	.99118
2.72	15.180	1.18128	.06587	7.5572	.87836	7.6231	.88213	.99136
2.73	15.333	1.18562	.06522	7.6338	.88274	7.6991	.89644	.99153
2.74	15.487	1.18997	.06457	7.7112	.88712	7.7758	.89074	.99170
2.75	15.643	1.19431	06393	7.7894	.89150	7.8533	.89505	.99186
2.76	15.800	1.19865	.06329	7.8683	.89588	7.9316	.89936	.99202
2.77	15.959	1.20300	.06266	7.9480	.90026	8.0106	.90367	.99218
2.78	16.119	1.20734	.06204	8.0285	.90463	8.0905	.90798	.99233
2.79	16.281	1.21168	.06142	8.1098	.90901	8.1712	.91229	.99248
2.80	16.445	1.21602	.06081	8.1919	.91339	8.2527	.91660	.99263
2.81	16.610	1.22037	.06020	8.2749	.91776	8.3351	92091	.99278
2.82	16.777	1.22471	.05961	8.3586	.92213	8.4182	.92522	.99292
2.83	16.945	1.22905	.05901	8.4432	.92651	8.5022	.92953	.99306
2.84	17.116	1.23340	.05843	8.5287	.93088	8.5871	.93385	.99320
2.85	17.288	1.23774	.05784	8.6150	.93525	8.6728	.93816	.99333
2.86	17.462	1.24208	.05727	8.7021	.93963	8.7594	.94247	.99346

x	e^x Value	e^x $\log^{10}$	e^{-x} (value)	sinh x Value	sinh x $\log^{10}$	cosh x Value	cosh x $\log^{10}$	tanh x (value)
2.87	17.637	1.24643	.05670	8.7902	.94400	8.8469	.94679	.99359
2.88	17.814	1.25077	.05613	8.8791	.94837	8.9352	95110	.99372
2.89	17.993	1.25511	.05558	8.9689	.95274	9.0244	.95542	.99384
2.90	18.174	1.25945	.05502	9.0596	.95711	9.1146	.95974	.99396
2.91	18.357	1.26380	.05448	9.1512	.96148	9.2056	.96405	.99408
2.92	18.541	1.26814	.05393	9.2437	.96584	9.2976	.96837	.99420
2.93	18.728	1.27248	.05340	9.3371	.97021	9.3905	.97269	.99431
2.94	18.916	1.27683	.05287	9.4315	.97458	9.4844	.97701	.99443
2.95	19.106	1.28117	.05234	9.5268	.97895	9.5791	.98133	.99454
2.96	19.298	1.28551	.05182	9.6231	.98331	9.6749	.98565	.99464
2.97	19.492	1.28985	.05130	9.7203	.98768	9.7716	.98997	.99475
2.98	19.688	1.29420	.05079	9.8185	.99205	9.8693	.99429	.99485
2.99	19.886	1.29854	.05029	9.9177	.99641	9.9680	.99861	.99496
3.00	20.086	1.30288	.04979	10.018	1.00078	10.068	1.00293	.99505
3.05	21.115	1.32460	.04736	10.534	1.02259	10.581	1.02454	0.99552
3.10	22.198	1.34631	.04505	11.076	1.04440	11.122	1.04616	0.99595
3.15	23.336	1.36803	.04285	11.647	1.06620	11.690	1.06779	0.99633
3.20	24.533	1.38974	.04076	12.246	1.08799	12.287	1.08943	0.99668
3.25	25.790	1.41146	.03877	12.876	1.10977	12.915	1.11108	0.99700
3.30	27.113	1.43317	.03688	13.538	1.13155	13.575	1.13273	0.99728
3.35	28.503	1.45489	.03508	14.234	1.15332	14.269	1.15439	0.99754
3.40	29.964	1.47660	.03337	14.965	1.17509	14.999	1.17605	0.99777
3.45	31.500	1.49832	.03175	15.734	1.19685	15.766	1.19772	0.99799
3.50	33.115	1.52003	.03020	16.543	1.21860	16.573	1.21940	0.99818
3.55	34.813	1.54175	.02872	17.392	1.24036	17.421	1.24107	0.99835
3.60	36.598	1.56346	.02732	18.286	1.26211	18.313	1.26275	0.99851
3.65	38.475	1.58517	02599	19.224	1.28385	19.250	1.28444	9.99865
3.70	40.447	1.60689	.02472	20.211	1.30559	20.236	1.30612	0.99878
3.75	42.521	1.62860	.02352	21.249	1.32733	21.272	1.32781	0.99889
3.80	44.701	1.65032	.02237	22.339	1.34907	22.362	1.34951	0.99900
3.85	46.993	1.67203	.02128	23.486	1.37081	23.507	1.37120	0.99909
3.90	49.402	1.69375	.02024	24.691	1.39254	24.711	1.39290	0.99918
3.95	51.935	1.71546	.01925	25.958	1.41427	25.977	1.41459	0.09926
4.00	54.598	1.73718	.01832	27.290	1.43600	27.308	1.43629	0.99933
4.10	60.340	1.78061	.01657	30.162	1.47946	30.178	1.47970	0.99945
4.20	66.686	1.82404	.01500	33.336	1.52291	33.351	1.52310	0.99955
4.30	73.700	1.86747	.01357	36.843	1.56636	36.857	1.56652	0.99963
4.40	81.451	1.91090	.01227	40.719	1.60980	40.732	1.60993	0.99970
4.50	90.017	1.95433	.01111	45.003	1.65324	45.014	1.65335	0.99975
4.60	99.484	1.99775	.01005	49.737	1.69668	49.747	1.69677	0.99980
4.70	109.95	2.04118	.00910	54.969	1.74012	54.978	1.74019	0.99983
4.80	121.51	2.08461	.00823	60.751	1.78355	60.759	1.78361	0.99986
4.90	134.29	2.12804	.00745	67.141	1.82699	67.149	1.82704	0.99989
5.00	148.41	2.17147	.00674	74.203	1.87042	74.210	1.87046	0.99991
5.10	164.02	2.21490	.00610	82.008	1.91389	82.014	1.91389	0.99993
5.20	181.27	2.25833	.00552	90.633	1.95729	90.639	1.95731	0.99994
5.30	200.34	2.30176	.00499	100.17	2.00074	100.17	2.00074	0.99995
5.40	221.41	2.34519	.00452	110.70	2.04415	110.71	2.04417	0.99996
5.50	244.69	2.38862	.00409	122.34	2.08758	122.35	2.08760	0.99997
5.60	270.43	2.43205	.00370	135.21	2.13101	135.22	2.13103	0.99997
5.70	298.87	2.47548	.00335	149.43	2.17444	149.44	2.17445	0.99998
5.80	330.30	2.51891	.00303	165.15	2.21787	165.15	2.21788	0.99998
5.90	365.04	2.56234	.00274	182.52	2.26130	182.52	2.26131	0.99998
6.00	403.43	2.60577	.00248	201.71	2.30473	201.72	2.30474	0.99999
6.25	518.01	2.71434	.00193	259.01	2.41331	259.01	2.41331	0.99999
6.50	665.14	2.82291	.00150	332.57	2.52188	332.57	2.52189	1.00000
6.75	854.06	2.93149	.00117	427.03	2.63046	427.03	2.63046	1.00000
7.00	1096.6	3.04006	.00091	548.32	2.73904	548.32	2.73903	1.00000
7.50	1808.0	3.25721	.00055	904.02	2.95618	904.02	2.95618	1.00000
8.00	2981.0	3.47436	.00034	1490.5	3.17333	1490.5	3.17333	1.00000
8.50	4914.8	3.69150	.00020	2457.4	3.39047	2457.4	3.39047	1.00000
9.00	8103.1	3.90865	.00012	4051.5	3.60762	4051.5	3.60762	1.00000
9.50	13360.	4.12580	.00007	6679.9	3.82477	6679.9	3.82477	1.00000
10.00	22026.	4.34294	.00005	11013.	4.04191	11013.	4.04191	1.00000

NATURAL TRIGONOMETRIC FUNCTIONS TO FOUR PLACES

X radians	X degrees	sin x	cos x	tan x	cot x	sec x	csc x		
.0000	0° 00′	.0000	1.0000	.0000	—	1.000	—	90° 00′	1.5708
.0029	10	.0029	1.0000	.0029	343.8	1.000	343.8	50	1.5679
.0058	20	.0058	1.0000	.0058	171.9	1.000	171.9	40	1.5650
.0087	30	.0087	1.0000	.0087	114.6	1.000	114.6	30	1.5621
.0116	40	.0116	.9999	.0116	85.94	1.000	85.95	20	1.5592
.0145	50	.0145	.9999	.0145	68.75	1.000	68.76	10	1.5563
.0175	1° 00′	.0175	.9998	.0175	57.29	1.000	57.30	89° 00′	1.5533
.0204	10	.0204	.9998	.0204	49.10	1.000	49.11	50	1.5504
.0233	20	.0233	.9997	.0233	42.96	1.000	42.98	40	1.5475
.0262	30	.0262	.9997	.0262	38.19	1.000	38.20	30	1.5446
.0291	40	.0291	.9996	.0291	34.37	1.000	34.38	20	1.5417
.0320	50	.0320	.9995	.0320	31.24	1.001	31.26	10	1.5388
.0349	2° 00′	.0349	.9994	.0349	28.64	1.001	28.65	88° 00′	1.5359
.0378	10	.0378	.9993	.0378	26.43	1.001	26.45	50	1.5330
.0407	20	.0407	.9992	.0407	24.54	1.001	24.56	40	1.5301
.0436	30	.0436	.9990	.0437	22.90	1.001	22.93	30	1.5272
.0465	40	.0465	.9989	.0466	21.47	1.001	21.49	20	1.5243
.0495	50	.0494	.9988	.0495	20.21	1.001	20.23	10	1.5213
.0524	3° 00′	.0523	.9986	.0524	19.08	1.001	19.11	87° 00′	1.5184
.0553	10	.0552	.9985	.0553	18.07	1.002	18.10	50	1.5155
.0582	20	.0581	.9983	.0582	17.17	1.002	17.20	40	1.5126
.0611	30	.0610	.9981	.0612	16.35	1.002	16.38	30	1.5097
.0640	40	.0640	.9980	.0641	15.60	1.002	15.64	20	1.5068
.0669	50	.0669	.9978	.0670	14.92	1.002	14.96	10	1.5039
.0698	4° 00′	.0698	.9976	.0699	14.30	1.002	14.34	86° 00′	1.5010
.0727	10	.0727	.9974	.0729	13.73	1.003	13.76	50	1.4981
.0756	20	.0756	.9971	.0758	13.20	1.003	13.23	40	1.4952
.0785	30	.0785	.9969	.0787	12.71	1.003	12.75	30	1.4923
.0814	40	.0814	.9967	.0816	12.25	1.003	12.29	20	1.4893
.0844	50	.0843	.9964	.0846	11.83	1.004	11.87	10	1.4864
.0873	5° 00′	.0872	.9962	.0875	11.43	1.004	11.47	85° 00′	1.4835
.0902	10	.0901	.9959	.0904	11.06	1.004	11.10	50	1.4806
.0931	20	.0929	.9957	.0934	10.71	1.004	10.76	40	1.4777
.0960	30	.0958	.9954	.0963	10.39	1.005	10.43	30	1.4748
.0989	40	.0987	.9951	.0992	10.08	1.005	10.13	20	1.4719
.1018	50	.1016	.9948	.1022	9.788	1.005	9.839	10	1.4690
.1047	6° 00′	.1045	.9945	.1051	9.514	1.006	9.567	84° 00′	1.4661
.1076	10	.1074	.9942	.1080	9.255	1.006	9.309	50	1.4632
.1105	20	.1103	.9939	.1110	9.010	1.006	9.065	40	1.4603
.1134	30	.1132	.9936	.1139	8.777	1.006	8.834	30	1.4573
.1164	40	.1161	.9932	.1169	8.556	1.007	8.614	20	1.4544
.1193	.50	.1190	.9929	.1198	8.345	1.007	8.405	10	1.4515
.1222	7° 00′	.1219	.9925	.1228	8.144	1.008	8.206	83° 00′	1.4486
.1251	10	.1248	.9922	.1257	7.953	1.008	8.016	50	1.4457
.1280	20	.1276	.9918	.1287	7.770	1.008	7.834	40	1.4428
.1309	30	.1305	.9914	.1317	7.596	1.009	7.661	30	1.4399
.1338	40	.1334	.9911	.1346	7.429	1.009	7.496	20	1.4370
.1367	50	.1363	.9907	.1376	7.269	1.009	7.337	10	1.4341
.1396	8° 00′	.1392	.9903	.1405	7.115	1.010	7.185	82° 00′	1.4312
.1425	10	.1421	.9899	.1435	6.968	1.010	7.040	50	1.4283
.1454	20	.1449	.9894	.1465	6.827	1.011	6.900	40	1.4254
.1484	30	.1478	.9890	.1495	6.691	1.011	6.765	30	1.4224
.1513	40	.1507	.9886	.1524	6.561	1.012	6.636	20	1.4195
.1542	50	.1536	.9881	.1554	6.435	1.012	6.512	10	1.4166
.1571	9° 00′	.1564	.9877	.1584	6.314	1.012	6.392	81° 00′	1.4137
.1600	10	.1593	.9872	.1614	6.197	1.013	6.277	50	1.4108
.1629	20	.1622	.9868	.1644	6.084	1.013	6.166	40	1.4079
.1658	30	.1650	.9863	.1673	5.976	1.014	6.059	30	1.4050
.1687	40	.1679	.9858	.1703	5.871	1.014	5.955	20	1.4021
.1716	50	.1708	.9853	.1733	5.769	1.015	5.855	10	1.3992
.1745	10° 00′	.1736	.9848	.1763	5.671	1.015	5.759	80° 00′	1.3963
.1774	10	.1765	.9843	.1793	5.576	1.016	5.665	50	1.3934
.1804	20	.1794	.9838	.1823	5.485	1.016	5.575	40	1.3904
.1833	30	.1822	.9833	.1853	5.396	1.017	5.487	30	1.3875
.1862	40	.1851	.9827	.1883	5.309	1.018	5.403	20	1.3846
.1891	50	.1880	.9822	.1914	5.226	1.018	5.320	10	1.3817
.1920	11° 00′	.1908	.9816	.1944	5.145	1.019	5.241	79° 00′	1.3788
.1949	10	.1937	.9811	.1974	5.066	1.019	5.164	50	1.3759
.1978	20	.1965	.9805	.2004	4.989	1.020	5.089	40	1.3730
.2007	30	.1994	.9799	.2035	4.915	1.020	5.016	30	1.3701
		cos x	sin x	cot x	tan x	csc x	sec x	x degrees	x radians

x radians	x degrees	sin x	cos x	tan x	cot x	sec x	csc x		
.2036	40	.2022	.9793	.2065	3.843	1.021	4.945	20	1.3672
.2065	50	.2051	.9787	.2095	4.773	1.022	4.876	10	1.3643
.2094	12° 00′	.2079	.9781	.2126	4.705	1.022	4.810	78° 00′	1.3614
.2123	10	.2108	.9775	.2156	4.638	1.023	4.745	50	1.3584
.2153	20	.2136	.9769	.2186	4.574	1.024	4.682	40	1.3555
.2182	30	.2164	.9763	.2217	4.511	1.025	4.620	30	1.3526
.2211	40	.2193	.9757	.2247	4.449	1.025	4.560	20	1.3497
.2240	50	.2221	.9750	.2278	4.390	1.026	4.502	10	1.3468
.2269	13° 00′	.2250	.9744	.2309	4.331	1.026	4.445	77° 00′	1.3439
.2298	10	.2278	.9737	.2339	4.275	1.027	4.390	50	1.3410
.2327	20	.2306	.9730	.2370	4.219	1.028	4.336	40	1.3381
.2356	30	.2334	.9724	.2401	4.165	1.028	4.284	30	1.3352
.2385	40	.2363	.9717	.2432	4.113	1.029	4.232	20	1.3323
.2414	50	.2391	.9710	.2462	4.061	1.030	4.182	10	1.3294
.2443	14° 00′	.2419	.9703	.2493	4.011	1.031	4.134	76° 00′	1.3265
.2473	10	.2447	.9696	.2524	3.962	1.031	4.086	50	1.3235
.2502	20	.2476	.9689	.2555	3.914	1.032	4.039	40	1.3206
.2531	30	.2404	.9681	.2586	3.867	1.033	3.994	30	1.3177
.2560	40	.2532	.9674	.2617	3.821	1.034	3.950	20	1.3148
.2589	50	.2560	.9667	.2648	3.776	1.034	3.906	10	1.3119
.2618	15° 00′	.2588	.9659	.2679	3.732	1.035	3.864	75° 00′	1.3090
.2647	10	.2616	.9652	.2711	3.689	1.036	3.822	50	1.3061
.2676	20	.2644	.9644	.2732	3.647	1.037	3.782	40	1.3032
.2705	30	.2672	.9636	.2773	3.606	1.038	3.742	30	1.3003
.2734	40	.2700	.9628	.2805	3.566	1.039	3.703	20	1.2974
.2763	50	.2728	.9621	2836	3.526	1.039	3.665	10	1.2945
.2793	16° 00′	.2756	.9613	.2867	3.487	1.040	3.628	74° 00′	1.2915
.2822	10	.2784	.9605	.2899	3.450	1.041	3.592	50	1.2886
.2851	20	.2812	.9596	.2931	3.412	1.042	3.556	40	1.2857
.2880	30	.2840	.9588	.2962	3.376	1.043	3.521	30	1.2828
.2909	40	.2868	.9580	.2994	3.340	1.044	3.487	20	1.2799
.2938	50	.2896	.9572	.3026	3.305	1.045	3.453	10	1.2770
.2967	17° 00′	.2924	.9563	.3057	3.271	1.046	3.420	73° 00′	1.2741
.2996	10	.2952	.9555	.3089	3.237	1.047	3.388	50	1.2712
.3025	20	.2979	.9546	.3121	4.204	1.048	3.356	40	1.2683
.3054	30	.3007	.9537	.3153	3.172	1.049	3.326	30	1.2654
.3083	40	.3035	.9528	.3185	3.140	1.049	3.295	20	1.2625
.3113	50	.3062	.9520	.3217	3.108	1.050	3.265	10	1.2595
.3142	18° 00′	.3090	.9511	.3249	3.078	1.051	3.236	72° 00′	1.2566
.3171	10	.3118	.9502	.3281	3.047	1.052	3.207	50	1.2537
.3200	20	.3145	.9492	.3314	3.018	1.053	3.179	40	1.2508
.3229	30	.3173	.9483	.3346	2.989	1.054	3.152	30	1.2479
.3258	40	.3201	.9474	.3378	2.960	1.056	3.124	20	1.2450
.3287	50	.3228	.9465	.3411	2.932	1.057	3.098	10	1.2421
.3316	19° 00′	.3256	.9455	.3443	2.904	1.058	3.072	71° 00′	1.2392
.3345	10	.3283	.9446	.3476	2.877	1.059	3.046	50	1.2363
.3374	20	.3311	.9436	.3508	2.850	1.060	3.021	40	1.2334
.3403	30	.3338	.9426	.3541	2.824	1.061	2.996	30	1.2305
.3432	40	.3365	.9417	.3574	2.798	1.062	2.971	20	1.2275
.3462	50	.3393	.9407	.3607	2.773	1.063	2.947	10	1.2246
.3491	20° 00′	.3420	.9397	.3640	2.747	1.064	2.924	70° 00′	1.2217
.3520	10	.3448	.9387	.3673	2.723	1.065	2.901	50	1.2188
.3599	20	.3475	.9377	.3706	2.699	1.066	2.878	40	1.2159
.3578	30	.3502	.9367	.3739	2.675	1.068	2.855	30	1.2130
.3607	40	.3529	.9356	.3772	2.651	1.069	2.833	20	1.2101
.3636	50	.3557	.9346	.3805	2.628	1.070	2.812	10	1.2072
.3665	21° 00′	.3584	.9336	.3839	2.605	1.071	2.790	69° 00′	1.2043
.3694	10	.3611	.9325	.3872	2.583	1.072	2.769	50	1.2014
.3723	20	.3638	.9315	.3906	2.560	1.074	2.749	40	1.1985
.3752	30	.3665	.9304	.3939	2.539	1.075	2.729	30	1.1956
.3782	40	.3692	.9293	.3973	2.517	1.076	2.709	20	1.1926
.3811	50	.3719	.9283	.4006	2.496	1.077	2.689	10	1.1897
.3840	22° 00′	.3746	.9272	.4040	2.475	1.079	2.669	68° 00′	1.1868
.3869	10	.3773	.9261	.4074	2.455	1.080	2.650	50	1.1839
.3898	20	.3800	.9250	.4108	2.434	1.081	2.632	40	1.1810
.3927	30	.3827	.9239	.4142	2.414	1.082	2.613	30	1.1781
.3956	40	.3854	.9228	.4176	2.394	1.084	2.595	20	1.1752
.3985	50	.3881	.9216	.4210	2.375	1.085	2.577	10	1.1723
.4014	23° 00′	.3907	.9205	.4245	2.356	1.086	2.559	67° 00′	1.1694
.4043	10	.3934	.9194	.4279	2.337	1.088	2.542	50	1.1665
		cos x	sin x	cot x	tan x	csc x	sec x	x degrees	x radians

x radians	x degrees	sin x	cos x	tan x	cot x	sec x	csc x		
.4072	20	.3961	.9182	.4314	2.318	1.089	2.525	40	1.1636
.4102	30	.3987	.9171	.4348	2.300	1.090	2.508	30	1.1606
4131	40	.4014	.9159	.4383	2.282	1.092	2.491	20	1.1577
.4160	50	4041	.9147	.4417	2.264	1.093	2.475	10	1.1548
.4189	24° 00′	.4067	.9135	.4452	2.246	1.095	2.459	66° 00′	1.1519
.4218	10	.4094	.9124	.4487	2.229	1.096	2.443	50	1.1490
.4247	20	.4120	.9112	.4522	2.211	1.097	2.427	40	1.1461
.4276	30	.4147	.9100	.4557	2.194	1.099	2.411	30	1.1432
.4305	40	.4173	.9088	.4592	2.177	1.100	2.396	20	1.1403
.4334	50	.4200	.9075	.4628	2.161	1.102	2.381	10	1.1374
.4363	25° 00′	.4226	.9063	.4663	2.145	1.103	2.366	65° 00′	1.1345
.4392	10	.4253	.9051	.4699	2.128	1.105	2.352	50	1.1316
.4422	20	.4279	.9038	.4734	2.112	1.106	2.337	40	1.1286
.4451	30	.4305	.9026	.4770	2.097	1.108	2.323	30	1.1257
.4480	40	.4331	.9013	.4806	2.081	1.109	2.309	20	1.1228
.4509	50	.4358	.9001	.4841	2.066	1.111	2.295	10	1.1199
.4538	26° 00′	.4384	.8988	.4877	2.050	1.113	2.281	64° 00′	1.1170
.4567	10	.4410	.8975	.4913	2.035	1.114	2.268	50	1.1141
.4596	20	.4436	.8962	.4950	2.020	1.116	2.254	40	1.1112
.4625	30	.4462	.8949	.4986	2.006	1.117	2.241	30	1.1083
.4654	40	.4488	.8936	.5022	1.991	1.119	2.228	20	1.1054
.4683	50	.4514	.8923	.5059	1.977	1.121	2.215	10	1.1025
.4712	27° 00′	.4540	.8910	.5095	1.963	1.122	2.203	63° 00′	1.0996
.4741	10	.4566	.8897	.5132	1.949	1.124	2.190	50	1.0966
.4771	20	.4592	.8884	.5169	1.935	1.126	2.178	40	1.0937
.4800	30	.4617	.8870	.5206	1.921	1.127	2.166	30	1.0908
.4829	40	.4643	.8857	.5243	1.907	1.129	2.154	20	1.0879
.4858	50	.4669	.8843	.5280	1.894	1.131	2.142	10	1.0850
.4887	28° 00′	.4695	.8829	.5317	1.881	1.133	2.130	62° 00′	1.0821
.4916	10	.4720	.8816	.5354	1.868	1.134	2.118	50	1.0792
.4945	20	.4746	.8802	.5392	1.855	1.136	2.107	40	1.0763
.4974	30	.4772	.8788	.5430	1.842	1.138	2.096	30	1.0734
.5003	40	.4797	.8774	.5467	1.829	1.140	2.085	20	1.0705
.5032	50	.4823	.8760	.5505	1.816	1.142	2.074	10	1.0676
.5061	29° 00′	.4848	.8746	.5543	1.804	1.143	2.063	61° 00′	1.0647
.5091	10	.4874	.8732	.5581	1.792	1.145	2.052	50	1.0617
.5120	20	.4899	.8718	.5619	1.780	1.147	2.041	40	1.0588
.5149	30	.4924	.8704	.5658	1.767	1.149	2.031	30	1.0559
.5178	40	.4950	.8689	.5696	1.756	1.151	2.020	20	1.0530
.5207	50	.4975	.8675	.5735	1.744	1.153	2.010	10	1.0501
.5236	30° 00′	.5000	.8660	.5774	1.732	1.155	2.000	60° 00′	1.0472
.5265	10	.5025	.8646	.5812	1.720	1.157	1.990	50	1.0443
.5294	20	.5050	.8631	.5851	1.709	1.159	1.980	40	1.0414
.5323	30	.5075	.8616	.5890	1.698	1.161	1.970	30	1.0385
.5352	40	.5100	.8601	.5930	1.686	1.163	1.961	20	1.0356
.5381	50	.5125	.8587	.5969	1.675	1.165	1.951	10	1.0327
.5411	31° 00′	.5150	.8572	.6009	1.664	1.167	1.942	59° 00′	1.0297
.5440	10	.5175	.8557	.6048	1.653	1.169	1.932	50	1.0268
.5469	20	.5200	.8542	.6088	1.643	1.171	1.923	40	1.0239
.5498	30	.5225	.8526	.6128	1.632	1.173	1.914	30	1.0210
.5527	40	.5250	.8511	.6168	1.621	1.175	1.905	20	1.0181
.5556	50	.5275	.8496	.6208	1.611	1.177	1.896	10	1.0152
.5585	32° 00′	.5299	8480	.6249	1.600	1.179	1.887	58° 00′	1.0123
.5614	10	.5324	.8465	.6289	1.590	1.181	1.878	50	1.0094
.5643	20	.5348	.8450	.6330	1.580	1.184	1.870	40	1.0065
.5672	30	.5373	.8434	.6371	1.570	1.186	1.861	30	1.0036
.5701	40	.5398	.8418	.6412	1.560	1.188	1.853	20	1.0007
.5730	50	.5422	.8403	.6453	1.550	1.190	1.844	10	.9977
.5760	33° 00′	.5446	.8397	.6494	1.540	1.192	1.836	57° 00′	.9948
.5789	10	.5471	.8371	.6536	1.530	1.195	1.828	50	.9919
.5818	20	.5495	.8355	.6577	1.520	1.197	1.820	40	.9890
.5847	30	.5519	.8339	.6619	1.511	1.199	1.812	30	.9861
.5876	40	.5544	.8323	.6661	1.501	1.202	1.804	20	.9832
.5905	50	.5568	.8307	.6703	1.492	1.204	1.796	10	.9803
.5934	34° 00′	.5592	.8290	.6745	1.483	1.206	1.788	56° 00′	.9774
.5963	10	.5616	.8274	.6787	1.473	1.209	1.781	50	.9745
.5992	20	.5640	.8258	.6830	1.464	1.211	1.773	40	.9716
.6021	30	.5664	.8241	.6873	1.455	1.213	1.766	30	.9687
.6050	40	.5688	.8225	.6916	1.446	1.216	1.758	20	.9657
.6080	50	.5712	.8208	.6959	1.437	1.218	1.751	10	.9628
		cos x	sin x	cot x	tan x	csc x	sec x	x degrees	x radians

x radians	x degrees	sin x	cos x	tan x	cot x	sec x	csc x		
.6109	35° 00′	.5736	.8192	.7002	1.428	1.221	1.743	55° 00′	.9599
.6138	10	.5760	.8175	.7046	1.419	1.223	1.736	50	.9570
.6167	20	.5783	.8158	.7089	1.411	1.226	1.729	40	.9541
.6196	30	.5807	.8141	.7133	1.402	1.228	1.722	30	.9512
.6225	40	.5831	.8124	.7177	1.393	1.231	1.715	20	.9483
.6254	50	.5854	.8107	.7221	1.385	1.233	1.708	10	.9454
.6283	36° 00′	.5878	.8090	.7265	1.376	1.236	1.701	54° 00′	.9425
.6312	10	.5901	.8073	.7310	1.368	1.239	1.695	50	.9396
.6341	20	.5925	.8056	.7355	1.360	1.241	1.688	40	.9367
.6370	30	.5948	.8039	.7400	1.351	1.244	1.681	30	.9338
.6400	40	.5972	.8021	.7445	1.343	1.247	1.675	20	.9308
.6429	50	.5995	.8004	.7490	1.335	1.249	1.668	10	.9279
.6458	37° 00′	.6018	.7986	.7536	1.327	1.252	1.662	53° 00′	.9250
.6487	10	.6041	.7969	.7581	1.319	1.255	1.655	50	.9221
.6516	20	.6065	.7951	.7627	1.311	1.258	1.649	40	.9192
.6545	30	.6088	.7934	.7673	1.303	1.260	1.643	30	.9163
.6574	40	.6111	.7916	.7720	1.295	1.263	1.636	20	.9134
.6603	50	.6134	.7898	.7766	1.288	1.266	1.630	10	.9105
.6632	38° 00′	.6157	.7880	.7813	1.280	1.269	1.624	52° 00′	.9076
.6661	10	.6180	.7862	.7860	1.272	1.272	1.618	50	.9047
.6690	20	.6202	.7844	.7907	1.265	1.275	1.612	40	.9018
.6720	30	.6225	.7826	.7954	1.257	1.278	1.606	30	.8988
.6749	40	.6248	.7808	.8002	1.250	1.281	1.601	20	.8959
.6778	50	.6271	.7790	.8050	1.242	1.284	1.595	10	.8930
.6807	39° 00′	.6293	.7771	.8098	1.235	1.287	1.589	51° 00′	.8901
.6836	10	.6316	.7753	.8146	1.228	1.290	1.583	50	.8872
.6865	20	.6338	.7735	.8195	1.220	1.293	1.578	40	.8843
.6894	30	.6361	.7716	.8243	1.213	1.296	1.572	30	.8814
.6923	40	.6383	.7698	.8292	1.206	1.299	1.567	20	.8785
.6952	50	.6406	.7679	.8342	1.199	1.302	1.561	10	.8756
.6981	40° 00′	.6428	.7660	.8391	1.192	1.305	1.556	50° 00′	.8727
.7010	10	.6450	.7642	.8441	1.185	1.309	1.550	50	.8698
.7039	20	.6472	.7623	.8491	1.178	1.312	1.545	40	.8668
.7069	30	.6494	.7604	.8541	1.171	1.315	1.540	30	.8639
.7098	40	.6517	.7585	.8591	1.164	1.318	1.535	20	.8610
.7127	50	.6539	.7566	.8642	1.157	1.322	1.529	10	.8581
.7156	41° 00′	.6561	.7547	.8693	1.150	1.325	1.524	49° 00′	.8552
.7185	10	.6583	.7528	.8744	1.144	1.328	1.519	50	.8523
.7214	20	.6604	.7509	.8796	1.137	1.332	1.514	40	.8494
.7243	30	.6626	.7490	.8847	1.130	1.335	1.509	30	.8465
.7272	40	.6648	.7470	.8899	1.124	1.339	1.504	20	.8436
.7301	50	.6670	.7451	.8952	1.117	1.342	1.499	10	.8407
.7330	42° 00′	.6691	.7431	.9004	1.111	1.346	1.494	48° 00′	.8378
.7359	10	.6713	.7412	.9057	1.104	1.349	1.490	50	.8348
7389	20	.6734	.7392	.9110	1.098	1.353	1.485	40	.8319
.7418	30	.6756	.7373	.9163	1.091	1.356	1.480	30	.8290
.7447	40	.6777	.7353	.9217	1.085	1.360	1.476	20	.8261
.7476	50	.6799	.7333	.9271	1.079	1.364	1.471	10	.8232
.7505	43° 00′	.6820	.7314	.9325	1.072	1.367	1.466	47° 00′	.8203
.7534	10	.6841	.7294	.9380	1.066	1.371	1.462	50	.8174
.7563	20	.6862	.7274	.9435	1.060	1.375	1.457	40	.8145
.7592	30	.6884	.7254	.9490	1.054	1.379	1.453	30	.8116
.7621	40	.6905	.7234	.9545	1.048	1.382	1.448	20	.8087
.7650	50	.6926	.7214	.9601	1.042	1.386	1.444	10	.8058
.7679	44° 00′	.6947	.7193	.9657	1.036	1.390	1.440	46° 00′	.8029
.7709	10	.6967	.7173	.9713	1.030	1.394	1.435	50	.7999
.7738	20	.6988	.7153	.9770	1.024	1.398	1.431	40	.7970
.7767	30	.7009	.7133	.9827	1.018	1.402	1.427	30	.7941
.7796	40	.7030	.7112	.9884	1.012	1.406	1.423	20	.7912
.7825	50	.7050	.7092	.9942	1.006	1.410	1.418	10	.7883
.7854	45° 00′	.7071	.7071	1.0000	1.0000	1.414	1.414	45° 00′	.7854
		cos x	sin x	cot x	tan x	csc x	sec x	x degrees	x radians

RELATION OF ANGULAR FUNCTIONS IN TERMS OF ONE ANOTHER

TRIGONOMETRIC FUNCTIONS

Function	$\sin\alpha$	$\cos\alpha$	$\tan\alpha$	$\cot\alpha$	$\sec\alpha$	$\csc\alpha$
$\sin\alpha$	$\sin\alpha$	$\pm\sqrt{1-\cos^2\alpha}$	$\dfrac{\tan\alpha}{\pm\sqrt{1+\tan^2\alpha}}$	$\dfrac{1}{\pm\sqrt{1+\cot^2\alpha}}$	$\dfrac{\pm\sqrt{\sec^2\alpha-1}}{\sec\alpha}$	$\dfrac{1}{\csc\alpha}$
$\cos\alpha$	$\pm\sqrt{1-\sin^2\alpha}$	$\cos\alpha$	$\dfrac{1}{\pm\sqrt{1+\tan^2\alpha}}$	$\dfrac{\cot\alpha}{\pm\sqrt{1+\cot^2\alpha}}$	$\dfrac{1}{\sec\alpha}$	$\dfrac{\pm\sqrt{\csc^2\alpha-1}}{\csc\alpha}$
$\tan\alpha$	$\dfrac{\sin\alpha}{\pm\sqrt{1-\sin^2\alpha}}$	$\dfrac{\pm\sqrt{1-\cos^2\alpha}}{\cos\alpha}$	$\tan\alpha$	$\dfrac{1}{\cot\alpha}$	$\pm\sqrt{\sec^2\alpha-1}$	$\dfrac{1}{\pm\sqrt{\csc^2\alpha-1}}$
$\cot\alpha$	$\dfrac{\pm\sqrt{1-\sin^2\alpha}}{\sin\alpha}$	$\dfrac{\cos\alpha}{\pm\sqrt{1-\cos^2\alpha}}$	$\dfrac{1}{\tan\alpha}$	$\cot\alpha$	$\dfrac{1}{\pm\sqrt{\sec^2\alpha-1}}$	$\pm\sqrt{\csc^2\alpha-1}$
$\sec\alpha$	$\dfrac{1}{\pm\sqrt{1-\sin^2\alpha}}$	$\dfrac{1}{\cos\alpha}$	$\pm\sqrt{1+\tan^2\alpha}$	$\dfrac{\pm\sqrt{1+\cot^2\alpha}}{\cot\alpha}$	$\sec\alpha$	$\dfrac{\csc\alpha}{\pm\sqrt{\csc^2\alpha-1}}$
$\csc\alpha$	$\dfrac{1}{\sin\alpha}$	$\dfrac{1}{\pm\sqrt{1-\cos^2\alpha}}$	$\dfrac{\pm\sqrt{1+\tan^2\alpha}}{\tan\alpha}$	$\pm\sqrt{1+\cot^2\alpha}$	$\dfrac{\sec\alpha}{\pm\sqrt{\sec^2\alpha-1}}$	$\csc\alpha$

Note: The choice of sign depends upon the quadrant in which the angle terminates.

HYPERBOLIC FUNCTIONS

Function	$\sinh x$	$\cosh x$	$\tanh x$
$\sinh x =$	$\sinh x$	$+\sqrt{\cosh^2 x-1}$	$\dfrac{\tanh x}{\sqrt{1-\tanh^2 x}}$
$\cosh x =$	$\sqrt{1+\sinh^2 x}$	$\cosh x$	$\dfrac{1}{\sqrt{1-\tanh^2 x}}$
$\tanh x =$	$\dfrac{\sinh x}{\sqrt{1+\sinh^2 x}}$	$\pm\dfrac{\sqrt{\cosh^2 x-1}}{\cosh x}$	$\tanh x$
$\operatorname{cosech} x =$	$\dfrac{1}{\sinh x}$	$\pm\dfrac{1}{\sqrt{\cosh^2 x-1}}$	$\dfrac{\sqrt{1-\tanh^2 x}}{\tanh x}$
$\operatorname{sech} x =$	$\dfrac{1}{\sqrt{1+\sinh^2 x}}$	$\dfrac{1}{\cosh x}$	$\sqrt{1-\tanh^2 x}$
$\coth x =$	$\dfrac{\sqrt{1+\sinh^2 x}}{\sinh x}$	$\pm\dfrac{\cosh x}{\sqrt{\cosh^2 x-1}}$	$\dfrac{1}{\tanh x}$

Function	$\operatorname{cosech} x$	$\operatorname{sech} x$	$\coth x$
$\sinh x =$	$\dfrac{1}{\operatorname{cosech} x}$	$\pm\dfrac{\sqrt{1-\operatorname{sech}^2 x}}{\operatorname{sech} x}$	$\dfrac{\pm 1}{\sqrt{\coth^2 x-1}}$
$\cosh x =$	$\pm\dfrac{\sqrt{\operatorname{cosech}^2 x+1}}{\operatorname{cosech} x}$	$\dfrac{1}{\operatorname{sech} x}$	$\pm\dfrac{\coth x}{\sqrt{\coth^2 x-1}}$
$\tanh x =$	$\dfrac{1}{\sqrt{\operatorname{cosech}^2 x+1}}$	$\pm\sqrt{1-\operatorname{sech}^2 x}$	$\dfrac{1}{\coth x}$
$\operatorname{cosech} x =$	$\operatorname{cosech} x$	$\pm\dfrac{\operatorname{sech} x}{\sqrt{1-\operatorname{sech}^2 zsx}}$	$\pm\dfrac{\sqrt{\coth^2 x-1}}{1}$
$\operatorname{sech} x =$	$\pm\dfrac{\operatorname{cosec} x}{\sqrt{\operatorname{cosech}^2 x+1}}$	$\operatorname{sech} x$	$\pm\dfrac{\sqrt{\coth^2 x-1}}{\coth x}$
$\coth x =$	$\sqrt{\operatorname{cosech}^2 x+1}$	$\pm\dfrac{1}{\sqrt{1-\operatorname{sech}^2 x}}$	$\coth x$

Whenever two signs are shown, choose + sign if x is positive, − sign if x is negative.

Derivatives*

In the following formulas u, v, w represent functions of x, while a, c, n represent fixed real numbers. All arguments in the trigonometric functions are measured in radians, and all inverse trigonometric and hyperbolic functions represent principal values.

1. $\dfrac{d}{dx}(a) = 0$

2. $\dfrac{d}{dx}(x) = 1$

3. $\dfrac{d}{dx}(au) = a\dfrac{du}{dx}$

4. $\dfrac{d}{dx}(u + v - w) = \dfrac{du}{dx} + \dfrac{dv}{dx} - \dfrac{dw}{dx}$

5. $\dfrac{d}{dx}(uv) = u\dfrac{dv}{dx} + v\dfrac{du}{dx}$

6. $\dfrac{d}{dx}(uvw) = uv\dfrac{dw}{dx} + vw\dfrac{du}{dx} + uw\dfrac{dv}{dx}$

7. $\dfrac{d}{dx}\left(\dfrac{u}{v}\right) = \dfrac{v\dfrac{du}{dx} - u\dfrac{dv}{dx}}{v^2} = \dfrac{1}{v}\dfrac{du}{dx} - \dfrac{u}{v^2}\dfrac{dv}{dx}$

8. $\dfrac{d}{dx}(u^n) = nu^{n-1}\dfrac{du}{dx}$

9. $\dfrac{d}{dx}(\sqrt{u}) = \dfrac{1}{2\sqrt{u}}\dfrac{du}{dx}$

10. $\dfrac{d}{dx}\left(\dfrac{1}{u}\right) = -\dfrac{1}{u^2}\dfrac{du}{dx}$

11. $\dfrac{d}{dx}\left(\dfrac{1}{u^n}\right) = -\dfrac{n}{u^{n+1}}\dfrac{du}{dx}$

12. $\dfrac{d}{dx}\left(\dfrac{u^n}{v^m}\right) = \dfrac{u^{n-1}}{v^{m+1}}\left(nv\dfrac{du}{dx} - mu\dfrac{dv}{dx}\right)$

13. $\dfrac{d}{dx}(u^n v^m) = u^{n-1}v^{m-1}\left(nv\dfrac{du}{dx} + mu\dfrac{dv}{dx}\right)$

14. $\dfrac{d}{dx}[f(u)] = \dfrac{d}{du}[f(u)] \cdot \dfrac{du}{dx}$

* Let $y = f(x)$ and $\dfrac{dy}{dx} = \dfrac{d[f(x)]}{dx} = f'(x)$ define respectively a function and its derivative for any value x in their common domain. The differential for the function at such a value x is accordingly defined as

$$dy = d[f(x)] = \dfrac{dy}{dx}dx = \dfrac{d[f(x)]}{dx}dx = f'(x)\,dx$$

Each derivative formula has an associated differential formula. For example, formula 6 above has the differential formula

$$d(uvw) = uv\,dw + vw\,du + uw\,dv$$

15. $\dfrac{d^2}{dx^2}[f(u)] = \dfrac{df(u)}{du} \cdot \dfrac{d^2u}{dx^2} + \dfrac{d^2f(u)}{du^2} \cdot \left(\dfrac{du}{dx}\right)^2$

16. $\dfrac{d^n}{dx^n}[uv] = \dbinom{n}{0} v \dfrac{d^nu}{dx^n} + \dbinom{n}{1}\dfrac{dv}{dx}\dfrac{d^{n-1}u}{dx^{n-1}} + \dbinom{n}{2}\dfrac{d^2v}{dx^2}\dfrac{d^{n-2}u}{dx^{n-2}}$

$$+ \cdots + \dbinom{n}{k}\dfrac{d^kv}{dx^k}\dfrac{d^{n-k}u}{dx^{n-k}} + \cdots + \dbinom{n}{n}u\dfrac{d^nv}{dx^n}$$

where $\dbinom{n}{r} = \dfrac{n!}{r'(n-r)'}$ the binomial coefficient, n non-negative integer and $\dbinom{n}{0} = 1$.

17. $\dfrac{du}{dx} = \dfrac{1}{\dfrac{dx}{du}}$ if $\dfrac{dx}{du} \neq 0$

18. $\dfrac{d}{dx}(\log_a u) = (\log_a e)\dfrac{1}{u}\dfrac{du}{dx}$

19. $\dfrac{d}{dx}(\log_e u) = \dfrac{1}{u}\dfrac{du}{dx}$

20. $\dfrac{d}{dx}(a^u) = a^u(\log_e a)\dfrac{du}{dx}$

21. $\dfrac{d}{dx}(e^u) = e^u\dfrac{du}{dx}$

22. $\dfrac{d}{dx}(u^v) = vu^{v-1}\dfrac{du}{dx} + (\log_e u)u^v\dfrac{dv}{dx}$

23. $\dfrac{d}{dx}(\sin u) = \dfrac{du}{dx}(\cos u)$

24. $\dfrac{d}{dx}(\cos u) = -\dfrac{du}{dx}(\sin u)$

25. $\dfrac{d}{dx}(\tan u) = \dfrac{du}{dx}(\sec^2 u)$

26. $\dfrac{d}{dx}(\cot u) = -\dfrac{du}{dx}(\csc^2 u)$

27. $\dfrac{d}{dx}(\sec u) = \dfrac{du}{dx}\sec u \cdot \tan u$

28. $\dfrac{d}{dx}(\csc u) = -\dfrac{du}{dx}\csc u \cdot \cot u$

29. $\dfrac{d}{dx}(\text{vers } u) = \dfrac{du}{dx}\sin u$

30. $\dfrac{d}{dx}(\text{arc sin } u) = \dfrac{1}{\sqrt{1-u^2}}\dfrac{du}{dx}$, $\quad \left(-\dfrac{\pi}{2} \leq \text{arc sin } u \leq \dfrac{\pi}{2}\right)$

31. $\dfrac{d}{dx}(\text{arc cos } u) = -\dfrac{1}{\sqrt{1-u^2}}\dfrac{du}{dx},$ $(0 \le \text{arc cos } u \le \pi)$

32. $\dfrac{d}{dx}(\text{arc tan } u) = \dfrac{1}{1+u^2}\dfrac{du}{dx},$ $\left(-\dfrac{\pi}{2} < \text{arc tan } u < \dfrac{\pi}{2}\right)$

33. $\dfrac{d}{dx}(\text{arc cot } u) = -\dfrac{1}{1+u^2}\dfrac{du}{dx},$ $(0 \le \text{arc cot } u \le \pi)$

34. $\dfrac{d}{dx}(\text{arc sec } u) = \dfrac{1}{u\sqrt{u^2-1}}\dfrac{du}{dx},$ $\left(0 \le \text{arc sec } u < \dfrac{\pi}{2}, -\pi \le \text{arc sec } u < -\dfrac{\pi}{2}\right)$

35. $\dfrac{d}{dx}(\text{arc csc } u) = -\dfrac{1}{u\sqrt{u^2-1}}\dfrac{du}{dx},$ $\left(0 < \text{arc csc } u \le \dfrac{\pi}{2}, -\pi < \text{arc csc } u \le -\dfrac{\pi}{2}\right)$

36. $\dfrac{d}{dx}(\text{arc vers } u) = \dfrac{1}{\sqrt{2u-u^2}}\dfrac{du}{dx},$ $(0 \le \text{arc vers } u \le \pi)$

37. $\dfrac{d}{dx}(\sinh u) = \dfrac{du}{dx}(\cosh u)$

38. $\dfrac{d}{dx}(\cosh u) = \dfrac{du}{dx}(\sinh u)$

39. $\dfrac{d}{dx}(\tanh u) = \dfrac{du}{dx}(\text{sech}^2 u)$

40. $\dfrac{d}{dx}(\coth u) = -\dfrac{du}{dx}(\text{csch}^2 u)$

41. $\dfrac{d}{dx}(\text{sech } u) = -\dfrac{du}{dx}(\text{sech } u \cdot \tanh u)$

42. $\dfrac{d}{dx}(\text{csch } u) = -\dfrac{du}{dx}(\text{csch } u \cdot \coth u)$

43. $\dfrac{d}{dx}(\sinh^{-1} u) = \dfrac{d}{dx}[\log(u + \sqrt{u^2+1})] = \dfrac{1}{\sqrt{u^2+1}}\dfrac{du}{dx}$

44. $\dfrac{d}{dx}(\cosh^{-1} u) = \dfrac{d}{dx}[\log(u + \sqrt{u^2-1})] = \dfrac{1}{\sqrt{u^2-1}}\dfrac{du}{dx},$ $(u > 1, \cosh^{-1} u > 0)$

45. $\dfrac{d}{dx}(\tanh^{-1} u) = \dfrac{d}{dx}\left[\dfrac{1}{2}\log\dfrac{1+u}{1-u}\right] = \dfrac{1}{1-u^2}\dfrac{du}{dx},$ $(u^2 < 1)$

46. $\dfrac{d}{dx}(\coth^{-1} u) = \dfrac{d}{dx}\left[\dfrac{1}{2}\log\dfrac{u+1}{u-1}\right] = \dfrac{1}{1-u^2}\dfrac{du}{dx},$ $(u^2 > 1)$

47. $\dfrac{d}{dx}(\text{sech}^{-1} u) = \dfrac{d}{dx}\left[\log\dfrac{1+\sqrt{1-u^2}}{u}\right] = -\dfrac{1}{u\sqrt{1-u^2}}\dfrac{du}{dx},$ $(0 < u < 1, \text{sech}^{-1} u > 0)$

48. $\dfrac{d}{dx}(\text{csch}^{-1} u) = \dfrac{d}{dx}\left[\log\dfrac{1+\sqrt{1+u^2}}{u}\right] = -\dfrac{1}{|u|\sqrt{1+u^2}}\dfrac{du}{dx}$

49. $\dfrac{d}{dq} \displaystyle\int_p^q f(x)\, dx = f(q),$ [p constant]

50. $\dfrac{d}{dp} \displaystyle\int_p^q f(x)\, dx = -f(p),$ [q constant]

51. $\dfrac{d}{da} \displaystyle\int_p^q f(x, a)\, dx = \int_p^q \dfrac{\partial}{\partial a}[f(x, a)]\, dx + f(q, a)\dfrac{dq}{da} - f(p, a)\dfrac{dp}{da}$

INTEGRATION

The following is a brief discussion of some integration techniques. A more complete discussion can be found in a number of good text books. However, the purpose of this introduction is simply to discuss a few of the important techniques which may be used, in conjunction with the integral table which follows, to integrate particular functions.

No matter how extensive the integral table, it is a fairly uncommon occurrence to find in the table the exact integral desired. Usually some form of transformation will have to be made. The simplest type of transformation, and yet the most general, is substitution. Simple forms of substitution, such as $y = ax$, are employed almost unconsciously by experienced users of integral tables. Other substitutions may require more thought. In some sections of the tables, appropriate substitutions are suggested for integrals which are similar to, but not exactly like, integrals in the table. Finding the right substitution is largely a matter of intuition and experience.

Several precautions must be observed when using substitutions:

1. Be sure to make the substitution in the dx term, as well as everywhere else in the integral.
2. Be sure that the function substituted is one-to-one and continuous. If this is not the case, the integral must be restricted in such a way as to make it true. See the example following.
3. With definite integrals, the limits should also be expressed in terms of the new dependent variable. With indefinite integrals, it is necessary to perform the reverse substitution to obtain the answer in terms of the original independent variable. This may also be done for definite integrals, but it is usually easier to change the limits.

Example:

$$\int \frac{x^4}{\sqrt{a^2 - x^2}}\, dx$$

Here we make the substitution $x = |a| \sin \theta$. Then $dx = |a| \cos \theta\, d\theta$, and

$$\sqrt{a^2 - x^2} = \sqrt{a^2 - a^2 \sin^2 \theta} = |a|\sqrt{1 - \sin^2 \theta} = |a \cos \theta|$$

Notice the absolute value signs. It is very important to keep in mind that a square root radical always denotes the positive square root, and to assure the sign is always kept positive. Thus $\sqrt{x^2} = |x|$. Failure to observe this is a common cause of errors in integration.

Notice also that the indicated substitution is not a one-to-one function, that is, it does not have a unique inverse. Thus we must restrict the range of θ in such a way as to make the function one-to-one. Fortunately, this is easily done by solving for θ

$$\theta = \sin^{-1} \frac{x}{|a|}$$

and restricting the inverse sine to the principal values, $-\dfrac{\pi}{2} \le \theta \le \dfrac{\pi}{2}$.

Thus the integral becomes

$$\int \frac{a^4 \sin^4 \theta |a| \cos \theta \, d\theta}{|a| |\cos \theta|}$$

Now, however, in the range of values chosen for θ, $\cos \theta$ is always positive. Thus we may remove the absolute value signs from $\cos \theta$ in the denominator. (This is one of the reasons that the principal values of the inverse trigonometric functions are defined as they are.)

Then the $\cos \theta$ terms cancel, and the integral becomes

$$a^4 \int \sin^4 \theta \, d\theta$$

By application of integral formulas 299 and 296, we integrate this to

$$-a^4 \frac{\sin^3 \theta \cos \theta}{4} - \frac{3a^4}{8} \cos \theta \sin \theta + \frac{3a^4}{8} \theta + C$$

We now must perform the inverse substitution to get the result in terms of x. We have

$$\theta = \sin^{-1} \frac{x}{|a|}$$

$$\sin \theta = \frac{x}{|a|}$$

Then

$$\cos \theta = \pm \sqrt{1 - \sin^2 \theta} = \pm \sqrt{1 - \frac{x^2}{a^2}} = \pm \frac{\sqrt{a^2 - x^2}}{|a|}.$$

Because of the previously mentioned fact that $\cos \theta$ is positive, we may omit the $\pm$ sign. The reverse substitution then produces the final answer

$$\int \frac{x^4}{\sqrt{a^2 - x^2}} \, dx = -\tfrac{1}{4} x^3 \sqrt{a^2 - x^2} - \tfrac{3}{8} a^2 x \sqrt{a^2 - x^2} + \frac{3a^4}{8} \sin^{-1} \frac{x}{|a|} + C.$$

Any rational function of x may be integrated, if the denominator is factored into linear and irreducible quadratic factors. The function may then be broken into partial fractions, and the individual partial fractions integrated by use of the appropriate formula from the integral table. See the section on partial fractions for further information.

Many integrals may be reduced to rational functions by proper substitutions. For example,

$$z = \tan \frac{x}{2}$$

will reduce any rational function of the six trigonometric functions of x to a rational function of z. (Frequently there are other substitutions which are simpler to use, but this one will always work. See integral formula number 484.)

Any rational function of x and $\sqrt{ax + b}$ may be reduced to a rational function of z by making the substitution

$$z = \sqrt{ax + b}.$$

Other likely substitutions will be suggested by looking at the form of the integrand.

The other main method of transforming integrals is integration by parts. This involves applying formula number 5 or 6 in the accompanying integral table. The critical factor in this method is the choice of the functions u and v. In order for the method to be successful, $v = \int dv$ and $\int v \, du$ must be easier to integrate than the original integral. Again, this choice is largely a matter of intuition and experience.

Example:

$$\int x \sin x \, dx$$

Two obvious choices are $u = x, dv = \sin x \, dx$, or $u = \sin x, dv = x \, dx$. Since a preliminary mental calculation indicates that $\int v \, du$ in the second choice would be more, rather than less, complicated than the original integral (it would contain x^2), we use the first choice.

$$u = x \qquad\qquad du = dx$$

$$dv = \sin x \, dx \qquad\qquad v = -\cos x$$

$$\int x \sin x \, dx = \int u \, dv = uv - \int v \, du = -x \cos x + \int \cos x \, dx$$

$$= \sin x - x \cos x$$

Of course, this result could have been obtained directly from the integral table, but it provides a simple example of the method. In more complicated examples the choice of u and v may not be so obvious, and several different choices may have to be tried. Of course, there is no guarantee that any of them will work.

Integration by parts may be applied more than once, or combined with substitution. A fairly common case is illustrated by the following example.

Example:

$$\int e^x \sin x \, dx$$

Let

$$u = e^x \qquad \text{Then} \quad du = e^x \, dx$$

$$dv = \sin x \, dx \qquad\qquad v = -\cos x$$

$$\int e^x \sin x \, dx = \int u \, dv = uv - \int v \, du = -e^x \cos x + \int e^x \cos x \, dx$$

In this latter integral,

$$\text{let} \quad u = e^x \qquad \text{Then} \quad du = e^x \, dx$$

$$dv = \cos x \, dx \qquad\qquad v = \sin x$$

$$\int e^x \sin x \, dx = -e^x \cos x + \int e^x \cos x \, dx = -e^x \cos x + \int u \, dv$$

$$= -e^x \cos x + uv - \int v \, du$$

$$= -e^x \cos x + e^x \sin x - \int e^x \sin x \, dx$$

This looks as if a circular transformation has taken place, since we are back at the same integral we started from. However, the above equation can be solved algebraically for the required integral:

$$\int e^x \sin x \, dx = \tfrac{1}{2}(e^x \sin x - e^x \cos x)$$

In the second integration by parts, if the parts had been chosen as $u = \cos x, dv = e^x \, dx$, we would indeed have made a circular transformation, and returned to the starting place.

In general, when doing repeated integration by parts, one should never choose the function u at any stage to be the same as the function v at the previous stage, or a constant times the previous v.

The following rule is called the extended rule for integration by parts. It is the result of $n + 1$ successive applications of integration by parts.

If

$$g_1(x) = \int g(x)\,dx, \qquad g_2(x) = \int g_1(x)\,dx,$$

$$g_3(x) = \int g_2(x)\,dx, \ldots, g_m(x) = \int g_{m-1}(x)\,dx, \ldots,$$

then

$$\int f(x) \cdot g(x)\,dx = f(x) \cdot g_1(x) - f'(x) \cdot g_2(x) + f''(x) \cdot g_3(x) - + \cdots$$

$$+ (-1)^n f^{(n)}(x) g_{n+1}(x) + (-1)^{n+1} \int f^{(n+1)}(x) g_{n+1}(x)\,dx.$$

A useful special case of the above rule is when $f(x)$ is a polynomial of degree n. Then $f^{(n+1)}(x) = 0$, and

$$\int f(x) \cdot g(x)\,dx = f(x) \cdot g_1(x) - f'(x) \cdot g_2(x) + f''(x) \cdot g_3(x) - + \cdots + (-1)^n f^{(n)}(x) g_{n+1}(x) + C$$

Example:

If $f(x) = x^2, g(x) = \sin x$

$$\int x^2 \sin x\,dx = -x^2 \cos x + 2x \sin x + 2 \cos x + C$$

Another application of this formula occurs if

$$f''(x) = af(x) \quad \text{and} \quad g''(x) = bg(x),$$

where a and b are unequal constants. In this case, by a process similar to that used in the above example for $\int e^x \sin x\,dx$, we get the formula

$$\int f(x)g(x)\,dx = \frac{f(x) \cdot g'(x) - f'(x) \cdot g(x)}{b - a} + C$$

This formula could have been used in the example mentioned. Here is another example.

Example:

If $f(x) = e^{2x}, g(x) = \sin 3x$, then $a = 4, b = -9$, and

$$\int e^{2x} \sin 3x\,dx = \frac{3 e^{2x} \cos 3x - 2 e^{2x} \sin 3x}{-9 - 4} + C = \frac{e^{2x}}{13}(2 \sin 3x - 3 \cos 3x) + C$$

The following additional points should be observed when using this table.

1. A constant of integration is to be supplied with the answers for indefinite integrals.
2. Logarithmic expressions are to base $e = 2.71828\ldots$, unless otherwise specified, and are to be evaluated for the absolute value of the arguments involved therein.
3. All angles are measured in radians, and inverse trigonometric and hyperbolic functions represent principal values, unless otherwise indicated.
4. If the application of a formula produces either a zero denominator or the square root of a negative number in the result, there is usually available another form of the answer which

avoids this difficulty. In many of the results, the excluded values are specified, but when such are omitted it is presumed that one can tell what these should be, especially when difficulties of the type herein mentioned are obtained.

5. When inverse trigonometric functions occur in the integrals, be sure that any replacements made for them are strictly in accordance with the rules for such functions. This causes little difficulty when the argument of the inverse trigonometric function is positive, since then all angles involved are in the first quadrant. However, if the argument is negative, special care must be used. Thus if $u > 0$,

$$\sin^{-1} u = \cos^{-1}\sqrt{1 - u^2} = \csc^{-1}\frac{1}{u}, \text{ etc.}$$

However, if $u < 0$,

$$\sin^{-1} u = -\cos^{-1}\sqrt{1 - u^2} = -\pi - \csc^{-1}\frac{1}{u}, \text{ etc.}$$

See the section on inverse trigonometric functions for a full treatment of the allowable substitutions.

6. In integrals 340–345 and some others, the right side includes expressions of the form

$$A \tan^{-1} [B + C \tan f(x)].$$

In these formulas, the $\tan^{-1}$ does not necessarily represent the principal value. Instead of always employing the principal branch of the inverse tangent function, one must instead use that branch of the inverse tangent function upon which $f(x)$ lies for any particular choice of x.

Example:

$$\int_0^{4\pi} \frac{dx}{2 + \sin x} = \frac{2}{\sqrt{3}}\tan^{-1}\frac{2\tan\frac{x}{2} + 1}{\sqrt{3}}\Bigg]_0^{4\pi}$$

$$= \frac{2}{\sqrt{3}}\left[\tan^{-1}\frac{2\tan 2\pi + 1}{\sqrt{3}} - \tan^{-1}\frac{2\tan 0 + 1}{\sqrt{3}}\right]$$

$$= \frac{2}{\sqrt{3}}\left[\frac{13\pi}{6} - \frac{\pi}{6}\right] = \frac{4\pi}{\sqrt{3}} = \frac{4\sqrt{3}\pi}{3}$$

Here

$$\tan^{-1}\frac{2\tan 2\pi + 1}{\sqrt{3}} = \tan^{-1}\frac{1}{\sqrt{3}} = \frac{13\pi}{6},$$

since $f(x) = 2\pi$; and

$$\tan^{-1}\frac{2\tan 0 + 1}{\sqrt{3}} = \tan^{-1}\frac{1}{\sqrt{3}} = \frac{\pi}{6},$$

since $f(x) = 0$.

7. B_n and E_n where used in Integrals represents the Bernoulli and Euler numbers as defined in tables of Bernoulli and Euler polynomials contained in certain mathematics reference and handbooks, as for example, Beyer, W. H., *Handbook of Mathematical Sciences*, 5th ed., CRC Press, Inc., West Palm Beach 1978, 577–583.

ELEMENTARY FORMS

1. $\int a \, dx = ax$

2. $\int a \cdot f(x) \, dx = a \int f(x) \, dx$

3. $\int \phi(y) \, dx = \int \frac{\phi(y)}{y'} \, dy,$ where $y' = \frac{dy}{dx}$

4. $\int (u + v) \, dx = \int u \, dx + \int v \, dx,$ where u and v are any functions of x

5. $\int u \, dv = u \int dv - \int v \, du = uv - \int v \, du$

6. $\int u \frac{dv}{dx} dx = uv - \int v \frac{du}{dx} dx$

7. $\int x^n \, dx = \frac{x^{n+1}}{n+1},$ except $n = -1$

8. $\int \frac{f'(x) \, dx}{f(x)} = \log f(x),$ $(df(x) = f'(x) \, dx)$

9. $\int \frac{dx}{x} = \log x$

10. $\int \frac{f'(x) \, dx}{2\sqrt{f(x)}} = \sqrt{f(x)},$ $(df(x) = f'(x) \, dx)$

11. $\int e^x \, dx = e^x$

12. $\int e^{ax} \, dx = e^{ax}/a$

13. $\int b^{ax} \, dx = \frac{b^{ax}}{a \log b},$ $(b > 0)$

14. $\int \log x \, dx = x \log x - x$

15. $\int a^x \log a \, dx = a^x,$ $(a > 0)$

16. $\int \frac{dx}{a^2 + x^2} = \frac{1}{a} \tan^{-1} \frac{x}{a}$

17. $\int \frac{dx}{a^2 - x^2} = \begin{cases} \dfrac{1}{a} \tanh^{-1} \dfrac{x}{a} \\ \text{or} \\ \dfrac{1}{2a} \log \dfrac{a + x}{a - x}, \quad (a^2 > x^2) \end{cases}$

18. $\int \frac{dx}{x^2 - a^2} = \begin{cases} -\dfrac{1}{a} \coth^{-1} \dfrac{x}{a} \\ \text{or} \\ \dfrac{1}{2a} \log \dfrac{x - a}{x + a}, \quad (x^2 > a^2) \end{cases}$

19. $\int \dfrac{dx}{\sqrt{a^2 - x^2}} = \begin{cases} \sin^{-1} \dfrac{x}{|a|} \\ \quad\text{or} \\ -\cos^{-1} \dfrac{x}{|a|}, \quad (a^2 > x^2) \end{cases}$

20. $\int \dfrac{dx}{\sqrt{x^2 \pm a^2}} = \log\left(x + \sqrt{x^2 \pm a^2}\right)$

21. $\int \dfrac{dx}{x\sqrt{x^2 - a^2}} = \dfrac{1}{|a|} \sec^{-1} \dfrac{x}{a}$

22. $\int \dfrac{dx}{x\sqrt{a^2 \pm x^2}} = -\dfrac{1}{a} \log\left(\dfrac{a + \sqrt{a^2 \pm x^2}}{x}\right)$

FORMS CONTAINING $(a + bx)$

For forms containing $a + bx$, but not listed in the table, the substitution $u = \dfrac{a + bx}{x}$ may prove helpful.

23. $\int (a + bx)^n \, dx = \dfrac{(a + bx)^{n+1}}{(n + 1)b}, \quad (n \neq -1)$

24. $\int x(a + bx)^n \, dx$

$$= \dfrac{1}{b^2(n + 2)}(a + bx)^{n+2} - \dfrac{a}{b^2(n + 1)}(a + bx)^{n+1}, \quad (n \neq -1, -2)$$

25. $\int x^2(a + bx)^n \, dx = \dfrac{1}{b^3}\left[\dfrac{(a + bx)^{n+3}}{n + 3} - 2a\dfrac{(a + bx)^{n+2}}{n + 2} + a^2\dfrac{(a + bx)^{n+1}}{n + 1}\right]$

26. $\int x^m(a + bx)^n \, dx = \begin{cases} \dfrac{x^{m+1}(a + bx)^n}{m + n + 1} + \dfrac{an}{m + n + 1}\displaystyle\int x^m(a + bx)^{n-1}\, dx \\ \qquad\qquad\text{or} \\ \dfrac{1}{a(n + 1)}\left[-x^{m+1}(a + bx)^{n+1}\right. \\ \qquad\qquad\qquad\left. + (m + n + 2)\displaystyle\int x^m(a + bx)^{n+1}\, dx\right] \\ \qquad\qquad\text{or} \\ \dfrac{1}{b(m + n + 1)}\left[x^m(a + bx)^{n+1} - ma\displaystyle\int x^{m-1}(a + bx)^n\, dx\right] \end{cases}$

27. $\int \dfrac{dx}{a + bx} = \dfrac{1}{b} \log(a + bx)$

28. $\int \dfrac{dx}{(a + bx)^2} = -\dfrac{1}{b(a + bx)}$

29. $\int \dfrac{dx}{(a + bx)^3} = -\dfrac{1}{2b(a + bx)^2}$

30. $\int \dfrac{x\, dx}{a + bx} = \begin{cases} \dfrac{1}{b^2}[a + bx - a\log(a + bx)] \\ \qquad\text{or} \\ \dfrac{x}{b} - \dfrac{a}{b^2}\log(a + bx) \end{cases}$

A-21

31. $\int \dfrac{x\,dx}{(a + bx)^2} = \dfrac{1}{b^2}\left[\log(a + bx) + \dfrac{a}{a + bx}\right]$

32. $\int \dfrac{x\,dx}{(a + bx)^n} = \dfrac{1}{b^2}\left[\dfrac{-1}{(n - 2)(a + bx)^{n-2}} + \dfrac{a}{(n - 1)(a + bx)^{n-1}}\right], \qquad n \neq 1, 2$

33. $\int \dfrac{x^2\,dx}{a + bx} = \dfrac{1}{b^3}\left[\dfrac{1}{2}(a + bx)^2 - 2a(a + bx) + a^2\log(a + bx)\right]$

34. $\int \dfrac{x^2\,dx}{(a + bx)^2} = \dfrac{1}{b^3}\left[a + bx - 2a\log(a + bx) - \dfrac{a^2}{a + bx}\right]$

35. $\int \dfrac{x^2\,dx}{(a + bx)^3} = \dfrac{1}{b^3}\left[\log(a + bx) + \dfrac{2a}{a + bx} - \dfrac{a^2}{2(a + bx)^2}\right]$

36. $\int \dfrac{x^2\,dx}{(a + bx)^n} = \dfrac{1}{b^3}\left[\dfrac{-1}{(n - 3)(a + bx)^{n-3}}\right.$

$\left. \qquad\qquad + \dfrac{2a}{(n - 2)(a + bx)^{n-2}} - \dfrac{a^2}{(n - 1)(a + bx)^{n-1}}\right], \qquad n \neq 1, 2, 3$

37. $\int \dfrac{dx}{x(a + bx)} = -\dfrac{1}{a}\log\dfrac{a + bx}{x}$

38. $\int \dfrac{dx}{x(a + bx)^2} = \dfrac{1}{a(a + bx)} - \dfrac{1}{a^2}\log\dfrac{a + bx}{x}$

39. $\int \dfrac{dx}{x(a + bx)^3} = \dfrac{1}{a^3}\left[\dfrac{1}{2}\left(\dfrac{2a + bx}{a + bx}\right)^2 + \log\dfrac{x}{a + bx}\right]$

40. $\int \dfrac{dx}{x^2(a + bx)} = -\dfrac{1}{ax} + \dfrac{b}{a^2}\log\dfrac{a + bx}{x}$

41. $\int \dfrac{dx}{x^3(a + bx)} = \dfrac{2bx - a}{2a^2x^2} + \dfrac{b^2}{a^3}\log\dfrac{x}{a + bx}$

42. $\int \dfrac{dx}{x^2(a + bx)^2} = -\dfrac{a + 2bx}{a^2x(a + bx)} + \dfrac{2b}{a^3}\log\dfrac{a + bx}{x}$

FORMS CONTAINING $c^2 \pm x^2$, $x^2 - c^2$

43. $\int \dfrac{dx}{c^2 + x^2} = \dfrac{1}{c}\tan^{-1}\dfrac{x}{c}$

44. $\int \dfrac{dx}{c^2 - x^2} = \dfrac{1}{2c}\log\dfrac{c + x}{c - x}, \qquad (c^2 > x^2)$

45. $\int \dfrac{dx}{x^2 - c^2} = \dfrac{1}{2c}\log\dfrac{x - c}{x + c}, \qquad (x^2 > c^2)$

46. $\int \dfrac{x\,dx}{c^2 \pm x^2} = \pm\dfrac{1}{2}\log(c^2 \pm x^2)$

47. $\int \dfrac{x\,dx}{(c^2 \pm x^2)^{n+1}} = \mp\dfrac{1}{2n(c^2 \pm x^2)^n}$

48. $\int \dfrac{dx}{(c^2 \pm x^2)^n} = \dfrac{1}{2c^2(n - 1)}\left[\dfrac{x}{(c^2 \pm x^2)^{n-1}} + (2n - 3)\int\dfrac{dx}{(c^2 \pm x^2)^{n-1}}\right]$

49. $\int \dfrac{dx}{(x^2 - c^2)^n} = \dfrac{1}{2c^2(n - 1)}\left[-\dfrac{x}{(x^2 - c^2)^{n-1}} - (2n - 3)\int\dfrac{dx}{(x^2 - c^2)^{n-1}}\right]$

50. $\int \dfrac{x\,dx}{x^2 - c^2} = \dfrac{1}{2}\log(x^2 - c^2)$

51. $\displaystyle\int \frac{x\,dx}{(x^2 - c^2)^{n+1}} = -\frac{1}{2n(x^2 - c^2)^n}$

FORMS CONTAINING $a + bx$ and $c + dx$

$$u = a + bx, \qquad v = c + dx, \qquad k = ad - bc$$

If $k = 0$, then $v = \dfrac{c}{a}u$

52. $\displaystyle\int \frac{dx}{u \cdot v} = \frac{1}{k} \cdot \log\left(\frac{v}{u}\right)$

53. $\displaystyle\int \frac{x\,dx}{u \cdot v} = \frac{1}{k}\left[\frac{a}{b}\log(u) - \frac{c}{d}\log(v)\right]$

54. $\displaystyle\int \frac{dx}{u^2 \cdot v} = \frac{1}{k}\left(\frac{1}{u} + \frac{d}{k}\log\frac{v}{u}\right)$

55. $\displaystyle\int \frac{x\,dx}{u^2 \cdot v} = \frac{-a}{bku} - \frac{c}{k^2}\log\frac{v}{u}$

56. $\displaystyle\int \frac{x^2\,dx}{u^2 \cdot v} = \frac{a^2}{b^2 ku} + \frac{1}{k^2}\left[\frac{c^2}{d}\log(v) + \frac{a(k - bc)}{b^2}\log(u)\right]$

57. $\displaystyle\int \frac{dx}{u^n \cdot v^m} = \frac{1}{k(m-1)}\left[\frac{-1}{u^{n-1} \cdot v^{m-1}} - (m + n - 2)b\int \frac{dx}{u^n \cdot v^{m-1}}\right]$

58. $\displaystyle\int \frac{u}{v}\,dx = \frac{bx}{d} + \frac{k}{d^2}\log(v)$

59. $\displaystyle\int \frac{u^m\,dx}{v^n} = \begin{cases} \dfrac{-1}{k(n-1)}\left[\dfrac{u^{m+1}}{v^{n-1}} + b(n - m - 2)\displaystyle\int \dfrac{u^m}{v^{n-1}}\,dx\right] \\[6pt] \qquad\qquad\text{or} \\[4pt] \dfrac{-1}{d(n - m - 1)}\left[\dfrac{u^m}{v^{n-1}} + mk\displaystyle\int \dfrac{u^{m-1}}{v^n}\,dx\right] \\[6pt] \qquad\qquad\text{or} \\[4pt] \dfrac{-1}{d(n-1)}\left[\dfrac{u^m}{v^{n-1}} - mb\displaystyle\int \dfrac{u^{m-1}}{v^{n-1}}\,dx\right] \end{cases}$

FORMS CONTAINING $(a + bx^n)$

60. $\displaystyle\int \frac{dx}{a + bx^2} = \frac{1}{\sqrt{ab}}\tan^{-1}\frac{x\sqrt{ab}}{a}, \qquad (ab > 0)$

61. $\displaystyle\int \frac{dx}{a + bx^2} = \begin{cases} \dfrac{1}{2\sqrt{-ab}}\log\dfrac{a + x\sqrt{-ab}}{a - x\sqrt{-ab}}, \qquad (ab < 0) \\[6pt] \qquad\qquad\text{or} \\[4pt] \dfrac{1}{\sqrt{-ab}}\tanh^{-1}\dfrac{x\sqrt{-ab}}{a}, \qquad (ab < 0) \end{cases}$

62. $\displaystyle\int \frac{dx}{a^2 + b^2 x^2} = \frac{1}{ab}\tan^{-1}\frac{bx}{a}$

63. $\displaystyle\int \frac{x\,dx}{a + bx^2} = \frac{1}{2b}\log(a + bx^2)$

64. $\displaystyle\int \frac{x^2\,dx}{a + bx^2} = \frac{x}{b} - \frac{a}{b}\int \frac{dx}{a + bx^2}$

65. $\displaystyle\int \frac{dx}{(a + bx^2)^2} = \frac{x}{2a(a + bx^2)} + \frac{1}{2a}\int \frac{dx}{a + bx^2}$

66. $\int \dfrac{dx}{a^2 - b^2 x^2} = \dfrac{1}{2ab} \log \dfrac{a + bx}{a - bx}$

67. $\int \dfrac{dx}{(a + bx^2)^{m+1}} = \begin{cases} \dfrac{1}{2ma} \dfrac{x}{(a + bx^2)^m} + \dfrac{2m-1}{2ma} \int \dfrac{dx}{(a + bx^2)^m} \\[2mm] \text{or} \\[2mm] \dfrac{(2m)!}{(m!)^2} \left[\dfrac{x}{2a} \displaystyle\sum_{r=1}^{m} \dfrac{r!(r-1)!}{(4a)^{m-r}(2r)!(a + bx^2)^r} + \dfrac{1}{(4a)^m} \int \dfrac{dx}{a + bx^2} \right] \end{cases}$

68. $\int \dfrac{x\,dx}{(a + bx^2)^{m+1}} = -\dfrac{1}{2bm(a + bx^2)^m}$

69. $\int \dfrac{x^2\,dx}{(a + bx^2)^{m+1}} = \dfrac{-x}{2mb(a + bx^2)^m} + \dfrac{1}{2mb} \int \dfrac{dx}{(a + bx^2)^m}$

70. $\int \dfrac{dx}{x(a + bx^2)} = \dfrac{1}{2a} \log \dfrac{x^2}{a + bx^2}$

71. $\int \dfrac{dx}{x^2(a + bx^2)} = -\dfrac{1}{ax} - \dfrac{b}{a} \int \dfrac{dx}{a + bx^2}$

72. $\int \dfrac{dx}{x(a + bx^2)^{m+1}} = \begin{cases} \dfrac{1}{2am(a + bx^2)^m} + \dfrac{1}{a} \int \dfrac{dx}{x(a + bx^2)^m} \\[2mm] \text{or} \\[2mm] \dfrac{1}{2a^{m+1}} \left[\displaystyle\sum_{r=1}^{m} \dfrac{a^r}{r(a + bx^2)^r} + \log \dfrac{x^2}{a + bx^2} \right] \end{cases}$

73. $\int \dfrac{dx}{x^2(a + bx^2)^{m+1}} = \dfrac{1}{a} \int \dfrac{dx}{x^2(a + bx^2)^m} - \dfrac{b}{a} \int \dfrac{dx}{(a + bx^2)^{m+1}}$

74. $\int \dfrac{dx}{a + bx^3} = \dfrac{k}{3a} \left[\dfrac{1}{2} \log \dfrac{(k + x)^3}{a + bx^3} + \sqrt{3} \tan^{-1} \dfrac{2x - k}{k\sqrt{3}} \right], \qquad \left(k = \sqrt[3]{\dfrac{a}{b}} \right)$

75. $\int \dfrac{x\,dx}{a + bx^3} = \dfrac{1}{3bk} \left[\dfrac{1}{2} \log \dfrac{a + bx^3}{(k + x)^3} + \sqrt{3} \tan^{-1} \dfrac{2x - k}{k\sqrt{3}} \right], \qquad \left(k = \sqrt[3]{\dfrac{a}{b}} \right)$

76. $\int \dfrac{x^2\,dx}{a + bx^3} = \dfrac{1}{3b} \log (a + bx^3)$

77. $\int \dfrac{dx}{a + bx^4} = \dfrac{k}{2a} \left[\dfrac{1}{2} \log \dfrac{x^2 + 2kx + 2k^2}{x^2 - 2kx + 2k^2} + \tan^{-1} \dfrac{2kx}{2k^2 - x^2} \right],$

$$\left(ab > 0, k = \sqrt[4]{\dfrac{a}{4b}} \right)$$

78. $\int \dfrac{dx}{a + bx^4} = \dfrac{k}{2a} \left[\dfrac{1}{2} \log \dfrac{x + k}{x - k} + \tan^{-1} \dfrac{x}{k} \right], \qquad \left(ab < 0, k = \sqrt[4]{-\dfrac{a}{b}} \right)$

79. $\int \dfrac{x\,dx}{a + bx^4} = \dfrac{1}{2bk} \tan^{-1} \dfrac{x^2}{k}, \qquad \left(ab > 0, k = \sqrt{\dfrac{a}{b}} \right)$

80. $\int \dfrac{x\,dx}{a + bx^4} = \dfrac{1}{4bk} \log \dfrac{x^2 - k}{x^2 + k}, \qquad \left(ab < 0, k = \sqrt{-\dfrac{a}{b}} \right)$

81. $\int \dfrac{x^2\,dx}{a + bx^4} = \dfrac{1}{4bk} \left[\dfrac{1}{2} \log \dfrac{x^2 - 2kx + 2k^2}{x^2 + 2kx + 2k^2} + \tan^{-1} \dfrac{2kx}{2k^2 - x^2} \right],$

$$\left(ab > 0, k = \sqrt[4]{\dfrac{a}{4b}} \right)$$

82. $\displaystyle\int \frac{x^2\,dx}{a + bx^4} = \frac{1}{4bk}\left[\log\frac{x - k}{x + k} + 2\tan^{-1}\frac{x}{k}\right], \qquad \left(ab < 0, k = \sqrt[4]{-\frac{a}{b}}\right)$

83. $\displaystyle\int \frac{x^3\,dx}{a + bx^4} = \frac{1}{4b}\log(a + bx^4)$

84. $\displaystyle\int \frac{dx}{x(a + bx^n)} = \frac{1}{an}\log\frac{x^n}{a + bx^n}$

85. $\displaystyle\int \frac{dx}{(a + bx^n)^{m+1}} = \frac{1}{a}\int \frac{dx}{(a + bx^n)^m} - \frac{b}{a}\int \frac{x^n\,dx}{(a + bx^n)^{m+1}}$

86. $\displaystyle\int \frac{x^m\,dx}{(a + bx^n)^{p+1}} = \frac{1}{b}\int \frac{x^{m-n}\,dx}{(a + bx^n)^p} - \frac{a}{b}\int \frac{x^{m-n}\,dx}{(a + bx^n)^{p+1}}$

87. $\displaystyle\int \frac{dx}{x^m(a + bx^n)^{p+1}} = \frac{1}{a}\int \frac{dx}{x^m(a + bx^n)^p} - \frac{b}{a}\int \frac{dx}{x^{m-n}(a + bx^n)^{p+1}}$

88. $\displaystyle\int x^m(a + bx^n)^p\,dx = \begin{cases} \dfrac{1}{b(np + m + 1)}\left[x^{m-n+1}(a + bx^n)^{p+1} \right. \\ \left. \qquad - a(m - n + 1)\displaystyle\int x^{m-n}(a + bx^n)^p\,dx\right] \\ \qquad\text{or} \\ \dfrac{1}{np + m + 1}\left[x^{m+1}(a + bx^n)^p \right. \\ \left. \qquad + anp\displaystyle\int x^m(a + bx^n)^{p-1}\,dx\right] \\ \qquad\text{or} \\ \dfrac{1}{a(m + 1)}\left[x^{m+1}(a + bx^n)^{p+1} \right. \\ \left. \qquad - (m + 1 + np + n)b\displaystyle\int x^{m+n}(a + bx^n)^p\,dx\right] \\ \qquad\text{or} \\ \dfrac{1}{an(p + 1)}\left[-x^{m+1}(a + bx^n)^{p+1} \right. \\ \left. \qquad + (m + 1 + np + n)\displaystyle\int x^m(a + bx^n)^{p+1}\,dx\right] \end{cases}$

FORMS CONTAINING $c^3 \pm x^3$

89. $\displaystyle\int \frac{dx}{c^3 \pm x^3} = \pm\frac{1}{6c^2}\log\frac{(c \pm x)^3}{c^3 \pm x^3} + \frac{1}{c^2\sqrt{3}}\tan^{-1}\frac{2x \mp c}{c\sqrt{3}}$

90. $\displaystyle\int \frac{dx}{(c^3 \pm x^3)^2} = \frac{x}{3c^3(c^3 \pm x^3)} + \frac{2}{3c^3}\int \frac{dx}{c^3 \pm x^3}$

91. $\displaystyle\int \frac{dx}{(c^3 \pm x^3)^{n+1}} = \frac{1}{3nc^3}\left[\frac{x}{(c^3 \pm x^3)^n} + (3n - 1)\int \frac{dx}{(c^3 \pm x^3)^n}\right]$

92. $\displaystyle\int \frac{x\,dx}{c^3 \pm x^3} = \frac{1}{6c}\log\frac{c^3 \pm x^3}{(c \pm x)^3} \pm \frac{1}{c\sqrt{3}}\tan^{-1}\frac{2x \mp c}{c\sqrt{3}}$

93. $\displaystyle\int \frac{x\,dx}{(c^3 \pm x^3)^2} = \frac{x^2}{3c^3(c^3 \pm x^3)} + \frac{1}{3c^3}\int \frac{x\,dx}{c^3 \pm x^3}$

94. $\displaystyle\int \frac{x\,dx}{(c^3 \pm x^3)^{n+1}} = \frac{1}{3nc^3}\left[\frac{x^2}{(c^3 \pm x^3)^n} + (3n - 2)\int \frac{x\,dx}{(c^3 \pm x^3)^n}\right]$

95. $\displaystyle\int \frac{x^2\,dx}{c^3 \pm x^3} = \pm\frac{1}{3}\log(c^3 \pm x^3)$

96. $\displaystyle\int \frac{x^2\,dx}{(c^3 \pm x^3)^{n+1}} = \mp\frac{1}{3n(c^3 \pm x^3)^n}$

97. $\displaystyle\int \frac{dx}{x(c^3 \pm x^3)} = \frac{1}{3c^3}\log\frac{x^3}{c^3 \pm x^3}$

98. $\displaystyle\int \frac{dx}{x(c^3 \pm x^3)^2} = \frac{1}{3c^3(c^3 \pm x^3)} + \frac{1}{3c^6}\log\frac{x^3}{c^3 \pm x^3}$

99. $\displaystyle\int \frac{dx}{x(c^3 \pm x^3)^{n+1}} = \frac{1}{3nc^3(c^3 \pm x^3)^n} + \frac{1}{c^3}\int \frac{dx}{x(c^3 \pm x^3)^n}$

100. $\displaystyle\int \frac{dx}{x^2(c^3 \pm x^3)} = -\frac{1}{c^3x} \mp \frac{1}{c^3}\int \frac{x\,dx}{c^3 \pm x^3}$

101. $\displaystyle\int \frac{dx}{x^2(c^3 \pm x^3)^{n+1}} = \frac{1}{c^3}\int \frac{dx}{x^2(c^3 \pm x^3)^n} \mp \frac{1}{c^3}\int \frac{x\,dx}{(c^3 \pm x^3)^{n+1}}$

FORMS CONTAINING $c^4 \pm x^4$

102. $\displaystyle\int \frac{dx}{c^4 + x^4} = \frac{1}{2c^3\sqrt{2}}\left[\frac{1}{2}\log\frac{x^2 + cx\sqrt{2} + c^2}{x^2 - cx\sqrt{2} + c^2} + \tan^{-1}\frac{cx\sqrt{2}}{c^2 - x^2}\right]$

103. $\displaystyle\int \frac{dx}{c^4 - x^4} = \frac{1}{2c^3}\left[\frac{1}{2}\log\frac{c + x}{c - x} + \tan^{-1}\frac{x}{c}\right]$

104. $\displaystyle\int \frac{x\,dx}{c^4 + x^4} = \frac{1}{2c^2}\tan^{-1}\frac{x^2}{c^2}$

105. $\displaystyle\int \frac{x\,dx}{c^4 - x^4} = \frac{1}{4c^2}\log\frac{c^2 + x^2}{c^2 - x^2}$

106. $\displaystyle\int \frac{x^2\,dx}{c^4 + x^4} = \frac{1}{2c\sqrt{2}}\left[\frac{1}{2}\log\frac{x^2 - cx\sqrt{2} + c^2}{x^2 + cx\sqrt{2} + c^2} + \tan^{-1}\frac{cx\sqrt{2}}{c^2 - x^2}\right]$

107. $\displaystyle\int \frac{x^2\,dx}{c^4 - x^4} = \frac{1}{2c}\left[\frac{1}{2}\log\frac{c + x}{c - x} - \tan^{-1}\frac{x}{c}\right]$

108. $\displaystyle\int \frac{x^3\,dx}{c^4 \pm x^4} = \pm\frac{1}{4}\log(c^4 \pm x^4)$

FORMS CONTAINING $(a + bx + cx^2)$

$$X = a + bx + cx^2 \quad \text{and} \quad q = 4ac - b^2$$

If $q = 0$, then $X = c\left(x + \dfrac{b}{2c}\right)^2$, and formulas starting with 23 should be used in place of these.

109. $\displaystyle\int \frac{dx}{X} = \frac{2}{\sqrt{q}}\tan^{-1}\frac{2cx + b}{\sqrt{q}}, \qquad (q > 0)$

110. $\displaystyle\int \frac{dx}{X} = \begin{cases} \dfrac{-2}{\sqrt{-q}}\tanh^{-1}\dfrac{2cx + b}{\sqrt{-q}} \\[2mm] \qquad\qquad \text{or} \\[2mm] \dfrac{1}{\sqrt{-q}}\log\dfrac{2cx + b - \sqrt{-q}}{2cx + b + \sqrt{-q}}, \qquad (q < 0) \end{cases}$

111. $\displaystyle\int \frac{dx}{X^2} = \frac{2cx + b}{qX} + \frac{2c}{q}\int \frac{dx}{X}$

112. $\displaystyle\int \frac{dx}{X^3} = \frac{2cx+b}{q}\left(\frac{1}{2X^2} + \frac{3c}{qX}\right) + \frac{6c^2}{q^2}\int \frac{dx}{X}$

113. $\displaystyle\int \frac{dx}{X^{n+1}} = \begin{cases} \dfrac{2cx+b}{nqX^n} + \dfrac{2(2n-1)c}{qn}\displaystyle\int \frac{dx}{X^n} \\[2mm] \text{or} \\[2mm] \dfrac{(2n)!}{(n!)^2}\left(\dfrac{c}{q}\right)^n\left[\dfrac{2cx+b}{q}\displaystyle\sum_{r=1}^n \left(\dfrac{q}{cX}\right)^r\dfrac{(r-1)!r!}{(2r)!}\right) + \displaystyle\int \frac{dx}{X}\right] \end{cases}$

114. $\displaystyle\int \frac{x\,dx}{X} = \frac{1}{2c}\log X - \frac{b}{2c}\int \frac{dx}{X}$

115. $\displaystyle\int \frac{x\,dx}{X^2} = \frac{bx+2a}{qX} - \frac{b}{q}\int \frac{dx}{X}$

116. $\displaystyle\int \frac{x\,dx}{X^{n+1}} = -\frac{2a+bx}{nqX^n} - \frac{b(2n-1)}{nq}\int \frac{dx}{X^n}$

117. $\displaystyle\int \frac{x^2}{X}dx = \frac{x}{c} - \frac{b}{2c^2}\log X + \frac{b^2-2ac}{2c^2}\int \frac{dx}{X}$

118. $\displaystyle\int \frac{x^2}{X^2}dx = \frac{(b^2-2ac)x+ab}{cqX} + \frac{2a}{q}\int \frac{dx}{X}$

119. $\displaystyle\int \frac{x^m\,dx}{X^{n+1}} = -\frac{x^{m-1}}{(2n-m+1)cX^n} - \frac{n-m+1}{2n-m+1}\cdot\frac{b}{c}\int \frac{x^{m-1}\,dx}{X^{n+1}}$
$$+ \frac{m-1}{2n-m+1}\cdot\frac{a}{c}\int \frac{x^{m-2}\,dx}{X^{n+1}}$$

120. $\displaystyle\int \frac{dx}{xX} = \frac{1}{2a}\log \frac{x^2}{X} - \frac{b}{2a}\int \frac{dx}{X}$

121. $\displaystyle\int \frac{dx}{x^2X} = \frac{b}{2a^2}\log \frac{X}{x^2} - \frac{1}{ax} + \left(\frac{b^2}{2a^2} - \frac{c}{a}\right)\int \frac{dx}{X}$

122. $\displaystyle\int \frac{dx}{xX^n} = \frac{1}{2a(n-1)X^{n-1}} - \frac{b}{2a}\int \frac{dx}{X^n} + \frac{1}{a}\int \frac{dx}{xX^{n-1}}$

123. $\displaystyle\int \frac{dx}{x^mX^{n+1}} = -\frac{1}{(m-1)ax^{m-1}X^n} - \frac{n+m-1}{m-1}\cdot\frac{b}{a}\int \frac{dx}{x^{m-1}X^{n+1}}$
$$- \frac{2n+m-1}{m-1}\cdot\frac{c}{a}\int \frac{dx}{x^{m-2}X^{n+1}}$$

FORMS CONTAINING $\sqrt{a+bx}$

124. $\displaystyle\int \sqrt{a+bx}\,dx = \frac{2}{3b}\sqrt{(a+bx)^3}$

125. $\displaystyle\int x\sqrt{a+bx}\,dx = -\frac{2(2a-3bx)\sqrt{(a+bx)^3}}{15b^2}$

126. $\displaystyle\int x^2\sqrt{a+bx}\,dx = \frac{2(8a^2-12abx+15b^2x^2)\sqrt{(a+bx)^3}}{105b^3}$

127. $\displaystyle\int x^m\sqrt{a+bx}\,dx = \begin{cases} \dfrac{2}{b(2m+3)}\left[x^m\sqrt{(a+bx)^3} - ma\displaystyle\int x^{m-1}\sqrt{a+bx}\,dx\right] \\[2mm] \text{or} \\[2mm] \dfrac{2}{b^{m+1}}\sqrt{a+bx}\displaystyle\sum_{r=0}^m \dfrac{m!(-a)^{m-r}}{r!(m-r)!(2r+3)}(a+bx)^{r+1} \end{cases}$

128. $\int \dfrac{\sqrt{a + bx}}{x} dx = 2\sqrt{a + bx} + a \int \dfrac{dx}{x\sqrt{a + bx}}$

129. $\int \dfrac{\sqrt{a + bx}}{x^2} dx = -\dfrac{\sqrt{a + bx}}{x} + \dfrac{b}{2} \int \dfrac{dx}{x\sqrt{a + bx}}$

130. $\int \dfrac{\sqrt{a + bx}}{x^m} dx = -\dfrac{1}{(m - 1)a}\left[\dfrac{\sqrt{(a + bx)^3}}{x^{m-1}} + \dfrac{(2m - 5)b}{2} \int \dfrac{\sqrt{a + bx}}{x^{m-1}} dx \right]$

131. $\int \dfrac{dx}{\sqrt{a + bx}} = \dfrac{2\sqrt{a + bx}}{b}$

132. $\int \dfrac{x\, dx}{\sqrt{a + bx}} = -\dfrac{2(2a - bx)}{3b^2}\sqrt{a + bx}$

133. $\int \dfrac{x^2\, dx}{\sqrt{a + bx}} = \dfrac{2(8a^2 - 4abx + 3b^2x^2)}{15b^3}\sqrt{a + bx}$

134. $\int \dfrac{x^m\, dx}{\sqrt{a + bx}} = \begin{cases} \dfrac{2}{(2m + 1)b}\left[x^m\sqrt{a + bx} - ma \int \dfrac{x^{m-1}\, dx}{\sqrt{a + bx}} \right] \\[2mm] \text{or} \\[2mm] \dfrac{2(-a)^m\sqrt{a + bx}}{b^{m+1}} \displaystyle\sum_{r=0}^{m} \dfrac{(-1)^r m!(a + bx)^r}{(2r + 1)r!(m - r)!a^r} \end{cases}$

135. $\int \dfrac{dx}{x\sqrt{a + bx}} = \dfrac{1}{\sqrt{a}} \log\left(\dfrac{\sqrt{a + bx} - \sqrt{a}}{\sqrt{a + bx} + \sqrt{a}} \right), \qquad (a > 0)$

136. $\int \dfrac{dx}{x\sqrt{a + bx}} = \dfrac{2}{\sqrt{-a}} \tan^{-1} \sqrt{\dfrac{a + bx}{-a}}, \qquad (a < 0)$

137. $\int \dfrac{dx}{x^2\sqrt{a + bx}} = -\dfrac{\sqrt{a + bx}}{ax} - \dfrac{b}{2a} \int \dfrac{dx}{x\sqrt{a + bx}}$

138. $\int \dfrac{dx}{x^n\sqrt{a + bx}} = \begin{cases} -\dfrac{\sqrt{a + bx}}{(n - 1)ax^{n-1}} - \dfrac{(2n - 3)b}{(2n - 2)a} \int \dfrac{dx}{x^{n-1}\sqrt{a + bx}} \\[2mm] \text{or} \\[2mm] \dfrac{(2n - 2)!}{[(n - 1)!]^2}\left[-\dfrac{\sqrt{a + bx}}{a} \displaystyle\sum_{r=1}^{n-1} \dfrac{r!(r - 1)!}{x^r(2r)!}\left(-\dfrac{b}{4a} \right)^{n-r-1} \right. \\[2mm] \qquad\qquad\qquad\qquad \left. + \left(-\dfrac{b}{4a} \right)^{n-1} \int \dfrac{dx}{x\sqrt{a + bx}} \right] \end{cases}$

139. $\int (a + bx)^{\pm\frac{n}{2}}\, dx = \dfrac{2(a + bx)^{\frac{2 \pm n}{2}}}{b(2 \pm n)}$

140. $\int x(a + bx)^{\pm\frac{n}{2}}\, dx = \dfrac{2}{b^2}\left[\dfrac{(a + bx)^{\frac{4 \pm n}{2}}}{4 \pm n} - \dfrac{a(a + bx)^{\frac{2 \pm n}{2}}}{2 \pm n} \right]$

141. $\int \dfrac{dx}{x(a + bx)^{\frac{m}{2}}} = \dfrac{1}{a} \int \dfrac{dx}{x(a + bx)^{\frac{m-2}{2}}} - \dfrac{b}{a} \int \dfrac{dx}{(a + bx)^{\frac{m}{2}}}$

142. $\displaystyle\int \frac{(a + bx)^{\frac{n}{2}}\,dx}{x} = b \int (a + bx)^{\frac{n-2}{2}}\,dx + a \int \frac{(a + bx)^{\frac{n-2}{2}}}{x}\,dx$

143. $\displaystyle\int f(x, \sqrt{a + bx})\,dx = \frac{2}{b} \int f\!\left(\frac{z^2 - a}{b}, z\right) z\,dz, \qquad (z = \sqrt{a + bx})$

FORMS CONTAINING $\sqrt{a + bx}$ and $\sqrt{c + dx}$

$$u = a + bx \qquad v = c + dx \qquad k = ad - bc$$

If $k = 0$, then $v = \dfrac{c}{a} u$, and formulas starting with 124 should be used in place of these.

144. $\displaystyle\int \frac{dx}{\sqrt{uv}} = \begin{cases} \dfrac{2}{\sqrt{bd}} \tanh^{-1} \dfrac{\sqrt{bduv}}{bv}, & bd > 0,\ k < o \\[2mm] \text{or} \\[2mm] \dfrac{2}{\sqrt{bd}} \tanh^{-1} \dfrac{\sqrt{bduv}}{du}, & bd > o,\ k > o. \\[2mm] \text{or} \\[2mm] \dfrac{1}{\sqrt{bd}} \log \dfrac{(bv + \sqrt{bduv})^2}{v}, & (bd > 0) \end{cases}$

145. $\displaystyle\int \frac{dx}{\sqrt{uv}} = \begin{cases} \dfrac{2}{\sqrt{-bd}} \tan^{-1} \dfrac{\sqrt{-bduv}}{bv} \\[2mm] \text{or} \\[2mm] -\dfrac{1}{\sqrt{-bd}} \sin^{-1}\!\left(\dfrac{2bdx + ad + bc}{|k|}\right), & (bd < 0) \end{cases}$

146. $\displaystyle\int \sqrt{uv}\,dx = \frac{k + 2bv}{4bd} \sqrt{uv} - \frac{k^2}{8bd} \int \frac{dx}{\sqrt{uv}}$

147. $\displaystyle\int \frac{dx}{v\sqrt{u}} = \begin{cases} \dfrac{1}{\sqrt{kd}} \log \dfrac{d\sqrt{u} - \sqrt{kd}}{d\sqrt{u} + \sqrt{kd}} \\[2mm] \text{or} \\[2mm] \dfrac{1}{\sqrt{kd}} \log \dfrac{(d\sqrt{u} - \sqrt{kd})^2}{v}, & (kd > 0) \end{cases}$

148. $\displaystyle\int \frac{dx}{v\sqrt{u}} = \frac{2}{\sqrt{-kd}} \tan^{-1} \frac{d\sqrt{u}}{\sqrt{-kd}}, \qquad (kd < 0)$

149. $\displaystyle\int \frac{x\,dx}{\sqrt{uv}} = \frac{\sqrt{uv}}{bd} - \frac{ad + bc}{2bd} \int \frac{dx}{\sqrt{uv}}$

150. $\displaystyle\int \frac{dx}{v\sqrt{uv}} = \frac{-2\sqrt{uv}}{kv}$

151. $\displaystyle\int \frac{v\,dx}{\sqrt{uv}} = \frac{\sqrt{uv}}{b} - \frac{k}{2b} \int \frac{dx}{\sqrt{uv}}$

152. $\displaystyle\int \sqrt{\frac{v}{u}}\,dx = \frac{v}{|v|} \int \frac{v\,dx}{\sqrt{uv}}$

153. $\displaystyle\int v^m \sqrt{u}\,dx = \frac{1}{(2m + 3)d}\left(2v^{m+1}\sqrt{u} + k \int \frac{v^m\,dx}{\sqrt{u}}\right)$

154. $\displaystyle\int \frac{dx}{v^m\sqrt{u}} = -\frac{1}{(m-1)k}\left(\frac{\sqrt{u}}{v^{m-1}} + \left(m - \frac{3}{2}\right)b\int \frac{dx}{v^{m-1}\sqrt{u}}\right)$

155. $\displaystyle\int \frac{v^m\,dx}{\sqrt{u}} = \begin{cases} \dfrac{2}{b(2m+1)}\left[v^m\sqrt{u} - mk\int \dfrac{v^{m-1}}{\sqrt{u}}\,dx\right] \\ \text{or} \\ \dfrac{2(m!)^2\sqrt{u}}{b(2m+1)!}\displaystyle\sum_{r=0}^{m}\left(-\dfrac{4k}{b}\right)^{m-r}\dfrac{(2r)!}{(r!)^2}v^r \end{cases}$

FORMS CONTAINING $\sqrt{x^2 \pm a^2}$

156. $\displaystyle\int \sqrt{x^2 \pm a^2}\,dx = \tfrac{1}{2}[x\sqrt{x^2 \pm a^2} \pm a^2 \log(x + \sqrt{x^2 \pm a^2})]$

157. $\displaystyle\int \frac{dx}{\sqrt{x^2 \pm a^2}} = \log(x + \sqrt{x^2 \pm a^2})$

158. $\displaystyle\int \frac{dx}{x\sqrt{x^2 - a^2}} = \frac{1}{|a|}\sec^{-1}\frac{x}{a}$

159. $\displaystyle\int \frac{dx}{x\sqrt{x^2 + a^2}} = -\frac{1}{a}\log\left(\frac{a + \sqrt{x^2 + a^2}}{x}\right)$

160. $\displaystyle\int \frac{\sqrt{x^2 + a^2}}{x}\,dx = \sqrt{x^2 + a^2} - a\log\left(\frac{a + \sqrt{x^2 + a^2}}{x}\right)$

161. $\displaystyle\int \frac{\sqrt{x^2 - a^2}}{x}\,dx = \sqrt{x^2 - a^2} - |a|\sec^{-1}\frac{x}{a}$

162. $\displaystyle\int \frac{x\,dx}{\sqrt{x^2 \pm a^2}} = \sqrt{x^2 \pm a^2}$

163. $\displaystyle\int x\sqrt{x^2 \pm a^2}\,dx = \tfrac{1}{3}\sqrt{(x^2 \pm a^2)^3}$

164. $\displaystyle\int \sqrt{(x^2 \pm a^2)^3}\,dx = \frac{1}{4}\left[x\sqrt{(x^2 \pm a^2)^3} \pm \frac{3a^2 x}{2}\sqrt{x^2 \pm a^2} \right.$
$$\left. + \frac{3a^4}{2}\log(x + \sqrt{x^2 \pm a^2})\right]$$

165. $\displaystyle\int \frac{dx}{\sqrt{(x^2 \pm a^2)^3}} = \frac{\pm x}{a^2\sqrt{x^2 \pm a^2}}$

166. $\displaystyle\int \frac{x\,dx}{\sqrt{(x^2 \pm a^2)^3}} = \frac{-1}{\sqrt{x^2 \pm a^2}}$

167. $\displaystyle\int x\sqrt{(x^2 \pm a^2)^3}\,dx = \tfrac{1}{5}\sqrt{(x^2 \pm a^2)^5}$

168. $\displaystyle\int x^2\sqrt{x^2 \pm a^2}\,dx = \frac{x}{4}\sqrt{(x^2 \pm a^2)^3} \mp \frac{a^2}{8}x\sqrt{x^2 \pm a^2} - \frac{a^4}{8}\log(x + \sqrt{x^2 \pm a^2})$

169. $\displaystyle\int x^3\sqrt{x^2 + a^2}\,dx = (\tfrac{1}{5}x^2 - \tfrac{2}{15}a^2)\sqrt{(a^2 + x^2)^3}$

170. $\displaystyle\int x^3\sqrt{x^2 - a^2}\,dx = \frac{1}{5}\sqrt{(x^2 - a^2)^5} + \frac{a^2}{3}\sqrt{(x^2 - a^2)^3}$

171. $\displaystyle\int \frac{x^2\,dx}{\sqrt{x^2 \pm a^2}} = \frac{x}{2}\sqrt{x^2 \pm a^2} \mp \frac{a^2}{2}\log(x + \sqrt{x^2 \pm a^2})$

172. $\int \dfrac{x^3\,dx}{\sqrt{x^2 \pm a^2}} = \dfrac{1}{3}\sqrt{(x^2 \pm a^2)^3} \mp a^2\sqrt{x^2 \pm a^2}$

173. $\int \dfrac{dx}{x^2\sqrt{x^2 \pm a^2}} = \mp \dfrac{\sqrt{x^2 \pm a^2}}{a^2 x}$

174. $\int \dfrac{dx}{x^3\sqrt{x^2 + a^2}} = -\dfrac{\sqrt{x^2 + a^2}}{2a^2 x^2} + \dfrac{1}{2a^3}\log\dfrac{a + \sqrt{x^2 + a^2}}{x}$

175. $\int \dfrac{dx}{x^3\sqrt{x^2 - a^2}} = \dfrac{\sqrt{x^2 - a^2}}{2a^2 x^2} + \dfrac{1}{2|a^3|}\sec^{-1}\dfrac{x}{a}$

176. $\int x^2\sqrt{(x^2 \pm a^2)^3}\,dx = \dfrac{x}{6}\sqrt{(x^2 \pm a^2)^5} \mp \dfrac{a^2 x}{24}\sqrt{(x^2 \pm a^2)^3} - \dfrac{a^4 x}{16}\sqrt{x^2 \pm a^2}$

$$\mp \dfrac{a^6}{16}\log(x + \sqrt{x^2 \pm a^2})$$

177. $\int x^3\sqrt{(x^2 \pm a^2)^3}\,dx = \dfrac{1}{7}\sqrt{(x^2 \pm a^2)^7} \mp \dfrac{a^2}{5}\sqrt{(x^2 \pm a^2)^5}$

178. $\int \dfrac{\sqrt{x^2 \pm a^2}\,dx}{x^2} = -\dfrac{\sqrt{x^2 \pm a^2}}{x} + \log(x + \sqrt{x^2 \pm a^2})$

179. $\int \dfrac{\sqrt{x^2 + a^2}}{x^3}\,dx = -\dfrac{\sqrt{x^2 + a^2}}{2x^2} - \dfrac{1}{2a}\log\dfrac{a + \sqrt{x^2 + a^2}}{x}$

180. $\int \dfrac{\sqrt{x^2 - a^2}}{x^3}\,dx = -\dfrac{\sqrt{x^2 - a^2}}{2x^2} + \dfrac{1}{2|a|}\sec^{-1}\dfrac{x}{a}$

181. $\int \dfrac{\sqrt{x^2 \pm a^2}}{x^4}\,dx = \mp \dfrac{\sqrt{(x^2 \pm a^2)^3}}{3a^2 x^3}$

182. $\int \dfrac{x^2\,dx}{\sqrt{(x^2 \pm a^2)^3}} = \dfrac{-x}{\sqrt{x^2 \pm a^2}} + \log(x + \sqrt{x^2 \pm a^2})$

183. $\int \dfrac{x^3\,dx}{\sqrt{(x^2 \pm a^2)^3}} = \sqrt{x^2 \pm a^2} \pm \dfrac{a^2}{\sqrt{x^2 \pm a^2}}$

184. $\int \dfrac{dx}{x\sqrt{(x^2 + a^2)^3}} = \dfrac{1}{a^2\sqrt{x^2 + a^2}} - \dfrac{1}{a^3}\log\dfrac{a + \sqrt{x^2 + a^2}}{x}$

185. $\int \dfrac{dx}{x\sqrt{(x^2 - a^2)^3}} = -\dfrac{1}{a^2\sqrt{x^2 - a^2}} - \dfrac{1}{|a^3|}\sec^{-1}\dfrac{x}{a}$

186. $\int \dfrac{dx}{x^2\sqrt{(x^2 \pm a^2)^3}} = -\dfrac{1}{a^4}\left[\dfrac{\sqrt{x^2 \pm a^2}}{x} + \dfrac{x}{\sqrt{x^2 \pm a^2}}\right]$

187. $\int \dfrac{dx}{x^3\sqrt{(x^2 + a^2)^3}} = -\dfrac{1}{2a^2 x^2\sqrt{x^2 + a^2}} - \dfrac{3}{2a^4\sqrt{x^2 + a^2}}$

$$+ \dfrac{3}{2a^5}\log\dfrac{a + \sqrt{x^2 + a^2}}{x}$$

188. $\int \dfrac{dx}{x^3\sqrt{(x^2 - a^2)^3}} = \dfrac{1}{2a^2 x^2\sqrt{x^2 - a^2}} - \dfrac{3}{2a^4\sqrt{x^2 - a^2}} - \dfrac{3}{2|a^5|}\sec^{-1}\dfrac{x}{a}$

189. $\int \dfrac{x^m}{\sqrt{x^2 \pm a^2}}\,dx = \dfrac{1}{m}x^{m-1}\sqrt{x^2 \pm a^2} \mp \dfrac{m-1}{m}a^2\int \dfrac{x^{m-2}}{\sqrt{x^2 \pm a^2}}\,dx$

190. $\int \dfrac{x^{2m}}{\sqrt{x^2 \pm a^2}}\, dx = \dfrac{(2m)!}{2^{2m}(m!)^2}\left[\sqrt{x^2 \pm a^2}\;\sum_{r=1}^{m}\dfrac{r!(r-1)!}{(2r)!}(\mp a^2)^{m-r}(2x)^{2r-1}\right.$

$$\left. + (\mp a^2)^m \log(x + \sqrt{x^2 \pm a^2})\right]$$

191. $\int \dfrac{x^{2m+1}}{\sqrt{x^2 \pm a^2}}\, dx = \sqrt{x^2 \pm a^2}\;\sum_{r=0}^{m}\dfrac{(2r)!(m!)^2}{(2m+1)!(r!)^2}(\mp 4a^2)^{m-r}x^{2r}$

192. $\int \dfrac{dx}{x^m \sqrt{x^2 \pm a^2}} = \mp\dfrac{\sqrt{x^2 \pm a^2}}{(m-1)a^2 x^{m-1}} \mp \dfrac{(m-2)}{(m-1)a^2}\int\dfrac{dx}{x^{m-2}\sqrt{x^2 \pm a^2}}$

193. $\int \dfrac{dx}{x^{2m}\sqrt{x^2 \pm a^2}} = \sqrt{x^2 \pm a^2}\;\sum_{r=0}^{m-1}\dfrac{(m-1)!\,m!\,(2r)!\,2^{2m-2r-1}}{(r!)^2(2m)!(\mp a^2)^{m-r}x^{2r+1}}$

194. $\int \dfrac{dx}{x^{2m+1}\sqrt{x^2 + a^2}} = \dfrac{(2m)!}{(m!)^2}\left[\dfrac{\sqrt{x^2 + a^2}}{a^2}\sum_{r=1}^{m}(-1)^{m-r+1}\dfrac{r!(r-1)!}{2(2r)!(4a^2)^{m-r}x^{2r}}\right.$

$$\left. + \dfrac{(-1)^{m+1}}{2^{2m}a^{2m+1}}\log\dfrac{\sqrt{x^2 + a^2} + a}{x}\right]$$

195. $\int \dfrac{dx}{x^{2m+1}\sqrt{x^2 - a^2}} = \dfrac{(2m)!}{(m!)^2}\left[\dfrac{\sqrt{x^2 - a^2}}{a^2}\sum_{r=1}^{m}\dfrac{r!(r-1)!}{2(2r)!(4a^2)^{m-r}x^{2r}}\right.$

$$\left. + \dfrac{1}{2^{2m}|a|^{2m+1}}\sec^{-1}\dfrac{x}{a}\right]$$

196. $\int \dfrac{dx}{(x-a)\sqrt{x^2 - a^2}} = -\dfrac{\sqrt{x^2 - a^2}}{a(x-a)}$

197. $\int \dfrac{dx}{(x+a)\sqrt{x^2 - a^2}} = \dfrac{\sqrt{x^2 - a^2}}{a(x+a)}$

198. $\int f(x, \sqrt{x^2 + a^2})\, dx = a\int f(a\tan u, a\sec u)\sec^2 u\, du, \qquad \left(u = \tan^{-1}\dfrac{x}{a},\, a > 0\right)$

199. $\int f(x, \sqrt{x^2 - a^2})\, dx = a\int f(a\sec u, a\tan u)\sec u\tan u\, du, \qquad \left(u = \sec^{-1}\dfrac{x}{a}\right.$

$$\left. a > 0\right)$$

FORMS CONTAINING $\sqrt{a^2 - x^2}$

200. $\int \sqrt{a^2 - x^2}\, dx = \dfrac{1}{2}\left[x\sqrt{a^2 - x^2} + a^2\sin^{-1}\dfrac{x}{|a|}\right]$

201. $\int \dfrac{dx}{\sqrt{a^2 - x^2}} = \begin{cases} \sin^{-1}\dfrac{x}{|a|} \\ \text{or} \\ -\cos^{-1}\dfrac{x}{|a|} \end{cases}$

202. $\int \dfrac{dx}{x\sqrt{a^2 - x^2}} = -\dfrac{1}{a}\log\left(\dfrac{a + \sqrt{a^2 - x^2}}{x}\right)$

203. $\int \dfrac{\sqrt{a^2 - x^2}}{x}\, dx = \sqrt{a^2 - x^2} - a\log\left(\dfrac{a + \sqrt{a^2 - x^2}}{x}\right)$

204. $\int \dfrac{x\, dx}{\sqrt{a^2 - x^2}} = -\sqrt{a^2 - x^2}$

205. $\int x\sqrt{a^2 - x^2}\, dx = -\dfrac{1}{3}\sqrt{(a^2 - x^2)^3}$

206. $\int \sqrt{(a^2 - x^2)^3}\, dx = \dfrac{1}{4}\left[x\sqrt{(a^2 - x^2)^3} + \dfrac{3a^2 x}{2}\sqrt{a^2 - x^2} + \dfrac{3a^4}{2}\sin^{-1}\dfrac{x}{|a|} \right]$

207. $\int \dfrac{dx}{\sqrt{(a^2 - x^2)^3}} = \dfrac{x}{a^2\sqrt{a^2 - x^2}}$

208. $\int \dfrac{x\, dx}{\sqrt{(a^2 - x^2)^3}} = \dfrac{1}{\sqrt{a^2 - x^2}}$

209. $\int x\sqrt{(a^2 - x^2)^3}\, dx = -\dfrac{1}{5}\sqrt{(a^2 - x^2)^5}$

210. $\int x^2\sqrt{a^2 - x^2}\, dx = -\dfrac{x}{4}\sqrt{(a^2 - x^2)^3} + \dfrac{a^2}{8}\left(x\sqrt{a^2 - x^2} + a^2\sin^{-1}\dfrac{x}{|a|} \right)$

211. $\int x^3\sqrt{a^2 - x^2}\, dx = \left(-\dfrac{1}{5}x^2 - \dfrac{2}{15}a^2 \right)\sqrt{(a^2 - x^2)^3}$

212. $\int x^2\sqrt{(a^2 - x^2)^3}\, dx = -\dfrac{1}{6}x\sqrt{(a^2 - x^2)^5} + \dfrac{a^2 x}{24}\sqrt{(a^2 - x^2)^3}$

$$+ \dfrac{a^4 x}{16}\sqrt{a^2 - x^2} + \dfrac{a^6}{16}\sin^{-1}\dfrac{x}{|a|}$$

213. $\int x^3\sqrt{(a^2 - x^2)^3}\, dx = \dfrac{1}{7}\sqrt{(a^2 - x^2)^7} - \dfrac{a^2}{5}\sqrt{(a^2 - x^2)^5}$

214. $\int \dfrac{x^2\, dx}{\sqrt{a^2 - x^2}} = -\dfrac{x}{2}\sqrt{a^2 - x^2} + \dfrac{a^2}{2}\sin^{-1}\dfrac{x}{|a|}$

215. $\int \dfrac{dx}{x^2\sqrt{a^2 - x^2}} = -\dfrac{\sqrt{a^2 - x^2}}{a^2 x}$

216. $\int \dfrac{\sqrt{a^2 - x^2}}{x^2}\, dx = -\dfrac{\sqrt{a^2 - x^2}}{x} - \sin^{-1}\dfrac{x}{|a|}$

217. $\int \dfrac{\sqrt{a^2 - x^2}}{x^3}\, dx = -\dfrac{\sqrt{a^2 - x^2}}{2x^2} + \dfrac{1}{2a}\log\dfrac{a + \sqrt{a^2 - x^2}}{x}$

218. $\int \dfrac{\sqrt{a^2 - x^2}}{x^4}\, dx = -\dfrac{\sqrt{(a^2 - x^2)^3}}{3a^2 x^3}$

219. $\int \dfrac{x^2\, dx}{\sqrt{(a^2 - x^2)^3}} = \dfrac{x}{\sqrt{a^2 - x^2}} - \sin^{-1}\dfrac{x}{|a|}$

220. $\int \dfrac{x^3\, dx}{\sqrt{a^2 - x^2}} = -\dfrac{2}{3}(a^2 - x^2)^{\frac{3}{2}} - x^2(a^2 - x^2)^{\frac{1}{2}} = -\dfrac{1}{3}\sqrt{a^2 - x^2}(x^2 + 2a^2)$

221. $\int \dfrac{x^3\, dx}{\sqrt{(a^2 - x^2)^3}} = 2(a^2 - x^2)^{\frac{1}{2}} + \dfrac{x^2}{(a^2 - x^2)^{\frac{1}{2}}} = -\dfrac{a^2}{\sqrt{a^2 - x^2}} + \sqrt{a^2 - x^2}$

222. $\int \dfrac{dx}{x^3\sqrt{a^2 - x^2}} = -\dfrac{\sqrt{a^2 - x^2}}{2a^2 x^2} - \dfrac{1}{2a^3}\log\dfrac{a + \sqrt{a^2 - x^2}}{x}$

223. $\int \dfrac{dx}{x\sqrt{(a^2 - x^2)^3}} = \dfrac{1}{a^2\sqrt{a^2 - x^2}} - \dfrac{1}{a^3}\log\dfrac{a + \sqrt{a^2 - x^2}}{x}$

224. $\int \dfrac{dx}{x^2\sqrt{(a^2 - x^2)^3}} = \dfrac{1}{a^4}\left[-\dfrac{\sqrt{a^2 - x^2}}{x} + \dfrac{x}{\sqrt{a^2 - x^2}} \right]$

225. $\int \dfrac{dx}{x^3\sqrt{(a^2 - x^2)^3}} = -\dfrac{1}{2a^2 x^2\sqrt{a^2 - x^2}} + \dfrac{3}{2a^4\sqrt{a^2 - x^2}}$

$$-\dfrac{3}{2a^5}\log\dfrac{a + \sqrt{a^2 - x^2}}{x}$$

226. $\displaystyle\int \frac{x^m}{\sqrt{a^2-x^2}}\,dx = -\frac{x^{m-1}\sqrt{a^2-x^2}}{m} + \frac{(m-1)a^2}{m}\int \frac{x^{m-2}}{\sqrt{a^2-x^2}}\,dx$

227. $\displaystyle\int \frac{x^{2m}}{\sqrt{a^2-x^2}}\,dx = \frac{(2m)!}{(m!)^2}\left[-\sqrt{a^2-x^2}\sum_{r=1}^{m}\frac{r!(r-1)!}{2^{2m-2r+1}(2r)!}a^{2m-2r}x^{2r-1}\right.$

$$\left. + \frac{a^{2m}}{2^{2m}}\sin^{-1}\frac{x}{|a|}\right]$$

228. $\displaystyle\int \frac{x^{2m+1}}{\sqrt{a^2-x^2}}\,dx = -\sqrt{a^2-x^2}\sum_{r=0}^{m}\frac{(2r)!(m!)^2}{(2m+1)!(r!)^2}(4a^2)^{m-r}x^{2r}$

229. $\displaystyle\int \frac{dx}{x^m\sqrt{a^2-x^2}} = -\frac{\sqrt{a^2-x^2}}{(m-1)a^2x^{m-1}} + \frac{m-2}{(m-1)a^2}\int \frac{dx}{x^{m-2}\sqrt{a^2-x^2}}$

230. $\displaystyle\int \frac{ax}{x^{2m}\sqrt{a^2-x^2}} = -\sqrt{a^2-x^2}\sum_{r=0}^{m-1}\frac{(m-1)!m!(2r)!2^{2m-2r-1}}{(r!)^2(2m)!a^{2m-2r}x^{2r+1}}$

231. $\displaystyle\int \frac{dx}{x^{2m+1}\sqrt{a^2-x^2}} = \frac{(2m)!}{(m!)^2}\left[-\frac{\sqrt{a^2-x^2}}{a^2}\sum_{r=1}^{m}\frac{r!(r-1)!}{2(2r)!(4a^2)^{m-r}x^{2r}}\right.$

$$\left. + \frac{1}{2^{2m}a^{2m+1}}\log\frac{a-\sqrt{a^2-x^2}}{x}\right]$$

232. $\displaystyle\int \frac{dx}{(b^2-x^2)\sqrt{a^2-x^2}} = \frac{1}{2b\sqrt{a^2-b^2}}\log\frac{(b\sqrt{a^2-x^2}+x\sqrt{a^2-b^2})^2}{b^2-x^2},$

$$(a^2 > b^2)$$

233. $\displaystyle\int \frac{dx}{(b^2-x^2)\sqrt{a^2-x^2}} = \frac{1}{b\sqrt{b^2-a^2}}\tan^{-1}\frac{x\sqrt{b^2-a^2}}{b\sqrt{a^2-x^2}}, \qquad (b^2 > a^2)$

234. $\displaystyle\int \frac{dx}{(b^2+x^2)\sqrt{a^2-x^2}} = \frac{1}{b\sqrt{a^2+b^2}}\tan^{-1}\frac{x\sqrt{a^2+b^2}}{b\sqrt{a^2-x^2}}$

235. $\displaystyle\int \frac{\sqrt{a^2-x^2}}{b^2+x^2}\,dx = \frac{\sqrt{a^2+b^2}}{|b|}\sin^{-1}\frac{x\sqrt{a^2+b^2}}{|a|\sqrt{x^2+b^2}} - \sin^{-1}\frac{x}{|a|}$

236. $\displaystyle\int f(x,\sqrt{a^2-x^2})\,dx = a\int f(a\sin u, a\cos u)\cos u\,du, \qquad \left(u=\sin^{-1}\frac{x}{a}, a>0\right)$

FORMS CONTAINING $\sqrt{a+bx+cx^2}$

$$X = a+bx+cx^2, q = 4ac-b^2, \text{and } k = \frac{4c}{q}$$

If $q=0$, then $\sqrt{X} = \sqrt{c}\left|x+\dfrac{b}{2c}\right|$

237. $\displaystyle\int \frac{dx}{\sqrt{X}} = \begin{cases} \dfrac{1}{\sqrt{c}}\log(2\sqrt{cX}+2cx+b) \\[2mm] \quad\text{or} \\[2mm] \dfrac{1}{\sqrt{c}}\sinh^{-1}\dfrac{2cx+b}{\sqrt{q}}, \quad (c>0) \end{cases}$

238. $\displaystyle\int \frac{dx}{\sqrt{X}} = -\frac{1}{\sqrt{-c}}\sin^{-1}\frac{2cx+b}{\sqrt{-q}}, \qquad (c<0)$

239. $\displaystyle\int \frac{dx}{X\sqrt{X}} = \frac{2(2cx+b)}{q\sqrt{X}}$

240. $\displaystyle\int \frac{dx}{X^2\sqrt{X}} = \frac{2(2cx + b)}{3q\sqrt{X}}\left(\frac{1}{X} + 2k\right)$

241. $\displaystyle\int \frac{dx}{X^n\sqrt{X}} = \begin{cases} \dfrac{2(2cx + b)\sqrt{X}}{(2n - 1)qX^n} + \dfrac{2k(n - 1)}{2n - 1}\displaystyle\int \dfrac{dx}{X^{n-1}\sqrt{X}} \\ \text{or} \\ \dfrac{(2cx + b)(n!)(n - 1)!4^n k^{n-1}}{q[(2n)!]\sqrt{X}} \displaystyle\sum_{r=0}^{n-1} \dfrac{(2r)!}{(4kX)^r(r!)^2} \end{cases}$

242. $\displaystyle\int \sqrt{X}\, dx = \frac{(2cx + b)\sqrt{X}}{4c} + \frac{1}{2k}\int \frac{dx}{\sqrt{X}}$

243. $\displaystyle\int X\sqrt{X}\, dx = \frac{(2cx + b)\sqrt{X}}{8c}\left(X + \frac{3}{2k}\right) + \frac{3}{8k^2}\int \frac{dx}{\sqrt{X}}$

244. $\displaystyle\int X^2\sqrt{X}\, dx = \frac{(2cx + b)\sqrt{X}}{12c}\left(X^2 + \frac{5X}{4k} + \frac{15}{8k^2}\right) + \frac{5}{16k^3}\int \frac{dx}{\sqrt{X}}$

245. $\displaystyle\int X^n\sqrt{X}\, dx = \begin{cases} \dfrac{(2cx + b)X^n\sqrt{X}}{4(n + 1)c} + \dfrac{2n + 1}{2(n + 1)k}\displaystyle\int X^{n-1}\sqrt{X}\, dx \\ \text{or} \\ \dfrac{(2n + 2)!}{[(n + 1)!]^2(4k)^{n+1}}\left[\dfrac{k(2cx + b)\sqrt{X}}{c}\displaystyle\sum_{r=0}^{n}\dfrac{r!(r + 1)!(4kX)^r}{(2r + 2)!} \right. \\ \left. \qquad\qquad\qquad\qquad\qquad\qquad\qquad + \displaystyle\int \dfrac{dx}{\sqrt{X}}\right] \end{cases}$

246. $\displaystyle\int \frac{x\, dx}{\sqrt{X}} = \frac{\sqrt{X}}{c} - \frac{b}{2c}\int \frac{dx}{\sqrt{X}}$

247. $\displaystyle\int \frac{x\, dx}{X\sqrt{X}} = -\frac{2(bx + 2a)}{q\sqrt{X}}$

248. $\displaystyle\int \frac{x\, dx}{X^n\sqrt{X}} = -\frac{\sqrt{X}}{(2n - 1)cX^n} - \frac{b}{2c}\int \frac{dx}{X^n\sqrt{X}}$

249. $\displaystyle\int \frac{x^2\, dx}{\sqrt{X}} = \left(\frac{x}{2c} - \frac{3b}{4c^2}\right)\sqrt{X} + \frac{3b^2 - 4ac}{8c^2}\int \frac{dx}{\sqrt{X}}$

250. $\displaystyle\int \frac{x^2\, dx}{X\sqrt{X}} = \frac{(2b^2 - 4ac)x + 2ab}{cq\sqrt{X}} + \frac{1}{c}\int \frac{dx}{\sqrt{X}}$

251. $\displaystyle\int \frac{x^2\, dx}{X^n\sqrt{X}} = \frac{(2b^2 - 4ac)x + 2ab}{(2n - 1)cqX^{n-1}\sqrt{X}} + \frac{4ac + (2n - 3)b^2}{(2n - 1)cq}\int \frac{dx}{X^{n-1}\sqrt{X}}$

252. $\displaystyle\int \frac{x^3\, dx}{\sqrt{X}} = \left(\frac{x^2}{3c} - \frac{5bx}{12c^2} + \frac{5b^2}{8c^3} - \frac{2a}{3c^2}\right)\sqrt{X} + \left(\frac{3ab}{4c^2} - \frac{5b^3}{16c^3}\right)\int \frac{dx}{\sqrt{X}}$

253. $\displaystyle\int \frac{x^n\, dx}{\sqrt{X}} = \frac{1}{nc}x^{n-1}\sqrt{X} - \frac{(2n - 1)b}{2nc}\int \frac{x^{n-1}\, dx}{\sqrt{X}} - \frac{(n - 1)a}{nc}\int \frac{x^{n-2}\, dx}{\sqrt{X}}$

254. $\displaystyle\int x\sqrt{X}\, dx = \frac{X\sqrt{X}}{3c} - \frac{b(2cx + b)}{8c^2}\sqrt{X} - \frac{b}{4ck}\int \frac{dx}{\sqrt{X}}$

255. $\displaystyle\int xX\sqrt{X}\, dx = \frac{X^2\sqrt{X}}{5c} - \frac{b}{2c}\int X\sqrt{X}\, dx$

256. $\displaystyle\int xX^n\sqrt{X}\,dx = \frac{X^{n+1}\sqrt{X}}{(2n+3)c} - \frac{b}{2c}\int X^n\sqrt{X}\,dx$

257. $\displaystyle\int x^2\sqrt{X}\,dx = \left(x - \frac{5b}{6c}\right)\frac{X\sqrt{X}}{4c} + \frac{5b^2 - 4ac}{16c^2}\int\sqrt{X}\,dx$

258. $\displaystyle\int\frac{dx}{x\sqrt{X}} = -\frac{1}{\sqrt{a}}\log\frac{2\sqrt{aX} + bx + 2a}{x}, \qquad (a > 0)$

259. $\displaystyle\int\frac{dx}{x\sqrt{X}} = \frac{1}{\sqrt{-a}}\sin^{-1}\left(\frac{bx + 2a}{|x|\sqrt{-q}}\right), \qquad (a < 0)$

260. $\displaystyle\int\frac{dx}{x\sqrt{X}} = -\frac{2\sqrt{X}}{bx}, \qquad (a = 0)$

261. $\displaystyle\int\frac{dx}{x^2\sqrt{X}} = -\frac{\sqrt{X}}{ax} - \frac{b}{2a}\int\frac{dx}{x\sqrt{X}}$

262. $\displaystyle\int\frac{\sqrt{X}\,dx}{x} = \sqrt{X} + \frac{b}{2}\int\frac{dx}{\sqrt{X}} + a\int\frac{dx}{x\sqrt{X}}$

263. $\displaystyle\int\frac{\sqrt{X}\,dx}{x^2} = -\frac{\sqrt{X}}{x} + \frac{b}{2}\int\frac{dx}{x\sqrt{X}} + c\int\frac{dx}{\sqrt{X}}$

FORMS INVOLVING $\sqrt{2ax - x^2}$

264. $\displaystyle\int\sqrt{2ax - x^2}\,dx = \frac{1}{2}\left[(x - a)\sqrt{2ax - x^2} + a^2\sin^{-1}\frac{x - a}{|a|}\right]$

265. $\displaystyle\int\frac{dx}{\sqrt{2ax - x^2}} = \begin{cases}\cos^{-1}\dfrac{a - x}{|a|} \\[1em] \text{or} \\[1em] \sin^{-1}\dfrac{x - a}{|a|}\end{cases}$

266. $\displaystyle\int x^n\sqrt{2ax - x^2}\,dx = \begin{cases}-\dfrac{x^{n-1}(2ax - x^2)^{\frac{3}{2}}}{n + 2} + \dfrac{(2n + 1)a}{n + 2}\displaystyle\int x^{n-1}\sqrt{2ax - x^2}\,dx \\[1.5em] \text{or} \\[1.5em] \sqrt{2ax - x^2}\left[\dfrac{x^{n+1}}{n + 2} - \displaystyle\sum_{r=0}^{n}\dfrac{(2n + 1)!(r!)^2 a^{n-r+1}}{2^{n-r}(2r + 1)!(n + 2)!n!}x^r\right] \\[1.5em] \hspace{6em} + \dfrac{(2n + 1)!a^{n+2}}{2^n n!(n + 2)!}\sin^{-1}\dfrac{x - a}{|a|}\end{cases}$

267. $\displaystyle\int\frac{\sqrt{2ax - x^2}}{x^n}\,dx = \frac{(2ax - x^2)^{\frac{3}{2}}}{(3 - 2n)ax^n} + \frac{n - 3}{(2n - 3)a}\int\frac{\sqrt{2ax - x^2}}{x^{n-1}}\,dx$

268. $\displaystyle\int\frac{x^n\,dx}{\sqrt{2ax - x^2}} = \begin{cases}\dfrac{-x^{n-1}\sqrt{2ax - x^2}}{n} + \dfrac{a(2n - 1)}{n}\displaystyle\int\frac{x^{n-1}}{\sqrt{2ax - x^2}}\,dx \\[1.5em] \text{or} \\[1.5em] -\sqrt{2ax - x^2}\displaystyle\sum_{r=1}^{n}\dfrac{(2n)!r!(r - 1)!a^{n-r}}{2^{n-r}(2r)!(n!)^2}x^{r-1} \\[1.5em] \hspace{6em} + \dfrac{(2n)!a^n}{2^n(n!)^2}\sin^{-1}\dfrac{x - a}{|a|}\end{cases}$

269. $\displaystyle \int \frac{dx}{x^n \sqrt{2ax - x^2}} = \begin{cases} \dfrac{\sqrt{2ax - x^2}}{a(1 - 2n)x^n} + \dfrac{n - 1}{(2n - 1)a} \displaystyle\int \frac{dx}{x^{n-1}\sqrt{2ax - x^2}} \\[2mm] \text{or} \\[2mm] -\sqrt{2ax - x^2} \displaystyle\sum_{r=0}^{n-1} \frac{2^{n-r}(n - 1)!\, n!\,(2r)!}{(2n)!\,(r!)^2 a^{n-r} x^{r+1}} \end{cases}$

270. $\displaystyle \int \frac{dx}{(2ax - x^2)^{\frac{3}{2}}} = \frac{x - a}{a^2\sqrt{2ax - x^2}}$

271. $\displaystyle \int \frac{x\, dx}{(2ax - x^2)^{\frac{3}{2}}} = \frac{x\cdot}{a\sqrt{2ax - x^2}}$

MISCELLANEOUS ALGEBRAIC FORMS

272. $\displaystyle \int \frac{dx}{\sqrt{2ax + x^2}} = \log\left(x + a + \sqrt{2ax + x^2}\right)$

273. $\displaystyle \int \sqrt{ax^2 + c}\, dx = \frac{x}{2}\sqrt{ax^2 + c} + \frac{c}{2\sqrt{a}}\log\left(x\sqrt{a} + \sqrt{ax^2 + c}\right), \quad (a > 0)$

274. $\displaystyle \int \sqrt{ax^2 + c}\, dx = \frac{x}{2}\sqrt{ax^2 + c} + \frac{c}{2\sqrt{-a}}\sin^{-1}\left(x\sqrt{-\frac{a}{c}}\right), \quad (a < 0)$

275. $\displaystyle \int \sqrt{\frac{1 + x}{1 - x}}\, dx = \sin^{-1}x - \sqrt{1 - x^2}$

276. $\displaystyle \int \frac{dx}{x\sqrt{ax^n + c}} = \begin{cases} \dfrac{1}{n\sqrt{c}}\log\dfrac{\sqrt{ax^n + c} - \sqrt{c}}{\sqrt{ax^n + c} + \sqrt{c}} \\[2mm] \text{or} \\[2mm] \dfrac{2}{n\sqrt{c}}\log\dfrac{\sqrt{ax^n + c} - \sqrt{c}}{\sqrt{x^n}}, \quad (c > 0) \end{cases}$

277. $\displaystyle \int \frac{dx}{x\sqrt{ax^n + c}} = \frac{2}{n\sqrt{-c}}\sec^{-1}\sqrt{-\frac{ax^n}{c}}, \quad (c < 0)$

278. $\displaystyle \int \frac{dx}{\sqrt{ax^2 + c}} = \frac{1}{\sqrt{a}}\log\left(x\sqrt{a} + \sqrt{ax^2 + c}\right), \quad (a > 0)$

279. $\displaystyle \int \frac{dx}{\sqrt{ax^2 + c}} = \frac{1}{\sqrt{-a}}\sin^{-1}\left(x\sqrt{-\frac{a}{c}}\right), \quad (a < 0)$

280. $\displaystyle \int (ax^2 + c)^{m+\frac{1}{2}}\, dx = \begin{cases} \dfrac{x(ax^2 + c)^{m+\frac{1}{2}}}{2(m + 1)} + \dfrac{(2m + 1)c}{2(m + 1)}\displaystyle\int (ax^2 + c)^{m-\frac{1}{2}}\, dx \\[2mm] \text{or} \\[2mm] x\sqrt{ax^2 + c}\displaystyle\sum_{r=0}^{m} \dfrac{(2m + 1)!\,(r!)^2 c^{m-r}}{2^{2m-2r+1}m!(m + 1)!(2r + 1)!}(ax^2 + c)^r \\[2mm] \quad + \dfrac{(2m + 1)!\,c^{m+1}}{2^{2m+1}m!(m + 1)!}\displaystyle\int \dfrac{dx}{\sqrt{ax^2 + c}} \end{cases}$

281. $\displaystyle \int x(ax^2 + c)^{m+\frac{1}{2}}\, dx = \frac{(ax^2 + c)^{m+\frac{3}{2}}}{(2m + 3)a}$

282. $\displaystyle\int \frac{(ax^2 + c)^{m+\frac{1}{2}}}{x}\,dx = \begin{cases} \dfrac{(ax^2 + c)^{m+\frac{1}{2}}}{2m+1} + c \displaystyle\int \dfrac{(ax^2 + c)^{m-\frac{1}{2}}}{x}\,dx \\ \text{or} \\ \sqrt{ax^2 + c}\ \displaystyle\sum_{r=0}^{m} \dfrac{c^{m-r}(ax^2 + c)^r}{2r+1} + c^{m+1} \displaystyle\int \dfrac{dx}{x\sqrt{ax^2 + c}} \end{cases}$

283. $\displaystyle\int \frac{dx}{(ax^2 + c)^{m+\frac{1}{2}}} = \begin{cases} \dfrac{x}{(2m-1)c(ax^2 + c)^{m-\frac{1}{2}}} + \dfrac{2m-2}{(2m-1)c} \displaystyle\int \dfrac{dx}{(ax^2 + c)^{m-\frac{1}{2}}} \\ \text{or} \\ \dfrac{x}{\sqrt{ax^2 + c}} \displaystyle\sum_{r=0}^{m-1} \dfrac{2^{2m-2r-1}(m-1)!\,m!\,(2r)!}{(2m)!\,(r!)^2 c^{m-r}(ax^2 + c)^r} \end{cases}$

284. $\displaystyle\int \frac{dx}{x^m\sqrt{ax^2 + c}} = -\frac{\sqrt{ax^2 + c}}{(m-1)cx^{m-1}} - \frac{(m-2)a}{(m-1)c}\int \frac{dx}{x^{m-2}\sqrt{ax^2 + c}}$

285. $\displaystyle\int \frac{1+x^2}{(1-x^2)\sqrt{1+x^4}}\,dx = \frac{1}{\sqrt{2}}\log\frac{x\sqrt{2} + \sqrt{1+x^4}}{1-x^2}$

286. $\displaystyle\int \frac{1-x^2}{(1+x^2)\sqrt{1+x^4}}\,dx = \frac{1}{\sqrt{2}}\tan^{-1}\frac{x\sqrt{2}}{\sqrt{1+x^4}}$

287. $\displaystyle\int \frac{dx}{x\sqrt{x^n + a^2}} = -\frac{2}{na}\log\frac{a + \sqrt{x^n + a^2}}{\sqrt{x^n}}$

288. $\displaystyle\int \frac{dx}{x\sqrt{x^n - a^2}} = -\frac{2}{na}\sin^{-1}\frac{a}{\sqrt{x^n}}$

289. $\displaystyle\int \sqrt{\frac{x}{a^3 - x^3}}\,dx = \frac{2}{3}\sin^{-1}\left(\frac{x}{a}\right)^{\frac{1}{2}}$

FORMS INVOLVING TRIGONOMETRIC FUNCTIONS

290. $\displaystyle\int (\sin ax)\,dx = -\frac{1}{a}\cos ax$

291. $\displaystyle\int (\cos ax)\,dx = \frac{1}{a}\sin ax$

292. $\displaystyle\int (\tan ax)\,dx = -\frac{1}{a}\log\cos ax = \frac{1}{a}\log\sec ax$

293. $\displaystyle\int (\cot ax)\,dx = \frac{1}{a}\log\sin ax = -\frac{1}{a}\log\csc ax$

294. $\displaystyle\int (\sec ax)\,dx = \frac{1}{a}\log(\sec ax + \tan ax) = \frac{1}{a}\log\tan\left(\frac{\pi}{4} + \frac{ax}{2}\right)$

295. $\displaystyle\int (\csc ax)\,dx = \frac{1}{a}\log(\csc ax - \cot ax) = \frac{1}{a}\log\tan\frac{ax}{2}$

296. $\displaystyle\int (\sin^2 ax)\,dx = -\frac{1}{2a}\cos ax\sin ax + \frac{1}{2}x = \frac{1}{2}x - \frac{1}{4a}\sin 2ax$

297. $\displaystyle\int (\sin^3 ax)\,dx = -\frac{1}{3a}(\cos ax)(\sin^2 ax + 2)$

298. $\displaystyle\int (\sin^4 ax)\,dx = \frac{3x}{8} - \frac{\sin 2ax}{4a} + \frac{\sin 4ax}{32a}$

299. $\displaystyle\int (\sin^n ax)\,dx = -\frac{\sin^{n-1} ax\cos ax}{na} + \frac{n-1}{n}\int (\sin^{n-2} ax)\,dx$

300. $\displaystyle\int (\sin^{2m} ax)\,dx = -\frac{\cos ax}{a} \sum_{r=0}^{m-1} \frac{(2m)!(r!)^2}{2^{2m-2r}(2r+1)!(m!)^2} \sin^{2r+1} ax + \frac{(2m)!}{2^{2m}(m!)^2} x$

301. $\displaystyle\int (\sin^{2m+1} ax)\,dx = -\frac{\cos ax}{a} \sum_{r=0}^{m} \frac{2^{2m-2r}(m!)^2(2r)!}{(2m+1)!(r!)^2} \sin^{2r} ax$

302. $\displaystyle\int (\cos^2 ax)\,dx = \frac{1}{2a}\sin ax \cos ax + \frac{1}{2}x = \frac{1}{2}x + \frac{1}{4a}\sin 2ax$

303. $\displaystyle\int (\cos^3 ax)\,dx = \frac{1}{3a}(\sin ax)(\cos^2 ax + 2)$

304. $\displaystyle\int (\cos^4 ax)\,dx = \frac{3x}{8} + \frac{\sin 2ax}{4a} + \frac{\sin 4ax}{32a}$

305. $\displaystyle\int (\cos^n ax)\,dx = \frac{1}{na}\cos^{n-1} ax \sin ax + \frac{n-1}{n}\int (\cos^{n-2} ax)\,dx$

306. $\displaystyle\int (\cos^{2m} ax)\,dx = \frac{\sin ax}{a} \sum_{r=0}^{m-1} \frac{(2m)!(r!)^2}{2^{2m-2r}(2r+1)!(m!)^2} \cos^{2r+1} ax + \frac{(2m)!}{2^{2m}(m!)^2} x$

307. $\displaystyle\int (\cos^{2m+1} ax)\,dx = \frac{\sin ax}{a} \sum_{r=0}^{m} \frac{2^{2m-2r}(m!)^2(2r)!}{(2m+1)!(r!)^2} \cos^{2r} ax$

308. $\displaystyle\int \frac{dx}{\sin^2 ax} = \int (\csc^2 ax)\,dx = -\frac{1}{a}\cot ax$

309. $\displaystyle\int \frac{dx}{\sin^m ax} = \int (\csc^m ax)\,dx = -\frac{1}{(m-1)a} \cdot \frac{\cos ax}{\sin^{m-1} ax} + \frac{m-2}{m-1}\int \frac{dx}{\sin^{m-2} ax}$

310. $\displaystyle\int \frac{dx}{\sin^{2m} ax} = \int (\csc^{2m} ax)\,dx = -\frac{1}{a}\cos ax \sum_{r=0}^{m-1} \frac{2^{2m-2r-1}(m-1)!m!(2r)!}{(2m)!(r!)^2 \sin^{2r+1} ax}$

311. $\displaystyle\int \frac{dx}{\sin^{2m+1} ax} = \int (\csc^{2m+1} ax)\,dx =$

$$-\frac{1}{a}\cos ax \sum_{r=0}^{m-1} \frac{(2m)!(r!)^2}{2^{2m-2r}(m!)^2(2r+1)! \sin^{2r+2} ax} + \frac{1}{a} \cdot \frac{(2m)!}{2^{2m}(m!)^2} \log \tan \frac{ax}{2}$$

312. $\displaystyle\int \frac{dx}{\cos^2 ax} = \int (\sec^2 ax)\,dx = \frac{1}{a}\tan ax$

313. $\displaystyle\int \frac{dx}{\cos^n ax} = \int (\sec^n ax)\,dx = \frac{1}{(n-1)a} \cdot \frac{\sin ax}{\cos^{n-1} ax} + \frac{n-2}{n-1}\int \frac{dx}{\cos^{n-2} ax}$

314. $\displaystyle\int \frac{dx}{\cos^{2m} ax} = \int (\sec^{2m} ax)\,dx = \frac{1}{a}\sin ax \sum_{r=0}^{m-1} \frac{2^{2m-2r-1}(m-1)!m!(2r)!}{(2m)!(r!)^2 \cos^{2r+1} ax}$

315. $\displaystyle\int \frac{dx}{\cos^{2m+1} ax} = \int (\sec^{2m+1} ax)\,dx =$

$$\frac{1}{a}\sin ax \sum_{r=0}^{m-1} \frac{(2m)!(r!)^2}{2^{2m-2r}(m!)^2(2r+1)! \cos^{2r+2} ax}$$

$$+ \frac{1}{a} \cdot \frac{(2m)!}{2^{2m}(m!)^2} \log (\sec ax + \tan ax)$$

316. $\displaystyle\int (\sin mx)(\sin nx)\,dx = \frac{\sin(m-n)x}{2(m-n)} - \frac{\sin(m+n)x}{2(m+n)}, \qquad (m^2 \neq n^2)$

317. $\displaystyle\int (\cos mx)(\cos nx)\,dx = \frac{\sin(m-n)x}{2(m-n)} + \frac{\sin(m+n)x}{2(m+n)}, \qquad (m^2 \neq n^2)$

318. $\displaystyle\int (\sin ax)(\cos ax)\,dx = \frac{1}{2a}\sin^2 ax$

319. $\int (\sin mx)(\cos nx)\,dx = -\dfrac{\cos(m-n)x}{2(m-n)} - \dfrac{\cos(m+n)x}{2(m+n)}, \qquad (m^2 \neq n^2)$

320. $\int (\sin^2 ax)(\cos^2 ax)\,dx = -\dfrac{1}{32a}\sin 4ax + \dfrac{x}{8}$

321. $\int (\sin ax)(\cos^m ax)\,dx = -\dfrac{\cos^{m+1} ax}{(m+1)a}$

322. $\int (\sin^m ax)(\cos ax)\,dx = \dfrac{\sin^{m+1} ax}{(m+1)a}$

323. $\int (\cos^m ax)(\sin^n ax)\,dx = \begin{cases} \dfrac{\cos^{m-1} ax \sin^{n+1} ax}{(m+n)a} \\[2mm] \qquad + \dfrac{m-1}{m+n}\int (\cos^{m-2} ax)(\sin^n ax)\,dx \\[2mm] \text{or} \\[2mm] -\dfrac{\sin^{n-1} ax \cos^{m+1} ax}{(m+n)a} \\[2mm] \qquad + \dfrac{n-1}{m+n}\int (\cos^m ax)(\sin^{n-2} ax)\,dx \end{cases}$

324. $\int \dfrac{\cos^m ax}{\sin^n ax}\,dx = \begin{cases} -\dfrac{\cos^{m+1} ax}{(n-1)a \sin^{n-1} ax} - \dfrac{m-n+2}{n-1}\int \dfrac{\cos^m ax}{\sin^{n-2} ax}\,dx \\[2mm] \text{or} \\[2mm] \dfrac{\cos^{m-1} ax}{a(m-n)\sin^{n-1} ax} + \dfrac{m-1}{m-n}\int \dfrac{\cos^{m-2} ax}{\sin^n ax}\,dx \end{cases}$

325. $\int \dfrac{\sin^m ax}{\cos^n ax}\,dx = \begin{cases} \dfrac{\sin^{m+1} ax}{a(n-1)\cos^{n-1} ax} - \dfrac{m-n+2}{n-1}\int \dfrac{\sin^m ax}{\cos^{n-2} ax}\,dx \\[2mm] \text{or} \\[2mm] -\dfrac{\sin^{m-1} ax}{a(m-n)\cos^{n-1} ax} + \dfrac{m-1}{m-n}\int \dfrac{\sin^{m-2} ax}{\cos^n ax}\,dx \end{cases}$

326. $\int \dfrac{\sin ax}{\cos^2 ax}\,dx = \dfrac{1}{a\cos ax} = \dfrac{\sec ax}{a}$

327. $\int \dfrac{\sin^2 ax}{\cos ax}\,dx = -\dfrac{1}{a}\sin ax + \dfrac{1}{a}\log \tan\left(\dfrac{\pi}{4} + \dfrac{ax}{2}\right)$

328. $\int \dfrac{\cos ax}{\sin^2 ax}\,dx = -\dfrac{1}{a\sin ax} = -\dfrac{\csc ax}{a}$

329. $\int \dfrac{dx}{(\sin ax)(\cos ax)} = \dfrac{1}{a}\log \tan ax$

330. $\int \dfrac{dx}{(\sin ax)(\cos^2 ax)} = \dfrac{1}{a}\left(\sec ax + \log \tan \dfrac{ax}{2}\right)$

331. $\int \dfrac{dx}{(\sin ax)(\cos^n ax)} = \dfrac{1}{a(n-1)\cos^{n-1} ax} + \int \dfrac{dx}{(\sin ax)(\cos^{n-2} ax)}$

332. $\int \dfrac{dx}{(\sin^2 ax)(\cos ax)} = -\dfrac{1}{a}\csc ax + \dfrac{1}{a}\log \tan\left(\dfrac{\pi}{4} + \dfrac{ax}{2}\right)$

333. $\int \dfrac{dx}{(\sin^2 ax)(\cos^2 ax)} = -\dfrac{2}{a}\cot 2ax$

334. $\displaystyle\int \frac{dx}{\sin^m ax \cos^n ax} = \begin{cases} -\dfrac{1}{a(m-1)(\sin^{m-1} ax)(\cos^{n-1} ax)} \\ \qquad + \dfrac{m+n-2}{m-1}\displaystyle\int \dfrac{dx}{(\sin^{m-2} ax)(\cos^n ax)} \\ \text{or} \\ \dfrac{1}{a(n-1)\sin^{m-1} ax \cos^{n-1} ax} \\ \qquad - \dfrac{m+n-2}{n-1}\displaystyle\int \dfrac{dx}{\sin^m ax \cos^{n-2} ax} \end{cases}$

335. $\displaystyle\int \sin(a+bx)\,dx = -\frac{1}{b}\cos(a+bx)$

336. $\displaystyle\int \cos(a+bx)\,dx = \frac{1}{b}\sin(a+bx)$

337. $\displaystyle\int \frac{dx}{1 \pm \sin ax} = \mp \frac{1}{a}\tan\left(\frac{\pi}{4} \mp \frac{ax}{2}\right)$

338. $\displaystyle\int \frac{dx}{1+\cos ax} = \frac{1}{a}\tan\frac{ax}{2}$

339. $\displaystyle\int \frac{dx}{1-\cos ax} = -\frac{1}{a}\cot\frac{ax}{2}$

*340. $\displaystyle\int \frac{dx}{a+b\sin x} = \begin{cases} \dfrac{2}{\sqrt{a^2-b^2}}\tan^{-1}\dfrac{a\tan\frac{x}{2}+b}{\sqrt{a^2-b^2}} \\ \text{or} \\ \dfrac{1}{\sqrt{b^2-a^2}}\log\dfrac{a\tan\frac{x}{2}+b-\sqrt{b^2-a^2}}{a\tan\frac{x}{2}+b+\sqrt{b^2-a^2}} \end{cases}$

*341. $\displaystyle\int \frac{dx}{a+b\cos x} = \begin{cases} \dfrac{2}{\sqrt{a^2-b^2}}\tan^{-1}\dfrac{\sqrt{a^2-b^2}\tan\frac{x}{2}}{a+b} \\ \text{or} \\ \dfrac{1}{\sqrt{b^2-a^2}}\log\left(-\dfrac{\sqrt{b^2-a^2}\tan\frac{x}{2}+a+b}{\sqrt{b^2-a^2}\tan\frac{x}{2}-a-b}\right) \end{cases}$

*342. $\displaystyle\int \frac{dx}{a+b\sin x + c\cos x}$

$= \begin{cases} \dfrac{1}{\sqrt{b^2+c^2-a^2}}\log\dfrac{b-\sqrt{b^2+c^2-a^2}+(a-c)\tan\frac{x}{2}}{b+\sqrt{b^2+c^2-a^2}+(a-c)\tan\frac{x}{2}}, & \text{if } a^2 < b^2+c^2, a\neq c \\ \text{or} \\ \dfrac{2}{\sqrt{a^2-b^2-c^2}}\tan^{-1}\dfrac{b+(a-c)\tan\frac{x}{2}}{\sqrt{a^2-b^2-c^2}}, & \text{if } a^2 > b^2+c^2 \\ \text{or} \\ \dfrac{1}{a}\left[\dfrac{a-(b+c)\cos x-(b-c)\sin x}{a-(b-c)\cos x+(b+c)\sin x}\right], & \text{if } a^2 = b^2+c^2, a\neq c. \end{cases}$

*See note 6 on page A-19.

*343. $\displaystyle\int \frac{\sin^2 x\, dx}{a + b\cos^2 x} = \frac{1}{b}\sqrt{\frac{a+b}{a}}\tan^{-1}\left(\sqrt{\frac{a}{a+b}}\tan x\right) - \frac{x}{b},$ $\quad (ab > 0,\text{ or } |a| > |b|)$

*344. $\displaystyle\int \frac{dx}{a^2\cos^2 x + b^2\sin^2 x} = \frac{1}{ab}\tan^{-1}\left(\frac{b\tan x}{a}\right)$

*345. $\displaystyle\int \frac{\cos^2 cx}{a^2 + b^2\sin^2 cx}\,dx = \frac{\sqrt{a^2 + b^2}}{ab^2 c}\tan^{-1}\frac{\sqrt{a^2+b^2}\,\tan cx}{a} - \frac{x}{b^2}$

346. $\displaystyle\int \frac{\sin cx \cos cx}{a\cos^2 cx + b\sin^2 cx}\,dx = \frac{1}{2c(b-a)}\log(a\cos^2 cx + b\sin^2 cx)$

347. $\displaystyle\int \frac{\cos cx}{a\cos cx + b\sin cx}\,dx = \int \frac{dx}{a + b\tan cx} =$

$$\frac{1}{c(a^2+b^2)}[acx + b\log(a\cos cx + b\,\text{si}$$

348. $\displaystyle\int \frac{\sin cx}{a\sin cx + b\cos cx}\,dx = \int \frac{dx}{a + b\cot cx} =$

$$\frac{1}{c(a^2+b^2)}[acx - b\log(a\sin cx + b\cos x)]$$

*349. $\displaystyle\int \frac{dx}{a\cos^2 x + 2b\cos x\sin x + c\sin^2 x} =$
$$\begin{cases}\dfrac{1}{2\sqrt{b^2-ac}}\log\dfrac{c\tan x + b - \sqrt{b^2-ac}}{c\tan x + b + \sqrt{b^2-ac}} \\ \qquad\qquad\qquad\qquad (b^2 > ac) \\[4pt] \text{or} \\[4pt] \dfrac{1}{\sqrt{ac-b^2}}\tan^{-1}\dfrac{c\tan x + b}{\sqrt{ac-b^2}}, \quad (b_2 < ac) \\[4pt] \text{or} \\[4pt] -\dfrac{1}{c\tan x + b}, \qquad (b^2 = ac)\end{cases}$$

350. $\displaystyle\int \frac{\sin ax}{1 \pm \sin ax}\,dx = \pm x + \frac{1}{a}\tan\left(\frac{\pi}{4} \mp \frac{ax}{2}\right)$

351. $\displaystyle\int \frac{dx}{(\sin ax)(1 \pm \sin ax)} = \frac{1}{a}\tan\left(\frac{\pi}{4} \mp \frac{ax}{2}\right) + \frac{1}{a}\log\tan\frac{ax}{2}$

352. $\displaystyle\int \frac{dx}{(1 + \sin ax)^2} = -\frac{1}{2a}\tan\left(\frac{\pi}{4} - \frac{ax}{2}\right) - \frac{1}{6a}\tan^3\left(\frac{\pi}{4} - \frac{ax}{2}\right)$

353. $\displaystyle\int \frac{dx}{(1 - \sin ax)^2} = \frac{1}{2a}\cot\left(\frac{\pi}{4} - \frac{ax}{2}\right) + \frac{1}{6a}\cot^3\left(\frac{\pi}{4} - \frac{ax}{2}\right)$

354. $\displaystyle\int \frac{\sin ax}{(1 + \sin ax)^2}\,dx = -\frac{1}{2a}\tan\left(\frac{\pi}{4} - \frac{ax}{2}\right) + \frac{1}{6a}\tan^3\left(\frac{\pi}{4} - \frac{ax}{2}\right)$

355. $\displaystyle\int \frac{\sin ax}{(1 - \sin ax)^2}\,dx = -\frac{1}{2a}\cot\left(\frac{\pi}{4} - \frac{ax}{2}\right) + \frac{1}{6a}\cot^3\left(\frac{\pi}{4} - \frac{ax}{2}\right)$

356. $\displaystyle\int \frac{\sin x\, dx}{a + b\sin x} = \frac{x}{b} - \frac{a}{b}\int \frac{dx}{a + b\sin x}$

357. $\displaystyle\int \frac{dx}{(\sin x)(a + b\sin x)} = \frac{1}{a}\log\tan\frac{x}{2} - \frac{b}{a}\int \frac{dx}{a + b\sin x}$

358. $\displaystyle\int \frac{dx}{(a + b\sin x)^2} = \frac{b\cos x}{(a^2 - b^2)(a + b\sin x)} + \frac{a}{a^2 - b^2}\int \frac{dx}{a + b\sin x}$

*See note 6 on page A-19.

359. $\int \dfrac{\sin x \, dx}{(a + b \sin x)^2} = \dfrac{a \cos x}{(b^2 - a^2)(a + b \sin x)} + \dfrac{b}{b^2 - a^2} \int \dfrac{dx}{a + b \sin x}$

*360. $\int \dfrac{dx}{a^2 + b^2 \sin^2 cx} = \dfrac{1}{ac\sqrt{a^2 + b^2}} \tan^{-1} \dfrac{\sqrt{a^2 + b^2} \tan cx}{a}$

*361. $\int \dfrac{dx}{a^2 - b^2 \sin^2 cx} = \begin{cases} \dfrac{1}{ac\sqrt{a^2 - b^2}} \tan^{-1} \dfrac{\sqrt{a^2 - b^2} \tan cx}{a}, & (a^2 > b^2) \\ \text{or} \\ \dfrac{1}{2ac\sqrt{b^2 - a^2}} \log \dfrac{\sqrt{b^2 - a^2} \tan cx + a}{\sqrt{b^2 - a^2} \tan cx - a}, & (a^2 < b^2) \end{cases}$

362. $\int \dfrac{\cos ax}{1 + \cos ax} dx = x - \dfrac{1}{a} \tan \dfrac{ax}{2}$

363. $\int \dfrac{\cos ax}{1 - \cos ax} dx = -x - \dfrac{1}{a} \cot \dfrac{ax}{2}$

364. $\int \dfrac{dx}{(\cos ax)(1 + \cos ax)} = \dfrac{1}{a} \log \tan \left(\dfrac{\pi}{4} + \dfrac{ax}{2} \right) - \dfrac{1}{a} \tan \dfrac{ax}{2}$

365. $\int \dfrac{dx}{(\cos ax)(1 - \cos ax)} = \dfrac{1}{a} \log \tan \left(\dfrac{\pi}{4} + \dfrac{ax}{2} \right) - \dfrac{1}{a} \cot \dfrac{ax}{2}$

366. $\int \dfrac{dx}{(1 + \cos ax)^2} = \dfrac{1}{2a} \tan \dfrac{ax}{2} + \dfrac{1}{6a} \tan^3 \dfrac{ax}{2}$

367. $\int \dfrac{dx}{(1 - \cos ax)^2} = -\dfrac{1}{2a} \cot \dfrac{ax}{2} - \dfrac{1}{6a} \cot^3 \dfrac{ax}{2}$

368. $\int \dfrac{\cos ax}{(1 + \cos ax)^2} dx = \dfrac{1}{2a} \tan \dfrac{ax}{2} - \dfrac{1}{6a} \tan^3 \dfrac{ax}{2}$

369. $\int \dfrac{\cos ax}{(1 - \cos ax)^2} dx = \dfrac{1}{2a} \cot \dfrac{ax}{2} - \dfrac{1}{6a} \cot^3 \dfrac{ax}{2}$

370. $\int \dfrac{\cos x \, dx}{a + b \cos x} = \dfrac{x}{b} - \dfrac{a}{b} \int \dfrac{dx}{a + b \cos x}$

371. $\int \dfrac{dx}{(\cos x)(a + b \cos x)} = \dfrac{1}{a} \log \tan \left(\dfrac{x}{2} + \dfrac{\pi}{4} \right) - \dfrac{b}{a} \int \dfrac{dx}{a + b \cos x}$

372. $\int \dfrac{dx}{(a + b \cos x)^2} = \dfrac{b \sin x}{(b^2 - a^2)(a + b \cos x)} - \dfrac{a}{b^2 - a^2} \int \dfrac{dx}{a + b \cos x}$

373. $\int \dfrac{\cos x}{(a + b \cos x)^2} dx = \dfrac{a \sin x}{(a^2 - b^2)(a + b \cos x)} - \dfrac{b}{a^2 - b^2} \int \dfrac{dx}{a + b \cos x}$

*374. $\int \dfrac{dx}{a^2 + b^2 - 2ab \cos cx} = \dfrac{2}{c(a^2 - b^2)} \tan^{-1} \left(\dfrac{a + b}{a - b} \tan \dfrac{cx}{2} \right)$

*375. $\int \dfrac{dx}{a^2 + b^2 \cos^2 cx} = \dfrac{1}{ac\sqrt{a^2 + b^2}} \tan^{-1} \dfrac{a \tan cx}{\sqrt{a^2 + b^2}}$

*376. $\int \dfrac{dx}{a^2 - b^2 \cos^2 cx} = \begin{cases} \dfrac{1}{ac\sqrt{a^2 - b^2}} \tan^{-1} \dfrac{a \tan cx}{\sqrt{a^2 - b^2}}, & (a^2 > h^2) \\ \text{or} \\ \dfrac{1}{2ac\sqrt{b^2 - a^2}} \log \dfrac{a \tan cx - \sqrt{b^2 - a^2}}{a \tan cx + \sqrt{b^2 - a^2}}, & (b^2 > a^2) \end{cases}$

377. $\int \dfrac{\sin ax}{1 \pm \cos ax} dx = \mp \dfrac{1}{a} \log (1 \pm \cos ax)$

*See note 6 on page A-19.

378. $\displaystyle \int \frac{\cos ax}{1 \pm \sin ax}\, dx = \pm \frac{1}{a}\log\,(1 \pm \sin ax)$

379. $\displaystyle \int \frac{dx}{(\sin ax)(1 \pm \cos ax)} = \pm \frac{1}{2a(1 \pm \cos ax)} + \frac{1}{2a}\log\tan\frac{ax}{2}$

380. $\displaystyle \int \frac{dx}{(\cos ax)(1 \pm \sin ax)} = \mp \frac{1}{2a(1 \pm \sin ax)} + \frac{1}{2a}\log\tan\left(\frac{\pi}{4} + \frac{ax}{2}\right)$

381. $\displaystyle \int \frac{\sin ax}{(\cos ax)(1 \pm \cos ax)}\, dx = \frac{1}{a}\log\,(\sec ax \pm 1)$

382. $\displaystyle \int \frac{\cos ax}{(\sin ax)(1 \pm \sin ax)}\, dx = -\frac{1}{a}\log\,(\csc ax \pm 1)$

383. $\displaystyle \int \frac{\sin ax}{(\cos ax)(1 \pm \sin ax)}\, dx = \frac{1}{2a(1 \pm \sin ax)} \pm \frac{1}{2a}\log\tan\left(\frac{\pi}{4} + \frac{ax}{2}\right)$

384. $\displaystyle \int \frac{\cos ax}{(\sin ax)(1 \pm \cos ax)}\, dx = -\frac{1}{2a(1 \pm \cos ax)} \pm \frac{1}{2a}\log\tan\frac{ax}{2}$

385. $\displaystyle \int \frac{dx}{\sin ax \pm \cos ax} = \frac{1}{a\sqrt{2}}\log\tan\left(\frac{ax}{2} \pm \frac{\pi}{8}\right)$

386. $\displaystyle \int \frac{dx}{(\sin ax \pm \cos ax)^2} = \frac{1}{2a}\tan\left(ax \mp \frac{\pi}{4}\right)$

387. $\displaystyle \int \frac{dx}{1 + \cos ax \pm \sin ax} = \pm \frac{1}{a}\log\left(1 \pm \tan\frac{ax}{2}\right)$

388. $\displaystyle \int \frac{dx}{a^2 \cos^2 cx - b^2 \sin^2 cx} = \frac{1}{2abc}\log\frac{b\tan cx + a}{b\tan cx - a}$

389. $\displaystyle \int x(\sin ax)\, dx = \frac{1}{a^2}\sin ax - \frac{x}{a}\cos ax$

390. $\displaystyle \int x^2(\sin ax)\, dx = \frac{2x}{a^2}\sin ax - \frac{a^2 x^2 - 2}{a^3}\cos ax$

391. $\displaystyle \int x^3(\sin ax)\, dx = \frac{3a^2 x^2 - 6}{a^4}\sin ax - \frac{a^2 x^3 - 6x}{a^3}\cos ax$

392. $\displaystyle \int x^m \sin ax\, dx = \begin{cases} -\dfrac{1}{a}x^m \cos ax + \dfrac{m}{a}\displaystyle\int x^{m-1}\cos ax\, dx \\[4pt] \text{or} \\[4pt] \cos ax \displaystyle\sum_{r=0}^{\left[\frac{m}{2}\right]} (-1)^{r+1}\frac{m!}{(m-2r)!}\cdot\frac{x^{m-2r}}{a^{2r+1}} \\[4pt] + \sin ax \displaystyle\sum_{r=0}^{\left[\frac{m-1}{2}\right]} (-1)^{r}\frac{m!}{(m-2r-1)!}\cdot\frac{x^{m-2r-1}}{a^{2r+2}} \end{cases}$

Note: $[s]$ means greatest integer $\leq s$; $[3\tfrac{1}{2}] = 3$, $[\tfrac{1}{2}] = 0$, etc.

393. $\displaystyle \int x(\cos ax)\, dx = \frac{1}{a^2}\cos ax + \frac{x}{a}\sin ax$

394. $\displaystyle \int x^2(\cos ax)\, dx = \frac{2x\cos ax}{a^2} + \frac{a^2 x^2 - 2}{a^3}\sin ax$

395. $\displaystyle \int x^3(\cos ax)\, dx = \frac{3a^2 x^2 - 6}{a^4}\cos ax + \frac{a^2 x^3 - 6x}{a^3}\sin ax$

396. $\displaystyle\int x^m(\cos ax)\,dx = \begin{cases} \dfrac{x^m \sin ax}{a} - \dfrac{m}{a}\displaystyle\int x^{m-1}\sin ax\,dx \\[4pt] \text{or} \\[4pt] \sin ax \displaystyle\sum_{r=0}^{\left[\frac{m}{2}\right]} (-1)^r \dfrac{m!}{(m-2r)!}\cdot\dfrac{x^{m-2r}}{a^{2r+1}} \\[4pt] \quad + \cos ax \displaystyle\sum_{r=0}^{\left[\frac{m-1}{2}\right]} (-1)^r \dfrac{m!}{(m-2r-1)!}\cdot\dfrac{x^{m-2r-1}}{a^{2r+2}} \end{cases}$

See note integral 392.

397. $\displaystyle\int \frac{\sin ax}{x}\,dx = \sum_{n=0}^{r} (-1)^n \frac{(ax)^{2n+1}}{(2n+1)(2n+1)!}$

398. $\displaystyle\int \frac{\cos ax}{x}\,dx = \log x + \sum_{n=1}^{r} (-1)^n \frac{(ax)^{2n}}{2n(2n)!}$

399. $\displaystyle\int x(\sin^2 ax)\,dx = \frac{x^2}{4} - \frac{x\sin 2ax}{4a} - \frac{\cos 2ax}{8a^2}$

400. $\displaystyle\int x^2(\sin^2 ax)\,dx = \frac{x^3}{6} - \left(\frac{x^2}{4a} - \frac{1}{8a^3}\right)\sin 2ax - \frac{x\cos 2ax}{4a^2}$

401. $\displaystyle\int x(\sin^3 ax)\,dx = \frac{x\cos 3ax}{12a} - \frac{\sin 3ax}{36a^2} - \frac{3x\cos ax}{4a} + \frac{3\sin ax}{4a^2}$

402. $\displaystyle\int x(\cos^2 ax)\,dx = \frac{x^2}{4} + \frac{x\sin 2ax}{4a} + \frac{\cos 2ax}{8a^2}$

403. $\displaystyle\int x^2(\cos^2 ax)\,dx = \frac{x^3}{6} + \left(\frac{x^2}{4a} - \frac{1}{8a^3}\right)\sin 2ax + \frac{x\cos 2ax}{4a^2}$

404. $\displaystyle\int x(\cos^3 ax)\,dx = \frac{x\sin 3ax}{12a} + \frac{\cos 3ax}{36a^2} + \frac{3x\sin ax}{4a} + \frac{3\cos ax}{4a^2}$

405. $\displaystyle\int \frac{\sin ax}{x^m}\,dx = -\frac{\sin ax}{(m-1)x^{m-1}} + \frac{a}{m-1}\int \frac{\cos ax}{x^{m-1}}\,dx$

406. $\displaystyle\int \frac{\cos ax}{x^m}\,dx = -\frac{\cos ax}{(m-1)x^{m-1}} - \frac{a}{m-1}\int \frac{\sin ax}{x^{m-1}}\,dx$

407. $\displaystyle\int \frac{x}{1\pm\sin ax}\,dx = \mp\frac{x\cos ax}{a(1\pm\sin ax)} + \frac{1}{a^2}\log(1\pm\sin ax)$

408. $\displaystyle\int \frac{x}{1+\cos ax}\,dx = \frac{x}{a}\tan\frac{ax}{2} + \frac{2}{a^2}\log\cos\frac{ax}{2}$

409. $\displaystyle\int \frac{x}{1-\cos ax}\,dx = -\frac{x}{a}\cot\frac{ax}{2} + \frac{2}{a^2}\log\sin\frac{ax}{2}$

410. $\displaystyle\int \frac{x+\sin x}{1+\cos x}\,dx = x\tan\frac{x}{2}$

411. $\displaystyle\int \frac{x-\sin x}{1-\cos x}\,dx = -x\cot\frac{x}{2}$

412. $\displaystyle\int \sqrt{1-\cos ax}\,dx = -\frac{2\sin ax}{a\sqrt{1-\cos ax}} = -\frac{2\sqrt{2}}{a}\cos\left(\frac{ax}{2}\right)$

413. $\displaystyle\int \sqrt{1+\cos ax}\,dx = \frac{2\sin ax}{a\sqrt{1+\cos ax}} = \frac{2\sqrt{2}}{a}\sin\left(\frac{ax}{2}\right)$

414. $\displaystyle\int \sqrt{1+\sin x}\,dx = \pm 2\left(\sin\frac{x}{2} - \cos\frac{x}{2}\right),$

$\left[\text{use } + \text{ if } (8k-1)\dfrac{\pi}{2} < x \le (8k+3)\dfrac{\pi}{2}, \text{ otherwise } -; k \text{ an integer}\right]$

415. $\int \sqrt{1 - \sin x}\, dx = \pm 2\left(\sin\dfrac{x}{2} + \cos\dfrac{x}{2}\right),$

[use $+$ if $(8k - 3)\dfrac{\pi}{2} < x \le (8k + 1)\dfrac{\pi}{2}$, otherwise $-$; k an integer]

416. $\int \dfrac{dx}{\sqrt{1 - \cos x}} = \pm \sqrt{2}\log\tan\dfrac{x}{4},$

[use $+$ if $4k\pi < x < (4k + 2)\pi$, otherwise $-$; k an integer]

417. $\int \dfrac{dx}{\sqrt{1 + \cos x}} = \pm \sqrt{2}\log\tan\left(\dfrac{x + \pi}{4}\right),$

[use $+$ if $(4k - 1)\pi < x < (4k + 1)\pi$, otherwise $-$; k an integer]

418. $\int \dfrac{dx}{\sqrt{1 - \sin x}} = \pm \sqrt{2}\log\tan\left(\dfrac{x}{4} - \dfrac{\pi}{8}\right),$

[use $+$ if $(8k + 1)\dfrac{\pi}{2} < x < (8k + 5)\dfrac{\pi}{2}$, otherwise $-$; k an integer]

419. $\int \dfrac{dx}{\sqrt{1 + \sin x}} = \pm \sqrt{2}\log\tan\left(\dfrac{x}{4} + \dfrac{\pi}{8}\right),$

[use $+$ if $(8k - 1)\dfrac{\pi}{2} < x < (8k + 3)\dfrac{\pi}{2}$, otherwise $-$; k an integer]

420. $\int (\tan^2 ax)\, dx = \dfrac{1}{a}\tan ax - x$

421. $\int (\tan^3 ax)\, dx = \dfrac{1}{2a}\tan^2 ax + \dfrac{1}{a}\log\cos ax$

422. $\int (\tan^4 ax)\, dx = \dfrac{\tan^3 ax}{3a} - \dfrac{1}{a}\tan x + x$

423. $\int (\tan^n ax)\, dx = \dfrac{\tan^{n-1} ax}{a(n - 1)} - \int (\tan^{n-2} ax)\, dx$

424. $\int (\cot^2 ax)\, dx = -\dfrac{1}{a}\cot ax - x$

425. $\int (\cot^3 ax)\, dx = -\dfrac{1}{2a}\cot^2 ax - \dfrac{1}{a}\log\sin ax$

426. $\int (\cot^4 ax)\, dx = -\dfrac{1}{3a}\cot^3 ax + \dfrac{1}{a}\cot ax + x$

427. $\int (\cot^n ax)\, dx = -\dfrac{\cot^{n-1} ax}{a(n - 1)} - \int (\cot^{n-2} ax)\, dx$

428. $\int \dfrac{x}{\sin^2 ax}\, dx = \int x(\csc^2 ax)\, dx = -\dfrac{x\cot ax}{a} + \dfrac{1}{a^2}\log\sin ax$

429. $\int \dfrac{x}{\sin^n ax}\, dx = \int x(\csc^n ax)\, dx = -\dfrac{x\cos ax}{a(n - 1)\sin^{n-1} ax}$

$$-\dfrac{1}{a^2(n - 1)(n - 2)\sin^{n-2} ax} + \dfrac{(n - 2)}{(n - 1)}\int \dfrac{x}{\sin^{n-2} ax}\, dx$$

430. $\int \dfrac{x}{\cos^2 ax}\, dx = \int x(\sec^2 ax)\, dx = \dfrac{1}{a}x\tan ax + \dfrac{1}{a^2}\log\cos ax$

431. $\int \dfrac{x}{\cos^n ax}\, dx = \int x(\sec^n ax)\, dx = \dfrac{x\sin ax}{a(n - 1)\cos^{n-1} ax}$

$$-\dfrac{1}{a^2(n - 1)(n - 2)\cos^{n-2} ax} + \dfrac{n - 2}{n - 1}\int \dfrac{x}{\cos^{n-2} ax}\, dx$$

432. $\int \dfrac{\sin ax}{\sqrt{1 + b^2 \sin^2 ax}} dx = -\dfrac{1}{ab} \sin^{-1} \dfrac{b \cos ax}{\sqrt{1 + b^2}}$

433. $\int \dfrac{\sin ax}{\sqrt{1 - b^2 \sin^2 ax}} dx = -\dfrac{1}{ab} \log (b \cos ax + \sqrt{1 - b^2 \sin^2 ax})$

434. $\int (\sin ax)\sqrt{1 + b^2 \sin^2 ax}\, dx = -\dfrac{\cos ax}{2a} \sqrt{1 + b^2 \sin^2 ax}$
$$-\dfrac{1 + b^2}{2ab} \sin^{-1} \dfrac{b \cos ax}{\sqrt{1 + b^2}}$$

435. $\int (\sin ax)\sqrt{1 - b^2 \sin^2 ax}\, dx = -\dfrac{\cos ax}{2a} \sqrt{1 - b^2 \sin^2 ax}$
$$-\dfrac{1 - b^2}{2ab} \log (b \cos ax + \sqrt{1 - b^2 \sin^2 ax})$$

436. $\int \dfrac{\cos ax}{\sqrt{1 + b^2 \sin^2 ax}} dx = \dfrac{1}{ab} \log (b \sin ax + \sqrt{1 + b^2 \sin^2 ax})$

437. $\int \dfrac{\cos ax}{\sqrt{1 - b^2 \sin^2 ax}} dx = \dfrac{1}{ab} \sin^{-1} (b \sin ax)$

438. $\int (\cos ax) \sqrt{1 + b^2 \sin^2 ax}\, dx = \dfrac{\sin ax}{2a} \sqrt{1 + b^2 \sin^2 ax}$
$$+ \dfrac{1}{2ab} \log (b \sin ax + \sqrt{1 + b^2 \sin^2 ax})$$

439. $\int (\cos ax) \sqrt{1 - b^2 \sin^2 ax}\, dx = \dfrac{\sin ax}{2a} \sqrt{1 - b^2 \sin^2 ax} + \dfrac{1}{2ab} \sin^{-1} (b \sin ax)$

440. $\int \dfrac{dx}{\sqrt{a + b \tan^2 cx}} = \dfrac{\pm 1}{c\sqrt{a - b}} \sin^{-1} \left(\sqrt{\dfrac{a - b}{a}} \sin cx \right), \qquad (a > |b|)$

$\left[\text{use } + \text{ if } (2k - 1)\dfrac{\pi}{2} < x \le (2k + 1)\dfrac{\pi}{2}, \text{ otherwise } - : k \text{ an integer}\right]$

FORMS INVOLVING INVERSE TRIGONOMETRIC FUNCTIONS

441. $\int (\sin^{-1} ax)\, dx = x \sin^{-1} ax + \dfrac{\sqrt{1 - a^2 x^2}}{a}$

442. $\int (\cos^{-1} ax)\, dx = x \cos^{-1} ax - \dfrac{\sqrt{1 - a^2 x^2}}{a}$

443. $\int (\tan^{-1} ax)\, dx = x \tan^{-1} ax - \dfrac{1}{2a} \log (1 + a^2 x^2)$

444. $\int (\cot^{-1} ax)\, dx = x \cot^{-1} ax + \dfrac{1}{2a} \log (1 + a^2 x^2)$

445. $\int (\sec^{-1} ax)\, dx = x \sec^{-1} ax - \dfrac{1}{a} \log (ax + \sqrt{a^2 x^2 - 1})$

446. $\int (\csc^{-1} ax)\, dx = x \csc^{-1} ax + \dfrac{1}{a} \log (ax + \sqrt{a^2 x^2 - 1})$

447. $\int \left(\sin^{-1} \dfrac{x}{a} \right) dx = x \sin^{-1} \dfrac{x}{a} + \sqrt{a^2 - x^2}, \qquad (a > 0)$

448. $\int \left(\cos^{-1} \dfrac{x}{a} \right) dx = x \cos^{-1} \dfrac{x}{a} - \sqrt{a^2 - x^2}, \qquad (a > 0)$

449. $\displaystyle\int \left(\tan^{-1}\frac{x}{a}\right) dx = x\tan^{-1}\frac{x}{a} - \frac{a}{2}\log(a^2 + x^2)$

450. $\displaystyle\int \left(\cot^{-1}\frac{x}{a}\right) dx = x\cot^{-1}\frac{x}{a} + \frac{a}{2}\log(a^2 + x^2)$

451. $\displaystyle\int x[\sin^{-1}(ax)]\,dx = \frac{1}{4a^2}[(2a^2x^2 - 1)\sin^{-1}(ax) + ax\sqrt{1 - a^2x^2}]$

452. $\displaystyle\int x[\cos^{-1}(ax)]\,dx = \frac{1}{4a^2}[(2a^2x^2 - 1)\cos^{-1}(ax) - ax\sqrt{1 - a^2x^2}]$

453. $\displaystyle\int x^n[\sin^{-1}(ax)]\,dx = \frac{x^{n+1}}{n+1}\sin^{-1}(ax) - \frac{a}{n+1}\int \frac{x^{n+1}\,dx}{\sqrt{1 - a^2x^2}}, \qquad (n \neq -1)$

454. $\displaystyle\int x^n[\cos^{-1}(ax)]\,dx = \frac{x^{n+1}}{n+1}\cos^{-1}(ax) + \frac{a}{n+1}\int \frac{x^{n+1}\,dx}{\sqrt{1 - a^2x^2}}, \qquad (n \neq -1)$

455. $\displaystyle\int x(\tan^{-1}ax)\,dx = \frac{1 + a^2x^2}{2a^2}\tan^{-1}ax - \frac{x}{2a}$

456. $\displaystyle\int x^n(\tan^{-1}ax)\,dx = \frac{x^{n+1}}{n+1}\tan^{-1}ax - \frac{a}{n+1}\int \frac{x^{n+1}}{1 + a^2x^2}\,dx$

457. $\displaystyle\int x(\cot^{-1}ax)\,dx = \frac{1 + a^2x^2}{2a^2}\cot^{-1}ax + \frac{x}{2a}$

458. $\displaystyle\int x^n(\cot^{-1}ax)\,dx = \frac{x^{n+1}}{n+1}\cot^{-1}ax + \frac{a}{n+1}\int \frac{x^{n+1}}{1 + a^2x^2}\,dx$

459. $\displaystyle\int \frac{\sin^{-1}(ax)}{x^2}\,dx = a\log\left(\frac{1 - \sqrt{1 - a^2x^2}}{x}\right) - \frac{\sin^{-1}(ax)}{x}$

460. $\displaystyle\int \frac{\cos^{-1}(ax)\,dx}{x^2} = -\frac{1}{x}\cos^{-1}(ax) + a\log\frac{1 + \sqrt{1 - a^2x^2}}{x}$

461. $\displaystyle\int \frac{\tan^{-1}(ax)\,dx}{x^2} = -\frac{1}{x}\tan^{-1}(ax) - \frac{a}{2}\log\frac{1 + a^2x^2}{x^2}$

462. $\displaystyle\int \frac{\cot^{-1}ax}{x^2}\,dx = -\frac{1}{x}\cot^{-1}ax - \frac{a}{2}\log\frac{x^2}{a^2x^2 + 1}$

463. $\displaystyle\int (\sin^{-1}ax)^2\,dx = x(\sin^{-1}ax)^2 - 2x + \frac{2\sqrt{1 - a^2x^2}}{a}\sin^{-1}ax$

464. $\displaystyle\int (\cos^{-1}ax)^2\,dx = x(\cos^{-1}ax)^2 - 2x - \frac{2\sqrt{1 - a^2x^2}}{a}\cos^{-1}ax$

465. $\displaystyle\int (\sin^{-1}ax)^n\,dx = \begin{cases} x(\sin^{-1}ax)^n + \dfrac{n\sqrt{1 - a^2x^2}}{a}(\sin^{-1}ax)^{n-1} \\ \qquad\qquad\qquad - n(n-1)\displaystyle\int (\sin^{-1}ax)^{n-2}\,dx \\[4pt] \qquad\qquad\text{or} \\ \displaystyle\sum_{r=0}^{\left[\frac{n}{2}\right]} (-1)^r\frac{n!}{(n-2r)!}x(\sin^{-1}ax)^{n-2r} \\ \qquad + \displaystyle\sum_{r=0}^{\left[\frac{n-1}{2}\right]} (-1)^r\frac{n!\sqrt{1 - a^2x^2}}{(n-2r-1)!a}(\sin^{-1}ax)^{n-2r-1} \end{cases}$

Note: $[s]$ means greatest integer $\leq s$. Thus $[3.5]$ means $3 : [5] = 5, [\frac{1}{2}] = 0.$

466. $\displaystyle\int (\cos^{-1} ax)^n\, dx = \begin{cases} x(\cos^{-1} ax)^n - \dfrac{n\sqrt{1 - a^2x^2}}{a}(\cos^{-1} ax)^{n-1} \\[2mm] \qquad\qquad\qquad -n(n-1)\displaystyle\int (\cos^{-1} ax)^{n-2}\, dx \\[2mm] \qquad\quad \text{or} \\[2mm] \displaystyle\sum_{r=0}^{\left[\frac{n}{2}\right]} (-1)^r \dfrac{n!}{(n - 2r)!} x(\cos^{-1} ax)^{n-2r} \\[2mm] \qquad\qquad \displaystyle\sum_{r=0}^{\left[\frac{n-1}{2}\right]} (-1)^r \dfrac{n!\sqrt{1 - a^2x^2}}{(n - 2r - 1)!a}(\cos^{-1} ax)^{n-2r-1} \end{cases}$

467. $\displaystyle\int \frac{1}{\sqrt{1 - a^2x^2}}(\sin^{-1} ax)\, dx = \frac{1}{2a}(\sin^{-1} ax)^2$

468. $\displaystyle\int \frac{x^n}{\sqrt{1 - a^2x^2}}(\sin^{-1} ax)\, dx = -\frac{x^{n-1}}{na^2}\sqrt{1 - a^2x^2}\,\sin^{-1} ax + \frac{x^n}{n^2a}$

$$+ \frac{n-1}{na^2}\int \frac{x^{n-2}}{\sqrt{1 - a^2x^2}}\sin^{-1} ax\, dx$$

469. $\displaystyle\int \frac{1}{\sqrt{1 - a^2x^2}}(\cos^{-1} ax)\, dx = -\frac{1}{2a}(\cos^{-1} ax)^2$

470. $\displaystyle\int \frac{x^n}{\sqrt{1 - a^2x^2}}(\cos^{-1} ax)\, dx = -\frac{x^{n-1}}{na^2}\sqrt{1 - a^2x^2}\,\cos^{-1} ax - \frac{x^n}{n^2a}$

$$+ \frac{n-1}{na^2}\int \frac{x^{n-2}}{\sqrt{1 - a^2x^2}}\cos^{-1} ax\, dx$$

471. $\displaystyle\int \frac{\tan^{-1} ax}{a^2x^2 + 1}\, dx = \frac{1}{2a}(\tan^{-1} ax)^2$

472. $\displaystyle\int \frac{\cot^{-1} ax}{a^2x^2 + 1}\, dx = -\frac{1}{2a}(\cot^{-1} ax)^2$

473. $\displaystyle\int x\sec^{-1} ax\, dx = \frac{x^2}{2}\sec^{-1} ax - \frac{1}{2a^2}\sqrt{a^2x^2 - 1}$

474. $\displaystyle\int x^n\sec^{-1} ax\, dx = \frac{x^{n+1}}{n+1}\sec^{-1} ax - \frac{1}{n+1}\int \frac{x^n\, dx}{\sqrt{a^2x^2 - 1}}$

475. $\displaystyle\int \frac{\sec^{-1} ax}{x^2}\, dx = -\frac{\sec^{-1} ax}{x} + \frac{\sqrt{a^2x^2 - 1}}{x}$

476. $\displaystyle\int x\csc^{-1} ax\, dx = \frac{x^2}{2}\csc^{-1} ax + \frac{1}{2a^2}\sqrt{a^2x^2 - 1}$

477. $\displaystyle\int x^n\csc^{-1} ax\, dx = \frac{x^{n+1}}{n+1}\csc^{-1} ax + \frac{1}{n+1}\int \frac{x^n\, dx}{\sqrt{a^2x^2 - 1}}$

478. $\displaystyle\int \frac{\csc^{-1} ax}{x^2}\, dx = -\frac{\csc^{-1} ax}{x} - \frac{\sqrt{a^2x^2 - 1}}{x}$

FORMS INVOLVING TRIGONOMETRIC SUBSTITUTIONS

479. $\displaystyle\int f(\sin x)\, dx = 2\int f\left(\frac{2z}{1 + z^2}\right)\frac{dz}{1 + z^2}, \qquad \left(z = \tan\frac{x}{2}\right)$

480. $\displaystyle\int f(\cos x)\, dx = 2\int f\left(\frac{1 - z^2}{1 + z^2}\right)\frac{dz}{1 + z^2}, \qquad \left(z = \tan\frac{x}{2}\right)$

*481. $\int f(\sin x)\, dx = \int f(u)\, \dfrac{du}{\sqrt{1 - u^2}}, \qquad (u = \sin x)$

*482. $\int f(\cos x)\, dx = - \int f(u)\, \dfrac{du}{\sqrt{1 - u^2}}, \qquad (u = \cos x)$

*483. $\int f(\sin x, \cos x)\, dx = \int f(u, \sqrt{1 - u^2})\, \dfrac{du}{\sqrt{1 - u^2}}, \qquad (u = \sin x)$

484. $\int f(\sin x, \cos x)\, dx = 2 \int f\left(\dfrac{2z}{1 + z^2}, \dfrac{1 - z^2}{1 + z^2}\right) \dfrac{dz}{1 + z^2}, \qquad \left(z = \tan \dfrac{x}{2}\right)$

LOGARITHMIC FORMS

485. $\int (\log x)\, dx = x \log x - x$

486. $\int x(\log x)\, dx = \dfrac{x^2}{2} \log x - \dfrac{x^2}{4}$

487. $\int x^2(\log x)\, dx = \dfrac{x^3}{3} \log x - \dfrac{x^3}{9}$

488. $\int x^n(\log ax)\, dx = \dfrac{x^{n+1}}{n + 1} \log ax - \dfrac{x^{n+1}}{(n + 1)^2}$

489. $\int (\log x)^2\, dx = x(\log x)^2 - 2x \log x + 2x$

490. $\int (\log x)^n\, dx = \begin{cases} x(\log x)^n - n \int (\log x)^{n-1}\, dx, & (n \neq -1) \\ \text{or} \\ (-1)^n n!\, x \sum\limits_{r=0}^{n} \dfrac{(-\log x)^r}{r!} \end{cases}$

491. $\int \dfrac{(\log x)^n}{x}\, dx = \dfrac{1}{n + 1}(\log x)^{n+1}$

492. $\int \dfrac{dx}{\log x} = \log(\log x) + \log x + \dfrac{(\log x)^2}{2 \cdot 2!} + \dfrac{(\log x)^3}{3 \cdot 3!} + \cdots$

493. $\int \dfrac{dx}{x \log x} = \log(\log x)$

494. $\int \dfrac{dx}{x(\log x)^n} = - \dfrac{1}{(n - 1)(\log x)^{n-1}}$

495. $\int \dfrac{x^m\, dx}{(\log x)^n} = - \dfrac{x^{m+1}}{(n - 1)(\log x)^{n-1}} + \dfrac{m + 1}{n - 1} \int \dfrac{x^m\, dx}{(\log x)^{n-1}}$

496. $\int x^m(\log x)^n\, dx = \begin{cases} \dfrac{x^{m+1}(\log x)^n}{m + 1} - \dfrac{n}{m + 1} \int x^m(\log x)^{n-1}\, dx \\ \text{or} \\ (-1)^n \dfrac{n!}{m + 1} x^{m+1} \sum\limits_{r=0}^{n} \dfrac{(-\log x)^r}{r!(m + 1)^{n-r}} \end{cases}$

497. $\int x^p \cos(b \ln x)\, dx = \dfrac{x^{p+1}}{(p + 1)^2 + b^2} \cdot [b \sin(b \ln x) + (p + 1) \cos(b \ln x)] + c$

498. $\int x^p \sin(b \ln x)\, dx = \dfrac{x^{p+1}}{(p + 1)^2 + b^2} \cdot [(p + 1) \sin(b \ln x) - b \cos(b \ln x)] + c$

499. $\int [\log(ax + b)]\, dx = \dfrac{ax + b}{a} \log(ax + b) - x$

---

*The square roots appearing in these formulas may be plus or minus, depending on the quadrant of x. Care must be used to give them the proper sign.

500. $\displaystyle\int \frac{\log(ax + b)}{x^2}\, dx = \frac{a}{b}\log x - \frac{ax + b}{bx}\log(ax + b)$

501. $\displaystyle\int x^m[\log(ax + b)]\, dx = \frac{1}{m + 1}\left[x^{m+1} - \left(-\frac{b}{a}\right)^{m+1}\right]\log(ax + b)$

$$-\frac{1}{m + 1}\left(-\frac{b}{a}\right)^{m+1}\sum_{r=1}^{m+1}\frac{1}{r}\left(-\frac{ax}{b}\right)^r$$

502. $\displaystyle\int \frac{\log(ax + b)}{x^m}\, dx = -\frac{1}{m - 1}\frac{\log(ax + b)}{x^{m-1}} + \frac{1}{m - 1}\left(-\frac{a}{b}\right)^{m-1}\log\frac{ax + b}{x}$

$$+\frac{1}{m - 1}\left(-\frac{a}{b}\right)^{m-1}\sum_{r=1}^{m-2}\frac{1}{r}\left(-\frac{b}{ax}\right)^r, (m > 2)$$

503. $\displaystyle\int \left[\log\frac{x + a}{x - a}\right]dx = (x + a)\log(x + a) - (x - a)\log(x - a)$

504. $\displaystyle\int x^m\left[\log\frac{x + a}{x - a}\right]dx = \frac{x^{m+1} - (-a)^{m+1}}{m + 1}\log(x + a) - \frac{x^{m+1} - a^{m+1}}{m + 1}\log(x - a)$

$$+\frac{2a^{m+1}}{m + 1}\sum_{r=1}^{\left[\frac{m+1}{2}\right]}\frac{1}{m - 2r + 2}\left(\frac{x}{a}\right)^{m-2r+2}$$

See note integral 392.

505. $\displaystyle\int \frac{1}{x^2}\left[\log\frac{x + a}{x - a}\right]dx = \frac{1}{x}\log\frac{x - a}{x + a} - \frac{1}{a}\log\frac{x^2 - a^2}{x^2}$

506. $\displaystyle\int (\log X)\, dx = \begin{cases}\left(x + \dfrac{b}{2c}\right)\log X - 2x + \dfrac{\sqrt{4ac - b^2}}{c}\tan^{-1}\dfrac{2cx + b}{\sqrt{4ac - b^2}}, \\ \hspace{4cm}(b^2 - 4ac < 0) \\[1em] \text{or} \\[1em] \left(x + \dfrac{b}{2c}\right)\log X - 2x + \dfrac{\sqrt{b^2 - 4ac}}{c}\tanh^{-1}\dfrac{2cx + b}{\sqrt{b^2 - 4ac}}, \\ \hspace{4cm}(b^2 - 4ac > 0) \\[1em] \text{where} \\ X = a + bx + cx^2\end{cases}$

507. $\displaystyle\int x^n(\log X)\, dx = \frac{x^{n+1}}{n + 1}\log X - \frac{2c}{n + 1}\int\frac{x^{n+2}}{X}\, dx - \frac{b}{n + 1}\int\frac{x^{n+1}}{X}\, dx$

$$\text{where } X = a + bx + cx^2$$

508. $\displaystyle\int [\log(x^2 + a^2)]\, dx = x\log(x^2 + a^2) - 2x + 2a\tan^{-1}\frac{x}{a}$

509. $\displaystyle\int [\log(x^2 - a^2)]\, dx = x\log(x^2 - a^2) - 2x + a\log\frac{x + a}{x - a}$

510. $\displaystyle\int x[\log(x^2 \pm a^2)]\, dx = \tfrac{1}{2}(x^2 \pm a^2)\log(x^2 \pm a^2) - \tfrac{1}{2}x^2$

511. $\displaystyle\int [\log(x + \sqrt{x^2 \pm a^2})]\, dx = x\log(x + \sqrt{x^2 \pm a^2}) - \sqrt{x^2 \pm a^2}$

512. $\displaystyle\int x[\log(x + \sqrt{x^2 \pm a^2})]\, dx = \left(\frac{x^2}{2} \pm \frac{a^2}{4}\right)\log(x + \sqrt{x^2 \pm a^2}) - \frac{x\sqrt{x^2 \pm a^2}}{4}$

513. $\displaystyle\int x^m[\log(x + \sqrt{x^2 \pm a^2})]\, dx = \frac{x^{m+1}}{m + 1}\log(x + \sqrt{x^2 \pm a^2})$

$$-\frac{1}{m + 1}\int\frac{x^{m+1}}{\sqrt{x^2 \pm a^2}}\, dx$$

514. $\displaystyle\int \frac{\log(x + \sqrt{x^2 + a^2})}{x^2}\,dx = -\frac{\log(x + \sqrt{x^2 + a^2})}{x} - \frac{1}{a}\log\frac{a + \sqrt{x^2 + a^2}}{x}$

515. $\displaystyle\int \frac{\log(x + \sqrt{x^2 - a^2})}{x^2}\,dx = -\frac{\log(x + \sqrt{x^2 - a^2})}{x} + \frac{1}{|a|}\sec^{-1}\frac{x}{a}$

516. $\displaystyle\int x^n \log(x^2 - a^2)\,dx = \frac{1}{n+1}\bigg[x^{n+1}\log(x^2 - a^2) - a^{n+1}\log(x - a)$

See note integral 392. $\displaystyle -(-a)^{n+1}\log(x + a) - 2\sum_{r=0}^{\left[\frac{n}{2}\right]} \frac{a^{2r}x^{n-2r+1}}{n - 2r + 1}\bigg]$

EXPONENTIAL FORMS

517. $\displaystyle\int e^x\,dx = e^x$

518. $\displaystyle\int e^{-x}\,dx = -e^{-x}$

519. $\displaystyle\int e^{ax}\,dx = \frac{e^{ax}}{a}$

520. $\displaystyle\int x\,e^{ax}\,dx = \frac{e^{ax}}{a^2}(ax - 1)$

521. $\displaystyle\int x^m e^{ax}\,dx = \begin{cases} \dfrac{x^m e^{ax}}{a} - \dfrac{m}{a}\displaystyle\int x^{m-1} e^{ax}\,dx \\[2mm] \qquad\quad \text{or} \\[2mm] e^{ax}\displaystyle\sum_{r=0}^{m} (-1)^r \dfrac{m!\,x^{m-r}}{(m-r)!\,a^{r+1}} \end{cases}$

522. $\displaystyle\int \frac{e^{ax}\,dx}{x} = \log x + \frac{ax}{1!} + \frac{a^2x^2}{2\cdot 2!} + \frac{a^3x^3}{3\cdot 3!} + \cdots$

523. $\displaystyle\int \frac{e^{ax}}{x^m}\,dx = -\frac{1}{m-1}\frac{e^{ax}}{x^{m-1}} + \frac{a}{m-1}\int \frac{e^{ax}}{x^{m-1}}\,dx$

524. $\displaystyle\int e^{ax}\log x\,dx = \frac{e^{ax}\log x}{a} - \frac{1}{a}\int \frac{e^{ax}}{x}\,dx$

525. $\displaystyle\int \frac{dx}{1 + e^x} = x - \log(1 + e^x) = \log\frac{e^x}{1 + e^x}$

526. $\displaystyle\int \frac{dx}{a + be^{px}} = \frac{x}{a} - \frac{1}{ap}\log(a + be^{px})$

527. $\displaystyle\int \frac{dx}{ae^{mx} + be^{-mx}} = \frac{1}{m\sqrt{ab}}\tan^{-1}\left(e^{mx}\sqrt{\frac{a}{b}}\right), \qquad (a > 0, b > 0)$

528. $\displaystyle\int \frac{dx}{ae^{mx} - be^{-mx}} = \begin{cases} \dfrac{1}{2m\sqrt{ab}}\log\dfrac{\sqrt{a}\,e^{mx} - \sqrt{b}}{\sqrt{a}\,e^{mx} + \sqrt{b}} \\[3mm] \qquad\qquad \text{or} \\[3mm] \dfrac{-1}{m\sqrt{ab}}\tanh^{-1}\left(\sqrt{\dfrac{a}{b}}\,e^{mx}\right), \qquad (a > 0, b > 0) \end{cases}$

529. $\displaystyle\int (a^x - a^{-x})\,dx = \frac{a^x + a^{-x}}{\log a}$

530. $\displaystyle\int \frac{e^{ax}}{b + ce^{ax}}\,dx = \frac{1}{ac}\log(b + ce^{ax})$

531. $\displaystyle\int \frac{x\,e^{ax}}{(1 + ax)^2}\,dx = \frac{e^{ax}}{a^2(1 + ax)}$

532. $\int x\,e^{-x^2}\,dx = -\tfrac{1}{2}e^{-x^2}$

533. $\int e^{ax}[\sin{(bx)}]\,dx = \dfrac{e^{ax}[a\sin{(bx)} - b\cos{(bx)}]}{a^2 + b^2}$

534. $\int e^{ax}[\sin{(bx)}][\sin{(cx)}]\,dx = \dfrac{e^{ax}[(b - c)\sin{(b - c)x} + a\cos{(b - c)x}]}{2[a^2 + (b - c)^2]}$

$$-\dfrac{e^{ax}[(b + c)\sin{(b + c)x} + a\cos{(b + c)x}]}{2[a^2 + (b + c)^2]}$$

535. $\int e^{ax}[\sin{(bx)}][\cos{(cx)}]\,dx = \begin{cases} \dfrac{e^{ax}[a\sin{(b - c)x} - (b - c)\cos{(b - c)x}]}{2[a^2 + (b - c)^2]} \\[3mm] \quad + \dfrac{e^{ax}[a\sin{(b + c)x} - (b + c)\cos{(b + c)x}]}{2[a^2 + (b + c)^2]} \\[3mm] \quad\quad\text{or} \\[3mm] \dfrac{e^{ax}}{\rho}[(a\sin{bx} - b\cos{bx})[\cos{(cx - \alpha)}] \\[3mm] \quad\quad\quad\quad - c(\sin{bx})\sin{(cx - \alpha)}] \\[3mm] \text{where} \\[3mm] \rho = \sqrt{(a^2 + b^2 - c^2)^2 + 4a^2c^2}, \\[2mm] \quad \rho\cos\alpha = a^2 + b^2 - c^2, \quad \rho\sin\alpha = 2ac \end{cases}$

536. $\int e^{ax}[\sin{(bx)}][\sin{(bx + c)}]\,dx$

$$= \dfrac{e^{ax}\cos c}{2a} - \dfrac{e^{ax}[a\cos{(2bx + c)} + 2b\sin{(2bx + c)}]}{2(a^2 + 4b^2)}$$

537. $\int e^{ax}[\sin{(bx)}][\cos{(bx + c)}]\,dx$

$$= \dfrac{-e^{ax}\sin c}{2a} + \dfrac{e^{ax}[a\sin{(2bx + c)} - 2b\cos{(2bx + c)}]}{2(a^2 + 4b^2)}$$

538. $\int e^{ax}[\cos{(bx)}]\,dx = \dfrac{e^{ax}}{a^2 + b^2}[a\cos{(bx)} + b\sin{(bx)}]$

539. $\int e^{ax}[\cos{(bx)}][\cos{(cx)}]\,dx = \dfrac{e^{ax}[(b - c)\sin{(b - c)x} + a\cos{(b - c)x}]}{2[a^2 + (b - c)^2]}$

$$+ \dfrac{e^{ax}[(b + c)\sin{(b + c)x} + a\cos{(b + c)x}]}{2[a^2 + (b + c)^2]}$$

540. $\int e^{ax}[\cos{(bx)}][\cos{(bx + c)}]\,dx$

$$= \dfrac{e^{ax}\cos c}{2a} + \dfrac{e^{ax}[a\cos{(2bx + c)} + 2b\sin{(2bx + c)}]}{2(a^2 + 4b^2)}$$

541. $\int e^{ax}[\cos{(bx)}][\sin{(bx + c)}]\,dx$

$$= \dfrac{e^{ax}\sin c}{2a} + \dfrac{e^{ax}[a\sin{(2bx + c)} - 2b\cos{(2bx + c)}]}{2(a^2 + 4b^2)}$$

542. $\int e^{ax}[\sin^n bx]\,dx = \dfrac{1}{a^2 + n^2b^2}\left[(a\sin bx - nb\cos bx)\,e^{ax}\sin^{n-1}bx\right.$

$$\left. + n(n - 1)b^2\int e^{ax}[\sin^{n-2}bx]\,dx\right]$$

543. $\int e^{ax}[\cos^n bx]\,dx = \dfrac{1}{a^2 + n^2b^2}\left[(a\cos bx + nb\sin bx)\,e^{ax}\cos^{n-1}bx\right.$

$$\left. + n(n - 1)b^2\int e^{ax}[\cos^{n-2}bx]\,dx\right]$$

544. $\int x^m e^x \sin x \, dx = \dfrac{1}{2} x^m e^x (\sin x - \cos x) - \dfrac{m}{2} \int x^{m-1} e^x \sin x \, dx$

$$+ \dfrac{m}{2} \int x^{m-1} e^x \cos x \, dx$$

545. $\int x^m e^{ax} [\sin bx] \, dx = \begin{cases} x^m e^{ax} \dfrac{a \sin bx - b \cos bx}{a^2 + b^2} \\ \qquad - \dfrac{m}{a^2 + b^2} \int x^{m-1} e^{ax} (a \sin bx - b \cos bx) \, dx \\ \qquad\qquad \text{or} \\ e^{ax} \displaystyle\sum_{r=0}^{m} \dfrac{(-1)^r m! x^{m-r}}{\rho^{r+1}(m - r)!} \sin [bx - (r + 1)\alpha] \\ \qquad\qquad \text{where} \\ \rho = \sqrt{a^2 + b^2}, \qquad \rho \cos \alpha = a, \qquad \rho \sin \alpha = b \end{cases}$

546. $\int x^m e^x \cos x \, dx = \dfrac{1}{2} x^m e^x (\sin x + \cos x)$

$$- \dfrac{m}{2} \int x^{m-1} e^x \sin x \, dx - \dfrac{m}{2} \int x^{m-1} e^x \cos x \, dx$$

547. $\int x^m e^{ax} \cos bx \, dx = \begin{cases} x^m e^{ax} \dfrac{a \cos bx + b \sin bx}{a^2 + b^2} \\ \qquad - \dfrac{m}{a^2 + b^2} \int x^{m-1} e^{ax} (a \cos bx + b \sin bx) \, dx \\ \qquad\qquad \text{or} \\ e^{ax} \displaystyle\sum_{r=0}^{m} \dfrac{(-1)^r m! x^{m-r}}{\rho^{r+1}(m - r)!} \cos [bx - (r + 1)\alpha] \\ \qquad\qquad \text{where} \\ \rho = \sqrt{a^2 + b^2}, \qquad \rho \cos \alpha = a, \qquad \rho \sin \alpha = b \end{cases}$

548. $\displaystyle\int e^{ax}(\cos^m x)(\sin^n x)\,dx = $

$$\dfrac{e^{ax}\cos^{m-1} x \sin^n x[a\cos x + (m+n)\sin x]}{(m+n)^2 + a^2}$$

$$-\dfrac{na}{(m+n)^2 + a^2}\int e^{ax}(\cos^{m-1} x)(\sin^{n-1} x)\,dx$$

$$+\dfrac{(m-1)(m+n)}{(m+n)^2 + a^2}\int e^{ax}(\cos^{m-2} x)(\sin^n x)\,dx$$

or

$$\dfrac{e^{ax}\cos^m x \sin^{n-1} x[a\sin x - (m+n)\cos x]}{(m+n)^2 + a^2}$$

$$+\dfrac{ma}{(m+n)^2 + a^2}\int e^{ax}(\cos^{m-1} x)(\sin^{n-1} x)\,dx$$

$$+\dfrac{(n-1)(m+n)}{(m+n)^2 + a^2}\int e^{ax}(\cos^m x)(\sin^{n-2} x)\,dx$$

or

$$\dfrac{e^{ax}(\cos^{m-1} x)(\sin^{n-1} x)(a\sin x\cos x + m\sin^2 x - n\cos^2 x)}{(m+n)^2 + a^2}$$

$$+\dfrac{m(m-1)}{(m+n)^2 + a^2}\int e^{ax}(\cos^{m-2} x)(\sin^n x)\,dx$$

$$+\dfrac{n(n-1)}{(m+n)^2 + a^2}\int e^{ax}(\cos^m x)(\sin^{n-2} x)\,dx$$

or

$$\dfrac{e^{ax}(\cos^{m-1} x)(\sin^{n-1} x)(a\cos x\sin x + m\sin^2 x - n\cos^2 x)}{(m+n)^2 + a^2}$$

$$+\dfrac{m(m-1)}{(m+n)^2 + a^2}\int e^{ax}(\cos^{m-2} x)(\sin^{n-2} x)\,dx$$

$$+\dfrac{(n-m)(n+m-1)}{(m+n)^2 + a^2}\int e^{ax}(\cos^m x)(\sin^{n-2} x)\,dx$$

549. $\displaystyle\int x\,e^{ax}(\sin bx)\,dx = \dfrac{x\,e^{ax}}{a^2 + b^2}(a\sin bx - b\cos bx)$

$$-\dfrac{e^{ax}}{(a^2+b^2)^2}[(a^2-b^2)\sin bx - 2ab\cos bx]$$

550. $\displaystyle\int x\,e^{ax}(\cos bx)\,dx = \dfrac{x\,e^{ax}}{a^2 + b^2}(a\cos bx + b\sin bx)$

$$-\dfrac{e^{ax}}{(a^2+b^2)^2}[(a^2-b^2)\cos bx + 2ab\sin bx]$$

551. $\displaystyle\int \dfrac{e^{ax}}{\sin^n x}\,dx = -\dfrac{e^{ax}[a\sin x + (n-2)\cos x]}{(n-1)(n-2)\sin^{n-1} x} + \dfrac{a^2 + (n-2)^2}{(n-1)(n-2)}\int \dfrac{e^{ax}}{\sin^{n-2} x}\,dx$

552. $\displaystyle\int \dfrac{e^{ax}}{\cos^n x}\,dx = -\dfrac{e^{ax}[a\cos x - (n-2)\sin x]}{(n-1)(n-2)\cos^{n-1} x} + \dfrac{a^2 + (n-2)^2}{(n-1)(n-2)}\int \dfrac{e^{ax}}{\cos^{n-2} x}\,dx$

553. $\displaystyle\int e^{ax}\tan^n x\,dx = e^{ax}\dfrac{\tan^{n-1} x}{n-1} - \dfrac{a}{n-1}\int e^{ax}\tan^{n-1} x\,dx - \int e^{ax}\tan^{n-2} x\,dx$

HYPERBOLIC FORMS

554. $\displaystyle\int (\sinh x)\,dx = \cosh x$

555. $\displaystyle\int (\cosh x)\,dx = \sinh x$

556. $\int (\tanh x)\,dx = \log \cosh x$

557. $\int (\coth x)\,dx = \log \sinh x$

558. $\int (\operatorname{sech} x)\,dx = \tan^{-1}(\sinh x)$

559. $\int \operatorname{csch} x\,dx = \log \tanh\left(\dfrac{x}{2}\right)$

560. $\int x(\sinh x)\,dx = x \cosh x - \sinh x$

561. $\int x^n(\sinh x)\,dx = x^n \cosh x - n \int x^{n-1}(\cosh x)\,dx$

562. $\int x(\cosh x)\,dx = x \sinh x - \cosh x$

563. $\int x^n(\cosh x)\,dx = x^n \sinh x - n \int x^{n-1}(\sinh x)\,dx$

564. $\int (\operatorname{sech} x)(\tanh x)\,dx = -\operatorname{sech} x$

565. $\int (\operatorname{csch} x)(\coth x)\,dx = -\operatorname{csch} x$

566. $\int (\sinh^2 x)\,dx = \dfrac{\sinh 2x}{4} - \dfrac{x}{2}$

567. $\int (\sinh^m x)(\cosh^n x)\,dx =$
$$\begin{cases} \dfrac{1}{m+n}(\sinh^{m+1} x)(\cosh^{n-1} x) \\[2mm] \qquad + \dfrac{n-1}{m+n}\int (\sinh^m x)(\cosh^{n-2} x)\,dx \\[2mm] \qquad\qquad \text{or} \\[2mm] \dfrac{1}{m+n}\sinh^{m-1} x \cosh^{n+1} x \\[2mm] \qquad - \dfrac{m-1}{m+n}\int (\sinh^{m-2} x)(\cosh^n x)\,dx, \quad (m+n \neq 0) \end{cases}$$

568. $\int \dfrac{dx}{(\sinh^m x)(\cosh^n x)} =$
$$\begin{cases} -\dfrac{1}{(m-1)(\sinh^{m-1} x)(\cosh^{n-1} x)} \\[2mm] \qquad - \dfrac{m+n-2}{m-1}\int \dfrac{dx}{(\sinh^{m-2} x)(\cosh^n x)}, \quad (m \neq 1) \\[2mm] \qquad\qquad \text{or} \\[2mm] \dfrac{1}{(n-1)\sinh^{m-1} x \cosh^{n-1} x} \\[2mm] \qquad + \dfrac{m+n-2}{n-1}\int \dfrac{dx}{(\sinh^m x)(\cosh^{n-2} x)}, \quad (n \neq 1) \end{cases}$$

569. $\int (\tanh^2 x)\,dx = x - \tanh x$

570. $\int (\tanh^n x)\,dx = -\dfrac{\tanh^{n-1} x}{n-1} + \int (\tanh^{n-2} x)\,dx, \quad (n \neq 1)$

571. $\int (\operatorname{sech}^2 x)\,dx = \tanh x$

572. $\int (\cosh^2 x)\,dx = \dfrac{\sinh 2x}{4} + \dfrac{x}{2}$

573. $\int (\coth^2 x)\,dx = x - \coth x$

574. $\int (\coth^n x)\,dx \doteq -\dfrac{\coth^{n-1} x}{n-1} + \int \coth^{n-2} x\,dx, \qquad (n \neq 1)$

575. $\int (\operatorname{csch}^2 x)\,dx = -\operatorname{ctnh} x$

576. $\int (\sinh mx)(\sinh nx)\,dx = \dfrac{\sinh (m+n)x}{2(m+n)} - \dfrac{\sinh (m-n)x}{2(m-n)}, \qquad (m^2 \neq n^2)$

577. $\int (\cosh mx)(\cosh nx)\,dx = \dfrac{\sinh (m+n)x}{2(m+n)} + \dfrac{\sinh (m-n)x}{2(m-n)}, \qquad (m^2 \neq n^2)$

578. $\int (\sinh mx)(\cosh nx)\,dx = \dfrac{\cosh (m+n)x}{2(m+n)} + \dfrac{\cosh (m-n)x}{2(m-n)}, \qquad (m^2 \neq n^2)$

579. $\int \left(\sinh^{-1} \dfrac{x}{a}\right) dx = x \sinh^{-1} \dfrac{x}{a} - \sqrt{x^2 + a^2}, \qquad (a > 0)$

580. $\int x\left(\sinh^{-1} \dfrac{x}{a}\right) dx = \left(\dfrac{x^2}{2} + \dfrac{a^2}{4}\right) \sinh^{-1} \dfrac{x}{a} - \dfrac{x}{4}\sqrt{x^2 + a^2}, \qquad (a > 0)$

581. $\int x^n(\sinh^{-1} x)\,dx = \dfrac{x^{n+1}}{n+1} \sinh^{-1} x - \dfrac{1}{n+1} \int \dfrac{x^{n+1}}{(1 + x^2)^{\frac{1}{2}}}\,dx, \qquad (n \neq -1)$

582. $\int \left(\cosh^{-1} \dfrac{x}{a}\right) dx = \begin{cases} x \cosh^{-1} \dfrac{x}{a} - \sqrt{x^2 - a^2}, & \left(\cosh^{-1} \dfrac{x}{a} > 0\right) \\[2mm] \qquad\qquad \text{or} \\[2mm] x \cosh^{-1} \dfrac{x}{a} + \sqrt{x^2 - a^2}, & \left(\cosh^{-1} \dfrac{x}{a} < 0\right), \qquad (a > 0) \end{cases}$

583. $\int x\left(\cosh^{-1} \dfrac{x}{a}\right) dx = \dfrac{2x^2 - a^2}{4} \cosh^{-1} \dfrac{x}{a} - \dfrac{x}{4}(x^2 - a^2)^{\frac{1}{2}}$

584. $\int x^n(\cosh^{-1} x)\,dx = \dfrac{x^{n+1}}{n+1} \cosh^{-1} x - \dfrac{1}{n+1} \int \dfrac{x^{n+1}}{(x^2 - 1)^{\frac{1}{2}}}\,dx, \qquad (n \neq -1)$

585. $\int \left(\tanh^{-1} \dfrac{x}{a}\right) dx = x \tanh^{-1} \dfrac{x}{a} + \dfrac{a}{2} \log (a^2 - x^2), \qquad \left(\left|\dfrac{x}{a}\right| < 1\right)$

586. $\int \left(\coth^{-1} \dfrac{x}{a}\right) dx = x \coth^{-1} \dfrac{x}{a} + \dfrac{a}{2} \log (x^2 - a^2), \qquad \left(\left|\dfrac{x}{a}\right| > 1\right)$

587. $\int x\left(\tanh^{-1} \dfrac{x}{a}\right) dx = \dfrac{x^2 - a^2}{2} \tanh^{-1} \dfrac{x}{a} + \dfrac{ax}{2}, \qquad \left(\left|\dfrac{x}{a}\right| < 1\right)$

588. $\int x^n\left(\tanh^{-1} x\right) dx = \dfrac{x^{n+1}}{n+1} \tanh^{-1} x - \dfrac{1}{n+1} \int \dfrac{x^{n+1}}{1 - x^2}\,dx, \qquad (n \neq -1)$

589. $\int x\left(\coth^{-1} \dfrac{x}{a}\right) dx = \dfrac{x^2 - a^2}{2} \coth^{-1} \dfrac{x}{a} + \dfrac{ax}{2}, \qquad \left(\left|\dfrac{x}{a}\right| > 1\right)$

590. $\int x^n(\coth^{-1} x)\,dx = \dfrac{x^{n+1}}{n+1} \coth^{-1} x + \dfrac{1}{n+1} \int \dfrac{x^{n+1}}{x^2 - 1}\,dx, \qquad (n \neq -1)$

591. $\int (\operatorname{sech}^{-1} x)\,dx = x \operatorname{sech}^{-1} x + \sin^{-1} x$

592. $\int x \operatorname{sech}^{-1} x\,dx = \dfrac{x^2}{2} \operatorname{sech}^{-1} x - \dfrac{1}{2}\sqrt{1 - x^2}$

593. $\int x^n \operatorname{sech}^{-1} x\,dx = \dfrac{x^{n+1}}{n+1} \operatorname{sech}^{-1} x + \dfrac{1}{n+1} \int \dfrac{x^n}{(1 - x^2)^{\frac{1}{2}}}\,dx, \qquad (n \neq -1)$

594. $\int \operatorname{csch}^{-1} x\,dx = x \operatorname{csch}^{-1} x + \dfrac{x}{|x|} \sinh^{-1} x$

595. $\int x \operatorname{csch}^{-1} x \, dx = \dfrac{x^2}{2} \operatorname{csch}^{-1} x + \dfrac{1}{2} \dfrac{x}{|x|} \sqrt{1 + x^2}$

596. $\int x^n \operatorname{csch}^{-1} x \, dx = \dfrac{x^{n+1}}{n+1} \operatorname{csch}^{-1} x + \dfrac{1}{n+1} \dfrac{x}{|x|} \int \dfrac{x^n}{(x^2+1)^{\frac{1}{2}}} \, dx, \qquad (n \neq -1)$

DEFINITE INTEGRALS

597. $\displaystyle\int_0^\infty x^{n-1} e^{-x} \, dx = \int_0^1 \left(\log \dfrac{1}{x}\right)^{n-1} dx = \dfrac{1}{n} \prod_{m=1}^\infty \dfrac{\left(1 + \dfrac{1}{m}\right)^n}{1 + \dfrac{n}{m}}$

$$= \Gamma(n), n \neq 0, -1, -2, -3, \ldots \quad \text{(Gamma Function)}$$

598. $\displaystyle\int_0^\infty t^n p^{-t} \, dt = \dfrac{n!}{(\log p)^{n+1}}, \qquad (n = 0, 1, 2, 3, \ldots \text{ and } p > 0)$

599. $\displaystyle\int_0^\infty t^{n-1} e^{-(a+1)t} \, dt = \dfrac{\Gamma(n)}{(a+1)^n}, \qquad (n > 0, a > -1)$

600. $\displaystyle\int_0^1 x^m \left(\log \dfrac{1}{x}\right)^n dx = \dfrac{\Gamma(n+1)}{(m+1)^{n+1}}, \qquad (m > -1, n > -1)$

601. $\Gamma(n)$ is finite if $n > 0$, $\Gamma(n+1) = n\Gamma(n)$

602. $\Gamma(n) \cdot \Gamma(1-n) = \dfrac{\pi}{\sin n\pi}$

603. $\Gamma(n) = (n-1)!$ if $n = $ integer > 0

604. $\Gamma(\frac{1}{2}) = 2 \displaystyle\int_0^\infty e^{-t^2} \, dt = \sqrt{\pi} = 1.7724538509 \cdots = (-\frac{1}{2})!$

605. $\Gamma(n + \frac{1}{2}) = \dfrac{1 \cdot 3 \cdot 5 \ldots (2n-1)}{2^n} \sqrt{\pi} \qquad n = 1, 2, 3, \ldots$

606. $\Gamma(-n + \frac{1}{2}) = \dfrac{(-1)^n 2^n \sqrt{\pi}}{1 \cdot 3 \cdot 5 \ldots (2n-1)} \qquad n = 1, 2, 3, \ldots$

607. $\displaystyle\int_0^1 x^{m-1}(1-x)^{n-1} \, dx = \int_0^\infty \dfrac{x^{m-1}}{(1+x)^{m+n}} \, dx = \dfrac{\Gamma(m)\Gamma(n)}{\Gamma(m+n)} = B(m, n)$

(Beta function)

608. $B(m, n) = B(n, m) = \dfrac{\Gamma(m)\Gamma(n)}{\Gamma(m+n)}$, where m and n are any positive real numbers.

609. $\displaystyle\int_a^b (x-a)^m (b-x)^n \, dx = (b-a)^{m+n+1} \dfrac{\Gamma(m+1) \cdot \Gamma(n+1)}{\Gamma(m+n+2)},$

$$(m > -1, n > -1, b > a)$$

610. $\displaystyle\int_1^\infty \dfrac{dx}{x^m} = \dfrac{1}{m-1}, \qquad [m > 1]$

611. $\displaystyle\int_0^\infty \dfrac{dx}{(1+x)x^p} = \pi \csc p\pi, \qquad [p < 1]$

612. $\displaystyle\int_0^\infty \dfrac{dx}{(1-x)x^p} = -\pi \cot p\pi, \qquad [p < 1]$

613. $\displaystyle\int_0^\infty \dfrac{x^{p-1} \, dx}{1+x} = \dfrac{\pi}{\sin p\pi}$

$$= B(p, 1-p) = \Gamma(p)\Gamma(1-p), \qquad [0 < p < 1]$$

614. $\displaystyle\int_0^\infty \dfrac{x^{m-1} \, dx}{1+x^n} = \dfrac{\pi}{n \sin \dfrac{m\pi}{n}}, \qquad [0 < m < n]$

615. $\displaystyle\int_0^\infty \frac{x^a\,dx}{(m+x^b)^c} = \frac{m^{\frac{a+1-bc}{b}}}{b}\left[\frac{\Gamma\!\left(\dfrac{a+1}{b}\right)\Gamma\!\left(c-\dfrac{a+1}{b}\right)}{\Gamma(c)}\right]$

$$\left(a>-1,\,b>0,\,m>0,\,c>\frac{a+1}{b}\right)$$

616. $\displaystyle\int_0^\infty \frac{dx}{(1+x)\sqrt{x}} = \pi$

617. $\displaystyle\int_0^\infty \frac{a\,dx}{a^2+x^2} = \frac{\pi}{2}$, if $a>0$; 0, if $a=0$; $-\dfrac{\pi}{2}$, if $a<0$

618. $\displaystyle\int_0^a (a^2-x^2)^{\frac{n}{2}}\,dx = \frac{1}{2}\int_{-a}^a (a^2-x^2)^{\frac{n}{2}}\,dx = \frac{1\cdot3\cdot5\ldots n}{2\cdot4\cdot6\ldots(n+1)}\cdot\frac{\pi}{2}\cdot a^{n+1}$ $(n\text{ odd})$

619. $\displaystyle\int_0^a x^m(a^2-x^2)^{\frac{n}{2}}\,dx = \begin{cases}\dfrac{1}{2}a^{m+n+1}B\!\left(\dfrac{m+1}{2},\dfrac{n+2}{2}\right)\\[2mm] \text{or}\\[2mm] \dfrac{1}{2}a^{m+n+1}\dfrac{\Gamma\!\left(\dfrac{m+1}{2}\right)\Gamma\!\left(\dfrac{n+2}{2}\right)}{\Gamma\!\left(\dfrac{m+n+3}{2}\right)}\end{cases}$

620. $\displaystyle\int_0^{\pi/2}(\sin^n x)\,dx = \begin{cases}\displaystyle\int_0^{\pi/2}(\cos^n x)\,dx\\[2mm] \text{or}\\[2mm] \dfrac{1\cdot3\cdot5\cdot7\ldots(n-1)}{2\cdot4\cdot6\cdot8\ldots(n)}\dfrac{\pi}{2}, \quad (n\text{ an even integer, }n\neq0)\\[2mm] \text{or}\\[2mm] \dfrac{2\cdot4\cdot6\cdot8\ldots(n-1)}{1\cdot3\cdot5\cdot7\ldots(n)}, \quad (n\text{ an odd integer, }n\neq1)\\[2mm] \text{or}\\[2mm] \dfrac{\sqrt{\pi}}{2}\dfrac{\Gamma\!\left(\dfrac{n+1}{2}\right)}{\Gamma\!\left(\dfrac{n}{2}+1\right)}, \quad (n>-1)\end{cases}$

621. $\displaystyle\int_0^\infty \frac{\sin mx\,dx}{x} = \frac{\pi}{2}$, if $m>0$; 0, if $m=0$; $-\dfrac{\pi}{2}$, if $m<0$

622. $\displaystyle\int_0^\infty \frac{\cos x\,dx}{x} = \infty$

623. $\displaystyle\int_0^\infty \frac{\tan x\,dx}{x} = \frac{\pi}{2}$

624. $\displaystyle\int_0^\pi \sin ax\cdot\sin bx\,dx = \int_0^\pi \cos ax\cdot\cos bx\,dx = 0,$ $(a\neq b;a,b\text{ integers})$

625. $\displaystyle\int_0^{\pi/a}[\sin(ax)][\cos(ax)]\,dx = \int_0^\pi[\sin(ax)][\cos(ax)]\,dx = 0$

626. $\displaystyle\int_0^\pi[\sin(ax)][\cos(bx)]\,dx = \frac{2a}{a^2-b^2}$, if $a-b$ is odd, or 0 if $a-b$ is even

627. $\displaystyle\int_0^\infty \frac{\sin x\cos mx\,dx}{x}$

$$= 0,\text{ if }m<-1\text{ or }m>1;\frac{\pi}{4},\text{ if }m=\pm1;\frac{\pi}{2},\text{ if }m^2<1$$

628. $\int_0^\infty \dfrac{\sin ax \sin bx}{x^2} dx = \dfrac{\pi a}{2}, \quad (a \le b)$

629. $\int_0^\pi \sin^2 mx \, dx = \int_0^\pi \cos^2 mx \, dx = \dfrac{\pi}{2}$

630. $\int_0^\infty \dfrac{\sin^2 (px)}{x^2} dx = \dfrac{\pi p}{2}$

631. $\int_0^\infty \dfrac{\sin x}{x^p} dx = \dfrac{\pi}{2\Gamma(p) \sin(p\pi/2)}, \quad 0 < p < 1$

632. $\int_0^\infty \dfrac{\cos x}{x^p} dx = \dfrac{\pi}{2\Gamma(p) \cos(p\pi/2)}, \quad 0 < p < 1$

633. $\int_0^\infty \dfrac{1 - \cos px}{x^2} dx = \dfrac{\pi p}{2}$

634. $\int_0^\infty \dfrac{\sin px \cos qx}{x} dx = \left\{ 0, \quad q > p > 0; \quad \dfrac{\pi}{2}, \ p > q > 0; \quad \dfrac{\pi}{4}, \ p = q > 0 \right\}$

635. $\int_0^\infty \dfrac{\cos (mx)}{x^2 + a^2} dx = \dfrac{\pi}{2|a|} e^{-|ma|}$

636. $\int_0^\infty \cos (x^2) \, dx = \int_0^\infty \sin (x^2) \, dx = \dfrac{1}{2}\sqrt{\dfrac{\pi}{2}}$

637. $\int_0^\infty \sin ax^n \, dx = \dfrac{1}{na^{1/n}} \Gamma(1/n) \sin \dfrac{\pi}{2n}, \quad n > 1$

638. $\int_0^\infty \cos ax^n \, dx = \dfrac{1}{na^{1/n}} \Gamma(1/n) \cos \dfrac{\pi}{2n}, \quad n > 1$

639. $\int_0^\infty \dfrac{\sin x}{\sqrt{x}} dx = \int_0^\infty \dfrac{\cos x}{\sqrt{x}} dx = \sqrt{\dfrac{\pi}{2}}$

640. (a) $\int_0^\infty \dfrac{\sin^3 x}{x} dx = \dfrac{\pi}{4}$ (b) $\int_0^\infty \dfrac{\sin^3 x}{x^2} dx \dfrac{3}{4} \log 3$

641. $\int_0^\infty \dfrac{\sin^3 x}{x^3} dx = \dfrac{3\pi}{8}$

642. $\int_0^\infty \dfrac{\sin^4 x}{x^4} dx = \dfrac{\pi}{3}$

643. $\int_0^{\pi/2} \dfrac{dx}{1 + a \cos x} = \dfrac{\cos^{-1} a}{\sqrt{1 - a^2}}, \quad (a < 1)$

644. $\int_0^\pi \dfrac{dx}{a + b \cos x} = \dfrac{\pi}{\sqrt{a^2 - b^2}}, \quad (a > b \ge 0)$

645. $\int_0^{2\pi} \dfrac{dx}{1 + a \cos x} = \dfrac{2\pi}{\sqrt{1 - a^2}}, \quad (a^2 < 1)$

646. $\int_0^\infty \dfrac{\cos ax - \cos bx}{x} dx = \log \dfrac{b}{a}$

647. $\int_0^{\pi/2} \dfrac{dx}{a^2 \sin^2 x + b^2 \cos^2 x} = \dfrac{\pi}{2ab}$

648. $\int_0^{\pi/2} \dfrac{dx}{(a^2 \sin^2 x + b^2 \cos^2 x)^2} = \dfrac{\pi(a^2 + b^2)}{4a^3 b^3}, \quad (a, b > 0)$

649. $\int_0^{\pi/2} \sin^{n-1} x \cos^{m-1} x \, dx = \dfrac{1}{2} B\left(\dfrac{n}{2}, \dfrac{m}{2}\right), \quad m$ and n positive integers

650. $\displaystyle\int_0^{\pi/2} (\sin^{2n+1}\theta)\,d\theta = \frac{2\cdot 4\cdot 6\ldots(2n)}{1\cdot 3\cdot 5\ldots(2n+1)},\qquad (n=1,2,3\ldots)$

651. $\displaystyle\int_0^{\pi/2} (\sin^{2n}\theta)\,d\theta = \frac{1\cdot 3\cdot 5\ldots(2n-1)}{2\cdot 4\ldots(2n)}\left(\frac{\pi}{2}\right),\qquad (n=1,2,3\ldots)$

652. $\displaystyle\int_0^{\pi/2} \frac{x}{\sin x}\,dx = 2\left\{\frac{1}{1^2}-\frac{1}{3^2}+\frac{1}{5^2}-\frac{1}{7^2}+\cdots\right\}$

653. $\displaystyle\int_0^{\pi/2} \frac{dx}{1+\tan^m x} = \frac{\pi}{4}$

654. $\displaystyle\int_0^{\pi/2} \sqrt{\cos\theta}\,d\theta = \frac{(2\pi)^{\frac{1}{2}}}{[\Gamma(\frac{1}{4})]^2}$

655. $\displaystyle\int_0^{\pi/2} (\tan^h\theta)\,d\theta = \frac{\pi}{2\cos\left(\dfrac{h\pi}{2}\right)},\qquad (0<h<1)$

656. $\displaystyle\int_0^{\infty} \frac{\tan^{-1}(ax)-\tan^{-1}(bx)}{x}\,dx = \frac{\pi}{2}\log\frac{a}{b},\qquad (a,b>0)$

657. The area enclosed by a curve defined through the equation $x^{\frac{b}{c}}+y^{\frac{b}{c}}=a^{\frac{b}{c}}$ where $a>0$, c a positive odd integer and b a positive even integer is given by

$$\frac{\left[\Gamma\left(\dfrac{c}{b}\right)\right]^2}{\Gamma\left(\dfrac{2c}{b}\right)}\left(\frac{2ca^2}{b}\right)$$

658. $\displaystyle I = \iiint_R x^{h-1}y^{m-1}z^{n-1}\,dv$, where R denotes the region of space bounded by

the co-ordinate planes and that portion of the surface $\left(\dfrac{x}{a}\right)^p + \left(\dfrac{y}{b}\right)^q + \left(\dfrac{z}{c}\right)^k = 1$,

which lies in the first octant and where $h, m, n, p, q, k, a, b, c$, denote positive real numbers is given by

$$\int_0^a x^{h-1}\,dx \int_0^{b\left[1-\left(\frac{x}{a}\right)^p\right]^{\frac{1}{q}}} y^m\,dy \int_0^{c\left[1-\left(\frac{x}{a}\right)^p-\left(\frac{y}{b}\right)^q\right]^{\frac{1}{k}}} z^{n-1}\,dz$$

$$= \frac{a^h b^m c^n}{pqk}\,\frac{\Gamma\left(\dfrac{h}{p}\right)\Gamma\left(\dfrac{m}{q}\right)\Gamma\left(\dfrac{n}{k}\right)}{\Gamma\left(\dfrac{h}{p}+\dfrac{m}{q}+\dfrac{n}{k}+1\right)}$$

659. $\displaystyle\int_0^{\infty} e^{-ax}\,dx = \frac{1}{a},\qquad (a>0)$

660. $\displaystyle\int_0^{\infty} \frac{e^{-ax}-e^{-bx}}{x}\,dx = \log\frac{b}{a},\qquad (a,b>0)$

661. $\displaystyle\int_0^{\infty} x^n e^{-ax}\,dx = \begin{cases} \dfrac{\Gamma(n+1)}{a^{n+1}}, & (n>-1, a>0) \\[2mm] \qquad\text{or} \\[2mm] \dfrac{n!}{a^{n+1}}, & (a>0, n\text{ positive integer}) \end{cases}$

662. $\displaystyle\int_0^{\infty} x^n \exp(-ax^p)\,dx = \frac{\Gamma(k)}{pa^k},\qquad \left(n>-1, p>0, a>0, k=\frac{n+1}{p}\right)$

663. $\displaystyle\int_0^{\infty} e^{-a^2 x^2}\,dx = \frac{1}{2a}\sqrt{\pi} = \frac{1}{2a}\Gamma\left(\frac{1}{2}\right),\qquad (a>0)$

664. $\displaystyle\int_0^{\infty} x\,e^{-x^2}\,dx = \frac{1}{2}$

665. $\displaystyle\int_0^\infty x^2 e^{-x^2}\, dx = \frac{\sqrt{\pi}}{4}$

666. $\displaystyle\int_0^\infty x^{2n} e^{-ax^2}\, dx = \frac{1\cdot 3\cdot 5 \ldots (2n-1)}{2^{n+1}a^n}\sqrt{\frac{\pi}{a}}$

667. $\displaystyle\int_0^\infty x^{2n+1} e^{-ax^2}\, dx = \frac{n!}{2a^{n+1}}, \qquad (a > 0)$

668. $\displaystyle\int_0^1 x^m e^{-ax}\, dx = \frac{m!}{a^{m+1}}\left[1 - e^{-a}\sum_{r=0}^m \frac{a^r}{r!}\right]$

669. $\displaystyle\int_0^\infty e^{\left(-x^2 - \frac{a^2}{x^2}\right)}\, dx = \frac{e^{-2a}\sqrt{\pi}}{2}, \qquad (a \geq 0)$

670. $\displaystyle\int_0^\infty e^{-nx}\sqrt{x}\, dx = \frac{1}{2n}\sqrt{\frac{\pi}{n}}$

671. $\displaystyle\int_0^\infty \frac{e^{-nx}}{\sqrt{x}}\, dx = \sqrt{\frac{\pi}{n}}$

672. $\displaystyle\int_0^\infty e^{-ax}(\cos mx)\, dx = \frac{a}{a^2 + m^2}, \qquad (a > 0)$

673. $\displaystyle\int_0^\infty e^{-ax}(\sin mx)\, dx = \frac{m}{a^2 + m^2}, \qquad (a > 0)$

674. $\displaystyle\int_0^\infty x\, e^{-ax}[\sin(bx)]\, dx = \frac{2ab}{(a^2 + b^2)^2}, \qquad (a > 0)$

675. $\displaystyle\int_0^\infty x\, e^{-ax}[\cos(bx)]\, dx = \frac{a^2 - b^2}{(a^2 + b^2)^2}, \qquad (a > 0)$

676. $\displaystyle\int_0^\infty x^n\, e^{-ax}[\sin(bx)]\, dx = \frac{n![(a+ib)^{n+1} - (a-ib)^{n+1}]}{2i(a^2 + b^2)^{n+1}}, \qquad (i^2 = -1, a > 0)$

677. $\displaystyle\int_0^\infty x^n\, e^{-ax}[\cos(bx)]\, dx = \frac{n![(a-ib)^{n+1} + (a+ib)^{n+1}]}{2(a^2 + b^2)^{n+1}}, \qquad (i^2 = -1, a > 0)$

678. $\displaystyle\int_0^\infty \frac{e^{-ax}\sin x}{x}\, dx = \cot^{-1} a, \qquad (a > 0)$

679. $\displaystyle\int_0^\infty e^{-a^2x^2}\cos bx\, dx = \frac{\sqrt{\pi}}{2a}\exp\left(-\frac{b^2}{4a^2}\right), \qquad (ab \neq 0)$

680. $\displaystyle\int_0^\infty e^{-t\cos\phi}\, t^{b-1}\sin(t\sin\phi)\, dt = [\Gamma(b)]\sin(b\phi), \qquad \left(b > 0, -\frac{\pi}{2} < \phi < \frac{\pi}{2}\right)$

681. $\displaystyle\int_0^\infty e^{-t\cos\phi}\, t^{b-1}[\cos(t\sin\phi)]\, dt = [\Gamma(b)]\cos(b\phi), \qquad \left(b > 0, -\frac{\pi}{2} < \phi < \frac{\pi}{2}\right)$

682. $\displaystyle\int_0^\infty t^{b-1}\cos t\, dt = [\Gamma(b)]\cos\left(\frac{b\pi}{2}\right), \qquad (0 < b < 1)$

683. $\displaystyle\int_0^\infty t^{b-1}(\sin t)\, dt = [\Gamma(b)]\sin\left(\frac{b\pi}{2}\right), \qquad (0 < b < 1)$

684. $\displaystyle\int_0^1 (\log x)^n\, dx = (-1)^n \cdot n!$

685. $\displaystyle\int_0^1 \left(\log\frac{1}{x}\right)^{\frac{1}{2}}\, dx = \frac{\sqrt{\pi}}{2}$

686. $\displaystyle\int_0^1 \left(\log\frac{1}{x}\right)^{-\frac{1}{2}}\, dx = \sqrt{\pi}$

687. $\displaystyle\int_0^1 \left(\log\frac{1}{x}\right)^n dx = n!$

688. $\displaystyle\int_0^1 x\log(1-x)\,dx = -\frac{3}{4}$

689. $\displaystyle\int_0^1 x\log(1+x)\,dx = \frac{1}{4}$

690. $\displaystyle\int_0^1 x^m(\log x)^n\,dx = \frac{(-1)^n n!}{(m+1)^{n+1}}, \qquad m > -1, n = 0, 1, 2, \ldots$

If $n \neq 0, 1, 2, \ldots$ replace $n!$ by $\Gamma(n+1)$.

691. $\displaystyle\int_0^1 \frac{\log x}{1+x}\,dx = -\frac{\pi^2}{12}$

692. $\displaystyle\int_0^1 \frac{\log x}{1-x}\,dx = -\frac{\pi^2}{6}$

693. $\displaystyle\int_0^1 \frac{\log(1+x)}{x}\,dx = \frac{\pi^2}{12}$

694. $\displaystyle\int_0^1 \frac{\log(1-x)}{x}\,dx = -\frac{\pi^2}{6}$

695. $\displaystyle\int_0^1 (\log x)[\log(1+x)]\,dx = 2 - 2\log 2 - \frac{\pi^2}{12}$

696. $\displaystyle\int_0^1 (\log x)[\log(1-x)]\,dx = 2 - \frac{\pi^2}{6}$

697. $\displaystyle\int_0^1 \frac{\log x}{1-x^2}\,dx = -\frac{\pi^2}{8}$

698. $\displaystyle\int_0^1 \log\left(\frac{1+x}{1-x}\right)\cdot\frac{dx}{x} = \frac{\pi^2}{4}$

699. $\displaystyle\int_0^1 \frac{\log x\,dx}{\sqrt{1-x^2}} = -\frac{\pi}{2}\log 2$

700. $\displaystyle\int_0^1 x^m\left[\log\left(\frac{1}{x}\right)\right]^n dx = \frac{\Gamma(n+1)}{(m+1)^{n+1}}, \qquad \text{if } m+1 > 0, n+1 > 0$

701. $\displaystyle\int_0^1 \frac{(x^p - x^q)\,dx}{\log x} = \log\left(\frac{p+1}{q+1}\right), \qquad (p+1 > 0, q+1 > 0)$

702. $\displaystyle\int_0^1 \frac{dx}{\sqrt{\log\left(\frac{1}{x}\right)}} = \sqrt{\pi}$, (same as integral 686)

703. $\displaystyle\int_0^\infty \log\left(\frac{e^x + 1}{e^x - 1}\right) dx = \frac{\pi^2}{4}$

704. $\displaystyle\int_0^{\pi/2} (\log\sin x)\,dx = \int_0^{\pi/2} \log\cos x\,dx = -\frac{\pi}{2}\log 2$

705. $\displaystyle\int_0^{\pi/2} (\log\sec x)\,dx = \int_0^{\pi/2} \log\csc x\,dx = \frac{\pi}{2}\log 2$

706. $\displaystyle\int_0^\pi x(\log\sin x)\,dx = -\frac{\pi^2}{2}\log 2$

707. $\displaystyle\int_0^{\pi/2} (\sin x)(\log\sin x)\,dx = \log 2 - 1$

708. $\displaystyle\int_0^{\pi/2} (\log \tan x)\, dx = 0$

709. $\displaystyle\int_0^{\pi} \log\,(a \pm b \cos x)\, dx = \pi \log\left(\frac{a + \sqrt{a^2 - b^2}}{2}\right), \qquad (a \geq b)$

710. $\displaystyle\int_0^{\pi} \log\,(a^2 - 2ab \cos x + b^2)\, dx = \begin{cases} 2\pi \log a, & a \geq b > 0 \\ 2\pi \log b, & b \geq a > 0 \end{cases}$

711. $\displaystyle\int_0^{\infty} \frac{\sin ax}{\sinh bx}\, dx = \frac{\pi}{2b} \tanh \frac{a\pi}{2b}$

712. $\displaystyle\int_0^{\infty} \frac{\cos ax}{\cosh bx}\, dx = \frac{\pi}{2b} \operatorname{sech} \frac{a\pi}{2b}$

713. $\displaystyle\int_0^{\infty} \frac{dx}{\cosh ax} = \frac{\pi}{2a}$

714. $\displaystyle\int_0^{\infty} \frac{x\, dx}{\sinh ax} = \frac{\pi^2}{4a^2}$

715. $\displaystyle\int_0^{\infty} e^{-ax}(\cosh bx)\, dx = \frac{a}{a^2 - b^2}, \qquad (0 \leq |b| < a)$

716. $\displaystyle\int_0^{\infty} e^{-ax}(\sinh bx)\, dx = \frac{b}{a^2 - b^2}, \qquad (0 \leq |b| < a)$

717. $\displaystyle\int_0^{\infty} \frac{\sinh ax}{e^{bx} + 1}\, dx = \frac{\pi}{2b} \csc \frac{a\pi}{b} - \frac{1}{2a}$

718. $\displaystyle\int_0^{\infty} \frac{\sinh ax}{e^{bx} - 1}\, dx = \frac{1}{2a} - \frac{\pi}{2b} \cot \frac{a\pi}{b}$

719. $\displaystyle\int_0^{\pi/2} \frac{dx}{\sqrt{1 - k^2 \sin^2 x}} = \frac{\pi}{2}\left[1 + \left(\frac{1}{2}\right)^2 k^2 + \left(\frac{1 \cdot 3}{2 \cdot 4}\right)^2 k^4 \right.$
$$\left. + \left(\frac{1 \cdot 3 \cdot 5}{2 \cdot 4 \cdot 6}\right)^2 k^6 + \cdots\right], \text{ if } k^2 < 1$$

720. $\displaystyle\int_0^{\pi/2} \sqrt{1 - k^2 \sin^2 x}\, dx = \frac{\pi}{2}\left[1 - \left(\frac{1}{2}\right)^2 k^2 - \left(\frac{1 \cdot 3}{2 \cdot 4}\right)^2 \frac{k^4}{3} \right.$
$$\left. - \left(\frac{1 \cdot 3 \cdot 5}{2 \cdot 4 \cdot 6}\right)^2 \frac{k^6}{5} - \cdots\right], \text{ if } k^2 < 1$$

721. $\displaystyle\int_0^{\infty} e^{-x} \log x\, dx = -\gamma = -0.5772157\ldots$

722. $\displaystyle\int_0^{\infty} e^{-x^2} \log x\, dx = -\frac{\sqrt{\pi}}{4}(\gamma + 2 \log 2)$

723. $\displaystyle\int_0^{\infty} \left(\frac{1}{1 - e^{-x}} - \frac{1}{x}\right) e^{-x}\, dx = \gamma = 0.5772157\ldots$ [Euler's Constant]

724. $\displaystyle\int_0^{\infty} \frac{1}{x}\left(\frac{1}{1 + x} - e^{-x}\right) dx = \gamma = 0.5772157\ldots$

For n even:

725. $\displaystyle\int \cos^n x\, dx = \frac{1}{2^{n-1}} \sum_{k=0}^{\frac{n}{2}-1} \binom{n}{k} \frac{\sin (n - 2k)\, x}{(n - 2k)} + \frac{1}{2^n} \binom{n}{\frac{n}{2}} x$

726. $\int \sin^n x \, dx = \dfrac{1}{2^{n-1}} \displaystyle\sum_{k=0}^{\frac{n}{2}-1} \binom{n}{k} \dfrac{\sin\left[(n-2k)\left(\frac{\pi}{2}-x\right)\right]}{2k-n} + \dfrac{1}{2^n} \binom{n}{\frac{n}{2}} x$

For n odd:

727. $\int \cos^n x \, dx = \dfrac{1}{2^{n-1}} \displaystyle\sum_{k=0}^{\frac{n-1}{2}} \binom{n}{k} \dfrac{\sin(n-2k)x}{(n-2k)}$

728. $\int \sin^n x \, dx = \dfrac{1}{2^{n-1}} \displaystyle\sum_{k=0}^{\frac{n-1}{2}} \binom{n}{k} \dfrac{\sin\left[n-2k)\left(\frac{\pi}{2}-x\right)\right]}{2k-n}$

DIFFERENTIAL EQUATIONS
SPECIAL FORMULAS

Certain types of differential equations occur sufficiently often to justify the use of formulas for the corresponding particular solutions. The following set of tables 1 to XIV covers all first, second, and nth order ordinary linear differential equations with constant coefficients for which the right members are of the form $P(x)e^{rx} \sin sx$ or $P(x)e^{rx} \cos sx$, where r and s are constants and $P(x)$, is a polynomial of degree n.

When the right member of a reducible linear partial differential equation with constant coefficients is not zero, particular solutions for certain types of right members are contained in tables XV to XXI. In these tables both F and P are used to denote polynomials, and it is assumed that no denominator is zero. In any formula the roles of x and y may be reversed throughout, changing a formula in which x dominates to one in which y dominates. Tables XIX, XX, XXI are applicable whether the equations are reducible or not.

The symbol $\binom{m}{n}$ stands for $\dfrac{m!}{(m-n)!n!}$ and is the $n + 1$ st coefficient in the expansion of $(a + b)^m$. Also $0! = 1$ by definition.

The tables as herewith given are those contained in the text *Differential Equations* by Ginn and Company (1955) and are published with their kind permission and that of the author, Professor Frederick H. Steen.

Solution of Linear Differential Equations with Constant Coefficients

Any linear differential equation with constant coefficients may be written in the form

$$p(D)y = R(x)$$

where D is the differential operation

$$Dy = \frac{dy}{dx}$$

$p(D)$ is a polynomial in D,
y is the dependent variable,
x is the independent variable,
$R(x)$ is an arbitrary function of x.

A power of D represents repeated differentiation, that is

$$D^n y = \frac{d^n y}{dx^n}$$

For such an equation, the general solution may be written in the form

$$y = y_c + y_p$$

where y_p is any particular solution, and y_c is called the *complementary function*. This complementary function is defined as the general solution of the *homogeneous equation*, which is the original differential equation with the right side replaced by zero, i.e.

$$p(D)y = 0$$

The complementary function y_c may be determined as follows:

1. Factor the polynomial $p(D)$ into real and complex linear factors, just as if D were a variable instead of an operator.
2. For each nonrepeated linear factor of the form $(D - a)$, where a is real, write down a term of the form

$$ce^{ax}$$

where c is an arbitrary constant.
3. For each repeated real linear factor of the form $(D - a)^n$, write down n terms of the form

$$c_1 e^{ax} + c_2 x e^{ax} + c_3 x^2 e^{ax} + \cdots + c_n x^{n-1} e^{ax}$$

where the c_i's are arbitrary constants.

4. For each non-repeated conjugate complex pair of factors of the form $(D - a + ib)(D - a - ib)$, write down 2 terms of the form

$$c_1 e^{ax} \cos bx + c_2 e^{ax} \sin bx$$

5. For each repeated conjugate complex pair of factors of the form $(D - a + ib)^n (D - a - ib)^n$, write down $2n$ terms of the form

$$c_1 e^{ax} \cos bx + c_2 e^{ax} \sin bx + c_3 x e^{ax} \cos bx + c_4 x e^{ax} \sin bx + \cdots$$
$$+ c_{2n-1} x^{n-1} e^{ax} \cos bx + c_{2n} x^{n-1} e^{ax} \sin bx$$

6. The sum of all the terms thus written down is the complementary function y_c.

To find the particular solution y_p, use the following tables, as shown in the examples. For cases not shown in the tables, there are various methods of finding y_p. The most general method is called *variation of parameters*. The following example illustrates the method:

Find y_p for $(D^2 - 4) y = e^x$.

This example can be solved most easily by use of equation 63 in the tables following. However it is given here as an example of the method of variation of parameters.

The complementary function is

$$y_c = c_1 e^{2x} + c_2 e^{-2x}$$

To find y_p, replace the constants in the complementary function with unknown functions,

$$y_p = u e^{2x} + v e^{-2x}$$

We now prepare to substitute this assumed solution into the original equation. We begin by taking all the necessary dervatives:

$$y_p = u e^{2x} + v e^{-2x}$$
$$y_p' = 2u e^{2x} - 2v e^{-2x} + u' e^{2x} + v' e^{-2x}$$

For each derivative of y_p except the highest, we set the sum of all the terms containing u' and v' to 0. Thus the above equation becomes

$$u' e^{2x} + v' e^{-2x} = 0 \quad \text{and} \quad y_p' = 2u e^{2x} - 2v e^{-2x}$$

Continuing to differentiate, we have

$$y_p'' = 4u e^{2x} + 4v e^{-2x} + 2u' e^{2x} - 2v' e^{-2x}$$

When we substitute into the original equation, all the terms not containing u' or v' cancel out. This is a consequence of the method by which y_p was set up.

Thus all that is necessary is to write down the terms containing u' or v' in the highest order derivative of y_p, multiply by the constant coefficient of the highest power of D in $p(D)$, and set it equal to $R(x)$. Together with the previous terms in u' and v' which were set equal to 0, this gives us as many linear equations in the first derivatives of the unknown functions as there are unknown functions. The first derivatives may then be solved for by algebra, and the unknown functions found by integration. In the present example, this becomes

$$u' e^{2x} + v' e^{-2x} = 0$$
$$2u' e^{2x} - 2v' e^{-2x} = e^x$$

We eliminate v' and u' separately, getting

$$4u' e^{2x} = e^x$$
$$4v' e^{-2x} = -e^x$$

Thus

$$u' = \tfrac{1}{4} e^{-x}$$
$$v' = -\tfrac{1}{4} e^{3x}$$

Therefore, by integrating

$$u = -\tfrac{1}{4} e^{-x}$$
$$v = -\tfrac{1}{12} e^{3x}$$

A constant of integration is not needed, since we need only one particular solution. Thus

$$y_p = u e^{2x} + v e^{-2x} = -\tfrac{1}{4} e^{-x} e^{2x} - \tfrac{1}{12} e^{3x} e^{-2x}$$
$$= -\tfrac{1}{4} e^x - \tfrac{1}{12} e^x = -\tfrac{1}{3} e^x$$

and the general solution is

$$y = y_c + y_p = c_1 e^{2x} + c_2 e^{-2x} - \tfrac{1}{3} e^x$$

The following examples illustrate the use of the tables.

Example 1. Solve $(D^2 - 4)y = \sin 3x$.

Substitution of $q = -4$, $s = 3$ in formula 24 gives

$$y_p = \frac{\sin 3x}{-9 - 4}$$

wherefore the general solution is

$$y = c_1 e^{2x} + c_2 e^{-2x} - \frac{\sin 3x}{13}$$

Example 2. Obtain a particular solution of $(D^2 - 4D + 5)y = x^2 e^{3x} \sin x$.

Applying formula 40 with $a = 2, b = 1, r = 3, s = 1$, $P(x) = x^2$, $s + b = 2$, $s - b = 0$, $a - r = -1$, $(a - r)^2 + (s + b)^2 = 5$, $(a - r)^2 + (s - b)^2 = 1$, we have

$$y_p = \frac{e^{3x} \sin x}{2} \left[\left(\frac{2}{5} - \frac{0}{1} \right) x^2 + \left(\frac{2(-1)2}{25} - \frac{2(-1)0}{1} \right) 2x \right.$$

$$\left. + \left(\frac{3 \cdot 1 \cdot 2 - 2^3}{125} - \frac{3 \cdot 1 \cdot 0 - 0}{1} \right) 2 \right]$$

$$- \frac{e^{3x} \cos x}{2} \left[\left(\frac{-1}{5} - \frac{-1}{1} \right) x^2 + \left(\frac{1 - 4}{25} - \frac{1 - 0}{1} \right) 2x \right.$$

$$\left. + \left(\frac{-1 - 3(-1)4}{125} - \frac{-1 - 3(-1)0}{1} \right) 2 \right]$$

$$= (\tfrac{1}{5}x^2 - \tfrac{4}{25}x - \tfrac{2}{125}) e^{3x} \sin x + (-\tfrac{2}{5}x^2 + \tfrac{28}{25}x - \tfrac{136}{125}) e^{3x} \cos x$$

The special formulas effect a very considerable saving of time in problems of this type.

Example 3. Obtain a particular solution of $(D^2 - 4D + 5)y = x^2 e^{2x} \cos x$. (Compare with Example 2.)

Formula 40 is not applicable here since for this equation $r = a$, $s = b$, wherefore the denominator $(a - r)^2 + (s - b)^2 = 0$. We turn instead to formula 44. Substituting $a = 2$, $b = 1$, $P(x) = x^2$ and replacing sin by cos, cos by $-\sin$, we obtain

$$y_p = \frac{e^{2x} \cos x}{4} (x^2 - \tfrac{2}{4}) + \frac{e^{2x} \sin x}{2} \int (x^2 - \tfrac{1}{2}) dx$$

$$= \left(\frac{x^2}{4} - \frac{1}{8} \right) e^{2x} \cos x + \left(\frac{x^3}{6} - \frac{x}{4} \right) e^{2x} \sin x$$

which is the required solution.

Example 4. Find z_p for $(D_x - 3 D_y) z = \ln(y + 3x)$.

Referring to Table XV we note that formula 69 (not 68) is applicable. This gives

$$z_p = x \ln(y + 3x)$$

It is easily seen that $- y/3 \ln(y + 3x)$ would serve equally well.

Example 5. Solve $(D_x + 2D_y - 4) z = y \cos(y - 2x)$.

Since R in formula 76 contains a polynomial in x, not y, we rewrite the given equation in the form

$$(D_y + \tfrac{1}{2}D_x - 2)z = \tfrac{1}{2} y \cos(y - 2x)$$

Then

$$z_c = e^{2y} F(x - \tfrac{1}{2}y) = e^{2y} f(2x - y)$$

and by the formula

$$z_p = -\tfrac{1}{2} \cos(y - 2x) \cdot \left(\frac{y}{2} + \frac{\tfrac{1}{2}}{2} \right)$$

$$= -\tfrac{1}{8} (2y + 1) \cos(y - 2x)$$

Example 6. Find z_p for $(D_x + 4D_y)^3 z = (2x - y)^2$.

Using formula 79, we obtain

$$z_p = \frac{\int \int \int u^2 du^3}{[2 + 4(-1)]^3} = \frac{u^5}{5 \cdot 4 \cdot 3 \cdot (-8)} = -\frac{(2x - y)^5}{480}$$

Example 7. Find z_p for $(D_x^3 + 5D_x^2 D_y - 7D_x + 4)z = e^{2x+3y}$.

By formula 87

$$z_p = \frac{e^{2x+3y}}{2^3 + 5 \cdot 2^2 \cdot 3 - 7 \cdot 2 + 4} = \frac{e^{2x+3y}}{58}$$

Example 8. Find z_p for

$$(D_x^4 + 6D_x^3 D_y + D_x D_y + D_y^2 + 9)z = \sin(3x + 4y)$$

A-67

Since every term in the left member is of even degree in the two operators D_x and D_y, formula 90 is applicable. It gives

$$z_p = \frac{\sin(3x + 4y)}{(-9)^2 + 6(-9)(-12) + (-12) + (-16) + 9}$$

$$= \frac{\sin(3x + 4y)}{710}$$

TABLE I: $(D - a)y = R$

R	y_p

1. e^{rx} $\dfrac{e^{rx}}{r - a}$

2. $\sin sx^*$ $-\dfrac{a \sin sx + s \cos sx}{a^2 + s^2} = -\dfrac{1}{\sqrt{a^2 + s^2}} \sin\left(sx + \tan^{-1}\dfrac{s}{a}\right)$

3. $P(x)$ $-\dfrac{1}{a}\left[P(x) + \dfrac{P'(x)}{a} + \dfrac{P''(x)}{a^2} + \cdots + \dfrac{P^{(n)}(x)}{a^n}\right]$

4. $e^{rx} \sin sx^*$ Replace a by $a - r$ in formula 2 and multiply by e^{rx}.

5. $P(x)e^{rx}$ Replace a by $a - r$ in formula 3 and multiply by e^{rx}.

6. $P(x)\sin sx^*$ $-\sin sx\left[\dfrac{a}{a^2 + s^2}P(x) + \dfrac{a^2 - s^2}{(a^2 + s^2)^2}P'(x) + \dfrac{a^3 - 3as^2}{(a^2 + s^2)^3}P''(x) + \cdots\right.$

$$\left. + \dfrac{a^k - \binom{k}{2}a^{k-2}s^2 + \binom{k}{4}a^{k-4}s^4 - \cdots}{(a^2 + s^2)^k}P^{(k-1)}(x) + \cdots\right]$$

$$-\cos sx\left[\dfrac{s}{a^2 + s^2}P(x) + \dfrac{2as}{(a^2 + s^2)^2}P'(x) + \dfrac{3a^2s - s^3}{(a^2 + s^2)^3}P''(x) + \cdots\right.$$

$$\left. + \dfrac{\binom{k}{1}a^{k-1}s - \binom{k}{3}a^{k-3}s^3 + \cdots}{(a^2 + s^2)^k}P^{(k-1)}(x) + \cdots\right]$$

7. $P(x)e^{rx}\sin sx^*$ Replace a by $a - r$ in formula 6 and multiply by e^{rx}.

8. e^{ax} xe^{ax}

9. $e^{ax}\sin sx^*$ $-\dfrac{e^{ax}\cos sx}{s}$

10. $P(x)e^{ax}$ $e^{ax}\displaystyle\int P(x)\,dx$

11. $P(x)e^{ax}\sin sx^*$ $\dfrac{e^{ax}\sin sx}{s}\left[\dfrac{P'(x)}{s} - \dfrac{P'''(x)}{s^3} + \dfrac{P^{v}(x)}{s^5} - \cdots\right] - \dfrac{e^{ax}\cos sx}{s}\left[P(x) - \dfrac{P''(x)}{s^2} + \dfrac{P^{iv}(x)}{s^4} - \cdots\right]$

* For $\cos sx$ in R replace "sin" by "cos" and "cos" by " $-$ sin" in y_p.

$$D^n = \frac{d^n}{dx^n} \qquad \binom{m}{n} = \frac{m!}{(m - n)!n!} \qquad 0! = 1$$

TABLE II: $(D - a)^2 y = R$

R	y_p

12. e^{rx} $\dfrac{e^{rx}}{(r - a)^2}$

13. $\sin sx^*$ $\dfrac{1}{(a^2 + s^2)^2}[(a^2 - s^2)\sin sx + 2as\cos sx] = \dfrac{1}{a^2 + s^2}\sin\left(sx + \tan^{-1}\dfrac{2as}{a^2 - s^2}\right)$

14. $P(x)$ $\dfrac{1}{a^2}\left[P(x) + \dfrac{2P'(x)}{a} + \dfrac{3P''(x)}{a^2} + \cdots + \dfrac{(n + 1)P^{(n)}(x)}{a^n}\right]$

15. $e^{rx}\sin sx^*$ Replace a by $a - r$ in formula 13 and multiply by e^{rx}.

16. $P(x)e^{rx}$ Replace a by $a - r$ in formula 14 and multiply by e^{rx}.

17. $P(x)\sin sx^*$ $\sin sx\left[\dfrac{a^2 - s^2}{(a^2 + s^2)^2}P(x) + 2\dfrac{a^3 - 3as^2}{(a^2 + s^2)^3}P'(x) + 3\dfrac{a^4 - 6a^2s^2 + s^4}{(a^2 + s^2)^4}P''(x) + \cdots\right.$

$$\left. + (k - 1)\dfrac{a^k - \binom{k}{2}a^{k-2}s^2 + \binom{k}{4}a^{k-4}s^4 - \cdots}{(a^2 + s^2)^k}P^{(k-2)}(x) + \cdots\right]$$

$$+ \cos sx\left[\dfrac{2as}{(a^2 + s^2)^2}P(x) + 2\dfrac{3a^2s - s^3}{(a^2 + s^2)^3}P'(x) + 3\dfrac{4a^3s - 4as^3}{(a^2 + s^2)^4}P''(x) + \cdots\right.$$

$$\left. + (k - 1)\dfrac{\binom{k}{1}a^{k-1}s - \binom{k}{3}a^{k-3}s^3 + \cdots}{(a^2 + s^2)^k}P^{(k-2)}(x) + \cdots\right]$$

18. $P(x)e^{rx}\sin sx^*$ Replace a by $a - r$ in formula 17 and multiply by e^{rx}.

19. e^{ax} $\frac{1}{2}x^2 e^{ax}$

20. $e^{ax}\sin sx^*$ $-\dfrac{e^{ax}\sin sx}{s^2}$

21. $P(x)e^{ax}$ $e^{ax}\displaystyle\iint P(x)\,dx\,dx$

22. $P(x)e^{ax}\sin sx^*$ $-\dfrac{e^{ax}\sin sx}{s^2}\left[P(x) - \dfrac{3P''(x)}{s^2} + \dfrac{5P^{iv}(x)}{s^4} - \dfrac{7P^{vi}(x)}{s^6} + \cdots\right]$

$$-\dfrac{e^{ax}\cos sx}{s^2}\left[\dfrac{2P'(x)}{s} - \dfrac{4P'''(x)}{s^3} + \dfrac{6P^{v}(x)}{s^5} - \cdots\right]$$

* For $\cos sx$ in R replace "sin" by "cos" and "cos" by " $-$ sin" in y_p.

TABLE III: $(D^2 + q)y = R$

R	y_p

23. e^{rx} $\dfrac{e^{rx}}{r^2 + q}$

24. $\sin sx^*$ $\dfrac{\sin sx}{-s^2 + q}$

25. $P(x)$ $\dfrac{1}{q}\left[P(x) - \dfrac{P''(x)}{q} + \dfrac{P^{iv}(x)}{q^2} - \cdots + (-1)^k \dfrac{P^{(2k)}(x)}{q^k} \cdots \right]$

26. $e^{rx}\sin sx^*$ $\dfrac{(r^2 - s^2 + q)e^{rx}\sin sx - 2\,rse^{rx}\cos sx}{(r^2 - s^2 + q)^2 + (2rs)^2} = \dfrac{e^{rx}}{\sqrt{(r^2 - s^2 + q)^2 + (2rs)^2}}\sin\left[sx - \tan^{-1}\dfrac{2rs}{r^2 - s^2 + q} \right]$

27. $P(x)e^{rx}$ $\dfrac{e^{rx}}{r^2 + q}\left[P(x) - \dfrac{2r}{r^2 + q}P'(x) + \dfrac{3r^2 - q}{(r^2 + q)^2}P''(x) - \dfrac{4r^3 - 4qr}{(r^2 + q)^3}P'''(x) \right.$
$$\left. + \cdots + (-1)^{k-1}\dfrac{\binom{k}{1}r^{k-1} - \binom{k}{3}r^{k-3}q + \binom{k}{5}r^{k-5}q^2 - \cdots}{(r^2 + q)^{k-1}}P^{(k-1)}(x) + \cdots \right]$$

28. $P(x)\sin sx^*$ $\dfrac{\sin sx}{(-s^2 + q)}\left[P(x) - \dfrac{3s^2 + q}{(-s^2 + q)^2}P''(x) + \dfrac{5s^4 + 10s^2q + q^2}{(-s^2 + q)^4}P^{iv}(x) + \cdots \right.$
$$\left. + (-1)^k\dfrac{\binom{2k+1}{1}s^{2k} + \binom{2k+1}{3}s^{2k-2}q + \binom{2k+1}{5}s^{2k-4}q^2 + \cdots}{(-s^2 + q)^{2k}}P^{(2k)}(x) + \cdots \right]$$
$$- \dfrac{s\cos sx}{(-s^2 + q)}\left[\dfrac{2P'(x)}{(-s^2 + q)} - \dfrac{4s^2 + 4q}{(-s^2 + q)^3}P'''(x) + \cdots \right.$$
$$\left. + (-1)^{k+1}\dfrac{\binom{2k}{1}s^{2k-2} + \binom{2k}{3}s^{2k-4}q + \cdots}{(-s^2 + q)^{2k-1}}P^{(2k-1)}(x) + \cdots \right]$$

TABLE IV: $(D^2 + b^2)y = R$

29. $\sin bx^*$ $-\dfrac{x\cos bx}{2b}$

30. $P(x)\sin bx^*$ $\dfrac{\sin bx}{(2b)^2}\left[P(x) - \dfrac{P''(x)}{(2b)^2} + \dfrac{P^{iv}(x)}{(2b)^4} - \cdots \right] - \dfrac{\cos bx}{2b}\int\left[P(x) - \dfrac{P''(x)}{(2b)^2} + \cdots \right]dx$

* For $\cos sx$ in R replace "sin" by "cos" and "cos" by "$-$ sin" in y_p.

TABLE V: $(D^2 + pD + q)y = R$

R	y_p

31. e^{rx} $\dfrac{e^{rx}}{r^2 + pr + q}$

32. $\sin sx^*$ $\dfrac{(q - s^2)\sin sx - ps\cos sx}{(q - s^2)^2 + (ps)^2} = \dfrac{1}{\sqrt{(q - s^2)^2 + (ps)^2}}\sin\left(sx - \tan^{-1}\dfrac{ps}{q - s^2} \right)$

33. $P(x)$ $\dfrac{1}{q}\left[P(x) - \dfrac{p}{q}P'(x) + \dfrac{p^2 - q}{q^2}P''(x) - \dfrac{p^3 - 2pq}{q^3}P'''(x) + \cdots \right.$
$$\left. + (-1)^n\dfrac{p^n - \binom{n-1}{1}p^{n-2}q + \binom{n-2}{2}p^{n-4}q^2 - \cdots}{q^n}P^{(n)}(x) \right]$$

34. $e^{rx}\sin sx^*$ Replace p by $p + 2r$, q by $q + pr + r^2$ in formula 32 and multiply by e^{rx}.

35. $P(x)e^{rx}$ Replace p by $p + 2r$, q by $q + pr + r^2$ in formula 33 and multiply by e^{rx}.

TABLE VI: $(D - b)(D - a)y = R$

36. $P(x)\sin sx^*$ $\dfrac{\sin sx}{b - a}\left[\left(\dfrac{a}{a^2 + s^2} - \dfrac{b}{b^2 + s^2} \right)P(x) + \left(\dfrac{a^2 - s^2}{(a^2 + s^2)^2} - \dfrac{b^2 - s^2}{(b^2 + s^2)^2} \right)P'(x) \right.$
$$\left. + \left(\dfrac{a^3 - 3as^2}{(a^2 + s^2)^3} - \dfrac{b^3 - 3bs^2}{(b^2 + s^2)^3} \right)P''(x) + \cdots \right]$$
$$+ \dfrac{\cos sx}{b - a}\left[\left(\dfrac{s}{a^2 + s^2} - \dfrac{s}{b^2 + s^2} \right)P(x) + \left(\dfrac{2as}{(a^2 + s^2)^2} - \dfrac{2bs}{(b^2 + s^2)^2} \right)P'(x) \right.$$
$$\left. + \left(\dfrac{3a^2s - s^3}{(a^2 + s^2)^3} - \dfrac{3b^2s - s^3}{(b^2 + s^2)^3} \right)P''(x) + \cdots \right]^\dagger$$

37. $P(x)e^{rx}\sin sx^*$ Replace a by $a - r$, b by $b - r$ in formula 36 and multiply by e^{rx}.

38. $P(x)e^{ax}$ $\dfrac{e^{ax}}{a - b}\left[\int P(x)dx + \dfrac{P(x)}{(b - a)} + \dfrac{P'(x)}{(b - a)^2} + \dfrac{P''(x)}{(b - a)^3} + \cdots + \dfrac{P^{(n)}(x)}{(b - a)^{n+1}} \right]$

* For $\cos sx$ in R replace "sin" by "cos" and "cos" by "$-$ sin" in y_p.
† For additional terms, compare with formula 6.

TABLE VII: $(D^2 - 2aD + a^2 + b^2)y = R$

R	y_p

39. $P(x) \sin sx$*

$$\frac{\sin sx}{2b}\left[\left(\frac{s+b}{a^2+(s+b)^2} - \frac{s-b}{a^2+(s-b)^2}\right)P(x) + \left(\frac{2a(s+b)}{[a^2+(s+b)^2]^2} - \frac{2a(s-b)}{[a^2+(s-b)^2]^2}\right)P'(x)\right.$$

$$\left. + \left(\frac{3a^2(s+b)-(s+b)^3}{[a^2+(s+b)^2]^3} - \frac{3a^2(s-b)-(s-b)^3}{[a^2+(s-b)^2]^3}\right)P''(x) + \cdots\right]$$

$$-\frac{\cos sx}{2b}\left[\left(\frac{a}{a^2+(s+b)^2} - \frac{a}{a^2+(s-b)^2}\right)P(x) + \left(\frac{a^2-(s+b)^2}{[a^2+(s+b)^2]^2} - \frac{a^2-(s-b)^2}{[a^2+(s-b)^2]^2}\right)P'(x)\right.$$

$$\left. + \left(\frac{a^3-3a(s+b)^2}{[a^2+(s+b)^2]^3} - \frac{a^3-3a(s-b)^2}{[a^2+(s-b)^2]^3}\right)P''(x) + \cdots\right]†$$

40. $P(x)e^{rx}\sin sx$*   Replace a by $a - r$ in formula 39 and multiply by e^{rx}.

41. $P(x)e^{ax}$ $\dfrac{e^{ax}}{b^2}\left[P(x) - \dfrac{P''(x)}{b^2} + \dfrac{P^{iv}(x)}{b^4} - \cdots\right]$

42. $e^{ax}\sin sx$* $\dfrac{e^{ax}\sin sx}{-s^2+b^2}$

43. $e^{ax}\sin bx$* $-\dfrac{xe^{ax}\cos bx}{2b}$

44. $P(x)e^{ax}\sin bx$* $\dfrac{e^{ax}\sin bx}{(2b)^2}\left[P(x) - \dfrac{P''(x)}{(2b)^2} + \dfrac{P^{iv}(x)}{(2b)^4} - \cdots\right] - \dfrac{e^{ax}\cos bx}{2b}\displaystyle\int\left[P(x) - \dfrac{P''(x)}{(2b)^2} + \dfrac{P^{iv}(x)}{(2b)^4} - \cdots\right]dx$

* For $\cos sx$ in R replace "sin" by "cos" and "cos" by "$-$ sin" in y_p.
† For additional terms, compare with formula 6.

TABLE VIII: $f(D)y = [D^n + a_{n-1}D^{n-1} + \cdots + a_1D + a_0]y = R$

R	y_p

45. e^{rx} $\dfrac{e^{rx}}{f(r)}$

46. $\sin sx$* $\dfrac{[a_0 - a_2s^2 + a_4s^4 - \cdots]\sin sx - [a_1s - a_3s^3 + a_5s^5 + \cdots]\cos sx}{[a_0 - a_2s^2 + a_4s^4 - \cdots]^2 + [a_1s - a_3s^3 + a_5s^5 - \cdots]^2}$

TABLE IX: $f(D^2)y = R$

47. $\sin sx$* $\dfrac{\sin sx}{f(-s^2)} = \dfrac{\sin sx}{a_0 - a_2s^2 + \cdots \pm s^{2n}}$

TABLE X: $(D - a)^ny = R$

48. e^{rx} $\dfrac{e^{rx}}{(r-a)^n}$

49. $\sin sx$* $\dfrac{(-1)^n}{(a^2+s^2)^n}\{[a^n - \binom{n}{2}a^{n-2}s^2 + \binom{n}{4}a^{n-4}s^4 - \cdots]\sin sx + [\binom{n}{1}a^{n-1}s - \binom{n}{3}a^{n-3}s^3 + \cdots]\cos sx\}$

50. $P(x)$ $\dfrac{(-1)^n}{a^n}\left[P(x) + \binom{n}{1}\dfrac{P'(x)}{a} + \binom{n+1}{2}\dfrac{P''(x)}{a^2} + \binom{n+2}{3}\dfrac{P'''(x)}{a^3} + \cdots\right]$

51. $e^{rx}\sin sx$*   Replace a by $a - r$ in formula 49 and multiply by e^{rx}.

52. $e^{rx}P(x)$ Replace a by $a - r$ in formula 50 and multiply by e^{rx}.

53. $P(x)\sin sx$* $(-1)^n\sin sx[A_nP(x) + \binom{n}{1}A_{n+1}P'(x) + \binom{n+1}{2}A_{n+2}P''(x) + \binom{n+2}{3}A_{n+2}P'''(x) + \cdots]$
$\qquad + (-1)^n\cos sx[B_nP(x) + \binom{n}{1}B_{n+1}P'(x) + \binom{n+1}{2}B_{n+2}P''(x) + \binom{n+2}{3}B_{n+2}P'''(x) + \cdots]$

$$A_1 = \frac{a}{a^2+s^2}, A_2 = \frac{a^2-s^2}{(a^2+s^2)^2}, \cdots, A_k = \frac{a^k - \binom{k}{2}a^{k-2}s^2 + \binom{k}{4}a^{k-4}s^4 - \cdots}{(a^2+s^2)^k}.$$

$$B_1 = \frac{s}{a^2+s^2}, B_2 = \frac{2as}{(a^2+s^2)^2}, \cdots, B_k = \frac{\binom{k}{1}a^{k-1}s - \binom{k}{3}a^{k-3}s^3 + \cdots}{(a^2+s^2)^k}.$$

54. $P(x)e^{rx}\sin sx$*   Replace a by $a - r$ in formula 53 and multiply by e^{rx}.

55. $e^{ax}P(x)$ $e^{ax}\displaystyle\int\int\cdots\int P(x)dx^n$

56. $P(x)e^{ax}\sin sx{*}$ $\dfrac{(-1)^{\frac{n-1}{2}}e^{ax}\sin sx}{s^n}\left[\binom{n}{n-1}\dfrac{P'(x)}{s}-\binom{n+2}{n-1}\dfrac{P'''(x)}{s^3}+\binom{n+4}{n-1}\dfrac{P^{v}(x)}{s^5}-\cdots\right]$

$$+\dfrac{(-1)^{\frac{n+1}{2}}e^{ax}\cos sx}{s^n}\left[\binom{n-1}{n-1}P(x)-\binom{n+1}{n-1}\dfrac{P''(x)}{s^2}+\binom{n+3}{n-1}\dfrac{P^{iv}(x)}{s^4}-\cdots\right]\ (n\ \text{odd})$$

$$\dfrac{(-1)^{\frac{n}{2}}e^{ax}\sin sx}{s^n}\left[\binom{n-1}{n-1}P(x)-\binom{n+1}{n-1}\dfrac{P''(x)}{s^2}+\binom{n+3}{n-1}\dfrac{P^{iv}(x)}{s^4}-\cdots\right]$$

$$+\dfrac{(-1)^{\frac{n}{2}}e^{ax}\cos sx}{s^n}\left[\binom{n}{n-1}\dfrac{P'(x)}{s}-\binom{n+2}{n-1}\dfrac{P'''(x)}{s^3}+\binom{n+4}{n-1}\dfrac{P^{v}(x)}{s^5}-\cdots\right]\ (n\ \text{even})$$

* For $\cos sx$ in R replace "sin" by "cos" and "cos" by "$-$ sin" in y_p.

TABLE XI: $(D-a)^n f(D)y=R$

57. e^{ax} $\dfrac{x^n}{n!}\cdot\dfrac{e^{ax}}{f(a)}$

* For $\cos sx$ in R replace "sin" by "cos" and "cos" by "$-$ sin" in y_p.

TABLE XII: $(D^2+q)^n y=R$

R	y_p
58. e^{rx}	$e^{rx}/(r^2+q)^n$
59. $\sin sx{*}$	$\sin sx/(q-s^2)^n$
60. $P(x)$	$\dfrac{1}{q^n}\left[P(x)-\binom{n}{1}\dfrac{P''(x)}{q}+\binom{n+1}{2}\dfrac{P^{iv}(x)}{q^2}-\binom{n+2}{3}\dfrac{P^{vi}(x)}{q^3}+\cdots\right]$
61. $e^{rx}\sin sx{*}$	$\dfrac{e^{rx}}{(A^2+B^2)^n}\{[A^n-\binom{n}{2}A^{n-2}B^2+\binom{n}{4}A^{n-4}B^4-\cdots]\sin sx-[\binom{n}{1}A^{n-1}B-\binom{n}{3}A^{n-3}B^3+\cdots]\cos sx\}$

$$A=r^2-s^2+q,\quad B=2rs$$

TABLE XIII: $(D^2+b^2)^n y=R$

62. $\sin bx{*}$ $(-1)^{\frac{n+1}{2}}\dfrac{x^n\cos bx}{n!(2b)^n}$ (n odd), $(-1)^{\frac{n}{2}}\dfrac{x^n\sin bx}{n!(2b)^n}$ (n even)

TABLE XIV: $(D^n-q)y=R$

63. e^{rx} $e^{rx}/(r^n-q)$

64. $P(x)$ $-\dfrac{1}{q}\left[P(x)\dfrac{P^{(n)}(x)}{q}+\dfrac{P^{(2n)}(x)}{q^2}+\cdots\right]$

65. $\sin sx{*}$ $-\dfrac{q\sin sx+(-1)^{\frac{n-1}{2}}s^n\cos sx}{q^2+s^{2n}}$ (n odd), $\dfrac{\sin sx}{(-s^2)^{n/2}-q}$ (n even)

66. $e^{rx}\sin sx{*}$ $\dfrac{Ae^{rx}\sin sr-Be^{rx}\cos sx}{A^2+B^2}=\dfrac{e^{rx}}{\sqrt{A^2+B^2}}\sin\left(sx-\tan^{-1}\dfrac{B}{A}\right)$

$$A=[r^n-\binom{n}{2}r^{n-2}s^2+\binom{n}{4}r^{n-4}s^4-\cdots]-q,\quad B=[\binom{n}{1}r^{n-1}s-\binom{n}{3}r^{n-3}s^3+\cdots]$$

* For $\cos sx$ in R replace "sin" by "cos" and "cos" by "$-$ sin" in y_p.

TABLE XV: $(D_x+mD_y)z=R$

R	z_p
67 e^{ax+by}	$\dfrac{e^{ax+by}}{a+mb}$
68. $f(ax+by)$	$\dfrac{\int f(u)du}{a+mb},\ u=ax+by$
69. $f(y-mx)$	$xf(y-mx)$
70. $\phi(x,y)f(y-mx)$	$f(y-mx)\int\phi(x,a+mx)dx$ $(a=y-mx$ after integration$)$

TABLE XVI: $(D_x + mD_y - k)z = R$

71. e^{ax+by}

$$\frac{e^{ax+by}}{a + mb - k}$$

72. $\sin(ax + by)$*

$$-\frac{(a + bm)\cos(ax + by) + k\sin(ax + by)}{(a + bm)^2 + k^2}$$

73. $e^{\alpha x+\beta y}\sin(ax + by)$*   Replace k in 72 by $k - \alpha - m\beta$ and multiply by $e^{\alpha x+\beta y}$

74. $e^{zx}f(ax + by)$

$$\frac{e^{kx}\int f(u)du}{a + mb}, u = ax + by$$

75. $f(y - mx)$

$$-\frac{f(y - mx)}{k}$$

76. $P(x)f(y - mx)$

$$-\frac{1}{k}f(y - mx)\left[P(x) + \frac{P'(x)}{k} + \frac{P''(x)}{k^2} + \cdots + \frac{P^{(n)}(x)}{k^n}\right]$$

77. $e^{kx}f(y - mx)$ $xe^{kx}f(y - mx)$

* For $\cos(ax + by)$ replace "sin" by "cos," and "cos" by "$-\sin$" in z_p.

$$D_x = \frac{\partial}{\partial x}; \qquad D_y = \frac{\partial}{\partial y}; \qquad D_x{}^k D_y{}^r = \frac{\partial^{k+r}}{\partial_x{}^k\partial_y{}^r}$$

TABLE XVII: $(D_x + mD_y)^n z = R$

R z_p

78. e^{ax+by}

$$\frac{e^{ax+by}}{(a + mb)^n}$$

79. $f(ax + by)$

$$\frac{\int\int\cdots\int f(u)du^n}{(a + mb)^n}, u = ax + by$$

80. $f(y - mx)$

$$\frac{x^n}{n!}f(y - mx)$$

81. $\phi(x, y)f(y + mx)$ $f(y - mx)\int\int\cdots\int\phi(x, a + mx)dx^n$ ($a = y - mx$ after integration)

TABLE XVIII: $(D_x + mD_y - k)^n z = R$

82. e^{ax+by}

$$\frac{e^{ax+by}}{(a + mb - k)^n}$$

83. $f(y - mx)$

$$\frac{(-1)^n f(y - mx)}{k^n}$$

84. $P(x)f(y - mx)$

$$\frac{(-1)^n}{k^n}f(y - mx)\left[P(x) + \binom{n}{1}\frac{P'(x)}{k} + \binom{n+1}{2}\frac{P''(x)}{k^2} + \binom{n+2}{3}\frac{P'''(x)}{k^3} + \cdots\right]$$

85. $e^{kx}f(ax + by)$

$$\frac{e^{kx}\int\int\cdots\int f(u)du^n}{(a + mb)^n}, u = ax + by$$

86. $e^{kx}f(y - mx)$

$$\frac{x^n}{n!}e^{kx}f(y - mx)$$

TABLE XIX: $[D_x{}^n + a_1 D_x{}^{n-1}D_y + a_2 D_x{}^{n-2}D_y{}^2 + \cdots + a^n D_y{}^n]z = R$

87. $e^{ax + by}$

$$\frac{e^{ax + by}}{a + a_1 a^{n-1}b + a_2 a^{n-2}b^2 + \cdots + a_n b^n}$$

88. $f(ax + by)$

$$\frac{\int\int\cdots\int f(u)du^n}{a^n + a_1 a^{n-1}b + a_2 a^{n-2}b^2 + \cdots + a^n b^n}, (u = ax + by)$$

TABLE XX: $F(D_x, D_y)z = R$

89. $e^{ax + by}$

$$\frac{e^{ax + by}}{F(a, b)}$$

TABLE XXI: $F(D_x{}^2, D_x D_y, D_y{}^2)z = R$

90. $\sin(ax + by)$*

$$\frac{\sin(ax + by)}{F(-a^2, -ab, -b^2)}$$

* For $\cos(ax + by)$ replace "sin" by "cos", and "cos" by "$-\sin$" in z_p.

DIFFERENTIAL EQUATIONS

Differential equation	Method of solution
Separation of variables $$f_1(x)\,g_1(y)\,dx + f_2(x)\,g_2(y)\,dy = 0$$	$$\int \frac{f_1(x)}{f_2(x)}\,dx + \int \frac{g_2(y)}{g_1(y)}\,dy = c$$
Exact equation $$M(x,y)dx + N(x,y)dy = 0$$ where $\partial M/\partial y = \partial N/\partial x$	$$\int M\,\partial x + \int \left(N - \frac{\partial}{\partial y}\int M\,\partial x \right) dy = c$$ where ∂x indicates that the integration is to be performed with respect to x keeping y constant.
Linear first order equation $$\frac{dy}{dx} + P(x)y = Q(x)$$	$$ye^{\int P dx} = \int Q e^{\int P dx}\,dx + c$$
Bernoulli's equation $$\frac{dy}{dx} + P(x)y = Q(x)y^n$$	$$ve^{(1-n)\int P dx} = (1-n)\int Q e^{(1-n)\int P dx}\,dx + c$$ where $v = y^{1-n}$. If $n = 1$, the solution is $$\ln y = \int (Q-P)\,dx + c$$
Homogeneous equation $$\frac{dy}{dx} = F\left(\frac{y}{x}\right)$$	$$\ln x = \int \frac{dv}{F(v)-v} + c$$ where $v = y/x$. If $F(v) = v$, the solution is $y = cx$
Reducible to homogeneous $$(a_1 x + b_1 y + c_1)\,dx + (a_2 x + b_2 y + c_2)\,dy = 0$$ $$\frac{a_1}{a_2} \neq \frac{b_1}{b_2}$$	Set $u = a_1 x + b_1 y + c_1$. $$v = a_2 x + b_2 y + c_2$$ Eliminate x and y and the equation becomes homogenous
Reducible to separable $$(a_1 x + b_1 y + c_1)\,dx + (a_2 x + b_2 y + c_2)\,dy = 0$$ $$\frac{a_1}{a_2} = \frac{b_1}{b_2}$$	Set $u = a_1 x + b_1 y$ Eliminate x or y and equation becomes separable

$y\,F(xy)\,dx + x\,G(xy)dy = 0$	$\ln x = \displaystyle\int \frac{G(v)\,dv}{v\,\{G(v) - F(v)\,\}} + c$ where $v = xy$. If $G(v) = F(v)$, the solution is $xy = c$.
Linear, homogeneous second order equation $\dfrac{d^2 y}{dx^2} + b\,\dfrac{dy}{dx} + cy = 0$ b,c are real constants	Let m_1, m_2 be the roots of $m^2 + bm + c = 0$. Then there are 3 cases: Case 1. m_1, m_2 real and distinct: $$y = c_1 e^{m_1 x} + c_2 e^{m_2 x}$$ Case 2. m_1, m_2 real and equal: $$y = c_1 e^{m_1 x} + c_2 x e^{m_1 x}$$ Case 3. $m_1 = p + qi,\ m_2 = p - qi$: $$y = e^{px}(c_1 \cos qx + c_2 \sin qx)$$ where $p = -b/2,\ q = \sqrt{4c - b^2}/2$
Linear, nonhomogeneous second order equation $\dfrac{d^2 y}{dx^2} + b\dfrac{dy}{dx} + cy = R(x)$ b, c are real constants	There are 3 cases corresponding to those immediately above: Case 1. $$y = c_1 e^{m_1 x} + c_2 e^{m_2 x}$$ $$+ \frac{e^{m_1 x}}{m_1 - m_2}\int e^{-m_1 x}\,R(x)\,dx$$ $$+ \frac{e^{m_2 x}}{m_2 - m_1}\int e^{-m_2 x}\,R(x)\,dx$$ Case 2. $$y = c_1 e^{m_1 x} + c_2 x e^{m_1 x}$$ $$+ x e^{m_1 x}\int e^{-m_1 x}\,R(x)\,dx$$ $$- e^{m_1 x}\int x e^{-m_1 x}\,R(x)\,dx$$ Case 3. $$y = e^{px}(c_1 \cos qx + c_2 \sin qx)$$ $$+ \frac{e^{px}\sin qx}{q}\int e^{-px}\,R(x)\cos qx\,dx$$ $$- \frac{e^{px}\cos qx}{q}\int e^{-px}\,R(x)\sin qx\,dx$$

DIFFERENTIAL EQUATIONS (Continued)

Euler or Cauchy equation $$x^2 \frac{d^2 y}{dx^2} + bx \frac{dy}{dx} + cy = S(x)$$	Putting $x = e^t$, the equation becomes $$\frac{d^2 y}{dt^2} + (b-1)\frac{dy}{dt} + cy = S(e^t)$$ and can then be solved as a linear second order equation.
Bessel's equation $$x^2 \frac{d^2 y}{dx^2} + x \frac{dy}{dx} + (\lambda^2 x^2 - n^2)y = 0$$	$$y = c_1 J_n(\lambda x) + c_2 Y_n(\lambda x)$$
Transformed Bessel's equation $$x^2 \frac{d^2 y}{dx^2} + (2p+1)x \frac{dy}{dx} + (\alpha^2 x^{2r} + \beta^2)y = 0$$	$$y = x^{-p} \left\{ c_1 J_{q/r}\left(\frac{\alpha}{r}x^r\right) + c_2 Y_{q/r}\left(\frac{\alpha}{r}x^r\right) \right\}$$ where $q = \sqrt{p^2 - \beta^2}$.
Legendre's equation $$(1-x^2)\frac{d^2 y}{dx^2} - 2x \frac{dy}{dx} + n(n+1)y = 0$$	$$y = c_1 P_n(x) + c_2 Q_n(x)$$

FOURIER SERIES

If $f(x)$ is a bounded periodic function of period 2L (i.e. $f(x+2L) = f(x)$), and satisfies the *Dirichlet conditions:*

A.   In any period $f(x)$ is continuous, except possibly for a finite number of jump discontinuities.
B.   In any period $f(x)$ has only a finite number of maxima and minima.

Then $f(x)$ may be represented by the *Fourier series*

$$\frac{a_0}{2} + \sum_{n=1}^{\infty} \left(a_n \cos \frac{n\pi x}{L} + b_n \sin \frac{n\pi x}{L} \right)$$

where a_n and b_n are as determined below. This series will converge to $f(x)$ at every point where $f(x)$ is continuous, and to

$$\frac{f(x^+) + f(x^-)}{2}$$

(i.e. the average of the left-hand and right-hand limits) at every point where $f(x)$ has a jump discontinuity.

$$a_n = \frac{1}{L} \int_{-L}^{L} f(x) \cos \frac{n\pi x}{L} \, dx, \quad n = 0, 1, 2, 3, \ldots,$$

$$b_n = \frac{1}{L} \int_{-L}^{L} f(x) \sin \frac{n\pi x}{L} \, dx, \quad n = 1, 2, 3, \ldots$$

We may also write

$$a_n = \frac{1}{L} \int_{a}^{a+2L} f(x) \cos \frac{n\pi x}{L} \, dx \text{ and } b_n = \frac{1}{L} \int_{a}^{a+2L} f(x) \sin \frac{n\pi x}{L} \, dx$$

where α is any real number. Thus if $\alpha = 0$,

$$a_n = \frac{1}{L} \int_{0}^{2L} f(x) \cos \frac{n\pi x}{L} \, dx, \quad n = 0, 1, 2, 3, \ldots,$$

$$b_n = \frac{1}{L} \int_{0}^{2L} f(x) \sin \frac{n\pi x}{L} \, dx, \quad n = 1, 2, 3, \ldots$$

2. If in addition to the above restrictions, $f(x)$), is even (i.e., $f(-x) = f(x)$ the Fourier series reduces to

$$\frac{a_0}{2} + \sum_{n=1}^{\infty} a_n \cos \frac{n\pi x}{L}$$

That is, $b_n = 0$. In this case, a simpler formula for a_n is

$$a_n = \frac{2}{L} \int_0^L f(x) \cos \frac{n\pi x}{L} dx, \quad n = 0, 1, 2, 3, \dots$$

3. If in addition to the restrictions in (1), $f(x)$ is an odd function (i.e., $f(-x) = -f(x)$), then the Fourier series reduces to

$$\sum_{n=1}^{\infty} b_n \sin \frac{n\pi x}{L}$$

That is, $a_n = 0$. In this case, a simpler formula for the b_n is

$$b_n = \frac{2}{L} \int_0^L f(x) \sin \frac{n\pi x}{L} dx, \quad n = 1, 2, 3, \dots$$

4. If in addition to the restrictions in (2) above, $f(x) = -f(L - x)$, then a_n will be 0 for all even values of n, including $n = 0$. Thus in this case, the expansion reduces to

$$\sum_{m=1}^{\infty} a_{2m-1} \cos \frac{(2m-1)\pi x}{L}$$

5. If in addition to the restrictions in (3) above, $f(x)$, $= f(L - x)$, then b_n will be 0 for all even values of n. Thus in this case, the expansion reduces to

$$\sum_{m=1}^{\infty} b_{2m-1} \sin \frac{(2m-1)\pi x}{L}$$

(The series in (4) and (5) are known as *odd-harmonic series*, since only the odd harmonics appear. Similar rules may be stated for even-harmonic series, but when a series appears in the even-harmonic form, it means that $2L$ has not been taken as the smallest period of $f(x)$. Since any integral multiple of a period is also a period, series obtained in this way will also work, but in general computation is simplified if $2L$ is taken to be the smallest period.)

6. If we write the Euler definitions for $\cos \theta$ and $\sin \theta$, we obtain the complex form of the Fourier Series known either as the "Complex Fourier Series" or the "Exponential Fourier Series" of $f(x)$. It is represented as

$$f(x) = \frac{1}{2} \sum_{n=-\infty}^{n=+\infty} c_n e^{i\omega_n x}$$

where

$$c_n = \frac{1}{L} \int_{-L}^{L} f(x) e^{-i\omega_n x} dx, \quad n = 0, \pm 1, \pm 2, \pm 3, \dots$$

with $\omega_n = \frac{n\pi}{L}, \quad n = 0, \pm 1, \pm 2, \dots$

The set of coefficients $\{c_n\}$ is often referred to as the Fourier spectrum.

7. If both sine and cosine terms are present and if $f(x)$ is of period $2L$ and expandable by a Fourier series, it can be represented as

$$f(x) = \frac{a_0}{2} + \sum_{n=1}^{\infty} c_n \sin\left(\frac{n\pi x}{L} + \phi_n\right), \text{ where } a_n = c_n \sin \phi_n,$$

$$b_n = c_n \cos \phi_n, \quad c_n = \sqrt{a_n^2 + b_n^2}, \quad \phi_n = \arctan\left(\frac{a_n}{b_n}\right)$$

It can also be represented as

$$f(x) = \frac{a_0}{2} + \sum_{n=1}^{\infty} c_n \cos\left(\frac{n\pi x}{L} + \phi_n\right), \text{ where } a_n = c_n \cos \phi_n,$$

$$b_n = -c_n \sin \phi_n, \quad c_n = \sqrt{a_n^2 + b_n^2}, \quad \phi_n = \arctan\left(-\frac{b_n}{a_n}\right)$$

where ϕ_n is chosen so as to make a_n, b_n, and c_n hold.

8. The following table of trigonometric identities should be helpful for developing Fourier Series.

	n	n even	n odd	$n/2$ odd	$n/2$ even
$\sin n\pi$	0	0	0	0	0
$\cos n\pi$	$(-1)^n$	$+1$	-1	$+1$	$+1$
$*\sin \dfrac{n\pi}{2}$		0	$(-1)^{(n-1)/2}$	0	0
$*\cos \dfrac{n\pi}{2}$		$(-1)^{n/2}$	0	-1	$+1$
$\sin \dfrac{n\pi}{4}$			$\dfrac{\sqrt{2}}{2}(-1)^{(n^2+4n+11)/8}$	$(-1)^{(n-2)/4}$	0

*A useful formula for $\sin \dfrac{n\pi}{2}$ and $\cos \dfrac{n\pi}{2}$ is given by

$$\sin \frac{n\pi}{2} = \frac{(i)^{n+1}}{2}[(-1)^n - 1] \text{ and } \cos \frac{n\pi}{2} = \frac{(i)^n}{2}[(-1)^n + 1], \text{ where } i^2 = -1.$$

AUXILIARY FORMULAS FOR FOURIER SERIES

$$1 = \frac{4}{\pi}\left[\sin \frac{\pi x}{k} + \frac{1}{3}\sin \frac{3\pi x}{k} + \frac{1}{5}\sin \frac{5\pi x}{k} + \cdots\right] \qquad [0 < x < k]$$

$$x = \frac{2k}{\pi}\left[\sin \frac{\pi x}{k} - \frac{1}{2}\sin \frac{2\pi x}{k} + \frac{1}{3}\sin \frac{3\pi x}{k} - \cdots\right] \qquad [-k < x < k]$$

$$x = \frac{k}{2} - \frac{4k}{\pi^2}\left[\cos \frac{\pi x}{k} + \frac{1}{3^2}\cos \frac{3\pi x}{k} + \frac{1}{5^2}\cos \frac{5\pi x}{k} + \cdots\right] \qquad [0 < x < k]$$

$$x^2 = \frac{2k^2}{\pi^3}\left[\left(\frac{\pi^2}{1} - \frac{4}{1}\right)\sin \frac{\pi x}{k} - \frac{\pi^2}{2}\sin \frac{2\pi x}{k} + \left(\frac{\pi^2}{3} - \frac{4}{3^3}\right)\sin \frac{3\pi x}{k}\right.$$
$$\left. - \frac{\pi^2}{4}\sin \frac{4\pi x}{k} + \left(\frac{\pi^2}{5} - \frac{4}{5^3}\right)\sin \frac{5\pi x}{k} + \cdots\right] \qquad [0 < x < k]$$

$$x^2 = \frac{k^2}{3} - \frac{4k^2}{\pi^2}\left[\cos \frac{\pi x}{k} - \frac{1}{2^2}\cos \frac{2\pi x}{k} + \frac{1}{3^2}\cos \frac{3\pi x}{k} - \frac{1}{4^2}\cos \frac{4\pi x}{k} + \cdots\right]$$
$$[-k < x < k]$$

$$1 - \frac{1}{3} + \frac{1}{5} - \frac{1}{7} + \cdots = \frac{\pi}{4}$$

$$1 + \frac{1}{2^2} + \frac{1}{3^2} + \frac{1}{4^2} + \cdots = \frac{\pi^2}{6}$$

$$1 - \frac{1}{2^2} + \frac{1}{3^2} - \frac{1}{4^2} + \cdots = \frac{\pi^2}{12}$$

$$1 + \frac{1}{3^2} + \frac{1}{5^2} + \frac{1}{7^2} + \cdots = \frac{\pi^2}{8}$$

$$\frac{1}{2^2} + \frac{1}{4^2} + \frac{1}{6^2} + \frac{1}{8^2} + \cdots = \frac{\pi^2}{24}$$

FOURIER EXPANSIONS FOR BASIC PERIODIC FUNCTIONS

$$f(x) = \frac{4}{\pi} \sum_{n=1,3,5\ldots} \frac{1}{n} \sin \frac{n\pi x}{L}$$

$$f(x) = \frac{2}{\pi} \sum_{n=1}^{\infty} \frac{(-1)^n}{n} \left(\cos \frac{n\pi c}{L} - 1 \right) \sin \frac{n\pi x}{L}$$

$$f(x) = \frac{c}{L} + \frac{2}{\pi} \sum_{n=1}^{\infty} \frac{(-1)^n}{n} \sin \frac{n\pi c}{L} \cos \frac{n\pi x}{L}$$

$$f(x) = \frac{2}{L} \sum_{n=1}^{\infty} \sin \frac{n\pi}{2} \frac{\sin(\frac{1}{2}n\pi c/L)}{\frac{1}{2}n\pi c/L} \sin \frac{n\pi x}{L}$$

$$f(x) = \frac{2}{\pi} \sum_{n=1}^{\infty} \frac{(-1)^{n+1}}{n} \sin \frac{n\pi x}{L}$$

$$f(x) = \frac{1}{2} - \frac{4}{\pi^2} \sum_{n=1,3,5\ldots} \frac{1}{n^2} \cos \frac{n\pi x}{L}$$

$$f(x) = \frac{8}{\pi^2} \sum_{n=1,3,5\ldots} \frac{(-1)^{(n-1)/2}}{n^2} \sin \frac{n\pi x}{L}$$

$$f(x) = \frac{1}{2} - \frac{1}{\pi} \sum_{n=1}^{\infty} \frac{1}{n} \sin \frac{n\pi x}{L}$$

$$f(x) = \tfrac{1}{2}(1 + a) + \frac{2}{\pi^2(1-a)} \sum_{n=1}^{\infty} \frac{1}{n^2} [(-1)^n \cos n\pi a - 1] \cos \frac{n\pi x}{L} \; ; \; \left(a = \frac{c}{2L}\right)$$

$$f(x) = \frac{2}{\pi} \sum_{n=1}^{\infty} \frac{(-1)^{n-1}}{n} \left[1 + \frac{\sin n\pi a}{n\pi(1-a)}\right] \sin \frac{n\pi x}{L} \; ; \; \left(a = \frac{c}{2L}\right)$$

$$f(x) = \tfrac{1}{2} - \frac{4}{\pi^2(1-2a)} \sum_{n=1,3,5\ldots} \frac{1}{n^2} \cos n\pi a \cos \frac{n\pi x}{L} \; ; \; \left(a = \frac{c}{2L}\right)$$

$$f(x) = \frac{2}{\pi} \sum_{n=1}^{\infty} \frac{(-1)^n}{n} \left[1 + \frac{1+(-1)^n}{n\pi(1-2a)} \sin n\pi a\right] \sin \frac{n\pi x}{L} \; ; \; \left(a = \frac{c}{2L}\right)$$

$$f(x) = \frac{4}{\pi} \sum_{n=1}^{\infty} \frac{1}{n} \sin \frac{n\pi}{4} \sin n\pi a \sin \frac{n\pi x}{L} \; ; \; \left(a = \frac{c}{2L}\right)$$

$$f(x) = \frac{9}{\pi^2} \sum_{n=1}^{\infty} \frac{1}{n^2} \sin \frac{n\pi}{3} \sin \frac{n\pi x}{L} \; ; \; \left(a = \frac{c}{2L}\right)$$

$$f(x) = \frac{32}{3\pi^2} \sum_{n=1}^{\infty} \frac{1}{n^2} \sin \frac{n\pi}{4} \sin \frac{n\pi x}{L} \; ; \; \left(a = \frac{c}{2L}\right)$$

$$f(x) = \frac{1}{\pi} + \frac{1}{2} \sin \omega t - \frac{2}{\pi} \sum_{n=2,4,6\ldots} \frac{1}{n^2-1} \cos n\omega t$$

Extracted from graphs and formulas, pages 372, 373, Differential Equations in Engineering Problems, Salvadori and Schwarz, published by Prentice-Hall, Inc., 1954.

THE FOURIER TRANSFORMS*

R. E. Gaskell

For a piecewise continuous function $F(x)$ over a finite interval $0 \leq x \leq \pi$, the *finite Fourier cosine transform* of $F(x)$ is

$$f_c(n) = \int_0^\pi F(x) \cos nx \, dx \quad (n = 0, 1, 2, \ldots) \tag{1}$$

If x ranges over the interval $0 \leq x \leq L$, the substitution $x' = \pi x/L$ allows the use of this definition, also. The inverse transform is written.

$$\bar{F}(x) = \frac{1}{\pi} f_c(0) - \frac{2}{\pi} \sum_{n=1}^\infty f_c(n) \cos nx \quad (0 < x < \pi) \tag{2}$$

where $F(x) = \dfrac{[F(x + o) + F(x - o)]}{2}$. We observe that $F(x) = F(x) =$ at points of continuity. The formula

$$f_c^{(2)}(n) = \int_0^\pi F''(x) \cos nx \, dx$$
$$= -n^2 f_c(n) - F'(0) + (-1)^n F'(\pi) \tag{3}$$

makes the finite Fourier cosine transform useful in certain boundary value problems.

Analogously, the *finite Fourier sine transform* of $F(x)$ is

$$f_s(n) = \int_0^\pi F(x) \sin nx \, dx \quad (n = 1, 2, 3, \ldots) \tag{4}$$

and

$$\bar{F}(x) = \frac{2}{\pi} \sum_{n=1}^\infty f_s(n) \sin nx \quad (0 < x < \pi) \tag{5}$$

Corresponding to (3) we have

$$f_s^{(2)}(n) = \int_0^\pi F''(x) \sin nx \, dx$$
$$= -n^2 f_s(n) - nF(0) - n(-1)^n F(\pi) \tag{6}$$

Fourier Transforms

If $F(x)$ is defined for $x \geq 0$ and is piecewise continuous over any finite interval, and if

$$\int_0^\infty F(x) \, dx$$

is absolutely convergent, then

$$f_c(\alpha) = \sqrt{\frac{2}{\pi}} \int_0^\infty F(x) \cos(\alpha x) \, dx \tag{7}$$

is the *Fourier cosine transform* of $F(x)$. Furthermore.

$$\bar{F}(x) = \sqrt{\frac{2}{\pi}} \int_0^\infty f_c(\alpha) \cos(\alpha x) \, d\alpha \tag{8}$$

If $\lim_{x \to \infty} \dfrac{d^n F}{dx^n} = 0$, an important property of the Fourier cosine transform

$$f_c^{(2r)}(\alpha) = \sqrt{\frac{2}{\pi}} \int_0^\infty \left(\frac{d^{2r} F}{dx^{2r}}\right) \cos(\alpha x) \, dx$$
$$= -\sqrt{\frac{2}{\pi}} \sum_{n=0}^{r-1} (-1)^n a_{2r-2n-1} \alpha^{2n} + (-1)^r \alpha^{2r} f_c(\alpha) \tag{9}$$

where $\lim_{x \to \infty} \dfrac{d^r F}{dx^r} = a_r$, makes it useful in the solution of many problems.

Under the same conditions.

$$f_s(\alpha) = \sqrt{\frac{2}{\pi}} \int_0^\infty F(x) \sin(\alpha x) \, dx \tag{10}$$

* From Beyer, W. H., Ed., *CRC Handbook of Mathematical Sciences,* 5th ed., CRC Press, Boca Raton, 1978, 592—598. With permission.

defines the *Fourier sine transform* of $F(x)$, and

$$\bar{F}(x) = \sqrt{\frac{2}{\pi}} \int_0^\infty f_s(\alpha) \sin(\alpha x)\,d\alpha \tag{11}$$

Corresponding to (9) we have

$$f_s^{(2r)}(\alpha) = \sqrt{\frac{2}{\pi}} \int_0^\infty \frac{d^{2r}F}{dx^{2r}} \sin(\alpha x)\,dx$$

$$= -\sqrt{\frac{2}{\pi}} \sum_{n=1}^r (-1)^n \alpha^{2n-1} a_{2r-2n} + (-1)^{r-1} \alpha^{2r} f_s(\alpha) \tag{12}$$

Similarly, if $F(x)$ is defined for $-\infty < x < \infty$, and if $\int_{-\infty}^\infty F(x)\,dx$ is absolutely convergent, then

$$f(\alpha) = \frac{1}{\sqrt{2\pi}} \int_{-\infty}^\infty F(x) e^{i\alpha x}\,dx \tag{13}$$

is the *Fourier transform* of $F(x)$, and

$$\bar{F}(x) = \frac{1}{\sqrt{2\pi}} \int_{-\infty}^\infty f(\alpha) e^{-i\alpha x}\,d\alpha \tag{14}$$

Also, if

$$\lim_{|x| \to \infty} \left| \frac{d^n F}{dx^n} \right| = 0 \quad (n = 1, 2, \ldots, r-1)$$

then

$$f^{(r)}(\alpha) = \frac{1}{\sqrt{2\pi}} \int_{-\infty}^\infty F^{(r)}(x) e^{i\alpha x}\,dx = (-i\alpha)^r f(\alpha) \tag{15}$$

Finite Sine Transforms

	$f_s(n)$	$F(x)$
1	$f_s(n) = \int_0^\pi F(x) \sin nx\,dx \ (n = 1, 2, \cdots)$	$F(x)$
2	$(-1)^{n+1} f_s(n)$	$F(\pi - x)$
3	$\dfrac{1}{n}$	$\dfrac{\pi - x}{\pi}$
4	$\dfrac{(-1)^{n+1}}{n}$	$\dfrac{x}{\pi}$
5	$\dfrac{1 - (-1)^n}{n}$	1
6	$\dfrac{2}{n^2} \sin \dfrac{n\pi}{2}$	$\begin{cases} x & \text{when } 0 < x < \pi/2 \\ \pi - x & \text{when } \pi/2 < x < \pi \end{cases}$
7	$\dfrac{(-1)^{n+1}}{n^3}$	$\dfrac{x(\pi^2 - x^2)}{6\pi}$
8	$\dfrac{1 - (-1)^n}{n^3}$	$\dfrac{x(\pi - x)}{2}$
9	$\dfrac{\pi^2(-1)^{n-1}}{n} - \dfrac{2[1 - (-1)^n]}{n^3}$	x^2
10	$\pi(-1)^n \left(\dfrac{6}{n^3} - \dfrac{\pi^2}{n} \right)$	x^3
11	$\dfrac{n}{n^2 + c^2} [1 - (-1)^n e^{c\pi}]$	e^{cx}
12	$\dfrac{n}{n^2 + c^2}$	$\dfrac{\sinh c(\pi - x)}{\sinh c\pi}$
13	$\dfrac{n}{n^2 - k^2} \ (k \neq 0, 1, 2, \cdots)$	$\dfrac{\sin k(\pi - x)}{\sin k\pi}$
14	$\begin{cases} \dfrac{\pi}{2} & \text{when } n = m \\ 0 & \text{when } n \neq m \end{cases} \ (m = 1, 2, \cdots)$	$\sin mx$

	$f_s(n)$	$F(x)$
15	$\dfrac{n}{n^2 - k^2}[1 - (-1)^n \cos k\pi]$ $(k \neq 1, 2, \cdots)$	$\cos kx$
16	$\begin{cases}\dfrac{n}{n^2 - m^2}[1 - (-1)^{n+m}] \\ \qquad\text{when } n \neq m = 1, 2, \cdots \\ 0 \qquad \text{when } n = m\end{cases}$	$\cos mx$
17	$\dfrac{n}{(n^2 - k^2)^2}$ $(k \neq 0, 1, 2, \cdots)$	$\dfrac{\pi \sin kx}{2k \sin^2 k\pi} - \dfrac{x \cos k(\pi - x)}{2k \sin k\pi}$
18	$\dfrac{b^n}{n}$ $(|b| \leq 1)$	$\dfrac{2}{\pi} \arctan \dfrac{b \sin x}{1 - b \cos x}$
19	$\dfrac{1 - (-1)^n}{n} b^n$ $(|b| \leq 1)$	$\dfrac{2}{\pi} \arctan \dfrac{2b \sin x}{1 - b^2}$

Finite Cosine Transforms

	$f_c(n)$	$F(x)$
1	$f_c(n) = \displaystyle\int_0^\pi F(x) \cos nx\, dx$ $(n = 0, 1, 2, \cdots)$	$F(x)$
2	$(-1)^n f_c(n)$	$F(\pi - x)$
3	0 when $n = 1, 2, \cdots$; $f_c(0) = \pi$	1
4	$\dfrac{2}{n} \sin \dfrac{n\pi}{2}$; $f_c(0) = 0$	$\begin{cases} 1 \text{ when } 0 < x < \pi/2 \\ -1 \text{ when } \pi/2 < x < \pi \end{cases}$
5	$-\dfrac{1 - (-1)^n}{n^2}$; $f_c(0) = \dfrac{\pi^2}{2}$	x
6	$\dfrac{(-1)^n}{n^2}$; $f_c(0) = \dfrac{\pi^2}{6}$	$\dfrac{x^2}{2\pi}$
7	$\dfrac{1}{n^2}$; $f_c(0) = 0$	$\dfrac{(\pi - x)^2}{2\pi} - \dfrac{\pi}{6}$
8	$3\pi^2 \dfrac{(-1)^n}{n^2} - 6 \dfrac{1 - (-1)^n}{n^4}$; $f_c(0) = \dfrac{\pi^4}{4}$	x^3
9	$\dfrac{(-1)^n e^c \pi - 1}{n^2 + c^2}$	$\dfrac{1}{c} e^{cx}$
10	$\dfrac{1}{n^2 + c^2}$	$\dfrac{\cosh c(\pi - x)}{c \sinh c\pi}$
11	$\dfrac{k}{n^2 - k^2}[(-1)^n \cos \pi k - 1]$ $(k \neq 0, 1, 2, \cdots)$	$\sin kx$
12	$\dfrac{(-1)^{n+m} - 1}{n^2 - m^2}$; $f_c(m) = 0$ $(m = 1, 2, \cdots)$	$\dfrac{1}{m} \sin mx$
13	$\dfrac{1}{n^2 - k^2}$ $(k \neq 0, 1, 2, \cdots)$	$-\dfrac{\cos k(\pi - x)}{k \sin k\pi}$
14	0 when $n = 1, 2, \cdots$; $f_c(m) = \dfrac{\pi}{2}$ $(m = 1, 2, \cdots)$	$\cos mx$

Fourier Sine Transforms*

	$F(x)$	$f_s(\alpha)$
1	$\begin{cases} 1 & (0 < x < a) \\ 0 & (x > a) \end{cases}$	$\sqrt{\dfrac{2}{\pi}}\left[\dfrac{1 - \cos \alpha}{\alpha}\right]$
2	$x^{p-1} (0 < p < 1)$	$\sqrt{\dfrac{2}{\pi}} \dfrac{\Gamma(p)}{\alpha^p} \sin \dfrac{p\pi}{2}$
3	$\begin{cases} \sin x & (0 < x < a) \\ 0 & (x > a) \end{cases}$	$\dfrac{1}{\sqrt{2\pi}}\left[\dfrac{\sin[a(1 - \alpha)]}{1 - \alpha} - \dfrac{\sin[a(1 + \alpha)]}{1 + \alpha}\right]$
4	e^{-x}	$\sqrt{\dfrac{2}{\pi}}\left[\dfrac{\alpha}{1 + \alpha^2}\right]$
5	$x e^{-x^2/2}$	$\alpha e^{-\alpha^2/2}$
6	$\cos \dfrac{x^2}{2}$	$\sqrt{2}\left[\sin \dfrac{\alpha^2}{2} C\left(\dfrac{\alpha^2}{2}\right) - \cos \dfrac{\alpha^2}{2} S\left(\dfrac{\alpha^2}{2}\right)\right]^*$

$F(x)$	$f_s(\alpha)$
7 $\sin \dfrac{x^2}{2}$	$\sqrt{2}\left[\cos\dfrac{\alpha^2}{2}\,C\!\left(\dfrac{\alpha^2}{2}\right)+\sin\dfrac{\alpha^2}{2}\,S\!\left(\dfrac{\alpha^2}{2}\right)\right]$*

* $C(y)$ and $S(y)$ are the Fresnel integrals

$$C(y)=\frac{1}{\sqrt{2\pi}}\int_0^y \frac{1}{\sqrt{t}}\cos t\,dt,$$

$$S(y)=\frac{1}{\sqrt{2\pi}}\int_0^y \frac{1}{\sqrt{t}}\sin t\,dt$$

*   More extensive tables of the Fourier sine and cosine transforms can be found in Fritz Oberhettinger, *Tabellen zur-Fourier Transformation*, Springer, 1957.

Fourier Cosine Transforms

$F(x)$	$f_c(\alpha)$
1 $\begin{cases} 1 & (0 < x < a) \\ 0 & (x > a) \end{cases}$	$\sqrt{\dfrac{2}{\pi}}\dfrac{\sin a\alpha}{\alpha}$
2 x^{p-1} $(0 < p < 1)$	$\sqrt{\dfrac{2}{\pi}}\dfrac{\Gamma(p)}{\alpha^p}\cos\dfrac{p\pi}{2}$
3 $\begin{cases} \cos x & (0 < x < a) \\ 0 & (x > a) \end{cases}$	$\dfrac{1}{\sqrt{2\pi}}\left[\dfrac{\sin[a(1-\alpha)]}{1-\alpha}+\dfrac{\sin[a(1+\alpha)]}{1+\alpha}\right]$
4 e^{-x}	$\sqrt{\dfrac{2}{\pi}}\left(\dfrac{1}{1+\alpha^2}\right)$
5 $e^{-x^2/2}$	$e^{-\alpha^2/2}$
6 $\cos\dfrac{x^2}{2}$	$\cos\left(\dfrac{\alpha^2}{2}-\dfrac{\pi}{4}\right)$
7 $\sin\dfrac{x^2}{2}$	$\cos\left(\dfrac{\alpha^2}{2}+\dfrac{\pi}{4}\right)$

Fourier Transforms*

$F(x)$	$f(\alpha)$
1 $\dfrac{\sin ax}{x}$	$\begin{cases} \sqrt{\dfrac{\pi}{2}} & \lvert\alpha\rvert < a \\ 0 & \lvert\alpha\rvert > a \end{cases}$
2 $\begin{cases} e^{iwx} & (p < x < q) \\ 0 & (x < p, x > q) \end{cases}$	$\dfrac{i}{\sqrt{2\pi}}\dfrac{e^{ip(w+\alpha)}-e^{iq(w+\alpha)}}{(w+\alpha)}$
3 $\begin{cases} e^{-cx+iwx} & (x > 0) \\ 0 & (x < 0) \end{cases}$ $(c > 0)$	$\dfrac{i}{\sqrt{2\pi}(w+\alpha+ic)}$
4 e^{-px^2} $R(p) > 0$	$\dfrac{1}{\sqrt{2p}}e^{-\alpha^2/4p}$
5 $\cos px^2$	$\dfrac{1}{\sqrt{2p}}\cos\left[\dfrac{\alpha^2}{4p}-\dfrac{\pi}{4}\right]$
6 $\sin px^2$	$\dfrac{1}{\sqrt{2p}}\cos\left[\dfrac{\alpha^2}{4p}+\dfrac{\pi}{4}\right]$
7 $\lvert x\rvert^{-p}$ $(0 < p < 1)$	$\sqrt{\dfrac{2}{\pi}}\dfrac{\Gamma(1-p)\sin\dfrac{p\pi}{2}}{\lvert\alpha\rvert^{(1-p)}}$
8 $\dfrac{e^{-a\lvert x\rvert}}{\sqrt{\lvert x\rvert}}$	$\dfrac{\sqrt{\sqrt{(a^2+\alpha^2)}+a}}{\sqrt{a^2+\alpha^2}}$
9 $\dfrac{\cosh ax}{\cosh \pi x}$ $(-\pi < a < \pi)$	$\sqrt{\dfrac{2}{\pi}}\dfrac{\cos\dfrac{a}{2}\cosh\dfrac{\alpha}{2}}{\cosh\alpha+\cos a}$
10 $\dfrac{\sinh ax}{\sinh \pi x}$ $(-\pi < a < \pi)$	$\dfrac{1}{\sqrt{2\pi}}\dfrac{\sin a}{\cosh\alpha+\cos a}$
11 $\begin{cases} \dfrac{1}{\sqrt{a^2-x^2}} & (\lvert x\rvert < a) \\ 0 & (\lvert x\rvert > a) \end{cases}$	$\sqrt{\dfrac{\pi}{2}}\,J_0(a\alpha)$

	$F(x)$	$f(\alpha)$
12	$\dfrac{\sin\left[b\sqrt{a^2+x^2}\right]}{\sqrt{a^2+x^2}}$	$\begin{cases} 0 & (\lvert\alpha\rvert > b) \\ \sqrt{\dfrac{\pi}{2}}\,J_0(a\sqrt{b^2-\alpha^2}) & (\lvert\alpha\rvert < b) \end{cases}$
13	$\begin{cases} P_n(x) & (\lvert x\rvert < 1) \\ 0 & (\lvert x\rvert > 1) \end{cases}$	$\dfrac{i^n}{\sqrt{\alpha}}\,J_{n+1}(\alpha)$
14	$\begin{cases} \dfrac{\cos\left[b\sqrt{a^2-x^2}\right]}{\sqrt{a^2-x^2}} & (\lvert x\rvert < a) \\ 0 & (\lvert x\rvert > a) \end{cases}$	$\sqrt{\dfrac{\pi}{2}}\,J_0(a\sqrt{a^2+b^2})$
15	$\begin{cases} \dfrac{\cosh\left[b\sqrt{a^2-x^2}\right]}{\sqrt{a^2-x^2}} & (\lvert x\rvert < a) \\ 0 & (\lvert x\rvert > a) \end{cases}$	$\sqrt{\dfrac{\pi}{2}}\,J_0(a\sqrt{\alpha^2-b^2})$

* More extensive tables of Fourier transforms can be found in W. Magnus and F. Oberhettinger, *Formulas and Theorems of the Special Functions of Mathematical Physics*. Chelsea, 1949, 116—120.

The following functions appear among the entries of the tables on transforms.

Function	Definition	Name
$Ei(x)$	$\displaystyle\int_{-\infty}^{x}\frac{e^v}{v}\,dv$; or sometimes defined as $-Ei(-x)=\displaystyle\int_{x}^{\infty}\frac{e^{-v}}{v}\,dv$	Sine, Cosine, and Exponential Integral tables pages 548–556
$Si(x)$	$\displaystyle\int_{0}^{x}\frac{\sin v}{v}\,dv$	Sine, Cosine, and Exponential Integral tables pages 548–556
$Ci(x)$	$\displaystyle\int_{\infty}^{x}\frac{\cos v}{v}\,dv$; or sometimes defined as negative of this integral	Sine, Cosine, and Exponential Integral tables pages 548–546
$erf(x)$	$\dfrac{2}{\sqrt{\pi}}\displaystyle\int_{0}^{x}e^{-v^2}\,dv$	Error function
$erfc(x)$	$1-erf(x)=\dfrac{2}{\sqrt{\pi}}\displaystyle\int_{x}^{\infty}e^{-v^2}\,dv$	Complementary function to error function
$L_n(x)$	$\dfrac{e^x}{n!}\dfrac{d^n}{dx^n}(x^n e^{-x}),\quad n=0,1,\cdots$	Laguerre polynomial of degree n

SERIES EXPANSION

The expression in parentheses following certain of the series indicates the region of convergence. If not otherwise indicated it is to be understood that the series converges for all finite values of x.

BINOMIAL

$$(x+y)^n = x^n + nx^{n-1}y + \frac{n(n-1)}{2!}x^{n-2}y^2$$
$$+ \frac{n(n-1)(n-2)}{3!}x^{n-3}y^3 + \cdots \quad (y^2 < x^2)$$

$$(1 \pm x)^n = 1 \pm nx + \frac{n(n-1)x^2}{2!} \pm \frac{n(n-1)(n-2)x^3}{3!} + \cdots \quad (x^2 < 1)$$

$$(1 \pm x)^{-n} = 1 \mp nx + \frac{n(n+1)x^2}{2!} \mp \frac{n(n+1)(n+2)x^3}{3!} + \cdots \quad (x^2 < 1)$$

$$(1 \pm x)^{-1} = 1 \mp x + x^2 \mp x^3 + x^4 \mp x^5 + \cdots \qquad (x^2 < 1)$$
$$(1 \pm x)^{-2} = 1 \mp 2x + 3x^2 \mp 4x^3 + 5x^4 \mp 6x^5 + \cdots \qquad (x^2 < 1)$$

REVERSION OF SERIES

Let a series be represented by

$$y = a_1 x + a_2 x^2 + a_3 x^3 + a_4 x^4 + a_5 x^5 + a_6 x^6 + \cdots \quad (a_1 \neq 0)$$

to find the coefficients of the series

$$x = A_1 y + A_2 y^2 + A_3 y^3 + A_4 y^4 + \cdots$$

$$A_1 = \frac{1}{a_1} \qquad A_2 = -\frac{a_2}{a_1^3} \qquad A_3 = \frac{1}{a_1^5}(2a_2^2 - a_1 a_3)$$

$$A_4 = \frac{1}{a_1^7}(5a_1 a_2 a_3 - a_1^2 a_4 - 5a_2^3)$$

$$A_5 = \frac{1}{a_1^9}(6a_1^2 a_2 a_4 + 3a_1^2 a_3^2 + 14a_2^4 - a_1^3 a_5 - 21a_1 a_2^2 a_3)$$

$$A_6 = \frac{1}{a_1^{11}}(7a_1^3 a_2 a_5 + 7a_1^3 a_3 a_4 + 84a_1 a_2^3 a_3 - a_1^4 a_6 - 28a_1^2 a_2^2 a_4 - 28a_1^2 a_2 a_3^2 - 42a_2^5)$$

$$A_7 = \frac{1}{a_1^{13}}(8a_1^4 a_2 a_6 + 8a_1^4 a_3 a_5 + 4a_1^4 a_4^2 + 120a_1^2 a_2^3 a_4$$
$$+ 180a_1^2 a_2^2 a_3^2 + 132a_2^6 - a_1^5 a_7$$
$$- 36a_1^3 a_2^2 a_5 - 72a_1^3 a_2 a_3 a_4 - 12a_1^3 a_3^3 - 330a_1 a_2^4 a_3)$$

TAYLOR

1. $f(x) = f(a) + (x - a)f'(a) + \dfrac{(x - a)^2}{2!}f''(a) + \dfrac{(x - a)^3}{3!}f'''(a)$
$$+ \cdots + \frac{(x - a)^n}{n!}f^{(n)}(a) + \cdots \text{ (Taylor's Series)}$$

(Increment form)

2. $f(x + h) = f(x) + hf'(x) + \dfrac{h^2}{2!}f''(x) + \dfrac{h^3}{3!}f'''(x) + \cdots$
$$= f(h) + xf'(h) + \frac{x^2}{2!}f''(h) + \frac{x^3}{3!}f'''(h) + \cdots$$

3. If $f(x)$ is a function possessing derivatives of all orders throughout the interval $a \leq x \leq b$, then there is a value X, with $a < X < b$, such that

$$f(b) = f(a) + (b - a)f'(a) + \frac{(b - a)^2}{2!}f''(a) + \cdots$$
$$+ \frac{(b - a)^{n-1}}{(n - 1)!}f^{(n-1)}(a) + \frac{(b - a)^n}{n!}f^{(n)}(X)$$

$$f(a + h) = f(a) + hf'(a) + \frac{h^2}{2!}f''(a) + \cdots + \frac{h^{n-1}}{(n - 1)!}f^{(n-1)}(a)$$
$$+ \frac{h^n}{n!}f^{(n)}(a + \theta h), \ b = a + h, 0 < \theta < 1.$$

or

$$f(x) = f(a) + (x - a)f'(a) + \frac{(x - a)^2}{2!}f''(a) + \cdots + (x - a)^{n-1}\frac{f^{(n-1)}(a)}{(n - 1)!} + R_n,$$

where

$$R_n = \frac{f^{(n)}[a + \theta \cdot (x - a)]}{n!}(x - a)^n, \quad 0 < \theta < 1.$$

The above forms are known as Taylor's series with the remainder term.

4. *Taylor's series for a function of two variables*

If $\left(h\dfrac{\partial}{\partial x} + k\dfrac{\partial}{\partial y}\right)f(x,y) = h\dfrac{\partial f(x,y)}{\partial x} + k\dfrac{\partial f(x,y)}{\partial y}$;

$$\left(h\frac{\partial}{\partial x} + k\frac{\partial}{\partial y}\right)^2 f(x,y) = h^2\frac{\partial^2 f(x,y)}{\partial x^2} + 2hk\frac{\partial^2 f(x,y)}{\partial x\,\partial y} + k^2\frac{\partial^2 f(x,y)}{\partial y^2}$$

etc., and if $\left(h\dfrac{\partial}{\partial x} + k\dfrac{\partial}{\partial y}\right)^n f(x,y)\Big|_{\substack{x=a \\ y=b}}$ with the bar and subscripts means that after differentiation we are to replace x by a and y by b,

$$f(a + h, b + k) = f(a,b) + \left(h\frac{\partial}{\partial x} + k\frac{\partial}{\partial y}\right)f(x,y)\Big|_{\substack{x=a \\ y=b}} + \cdots$$
$$+ \frac{1}{n!}\left(h\frac{\partial}{\partial x} + k\frac{\partial}{\partial y}\right)^n f(x,y)\Big|_{\substack{x=a \\ y=b}} + \cdots$$

MACLAURIN

$$f(x) = f(0) + xf'(0) + \frac{x^2}{2!}f''(0) + \frac{x^3}{3!}f'''(0) + \cdots + x^{n-1}\frac{f^{(n-1)}(0)}{(n - 1)!} + R_n,$$

where

$$R_n = \frac{x^n f^{(n)}(\theta x)}{n!}, \quad 0 < \theta < 1.$$

EXPONENTIAL

$$e = 1 + \frac{1}{1!} + \frac{1}{2!} + \frac{1}{3!} + \frac{1}{4!} + \cdots$$

$$e^x = 1 + x + \frac{x^2}{2!} + \frac{x^3}{3!} + \frac{x^4}{4!} + \cdots \qquad \text{(all real values of } x\text{)}$$

$$a^x = 1 + x \log_e a + \frac{(x \log_e a)^2}{2!} + \frac{(x \log_e a)^3}{3!} + \cdots$$

$$e^x = e^a \left[1 + (x - a) + \frac{(x - a)^2}{2!} + \frac{(x - a)^3}{3!} + \cdots \right]$$

LOGARITHMIC

$$\log_e x = \frac{x - 1}{x} + \frac{1}{2} \left(\frac{x - 1}{x} \right)^2 + \frac{1}{3} \left(\frac{x - 1}{x} \right)^3 + \cdots \qquad \left(x > \tfrac{1}{2} \right)$$

$$\log_e x = (x - 1) - \tfrac{1}{2}(x - 1)^2 + \tfrac{1}{3}(x - 1)^3 - \cdots \qquad (2 \geq x > 0)$$

$$\log_e x = 2 \left[\frac{x - 1}{x + 1} + \frac{1}{3} \left(\frac{x - 1}{x + 1} \right)^3 + \frac{1}{5} \left(\frac{x - 1}{x + 1} \right)^5 + \cdots \right] \qquad (x > 0)$$

$$\log_e (1 + x) = x - \tfrac{1}{2}x^2 + \tfrac{1}{3}x^3 - \tfrac{1}{4}x^4 + \cdots \qquad (-1 < x \leq 1)$$

$$\log_e (n + 1) - \log_e (n - 1) = 2 \left[\frac{1}{n} + \frac{1}{3n^3} + \frac{1}{5n^5} + \cdots \right]$$

$$\log_e (a + x) = \log_e a + 2 \left[\frac{x}{2a + x} + \frac{1}{3} \left(\frac{x}{2a + x} \right)^3 + \frac{1}{5} \left(\frac{x}{2a + x} \right)^5 + \cdots \right]$$
$$(a > 0, -a < x < +\infty)$$

$$\log_e \frac{1 + x}{1 - x} = 2 \left[x + \frac{x^3}{3} + \frac{x^5}{5} + \cdots + \frac{x^{2n-1}}{2n - 1} + \cdots \right], \qquad -1 < x < 1$$

$$\log_e x = \log_e a + \frac{(x - a)}{a} - \frac{(x - a)^2}{2a^2} + \frac{(x - a)^3}{3a^3} - + \cdots, \qquad 0 < x \leq 2a$$

TRIGONOMETRIC

$$\sin x = x - \frac{x^3}{3!} + \frac{x^5}{5!} - \frac{x^7}{7!} + \cdots \qquad \text{(all real values of } x)$$

$$\cos x = 1 - \frac{x^2}{2!} + \frac{x^4}{4!} - \frac{x^6}{6!} + \cdots \qquad \text{(all real values of } x)$$

$$\tan x = x + \frac{x^3}{3} + \frac{2x^5}{15} + \frac{17x^7}{315} + \frac{62x^9}{2835} + \cdots + \frac{(-1)^{n-1} 2^{2n} (2^{2n} - 1) B_{2n}}{(2n)!} x^{2n-1} + \cdots,$$
$$\left[x^2 < \frac{\pi^2}{4}, \text{ and } B_n \text{ represents the } n \text{th Bernoulli number.} \right]$$

$$\cot x = \frac{1}{x} - \frac{x}{3} - \frac{x^3}{45} - \frac{2x^5}{945} - \frac{x^7}{4725} - \cdots - \frac{(-1)^{n+1} 2^{2n}}{(2n)!} B_{2n} x^{2n-1} - \cdots,$$
$$[x^2 < \pi^2, \text{ and } B_n \text{ represents the } n \text{th Bernoulli number.}]$$

$$\sec x = 1 + \frac{x^2}{2} + \frac{5}{24} x^4 + \frac{61}{720} x^6 + \frac{277}{8064} x^8 + \cdots + \frac{(-1)^n}{(2n)!} E_{2n} x^{2n} + \cdots,$$
$$\left[x^2 < \frac{\pi^2}{4}, \text{ and } E_n \text{ represents the } n \text{th Euler number.} \right]$$

$$\csc x = \frac{1}{x} + \frac{x}{6} + \frac{7}{360} x^3 + \frac{31}{15,120} x^5 + \frac{127}{604,800} x^7 + \cdots$$
$$+ \frac{(-1)^{n+1} 2 (2^{2n-1} - 1)}{(2n)!} B_{2n} x^{2n-1} + \cdots,$$
$$[x^2 < \pi^2, \text{ and } B_n \text{ represents } n \text{th Bernoulli number.}]$$

$$\sin x = x \left(1 - \frac{x^2}{\pi^2} \right) \left(1 - \frac{x^2}{2^2 \pi^2} \right) \left(1 - \frac{x^2}{3^2 \pi^2} \right) \cdots \qquad (x^2 < \infty)$$

$$\cos x = \left(1 - \frac{4x^2}{\pi^2} \right) \left(1 - \frac{4x^2}{3^2 \pi^2} \right) \left(1 - \frac{4x^2}{5^2 \pi^2} \right) \cdots \qquad (x^2 < \infty)$$

$$\sin^{-1} x = x + \frac{x^3}{2 \cdot 3} + \frac{1 \cdot 3}{2 \cdot 4 \cdot 5} x^5 + \frac{1 \cdot 3 \cdot 5}{2 \cdot 4 \cdot 6 \cdot 7} x^7 + \cdots \qquad \left(x^2 < 1, -\frac{\pi}{2} < \sin^{-1} x < \frac{\pi}{2} \right)$$

$$\cos^{-1} x = \frac{\pi}{2} - \left(x + \frac{x^3}{2 \cdot 3} + \frac{1 \cdot 3}{2 \cdot 4 \cdot 5} x^5 + \frac{1 \cdot 3 \cdot 5 x^7}{2 \cdot 4 \cdot 6 \cdot 7} + \cdots \right) \qquad (x^2 < 1, 0 < \cos^{-1} x < \pi)$$

$$\tan^{-1} x = x - \frac{x^3}{3} + \frac{x^5}{5} - \frac{x^7}{7} + \cdots \qquad (x^2 < 1)$$

$$\tan^{-1} x = \frac{\pi}{2} - \frac{1}{x} + \frac{1}{3x^3} - \frac{1}{5x^5} + \frac{1}{7x^7} - \cdots \qquad (x > 1)$$

$$\tan^{-1} x = -\frac{\pi}{2} - \frac{1}{x} + \frac{1}{3x^3} - \frac{1}{5x^5} + \frac{1}{7x^7} - \cdots \qquad (x < -1)$$

$$\cot^{-1} x = \frac{\pi}{2} - x + \frac{x^3}{3} - \frac{x^5}{5} + \frac{x^7}{7} - \cdots \qquad (x^2 < 1)$$

$$\log_e \sin x = \log_e x - \frac{x^2}{6} - \frac{x^4}{180} - \frac{x^6}{2835} - \cdots \qquad (x^2 < \pi^2)$$

$$\log_e \cos x = -\frac{x^2}{2} - \frac{x^4}{12} - \frac{x^6}{45} - \frac{17x^8}{2520} - \cdots \qquad \left(x^2 < \frac{\pi^2}{4}\right)$$

$$\log_e \tan x = \log_e x + \frac{x^2}{3} + \frac{7x^4}{90} + \frac{62x^6}{2835} + \cdots \qquad \left(x^2 < \frac{\pi^2}{4}\right)$$

$$e^{\sin x} = 1 + x + \frac{x^2}{2!} - \frac{3x^4}{4!} - \frac{8x^5}{5!} - \frac{3x^6}{6!} + \frac{56x^7}{7!} + \cdots$$

$$e^{\cos x} = e\left(1 - \frac{x^2}{2!} + \frac{4x^4}{4!} - \frac{31x^6}{6!} + \cdots\right)$$

$$e^{\tan x} = 1 + x + \frac{x^2}{2!} + \frac{3x^3}{3!} + \frac{9x^4}{4!} + \frac{37x^5}{5!} + \cdots \qquad \left(x^2 < \frac{\pi^2}{4}\right)$$

$$\sin x = \sin a + (x - a)\cos a - \frac{(x - a)^2}{2!}\sin a$$
$$- \frac{(x - a)^3}{3!}\cos a + \frac{(x - a)^4}{4!}\sin a + \cdots$$

VECTOR ANALYSIS

Definitions

Any quantity which is completely determined by its magnitude is called a *scalar*. Examples of such are mass, density, temperature, etc. Any quantity which is completely determined by its magnitude and direction is called a *vector*. Examples of such are velocity, acceleration, force, etc. A vector quantity is represented by a directed line segment, the length of which represents the magnitude of the vector. A vector quantity is usually represented by a boldfaced letter such as $\mathbf{V}$. Two vectors $\mathbf{V}_1$ and $\mathbf{V}_2$ are equal to one another if they have equal magnitudes and are acting in the same directions. A negative vector, written as $-\mathbf{V}$, is one which acts in the opposite direction to $\mathbf{V}$, but is of equal magnitude to it. If we represent the magnitude of $\mathbf{V}$ by v, we write $|\mathbf{V}| = v$. A vector parallel to $\mathbf{V}$, but equal to the reciprocal of its magnitude is written as $\mathbf{V}^{-1}$ or as $\frac{1}{\mathbf{V}}$.

The *unit vector* $\frac{\mathbf{V}}{v}$ $(v \neq 0)$ is that vector which has the same direction as $\mathbf{V}$, but has a magnitude of unity (sometimes represented as $\mathbf{V}_0$ or $\hat{\mathbf{v}}$).

Vector Algebra

The vector sum of $\mathbf{V}_1$ and $\mathbf{V}_2$ is represented by $\mathbf{V}_1 + \mathbf{V}_2$. The vector sum of $\mathbf{V}_1$ and $-\mathbf{V}_2$, or the difference of the vector $\mathbf{V}_2$ from $\mathbf{V}_1$ is represented by $\mathbf{V}_1 - \mathbf{V}_2$.

If r is a scalar, then $r\mathbf{V} = \mathbf{V}r$, and represents a vector r times the magnitude of $\mathbf{V}$, in the same direction as $\mathbf{V}$ if r is positive, and in the opposite direction if r is negative. If r and s are scalars, $\mathbf{V}_1$, $\mathbf{V}_2$, $\mathbf{V}_3$, vectors, then the following rules of scalars and vectors hold:

$$\mathbf{V}_1 + \mathbf{V}_2 = \mathbf{V}_2 + \mathbf{V}_1$$
$$(r + s)\mathbf{V}_1 = r\mathbf{V}_1 + s\mathbf{V}_1; \qquad r(\mathbf{V}_1 + \mathbf{V}_2) = r\mathbf{V}_1 + r\mathbf{V}_2$$
$$\mathbf{V}_1 + (\mathbf{V}_2 + \mathbf{V}_3) = (\mathbf{V}_1 + \mathbf{V}_2) + \mathbf{V}_3 = \mathbf{V}_1 + \mathbf{V}_2 + \mathbf{V}_3$$

Vectors in Space

A plane is described by two distinct vectors $\mathbf{V}_1$ and $\mathbf{V}_2$. Should these vectors not intersect each other, then one is displaced parallel to itself until they do (fig. 1.) Any other vector $\mathbf{V}$ lying in this plane is given by

$$\mathbf{V} = r\mathbf{V}_1 + s\mathbf{V}_2$$

A *position vector* specifies the position in space of a point relative to a fixed origin. If therefore $\mathbf{V}_1$ and $\mathbf{V}_2$ are the position vectors of the points A and B, relative to the origin O, then any point P on the line AB has a position vector $\mathbf{V}$ given by

$$\mathbf{V} = r\mathbf{V}_1 + (1 - r)\mathbf{V}_2$$

The scalar "r" can be taken as the parametric representation of P since $r = 0$ implies $P = B$ and $r = 1$ implies $P = A$. (fig. 2). If P divides the line AB in the ratio $r:s$ then

$$\mathbf{V} = \left(\frac{r}{r + s}\right)\mathbf{V}_1 + \left(\frac{s}{r + s}\right)\mathbf{V}_2$$

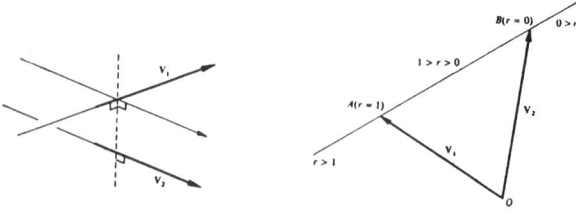

Fig. 1.   Fig. 2.

The vectors V_1, V_2, V_3, . . . , V_n are said to be *linearly dependent* if there exist scalars r_1, r_2, r_3, . . . , r_n, not all zero, such that

$$r_1V_1 + r_2V_2 + \cdots + r_nV_n = 0$$

A vector V is linearly dependent upon the set of vectors V_1, V_2, V_3, . . . , V_n if

$$V = r_1V_1 + r_2V_2 + r_3V_3 + \cdots + r_nV_n$$

Three vectors are linearly dependent if and only if they are co-planar.

All points in space can be uniquely determined by linear dependence upon three *base vectors* i.e. three vectors any one of which is linearly independent of the other two. The simplest set of base vectors are the unit vectors along the coordinate Ox, Oy and Oz axes. These are usually designated by i, j and k respectively.

If V is a vector in space, and a, b and c are the respective magnitudes of the projections of the vector along the axes then

$$V = ai + bj + ck$$

and

$$v = \sqrt{a^2 + b^2 + c^2}$$

and the direction cosines of V are

$$\cos \alpha = a/v, \quad \cos \beta = b/v, \quad \cos \gamma = c/v.$$

The law of addition yields

$$V_1 + V_2 = (a_1 + a_2)i + (b_1 + b_2)j + (c_1 + c_2)k$$

The Scalar, Dot, or Inner Product of Two Vectors V_1 and V_2

This product is represented as $V_1 \cdot V_2$ and is defined to be equal to $v_1v_2 \cos \theta$, where θ is the angle from V_1 to V_2, i.e.,

$$V_1 \cdot V_2 = v_1v_2 \cos \theta$$

The following rules apply for this product:

$$V_1 \cdot V_2 = a_1a_2 + b_1b_2 + c_1c_2 = V_2 \cdot V_1$$

It should be noted that this verifies that scalar multiplication is commutative.

$$(V_1 + V_2) \cdot V_3 = V_1 \cdot V_3 + V_2 \cdot V_3$$
$$V_1 \cdot (V_2 + V_3) = V_1 \cdot V_2 + V_1 \cdot V_3$$

If V_1 is perpendicular to V_2 then $V_1 \cdot V_2 = 0$, and if V_1 is parallel to V_2 then $V_1 \cdot V_2 = v_1v_2 = rw_1^2$

In particular

$$i \cdot i = j \cdot j = k \cdot k = 1,$$

and

$$i \cdot j = j \cdot k = k \cdot i = 0$$

The Vector or Cross Product of Vectors V_1 and V_2

This product is represented as $V_1 \times V_2$ and is defined to be equal to $v_1v_2(\sin \theta)1$, where θ is the angle from V_1 to V_2 and 1 is a unit vector perpendicular to the plane of V_1 and V and so directed that a right-handed screw driven in the direction of 1 would carry V_1 into V_2, i.e.,

$$V_1 \times V_2 = v_1v_2(\sin \theta)1$$

and

$$\tan \theta = \frac{|V_1 \times V_2|}{V_1 \cdot V_2}$$

The following rules apply for vector products:

$$V_1 \times V_2 = -V_2 \times V_1$$
$$V_1 \times (V_2 + V_3) = V_1 \times V_2 + V_1 \times V_3$$
$$(V_1 + V_2) \times V_3 = V_1 \times V_3 + V_2 \times V_3$$
$$V_1 \times (V_2 \times V_3) = V_2(V_3 \cdot V_1 - V_3(V_1 \cdot V_2)$$
$$i \times i = j \times j = k \times k = 0.1 \text{ (zero vector)}$$
$$= 0$$

$$i \times j = k, \qquad j \times k = i, \qquad k \times i = j$$

A-88

If $\mathbf{V}_1 = a_1\mathbf{i} + b_1\mathbf{j} + c_1\mathbf{k}$, $\qquad \mathbf{V}_2 = a_2\mathbf{i} + b_2\mathbf{j} + c_2\mathbf{k}$, $\qquad \mathbf{V}_3 = a_3\mathbf{i} + b_3\mathbf{j} + c_3\mathbf{k}$,

then

$$\mathbf{V}_1 \times \mathbf{V}_2 = \begin{vmatrix} \mathbf{i} & \mathbf{j} & \mathbf{k} \\ a_1 & b_1 & c_1 \\ a_2 & b_2 & c_2 \end{vmatrix} = (b_1c_2 - b_2c_1)\mathbf{i} + (c_1a_2 - c_2a_1)\mathbf{j} + (a_1b_2 - a_2b_1)\mathbf{k}$$

It should be noted that, since $\mathbf{V}_1 \times \mathbf{V}_2 = -\mathbf{V}_2 \times \mathbf{V}_1$, the vector product is not commutative.

Scalar Triple Product

There is only one possible interpretation of the expression $\mathbf{V}_1 \cdot \mathbf{V}_2 \times \mathbf{V}_3$ and that is $\mathbf{V}_1 \cdot (\mathbf{V}_2 \times \mathbf{V}_3)$ which is obviously a scalar.

Further $\mathbf{V}_1 \cdot (\mathbf{V}_2 \times \mathbf{V}_3) = (\mathbf{V}_1 \times \mathbf{V}_2) \cdot \mathbf{V}_3 = \mathbf{V}_2 \cdot (\mathbf{V}_3 \times \mathbf{V}_1)$

$$= \begin{vmatrix} a_1 & b_1 & c_1 \\ a_2 & b_2 & c_2 \\ a_3 & b_3 & c_3 \end{vmatrix}$$

$$= v_1v_2v_3 \cos \phi \sin \theta,$$

Where θ is the angle between $\mathbf{V}_2$ and $\mathbf{V}_3$ and ϕ is the angle between $\mathbf{V}_1$ and the normal to the plane of $\mathbf{V}_2$ and $\mathbf{V}_3$.

This product is called the *scalar triple product* and is written as $[\mathbf{V}_1\mathbf{V}_2\mathbf{V}_3]$.

The determinant indicates that it can be considered as the volume of the parallelepiped whose three determining edges are $\mathbf{V}_1$, $\mathbf{V}_2$ and $\mathbf{V}_3$.

It also follows that cyclic permutation of the subscripts does not change the value of the scalar triple product so that

$$[\mathbf{V}_1\mathbf{V}_2\mathbf{V}_3] = [\mathbf{V}_2\mathbf{V}_3\mathbf{V}_1] = [\mathbf{V}_3\mathbf{V}_1\mathbf{V}_2]$$

but $\quad [\mathbf{V}_1\mathbf{V}_2\mathbf{V}_3] = -[\mathbf{V}_2\mathbf{V}_1\mathbf{V}_3]$ etc. $\qquad$ and $\quad [\mathbf{V}_1\mathbf{V}_1\mathbf{V}_2] \equiv 0$ etc.

Given three non-coplanar reference vectors $\mathbf{V}_1$, $\mathbf{V}_2$ and $\mathbf{V}_3$, the *reciprocal system is* given by $\mathbf{V}_1^*$, $\mathbf{V}_2^*$ and $\mathbf{V}_3^*$, where

$$1 = v_1v_1^* = v_2v_2^* = v_3v_3^*$$

$$0 = v_1v_2^* = v_1v_3^* = v_2v_1^* \quad \text{etc.}$$

$$\mathbf{V}_1^* = \frac{\mathbf{V}_2 \times \mathbf{V}_3}{[\mathbf{V}_1\mathbf{V}_2\mathbf{V}_3]}, \qquad \mathbf{V}_2^* = \frac{\mathbf{V}_3 \times \mathbf{V}_1}{[\mathbf{V}_1\mathbf{V}_2\mathbf{V}_3]}, \qquad \mathbf{V}_3^* = \frac{\mathbf{V}_1 \times \mathbf{V}_2}{[\mathbf{V}_1\mathbf{V}_2\mathbf{V}_3]}$$

The system $\mathbf{i}$, $\mathbf{j}$, $\mathbf{k}$ is its own reciprocal.

Vector Triple Product

The product $\mathbf{V}_1 \times (\mathbf{V}_2 \times \mathbf{V}_3)$ defines the *vector triple product*. Obviously, in this case, the brackets are vital to the definition.

$$\mathbf{V}_1 \times (\mathbf{V}_2 \times \mathbf{V}_3) = (\mathbf{V}_1 \cdot \mathbf{V}_3)\mathbf{V}_2 - (\mathbf{V}_1 \cdot \mathbf{V}_2)\mathbf{V}_3$$

$$= \begin{vmatrix} \mathbf{i} & \mathbf{j} & \mathbf{k} \\ a_1 & b_1 & c_1 \\ \begin{vmatrix} b_2 & c_2 \\ b_3 & c_3 \end{vmatrix} & \begin{vmatrix} c_2 & a_2 \\ c_3 & a_3 \end{vmatrix} & \begin{vmatrix} a_2 & b_2 \\ a_3 & b_3 \end{vmatrix} \end{vmatrix}$$

i.e. it is a vector, perpendicular to $\mathbf{V}_1$, lying in the plane of $\mathbf{V}_2$, $\mathbf{V}_3$.

Similarly $\qquad (\mathbf{V}_1 \times \mathbf{V}_2) \times \mathbf{V}_3 = \begin{vmatrix} \mathbf{i} & \mathbf{j} & \mathbf{k} \\ \begin{vmatrix} b_1 & c_1 \\ b_2 & c_2 \end{vmatrix} & \begin{vmatrix} c_1 & a_1 \\ c_2 & a_2 \end{vmatrix} & \begin{vmatrix} a_1 & b_1 \\ a_2 & b_2 \end{vmatrix} \\ a_3 & b_3 & c_3 \end{vmatrix}$

$$\mathbf{V}_1 \times (\mathbf{V}_2 \times \mathbf{V}_3) + \mathbf{V}_2 \times (\mathbf{V}_3 \times \mathbf{V}_1) + \mathbf{V}_3 \times (\mathbf{V}_1 \times \mathbf{V}_2) \equiv 0$$

If $\mathbf{V}_1 \times (\mathbf{V}_2 \times \mathbf{V}_3) = (\mathbf{V}_1 \times \mathbf{V}_2) \times \mathbf{V}_3$ then $\mathbf{V}_1$, $\mathbf{V}_2$, $\mathbf{V}_3$ form an *orthogonal set*. Thus $\mathbf{i}$, $\mathbf{j}$, $\mathbf{k}$ form an orthogonal set.

Geometry of the Plane, Straight Line and Sphere

The position vectors of the fixed points A, B, C, D relative to O are $\mathbf{V}_1$, $\mathbf{V}_2$, $\mathbf{V}_3$, $\mathbf{V}_4$ and the position vector of the variable point P is $\mathbf{V}$.

The vector form of the equation of the straight line through A parallel to $\mathbf{V}_2$ is

$$\mathbf{V} = \mathbf{V}_1 + r\mathbf{V}_2$$

or $\qquad (\mathbf{V} - \mathbf{V}_1) = r\mathbf{V}_2$

or $\qquad (\mathbf{V} - \mathbf{V}_1) \times \mathbf{V}_2 = 0$

while that of the plane through A perpendicular to $\mathbf{V}_2$ is

$$(\mathbf{V} - \mathbf{V}_1) \cdot \mathbf{V}_2 = 0$$

The equation of the line AB is

$$\mathbf{V} = r\mathbf{V}_1 + (1 - r)\mathbf{V}_2$$

and those of the bisectors of the angles between $\mathbf{V}_1$ and $\mathbf{V}_2$ are

$$\mathbf{V} = r\left(\frac{\mathbf{V}_1}{v} \pm \frac{\mathbf{V}_2}{v_2}\right)$$

or $\qquad \mathbf{V} = r(\hat{\mathbf{v}}_1 \pm \hat{\mathbf{v}}_2)$

The perpendicular from C to the line through A parallel to $\mathbf{V}_2$ has as its equation

$$\mathbf{V} = \mathbf{V}_1 - \mathbf{V}_3 - \hat{\mathbf{v}}_2 \cdot (\mathbf{V}_1 - \mathbf{V}_3)\hat{\mathbf{v}}_2.$$

The condition for the intersection of the two lines,

$$\mathbf{V} = \mathbf{V}_1 + r\mathbf{V}_3$$

and $\qquad \mathbf{V} = \mathbf{V}_2 + s\mathbf{V}_4$

is $\qquad [(\mathbf{V}_1 - \mathbf{V}_2)\mathbf{V}_3\mathbf{V}_4] = 0.$

The common perpendicular to the above two lines is the line of intersection of the two planes

$$[(\mathbf{V} - \mathbf{V}_1)\mathbf{V}_3(\mathbf{V}_3 \times \mathbf{V}_4)] = 0$$

and $\qquad [(\mathbf{V} - \mathbf{V}_2)\mathbf{V}_4(\mathbf{V}_3 \times \mathbf{V}_4)] = 0$

and the length of this perpendicular is

$$\frac{[(\mathbf{V}_1 - \mathbf{V}_2)\mathbf{V}_3\mathbf{V}_4]}{|\mathbf{V}_3 \times \mathbf{V}_4|}.$$

The equation of the line perpendicular to the plane ABC is

$$\mathbf{V} = \mathbf{V}_1 \times \mathbf{V}_2 + \mathbf{V}_2 \times \mathbf{V}_3 + \mathbf{V}_3 \times \mathbf{V}_1$$

and the distance of the plane from the origin is

$$\frac{[\mathbf{V}_1\mathbf{V}_2\mathbf{V}_3]}{|(\mathbf{V}_2 - \mathbf{V}_1) \times (\mathbf{V}_3 - \mathbf{V}_1)|}.$$

In general the vector equation

$$\mathbf{V} \cdot \mathbf{V}_2 = r$$

defines the plane which is perpendicular to $\mathbf{V}_2$, and the perpendicular distance from A to this plane is

$$\frac{r - \mathbf{V}_1 \cdot \mathbf{V}_2}{v_2}.$$

The distance from A, measured along a line parallel to $\mathbf{V}_3$, is

$$\frac{r - \mathbf{V}_1 \cdot \mathbf{V}_2}{\mathbf{V}_2 \cdot \hat{\mathbf{v}}_3} \quad \text{or} \quad \frac{r - \mathbf{V}_1 \cdot \mathbf{V}_2}{v_2 \cos \theta}$$

where θ is the angle beween $\mathbf{V}_2$ and $\mathbf{V}_3$.
(If this plane contains the point C then $r = \mathbf{V}_3 \cdot \mathbf{V}_2$ and if it passes through the origin then $r = 0$.)

Given two planes $\qquad \mathbf{V} \cdot \mathbf{V}_1 = r$

$$\mathbf{V} \cdot \mathbf{V}_2 = s$$

then any plane through the line of intersection of these two planes is given by

$$\mathbf{V} \cdot (\mathbf{V}_1 + \lambda\mathbf{V}_2) = r + \lambda s$$

where λ is a scalar parameter. In particular $\lambda = \pm r_1/r_2$ yields the equation of the two planes bisecting the angle between the given planes.
The plane through A parallel to the plane of $\mathbf{V}_2$, $\mathbf{V}_3$ is

$$\mathbf{V} = \mathbf{V}_1 + r\mathbf{V}_2 + s\mathbf{V}_3$$

or $\quad (\mathbf{V} - \mathbf{V}_1) \cdot \mathbf{V}_2 \times \mathbf{V}_3 = 0$

or $\quad [\mathbf{V}\mathbf{V}_2\mathbf{V}_3] - [\mathbf{V}_1\mathbf{V}_2\mathbf{V}_3] = 0$

so that the expansion in rectangular Cartesian coordinates yields

$$\begin{vmatrix} (x - a_1) & (y - b_1) & (z - c_1) \\ a_2 & b_2 & c_2 \\ a_3 & b_3 & c_3 \end{vmatrix} = 0 \qquad (\mathbf{V} \equiv x\mathbf{i} + y\mathbf{j} + z\mathbf{k})$$

which is obviously the usual linear equation in x, y and z.
The plane through AB parallel to $\mathbf{V}_3$ is given by

$$[(\mathbf{V} - \mathbf{V}_1)(\mathbf{V}_1 - \mathbf{V}_2)\mathbf{V}_3] = 0$$

or $\quad [\mathbf{V}\mathbf{V}_2\mathbf{V}_3] - [\mathbf{V}\mathbf{V}_1\mathbf{V}_3] - [\mathbf{V}_1\mathbf{V}_2\mathbf{V}_3] = 0$

The plane through the three points A, B and C is

$$\mathbf{V} = \mathbf{V}_1 + s(\mathbf{V}_2 - \mathbf{V}_1) + t(\mathbf{V}_3 - \mathbf{V}_1)$$

or $\quad\quad \mathbf{V} = r\mathbf{V}_1 + s\mathbf{V}_2 + t\mathbf{V}_3 \quad\quad (r + s + t \equiv 1)$

or $\quad\quad [(\mathbf{V} - \mathbf{V}_1)(\mathbf{V}_1 - \mathbf{V}_2)(\mathbf{V}_2 - \mathbf{V}_3)] = 0$

or $\quad\quad [\mathbf{V}\mathbf{V}_1\mathbf{V}_2] + [\mathbf{V}\mathbf{V}_2\mathbf{V}_3] + [\mathbf{V}\mathbf{V}_3\mathbf{V}_1] - [\mathbf{V}_1\mathbf{V}_2\mathbf{V}_3] = 0$

For four points A, B, C, D to be coplanar, then

$$r\mathbf{Y}_1 + s\mathbf{V}_2 + t\mathbf{V}_3 + u\mathbf{V}_4 \equiv 0 \equiv r + s + t + u$$

The following formulae relate to a sphere when the vectors are taken to lie in three dimensional space and to a circle when the space is two dimensional. For a circle in three dimensions take the intersection of the sphere with a plane.

The equation of a sphere with center O and radius OA is

$$\mathbf{V} \cdot \mathbf{V} = v_1^2 \quad\quad (\text{not } \mathbf{V} = \mathbf{V}_1)$$

or $\quad\quad (\mathbf{V} - \mathbf{V}_1) \cdot (\mathbf{V} + \mathbf{V}_1) = 0$

while that of a sphere with center B radius v_1 is

$$(\mathbf{V} - \mathbf{V}_2) \cdot (\mathbf{V} - \mathbf{V}_2) = v_1^2$$

or $\quad\quad \mathbf{V} \cdot (\mathbf{V} - 2\mathbf{V}_2) = v_1^2 - v_2^2$

If the above sphere passes through the origin then

$$\mathbf{V} \cdot (\mathbf{V} - 2\mathbf{V}_2) = 0$$

(note that in two dimensional polar coordinates this is simply)

$$r = 2a \cdot \cos \theta$$

while in three dimensional Cartesian coordinates it is

$$x^2 + y^2 + z^2 - 2(a_2x + b_2y + c_2x) = 0.$$

The equation of a sphere having the points A and B as the extremities of a diameter is

$$(\mathbf{V} - \mathbf{V}_1) \cdot (\mathbf{V} - \mathbf{V}_2) = 0.$$

The square of the length of the tangent from C to the sphere with center B and radius v_1 is given by

$$(\mathbf{V}_3 - \mathbf{V}_2) \cdot (\mathbf{V}_3 - \mathbf{V}_2) = v_1^2$$

The condition that the plane $\mathbf{V} \cdot \mathbf{V}_3 = s$ is tangential to the sphere $(\mathbf{V} - \mathbf{V}_2) \cdot (\mathbf{V} - \mathbf{V}_2) = v_1^2$ is

$$(s - \mathbf{V}_3 \cdot \mathbf{V}_2) \cdot (s - \mathbf{V}_3 \cdot \mathbf{V}_2) = v_1^2 v_3^2.$$

The equation of the tangent plane at D, on the surface of sphere $(\mathbf{V} - \mathbf{V}_2) \cdot (\mathbf{V} - \mathbf{V}_2) = v_1^2$, is

$$(\mathbf{V} - \mathbf{V}_4) \cdot (\mathbf{V}_4 - \mathbf{V}_2) = 0$$

or $\quad\quad \mathbf{V} \cdot \mathbf{V}_4 - \mathbf{V}_2 \cdot (\mathbf{V} + \mathbf{V}_4) = v_1^2 - v_2^2$

The condition that the two circles $(\mathbf{V} - \mathbf{V}_2) \cdot (\mathbf{V} - \mathbf{V}_2) = v_1^2$ and $(\mathbf{V} - \mathbf{V}_4) \cdot (\mathbf{V} - \mathbf{V}_4) = v_3^2$ intersect orthogonally is clearly

$$(\mathbf{V}_2 - \mathbf{V}_4) \cdot (\mathbf{V}_2 - \mathbf{V}_4) = v_1^2 + v_3^2$$

The polar plane of D with respect to the circle

$$(\mathbf{V} - \mathbf{V}_2) \cdot (\mathbf{V} - \mathbf{V}_2) = v_1^2 \quad \text{is}$$

$$\mathbf{V} \cdot \mathbf{V}_4 - \mathbf{V}_2 \cdot (\mathbf{V} + \mathbf{V}_4) = v_1^2 - v_2^2$$

Any sphere through the intersection of the two spheres $(\mathbf{V} - \mathbf{V}_2) \cdot (\mathbf{V} - \mathbf{V}_2) = v_1^2$ and $(\mathbf{V} - \mathbf{V}_4) \cdot (\mathbf{V} - \mathbf{V}_1) = v_3^2$ is given by

$$(\mathbf{V} - \mathbf{V}_2) \cdot (\mathbf{V} - \mathbf{V}_2) + \lambda(\mathbf{V} - \mathbf{V}_4) \cdot (\mathbf{V} - \mathbf{V}_4) = v_1^2 + \lambda v_3^2$$

while the radical plane of two such spheres is

$$\mathbf{V} \cdot (\mathbf{V}_2 - \mathbf{V}_4) = -\tfrac{1}{2}(v_1^2 - v_2^2 - v_3^2 + v_4^2)$$

Differentiation of Vectors

If $\mathbf{V}_1 = a_1\mathbf{i} + b_1\mathbf{j} + c_1\mathbf{k}$, and $\mathbf{V}_2 = a_2\mathbf{i} + b_2\mathbf{j} + c_2\mathbf{k}$, and if $\mathbf{V}_1$ and $\mathbf{V}_2$ are functions of the scalar t, then

$$\frac{d}{dt}(\mathbf{V}_1 + \mathbf{V}_2 + \cdots) = \frac{d\mathbf{V}_1}{dt} + \frac{d\mathbf{V}_2}{dt} + \cdots,$$

where $\quad \dfrac{d\mathbf{V}_1}{dt} = \dfrac{da_1}{dt}\mathbf{i} + \dfrac{db_1}{dt}\mathbf{j} + \dfrac{dc_1}{dt}\mathbf{k}$, etc.

$$\frac{d}{dt}(\mathbf{V}_1 \cdot \mathbf{V}_2) = \frac{d\mathbf{V}_1}{dt} \cdot \mathbf{V}_2 + \mathbf{V}_1 \cdot \frac{d\mathbf{V}_2}{dt}$$

$$\frac{d}{dt}(\mathbf{V}_1 \times \mathbf{V}_2) = \frac{d\mathbf{V}_1}{dt} \times \mathbf{V}_2 + \mathbf{V}_1 \times \frac{d\mathbf{V}_2}{dt}$$

$$\mathbf{V} \cdot \frac{d\mathbf{V}}{dt} = v \cdot \frac{dv}{dt}$$

In particular, if **V** is a vector of constant length then the right hand side of the last equation is identically zero showing that **V** is perpendicular to its derivative.

The derivatives of the triple products are

$$\frac{d}{dt}[V_1V_2V_3] = \left[\left(\frac{dV_1}{dt}\right)V_2V_3\right] + \left[V_1\left(\frac{dV_2}{dt}\right)V_3\right] + \left[V_1V_2\left(\frac{dV_3}{dt}\right)\right]$$

and $\dfrac{d}{dt}\left\{V_1 \times (V_2 \times V_3)\right\} = \left(\dfrac{dV_1}{dt}\right) \times (V_2 \times V_3) + V_1$

$$\times \left(\left(\frac{dV_2}{dt}\right) \times V_3\right) + V_1 \times \left(V_2 \times \left(\frac{dV_3}{dt}\right)\right)$$

Geometry of Curves in Space

s = the *length of arc*, measured from some fixed point on the curve (fig. 3).

V_1 = the position vector of the point A on the curve

$V_1 + \delta V_1$ = the position vector of the point P in the neighborhood of A

$\hat{t}$ = the *unit tangent* to the curve at the point A, measured in the direction of s increasing.

The *normal plane* is that plane which is perpendicular to the unit tangent. The principal normal is defined as the intersection of the normal plane with the plane defined by V_1 and $V_1 + \delta V_1$ in the limit as $\delta V_1 - 0$.

$\hat{n}$ = the *unit normal* (principal) at the point A. The plane defined by $\hat{t}$ and $\hat{n}$ is called the *osculating plane* (alternatively plane of curvature or local plane).

ρ = the radius of curvature at A

$\delta\theta$ = the angle subtended at the origin by δV_1.

$$\kappa = \frac{d\theta}{ds} = \frac{1}{\rho}$$

$\hat{b}$ = the *unit binormal* i.e. the unit vector which is parallel to $\hat{t} \times \hat{n}$ at the point A:

λ = the *torsion* of the curve at A

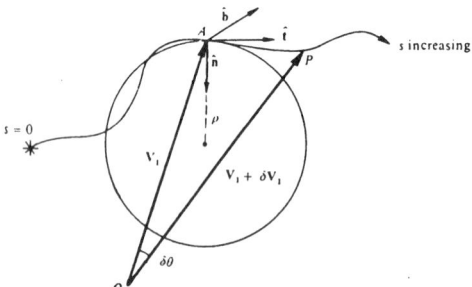

Figure 3.

Frenet's Formulae:

$$\frac{d\hat{t}}{ds} = \kappa\hat{n}$$

$$\frac{d\hat{n}}{ds} = -\kappa\hat{t} + \lambda\hat{b}$$

$$\frac{d\hat{b}}{ds} = -\lambda\hat{n}$$

The following formulae are also applicable:

Unit tangent $\qquad\qquad\qquad \hat{t} = \dfrac{dV_1}{ds}$

Equation of the tangent $\qquad (V - V_1) \times \hat{t} = 0$

or $\qquad V = V_1 + q\hat{t}$

Unit normal $\qquad\qquad\qquad \hat{n} = \dfrac{1}{\kappa}\dfrac{d^2V_1}{ds^2}$

Equation of the normal plane $\qquad (V - V_1) \cdot \hat{t} = 0$

Equation of the normal $\qquad (V - V_1) \times \hat{n} = 0$

or $\qquad V = V_1 + r\hat{n}$

Unit binormal $\qquad\qquad \hat{b} = \hat{t} \times \hat{n}$

Equation of the binormal $\qquad (V - V_1) \times \hat{b} = 0$

or $\qquad V = V_1 + u\hat{b}$

or $\qquad V = V_1 + w\dfrac{dV_1}{ds} \times \dfrac{d^2V_1}{ds^2}$

Equation of the osculating plane:

$$[(V - V_1)\hat{t}\hat{n}] = 0$$

or $\qquad \left[(V - V_1)\left(\dfrac{dV_1}{ds}\right)\left(\dfrac{d^2V_1}{ds^2}\right)\right] = 0$

A *geodetic line* on a surface is a curve, the osculating plane of which is everywhere normal to the surface.

The differential equation of the geodetic is

$$[\dot{n}\,dV_1\,d^2V_1] = 0$$

Differential Operators—Rectangular Coordinates

$$dS = \frac{\partial S}{\partial x}\cdot dx + \frac{\partial S}{\partial y}\cdot dy + \frac{\partial S}{\partial z}\cdot dz$$

By definition

$$\nabla \equiv \text{del} \equiv \mathbf{i}\,\frac{\partial}{\partial x} + \mathbf{j}\,\frac{\partial}{\partial y} + \mathbf{k}\,\frac{\partial}{\partial z}$$

$$\nabla^2 \equiv \text{Laplacian} \equiv \frac{\partial^2}{\partial x^2} + \frac{\partial^2}{\partial y^2} + \frac{\partial^2}{\partial z^2}$$

If S is a scalar function, then

$$\nabla S \equiv \text{grad } S \equiv \frac{\partial S}{\partial x}\mathbf{i} + \frac{\partial S}{\partial y}\mathbf{j} + \frac{\partial S}{\partial z}\mathbf{k}$$

Grad S defines both the direction and magnitude of the maximum rate of increase of S at any point. Hence the name *gradient* and also its vectorial nature. ∇S is independent of the choice of rectangular coordinates.

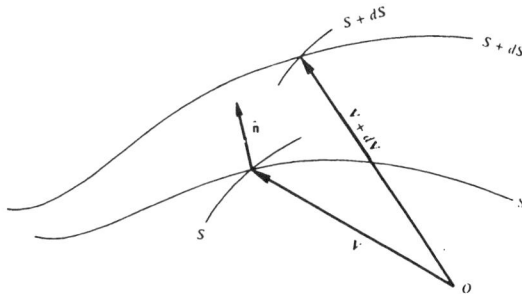

Figure 4

$$\nabla S = \frac{\partial S}{\partial n}\,\dot{n}$$

where $\dot{n}$ is the unit normal to the surface $S = $ constant, in the direction of S increasing. The total derivative of S at a point having the position vector $\mathbf{V}$ is given by (fig. 4)

$$dS = \frac{\partial S}{\partial n}\,\dot{n}\cdot d\mathbf{V}$$
$$= d\mathbf{V}\cdot\nabla S$$

and the directional derivative of S in the direction of $\mathbf{U}$ is

$$\mathbf{U}\cdot\nabla S = \mathbf{U}\cdot(\nabla S) = (\mathbf{U}\cdot\nabla)S$$

Similarly the directional derivative of the vector $\mathbf{V}$ in the direction of $\mathbf{U}$ is

$$(\mathbf{U}\cdot\nabla)\mathbf{V}$$

The *distributive* law holds for finding a gradient. Thus if S and T are scalar functions
$$\nabla(S + T) = \nabla S + \nabla T$$
The *associative* law becomes the rule for differentiating a product:
$$\nabla(ST) = S\nabla T + T\nabla S$$
If $\mathbf{V}$ is a vector function with the magnitudes of the components parallel to the three coordinate axes V_x, V_y, V_z, then

$$\nabla\cdot\mathbf{V} \equiv \text{div }\mathbf{V} \equiv \frac{\partial V_x}{\partial x} + \frac{\partial V_y}{\partial y} + \frac{\partial V_z}{\partial z}$$

The divergence obeys the distributive law. Thus, if $\mathbf{V}$ and $\mathbf{U}$ are vectors functions, then

$$\nabla\cdot(\mathbf{V} + \mathbf{U}) = \nabla\cdot\mathbf{V} + \nabla\cdot\mathbf{U}$$
$$\nabla\cdot(S\mathbf{V}) = (\nabla S)\cdot\mathbf{V} + S(\nabla\cdot\mathbf{V})$$
$$\nabla\cdot(\mathbf{U}\times\mathbf{V}) = \mathbf{V}\cdot(\nabla\times\mathbf{U}) - \mathbf{U}\cdot(\nabla\times\mathbf{V})$$

As with the gradient of a scalar, the divergence of a vector is invariant under a transformation from one set of rectangular coordinates to another.

$$\nabla\times\mathbf{V} \equiv \text{curl }\mathbf{V}\ (\text{sometimes }\nabla\wedge\mathbf{V}\text{ or rot }\mathbf{V})$$

$$\equiv \left(\frac{\partial V_z}{\partial y} - \frac{\partial V_y}{\partial z}\right)\mathbf{i} + \left(\frac{\partial V_x}{\partial z} - \frac{\partial V_z}{\partial x}\right)\mathbf{j} + \left(\frac{\partial V_y}{\partial x} - \frac{\partial V_x}{\partial y}\right)\mathbf{k}$$

$$= \begin{vmatrix} \mathbf{i} & \mathbf{j} & \mathbf{k} \\ \dfrac{\partial}{\partial x} & \dfrac{\partial}{\partial y} & \dfrac{\partial}{\partial z} \\ V_x & V_y & V_z \end{vmatrix}$$

A-93

The *curl* (or *rotation*) of a vector is a vector which is invariant under a transformation from one set of rectangular coordinates to another.

$$\nabla \times (\mathbf{U} + \mathbf{V}) = \nabla \times \mathbf{U} + \nabla \times \mathbf{V}$$
$$\nabla \times (S\mathbf{V}) = (\nabla S) \times \mathbf{V} + S(\nabla \times \mathbf{V})$$
$$\nabla \times (\mathbf{U} \times \mathbf{V}) = (\mathbf{V} \cdot \nabla)\mathbf{U} - (\mathbf{U} \cdot \nabla)\mathbf{V} + \mathbf{U}(\nabla \cdot \mathbf{V}) - \mathbf{V}(\nabla \cdot \mathbf{U})$$

$$\operatorname{grad} (\mathbf{U} \cdot \mathbf{V}) = \nabla(\mathbf{U} \cdot \mathbf{V})$$
$$= (\mathbf{V} \cdot \nabla)\mathbf{U} + (\mathbf{U} \cdot \nabla)\mathbf{V} + \mathbf{V} \times (\nabla \times \mathbf{U}) + \mathbf{U} \times (\nabla \times \mathbf{V})$$

If
$$\mathbf{V} = V_x\mathbf{i} + V_y\mathbf{j} + V_z\mathbf{k}$$
$$\nabla \cdot \mathbf{V} = \nabla V_x \cdot \mathbf{i} + \nabla V_y \cdot \mathbf{j} + \nabla V_z \cdot \mathbf{k}$$
and
$$\nabla \times \mathbf{V} = \nabla V_x \times \mathbf{i} + \nabla V_y \times \mathbf{j} + \nabla V_z \times \mathbf{k}$$

The operator ∇ can be used more than once. The number of possibilities where ∇ is used twice are

$$\nabla \cdot (\nabla\theta) \equiv \operatorname{div} \operatorname{grad} \theta$$
$$\nabla \times (\nabla\theta) \equiv \operatorname{curl} \operatorname{grad} \theta$$
$$\nabla(\nabla \cdot \mathbf{V}) \equiv \operatorname{grad} \operatorname{div} \mathbf{V}$$
$$\nabla \cdot (\nabla \times \mathbf{V}) \equiv \operatorname{div} \operatorname{curl} \mathbf{V}$$
$$\nabla \times (\nabla \times \mathbf{V}) \equiv \operatorname{curl} \operatorname{curl} \mathbf{V}$$

Thus: $\operatorname{div} \operatorname{grad} S \equiv \nabla \cdot (\nabla S) \equiv \text{Laplacian } S \equiv \nabla^2 S$

$$\equiv \frac{\partial^2 S}{\partial x^2} + \frac{\partial^2 S}{\partial y^2} + \frac{\partial^2 S}{\partial z^2}$$

$\operatorname{curl} \operatorname{grad} S \equiv 0;$ $\operatorname{curl} \operatorname{curl} \mathbf{V} \equiv \operatorname{grad} \operatorname{div} \mathbf{V} - \nabla^2\mathbf{V};$

$$\operatorname{div} \operatorname{curl} \mathbf{V} \equiv \qquad 0$$

Taylor's expansion in three dimensions can be written

$$f(\mathbf{V} + \mathbf{\varepsilon}) = e^{\varepsilon \cdot \nabla} f(\mathbf{V})$$

where $\mathbf{V} = x\mathbf{i} + y\mathbf{j} + z\mathbf{k}$

and $\mathbf{\varepsilon} = h\mathbf{i} + l\mathbf{j} + m\mathbf{k}$

(note the analogy with $f_p = e^{phD}f_0$ in finite difference methods).

Orthogonal Curvilinear Coordinates

If at a point P there exist three uniform point functions u, v and w so that the surfaces $u = $ const., $v = $ const., and $w = $ const., intersect in three distinct curves through P then the surfaces are called the *coordinate surfaces* through P. The three lines of intersection are referred to as the *coordinate lines* and their tangents a, b, and c as the *coordinate axes*. When the coordinate axes form an orthogonal set the system is said to define *orthogonal curvilinear coordinates* at P.

Consider an infinitesimal volume enclosed by the surfaces u, v, w, $u + du$, $v + dv$, and $w + dw$ (fig. 5).

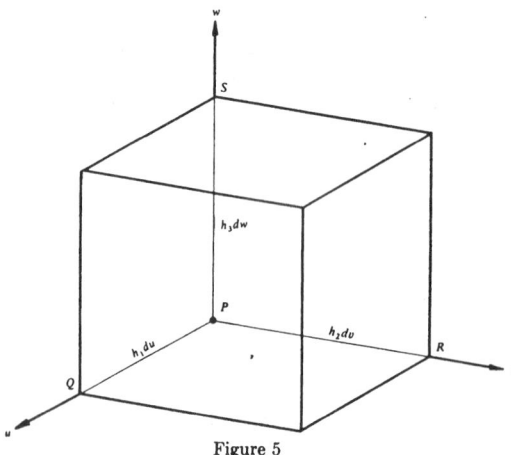

Figure 5

The surface $PRS \equiv u = $ const., and the face of the curvilinear figure immediately opposite this is $u + du = $ const. etc.

In terms of these surface constants

$$P = P(u,v,w)$$
$$Q = Q(u + du,v,w) \qquad \text{and} \qquad PQ = h_1 du$$

$$R = R(u, v + dv, w) \qquad PR = h_2 dv$$
$$S = S(u, v, w + dw) \qquad PS = h_3 dw$$

where h_1, h_2, and h_3 are functions of u, v, and w.

In rectangular Cartesians $\mathbf{i}$, $\mathbf{j}$, $\mathbf{k}$

$$h_1 = 1, \qquad h_2 = 1, \qquad h_3 = 1.$$
$$\frac{\hat{\mathbf{a}}}{h_1}\frac{\partial}{\partial u} = \mathbf{i}\,\frac{\partial}{\partial x}, \qquad \frac{\hat{\mathbf{b}}}{h_2}\frac{\partial}{\partial v} = \mathbf{j}\,\frac{\partial}{\partial y}, \qquad \frac{\hat{\mathbf{c}}}{h_3}\frac{\partial}{\partial w} = \mathbf{k}\,\frac{\partial}{\partial z}.$$

In cylindrical coordinates $\hat{\mathbf{r}}$, $\hat{\boldsymbol{\phi}}$, $\hat{\mathbf{k}}$

$$h_1 = 1, \qquad h_2 = r \qquad h_3 = 1.$$
$$\frac{\hat{\mathbf{a}}}{h_1}\frac{\partial}{\partial u} = \hat{\mathbf{r}}\,\frac{\partial}{\partial r}, \qquad \frac{\hat{\mathbf{b}}}{h_2}\frac{\partial}{\partial v} = \frac{\hat{\boldsymbol{\phi}}}{r}\,\frac{\partial}{\partial \phi} \qquad \frac{\hat{\mathbf{c}}}{h_3}\frac{\partial}{\partial w} = \hat{\mathbf{k}}\,\frac{\partial}{\partial z}$$

In spherical coordinates $\hat{\mathbf{r}}$, $\hat{\boldsymbol{\theta}}$, $\hat{\boldsymbol{\phi}}$

$$h_1 = 1, \qquad h_2 = r, \qquad h_3 = r\sin\theta$$
$$\frac{\hat{\mathbf{a}}}{h_1}\frac{\partial}{\partial u} = \hat{\mathbf{r}}\,\frac{\partial}{\partial r}, \qquad \frac{\mathbf{b}}{h_2}\frac{\partial}{\partial v} = \frac{\hat{\boldsymbol{\phi}}}{r}\,\frac{\partial}{\partial \theta}, \qquad \frac{\hat{\mathbf{c}}}{h_3}\frac{\partial}{\partial w} = \frac{\hat{\boldsymbol{\phi}}}{r\sin\theta}\,\frac{\partial}{\partial \phi}$$

The general expressions for grad, div and curl together with those for ∇^2 and the directional derivative are, in orthogonal curvilinear coordinates, given by

$$\nabla S = \frac{\hat{\mathbf{a}}}{h_1}\frac{\partial S}{\partial u} + \frac{\hat{\mathbf{b}}}{h_2}\frac{\partial S}{\partial v} + \frac{\hat{\mathbf{c}}}{h_3}\frac{\partial S}{\partial w}$$

$$(\mathbf{V}\cdot\nabla)S = \frac{V_1}{h_1}\frac{\partial S}{\partial u} + \frac{V_2}{h_2}\frac{\partial S}{\partial v} + \frac{V_3}{h_3}\frac{\partial S}{\partial w}$$

$$\nabla\cdot\mathbf{V} = \frac{1}{h_1 h_2 h_3}\left\{\frac{\partial}{\partial u}(h_2 h_3 V_1) + \frac{\partial}{\partial v}(h_3 h_1 V_2) + \frac{\partial}{\partial w}(h_1 h_2 V_3)\right\}$$

$$\nabla\times\mathbf{V} = \frac{\hat{\mathbf{a}}}{h_2 h_3}\left\{\frac{\partial}{\partial v}(h_3 V_3) - \frac{\partial}{\partial w}(h_2 V_2)\right\} + \frac{\hat{\mathbf{b}}}{h_3 h_1}\left\{\frac{\partial}{\partial w}(h_1 V_1) - \frac{\partial}{\partial u}(h_3 V_3)\right\}$$
$$+ \frac{\hat{\mathbf{c}}}{h_1 h_2}\left\{\frac{\partial}{\partial u}(h_2 V_2) - \frac{\partial}{\partial v}(h_1 V_1)\right\}$$

$$\nabla^2 S = \frac{1}{h_1 h_2 h_3}\left\{\frac{\partial}{\partial u}\left(\frac{h_2 h_3}{h_1}\frac{\partial S}{\partial u}\right) + \frac{\partial}{\partial v}\left(\frac{h_3 h_1}{h_2}\frac{\partial S}{\partial v}\right) + \frac{\partial}{\partial w}\left(\frac{h_1 h_2}{h_3}\frac{\partial S}{\partial w}\right)\right\}$$

FORMULAS OF VECTOR ANALYSIS

	Rectangular coordinates	Cylindrical coordinates	Spherical coordinates
Conversion to rectangular coordinates		$x = r\cos\varphi \quad y = r\sin\varphi \quad z = z$	$x = r\cos\varphi\sin\theta \quad y = r\sin\varphi\sin\theta$ $z = r\cos\theta$
Gradient	$\nabla\phi = \dfrac{\partial\phi}{\partial x}\mathbf{i} + \dfrac{\partial\phi}{\partial y}\mathbf{j} + \dfrac{\partial\phi}{\partial z}\mathbf{k}$	$\nabla\phi = \dfrac{\partial\phi}{\partial r}\mathbf{r} + \dfrac{1}{r}\dfrac{\partial\phi}{\partial\varphi}\boldsymbol{\phi} + \dfrac{\partial\phi}{\partial z}\mathbf{k}$	$\nabla\phi = \dfrac{\partial\phi}{\partial r}\mathbf{r} + \dfrac{1}{r}\dfrac{\partial\phi}{\partial\theta}\boldsymbol{\theta} + \dfrac{1}{r\sin\theta}\dfrac{\partial\phi}{\partial\varphi}\boldsymbol{\phi}$
Divergence	$\nabla\cdot\mathbf{A} = \dfrac{\partial A_x}{\partial x} + \dfrac{\partial A_y}{\partial y} + \dfrac{\partial A_z}{\partial z}$	$\nabla\cdot\mathbf{A} = \dfrac{1}{r}\dfrac{\partial(rA_r)}{\partial r} + \dfrac{1}{r}\dfrac{\partial A_\varphi}{\partial\varphi}$ $+ \dfrac{\partial A_z}{\partial z}$	$\nabla\cdot\mathbf{A} = \dfrac{1}{r^2}\dfrac{\partial(r^2 A_r)}{\partial r} + \dfrac{1}{r\sin\theta}\dfrac{\partial(A_\theta\sin\theta)}{\partial\theta}$ $+ \dfrac{1}{r\sin\theta}\dfrac{\partial A_\varphi}{\partial\varphi}$
Curl	$\nabla\times\mathbf{A} = \begin{vmatrix} \mathbf{i} & \mathbf{j} & \mathbf{k} \\ \dfrac{\partial}{\partial x} & \dfrac{\partial}{\partial y} & \dfrac{\partial}{\partial z} \\ A_x & A_y & A_z \end{vmatrix}$	$\nabla\times\mathbf{A} = \begin{vmatrix} \dfrac{1}{r}\mathbf{r} & \boldsymbol{\phi} & \dfrac{1}{r}\mathbf{k} \\ \dfrac{\partial}{\partial r} & \dfrac{\partial}{\partial\varphi} & \dfrac{\partial}{\partial z} \\ A_r & rA_\varphi & A_z \end{vmatrix}$	$\nabla\times\mathbf{A} = \begin{vmatrix} \dfrac{\mathbf{r}}{r^2\sin\theta} & \dfrac{\boldsymbol{\theta}}{r\sin\theta} & \dfrac{\boldsymbol{\phi}}{r} \\ \dfrac{\partial}{\partial r} & \dfrac{\partial}{\partial\theta} & \dfrac{\partial}{\partial\varphi} \\ A_r & rA_\theta & rA_\varphi\sin\theta \end{vmatrix}$
Laplacian	$\nabla^2\phi = \dfrac{\partial^2\phi}{\partial x^2} + \dfrac{\partial^2\phi}{\partial y^2} + \dfrac{\partial^2\phi}{\partial z^2}$	$\nabla^2\phi = \dfrac{1}{r}\dfrac{\partial}{\partial r}\left(r\dfrac{\partial\phi}{\partial r}\right) + \dfrac{1}{r^2}\dfrac{\partial^2\phi}{\partial\varphi^2}$ $+ \dfrac{\partial^2\phi}{\partial z^2}$	$\nabla^2\phi = \dfrac{1}{r^2}\dfrac{\partial}{\partial r}\left(r^2\dfrac{\partial\phi}{\partial r}\right) + \dfrac{1}{r^2\sin\theta}\dfrac{\partial}{\partial\theta}\left(\sin\theta\dfrac{\partial\phi}{\partial\theta}\right)$ $+ \dfrac{1}{r^2\sin^2\theta}\dfrac{\partial^2\phi}{\partial\varphi^2}$

Transformation of Integrals

s = the distance along some curve "C" in space and is measured from some fixed point.
S = a surface area
V = a volume contained by a specified surface
$\hat{t}$ = the unit tangent to C at the point P
$\hat{n}$ = the unit outward pointing normal
F = some vector function
ds = the vector element of curve ($= \hat{t}\,ds$)
dS = the vector element of surface ($= \hat{n}\,dS$)

Then
$$\int_{(c)} \mathbf{F} \cdot \hat{t}\,ds = \int_{(c)} \mathbf{F} \cdot d\mathbf{s}$$

and when
$$\mathbf{F} = \nabla\phi$$

$$\int_{(c)} (\nabla\phi) \cdot \hat{t}\,ds = \int_{(c)} d\phi$$

Gauss' Theorem (Green's Theorem)

When S defines a closed region having a volume V

$$\iiint_{(v)} (\nabla \cdot \mathbf{F})\,dV = \iint_{(s)} (\mathbf{F} \cdot \hat{n})\,dS = \iint_{(s)} \mathbf{F} \cdot d\mathbf{S}$$

also
$$\iiint_{(v)} (\nabla\phi)\,dV = \iint_{(s)} \phi\hat{n}\,dS$$

and
$$\iiint_{(v)} (\nabla \times \mathbf{F})\,dV = \iint_{(s)} (\hat{n} \times \mathbf{F})\,dS$$

Stokes' Theorem

When C is closed and bounds the open surface S.

$$\iint_{(s)} \hat{n} \cdot (\nabla \times \mathbf{F})\,dS = \int_{(c)} \mathbf{F} \cdot d\mathbf{s}$$

also
$$\iint_{(s)} (\hat{n} \times \nabla\phi)\,dS = \int_{(c)} \phi\,d\mathbf{s}$$

Green's Theorem

$$\iint_{(s)} (\nabla\phi \cdot \nabla\theta)\,dS = \iint_{(s)} \phi\hat{n} \cdot (\nabla\theta)\,dS = \iiint_{(v)} \phi(\nabla^2\theta)\,dV$$

$$= \iint_{(s)} \theta \cdot \hat{n}(\nabla\phi)\,dS = \iiint_{(v)} \theta(\nabla^2\phi)\,dV$$

MOMENT OF INERTIA FOR VARIOUS BODIES OF MASS

The mass of the body is indicated by m

Body	Axis	Moment of inertia	Body	Axis	Moment of inertia
Uniform thin rod	Normal to the length, at one end	$m\dfrac{l^2}{3}$	Spherical shell, very thin, mean radius, r	Any diameter	$m\dfrac{2}{3}r^2$
Uniform thin rod	Normal to the length, at the center	$m\dfrac{l^2}{12}$	Right circular cylinder of radius r, length l	The longitudinal axis of the solid	$m\dfrac{r^2}{2}$
Thin rectangular sheet, sides a and b	Through the center parallel to b	$m\dfrac{a^2}{12}$	Right circular cylinder of radius r, length l	Transverse diameter	$m\left(\dfrac{r^2}{4} + \dfrac{l^2}{12}\right)$
Thin rectangular sheet, sides a and b	Through the center perpendicular to the sheet	$m\dfrac{a^2 + b^2}{12}$	Hollow circular cylinder, length l, radii r_1 and r_2	The longitudinal axis of the figure	$m\dfrac{(r_1^2 + r_2^2)}{2}$
Thin circular sheet of radius r	Normal to the plate through the center	$m\dfrac{r^2}{2}$	Thin cylindrical shell, length l, mean radius, r	The longitudinal axis of the figure	mr^2
Thin circular sheet of radius r	Along any diameter	$m\dfrac{r^2}{4}$	Hollow circular cylinder, length l, radii r_1 and r_2	Transverse diameter	$m\left[\dfrac{r_1^2 + r_2^2}{4} + \dfrac{l^2}{12}\right]$
Thin circular ring. Radii r_1 and r_2	Through center normal to plane of ring	$m\dfrac{r_1^2 + r_2^2}{2}$	Hollow circular cylinder, length l, very thin, mean radius	Transverse diameter	$m\left(\dfrac{r^2}{2} + \dfrac{l^2}{12}\right)$
Thin circular ring. Radii r_1 and r_2	Any diameter	$m\dfrac{r_1^2 + r_2^2}{4}$	Elliptic cylinder, length l, transverse semiaxes a and b	Longitudinal axis	$m\left(\dfrac{a^2 + b^2}{4}\right)$
Rectangular parallelopiped, edges a, b, and c	Through center perpendicular to face ab, (parallel to edge c)	$m\dfrac{a^2 + b^2}{12}$	Right cone, altitude h, radius of base r	Axis of the figure	$m\dfrac{3}{10}r^2$
Sphere, radius r	Any diameter	$m\dfrac{2}{5}r^2$	Spheroid of revolution, equatorial radius r	Polar axis	$m\dfrac{2r^2}{5}$
Spherical shell, external radius, r_1, internal radius r_2	Any diameter	$m\dfrac{2}{5}\dfrac{(r_1^5 - r_2^5)}{(r_1^3 - r_2^3)}$	Ellipsoid, axes $2a$, $2b$, $2c$	Axis $2a$	$m\dfrac{(b^2 + c^2)}{5}$

Bessel Functions*

1. Bessel's differential equation for a real variable x is

$$x^2 \frac{d^2 y}{dx^2} + x \frac{dy}{dx} + (x^2 - n^2)y = 0$$

* From Beyer, W. H., Ed., *CRC Handbook of Mathematical Sciences*, 5th ed., CRC Press, Boca Raton, 1978, 500—503. With permission.

2. When n is not an integer, two independent solutions of the equation are $J_n(x)$, $J_{-n}(x)$, where

$$J_n(x) = \sum_{k=0}^{\infty} \frac{(-1)^k}{k!\Gamma(n + k + 1)} \left(\frac{x}{2}\right)^{n+2k}$$

3. If n is an integer $J_{-n}(x) = (-1)^n J_n(x)$, where

$$J_n(x) = \frac{x^n}{2^n n!} \left\{ 1 - \frac{x^2}{2^2 \cdot 1!(n + 1)} + \frac{x^4}{2^4 \cdot 2!(n + 1)(n + 2)} \right.$$
$$\left. + \frac{x^6}{2^6 \cdot 3!(n + 1)(n + 2)(n + 3)} + \cdots \right\}$$

4. For $n = 0$ and $n = 1$, this formula becomes

$$J_0(x) = 1 - \frac{x^2}{2^2(1!)^2} + \frac{x^4}{2^4(2!)^2} - \frac{x^6}{2^6(3!)^2} + \frac{x^8}{2^8(4!)^2} - \cdots$$

$$J_1(x) = \frac{x}{2} - \frac{x^3}{2^3 \cdot 1!2!} + \frac{x^5}{2^5 \cdot 2!3!} - \frac{x^7}{2^7 \cdot 3!4!} + \frac{x^9}{2^9 \cdot 4!5!} - \cdots$$

5. When x is large and positive, the following asymptotic series may be used

$$J_0(x) = \left(\frac{2}{\pi x}\right)^{\frac{1}{2}} \left\{ P_0(x) \cos\left(x - \frac{\pi}{4}\right) - Q_0(x) \sin\left(x - \frac{\pi}{4}\right) \right\}$$

$$J_1(x) = \left(\frac{2}{\pi x}\right)^{\frac{1}{2}} \left\{ P_1(x) \cos\left(x - \frac{3\pi}{4}\right) - Q_1(x) \sin\left(x - \frac{3\pi}{4}\right) \right\}$$

where

$$P_0(x) \sim 1 - \frac{1^2 \cdot 3^2}{2!(8x)^2} + \frac{1^2 \cdot 3^2 \cdot 5^2 \cdot 7^2}{4!(8x)^4} - \frac{1^2 \cdot 3^2 \cdot 5^2 \cdot 7^2 \cdot 9^2 \cdot 11^2}{6!(8x)^6} + \cdots$$

$$Q_0(x) \sim - \frac{1^2}{1!8x} + \frac{1^2 \cdot 3^2 \cdot 5^2}{3!(8x)^3} - \frac{1^2 \cdot 3^2 \cdot 5^2 \cdot 7^2 \cdot 9^2}{5!(8x)^5} + - \cdots$$

$$P_1(x) \sim 1 + \frac{1^2 \cdot 3 \cdot 5}{2!(8x)^2} - \frac{1^2 \cdot 3^2 \cdot 5^2 \cdot 7 \cdot 9}{4!(8x)^4} + \frac{1^2 \cdot 3^2 \cdot 5^2 \cdot 7^2 \cdot 9^2 \cdot 11 \cdot 13}{6!(8x)^6} - + \cdots$$

$$Q_1(x) \sim \frac{1 \cdot 3}{1!8x} - \frac{1^2 \cdot 3^2 \cdot 5 \cdot 7}{3!(8x)^3} + \frac{1^2 \cdot 3^2 \cdot 5^2 \cdot 7^2 \cdot 9 \cdot 11}{5!(8x)^5} - \cdots$$

[In $P_1(x)$ the signs alternate from $+$ to $-$ after the first term]

6. If $x > 25$, it is convenient to use the formulas
$$J_0(x) = A_0(x) \sin x + B_0(x) \cos x$$
$$J_1(x) = B_1(x) \sin x - A_1(x) \cos x$$

where

$$A_0(x) = \frac{P_0(x) - Q_0(x)}{(\pi x)^{\frac{1}{2}}} \quad \text{and} \quad A_1(x) = \frac{P_1(x) - Q_1(x)}{(\pi x)^{\frac{1}{2}}}$$

$$B_0(x) = \frac{P_0(x) + Q_0(x)}{(\pi x)^{\frac{1}{2}}} \quad \text{and} \quad B_1(x) = \frac{P_1(x) + Q_1(x)}{(\pi x)^{\frac{1}{2}}}$$

7. The zeros of $J_0(x)$ and $J_1(x)$
If j_{0s} and j_{1s} are the sth zeros of $J_0(x)$ and $J_1(x)$ respectively, and if $a = 4_s - 1$, $b = 4_s + 1$

$$j_{0,s} \sim \frac{1}{4}\pi a \left\{ 1 + \frac{2}{\pi^2 a^2} - \frac{62}{3\pi^4 a^4} + \frac{15,116}{15\pi^6 a^6} - \frac{12,554,474}{105\pi^8 a^8} + \frac{8,368,654,292}{315\pi^{10}a^{10}} - + \cdots \right\}$$

$$j_{1,s} \sim \frac{1}{4}\pi b \left\{ 1 - \frac{6}{\pi^2 b^2} + \frac{6}{\pi^4 b^4} - \frac{4716}{5\pi^6 b^6} + \frac{3,902,418}{35\pi^8 b^8} - \frac{895,167,324}{35\pi^{10}b^{10}} + \cdots \right\}$$

$$J_1(j_{0,s}) \sim \frac{(-1)^{s+1} 2^{\frac{1}{2}}}{\pi a^{\frac{1}{2}}} \left\{ 1 - \frac{56}{3\pi^4 a^4} + \frac{9664}{5\pi^6 a^6} - \frac{7,381,280}{21\pi^8 a^8} + \cdots \right\}$$

$$J_0(j_{1,s}) \sim \frac{(-1)^s 2^{\frac{1}{2}}}{\pi b^{\frac{1}{2}}} \left\{ 1 + \frac{24}{\pi^4 b^4} - \frac{19,584}{10\pi^6 b^6} + \frac{2,466,720}{7\pi^8 b^8} - \cdots \right\}$$

8. Table of zeros for $J_0(x)$ and $J_1(x)$

Roots α_n	$J_1(\alpha_n)$	Roots β_n	$J_0(\beta_n)$
2.4048	0.5191	0.0000	1.0000
5.5201	−0.3403	3.8317	−0.4028
8.6537	0.2715	7.0156	0.3001
11.7915	−0.2325	10.1735	−0.2497
14.9309	0.2065	13.3237	0.2184
18.0711	−0.1877	16.4706	−0.1965
21.2116	0.1733	19.6159	0.1801

$$J_1(\alpha_n) = 0 \qquad J_0(\beta_n) = 0$$

9. Recurrence formulas

$$J_{n-1}(x) + J_{n+1}(x) = \frac{2n}{x} J_n(x) \qquad\qquad nJ_n(x) + xJ_n'(x) = xJ_{n-1}(x)$$

$$J_{n-1}(x) - J_{n+1}(x) = 2J_n'(x) \qquad\qquad nJ_n(x) - xJ_n'(x) = xJ_{n+1}(x)$$

10. If J_n is written for $J_n(x)$ and $J_n^{(k)}$ is written for $\dfrac{d^k}{dx^k}\{J_n(x)\}$, then the following derivative relationships are important

$$J_0^{(r)} = -J_1^{(r-1)}$$

$$J_0^{(2)} = -J_0 + \frac{1}{x} J_1 = \frac{1}{2}(J_2 - J_0)$$

$$J_0^{(3)} = \frac{1}{x} J_0 + \left(1 - \frac{2}{x^2}\right)J_1 = \frac{1}{4}(-J_3 + 3J_1)$$

$$J_0^{(4)} = \left(1 - \frac{3}{x^2}\right)J_0 - \left(\frac{2}{x} - \frac{6}{x^3}\right)J_1 = \frac{1}{8}(J_4 - 4J_2 + 3J_0), \text{ etc.}$$

11. Half order Bessel functions

$$J_{\frac{1}{2}}(x) = \sqrt{\frac{2}{\pi x}} \sin x$$

$$J_{-\frac{1}{2}}(x) = \sqrt{\frac{2}{\pi x}} \cos x$$

$$J_{n+\frac{1}{2}}(x) = -x^{n+\frac{1}{2}} \frac{d}{dx}\{x^{-(n+\frac{1}{2})}J_{n+\frac{1}{2}}(x)\}$$

$$J_{n-\frac{1}{2}}(x) = x^{-(n+\frac{1}{2})} \frac{d}{dx}\{x^{n+\frac{1}{2}}J_{n+\frac{1}{2}}(x)\}$$

n	$\left(\dfrac{\pi x}{2}\right)^{\frac{1}{2}} J_{n+\frac{1}{2}}(x)$	$\left(\dfrac{\pi x}{2}\right)^{\frac{1}{2}} J_{-(n+\frac{1}{2})}(x)$
0	$\sin x$	$\cos x$
1	$\dfrac{\sin x}{x} - \cos x$	$-\dfrac{\cos x}{x} - \sin x$
2	$\left(\dfrac{3}{x^2} - 1\right)\sin x - \dfrac{3}{x}\cos x$	$\left(\dfrac{3}{x^2} - 1\right)\cos x + \dfrac{3}{x}\sin x$
3	$\left(\dfrac{15}{x^3} - \dfrac{6}{x}\right)\sin x - \left(\dfrac{15}{x^2} - 1\right)\cos x$	$-\left(\dfrac{15}{x^3} - \dfrac{6}{x}\right)\cos x - \left(\dfrac{15}{x^2} - 1\right)\sin x$

etc.

12. Additional solutions to Bessel's equation are

$Y_n(x)$ (also called Weber's function, and sometimes denoted by $N_n(x)$)

$H_n^{(1)}(x)$ and $H_n^{(2)}(x)$ (also called Hankel functions)

These solutions are defined as follows

$$Y_n(x) = \begin{cases} \dfrac{J_n(x)\cos(n\pi) - J_{-n}(x)}{\sin(n\pi)} & n \text{ not an integer} \\[3mm] \lim_{v \to n} \dfrac{J_v(x)\cos(v\pi) - J_{-v}(x)}{\sin(v\pi)} & n \text{ an integer} \end{cases}$$

$$H_n^{(1)}(x) = J_n(x) + iY_n(x)$$

$$H_n^{(2)}(x) = J_n(x) - iY_n(x)$$

The additional properties of these functions may all be derived from the above relations and the known properties of $J_n(x)$.

13. Complete solutions to Bessel's equation may be written as

$$c_1 J_n(x) + c_2 J_{-n}(x) \qquad \text{if } n \text{ is not an integer}$$

or

$$\left. \begin{array}{l} c_1 J_n(x) + c_2 Y_n(x) \\ \\ c_1 H_n^{(1)}(x) + c_2 H_n^{(2)}(x) \end{array} \right\} \text{for any value of } n$$

or

14. The modified (or hyperbolic) Bessel's differential equation is

$$x^2 \frac{d^2 y}{dx^2} + x \frac{dy}{dx} - (x^2 + n^2)y = 0$$

15. When n is not an integer, two independent solutions of the equation are $I_n(x)$ and $I_{-n}(x)$, where

$$I_n(x) = \sum_{k=0}^{\infty} \frac{1}{k!\,\Gamma(n+k+1)} \left(\frac{x}{2}\right)^{n+2k}$$

16. If n is an integer,

$$I_n(x) = I_{-n}(x) = \frac{x^n}{2^n n!} \left\{ 1 + \frac{x^2}{2^2 \cdot 1!(n+1)} + \frac{x^4}{2^4 \cdot 2!(n+1)(n+2)} \right.$$
$$\left. + \frac{x^6}{2^6 \cdot 3!(n+1)(n+2)(n+3)} + \cdots \right\}$$

17. For $N = 0$ and $n = 1$, this formula becomes

$$I_0(x) = 1 + \frac{x^2}{2^2(1!)^2} + \frac{x^4}{2^4(2!)^2} + \frac{x^6}{2^6(3!)^2} + \frac{x^8}{2^8(4!)^2} + \cdots$$

$$I_1(x) = \frac{x}{2} + \frac{x^3}{2^3 \cdot 1!2!} + \frac{x^5}{2^5 \cdot 2!3!} + \frac{x^7}{2^7 \cdot 3!4!} + \frac{x^9}{2^9 \cdot 4!5!} + \cdots$$

18. Another solution to the modified Bessel's equation is

$$K_n(x) = \begin{cases} \dfrac{1}{2}\,\pi\,\dfrac{I_{-n}(x) - I_n(x)}{\sin(n\pi)} & n \text{ not an integer} \\ \\ \lim_{v \to n} \dfrac{1}{2}\,\pi\,\dfrac{I_{-v}(x) - I_v(x)}{\sin(v\pi)} & n \text{ an integer} \end{cases}$$

This function is linearly independent of $I_n(x)$ for all values of n. Thus the complete solution to the modified Bessel's equation may be written as

$$c_1 I_n(x) + c_2 I_{-n}(x) \qquad n \text{ not an integer}$$

or

$$c_1 I_n(x) + c_2 K_n(x) \qquad \text{any } n$$

19. The following relations hold among the various Bessel functions:

$$I_n(z) = i^{-m} J_m(iz)$$

$$Y_n(iz) = (i)^{n+1} I_n(z) - \frac{2}{\pi} i^{-n} K_n(z)$$

Most of the properties of the modified Bessel function may be deduced from the known properties of $J_n(x)$ by use of these relations and those previously given.

20. Recurrence formulas

$$I_{n-1}(x) - I_{n+1}(x) = \frac{2n}{x}\,I_n(x) \qquad I_{n-1}(x) + I_{n+1}(x) = 2I_n'(x)$$

$$I_{n-1}(x) - \frac{n}{x}\,I_n(x) = I_n'(x) \qquad I_n'(x) = I_{n+1}(x) + \frac{n}{x}\,I_n(z)$$

The Gamma Function*

Definition: $\Gamma\,(n = \int_0^{\infty} t^{n-1}\,e^{-t}\,dt \; n > 0$

Recursion Formula: $\Gamma(n+1 = n\Gamma(n)$
$\Gamma(n) + 1) = n!$ if $n = 0,1,2, \ldots$ where $0! = 1$
For $n < 0$ the gamma function can be defined by using

---

* From Beyer, W. H., Ed., *CRC Handbook of Mathematical Sciences*, 5th ed., CRC Press, Boca Raton, 1978, 484—485. With permission.

$$\Gamma(n) = \frac{\Gamma(n + 1)}{n}$$

Graph:

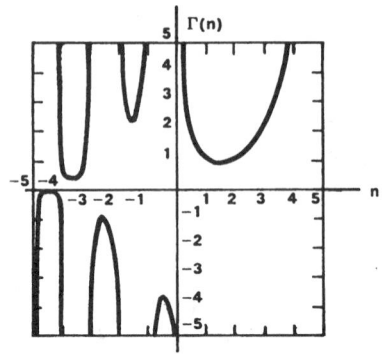

Special Values:

$$\Gamma(\tfrac{1}{2}) = \sqrt{\pi}$$

$$\Gamma(m + \tfrac{1}{2}) = \frac{1 \cdot 3 \cdot 5 \cdots (2m - 1)}{2^m} \sqrt{\pi} \qquad m = 1,2,3,\ldots$$

$$\Gamma(-m + \tfrac{1}{2}) = \frac{(-1)^m 2^m \sqrt{\pi}}{1 \cdot 3 \cdot 5 \cdots (2m - 1)} \qquad m = 1,2,3,\ldots$$

Definition:

$$\Gamma(x + 1) = \lim_{k \to \infty} \frac{1 \cdot 2 \cdot 3 \cdots k}{(x + 1)(x + 2) \cdots (x + k)} \, k^x$$

$$\frac{1}{\Gamma(x)} = x e^{\gamma x} \prod_{m = 1}^{\infty} \left\{ \left(1 + \frac{x}{m}\right) e^{-x/m} \right\}$$

This is an infinite product representation for the gamma function where γ is Euler's constant.

Properties:

$$\Gamma'(1) = \int_{0}^{\infty} e^{-x} \ln x \, dx = -\gamma$$

$$\frac{\Gamma'(x)}{\Gamma(x)} = -\gamma + \left(\frac{1}{1} - \frac{1}{x}\right) + \left(\frac{1}{2} - \frac{1}{x+1}\right) + \ldots + \left(\frac{1}{n} - \frac{1}{x+n-1}\right) + \ldots$$

$$\Gamma(x + 1) = \sqrt{2\pi x} \, x^x e^{-x} \left\{ 1 + \frac{1}{12x} + \frac{1}{288x^2} - \frac{139}{51,840x^3} + \ldots \right\}$$

This is called *Stirling's asymptotic series.*

If we let $x = n$ a positive integer, then a useful approximation for $n!$ where n is large (e.g., $n > 10$) is given by *Stirling's formula*

$$n! \approx \sqrt{2\pi n} \, n^n e^{-n}$$

The Gamma Function*

$$\text{Values of } \Gamma(n) = \int_0^\infty e^{-x} x^{n-1} \, dx; \; \Gamma(n+1) = n\Gamma(n)$$

n	$\Gamma(n)$	n	$\Gamma(n)$	n	$\Gamma(n)$	n	$\Gamma(n)$
1.00	1.00000	1.25	.90640	1.50	.88623	1.75	.91906
1.01	.99433	1.26	.90440	1.51	.88659	1.76	.92137
1.02	.98884	1.27	.90250	1.52	.88704	1.77	.92376
1.03	.98355	1.28	.90072	1.53	.88757	1.78	.92623
1.04	.97844	1.29	.89904	1.54	.88818	1.79	.92877
1.05	.97350	1.30	.89747	1.55	.88887	1.80	.93138
1.06	.96874	1.31	.89600	1.56	.88964	1.81	.93408
1.07	.96415	1.32	.89464	1.57	.89049	1.82	.93685
1.08	.95973	1.33	.89338	1.58	.89142	1.83	.93969
1.09	.95546	1.34	.89222	1.59	.89243	1.84	.94261
1.10	.95135	1.35	.89115	1.60	.89352	1.85	.94561
1.11	.94740	1.36	.89018	1.61	.89468	1.86	.94869
1.12	.94359	1.37	.88931	1.62	.89592	1.87	.95184
1.13	.93993	1.38	.88854	1.63	.89724	1.88	.95507
1.14	.93642	1.39	.88785	1.64	.89864	1.89	.95838
1.15	.93304	1.40	.88726	1.65	.90012	1.90	.96177
1.16	.92980	1.41	.88676	1.66	.90167	1.91	.96523
1.17	.92670	1.42	.88636	1.67	.90330	1.92	.96877
1.18	.92373	1.43	.88604	1.68	.90500	1.93	.97240
1.19	.92089	1.44	.88581	1.69	.90678	1.94	.97610
1.20	.91817	1.45	.88566	1.70	.90864	1.95	.97988
1.21	.91558	1.46	.88560	1.71	.91057	1.96	.98374
1.22	.91311	1.47	.88563	1.72	.91258	1.97	.98768
1.23	.91075	1.48	.88575	1.73	.91466	1.98	.99171
1.24	.90852	1.49	.88595	1.74	.91683	1.99	.99581
						2.00	1.00000

* For large positive values of x, $\Gamma(x)$ approximates Stirling's asymptotic series

$$x^x e^{-x} \sqrt{\frac{2\pi}{x}} \left[1 + \frac{1}{12x} + \frac{1}{288x^2} - \frac{139}{51840x^3} - \frac{571}{2488320x^4} + \cdots \right]$$

The Beta Function*

Definition: $B(m,n) = \displaystyle\int_0^1 t^{m-1} (1-t)^{m-1} \, dt \quad m > 0, n > 0$

Relationship with
Gamma Function: $B(m,n) = \dfrac{\Gamma(m)\Gamma(n)}{\Gamma(m+n)}$

Properties: $\qquad\qquad\qquad\qquad B(m,n) = B(n,m)$

$$B(m,n) = 2\int_0^{\pi/2} \sin^{2m-1}\theta \, \cos^{2n-1}\theta \, d\theta$$

$$B(m,n) = \int_0^\infty \frac{t^{m-1}}{(1+t)^{m+n}} \, dt$$

$$B(m,n) = r^n(r+1)^m \int_0^1 \frac{t^{m-1}(1-t)^{n-1}}{(r+t)^{m+n}} \, dt$$

The Error Function

Definition: $erf\, x = \dfrac{2}{\sqrt{\pi}} \displaystyle\int_0^x e^{-t^2} \, dt$

Series: $erf\, x = \dfrac{2}{\sqrt{\pi}} \left(x - \dfrac{x^3}{3} + \dfrac{1}{2!}\dfrac{x^5}{5} - \dfrac{1}{3!}\dfrac{x^7}{7} + \cdots \right)$

Property: $erf\, x = -erf(-x)$

Relationship with Normal Probability Function $f(t)$: $\displaystyle\int_0^x f(t) \, dt = \frac{1}{2} erf\left(\frac{x}{\sqrt{2}}\right)$

To evaluate $erf\,(2.3)$, one proceeds as follows: Since $\dfrac{x}{\sqrt{2}} = 2.3$, one finds $x = (2.3)(\sqrt{2}) = 3.25$. In the normal probability function table (page A-104), one finds the entry 0.4994 opposite the value 3.25. Thus $erf\,(2.3) = 2(0.4994) = 0.9988$.

* From Beyer, W. H., Ed., *CRC Handbook of Mathematical Sciences*, 5th ed., CRC Press, Boca Raton, 1978, 499. With permission.

$$erfc\ z = 1 - erf\ z = \frac{2}{\sqrt{\pi}} \int_z^\infty e^{-t^2}\,dt$$

is known as the complementary error function.

Orthogonal Polynomials*

I

Name: Legendre   *Symbol:* $P_n(x)$ *Interval:* $[-1,1]$
Differential Equation: $(1 - x^2)\,y'' - 2\,xy' + n(n + 1)\,y = 0$
$$y = P_n(x)$$

Explicit Expression: $P_n(x) = \dfrac{1}{2^n} \displaystyle\sum_{m=0}^{[n/2]} (-1)^m \binom{n}{m}\binom{2n - 2m}{n} x^{n-2m}$

Recurrence Relation: $(n + 1)\,P_{n+1}(x) = (2n + 1)\,x\,P_n(x) - nP_{n-1}(x)$
Weight: 1   *Standardization:* $P_n(1) = 1$

Norm: $\displaystyle\int_{-1}^{+1} [P_n(x)]^2\,dx = \dfrac{2}{2n + 1}$

Rodrigues' Formula: $P_n(x) = \dfrac{(-1)^n}{2^n n!} \dfrac{d^n}{dx^n}\{(1 - x^2)^n\}$

Generating Function: $R^{-1} = \displaystyle\sum_{n=0}^\infty P_n(x)\,z^n;\ -1 < x < 1,\ |z| < 1,$
$$R = \sqrt{1 - 2xz + z^2}$$

Inequality: $|P_n(x)| \leq 1,\ -1 \leq x \leq 1.$

II

Name: Tschebysheff, First Kind   *Symbol:* $T_n(x)$ *Interval:* $[-1,1]$
Differential Equation: $(1 - x^2)y - xy' + n^2 y = 0$
$$y = T_n(x)$$

Explicit Expression: $\dfrac{n}{2} \displaystyle\sum_{m=0}^{[n/2]} (-1)^m \dfrac{(n - m - 1)!}{m!(n - 2m)!} (2x)^{n-2m} = \cos(n \arccos x) = T_n(x)$

Recurrence Relation: $T_{n+1}(x) = 2xT_n(x) - T_{n-1}(x)$
Weight: $(1 - x^2)^{-1/2}$ *Standardization:* $T_n(1) = 1$

Norm: $\displaystyle\int_{-1}^{+1} (1 - x^2)^{-1/2}[T_n(x)]^2\,dx = \begin{cases} \pi/2, & n \neq 0 \\ \pi, & n = 0 \end{cases}$

Rodrigues' Formula: $\dfrac{(-1)^n(1 - x^2)^{1/2} \sqrt{\pi}}{2^{n+1}\Gamma(n + \frac{1}{2})} \dfrac{d^n}{dx^n}\{(1 - x^2)^{n-(1/2)}\} = T_n(x)$

Generating Function: $\dfrac{1 - xz}{1 - 2xz + z^2} = \displaystyle\sum_{n=0}^\infty T_n(x)\,z^n,\ -1 < x < 1,\ |z| < 1$

Inequality: $|T_n(x)| \leq 1,\ -1 \leq x \leq 1.$

III

Name: Tschebysheff, Second Kind   *Symbol:* $U_n(x)$ *Interval:* $[-1,1]$
Differential Equation: $(1 - x^2)\,y'' - 3\,xy' + n(n + 2)y = 0$
$$y = U_n(x)$$

Explicit Expression: $U_n(x) = \displaystyle\sum_{m=0}^{[n/2]} (-1)^m \dfrac{(m - n)!}{m!(n - 2m)!} (2x)^{n-2m}$

$$U_n(\cos\theta) = \dfrac{\sin[(n + 1)\theta]}{\sin\theta}$$

* From Beyer, W. H., Ed., *CRC Handbook of Mathematical Sciences,* 5th ed., CRC Press, Boca Raton, 1978, 557—560. With permission.

Recurrence Relation: $U_{n+1}(x) = 2xU_n(x) - U_{n-1}(x)$

Weight: $(1 - x^2)^{1/2}$ *Standardization:* $U_n(1) = n + 1$

Norm: $\displaystyle\int_{-1}^{+1} (1 - x^2)^{1/2}[U_n(x)]^2\,dx = \frac{\pi}{2}$

Rodrigues' Formula: $\displaystyle U_n(x) = \frac{(-1)^n(n + 1)\sqrt{\pi}}{(1 - x^2)^{1/2}2^{n+1}\Gamma(n + \frac{3}{2})}\frac{d^n}{dx^n}\{(1 - x^2)^{n+(1/2)}\}$

Generating Function: $\displaystyle\frac{1}{1 - 2xz + z^2} = \sum_{n=0}^{\infty} U_n(x)z^n, \; -1 < x < 1, \; |z| < 1$

Inequality: $|U_n(x)| \leq n + 1, \; -1 \leq x \leq 1.$

IV

Name: Jacobi   *Symbol:* $P_n^{(\alpha,\beta)}(x)$ *Interval:* $[-1,1]$

Differential Equation:

$$(1 - x^2)y'' + [\beta - \alpha - (\alpha + \beta + 2)x]y' + n(n + \alpha + \beta + 1)y = 0$$
$$y = P_n^{(\alpha,\beta)}(x)$$

Explicit Expression: $\displaystyle P_n^{(\alpha,\beta)}(x) = \frac{1}{2^n}\sum_{m=0}^{n}\binom{n + \alpha}{m}\binom{n + \beta}{n - m}(x - 1)^{n-m}(x + 1)^m$

Recurrence Relation: $2(n + 1)(n + \alpha + \beta + 1)(2n + \alpha + \beta)P_{n+1}^{(\alpha,\beta)}(x)$
$$= (2n + \alpha + \beta + 1)[(\alpha^2 - \beta^2) + (2n + \alpha + \beta + 2)$$
$$\times (2n + \alpha + \beta)x]P_n^{(\alpha,\beta)}(x)$$
$$- 2(n + \alpha)(n + \beta)(2n + \alpha + \beta + 2)P_{n-1}^{(\alpha,\beta)}(x)$$

Weight: $(1 - x)^\alpha(1 + x)^\beta; \; \alpha, \beta > 1$ *Standardization:* $P_n^{(\alpha,\beta)}(x) = \binom{n + \alpha}{n}$

Norm: $\displaystyle\int_{-1}^{+1}(1 - x)^\alpha(1 + x)^\beta[P_n^{(\alpha,\beta)}(x)]^2\,dx = \frac{2^{\alpha+\beta+1}\Gamma(n + \alpha + 1)\Gamma(n + \beta + 1)}{(2n + \alpha + \beta + 1)n!\,\Gamma(n + \alpha + \beta + 1)}$

Rodrigues' Formula: $\displaystyle P_n^{(\alpha,\beta)}(x) = \frac{(-1)^n}{2^n n!(1 - x)^\alpha(1 + x)^\beta}\frac{d^n}{dx^n}\{(1 - x)^{n+\alpha}(1 + x)^{n+\beta}\}$

Generating Function: $\displaystyle R^{-1}(1 - z + R)^{-\alpha}(1 + z + R)^{-\beta} = \sum_{n=0}^{\infty}2^{-\alpha-\beta}P_n^{(\alpha,\beta)}(x)z^n,$

$$R = \sqrt{1 - 2xz + z^2}, \; |z| < 1$$

Inequality: $\displaystyle \max_{-1 \leq x \leq 1}|P_n^{(\alpha,\beta)}(x)| = \begin{cases} \dbinom{n + q}{n} \sim n^q \text{ if } q = \max(\alpha,\beta) \geq -\frac{1}{2} \\ |P_n^{(\alpha,\beta)}(x')| \sim n^{-1/2} \text{ if } q < -\frac{1}{2} \\ x' \text{ is one of the two maximum points nearest} \\ \dfrac{\beta - \alpha}{\alpha + \beta + 1} \end{cases}$

V

Name: Generalized Laguerre   *Symbol:* $L_n^{(\alpha)}(x)$ *Interval:* $[0, \infty]$

Differential Equation: $xy'' + (\alpha + 1 - x)y' + ny = 0$
$$y = L_n^{(\alpha)}(x)$$

Explicit Expression: $\displaystyle L_n^{(\alpha)}(x) = \sum_{m=0}^{n}(-1)^m\binom{n + \alpha}{n - m}\frac{1}{m!}x^m$

Recurrence Relation: $(n + 1)L_{n+1}^{(\alpha)}(x) = [(2n + \alpha + 1) - x]L_n^{(\alpha)}(x) - (n + \alpha)L_{n-1}^{(\alpha)}(x)$

Weight: $x^\alpha e^{-x}, \; \alpha > -1$ *Standardization:* $L_n^{(\alpha)}(x) = \dfrac{(-1)^n}{n!}x^n + \cdots$

Norm: $\displaystyle\int_0^{\infty} x^\alpha e^{-x}[L_n^{(\alpha)}(x)]^2\,dx = \frac{\Gamma(n + \alpha + 1)}{n!}$

Rodrigues' Formula: $\displaystyle L_n^{(\alpha)}(x) = \frac{1}{n!\,x^\alpha e^{-x}}\frac{d^n}{dx^n}\{x^{n+\alpha}e^{-x}\}$

Generating Function: $(1 - z)^{-\alpha - 1} \exp\left(\dfrac{xz}{z - 1}\right) = \displaystyle\sum_{n=0}^{\infty} L_n^{(\alpha)}(x) z^n$

Inequality: $|L_n^{(\alpha)}(x)| \leq \dfrac{\Gamma(n + \alpha + 1)}{n!\,\Gamma(\alpha + 1)} e^{x/2}; \quad \begin{array}{l} x \geq 0 \\ \alpha > 0 \end{array}$

$|L_n^{(\alpha)}(x)| \leq \left[2 - \dfrac{\Gamma(\alpha + n + 1)}{n!\,\Gamma(\alpha + 1)}\right] e^{x/2}; \quad \begin{array}{l} x \geq 0 \\ -1 < \alpha < 0 \end{array}$

Orthogonal Polynomials

Name: Hermite Symbol: $H_n(x)$ Interval: $[-\infty, \infty]$

Differential Equation: $y'' - 2xy' + 2ny = 0$

Explicit Expression: $H_n(x) = \displaystyle\sum_{m=0}^{[n/2]} \dfrac{(-1)^m\, n!\, (2x)^{n-2m}}{m!\,(n-2m)!}$

Recurrence Relation: $H_{n+1}(x) = 2x\,H_n(x) - 2n H_{n-1}(x)$

Weight: e^{-x^2} Standardization: $H_n(1) = 2^n x^n + \ldots$

Norm: $\displaystyle\int_{-\infty}^{\infty} e^{-x^2} [H_n(x)]^2\, dx = 2^n n! \sqrt{\pi}$

Rodriques' Formula: $H_n(x) = (-1)^n e^{x^2} \dfrac{d^n}{dx^n}\left(e^{-x^2}\right)$

Generating Function: $e^{-z^2 + 2zx} = \displaystyle\sum_{n=0}^{\infty} H_n(x)\, \dfrac{z^n}{n!}$

Inequality: $|H_n(x)| < e^{\frac{x^2}{2}}\, k\, 2^{n/2} \sqrt{n!}\quad k \approx 1.086435$

NORMAL PROBABILITY FUNCTION

Areas under the Standard Normal Curve from 0 to z

z	0	1	2	3	4	5	6	7	8	9
0.0	.0000	.0040	.0080	.0120	.0160	.0199	.0239	.0279	.0319	.0359
0.1	.0398	.0438	.0478	.0517	.0557	.0596	.0636	.0675	.0714	.0754
0.2	.0793	.0832	.0871	.0910	.0948	.0987	.1026	.1064	.1103	.1141
0.3	.1179	.1217	.1255	.1293	.1331	.1368	.1406	.1443	.1480	.1517
0.4	.1554	.1591	.1628	.1664	.1700	.1736	.1772	.1808	.1844	.1879
0.5	.1915	.1950	.1985	.2019	.2054	.2088	.2123	.2157	.2190	.2224
0.6	.2258	.2291	.2324	.2357	.2389	.2422	.2454	.2486	.2518	.2549
0.7	.2580	.2612	.2652	.2673	.2704	.2734	.2764	.2794	.2823	.2852
0.8	.2881	.2910	.2939	.2967	.2996	.3023	.3051	.3078	.3106	.3133
0.9	.3159	.3186	.3212	.3238	.3264	.3289	.3315	.3340	.3365	.3389
1.0	.3413	.3438	.3461	.3485	.3508	.3531	.3554	.3577	.3599	.3621
1.1	.3643	.3665	.3686	.3708	.3729	.3749	.3770	.3790	.3810	.3830
1.2	.3849	.3869	.3888	.3907	.3925	.3944	.3962	.3980	.3997	.4015
1.3	.4032	.4049	.4066	.4082	.4099	.4115	.4131	.4147	.4162	.4177
1.4	.4192	.4207	.4222	.4236	.4251	.4265	.4279	.4292	.4306	.4319
1.5	.4332	.4345	.4357	.4370	.4382	.4394	.4406	.4418	.4429	.4441
1.6	.4452	.4463	.4474	.4484	.4495	.4505	.4515	.4525	.4535	.4545
1.7	.4554	.4564	.4573	.4582	.4591	.4599	.4608	.4616	.4625	.4633
1.8	.4641	.4649	.4656	.4664	.4671	.4678	.4686	.4693	.4699	.4706
1.9	.4713	.4719	.4726	.4732	.4738	.4744	.4750	.4756	.4761	.4767
2.0	.4772	.4778	.4783	.4788	.4793	.4798	.4803	.4808	.4812	.4817
2.1	.4821	.4826	.4830	.4834	.4838	.4842	.4846	.4850	.4854	.4857
2.2	.4861	.4864	.4868	.4871	.4875	.4878	.4881	.4884	.4887	.4890
2.3	.4893	.4896	.4898	.4901	.4904	.4906	.4909	.4911	.4913	.4916
2.4	.4918	.4920	.4922	.4925	.4927	.4929	.4931	.4932	.4934	.4936
2.5	.4938	.4940	.4941	.4943	.4945	.4946	.4948	.4949	.4951	.4952
2.6	.4953	.4955	.4956	.4957	.4959	.4960	.4961	.4962	.4963	.4964
2.7	.4965	.4966	.4967	.4968	.4969	.4970	.4971	.4972	.4973	.4974
2.8	.4974	.4975	.4976	.4977	.4977	.4978	.4979	.4979	.4980	.4981
2.9	.4981	.4982	.4982	.4983	.4984	.4984	.4985	.4985	.4986	.4986
3.0	.4987	.4987	.4987	.4988	.4988	.4989	.4989	.4989	.4990	.4990
3.1	.4990	.4991	.4991	.4991	.4992	.4992	.4992	.4992	.4993	.4993
3.2	.4993	.4993	.4994	.4994	.4994	.4994	.4994	.4995	.4995	.4995
3.3	.4995	.4995	.4995	.4996	.4996	.4996	.4996	.4996	.4996	.4997
3.4	.4997	.4997	.4997	.4997	.4997	.4997	.4997	.4997	.4997	.4998
3.5	.4998	.4998	.4998	.4998	.4998	.4998	.4998	.4998	.4998	.4998
3.6	.4998	.4998	.4999	.4999	.4999	.4999	.4999	.4999	.4999	.4999
3.7	.4999	.4999	.4999	.4999	.4999	.4999	.4999	.4999	.4999	.4999
3.8	.4999	.4999	.4999	.4999	.4999	.4999	.4999	.4999	.4999	.4999
3.9	.5000	.5000	.5000	.5000	.5000	.5000	.5000	.5000	.5000	.5000

F(z) below refers to area under Standard Normal Curve from $-\infty$ to z

z	1.282	1.645	1.960	2.326	2.576	3.090
F(z)	.90	.95	.975	.99	.995	.999
2[1 − F(z)]	.20	.10	.05	.02	.01	.002

PERCENTAGE POINTS, STUDENT'S t-DISTRIBUTION

This table gives values of t such that

$$F(t) = \int_{-\infty}^{t} \frac{\Gamma\left(\frac{n+1}{2}\right)}{\sqrt{n\pi}\ \Gamma\left(\frac{n}{2}\right)} \left(1 + \frac{x^2}{n}\right)^{-\frac{n+1}{2}} dx$$

for n, the number of degrees of freedom, equal to 1, 2, . . ., 30, 40, 60, 120, ∞; and for $F(t) = 0.60, 0.75, 0.90, 0.95, 0.975, 0.99, 0.995,$ and 0.9995. The t-distribution is symmetrical, so that $F(-t) = 1 - F(t)$

F n	.60	.75	.90	.95	.975	.99	.995	.9995
1	.325	1.000	3.078	6.314	12.706	31.821	63.657	636.619
2	.289	.816	1.886	2.920	4.303	6.965	9.925	31.598
3	.277	.765	1.638	2.353	3.182	4.541	5.841	12.924
4	.271	.741	1.533	2.132	2.776	3.747	4.604	8.610
5	.267	.727	1.476	2.015	2.571	3.365	4.032	6.869
6	.265	.718	1.440	1.943	2.447	3.143	3.707	5.959
7	.263	.711	1.415	1.895	2.365	2.998	3.499	5.408
8	.262	.706	1.397	1.860	2.306	2.896	3.355	5.041
9	.261	.703	1.383	1.833	2.262	2.821	3.250	4.781
10	.260	.700	1.372	1.812	2.228	2.764	3.169	4.587
11	.260	.697	1.363	1.796	2.201	2.718	3.106	4.437
12	.259	.695	1.356	1.782	2.179	2.681	3.055	4.318
13	.259	.694	1.350	1.771	2.160	2.650	3.012	4.221
14	.258	.692	1.345	1.761	2.145	2.624	2.977	4.140
15	.258	.691	1.341	1.753	2.131	2.602	2.947	4.073
16	.258	.690	1.337	1.746	2.120	2.583	2.921	4.015
17	.257	.689	1.333	1.740	2.110	2.567	2.898	3.965
18	.257	.688	1.330	1.734	2.101	2.552	2.878	3.922
19	.257	.688	1.328	1.729	2.093	2.539	2.861	3.883
20	.257	.687	1.325	1.725	2.086	2.528	2.845	3.850
21	.257	.686	1.323	1.721	2.080	2.518	2.831	3.819
22	.256	.686	1.321	1.717	2.074	2.508	2.819	3.792
23	.256	.685	1.319	1.714	2.069	2.500	2.807	3.767
24	.256	.685	1.318	1.711	2.064	2.492	2.797	3.745
25	.256	.684	1.316	1.708	2.060	2.485	2.787	3.725
26	.256	.684	1.315	1.706	2.056	2.479	2.779	3.707
27	.256	.684	1.314	1.703	2.052	2.473	2.771	3.690
28	.256	.683	1.313	1.701	2.048	2.467	2.763	3.674
29	.256	.683	1.311	1.699	2.045	2.462	2.756	3.659
30	.256	.683	1.310	1.697	2.042	2.457	2.750	3.646
40	.255	.681	1.303	1.684	2.021	2.423	2.704	3.551
60	.254	.679	1.296	1.671	2.000	2.390	2.660	3.460
120	.254	.677	1.289	1.658	1.980	2.358	2.617	3.373
∞	.253	.674	1.282	1.645	1.960	2.326	2.576	3.291

* This table is abridged from the "Statistical Tables" of R. A. Fisher and Frank Yates published by Oliver & Boyd. Ltd., Edinburgh and London, 1938. It is here published with the kind permission of the authors and their publishers.

PERCENTAGE POINTS, CHI-SQUARE DISTRIBUTION

This table gives values of χ^2 such that

$$F(\chi^2) = \int_0^{\chi^2} \frac{1}{2^{\frac{n}{2}} \Gamma\left(\frac{n}{2}\right)} x^{\frac{n-2}{2}} e^{-\frac{x}{2}} dx$$

for n, the number of degrees of freedom, equal to 1, 2, . . . , 30. For $n > 30$, a normal approximation is quite accurate. The expression $\sqrt{2\chi^2} - \sqrt{2n-1}$ is approximately normally distributed as the standard normal distribution. Thus χ^2_α, the α-point of the distribution, may be computed by the formula

$$\chi^2_\alpha = \tfrac{1}{2}[x_\alpha + \sqrt{2n-1}]^2,$$

where x_α is the α-point of the cumulative normal distribution. For even values of n, $F(\chi^2)$ can be written as

$$1 - F(\chi^2) = \sum_{x=0}^{x'-1} \frac{e^{-\lambda}\lambda^x}{x!}$$

with $\lambda = \tfrac{1}{2}\chi^2$ and $x' = \tfrac{1}{2}n$. Thus the cumulative Chi-Square distribution is related to the cumulative Poisson distribution.

Another approximate formula for large n

$$\chi_\alpha^2 = n\left(1 - \frac{2}{9n} + z_\alpha\sqrt{\frac{2}{9n}}\right)^3$$

n = degrees of freedom

z_α = the normal deviate, (the value of x for which $F(x)$ = the desired percentile).

x	1.282	1.645	1.960	2.326	2.576	3.090
$F(x)$	.90	.95	.975	.99	.995	.999

$\chi_{.99}^2 = 60[1 - 0.00370 + 2.326(0.06086)]^3 = 88.4$ is the 99th percentile for 60 degrees of freedom.

$$F(\chi^2) = \int_0^{\chi^2} \frac{1}{2^{\frac{n}{2}}\,\Gamma\left(\frac{n}{2}\right)} x^{\frac{n-2}{2}} e^{-\frac{x}{2}}\,dx$$

n \ F	.005	.010	.025	.050	.100	.250	.500	.750	.900	.950	.975	.990	.995
1	.0000393	.000157	.000982	.00393	.0158	.102	.455	1.32	2.71	3.84	5.02	6.63	7.88
2	.0100	.0201	.0506	.103	.211	.575	1.39	2.77	4.61	5.99	7.38	9.21	10.6
3	.0717	.115	.216	.352	.584	1.21	2.37	4.11	6.25	7.81	9.35	11.3	12.8
4	.207	.297	.484	.711	1.06	1.92	3.36	5.39	7.78	9.49	11.1	13.3	14.9
5	.412	.554	.831	1.15	1.61	2.67	4.35	6.63	9.24	11.1	12.8	15.1	16.7
6	.676	.872	1.24	1.64	2.20	3.45	5.35	7.84	10.6	12.6	14.4	16.8	18.5
7	.989	1.24	1.69	2.17	2.83	4.25	6.35	9.04	12.0	14.1	16.0	18.5	20.3
8	1.34	1.65	2.18	2.73	3.49	5.07	7.34	10.2	13.4	15.5	17.5	20.1	22.0
9	1.73	2.09	2.70	3.33	4.17	5.90	8.34	11.4	14.7	16.9	19.0	21.7	23.6
10	2.16	2.56	3.25	3.94	4.87	6.74	9.34	12.5	16.0	18.3	20.5	23.2	25.2
11	2.60	3.05	3.82	4.57	5.58	7.58	10.3	13.7	17.3	19.7	21.9	24.7	26.8
12	3.07	3.57	4.40	5.23	6.30	8.44	11.3	14.8	18.5	21.0	23.3	26.2	28.3
13	3.57	4.11	5.01	5.89	7.04	9.30	12.3	16.0	19.8	22.4	24.7	27.7	29.8
14	4.07	4.66	5.63	6.57	7.79	10.2	13.3	17.1	21.1	23.7	26.1	29.1	31.3
15	4.60	5.23	6.26	7.26	8.55	11.0	14.3	18.2	22.3	25.0	27.5	30.6	32.8
16	5.14	5.81	6.91	7.96	9.31	11.9	15.3	19.4	23.5	26.3	28.8	32.0	34.3
17	5.70	6.41	7.56	8.67	10.1	12.8	16.3	20.5	24.8	27.6	30.2	33.4	35.7
18	6.26	7.01	8.23	9.39	10.9	13.7	17.3	21.6	26.0	28.9	31.5	34.8	37.2
19	6.84	7.63	8.91	10.1	11.7	14.6	18.3	22.7	27.2	30.1	32.9	36.2	38.6
20	7.43	8.26	9.59	10.9	12.4	15.5	19.3	23.8	28.4	31.4	34.2	37.6	40.0
21	8.03	8.90	10.3	11.6	13.2	16.3	20.3	24.9	29.6	32.7	35.5	38.9	41.4
22	8.64	9.54	11.0	12.3	14.0	17.2	21.3	26.0	30.8	33.9	36.8	40.3	42.8
23	9.26	10.2	11.7	13.1	14.8	18.1	22.3	27.1	32.0	35.2	38.1	41.6	44.2
24	9.89	10.9	12.4	13.8	15.7	19.0	23.3	28.2	33.2	36.4	39.4	43.0	45.6
25	10.5	11.5	13.1	14.6	16.5	19.9	24.3	29.3	34.4	37.7	40.6	44.3	46.9
26	11.2	12.2	13.8	15.4	17.3	20.8	25.3	30.4	35.6	38.9	41.9	45.6	48.3
27	11.8	12.9	14.6	16.2	18.1	21.7	26.3	31.5	36.7	40.1	43.2	47.0	49.6
28	12.5	13.6	15.3	16.9	18.9	22.7	27.3	32.6	37.9	41.3	44.5	48.3	51.0
29	13.1	14.3	16.0	17.7	19.8	23.6	28.3	33.7	39.1	42.6	45.7	49.6	52.3
30	13.8	15.0	16.8	18.5	20.6	24.5	29.3	34.8	40.3	43.8	47.0	50.9	53.7

PERCENTAGE POINTS, *F*-DISTRIBUTION

This table gives values of F such that

$$F(F) = \int_0^F \frac{\Gamma\left(\frac{m+n}{2}\right)}{\Gamma\left(\frac{m}{2}\right)\Gamma\left(\frac{n}{2}\right)} m^{\frac{m}{2}} n^{\frac{n}{2}} x^{\frac{m-2}{2}} (n + mx)^{-\frac{m+n}{2}}\,dx$$

for selected values of m, the number of degrees of freedom of the numerator of F; and for selected values of n, the number of degrees of freedom of the denominator of F. The table also provides values corresponding to $F(F) = .10, .05, .025, .01, .005, .001$ since $F_{1-\alpha}$ for m and n degrees of freedom is the reciprocal of F_α for n and m degrees of freedom. Thus

$$F_{.05}(4, 7) = \frac{1}{F_{.95}(7, 4)} = \frac{1}{6.09} = .164 \;.$$

$$F(F) = \int_0^F \frac{\Gamma\left(\frac{m+n}{2}\right)}{\Gamma\left(\frac{m}{2}\right)\Gamma\left(\frac{n}{2}\right)} m^{\frac{m}{2}} n^{\frac{n}{2}} x^{\frac{m}{2}-1} (n+mx)^{-\frac{m+n}{2}}\, dx = .90$$

m \ n	1	2	3	4	5	6	7	8	9	10	12	15	20	24	30	40	60	120	∞
1	39.86	49.50	53.59	55.83	57.24	58.20	58.91	59.44	59.86	60.19	60.71	61.22	61.74	62.00	62.26	62.53	62.79	63.06	63.33
2	8.53	9.00	9.16	9.24	9.29	9.33	9.35	9.37	9.38	9.39	9.41	9.42	9.44	9.45	9.46	9.47	9.47	9.48	9.49
3	5.54	5.46	5.39	5.34	5.31	5.28	5.27	5.25	5.24	5.23	5.22	5.20	5.18	5.18	5.17	5.16	5.15	5.14	5.13
4	4.54	4.32	4.19	4.11	4.05	4.01	3.98	3.95	3.94	3.92	3.90	3.87	3.84	3.83	3.82	3.80	3.79	3.78	3.76
5	4.06	3.78	3.62	3.52	3.45	3.40	3.37	3.34	3.32	3.30	3.27	3.24	3.21	3.19	3.17	3.16	3.14	3.12	3.10
6	3.78	3.46	3.29	3.18	3.11	3.05	3.01	2.98	2.96	2.94	2.90	2.87	2.84	2.82	2.80	2.78	2.76	2.74	2.72
7	3.59	3.26	3.07	2.96	2.88	2.83	2.78	2.75	2.72	2.70	2.67	2.63	2.59	2.58	2.56	2.54	2.51	2.49	2.47
8	3.46	3.11	2.92	2.81	2.73	2.67	2.62	2.59	2.56	2.54	2.50	2.46	2.42	2.40	2.38	2.36	2.34	2.32	2.29
9	3.36	3.01	2.81	2.69	2.61	2.55	2.51	2.47	2.44	2.42	2.38	2.34	2.30	2.28	2.25	2.23	2.21	2.18	2.16
10	3.29	2.92	2.73	2.61	2.52	2.46	2.41	2.38	2.35	2.32	2.28	2.24	2.20	2.18	2.16	2.13	2.11	2.08	2.06
11	3.23	2.86	2.66	2.54	2.45	2.39	2.34	2.30	2.27	2.25	2.21	2.17	2.12	2.10	2.08	2.05	2.03	2.00	1.97
12	3.18	2.81	2.61	2.48	2.39	2.33	2.28	2.24	2.21	2.19	2.15	2.10	2.06	2.04	2.01	1.99	1.96	1.93	1.90
13	3.14	2.76	2.56	2.43	2.35	2.28	2.23	2.20	2.16	2.14	2.10	2.05	2.01	1.98	1.96	1.93	1.90	1.88	1.85
14	3.10	2.73	2.52	2.39	2.31	2.24	2.19	2.15	2.12	2.10	2.05	2.01	1.96	1.94	1.91	1.89	1.86	1.83	1.80
15	3.07	2.70	2.49	2.36	2.27	2.21	2.16	2.12	2.09	2.06	2.02	1.97	1.92	1.90	1.87	1.85	1.82	1.79	1.76
16	3.05	2.67	2.46	2.33	2.24	2.18	2.13	2.09	2.06	2.03	1.99	1.94	1.89	1.87	1.84	1.81	1.78	1.75	1.72
17	3.03	2.64	2.44	2.31	2.22	2.15	2.10	2.06	2.03	2.00	1.96	1.91	1.86	1.84	1.81	1.78	1.75	1.72	1.69
18	3.01	2.62	2.42	2.29	2.20	2.13	2.08	2.04	2.00	1.98	1.93	1.89	1.84	1.81	1.78	1.75	1.72	1.69	1.66
19	2.99	2.61	2.40	2.27	2.18	2.11	2.06	2.02	1.98	1.96	1.91	1.86	1.81	1.79	1.76	1.73	1.70	1.67	1.63
20	2.97	2.59	2.38	2.25	2.16	2.09	2.04	2.00	1.96	1.94	1.89	1.84	1.79	1.77	1.74	1.71	1.68	1.64	1.61
21	2.96	2.57	2.36	2.23	2.14	2.08	2.02	1.98	1.95	1.92	1.87	1.83	1.78	1.75	1.72	1.69	1.66	1.62	1.59
22	2.95	2.56	2.35	2.22	2.13	2.06	2.01	1.97	1.93	1.90	1.86	1.81	1.76	1.73	1.70	1.67	1.64	1.60	1.57
23	2.94	2.55	2.34	2.21	2.11	2.05	1.99	1.95	1.92	1.89	1.84	1.80	1.74	1.72	1.69	1.66	1.62	1.59	1.55
24	2.93	2.54	2.33	2.19	2.10	2.04	1.98	1.94	1.91	1.88	1.83	1.78	1.73	1.70	1.67	1.64	1.61	1.57	1.53
25	2.92	2.53	2.32	2.18	2.09	2.02	1.97	1.93	1.89	1.87	1.82	1.77	1.72	1.69	1.66	1.63	1.59	1.56	1.52
26	2.91	2.52	2.31	2.17	2.08	2.01	1.96	1.92	1.88	1.86	1.81	1.76	1.71	1.68	1.65	1.61	1.58	1.54	1.50
27	2.90	2.51	2.30	2.17	2.07	2.00	1.95	1.91	1.87	1.85	1.80	1.75	1.70	1.67	1.64	1.60	1.57	1.53	1.49
28	2.89	2.50	2.29	2.16	2.06	2.00	1.94	1.90	1.87	1.84	1.79	1.74	1.69	1.66	1.63	1.59	1.56	1.52	1.48
29	2.89	2.50	2.28	2.15	2.06	1.99	1.93	1.89	1.86	1.83	1.78	1.73	1.68	1.65	1.62	1.58	1.55	1.51	1.47
30	2.88	2.49	2.28	2.14	2.05	1.98	1.93	1.88	1.85	1.82	1.77	1.72	1.67	1.64	1.61	1.57	1.54	1.50	1.46
40	2.84	2.44	2.23	2.09	2.00	1.93	1.87	1.83	1.79	1.76	1.71	1.66	1.61	1.57	1.54	1.51	1.47	1.42	1.38
60	2.79	2.39	2.18	2.04	1.95	1.87	1.82	1.77	1.74	1.71	1.66	1.60	1.54	1.51	1.48	1.44	1.40	1.35	1.29
120	2.75	2.35	2.13	1.99	1.90	1.82	1.77	1.72	1.68	1.65	1.60	1.55	1.48	1.45	1.41	1.37	1.32	1.26	1.19
∞	2.71	2.30	2.08	1.94	1.85	1.77	1.72	1.67	1.63	1.60	1.55	1.49	1.42	1.38	1.34	1.30	1.24	1.17	1.00

$F = \dfrac{s_1^2}{s_2^2} = \dfrac{S_1}{m}\Big/\dfrac{S_2}{n}$, where $s_1^2 = S_1/m$ and $s_2^2 = S_2/n$ are independent mean squares estimating a common variance σ^2 and based on m and n degrees of freedom, respectively.

$$F(F) = \int_0^F \frac{\Gamma\left(\frac{m+n}{2}\right)}{\Gamma\left(\frac{m}{2}\right)\Gamma\left(\frac{n}{2}\right)} m^{\frac{m}{2}} n^{\frac{n}{2}} x^{\frac{m}{2}-1} (n+mx)^{-\frac{m+n}{2}}\, dx = .95$$

m \ n	1	2	3	4	5	6	7	8	9	10	12	15	20	24	30	40	60	120	∞
1	161.4	199.5	215.7	224.6	230.2	234.0	236.8	238.9	240.5	241.9	243.9	245.9	248.0	249.1	250.1	251.1	252.2	253.3	254 3
2	18.51	19.00	19.16	19.25	19.30	19.33	19.35	19.37	19.38	19.40	19.41	19.43	19.45	19.45	19.46	19.47	19.48	19.49	19.50
3	10.13	9.55	9.28	9.12	9.01	8.94	8.89	8.85	8.81	8.79	8.74	8.70	8.66	8.64	8.62	8.59	8.57	8.55	8.53
4	7.71	6.94	6.59	6.39	6.26	6.16	6.09	6.04	6.00	5.96	5.91	5.86	5.80	5.77	5.75	5.72	5.69	5.66	5.63
5	6.61	5.79	5.41	5.19	5.05	4.95	4.88	4.82	4.77	4.74	4.68	4.62	4.56	4.53	4.50	4.46	4.43	4.40	4.36
6	5.99	5.14	4.76	4.53	4.39	4.28	4.21	4.15	4.10	4.06	4.00	3.94	3.87	3.84	3.81	3.77	3.74	3.70	3.67
7	5.59	4.74	4.35	4.12	3.97	3.87	3.79	3.73	3.68	3.64	3.57	3.51	3.44	3.41	3.38	3.34	3.30	3.27	3.23
8	5.32	4.46	4.07	3.84	3.69	3.58	3.50	3.44	3.39	3.35	3.28	3.22	3.15	3.12	3.08	3.04	3.01	2.97	2.93
9	5.12	4.26	3.86	3.63	3.48	3.37	3.29	3.23	3.18	3.14	3.07	3.01	2.94	2.90	2.86	2.83	2.79	2.75	2.71
10	4.96	4.10	3.71	3.48	3.33	3.22	3.14	3.07	3.02	2.98	2.91	2.85	2.77	2.74	2.70	2.66	2.62	2.58	2.54
11	4.84	3.98	3.59	3.36	3.20	3.09	3.01	2.95	2.90	2.85	2.79	2.72	2.65	2.61	2.57	2.53	2.49	2.45	2.40
12	4.75	3.89	3.49	3.26	3.11	3.00	2.91	2.85	2.80	2.75	2.69	2.62	2.54	2.51	2.47	2.43	2.38	2.34	2.30
13	4.67	3.81	3.41	3.18	3.03	2.92	2.83	2.77	2.71	2.67	2.60	2.53	2.46	2.42	2.38	2.34	2.30	2.25	2.21
14	4.60	3.74	3.34	3.11	2.96	2.85	2.76	2.70	2.65	2.60	2.53	2.46	2.39	2.35	2.31	2.27	2.22	2.18	2.13
15	4.54	3.68	3.29	3.06	2.90	2.79	2.71	2.64	2.59	2.54	2.48	2.40	2.33	2.29	2.25	2.20	2.16	2.11	2.07
16	4.49	3.63	3.24	3.01	2.85	2.74	2.66	2.59	2.54	2.49	2.42	2.35	2.28	2.24	2.19	2.15	2.11	2.06	2.01
17	4.45	3.59	3.20	2.96	2.81	2.70	2.61	2.55	2.49	2.45	2.38	2.31	2.23	2.19	2.15	2.10	2.06	2.01	1.96
18	4.41	3.55	3.16	2.93	2.77	2.66	2.58	2.51	2.46	2.41	2.34	2.27	2.19	2.15	2.11	2.06	2.02	1.97	1.92
19	4.38	3.52	3.13	2.90	2.74	2.63	2.54	2.48	2.42	2.38	2.31	2.23	2.16	2.11	2.07	2.03	1.98	1.93	1.88
20	4.35	3.49	3.10	2.87	2.71	2.60	2.51	2.45	2.39	2.35	2.28	2.20	2.12	2.08	2.04	1.99	1.95	1.90	1.84
21	4.32	3.47	3.07	2.84	2.68	2.57	2.49	2.42	2.37	2.32	2.25	2.18	2.10	2.05	2.01	1.96	1.92	1.87	1.81
22	4.30	3.44	3.05	2.82	2.66	2.55	2.46	2.40	2.34	2.30	2.23	2.15	2.07	2.03	1.98	1.94	1.89	1.84	1.78
23	4.28	3.42	3.03	2.80	2.64	2.53	2.44	2.37	2.32	2.27	2.20	2.13	2.05	2.01	1.96	1.91	1.86	1.81	1.76
24	4.26	3.40	3.01	2.78	2.62	2.51	2.42	2.36	2.30	2.25	2.18	2.11	2.03	1.98	1.94	1.89	1.84	1.79	1.73
25	4.24	3.39	2.99	2.76	2.60	2.49	2.40	2.34	2.28	2.24	2.16	2.09	2.01	1.96	1.92	1.87	1.82	1.77	1.71
26	4.23	3.37	2.98	2.74	2.59	2.47	2.39	2.32	2.27	2.22	2.15	2.07	1.99	1.95	1.90	1.85	1.80	1.75	1.69
27	4.21	3.35	2.96	2.73	2.57	2.46	2.37	2.31	2.25	2.20	2.13	2.06	1.97	1.93	1.88	1.84	1.79	1.73	1.67
28	4.20	3.34	2.95	2.71	2.56	2.45	2.36	2.29	2.24	2.19	2.12	2.04	1.96	1.91	1.87	1.82	1.77	1.71	1.65
29	4.18	3.33	2.93	2.70	2.55	2.43	2.35	2.28	2.22	2.18	2.10	2.03	1.94	1.90	1.85	1.81	1.75	1.70	1.64
30	4.17	3.32	2.92	2.69	2.53	2.42	2.33	2.27	2.21	2.16	2.09	2.01	1.93	1.89	1.84	1.79	1.74	1.68	1.62
40	4.08	3.23	2.84	2.61	2.45	2.34	2.25	2.18	2.12	2.08	2.00	1.92	1.84	1.79	1.74	1.69	1.64	1.58	1.51
60	4.00	3.15	2.76	2.53	2.37	2.25	2.17	2.10	2.04	1.99	1.92	1.84	1.75	1.70	1.65	1.59	1.53	1.47	1.39
120	3.92	3.07	2.68	2.45	2.29	2.17	2.09	2.02	1.96	1.91	1.83	1.75	1.66	1.61	1.55	1.50	1.43	1.35	1.25
∞	3.84	3.00	2.60	2.37	2.21	2.10	2.01	1.94	1.88	1.83	1.75	1.67	1.57	1.52	1.46	1.39	1.32	1.22	1.00

$F = \dfrac{s_1^2}{s_2^2} = \dfrac{S_1}{m}\Big/\dfrac{S_2}{n}$, where $s_1^2 = S_1/m$ and $s_2^2 = S_2/n$ are independent mean squares estimating a common variance σ^2 and based on m and n degrees of freedom, respectively.

$$F(F) = \int_0^F \frac{\Gamma\left(\dfrac{m+n}{2}\right)}{\Gamma\left(\dfrac{m}{2}\right)\Gamma\left(\dfrac{n}{2}\right)}\, m^{\frac{m}{2}} n^{\frac{n}{2}} x^{\frac{m}{2}-1}(n+mx)^{-\frac{m+n}{2}}\, dx = .975$$

n \ m	1	2	3	4	5	6	7	8	9	10	12	15	20	24	30	40	60	120	∞
1	647.8	799.5	864.2	899.6	921.8	937.1	948.2	956.7	963.3	968.6	976.7	984.9	993.1	997.2	1001	1006	1010	1014	1018
2	38.51	39.00	39.17	39.25	39.30	39.33	39.36	39.37	39.39	39.40	39.41	39.43	39.45	39.46	39.46	39.47	39.48	39.49	39.50
3	17.44	16.04	15.44	15.10	14.88	14.73	14.62	14.54	14.47	14.42	14.34	14.25	14.17	14.12	14.08	14.04	13.99	13.95	13.90
4	12.22	10.65	9.98	9.60	9.36	9.20	9.07	8.98	8.90	8.84	8.75	8.66	8.56	8.51	8.46	8.41	8.36	8.31	8.26
5	10.01	8.43	7.76	7.39	7.15	6.98	6.85	6.76	6.68	6.62	6.52	6.43	6.33	6.28	6.23	6.18	6.12	6.07	6.02
6	8.81	7.26	6.60	6.23	5.99	5.82	5.70	5.60	5.52	5.46	5.37	5.27	5.17	5.12	5.07	5.01	4.96	4.90	4.85
7	8.07	6.54	5.89	5.52	5.29	5.12	4.99	4.90	4.82	4.76	4.67	4.57	4.47	4.42	4.36	4.31	4.25	4.20	4.14
8	7.57	6.06	5.42	5.05	4.82	4.65	4.53	4.43	4.36	4.30	4.20	4.10	4.00	3.95	3.89	3.84	3.78	3.73	3.67
9	7.21	5.71	5.08	4.72	4.48	4.32	4.20	4.10	4.03	3.96	3.87	3.77	3.67	3.61	3.56	3.51	3.45	3.39	3.33
10	6.94	5.46	4.83	4.47	4.24	4.07	3.95	3.85	3.78	3.72	3.62	3.52	3.42	3.37	3.31	3.26	3.20	3.14	3.08
11	6.72	5.26	4.63	4.28	4.04	3.88	3.76	3.66	3.59	3.53	3.43	3.33	3.23	3.17	3.12	3.06	3.00	2.94	2.88
12	6.55	5.10	4.47	4.12	3.89	3.73	3.61	3.51	3.44	3.37	3.28	3.18	3.07	3.02	2.96	2.91	2.85	2.79	2.72
13	6.41	4.97	4.35	4.00	3.77	3.60	3.48	3.39	3.31	3.25	3.15	3.05	2.95	2.89	2.84	2.78	2.72	2.66	2.60
14	6.30	4.86	4.24	3.89	3.66	3.50	3.38	3.29	3.21	3.15	3.05	2.95	2.84	2.79	2.73	2.67	2.61	2.55	2.49
15	6.20	4.77	4.15	3.80	3.58	3.41	3.29	3.20	3.12	3.06	2.96	2.86	2.76	2.70	2.64	2.59	2.52	2.46	2.40
16	6.12	4.69	4.08	3.73	3.50	3.34	3.22	3.12	3.05	2.99	2.89	2.79	2.68	2.63	2.57	2.51	2.45	2.38	2.32
17	6.04	4.62	4.01	3.66	3.44	3.28	3.16	3.06	2.98	2.92	2.82	2.72	2.62	2.56	2.50	2.44	2.38	2.32	2.25
18	5.98	4.56	3.95	3.61	3.38	3.22	3.10	3.01	2.93	2.87	2.77	2.67	2.56	2.50	2.44	2.38	2.32	2.26	2.19
19	5.92	4.51	3.90	3.56	3.33	3.17	3.05	2.96	2.88	2.82	2.72	2.62	2.51	2.45	2.39	2.33	2.27	2.20	2.13
20	5.87	4.46	3.86	3.51	3.29	3.13	3.01	2.91	2.84	2.77	2.68	2.57	2.46	2.41	2.35	2.29	2.22	2.16	2.09
21	5.83	4.42	3.82	3.48	3.25	3.09	2.97	2.87	2.80	2.73	2.64	2.53	2.42	2.37	2.31	2.25	2.18	2.11	2.04
22	5.79	4.38	3.78	3.44	3.22	3.05	2.93	2.84	2.76	2.70	2.60	2.50	2.39	2.33	2.27	2.21	2.14	2.08	2.00
23	5.75	4.35	3.75	3.41	3.18	3.02	2.90	2.81	2.73	2.67	2.57	2.47	2.36	2.30	2.24	2.18	2.11	2.04	1.97
24	5.72	4.32	3.72	3.38	3.15	2.99	2.87	2.78	2.70	2.64	2.54	2.44	2.33	2.27	2.21	2.15	2.08	2.01	1.94
25	5.69	4.29	3.69	3.35	3.13	2.97	2.85	2.75	2.68	2.61	2.51	2.41	2.30	2.24	2.18	2.12	2.05	1.98	1.91
26	5.66	4.27	3.67	3.33	3.10	2.94	2.82	2.73	2.65	2.59	2.49	2.39	2.28	2.22	2.16	2.09	2.03	1.95	1.88
27	5.63	4.24	3.65	3.31	3.08	2.92	2.80	2.71	2.63	2.57	2.47	2.36	2.25	2.19	2.13	2.07	2.00	1.93	1.85
28	5.61	4.22	3.63	3.29	3.06	2.90	2.78	2.69	2.61	2.55	2.45	2.34	2.23	2.17	2.11	2.05	1.98	1.91	1.83
29	5.59	4.20	3.61	3.27	3.04	2.88	2.76	2.67	2.59	2.53	2.43	2.32	2.21	2.15	2.09	2.03	1.96	1.89	1.81
30	5.57	4.18	3.59	3.25	3.03	2.87	2.75	2.65	2.57	2.51	2.41	2.31	2.20	2.14	2.07	2.01	1.94	1.87	1.79
40	5.42	4.05	3.46	3.13	2.90	2.74	2.62	2.53	2.45	2.39	2.29	2.18	2.07	2.01	1.94	1.88	1.80	1.72	1.64
60	5.29	3.93	3.34	3.01	2.79	2.63	2.51	2.41	2.33	2.27	2.17	2.06	1.94	1.88	1.82	1.74	1.67	1.58	1.48
120	5.15	3.80	3.23	2.89	2.67	2.52	2.39	2.30	2.22	2.16	2.05	1.94	1.82	1.76	1.69	1.61	1.53	1.43	1.31
∞	5.02	3.69	3.12	2.79	2.57	2.41	2.29	2.19	2.11	2.05	1.94	1.83	1.71	1.64	1.57	1.48	1.39	1.27	1.00

$F = \dfrac{s_1^2}{s_2^2} = \dfrac{S_1}{m}\Big/\dfrac{S_2}{n}$, where $s_1^2 = S_1/m$ and $s_2^2 = S_2/n$ are independent mean squares estimating a common variance σ^2 and based on m and n degrees of freedom, respectively.

$$F(F) = \int_0^F \frac{\Gamma\left(\dfrac{m+n}{2}\right)}{\Gamma\left(\dfrac{m}{2}\right)\Gamma\left(\dfrac{n}{2}\right)}\, m^{\frac{m}{2}} n^{\frac{n}{2}} x^{\frac{m}{2}-1}(n+mx)^{-\frac{m+n}{2}}\, dx = .99$$

n \ m	1	2	3	4	5	6	7	8	9	10	12	15	20	24	30	40	60	120	∞
1	4052	4999.5	5403	5625	5764	5859	5928	5982	6022	6056	6106	6157	6209	6235	6261	6287	6313	6339	6366
2	98.50	99.00	99.17	99.25	99.30	99.33	99.36	99.37	99.39	99.40	99.42	99.43	99.45	99.46	99.47	99.47	99.48	99.49	99.50
3	34.12	30.82	29.46	28.71	28.24	27.91	27.67	27.49	27.35	27.23	27.05	26.87	26.69	26.60	26.50	26.41	26.32	26.22	26.13
4	21.20	18.00	16.69	15.98	15.52	15.21	14.98	14.80	14.66	14.55	14.37	14.20	14.02	13.93	13.84	13.75	13.65	13.56	13.46
5	16.26	13.27	12.06	11.39	10.97	10.67	10.46	10.29	10.16	10.05	9.89	9.72	9.55	9.47	9.38	9.29	9.20	9.11	9.02
6	13.75	10.92	9.78	9.15	8.75	8.47	8.26	8.10	7.98	7.87	7.72	7.56	7.40	7.31	7.23	7.14	7.06	6.97	6.88
7	12.25	9.55	8.45	7.85	7.46	7.19	6.99	6.84	6.72	6.62	6.47	6.31	6.16	6.07	5.99	5.91	5.82	5.74	5.65
8	11.26	8.65	7.59	7.01	6.63	6.37	6.18	6.03	5.91	5.81	5.67	5.52	5.36	5.28	5.20	5.12	5.03	4.95	4.86
9	10.56	8.02	6.99	6.42	6.06	5.80	5.61	5.47	5.35	5.26	5.11	4.96	4.81	4.73	4.65	4.57	4.48	4.40	4.31
10	10.04	7.56	6.55	5.99	5.64	5.39	5.20	5.06	4.94	4.85	4.71	4.56	4.41	4.33	4.25	4.17	4.08	4.00	3.91
11	9.65	7.21	6.22	5.67	5.32	5.07	4.89	4.74	4.63	4.54	4.40	4.25	4.10	4.02	3.94	3.86	3.78	3.69	3.60
12	9.33	6.93	5.95	5.41	5.06	4.82	4.64	4.50	4.39	4.30	4.16	4.01	3.86	3.78	3.70	3.62	3.54	3.45	3.36
13	9.07	6.70	5.74	5.21	4.86	4.62	4.44	4.30	4.19	4.10	3.96	3.82	3.66	3.59	3.51	3.43	3.34	3.25	3.17
14	8.86	6.51	5.56	5.04	4.69	4.46	4.28	4.14	4.03	3.94	3.80	3.66	3.51	3.43	3.35	3.27	3.18	3.09	3.00
15	8.68	6.36	5.42	4.89	4.56	4.32	4.14	4.00	3.89	3.80	3.67	3.52	3.37	3.29	3.21	3.13	3.05	2.96	2.87
16	8.53	6.23	5.29	4.77	4.44	4.20	4.03	3.89	3.78	3.69	3.55	3.41	3.26	3.18	3.10	3.02	2.93	2.84	2.75
17	8.40	6.11	5.18	4.67	4.34	4.10	3.93	3.79	3.68	3.59	3.46	3.31	3.16	3.08	3.00	2.92	2.83	2.75	2.65
18	8.29	6.01	5.09	4.58	4.25	4.01	3.84	3.71	3.60	3.51	3.37	3.23	3.08	3.00	2.92	2.84	2.75	2.66	2.57
19	8.18	5.93	5.01	4.50	4.17	3.94	3.77	3.63	3.52	3.43	3.30	3.15	3.00	2.92	2.84	2.76	2.67	2.58	2.49
20	8.10	5.85	4.94	4.43	4.10	3.87	3.70	3.56	3.46	3.37	3.23	3.09	2.94	2.86	2.78	2.69	2.61	2.52	2.42
21	8.02	5.78	4.87	4.37	4.04	3.81	3.64	3.51	3.40	3.31	3.17	3.03	2.88	2.80	2.72	2.64	2.55	2.46	2.36
22	7.95	5.72	4.82	4.31	3.99	3.76	3.59	3.45	3.35	3.26	3.12	2.98	2.83	2.75	2.67	2.58	2.50	2.40	2.31
23	7.88	5.66	4.76	4.26	3.94	3.71	3.54	3.41	3.30	3.21	3.07	2.93	2.78	2.70	2.62	2.54	2.45	2.35	2.26
24	7.82	5.61	4.72	4.22	3.90	3.67	3.50	3.36	3.26	3.17	3.03	2.89	2.74	2.66	2.58	2.49	2.40	2.31	2.21
25	7.77	5.57	4.68	4.18	3.85	3.63	3.46	3.32	3.22	3.13	2.99	2.85	2.70	2.62	2.54	2.45	2.36	2.27	2.17
26	7.72	5.53	4.64	4.14	3.82	3.59	3.42	3.29	3.18	3.09	2.96	2.81	2.66	2.58	2.50	2.42	2.33	2.23	2.13
27	7.68	5.49	4.60	4.11	3.78	3.56	3.39	3.26	3.15	3.06	2.93	2.78	2.63	2.55	2.47	2.38	2.29	2.20	2.10
28	7.64	5.45	4.57	4.07	3.75	3.53	3.36	3.23	3.12	3.03	2.90	2.75	2.60	2.52	2.44	2.35	2.26	2.17	2.06
29	7.60	5.42	4.54	4.04	3.73	3.50	3.33	3.20	3.09	3.00	2.87	2.73	2.57	2.49	2.41	2.33	2.23	2.14	2.03
30	7.56	5.39	4.51	4.02	3.70	3.47	3.30	3.17	3.07	2.98	2.84	2.70	2.55	2.47	2.39	2.30	2.21	2.11	2.01
40	7.31	5.18	4.31	3.83	3.51	3.29	3.12	2.99	2.89	2.80	2.66	2.52	2.37	2.29	2.20	2.11	2.02	1.92	1.80
60	7.08	4.98	4.13	3.65	3.34	3.12	2.95	2.82	2.72	2.63	2.50	2.35	2.20	2.12	2.03	1.94	1.84	1.73	1.60
120	6.85	4.79	3.95	3.48	3.17	2.96	2.79	2.66	2.56	2.47	2.34	2.19	2.03	1.95	1.86	1.76	1.66	1.53	1.38
∞	6.63	4.61	3.78	3.32	3.02	2.80	2.64	2.51	2.41	2.32	2.18	2.04	1.88	1.79	1.70	1.59	1.47	1.32	1.00

$F = \dfrac{s_1^2}{s_2^2} = \dfrac{S_1}{m}\Big/\dfrac{S_2}{n}$, where $s_1^2 = S_1/m$ and $s_2^2 = S_2/n$ are independent mean squares estimating a common variance σ^2 and based on m and n degrees of freedom, respectively.

$$F(F) = \int_0^F \frac{\Gamma\left(\frac{m+n}{2}\right)}{\Gamma\left(\frac{m}{2}\right)\Gamma\left(\frac{n}{2}\right)} m^{\frac{m}{2}} n^{\frac{n}{2}} x^{\frac{m}{2}-1} (n+mx)^{-\frac{m+n}{2}} dx = .995$$

n \ m	1	2	3	4	5	6	7	8	9	10	12	15	20	24	30	40	60	120	∞
1	16211	20000	21615	22500	23056	23437	23715	23925	24091	24224	24426	24630	24836	24940	25044	25148	25253	25359	25465
2	198.5	199.0	199.2	199.2	199.3	199.3	199.4	199.4	199.4	199.4	199.4	199.4	199.4	199.5	199.5	199.5	199.5	199.5	199.5
3	55.55	49.80	47.47	46.19	45.39	44.84	44.43	44.13	43.88	43.69	43.39	43.08	42.78	42.62	42.47	42.31	42.15	41.99	41.83
4	31.33	26.28	24.26	23.15	22.46	21.97	21.62	21.35	21.14	20.97	20.70	20.44	20.17	20.03	19.89	19.75	19.61	19.47	19.32
5	22.78	18.31	16.53	15.56	14.94	14.51	14.20	13.96	13.77	13.62	13.38	13.15	12.90	12.78	12.66	12.53	12.40	12.27	12.14
6	18.63	14.54	12.92	12.03	11.46	11.07	10.79	10.57	10.39	10.25	10.03	9.81	9.59	9.47	9.36	9.24	9.12	9.00	8.88
7	16.24	12.40	10.88	10.05	9.52	9.16	8.89	8.68	8.51	8.38	8.18	7.97	7.75	7.65	7.53	7.42	7.31	7.19	7.08
8	14.69	11.04	9.60	8.81	8.30	7.95	7.69	7.50	7.34	7.21	7.01	6.81	6.61	6.50	6.40	6.29	6.18	6.06	5.95
9	13.61	10.11	8.72	7.96	7.47	7.13	6.88	6.69	6.54	6.42	6.23	6.03	5.83	5.73	5.62	5.52	5.41	5.30	5.19
10	12.83	9.43	8.08	7.34	6.87	6.54	6.30	6.12	5.97	5.85	5.66	5.47	5.27	5.17	5.07	4.97	4.86	4.75	4.64
11	12.23	8.91	7.60	6.88	6.42	6.10	5.86	5.68	5.54	5.42	5.24	5.05	4.86	4.76	4.65	4.55	4.44	4.34	4.23
12	11.75	8.51	7.23	6.52	6.07	5.76	5.52	5.35	5.20	5.09	4.91	4.72	4.53	4.43	4.33	4.23	4.12	4.01	3.90
13	11.37	8.19	6.93	6.23	5.79	5.48	5.25	5.08	4.94	4.82	4.64	4.46	4.27	4.17	4.07	3.97	3.87	3.76	3.65
14	11.06	7.92	6.68	6.00	5.56	5.26	5.03	4.86	4.72	4.60	4.43	4.25	4.06	3.96	3.86	3.76	3.66	3.55	3.44
15	10.80	7.70	6.48	5.80	5.37	5.07	4.85	4.67	4.54	4.42	4.25	4.07	3.88	3.79	3.69	3.58	3.48	3.37	3.26
16	10.58	7.51	6.30	5.64	5.21	4.91	4.69	4.52	4.38	4.27	4.10	3.92	3.73	3.64	3.54	3.44	3.33	3.22	3.11
17	10.38	7.35	6.16	5.50	5.07	4.78	4.56	4.39	4.25	4.14	3.97	3.79	3.61	3.51	3.41	3.31	3.21	3.10	2.98
18	10.22	7.21	6.03	5.37	4.96	4.66	4.44	4.28	4.14	4.03	3.86	3.68	3.50	3.40	3.30	3.20	3.10	2.99	2.87
19	10.07	7.09	5.92	5.27	4.85	4.56	4.34	4.18	4.04	3.93	3.76	3.59	3.40	3.31	3.21	3.11	3.00	2.89	2.78
20	9.94	6.99	5.82	5.17	4.76	4.47	4.26	4.09	3.96	3.85	3.68	3.50	3.32	3.22	3.12	3.02	2.92	2.81	2.69
21	9.83	6.89	5.73	5.09	4.68	4.39	4.18	4.01	3.88	3.77	3.60	3.43	3.24	3.15	3.05	2.95	2.84	2.73	2.61
22	9.73	6.81	5.65	5.02	4.61	4.32	4.11	3.94	3.81	3.70	3.54	3.36	3.18	3.08	2.98	2.88	2.77	2.66	2.55
23	9.63	6.73	5.58	4.95	4.54	4.26	4.05	3.88	3.75	3.64	3.47	3.30	3.12	3.02	2.92	2.82	2.71	2.60	2.48
24	9.55	6.66	5.52	4.89	4.49	4.20	3.99	3.83	3.69	3.59	3.42	3.25	3.06	2.97	2.87	2.77	2.66	2.55	2.43
25	9.48	6.60	5.46	4.84	4.43	4.15	3.94	3.78	3.64	3.54	3.37	3.20	3.01	2.92	2.82	2.72	2.61	2.50	2.38
26	9.41	6.54	5.41	4.79	4.38	4.10	3.89	3.73	3.60	3.49	3.33	3.15	2.97	2.87	2.77	2.67	2.56	2.45	2.33
27	9.34	6.49	5.36	4.74	4.34	4.06	3.85	3.69	3.56	3.45	3.28	3.11	2.93	2.83	2.73	2.63	2.52	2.41	2.25
28	9.28	6.44	5.32	4.70	4.30	4.02	3.81	3.65	3.52	3.41	3.25	3.07	2.89	2.79	2.69	2.59	2.48	2.37	2.29
29	9.23	6.40	5.28	4.66	4.26	3.98	3.77	3.61	3.48	3.38	3.21	3.04	2.86	2.76	2.66	2.56	2.45	2.33	2.24
30	9.18	6.35	5.24	4.62	4.23	3.95	3.74	3.58	3.45	3.34	3.18	3.01	2.82	2.73	2.63	2.52	2.42	2.30	2.18
40	8.83	6.07	4.98	4.37	3.99	3.71	3.51	3.35	3.22	3.12	2.95	2.78	2.60	2.50	2.40	2.30	2.18	2.06	1.93
60	8.49	5.79	4.73	4.14	3.76	3.49	3.29	3.13	3.01	2.90	2.74	2.57	2.39	2.29	2.19	2.08	1.96	1.83	1.69
120	8.18	5.54	4.50	3.92	3.55	3.28	3.09	2.93	2.81	2.71	2.54	2.37	2.19	2.09	1.98	1.87	1.75	1.61	1.43
∞	7.88	5.30	4.28	3.72	3.35	3.09	2.90	2.74	2.62	2.52	2.36	2.19	2.00	1.90	1.79	1.67	1.53	1.36	1.00

$F = \dfrac{s_1^2}{s_2^2} = \dfrac{S_1}{m} \Big/ \dfrac{S_2}{n}$, where $s_1^2 = S_1/m$ and $s_2^2 = S_2/n$ are independent mean squares estimating a common variance σ^2 and based on m and n degrees of freedom, respectively.

$$F(F) = \int_0^F \frac{\Gamma\left(\frac{m+n}{2}\right)}{\Gamma\left(\frac{m}{2}\right)\Gamma\left(\frac{n}{2}\right)} m^{\frac{m}{2}} n^{\frac{n}{2}} x^{\frac{m}{2}-1} (n+mx)^{-\frac{m+n}{2}} dx = .999$$

n \ m	1	2	3	4	5	6	7	8	9	10	12	15	20	24	30	40	60	120	∞
1	4053*	5000*	5404*	5625*	5764*	5859*	5929*	5981*	6023*	6056*	6107*	6158*	6209*	6235*	6261*	6287*	6313*	6340*	6366*
2	998.5	999.0	999.2	999.2	999.3	999.3	999.4	999.4	999.4	999.4	999.4	999.4	999.4	999.5	999.5	999.5	999.5	999.5	999.5
3	167.0	148.5	141.1	137.1	134.6	132.8	131.6	130.6	129.9	129.2	128.3	127.4	126.4	125.9	125.4	125.0	124.5	124.0	123.5
4	74.14	61.25	56.18	53.44	51.71	50.53	49.66	49.00	48.47	48.05	47.41	46.76	46.10	45.77	45.43	45.09	44.75	44.40	44.05
5	47.18	37.12	33.20	31.09	29.75	28.84	28.16	27.64	27.24	26.92	26.42	25.91	25.39	25.14	24.87	24.60	24.33	24.06	23.79
6	35.51	27.00	23.70	21.92	20.81	20.03	19.46	19.03	18.69	18.41	17.99	17.56	17.12	16.89	16.67	16.44	16.21	15.99	15.75
7	29.25	21.69	18.77	17.19	16.21	15.52	15.02	14.63	14.33	14.08	13.71	13.32	12.93	12.73	12.53	12.33	12.12	11.91	11.70
8	25.42	18.49	15.83	14.39	13.49	12.86	12.40	12.04	11.77	11.54	11.19	10.84	10.48	10.30	10.11	9.92	9.73	9.53	9.33
9	22.86	16.39	13.90	12.56	11.71	11.13	10.70	10.37	10.11	9.89	9.57	9.24	8.90	8.72	8.55	8.37	8.19	8.00	7.81
10	21.04	14.91	12.55	11.28	10.48	9.92	9.52	9.20	8.96	8.75	8.45	8.13	7.80	7.64	7.47	7.30	7.12	6.94	6.76
11	19.69	13.81	11.56	10.35	9.58	9.05	8.66	8.35	8.12	7.92	7.63	7.32	7.01	6.85	6.68	6.52	6.35	6.17	6.00
12	18.64	12.97	10.80	9.63	8.89	8.38	8.00	7.71	7.48	7.29	7.00	6.71	6.40	6.25	6.09	5.93	5.76	5.59	5.42
13	17.81	12.31	10.21	9.07	8.35	7.86	7.49	7.21	6.98	6.80	6.52	6.23	5.93	5.78	5.63	5.47	5.30	5.14	4.97
14	17.14	11.78	9.73	8.62	7.92	7.43	7.08	6.80	6.58	6.40	6.13	5.85	5.56	5.41	5.25	5.10	4.94	4.77	4.60
15	16.59	11.34	9.34	8.25	7.57	7.09	6.74	6.47	6.26	6.08	5.81	5.54	5.25	5.10	4.95	4.80	4.64	4.47	4.31
16	16.12	10.97	9.00	7.94	7.27	6.81	6.46	6.19	5.98	5.81	5.55	5.27	4.99	4.85	4.70	4.54	4.39	4.23	4.06
17	15.72	10.66	8.73	7.68	7.02	6.56	6.22	5.96	5.75	5.58	5.32	5.05	4.78	4.63	4.48	4.33	4.18	4.02	3.85
18	15.38	10.39	8.49	7.46	6.81	6.35	6.02	5.76	5.56	5.39	5.13	4.87	4.59	4.45	4.30	4.15	4.00	3.84	3.67
19	15.08	10.16	8.28	7.26	6.62	6.18	5.85	5.59	5.39	5.22	4.97	4.70	4.43	4.29	4.14	3.99	3.84	3.68	3.51
20	14.82	9.95	8.10	7.10	6.46	6.02	5.69	5.44	5.24	5.08	4.82	4.56	4.29	4.15	4.00	3.86	3.70	3.54	3.38
21	14.59	9.77	7.94	6.95	6.32	5.88	5.56	5.31	5.11	4.95	4.70	4.44	4.17	4.03	3.88	3.74	3.58	3.42	3.26
22	14.38	9.61	7.80	6.81	6.19	5.76	5.44	5.19	4.99	4.83	4.58	4.33	4.06	3.92	3.78	3.63	3.48	3.32	3.15
23	14.19	9.47	7.67	6.69	6.08	5.65	5.33	5.09	4.89	4.73	4.48	4.23	3.96	3.82	3.68	3.53	3.38	3.22	3.05
24	14.03	9.34	7.55	6.59	5.98	5.55	5.23	4.99	4.80	4.64	4.39	4.14	3.87	3.74	3.59	3.45	3.29	3.14	2.97
25	13.88	9.22	7.45	6.49	5.88	5.46	5.15	4.91	4.71	4.56	4.31	4.06	3.79	3.66	3.52	3.37	3.22	3.06	2.89
26	13.74	9.12	7.36	6.41	5.80	5.38	5.07	4.83	4.64	4.48	4.24	3.99	3.72	3.59	3.44	3.30	3.15	2.99	2.82
27	13.61	9.02	7.27	6.33	5.73	5.31	5.00	4.76	4.57	4.41	4.17	3.92	3.66	3.52	3.38	3.23	3.08	2.92	2.75
28	13.50	8.93	7.19	6.25	5.66	5.24	4.93	4.69	4.50	4.35	4.11	3.86	3.60	3.46	3.32	3.18	3.02	2.86	2.69
29	13.39	8.85	7.12	6.19	5.59	5.18	4.87	4.64	4.45	4.29	4.05	3.80	3.54	3.41	3.27	3.12	2.97	2.81	2.64
30	13.29	8.77	7.05	6.12	5.53	5.12	4.82	4.58	4.39	4.24	4.00	3.75	3.49	3.36	3.22	3.07	2.92	2.76	2.59
40	12.61	8.25	6.60	5.70	5.13	4.73	4.44	4.21	4.02	3.87	3.64	3.40	3.15	3.01	2.87	2.73	2.57	2.41	2.23
60	11.97	7.76	6.17	5.31	4.76	4.37	4.09	3.87	3.69	3.54	3.31	3.08	2.83	2.69	2.55	2.41	2.25	2.08	1.89
120	11.38	7.32	5.79	4.95	4.42	4.04	3.77	3.55	3.38	3.24	3.02	2.78	2.53	2.40	2.26	2.11	1.95	1.76	1.54
∞	10.83	6.91	5.42	4.62	4.10	3.74	3.47	3.27	3.10	2.96	2.74	2.51	2.27	2.13	1.99	1.84	1.66	1.45	1.00

* Multiply these entries by 100.

Index

The most efficient way to use this index is to look for the pertinent *property* (e.g., vapor pressure, entropy), *process* (e.g., disposal of chemicals, calibration), or *general concept* (e.g., units, radiation). Most primary entries are subdivided into several secondary entries, e.g., under heat capacity there are 17 secondary entries such as air, metals, water, etc. Primary entries will be found for certain *classes of substances*, such as alloys, elements, organic compounds, refrigerants, semiconductors, etc. Primary entries are also given for the individual chemical elements and for a few compounds such as water and carbon dioxide. However, only the most important tables are listed under these substances. Therefore, the user will find in most cases that it is best to look first for the property of interest, then examine the table or tables that are referenced.

The reference given for each index term is the inclusive pages of the pertinent table (e.g., **8**-45 to 55). The introduction to each table describes the method of ordering the substances within that table.

The Editor would be grateful for comments and suggestions on this index.

INDEX

A

AAS, definition, **12**-1 to 6
Abbreviations
   amino acids, **7**-3
   physical quantities, **2**-1 to 21
   scientific terms, **2**-29 to 36
   units, **1**-24 to 28
Absorption
   infrared, by earth's atmosphere, **14**-24
   light, by metals and semiconductors,
      12-133 to 156
   microwave power, by water, **6**-15
   sound, by air, **14**-40
Abundance, isotopic, **1**-15 to 17, **11**-50 to 158
Acceleration due to gravity
   at poles and equator, **14**-6 to 7
   at various latitudes, **14**-9
   on the sun, moon, and planets, **14**-2 to 3
   standard value, **1**-1 to 11
Acid dissociation constant
   amino acids, **7**-1 to 2
   biological buffers, **7**-9
   indicators, **8**-16 to 18
   inorganic acids and bases, **8**-44 to 45
   organic acids and bases, **8**-46 to 56
   purine and pyrimidine bases, **7**-4
Acid rain, pH measurement, **8**-41 to 42
Acids
   activity coefficients, **5**-97 to 98, **5**-99 to
      103
   decinormal solutions, **8**-5 to 6
   enthalpy of dilution, **5**-104
   fatty, **7**-6
   indicators, **8**-16 to 18
   inorganic, dissociation constant, **8**-44 to
      45
   organic, dissociation constant, **8**-46 to 56
Acoustics
   human hearing, **14**-44 to 45
   musical scales, **14**-42 to 43
   noise levels, **14**-44 to 45
   sound velocity, **14**-38 to 39, **14**-40
Acronyms, definitions, **2**-29 to 36
Actinium: see also Elements
   electron configuration, **1**-18 to 19
   heat capacity, **4**-125
   history, occurrence, uses, **4**-1 to 36
   ionization energy, **10**-175 to 177
   isotopes and their properties, **11**-50 to 158
   physical properties, **4**-124
   thermal properties, **12**-197 to 198
Activation energy
   atmospheric reactions, **5**-106 to 120
   diffusion in semiconductors, **12**-108 to
      116
Activity coefficients of acids, bases, and salts,
      5-97 to 98, **5**-99 to 103
AES, definition, **12**-1 to 6
AFM, definition, **12**-1 to 6

Air
   absorption of sound, **14**-40
   cryogenic properties, **6**-138
   density, **6**-1 to 2
   diffusion of gases in, **6**-189 to 190
   enthalpy, **6**-1 to 2
   entropy, **6**-1 to 2
   heat capacity, **6**-1 to 2
   index of refraction, **10**-220
   mean free path, **6**-46
   permittivity (dielectric constant), **6**-172 to
      173
   thermal conductivity, **6**-185 to 186
   thermodynamic properties, **6**-1 to 2
   U.S. Standard Atmosphere, **14**-16 to 22
   vapor pressure, **6**-1 to 2
   velocity of sound, **14**-40
   velocity of sound, as function of
      temperature, **14**-41
   viscosity, **6**-178 to 179
Airborne contaminants, threshold limits,
      16-32 to 42
Albedo
   planets, **14**-2 to 3
   satellites of the planets, **14**-4 to 5
ALI for radionuclides, **16**-49 to 52
Alkali halides, secondary electron emission,
      12-131 to 132
Alkali metals: see entries for Lithium,
      Sodium, etc.
Alloys
   composition, **12**-117 to 125
   electrical resistivity, **12**-48 to 51
   magnetic properties, **12**-117 to 125
   phase diagrams, **12**-178 to 195
   superconducting properties, **12**-64 to 90
   thermal conductivity, **12**-201
Alloys, commercial
   electrical resistivity, **12**-210
   mechanical and thermal properties,
      12-210
   thermal conductivity, **12**-201
Alphabets: Greek, Russian, Hebrew, **2**-37
Aluminum: see also Elements
   electrical resistivity, **12**-45 to 47
   electron configuration, **1**-18 to 19
   heat capacity, **4**-125, **12**-196
   history, occurrence, uses, **4**-1 to 36
   ionization energy, **10**-175 to 177
   isotopes and their properties, **11**-50 to 158
   magnetic susceptibility, **4**-133 to 138
   molten, density, **4**-129 to 132
   physical properties, **4**-124
   resistance of wires, **15**-27
   thermal conductivity, **12**-199 to 200
   thermal properties, **12**-197 to 198
   thermodynamic properties, **5**-1 to 3
   vapor pressure, **6**-64 to 91
   wire tables, **15**-27
Americium: see also Elements

electron configuration, **1**-18 to 19
history, occurrence, uses, **4**-1 to 36
ionization energy, **10**-175 to 177
isotopes and their properties, **11**-50 to 158
Amino acids
   abbreviations and symbols, **7**-3
   dissociation constants, **7**-1 to 2
   physical properties, **7**-1 to 2
   solubility, **7**-1 to 2
   structure, **7**-3
Ampere
   definition, **1**-24 to 28
   maintained value, **1**-1 to 11
Analytical procedures
   organic reagents, **8**-8 to 15
   preparation of reagents, **8**-1 to 4
   reduction of weighings, **8**-119
   solids and surfaces, **12**-1 to 6
   volumetric calibrations, **8**-119
Ångstrom, definition, **1**-24 to 28
Antiferroelectric crystals, Curie temperature,
      12-62
Antiferromagnetic elements and compounds,
      12-117 to 125
Antiferromagnetic materials, Faraday
      rotation, **12**-157 to 173
Antimony: see also Elements
   electrical resistivity, **12**-45 to 47
   electron configuration, **1**-18 to 19
   heat capacity, **4**-125
   history, occurrence, uses, **4**-1 to 36
   ionization energy, **10**-175 to 177
   isotopes and their properties, **11**-50 to 158
   magnetic susceptibility, **4**-133 to 138
   molten, density, **4**-129 to 132
   physical properties, **4**-124
   thermal properties, **12**-197 to 198
   vapor pressure, **6**-64 to 91
Appearance potential, molecules, **10**-178 to
      195
APS, definition, **12**-1 to 6
Aqueous solutions
   activity coefficients, **5**-97 to 98, **5**-99 to
      103
   concentrative properties, **8**-57 to 83
   conductivity, **5**-94, **5**-95 to 97
   density, **8**-57 to 83
   diffusion of ions, **5**-95 to 97
   diffusion of non-electrolytes, **6**-192
   enthalpy, **5**-105
   freezing point depression, **8**-57 to 83
   heat capacity, **5**-85 to 88
   hydrohalogen acids, conductivity, **5**-92 to
      93
   index of refraction, **8**-57 to 83
   sodium chloride, volumetric properties,
      6-8
   solubility product constant, **8**-111 to 114
   thermodynamic properties, **5**-85 to 88
   vapor pressure, **6**-103

viscosity, **8**-57 to 83
Argon: see also Elements
  critical constants, **6**-48 to 60
  cryogenic properties, **6**-138
  electron configuration, **1**-18 to 19
  entropy, **5**-1 to 3
  history, occurrence, uses, **4**-1 to 36
  ionization energy, **10**-175 to 177
  isotopes and their properties, **11**-50 to 158
  magnetic susceptibility, **4**-133 to 138
  mean free path, **6**-46
  permittivity (dielectric constant), **6**-149 to 171, **6**-172 to 173
  physical properties, **4**-124
  solubility in water, **8**-86 to 89
  thermal conductivity, **6**-16 to 24, **6**-185 to 186
  thermodynamic properties, **6**-16 to 24
  thermodynamic properties at high temperature, **5**-61 to 84
  van der Waals constants, **6**-45
  vapor pressure, **6**-64 to 91, **6**-92 to 102
  viscosity, **6**-16 to 24, **6**-178 to 179
Arsenic: see also Elements
  critical constants, **6**-48 to 60
  electron configuration, **1**-18 to 19
  heat capacity, **4**-125
  history, occurrence, uses, **4**-1 to 36
  ionization energy, **10**-175 to 177
  isotopes and their properties, **11**-50 to 158
  magnetic susceptibility, **4**-133 to 138
  molten, density, **4**-129 to 132
  physical properties, **4**-124
  vapor pressure, **6**-64 to 91
Astatine: see also Elements
  electron configuration, **1**-18 to 19
  history, occurrence, uses, **4**-1 to 36
  ionization energy, **10**-175 to 177
  isotopes and their properties, **11**-50 to 158
  vapor pressure, **6**-64 to 91
Astronomical constants, **14**-1
Astronomical unit (AU), **14**-1
Atmosphere
  carbon dioxide concentration, **14**-25 to 26
  cosmic ray background, **11**-176 to 180
  electrical conductance, **14**-30 to 37
  infrared absorption, **14**-24
  ion mobility, **14**-30 to 37
  kinetic and photochemical data, **5**-106 to 120
  mass, **14**-6 to 7
  planetary, composition, **14**-2 to 3
  potential, electrical, **14**-30 to 37
  standard (unit), **1**-24 to 28
  U.S. Standard, various properties, **14**-16 to 22
Atmospheric electricity, **14**-30 to 37
Atomic mass unit (amu), **1**-1 to 11, **1**-24 to 28
Atomic masses, **1**-15 to 17, **11**-50 to 158
Atomic radius, rare earth elements, **4**-114 to 123
Atomic spectra
  elements, line spectra, **10**-1 to 87

ionization energies, **10**-175 to 177
  transition probabilities, **10**-88 to 146
  wavelengths, **10**-1 to 87
Atomic transition probability, **10**-88 to 146
Atomic weights, **1**-12 to 14
Atoms
  electron affinity, **10**-147 to 159
  electron binding energy, **10**-200 to 205
  electron configuration, **1**-18 to 19
  ionization energies, **10**-175 to 177
  masses, **1**-15 to 17
  photon cross sections, **10**-207 to 211
  polarizability, **10**-160 to 174
  spectra, **10**-1 to 87
  x-ray energy levels, **10**-196 to 199
ATR, definition, **12**-1 to 6
Autoignition temperature, **16**-16 to 31
Avogadro constant, **1**-1 to 11
Azeotropes, **6**-174 to 177

B

Bands, electromagnetic (classification), **10**-212 to 213
Barium: see also Elements
  electrical resistivity, **12**-45 to 47
  electron configuration, **1**-18 to 19
  heat capacity, **4**-125
  history, occurrence, uses, **4**-1 to 36
  ionization energy, **10**-175 to 177
  isotopes and their properties, **11**-50 to 158
  magnetic susceptibility, **4**-133 to 138
  molten, density, **4**-129 to 132
  physical properties, **4**-124
  thermal properties, **12**-197 to 198
  vapor pressure, **6**-64 to 91
Barn, definition, **1**-24 to 28
Baryons, summary of properties, **11**-1 to 49
Bases
  activity coefficients, **5**-97 to 98, **5**-99 to 103
  decinormal solutions, **8**-5 to 6
  indicators, **8**-16 to 18
  inorganic, dissociation constant, **8**-44 to 45
  organic, dissociation constant, **8**-46 to 56
  purine and pyrimidine, **7**-4
Becquerel, definition, **1**-24 to 28
Beilstein references for organic compounds, **3**-1 to 330
Berkelium: see also Elements
  electron configuration, **1**-18 to 19
  history, occurrence, uses, **4**-1 to 36
  isotopes and their properties, **11**-50 to 158
Beryllium: see also Elements
  electrical resistivity, **12**-45 to 47
  electron configuration, **1**-18 to 19
  heat capacity, **4**-125
  history, occurrence, uses, **4**-1 to 36
  ionization energy, **10**-175 to 177
  isotopes and their properties, **11**-50 to 158
  magnetic susceptibility, **4**-133 to 138
  molten, density, **4**-129 to 132

physical properties, **4**-124
  thermal properties, **12**-197 to 198
  thermodynamic properties, **5**-1 to 3
  vapor pressure, **6**-64 to 91
Bessel functions, **A**-96 to 99
BET, definition, **12**-1 to 6
Beta function, **A**-101
Billion, billiard (definition), **1**-41
Binding energy
  electrons in atoms, **10**-200 to 205
  in molecules, **9**-51 to 74
Binomial series, **A**-84 to 87
Biochemical nomenclature, references, **2**-22
Biological buffers, **7**-9
Biological materials and tissues
  effect of cosmic rays, **11**-176 to 180
  pH, **7**-10
Biosphere, mass of, **14**-6 to 7
Bismuth: see also Elements
  electrical resistivity, **12**-45 to 47
  electron configuration, **1**-18 to 19
  heat capacity, **4**-125
  history, occurrence, uses, **4**-1 to 36
  ionization energy, **10**-175 to 177
  isotopes and their properties, **11**-50 to 158
  magnetic susceptibility, **4**-133 to 138
  molten, density, **4**-129 to 132
  physical properties, **4**-124
  thermal properties, **12**-197 to 198
  vapor pressure, **6**-64 to 91
Black body radiation, **10**-214 to 215
Blood, pH of, **7**-10
Bohr magneton, **1**-1 to 11
Bohr radius, **1**-1 to 11
Bohrium (element 107), **4**-1 to 36, **11**-50 to 158
Boiling point
  correction to standard pressure, **15**-19
  cryogenic fluids, **6**-138
  D_2O, **6**-4
  elements, **4**-124
  elevation of, **15**-20
  halocarbons, **6**-140 to 142
  index for organic compounds, **3**-751 to 763
  inorganic compounds, **4**-37 to 100, **6**-48 to 60, **6**-105 to 120
  metals, **12**-197 to 198
  organic compounds, **3**-1 to 330, **6**-48 to 60, **6**-105 to 120
  organic compounds, index, **3**-751 to 763
  pressure dependence, **15**-19
  rare earth elements, **4**-114 to 123
  solvents, **15**-14 to 18
  water, as function of pressure, **6**-12
Boltzmann constant, **1**-1 to 11
Bond lengths
  diatomic molecules, **9**-81 to 86
  gas-phase molecules, **9**-17 to 43
  organic crystals, **9**-1 to 14
  organometallic compounds, **9**-15 to 16
Bonds, chemical
  lengths and angles, **9**-1 to 14, **9**-15 to 16, **9**-17 to 43

strengths, **9**-51 to 74
stretching force constants, **9**-76
Born-Haber cycle, **12**-22 to 35
Boron: see also Elements
electron configuration, **1**-18 to 19
heat capacity, **4**-125
history, occurrence, uses, **4**-1 to 36
ionization energy, **10**-175 to 177
isotopes and their properties, **11**-50 to 158
magnetic susceptibility, **4**-133 to 138
molten, density, **4**-129 to 132
physical properties, **4**-124
thermodynamic properties, **5**-1 to 3
vapor pressure, **6**-64 to 91
Bosons, summary of properties, **11**-1 to 49
Brass
phase diagram, **12**-178 to 195
resistance of wires, **15**-27
thermal conductivity, **12**-201
wire tables, **15**-27
Bravais lattices, **12**-7 to 13
Breakdown voltage, **15**-30 to 35
Bromine: see also Elements
critical constants, **6**-48 to 60
electron configuration, **1**-18 to 19
heat capacity, **4**-125
history, occurrence, uses, **4**-1 to 36
ionization energy, **10**-175 to 177
isotopes and their properties, **11**-50 to 158
physical properties, **4**-124
thermodynamic properties, **5**-1 to 3
thermodynamic properties at high
temperature, **5**-61 to 84
vapor pressure, **6**-64 to 91, **6**-92 to 102
Buffers
biological, **7**-9
for round values of pH, **8**-43
for seawater measurements, **8**-41 to 42
standard solutions, **8**-35 to 40, **8**-41 to 42

C

Cadmium: see also Elements
electrical resistivity, **12**-45 to 47
electron configuration, **1**-18 to 19
heat capacity, **4**-125
history, occurrence, uses, **4**-1 to 36
ionization energy, **10**-175 to 177
isotopes and their properties, **11**-50 to 158
magnetic susceptibility, **4**-133 to 138
molten, density, **4**-129 to 132
physical properties, **4**-124
thermal properties, **12**-197 to 198
thermodynamic properties, **5**-1 to 3
vapor pressure, **6**-64 to 91
Calcium: see also Elements
electrical resistivity, **12**-45 to 47
electron configuration, **1**-18 to 19
heat capacity, **4**-125
history, occurrence, uses, **4**-1 to 36
ionization energy, **10**-175 to 177
isotopes and their properties, **11**-50 to 158
magnetic susceptibility, **4**-133 to 138
molten, density, **4**-129 to 132

physical properties, **4**-124
thermal properties, **12**-197 to 198
thermodynamic properties, **5**-1 to 3
vapor pressure, **6**-64 to 91
Calibration
boiling points, to standard pressure, **15**-19
conductivity cells, **5**-91
index of refraction, **10**-219
infrared and far infrared frequencies,
10-236 to 240
infrared frequencies, **10**-228 to 235
relative humidity, **15**-26
temperature scale, **1**-20 to 21
thermocouples, **15**-1 to 13
vapor pressure, **6**-104
volumetric, **8**-119
weighings in air, **8**-119
Californium: see also Elements
electron configuration, **1**-18 to 19
history, occurrence, uses, **4**-1 to 36
isotopes and their properties, **11**-50 to 158
Calorie, value of, **1**-29 to 40
Candela, definition, **1**-24 to 28
Carbohydrate names and symbols, **7**-7 to 8
Carbon: see also Elements
dielectric constant, **12**-52 to 60
electron configuration, **1**-18 to 19
heat capacity, **4**-125
history, occurrence, uses, **4**-1 to 36
ionization energy, **10**-175 to 177
isotopes and their properties, **11**-50 to 158
magnetic susceptibility, **4**-133 to 138
physical properties, **4**-124
thermodynamic properties, **5**-1 to 3
vapor pressure, **6**-64 to 91
Carbon dioxide
atmospheric concentration (historical),
14-25 to 26
critical constants, **6**-48 to 60
infrared laser frequencies, **10**-228 to 235
mean free path, **6**-46
permittivity (dielectric constant), **6**-149 to
171, **6**-172 to 173
solubility in water, **8**-86 to 89
solubility in water at various pressures,
8-90
speed of sound in, **14**-38 to 39
standard thermodynamic properties, **5**-4
to 60
sublimation pressure, **6**-61 to 63
thermal conductivity, **6**-185 to 186
thermodynamic properties at high
temperature, **5**-61 to 84
van der Waals constants, **6**-45
vapor pressure, **6**-92 to 102, **6**-104
viscosity, **6**-178 to 179
Carcinogenic chemicals, list, **16**-53 to 57
Carrier gas properties, for chromatography,
8-120
CARS, definition, **12**-1 to 6
CAS Registry Numbers
inorganic compounds, **4**-37 to 100
inorganic compounds, index, **4**-107 to 113
organic compounds, **3**-1 to 330

organic compounds, index, **3**-709 to 736
Celsius temperature
conversion to other scales, **1**-41
definition, **1**-24 to 28
Ceramics
breakdown voltage, **15**-30 to 35
permittivity (dielectric constant), **12**-52 to
60
phase diagrams, **12**-178 to 195
thermal conductivity, **12**-204 to 205
Cerenkov light, in cosmic ray showers,
11-176 to 180
Cerium: see also Elements
electrical resistivity, **12**-45 to 47
electron configuration, **1**-18 to 19
heat capacity, **4**-125
history, occurrence, uses, **4**-1 to 36
ionization energy, **10**-175 to 177
isotopes and their properties, **11**-50 to 158
magnetic susceptibility, **4**-133 to 138
molten, density, **4**-129 to 132
physical properties, **4**-124
thermal properties, **12**-197 to 198
vapor pressure, **6**-64 to 91
Cesium: see also Elements
electrical resistivity, **12**-45 to 47
electron configuration, **1**-18 to 19
heat capacity, **4**-125
history, occurrence, uses, **4**-1 to 36
ionization energy, **10**-175 to 177
isotopes and their properties, **11**-50 to 158
magnetic susceptibility, **4**-133 to 138
molten, density, **4**-129 to 132
physical properties, **4**-124
thermal properties, **12**-197 to 198
thermodynamic properties, **5**-1 to 3
vapor pressure, **6**-64 to 91
CFC's, various properties, **6**-140 to 142
Characterization of materials, **12**-1 to 6
Charge
electron, **1**-1 to 11
fundamental particles, **11**-1 to 49
Chemical Abstracts Service nomenclature,
2-22
Chemical Abstracts Service Registry
Numbers: see CAS Registry
Numbers
Chemical carcinogens, **16**-53 to 57
Chemical composition of human body, **7**-11
Chemical kinetics
atmospheric reactions, **5**-106 to 120
conversion factors, **1**-46
Chemical nomenclature, **2**-22
Chemical shifts, NMR
for ^{13}C, **9**-96
for protons, **9**-95
of solvents for NMR, **8**-122
Chemicals, safe handling and disposal, **16**-1
to 15
Chi-square distribution, **A**-105 to 106
Chlorine: see also Elements
critical constants, **6**-48 to 60
electron configuration, **1**-18 to 19
heat capacity, **4**-125

history, occurrence, uses, **4**-1 to 36
ionization energy, **10**-175 to 177
isotopes and their properties, **11**-50 to 158
physical properties, **4**-124
thermodynamic properties, **5**-1 to 3
thermodynamic properties at high
    temperature, **5**-61 to 84
vapor pressure, **6**-64 to 91, **6**-92 to 102
Chlorofluorocarbon refrigerants, **6**-140 to 142
Chromatography, carrier gas properties,
    8-120
Chromium: see also Elements
electrical resistivity, **12**-45 to 47
electron configuration, **1**-18 to 19
heat capacity, **4**-125
history, occurrence, uses, **4**-1 to 36
ionization energy, **10**-175 to 177
isotopes and their properties, **11**-50 to 158
magnetic susceptibility, **4**-133 to 138
molten, density, **4**-129 to 132
physical properties, **4**-124
thermal properties, **12**-197 to 198
vapor pressure, **6**-64 to 91
Clausius-Mosotti equation, **12**-17 to 18
Cobalt: see also Elements
electrical resistivity, **12**-45 to 47
electron configuration, **1**-18 to 19
heat capacity, **4**-125
history, occurrence, uses, **4**-1 to 36
ionization energy, **10**-175 to 177
isotopes and their properties, **11**-50 to 158
magnetic susceptibility, **4**-133 to 138
molten, density, **4**-129 to 132
physical properties, **4**-124
thermal properties, **12**-197 to 198
vapor pressure, **6**-64 to 91
CODATA fundamental constants, **1**-1 to 11
CODATA Key Values for Thermodynamics,
    5-1 to 3
Coercivity, magnetic materials, **12**-117 to 125
Collision diameter of gases, **6**-46
Collision frequency, common gases, **6**-46
Combustion, heat of, **5**-89
Composition
atmosphere, **14**-2 to 3
earth's crust, **14**-14
glasses, **12**-206 to 209
human body, **7**-11
magnetic materials, **12**-117 to 125
planetary atmospheres, **14**-2 to 3
sea water, **14**-14
U.S. Standard Atmosphere, **14**-16 to 22
Compressibility
ice, **6**-6 to 7
liquids, **6**-136 to 137
semiconductors, **12**-97 to 107
sodium chloride solutions, **6**-8
Compton wavelength (electron, proton,
    neutron), **1**-1 to 11
Concentrative properties of aqueous
    solutions, **8**-57 to 83
Conductance: see Conductivity, electrical
Conductivity, electrical
calibration standards, **5**-91

earth's atmosphere, **14**-30 to 37
electrolyte solutions, **5**-94
hydrohalogen acids, **5**-92 to 93
ions, at infinite dilution, **5**-95 to 97
potassium chloride solutions, **5**-91
seawater, **14**-12 to 13
standard solutions, **5**-91
water, **5**-90
Conductivity, thermal: see Thermal
    conductivity
Constantan
thermal conductivity, **12**-201
wire tables, **15**-27
Construction materials
density, **15**-29
thermal conductivity, **12**-204 to 205
Conversion factors
chemical kinetics, **1**-46
electrical resistivity, **1**-45
energy, **1**-42
gas constant, **1**-49
general, **1**-29 to 40
ionizing radiation, **1**-47 to 48
pressure, **1**-43
SI units, **1**-29 to 40
temperature scales, **1**-41
temperature, to ITS-90, **1**-22 to 23
thermal conductivity, **1**-44
Conversion formulas for concentration of
    solutions, **8**-20
Copper: see also Elements
electrical resistivity, **12**-45 to 47
electron configuration, **1**-18 to 19
heat capacity, **4**-125, **12**-196
history, occurrence, uses, **4**-1 to 36
ionization energy, **10**-175 to 177
isotopes and their properties, **11**-50 to 158
magnetic susceptibility, **4**-133 to 138
molten, density, **4**-129 to 132
physical properties, **4**-124
resistance of wires, **15**-27
thermal conductivity, **12**-199 to 200
thermal properties, **12**-197 to 198
thermodynamic properties, **5**-1 to 3
vapor pressure, **6**-64 to 91
wire tables, **15**-27
Copper-constantan thermocouple tables, **15**-1
    to 13
Correlation charts for infrared spectra, **9**-87 to
    91
Cosecant function, **A**-7 to 10
Cosine function, **A**-7 to 10
Cosmic rays, **11**-176 to 180
Cotangent function, **A**-7 to 10
Coulomb, definition, **1**-24 to 28
Critical constants
cryogenic fluids, **6**-138
elements, **4**-124, **6**-48 to 60
H_2O and D_2O, **6**-4
halocarbons, **6**-140 to 142
inorganic compounds, **6**-48 to 60
organic compounds, **6**-48 to 60
Cross section, x-ray and gamma-ray, **10**-207
    to 211

Crust
composition, **14**-14
density, pressure, gravity, **14**-10
Cryogenic fluids
density, **6**-16 to 24, **6**-138
liquid helium properties, **6**-139
vapor pressure, **6**-92 to 102
various properties, **6**-138
Cryoscopic constant, **15**-21
Crystal elastic constants, **12**-37 to 44
Crystal ionic radii, **12**-14 to 16
Crystal lattice energy, **12**-22 to 35, **12**-36
Crystal structure
elements, **4**-148 to 158, **12**-19 to 21
inorganic compounds, **4**-37 to 100, **4**-148
    to 158
magnetic materials, **12**-117 to 125
minerals, **4**-141 to 147, **4**-148 to 158
rare earth elements, **4**-114 to 123
semiconductors, **12**-97 to 107
superconductors, **12**-64 to 90, **12**-91 to 93
Crystal symmetry, **12**-7 to 13
Curie temperature
antiferroelectric crystals, **12**-62
ferroelectric crystals, **12**-61 to 62
magnetic materials, **12**-117 to 125
rare earth elements, **4**-114 to 123
Curie, definition, **16**-48
Curium: see also Elements
electron configuration, **1**-18 to 19
history, occurrence, uses, **4**-1 to 36
isotopes and their properties, **11**-50 to 158

D

D_2O
boiling point, **6**-4
critical constants, **6**-4
density, **6**-9
dissociation constant, **8**-85
fixed point properties, **6**-4
ion product, **8**-85
thermal conductivity, **6**-4, **6**-185 to 186
triple point constants, **6**-4
vapor pressure, **6**-4
viscosity, **6**-178 to 179
Debye equation, **6**-15
Debye temperature
rare earth elements, **4**-114 to 123
semiconductors, **12**-97 to 107
Decay mode
fundamental particles, **11**-1 to 49
nuclides, **11**-50 to 158
Decinormal solutions
oxidation and reduction reagents, **8**-7
salts, acids, and bases, **8**-5 to 6
Definitions
abbreviations and acronyms, **2**-29 to 36
scientific terms, **2**-38 to 59
Density
air, **6**-1 to 2
aqueous solutions, **8**-57 to 83
atmosphere, as function of altitude, **14**-16
    to 22

carrier gases for chromatography, **8**-120
commercial metals and alloys, **12**-210
common fluids, as function of temperature and pressure, **6**-16 to 24
construction materials, **15**-29
cryogenic fluids, **6**-16 to 24, **6**-138
D_2O, **6**-4, **6**-9
earth, **14**-6 to 7
earth, as function of depth, **14**-10
elements, **4**-129 to 132
ice, **6**-6 to 7
inorganic compounds, **4**-37 to 100
liquid elements and salts, **4**-129 to 132
liquids, pressure and temperature dependence, **6**-136 to 137
mercury, **6**-143
metallic elements, **12**-197 to 198
minerals, **4**-141 to 147
miscellaneous materials, **15**-29
molten salts, **4**-129 to 132
organic compounds, **3**-1 to 330
planets, **14**-2 to 3
polymer melts, **13**-13 to 17
rare earth elements, **4**-114 to 123
rocks, **15**-29
satellites, **14**-4 to 5
seawater, **14**-12 to 13
semiconductors, **12**-97 to 107
sodium chloride solutions, **6**-8
solvents, **15**-14 to 18
steam, **6**-13 to 14
water, **6**-3, **6**-4, **6**-5, **6**-13 to 14
water, supercooled, **6**-6 to 7
wood, **15**-29
Depression of the freezing point, **8**-57 to 83, **15**-21, **15**-22
Derivatives, **A**-12 to 15
Detectors, infrared, **10**-216
Deuterium
solubility in water, **8**-86 to 89
viscosity, **6**-178 to 179
Dew point and relative humidity, **15**-23
Diamagnetic susceptibility
elements, **4**-133 to 138
inorganic compounds, **4**-133 to 138
organic compounds, **3**-764 to 767
Diamond
dielectric constant, **12**-52 to 60
optical properties, **12**-133 to 156
phase diagram, **12**-178 to 195
thermal conductivity, **12**-199 to 200
Diatomic molecules
bond lengths, **9**-81 to 86
bond strengths, **9**-51 to 74
electron affinity, **10**-147 to 159
force constants, **9**-76
polarizability, **10**-160 to 174
spectroscopic constants, **9**-81 to 86
vibrational frequencies, **9**-81 to 86
Dielectric constant
common fluids, as function of temperature and pressure, **6**-16 to 24
cryogenic fluids, temperature and pressure dependence, **6**-16 to 24

crystals, **12**-52 to 60
gases, **6**-172 to 173
glass, **12**-63
ice, **6**-6 to 7
liquids, **6**-149 to 171
plastics, **13**-12
quartz, **12**-63
rubbers, **13**-12
semiconductors, **12**-97 to 107
solids, **12**-52 to 60
solvents, **8**-121
vacuum, **1**-1 to 11
water, **6**-3
water, frequency dependence, **6**-15
Dielectric strength, insulating materials, **15**-30 to 35
Differential equations, **A**-65 to 75
Diffusion
gases, **6**-189 to 190
gases in water, **6**-191
in air, **6**-189 to 190
ions in solution, **5**-95 to 97
liquids, **6**-192
semiconductors, **12**-108 to 116
Dipole moment
electric, of molecules, **9**-44 to 50
magnetic, of nuclides, **11**-50 to 158
solvents, **15**-14 to 18
Discharges, electrical, in the atmosphere, **14**-30 to 37
Disposal of laboratory chemicals, **16**-1 to 15
Dissociation constant
acid-base indicators, **8**-16 to 18
amino acids, **7**-1 to 2
biological buffers, **7**-9
D_2O, **8**-85
inorganic acids and bases, **8**-44 to 45
inorganic salts in water, **8**-111 to 114
organic acids and bases, **8**-46 to 56
purine and pyrimidine bases, **7**-4
water, **8**-84, **8**-85
Dissociation energy of chemical bonds, **9**-51 to 74
DSC, definition, **12**-1 to 6
DTA, definition, **12**-1 to 6
Dubnium (element 105), **4**-1 to 36, **11**-50 to 158
Dysprosium: see also Elements
electrical resistivity, **12**-45 to 47
electron configuration, **1**-18 to 19
heat capacity, **4**-125
history, occurrence, uses, **4**-1 to 36
ionization energy, **10**-175 to 177
isotopes and their properties, **11**-50 to 158
magnetic susceptibility, **4**-133 to 138
molten, density, **4**-129 to 132
physical properties, **4**-124
thermal properties, **12**-197 to 198
vapor pressure, **6**-64 to 91

E

Earth
age, **14**-6 to 7

area of land and oceans, **14**-6 to 7
atmospheric composition, **14**-2 to 3
composition of crust, **14**-14
density as function of depth, **14**-10
dimensions, **14**-1, **14**-6 to 7
gravity in interior, **14**-10
mass and density, **14**-6 to 7
orbital and rotational parameters, **14**-2 to 3, **14**-6 to 7
pressure in interior, **14**-10
Ebullioscopic constant, **15**-20
ECR, definition, **12**-1 to 6
EELS, definition, **12**-1 to 6
Einsteinium: see also Elements
electron configuration, **1**-18 to 19
history, occurrence, uses, **4**-1 to 36
isotopes and their properties, **11**-50 to 158
Elastic constants, single crystal, **12**-37 to 44
Elastic modulus
commercial metals and alloys, **12**-210
rare earth elements, **4**-114 to 123
Elasto-optic constants, **12**-157 to 173
Electrical conductance: see Conductivity, electrical
Electrical resistivity: see Resistivity
Electro-optic constants, **12**-157 to 173
Electrochemical series, **8**-21 to 31
Electrode potential
general table, **8**-21 to 31
ion radicals, **8**-32 to 34
Electrolytes
activity coefficients, **5**-97 to 98, **5**-99 to 103
diffusion of ions, **5**-95 to 97
electrical conductivity, **5**-94, **5**-95 to 97
enthalpy of solution, **5**-105
freezing point lowering, **8**-57 to 83, **15**-22
Electromagnetic radiation, classification of bands, **10**-212 to 213
Electron
charge, **1**-1 to 11
in cosmic ray showers, **11**-176 to 180
magnetic moment, **1**-1 to 11, **11**-1 to 49
range in various materials, **16**-48
Electron affinity, atoms and molecules, **10**-147 to 159
Electron binding energy of the elements, **10**-200 to 205
Electron configuration
neutral atoms, **1**-18 to 19
rare earth elements, **4**-114 to 123
Electron volt, **1**-1 to 11, **1**-24 to 28
Electron work function of the elements, **12**-130
Electronegativity values, **9**-75
Electrons, secondary, emission by metals, **12**-131 to 132
Elementary charge, **1**-1 to 11
Elements
abundance of isotopes, **1**-15 to 17, **11**-50 to 158
atomic mass, **1**-15 to 17, **11**-50 to 158
atomic spectrum, **10**-1 to 87
atomic transition probability, **10**-88 to 146

atomic weight, **1**-12 to 14
boiling point, **4**-124
critical temperature, **4**-124
crystal ionic radii, **12**-14 to 16
crystal structure, **4**-148 to 158, **12**-19 to 21
density, **4**-129 to 132
electrical resistivity, **12**-45 to 47
electron affinity, **10**-147 to 159
electron binding energy, **10**-200 to 205
electron configuration, **1**-18 to 19
electronegativity, **9**-75
enthalpy of fusion, **6**-121 to 135
enthalpy of vaporization, **6**-105 to 120
gamma-ray cross sections, **10**-207 to 211
gamma-ray emission, **11**-50 to 158
general information, **4**-1 to 36
heat capacity, **4**-125
historical information, **4**-1 to 36
in seawater, **14**-14
in the earth's crust, **14**-14
in the human body, **7**-11
line spectrum, **10**-1 to 87
magnetic susceptibility, **4**-133 to 138
melting point, **4**-124
periodic table, inside front cover
photon attenuation coefficients, **10**-207 to 211
polarizability, **10**-160 to 174
radii of ions, **12**-14 to 16
reference states, **5**-4 to 60
semiconducting properties, **12**-97 to 107
superconducting properties, **12**-64 to 90
thermal conductivity, **6**-185 to 186, **12**-197 to 198, **12**-199 to 200
thermal properties, **12**-197 to 198
thermodynamic properties, **5**-4 to 60
transition probabilities, **10**-88 to 146
vapor pressure, **6**-64 to 91
vapor pressure at high temperature, **4**-126 to 128
work function, **12**-130
x-ray atomic energy levels, **10**-196 to 199
x-ray cross sections, **10**-207 to 211
Elevation of the boiling point, **15**-20
Emission, secondary electrons, **12**-131 to 132
Emissivity, rare earth metals, **4**-114 to 123
ENDOR, definition, **12**-1 to 6
Energy
conversion factors, **1**-42
crystal lattice, **12**-22 to 35, **12**-36
spectrum of cosmic rays, **11**-176 to 180
Energy, activation
chemical reactions, **5**-106 to 120
diffusion in semiconductors, **12**-108 to 116
Energy gap
semiconductors, **12**-97 to 107
superconductors, **12**-91 to 93
Energy levels, x-ray, **10**-196 to 199
Energy states of solids, **12**-1 to 6
Enthalpy
air, **6**-1 to 2

common fluids, as function of temperature and pressure, **6**-16 to 24
crystal lattices, **12**-22 to 35
steam, **6**-13 to 14
water, **6**-13 to 14
Enthalpy of combustion, **5**-89
Enthalpy of dilution, acids, **5**-104
Enthalpy of formation
aqueous systems, **5**-85 to 88
CODATA Key Values, **5**-1 to 3
free radicals, **9**-51 to 74
gaseous atoms, **9**-51 to 74
high temperature, **5**-61 to 84
inorganic compounds, **5**-4 to 60
ions, **10**-178 to 195
organic compounds, **5**-4 to 60
semiconductors, **12**-97 to 107
standard state values, **5**-4 to 60
Enthalpy of fusion
cryogenic fluids, **6**-138
elements, **6**-121 to 135
ice, **6**-6 to 7
inorganic compounds, **6**-121 to 135
metals, **12**-197 to 198
organic compounds, **6**-121 to 135
rare earth elements, **4**-114 to 123
Enthalpy of solution, electrolytes, **5**-105
Enthalpy of vaporization
cryogenic fluids, **6**-138
elements, **6**-105 to 120
ice, **6**-6 to 7
inorganic compounds, **6**-105 to 120
organic compounds, **6**-105 to 120
rare earth elements, **4**-114 to 123
water, **6**-3
Entropy
air, **6**-1 to 2
aqueous systems, **5**-85 to 88
CODATA Key Values, **5**-1 to 3
common fluids, as function of temperature and pressure, **6**-16 to 24
high temperature, **5**-61 to 84
rare earth elements, **4**-114 to 123
standard state values, **5**-4 to 60
steam, **6**-13 to 14
water, **6**-13 to 14
EPMA, definition, **12**-1 to 6
EPR, definition, **12**-1 to 6
Equation of state
Tait (for polymer melts), **13**-13 to 17
van der Waals, **6**-45
virial, **6**-25 to 44
Equilibrium constant of formation, **5**-61 to 84
Equivalent conductance: see Conductivity, electrical
Erbium: see also Elements
electrical resistivity, **12**-45 to 47
electron configuration, **1**-18 to 19
heat capacity, **4**-125
history, occurrence, uses, **4**-1 to 36
ionization energy, **10**-175 to 177
isotopes and their properties, **11**-50 to 158
magnetic susceptibility, **4**-133 to 138

molten, density, **4**-129 to 132
physical properties, **4**-124
thermal properties, **12**-197 to 198
vapor pressure, **6**-64 to 91
Error function, **A**-101 to 102
ESCA, definition, **12**-1 to 6
ESD, definition, **12**-1 to 6
Euler's constant, value of, **A**-1
Europium: see also Elements
electrical resistivity, **12**-45 to 47
electron configuration, **1**-18 to 19
heat capacity, **4**-125
history, occurrence, uses, **4**-1 to 36
ionization energy, **10**-175 to 177
isotopes and their properties, **11**-50 to 158
magnetic susceptibility, **4**-133 to 138
molten, density, **4**-129 to 132
physical properties, **4**-124
thermal properties, **12**-197 to 198
vapor pressure, **6**-64 to 91
EXAFS, definition, **12**-1 to 6
EXELFS, definition, **12**-1 to 6
Expansion coefficient
commercial metals and alloys, **12**-210
liquid helium, **6**-139
metals, **12**-197 to 198
rare earth elements, **4**-114 to 123
semiconductors, **12**-97 to 107
sodium chloride solutions, **6**-8
Explosion hazards of laboratory chemicals, **16**-1 to 15
Explosive limits, **16**-16 to 31
Exponential function, **A**-2 to 6
Exponential series, **A**-84 to 87
Exposure limits
airborne contaminants, **16**-32 to 42
ionizing radiation, **16**-48
radionuclides, **16**-49 to 52
Extinction coefficient, in metals and semiconductors, **12**-133 to 156

F

Fahrenheit temperature, conversion to other scales, **1**-41
Farad, definition, **1**-24 to 28
Faraday constant, **1**-1 to 11
Faraday effect, **12**-157 to 173
Fatty acids, physical properties, **7**-6
F-distribution, **A**-106 to 109
Fehling's solution, preparation, **8**-1 to 4
FEM, definition, **12**-1 to 6
Fermium: see also Elements
electron configuration, **1**-18 to 19
history, occurrence, uses, **4**-1 to 36
isotopes and their properties, **11**-50 to 158
Ferrimagnetic materials
Faraday rotation, **12**-157 to 173
organic, **12**-126 to 129
Ferrites, magnetic properties, **12**-117 to 125
Ferroelectric crystals
Curie temperature, **12**-61 to 62
Kerr constants, **12**-157 to 173

Ferromagnetic materials
    Faraday rotation, **12**-157 to 173
    organic, **12**-126 to 129
    various properties, **12**-117 to 125
Ferromagnetic moment, rare earth elements, **4**-114 to 123
FIM, definition, **12**-1 to 6
Fine structure constant, **1**-1 to 11
First radiation constant, **1**-1 to 11
Fixed point properties
    cryogenic fluids, **6**-138
    water and heavy water, **6**-4
Flammability
    chemical substances, general, **16**-16 to 31
    laboratory chemicals, **16**-1 to 15
    organic solvents, **15**-14 to 18
Flash point: see Flammability
Flattening factor for the earth, **14**-1
Fluorine: see also Elements
    critical constants, **6**-48 to 60
    electron configuration, **1**-18 to 19
    heat capacity, **4**-125
    history, occurrence, uses, **4**-1 to 36
    ionization energy, **10**-175 to 177
    isotopes and their properties, **11**-50 to 158
    physical properties, **4**-124
    thermodynamic properties, **5**-1 to 3
    thermodynamic properties at high temperature, **5**-61 to 84
    vapor pressure, **6**-64 to 91, **6**-92 to 102
Fluorocarbon refrigerants, **6**-140 to 142
Foods, pH, **7**-10
Force constants for bond stretching, **9**-76
Formation, heat of: see Enthalpy of formation
Formula index of organic compounds, **3**-631 to 708
Fossils, age of, **14**-6 to 7
Fourier series, **A**-75 to 79
Fourier transforms, **A**-80 to 84
Francium: see also Elements
    electron configuration, **1**-18 to 19
    history, occurrence, uses, **4**-1 to 36
    ionization energy, **10**-175 to 177
    isotopes and their properties, **11**-50 to 158
    vapor pressure, **6**-64 to 91
Free energy: see Thermodynamic properties
Free radicals
    dipole moment, **9**-44 to 50
    enthalpy of formation, **9**-51 to 74
    ionization energy, **10**-178 to 195
    reaction rates, **5**-106 to 120
    vibrational frequencies, **9**-77 to 80
Freezing point: see also Melting point
    depression of, **8**-57 to 83, **15**-21
    pressure dependence, **6**-47
    seawater, **14**-12 to 13
Freezing point depression
    aqueous solutions, **8**-57 to 83
    cryoscopic constants for various liquids, **15**-21
    electrolytes, **15**-22
Frequency
    electromagnetic radiation bands, **10**-212 to 213

human hearing range, **14**-44 to 45
musical scales, **14**-42 to 43
NMR resonances, **9**-92 to 94
standards, infrared, **10**-228 to 235, **10**-236 to 240
FTIR, definition, **12**-1 to 6
Fundamental constants, **1**-1 to 11
Fundamental particles, **11**-1 to 49
Fundamental vibrations of molecules, **9**-77 to 80
Fusion: see Enthalpy of fusion

G

g-Factor of the electron, **1**-1 to 11
Gadolinium: see also Elements
    electrical resistivity, **12**-45 to 47
    electron configuration, **1**-18 to 19
    heat capacity, **4**-125
    history, occurrence, uses, **4**-1 to 36
    ionization energy, **10**-175 to 177
    isotopes and their properties, **11**-50 to 158
    magnetic susceptibility, **4**-133 to 138
    molten, density, **4**-129 to 132
    physical properties, **4**-124
    thermal properties, **12**-197 to 198
    vapor pressure, **6**-64 to 91
Gallium: see also Elements
    electrical resistivity, **12**-45 to 47
    electron configuration, **1**-18 to 19
    heat capacity, **4**-125
    history, occurrence, uses, **4**-1 to 36
    ionization energy, **10**-175 to 177
    isotopes and their properties, **11**-50 to 158
    magnetic susceptibility, **4**-133 to 138
    molten, density, **4**-129 to 132
    physical properties, **4**-124
    thermal properties, **12**-197 to 198
    vapor pressure, **6**-64 to 91
Gamma function, **A**-99 to 101
Gamma rays
    cross sections, for the elements, **10**-207 to 211
    energy, of nuclides, **11**-50 to 158
    in cosmic ray showers, **11**-176 to 180
    photon attenuation coefficients, **10**-207 to 211
    protection against, **16**-48
Gas chromatography, carrier gas properties, **8**-120
Gas constant, **1**-1 to 11, **1**-49
Gases
    average velocity, **6**-46
    breakdown voltage, **15**-30 to 35
    collision diameter, **6**-46
    dielectric constant, **6**-172 to 173
    diffusion, **6**-189 to 190
    diffusion in water, **6**-191
    dipole moment, **9**-44 to 50
    mean free path, **6**-46
    permittivity, **6**-172 to 173
    solubility in water, **8**-86 to 89
    speed of sound in, **14**-38 to 39
    thermal conductivity, **6**-185 to 186

threshold limits, **16**-32 to 42
van der Waals constants, **6**-45
Verdet constants, **12**-157 to 173
virial coefficients, **6**-25 to 44
viscosity, **6**-178 to 179
Gauges, of wires, **15**-27
Gaussian gravitational constant, **14**-1
GDMS, definition, **12**-1 to 6
Genetic code, **7**-5
Geological time scale, **14**-8
Geophysical constants, **14**-6 to 7
Germanium: see also Elements
    dielectric constant, **12**-52 to 60
    electron configuration, **1**-18 to 19
    heat capacity, **4**-125, **12**-196
    history, occurrence, uses, **4**-1 to 36
    ionization energy, **10**-175 to 177
    isotopes and their properties, **11**-50 to 158
    magnetic susceptibility, **4**-133 to 138
    molten, density, **4**-129 to 132
    optical properties, **12**-133 to 156
    physical properties, **4**-124
    semiconducting properties, **12**-97 to 107
    thermal conductivity, **12**-199 to 200
    thermodynamic properties, **5**-1 to 3
    vapor pressure, **6**-64 to 91
Gibbs energy of formation
    aqueous systems, **5**-85 to 88
    high temperature, **5**-61 to 84
    standard state values, **5**-4 to 60
Glass transition temperature of polymers, **13**-4 to 11
Glasses
    composition, **12**-206 to 209
    density, **15**-29
    dielectric constant, **12**-63
    index of refraction, **10**-217
    loss factor, **12**-63
    resistivity, **12**-63
    speed of sound in, **14**-38 to 39
    thermal conductivity, **12**-202 to 203, **12**-206 to 209
    transmittance, **10**-217
    Verdet constants, **12**-157 to 173
Global warming
    atmospheric carbon dioxide concentration, **14**-25 to 26
    mean temperatures, global, **14**-29
    mean temperatures, U. S., **14**-27 to 28
Gloves, resistance to chemicals, **16**-1 to 15
Glucose, aqueous solution properties, **8**-57 to 83
Gold: see also Elements
    electrical resistivity, **12**-45 to 47
    electron configuration, **1**-18 to 19
    heat capacity, **4**-125, **12**-196
    history, occurrence, uses, **4**-1 to 36
    ionization energy, **10**-175 to 177
    isotopes and their properties, **11**-50 to 158
    magnetic susceptibility, **4**-133 to 138
    molten, density, **4**-129 to 132
    physical properties, **4**-124
    thermal conductivity, **12**-199 to 200
    thermal properties, **12**-197 to 198

vapor pressure, **6**-64 to 91
Graphite
    heat capacity, **12**-196
    heat of combustion, **5**-89
    phase diagram, **12**-178 to 195
    sublimation pressure, **6**-64 to 91
    thermal conductivity, **12**-199 to 200
Gravitational constant, **1**-1 to 11, **14**-1
Gravitational potential, **14**-2 to 3
Gravity, acceleration of
    at poles and equator, **14**-6 to 7
    at various latitudes, **14**-9
    in interior of earth, **14**-10
    standard value, **1**-1 to 11
Gray, definition, **1**-24 to 28, **16**-48
Greek alphabet, **2**-37

H

Hafnium: see also Elements
    electrical resistivity, **12**-45 to 47
    electron configuration, **1**-18 to 19
    heat capacity, **4**-125
    history, occurrence, uses, **4**-1 to 36
    ionization energy, **10**-175 to 177
    isotopes and their properties, **11**-50 to 158
    magnetic susceptibility, **4**-133 to 138
    molten, density, **4**-129 to 132
    physical properties, **4**-124
    thermal properties, **12**-197 to 198
    vapor pressure, **6**-64 to 91
Hall coefficient, rare earth elements, **4**-114 to 123
Hall density, superconductors, **12**-91 to 93
Hall resistance, quantized, **1**-1 to 11
Halocarbon refrigerants, **6**-140 to 142
Hardness
    ceramics, **12**-211 to 212
    minerals, **4**-141 to 147, **12**-211 to 212
    scales, comparison, **12**-211 to 212
    semiconductors, **12**-97 to 107
    various materials, **12**-211 to 212
Hartree energy, **1**-1 to 11
Hassium (element 108), **4**-1 to 36, **11**-50 to 158
Hazardous chemicals, handling and disposal, **16**-1 to 15
Hearing, characteristics, **14**-44 to 45
Heat capacity
    air, **6**-1 to 2
    aqueous systems, **5**-85 to 88
    carrier gases for chromatography, **8**-120
    common fluids, as function of temperature and pressure, **6**-16 to 24
    cryogenic fluids, **6**-138
    elements, **4**-125
    high temperature, **5**-61 to 84
    ice, **6**-6 to 7
    liquid helium, **6**-139
    mercury, **6**-144
    metals, **12**-196, **12**-197 to 198
    rare earth elements, **4**-114 to 123
    seawater, **14**-12 to 13
    semiconductors, **12**-97 to 107

    solids, **12**-196
    solvents, **15**-14 to 18
    standard state values, **5**-4 to 60
    steam, **6**-13 to 14
    water, **6**-3
Heat conductivity: see Thermal conductivity
Heat of combustion, **5**-89
Heat of dilution: see Enthalpy of dilution
Heat of formation: see Enthalpy of formation
Heat of fusion: see Enthalpy of fusion
Heat of solution: see Enthalpy of solution
Heat of vaporization: see Enthalpy of vaporization
Hebrew alphabet, **2**-37
Helium: see also Elements
    critical constants, **6**-48 to 60
    cryogenic properties, **6**-138, **6**-139
    density, **6**-139
    electron configuration, **1**-18 to 19
    enthalpy of vaporization, **6**-139
    entropy, **5**-1 to 3
    expansion coefficient, **6**-139
    heat capacity, **6**-139
    history, occurrence, uses, **4**-1 to 36
    ionization energy, **10**-175 to 177
    isotopes and their properties, **11**-50 to 158
    liquid properties, **6**-139
    magnetic susceptibility, **4**-133 to 138
    mean free path, **6**-46
    permittivity (dielectric constant), **6**-139, **6**-149 to 171, **6**-172 to 173
    physical properties, **4**-124
    solubility in water, **8**-86 to 89
    surface tension, **6**-139
    thermal conductivity (gas), **6**-185 to 186
    thermal conductivity (liquid), **6**-139
    thermal conductivity (two-phase), **6**-16 to 24
    thermodynamic properties, **6**-16 to 24
    van der Waals constants, **6**-45
    vapor pressure, **6**-64 to 91, **6**-92 to 102, **6**-139
    viscosity (gas), **6**-178 to 179
    viscosity (liquid), **6**-139
    viscosity (two-phase), **6**-16 to 24
Henry, definition, **1**-24 to 28
Henry's Law constant, **8**-91 to 101
Hermite polynomials, A-102 to 104
Hertz, definition, **1**-24 to 28
Holmium: see also Elements
    electrical resistivity, **12**-45 to 47
    electron configuration, **1**-18 to 19
    heat capacity, **4**-125
    history, occurrence, uses, **4**-1 to 36
    ionization energy, **10**-175 to 177
    isotopes and their properties, **11**-50 to 158
    magnetic susceptibility, **4**-133 to 138
    molten, density, **4**-129 to 132
    physical properties, **4**-124
    thermal properties, **12**-197 to 198
    vapor pressure, **6**-64 to 91
Human body
    chemical composition, **7**-11
    pH of fluids, **7**-10

Humidity, relative
    from wet and dry bulb temperatures, **15**-24
    relation to dew point, **15**-23
    solutions for calibration, **15**-26
    solutions for constant humidity, **15**-25
Hydrocarbons
    heat of combustion, **5**-89
    thermophysical properties, **6**-16 to 24
Hydrogen: see also Elements
    critical constants, **6**-48 to 60
    cryogenic properties, **6**-138
    electron configuration, **1**-18 to 19
    enthalpy of fusion, **6**-121 to 135
    enthalpy of vaporization, **6**-105 to 120
    heat of combustion, **5**-89
    history, occurrence, uses, **4**-1 to 36
    ionization energy, **10**-175 to 177
    isotopes and their properties, **11**-50 to 158
    magnetic susceptibility, **4**-133 to 138
    mean free path, **6**-46
    permittivity (dielectric constant), **6**-149 to 171, **6**-172 to 173
    physical properties, **4**-124
    solubility in water, **8**-86 to 89
    thermal conductivity, **6**-185 to 186
    thermodynamic properties, **5**-1 to 3, **6**-16 to 24
    thermodynamic properties at high temperature, **5**-61 to 84
    van der Waals constants, **6**-45
    vapor pressure, **6**-64 to 91, **6**-92 to 102
    viscosity, **6**-178 to 179
Hydrosphere, mass of, **14**-6 to 7
Hyperbolic functions, A-2 to 6, A-11
Hysteresis, in magnetic materials, **12**-117 to 125

I

Ice
    compressibility, **6**-6 to 7
    crystal structure, **4**-148 to 158
    density, **6**-6 to 7
    dielectric constant, **6**-6 to 7, **12**-52 to 60
    heat capacity, **6**-6 to 7
    melting point, pressure dependence, **6**-12, **6**-47
    phase diagram, **12**-178 to 195
    phase transitions, **6**-6 to 7, **6**-12
    thermal conductivity, **6**-6 to 7, **12**-202 to 203
    thermal expansion coefficient, **6**-6 to 7
    vapor pressure, **6**-9
ICPMS, definition, **12**-1 to 6
Ignition temperature
    chemical substances, general, **16**-16 to 31
    laboratory chemicals, **16**-1 to 15
    solvents, **15**-14 to 18
Index of refraction
    air, **10**-220
    aqueous solutions, **8**-57 to 83
    glass, **10**-217
    inorganic liquids, **4**-139 to 140

liquids, for calibration, **10**-219
metals, **12**-133 to 156
minerals, **4**-141 to 147
organic compounds, **3**-1 to 330
semiconductors, **12**-97 to 107, **12**-133 to 156
water, **10**-218
Indicators
  acid-base, **8**-16 to 18
  fluorescent, **8**-19 to 20
  pH, **8**-16 to 18, **8**-19 to 20
  pK, **8**-16 to 18
  preparation, **8**-1 to 4, **8**-16 to 18
Indium: see also Elements
  electrical resistivity, **12**-45 to 47
  electron configuration, **1**-18 to 19
  heat capacity, **4**-125
  history, occurrence, uses, **4**-1 to 36
  ionization energy, **10**-175 to 177
  isotopes and their properties, **11**-50 to 158
  magnetic susceptibility, **4**-133 to 138
  molten, density, **4**-129 to 132
  physical properties, **4**-124
  thermal properties, **12**-197 to 198
  vapor pressure, **6**-64 to 91
Infrared detectors, **10**-216
Infrared spectrum
  carbon dioxide laser, **10**-228 to 235
  characteristic group frequencies, **9**-87 to 91
  correlation charts, **9**-87 to 91
  earth's atmosphere, **14**-24
  frequency standards, **10**-236 to 240
  vibrational frequencies of molecules, **9**-77 to 80
Inorganic compounds
  activity coefficients, **5**-97 to 98, **5**-99 to 103
  boiling point, **6**-105 to 120
  bond lengths and angles, **9**-17 to 43
  characteristic infrared frequencies, **9**-87 to 91
  crystal lattice energy, **12**-22 to 35
  crystal structure, **4**-37 to 100, **4**-148 to 158
  dielectric constant, **12**-52 to 60
  dipole moment, **9**-44 to 50
  dissociation constant in water, **8**-111 to 114
  electrical conductivity, **5**-94
  enthalpy of formation, **5**-4 to 60
  enthalpy of fusion, **6**-121 to 135
  enthalpy of solution, **5**-105
  enthalpy of vaporization, **6**-105 to 120
  entropy, **5**-4 to 60
  Gibbs energy of formation, **5**-4 to 60
  heat capacity, **5**-4 to 60
  index of refraction, **4**-139 to 140
  magnetic susceptibility, **4**-133 to 138
  nomenclature, **2**-23 to 24
  permittivity, **12**-52 to 60
  physical properties, **4**-37 to 100
  polarizability, **10**-160 to 174
  reagents for determination, **8**-8 to 15

solubility as a function of temperature, **8**-102 to 110
solubility product constant, **8**-111 to 114
solubility, qualitative rules, **8**-116 to 118
standard thermodynamic properties, **5**-4 to 60
surface tension, **6**-145 to 148
INS, definition, **12**-1 to 6
Insulation, thermal conductivity of, **12**-204 to 205
Insulators, breakdown voltage, **15**-30 to 35
Integral tables, **A**-20 to 65
Integration, methods and techniques, **A**-15 to 19
Interatomic distances
  diatomic molecules, **9**-81 to 86
  gas-phase molecules, **9**-17 to 43
  organic crystals, **9**-1 to 14
  organometallic compounds, **9**-15 to 16
International System of Units (SI), **1**-24 to 28
International Temperature Scale (ITS-90)
  conversion from IPTS-68 and IPTS-48, **1**-22 to 23
  definition and fixed points, **1**-20 to 21
International Union of Pure and Applied Chemistry: see IUPAC
Iodine: see also Elements
  critical constants, **6**-48 to 60
  dielectric constant, **12**-52 to 60
  electron configuration, **1**-18 to 19
  heat capacity, **4**-125
  history, occurrence, uses, **4**-1 to 36
  ionization energy, **10**-175 to 177
  isotopes and their properties, **11**-50 to 158
  magnetic susceptibility, **4**-133 to 138
  physical properties, **4**-124
  thermodynamic properties, **5**-1 to 3
  thermodynamic properties at high temperature, **5**-61 to 84
  vapor pressure, **6**-64 to 91
Ion product
  D_2O, **8**-85
  water, **8**-84, **8**-85
Ionic radii
  in crystals, **12**-14 to 16
  rare earth elements, **4**-114 to 123
Ionization constant
  D_2O, **8**-85
  inorganic acids and bases, **8**-44 to 45
  inorganic compounds in water, **8**-111 to 114
  organic acids and bases, **8**-46 to 56
  water, **8**-84, **8**-85
Ionization energy
  atoms and ions, **10**-175 to 177
  molecules, **10**-178 to 195
  rare earth elements, **4**-114 to 123
Ionization potential: See Ionization energy
Ions
  aerosol, **14**-30 to 37
  diffusion in aqueous solutions, **5**-95 to 97
  electrical conductivity in aqueous solutions, **5**-95 to 97
  enthalpy of formation, **10**-178 to 195

heat capacity, aqueous solutions, **5**-85 to 88
in the atmosphere, **14**-30 to 37
magnetic properties, **12**-117 to 125
nomenclature, **2**-23 to 24
polarizability, **12**-17 to 18
radii, in crystals, **12**-14 to 16
thermodynamic properties, aqueous solutions, **5**-85 to 88
Iridium: see also Elements
  electrical resistivity, **12**-45 to 47
  electron configuration, **1**-18 to 19
  heat capacity, **4**-125
  history, occurrence, uses, **4**-1 to 36
  ionization energy, **10**-175 to 177
  isotopes and their properties, **11**-50 to 158
  magnetic susceptibility, **4**-133 to 138
  molten, density, **4**-129 to 132
  physical properties, **4**-124
  thermal properties, **12**-197 to 198
  vapor pressure, **6**-64 to 91
Iron: see also Elements
  electrical resistivity, **12**-45 to 47
  electron configuration, **1**-18 to 19
  heat capacity, **4**-125, **12**-196
  history, occurrence, uses, **4**-1 to 36
  ionization energy, **10**-175 to 177
  isotopes and their properties, **11**-50 to 158
  magnetic susceptibility, **4**-133 to 138
  molten, density, **4**-129 to 132
  physical properties, **4**-124
  thermal conductivity, **12**-199 to 200
  thermal properties, **12**-197 to 198
  vapor pressure, **6**-64 to 91
Iron-constantan thermocouple tables, **15**-1 to 13
Irradiance of the sun, **14**-15
Isoelectric point, amino acids, **7**-1 to 2
Isotopes, summary of properties, **11**-50 to 158
Isotopic abundance, **1**-15 to 17, **11**-50 to 158
ITS-90
  conversion from IPTS-68 and IPTS-48, **1**-22 to 23
  definition and fixed points, **1**-20 to 21
IUPAC
  atomic weights, **1**-12 to 14
  nomenclature for carbohydrates, **7**-7 to 8
  nomenclature for organic substituent groups, **2**-25 to 28
  nomenclature for polymers, **13**-1 to 2
  symbols for physical quantities, **2**-1 to 21
  vapor pressure calibration data, **6**-104
Jacobi polynomials, **A**-102 to 104
Josephson ratio, **1**-1 to 11
Joule, definition, **1**-24 to 28
Jupiter, orbital data and dimensions, **14**-2 to 3

K

Kelvin, definition, **1**-24 to 28
Kerr constants, **12**-157 to 173
Kilogram, definition, **1**-24 to 28
Kinetics
  atmospheric reactions, **5**-106 to 120

conversion factors, 1-46

Krypton: see also Elements
   critical constants, 6-48 to 60
   cryogenic properties, 6-138
   electron configuration, 1-18 to 19
   entropy, 5-1 to 3
   history, occurrence, uses, 4-1 to 36
   ionization energy, 10-175 to 177
   isotopes and their properties, 11-50 to 158
   magnetic susceptibility, 4-133 to 138
   mean free path, 6-46
   permittivity (dielectric constant), 6-149 to
       171, 6-172 to 173
   physical properties, 4-124
   solubility in water, 8-86 to 89
   thermal conductivity, 6-185 to 186
   van der Waals constants, 6-45
   vapor pressure, 6-64 to 91, 6-92 to 102
   viscosity, 6-178 to 179

L

Laboratory reagents, preparation of, 8-1 to 4
Laguerre polynomials, A-102 to 104
Lanthanum: see also Elements
   electrical resistivity, 12-45 to 47
   electron configuration, 1-18 to 19
   heat capacity, 4-125
   history, occurrence, uses, 4-1 to 36
   ionization energy, 10-175 to 177
   isotopes and their properties, 11-50 to 158
   magnetic susceptibility, 4-133 to 138
   molten, density, 4-129 to 132
   physical properties, 4-124
   thermal properties, 12-197 to 198
   vapor pressure, 6-64 to 91
Lasers
   characteristics of various types, 10-221 to
       227
   infrared, frequencies, 10-228 to 235
Lattice constants
   elements, 4-148 to 158, 12-19 to 21
   inorganic compounds, 4-148 to 158
   minerals, 4-148 to 158
   rare earth elements, 4-114 to 123
   semiconductors, 12-97 to 107
Lattice energy, 12-22 to 35, 12-36
Lawrencium: see also Elements
   electron configuration, 1-18 to 19
   history, occurrence, uses, 4-1 to 36
   isotopes and their properties, 11-50 to 158
Lead: see also Elements
   electrical resistivity, 12-45 to 47
   electron configuration, 1-18 to 19
   heat capacity, 4-125, 12-196
   history, occurrence, uses, 4-1 to 36
   ionization energy, 10-175 to 177
   isotopes and their properties, 11-50 to 158
   magnetic susceptibility, 4-133 to 138
   molten, density, 4-129 to 132
   physical properties, 4-124
   thermal conductivity, 12-199 to 200
   thermal properties, 12-197 to 198
   thermodynamic properties, 5-1 to 3

vapor pressure, 6-64 to 91
LEED, definition, 12-1 to 6
Legendre polynomials, A-102 to 104
Leptons, summary of properties, 11-1 to 49
Lifetime
   fundamental particles, 11-1 to 49
   in laser systems, 10-221 to 227
   nuclides, 11-50 to 158
Light, speed of, 1-1 to 11
Lightning, 14-30 to 37
LIMS, definition, 12-1 to 6
Line spectra of the elements, 10-1 to 87
Line strengths in atomic spectra, 10-88 to 146
Line width, x-ray lines, 10-206
Liquid air, thermodynamic properties, 6-1 to
   2
Liquid helium properties, 6-139
Liquids
   breakdown voltage, 15-30 to 35
   dielectric constant, 6-149 to 171
   diffusion, 6-192
   flammability, 15-14 to 18
   index of refraction, 4-139 to 140, 10-219
   Kerr constants, 12-157 to 173
   permittivity, 6-149 to 171
   speed of sound in, 14-38 to 39
   surface tension, 6-145 to 148
   thermal conductivity, 6-187 to 188
   Verdet constants, 12-157 to 173
   viscosity, 6-180 to 184
Liter, definition, 1-24 to 28
Lithium: see also Elements
   electrical resistivity, 12-45 to 47
   electron configuration, 1-18 to 19
   heat capacity, 4-125
   history, occurrence, uses, 4-1 to 36
   ionization energy, 10-175 to 177
   isotopes and their properties, 11-50 to 158
   magnetic susceptibility, 4-133 to 138
   molten, density, 4-129 to 132
   physical properties, 4-124
   thermal properties, 12-197 to 198
   thermodynamic properties, 5-1 to 3
   vapor pressure, 6-64 to 91
Lithosphere, mass of, 14-6 to 7
Logarithmic series, A-84 to 87
Log P, 16-43 to 47
Loss factor, glasses, 12-63
Loss tangent, 6-15
Loudness level, definition, 14-44 to 45
Low temperature liquid baths, 15-26
Lumen, definition, 1-24 to 28
Lutetium: see also Elements
   electrical resistivity, 12-45 to 47
   electron configuration, 1-18 to 19
   heat capacity, 4-125
   history, occurrence, uses, 4-1 to 36
   ionization energy, 10-175 to 177
   isotopes and their properties, 11-50 to 158
   magnetic susceptibility, 4-133 to 138
   molten, density, 4-129 to 132
   physical properties, 4-124
   thermal properties, 12-197 to 198
   vapor pressure, 6-64 to 91

Lux, definition, 1-24 to 28

M

Maclaurin series, A-84 to 87
Madelung constant, 12-36
Magnesium: see also Elements
   electrical resistivity, 12-45 to 47
   electron configuration, 1-18 to 19
   heat capacity, 4-125
   history, occurrence, uses, 4-1 to 36
   ionization energy, 10-175 to 177
   isotopes and their properties, 11-50 to 158
   magnetic susceptibility, 4-133 to 138
   molten, density, 4-129 to 132
   thermal conductivity, 12-199 to 200
   thermal properties, 12-197 to 198
   thermodynamic properties, 5-1 to 3
   vapor pressure, 6-64 to 91
Magnetic materials, composition and
   properties, 12-117 to 125
Magnetic moment
   electron, 1-1 to 11, 11-1 to 49
   fundamental particles, 11-1 to 49
   NMR, for important nuclei, 9-92 to 94
   nuclides, 11-50 to 158
   proton, 1-1 to 11
   rare earth elements, 4-114 to 123
Magnetic properties
   alloys, 12-117 to 125
   organic magnets, 12-126 to 129
   rare earth elements, 4-114 to 123
   superconductors, 12-64 to 90, 12-91 to 93
Magnetic susceptibility
   elements, 4-133 to 138
   inorganic compounds, 4-133 to 138
   organic compounds, 3-764 to 767
   rare earth elements, 4-114 to 123
   semiconductors, 12-97 to 107
   various materials, 12-117 to 125
Magnetism, symbols and units, 12-117 to 125
Magneton (nuclear, Bohr), 1-1 to 11
Magneto-optic constants, 12-157 to 173
Magnetostriction, 12-117 to 125
Manganese: see also Elements
   electrical resistivity, 12-45 to 47
   electron configuration, 1-18 to 19
   heat capacity, 4-125
   history, occurrence, uses, 4-1 to 36
   ionization energy, 10-175 to 177
   isotopes and their properties, 11-50 to 158
   magnetic susceptibility, 4-133 to 138
   molten, density, 4-129 to 132
   physical properties, 4-124
   thermal properties, 12-197 to 198
   vapor pressure, 6-64 to 91
Mars, orbital data and dimensions, 14-2 to 3
Mass
   atmosphere, oceans, and crust, 14-6 to 7
   atomic mass unit, 1-1 to 11
   atomic, of nuclides, 11-50 to 158
   earth, moon, and sun, 14-1
   electron, proton, neutron, 1-1 to 11
   fundamental particles, 11-1 to 49

planets, **14**-2 to 3
planets, relative to sun, **14**-1
satellites, **14**-4 to 5
sun, **14**-2 to 3
Mass spectral peaks of common solvents, **8**-123 to 128
Materials characterization, techniques, **12**-1 to 6
Mathematical constants, **A**-1
Mean free path
    as function of altitude, **14**-16 to 22
    common gases, **6**-46
Mechanical properties
    commercial metals and alloys, **12**-210
    rare earth elements, **4**-114 to 123
Meitnerium (element 109), **4**-1 to 36, **11**-50 to 158
Melting point
    amino acids, **7**-1 to 2
    commercial metals and alloys, **12**-210
    cryogenic fluids, **6**-138
    D_2O, **6**-4
    depression of, **15**-21
    elements, **4**-124
    fatty acids, **7**-6
    halocarbons, **6**-140 to 142
    ice, as function of pressure, **6**-12
    index for organic compounds, **3**-737 to 750
    inorganic compounds, **4**-37 to 100, **6**-121 to 135
    metals, **12**-197 to 198
    organic compounds, **3**-1 to 330, **6**-121 to 135
    organic compounds, index, **3**-737 to 750
    pressure dependence, **6**-47
    rare earth elements, **4**-114 to 123
    semiconductors, **12**-97 to 107
    solvents, **15**-14 to 18
Mendelevium: see also Elements
    electron configuration, **1**-18 to 19
    history, occurrence, uses, **4**-1 to 36
    isotopes and their properties, **11**-50 to 158
Merck Index references for organic compounds, **3**-1 to 330
Mercury: see also Elements
    compressibility, **6**-144
    critical constants, **6**-48 to 60
    density, **6**-143
    electrical resistivity, **12**-45 to 47
    electron configuration, **1**-18 to 19
    heat capacity, **4**-125, **6**-144
    history, occurrence, uses, **4**-1 to 36
    ionization energy, **10**-175 to 177
    isotopes and their properties, **11**-50 to 158
    magnetic susceptibility, **4**-133 to 138
    physical properties, **4**-124
    sound velocity, **6**-144
    specific volume, **6**-143
    thermal conductivity, **6**-187 to 188
    thermal expansion, **6**-144
    thermal properties, **12**-197 to 198
    thermodynamic properties, **5**-1 to 3
    vapor pressure, **6**-64 to 91, **6**-104

Mercury (planet), orbital data and dimensions, **14**-2 to 3
Mesons, summary of properties, **11**-1 to 49
Metal oxides, secondary electron emission, **12**-131 to 132
Metals
    commercial, mechanical and thermal properties, **12**-210
    crystal structure, **12**-19 to 21
    elastic constants, **12**-37 to 44
    electrical resistivity, **12**-45 to 47
    extinction coefficient, **12**-133 to 156
    heat capacity, **12**-196
    index of refraction, **12**-133 to 156
    optical properties, **12**-133 to 156
    reflection coefficient, **12**-133 to 156
    secondary electron emission, **12**-131 to 132
    speed of sound in, **14**-38 to 39
    sublimation pressure, **6**-64 to 91
    superconducting properties, **12**-64 to 90
    thermal conductivity, **12**-199 to 200
    thermal properties, **12**-197 to 198
Meter, definition, **1**-24 to 28
Microwave bands, classification, **10**-212 to 213
Million, milliard (definition), **1**-41
Minerals
    chemical formulas, **4**-141 to 147, **4**-148 to 158
    crystal structure, **4**-148 to 158
    elastic constants, **12**-37 to 44
    hardness, **12**-211 to 212
    index of refraction, **4**-141 to 147
    physical constants, **4**-141 to 147
    semiconducting properties, **12**-97 to 107
    solubility as a function of temperature, **8**-102 to 110
    thermal conductivity, **12**-204 to 205
Mobility
    in semiconductors, **12**-97 to 107
    of atmospheric ions, **14**-30 to 37
Mole, definition, **1**-24 to 28
Molecular weight
    amino acids, **7**-1 to 2
    inorganic compounds, **4**-37 to 100
    organic compounds, **3**-1 to 330
Molecules
    appearance potential, **10**-178 to 195
    bond lengths, **9**-1 to 14, **9**-15 to 16
    bond lengths and angles, **9**-17 to 43
    bond strengths, **9**-51 to 74
    electron affinity, **10**-147 to 159
    force constants, **9**-76
    fundamental vibrational frequencies, **9**-77 to 80
    ionization energy, **10**-178 to 195
    polarizability, **10**-160 to 174
Molybdenum: see also Elements
    electrical resistivity, **12**-45 to 47
    electron configuration, **1**-18 to 19
    heat capacity, **4**-125
    history, occurrence, uses, **4**-1 to 36
    ionization energy, **10**-175 to 177

isotopes and their properties, **11**-50 to 158
    magnetic susceptibility, **4**-133 to 138
    molten, density, **4**-129 to 132
    physical properties, **4**-124
    thermal properties, **12**-197 to 198
    vapor pressure, **6**-64 to 91
Moment of inertia, formulas for, **A**-96
Moon
    orbital constants and other parameters, **14**-2 to 3, **14**-4 to 5
    ratio of mass to earth's mass, **14**-1
Muon
    in cosmic ray showers, **11**-176 to 180
    summary of properties, **11**-1 to 49
Musical scales, **14**-42 to 43

N

NAA, definition, **12**-1 to 6
Néel temperature
    magnetic materials, **12**-117 to 125
    rare earth elements, **4**-114 to 123
Neodymium: see also Elements
    electrical resistivity, **12**-45 to 47
    electron configuration, **1**-18 to 19
    heat capacity, **4**-125
    history, occurrence, uses, **4**-1 to 36
    ionization energy, **10**-175 to 177
    isotopes and their properties, **11**-50 to 158
    magnetic susceptibility, **4**-133 to 138
    molten, density, **4**-129 to 132
    physical properties, **4**-124
    thermal properties, **12**-197 to 198
    vapor pressure, **6**-64 to 91
Neon: see also Elements
    critical constants, **6**-48 to 60
    cryogenic properties, **6**-138
    electron configuration, **1**-18 to 19
    entropy, **5**-1 to 3
    history, occurrence, uses, **4**-1 to 36
    ionization energy, **10**-175 to 177
    isotopes and their properties, **11**-50 to 158
    magnetic susceptibility, **4**-133 to 138
    mean free path, **6**-46
    permittivity (dielectric constant), **6**-149 to 171, **6**-172 to 173
    physical properties, **4**-124
    solubility in water, **8**-86 to 89
    thermal conductivity, **6**-185 to 186
    van der Waals constants, **6**-45
    vapor pressure, **6**-64 to 91, **6**-92 to 102
    viscosity, **6**-178 to 179
Neptune, orbital data and dimensions, **14**-2 to 3
Neptunium: see also Elements
    electron configuration, **1**-18 to 19
    history, occurrence, uses, **4**-1 to 36
    ionization energy, **10**-175 to 177
    isotopes and their properties, **11**-50 to 158
    magnetic susceptibility, **4**-133 to 138
    physical properties, **4**-124
Neutrino, summary of properties, **11**-1 to 49
Neutron
    mass, **1**-1 to 11

range in paraffin, **16**-48
scattering and absorption, **11**-159 to 175
summary of properties, **11**-1 to 49
Neutron cross sections, **11**-159 to 175
Neutron resonance integrals, **11**-159 to 175
Newton, definition, 1-24 to 28
Nichrome, wire tables, **15**-27
Nickel: see also Elements
   electrical resistivity, **12**-45 to 47
   electron configuration, **1**-18 to 19
   heat capacity, **4**-125
   history, occurrence, uses, **4**-1 to 36
   ionization energy, **10**-175 to 177
   isotopes and their properties, **11**-50 to 158
   magnetic susceptibility, **4**-133 to 138
   molten, density, **4**-129 to 132
   physical properties, **4**-124
   thermal properties, **12**-197 to 198
   vapor pressure, **6**-64 to 91
Niobium: see also Elements
   electrical resistivity, **12**-45 to 47
   electron configuration, **1**-18 to 19
   heat capacity, **4**-125
   history, occurrence, uses, **4**-1 to 36
   ionization energy, **10**-175 to 177
   isotopes and their properties, **11**-50 to 158
   magnetic susceptibility, **4**-133 to 138
   physical properties, **4**-124
   thermal properties, **12**-197 to 198
   vapor pressure, **6**-64 to 91
Nitrogen: see also Elements
   critical constants, **6**-48 to 60
   cryogenic properties, **6**-138
   electron configuration, **1**-18 to 19
   enthalpy of vaporization, **6**-105 to 120
   history, occurrence, uses, **4**-1 to 36
   ionization energy, **10**-175 to 177
   isotopes and their properties, **11**-50 to 158
   magnetic susceptibility, **4**-133 to 138
   mean free path, **6**-46
   permittivity (dielectric constant), **6**-149 to 171, **6**-172 to 173
   physical properties, **4**-124
   solubility in water, **8**-86 to 89
   thermal conductivity, **6**-16 to 24, **6**-185 to 186
   thermodynamic properties, **5**-1 to 3, **6**-16 to 24
   thermodynamic properties at high temperature, **5**-61 to 84
   van der Waals constants, **6**-45
   vapor pressure, **6**-64 to 91, **6**-92 to 102
   viscosity, **6**-16 to 24, **6**-178 to 179
NMR spectrum
   characteristic ^{13}C chemical shifts, **9**-96
   characteristic shifts for protons, **9**-95
   chemical shifts of solvents, **8**-122
   nuclear moments and resonance frequencies, **9**-92 to 94
Nobelium: see also Elements
   electron configuration, **1**-18 to 19
   history, occurrence, uses, **4**-1 to 36
   isotopes and their properties, **11**-50 to 158
Noise thresholds, **14**-44 to 45

Nomenclature
   carbohydrates, **7**-7 to 8
   chemical, references, **2**-22
   inorganic ions and ligands, **2**-23 to 24
   minerals, **4**-141 to 147
   organic substituent groups and ring systems, **2**-25 to 28
   physical quantities, **2**-1 to 21
   polymers, **13**-1 to 2
NQR, definition, **12**-1 to 6
NRA, definition, **12**-1 to 6
Nuclear magnetic resonance: see NMR
Nuclear magneton, **1**-1 to 11
Nuclear spins and moments
   for all nuclides, **11**-50 to 158
   for important nuclei in NMR, **9**-92 to 94
Nucleic acids
   genetic code, **7**-5
   purine and pyrimidine bases, **7**-4
Nuclides, summary of properties, **11**-50 to 158

O

Oceans
   abundance of chemical elements, **14**-14
   pressure as a function of depth and latitude, **14**-11
Octanol-water partition coefficients, **16**-43 to 47
Ohm
   definition, **1**-24 to 28
   maintained value, **1**-1 to 11
Optical materials
   elasto-, electro-, and magneto-optic constants, **12**-157 to 173
   harmonic generation, **12**-174 to 177
   index of refraction, **10**-217
   nonlinear constants, **12**-174 to 177
Optical properties
   glass, **10**-217
   metals, **12**-133 to 156
   semiconductors, **12**-133 to 156, **12**-174 to 177
   various materials, **12**-157 to 173, **12**-174 to 177
Organic compounds
   boiling point index, **3**-751 to 763
   bond lengths (in crystals), **9**-1 to 14
   bond lengths and angles (in gas phase), **9**-17 to 43
   bond strengths, **9**-51 to 74
   characteristic ^{13}C chemical shifts, **9**-96
   characteristic proton chemical shifts, **9**-95
   classes, definitions, **2**-38 to 59
   dipole moment, **9**-44 to 50
   enthalpy of fusion, **6**-121 to 135
   enthalpy of vaporization, **6**-105 to 120
   heat of combustion, **5**-89
   infrared correlation charts, **9**-87 to 91
   magnetic susceptibility, **3**-764 to 767
   mass spectral peaks, **8**-123 to 128
   melting point index, **3**-737 to 750
   nomenclature, **2**-25 to 28

physical properties, **3**-1 to 330
polarizability, **10**-160 to 174
solubility, aqueous, **8**-91 to 101
sublimation pressure, **6**-61 to 63
superconducting properties, **12**-94 to 96
surface tension, **6**-145 to 148
thermal conductivity, **6**-187 to 188
thermodynamic properties, **5**-4 to 60
Organic magnets, **12**-126 to 129
Organometallic compounds, bond lengths, **9**-15 to 16
Orthogonal polynomials, **A**-102 to 104
Oscillator strengths in atomic spectra, **10**-88 to 146
Osmium: see also Elements
   electrical resistivity, **12**-45 to 47
   electron configuration, **1**-18 to 19
   heat capacity, **4**-125
   history, occurrence, uses, **4**-1 to 36
   ionization energy, **10**-175 to 177
   isotopes and their properties, **11**-50 to 158
   magnetic susceptibility, **4**-133 to 138
   molten, density, **4**-129 to 132
   physical properties, **4**-124
   thermal properties, **12**-197 to 198
   vapor pressure, **6**-64 to 91
Oxidation and reduction reagents, decinormal solutions, **8**-7
Oxidation-reduction potentials
   general table, **8**-21 to 31
   ion radicals, **8**-32 to 34
Oxygen: see also Elements
   critical constants, **6**-48 to 60
   cryogenic properties, **6**-138
   electron configuration, **1**-18 to 19
   enthalpy of fusion, **6**-121 to 135
   enthalpy of vaporization, **6**-105 to 120
   history, occurrence, uses, **4**-1 to 36
   ionization energy, **10**-175 to 177
   isotopes and their properties, **11**-50 to 158
   magnetic susceptibility, **4**-133 to 138
   mean free path, **6**-46
   permittivity (dielectric constant), **6**-149 to 171, **6**-172 to 173
   physical properties, **4**-124
   solubility in water, **8**-86 to 89
   thermal conductivity, **6**-16 to 24, **6**-185 to 186
   thermodynamic properties, **5**-1 to 3, **6**-16 to 24
   van der Waals constants, **6**-45
   vapor pressure, **6**-64 to 91, **6**-92 to 102
   viscosity, **6**-16 to 24, **6**-178 to 179

P

Palladium: see also Elements
   electrical resistivity, **12**-45 to 47
   electron configuration, **1**-18 to 19
   heat capacity, **4**-125
   history, occurrence, uses, **4**-1 to 36
   ionization energy, **10**-175 to 177
   isotopes and their properties, **11**-50 to 158
   magnetic susceptibility, **4**-133 to 138

molten, density, **4**-129 to 132
  physical properties, **4**-124
  thermal properties, **12**-197 to 198
  vapor pressure, **6**-64 to 91
Paramagnetic moment, rare earth elements,
      4-114 to 123
Paramagnetic susceptibility, elements and
      inorganic compounds, **4**-133 to 138
Particle size, **15**-28
Particles, fundamental, summary of
      properties, **11**-1 to 49
Pascal, definition, **1**-24 to 28
Pauling electronegativity scale, **9**-75
Pearson symbols, **12**-7 to 13, **12**-19 to 21
Periodic table of the elements, inside front
      cover
Permeability, magnetic alloys, **12**-117 to 125
Permeability of vacuum, **1**-1 to 11
Permittivity (dielectric constant)
  cryogenic fluids, temperature and
      pressure dependence, **6**-16 to 24
  crystals, **12**-52 to 60
  gases, **6**-172 to 173
  glass, **12**-63
  ice, **6**-6 to 7
  liquid helium, **6**-139
  liquids, **6**-149 to 171
  plastics, **13**-12
  quartz, **12**-63
  rubbers, **13**-12
  semiconductors, **12**-97 to 107
  solids, **12**-52 to 60
  solvents, **8**-121
  vacuum, **1**-1 to 11
  water, **6**-3
  water, frequency dependence, **6**-15
Peroxide formation by laboratory chemicals,
      16-1 to 15
pH
  acid-base indicators, **8**-16 to 18
  biological buffers, **7**-9
  biological materials and tissues, **7**-10
  blood, **7**-10
  definition of pH scale, **8**-35 to 40
  fluorescent indicators, **8**-19 to 20
  foods, **7**-10
  measurement in natural waters, **8**-41 to 42
  seawater, **8**-41 to 42
  solutions giving round values, **8**-43
  standards, **8**-35 to 40
Phase diagrams, **12**-178 to 195
Phase transitions
  enthalpy of fusion, **6**-121 to 135
  enthalpy of vaporization, **6**-105 to 120
  ice, **6**-6 to 7, **6**-12
  polymers, glass to crystal, **13**-4 to 11
  rare earth elements, **4**-114 to 123
Phon, definition, **14**-44 to 45
Phonon-electron coupling, rare earth
      elements, **4**-114 to 123
Phosphorus: see also Elements
  critical constants, **6**-48 to 60
  dielectric constant, **12**-52 to 60
  electron configuration, **1**-18 to 19

heat capacity, **4**-125
history, occurrence, uses, **4**-1 to 36
ionization energy, **10**-175 to 177
isotopes and their properties, **11**-50 to 158
magnetic susceptibility, **4**-133 to 138
physical properties, **4**-124
thermodynamic properties, **5**-1 to 3
vapor pressure, **6**-64 to 91
Photochemical data, **5**-106 to 120
Photoelastic constants, **12**-157 to 173
Photon attenuation coefficients, **10**-207 to
      211
Physical constants, fundamental, **1**-1 to 11
Physical properties
  amino acids, **7**-1 to 2
  inorganic compounds, **4**-37 to 100
  minerals, **4**-141 to 147
  organic compounds, **3**-1 to 330
  semiconductors, **12**-97 to 107
  solvents, **15**-14 to 18
Physical quantities
  definitions, **2**-38 to 59
  terminology and symbols, **2**-1 to 21
Pi, value of, **A**-1
Pion, summary of properties, **11**-1 to 49
Pitch, in musical scales, **14**-42 to 43
PIXE, definition, **12**-1 to 6
pK
  acid-base indicators, **8**-16 to 18
  amino acids, **7**-1 to 2
  biological buffers, **7**-9
  inorganic acids and bases, **8**-44 to 45
  organic acids and bases, **8**-46 to 56
  purine and pyrimidine bases, **7**-4
Planck constant, **1**-1 to 11
Planets
  atmospheric composition, **14**-2 to 3
  general properties, **14**-2 to 3
  orbital parameters, **14**-2 to 3
  satellites, **14**-4 to 5
Plastics
  breakdown voltage, **15**-30 to 35
  density, **15**-29
  dielectric constant, **13**-12
  speed of sound in, **14**-38 to 39
  thermal conductivity, **12**-204 to 205
Platinum: see also Elements
  electrical resistivity, **12**-45 to 47
  electron configuration, **1**-18 to 19
  heat capacity, **4**-125
  history, occurrence, uses, **4**-1 to 36
  ionization energy, **10**-175 to 177
  isotopes and their properties, **11**-50 to 158
  magnetic susceptibility, **4**-133 to 138
  molten, density, **4**-129 to 132
  physical properties, **4**-124
  resistance of wires, **15**-27
  thermal conductivity, **12**-199 to 200
  thermal properties, **12**-197 to 198
  vapor pressure, **6**-64 to 91
  wire tables, **15**-27
Platinum-rhodium thermocouple tables, **15**-1
      to 13
Pluto, orbital data and dimensions, **14**-2 to 3

Plutonium: see also Elements
  electron configuration, **1**-18 to 19
  heat capacity, **4**-125
  history, occurrence, uses, **4**-1 to 36
  ionization energy, **10**-175 to 177
  isotopes and their properties, **11**-50 to 158
  magnetic susceptibility, **4**-133 to 138
  molten, density, **4**-129 to 132
  physical properties, **4**-124
  thermal properties, **12**-197 to 198
  vapor pressure, **6**-64 to 91
Point groups of small molecules, **9**-77 to 80
Polarizability
  atoms and ions in solids, **12**-17 to 18
  free atoms and molecules, **10**-160 to 174
Pollutants
  airborne, limits in the workplace, **16**-32 to
      42
  Henry's Law constants, **8**-91 to 101
  octanol-water partition coefficients, **16**-43
      to 47
  solubility, **8**-91 to 101
Polonium: see also Elements
  electrical resistivity, **12**-45 to 47
  electron configuration, **1**-18 to 19
  heat capacity, **4**-125
  history, occurrence, uses, **4**-1 to 36
  ionization energy, **10**-175 to 177
  isotopes and their properties, **11**-50 to 158
  physical properties, **4**-124
  thermal properties, **12**-197 to 198
  vapor pressure, **6**-64 to 91
Polymers
  breakdown voltage, **15**-30 to 35
  density of melts, **13**-13 to 17
  dielectric constant, **13**-12
  glass transition temperature, **13**-4 to 11
  molar volume, **13**-13 to 17
  nomenclature, **13**-1 to 2
  solvents for, **13**-3
Polynomials, orthogonal, **A**-102 to 104
Potassium: see also Elements
  electrical resistivity, **12**-45 to 47
  electron configuration, **1**-18 to 19
  heat capacity, **4**-125
  history, occurrence, uses, **4**-1 to 36
  ionization energy, **10**-175 to 177
  isotopes and their properties, **11**-50 to 158
  magnetic susceptibility, **4**-133 to 138
  molten, density, **4**-129 to 132
  physical properties, **4**-124
  thermal properties, **12**-197 to 198
  thermodynamic properties, **5**-1 to 3
  thermodynamic properties at high
      temperature, **5**-61 to 84
  vapor pressure, **6**-64 to 91
Potential
  electrical, in the atmosphere, **14**-30 to 37
  oxidation-reduction, **8**-21 to 31
  oxidation-reduction, of ion radicals, **8**-32
      to 34
Praseodymium: see also Elements
  electrical resistivity, **12**-45 to 47
  electron configuration, **1**-18 to 19

heat capacity, 4-125
history, occurrence, uses, 4-1 to 36
ionization energy, 10-175 to 177
isotopes and their properties, 11-50 to 158
magnetic susceptibility, 4-133 to 138
molten, density, 4-129 to 132
physical properties, 4-124
thermal properties, 12-197 to 198
vapor pressure, 6-64 to 91
Precipitation of metal ions, 16-1 to 15
Pressure
    atmosphere, as function of altitude, 14-16
        to 22
    conversion factors, 1-43
    earth, as function of depth, 14-10
    effect on boiling point, 15-19
    effect on freezing point, 6-47
    ocean, as function of depth, 14-11
    planetary atmospheres, 14-2 to 3
Probability function, A-104
Promethium: see also Elements
    electrical resistivity, 12-45 to 47
    electron configuration, 1-18 to 19
    heat capacity, 4-125
    history, occurrence, uses, 4-1 to 36
    ionization energy, 10-175 to 177
    isotopes and their properties, 11-50 to 158
    physical properties, 4-124
    thermal properties, 12-197 to 198
Protactinium: see also Elements
    electrical resistivity, 12-45 to 47
    electron configuration, 1-18 to 19
    heat capacity, 4-125
    history, occurrence, uses, 4-1 to 36
    ionization energy, 10-175 to 177
    isotopes and their properties, 11-50 to 158
    magnetic susceptibility, 4-133 to 138
    physical properties, 4-124
    thermal properties, 12-197 to 198
Proton
    in cosmic ray showers, 11-176 to 180
    magnetic moment, 1-1 to 11
    mass, 1-1 to 11
    summary of properties, 11-1 to 49
PSD, definition, 12-1 to 6
psia and psig, definition, 1-43
Purine bases, properties of, 7-4
Pyrimidine bases, properties of, 7-4
Pyrophoric chemicals, safe handling, 16-1 to
    15

Q

Quadrupole moments
    all nuclides, 11-50 to 158
    important nuclei for NMR, 9-92 to 94
Quartz
    crystallographic data, 4-148 to 158
    dielectric constant, 12-63
    loss factor, 12-63
    optical properties, 4-141 to 147
    phase diagram, 12-178 to 195
    thermal conductivity, 12-202 to 203

R

Rad, definition, 16-48
Radiation
    black body, 10-214 to 215
    electromagnetic, classification, 10-212 to
        213
    microwave, classification of bands,
        10-212 to 213
Radiation, ionizing
    conversion factors, 1-47 to 48
    from nuclear decay, 11-50 to 158
    permissible intake of radionuclides, 16-49
        to 52
    protection against, 16-48
Radiation, solar
    by month and latitude, 14-23
    by wavelength, 14-15
    flux, solar constant, 14-2 to 3
Radiative transition probability, 10-88 to 146
Radicals, free: see Free radicals
Radicals, nomenclature, 2-23 to 24, 2-25 to
    28
Radionuclides, permissible intake, 16-49 to
    52
Radium: see also Elements
    electron configuration, 1-18 to 19
    heat capacity, 4-125
    history, occurrence, uses, 4-1 to 36
    ionization energy, 10-175 to 177
    isotopes and their properties, 11-50 to 158
    physical properties, 4-124
    thermal properties, 12-197 to 198
    vapor pressure, 6-64 to 91
Radius of ions in crystals, 12-14 to 16
Radon: see also Elements
    critical constants, 6-48 to 60
    electron configuration, 1-18 to 19
    history, occurrence, uses, 4-1 to 36
    ionization energy, 10-175 to 177
    isotopes and their properties, 11-50 to 158
    physical properties, 4-124
    solubility in water, 8-86 to 89
    vapor pressure, 6-64 to 91, 6-92 to 102
Rankine temperature, conversion to other
    scales, 1-41
Rare earth crystals, Verdet constants, 12-157
    to 173
Rare earth metals, general properties, 4-114
    to 123
Rate constants
    atmospheric reactions, 5-106 to 120
    conversion factors, 1-46
RBS, definition, 12-1 to 6
Reagents
    decinormal solutions, 8-5 to 6, 8-7
    disposal of, 16-1 to 15
    formulas for concentration, 8-20
    organic, for analysis, 8-8 to 15
    preparation, 8-1 to 4
Reduction potentials
    general table, 8-21 to 31
    ion radicals, 8-32 to 34
Reference states of elements, 5-4 to 60

Reflection coefficient, metals and semicon-
    ductors, 12-133 to 156
Refractive index: see Index of refraction
Refractory materials
    hardness, 12-211 to 212
    thermal conductivity, 12-204 to 205
Refrigerants, various properties, 6-140 to 142
Relative humidity
    from wet and dry bulb temperatures,
        15-24
    relation to dew point, 15-23
    solutions for calibration, 15-26
    solutions for constant humidity, 15-25
Relaxation time, in water, 6-15
Rem, definition, 16-48
Remanence, magnetic materials, 12-117 to
    125
Resistance of wires, 15-27
Resistivity, electrical
    alloys, 12-48 to 51
    commercial metals and alloys, 12-210
    conversion factors, 1-45
    elements, 12-45 to 47
    glasses, 12-63
    pure metals, 12-45 to 47
    quartz, 12-63
    rare earth elements, 4-114 to 123
    semiconducting minerals, 12-97 to 107
    semiconductors, 12-97 to 107
    superconductors, 12-91 to 93
Respirators, for laboratory use, 16-1 to 15
RHEED, definition, 12-1 to 6
Rhenium: see also Elements
    electrical resistivity, 12-45 to 47
    electron configuration, 1-18 to 19
    heat capacity, 4-125
    history, occurrence, uses, 4-1 to 36
    ionization energy, 10-175 to 177
    isotopes and their properties, 11-50 to 158
    magnetic susceptibility, 4-133 to 138
    molten, density, 4-129 to 132
    physical properties, 4-124
    thermal properties, 12-197 to 198
    vapor pressure, 6-64 to 91
Rhodium: see also Elements
    electrical resistivity, 12-45 to 47
    electron configuration, 1-18 to 19
    heat capacity, 4-125
    history, occurrence, uses, 4-1 to 36
    ionization energy, 10-175 to 177
    isotopes and their properties, 11-50 to 158
    magnetic susceptibility, 4-133 to 138
    molten, density, 4-129 to 132
    physical properties, 4-124
    thermal properties, 12-197 to 198
    vapor pressure, 6-64 to 91
Ring systems, nomenclature, 2-25 to 28
Rochelle salts, 12-61 to 62
Rocks
    age, 14-6 to 7
    density, 15-29
    thermal conductivity, 12-204 to 205
Roentgen, definition, 16-48
Rotational constants, diatomic molecules,
    9-81 to 86

Rounding of numbers, **1**-29 to 40
Rubbers
    breakdown voltage, **15**-30 to 35
    density, **15**-29
    dielectric constant, **13**-12
    speed of sound in, **14**-38 to 39
    thermal conductivity, **12**-204 to 205
Rubidium: see also Elements
    electrical resistivity, **12**-45 to 47
    electron configuration, **1**-18 to 19
    heat capacity, **4**-125
    history, occurrence, uses, **4**-1 to 36
    ionization energy, **10**-175 to 177
    isotopes and their properties, **11**-50 to 158
    magnetic susceptibility, **4**-133 to 138
    molten, density, **4**-129 to 132
    physical properties, **4**-124
    thermal properties, **12**-197 to 198
    thermodynamic properties, **5**-1 to 3
    vapor pressure, **6**-64 to 91
Russian alphabet, **2**-37
Ruthenium: see also Elements
    electrical resistivity, **12**-45 to 47
    electron configuration, **1**-18 to 19
    heat capacity, **4**-125
    history, occurrence, uses, **4**-1 to 36
    ionization energy, **10**-175 to 177
    isotopes and their properties, **11**-50 to 158
    magnetic susceptibility, **4**-133 to 138
    molten, density, **4**-129 to 132
    physical properties, **4**-124
    thermal properties, **12**-197 to 198
    vapor pressure, **6**-64 to 91
Rutherfordium (element 104)
    electron configuration, **1**-18 to 19
    history, occurrence, uses, **4**-1 to 36
    isotopes and their properties, **11**-50 to 158
Rydberg constant, **1**-1 to 11

S

Sackur-Tetrode constant, **1**-1 to 11
Safety
    chemical carcinogens, **16**-53 to 57
    flammability of chemicals, **16**-16 to 31
    laboratory practices, **16**-1 to 15
    radiation, **16**-48, **16**-49 to 52
SALI, definition, **12**-1 to 6
Salinity scale for seawater, **14**-12 to 13
Salts
    activity coefficients, **5**-97 to 98, **5**-99 to 103
    decinormal solutions, **8**-5 to 6
    electrical conductivity, **5**-94
    enthalpy of solution, **5**-105
    molten, density of, **4**-129 to 132
    solubility as a function of temperature, **8**-102 to 110, **8**-115
    vapor pressure of aqueous solutions, **6**-103
Samarium: see also Elements
    electrical resistivity, **12**-45 to 47
    electron configuration, **1**-18 to 19
    heat capacity, **4**-125

    history, occurrence, uses, **4**-1 to 36
    ionization energy, **10**-175 to 177
    isotopes and their properties, **11**-50 to 158
    magnetic susceptibility, **4**-133 to 138
    molten, density, **4**-129 to 132
    physical properties, **4**-124
    thermal properties, **12**-197 to 198
    vapor pressure, **6**-64 to 91
SANS, definition, **12**-1 to 6
Saturn, orbital data and dimensions, **14**-2 to 3
Scandium: see also Elements
    electrical resistivity, **12**-45 to 47
    electron configuration, **1**-18 to 19
    heat capacity, **4**-125
    history, occurrence, uses, **4**-1 to 36
    ionization energy, **10**-175 to 177
    isotopes and their properties, **11**-50 to 158
    magnetic susceptibility, **4**-133 to 138
    molten, density, **4**-129 to 132
    physical properties, **4**-124
    thermal properties, **12**-197 to 198
    vapor pressure, **6**-64 to 91
Seaborgium (element 106), **4**-1 to 36, **11**-50 to 158
Seawater

    composition (elemental), **14**-14
    composition (ions), **14**-12 to 13
    density, **14**-12 to 13
    electrical conductivity, **14**-12 to 13
    freezing point, **14**-12 to 13
    pH measurement, **8**-41 to 42
    pressure as a function of depth, **14**-11
    salinity scale, **14**-12 to 13
    specific heat, **14**-12 to 13
    speed of sound in, **14**-38 to 39
    viscosity, **14**-12 to 13
Secant function, **A**-7 to 10
Second, definition, **1**-24 to 28
Second radiation constant, **1**-1 to 11
Secondary electron emission, **12**-131 to 132
Selenium: see also Elements
    critical constants, **6**-48 to 60
    electron configuration, **1**-18 to 19
    heat capacity, **4**-125
    history, occurrence, uses, **4**-1 to 36
    ionization energy, **10**-175 to 177
    isotopes and their properties, **11**-50 to 158
    magnetic susceptibility, **4**-133 to 138
    molten, density, **4**-129 to 132
    physical properties, **4**-124
    vapor pressure, **6**-64 to 91
SEM, definition, **12**-1 to 6
Semiconductors
    crystal structure, **12**-19 to 21, **12**-97 to 107
    diffusion in, **12**-108 to 116
    effective mass, **12**-97 to 107
    elastic constants, **12**-37 to 44
    electrical properties, **12**-97 to 107
    extinction coefficient, **12**-133 to 156
    index of refraction, **12**-133 to 156
    minerals, resistivity of, **12**-97 to 107
    optical properties, **12**-133 to 156, **12**-157 to 173

    physical properties, **12**-97 to 107
    reflection coefficient, **12**-133 to 156
    thermal conductivity, **12**-97 to 107, **12**-199 to 200
Series expansions, **A**-84 to 87
Shielding, from radiation, **10**-207 to 211, **16**-48
SI units
    conversion factors to, **1**-29 to 40
    definitions and symbols, **1**-24 to 28
    prefixes, **1**-24 to 28
Siemens, definition, **1**-24 to 28
Sievert, definition, **1**-24 to 28, **16**-48
Silicon: see also Elements
    dielectric constant, **12**-52 to 60
    electron configuration, **1**-18 to 19
    heat capacity, **4**-125, **12**-196
    history, occurrence, uses, **4**-1 to 36
    ionization energy, **10**-175 to 177
    isotopes and their properties, **11**-50 to 158
    magnetic susceptibility, **4**-133 to 138
    molten, density, **4**-129 to 132
    optical properties, **12**-133 to 156
    physical properties, **4**-124
    semiconducting properties, **12**-97 to 107
    thermal conductivity, **12**-199 to 200
    thermodynamic properties, **5**-1 to 3
    thermodynamic properties at high temperature, **5**-61 to 84
    vapor pressure, **6**-64 to 91
Silver: see also Elements
    electrical resistivity, **12**-45 to 47
    electron configuration, **1**-18 to 19
    heat capacity, **4**-125, **12**-196
    history, occurrence, uses, **4**-1 to 36
    ionization energy, **10**-175 to 177
    isotopes and their properties, **11**-50 to 158
    magnetic susceptibility, **4**-133 to 138
    molten, density, **4**-129 to 132
    physical properties, **4**-124
    resistance of wires, **15**-27
    thermal conductivity, **12**-199 to 200
    thermal properties, **12**-197 to 198
    thermodynamic properties, **5**-1 to 3
    vapor pressure, **6**-64 to 91
    wire tables, **15**-27
SIMS, definition, **12**-1 to 6
Sine function, **A**-7 to 10
SLAM, definition, **12**-1 to 6
SMOW (standard mean ocean water), density, **6**-5
SNMS, definition, **12**-1 to 6
Sodium: see also Elements
    electrical resistivity, **12**-45 to 47
    electron configuration, **1**-18 to 19
    heat capacity, **4**-125, **12**-196
    history, occurrence, uses, **4**-1 to 36
    ionization energy, **10**-175 to 177
    isotopes and their properties, **11**-50 to 158
    magnetic susceptibility, **4**-133 to 138
    molten, density, **4**-129 to 132
    physical properties, **4**-124
    thermal properties, **12**-197 to 198
    thermodynamic properties, **5**-1 to 3

vapor pressure, **6**-64 to 91
Sodium chloride
    activity coefficients, **5**-97 to 98, **5**-99 to 103
    aqueous solutions, concentrative properties, **8**-57 to 83
    aqueous solutions, relative humidity, **15**-26
    aqueous solutions, volumetric properties, **6**-8
    density of aqueous solutions, **6**-8
    enthalpy of solution, **5**-105
    standard thermodynamic properties, **5**-4 to 60
Solar constant, **14**-2 to 3
Solar radiation
    by month and latitude, **14**-23
    by wavelength, **14**-15
Solar spectrum, **14**-15
Solar system, **14**-2 to 3
Solder
    phase diagram, **12**-178 to 195
    thermal conductivity, **12**-201
Solids, characterization and analysis, **12**-1 to 6
Solubility
    amino acids, **7**-1 to 2
    carbon dioxide in water, **8**-90
    fatty acids in water, **7**-6
    gases in water, **8**-86 to 89
    inorganic compounds, **4**-37 to 100
    inorganic compounds, as function of temperature, **8**-102 to 110, **8**-115
    inorganic compounds, qualitative rules, **8**-116 to 118
    inorganic compounds, sparingly soluble, **8**-111 to 114
    octanol-water partition coefficients, **16**-43 to 47
    organic compounds, **3**-1 to 330
    organic compounds in water, **8**-91 to 101
    purine and pyrimidine bases, **7**-4
    salts in water, **8**-115
Solubility product constant, **8**-111 to 114
Solutions
    aqueous, concentrative properties, **8**-57 to 83
    decinormal, oxidation reagents, **8**-7
    decinormal, salts, **8**-5 to 6
    density, **8**-57 to 83
    diffusion of ions, **5**-95 to 97
    enthalpy, for common electrolytes, **5**-105
    for round values of pH, **8**-43
    formulas for concentration, **8**-20
    freezing point depression, **8**-57 to 83
    index of refraction, **8**-57 to 83
    ionic conductivity, **5**-95 to 97
    viscosity, **8**-57 to 83
Solvents
    dielectric constant, **8**-121
    dipole moment, **15**-14 to 18
    flammability, **15**-14 to 18
    for NMR, chemical shifts, **8**-122
    for polymers, **13**-3

for ultraviolet spectrophotometry, **8**-121
    heat capacity, **15**-14 to 18
    mass spectral peaks, **8**-123 to 128
    physical properties, **15**-14 to 18
    threshold limit in air, **15**-14 to 18
    vapor pressure, **15**-14 to 18
    wavelength cutoff (UV), **8**-121
Sound level, in human hearing, **14**-44 to 45
Sound velocity
    air, as function of frequency, **14**-40
    air, as function of humidity, **14**-40
    air, as function of temperature, **14**-41
    atmosphere, as function of altitude, **14**-16 to 22
    fluids, **6**-16 to 24
    mercury, **6**-144
    various solids, liquids, and gases, **14**-38 to 39
    water and seawater, **14**-38 to 39
Space group
    elements, **12**-19 to 21
    notation, **12**-7 to 13
Specific gravity: see Density
Specific heat: see Heat capacity
Specific volume: see also Density
    mercury, **6**-143
    sodium chloride solutions, **6**-8
    water, **8**-119
Spectroscopic constants, diatomic molecules, **9**-81 to 86
Spectrum, infrared
    calibration frequencies, **10**-228 to 235, **10**-236 to 240
    correlation charts, **9**-87 to 91
    fundamental vibrational frequencies, **9**-77 to 80
Spectrum, line, of the elements, **10**-1 to 87
Speed of light, **1**-1 to 11
Speed of sound
    air, as function of frequency, **14**-40
    air, as function of humidity, **14**-40
    air, as function of temperature, **14**-41
    atmosphere, as function of altitude, **14**-16 to 22
    fluids, **6**-16 to 24
    various solids, liquids, and gases, **14**-38 to 39
    water and seawater, **14**-38 to 39
Spin
    fundamental particles, **11**-1 to 49
    nuclides, **11**-50 to 158
    nuclides of NMR interest, **9**-92 to 94
    ordering in magnetic materials, **12**-126 to 129
SPM, definition, **12**-1 to 6
SSMS, definition, **12**-1 to 6
Standard Atmosphere (U.S.), **14**-16 to 22
Standard atomic weights, **1**-12 to 14
Standard density of water, **6**-5
Standard solutions, for pH measurement, **8**-35 to 40, **8**-41 to 42
Standards
    CODATA thermodynamic values, **5**-1 to 3

index of refraction, **10**-219
    infrared absorption frequencies, **10**-236 to 240
    infrared laser frequencies, **10**-228 to 235
    temperature, **1**-20 to 21
    vapor pressure, **6**-104
Steam, thermodynamic properties (see also Water), **6**-13 to 14
Steel
    mechanical properties, **12**-210
    thermal conductivity, **12**-201
Stefan-Boltzmann constant, **1**-1 to 11
STEM, definition, **12**-1 to 6
STM, definition, **12**-1 to 6
Stratosphere
    chemical reactions, **5**-106 to 120
    properties, **14**-16 to 22
Strontium: see also Elements
    electrical resistivity, **12**-45 to 47
    electron configuration, **1**-18 to 19
    heat capacity, **4**-125
    history, occurrence, uses, **4**-1 to 36
    ionization energy, **10**-175 to 177
    isotopes and their properties, **11**-50 to 158
    magnetic susceptibility, **4**-133 to 138
    molten, density, **4**-129 to 132
    thermal properties, **12**-197 to 198
    vapor pressure, **6**-64 to 91
Structure
    amino acids, **7**-3
    bond lengths in organic crystals, **9**-1 to 14
    bond lengths in organometallic compounds, **9**-15 to 16
    characteristic ^{13}C chemical shifts, **9**-96
    characteristic infrared frequencies, **9**-87 to 91
    characteristic proton chemical shifts, **9**-95
    crystal, of elements, **12**-19 to 21
    crystal, of superconductors, **12**-64 to 90, **12**-91 to 93
    force constants, **9**-76
    formulas for organic compounds, **3**-331 to 585
    fundamental vibrational frequencies, **9**-77 to 80
    gas-phase molecules, **9**-17 to 43
    geometry of small molecules, **9**-77 to 80
    solids, characterization techniques, **12**-1 to 6
Student's *t*-distribution, A-105
Sublimation pressure of solids, **6**-61 to 63
Sugars
    aqueous solution properties, **8**-57 to 83
    nomenclature, **7**-7 to 8
Sulfur: see also Elements
    critical constants, **6**-48 to 60
    dielectric constant, **12**-52 to 60
    electron configuration, **1**-18 to 19
    heat capacity, **4**-125
    history, occurrence, uses, **4**-1 to 36
    ionization energy, **10**-175 to 177
    isotopes and their properties, **11**-50 to 158
    magnetic susceptibility, **4**-133 to 138
    molten, density, **4**-129 to 132

physical properties, **4**-124
thermodynamic properties, **5**-1 to 3
thermodynamic properties at high temperature, **5**-61 to 84
vapor pressure, **6**-64 to 91
Sulfuric acid
activity coefficients, **5**-97 to 98, **5**-99 to 103
concentrative properties, **8**-57 to 83
constant humidity solutions, **15**-25
vapor pressure, **6**-64 to 91
Sun
mass, dimensions, and other properties, **14**-2 to 3
radiative properties, **14**-2 to 3
spectral irradiance, **14**-15
Superconductors
electrical and magnetic properties, **12**-64 to 90, **12**-91 to 93
organic, **12**-94 to 96
rare earth elements, **4**-114 to 123
transition temperature, **12**-64 to 90, **12**-91 to 93, **12**-94 to 96
Superconductors, high temperature
general properties, **12**-91 to 93
phase diagram, **12**-178 to 195
Supercooled water, **6**-6 to 7
Surface characterization and analysis, **12**-1 to 6
Surface tension
liquid helium, **6**-139
liquid rare earth metals, **4**-114 to 123
various liquids, **6**-145 to 148
water, **6**-3
Susceptibility: see Magnetic susceptibility
Symbols
amino acids, **7**-3
carbohydrates, **7**-7 to 8
magnetism, **12**-117 to 125
physical quantities, **2**-1 to 21, **2**-29 to 36
SI units, **1**-24 to 28
units, **2**-29 to 36
Symmetry of crystals, **12**-7 to 13
Synonym index
inorganic compounds, **4**-101 to 106
organic compounds, **3**-586 to 630

T

Tangent function, A-7 to 10
Tantalum: see also Elements
electrical resistivity, **12**-45 to 47
electron configuration, **1**-18 to 19
heat capacity, **4**-125
history, occurrence, uses, **4**-1 to 36
ionization energy, **10**-175 to 177
isotopes and their properties, **11**-50 to 158
magnetic susceptibility, **4**-133 to 138
molten, density, **4**-129 to 132
physical properties, **4**-124
thermal properties, **12**-197 to 198
vapor pressure, **6**-64 to 91
Taylor series, A-84 to 87

Technetium: see also Elements
electron configuration, **1**-18 to 19
heat capacity, **4**-125
history, occurrence, uses, **4**-1 to 36
ionization energy, **10**-175 to 177
isotopes and their properties, **11**-50 to 158
magnetic susceptibility, **4**-133 to 138
physical properties, **4**-124
thermal properties, **12**-197 to 198
vapor pressure, **6**-64 to 91
Tellurium: see also Elements
dielectric constant, **12**-52 to 60
electron configuration, **1**-18 to 19
history, occurrence, uses, **4**-1 to 36
ionization energy, **10**-175 to 177
isotopes and their properties, **11**-50 to 158
magnetic susceptibility, **4**-133 to 138
molten, density, **4**-129 to 132
physical properties, **4**-124
vapor pressure, **6**-64 to 91
Temperature
atmosphere, as function of altitude, **14**-16 to 22
baths for low temperature, **15**-26
calibration, ITS-90, **1**-20 to 21
celsius and absolute, definitions, **1**-24 to 28
conversion between scales, **1**-41
conversion to ITS-90, **1**-22 to 23
glass transition, in polymers, **13**-4 to 11
International Temperature Scale (ITS-90), **1**-20 to 21
mean global, **14**-29
mean United States, **14**-27 to 28
planetary atmospheres, **14**-2 to 3
superconducting transition, **12**-64 to 90
thermocouple tables, **15**-1 to 13
Tensile strength
commercial metals and alloys, **12**-210
rare earth elements, **4**-114 to 123
Terbium: see also Elements
electrical resistivity, **12**-45 to 47
electron configuration, **1**-18 to 19
heat capacity, **4**-125
history, occurrence, uses, **4**-1 to 36
ionization energy, **10**-175 to 177
isotopes and their properties, **11**-50 to 158
magnetic susceptibility, **4**-133 to 138
molten, density, **4**-129 to 132
physical properties, **4**-124
thermal properties, **12**-197 to 198
vapor pressure, **6**-64 to 91
Terminology
inorganic ions and ligands, **2**-23 to 24
organic substituent groups and ring systems, **2**-25 to 28
physical quantities, **2**-1 to 21
polymers, **13**-1 to 2
scientific terms, definitions, **2**-38 to 59
Tesla, definition, **1**-24 to 28
TGA, definition, **12**-1 to 6
Thallium: see also Elements
electrical resistivity, **12**-45 to 47
electron configuration, **1**-18 to 19

heat capacity, **4**-125
history, occurrence, uses, **4**-1 to 36
ionization energy, **10**-175 to 177
isotopes and their properties, **11**-50 to 158
magnetic susceptibility, **4**-133 to 138
molten, density, **4**-129 to 132
physical properties, **4**-124
thermal properties, **12**-197 to 198
vapor pressure, **6**-64 to 91
Thermal conductivity
alloys, **12**-201
argon, liquid and gas, **6**-16 to 24
atmosphere, as function of altitude, **14**-16 to 22
carrier gases for chromatography, **8**-120
ceramics, **12**-204 to 205
commercial metals, **12**-201
commercial metals and alloys, **12**-210
common fluids, as function of temperature and pressure, **6**-16 to 24
construction materials, **12**-204 to 205
conversion factors, **1**-44
cryogenic solids, **12**-202 to 203
crystalline solids, **12**-202 to 203
D_2O, **6**-4
dielectric crystals, **12**-202 to 203
gases, at atmospheric pressure, **6**-185 to 186
glasses, **12**-206 to 209
helium, liquid, **6**-139
helium, liquid and gas, **6**-16 to 24
ice, **6**-6 to 7, **12**-202 to 203
insulation, **12**-204 to 205
liquids, **6**-187 to 188
mercury, **6**-187 to 188
metals, **12**-197 to 198, **12**-199 to 200
methane, liquid and gas, **6**-16 to 24
minerals, **12**-204 to 205
nitrogen, liquid and gas, **6**-16 to 24
oxygen, liquid and gas, **6**-16 to 24
plastics, **12**-204 to 205
quartz, **12**-202 to 203
rare earth elements, **4**-114 to 123
refractory materials, **12**-204 to 205
rocks, **12**-204 to 205
rubber, **12**-204 to 205
semiconductors, **12**-97 to 107, **12**-199 to 200
superconductors, **12**-91 to 93
water, **6**-3, **6**-4, **6**-187 to 188
wood, **12**-204 to 205
Thermal expansion coefficient
commercial metals and alloys, **12**-210
ice, **6**-6 to 7
liquids, **6**-136 to 137
metals, **12**-197 to 198
rare earth elements, **4**-114 to 123
semiconductors, **12**-97 to 107
sodium chloride solutions, **6**-8
Thermal neutron cross sections, **11**-159 to 175
Thermocouple calibration tables, **15**-1 to 13
Thermodynamic properties (see also Enthalpy, Heat capacity, etc.)

air, **6**-1 to 2
aqueous systems, **5**-85 to 88
argon, **6**-16 to 24
CODATA Key Values, **5**-1 to 3
common fluids, as function of temperature and pressure, **6**-16 to 24
ethane, **6**-16 to 24
helium, **6**-16 to 24
high temperature, **5**-61 to 84
hydrogen, **6**-16 to 24
inorganic compounds, **5**-4 to 60
nitrogen, **6**-16 to 24
organic compounds, **5**-4 to 60
oxygen, **6**-16 to 24
propane, **6**-16 to 24
rare earth elements, **4**-114 to 123
standard state values, **5**-4 to 60
steam, **6**-13 to 14
temperature dependence, **5**-61 to 84
water, **6**-13 to 14
Thermometers, wet and dry bulb, **15**-24
Thorium: see also Elements
electrical resistivity, **12**-45 to 47
electron configuration, **1**-18 to 19
heat capacity, **4**-125
history, occurrence, uses, **4**-1 to 36
ionization energy, **10**-175 to 177
isotopes and their properties, **11**-50 to 158
magnetic susceptibility, **4**-133 to 138
physical properties, **4**-124
thermal properties, **12**-197 to 198
thermodynamic properties, **5**-1 to 3
vapor pressure, **6**-64 to 91
Threshold limits
airborne contaminants, **16**-32 to 42
halocarbon refrigerants, **6**-140 to 142
solvents, **15**-14 to 18
Thulium: see also Elements
electrical resistivity, **12**-45 to 47
electron configuration, **1**-18 to 19
heat capacity, **4**-125
history, occurrence, uses, **4**-1 to 36
ionization energy, **10**-175 to 177
isotopes and their properties, **11**-50 to 158
magnetic susceptibility, **4**-133 to 138
molten, density, **4**-129 to 132
physical properties, **4**-124
thermal properties, **12**-197 to 198
vapor pressure, **6**-64 to 91
Thunderstorm electricity, **14**-30 to 37
Time
astronomical units, **14**-1
geological scale, **14**-8
Tin: see also Elements
electrical resistivity, **12**-45 to 47
electron configuration, **1**-18 to 19
heat capacity, **4**-125
history, occurrence, uses, **4**-1 to 36
ionization energy, **10**-175 to 177
isotopes and their properties, **11**-50 to 158
magnetic susceptibility, **4**-133 to 138
molten, density, **4**-129 to 132
physical properties, **4**-124
thermal conductivity, **12**-199 to 200

thermal properties, **12**-197 to 198
thermodynamic properties, **5**-1 to 3
vapor pressure, **6**-64 to 91
Titanium: see also Elements
electrical resistivity, **12**-45 to 47
electron configuration, **1**-18 to 19
heat capacity, **4**-125, **12**-196
history, occurrence, uses, **4**-1 to 36
ionization energy, **10**-175 to 177
isotopes and their properties, **11**-50 to 158
magnetic susceptibility, **4**-133 to 138
molten, density, **4**-129 to 132
physical properties, **4**-124
thermal properties, **12**-197 to 198
thermodynamic properties, **5**-1 to 3
vapor pressure, **6**-64 to 91
Toxicity of anions and cations, **16**-1 to 15
Transition probability, atomic, **10**-88 to 146
Transition temperature
glass, in polymers, **13**-4 to 11
superconductors, **12**-64 to 90, **12**-91 to 93, **12**-94 to 96
Transport properties: see Thermal conductivity, Viscosity, Diffusion
Trigonometric functions, A-7 to 10, A-11
Trigonometric series, A-84 to 87
Trillion, definition, **1**-41
Triple point constants
carbon dioxide, **6**-61 to 63
cryogenic fluids, **6**-138
D$_2$O, **6**-4
various compounds, **6**-61 to 63
water, **6**-4
Tschebysheff polynomials, A-102 to 104
Tungsten: see also Elements
electrical resistivity, **12**-45 to 47
electron configuration, **1**-18 to 19
heat capacity, **4**-125, **12**-196
history, occurrence, uses, **4**-1 to 36
ionization energy, **10**-175 to 177
isotopes and their properties, **11**-50 to 158
magnetic susceptibility, **4**-133 to 138
molten, density, **4**-129 to 132
physical properties, **4**-124
resistance of wires, **15**-27
thermal conductivity, **12**-199 to 200
thermal properties, **12**-197 to 198
vapor pressure, **6**-64 to 91
wire tables, **15**-27

U

Ultraviolet spectrophotometry, solvents for, **8**-121
Units
conversion factors, **1**-29 to 40
definitions, **2**-38 to 59
ionizing radiation, **1**-47 to 48, **16**-48
magnetic quantities, **12**-117 to 125
SI, definitions and symbols, **1**-24 to 28
UPS, definition, **12**-1 to 6
Uranium: see also Elements
electrical resistivity, **12**-45 to 47
electron configuration, **1**-18 to 19

heat capacity, **4**-125
history, occurrence, uses, **4**-1 to 36
ionization energy, **10**-175 to 177
isotopes and their properties, **11**-50 to 158
magnetic susceptibility, **4**-133 to 138
molten, density, **4**-129 to 132
physical properties, **4**-124
thermal properties, **12**-197 to 198
thermodynamic properties, **5**-1 to 3
vapor pressure, **6**-64 to 91
Uranus, orbital data and dimensions, **14**-2 to 3

V

Van der Waals constants for gases, **6**-45
Vanadium: see also Elements
electrical resistivity, **12**-45 to 47
electron configuration, **1**-18 to 19
heat capacity, **4**-125
history, occurrence, uses, **4**-1 to 36
ionization energy, **10**-175 to 177
isotopes and their properties, **11**-50 to 158
magnetic susceptibility, **4**-133 to 138
molten, density, **4**-129 to 132
physical properties, **4**-124
thermal properties, **12**-197 to 198
vapor pressure, **6**-64 to 91
Vapor pressure
air, **6**-1 to 2
aqueous salt solutions, **15**-26
at temperatures below 300 K, **6**-92 to 102
calibration data, **6**-104
carbon dioxide, **6**-61 to 63, **6**-92 to 102, **6**-104
cryogenic fluids, **6**-92 to 102
elements, **6**-64 to 91
elements, high temperature, **4**-126 to 128
general table, **6**-64 to 91
helium, **1**-20 to 21, **6**-92 to 102, **6**-139
ice, **6**-9, **6**-104
inorganic compounds, **6**-64 to 91
IUPAC recommended data, **6**-104
metals, at high temperatures, **4**-126 to 128
organic compounds, **6**-64 to 91
rare earth elements, **4**-114 to 123
rare gases, **6**-92 to 102
salt solutions, **6**-103
solids, **6**-61 to 63
solvents, **15**-14 to 18
water, **6**-3, **6**-10 to 11, **6**-104
water, over salt solutions, **15**-25, **15**-26
Vaporization: see Enthalpy of vaporization
Vector analysis, A-87 to 96
Velocity of light, **1**-1 to 11
Velocity of sound
air, as function of frequency, **14**-40
air, as function of humidity, **14**-40
air, as function of temperature, **14**-41
atmosphere, as function of altitude, **14**-16 to 22
fluids, **6**-16 to 24
various solids, liquids, and gases, **14**-38 to 39

Velocity, mean, in gases, **6**-46
Venus, orbital data and dimensions, **14**-2 to 3
Verdet constants, **12**-157 to 173
Vibrational force constants, **9**-76
Vibrational frequencies of molecules, **9**-77 to 80, **9**-81 to 86
Vibrational-rotational spectra, for frequency calibration, **10**-236 to 240
Virial coefficients of gases, **6**-25 to 44
Viscosity
    aqueous solutions, **8**-57 to 83
    argon, liquid and gas, **6**-16 to 24
    atmosphere, as function of altitude, **14**-16 to 22
    carrier gases for chromatography, **8**-120
    common fluids, as function of temperature and pressure, **6**-16 to 24
    gases, at atmospheric pressure, **6**-178 to 179
    helium, liquid, **6**-139
    helium, liquid and gas, **6**-16 to 24
    liquids, **6**-180 to 184
    methane, liquid and gas, **6**-16 to 24
    nitrogen, liquid and gas, **6**-16 to 24
    oxygen, liquid and gas, **6**-16 to 24
    seawater, **14**-12 to 13
    water, as function of temperature, **6**-3
Volt
    definition, **1**-24 to 28
    maintained value, **1**-1 to 11

W

Water
    azeotropic mixtures, **6**-174 to 177
    boiling point, as function of pressure, **6**-12
    compressibility, **6**-136 to 137
    critical constants, **6**-4
    density, **6**-3, **6**-4, **6**-5
    density, as function of pressure, **6**-136 to 137
    density (supercooled), **6**-6 to 7
    dielectric constant, **6**-3, **6**-149 to 171, **6**-172 to 173
    dielectric constant, as function of frequency, **6**-15
    diffussion of gases, **6**-191
    dissociation constant, **8**-84, **8**-85
    electrical conductivity, **5**-90
    enthalpy of fusion, **6**-121 to 135
    enthalpy of vaporization, **6**-3
    fixed point properties, **6**-4
    freezing point, pressure dependence, **6**-47
    heat capacity, **6**-3
    index of refraction, **10**-218
    ion product, **8**-84, **8**-85
    octanol-water partition coefficients, **16**-43 to 47
    permittivity (dielectric constant), **6**-3, **6**-149 to 171, **6**-172 to 173
    permittivity, as function of frequency, **6**-15
    pH measurement, **8**-41 to 42

speed of sound in, **14**-38 to 39
surface tension, **6**-3
thermal conductivity, **6**-4, **6**-185 to 186
thermal expansion coefficient, **6**-136 to 137
thermodynamic properties, **6**-13 to 14
thermodynamic properties at high temperature, **5**-61 to 84
triple point constants, **6**-4
van der Waals constants, **6**-45
vapor pressure, **6**-10 to 11
vapor pressure over salt solutions, **15**-25
viscosity, **6**-3, **6**-178 to 179
volume of one gram, **8**-119
Water (D$_2$O)
    boiling point, **6**-4
    critical constants, **6**-4
    density, **6**-9
    dissociation constant, **8**-85
    fixed point properties, **6**-4
    ion product, **8**-85
    thermal conductivity, **6**-4, **6**-185 to 186
    triple point constants, **6**-4
    vapor pressure, **6**-4
    viscosity, **6**-178 to 179
Watt, definition, **1**-24 to 28
Wavelengths
    atomic spectra, **10**-1 to 87
    correction to vacuum, **10**-220
    electromagnetic radiation bands, **10**-212 to 213
    laser sources, **10**-221 to 227
Weber, definition, **1**-24 to 28
Weighings, reduction from air to vacuum, **8**-119
Width, x-ray lines, **10**-206
Wien displacement law constant, **1**-1 to 11
Wire tables, **15**-27
Wood
    density, **15**-29
    speed of sound in, **14**-38 to 39
    thermal conductivity, **12**-204 to 205
Work function, of the elements, **12**-130

X

Xenon: see also Elements
    critical constants, **6**-48 to 60
    cryogenic properties, **6**-138
    electron configuration, **1**-18 to 19
    entropy, **5**-1 to 3
    history, occurrence, uses, **4**-1 to 36
    ionization energy, **10**-175 to 177
    isotopes and their properties, **11**-50 to 158
    magnetic susceptibility, **4**-133 to 138
    mean free path, **6**-46
    permittivity (dielectric constant), **6**-149 to 171, **6**-172 to 173
    physical properties, **4**-124
    solubility in water, **8**-86 to 89
    thermal conductivity, **6**-185 to 186
    van der Waals constants, **6**-45
    vapor pressure, **6**-64 to 91, **6**-92 to 102
    viscosity, **6**-178 to 179

X-rays
    atomic energy levels, **10**-196 to 199
    attenuation coefficients, **10**-207 to 211
    cross sections, for the elements, **10**-207 to 211
    natural line width, **10**-206

Y

Young's modulus, rare earth elements, **4**-114 to 123
Ytterbium: see also Elements
    electrical resistivity, **12**-45 to 47
    electron configuration, **1**-18 to 19
    heat capacity, **4**-125
    history, occurrence, uses, **4**-1 to 36
    ionization energy, **10**-175 to 177
    isotopes and their properties, **11**-50 to 158
    magnetic susceptibility, **4**-133 to 138
    molten, density, **4**-129 to 132
    physical properties, **4**-124
    thermal properties, **12**-197 to 198
    vapor pressure, **6**-64 to 91
Yttrium: see also Elements
    electrical resistivity, **12**-45 to 47
    electron configuration, **1**-18 to 19
    heat capacity, **4**-125
    history, occurrence, uses, **4**-1 to 36
    ionization energy, **10**-175 to 177
    isotopes and their properties, **11**-50 to 158
    magnetic susceptibility, **4**-133 to 138
    molten, density, **4**-129 to 132
    physical properties, **4**-124
    thermal properties, **12**-197 to 198
    vapor pressure, **6**-64 to 91

Z

Zeotropes, **6**-174 to 177
Zinc: see also Elements
    electrical resistivity, **12**-45 to 47
    electron configuration, **1**-18 to 19
    heat capacity, **4**-125
    history, occurrence, uses, **4**-1 to 36
    ionization energy, **10**-175 to 177
    isotopes and their properties, **11**-50 to 158
    magnetic susceptibility, **4**-133 to 138
    molten, density, **4**-129 to 132
    physical properties, **4**-124
    thermal properties, **12**-197 to 198
    thermodynamic properties, **5**-1 to 3
    vapor pressure, **6**-64 to 91
Zirconium: see also Elements
    electrical resistivity, **12**-45 to 47
    electron configuration, **1**-18 to 19
    heat capacity, **4**-125
    history, occurrence, uses, **4**-1 to 36
    ionization energy, **10**-175 to 177
    isotopes and their properties, **11**-50 to 158
    magnetic susceptibility, **4**-133 to 138
    molten, density, **4**-129 to 132
    physical properties, **4**-124
    thermal properties, **12**-197 to 198
    vapor pressure, **6**-64 to 91